Jane's

ALL THE WORLD'S AIRCRAFT

Editor-in-Chief: Paul Jackson MRAeS
Deputy Editor: Kenneth Munson AMRAeS
Assistant Editor: Lindsay Peacock

Ninety-fourth year of issue
2003-2004

Total number of entries 1,589   New and updated entries 1,454
Total number of images 1,794   New images 880

Visit jawa.janes.com and view the list of latest updates that have been added to the online version of *Jane's All the World's Aircraft* subsequent to this print edition.

Bookmark jawa.janes.com today!
Jane's All the World's Aircraft online site gives you details of the additional information that is unique to online subscribers and the many benefits to upgrading to an online subscription. Don't delay, visit jawa.janes.com today and view the list of latest updates to this online service.

ISBN 0 7106 2537 5
"Jane's" is a registered trade mark

Contents

This Edition has been compiled by:

Susan Bushell	AIRCRAFT: LIGHT AVIATION
Bill Gunston, OBE, FRAeS	GLOSSARY; AERO-ENGINES TABLES
Paul Jackson, MRAeS	AIRCRAFT: FRANCE (part), INTERNATIONAL (part), RUSSIAN FEDERATION (part), UKRAINE, UNITED KINGDOM (part) and UZBEKISTAN
Mike Jerram	AIRCRAFT: GENERAL AVIATION
Kenneth Munson, AMRAeS, ARHistS	AIRCRAFT: CIVIL AND MILITARY (part); LIGHTER THAN AIR
Lindsay Peacock	AIRCRAFT: INTERNATIONAL (military) and UNITED STATES OF AMERICA (military)

CONTENTS

Jane's All the World's Aircraft website: jawa.janes.com

How to use *Jane's All the World's Aircraft*

Described on the following pages are all known powered aircraft, of which details have been received, currently in, or anticipating, commercial production in all countries of the World, apart from those built at home, entirely from plans, and rapidly dismantled, ultralight recreational machines. Exceptions to these criteria are made in the cases of 'one off' aircraft of technical interest – for example those undertaking pioneering work with NASA or designed as a practical exercise by aeronautical universities. Many other of the World's aircraft remaining in service, but no longer being built, will be found in *Jane's Aircraft Upgrades*.

Entries in this book are in alphabetical order (a) by country (including International) and then (b) by manufacturer's name. In the cases of manufacturers producing a diversity of aircraft, those of obvious military potential are presented first. However, it should be noted that a few of the larger constructors are divided into operating divisions, within which individual aircraft types are arranged. In the case of the multinational EADS, aircraft descriptions are retained in the French, German and Spanish sections.

Company entries begin with a brief introduction, including postal address, telephone/fax numbers, e-mail and website addresses and the names of some significant executives. The last-mentioned listing is by no means exhaustive and further details will be found in *Jane's International ABC Aerospace Directory* and *Jane's International Defence Directory*. Subcontractors and suppliers to the aerospace industry appear in *Jane's Aircraft Component Manufacturers,* which also provides rapid references to the subcontractors involved in certain prominent aircraft programmes.

For ease of access to information, entries on individual types of aircraft are subdivided under the following headings:

TYPE: A brief description of the aircraft's function. A full list of categories appears under the heading Type Classifications.

PROGRAMME: A record of key events in an aircraft's production history.

CURRENT VERSIONS: Where applicable, details of available models (marks) and cross-reference to earlier versions now out of production.

CUSTOMERS: Present total on order and produced, often in tabular form, for military aircraft and those civil types for which such a list would not be of prohibitive length.

COSTS: Price per unit or programme price, plus any other disclosed information on R&D expenditure.

DESIGN FEATURES: Where appropriate, opens with a broad statement of design objectives and the means by which they were achieved. This is followed by details such as aerofoil section and helicopter rotor speeds.

FLYING CONTROLS: Here is described the method of controlling the aircraft, it being assumed that the reader has a basic understanding of the function of ailerons, flaps, rudders, trim tabs and the other conventional manoeuvring surfaces (refer to 'conventional and manual' in the Glossary). Descriptions are concerned with the method by which the controls are operated (manual/powered) and

appropriate control inputs determined (autopilot/fly-by-wire, for example).

STRUCTURE: Configuration, materials and any special manufacturing methods; details of subcontractors or partners producing significant elements of the airframe.

LANDING GEAR: Includes tyre sizes and pressures for wheeled aircraft, as well as ground turning circle. Braking parachutes, where fitted.

POWER PLANT: Number and power of engines; helicopter transmission ratings; fuel capacity. Brief additional details are provided in the Aero-Engines listing; fuller descriptions of turbine power plants are in *Jane's Aero-Engines*.

ACCOMMODATION: Seating arrangements, access, environmental control and, for transport aircraft, cargo loading capacity; type of ejection seat, if fitted.

SYSTEMS: Power generation provisions, de-/anti-icing equipment, pressurisation/air conditioning and similar equipment.

AVIONICS: The entry is subdivided into communications, radar, flight aids, instruments, mission equipment (mostly military or law-enforcement) and self-defence (military).

EQUIPMENT: Cargo-handling aids, spraying/firefighting apparatus, lighting, ballistic recovery parachutes and similar items.

ARMAMENT: Fixed and air-dropped/launched weapons listed by the manufacturer as actual or potential armament. Not all items may have been cleared for carriage and not all operators will use those which have. Refer also to the Missiles listing, where appropriate.

DIMENSIONS, EXTERNAL: Includes door sizes and certain ground clearances.

DIMENSIONS, INTERNAL: Includes areas and volumes where relevant.

AREAS: Wings, fixed tail surfaces and control surfaces.

WEIGHTS AND LOADINGS: As supplied by the manufacturer; individual aircraft may vary.

PERFORMANCE: Observations as above; all speeds assumed TAS unless stated otherwise.

OPERATIONAL NOISE LEVELS: Internationally recognised measurements of landing and take-off sound at airports.

Measurements of all types are given in both SI (metric) and Imperial units, the more common conversion factors for which are in the Glossary. Performance details are quoted in good faith and certain critical conversions 'rounded' to give a margin of safety, although *Jane's* does not purport to be an alternative to the manufacturer's operating notes.

In addition to the main section on aeroplanes and helicopters are others detailing Lighter than Air craft, Air-Launched Missiles, Aero-Engines and Propellers, the first also arranged alphabetically by country. Lighter than Air covers commercially and militarily operated airships, and vehicles of special technical interest, but not recreational balloons. Air-Launched Missiles is a rapid reference which seeks not to duplicate the separate and vastly more detailed *Jane's Air-Launched Weapons,* indicating instead how the potential of aircraft described is augmented by the weapons they carry. Aero-Engines are similarly arranged in alphabetical order of manufacturer, irrespective of their country of origin, within the classifications of piston, turboprop, turboshaft and jet engine. Propellers, introduced in 2003, lists some of the main airscrew manufacturers and provides explanations of their designation systems.

Indexes are arranged to speed reference to both current and past aircraft. The first index deals with aircraft in this volume only and is followed by a second covering the past 10 editions of the book. Lighter than Air and the former comprehensive engines section have their separate indexes, but the Air-Launched Missiles and Aero-Engines listings are self-indexing, with multiple cross-references.

Each entry for a company or aircraft is annotated to indicate whether or not it has been altered since the last edition was published.

To help users evaluate the data in this edition, the following descriptions have been used:-

● **VERIFIED** The editor has made a detailed examination of the entry's content and checked its relevance and accuracy for publication in the new edition to the best of his ability.

● **UPDATED** During the verification process, significant changes to content have been made to reflect the latest position known to *Jane's* at the time of publication.

● **NEW ENTRY** Information on new equipment and/or systems appearing for the first time in the title.

● **NEW** New images are identified as *NEW* and some are followed by a seven digit number for ease of identification by our image library.

Full listings of all entries, with an indication of their current status, are provided in the indexes.

Total number of entries 1,589   New and updated entries 1,454
Total number of images 1,794   New images 880

Visit jawa.janes.com and view the list of latest updates that have been added to the online version of *Jane's All the World's Aircraft* subsequent to this print edition.

Copyright enquiries
Contact: Keith Faulkner, Tel/Fax: +44 (0) 1342 305032, e-mail: keith.faulkner@janes.co.uk

British Library Cataloguing-in-Publication Data.
A catalogue record for this book is available from the British Library.

Printed and bound in Great Britain by Bath Press, Bath and Glasgow

EDITORIAL AND ADMINISTRATION

Director: Ian Kay, e-mail: Ian.Kay@janes.co.uk

Managing Editor: Simon Michell, e-mail: Simon.Michell@janes.co.uk

Group Content Director: Anita Slade, e-mail: Anita.Slade@janes.co.uk

Content Editing Manager: Jo Agius, e-mail: Jo.Agius@janes.co.uk

Pre-Press Manager: Christopher Morris, e-mail: Christopher.Morris@janes.co.uk

Team Leaders: Sharon Marshall, e-mail: Sharon.Marshall@janes.co.uk
Neil Grace, e-mail: Neil.Grace@janes.co.uk

Production Editor: Diana Burns, e-mail: Diana.Burns@janes.co.uk
Tracy Johnson, e-mail: Tracy.Johnson@janes.co.uk

Production Controller: Victoria Powell, e-mail: Victoria.Powell@janes.co.uk

Content Update: Jacqui Beard, Information Collection Co-Ordinator
Tel: (+44 20) 87 00 38 08 Fax: (+44 20) 87 00 39 59
e-mail: yearbook@janes.co.uk

Jane's Information Group Limited, Sentinel House, 163 Brighton Road, Coulsdon, Surrey
CR5 2YH, UK
Tel: (+44 20) 87 00 37 00 Fax: (+44 20) 87 00 37 88
e-mail: jawa@janes.co.uk

SALES OFFICE

Send Europe, Middle East and Africa enquiries to: *Mike Gwynn – Head of Information Sales*
Jane's Information Group Limited, Sentinel House, 163 Brighton Road, Coulsdon, Surrey
CR5 2YH, UK
Tel: (+44 20) 87 00 37 00 Fax: (+44 20) 87 63 10 06
e-mail: info@janes.co.uk

Send USA enquiries to: *Robert Loughman – Sales Director*
Jane's Information Group Inc, 110 N Royal Street, Suite 200, Alexandria 22314, USA
Tel: (+1 703) 683 37 00 Fax: (+1 703) 836 02 97 Telex: 6819193
Tel: (+1 800) 824 07 68 Fax: (+1 800) 836 02 971
e-mail: info@janes.com

Send Asia enquiries to: *David Fisher – Group Business Manager*
Jane's Information Group Asia, 5 Shenton Way , #01-01 UIC Building, Singapore 068808
Tel: (+65) 6410 1240 Fax: (+65) 6226 1185
e-mail: info@janes.com.sg

Send Australia/New Zealand enquiries to: *Pauline Roberts – Business Manager*
Jane's Information Group, PO Box 3502, Rozelle Delivery Centre, New South Wales 2039,
Australia
Tel: (+61 2) 85 87 79 00 Fax: (+61 2) 85 87 79 01
e-mail: info@janes.thomson.com.au

ADVERTISEMENT SALES OFFICES

(Head Office)
Jane's Information Group
Sentinel House, 163 Brighton Road,
Coulsdon, Surrey CR5 2YH, UK
Tel: (+44 20) 87 00 37 00
Fax: (+44 20) 87 00 38 59/37 44
e-mail: defadsales@janes.co.uk

Richard West, Senior Key Accounts Manager
Tel: (+44 1892) 72 55 80 Fax: (+44 1892) 72 55 81
e-mail: richard.west@janes.co.uk

Joni Beeden, Advertising Sales Executive
Tel: (+44 20) 87 00 39 63 Fax: (+44 20) 87 00 38 59/37 44
e-mail: joni.beeden@janes.co.uk

(USA/Canada office)
Jane's Information Group
110 N Royal Street, Suite 200,
Alexandria, Virginia 22314, USA
Tel: (+1 703) 683 37 00
Fax: (+1 703) 836 55 37
e-mail: defadsales@janes.com

USA and Canada
Katie Taplett, US Advertising Sales Director
Tel: (+1 703) 683 37 00 Fax: (+1 703) 836 55 37
e-mail: katie.taplett@janes.com

Northern USA and Eastern Canada
Harry Carter, Northeast Region Advertising Sales Manager
Tel: (+1 703) 683 37 00 Fax: (+1 703) 836 55 37
e-mail: harry.carter@janes.com

Southeastern USA
Kristin D Schulze, Advertising Sales Manager
PO Box 270190, Tampa, Florida 33688-0190
Tel: (+1 813) 961 81 32 Fax: (+1 813) 961 96 42
e-mail: kristin@intnet.net

Western USA and Western Canada
Richard L Ayer
127 Avenida Del Mar, Suite 2A, San Clemente, California 92672
Tel: (+1 949) 366 84 55 Fax: (+1 949) 366 92 89
e-mail: ayercomm@earthlink.com

Australia: *Richard West* (see UK Head Office)

Benelux: *Joni Beeden* (see UK Head Office)

Brazil: *Katie Taplett* (see USA address)

Corporate Accounts: Simon Kay
33 St John's Street, Crowthorne, Berkshire RG45 7NQ, UK
Tel: (+44 1344) 77 71 23 Mobile: (+44 7702) 54 96 84
Fax: (+44 1344) 77 58 85
e-mail: simon.kay@btclick.com

Eastern Europe: MCW Media & Consulting Wehrstedt
Dr Uwe H Wehrstedt
Hagenbreite 9, D-06463 Ermsleben, Germany
Tel: (+49) 0700/WEHRSTEDT / (+49) 03 47 43/620 90
Fax: (+49) 03 47 43/620 91
e-mail: info@Wehrstedt.org

France: Patrice Février
BP 418, 35 avenue MacMahon,
F-75824 Paris Cedex 17, France
Tel: (+33 1) 45 72 33 11 Fax: (+33 1) 45 72 17 95
e-mail: patrice.fevrier@wanadoo.fr

Germany and Austria: *MCW Media & Consulting Wehrstedt* (see Eastern Europe)

Greece: *Joni Beeden* (see UK Head Office)

Hong Kong: *Joni Beeden* (see UK Head Office)

India: *Joni Beeden* (see UK Head Office)

Israel: Oreet – International Media
15 Kinneret Street, IL-51201 Bene Berak, Israel
Tel: (+972 3) 570 65 27 Fax: (+972 3) 570 65 27
e-mail: admin@oreet-marcom.com
Defence: Liat Shaham
e-mail: liat_s@oreet-marcom.com

Italy and Switzerland: Ediconsult Internazionale Srl
Piazza Fontane Marose 3, I-16123 Genoa, Italy
Tel: (+39 010) 58 36 84 Fax: (+39 010) 56 65 78
e-mail: genova@ediconsult.com

Middle East: *Joni Beeden* (see UK Head Office)

Pakistan: *Joni Beeden* (see UK Head Office)

Russia: Vladimir N Usov, PO Box 98, Nizhniy Tagil,
Sverdlovsk Region, 622018, Russia
Tel/Fax: +007 3435 230268
e-mail: uvn125@uraltelecom.ru

Scandinavia: The Falsten Partnership
PO Box 21175, London N16 6ZG, UK
Tel: (+44 20) 88 06 23 01 Fax: (+ 44 20) 88 06 81 37
e-mail: sales@falsten.com

Singapore: *Richard West/Joni Beeden* (see UK Head Office)

South Africa: *Richard West* (see UK Head Office)

South Korea: Infonet Group Inc
Sanbu Rennaissance Tower 902, 456 Gongdukdong, Mapogu, Seoul, South Korea
Contact: Mr Jongseog Lee
Tel: (+82 2) 716 99 22
Fax: (+82 2) 716 95 31
e-mail: jslee@infonetgroup.co.kr

Spain: Via Exclusivas SL
Contact: Julio de Andres
Viriato 69SC, E-28010 Madrid, Spain
Tel: (+34 91) 448 76 22 Fax: (+34 91) 446 02 14
e-mail: j.a.deandres@viaexclusivas.com

Turkey: *Richard West* (see UK Head Office)

ADVERTISING COPY
Linda Letori (Jane's UK Head Office)
Tel: (+44 20) 87 00 37 42 Fax: (+44 20) 87 00 38 59/37 44
e-mail: linda.letori@janes.co.uk

For North America, South America and Caribbean only:
Shanee Johnson (Jane's USA/Canada Office)
Tel: (+1 703) 683 37 00 Fax: (+1 703) 836 55 37
e-mail: shanee.johnson@janes.com

Jane's Electronic Solutions

Access over 200 sources of Jane's news, analysis and reference covering defence, risk, security, aerospace, transport and law enforcement information with electronic solutions designed to meet the demands of your information needs.

Jane's electronic solutions offer you the choice of how you want to receive the data. Whether you want to integrate data into your organisation's network, access the data online or on CD-ROM, Jane's can provide you with critical information, quickly and easily in a format that suits your organisation.

www.janes.com

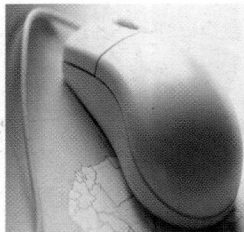

Accessible by IP address, for networking within organisations or by unique username and password, janes.com enables you to access Jane's information wherever you are in the world.

Why subscribe to janes.com:

- **Search function**
  Allows you to explore within the contents of all Jane's datasets by keyword and/or fielded search terms.

- **Image Searching**
  Allows you to search captions within Jane's data with results in the form of a thumbnail and caption.

- **Latest News Extra**
  Search for the latest news globally across all areas of defence, transport, aerospace, security and business.

- **Active Interlinking**
  Allows you to navigate via hyperlinks in records to other related product entries throughout Jane's content to which you subscribe, reducing your research time significantly.

- **Browse function**
  Allows you to view the available contents of Jane's datasets by Country, Image, Market, News/Analysis, Operational Guide, Organisation, and Systems & Equipment.

Jane's Libraries

Jane's Libraries offer you and your organisation comprehensive datasets in the areas of defence, risk, security, aerospace, transport and law enforcement. Each Library groups relevant information and graphics that can be cross-searched to ensure you find every reference you are looking for.

With Jane's Libraries you can:

- Pinpoint the information you need using a variety of basic and advanced searches.

- Export text easily into ASCII, dbase or comma-delimited formats for use in your own reports and presentations.

- Download JPEG photographs and technical line drawings for your own internal use.

- Print all text and graphics straight from the Libraries.

- Network the Libraries within your organisation to ensure easy access by all.

Libraries available:

Jane's Sea and Systems Library
Jane's Air and Systems Library
Jane's Land and Systems Library
Jane's Police and Security Library
Jane's Market Intelligence Library
Jane's Geopolitical Library
Jane's Transport Library
Jane's Defence Equipment Library
Jane's Defence Magazines Library

For information on how a Jane's Library could help your organisation please contact your local Jane's sales office.

Jane's Data Service

Jane's information on your intranet or controlled military network

Jane's Data Services brings together more than 200 sources of near-realtime and technical reference information serving defence, intelligence, space, transportation and law enforcement professionals.

Jane's Data Service provides you with:

- Full integration of data into your own secure environment

- Flexibility of choice with your selection of data

- Frequent updates via CD-ROM or FTP

- Re-usable data for internal presentations and reports

- Full integration with other data sources for cross-databank searching

- High-quality JPEG images

http://consultancy.janes.com

As a decision maker, you shoulder a heavy burden. Yes, you can access more information faster and more easily than ever before, but more information does not always help you to make better choices. So, how do you maintain your advantage in such a world?

You could build a sophisticated global network and collect, weigh and analyse the information it gathers. Or, you could simply turn to Jane's Consultancy.

What we can do for you:

- Market Intelligence
- Systems Analysis

- Military Assessments
- Security Risks and Red Teaming
- Expert Testimony

www.janes.com

Alphabetical list of advertisers

Users' Charter

This publication is brought to you by Jane's Information Group, a global company with more than 100 years of innovation and unrivalled reputation for impartiality, accuracy and authority.

Our collection and output of information and images is not dictated by any political or commercial affiliation. Our reportage is undertaken without fear of, or favour from, any government, alliance, state or corporation.

We publish information that is collected overtly from unclassified sources, although much could be regarded as extremely sensitive or not publicly accessible.

Our validation and analysis aims to eradicate misinformation or disinformation as well as factual errors; our objective is always to produce the most accurate and authoritative data.

In the event of any significant inaccuracies, we undertake to draw these to the readers' attention to preserve the highly valued relationship of trust and credibility with our customers worldwide.

If you believe that these policies have been breached by this title, you are invited to contact the editor.

A copy of Jane's Information Group's Code of Conduct for its editorial teams is available from the publisher.

Quality Policy

Jane's Information Group is the world's leading unclassified information integrator for military, government and commercial organisations worldwide. To maintain this position, the Company will strive to meet and exceed customers' expectations in the design, production and fulfilment of goods and services.

Information published by Jane's is renowned for its accuracy, authority and impartiality, and the Company is committed to seeking ongoing improvement in both products and processes.

Jane's will at all times endeavour to respond directly to market demands and will also ensure that customer satisfaction is measured and employees are encouraged to question and suggest improvements to working practices.

Jane's will continue to invest in its people through training and development, to meet the Investor in People standards and changing customer requirements.

Jane's

While Boeing has spent the last year hesitating over its next move, rival Airbus has forged ahead with design of the A380 high-capacity airliner; flown the 117-seat A318; and brought out a second A340 variant, the -500, to complement the -600, delivered to its first customer in July 2002. Airbus' two most recent world-beaters claim to have – for airliners – the longest range (A340-500, nearest) and longest fuselage (A340-600)

NEW/0543352

Executive Overview: We Apologise for Late Departure of the Next Century of Flight

"Annual income twenty pounds, annual expenditure nineteen nineteen six, result happiness. Annual income twenty pounds, annual expenditure twenty pounds ought and six, result misery."

Mr Micawber (Charles Dickens, *David Copperfield*).

A valued benefit conferred upon the writers of annual overviews, such as this, is the incomparable gift of hindsight. Even more precious still is the further interval granted before the writer is again required to practise the arcane arts of analysis and prognostication. Thus, two years after the terrorist attacks on New York and Washington, and one year after this Executive Overview noted the start of a recovery in aerospace business, a more considered analysis is forced to offer a less sanguine appraisal. True, prospects for aviation are better than in late 2001, but the climb-out from recession noted 12 months ago is now seen to be the equally transient mirror reaction to the sharp downturn after September 11. The underlying recession, which *Jane's* noted had begun well before the terrorist attacks, is still with us and has been made slightly worse by those terrible events.

The title of last year's Foreword, "The art or science of riding two cycles simultaneously" proved to encapsulate more prescience than its author could reasonably have claimed. Then, the expectation was of a formation recovery from both downturns. Now, it is evident that a return to the relative prosperity of 2000 will not be achieved by the end of the current year. Indeed, some air transport analysts are predicting four years from September 2001 before airlines have the confidence to begin placing orders at late-1990s volumes; and a further interval until aircraft makers and their subcontractors feel that benefit in their backlog ledgers.

Nothing more graphically illustrates the major airliner manufacturers' predicament than the fact that, in composing the tables for orders and deliveries in the past year, *Jane's* has had to rummage in a dusty corner of the printer's fount (in truth, a no less neglected corner of the computer keyboard) for the *minus sign*. While, in recent months, new orders from strange-sounding airlines in faraway places have been trumpeted at a volume more normally associated with two-digit contracts from national flag-carriers, less has been heard of the cancellations quietly agreed behind the scenes.

For Boeing, this has meant the 757 order book marking time throughout 2002, while the Boeing 767 total has shrunk by two. Of course, production is continuing, with the consequence that the backlog — rightly or wrongly perceived as a measure of the company's worth — is being consumed from both ends. Standing back to look at the

figures, Boeing announced 251 orders in 2002, although the net rise after cancellations was a mere 176, compared to 381 delivered. The latter figure, itself, sits ill alongside 527 in 2001 and a peak of 620 the previous year, but is on a par with 375 during the 1997 downturn. As with the Dickensian example cited above, the 'expenditure' of production is temporarily exceeding the 'income' of orders, although the predicted misery can be avoided in the short term by recourse to savings amassed in the Bank of Backlog.

Airbus, while able to cock a snook at its transatlantic rival, with 300 gross orders and 233 net (but only 303 deliveries), remains in no doubt that 2003 "will be an extremely difficult year". It is of note that, as with the Boeing 757/767 figures, it is the middle of the passenger capacity range that is suffering most from customer indifference, with no new orders for either the A300 or A310. In view of the long-standing rivalry between Airbus and Boeing, the former's convincing 57 per cent share of the post-cancellations market for 2002 is but a Pyrrhic victory.

Furthermore, the fact that two European low-cost airlines, easyJet and Ryanair, were responsible for 40 per cent of 2002's Airbus/Boeing orders suggests that the manufacturers' predictions of "similar" results in 2003 will be, actually, lower because of the probable absence of equivalent special cases (and had easyJet extended its existing Boeing 737 fleet instead of changing its favours to Airbus, the year-end picture would have been radically different). In terms of deliveries, both manufacturers are predicting 280-300 in the current year, a fact which Airbus claims is bringing ever closer the day of its total eclipse of Boeing.

What is Boeing doing?

Is the US giant dithering and losing its grip, or is it manoeuvring to snatch back its dominance from behind a carefully laid smokescreen? Two years ago, *Jane's* applauded the unveiling of the Boeing Sonic Cruiser as a bold move to advance air travel, while avoiding a damaging battle with Airbus in the 'Super Jumbo' category already claimed by the A380. Even before the recent slump in airliner orders, there were well-founded doubts that the economics of transonic transport did not add up, yet the initiative was no less praiseworthy for all that.

By late 2002, it had become clear that at least two projects were on the drawing boards, the second a more conventional aircraft dubbed the Super Efficient. In December, the Sonic Cruiser went the way of Boeing's proposed Concorde-challenger of the early 1970s to be replaced in the shop window by what is now being called the 7E7 and some are already prophesying will exchange its 'E' (for efficient) for an '8' within the next year, and could be in airline service by 2008.

Resembling a 777 scaled down in all dimensions, the 7E7 is talked of in terms of an Airbus A330 rival, but with 210- and 250-seat capacity versions, which puts it in the A300/A310 (or B757/B767) sector. These four aircraft, it will be recalled, are the slowest-selling part of both manufacturers' ranges. And within another meaning of the word 'range' lies a clue to Boeing's thinking. The 7E7 will cruise at an unremarkable (compared to the 'Cruiser) M0.85+, but the challenge to its rivals comes with its long reach: up to 8,000 n miles (14,800 km; 9,200 miles).

The Sonic Cruiser may have camouflaged Boeing's real intentions with the 7E7, but is

Without two major sales to low-cost operators of the Airbus A320 (illustrated) and Boeing 737 families, the airliner contract totals for 2002 would have been even more bleak *NEW*/0543356

the 'E' just further obfuscation? The Super Efficient epithet caused further head-scratching when artist's impressions showed a very conventional-looking airliner lacking the aerodynamic ploys currently in vogue to enhance 'slipperiness'. The lower operating costs implied in the name seemed to be denied by the immutable laws of physics, yet Boeing persists with claims of 23 per cent better fuel consumption per passenger than the A330-200 and cites materials development begun for the 'Cruiser as having valuable applications in the 7E7. Granting some efficiencies from structural weight savings and aerodynamically clever wingtips, this still leaves most of the percentage improvements to come from engines — which will, presumably, be available for purchase by competitors.

Nevertheless, Boeing's strategic thinking seems to be that, short-term cycles aside, the number of people flying is increasing and more airliner seats are needed. The Sonic Cruiser relied on the premise that a few of those people would be prepared to pay extra for a superior form of transport; the 7E7 is based on the belief that a larger number — even as far down the pecking order as the budget traveller — are disinclined to undertake part of their journey in an A380-size mega-jet; then fight their way across a crowded hub airport; and repeat the check-in, security and waiting nausea in order to board a 737 or A320 to continue to their destination.

Budget airlines are enjoying good business with direct flights to where passengers want to go, but usually over hundreds, not thousands, of miles. While detractors may claim never to have heard of the out-of-the-way airports billed as destinations, these supposed backwaters have an uncanny knack of being a short bus or taxi ride away from a ski slope, spa resort, picturesque city or other attraction. As a case in point, the touristic lure of Lake Constance was recently recognised by two low-cost airlines and enabled the Editor to attend the Friedrichshafen air show at one third the travel cost and aggravation previously experienced in attending the event. This latest foray took advantage of a direct flight from the UK right in to the show airfield, where the alternatives were previously wallet-crippling air travel via Frankfurt; or by air to Zurich, followed by car hire and self-drive, including a ferry crossing.

Boeing, while agreeing with Airbus that more passengers will be travelling in the years ahead, sees not larger airliners, but a larger fleet of mid-size airliners as the answer. The European company appears to be neglecting to renew the centre (A300 and A310 replacement) of its portfolio and, instead is extending at both ends (A318 and A380). Longer range is the means whereby passengers in new middle-sized Boeings will avoid the hub airports into whose tender mercies the A380 will deliver them, already travel-weary, at only the mid-point of their journey. Suddenly, Boeing seems like the champion of the common man, whereas Airbus and its A380 can be construed as pandering to the 'herd them on board' school of airline management.

Among the nether-regionals

At the next level down from intercontinental transport, regional airliner orders have plummeted over the past year, continuing a trend signalled previously, but confirming beyond any doubt that the buying boom which built up throughout the 1990s is at an end. Regional jet orders for the four major manufacturers amounted to a paltry 86 (the lowest since 1993) and were completely overshadowed by 207 cancellations, the year-end result being a net reduction of 121. The turboprop sector continued shrinking, with 26 gross orders, the only good news here being that the low figure of two cancellations indicated that order books contained no measure of ill-founded optimism. For regionals as a whole, 112 gross orders compares badly with 400 in 2001 and is more than offset by 369 deliveries (including 60 turboprops).

Already, the figure of 'four major manufacturers' mentioned above has been halved. BAE Systems had previously announced its departure from the RJ market in 2001 and flew the last of the 146/RJX family in April 2002. Fairchild Dornier suffered financial meltdown during the year, apparently taking with it the rolled-out, but unflown, prototype of the 728 and leaving AvCraft Aviation to inherit the 328JET programme and formulate plans for its comeback. Not unexpectedly, there have been no recent echoes of the once-regular reports from the Netherlands of one plan or another to resuscitate the Fokker line of regionals.

With a foot in both the jet and turboprop camps, Bombardier has emerged from 2002 if not well, then least scathed, despite a two-month shutdown and continued questions over the viability of the Dash 8 line (10 orders in 2002) and, especially, a payback for the development investment made in the Q400 version. Its 700th CRJ airliner was delivered last October, while Embraer handed over the 600th of its regional jets in May 2002. The Brazilian company is expected to make a strong showing in the near future as the 170/190 family enters production on the back of an intensive flight trials programme which witnessed six prototypes taking to the skies in as many months. ATR, locked into the turboprop market, ticks over steadily on an order and delivery figure in the late 'teens, while Raytheon's Beech 1900 appears to have turned off the runway and on to the taxiway of history.

Set against this gloomy picture, the airlines operating regional jets are more content with their lot. All the top 10 US operators returned double-digit growth last year, but through doing better with what is currently available, rather than placing new orders for what is not. Larger airlines have been handing down unprofitable routes and aircraft to the regionals which, apparently, are making a go of them. As a result of this convoluted arithmetic, there are now 27 per cent more airliners operating US regional services, although the number of airframes has remained, essentially, static. Making sense of the turboprop sales picture, cited previously, the percentage of US regional airports served by jets has increased 13 points in a year, even though domestic traffic (all operators) fell 3.8 per cent in 2002.

The mob demand a protection racket

When a company begins to demand protection or special treatment from the government it is an opportune time to check how many of its shares one is holding. Recent months have witnessed a small crop of aerospace manufacturers in several countries calling for shelter from the icy blast of the open market. Leading these is BAE Systems, which prefixed its appearance at the 2002 Farnborough International aerospace show with a public demand to the UK government to order the latest version of Hawk for the RAF training programme without competition, further debate or synchronisation with European air arms with a similar need. In effect, BAE wants to be the 'national champion', guaranteed as the first stop when Her Majesty's armed forces go shopping.

How far that privileged status is deserved is evident from perusal of the *Jane's* entry for the Nimrod MRA. Mk 4 maritime patrol jet. Neither customer nor supplier can escape criticism in this sorry saga, but the latest scene of the farce, screened in mid-2002, has

Expenditure on development of the Fairchild Dornier 728/928 outstripped income from the 328JET (illustrated) during 2002, with the inevitable result. The latter will be saved; the former's fate appears to be sealed
NEW/0543357

US helicopter makers are worried by European encroachment in the domestic civil market and fear military may be next. Now flying in production form, the Bell/Agusta AB 139 is a product of the "join them" business strategy
NEW/0543360

national champion, BAE Systems appears to have suffered the rare indignity of being on the receiving end of hostile briefings of the media by government officials in the run-up to its qualified victory in the competition to build two new aircraft carriers for the Royal Navy. Hard work will be needed before mutual trust is restored between manufacturer and customer.

US helicopter makers are also calling for protection from the predations of foreign companies after a lacklustre year in 2002. With 320 non-military sales, Eurocopter has claimed 60 per cent of the world market for turbine-powered machines and Agusta has added some 50 of its own European-built machines. Bell came a poor second on 87, MD scoring 15 and Sikorsky trailing on 13. Only when the lighter Robinsons, Enstroms and Schweizers are taken into account do the numbers improve, even though the total monetary value is less influenced. The US helicopter makers' call is for a ban on foreign military helicopter imports (a civil embargo being impossible to implement), implicit in which is a fear that this last bastion of sales security is under attack because the Europeans are seeking US-based Trojan Horse companies to manufacture the NH90 and EH101 under licence. At least one US helicopter manufacturer has even resorted to boosting a published table of its 2002 production with numbers of upgrades and its re-sales of second-hand trade-ins to arrive at a more respectable-looking total.

During candid moments, US observers admit that their industry has not matched the recent European investment in helicopter technology and is in need of some form of consolidation along the path already trodden by the USA's aeroplane and systems manufacturers. In fact, there appears to have been nothing in the past 30 years to match the stimuli given to fixed-wing aviation of the sort typified by the JSF programme. With NASA having opted out of rotorcraft research in 2003, US companies are failing to grasp the baton themselves and are looking to Washington to set the pace — or, at least, put up half the money needed for a development 'push' to begin in 2004 with US$50 million of public money. This will be used to research improvements in noise, vibration, structures, safety and flight in extreme IFR conditions, so maximising the

a story-line which would be rejected as implausible if written as a drama. In its fabrication of new wings to be attached to 18 existing fuselages, BAE appears to have made fullest use of all the latest tools — dynamic young management, high-powered design computers and state-of-the-art machines which ensure a Eurofighter Typhoon part, made in Spain or Italy, will fit exactly to another component from Germany or the UK. In all, 18 perfectly identical sets of wings based on measurement of the pattern Nimrod.

No one, it seems, consulted the old retainers who, perhaps now retired, *hand-built* the original Nimrods three decades ago on a production line which had changed little since Avro Manchesters and Lancasters

rolled down it en route to the aerodromes of Lincolnshire. In those days, one size most certainly did not fit all, with the inevitable present consequence that, as one BAE employee anonymously told a national newspaper, "Engineers have been trimming the wings to try to get them to fit. The whole thing is a fiasco." If ever there was proof that a wise company respects its history and should listen to the accumulated wisdom of those who participated, then this is it.

Taken in conjunction with related problems regarding the strength of the new wing's structure, it is little wonder that the aircraft's in-service date has slipped four years — even before flight tests and airborne trials of the complex avionics have taken place. In truth, instead of becoming a

Struggling to find industry information?

helicopter's natural advantages. Excellent aims; but why are the capitalists of the Land of the Free refusing to take the initiative without a fistful of taxpayers' dollars?

While the US rotary-wing community is urging its government to follow European lines, the European lightplane manufacturing leader, Socata, is clamouring for the EC to provide defence from American manufacturers who have the unfair advantage of a supportive government (tell *that* to Ayres, Commander, Lancair, Micco, Mooney and Stoddard Hamilton). Socata predicts the "extinction" of the European lightplane industry (for which, read 'government-supported French lightplane industry') in 15 years, unless Brussels shows willing and rescues companies such as Apex (CAP, Robin and BUL) and Reims Aviation from their present financial difficulties.

Among the means for assisting light aviation, says Socata, are relaxation of the prohibition of single-engine VFR flight (a boon to the TBM 700 and Pilatus PC-12, among others); closer harmonisation of certification rules between JAA members and with the FAA; unified European airspace below FL290; and more official investment in development initiatives which will benefit general aviation. While pilots would unanimously raise a non-alcoholic glass to such measures, we would caution that urging politicians to 'do something' is a double-edged weapon, best wielded with great care. The last big JAA initiative for general aviation was the pan-European private pilot's licence of 2000 which damaged private flying to such a degree that several member states were forced to take urgent, disparate and unilateral action to avert further alienation of potential pilots.

Excuse me; who's in charge here?

While the West may fret at alleged muddle and lack of support within its own aerospace industry and armed forces, a look at recent events in the CIS provides a means of placing matters in perspective. At each turn in the convoluted saga of the Airbus A400M military airlifter, it has been possible to point to the comparable and more advanced Antonov An-70 programme progressing, albeit slowly, with combined support from Russia and Ukraine towards service entry. At least that was the position until January, when the C-in-C of the Russian Air Forces, Col Gen Vladimir Mikhailov, revealed at a press conference that "an error has been made and [we must] choose another aircraft".

The staggering announcement is not without its merits for, as an Antonov An-12 replacement, the An-70 has almost twice the desired payload, and is thus in the category of the Ilyushin Il-76 — which does not need replacement at the moment. The correct size is typified by the Tupolev Tu-330, which has been long waiting in the wings, and Ilyushin Il-214, being developed jointly with Indian finance (47 per cent of development costs, no less). One must assume that the political

Socata has warned that the Tampico, Trinidad and TBM 700, among others, will be extinct in 15 years if the EC does not show greater commitment to the European lightplane industry   *NEW*/0543359

merits of a post-USSR liaison between Russia and Ukraine took precedence over military considerations at the An-70's genesis, yet such an unequivocal condemnation of a project blessed less than a year before by President Putin must be viewed as a serious bid for the world insubordination record.

As if to reinforce his claim, Gen Mikhailov returned to the fray in the following month with a statement on the MiG-AT advanced trainer. "One may say with confidence that after the state certification tests, the MiG-AT will be examined as a training aeroplane for air force instructional establishments," he declared. This will have come as news to the official committee which, on 10 April 2002, declared the rival Yak-130 as the preferred type for re-equipment of military flying schools, but encouraged MiG to continue with its project "for export". A similar non-decision has kept the Mil Mi-28 snapping at the heels of its supposed victor, the Kamov Ka-50, since the latter was pronounced the army's choice for a new combat helicopter on 5 October 1994.

It is difficult to imagine such a confusing situation acting as a stimulus to a prospective — and desperately needed — export customer who could find his investment in a

programme suddenly turning to dust. For a parallel, one would have to picture the current C-in-C of the US Air Force declaring a resumption of the F-22 Raptor versus Northrop Grumman YF-23 fly-off and announcing that Boeing's X-32 will be the new Joint Strike Fighter, whatever the Pentagon and Lockheed Martin might say.

Russia is now committed to its own 'JSF' in the form of the PACFA project led by Sukhoi. This 'fifth generation' multirole fighter received a surprise boost in January when India announced its intention of taking up a Russian offer to participate. As with the Il-214 transport, this move will inject cash into a programme which would otherwise run the risk of financial stagnation. India has been a favoured customer for Soviet and Russian combat aircraft for four decades, but this proposal represents a step up to the highest levels of joint development.

In conjunction with current plans for Indian licensed production of the Sukhoi Su-30M it also raises questions about the proposed large-scale manufacture of the indigenous LCA, an airframe which appears to be competing against its Kaveri engine in the aeronautical equivalent of a slow bicycle race. The long awaited first order for LCAs has just been placed, yet it is for a meagre 16 aircraft, instead of the 140 or so which the

The Sukhoi RRJ is leading candidate to fill a position supposedly allocated to the Tupolev Tu-334 more than a decade ago. Duplication and under-funding are getting Russia nowhere   *NEW*/0543405

IAF traditionally requests of a new type. Adding to the measure of caution, these 16 will have interim General Electric engines, like the prototypes.

Civil aviation in Russia is looking forward to happier days, with the indication that 2002's figures of a 6 per cent rise in passenger numbers have about them the joyous ring of a world trend being bucked. Continued growth is anticipated in the internal travel market and there are plans for more state subsidies of non-profitable, but socially essential routes, as in the days of the 'old' Aeroflot. Already, in the Far East region, operators with new, Russian-built aircraft are to see their annual subsidy increased from Rb80 million to Rb300 million this year.

Just as regional airliner orders in the West have slumped, as previously noted, the Russian market seems to be ready for immediate take-off. This has prompted a government-sponsored competition for a new regional airliner, which is expected to enter service in 2007, once more immediate attention (in other words, funds) has been devoted to sorting out the present tangle of existing programmes. The Sukhoi RRJ, with its Boeing backing, must be regarded as a favourite, but strong opposition will be mounted by the Tupolev Tu-414 and/or Tu-324, with contenders from Myasishchev (M-70) and Ukraine's Antonov (An-74-300 and new An-148) also running.

Attempts by a Russian consortium to obtain the stillborn Fairchild Dornier 728 programme may be regarded as showing initiative, but little appreciation of the larger picture of obtaining local design expertise. As for Tupolev's Tu-324, this was supposed to have begun regional jet services in 1995, but there is still only one prototype flying and its future would appear to be overshadowed by the latest contest. The traditions of the Soviet era die hard, but Russia must realise that it now has neither the funds nor manpower to run competing programmes farther than the drawing board.

Between now and 2010, Russian civil transport needs for new equipment are estimated as 200 airliners, 300 utility aeroplanes, 80 cargo carriers and up to 350 helicopters. This can be compared with a 2002 production of 15 aeroplanes and 45 helicopters, only four and five, respectively, having been bought by Russian customers, the rest exported. The output of fixed-wing types from factories last year makes pitiful reading: two Tu-214s, two Tu-204s, one Tu-154M, one An-74, three An-38s, one M-101 and five An-3 turboprop conversions from An-2.

Demand from local operators in 2003 is estimated as 16 passenger aeroplanes, followed by 27, 45, then an average of 80 per annum between 2006 and 2010. These will go some way towards swelling the commercial and general aviation fleet (excluding the FLA RF private flying movement) of 653 long-range airliners, 810 regionals, 1,990 business and general purpose aeroplanes, 530 specialised cargo types and 1,915 helicopters, or 5,898 in all.

The 'war dividend' proves elusive

Western aerospace manufacturers are also looking forward to one door opening as another, if not altogether closing, certainly sticks half-way. It is not without its irony — and a certain symmetry — that both trends were precipitated by a single terrorist event. Boeing's medium-term forecast is for lowered revenue in 2003, but a better return for its shareholders next year as benefits are felt from increased military spending. Not all in the business straddle both military and civil markets, yet the industry-wide picture may be seen as broadly following the Boeing outlook.

It would be unwise to presume the symmetry will be perfect. Although the US defence budget is heading towards US$400 billion in 2004, and targeting a further 25 per cent rise before the end of the decade, few lights have been noted burning late into the night in the Pentagon offices responsible for ordering combat systems which would appear in future editions of *Jane's All the World's Aircraft*. Quite the reverse. The draft 2004 budget calls for *cuts* in procurement of Lockheed Martin F-35 JSFs, Boeing F/A-18 Super Hornets and Lockheed Martin F/A-22 Raptors in order to free funds for missile defence, unmanned air vehicles, surveillance satellites and special operations forces, among others.

Nor will general aviation profit much from the present emergency. Post September 11, 'homeland defence' became the new buzz-word, as manufacturers of jet-powered kitplanes offered fighter versions of their products to defend the USA — flown by whom, against what, not being adequately elaborated. Less improbable schemes there also were, but apart from a promise of possibly a dozen orders for an Australian-built utility lightplane, the impact on the contents of this publication has been imperceptible. The new US Department of Homeland Security is (rightly) more interested in buying X-ray scanners and bugging devices than forming its own air arm.

The keystone (in more senses than one) of the EU's intervention forces is the Airbus A400M military transport, which has made no further progress towards the cutting of metal since the preceding Executive Overview was written 12 months ago. A German announcement, in December 2002, of a reduced purchase commitment is expected to clear the way to a production go-ahead just as this is published, but with one fundamental change in the programme which has, so far, generated no comment.

Military aviation in Europe appears to have in store what the well-known Chinese curse describes as 'interesting times'. France's attempt, during the overture to the recent US-led attack on Iraq, to complete its self-appointed task of severing European defence interests from those of the USA will have consequences not yet clearly appreciated. The French move to cripple US-led NATO and replace it by the European Union defence entity may be successful, but Paris has yet to explain the practical use of a military alliance which suffers hopelessly inadequate funding, inept political direction and a leading member who professes a steadfast moral opposition to the use of military force under any circumstances.

According to Airbus, the engine choice, between two contenders, for the A400M will now be the customer's (that is to say, the single agency acting for all). This would appear to overturn the previous, clear-cut arrangement whereby the contractor was to produce an aircraft to meet the customer's specification and make recompense from his own pocket for any shortfall. Now, who is unequivocally responsible if A400M fails to meet its performance guarantees? Translated into legal-speak, the phrase, "It's all your fault for messing with the

engines," can be made into a court case to last years.

In more positive vein, we now have, at least, the ability to view airborne production examples of the Eurofighter Typhoon and Eurocopter Tigre/Tiger, which serve as symbols of more productive European collaboration — not to mention as dire warnings of the inordinate period of time required to bring any such programme to fruition. Tiger began in the mid-1970s and, after two failed attempts at defining a Franco-German specification, was eventually launched as a programme in 1984. Exactly 20 years later, personnel of its first regiment will just be learning how to operate it.

Regarding Eurofighter, it is only necessary to recall that the original German specification was designated TKF-90, indicating a 1990 operational date; and that the Royal Air Force has spent a quarter of the time since the Wright Brothers first flew preparing for the aircraft, having issued the original specification in May 1979. Competitors should not sneer; the Dassault Rafale was conceived to ACT-88, indicating intended 1988 service-entry.

Production status notwithstanding, European fighters have had a year of misfortune, including Austria's on-then-off purchase of 24 Typhoons; Brazilian postponement of a contest including Dassault Mirage 2000 and Saab Gripen; the Czech cancellation of 24 Gripens; Hungary's renegotiation of its Gripen agreement for a lesser quantity; the Polish selection of F-16s instead of a European fighter; and the not altogether unexpected UK move to abandon its proposed third batch of Typhoons.

For the future, one matter occupying an increasing amount of time will be the 'Eurotraining' venture to establish a contractor-run, military training scheme involving as many as possible of the European air forces, and starting in around 2010. Central to the venture will be an advanced jet trainer, and although BAE Systems will promote the Hawk in an upgraded guise, it is likely that most partners will be seeking a more modern design. The heir apparent is EADS' Mako, currently progressing slowly towards prototype status on the untenable premise that its fortunes are being determined by the air force of United Arab Emirates.

Further advanced is the Italian challenger, the Aermacchi M-346, now an Alenia product since that company's acquisition of Macchi and its extensive stable of trainers last December. The M-346 is based around the Yakovlev Yak-130, but Aermacchi is now sufficiently confident of the aircraft's maturity to begin downplaying the relationship to an already flying aircraft and stressing the amount of redesign it has incorporated. Eurotraining will be hard-fought, not only as it will coincidentally launch a light attack fighter for export, but also because the more politically minded may see it as the foundation stone of a single

After 2002, there are many new names in the world of business jets, but few genuinely new designs from the large manufacturers. Cessna's Citation CJ3 was announced last September and has obvious antecedents
NEW/0543358

European Union air force and the setter of all manner of precedents.

Business not as usual

Most business aircraft manufacturers have reduced their production rates since late 2001 although, in truth, the downwards trend in orders began somewhat earlier. 'Steady' is the best prognosis for the current year, perhaps with piston-engined types rallying the numbers, but having proportionally less effect on the balance sheet.

Members of the General Aviation Manufacturers' Association delivered 2,539 aircraft in 2002, ranging from large corporate jets to light Cessnas, Pipers, Socatas and Pitts Specials, reflecting an overall decline of 15 per cent in relation to the previous year's 2,994. Of the former total, 2,214 were US-built. No comparable data are available for the multitude of lightplanes, kitbuilts and other aircraft delivered by smaller manufacturers not in the Association, but there is nothing to suggest that they have had a better time.

Business jets account for 683 of the 2002 deliveries, and turboprops for 280, the piston-powered sector featuring 1,446 singles and 130 twins. Most notable in the first-mentioned sector is the almost total collapse of the market for ultra-large business jets: the Boeing BBJ and Airbus ACJ. Turboprops are also faring worse than average, as in the regional airliner market, and second-hand machines of this type are languishing in showrooms in higher than average numbers.

That said, Cessna and Bombardier/Learjet have been bold enough to launch two new business jets each in the past year and Gulfstream has followed with one. On closer examination, however, all but the Cessna Mustang prove to be adaptations of existing designs with little in the way of radical changes. Both Bombardier/Canadair and

Gulfstream took the opportunity of the September 2002 NBAA Convention to rename some, or all of their portfolio to present a more uniform line-up.

The Mustang is Cessna's move to prevent newcomers muscling in on the lightest end of the business jet market. Hitherto, financial constraints and — it must be admitted — design inexperience have thinned the attackers' ranks without the company spending a cent, but with challengers such as the Eclipse 500 now a flying reality and claiming 1,357 firm (non-refundable deposit) orders, the time for action is at hand. Not far behind Eclipse, thanks to a late 2002 intervention by Swiss backers, is Safire, whose S-26 is due to fly in early 2004, powered by a pair of Williams FJ33 small turbofans, and with 722 deposits banked by early 2003. Eclipse flew just once on Williams EJ22 power before rejecting the engine as unsuitable for its needs, having previously signed an exclusive supplier agreement with Williams.

Accordingly, certification and availability of the Eclipse will be in 2006, when its rivals also expect to be opening shop. Cessna's slightly bigger Mustang will launch Pratt & Whitney's PW600 small turbofan range and cost US$2.6 million, compared to just over US$1 million for the S-26 and US$950,000 for early-production Eclipses. All are six-seaters, with two on the flight deck and a 'club-four' cabin. Recently, Diamond of Austria has revealed plans to enter this market, while those with a bent for do-it-yourself can obtain a comparable Aerocomp kit.

Some in the industry shun these 'ultra light' or personal jets, regarding them as outside the realm of genuine business transports. With Cessna taking an interest, and estimates of a market for 8,000 such machines over the next 10 to 15 years, that may have to change. While Cessna presumes many Mustang owners will have traded up from turboprops, none denies that the latter are correctly counted in the business category. Newcomers this year include the Extra 500, representing the logical addition of a PT6 turboprop to the Extra 400; and Adam 500 production version of the M-309, which also promises a light jet derivative.

Nor, too, should the advance of Diesel be forgotten. Here, Europe has stolen a march on America, with Diesel Air, SMA, Thielert and Wilksch engines available, among others, for lightplane power. Socata has been strangely reluctant to launch its long-promised range of Diesel lightplanes, but Diamond has stepped into the breach with the latest variant of DA 40 Star and a new twin, the recently flown DA 42 Twin Star. Toyota is rumoured to be developing a Diesel-powered aircraft engine, even though the enigmatic TAA-1 trials aircraft, which Burt Rutan's Scaled Composites company built for it last year, is conventionally powered.

In parallel with new piston engine technology, research is known to be proceeding on new designs of propeller including, by the University of Miami, one with blades kinked forward from half-span outwards to increase efficiency and reduce noise. Trials by NASA have also demonstrated the advantages of a helicopter blade with a variable-droop leading edge, equivalent to 25 per cent chord, to reduce drag and pitching moment coefficients, leading to greater speed, and less vibration.

It was such fundamental considerations which occupied the methodical brothers, Orville and Wilbur Wright, as they prepared for their historic flight exactly a century ago, and it is a lesson to us all that we still have not perfected these basics in the intervening period. To mark the anniversary, we have added a special First Flights section to this edition covering the period since 17 December 1903 and mentioning at least some of the aircraft which have made a special contribution to the inexorable progress of aviation. Even those which were failures (and who has not seen the multi-winged Gerhardt Cycloplane dramatically collapse in a heap in front of the newsreel camera?) contributed knowledge — if only of what was the wrong way to do it. Readers will, doubtless, offer alternative candidates for this Hangar of Fame, but to quote the first rule of newspaper competitions, "The (Deputy) Editor's decision is final".

The second century of heavier than air, powered flight begins with expectations

Aviation history has been less than kind to the self-collapsing, man-powered Gerhardt Cycloplane, but even the most public of last century's failures has contributed, in a perverse way, to the fund of aeronautical knowledge   NEW/0543406

lowered by financial uncertainty and conflict. However, history shows that the world of aerospace has always been at its best when it was fully briefed and prepared for the next opportunity — or, to quote Mr Micawber once again, standing by "in case anything turned up".

Paul Jackson
Pulham Market, Norfolk, March 2003

Paul Jackson, MRAeS

Editor-in-Chief,
Jane's All the World's Aircraft

The chance gift of an aircraft book at the age of six was the Genesis of a passion for all matters aeronautical that has absorbed Paul Jackson for almost half a century. Editor of an aviation society newsletter in his native Yorkshire shortly after leaving school, he contributed to amateur and professional aviation journals, on both historical and current matters, for a further decade before becoming a freelance aerospace writer and editor. In 1987, he was invited to join the compiling team of *Jane's All the World's Aircraft*, being appointed its editor in 1994. His hobbies are flying and maintaining a veteran lightplane and aviation photography.

Notes and Acknowledgements

Following an unprecedented number of detail changes in the format of last year's *All the World's Aircraft*, the current edition confines itself to but one change of any import, namely the addition of a new appendix containing details of some of the world's propeller manufacturers and their designation systems. As is so often the case, the Genesis of this small section lay in a desire to add a measure of standardisation to the POWER PLANT paragraphs of certain aircraft descriptions, but it soon became apparent that no one source existed — within or without Jane's — to provide a baseline. Thereby did this appendix argue itself into existence.

Despite what we have recorded above in regard to continued advances in propeller technology, the airscrew is still, at one end, a simple device well within the capabilities of a cottage industry, and there are, undoubtedly, manufacturers whose efforts we have neglected through ignorance, rather than indifference. As with all other sections of this work, it is inevitable that the propellers' section will grow in the coming years.

As has become traditional, the First Flights and Aerospace Calendar continue to record milestones of the previous twelve months with the intention of providing a permanent catalogue of aviation development as each year is added. A little late re-editing has been necessary here, for some manufacturers appear to have adopted the spirit of the Wright Brothers too literally and announced 'first flights' which are no more than the usual short 'hop' which traditionally comes between fast taxying runs and (at least) the first airborne circuit of the airfield. These traps for the unwary occupy the same category as 'official' first flights which are well publicised events coming shortly after a furtive sortie, the date of which is difficult to obtain, to ensure all will be well for the prearranged day. Future historians beware.

Readers will also note that the order of entries in the US section has been changed following the announcement by Raytheon Aircraft Company that it is to give greater prominence to its Hawker and Beech product lines. We are delighted to restore the title of the Beech company and its 'Beechcraft' trading name to its rightful position in the 'Bs' and to insert Hawker a little further along, even though it has only the most tenuous connection with the maker of Harts, Hurricanes and Hunters.

We have also been taken to task for our sloth in reflecting some mergers and new associations. A case in point is AgustaWestland, which some view as a single company deserving of one entry in the International section containing all its products. However, there are some loose ends here, in the form of Bell/Agusta (which seems to be nothing to do with Westland) and EHI (logically subsumed by the larger identity, but still extant). Our treatment of EADS is not dissimilar, for it is clear that the melting pot still has not reached the critical temperature and the company's components are still describing themselves as 'EADS hyphen something'.

It may be prudent to give advance notice of some changes which are, even now, being incorporated in the on-line and CD-ROM versions of *All the World's Aircraft* and will be seen in the book from 2004. To ease research by our 'electronic' readers, we shall no longer be publishing short 'obituaries' of companies and aircraft withdrawn from the current edition, nor making referrals to the edition containing the last complete entry. Instead, the final mention will comprise the full entry appended with an explanation of the reason for the forthcoming move to ''online only''. It will also be necessary to delete some of our 'signpost' entries for aircraft or companies which do not appear in the place which the reader might suppose. These only serve to confuse the electronic search process, so we ask our 'paper' readers to turn to the index as a matter of prime recourse if something can not be found in what might appear to be its proper place.

Attending to these matters, as always, is the editorial team of Susan Bushell, Mike Jerram, Kenneth Munson and Lindsay Peacock, without whose support the immense task of updating this book would not be possible. James Goulding has supplied new and updated general arrangement drawings, with Mike Badrocke providing most of the cutaways. Thanks go, too, to Bill Gunston for the aero-engines section and glossary, and to Maurice Allward who compiles the index pages at the extreme rear. The last-mentioned task has been lightened this edition with the expiry of the ten-year indexes for private aircraft, sailplanes and hang gliders sections, which were reluctantly removed a decade ago.

Colleagues have been instrumental in the compilation of an edition containing 47 new companies and 75 new aircraft, bringing the total contents to 565 and 965, respectively after allowance for the inevitable deletions. Of those entries carried forward from the previous edition, 96 per cent of aircraft (855 of 890) and 86 per cent of companies (446 of 518) have been updated with new information — and that does not include outdated text removed or new illustrations added without a change of text.

The main body of the book contains 1,794 illustrations, of which 957 show complete aircraft, 72 per cent of them new in the past 12 months. Of the others, there are 198 close-ups of notable features, 52 artist's impressions (now almost exclusively computer-generated), 18 cutaways, 91 diagrams and 478 general arrangement drawings (including 61 new and 36 modified since the last edition).

Outside help from aviation colleagues, both amateur and professional, is always welcome. Particular thanks for illustrative material and supporting text has been well earned by Yefim Gordon, Robert Hewson and Ryszard Jaxa-Malachowski. Additionally, Peter Cooper, Kensuke Ebata, John Fricker, Jean-Louis Gaynecoetche, Arnold Nayler, Rod Simpson, Jim Thorn (Australian Aviation), Mike Vines (Photo Link) and Sebastian Zacharias have all shown long-term support.

On the inside, managing editor Simon Michell is kept busy smoothing the many bumps which would otherwise impede our progress; while Diana Burns and Tracy Johnson, our database editors, cheerfully keep on top of the mountain of update material which descends upon their desks each month of the year. The production process is managed by Jo Agius, Neil Grace and Sharon Marshall and tangible pages conjured out of thin, electronic air by Lynette Murphy and Hayley Austin. The results of their labours beckon.

Type Classifications

To assist comparison between different makes of aircraft undertaking broadly similar roles, the 'TYPE' classifications which introduce each aircraft description are standardised and arranged into 13 classes.

Of these, seven are of a sufficiently interrelated nature to benefit from cross-comparison as one group. Accordingly, the contents of Classes 4 to 10 are presented in a single table, in which approximate equivalence between classes is indicated by an entry on the same line, but in a different column. For example, a military strategic transport is in the same capacity bracket as a civil outsize freighter; or the prospective purchaser of a business jet may wish to be reminded that a business jet amphibian is also available. Large and high-performance aircraft are in the table's upper left areas; single-seat lightplanes in the lower right. A few lightplanes are optimised for easy conversion to ultralight category.

Classification is according to *Jane's* own criteria; in some instances this may differ from a manufacturer's description. Exact capacities and engine types and numbers for large aircraft are omitted in these shorthand descriptions. However, as size diminishes, such aspects assume greater importance and, accordingly, are quoted.

Except where the type description immediately and obviously precludes it (for example, 'Strategic transport'), aircraft are assumed to be monoplanes powered by a single piston engine and propeller. Multiple engines, turboprop and jet propulsion are specifically mentioned. In deference to common usage, 'jet' includes turbojets and turbofans, as more properly described in the 'POWER PLANT' paragraph.

All aircraft are Class A (or equivalent) certified/certifiable and factory-assembled; where clarification is necessary, the term 'lightplane' is employed. Ultralights and/or aircraft built from kits are specifically noted as such. However, in view of widely differing national legislation on ultralights (from prohibited, up to 540 kg; 1,190 lb maximum take-off weight) it must be noted that the term has several interpretations; the most liberal is used in this book.

The more versatile makes of aircraft cannot be given justice in deliberately short type descriptions. Readers should be aware that the ability to undertake surveillance, geological survey, VIP transport, water bombing and many other duties can be easily bestowed on medium transports and some other types of aircraft. Crop sprayers can be reconfigured in moments for firefighting and oil pollution dispersal. The full story will be found under the 'CURRENT VERSIONS' heading.

As an aid to analysis, totals for each category are given in parentheses.

Class 1: Bomber and surveillance (25)
These are military or paramilitary aircraft of widely differing size and performance.

Embraer R-99A   *NEW*/0143348

Strategic bomber (1)
Tupolev Tu-160 (Russian Federation)
Maritime reconnaissance four-jet (2)
BAE Systems Nimrod MRA. Mk 4 (United Kingdom)
Kawasaki MP-X (Japan)
Maritime surveillance twin-jet (2)
Airbus MPA320 (International)
Boeing 737 MMA (USA)
Maritime surveillance twin-turboprop (4)
Airtech CN-235 MP Persuader and CN-235 MPA (International)
ATR 42 Surveyor (International)
CASA C-212 Patrullero (Spain)
PZL (Antonov) M-28 Bryza (Poland)
Airborne early-warning and control system (3)
Airbus A310 AEW&C (International)
Boeing 737 AEW&C (USA)
Northrop Grumman E-2C Hawkeye (USA)
Airborne ground surveillance system (3)
Airbus A321 AGS (International)
Boeing 767 Military Versions (USA)
Northrop Grumman E-8 Joint STARS (USA)
Multisensor surveillance twin-turboprop (4)
BNG BN2T-4S Defender 4000 (United Kingdom)
DHC-8 Dash 8M and Triton (Canada)
Ilyushin Il-140 (Russian Federation)
Vulcanair P.68 Observer (Italy)

Multisensor surveillance turboprop (1)
Pilatus PC-12M Eagle (Switzerland)
Multisensor surveillance lightplane (5)
Creative Flight MPA (Canada)
Diamond MPX (Austria)
SAI G97V Spotter (Italy)
Schweizer SA 2-37 (USA)
Seabird SB7L-360 Seeker (Australia)

Class 2: Fighter and trainer (71)

CAC J-10   *NEW*/0524667

Air superiority fighter (4)
AIDC F-CK-1 Ching-Kuo (Taiwan)
SAC (Sukhoi Su-27) J-11 (China)
Sukhoi Su-27 (Russian Federation)
Sukhoi Su-30 (Su-27PU) (Russian Federation)
Multirole fighter (23)
ADA Light Combat Aircraft (LCA) (India)
BAE Systems Hawk 200 Series (UK)
Boeing F-15E Eagle (USA)
Boeing F/A-18E/F Super Hornet (USA)
CAC J-7 (China)
CAC J-10 (China)
Dassault Mirage 2000 (France)
Dassault Rafale (France)
Eurofighter Typhoon (International)
KAI F-X (Korea, South)
KAI (Lockheed Martin) F-16C/D Fighting Falcon (Korea, South)
Lockheed Martin F-16 Fighting Falcon (USA)
Lockheed Martin (645) F-22 Raptor (USA)
Lockheed Martin F-35 Joint Strike Fighter (USA)
MiG-29 (Russian Federation)
Saab JAS 39 Gripen (Sweden)
SAC J-8 II (China)
SAC F-8 IIM (China)
Sukhoi Su-30M (Russian Federation)
Sukhoi Su-33 (Su-27K) (Russian Federation)
Sukhoi Su-33UB (Su-27KUB) (Russian Federation)
Sukhoi Su-35 and Su-37 (Su-27M) (Russian Federation)
Sukhoi PACFA (Russian Federation)
Attack fighter (9)
AMX AMX (International)
CAC FC-1 (China)
HAL (SEPECAT) Jaguar International (India)
IAMI (HESA) Azarakhsh (Iran)
Mitsubishi F-2 (Japan)
Sukhoi Su-25 and Su-28 (Russian Federation)
Sukhoi Su-25TM (Su-39) (Russian Federation)
Sukhoi Su-27IB (Su-32 and Su-34) (Russian Federation)
XAC JH-7 (China)
Light fighter/advanced jet trainer (2)
Aero L 159 (Czech Republic)
EADS Mako (Germany)
Advanced jet trainer (5)
Boeing/BAE Systems T-45 Goshawk (International)
GAIC JJ-7 (China)
GAIC FTC-2000 (China)
IACI (SAHA) Safak (Iran)
Kawasaki T-4 (Japan)
Advanced jet trainer/light attack jet (8)
Aermacchi M-346 (Italy)
Aero L-39 MS and L-59 (Czech Republic)
Avioane IAR-99 Soim (Romania)
BAE Systems Hawk 50, 60 and 100 Series (United Kingdom)
HAIG L-15 (China)
KAI T-50 and A-50 Golden Eagle (Korea, South)
MiG-AT (Russian Federation)
Yakovlev Yak-130 (Russian Federation)
Basic jet trainer (3)
IAMI (HESA) Shahbaz (Iran)
IRIAF Tazarve (Iran)
ZRPSL Bielik and Fenix (Poland)

Basic jet trainer/light attack jet (5)
Aermacchi MB-339 (Italy)
Aero L-39 and L-139 Albatros (Czech Republic)
HAL HJT-36 (India)
LMAASA AT-63 Pampa NG (Argentina)
NAMC K-8 Karakorum 8 (China)
Basic turboprop trainer/attack lightplane (2)
Embraer EMB-314 Super Tucano (Brazil)
EADS PZL PZL-130 Orlik (Poland)
Basic turboprop trainer (6)
Beech T-6A Texan II (USA)
Fuji T-3Kai/T-7 (Japan)
KAI KT-1 Woong-Bee (Korea, South)
Pilatus PC-7 Mk II M Turbo Trainer (Switzerland)
Pilatus PC-9M Advanced Turbo Trainer (Switzerland)
Pilatus PC-21 (Switzerland)
Basic prop trainer/attack lightplane (1)
Aermacchi SF-260 (Italy)
Basic prop trainer (3)
ENAER (ECH-51) T-35 Pillán (Chile)
IAMI (HESA) Parastu (Iran)
Sukhoi Su-49 (Russian Federation)

Class 3: Miscellaneous and/or government (6)
Aircraft of diverse or multiple duties employed generally, but not exclusively, by the state.

Missile defence system (1)
Boeing AL-1A (USA)
Multirole twin-jet (1)
KAC (Raytheon) Hawker 800 (Japan)
High-altitude platform (2)
Scaled 281 Proteus (USA)
Scaled 318 (USA)
Technology demonstrator (2)
Aerocopter Humming (USA)
Boeing Bird of Prey (USA)
Ornithopter (1)
UTIAS 'Big Flapper' (Canada)

Class 4: Transport (33)

Sukhoi Su-80   *NEW*/0525056

Generally of a military nature, often with rear loading ramps. The larger aircraft are usually, but not exclusively, jet-powered. Light transports are those not exceeding 5,670 kg (12,500 lb).

Strategic transport (3)
Airbus A400M (International)
Antonov An-70 (Ukraine)
Boeing C-17A Globemaster III (USA)
Tanker-transport (2)
Airbus MultiRole Tanker-Transport (MRTT) (International)
Boeing 767 Military Versions (USA)
Medium transport/multirole (7)
Boeing Advanced Medium Transport (USA)
Ilyushin/IAPO/HAL Il-214 (International)
Kawasaki C-X (Japan)
Lockheed Martin 382U/V Hercules (USA)
Lockheed Martin Advanced Mobility Aircraft (AMA) (USA)
Multirole Transport Aircraft (International)
SAC Y-8 (China)
Twin-turboprop transport (17)
Airtech CN-235 (International)
Alenia/Lockheed Martin C-27J Spartan (Italy)
Antonov An-32 (Ukraine)
Antonov An-38 (Ukraine)
Be-132 (Russian Federation)
CASA C-212 Series 400 (Spain)
CASA C-295 (Spain)
CSIR Saras (India)
Dirgantara (CASA) NC-212-200 (Indonesia)
HAL (Dornier) 228 (India)
IAMI (HESA) Ir.an 140 (Iran)
Ilyushin Il-112 (Russian Federation)
LZ L 610 G (Czech Republic)
MiG-110 (Russian Federation)

TYPE CLASSIFICATIONS

PZL M-28 Skytruck (Poland)
Sukhoi Su-80 (Russian Federation)
TAI (Airtech) CN-235M (Turkey)
Twin-turboprop light transport (4)
AII AVA-404 (Iran)
Beriev Be-132 (Russian Federation)
HAI Y-12 (China)
Skydesign SB100 Skylander (France)

Class 5: Airliner and freighter (63)

Antonov An-140 *NEW*/0525074

Civilian passenger and cargo aircraft. Jet power is implied for all large airliners; number of engines given only for medium-size aircraft; regional jets are all twins.

Outsize freighter (3)
Antonov An-124 (Ukraine)
Antonov An-225 Mriya (Ukraine)
Lockheed Martin Aeronaft (USA)
High-capacity airliner (1)
Airbus A380 (International)
Wide-bodied airliner (8)
Airbus A300-600 (International)
Airbus A310 (International)
Airbus A330 (International)
Airbus A340 (International)
Boeing 747 (USA)
Boeing 767 (USA)
Boeing 777 (USA)
Ilyushin Il-96-300 (Russian Federation)
New concept airliner (3)
Boeing Sonic Cruiser (USA)
Boeing Super Efficient Airliner (USA)
Boeing Blended Wing Body (USA)
Four-jet freighter (2)
Boeing BC-17X Globemaster (USA)
Boeing 747-400F (USA)
Tri-jet airliner (2)
Tupolev Tu-154M (Russian Federation)
Tupolev Tu-156 (Russian Federation)
Twin-jet transport (2)
Antonov An-72 and An-74 (Ukraine)
Antonov An-74-300 and An-174 (Ukraine)
Twin-jet airliner (12)
Airbus A318 (International)
Airbus A319 (International)
Airbus A320 (International)
Airbus A321 (International)
Boeing 717 (USA)
Boeing 737 (USA)
Boeing 757 (USA)
Boeing 757-300 (USA)
JADC YSX (Japan)
Tupolev Tu-204 and Tu-214 (Russian Federation)
Tupolev Tu-204K (Russian Federation)
Tupolev Tu-334, Tu-336 and Tu-354 (Russian Federation)
Twin-jet freighter (3)
Airbus A300-600 Convertible and A300-600 Freighter (International)
Boeing 767-300 General Market Freighter (USA)
Tupolev Tu-330 (Russian Federation)
Regional jet airliner (14)
Antonov An-148 (Ukraine)
Bombardier CRJ200 and Challenger 800 (Canada)
Bombardier CRJ700 (Canada)
Bombardier CRJ900 (Canada)
Embraer ERJ-145 (Brazil)
Embraer ERJ-135 (Brazil)
Embraer ERJ-140 (Brazil)
Embraer 170 and 190 (Brazil)
Fairchild Dornier 328JET (Germany)
Fairchild Dornier 528, 728 and 928 (Germany)
Myasishchev M-70 (Russian Federation)
Sukhoi RRJ (Russian Federation)
Tupolev Tu-324 (Russian Federation)
Tupolev Tu-414 (Russian Federation)
Twin-turboprop airliner (11)
Antonov An-140 (Russian Federation)
ATR 42 (International)
ATR 72 (International)
Beech 1900D (USA)
DHC-8 Dash 8 Q100 and Q200 (Canada)
DHC-8 Dash 8 Q300 (Canada)
DHC-8 Dash 8 Q400 (Canada)

Dirgantara N-250 (Indonesia)
Embraer EMB-120 Brasilia (Brazil)
Ilyushin Il-114 (Russian Federation)
XAC MA-60 (China)
Twin-turboprop freighter (2)
AUC FF-1080-200 Freight Feeder (USA)
IAI C-5WA Airtruck (Israel)

Class 6: Business (57)

Canadair Challenger 604 *NEW*/0137420

The established configuration of a business jet being a twin, only single- and tri-jet configurations are specifically noted. Detailed strata devised by participants in the business jet market are not reproduced here. Instead, the main category is subdivided into small business jets with accommodation for six or fewer passengers; long-range (sub-airliner size) business jets with the 4,000 n mile (7,400 km; 4,600 mile) range necessary to fly from the US eastern seaboard to Western Europe; and the central core of twin-jets with upwards of seven seats.

Supersonic business jet (1)
Gulfstream SBJ (USA)
Long-range business jet (4)
Bombardier BD-700 Global Express (Canada)
Bombardier Global 5000 (Canada)
Gulfstream IV, G300 and G400 (USA)
Gulfstream V, G500 and G550 (USA)
Large business jet (2)
Boeing Business Jet (BBJ) (USA)
Boeing Business Jet 2 (BBJ 2) (USA)
Long-range business tri-jet (2)
Dassault Falcon 900 (France)
Dassault Falcon 7X (France)
Business tri-jet (1)
Dassault Falcon 50 (France)
Business jet (20)
Aerostar FJ-100 (USA)
Beech Beechjet 400A (USA)
Bombardier BD-100 Challenger 300 (Canada)
Canadair CL-600 Challenger (Canada)
Cessna 550 Citation Bravo (USA)
Cessna 560 Citation Encore (USA)
Cessna 560XL Citation Excel (USA)
Cessna 680 Citation Sovereign (USA)
Cessna 750 Citation X (USA)
Dassault Falcon 2000 (France)
Embraer Legacy (Brazil)
Hawker 800 (USA)
Hawker Horizon (USA)
Hawker 450 (USA)
Gulfstream G100 (USA)
Gulfstream G200 (USA)
Learjet 31 (USA)
Learjet 40 (USA)
Learjet 45 (USA)
Learjet 60 (USA)
Light business jet (11)
Adam A700 (USA)
Ameur B1 Fanjet (France)
Beech Premier (USA)
Cessna Citation Mustang (USA)
Cessna 525 Citation CJ1 (USA)
Cessna 525A Citation CJ2 (USA)
Cessna 525B Citation CJ3 (USA)
CMC Leopard (United Kingdom)
Eclipse Eclipse 500 (USA)
Safire S-26 (USA)
Sino Swearingen SJ30-2 (USA)
Light business jet/kitbuilt (1)
Maverick Leader (USA)
Business mono-jet (1)
Visionaire VA-10 Vantage (USA)
Business mono-jet kitbuilt (1)
Aerocomp CA-J Comp Air Jet (USA)
Business twin-turboprop (2)
Beech King Air C90B (USA)
Piaggio P.180 Avanti (Italy)
Aerobatic business turboprop (1)
Grob G 140TP (Germany)
Business turboprop (5)
Extra 500 (Germany)
FACL F1 (United Kingdom)
Pilatus PC-12 (Switzerland)

Piper PA-46-500TP Malibu Meridian (USA)
Socata TBM 700 (France)
Business turboprop kitbuilt (1)
Comp Air 12 (USA)
Business twin-prop (2)
Adam A500 CarbonAero (USA)
High Performance TT62 (Germany)
Business prop (2)
Extra 400 (Germany)
Piper PA-46-350P Malibu Mirage (USA)

Class 7: Utility (70)

Air Tractor AT-502B *NEW*/0121677

Restricted to single- and twin-engine passenger or passenger/light freight aircraft with accommodation, typically, for four or six persons. An intermediate category bridging commercial and private ownership. Crop sprayers are included.

Utility turboprop twin (7)
Atlas Liftmaster (USA)
AvtekAir 9000T (USA)
Beech King Air B200 (USA)
Beech King Air 350 (USA)
Ilyushin Il-100 (Russian Federation)
Piaggio P.166 (Italy)
Reims F406 Caravan II (France)
Light utility twin-prop transport (8)
Avia Accord-201 (Russia)
BNG BN2B Islander (United Kingdom)
Romaero (BNG) Islander and Defender (Romania)
TAAL (Partenavia) P.68 Series (India)
TKEF 44 Angel (USA)
VulcanAir P.68 (Italy)
VulcanAir VA 300 (Italy)
Wolfsberg-Letov Raven 257 (Czech Republic)
Six-seat utility twin (3)
Beech Baron 58 (USA)
Neiva EMB-810D (Brazil)
Piper PA-34-220T Seneca V (USA)
Four-seat utility twin (4)
Aviaton Merkury (Russian Federation)
Diamond Twin Star (Austria)
Dubna-4 (Russian Federation)
Piper PA-44-180 Seminole (USA)
Light utility turboprop (10)
Aerocourier Aerocourier (USA)
Aeroprogress/ROKS-Aero T-101 Grach (Russian Federation)
Explorer Explorer (USA)
Ibis Ae 270 Ibis (International)
Intracom GM-17 Viper (International)
Khrunichev T-507 Skvorets (Russian Federation)
Myasishchev M-101T Gzhel (Russian Federation)
PAC 750XL (New Zealand)
Technoavia SMG-92 Turbo Finist (Russian Federation)
VulcanAir VF 600W (Italy)
Light utility transport (5)
El Gavilán EL-1 Gavilán 358 and 508T (Colombia)
Gippsland GA-8 Airvan (Australia)
Sherpa Sherpa (USA)
Technoavia SM-92 Finist (Russian Federation)
Zlin Z 400 Rhino (Czech Republic)
Utility biplane (1)
SAMC Y-5 (China)
Six-seat utility transport (7)
Beech Bonanza A36 (USA)
Beech Bonanza B36TC (USA)
Cessna 206 Stationair and T206 Turbo Stationair (USA)
Found FBA-2C1 Bush Hawk (Canada)
Khrunichev T-415 Snegir (Russian Federation)
Piper PA-32R-301 Saratoga II HP (USA)
Piper PA-32R-301T Saratoga II TC (USA)
Agricultural sprayer (23)
Agrolot PZL-126P Mrówka 2001 (Poland)

Air Tractor AT-401 Air Tractor (USA)
Air Tractor AT-402 Turbo Air Tractor (USA)
Air Tractor AT-502 Turbo (USA)
Air Tractor AT-602 (USA)
Air Tractor AT-802 (USA)
Ayres Turbo-Thrush S2R (USA)
Ayres 660 Turbo-Thrush (USA)
CZAW (Zenair) CH 701-AG STOL (Czech Republic)
EADS PZL PZL-106BT Turbo-Kruk (Poland)
Gippsland GA-200 Fatman (Australia)
Khrunichev T-419 Strezoka (Russian Federation)
MVEN-1 Fermer (Russian Federation)
NEIVA EMB-202 Ipanema (Brazil)
PAC Cresco (New Zealand)
PZL M-18 Dromader (Poland)
Saviat E-4 (Russian Federation)
Sukhoi Su-38L (Russian Federattion)
TAI ZIU (Turkey)
Unikomtranso Don (Russian Federation)
Vitek Ezhik (Russian Federation)
Weatherly 620 (USA)
Yakovlev Yak-SKh (Russian Federation)
Agricultural sprayer ultralight (1)
Aerolites Aeromaster (USA)

Class 8: Amphibian (16)

ShinMaywa US-1A   *NEW*/0120432

All such aircraft (including the occasional flying boat) are included here for ease of reference, even those assembled from kits or classified as ultralights.

Amphibious airliner (1)
Beriev Be-210 (Russian Federation)
Four-turboprop amphibian (1)
ShinMaywa US-1A (Japan)
Twin-turboprop amphibian (1)
Canadair CL-415 Superscooper (Canada)
Twin-jet amphibian (1)
Beriev (Betair) Be-200 (Russian Federation)
Six-seat utility twin-prop amphibian (1)
Beriev Be-103 (Russian Federation)
Six seat amphibian (2)
Aero-Volga L-6 (Russian Federation)
Warrior Centaur (United Kingdom)
Six-seat amphibian/kitbuilt (1)
Seastar Seastar (USA)
Four-seat amphibian (2)
Aquastar TA16 Seafire (USA)
Chernov Che-25 (Russian Federation)
Two-/four-seat amphibian kitbuilt (1)
SG Sea Storm (Italy)
Three-seat amphibian/kitbuilt (1)
Gidroplan Che-22 Korvet (Russian Federation)
Three-seat amphibian kitbuilt (1)
Aeroprakt-24 (Ukraine)
Two-seat amphibian kitbuilt (1)
Quikkit Glass Goose (USA)
Two-seat amphibian/kitbuilt (1)
EDRA Helicentro Paturi (Brazil)
Two-seat amphibian ultralight kitbuilt (1)
Aerolites Aeroskiff (USA)

Class 9: Lightplane (factory built) (126)

Extra 300L   *NEW*/0132416

Single-prop aircraft seating up to five persons and available only in complete form, unless otherwise stated.

Two-seat jet sportplane (2)
ATG-1 Javelin (USA)
Messerschmitt Me 262 (USA)
Five-seat lightplane/turboprop (1)
Maule M-7 Series (USA)
Five-seat lightplane (1)
NLA AC-500 Aircar (China)
Four-seat lightplane (35)

Cessna 172 Skyhawk (USA)
Cessna 182 Skylane (USA)
Cessna T-182T Turbo Skylane (USA)
Cirrus SR20 (USA)
Cirrus SR22 (USA)
Commander 115 (USA)
Diamond DA 40-TDI Star (Austria)
Diamond DA 40-180 Star (Canada)
EADS PZL PZL-104M Wilga 2000 (Poland)
EADS PZL PZL-110 Koliber 160 (Poland)
EADS PZL PZL-111 Koliber 235A (Poland)
FACI F-3 (Iran)
FFA Bravo 3000 (Switzerland)
III F.300 (Italy)
IL I-23 Manager (Poland)
Ilyushin Il-103 (Russian Federation)
Lancair Columbia (USA)
Luscombe 185-11E (USA)
Mooney M20M Bravo (USA)
Mooney M20R Ovation 2 and M20S Eagle (USA)
Mylius MY-104/200 (Germany)
Piper PA-28-161 Warrior III (USA)
Piper PA-28-181 Archer III (USA)
Piper PA-28-201 Arrow (USA)
Robin DR 400 Dauphin (France)
Robin DR 400/160 Major (France)
Robin DR 400/180 Régent (France)
Robin DR 500 Président (France)
Scaled TAA (USA)
Socata TB 9 Tampico and TB10 and TB 200 Tobago (France)
Socata TB 20 and TB 21 Trinidad (France)
Solaris Sigma and Alpha (USA)
Tiger AG-5B Tiger (USA)
Yakovlev Yak-18T (Russian Federation)
Zlin Z 143 (Czech Republic)
Three-seat lightplane (1)
PAC (AMF) Mushshak and Super Mushshak (Pakistan)
Three-seat biplane (1)
Waco Classic YMF Super (USA)
Primary prop trainer/sportplane (12)
Aerostar (Yakovlev) Iak-52 (Romania)
AII AVA-202 (Iran)
BHEL LT-IIM Swati (India)
EADS PZL PZL-112 Koliber Junior (Poland)
III D-130 and D-200 (Italy)
NAL/TAAL Hansa-3 (India)
PAC Airtrainer CT-4E (New Zealand)
PZL M-26 Iskierka (Poland)
Robin HR 200 (France)
Robin R 2120 Alpha (France)
Slingsby T67 Firefly (United Kingdom)
Zlin Z 242 L (Czech Republic)
Aerobatic two-seat turboprop sportplane (1)
Endeavour Phoenix (Australia)
Aerobatic two-seat sportplane (10)
CAP 10 (France)
CAP 222 (France)
Extra 200 (Germany)
Extra 300 and 330 (Germany)
Kovaks K-51 Peregrino (Brazil)
Mylius MY-103 (Germany)
Staudacher S-300 (USA)
Sukhoi Su-29 (Russian Federation)
Yakovlev Yak-54 (Russian Federation)
Zivko Edge 540T (USA)
Aerobatic two-seat sportplane/kitbuilt (2)
Dyn'Aero CR100, CR110 and CR120 (France)
MSW Votec 322 and 332 (Switzerland)
Aerobatic two-seat lightplane (4)
Aeromot AMT-600 Guri (Brazil)
Grob G 115 (Germany)
Grob G 120A (Germany)
Micco SP20 and SP26 (USA)
Two-seat lightplane (38)
3Xtrim 550 (Poland)
AMD Zenith CH 2000 Alarus (USA)
American Champion 7ACA Champ (USA)
American Champion 7ECA Citabria Aurora (USA)
American Champion 7GCAA Citabria Adventure (USA)
American Champion 7GCBC Citabria Explorer (USA)
American Champion 8GCBC Scout (USA)
American Champion 8KCAB Super Decathlon (USA)
Aquila A210 (Germany)
Association Zephyr Alizé (France)
Aviat Husky A-1 (USA)
Aviat Husky Pup (USA)
Aviatika-MAI-910 Interfly (Russian Federation)
BKLAIC AD-200 (China)
Cub Crafters PA-18-180 Top Cub (USA)
Diamond DA 20-A1 Katana (Canada)
Diamond DA 20-C1 Katana (Canada)
Dorna D139-PT1 Blue Bird (Iran)
Dubna-2 (Russian Federation)
Eagle Eagle 150 (Australia)
Elitar IE-201 Senator (Russia)

EV-AT Harmony (Czech Republic)
Flight Design CT (Germany)
HK Wega (Germany)
IAR IAR-46 (Romania)
III Sky Arrow (Italy)
III Speed Arrow (Italy)
Impulse Impulse 100 (Germany)
Issoire APM-20 Lionceau and APM-21 Lion (France)
Jurca MJ5 Sirocco (France)
Khrunichev T-21 Sapsan (Russian Federation)
Kinetic Mountain Goat (USA)
Liberty XL-2 (USA)
MAI-217 (Russia)
OMF-100-160 Symphony (Germany)
Renaissance 8F (USA)
Robin R 2160 (France)
SSVOBB Impuls (Netherlands)
Motor glider (7)
ABS RF-9 (Switzerland)
Aeromot Ximango (Brazil)
Diamond HK 36 (Austria)
Scheibe SF 25C Falke (Germany)
Stemme S 10 and S 15 (Germany)
Stemme S 6 and S 8 (Germany)
Technoflug TFK-2 Carat (Germany)
Aerobatic two-seat biplane (3)
Aviat Pitts S-2C (USA)
Kaiser Magic (Germany)
SSH (Bücker) T-131 Jungmann (Poland)
Aerobatic single-seat sportplane (5)
CAP 232 (France)
Sukhoi Su-31 (Russia)
Technoavia SP-55M (Russia)
Yakovlev Yak-3M and Yak-9U-M (Russian Federation)
Zivko Edge 540
Aerobatic single-seat biplane (1)
SSH (Bücker) Jungmeister (Poland)
Glider tug (2)
Robin DR 400/180R Remo 180 (France)
Robin DR 400/200R Remo 200 (France)

Class 10: Utility kitbuilt (39)

Leza AirCam   *NEW*/0533700

This class reflects the growing number of kitbuilt aircraft intended for more than recreational use. The largest has 10 seats and turboprop power.

Two-seat jet sportplane kitbuilt (1)
Viper ViperJet (USA)
Two-seat kitbuilt twin (4)
Leza Air Cam (USA)
SlipStream SkyBlaster (USA)
SlipStream SkyQuest (USA)
VSTOL SST 2000 Pairadigm (USA)
Utility kitbuilt (4)
Aerocomp Comp Air 7 (USA)
Aerocomp Comp Air 8 (USA)
Aerocomp Comp Air 10 (USA)
Khrunichev T-411 Aist (Russian Federation)
Utility biplane kitbuilt (1)
Griffon Lionheart (USA)
Six-seat kitbuilt (5)
ADI Super Stallion (USA)
Aerocomp Comp Air 6 (USA)
Aviabellanca Skyrocket III (USA)
Barr BarrSix (USA)
Murphy Super Rebel and Moose (Canada)
Four-seat kitbuilt twin (2)
Aeroprakt-28 (Ukraine)
Hensley H-1 Wolf (USA)
Four-seat kitbuilt (19)
Aerocomp Comp Air 4 (USA)
AMD Zodiac CH 640 (USA)
Aviation Development Alaskan Bushmaster (USA)
CLASS Bush Caddy (Canada)
Culp's Monoculp (USA)
DFL Foxtrot-4 (USA)
Dyn'Aero MCR4S (France)
Express Express (USA)
Four Winds 192, 210 and 250T (USA)
Jabiru J200 and J400 (Australia)
Lambert Mission M212 (Belgium)
Lancair Lancair IV, IVP and Sentry (USA)
Lancair Lancair ES and Super ES (USA)
Pottier P 200 Series (France)

Pulsar Cruiser and Super Cruiser (USA)
St Just Super Cyclone (Canada)
Ullman Panther (USA)
Wag-Aero 2+2 Sportsman (USA)
Zenith Super STOL CH 801 (USA)

Three-seat kitbuilt (2)
Aerocomp Comp Air 3 (USA)
Murphy Rebel (Canada)

Aerobatic three-seat biplane/kitbuilt (1)
Culp's Special (USA)
Bel-Aire 2000 (USA)

Class 11: Kitplanes and/or ultralights (190)

Interplane Skyboy   *NEW*/0137214

One- or two-seat, single-prop machines intended for private ownership. 'Kitbuilt' aircraft are in the lightplane category, unless described as ultralights (and flyable as such in at least some countries). Some aircraft are available only in complete form, while kit manufacturers may offer the option of fly-away aircraft to those requiring the aeroplane, but not the assembly work. These are indicated as follows:

Single-seat kitbuilt (1)
Fisher Avenger (USA)

Single-seat sportplane kitbuilt (4)
Bede BD-17 Nugget (USA)
Flug Werk Fw 190 (Germany)
Mini-Imp Mini-Imp (USA)
TBG Thunder Mustang (USA)

Single-seat ultralight/motor glider (1)
Noin Exel (France)

Single-seat ultralight (1)
Kolb Firefly (USA)

Single-seat ultralight kitbuilt (10)
Aerolites Bearcat (USA)
Avid Champion and Buzz (USA)
Blue Yonder E-Z King Cobra (Canada)
Carlson Skycycle (USA)
Flightstar Spyder and Formula (USA)
Flying K Sky Raider (USA)
Fly Synthesis Wallaby (Italy)
Joplin ½ Tun (USA)
Murphy JDM-8 (Canada)
SupaPup SupaPup Mk 4 (Australia)

Single-seat ultralight/kitbuilt (5)
Interplane Griffon 103 (Czech Republic)
Rans S-17 Stinger (USA)
Sapphire LSA Mk II (Australia)
Silence Silence (Germany)
Skystar Kitfox Lite (USA)

Single-seat ultralight biplane (1)
(MAI) Aviatika-MAI-890 (Russian Federation)

Single-seat ultralight biplane kitbuilt (1)
Fisher Youngster and Youngster V (USA)

Single-seat motor glider/kitbuilt (1)
TeST TST-9 Junior (Czech Republic)

Tandem-seat kitbuilt (10)
AeroKuhlmann SCUB (France)
American Homebuilts' Vaquero John Doe (USA)
Blue Yonder E-Z Flyer (Canada)
Carlson CA6 Criquet (USA)
Custom Flight North Star (Canada)
Partenair S45 Mystère (Canada)
Rans S-7 Courier (USA)
Van's RV-4 (USA)
Van's RV-8 and RV-8A (USA)
Wag-Aero Sport Trainer (USA)

Tandem-seat lightplane/kitbuilt (1)
Storch Storch SS Mk 4 (Australia)

Tandem-seat sportplane kitbuilt (9)
Aceair Aeriks A-200 (Switzerland)
HB Flugtechnik HB 204 Tornado (Austria)
Kimball McCulloccoupe (USA)
Kimball Model 15 Raptor A S (USA)
Legend Legend (USA)
Super-Chipmunk Super-Chipmunk (Canada)
Supermarine Spitfire (Australia)
Team Rocket F1 (USA)
Warner Sportster (USA)

Tandem-seat turboprop sportplane kitbuilt (2)
Legend Turbine Legend (USA)
Turbine Design TD-2 Tempest (USA)

Tandem-seat turboprop sportplane/kitbuilt (1)
Cameron P-51G (USA)

Tandem-seat ultralight (5)
DAR-21 Vector (Bulgaria)
III Sky Arrow (Italy)
Kolb Sling Shot (USA)
Let-Mont UL Piper (Czech Republic)
Sauper J-300 Series 3 (France)

Tandem-seat ultralight kitbuilt (8)
Flying K Sky Raider II (USA)
Joplin Tundra (USA)
Kolb Kolbra (USA)
Legend Lite Skywatch SS-11 (Canada)
Rans S-18 Stinger II (USA)
Reality Easy Raider (United Kingdom)
RMT 03 Bateleur (Germany)
US Light Aircraft Hornet (USA)

Tandem-seat ultralight/kitbuilt (6)
Aeroprakt-20 (Ukraine)
Aeroprakt-20R-912 Sky Cruiser (Ukraine)
CFM Shadow Series D and E (United Kingdom)
CFM Streak (United Kingdom)
CFM SA-II Star Streak (United Kingdom)
TL-232 Condor (Czech Republic)

Tandem-seat biplane kitbuilt (2)
Aviat (Christen) Eagle II (USA)
Culp's Sopwith Pup (USA)

Tandem-seat biplane/kitbuilt (1)
Kimball (Pitts) Model 12 (USA)

Tandem-seat ultralight biplane (1)
B&F FK 12 Comet (Germany)

Tandem-seat ultralight biplane kitbuilt (3)
Fisher R-80 and RS-80 Tiger Moth (USA)
Murphy Renegade II (Canada)
Murphy Renegade Spirit (Canada)

Tandem-seat ultralight twin/kitbuilt (1)
Aeroprakt-26 (Ukraine)

Side-by-side kitbuilt (23)
Aerocomp Merlin GT (USA)
AirStart Griffin (Canada)
Avid Magnum (USA)
DFL Tango-2 (USA)
Dyn'Aero MCR01 Ban Bi (France)
Glasair Super II (USA)
Glasair III (USA)
GlaStar GlaStar (USA)
HB Flugtechnik HB 207 Alfa (Austria)
Lancair Legacy (USA)
NuVenture Venture (USA)
Paxman's Northern Lite Viper P3/A (Canada)
Prosport Freebird Classic (USA)
Pulsar Super Pulsar 100 (USA)
Pulsar Sport 150 (USA)
Rans S-16 Shekari (USA)
Skystar Kitfox (USA)
Ultravia Pelican (Canada)
Van's RV-7 and RV-7A (USA)
Van's RV-9 and RV-9A (USA)
Venture Thorp T-211 (USA)
Wag-Aero Wag-A-Bond
Zenith Super Zodiac CH 601XL (USA)

Side-by-side sportplane/kitbuilt (1)
Team Rocket F2 (USA)

Side-by-side lightplane/kitbuilt (6)
Aero AT-3 (Poland)
Aeropract-27 (Russian Federation)
Goair GT-1 Trainer
Sequoia Falco F.8L (USA)
SG Storm (Italy)
Tecnam P92 Echo (Italy)

Side-by-side lightplane/motor glider kitbuilt (1)
Europa Europa (United Kingdom)

Side-by-side ultralight (31)
3Xtrim 450 (Poland)
Aeros UL-2000 Flamingo (Czech Republic)
Aerospool WT9 Dynamic (Slovak Republic)
Aerostyle Breezer (Germany)
ATEC Zephyr (Czech Republic)
B&F FK 9 Mark 3 (Germany)
B&F FK 14 Polaris (Germany)
Bussard Raptor (Germany)
BUL Zùlù (France)
DAR-23 Accent (Bulgaria)
Delta Pegass (Czech Republic)
Ekoflug MXP-740 (Slovak Republic)
Elitar IE-101 Elitar (Russian Federation)
Euro ALA Jet Fox 97 (Italy)
Flight Design CT (Germany)
Fly Synthesis Texan (Italy)
ICP Amigo (Italy)
ICP Savannah (Italy)
Ikar Ai-10 Ikar (Ukraine)
Impulse Impulse 100 (Germany)
Packo Larus (Czech Republic)
Pipistrel Virus (Slovenia)
PJB Vega (France)
PJB Spectra (France)
Remos G-3 Mirage (Germany)
Remos G-4 Velocity (Germany)
SAI G97 Spotter (Italy)
SG Rally (Italy)
Tecnam P96 Golf (Italy)

TL-96 Star and TL-200 Sting Carbon (Czech Republic)
USA SF 45 SA Spirit (Germany)

Side-by-side ultralight kitbuilt (24)
Alpi Pioneer 200 (Italy)
Alpi Pioneer 300 (Italy)
Avid Flyer Mark IV (USA)
Best Off Sky Ranger (France)
Dyn'Aero MCR01 Ban Bi (France)
Fisher Dakota Hawk (USA)
Flightstar IISL and IISC (USA)
Kaiser Whisper (Germany)
Murphy Maverick (Canada)
Noin Choucas (France)
PC Flight Pretty Flight (Germany)
Precision Tech Ferguson F-IIB Fergy (USA)
Raj Hamsa X-Air (India)
Raj Hamsa Hanuman (India)
Rans S-12XL Airaile and S-12S Super Airaile (USA)
Skygear Comet (Korea, South)
Skystar Kitfox Lite² (USA)
SlipStream Dragonfly (USA)
SlipStream Genesis and Revelation (USA)
SlipStream Scepter (USA)
Sonex Sonex (USA)
Vol Mediterrani VM-1 Esqual (Spain)
Zenith Zodiac CH 601HD (USA)
Zenith STOL CH 701 (USA)

Side-by-side ultralight/kitbuilt (25)
Aeroprakt-22 (Ukraine)
Aeropro Eurofox (Slovak Republic)
Aerostar 01 (Romania)
Ameur Altania (France)
EV-AT teamEurostar (Czech Republic)
Fantasy Air Allegro and Arius (Czech Republic)
Fläming Air FSU Saphir (Germany)
Fläming Air Trener Baby (Germany)
Fly Synthesis Storch (Italy)
Hughes Australian Light Wing GR-912 (Australia)
Hughes Australian Light Wing Sport 2000 (Australia)
Humbert Tétras (France)
Ikarus C 42 (Germany)
Interplane Skyboy (Czech Republic)
Jabiru Jabiru (Australia)
Kappa KP-2U Sova (Czech Republic)
Let-Mont UL Tulák (Czech Republic)
Skycraft ARV Super2 (USA)
TeST TST-5 Variant (Czech Republic)
TeST TST-6 Duo (Czech Republic)
ULBI Wild Thing (Germany)
Urban UFM-10 Samba (Czech Republic)
Urban UFM-13 Lambáda (Czech Republic)
WD D4 BK Fascination (Germany)
WD D5 Evolution (Germany)

Side-by-side ultralight biplane (1)
(MAI) Aviatika-MAI-890

Side-by-side ultralight/motor glider (2)
Noin Excel (France)
Pipistrel Sinus (Slovenia)

Side-by-side ultralight motor glider kitbuilt (1)
Sonex Xenos (USA)

Side-by-side ultralight/motor glider kitbuilt (1)
Tomair Cobra Arrow (Australia)

Side-by-side ultralight twin (1)
Airox F-31 (Czech Republic)

Class 12: Rotary wing (147)

Bell 427   *NEW*/0132964

Includes tiltrotors, autogyros and lifting platforms. Light utility helicopters are those not exceeding 5,000 kg (11,023 lb) maximum take-off weight.

Attack helicopter (15)
Agusta A 129 Mangusta (Italy)
Agusta A 129 International (Italy)
Bell 446, SuperCobra and King Cobra (USA)
Boeing AH-64 Apache (USA)
Boeing Sikorsky RAH-66 Comanche (USA)
Denel AH-2 Rooivalk (South Africa)
Eurocopter 665 Tiger/Tigre (International)
Fuji (Boeing) AH-64D (Japan)
HAI Z-10 (China)

HAL LAH (India)
IHSRC (PANHA) 2091 (Iran)
Kamov Ka-50 Chernaya Akula (Russian Federation)
Mil Mi-28 (Russian Federation)
TAI (Bell) AH-1Z KingCobra (Turkey)
Westland WAH-64D Apache Longbow (United Kingdom)

Armed observation helicopter (2)
HAL Lancer (India)
Kawasaki OH-1 (Japan)

Naval combat helicopter (5)
CHAIG Z-8 (China)
Kaman (K894) SH-2G Super Seasprite (USA)
Mitsubishi (Sikorsky) SH-60 (Japan)
PZL-Swidnik Sep (Poland)
Sikorsky S-70B (USA)

Light observation helicopter (1)
HAL Light Observation Helicopter (LOH) (India)

Heavy lift helicopter (1)
Mil Mi-26 (Russian Federation)

Medium lift helicopter (3)
Boeing 114 and 414 (USA)
Kawasaki (Boeing) CH-47J Chinook (Japan)

Light lift helicopter (1)
Kaman K-1200 K-MAX (USA)

Medium transport helicopter (7)
Euromil Mi-38 (International)
Kamov Ka-29 (Russian Federation)
Kamov Ka-60 Kasatka (Russian Federation)
Kazan/Mil Mi-17 (Russian Federation)
Mil Mi-38 (Russian Federation)
Mitsubishi (Sikorsky) UH-60 (Japan)
Sikorsky S/H-92 (USA)

Multirole medium helicopter (19)
Agusta-Bell 412 and Griffon (Italy)
Bell 412 (Canada)
Dirgantara (Eurocopter) NAS-332 Super Puma (Indonesia)
Dirgantara (Bell) NBell-412 (Indonesia)
EH Industries EH101 (International)
Eurocopter Super Puma Mk I and Cougar Mk I (International)
Eurocopter Super Puma Mk II and Cougar Mk II (International)
IHSRC (PANHA) Shabaviz 2-75 (Iran)
IRGC Transport Helicopter (Iran)
KAI (Sikorsky) UH-60P (Korea, South)
Kamov Ka-32 (Russian Federation)
Kamov Ka-62 (Russian Federation)
Mil Mi-17 (Mi-8M), Mi-19, Mi-171 and Mi-172 (Russian Federation)
NH Industries NH 90 (International)
PZL-Swidnik W-3 Sokół (Poland)
Sikorsky S-70A (USA)
Sikorsky S-70C (USA)
Sikorsky S-76C+ (USA)
TAI (Eurocopter) Cougar Mk 1 (Turkey)

Light utility helicopter (44)
AVA-505 Thunder (Iran)
Agusta A 109 (Italy)
Agusta A 119 Koala (Italy)
Bell 206B-3 JetRanger III (Canada)
Bell 206L-4 LongRanger IV (Canada)
Bell 407 (Canada)
Bell 427 (Canada)
Bell 430 (Canada)
Bell/Agusta AB 139 (International)
CHAIG Z-11 (China)
CHRDI Z-X (China)
Dirgantara (Eurocopter) NBO-105 (Indonesia)
Enstrom 480 and TH-28 (USA)
Eurocopter AS 350 Ecureuil/AStar and AS 550 Fennec (International)
Eurocopter AS 355 Ecureuil 2/TwinStar and AS 555 Fennec (International)
Eurocopter AS 365N Dauphin 2 and AS 565 Panther (France)
Eurocopter BO 105 (International)
Eurocopter EC 130 (International)
Eurocopter EC 135 and EC 635 (France)
Eurocopter EC 155B (International)
Eurocopter EC 120 B Colibri (International)
Eurocopter BK 117 (International)
Eurocopter EC 145 (International)
FH-1100 FHoenix (USA)

HAI (Eurocopter) Z-9 Haitun (China)
HAL Dhruv (India)
HAL Cheetah (India)
HAL Chetak (India)
IHSRC (PANHA) 2061 (Iran)
IHSRC (PANHA) 287 (Iran)
IRGC Shahed 274 (Iran)
KAI-Bell SB 427 (Korea, South)
KAI KMH (Korea, South)
KAL KMH (Korea, South)
Kamov Ka-215 (Russian Federation)
Kamov Ka-226A Sergei (Russian Federation)
Kawasaki (MDH) MD 500D (Japan)
Kazan Ansat (Russian Federation)
MD 500 and MD 530 (USA)
MD 520N (USA)
MD 600N (USA)
MD Explorer (USA)
PZL-Swidnik SW-4 (Poland)
Westland Lynx (United Kingdom)

AEW helicopter (1)
Kamov Ka-31 (Russian Federation)

Four-seat helicopter (4)
Mil Mi-34 (Russian Federation)
Mil Mi-234 (Russian Federation)
Robinson R44 (USA)
Schweizer 330 and 333 (USA)

Three-seat helicopter (6)
Aviaimpex Yanhol (Ukraine)
Enstrom F28 and 280 (USA)
Kazan Aktai (Russian Federation)
MK Helicopter MK III (Germany)
Mil Mi-60 (Russian Federation)
Schweizer 300C (USA)

Two-seat helicopter (7)
Aerokopter ZA-6 San'ka (Ukraine)
Aerokopter AK-1 (Ukraine)
Brantly B-2B (USA)
Dragon Fly 333 (Italy)
IL IS-2 (Poland)
IHSRC (PANHA) Sanjaqak (Iran)
Robinson R22 (USA)

Four-seat helicopter kitbuilt (2)
Hillberg 1-04 Baby Huey (USA)
VAT Dragonfly (USA)

Two-seat helicopter kitbuilt (3)
Hillberg EH 1-02 TandemMouse (USA)
JAG JAG (USA)
Rotorway Exec 162F (USA)

Two-seat ultralight helicopter (1)
NA Design NA 40 Bongo (Czech Republic)

Two-seat ultralight helicopter kitbuilt (3)
ASII Ultrasport 496 (USA)
Helisport CH-7 Kompress (Italy)
Masquito M80 (Belgium)

Single-seat helicopter kitbuilt (2)
Eagle Helicycle (USA)
Hillberg EH 1-01 RotorMouse (USA)

Single-seat ultralight helicopter kitbuilt (1)
ASII Ultrasport 254 and 331 (USA)

Utility autogyro (1)
Groen Hawk 6 and Hawk 8 (USA)

Four-seat autogyro (1)
Groen Hawk 4 and Jet Hawk 4T (USA)

Three-seat autogyro (1)
IAPO A-002 (Russian Federation)

Two-seat autogyro kitbuilt (2)
Air Command Commander Elite (USA)
RAF 2000 (Canada)

Two-seat autogyro ultralight (1)
Magni M 19 Shark (Italy)

Single-seat autogyro ultralight (2)
Magni M 20 Talon (Italy)
MAI-205 (Russia)

New-concept rotorcraft (4)
Boeing X-50A Dragonfly Canard Rotor/Wing (USA)
Modus Vertijet and Verticraft (Taiwan)
Modus Military Scout (Taiwan)
Piasecki Vectored Thrust Ducted Propeller (USA)

Lifting vehicle (3)
AD & D Hummingbird (Israel)
PAM 100B Individual Lifting Vehicle (USA)
Urban Aero CityHawk, X-Hawk and TurboHawk (Israel)

Exoskelitor flying vehicle (3)
Millennium Jet Solotrek (USA)
Polyot Yula (Russian Federation)
Yanagisawa Gen H-4 (Japan)

Light utility tiltrotor (1)
Bell/Agusta BA609 (International)

Multimission tiltrotor (3)
Bell Quad Tiltrotor (USA)
Bell Boeing V-22 Osprey (USA)
Eurotilt (International)

Convertiplane (2)
Carter CarterCopter CC1 (USA)
Carter Heliplane Transport (USA)

Class 13: Lighter than air (42)

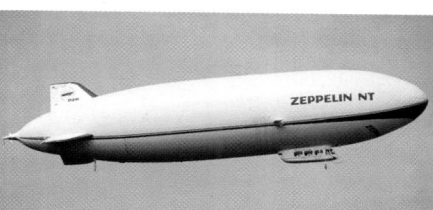

Zeppelin N 07   *NEW*/0143726

All are (at least broadly) cylindrical dirigibles, unless otherwise specified.

Helium rigid (1)
Aeros ML (USA)

Helium semi-rigid (11)
AMSAT Mantha (France)
CargoLifter Joey (Germany)
CargoLifter CL 160 (Germany)
EADS First 1A (France)
RosAeroSystems PD-300 (Russian Federation)
RosAeroSystems MD-900 (Russian Federation)
RosAeroSystems DPD-5000 (Russian Federation)
RosAeroSystems DTs-N1 (Russian Federation)
UPship 001 (USA)
Voliris 900 (France)
Zeppelin N 07 (Germany)

Helium semi-rigid hybrid (1)
Vozdukh B-150 (Russian Federation)

Helium non-rigid (15)
ABC Lightship A-60+ (USA)
ABC Lightship A-100 (USA)
ABC Lightship/Spector Series (USA)
AERALL AVEA (France)
Aeros 40A (USA)
Aeros 40B Sky Dragon (USA)
Aeros 40C (USA)
Aerostatica -200 (Russian Federation)
Aerostatica -300 (Russian Federation)
ATG AT-10 (United Kingdom)
GSI Skyship 500HL (USA)
GSI Skyship 600 (USA)
Pan Atlantic Cargo Airship System (Canada)
Rosaerosystems Au-11 (Russian Federation)
Rosaerosystems Au-12 (Russian Federation)

Helium non-rigid special shape (2)
21st Century SPAS-13 (Canada)
21st Century SPAS-70 (Canada)

Helium non-rigid hybrid (2)
AHA Hornet (USA)
AHA Light Utility (USA)

Hybrid non-rigid air vehicle (1)
ATG SkyCat (United Kingdom)

Hybrid rigid air vehicle (1)
Ohio Dynalifter (USA)

Hot air non-rigid (6)
Aerostatica-200 (Russia)
Aerostatica-300 (Russia)
Cameron DP Series (United Kingdom)
Gefa-Flug AS 105 GD (Germany)
Kubicek AV-2 (Czech Republic)
Skymedia SMA (USA)

Helium balloon (2)
CargoLifter CL 75 AirCrane (Germany)
QinetiQ QinetiQ 1 (UK)

TYPE CLASSIFICATIONS

COMPARISON: CLASSESS 4 to 10

Transport	Airliner/freighter	Business	Utility	Amphibian (all)	Lightplane (factory-built)	Utility kitbuilt
		Supersonic business jet				
	Outsize freighter					
Strategic transport	High-capacity airliner					
	Wide-bodied airliner					
	New concept airliner					
Tanker-transport	Four-jet freighter					
Medium transport/ multirole				Four-turboprop amphibian		
	Twin-jet transport					
	Twin-jet airliner	Long-range business jet		Amphibious airliner		
	Twin-jet freighter					
	Regional jet airliner	Large business jet		Twin-jet amphibian		
		Long-range business tri-jet				
		Business tri-jet				
		Business jet				
		Light business jet			Two-seat jet sportplane	Two-seat jet sportplane kitbuilt
		Business mono-jet				
	Twin-turboprop airliner					
Twin-turboprop transport	Twin-turboprop freighter					
Twin-turboprop light transport		Business twin-turboprop	Utility turboprop twin	Twin-turboprop amphibian		
		Business twin-prop	Light utility twin-prop transport			
			Six-seat utility twin	Six-seat utility twin-prop amphibian		
			Four-seat utility twin			Two-seat kitbuilt twin
		Business turboprop	Light utility turboprop		Five-seat lightplane/ turboprop	
		Business prop	Light utility transport			Utility kitbuilt Utility/kitbuilt
			Utility biplane			Utility biplane kitbuilt
			Six-seat utility transport	Six-seat amphibian Six-seat amphibian/ kitbuilt		Six-seat kitbuilt
					Five-seat lightplane	
				Four-seat amphibian	Four-seat lightplane	
						Four-seat kitbuilt twin
				Two-/four seat amphibian kitbuilt		Four-seat kitbuilt
				Three-seat amphibian kitbuilt	Three-seat lightplane	Three-seat kitbuilt
						Three-seat biplane kitbuilt
					Three-seat biplane	
					Primary prop trainer/ sportplane	

Transport	Airliner/freighter	Business	Utility	Amphibian (all)	Lightplane (factory-built)	Utility kitbuilt
					Aerobatic two-seat turboprop sport-plane	
					Aerobatic two-seat sportplane	
					Aerobatic two-seat lightplane	
				Two-seat amphibian	Two-seat lightplane	
				Two-seat amphibian kitbuilt		
					Motor glider	
					Two-seat lightplane ultralight	
				Two-seat amphibian ultralight kitbuilt		
					Aerobatic two-seat biplane	
					Aerobatic single-seat sportplane	
					Glider tug	
			Agricultural sprayer			

First Flights

Some of the first flights made during the period January to December 2002

The first three production Eurofighter Typhoons flew within days of each other during April 2002, Germany's contribution being two-seat version 9803

Australia

Tomair Cobra Arrow (19-3700)   18 May

Austria

Diamond Aircraft DA 42 Twin Star
  (OE-VPS)   9 Dec

Brazil

Embraer 170 (PP-XJE)   19 Feb

Embraer 170, second prototype (PP-XJC)   9 Apr
Embraer 170, third prototype (PP-XJB)   25 May
Embraer 170, fourth prototype (PP-XJF)   19 Jun
Embraer 170, fifth prototype (PP-XJA)   14 Jul
Embraer 170, sixth prototype (PP-XJS)   27 Jul
Embraer ERJ-145XR first production
  (c/n 145590, PP-XHF/N18101)   not known

Canada

21st Century Airships SPAS-R1 (C-GJKI)   30 Jun
Bombardier BD-100 Challenger 300, fourth
  prototype (C-GJCV)   5 Apr

Canadair 415MP, first 415GR for Greece
  (C-GHVX)   6 Mar

Creative Flight Aerocat (formerly MPA),
  first flight in twin-engined configuration
  (C-GYCC)   5 Sep

China

CAC J-10, first (pre-?) production   28 Jun

Czech Republic

Aero L159B (6073)   1 Jun
Zlin Z 400 Rhino (OK-HZL)   22 May

The Airbus range of airliners got smaller in 2002 with maiden flight of the doubly shortened A318 on 15 January

France

Dyn'Aero MCR01 Ban Bi, first flight with Jabiru engine	19 Apr
Sauper Aviation Papango (F-WWNO)	30 Nov
Socata TBM 700C2 (F-WWRL)	Feb

Germany

CargoLifter Scala	9 Jul
Extra 500 (D-EKEW)	26 Apr
Grob G 140TP (D-ETPG)	12 Dec

NEW/0137731

WD D5 Evolution (D-MPMM)	16 Feb

India

ADA LCA, second prototype (TD2) (KH2002)	6 Jun
HAL Dhruv, fifth (PTC-2 civil) prototype (VT-XLH)	6 Mar
IITB PADD Micro airship (unregistered)	23 Jan

International

Airbus A318-100 (F-WWIA)	15 Jan
Airbus A318-100, first flight after refit with CFM56 engines (F-WWIA)	29 Aug
Airbus A318-100, second prototype (F-WWIB)	3 Jun
Airbus A340, first flight with CFM56-5C/P engine	19 Nov
Airbus A340-500 (F-WWTE)	11 Feb
Bell/Agusta AB139, first production (I-ANEW)	24 Jun
Eurocopter EC 135 ACT/FHS fly-by-light testbed (D-HECV)	28 Jan
Eurocopter Tiger/Tigre, first production UHT (9826)	2 Aug
Eurocopter AWH (converted EC 155) (F-WQEY)	15 Oct
Eurofighter Typhoon IPA 1/PT001 (first UK-built production two-seater)	14 Apr
Eurofighter Typhoon IPA 2/PT002 (first, and first Italian-built, production two-seater) (MMX614)	5 Apr
Eurofighter Typhoon IPA 3/PT003 (first German-built production two-seater) (9803)	8 Apr

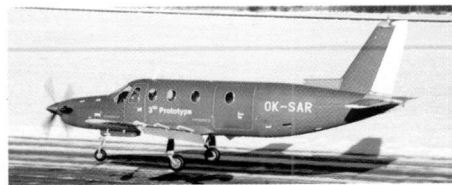

NEW/0123951

Ibis Aerospace Ae 270 Ibis, third (second flying) prototype (OK-SAR)	11 Jan

Japan

Fuji T-3Kai/T-7, first production (26-5901)	July

Korea, South

KAI T-50 Golden Eagle, first prototype (001)	20 Aug
KAI T-50 Golden Eagle, second prototype (002)	8 Nov

Russia

Beriev Be-200ChS, first production (RA-21512)	27 Aug

NEW/0543372

Irkut A-002 (unregistered)	21 Apr

Little was known of the Pilatus PC-21 turboprop trainer until shortly before its first flight on 1 July 2002
NEW/0543331

MVEN MVEN-1 Fermer	9 Nov
Tupolev Tu-214VSSN (RA-64504)	22 Jun

Sweden

Saab JAS 39C Gripen (39-6)	29 Apr
Saab JAS 39C Gripen, first production (39-7)	14 Aug

Switzerland

Aceair Aeriks A-200 (HB-YKS)	29 May
Pilatus PC-21 (HB-HZA)	1 Jul

UK

NEW/0543373

ATG AT-10 airship (G-OATG)	28 Mar
BAE Systems Harrier GR Mk 7A (ZD318)	20 Sep
BAE Systems Hawk NDA (ZJ951)	5 Aug
GKN Westland Super Lynx 300, first for Royal Malaysian Navy (ZJ804)	28 Mar

USA

NEW/0543374

Adam A500 (N500AX)	11 Jul
Aviat Husky Pup (N180HY)	7 Mar
Bell AH-1Z, second prototype (third to fly)	4 Oct
Boeing 707, first flight as MC2A-X testbed *Paul Revere* (N404PA)	18 Apr
Boeing 747 testbed, first flight with (one) GE90-115B engine (N747GE)	18 Sep
Boeing 747-400ER (N747ER)	31 Jul
Boeing 757-300, first flight with PW2040 engines (N753JM)	20 Feb

NEW/0525360

Boeing YAL-1A Airborne Laser (ABL) (converted 747F, 00-0001)	18 Jul
Cessna 680 Citation Sovereign (N680CS)	27 Feb
Cessna 680 Citation Sovereign, first pre-production (N681CS)	27 Jun
Eclipse Aviation Eclipse 500 (N500EA)	26 Aug
Gulfstream Aerospace Gulfstream V-SP, (now G550), first production (N5SP)	18 Jul
Hawker Horizon, second prototype (N802HH)	10 May
Learjet 40 (converted prototype Learjet 45) (N40LX)	31 Aug
Learjet 40, first production (N40LJ)	5 Sep
Lockheed Martin F/A-22 Raptor, second (first to fly) production-representative test vehicle (99-4011)	16 Sep
Me 262 Project Me 262 replica (N262AZ)	20 Dec
Millennium Jet (now Trek) SoloTrek XFV (N102MJ)	15 Jan

NEW/0543377

NASA (Northrop Grumman) Active Aeroelastic Wing, first information-gathering flight (N853NA; converted F/A-18A Hornet, 161744)	15 Nov
Rockwell Sabreliner 60, Williams EJ22 engine testbed for Eclipse 500 (N15HF)	30 May

NEW/0543376

Scaled Composites (Toyota) TAA-1 (N72TA)	31 May
Sikorsky MH-60R Seahawk	4 Apr
Sikorsky S-92/H, third prototype, first flight in military guise (N392SA)	22 Mar
Van's RV-9 (N179RV)	4 Mar

Despite a successful initial flight on 26 August 2002, the promising Eclipse 500 was grounded for a change of engine type
NEW/0543375

A Century of First Flights

A selection of the first flights made during the period 1903 to 2002

The Blériot Type XI of 1909 was the first aeroplane to cross the English Channel, coincidentally in the same year that *Jane's All the World's Air-Ships* was first published. This airworthy replica is flown by the Rhinebeck Aerodrome Museum of New York, USA *(Jane's/Paul Jackson)*   *NEW*/0543349

For most of the history of heavier-than-air powered flight, *Jane's All the World's Aircraft* has recorded the progression of winged aviation from Orville Wright's 37 m (120 ft) first 'hop' to non-stop circumnavigation of the Globe. The list which follows captures the flavour of this century of flight by highlighting some of the significant aircraft which have contributed to the advancement of aviation, whether by technical merit or pure luck; through shouldering part of mankind's burden or sowing the seeds of sorrow. Illustrations are principally from the Kenneth Munson archive.

1903
17 Dec	Wright Flyer I	USA

1904
23 May	Wright Flyer II	USA

1905
23 Jun	Wright Flyer III	USA

1906
3 Mar	Vuia No. 1 monoplane	Romania
13 Sep	Santos-Dumont 14*bis*	France

1907
29 Sep	Breguet-Richet helicopter	France
20 Sep	Henri Farman No. 1	France

1908
8 Jun	Roe I biplane	UK
21 Jun	AEA (Curtiss) *June Bug*	USA
16 Oct	(Cody) British Army Aeroplane No. 1	UK

1909
23 Jan	Blériot Type XI	France
27 Jul	Antoinette VII	France

1910
28 Mar	Fabre Hydravion	France
29 Jul	Bristol Boxkite	UK

One of the first British aircraft built in modest quantities, the Bristol Boxkite
NEW/0543311

1911
30 Jun	Curtiss A-1	USA
18 Sep	Short Triple Twin	UK

1912
3 Mar	Avro 500/Type E	UK

1913
2 Mar	Sikorskii *Bolshoi Baltiski*	Russia

1914
23 Feb	Bristol Scout	UK

Samuel F Cody, though American-born, was first to fly an aeroplane in the UK, on the site now hosting the Farnborough International air show   *NEW*/0543310

Stick-and-string gave way to the first all-metal construction in the Junkers J1 'Tin Donkey'
NEW/0543312

Boeing 40, designed for the pioneering trans-USA air mail service
NEW/0543314

1915

12 Dec	Junkers J1 'Blechesel'	Germany
17 Dec	Handley Page O/100	UK

1916

28 May	Sopwith Triplane	UK
9 Sep	Bristol F.2A	UK
21 Nov	Breguet XIV	France

1917

14 Jun	Nieuport 28	France
21 Oct	Curtiss HS-1	USA
30 Nov	Vickers Vimy	UK

1925

22 Feb	de Havilland D.H.60 Moth	UK
7 Jul	Boeing 40	USA
4 Sep	Fokker F.VIIa-3m	Netherlands
24 Nov	Tupolev (Petlyakov) ANT-4/TB-1	USSR

1926

19 Mar	Fairey IIIF	UK
11 Jun	Ford 4-AT	USA
30 Sep	de Havilland D.H.66 Hercules	UK

1927

28 Apr	Ryan NYP *Spirit of St Louis*	USA
17 May	Bristol Bulldog	UK

Former bomber, the Vickers Vimy made long-distance pioneer flights in the first years of peace
NEW/0543329

Charles Lindbergh crossed the Atlantic in the Ryan NYP, 20-21 May 1927
NEW/0543350

1918

27 Apr	Sopwith Salamander	UK
4 May	Junkers CLI	Germany

1919

25 Jun	Junkers F 13	Germany
4 Dec	Handley Page W 8	UK

1920

24 Nov	Dornier Delphin	Germany
12 Dec	Blériot-Spad 33	France

1921

16 Aug	Dornier Libelle	Germany

1922

26 Mar	de Havilland D.H.34	UK
6 Nov	Dornier Do J Wal	Germany
18 Nov	Dewoitine D 1	France
28 Nov	Fairey Flycatcher	UK

1928

7 Jan	Polikarpov Po-2 (U-2)	USSR
25 Jun	Boeing XF4B-1	USA
19 Dec	Pitcairn PCA 1 autogyro	USA

1929

25 Jul	Dornier Do X	Germany
25 Sep	PZL P-1	Poland
6 Nov	Junkers G 38	Germany

1930

29 Apr	Polikarpov I-5	USSR
6 May	Boeing 200 Monomail	USA
14 Nov	Handley Page H.P.42	UK
22 Dec	Tupolev ANT-6/TB-3	USSR

Handley Page H.P.42s flew Imperial Airways' services to Europe and the Empire
NEW/0543315

Flying boats, such as the Dornier Wal, were prominent long-distance airliners between the wars
NEW/0543313

1923

9 Jan	Cierva C4 Autogiro	Spain
30 Jul	de Havilland D.H.50	UK
23 Aug	Polikarpov I-1 (Il-400)	USSR

1924

4 Nov	Canadian Vickers Vedette	Canada
7 Sep	Dornier Komet III	Germany

1931

25 Mar	Hawker Fury I	UK
26 May	Consolidated P2Y	USA
26 Oct	de Havilland D.H.82A Tiger Moth	UK

1932

13 Aug	Gee Bee R-1 Super Sportster	USA
4 Nov	Beechcraft 17	USA

1933

8 Feb	Boeing 247	USA
22 Jun	Tupolev ANT-25	USSR
1 Jul	Douglas DC-1	USA
31 Dec	Polikarpov I-16/TsKB-12	USSR

Prototype Douglas DC-1, predecessor of the DC-3/C-47 which still earns its living in the 21st Century *NEW*/0543316

1934

10 Jan	Boeing P-26A	USA
17 Apr	Fairey TSR II (Swordfish prototype)	UK
12 Sep	Gloster SS37 (Gladiator prototype)	UK

1935

28 May	Messerschmitt Bf 109 V1	Germany
17 Sep	Junkers Ju 87 V1	Germany
6 Nov	Hawker Hurricane	UK
17 Dec	Douglas DST	USA

Messerschmitt Bf 109, first of an estimated world record of 33,000 built *NEW*/0543317

1936

5 Mar	Supermarine Spitfire	UK
19 May	Consolidated XPBY-1	USA
26 Jun	Focke-Wulf Fw 61	Germany
3 Jul	Short S.23 (*Canopus*)	UK
2 Dec	Boeing YB-17/Y1B-17	USA

1937

7 May	Lockheed XC-35	USA
24 Dec	Macchi C 200 Saetta	Italy

1938

6 Apr	Bell XP-39	USA
2 Oct	Dewoitine 520	France
14 Oct	Curtiss XP-40 Warhawk	USA
31 Dec	Boeing 307 Stratoliner	USA

1939

1 Apr	Mitsubishi A6M1 Zero-Sen	Japan
1 Jun	Focke-Wulf Fw 190 V1	Germany
27 Aug	Heinkel He 178	Germany
30 Dec	Ilyushin TsKB-55/BSh-2 (Il-2 prototype)	USSR

1940

13 May	Sikorsky VS-300	USA
29 May	Vought XF4U-1 Corsair	USA
19 Aug	North American B-25	USA
26 Oct	North American NA-73X (Mustang prototype)	USA
25 Nov	de Havilland D.H.98 Mosquito	UK

Practical helicopters began with the Sikorsky VS-300 *NEW*/0543319

1941

9 Jan	Avro Manchester III (Lancaster prototype)	UK
2 Apr	Heinkel He 280 V1	Germany
18 Apr	Messerschmitt Me 262 V1	Germany
15 May	Gloster E.28/39	UK
13 Aug	Messerschmitt Me 163 V1	Germany

1942

20 Mar	Mitsubishi J2M1 Raiden	Japan
2 Sep	Hawker Tempest Mk V	UK
21 Sep	Boeing XB-29 Superfortress	USA
1 Oct	Bell XP-59A Airacomet	USA
27 Dec	Kawanishi N1K1-J Shiden	Japan

1943

9 Jan	Lockheed L-49 Constellation	USA
5 Mar	Gloster Meteor Mk I	UK
15 Jun	Arado Ar 234 V1 Blitz	Germany
29 Sep	de Havilland D.H.100 Vampire	UK
26 Oct	Dornier Do 335 V1 Pfeil	Germany

1944

9 Jan	Lockheed XP-80 Shooting Star	USA
6 May	Blohm und Voss BV 40	Germany
1 Sep	Hawker Fury	UK
15 Nov	Boeing XC-97 Stratofreighter	USA
6 Dec	Heinkel He 162 V1 Salamander	Germany

America's first practical jet fighter, the P-80 (later T-33) *NEW*/0543318

1945

14 Jun	Avro Tudor I	UK
22 Jun	Vickers Viking	UK
2 Dec	Bristol 170 Freighter	UK
8 Dec	Bell 47	USA

1946

24 Apr	Mikoyan-Gurevich MiG-9	USSR
24 Apr	Yakovlev Yak-15	USSR
22 May	de Havilland Canada DHC-1 Chipmunk	Canada
9 Dec	Bell X-1	USA

1947

8 Jul	Boeing Stratocruiser	USA
31 Aug	Antonov SKh-1 (An-2 prototype)	USSR
1 Oct	North American XP-86 Sabre	USA
17 Dec	Boeing XB-47 Stratojet	USA

1948

23 Jan	de Havilland Australia DHA-3 Drover	Australia
16 Jul	Vickers Viscount	UK
1 Sep	Saab 29 Tunnan	Sweden

1949

13 May	English Electric A.1 (Canberra prototype)	UK
27 Jul	de Havilland D.H.106 Comet 1	UK
4 Sep	Bristol Brabazon I	UK
7 Nov	Sikorsky S-55	USA

Canberra: first British jet licence-built in the USA (as the Martin B-57) *NEW*/0543320

1950

20 Jan	North America XT-28A	USA
19 Jun	Hawker P.1081	UK

1951

23 Feb	Dassault MD 452 Mystère	France
18 May	Vickers Valiant	UK
20 Jul	Hawker Hunter	UK

Saro's unfortunate Princess arrived just in time to greet the demise of the flying boat   *NEW*/0543321

1952

15 Apr	Boeing YB-52 Stratofortress	USA
23 Jul	Fouga Magister	France
20 Aug	Saunders-Roe S.R.45 Princess	UK
30 Aug	Avro Vulcan	UK
3 Nov	Saab 32 Lansen	Sweden

1953

18 May	Douglas DC-7	USA
25 May	North American YF-100A Super Sabre	USA
14 Jun	Blackburn Beverley	UK

1954

15 Jul	Boeing 367-80 (707 prototype)	USA
2 Aug	Rolls-Royce Thrust Measuring Rig	UK
4 Aug	English Electric P.1A (Lightning prototype)	UK
6 Oct	Fairey Delta 2	UK

1955

12 Mar	Sud-Est SE 3130 Alouette II	France
27 May	Sud-Est SE 210 Caravelle	France
15 Jun	Tupolev Tu-104	USSR
25 Oct	Saab 35 Draken	Sweden
24 Nov	Fokker F27 Friendship	Netherlands

1956

24 Jul	Dassault Etendard IV	France
9 Aug	Fiat G91	Italy
11 Nov	Convair XB-58 Hustler	USA
17 Nov	Dassault Mirage III	France

Mirage III: launch pad for French military aircraft exports   *NEW*/0543322

1957

16 May	Saunders-Roe SR.53	UK
6 Nov	Fairey Rotodyne	UK
6 Dec	Lockheed L-188 Electra	USA
10 Dec	Aermacchi MB-326	Italy

1958

22 Apr	Boeing Vertol 107	USA
27 May	McDonnell F4H-1 Phantom II	USA
30 May	Douglas DC-8	USA

1959

31 Mar	Millicer Airtourer	Australia
30 Jul	Northrop N-156C Freedom Fighter	USA
15 Nov	Canadair CL-44	Canada
20 Dec	Antonov An-24	USSR

The Hawker P.1127 was forerunner of today's Harrier   *NEW*/0543323

1960

| 15 Jan | Yeoman YA1 Cropmaster | Australia |
| 24 Jun | Avro 748 | UK |

1961

| 13 Mar | Hawker Siddeley P.1127 | UK |
| 21 Sep | Boeing Vertol YHC-1B/YCH-47A Chinook | USA |

1962

6 Feb	DINFIA FA 1 Guarani I	Argentina
19 Sep	Aero Spacelines B-377PG Pregnant Guppy	USA
16 Oct	Boeing Vertol CH-46A Sea Knight	USA
24 Dec	Nord Aviation 262	France

1963

9 Feb	Boeing 727	USA
20 Aug	BAC One-Eleven	UK
6 Dec	Piper PA-32-260 Cherokee SIX	USA
17 Dec	Lockheed C-141A StarLifter	USA

1964

27 Sep	BAC TSR-2	UK
14 Oct	Sikorsky CH-53A Sea Stallion	USA
2 Nov	Hawker Siddeley Trident 1E	UK
21 Dec	General Dynamics F-111A	USA
22 Dec	Lockheed SR-71A	USA

1965

25 Feb	Douglas DC-9	USA
27 Feb	Antonov An-22 Antheus	USSR
7 May	Canadair CL-84	Canada
12 Jun	Britten-Norman BN-2 Islander	UK

Canadair CL-84: a tilting wing idea whose time may, at last, have come
NEW/0543324

1966

| 10 Jan | Bell 206A JetRanger | USA |
| 31 Aug | Hawker Siddeley Harrier | UK |

1967

10 Feb	Dornier Do 31 E	Germany
9 Apr	Boeing 737	USA
9 May	Fokker F28 Fellowship	Netherlands
23 Oct	Canadair CL-215	Canada

1968

30 Jun	Lockheed C-5A Galaxy	USA
8 Sep	Sepecat Jaguar	International
4 Nov	Aero L-39 Albatros	Czechoslovakia
31 Dec	Tupolev Tu-144	USSR

1969

| 9 Feb | Boeing 747 | USA |
| 2 Mar | BAC/Sud-Aviation Concorde | International |

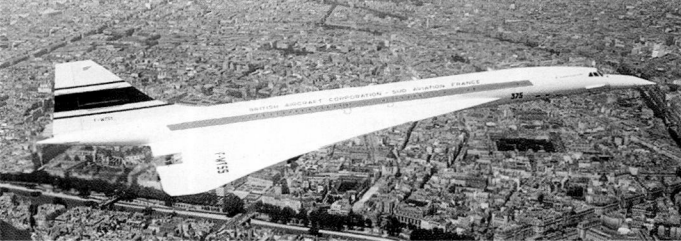

Concorde's impending withdrawal from service was announced in April 2003
NEW/0543378

1970

2 Jul	Saab 37 Viggen	Sweden
29 Aug	McDonnell Douglas DC-10	USA
16 Nov	Lockheed L-1011 TriStar	USA
21 Dec	Grumman F-14A Tomcat	USA

1971

21 Mar	Westland WG 13 Lynx	UK
14 Jul	VFW-Fokker VFW 614	Germany
20 Jul	Mitsubishi XT-2	Japan
4 Aug	Agusta A 109	Italy

Tornado became an exemplar of European collaboration *NEW*/0543326

Voyager made the first unrefuelled aeroplane world circumnavigation *NEW*/0543351

1972

23 Feb	Victa CT4 Airtrainer	Australia
10 May	Fairchild A-10A Thunderbolt II	USA
27 Jul	McDonnell Douglas F-15A Eagle	USA
28 Oct	Airbus A300B1	International

1973

26 Jul	Sikorsky S-69	USA
26 Oct	Dassault-Breguet/Dornier Alpha Jet	International

1974

20 Jan	General Dynamics YF-16	USA
27 Jun	Aerospatiale AS 350 Ecureuil	France
14 Aug	Panavia 200 MRCA (Tornado prototype)	International

1975

24 Jan	Aerospatiale SA 365C Dauphin 2	France
26 Aug	McDonnell Douglas YC-15	USA
30 Sep	Hughes YAH-64	USA

1976

12 Aug	Aermacchi MB 339	Italy
22 Dec	Ilyushin Il-86	USSR

1977

13 Aug	NASA/Rockwell Space Shuttle Orbiter	USA
22 Dec	Antonov An-72	USSR

1978

10 Mar	Dassault-Breguet Mirage 2000	France
8 Nov	Canadair CL-600 Challenger	Canada
18 Nov	McDonnell Douglas F/A-18 Hornet	USA

1979

21 Jul	Bell 214ST	USA
16 Nov	Edgley EA-7 Optica	UK

1980

23 Jun	Mudry CAP 21	France
16 Aug	Embraer EMB-312 Tucano	Brazil
6 Nov	MacCready Solar Challenger	USA
14 Nov	Socata TB 20 Trinidad	France

1984

22 Jun	Voyager (Rutan) Voyager	USA
16 Aug	ATR 42	International
6 Oct	FMA IA 63 Pampa	Argentina
14 Dec	Grumman X-29A	USA

1985

12 Feb	Valmet L-80 TP	Finland
24 Jun	Sikorsky S-76 SHADOW	USA
29 Jul	Kawasaki XT-4	Japan

1986

15 Feb	Beechcraft Starship 1	USA
4 Jul	Dassault Rafale A	France
23 Sep	Piaggio P.180 Avanti	Italy

1987

24 Jun	E-Systems/Grob/Garrett Egrett-1	International
9 Oct	EH Industries EH 101	International

1988

14 Jun	Schweizer 330 Sky Knight	USA
28 Sep	Ilyushin Il-96-300	USSR
9 Dec	Saab JAS 39 Gripen	Sweden
21 Dec	Antonov An-225 Mriya	USSR

1989

2 Jan	Tupolev Tu-204	USSR
19 Mar	Bell Boeing V-22 Osprey	USA
28 May	AIDC Ching-Kuo	Taiwan
17 Jul	Northrop B-2A Spirit	USA

Tucano is one of an increasing number of Brazilian successes *NEW*/0543327

1981

28 Mar	Dornier 228-100	Germany
2 Sep	Cameron D-50 airship	UK
26 Sep	Boeing 767	USA
17 Dec	Hughes OH-6A NOTAR	USA

1982

28 Jan	Hughes 500E	USA
9 Dec	Cessna 208 Caravan	USA

1983

25 Jan	Saab-Fairchild 340	Sweden
20 Jun	de Havilland Canada DHC-8 Dash 8	Canada
27 Jul	Embraer EMB-120 Brasilia	Brazil
15 Sep	Agusta A 129 Mangusta	Italy

Second of the 'stealth' generation, the B-2A Spirit *NEW*/0126863

1990
11 Feb	Atlas XH-2 Rooivalk	South Africa
29 Mar	Ilyushin Il-114	USSR
27 Apr	Aero Mercantil EL 1 Gavilán	Colombia
29 Sep	Lockheed YF-22A	USA

1991
4 Apr	De Chevigny/Wilson Explorer	International
31 May	Pilatus PC-12	Switzerland
15 Sep	McDonnell Douglas C-17 Globemaster III	USA
23 Dec	Kaman K-MAX	USA

1992
20 Aug	Hindustan Aeronautics ALH	India
2 Nov	Airbus A330	International

1993
4 Mar	Dassault Falcon 2000	France
11 Mar	Airbus A321	International
24 Jul	American Sportscopter Ultrasport 254	USA
30 Aug	HAMC Y-12 (IV)	China

1994
27 Mar	Eurofighter 2000	International
17 May	Ilyushin Il-103	Russia
12 Jun	Boeing 777	USA

1995
3 Mar	Gippsland GA-8 Airvan	Australia
31 Mar	Cirrus Design SR20	USA
7 Oct	Mitsubishi F-2	Japan

1996
4 Jan	Boeing Sikorsky YRAH-66 Comanche	USA
25 Apr	Yakovlev Yak-130	Russia
6 Aug	Kawasaki XOH-1	Japan
26 Oct	PZL-Swidnik SW-4	Poland

1997
8 Jan	ABC Lightship A-1-50	USA
15 Jul	Beriev Be-103	Russia
5 Nov	Diamond DA 40 Katana	Austria
11 Dec	Bell 427	Canada/USA

1998
20 Jan	Fairchild Dornier 328JET	Germany

24 Mar	Chengdu J-10	China
11 May	TAAL Hansa-3	India
4 Jul	Embraer ERJ-135	Brazil

1999
8 Feb	Tupolev Tu-334	Russia
27 Apr	Cessna Citation CJ2	USA
17 Aug	Kazan Ansat	Russia
18 Oct	CargoLifter Joey	Germany

2000
29 Feb	MiG 1-44	Russia
26 Jun	Turkish Aerospace Industries ZIU	Turkey
25 Jul	Ibis Aerospace Ae 270 Ibis	International
18 Sep	Boeing X-32A	USA

Turkey's ZIU agricultural sprayer programme is suspended *NEW*/0092518

2001
4 Jan	Aeronautical Development Agency LCA	India
31 May	Aerostar 01	Romania
11 Aug	Raytheon Hawker Horizon	USA

2002
26 Apr	Extra 500	Germany
22 May	Zlin Z 400 Rhino	Czech Republic
1 Jul	Pilatus PC-21	Switzerland
20 Aug	KAI T-50 Golden Eagle	South Korea

First KAI T-50 Golden Eagle 0137979

Aerospace Calendar

Some significant aerospace events during the period January to December 2002

The second production-representative Lockheed Martin F-22A Raptor became the first to this standard to fly when it took off at Marietta, Georgia, on 16 September 2002
NEW/0522852

2002	Military aviation	Civil aviation
3 Jan		Antonov An-225 Mriya made first commercial charter flight (from Stuttgart, Germany, to Thumrait, Oman)
14 Jan		Agusta and Denel signed agreement on licensed production of the A 109 and A 119 in South Africa
15 Jan		First flight of prototype Airbus A318
23 Jan		First flight, in India, of unmanned PADD Micro airship prototype
28 Jan		Extended range Fairchild Dornier 728-200 announced
29 Jan		FAA certification awarded to Grob G 120A
1 Feb		Micco SP26 awarded aerobatic certification
7 Feb	First Bombardier Global Express for the UK's ASTOR programme is delivered to Greenville, USA, for outfitting	
11 Feb		Aerocomp announced it is working on the CA-J Comp Air Jet jet-powered kitbuilt aircraft First flight of first Airbus A340-500
19 Feb		First flight of the Embraer 170 regional jet airliner
25 Feb		First flight of WD Flugzeugleichtbau D5 Evolution
Feb		Turbomeca delivered its 5,000th production Arriel turboshaft engine, which was installed in a Squirrel helicopter operated by Rocky Mountain Helicopters of the USA Announcement regarding formation of Shanghai Sikorsky Helicopter Company by Sikorsky and Shanghai Little Eagle Science and Technology Company
7 Mar		First flight of Aviat Husky Pup
16 Mar		First CargoLifter CL 75 AirCrane sale announced, to Heavy Lift Canada Ltd
22 Mar	First production Eurocopter Tiger rolled out at Donauwörth	
28 Mar		First flight of ATG AT-10 non-rigid airship
Mar		Assembly begun of Diamond DA 42 Twin Star
2 Apr		First Cirrus SR20 manufactured to Canadian certification standard delivered
3 Apr	First helicopter from second Boeing AH-64D Apache multiyear procurement batch delivered to US Army	
5 Apr	First flight of first production Eurofighter Typhoon in Italy (actually second production example)	

2002	Military aviation	Civil aviation
7 Apr		Public debut of Turbine Design TD-2 Tempest kitbuilt sportplane at Sun 'n' Fun show Public debut of Avid Buzz kitbuilt ultralight at Sun 'n' Fun show

NEW/0543328

2002	Military aviation	Civil aviation
10 Apr	Fourth Eurofighter Typhoon development aircraft made first guided AMRAAM firing against a drone target Yakovlev Yak-130 announced as winner of Russian Air Force trainer competition	High Performance Aircraft founded in Germany to produce TT62 light twin propeller aircraft
18 Apr	First flight of 'Paul Revere' Boeing 707 testbed for new Multimission Command & Control Aircraft (MC2A) concept	AASI buys Mooney lightplane company, subsequently changing name to match the acquisition
21 Apr		Prototype Irkut A-002 autogyro flown for the first time
24 Apr		Eurocopter EC 145 deliveries began with the first aircraft for France's Sécurité Civile
26 Apr	Sukhoi announced as lead developer of new Russian combat aircraft known as PACFA	First flight of prototype Extra 500 business turboprop aircraft First flight of 394th and last example of BAE Systems Avro RJ family of regional jet airliners

2002	Military aviation	Civil aviation
4 May		Thielert TAE 125 turbocharged diesel aero-engine received JAR-E certification

NEW/0543344

6 May		The 200th Learjet 45 is delivered to JM Aviation Holdings by Bombardier
15 May	Second EMD Boeing CH-47F Chinook delivered to US Army for flight test duties	
16 May	Turkey signed contract covering purchase of four Boeing 737 AEW&C aircraft, with option on two more	
17 May	First Boeing AH-64D Apache delivered to Singapore Air Force	
22 May		First flight of prototype Zlin Z 400 Rhino

NEW/0543345

28 May		Embraer delivered 600th example of the ERJ regional jet airliner family, an ERJ-145 to Swiss of Europe
29 May	Bell Boeing V-22 Osprey resumed flight test operations after 17-month grounding	JAA certification awarded to Airbus A340-600
31 May		First flight of Aceair Aeriks 200 First flight of Scaled TAA proof-of-concept prototype at Mojave, California
May	Sukhoi PACFA development agreement signed	
6 Jun	First flight of second prototype ADA Light Combat Aircraft	
7 Jun		Cessna simultaneously rolled out the 500th Citation 525 CJ1 and the 100th Citation CJ2
10 Jun		Roll-out of prototype Boeing 747-400ER
14 Jun	Boeing delivered the 1,000th F/A-18E/F Super Hornet (an F/A-18F)	
20 Jun		First flight of first production Bell/Agusta AB 139 helicopter
21 Jun	Delivery to USAF of first F-15E Eagle of follow-on batch of 10 (Lot 12 aircraft), the overall 227th of the type	Mooney announced maiden flight of first aircraft built since takeover by former AASI
25 Jun	First flight of first EMD Boeing CH-47F Chinook	
27 Jun	Initial order for Alenia C-27J: Italy contracted for five of planned 12 for Air Force	
1 Jul	First flight of Pilatus PC-21 basic trainer First flight of Aero Vodochody L159B two-seat advanced light combat aircraft	

2002	Military aviation	Civil aviation
5 Jul	Five aerospace companies agree on joint study for 'Eurotraining' European military pilot training scheme under discussion by several air forces	

NEW/0543346

13 Jul		Eclipse Aviation Eclipse 500 rolled out at Albuquerque, New Mexico
18 Jul	First flight of Boeing YAL-1A Airborne Laser following installation of nose turret and laser targeting pod at Wichita	Airbus completed its 3,000th aircraft, an A320 for JetBlue Airways
20 Jul	Pilatus PC-21 trainer made public debut at Royal International Air Tattoo at RAF Fairford	
21 Jul		On the eve of the Farnborough International Air Show Bombardier announced intention to add Learjet 40 and Learjet 45^{XR} to its range of executive jet aircraft
22 Jul	Bahrain announced selection of BAE Systems Hawk jet trainer Bell/Agusta confirmed selection of AB 139 for US Coast Guard's 'Deepwater' requirement as the type makes its public debut at Farnborough	

NEW/0526948

First Dash 600 version of Airbus A340 handed-over to a customer (Virgin Atlantic) at Heathrow and immediately repositions to Farnborough Show

23 Jul	Lockheed Martin and AgustaWestland announced 10-year agreement to jointly market, produce and support the US101 medium-lift helicopter in the USA United Kingdom adopted name 'Typhoon' (formerly reserved for export marketing only) for the RAF's Eurofighters on the day that the first four-aircraft formation flight displays at Farnborough Initial production order for first two Aero Vodochody L159B for Czech Air Force signed at Farnborough	
24 Jul	Announcement of the 50,000th hour flown by the Lockheed Martin C-130J Hercules First operational deployment by F/A-18E Super Hornet (aboard USS Abraham Lincoln)	
25 Jul		Mooney announced the first sale of an aircraft since its rescue from bankruptcy by the former AASI company
31 Jul		First flight of prototype Boeing 747-400ER

2002	Military aviation	Civil aviation
1 Aug		First flight of Scaled 318 research platform
2 Aug	First flight of first production Eurocopter 665 Tiger/Tigre	

NEW/0521565

2002	Military aviation	Civil aviation
16 Aug	First BAE Systems Nimrod MRA. Mk 4 rolled out	
20 Aug	Initial flight of first prototype KAI T-50 Golden Eagle jet trainer	
26 Aug	First flight of third engineering development Bell AH-1Z SuperCobra	First flight of prototype Eclipse Aviation Eclipse 500 personal jet
31 Aug		First flight of prototype Learjet 40
5 Sep		First flight of first production Learjet 40
7 Sep		First flight of prototype Lancair Sentry
		Beriev Be-200 set six climb-with-payload world records for jet flying boats
8 Sep		Gulfstream Aerospace renamed its range of executive jets to include Gulfstream G100, G300, G400, G500 and G550 versions. Also announced is G150, a wider-bodied G100 to be available in 2005
9 Sep		The 100th Cessna Citation CJ2 was delivered

NEW/0543347

2002	Military aviation	Civil aviation
18 Oct	Boeing revealed existence of, and first exhibited in public, Bird of Prey technology demonstrator that was tested in the mid-1990s	
25 Oct	Italian Army received first EES-standard Agusta A 129 Mangusta combat helicopter	250th EC 135 helicopter was shipped from the Eurocopter factory
27 Oct	First three Grob 120A-1 primary trainers handed over to Israel Defence Force	

NEW/0533392

2002	Military aviation	Civil aviation
10 Sep		Bombardier announced business jet name changes at NBAA: Continental became Challenger 300; Corporate Jetliner became Challenger 800
		Cessna announced and unveiled mockups of Mustang and Citation CJ3 business jets at NBAA
		Public debut of Sino Swearingen SJ30-2 business jet's conforming prototype at NBAA Convention at Orlando, Florida
		Formal launch of Learjet 45XR at NBAA
		Public debut and formal launch of Learjet 40 at NBAA
		Raytheon announced the return of Beech and Hawker as marketing names for its general aviation fleet
		The 600th Beechjet 400 is displayed at NBAA
		CAP Aviation entered voluntary receivership
11 Sep		Socata formally announced 700C version of TBM 700 at NBAA and confirmed almost immediate discontinuation of TBM 700B
16 Sep	First flight of Lockheed Martin F-22A Raptor production representative test vehicle at Marietta, Georgia	
3 Oct	Eurocopter EC 725 made first flight on the power of Makila 2A turboshaft engines	
4 Oct	First flight of second engineering development Bell AH-1Z SuperCobra	
14 Oct		Cessna delivered the 200th Citation X business jet

2002	Military aviation	Civil aviation
30 Oct		Bombardier delivered the 700th Canadair Regional Jet to Air Nostrum of Spain
31 Oct	First Boeing 737-700 AEW&C aircraft for Royal Australian Air Force rolled out at Renton, Washington	
6 Nov	Combat debut of F/A-18E Super Hornet; bombing mission against SAM sites in Iraq by VFA-115 squadron	
8 Nov	First flight of second prototype KAI T-50 Golden Eagle jet trainer	300th Eurocopter EC 120B is delivered to an Australian customer
15 Nov	First official data-gathering flight of NASA F/A-18A test aircraft following installation of Active Aeroelastic Wing	
25 Nov	Greece announced intent to purchase total of 42 NH Industries NH90 helicopters and also revealed plan to acquire at least 12 Boeing AH-64 Apache attack helicopters	
3 Dec		500th and last Gulfstream Aerospace Gulfstream IV rolled out; production line turns to Gulfstream G300 and G400
6 Dec		Bell/Agusta BA609 tiltrotor prototype begins ground running

NEW/0543333

2002	Military aviation	Civil aviation
9 Dec	First production German Eurofighter ''presented'' to Luftwaffe in advance of planned delivery in January 2003	Dash 500 version of Airbus A340 received JAA certification
		First flight of prototype Diamond DA 42 Twin Star

2002	**Military aviation**	**Civil aviation**
10 Dec	French Gendarmerie received its first Eurocopter EC 145s	
12 Dec		The 1,000th Cessna 208B Grand Caravan was delivered to a Thai-based operator

NEW/0543348

18 Dec		The 100th Bombardier Global Express long-range executive jet was handed over to Stanford Financial Group of Texas
19 Dec		Finmeccanica gained control of Aeronautica Macchi by obtaining a further 67 per cent of shares
20 Dec		Maiden flight of first reproduction Messerschmitt Me 262 in USA
		Bell/Agusta AB 139 began certification flying in the USA
		1,000th Cessna 208B Grand Caravan was delivered
		Russia's IAPO changes name to Irkut Scientific Production Corporation JSC
27 Dec	Poland announces decision to order 48 Lockheed Martin F-16s	

Rebelling against conformity, Murphy Rebel N490R has an unorthodox presentation (*Jane's/Paul Jackson*) **NEW**/0543342

As an 'ultralight vehicle' Golden Circle T-Bird II A11RJE is not required by the US FAA to have a regular N-registration, but is in the voluntary series managed by the Aero Sports Connection. Parallel series are managed by the Experimental Aircraft Association ('E' followed by two numbers and three letters) and US Ultralight Association (two numbers and three letters) (*Jane's/Paul Jackson*)   **NEW**/0543339

VQ-T	Turks & Caicos Islands	YJ-	Vanuatu	
VT-	India	YK-	Syria	
V2-	Antigua & Barbuda	YL-	Latvia	
V3-	Belize	YN-	Nicaragua	
V4-	St Kitts & Nevis	YR-	Romania	
V5-	Namibia	YS-	El Salvador	
V6-	Micronesia	YU-	Yugoslavia	
V7-	Marshall Islands	YV-	Venezuela	
V8-	Brunei	Z-	Zimbabwe	
XA- to XC-	Mexico	ZA-	Albania	
XT-	Burkina Faso	ZK- to ZM-	New Zealand	
XU-	Cambodia	ZP-	Paraguay	
XY-, XZ-	Myanmar	ZS- to ZU-	South Africa	
YA-	Afghanistan	Z3-	Macedonia	
YI-	Iraq	3A-	Monaco	
		3B-	Mauritius	
		3C-	Equatorial Guinea	
		3D-	Swaziland	
		3X-	Guinea	
		4K-	Azerbaijan	
		4L-	Georgia	

4R-	Sri Lanka
4X-	Israel
5A-	Libya
5B-	Cyprus
5H-	Tanzania
5N-	Nigeria
5R-	Madagascar
5T-	Mauritania
5U-	Niger
5V-	Togo
5W-	Samoa
5X-	Uganda
5Y-	Kenya
60-	Somalia
6V-, 6W-	Senegal
6Y-	Jamaica
7O-	Yemen
7P-	Lesotho
7Q-	Malawi
7T-	Algeria
8P-	Barbados
8Q-	Maldives
8R-	Guyana
9A-	Croatia
9G-	Ghana
9H-	Malta
9J-	Zambia
9K-	Kuwait
9L-	Sierra Leone
9M-	Malaysia
9N-	Nepal
9Q-	Congo (Democratic Republic)
9U-	Burundi
9V-	Singapore
9XR-	Rwanda
9Y-	Trinidad & Tobago

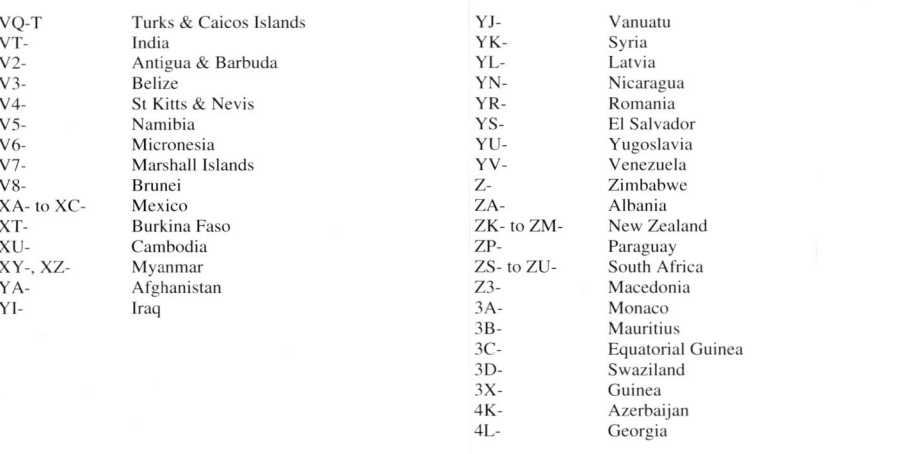

Aide Memoire. Hoffmann H.36 Dimona powered glider (*Jane's/Paul Jackson*)   **NEW**/0543340

Registration prefixes: by country

Afghanistan	YA-
Albania	ZA-
Algeria	7T-
Andorra	C3-
Angola	D2-
Anguilla	VP-LA
Antigua & Barbuda	V2-
Argentina	LQ-, LV-
Armenia	EK-
Aruba	P4-
Australia	VH-
Austria	OE-
Azerbaijan	4K-
Bahamas	C6-
Bahrain	A9C-
Bangladesh	S2-
Barbados	8P-
Belarus	EW-
Belgium	OO-, OQ-
Belize	V3-
Benin	TY-
Bermuda	VP-B
Bhutan	A5-
Bolivia	CP-
Bosnia-Herzegovina	T9-
Botswana	A2-
Brazil	PP- to PU-
Brunei	V8-
Bulgaria	LZ-
Burkina Faso	XT-
Burundi	9U-
Cambodia	XU-
Cameroon	TJ-
Canada	C-, CF-
Cape Verde	D4-
Cayman Islands	VP-C
Central African Republic	TL-
Chad	TT-
Chile	CC-
China	B-
(Hong Kong	B-H)
(Macao	B-M)
Colombia	HK-
Comoros	D6-
Congo (Democratic Republic)	9Q-
Congo (Republic)	TN-
Costa Rica	TI-
Côte d'Ivoire	TU-
Croatia	9A-
Cuba	CU-
Cyprus	5B-
Czech Republic	OK-
Denmark	OY-
Djibouti	J2-
Dominica	J7-
Dominican Republic	HI-
Ecuador	HC-
Egypt	SU-
Equatorial Guinea	3C-
Eritrea	E3-
Estonia	ES-
Ethiopia	ET-
Falkland Islands	VP-F
Fiji	DQ-
Finland	OH-
France and colonies	F-
Gabon	TR-
Gambia	C5-
Georgia	4L-
Germany	D-
Ghana	9G-
Gibraltar	VP-G
Greece (ultralights)	HE-
Grenada	J3-
Guatemala	TG-
Guinea	3X-
Guinea-Bissau	J5-
Guyana	8R-
Haiti	HH-

Honduras	HR-
Hungary	HA-
Iceland	TF-
India	VT-
Indonesia	PK-
Iran	EP-
Iraq	YI-
Ireland	EI-, EJ-
Israel	4X-
Italy	I-
Jamaica	6Y-
Japan	JA
Jordan	JY-
Kazakhstan	UN-
Kenya	5Y-
Kiribati	T3-
Korea, North	P-
Korea, South	HL-
Kuwait	9K-
Kyrgyzstan	EX-
Laos	RDPL-
Latvia	YL-
Lebanon	OD-
Lesotho	7P-
Liberia	pending
Libya	5A-
Liechtenstein	HB-
Lithuania	LY-
Luxembourg	LX-
Macedonia	Z3-
Madagascar	5R-
Malawi	7Q-
Malaysia	9M-
Maldives	8Q-
Mali	TZ-
Malta	9H-
Marshall Islands	V7-
Mauritania	5T-
Mauritius	3B-
Mexico	XA- to XC-
Micronesia	V6-
Moldova	ER-
Monaco	3A-
Mongolia	JU-
Montserrat	VP-LM
Morocco	CN-
Mozambique	C9-
Myanmar	XY-, XZ-
Namibia	V5-
Nauru	C2-
Nepal	9N-
Netherlands	PH-
Netherlands Antilles	PJ-
New Zealand	ZK- to ZM-
Nicaragua	YN-
Niger	5U-
Nigeria	5N
Norway	LN-
Oman	A4O-
Pakistan	AP-
Palestine	SU-YA
Panama	HP-
Papua New Guinea	P2-
Paraguay	ZP-
Peru	OB-
Philippines	RP-
Poland	SP-
Portugal	CR-, CS-
Qatar	A7-
Romania	YR-
Russian Federation	RA-, RF-
Rwanda	9XR-
El Salvador	YS-
Samoa	5W-
San Marino	T7-
São Tomé e Príncipé	S9-
Saudi Arabia	HZ-
Senegal	6V-, 6W-
Seychelles	S7-
Sierra Leone	9L-
Singapore	9V-

Slovak Republic	OM-
Slovenia	S5-
Solomon Islands	H4-
Somalia	6O-
South Africa	ZS- to ZU-
Spain	EC-
Sri Lanka	4R-
St Helena & Ascension	VQ-H
St Kitts & Nevis	V4-
St Lucia	J6-
St Vincent & Grenadines	J8-
Sudan	ST-
Suriname	PZ-
Swaziland	3D-
Sweden	SE-
Switzerland	HB-
Syria	YK-
Taiwan	B-
Tajikistan	EY-
Tanzania	5H-
Thailand	HS-
Togo	5V-
Tonga	A3-
Trinidad & Tobago	9Y-
Tunisia	TS-
Turkey	TC-
Turkmenistan	EZ-
Turks & Caicos	VQ-T
Tuvalu	T2-
Uganda	5X-
Ukraine	UR-
United Arab Emirates	A6-
United Kingdom	G-
United States	N
Uruguay	CX-
Uzbekistan	UK-
Vanuatu	YJ-
Vatican City	HV-
Venezuela	YV-
Vietnam	VN-
Virgin Islands	VP-L
Yemen	7O-
Yugoslavia	YU-
Zambia	9J-
Zimbabwe	Z-

His and his. Geoffrey T Leedham and Richard Morris are both owners of UK-registered Europas (*Jane's/Paul Jackson*) NEW/0543343

Glossary of Aerospace Terms in Jane's All the World's Aircraft

AFRC: United States Air Force Reserve Command operates weather reconnaissance Lockheed Martin WC-130Hs including this aircraft of 53rd WRS 'Hurricane Hunters' (*Jane's/Paul Jackson*)

NEW/0543334

AAM Air-to-air missile.
AAR Air-to-air refuelling.
AATH Automatic approach to hover.
AB Aktiebolag (Swedish company constitution).
absolute ceiling Greatest altitude attainable by aircraft in level flight.
AC Alternating current.
ACC Air Combat Command (US).
ACE Actuator control electronics.
ACLS Automatic carrier landing system.
ACMI Air combat manoeuvring instrumentation.
ACN Aircraft classification number (ICAO system for aircraft pavements).
ADC Air data computer.
ADF Medium-frequency automatic direction-finding (equipment).
ADG Accessory-drive generator.
ADI Attitude/director indicator.
aerofoil Any solid body so shaped that, as a fluid medium (air or hot gas) moves past it, it experiences a useful force perpendicular to the direction of relative motion; thus, a wing generates lift, while a turbine blade generates torque on a shaft.
aeroplane (N America, airplane) Heavier-than-air aircraft with propulsion and a wing that does not rotate in order to generate lift.
AEW Airborne early warning.
AFB Air Force Base (USAF).
AFCS Automatic flight control system.
AFRC Air Force Reserve Command (US).
AFRP Aramid fibre-reinforced plastics.

NEW/0543353

afterburning Temporarily augmenting the thrust of a turbofan or turbojet by burning additional fuel in the jetpipe.
AGM Air-to-ground missile.
Ah Ampère-hours.
AHRS Attitude/heading reference system.
AIDS Airborne integrated data system.
airbrake Passive device extended from aircraft to increase drag. Most common form is hinged flap(s) or plate(s), mounted in locations where operation causes no

significant deterioration in stability and control at any attainable airspeed.
aircraft All manmade vehicles for off-surface navigation within the atmosphere, including helicopters and balloons. For practical purposes, air-cushion vehicles and wing-in-ground-effect vehicles are excluded from the classification.
airship Power-driven lighter-than-air aircraft. Traditional classes are: blimp, a small non-rigid; non-rigid, in which envelope is essentially devoid of rigid members and maintains shape by inflation pressure; semi-rigid, non-rigid with strong axial keel acting as beam to support load; and rigid, in which envelope is itself stiff in local bending or supported within or around rigid framework.
airstair Retractable stairway built into aircraft.
AIS Advanced instrumentation subsystem.
ALARM Air-launched anti-radiation missile.
ALCM Air-launched cruise missile.
Allithium Aluminium-lithium alloy.
ALLTV All-light level television.
AM Amplitude modulation.
AMC Air Mobility Command (US).
AMRAAM Advanced Medium-Range AAM.
ANG Air National Guard (US).
anhedral Downward slope of wing or tailplane from root to tip.
anti-balance tab Hinged surface on trailing-edge of stabilator and operating in same direction, so as to dampen its movement.
ANVIS Aviator's night vision system.
AO Aktsionernoye Obshchestvo (Co Ltd; Russian company constitution).
AoA Angle of attack (see 'attack' below).
AOOT Aktsionernoye Obshchestvo Oktrytogo Tipa (Russian company constitution).
approach noise Measured 1 n mile from downwind end of runway with aircraft passing overhead at 113 m (370 ft).
APR Auxiliary power reserve.
APU Auxiliary power unit (part of aircraft).
ARINC Aeronautical Radio Inc, US company whose electronic box sizes (racking sizes) are the international standard.
ARM Anti-radiation missile.
ArNG Army National Guard (US).
ASE (1) Automatic stabilisation equipment; (2) Aircraft survivability equipment.
ASI Airspeed indicator.
ASM Air-to-surface missile.
aspect ratio Measure of wing (or other aerofoil) slenderness seen in plan view, usually defined as the square of the span divided by gross area.
ASPJ Advanced self-protection jammer.
AST Air Staff Target (UK).
ASTOVL Advanced STOVL.
ASUW Anti-surface unit warfare.
ASV Anti-surface vessel.
ASW Anti-submarine warfare.
ATC Air traffic control.
ATDS Airborne tactical data system.
ATHS Airborne target handover (US, handoff) system.
ATPCS Automatic take-off power control system.

ATR Airline Transport Radio ARINC 404 black box racking standards.
attack, angle of (alpha) Angle at which airstream meets aerofoil (angle between mean chord and free-stream direction). Not to be confused with angle of incidence (which see).
augmented Boosted by afterburning (turbofan).
autogyro Rotary-wing aircraft propelled by a propeller (or other thrusting device) and lifted by a freely running autorotating rotor.
AUW All-up weight (term meaning total weight of aircraft under defined conditions, or at a specific time during flight). Not to be confused with MTOW (which see).
avionics Aviation electronics.
AWACS Airborne warning and control system (aircraft category).
axisymmetric intakes Twin, circular engine air intakes mounted astride the spinner of New Piper light aircraft.
axisymmetric nozzle Circular jet-engine nozzle capable of unrestricted vectoring movement (within a cone specified by mechanical limitation) to enhance aircraft manoeuvrability.

ballistic parachute Emergency recovery parachute installed in (generally light) aircraft and capable of supporting both machine and occupants.
band See radar frequency.
bar Non-SI unit of pressure adopted by this yearbook pending wider acceptance of Pa. 1 bar = 10^5 Pa. ISA pressure at S/L is 1013.2 mb or just over 1 bar. ICAO has standardised hectoPascal for atmospheric pressure, in which ISA S/L pressure is 101.32 hPa.
basic operating weight MTOW minus payload (thus, including crew, fuel and oil, bar stocks, cutlery and so on).
bearingless rotor Rotor in which flapping, lead/lag and pitch change movements are provided by the flexibility of the structural material and not by bearings. No rotor is truly rigid.
BITE Built-in test equipment.
bladder tank Fuel (or other fluid) tank of flexible material.
BLC Boundary-layer control.
bleed air Hot high-pressure air extracted from gas turbine engine compressor or combustor and taken through valves and pipes to perform useful work such as pressurisation, driving machinery or anti-icing by heating surfaces.
blown flap Flap across which bleed air is discharged at high (often supersonic) speed to prevent flow breakaway.
BOW Basic operating weight (which see).
BPR Bypass ratio.
BTU Non-SI energy unit (British Thermal Unit) = 0.9478 J.
bulk cargo All cargo not packed in containers or on pallets.
bus Busbar, main terminal in electrical system to which battery or generator power is supplied.
BV Besloten Vennootschap (Netherlands company constitution).
BVR Beyond visual range.
BWB Blended wing/body.

bypass ratio Air flow through fan duct (not passing through core) divided by airflow through core.
byte Group of bits of information forming unit in computer processing.

C³ Command, control and communications.
CAA Civil Aviation Authority (UK).
cabane Structure, usually of braced struts, to support load above fuselage or wing. May carry parasol wing, engine nacelle or upper wing of most biplanes.
cabin altitude Height above S/L at which ambient pressure is same as inside cabin.

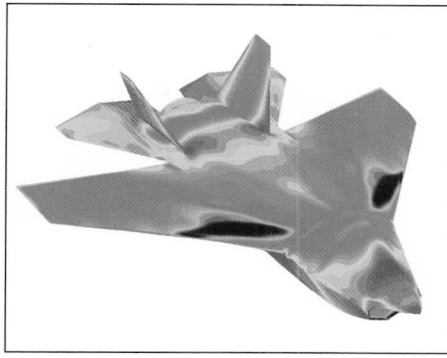

NEW/0106241

CAD/CAM Computer-assisted design/computer-assisted manufacture.
CAHI Central Aero and Hydrodynamics Institute of the Russian Federation; also transliterated as TsAGI.
canards Foreplanes, fixed or controllable aerodynamic surfaces ahead of CG.
capacity The volume swept out on each stroke by the pistons of a piston engine. It is expressed in cc (cubic centimetres) for small engines and in litres (1 litre = 1,000 cc) for larger ones. Also known as displacement or swept volume.
carbon fibre Fine filament of carbon/graphite used as strength element in composites.
CAS (1) Calibrated airspeed, ASI calibrated to allow for air compressibility according to ISA S/L; (2) close air support.
casevac Casualty evacuation.
Cat Category. Meanings include runway visibility and decision height minima for ILS. See diagram.
CATIA Computer-aided three-dimensional interactive analysis; Anglicised form of French CAD proprietary system (*Conception assistée tridimensionelle interactive d'applications*).
CBU Cluster bomb unit.
CCV Control-configured vehicle.
CEAM Centre d'Expériences Aériennes Militaires.
CEAT Centre d'Essais Aéronautiques de Toulouse.
Ceconite Manmade covering material for light aircraft; trade name.
CEO Chief executive officer.
CEV Centre d'Essais en Vol.
CFE Conventional Forces in Europe.
CFRP Carbon fibre-reinforced plastics.
CG Centre of gravity.
chaff Thin slivers of radar-reflective material cut to length appropriate to wavelengths of hostile radars and scattered in bundles to protect friendly aircraft.
chord Distance from leading-edge to trailing-edge measured parallel to longitudinal axis.
CIS Commonwealth of Independent [ex-USSR] States. See also RFAS.
CKD Component knocked down, for assembly elsewhere.
clean (1) In flight configuration with landing gear, flaps, slats and so on retracted; (2) Without any optional external stores.

c/n Constructor's (or construction) number; manufacturer's serial number.
C of A Certificate of Airworthiness; awarded to each individual aircraft (compare Type Certificate).
COIN Counter-insurgency.
collective pitch Controls pitch of all blades of helicopter main rotor in unison.
combi Civil aircraft carrying both freight and passengers on main deck.
comint Communications intelligence.
composite Material made of two constituents, such as filaments or short whiskers plus adhesive forming binding matrix.
constant-speed Variable-pitch propeller governed by a CSU so that its rotational speed is held constant.
contrarotating Propellers on same axis turning in opposite directions (compare C/R).
conventional and manual Aeroplane manoeuvring surfaces mechanically linked to pilot's hand and foot controls, unassisted (except, optionally, by aerodynamic or mass balances) and comprising ailerons on the outboard wing, rudder(s) to the rear of fixed tailfin(s) and elevators to the rear of a fixed (but optionally, incidence angle trimmable) tailplane. The description optionally includes leading-edge slats, flaps on inboard trailing-edges and trim tabs, all of which are mentioned separately, if installed. Ailerons which droop in unison with flaps (and thus are not the primary means of lowering stalling speed) are regarded as conventional. Control systems not conforming to the above – in that they have foreplanes, one or more all-moving tail surfaces, or flaperons, and those with mechanical/electronic assistance or interception of the pilot's movements – are described in appropriate detail.
convertible Transport aircraft able to be equipped to carry passengers or cargo, but not both simultaneously.
COO Chief operating officer.
core Gas generator portion of turbofan comprising compressor(s), combustion chamber and turbine(s).
C/R Counter-rotating; propellers of multi-engined aircraft turning in opposite directions on different axes (compare contrarotating).
CRT Cathode-ray tube.
cruising speed Flight speed on less than full engine power; maximum is normally at 75%, if not otherwise specified, but some manufacturers use higher throttle settings.
CSAS Command and stability augmentation system (part of AFCS).
CSD Constant-speed drive (output shaft speed held steady, no matter how input may vary). Used to enable an alternator to generate AC at a fixed frequency.
CTOL Conventional take-off and landing (compare V/STOL).
CVR Cockpit voice recorder.
cyclic pitch Controls variation of pitch as helicopter rotor blade makes each revolution.

Dacron Artificial fabric for light aircraft covering; trade name.
DADC Digital air data computer.
DADS Digital air data system.
DA/FD Digital autopilot/flight director.
DARPA Defense Advanced Research Projects Agency (US) (briefly ARPA before February 1996).
databus Electronic highway for passing digital data between aircraft sensors and system processors, usually MIL-STD-1553B or ARINC 419 (one-way) and 619 (two-way) systems.
dB Decibel.
DC Direct current.
DECU Digital engine (or electronic) control unit.
dem/val Demonstration/validation.
DERA Defence Evaluation and Research Agency (UK). Divided, 2001, as QinetiQ and Defence Science and Technology Laboratory.
derated Engine restricted to power less than potential maximum (usually such engine is flat rated, which see).

design weight Different authorities have different definitions; weight chosen as typical of mission but usually much less than MTOW.
DF Direction-finder, or direction-finding.
DGAC Direction Générale à l'Aviation Civile. French certification authority.

NEW/0110630

dihedral Upward slope of wing from root (or intermediate point) to tip.
disposable load Sum of masses that can be loaded or unloaded, including payload, crew, removable equipment, usable fuel and other consumables; MTOW minus OWE.
DME UHF distance-measuring equipment; gives slant distance to a beacon; DME element of Tacan.
DoD Department of Defense.
dog-tooth A sharp discontinuity in the leading-edge of a wing or tail surface resulting from an increase in chord (see also sawtooth).
Doppler Short for Doppler radar – radar using fact that received frequency is a function of relative velocity between transmitter or reflecting surface and receiver; used for measuring speed over ground or for detecting aircraft or moving vehicles against static ground or sea.
double-slotted flap One having an auxiliary aerofoil ahead of main surface to increase maximum lift.

EAA Experimental Aircraft Association (divided into local branches called Chapters).
EAS Equivalent airspeed, RAS minus correction for compressibility.
ECCM Electronic counter-countermeasures.
ECM Electronic countermeasures.
ECS Environmental control system.
EEZ Economic exclusion (or exclusive-economic) zone.
EFIS Electronic flight instrument(ation) system, in which large multifunction CRT displays replace traditional instruments.
EGPWS Enhanced ground proximity warning system
EGT Exhaust gas temperature.
ehp Equivalent horsepower, measure of propulsive power of turboprop made up of shp plus addition due to residual thrust from jet.
EICAS Engine indication (and) crew alerting system.
ekW Equivalent kilowatts, SI measure of propulsive power of turboprop (see ehp).
elevon Wing trailing-edge control surface combining functions of aileron and elevator.
ELF Extreme low frequency.
elint electronics intelligence.
ELT Emergency locator transmitter, to help rescuers home on to a disabled or crashed aircraft.
EMD Engineering and manufacturing development.
EMP Electromagnetic pulse of nuclear or electronic origin.
EO Electro-optical.
EPNL Effective perceived noise level.
EPNdB Effective perceived noise decibel, SI unit of EPNL.
EPU Emergency power unit (part of aircraft, not used for propulsion).
ERU Ejector release unit.
ESM (1) Electronic surveillance (or support) measures; (2) Electronic signal monitoring.
ETOPS Extended-range twin (engine) operations (thus sometimes given as EROPS), routeing not more than a given flight time (120, 180 or 240 minutes) from a usable alternative airfield.
EW Electronic warfare.

FAA Federal Aviation Administration.
FAC Forward air control (or controller).
factored Multiplied by an agreed number to take account of extreme adverse conditions, errors, design deficiencies or other inaccuracies.
FADEC Full-authority digital engine (or electronic) control.
FAI Fédération Aéronautique Internationale.
fail-operational System which continues to function after any single fault has occurred.
fail-safe Structure or system which survives failure (in case of system, may no longer function normally).
FAR Federal Aviation Regulations.
FAR Pt 23 Defines the airworthiness of private and air taxi aeroplanes of 5,670 kg (12,500 lb) MTOW and below.
FAR Pt 25 Defines the airworthiness of public transport aeroplanes exceeding 5,670 kg (12,500 lb) MTOW.
FBL Fly-by-light (which see).
FBW Fly-by-wire (which see).

Decision Height: 200 ft (60 m) | 100 ft (30 m)

Runway visual range

150 ft (50 m)
700 ft (200 m)
1,200 ft (400 m)
2,600 ft (800 m)

Cat. I | **Cat. II** | **Cat. IIIa** | **Cat. IIIb**

Categories of ILS minima

NEW/0131472

FCS Flight control system.
FDR Flight data recorder (which see).
FDS Flight director system.
feathering Setting propeller blades at pitch aligned with slipstream to minimise drag.
fence A chordwise projection on the surface of a wing, used to modify the distribution of pressure.
Fenestron Helicopter tail rotor with many slender blades rotating in short duct (registered name).
ferry range Extreme safe range with zero payload.
FFAR Folding-fin (or free-flight) aircraft rocket.
field length Measure of distance needed to land and/or take off; many different measures for particular purposes, each precisely defined.
fixed-pitch Propeller with blades fixed to the hub.
FL Flight level. Notional altitude for air traffic control purposes which assumes ISA pressure (1013.25 mb; 29.92 in Hg) at S/L. Expressed in hundreds of feet; thus FL255 indicates approximately 25,500 ft

NEW/0131450

flap A surface carried on the leading- or trailing-edge of a wing and able to move relative to it. The simplest leading-edge flap and so-called plain (trailing-edge) flap is formed by hinging the entire edge of the wing. The Krueger is a leading-edge flap forming part of the wing undersurface, swung down and forwards on arms to give a bluff leading-edge. A split flap is formed by hinging only the undersurface of the trailing-edge. A slotted flap is a hinged trailing-edge which moves aft as well as down on tracks to leave a narrow slot ahead of it; hence double- and triple-slotted. A Fowler flap is a complete auxiliary aerofoil mounted on tracks under a fixed trailing-edge; initially it moves aft, to emerge behind the fixed part of the wing, and at the end of its travel it rotates down. A Gouge flap has an upper surface forming part of a cylinder, and rotates (on rails or brackets) about that cylinder's centre.
flaperon Wing trailing-edge surface combining functions of flap and aileron.
FLAR Federatsii Lyubitelei Aviatsii Rossii, Russian PFA
flat-four Piston engine having four horizontally opposed cylinders; thus, flat-twin, flat-six and so on.
flight-adjustable pitch Propeller with blades that can be changed in pitch during flight to a limited extent (eg one way only). Compare variable pitch.
flat rated Propulsion engine capable of giving full thrust or power for take-off at an airfield well above S/L and/or at high ambient temperature (thus, probably derated at S/L).
flight data recorder Crash-protected recorder of dynamic/static pressure, air temperature, control-surface and slat/flap positions, 3-axis accelerations, engine parameters and possibly other variables.
FLIR Forward-looking infra-red.
fly-by-light Flight control system in which signals pass between computers and actuators along fibre-optic leads.
fly-by-wire Flight control system with electrical signalling, without mechanical interconnection between cockpit flying controls and control surfaces.
FM Frequency modulation.
FMCS Flight management computer system.
FMS (1) Foreign military sales (US DoD); (2) Flight management system.
FOD Foreign-object damage.
footprint (1) A precisely delineated boundary on the surface around an airfield, inside which the perceived noise of an aircraft exceeds a specified level during take-off and/or landing; (2) Dispersion of weapon or submunition impact points.
foreplanes Pivoted canard surfaces forming part of the primary flight control system with authority in pitch and possibly also in roll. See also canards.
FOV Field of view.
Fowler flap See flap.
FRADU Fleet Requirements and Air Direction Unit (UK).
frequency See radar frequency.
frequency agile (frequency hopping) Making a transmission harder to detect by switching automatically to a succession of frequencies.
FSD Full-scale development.
FSED Full-scale engineering development.
FY Fiscal year; in US government affairs, runs from 1 October to 30 September (FY02 begins 1 October 2001); in Japan, from 1 April (FY13 or FY01 began 1 April 2001).

g Acceleration due to mean Earth gravity, that is of a body in free-fall; or acceleration due to rapid change of direction of flight path.
gallons Non-SI measure; 1 Imp gallon (UK) = 4.546 litres, 1 US gallon = 3.785 litres.
GFRP Glass fibre-reinforced plastics.
'glass cockpit' Cockpit in which dial instruments are replaced by multifunction electronic displays.
glass fibre Spun molten glass; see GFRP.
glideslope Element giving vertical (height) guidance in ILS.
glove (1) Fixed portion of wing inboard of variable sweep wing; (2) additional aerofoil profile added around normal wing for test purposes.
GmbH Gesellschaft mit beschränkter Haftpflicht (or Haftung) (German company constitution).
GPS Global Positioning System, US military/civil satellite-based precision navaid.
GPU Ground power unit (not part of aircraft).
GPWS Ground-proximity warning system.
green aircraft Aircraft flyable but unpainted, unfurnished and basically equipped.
gross wing area See wing area.
ground-adjustable pitch Propeller with blades that can be adjusted in pitch by an engineer on the ground. Compare flight adjustable and variable pitch.
GS See glideslope.
gunship Aircraft designed for battlefield attack; helicopter gunships normally with slim body carrying pilot and weapon operator only.
GUP Gosudarstvennoye Unitarnoye Predpriyatie (Russian State Unitary Enterprise).

h Hour(s).
handed Rotating in opposite directions.
hardened Protected as far as possible against nuclear explosion.
hardpoint Reinforced part of aircraft to which external load can be attached, for example weapon or tank pylon.
HDU Hose-drum unit.
head-down display On the cockpit instrument panel (as distinct from a HUD).
head-level display Immediately below HUD.
hectopascal Unit of pressure (hPa) equal to 100 Pascals.
helicopter Rotary-wing aircraft both lifted and propelled by one or more power-driven rotors turning about substantially vertical axes.
HF High frequency.
HIFR Helicopter in-flight refuelling.
HIRF High-intensity radiated field(s).
HMD Helmet-mounted display; hence HMS = sight.
HOCAC Hands on cyclic and collective.
homebuilt Aircraft built/assembled from plans or kits.
hot and high Adverse combination of airfield height and high ambient temperature, which lengthens required take-off distance.

NEW/0543354

HOTAS Hands on throttle and stick.
hot refuelling Replenishment of fuel while engine(s) running.
hovering ceiling Ceiling of helicopter (corresponding to air density at which maximum rate of climb is zero), either IGE or OGE.
HP High pressure (HPC, compressor; HPT, turbine).
hp Horsepower, non-SI unit of power.
HSI Horizontal situation indicator.
HUD Head-up display (bright numbers and symbols projected on pilot's aiming sight glass and focused on infinity so that pilot can simultaneously read display and look ahead). The term is increasingly rendered in the USA as "heads up", which is incorrect.
Hz Hertz, cycles per second.

IAS Indicated airspeed, airspeed indicator reading corrected for instrument error.
IATA International Air Transport Association.
ICAO International Civil Aviation Organisation.
IDG Integrated-drive generator.

IFF Identification friend or foe.
IFR (1) Instrument flight rules (compare VFR); (2) in-flight refuelling.
IGE In ground effect: helicopter performance with theoretical flat horizontal surface just below it (for example mountain).
IIR Imaging infra-red.
ILS Instrument landing system. See Cat.
Imperial gallon 1.20095 US gallons; 4.546 litres.
IMS Integrated multiplex system.
INAS Integrated nav/attack system.
Inc Incorporated (company constitution).
incidence The angle at which the wing is set in relation to the fore/aft axis. Often wrongly used to mean angle of attack (which see).
inertial navigation Measuring all accelerations imparted to a vehicle and, by integrating these with respect to time, calculating speed at every instant (in all three planes) and, by integrating a second time, calculating total change of position in relation to starting point.
INS Inertial navigation system.
integral construction Machined from solid instead of assembled from separate parts.
integral tank Fuel (or other liquid) tank formed by sealing part of structure.
intercom Wired telephone system for communication within aircraft.
inverter Electric or electronic device for inverting (reversing polarity of) alternate waves in AC power to produce DC.
IOC Initial operational capability.
IR Infra-red.
IRCM Infra-red countermeasures.
IRLS Infra-red linescan (builds TV-type picture showing cool and hot regions as contrasting shades).
IRS Inertial reference system.
IRST Infra-red search and track.
ISA International Standard Atmosphere (1013.25 mb, 1,225 g/m^3 and 15°C at mean sea level; lapse rate 1.98°C per 1,000 ft up to −56.5°C at 36,090 ft).

J Joule, SI unit of energy.
JAA Joint Aviation Authorities.
JAR Joint Aviation Requirements, agreed by all major EC countries (JAR 25 equivalent to FAR Pt 25).
JAR-VLA JAR classification for Very Light Aircraft (MTOW limit of 750 kg; 1,653 lb).
JASDF Japan Air Self-Defence Force.
JATO Jet-assisted take-off (actually means rocket-assisted).
JCAB Japan Civil Airworthiness Board.
JDA Japan Defence Agency.
JGSDF Japan Ground Self-Defence Force.
JMSA Japan Maritime Safety Agency.
JMSDF Japan Maritime Self-Defence Force.
joined wing Tandem wing layout in which forward and aft wings are swept so that the outer sections meet.
joule SI unit of energy, = 1 Nm = 1 Ws.
JPATS Joint Primary Aircraft Training System (Raytheon T-6A Texan II).
JSC Joint stock company.
JSF Joint Strike Fighter.
J-STARS US Air Force/Navy Joint Surveillance and Target Attack Radar System in Northrop Grumman E-8C.
JTIDS Joint Tactical Information Distribution System (NATO Link 16).
Junkers aileron Control surface (sometimes flaperon) suspended from mountings to the rear of wing trailing edge.

G-BWI

NEW/0543335

kbit One thousand bits of memory.
Kevlar Aramid fibre used as basis of high-strength composites material.
kg Kilogramme (2.20462 lb).
kitbuilt Prefabricated aircraft for amateur assembly.
KK Kabushiki Kaisha (Japanese company constitution).
km/h Kilometres per hour.
kN KiloNewtons (N×10³). See N.
knot 1 n mile per hour (1.852 km/h; 1.15078 mph).
Krueger flap Hinges down and then forward from below the leading-edge.
kVA Kilovolt-ampères.
kW Kilowatt, SI measure of all forms of power (not just electrical).

LAMPS Light airborne multipurpose system.
LANTIRN Low-altitude navigation and targeting infra-red, night.
LAPES Low-altitude parachute extraction system.
LBA Luftfahrtbundesamt (German civil aviation authority).
lb Pound, non-SI unit of weight: 0.453592 kg.
lb st Pounds of static thrust.
LCD Liquid crystal display, used for showing instrument information.
LCN Load classification number, measure of 'flotation' of aircraft landing gear linking aircraft weight, weight distribution, tyre numbers, pressures and disposition.
LED Light-emitting diode.
lift dumper Spoiler designed to open on landing to reduce lift and thus increase effectiveness of wheel braking.
LINS Laser inertial navigation system.
litre SI unit of volume (0.264177 US gallon; 0.219975 Imp gallon).
LLTV Low-light TV (thus, LLLTV, low-light level); see ALLTV.
LO Low-observables, which see (stealth).
load factor (1) Percentage of maximum payload; (2) design factor (g limit) for airframe.
LOC Localiser (which see).
localiser Element giving steering guidance in ILS.
LOH Light observation helicopter.
loiter Fly for maximum endurance, at much less than normal cruise speed.
longerons Principal fore-and-aft structural members (for example in fuselage).
Loran Long-range navigation; family of hyperbolic navaids based on ground radio emissions, now mainly Loran C.
LOROP Long-range oblique photography.
LOS Line of sight.
low-observables Materials, structures and techniques designed to minimise aircraft signatures of all kinds.
lox Liquid oxygen.
LP Low pressure (LPC, compressor; LPT, turbine).
LRIP Low-rate initial production.
LRMTS Laser ranger and marked-target seeker.
LRU Line-replaceable unit.
Ltd Limited (company constitution).
LV Low volume (crop-spraying intensity).

m Metre(s), SI unit of length (3.28084 feet).
M or Mach number The ratio of the speed of a body to the speed of sound (340 m; 1,116 ft/s in air at 15°C) under the same ambient conditions.
MAD Magnetic anomaly detector.
mass balance Mass attached to flight control surface, typically ahead of hinge axis, internally or externally, to reduce or eliminate coupling with airframe flutter modes.
mass flow Mass of air passing per second (usually at T-O, S/L).
MAWS Missile-approach warning system.
mb Millibar, bar × 10⁻³.
Mbyte One million (10⁶) bytes.
MCM Mine countermeasures.
medevac Medical evacuation.
MFD Multifunction (electronic) display.
MHz Megahertz: 1 million (10⁶) Hertz.
microlight See ultralight.
MIDS Multifunction information distribution system.
MKR Marker beacon receiver.
MLS Microwave landing system.
MLU Mid-life update.
MLW Maximum landing weight.
mm Millimetres, metres × 10⁻³.
M$_{MO}$ Maximum operating Mach number.
MMS Mast-mounted sight.
MoD Ministry of Defence.
monocoque Structure with strength in outer shell, devoid of internal bracing (semi-monocoque, with some internal supporting structure).
MoU Memorandum of Understanding.
MPA Maritime patrol aircraft.
mph Miles per hour.
MSIP Multistaged improvement program (US).
MTBF Mean time between failures.
MTBR Mean time between removals.
MTI Moving-target indication (radar).
MTOW Maximum take-off weight (minus taxi/run-up fuel).

MYP Multiyear procurement (US).
MZFW Maximum zero-fuel weight.

N Newton, SI unit of force, = 0.22480455 lb force.
NACES Navy aircrew common ejection seat (US).
NAS Naval Air Station (US).
NASA National Aeronautics and Space Administration (US).
NASC Naval Air Systems Command (also several other aerospace meanings) (US).
NATC Naval Air Training Command or Test Center (also several other aerospace meanings) (US).
NATO North Atlantic Treaty Organisation.
nav/com Navigation and communications receiver.
NBAA National Business Aircraft Association (US).
NBC Nuclear, biological, chemical (warfare).
NDT Non-destructive testing.
Newton See N.
NFO Naval flight officer; second crew member in US Navy aircraft; compare WSO.
Ni/Cd Nickel/cadmium.
Nib Forward-pointing extension at inner end of fixed glove on VG aircraft or leading-edge root extension on light aircraft.
n mile nautical mile, 1.852 km, 1.15078 miles.
NOE Nap-of-the-Earth (low flying in military aircraft, using natural cover of hills, trees and so on).
NOS Night observation surveillance.
NV Naamloze Vennootschap (Belgian/Netherlands company constitution).
NVG Night vision goggles.
NVS Noise vibration suppression.

OAO Otkrytolye Aktsionernoye Obshchestvo (JSC; Russian company constitution).
OAT Outside air temperature.
OBIGGS Onboard inert gas generating system.
OBOGS Onboard oxygen generating system.
OCU (1) Operational Conversion Unit; (2) operational capabilities upgrade.
OEI One engine inoperative.
OEU Operational Evaluation Unit.
offset Workshare granted to a customer nation to offset the cost of an imported system.
OGE Out of ground effect; helicopter hovering, far above nearest surface.
OKB Opytnyi Konstruktorskoye Byuro (Russian experimental design bureau)
Omega Long-range hyperbolic radio navaid.
omni Generalised word meaning equal in all directions (as in omnirange, omniflash beacon).
on condition maintenance According to condition rather than at fixed intervals.
OOO Obshchestvo Ogranichennoye Otvetstvennostyu (Russian company constitution).
opeval Operational evaluation.
OTH Over-the-horizon (OTHT adds targeting).
OTPI On-top position indicator (indicates overhead of submarine in ASW).
OWE Operating weight empty. MTOW minus payload, usable fuel and oil and other consumables (thus, includes crew).

PA system Public or passenger address.
pallet (1) for freight, rigid platform for handling by forklift or conveyor; (2) for missile, interface mounting and electronics box outside aircraft.
Pascal SI unit of pressure =1 Nm⁻² (one Newton per square metre).
payload Disposable load generating revenue (passengers, cargo, mail and other paid items); in military aircraft, loosely used to mean total load carried of weapons, cargo or other mission equipment.
Performance Aircraft capabilities after S/L take-off at MTOW in ISA with normal full tankage, except as otherwise specified, and with landing data at MLW (where different).
PFA Popular Flying Association (UK).
PFCS Primary flight computer system.
PGM Precision-guided munition.
phased array Radar in which the beam is scanned electronically in one or both axes without moving the antenna.
Pirate Passive infra-red airborne tracking equipment.
PLA Prelaunch activities.
plane A lifting surface (for example wing, tailplane).
plc Public limited company (company constitution).
plug door Door larger than its frame in pressurised fuselage, either opening inwards or arranged to retract parts before opening outwards.
plume The region of hot air and gas emitted by a helicopter jetpipe.
ply Indication (ply rating) of tyre strength in a specific application; not necessarily the actual number of carcass plies in the tyre.
pneumatic de-icing Covered with flexible surfaces alternately pumped up and deflated to throw off ice.
port Left side, looking forward.

power-by-wire Using electric power alone (not electro-hydraulic) to drive control surfaces and perform other mechanical tasks.
power loading Aircraft weight (usually MTOW) divided by total propulsive power or thrust at T-O. For helicopters, based on transmission rating rather than total engine power.
power train A complete mechanical drive system, for example the sequence of gearwheels, clutches and shafts transmitting power from one or more engines to the rotors of a helicopter.
PPV Pre-production verification.
prepreg Glass fibre cloth or rovings pre-impregnated with resin to simplify layup.
pressure fuelling Fuelling via a leakproof connection through which fuel passes at high rate under pressure.
primary flight controls Those used to control trajectory of aircraft (thus, not trimmers, tabs, flaps, slats, airbrakes or lift dumpers, and so on).
primary flight display Single screen bearing all data for aircraft flight-path control.

NEW/0052900

propfan A family of new-technology propellers characterised by multiple scimitar-shaped blades with thin sharp-edged profile. Single and contrarotating examples promise to extend propeller efficiency up to an aircraft Mach number of about 0.8.
proprotor Large propeller, tilting for forward or vertical flight.
PT Pesawat Terbang (Indonesian company constitution).
Pty Proprietary (company constitution).
pulse Doppler Radar sending out pulses and measuring frequency-shift to detect returns only from moving target (s) seen against background clutter.
pylon Structure linking aircraft to external load (engine nacelle, drop tank, bomb, and so on).

Radar frequency Operating bands of airborne radars are given according to frequency. That part of the electromagnetic spectrum appropriate to above-surface short-range communication and radar (but not OTH) used in aviation is given in the adjacent table with an approximate cross-reference to previously used wavelength bands.
radius The approximate distance an aircraft can fly from base and return without intermediate landing.
RAI Registro Aeronautico Italiano (Italian civil aviation authority).
RAM Radar absorbent material.
ram pressure Increased pressure in forward-facing aircraft inlet, generated by converting relative kinetic energy to pressure.
ramp weight Maximum weight at start of flight (MTOW plus taxi/run-up fuel).
range Too many definitions to list, but essentially the distance an aircraft can fly (or is permitted to fly) with specified load and usually while making allowance for specified additional manoeuvres (diversions, standoff, go-around and so on).
RANSAC Range surveillance aircraft.
RAS Rectified airspeed, IAS corrected for position error.
raster Generation of large-area display, for example TV screen, by close-spaced horizontal lines scanned either alternately or in sequence.
RAT Ram air turbine.
rating Any of several values of thrust or shaft power which an engine is qualified (usually also guaranteed) to develop under specified conditions.
RCS Radar cross-section; apparent size of echo.
redundant Provided with spare capacity or data channels and thus able to survive failures.
reversion Ability to switch to manual control following failure of a powered system.
RFAS Russian Federation and Associated States (CIS).
RFP Request(s) for proposals.
rigid rotor See bearingless rotor.
RMI Radio magnetic indicator; combines compass and navaid bearings:

The Electromagnetic Spectrum

Frequency	Wavelength	General Designation	NATO Band	US Band
30-3 kHz	10,000-100 km	ELF		
3-30 kHz	100-10 km	VLF		
30-300 kHz	10-1 km	LF		
300 kHz-3 MHz	1,000-100 m	MF		
3-30 MHz	100-10 m	HF	A	
30-230 MHz	10-1.3 m	VHF	A	
230-250 MHz	1.3-1.2 m	VHF	A	P
250-300 MHz	1.2-1 m	VHF	B	P
300-500 MHz	100-60 cm	UHF	B	P
500-1,000 MHz	60-30 cm	UHF	C	P
1-2 GHz	30-15 cm	UHF	D	L
2-3 GHz	15-10 cm	UHF	E	S
3-4 GHz	10-7.5 cm	SHF	F	S
4-6 GHz	7-5.5 cm	SHF	G	C
6-8 GHz	5-7.5 cm	SHF	H	C
8-10 GHz	3.75-3 cm	SHF	I	X
10-12.5 GHz	3-2.5 cm	SHF	J	X
12.5-18 GHz	2.5-1.6 cm	SHF	J	Ku
18-20 GHz	1.6-1.5 cm	SHF	J	K
20-26.5 GHz	1.5-1.1 cm	SHF	K	K
26.5-30 GHz	1.1-1 cm	SHF	K	Ka
30-40 GHz	10-7.5 mm	EHF	K	Ka
40-60 GHz	7.5-5 mm	EHF	L	mm
60-100 GHz	5-3 mm	EHF	M	mm
100-300 GHz	3-1 mm	EHF		

Notes: Three overlapping descriptive systems are used in the West.
General designations are Extremely Low, Very Low, Low, Medium, High, Very High, Ultra High, Super High and Extremely High Frequency.
Frequencies are measured in kilo (1,000), mega (1,000,000) and giga (1,000,000,000) cycles per second (Hertz); wavelengths measured in kilometres, metres, centimetres and millimetres.
'NATO' bands describe radar and electronic warfare equipment; 'US' bands are used for radar and satellite communications. The latter's bounds are slightly 'elastic'.
Aircraft-to-ground voice communications for air traffic control and similar purposes (including ground radio beacons) uses 108-136 MHz in the VHF band and 225-400 MHz in the V/UHF bands, the latter principally military, and not entirely accurately termed 'UHF'.

R/Nav Calculates position, distance and time from groups of airways beacons.

RON Research octane number of fuel.

roving Multiple strands of fibre, as in a rope (but usually not twisted).

rpm Revolutions per minute.

RPV Remotely piloted vehicle (pilot in other aircraft or on ground); contrast UAV.

RSA Réseau du Sport de l'Air.

ruddervators Flying control surfaces, usually a V tail, that control both yaw and pitch attitude.

RVSM Reduced vertical separation minimum. Halved (1,000 ft) air traffic control separation between FL290 and FL410.

RWR Radar warning receiver.

s Second(s)

SA Société Anonyme (France, Romania), Sociedad Anónima (Brazil, Spain) or Spółka Akeyjna (Poland) (company constitution).

safe-life A term denoting that a component has proved by testing that it can be expected to continue to function safely for a precisely defined period before replacement.

SAM Surface-to-air missile.

SAR (1) Search and rescue; (2) synthetic aperture radar.

SAS Stability augmentation system.

satcom Satellite communications.

sawtooth Same as dog-tooth.

SCAS Stability and control augmentation system.

Sdn Bhd Sendirian Berhad (Malaysian company constitution).

SEAD Suppression of enemy air defence(s).

second-source Production of identical item by second factory or company.

semi-active Homing on to radiation reflected from target illuminated by radar or laser energy beamed from elsewhere (for example, from launch aircraft).

sensitive altimeter Altitude indicator of mechanical type, having acute sensitivity.

service ceiling Usually height equivalent to air density at which maximum attainable rate of climb is 100 ft/min.

servo A device which acts as a relay, usually augmenting the pilot's efforts to move a control surface, or the like.

SFAR Special Federal Aviation Regulation(s).

sfc Specific fuel consumption (which see).

shaft Connection between gas-turbine and compressor or other driven unit. Two-shaft engine has second shaft, rotating at different speed, surrounding the first (thus, HP surrounds inner LP or fanshaft).

shipment One item or consignment delivered (by any means of transport) to customer.

shp Shaft horsepower, measure of power transmitted via rotating shaft.

shroud Many meanings, including: (1) a fixed circular duct surrounding a fan or propfan; (2) a ring formed by lateral projections on a rotor (for example fan) blade (part-span or at the tip); (3) a portion of a wing or other fixed aerofoil projecting aft over the leading-edge of a hinged or otherwise movable surface such as a flap, aileron or elevator.

sideline noise EPNdB measure of aircraft landing and taking off, at point 0.25 n mile (2- or 3-engined) or 0.35 n mile (4-engined) from runway centreline.

sidestick Control column in the form of a short handgrip beside the pilot.

SIF Selective identification facility.

sigint Signals intelligence.

signature Characteristic 'fingerprint' of all acoustic or electromagnetic radiation (radar, IR, and so on).

NEW/0543355

single-aisle Passenger cabin has seats on each side of a single aisle along or near the centre.

single-shaft Gas-turbine in which all compressors and turbines are fixed to common shaft.

S/L Sea level.

SLAR Side-looking airborne radar.

slat Auxiliary curved or mini-aerofoil surface designed to prevent flow breakaway from a wing or tail. On a tail leading-edge it may be fixed, leaving a narrow slot. On a wing it is almost always retractable, normally flush with the wing profile but extended (under power or by aerodynamic lift) to leave a narrow slot for take-off, low-speed loiter or landing.

slot, slotted See slat.

snap-down Air-to-air interception of low-flying aircraft by AAM fired from fighter at a higher altitude.

soft target Not armoured or hardened.

SONAR, sonar Sound navigation and ranging.

SpA Società per Azioni (Italian company constitution).

specific fuel consumption Rate at which fuel is consumed divided by power or thrust developed, and thus a measure of engine efficiency. For jet engines (air-breathing, not rockets) unit is mg/Ns, milligrams per Newton-second; for shaft engines unit is μg/J, micrograms (millionths of a gram) per Joule (SI unit of work or energy).

spoiler Plank-like surface normally recessed into top of wing, hinged up under power to reduce (spoil) lift and increase drag. Used asymmetrically for lateral control.

spoileron Small spoiler augmenting ailerons.

sportplane Light aircraft design in which performance takes precedence over utility.

Sp. z o.o Spółka z ograniczoną odpowiedzialnością (Polish company constitution)

SRL Società Reponsibilita Limitata (Italian company constitution).

SSB Single-sideband (radio).

SSR Secondary surveillance radar.

SST Supersonic transport.

st Static thrust.

stabilator One-piece, all-moving horizontal tail, combining functions of horizontal stabiliser and elevator.

stabilizer Tailplane (US); vertical stabilizer = fin.

stall Sudden near-total loss of lift of a wing because AoA has exceeded a critical value.

stall strips Sharp-edged strips on wing leading-edge to induce stall to initiate at that point.

stalling speed Airspeed at which aircraft stalls at $1g$.

starboard Right side, looking forward.

static inverter Solid-state (not rotary machine) inverter of alternating wave-form to produce DC from AC.

STC Supplementary Type Certificate.

stealth See low-observables.

stick-pusher Stall-protection device that forces pilot's control column forward as stalling angle of attack is neared.

stick-shaker Stall-warning device that noisily shakes pilot's control column as stalling angle of attack is neared.

STOL Short take-off and landing. (Several definitions, stipulating allowable horizontal distance to clear screen height of 35 or 50 ft or various SI measures.)

store Object carried as part of payload on external attachment (for example bomb, drop tank) (see photograph above).

STOVL Short take-off, vertical landing.

strobe light High-intensity flashing beacon.

supercritical wing Wing of relatively deep, flat-topped profile generating lift right across upper surface instead of concentrated close behind leading-edge.

sweepback Backwards inclination of wing or other aerofoil, seen from above, measured relative to fuselage or other reference axis, usually measured at quarter-chord (25 per cent) or at leading-edge.

t Tonne, 1 Megagram, 1,000 kg.

tab Small auxiliary surface hinged (flight-adjustable) or attached in a fixed position (ground-adjustable) to trailing-edge of control surface for trimming, balancing (reducing hinge moment: force needed to operate main surface) or in other way assisting pilot. Compare anti-balance tab.

tabbed flap Fitted with narrow-chord tab along trailing-edge which deflects to greater angle than main surface.

Tacan Tactical air navigation, UHF navaid giving bearing and distance to ground beacons; distance element (see DME) can be paired with civil VOR.

TACCO Tactical commander, ASW aircraft.

taileron Left and right tailplanes used as primary control surfaces in both pitch and roll.

tailplane Horizontal stabiliser; main horizontal tail surface, originally fixed and carrying hinged elevator(s) but today often a single 'slab' serving as control surface (see stabiliser, stabilator).

TANS Tactical air navigation system; Decca Navigator or Doppler-based computer, control and display unit.

TAS True airspeed, EAS corrected for density (often very large factor) appropriate to aircraft altitude.

TBO Time between overhauls.

t/c ratio Ratio of the thickness (aerodynamic depth) of a wing or other surface to its chord, both measured at the same place parallel to the fore-and-aft axis.

TCAS Traffic-alert and collision-avoidance system.

Tercom Terrain-comparison (or contour-matching), navigation aid which compares relief of terrain with profile stored in memory.

TFR Terrain-following radar (for low-level attack).

thickness Depth of wing or other aerofoil; maximum perpendicular distance between upper and lower surfaces.

thrust vectoring Rotation of a vehicle's thrust axis to control its trajectory or support its weight.

TIALD Thermal imaging and laser designation (pod).

tiltrotor Aircraft with fixed wing and rotors that tilt up for hovering and forward for fast flight.

NEW/0543336

Tyre classification systems				
Classification name	Example	Nominal diameter	Nominal section width	Nominal rim diameter
Type III	8.50-10		8½ in	10 in
Type VII	49×17	49 in	17 in	
Three Part	49×19.0-20	49 in	19.0 in	20 in
Radial	32×8.8R16	32 in	8.8 in	16 in
Metric	670×210-12	670 mm	210 mm	12 in

Example microlight/ultralight maxima						
Country	Name	Seat(s)	Empty weight	MTOW	Fuel	Vso
Australia	Ultralight	1-2		540 kg		
France	ULM[1]	1		300 kg[2]		35 kt[3]
	ULM[1]	2		450 kg[2]		35 kt[3]
UK	Microlight	1-2		450 kg		35 kt
USA	Ultralight	1	254 lb		5 USg	

[1] Ultra léger motorisé; criteria accepted by several European countries, including Germany
[2] Plus 10 per cent for seaplanes and amphibians
[3] Plus 5 per cent for seaplanes and amphibians

T-O Take-off.

T-O noise EPNdB measure of aircraft taking off, at point directly under flight path 3.5 n miles from brakes-release.

TOGW Take-off gross weight (not necessarily MTOW)

ton Imperial (long) ton = 1.016 t or 2,240 lb, US (short) ton = 0.9072 t or 2,000 lb.

track Distance between centres of contact areas of main landing wheels measured left/right across aircraft (with bogies, distance between centres of contact areas of each bogie).

transceiver Radio transmitter/receiver.

transformer-rectifier Device for converting AC to DC at a different voltage.

transponder Radio transmitter triggered automatically by a particular received signal, as in secondary surveillance radar (SSR).

TRU Transformer/rectifier unit.

TsENTROSPAS (in Russian Federation) Ministry for Civil Defence, Emergencies and Elimination of the Consequences of Natural Disasters.

turbofan Gas-turbine jet engine generating most thrust by a large-diameter cowled fan, with small part added by jet from core.

turbojet Simplest form of gas turbine comprising compressor, combustion chamber, turbine and propulsive nozzle.

turboprop Gas turbine in which as much energy as possible is taken from gas jet and used to drive reduction gearbox and propeller.

turboshaft Gas turbine in which as much energy as possible is taken from gas jet and used to drive high-speed shaft (which in turn drives external load such as helicopter transmission).

twist Progressive change of angle of incidence of a wing, rotor blade or other aerofoil from root to tip.

Type Certificate Airworthiness licence granted to enable a manufacturer to produce and market a specified type of aircraft (compare C of A).

tyre sizes Five systems of classification are in current use; Type I, consisting of a single figure indicating nominal diameter in inches, is obsolete. See adjacent table and also 'ply'

UAV Unmanned (or uninhabited) aerial vehicle; contrast RPV.

UHCA Ultra-high capacity airliner.

UHF Ultra-high frequency.

ultralight Light aircraft with parameters below specified national limits, qualifying for less rigorous licensing; also known as microlight. See table.

ULV Ultra-low volume (crop-spraying intensity).

unfactored Performance level expected of average pilot, in average aircraft, without additional safety factors.

upper surface blowing Turbofan jet expelled over upper surface of wing to increase lift.

usable fuel Total mass of fuel consumable in flight, usually 95 to 98 per cent of system capacity.

useful load Usable fuel and other consumables plus payload.

US gallon 0.83267 Imperial gallon; 3.785 litres.

UV Ultra-violet.

'V-speeds' Shorthand notation of significant speeds within an aircraft's flight envelope (see table).

variable geometry Capable of grossly changing shape in flight, especially by varying sweep of wings.

variable pitch Propeller with its blades held in rotary bearings in the hub, so that pitch (of all blades in unison) can be altered in flight. See constant speed; compare ground- and flight-adjustable pitch.

VDU Video (or visual) display unit.

vectored Capable of being pointed in different directions.

vertrep Vertical replenishment.

VFR Visual flight rules.

VHF Very high frequency.

VLF Very low frequency (area-coverage navaid).

VMS Vehicle management system.

VOR VHF omnidirectional range (network of VHF radio beacons each providing to/from bearing).

vortex generators Small blades attached to wing and tail surfaces to energise local airflow and improve control.

vortillon Short-chord fence (particularly on MD-80 series) ahead of and below leading-edge.

VSI Vertical speed (climb/descent) indicator.

V/STOL Vertical/short take-off and landing.

washout Inbuilt twist of wing or rotor blade reducing angle of incidence towards the tip.

watt SI unit of power, equal to 1 Js^{-1} (one Joule per second).

WDNS Weapon delivery and navigation system.

wet Housing fuel; wet wing often has extra connotation of integral tankage. Wet pylon can accommodate external fuel tank.

wheelbase Minimum distance from nosewheel or tailwheel (centre of contact area) to line joining mainwheels (centres of contact areas).

NEW/0131451

wide-body Passenger aircraft with cabin wide enough to have two longitudinal aisles between seats.

wing area Total projected area of clean wing (no projecting flaps, slats and so on) including all control surfaces and area of fuselage bounded by leading- and trailing-edges projected to centreline (inapplicable to slender-delta aircraft with extremely large leading-edge sweep angle). Described in *Jane's* as gross wing area; net area excludes projected areas of fuselage, nacelles, and so on.

wing loading Aircraft weight (usually MTOW) divided by wing area.

winglet Small auxiliary aerofoil, usually sharply upturned and often sweptback, at tip of wing.

WSO Weapon(s) system(s) officer.

yoke Pilot's flight control interface for pitch and roll axes in the form of a stick (control column) to the top of which is laterally pivoted a pair of handgrips in the form of a Y.

ZAO Zakrytoe Aktsionernoye Obshchestvo (Russian company constitution).

zero-fuel weight MTOW minus usable fuel and other consumables, in most aircraft imposing severest stress on wing and defining limit on payload.

zero/zero seat Ejection seat designed for use even at zero speed on ground.

ZFW Zero-fuel weight.

V-speeds definitions

V_1 Decision speed, up to which it should be possible to abort a take-off after failure of the critical engine and stop safely within the remaining runway length. After reaching V_1 the take-off must be continued.

V_2 Minimum take-off safety speed.

V_A Design manoeuvring speed. The speed below which abrupt and extreme control movements are possible (though not advised) without exceeding the airframe's limiting load factors.

V_B Design speed for maximum gust intensity.

V_C Design cruising speed.

V_D Design diving speed.

V_{DF} Maximum demonstrated diving speed. Also M_{DF}, maximum demonstrated Mach No. in a dive.

V_E Maximum speed at which landing gear (or other item) may be extended or retracted (cycled).

V_{FE} Maximum flap extension speed (top of white arc on ASI).

V_H Maximum level-flight speed with maximum continuous power.

V_{LE} Maximum speed with landing gear extended.

V_{MCA} Minimum control speed (air). Minimum speed at which directional control of a multi-engined aircraft can be maintained after failure of critical engine (in effect, the lowest speed at which the aircraft possesses sufficient rudder authority to counteract the yaw induced by asymmetric thrust).

V_{MU} Minimum unstick speed.

V_{NE} Never-exceed speed.

V_{MO} Maximum operating speed. Also M_{MO}, maximum operating Mach No.

V_{NO} Normal operating speed. The maximum structural cruising speed allowable for normal operating conditions.

V_R Rotation speed, at which to raise the nose for take-off.

V_{RA} Rough-air speed. Maximum recommended airspeed for penetrating turbulent air.

V_{REF} Any reference or 'bug' speed, typically quoted for approach speeds.

V_{S0} Stalling speed at maximum take-off weight, in landing configuration with flaps and landing gear down, at sea level, ISA conditions (bottom of white arc on ASI). Also V_S, stalling speed 'clean', and V_{S1}, stalling speed for a given configuration other than 'clean'.

V_{SSE} Minimum speed for deliberate shutting down of one engine for purposes of asymmetric flight training.

V_X Best angle of climb speed on all engines. Sometimes (UK usage) $V_\%$.

V_{XSE} Best engine-out angle of climb speed.

V_Y Best rate of climb speed on all engines.

V_{YSE} Best engine-out rate of climb speed. Sometimes $V_{\%SE}$.

V_{ZRC} Zero rate of climb speed (on one engine, where drag of inoperative engine reduces climb gradient to zero).

FREE ENTRY/CONTENT IN THIS PUBLICATION

Having your products and services represented in our titles means that they are being seen by the professionals who matter – both by those involved in procurement and by those working for the companies that are likely to affect your business. We therefore feel that it is very much in the interests of your organisation, as well as Jane's, to ensure your data is current and accurate.

■ **Don't forget** – You may be missing out on business if your entry in a Jane's book, CD-ROM or Online product is incorrect because you have not supplied the latest information to us.

■ **Ask yourself** – Can you afford not to be represented in Jane's printed and electronic products? And if you are listed, can you afford for your information to be out of date?

■ **And most importantly** – The best part of all is that your entries in Jane's products are TOTALLY FREE OF CHARGE.

Please provide (using a photocopy of this form) the information on the following categories where appropriate:

1. Organisation name: _____

2. Division name: _____

3. Location address: _____

   _____

4. Mailing address if different: _____

   _____

5. Telephone (please include switchboard and main department contact numbers, for example Public Relations, Sales, and so on):

   _____

6. Facsimile: _____

7. E-mail: _____

8. Web sites: _____

9. Contact name and job title: _____

   _____

10. A brief description of your organisation's activities, products and services: _____

   _____

   _____

11. Jane's publications in which you would like to be included: _____

   _____

Please send this information to:
Jacqui Beard, Information Collection, Jane's Information Group,
Sentinel House, 163 Brighton Road, Coulsdon, Surrey, CR5 2YH, UK
Tel: (+44 20) 87 00 38 08
Fax: (+44 20) 87 00 39 59
E-mail: yearbook@janes.co.uk

Copyright enquiries:
Contact: Keith Faulkner
Tel/Fax: (+44 1342) 30 50 32
E-mail: keith.faulkner@janes.co.uk

Please tick this box if you do not wish your organisation's staff to be included in Jane's mailing lists ☐

JAWA

ARGENTINA

LMAASA

LOCKHEED MARTIN AIRCRAFT ARGENTINA SA

Avenida Fuerza Aérea Argentina Km 5500, CP5010 Córdoba
Tel: (+54 351) 466 87 33
Fax: (+54 351) 466 87 34
e-mail: busdev@lmaasa.com
PRESIDENT: James A Taylor
GENERAL MANAGER, PROGRAMS AND MANUFACTURE:
Alberto O Buthet
INTERNATIONAL BUSINESS DEVELOPMENT DIRECTOR:
Bernard R Kelleher

Original FMA (Military Aircraft Factory) came into operation 10 October 1927 as central organisation for aeronautical research and production; underwent several name changes (see 1987-88 and earlier *Jane's*) before reverting, in 1968, to original title as component of Aérea de Material Córdoba (AMC) of Argentine Air Force. FMA converted into a joint stock company from April 1992, Air Force buying 30 per cent of the shares to establish itself as holding and management authority; control of FMA handed over to Planning Secretary of MoD on 20 December 1993. MoD and Lockheed Aircraft Argentina SA signed concession agreement on 15 December 1994, allocating management of FMA to Lockheed Martin, now Lockheed Martin Aircraft Argentina SA, from 1 July 1995.

Principal activities are aircraft design, manufacture, upgrading, maintenance and repair, current tasks including maintenance and modification of A-4AR Fighting Hawks, C-130 Hercules, IA 58 Pucarás and IA 63 Pampas and manufacture of a further batch of 12 Pampas. Further data on out-of-production aircraft in *Jane's Aircraft Upgrades*. Argentine Air Force maintenance contract renewed in January 2001 for five years at estimated value of US$230 million.

Laboratories, factories and other aeronautical division buildings occupy total covered area of 220,000 m² (2,368,075 sq ft); had workforce of more than 1,000 in 2001. Córdoba facility also accommodates Centro de Ensayos en Vuelo (Flight Test Centre), a separate division also controlled by Argentine Air Force, at which all aircraft produced in Argentina undergo certification testing.

UPDATED

LMAASA AT-63 PAMPA

TYPE: Basic jet trainer/light attack jet.
PROGRAMME: First-generation IA 63 Pampa initiated by Fuerza Aérea Argentina (FAA; Argentine Air Force) 1979, eventual configuration being selected over six other designs early 1980, with Dornier of Germany providing technical assistance (including manufacture of prototypes' wings and tailplanes); two static/fatigue test airframes and three flying prototypes built (first flight, by EX-01, 6 October 1984); first flight of production Pampa October 1987; 14 of initial batch of 18 (including three prototypes) delivered to FAA (10 survivors currently serving I Escuadron of 4 Grupo de Caza at El Plumerillo, Mendoza) from 1988; expected follow-on order for 46 did not materialise, and Argentine Navy requirement for 12 long remained in abeyance, but one new aircraft (E-816), assembled from existing and new components, delivered to FAA on 28 September 1999.

'New-generation' Pampa NG revealed late 1997 and offered to Argentine Air Force and Navy.

Resumption of production announced 29 June 2000 for a further batch of 12 Pampas in new AT-63 (with hyphen) configuration, deliveries beginning in 2003 and completing in June 2005; contract (2001) includes option for further 12. This reaffirmed in January 2001 to be part of five-year FAA support contract placed with LMAASA. US$230 million contract also includes upgrade of existing Pampas. Argentine Navy interest renewed in 2000; initial batch of eight in prospect, but no order yet announced (mid-2002).

CURRENT VERSIONS: **IA 63 Pampa:** Standard Argentine Air Force version; last described in 1998-99 *Jane's*.

Naval version: Strengthened landing gear, uprated engine and some changed avionics.

Pampa 2000 International: Details last appeared in 1997-98 *Jane's*.

Pampa NG A: Proposed advanced trainer with updated avionics. Essentially became AT-63 (initial version).

Pampa NG B: Proposed combat-capable version; uprated engine and avionics. Features to be incorporated in later upgrade of AT-63.

AT-63: Attack-trainer. Announced 2001 as new standard version for Argentine Air Force and potential export. Two prototypes modified from existing IA 63s in

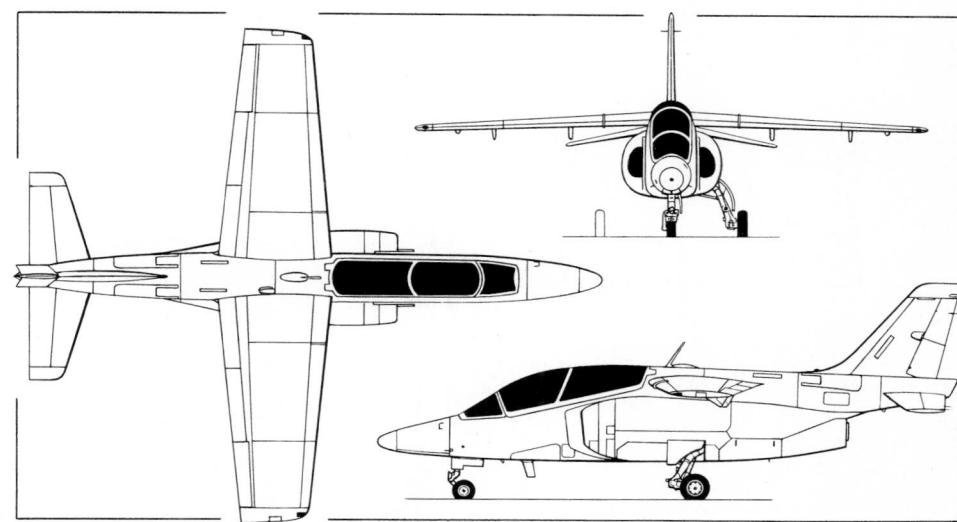

LMAASA AT-63 Pampa NG two-seat basic and advanced jet trainer *(Jane's/Dennis Punnett)*

2001-02; first flight planned for November 2002. Initial version (Pampa Phase II) features new processor for DECU, Elbit avionics suite (MIL-STD-1553B databus, mission computer, RLG INS/GPS, integrated weapons system, liquid crystal MFD in each cockpit, and front cockpit HUD). Advanced version (Pampa Phase III) proposed for 2005 with structurally enhanced (+7/–3 *g*) wing, 18.9 kN (4,250 lb st) TFE731-40R engine, missionised avionics with enhanced tactical capabilities, nose-mounted laser range-finder, fin-mounted RWR, conformal chaff/flare dispensers, two additional hardpoints (total seven) for AAMs and strengthened landing gear for increased MTOW.

AT-63 version also proposed with Lockheed Martin AN/APG-67(V)4 multimode radar in response to Colombian interest; announced at FIDAE, Chile, April 2002. Also reported that Singapore plans to evaluate AT-63 in early 2003 as S.211 replacement.

Description applies to Phase II Pampa.

DESIGN FEATURES: Intended for cost-effective pilot training in mission management techniques, advanced fighter lead-in training and extended-range anti-drug patrol missions. High degree of commonality with the original IA 63 Pampa, providing a customised low life-cycle cost fleet. Fail-safe service life 8,000 hours.

Non-swept shoulder-mounted wings and anhedral tailplane, sweptback fin and rudder; single engine with twin lateral air intakes. Wing section Dornier DoA-7/-8 advanced transonic; leading-edge sweep 5° 24′; thickness/chord ratio 14.5 per cent at root, 12.5 per cent at tip; anhedral 3°.

FLYING CONTROLS: Conventional hydraulically powered ailerons, rudder, all-moving tailplane, single-slotted Fowler flaps, and door-type airbrake, deployable at all speeds, on each side of upper rear fuselage; primary surfaces have Liebherr tandem actuators and electromechanical trim.

STRUCTURE: Conventional all-metal semi-monocoque/stressed skin; two-spar wing box forms integral fuel tank. Modified fuselage nosecone, tailcone and fin-tip in AT-63.

LANDING GEAR: SHL (Israel) retractable tricycle type, with hydraulic extension/retraction and emergency free-fall extension. Oleo-pneumatic shock-absorbers. Single Messier-Bugatti wheel on each unit with Goodrich (main) or Continental (nose) low-pressure tyre; nosewheel offset 10 cm (3.9 in) to starboard. Tyre sizes 6.50-10 (10 ply) on mainwheels, 380×150 (4/6 ply) on nosewheel, with respective pressures of 6.55 bar (95 lb/sq in) and 4.00 bar (58 lb/sq in). Nosewheel retracts rearward, mainwheels inward into underside of engine air intake trunks. Messier-Bugatti mainwheel hydraulic disc brakes incorporate anti-skid device; nosewheel steering (±47°). Gear designed for operation from unprepared surfaces.

POWER PLANT: One 15.57 kN (3,500 lb st) Honeywell TFE731-2C-2N turbofan installed in rear fuselage. Single-point pressure refuelling, plus gravity point in upper surface of each wing. Standard internal fuel capacity of 968 litres (255 US gallons; 213 Imp gallons) in integral wing tank of 550 litres (145 US gallons; 121 Imp gallons) and 418 litre (110 US gallon; 92.0 Imp gallon) flexible fuselage tank with a negative *g* chamber permitting up to 10 seconds of inverted flight. Additional 415 litres (109 US gallons; 91.0 Imp gallons) can be carried in auxiliary tanks inside outer wing panels, to give a maximum internal capacity of 1,383 litres (364 US gallons; 304 Imp gallons). Provision on centre underwing stations for two external drop tanks, each of 317 litres (83.7 US gallons; 69.7 Imp gallons). Total fuel capacity 2,017 litres (533 US gallons; 444 Imp gallons).

ACCOMMODATION: Tandem, rear seat elevated, on UPC (Stencel) S-III-S3IA63 zero/zero ejection seats. Ejection procedure can be preselected for separate single ejections, or for both seats to be fired from front or rear cockpit. HOTAS operation; dual controls standard. One-piece wraparound windscreen. One-piece canopy, with internal screen, is hinged at rear and opens upward. Entire accommodation pressurised and air conditioned.

SYSTEMS: Honeywell environmental control system, maximum differential 0.30 bar (4.4 lb/sq in), supplied by high- or low-pressure engine bleed air, provides a 1,980 m

IA 63 Pampa of 4 Grupo de Caza, Argentine Air Force   *NEW*/0118292

(6,500 ft) cockpit environment up to flight level 5,730 m (18,800 ft) and also provides ram air for negative *g* system and canopy seal. Oxygen system supplied by 10 litre (0.35 cu ft) lox converter. Engine air intakes anti-iced by engine bleed air.

Two independent hydraulic systems, each at pressure of 207 bar (3,000 lb/sq in), each supplied by engine-driven pump. Each system incorporates a bootstrap reservoir pressurised at 4 bar (58 lb/sq in). No.1 system, with flow rate of 16 litres (4.2 US gallons; 3.5 Imp gallons)/min, actuates primary flight controls, airbrakes, landing gear and wheel brakes; No. 2 system, with flow rate of 8 litres (2.1 US gallons; 1.75 Imp gallons)/min, actuates primary flight controls, wing flaps, emergency and parking brakes, and nosewheel steering. Honeywell ram air turbine provides emergency hydraulic power for No. 2 system if engine shuts down in flight and pressure in this system drops below minimum.

Electrical system (28 V DC) supplied by Lear Siegler 400 A 11.5 kW engine-driven starter/generator; secondary supply (115/26 V AC power at 400 Hz) from two Flite-Tronics 450 VA static inverters and two SAFT 27 Ah sealed lead batteries. Thirty minutes of emergency electrical power available in case of in-flight engine shutdown.

AVIONICS: *Comms:* Two VHF/UHF transceivers and direction-finder, intercom and IFF or ATC transponder.

Radar: Lockheed Martin AN/APG-67(V)4 multimode radar optional (not fitted in Argentine Air Force aircraft).

Flight: VOR/ILS with marker beacon receiver; DME or Tacan optional; autonomous ESIS/air data computer, HSI, ADF, ADI; Honeywell HG 764 laser INS with GPS; radar altimeter; air data computer.

Instrumentation: Single 127 mm (5 in) HUD (multimode: UFCP, PDU, PSVS and camera); colour HUD camera/airborne videotape recorder; video repeater; 12.7 × 17.8 cm (5 × 7 in) liquid crystal MFD in each

cockpit for Argentine Air Force; second, similar MFD in each cockpit (for EICAS) for export version; Multirole central computer and MIL-STD-1553B digital databus.

Mission: New integrated weapon delivery system. Mission computer/symbol generator/integrated comms; weapon management system/data transfer unit; steerable laser ring finder optional; radar warning receiver optional; chaff/flare dispenser.

ARMAMENT: Five stations for external stores, stressed for 440 kg (970 lb) on centre fuselage and each inboard underwing station, 290 kg (639 lb) on each outboard underwing station, all at +5.5/−2 *g*. Phase III aircraft will have further pair of pylons further outboard, each rated at 170 kg (375 lb), for total of seven, plus uprated inboard wing pylons. Several external stores configurations including Mk 81 and Mk 82 bombs; LAU-32, LAU-51 and LAU-10 rocket pods; 30 mm gun pod (centreline), twin machine gun pods; and CBLS 200 practice bomb carriers.

DIMENSIONS, EXTERNAL:

Wing span	9.69 m (31 ft 9½ in)
Wing aspect ratio	6.0
Length overall	10.93 m (35 ft 10¼ in)
Height overall	4.29 m (14 ft 1 in)
Tailplane span	4.58 m (15 ft 0 ¼ in)
Wheel track	2.66 m (8 ft 8¾ in)
Wheelbase	4.42 m (14 ft 6 in)

AREAS:

Wings, gross	15.63 m² (168.3 sq ft)
Ailerons (total)	0.89 m² (9.58 sq ft)
Trailing-edge flaps (total)	2.93 m² (31.54 sq ft)
Fin	1.86 m² (20.02 sq ft)
Rudder	0.655 m² (7.05 sq ft)
Tailplane	4.35 m² (46.87 sq ft)

WEIGHTS AND LOADINGS (estimated):

Weight empty	2,820 kg (6,217 lb)
Max fuel weight:	
(A) fuselage	327 kg (721 lb)
(B) wings, internal inboard	430 kg (948 lb)
(C) wings, internal outboard	325 kg (717 lb)
underwing drop tanks	496 kg (1,093 lb)
Max external stores load	1,900 kg (4,189 lb)
Max T-O weight	5,000 kg (11,023 lb)
Max wing loading	319.9 kg/m² (65.52 lb/sq ft)
Max power loading	321 kg/kN (3.15 lb/lb st)

PERFORMANCE (estimated at 3,764 kg; 8,300 lb clean T-O weight with normal internal fuel, except where indicated):

Max level speed at 7,985 m (25,900 ft)	440 kt (814 km/h; 506 mph)
Max operating speed (Vᴍᴏ)	M0.8
Econ cruising speed at 9,145 m (30,000 ft)	350 kt (648 km/h; 402 mph)
Stalling speed at S/L, 50% normal internal fuel:	
flaps up	104 kt (193 km/h; 120 mph)
flaps down	82 kt (152 km/h; 95 mph)
Max rate of climb at S/L	1,560 m (5,118 ft)/min
Service ceiling	12,900 m (42,320 ft)
T-O run	430 m (1,410 ft)
Landing run at 3,497 kg (7,710 lb)	460 m (1,510 ft)

Radius of action:
air-to-air (hi-hi), T-O weight of 4,300 kg (9,480 lb) with 254 kg (560 lb) external load, 5 min allowance for dogfight, normal internal fuel, 30 min reserves
380 n mile (703 km, 437 miles)
air-to-ground (hi-lo-lo-hi), 30 n mile dash out/in, T-O weight of 5,000 kg (11,023 lb) with 1,000 kg (2,205 lb) external load, max internal fuel, 5 min allowance for weapon delivery, plus 30 min reserves
236 n miles (127 km, 205 miles)
Ferry range 9,145 m (30,000 ft), ISA with 15 min reserves
1,140 n mile (2,111 km; 1,311 miles)
g limits +6/−3

UPDATED

'First-generation' IA 63 Pampa tandem-seat jet trainer

NEW/0131688

AUSTRALIA

BAE

BAE SYSTEMS AUSTRALIA LIMITED
Delivery of Australian-assembled Hawk jet trainers was completed in 2001; see 2002-03 *Jane's* for details.

UPDATED

EAGLE

EAGLE AIRCRAFT PTY LTD
PO Box 586, Fremantle, Western Australia 6959
Tel: (+61 8) 94 14 11 74
Fax: (+61 8) 94 14 11 75
e-mail: sales@eagleair.com.au
Web: http://www.eagleair.com.au
CEO: Nor Manshor Gafar
MARKETING MANAGER: Ron Scherpenzeel
MARKETING ASSISTANT: Julia Kemmer

US DISTRIBUTOR:
HGL Aero Inc, 10504 SW Indianola Road, Augusta Airport, Augusta, Kansas 67010
Tel: (+1 316) 733 60 15
Fax: (+1 316) 733 62 50
e-mail: jdavis@hglaero.com
Web: http://www.hglaero.com
SALES MANAGER: James Davis

Australian company Eagle Aircraft was purchased by the Malaysian Government in 1990. In December 2001 it announced that production would move to Malaysia, the Henderson, Western Australia, plant closing at the end of February 2002. Manufacture is now entirely by CTRM (which see in Malaysian section).

UPDATED

EAGLE EAGLE 150
TYPE: Two-seat lightplane.
PROGRAMME: Launched 1981 with objective of producing first all-composites light aircraft in Australia; single-seat POC (proof-of-concept) aircraft now displayed at Power House Museum, Sydney; construction of two-seat preproduction prototype Eagle X started fourth quarter 1987; first flight (VH-XEG) second quarter 1988, with 58 kW (78 hp) Aeropower engine, replaced later by 74.5 kW (100 hp) Continental O-200; 200 hour test programme, meeting all original design criteria, completed by October 1988; plans for 1989 production start aborted (see 1991-92 *Jane's*); component manufacture began March 1991; production prototype (VH-XEP) made first flight 6 November 1992; weight-restricted certification by Australian CASA 21 September 1993; European JAR-VLA certificate not awarded at that time and aircraft consequently redesigned with changes including IO-240-B engine; increased span and chord on foreplane and mainplane flaps; redesigned and repositioned wing cuffs, with leading-edge extension

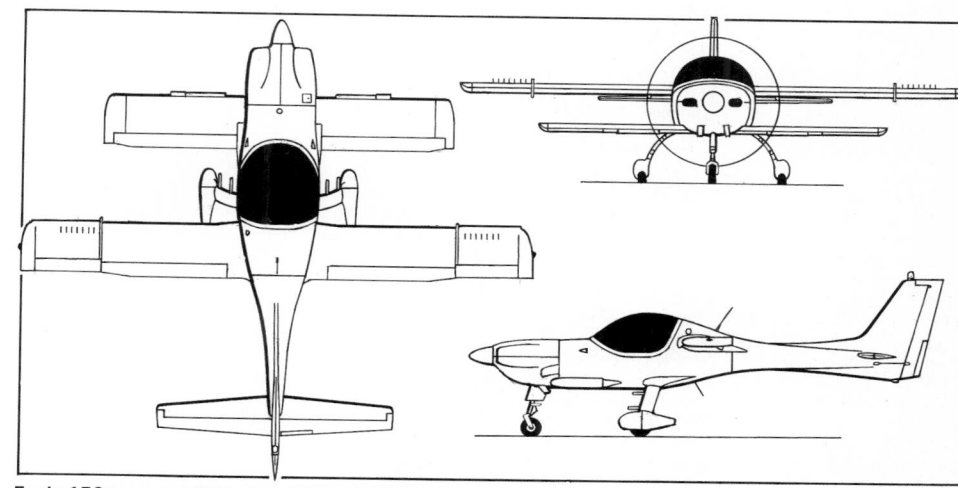

Eagle 150 two-seat light aircraft *(Jane's/Mike Keep)*   0092121

outboard; vortex generators and repositioned horizontal stabiliser.

Series production in Australia, as Eagle X-TS, launched August 1993; first flight (VH-AHH) 23 October 1993 and first customer delivery December 1993 (VH-FPO to Department of Conservation and Land Management in Western Australia). Initial production in Australia only, but using some components manufactured by Eagle Aircraft (Malaysia) of which Eagle Aircraft Pty Ltd is a wholly owned subsidiary. First aircraft completed from Malaysian components (VH-PMI, c/n MA 00001) made maiden flight 27 March 2001. In December 2001 it was announced that production was moving to Malaysia and operations in Australia were being run down; this had been achieved by late February 2002.

First Series 150 (converted production prototype VH-XEP) rolled out August 1997; certified on 13 November 1997 by CASA (Civil Aviation Safety Authority of Australia) to JAR-VLA standards at MTOW of 640 kg (1,411 lb); this certification also valid in Malaysia. FAA certification achieved 11 February 1999, followed by New Zealand approval in mid-1999; also approved in Thailand; JAA certification expected in 2002. First night VFR instrumented aircraft certified and delivered in December 2000.

CURRENT VERSIONS: **Series 100:** Retrospective designation of former Eagle X-TS: production completed. Total of 10 built, of which five converted or undergoing conversion to Series 150 by late April 1998. Described in Malaysia section in 1998-1999 *Jane's*.

Series 150: Current production version, designated **150A** with IO-240-A engine and **150B** with IO-240-B. Engines have same maximum power of 93.2 kW (125 hp), but B version has increased mid-range power of almost 29 per cent to 69.4 kW (93 hp) and is 3.7 per cent quieter than A model. Available in Basic, Training and Executive variants, as below. US promotion devoted to 150B version only.

Basic: Standard Series 150 version; *as described.*

Training: As Basic, with addition of tachometer, turn co-ordinator, directional gyro, digital clock, cockpit internal lighting, transponder, nav/com/VOR, navigation and landing lights and Eagle external pinstriping and decals.

Executive: As Training, plus GPS moving map display, mainwheel spats, leather upholstery, cabin heater, baggage bins, sun visor and chart pocket.

Eagle ARV: Optionally piloted surveillance version (Aerial Reconnaissance Vehicle), co-developed by CTRM with BAE Systems Controls of California, which supplies complete modification kit including flight management system, sensor payload, datalink and ground control station. First three-aircraft system delivered to Royal Malaysian Air Force in late 2001. Further details in *Jane's Unmanned Aerial Vehicles and Targets.*

CUSTOMERS: Ten Series 100/Eagle X-TS built, including two for Malaysia (c/n 0003 and 0005) and one for John Roncz in USA (c/n 0010): five conversions (c/n 0002/3/5/7/8) to Series 150.

Series 150s operators include: Aerostaff (one), Australian Flying Training School (one), DTIL New Zealand (one), Guernsey Aviation USA (two), HGL Aero of Kansas, USA (28), Horizon Airways (one), International Aviation and Travel Academy USA (two), Malacca Flying Club Malaysia (three), Perak Aero Club Malaysia (one), Phoenix Aviation (two), Royal Queensland Aero Club (two), Tanjung Flying Club (one), Troy Aviation (one) and Victoria Civil Aviation Academy (three).

A pair of US-registered Eagle Series 150s   NEW/0131695

Eagle 150 instrument panel   0113154

By February 2002, production (including Eagle X) totalled 44 in Australia (of which only two registered in 2001) and three in Malaysia.

COSTS: Basic Series 150 A$150,000 (US$104,900), Sports US$119,900, Training US$121,400 (all 1999). Basic Eagle 150B A$159,000 (US$119,000), Training A$187,500 (US$124,400), Executive A$198,500 (US$129,500) (all 2000). Total maintenance cost US$9.98 per hour over 2,000 hours' utilisation (2000).

DESIGN FEATURES: Intended primarily for *ab initio* training, recreational flying and surveillance. 'Tri-surface' configuration, with high-mounted mainplane, large low-mounted foreplane, and tailplane. Stall strips on foreplane ensure that it stalls before the mainplane.

Mainplane of tailored Roncz aerofoil section, thickness/chord ratio 16 per cent; no sweep; no dihedral.

FLYING CONTROLS: Conventional and manual. Slotted ailerons on mainplane, elevators on tailplane, and rudder, all with normal manual/mechanical actuation. Pushrods on ailerons and elevators, cables on rudder. Electric pitch trim (tab in starboard elevator), manual roll trim; rudder tab. Electrically actuated single-slotted flaps on foreplane (full span) and mainplane (part span). Leading-edge stall strips, vortex generators and fences on mainplane, stall strips on foreplane. Flap and airflow control systems designed to achieve low stalling speed with relatively high wing loading, to provide good ride quality in turbulence.

STRUCTURE: Except for metal engine mounts and flight control rods, Model 150 built entirely of composites, with 90 per cent of structure bonded/assembled by Composites Technology Research Malaysia (CTRM). Wings, fuselage and all control surfaces are Nomex honeycomb or high-density foams, sandwiched between multiple layers of carbon fibre; Kevlar reinforcement around wing leading-edges and shoulder of mainplane; cockpit is impact-resistant capsule of multilayered Kevlar and carbon fibre; carbon fibre spars. Entire structure uses Eagle-designed vinylester resins for strength/longevity/impact resistance and to minimise 'wet environment' problems inherent in standard epoxies.

LANDING GEAR: Non-retractable tricycle type, with glass fibre/epoxy self-sprung main legs and oleo nose leg. Cleveland 5.00-5 wheel on each unit, with optional speed fairings. Cleveland hydraulic single-disc brakes on mainwheels. Castoring nosewheel, size 11×4.00. Minimum ground turning radius 5.21 m (17 ft 1 in). Brakes applied by pressure on rudder pedals. Twin landing lights in nosewheel fairing.

POWER PLANT: One 93.2 kW (125 hp) Teledyne Continental IO-240-A or IO-240-B7B flat-four engine (see Current Versions above); respectively driving a McCauley IA135BRM7054 or IA135CRM7057 two-blade fixed-pitch metal propeller. Trials with MTV-7-D/175-112 three-blade propeller under way in early 2001. Fuel capacity 100 litres (26.4 US gallons: 22.0 Imp gallons), of which 97 litres (25.6 US gallons: 21.3 Imp gallons) are usable.

ACCOMMODATION: Two seats side by side with Y-shape control column operable from either seat. Adjustable rudder pedals. Bubble canopy hinged at front and opens upward. Four-point safety harnesses. Hat shelf and two baggage compartments.

SYSTEMS: Hydraulic system for manual brake actuation only; 12 V DC electrical system with 60 A alternator.

AVIONICS: Standard avionics suite as described below. Optional equipment as detailed under Current Versions. Night VFR package also available.

Comms: VHF com, intercom and ELT standard.

Instrumentation: Airspeed indicator, altimeter, tachometer, oil pressure/temperature gauge, fuel pressure gauge, fuel quantity gauge, voltmeter/ammeter, OAT gauge, magnetic compass, pitch trim indicator, engine hours meter and flight hours meter.

DIMENSIONS, EXTERNAL:

Wing span	7.16 m (23 ft 6 in)
Wing chord, constant	0.74 m (2 ft 5 in)
Wing aspect ratio	9.9
Foreplane span	4.88 m (16 ft 0 in)
Foreplane chord, constant	0.74 m (2 ft 5 in)
Length overall	6.45 m (21 ft 2 in)
Height overall	2.31 m (7ft 7 in)
Tailplane span	3.30 m (10 ft 10 in)
Wheel track	1.93 m (6 ft 4 in)
Wheelbase	1.40 m (4 ft 7 in)
Propeller diameter	1.78 m (5 ft 10 in)

DIMENSIONS, INTERNAL:

Cabin: Length	1.37 m (4 ft 6 in)
Max width	1.07 m (3 ft 6 in)
Max height	0.86 m (2 ft 10 in)

AREAS:

Wings, gross	5.20 m² (56.0 sq ft)
Foreplanes, gross	3.62 m² (39.0 sq ft)
Wing flaps (total)	0.90 m² (9.68 sq ft)
Foreplane flaps (total)	0.90 m² (9.68 sq ft)
Tailplane	1.49 m² (16.00 sq ft)

WEIGHTS AND LOADINGS:

Weight empty	429 kg (946 lb)
Baggage capacity: shelf	9 kg (20 lb)
compartments (total)	36 kg (80 lb)
Max T-O weight	650 kg (1,433 lb)
Max wing/foreplane loading	73.7 kg/m² (15.09 lb/sq ft)
Max power loading	6.97 kg/kW (11.45 lb/hp)

PERFORMANCE:

Max level speed at S/L	130 kt (241 km/h; 150 mph)
Cruising speed at 75% power at 660 m (2,000 ft)	125 kt (232 km/h; 144 mph)
Stalling speed at S/L: flaps up	52 kt (97 km/h; 60 mph)
flaps down	43 kt (80 km/h; 50 mph)
Max rate of climb at S/L	322 m (1,055 ft)/min
Service ceiling	4,575 m (15,000 ft)
T-O to 15 m (50 ft)	349 m (1,145 ft)
Landing from 15 m (50 ft)	365 m (1,200 ft)
Range with max fuel	520 n miles (963 km; 598 miles)
Endurance at 60% power	5 h
g limits: JAR-VLA	+3.8/–1.9
ultimate	+8.55/–4.27
	UPDATED

ENDEAVOUR

ENDEAVOUR AEROSPACE

PO Box 582, Werribee, Victoria 3030
Tel: (+61 0402) 29 37 35
Fax: (+61 397) 34 78 87
e-mail: endaero@optusnet.com.au
MANAGING DIRECTOR: Mark Graham
ENGINEERING DIRECTOR: Richard Bartholomeusz

Company formed in 2000 to create an Australian general aviation export business. Acquired rights to complete development of, and export, Hawker de Havilland Wamira turboprop trainer; aircraft will be offered at a competitive price, with emphasis on US private market. Financial backing being obtained in 2002-03.

NEW ENTRY

ENDEAVOUR PHOENIX

TYPE: Aerobatic two-seat turboprop sportplane.

PROGRAMME: HDH A10 Wamira programme begun in 1982 to meet Royal Australian Air Force requirement for turboprop pilot trainer; redefined as A10B in May 1985 but rejected in December 1985 in favour of Pilatus PC-9. Prototype then substantially complete and HDH continued development with aim of late 1986 first flight. However, venture was cancelled and A$80+ million costs written off.

Rights acquired by Endeavour, which is to produce sportplane and touring versions for export market, while retaining military potential. Following implementation of

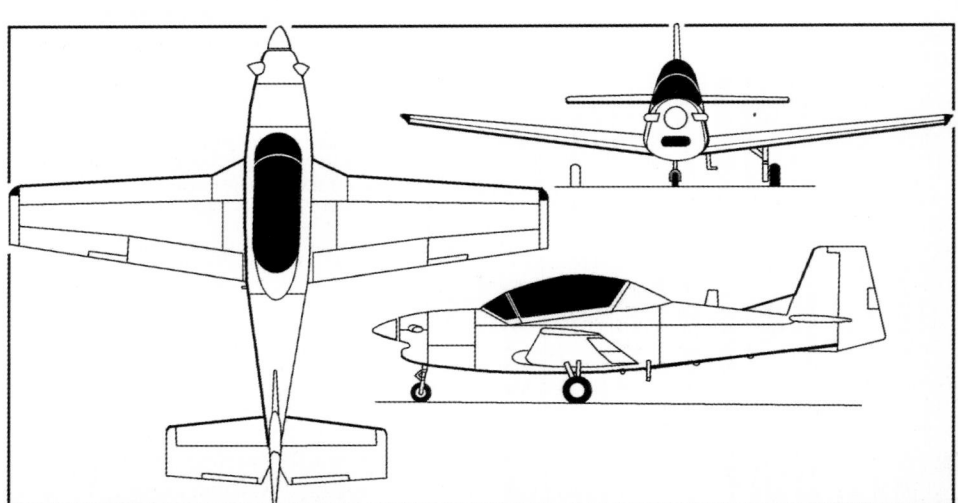

Tandem-seat version of Endeavour Phoenix   *NEW*/0527121

funding plan, estimated two-year programme to complete prototype/demonstrator under Experimental certification.

CURRENT VERSIONS: **Tandem:** Sportplane or intermediate military trainer conforming to current preference.

Side-by-side: Original Wamira configuration. Tourer or basic military trainer.

Both are available with choice of engine power and are suitable for pilot training up to first 100 hours of a military syllabus. Version commonality is 80 per cent.

DESIGN FEATURES: Low-wing, high-performance, aerobatic turboprop; mid-mounted tailplane. Tapered wing planform with root nibs and dihedral. Enlarged rudder on higher-powered version. Sweptback fin with small fillet; tapered tailplane.

FLYING CONTROLS: Conventional and manual. Mass-balanced ailerons with tab each side, port tab electrically actuated. Electric trim tabs in elevators (two) and rudder. Pushrod actuation throughout. Electrically actuated flaps.

STRUCTURE: Aluminium 2024 semi-monocoque with two-spar wing; single-spar rudder; tailplane as wings; some composites fairings. Comprehensive corrosion-proofing.

LANDING GEAR: Tricycle type; retractable. Mainwheels, with 7.00-8 tyres, retract inwards electromechanically and have no doors; nosewheel, 5.00-5, retracts rearward. All tyre pressures 2.9 bar (42 lb/sq in). Manual emergency extension. Oleo-pneumatic shock-absorber on each leg.

POWER PLANT: One Pratt and Whitney Canada PT6A-25C turboprop, flat-rated to either 559 kW (750 shp) or 634 kW (850 shp), driving a Hartzell HC-B3TN-T10173-11R constant-speed, fully feathering, reversible-pitch, three-blade propeller. Single fuel tank in lower centre fuselage, capacity 580 litres (153 US gallons; 128 Imp gallons).

ACCOMMODATION: Two pilots in tandem (single-piece canopy) or side-by-side (two-piece canopy) with raised

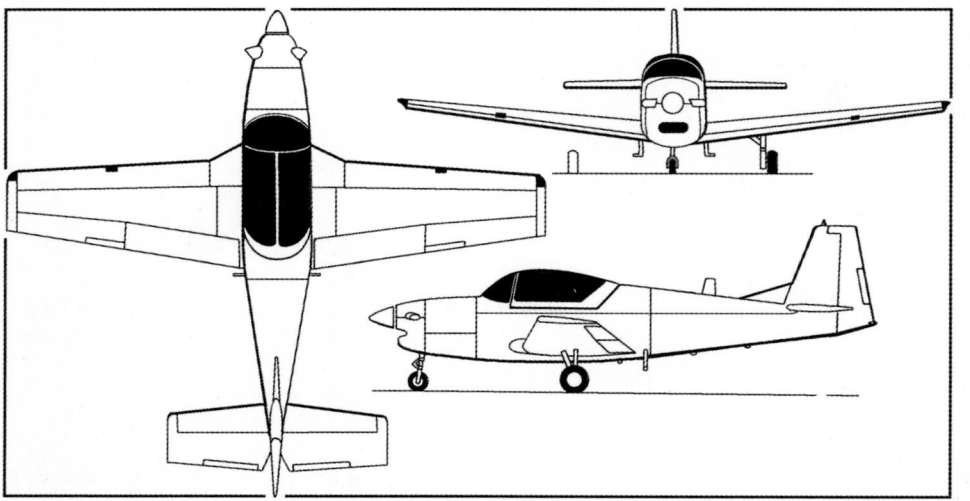

Side-by-side Phoenix more closely resembles the HDH Wamira   *NEW*/0527115

rear seat. Separate windscreen; canopies slide rearwards. Partial pressurisation. Baggage space behind seats, with external access.

SYSTEMS: Hydraulic system for brakes; no pneumatic system. 28 V DC electrical system incorporating 28 V, 200A starter/generator, with two 125 VA static invertors for AC supply. Crew oxygen system. Environmental control system.

DIMENSIONS, EXTERNAL:

Wing span	11.00 m (36 ft 1 in)
Wing chord: at root	3.03 m (9 ft 11¼ in)
at tip	1.21 m (3 ft 11½ in)
Wing aspect ratio	6.1
Length overall	10.20 m (33 ft 5½ in)
Height overall	3.51 m (11 ft 6¼ in)
Tailplane span	4.50 m (14 ft 9¼ in)
Wheel track	4.00 m (13 ft 1½ in)
Wheelbase	3.10 m (10 ft 2 in)
Propeller diameter	2.29 m (7 ft 6 in)

AREAS:

Wings, gross reference	20.00 m² (215.3 sq ft)

WEIGHTS AND LOADINGS (A: 559 kW; 750 shp engine, B: 634 kW; 850 shp engine):

Weight empty: A	1,636 kg (3,607 lb)
B	1,698 kg (3,744 lb)
Max ramp weight: A	2,213 kg (4,878 lb)
B	2,303 kg (5,077 lb)

PERFORMANCE (estimated; A, B as above):
Never-exceed speed (VNE)

	M0.575 (320 kt; 593 mph; 368 mph) EAS

Max level speed: at FL150:

A	238 kt (441 km/h; 274 mph)
B	248 kt (459 km/h; 285 mph)
at S/L: A	230 kt (426 km/h; 265 mph)
B	238 kt (441 km/h; 274 mph)

Max cruising speed: at FL150:

A	229 kt (424 km/h; 264 mph)
B	244 kt (452 km/h; 281 mph)
at S/L: A	225 kt (417 km/h; 259 mph)
B	227 kt (420 km/h; 261 mph)
Stalling speed: flaps up: A	69 kt (128 km/h; 80 mph)
B	71 kt (132 km/h; 82 mph)
flaps down: A	59 kt (110 km/h; 68 mph)
B	60 kt (112 km/h; 69 mph)
Max rate of climb at S/L: A	885 m (2,905 ft)/min
B	1,012 m (3,320 ft)/min
Time to FL150: A	6 min 40 s
B	6 min 5 s
Max certified ceiling	9,570 m (31,400 ft)
T-O to 15 m (50 ft): A	363 m (1,190 ft)
B	397 m (1,300 ft)
g limits	+7.0/−3.5

NEW ENTRY

GIPPSLAND

GIPPSLAND AERONAUTICS PTY LTD

Latrobe Valley Airport, PO Box 881, Traralgon, Morwell, Victoria 3840
Tel: (+61 3) 51 74 30 86
Fax: (+61 3) 51 74 09 56
e-mail: sales@gipplandaeronautics.com.au
Web: http://www.gippslandaeronautics.com.au
DIRECTORS:
   George Morgan
   Peter Furlong

Involved since 1971 in design and modification programmes for a wide range of aircraft, from wooden homebuilts to pressurised turboprops. Conversion of five Piper Pawnees to two-seat configuration in second half of 1980s led eventually to a totally new design, the GA-200 crop-sprayer. Since 1997, Gippsland has been seeking partners in Asia and South America for licensed production of the GA-200. Second Gippsland aircraft, GA-8 Airvan, first flew in 1995 and was delivered to customers from December 2000 onwards. First export Airvan flew in 2001.

Other activities include main spar life extension of Piper Pawnee and Navajo/Chieftain, of which combined total of 50 refurbished by 2001.

UPDATED

GIPPSLAND GA-200 FATMAN

TYPE: Agricultural sprayer.

PROGRAMME: Precursor was two-seat conversion of five Piper PA-25-235 Pawnees to -235/A8 or -235/A9 Fatman standard in 1986-90. Prototype GA-200 (VH-BCE) registered 1991; this and second aircraft used airframes from damaged Pawnees; No 3 (VH-SKG) was first new-build, in 1992. Designation derived from hopper capacity in US gallons. Full Australian CAA certification in both Normal and Agricultural categories (to CAO 101.16 and 101.22 and FAR Pt 23 Amendment 23.36) awarded 1 March 1991; FAR Pt 23 certification awarded 15 October 1997 in Restricted category. Certification also achieved in Brazil, Canada and elsewhere. Production at Traralgon, Victoria, since 1993.

CURRENT VERSIONS: **GA-200:** Initial version; 194 kW (260 hp) Textron Lycoming O-540-H2A5 engine. Hopper capacity 776 litres (205 US gallons; 170 Imp gallons). Replaced by GA-200C.

GA-200 Ag-trainer: Training version, with dual controls, dual rudder pedals and smaller hopper. Total of three built, plus unspecified number of conversions from GA-200s.

Gippsland GA-200C Fatman agricultural aircraft *(Jane's/Paul Jackson)*   **NEW**/0131701

GA-200B: Unofficial designation of GA-200 with extended wingtips. Most built up to 1998 were to this standard.

GA-200C: *As described.* Uprated engine beneath cowling of revised shape; constant-speed propeller. In early 1998, 21st production (23rd overall) aircraft completed as prototype GA-200C; all subsequent aircraft to this standard.

CUSTOMERS: Delivery of 44th (including prototypes) effected in November 2001; 45th and 46th then substantially complete. Exports to China (nine), New Zealand (eight, beginning 1994), South Africa (one, in March 1999) and USA (five, assembled locally from Australian-manufactured airframe components and US engines, wheels, brakes, instruments and avionics, plus at least two more, exported in early 2000).

DESIGN FEATURES: Purpose-designed crop-sprayer. Braced low wing, with large integral hopper forward of cockpit; crash-resistant, corrosion-proofed structure; gap-sealed ailerons; detachable wingtips. Although initially based on Piper Pawnee, GA-200 in current form is substantially different. Wing dihedral 7° from roots. Flaps and ailerons non-handed. Horn-balanced elevators and rudder.

FLYING CONTROLS: Conventional and manual, cable actuated. Single-slotted trailing-edge wing flaps can be deployed to tighten turning radius during agricultural operations; T-O setting 15°, maximum 38°. Interconnect system applies bias to elevator trim spring when flaps extended, to avoid pitch trim changes. Fixed tab on rudder.

STRUCTURE: Fuselage of welded SAE 4130 chromoly steel tube with removable metal side and top panels; wings (braced by overwing inverted V strut of aluminium each side) and wire-braced tail surfaces conventional all-metal, but wing spars constructed from sheet metal to expedite repairs; wingtips detachable. Rear fuselage and tail surfaces are fabric covered.

LANDING GEAR: Non-retractable, with 6 in diameter Cleveland P/N 40-84A mainwheels mounted on tubular steel side Vs; wire-cutter on leading tube of landing gear legs; rubber cord shock-absorption with hydraulic dampers and Cleveland hydraulic disc brakes. Mainwheel tyres size 8.50×6 (6 ply). Scott 3200 steerable/castoring tailwheel, size 5.00-5 mounted on multileaf flat springs.

POWER PLANT: One 224 kW (300 hp) Textron Lycoming IO-540-K1A5 flat-six engine, driving an HC-C2YR-1BF/F8475R constant-speed, two-blade metal propeller. Fuel in integral tank in each wing, combined usable capacity 200 litres (52.8 US gallons; 44.0 Imp gallons), plus small 14 litre (3.7 US gallon; 3.1 Imp gallon) header tank in upper front fuselage. Oil capacity 11.4 litres (3.0 US gallons; 2.5 Imp gallons).

ACCOMMODATION: Two energy-absorbing seats, side by side (smaller, right-hand seat for loader/driver); dual controls and second set of rudder pedals in right-hand seat in Ag-trainer. Four-point 25 g restraint harness(es). Cockpit doors open upward, each with bulged window to improve shoulder room and outward view.

SYSTEMS: 14 V 55 A automotive alternator and automotive or R-35 aviation battery for electrical power; 50 A circuit breaker switch serves as master switch.

AVIONICS: *Instrumentation:* Basic VFR instruments only.

EQUIPMENT: One 1,050 litre (277 US gallon; 231 Imp gallon) capacity hopper in forward fuselage (approximately 76 litres; 20.0 US gallons; 16.7 Imp gallons less in Ag-trainer), with minimum load dump time of five seconds. Multirole door/hopper outlet, eliminating need to change outlet when changing between solids and liquids, can also be used for fire bombing or laying of fire retardants. Spreader vanes can be added to increase swath width. 100 W landing light in each wingtip; 28 V night working lights system (two retractable underwing 600 W lights, powered by separate 28 V 55 A alternator) also available.

DIMENSIONS, EXTERNAL:

Wing span, standard version	11.985 m (39 ft 3¾ in)
Wing chord, constant	1.60 m (5 ft 3 in)
Wing aspect ratio	7.3
Length overall (flying attitude)	7.48 m (24 ft 6½ in)
Height (static) over cockpit canopy	2.33 m (7 ft 7¾ in)
Tailplane span	2.90 m (9 ft 6¼ in)
Wheel track	2.335 m (7 ft 8 in)
Propeller diameter	2.13 m (7 ft 0 in)

AREAS:

Wings, gross	19.60 m² (211.0 sq ft)

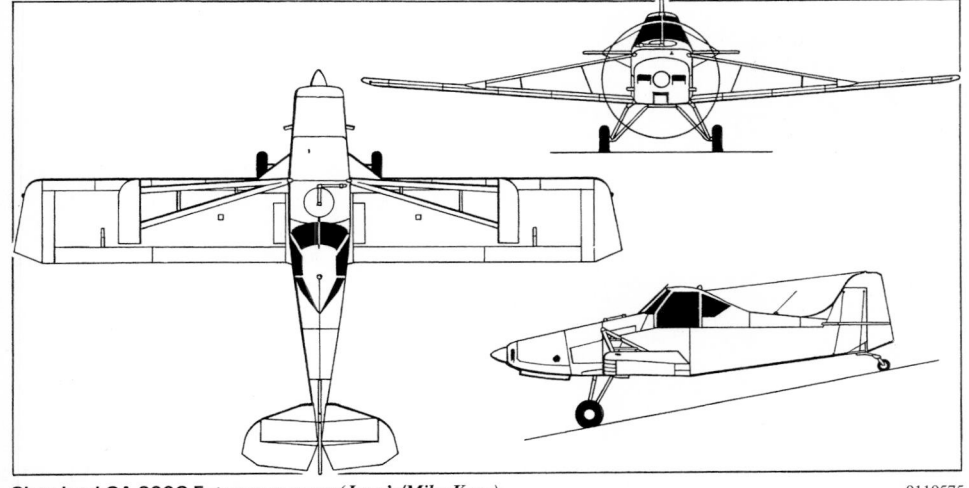

Gippsland GA-200C Fatman sprayer *(Jane's/Mike Keep)*   0110575

WEIGHTS AND LOADINGS (GA-200C, agricultural):

Operating weight empty	868 kg (1,914 lb)
Max T-O weight	1,995 kg (4,400 lb)
Max wing loading	101.8 kg/m² (20.85 lb/sq ft)
Max power loading	8.92 kg/kW (14.67 lb/hp)

PERFORMANCE (GA-200C, agricultural):

Long-range cruising speed (clean)	
	110 kt (204 km/h; 127 mph)
Stalling speed at 1,656 kg (3,650 lb):	
flaps up	57 kt (106 km/h; 66 mph) IAS
flaps down	51 kt (95 km/h; 59 mph) IAS
Max rate of climb (clean) at S/L	149 m (490 ft)/min
T-O run	580 m (1,903 ft)

UPDATED

GIPPSLAND GA-8 AIRVAN

TYPE: Light utility transport.

PROGRAMME: Design completed and prototype construction started early 1994; first flight 3 March 1995 (aircraft unregistered, but marked 'GA-8' on the fin); 186 kW (250 hp) Textron Lycoming O-540; prototype destroyed 7 February 1996 during spinning trials; second airframe completed to major component stage for static testing; third airframe (also unmarked; 'AIRVAN' on fin; VH-ZGI not worn) was second flying prototype, first flown in August 1996 with an interim 224 kW (300 hp) Textron Lycoming IO-540K-1A5 engine and three-blade propeller, pending installation of intended (but temporarily discontinued) production standard IO-580 in 1997; total of 350 hours flight testing completed by two prototypes by November 1998; second prototype re-registered VH-XGA in January 1999.

Provisional type certification achieved in early 1999, with full certification to FAR Pt 23 then scheduled for third quarter of 1999, but not achieved until 18 December 2000. First delivery (VH-RYT) to Air Fraser Island Air 22 December 2000. Initial export aircraft (seventh overall) first flew 24 November 2001, and delivered to Maya Island Air of Belize in following month. International debut (VH-BJY) at Asian Aerospace show in Singapore February 2002. Certification then being sought in Canada, South Africa, UK and USA. In early 2002 Gippsland was investigating the possibility of overseas assembly of the Airvan, and was holding discussions with potential sub-assembly manufacturers in China, Indonesia and South Africa. Initial production rate 17 to 20 per year, rising to 50 in 2005.

CURRENT VERSIONS: **GA-8 Airvan:** Initial production version; IO-540 engine. To be recertified subsequently with IO-580 and turbocharged TIO-540 for hot-and-high operations. Diesel version powered by Thielert TAE 135 turbo-diesel under study in early 2002.

GA-10T Taska: Stretched, 10-seat turboprop version to be powered by a 313 kW (420 shp) Rolls-Royce 250. Under development in early 2002 as joint venture with Helitech of Brisbane, aimed at paramilitary and special missions applications and to compete against Pacific Aerospace 750XL (which see in New Zealand section) for Australian Army requirement.

CUSTOMERS: Twenty-five sold by February 2002. Australian customers include Wrightsair, Alligator Airways (two) and Air Fraser Island. Export orders had been received from Advanced Wing Technologies (AWT) of Canada (six), Maya Island Air of Belize (six), an undisclosed operator in Surinam and Satmarindo, the manufacturer's Indonesian agent. AWT (which see) has sales rights in North America and plans to produce stretched GA-8 under licence.

COSTS: A$550,000, plus tax (2001).

DESIGN FEATURES: Strut-braced, high-wing monoplane with sweptback vertical tail and fixed tricycle landing gear; designed to operate from unprepared strips. Fin and rudder modified, and ventral finlet added on second prototype, late 1996.

Wing aerofoil modified V-35; dihedral 2° 30′; incidence 2°; twist 1.6°.

FLYING CONTROLS: Conventional and manual. Trimmable tailplane; slotted flaps, deflections 14 and 38°.

STRUCTURE: Light alloy; two-spar wing based on GA-200 Fatman unit; composites cowlings and fairings; entire structure designed for easy manufacture, maintenance and repair.

LANDING GEAR: Non-retractable tricycle type; single-piece tubular spring main gear, steerable steel spring/oleo nose leg; Cleveland hydraulic disc brakes; mainwheels 8.50-6 (6 ply); nosewheel 6.00-6 (6 ply). Wipline float installation under development in Canada by AWT.

POWER PLANT: One 224 kW (300 hp) Textron Lycoming IO-540-K1A5 flat-six engine driving a three-blade Hartzell F8475R constant-speed propeller. Fuel capacity 340 litres (89.8 US gallons; 74.8 Imp gallons) in two wing tanks, of

Gippsland GA-8 Airvan (*Jane's/Paul Jackson*)   *NEW*/0131702

which 332 litrés (87.7 US gallons; 73.0 Imp gallons) usable. Oil capacity 9 litres (2.4 US gallons; 2.0 Imp gallons).

ACCOMMODATION: Pilot and up to seven passengers or equivalent cargo; full FAR Pt 23 (26 g) crashworthy seats. Sliding door each side of flight deck; flight openable forward-sliding cargo door on port side aft of wing. Cabin is heated and ventilated. Baggage compartment, with security net, at rear of cabin.

SYSTEMS: Split bus electrical system powered by 14 V 95 A engine-driven alternator. Maintenance-free battery and external power receptacle with overvolt protection standard.

AVIONICS: Standard Bendix/King avionics suite.

Comms: Dual KY 97A-61 VHF; KT 76A-12 transponder; KMA 24H audio control/intercom.

Flight: KR 87 ADF; altitude encoder.

Instrumentation: KMD 150 moving map, airspeed indicator, artificial horizon, altimeter with hectoPascal subscale, electric turn co-ordinator, directional gyro, vertical speed indicator, OAT gauge, magnetic compass, electronic digital fuel flow and totaliser gauge, EGT gauge, tachometer, fuel supply gauge, oil temperature/pressure gauges, CHT gauge, and manifold pressure gauge.

EQUIPMENT: Standard equipment includes automatic fuel management system, windscreen demisting, instrument panel lighting, map and cabin reading lights, navigation lights, landing/taxiing lights, low-profile red beacon lights; pilot's and co-pilot's five-point inertia reel shoulder harnesses; six three-point inertia reel harnesses for passengers; cabin soundproofing; and entrance step.

DIMENSIONS, EXTERNAL:

Wing span	12.28 m (40 ft 3¾ in)
Wing chord, constant	1.60 m (5 ft 3 in)
Wing aspect ratio	7.9
Length overall	8.95 m (29 ft 4¼ in)
Fuselage max width	1.37 m (4 ft 6 in)
Height overall	3.89 m (12 ft 9 in)
Tailplane span	4.17 m (13 ft 8 in)
Wheel track	2.79 m (9 ft 2 in)
Wheelbase	2.30 m (7 ft 6½ in)
Cargo/passenger door: Height	1.08 m (3 ft 6½ in)
Width	1.07 m (3 ft 6 in)
Height to sill	0.85 m (2 ft 9½ in)

DIMENSIONS, INTERNAL:

Cabin: Length, incl flight deck	4.01 m (13 ft 2 in)
Max width	1.27 m (4 ft 2 in)

Max height	1.19 m (3 ft 11 in)
Floor area: incl flight deck	5.02 m² (54.0 sq ft)
excl flight deck	3.44 m² (37.0 sq ft)
Volume, incl flight deck	5.1 m³ (180 cu ft)

AREAS:

Wings, gross	19.32 m² (208.0 sq ft)
Ailerons (total)	0.80 m² (8.60 sq ft)
Trailing-edge flaps (total)	0.80 m² (8.60 sq ft)
Fin	1.35 m² (14.50 sq ft)
Rudder	0.73 m² (7.90 sq ft)
Tailplane	2.41 m² (25.90 sq ft)
Elevators (total)	1.76 m² (18.90 sq ft)

WEIGHTS AND LOADINGS:

Weight empty	997 kg (2,198 lb)
Baggage capacity	90 kg (200 lb)
Max fuel	270 kg (595 lb)
Max T-O and landing weight:	
initial certification	1,814 kg (4,000 lb)
target	1,905 kg (4,200 lb)
Max wing loading:	
initial certification	93.9 kg/m² (19.23 lb/sq ft)
target	98.6 kg/m² (20.19 lb/sq ft)
Max power loading:	
first prototype	9.74 kg/kW (16.00 lb/hp)
initial certification	8.12 kg/kW (13.33 lb/hp)
target	8.52 kg/kW (14.00 lb/hp)

PERFORMANCE:

Never-exceed speed (VNE)	185 kt (342 km/h; 212 mph)
Max cruising speed at 1,525 m (5,000 ft)	
	130 kt (241 km/h; 150 mph)
Cruising speed at 3,050 m (10,000 ft)	
	120 kt (222 km/h; 138 mph)
Stalling speed: flaps up	60 kt (112 km/h; 69 mph)
flaps down	52 kt (97 km/h; 60 mph)
Max rate of climb at S/L	240 m (788 ft)/min
Service ceiling, estimated	6,100 m (20,000 ft)
T-O run	305 m (1,000 ft)
T-O to 15 m (50 ft)	549 m (1,800 ft)
Landing from 15 m (50 ft)	366 m (1,200 ft)
Landing run	396 m (1,299 ft)
Range: with max fuel:	
normal cruise	650 n miles (1,203 km; 748 miles)
at 67% power	730 n miles (1,352 km; 840 miles)
with max payload	100 n miles (185 km; 115 miles)
Endurance at 67% power, no reserves	6 h

UPDATED

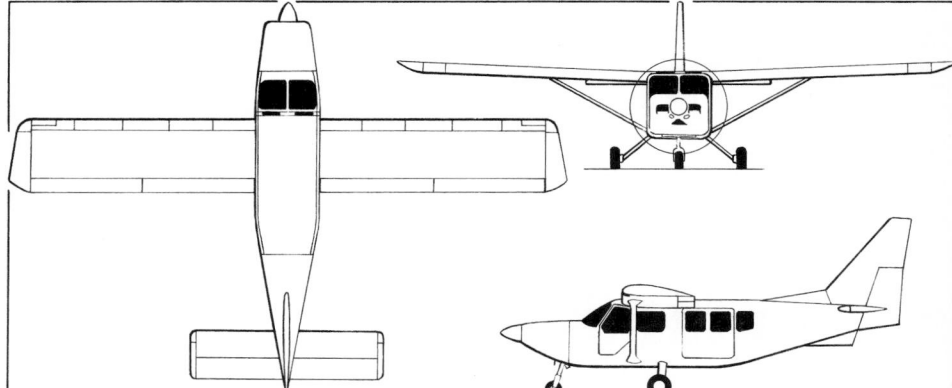

Gippsland GA-8 Airvan (Textron Lycoming IO-540 engine) (*Jane's/James Goulding*)   0110574

GOAIR

GOAIR PRODUCTS

675 Drover Road, Bankstown Airport, New South Wales 2200
Tel: (+61 2) 97 96 34 26

Fax: (+61 2) 97 91 03 54
e-mail: goair@bigpond.com
MANAGING DIRECTOR: Phil Goard

Company machines, welds and manufactures aircraft components. During the 1990s it designed, built and flight-

tested the Goair Trainer, production of which is now under way.

In 2001, Goair completed a replica of the 1920s Westland Widgeon.

UPDATED

GOAIR GT-1 TRAINER

TYPE: Side-by-side lightplane/kitbuilt.

PROGRAMME: Four years in design; public debut (VH-BBR, then unflown) at Avalon Air Show, March 1995; first flight July 1995; some 120 hours of flight testing completed by November 1998; second aircraft (pre-production) displayed without markings at Aviex 2000 at Sydney/ Bankstown and subsequently registered (VH-AYS) on 22 January 2001; had flown 30 hours by late 2001; certification to JAR-VLA expected in late 2001; will also be marketed as a kitbuilt.

CUSTOMERS: Four aircraft built or under construction by early 2002, of which two had flown.

COSTS: US$70,000 (2000).

DESIGN FEATURES: Conventional low-wing, fixed landing gear monoplane; design goals strength, simplicity, low costs, docile stall characteristics and ease of maintenance. Production aircraft differs from prototype in having a forward-sliding canopy; cabin 7.6 cm (3 in) wider; larger ailerons and slotted flaps; and leaf spring main landing gear legs.

FLYING CONTROLS: Conventional and manual; elevators and ailerons are pushrod operated, rudder is cable operated.

STRUCTURE: All metal.

LANDING GEAR: Non-retractable tricycle type with spring steel main legs and rubber-in-compression nose leg. Wheel size (all) 5.00×5. Hydraulic toe brakes standard.

POWER PLANT: One 88 kW (118 hp) Textron Lycoming O-235-C2A (in first prototype) or -C1 (in preproduction aircraft) flat-four engine, driving a two-blade metal propeller. Fuel contained in integral tanks in leading-edges of wings, combined capacity 130 litres (34.3 US gallons; 28.6 Imp gallons).

ACCOMMODATION: Two persons, side by side under forward-sliding bubble canopy; dual controls (sticks) standard, control wheels optional; seats adjust fore and aft; four-point harnesses standard. Baggage compartment behind seats.

DIMENSIONS, EXTERNAL:
Wing span	8.69 m (28 ft 6 in)
Wing chord at root	1.52 m (5 ft 0 in)
Wing aspect ratio: prototype	6.9
production	7.2
Length overall: prototype	6.25 m (20 ft 6 in)
production	7.16 m (23 ft 6 in)

Height overall: prototype	2.03 m (6 ft 8 in)
production	2.29 m (7 ft 6 in)

DIMENSIONS, INTERNAL:
Cabin max width	1.14 m (3 ft 9 in)
Baggage compartment volume	0.17 m³ (6.0 cu ft)

AREAS:
Wings, gross	10.50 m² (113.0 sq ft)

WEIGHTS AND LOADINGS:
Weight empty	408 kg (900 lb)
Max T-O weight	748 kg (1,650 lb)
Max wing loading	71.3 kg/m² (14.60 lb/sq ft)
Max power loading	8.51 kg/kW (13.98 lb/hp)

PERFORMANCE:
Max level speed	115 kt (213 km/h; 132 mph)
Cruising speed	100 kt (185 km/h; 115 mph)
Stalling speed	45 kt (84 km/h; 52 mph)
Max rate of climb at S/L	244 m (800 ft)/min
T-O run	366 m (1,200 ft)
Landing run	198 m (650 ft)
Endurance	5 h

UPDATED

Prototype Goair Trainer (*Gerard Frawley/Australian Aviation*)

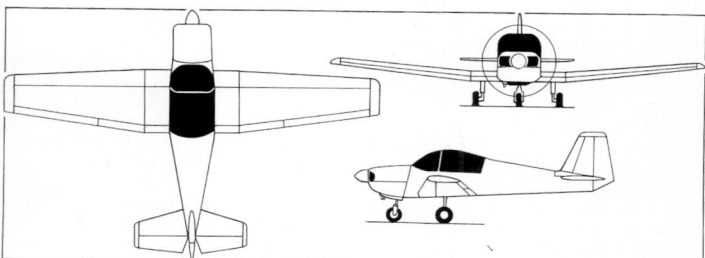

Goair Trainer (115 hp Textron Lycoming O-235) in prototype configuration (*Jane's/Paul Jackson*)

HUGHES

HOWARD HUGHES ENGINEERING PTY LTD

PO Box 89, Lot 8, Southern Cross Drive, Ballina, New South Wales 2478
Tel: (+61 2) 66 86 86 58
Fax: (+61 2) 66 86 83 43
e-mail: alw@spot.com.au
Web: http://www.lightwing.com.au
MANAGING DIRECTOR: Howie Hughes

Company formed 1984 to design and produce the Hughes Australian Light Wing series which has developed from GR-582 via GA-55 (see 1994-95 *Jane's*) to GR-912 and the associated Sport 2000. Most production, undertaken at Ballina Airport, is for local use and either full certification or registry under AUF rules; around 10 aircraft per year are built.

UPDATED

HUGHES AUSTRALIAN LIGHT WING GR-912

TYPE: Side-by-side ultralight/kitbuilt.

PROGRAMME: Prototype Australian Light Wing (ALW) made first flight June 1986; initially with 47.8 kW (64.1 hp) Rotax 582 UL; many sold conforming to CAO 95 Pt 25 requirements; also produced under homebuilt or CAO 101 Pt 28 regulations in non-retractable tailwheel and twin-float configurations; first amphibian model delivered in 1990. Type certificate issued 25 September 1998.

CURRENT VERSIONS: **GR-912:** Tailwheel version. *As described.*

GR-912-T: Tricycle landing gear version; marketing designation Sport 2000; see following entry.

Hughes GR-912 tailwheel version (*Jane's/Paul Jackson*)   *NEW*/0131711

CUSTOMERS: 160 aircraft (all models) delivered by 2001; most registered under AUF rules. Claimed 29 per cent share of Australian two-seat light trainer market with GR-912.

COSTS: A$85,000 factory built (2001).

DESIGN FEATURES: Modified Clark Y wing section.

FLYING CONTROLS: Conventional and manual. Optional flaps.

STRUCTURE: Strut-braced metal wings; welded steel tube fuselage and tail; tail surfaces are wire-braced; composites and polyfibre covering.

LANDING GEAR: Non-retractable tailwheel type; optional amphibious floats with retractable mainwheels and non-retractable tailwheels (latter doubling as water rudders).

POWER PLANT: One 73.5 kW (98.6 hp) Rotax 912 ULS flat-four driving a two- or three-blade propeller; Aerofiber wooden two-blade fixed-pitch propeller recommended. Fuel capacity 60 litres (15.8 US gallons; 13.2 Imp gallons).

DIMENSIONS, EXTERNAL:
Wing span	9.60 m (31 ft 6 in)
Length overall	5.70 m (18 ft 8½ in)
Height overall	1.90 m (6 ft 2¼ in)

AREAS:
Wings, gross	14.86 m² (160.0 sq ft)

WEIGHTS AND LOADINGS:
Weight empty	295 kg (650 lb)
Max T-O weight	520 kg (1,146 lb)

PERFORMANCE:
Max level speed	95 kt (176 km/h; 109 mph)
Max cruising speed	80 kt (148 km/h; 92 mph)
Loiter speed	50 kt (93 km/h; 58 mph)
Stalling speed, power off	35 kt (65 km/h; 41 mph)
Max rate of climb at S/L	244 m (800 ft)/min
T-O run	100 m (330 ft)
Landing run	80 m (263 ft)
Range with max fuel	216 n miles (400 km; 248 miles)

UPDATED

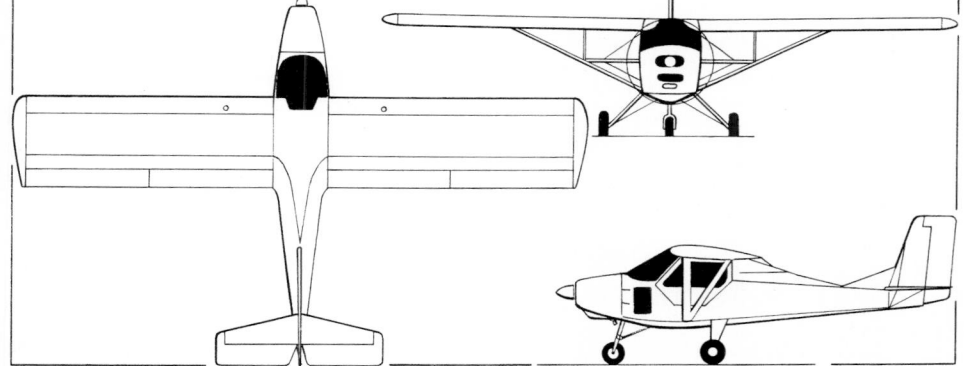

Hughes Australian Light Wing Sport 2000, showing optional 'Heliview' cockpit with side window at floor height (Rotax 912 UL flat-four) (*Jane's/James Goulding*)   0010201

HUGHES AUSTRALIAN LIGHT WING SPORT 2000

TYPE: Side-by-side ultralight/kitbuilt.

PROGRAMME: Developed version of GR-912 (which see); certified in Australian Primary category. Uses original Light Wing type certificate.

COSTS: A$86,500 factory built; A$45,900 kit, including alternative 59.6 kW (79.9 hp) Werner engine (2000).

FLYING CONTROLS: Conventional and manual. Horn-balanced tail surfaces; elevator pushrod-operated; aileron and rudder cable-operated.

STRUCTURE: Welded steel tube fuselage; riveted aluminium alloy wings; mainly fabric covered, with non-structural glass fibre fairings.

LANDING GEAR: Non-retractable tricycle type.

POWER PLANT: One 59.6 kW (79.9 hp) Rotax 912 UL flat-four driving an Aerofiber two-blade fixed-pitch wooden propeller. Fuel capacity 62 litres (16.4 US gallons; 13.6 Imp gallons).

DIMENSIONS, EXTERNAL:	
Wing span	9.50 m (31 ft 2 in)
Length overall	5.68 m (18 ft 7½ in)
Height overall	2.20 m (7 ft 2½ in)
AREAS:	
Wings, gross	15.14 m² (163.0 sq ft)
WEIGHTS AND LOADINGS:	
Weight empty	300 kg (661 lb)
Max T-O weight	480 kg (1,058 lb)
PERFORMANCE:	
Cruising speed at 75% power	72 kt (133 km/h; 83 mph)

Stalling speed	35 kt (65 km/h; 41 mph)
T-O and landing from 15 m (50 ft)	300 m (985 ft)
Range with max fuel	300 n miles (555 km; 345 miles)
g limits	+6/−4

UPDATED

JABIRU

JABIRU AIRCRAFT PTY LTD

PO Box 5186, Bundaberg West, Queensland 4670
Tel: (+61 7) 41 55 17 78
Fax: (+61 7) 41 55 26 69
e-mail: jabiru@tpgi.com.au
Web: http://www.jabiru.net.au
JOINT MANAGING DIRECTOR: Phil Ainsworth

EUROPEAN AGENT:
ST Aviation Ltd, Technology House, High Street, Downham Market, Norfolk PE38 9HH, UK
Tel: (+44 1366) 38 55 58
Fax: (+44 1366) 38 55 59
Web: http://www.jabiru.co.uk

Jabiru formed 1988 by Rodney Stiff and Phil Ainsworth to produce Jabiru LSA 55/2K, but branched into engine design and manufacture when Italian-built KFM engine, which powered this variant, was withdrawn from market. Now produces 1,600 and 2,200 cc flat-four and 3,300 cc flat-six engines for its own and other manufacturers' aircraft. A factory-built SLA version of Jabiru, the UL 450, was launched in 2000. The company and aircraft are named for Australia's sole indigenous stork. Production averages 100 aircraft and 360 engines per year. A prototype of a four-seat Jabiru, the J400, first flew in March 2001 and customer deliveries began almost immediately.

UPDATED

JABIRU JABIRU

TYPE: Side-by-side ultralight/kitbuilt.

PROGRAMME: Design of original Jabiru LSA, started 1987; prototype (first of two, VH-JCX and VH-JQX) made first flight August 1989; first customer delivery April 1991; factory-built version certified to CAO 101 Pt 55 on 1 October 1991; 20 built with KFM engine before change to 44.7 kW (60 hp) Jabiru 1600 flat-four; kitplane construction certification achieved in Australia, New Zealand, South Korea, UK and USA by early 1997; marketed in US as amateur-built kit, achieving FAR Pt 21.191 (g) certification on 8 February 1996 (as well as Australian CAO 19 equivalent); marketing office opened at Aiken, South Carolina, in anticipation of substantial US sales of at least 100 per year but later closed in favour of network of dealerships. Total of 54 built with 1600 engine between April 1993 and March 1996. More powerful ST version introduced in 1994. Joint venture announced in 1998 with CDE Aviation Company of Sri Lanka, subsidiary of Lionair, to establish Jabiru Asia for manufacture of Jabiru ST3 for Asian markets, especially India. Jabiru factory at Koggala, in southern Sri Lanka, will have production capacity for 40 aircraft per year, using engines and avionics imported from Australia, and airframes manufactured locally.

CURRENT VERSIONS: **Jabiru LSA:** Light Sport Aircraft; factory-built; registered under AUF (Australian Ultralight Federation) rules. 'Short span; short fuselage' version: span 8.03 m (26 ft 4 in), length 5.03 m (16 ft 6 in), wing area 7.90 m² (85.0 sq ft).

Jabiru ST: Certified version; in factory production from mid-1994; first was 50th production Jabiru. In mid-1998, the 17th production ST was registered VH-JRU to Jabiru Aircraft as the first **ST3** short span and fuselage.

Jabiru ST certified version (*Jane's/Paul Jackson*)   *NEW*/0131712

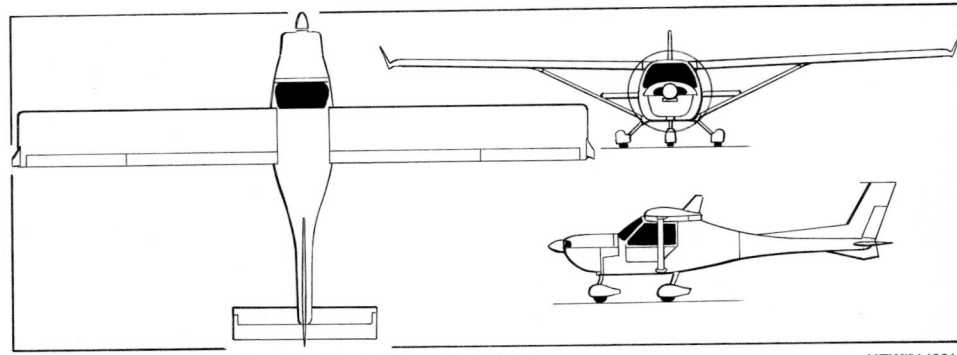

Jabiru Calypso two-seat ultralight (*Jane's/James Goulding*)   *NEW*/0143319

Jabiru SK: Quick-build kit version of ST (components are manufactured alongside production aircraft); first was 66th production Jabiru; quoted home-build time is 600 hours.

Jabiru SP4: Kit, combining longer fuselage of UL 450 with Jabiru 2200 engine and short-span wings of SK; originally designated SP. MTOW 470 kg (1,036 lb); stalling speed 39 kt (73 km/h; 45 mph); optimised for Canadian, Chilean and South African markets, among others. Fuel capacity 70 litres (18.5 US gallons; 15.4 Imp gallons).

Jabiru SP6: Company designation of SP4 kit with 89.5 kW (120 hp) Jabiru 3300 engine. MTOW 500 kg (1,102 lb). Cruising speed 117 kt (217 km/h; 135 mph). Fuel capacity as SP4, or 85 litres (22.5 US gallons; 18.7 Imp gallons) optional.

Jabiru SP-T: Tailwheel version of SP. Enlarged rudder; mainwheel legs moved forward; steerable tailwheel and differential braking. Prototype built by Peter Kayne of Narrogin, Western Australia.

Jabiru SP 480: Proposed version of SP for JAR-VLA certification, with maximum take-off weight of 480 kg (1,058 lb); under development in late 1999.

Jabiru Calypso: Initially factory-built; now available only as kit; previously known as **UL 450;** SLA (Small Light Aeroplane to UK's now-replaced BCAR Section S), first aircraft was built in Ireland in 1998. Launched in UK in 2000 and meeting 35 kt (65 km/h; 41 mph) stalling speed requirement. Differences from SK include wing span increased by 1.37 m (4 ft 6 in); fuselage by 0.61 m (2 ft 0 in); flaps by total of 1.83 m (6 ft 0 in); and reduced ventral fin. Also available with tailwheel as **Calypso T.** Meets FAA LSA. 150 plus sold by end 2001.

J200 and **J400:** Stretched version; described separately.

CUSTOMERS: Production of complete and kit aircraft totalled some 520 by December 2001. Sold to 22 countries.

COSTS: Calypso: US$18,000; SP US$17,500 (2002) both minus engine. Fly-away Calypso 2200 US$59,995 (April 2002).

DESIGN FEATURES: Designed to Australian CAO 101 Pt 55 and British BCAR Section S, with reference to FAR Pt 21 and JAR-VLA. Ultralight or certified aircraft, suitable for factory or amateur assembly. Unswept wing (braced) and tailplane, swept fin; dorsal and ventral fins; all flying surfaces square-tipped; wings detachable for storage and transportation.

Wing section NASA 4412; drooped wingtips; dihedral 1° 15'; incidence 2° 30'; no twist.

FLYING CONTROLS: Conventional and manual. In-flight adjustable pitch trim; wide-span slotted flaps.

STRUCTURE: All-GFRP except metal wing struts, nosewheel assembly and engine mount.

LANDING GEAR: Non-retractable tricycle type, with 15° steerable nosewheel coupled to rudder. Cantilever self-sprung GFRP mainwheel legs; rubber block nosewheel shock-absorption; wheel fairings standard over 4.00-6 or 5.00-6 tyres; Bigfoot 6.00-6 tyres optional. Nosewheel 2.60-4, 4.00-4 or 5.00-6; tailwheel 250×50.

POWER PLANT: One 59.7 kW (80 hp) Jabiru 2200A/J flat-four engine, driving a two-blade, fixed-pitch, wood/composites propeller. Fuel capacity 65 litres (17.2 US gallons; 14.3 Imp gallons).

SYSTEMS: 12 V DC electrical system with 100 W capacity; 20 Ah battery.

AVIONICS: *Comms:* VHF antenna (inside fin) and intercom standard.

Instrumentation: Basic VFR; vacuum instruments optional.

Tailwheel Jabiru SP-T (*Jane's/Paul Jackson*)   *NEW*/0530188

Data below refer to Jabiru Calypso with Jabiru 2200 engine.

DIMENSIONS, EXTERNAL:
Wing span: normal	9.40 m (30 ft 10 in)
with winglets	9.56 m (31 ft 41/2 in)
Length overall	5.64 m (18 ft 6 in)
Height overall	2.01 m (6 ft 7 in)

DIMENSIONS, INTERNAL:
Cockpit max width at shoulder	1.07 m (3 ft 6 in)

AREAS:
Wings, gross	9.29 m² (100.0 sq ft)

WEIGHTS AND LOADINGS:
Weight empty, equipped	242 kg (534 lb)
Max T-O weight	450 kg (992 lb)

PERFORMANCE:
Never-exceed speed (VNE)	120 kt (222 km/h; 138 mph)
Max level speed	110 kt (204 km/h; 127 mph)
Cruising speed at 75% power	
	100 kt (185 km/h; 115 mph)
Stalling speed, flaps down	35 kt (65 km/h; 41 mph)
Max rate of climb at S/L	305 m (1,000 ft)/min
T-O run	100 m (330 ft)
Landing run	168 m (555 ft)
Range with max fuel	430 n miles (796 km; 494 miles)

UPDATED

One of the first Jabiru J200s completed in Australia *(Jane's/Paul Jackson)*   *NEW*/0530187

JABIRU J200 and J400

TYPE: Four-seat kitbuilt.

PROGRAMME: Stretched development of Jabiru, described in previous entry. Prototype (19-3512), in ultralight configuration, first flew 16 March 2001. Maiden flight of first customer-built J200, 13 December 2001.

CURRENT VERSIONS: **J200:** Two-seat ultralight with same dimensions as J400. **200A** has maximum allowable T-O weight; **200B** limited to 544 kg (1,200 lb) to meet AUF ultralight rules.

J250: Announced at Sun 'n' Fun, April 2002; meets FAA LSA kitbuilt rules. Wing span increased 1.65 m (5 ft 5 in) to 9.75 m (32 ft 0 in) and area 3.89 m² (41.9 sq ft) to 11.89 m² (128.0 sq ft) to meet stall requirements of 44 kt (82 km/h; 51 mph) 'clean' and 39 kt (73 km/h; 45 mph) flaps down. Empty weight 326 kg (720 lb); MTOW 559 kg (1,232 lb).

J400: Four-seat lightplane.

CUSTOMERS: 26 kits sold by 17 December 2001, including export to South Africa.

COSTS: Kit US$24,500 without engine; US$35,300 with engine (2002).

Data generally as for Jabiru Jabiru, except that below.

DESIGN FEATURES: Stretched fuselage with second pair of side windows. Wings of Jabiru LSA, with fuel in wings.

POWER PLANT: One 89.5 kW (120 hp) Jabiru 3300 flat-six, driving two-blade fixed-pitch propeller. Fuel capacity 100 litres (26.4 US gallons; 22.0 Imp gallons) in wings.

ACCOMMODATION: Four persons; or two, if flown to ultralight regulations.

DIMENSIONS, EXTERNAL:
Wing span	8.10 m (26 ft 7 in)
Wing aspect ratio	8.2
Length overall	6.55 m (21 ft 5¾ in)

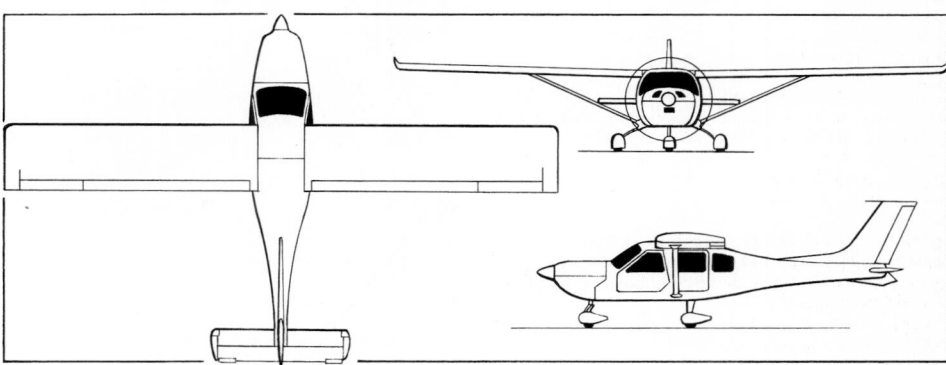

Two/four-seat Jabiru J200/J400 *(Jane's/James Goulding)*   *NEW*/0143320

Height overall	2.20 m (7 ft 2½ in)
Tailplane span	2.66 m (8 ft 8¾ in)
Wheel track	1.80 m (5 ft 10¾ in)
Propeller diameter	1.52 m (5 ft 0 in)

DIMENSIONS, INTERNAL:
Cabin: Max width	1.12 m (3 ft 8 in)
Max height	1.09 m (3 ft 7 in)

AREAS:
Wings, gross	8.00 m² (86.1 sq ft)

WEIGHTS AND LOADINGS:
Weight empty: J200A	300 kg (661 lb)
J200B/J400	310 kg (683 lb)
Max T-O weight: J200B	544 kg (1,200 lb)
J200A/J400	700 kg (1,543 lb)
Max wing loading: J200B	68.0 kg/m² (13.92 lb/sq ft)
J200A/J400	87.5 kg/m² (17.92 lb/sq ft)
Max power loading: J200B	6.08 kg/kW (10.00 lb/hp)
J200A/J400	7.82 kg/kW (12.86 lb/hp)

PERFORMANCE:
Never-exceed speed (VNE)	138 kt (255 km/h; 158 mph)
Max level speed	130 kt (240 km/h; 150 mph)
Cruising speed	120 kt (222 km/h; 138 mph)
Manoeuvring speed	90 kt (166 km/h; 103 mph)
Stalling speed, power off: flaps up:	
J200B	51 kt (94 km/h; 58 mph)
J200A/J400	60 kt (111 km/h; 69 mph)
flaps down: J200B	45 kt (83 km/h; 51 mph)
J200A/J400	48 kt (88 km/h; 55 mph)
Max rate of climb at S/L: J200	457 m (1,500 ft)/min
Service ceiling: all	4,570 m (15,000 ft)
T-O and landing run	200 m (660 ft)
Range	520 n miles (963 km; 598 miles)
Endurance	4 h 25 min
g limits	+3.8/−1.9

UPDATED

SAPPHIRE

SAPPHIRE AIRCRAFT AUSTRALIA PTY LTD

13 Ancura Court, Wattle Grove, New South Wales 2173
Tel: (+61 2) 98 25 33 75
Fax: (+61 4) 18 30 19 16
e-mail: sapphireaircraft@bigfoot.com
Web: http://www.users.bigpond.com/stevendumesny
SALES MANAGER: Steven Dumesny

Sapphire markets an updated version of the Winton Sapphire, designed to CAO 19 by Scott Winton and last described (briefly) in the 1992-93 *Jane's.* A two-seat version, nominally designated Sapphire Scenic, and a self-launching glider, are under development.

UPDATED

SAPPHIRE LSA Mk II

TYPE: Single-seat ultralight/kitbuilt.

PROGRAMME: Introduced in 1998. First kit delivered in May 1999 and flown in September 1999.

CURRENT VERSIONS: **Sapphire LSA:** Available as factory-built version meeting Australian CAO 95 Pt 25 certification standards, or as kitbuilt to CAO 95 Pt 10. Three kits available: Basic, comprising main airframe components; Intermediate, with control systems installed; and Advanced, including engine, propeller and instruments.

Sapphire Scenic: Proposed two-seat version under development; will have four-stroke engine, MTOW of 500 kg (1,102 lb), 94 kt (174 km/h; 108 mph) cruise. Cost approximately A$35,000, plus tax (2002).

CUSTOMERS: First kit sold May 1999. Total of 36 flying by February 2001, including two factory-built.

COSTS: Factory-built flyaway A$26,100; Basic kit A$12,900; Intermediate kit A$19,000; Advanced kit A$22,300, all excluding tax (2002). Operating cost less than A$20 per hour.

Kitbuilt Sapphire LSA Mk II *(Jane's/Paul Jackson)*   *NEW*/0131713

DESIGN FEATURES: Pod-and-boom fuselage with strut-braced shoulder-mounted wings; open cockpit with large windscreen. Baggage box, volume 25 litres (0.88 cu ft), in fuselage pod.

FLYING CONTROLS: Conventional and manual. Ailerons and trimmable all-flying tailplane operated by pushrods; rudder spring-coupled to tailwheel; quarter-span flaps; deflections 5° up (for fast cruising) and 10 and 20° down.

STRUCTURE: Mostly composites; GFRP fuselage pod reinforced with internal bulkheads; aluminium tailboom; wings of GFRP with foam ribs; fin is of GFRP; horizontal tail and rudder of aluminium tube with fabric covering. Wings and horizontal tail detach for storage or transport;

quoted disassembly/assembly times 10 minutes and 5 minutes respectively.

LANDING GEAR: Tailwheel type; fixed. One-piece laminated GFRP main legs. Mainwheel size 3.50-5. Cable-operated brakes. Wheel fairings optional.

POWER PLANT: One 29.5 kW (39.6 hp) Rotax 447 UL-1V or 34.0 kW (45.6 hp) Rotax 503 UL-1V two-cylinder two-stroke, driving a Sweetapple two-blade wooden pusher propeller via 2.24:1 reduction gearing. Integral fuel tanks in wings, standard capacity 38 litres (10.0 US gallons; 8.4 Imp gallons), optional 58 litres (15.3 US gallons; 12.8 Imp gallons).

EQUIPMENT: Fully enclosed canopy and strut fairings optional.

DIMENSIONS, EXTERNAL:		WEIGHTS AND LOADINGS:		Stalling speed: power on	34 kt (63 km/h; 40 mph)
Wing span	8.84 m (29 ft 0 in)	Weight empty	158 kg (348 lb)	power off	36 kt (67 km/h; 42 mph)
Length overall	5.00 m (16 ft 4¾ in)	Max T-O weight	320 kg (705 lb)	Max rate of climb at S/L	305 m (1,000 ft)/min
Height overall	1.25 m (4 ft 1¼ in)	PERFORMANCE:		Service ceiling	3,660 m (12,000 ft)
Propeller diameter	1.27 m (4 ft 2 in)	Never-exceed speed (VNE)	98 kt (181 km/h; 112 mph)	T-O run	55 m (180 ft)
DIMENSIONS, INTERNAL:		Cruising speed at 75% power:		Landing run	100 m (330 ft)
Cabin max width	0.61 m (2 ft 0 in)	without fairings	75 kt (139 km/h; 86 mph)	Range	217 n miles (402 km; 250 miles)
AREAS:		with fairings		Endurance	3 h
Wings, gross	9.13 m² (98.3 sq ft)		75 to 85 kt (139 to 157 km/h; 86 to 98 mph)	g limits	+8/−4
					UPDATED

SEABIRD

SEABIRD AVIATION AUSTRALIA PTY LTD

Hervey Bay Airport, PO Box 618, Pialba, Hervey Bay,
  Queensland 4655
Tel: (+61 7) 41 25 31 44
Fax: (+61 7) 41 25 31 23
e-mail: seabird@australia.aunz.com
CHAIRMAN AND MANAGING DIRECTOR: Donald C Adams
PRODUCTION DIRECTOR: Peter Adams

Seabird Aviation was founded in 1983 to develop the Seeker
observation aircraft. Despite having built Seekers for
certification, Seabird's business plan was always to franchise
Seeker production capability, rather than sell individual
aircraft; entire manufacturing equipment for a Seeker
production line can be forwarded in two standard-size ISO
shipping containers. On 25 July 2000, Seabird signed an
agreement with Evektor-Aerotechnik (EV-AT) of the Czech
Republic (which see) leading to the formation of a joint
venture company, Evektor-Seabird Pty Ltd, to manufacture
and market the Seeker.

UPDATED

Model of Seabird SB9 Stormer   *NEW*/0531267

SEABIRD SB7L-360 SEEKER

TYPE: Multisensor surveillance lightplane.
PROGRAMME: Design began January 1985, construction
  January 1988; first flight of first (SB5N) prototype (VH-
  SBI, c/n 001, with Norton rotary engine) 1 October 1989,
  and of SB5E second prototype (VH-SBU, c/n 003, with
  Emdair engine) 11 January 1991 (c/n 002 was structural
  test airframe); fourth aircraft (c/n 004, VH-ZIG) is SB7L
  Seeker prototype, making first flight 6 June 1991;
  definitive SB7L-360A (VH-DPT; first flight early 1993)
  has Textron Lycoming O-360 engine for higher
  performance and received Australian CAA certification to
  FAR Pt 23 on 24 January 1994. Operational testing of
  Seeker included use as an airborne surveillance platform in
  October 1995 trials by the Australian Army. Components
  for an initial production batch of five aircraft completed by
  December 1997. Project then frozen and franchised
  producer sought, being achieved in July 2000, on signature
  of agreement with Evektor-Aerotechnik.
CURRENT VERSIONS: **SB7L:** *As described.*
  SB9 Stormer: Projected light attack version with
  Rolls-Royce 250 turbine engine; tandem seats for two
  crew; chin turret below nose-mounted sensor package;
  pylon for gun pod, bombs or rocket launchers beneath each
  wing. MTOW 1,680 kg (3,703 lb), including 725 kg
  (1,598 lb) of sensors and weapons; range 550 n miles
  (1,018 km; 632 miles).
CUSTOMERS: Orders for 13 received by July 2001.
DESIGN FEATURES: Braced high-wing monoplane with pod and
  boom fuselage, extensively glazed cabin and slightly
  sweptback vertical tail. Ventral and auxiliary fins added
  early 1993. Primary applications are observation/
  reconnaissance, offering helicopter-like view from cockpit
  and good low-speed handling and loiter capabilities;
  training and agricultural use also foreseen.

Fourth Seabird Seeker, employed on power line patrols by Australia's Country Energy in 2003
(Jane's/Paul Jackson)   *NEW*/0530177

  Wing section NACA 63₂-215 (modified); wedge tips;
twist 3°, incidence 4° at root, 1° at tip; dihedral 2° 30′ from
roots.
FLYING CONTROLS: Conventional and manual. Rod-actuated
  slotted ailerons, one-piece horn-balanced elevator (with
  trim tab) and horn-balanced rudder; slotted flaps on wing
  trailing-edge; fixed incidence tailplane. Inboard stall strips;
  vortex generators on wing upper surface near leading-edge
  and lower surface forward of ailerons; elevator bias trim.
STRUCTURE: Front fuselage mainly of 4130 chromoly steel
  tube with Kevlar non-load-bearing skin; aluminium alloy
  tubular tailboom; wings and tail unit conventional
  aluminium alloy stressed skin structures, former with
  single bracing strut and jury strut each side.

LANDING GEAR: Non-retractable, with Cleveland 8.00-6
  mainwheels and fairings on cantilever spring steel legs;
  fully castoring Scott 8 in tailwheel with oil/nitrogen oleo
  strut and 210×65 McCreary tyre. Mainwheel tyre pressure
  1.38 bar (20 lb/sq in); Cleveland disc brakes. Float gear to
  be developed.
POWER PLANT: One Textron Lycoming O-360-B2C flat-four
  engine, derated to 125 kW (168 hp) from 134 kW (180 hp)
  with lower compression ratio of 7.5:1 for Mogas fuel. Low
  propeller tip speed for minimal noise; Bishton BB177 two-
  blade fixed-pitch wood/composites pusher propeller. Fuel
  in two 96 litre (25.4 US gallon; 21.1 Imp gallon) integral
  wing tanks, each with overwing gravity filling point. At
  maximum T-O weight of 897 kg (1,977 lb) with two 77 kg
  (170 lb) occupants, maximum fuel load 152 litres (40.2 US
  gallons; 33.4 Imp gallons); higher MTOW of 925 kg
  (2,040 lb) would permit this to be increased to 182 litres
  (48.0 US gallons; 40.0 Imp gallons). Provision for
  auxiliary fuel tanks on underwing hardpoints. Oil capacity
  7 litres (1.9 US gallons; 1.5 Imp gallons).
ACCOMMODATION: Side-by-side seats, adjustable fore and aft,
  for pilot and co-pilot or observer/passenger in extensively
  glazed cabin. Dual controls and four-point inertia reel
  seatbelts standard. Split (upward/downward-hinged) door
  each side, both removable. Ram air cabin ventilation.
  Space for 43 kg (95 lb) of baggage aft of seats.
SYSTEMS: 28 V electrical system, with 70 A alternator and Gill
  246 battery, for engine start, instruments and lighting.
AVIONICS: *Comms:* Bendix/King KY 97A VHF, KT 76A
  transponder, AR 850 encoder and Sigtronics SPA 400
  intercom standard.
    Flight: Bendix/King nav/com, ADF, second com and
  ELT optional.
EQUIPMENT: Hardpoint for 60 kg (132 lb) of external stores
  beneath each wing. Quick-change photo/survey modules,
  stretcher or 100 litre (26.4 US gallon; 22.0 Imp gallon)
  spraytank optional in place of right-hand seat.
DIMENSIONS, EXTERNAL:
  Wing span   11.07 m (36 ft 4 in)
  Wing chord, constant   1.22 m (4 ft 0 in)
  Wing aspect ratio   9.4

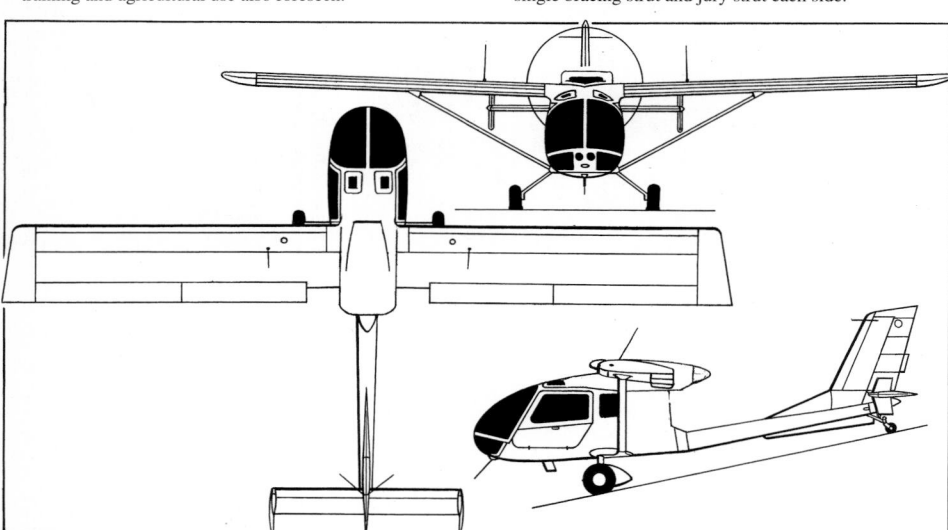

Seabird SB7L Seeker (Jane's/Mike Keep)   0093657

Length overall	7.01 m (23 ft 0 in)	Baggage compartment volume	0.42 m³ (15.0 cu ft)	Stalling speed, flaps down	50 kt (93 km/h; 58 mph) IAS	
Fuselage: max width	1.14 m (3 ft 9 in)	AREAS:		Rate of climb: at S/L	288 m (944 ft)/min	
Height overall, propeller vertical	2.49 m (8 ft 2 in)	Wings, gross	13.05 m² (140.5 sq ft)	at 1,830 m (6,000 ft)	173 m (567 ft)/min	
Tailplane span	2.90 m (9 ft 6 in)	WEIGHTS AND LOADINGS:		Service ceiling	4,648 m (15,250 ft)	
Wheel track	2.03 m (6 ft 8 in)	Basic weight empty	604 kg (1,332 lb)	T-O run	265 m (870 ft)/min	
Wheelbase	4.75 m (15 ft 7 in)	Max fuel	115 kg (254 lb)	Landing run	199 m (654 ft)	
Propeller diameter	1.77 m (5 ft 9¾ in)	Max T-O weight	896 kg (1,977 lb)	Range with reserves:		
Cabin doors (two each): Height	0.83 m (2 ft 8¾ in)	Max wing loading	68.7 kg/m² (14.07 lb/sq ft)	at min patrol speed	476 n miles (881 km; 547 miles)	
Width	0.98 m (3 ft 2½ in)	Max power loading	7.16 kg/kW (11.77 lb/hp)	at 65% power	470 n miles (870 km; 540 miles)	
Height to sill	0.98 m (3 ft 2½ in)	PERFORMANCE:		Endurance, with reserves: at min patrol speed	7 h 15 min	
DIMENSIONS, INTERNAL:		Never-exceed speed (V_{NE})		at 65% power	4 h 30 min	
Cabin, incl baggage space:			129 kt (238 km/h; 148 mph) CAS		**UPDATED**	
Length	2.21 m (7 ft 3 in)	Cruising speed at 75% power				
Max width	1.12 m (3 ft 8 in)		112 kt (207 km/h; 129 mph)			
Max height	1.09 m (3 ft 7 in)	Minimum patrol speed	65 kt (120 km/h; 75 mph) CAS			

STORCH

STORCH AVIATION AUSTRALIA PTY LTD

113 Koree Island Road, Beechwood, New South Wales 2446
Tel/Fax: (+61 2) 65 85 64 58
Fax: (+61 2) 65 85 66 22
e-mail: slepcev@nor.com.au
Web: http://www.storch.com.au
DESIGNER: Nestor Slepcev

Nestor Slepcev designed, built and flew two earlier aircraft before producing the Storch Mk 4. By 2002, 100 of the type had been produced. In August 1996, Mr Slepcev landed a Storch Mk 4 on Gran Sasso, Italy, in re-enactment of the rescue of Benito Mussolini, 53 years earlier.

UPDATED

STORCH STORCH SS Mk 4

TYPE: Tandem-seat lightplane/kitbuilt.
PROGRAMME: First flight (VH-ZOR) 1991, then with single seat and powered by 59.6 kW (79.9 hp) Rotax 912 engine; two-seat version first flown 1994; JAR-VLA certification achieved in Australia, 14 October 1999.
CURRENT VERSIONS:: **Storch 'Ultra':** Ultralight; Australian Primary Category; CAO 19 (amateur build).
  Storch 'Muster': Utility version; envisaged uses include cattle mustering. Australian CAO 24 or certified categories.
  Storch 'Criquet': Powered by Rotec Fireball 7 seven-cylinder radial engine of 82.0 kW (110 hp), thus resembling Morane-Saulnier MS 502 Criquet. Prototype VH-AJH public debut at Avalon Air Show, February 2001.
  Super Storch Mk 4: Higher-powered version with 119 kW (160 hp) Textron Lycoming O-320 engine and top speed of 110 kt (203 km/h; 126 mph). 13 in Australia by April 2002.
  Storch SS-FP: Floatplane; prototype, with full Lotus floats and Rotax 912S2 engine, registered VH-ANB in November 2001.
  Storch SS-M: M-14P radial engine (265 kW; 355 hp) reportedly installed in prototype VH-AYQ (100th Storch), which registered in November 2001.
CUSTOMERS: More than 100 sold by early 2002 (including some 50 factory-built) to customers in 19 countries in Africa, Australasia, Europe, North America and Pacific Rim.
COSTS: US$58,000, factory-complete (2000).
DESIGN FEATURES: Three-quarters scale version of Second World War German Fieseler Fi 156 Storch STOL liaison and observation aircraft. Wings are removable for

Storch 'Criquet' three-quarters scale Morane-Saulnier MS 502 *(Jane's/Paul Jackson)*   NEW/0131714

transport or storage. Quoted build time for kitbuilt version approximately 600 hours.
FLYING CONTROLS: Conventional and manual. Actuation by pushrods and cables; ground-adjustable tab on each aileron. Flaps and full span leading-edge slats. Horn-balanced tail surfaces. Electronic flight-adjustable tailplane incidence for pitch trim in Ultralight; VLA has fixed tailplane with trim tab in port elevator.
STRUCTURE: Welded 4130 chromoly steel tube fabric-covered fuselage and tail surfaces; metal engine cowling. Strut-braced metal wings with stamped aluminium ribs and D-section leading-edge; latest versions aluminium-covered, except fabric flaps and ailerons. Composites wheel fairings.
LANDING GEAR: Tailwheel type; fixed. Original compression spring suspension replaced by bungee. Mainwheel size 8.00×6-6; Maule tailwheel, diameter 15 cm (6 in); hydraulic disc brakes. Float installation under development in late 1999.
POWER PLANT: Standard factory-built VLA version has one 73.5 kW (98.6 hp) Rotax 912 ULS flat-four driving two-blade Bruce de Chastel wooden propeller; 84.6 kW (113.4 hp) turbocharged Rotax 914 optional. Fuel in two wing tanks, combined capacity 76 litres (20.1 US gallons; 16.7 Imp gallons); belly tank, capacity 100 litres (26.4 US gallons; 22.0 Imp gallons) optional.

ACCOMMODATION: Two, in tandem, with dual controls.

DIMENSIONS, EXTERNAL:	
Wing span	10.00 m (32 ft 9¾ in)
Length overall	6.80 m (22 ft 3¾ in)
Propeller diameter	1.88 m (6 ft 2 in)
DIMENSIONS, INTERNAL:	
Cabin: Length	1.50 m (4 ft 11 in)
Max width	0.80 m (2 ft 7½ in)
Max height	1.00 m (3 ft 3¼ in)
AREAS:	
Wings, gross	16.00 m² (172.2 sq ft)
WEIGHTS AND LOADINGS (Storch 'Ultra'):	
Weight empty	345 kg (761 lb)
Max T-O weight	544 kg (1,200 lb)
PERFORMANCE:	
Max level speed	78 kt (144 km/h; 90 mph)
Cruising speed	70 kt (130 km/h; 81 mph)
Stalling speed, flaps down, power on	
	22 kt (41 km/h; 26 mph)
Max rate of climb at S/L	213 m (700 ft)/min
Service ceiling	4,575 m (15,000 ft)
T-O and landing run	9 to 15 m (30 to 50 ft)
Endurance: standard fuel	3 h 0 min
optional fuel	7 h 0 min
g limits	+6/−3

UPDATED

SUPAPUP

SUPAPUP AIRCRAFT

PO Box 630, Oakbank, South Australia 5243
Tel: (+61 8) 83 88 43 49
Fax: (+61 8) 83 88 46 44
e-mail: supapup@senet.com.au
Web: http://www.supapup.com
MANAGING DIRECTOR: John Cotton

Company previously named Aerosport Pty Ltd. Early versions (Mks 1, 2 and 3) of SupaPup hand-built at Hahndorf, South Australia, until cost of compliance with local airworthiness directives forced closure of company. Aerosport redesigned aircraft as Mk 4 in 1992 and returned to business with relaunched production line for kits. Most are operated under ultralight rules (homebuilt, uncertified; CAO 95 Pt 10).

VERIFIED

SUPAPUP SUPAPUP Mk 4

TYPE: Single-seat ultralight kitbuilt.
PROGRAMME: Original Aerosport company developed and produced three earlier versions of the SupaPup from the mid-1980s onwards; SupaPup Mk 4 first flown in early

SupaPup SupaPup Mk 4 kitbuilt demonstrating wing folding   NEW/0087865

1998. Some registered as lightplanes; most as ultralights with 300 kg (661 lb) MTOW limitation under CAO 95 Pt 10.

COSTS: Kit, less engine A\$17,900 or US\$8,000 (2002).

DESIGN FEATURES: Strut-braced, high-wing cabin monoplane; wings fold rearwards for storage or transportation without disconnection of fuel lines or control runs. Quoted build time 250 hours.
Wing dihedral 2°.

FLYING CONTROLS: Conventional and manual.

STRUCTURE: Welded steel tube fuselage covered in Stitts fabric; glass fibre engine cowling and wingtips.

LANDING GEAR: Tailwheel type; fixed.

POWER PLANT: One 34.0 kW (45.6 hp) Rotax 503 UL-1V two-cylinder two-stroke or 40.3 kW (54 hp) Jabiru 1600 four-cylinder four-stroke. Jabiru 2200 and Rotax 582 and 618 also suitable. Fuel contained in wing tanks, combined capacity 54 litres (14.3 US gallons; 11.9 Imp gallons).

EQUIPMENT: Optional cargo pod under development.

DIMENSIONS, EXTERNAL:
Wing span	7.925 m (26 ft 0 in)
Length overall	5.70 m (18 ft 8½ in)

DIMENSIONS, INTERNAL:
Cabin max width	0.61 m (2 ft 0 in)

AREAS:
Wings, gross	9.50 m² (102.3 sq ft)

WEIGHTS AND LOADINGS:
Weight empty	180 to 200 kg (397 to 441 lb)
Baggage capacity	20 kg (44 lb)
Max T-O weight	340 kg (750 lb)

PERFORMANCE:
Never-exceed speed (VNE)	130 kt (240 km/h; 149 mph)
Max level speed	100 kt (185 km/h; 115 mph)
Cruising speed	95 kt (176 km/h; 109 mph)
Stalling speed	35 kt (65 km/h; 41 mph)
Max rate of climb at S/L	335 m (1,100 ft)/min
T-O to 15 m (50 ft)	200 m (660 ft)
Endurance with 10% reserves	4 h 30 min

UPDATED

SUPERMARINE

SUPERMARINE AIRCRAFT PTY LTD

Archerfield, Queensland
Tel: (+61 7) 32 55 55 25
e-mail: sales@supermarineaircraft.com
Web: http://www.supermarineaircraft.com

Michael (Mike) O'Sullivan founded Supermarine Aircraft Pty Ltd in 1992 to develop a kitbuilt scale replica of the Supermarine Spitfire Mk V. A partnership was formed with John McCarron to finance production.

VERIFIED

SUPERMARINE SPITFIRE

TYPE: Sportplane kitbuilt.

PROGRAMME: Prototype Mk 25 first flew in November 1996. Designations follow from Mk 24 version produced by original Supermarine company.

CURRENT VERSIONS: **Spitfire Mk 25:** Single-seat version to 75 per cent scale; 11 built by 2001.
Spitfire Mk 26: Two-seat version, with accommodation for pilot and passenger in tandem under single canopy, or pilot and additional fuel or baggage. Five completed by 2001. *Data below refer to this version.*

CUSTOMERS: Total of 20 of both versions sold by July 2001, at which time eight were flying.

COSTS: Complete basic kit from A\$85,500 not including engine, instruments, shipping or taxes (2001). Also available factory-built.

DESIGN FEATURES: Scale replica of Supermarine Spitfire Mk V. Wings de-rig for storage or road transport. Supplied in five subkits, or as complete kit; main airframe components are jig-assembled at factory for completion by customer; quoted build time 800 to 900 hours. Complies with FAA and other authorities' 51 per cent homebuilding rules.

Supermarine Spitfire Mk 25 (*Jane's/Paul Jackson*)   NEW/0530190

FLYING CONTROLS: Conventional and manual; electric split flaps; electric pitch trim.

STRUCTURE: Aluminium throughout, except for glass fibre cowling and non-structural fairings. Two-spar wing.

LANDING GEAR: Tailwheel type; electrically retractable mainwheels.

POWER PLANT: One 149 kW (200 hp) Jabiru eight-cylinder four-stroke, driving a two- or three-blade propeller. Constant-speed, four-blade propeller under development. Fuel capacity 115 litres (30.4 US gallons; 25.3 Imp gallons).

DIMENSIONS, EXTERNAL:
Wing span	8.43 m (27 ft 8 in)
Length overall	7.01 m (23 ft 0 in)
Height overall	1.98 m (6 ft 6 in)

AREAS:
Wings, gross	11.33 m² (122.0 sq ft)

WEIGHTS AND LOADINGS:
Weight empty	380 kg (838 lb)
Max T-O weight	650 kg (1,433 lb)
Max wing loading	59.8 kg/m² (12.25 lb/sq ft)
Max power loading (Jabiru)	4.36 kg/kW (7.17 lb/hp)

PERFORMANCE:
Never-exceed speed (VNE)	180 kt (333 km/h; 207 mph)
Cruising speed at 75% power	150-160 kt (278-296 km/h; 173-184 mph)
Manoeuvring speed	125 kt (232 km/h; 144 mph)
Stalling speed: flaps up	44 kt (82 km/h; 51 mph)
landing configuration	42 kt (78 km/h; 49 mph)
Max rate of climb at S/L	914 m (3,000 ft)/min
Service ceiling	3,660 m (12,000 ft)
T-O run	335 m (1,100 ft)
Landing run	455 m (1,500 ft)
Max range	434 n miles (804 km; 500 miles)
Endurance at econ cruising speed, with reserves 3 h 0 min	
g limits	+6/-3

UPDATED

TOMAIR

TOMAIR AIRCRAFT SALES

Tel: (+61 3) 97 20 31 33
Fax: (+61 3) 97 20 33 49
e-mail: cobra@tomair.com.au
Web: http://www.tomair.com.au
MANAGING DIRECTOR: Tom Wickers

In addition to the Cobra Arrow, Tomair offers a scale replica of an early model ('razorback') North American P-51 Mustang, the prototype of which was nearing completion in early 2002, and conversion kits for installation of Subaru motorcar engines in light aircraft.

UPDATED

TOMAIR COBRA ARROW

TYPE: Side-by-side ultralight/motor glider kitbuilt.

PROGRAMME: Non-flying exhibition airframe, then incomplete, displayed at Avalon Air Show, February 2001; prototype (19-3700) was removed from its moulds in August 2001 and was fitted out in late 2001/early 2002. First flight 18 May 2002. Conventional tailplane and elevators replaced in July 2002 by stabilator.

CURRENT VERSIONS: **Arrow Sport:** Standard version, *as described.* Available to Ultralight or Certified rules.
Arrow M/G: Motor glider version with 3.66 m (12 ft) increase in wing span.

CUSTOMERS: Four kits sold by February 2001.

COSTS: Arrow Sport kit, less engine, instruments and retractable landing gear, A\$35,000 (2002).

DESIGN FEATURES: Derived from US Polliwagen homebuilt, described in 1983-84 and earlier editions of *Jane's.* Designed to meet AUF 51 per cent amateur-built rules. Conventional low-wing monoplane; sweptback T tail. Kit comprises prejoined fuselage mouldings with cowling, canopy and main spar carry-through installed; tail surfaces; wing skins with main spar installed; laser-cut profiles for rib shapes; carbon fibre instrument panel; plus seats, fuel cells, control columns, rudder pedals and hardware. Quoted build time 400 hours.

Prototype Cobra Arrow taking off for its maiden flight   NEW/0137985

FLYING CONTROLS: Manual. Ailerons, rudder and stabilator. Flaps.

STRUCTURE: Mostly of pre-moulded composites.

LANDING GEAR: Option of tailwheel type, fixed; or tricycle type, retractable.

POWER PLANT: Prototype has one 74.6 kW (100 hp) Subaru EA81. Other suitable engines include JPX 4TX75, Rotax 582, 912 and 914, Jabiru 2200, Textron Lycoming O-200 and O-360 and 74.6 kW (100 hp) Honeywell turboprop. Fuel capacity 71.9 litres (19.0 US gallons; 15.8 Imp gallons).

ACCOMMODATION: Two persons side by side under enclosed canopy. Dual controls (sticks) standard.

DIMENSIONS, EXTERNAL:
Wing span: Arrow Sport	8.53 m (28 ft 0 in)
Arrow M/G	12.19 m (40 ft 0 in)
Length overall	5.79 m (19 ft 0 in)
Height overall (nosewheel)	1.83 m (6 ft 0 in)
Wheelbase: fixed landing gear	2.34 m (7 ft 8 in)
retractable landing gear	1.98 m (6 ft 6 in)

WEIGHTS AND LOADINGS:
Weight empty	250 kg (551 lb)
Max T-O weight: Ultralight	544 kg (1,200 lb)
Certified	700 kg (1,543 lb)
Max power loading (prototype)	6.63 kg/kN (10.90 lb/hp)

PERFORMANCE (estimated, with Subaru engine):
Max level speed	200 kt (370 km/h; 230 mph)
Cruising speed at 2,590 m (8,500 ft)	152 kt (282 km/h; 175 mph)
Stalling speed, flaps down	38 kt (71 km/h; 44 mph)
T-O and landing run	122 m (400 ft)
Range	782 n miles (1,448 km; 900 miles)
g limits	+10/-6

UPDATED

AUSTRIA

DIAMOND

DIAMOND AIRCRAFT INDUSTRIES GmbH

N A Otto-Strasse 5, A-2700 Wiener Neustadt
Tel: (+43 2622) 267 00
Fax: (+43 2622) 267 80
e-mail: sales@diamond-air.at
Web: http://www.diamondair.com
PRESIDENT: Christian Dries
MANAGING DIRECTORS: Michael Feinig, Wolfgang Grumeth
TECHNICAL MANAGER: Martin Volck
PURCHASING MANAGER: Helmuth Stueckler

Company formed 1981 in Friesach, Carinthia, southern Austria, as Hoffman Flugzeugbau GmbH; re-formed after August 1984 bankruptcy as Hoffman Aircraft Ltd, subsidiary of Simmering-Graz-Pauker AG; relocated to new facility at Wiener Neustadt 1987; name HOAC Austria Flugzeugwerk Wiener Neustadt GmbH adopted 1990 after 1989 management buyout; renamed Diamond Aircraft Industries GmbH March 1996 and achieved JAR 21 design approval in 2001.

Factory expanded to 5,200 m² (56,000 sq ft) of floor space in 1996 to accommodate wider model range, increase production levels and provide new research and maintenance facilities. Workforce 118. Current products include the HK 36 Super Dimona/Katana Xtreme and refurbished DV 20 Katana 100. Diamond Aircraft Corporation produces the two-seat DA 20 in Canada (which see); the four-seat DA 40 Star became available in 2000 and deliveries of production aircraft accelerated in 2001. DA 42 TwinStar launched in 2002.

Diamond announced in May 2002 that it is to expand aircraft production into Germany, initially by building Super Dimonas at plant funded in part by Mecklenburg-Vorpommern regional government. Having achieved sales of 1,700 aircraft up to the end of 2000, Diamond gained a further 176 (including Canada) in 2001 and was projecting 249, 395 and 489 in the years up to 2004.

UPDATED

DIAMOND HK 36

European marketing name: Super Dimona
North American marketing name: Katana Xtreme
TYPE: Motor glider.
PROGRAMME: Original Hoffman H-36 Dimona (Diamond) was designed by Austrian-German team, first flight by first of three prototypes taking place in Germany 9 October 1980; production, in Austria, started May 1981; Dimona Mk II introduced May 1985 (smaller wingtip fairings, modified cowling, better propeller pitch control, stronger main gear, sprung steerable tailwheel); production of H-36 totalled over 275 by 1989.

Development of HK 36R Super Dimona (considerably redesigned by Dieter Kohler − hence K in designation) started March 1987; first flight, with 67.1 kW (90 hp) Limbach L 2400 engine, October 1989, and one or two more completed with Limbach engines, but Rotax 912 adopted as standard power plant. Production began April 1990 and Austrian certification to JAR 22 awarded 15 May 1990. Manufacture totalled 114 by 1995, when supplanted by current versions. 'Short-wing' LF 2000 was proof-of-concept for DA 20 Katana; TS and TTS versions introduced 1995; these and current TC/TTC featured winglets; prototypes (OE-9415 and OE-9416) built 1995. In 1997, export Dimonas for the North American market were renamed Katana Xtreme.
CURRENT VERSIONS: **HK 36TC:** Tricycle landing gear version of now discontinued HK 36TS. Some 20 TS/TC built in 1996, four in 1998 and eight for Indian Air Force in early 2000. Additionally, **Eco** Super Dimona HB-2335, operated on weather research missions by Met Air of Switzerland, fitted with two underwing sensor pods.

HK 36TTC: As HK 36TC but with Rotax 914. Majority of production is this version. **Eco** version described separately.

MPX: Described separately.

Late production (197th) Diamond HK 36TTC Super Dimona motor glider (*Jane's/Paul Jackson*)   NEW/0131704

CUSTOMERS: Two prototypes and 201 of T/TT series built by early 2001, when manufacture temporarily halted to concentrate on Diamond Star. Total includes at least 120 TTCs and 35 TCs. Sold to customers in Austria, Canada, France, Germany, Netherlands, UK, USA and elsewhere. Production totals for previous versions given under Programme.
COSTS: TC, Sch94,370; TTC, Sch99,850 (2000).
DESIGN FEATURES: Conventional motor glider, with T tail and low/mid wing. Wings can be detached and folded back alongside fuselage for transportation and storage. Improvements of T/TT series (from 1996) include carbon fibre main spar. Compared with H-36 Dimona, HK 36 has modified inboard leading-edge, sweptback wingtips and Schempp-Hirth (instead of DFS) type airbrakes; better access to engine and control system via completely removable cowlings; improved spin characteristics; stronger main landing gear and improved tailwheel springing; larger canopy with new mechanism; wingroot fairings and rear fuselage redesigned.

Wortmann wing sections: FX-63-137 at root, FX-71 at tip fairings
FLYING CONTROLS: Conventional and manual. Longitudinal stability improved in Super Dimona over original by increasing elevator chord and adding trim tab; airbrakes in wing upper surface.
STRUCTURE: Mainly of GFRP; carbon fibre wing spar and main bulkhead. GFRP mainwheel legs and fairings.
LANDING GEAR: Non-retractable tricycle type, with free-castoring nosewheel, ±30°. Wheel fairings on all three legs.
POWER PLANT: TC has one 59.6 kW (79.9 hp) Rotax 912A-3 flat-four with 2.27:1 reduction drive to an MT composites, two-blade, three-position constant-speed and feathering propeller. TTC has one 84.6 kW (113.4 hp) Rotax 914F-3. Fuel tank in fuselage, capacity 77 litres (20.3 US gallons; 16.9 Imp gallons).
ACCOMMODATION: Two seats side by side. Cockpit canopy hinged at rear to open upward. Baggage space aft of seats.
DIMENSIONS, EXTERNAL:

Wing span	16.33 m (53 ft 7 in)
Wing aspect ratio	17.4
Length overall	7.28 m (23 ft 10½ in)
Width, wings folded	2.20 m (7 ft 2½ in)
Height over tail	2.40 m (7 ft 10½ in)

AREAS:

Wings, gross	15.30 m² (164.7 sq ft)

WEIGHTS AND LOADINGS:

Weight empty: TC	555 kg (1,224 lb)
TTC	560 kg (1,235 lb)
Max T-O weight: both	770 kg (1,697 lb)
Max towing load: TC	370 kg (815 lb)
TTC	600 kg (1,322 lb)
Max wing loading: both	50.3 kg/m² (10.31 lb/sq ft)
Max power loading: TC	12.92 kg/kW (21.23 lb/hp)
TTC	9.10 kg/kW (14.95 lb/hp)

PERFORMANCE, POWERED:
Never-exceed speed (V$_{NE}$):

both	141 kt (261 km/h; 162 mph)

Max cruising speed: TC	108 kt (200 km/h; 124 mph)
TTC	119 kt (220 km/h; 137 mph)
Cruising speed at 65% power:	
TC	90 kt (167 km/h; 104 mph)
TTC	106 kt (196 km/h; 122 mph)
Max rate of climb at S/L: TC	240 m (787 ft)/min
TTC	324 m (1,063 ft)/min
T-O run: TC	201 m (659 ft)
TTC	182 m (600 ft)
T-O to 15 m (50 ft): TC	338 m (1,110 ft)
TTC	274 m (900 ft)
g limits: both	+5.3/−2.65

PERFORMANCE, UNPOWERED:

Best glide ratio: both	27
Min sink rate: both	1.17 m (3.84 ft)/s

UPDATED

DIAMOND MPX

TYPE: Multisensor surveillance lightplane.
CURRENT VERSIONS: **MPX:** Generic designation for sensor-equipped Dimona. Developed by Diamond Aircraft Corporation in association with Wescam Inc of Ontario; multimission utility aircraft version of Diamond HK 36TTC Katana Xtreme. MPX (Multipurpose Xtreme) incorporates hardpoint under each wing at approximately mid-span for quick-release sensor pods. Each pod can accommodate up to 45 kg (100 lb) of equipment, including gyrostabilised video surveillance cameras (such as Wescam Model 16 ENG broadcast camera system) and IR sensors, with provision for onboard data- and image collection, or downlink telemetry. An independent 28 V 40 A DC power supply is provided for surveillance equipment, and a cockpit equipment bay, capacity 0.4 m³ (14 cu ft) or 30 kg (66 lb), accommodates rack-mounted electronics or instruments for in-flight monitoring. Standard fuel capacity of 110 litres (29.1 US gallons; 24.2 Imp gallons) provides endurance up to 7 hours.

HK 36TTC-Eco: Specific designation for aircraft currently in service. Certified in FAA Utility category 29 March 1999 and Restricted category (aerial photography only) 21 December 2000; latter permits increased MTOW of 929 kg (2,050 lb).

Prototype OE-9481 currently operated (since 1998) in Switzerland by Met Air atmospheric sampling company as HB-2335. Five conversions undertaken in Canada for US market: N842WS, N845WS and N846WS operated by Diamond Aircraft; and N843WS and N844WS of Wescam Air Operations, Van Nuys, California.

UPDATED

DIAMOND DV 20 KATANA and KATANA 100

See Canadian section under DA 20.

DIAMOND DA 40-TDI STAR

TYPE: Four-seat lightplane.
PROGRAMME: Formally launched 23 April 1997 at the Aero 97 show at Friedrichshafen, Germany, when mockup displayed; initially retained name of Katana; proof-of-concept prototype DA 40-V1 (OE-VPC) first flew on 5 November 1997, powered by a Rotax 914 engine; a Teledyne Continental IO-240-engined DA 40-V2 (OE-VPE) followed shortly thereafter; neither event was publicised; a third prototype, DA 40-V3, with Textron Lycoming IO-360 engine, flew in late 1998; fourth, near-production standard, prototype (OE-VPM) joined the test programme in mid-1999 followed by three more prototypes comprising DA40-P5 (c/n 40005/OE-VPQ), -P6 (40006/OE-VPB), -P7 (40007/OE-VPW), these having enlarged ventral strake and cutaway rudder.

First production aircraft (40008/OE-KPN) registered June 2000; second exhibited at Farnborough in following month. Lycoming-powered variant was initial production version, for which JAR 23 certification was achieved 25

Diamond DA 40-TDI development aircraft (*Jane's/Paul Jackson*)   NEW/0143326

October 2000; FAA certification, 9 April 2001, followed by JAA IFR certification on 27 April 2001, and FAA IFR certification on 23 August 2001. First deliveries to US customers (seven aircraft) took place during EAA AirVenture 2001 at Oshkosh in July 2001.

Austrian production temporarily halted in early 2002 upon transfer of Lycoming-engined variant to Canada. Thielert-engined version now sole European-built DA 40.

CURRENT VERSIONS: **DA 40-180:** Original version with Lycoming IO-360 engine; now produced by Diamond Canada (which see).

DA 40-TDI: Powered by Thielert TAE 125 turbo-diesel; converted prototype (OE-XPP) first flew 22 November 2001; certification target then October 2002. Development aircraft OE-VTA (80th Austrian-built). Initial batch will comprise 75 aircraft.

CUSTOMERS: More than 280 orders and options by September 2000. At least 80 from Austrian production had been registered by early 2002 to customers in Austria, Germany, Netherlands, Sweden, Switzerland, UK, USA and elsewhere.

COSTS: DA 40-TDI €186,700 basically (N-VFR) equipped; full standard avionics (IFR) €29,100 extra; all excluding tax (2002).

DESIGN FEATURES: Four-seat development of DA 20-C1; to be certified to FAR/JAR 23 or equivalents in Europe, North America, Russian Federation, South Africa and Turkey. DA 20-C1 wings attach to new, wider centre-section to increase span; downturned tips to horizontal tail; single strake/tail bumper. Enlarged cabin increases fuselage length by 0.67 m (2 ft 2¼ in).

Wing section is Diamond-modified Wortmann FX 63-137/20; washout 1°.

FLYING CONTROLS: Conventional and manual. Ailerons and elevators operated by pushrods with low-friction bearings, rudder by cables; manual pitch trim standard, electric pitch trim optional via control stick-mounted switch; ground-adjustable trim tab on rudder; electrically actuated slotted flaps.

STRUCTURE: Fuselage, with integral vertical fin, is mostly of GFRP construction with local CFRP reinforcement in high-stress areas, comprising two half-shells bonded together. Two-spar wings with GFRP/PVC foam/GFRP sandwich skins; horizontal tail and rudder surfaces similarly covered.

LANDING GEAR: Fixed tricycle type, with cantilever spring steel leg on each main unit and elastomeric suspension on nose leg. Mainwheel size 6.00-6; nosewheel 5.00-5. Steering is provided by differential braking of mainwheel and friction-damped castoring nosewheel. Speed fairings on all three wheels.

POWER PLANT: One 99 kW (133 hp) Thielert TAE 125 turbo-diesel, driving an MTV-12-B/180-17 three-blade constant-speed propeller. Fuel in two wing tanks, standard capacity 110 litres (29.0 US gallons; 24.2 Imp gallons); 155 litres (40.9 US gallons; 34.1 Imp gallons) optional.

ACCOMMODATION: Four persons in two side-by-side pairs. Single-piece canopy lifts up and forwards for access to front seats; rear occupants board through an upward-opening door to port. Cabin has 26 g composites seats with three-point automatic safety harnesses and roll-over bar. Baggage compartment behind rear seats includes tubular extension into rear fuselage for extra-long items. Rigid baggage compartment extension for long objects such as skis.

SYSTEMS: Full standard fit includes Bendix/King KAP 140 dual-axis autopilot. 28 V electrical system includes 70 A alternator; 28 V 35 Ah battery.

AVIONICS: *Comms:* Baseline Garmin GNS 430 com/nav/GPS and GTX 327 transponder. IFR standard adds GMA 430 audio selector with marker, second GNS 430.

Flight: Baseline Garmin GI 106 VOR/ILS. IFR adds second GI 106. Further options include KN 62A DME, KCS 55A slaved compass system, GNS 530 (replacing GNS 430) and Goodrich WX-500 Stormscope.

Optional Brightline 'glass cockpit' of TwinStar   *NEW*/0137410

EQUIPMENT: Landing light, taxying light, position lights, instrument lighting, overhead cabin light and pitot/static heat all standard. Optional equipment includes glider towing kit, with retractor winch.

Data apply to DA 40-TDI.

DIMENSIONS, EXTERNAL:
Wing span	11.94 m (39 ft 2 in)
Wing aspect ratio	10.6
Length overall	8.01 m (26 ft 3¼ in)
Height overall	2.00 m (6 ft 6¾ in)

DIMENSIONS, INTERNAL:
Baggage compartment: Length	1.08 m (3 ft 6½ in)
Max width	0.35 m (1 ft 1¾ in)
Max height	0.80 m (2 ft 7½ in)

AREAS:
Wings, gross	13.50 m² (145.3 sq ft)

WEIGHTS AND LOADINGS:
Weight empty	750 kg (1,653 lb)
Baggage capacity	35 kg (77 lb)
Max T-O weight	1,150 kg (2,535 lb)
Max landing weight	1,092 kg (2,407 lb)
Max wing loading	85.2 kg/m² (17.45 lb/sq ft)
Max power loading	11.60 kg/kW (19.06 lb/hp)

PERFORMANCE:
Never-exceed speed (VNE)	
	178 kt (329 km/h; 204 mph) IAS
Max level speed at FL100	154 kt (285 km/h; 177 mph)
Cruising speed at FL100:	
at 70% power	132 kt (244 km/h; 152 mph)
at 50% power	110 kt (204 km/h; 127 mph)
at 40% power	89 kt (165 km/h; 102 mph)
Stalling speed	49 kt (91 km/h; 57 mph)
Max rate of climb at S/L	238 m (780 ft)/min
Max certified altitude	5,000 m (16,400 ft)
T-O run	357 m (1,175 ft)
T-O to 15 m (50 ft)	430 m (1,410 ft)
Range, 45 min reserves:	
standard fuel	750 n miles (1,389 km; 863 miles)
optional fuel	1,100 n miles (2,037 km; 1,265 miles)

UPDATED

DIAMOND DA 42 TWINSTAR

TYPE: Four-seat utility twin.

PROGRAMME: Prototype (D-GENI worn, but not officially registered as such); shown statically at Berlin, 6 to 12 May 2002; first flight then projected for July 2002. Although long reported, company decision to build was only made in September 2001, and assembly began March 2002. Re-registered as OE-VPS and made first flight 9 December 2002.

CUSTOMERS: At time of launch, 7 May 2002; Lufthansa training school had signed letter of intent for 40.

COSTS: €359,800 basically equipped (2002).

DESIGN FEATURES: Twin-engined version of DA 40 Star. Low-mounted wing with winglets; T tail; engine cowlings of large frontal area. Downturned tailplane tips. Enlarged rudder and fin dorsal fillet; underfin faired into rudder.

FLYING CONTROLS: Conventional and manual. Flight-adjustable tabs on elevator (central) and rudder; ground-adjustable tab on port aileron. Slotted flaps outboard of engines; split flaps inboard; both electrically actuated.

STRUCTURE: Composites throughout.

LANDING GEAR: Retractable tricycle type. Nosewheel, with 5.00-5 tyre, retracts rearward and covered by single, starboard-hinged door; mainwheels, with 15×6.00-6 tyres and trailing-link configuration, retract inboard, having single part-door on legs, leaving wheel exposed after retraction. Oleo-pneumatic shock-absorbers on all legs. Hydraulic brakes.

POWER PLANT: Two 99 kW (133 hp) Thielert TAE 125 turbocharged diesel engines, each driving an MT-Propeller MTV-6-A/187-129 three-blade, variable-pitch (hydraulic) propeller. Fuel tanks in wings, total normal capacity 200 litres (52.8 US gallons; 44.0 Imp gallons); optional capacity 280 litres (74.0 US gallons; 61.6 Imp gallons).

ACCOMMODATION: Four persons in individual seats; two side-by-side pairs. Cabin doors as for DA 40 Star. Baggage compartment in nose; door port side.

SYSTEMS: Bendix/King KAP 140 dual-axis autopilot.

AVIONICS: *Comms:* Twin Garmin GNS 430 com/nav/GPS; GTX 327 transponder; Garmin GMA 340 audio selector panel. Optional GNS 530.

Flight: Twin GI 106 VOR/ILS. Optional KR 87 ADF and Goodrich WX-500 Stormscope.

Instrumentation: Conventional displays standard. Optional Brightline three-screen 'glass cockpit' (as fitted to prototype).

DIMENSIONS, EXTERNAL:
Wing span	13.42 m (44 ft 0¼ in)
Wing aspect ratio	10.9
Length overall	8.50 m (27 ft 10¾ in)

Diamond DA 42 TwinStar prototype prior to its December 2002 first flight   *NEW*/0137409

Height overall	2.60 m (8 ft 6¼ in)
Propeller diameter	1.87 m (6 ft 1½ in)

AREAS:

Wings, gross	16.46 m² (177.2 sq ft)

WEIGHTS AND LOADINGS:

Weight empty	1,030 kg (2,271 lb)
Max T-O weight	1,650 kg (3,637 lb)
Max wing loading	100.2 kg/m² (20.53 lb/sq ft)
Max power loading	8.32 kg/kW (13.67 lb/hp)

PERFORMANCE:

Cruising speed: max at 95% power at FL125

    203 kt (376 km/h; 234 mph)

econ at 60% power at FL100

	168 kt (311 km/h; 193 mph)
Max rate of climb: at S/L	527 m (1,730 ft)/min
at FL100	503 m (1,650 ft)/min
Max certified altitude	6,095 m (20,000 ft)
T-O run	290 m (955 ft)
T-O to 15 m (50 ft)	510 m (1,675 ft)

Range at econ cruising speed:

normal fuel	1,061 n miles (1,965 km; 1,221 miles)
optional fuel	1,485 n miles (2,750 km; 1,708 miles)

NEW ENTRY

DIAMOND FUTURE DEVELOPMENTS

Immediately on revealing the Diamond TwinStar in May 2002, the company announced that its next project is a five-seater powered by a single Thielert turbo-diesel engine of 228 kW (306 hp). Deliveries are due to begin in 2004.

UPDATED

HB FLUGTECHNIK

HB FLUGTECHNIK GmbH

Dr Adolf Scharf Strasse 42, Postfach 74, A-4053 Haid-Ansfelden

Tel: (+43 7229) 791 04 and 791 17
Fax: (+43 7229) 791 04 15 and 791 17 15
e-mail: office@hb-aviation.com
Web: http://www.hb-aviation.com
MANAGING DIRECTOR: Ing Heino Brditschka
TECHNICAL DIRECTOR: Georg Passenbrunner

DISTRIBUTOR:

Aero-Service Thüringen GmbH
Mühlberger Strasse 13, D-99869 Wandersleben, Germany
Tel: (+49 36202) 823 95
Fax: (+49 36202) 823 86
e-mail: info@aeroserv.de
Web: http://www.aeroserv.de
DIRECTOR: Karl-Heinz Maletschek

Company name indicates Heino Brditschka, designer of motor gliders, including the HB 23 and HB 202, last described in the 1991-92 *Jane's*. Both had the unconventional mid-fuselage propeller, to which HB Flugtechnik has returned for its most recent project, the Tornado, which will now be available to kitbuilders. First flight of this aircraft was rescheduled from 2001 to 2002, while company concentrates on marketing of HB 207.

HB provides a wide variety of other aviation services, including overhaul and repair, aerial photography and advertising, flying training, aircraft charter and support of amateur constructors. By 2000, distribution of HB 207 kits had been assigned to Aero-Service Thüringen.

UPDATED

HB 207RG Alfa two-seat lightplane undertaking US promotion at Sun 'n' Fun, April 2002 *(Jane's/Paul Jackson)*

NEW/0132942

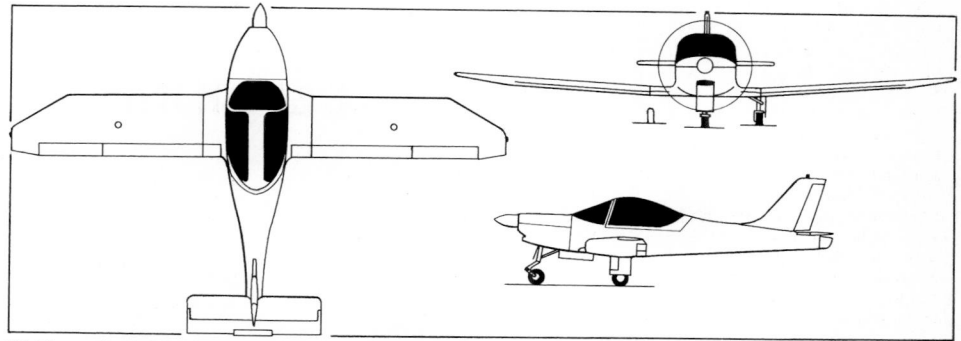

HB Flugtechnik HB-207RG Alfa (Volkswagen engine) *(Jane's/Paul Jackson)*   0092124

HB FLUGTECHNIK HB 207 ALFA

TYPE: Side-by-side kitbuilt.
PROGRAMME: First flight (HB 207RG OE-CHC) 14 March 1995. Second prototype (HB 207 OE-CAA) made public debut, unmarked and then unflown, at AERO 97 at Friedrichshafen April 1997. Certified to JAR-VLA.
CURRENT VERSIONS: **HB 207:** Fixed landing gear version.
    HB 207RG: Retractable landing gear version.
CUSTOMERS: Total 70 sold in Austria, Brazil, Czech Republic, France, Germany, Sweden, Switzerland, United Kingdom and USA by April 2001.
COSTS: Standard aircraft fast-build kit, no engine, propeller or avionics €21,100 (2002).
DESIGN FEATURES: Quoted build time 1,000 hours. Constant chord low wings with raked tips; slightly sweptback fin and rudder; one-piece elevator aft of vertical tail. Airframe features guide derigging and rigging for road transport or storage.
FLYING CONTROLS: Conventional and manual. Ailerons and elevator actuated by pushrods; rudder by cables. Large central trim tab in elevator. Manual (optionally electric) operation for flaps.
STRUCTURE: Mainly metal primary structure with GFRP skin; moving surfaces have Ceconite covering.
LANDING GEAR: HB 207RG has retractable tricycle type, with size 4.00-4 tyre on each unit. Main units have brakes and rubber-in-compression shock-absorption, and retract inward; nosewheel retracts rearward. HB 207 has fixed tricycle type with streamlined fairings and spats on each unit.
POWER PLANT: Standard engine is an 82.0 kW (110 hp) VW-Porsche HB 2400 G/2 flat-four based on that of Porsche 911 sports car; alternatives include 59.6 kW (79.9 hp) Rotax 912 A, and Rotax 914, or Limbach or Textron Lycoming engines up to 74.6 kW (100 hp). Geared drive to choice of propellers: two-blade wooden fixed-pitch, three-blade adjustable- or variable-pitch, or (as on prototypes) five-blade adjustable- or variable-pitch. Fuel tank in each wing, combined capacity 100 litres (26.4 US gallons; 22.0 Imp gallons).
ACCOMMODATION: Rearward-sliding fully transparent canopy.
SYSTEMS: Electrical system: 200 A 15 Ah (optionally 20 Ah) battery; 14 V 75 A alternator.
AVIONICS: *Comms:* Honeywell 760-channel KX 155 transceiver, KT 76A transponder and ELT optional.

Flight: Honeywell KX 125 VOR with KI 208 indicator optional.
Instrumentation: VFR flight and engine transmission instrumentation standard.

DIMENSIONS, EXTERNAL:

Wing span	9.00 m (29 ft 6¼ in)
Wing aspect ratio	8.5
Length overall	5.95 m (19 ft 6¼ in)
Height overall	1.95 m (6 ft 4¼ in)

DIMENSIONS, INTERNAL:

Cabin: Length	1.94 m (6 ft 4½ in)
Max width	1.10 m (3 ft 7¼ in)
Max height	0.97 m (3 ft 2¼ in)

AREAS:

Wings, gross	9.50 m² (102.3 sq ft)
Vertical tail surfaces (total)	0.65 m² (7.00 sq ft)
Horizontal tail surfaces (total)	1.68 m² (18.08 sq ft)

WEIGHTS AND LOADINGS (U: Utility, N: Normal category):

Weight empty: U, N	450 kg (992 lb)
Max T-O weight: U	640 kg (1,411 lb)
N	700 kg (1,543 lb)
Max wing loading: U	67.4 kg/m² (13.80 lb/sq ft)
N	73.7 kg/m² (15.09 lb/sq ft)

Max power loading:

VW-Porsche HB 2400 G/2	8.54 kg/kW (14.03 lb/hp)
Rotax 912 A	11.75 kg/kW (19.31 lb/hp)

PERFORMANCE (HB 2400 engine, U and N as above):

Never-exceed speed (V_{NE}):

U, N	166 kt (309 km/h; 192 mph)
Cruising speed: U	132 kt (245 km/h; 152 mph)
N	130 kt (240 km/h; 149 mph)
Stalling speed: flaps up: U	51 kt (93 km/h; 58 mph)
N	53 kt (98 km/h; 61 mph)
30° flap: U	46 kt (84 km/h; 53 mph)
N	47 kt (87 km/h; 55 mph)
Max rate of climb at S/L: U	330 m (1,082 ft)/min
N	300 m (984 ft)/min
T-O run: U	190 m (625 ft)
N	210 m (690 ft)
T-O to 15 m (50 ft): U	300 m (985 ft)
N	320 m (1,050 ft)
Range: U, N	647 n miles (1,200 km; 745 miles)
g limits: U	+4.4/−2.2
N	+3.8/−1.9

UPDATED

HB FLUGTECHNIK HB 204 TORNADO

TYPE: Tandem-seat sportplane kitbuilt.
PROGRAMME: Developed from previous, unbuilt, design; unflown airframe, based on components from first two aircraft, exhibited at Aero '99 at Friedrichshafen in April

Prototype HB 204 Tornado kitbuilt sportplane *(Jane's/Paul Jackson)*   *NEW*/0131715

1999; prototype completed by early 2001 and painted with spurious registration OE-VPP for Aero '01; first flight repeatedly postponed and had not been achieved by January 2002.

DESIGN FEATURES: Unswept wing with wingtip pods housing landing lights; cheek engine air intakes aft of cabin; pusher propeller mounted at mid-fuselage, rotating in triangular-shaped cutout.

FLYING CONTROLS: Conventional and mechanical. Trim tab in starboard elevator; horn-balanced rudder; two-section Fowler-type flaps.

STRUCTURE: Primarily composites, with welded steel tube fuselage frame and alloy control surfaces.

LANDING GEAR: Retractable tricycle type; mainwheels retract inwards, nosewheel forwards.

POWER PLANT: One 147 kW (197 hp) Porsche 911 Carrera aircooled flat-four, driving a mid-mounted HB five-blade, constant-speed pusher propeller at up to 2,000 rpm, via an extension shaft.

ACCOMMODATION: Two in tandem under one-piece sideways-opening canopy with fixed windscreen.

DIMENSIONS, EXTERNAL:

Wing span	7.45 m (24 ft 5¼ in)
Wing aspect ratio	6.3
Length overall	6.60 m (21 ft 7¾ in)
Height overall	2.55 m (8 ft 4½ in)

AREAS:

Wings, gross	8.85 m² (95.3 sq ft)
Fin and rudder	1.05 m² (11.30 sq ft)
Tailplane and elevator	1.86 m² (20.02 sq ft)

WEIGHTS AND LOADINGS:

Weight empty	520 kg (1,146 lb)
Max T-O weight	750 kg (1,653 lb)
Max wing loading	84.7 kg/m² (17.36 lb/sq ft)
Max power loading	5.03 kg/kW (8.27 lb/hp)

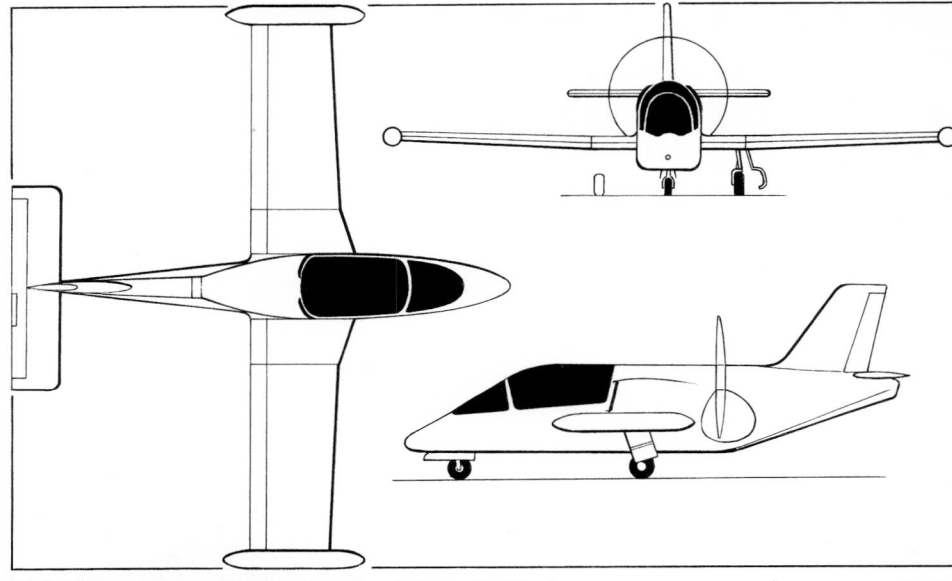

HB 204 Tornado two-seat sportplane (*Jane's/James Goulding*)   0114481

PERFORMANCE (estimated):

Never-exceed speed (VNE)	217 kt (403 km/h; 250 mph)	Max rate of climb at S/L	720 m (2,362 ft)/min
Cruising speed	189 kt (350 km/h; 217 mph)	T-O run	171 m (561 ft)
Stalling speed: flaps up	56 kt (103 km/h; 64 mph)	T-O to 15 m (50 ft)	235 m (771 ft)
flaps down	46 kt (84 km/h; 53 mph)	Max range	864 n miles (1,600 km; 994 miles)
		g limits	+6/−3

UPDATED

BELGIUM

LAMBERT

LAMBERT AIRCRAFT ENGINEERING BVBA

Lupinestraat 5, B-8500 Kortrijk
Tel/Fax: (+32 56) 21 33 47
e-mail: info@lambert-aircraft.com
Web: http://www.lambert-aircraft.com
DIRECTOR: Filip Lambert

In late 1995 Filip Lambert, a student at the College of Aeronautics, Cranfield, UK, was named one of three joint winners of the Royal Aeronautical Society Light Aircraft Design Competition with his Mission M212-100, described below. The prototype was fabricated in Belgium with the assistance of Steven Lambert and the intention of undertaking test flying at Cranfield. No decision has yet been announced as to where production will take place.

UPDATED

LAMBERT MISSION M212

TYPE: Four-seat kitbuilt.

PROGRAMME: Design started November 1992; fuselage mockup completed July 1993; construction of proof-of-concept prototype began at Geluveld, Belgium, January 1996; structural test programme completed 23 December 1998; prototype, M212-200 four-seater, registered G-XFLY in February 2000, was structurally complete by March 2001; all composites work and landing gear completed November 2001; power plant and fuel system work began March 2002. Public debut in UK at PFA Rally, Cranfield, 21-23 June 2002.

CURRENT VERSIONS: Family of five aircraft planned around common core design, including 2+2, aerobatic and four-seat tourer versions, all derived with minimum of modifications to basic two-seat aircraft. Initial versions will be **M212-100** with two seats and **M212-200** (the prototype), with four seats. Available factory-built or as kit, to be certified to JAR-VLA and JAR-23, respectively.

Proposed future variants include **M212-300** two-seat basic aerobatic trainer powered by a 171.5 kW (230 hp) Zoche ZO-02A engine driving a constant-speed propeller, and **M212-400** four-seat tourer based on M212-300 but with increased fuel and baggage capacity, optional retractable landing gear, maximum take-off weight 1,100 kg (2,425 lb) and range of over 1,000 n miles (1,852 km; 1,151 miles).

CUSTOMERS: Options on 96 kits held in 2001; customers in the Benelux countries, France and, mostly, the UK.

COSTS: Kit, including engine, propeller and VFR avionics, £85,000 (2001).

DESIGN FEATURES: Designed using CAD and wind tunnel research facilities; will meet FAR Pt 23 and JAR 23 certification criteria. Low-wing monoplane; wings, which have laminar flow, moderate taper and upturned tips, can be folded in under 5 minutes by one person without tools. Quoted kit build time 600 hours.

Prototype Lambert Mission M212 making its public debut at the PFA Rally, Cranfield, UK, 21 June 2002 (*Jane's/Paul Jackson*)   *NEW*/0137966

FLYING CONTROLS: Conventional and manual. Cable and pushrod actuation. Electrically-actuated single-slotted flaps.

STRUCTURE: All-composites; designed for ease of assembly, robustness, low maintenance and long life.

LANDING GEAR: Fixed tricycle type. Mainwheels size 6.00-6, nosewheel 5.00-5 or 6.00-6; Cleveland brakes. Steerable nosewheel, maximum steering angle ±50°. Potential for development of tailwheel and retractable versions.

POWER PLANT: One 112 kW (150 hp) Zoche ZO-01A radial four-cylinder, two-stroke diesel engine with multifuel capability, driving a three-blade constant-speed propeller. However, prototype fitted with second-hand 112 kW (150 hp) Textron Lycoming O-320-E2D flat-four; MT-

Propeller MTV-18-C/175-17d three-blade, constant-speed (electric) propeller; and Gomolzig twin-muffler exhaust. Fuel capacity 120 litres (31.7 US gallons; 26.4 Imp gallons). In developed form as four-seat tourer, suitable for engines in 172 to 194 kW (230 to 260 hp) power range.

ACCOMMODATION: Two persons, side by side, or four in 2+2 configuration, under one-piece, forward-hinged canopy. Seats and rudder pedals adjustable; dual controls standard. *All data are provisional.*

DIMENSIONS, EXTERNAL:

Wing span	9.80 m (32 ft 1¾ in)
Wing aspect ratio	8.0
Length overall	7.40 m (24 ft 3½ in)
Height overall	2.90 m (9 ft 6 in)

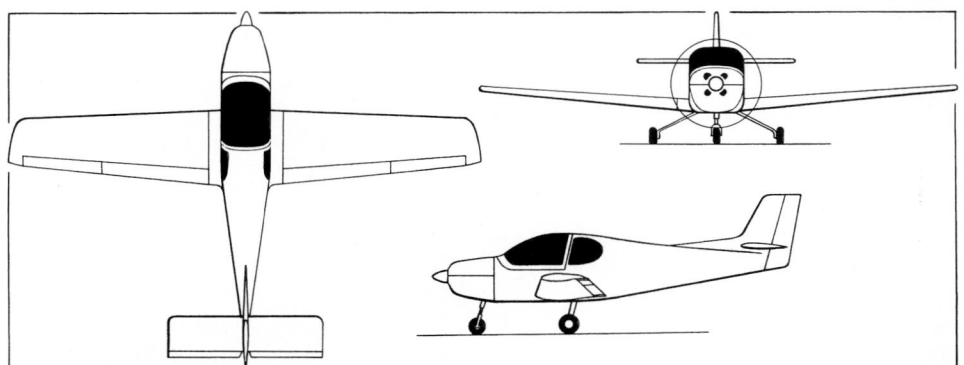

Lambert Mission M212-100 (Zoche ZO-01A diesel engine) (*Jane's/James Goulding*)   0044443

Tailplane span	3.20 m (10 ft 6 in)					
Wheel track	2.60 m (8 ft 6¼ in)					

DIMENSIONS, INTERNAL:
Cabin: Length	2.50 m (8 ft 2½ in)
Max width	1.12 m (3 ft 8 in)
Max height	1.25 m (4 ft 1¼ in)

Baggage compartment volume:
M212-100	0.85 m³ (30.0 cu ft)
M212-200	0.10 m³ (3.5 cu ft)

AREAS:
Wings, gross	12.00 m² (129.2 sq ft)
Ailerons (total)	0.60 m² (6.46 sq ft)
Flaps (total)	1.08 m² (11.62 sq ft)
Elevators (total)	2.65 m² (28.52 sq ft)

WEIGHTS AND LOADINGS (A: M212-100, B: M212-200):
Weight empty: A	460 kg (1,014 lb)
B	485 kg (1,069 lb)
Max T-O weight: A	750 kg (1,653 lb)
B	900 kg (1,984 lb)
Max wing loading: A	62.5 kg/m² (12.80 lb/sq ft)
B	75.0 kg/m² (15.36 lb/sq ft)
Max power loading: A	6.70 kg/kW (11.00 lb/hp)
B	8.04 kg/kW (13.20 lb/hp)

PERFORMANCE:
Never-exceed speed (V$_{NE}$):	
A, B	183 kt (338 km/h; 210 mph)
Max level speed at S/L: A	138 kt (256 km/h; 159 mph)
B	136 kt (252 km/h; 157 mph)

Max cruising speed at 2,440 m (8,000 ft):	
at 75% power: A	134 kt (248 km/h; 154 mph)
B	131 kt (243 km/h; 151 mph)
at 60% power: A	121 kt (224 km/h; 139 mph)
B	118 kt (219 km/h; 136 mph)
Stalling speed, flaps down: A	45 kt (84 km/h; 52 mph)
B	49 kt (91 km/h; 57 mph)
Max rate of climb at S/L: A	378 m (1,240 ft)/min
B	293 m (960 ft)/min
T-O run: A	175 m (575 ft)
B	224 m (735 ft)
Range with max fuel: A, B:	
at 75% power	640 n miles (1,185 km; 736 miles)
at 60% power	800 n miles (1,481 km; 920 miles)

UPDATED

MASQUITO

MASQUITO AIRCRAFT n.v.
Reigersbaan 31, B-1760 Roosdaal-Strijtem
Tel: (+32 54) 34 30 08
Fax: (+32 54) 34 30 09
e-mail: info@masquito.be
Web: http://www.masquito.be
FOUNDERS: Paul Maschelein
　　Stefaan Maschelein
　　John Pescod

This company was established to produce the Masquito M80 ultralight helicopter, two prototypes of which have been completed. Production deliveries were to have begun in late 2002 or early 2003.

UPDATED

MASQUITO M80
TYPE: Two-seat ultralight helicopter kitbuilt.
PROGRAMME: Design started November 1994; construction of prototype began December 1995; first flight (G-MASZ) May 1996; 25 hours of test flying completed by April 1997. Prototype reconfigured as M80 with Jabiru engine and began ground trials in late 1997. Two preproduction prototypes (G-MASX and G-MASY) registered June 1998 in M80 configuration; one displayed unflown at the PFA International Air Rally at Cranfield in July 1999; but neither had flown by 2001; CAA go-ahead for ground running given mid-2001; this completed on 26 April 2002, confirming engine integrity. UK certification will be to BCAR-VLH requirements; French CNSK 2 certificate will then be issued automatically; US certification of experimental version will follow.
CURRENT VERSIONS: **M58:** Prototype only; initially with 47.7 kW (64 hp) Rotax 582 engine.
　　M80: Production version; Jabiru engine. *As described.*
COSTS: Kit €65,000 to €70,000 (2002), including engine and basic instruments.
DESIGN FEATURES: Conforms to FAR Pt 27 crashworthiness requirements. Two-blade, teetering main rotor and two-blade tail rotor; main rotor blades of modified Wortmann S aerofoil section with 13 per cent thickness/chord ratio and 8° 15′ linear twist. Hiller servo rotor. Main rotor speed 690 rpm; 2 kg (4.4 lb) lead weight in each tip to increase inertia and improve stability at the hover. Tail rotor 3,200 rpm. Ventral fin offset 2°. Composites 'virtual hinge' main rotor head with elastomeric collective pitch thrust bearings. Prototype had Rotax gearbox for primary reduction and Gates powerdrive toothed aramid fibre belts for secondary reduction; production version employs toothed drivebelt, centrifugal clutch and flywheel as primary transmission, with flexible driveshaft for tail rotor.
FLYING CONTROLS: Conventional and manual, with Hiller servo rotor for cyclic pitch control.
STRUCTURE: Primary structure of L-section Ti3A12, 5V alloy, riveted and glued; tailboom of 8.9 cm (3½ in) carbon fibre tube. Main rotor blades have unidirectional composites spar with PVC foam trailing-edge core and bidirectional composites skin.
LANDING GEAR: Two fixed skids of 6061 alloy mounted on unidirectional carbon fibre supports.
POWER PLANT: One modified Jabiru 2200 flat-four, nominal power output 70.8 kW (95 hp) at sea level, derated to 59.7 kW (80 hp) and equipped with dual ignition, electronic fuel injection, rpm governor, engine overspeed protection, larger air intake ducts and tuned exhaust;

Masquito two-seat ultralight helicopter (*Jane's/Paul Jackson*) 　　　　NEW/0137967

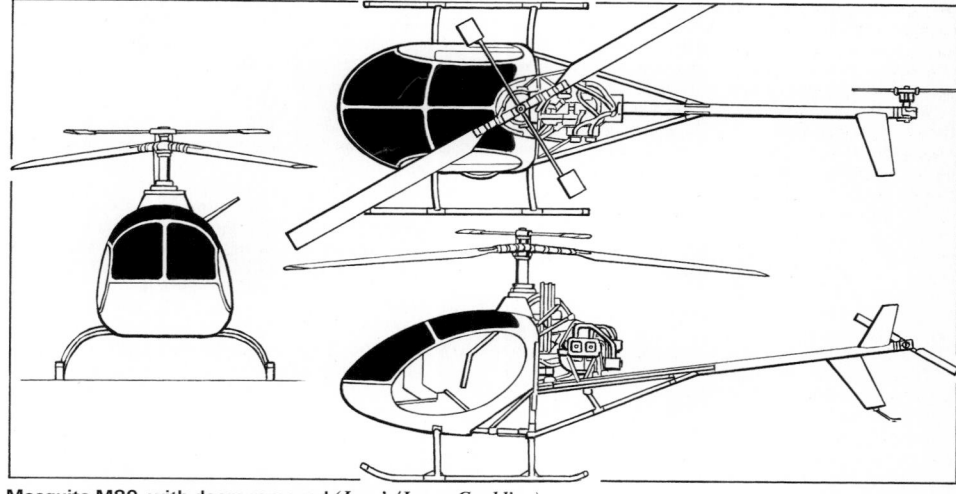

Masquito M80, with doors removed (*Jane's/James Goulding*) 　　　　0092125

transmission rating in excess of 74.6 kW (100 hp). Fuel capacity 62 litres (16.4 US gallons; 13.6 Imp gallons).

DIMENSIONS, EXTERNAL:
Main rotor diameter	5.40 m (17 ft 8½ in)
Tail rotor diameter	1.00 m (3 ft 3¼ in)
Length: overall, rotors turning	6.16 m (20 ft 2½ in)
fuselage	4.85 m (15 ft 11 in)
Height overall	2.25 m (7 ft 4½ in)
Skid track	1.70 m (5 ft 7 in)

DIMENSIONS, INTERNAL:
Cockpit max width	1.30 m (4 ft 3¼ in)

AREAS:
Main rotor disc	22.90 m² (246.49 sq ft)
Tail rotor disc	0.79 m² (8.45 sq ft)

WEIGHTS AND LOADINGS:
Weight empty, equipped	230 kg (507 lb)
Max T-O and landing weight	450 kg (992 lb)

PERFORMANCE (estimated):
Never-exceed speed (V$_{NE}$)	97 kt (180 km/h; 111 mph)
Max cruising speed	80 kt (148 km/h; 92 mph)
Max rate of climb at S/L	335 m (1,100 ft)/min
Hovering ceiling IGE	1,830 m (6,000 ft)
Range with max internal fuel, no reserves	
	324 n miles (600 km; 372 miles)
Endurance	4 h

UPDATED

WOLFSBERG

WOLFSBERG AIRCRAFT CORPORATION NV
Woudstraat 23, B-3600 Genk
Tel: (+32 89) 38 08 31
Fax: (+32 89) 38 61 41
DIRECTOR AND CHIEF DESIGNER: Alec N Clark

Formerly Triloader Aircraft Corporation NV (see entry in 1996-97 edition), which intended to develop the Clark-Norman Triloader, described in the UK section of the 1996-97 *Jane's*. Wolfsberg Aircraft Corporation NV is now developing the Raven 257 small multirole aircraft. The programme was previously managed by Wolfsberg-Evektor (which see) in the Czech Republic; the prototype, which flew on 28 July 2000, was built by Evektor-Aerotechnik (EV-AT) in Kunovice, with assistance from Letov Air. However, on 21 November 2000 it was announced that development

(including four preproduction aircraft for certification trials) and production of the Raven 257 had been assigned to Letov, under which heading it is described in the Czech section. Manufacture of wings, originally subcontracted to EV-AT, was reassigned to Letov.

In 2001, Wolfsberg announced preliminary details of the proposed **Sparrow**, which will compete with the Cessna 206 Stationair, albeit in twin-engine configuration.

UPDATED

BOSNIA-HERZEGOVINA

SOKO

SOKO AIR MOSTAR

This company appears not to have proceeded with production of the Soko 2 ultralight, last described in the 2001-02 *Jane's*. Details of the firm appeared in the 2002-03 edition.

UPDATED

BRAZIL

AEROMOT

AEROMOT INDÚSTRIA MECANICO-METALURGICA LTDA

Caixa Postale 8031, Avenida das Indústrias 1210, 90200-290 Porto Alegre, RS
Tel: (+55 51) 371 16 44
Fax: (+55 51) 371 16 55
e-mail: industria@aeromot.com.br
Web (1): http://www.ximango.com
Web (2): http://www.aeromot.com
PRESIDENT: Claudio B Viana
MANAGING DIRECTOR: João Claudio Jotz
COMMERCIAL DIRECTOR: Vitor J P Neves
MARKETING CONTACT: Patricia Munstock

Aeromot Indústria is part of Aeromot Group with Aeromot Aeronaves e Motores SA (parent company, founded 1967) and Aeroeletrônica Indústria de Componentes Aviónicos SA (established 1981); former sells aircraft and spares and provides maintenance; latter certifies and manufactures avionic equipment for civil and military aircraft, including 11 items for Embraer Tucano and 13 for Italian-Brazilian AMX.

Aeromot Indústria initially designed, certified and built seats for aircraft built by Embraer; it later produced several structural parts for Embraer aircraft, and designed and certified seats for Airbus, Boeing, Fokker and McDonnell Douglas (now Boeing) commercial transports. In 1985 it purchased assets of Fournier motor glider factory in France, including sole manufacturing rights for RF-10; has since incorporated several improvements to basic model, including a turbocharged engine and a trainer variant. A lightplane version, with reduced wing span, first flew in 1999.

Factory has shop floor area of 3,100 m² (33,375 sq ft) and workforce of approximately 150. Future plans include eventual manufacture under licence of high-performance foreign sailplanes and of all-composites four-seat light aircraft.

VERIFIED

AEROMOT XIMANGO

TYPE: Motor glider.
PROGRAMME: Brazilian production version of French Aérostructure (Fournier) RF-10; first flight (French prototype) 6 March 1981; see Sailplanes section of 1990-91 *Jane's* for French production history. All production rights sold to Aeromot July 1985; Brazilian CTA certification of AMT-100 granted 5 June 1986 and French (DGAC) 10 October 1990; AMT-200, developed by Aeromot, made first flight July 1992; was certified 3 February 1993 in Brazil, on 29 December 1993 in USA (to FAR Pt 22), during 1995 in UK and on 24 November 1998 in Canada. Certification of AMT-200 also achieved in Australia, Colombia, France, Germany, Japan, Netherlands and New Zealand.
CURRENT VERSIONS: **AMT-100 Ximango:** Initial production version; powered by one 59.7 kW (80 hp) Limbach L 2000 EO1 flat-four engine driving a Hoffmann HO-V62RL/160BT two-blade three-position variable-pitch propeller. Total 43 built, 1988-93.

Winglet-equipped Aeromot AMT-200S Super Ximango S two-seat motor glider (*Jane's/Paul Jackson*) 0062405

AMT-100P and -100R Ximango: Military/police (100P) and observation (100R) versions. Two side windows below canopy; underfuselage pod for 100 kg (220 lb) of weapons and avionics or surveillance equipment. In service with Brazilian military police and other law enforcement agencies.
AMT-200 Super Ximango: Generally similar to AMT-100 except for Rotax 912A power plant. Main production variant since 1995. Improvements introduced in 1997 include redesigned instrument panel; new canopy locking mechanism; fully enclosing main landing gear doors; tailwheel fairing; and metal fuel tanks.
AMT-200S Super Ximango S: Version intended for pilot training and soaring. Generally similar to AMT-200 except for Rotax 912 S4 engine and optional removable winglets. Prototype first flew in September 1999; deliveries from December 1999, to customers in Australia (one), South Africa (two), UK (two) and USA (three).

Selected from at least 13 types evaluated for the USAF's Introductory Flight Training Program (IFTP), replacing grounded Slingsby T-3A Fireflies in flight screening/primary training role; initial order for 14 aircraft announced October 2001, with delivery of first five scheduled for completion by March 2002.
Detailed description applies to the AMT-200 except where indicated.
AMT-300 Turbo Ximango Shark: Prototype (PT-ZAM, converted from an early AMT-100) flew July 1997; based on AMT-200, but with Rotax 914F turbocharged engine, redesigned cockpit, new engine cowling and winglets. Brazilian certification achieved 31 March 1999, followed by FAA approval 19 July 1999; deliveries began in May 1999 with two aircraft for US market delivered to Ximango US at Spruce Creek, Florida.
AMT-300R Reboque: Glider tug version of AMT-300.
AMT-600 Guri: Trainer and sport aerobatic version. Described separately.
CUSTOMERS: Initial order for eventual 100 from Brazilian Civil Aeronautical Department; also produced for military/paramilitary roles such as observation, patrol and counter-insurgency. Total of 130 AMT-100s, AMT-200s and AMT-300s produced by mid-2001, including deliveries to

customers in Argentina (two), Australia (seven), Belgium (one), Brazil (63), Canada (one), Colombia (one), Dominican Republic (one), France (seven), Germany (two), Japan (four), South Africa (two), UK (six) and USA (27). AMT-300 production began at No. 106, but continues in parallel with AMT-200.
COSTS: US$115,000 basic for AMT-200S; US$125,000 basic for AMT-300 (both 2000).
DESIGN FEATURES: Typical motor glider; low, tapered wing with fixed incidence T tail. Wings detachable for transportation and storage; outboard half of wing can be folded inward without disconnecting aileron controls.

Wing section NACA 64₃-618; dihedral 2° 30′.
FLYING CONTROLS: Conventional and manual; ailerons and elevators operated by pushrods, rudder by cables; trim tab actuated by cable on AMT-100/200; trim control on AMT-200S/300 effected by spring device on elevator pushrod. Schempp-Hirth airbrakes in wing upper surface.
STRUCTURE: All-GFRP/foam except for carbon fibre main spar and light alloy airbrakes.
LANDING GEAR: Mechanically retractable mainwheels (tyre size 330×130), with hydraulic suspension and Cleveland hydraulic disc brakes with differential action; steerable tailwheel with size 210×65 tyre.
POWER PLANT: One 59.6 kW (79.9 hp) Rotax 912A flat-four with Hoffmann HO-V62R/170FA propeller. Fuel in two main tanks in wings, combined capacity 90 litres (23.8 US gallons; 19.75 Imp gallons). AMT-200S has one 73.5 kW (98.6 hp) Rotax 914 S4, driving a three-position variable-pitch Hoffmann propeller. AMT-300 has one 84.6 kW (113.4 hp) Rotax 914 F3 turbocharged flat-four driving an MT/MTV-21 constant-speed feathering propeller; fuel capacity 90 litres (23.8 US gallons; 19.8 Imp gallons).
ACCOMMODATION: Two seats side by side. One-piece canopy hinged at rear to open upward. Dual controls standard.
SYSTEMS: Electric starter and 12 V 30 A alternator.

DIMENSIONS, EXTERNAL:

Wing span: 200, 200S	17.47 m (57 ft 3¾ in)
300	17.70 m (58 ft 0¾ in)
Wing aspect ratio: 200, 200S	16.3
300	16.7
Width, wings folded	10.15 m (33 ft 3½ in)
Length overall: 200, 200S	8.05 m (26 ft 5 in)
300	8.08 m (26 ft 6 in)
Height overall: 200, 200S, 300	1.93 m (6 ft 4 in)
Propeller diameter: 200, 200S	1.70 m (5 ft 7 in)
300	1.65 m (5 ft 5 in)

DIMENSIONS, INTERNAL:

Cockpit: Length	1.85 m (6 ft 0¾ in)
Max width	1.15 m (3 ft 9¼ in)
Max height	1.00 m (3 ft 3¼ in)

AREAS:

Wings, gross: 200, 200S	18.70 m² (201.3 sq ft)
300	18.75 m² (201.8 sq ft)
Ailerons (total)	2.72 m² (29.28 sq ft)
Airbrakes (total)	1.16 m² (12.49 sq ft)
Fin	1.44 m² (15.50 sq ft)
Rudder	0.69 m² (7.43 sq ft)
Tailplane	1.95 m² (20.99 sq ft)
Elevator	0.95 m² (10.23 sq ft)

WEIGHTS AND LOADINGS:

Weight empty: 200	620 kg (1,367 lb)
200S	625 kg (1,378 lb)

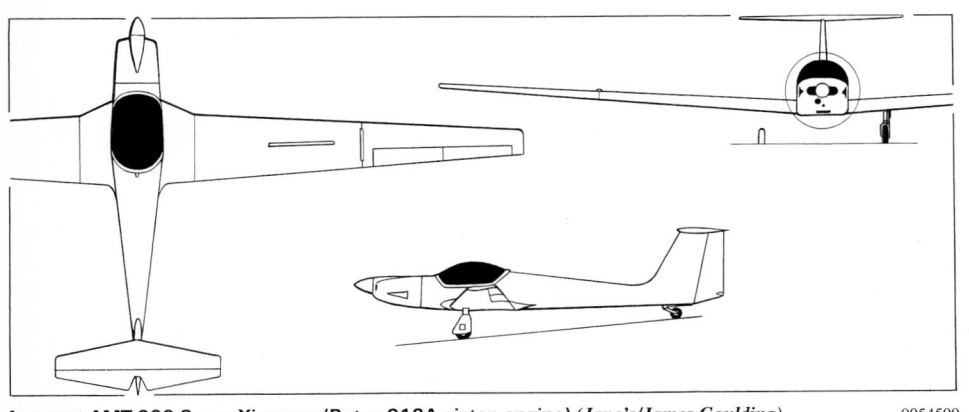

Aeromot AMT-200 Super Ximango (Rotax 912A piston engine) (*Jane's/James Goulding*)   0054599

300	630 kg (1,389 lb)
300R	620 kg (1,367 lb)
Max T-O weight: 200, 200S, 300	850 kg (1,874 lb)
Max baggage: 300	10 kg (22 lb)
Max wing loading: 200, 200S	45.5 kg/m² (9.31 lb/sq ft)
300	45.3 kg/m² (9.28 lb/sq ft)
Max power loading: 200	14.26 kg/kW (23.42 lb/hp)
200S	11.56 kg/kW (19.00 lb/hp)
300	9.92 kg/kW (16.29 lb/hp)

PERFORMANCE, POWERED:

Never-exceed speed (VNE):	
200, 200S, 300	133 kt (245 km/h; 153 mph)
Max cruising speed: 200	110 kt (205 km/h; 127 mph)
200S	119 kt (220 km/h; 137 mph)
300	124 kt (230 km/h; 143 mph)
Normal cruising speed: 300	122 kt (225 km/h; 140 mph)
Stalling speed: 300	42 kt (78 km/h; 48 mph)
Max rate of climb at S/L: 200	156 m (512 ft)/min
200S	180 m (591 ft)/min
300	198 m (650 ft)/min
Service ceiling: 200, 200S	more than 4,900 m (16,076 ft)
300	more than 8,700 m (28,543 ft)
T-O run, hard runway: 200	226 m (745 ft)
200S	210 m (689 ft)
300	174 m (570 ft)
T-O to 15 m (50 ft), hard runway:	
200	305 m (1,001 ft)
300	296 m (970 ft)
Landing from 15 m (50 ft)	less than 300 m (984 ft)
Landing run	less than 150 m (492 ft)
Range with max fuel:	
200: best power	594 n miles (1,100 km; 683 miles)
best economy	756 n miles (1,400 km; 870 miles)
200S: best power	575 n miles (1,065 km; 661 miles)
best economy	783 n miles (1,450 km; 901 miles)
300: best power	548 n miles (1,015 km; 630 miles)
best economy	810 n miles (1,500 km; 932 miles)
Max endurance:	
200: best power	5 h 30 min
best economy	12 h
200S: best power	4 h 54 min
best economy	12 h
300: best power	4 h 30 min
best economy	12 h
g limits: 200, 200S, 300	+5.3/−2.65

PERFORMANCE, UNPOWERED:

Best glide ratio at 58 kt (108 km/h; 67 mph):	
200, 200S	31.1
300	32.1
Min rate of sink at 52 kt (97 km/h; 60 mph):	
200, 200S	0.93 m (3.05 ft)/s
300	0.96 m (3.15 ft)/s
Stalling speed: 200, 200S, 300	42 kt (78 km/h; 48 mph)

VERIFIED

Prototype Aeromot AMT-600 Guri   *NEW*/0131751

AEROMOT AMT-600 GURI
English name: Boy

TYPE: Aerobatic two-seat lightplane.

PROGRAMME: Design began July 1998; prototype (PP-XBS) first flown 13 July 1999; total of more than 110 flight hours logged by mid-2000; Brazilian CTA and FAA certification was scheduled before end of 2001.

CURRENT VERSIONS: **AMT-600 Guri:** Initial production version, as described.

Data below are provisional, and refer to prototype.

Advanced Trainer: Projected derivative with 134 kW (180 hp) Lycoming engine, retractable landing gear and IFR avionics.

CUSTOMERS: Brazilian state aero clubs seen as initial target market. No production aircraft had been registered by early 2002.

DESIGN FEATURES: Generally as Ximango, but shortened wings (of same aerofoil section) and tricycle landing gear. Wings can be removed for storage.

FLYING CONTROLS: Conventional and manual; ailerons and elevator operated by pushrods, rudder by cables; ground-adjustable trim tab on rudder and starboard aileron. Control surface maximum deflections: elevator +22/−25°, ailerons +15/−27°, rudder 28°, flaps 45°.

STRUCTURE: Generally as for Ximango. Wing includes mainspar, plus rear-spar of Z shape to accommodate ailerons and flaps; two ribs in each panel, at centre-section break and tip, but three ribs per side at centre-section to strengthen walkways. Ailerons and flaps have leading- and trailing-edge spars, each with six glass fibre ribs stiffened with foam and aluminium skirt. Two fuselage bulkheads only: one horizontal, one vertical, for attachment of tail surfaces. Plywood for local strengthening in areas of equipment installations.

LANDING GEAR: Non-retractable tricycle type, with hydropneumatic trailing-link suspension on main units and rubber-in-compression suspension on nose leg. Oldi wheels and brakes. Mainwheel tyre size 6.00×6, nosewheel 5.00×5.

POWER PLANT: One 85.8 kW (115 hp) Textron Lycoming O-235-NBR flat-four, driving a two-blade Sensenich 72CK-0-50 aluminium propeller. Fuel contained in two wing tanks, combined capacity 90 litres (23.8 US gallons; 19.8 Imp gallons), with filler port in upper surface of each wing.

ACCOMMODATION: Two persons, side by side, under single-piece, upward- and rearward-hinged canopy. Dual controls standard. Boarding step on each side forward of wing on prototype will be relocated aft of wing on production aircraft.

AVIONICS: VFR avionics, including Garmin GPS 100; IFR avionics optional.

DIMENSIONS, EXTERNAL:

Wing span	10.50 m (34 ft 5½ in)
Wing aspect ratio	8.0
Length overall	8.07 m (26 ft 5¾ in)
Height overall	2.65 m (8 ft 8¼ in)
Tailplane span	2.60 m (8 ft 6¼ in)
Aileron span	1.99 m (6 ft 6½ in)
Flap span	2.44 m (8 ft 0 in)
Wheel track	2.99 m (9 ft 9¾ in)
Wheelbase	1.75 m (5 ft 9 in)
Propeller diameter	1.83 m (6 ft 0 in)

DIMENSIONS, INTERNAL:

Cabin max width	0.86 m (2 ft 10 in)

AREAS:

Wings, gross	13.79 m² (148.4 sq ft)
Ailerons (total)	0.90 m² (9.69 sq ft)
Flaps (total)	1.40 m² (15.07 sq ft)
Rudder	0.63 m² (6.78 sq ft)
Tailplane	1.41 m² (15.18 sq ft)
Elevator	0.76 m² (8.18 sq ft)

WEIGHTS AND LOADINGS:

Weight empty	675 kg (1,488 lb)
Max T-O weight	900 kg (1,984 lb)
Max wing loading	65.3 kg/m² (13.37 lb/sq ft)
Max power loading	10.49 kg/kW (17.23 lb/hp)

PERFORMANCE:

Never-exceed speed (VNE)	135 kt (250 km/h; 155 mph)
Cruising speed	108 kt (200 km/h; 124 mph)
Stalling speed	49 kt (90 km/h; 56 mph)
Max rate of climb at S/L	216 m (709 ft)/min
Service ceiling	5,300 m (17,380 ft)
T-O run	130 m (427 ft)
Max range	783 n miles (1,450 km; 901 miles)
Endurance	6 h 0 min
g limits	+3.8/−1.52

UPDATED

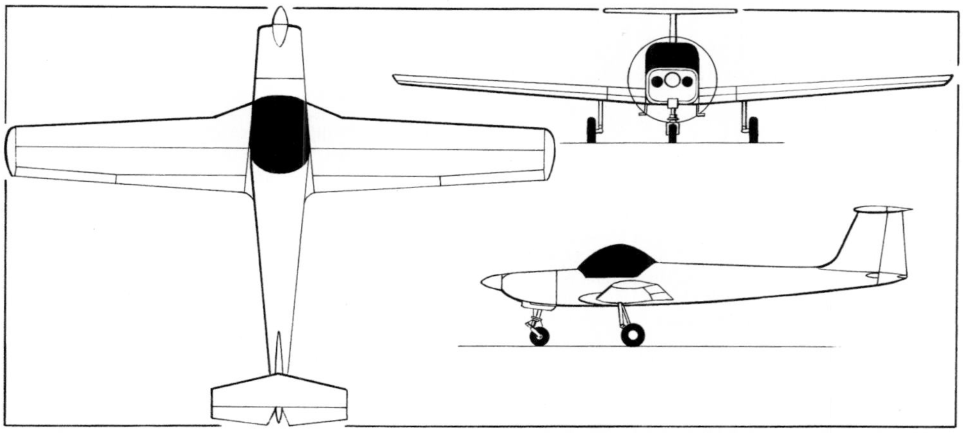

Aeromot AMT-600 Guri with raised canopy intended for production version (*Jane's/James Goulding*) 0121661

EDRA HELICENTRO

EDRA HELICENTRO, PEÇAS e MANUTENÇÃO LTDA
Rodovia Estadual SP-191 Km 87.5, 13537-000 Ipeúna, São Paulo
Tel: (+55 19) 576 12 92
Fax: (+55 19) 576 13 92
e-mail: mnd@edraescola.com.br
Web: http://www.edraescola.com.br

WORLD DISTRIBUTOR:
Amphibian Airplanes of Canada Ltd
Box 1100, Squamish, British Columbia, V0N 3G0, Canada
Tel: (+1 604) 898 53 27 and 32 00
Fax: (+1 604) 898 20 09
e-mail: harusch@uniserve.com
Web: http://www.seastaramphibian.com

Following cessation of manufacture by the Billie company in France, the Pétrel amphibian is being produced in Brazil for local and export markets. Marketing outside South America is undertaken by Amphibian Airplanes of Canada, which built the first North American SeaStar in 1998. Components are imported and augmented by North American standard items before sale as kits.

EDRA Helicentro is one of three components of the EDRA group, founded in 1994 and with interests in pilot training, aircraft maintenance, air taxi services and aircraft dealership (Schweizer helicopters for Brazil).

UPDATED

EDRA HELICENTRO PATURI
English name: Masked Duck
Export marketing name: SeaStar

TYPE: Two-seat amphibian/kitbuilt.

PROGRAMME: Derived from Claude Tisserand's Hydroplum II, prototype of which first flew 1 November 1986; SMAN bought design rights in 1987; construction of Pétrel prototype began January 1989; first flight July 1989. At least 70 built in France; rights passed to Billie Aero Marine. Production transferred to Brazil by 1997.

CUSTOMERS: Total of 135 produced by Edra by late 1999. Production rate five kits and one factory-built aircraft per month in 1999. Customers in Europe, Africa, Canada, Australia, USA and Brazil.

COSTS: Kit, less engine, US$18,500 with basic items only, or US$26,000 for complete components set; fly-away US$43,000 (2002).

DESIGN FEATURES: Experimental category pusher-engined biplane; road-towable on custom-designed trailer with assembly/disassembly time of 10 minutes. Quoted kit build time 500 hours (fast-build kit).

EDRA Helicentro Paturi, marketed abroad as the SeaStar (*Jane's/Paul Jackson*)   0084581

Length overall	6.47 m (21 ft 2¾ in)
Height overall	2.26 m (7 ft 5 in)
Wheel track	1.78 m (5 ft 10 in)
Wheelbase	2.10 m (6 ft 10¾ in)
Propeller diameter	1.65 m (5 ft 5 in)

DIMENSIONS, INTERNAL:
Cabin max width	1.13 m (3 ft 8½ in)

AREAS:
Wings, gross	16.50 m² (177.6 sq ft)

WEIGHTS AND LOADINGS:
Weight empty	300 kg (661 lb)
Max T-O weight	600 kg (1,323 lb)
Max wing loading	36.4 kg/m² (7.45 lb/sq ft)
Max power loading	10.08 kg/kW (16.56 lb/hp)

PERFORMANCE (Rotax 912 ULS):
Never-exceed speed (V$_{NE}$)	98 kt (180 km/h; 112 mph)
Max level speed	92 kt (170 km/h; 106 mph)
Max cruising speed	86 kt (160 km/h; 99 mph)
Manoeuvring speed	81 kt (150 km/h; 92 mph)
Stalling speed	35 kt (65 km/h; 41 mph)
Max rate of climb at S/L	152 m (500 ft)/min
Service ceiling	3,660 m (12,000 ft)
T-O run: on land	200 m (660 ft)
on water	150 m (495 ft)
Range with max fuel	600 n miles (1,111 km; 690 miles)
Radius of action, auxiliary fuel	405 n miles (750 km; 466 miles)
Endurance	2 h 30 min
g limits	+4/−2

UPDATED

Equal-span, constant-chord wings, upper unit mounted on cabane; V-type interplane struts, with diagonal strut brace to fuselage from upper plane. Single-step hull. Boom-mounted empennage, wire-braced. Floats at midspan of lower plane. (French-built Pétrels have shorter lower span with floats at tips.)

Wing section NACA 2412; dihedral 2° 13′ on upper wings, 3° 26′ on lower; sweepback 4° upper, 2° lower.

FLYING CONTROLS: Conventional and manual. Ailerons on upper wing only. No flaps. Actuation by wires (rudder) and pushrods. Electric trim.

STRUCTURE: Moulded monocoque single-step hull of epoxy/carbon fibre foam with carbon fibre tailboom; wings have 2014-T6 aluminium alloy tubular main spar with PVC foam ribs and glass fibre/epoxy leading-edge and tips covered with fabric and braced by single diagonal strut and V interplane struts of 6061-T6 aluminium. Wings fold for ground transportation. Tail surfaces of glass fibre spars and PVC foam ribs, with fabric and glass fibre covering, stainless steel wire braced.

LANDING GEAR: Retractable tricycle type; based on Mooney design, with Johnson bar actuation; main units retract upwards into hull and undersurface of lower wing; nosewheel retracts upwards and forwards, tyre remaining partially exposed to serve as docking bumper when operating on water. Nosewheel maximum steering angle 80°. Hydraulic disc brakes on main units. No water rudders.

POWER PLANT: One 59.6 kW (79.9 hp) Rotax 912 UL or 73.5 kW (98.6 hp) Rotax 912 ULS four-stroke piston engine driving an Airplast 175 three-blade pusher propeller. Optional Airplast PV 50 in-flight adjustable pitch propeller with electric or manual control. Fuel capacity 50 litres (13.2 US gallons; 11.0 Imp gallons), of which 45 litres (11.9 US gallons; 9.9 Imp gallons) usable. Optional auxiliary tank, capacity 106 litres (28.0 US gallons; 23.3 Imp gallons).

ACCOMMODATION: Two, side by side in open or enclosed cockpit, with windscreen or single-piece forward-hinging canopy. Dual controls.

EQUIPMENT: Optional BRS ballistic parachute. Bilge pump.

DIMENSIONS, EXTERNAL:
Wing span (both)	8.25 m (27 ft 8 in)

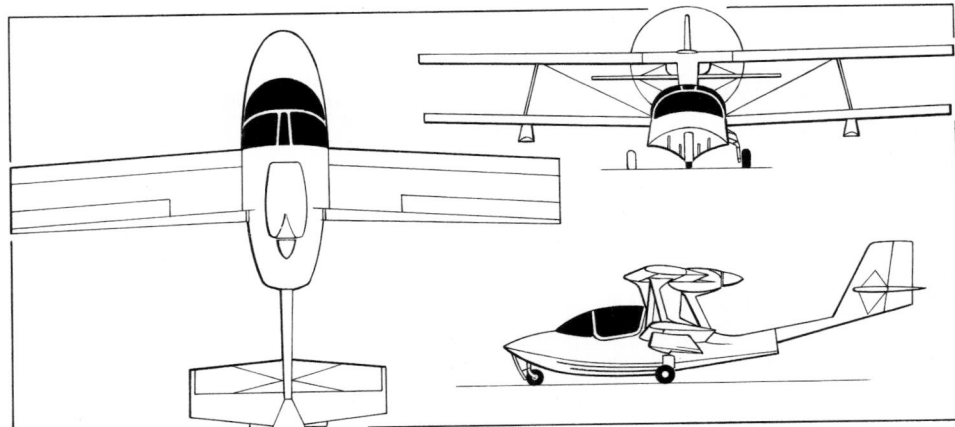

Paturi amphibious two-seat lightplane (*Jane's/James Goulding*)   0100404

EMBRAER

EMPRESA BRASILEIRA DE AERONÁUTICA SA

Av Brig Faria Lima 2170, Caixa Postal 343, 12227-901 São José dos Campos, SP
Tel: (+55 12) 39 27 10 00
Fax: (+55 12) 39 21 23 94
Web: http://www.embraer.com
PRESIDENT AND CEO: Maurício Novis Botelho
EXECUTIVE VICE-PRESIDENT, INDUSTRIAL: Satoshi Yokota
EXECUTIVE VICE-PRESIDENT, COMMERCIAL:
    Frederico Pinheiro Fleury Curado
EXECUTIVE VICE-PRESIDENT, PLANNING AND ORGANISATIONAL DEVELOPMENT: Horácio Aragonés Forjaz
EXECUTIVE VICE-PRESIDENT, FINANCE:
    Antonio Luis Pizarro Manso
VICE-PRESIDENT, COMMERCIAL (Military Marketing): Romualdo Monteiro de Barros
VICE-PRESIDENT, CORPORATE JETS: Sam Hill
PRESS OFFICER: Marcia Benevides

EUROPEAN OFFICE
    Embraer Europe Corporate Jets
    Aéroport du Bourget, Zone d'Aviation d'Affaires, BP 74, F-93352 Le Bourget Cedex, France
    Tel: (+33 1) 49 38 44 44
    Fax: (+33 1) 49 38 44 45
    e-mail: corporatejets@embraer.com
    MANAGER: Neil Patton

Created 19 August 1969, Embraer began operating on 2 January 1970. Its 250,000 m² (2,691,000 sq ft) factory at Faria Lima was officially joined on 15 January 2001 by the Eugênio de Mello plant, comprising 147,650 m² (1,589,300 sq ft) of covered space. Both facilities are near São José dos Campos. Fifth company plant, at Gavião Peixoto (300 km; 186 miles northwest of São Paulo) opened 11 June 2002; to expand over five years to 3,000,000 m² (32 million sq ft) and undertake final assembly of Legacy, Super Tucano, AMX-T and special missions versions of EMB-145.

Total Embraer workforce was 11,264 on 1 April 2002. Neiva (which see) became a subsidiary in March 1980. Embraer and subsidiaries have delivered nearly 5,500 aircraft.

Privatisation of Embraer was undertaken on 7 December 1994 with the auction of 55.4 per cent of voting stock. A consortium, led by Bozano Simonsen bank, acquired 45.44 per cent of auctioned stock, assumed a controlling interest in the company, and provided an extra US$36 million in capitalisation. A further 10 per cent of voting stock was offered to the public within 60 days, with Embraer employees and the Brazilian government retaining 10 and 18.4 per cent respectively. In October 1999 a consortium of French aerospace companies, comprising Aerospatiale Matra (now EADS France), Dassault, SNECMA and Thomson-CSF (now Thales), acquired 20 per cent of Embraer's voting shares. Embraer also embarked on a joint venture with Liebherr International of Germany to establish Embraer-Liebherr Equipamentos do Brasil SA (ELEB) to create additional opportunities for the company's landing gear and hydraulics components business. Embraer's controlling group – with 60 per cent of voting shares – comprises pension funds Previ and Sistel, plus investment house Companhia Bozano Simonsen.

Principal current own-design manufacturing programmes are EMB-120 Brasilia commuter transport, ERJ-135/ERJ-140/ERJ-145 and ERJ-170/190 regional jet families and EMB-312/EMB-314 Tucano turboprop military trainer. Subsidiary Neiva (which see) manufactures EMB-202 Ipanema agricultural aircraft and, under licence from Piper, PA-32-301 Saratoga (as EMB-720D Minuano) and PA-34-220T (as EMB-810D Seneca IV).

In subcontract field, first deliveries made in 1994 of wingtips and vertical fin fairings for Boeing 777 under 1991 contract.

Embraer is a risk-sharing partner in the Sikorsky S-92 Helibus programme; under a June 1995 contract valued at US$170 million, it will supply 730 sets of S-92 fuel sponsons. In 1995 Embraer signed a co-operation agreement with PZL Warszawa-Okecie of Poland under which Embraer subsidiary Neiva SA would market PZL light aircraft in Brazil and PZL subsidiary Skypol would acquire Brasilias for its freight and charter services. This appears to have lapsed.

On 3 April 2002 Embraer, Dassault Aviation, SNECMA Moteurs and Thales Airborne Systems formed the Mirage 2000BR Consortium to develop a version of the Dassault Mirage 2000-5 Mk 2 as a contender for the Brazilian Air Force's F-X BR programme. If selected, FAB Mirage 2000BRs would be assembled at a new Embraer facility at Gavião Peixoto.

Embraer delivered 177 aircraft (45 ERJ-135s, 112 ERJ-145s, three EMB-135s and 17 lightplanes) in 2000, and 174 aircraft (two EMB-120s, 27 ERJ-135s, 22 ERJ-140s, 104 ERJ-145, seven EMB-135, one EMB-145 and 11 light aircraft) in 2001. Planned target for 2002 is 135 deliveries rising to 145 in 2003. Firm order backlog valued at US$10.7 billion at 31 December 2001 including 60 regional jets and 25 Legacy business jets, giving backlog of 364 ERJ-135/140/145s, 112 Embraer 170/190s and 68 Legacys.

EMBRAER DELIVERIES
(at December 2001)

Type	Deliveries
EMB-110 Bandeirante	469
EMB-111 Bandeirante Patrulha	31
EMB-120 Brasilia	352
EMB-121 Xingu	105
ERJ-135	98
ERJ-140	22
ERJ-145	397
EMB-200/201/202 Ipanema	826
EMB-312 Tucano	650
EMB-326GB Xavante	182
AMX	56
Light aircraft (incl Neiva)	2,470
Total	**5,658**

UPDATED

EMBRAER EMB-314 SUPER TUCANO
English name: Super Toucan
Brazilian Air Force designations: A-29 and AT-29
TYPE: Basic turboprop trainer/attack lightplane.

PROGRAMME: Design of original EMB-312 Tucano started January 1978. Ministry of Aeronautics contract received 6 December that year for two flying prototypes and two static/fatigue test airframes; first prototype (Brazilian Air Force serial number 1300) made first flight 16 August 1980, second (1301) on 10 December 1980; third to fly (PP-ZDK, on 16 August 1982) was to production standard. Total of 650 built by 1998; further details in 2000-01 and previous *Jane's* and in *Jane's Aircraft Upgrades*.

Development of EMB-314 began January 1991; announced (as EMB-312H) at Paris Air Show June 1991; Embraer development aircraft PT-ZTW (c/n 312161, previously used as prototype for TPE331-powered Tucano adopted by Royal Air Force) modified as Tucano H proof-of-concept (POC) prototype, making first flight in this form 9 September 1991. This aircraft toured US Air Force/Navy bases August and September 1992 as preliminary to Super Tucano entry in JPATS competition; Embraer teamed with Northrop May 1992 to bid Super Tucano for JPATS, but was unsuccessful. Provisional Brazilian type certification granted August 1994 after 500 hour, 396 sortie test and certification programme.

CURRENT VERSIONS: **EMB-314:** Two EMB-312H prototypes (PT[later PP]-ZTV, c/n 312454, first flight 15 May 1993, and PP-ZTF, c/n 312455, first flight 14 October 1993) tailored to US JPATS requirements as EMB-312HJ. Designation subsequently changed to EMB-314 to reflect extensive modifications to structure and systems.

ALX (EMB-314M): Brazilian Air Force (FAB) version, for border patrol missions under its SIVAM (*SIstema de Vigilancia da AMazonia*) programme. FAB finalised specification in early 1994; trials to validate projected flight characteristics, using POC aircraft and both Super Tucano prototypes, completed 1994. US$50 million development contract signed 18 August 1995 for two prototypes (one single-seat) to be modified from Super Tucano prototypes. The first flew in May 1996 and is being used for external stores compatibility and handling qualities testing; the second, which flew in early 1997, for testing the advanced weapons systems. Total 650 hours accumulated by early 2002, with up to 300 more required for completion.

FAB commitment to purchase 99 ALXs, of which 50 will be two-seat **AT-29**, 30 of these replacing AT-26 Xavante with 2°/5° Grupo of Training Command at Natal AFB, remainder configured for night intruder role and expected to serve with 1°/3° Grupo at Boa Vista and 2°/3° Grupo at Porto Velho; 49 will be single-seat **A-29**. Prototype 5700 (the former EMB-314 PP-ZTV), designated **YA-29**, rolled out 28 May 1999; IOC originally intended in May 2001 but FAB formal order only placed (76 firm, plus 23 options) on 8 August 2001; production started February 2002; deliveries to begin in December 2003. Elbit selected December 1996 to supply mission avionics, including ventral FLIR Systems AN-AAQ-22

turret, GPS/INS, radalt, Mode S transponder, DME, ILS, ADF, VOR, RWR, MAWS and chaff/flare dispenser. Export variants of both versions will be offered for border patrol/COIN missions and for basic/advanced pilot training. Can be flown as single-seat attack aircraft with fuel tank in rear cockpit.

Dominican Republic Air Force ordered 10 Super Tucanos on 20 August 2001.

Estimated market for 500 aircraft over 10 years.

DESIGN FEATURES: Meets requirements of FAR Pt 23 Appendix A, and MIL and CAA Section K specifications. Low-mounted wings, stepped cockpits in tandem, fully aerobatic. Small fillet forward of tailplane root each side.

EMB-314 differs from EMB-312 mainly in having more powerful engine, reprofiled wing and plugs of 0.37 m (1 ft 2½ in) forward and 1.00 m (3 ft 3¼ in) aft of cockpit to accommodate longer engine and retain CG and stability. Other changes include strengthened airframe for higher *g* loads and longer fatigue life; ventral strakes; five weapons hardpoints; NVG-compatible 'glass cockpit' with HOTAS controls. Able to cover whole primary and half of advanced pilot training syllabus, and fly precision weapons delivery and target towing missions.

Wing section NACA 63-415 at root; NACA 63A-212 at tip.

FLYING CONTROLS: Conventional and manual. Primary surfaces internally balanced; electrically actuated trim tab in, and small geared tab on, each Frise aileron; electromechanically actuated spring tab in rudder and port elevator. Electrically actuated single-slotted Fowler flaps on wing trailing-edges. Fixed incidence tailplane. Ventral airbrake.

STRUCTURE: Conventional all-metal construction from 2024 series aluminium alloys; continuous three-spar wing box forms integral fuel tankage. Steel flap tracks.

LANDING GEAR: Hydraulically retractable tricycle type, with single wheel and Piper oleo-pneumatic shock-absorber on each unit. Accumulator for emergency extension in the event of hydraulic system failure. Hydraulic steering for nose unit. Rearward-retracting steerable nose unit; main units retract inward into wings. Parker Hannifin 40-130 mainwheels, Oldi-DI-1.555-02-OL nosewheel. Tyre sizes 6.50-10 (8 ply) tubeless on mainwheels, 5.00-5 (6 ply) tubeless on nosewheel. Tyre pressures (±0.21 bar; 3 lb/ sq in in each case) are 5.17 bar (75 lb/sq in) on mainwheels, 4.48 bar (65 lb/sq in) on nosewheel. Parker Hannifin 30-95A hydraulic mainwheel brakes.

POWER PLANT: *ALX:* One 1,193 kW (1,600 shp) Pratt & Whitney Canada PT6A-68-3 turboprop, with FADEC, driving a Hartzell five-blade, constant-speed, fully feathering, reversible-pitch propeller.

EMB-314: One 969 kW (1,300 shp) PT6A-68A.

Single-lever combined control for engine throttling and propeller pitch adjustment. Two integral fuel tanks in each wing, total capacity 694 litres (183.3 US gallons; 152.7 Imp gallons). Fuel tanks lined with anti-detonation plastics foam. Optional self-sealing 303 litre (80.0 US gallon; 66.6 Imp gallon) tank in rear cockpit. Single-point pressure refuelling. Fuel system allows nominally for up to 30 seconds of inverted flight. Provision for three ferry fuel tanks (centreline and inboard wing pylons), each total capacity 330 litres (87.1 US gallons; 72.6 Imp gallons).

ACCOMMODATION: Two in tandem, on Martin-Baker Mk 10 LCX zero-zero ejection seats, in air conditioned and

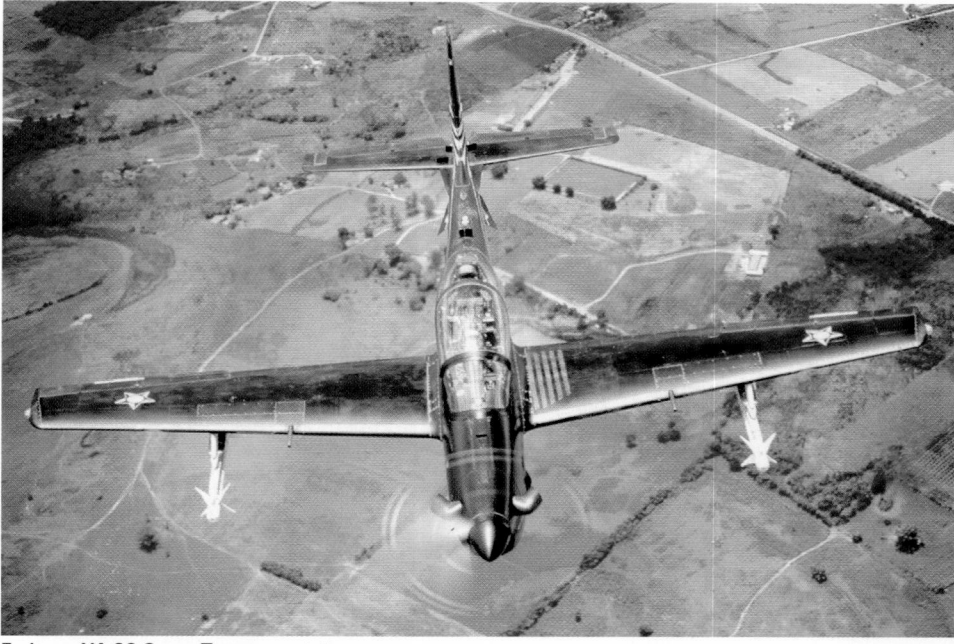
Embraer YA-29 Super Tucano prototype *NEW*/0137411

Front cockpit of Embraer Super Tucano *NEW*/0131837

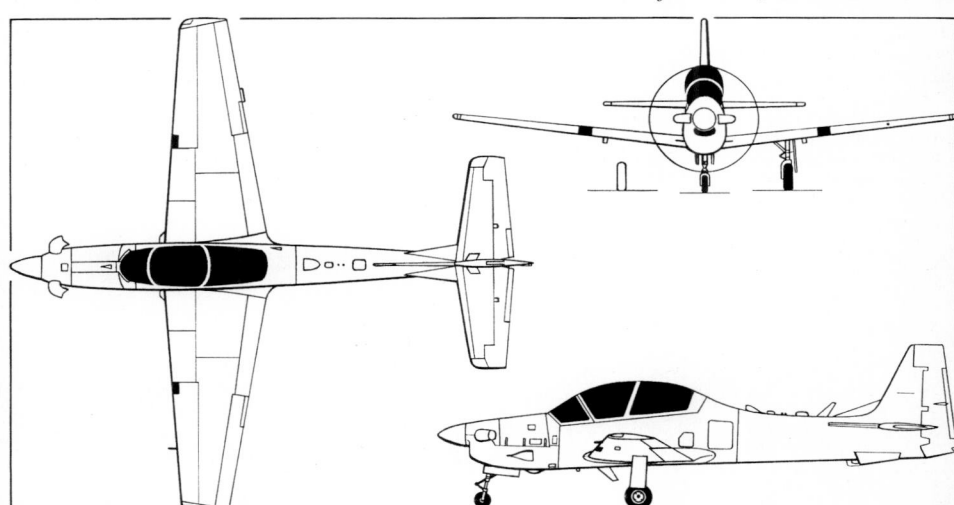
AT-29 advanced trainer version of Embraer Super Tucano (*Jane's/James Goulding*) 0056316

For details of the latest updates to *Jane's All the World's Aircraft* online and to discover the additional information available exclusively to online subscribers please visit
jawa.janes.com

pressurised cockpit. One-piece fully transparent vacuum-formed canopy, opening sideways to starboard, with internal and external jettison provisions. Rear seat elevated 25 cm (9.9 in). Dual controls standard. Baggage compartment in rear fuselage, with access via door on port side.

SYSTEMS: Two-axis autopilot with embedded mission planning capability. Freon air cycle conditioning system, with engine-driven compressor. Single hydraulic system, consisting basically of (a) control unit, including reservoir with usable capacity of 1.9 litres (0.5 US gallon; 0.42 Imp gallon); (b) an engine-driven pump with nominal pressure of 131 bar (1,900 lb/sq in) and nominal flow rate of 4.6 litres (1.22 US gallons; 1.01 Imp gallons)/min at 3,800 rpm; (c) landing gear and gear door actuators; (d) filter; (e) shutoff valve; and (f) hydraulic fluid to MIL-H-5606. Under normal operation, hydraulic system actuates landing gear extension/retraction and control of gear doors. Landing gear extension can be performed under emergency operation; emergency retraction also possible during landing and T-O with engine running. Reservoir and system are suitable for aerobatics. No pneumatic system. 28 V DC electrical power provided by a 6 kW starter/generator, 26 Ah battery and, for 115 V and 26 V AC power at 400 Hz, a 250 VA inverter. Onboard oxygen generation system. Canopy and propeller de-icing.

AVIONICS: *Comms:* Standard Rockwell Collins equipment. Integrated nav/com. Tactical VHF/UHF with datalink provisions; transponder.

Flight: Laser INS/dual GPS; twin VOR; ILS; DME; ADF; autopilot.

Instrumentation: Two 152 × 203 mm (6 × 8 in) CMFDs in each cockpit; HUD with up-front control panel in front cockpit. NVG Gen III compatible internal/external lighting. Optional HMD.

Mission: Provision for datalink; video camera and recorder. Embedded mission planning capability. FLIR. Chaff/flare dispensers; MAWS and RWR.

EQUIPMENT: Landing light in each wing leading-edge; taxying lights on nosewheel unit. Optional Kevlar cockpit armour.

ARMAMENT: One 12.7 mm machine gun and 200 rounds mounted in each wing. Provision for a variety of ordnance including two Giat NC621 20 mm cannon pods, Mk 81/82 bombs, MAA-1 Piranha AAMs, BLG-252 cluster bombs, SBAT-70/19 or LAU-68A/G rocket pods or MLBs on underwing stations.

DIMENSIONS, EXTERNAL:

Wing span	11.14 m (36 ft 6½ in)
Wing chord: at root	2.30 m (7 ft 6½ in)
at tip	1.07 m (3 ft 6⅛ in)
Wing aspect ratio	6.4
Length overall	11.42 m (37 ft 5¾ in)
Fuselage: Length (excl rudder)	10.53 m (34 ft 6½ in)
Max depth	1.86 m (6 ft 1¼ in)
Height overall (static)	3.90 m (12 ft 9½ in)
Tailplane span	4.66 m (15 ft 3½ in)
Wheel track	3.76 m (12 ft 4 in)
Wheelbase	3.36 m (11 ft 0¼ in)
Propeller ground clearance (static)	0.345 m (1 ft 1½ in)
Baggage compartment door:	
Height	0.60 m (1 ft 11½ in)
Width	0.54 m (1 ft 9¼ in)
Height to sill	1.25 m (4 ft 1¼ in)

DIMENSIONS, INTERNAL:

Cockpits: Combined length	2.90 m (9 ft 6⅛ in)
Max height	1.55 m (5 ft 1 in)
Max width	0.85 m (2 ft 9½ in)
Baggage compartment volume	0.17 m³ (6.0 cu ft)

AREAS:

Wings, gross	19.40 m² (208.8 sq ft)
Ailerons (total)	1.97 m² (21.20 sq ft)
Trailing-edge flaps (total)	2.58 m² (27.77 sq ft)
Fin, incl dorsal fin	2.29 m² (24.65 sq ft)
Rudder, incl tab	1.38 m² (14.85 sq ft)
Tailplane, incl fillets	4.77 m² (51.34 sq ft)
Elevators, incl tab	2.00 m² (21.53 sq ft)

WEIGHTS AND LOADINGS (EMB-314, except where indicated):

Basic weight empty	2,420 kg (5,335 lb)
Max external load	1,500 kg (3,307 lb)
Max internal fuel load (usable)	538 kg (1,186 lb)
Max T-O weight: EMB-314, clean	3,190 kg (7,033 lb)
ALX	3,600 kg (7,936 lb)
Max ramp weight	3,210 kg (7,077 lb)
Max zero-fuel weight: EMB-314	2,670 kg (5,886 lb)
ALX	3,150 kg (6,945 lb)
Max wing loading	164.4 kg/m² (33.68 lb/sq ft)
Max power loading	3.29 kg/kW (5.41 lb/shp)

PERFORMANCE (EMB-314, at max clean T-O weight except where indicated):

Max level speed: EMB-314 at 6,100 m (20,000 ft)	301 kt (557 km/h; 346 mph)
ALX with external stores	245 kt (454 km/h; 282 mph)
Max cruising speed at 6,100 m (20,000 ft)	286 kt (530 km/h; 329 mph)
Econ cruising speed at 6,100 m (20,000 ft)	228 kt (422 km/h; 262 mph)
Stalling speed, power off:	
flaps and landing gear up	85 kt (157 km/h; 98 mph) EAS
flaps and landing gear down	78 kt (145 km/h; 90 mph) EAS
Max rate of climb at S/L	895 m (2,925 ft)/min
Service ceiling	10,670 m (35,000 ft)
T-O run	350 m (1,150 ft)
T-O to 15 m (50 ft)	550 m (1,805 ft)
Landing from 15 m (50 ft)	860 m (2,820 ft)
Landing run	550 m (1,805 ft)
Range at 9,150 m (30,000 ft) with max fuel, 30 min reserves	847 n miles (1,568 km; 974 miles)
Ferry range at 7,620 m (25,000 ft) with underwing tanks and 30 min reserves	1,495 n miles (2,768 km; 1,720 miles)
Endurance on internal fuel at econ cruising speed at 7,620 m (25,000 ft), 30 min reserves	6 h 30 min
g limits: fully Aerobatic category at 2,770 kg (6,107 lb)	+7/–3.5
at 2,770 kg (6,107 lb) with external stores	+4/–2.2

UPDATED

EMBRAER ERJ-135 and ERJ-140

Follow entry for ERJ-145, of which they are variants.

EMBRAER ERJ-145

Brazilian Air Force designations: R-99A and R-99B

TYPE: Regional jet airliner.

PROGRAMME: Development plans revealed 12 June 1989, aimed at first flight late 1991 and first deliveries mid-1993; programme delayed by company cutbacks, complete redesign of wing and other changes, as described in 2000-01 and earlier *Jane's*.

First metal cut for prototype, and tooling fabrication, in second quarter 1993; assembly of prototype (PT-ZJA) began October 1994; fuselage sections mated January 1995; first flight 11 August 1995 ahead of formal roll-out and 'official' first flight a week later; first of three pre-series aircraft (PT-ZJB) first flown 17 November 1995; second ('ZJC) flew on 14 February and third ('ZJD) 2 April 1996; FAA and Brazilian CTA certification (to FAR/JAR 25, FAR Pt 36, ICAO Annex 16 and FAR Pt 121) achieved 16 December 1996. Single prototype and three pre-series aircraft undertook a 1,600 hour, 13 month development flight testing and certification programme. Deliveries began on 19 December 1996 with two aircraft (N15925 and N15926) to US launch customer Continental Express. Designation changed from EMB-145 to ERJ-145 in October 1997 to reflect 'Regional Jet' terminology. Certified by the aviation authorities of 27 countries by September 1998.

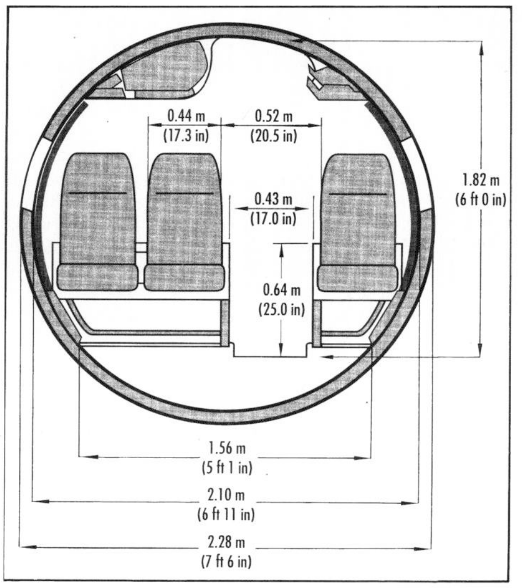

ERJ-145 fuselage cross-section 0011037

In addition to its suitability for executive transport and corporate shuttle roles, Embraer foresees military potential for the ERJ-145 as a tanker aircraft for small combat units, and as an AEW, elint, comint, sigint or battlefield surveillance platform. See Current Versions.

Two static test airframes: first (802) completed trials on 30 August 1996; second (803) began 10 year programme in December 1996.

CURRENT VERSIONS: **ERJ-145ER** (Extended Range): Initial version.

ERJ-145LR (Long Range): Introduced in 1998 for full passenger payload range of 1,640 n miles (3,037 km; 1,887 miles); increases in fuel capacity and all operating weights; and uprated AE 3007A1 turbofans providing 15 per cent more thermodynamic power, but flat rated to 33.1 kN (7,430 lb st), for improved climb and hot weather cruise performance.

ERJ-145XR (Extra Long Range): Announced at Farnborough International Air Show 25 July 2000 with launch order for 75, plus 100 options, from Continental Express subsequently increased to 104, plus 100 options. Features strengthened fuselage, wings and horizontal stabiliser, winglets, uprated 39.6 kN (8,910 lb st) AE 3007A1E turbofans providing lower specific fuel consumption, improved hot-and-high operation and higher single-engine ceiling; and auxiliary fuel tank in wing/fuselage fairing, capacity 1,000 litres; (264 US gallons; 220 Imp gallons); range 2,000 n miles (3,704 km; 2,301 miles). Prototype (PT-ZJB, modified from first pre-series ERJ-145) first flown 29 June 2001; first production aircraft (c/n 590) flew early April 2002; two aircraft took part in a 400 hour test programme, culminating in certification scheduled for September 2002 and first deliveries shortly thereafter.

ERJ-135: Short-fuselage, 37-seat version; described separately.

ERJ-140: Mid-size, 44-seat version; described separately.

EMB-145AEW&C: Formerly EMB-145SA; Brazilian Air Force designation R-99A. Airborne early warning and remote sensing version of ERJ-145LR developed for Brazilian government's *SIstema de Vigilancia de AMazonia* (SIVAM) programme for which Raytheon is prime contractor; initial requirement for five; contract signature March 1997; deliveries for completion by May 2002. Selected on 15 December 1998 by Greek Air Force for its four-aircraft AEW requirement, with delivery from 2002; contract, signed 1 July 1999, valued at US$500

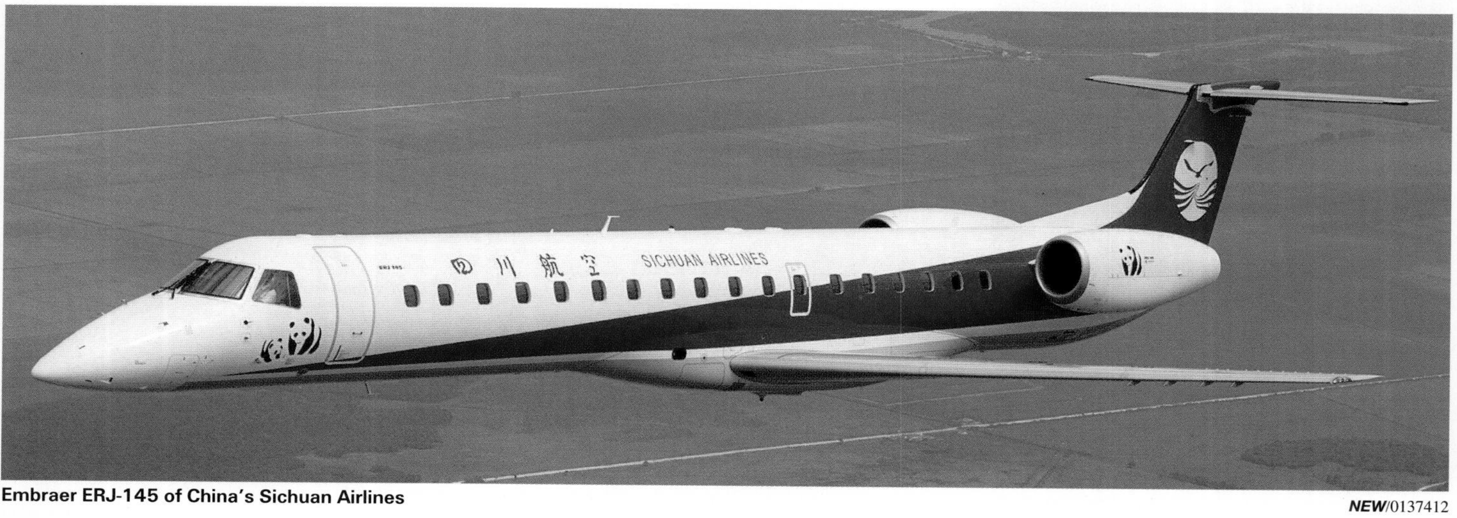

Embraer ERJ-145 of China's Sichuan Airlines

NEW/0137412

Brazilian Air Force Embraer R-99A early warning and control aircraft

NEW/0143348

million. Announced late 1996 and features a strengthened fuselage, ventral strakes, more powerful APU, increased fuel capacity (three extra tanks at extreme rear of cabin, plus jettison capability), extending endurance to more than 8 hours; enhanced electrical system, five seats for relief crew and four (with provision for additional two) operators' consoles, including tactical co-ordinator. Flight crew of two.

Mission systems comprise civil version of Ericsson PS-890 Erieye side-looking airborne radar with antenna housed in long overfuselage fairing, optimised for lower-speed targets typically encountered in border incursions; five Erieye systems purchased at cost of US$145 million in 1997 for installation in EMB-145SAs, with first system delivery scheduled for 1999. Onboard command and control system, and BAE Systems North America Comms-Non Comms system. Radar is pulse Doppler type, operating in E/F-band, offering coverage from very low level up to about 25,000 m (82,000 ft) and at ranges exceeding 162 n miles (300 km; 186 miles). Datalink; GPS; secure communications. Second airframe (PP-XSA/6702) was first to fly, on 22 May 1999, ahead of formal roll-out on 28 May. Service entry with 2°/6°Grupo at Anapolis AFB scheduled for June 2001, following systems integration by Raytheon in USA.

Greek selection of Erieye-equipped EMB-145 announced 1 July 1999; four aircraft ordered, of which first two delivered on 24 September 2001.

Mexican government order for one EMB-145AEW&C announced 1 March 2001; equipment includes comint system installed by Raytheon.

EMB-145RS: Brazilian Air Force designation R-99B. Remote sensing version, of which three ordered for FAB's SIVAM programme for delivery commencing first quarter 2001. Similar to AEW variant, and with ventral strakes, but different mission systems for primary roles in natural resources exploitation, environmental and river pollution control, economic activities, ground occupation monitoring and illegal activities surveillance. Main sensor is version of Canadian MacDonald Dettwiler IRIS (Integrated Radar Imaging System) synthetic aperture radar, installed in underfuselage bulge with auxiliary antennas beneath wingroots, operating in D-band interferometric mode and capable of generating 3-D imagery. Other main sensors include Star Safire FLIR mounted behind nosewheel bay, Daedalus ultraviolet/visible/infra-red linescanner and BAE Systems North America Comms-Non Comms system. Roll-out (PP-XRT/6751) November 1999, with first delivery to 2°/6° Grupo at Anapolis AFB was due in January 2001 and final delivery in May 2002.

EMB-145AGS: Airborne ground sensor version, under study during 2000; equipped with a mission package comprising Airborne Platform Subsystem (APS), Airborne Mission Equipment Subsystem (AMES) and Ground Exploitation Station Subsystem (GESS) including HF, UHF, VHF, ELINT and IMINT equipment, providing a self-deployable and cost-effective surface reconnaissance system.

EMB-145MP and **EMB-145MP/ASW:** Maritime patrol and anti-submarine warfare versions, under development by 2000; equipped with surveillance radar with multiple target track-while-scan mode, autodetection, FLIR interface, digital map, incorporated tactical aids, SAR/ISAR mode allowing real-time imaging, adaptive processing for different sea states, and simultaneous side and range views; high-altitude and resolution FLIR; ESM suite; COMINT/ELINT; MAD; IFF/SSR and acoustics.

Mexican government order for two EMB-145MPs announced 1 March 2001. Equipment includes SeaVue radar and AN/APX-114 IFF interrogator to be installed by Raytheon at Greenville.

CUSTOMERS: Total of 570 firm orders for EMB-145 commercial variant by 15 June. Four hundredth ERJ series delivery was an ERJ-145 (HB-JAL) to Crossair on 22 March 2001; 500th was an ERJ-145 (N2933K) delivered to Chatauqua Airlines of Indiana on 21 September 2001; 600th to Swiss (as HB-JAY) on 28 May 2002. Total of 112

ERJ-145s delivered in 2000, and 104 in 2001. See table. Belgian Air Force took delivery of two ERJ-145LRs in VIP configuration (CE-03 and CE-04) on 11 December 2001 and 21 January 2002 respectively.

EMBRAER ERJ-135/140/145 COMMERCIAL ORDERS
(at 15 June 2002)

Customer	Variant	Orders
Air Caraibes	145	2
Air Moldova	145	2
Alitalia Express	145	8
American Eagle	135	40
	140	139
	145	56
Axon Airlines	145	3
bmi (British Midland)	135	4
	145	11
Brymon Airways	145	7
Cirrus Airlines	145	1
City Airline AB/	135	2
Skyways	145	4
Continental Express	135	30
	145	245
Crossair (Swiss)	145	25
Flandre Air	135	7
	145	7
GATX	145	1
KLM Excel	145	3
LOT Polish	145	16
Luxair	145	9
Manx Airlines	145	23
Mesa Airlines	145	36
Midwest	140	20
Occitania	135	1
Pan Européenne	135	1
Portugalia	145	8
Proteus Airlines	135	3
	145	11
Regional Airlines	135	5
(France)	145	17
Regional Air Lines	135	5
(Morocco)		
Rheintalflug	145	4
Rio-Sul	145	16
Sichuan Airlines	145	5
South African Airlink	135	30
Trans States Airlines	145	12
Wexford Management	140	15
	145	38
Total		**872**

COSTS: Estimated development costs US$300 million; unit cost US$15.5 million (1997). AEW version (PS-890 radar) US$50 million to US$60 million (1996).

Following description applies to ERJ-145ER except where indicated.

DESIGN FEATURES: Stretched EMB-120 Brasilia fuselage (with tailcone adapted for rear-mounted engine installation), allied to new-design wing with Embraer supercritical section; CBA-123 nose and cabin; T tailplane.

Wing sweepback 22° 43′ 48″ at quarter-chord.

FLYING CONTROLS: Conventional and assisted. Ailerons and two-section rudder hydraulically actuated, with artificial feel; mechanically actuated elevator with automatic and spring tab. Four-segment in-flight and ground spoilers; two pairs of electrically actuated double-slotted flaps.

STRUCTURE: Fuselage as for Brasilia; two-spar wing with integral fuel tanks, plus auxiliary third spar supporting landing gear; T tail unit with aluminium main boxes; wing and tailplane leading-edges aluminium, fin leading-edge composites sandwich. Gamesa (Spain) builds wings, wing/body fairings, main landing gear doors and engine nacelles; rear fuselage section 1, including engine pylons

and passenger/service/baggage doors, plus centre-fuselage section 1, including doors, by Sonaca (Belgium); fin, tailplane and elevators by ENAER (Chile); engine nacelles and thrust reversers by International Nacelle Systems; nose radome by Norton; passenger cabin and baggage compartment interiors by C & D Interiors (USA). Structure designed for an economical service life of 60,000 flights.

LANDING GEAR: Twin-wheel main legs retract inward into wing/fuselage fairings; twin-wheel nose unit retracts forward. EDE/Liebherr landing gear system, with EDE responsible for whole system and Liebherr for development and production of nose unit. Goodrich wheels and carbon brakes. Tyre sizes 30×9.5-14 (16 ply) tubeless (main), 19.5×6.75-8 (8 ply) tubeless (nose); tyre pressure 8.60 to 9.00 bar (125 to 130 lb/sq in). Minimum ground turning radius at nosewheel 12.51 m (41 ft 0 in). Minimum turning circle 29.22 m (95 ft 10½ in).

POWER PLANT: Two turbofans pylon-mounted on rear cone of fuselage: ERJ-145ER has two 31.3 kN (7,040 lb st), FADEC-equipped Rolls-Royce AE 3007As; and ERJ-145LR has two AE 3007A1s flat rated at 33.1 kN (7,430 lb st), which are also optional on the ERJ-145ER, both with FADEC. ERJ-145XR has two 39.6 kN (8,910 lb st) AE 3007A1Es. For fuel, see Weights and Loadings. Parker Hannifin fuel system. Clamshell-type thrust reversers optional.

ACCOMMODATION: Two pilots, flight observer and cabin attendant. Standard accommodation for 50 passengers, three-abreast at seat pitch of 79 cm (31 in). Carry-on baggage wardrobe, galley and cabin attendant's seat at front of cabin; lavatory and main baggage compartment at rear of cabin. Cabinet plus overhead bins carry-on baggage capacity 358 kg (789 lb); underseat capacity 450 kg (992 lb); main baggage compartment capacity 1,200 kg (2,646 lb). Additional baggage cabinet or galley capacity can be provided by removing one or two single forward passenger seats. Outward-opening plug-type door, incorporating airstair, at front on port side, identical to that of EMB-120; upward-sliding baggage door at rear on port side; sideways-opening service door at front on starboard side; inward-opening emergency exit above wing on each side. Entire accommodation, including baggage compartments, pressurised and air conditioned.

SYSTEMS: Liebherr Aerospace pressurisation system (maximum differential 0.54 bar; 7.8 lb/sq in) maintains 2,440 m (8,000 ft) cabin altitude to 11,275 m (37,000 ft). Hamilton Sundstrand air conditioning and bleed air systems (wing and tailplane leading-edges and engine intakes anti-iced by engine bleed air); electric anti-icing system for windscreen and static and pitot tubes and sensors. Lucas electrical power generation system. APIC 18.6 kW (25 shp) APS-500 APU. Honeywell air turbine starter. Parker Hannifin flight control and steering systems. Hydro-Aire brake-by-wire control system. EROS oxygen system.

AVIONICS: Honeywell Primus 1000 as core system.

Comms: Dual Primus II radios and radio management units.

Radar: Primus 1000 colour weather radar.

Flight: Dual digital air data computers, dual AHRS, TCAS and GPWS standard. FMS/GPS optional. Flight Dynamics HUD selected April 1998 for certification in 2000, providing Cat. III landing capability.

Instrumentation: EFIS panel comprising five 280 × 180 mm (11 × 7 in) displays, two PFDs, two MFDs and IECAS.

DIMENSIONS, EXTERNAL:

Wing span: ERJ-145ER/LR	20.04 m (65 ft 9 in)
ERJ-145XR over winglets	21.01 m (68 ft 11 in)
Wing chord: at root	4.09 m (13 ft 5 in)
at tip	1.04 m (3 ft 5 in)
Wing aspect ratio	7.9
Length overall	29.87 m (98 ft 0 in)
Fuselage: Length	27.93 m (91 ft 7½ in)
Max diameter	2.28 m (7 ft 5¾ in)
Height overall	6.76 m (22 ft 2 in)
Tailplane span	7.55 m (24 ft 9 in)
Wheel track (c/l of shock-struts)	4.10 m (13 ft 5½ in)

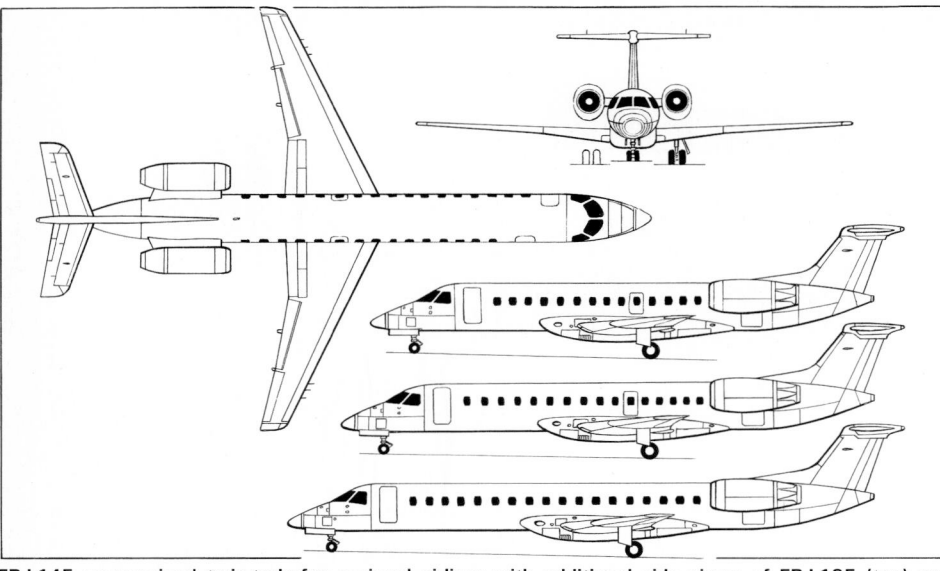

ERJ-145 rear-engined twin-turbofan regional airliner with additional side views of ERJ-135 (top) and ERJ-140 (centre) *(Jane's/James Goulding)*   *NEW*/0143318

Wheelbase	14.45 m (47 ft 5 in)
Passenger door (fwd, port): Height	1.70 m (5 ft 7 in)
Width	0.71 m (2 ft 4 in)
Height to sill (max)	1.63 m (5 ft 4 in)
Baggage door (rear, port): Height	1.12 m (3 ft 8 in)
Width	1.00 m (3 ft 3¼ in)
Height to sill (max)	1.76 m (5 ft 9¼ in)
Service door (rear, stbd): Height	1.42 m (4 ft 8 in)
Width	0.62 m (2 ft 0½ in)
Height to sill	1.60 m (5 ft 3 in)
Emergency exits (overwing, each):	
Height	0.92 m (3 ft 0¼ in)
Width	0.51 m (1 ft 8 in)

DIMENSIONS, INTERNAL:

Cabin (excl flight deck and baggage compartment, incl lavatory):	
Length	16.49 m (54 ft 1¼ in)
Max width	2.10 m (6 ft 10¾ in)
Max height	1.83 m (6 ft 0 in)
Max aisle width	0.52 m (1 ft 8½ in)
Floor area	25.7 m² (277 sq ft)
Volume	53.0 m³ (1,872 cu ft)
Baggage compartment: Length	3.26 m (10 ft 8¼ in)
Baggage volume:	
wardrobe and stowage compartment	1.4 m³ (49 cu ft)
overhead bins	1.9 m³ (67 cu ft)
underseat	2.3 m³ (80 cu ft)
baggage compartment	9.2 m³ (325 cu ft)

AREAS:

Wings, gross	51.18 m² (550.9 sq ft)
Ailerons (total)	1.70 m² (18.30 sq ft)
Trailing-edge flaps (total)	8.36 m² (89.99 sq ft)
Spoilers (total)	2.32 m² (24.97 sq ft)
Fin	5.07 m² (54.57 sq ft)
Rudder	2.13 m² (22.93 sq ft)
Tailplane	11.20 m² (120.55 sq ft)
Elevators (total, incl tabs)	3.34 m² (35.95 sq ft)

WEIGHTS AND LOADINGS:

Operating weight empty:	
ERJ-145ER	11,700 kg (25,794 lb)
ERJ-145LR	11,800 kg (26,015 lb)
ERJ-145XR	12,520 kg (27,602 lb)
Max fuel: ERJ-145ER	4,173 kg (9,200 lb)
ERJ-145LR	5,187 kg (11,435 lb)
ERJ-145XR	5,987 kg (13,199 lb)
Max payload: ERJ-145ER	5,400 kg (11,905 lb)
ERJ-145LR	6,100 kg (13,448 lb)
ERJ-145XR	5,980 kg (13,184 lb)
Max T-O weight: ERJ-145ER	20,600 kg (45,415 lb)
ERJ-145LR	22,000 kg (48,500 lb)
ERJ-145XR	24,000 kg (52,910 lb)
Max ramp weight: ERJ-145ER	20,700 kg (45,635 lb)
ERJ-145LR	22,100 kg (48,721 lb)
ERJ-145XR	24,100 kg (53,131 lb)
Max landing weight: ERJ-145ER	18,700 kg (41,226 lb)
ERJ-145LR	19,300 kg (42,549 lb)
ERJ-145XR	20,000 kg (44,092 lb)
Max zero-fuel weight:	
ERJ-145ER	17,100 kg (37,698 lb)
ERJ-145LR	17,900 kg (39,463 lb)
ERJ-145XR	18,500 kg (40,785 lb)
Max wing loading:	
ERJ-145ER	402.5 kg/m² (82.44 lb/sq ft)
ERJ-145LR	429.8 kg/m² (88.04 lb/sq ft)
Max power loading:	
ERJ-145ER	329 kg/kN (3.23 lb/lb st)
ERJ-145LR	333 kg/kN (3.26 lb/lb st)
ERJ-145XR	303 kg/kN (2.97 lb/lb st)

PERFORMANCE:

High cruising speed: all	
	M0.78 (450 kt; 833 km/h; 518 mph)

Time to climb to 10,670 m (35,000 ft)	20 min
Service ceiling: all	11,275 m (37,000 ft)
Service ceiling, OEI: ERJ-145ER/LR	6,100 m (20,000 ft)
FAR T-O field length at S/L:	
ERJ-145ER/LR	1,970 m (6,465 ft)
ERJ-145XR	2,126 m (6,975 ft)
FAR landing field length, S/L, at typical landing weight:	
ERJ-145ER/LR	1,300 m (4,265 ft)
ERJ-145XR	1,440 m (4,724 ft)
Range, 50 passengers, 100 n miles (185 km; 115 mile) diversion, 45 min reserves	
ERJ-145ER	1,600 n miles (2,963 km; 1,841 miles)
ERJ-145XR	2,000 n miles (3,704 km; 2,301 miles)

UPDATED

EMBRAER ERJ-135

TYPE: Regional jet airliner.

PROGRAMME: Launched 16 September 1997; two pre-series ERJ-145s (001/PT-ZJA and 002/PT-ZJC) modified to create two prototype ERJ-135s; roll-out (PT-ZJA) 12 May 1998, followed by first flight 4 July 1998; public debut at Farnborough Air Show September 1998; second aircraft (PT-ZJC), flown 24 September 1998, for systems testing before conversion to production standard in March 1999. Brazilian Centro Técnico Aerospacial (CTA) certification achieved in June 1999; FAA certification 16 July 1999; JAA approval was expected in October 1999. First delivery 23 July 1999 to Continental Express; other early aircraft to American Eagle.

CURRENT VERSIONS: **ERJ-135:** Regional airliner, *as described.*

Legacy: Corporate version; described separately.

CUSTOMERS: Total of 128 firm commercial orders by 15 June 2002. See table. Additionally, one VIP-configured ERJ-135LR handed over to Greek Air Force on 7 January 2000 and two, also in VIP configuration, to the Belgian Air Force on 4 June and in August 2001, for operation by No. 21 Squadron at Melsbroek with two similarly configured ERJ-145s. Deliveries have been 16 in 1999, 45 in 2000 and 27 in 2001.

Programme based on estimates of 500 sales.

COSTS: Development cost US$100 million, of which 40 per cent provided by risk-sharing partners. Unit cost US$11.8 million.

DESIGN FEATURES: Shares 96 per cent commonality with ERJ-145 including engines, wings, tail surfaces, flight deck and main systems; fuselage shortened by 3.53 m (11 ft 7 in) by removal of two frames (4.84 m; 15 ft 10½ in ahead of wing and 3.07 m; 10 ft 0¾ in at rear) and substitution of two shorter frames (2.85 m; 9 ft 4¼ in and 1.53 m; 5 ft 0¼ in, respectively).

FLYING CONTROLS: As for ERJ-145.

STRUCTURE: As for ERJ-145.

LANDING GEAR: As for ERJ-145.

POWER PLANT: ERJ-135ER has two 31.3 kN (7,040 lb st) Rolls-Royce AE 3007A turbofans. ERJ-135LR has two Rolls-Royce AE 3007A1 turbofans.

ACCOMMODATION: Standard accommodation for 37 passengers in three-abreast configuration.

SYSTEMS: As for ERJ-145.

DIMENSIONS, EXTERNAL: As for ERJ-145 except:

Length: overall	26.33 m (86 ft 4½ in)
fuselage	24.39 m (80 ft 0¼ in)
Wheelbase	12.43 m (40 ft 9¼ in)

DIMENSIONS, INTERNAL:

Cabin (excl flight deck and baggage compartment, incl lavatory): Length	12.95 m (42 ft 5¾ in)
Baggage compartment: Length	3.34 m (10 ft 11½ in)
Baggage volume:	
wardrobe and stowage compartment	1.0 m³ (35 cu ft)
overhead bins	1.4 m³ (49 cu ft)
underseat	1.7 m³ (60 cu ft)
Galley volume	0.99 m³ (35 cu ft)

Embraer ERJ-135 VIP transport of Belgian Air Force *(Jean-Louis Gaynecoetche)*   *NEW*/0137341

Flight deck of Embraer Legacy and ERJ-135   *NEW*/0131839

WEIGHTS AND LOADINGS:
Operating weight empty:
ERJ-135ER	11,200 kg (24,692 lb)
ERJ-135LR	11,300 kg (24,912 lb)
Max fuel: ERJ-135ER	4,173 kg (9,200 lb)
ERJ-135LR	5,187 kg (11,435 lb)
Max payload: ERJ-135ER	4,400 kg (9,700 lb)
ERJ-135LR	4,700 kg (10,362 lb)
Max T-O weight: ERJ-135ER	19,000 kg (41,888 lb)
ERJ-135LR	20,000 kg (44,092 lb)

Max landing weight:
ERJ-135ER, ERJ-135LR	18,500 kg (40,785 lb)
Max ramp weight: ERJ-135ER	19,100 kg (42,108 lb)
ERJ-135LR	20,100 kg (44,313 lb)
Max zero-fuel weight: ERJ-135ER	15,600 kg (34,392 lb)
ERJ-135LR	16,000 kg (35,274 lb)

Max wing loading:
ERJ-135ER	371.2 kg/m² (76.04 lb/sq ft)
ERJ-135LR	390.8 kg/m² (80.04 lb/sq ft)

Max power loading:
ERJ-135ER	304 kg/kN (2.98 lb/lb st)
ERJ-135LR	319 kg/kN (3.13 lb/lb st)

PERFORMANCE (estimated):
Max cruising speed	M0.78 (450 kt; 833 km/h; 518 mph)
Time to 10,670 m (35,000 ft)	20 min
Service ceiling	11,275 m (37,000 ft)
T-O field length at S/L	1,700 m (5,577 ft)
Landing field length, S/L, at typical landing weight	1,360 m (4,460 ft)

Range with 37 passengers, 100 n mile (185 km; 115 mile) diversion, 45 min reserves
1,700 n miles (3,148 km; 1,956 miles)

UPDATED

Embraer Legacy twin-turbofan business aircraft   *NEW*/0137438

Legacy Executive interior   *NEW*/0137437

EMBRAER ERJ-140

TYPE: Regional jet airliner.
PROGRAMME: Launched 30 September 1999 at the European Regional Airline Association annual meeting in Paris; first flight of prototype, modified from prototype ERJ-135 c/n 801/PT-ZJA, 27 June 2000; public debut at Farnborough International Air Show July 2000; Brazilian CTA and FAA certification achieved in June and July 2001 respectively; first delivery (PP-XGF/N800AE) to American Eagle late July 2001.
CURRENT VERSIONS: **ERJ-140ER:** Standard version, *as described.*
ERJ-140LR: Long-range version.
CUSTOMERS: Launch customer American Eagle announced order for 66 on 27 September 2000, subsequently increased to 139, through transfer of EMB-135 orders and exercising of options beginning with August 2001 deliveries. Total of 174 orders by 15 June 2002. See table.
COSTS: US$15.2 million (1999).
Descriptions for the ERJ-135 and ERJ-145 apply also to the ERJ-140 except as follows:
DESIGN FEATURES: Shares 98 per cent commonality with ERJ-135/145, including engines, wings, tail surfaces, flight deck and main systems; ERJ-135 fuselage stretched by 2.30 m (7 ft 6½ in) by removal of two frames (2.85 m; 9 ft 4¼ in ahead of wing and 1.35 m; 4 ft 5 in at rear) and substitution of two longer frames (3.94 m; 12 ft 11 in and 2.56 m; 8 ft 4¼ in respectively).
ACCOMMODATION: Standard accommodation for 44 passengers, three-abreast at seat pitch of 39 cm (31 in). Flight attendant seat on port side immediately aft of flight deck standard; attendant seat in centre of aisle at rear of cabin optional. Wardrobe/carry-on baggage cabinet and galley at front of cabin, lavatory at rear.
DIMENSIONS, EXTERNAL:
Length: overall	28.47 m (93 ft 5 in)
fuselage	26.52 m (87 ft 0 in)
Wheelbase	13.51 m (44 ft 4 in)

DIMENSIONS, INTERNAL:
Baggage volume:
Wardrobe and stowage compartment
0.93 m³ (32.84 cu ft)
WEIGHTS AND LOADINGS:
Basic operating weight: ER, LR	11,770 kg (25,948 lb)
Max fuel: ER	4,173 kg (9,200 lb)
LR	5,187 kg (11,435 lb)
Max T-O weight: ER	20,100 kg (44,312 lb)
LR	21,100 kg (46,517 lb)
Max landing weight: ER, LR	18,700 kg (41,226 lb)
Max ramp weight: ER	20,200 kg (44,533 lb)
LR	21,200 kg (46,738 lb)
Max zero-fuel weight: ER, LR	17,100 kg (37,699 lb)
Max wing loading: ER	392.7 kg/m² (80.44 lb/sq ft)
LR	412.3 kg/m² (84.44 lb/sq ft)
Max power loading: ER	321 kg/kN (3.15 lb/lb st)
LR	320 kg/kN (3.14 lb/lb st)

PERFORMANCE:
Max cruising speed:
ER, LR	Mach 0.78 (450 kt; 833 km/h; 518 mph)
T-O field length: ER	1,720 m (5,643 ft)
LR	1,320 m (4,331 ft)
Time to climb to 10,670 m (35,000 ft): ER, LR	22 min
Service ceiling: ER, LR	11,278 m (37,000 ft)

Range with 44 passengers at long-range cruising speed 100 n mile (185 km; 115 mile) alternate and 45 min reserves:
ER	1,230 n miles (2,278 km; 1,415 miles)
LR	1,630 n miles (3,018 km; 1,875 miles)

UPDATED

EMBRAER LEGACY

TYPE: Business jet.
PROGRAMME: Announced on eve of Farnborough International Air Show 23 July 2000; first flight of prototype (PP-XJO), converted from second ERJ-135 prototype (PT-ZJC), 31 March 2001; Brazilian CTA certification achieved 10 December 2001, with FAA/JAA approval expected within several months.
CURRENT VERSIONS: Offered in **Executive, Corporate Shuttle** and **Regional** variants.
CUSTOMERS: Total of 73 firm orders and 94 options by 15 June 2002 from customers in Europe, Africa, Middle East, Latin America and North America. Launch customer Swift Aviation of Phoenix, Arizona, ordered 25, with 25 options in July 2000; other announced customers include an unnamed major energy company based in Houston, Texas, which ordered one in corporate shuttle configuration in April 2001, and the Greek Air Force, which ordered one in Executive configuration for delivery in December 2001. Chicago-based Indigo ordered 25 Corporate Shuttle versions and placed options on 50 more in December 2001; market estimated at 240 aircraft over a ten-year period.

EMBRAER LEGACY ORDERS
(at 15 June 2002)

	Orders	Options	Total
Executive			
Brazil	34	35	69
Switzerland	6	3	9
USA	1		1
Shuttle			
USA	25	50	75
Regional			
USA	7	6	13
Totals	**73**	**94**	**167**

COSTS: US$20.275 million, outfitted (2002).
Description for ERJ-135 applies also to the Legacy, except the following.
DESIGN FEATURES: Compared with ERJ-135 has belly and aft fuel tanks in extended underwing fairing, plus winglets.
STRUCTURE: Airframe manufactured as ERJ-135 and modified to Legacy configuration; Embraer performs interior completion, using components supplied by Nordam.
POWER PLANT: Legacy Executive has two 33.0 kN (7,426 lb st) Rolls-Royce AE 3007A1P turbofans; Legacy Corporate Shuttle has 31.4 kN (7,057 lb st) AE 3007A1/3s.
ACCOMMODATION: Typical accommodation for 10 passengers in Executive configuration, or 20 in two-abreast arrangement at 91 cm (36 in) seat pitch in Corporate Shuttle layout. Baggage compartment at rear of cabin.

Embraer ERJ-140 regional jet airliner *(Jane's/Paul Jackson)*   *NEW*/0137350

AVIONICS: Honeywell Primus 1000 as core system.
DIMENSIONS, INTERNAL (A: Executive, B: Corporate Shuttle):
  Cabin volume: B   40 m³ (1,413 cu ft)
WEIGHTS AND LOADINGS (A, B as above):
  Weight empty, equipped: B   11,930 kg (26,301 lb)
  Basic operating weight: A   12,990 kg (28,638 lb)
              B   12,510 kg (27,580 lb)
  Baggage capacity   1,000 kg (2,205 lb)
  Max fuel: A   8,030 kg (17,703 lb)
            B   5,135 kg (11,321 lb)
  Max payload: A   3,010 kg (6,636 lb)
              B   3,490 kg (7,694 lb)
  Payload with max fuel: A   970 kg (2,138 lb)
                        B   2,355 kg (5,192 lb)
  Fuel with max payload: A   5,990 kg (13,206 lb)
                        B   4,000 kg (8,818 lb)
  Max T-O weight: A   21,990 kg (48,479 lb)
                B   20,000 kg (44,092 lb)
  Max landing weight: A, B   18,500 kg (40,785 lb)
  Max ramp weight: A   22,060 kg (48,634 lb)
                B   20,070 kg (44,247 lb)
  Max zero-fuel weight: A, B   16,000 kg (35,274 lb)
  Max wing loading: A   429.7 kg/m² (88.00 lb/sq ft)
                B   390.8 kg/m² (88.04 lb/sq ft)
  Max power loading: A   332 kg/kN (3.26 lb/lb st)
                B   318 kg/kN (3.12 lb/lb st)
PERFORMANCE (estimated):
  Cruising speed at 11,885 m (39,000 ft):
    A, B   M 0.78 (447 kt; 828 km/h; 514 mph)
  T-O field length: A   1,924 m (6,312 ft)
                  B   1,708 m (5,604 ft)
  Landing field length: A, B   1,360 m (4,462 ft)
  Range with NBAA IFR reserves:
    A with 10 passengers
                      3,200 n miles (5,926 km; 3,682 miles)
    B with 20 passengers
                      1,834 n miles (3,396 km; 2,110 miles)
                                              UPDATED

EMBRAER 170 and 190

TYPE: Regional jet airliner.
PROGRAMME: Announced (officially 'pre-launch') in February
  1999; designations then ERJ-170 and ERJ-190; engine
  selection May 1999; first orders announced 14 June 1999;
  risk-sharing partners (see 'Structure', below) revealed at
  European Regional Airline Association annual meeting in
  Paris, 30 September 1999. Joint definition phase leading to
  ERJ-170 design freeze, completed in April 2000; first
  metal cut for first of six pre-series ERJ-170s 14 July 2000;
  ERJ-170 (c/n 0001/PP-XJE) first rolled out 29 October
  2001, first flight 19 February 2002; followed by c/n 0002/
  PP-XJC (to be HB-JCA) in Crossair (to become Swiss)
  colours on 9 April 2002, by which time the first aircraft had
  completed 40 hours of test flying; c/n 0003/PP-XJB on 25
  May 2002; and c/n 0004/PP-XJF on 19 June 2002.
  Announced at roll-out that ERJ prefix was discontinued
  and that ERJ-190-200 was to be known as Embraer 195;
  intermediate Embraer 175 simultaneously revealed.
    Six aircraft (0001 to 0006) scheduled to have been
  completed during first half of 2002 and conduct 1,800-hour
  flight test programme, culminating in certification in first
  quarter 2003; initial production rate up to two per month in
  2002, rising to three per month in 2003 and four per month
  in 2004, with maximum capacity of six per month. Further
  two static and fatigue test airframes (801 and 802). Final
  assembly of the first of two pre-series Embraer 195s will
  begin in late 2002; first flight in third quarter of 2003; first
  delivery (to Swiss, formerly Crossair) in December 2004;
  first of two pre-series Embraer 190s will fly in 2004 with
  certification and deliveries in December 2005.
CURRENT VERSIONS: **Embraer 170:** Baseline version with 70
  seats; available in Standard and Long-Range versions; first
  delivery December 2002.
    Embraer 175: Longer version with 1.77 m (5 ft 9¾ in)
  fuselage stretch by means of two plugs to accommodate 78
  to 86 passengers; offered in Standard and Long-Range
  versions; first delivery scheduled for June 2004. No launch
  customer announced by late 2001, but British Airways and
  Qantas seen as potential operators.
    Embraer 190: Further stretch by 6.25 m (20 ft 6 in) to
  accommodate up to 104 passengers; wing span increased
  by 2.56 m (8 ft 4¾ in); GE CF34-8E-10 engines;

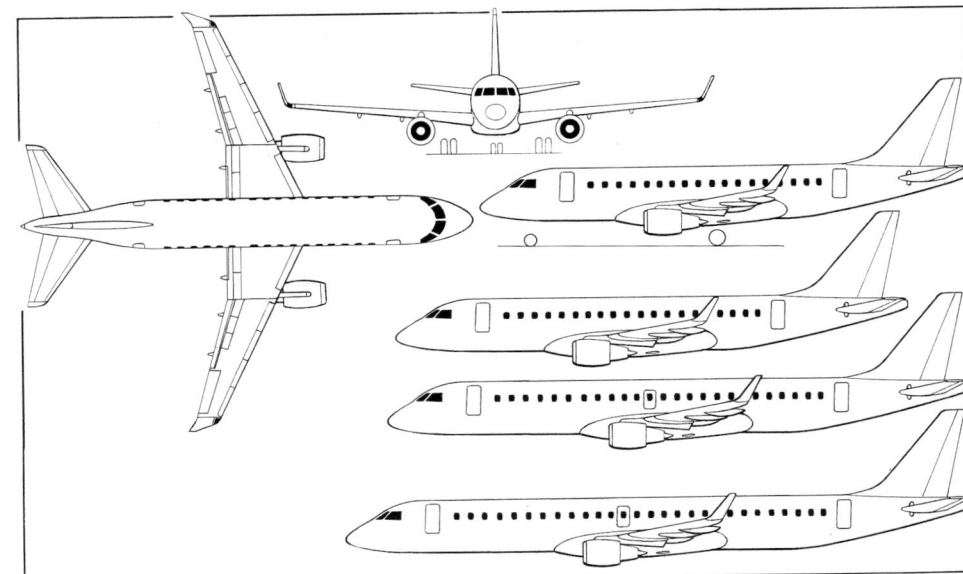

Provisional arrangement of Embraer 170 regional jet, with additional side views of Embraer 175, Embraer 190 and Embraer 195 (bottom) *(Jane's/James Goulding)*   *NEW*/0526909

Embraer 170 first prototype   *NEW*/0137413

strengthened landing gear; available in Standard and Long-
Range versions. First delivery December 2005.
    Embraer 195: Formerly ERJ-190-200. Further stretch
of Embraer 190-200 by 2.41 m (7 ft 11 in) to accommodate
up to 110 passengers; available in Standard and Long-
Range versions; first delivery December 2004.
    Corporate: Proposed variant of Embraer 170 Long-
Range with additional fuel tanks in baggage compartment
area to extend range to more than 4,000 n miles (7,408 km;
4,603 miles).
CUSTOMERS: See table. Launch customers Crossair (now
Swiss) and Regional Airlines of France announced at Paris
Air Show 14 June 1999; Crossair ordered 30 Embraer 170s
and 30 Embraer 195s, with options on a further 100
Embraer 170/195s, for delivery (Embraer 170) from
December 2002. Total of 112 firm orders (see table) and
212 options by April 2002. Alitalia negotiating for six in
mid-2002.

EMBRAER 170/190 ORDERS
(at 15 June 2002)

Customer	Type	Orders
Air Caraibes	170	2
Swiss	170	30
	195	30
GECAS	170	50
Total		**112**

Note: Further 202 options held.

COSTS: Embraer 170 US$21 million; Embraer 190, US$24
  million (both 1999). Development US$850 million for
  whole family (2001).
DESIGN FEATURES: Design goals include low weight,
  simplicity of operation, high reliability, ease and economy
  of maintenance and ability to operate from same airports as
  ERJ-135/145. Low-wing airliner of conventional
  appearance, with podded engine below each wing and (on
  Embraer 170) winglets; airframe designed for an economic
  life of 60,000 to 80,000 cycles.
FLYING CONTROLS: Fly-by-wire. Ailerons, rudder and all-
  moving tailplane. Double-slotted flaps, five-section
  leading-edge slats and five-section spoilers on each wing.
STRUCTURE: Embraer is responsible for radome, forward
  fuselage, centre fuselage II, wing-to-fuselage fairing and
  final assembly; risk-sharing partners are C & D Interiors
  (cabin interior); Gamesa (rear fuselage and horizontal and
  vertical tail surfaces); General Electric (power plant and
  nacelles); Hamilton Sundstrand (tailcone, APU and air
  management and electrical systems); Honeywell
  (avionics); Kawasaki (wing stub, fixed leading- and
  trailing-edge assemblies. flaps, spoilers, control surfaces
  and engine pylons); Latécoère (centre fuselage I and III);
  Liebherr (landing gear); Parker Hannifin (hydraulic, flight
  control and fuel systems); and Sonaca (wing slats).
LANDING GEAR: Retractable tricycle type. Twin wheels on
  each unit.
POWER PLANT: Embraer 170/175 series has two 62.28 kN
  (14,000 lb st) General Electric CF34-8E turbofans; 190/
  195 series has two 82.29 kN (18,500 lb st) CF34-10E
  turbofans; both engines have FADEC.
ACCOMMODATION: Total of 70 to 76 (Embraer 170), 78 to 86
  (Embraer 175), 96 to 104 (Embraer 190) or 108 to 110
  (Embraer 195) passengers, four abreast at 81 cm (32 in)
  seat pitch. Optional first class cabin on Embraer 170 and
  Embraer 190 with three-abreast seating.
AVIONICS: Honeywell Primus Epic five-tube EFIS. Thales
  Avionics integrated electronic standby instrument (IESI)
  system standard.
DIMENSIONS, EXTERNAL (all versions, except where stated):
  Wing span over winglets: 170   26.00 m (85 ft 3½ in)
                  190, 195   28.72 m (94 ft 2¾ in)
  Length overall: 170   29.90 m (98 ft 1¼ in)
                  175   31.68 m (103 ft 11¼ in)
                  190   36.24 m (118 ft 10¾ in)
                  195   38.65 m (126 ft 9½ in)
  Height overall: 170   9.67 m (31 ft 8¾ in)
                  175   9.73 m (31 ft 11 in)
                  190, 195   10.28 m (33 ft 8¾ in)
  Fuselage height: all   3.35 m (11 ft 0 in)

Artist's impression of Embraer 195, longest of the family   *NEW*/0137342

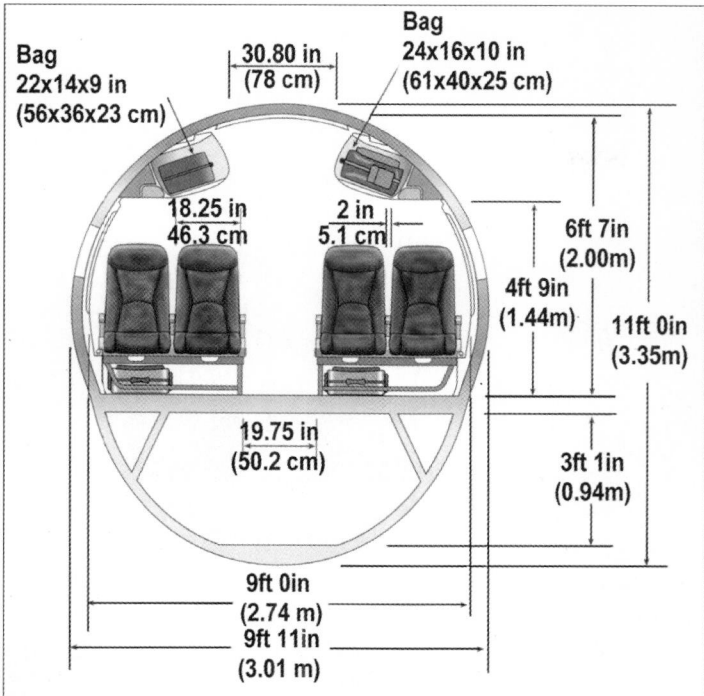

Bag
22x14x9 in
(56x36x23 cm)

Bag
24x16x10 in
(61x40x25 cm)

30.80 in
(78 cm)

18.25 in
46.3 cm

2 in
5.1 cm

6ft 7in
(2.00m)

4ft 9in
(1.44m)

11ft 0in
(3.35m)

19.75 in
(50.2 cm)

3ft 1in
(0.94m)

9ft 0in
(2.74 m)

9ft 11in
(3.01 m)

Common flight deck of Embraer 170/190   NEW/0131838

Embraer 170/190 fuselage cross-section
NEW/0137414

Fuselage width: all	3.01 m (9 ft 10½ in)		
Tailplane span: all	10.00 m (32 ft 9¾ in)		

DIMENSIONS, INTERNAL:
Cabin (excl flight deck):
Length: 170 — 19.38 m (63 ft 7 in)
175 — 21.16 m (69 ft 5 in)
190 — 25.70 m (84 ft 3¾ in)
195 — 28.13 m (92 ft 3½ in)
Max width: all — 2.74 m (8 ft 11¾ in)
Max height: all — 2.00 m (6 ft 6¾ in)
Aisle width: 170 — 0.46 m (1 ft 6 in)
175, 190, 195 — 0.50 m (1 ft 7¾ in)

WEIGHTS AND LOADINGS:
Basic operating weight:
170 — 20,150 kg (44,423 lb)
175 — 21,150 kg (46,627 lb)
190 — 26,200 kg (57,761 lb)
195 — 27,100 kg (59,745 lb)
Max payload: 170 — 9,100 kg (20,062 lb)
175 — 10,550 kg (23,259 lb)
190 — 12,400 kg (27,337 lb)
195 — 12,700 kg (27,999 lb)
Max fuel: 170 — 9,470 kg (20,878 lb)
190, 195 — 13,000 kg (28,660 lb)
Max T-O weight:
170 Standard — 35,450 kg (78,153 lb)
170 Long-Range — 36,850 kg (81,240 lb)
175 Standard — 35,990 kg (79,344 lb)
175 Long-Range — 37,500 kg (82,673 lb)
190 Standard — 45,990 kg (101,390 lb)
190 Long-Range — 48,500 kg (106,924 lb)
195 Standard — 46,990 kg (103,595 lb)
190 Long-Range — 48,990 kg (108,004 lb)
Max landing weight: 170 — 32,450 kg (71,540 lb)
175 — 33,350 kg (73,524 lb)
190 — 42,500 kg (93,696 lb)
195 — 44,500 kg (98,106 lb)
Max power loading:
170 Standard — 284 kg/kN (2.79 lb/lb st)
170 Long-Range — 296 kg/kN (2.90 lb/lb st)
190 Standard — 279 kg/kN (2.74 lb/lb st)
190 Long-Range — 295 kg/kN (2.89 lb/lb st)
195 Standard — 285 kg/kN (2.80 lb/lb st)
195 Long-Range — 298 kg/kN (2.92 lb/lb st)

PERFORMANCE (estimated):
Max cruising speed M0.80 (470 kt; 870 km/h; 541 mph)
Max speed below 3,050 m (10,000 ft):
170 — 300 kt (556 km/h: 345 mph)
Time to climb to FL300: 170, 195 — 17 min
175 — 19 min
190 — 16 min
T-O field length: 170 — 1,676 m (5,500 ft)
175 — 1,995 m (6,545 ft)
190 — 1,986 m (6,515 ft)
195 — 2,046 m (6,715 ft)
Landing field length at S/L, ISA, at typical landing weight:
170 — 1,250 m (4,105 ft)
175 — 1,305 m (4,285 ft)
190 — 1,318 m (4,325 ft)
195 — 1,366 m (4,485 ft)
Range with max passengers at long-range cruising speed:
170 Long-Range 2,100 n miles (3,889 km; 2,416 miles)
175 Standard 1,298 n miles (2,405 km; 1,494 miles)
175 Long-Range 1,598 n miles (2,960 km; 1,839 miles)
190 Long-Range 2,300 n miles (4,259 km; 2,646 miles)
195 Standard 1,400 n miles (2,592 km; 1,611 miles)
195 Long-Range 1,800 n miles (3,333 km; 2,071 miles)

UPDATED

OTHER AIRCRAFT

Refer to International section for **AMX, AMX-T** and other variants. Details of the **EMB-202 Ipanema** and **Piper** aircraft built under licence granted to Embraer can be found under the Neiva entry in this section and in the US section of this and earlier editions of *Jane's*.

VERIFIED

Second prototype Embraer 170 in Crossair colours, prior to adopting Swiss livery   NEW/0143349

HELIBRAS

HELICÓPTEROS DO BRASIL SA
(Subsidiary of Eurocopter)

Distritio Industrial de Itajubá, Rua Santos Dumont 200,
   Caixa Postal 184, 37500-000 Itajubá, Minas Gerais
Tel: (+55 35) 36 23 20 00
Fax: (+55 35) 36 23 20 01
Web: http://www.helibras.com.br

COMMERCIAL DEPARTMENT
Aeroporto Campo de Marte, Avenida Santos-Dumont 1979,
   CEP 02012-010 Saõ Paulo, SP
Tel: (+55 11) 69 90 37 00
Fax: (+55 11) 62 21 55 35
PRESIDENT: Jean Raquin
COMMERCIAL MANAGER: Fabrice Cagnat
MARKETING MANAGER: Luis Henrique Testa

Formed 1978; owned jointly by Grupo Bueninvest (30 per
cent), MGI Participações (25 per cent) and Eurocopter France
(45 per cent). First assembly hall inaugurated 28 March 1980;
facility occupies 12,000 m² (129,200 sq ft); new facilities,
totalling 3,800 m² (40,900 sq ft) opened at São Paulo in
second quarter of 1998 for maintenance, spares, training,
sales and marketing. Total of 300 personnel employed in
2000.

Assembly, marketing and overhaul, under Eurocopter
licence, of single- and twin-engined Ecureuil/Fennec
helicopters and twin-engined Dauphin/Panther; markets in
Brazil and overhauls civil Eurocopter AS 350/355 Esquilo
(Fennec), EC 130, EC 135, AS 332 Super Puma; Eurocopter/
Kawasaki BK 117C-1 (EC 145); and Eurocopter/CATIC/ST
Aero EC 120 (see International section).

Total of more than 400 helicopters, including some 300
Esquilos, sold by January 2002, representing 50 per cent of
the turbine-powered helicopter fleet in Brazil); to more than
90 customers, including 160 to Brazilian police forces, Army,

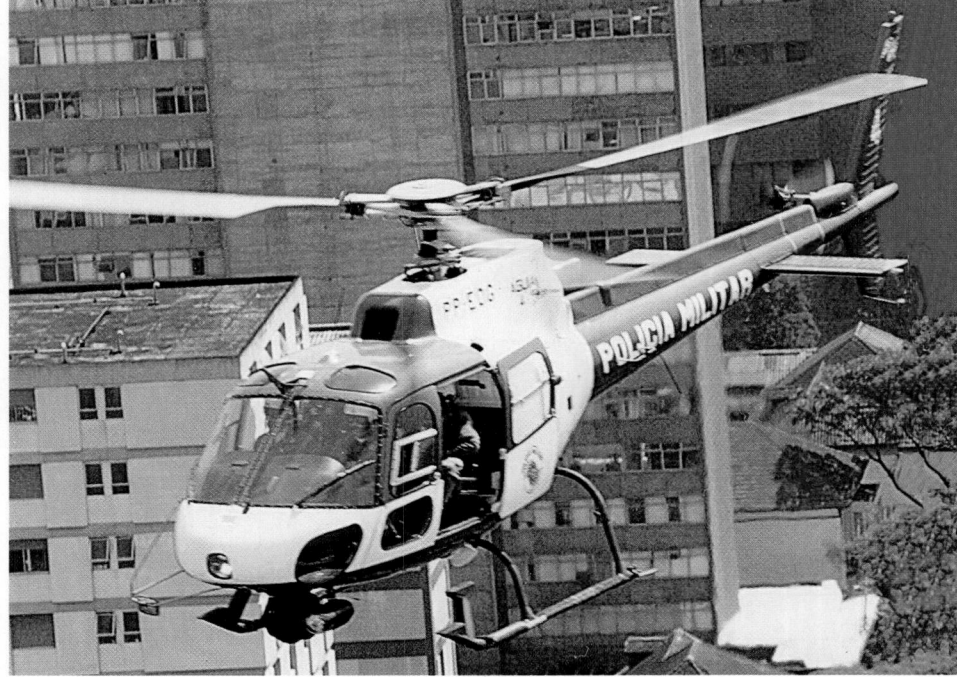

Helibras-supplied AS 350 Esquilo of São Paulo Police   *NEW*/0131689

Navy and Air Force; nearly 10 per cent of overall total
exported to Argentina, Bolivia, Chile, Paraguay and
Venezuela. Supply of new EC 120B Colibri to local owners

began in 1999. Details of local Esquilo versions last appeared
in 2000-01 *Jane's.*

UPDATED

KOVACS

JOSEPH KOVACS

Rua Maria de Lourdes Barbosa 32, 12244-690 São José dos
   Campos, SP
Tel: (+55 12) 97 19 78 17
e-mail: avkovacs@vol.com.br
DESIGNER: Joseph Kovács
MARKETING: André Kovács

Having produced a prototype of the K-51 Peregrino two-seat
aerobatic sportplane, Mr Kovacs is now working on design of
the K-55 aerobatic competition monoplane and K-32 STOL,
ultralight two-seater, which will incorporate variable
geometry.

UPDATED

KOVACS K-51 PEREGRINO
English name: Peregrine
TYPE: Aerobatic two-seat sportplane.
PROGRAMME: First outlined in 1983; design began 1988.
   Prototype (PP-XLI) first flew 28 November 1998, at which
   time a commercial manufacturer was being sought to
   launch production; kitbuilt version planned. However, by
   late 2002, no further aircraft of the type were known to
   have been built.
DESIGN FEATURES: Conventional low-wing design, similar to
   Embraer Tucano, but (initially, at least) with tailwheel.
   Intended for civil or military training, including aerobatics,
   and with provision for higher-powered engines; suitable
   for glider towing.
      Wing section NACA 64$_{15}$ A-313 ½ (a = 0.8) at root and
   NACA 64A-212 (modified) at tip for optimum
   performance in inverted flight and spins. Dihedral 5°;
   incidence 1° at root and 0° at tip.
FLYING CONTROLS: Conventional and manual. All
   manoeuvring surfaces horn balanced. Fully balanced
   slotted ('semi-Frise') ailerons interconnected with flaps to
   enhance aerobatic performance; slotted flaps adopt slight
   reflex angle for cruising flight. Trim tab in port elevator.
STRUCTURE: Wood, metal tube and composites. Wing has
   single-piece wood box spar with trellis ribs, rear auxiliary
   spar, plywood covering on lower surface and 10 mm
   (0.4 in) glass fibre honeycomb on upper surface; wooden
   ailerons and flaps covered with 3 mm (⅛ in) glass fibre/
   Nomex honeycomb sheet. Centre fuselage of welded steel
   tube with semi-structural plywood skin panels; two-cell
   semi-monocoque wooden empennage with plywood
   covering and integral fin. Rudder and elevators covered
   with 3 mm (⅛ in) glass fibre/Nomex honeycomb.
LANDING GEAR: Tailwheel type; fixed. Machined, conically
   tapered cantilever spring steel main gear legs constructed
   of five tubes of different lengths; glass fibre speed fairings.
   Cleveland mainwheels and hydraulic brakes; Goodyear
   tyre size 5.00-5 (6-ply). Steerable R & K tailwheel, size
   225×6. Turning circle 5.4 m (18 ft). Tricycle landing gear
   optional.

The Kovács K-51 Peregrino photographed during its maiden flight   0041490

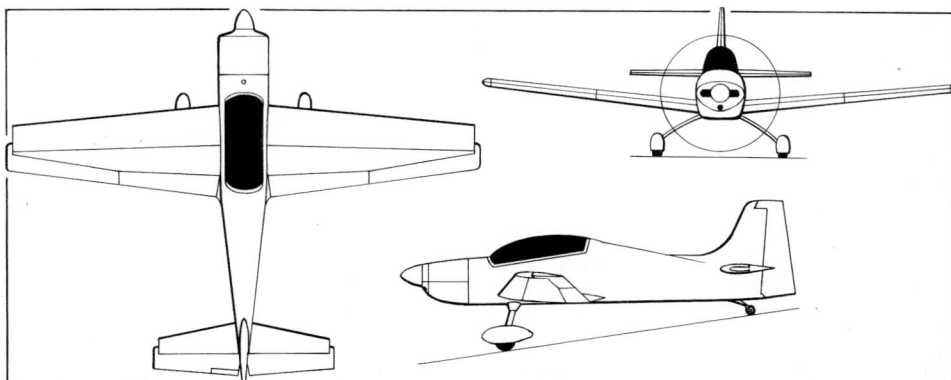

Kovács K-51 Peregrino aerobatic trainer (*Jane's/James Goulding*)   0100409

POWER PLANT: Prototype has one 149 kW (200 hp) Textron
   Lycoming IO-360-C1C-6 flat-four, driving a McCauley
   two-blade constant-speed metal propeller. Airframe
   designed to accommodate engines in 119 to 150 kW (160
   to 200 hp) class. Fuel for aerobatic flying in 83 litre (21.9
   US gallon; 18.3 Imp gallon) tank in forward fuselage; non-
   aerobatic tank beneath baggage compartment, capacity
   97 litres (25.6 US gallons; 21.3 Imp gallons). Total fuel
   180 litres (47.6 US gallons; 39.6 Imp gallons). Fuel and oil
   systems modified for long periods of inverted flight.
ACCOMMODATION: Two persons, in tandem, under starboard-
   hinged, single-piece canopy. Adjustable seats and dual

controls; rear seat, raised 15 cm (6 in), is occupied when
flown solo. Baggage compartment behind rear seat.
SYSTEMS: Electrical system 28 V DC.
AVIONICS: *Comms:* Honeywell KT 76A transponder.
   Flight: Garmin GNC 250 GPS.
   Instrumentation: Standard VFR.
DIMENSIONS, EXTERNAL:

Wing span	8.50 m (27 ft 10¾ in)
Wing chord: at root	1.70 m (5 ft 7 in)
at tip	0.84 m (2 ft 9 in)
Wing aspect ratio	6.5
Length overall	7.03 m (23 ft 0¾ in)

Height overall	2.05 m (6 ft 8¾ in)			Stalling speed: flaps up: A	55 kt (102 km/h; 64 mph)
Tailplane span	3.20 m (10 ft 6 in)			U	60 kt (112 km/h; 69 mph)
Wheel track	2.20 m (7 ft 2½ in)			flaps down: A	46 kt (86 km/h; 53 mph)
Wheelbase	4.38 m (14 ft 4½ in)			U	54 kt (100 km/h; 63 mph)
Propeller diameter	1.93 m (6 ft 4 in)			Max rate of climb at S/L: A	580 m (1,900 ft)/min

WEIGHTS AND LOADINGS (A: Aerobatic, U: Utility with 149 kW; 200 hp engine):

Height overall	2.05 m (6 ft 8¾ in)	
Tailplane span	3.20 m (10 ft 6 in)	
Wheel track	2.20 m (7 ft 2½ in)	
Wheelbase	4.38 m (14 ft 4½ in)	
Propeller diameter	1.93 m (6 ft 4 in)	
DIMENSIONS, EXTERNAL:		
Cockpit: Length	2.13 m (7 ft 0 in)	
Max width	0.73 m (2 ft 4¾ in)	
Baggage compartment volume	0.20 m³ (7.0 cu ft)	
AREAS:		
Wings, gross	11.04 m² (118.8 sq ft)	
Ailerons (total)	1.08 m² (11.63 sq ft)	
Trailing-edge flaps (total)	1.16 m² (12.49 sq ft)	
Fin	1.22 m² (13.13 sq ft)	
Rudder	0.65 m² (7.00 sq ft)	
Tailplane	2.40 m² (25.83 sq ft)	
Elevators, incl tab	0.95 m² (10.23 sq ft)	

WEIGHTS AND LOADINGS (A: Aerobatic, U: Utility with 149 kW; 200 hp engine):

Weight empty, equipped: A, U	600 kg (1,323 lb)
Max T-O weight: A	840 kg (1,852 lb)
U	940 kg (2,072 lb)
Max wing loading: A	76.1 kg/m² (15.59 lb/sq ft)
U	85.1 kg/m² (17.44 lb/sq ft)
Max power loading: A	5.64 kg/kW (9.26 lb/hp)
U	6.31 kg/kW (10.37 lb/hp)

PERFORMANCE (A, U as above, 119 kW; 160 hp engine):

Never-exceed speed (VNE):	
A, U	200 kt (370 km/h; 230 mph)
Max level speed at S/L: A	182 kt (337 km/h; 209 mph)
U	175 kt (324 km/h; 201 mph)
Max cruising speed at 75% power at 2,400 m (7,880 ft):	
A	173 kt (320 km/h; 199 mph)
U	170 kt (315 km/h; 196 mph)

Stalling speed: flaps up: A	55 kt (102 km/h; 64 mph)
U	60 kt (112 km/h; 69 mph)
flaps down: A	46 kt (86 km/h; 53 mph)
U	54 kt (100 km/h; 63 mph)
Max rate of climb at S/L: A	580 m (1,900 ft)/min
U	415 m (1,362 ft)/min
Service ceiling: A	8,230 m (27,000 ft)
U	7,075 m (23,220 ft)
T-O run: A	151 m (495 ft)
U	247 m (810 ft)
T-O to 15 m (50 ft): A	250 m (820 ft)
U	328 m (1,080 ft)
Landing from 15 m (50 ft): A	451 m (1,480 ft)
U	475 m (1,560 ft)
Range at max cruising speed and altitude, no reserves:	
U	890 n miles (1,648 km; 1,024 miles)
g limits	+6/−3

UPDATED

NEIVA

INDÚSTRIA AERONÁUTICA NEIVA SA
(Subsidiary of Embraer)

Caixa Postale 1011, Avenida Alcides Cagliari 2281, 18608-900 Botucatu-SP
Tel: (+55 14) 821 21 22
Fax: (+55 14) 822 12 85
e-mail: neiva@laser.com.br
PRESIDENT AND DIRECTOR: Mauricio Novis Botelho
MANAGING DIRECTOR: Paolo Urbanavicius
SALES MANAGERS: Luiz Fabiano
  Zaccarelli Cunha

Formed October 1953 by José Carlos de Barros Neiva; produced Paulistinha, Regente, Universal and related lightplanes of own design. Became wholly owned Embraer subsidiary 10 March 1980; factory area 18,550 m² (199,700 sq ft), workforce 900 in 2002, following October 2001 cut of 270 staff. Previously played major role in Embraer general aviation programmes, including complete production of EMB-711ST Corisco Turbo, EMB-710 Carioca, EMB-720D Minuano (of which some 300 delivered), EMB-721 Sertanejo and EMB-810D Seneca IV (licence-built Piper models); also responsible since 1980 for total production of EMB-202 and now discontinued EMB-201A agricultural sprayers. Low-volume production of EMB-120 Brasilia was transferred from Embraer in 2001. Built 17 light aircraft in 1999; same in 2000; and 11 in 2001.

UPDATED

EMBRAER EMB-120 BRASILIA
Brazilian Air Force designation: VC-97
TYPE: Twin-turboprop airliner.
PROGRAMME: Design began September 1979; three flying prototypes, two static/fatigue test aircraft and one pre-series demonstrator built (first flight, by PT-ZBA, 27 July 1983); Brazilian CTA certification with original PW115 engines on 10 May 1985 followed by FAA (FAR Pt 25) type approval 9 July 1985, British/French/German approval 1986 and Australian in April 1990; deliveries began June 1985 (to Atlantic Southeast Airlines, USA, entering service that October); first order for corporate version (United Technologies Corporation, USA) received August 1986, delivered following month.

From October 1986 (c/n 120028), all Brasilias delivered incorporate composites materials equivalent to 10 per cent of aircraft basic empty weight; also since late 1986 has been available in hot-and-high version (certified 26 August 1986) with PW118A engines which maintain maximum output up to ISA + 15°C at S/L (first customer Skywest of USA); on 4 January 1989, first prototype began flight trials with Honeywell TPE331-12B turboprop on port side of rear fuselage, as testbed for engine installation of subsequently cancelled Embraer/FMA CBA-123; 300th Brasilia delivered 4 September 1995; more than five million flying hours by September 1998, at which time Brasilias had carried more than 60 million passengers.

Extended-range EMB-120ER (now standard version) announced June 1991, certified by CTA February 1992.
CURRENT VERSIONS: **EMB-120:** Launch production version, with 1,118.5 kW (1,500 shp) PW115 engines and Hamilton Sundstrand 14RF four-blade propellers (replaced at early stage by higher output PW115s of 1,193 kW; 1,600 shp). Details in 1996-97 and earlier *Jane's.*

EMB-120RT (Reduced Take-off): Initial production version; 1,342 kW (1,800 shp) PW118 engines for better field performance; most early aircraft now retrofitted to RT standard. Also, from late 1986, in hot/high version with PW118As (see Programme above). See 1996-97 and earlier editions for details.

EMB-120FC: 'Full Cargo' conversion of EMB-120RT developed in USA by IASG and approved by Embraer. Prototype (N453UE) converted June 2000 and exhibited at Farnborough in following month. Cargo volume 31.0 m³ (1,095 cu ft); maximum payload 3,700 kg (8,157 lb); empty weight 7,200 kg (15,873 lb); maximum T-O weight

First Embraer EMB-120FC Brasilia freighter conversion *(Jane's/Paul Jackson)*   0100513

12,000 kg (26,455 lb). Initial 10 kits ordered by North South Airways.

EMB-120 Combi: Mixed configuration version of ER with quick-release seats, 9 g movable rear bulkhead and cargo restraint net; retains lavatory and galley; typical capacity 30 passengers and 700 kg (1,543 lb) of cargo or 19 passengers and 1,100 kg (2,425 lb) cargo.

EMB-120QC (Quick-Change): Convertible in 40 minutes from 30-passenger layout to 3,500 kg (7,716 lb) all-cargo configuration with floor and sidewall protection, fire detection system, smoke curtain aft of flight deck, 9 g movable rear bulkhead and cargo restraint net; conversion from cargo to passenger interior takes 50 minutes. First customer (14 May 1993 delivery), Total Linhas Aereas of Brazil.

EMB-120ER Brasilia Advanced: First delivery to Skywest Airlines in May 1993; standard production version from 1994, previously referred to as EMB-120X or Improved Brasilia. Additional range obtained by increasing maximum T-O weight without major structural change; this also allows for increased standard passenger-plus-baggage weight (97.5 kg; 215 lb per person instead of 91 kg; 200 lb) and obligatory fitting of TCAS and flight data recorder. Incorporates several modifications and some redesign aimed at maximising both passenger comfort and dispatchability while simultaneously reducing operational

and maintenance costs. These include redesigned, interchangeable leading-edges for all flying surfaces; deletion of the fin de-icing boot; improved door seals and redesigned interior panel joints to reduce passenger cabin noise; more comfortable Sicma-built pilot seats; redesigned flight deck door; and new-design overhead bins with wider doors and increased capacity. Other features include new passenger cabin lighting; redesigned flight deck and cabin ventilation systems; new windscreen frame fairings to improve resistance to condensation; new flight deck sun visors; improved flap system; and an increase in baggage/cargo compartment capacity to 700 kg (1,543 lb).

Existing Brasilias can be retrofitted to ER standard (first two customers for retrofit Delta Air Transport and Luxair); initial production deliveries (two) to Great Lakes Aviation in USA, December 1994.

EMB-120 Cargo: All-cargo version of ER with 4,000 kg (8,818 lb) payload capacity; floor and sidewall protection, fire detection system, smoke curtain between flight deck and cargo cabin, and cargo restraint net.

Detailed description applies to EMB-120ER except where indicated.

EMB-120K: Proposed maritime surveillance or ASW version for Brazilian Air Force; would be equipped with in-flight refuelling, giving 15 hour endurance; under study in 1999.

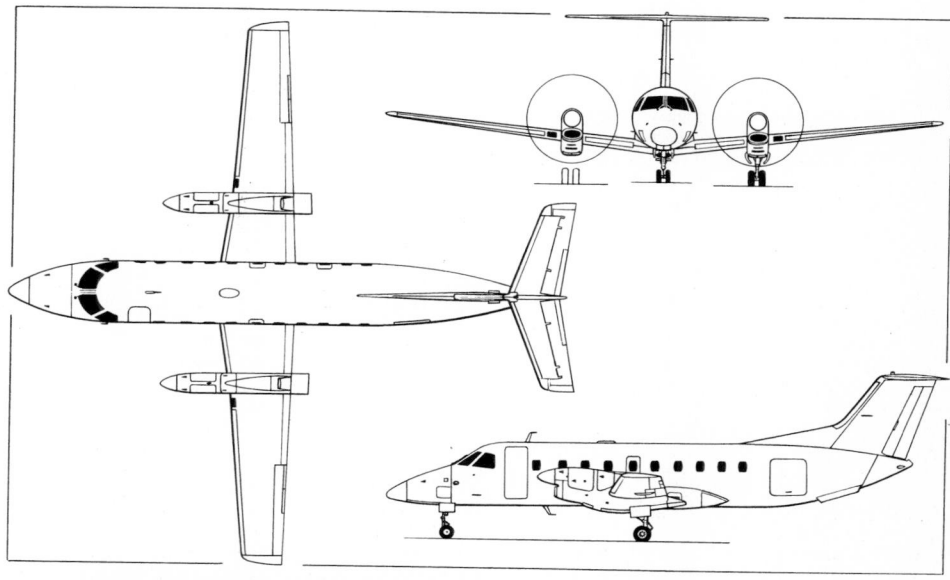

Embraer EMB-120 Brasilia twin-turboprop transport *(Jane's/Dennis Punnett)*

VC-97: VIP transport model for Brazilian Air Force's 6° Esquadrao de Transporte Aérea and Grupo de Transporte Especial, both at Brasilia.

CUSTOMERS: See table. Embraer delivered 352nd (including prototypes) aircraft to SkyWest (as N586SW) in September 1999. Manufacture then transferred to Neiva, which delivered two in 2001 (none in 2000) to Avior Express of Venezuela.

EMBRAER EMB-120 BRASILIA CUSTOMERS
(at 15 June 2002)

Customer	Qty
Air Aruba	1
Air Littoral	10
Air Midwest	14
ASA	62
Avior	2
Bop Air	2
Brazilian Air Force (VIP)	4
(SIVAM)	1
Comair	40
CSE Aviation	2
DAT	10
DLT	12
Ematec	1
Esquel	4
Flight West	2
Great Lakes	5
Interbrasil	3
Luxair	3
Mesa Air	2
Nordeste	1
Norsk Air	4
Ontario	5
Pantanal	2
Passaredo	3
Penta	1
Rico	1
Rio Sul	12
Skywest	70
TAT	2
Texas Air	34
Total	1
UTC	1
Westair	35
Embraer (retained)	2
Total	**354**

DESIGN FEATURES: Low-mounted unswept wings, circular-section pressurised fuselage, all-swept T tail unit. Fixed incidence tailplane.

Wing section NACA 23018 (modified) at root, NACA 23012 at tip; incidence 2°; at 66 per cent chord, wings have 6° 30′ dihedral.

FLYING CONTROLS: Conventional and assisted. Internally balanced ailerons, and horn-balanced elevators, actuated mechanically (ailerons by dual irreversible actuators); serially hinged two-segment rudder actuated hydraulically by Bertea CSD unit; trim tabs in ailerons (one port, two starboard) and each elevator. Hydraulically actuated, electrically controlled, double-slotted Fowler trailing-edge flap inboard and outboard of each engine nacelle, with small plain flap beneath each nacelle; no slats, slots, spoilers or airbrakes. Small fence on each outer wing between outer flap and aileron; twin ventral strakes under rear fuselage.

STRUCTURE: Kevlar-reinforced glass fibre for wing and tailplane leading-edges and tips, wingroot fairings, dorsal fin, fuselage nosecone and (when no APU fitted) tailcone; Fowler flaps of carbon fibre. Remainder conventional semi-monocoque/stressed skin structure of 2024/7050/7475 aluminium alloys with chemically milled skins. Fuselage pressurised between flat bulkhead forward of flight deck and hemispherical bulkhead aft of baggage compartment, and meets damage tolerance requirements of FAR Pt 25 (Transport category) up to Amendment 25-54. Wing is single continuous three-spar fail-safe structure, attached to underfuselage frames; tail surfaces also three-spar.

LANDING GEAR: Retractable tricycle type, with Goodrich twin wheels and oleo-pneumatic shock-absorber on each unit (main units 12 in, nose unit 8 in). Hydraulic actuation; all units retract forward (main units into engine nacelles). Hydraulically powered nosewheel steering. Mainwheel tyres 24×7.25-12 (12 ply) tubeless; nose tyre 18×5.5 (8 ply) tubeless; pressure 6.90 to 7.58 bar (100 to 110 lb/sq in) on main units, 4.14 to 4.83 bar (60 to 70 lb/sq in) on nose unit. Goodrich carbon brakes standard (steel optional). Hydro Aire anti-skid system standard; autobrake optional. Minimum ground turning radius 15.76 m (51 ft 8½ in). Nosewheel guard optional for operation from unpaved surfaces.

POWER PLANT: Two Pratt & Whitney Canada PW118 or PW118A turboprops, each rated at 1,342 kW (1,800 shp) for T-O and maximum continuous power, and driving a Hamilton Sundstrand 14RF-9 four-blade constant-speed reversible-pitch autofeathering propeller with glass fibre blades containing aluminium spars. Fuel in two-cell 1,670 litre (441 US gallon; 367.2 Imp gallon) integral tank in each wing; total capacity 3,340 litres (882 US gallons; 734.4 Imp gallons), of which 3,308 litres (874 US gallons; 728 Imp gallons) are usable. Single-point pressure refuelling (beneath outer starboard wing), plus gravity point in upper surface of each wing. Oil capacity 9 litres (2.4 US gallons; 2.0 Imp gallons).

ACCOMMODATION: Two-pilot flight deck. Main cabin accommodates attendant and 30 passengers in three-abreast seating at 79 cm (31 in) pitch, with overhead lockable baggage racks, in pressurised and air conditioned environment. Passenger seats of carbon fibre and Kevlar, floor and partitions of carbon fibre and Nomex sandwich, side panels and ceiling of glass fibre/Kevlar/Nomex/carbon fibre sandwich. Provisions for wardrobe, galley and lavatory. Quick-change interior available optionally (first customer, Total Linhas Aéreas in 1993), for 30 passengers or 3,500 kg (7,716 lb) of cargo. Downward-opening main passenger door, with airstairs, forward of wing on port side. Type II emergency exit on starboard side at rear. Overwing Type III emergency exit on each side. Pressurised baggage compartment aft of passenger cabin, with large door on port side.

Also available with all-cargo interior; executive or military transport interior; or in mixed-traffic version with 24 or 26 passengers (lavatory omitted in latter case), and 900 kg (1,984 lb) of cargo in enlarged rear baggage compartment.

SYSTEMS: Honeywell air conditioning/pressurisation system (differential 0.48 bar; 7 lb/sq in), with dual packs of recirculation equipment. Duplicated hydraulic systems (pressure 207 bar; 3,000 lb/sq in), each powered by an engine-driven pump, for landing gear, flap, rudder and brake actuation, and nosewheel steering. Emergency standby electric pumps on each system, plus single standby hand pump, for landing gear extension. Main electrical power supplied by two 28 V 400 A DC starter/generators; two 28 V 100 A DC auxiliary brushless generators for secondary and/or emergency power; one 24 V 40 Ah Ni/Cd battery for assisted starting and emergency power. Main and standby 450 VA static inverters for 26/115 V AC power at 400 Hz. Single high-pressure (127.5 bars; 1,850 lb/sq in) oxygen cylinder for crew; individual chemical oxygen generators for passengers. Pneumatic de-icing for wing and tail leading-edges; electrically heated windscreens, propellers and pitot tubes; bleed air de-icing of engine air intakes. Optional Honeywell GTCP36-150(A) APU in tailcone, for electrical and pneumatic power supply.

AVIONICS: Rockwell Collins Pro Line II digital package as core system.

Comms: Dual VHF-22 com transceivers, one TDR-90 transponder, Dorne & Margolin DMELT-81 emergency locator transmitter, Fairchild voice recorder, dual Avtech audio/interphones and Avtech PA and cabin interphone all standard. Third VHF com, second transponder, Motorola Selcal and flight entertainment music all optional. Alternative Bendix/King avionics package available to special order.

Radar: Rockwell Collins WXR-270 weather radar standard; WXR-300 optional.

Flight: Dual VIR-32 VHF nav receivers, one ADF-60A, DME-41 and CLT-22/32/62/92 control heads. Optional single/dual JET RNS-8000 3D or Racal Avionics RN 5000 nav and second DME. Optional equipment includes APS-65 digital autopilot, single/dual FCS-65 digital flight director, single/dual BAE Systems Canada CMA-771 Alpha VLF/Omega, MLS, GPWS, flight recorder.

Instrumentation: Dual AHRS-85 digital strapdown AHRS, dual ADI-84, dual EHSI-74, dual RMI-36 and JET standby altitude indicator standard. Optional equipment includes dual EFIS-86 electronic flight instrumentation systems, MFD-85 multifunction display, dual ALT-55 radio altimeters and altitude alerter/preselect.

DIMENSIONS, EXTERNAL:

Wing span	19.78 m (64 ft 10¾ in)
Wing chord: at root	2.81 m (9 ft 2¾ in)
at tip	1.40 m (4 ft 7 in)
Wing aspect ratio	9.9
Length overall	20.07 m (65 ft 10¼ in)
Fuselage: Length	18.73 m (61 ft 5½ in)
Max diameter	2.28 m (7 ft 5¾ in)
Height overall	6.35 m (20 ft 10 in)
Elevator span	6.94 m (22 ft 9¼ in)
Wheel track (c/l of shock-struts)	6.58 m (21 ft 7 in)
Wheelbase	6.97 m (22 ft 10½ in)
Propeller diameter	3.20 m (10 ft 6 in)
Propeller ground clearance (min)	0.52 m (1 ft 8½ in)
Passenger door (fwd, port): Height	1.70 m (5 ft 7 in)
Width	0.77 m (2 ft 6¼ in)
Height to sill	1.47 m (4 ft 10 in)
Cargo door (rear, port): Height	1.36 m (4 ft 5½ in)
Width	1.30 m (4 ft 3¼ in)
Height to sill	1.67 m (5 ft 5¾ in)
Emergency exit (rear, stbd): Height	1.37 m (4 ft 6 in)
Width	0.51 m (1 ft 8 in)
Height to sill	1.56 m (5 ft 1½ in)
Emergency exits (overwing, each):	
Height	0.91 m (3 ft 0 in)
Width	0.51 m (1 ft 8 in)
Emergency exits (flight deck side windows, each):	
Min height	0.48 m (1 ft 7 in)
Min width	0.51 m (1 ft 8 in)

DIMENSIONS, INTERNAL:

Cabin, excl flight deck and baggage compartment:	
Length	9.38 m (30 ft 9¼ in)
Max width	2.10 m (6 ft 10¾ in)
Max height	1.76 m (5 ft 9¼ in)
Floor area	15.0 m² (161 sq ft)
Volume	27.4 m² (968 cu ft)
Rear baggage compartment volume:	
30-passenger version	6.4 m² (226 cu ft)
all-cargo version	2.7 m² (95 cu ft)
passenger/cargo version	11.0 m² (388 cu ft)
Cabin, incl flight deck and baggage compartment:	
total volume	approx 41.8 m³ (1,476 cu ft)
Max available cabin volume (all-cargo version)	31.1 m³ (1,098 cu ft)

AREAS:

Wings, gross	39.43 m² (424.4 sq ft)
Ailerons (total)	2.88 m² (31.00 sq ft)
Trailing-edge flaps (total)	3.23 m² (34.77 sq ft)
Fin, incl dorsal fin	5.74 m² (61.78 sq ft)
Rudder	2.59 m² (27.88 sq ft)
Tailplane	6.10 m² (65.66 sq ft)
Elevators, (total, incl tabs)	3.90 m² (41.98 sq ft)

WEIGHTS AND LOADINGS:

Weight empty, equipped	7,150 kg (15,763 lb)*
Operating weight empty	7,560 kg (16,667 lb)
Max usable fuel	2,660 kg (5,864 lb)
Max payload	3,340 kg (7,363 lb)*
Max T-O weight	11,990 kg (26,433 lb)
Max ramp weight	12,080 kg (26,632 lb)
Max landing weight	11,700 kg (25,794 lb)
Max zero-fuel weight	10,900 kg (24,030 lb)
Max wing loading	304.1 kg/m² (62.28 lb/sq ft)
Max power loading	4.47 kg/kW (7.34 lb/shp)

** 4 kg (8.8 lb) payload decrease/empty weight increase with PW118A engines*

PERFORMANCE:

Max operating speed (VMO)	272 kt (504 km/h; 313 mph) EAS
Max level speed at 6,100 m (20,000 ft)	327 kt (606 km/h; 377 mph)
Max cruising speed at 7,620 m (25,000 ft)	
PW118	300 kt (555 km/h; 345 mph)
PW118A	313 kt (580 km/h; 360 mph)
Long-range cruising speed at 7,620 m (25,000 ft)	270 kt (500 km/h; 311 mph)
Stalling speed, power off:	
flaps up	120 kt (223 km/h; 138 mph) CAS
flaps down	89 kt (165 km/h; 103 mph) CAS
Max rate of climb at S/L	762 m (2,500 ft)/min
Rate of climb at S/L, OEI	168 m (550 ft)/min
Service ceiling: PW118	8,840 m (29,000 ft)
PW118A	9,750 m (32,000 ft)
Service ceiling, OEI, AUW of 11,200 kg (24,690 lb):	
PW118 or PW118A	4,600 m (15,100 ft)
FAR Pt 25 T-O field length	1,550 m (5,085 ft)
FAR Pt 135 landing field length, MLW at S/L	1,390 m (4,560 ft)

Range at 9,150 m (30,000 ft), reserves for 100 n mile (185 km; 115 mile) diversion and 45 min hold:

with max (30-passenger) payload (2,721 kg; 6,000 lb):

PW118	850 n miles (1,575 km; 979 miles)
PW118A	800 n miles (1,482 km; 921 miles)

with max fuel:

PW118, 20 passengers (1,785 kg; 3,935 lb payload) 1,629 n miles (3,017 km; 1,874 miles)

PW118A, 20 passengers (1,785 kg; 3,935 lb payload) 1,570 n miles (2,908 km; 1,806 miles)

OPERATIONAL NOISE LEVELS (FAR Pt 36, BCAR-N and ICAO Annex 16):

T-O	81.2 EPNdB
Approach	92.3 EPNdB
Sideline	83.5 EPNdB

NEW ENTRY

NEIVA EMB-202 IPANEMA

TYPE: Agricultural sprayer.

PROGRAMME: Embraer design; first flight (EMB-200 prototype PP-ZIP) 31 July 1970; certified 14 December 1971; see 1977-78 and earlier editions for EMB-200/200A (49/24 built), EMB-201 (200 built) and EMB-201R (three built); and 1992-93 edition for EMB-201A (402 built). Manufacture transferred to Neiva in March 1980; many detail improvements (listed in 1991-92 and earlier *Jane's*) introduced late 1988. Available from 2002 onwards as airframe only, giving option of customer engine installation.

On 10 October 2002, Embraer revealed the prototype (PP-XHQ) of an alcohol-fuelled (Lycoming) Ipanema.

CURRENT VERSIONS: **EMB-202 Ipanema:** Current version from 1992; generally similar to EMB-201A except for increased wing span, larger hopper and choice of Lycoming or Continental engine. First delivery 2 October 1992.

CUSTOMERS: 678 of earlier versions built (see Programme). Total of 850 built by October 2002; registration allocated to 872nd in November 2002. Total 15 produced in 2000

Prototype of alcohol-fuelled Neiva EMB-202 Ipanema

NEW/0527116

(latest information). In August 2001, five ordered by Dominican Republic. Production of alcohol-fuelled version to be 40 per year.

COSTS: US$200,000; alcohol fuel retrofit US$20,000 (both 2002).

DESIGN FEATURES: Designed to Brazilian Ministry of Agriculture specifications. Typical agricultural aircraft features, including hopper ahead of pilot and high cockpit for good view over nose; low-mounted unbraced wings, with cambered leading-edges (detachable) and tips; rectangular-section fuselage; slightly sweptback vertical tail.

Wing section NACA 23015 (modified); incidence 3°; dihedral 7° from roots.

FLYING CONTROLS: Conventional and manual. Frise ailerons; trim tab in starboard elevator, ground-adjustable tab on rudder. Slotted flaps on wing trailing-edges.

STRUCTURE: All-metal safe-life fuselage frame of welded 4130 steel tube, with removable skin panels of 2024 aluminium alloy and glass fibre and specially treated against chemical corrosion. All-metal wings (single-spar) and tail surfaces (two-spar).

LANDING GEAR: Non-retractable mainwheels and tailwheel, with rubber shock-absorbers in main units. Tailwheel has tapered spring shock-absorber. Mainwheels and tyres size 8.50-10. Tailwheel diameter 250 mm (10 in). Tyre pressures: main, 2.07 to 2.41 bar (30 to 35 lb/sq in); tailwheel, 3.79 bar (55 lb/sq in). Hydraulic disc brakes on mainwheels.

POWER PLANT: One 224 kW (300 hp) Textron Lycoming IO-540-K1J5D flat-six engine, driving a Hartzell two-blade (optionally three-blade) constant-speed metal propeller. Optional engine is Teledyne Continental 224 kW (300 hp) IO-550-D with McCauley two-blade constant-speed propeller. Integral fuel tanks in each wing leading-edge, with total capacity of 292 litres (77.1 US gallons; 64.2 Imp gallons), of which 264 litres (69.75 US gallons; 58.1 Imp gallons) are usable. Refuelling point on top of each tank. Oil capacity 12 litres (3.2 US gallons; 2.6 Imp gallons).

ACCOMMODATION: Single horizontally/vertically adjustable seat in fully enclosed cabin with bottom-hinged, triple-lock, jettisonable window/door on each side and two overhead windows. Ventilation airscoop at top of canopy front edge; second airscoop on fin leading-edge pressurises interior of rear fuselage. Inertial shoulder harness standard.

SYSTEMS: 28 V DC electrical system supplied by 24 Ah batteries and a Bosch KI 28 V 35 A alternator. Power receptacle for external battery (AN-2552-3A type) on port side of forward fuselage.

AVIONICS: *Comms:* Optional portable VHF transceiver.
Flight: Optional portable Garmin GPS.

EQUIPMENT: Hopper for agricultural chemicals, capacity 950 litres (251 US gallons; 209 Imp gallons) liquid or 750 kg (1,653 lb) dry. Dusting system below centre of fuselage. Spraybooms or Micronair atomisers aft of or above wing trailing-edges respectively. Options include ram air pressure generator for use with liquid spray system; improved lightweight spraybooms; smaller/lighter Micronair AU5000 rotary atomisers; trapezoidal spreader with adjustable inlet to improve application of dry chemicals; and, from 2002, electrostatic spray system.

DIMENSIONS, EXTERNAL:

Wing span	11.69 m (38 ft 4¼ in)
Wing chord, constant	1.71 m (5 ft 7½ in)
Wing aspect ratio	6.9
Length overall (tail up)	7.43 m (24 ft 4½ in)
Height overall (tail down)	2.20 m (7 ft 2½ in)
Fuselage: Max width	0.93 m (3 ft 0½ in)
Tailplane span	3.66 m (12 ft 0 in)
Wheel track	2.20 m (7 ft 2½ in)
Wheelbase	5.20 m (17 ft 7¼ in)
Propeller diameter:	
Hartzell two- and three-blade	2.13 m (7 ft 0 in)
McCauley two-blade	2.18 m (7 ft 2 in)

DIMENSIONS, INTERNAL:

Cockpit: Length	1.20 m (3 ft 11¼ in)
Max width	0.85 m (2 ft 9½ in)
Max height	1.34 m (4 ft 4¾ in)

AREAS:

Wings, gross	19.94 m² (214.6 sq ft)
Ailerons (total)	1.60 m² (17.22 sq ft)
Trailing-edge flaps (total)	2.30 m² (24.76 sq ft)
Fin	0.58 m² (6.24 sq ft)
Rudder	0.63 m² (6.78 sq ft)
Tailplane	3.17 m² (34.12 sq ft)
Elevators (total, incl tab)	1.50 m² (16.15 sq ft)

WEIGHTS AND LOADINGS (N: Normal, R: Restricted category):

Weight empty: N, R	1,020 kg (2,249 lb)
Max payload: N, R	750 kg (1,653 lb)
Max T-O and landing weight: N	1,550 kg (3,417 lb)
R	1,800 kg (3,968 lb)

Max wing loading: N		77.7 kg/m² (15.92 lb/sq ft)
R		90.3 kg/m² (18.49 lb/sq ft)
Max power loading: N		6.92 kg/kW (11.39 lb/hp)
R		8.03 kg/kW (13.23 lb/hp)

PERFORMANCE (clean):

Never-exceed speed (VNE):		
N		147 kt (272 km/h; 169 mph)
Max level speed at S/L:		
N		124 kt (230 km/h; 143 mph)
R		121 kt (225 km/h; 140 mph)
Max cruising speed at 75% power at 1,830 m (6,000 ft):		
N		115 kt (213 km/h; 132 mph)
R		111 kt (206 km/h; 128 mph)
Stalling speed, power off (N):		
flaps up		56 kt (103 km/h; 64 mph)
8° flap		54 kt (100 km/h; 62 mph)
30° flap		50 kt (92 km/h; 57 mph)
Stalling speed, power off (R):		
flaps up		60 kt (110 km/h; 68 mph)
8° flap		58 kt (107 km/h; 66 mph)
30° flap		53 kt (99 km/h; 61 mph)
Max rate of climb at S/L, 8° flap:		
N		283 m (930 ft)/min
R		201 m (660 ft)/min
Service ceiling, 8° flap: R		3,470 m (11,380 ft)
T-O run at S/L, asphalt runway:		
N		200 m (655 ft)
R		354 m (1,160 ft)
T-O to 15 m (50 ft), conditions as above:		
N		332 m (1,090 ft)
R		564 m (1,850 ft)
Landing from 15 m (50 ft) at S/L, 30° flap, asphalt		
runway: N		412 m (1,355 ft)
R		476 m (1,565 ft)
Landing run, conditions as above: N		153 m (505 ft)
R		170 m (560 ft)
Range at 1,830 m (6,000 ft), no reserves:		
N		506 n miles (938 km; 583 miles)
R		474 n miles (878 km; 545 miles)

UPDATED

NEIVA EMB-810D

TYPE: Six-seat utility twin.

PROGRAMME: Licence-built Piper PA-34 Seneca IV, which see in the US section of the 1996-97 *Jane's* for a detailed description; approximate data (Seneca V) in this edition.

CUSTOMERS: Some 876 produced in Brazil by late 2000, including two in that year; none further by January 2002.

UPDATED

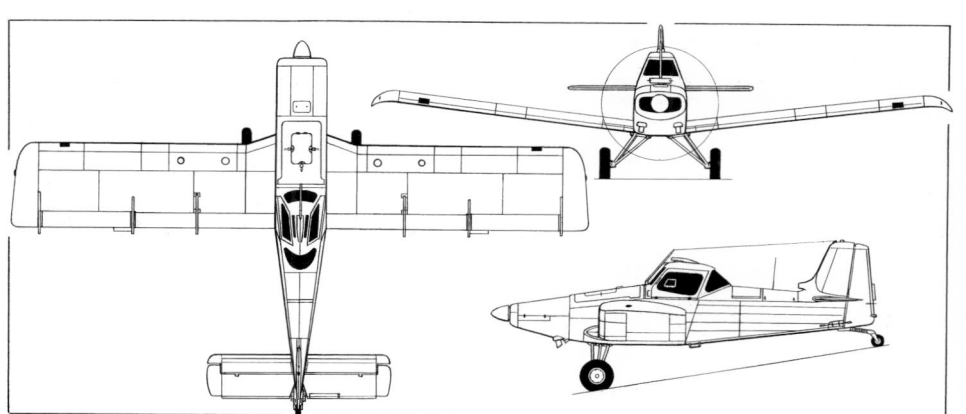

Neiva EMB-202 Ipanema (one Textron Lycoming IO-540) *(Jane's/Dennis Punnett)*

Neiva-built EMB-810

BULGARIA

DAR

DAR AIRCRAFT LTD
Mladost 103-A-49, Sofia 1797
Tel/Fax: (+35 927) 571 65
e-mail: dar-info@dir.bg
Web: http://dar.dir.bg
MANAGER AND CO-OWNER: Tony Ilief
CHIEF DESIGNER AND CO-OWNER: Vesselin Valkanov

Two types of ultralight are offered by this new company.

NEW ENTRY

DAR-21 VECTOR
TYPE: Tandem-seat ultralight.
PROGRAMME: Designed by Vesselin Valkanov. Prototype (LZ-DAR) first flew 20 May 2000, initially powered by a 39.0 kW (52.3 hp) Hirth 2704.
COSTS: US$50,000 (agricultural version, 2002).
DESIGN FEATURES: Conventional high-wing monoplane of angular appearance; constant-chord wing and tailplane, both strut-braced; sweptback fin; good all-round view from cockpit. Horizontal wingtip vanes and leading-edge slots outboard.
FLYING CONTROLS: Conventional and manual. Flaps (0, 15 and 30°). Trim tabs on starboard aileron and rudder; flight-adjustable (electric) tab on single-piece elevator. Aileron and rudder actuation by cables; elevator by pushrods. No aerodynamic balances.
STRUCTURE: Generally of 1050, 2024, 3130 and 6164 aluminium alloy sheet and type with milled and turned steel elements in landing gear. Composites engine cowling. Single-spar wing.
LANDING GEAR: Tailwheel type; fixed. Bungee shock-absorbers. Mainwheel size 6.00 standard; 8.00 optional. Steerable tailwheel.
POWER PLANT: One 47.8 kW (64.1 hp) Rotax 582 UL two-stroke with 2.58:1 reduction gear driving a Powerfin three-blade, composites propeller. Fuel tank in each wing and fuselage, each 30 litres (7.9 US gallons; 6.6 Imp gallons); total fuel 90 litres (23.8 US gallons; 19.8 Imp gallons).
ACCOMMODATION: Two pilots in tandem in enclosed, extensively glazed cockpit, with single door, starboard, to seats and baggage area to rear.
EQUIPMENT: Optional Spray Miser spraying bars mounted at mainwheel leg roots.

DIMENSIONS, EXTERNAL:
Wing span	8.00 m (26 ft 3 in)
Length overall	5.75 m (18 ft 10½ in)
Height overall	1.80 m (5 ft 10¾ in)

DIMENSIONS, INTERNAL:
Cockpit max width	0.63 m (2 ft 0¾ in)

AREAS:
Wings, gross	10.40 m² (111.9 sq ft)

WEIGHTS AND LOADINGS:
Weight empty	230 kg (507 lb)
Max T-O weight	450 kg (992 lb)

PERFORMANCE:
Max level speed	76 kt (140 km/h; 87 mph)
Max cruising speed	67 kt (125 km/h; 78 mph)
Stalling speed, flaps down	33 kt (60 km/h; 38 mph)
Endurance	9 h

NEW ENTRY

Prototype DAR-21 Vector   *NEW*/0527113

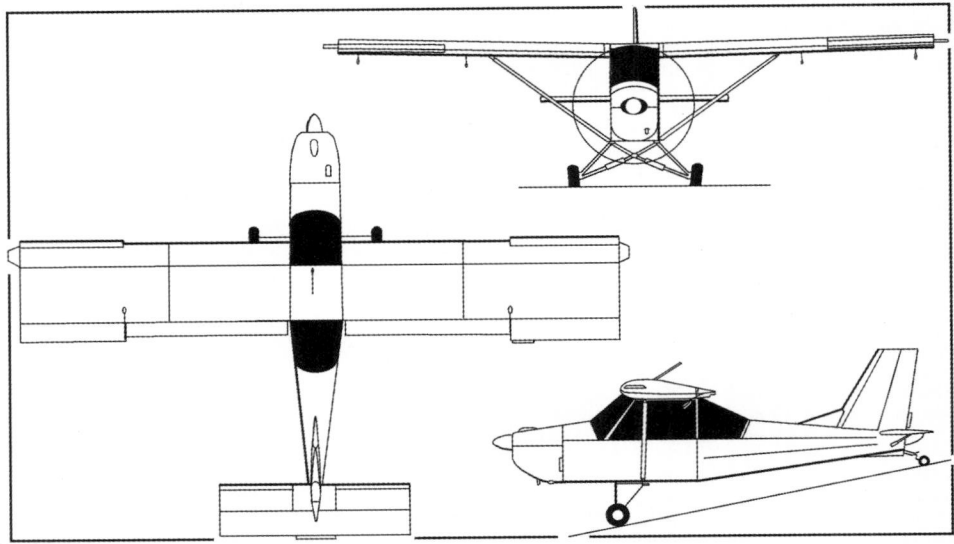
DAR-21 Vector tandem-seat ultralight (*Jane's/Paul Jackson*)   *NEW*/0527119

DAR-23 ACCENT
TYPE: Side-by-side ultralight.
PROGRAMME: Prototype (LZ-DAS) flying by 2002.

DESIGN FEATURES: Cabin pod underslung from boom carrying uncowled engine, wings and empennage. Wings braced to cabin; cabin braced to boom ahead of windscreen.
FLYING CONTROLS: Conventional and manual. Plain flaps. Flight-adjustable elevator tab. No aerodynamic balances. Ailerons, elevator and flaps actuated by pushrods; rudder by cables.
STRUCTURE: Metal boom and composites cabin. Metal wings fabric-covered.
LANDING GEAR: Tricycle type; fixed. Tubular steel legs. Mainwheels Azusa 6.00. Cable mainwheel brakes.
POWER PLANT: One 34.0 kW (45.6 hp) Rotax 503 UL two-stroke driving a Powerfin two-blade, composites propeller via 2.58:1 reduction gear. Fuel capacity 30 litres (7.9 US gallons; 6.6 Imp gallons).

DIMENSIONS, EXTERNAL:
Wing span	8.00 m (26 ft 3 in)
Length overall	5.00 m (16 ft 4¾ in)
Height at wing	1.80 m (5 ft 10¾ in)

DIMENSIONS, INTERNAL:
Cockpit max width	0.90 m (2 ft 11½ in)

AREAS:
Wings, gross	10.40 m² (111.9 sq ft)

WEIGHTS AND LOADINGS:
Weight empty	180 kg (397 lb)
Max T-O weight	380 kg (837 lb)

PERFORMANCE:
Max level speed	67 kt (125 km/h; 78 mph)
Max cruising speed	51 kt (95 km/h; 59 mph)
Stalling speed, flaps down	27 kt (50 km/h; 32 mph)
Endurance	3 h

NEW ENTRY

Prototype DAR-23 Accent   *NEW*/0527114

CANADA

AIRSTART

AIRSTART CORPORATION

Hangar 11, YXD 11760 109th Street, Edmonton, Alberta
T5G 2T8
Tel: (+1 780) 944 92 10
Fax: (+1 780) 461 05 84
e-mail: eric.robinson@business-aid.net
Web: http://www.airstart1.com
POSTAL ADDRESS: PO Box 34013, Kingsway Garden Mall,
Edmonton, Alberta T5G 3G4
VICE-PRESIDENT, MANUFACTURING: Eric A Robinson
VICE-PRESIDENT, SALES AND MARKETING:
William (Bill) Crampton

Following refinancing, the former Canada Air RV company
was reorganised. Manufacturing is now undertaken by AC
Millennium Corporation, and Sales and Marketing by
AirStart Corporation. Component supply is handled by AN
Hardware. The Griffin is at present available only in North
America; research and development continues under the
aegis of ARV Development Corporation.

UPDATED

AIRSTART GRIFFIN

TYPE: Side-by-side kitbuilt.
PROGRAMME: Design work started January 1989; construction
of first prototype Griffin I began September 1989; first
flight 1993. Construction of Mark II began 1995 with first
flight that year; Mark III first flown 1997 followed by Mark
IV first flight October 1998. First kit delivered 4 January
1996.
CURRENT VERSIONS: **Griffin III:** Lower-powered version.
   Griffin IV: Baseline version, *as described.*
CUSTOMERS: 72 ordered and eight flying by mid-2001
including three in Canada and two in USA.
COSTS: Kit C$28,000 excluding engine (2001). Estimated
design/programme cost C$3.5 million to date.
DESIGN FEATURES: High-wing monoplane. Good all-round
visibility, including roof windows, enhanced by
instrument panel positioned low in cockpit.
FLYING CONTROLS: Manual. Full-span continuously
positionable flaperons deflect to 20° and also reflex slightly
for cruise drag reduction. Electrically operated full-span
trim on port elevator.
STRUCTURE: All-aluminium semi-monocoque fuselage and
wings; elevator and rudder are fabric-covered. Single-strut
laminar flow high-aspect ratio wing; aerofoil section UA2-
180, thickness/chord ratio 19 per cent. Wing dihedral 1°
30′. Low-mounted tailplane.
LANDING GEAR: Choice of tricycle or tailwheel layout. Sprung
aluminium mainwheel legs. Tricycle model has oleo-
pneumatic nosewheel suspension. Tailwheel version has
steerable Matco 11 cm (4½ in) tailwheel. Matco disc
brakes. Mainwheels and nosewheel Matco 6.00-6, all tyres
pressurised to 2.9 bar (42 lb/sq in).
POWER PLANT: *Griffin IV:* One 119 kW (160 hp) Textron
Lycoming O-320 flat-four engine, driving a two-
blade fixed-pitch wooden propeller in demonstrator;
engines in range 74.6 to 119 kW (100 to 160 hp) can be
fitted.
   Griffin III: 74.6 kW (100 hp) CAM 100 piston engine.
   Fuel capacity 140 litres (37.0 US gallons; 30.8 Imp
gallons) in two wing tanks, of which 138 litres (36.5 US
gallons; 30.4 Imp gallons) usable. Oil capacity 8 litres (2.1
US gallons; 1.8 Imp gallons).
ACCOMMODATION: Pilot and passenger in side-by-side seating
with dual controls. Upward-opening door on each side of
fuselage. Baggage stowage behind seats.

DIMENSIONS, EXTERNAL:
Wing span	10.82 m (35 ft 6 in)
Wing chord: at root	1.52 m (5 ft 0 in)
at tip	0.76 m (2 ft 6 in)
Wing aspect ratio	11.0
Length overall	6.25 m (20 ft 6 in)
Fuselage max width	1.09 m (3 ft 7 in)
Height overall: tricycle	2.44 m (8 ft 0 in)
Tailplane span	2.54 m (8 ft 4 in)
Wheel track	2.44 m (8 ft 0 in)
Wheelbase: tricycle	2.44 m (8 ft 0 in)
Propeller diameter	1.83 m (6 ft 0 in)
Propeller ground clearance: tricycle	0.76 m (2 ft 6 in)
Passenger doors (each): Height	0.76 m (2 ft 6 in)
Width	0.76 m (2 ft 6 in)

DIMENSIONS, INTERNAL:
Cabin: Length	1.07 m (3 ft 6 in)
Max width	1.05 m (3 ft 5¼ in)
Max height	1.07 m (3 ft 6 in)
Floor area	1.11 m² (12.0 sq ft)
Baggage hold volume	0.51 m³ (18.0 cu ft)

AREAS:
Wings, gross	10.68 m² (115.0 sq ft)
Rudder	0.55 m² (5.90 sq ft)
Tailplane	1.49 m² (16.00 sq ft)
Elevator	0.79 m² (8.49 sq ft)

WEIGHTS AND LOADINGS (Griffin IV with 119 kW; 160 hp
engine, Griffin III with 74.6 kW; 100 hp engine):
Weight empty	381 kg (840 lb)
Baggage capacity	91 kg (200 lb)
Max T-O and landing weight:	
Griffin IV	787 kg (1,735 lb)
Griffin III	680 kg (1,500 lb)
Max zero-fuel weight	682 kg (1,503 lb)
Max wing loading:	
Griffin IV	73.7 kg/m² (15.09 lb/sq ft)
Griffin III	63.7 kg/m² (13.04 lb/sq ft)
Max power loading:	
Griffin IV	6.60 kg/kW (10.84 lb/hp)
Griffin III	9.13 kg/kW (15.00 lb/hp)

PERFORMANCE (Griffin IV with 119 kW; 160 hp engine;
Griffin III with 74.6 kW; 100 hp engine):
Never-exceed speed (VNE):	
Griffin IV	150 kt (277 km/h; 172 mph)
Max operating speed (VMO):	
Griffin IV	141 kt (261 km/h; 162 mph)
Griffin III	115 kt (214 km/h; 133 mph)
Max cruising speed at 75% power:	
Griffin IV	132 kt (244 km/h; 152 mph)
Griffin III	105 kt (195 km/h; 121 mph)
Econ cruising speed:	
Griffin IV	114 kt (211 km/h; 131 mph)
Stalling speed: both	40 kt (73 km/h; 45 mph)
Max rate of climb at S/L	259 m (850 ft)/min
Service ceiling	5,485 m (18,000 ft)
T-O run	107 m (350 ft)
Landing run	137 m (450 ft)
Range with max fuel	828 n miles (1,533 km; 952 miles)

VERIFIED

AWT

ADVANCED WING TECHNOLOGIES CORPORATION

1080-885 West Georgia Street, Vancouver, British Columbia
V6C 3E8
Tel: (+1 604) 685 92 66
Fax: (+1 604) 685 65 01
e-mail: tgill@advancedwing.com
Web: http://www.advancedwing.com
PRESIDENT AND CEO: Sacha S Gill
DIRECTOR OF SALES AND MARKETING: Gerardo Martinez
DIRECTOR OF CORPORATE RELATIONS: Tej S Gill

WORKS: 227-7080 River Road, Richmond, British Columbia,
V6X 1X5
Tel: (+1 604) 244 85 12
Fax: (+1 604) 244 85 13
e-mail: sales@advancedwing.com
MANAGER: Manny Rodrigues
SALES MANAGER: Dave Crerar

Company is division of AWT Holdings Ltd; founded in 1985
by John Hill and Bill Foyle to develop and manufacture an
advanced design of wing for retrofit to DHC-2 Beaver
bushplane (see *Jane's Aircraft Upgrades*). Present structure
of company dates from 1994, when investment for three-
stage expansion secured from Grosvenor Square Business
Capital Inc. In May 2000, AWT began proceedings for the
acquisition of tooling company ADC Engineering
Technology Inc of Carson, California.

By 2001, six Beaver wings had been delivered and further
12 were on order. AWT also overhauls and modifies Beavers
at its 2,415 m² (26,000 sq ft) plant at Richmond, offering
floats of greater capacity to take advantage of the new Beaver
wing; a fuselage stretch; larger cargo door; and an engine
upgrade. Aircraft of this type fitted with the new wing can
also accommodate an additional seat behind existing seats.
All products are certified by Transport Canada and the FAA.
Total of 24 staff were employed in 2001.

Second phase of AWT's expansion plan was to have
involved Canadian assembly of El Gavilán EL-1 Gavilán
bushplane (which see in Colombian section), of which
company acquired Type Certificate. By 2001, however,
promotion had switched to similar Australian Gippsland
GA-8 Airvan (which see) for which initial order for six
placed in same year, with deliveries due to begin in 2002.

Third phase will be design and production of **CA-21
Airwolf,** a 3,628 kg (8,000 lb) MTOW utility aircraft,
combining a stretched Airvan fuselage, the AWT advanced
wing and a turboprop engine.

VERIFIED

BELL

BELL HELICOPTER TEXTRON CANADA (Division of Textron Canada Ltd)

12800 Rue de l'Avenir, Mirabel, Québec J7J 1R4
Tel: (+1 514) 437 34 00
Fax: (+1 514) 437 60 10
e-mail: bhtchr@bellhelicopter.textron.com
PRESIDENT: Paul Costanzo

Memorandum of Understanding to start helicopter industry in
Canada signed 7 October 1983; 38,900 m² (418,725 sq ft)
factory opened late 1985 and employs 1,700 people; US civil
production of 206B JetRanger transferred to Canada in 1986,
206L LongRanger by early 1987; then 212 in 1988 and 412 in
1989. The 230 was introduced and certified in Canada in 1992,
but has been replaced by the 430. Newest products are the 407
and 427. About half of each helicopter made in Canada
(dynamic systems supplied by Bell Fort Worth). Total of 233
commercial helicopters delivered in 1997, followed by 197 in
1998, 146 in 1999, 145 in 2000, and 122 in 2001.

Bell helicopter 'families' do not follow a chrono-
numerical pattern; accordingly, entries in this section are
grouped in the following order:
   Bell 206, 407 and 427
   Bell 412
   Bell 430

UPDATED

BELL 206B-3 JETRANGER III
US Army designation: TH-67 Creek

TYPE: Light utility helicopter.
PROGRAMME: Model 206 originally offered (unsuccessfully)
to US Army as OH-4, having first flown on 8 December
1962. Civil version flew 10 January 1966, after extensive
redesign, and entered service as 206A JetRanger; followed
by 206A-1 and 206B JetRanger II. Delivery of current
206B-3 JetRanger III began Summer 1977; transferred to
Mirabel, Canada, 1986. Also built in Italy by Agusta.
Upgraded version, under consideration in 1999, would
feature improved engine, transmission and dynamic
components, with target selling price under US$900,000.
However, by late 2001, Bell was weighing possibility of
closing JetRanger production line.
CURRENT VERSIONS: **206B-3 JetRanger III:** Current civil
production version.
Following description refers to JetRanger III.
   TH-67 Creek (Bell designation TH-206): Selected
March 1993 as US Army NTH (New Training Helicopter)
choice to replace UH-1 at pilot training school, Fort
Rucker, Alabama (223 Aviation Regiment at Cairns AAF
and 212 Aviation Regiment at Lowe AHP). Instructor and
pupil in front seats; plans abandoned for second pupil to be
seated at the rear, observing flight instruments by closed-
circuit TV screen mounted on back of right-hand front
seat. Powered by Rolls-Royce (formerly Allison) 250-
C20JN engine. First batch included nine cockpit
procedures trainers outfitted by Frasca International;
three configurations: VFR, IFR, and VFR with IFR
provision.

   TH-67 features also include dual controls, crashworthy
seats, five-point seat restraints, heavy-duty battery, particle
separator, bleed air heater, heavy-duty skid shoes and
enlarged instrument panel. IFR version additionally
possesses force trim system, auxiliary electrical system
and is FAA certified for dual pilot operation. Also supplied
to Taiwan Army, based at Gwe-Ren.
CUSTOMERS: TH-206 (TH-67 Creek) declared winner of US
Army NTH competition March 1993; initial 102 ordered in
IFR configuration; second batch of 35 VFR helicopters
ordered February 1994; deliveries began 15 October 1993
with N67001 and N67014 (all TH-67s have civilian
registrations); 45 aircraft (and six procedures trainers)
delivered in time for first training course to open 5 May
1994; initial orders 137 TH-67s and nine cockpit
procedures trainers; all delivered by late 1995. Further
orders placed on behalf of Taiwanese Army, which
received 30 TH-67s in 1997-99. Deliveries in 1999
included six to the Bulgarian Air Force. Further 35 ordered
for US Army for delivery between September 2001 and
May 2002.

   Over 7,740 JetRangers produced by Bell and licensees,
including 4,400 Bell 206Bs and 2,200 military OH-58
series; and 900 in Italy. Bell Canada delivered 28 206Bs in
1999; and 14 in both 2000 and 2001.

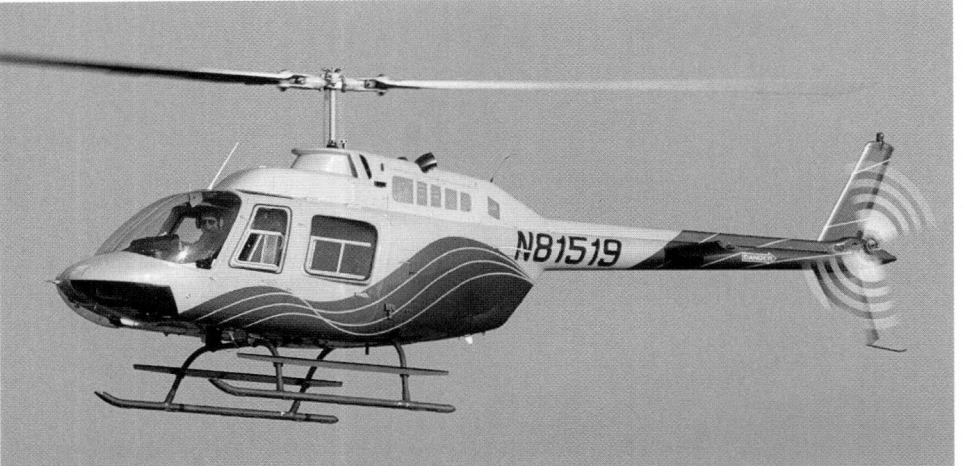

Bell 206B JetRanger III (Rolls-Royce 250 turboshaft)   *NEW*/0132967

COSTS: US Army initial NTH contract US$84.9 million. Civilian 206B-3, typically equipped, US$730,000 with C20J engine or US$765,000 with C20R (1999). Average direct operating cost US$191 per hour (1997).

DESIGN FEATURES: Typical light utility helicopter with skid landing gear, high-mounted tailboom and horizontal stabiliser. Two-blade teetering main rotor with preconed and underslung bearings; blades retained by grip, pitch change bearing and torsion-tension strap assembly; two-blade tail rotor; main rotor rpm 374 to 394.

FLYING CONTROLS: Hydraulic fully powered cyclic and collective controls and foot-powered tail rotor control; tailplane with highly cambered inverted aerofoil section and stall strip produces appropriate nose-up and nose-down attitude during climb and descent; optional autostabiliser, autopilot and IFR systems.

STRUCTURE: Conventional light alloy structure with two floor beams and bonded honeycomb sandwich floor; transmission mounted on two beams and deck joined to floor by three fuselage frames; main rotor blades have extruded aluminium D-section leading-edge with honeycomb core behind, covered by bonded skin; tail rotor blades have bonded skin without honeycomb core.

LANDING GEAR: Aluminium alloy tubular skids bolted to extruded cross-tubes. Tubular steel skid on ventral fin to protect tail rotor in tail-down landing. Special high skid gear (0.25 m; 10 in greater ground clearance) available for use in areas with high brush. Pontoons or stowed floats, capable of in-flight inflation, available as optional kits.

POWER PLANT: One Rolls-Royce 250-C20J turboshaft, rated at 313 kW (420 shp) for T-O; 276 kW (370 shp) max continuous. Optionally, one Rolls-Royce 250-C20R/4 rated at 336 kW (450 shp) for T-O. Transmission rating 236 kW (317 shp) for T-O; 201 kW (270 shp) continuous. Rupture-resistant fuel tank below and behind rear passenger seat, usable capacity 344 litres (91.0 US gallons; 75.75 Imp gallons). Refuelling point on starboard side of fuselage, aft of cabin. Oil capacity 5.68 litres (1.5 US gallons; 1.25 Imp gallons).

ACCOMMODATION: Two seats side by side in front and three-seat rear bench. Dual controls optional. Two forward-hinged doors on each side, made of formed aluminium alloy with transparent panels (bulged on rear pair). Baggage compartment aft of rear seats, capacity 113 kg (250 lb), with external door on port side.

SYSTEMS: Hydraulic system, pressure 41.5 bar (600 lb/sq in), for cyclic, collective and directional controls. Maximum flow rate 7.57 litres (2.0 US gallons; 1.65 Imp gallons)/min. Open reservoir. Electrical supply from 150 A starter/generator. One 28 V 17 Ah Ni/Cd battery; auxiliary 13 Ah battery optional. Optional ECS.

AVIONICS: *Comms:* VHF communications, intercom and speaker system.
Flight: VOR, ADF, DME and R/Nav optional.

EQUIPMENT: Standard equipment includes cabin fire extinguisher, first aid kit, door locks, night lighting, and dynamic flapping restraints. Optional items include clock, engine hour meter, turn and slip indicator, custom seating, internal stretcher kit, rescue hoist, cabin heater, camera access door, high-intensity night lights, engine fire detection system, and external cargo hook of 680 kg (1,500 lb) capacity.

DIMENSIONS, EXTERNAL:
Main rotor diameter	10.16 m (33 ft 4 in)
Main rotor blade chord	0.33 m (1 ft 1 in)
Tail rotor diameter	1.65 m (5 ft 5 in)
Tail rotor blade chord	0.12 m (4¾ in)
Distance between rotor centres	5.96 m (19 ft 6½ in)
Length: overall, rotors turning	11.82 m (38 ft 9½ in)
fuselage, incl tailskid	9.50 m (31 ft 2 in)
Height: over tailfin	2.54 m (8 ft 4 in)
to top of rotor head:	
standard skids	2.89 m (9 ft 6 in)
high skids	3.17 m (10 ft 4¾ in)
emergency floats	3.20 m (10 ft 6 in)
Stabiliser span	1.97 m (6 ft 5¾ in)

Width over landing gear:
standard skids	1.95 m (6 ft 4¾ in)
high skids	2.07 m (6 ft 9½ in)
emergency floats	2.68 m (8 ft 9½ in)
Forward cabin doors (each): Height	1.02 m (3 ft 4 in)
Width	0.61 m (2 ft 0 in)
Rear cabin doors (each): Height	1.02 m (3 ft 4 in)
Width	0.91 m (3 ft 0 in)

DIMENSIONS, INTERNAL:
Cabin: Length of seating area	2.13 m (7 ft 0 in)
Max width	1.27 m (4 ft 2 in)
Max height	1.28 m (4 ft 3 in)
Volume	1.1 m³ (40 cu ft)
Baggage compartment:	
Max width	1.10 m (3 ft 7¼ in)
Max height	0.55 m (1 ft 9½ in)
Max length	0.94 m (3 ft 1¼ in)
Volume	approx 0.45 m³ (16 cu ft)

AREAS:
Main rotor blades (each)	1.68 m² (18.05 sq ft)
Tail rotor blades (each)	0.11 m² (1.18 sq ft)
Main rotor disc	81.07 m² (872.7 sq ft)
Tail rotor disc	2.14 m² (23.05 sq ft)
Stabiliser	0.90 m² (9.65 sq ft)

WEIGHTS AND LOADINGS:
Weight empty: standard civil	772 kg (1,702 lb)
TH-67: VFR	852 kg (1,879 lb)
IFR	911 kg (2,009 lb)
Operating weight, TH-67: VFR	1,369 kg (3,019 lb)
IFR	1,428 kg (3,149 lb)
Max payload: internal	635 kg (1,400 lb)
external	680 kg (1,500 lb)
Max T-O weight: internal load	1,451 kg (3,200 lb)
external load	1,519 kg (3,350 lb)
Max disc loading:	
internal load	17.9 kg/m² (3.67 lb/sq ft)
external load	18.7 kg/m² (3.84 lb/sq ft)
Transmission loading at max T-O weight and power:	
internal load	6.15 kg/kW (10.09 lb/shp)
external load	6.43 kg/kW (10.57 lb/shp)

PERFORMANCE (at internal load max T-O weight, ISA):
Never-exceed speed (VNE) at S/L	122 kt (225 km/h; 140 mph)
Max and econ cruising speed at S/L	115 kt (214 km/h; 133 mph)
Max rate of climb at S/L	390 m (1,280 ft)/min
Vertical rate of climb at S/L	91 m (300 ft)/min
Service ceiling	4,115 m (13,500 ft)
Hovering ceiling: IGE	4,025 m (13,200 ft)
OGE	1,615 m (5,300 ft)

Range with max fuel, no reserves:
at S/L	374 n miles (692 km; 430 miles)
at 1,525 m (5,000 ft)	395 n miles (732 km; 455 miles)
Range, TH-67 VFR with normal 314 litres (83.0 US gallons; 69.0 Imp gallons) fuel	
	327 n miles (605 km; 376 miles)
Endurance	4 h 30 min

OPERATIONAL NOISE LEVELS (FAR Pt 36):
T-O	88.7 EPNdB
Flyover	85.4 EPNdB
Approach	90.6 EPNdB

UPDATED

BELL 206L-4 LONGRANGER IV

TYPE: Light utility helicopter.

PROGRAMME: Stretched JetRanger (see previous entry). LongRanger announced 25 September 1973; first flight 11 September 1974; initial versions were 206L and 206L-1 LongRanger II, of which 790 built; limited production (17) of 206L-2; replaced by 206L-3 LongRanger III in 1982, of which 612 built, production having transferred to Canada in January 1987.

CURRENT VERSIONS: **LongRanger IV:** Announced March 1992 as new current standard model; transmission uprated to absorb 365 kW (490 shp) instead of 324 kW (435 shp) from same engine; gross weight raised from 1,882 kg (4,150 lb) to 2,018 kg (4,450 lb); certified late 1992, delivered from December that year.

TwinRanger: Twin-engined version; model **206LT;** 13 built on LongRanger IV line between 1993 and 1997. No longer available; see 1998-99 *Jane's* for description.

Gemini ST: Twin-engined rebuild of LongRanger III/IV developed by Tridair and Soloy in USA: see *Jane's Aircraft Upgrades.*

CUSTOMERS: Over 1,690 LongRangers produced by late 2001, including 275 LongRanger IVs and 206LTs. Total of 12 delivered in 1999, 27 in 2000 and 10 in 2001.

COSTS: US$1.12 million, typically equipped (1999). Average direct operating cost US$293 per hour (1997).

DESIGN FEATURES: As JetRanger, but cabin length increased to make room for club seating and extra window; Bell Noda-Matic transmission to reduce vibration; vertical stabilisers added to horizontal tail surfaces. Improvements introduced on LongRanger II include new freewheel unit, modified shafting and increased-thrust tail rotor; main rotor rpm 394; rotor brake optional.

FLYING CONTROLS: As JetRanger, but with endplate fins on tailplane; single-pilot IFR with Rockwell Collins AP-107H autopilot; optional SFENA autopilot with stabilisation and holds for heading, height and approach.

STRUCTURE: As JetRanger.

LANDING GEAR: As JetRanger.

POWER PLANT: One 485 kW (650 shp) Rolls-Royce 250-C30P turboshaft (maximum continuous rating 415 kW; 557 shp). Transmission rated at 365 kW (490 shp) for take-off, with a continuous rating of 276 kW (370 shp); 340 kW (456 shp) transmission optional. Rupture-resistant fuel system, comprising three interconnected cells, total usable capacity 419 litres (111 US gallons; 92.0 Imp gallons).

ACCOMMODATION: Redesigned rear cabin, more spacious than JetRanger. With a crew of two, standard cabin layout accommodates five passengers in two canted rearward-facing seats and three forward-facing seats. Optional DeLuxe and Custom DeLuxe interiors with fabric, fabric/vinyl, all vinyl or fabric/leather seats. Port forward passenger seat has folding back to allow loading of a 2.44 × 0.91 × 0.30 m (8 × 3 × 1 ft) container, making possible carriage of such items as survey equipment, skis and other long components. Double doors on port side of cabin provide opening 1.55 m (5 ft 1 in) wide, for straight-in loading of stretcher patients or utility cargo; in ambulance or rescue role two stretcher patients and two ambulatory patients/attendants can be carried. Dual controls optional.

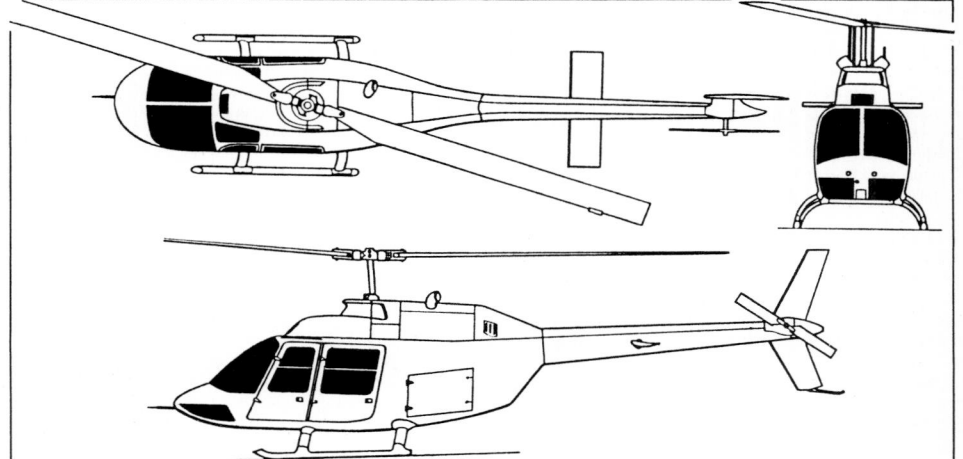

Bell 206B-3 JetRanger five-seat helicopter (*Jane's/Paul Jackson*)

Max T-O weight: normal	2,018 kg (4,450 lb)
external load	2,064 kg (4,550 lb)
Max disc loading: normal	20.7 kg/m² (4.23 lb/sq ft)
external load	19.3 kg/m² (3.95 lb/sq ft)
Transmission loading at max T-O weight and power:	
internal load	5.52 kg/kW (9.08 lb/shp)
external load	5.65 kg/kW (9.29 lb/shp)

PERFORMANCE (at max normal T-O weight, ISA):

Never-exceed speed (VNE):	
at S/L	130 kt (241 km/h; 150 mph)
at 1,525 m (5,000 ft)	133 kt (246 km/h; 153 mph)
Max cruising speed: at S/L	110 kt (204 km/h; 127 mph)
at 1,525 m (5,000 ft)	111 kt (206 km/h; 128 mph)
Max rate of climb at S/L	408 m (1,340 ft)/min
Service ceiling at max cruise power	3,050 m (10,000 ft)
Hovering ceiling: IGE	3,050 m (10,000 ft)
OGE	1,980 m (6,500 ft)
Range with max fuel, no reserves:	
at S/L	324 n miles (600 km; 372 miles)
at 1,525 m (5,000 ft)	357 n miles (661 km; 411 miles)
Endurance	3 h 42 min

OPERATIONAL NOISE LEVELS (FAR Pt 36):

T-O	88.4 EPNdB
Flyover	85.2 EPNdB
Approach	90.7 EPNdB

UPDATED

Bell 206L LongRanger   *NEW*/0132966

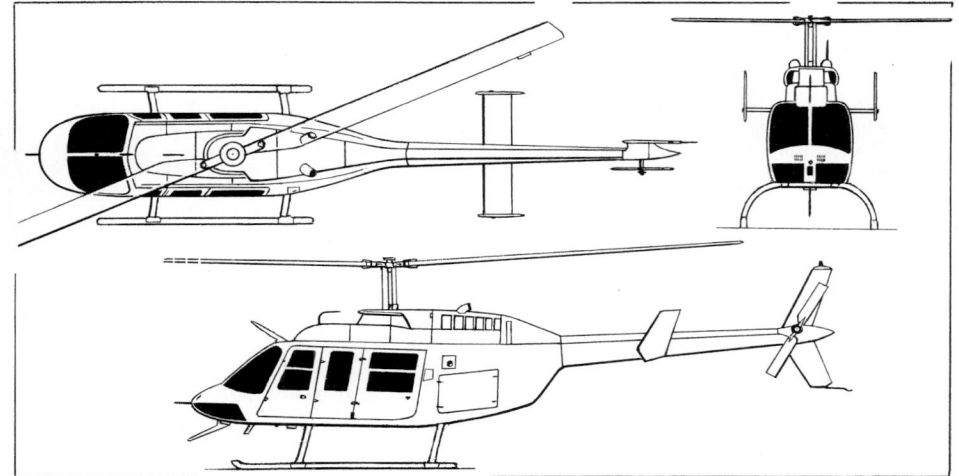

Bell 206L-4 LongRanger IV general purpose light helicopter *(Jane's/Dennis Punnett)*

SYSTEMS: Hydraulic system; 28 V DC electrical power from 180 A starter/generator and 17 Ah battery. Engine bleed air ECS optional.

AVIONICS: *Comms:* Bendix/King suite includes dual nav/com and transponder.

Flight: ADF, DME and marker beacon receiver. Honeywell R/Nav, radio altimeter and encoding altimeter optional.

EQUIPMENT: Optional kits include emergency flotation gear, 907 kg (2,000 lb) cargo hook, rescue hoist, Nightsun searchlight (requires high skid gear).

DIMENSIONS, EXTERNAL:

Main rotor diameter	11.28 m (37 ft 0 in)
Main rotor blade chord	0.33 m (1 ft 1 in)
Tail rotor diameter	1.65 m (5 ft 5 in)
Tail rotor blade chord	0.135 m (5¼ in)
Length: overall, rotors turning	13.02 m (42 ft 8½ in)
fuselage, incl tailskid	9.82 m (32 ft 2½ in)
Height: over tailfin	3.12 m (10 ft 2¾ in)
to top of rotor head	3.14 m (10 ft 3¾ in)
Fuselage: Max width	1.32 m (4 ft 4 in)
Stabiliser span	1.98 m (6 ft 6 in)
Width over skids	2.34 m (7 ft 8 in)
Forward cabin doors (each): Height	1.04 m (3 ft 5 in)
Max width	0.61 m (2 ft 0 in)
Centre cabin door (port): Height	1.04 m (3 ft 5 in)
Width	0.64 m (2 ft 1¼ in)
Rear cabin doors (each): Height	1.04 m (3 ft 5 in)
Width	0.91 m (3 ft 0 in)
Baggage door: Height	0.55 m (1 ft 9½ in)
Width	0.94 m (3 ft 1¼ in)

DIMENSIONS, INTERNAL:

Cabin: Length	2.74 m (9 ft 0 in)
Max width and height	as JetRanger
Volume	2.4 m³ (83 cu ft)

Cabin cargo volume (all passenger seats removed)	2.83 m³ (100 cu ft)
Baggage compartment volume	0.45 m³ (16.0 cu ft)

AREAS:

Main rotor disc	99.89 m² (1,075.2 sq ft)
Tail rotor disc	2.13 m² (22.97 sq ft)

WEIGHTS AND LOADINGS:

Weight empty, standard	1,047 kg (2,309 lb)
Max external load	907 kg (2,000 lb)

BELL 407

TYPE: Light utility helicopter.

PROGRAMME: Design definition launched in 1993 as Bell Light Helicopter to supplement, and eventually replace, JetRanger and LongRanger; concept demonstrator Model 407 (N407LR) first flown 21 April 1994 (standard Bell 206L-3 modified with tailboom and dynamic system of military OH-58D, plus sidewall fairings to simulate broader fuselage); programme first revealed at Heli-Expo '95, Las Vegas, January 1995. Two prototype/preproduction 407s (C-GFOS and C-FORS) first flown on 29 June and 13 July 1995, respectively; first production airframe (C-FWQY/N407BT) flown 10 November 1995. Transport Canada certification 9 February 1996, with FAA certification following on 23 February; first customer delivery at Heli-Expo '96, Dallas, in February. MoU of June 1996 provided for licensed assembly and marketing by Dirgantara (formerly IPTN) of Indonesia.

CUSTOMERS: Launch customers Petroleum Helicopters, Niagara Helicopters and Greenland Air. The 500th production Bell 407 was delivered in October 2001 to Pabst Air, Germany; fleet time then totalled more than 520,000 hours. Total of 62 delivered in 1999, 62 in 2000 and 47 in 2001.

COSTS: US$1.37 million (1999) flyaway. Average direct operating cost US$307 per hour (1999). Development cost estimated as US$50 million, of which US$9 million provided by Canadian government.

DESIGN FEATURES: As for JetRanger and LongRanger. Based on Bell 206L-4 LongRanger fuselage with cabin widened by 17.8 cm (7 in); 35 per cent larger cabin window area; Litton LCD active matrix cockpit displays; all-composites four-blade main rotor based on that of OH-58D, with soft-mounted pylon isolation system; single Rolls-Royce 250-C47 turboshaft; enlarged vertical stabilisers. 'Ring-fin' shrouded rotor tested in 1995 as possible future option. Rotor speed 413 rpm.

STRUCTURE: Generally as for LongRanger, but with carbon fibre tailboom and flush-fitting cabin door windows and skylights.

LANDING GEAR: LongRanger unit modified to match new rotor's ground-resonance characteristics, and with rear cross-tube pivoted in centre between outboard damping pads.

POWER PLANT: One Rolls-Royce 250-C47B turboshaft rated at 606 kW (813 shp) for T-O, 523 kW (701 shp) maximum continuous; transmission rating 503 kW (674 shp) for T-O,

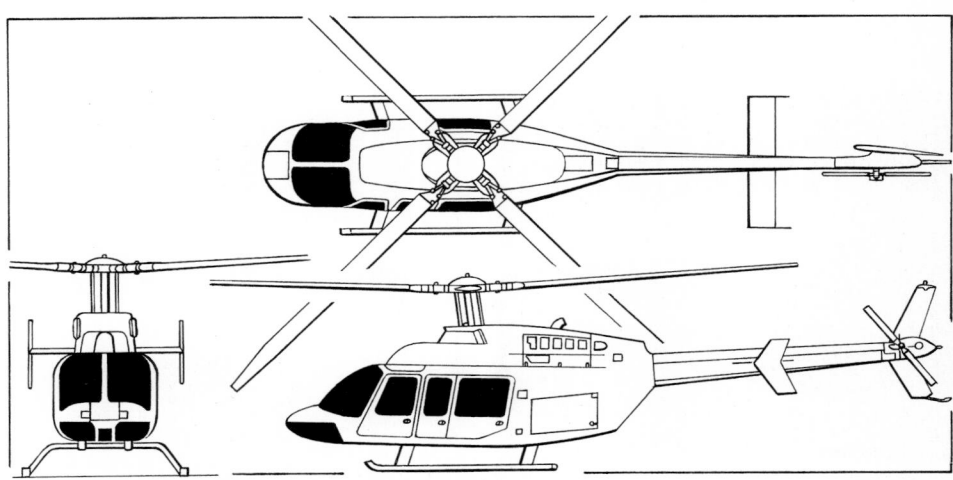

Bell 407 seven-seat light helicopter *(Jane's/James Goulding)*   0103545

Bell 407 seven-seat light helicopter

NEW/0132965

470 kW (630 shp) continuous operation. Single-channel FADEC standard. Standard usable fuel capacity 484 litres (128 US gallons; 106.6 Imp gallons); optional auxiliary fuel tank in aft baggage compartment, usable capacity 75 litres (20.0 US gallons; 16.5 Imp gallons). 'Quiet Cruise' system, in development in early 1997, reduces flyover noise level by using FADEC to reduce rotor rpm to 90 to 93 per cent in the cruise, with accompanying reduction in V_{NE} to 110 kt (204 km/h; 126 mph).

ACCOMMODATION: With a crew of two, standard cabin layout accommodates five passengers in two rearward-facing seats with centre armrest/console, and three forward-facing seats, all fabric-covered. Optional utility cabin has vinyl-covered seats; corporate interior features two extra-wide forward-facing seats in the outboard positions and an occasional-use seat in the centre.

AVIONICS: Bendix/King suite with various options.

DIMENSIONS, EXTERNAL:
Main rotor diameter	10.67 m (35 ft 0 in)
Main rotor blade chord	0.27 m (10¾ in)
Tail rotor diameter	1.65 m (5 ft 5 in)
Tail rotor blade chord	0.16 m (6½ in)
Length: overall, rotors turning	12.74 m (41 ft 9½ in)
rotor in X configuration	11.16 m (36 ft 7¼ in)*
fuselage	10.58 m (34 ft 8½ in)
Stabiliser span over endplate fins	2.22 m (7 ft 3½ in)
Height: over tailfin	3.10 m (10 ft 2 in)
overall: low skids	3.60 m (11 ft 9½ in)
high skids	3.81 m (12 ft 6 in)
Width over skids	2.29 m (7 ft 6 in)
Rear cabin door:	
Width: port	1.55 m (5 ft 1 in)
starboard	0.91 m (3 ft 0 in)
Height to sill (standard skids)	0.51 m (1 ft 8 in)

Blades at 45° to fuselage centreline

DIMENSIONS, INTERNAL:
Cabin: Max width	1.37 m (4 ft 6 in)
Max height	1.00 m (3 ft 3¼ in)
Baggage compartment volume	0.45 m³ (16.0 cu ft)

AREAS:
Main rotor disc	89.38 m² (962.1 sq ft)
Tail rotor disc	2.08 m² (22.34 sq ft)

WEIGHTS AND LOADINGS:
Weight empty, equipped	1,187 kg (2,617 lb)
Max payload	1,089 kg (2,402 lb)
Max hook capacity	1,200 kg (2,646 lb)
Max T-O weight: internal load:	
standard	2,268 kg (5,000 lb)
optional	2,381 kg (5,250 lb)
external load	2,721 kg (6,000 lb)
Max disc loading	27.9 kg/m² (5.72 lb/sq ft)

Max disc loading: internal load:
standard	25.4 kg/m² (5.20 lb/sq ft)
optional	26.6 kg/m² (5.46 lb/sq ft)
external load	30.4 kg/m² (6.24 lb/sq ft)

Transmission loading at max T-O weight and power:
internal load: standard	4.52 kg/kW (7.42 lb/shp)
optional	4.74 kg/kW (7.79 lb/shp)
external load	5.42 kg/kW (8.90 lb/shp)

PERFORMANCE (at internal load MTOW, ISA):
Never-exceed speed (V_{NE})	140 kt (259 km/h; 161 mph)
Max cruising speed:	
at S/L	128 kt (237 km/h; 147 mph)
at 1,220 m (4,000 ft)	131 kt (243 km/h; 151 mph)
Long-range cruising speed:	
at S/L	112 kt (207 km/h; 129 mph)
at 1,220 m (4,000 ft)	115 kt (213 km/h; 132 mph)
Max certified T-O height	5,180 m (17,000 ft)
Max certified altitude	6,100 m (20,000 ft)
Hovering ceiling: IGE	3,720 m (12,200 ft)
OGE	3,170 m (10,400 ft)
Max range	330 n miles (611 km; 379 miles)
Endurance	3 h 42 min

UPDATED

BELL 427

TYPE: Light utility helicopter.

PROGRAMME: Launched as New Light Twin (NLT) in February 1996 on signature of collaborative agreement with Samsung Aerospace Industries of South Korea. Prototype assembly began early 1997; first flight (C-GBLL) 11 December 1997; second prototype (C-FCSS) completed February 1998; two prototypes undertook flight test programme, gaining Transport Canada certification on 19 November 1999; FAA VFR certification achieved in January 2000, followed by Dual Pilot IFR (DPIFR) Category A certification on 24 May 2000. JAA certification was expected in early 2002. First production aircraft (C-GDEJ) flown June 1998; compared with prototypes, production 427 has longer exhausts and revised upper surface contours. Alternative FAR Pt 29 version, with increased T-O weight, will also be available.

CURRENT VERSIONS: **Bell 427:** *As described.*

SB427: South Korean-assembled version; see KAI entry in that national section.

Super Kiowa: Proposed armed scout offered by Bell.

CUSTOMERS: Orders for 22 placed during the mockup's first public display at Farnborough Air Show 1996; 85 on order by May 2000 from 50 customers, including five sold by Samsung to CitiAir in South Korea. Deliveries began 2000. Total of five delivered during 2000 and 15 in 2001. Expected to be offered to South African Air Force as

Bell 427 instrument panel *(Jane's/Paul Jackson)*

0062711

Alouette III replacement from 2003, possibly with local assembly. MoU signed in 1999 with Elbit Systems Inc for joint pre-design phase of military light helicopter using Bell 427 as baseline platform.

COSTS: Projected unit cost of US$2.2 million in 1999 and US$2.3 million in 2000 (both IFR). VFR version available from 2000 at US$2.675 million (dual pilot, without AFCS) or US$2.8 million (single pilot, with AFCS). Direct operating cost US$438 per hour (1999).

DESIGN FEATURES: Similar in appearance to Bell 407, with cabin stretch of 33 cm (13 in), but is all-new design, incorporating twin-engine safety margins. Flight dynamics based on four-blade rotor system of Bell OH-58D Kiowa (see US section), allied to tail rotor of Bell 407; folding main blades. Main rotor rpm 395; tail rotor rpm 2,375. Purpose-designed 'flat pack' main transmission, with direct input from both engines, has only four gear meshes to simplify design and operation. Transmission attached to airframe by four liquid-inertia vibration eliminators. First Bell helicopter designed entirely with use of computer (Dassault CATIA programme).

FLYING CONTROLS: Sextant AFDS 95-1 AFCS, with two- to four-axis autopilot computer, flight director computer and associated equipment, to be certified by 2000.

STRUCTURE: Generally as for Bell 206, but extensive use of carbon/epoxy composites reduces airframe parts count by some 33 per cent. Cabin floor and roof are flat panels for ease of manufacture; minimal use of curved panels elsewhere. Composites main and tail rotors; main blades have nickel-plated stainless steel leading-edges. Soft-in-plane hub of main rotor employs a composites flexbeam yoke and elastomeric joints, eliminating lubrication and maintenance requirements. Brake and main rotor blade folding optional. Composites cabins and rolled aluminium tailbooms built by Samsung; assembly in Canada, except for sales to Korea and China; Hexcel honeycomb as stiffener.

LANDING GEAR: Twin skids with dynamically tuned cross tubes to reduce ground resonance. Low skids standard; optional high skids and emergency floats.

POWER PLANT: Two Pratt & Whitney Canada PW207D turboshafts with FADEC, each rated at 529 kW (710 shp) for T-O (5 minutes) or 466 kW (625 shp) maximum continuous; OEI ratings 611 kW (820 shp) for 30 seconds,

Bell 407 panel, showing standard and optional instruments
0010221

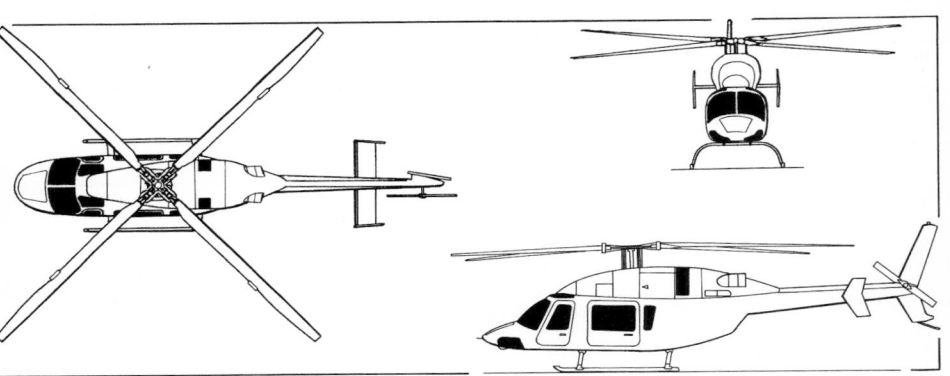

Provisional general arrangement of Bell 427 *(Jane's/Paul Jackson)*
0062757

Bell 427 (P&WC PW207D turboshafts)

NEW/0132964

582 kW (780 shp) for 2 minutes, 559 kW (750 shp) for 30 minutes or 529 kW (710 shp) maximum continuous. Twin-engine transmission rating, T-O and maximum continuous, 597 kW (800 shp). OEI transmission rating, 485 kW (650 shp) for 30 seconds; 451 kW (605 shp) for 2 minutes; 343 kW (460 shp) maximum continuous.

Fuel contained in three crash-resistant tanks; two forward, one aft, total usable capacity 770 litres (203.5 US gallons; 169 Imp gallons). One forward fuel tank can be removed in EMS configurations to provide additional stretcher space in cabin or to permit stretcher to extend into port side of cockpit.

ACCOMMODATION: Standard accommodation is for two crew in cockpit, on 20 *g* energy-attenuating seats, and six passengers in cabin on two rows of three seats in club configuration (all-forward-facing seats optional); all seats equipped with inertia-reel shoulder harnesses. Alternative configurations include corporate club-four seating with refreshment/entertainment console between each pair of seats. Optional EMS interiors provide for carriage of one or two stretchers with up to two medical attendants, affording either full patient or head-only in-flight access, with single- or two-person crew. In cargo configuration, with all passenger seats removed, an optional removable flat cargo floor can be installed, equipped with integral tie-downs. Two forward-hinged doors each side; cabin doors, both sides, are forward-hinged, but port unit can be replaced by optional rearward-sliding door for cargo handling. External door on starboard side to rear baggage hold.

SYSTEMS: Hydraulic system, operating pressure 86 bar (1,250 lb/sq in), provides boost power for main and tail rotor controls; 28 V DC electrical power from 17 Ah Ni/Cd battery and two engine-mounted 17 Ah starter/generators; 28 Ah battery and 200 A starter/generator optional. Air conditioning optional.

AVIONICS: Rogerson-Kratos NeoAV two-screen LCD integrated instrument display system (IIDS) for monitoring engine instruments, fuel quantity, hydraulic and electrical systems and weight and balance functions. Rogerson Kratos NeoAV EFIS optional, featuring GPS interface, area navigation map, non-precision approach capability and weather radar display. Bendix/King nav/com avionics suite to customer's choice.

EQUIPMENT: Optional kits include engine air particle separator, cargo hook, cargo floor, sliding cabin door, external rescue hoist, NightSun searchlight, EMS installation and wire strike protection system.

DIMENSIONS, EXTERNAL:
Main rotor diameter	11.28 m (37 ft 0 in)
Main rotor blade chord	0.27 m (10½ in)
Tail rotor diameter	1.73 m (5 ft 8 in)
Tail rotor blade chord	0.18 m (7¼ in)
Length: overall, rotors turning	13.07 m (42 ft 10¾ in)
overall, rotor in X configuration	11.55 m (37 ft 10¾ in)*
fuselage, incl tailskid	10.94 m (35 ft 10¾ in)

Height over tailfin	3.49 m (11 ft 5¼ in)
Ground clearance: low skids	0.41 m (1 ft 4¼ in)
high skids	0.67 m (2 ft 2½ in)
Width: overall, rotor in X configuration	8.24 m (27 ft 0½ in)*
over skids	2.36 m (7 ft 9 in)
Cabin door: Height	1.07 m (3 ft 6 in)
Width	1.24 m (4 ft 0¾ in)

Blades at 45° to fuselage centreline

DIMENSIONS, INTERNAL:
Cabin: Max height	1.30 m (4 ft 3 in)
Baggage hold volume	0.76 m³ (27.0 cu ft)

AREAS:
Main rotor disc	99.89 m² (1,075.2 sq ft)
Tail rotor disc	2.34 m² (25.2 sq ft)

WEIGHTS AND LOADINGS (provisional):
Weight empty	1,743 kg (3,842 lb)
Max payload	1,021 kg (2,250 lb)
Max baggage weight	113 kg (250 lb)
Max T-O weight: internal load:	2,880 kg (6,350 lb)
external load	2,948 kg (6,500 lb)
Cargo hook capacity	1,361 kg (3,000 lb)
Max cargo floor loading	366.2 kg/m² (75 lb/sq ft)
Max disc loading	29.52 kg/m² (6.05 lb/sq ft)
Transmission loading at max T-O weight and power:	
internal load	4.56 kg/kW (7.50 lb/shp)
external load	4.95 kg/kW (8.13 lb/shp)

PERFORMANCE:
Max cruising speed at S/L	138 kt (256 km/h; 159 mph)
Econ cruising speed at S/L	133 kt (246 km/h; 153 mph)
Service ceiling	
max continuous power	5,547 m (18,200 ft)
OEI (30 min)	3,050 m (10,000 ft)
Hovering ceiling: IGE	4,938 m (16,200 ft)
OGE	1,830 m (6,000 ft)
Range with max fuel, econ cruising speed, no reserves	380 n miles (703 km; 437 miles)
Endurance	4 h 0 min

UPDATED

BELL 412

Canadian Forces designation: CH-146 Griffon
UK armed forces designation: Griffin HT. Mk 1

TYPE: Multirole medium helicopter.

PROGRAMME: Original 412 announced 8 September 1978 (see earlier editions of *Jane's*); FAR Pt 29 VFR approval received 9 January 1981, IFR 13 February 1981; 213 built in USA; production (SP version) transferred to Canada February 1989; first delivery (civil) 18 January 1981. Production licences obtained by Dirgantara of Indonesia and Agusta of Italy (which see).

CURRENT VERSIONS: **412SP:** Special Performance version with increased maximum T-O weight, new seating options

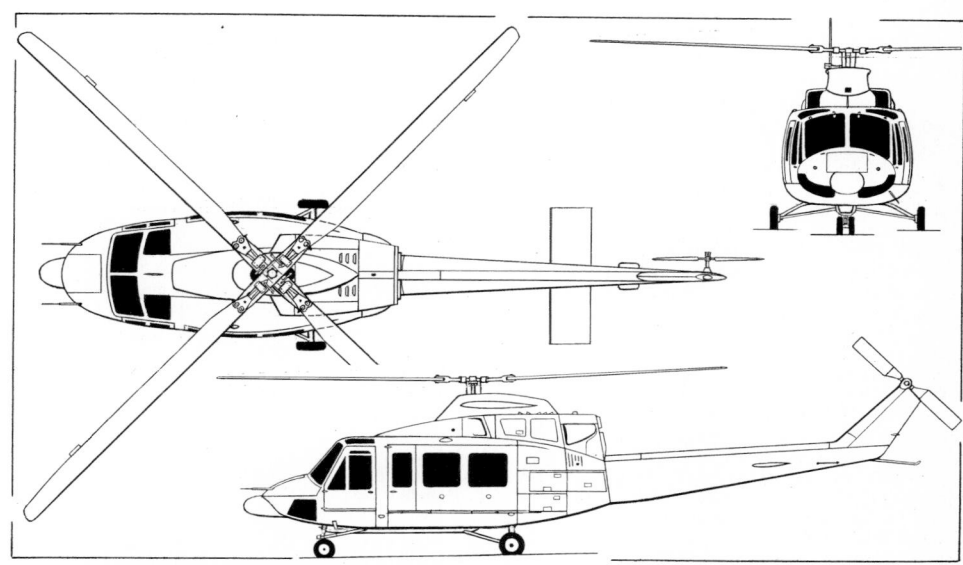

Bell 412EP (P&WC PT6T turboshaft) *(Jane's/Dennis Punnett)*

Bell 412EP Multirole helicopter   *NEW*/0132963

and 55 per cent greater standard fuel capacity. Superseded by 412HP early 1991. Details in 1991-92 *Jane's*.

Military 412: Announced by Bell June 1986; fitted with Lucas Aerospace chin turret and Honeywell Head Tracker helmet sight similar to that in AH-1S; turret carries 875 rounds, weighs 188 kg (414 lb) and can be removed in under 30 minutes; firing arcs ±110° in azimuth, +15° and −45° in elevation; other armament includes twin dual FN Herstal 7.62 mm gun pods, single FN Herstal 0.50 in pod, pods of seven or nineteen 2.75 in rockets, M240E1 pintle-mounted door guns, FN Herstal four-round 70 mm rocket launcher and a 0.50 in gun or two Giat M621 20 mm cannon pods.

412HP: Improved transmission giving better OGE hover; FAR Pt 29 certification 5 February 1991, first delivery (c/n 36020) later that month.

412EP (Enhanced Performance): PT6T-3D engine, dual digital automatic flight control system (DDAFCS), three-axis in basic aircraft but customer option for four-axis and EFIS. Category A certification was imminent in late 1998. Also customer option for SAR fit. Now standard current model.

Detailed description applies to Bell 412EP.

412CF (CH-146) Griffon: Canadian Forces C$700 million contract for 100 CH-146s (modified Bell 412EP) placed in 1992. Duties include armed support, troop/cargo transport, medevac, ASW, SAR and patrol; first flight (146000) 30 April 1994; deliveries began 14 October 1994; completed early 1998. Generally as commercial Bell 412EP except for avionics and mission equipment; see 1998-99 and earlier *Jane's*. Empty weight 3,402 kg (7,500 lb); maximum weight as civil version.

412EP Sentinel: First of two modified in 1998 by Heli-Dyne Systems with quick-change ASV and ASW mission packages; intended for Ecuadorean Navy, but order cancelled and aircraft became demonstrator. ASV equipment comprises Honeywell RDR-1500B chin radar, Hughes Starburst searchlight, radar warning receiver, Wescam sensor turret and possibly Penguin Mk 2 Mod 7 ASMs; ASW fit is L3 Ocean Systems AN/AQS-18A dipping sonar and Raytheon Mk 46 torpedo.

412 RSAF: First three of 16 ordered by Royal Saudi Air Force built in 2001 for manufacturer's trials. Equipment standard not disclosed, but sufficiently different from 412EP to warrant separate c/n sequence beginning 33501.

412 Plus: Projected improved version under study in 1999 with MTOW increased to 5,647 kg (12,450 lb), uprated PT6C engines, new dynamic components and Rogerson-Kratos avionics. Development terminated in early 2001.

NBell-412: Indonesia's Dirgantara (which see) has licence to produce up to 100 Model 412SPs.

CUSTOMERS: Some 610 Bell 412s of all versions built in North America by early 2002, including 26 delivered in 1999 and 24 in 2000. Shipments in 2001 comprised 22, including five to El Salvador.

Military deliveries include Venezuelan Air Force (two), Botswana Defence Force (three), Public Security Flying Wing of Bahrain Defence Force (two), Sri Lankan armed forces (four), Nigerian Police Air Wing (two), Mexican government (two VIP transports), South Korean Coast Guard (one), Honduras (10), Royal Norwegian Air Force

(19, of which 18 assembled by Helikopter Service, Stavanger, to replace UH-1Bs of 339 Squadron at Bardufoss and 720 Squadron at Rygge). Three 412EPs delivered to Slovenian Territorial Forces in 1995, for border patrol and rescue duties; four ordered by Philippine Air Force late in 1996, comprising two for VVIP transport and two SAR; first of nine 412EPs entered service in April 1997 with civilian-operated Defence Helicopter Flying School at RAF Shawbury, UK, within which they constitute No. 60 (Reserve) Squadron, RAF. Four in SAR/ utility fit delivered to Venezuelan Navy, 1999.

COSTS: Bell 412EP, VFR-equipped US$4.895 million (1999); Bell 412EP, IFR-equipped US$5.12 million (1999). Average direct operating cost US$746 per hour (1997).

DESIGN FEATURES: Four-blade main rotor with blades retained within central metal star fitting by single elastomeric bearings; shorter rotor mast than 212; blades can be folded; rotor brake standard; two-blade tail rotor; main rotor rpm 314.

FLYING CONTROLS: Fully powered hydraulic controls; gyroscopic stabiliser bar above main rotor; automatic tailplane incidence control.

STRUCTURE: Generally of conventional light metal. Main rotor blade spar unidirectional glass fibre with 45° wound torque casing of glass fibre cloth; Nomex rear section core with trailing-edge of unidirectional glass fibre; leading-edge protected by titanium abrasion strip and replaceable stainless steel cap at tip; lightning protection mesh embedded; provision for electric de-icing heater elements; main rotor hub of steel and light alloy; all-metal tail rotor.

LANDING GEAR: High skid, emergency pop-out float or non-retractable tricycle gear optional. Spats optional for last-named.

POWER PLANT: Pratt & Whitney Canada PT6T-3D Turbo Twin-Pac, rated at 1,342 kW (1,800 shp) for T-O and 1,193 kW (1,600 shp) maximum continuous. OEI ratings 850 kW (1,140 shp) for 2½ minutes, or 723 kW (970 shp) for 30 minutes. Transmission rating 1,022 kW (1,370 shp) for T-O, 828 kW (1,110 shp) maximum continuous; OEI rating 850 kW (1,140 shp). Optional 30 kW (40 shp) for accessory drives from main gearbox.

Seven interconnected rupture-resistant fuel cells, with automatic shutoff valves (breakaway fittings), have a combined usable capacity of 1,249 litres (330 US gallons; 275 Imp gallons). Two 76 or 310.5 litre (20.0 or 82.0 US gallon; 16.7 or 68.3 Imp gallon) auxiliary fuel tanks, in any combination, can increase maximum total capacity to 1,870 litres (494 US gallons; 411.6 Imp gallons). Single-point refuelling on starboard side of cabin.

ACCOMMODATION: Pilot and up to 14 passengers: one in front port seat and 13 in cabin. Dual controls optional. Accommodation heated and ventilated.

SYSTEMS: Dual hydraulic systems, pressure 69 bar (1,000 lb/ sq in) each. 28 V DC electrical system supplied by two completely independent 450 VA inverters. 40 Ah Ni/Cd battery.

AVIONICS: *Comms:* Optional IFR avionics include dual Bendix/King Gold Crown III.

Radar: Weather radar optional.

Flight: Dual KNR 660A VOR/LOC/RMI receivers, KDF 800 ADF, KMD 700A DME, KXP 750A transponder and KGM 690 marker beacon/glideslope receiver.

Optional Honeywell AFCS.

EQUIPMENT: Optional equipment includes cargo sling and rescue hoist.

DIMENSIONS, EXTERNAL:

Main rotor diameter	14.02 m (46 ft 0 in)
Main rotor blade chord: at root	0.40 m (1 ft 4 in)
at tip	0.22 m (8½ in)
Tail rotor diameter	2.62 m (8 ft 7 in)
Tail rotor blade chord	0.29 m (11½ in)
Length: overall, rotors turning	17.12 m (56 ft 2 in)
fuselage, excl rotors	12.70 m (41 ft 8 in)
Height: to top of rotor head	3.48 m (11 ft 5 in)
overall, tail rotor turning	4.57 m (15 ft 0 in)
Stabiliser: span	2.87 m (9 ft 5 in)
chord	0.79 m (2 ft 7 in)
Width over skids	2.84 m (9 ft 4 in)
Rear sliding doors (each): Height	1.24 m (4 ft 1 in)
Width	1.88 m (6 ft 2 in)
Height to sill	0.76 m (2 ft 6 in)
Baggage compartment door: Height	0.53 m (1 ft 9 in)
Width	1.71 m (2 ft 4 in)
Emergency exits (centre cabin windows, each):	
Height	0.76 m (2 ft 6 in)
Width	0.97 m (3 ft 2 in)

DIMENSIONS, INTERNAL:

Baggage compartment volume	0.79 m³ (28.0 cu ft)

AREAS:

Main rotor disc	154.40 m² (1,661.9 sq ft)
Tail rotor disc	5.38 m² (57.86 sq ft)

WEIGHTS AND LOADINGS:

Weight empty, standard equipped	3,079 kg (6,789 lb)
Max external hook load	2,041 kg (4,500 lb)
Max T-O and landing weight, internal or external load	5,397 kg (11,900 lb)
Max disc loading	35.0 kg/m² (7.16 lb/sq ft)
Transmission loading at max T-O weight and power	5.29 kg/kW (8.69 lb/shp)

PERFORMANCE:

Never-exceed speed (VNE)	140 kt (259 km/h; 161 mph)
Max cruising speed: at S/L	122 kt (226 km/h; 140 mph)
at 1,525 m (5,000 ft)	124 kt (230 km/h; 143 mph)
Long-range cruising speed at 1,525 m (5,000 ft)	130 kt (241 km/h; 150 mph)
Service ceiling, OEI, 30 min power rating	1,920 m (6,300 ft)
Hovering ceiling: IGE	3,110 m (10,200 ft)
OGE	1,585 m (5,200 ft)
Range at 1,525 m (5,000 ft), long-range cruising speed, standard fuel, no reserves	402 n miles (744 km; 462 miles)
Endurance	3 h 42 min

OPERATIONAL NOISE LEVELS (FAR Pt 36) (412 EP):

T-O	92.8 EPNdB
Flyover	93.4 EPNdB
Approach	95.6 EPNdB

UPDATED

BELL 430

TYPE: Light utility helicopter.

PROGRAMME: Preliminary design 1991; four-blade rotor, higher-powered and stretched variant of Bell 230; programme launched February 1992; two prototypes, modified from Bell 230 airframes; first prototype (C-GBLL; wheel-equipped) flown 25 October 1994; second prototype (C-GEXP; skid-equipped, with complete avionics suite) flown 19 December 1994; first flight of production 430 (C-GRND) in 1995; deliveries began 25 June 1996 after Canadian type approval on 23 February. Single-pilot IFR and Category A certification anticipated early 1999. MoU of June 1996 with Dirgantara (formerly IPTN) for licensed assembly and marketing in Indonesia.

Second production aircraft, N4300 circumnavigated the world in a record time of 17 days 6 hours 14 minutes, landing back at Fairoaks, UK, on 3 September 1996.

Neiman Marcus Special Edition introduced in 2001 for Neiman Marcus store's 75th Christmas catalogue, featuring three-colour custom metallic exterior paint scheme with NM signature, Italian leather and rosewood interior, sculpted carpets with NM logo, passenger refreshment centre, Blaupunkt AM/FM/CD entertainment centre, JetMap cabin information system with moving map displays on two ceiling-mounted monitors, Motorola cellphone and two computer ports.

CUSTOMERS: First delivery 25 June 1996, when sixth production aircraft (N6282X) handed over to IPTN (now Dirgantara) of Indonesia in eight-seat executive configuration. Thirteen delivered in 1996; followed by eight, 15 and 18 in 1997-99 11 in 2000 and 14 in 2001. Total of 80 in service by October 2001, when fleet time totalled more than 55,000 flight hours. Recent customers include STAT MedEvac of Pittsburgh, which took delivery of an EMS-equipped Bell 430 (N430Q) in January 2002.

COSTS: Programme US$18 million, 35 per cent financed by Canadian Defence Industry Productivity Program (DIPP) and repayable as royalty on each sale. Unit costs (1999) US$4.09 million, dual pilot, no autopilot; US$4.25 million, single pilot and autopilot; US$4.295 million, dual pilot and autopilot; US$4.395 million, single/dual pilot and autopilot; all versions equipped for IFR. Average direct operating cost US$565 per hour (1997).

DESIGN FEATURES: Optimised for high cruising speed with (retractable) wheel landing gear, although traditional skids optional; inclined towards executive transport market. Bell 230 fuselage lengthened by 0.46 m (1 ft 6 in) plug; Bell 680 all-composites four-blade bearingless, hingeless main rotor; approximately 10 per cent power increase over Bell 230; uprated transmission; and optional EFIS. Short-span sponson each side of fuselage houses mainwheel units and fuel tanks, and serves as work platform.

FLYING CONTROLS: Fully powered hydraulic, with elastomeric pitch change and flapping bearings; fixed tailplane with leading-edge slats and endplate fins; strakes under sponsons; single-pilot IFR system without auto-stabilisation.

STRUCTURE: Semi-monocoque fuselage of light alloy, with limited use of light alloy honeycomb panels. Fail-safe structure in critical areas. One-piece nosecone tilts forward and down for access to avionics and equipment bay. Short span cantilever sponson set low on each side of fuselage, serving as main landing gear housings, fuel tanks and work platforms. Section NACA 0035. Dihedral 3° 12′. Incidence 5°. Sweepback at quarter-chord 3° 30′.

Fixed vertical fin in sweptback upper and lower sections. Tailplane, with slotted leading-edge and endplate fins, mounted midway along rear fuselage. Small skid below ventral fin for protection in tail-down landing. Four-blade main rotor with stainless steel spars and leading-edges, Nomex honeycomb trailing-edge with glass fibre skin, and glass fibre safety straps; tail rotor blades stainless steel. Rotors shaft-driven through gearbox with two spiral bevel reductions and one planetary reduction. Main blade and hub life, 10,000 hours.

LANDING GEAR: Tubular skid type on Utility version. Executive version has hydraulically retractable tricycle gear, single mainwheels retracting forward into sponsons; forward-retracting nosewheel fully castoring and self-centring; hydraulic disc brakes on main units. Mainwheel tyre size 18×5.5, nosewheel 5.00-5. Emergency floats optional.

POWER PLANT: Two Rolls-Royce 250-C40B turboshafts, each rated at 584 kW (783 shp) for T-O and 521 kW (699 shp) maximum continuous. OEI ratings 701 kW (940 shp) for 30 seconds, 656 kW (880 shp) for 2 minutes, 623 kW (835 shp) for 30 minutes and 602 kW (808 shp) continuous. Chandler Evans FADEC. Transmission rating 779 kW (1,045 shp) for 5 minutes for T-O, 737 kW (989 shp) maximum continuous. Power train TBO, 5,000 hours. Usable fuel capacity 935 litres (247 US gallons; 206 Imp gallons) in skid version, 710 litres (187.5 US gallons; 156 Imp gallons) in wheeled version; provision in both versions for 182 litre (48 US gallon; 40 Imp gallon) auxiliary tank. Fuel system is rupture resistant, with self-sealing breakaway fittings.

ACCOMMODATION: Standard layout has forward-facing seats for nine persons (two-two-two-three) including pilot. Options include 10-seat layout (two-two-three-three); eight-seat executive (rear six in club layout), six-seat executive (rear four in club layout with console between each pair); and five- and four-seat executive with one or two refreshment cabinets; seat pitches vary between 86 cm (34 in) and 91 cm (36 in). Pilots on crashworthy (energy attenuating) seats, which are optional for passengers. Customised emergency medical service (EMS) versions also available, configured for pilot-only operation plus one or two pivotable stretchers and four or three medical attendants/sitting casualties respectively. Two forward-opening doors each side; EMS version has optional stretcher door between forward and rear doors on port side. Entire interior ram air ventilated and soundproofed. Dual controls optional.

SYSTEMS: Dual hydraulic systems (dual for main rotor collective and cyclic, single for tail rotor). Dual 28 V DC electrical systems, powered by two 30 V 200 A engine-mounted starter/generators (derated to 180 A) and a 24 V 28 Ah Ni/Cd battery.

AVIONICS: Comms: Bendix/King Gold Crown III.
Flight: Honeywell KFC 500 AFCS. GPS optional.
Instrumentation: Rogerson-Kratos LCD integrated instrument display system (IIDS) comprising two active matrix LCDs displaying engine and system parameters; optional Rogerson-Kratos EFIS.

Bell 430 cockpit test-rig, showing Rogerson-Kratos IIDS display in centre   0131800

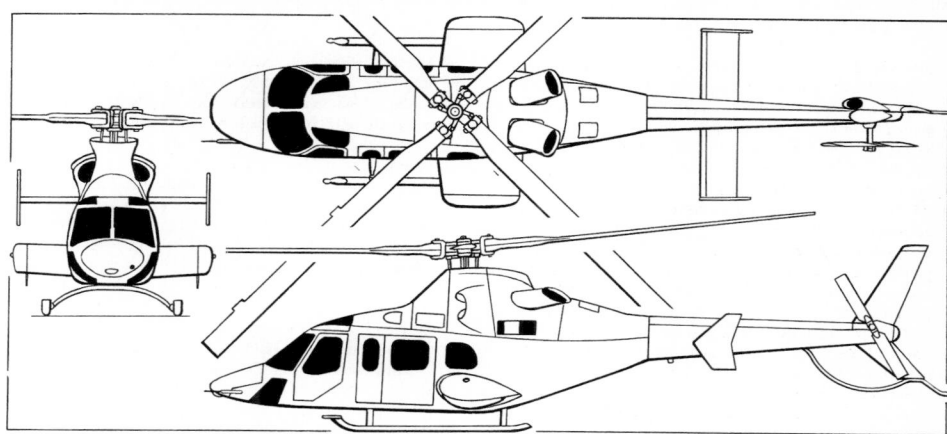
Skid version of Bell 430 nine-seat helicopter (two Rolls-Royce 250 turboshafts) (Jane's/James Goulding)   0103543

EQUIPMENT: Standard equipment includes rotor and cargo tiedowns, ground handling wheels for skid version, retractable 450 W search/landing light. Options include dual controls, auxiliary fuel tankage, force/feel trim system, more comprehensive nav/com avionics, 272 kg (600 lb) capacity rescue hoist, 1,587 kg (3,500 lb) capacity cargo hook, emergency flotation gear, heated windscreen, particle separator and snow baffles.

DIMENSIONS, EXTERNAL:

Main rotor diameter	12.80 m (42 ft 0 in)
Main rotor blade chord	0.36 m (1 ft 2¼ in)
Tail rotor diameter	2.10 m (6 ft 10½ in)
Tail rotor blade chord	0.25 m (10 in)
Length: fuselage (incl tailskid)	13.44 m (44 ft 1¼ in)
overall, rotors turning	15.30 m (50 ft 2½ in)
Fuselage max width over sponsons	3.45 m (11 ft 4 in)
Height to top of rotor head:	
standard skids	4.03 m (13 ft 2½ in)
high skids	4.24 m (13 ft 10¾ in)
wheels	3.72 m (12 ft 2½ in)
Skid track	2.54 m (8 ft 4 in)
Wheel track	2.78 m (9 ft 1½ in)
Wheelbase	4.17 m (13 ft 8¼ in)
Passenger doors: forward:	
Height	1.34 m (4 ft 4¾ in)
Width	0.88 m (2 ft 10¾ in)
aft:	
Height	1.22 m (4 ft 0 in)
Width	0.91 m (3 ft 0 in)
Baggage door: Height	0.60 m (1 ft 11½ in)
Width	0.85 m (2 ft 9½ in)

DIMENSIONS, INTERNAL:

Cabin: Length, excl cockpit	2.87 m (9 ft 5 in)
Max width	1.27 m (4 ft 2 in)
Max height	1.45 m (4 ft 9 in)
Volume	4.5 m³ (158 cu ft)
Baggage compartment volume	1.0 m³ (37 cu ft)

AREAS:

Main rotor disc	128.71 m² (1,385.4 sq ft)
Tail rotor disc	3.45 m² (37.12 sq ft)

WEIGHTS AND LOADINGS (A: wheeled version; B: skid version):

Weight empty: A	2,423 kg (5,342 lb)
B	2,407 kg (5,308 lb)
Max external load: A, B	1,587 kg (3,500 lb)
Max T-O weight, all conditions	4,218 kg (9,300 lb)
Max disc loading	32.8 kg/m² (6.71 lb/sq ft)
Transmission loading at max T-O weight and power	5.42 kg/kW (8.90 lb/shp)

PERFORMANCE:

Never-exceed speed (VNE):	
A, B	150 kt (277 km/h; 172 mph)
Max level speed: A	139 kt (257 km/h; 160 mph)
B	135 kt (250 km/h; 155 mph)
Long-range cruising speed:	
A	131 kt (243 km/h; 151 mph)
B	128 kt (237 km/h; 147 mph)
Service ceiling: A, B	5,590 m (18,340 ft)
Hovering ceiling, IGE: A, B	3,565 m (11,700 ft)
OGE: A, B	1,890 m (6,200 ft)

Bell 430 nine-seat utility helicopter   NEW/0132962

Max range, standard fuel, no reserves:
A	275 n miles (509 km; 316 miles)
B	353 n miles (653 km; 406 miles)
Max endurance	3 h 35 min

OPERATIONAL NOISE LEVELS (FAR Pt 36, Stage 2):
T-O	92.4 EPNdB
Sideline	91.6 EPNdB
Approach	93.8 EPNdB

UPDATED

BELL/AGUSTA AB139
A description of this joint venture helicopter appears under the Bell/Agusta entry in the International section.

BELL JRX
At Heli-Expo in Orlando, Florida in February 2002, Bell revealed brief details of this concept study for a new light turbine helicopter as part of its Vision 2020 production plan. The JRX is a possible entry-level successor to the JetRanger, featuring a four-blade main rotor, wider cabin, state-of-the-art cockpit, single-piece windscreen, larger cabin windows and improved operating economics.

NEW ENTRY

BLUE YONDER
BLUE YONDER AVIATION
Box 12, Suite 9, RR5, Calgary, Alberta T2P 2G6
Tel: (+1 403) 936 57 67
Fax: (+1 403) 936 51 08
e-mail: mailto:ezflyer@sprint.ca
Web: http://pages.sprint.ca/ezflyercom/files
PRESIDENT: Wayne Winters

Blue Yonder designed the E-Z Flyer in 1991; built under licence by Merlin in USA until bankruptcy of December 1995; production tooling transferred to Canada and manufacture of it (and Merlin GT) resumed; Aerocomp (which see) is US agent, although Blue Yonder responsible for global sales. The company has added the E-Z KingCobra to its product line.

UPDATED

BLUE YONDER E-Z FLYER
TYPE: Tandem-seat kitbuilt.
PROGRAMME: First kitbuilt E-Zs delivered by Merlin mid-1995. Programme transferred in 1998 to Blue Yonder, which has also produced a twin-engined prototype, designated E-Z Flyer Twin.
CURRENT VERSIONS: **E-Z Flyer:** *Basic version; as described.*
 E-Z Flyer Twin: As E-Z Flyer, but with two Rotax engines in range 37.0 to 73.5 kW (49.6 to 98.6 hp). Wing span 10.82 m (35 ft 6 in); maximum T-O weight 635 kg (1,400 lb). Prototype (C-ITEZ) only to date.
CUSTOMERS: Total 39 kits sold by late 2001.
COSTS: US$14,975 including engine (2002).
DESIGN FEATURES: Open tube fuselage married to high-mounted Aerocomp Merlin GT wing and tail sections.
FLYING CONTROLS: Manual. Non-drooping Junkers-type ailerons.
STRUCTURE: Fuselage is welded 4130 chromoly. Wings are aluminium D-cell construction with aluminium and Styrofoam constructed ribs; Stits Poly-Fiber covering.
LANDING GEAR: Conventional tricycle. Hegar rims with brakes; solid spring steel legs; 18 in tundra tyres on mainwheels.
POWER PLANT: One 37.0 kW (49.6 hp) Rotax 503 or 47.8 kW (64.1 hp) Rotax 582 engine, driving a two- or three-blade pusher propeller. Fuel capacity 34 litres (9.0 US gallons; 7.5 Imp gallons).
DIMENSIONS, EXTERNAL:
Wing span	9.45 m (31 ft 0 in)
Length overall	6.40 m (21 ft 0 in)
Height overall	1.83 m (6 ft 0 in)

AREAS:
Wings, gross, incl extended tips	16.35 m² (176.0 sq ft)

WEIGHTS AND LOADINGS:
Weight empty	225 kg (495 lb)
Max T-O weight	607 kg (1,340 lb)

Blue Yonder E-Z Flyer two-seat kitbuilt

PERFORMANCE:
Max cruising speed	56 kt (105 km/h; 65 mph)
Stalling speed	35 kt (65 km/h; 40 mph)
Max rate of climb at S/L	137 m (450 ft)/min
T-O to 15 m (50 ft)	99 m (325 ft)
Landing from 15 m (50 ft)	61 m (200 ft)

UPDATED

BLUE YONDER E-Z KINGCOBRA
TYPE: Single-seat ultralight kitbuilt.
PROGRAMME: Designed by Blue Yonder for Dr Jack Barlass; prototype (C-IFWW) first flown late 1998. Aircraft and design rights later purchased from Barlass estate.
CUSTOMERS: Prototype only by early 2002.
COSTS: Kit US$18,645 including engine; firewall back US$12,660 (2002).
DESIGN FEATURES: Low-wing monoplane; uses (shortened) wing from Merlin and E-Z Flyer allied to new tail unit. Options include electric starter.
FLYING CONTROLS: Manual. Junkers-type ailerons.
STRUCTURE: Fabric-covered 4130 chromoly welded steel tube fuselage and fabric-covered aluminium D-cell cantilever wings.

LANDING GEAR: Tailwheel type. Hegar 46 cm (18 in) wheels and tundra tyres standard. Hydraulic brakes.
POWER PLANT: One 47.8 kW (64.1 hp) Rotax 582 UL driving two-blade, wooden propeller; design will accept engines up to 73.5 kW (98.6 hp). Fuel capacity 57 litres (15.0 US gallons; 12.5 Imp gallons).
DIMENSIONS, EXTERNAL:
Wing span	8.23 m (27 ft 0 in)
Length overall	6.40 m (21 ft 0 in)
Height overall	1.83 m (6 ft 0 in)

AREAS:
Wings, gross	14.68 m² (158.0 sq ft)

WEIGHTS AND LOADINGS:
Weight empty	246 kg (543 lb)
Max T-O weight	544 kg (1,200 lb)

PERFORMANCE:
Never-exceed speed (V_{NE})	104 kt (193 km/h; 120 mph)
Normal cruising speed	78 kt (145 km/h; 90 mph)
Stalling speed	27 kt (49 km/h; 30 mph)
Service ceiling	4,265 m (14,000 ft)
T-O run	19 m (60 ft)
Landing run	46 m (150 ft)
Range	400 n miles (740 km; 460 miles)

NEW ENTRY

BOMBARDIER
BOMBARDIER AEROSPACE
400 Chemin de la Côte Vertu West, Dorval, Québec H4S 1Y9
Tel: (+1 514) 855 50 00
Fax: (+1 514) 855 79 03
Web: http://www.aero.bombardier.com
DESIGN DIVISIONS:
 Canadair: follows Bombardier entry
 de Havilland: follows Canadair entry
 Learjet: see US section
 Shorts: see 1996-97 edition and *Jane's Aircraft Upgrades*
PRESIDENT AND COO: Pierre Beaudoin
EXECUTIVE VICE-PRESIDENT, ENGINEERING AND PRODUCT DEVELOPMENT: John Holding
EXECUTIVE VICE-PRESIDENT, PROGRAMS AND STRATEGIC PLANNING: Steven A Ridolfi
VICE-PRESIDENT, OPERATIONS: Ken Brunrle
VICE-PRESIDENT, FINANCE: James Stewart
PRESIDENT, REGIONAL AIRCRAFT: John Giraudy
PRESIDENT, AMPHIBIOUS AIRCRAFT: Thomas Appleton

Bombardier Inc, a diversified Canadian corporation with 79,000 employees, formed Bombardier Aerospace in 1986, subsequently combining design and manufacturing activities of Canadair (1986), de Havilland (1992), Learjet (1990) and Shorts (1989). Sales, marketing and support are conducted by the Amphibious Aircraft, Regional Aircraft and Business Aircraft units.

Bombardier Aerospace also designs and manufactures components for Airbus and Boeing.

Bombardier Aerospace's revenue for the year ended 31 January 2001 totalled C$9.17 billion. Deliveries in 2001 were 370 aircraft, the same as 2000, but a reduction of 40 over target figure projected at the start of the year. Total for 2001 comprised 206 regional aircraft, 162 business jets and two amphibians. Projected deliveries in 2002 totalled 334, comprising 190 regional aircraft, 140 business jets and four amphibians. Previous deliveries in company's financial years were 146, 168, 178, 227 and 292 in 1995-99.

UPDATED

BOMBARDIER BD-700 GLOBAL EXPRESS
TYPE: Long-range business jet.
PROGRAMME: Announced 28 October 1991 at NBAA Convention; full-scale cabin mockup exhibited at NBAA Convention September 1992; conceptual design started early 1993. Programme launched 20 December 1993; high-speed configuration frozen June 1994; low-speed configuration established August 1994.
 Ground test programme using static test airframe c/n 0001 began August 1996; prototype C-FBGX (engineering designation BD-700-1A10) rolled out 26 August 1996; first flight 13 October 1996; public debut at NBAA Convention at Orlando, Florida, November 1996; prototype and three other aircraft undertook 2,000-hour, 18-month flight test programme based at Bombardier's

flight test centre in Wichita, Kansas; second aircraft, (C-FHGX), which is used for systems evaluation and testing, flew 3 February 1997, third (C-FJGX), which is used for avionics and autopilot testing, 22 April 1997; fourth (C-FKGX), first flown 8 September 1997 and the first to be fully outfitted, was used for function and reliability testing.
 Transport Canada certification 31 July 1998; FAA certification 13 November 1998; JAA certification 7 May 1999; German LBA certification 26 May 1999; first customer delivery of completed aircraft 8 July 1999 to AirFlite Inc of Long Beach, California, which operates the aircraft on behalf of Toyota Motor Sales USA; 50th 'green' airframe delivered to Montréal completion centre 7 June 2000. Transport Canada and JAA RVSM approval granted 16 January 2001, followed by FAA RVSM approval on 29 January. Thales Avionics head-up flight display system (HFDS) achieved Transport Canada certification on 14 September 2001 and FAA approval on 4 October 2001. Following exploratory discussions with potential suppliers, development of an enhanced vision system (EVS) is expected to begin in 2002. Total of 57 aircraft in customer service by 1 January 2002, by which date in-service aircraft had accumulated 23,573 flying hours with a despatch rate of 98.87 per cent.
CURRENT VERSIONS: **Corporate Transport:** *As described.*
 Airliner: Market studies were being conducted in early 1999 for a modified version of the Global Express, seating 12 to 16 business class passengers in a three-abreast layout,

Bombardier Global Express long-range business jet

for use on scheduled long-haul passenger flights between secondary airports.

Global Express-ASTOR: On 15 June 1999 Raytheon Systems Limited was chosen as the preferred bidder for the UK Ministry of Defence's Airborne Stand-Off Radar (ASTOR) programme, for which five Global Express airframes will be modified by Bombardier's Belfast division to provide the airborne platform for radar and communications systems. Contract value £800 million (1999). Preliminary Design Review (PDR) scheduled for completion in March 2001; 200-hour flight test programme of an aerodynamically representative airframe, modified from Global Express prototype C-FBGX, began on 3 August 2001, leading to Critical Design Review (CDR) and design freeze in second quarter 2002. First production airframe (C-GJR6/ZJ690) delivered 7 February 2002 to Raytheon Systems Limited at Greenville, Texas, for airframe modification and systems integration, to be followed in 2003 by flight tests and operational acceptance trials before delivery to UK Ministry of Defence in 2004. In-service date is 2005, with aircraft based at RAF Waddington, Lincolnshire.

CUSTOMERS: More than 120 firm orders by October 2000. Announced customers include Bombardier's Flexjet fractional ownership programme, which has ordered 22 for delivery from 2000, the Royal Malaysian Air Force, which has taken delivery of one for VIP duties; Dogus Air of Turkey, which ordered one for delivery in 2001 and the Japanese Civil Aviation Bureau (JCAB), which has ordered two for flight inspection and airways calibration duties, the first of which (c/n 9034) was handed over to JCAB's prime contractor Itochu Corporation on 4 May 2001 after outfitting by Marshall Aerospace at Cambridge, UK. Estimated market for 500 to 800 long-range business jets over 15 years; Bombardier anticipates capturing 50 per cent of the market, breaking even at approximately 100; target production rate 34 per year; total of 35 delivered in 2000, and 21 in 2001. Total of 47 in service at 1 November 2001; flight time then 22,668 hours.

COSTS: Development costs C$800 million; half carried by Bombardier, balance by risk-sharing partners. Unit cost approximately US$40.66 million outfitted (2000).

DESIGN FEATURES: Design goal was longest possible range at highest speed from short runway with 99.5 per cent despatch reliability; wide-body fuselage, combining Challenger cabin cross-section with cabin length of Regional Jet; all-new, 'third-generation supercritical' wings with leading-edge slats and winglets.

Wing sweep 35° at quarter-chord, thickness/chord ratio 11 per cent, dihedral 2° 30′, root incidence 2° 30′. Wing, high-lift devices and wing/fuselage interface and area-ruled rear fuselage/engine pylon junction contours developed with extensive use of computational fluid dynamics (CFD). Rear-mounted engines. Sweptback T tail with 38° sweep and 5° anhedral on tailplane, 45° sweep on fin.

FLYING CONTROLS: Conventional and mechanical. Fully powered primary flying controls with variable artifical feel and emergency back-up via ram air turbine following triple hydraulic failure; dual sidestick controllers; duplicated cable runs with automatic disconnect in the event of control surface jamming; dual power control units on ailerons (maximum deflections +26.5/–23°) and elevators (maximum deflections +24/–19°), triple units on rudder (maximum deflection 37° left/right). Eight-section (total) leading-edge slats (maximum deflection 20°) and six-section (total) single-slotted Fowler flaps (maximum deflection 30°) are signalled by dual electronic control units and operated by dual-motor power units connected by rigid driveshafts to ball-screw actuators. Electrically signalled, hydraulically actuated multifunction spoilers (four per side, outboard, operating differentially to assist ailerons and improve roll response, and symmetrically for speed brake or lift dump functions) and ground spoilers (two per side, inboard), maximum deflection +40°. Horizontal stabiliser incidence adjustable for pitch trim (+13/–2°) via dual-channel electrically driven screw actuator; roll trim accomplished by electric trim actuator

located at aileron feel simulator unit; yaw trim accomplished by electric trim actuator at summing unit in fin. Dual yaw damper stability augmentation system and stick shaker/pusher stall protection system standard.

STRUCTURE: Semi-monocoque fuselage with chemically milled C-188 A1 aluminium alloy skin riveted over alloy frames and stringers to form damage-tolerant structure; main two-spar torsion box wing structure mostly of alloy construction, with machined alloy spars and ribs and polyurethane-coated machined alloy skin panels; two-spar winglets of mixed alloy/composites construction; multispar fin is alloy; ailerons, flaps, spoilers, rudder, two-spar tailplane, elevators, wing/fuselage fairings, flap track fairings, main landing gear bay, upper and lower engine nacelle doors and cabin floor panels are of composites construction.

Bombardier's Canadair division is design authority and manufactures nose section; de Havilland manufactures rear fuselage, engine pylons and vertical stabiliser and is responsible for final assembly at Downsview; Mitsubishi supplies wings and centre fuselage; Short Brothers designed and manufactures forward fuselage, engine nacelles, horizontal stabiliser and other composites components; Bombardier is responsible for interior completions at its Montréal and Tucson, Arizona, facilities. Other participants in the programme are: Honeywell Aerospace (APU), Ametek Aerospace (data acquisition unit, engine vibration monitoring system, fuel flow transmitters and engine thermocouples), Rolls-Royce Deutschland (power plant), Hella (lighting systems), Honeywell (avionics), Liebherr-Aerospace Toulouse (air management system), Lucas Aerospace (electrical systems), Messier-Dowty International (landing gear), Parker Bertea Aerospace (flight controls and fuel and hydraulic systems), Raytheon E-Systems (pitch feel systems), Sextant Avionique (flight control system) and Hamilton Sundstrand (slat/flap actuation system and ram air turbine).

LANDING GEAR: Hydraulically retractable tricycle type with Messier-Dowty oleo-pneumatic shock-absorber and twin wheels on each unit; main units retract inward into wing, nosewheel forwards. Goodyear tyres, mainwheels tyre size H38×12.0-19 (20 ply) tubeless, maximum pressure 11.45 bar (166 lb/sq in); nosewheel tyres 21×7.25-10 (12 ply) tubeless (deflector-type), maximum pressure 9.93 bar (144 lb/sq in). Carbon brakes with dual Goodrich/HydroAire hydraulic digital brake-by-wire/modulated anti-skid system providing pilot-selectable, three-level autobrake capability. Steerable nosewheel, maximum steering angle ±75°; minimum ground turning radius 20.73 m (68 ft 0 in).

POWER PLANT: Two rear-mounted 65.6 kN (14,750 lb st) Rolls-Royce Deutschland BR710A2-20 turbofans, flat rated to ISA + 20°C, with FADEC. International Nacelle Systems (Shorts/Hurel-Dubois joint venture) hydraulically actuated two-petal target-type thrust reversers.

Fuel contained in two integral wing tanks, each of 8,479 litres (2,240 US gallons; 1,865 Imp gallons) capacity, centre-section tank, capacity 6,117 litres (1,616 US gallons; 1,346 Imp gallons), and auxiliary tank in aft fuselage, capacity 1,234 litres (326 US gallons; 271 Imp gallons), giving total standard capacity of 24,310 litres (6,422 US gallons; 5,347 Imp gallons). Fuel from centre-section and auxiliary tanks is transferred to wing tanks from where two AC main pumps and DC back-up pump feed to engines; automatic fuel management system balances quantities in port and starboard wing tanks. Gravity and pressure refuelling; single-point pressure fuelling/defuelling coupling in starboard wing/fuselage fairing. Oil capacity 20 litres (5.3 US gallons; 4.4 Imp

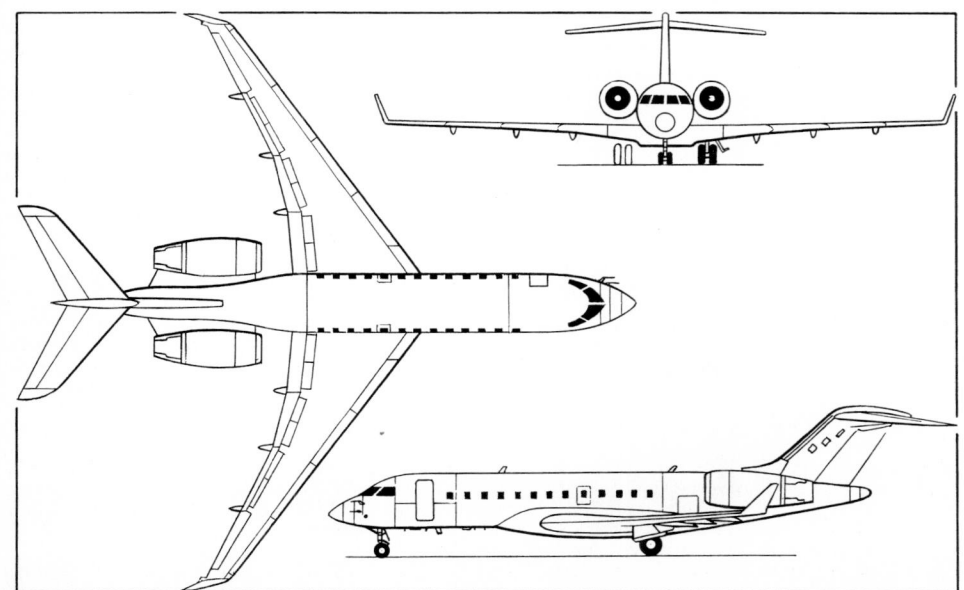

Bombardier BD-700 Global Express (*Jane's/James Goulding*)

Global Express cabin interior   *NEW*/0137416

gallons) with oil replenishment tank, capacity 5.7 litres (1.5 US gallons; 1.2 Imp gallons) permitting remote oil servicing from the cockpit.

ACCOMMODATION: Crew of three or four (including cabin attendant) and eight to 19 passengers depending on interior fit. Customised cabin interior according to customer requirements. Typical arrangement comprises three-compartment cabin with lavatory at rear, crew rest area, galley, small lavatory and wardrobe forward, and provision for 'office in the sky', stateroom or conference area. Flight-accessible baggage compartment at rear of cabin with external plug-type door forward of port engine intake. Accommodation is heated, air conditioned and pressurised; predicted cabin noise level 52 dB. Thirteen cabin windows per side, each 40 cm (1 ft 3¾ in) high × 27.4 cm (10¾ in) wide; one window over wing on starboard side doubles as plug-type emergency exit. Electrically operated airstair door at front of cabin on port side.

SYSTEMS: Integrated air management system by Liebherr-Aerospace Toulouse provides engine bleed, wing anti-ice, air conditioning, cabin pressurisation and avionics ventilation. Digitally controlled dual cooling pack system with ozone converters and bleed air filters provides cabin air circulation at standard rate 1.81 m³ (64 cu ft)/min/person, maximum rate 2.29 m³ (81 cu ft)/min/person with crew-selectable 100 per cent fresh air or recirculation and three air sources for cabin pressure control; maximum pressure differential 0.66 bar (9.64 lb/sq in) maintains a 1,525 m (5,000 ft) cabin altitude to 12,500 m (41,000 ft) and a 2,200 m (7,220 ft) cabin altitude to maximum operating altitude of 15,545 m (51,000 ft). Engine bleed air anti-icing for wing leading-edge fixed surfaces and slats; tail surfaces unprotected; bleed management system automatically switches between low- and high-pressure compressor air to improve engine efficiency. Oxygen system comprises four 1,417 litre (50 cu ft) oxygen cylinders pressurised to 127.6 bar (1,850 lb/sq in) for passenger and crew use.

Lucas/Leach electrical power generation and distribution system comprises two 40 kVA variable frequency generators on each engine, supplying primary 115/200 V three-phase AC electrical power at 324 to 596 Hz; alternative AC power provided by 45 kVA APU-mounted generator and emergency power by 9 kVA air-driven generator, the latter automatically deployed in the event of power loss; electrical management system automatically performs priority-based load-shedding and reconfiguration in event of failure. Four 150 A TRUs convert AC to 28 V DC; emergency DC provided by 25 Ah and 42 Ah low-maintenance Ni/Cd batteries. Provision for external AC and DC power connection. Triple logic-controlled AC power centre performs primary AC power distribution and high-power secondary distribution via solid-state switches and 'smart'-contactors; low-power AC distributed through thermal circuit breakers in the cockpit. Triple logic-controlled DC power centre provides non-interruptible primary DC power distribution, emergency bus supplies and normal DC supplies to four secondary power distribution assemblies (SPDAs) throughout the aircraft to provide remote logic-controlled power to all DC loads. Two CDUs in the cockpit allow for remote sensing/setting and resetting of circuit breakers.

Tailcone-mounted Honeywell RE220(GX) APU provides electrical power (45 kVA ground; 40 kVA flight),

as well as bleed air and main engine starting; APU is certified for operation up to 13,715 m (45,000 ft), in-flight starting up to 11,280 m (37,000 ft) and engine starting up to 9,145 m (30,000 ft).

Triple-redundant hydraulic systems at pressure of 207 bar (3,000 lb/sq in), with bootstrap reservoirs.

Walter Kidde Aerospace integrated aircraft fire detection and extinguishing system provides continuous fire detection monitoring in engine nacelles, APU compartment, main landing gear bays and cabin; dual extinguishers provide two-shot fire suppression in main engine and APU bays. Aircraft serviceability monitored by CAIMS (central aircraft information and maintenance system) with facilities including in-flight display.

AVIONICS: Honeywell Primus 2000XP as core system.

Comms: Dual VHF (third optional); dual Rockwell Collins HF; dual transponders; dual radio management systems; Coltech five-channel Selcal; Honeywell digital FDR and CVR; ELT; satcom optional, with antenna mounted in fin cap; Teledyne Magnastar Office in the Sky datalink optional.

Radar: Colour weather radar with dual controllers.

Flight: Dual flight management systems (third optional) with dual Cat. II autopilots and triple digital air data computers providing fail-safe AFCS; triple laser gyro inertial reference systems; GPS with option for second sensor; ADF; VOR/ILS; DME; TCAS II; Honeywell EGPWS with terrain database integrated into Primus 2000XP system for EFIS display.

Instrumentation: Dual EFIS comprising six 203 × 178 mm (8 × 7 in) CRT multifunction displays, for PFD and EICAS functions; dual Rockwell Collins digital radio altimeter; combined standby airspeed/altimeter, standby artificial horizon and stowable standby heading indicator. Thales HFDS with Cat. II landing capability and lightning sensor system optional.

DIMENSIONS, EXTERNAL:

Wing span over winglets	28.65 m (94 ft 0 in)
Wing chord: at root	6.43 m (21 ft 1 in)
at tip	1.24 m (4 ft 1 in)
Wing aspect ratio	8.6
Length: overall	30.30 m (99 ft 5 in)
fuselage	26.31 m (86 ft 4 in)
Diameter of fuselage, constant	2.69 m (8 ft 10 in)
Height overall	7.57 m (24 ft 10 in)
Tailplane span	9.68 m (31 ft 9 in)

Wheel track (c/l of shock-absorbers)	4.06 m (13 ft 4 in)
Wheelbase	12.78 m (41 ft 11 in)
Passenger door: Height	1.83 m (6 ft 0 in)
Width:	0.91 m (3 ft 0 in)
Baggage door: Height	0.84 m (2 ft 9 in)
Width	1.09 m (3 ft 7 in)
Emergency exit: Height	0.99 m (3 ft 3 in)
Width	0.51 m (1 ft 8 in)

DIMENSIONS, INTERNAL:

Cockpit: volume	3.99 m³ (141.0 cu ft)
Cabin (excl flight deck): Length	14.73 m (48 ft 4 in)
Width at floor	2.11 m (6 ft 11 in)
Max width	2.49 m (8 ft 2 in)
Max height	1.90 m (6 ft 3 in)
Floor area	31.1 m² (335 sq ft)
Volume, incl baggage compartment	
	60.6 m³ (2,140 cu ft)

AREAS:

Wings, basic	94.95 m² (1,022.0 sq ft)
Horizontal tail surfaces (total)	22.76 m² (245.0 sq ft)
Vertical tail surfaces (total)	17.28 m² (186.0 sq ft)

WEIGHTS AND LOADINGS:

Operating weight empty	22,816 kg (50,300 lb)
Max payload	2,585 kg (5,700 lb)
Payload with max fuel	725 kg (1,600 lb)
Fuel with max payload	17,804 kg (39,250 lb)
Max fuel weight	19,663 kg (43,350 lb)
Max ramp weight: standard	43,205 kg (95,250 lb)
optional	43,638 kg (96,250 lb)
Max T-O weight: standard	43,091 kg (95,000 lb)
optional	43,545 kg (96,000 lb)
Max landing weight	35,652 kg (78,600 lb)
Max zero-fuel weight	25,401 kg (56,000 lb)
Max wing loading: standard	453.8 kg/m² (92.95 lb/sq ft)
optional	458.6 kg/m² (93.93 lb/sq ft)
Max power loading: standard	328 kg/kN (3.22 lb/lb st)
optional	332 kg/kN (3.25 lb/lb st)

PERFORMANCE

Max level speed (VMO): S/L to 2,440 m (8,000 ft)	
	300 kt (555 km/h; 345 mph) CAS
2,440 m (8,000 ft) to 9,410 m (30,870 ft)	
	340 kt (629 km/h; 391 mph) CAS
above 9,410 m (30,870 ft)	M0.89
High cruising speed M0.88 (505 kt; 935 km/h; 581 mph)	
Normal cruising speed	
	M0.85 (488 kt; 904 km/h; 562 mph)
Long-range cruising speed at 13,715 m (45,000 ft)	
	M0.80 (459 kt; 850 km/h; 528 mph)
Max rate of climb at S/L	1,097 m (3,600 ft)/min
Initial cruising altitude	13,105 m (43,000 ft)
Time to climb to initial cruising altitude	30 min
Max certified altitude	15,545 m (51,000 ft)
T-O to 11 m (35 ft)	1,774 m (5,820 ft)
Landing from 15 m (50 ft) at max landing weight	
	814 m (2,670 ft)
Runway LCN	55
Range with max fuel and eight passengers, NBAA IFR reserves:	
at M0.85	6,010 n miles (11,130 km; 6,916 miles)
at M0.87	5,276 n miles (9,771 km; 6,071 miles)
Range with max payload:	
at M0.85	5,187 n miles (9,606 km; 5,969 miles)
at M0.87	4,596 n miles (8,511 km; 5,289 miles)
Design g limit	+2.5

OPERATIONAL NOISE LEVELS:

T-O	82.4 EPNdB
Approach	89.8 EPNdB
Sideline	88.6 EPNdB

UPDATED

BOMBARDIER BD-100 CHALLENGER 300

TYPE: Business jet.

PROGRAMME: Design study, then known as 'Bombardier Model 70', revealed at the Paris Air Show in June 1997; formally announced at NBAA Convention at Las Vegas 18 October 1998; launched at Paris Air Show 13 June 1999; initially named Continental; engineering designation BD-100-1A10; first metal cut 21 October 1999 following completion of joint definition phase; AS907 engine first flown 29 January 2000, engine certification expected in March 2001; wing/fuselage mating of first aircraft

Global Express ASTOR aerodynamic prototype C-FBGX   *NEW*/0137417

Bombardier Challenger 300 flight deck   0126928   **Bombardier Challenger 300 cabin mockup**   0126929

achieved 19 November 2000; first flight (c/n 20001/C-GJCJ) from the Bombardier Flight Test Center at Wichita's Mid-Continent Airport 14 August 2001, followed by second aircraft (c/n 20002/C-GJCF) on 9 October. These and three further aircraft (c/n 20003/C-GIPX, dedicated to avionics test and flown 6 December 2001; c/n 20004/C-GJCV for systems testing and the first to be fully outfitted with standard interior, flew 5 April 2002; and c/n 20005/C-GIPZ, for function and reliability testing, originally due to fly in May 2002 but had not done so by mid-December), will participate in the flight test and certification programme scheduled to last for more than 1,500 flight hours, culminating in Transport Canada 525 approval in the third quarter of 2002 and US Federal Aviation Administration FAR Pt 25 and European JAR 25 certification, with RVSM approval, shortly afterwards. Total 300 hours by three aircraft had been accumulated up to 5 April 2002 and 925 hours in 400 flights up to 4 September 2002.

Public debut at NBAA Convention, New Orleans, 11 December 2001 (formal presentation 12 December). Re-named Challenger 300 on 8 September 2002, immediately prior to NBAA Convention at Orlando, Florida. Customer deliveries scheduled to begin in late 2002, with up to five aircraft expected to be handed over to operators by the end of the year. Target production up to 30 in 2003, rising to 40 to 50 in 2004 and the maximum planned rate of 60 per year from 2005.

CUSTOMERS: Two orders signed at time of launch, by customers in Germany and United Arab Emirates; total of 115 firm orders received by 1 October 2001, including 25 for Bombardier's Flexjet fractional ownership programme and five for its Middle East and Arab nations distributor TAG Aeronautics. Bombardier anticipates gaining 30 per cent of estimated 1,200-aircraft market in this class by 2012, with fractional ownership operations especially targeted.

COSTS: Development cost C\$500 million (1998); break-even estimated at 300th aircraft. Unit cost US\$14.25 million typically equipped (1998). Direct operating cost estimated at US\$940.94 per hour (2001).

DESIGN FEATURES: Design goals included coast-to-coast range across USA with eight passengers in cabin with stand-up headroom and take-off field length less than 1,525 m (5,000 ft). General configuration is as shown in the accompanying illustrations; supercritical wing with winglets, sweepback 27° at quarter-chord.

FLYING CONTROLS: Conventional. Ailerons manually actuated via cables, pulleys and pushrods, each with a geared tab and fixed tab, plus trim tab on port aileron only; maximum aileron deflections +23/−19°. Horn-balanced elevators, maximum deflections +23/−18°, and single rudder panel, maximum deflection ±30°, hydraulically actuated by cables and pulleys with manual reversion, each with dual PCUs; variable incidence tailplane for pitch trim, maximum travel +2/−13°. Hydraulically actuated Fowler flaps, maximum deflection 30°; each wing has two-segment multifunction spoiler outboard, maximum deflection 45°, and two-segment ground spoiler/lift dumper inboard, maximum deflection 60°; yaw damper standard.

STRUCTURE: Primarily light alloy, with composites for some non-structural fairings; fuselage of semi-monocoque construction with frames and stringers; two-spar wing; three-spar fin. Programme suppliers include: AIDC Taiwan (rear fuselage and tail unit); Canadair (cockpit, forward fuselage and primary flight controls); DeCrane Aircraft (cabin interior) ECE (electrical system and cockpit lighting); Fischer Austria (wing-to-fuselage fairings); GKN Westland (engine nacelles); Goodrich (wheels and brakes); Hawker de Havilland Australia (tailcone and APU installation kit); Hella (lighting); Honeywell (power plant and APU); Hurel-Dubois (thrust reversers); Intertechnique (fuel system); Liebherr Aerospace-Toulouse (environmental control and anti-icing systems); Liebherr Aerospace Lindenberg (flap control system); Messier-Dowty (landing gear); Mitsubishi Heavy Industries (wing); Moog (secondary flight controls); NLX (flight training device and level C/D flight simulator); Parker Aerospace (hydraulic system); PPG Industries (cockpit windscreens and cabin windows); Rockwell Collins (avionics); Scott Aviation (oxygen system); Shorts (centre fuselage); and Walter Kidde (fire detection and suppression system). Final assembly will be at Bombardier's Learjet facility in Wichita, with interior completion in Tucson.

LANDING GEAR: Hydraulically retractable tricycle type by Messier-Dowty, with two wheels on each unit; trailing-link-type main units retract inwards, nosewheel forwards. Steerable nosewheel, maximum deflection ±65°. Mainwheel tyre size 26.5×8.0-18, nosewheel tyre size 18×5.5-10. Goodrich carbon composites multiple disc brakes. Turning radius 17.68 m (58 ft 0 in).

POWER PLANT: Two Honeywell AS907 turbofans with FADEC, each with thermodynamic rating of 35.81 kN (8,050 lb st), flat-rated to 28.91 kN (6,500 lb st) with APR at ISA+15°C. All fuel contained in two integral wing tanks, combined capacity 7,684 litres (2,030 US gallons; 1,690 Imp gallons). Gravity fuelling point in top of each wing, near leading-edge, plus single-point pressure refuelling/defuelling port in starboard wingroot near leading-edge. Target-type reversers standard.

ACCOMMODATION: Two crew flight deck; cabin, with flat floor, accommodates eight passengers in standard 'double club' arrangement on tracking, swivelling and reclining 16 g seats with retractable headrests, cabin management controls, cupholders and shoulder harnesses; three-seat 16 g take-off and landing-certified divan with extending backrest optional as interchange for two club seats. Standard cabin equipment includes fold-out work tables; one 110 V electrical outlet per club seat group; hot drinks dispensers; DVD/CD player with hi-fi grade audio speakers and two 381 mm (15 in) flat screen monitors; Airshow 400 system; Magnastar 2000 in-flight telephone with two handsets and switchable locations; extended-life LED lighting; forward galley with microwave oven; forward passenger wardrobe and crew coat closet; aft lavatory and vanity unit with hot and cold water and removable waste tank, and flight-accessible baggage compartment. Free-fall opening/power-assisted-closure, semi-plug-type airstair cabin door, on port side immediately aft of flight deck, also serves as Type I emergency exit; single plug-type overwing Type III emergency exit on starboard side, between rearmost pair of club seats. External baggage door aft of port wing trailing-edge. Cabin and baggage compartment pressurised, air-conditioned and heated.

SYSTEMS: Two independent phosphate-ester hydraulic systems with one engine-driven pump and one DC motor pump per system, pressure 207 bar (3,000 lb/sq in), plus one auxiliary system powered by an accumulator. Pressurisation system, differential 0.60 bar (8.78 lb/sq in), with auxiliary system providing pressurisation up to 10,670 m (35,000 ft). 28 V DC electrical system comprises three 400 Ah DC brushless generators (one each on

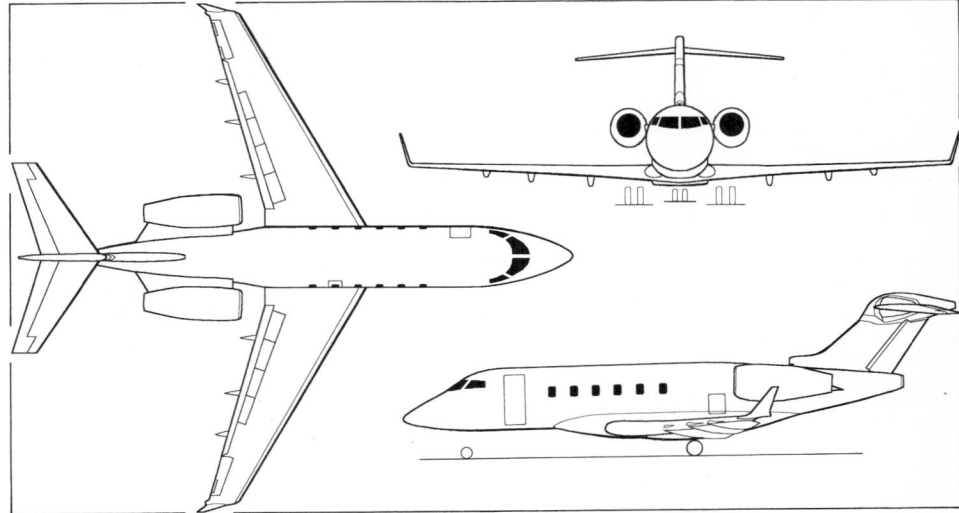

Provisional general arrangement of the Bombardier Challenger 300 *(Jane's/Paul Jackson)*   0085636

For details of the latest updates to *Jane's All the World's Aircraft* online and to discover the additional information available exclusively to online subscribers please visit

jawa.janes.com

First three BD-100 Challenger 300 prototypes *NEW*/0137418

the engines and one on the APU) and two 24 V 44 Ah Ni/Cd batteries which provide power for APU starting, in-flight emergency power and ground power. APU generator can carry load of a failed engine generator, and one battery can supply power for APU starting. Oxygen system, capacity to suit customer requirements, with demand-type masks for crew and drop-down masks for passengers.

Engine bleed air automatically controlled anti-icing for wing leading-edges and nacelle lips; electrically anti-iced windscreen; heated angle-of-attack vanes and pitot probes. Honeywell tailcone-mounted RE220 APU, with FADEC, will be certified for operation up to 11,280 m (37,000 ft) and in-flight starting to 9,150 m (30,000 ft).

AVIONICS: Rockwell Collins Pro Line 21 as core system.

Comms: Dual VHF com with 8.33 kHz frequency spacing capability; dual integrated radio control and display units; dual transponders, all standard. Third VHF com; dual HF com, satcom, VHF/satcom datalink capability, Selcal and ELT optional.

Radar: Dual-scan digital weather radar with optional turbulence detection.

Flight: Standard equipment includes dual ILS/VOR/markers, AHRS and air data computers; single ADF, DME, FMS/CDU, GPS sensor, EGPWS, TCAS II, EICAS, radio altimeter, CVR and flight deck aural warning system. Second ADF, DME, FMS/CDU and GPS, three-dimensional flight plan maps, FDR and lightning sensor optional.

Instrumentation: EFIS with four 305 × 254 mm (12 × 10 in) colour LCDs providing liquid PFD and MFD functions for pilot and co-pilot.

DIMENSIONS, EXTERNAL:

Wing span over winglets	19.46 m (63 ft 10 in)
Length overall	20.93 m (68 ft 8 in)
Height overall	6.20 m (20 ft 4 in)
Fuselage max diameter	2.34 m (7 ft 8 in)
Tailplane span	8.45 m (27 ft 8½ in)
Wheel track	3.20 m (10 ft 6 in)
Wheelbase	8.47 m (27 ft 9½ in)
Passenger door: Height	1.89 m (6 ft 2½ in)
Width	0.76 m (2 ft 6 in)
Baggage door: Height	0.76 m (2 ft 6 in)
Width	0.61 m (2 ft 0 in)
Height to sill	1.61 m (5 ft 3½ in)
Emergency exit: Height	0.91 m (3 ft 0 in)
Width	0.52 m (1 ft 8½ in)

DIMENSIONS, INTERNAL:

Cabin (excl cockpit):	
Length	8.71 m (28 ft 7 in)
Width: at centreline	2.18 m (7 ft 2 in)
at floor	1.55 m (5 ft 1 in)
Max height	1.85 m (6 ft 1 in)

Fourth Bombardier Challenger 300 taking off for the first time on 5 April 2002 *NEW*/0137388

Floor area	13.5 m² (146 sq ft)	Payload: max	1,360 kg (3,000 lb)
Volume	25.4 m³ (896 cu ft)	with max fuel	725 kg (1,600 lb)
Baggage compartment volume	3.0 m³ (106 cu ft)	Max fuel weight	6,214 kg (13,700 lb)
AREAS:		Max T-O weight	17,010 kg (37,500 lb)
Wings, net	48.49 m² (522.0 sq ft)	Max ramp weight	17,078 kg (37,650 lb)
Ailerons, total	0.93 m² (10.00 sq ft)	Max landing weight	15,308 kg (33,750 lb)
Trailing-edge flaps, total	7.56 m² (81.40 sq ft)	Max zero-fuel weight	11,498 kg (25,350 lb)
Spoilers, total	3.62 m² (39.00 sq ft)	Fuel with max payload	5,579 kg (12,300 lb)
Rudder, incl tabs	1.89 m² (20.40 sq ft)	Max wing loading	350.8 kg/m² (71.84 lb/sq ft)
Tailplane	3.92 m² (42.20 sq ft)	Max power loading	294 kg/kN (2.88 lb/lb st)
Elevators	3.81 m² (41.00 sq ft)	PERFORMANCE (estimated):	
WEIGHTS AND LOADINGS (provisional):		Max level speed	476 kt (882 km/h; 548 mph)
Operating weight empty	10,138 kg (22,350 lb)	High cruising speed	
Outfitting allowance	1,315 kg (2,900 lb)		M0.82 or 470 kt (870 km/h; 541 mph)

Challenger 300 wearing its new name for the first time at NBAA Convention, Orlando, Florida, September 2002 *(Jane's/Paul Jackson)* *NEW*/0533392

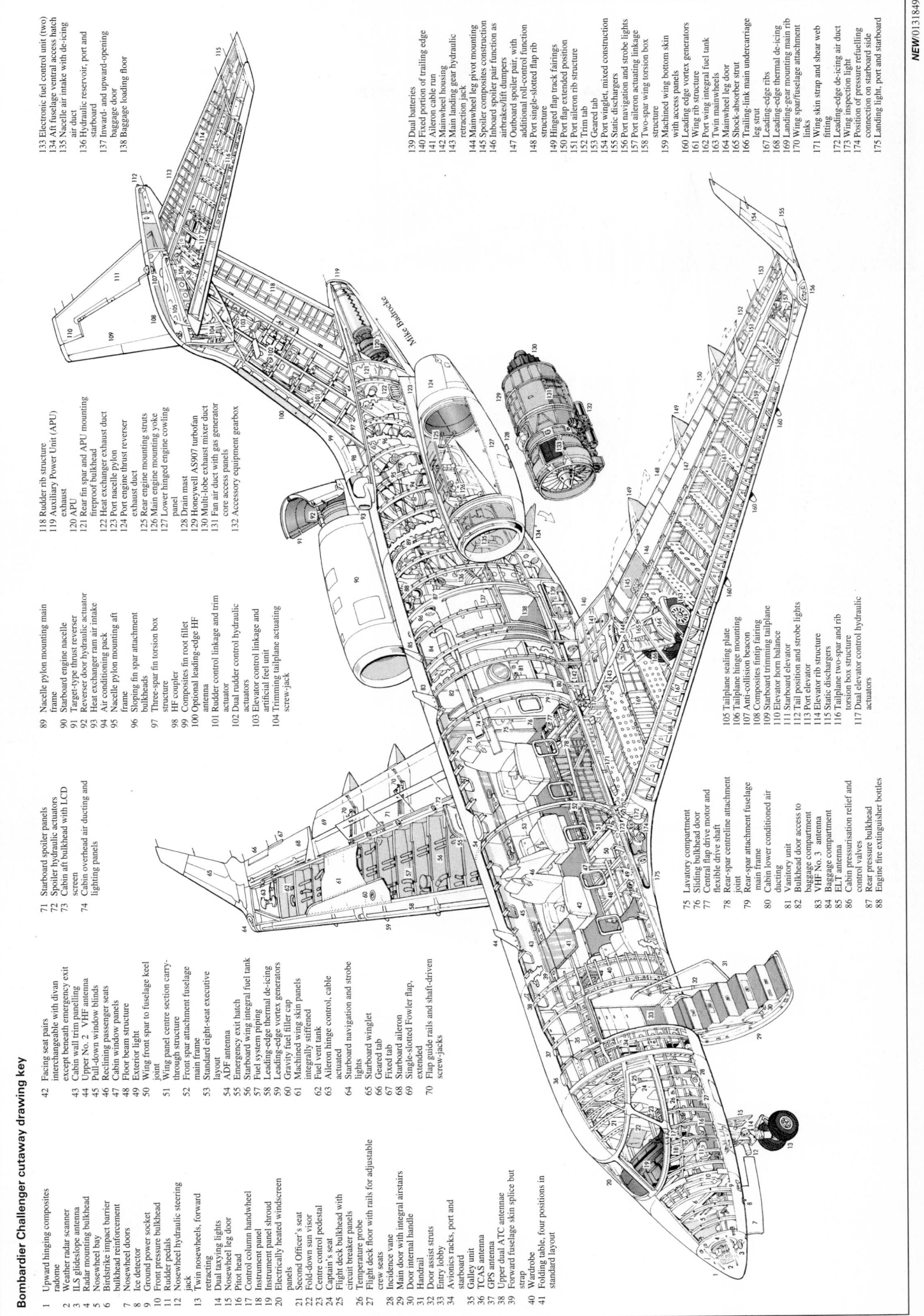

Mike Badrocke

NEW/0131849

Bombardier Challenger cutaway drawing key

1 Upward hinging composites radome
2 Weather radar scanner
3 ILS glideslope antenna
4 Radar mounting bulkhead
5 Nosewheel bay
6 Birdstrike impact barrier bulkhead reinforcement
7 Nosewheel doors
8 Ice detector
9 Ground power socket
10 Front pressure bulkhead
11 Rudder pedals
12 Nosewheel hydraulic steering jack
13 Twin nosewheels, forward retracting
14 Dual taxiing lights
15 Nosewheel leg door
16 Pitot head
17 Control column handwheel
18 Instrument panel
19 Instrument panel shroud
20 Electrically heated windscreen panels
21 Second Officer's seat
22 Fold-down sun visor
23 Centre control pedestal
24 Captain's seat
25 Flight deck bulkhead with circuit breaker panels
26 Temperature probe
27 Flight deck floor with rails for adjustable crew seats
28 Incidence vane
29 Main door with integral airstairs
30 Door internal handle
31 Handrail
32 Door assist struts
33 Entry lobby
34 Avionics racks, port and starboard
35 Galley unit
36 TCAS antenna
37 GPS antenna
38 Upper dual ATC antennae
39 Forward fuselage skin splice but strap
40 Wardrobe
41 Folding table, four positions in standard layout
42 Facing seat pairs interchangeable with divan except beneath emergency exit
43 Cabin wall trim panelling
44 Upper No. 2 VHF antenna
45 Pull-down window blinds
46 Reclining passenger seats
47 Cabin window panels
48 Floor beam structure
49 Exterior light
50 Wing front spar to fuselage keel joint
51 Wing panel centre section carry-through structure
52 Front spar attachment fuselage main frame
53 Standard eight-seat executive layout
54 ADF antenna
55 Emergency exit hatch
56 Starboard wing integral fuel tank
57 Fuel system piping
58 Leading-edge thermal de-icing
59 Leading-edge vortex generators
60 Gravity fuel filler cap
61 Machined wing skin panels integrally stiffened
62 Fuel vent tank
63 Aileron hinge control, cable actuated
64 Starboard navigation and strobe lights
65 Starboard winglet
66 Geared tab
67 Fixed tab
68 Starboard aileron
69 Single-slotted Fowler flap, extended
70 Flap guide rails and shaft-driven screw-jacks
71 Starboard spoiler panels
72 Spoiler hydraulic actuators
73 Cabin aft bulkhead with LCD screen
74 Cabin overhead air ducting and lighting panels
75 Lavatory compartment
76 Sliding bulkhead door
77 Central flap drive motor and flexible drive shaft
78 Rear-spar centreline attachment joint
79 Rear-spar attachment fuselage main frame
80 Cabin lower conditioned air ducting
81 Vanitory unit
82 Bulkhead door access to baggage compartment
83 VHF No. 3 antenna
84 Baggage compartment
85 ELT antenna
86 Cabin pressurisation relief and control valves
87 Rear pressure bulkhead
88 Engine fire extinguisher bottles
89 Nacelle pylon mounting main frame
90 Starboard engine nacelle
91 Target-type thrust reverser
92 Reverser door hydraulic actuator
93 Heat exchanger ram air intake
94 Air conditioning pack
95 Nacelle pylon mounting aft frame
96 Sloping fin spar attachment bulkheads
97 Three-spar fin torsion box structure
98 HF coupler
99 Composites fin root fillet
100 Optional leading-edge HF antenna
101 Rudder control linkage and trim actuator
102 Dual rudder control hydraulic actuators
103 Elevator control linkage and artificial feel unit
104 Trimming tailplane actuating screw-jack
105 Tailplane sealing plate
106 Tailplane hinge mounting
107 Anti-collision beacon
108 Composites fintip fairing
109 Starboard trimming tailplane
110 Elevator horn balance
111 Starboard elevator
112 Tail position and strobe lights
113 Port elevator
114 Elevator rib structure
115 Static dischargers
116 Tailplane two-spar and rib torsion box structure
117 Dual elevator control hydraulic actuators
118 Rudder rib structure
119 Auxiliary Power Unit (APU)
120 APU
121 Rear fin spar and APU mounting fireproof bulkhead
122 Heat exchanger exhaust duct
123 Port nacelle pylon
124 Port engine thrust reverser exhaust duct
125 Rear engine mounting struts
126 Main engine mounting yoke
127 Lower hinged engine cowling panel
128 Drain mast
129 Honeywell AS907 turbofan
130 Multi-lobe exhaust mixer duct
131 Fan air duct with gas generator core access panels
132 Accessory equipment gearbox
133 Electronic fuel control unit (two)
134 Aft fuselage ventral access hatch
135 Nacelle air intake with de-icing air duct
136 Hydraulic reservoir, port and starboard
137 Inward- and upward-opening baggage door
138 Baggage loading floor
139 Dual batteries
140 Fixed portion of trailing edge
141 Aileron cable run
142 Mainwheel housing
143 Main landing gear hydraulic retraction jack
144 Mainwheel leg pivot mounting
145 Spoiler composites construction
146 Inboard spoiler pair function as airbrakes/lift dumpers
147 Outboard spoiler pair, with additional roll-control function
148 Port single-slotted flap rib structure
149 Hinged flap track fairings
150 Port flap extended position
151 Port aileron rib structure
152 Trim tab
153 Geared tab
154 Port winglet, mixed construction
155 Static dischargers
156 Port navigation and strobe lights
157 Port aileron actuating linkage
158 Two-spar wing torsion box structure
159 Machined wing bottom skin with access panels
160 Leading edge vortex generators
161 Wing rib structure
162 Port wing integral fuel tank
163 Twin mainwheels
164 Mainwheel leg door
165 Shock-absorber strut
166 Trailing-link main undercarriage leg strut
167 Leading-edge ribs
168 Leading-edge thermal de-icing
169 Leading-gear mounting main rib
170 Wing spar/fuselage attachment links
171 Wing skin strap and shear web fitting
172 Leading-edge de-icing air duct
173 Wing inspection light
174 Position of pressure refuelling connection on starboard side
175 Landing light, port and starboard

Computer-generated image of Bombardier Global 5000 business jet   *NEW*/0137343

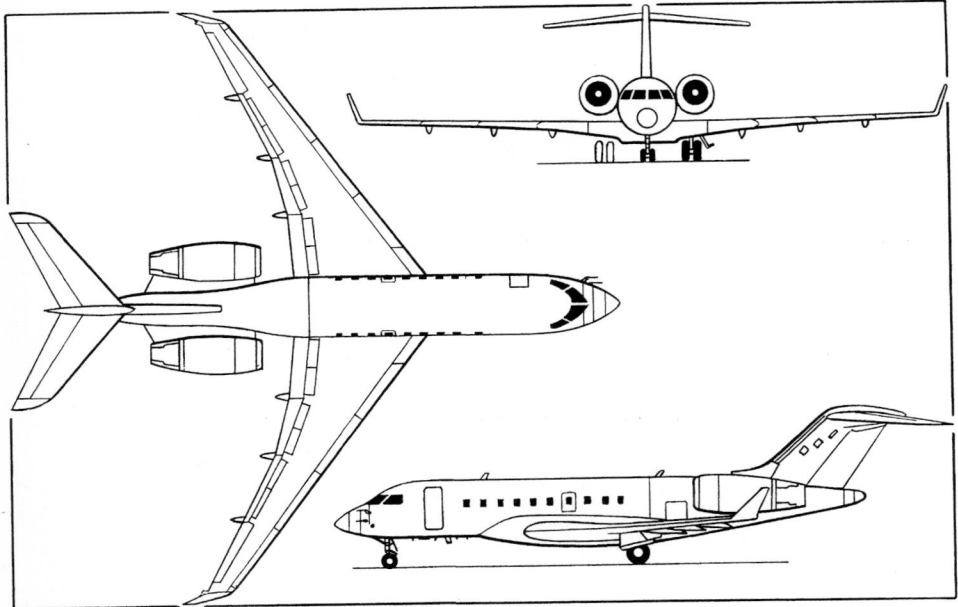

Bombardier Global 5000 general arrangement (*Jane's/James Goulding*)   0126930

Normal cruising speed	
M0.80 or 459 kt (850 km/h; 528 mph)	
Max rate of climb at S/L	1,097 m (3,600 ft)/min
Rate of climb at S/L, OEI	205 m (673 ft)/min
Initial cruising altitude	12,500 m (41,000 ft)
Max certified ceiling	13,715 m (45,000 ft)

T-O balanced field length	1,509 m (4,950 ft)
Landing run at max landing weight	792 m (2,600 ft)
Range with eight passengers, NBAA IFR reserves	
	3,100 n miles (5,741 km; 3,567 miles)

OPERATIONAL NOISE LEVELS:

T-O	89 EPNdB
Approach	98 EPNdB
Sideline	94 EPNdB

UPDATED

BOMBARDIER GLOBAL 5000

TYPE: Long-range business jet.

PROGRAMME: Market and design studies began 1999; announced 25 October 2001; formal launch 5 February 2002; first flight anticipated in first quarter 2003; certification first quarter of 2004; service entry late 2004.

CUSTOMERS: Launch customer TAG Aeronautics ordered five on 5 November 2001; letters of intent for 15 aircraft at time of launch; Sino Private Aviation of Hong Kong ordered one aircraft during Asian Aerospace 2002 in Singapore on 28 February 2002. Potential market for 750 aircraft in class by 2010.

COSTS: US$33 million 'green' (2002).

DESIGN FEATURES: Based on Global Express, with 1.22 m (4 ft 0 in) reduction in fuselage length. Up to 19 passengers.
Data generally as for Global Express, except the following.

DIMENSIONS, EXTERNAL:
Length overall	29.51 m (96 ft 10 in)

DIMENSIONS, INTERNAL:
Cabin (excl flight deck):	
Length	13.92 m (45 ft 8 in)
Volume	57.77 m³ (2,040 cu ft)

WEIGHTS AND LOADINGS:
Basic operating weight	22,838 kg (50,350 lb)
Payload: max	2,563 kg (5,650 lb)
with max fuel	726 kg (1,600 lb)
Max fuel weight	16,329 kg (36,000 lb)
Fuel with max payload	14,492 kg (31,950 lb)
Max T-O weight	39,780 kg (87,700 lb)
Max ramp weight	39,893 kg (87,950 lb)
Max landing weight	35,652 kg (78,600 lb)
Max zero-fuel weight	25,401 kg (56,000 lb)
Max wing loading	419.0 kg/m² (85.81 lb/sq ft)
Max power loading	303 kg/kN (2.97 lb/lb st)

PERFORMANCE:
Cruising speed: max	M0.88 (505 kt; 935 km/h; 581 mph)
normal	M0.85 (488 kt; 904 km/h; 562 mph)
Initial cruising altitude	13,105 m (43,000 ft)
Time to climb to initial cruising altitude	25 min
Max certified altitude	15,545 m (51,000 ft)
Balanced field length	1,525 m (5,000 ft)
Landing run	825 m (2,700 ft)
Range, NBAA IFR reserves, three crew:	
at M0.85	4,800 n miles (8,889 km; 5,523 miles)
at M0.88	3,700 n miles (6,852 km; 4,257 miles)

UPDATED

BOMBARDIER AEROSPACE CANADAIR OPERATIONS

400 Chemin de la Côte Vertu West, Dorval, Québec H4S 1Y9
POSTAL ADDRESS: PO Box 6087, Station Centreville, Montréal , Québec H3C 3G9
Tel: (+1 514) 855 50 00
Fax: (+1 514) 855 79 03
Web: http://www.aero.bombardier.com
VICE-PRESIDENT AND GENERAL MANAGER, OPERATIONS, CANADAIR: Serge Perron
VICE-PRESIDENT, ST LAURENT PLANT, OPERATIONS, CANADAIR: Jean Séguin
VICE-PRESIDENT, DORVAL PLANT, OPERATIONS, CANADAIR: Réal Gervais
INFORMATION OFFICER: Anne-Marie Laroche

Acquired by Bombardier Inc 23 December, 1986, Canadair has manufactured more than 5,000 aircraft since 1944. Mirabel plant, adjacent to Montréal International Airport, until recently comprised 78,975 m² (850,080 sq ft) of floor space for manufacture and assembly of the Challenger and Regional Jet. However, 27,870 m² (300,000 sq ft) final assembly building for CRJ700 and CRJ900 formally opened on 22 October 2001, having begun operations in previous August. Canadair 415 assembly is in a 4,750 m² (51,100 sq ft) plant at North Bay, Ontario. Parts, components and spare parts for various aircraft, including Challenger, Regional Jet, Regional Jet Series 700, Global Express and Canadair 415, are manufactured at St Laurent plant, 248,620 m² (2,676,110 sq ft) in Québec. Structural components for other aircraft builders, such as Boeing and Aerospatiale Matra, are also manufactured in this plant.

Global Express is completed at the 2,820 m² (30,350 sq ft) Bombardier Completion Center, Montréal, beside the Bombardier Aerospace headquarters at Dorval. Total workforce in the Montréal area is more than 11,500.

UPDATED

Canadair Challenger 604 business jet   *NEW*/0137420

CANADAIR CL-600 CHALLENGER
Canadian Forces designations: CC-144, CC-144B and CE-144A

TYPE: Business jet.

PROGRAMME: First flight of first of three prototypes (C-GCGR-X) 8 November 1978; first flight production Challenger 600 with AlliedSignal ALF 502L-2 turbofans 21 September 1979; first customer delivery 30 December 1980; first flight Challenger 601 with GE CF34s 10 April 1982; first 601-1A delivered 6 May 1983; first 601-3A 6 May 1987 and first 601-3A/ER 19 May 1989; first 601-3R 14 July 1993; first 604 25 January 1996. Challenger certified for operation in 40 countries by 1998. By 1 January 2002 the Challenger fleet had flown 2,096,494 hours, with a despatch reliability of 99.6 per cent. 500th Challenger rolled out 'green' 25 May 2000 and handed over (as N816CC) 1 September 2000.

CURRENT VERSIONS: **Challenger 600:** Total 84 built after certification in 1980 (76 since retrofitted with winglets); 12 delivered to Canadian Department of National Defence as CC-144 (three) and CE-144A (three) (see 1989-90 and earlier *Jane's*), plus three for coastal patrol, two for general transport and one test aircraft. Production completed with final delivery on 22 June 1983.

Challenger 601-1A: First production version to have CF34 engines (see 1990-91 *Jane's*); first flight 17 September 1982. Deliveries (66, including four CC-144Bs) between 6 May 1983 and 29 May 1987.

Canadair Challenger 604 flight deck with PrecisionPlus upgraded Rockwell Collins Pro Line 4 integrated avionics   0126924

Challenger 601-3A: Version with 'glass' cockpit and CF34-3A engines; first flight 28 September 1986; Canadian and US certification 21 and 30 April 1987; also certified for Cat. II and in 22 other countries; improvements include CF34-3A engines flat rated to 21°C, and fully integrated digital flight guidance and flight management systems. Total of 134 delivered between 6 May 1987 and 29 October 1993.

Challenger 601-3R: Extended-range option available on new 601-3As since 1989 (c/n 5135 and onwards) and as retrofit to 601-1As and 601-3As; range increased to 3,585 n miles (6,639 km; 4,125 miles) with NBAA IFR reserves; first flight 8 November 1988; Canadian certification 16 March 1989; tail fairing replaced with conformal tailcone fuel tank which extends fuselage length by 46 cm (1 ft 6 in) and adds 118 kg (260 lb) to operating weight empty; maximum ramp weight increased by 680 kg (1,500 lb). Optional gross weight increase of 227 kg (500 lb). Total of 92 modification kits supplied between March 1989 and October 1993. Challenger 601-3ER, incorporating extended-range modifications, CF34-3A1 engines and 20,457 kg (45,100 lb) max T-O weight, was standard production version from 14 July 1993 (first delivery); 59 new-build aircraft delivered by early 1996; no further production.

Challenger 604: Has range of 4,077 n miles (7,550 km; 4,691 miles) at M0.74 and is powered by General Electric CF34-3B engines each rated at 38.8 kN (8,729 lb st) T-O power at ISA + 15°C. Prototype (C-FTBZ) modified on the production line from a Challenger 601-3R; engineering designation CL-600-2B16; first flight (with CF34-3A engines) 18 September 1994; first flight with definitive CF34-3B engines 17 March 1995. Exploits systems developed in Regional Jet programme. Rockwell Collins Pro Line 4 EFIS; extra 1,242 litres (328 US gallons; 273 Imp gallons) of fuel in aft equipment bay, forward fuselage tank and tail tank. Automatic aft-CG control to reduce trim drag for longer range. New landing gear, carbon brakes and anti-skid system; strengthened tail unit; new wing-to-fuselage and underbelly fairings. Maximum T-O weight 21,863 kg (48,200 lb). Transport Canada certification achieved 20 September 1995; FAA certification 2 November 1995; 100th delivery to a customer was made in mid-1999.

From June 2001, Challenger 604s have been delivered with upgraded PrecisionPlus Collins Pro Line 4 avionics, intended to reduce pilot workload and make the aircraft more compatible with future air traffic environments. Standard PrecisionPlus features include: automatic look-up and display of take-off, approach, landing and missed-approach speeds, eliminating the need to refer to manual charts; automatic look-up and display of thrust setting (N_1) for take-off, climb, cruise and go-around; blending of actual observed wind and entered wind to improve the prediction of flight time and fuel requirements; position reporting in non-radar environments such as the North Atlantic; improved polar navigation, enabling the crew to navigate and steer the aircraft at latitudes over 89°; full-time DME reporting on the pilot's MFD; EICAS improvements including the addition of metric fuel indication capability, logic enhancements and FMS performance enhancements; and full integration with the Flight Dynamics HUD and Safe Flight AutoPower autothrottle system. Optional features include: flight plan map feature providing an intuitive, three-dimensional graphic representation of the programmed flight plan and predicted flight path for the pilot's and co-pilot's MFDs; long-range cruise feature allowing pilots to select a cruise speed computed by the FMS for either maximum range or maximum speed; search pattern feature offering automatic generation of waypoints; and expanded FDR to meet FAA FAR Pt 135.152 requirements. The PrecisionPlus avionics upgrade is also available for retrofit to earlier Challenger 604s.

Detailed description applies to Challenger 604.

Special Missions: One Challenger 604 delivered in late 2000 to (South) Korean National Maritime Police with unspecified sensor and communications suite. First of two maritime surveillance 604s was due to enter service with Royal Danish Air Force in 2002.

CUSTOMERS: See under individual headings in Current Versions. Total of 575 Challengers of all versions delivered (including to completion centres) by 31 January 2002, including 231 Challenger 604s. Recent customers include the Royal Jordanian Air Force, which ordered two in VIP configuration for delivery during 2000; the Australian government, which ordered three on 16 August 2000 for delivery in 2001 to Qantas Airways, which will operate them on behalf of the Royal Australian Air Force for transport of senior government officials; REGA Swiss Air-Ambulance Ltd, which ordered three on 26 September 2001 for delivery in June and November 2002, and Shandong Airlines of China, launch operator of Bombardier's Flexjet Asia-Pacific fractional ownership programme, which has ordered four. Annual deliveries have included 33 in 1997, 36 in 1998, 40 in 1999, 38 in 2000, and 41 in 2001.

COSTS: Unit cost (604), US$22.5 million, typically equipped (2000).

DESIGN FEATURES: Advanced wing section; quarter-chord sweep 25°; thickness/chord ratio 14 per cent at root, 12 per cent at leading-edge sweep break and 10 per cent at tip; dihedral 2° 33'; incidence at root 3° 30'; fuselage circular cross-section, pressurised.

FLYING CONTROLS: Conventional, fully powered hydraulic controls; electrically actuated variable incidence tailplane; two-segment spoilers (outboard airbrake panels, inboard lift dumpers); two-segment double-slotted flaps.

STRUCTURE: Two-spar wing torsion box; chemically milled fuselage skin panels with riveted frames and stringers form damage-tolerant structure; multispar fin and tailplane.

LANDING GEAR: Hydraulically retractable tricycle type, with twin wheels and Dowty oleo-pneumatic shock-absorber on each unit. Mainwheels retract inward into wing centre-section, nose unit forward. Nose unit steerable and self-centring. Mainwheels have H27×8.5-14 (16 ply) tubeless tyres, pressure 12.07 bar (175 lb/sq in); nosewheels have Goodrich 18×4.4 (12 ply) tubeless (deflector type) tyres, pressure 10.00 bar (145 lb/sq in). ABS (Aircraft Braking Systems) hydraulically operated multiple-disc carbon brakes with fully modulated anti-skid system. Minimum ground turning radius 12.19 m (40 ft 0 in).

POWER PLANT: Two General Electric CF34-3B1 turbofans, each rated at 41.0 kN (9,220 lb st) with automatic power reserve, or 38.8 kN (8,729 lb st) without APR, pylon-mounted on rear fuselage and fitted with cascade-type fan-air thrust reversers. Nacelles and thrust reversers by Shorts. Integral fuel tank in centre-section, capacity 2,839 litres (750 US gallons; 624 Imp gallons), one in each wing (each 2,725 litres; 720 US gallons; 600 Imp gallons) and auxiliary tanks (combined capacity 1,181 litres; 312 US gallons; 260 Imp gallons) beneath cabin floor. Saddle tanks, total capacity 999 litres (264 US gallons; 220 Imp gallons); tank in tailcone, capacity 745 litres (197 US gallons; 164 Imp gallons). Total fuel capacity 11,214 litres (2,963 US gallons; 2,468 Imp gallons). Pressure and gravity fuelling and defuelling. Oil capacity 13.6 litres (3.6 US gallons; 3.0 Imp gallons).

ACCOMMODATION: Two-pilot flight deck with dual controls. Blind-flying instrumentation standard. Cabin interiors to customer's specifications; maximum of 19 passenger seats and three crew approved. Typical installations include lavatory, buffet, bar and wardrobe. Medevac version can carry up to seven stretcher patients, infant incubator, full complement of medical staff and comprehensive intensive care equipment. Baggage compartment, with own loading door, accessible in flight. Downward-opening, power-assisted door on port side, forward of wing. Overwing emergency exit on starboard side. Entire accommodation heated, pressurised and air conditioned. Optional extended cabin interior increases cabin length by 0.51 m (1 ft 8 in) and provides two additional cabin windows by removing rear closet and moving the lavatory and baggage compartment bulkheads, with corresponding 0.20 m³ (7 cu ft) reductions in baggage capacity. Ultra Electronics active noise vibration control (ANVC) system optional.

SYSTEMS: Honeywell pressurisation and air conditioning systems, maximum pressure differential 0.63 bar (9.1 lb/sq in). Three independent hydraulic systems, each of 207 bar (3,000 lb/sq in). No. 1 system powers flight controls (via servo-actuators positioned by cables and pushrods); No. 2 system for flight controls and brakes; No. 3 system for flight controls, landing gear extension/retraction, brakes and nosewheel steering. Nos. 1 and 2 systems each powered by an engine-driven pump, supplemented by an AC electric pump; No. 3 system by two AC pumps. Two 30 kVA engine-driven generators supply primary 115/200 V three-phase AC electric power at 400 Hz. Four transformer-rectifiers to convert AC power to 28 V DC; one primary 24 V 17 Ah Ni/Cd battery and one auxiliary 24 V 43 Ah battery. Alternative primary power provided

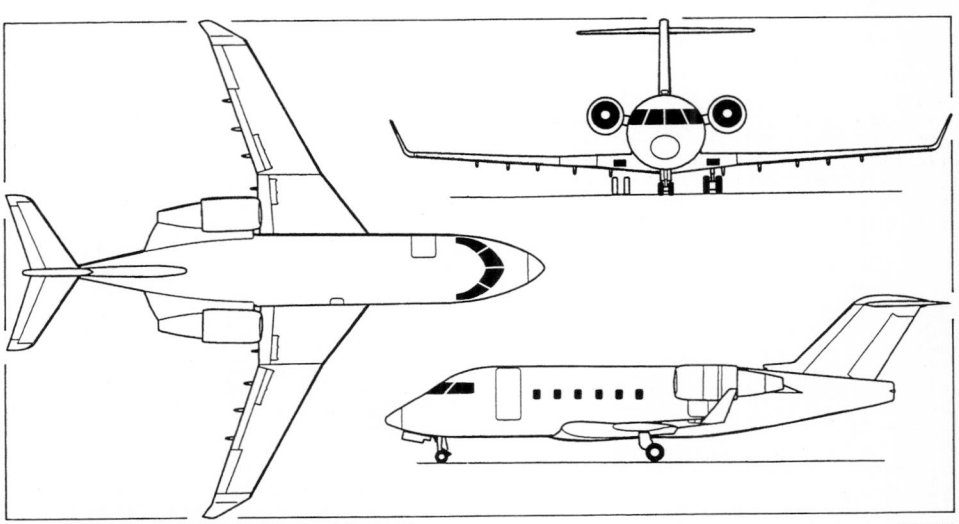

Canadair Challenger 604 *(Jane's/Paul Jackson)*   0085635

by APU and/or an air-driven generator, latter deployed automatically in flight if engine-driven generators and APU are inoperative. Stall warning system, with stick shakers and stick pusher. Honeywell GTCP-100E gas-turbine APU for engine start, ground air conditioning and other services. Electric anti-icing of windscreen, flight deck side windows and pitot heads; Hamilton Sundstrand bleed air anti-icing of wing leading-edges, engine intake cowls and guide vanes. Gaseous oxygen system, pressure 127.5 bar (1,850 lb/sq in). Continuous-element fire detectors in each engine nacelle, APU and main landing gear bays; two-shot extinguishing system for engines, single-shot system for APU.

AVIONICS: Rockwell Collins Pro Line 4 nav/com.

Comms: Dual VHF; dual ATC transponders; dual HF; cockpit voice recorder.

Radar: Rockwell Collins WXP-4220 colour digital weather radar with turbulence detection.

Flight: Dual VHF nav with provision for third; dual DME; dual ADF; dual Litton TN-101 laser inertial reference systems (LIRS) with full provision for third; dual flight management system with provision for third; digital automatic flight control system, with dual-channel autopilot and flight director; Mach trim and auto trim; digital air data system. Flight Dynamics HGS 2150 HUD received FAA approval on 4 April 2000 and is optional. Space provisions for flight data recorder, ELT, dual GPS, EGPWS, AFIS, TCAS and Safe Flight Instrument Corporation AutoPower enhanced autothrottle system.

Instrumentation: Rockwell Collins digital avionics include Pro Line 4 six-tube EFIS with 184 × 184 mm (7¼ × 7¼ in) CRT displays which include two-tube EICAS display (MFD); standby instruments (artificial horizon, airspeed indicator, compass and altimeter). Systems certified for Cat. II operations.

EQUIPMENT (Medevac version): Includes cardiopulmonary resuscitation unit; physio control lifepack comprising heart defibrillator, ECG and cardioscope; ophthalmoscope; respirators and resuscitators; infant monitor; X-ray viewer; cardiostimulator; foetal heart monitor; and anti-shock suit.

DIMENSIONS, EXTERNAL:

Wing span over winglets	19.61 m (64 ft 4 in)
Wing chord: at root	3.99 m (13 ft 1 in)
at tip	1.27 m (4 ft 2 in)
Wing aspect ratio (excl winglets)	8.0
Length overall	20.85 m (68 ft 5 in)
Fuselage: Max diameter	2.69 m (8 ft 10 in)
Length	18.77 m (61 ft 7 in)
Height overall	6.30 m (20 ft 8 in)
Tailplane span	6.20 m (20 ft 4 in)
Wheel track (c/l of shock-struts)	3.18 m (10 ft 5 in)
Wheelbase	7.99 m (26 ft 2½ in)
Passenger door (port, fwd): Height	1.78 m (5 ft 10 in)
Width	0.94 m (3 ft 1 in)
Height to sill	1.63 m (5 ft 4 in)
Baggage door (port, rear): Height	0.84 m (2 ft 9 in)
Width	0.71 m (2 ft 4 in)
Height to sill	1.73 m (5 ft 8 in)
Overwing emergency exit (stbd):	
Height	0.91 m (3 ft 0 in)
Width	0.51 m (1 ft 8 in)

DIMENSIONS, INTERNAL:

Cabin: Length, incl galley, lavatory and baggage area,	
excl flight deck	8.66 m (28 ft 5 in)
Max width	2.49 m (8 ft 2 in)
Width at floor level	2.18 m (7 ft 2 in)
Max height	1.85 m (6 ft 1 in)
Floor area	18.8 m² (202 sq ft)
Volume	32.6 m³ (1,150 cu ft)

AREAS:

Wings, gross (excl winglets)	48.31 m² (520.0 sq ft)
Ailerons (total)	1.39 m² (15.0 sq ft)
Trailing-edge flaps (total)	7.80 m² (84.0 sq ft)
Fin	9.18 m² (98.8 sq ft)
Rudder	2.03 m² (21.9 sq ft)
Tailplane	6.45 m² (69.4 sq ft)
Elevators (total)	2.15 m² (23.1 sq ft)

WEIGHTS AND LOADINGS:

Manufacturer's weight empty	9,806 kg (21,620 lb)
Operating weight empty	12,079 kg (26,630 lb)
Max fuel	9,072 kg (20,000 lb)
Max payload	2,435 kg (5,370 lb)
Payload with max fuel: standard	485 kg (1,070 lb)
optional	757 kg (1,670 lb)
Max T-O weight: standard	21,591 kg (47,600 lb)
optional	21,863 kg (48,200 lb)
Max ramp weight: standard	21,636 kg (47,700 lb)
optional	21,908 kg (48,300 lb)
Max landing weight	17,236 kg (38,000 lb)
Max zero-fuel weight	14,515 kg (32,000 lb)
Max wing loading:	
standard	446.9 kg/m² (91.54 lb/sq ft)
optional	452.6 kg/m² (92.69 lb/sq ft)
Max power loading:	
standard	263 kg/kN (2.58 lb/lb st)
optional	267 kg/kN (2.61 lb/lb st)

PERFORMANCE (at standard max T-O weight, except where indicated):

High-speed cruising speed	
	M0.82 (470 kt; 870 km/h; 541 mph)

Interior of a Challenger 604   NEW/0137419

Normal cruising speed	
	M0.80 (459 kt; 851 km/h; 529 mph)
Long-range cruising speed	
	M0.74 (425 kt; 787 km/h; 489 mph)
Time to initial cruising altitude	21 min
Initial cruising altitude	11,460 m (37,600 ft)
Max certified altitude	12,500 m (41,000 ft)
Service ceiling, OEI: at mid-cruise weight 17,373 kg	
(38,300 lb)	6,920 m (22,700 ft)
at max T-O weight	5,170 m (16,960 ft)
Balanced T-O field length (ISA at S/L) 1,737 m (5,700 ft)	
Landing distance at S/L at max landing weight	
	846 m (2,775 ft)
Range with max fuel and five passengers, NBAA IFR reserves (200 n mile; 370 km; 230 mile alternate):	
long-range cruising speed	
	4,077 n miles (7,550 km; 4,691 miles)
normal cruising speed	
	3,769 n miles (6,980 km; 4,337 miles)
Design *g* limit	+2.5

OPERATIONAL NOISE LEVELS:

T-O	80.9 EPNdB
Sideline	86.2 EPNdB
Approach	90.3 EPNdB

UPDATED

BOMBARDIER CRJ200 and CHALLENGER 800

TYPE: Regional jet airliner.

PROGRAMME: Design studies began in third quarter of 1987; basic configuration frozen June 1988; engineering designation CL-600-2B19; formal programme go-ahead given 31 March 1989; extended-range CRJ100ER announced September 1990. Three development aircraft built (c/n 7001-7003), plus static test airframe (c/n 7991) and forward fuselage test article (7992); first flight of 7001 (C-FCRJ) 10 May 1991; 7002 (C-FNRJ) first flew 2 August 1991 and 7003 on 17 November 1991; all three in 1,400-hour flight test programme in Wichita, USA. CF34-3A1 engine obtained its US type certificate 24 July 1991. Transport Canada type approval (CRJ100 and CRJ100ER) 31 July 1992. Japanese Civil Aviation Bureau certification 23 May 2000.

First delivery aircraft (c/n 7004) flew 4 July 1992, and to Lufthansa CityLine of Germany (as D-ARJA) 29 October 1992; European JAA and US FAA certification 14 and 21 January 1993 respectively; long-range CRJ100LR certified 29 April 1994; CRJ200 with CF34-3B1 engines announced in 1995. Replaced CRJ100 after 226 of the latter had been delivered. Total fleet time at December 2001 (not including CRJ700 or corporate aircraft) was 4,234,666 flight hours

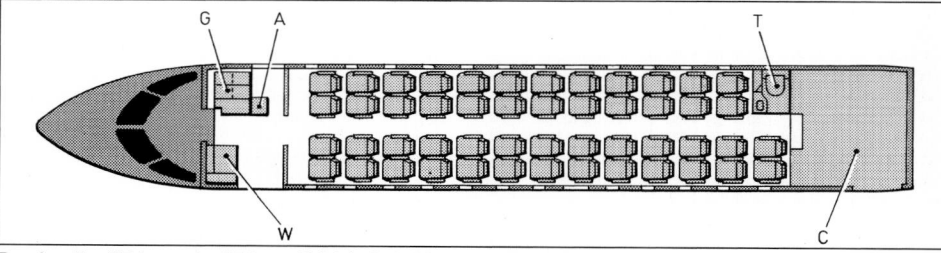

Bombardier CRJ standard 79 cm (31 in) pitch 50-seat layout (*Jane's/Mike Keep*)
A: attendant's seat, C: cargo, G: galley, T: toilet, W: wardrobe

Bombardier CRJ 200 of Australia's Kendell Airlines (*Jane's/Paul Jackson*)   NEW/0137351

Bombardier Challenger 800 business jet

NEW/0533371

CRJ100/200/440 PRODUCTION
(at 31 December 2003)

Customer	Variant	Ordered	Del'd	Backlog
Adria Airways	200	4	4	
Air Canada	100	24	24	
	200	2	2	
Air Dolomiti	200	6	5	1
Air Littoral	100	19	19	
Air Nostrum	200	21	18	3
Air Wisconsin	200	64	29	35
American Eagle	700	25	8	17
Atlantic Coast				
Airlines	200	121	74	47
Atlantic Southeast				
(ASA)	200	45	45	
	700	12	9	3
Brit Air	100	20	20	
	700	12	7	5
British European	200	4	4	
Cameroon Airlines	700	1		1
China Yunnan	200	6	6	
Cimber Air	200	2	2	
COMAIR	100	110	110	
	700	20	6	14
DAC Air	200	2	2	
Delta Connection	200	84	53	31
	700	26		26
GECAS	200	16	5	11
	700	25		25
	900	10		10
Horizon Air	700	30	16	14
Japan Air Lines	200	6	4	2
(JAir)				
Kendell/Ansett	200	12	12	
Lauda Air	100	8	8	
Lufthansa CityLine	100	35	35	
	200	10	10	
	700	20	11	9
Lufthansa/	200	15	7	8
Eurowings				
Maersk	200	11	11	
	700	5	5	
Malev	200	4	2	2
Mesa Air	200	32	32	
	700	20	6	14
	900	20		20
Midway	200	24	24	
Northwest Airlines	200	54	33	21
	440	75	18	57
Saeaga Airlines	200	1	1	
Shandong Airlines	200	5	5	
	700	2		2
Shanghai Airlines	200	3	3	
SkyWest	100	10	10	
	200	90	49	41
South African	200	6	6	
Express				
Southern Winds	200	2	2	
The Fair Inc	200	2	1	1
Tyrolean	200	15	12	3
Challenger 800	SE	17	17	
Totals		**1,215**	**792**	**423**

and 3,747,204 cycles, with 98.4 per cent despatch reliability rate. 200th aircraft delivered (to Lufthansa) 24 October 1997; 300th to Atlantic Coast Airlines in April 1999. 400th to Delta Connection/SkyWest in July 2000. 500th to Atlantic Coast Airliners 26 April 2001, and 600th to Atlantic Southeast Airlines 29 January 2002, and 700th to Air Nostrum 30 October 2002. Production of CRJ200 running at 9.5 per month in 2000, rising to 12.5 per month by late 2001, and 14.5 per month by 2003, with annual targets of 165 in 2003 and 174 in 2004.

CURRENT VERSIONS: **CRJ100:** Original standard aircraft.

CRJ100ER: Replaced by CRJ200ER.

CRJ100LR: Announced March 1994; launch customer, Lauda Air of Austria; replaced by CRJ200LR.

CRJ200: Standard aircraft; designed to carry 50 passengers over 985 n mile (1,824 km; 1,133 mile) range CF34-3B1 engines with 2.8 per cent lower specific fuel consumption than CF34-3A1 of CRJ100, increasing initial cruise altitude by 213 m (700 ft), cruising speed by 2.5 kt (4.5 km/h; 3 mph), and range typically by 1.5 per cent; Class C baggage compartment as standard. First delivery, to Tyrolean Airways as OE-LCF, 15 January 1996. Further improvements in development for introduction on CRJ200 variants during early 1996 included 3 kt (5.5 km/h; 3.5 mph) reduction in V_2 speed to provide 91 m (300 ft) reduction in T-O run at maximum T-O weight; 1 kt (1.8 km/h; 1.2 mph) reduction in V_{REF} to provide 15 m (50 ft) reduction in landing run at typical landing weights; new 8° flap setting to improve second-segment climb performance; and GPS integrated with an upgraded FMS.

CRJ200ER: Extended-range capability with optional increase in maximum T-O weight to 23,133 kg (51,000 lb) and optional additional fuel capacity, for range of 1,645 n miles (3,046 km; 1,893 miles).

CRJ200LR: Longer-range version of CRJ200ER (more than 2,005 n miles; 3,713 km; 2,307 miles); maximum T-O weight increased by 907 kg (2,000 lb) to 24,040 kg (53,000 lb).

CRJ200B, CRJ200B ER and CRJ200B LR: As above, but with optional hot-and-high CF34-3B1 engines providing normal T-O thrust up to ISA+7.8°C (ISA+6.1°C for standard engines), and APR thrust up to ISA+15°C (ISA+6.1°C for standard engines).

CRJ440: Version seating 44 passengers in standard configuration. Launch customer Northwest Airlines has ordered 75.

CRJ700: *Described separately.*

Corporate Jetliner: Company shuttle version with more spacious cabin accommodation for 18 to 30 passengers. One delivered June 1993 to Xerox Corporation. Five ordered by the People's Republic of China in January 1997; operated on behalf of PRC government by China United Airlines crews; contract value C$116 million, including outfitting, pilot and maintenance staff training and spares. Supplanted from September 2002 by corporate version of Challenger 800 (see below).

Challenger 800: Corporate version developed in consultation with launch customer TAG Aeronautics Ltd to meet requirement for non-stop flights, London to Jeddah or equivalent, with three crew and five passengers; or between Middle East city pairs with 15 passengers. First flown 26 May 1995 and formally announced at Paris Air Show in the following month; initially designated **Canadair Special Edition**; first delivery (N877SE) to TAG during Dubai International Aerospace Show in November 1995; second TAG aircraft delivered November 1997. Accommodation for up to 19 passengers

in customised cabin; additional 1,814 kg (4,000 lb) of fuel carried in two auxiliary tanks behind main cabin, extending range to more than 3,000 n miles (5,556 km; 3,452 miles); maximum T-O weight 24,040 kg (53,000 lb); first aircraft powered by standard CF34-3A1 turbofans, but subsequent examples are equipped with CF34-3B1s increasing range to 3,120 n miles (5,778 km; 3,590 miles); Rockwell Collins Pro Line 4 avionics as on RJ, but with third FMS, third VHF, dual Collins HF and Selcal. Manufactured to special order only. Recent customers include Poly Technologies Inc, which ordered two on 16 August 2001 for operation by China Ocean Aviation Group, with deliveries scheduled for 2002. Renamed Challenger 800 on eve of NBAA Convention at Orlando, Florida, 8 September 2002.

CUSTOMERS: See table.

COSTS: Programme development costs C$275 million. Atlantic Coast Airlines order for 10 CRJ200ERs valued at US$200 million (September 1998). Order for two Special Editions valued at US$54.2 million (August 2001).

DESIGN FEATURES: Evolved from Challenger (which see), designed expressly for regional airline operating environment. Advanced transonic wing design, with winglets for high-speed operations; fuel-efficient GE turbofans; options include higher design weights, additional fuel capacity, more comprehensive avionics, and maximum certified altitude raised to 12,500 m (41,000 ft).

Wings, designed with computational fluid dynamics (CFD), have 13.2 per cent (root) and 10 per cent (tip) thickness/chord ratios, 2° 20' dihedral, 3° 25' root incidence and 24° 45' quarter-chord sweepback.

FLYING CONTROLS: Conventional and power-assisted. Primary controls with cables and push/pull rods for multiple redundancy; hydraulically actuated ailerons, elevators and rudder with at least two hydraulic power control unit actuators per surface (three on rudder and elevator); ailerons and elevators fitted with flutter dampers (dual on elevators); rudder with dual-channel control yaw damping; artificial feel and electric trim for roll and yaw; electronically controlled, variable incidence T tailplane for pitch trim and electronically controlled artificial pitch feel. Double-slotted electromechanical flaps with electronically controlled Datron electric motors; BAE fly-by-wire spoiler and spoileron system, four spoilers each side, with inner two functioning as ground spoilers, outer two comprising one flight spoiler and one spoileron, both also providing lift dumping on touchdown. Avionics suite includes engine indication and crew alerting system (EICAS).

STRUCTURE: Semi-monocoque fuselage is damage tolerant FAR/JAR 25 certified airframe with chemically milled skins; flat pressure bulkheads forward of flight deck and aft of baggage compartment; extensive use of advanced composites in secondary structures (passenger compartment floor, wing/fuselage fairings, nacelle doors, wing access door covers, winglets, tailcone, avionics access doors and landing gear doors); comprehensive anti-corrosion treatment and drainage. Wing is one-piece unit mounted to underside of fuselage; two-spar box joined by ribs, covered top and bottom with integrally stiffened skin panels (three upper and three lower each side) for smooth flow; machined or built-up spars and shearweb-type ribs. Short Brothers (UK) manufactures fuselage central section, fore and aft fuselage plugs, wing flaps, ailerons, spoilerons and inboard spoilers.

LANDING GEAR: Hydraulically retractable tricycle type, manufactured by Dowty. Inward-retracting main units each have 15 in Aircraft Braking System (ABS) wheels

with H29×9.0-15 (16 ply) Goodyear tubeless tyres, pressure 11.17 bar (162 lb/sq in) unladen. Nose unit has 18×4.4 (12 ply) tyres (deflector type) and Dowty Canada steer-by-wire steering; unladen tyre pressure 8.62 bar (125 lb/sq in). Aircraft Braking System steel multidisc brakes and fully modulated Hydro Aire Mk III anti-skid system. Minimum taxiway width for 180° turn (with 3.35 m; 11 ft 0 in safety margin) is 22.86 m (75 ft 0 in).

POWER PLANT: Two General Electric CF34-3B1 turbofans, each rated at 41.0 kN (9,220 lb st) with APR and 38.8 kN (8,729 lb st) without. Nacelles produced by Short Brothers. Pneumatically actuated thrust reversers. Fuel in two integral wing tanks, combined capacity 5,300 litres (1,400 US gallons; 1,166 Imp gallons); increasable to 8,080 litres (2,135 US gallons; 1,778 Imp gallons) with optional centre-wing tank. Pressure refuelling point in starboard leading-edge wingroot; transfer rate 474 litres (125 US gallons; 104 Imp gallons)/min at 3.45 bar (50 lb/sq in); two gravity points on starboard wing (one for centre tank) and one on port wing.

ACCOMMODATION: Two-pilot flight deck; one or two cabin attendants. Main cabin seats up to 50 passengers in standard configuration, four-abreast at 79 cm (31 in) pitch, with centre aisle; maximum capacity 52 seats. Various configurations, from 15 to 50 seats, available for corporate version. Downward-opening front passenger door with integral airstairs on port side; plug-type forward emergency exit/service door opposite on starboard side (inoperative on Challenger 800). Inward-opening baggage door on port side at rear. Overwing Type III emergency exit each side (port side door inoperative on Challenger 800). Entire accommodation pressurised, including rear baggage compartment.

SYSTEMS: Cabin pressurisation and air conditioning system (maximum differential 0.57 bar; 8.3 lb/sq in). Primary flight control systems powered by hydraulic servo-actuators with distinct, alternate paths cable and pushrod systems. Electric trim and dual yaw dampers. Three fully independent 207 bar (3,000 lb/sq in) hydraulic systems. Three-phase 115 V AC electrical primary power at 400 Hz supplied by two 30 kVA engine-driven generators; alternative power provided by APU and air-driven generator. Conversion to 28 V DC by five transformer-rectifier units. Main (Ni/Cd) battery 17 Ah, APU battery 43 Ah. Honeywell GTCP 36-150 (RJ) APU and two-pack air conditioning system in rear of fuselage. Wing leading-edges and engine intake cowls anti-iced by engine bleed air. Electric anti-icing of windscreen and cockpit side windows, pitot heads, air data vanes, static sources and sensors. Ice detection system standard.

AVIONICS: *Comms:* Dual VHF nav/com radios. Options include HF radio, single Selcal and 8.33 kHz VHF.

Radar: Rockwell Collins digital weather radar system; split-scan weather radar and radar with turbulence mode optional.

Flight: Dual flight management systems optional. GPWS, windshear detection system and TCAS. EGPWS optional in place of GPWS. Lockheed Martin Fairchild flight data recorder. Dual FMS 4200 and dual IRS in Corporate Jetliner and Challenger 800.

Instrumentation: Rockwell Collins Pro Line 4 integrated all-digital suite, including dual primary flight displays, dual multifunction displays, dual EICAS, dual AFCS, dual AHRS, dual air data system and Cat. II capability with Cat. IIIa optional using head-up guidance system. Dual inertial reference system optional in lieu of AHRS. Flight Dynamics Inc HGS 2100 HUD approved by Transport Canada November 1995, permitting Cat. IIIa operation.

DIMENSIONS, EXTERNAL: As for Challenger 604 except:

Wing span over winglets	21.21 m (69 ft 7 in)
Wing chord: at fuselage c/l	5.13 m (16 ft 10 in)
at tip	1.27 m (4 ft 2 in)
Wing aspect ratio (excl winglets)	8.9
Length: overall	26.77 m (87 ft 10 in)
fuselage	24.38 m (80 ft 0 in)
Height overall	6.22 m (20 ft 5 in)
Fuselage max diameter	2.69 m (8 ft 10 in)
Wheel track	3.17 m (10 ft 5 in)
Wheelbase	11.41 m (37 ft 5 in)
Passenger/crew door (fwd, port):	
Height	1.78 m (5 ft 10 in)
Width	0.91 m (3 ft 0 in)
Service door (stbd, fwd): Height	1.22 m (4 ft 0 in)
Width	0.61 m (2 ft 0 in)
Height to sill (crew/service)	1.63 m (5 ft 4 in)
Baggage door (port, rear): Height	1.09 m (3 ft 7 in)
Width	0.84 m (2 ft 9 in)
Height to sill	1.63 m (5 ft 4 in)
Emergency exit (overwing, stbd):	
Height	0.96 m (3 ft 2 in)
Width	0.51 m (1 ft 8 in)
Turning circle	22.86 m (75 ft 0 in)

DIMENSIONS, INTERNAL: As for Challenger 604 except:

Cabin (incl baggage compartment, excl flight deck):	
Length	14.76 m (48 ft 5 in)
Max height	1.85 m (6 ft 1 in)
Width: at centreline	2.57 m (8 ft 5 in)
at floor level	2.18 m (7 ft 2 in)
Floor area: except 800	32.1 m² (346 sq ft)
800	30.3 m² (326 sq ft)
Volume: except 800	57.1 m³ (2,015 cu ft)
800	53.8 m³ (1,900 cu ft)
Stowage volume:	
main (rear) baggage compartment	9.0 m³ (318 cu ft)
wardrobes/bins/underseat (total)	4.7 m³ (165 cu ft)

AREAS:

Wings: gross (excl winglets)	54.54 m² (587.1 sq ft)
net	48.35 m² (520.4 sq ft)
Ailerons (total)	1.93 m² (20.8 sq ft)
Trailing-edge flaps (total)	10.60 m² (114.1 sq ft)
Spoilers (total)	2.26 m² (24.3 sq ft)
Winglets (total)	1.38 m² (14.9 sq ft)
Fin	9.18 m² (98.8 sq ft)
Rudder	2.03 m² (21.9 sq ft)
Tailplane	9.44 m² (101.6 sq ft)
Elevators (total)	2.84 m² (30.52 sq ft)

WEIGHTS AND LOADINGS:

Manufacturer's weight empty:	
200	13,236 kg (29,180 lb)
200ER, 200LR	13,243 kg (29,195 lb)
800 Corporate	11,703 kg (25,800 lb)
Operating weight empty: 200	13,730 kg (30,270 lb)
200ER, 200LR	13,835 kg (30,500 lb)
800 Corporate	14,424 kg (31,800 lb)
800	15,377 kg (33,900 lb)
Max payload (structural): 200	5,411 kg (11,930 lb)
200ER, 200LR	6,124 kg (13,500 lb)
800 Corporate	5,533 kg (12,200 lb)
800	2,476 kg (5,460 lb)
Max fuel: 200	4,254 kg (9,380 lb)
200ER, 200LR, 800 Corporate	6,489 kg (14,305 lb)
800	8,303 kg (18,305 lb)
Payload with max fuel: 200	3,651 kg (8,050 lb)

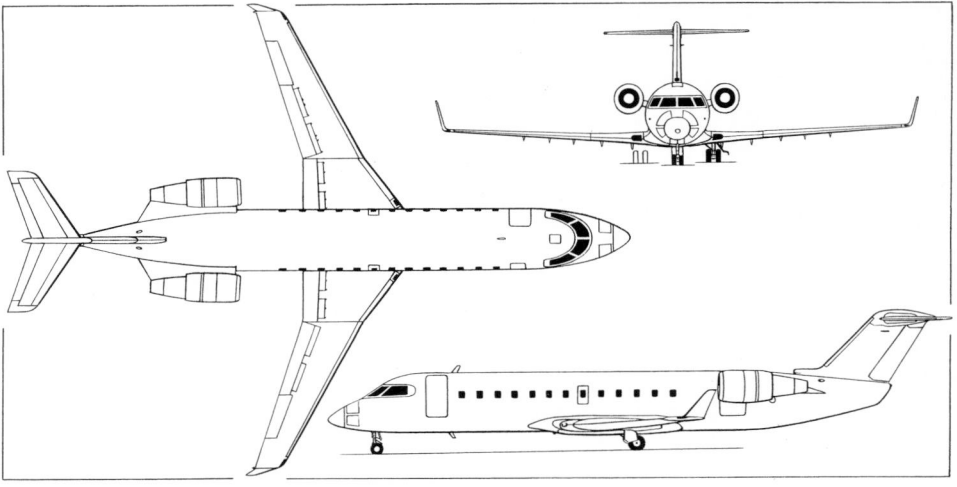

Bombardier CRJ200 (two General Electric CF34-3B1 turbofans)

200ER	2,923 kg (6,445 lb)
200LR	3,831 kg (8,445 lb)
800 Corporate	2,333 kg (5,145 lb)
800	410 kg (905 lb)
Max T-O weight: 200	21,523 kg (47,450 lb)
200ER, 800 Corporate	23,133 kg (51,000 lb)
200LR, 800 Corporate (optional), 800	24,040 kg (53,000 lb)
Max ramp weight: 200	21,636 kg (47,700 lb)
200ER	23,246 kg (51,250 lb)
200LR, 800	24,154 kg (53,250 lb)
Max zero-fuel weight: 800	17,917 kg (39,500 lb)
200	19,141 kg (42,200 lb)
200ER, 200LR, 800 Corporate	19,958 kg (44,000 lb)
Max landing weight: 200	20,275 kg (44,700 lb)
200ER, 200LR, 800 Corporate, 800	21,319 kg (47,000 lb)
Max wing loading: 200	394.6 kg/m² (80.82 lb/sq ft)
200ER	424.1 kg/m² (86.87 lb/sq ft)
200LR, 800 Corporate (optional), 800	440.8 kg/m² (90.27 lb/sq ft)
Max power loading (APR rating):	
200	263 kg/kN (2.57 lb/lb st)
200ER, 800 Corporate	282 kg/kN (2.77 lb/lb st)
200LR, 800 Corporate (optional), 800	293 kg/kN (2.87 lb/lb st)

PERFORMANCE:

Max operating speed:	
above 9,570 m (31,400 ft)	M0.85
below 7,740 m (25,400 ft)	335 kt (621 km/h; 386 mph)
High-speed cruising speed: CRJ200 at 11,275 m (37,000 ft)	M0.81 or 464 kt (859 km/h; 534 mph)
800 Corporate, 800	M0.80 or 459 kt (850 km/h; 528 mph)
Normal cruising speed: CRJ200 at 11,275 m (37,000 ft)	M0.74 or 424 kt (785 km/h; 488 mph)
800 Corporate, 800	M0.77 or 442 kt (819 km/h; 509 mph)
Long-range cruising speed, 800 Corporate, 800	M0.74 or 424 kt (785 km/h; 488 mph)
Approach speed, 45° flap, AUW of 19,504 kg (43,000 lb)	135 kt (250 km/h; 155 mph)
Max rate of climb at 457 m (1,500 ft), 250 kt CAS/M0.74	
climb schedule: 200	1,128 m (3,700 ft)/min
200LR	1,036 m (3,400 ft)/min
800	1,034 m (3,395 ft/min)
Max certified altitude	12,500 m (41,000 ft)
FAR T-O field length at S/L, ISA:	
200	1,527 m (5,010 ft)
200ER	1,768 m (5,800 ft)

200LR	1,917 m (6,290 ft)
800 Corporate	1,765 m (5,790 ft)
800	1,918 m (6,295 ft)
FAR landing field length at S/L, ISA, at max landing weight: 200	1,423 m (4,670 ft)
200ER, 200LR	1,478 m (4,850 ft)
800 Corporate, SE	887 m (2,910 ft)
Range with max payload at long-range cruising speed, FAR Pt 121 reserves:	
200	965 n miles (1,787 km; 1,110 miles)
200ER	1,645 n miles (3,046 km; 1,893 miles)
200LR	2,005 n miles (3,713 km; 2,307 miles)
Corporate (30 seats), NBAA IFR reserves	2,017 n miles (3,735 km; 2,321 miles)
800 with 3,674 kg (8,100 lb) payload	1,541 n miles (2,853 km; 1,773 miles)
Range with max fuel:	
800 Corporate	2,250 n miles (4,167 km; 2,589 miles)
800	3,120 n miles (5,778 km; 3,590 miles)

OPERATIONAL NOISE LEVELS: (CRJ200, FAR Pt 36):

CRJ200:	
T-O	89 EPNdB
Approach	98 EPNdB
Sideline	94 EPNdB
800 Corporate, 800:	
T-O	79 EPNdB
Approach	92 EPNdB
Sideline	82 EPNdB

UPDATED

BOMBARDIER CRJ700

TYPE: Regional jet airliner.

PROGRAMME: Design and market evaluation began in 1995 in consultation with 15-member advisory panel of airline operators; stretched derivative of Regional Jet; originally designated CRJ-X; engineering designation CL-600-2C10. GE CF34-8C1 engines selected February 1995; low-speed wind tunnel testing began early 1995 at Institute of Aeronautical Research, Ottawa; high-speed wind tunnel testing began November 1995 at Rockwell facilities in California. More than 650 hours of wind tunnel testing completed by September 1996. Bombardier Board gave approval for launch on 21 January 1997, at which time orders included four from launch customer Brit Air of France. Aerodynamic configuration frozen 14 March 1997; design frozen 17 July 1998. New version of CF34 engine first ground run in February 1998, followed by

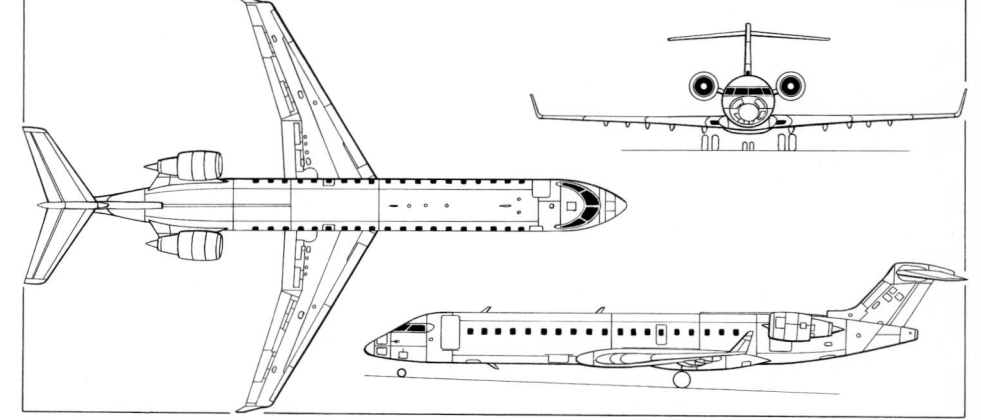

Bombardier CRJ700 regional jet airliner (*Jane's/James Goulding*)

0131848

CRJ700 in the colours of Lufthansa   *NEW*/0137403

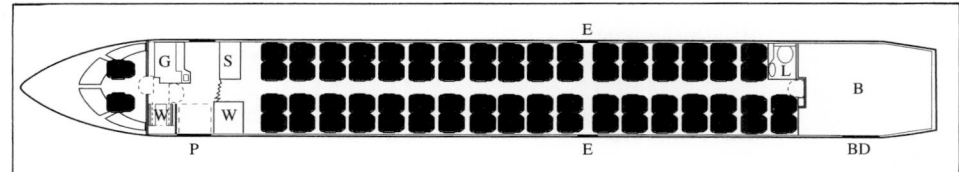

Typical CRJ700 seating plan for 70 passengers at 79 cm (31 in) pitch, with 15.7 m³ (555 cu ft) of baggage
B: baggage, BD: baggage door, E: emergency exit, G: galley, L: lavatory, P: entrance, S: stowage, W: wardrobe

flight tests on GE-owned Boeing 747 testbed in first quarter of 1999 and certification in November 1999. First flight (C-FRJX c/n 10001) 27 May 1999; official roll-out 28 May 1999; public debut (c/n 10004) at Farnborough International Air Show 23 July 2000; Transport Canada certification achieved 22 December 2000; first delivery, to Brit Air of France, in February 2001, followed by deliveries to Horizon Air and Lufthansa CityLine in May.

Four flying aircraft and two static test airframes participated in the certification programme; one static test airframe was used for Complete Aircraft Static Test (CAST), while the other underwent Durability and Damage Tolerance Test (DDTT), and was subjected to the equivalent of 160,000 flight cycles. Test programme was conducted at the Bombardier Flight Test Centre at Wichita, Kansas, totalling 1,600 flight hours on four aircraft: C-FRJX (handling and performance evaluation), 10002/C-FJFC (systems), 10003/C-FBKA (avionics) and 10004/C-FCRJ, the first to be fully furnished (function and reliability testing), which initially flew on 16 December 1999.

CURRENT VERSIONS: **CRJ500:** 'Shrink' derivative, under consideration in mid-2001, combining airframe and engines of CRJ700 with 50-passenger cabin.

CRJ700: 68-seat version in standard and extended range (ER) weight options.

CRJ701: 70-seat version in standard and extended range (ER) weight options.

CRJ702: 72- to 78-seat version in standard and extended range (ER) weight options.

CRJ900: further stretch, described separately.

CUSTOMERS: See table in CRJ200 entry. Total 198 firm orders, of which 68 delivered, by 1 January 2003.

COSTS: Development cost C$645 million, of which C$440 million provided by Bombardier and balance by risk-sharing partners. Unit cost US$24.7 million (1998). Break-even point is at 200th of 400 aircraft production run.

DESIGN FEATURES: Commonality (including crew training and type rating) with CRJ100/200; direct operating cost per seat-mile to be 20 per cent lower than that of CRJ100/200 and lowest of any airliner in its class. Fuselage stretched 4.72 m (15 ft 6 in) by plugs fore and aft of centre-section to seat 70 passengers; rear pressure bulkhead moved aft by 1.29 m (4 ft 3 in); cabin 6.02 m (19 ft 9 in) longer; APU moved to tailcone; cabin floor lowered by 2.5 cm (1 in) and ceiling raised by 1.3 cm (1/2 in) to provide 1.89 m (6 ft 21/4 in) headroom; cabin windows raised 11.5 cm (41/2 in); new underfloor baggage compartment, volume 3.09 m³ (109 cu ft) to facilitate ramp check-in; wing span increased by 1.83 m (6 ft 0 in) by wingroot plug; wing leading-edge extended and equipped with high-lift devices; larger horizontal tail surfaces; new pitch control system; main landing gear lengthened; new wheels, tyres, brakes; air conditioning and anti-icing systems upgraded; new engines; new underfloor baggage compartment on forward port side; overhead stowage bins redesigned.

Incorporation of CRJ900's strengthened wing and avionics software upgrades under consideration in mid-2001 to increase commonality between CRJ700 and CRJ900, affording a 200 n mile (370 km; 230 mile) increase in range for CRJ700.

FLYING CONTROLS: Sextant and Menasco primary and secondary flight control system; flying controls are actuated via a dual network of cables, pulleys and pushrods which operate hydraulic power drive units.

STRUCTURE: Programme participants include Avcorp (vertical and horizontal stabilisers); Bombardier Canadair operations (wing, cockpit, rudder and doors, electrical system, primary flight controls, plus final assembly and interior completion); C & D Interiors (cabin interior), General Electric (power plant), GKN Westland (tailcone and doors), Hella Aerospace GmbH (lighting system), Honeywell (APU), Intertechnique (fuel system), Liebherr Aerospace Toulouse (air management system), Menasco

Aerospace (landing gear), Mitsubishi Heavy Industries (aft fuselage), Parker Abex (hydraulic system), Rockwell Collins (avionics), Sextant Avionique (flight control system), Shorts (nacelles and thrust reversers) and Hamilton Sundstrand (flaps, leading-edge slats and electrical system).

LANDING GEAR: Hydraulically retractable tricycle type by Menasco with twin wheels on each unit. Mainwheel tyre size H36×12.0-8, pressure 10.55 bar (153 lb/sq in); nosewheel tyre size H20.5×6.75-10, pressure 9.24 bar (134 lb/sq in).

POWER PLANT: Two General Electric CF34-8C1 turbofans with dual-channel FADEC. Engine rating 56.4 kN (12,670 lb st), or 61.3 kN (13,790 lb st) with automatic power reserve, flat rated to ISA + 15°C. Intertechnique fuel management system with Ratier-Figeac controls. Fuel capacity 11,488 litres (3,035 US gallons; 2,527 Imp gallons).

ACCOMMODATION: Two-pilot flight deck. Main cabin seats 70 passengers, four-abreast at 79 cm (31 in) pitch. Baggage compartment and lavatory at rear of cabin; underfloor baggage compartment; various combinations of galleys, wardrobes and lavatory at front of cabin according to seating capacity. Passenger door, forward emergency exit/service door, overwing emergency exits and baggage door as for Regional Jet.

SYSTEMS: Hamilton Sundstrand electrical generation system comprising two 40 kVA integrated drive generators; tailcone-mounted Honeywell APU approved for operation up to 12,500 m (41,000 ft); Liebherr air management system; Intertechnique fuel system; Walter Kidde fire detection system; Goodrich portable water system; Hella lighting system; Parker/Abex Hydraulics hydraulic systems, and Teleflex anti-icing system.

AVIONICS: *Radar:* Rockwell Collins digital weather radar.

Flight: Rockwell Collins AHRS and TCAS.

Instrumentation: Rockwell Collins Pro Line 4 EFIS with six 127 × 178 mm (5 × 7 in) CRT displays, including dual PFD, dual MFD and dual EICAS; Flight Dynamics HGS 2000 head-up guidance system. Autopilot, FMS and centralised avionics maintenance functions are provided by integrated avionics processing system (IAPS). Windshear detection and recovery system standard. Equipped for Cat. II landings.

Following data are provisional.

DIMENSIONS, EXTERNAL:

Wing span	23.24 m (76 ft 3 in)
Wing aspect ratio	7.4
Length overall	32.51 m (106 ft 8 in)

Cabin of CRJ700   *NEW*/0137404

The Bombardier CRJ700 flight deck is based on that of the CRJ¹⁰⁰/₂₀₀
0131796

Max diameter of fuselage	2.69 m (8 ft 10 in)
Height overall	7.57 m (24 ft 10 in)
Tailplane span	8.53 m (28 ft 0 in)
Turning circle	22.86 m (75 ft 0 in)
Passenger door (port, forward):	
Height	1.78 m (5 ft 10 in)
Width	0.91 m (3 ft 0 in)
Height to sill	1.73 m (5 ft 8 in)
Baggage door (port, aft):	
Height	0.84 m (2 ft 9 in)
Width	1.09 m (3 ft 7 in)
Height to sill	2.31 m (7 ft 7 in)
Baggage door (port, forward):	
Height	0.51 m (1 ft 8 in)
Width	1.07 m (3 ft 6 in)
Height to sill	1.28 m (4 ft 21/2 in)
Service door (starboard, forward):	
Height	1.22 m (4 ft 0 in)
Width	0.61 m (2 ft 0 in)
Height to sill	1.73 m (5 ft 8 in)
DIMENSIONS, INTERNAL:	
Cabin (excl flight deck):	
Length	20.78 m (68 ft 2 in)
Max width: at centreline	2.57 m (8 ft 5 in)
at floor level	2.13 m (7 ft 0 in)
Max height	1.89 m (6 ft 2¼ in)
Floor area, excl cockpit	42.8 m² (461 sq ft)
Volume	75.95 m³ (2,682 cu ft)
Baggage volume (total)	23.3 m³ (824 cu ft)
AREAS:	
Wings, net	68.63 m² (738.7 sq ft)
Horizontal tail surfaces (total)	20.74 m² (223.3 sq ft)
Vertical tail surfaces (total)	13.36 m² (143.8 sq ft)
WEIGHTS AND LOADINGS:	
Operating weight empty	19,269 kg (42,480 lb)
Max payload	8,528 kg (18,800 lb)
Payload with max fuel: standard	4,364 kg (9,620 lb)
ER	5,384 kg (11,870 lb)
Max fuel weight	9,017 kg (19,880 lb)
Max T-O weight: standard	32,999 kg (72,750 lb)
ER	34,019 kg (75,000 lb)
Max ramp weight: standard	33,112 kg (73,000 lb)
ER	34,132 kg (75,250 lb)
Max landing weight	30,390 kg (67,000 lb)
Max zero-fuel weight	28,259 kg (62,300 lb)
Max wing loading: standard	480.8 kg/m² (98.48 lb/sq ft)
ER	495.7 kg/m² (101.53 lb/sq ft)
Max power loading: standard	293 kg/kN (2.87 lb/lb st)
ER	302 kg/kN (2.96 lb/lb st)
PERFORMANCE:	
Cruising Mach No.:	
high speed	0.825 (470 kt; 870 km/h; 541 mph)
normal	0.77 (442 kt; 818 km/h; 508 mph)
Max certified altitude	12,500 m (41,000 ft)
T-O field length at S/L, ISA:	
standard	1,564 m (5,130 ft)
ER	1,648 m (5,406 ft)
Landing field length at S/L, ISA, at max landing weight:	
standard and ER	1,478 m (4,850 ft)
Range, 70 passengers, at M0.77, IFR reserves:	
standard	1,687 n miles (3,124 km; 1,941 miles)
ER	1,984 n miles (3,674 km; 2,283 miles)
OPERATIONAL NOISE LEVELS (FAR Pt 36):	
T-O at max T-O weight	89 EPNdB
Approach at max landing weight	98 EPNdB
Sideline	94 EPNdB
	UPDATED

Prototype Bombardier CRJ900 wearing house colours   0130587

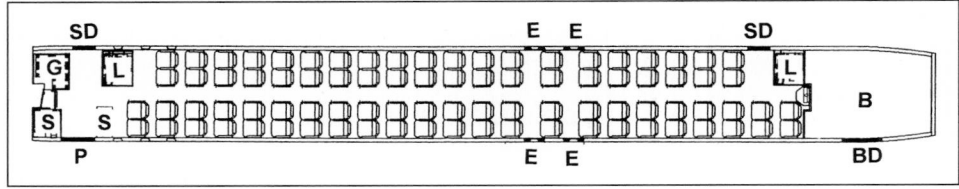
Bombardier CRJ900 seating 86 passengers
B: baggage, BD: baggage door, E: Type III exit, G: galley, L: lavatory, P: Type I passenger door, S: Stowage, SD: Type I service door   0105026

BOMBARDIER CRJ900

TYPE: Regional jet airliner.

PROGRAMME: Announced October 1999; stretched derivative of CRJ700; wind tunnel testing began November 1999; partner/supplier selection finalised second quarter 2000; interior mockup completed March 2000; formal launch at Farnborough International Air Show 24 July 2000; prototype, modified from CRJ700 prototype C-FRJX, with fuselage plugs but retaining CRJ700 wings, landing gear and engines, first flown 21 February 2001; public debut at Paris Air Show 14 June 2001; first production aircraft, c/n 15001/C-GRNH, first flown 20 October 2001, at which time prototype had completed 346 hours in 142 flights; certification third quarter 2002; first customer delivery (c/n 15002 for Mesa Air) 3 February 2003.

CURRENT VERSIONS: **900**: Standard version.
  900ER: Extended-range version.
  900ER European: As 900ER but with maximum T-O weight limited to 36,995 kg (81,560 lb) to minimise weight-related charges when operating in European airspace.

CUSTOMERS: Firm orders for 30 by 1 January 2003, comprising GE Capital Aviation Services (10), and Mesa (20). Estimated market for 800 aircraft in CRJ900 class over 20-year period.

COSTS: Development cost C$200 million; unit cost C$29 million (2000).

DESIGN FEATURES: Compared to CRJ700, has fuselage stretched – by means of 2.29 m (7 ft 6 in) plug forward of centre-section and 1.57 m (5 ft 2 in) plug aft of centre-section – to accommodate 86 to 90 passengers in four-abreast configuration with fore and aft lavatories; 5 to 10 per cent higher thrust engines; strengthened main landing gear with upgraded wheels and brakes; strengthened wing; two additional overwing emergency exits; increased volume in forward underfloor baggage hold, and an additional underfloor baggage door and aft service door on starboard side. Common crew qualification with CRJ$^{200}/_{700}$ series.

STRUCTURE: Programme partners generally as for CRJ700 except: Gamesa (vertical and horizontal stabilisers), Hamilton Sundstrand (flaps), and Shorts (mid-fuselage section). Final assembly and completion by Bombardier at new CRJ$^{700}/_{900}$ facility at Montréal-Mirabel Airport, scheduled to open in second quarter 2001.

LANDING GEAR: As for CRJ700.

POWER PLANT: Two General Electric CF34-8C5 turbofans. Engine rating 58.4 kN (13,123 lb st) or 63.4 kN (14,255 lb st) with automatic power reserve, flat rated to ISA +15°C. Fuel capacity 11,148 litres (2,945 US gallons; 2,452 Imp gallons).

ACCOMMODATION: Standard dual-class accommodation for 86 passengers in four-abreast configuration at 79 cm (31 in) seat pitch with fore and aft lavatories and forward galley; alternative configurations include high density with accommodation for 90 passengers at 79 cm (31 in) seat pitch, dual-class with 15 business class seats three-abreast at 86 cm (34 in) seat pitch in forward section and 60 economy class four-abreast at 79 cm (31 in) seat pitch at rear, and dual class with 55 business class seats in four-abreast configuration at 84 cm (33 in) seat pitch in forward section and 24 economy class at 79 cm (31 in) seat pitch at rear. Standard additional floor beam facilitates offset seat rail for three-abreast seating throughout cabin.

AVIONICS: As for CRJ700.
Following data are provisional.
DIMENSIONS, EXTERNAL: As for CRJ700 except

Wing span	23.24 m (76 ft 3 in)
Length overall	36.19 m (118 ft 9 in)
Height overall	7.49 m (24 ft 7 in)

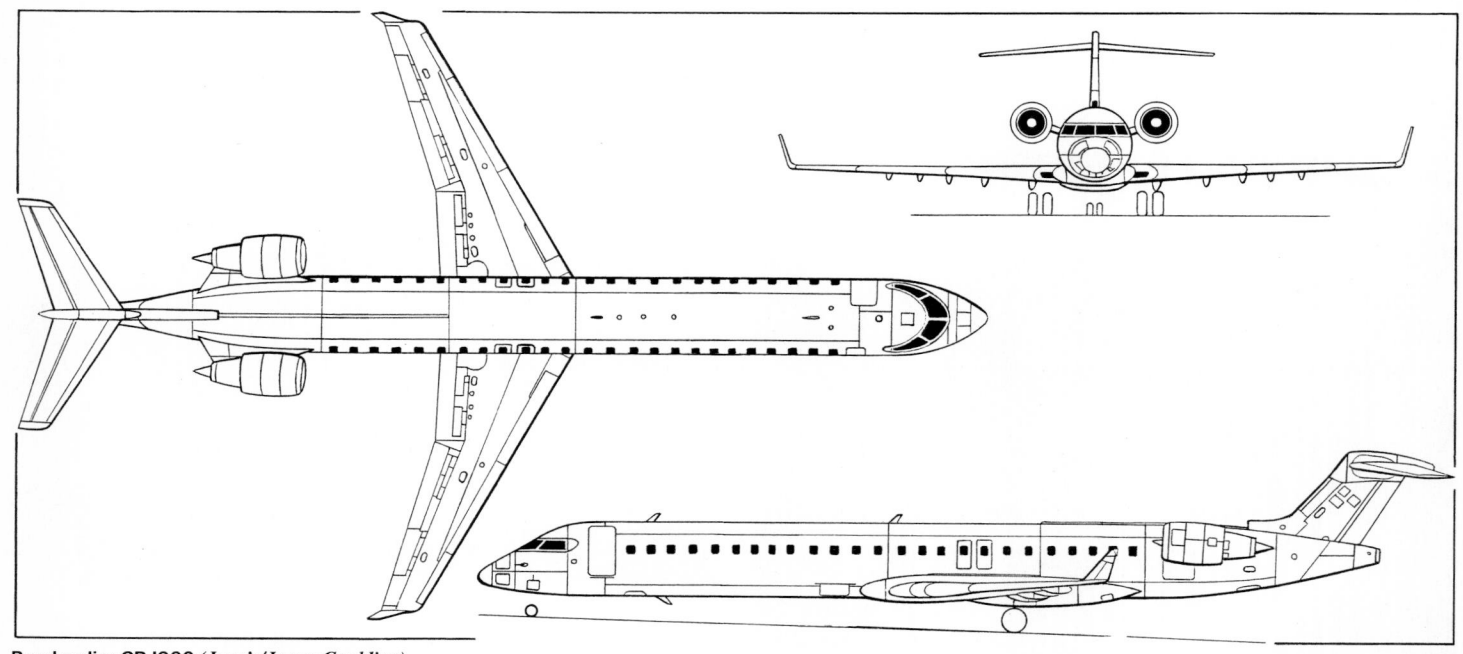
Bombardier CRJ900 (*Jane's/James Goulding*)   0131847

Wheelbase	14.73 m (48 ft 4 in)
Aft service door (starboard):	
Height	1.22 m (4 ft 0 in)
Width	0.61 m (2 ft 0 in)
Height to sill	1.73 m (5 ft 8 in)
Emergency exits (four, overwing):	
Height	0.91 m (3 ft 0 in)
Width	0.51 m (1 ft 8 in)

DIMENSIONS, INTERNAL:

Cabin (excl flight deck):	
Baggage volume: checked	16.81 m³ (593.5 cu ft)
total	25.57 m³ (903 cu ft)

WEIGHTS AND LOADINGS:

Operating weight empty	21,546 kg (47, 500 lb)
Max payload	10,206 kg (22,500 lb)
Payload with max fuel: 900	6,260 kg (13,800 lb)
900ER	7,167 kg (15,800 lb)
Max fuel weight	8,822 kg (19,450 lb)
Max T-O weight: 900	36,514 kg (80,500 lb)
900ER	37,421 kg (82,500 lb)
900ER European	36,995 kg (81,560 lb)
Max ramp weight: 900	36,627 kg (80,750 lb)
900ER	37,716 kg (83,150 lb)
Max landing weight	33,339 kg (73,500 lb)
Max zero-fuel weight	31,751 kg (70,000 lb)
Max wing loading: 900	532.1 kg/m² (108.98 lb/sq ft)
900ER	545.3 kg/m² (111.68 lb/sq ft)
900ER European	539.1 kg/m² (110.41 lb/sq ft)
Max power loading: 900	313 kg/kN (3.07 lb/lb st)
900ER	320 kg/kN (3.14 lb/lb st)
900ER European	317 kg/kN (3.11 lb/lb st)

PERFORMANCE:

Cruising Mach No:	
high speed	0.81 (464 kt; 859 km/h; 534 mph)
for max range	0.80 (460 kt; 852 km/h; 529 mph)
Max certified altitude: 900, 900ER	12,500 m (41,000 ft)
Service ceiling, OEI: 900	4,968 m (16,300 ft)
900ER	4,755 m (15,600 ft)
FAR T-O field length:	
900	1,878 m (6,160 ft)
900ER	1,970 m (6,465 ft)
FAR landing field length at S/L, ISA, at max landing weight:	
900, 900ER	1,565 m (5,135 ft)
Range, 86 passengers, at M0.77:	
900	1,498 n miles (2,774 km; 1,723 miles)
900ER	1,732 n miles (3,207 km; 1,993 miles)

UPDATED

CANADAIR 415 SUPERSCOOPER

TYPE: Twin-turboprop amphibian.

PROGRAMME: Introduced as product follow-on to piston-engined CL-215; given new designation Canadair 415 in 1991 to distinguish new production turboprop model from CL-215T retrofit, but engineering designation retained and new-build turboprop versions are CL-215-6B11. Launched officially 16 October 1991 with firm orders from France and (August 1992) Québec; first flight (C-GSCT) 6 December 1993, although preceded by initial CL-215T conversion (C-FASE), flown on 8 June 1989. Canadian certification 24 June 1994 in Restricted and Utility categories, FAA approval 14 October 1994 in Restricted category. RAI (Italy) approval 27 October 1994 in Restricted category. Fleet had achieved 34,027 flying hours and 126,318 scooping sorties by November 1999. Final assembly relocated to North Bay, Ontario, in November 1998. Production suspended in October 2001 pending new orders (but see 415MP entry below).

CURRENT VERSIONS: Standard **415 SuperScooper** and first production units are in **firefighting** configuration; **415M** can be modified for **maritime**, **SAR** and **special missions**.

Prototype Canadair 415MP (two P&WC PW123 turboprops) *NEW*/0137421

415MP: Multipurpose version incorporating FLIR, SLAR and nose-mounted search radar for missions such as search and rescue, coastal and border patrol and environmental monitoring, while retaining firefighting capability. First flight (C-GHVX; 55th aircraft, destined to become 415GR for Greece) 6 March 2002, preceding a 200-hour flight test programme.

415GR: Ordered by Greece, January 1999; based on 415MP; increased weights; boat handling and cargo hoisting provisions. Two (plus an optional third) will be configured for combat search-and-rescue (C-SAR) role, for which, during 2000, SAAB Nyge Aero of Sweden was awarded a three-year contract to install MSS 5000 mission equipment including SLAR (each side), wing-mounted FLIR Systems SeaFLIR, nose-mounted Honeywell Primus 660 weather/search radar, digital cameras, autopilot and provision for Have Quick secure radios and rescue beacon receivers. The aircraft will also have an enlarged cargo door to facilitate deployment of an inflatable rescue boat. The Hellenic Air Force C-SAR aircraft will be based at Elefsis AB; delivery due in May 2003.

CUSTOMERS: See table. First French Canadair 415 delivered to CEV experimental unit 8 February 1995, but trials revealed need for modifications and acceptance delayed until 13 June 1995; deliveries completed June 1997. Ontario provincial government announced order for nine on 2 April 1998. Government of Malaysia expressed interest in acquiring two following demonstration flights in February 2002. Total 59 built by October 2001 production suspension; 60th registered in January 2002.

CANADAIR 415 CUSTOMERS
(at July 2000)

Customer	Qty	First aircraft	Delivered
Canada: Ontario govmt	9	C-GAO1	29 Apr 98
Québec govmt	8	C-GQBA	Jul 95
Croatia[1]	3	9A-CAG	Feb 97
France: Sécurité Civile	12	F-ZBFS	8 Feb 95
Greek govmt[3]	10	2039	Jan 99
Italy: SISAM[2]	15	I-DPCD	27 Jan 95
Total	**57**		

[1] Requirement for six

[2] Societa Italiana Servizi Aerei Mediterranei

[3] Plus five options; order total includes two 415GR

COSTS: Approximately US$23 million (2000).

DESIGN FEATURES: Retains well-proven basic airframe of piston-engined CL-215 (thick wing with zero dihedral and 2° incidence; row of vortex generators on each wing outboard of fence; long stall strip inboard of starboard fence; leading-edge strakes and fences beside engine nacelles; water scoops behind planing step; anti-spray channels in planing bottom chine) but incorporates upgrading modifications and improvements including higher operating weights for increased firefighting productivity; pressure refuelling; wing endplates for lateral stability; finlets and tailplane/fin bullet to recover longitudinal and directional stability affected by relocated thrust line, increased power and new propellers; powered rudder, ailerons and elevators; new electrical system; new 'glass' cockpit with air conditioning; enlarged four-tank firefighting drop system.

FLYING CONTROLS: Conventional and power-assisted. Hydraulically actuated ailerons, elevators and rudder, standard; manual reversion in event of hydraulic failure; geared tab in each aileron, spring tab in rudder and each elevator, plus trim tab in port aileron and port elevator. Hydraulically operated single-slotted flaps, each supported by four external hinges.

STRUCTURE: No-dihedral, no-twist high wing, of constant chord; one-piece structure with two conventional spars, extruded spanwise stringers and interspar ribs and aluminium alloy skins. All-metal, fail-safe, single-step boat-hull fuselage with numerous watertight compartments. Tail surfaces of aluminium alloy sheet and extrusions, with honeycomb panels on control surfaces.

LANDING GEAR: Hydraulically retractable tricycle type. Self-centring twin-wheel nose unit retracts rearward into hull and is fully enclosed by conformal doors. Nosewheel steering standard. Main gear legs retract into wells in sides of hull. Plate mounted on each main gear assembly encloses bottom of wheel well. Mainwheel tyres 15.00-16 (16 ply) tubeless, pressure 5.31 bar (77 lb/sq in); nosewheel tyres 6.50-10 (10 ply) tubed, pressure 6.55 bar (95 lb/sq in). Hydraulic disc brakes. Non-retractable stabilising floats, each carried near wingtip on pylon cantilevered from wing box structure, with breakaway provision.

POWER PLANT: Two 1,775 kW (2,380 shp) Pratt & Whitney Canada PW123AF turboprops, on damage-tolerant mounts capable of withstanding a breach of the compressor/turbine casing, each driving a Hamilton Sundstrand 14SF-19 four-blade constant-speed fully feathering reversible-pitch propeller. Two fuel tanks, each of eight identical flexible cells, in wing spar box, with total usable capacity of 5,796 litres (1,531 US gallons; 1,275 Imp gallons). Single-point pressure refuelling (rear fuselage, starboard side), plus gravity points in wing upper surface.

ACCOMMODATION: Normal crew of two side by side on flight deck, with dual controls. Additional station in maritime patrol/SAR versions for third cockpit member, mission specialist and two observers. For water bomber cabin installation, see Equipment paragraph. Combi layout offers cargo at front, full firefighting capability, plus 11 seats at rear. Other quick-change interiors available for utility/paratroop (up to 14 troop-type folding canvas seats in cabin) or other special missions according to customer's requirements. Flush doors to main cabin on port side of fuselage forward and aft of wings. Optional aft cargo door, height 1.33 m (4 ft 4½ in), width 1.46 m (4 ft 9½ in). Emergency exit on starboard side aft of wing trailing-edge. Crew emergency hatch in flight deck roof on starboard side. Mooring hatch in upper surface of nose. Provision for additional cabin windows.

SYSTEMS: Vapour cycle air conditioning system and combustion heater. Hydraulic system, pressure 207 bar (3,000 lb/sq in), utilises two engine-driven pumps (maximum flow rate 45.5 litres; 12.0 US gallons; 10.0 Imp gallons/min) to actuate nosewheel steering, landing gear, flaps, water drop doors, pickup probes, flight controls, main gear unlocking and wheel brakes. Hydraulic fluid (MIL-H-83282) in air/oil reservoir slightly pressurised by engine bleed air. Electrically driven third pump provides hydraulic power for emergency actuation of landing gear and brakes and closure of water doors. Electrical system includes two 800 VA 115 V 400 Hz static inverters, two 28 V 400 A DC engine-driven starter/generators and two 40 Ah Ni/Cd batteries. Pneumatic/electric intake de-icing system; airframe ice protection system optional.

Canadair 415 SuperScooper twin-turboprop general purpose amphibian *(Jane's/Dennis Punnett)*

AVIONICS: Dual Honeywell Primus 2 digital integrated VHF nav/com.

Comms: Global VHF/UHF/AM/FM and Rockwell Collins HF radios with central control heads, ELT and dual transponders.

Radar: Search/weather radar optional.

Flight: Dual ADF, VOR/ILS, marker beacon receivers and single DME. Optional Garmin GPS and Goodrich Stormscope.

Instrumentation: Honeywell EDZ-605 EFIS with three-tube Integrated Instrument Display System for EADI and EHSI; dual Litef/Honeywell AHRS, dual air data computers, Honeywell radio altimeter.

EQUIPMENT (firefighter): Four integral water tanks in main fuselage compartment, near CG (combined capacity 6,137 litres; 1,621 US gallons; 1,350 Imp gallons), plus eight inward-facing seats in forward cabin. Tanks filled by two hydraulically actuated scoops aft of hull step, fillable also on ground by hose adaptor on each side of fuselage. Four independently openable water doors in hull bottom. Onboard foam concentrate reservoirs (capacity 680 kg; 1,500 lb) and mixing system. Improved drop pattern and drop door sequencing compared with CL-215. Optional spray kit can be coupled with firefighting tanks for large-scale spraying of oil dispersants and insecticides. In a typical firefighting mission, with a water source 6 n miles (11 km; 7 miles) from the fire, aircraft can remain on station for 3 hours, dropping 55,267 litres (14,600 US gallons; 12,157 Imp gallons)/h. Water tanks can be scoop-filled completely (ISA at S/L, zero wind) in 12 seconds over a water distance of 1,341 m (4,400 ft); partial water loads can be scooped on smaller bodies of water. Minimum safe water depth for scooping is only 1.40 m (4 ft 7 in).

EQUIPMENT (other versions): Stretcher kits, passenger or troop seats, cargo tiedowns, searchlight and other equipment according to mission and customer requirements. Provision for underwing pylon attachment points for variety of stores. Canadair 415 can be equipped with maritime surveillance radar and electro-optical sensors, precision navigation and communications equipment and autopilot.

DIMENSIONS, EXTERNAL:

Wing span	28.63 m (93 ft 11 in)
Wing chord, constant	3.54 m (11 ft 7½ in)
Wing aspect ratio	8.2
Length overall	19.82 m (65 ft 0½ in)
Beam (max)	2.59 m (8 ft 6 in)
Length/beam ratio	7.5
Height overall: on land	8.98 m (29 ft 5½ in)
on water	6.88 m (22 ft 7 in)
Draught: wheels up	1.12 m (3 ft 8 in)
wheels down	2.03 m (6 ft 8 in)
Tailplane span	10.97 m (36 ft 0 in)
Wheel track	5.28 m (17 ft 4 in)
Wheelbase	7.23 m (23 ft 9 in)
Propeller diameter	3.97 m (13 ft 0¼ in)
Propeller fuselage clearance	0.59 m (1 ft 11¼ in)
Propeller water clearance	1.30 m (4 ft 3¼ in)
Propeller ground clearance	2.77 m (9 ft 1 in)
Forward door: Height	1.37 m (4 ft 6 in)*
Width	1.03 m (3 ft 4 in)
Height to sill	1.68 m (5 ft 6 in)
Rear door: Height	1.12 m (3 ft 8 in)
Width	1.02 m (3 ft 4 in)
Height to sill	1.83 m (6 ft 0 in)
Water drop door: Length	1.60 m (5 ft 3 in)
Width	0.81 m (2 ft 8 in)
Emergency exit: Height	0.91 m (3 ft 0 in)
Width	0.51 m (1 ft 8 in)

*incl 25 cm (10 in) removable sill

DIMENSIONS, INTERNAL:

Cabin, excl flight deck: Length	9.38 m (30 ft 9½ in)
Max width	2.39 m (7 ft 10 in)
Max height	1.90 m (6 ft 3 in)
Floor area	19.7 m² (212 sq ft)
Volume	35.6 m³ (1,257 cu ft)

AREAS:

Wings, gross	100.33 m² (1,080.0 sq ft)
Ailerons (total)	8.05 m² (86.60 sq ft)
Flaps (total)	22.39 m² (241.00 sq ft)
Fin	11.22 m² (120.75 sq ft)
Rudder, incl tabs	6.02 m² (64.75 sq ft)
Tailplane	20.55 m² (221.20 sq ft)
Elevators (total, incl tabs)	7.88 m² (84.80 sq ft)

WEIGHTS AND LOADINGS (A: firefighter, B: utility, land or water based):

Typical operating weight empty:	
A	12,882 kg (28,400 lb)
B	12,733 kg (28,072 lb)
Max internal fuel weight: A, B	4,649 kg (10,250 lb)
Max payload: A (disposable)	6,123 kg (13,500 lb)
B	3,777 kg (8,328 lb)
Max ramp weight: A (land)	19,958 kg (44,000 lb)
A (water), B	17,236 kg (38,000 lb)
Max T-O weight: A (land)	19,890 kg (43,850 lb)
A (water), B (land and water)	17,168 kg (37,850 lb)
Max touchdown weight for water scooping:	
A	16,420 kg (36,200 lb)
Max flying weight after water scooping:	
A	21,319 kg (47,000 lb)
Max landing weight:	
A, B (land and water)	16,783 kg (37,000 lb)
Max zero-fuel weight: A	19,504 kg (43,000 lb)
B	16,511 kg (36,400 lb)
Max wing loading:	
A (after scoop)	212.5 kg/m² (43.52 lb/sq ft)
A (land)	198.2 kg/m² (40.60 lb/sq ft)
B (land and water)	171.1 kg/m² (35.05 lb/sq ft)
Max power loading:	
A (after scoop)	6.01 kg/kW (9.87 lb/shp)
A (land)	5.60 kg/kW (9.21 lb/shp)
B (land and water)	4.84 kg/kW (7.95 lb/shp)

PERFORMANCE (at weights shown):

Max cruising speed at 1,525 m (5,000 ft)	197 kt (365 km/h; 227 mph) TAS
Long-range cruising speed at 3,050 m (10,000 ft), AUW of 14,741 kg (32,500 lb)	145 kt (269 km/h; 167 mph)
Patrol speed at S/L, AUW of 14,741 kg (32,500 lb)	130 kt (241 km/h; 150 mph)
Stalling speed: 15° flap, AUW of 21,319 kg (47,000 lb)	80 kt (149 km/h; 93 mph)
Max rate of climb at S/L, AUW of 21,319 kg (47,000 lb)	396 m (1,300 ft)/min
T-O distance at S/L, ISA:	
land, AUW of 19,890 kg (43,850 lb)	844 m (2,770 ft)
land, AUW of 17,168 kg (37,850 lb)	799 m (2,620 ft)
water, AUW of 17,168 kg (37,850 lb)	814 m (2,670 ft)
Landing distance at S/L, ISA:	
land, AUW of 16,783 kg (37,000 lb)	674 m (2,210 ft)
water, AUW of 16,783 kg (37,000 lb)	665 m (2,180 ft)
Scooping distance at S/L, ISA (incl safe clearance heights)	1,341 m (4,400 ft)
Ferry range with 499 kg (1,100 lb) payload	1,310 n miles (2,426 km; 1,507 miles)
Design *g* limits (15° flap)	+3.25/−1

UPDATED

BOMBARDIER AEROSPACE DE HAVILLAND

Garratt Boulevard, Downsview, Ontario M3K 1Y5
Tel: (+1 416) 633 73 10
Fax: (+1 416) 375 45 46
VICE-PRESIDENT, CANADIAN OPERATIONS: Serge Perron
VICE-PRESIDENT AND GENERAL MANAGER, OPERATIONS: Alain Dugas
VICE-PRESIDENT, DASH 8 SERIES 100/200/300 OPERATIONS: Rob Comand
VICE-PRESIDENT, DASH 8 SERIES 400 OPERATIONS: Bruce Vannus
VICE-PRESIDENT, GLOBAL EXPRESS OPERATIONS: Trevor Anderson
VICE-PRESIDENT, FINANCE: Colin Fernie
VICE-PRESIDENT, HUMAN RESOURCES: Bernard Cormier
MANAGER, PUBLIC RELATIONS: Colin Fisher

Established 1928 as The de Havilland Aircraft of Canada Ltd, subsidiary of The de Havilland Aircraft Company Ltd, both absorbed 1961 by Hawker Siddeley Group; ownership transferred to Canadian government 26 June 1974; purchased by Boeing Company 31 January 1986 and made a division of Boeing of Canada Ltd; Boeing's intention to sell announced July 1990.

Sale to Bombardier Inc (51 per cent) and government of Ontario (49 per cent), signed 22 January 1992, supported by help from Ontario and federal governments, with Canadian Export Development Corporation to provide sales financing for Dash 8; all government support conditionally repayable. Bombardier acquired remaining 49 per cent of de Havilland in January 1997.

De Havilland manufactures Dash 8Q series, including spare parts and components, and also some components of the Bombardier Global Express; performs final assembly of Dash 8Q and Global Express; designed and builds wings for Learjet 45.

VERIFIED

DHC-8 DASH 8 Q100 and Q200

TYPE: Twin-turboprop airliner.

PROGRAMME: Launched 1980; first flight of first prototype (C-GDNK) 20 June 1983, second prototype (C-GGMP) 26 October 1983, third 22 December 1983; fourth aircraft, first with production P&WC PW120 engines, 3 April 1984. Certified to Canadian DoT, FAR Pts 25 and 36 and SFAR No. 27 on 28 September 1984, followed by FAA type approval; also certified in Australia, Austria, Brazil, Cameroon, China, Colombia, Germany, Ireland, Italy, Maldives, Netherlands, New Zealand, Norway, Papua New Guinea, South Africa, Taiwan, UK and United Arab Emirates. First delivery (NorOntair) 23 October 1984 followed by service entry in December; 500th Dash 8 (a Series 200, N355PH) delivered 21 November 1997 to Horizon Air. 600th of type delivered 6 March 2001 (see Q300). Total Dash 8 fleet time at December 2001 (not including Q400) was 10,906,707 hours and 13,196,656 cycles, with 98.8 per cent despatch reliability rate.

In October 2002, Bombardier announced six to eight week suspension (beginning in following month) of Dash 8 production because of reduced orders (estimated 29-aircraft backlog at that time).

CURRENT VERSIONS: **Dash 8 Q:** Redesigned interior and noise and vibration suppression system (NVS) standard from second quarter of 1996, reducing cabin noise levels by 12 dB; all current production aircraft are equipped with NVS and known as Dash 8 Qs. Further Noise Technology Demonstration Program, under way in mid-1999, is expected to result in additional vibration/noise damping in fuselage and floor areas. See below for specific versions.

Dash 8 Series 100: Initial version, with choice of PW120A or PW121 engines; described in 1992-93 and earlier editions.

Dash 8 Series 100A: Introduced 1990; PW120A (or optional PW121) engines and restyled interior with 6.35 cm (2.5 in) more headroom in aisle; first delivery to Pennsylvania Airlines July 1990. Specifications in 1997-98 and earlier *Jane's*.

Dash 8 Series 100B: Improved version from 1992; PW121 engines enhance airfield and climb performance. Specifications in 1997-98 and earlier *Jane's*.

Dash 8 Q100: Introduced 1998; PW 120A or optional PW 121 in basic weight version, PW 121 standard in high gross weight (HGW) version.

Dash 8 Series 200A: Increased speed/payload version of Series 100A with PW 123C engines. Transport Canada certification March 1995; first delivery 19 April 1995. Specifications in 1998-99 and earlier editions.

Dash 8 Series 200B: As 200A, but with PW123D engines for full power at higher ambient temperatures. Specifications in 1998-99 and earlier editions.

Dash 8 Q200: Introduced 1998; increased speed/payload version of Q100; increased OEI capability and greater commonality with Series 300. Same airframe as Q100, but PW 123C/D engines give 30 kt (56 km/h; 35 mph) increase in cruising speed, allowing airlines to increase frequencies or operational radius. PW 123D engine offers full power at higher ambient temperatures for improved hot-and-high airfield performance.

Detailed description applies to Q200 except where indicated.

Dash 8M and **Dash 8 Q300 and Q400;** *described separately.*

CUSTOMERS: By 28 February 2002, orders received for 299 Srs 100 subvariants and 94 Srs 200s, of which 298 and 94 had then been delivered (including military Dash 8Ms). See table.

DHC-8 Dash 8 Q100 of Southern Australia Airlines (*Jane's/Paul Jackson*)   NEW/0137352

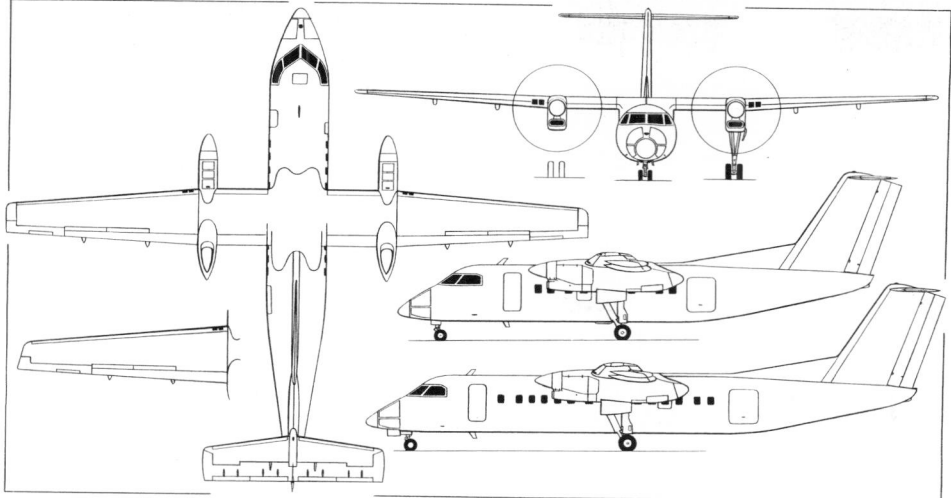

Rudder	4.31 m² (46.4 sq ft)
Tailplane	8.97 m² (96.5 sq ft)
Elevators (total)	4.97 m² (53.5 sq ft)

WEIGHTS AND LOADINGS:

Operating weight empty: Q100	10,406 kg (22,941 lb)
Q200	10,486 kg (23,117 lb)
Max usable fuel: standard	2,576 kg (5,678 lb)
optional	4,647 kg (10,244 lb)
Max payload: Q100	4,109 kg (9,059 lb)
Q200	4,211 kg (9,283 lb)
Payload with max fuel: Q100	3,484 kg (7,681 lb)
Q200	3,404 kg (7,505 lb)
Max T-O weight: Q100	15,649 kg (34,500 lb)
Q100 HGW, Q200	16,465 kg (36,300 lb)
Max landing weight: Q100	15,377 kg (33,900 lb)
Q200	15,649 kg (34,500 lb)
Max zero-fuel weight:	
Q100, Q200 JAA	14,515 kg (32,000 lb)
Q200	14,696 kg (32,400 lb)
Max wing loading: Q100	287.9 kg/m² (58.97 lb/sq ft)
Q100 HGW, Q200	303.0 kg/m² (62.05 lb/sq ft)
Max power loading: Q100	5.25 kg/kW (8.63 lb/shp)
Q100 HGW, Q200	5.14 kg/kW (8.44 lb/shp)

PERFORMANCE (at 95% standard MTOW, except where indicated):

Max cruising speed: Q100	265 kt (491 km/h; 305 mph)
Q100 HGW	270 kt (500 km/h; 311 mph)
Q200	290 kt (537 km/h; 334 mph)
Stalling speed, flaps down: all	72 kt (134 km/h; 83 mph)
Max rate of climb at S/L: all	450 m (1,475 ft)/min
Max certified altitude: all	7,620 m (25,000 ft)
Service ceiling: Q100	4,503 m (14,775 ft)
Q100 HGW	5,105 m (16,750 ft)
Q200	4,938 m (16,200 ft)
FAR Pt 25 T-O field length:	
Q100 (PW 121)	991 m (3,250 ft)
Q200	1,000 m (3,280 ft)
FAR Pt 25 landing field length at max landing weight:	
Q100	785 m (2,575 ft)
Q200	780 m (2,560 ft)
Range with 37 passengers:	
Q100	1,020 n miles (1,889 km; 1,173 miles)
Q200	925 n miles (1,713 km; 1,064 miles)

OPERATIONAL NOISE LEVELS: (FAR Pt 36 Stage 3 and ICAO Annex 16):

T-O	80.5 EPNdB
Sideline	85.6 EPNdB
Approach	94.8 EPNdB

UPDATED

DHC-8 Dash 8 Q100, with additional side view (bottom) and wingtip of Q300 (*Jane's/Dennis Punnett*)

COSTS: Unit cost approximately US$13 million (1998).

DESIGN FEATURES: T tail, swept fin with large dorsal fin. Wing has constant chord inboard section and tapered outer panels; thickness/chord ratio 18 per cent at root, 13 per cent at tip; dihedral 2° 30′ on outer panels; inboard leading-edges drooped; 3° washout at wingtips.

FLYING CONTROLS: Conventional and power-assisted. Fixed tailplane; horn-balanced elevator with four tabs; mechanically actuated horn-balanced ailerons with inset tabs; hydraulically actuated roll spoilers/lift dumpers forward of each outer flap; two-segment serially hinged rudder, hydraulically actuated; yaw damper; stall strips on leading-edges outboard of engines; two-section slotted Fowler flaps. Digital AFCS.

STRUCTURE: Fuselage near-circular section, flush riveted and pressurised; adhesive bonded stringers and cutout reinforcements; wing leading-edge, radome, nose bay, wing/fuselage and wingtip fairings, dorsal fin, fin leading-edge, fin/tailplane fairings, tailplane leading-edges, elevator tips, flap shrouds, flap trailing-edges and other components of Kevlar and Nomex; wing has tip-to-tip torsion box. Wheel doors of Kevlar and other composites.

LANDING GEAR: Retractable tricycle type, by Dowty Aerospace, with twin wheels on each unit. Steer-by-wire nose unit retracts forward, main units rearward into engine nacelles. Goodrich mainwheels and brakes; Hydro-Aire Mk 3 anti-skid system. Tyres 26.5×8.0-13 (12 ply) or high flotation H31×9.75-13 (10 ply) on main units; 18×5.5 (10 ply) or 22×6.50-10 (12 ply) on nose unit. Standard tyre pressures: main 9.03 bar (131 lb/sq in), nose 5.52 bar (80 lb/sq in). Low-pressure tyres optional, pressure 5.31 bar (77 lb/sq in) on main units, 3.31 bar (48 lb/sq in) on nose unit.

POWER PLANT: *Q100:* Two 1,491 kW (2,000 shp) Pratt & Whitney Canada PW 120A turboprops, driving Hamilton Sundstrand 14SF-7 four-blade constant-speed fully feathering aluminium/glass fibre propellers with reverse pitch, standard on basic weight version. Two 1,603 kW (2,150 shp) PW 121s optional on this version and standard on High Gross Weight version.

Q200: Two 1,603 kW (2,150 shp) PW 123C/D turboprops, driving Hamilton Sundstrand 14SF-23 propellers; PW 123C is flat-rated for full power at up to 26°C; PW 123D maintains same power at up to 45°C.

Standard usable fuel capacity (in-wing tanks) 3,160 litres (835 US gallons; 695 Imp gallons); optional auxiliary tank system increases this to 5,700 litres (1,506 US gallons; 1,254 Imp gallons). Pressure refuelling point in rear of starboard engine nacelle; overwing gravity point in each outer wing panel. Oil capacity 21 litres (5.5 US gallons; 4.6 Imp gallons) per engine.

ACCOMMODATION: Crew of two on flight deck, plus cabin attendant. Dual controls standard. Standard commuter layout provides four-abreast seating, with central aisle, for 37 passengers at 79 cm (31 in) pitch, plus buffet, lavatory and large rear baggage compartment. Wardrobe at front of passenger cabin, in addition to overhead lockers and underseat stowage, provides additional carry-on baggage capacity. Alternative 39-passenger, passenger/cargo or corporate layouts available at customer's option. Movable bulkhead to facilitate conversion to mixed traffic. Port side airstair door at front for crew and passengers; large inward-opening port side door aft of wing for cargo loading. Emergency exit each side, in line with wing leading-edge, and opposite passenger door on starboard side. Entire accommodation pressurised and air conditioned.

SYSTEMS: Pressurisation system with maximum differential 0.38 bar (5.5 lb/sq in). Normal hydraulic installation comprises two independent systems, each having an engine-driven variable displacement pump and an electrically driven standby pump; accumulator and hand pump for emergency use. Electrical system DC power provided by two starter/generators, two transformer-rectifier units, and two Ni/Cd batteries. Variable frequency

AC power provided by two engine-driven AC generators and three static inverters. Ground power receptacles in port side of nose (DC) and rear of starboard nacelle (AC). Rubber-boot de-icing of wing, tailplane and fin leading-edges and nacelle intakes by pneumatic system; electric de-icing of propeller blades, pitot static ports, stall warning transducer, engine intake adaptor and elevator horn leading-edge. APU standard in corporate version. Simmonds fuel monitoring system.

AVIONICS: Bendix/King Gold Crown III com/nav; Rockwell Collins equipment alternative option.
Comms: KTR 908 VHF com and KXP 756 transponder. Avtech audio integrating system. Telephonics PA system.
Radar: Primus P660 colour weather radar.
Flight: KNR 806 ADF, KDM 706A DME. Bendix/King SPZ-8000 dual-channel digital AFCS with integrated fail-operational flight director/autopilot system, dual digital air data system. Optional FMS and GPS.
Instrumentation: Bendix/King EFIS. Hughes/Flight Dynamics HGS 2000 head-up guidance system for Cat. IIIa operation optional.

DIMENSIONS, EXTERNAL:

Wing span	25.91 m (85 ft 0 in)
Wing aspect ratio	12.4
Length overall	22.25 m (73 ft 0 in)
Fuselage: Max diameter	2.69 m (8 ft 10 in)
Height overall	7.49 m (24 ft 7 in)
Elevator span	7.92 m (26 ft 0 in)
Wheel track (c/l of shock-struts)	7.87 m (25 ft 10 in)
Wheelbase	7.95 m (26 ft 1 in)
Propeller diameter	3.96 m (13 ft 0 in)
Propeller ground clearance	0.94 m (3 ft 1 in)
Propeller fuselage clearance	0.76 m (2 ft 6 in)
Passenger/crew door (fwd, port):	
Height	1.66 m (5 ft 5½ in)
Width	0.76 m (2 ft 6 in)
Height to sill	1.09 m (3 ft 7 in)
Baggage door (rear, port): Height	1.52 m (5 ft 0 in)
Width	1.27 m (4 ft 2 in)
Height to sill	1.09 m (3 ft 7 in)

DIMENSIONS, INTERNAL:

Cabin (excl flight deck): Length	9.14 m (30 ft 0 in)
Max width	2.49 m (8 ft 2 in)
Width at floor	2.03 m (6 ft 8 in)
Max height	1.96 m (6 ft 5 in)
Floor area	23.60 m² (254.0 sq ft)
Volume	37.6 m³ (1,328 cu ft)
Baggage compartment volume	8.5 m³ (300 cu ft)

AREAS:

Wings, gross	54.35 m² (585.0 sq ft)
Ailerons (total)	1.12 m² (12.1 sq ft)
Fin	9.81 m² (105.6 sq ft)

DHC-8 DASH 8 Q300

TYPE: Twin-turboprop airliner.

PROGRAMME: Dash 8 Series 300 announced mid-1985 as stretch of Series 200; launched March 1986; first flight (modified Series 100 prototype C-GDNK) 15 May 1987; Canadian DoT certification 14 February 1989; first delivery (Time Air) 27 February 1989; FAA type approval 8 June 1989; now also certified in Argentina, Australia, Austria, Brazil, China, Colombia, Germany, Ireland, Italy, Jordan, Malaysia, Maldives, Netherlands, Norway, South Africa, Taiwan, Thailand and Zambia. Low-noise Dash 8 Q (see Series 200 for details) became standard version from 1996. A Q300 delivered to Air Nippon on 6 March 2001 was the 600th Dash 8. Production temporarily suspended in November/December 2002.

CURRENT VERSIONS: **Series 300, 300A, 300B and 300E:** Initial versions, described in 1998-99 and earlier editions.
Q300: Introduced 1998; differs from Q200 in having extended wingtips; 3.43 m (11 ft 3 in) two-plug fuselage extension giving standard seating for 50 at 81 cm (32 in) pitch or 56 at 74 cm (29 in) pitch, plus second cabin attendant; also larger galley, galley service door, additional wardrobe, larger lavatory, dual air conditioning packs and optional Turbomach T-40 APU; powered by 1,775 kW (2,380 shp) P&WC PW123s driving Hamilton Sundstrand 14SF-23 four-blade propellers standard in basic version; 1,864 kW (2,500 shp) PW 123Bs optional in basic and standard in high gross weight (HGW) versions provide

Dash 8 Q300 twin-turboprop airliner (*Jane's/Paul Jackson*)

NEW/0137353

increased mechanical power for improved take-off performance in low and cold conditions; optional 1,775 kW (2,380 shp) PW 123Es provide 5 per cent increase in thermodynamic power up to 40°C (96°F) for improved hot-and-high performance; fuel capacity as Q200; tyre pressures increased (mainwheels 6.69 bar; 97 lb/sq in, nosewheels 4.14 bar; 60 lb/sq in). NVS system standard on all aircraft produced from second quarter 1996.

CUSTOMERS: Firm orders for 209 by 28 February 2002, of which 189 then delivered. See table.

COSTS: Air Senegal order valued at US$14.3 million (2000); Petroleum Air Services contract for two valued at US$30.04 million (2001).

DIMENSIONS, EXTERNAL: As for Q200 except:

Wing span	27.43 m (90 ft 0 in)
Wing aspect ratio	13.4
Length overall	25.68 m (84 ft 3 in)
Wheelbase	10.01 m (32 ft 10 in)

DIMENSIONS, INTERNAL: As for Q200 except:

Cabin (excl flight deck): Length	12.65 m (41 ft 6 in)
Floor area	30.57 m² (329.0 sq ft)
Volume	52.0 m³ (1,838 cu ft)
Baggage compartment volume:	
with 50 passengers	9.1 m³ (320 cu ft)
with 56 passengers	7.9 m³ (280 cu ft)

AREAS:

Wings, gross	56.21 m² (605.0 sq ft)
Ailerons (total)	1.87 m² (20.18 sq ft)
Tail surfaces	as for Series 100A/200A

WEIGHTS AND LOADINGS:

Operating weight empty:	
basic	11,709 kg (25,814 lb)
HGW	11,791 kg (25,995 lb)
Max usable fuel:	
standard	2,576 kg (5,679 lb)
optional	4,646 kg (10,243 lb)
Max payload: basic	5,166 kg (11,386 lb)
HGW	6,126 kg (13,505 lb)
Payload with max fuel	5,138 kg (11,327 lb)
Max T-O weight: basic	18,642 kg (41,100 lb)
HGW	19,504 kg (43,000 lb)
Max landing weight: basic	18,144 kg (40,000 lb)
HGW	19,050 kg (42,000 lb)
Max zero-fuel weight: basic	16,873 kg (37,200 lb)
HGW	17,917 kg (39,500 lb)
Max wing loading: basic	331.7 kg/m² (67.93 lb/sq ft)
HGW	347.0 kg/m² (71.07 lb/sq ft)
Max power loading: basic	5.25 kg/kW (8.63 lb/shp)
HGW	5.49 kg/kW (9.03 lb/shp)

PERFORMANCE (at 95% standard MTOW except where indicated):

Max cruising speed: basic	287 kt (532 km/h; 330 mph)
HGW	285 kt (528 km/h; 328 mph)
Stalling speed, flaps down	77 kt (141 km/h; 88 mph)
Max rate of climb at S/L	549 m (1,800 ft)/min
Rate of climb at S/L, OEI	137 m (450 ft)/min
Max certified altitude	7,620 m (25,000 ft)
Service ceiling, OEI: basic	4,145 m (13,600 ft)
HGW	3,734 m (12,250 ft)
FAR Pt 25 T-O field length at S/L, ISA, 15° flap:	
basic	1,097 m (3,600 ft)
HGW	1,180 m (3,870 ft)
FAR Pt 25 landing field length at max landing weight:	
basic	1,010 m (3,315 ft)
HGW	1,040 m (3,415 ft)
Range, HGW with 50 passengers:	
standard fuel	841 n miles (1,557 km; 967 miles)
auxiliary fuel	1,098 n miles (2,033 km; 1,263 miles)

OPERATIONAL NOISE LEVELS: (FAR Pt 36):

T-O	79.5 EPNdB
Sideline	87 EPNdB
Approach	93.3 EPNdB

UPDATED

DHC-8 DASH 8 Q400

TYPE: Twin-turboprop airliner.

PROGRAMME: Launched June 1995 as stretch of Series 300; component manufacture began December 1996; engine first test flown in January 1997, mounted on nose of Pratt & Whitney Canada Boeing 720B testbed; roll-out 21 November 1997, first flight (C-FJJA) 31 January 1998; five aircraft participated in the 1,900 hour, 1,400 sortie flight test programme, based at the Bombardier Flight Test Centre in Wichita, Kansas, (see 2000-01 edition), leading to Transport Canada CAR 525 Amendment 86 certification on 14 June 1999, JAA approval in December 1999 and FAA FAR Pt 25 approval on 8 February 2000. JAA approval for steep approach (5.5° glideslope) operations at London City Airport achieved 11 October 2001. First delivery (OY-KCA) to SAS Commuter 20 January 2000, followed by service entry 7 February on Copenhagen, Denmark, to Poznan, Poland route. Q400 fleet flight time totalled 19,804 hours and 24,697 cycles by March 2001. Mitsubishi joined programme as risk-sharing partner in October 1995, with responsibility for design, development and manufacture of fuselage and tail sections.

Production temporarily suspended in November/ December 2002.

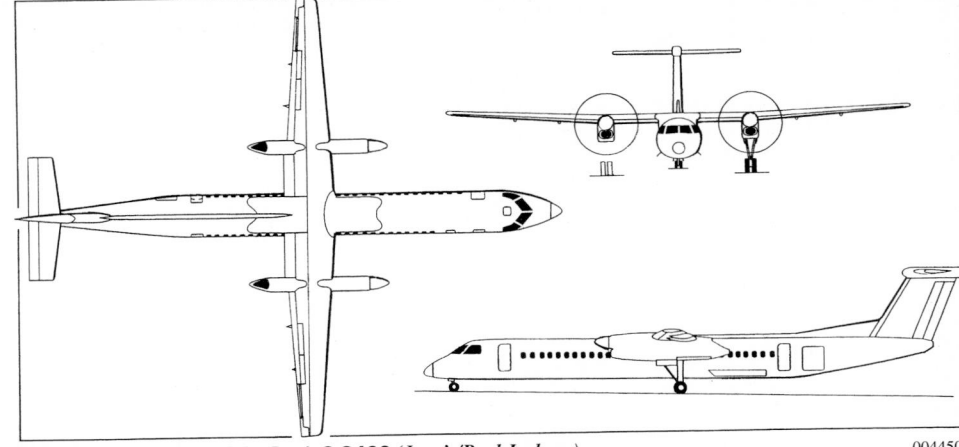

General arrangement of the Dash 8 Q400 (*Jane's/Paul Jackson*)   0044508

CUSTOMERS: Total of 74 firm orders by 28 February 2002, at which time 53 had been delivered; see table. Launch customer was Great China Airlines (now UNI Airways), which ordered six (since cancelled) in February 1996. US launch customer Horizon Air, which ordered 15 with 15 options, in June 1999, took delivery of its first aircraft on 8 January 2001. First for Changan handed over 27 October 2000. British European received first on 23 November 2001; Widerøe deliveries began 16 November 2001.

COSTS: Five for Japan Air Commuter valued at US$105 million (2001). Direct operating cost in US currency, based on 70 passengers over 200 n miles (370 km; 230 mile) stage length: in USA 8.06 cents per seat-mile; in Europe 9.38 cents per seat-kilometre (both 2002).

DESIGN FEATURES: Compared with Q300, fuselage stretched 6.83 m (22 ft 5 in) to seat up to 78. Other new features include engines; propeller; tapered inboard wing section increasing propeller/fuselage clearance to 1.09 m (3 ft 7 in); revised wing/fuselage fairings; ailerons, elevators and fin cap fairing; landing gear; baggage/service doors; upgraded avionics. Airframe designed for crack-free life of 40,000 flight hours/80,000 flight cycles and an economic life of 80,000 flight hours/160,000 flight cycles. Common type rating with Q100/200/300.

STRUCTURE: Generally as for Q100 and Q200. De Havilland Canada manufactures cockpit section and wing and performs final assembly; other participants in programme include AlliedSignal (electrical power system); Goodrich (brakes); Dowty Aerospace (propellers); Menasco (landing gear); Micro Technica (flap system); Mitsubishi (fuselage and tail sections); Parker Bertea Aerospace (hydraulics and fuel system); Pratt & Whitney Canada (engines); Sextant Avionique (avionics system); and Shorts (engine nacelles).

POWER PLANT: Two Pratt & Whitney Canada PW150A turboprops with FADEC, each flat rated to 3,781 kW (5,071 shp) at 37.4°C, driving Dowty R408 six-blade, slow-turning (1,020 rpm for T-O; 850 rpm cruise) composites propellers giving 15 per cent reduction in T-O rpm and 6 to 19 per cent reduction in cruise rpm over Q100/300. Fuel capacity 6,708 litres (1,772 US gallons; 1,475 Imp gallons).

ACCOMMODATION: Variety of cabin configurations providing four-abreast seating, with central aisle, for 70 passengers at 84 cm (33 in) pitch. 74 passengers at 79 cm (31 in) pitch, 78 passengers at 76 cm (30 in), or two-class layout for 10 business class passengers at 86 cm (34 in) pitch and 62 economy class passengers at 79 cm (31 in); NVS system standard.

AVIONICS: Prime contractor, Sextant Avionique.
Comms: Dual VHF nav/com and mode S transponder; solid-state cockpit voice recorder; ELT antenna.
Radar: Weather radar.
Flight: GPWS, radar altimeter, ADF and DME. Provision for FMS, GPS and TCAS II.
Instrumentation: EFIS employs five 152 × 203 mm (6 × 8 in) LCDs. Optional Flight Dynamics HGS4100 (later HGS4200) head-up guidance system.

DIMENSIONS, EXTERNAL:

Wing span	28.42 m (93 ft 3 in)
Wing aspect ratio	12.8
Length overall	32.84 m (107 ft 9 in)
Fuselage max diameter	2.69 m (8 ft 10 in)

Height overall	8.36 m (27 ft 5 in)
Tailplane span	9.27 m (30 ft 5 in)
Wheel track	8.79 m (28 ft 10 in)
Wheelbase	13.94 m (45 ft 9 in)
Propeller diameter	4.11 m (13 ft 6 in)
Propeller ground clearance	0.97 m (3 ft 2 in)
Propeller fuselage clearance	1.10 m (3 ft 7¼ in)
Passenger/crew door (port, fwd):	
Height	1.66 m (5 ft 5½ in)
Width	0.76 m (2 ft 6 in)
Height to sill	1.17 m (3 ft 10 in)
Passenger/crew door (port, rear):	
Height	1.73 m (5 ft 8 in)
Width	0.71 m (2 ft 4 in)
Height to sill	1.55 m (5 ft 1 in)
Baggage door (port, rear):	
Height	1.55 m (5 ft 1 in)
Width	1.27 m (4 ft 2 in)
Height to sill	1.55 m (5 ft 1 in)
Baggage door (stbd, fwd):	
Height	1.37 m (4 ft 6 in)
Width	0.61 m (2 ft 0 in)
Height to sill	1.17 m (3 ft 10 in)
Service door (stbd, aft):	
Height	1.42 m (4 ft 8 in)
Width	0.69 m (2 ft 3 in)
Height to sill	1.55 m (5 ft 1 in)

DIMENSIONS, INTERNAL:

Cabin (excl flight deck):	
Length	18.80 m (61 ft 8 in)
Max width	2.49 m (8 ft 2 in)
Max height	1.95 m (6 ft 5 in)
Floor area	43.6 m² (469 sq ft)
Volume	77.6 m³ (2,740 cu ft)
Baggage volume:	
forward	2.58 m³ (91.0 cu ft)
aft	11.6 m³ (411 cu ft)
total	14.2 m³ (502 cu ft)

AREAS:

Wings, gross	63.08 m² (679.0 sq ft)
Ailerons (total)	1.87 m² (20.18 sq ft)
Vertical tail surfaces (total)	14.12 m² (152.0 sq ft)
Horizontal tail surfaces (total)	16.72 m² (180.0 sq ft)

WEIGHTS AND LOADINGS:

Operating weight empty	17,108 kg (37,717 lb)
Max payload	8,747 kg (19,283 lb)
Payload with max fuel	6,831 kg (15,059 lb)
Max T-O weight: basic	27,329 kg (60,250 lb)
intermediate	27,995 kg (61,720 lb)
HGW	29,256 kg (64,500 lb)
Max landing weight	28,009 kg (61,750 lb)
Max zero-fuel weight	25,855 kg (57,000 lb)
Max wing loading: basic	433.2 kg/m² (88.73 lb/sq ft)
intermediate	443.8 kg/m² (90.90 lb/sq ft)
HGW	463.8 kg/m² (94.99 lb/sq ft)
Max power loading: basic	3.62 kg/kW (5.94 lb/shp)
intermediate	3.70 kg/kW (6.09 lb/shp)
HGW	3.87 kg/kW (6.36 lb/shp)

PERFORMANCE (estimated, at 95% of standard MTOW, except where indicated):

Max cruising speed	350 kt (648 km/h; 403 mph)
Max certified altitude: standard	7,620 m (25,000 ft)
optional	8,230 m (27,000 ft)

Dash 8 Q400 in 74-seat configuration
A: airstair/Type I exit, A2: passenger door/Type I exit, B: baggage, BD: baggage door, F: folding seat, G: galley, L: lavatory, S: service door/Type I exit, TI/TII: Type I/Type II exits, W: wardrobe   0044509

Dash 8 Q400 demonstrator *(Jane's/Paul Jackson)*

NEW/0137354

BOMBARDIER Q SERIES ORDER BOOK
(at 1 January 2003)

Customer	Variant	Ordered	Delivered	Backlog
Abu Dhabi Aviation	200	2	2	
AGES/Worldwide	300	6	6	
All Nippon Airways	400	4	0	4
Air Atlantic	100	15	15	
AirBC	100	12	12	
	300	6	6	
Air Creebec	100	1	1	
Air Dolomiti	300	3	3	
Air Maldives	200	1	1	
Air Manitoba	100	1	1	
Air Nippon Network/ ANK	300	5	3	2
Air Niugini	200	2	2	
Air Nostrum	300	29	19	10
Air Nova	100	6	6	
Air Ontario	100	30	30	
	300	6	6	
Air Senegal International	300	1	1	
Air Wisconsin	100	5	5	
	300	7	7	
Alberta Government	100	1	1	
ALM	300	2	2	
Amakusa	100	1	1	
America West	100	12	12	
Ansett New Zealand	100	2	2	
Augsburg Airways	200	2	2	
	300	7	7	
	400	5	5	
Australian Airlines	100	4	4	
Aviaco/Furlong	300	4	4	
Avline	300	4	4	
BPX Colombia	200	1	1	
Bahamasair	300	3	3	
British European	200	3	3	
	300	4	4	
	400	4	4	
Brymon Airways	300	10	10	
BWIA	300	4	4	
Canadian DND	100	6	6	
CCAir	100	2	2	
Changan	400	3	3	
City Express	100	4	4	
Contact Air	100	2	2	
	300	4	4	
DAC Air	300	1	1	
Eastern Australia	200	1	1	
Eastern Metro Express	100	8	8	
Elveden Investments	100	3	3	
	300	1	1	
Fairways Corp	100	1	1	
GPA Jetprop	100	16	16	
	300	14	14	
Hamburg Airlines	100	2	2	
	300	2	2	
Horizon Air	100	21	21	
	200	28	28	
	400	15	15	
HydroQuebec	400	2	2	
Interot	100	2	2	
	300	1	1	
JAS/Japan Air Commuter	400	5	2	3
LADS	200	1	1	
LIAT	100	5	5	
	300	3	3	
MarkAir	300	2	2	
Mesa Air	200	12	12	

Customer	Variant	Ordered	Delivered	Backlog
Mexican Navy	200	1	1	
Midroc Aviation	200	2	2	
Mobil Oil	100	1	1	
National Jet	200	6	6	
	300	2	2	
Norfolk Airlines	100	1	1	
NorOntair	100	2	2	
Northwest Airlines	100	25	25	
Norwegian CAA	100	1	1	
Oriental Air Bridge	200	2	2	
Palestinian	300	2	2	
Petroleum Air Service	300	3	2	1
Qantas Airways	200	1	1	
	300	8	5	3
Rheintalflug	100	3	3	
	300	3	3	
Royal Wings	300	1	1	
Ryukyu Air Commuter	100	4	3	1
South African Express	300	12	12	
SAS Commuter	400	28	28	
Saeaga Airlines	200	1	1	
	300	1	1	
Saudi Aramco	200	3	3	
Schreiner Airways	100	1	1	
	300	2	2	
Talair	100	2	2	
TAVAJ	200	2	2	
Time Air	100	4	4	
	300	10	10	
Transport Canada	100	2	2	
Tyrolean	100	11	11	
	300	19	19	
	400	8	6	2
UNI Airways	100	4	4	
	200	1	1	
	300	14	14	
USAF/Sierra	100	2	2	
USAir Express	100	56	56	
	200	19	19	
Wideroe's	100	15	15	
	300	4	4	
	400	3		3
Zhejiang Airlines	300	2	2	
Undisclosed	100	3	3	
	200	3	3	
	300	1	1	
	400	1	1	
Totals		**684**	**655**	**29**

Notes: Operators with leased Dash 8 aircraft include, but are not necessarily restricted to, the following: Air Alliance, ERA, Lloyd Aviation, Lufthansa CityLine, Sabena, TABA.
* US Airways operators include Piedmont Airlines and Allegheny Commuter.
UNI Airways received initial aircraft when named Great China Airlines.
British European received initial aircraft when named Jersey European Airways.

SUMMARY

	Delivered	Backlog	Total
Series 100	298	1	299
Series 200	94		94
Series 300	197	16	213
Series 400	66	12	78
Totals	**655**	**29**	**684**

Service ceiling, OEI:
standard   6,250 m (20,500 ft)
optional   5,790 m (19,000 ft)
FAR T-O field length at S/L, MTOW   1,193 m (3,915 ft)
FAR landing field length at S/L, MLW 1,287 m (4,221 ft)
Max range, 70 passengers, IFR reserves
                          1,360 n miles (2,518 km; 1,565 miles)
OPERATIONAL NOISE LEVELS (estimated, FAR Pt 36):
T-O   78.3 EPNdB
Sideline   84 EPNdB
Approach   94.8 EPNdB
                                             UPDATED

DHC-8 DASH 8M and TRITON
Canadian Forces designations: CC-142 and CT-142
US Air Force designation: E-9A
TYPE: Multisensor surveillance twin-turboprop.
CURRENT VERSIONS: **CC-142** and **CT-142**: Passenger/cargo transport and navigation trainer versions of Dash 8M-100 respectively; both have long-range fuel tanks, low-pressure tyres, high-strength floors and mission avionics; Canadian Department of National Defence operates two CC-142s and four CT-142s, latter with mapping radar in extended nose.

E-9A: US Air Force Dash 8-100 as missile range control aircraft; relays telemetry, voice and drone and fighter control data and simultaneously observes range with radar; avionics include large electronically steered phased-array radar in fuselage side and AN/APS-128D surveillance radar in ventral dome.

Aerial Survey/Coastal Patrol: LADS Corporation of Adelaide, Australia, operates a Series 200 equipped with a Visions System Group laser airborne depth sounder (LADS) for hydrographic survey of shallow coastal

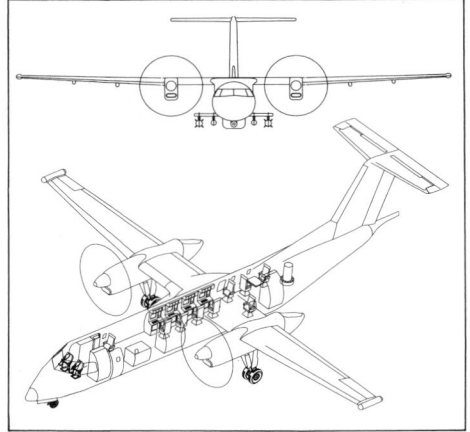

Interior and weapon positions of the DHC-8 Triton 300 maritime patrol aircraft

waters. Canadian Coast Guard aircraft are equipped with environmental monitoring equipment installed in fairing on starboard side of fuselage.

Surveillance Australia of Adelaide took delivery in mid-2000 of first of two Q200s that it operates for maritime patrol on behalf of the Australian Customs Service's Coastwatch. The Q200s, which supplement three earlier Series 200s Surveillance Australia has operated since 1996, were modified for the maritime patrol role by Field

Aviation Inc of Toronto, which installed a belly-mounted Raytheon SV1022 search radar, Wescam 16DS turret containing FLIR and daylight TV camera, large observation window inserts at each of the mid-fuselage emergency exits, and a systems operator's console in the main cabin. In Coastwatch configuration the Q200 can transit 300 n miles (556 km; 345 miles) at maximum cruising speed, fly a low-level search track of about 1,000 n miles (1,852 km; 1,150 miles) covering some 260,000 km² (100,390 sq miles) per mission, and return to base with adequate fuel reserves.

Triton: Maritime surveillance version; promotion began in 1997 of the Triton 300, based on the Series 300 airframe, but with weapon pylons on the lower fuselage and provision for ventral radar, MAD tail and wingtip ESM.
CUSTOMERS: Norwegian CAA (one Dash 8M-100), Canadian Department of Transport (two Dash 8M-100), Canadian Forces (two CC-142, four CT-142, all with No. 402 Squadron at Winnipeg), USAF (two E-9A at Tyndall AFB, Florida); all 11 included in Series 100/100A delivery total.
Data follow for Triton 300 where different from Dash 8 Series 300.
WEIGHTS AND LOADINGS:
Max fuel weight   4,647 kg (10,245 lb)
Max T-O weight   19,504 kg (43,000 lb)
Max ramp weight   19,595 kg (43,200 lb)
Max landing weight   19,051 kg (42,000 lb)
Max zero-fuel weight   17,916 kg (39,500 lb)
PERFORMANCE:
Econ cruising speed   240 kt (444 km/h; 276 mph)
Patrol speed   135 kt (250 km/h; 155 mph)
Endurance   9 h
                                             UPDATED

DHC-8 Dash 8 Q200 modified for surveillance duties with the Australian Coast Guard *(Jane's/Paul Jackson)*   *NEW*/0137355

CAG

CANADIAN AEROSPACE GROUP INTERNATIONAL INC
5353 John Lucas Drive, Burlington, Ontario L7L 6A8
Tel: (+1 905) 331 03 55
Fax: (+1 905) 331 03 56
e-mail: pnelso2@ibm.net
Web: http://www.canaero.com
CHAIRMAN: Joseph Rotolo
PRESIDENT: Philip A Nelson

Under a joint venture begun in 1997, CAG subsidiary Panda Aircraft Corporation of North Bay, Ontario, planned to receive bare HAMC Y-12 (IV) (see Chinese section) airframes shipped from China and equip them with 559 kW (750 shp) PT6A-34 engines, instrumentation and cabin furnishing. Aircraft would have been marketed for approximately US\$3 million, under the export name **Twin Panda**. The initial agreement covered 50 Y-12s in the first three years with options for up to 150 more later. The project was abandoned in April 1999 and Panda Aircraft was

required to repay a C\$917,000 government grant. A C\$1,625,000 grant was later obtained for relocation of manufacturing to Québec, and the programme revived. Deliveries of the PT6A-34-powered version were then targeted to begin in January 2001 to Fernani Trade Lease and Finance of Brazil, which has ordered 30. The Brazilian contract, reportedly worth C\$94.5 million, was said to be near to agreement in July 2001, but had not been confirmed by January 2002.

Under a second agreement, announced in April 1998, CAG obtained for its shareholder company, Can-Aero Leasing, the rights to assemble and complete at St Hubert, Québec, (including new engine, landing gear and avionics installations) up to 240 NAMC N-5A agricultural aircraft, now known as the **Hongdu AG Dragon**; Can-Aero was to form AG Dragon Aircraft Corporation for this purpose, but this appears not to have occurred. In return, NAMC has the option of producing in China the Windeagle light aircraft and/or the **Monitor Jet**, a turbofan trainer based on the Bede BD-10 but entirely redesigned by a CAG team and also offered as a UAV with SL Ventures guidance system. In December 1999, co-production of the Monitor Jet was

offered to Portugal as part of Sikorsky's bid to sell that country S-92 Helibus helicopters. Monitor Jet Corporation continued promotion of the aircraft in early 2002.

Windeagle Aircraft Corporation of Canada, another CAG company, acquired from Composite Aircraft Corporation of Texas, USA, all rights to the Windecker Eagle all-composites light aircraft, last described (under Composite Aircraft Company) in the 1984-85 edition of *Jane's*. CAG planned to resume production of the aircraft, renamed **Windeagle** and powered by a 212 kW (285 hp) Teledyne Continental IO-520C engine, in 1997, at a price of US\$198,000. Plans were delayed when CAG left North Bay. Proposed future developments include a turboprop version powered by a 335 kW (450 shp) Pratt and Whitney Canada PT6A and flight tests with a 447 kW (600 hp) Orenda V-8 engine.

CAG also acquired rights to the Timberwolf, a Windeagle derivative optimised for military training. This was last described in the 1995-96 *Jane's* under the Sino-Canadian partnership Venga/Baoshan. The aircraft is not currently being promoted.
                                             VERIFIED

CLASS

CANADIAN LIGHT AIRCRAFT SALES AND SERVICES INC

880 Chemin St-Féréol, Les Cèdres, Québec J7T 1N3
Tel: (+1 450) 452 47 72 and (+1 888) 977 14 47
Fax: (+1 450) 452 26 94
e-mail: marla@bushcaddy.com
Web: http://www.bushcaddy.com
PARTNERS: Sean Gilmore
Marlene Gill
Ronald Strauss

EUROPEAN, MIDDLE EAST AND NORTH AFRICAN AGENTS
AMA Aeromax Affairs, Aérodrome St Laurent, 4 route de Sémignan, F-33112 St Laurent Médoc, France

CLASS took over production of the previously named Cadi from designer Jean Edues Potvin and has recently expanded the range to provide six versions of the Bush Caddy.

UPDATED

CLASS BUSH CADDY

TYPE: Four-seat kitbuilt.
PROGRAMME: Designed by Jean Edues Potvin, the original Cadi was first flown in first half of 1993. Kits of the refined Cadi were marketed from 1994. The re-engineered Bush Caddy appeared in 2000. Several new versions have been added to the range since then, using both Rotax and Lycoming power plants.
CURRENT VERSIONS: **R80:** Smallest aircraft in range and classified as ultralight in USA, Canada and Australia; two seats; designed for engines up to 59.7 kW (80 hp). Can be purchased as a kit or as a factory-completed aircraft. Voyageur option can be converted from tailwheel to tricycle layout in a few hours, as hardpoints for both variants are provided as standard. 'R' in designation indicates Rotax engine.
Description below refers to R80, unless otherwise stated.
R60: Ultralight version of the above, intended for European market; due for launch during 2003.
R120: Engines in the 48.5 to 89.5 kW (65 to 120 hp) range; two seats. Introduced at AirVenture, Oshkosh, July 2001. Voyageur option available as for R80.
L160: Accepts engines up to 119 kW (160 hp). Seats two-plus-one. 'L' in designation indicates Lycoming engine.
L162: Two-plus-two version of L160.
L164: Four-seat version, with 134 kW (180 hp) engine. Displayed statically at AirVenture 2001. Prototype (C-GJDK) first flew 8 September 2001.
CUSTOMERS: Total 85 of all versions sold by mid-2001, of which 62 were then flying.
COSTS: Basic kit costs (excluding engine, propeller and internal finishing): R80, US$14,036; R120, US$15,089; R120 Voyageur, US$14,115; L160, US$17,897; L164, US$20,730 (2002). Factory-finished R80, US$45,500 (all 2002).
DESIGN FEATURES: Conventional lightplane; high, constant chord wing braced by V struts each side; sweptback cantilever tail surfaces.
Kit includes many preformed and welded parts. Quoted build time 1,000 hours. Fast-build kits are available in which wings and fuselage internal structures are complete and need only covering.
FLYING CONTROLS: Conventional and manual. No horn balances on control surfaces. Flaps standard on L160 and higher-powered versions and optional for R120.
STRUCTURE: All-metal construction throughout. Two-spar wing has a modified Piper Super Cub aerofoil with flat lower surface and has twin 7.6 cm (3 in) extruded aluminium bracing struts and composites wingtips. Monocoque fuselage contains 6061-T6 bulkheads covered with sheet aluminium and has a composites cowling.
LANDING GEAR: Tailwheel type, fixed; R120 has quick-change option from tailwheel to tricycle. Bungee cord suspension. Optional floats, amphibious floats and skis can be fitted. Cleveland mainwheels are 7.00-6, with McCreary tyres and Matco hydraulic brakes. Castoring tailwheel, size 2.50-4.
POWER PLANT: *R80:* One 59.6 kW (79.9 hp) Rotax 912 UL driving a two-blade Sensenich metal fixed-pitch propeller; options include 73.5 kW (98.6 hp) Rotax 912 ULS and 84.6 kW (113.4 hp) Rotax 914 UL, and two- and three-blade Warp Drive propellers. Fuel capacity 91 litres (24.0 US gallons; 20.0 Imp gallons) in two wingroot tanks.
R120: 73.5 kW (98.6 hp) Rotax 912 ULS standard; engines up to 89.5 kW (120 hp) optional. Standard fuel capacity 100 litres (26.4 US gallons; 22.0 Imp gallons); optional increase to 155 litres (40.8 US gallons; 34.0 Imp gallons).
L160 and *L162:* One 119 kW (160 hp) Lycoming O-320 standard. Fuel capacity 173 litres (45.6 US gallons; 38.0 Imp gallons).
L164: One 134 kW (180 hp) Lycoming IO-360 standard. Fuel capacity as L160/L162.
ACCOMMODATION: Pilot and passenger in fully adjustable seats in R80 and R120; bench seat in rear of L160 and L162 for one and two extra passengers, respectively; L164 is full four-seat version. Wraparound Lexan windscreen. Baggage locker behind cabin, with door on port side.

DIMENSIONS, EXTERNAL:

Wing span: R80, R120	9.75 m (32 ft 0 in)
L160, L162, L164	10.97 m (36 ft 0 in)
Wing chord, constant: all	1.60 m (5 ft 3 in)
Wing aspect ratio: R80, R120	6.1
L160, L162, L164	6.9
Length overall: R80, R120	6.73 m (22 ft 1 in)
L160, L162	7.26 m (23 ft 10 in)
L164	7.75 m (25 ft 5 in)
Tailplane span	2.62 m (8 ft 7 in)
Wheel track	2.13 m (7 ft 0 in)
Wheelbase	4.44 m (14 ft 7 in)

DIMENSIONS, INTERNAL:

Cabin: Max width: R80, R120, L160	1.12 m (3 ft 8 in)
L164	1.19 m (3 ft 11 in)
Max height: R80, R120	1.12 m (3 ft 8 in)
L160	1.17 m (3 ft 10 in)
L164	1.33 m (4 ft 4½ in)
Baggage hold volume	0.42 m³ (15.0 cu ft)

AREAS:

Wings, gross: R80, R120	15.61 m² (168.0 sq ft)
L160, L162, L164	17.56 m² (189.0 sq ft)

WEIGHTS AND LOADINGS:

Weight empty: R80	294 kg (648 lb)
R120	373 kg (823 lb)
L160	498 kg (1,098 lb)
L164	562 kg (1,239 lb)
Baggage capacity	34 kg (75 lb)
Max T-O weight: R80	559 kg (1,232 lb)
R120	680 kg (1,500 lb)
L160, L162	998 kg (2,200 lb)
L164	1,134 kg (2,500 lb)
Max wing loading: R80	35.8 kg/m² (7.33 lb/sq ft)
R120	43.6 kg/m² (8.93 lb/sq ft)
L160, L162	56.8 kg/m² (11.64 lb/sq ft)
L164	64.6 kg/m² (13.23 lb/sq ft)
Max power loading: R80	9.38 kg/kW (15.42 lb/hp)
R120	9.26 kg/kW (15.21 lb/hp)
L160, L162	8.37 kg/kW (13.75 lb/hp)
L164	8.45 kg/kW (13.89 lb/hp)

PERFORMANCE:

Never-exceed speed (VNE):	
R80, R120	113 kt (209 km/h; 130 mph)
L160	130 kt (241 km/h; 150 mph)
L164	139 kt (257 km/h; 160 mph)
Cruising speed at 75% power:	
R80	83 kt (153 km/h; 95 mph)
R120	87 kt (161 km/h; 100 mph)
L160	100 kt (185 km/h; 115 mph)
L164	117 kt (217 km/h; 135 mph)
Econ cruising speed at 60% power:	
R80, R120	78 kt (145 km/h; 90 mph)
L160	91 kt (169 km/h; 105 mph)
L164	104 kt (193 km/h; 120 mph)
Stalling speed, power off: R80	30 kt (55 km/h; 34 mph)
no flaps: R120, L160, L162, L164	37 kt (68 km/h; 42 mph)
flaps down: R120	28 kt (52 km/h; 32 mph)
L160, L162, L164	33 kt (62 km/h; 38 mph)
Max rate of climb at S/L: R80, L160	259 m (850 ft)/min
R120	274 m (900 ft)/min
L164	366 m (1,200 ft)/min
Service ceiling: all	3,960 m (13,000 ft)
T-O run: R80, R120	61 m (200 ft)
L160	110 m (360 ft)
L164	76 m (250 ft)
Endurance	more than 4 h

UPDATED

CLASS Bush Caddy L164 prototype (*Jane's/Paul Jackson*)   NEW/0137952

CREATIVE FLIGHT

CREATIVE FLIGHT

Elico House, 140 Industrial Park Road, PO Box 596, Haliburton, Ontario K0M 1S0
Tel: (+1 705) 457 21 92
Fax: (+1 705) 457 96 51
e-mail: info@creativeflight.com
Web: http://www.creativeflight.com
INFORMATION: Kirk Creelman

NEW ZEALAND OFFICE:
PO Box 13697, Johnsonville, Wellington
Fax: (+64 4) 233 98 44

UPDATED

CREATIVE FLIGHT MPA

TYPE: Multisensor surveillance lightplane.
PROGRAMME: Project began 1997. Launched at AirVenture, Oshkosh, July 2000. Development prototype (C-GYCC) first flown 15 July 2001; exhibited at Oshkosh 24 to 30 July 2001.
CURRENT VERSIONS: **Two-seat:** Single-engine version.
Four-seat: Twin-engine version, retaining same dimensions.

Prototype Creative Flight MPA (*Jane's/Paul Jackson*)   NEW/0131706

DESIGN FEATURES: Designation indicates 'multi platform aircraft'. Kitplane meeting '51 per cent' rule for amateur construction; easily completed in single- or twin-engine configuration and with amphibian and seaplane landing gear.

Cabin pod, with broad outward visibility, mounted on cranked wings of two-stage taper and with sweptback tips. Engine(s) on arched cabane with central, vertical reinforcement strut. Twin tailbooms and high tailplane; forward boom sections contain main landing gear and attachment points for floats with cutouts for mainwheels when in amphibian mode.

FLYING CONTROLS: Conventional and manual. Horn balances on elevator and rudders; flight-adjustable tab on starboard rudder and port aileron. Fowler flaps. Third flap, between booms, is independently actuated; potential advantages during take-off and landing are being assessed during trials.

STRUCTURE: Generally of glass fibre with foam cores, but carbon fibre wing spars and cabane.

LANDING GEAR: Tricycle type; retractable. Mainwheels 6.00-6; castoring, forward retracting nosewheel, size 5.00-5; differential hydraulic mainwheel brakes. Adjustable air bag suspension. Optional floats, with or without wheeled gear.

POWER PLANT: One or two 89 kW (120 hp) piston engines, driving three-blade propeller of constant-speed, flight- or ground-adjustable pitch or fixed-pitch type. Prototype has single 89 kW (120 hp) Jabiru 3300 flat-six and ground-adjustable pitch propeller.

ACCOMMODATION: Two or four persons, according to version. Rear seats may be replaced by freight. Door each side of cabin. Dual controls.

DIMENSIONS, EXTERNAL:

Wing span	10.12 m (33 ft 2½ in)
Wing aspect ratio	6.7
Length overall	8.05 m (26 ft 5 in)
Height overall	2.29 m (7 ft 6 in)

DIMENSIONS, INTERNAL:

Cabin: Length	2.77 m (9 ft 1 in)
Max width	1.48 m (4 ft 9 in)
Max height	1.14 m (3 ft 9 in)

AREAS:

Wings, gross	15.33 m² (165.0 sq ft)

WEIGHTS AND LOADINGS (S: single-engine, T: twin-engine):

Weight empty: S	476 kg (1,050 lb)
T	612 kg (1,350 lb)
Max T-O weight: S	757 kg (1,670 lb)
T	1,124 kg (2,480 lb)

Max wing loading: S	49.41 kg/m² (10.12 lb/sq ft)
T	73.38 kg/m² (15.03 lb/sq ft)
Max power loading: S	8.47 kg/kW (13.92 lb/hp)
T	6.29 kg/kW (10.33 lb/hp)

PERFORMANCE (S, T as above):

Max level speed: S	139 kt (257 km/h; 160 mph)
T	169 kt (313 km/h; 195 mph)
Cruising speed at 75% power:	
S	130 kt (241 km/h; 150 mph)
T	161 kt (298 km/h; 185 mph)
Stalling speed: S	40 kt (73 km/h; 45 mph)
T	48 kt (89 km/h; 55 mph)
Max rate of climb at S/L: S	305 m (1,000 ft)/min
T	488 m (1,600 ft)/min
T-O run: S	198 m (650 ft)
T	183 m (600 ft)
Range: S	695 n miles (1,287 km; 800 miles)
T	825 n miles (1,528 km; 950 miles)

UPDATED

CUSTOM FLIGHT

CUSTOM FLIGHT COMPONENTS LTD

RR1, Perkinsfield, Ontario L0L 2J0
Tel: (+1 705) 526 96 26
Fax: (+1 705) 526 25 29
e-mail: nstar@csolve.net
Web: customflightltd.com
PRESIDENT: Morgan Williams
EXECUTIVE VICE-PRESIDENT: Jeff Owen

Custom Flight Components produces the North Star and more basic Bright Star two-seat kitplanes, both modelled on the Piper Super Cub. The North Star wing can also be bought as a separate kit and retrofitted to Piper Super Cub and Aeronca Champ and Citabria aircraft. Completions continue at low rate.

Company plans include an eight-seat aircraft to be named Galaxie.

UPDATED

CUSTOM FLIGHT NORTH STAR

TYPE: Tandem-seat kitbuilt.

PROGRAMME: Re-creation of Piper Super Cub, but re-engineered for additional structural strength. Prototype North Star (C-FGYD) first flew in 1991 and made its public debut that year. Offered in kit form from 1992 onwards.

CURRENT VERSIONS: **North Star:** Standard version, *as described below.*

Bright Star: Basic version with no flaps, single one-piece door, leaf spring steel main landing gear, square-topped fin and fuel capacity of 87 litres (23.0 US gallons; 19.2 Imp gallons); introduced in 1996, but first aircraft had not flown by late 2001; optional tricycle or tailwheel versions. Accepts engines from 74.6 kW to 134 kW (100 to 180 hp).

CUSTOMERS: At least 12 (of which six completed) in Canada; five (three completed) in USA. Production aircraft Nos. 7, 8 and 16 completed in Ontario by D E Blackburn, A King and R M Harkness in 1998. Nos. 3 (Subaru engine) and 15, built by W Smith, Ontario, and J Y Gratton, Québec, followed in the first half of 1999. Total of at least 23 kits sold, of which at least eight registered, including No. 23 to Donald Bloom of Coronado, California, in October 2001.

COSTS: North Star: C$34,476; Bright Star: C$19,602; both excluding engine (2001).

DESIGN FEATURES: Strut-braced mainplanes; wire-braced tailplanes; raked wingtips. Easy maintenance/inspection covers. Quoted build time 1,000 hours.

Compared with Super Cub, has landing gear legs 8 cm (3 in) taller; flatter spine; squared wingtips, with ailerons moved outboard by 0.61 m (2 ft 0 in); aileron and flap chord increased 10 cm (4 in); flap span increased 0.60 m (2 ft 0 in); cockpit 6 cm (2½ in) wider.

Aerofoil USA 35B.

FLYING CONTROLS: Conventional and manual. Horn-balanced rudder and elevators. Four-position flaps.

STRUCTURE: Welded 4130 chromoly tube, fabric-covered fuselage. Tubes injected with oil as anti-corrosion measure. Wing of prototype constructed of wood, but kit aircraft have I-beam spar, aluminium ribs and 2024-T3 leading-edges.

LANDING GEAR: Tailwheel configuration (fixed) with hydraulic/coil-spring suspension. Steerable tailwheel. Optional floats and skis. Mainwheel tyres 6.00-6 or 8.50-6; tailwheel 2.80-4 in.

POWER PLANT: One 112 kW (150 hp) Textron Lycoming O-320-E3D flat-four driving a McCauley two-blade fixed-pitch propeller. Fuel capacity 197 litres (52.0 US gallons; 43.3 Imp gallons) in two wing tanks.

ACCOMMODATION: Pilot and passenger in tandem within enclosed cockpit. Door on each side – upper half Perspex and lower half fabric-covered steel tubing. Baggage compartment reached through separate door on starboard side of fuselage.

Data estimated for Bright Star.

DIMENSIONS, EXTERNAL:

Wing span	11.07 m (36 ft 4 in)
Wing chord, average	1.60 m (5 ft 3 in)
Length overall: North Star	6.86 m (22 ft 6 in)
Bright Star	7.16 m (23 ft 6 in)
Height overall: North Star	2.08 m (6 ft 10 in)
Bright Star	2.29 m (7 ft 6 in)
Propeller diameter	2.08 m (6 ft 10 in)

Cabin door (each): Height	0.89 m (2 ft 11 in)
Width	1.17 m (3 ft 10 in)

AREAS:

Wings, gross	17.65 m² (190.0 sq ft)

WEIGHTS AND LOADINGS:

Weight empty: North Star	531 kg (1,170 lb)
Bright Star	476 kg (1,050 lb)
Baggage capacity	41 kg (90 lb)
Max T-O weight: North Star	997 kg (2,200 lb)
Bright Star	816 kg (1,800 lb)
Max wing loading: North Star	56.3 kg/m² (11.53 lb/sq ft)
Bright Star	46.3 kg/m² (9.47 lb/sq ft)
Max power loading (112 kW; 150 hp engine):	
North Star	8.93 kg/kW (14.67 lb/hp)
Bright Star	7.30 kg/kW (12.00 lb/hp)

PERFORMANCE:

Max level speed	122 kt (225 km/h; 140 mph)
Max cruising speed:	
North Star: landplane	100 kt (185 km/h; 115 mph)
floatplane	85 kt (158 km/h; 98 mph)
Bright Star	87 kt (161 km/h; 100 mph)
Normal cruising speed:	
North Star	85 kt (158 km/h; 98 mph)
Stalling speed, power off:	
flaps up	35 kt (65 km/h; 40 mph)
flaps down	31 kt (57 km/h; 35 mph)
Stalling speed, full power:	
flaps up	31 kt (57 km/h; 35 mph)
flaps down	22 kt (41 km/h; 25 mph)
Max rate of climb at S/L:	
North Star	335 m (1,100 ft)/min
Bright Star	305 m (1,000 ft)/min
Service ceiling	4,880 m (16,000 ft)
T-O run (half fuel): North Star	55 m (180 ft)
Bright Star	92 m (300 ft)
Landing run: North Star	61 m (200 ft)
Bright Star	84 m (275 ft)
Range	521 n miles (965 km; 600 miles)

UPDATED

North Star, with flaps fully extended   0110200

Prototype Custom Flight North Star two-seat lightplane   0131799

DIAMOND

DIAMOND AIRCRAFT CORPORATION

1560 Crumlin Sideroad, London, Ontario N5V 1S2
Tel: (+1 519) 457 40 00
Fax: (+1 519) 457 40 21
e-mail: sales@diamondair.com
Web: http://www.diamondair.com
CEO: Christian Dries
MARKETING DIRECTOR: Jeff Owen
TECHNICAL DIRECTOR: Peter Maurer

Diamond Aircraft Corporation incorporated 12 January 1992 and established 24,625 m² (265,000 sq ft) plant at London, Ontario in June 1994, to build DV 20 Katana light aircraft.

An experimental design, the DA 20-B1 (C-GKAT), was completed in June 1996, followed by the DA 20-C1 prototype in February 1997. Series production of the DA 20-C1 is currently under way, augmented from March 2002 onwards by the Diamond DA 40-180 Star (which see in Austrian section). Workforce reduced from about 220 to 60 in March 2000, with reduction in production rate of DA 20-C1s, pending reorganisation of company.

UPDATED

DIAMOND DA 20-A1 KATANA
English name: Samurai Sword

TYPE: Two-seat lightplane.

PROGRAMME: See following entry for early details of Katana programme. Initial Canadian aircraft (C-FSQN) first flight 29 June 1994. First recipient, in March 1995, was Missouri State University; large orders include 42 for Spartan Flight School of Tulsa, Oklahoma. Diamond had built 331 Katana A1s in Canada up to October 1998, when manufacture of the variant was suspended. At least 20 were supplied to Diamond in Austria in 1998, being placed in storage to meet future European orders for Rotax-powered aircraft.

DV 20-A1 converted to Katana 100 (*Jane's/Paul Jackson*)   0089474

CURRENT VERSIONS: **DA 20-A1 Katana:** New-build version.

Katana 100: Rebuild and upgrade offered by Diamond in Austria from 1999 and Diamond Aircraft in Canada from late 2000. Refitted with 73.5 kW (98.6 hp) Rotax 912S3 engine. Prototype OE-VPP (DV 20) flew December 1999.

Further details generally as in following entry on modified C1 variant, but detailed description of -A1 in Austrian section of 1997-98 and earlier editions. Specific -A1 details below.

DESIGN FEATURES: Incidence 3°.

POWER PLANT: One 59.6 kW (79.9 hp) Rotax 912A-3 flat-four engine; 2.27 reduction gear to two-blade constant-speed Hoffmann HO-V72F-M/S propeller with composites blades. Fuel capacity 79 litres (20.9 US gallons; 17.4 Imp gallons), normal; 55 litres (14.5 US gallons; 12.1 Imp gallons), optional. Fuel grades 100LL or mogas (95 octane leaded/unleaded).

WEIGHTS AND LOADINGS (DA 20-A1 and Katana 100, where different):

Weight empty: A1	494 kg (1,090 lb)
100	510 kg (1,124 lb)
Max fuel	57 kg (125 lb)
Max T-O weight: A1	730 kg (1,610 lb)
100	750 kg (1,653 lb)
Max wing loading: A1	62.9 kg/m² (12.89 lb/sq ft)
100	64.66 kg/m² (13.24 lb/sq ft)
Max power loading: A1	12.25 kg/kW (20.13 lb/hp)
100	10.20 kg/kW (16.77 lb/hp)

PERFORMANCE:

Never-exceed speed (V_{NE})	161 kt (298 km/h; 185 mph)
Cruising speed at 75% power:	
A1	119 kt (220 km/h; 137 mph)
100	125 kt (231 km/h; 144 mph)
Stalling speed: flaps up	41 kt (76 km/h; 48 mph)
flaps down	37 kt (69 km/h; 43 mph)
Max rate of climb at S/L: A1	207 m (680 ft)/min
100	218 m (715 ft)/min
Service ceiling: A1	4,270 m (14,000 ft)
100	more than 4,000 m (13,120 ft)
T-O run: 100	326 m (1,070 ft)
T-O to 15 m (50 ft): A1	488 m (1,600 ft)
100	454 m (1,490 ft)
Landing from 15 m (50 ft): A1	454 m (1,490 ft)
Range with max fuel:	
A1	523 n miles (968 km; 601 miles)
100, with 45 min reserves	
	500 n miles (927 km; 576 miles)
g limits	+4.4/−2.2

OPERATIONAL NOISE LEVELS:

T-O	65.2 EPNdB

UPDATED

DIAMOND DA 20-C1

English name: Samurai Sword

TYPE: Two-seat lightplane.

PROGRAMME: Original Austrian-built Katana revealed following first flight on 16 March 1991 of proof-of-concept LF 2000 (OE-VPX); followed by LF 2 Katana predecessor (OE-CPU, first flight December 1991); DV 20 prototype (OE-AKL, first flight 17 December 1992) developed from this by Dries and Volck and was representative of production aircraft; Austrian and German certification 26 April 1993; JAR-VLA certification received later in 1993; now also certified by Canada, UK and USA (FAA). Second production line opened in Canada, initially for the DA 20-A1; European-built aircraft designated DV 20. European manufacture of the DV 20 ended in mid-1996.

The current C1 version, introducing minor airframe improvements, was announced at Aero '97 at Friedrichshafen, shortly after first flight of the Canadian prototype (C-FDVA); production of this version began in early 1998. Canadian variant introduced some 40 minor changes, including wider cabin, electric trim and doubled electrical capacity, but is unnamed. Exports of -C1s to USA began June 1998 with seventh production aircraft.

CURRENT VERSIONS: **DA 20-C1:** *As described.* Relaunched in two versions from early 2001: **Evolution** basically equipped trainer, capable of 135 kt (250 km/h; 155 mph) cruise at 1,830 m (6,000 ft) and **Eclipse**, featuring carpets, leather seats and other interior features for private owners.

DA 20-A2: Intended to have DA 20-C's improved airframe, but retain DA 20-A's 59.7 kW (80 hp) Rotax 912 flat-four engine. Not launched.

CUSTOMERS: Total of more than 650 DV/DA 20s of all versions delivered by end 2001 from Austrian and Canadian production. European production of DV 20s terminated with the 160th aircraft in July 1996. Total of 166 C1s registered up to December 2001. Most to USA, but some to Canada and one each to New Zealand (in March 1999) and UK (July 1999). Katanas of all versions operated by training schools in Australia (one), Austria (six), Canada (11), Czech Republic (one), Finland (one), Germany (six), New Zealand (one), Sweden (one), South Africa (one), Switzerland (one), UK (two) and USA (32).

DESIGN FEATURES: Conventional, mid-wing, T-tail monoplane of composites construction.

Wing section Wortmann FX-63-137/20; dihedral 4° at root, 27° at tip; sweepback 1°; incidence, 2° 30′. Fin sweepback 57°; tailplane incidence −2°.

FLYING CONTROLS: Conventional and manual. Ailerons and elevators are operated by push-pull tubes, rudder by cables; electrically actuated spring bias provides elevator trimming; electrically actuated flaps.

STRUCTURE: Fuselage, with integral vertical fin, is mostly of GFRP construction with local CFRP reinforcement in high-stress areas, comprising two half-shells which are bonded together and incorporate transverse bulkheads in the cabin/centre-section area and three ring bulkheads in the tailcone. Wings comprise I-section spar of GFRP, with ribs providing mounting surfaces for control tubes and bellcranks; CFRP spar caps; centre section shear web is foam and fibre; wing skins are of GFRP/PVC foam/GFRP sandwich construction, as are the rudder and horizontal tail surfaces. Flaps and ailerons of GFRP.

LANDING GEAR: Fixed tricycle type, with cantilever self-sprung aluminium leg on each main unit and elastomeric suspension on nose leg; latter offset to starboard by 8 cm (3 in). Steering is provided by differential braking of the mainwheels and friction-damped castoring nosewheel. Optional speed fairings on all three wheels.

POWER PLANT: One 93 kW (125 hp) Teledyne Continental IO-240-B flat-four driving a Sensenich two-blade wood and composites propeller. Fuel capacity 92.7 litres (24.5 US gallons; 20.4 Imp gallons), of which 91 litres (24.0 US

Diamond DA 20-C1 (*Jane's/James Goulding*)   0044445

Diamond DA 20-C1 Eclipse registered in the USA (*Jane's/Paul Jackson*)   *NEW*/0131707

Typical instrument panel of Eclipse   0109126

Evolution typical instrument panel   0109127

Instrument panel of the DA 40-180 Diamond Star (*Jane's/Paul Jackson*)   0092046

Early production Diamond DA 40 Star from 2002 Canadian manufacture (*Jane's/Paul Jackson*)   *NEW*/0132900

gallons; 20.0 Imp gallons usable); oil capacity 5.7 litres (1.5 US gallons; 1.25 Imp gallons).

ACCOMMODATION: Two seats side by side, with baggage space to rear; leather seats optional. Canopy hinged at rear to open upward.

SYSTEMS: Electrical system includes 12 V 240 Ah battery.

AVIONICS: *Comms:* Eclipse has Garmin GMA 340 audio panel, GNS 430 comm/nav/GPS, GI 106, and GTX 327 transponder, all as standard; other avionics to customer's choice.

Flight: Optional S-TEC 30 two-axis autopilot and slaved HSI in Eclipse.

EQUIPMENT: Options include fire extinguisher, first aid kit, baggage harness and landing/position lights.

DIMENSIONS, EXTERNAL:

Wing span	10.87 m (35 ft 8 in)
Wing aspect ratio	10.2
Length overall	7.165 m (23 ft 6 in)
Height overall	2.18 m (7 ft 2 in)
Tailplane span	2.65 m (8 ft 8¼ in)
Wheel track	1.90 m (6 ft 2¾ in)
Wheelbase	1.72 m (5 ft 7¾ in)
Propeller diameter	1.75 m (5 ft 9 in)

AREAS:

Wing area, gross	11.61 m² (125.0 sq ft)

WEIGHTS AND LOADINGS:

Weight empty	529 kg (1,166 lb)
Max T-O weight: JAA	750 kg (1,653 lb)
FAA	780 kg (1,720 lb)
Max wing loading (JAA)	64.6 kg/m² (13.22 lb/sq ft)
Max power loading (JAA)	8.03 kg/kW (13.20 lb/hp)

PERFORMANCE:

Never-exceed speed (VNE)	164 kt (303 km/h; 188 mph)
Cruising speed at 75% power at 1,830 m (6,000 ft):	
Evolution	135 kt (250 km/h; 155 mph)
Eclipse	140 kt (259 km/h; 161 mph)
Manoeuvring speed	106 kt (196 km/h; 122 mph)
Stalling speed:	
in landing configuration	44 kt (82 km/h; 51 mph)
flaps down	34 kt (63 km/h; 40 mph)
Max rate of climb at S/L	305 m (1,000 ft)/min
Service ceiling	5,365 m (17,600 ft)
T-O run	337 m (1,105 ft)
T-O to 15 m (50 ft)	448 m (1,470 ft)
Landing from 15 m (50 ft): Evolution	377 m (1,235 ft)
Eclipse	390 m (1,280 ft)
Landing run: Evolution	168 m (550 ft)
Eclipse	177 m (581 ft)
Range, 45 min reserves: Eclipse	
	410 n miles (759 km; 471 miles)

UPDATED

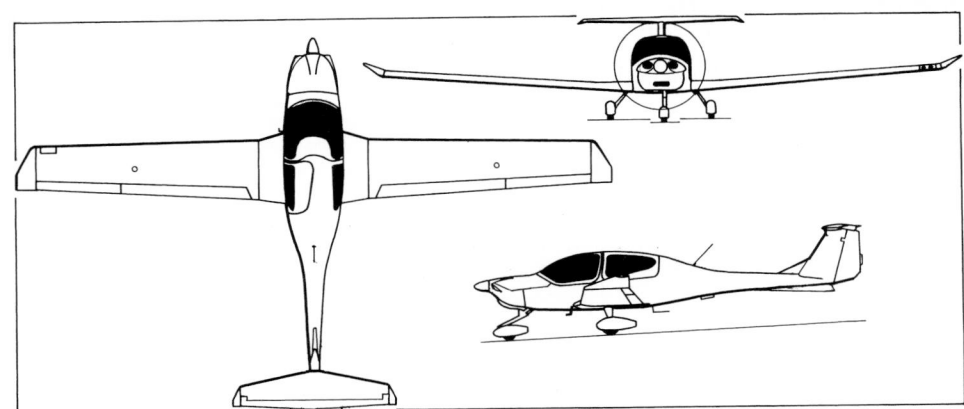

Diamond DA 40-180 Star (Lycoming IO-360) (*Jane's/Paul Jackson*)   0092123

DIAMOND DA 40-180 STAR

TYPE: Four-seat lightplane.

PROGRAMME: Development described under Diamond in Austrian section (which see). Production of Lycoming-engined variant transferred to Canada in 2002.

CURRENT VERSIONS: **DA 40-180:** Baseline version with Lycoming IO-360 engine. Preparation for Canadian assembly (which see) began September 2001, and first from this source (C-GDFC, c/n 40201) delivered on 28 March 2002 to Academy Air Services, Ontario.

DA 40-TDI: Powered by 99 kW (133 hp) Thielert turbo-diesel; Austrian-built.

COSTS: Baseline version €203,800 N-VFR-equipped (2002). Full standard (IFR) avionics €37,600 extra. Prices exclude tax.

POWER PLANT: One 134 kW (180 hp) Textron Lycoming IO-360-M1A flat-four with Lasar electronic ignition and self-adapting inlet cooling, driving an MTV-12-B/180-17 three-blade constant-speed (hydraulic) propeller. Fuel in two wing tanks, standard capacity 155 litres (40.9 US gallons; 34.1 Imp gallons); optional 200 litres (52.8 US gallons; 44.0 Imp gallons).

WEIGHTS AND LOADINGS:

Weight empty	740 kg (1,631 lb)
Baggage capacity	35 kg (77 lb)
Max T-O weight	1,150 kg (2,535 lb)
Max landing weight	1,092 kg (2,407 lb)
Max wing loading	85.2 kg/m² (17.45 lb/sq ft)
Max power loading	8.58 kg/kW (14.10 lb/hp)

PERFORMANCE:

Never-exceed speed (VNE)	
	178 kt (329 km/h; 204 mph) IAS
Max level speed	147 kt (272 km/h; 169 mph)
Cruising speed: at 75% power at FL65	
	137 kt (254 km/h; 158 mph)
at 55% power at FL100	122 kt (225 km/h; 140 mph)
Stalling speed	49 kt (91 km/h; 57 mph)
Max rate of climb at S/L	288 m (945 ft)/min
Max certified altitude	5,000 m (16,400 ft)
T-O run	310 m (1,020 ft)
T-O to 15 m (50 ft)	480 m (1,575 ft)
Range, 45 min reserves:	
standard fuel	570 n miles (1,055 km; 655 miles)
optional fuel	740 n miles (1,370 km; 851 miles)

NEW ENTRY

EUROCOPTER CANADA

EUROCOPTER CANADA LTD
(Subsidiary of Eurocopter)

HEAD OFFICE: PO Box 250, 1100 Gilmore Road, Fort Erie, Ontario L2A 5M9

Tel: (+1 905) 871 77 72
Fax: (+1 905) 871 33 20 and 35 99
e-mail: euro@eurocopter.ca
Web: http://www.eurocopter.ca

PRESIDENT AND CEO: Edouard Gaillat
VICE-PRESIDENT, FINANCE: Cyrille Fourny
DIRECTOR, OPERATIONS: Benoit Marcoux
DIRECTOR, CUSTOMER SUPPORT AND SALES: Gordon Kay
MANAGER, MARKETING SERVICES AND CUSTOMER RELATIONS: Diane Bourdeau

Started as MBB Helicopter Canada on 30 March 1984, Eurocopter Canada Limited (ECL) now employs 150 people at its 9,290 m² (100,000 sq ft) facility at Fort Erie. The company holds the world product mandate and design authority for the BO 105 LS A-3 and provides products and services for the complete Eurocopter range (which see in International section), including the new EC 135 and EC 120 B Colibri. Eurocopter Canada is certified for ISO 9001; its production department is certified for AQA P1, JAR 145 and FAR Pt 45 requirements; the engineering department makes use of the CATIA design system and is responsible for certification of modifications and repairs for the Eurocopter range. Pilot and engineer training is also carried out by the company and offered as part of the purchase package. Warehouses at Fort Erie, Montréal and Vancouver provide parts support for over 250 Eurocopter helicopters at present

operational in Canada and also supply approximately 20 new and used helicopters per year. Previously produced BO 105 LS utility helicopter (see 2002-03 and previous *Jane's*).

Current activities also involve assembly of the EC 120 B Colibri, AS 350/355 Ecureuil and EC 135. Turnover in 1999 was C$62 million and in 2000, C$60 million.

UPDATED

EUROCOPTER BO 105 LS

By early 2002, the Canadian-built BO 105 had been out of production for three years, although a stock of airframes is available to meet further demand. Refer to Eurocopter in International section for a description.

UPDATED

FOUND

FOUND AIRCRAFT CANADA INC

RR# 2, Site 12, Box 10, Georgian Bay Airport, Parry Sound, Ontario P2A 2W8
Tel: (+1 705) 378 05 30
Fax: (+1 705) 378 05 94

e-mail: found@zeuter.com
Web: http://www.foundair.com

The present company of this name was establised to reinstate production of the FBA-2 Bush Hawk, beginning deliveries in mid-2000. An XP version was introduced in 2001.

UPDATED

FOUND FBA-2C BUSH HAWK

TYPE: Six-seat utility transport.

PROGRAMME: Original Found FBA-2C first flew on 9 May 1962 and was last described in the 1967-68 *Jane's*. Production numbered 27, plus five improved Centenniel 100s (1969-70 edition) before company ceased trading.

Found FBA-2C Bush Hawk *(Jane's/Paul Jackson)*   *NEW*/0131717

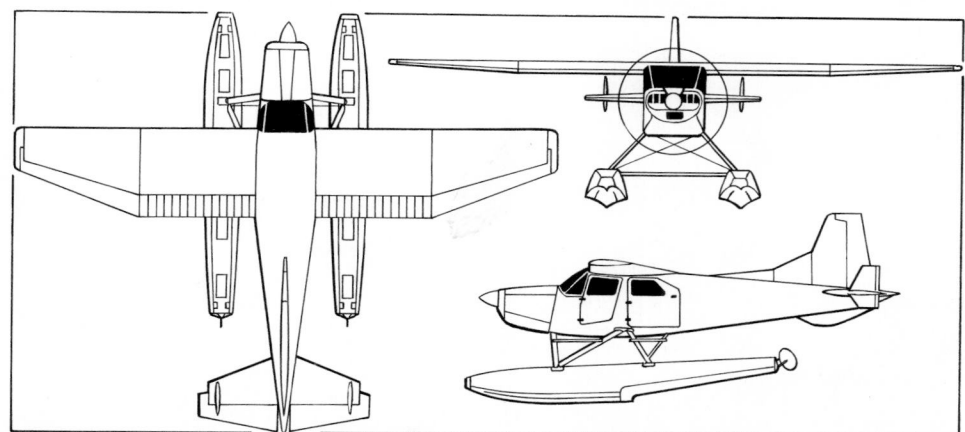

Found FBA-2C Bush Hawk six-seat utility transport *(Jane's/James Goulding)*   0099158

Prototype of current FBA-2C1 (C-FSVD), converted from an existing aircraft with IO-540-D engine, and other improvements, flew at Georgian Bay Airport, Ontario, in November 1996 and completed test flying, on wheels and floats, during 1998 and 1999. New production was due to start in July 1997 but programme slipped; preproduction aircraft C-GDWO (c/n 28; IO-540-L engine) made first flight October 1998 and certified 5 March 1999; was followed by C-GDWQ in 1999; initial production Bush Hawk C-GDWS (c/n 30) flew 2 March 2000. FAA certificate A7EA reinstated 10 March 2000, covering both FBA-2C and FBA-2C1; uprated (224 kW; 300 hp) version added 1 May 2001.

CURRENT VERSIONS: **FBA-2C Bush Hawk-260**: 194 kW (260 hp) version. Not currently available, due to Hawk-XP development taking priority.

FBA-2C1 Bush Hawk-300: 224 kW (300 hp) version. *Description and data apply to Bush Hawk-300 unless otherwise stated.*

Bush Hawk-XP: Improved version; demonstrator C-GDWC (c/n 31) built 2000; available from 2001.

CUSTOMERS: Eleven on order by end-1999; customers include Reliance Airways, Northwest Territories; Nakina Outpost Camps; and Air Service, Ontario. Initial production

aircraft exported to USA 24 May 2001. By end 2001, five had been built and production slots to mid-2003 sold.

COSTS: Bush Hawk-XP C$250,000; hourly operating cost C$54.13 (2002).

DESIGN FEATURES: Fully corrosion-proofed, all-metal construction for operation in exceptionally poor weather over all terrain.

High wing with constant chord to mid-span; outer panels unswept, with sharply swept forward trailing-edges. Sweptback fin with two-stage fillet; mid-mounted tailplane. Floatplane directional stability enhanced by underfin and auxiliary fins above and below tailplane at three-quarters span.

FLYING CONTROLS: Conventional and manual. Horn-balanced rudder and elevators. Flaps.

STRUCTURE: Aluminium-covered steel tube front fuselage and semi-monocoque rear fuselage. Deep-section, low-drag, all-aluminium wing.

LANDING GEAR: Fixed. Prototype, preproduction and first production aircraft have tailwheel configuration; all following landplanes have tricycle layout. Sprung steel main legs with urethane shock-absorbers. Mainwheels 8.00-6, but 66 cm (26 in) 8.50-10 and 76 cm (30 in) Tundra tyres being evaluated. Steerable tailwheel. Other

options include Wipline 3450 floats or amphibious floats or Aerocet 3500L floats and Federal C-3200 or C-3600 retractable skis.

POWER PLANT: One 194 kW (260 hp) Textron Lycoming IO-540-D4A5 or 224 kW (300 hp) IO-540-L1C5 flat-six engine, driving a Hartzell HC-C3YR-IRF constant-speed three-blade metal propeller. Fuel in two wing tanks, total capacity 379 litres (100 US gallons; 83.3 Imp gallons) of which 371 litres (98.0 US gallons; 81.6 Imp gallons) usable. Oil capacity 10 litres (2.75 US gallons; 2.3 Imp gallons).

ACCOMMODATION: Pilot and co-pilot, plus four passengers in cabin, with crew entrance through forward-hinged door on each side of fuselage. Rear hammock-style seat rolls up for storage in cabin roof when used in freight role; optional individual folding seats available. Rear cabin typically holds four 204 litre (54.0 US gallon; 45.0 Imp gallon) drums and has large, forward-hinged door on each side; door sills are flush with floor; doors swing 180° to allow easy loading. Alternatively, one stretcher and attendant in rear cabin. Camera hatch can be fitted in cabin floor.

SYSTEMS: 24 V 10 Ah battery, 28 V 70 A alternator.

AVIONICS: *Comms:* Optional Bendix/King KT-76A Mode C transponder.
Flight: KLX-135A GPS optional.

Data follows for Bush Hawk-300 landplane.

DIMENSIONS, EXTERNAL:
Wing span	10.97 m (36 ft 0 in)
Wing aspect ratio	7.2
Length overall: landplane	8.08 m (26 ft 6 in)
floatplane	8.74 m (28 ft 8 in)
Height overall: landplane	2.51 m (8 ft 3 in)
floatplane	4.17 m (13 ft 8 in)
Propeller diameter	2.13 m (7 ft 0 in)
Cabin door: Width	0.67 m (2 ft 2½ in)

DIMENSIONS, INTERNAL:
Cabin (including baggage compartment):	
Length	4.01 m (13 ft 2 in)
Max width	1.12 m (3 ft 8¼ in)
Max height	1.27 m (4 ft 2 in)
Volume	3.40 m³ (120 cu ft)

AREAS:
Wings, gross	16.72 m² (180.0 sq ft)

WEIGHTS AND LOADINGS:
Weight empty: landplane	862 kg (1,900 lb)
floatplane	1,021 kg (2,250 lb)
Baggage capacity	113 kg (250 lb)
Max T-O weight: landplane	1,587 kg (3,500 lb)
floatplane	1,678 kg (3,700 lb)
Max wing loading: landplane	94.9 kg/m² (19.44 lb/sq ft)
floatplane	100.4 kg/m² (20.56 lb/sq ft)
Max power loading: landplane	7.10 kg/kW (11.67 lb/hp)
floatplane	7.51 kg/kW (12.33 lb/hp)

PERFORMANCE:
Cruising speed at 75% power:	
landplane	150 kt (278 km/h; 173 mph)
floatplane	130 kt (241 km/h; 150 mph)
Stalling speed, power off:	
flaps up: landplane	57 kt (106 km/h; 66 mph)
floatplane	58 kt (108 km/h; 67 mph)
flaps down: landplane	52 kt (97 km/h; 60 mph)
floatplane	54 kt (100 km/h; 62 mph)
Max rate of climb at S/L:	
landplane	341 m (1,120 ft)/min
floatplane	320 m (1,050 ft)/min
Service ceiling: landplane	5,485 m (18,000 ft)
floatplane	4,570 m (15,000 ft)
T-O run: landplane	229 m (750 ft)
floatplane	366 m (1,200 ft)
T-O to 15 m (50 ft): landplane	406 m (1,330 ft)
floatplane	580 m (1,900 ft)
Range with max fuel:	
landplane	880 n miles (1,629 km; 1,012 miles)
floatplane	710 n miles (1,315 km; 817 miles)
Endurance	8 h 30 min

UPDATED

LEGEND LITE

LEGEND LITE INC

General Delivery, New Hamburg, Ontario N0B 2G0
Tel: (+1 519) 662 29 80
e-mail: teamskywatch@skywatchss11.on.ca
Web: http://www.skywatchss11.on.ca
CEO: Rob McIntosh

Freedom Lite company began production of the Skywatch SS-11 in 1996; changed name to Legend Lite during 2000.
VERIFIED

LEGEND LITE SKYWATCH SS-11/A

TYPE: Tandem-seat ultralight kitbuilt.
PROGRAMME: Based on the Beaver RX-650 (see 1992-93 edition of *Jane's*) which first flew 1991. In early 2002, some 450 Beaver lightplanes remained in Canada, including 22 RX-650s. First flight of SS-11, mid-1995.
CURRENT VERSIONS: Standard version *as described*; 37.0 kW (49.6 hp) Rotax 503 and 55.0 kW (73.8 hp) Rotax 618 engines optional.
CUSTOMERS: Nine flying by end-2000; no further additions known in 2001.

COSTS: Standard kits: with Rotax 503 US$18,980 (1999); with Rotax 582 US$16,995 (2002).
DESIGN FEATURES: Simple, pod-and-boom configuration, with braced wing and tailplane, plus pusher propeller. Optional folding of wings and horizontal tail surfaces. Parts manufacture 90 per cent computer numeric controlled (CNC). Quoted build time 200 hours.
FLYING CONTROLS: Conventional and manual. No aerodynamic balances.
STRUCTURE: Metal airframe with composites fuselage fairing and fabric covering on flying surfaces.

For details of the latest updates to *Jane's All the World's Aircraft* online and to discover the additional information available exclusively to online subscribers please visit
jawa.janes.com

LANDING GEAR: Fixed tricycle type; steerable nosewheel. Optional 4.57 m (15 ft 0 in) composites amphibian floats.
POWER PLANT: One 47.8 kW (64.1 hp) Rotax 582 UL. Standard fuel capacity 51 litres (13.5 US gallons; 11.2 Imp gallons).

DIMENSIONS, EXTERNAL:
Wing span	10.06 m (33 ft 0 in)
Length overall, wings folded	6.55 m (21 ft 6 in)
Height overall, wings folded	2.18 m (7 ft 2 in)

AREAS:
Wings, gross	14.40 m² (155.0 sq ft)

WEIGHTS AND LOADINGS:
Weight empty	190 kg (420 lb)
Max T-O weight	430 kg (950 lb)

PERFORMANCE:
Cruising speed	61-78 kt (113-145 km/h; 70-90 mph)
Stalling speed	26 kt (47 km/h; 29 mph)
Max rate of climb at S/L	238 m (780 ft)/min
T-O run	61 m (200 ft)
Landing run	77 m (250 ft)
Range	217 n miles (402 km; 250 miles)

UPDATED

Legend Lite Skywatch SS-11 two-seat ultralight

0044474

MURPHY

MURPHY AIRCRAFT MANUFACTURING LTD

Unit 1, 8155 Aitken Road, Chilliwack, British Columbia V2R 4H5
Tel: (+1 604) 792 58 55
Fax: (+1 604) 792 70 06
e-mail: mursales@murphyair.com
Web: http://www.murphyair.com
PRESIDENT: Darryl Murphy
SALES MANAGER: Colleen Dyck
MARKETING MANAGER: Dave Walker

Formed in 1985, the company moved from initial 745 m² (8,000 sq ft) plant to new 4,925 m² (53,000 sq ft) premises in 1995 to accommodate increase in business. In mid-2000, Murphy opened a fast-build factory in the Philippines, which employs 40 staff to produce Super Rebels and float kits; by mid-2001, production was due to have reached five aircraft per month. Subassemblies are built from Canadian parts which are then shipped back.

At Sun 'n' Fun 2001, Murphy displayed two new aircraft, the JDM-8 single-seat ultralight and SR 3500 Super Rebel. By January 2002, 176 Murphy aircraft were current in Canada.

UPDATED

MURPHY JDM-8

TYPE: Single-seat ultralight kitbuilt.
PROGRAMME: Initially shown in uncompleted form at Oshkosh in 2000. Prototype (C-IGTD) first flown 30 March 2001 and made public debut at Sun 'n' Fun the following month as 'conceptual' design to evaluate market response. Launched at AirVenture, Oshkosh, July 2001.
CUSTOMERS: One prototype only had been registered by January 2002.
COSTS: US$6,995, excluding engine, for first 30 kits (2002).
DESIGN FEATURES: Murphy's first low-wing design; meets FAR Pt 103. Produced with different wing and engine sizes to satisfy Canadian, European and US regulations. Wings fold upwards for storage; open cockpit.
FLYING CONTROLS: Conventional and manual. No aerodynamic balancing.
STRUCTURE: Semi-monocoque fuselage and constant-chord wings covered with aluminium sheet. Ailerons have aluminium ribs; aluminium tube empennage with wire bracing; control surfaces fabric-covered.
LANDING GEAR: Tricycle type; fixed. Strut-braced mainwheels.
POWER PLANT: One 22.0 kW (29.6 hp) air-cooled Rotax 277 engine driving a two-blade, wooden in prototype; design will accept engines up to 59.7 kW (80 hp) for the Canadian/European market and in the range 18.6 to

Prototype JDM-8 *(Jane's/Paul Jackson)*

NEW/0131718

22.4 kW (25 to 30 hp) for the US market. Fuel capacity in Canadian/European version 75.7 litres (20.0 gallons; 16.7 Imp gallons); in US version 18.9 litres (5.0 US gallons; 4.2 Imp gallons).

DIMENSIONS, EXTERNAL (C/E: Canadian/European version, US: FAR Pt 103 version):
Wing span: C/E	6.10 m (20 ft 0 in)
US	7.32 m (24 ft 0 in)
Width, wings folded	2.44 m (8 ft 0 in)
Length overall	5.87 m (19 ft 3 in)
Height overall: to top of fin	1.19 m (3 ft 11 in)
with wings folded	3.28 m (10 ft 9 in)

DIMENSIONS, INTERNAL:
Cabin max width	0.66 m (2 ft 2 in)

AREAS:
Wings, gross: C/E	11.15 m² (120.0 sq ft)
US	9.29 m² (100.0 sq ft)

WEIGHTS AND LOADINGS:
Weight empty: C/E	125-159 kg (275-350 lb)
US	113-115 kg (250-254 lb)
Max T-O weight: C/E	318 kg (700 lb)
US	227 kg (500 lb)

PERFORMANCE:
Max operating speed: US	54 kt (101 km/h; 63 mph)
Stalling speed: C/E	27 kt (49 km/h; 30 mph)
US	23 kt (42 km/h; 26 mph)
g limits: C/E	±4.0
US	+3.8/−1.8

UPDATED

MURPHY RENEGADE II

TYPE: Tandem-seat ultralight biplane kitbuilt.
PROGRAMME: Prototype Renegade II made first flight May 1985; first kits became available September that year. Ready assembled aircraft, plans and kits are available.
CUSTOMERS: By mid-2000 (latest information supplied), 350 Renegade IIs and Renegade Spirits had been completed and flown.
COSTS: Quick-build kit US$13,500; full materials kit US$9,500; partial parts kit US$3,600; plans US$325 (2002 prices).
DESIGN FEATURES: Sporting biplane of classic configuration. Conforms to TP101.41 ultralight regulations in Canada and some European countries. Optional wingtip extensions (for countries such as UK requiring lower wing loadings) increase gross area to 15.83 m² (170.4 sq ft). Optional extended wings and rounded tips; optional floats and parachute.
NACA 23012 wing section; 10° sweepback on upper wings; 3° dihedral on non-swept lower wings. Angle of incidence for both wings of 4°.
Kits in four standards ranging from partial parts kit to 300/400 working hour quick-build kit with all parts premanufactured plus partly assembled fuselage, wings and engine mounting.
STRUCTURE: Fabric-covered aluminium tubing; rear fuselage top-decking of preformed glass fibre; glass fibre wingtips optional; wire-braced tail assembly.
LANDING GEAR: Tailwheel type; fixed. Tailwheel steerable. Bungee cord suspension system.
POWER PLANT: One 34.0 kW (45.6 hp) Rotax 503 UL-1V two-cylinder two-stroke engine standard; 37.0 kW (49.6 hp) twin-carburettor Rotax 503 UL-2V optional. Alternative engines up to 59.6 kW (80 hp) can be fitted, including Rotax 582 and Teledyne Continental O-65, O-75 and O-85. Fuel capacity 53 litres (14.0 US gallons; 11.6 Imp gallons).

DIMENSIONS, EXTERNAL:
Wing span (normal): upper	6.48 m (21 ft 3 in)
lower	6.05 m (19 ft 10 in)
Length overall	5.61 m (18 ft 5 in)
Height overall	2.08 m (6 ft 10 in)

AREAS:
Wings, gross	14.29 or 15.83 m² (153.8 or 170.4 sq ft)

WEIGHTS AND LOADINGS:
Weight empty	170-236 kg (375-520 lb)
Max T-O weight, Rotax 503	431 kg (950 lb)

PERFORMANCE (Rotax 503 engine and two crew):
Max level speed at 305 m (1,000 ft)	
	70 kt (129 km/h; 80 mph)

Murphy JDM-8 single-seat ultralight *(Jane's/James Goulding)*

NEW/0132945

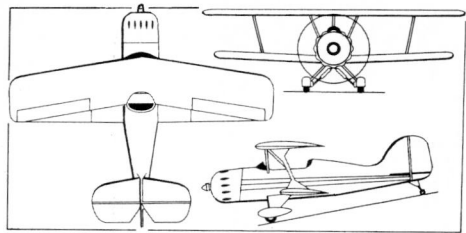

Murphy Renegade Spirit 0044475

Cruising speed at 75% power	56 kt (105 km/h; 65 mph)
Stalling speed, power off	33 kt (61 km/h; 38 mph)
Max rate of climb at S/L	152 m (500 ft)/min
T-O and landing run	92 m (300 ft)
Range at 75% power	197 n miles (365 km; 227 miles)

UPDATED

MURPHY RENEGADE SPIRIT

TYPE: Tandem-seat ultralight biplane kitbuilt.
PROGRAMME: First flight (C-IJLW) 6 May 1987; kit production began a month later.
DESIGN FEATURES: Variant of Renegade II. Alternative power plants; internal and external refinements.
COSTS: Fast-build kit US$15,000; materials kit US$10,500; partial parts kit US$4,000; plans US$325 (2002).
POWER PLANT: One 47.8 kW (64.1 hp) Rotax 582 in-line engine standard, in radial-type cowling; 59.7 kW (80 hp) Rotax 912 flat-four or Alvis rotary engine optional.
DIMENSIONS, EXTERNAL: As for Renegade II
AREAS: As for Renegade II
WEIGHTS AND LOADINGS:

Weight empty	190-227 kg (420-500 lb)
Max T-O weight	431 kg (950 lb)

PERFORMANCE (two crew and Rotax 582 engine):
As for Renegade II except:

Max level speed	78 kt (145 km/h; 90 mph)
Cruising speed at 75% power	63 kt (116 km/h; 72 mph)
Stalling speed	35 kt (64 km/h; 40 mph)
Max rate of climb at S/L	152 m (500 ft)/min
T-O and landing run	107 m (350 ft)
Range at 75% power	206 n miles (381 km; 237 miles)

UPDATED

MURPHY REBEL

TYPE: Three-seat kitbuilt.
PROGRAMME: Construction of prototype started October 1989; first flight 1990; building of production aircraft/kits began in October 1990.
CURRENT VERSIONS: **Rebel:** Standard version, *as described.*
Rebel Elite: Modified tricycle landing gear version, with metal-covered split flaps and fin of increased chord; O-320 or optional 134 kW (180 hp) Lycoming O-360 engine; and increased maximum T-O weight to 816 kg (1,800 lb). First flight (C-FWSF) 29 February 1996.
Super Rebel and Maverick: Described separately.
CUSTOMERS: Total of 700 complete aircraft or kits (including 82 Elites) ordered by 1 January 2000 (latest information supplied); over 400 flying.

Murphy Renegade Spirit registered in the French ultralight series *(Jane's/Paul Jackson)* NEW/0131696

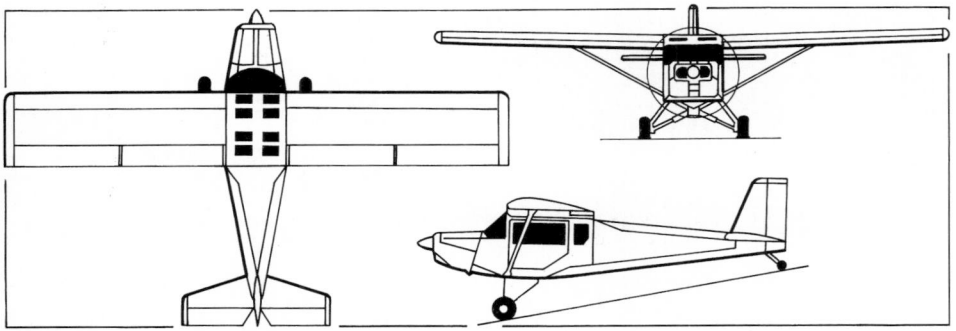

Murphy Rebel floatplane *(Jane's/Paul Jackson)* NEW/0131698

COSTS: Rebel kit without engine US$15,500; Elite kit without engine US$21,000 tailwheel, US$22,000 tricycle (2002). Available in three sub-component kits.
DESIGN FEATURES: High-wing, strut-braced cabin monoplane for recreational, training, cross-country, border patrol and other roles; designed to conform to FAR Pt 23 and JAR (Utility category); STOL performance. Optional fin fillet required for floatplane version. Wings detachable and horizontal tail folds upward for transportation and storage. Quoted build time 800 to 1,000 hours.
NACA 4415 (modified) wing section.
FLYING CONTROLS: Conventional and manual. Flaps. Elite has split flaps, metal-covered ailerons and cantilevered horizontal stabiliser.
STRUCTURE: Aluminium alloy, semi-monocoque fuselage; flaperons fabric covered; glass fibre wingtips. Tailplane also strut-braced. Rebel Elite wing leading-edge 60 per cent thicker than Rebel.
LANDING GEAR: Non-retractable tailwheel type with mainwheel brakes. Optional tricycle gear (Rebel Elite), skis, or Murphy 1800 straight or wheeled or Montana 2100 wheeled floats. 6.00-6 wheels with 18 in high-profile tyres are standard. Bungee suspension system is standard; optional aluminium spring gear.
POWER PLANT: One 59.6 kW (79.9 hp) Rotax 912 UL, or 86.5 kW (116 hp) Textron Lycoming O-235-N2C engine; 2.27:1 reduction gear with Rotax engine and direct drive with O-235. Elite has 119 kW (160 hp) Lycoming O-320; or optional 134 kW (180 hp) Lycoming O-360; ground-adjustable wooden two-blade propeller; three-blade propeller optional. Standard fuel capacity 167 litres (44.0 US gallons; 36.6 Imp gallons) in wing tanks; optional extra 220 litre (58.0 US gallon; 48.3 Imp gallon) tank.
ACCOMMODATION: Pilot and two passengers.
DIMENSIONS, EXTERNAL:

Wing span: Rebel	9.14 m (30 ft 0 in)
Elite	9.25 m (30 ft 4 in)
Wing chord, constant	1.52 m (5 ft 0 in)
Wing aspect ratio	6.0
Length overall: Rebel	6.50 m (21 ft 4 in)
Elite	6.78 m (22 ft 3 in)
Fuselage max width	1.12 m (3 ft 8 in)
Height overall: Rebel, Elite (tailwheel)	2.03 m (6 ft 8 in)
Elite (tricycle)	2.39 m (7 ft 10 in)
Tailplane span	2.79 m (9 ft 2 in)
Wheel track: Rebel	2.18 m (7 ft 2 in)
Elite	2.32 m (7 ft 7½ in)
Wheelbase: Rebel, Elite (tailwheel)	4.93 m (16 ft 2 in)
Elite (tricycle)	1.70 m (5 ft 7 in)
Propeller diameter	1.78 m (5 ft 10 in)

AREAS:

Wings, gross: Rebel	13.94 m² (150.0 sq ft)
Elite	14.12 m² (152.0 sq ft)

WEIGHTS AND LOADINGS (A: Rotax 912, B: O-235, E: Elite, O-360):

Weight empty: A	295-317 kg (650-700 lb)
B	374-408 kg (825-900 lb)
E	445 kg (980 lb)
Baggage capacity	45.5-68 kg (100-150 lb)
Max T-O weight: A	657 kg (1,450 lb)
B	748 kg (1,650 lb)
E	816 kg (1,800 lb)
Max wing loading: A	47.2 kg/m² (9.67 lb/sq ft)
B	53.7 kg/m² (11.00 lb/sq ft)
E	57.8 kg/m² (11.84 lb/sq ft)
Max power loading: A	11.03 kg/kW (18.13 lb/hp)
B	8.66 kg/kW (14.22 lb/hp)
E	6.09 kg/kW (10.00 lb/hp)

Privately modified STOL version of Murphy Rebel fitted with non-standard three-blade propeller and fin fillet *(W J Bushell)* NEW/0131697

Murphy Rebel three-seat kitbuilt *(Jane's/Paul Jackson)* 0064886

Murphy Super Rebel   *NEW*/0089040

PERFORMANCE (A, B, E as above):
  Never-exceed speed (VNE):
    A   124 kt (230 km/h; 143 mph)
    B   131 kt (243 km/h; 151 mph)
    E   136 kt (252 km/h; 157 mph)
  Max level speed: A   87 kt (162 km/h; 100 mph)
    B   108 kt (201 km/h; 125 mph)
    E   126 kt (233 km/h; 145 mph)
  Cruising speed at 65% power:
    A   69 kt (129 km/h; 80 mph)
    B   91 kt (169 km/h; 105 mph)
    E   115 kt (212 km/h; 132 mph)
  Stalling speed, power off, flaps up:
    A   35 kt (65 km/h; 40 mph)
    B   39 kt (71 km/h; 44 mph)
    E   40 kt (74 km/h; 46 mph)
  Stalling speed, power off, flaps down:
    A   32 kt (58 km/h; 36 mph)
    B   35 kt (65 km/h; 40 mph)
    E   37 kt (68 km/h; 42 mph)
  Max rate of climb at S/L: A   164 m (500 ft)/min
    B   244 m (800 ft)/min
    E   457 m (1,500 ft)/min
  Service ceiling: B   3,960 m (13,000 ft)
  T-O run: A   137 m (450 ft)
    B   122 m (400 ft)
    E   143 m (470 ft)
  T-O to 15 m (50 ft): A, B   244 m (800 ft)
  Landing run: A   92 m (300 ft)
    B, E   122 m (400 ft)
  Range with max fuel:
    A   764 n miles (1,416 km; 880 miles)
    B   691 n miles (1,281 km; 796 miles)
    E   593 n miles (1,099 km; 683 miles)
  Endurance: B   7 h 36 min
    E   5 h 12 min
  g limits   +3.8/−2.5
                              (+5.7/−3.8 ultimate)
                                         UPDATED

MURPHY SUPER REBEL and MOOSE

TYPE: Six-seat kitbuilt.
PROGRAMME: Announced April 1995; prototype Super Rebel
  (then designated SR 2500) SR 2500 (C-GKSR), with
  initial MTOW of 1,134 kg (2,500 lb), first flew November
  1995; parts shipments began in late 1995 with tail sections.
  Tricycle version first flew May 1997. Increased gross
  weight Moose (originally SR 3500) introduced at Sun 'n'
  Fun 2001; prototype converted from Super Rebel
  demonstrator C-GBZD.
CURRENT VERSIONS: **Super Rebel**: Standard version.
  Moose: Increased gross weight version, first flown 2
  April 2001; (VOKBM engine); choice of two power
  plants. Tailwheel version only.
CUSTOMERS: Over 130 aircraft sold.
COSTS: *Super Rebel*: Basic kit less engine, propeller,
  instruments and avionics: tailwheel version US$26,500;
  tricycle version US$28,000 (2002).
  Moose: Basic kit, less engine, propeller, instruments and
  avionics, US$32,000 (2002). Fast-build kit extra
  US$16,000.
DESIGN FEATURES: Development of Murphy Rebel (which
  see). All-metal construction. Quoted build time 1,400 to
  1,800 hours. Fast-build kits have quoted build time of 800-
  1,000 hours.
FLYING CONTROLS: Conventional and manual. Horn-balanced
  tail surfaces. Electric trim and three-stage (0, 18, 33°)
  flaps; ailerons deflect 20° up and 12° down; ailerons droop
  12° when 33° flap deployed.
STRUCTURE: Wing is identical in shape to that of Rebel and
  Elite; aerofoil modified NACA 4415. Three-spar wing;
  leading-edge of 0.032 sheet aluminium.
LANDING GEAR: Tailwheel configuration is standard; tricycle
  also available. Optional hardpoints enable rapid

conversion. Main and nose tyres 8.00-6 in; tailwheel has 9
in tyre. Dual brakes. Optional 3600 series floats.
POWER PLANT: *Super Rebel*: Prototype has 186 kW (250 hp)
  Textron Lycoming O-540-A4A5 driving two-blade
  constant-speed Hartzell propeller, but design
  accommodates engines from 134 to 224 kW (180 to
  300 hp). Fuel in two wing tanks, total capacity 227 litres
  (60.0 US gallons; 50.0 Imp gallons). Optional larger tanks
  increase capacity to 303 litres (80.0 US gallons; 66.6 Imp
  gallons).
  Moose: Choice of 186 kW (250 hp) Textron Lycoming
  IO-540 or 265 kW (355 hp) VOKBM M-14P radial, latter
  driving two-blade wooden propeller. Fuel capacity as SR
  2500.
ACCOMMODATION: Pilot and up to five passengers in two
  pairs of side-by-side seats plus optional jump seat;
  dual controls. Rear bench seat can be removed to
  provide cargo capacity. Baggage compartment is reached
  by large cargo door on port side of fuselage, behind
  passenger door.
DIMENSIONS, EXTERNAL:
  Wing span   10.97 m (36 ft 0 in)
  Wing chord, constant   1.52 m (5 ft 0 in)
  Wing aspect ratio   7.1
  Length overall   7.01 m (23 ft 0 in)
  Fuselage max width   1.12 m (3 ft 8 in)
  Height overall: tailwheel   1.98 m (6 ft 6 in)
    tricycle   2.67 m (8 ft 9 in)
  Tailplane span   3.30 m (10 ft 10 in)
  Wheel track   2.29 m (7 ft 6 in)
  Wheelbase: tailwheel   5.94 m (19 ft 6 in)
    tricycle   2.03 m (6 ft 8 in)
  Propeller diameter: Super Rebel   2.13 m (7 ft 0 in)
    Moose with M-14P engine   2.39 m (7 ft 10 in)
  Passenger door: Height   1.07 m (3 ft 6 in)
    Width   1.07 m (3 ft 6 in)

  Baggage door: Height   0.83 m (2 ft 8½ in)
    Width   0.93 m (3 ft 0½ in)
DIMENSIONS, INTERNAL:
  Cabin: Max width   1.12 m (3 ft 8 in)
    Length   2.62 m (8 ft 7 in)
    Height   1.30 m (4 ft 3 in)
AREAS:
  Wings, gross   16.91 m² (182.0 sq ft)
  Ailerons (total)   2.37 m² (25.50 sq ft)
  Flaps (total)   2.09 m² (22.50 sq ft)
WEIGHTS AND LOADINGS (Super Rebel with 186 kW; 250 hp
  engine, Moose with 265 kW; 355 hp engine):
  Weight empty: Super Rebel   748 kg (1,650 lb)
    Moose   816 kg (1,800 lb)
  Baggage capacity   113 kg (250 lb)
  Max T-O weight: Super Rebel 1,361 kg (3,000 lb)
    Moose   1,587 kg (3,500 lb)
  Max wing loading:
    Super Rebel   80.5 kg/m² (16.48 lb/sq ft)
    Moose   93.9 kg/m² (19.23 lb/sq ft)
  Max power loading:
    Super Rebel   7.30 kg/kW (12.00 lb/hp)
    Moose   6.00 kg/kW (9.86 lb/hp)
PERFORMANCE (engines as before):
  Never-exceed speed (VNE):
    Super Rebel   153 kt (284 km/h; 177 mph)
    Moose   164 kt (304 km/h; 189 mph)
  Max level speed:
    Super Rebel   139 kt (257 km/h; 160 mph)
    Moose   152 kt (282 km/h; 175 mph)
  Max cruising speed at 70% power:
    Super Rebel   126 kt (233 km/h; 145 mph)
    Moose   135 kt (249 km/h; 155 mph)
  Stalling speed, power off:
    flaps up: Super Rebel   46 kt (84 km/h; 52 mph)
      Moose   49 kt (91 km/h; 56 mph)
    flaps down: Super Rebel   40 kt (74 km/h; 46 mph)
      Moose   44 kt (81 km/h; 50 mph)
  Max rate of climb at S/L:
    Super Rebel   335 m (1,100 ft)/min
    Moose   457 m (1,500 ft)/min
  Service ceiling   4,575 m (15,000 ft)
  T-O run: Super Rebel, Moose 183 m (600 ft)
  Landing run: Super Rebel   152 m (500 ft)
    Moose   183 m (600 ft)
  Range: with max standard fuel:
    Super Rebel   537 n miles (996 km; 619 miles)
    Moose   521 n miles (965 km; 600 miles)
    with optional larger tanks: Super Rebel
                      729 n miles (1,350 km; 839 miles)
  Endurance: Super Rebel   4 h 15 min
    Moose   4 h 0 min
  g limits (ultimate): both   +5.7/−3.8
                                         UPDATED

MURPHY MAVERICK

TYPE: Side-by-side ultralight kitbuilt.
PROGRAMME: Original design shelved in favour of Rebel, but
  revived in 1994 to meet Japanese ultralight restrictions;
  prototype/demonstrator G-MYSS used to gain UK BCAR
  Section S type approval in 1995.

Murphy Moose with Lycoming engine (*Jane's/Paul Jackson*)   *NEW*/0132941

Murphy Maverick two-seat ultralight   0089050

CUSTOMERS: 34 completed and flown by late 1999; at which time 119 kits had been sold (latest data supplied). No further sales known.

COSTS: Kit: US$13,500 without engine (2002).

DESIGN FEATURES: Derivative of Rebel, with 40 per cent commonality of parts, including optional wingtip extensions.

FLYING CONTROLS: As Rebel, except cable-operated ailerons only instead of full-span flaperons.

STRUCTURE: Similar to Rebel, but with weight-saving features. Wings omit full-span stringers and are fabric covered; glass fibre engine cowling; tail surfaces also fabric covered.

LANDING GEAR: As Rebel.

POWER PLANT: One two-cylinder two-stroke engine (39.5 kW; 53 hp Rotax 503 DC or 47.8 kW; 64.1 hp Rotax 582 UL); GSC two-blade propeller. Testing with Hpower HKS-700E four-stroke engine completed in November 1999. Fuel capacity 18.9 litres (5.0 US gallons; 4.2 Imp gallons) standard, 53 litres (14.0 US gallons; 11.7 Imp gallons) in optional wing tanks.

DIMENSIONS, EXTERNAL:

Wing span: standard	8.97 m (29 ft 5 in)
with extended tips	9.88 m (32 ft 5 in)
Length overall, tail up	6.30 m (20 ft 8 in)

AREAS:

Wings, gross: standard	13.66 m² (147.0 sq ft)
with extended tips	15.05 m² (162.0 sq ft)

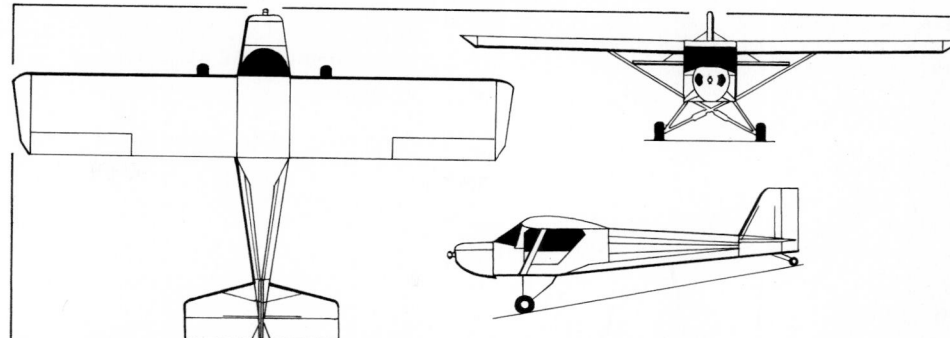

Murphy Maverick two-seat ultralight (*Jane's/Paul Jackson*)   0064887

WEIGHTS AND LOADINGS (standard span and Rotax 503 DC engine):

Weight empty	179-191 kg (395-420 lb)
Max T-O weight	431 kg (950 lb)

PERFORMANCE (standard span and Rotax 503 DC engine):

Max level speed	82 kt (153 km/h; 95 mph)
Cruising speed at 75% power	70 kt (129 km/h; 80 mph)
Stalling speed	28 kt (52 km/h; 32 mph)
Max rate of climb at S/L	183 m (600 ft)/min
T-O run	46 m (150 ft)
Landing run	61 m (200 ft)
Range with max fuel at 75% power	243 n miles (450 km; 280 miles)
g limits	+5.7/−3.8

UPDATED

PARTENAIR

PARTENAIR DESIGN INC

4 Chemin de l'Aeroport, Hangar H-4, St Jean-sur-le-Richelieu, Québec J3B 7B5
Tel: (+1 450) 357 00 15
Fax: (+1 450) 357 97 91
e-mail: info@partenairdesign.com
Web: http://www.partenairdesign.com
OWNER: Saleem Saleh

Company formed in 1998 to develop the S44 Mystère, designed by Saleem Saleh and Frédéric Amblard, which in turn has evolved into the S45.

UPDATED

PARTENAIR S45 MYSTERE

English name: Mystery

TYPE: Tandem-seat kitbuilt.

PROGRAMME: Design work on series began in 1991. Prototype (C-FZHP) first flew as Rotax 912-powered S44 on 16 November 1996, and as S45 in 1998; total 50 hours flown by July 2001, when second prototype (C-GGYY), with Textron Lycoming engine, exhibited complete, but unflown, at AirVenture, Oshkosh. First flight eventually accomplished 4 October 2001.

COSTS: Basic kit, excluding engine, propeller and avionics, US$26,000 (2001); can be purchased in three sub-kits.

DESIGN FEATURES: Configured to eliminate pitch-up associated with high thrust-line; horizontal stabiliser immediately above propeller slipstream counteracts thrust increases by exerting proportional pitch-down and additionally has increased effectiveness at lower airspeeds. Streamlined, pod-and-boom configuration, with T tail and sweptback fin with large fillet. Wings have constant-chord inboard sections and tapered outer panels. Engine and pusher propeller behind rear seat. Quoted build time 800 hours. Provision for 4 or 6° wing dihedral to be incorporated during manufacture.

FLYING CONTROLS: Conventional and manual. Aileron movements +30/−15°. Double slotted flaps, electrically actuated; positions 10/25/45°. Spade-type balance below each aileron. Flight-adjustable tab in elevator.

STRUCTURE: Generally of composites.

LANDING GEAR: Tricycle type; fixed. Cantilever sprung main legs; trailing link nosewheel. Speed fairings on all wheels. Hydraulic brakes. Mainwheels 5.00-5; nosewheel 11×4.00-5.

POWER PLANT: One 119.3 kW (160 hp) Textron Lycoming IO-320 flat-four, driving a PAC three-blade metal propeller. Fuel capacity 208 litres (55.0 US gallons; 45.8 Imp gallons).

ACCOMMODATION: Two persons in tandem, beneath two-piece canopy/windscreen. Dual controls.

Second prototype Partenair Mystère (*Jane's/Paul Jackson*)   *NEW*/0137222

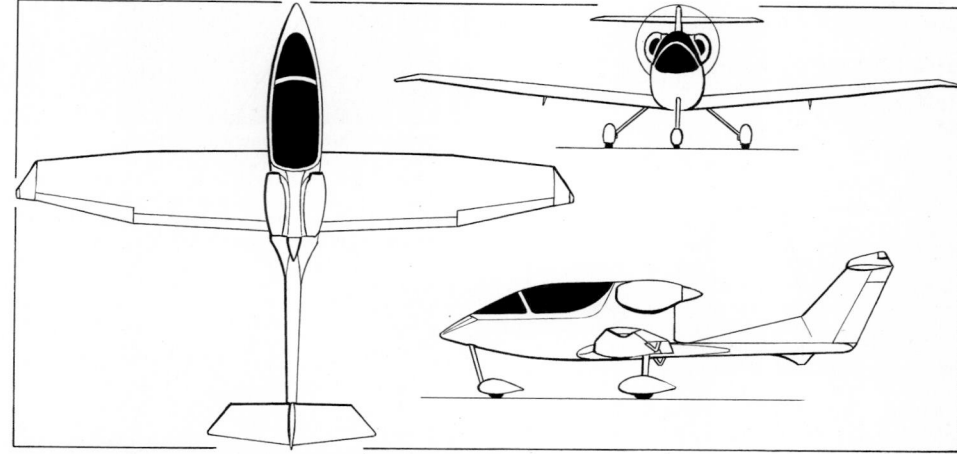

Partenair S45 Mystère tandem-seat kitbuilt (*Jane's/James Goulding*)   *NEW*/0526897

DIMENSIONS, EXTERNAL:

Wing span	7.13 m (28 ft 6 in)
Wing aspect ratio	7.6
Length overall	7.32 m (24 ft 0 in)
Height overall	2.44 m (8 ft 0 in)

DIMENSIONS, INTERNAL:

Cabin max width	0.76 m (2 ft 6 in)

AREAS:

Wings, gross	9.94 m² (107.0 sq ft)

WEIGHTS AND LOADINGS:

Weight empty	522 kg (1,150 lb)
Baggage capacity	22 kg (50 lb)
Max T-O weight	861 kg (1,900 lb)
Max wing loading	86.70 kg/m² (17.76 lb/sq ft)
Max power loading	7.23 kg/kW (11.88 lb/hp)

PERFORMANCE:

Never-exceed speed (VNE)	182 kt (338 km/h; 210 mph)
Max cruising speed	148 kt (274 km/h; 170 mph)
Stalling speed	48 kt (89 km/h; 55 mph)
Max rate of climb at S/L	518 m (1,700 ft)/min
T-O run	153 m (500 ft)
Max range	1,000 n miles (1,852 km; 1,150 miles)

UPDATED

PAXMAN

PAXMAN'S NORTHERN LITE AEROCRAFT

PO Box 1155, Glenwood, Alberta T0K 2R0
Tel: (+1 403) 626 30 94
Fax: (+1 403) 626 34 90
CEO: Elbert D Paxman

There have been no recent completions of Viper kits, although the aircraft remains available to order.

UPDATED

PAXMAN'S NORTHERN LITE VIPER P3/A

TYPE: Side-by-side kitbuilt.

PROGRAMME: Prototype (C-FVPR) first flew 1994.

CUSTOMERS: Two flying by end 2001.

COSTS: Kit US$17,500 (2002); with engine and instruments. Plans only, US$300.

DESIGN FEATURES: Low-wing monoplane; constant chord wings; broad, sweptback fin. Wood and fabric construction; quoted 600 hour build time. Wings detach for storage.

FLYING CONTROLS: Conventional and manual. Flaps.

STRUCTURE: Steel tube covered with wood and fabric; glass fibre engine cowling.

LANDING GEAR: Tailwheel type; fixed. Sprung steel main legs; brakes on mainwheels.

ACCOMMODATION: Dual controls. Dual gull-wing doors; baggage compartment behind seats.

POWER PLANT: One 74.6 kW (100 hp) Subaru EA81; alternative engines between 48.5 and 85.8 kW (65 and 115 hp) can be fitted. Fuel capacity 64 litres (17.0 US gallons; 14.2 Imp gallons).

DIMENSIONS, EXTERNAL:

Wing span	8.08 m (26 ft 6 in)
Wing aspect ratio	5.9
Length overall	5.94 m (19 ft 6 in)
Height overall	1.88 m (6 ft 2 in)
Wheel track	2.13 m (7 ft 0 in)

DIMENSIONS, INTERNAL:

Cabin: Length	1.09 m (3 ft 7 in)
Max width	1.04 m (3 ft 5 in)
Max height	1.02 m (3 ft 4 in)

AREAS:

Wings, gross	11.08 m² (119.3 sq ft)

WEIGHTS AND LOADINGS (74.6 kW; 100 hp engine):

Weight empty	295 kg (650 lb)
Baggage capacity	22 kg (50 lb)
Max T-O weight	589 kg (1,300 lb)
Max wing loading	53.2 kg/m² (10.90 lb/sq ft)
Max power loading	7.91 kg/kW (13.00 lb/hp)

PERFORMANCE:

Never-exceed speed (VNE)	126 kt (233 km/h; 145 mph)
Max level speed	113 kt (209 km/h; 130 mph)
Cruising speed	100 kt (185 km/h; 115 mph)
Stalling speed	33 kt (62 km/h; 38 mph)
Max rate of climb at S/L	457 m (1,500 ft)/min
T-O and landing run	91 m (300 ft)
Range	477 n miles (885 km; 550 miles)
g limits	+4.4/−2.2

UPDATED

PHOENIX

PHOENIX FANJET CORPORATION

Suite 804, 7015 Macleod Trail South, Calgary, Alberta T2H 2K6
Tel: (+1 403) 255 28 10 or (+1 877) FJ44JET
Fax: (+1 403) 255 26 49
e-mail: sales@phoenixfanjet.com
Web: http://www.phoenixfanjet.com

CEO: Don Jewett
PRESIDENT: C Raymond Johnson
SALES DIRECTOR: Tom Heath

Alberta Aerospace, parent company of Phoenix and its SigmaJet and MagnaJet, ceased operations in November 2001. Data last appeared in the 2002-03 *Jane's*.

UPDATED

PHOENIX SIGMAJET and MAGNAJET

Development has been terminated. See 2002-03 and earlier editions.

UPDATED

RAF

ROTARY AIR FORCE MARKETING INC

PO Box 1236, 1107-9th Street W, Kindersley, Saskatchewan S0L 1S0
Tel: (+1 306) 463 60 30
Fax: (+1 306) 463 60 32
e-mail: info@raf2000.com
Web: http://www.raf2000.com
SALES MANAGER: Don LaFleur

Incorporated in 1987, Rotary Air Force employs 16 people. For details of the single-seat RAF 1000/GT, see the 1992-93 edition of *Jane's*.

In May 2001, RAF announced that it would be expanding its activities into commercial applications, including agricultural spraying.

UPDATED

RAF 2000

TYPE: Two-seat autogyro kitbuilt.
PROGRAMME: Introduced 1990; conforms to 51 per cent homebuilt rules.
CURRENT VERSIONS: **2000 STD-SE**: Basic version, *as described.*

2000 GTX-SE: Top-of-range model; kit includes rotor brake, heater, dual controls and adjustable pitch and roll trim assembly.
CUSTOMERS: RAF autogyros sold in Argentina, Australia, Austria, Brazil, Canada, Chile, China, Ecuador, Germany, Greece, Hungary, Ireland, Italy, Japan, Kazakhstan, Mexico, Netherlands, New Caledonia, New Zealand, Norway, Portugal, Puerto Rico, Russia, South Africa, Spain, Thailand, UK and USA. At least 470 believed completed and flown by end 2000.
COSTS: 2000 STD-SE kit price US$20,615, including engine; 2000 GTX-SE US$22,500 (2002).
DESIGN FEATURES: Conventional autogyro. Suitable for training, crop-spraying, power line inspection and aerial photography. Quoted build time 150-250 hours.
STRUCTURE: Composites RAF rotor blade features aluminium spar and foam filler. Patented 5 × 10 cm (2 × 4 in) rigid rotor mast. Removable doors. Mast folds for storage.
LANDING GEAR: Tricycle configuration; fixed; optional speed fairings.
POWER PLANT: One 97 kW (130 hp) Subaru EJ22 16-valve four-cylinder liquid-cooled engine driving a ground-adjustable three-blade Warp Drive composites propeller through RAF cog belt reduction gear, ratio 2.10:1. Fuel capacity 87 litres (23.0 US gallons; 19.2 Imp gallons) of premium unleaded Mogas, of which 79 litres (21.0 US gallons; 17.5 Imp gallons) are usable. Fuel consumption 18.2 litres (4.8 US gallons; 4.0 Imp gallons)/h at 75 per cent power.
ACCOMMODATION: Pilot and passenger side by side in enclosed cabin.
SYSTEMS: 35 A alternator, electric starter.

RAF 2000 GTX-SE autogyro (*Jane's/Paul Jackson*)   NEW/0131744

EQUIPMENT: Agricultural version has 114 litre (30.0 US gallon; 25.0 Imp gallon) tank fitted between main landing gear legs, supplying a 24-nozzle spraybar fitted behind and below the engine. Spray width 6.7 to 7.3 m (22 to 24 ft).

DIMENSIONS, EXTERNAL:

Rotor diameter	9.14 m (30 ft 0 in)
Rotor blade chord	0.22 m (8½ in)
Fuselage: Length	4.11 m (13 ft 6 in)
Max width	1.08 m (3 ft 6½ in)
Width over wheels	1.59 m (5 ft 2½ in)
Height overall: STD-SE	2.50 m (8 ft 2½ in)
GTX-SE	2.58 m (8 ft 5½ in)
Propeller diameter	1.73 m (5 ft 8 in)
Wheel track	1.55 m (5 ft 1 in)
Wheelbase	3.91 m (12 ft 10 in)

DIMENSIONS, INTERNAL:

Cabin: Length	1.50 m (4 ft 11 in)
Max width	1.09 m (3 ft 7 in)
Max height	1.32 m (4 ft 4 in)

AREAS:

Rotor disc	65.67 m² (706.9 sq ft)

WEIGHTS AND LOADINGS:

Weight empty: STD-SE	345 kg (760 lb)
GTX-SE	331 kg (730 lb)
Max payload: STD-SE	261 kg (575 lb)
GTX-SE	317 kg (700 lb)
Max T-O weight	698 kg (1,540 lb)
Max disc loading	10.6 kg/m² (2.18 lb/sq ft)
Max power loading	7.21 kg/kW (11.85 lb/hp)

PERFORMANCE, POWERED (two occupants):

Max operating speed (VMO)	87 kt (161 km/h; 100 mph)
Cruising speed: STD-SE	61 kt (113 km/h; 70 mph)
GTX-SE	65 kt (121 km/h; 75 mph)
Min flying speed	17-26 kt (32-48 km/h; 20-30 mph)

Unstick speed:

STD-SE	26-35 kt (48-64 km/h; 30-40 mph)
GTX-SE	39-43 kt (72-80 km/h; 45-50 mph)
Max rate of climb at S/L: STD-SE	213 m (700 ft)/min
GTX-SE	305 m (1,000 ft)/min
Service ceiling	3,050 m (10,000 ft)
T-O run	23-107 m (75-350 ft)
Landing run	3 m (10 ft)

Range with max fuel:

STD-SE	278 n miles (515 km; 320 miles)
GTX-SE	182 n miles (338 km; 210 miles)
Endurance (with 30 min reserves): STD-SE	4 h 0 min
GTX-SE	3 h 0 min

PERFORMANCE, UNPOWERED:

Glide ratio	4:1

UPDATED

ST JUST

ST JUST AVIATION INC

Suite 204, 1310 Gay-Lussac, Boucherville, Québec J4B 7G4
Tel: (+1 450) 641 86 86
Fax: (+1 450) 641 84 21
e-mail: st-justaviation@sympatico.ca
Web: http://www.cycloneaviation.com
PRESIDENT: Guy Cantin
DIRECTOR: Albert Beaudry

Original Avionerrie du Lac St John formed in 1991 to provide parts for Cessna 185 owners after that company removed the aircraft from its product line; later progressed to manufacturing kits of the Cyclone. Sold to St Just Aviation in 1997.

VERIFIED

ST JUST SUPER CYCLONE

TYPE: Four-seat kitbuilt.
PROGRAMME: Introduced in 1997; based on reverse engineering of the Cessna 180/185 (which see in *Jane's Aircraft Upgrades*).
CUSTOMERS: Total 67 sold and 36 flying by July 2001.
COSTS: Basic kit US$39,900 (2002). Fast-build kits US$3,000 extra.
DESIGN FEATURES: Compared to Cessna 180/185, Super Cyclone has span increased by addition of 30 cm (1 ft) at each wingroot, which increases flap length, lowers stalling speed, enhances stability and increases maximum T-O weight. Quoted build time 2,000 hours.
FLYING CONTROLS: Conventional and manual. Flaps.
STRUCTURE: Single-strut braced, two-spar wing with constant-chord centre-section and tapered outer sections. Wing and empennage constructed of 2024 T-3 aluminium;

Seaplane version of St Just Super Cyclone   *NEW*/0137973

composites engine cowling. Large fin fillet. Low-mounted tailplane.

LANDING GEAR: Tailwheel type, with sprung steel main legs; Federal wheel/skis and floats can be fitted. Mainwheels 6.00-6.

POWER PLANT: One 224 kW (300 hp) Teledyne Continental IO-520, flat-six engine, driving a three-blade, constant-speed McCauley propeller. Optional alternative engines between 149 and 261 kW (200 and 350 hp). Fuel capacity 341 litres (90.0 US gallons; 75.0 Imp gallons), of which 318 litres (84.0 US gallons; 70.0 Imp gallons) are usable.

ACCOMMODATION: Pilot and three passengers in two pairs of seats with upward-hinged door on each side of cabin. Additional baggage door on port side of fuselage behind cabin. Compared to Cessna 185, has a third cabin window each side.

AVIONICS: Full IFR panel available.

DIMENSIONS, EXTERNAL:

Wing span	11.58 m (38 ft 0 in)
Wing aspect ratio	7.2
Length overall	7.92 m (26 ft 0 in)
Height overall	2.41 m (7 ft 11 in)
Propeller diameter	2.18 m (7 ft 2 in)

DIMENSIONS, INTERNAL:

Cabin max width	1.07 m (3 ft 6 in)

AREAS:

Wings, gross	17.74 m² (191.0 sq ft)

WEIGHTS AND LOADINGS:

Weight empty	839 kg (1,850 lb)
Baggage capacity	68 kg (150 lb)
Max payload	680 kg (1,500 lb)
Max T-O weight	1,587 kg (3,500 lb)
Max wing loading	89.5 kg/m² (18.32 lb/sq ft)
Max power loading	7.10 kg/kW (11.67 lb/hp)

PERFORMANCE:

Max operating speed	152 kt (281 km/h; 175 mph)
Max cruising speed	142 kt (263 km/h; 163 mph)
Stalling speed: flaps up	37 kt (68 km/h; 42 mph)
flaps down	33 kt (62 km/h; 38 mph)
Max rate of climb at S/L	488 m (1,600 ft)/min
Service ceiling	4,570 m (15,000 ft)
T-O run	69 m (225 ft)
Landing run	122 m (400 ft)
Range with max fuel	788 n miles (1,460 km; 907 miles)
g limits	+3.8/−1.5

UPDATED

Cockpit of St Just Super Cyclone   *NEW*/0137974

Fuselage of St Just Super Cyclone under construction (*Jane's/Paul Jackson*)   0126908

SUPER-CHIPMUNK

SUPER-CHIPMUNK INC

109, Route 201, St Louis-de-Gonzague, Québec J0S 1T0
Tel: (+1 450) 373 07 06
Fax: (+1 450) 373 31 44
e-mail: info@super-chipmunk.com
Web: http://www.super-chipmunk.com
PRESIDENT: Gilles Leger

This company was launched to manufacture a modern reproduction of the de Havilland Canada DHC-1 Chipmunk, externally similar rear of the firewall, but with all-new structural design. Company owns Supplementary Type Certificate (STC) for original Super Chipmunk conversion of ex-military aircraft. Engine modification kits for existing Chipmunks are available. Promotion was continuing in 2002. The prototype is officially registered as a Leger Super Chipmunk.

UPDATED

SUPER-CHIPMUNK SUPER-CHIPMUNK

TYPE: Tandem-seat sportplane kitbuilt.

PROGRAMME: Original Chipmunk described in *Jane's Aircraft Upgrades*. Earlier Super Chipmunk modification of surplus military aircraft, with Teledyne Continental IO-360 engine, undertaken by Jean-Paul Huneault (see 1970-71 *Jane's*), achieved only seven sales.

Prototype of current Super-Chipmunk (C-GLSC), with similar power plant, constructed in 1998-99 from DHC-1 wings and rear fuselage, but with steel tube centre-section providing additional accommodation for pilots of above average height and/or girth. Shown, incomplete, at Oshkosh in 1998; first flight 2 September 1999. Debut at Sun 'n' Fun, Lakeland, Florida, April 2000, when all-new fuselage of third aircraft also exhibited. Lost on 14 July 2002 when wing reportedly detached in flight. Complete airframe will be available for assembly from kit.

CUSTOMERS: First production aircraft (002) bought by André Gauthier, but had not been registered in Canada by early 2002. No further information received.

COSTS: Kit US$33,900 (2000).

DESIGN FEATURES: Primary trainer adapted for civilian sporting use. Low wing and mid-mounted tailplane, both tapered; elliptical fin/rudder. Super-Chipmunk features

new internal structure and Canadian DHC-1 standard single-piece canopy and lack of strakes ahead of tailplane.

FLYING CONTROLS: Conventional and manual. Horn-balanced rudder and elevators. Flaps. Flight-adjustable trim tab on starboard elevator; ground-adjustable tab on rudder.

STRUCTURE: Metal throughout, except fabric-covered wings (except roots and leading edges), rudder, ailerons, elevator and flaps and composites engine cowling, wheel fairings and wingtips. Kit includes pre-welded 4130 steel fuselage frame and T6736 centre-section, plus aluminium (6061T6 and 2024T3) rear fuselage.

LANDING GEAR: Tailwheel type; fixed. Mainwheels size 6.00-6. Mainwheel brakes.

POWER PLANT: One 157 kW (210 hp) Teledyne Continental IO-360 flat-six driving a Hartzell F8459A-4 two-blade propeller. Alternative engines between 112 and 224 kW (150 and 300 hp) with fixed-pitch or constant-speed propellers. Fuel in wingroot tanks, combined capacity 114 litres (30.0 US gallons; 25.0 Imp gallons) normal, 189 litres (50.0 US gallons; 41.7 Imp gallons) optional. Oil capacity 9.5 litres (2.5 US gallons; 2.0 Imp gallons).

ACCOMMODATION: Two in tandem, beneath single-piece, rearward-sliding canopy with fixed windscreen. Single cockpit.

DIMENSIONS, EXTERNAL:

Wing span	10.57 m (34 ft 8 in)
Length overall	7.75 m (25 ft 5 in)
Height overall	2.18 m (7 ft 2 in)
Tailplane span	3.63 m (11 ft 11 in)
Wheel track	2.82 m (9 ft 3 in)
Wheelbase	5.23 m (17 ft 2 in)

DIMENSIONS, INTERNAL:

Cockpit: Max height	1.40 m (4 ft 7 in)
Width: max	0.91 m (3 ft 0 in)
at shoulders	0.69 m (2 ft 3 in)

AREAS:

Wings, gross	16.07 m² (173.0 sq ft)
Ailerons (total)	1.29 m² (13.90 sq ft)
Flaps (total)	2.04 m² (22.00 sq ft)
Fin	0.55 m² (5.90 sq ft)
Rudder	0.63 m² (6.80 sq ft)
Tailplane	1.58 m² (17.00 sq ft)
Elevators (total)	1.41 m² (15.20 sq ft)

Prototype Super-Chipmunk reproduction of de Havilland DHC-1 (*Jane's/Paul Jackson*)   *NEW*/0131745

WEIGHTS AND LOADINGS:	
Weight empty	693 kg (1,527 lb)
Baggage capacity	40 kg (90 lb)
Max T-O weight	1,088 kg (2,400 lb)
Max wing loading	67.7 kg/m² (13.87 lb/sq ft)
Max power loading	6.96 kg/kW (11.43 lb/hp)

PERFORMANCE:	
Never-exceed speed (VNE)	173 kt (321 km/h; 200 mph)
Max level speed	156 kt (290 km/h; 180 mph)
Cruising speed at 75% power	
	139 kt (257 km/h; 160 mph)
Max rate of climb at S/L	610 m (2,000 ft)/min

Service ceiling	5,245 m (17,200 ft)
T-O to 15 m (50 ft)	204 m (670 ft)
Landing from 15 m (50 ft)	283 m (930 ft)
Max range with optional fuel	
	521 n miles (965 km; 600 miles)
g limits	+9/−6

UPDATED

ULTRAVIA

ULTRAVIA AERO INTERNATIONAL INC
152-A Industrial, Gatineau, Québec J8P 7G7
Tel: (+1 819) 669 31 44
Fax: (+1 819) 669 84 06
e-mail: pelican@ultravia.ca
Web: http://www.ultravia.ca

The Pelican PL was added to the Ultravia line in 1991, followed by the Pelican Sport in 1998; for details of earlier Ultravia models, including the Pelican Club, see the 1992-93 edition of *Jane's All the World's Aircraft.* In April 2002, New Kolb Aircraft of USA was appointed distributor.

NEW ENTRY

ULTRAVIA PELICAN
TYPE: Side-by-side kitbuilt.
PROGRAMME: Original Pelican Club first flown 1984 and details appear in the 1992/93 edition of *Jane's All the World's Aircraft.* The Pelican PL was introduced in 1991 and the Sport in 1998. Meets Transport Canada TP 10141 regulations.
CURRENT VERSIONS: **Pelican PL:** Standard ultralight version.
   Pelican Sport: Redesigned wing with higher-lift aerofoil, longer span and STOL kit. Comes in three subvariants, depending on operating area: Sport 450 (Europe), Sport 500 (Brazil) and sport 600 (US, Canada and Mexico).
Description applies to Pelican Sport 600 unless otherwise indicated.
CUSTOMERS: Over 250 Pelican PLs and 60 Pelican Sports completed and flown by December 2002.
COSTS: PL: standard US$19,800; fast-build US$25,600; Sport: standard US$21,500; fast-build US$27,750 (all 2003).
DESIGN FEATURES: High-wing of constant chord, single bracing strut each side; sweptback fin, with constant-chord tailplane at base. Cantilever landing gear.
   Fast-build kit, with 49 per cent of assembly completed, is also available. Quoted build time of standard kit 1,000 hours.
FLYING CONTROLS: Conventional and manual. Five position (0 to 45°) flaps. Pelican Sport ailerons droop 10° when full flap deployed. Mechanically operated elevator trim tab, ground-adjustable rudder and aileron tabs. Horn-balanced rudder.
STRUCTURE: Composites, vacuum-moulded fuselage with rigid PVC core. Single-strut single-spar all-metal wing

Kolb Sport 600, an Ultravia Pelican marketed in the USA *(Jane's/Paul Jackson)*   *NEW*/0533693

with aluminium-covered control surfaces. Cantilever horizontal tail surfaces. Landing gear legs of 4130 steel.
LANDING GEAR: Fixed tricycle versions standard. Steerable nosewheel has bungee suspension and uses Azusa 15 cm (6 in) wheel. Optional tailwheel version uses a Marco WLT-6 or Kolb 10 cm (4 in) tailwheel. Cleveland 5.00-5 mainwheels with disc brakes on both versions. Optional floats and speed fairings.
POWER PLANT: Typically, one 73.5 kW (98.6 hp) Rotax 912 ULS four-cylinder four-stroke driving a three-blade Warp Drive ground-adjustable propeller. Other engines can be fitted, including the 59.6 kW (79.9 hp) Rotax 912 UL and 84.6 kW (113.4 hp) Rotax 914 UL. Fuel capacity 91 litres (24.0 US gallons; 20.0 Imp gallons) in two wing tanks, of which 87 litres (23.0 US gallons; 19.2 Imp gallons) usable.
ACCOMMODATION: Pilot and passenger in side-by-side seating. Baggage area behind seats.
SYSTEMS: 12 V 250 W generator.
DIMENSIONS, EXTERNAL:

Wing span: PL	8.99 m (29 ft 6 in)
Sport	9.75 m (32 ft 0 in)
Length overall: PL	6.02 m (19 ft 9 in)
Sport	6.07 m (19 ft 11 in)
Height overall, nosewheel configuration:	
PL	2.54 m (8 ft 4 in)
Sport	2.59 m (8 ft 6 in)

AREAS:

Wings, gross: PL	10.03 m² (108.0 sq ft)
Sport	10.89 m² (117.3 sq ft)

WEIGHTS AND LOADINGS:

Weight empty: PL	363-386 kg (800-850 lb)
Sport	327-340 kg (720-750 lb)
Max T-O weight: PL	635 kg (1,400 lb)
Sport	599 kg (1,320 lb)

PERFORMANCE (Rotax 912 ULS engine):

Never-exceed speed (VNE):	
PL	142 kt (263 km/h; 163 mph)
Sport	135 kt (250 km/h; 155 mph)
Normal cruising speed: both	115 kt (213 km/h; 132 mph)
Stalling speed, flaps down:	
PL	44 kt (82 km/h; 51 mph)
Sport	39 kt (73 km/h; 45 mph)
Max rate of climb at S/L: PL	366 m (1,200 ft)/min
Sport	411 m (1,350 ft)/min
Service ceiling: both	4,875 m (16,000 ft)
T-O run: PL	183 m (600 ft)
Sport	91 m (300 ft)
Landing run: Sport	152 m (500 ft)
Range with max fuel:	
PL	695 n miles (1,287 km; 800 miles)
Sport	608 n miles (1,126 km; 700 miles)

NEW ENTRY

UTIAS

UNIVERSITY OF TORONTO INSTITUTE FOR AEROSPACE STUDIES
4295 Dufferin Street, Toronto, Ontario M3H 5T6
Tel: (+1 416) 667 77 57
Fax: (+1 416) 667 77 99
Web: http://www.ornithopter.net
RESEARCHER: Prof James D DeLaurier

NEW ENTRY

UTIAS 'BIG FLAPPER'
TYPE: Ornithopter.
PROGRAMME: Early research resulted 1985 in quarter-scale, hand-launched, engine-powered and radio-controlled flying model; this progressively improved into more efficient model flown 4 September 1991 and recognised by FAI as first successful engine-powered ornithopter. Feasibility study 1993-94 for full-scale prototype, construction of which began 1995; this aircraft (C-GPTR) began taxi tests October 1996 and by 1999 had demonstrated ability to accelerate under own power to more than 43 kt (80 km/h; 50 mph) and make a number of brief hops. Further activity suspended since third quarter 2001 pending additional funding. In addition to Toronto University, Project Ornithopter currently has some 16 sponsors including Canadian National Research Council, the Ken Molson Foundation and Toronto Aerospace Museum. Criterion for success is considered to be sustained flight of at least 11 seconds.
COSTS: Development costs of full-size prototype £140,000 up to mid-2001.
DESIGN FEATURES: Conventional fuselage and tail unit.
   Flapping wings are hinged to a centre-section which is

UTIAS 'Big Flapper' during taxi tests in August 2001   *NEW*/0527147

moved up and down by pylons connected to the drive train and pivot on the ends of vertical struts mounted on short outriggers attached to the base of the fuselage. Thus, when centre-section pushes up, wings flap down, and vice versa.

Wings deflect 53° 36′ up and down, and are designed to flap 1.3 times per second.
FLYING CONTROLS: All of thrust and most of lift created by mechanical flapping of wings. Thrust due mainly to

low-pressure region around wing leading-edges, which creates suction. Wings also able to twist passively ('aeroelastic tailoring') to prevent airflow separation at trailing-edges. Rudder provides control in roll as well as yaw, due to wing opposite direction of yaw gaining lift while that of other wing loses lift.

STRUCTURE: Fuselage frame of aluminium tube; wings of Kevlar, carbon fibre and epoxy resin; Dacron overall covering.

LANDING GEAR: Tricycle type, fixed, with cantilever, self-sprung mainwheel bows; shock-absorption on nosewheel leg. Gear designed to give aircraft a nose-down attitude on ground, to minimise bounce during take-off run.

POWER PLANT: One 17.9 kW (24 hp) König three-cylinder, two-stroke, fan-cooled engine, with 60:1 reduction gear drive to power wing centre-section. As latter requires more power on downstrokes than on upstrokes, engine stores energy on flywheel between downstrokes.

ACCOMMODATION: Pilot only.

DIMENSIONS, EXTERNAL (approx):
Wing span	12.3 m (40 ft 6 in)
Length overall	7.3 m (24 ft 0 in)
Height overall	2.7 m (9 ft 0 in)

WEIGHTS AND LOADINGS:
Weight empty	318 kg (700 lb)

PERFORMANCE (estimated):
Unstick speed	50 kt (92 km/h; 57 mph)
Cruising speed	44 kt (82 km/h; 51 mph)

NEW ENTRY

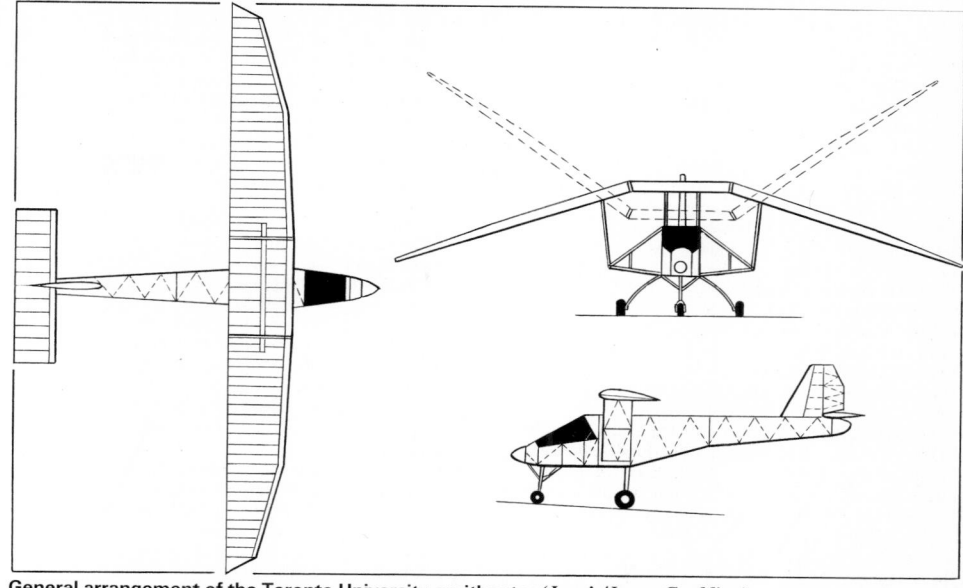

General arrangement of the Toronto University ornithopter (*Jane's/James Goulding*) *NEW*/0526866

ZENAIR

ZENAIR LTD

Huronia Airport, Midland, Ontario L4R 4K8
Tel: (+1 705) 526 28 71
Fax: (+1 705) 526 80 22
e-mail: zenair@zenair.com
Web: http://www.zenair.com
PRESIDENT AND DESIGNER: Christophe Heintz
VICE-PRESIDENT: Mathieu Heintz
GENERAL MANAGER: Bruce Barker

Founded in 1974, Zenair designs and develops light aircraft. Christophe Heintz was formerly chief engineer of Avions Pierre Robin in France. In 1992 Zénith Aircraft Company was formed to manufacture and market all models except the CH 2000 in the USA (which see), and in September 1996 reached full production status. In June 1999, Zenith announced that it was transferring production of the CH 2000 to a new, purpose-built 2,800 m² (30,000 sq ft) factory at Eastman, Georgia; Aircraft Manufacturing and Development Company (which see in US section) was formed to produce

the aircraft and its derivative, the kitbuilt CH 640; the transfer is now complete.

Zenair has licensed representatives in Africa, South America, Belgium, Czech Republic (see entry for CZAW), France, Germany, Israel, Italy (see ICP), Spain and the UK.

Zenith Aircraft Company (Mexico, Missouri, USA) licenses kit production of Zenair designs. In addition to light aircraft, company constructs metal floats and amphibious floats.

UPDATED

CHILE

ENAER

EMPRESA NACIONAL DE AERONÁUTICA DE CHILE

Avenida José Miguel Carrera 11087, El Bosque, Santiago
Tel: (+56 2) 528 27 35, 28 23 and 25 99
Fax: (+56 2) 528 26 99
e-mail: enaercom@interaccess.cl
Web: http://www.enaer.cl
PRESIDENT: Gen Patricio Rios
MANAGING DIRECTOR: Brig Gen Alfredo Guzmán
TECHNICAL DIRECTOR: Rodolfo Pinto
COMMERCIAL DIRECTOR: Hector Monje
SALES DIRECTOR: Manuel Vargas
MARKETING MANAGER: Felipe Contardo
COMMUNICATIONS MANAGER: Cecilia Sola

State-owned company, formed 1984 from IndAer industrial organisation set up 1980 by Chilean Air Force (FACh); aircraft manufacture started 1979 with assembly of 27 Piper PA-28 Dakota lightplanes for Chilean Air Force and flying clubs. Covered plant area 47,000 m² (505,900 sq ft); workforce of 1,615 in mid-2001.

Recent main programmes have included Pantera upgrade of FACh Mirage 50s, Tigre upgrade of FACh F-5Es and co-production of CASA C-101 as A-36 Halcón. Also modified, with IAI assistance, a Chilean Air Force Boeing 707-320B to combined tanker/AEW configuration with similar nose and fuselage antenna fairings to those of original FACh Phalcon plus hose and drogue aerial refuelling system. Links with IAI recently widened to encompass flight deck upgrades of early model C-130 Hercules (first of two C-130Bs redelivered to Uruguayan Air Force in early 2002); and sensor upgrades for Latin American operators of Bell 412 and other aircraft. Currently undertaking sole-source fin/tailplane manufacture of ERJ-135/140/145 (560 shipsets ordered) as Embraer risk-sharing partner, and subcontract manufacture of part of tail unit of CASA CN-235 and C-295; is also manufacturing components for Dassault Falcon 900 and Falcon 2000.

Own-design programmes have comprised T-35 piston-engined Pillán, T-35DT Turbo Pillán military trainer (1996-97 and earlier *Jane's*) and Ñamcu light aircraft. Last-named became centred in Netherlands under Euro-ENAER (which see) as the EE-10 Eaglet; ENAER was to have been South American distributor for this aircraft, but had not begun to supply structures by early 2002, when Euro-ENAER declared bankrupt.

UPDATED

T-35DT Turbo Pillán with underwing FLIR turret, displayed in April 2002 (*Jane's/Robert Hewson*) *NEW*/0118371

ENAER (ECH-51) T-35 PILLÁN

English name: Devil
Spanish Air Force designation: E.26 Tamiz (Grader)
TYPE: Basic prop trainer.
PROGRAMME: First two prototypes (first flight 6 March 1981) developed by Piper; followed by three Piper kits for ENAER assembly (first flight, by FACh s/n 101, 30 January 1982); then slight redesign, as recorded in earlier editions; ENAER series production started September 1984; first flight of production T-35A, 28 December 1984. Deliveries to FACh began 31 July 1985; original export deliveries completed 1991. Production resumed in November 1998 to build 12 (all now delivered) for Dominican Republic and again in 2002 to meet an Ecuadorean Navy order for four. Promotion continued at FIDAE Air Show in April 2002.

CURRENT VERSIONS: **T-35A:** Primary trainer version for Chilean Air Force. In service with Escuela de Aviación 'Capitán Avalos' at El Bosque AB, Santiago.

Detailed description applies to the above version except where indicated.

T-35B: Chilean Air Force and Dominican Republic instrument trainer, with more comprehensive instrumentation.

T-35C: Primary trainer for Spanish Air Force (designation E.26 Tamiz), assembled by CASA from ENAER kits.

T-35D: Instrument trainer for Panama and Paraguay.

T-35S: Single-seat prototype (CC-PZB); illustrated in 1991-92 *Jane's*.

T-35DT Turbo Pillán: Turboprop version; described in 1996-97 and earlier editions. No orders announced by

April 2002, when shown at FIDAE Air Show after mid-March maiden flight with new 'glass cockpit' (203 × 254 mm; 8 × 10 in ARNAV ICDS 2000 display) and offered with IAI Tamam POP or FLIR Systems 7500 infra-red sensor in turret under port wing as border surveillance aircraft.

Pillán 2000: Version proposed in 1998, retaining the existing T-35 fuselage, tail unit and piston engine, combined with a lighter weight but increased span wing designed by the Russian companies Technoavia and Tyazhpromexport. No further news since that time.

CUSTOMERS: Total of 123 (excluding prototypes and trials aircraft) built by 2001 for air forces of Chile (34 T-35A, 14 T-35B), Dominican Republic (12 T-35B), Panama (10 T-35D), Paraguay (13 T-35D) and Spain (40 T-35C/E.26). Further four delivered to Ecuadorean Navy in 2002; total 127. Surplus FACh aircraft also sold to El Salvador (five), Guatemala (five) and Panama (two). Pillán was short-listed (unsuccessfully) as potential replacement for Israel Defence Force Piper Super Cubs.

COSTS: US$1.3 million, T-35DT (2002).

DESIGN FEATURES: Low-risk, low-cost and international marketability achieved by reliance on Piper for development, some structural components and FAA compliance. Based on Piper Cherokee series (utilising many components of PA-28 Dakota and PA-32 Saratoga); cleared to FAR Pt 23 (Aerobatic category) and military standards for basic, intermediate and instrument flying training. Low-wing, tandem-seat design; sweptback vertical tail, non-swept horizontal surfaces.

Wing section NACA 65_2-415 on constant chord inboard panels. NACA 65_2-415 (modified) at tips; incidence 2° at root, $-0°$ 30' at tip; dihedral 7° from roots.

FLYING CONTROLS: Conventional and manual. Mass-balanced elevators and rudder; single-slotted wing flaps, aileron trim tab (port) and tailplane/elevator trim all electrically actuated; variable incidence tailplane.

STRUCTURE: Main structure of aluminium alloy and steel, with riveted skins, except for glass fibre engine cowling, wingtips and tailplane tips. Single-spar fail-safe wings, with components mainly from PA-28-236 Dakota (leading-edges) and PA-32R-301 Saratoga (trailing-edges), modified for shorter span; vertical tail virtually identical with Dakota; tailplane uses some standard components from Dakota and PA-31 (Navajo/Cheyenne); tailcone from Cherokee components, modified for narrower fuselage.

LANDING GEAR: Hydraulically retractable tricycle type, with single wheel on each unit. Main units retract inward, steerable nosewheel rearward. Piper oleo-pneumatic shock-absorber in each unit. Emergency free-fall extension. Cleveland mainwheels and McCreary tyres size 6.00-6 (8 ply), nosewheel and tyre size 5.00-5 (6 ply). Tyre pressures: 2.62 bar (38 lb/sq in) on mainwheels, 2.41 bar (35 lb/sq in) on nosewheel. Single-disc air-cooled hydraulic brake on each mainwheel. Parking brake. Minimum ground turning radius 6.20 m (20 ft 4 in).

POWER PLANT: One 224 kW (300 hp) Textron Lycoming IO-540-K1K5 flat-six engine, driving a Hartzell HC-C3YR-4BF/FC7663R three-blade constant-speed metal propeller. Fuel in two integral aluminium tanks in wing leading-edges, total capacity 291.5 litres (77.0 US gallons; 64.1 Imp gallons), of which 278 litres (73.4 US gallons; 61.1 Imp gallons) are usable. Overwing gravity refuelling point on each wing. Oil capacity 11.4 litres (3.0 US gallons; 2.5 Imp gallons). Fuel and oil systems permit up to 40 seconds of inverted flight.

ACCOMMODATION: Two vertically adjustable seats, with seat belts and shoulder harnesses, in tandem beneath one-piece transparent jettisonable canopy which opens sideways to starboard. One-piece acrylic windscreen and one-piece window in glass fibre fairing aft of canopy. Rear (instructor's) seat 22 cm (8.7 in) higher. Dual controls standard. Baggage compartment aft of rear cockpit, with external access on port side. Cockpits ventilated; cockpit heating and canopy demisting by engine bleed air.

SYSTEMS: Electrically operated hydraulic system, at 124 bar (1,800 lb/sq in) pressure for landing gear retraction and 44.8 bar (650 lb/sq in) for gear extension; separate system at 20.7 bar (300 lb/sq in) for wheel brakes. Electrical system is 24 V DC, powered by 28 V 70 A engine-driven Prestolite alternator and 24 V 15.5 Ah lead-acid battery, with inverter for AC power at 400 Hz to operate RMIs and attitude indicators. External power socket. No oxygen or de-icing provisions.

AVIONICS: Standard basic avionics by Rockwell Collins; optional items by Bendix/King.

Comms: Two VHF-251 com transceivers, two AMR-350 audio selector panels, TDR-950 transponder and Isocom interphone standard; Bendix/King KX 165 nav/com/glideslope, KMA 24H audio control/MKR and KT 76A transponder optional.

Flight: VIR-351 VOR with IND-350A, ADF-650A with IND-650 indicator and TOR-950 IFF standard; options include Bendix/King KR 87 ADF with KA 44B indicator, KN 63 DME with KDI 572 or IND 450 indicator, KDI 573 DME remote indicator, KR 21 or KML 351 marker beacon receiver, KA 40 marker beacon remote, KI 525A HSI and KNI 582 RMI indicator.

DIMENSIONS, EXTERNAL:

Wing span	8.84 m (29 ft 0 in)
Wing aspect ratio	5.7
Length overall	8.00 m (26 ft 3 in)
Height overall	2.64 m (8 ft 8 in)
Tailplane span	3.05 m (10 ft 0 in)
Wheel track	3.02 m (9 ft 11 in)
Wheelbase	2.09 m (6 ft 10¼ in)
Propeller diameter	1.93 m (6 ft 4 in)

DIMENSIONS, INTERNAL:

Cockpit: Length	3.24 m (10 ft 7½ in)
Max width	1.04 m (3 ft 5 in)
Max height	1.48 m (4 ft 10¼ in)

AREAS:

Wings, gross	13.69 m² (146.3 sq ft)

WEIGHTS AND LOADINGS:

Weight empty, equipped	930 kg (2,050 lb)
Fuel weight	210 kg (462 lb)
Max aerobatic T-O weight	1,315 kg (2,900 lb)
Max T-O and landing weight	1,338 kg (2,950 lb)
Max wing loading	97.73 kg/m² (20.03 lb/sq ft)
Max power loading	5.98 kg/kW (9.83 lb/hp)

PERFORMANCE:

Never-exceed speed (V_{NE})	241 kt (446 km/h; 277 mph)
Max level speed at S/L	168 kt (311 km/h; 193 mph)
Cruising speed:	
at 75% power at 2,680 m (8,800 ft)	144 kt (266 km/h; 166 mph) IAS
at 55% power at 5,120 m (16,800 ft)	138 kt (255 km/h; 159 mph) IAS
Stalling speed: flaps up	67 kt (125 km/h; 78 mph)
flaps down	62 kt (115 km/h; 72 mph)
Max rate of climb at S/L	465 m (1,525 ft)/min
Time to: 1,830 m (6,000 ft)	4 min 42 s
3,050 m (10,000 ft)	8 min 48 s
Service ceiling	5,840 m (19,160 ft)
Absolute ceiling	6,250 m (20,500 ft)
T-O run	287 m (940 ft)
T-O to 15 m (50 ft)	494 m (1,620 ft)
Landing from 15 m (50 ft)	509 m (1,670 ft)
Landing run	238 m (780 ft)
Range with 45 min reserves:	
at 75% power at 2,440 m (8,000 ft)	590 n miles (1,093 km; 679 miles)
at 55% power at 3,660 m (12,000 ft)	650 n miles (1,204 km; 748 miles)
Range, no reserves:	
at 75% power at 2,440 m (8,000 ft)	680 n miles (1,260 km; 783 miles)
at 55% power at 3,660 m (12,000 ft)	735 n miles (1,362 km; 846 miles)
Endurance at S/L: at 75% power	4 h 24 min
at 55% power	5 h 36 min
g limits	+6/−3

UPDATED

CHINA, PEOPLE'S REPUBLIC

AVIC

CHINA AVIATION INDUSTRY CORPORATION (Zhongguo Hangkong Gongye Zonggongsi)

67 Jiao Nan Street (PO Box 33), Beijing 100009
Tel: (+86 10) 64 09 31 14
Fax: (+86 10) 64 01 36 48
Web (1): http://www.avic1.com.cn
Web (2): http://www.avic2.com.cn
PRESIDENTS:
 Liu Gaozhuo (AVIC I)
 Zhang Yanzhong (AVIC II)
VICE-PRESIDENTS:
 Yang Yuzhong (AVIC I)
 Shi Jianzhong (AVIC II)
DIRECTORS-GENERAL OF MARKETING:
 Tang Xiaoping (AVIC I)
 Cui Degang (AVIC II)

INTERNATIONAL MARKETING:
CATIC (Zhongguo Hangkong Jishu Jinchukou Zonggongsi: China National Aero-Technology Import and Export Corporation)
CATIC Plaza, 18 Beichen Dong Street, Chaoyang District, Beijing 100101
Tel: (+86 10) 64 94 22 55, 64 94 03 70 and 64 94 10 90
Fax: (+86 10) 64 94 06 58 and 64 94 10 88
e-mail: webmaster@catic.com.cn
Web: http://www.catic.com.cn
PRESIDENT: Yang Chunshu
VICE-PRESIDENT, EXPORTS: Ding Shiqing
DIRECTOR, PUBLIC RELATIONS: Bi Jianfa

Present Chinese aviation industry created in 1951 and has since manufactured some 14,000 aircraft (including more than 10,000 military), more than 50,000 aero-engines and 10,000 air-to-air and tactical missiles. Some 700 aircraft have been exported, approximately 10 per cent of them civil types.

Former Ministry of Aero-Space Industry abolished 1993 and AVIC created on 26 June 1993 as economic entity to develop market economy and expand international collaboration in aviation programmes. CATIC Group formed 26 August 1993, with CATIC (founded January 1979) as its core company, to be responsible for import and export of aero and non-aero products, subcontract work and joint ventures.

Xian, Chengdu, Shanghai, Shenyang, Harbin and other factories carry out subcontract work on Airbus A300/310/318/320; ATR 42; Boeing 737/747/757; de Havilland Dash 8Q; and Bombardier 415. Licensed manufacture of Sukhoi Su-27s is undertaken at Shenyang. Co-development with UK/France/Italy and Singapore of AE-100 regional airliner abandoned in 1998 and replaced by AVIC I's ARJ21 project for 72/99-seat-family of aircraft (see ACAC entry in this section); AVIC II programme for smaller regional jet now to be met by co-production of Embraer ERJ-145, as described in HAIG (Harbin) entry.

Total workforce of aerospace industry was reduced to about 500,000 in 1998, when about 34,000 workers were made redundant and some 14,000 others transferred to non-aerospace activities. AVIC's President announced plans at Airshow China in November 1998 to restructure aircraft industry into two ''competing and co-operating'' groups in the near future. Draft organisation plan submitted to State Council in February 1999, resulting in division of AVIC into two separate companies (AVIC I and II) with effect from July 1999. Workforces in 2002 were 281,000 and 210,000 respectively, with corresponding assets of 34.9 billion and 31.5 billion yuan. AVIC I comprises 73 enterprises and 31 research institutes; corresponding figures for AVIC II are 79 and three. Principal organisations in the new structure, and their major indigenous programmes, are as follows. AVIC I planned to separate its military programmes in 2003; AVIC II stated in mid-2002 to have government approval to float its non-military business on Hong Kong Stock Exchange.

AVIC I
AVIC I Commercial Aircraft Company (ARJ21)
Beijing Aviation Simulator
Chengdu Aircraft Industry Group (J-7/F-7, FC-1 and J-10)
China Air-to-Air Missile Research Institute
Guizhou Aviation Industry Group (JJ-7/FT-7; FTC 2000; jet engines; missiles)
Shanghai Aviation Industry Group (airliner subcontracts)
Shenyang Aircraft Industry Group (J-8/F-8 and J-11/Su-27; civil subcontracts)
Xian Aero-Engine (WP8 and WS9)
Xian Aircraft Industry Group (JH-7/FBC-1 and MA-60)

AVIC II
Changhe Aircraft Industry Group (Z-8 and Z-11)
Chengdu Engine (WP7 and WP13)
China Helicopter Design and Research Institute ('Z-X')
Dongan Engine (WJ5)
Harbin Aircraft Industry Group (Hafei Z-9, Z-10 (?) and Y-12)
Hongdu Aviation Industry Group (NAMC K-8 and N-5)
Shaanxi Aircraft Company (Y-8)
Shijiazhuang Aircraft Industry Corporation (Y-5)
South Aero-Engine (HS5, WJ6 and WZ8)

UPDATED

ACAC

AVIC I COMMERCIAL AIRCRAFT COMPANY

PARTICIPATING COMPANIES:
  Chengdu Aircraft Industry Group (CAC)
  Shanghai Aircraft Research Institute (SARI)
  Shanghai Aircraft Industry Company (SAIC)
  Shenyang Aircraft Corporation (SAC)
  Xian Aircraft Design and Research Institute (XADRI)
  Xian Aircraft Company (XAC)
CHAIRMAN: Yang Yuzhong
PRESIDENT AND VICE-CHAIRMAN: Tang Xiaoping

Six Chinese organisations formed a consortium in 1998 to study the prospects of launching, with risk-sharing foreign industrial partners, a programme for a new regional jet in the 70-seat class. Initial concepts were focused on 58-seat (NRJ 58) and 76-seat (NRJ 76) variants, as described under NRJ heading in earlier editions. However, the latter project received the designation ARJ21 (Advanced Regional Jet, 21st Century) in 2001 and ACAC formed in October 2002 to manage this programme.

NEW ENTRY

ACAC ARJ21

TYPE: Regional jet airliner.
PROGRAMME: ARJ21 project officially launched by AVIC I on 7 November 2000 as US$700 million programme under the auspices of a newly created New Regional Jet Programme Management Company. This was allocated initial investments in 2001 from AVIC I (20 million yuan); CAC, SAIC and XAC (each 5 million yuan); and SAC (3 million yuan). Further investment is being sought from other Chinese and foreign companies. The parent company is responsible for defining airframe configuration; defining and subcontracting work-shares; schedule and quality control; final assembly, certification, marketing and customer support. Programme go-ahead announced October 2002 with formation of ACAC and allocation of 5 billion yuan (US$602 million) start-up funding; Boeing to provide engineering support; production to be at Shanghai. Selection of General Electric CF34 announced in November 2002. First flight targeted for 2006.

Current configuration of ARJ21 airliner   *NEW*/0137945

CURRENT VERSIONS: **Standard version:** For 72 passengers.
  Stretched version: To seat 99 passengers.
CUSTOMERS: Estimated sale of 300 (domestic and export) over 20 years.
DESIGN FEATURES: Twin-turbofan design; rear-mounted engines and T tail. Antonov (Ukraine) to design supercritical wing.
POWER PLANT: Two 80.1 kN (18,000 lb st) class rear-mounted General Electric CF34-10A turbofans. Standard fuel capacity 12,719 litres (3,360 US gallons; 2,798 Imp gallons).
AVIONICS: To be selected. Candidate suppliers include Honeywell, Rockwell Collins and Thales.
Preliminary data as follows (A: 72-passenger, B: 99-passenger):
DIMENSIONS, EXTERNAL:

Wing span: A, B		27.39 m (89 ft 10¼ in)
Length overall: A		31.74 m (104 ft 1½ in)
B		35.11 m (115 ft 2¼ in)
Height overall: A		8.795 m (28 ft 10¼ in)
B		8.815 m (28 ft 11 in)

DIMENSIONS, INTERNAL:

Cabin max width: A, B		3.145 m (10 ft 3¾ in)

WEIGHTS AND LOADINGS:

Max payload: A	8,280 kg (18,254 lb)
B	10,589 kg (23,345 lb)
Max T-O weight: A standard	36,547 kg (80,572 lb)
A option	39,558 kg (87,210 lb)
B standard	40,281 kg (88,804 lb)
B option	42,758 kg (94,267 lb)
Max landing weight: A standard	33,989 kg (74,933 lb)
A option	36,789 kg (81,106 lb)
B standard	37,461 kg (82,587 lb)
B option	39,765 kg (87,667 lb)

PERFORMANCE (estimated):

Normal operating speed: A, B	M0.78
Max operating altitude: A, B	10,670 m (35,000 ft)
T-O field length: A	1,376 m (4,515 ft)
B	1,612 m (5,290 ft)
Landing field length: A	1,415 m (4,640 ft)
B	1,491 m (4,890 ft)
Design range: A and B standard	
	1,200 n miles (2,222 km; 1,380 miles)
A option	2,000 n miles (3,704 km; 2,301 miles)
B option	1,800 n miles (3,333 km; 2,071 miles)

NEW ENTRY

BKLAIC

BEIJING KEYUAN LIGHT AIRCRAFT INDUSTRIAL COMPANY LTD

7 Keyuan South Road, Zhongguanchun, Haidian, Beijing 100080

Tel: (+86 10) 62 57 29 22, 62 57 28 22 and 62 55 79 43
Fax: (+86 10) 62 57 29 22
CHAIRMAN: Yuan Yong Min
VICE GENERAL MANAGER: Fangjun Dong

This company produced the AD-200 tandem two-seat light aircraft last described in the 2001-02 *Jane's*. It was also responsible for the Zhong Hua hot-air airship (see Lighter Than Air section, 2000-01 and previous editions). No indication of more recent activities has been received.

UPDATED

BUAA

BEIJING UNIVERSITY OF AERONAUTICS AND ASTRONAUTICS (Beijing Hangkong Hangtian Daxue)

37 Xue Yuan Road, Haidian District, Beijing 100083
Tel: (+86 10) 82 31 75 91
Fax: (+86 10) 82 31 75 90
e-mail: webmaster@buaa.edu.cn
Web: http://www.buaa.edu.cn/main/eindex.html

Beijing University's Aeronautics and Astronautics department has designed several microlight aircraft during the past two decades, under the series name Mifeng (Bee). Its most recent known manned aircraft project is the M-16 small autogyro, which made its international debut in November 2000.

UPDATED

BUAA MIFENG M-16

English name: Bee
PROGRAMME: Prototype of this single-seat ultralight autogyro completed August 1998. Seen publicly at Airshow China, Zhuhai, November 2000, but no indication of production plans (if any) by mid-2002. See 2002-03 and earlier editions for description and illustrations.

UPDATED

CAC

CHENGDU AIRCRAFT INDUSTRY GROUP (Chengdu Feiji Gongye Gongsi) (Subsidiary of AVIC I)

Huangtianba (PO Box 800), Chengdu, Sichuan 610092
Tel: (+86 28) 740 10 33
Fax: (+86 28) 740 49 84
e-mail: cacoa@mail.cac.com.cn
Web: http://:www.cac.com.cn
CHAIRMAN: Yang Tingkuo
PRESIDENT: Yang Baoshu
DEPUTY GENERAL MANAGER: Li Shaoming
DIRECTOR, INTERNATIONAL CO-OPERATION DIVISION:
  Wang Yinggong

Major centre for fighter development and production, founded 1958; has since built over 2,000 fighters of more than 10 models or variants; current facility occupies site area of 462 ha (1,142 acres) and had 1995 workforce (latest figure known) of nearly 20,000. Production continues mainly to concern J-7/F-7 fighter series (several models), but new fighters are currently under development (see J-10 and FC-1 entries which follow). Chengdu also thought to be in concept stage of an advanced combat aircraft, possibly to the same outline requirement as the 'XXJ' described under the SAC (Shenyang) entry. In addition, CAC was named in 2001 as a risk-sharing partner in the ARJ21 regional jet programme

described under the ACAC heading in this section.
  Subcontract work includes passenger doors for the Airbus A320; production of fuselage components for the Dassault Falcon 2000EX is due to begin in 2003 under an agreement signed in early 2002. Non-aerospace products, which previously accounted for about 10 per cent of current output, were targeted to have reached 40 per cent by 2000; it is not known whether this was achieved.

UPDATED

CAC J-7

Chinese name: Jianjiji-7 (Fighter aircraft 7)
Westernised designation: F-7
TYPE: Multirole fighter.
PROGRAMME: Soviet licence to manufacture MiG-21F-13 and its R-11F-300 engine granted 1961, when some pattern aircraft and CKD kits delivered, but technical documentation not completed; assembly of first J-7 using Chinese-made components began early 1964; original plan, for Chengdu and Guizhou to become main airframe/engine production centres, backed up by Shenyang until these were fully productive, delayed by cultural revolution.
  First flight of Shenyang-built J-7, 17 January 1966; Chengdu production of J-7 I began June 1967 (first flight 16 June 1969); development of J-7 II began 1975, followed by first flight 30 December 1978 and production approval September 1979; development of F-7M and J-7 III started

1981; J-7 III first flight 26 April 1984; F-7M (first flight 31 August 1983) revealed publicly October 1984, production go-ahead December 1984, named Airguard early 1986; first F-7P deliveries to Pakistan 1988; first F-7MPs to Pakistan mid-1989; F-7MG public debut November 1996; J-7FS revealed September 1998 and F-7MF November 2000.
CURRENT VERSIONS (domestic): **J-7:** Initial Shenyang licence version using Chinese-made components; few only.
  J-7 I: Initial Chengdu version (1967), with variable intake shock cone and second 30 mm gun; not accepted in large numbers.
  J-7 II: Modified and improved J-7 I, with WP7B turbojet of increased thrust (43.2 kN; 9,700 lb st dry, 59.8 kN; 13,450 lb st with afterburning); 720 litre (190 US gallon; 158 Imp gallon) centreline drop tank for increased range; brake-chute relocated at base of rudder to improve landing performance and shorten run; rear-hinged canopy, jettisoned before ejection seat deploys; new Chengdu Type II seat operable at zero height and speeds down to 135 kt (250 km/h; 155 mph); and new Lanzhou compass system. Small batch production (typically, 14 in 1989), notwithstanding advent of J-7 III and J-7E. Still the major PLAAF variant.
  J-7 III: Chinese equivalent of MiG-21MF, with blown flaps and all-weather, day/night capability. Main improvements are change to WP13 engine with greater power; additional fuel in deeper dorsal spine; JL-7 (J-band)

'Red 139', the Chengdu J-7FS technology demonstrator prototype   0116932

Canards and rectangular chin intake characterise the F-7MF (*Photo Link*) 0100788

interception radar, with correspondingly larger nose intake and centrebody radome; sideways-opening (to starboard) canopy, with centrally located rearview mirror; improved HTY-4 low-speed/zero height ejection seat; more advanced fire-control system; twin-barrel 23 mm gun under fuselage (with HK-03D optical gunsight); broader-chord vertical tail surfaces, incorporating antennas for LJ-2 omnidirectional RWR in hemispherical fairing each side at base of rudder; increased weapon/stores capability (four underwing stations), similar to that of F-7M; and new or additional avionics (which see). Joint development by Chengdu and Guizhou (GAIC); entered PLA Air Force and Navy service from 1992, but reportedly limited use only.

J-7 IV: Further (early 1990s) attempt to improve upon J-7 III; 71.6 kN (16,093 lb st) WP13F1 engine, JL-7A interception radar, RWR antennas atop vertical fin; HUD, Tacan and ADC; weapons upgrade to include PL-8 AAMs. Believed to equip two night fighter regiments.

J-7E: Upgraded version of J-7 II with modified, double-delta wing, retaining existing leading-edge sweep angle of 57° inboard but reduced sweep of only 42° outboard; span increased by 1.17 m (3 ft 10 in) and area by 1.88 m² (20.2 sq ft), giving 8.17 per cent more wing area; four underwing stations instead of two, outer pair each plumbed for 480 litre (127 US gallon; 106 Imp gallon) drop tank; new WP7F version of WP7 engine, rated at 44.1 kN (9,921 lb st) dry and 63.7 kN (14,330 lb st) with afterburning; armament generally as listed for F-7M, but capability extended to include PL-8 (Python 3) air-to-air missiles; *g* limits of 8 (up to M0.8) and 6.5 (above M0.8); avionics include head-up display and air data computer. Believed to have made first flight in April 1990 and entered service 1993. In production.

J-7EB: Version of J-7E equipping PLA Air Force 'August 1st' aerobatic team (nine in 1998 display season); fitted with smoke canisters for display purposes.

J-7FS: Technology demonstrator, modified from standard J-7 II, with new chin-mounted intake and central splitter plate under reconfigured ogival nosecone; more powerful (73.4 to 78.3 kN; 16,502 to 17,604 lb thrust class with afterburning) Liyang (LMC) WP13F II turbojet. Began 22 month flight test programme on 8 June 1998. Enlarged nose avionics bay able to accept 60 cm (23.6 in) diameter multimode pulse Doppler radar believed to be under development in China. Development work formed basis for new F-7MF export variant (which see). Planned future changes were to include wing modifications based on J-7E/F-7MG double-delta configuration.

J-7G: Recently reported as further upgrade of J-7E and expected to make maiden flight in mid-2002.

JJ-7: Tandem two-seat operational trainer, based on J-7 II and MiG-21US; developed at Guizhou and described under GAIC entry.

CURRENT VERSIONS (export): **F7-3:** Export designation of J-7 III. No orders known.

F-7A: Export counterpart of J-7 I, supplied to Albania and Tanzania.

F-7B: Export version of J-7 II, with R550 Magic missile capability; supplied to Egypt and Iraq in 1982-83 and also to Sudan. Some supplied to Air Force of Zimbabwe (known locally as F-7 II) appear to be of this version.

F-7BS: Hybrid version supplied to Sri Lanka 1991: has F-7B fuselage/tail and Chinese avionics (no HUD), combined with four-pylon wings of F-7M. Equips No. 5 Squadron of SLAF. Zimbabwe also has some four-pylon aircraft with its No. 5 Squadron; these known locally as the F-7 IIN.

F-7M Airguard: Upgraded export version, developed from J-7 II; new avionics imported from May 1979 included Marconi HUDWAC (head-up display and weapon aiming computer); new ranging radar, air data computer, radar altimeter and IFF; more secure com radio; improved electrical power generation system for the new avionics; two additional underwing stores points; improved WP7B(BM) engine; birdproof windscreen; strengthened landing gear; ability to carry PL-7 air-to-air missiles; nose probe relocated from beneath intake to top lip of intake, offset to starboard. Exported to Iran and Myanmar. Reported in early 1999 that Zimbabwe planned to purchase 12, but no order yet confirmed.

Description applies to F-7M version, except where indicated.

F-7MB: Variant of F-7M; mentioned in 1996 F-7MG brochure. Customer believed to be Bangladesh (16); one planned to be equipped with reconnaissance pod.

F-7MF: Latest known variant; debut at Airshow China, November 2000; further development of J-7FS and F-7MG for export market. Larger 'solid' nose; shorter, rectangular intake located farther back under nose; small, shoulder-mounted canards just forward of wingroot leading-edge; WP13F engine; 1553B databus; avionics include 43 n mile (80 km; 50 mile) range pulse Doppler radar (possibly IAI Elta EL/M-2032), single HUD and dual HDDs; 3,000 kg (6,614 lb) external stores load.

Wind tunnel testing completed; first flight targeted for late 2001/early 2002. Performance expectations include

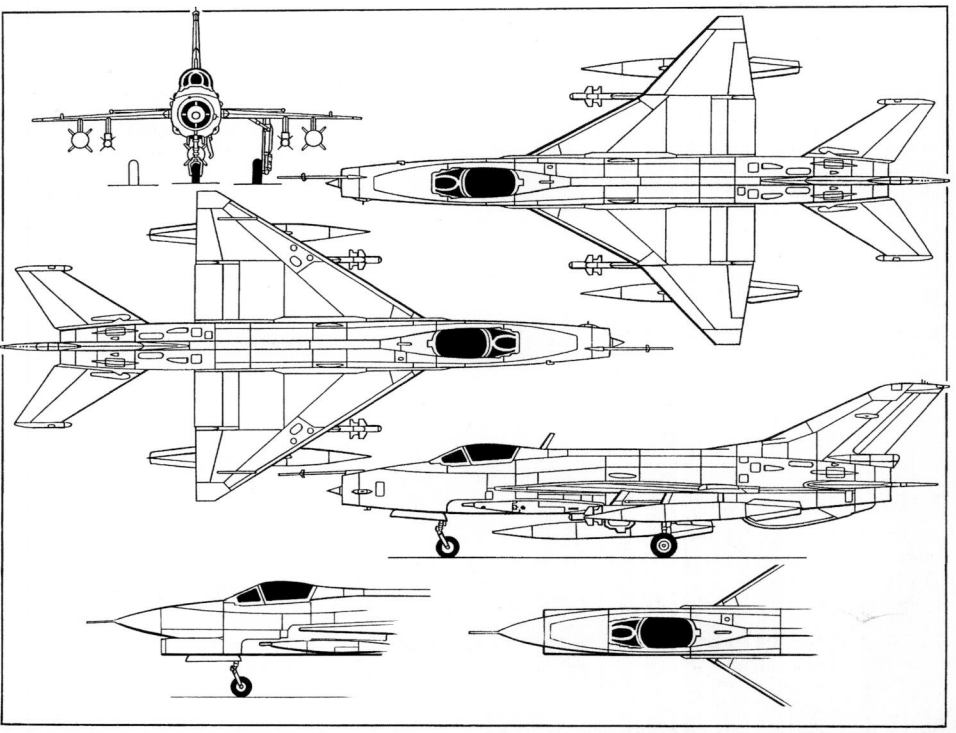

CAC F-7M Airguard single-seat fighter and close support aircraft; upper plan view shows modified outer wings of J-7E and F-7MG; lower scrap views of J-7FS nose in profile and plan (*Jane's/Mike Keep*)   0126688

CAC F-7Ps of the Pakistan Air Force
0116933

CAC J-7EB of the PLAAF 'August 1st' aerobatic team (*Robert Hewson*)   0105846

M1.8 top speed, 16,000 m (52,500 ft) ceiling, 650 m (2,135 ft) T-O run and 1,403 n mile (2,600 km; 1,615 mile) ferry range. Possibly testbed for some features of J-10.

F-7MG: Improved version of F-7M (G suffix indicates *gai:* modified), combining double-delta wings of J-7E with Grifo MG radar, other upgraded avionics, uprated (WP13F) engine and leading/trailing-edge manoeuvring flaps. Said to have 45 per cent better manoeuvrability than F-7M. Public debut (aircraft 0142 and 0144) at China Air Show, Zhuhai, November 1996; Pakistan (see F-7PG) only customer so far.

F-7MP: Further modified variant of F-7P; improved cockpit layout and navigation system incorporating Rockwell Collins AN/ARN-147 VOR/ILS receiver, AN/ARN-149 ADF and Pro Line II digital DME-42. Avionics (contract for up to 100 sets) delivered to China from early 1989. FIAR (now Galileo) Grifo MG fire-control radar (range of more than 30 n miles; 55 km; 34 miles) for F-7P and MP ordered 1993, to replace Skyranger; flight trials began May 1996 and completed in 1997.

F-7N: Variant of F-7M; mentioned in 1996 F-7MG brochure, but no details given. Possibly an alternative designation for Zimbabwe F-7 IIN (see F-7BS paragraph above).

F-7P Airguard: Variant of F-7M (briefly called Skybolt), embodying 24 modifications to meet specific requirements of Pakistan Air Force, including ability to carry four air-to-air missiles (Sidewinders) instead of two and fitment of Martin-Baker Mk 10L ejection seat. Delivered 1988-91.

F-7PG: Pakistan Air Force designation of F-7MG; 57 (plus six FT-7PGs) ordered in late 2000. Deliveries to Nos. 17 and 23 Squadrons at Samungli, replacing Shenyang F-6s, completed by second quarter 2002; Grifo radars produced by KARF factory of PAC (which see). Possibility of follow-on order for up to 25 more (including FT-7PG trainers).

FT-7: Export designation of JJ-7 two-seat trainer (see GAIC entry for details).

CUSTOMERS: Several hundred built for Chinese air forces; more than 400 exported to Albania (12 F-7A), Bangladesh (16 F-7MB), Egypt (approximately 90 F-7B?), Iran (18 or more F-7M), Iraq (approximately 90 F-7B?), Myanmar (24 F-7M), Pakistan (20 F-7P and 100 F-7MP, all designated F-7P by PAF; followed by 57 F-7PGs), Sri Lanka (four F-7BS), Sudan (22 F-7B); Tanzania (16 F-7A) and Zimbabwe (approximately 12 F-7B/F-7IIN variants). Pakistan Air Force F-7P squadrons are No. 2 at Masroor, Nos. 18 and 20 at Rafiqui and No. 25 at Mianwali; No. 17 first with F-7PG (27 March 2002), followed by No. 23, both at Quetta.

DESIGN FEATURES: Typical mid-1950s design of fighter, incorporating diminutive delta wing (double-delta on J-7E/EB and F-7MG/PG), with clipped tips to mid-

F-7MG demonstrator about to land

0106477

mounted wings, plus all-moving horizontal tail; circular-section fuselage with dorsal spine; nose intake with conical centrebody; swept tail, with large vertical surfaces and ventral fin.

Wing anhedral 2° from roots; incidence 0°; thickness/chord ratio approximately 5 per cent at root, 4.2 per cent at tip; quarter-chord sweepback 49° 6' 36" (reduced on J-7E/EB and F-7MG/PG outer panels); no wing leading-edge camber.

FLYING CONTROLS: Manual operation, with autostabilisation in pitch and roll; hydraulically boosted inset ailerons; plain trailing-edge flaps, actuated hydraulically; forward-hinged door-type airbrake each side of underfuselage below wing leading-edge; third, forward-hinged airbrake under fuselage forward of ventral fin; airbrakes actuated hydraulically; hydraulically boosted rudder and all-moving, trimmable tailplane. Leading/trailing-edge manoeuvring flaps on J-7E/EB and F-7MG/PG.

STRUCTURE: All-metal; wings have two primary spars and auxiliary spar; semi-monocoque fuselage, with spine housing control pushrods, avionics, single-point refuelling cap and fuel tank; blister fairings on fuselage above and below each wing to accommodate retracted mainwheels.

LANDING GEAR: Inward-retracting mainwheels, with 600×200 tyres (pressure 11.50 bar; 167 lb/sq in) and LS-16 disc brakes; forward-retracting nosewheel, with 500×180 tyre (pressure 7.00 bar; 102 lb/sq in) and LS-15 double-acting brake. Nosewheel steerable ±47°. Minimum ground turning radius 7.04 m (23 ft 1¼ in). Tail braking parachute at base of vertical tail.

POWER PLANT: *F-7M:* One LMC (Liyang) WP7B(BM) turbojet (43.2 kN; 9,700 lb st dry, 59.8 kN; 13,448 lb st with afterburning).

J-7 III: LMC WP13 turbojet (40.2 kN; 9,039 lb st dry, 64.7 kN; 14,550 lb st with afterburning).

J-7E/EB: See Current Versions.

J-7FS: See Current Versions.

F-7MF/MG/PG: WP13F (44.1 kN; 9,921 lb st dry, 64.7 kN; 14,550 lb st with afterburning).

Total F-7M internal fuel capacity 2,385 litres (630 US gallons; 525 Imp gallons), contained in six flexible tanks in fuselage and two integral tanks in each wing. Provision for carrying a 500 or 800 litre (132 or 211 US gallon; 110 or 176 Imp gallon) centreline drop tank, and/or a 500 litre drop tank on each outboard underwing pylon. Maximum internal/external fuel capacity 4,185 litres (1,105 US gallons; 921 Imp gallons).

ACCOMMODATION: Pilot only, on CAC zero-height/low-speed ejection seat operable between 70 and 459 kt (130 and 850 km/h; 81 and 528 mph) IAS. Martin-Baker Mk 10L seat in F-7P/MP/PG. One-piece canopy, hinged at rear to open upward. J-7 III/F7-3 canopy opens sideways to starboard.

SYSTEMS: Improved electrical system in F-7M, using three static inverters, to cater for additional avionics. Jianghuai YX-3 oxygen system.

AVIONICS: *Comms:* BAE Systems AD 3400 UHF/VHF multifunction com, Chinese Type 602 IFF transponder; Type 605A ('Odd Rods' type) IFF in J-7 III.

Radar: BAE Systems Type 226 Skyranger ranging radar in F-7M; Chinese JL-7 fire-control radar in J-7 III. Galileo (formerly FIAR) Grifo MG in F-7P/MP/MG/PG; (look-down, shoot-down and track-while-scan capability).

Flight: Navigation function of BAE Systems HUDWAC includes approach mode. WL-7 radio compass, XS-6A marker beacon receiver, Type 0101 HR A/2 radar

CONFIGURATION Stores	O-Board	I-Board	C-Line	I-Board	O-Board
2 x PL-7 (or R550) Missiles		○		○	
2 x AIM-9P Missiles		○		○	
4 x AIM-9P Missiles	○	○		○	○
4 x AIM-9P Missiles + 500L Tank	○	○	●	○	○
2 x AIM-9P Missiles + 3 x 500L Tanks	●	○	●	○	●
2 x PL-7 (or R550) Missiles + 3 x 500L Tanks	●	○	●	○	●
4 x HF-7C Rockets Pods	○	○		○	○
2 x HF-7C Rockets Pods + 3 x 500L Tanks	●	○	●	○	●
4 x HF-7C Rockets Pods + 800L Tank	○	○	●	○	○
2 x 500 kg Bombs		○		○	
2 x 500 kg Bombs + 800L Tank		○	●	○	
4 x MK-82 Bombs	○	○		○	○
2 x MK-82 Bombs + 3 x 500L Tanks	●	○	●	○	●
4 x MK-82 Bombs + 800L Tank	○	○	●	○	○
2 x MK-82 Bombs + 2 x HF-7C Rockets Pods + 800L Tank	○	○	●	○	○
2 x 500L Tanks + 800L Tank	●		●		●

● Drop tank   ○ Weapon

F-7MG weapon options

Cockpit layout of the F-7MG

altimeter and BAE Systems air data computer in F-7M. Beijing Aeronautical Instruments Factory KJ-11 twin-channel autopilot and FJ-1 flight data recorder in J-7 III. F-7MG/PG suite includes VOR/DME/INS and Tacan.

Instrumentation: BAE Systems Type 956 HUDWAC (head-up display and weapon aiming computer) in F-7M provides pilot with displays for instrument flying, with air-to-air and air-to-ground weapon aiming symbols integrated with flight-instrument symbology. It can store 32 weapon parameter functions, allowing for both current and future weapon variants. In air-to-air combat its four modes (missiles, conventional gunnery, snapshoot gunnery, dogfight) and standby aiming reticle allow for all eventualities. VCR and infra-red cockpit lighting in F-7MG/PG, for which licence-built Russian helmet sight, slaved to PL-9 AAM, is also in production.

Self-defence: Skyranger ECCM in F-7M. Chinese LJ-2 RWR and GT-4 ECM jammer in J-7 III.

ARMAMENT (F-7M): Two 30 mm Type 30-1 belt-fed cannon, with 60 rds/gun, in fairings under front fuselage just forward of wingroot leading-edges. Two hardpoints under each wing, of which outer ones are wet for carriage of drop tanks. Centreline pylon used for drop tank only. Each inboard pylon capable of carrying a PL-2, -2A, -5B, -7 or -8 (Python 3) missile (and PL-9 on F-7MG/PG) or, at customer's option, an R550 Magic; one 18-tube pod of Type 57-2 (57 mm) air-to-air and air-to-ground rockets; one Type 90-1 (90 mm) seven-tube pod of air-to-ground rockets; or a 50, 150, 250 or 500 kg bomb. Each outboard pylon can carry one of above rocket pods, a 50 or 150 kg bomb, or a 500 litre drop tank.

ARMAMENT (J-7 III): One 23 mm Type 23-3 twin-barrel gun in ventral pack. Five external stores stations can carry two to four PL-2 or PL-5B air-launched missiles; two or four Qingan HF-16B 12-round launchers for Type 57-2 or seven-round pods of Type 90-1 rockets; or two 500 kg, four 250 kg or ten 100 kg bombs, in various combinations with 500 litre (one centreline and/or one under each wing) or 800 litre (underfuselage station only) drop tanks.

DIMENSIONS, EXTERNAL:
Wing span: except J-7E/F-7MG 7.15 m (23 ft 5½ in)
J-7E/F-7MG 8.32 m (27 ft 3½ in)
Wing chord: at root 5.51 m (18 ft 0¾ in)
at tip (except J-7E/F-7MG) 0.46 m (1 ft 6¼ in)
Wing aspect ratio: except J-7E/F-7MG 2.2
J-7E/F-7MG 2.8
Length overall: excl nose probe 13.945 m (45 ft 9 in)
incl nose probe 14.885 m (48 ft 10 in)
Fuselage: Length 12.175 m (39 ft 11½ in)
Max diameter 1.34 m (4 ft 4¾ in)
Height overall 4.105 m (13 ft 5½ in)
Tailplane span 3.74 m (12 ft 3¼ in)
Wheel track 2.69 m (8 ft 10 in)
Wheelbase 4.805 m (15 ft 9¼ in)

AREAS:
Wings, gross: except J-7E/F-7MG 23.00 m² (247.6 sq ft)
J-7E/F-7MG 24.88 m² (267.8 sq ft)
Ailerons (total): except J-7E/F-7MG 1.18 m² (12.70 sq ft)
Trailing-edge flaps (total) 1.87 m² (20.13 sq ft)
Fin 3.48 m² (37.46 sq ft)
Rudder 0.97 m² (10.44 sq ft)
Tailplane 3.94 m² (42.41 sq ft)

WEIGHTS AND LOADINGS:
Weight empty: F-7M 5,275 kg (11,629 lb)
F-7MF 5,500 kg (12,125 lb)
F-7MG 5,292 kg (11,667 lb)
Normal T-O weight with two PL-2 or PL-7 air-to-air missiles: F-7M 7,531 kg (16,603 lb)
J-7 III 8,150 kg (17,967 lb)
F-7MG 7,540 kg (16,623 lb)
Max T-O weight: F-7MG 9,100 kg (20,062 lb)
Wing loading at normal T-O weight:
F-7M 327.4 kg/m² (67.06 lb/sq ft)
J-7 III 354.3 kg/m² (72.56 lb/sq ft)
Max wing loading: F-7MG 365.8 kg/m² (74.91 lb/sq ft)
Power loading at normal T-O weight:
F-7M, J-7 III 126 kg/kN (1.23 lb/lb st)
Max power loading: F-7MG 141 kg/kN (1.38 lb/lb st)

PERFORMANCE (F-7M at normal T-O weight with two PL-2 or PL-7 air-to-air missiles, except where indicated):
Never-exceed speed (VNE) above 12,500 m (41,010 ft) M2.35 (1,346 kt; 2,495 km/h; 1,550 mph)
Max level speed between 12,500 and 18,500 m (41,010-60,700 ft) M2.05 (1,175 kt; 2,175 km/h; 1,350 mph)
Unstick speed 167-178 kt (310-330 km/h; 193-205 mph)
Touchdown speed 162-173 kt (300-320 km/h; 186-199 mph)
Max rate of climb at S/L 10,800 m (35,435 ft)/min
Acceleration from M0.9 to M1.2 at 5,000 m (16,400 ft) 35 s
Max sustained turn rate: M0.7 at S/L 14.7°/s
M0.8 at 5,000 m (16,400 ft) 9.5°/s
Service ceiling 18,200 m (59,720 ft)
Absolute ceiling 18,700 m (61,360 ft)
T-O run 700-950 m (2,300-3,120 ft)
Landing run with brake-chute 600-900 m (1,970-2,955 ft)
Typical mission profiles:
combat air patrol at 11,000 m (36,080 ft) with two air-to-air missiles and three 500 litre drop tanks, incl 5 min combat 45 min
long-range interception at 11,000 m (36,080 ft) at 351 n miles (650 km; 404 miles) from base, incl M1.5 dash and 5 min combat, stores as above
hi-lo-hi interdiction radius, out and back at 11,000 m (36,080 ft), with three 500 litre drop tanks and two 150 kg bombs 324 n miles (600 km; 373 miles)
lo-lo-lo close air support radius with four rocket pods, no external tanks 200 n miles (370 km; 230 miles)
Range: two PL-7 missiles and three 500 litre drop tanks 939 n miles (1,740 km; 1,081 miles)
self-ferry with one 800 litre and two 500 litre drop tanks, no missiles 1,203 n miles (2,230 km; 1,385 miles)
g limit +8

PERFORMANCE (J-7 III at normal T-O weight):
Max operating Mach No. (MMO) 2.1
Unstick speed with afterburning 173 kt (320 km/h; 199 mph)
Touchdown speed with flap blowing 146 kt (270 km/h; 168 mph)
Min level flight speed 140 kt (260 km/h; 162 mph)
Max rate of climb at S/L 9,000 m (29,525 ft)/min
Service ceiling 18,000 m (59,060 ft)
Acceleration from M1.2 to M1.9 at 13,000 m (42,660 ft) 3 min 27 s
Air turning radius at 5,000 m (16,400 ft) at M1.2 5,093 m (16,710 ft)
T-O run with afterburning 800 m (2,625 ft)
Landing run with flap blowing, drag-chute and brakes deployed 550 m (1,805 ft)
Range: on internal fuel 518 n miles (960 km; 596 miles)
with 800 litre belly tank 701 n miles (1,300 km; 807 miles)
with 800 litre belly tank and two 500 litre underwing tanks 1,025 n miles (1,900 km; 1,180 miles)
g limits: up to M0.8 +8.5
above M0.8 +7

PERFORMANCE (F-7MG):
Max operating Mach No. (MMO) 2.0
Max level speed 648 kt (1,200 km/h; 745 mph) IAS
Min level speed 114 kt (210 km/h; 131 mph) IAS
Max rate of climb at S/L 11,700 m (38,386 ft)/min
Max instantaneous turn rate 25.2°/s
Sustained turn rate: at 1,000 m (3,280 ft) 16°/s
at 5,000 m (16,400 ft) 11°/s
at 8,000 m (26,240 ft) 8°/s
Service ceiling 17,500 m (57,420 ft)
Theoretical ceiling 18,000 m (59,060 ft)
T-O run, with afterburning, and landing run 600-700 m (1,970-2,300 ft)
Operational radius:
air superiority (hi-hi-hi) with two AIM-9P AAMs and three 500 litre drop tanks, incl 5 min combat with afterburner 459 n miles (850 km; 528 miles)
air-to-ground attack (lo-lo-hi) with two Mk 82 bombs and two 500 litre drop tanks 297 n miles (550 km; 342 miles)
Ferry range 1,187 n miles (2,200 km; 1,367 miles)
g limits +8/–3

UPDATED

Full-size mockup of the FC-1/Super-7 ***(Yihong Chang)*** ***NEW***/0137976

CAC FC-1

Export designation: Super-7

TYPE: Attack fighter.

PROGRAMME: Fighter China (FC) programme launched 1991 following cancellation of US participation in development of Chengdu Super-7 (see 1995-96 *Jane's*). Some design assistance from MiG OKB, possibly based on (then-designated MiG-33) mid-1980s project for single-engined variant of the MiG-29. (Sources at MiG experimental bureau quoted as saying that FC-1 was designed there to a military specification as Izd (*Izdeliye*: article) 33 and later offered for Chinese production following cancellation of Russian requirement.)

Two static test airframes were due to begin testing 1996, though not yet known to have done so; full-scale mockup completed 2001. First prototype under construction by then, with oft-postponed maiden flight now targeted for second half of 2003. Avionics competition still open in 2002; contenders include IAI Elta, SAGEM and Thales. Choice of a multimode pulse Doppler radar apparently still lies between Phazotron of Russia (Kopyo), Galileo (originally FIAR) of Italy (Grifo MG), and Thales (RC 400). Selection had not been made by mid-2002, due to Western misgivings about technology transfer; China may look elsewhere if clearance is much further delayed.

Chinese domestic requirement for the FC-1 is said to be for several hundred aircraft, with Pakistan wanting about 150, but a Pakistani order is conditional upon PLAAF commitment, which was still awaited in mid- 2002. A letter of intent for joint development by China and Pakistan was reportedly signed in February 1998, followed by signature of a joint development and production agreement in June 1999.

CUSTOMERS: Envisaged for air forces of China and Pakistan initially, but costed to be competitive in wider export market. Seen as potential replacement for Shenyang J-6, Chengdu J-7, Nanchang Q-5, Northrop F-5 and Dassault Mirage III/5.

COSTS: Development investment of some US$150 million by mid-2001; estimated unit cost approximately US$2 million (2001).

DESIGN FEATURES: Agile light fighter. Mid-mounted delta wing with narrow wingroot strakes at leading-edge; leading-edge manoeuvring flaps; single turbofan engine; side-mounted twin intakes, with splitter plates; large intake trunks provide space for considerable internal fuel capacity. Large main fin with dorsal fairing; two smaller, uncanted ventral fins.

FLYING CONTROLS: Conventional hydraulic servo-operated control of ailerons, rudder and all-moving tailplane initially, with single analogue fly-by-wire system for back-up; provision for FBW to become primary system later. Trailing-edge and leading-edge flaps.

STRUCTURE: Primary structure conventional aluminium and steel alloy semi-monocoque. Some components may be manufactured in Pakistan.

LANDING GEAR: Retractable tricycle type, with single wheel and oleo shock-absorber on each unit. Mainwheels retract upward into engine intake trunks; nosewheel retracts rearward.

POWER PLANT: One Klimov RD-93 (RD-33 derivative) turbofan (81.4 kN; 18,300 lb st with afterburning), possibly to be licence built by Liyang Machinery Corporation (LMC) for production aircraft. Could have alternative Western engine at customer's option. Substantial internal fuel capacity. Provision for external fuel tanks.

ACCOMMODATION: Single seat (Martin-Baker zero/zero Mk 10 in any aircraft for Pakistan) under one-piece canopy. Two-seat training versions also planned.

AVIONICS: Still to be selected in mid-2002 (see Programme above). Prototype to have IAI Elta EL/M-2032 fire-control radar.

ARMAMENT: Underfuselage centreline station for 23 mm GSh-23-2 twin-barrel cannon or other store; two attachments under each wing and one at each wingtip. Weapons expected to include advanced AAMs, ASMs, bombs, gun and rocket pods, or other stores. Mockup seen in early 2002 mounting what was thought to be the AMRAAM-class LETRI (607 Institute) SD-10 BVR, active radar homing AAM (lock-on range 20 km; 12.4 miles).

DIMENSIONS, EXTERNAL:
Wing span, to c/l of AAMs 9.00 m (29 ft 6¼ in)
Length overall 14.00 m (45 ft 11¼ in)
Height overall 5.10 m (16 ft 8¾ in)
Wheel track 2.30 m (7 ft 6½ in)
Wheelbase 5.14 m (16 ft 10¼ in)

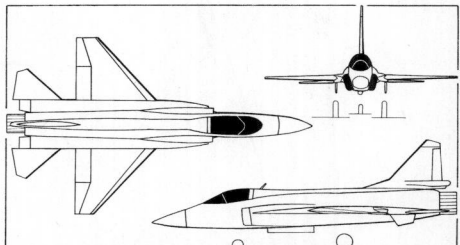

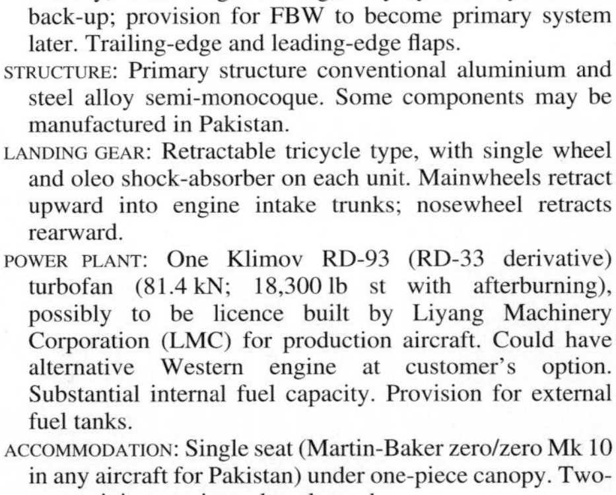
Provisional drawing of the CAC FC-1 multirole fighter ***(Jane's/Paul Jackson)*** 0062693

WEIGHTS AND LOADINGS:
Max external stores load 3,800 kg (8,378 lb)
T-O weight: normal (with wingtip AAMs only)
9,100 kg (20,062 lb)
max 12,700 kg (27,998 lb)
Max power loading 156 kg/kN (1.53 lb/lb st)
PERFORMANCE (estimated):
Max level speed at altitude, clean M1.6
Service ceiling 16,500 m (54,140 ft)
T-O run 500 m (1,640 ft)
Landing run 700 m (2,300 ft)
Combat radius:
fighter 648 n miles (1,200 km; 745 miles)
ground attack 378 n miles (700 km; 435 miles)
Max range on internal fuel
864 n miles (1,600 km; 994 miles)
Max ferry range 1,619 n miles (3,000 km; 1,864 miles)
g limit +8

UPDATED

Anonymously photographed J-10 apparently belonging to a PLA Air Force squadron *NEW*/0524668

CAC J-10

Chinese name: Jianjiji-10 (Fighter aircraft 10)
Westernised designation: F-10

TYPE: Multirole fighter.

PROGRAMME: Reports of the existence of this new Chinese fighter began to emerge in October 1994, following the detection by a US intelligence satellite of a prototype at Chengdu. Said to be in the weight and performance class of the Eurofighter Typhoon and Dassault Rafale, the J-10 bears a close external resemblance to the cancelled IAI Lavi (see 1988-89 *Jane's*), despite Israeli government statements that fears of unauthorised technology transfer are unfounded. A photograph, released in 1996 by the People's Liberation Army, of a wind tunnel model of the J-10 showed it to be outwardly identical to the Lavi in all essential respects, apart from slightly raised foreplanes and the addition of wingroot trailing-edge extensions.

This was confirmed in January 2001 when an unauthorised source posted a photograph of prototype 1001 on the Worldwide Web (although it was rapidly removed). The photograph showed the artist's impression first published in the 1997-98 *Jane's* to be accurate in all

respects, except that it was larger, and fin chord was some 20 per cent greater from root to tip.

It now appears that go-ahead for what became the J-10 was given in September 1988, development by No. 611 Research Institute beginning in the following month. An all-metal mockup was completed by late 1993, but poor performance predictions caused some redesign and consequent delays in the programme.

In late 1995, Russian sources indicated that the first flight, then expected during the early part of 1996, would be powered by a single AL-31FN turbofan; 1001 is now thought to have achieved this milestone in mid-1996. Russian avionics manufacturer Phazotron offered its Zhuk (Beetle) radar (already selected for the F-8 IIM upgrade programme) and the more capable Zhemchug (Pearls) and RP-35 as alternatives to the Elta EL/M-2035 or a derivative.

In late 1997, the slightly modified prototype 1002 was lost in a fatal accident. Two further prototypes (1003 and 1004) had been completed, and designation J-10 bestowed, during that year; the officially announced 'first' flight date of 23 March 1998 apparently referred to resumption of

flight testing by one of these aircraft. According to China-based sources, 10 AL-31FN engines had been imported by end 1997 and four J-10 prototypes completed. An unofficial CAC source in mid-1999 stated that two prototypes (evidently 1001 and 1003) were then flying, with four others undergoing static test, still being assembled or only just completed; of these, 1004 and/or 1005 may have been static/fatigue test aircraft. Chinese sources at Zhuhai in November 2000 stated that between five and eight J-10s then built and more than 140 test flights made; nine aircraft identified by mid-2002, of which 1007 to 1009 may be pre-production or initial production aircraft, one of which reportedly having made a 'first' flight on 28 June 2002. Service entry should be in about 2005 and some may be deployed in the two aircraft carriers that are due to be built by then.

CUSTOMERS: Reports have suggested a PLAAF requirement for up to 300, but achievement of this may depend upon China's progress with licensed manufacture of the Sukhoi Su-27 (see entry for Shenyang (SAC) J-11).

DESIGN FEATURES: Tail-less delta wing and close-coupled foreplanes; single sweptback vertical tail with outward-canted ventral fins; single ventral engine air intake.

FLYING CONTROLS: All-moving foreplanes; inboard and outboard elevons; single-piece rudder; wing leading-edge manoeuvring flaps.

LANDING GEAR: Retractable tricycle type. Main units retract forwards, nose unit rearwards.

POWER PLANT: One Saturn/Lyulka AL-31FN turbofan (125.5 kN; 28,219 lb st with afterburning). Fuel capacity reported to be in region of 3,175 litres (839 US gallons; 698.5 Imp gallons) in wings and 1,775 litres (469 US gallons; 390.5 Imp gallons), giving total internal capacity of 4,950 litres (1,308 US gallons; 1,089 Imp gallons). Provision for auxiliary fuel tanks on centreline and inboard underwing stations.

ACCOMMODATION: Pilot only, on zero/zero ejection seat.

AVIONICS: Reportedly equipped with quadruplex digital fly-by-wire FCS, HOTAS controls, a wide field of view HUD, three MFDs (weapons, RWR/navigation and radar information) and a helmet-mounted weapon sight.

ARMAMENT: Recently released photographs indicate 11 external stores points, including one on centreline, tandem pairs on fuselage sides and three under each wing, the outboard wing stations each carrying PL-8 (Python 3) or later AAMs. Other potential weapons could include Vympel R-73 and R-77 AAMs; C-801 or C-802 ASMs; and laser-guided or free-fall bombs. The J-10 is believed to carry an internally mounted 23 mm cannon and, probably, a Chinese-developed IR/laser navigation and targeting pod.

Following estimated data have not been confirmed.

Clandestinely photographed J-10 landing at a PLAAF base *NEW*/0524667

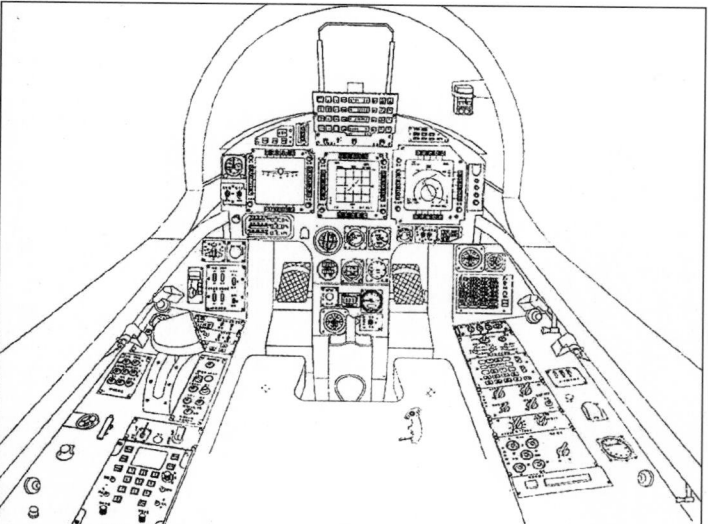

Provisional general arrangement of J-10 fighter *(Michael Badrocke)* *NEW*/0137950

Uncorroborated drawing depicting possible J-10 cockpit layout
NEW/0096948

DIMENSIONS, EXTERNAL:
Wing span	8.78 m (28 ft 9¾ in)
Length overall	14.57 m (47 ft 9½ in)
Height overall	4.80 m (15 ft 9 in)

AREAS:
Wings, gross	33.0 m² (355.2 sq ft)

WEIGHTS AND LOADINGS:
Weight empty	6,940 kg (15,300 lb)

Max external stores load	7,300 kg (16,095 lb)
Max T-O weight: clean	10,000 kg (22,046 lb)
with max external stores	19,227 kg (42,388 lb)

PERFORMANCE:
Max level speed: at altitude	M1.85
at S/L	M1.2
Service ceiling	18,000 m (59,050 ft)
T-O run	350 m (1,150 ft)

Combat radius	250-300 n miles (463-555 km; 287-345 miles)
Range	1,000 n miles (1,850 km; 1,150 miles)
g limit	+9

UPDATED

CHAIG

CHANGHE AIRCRAFT INDUSTRY GROUP
(Changhe Feiji Gongye Gongsi)
(Subsidiary of AVIC II)
PO Box 109, Jingdezhen, Jiangxi 333002
Tel: (+86 798) 844 20 19 and 844 24 30
Fax: (+86 798) 844 71 69 and 844 14 60
Web: http://www.changhe.com
PRESIDENT: Yang Jinhuai

CHAIG (formerly Changhe Aircraft Factory), which occupies a 433 ha (1,070 acre) site at Jingdezhen, has a workforce of more than 10,000 and began producing coaches and commercial road vehicles in 1974. These and other automotive products still account for much of output, but batch-produced helicopters have included the Z-8 and Z-11. Reports from the November 2000 Airshow China indicated that CHAIG was in talks with Western manufacturers concerning a possible re-engining programme for the Z-8, limited production and/or upgrading of which may be continuing.

CHAIG is responsible for manufacture of the tailcone, vertical fin and horizontal stabiliser of the Sikorsky S-92 helicopter (see US section). The tail for the first S-92 was delivered to Sikorsky in May 1997.

UPDATED

CHAIG Z-8
Chinese name: Zhishengji-8 (Vertical take-off aircraft 8)
TYPE: Naval combat helicopter.
PROGRAMME: Design work begun 1976, but suspended from 1979 to mid-1984; initial flight of first prototype 11 December 1985, second prototype October 1987; domestic type approval awarded 8 April 1989; first Z-8 handed over to PLA Naval Air Force for service trials 5 August 1989; initial production approved; final design approval granted 12 November 1994. Applications include troop transport, ASW/ASV, search and rescue, minelaying/sweeping, aerial survey and firefighting.
CURRENT VERSIONS: **Z-8:** Standard version, *as described.*
Z-8A: Reported designation of upgraded version, with Turbomeca Makila 2A engines matched to Turmo gearbox. Pratt & Whitney Canada also a candidate for any re-engining programme. Two Z-8As reportedly delivered to PLA Army Aviation for evaluation in 2001; camouflaged; lack nose radome and side-mounted floats.
CUSTOMERS: Up to 20 delivered to PLA Navy by end of 1999. No firm evidence of production status since then, but type still being promoted by CATIC in 2002. Sole operating unit is Shipborne Helicopter Group within East Sea Fleet at Dachang (Shanghai).
DESIGN FEATURES: Chinese equivalent of Aerospatiale Super Frelon (see 1982-83 *Jane's All the World's Aircraft*), of which 16 supplied to PLA Navy in 1977-78. Six-blade main rotor and five-blade tail rotor; boat-hull fuselage with watertight compartments inside planing bottom; stabilising float at rear each side, attached to small stubwing; small, strut-braced fixed horizontal stabiliser on starboard side of tail rotor pylon. Search radar in nose 'thimble' on SAR version.
FLYING CONTROLS: Pitch control fitting at root of each main rotor blade; drag and flapping hinges for each blade mounted on rotor head starplates; each main blade also has

Z-8 multirole civil and military utility helicopter   0121073

a hydraulic drag damper. Fully redundant flight control system, with Dong Fang KJ-8 autopilot.
STRUCTURE: Stressed skin metal fuselage, with riveted watertight compartments; gearboxes manufactured by Zhongnan Transmission Machinery Factory.
LANDING GEAR: Non-retractable tricycle type, with twin wheels and low-pressure oleo-pneumatic shock-absorber on each unit. Small tripod tailskid under rear of tailboom. Boat hull and side floats permit emergency water landings and take-offs.
POWER PLANT: Three Changzhou (CLXMW) WZ6 turboshafts, each with maximum emergency rating of 1,156 kW (1,550 shp) and 20 per cent power reserve at S/L, ISA. Two engines side by side in front of main rotor shaft and one aft of shaft. Transmission rated at 3,072 kW (4,120 shp).
Standard internal fuel capacity 3,900 litres (1,030 US gallons; 858 Imp gallons), in flexible tanks under floor of centre-fuselage. Auxiliary fuel tanks can be carried inside cabin for extended range or self-ferry missions, increasing total capacity to 5,800 litres (1,532 US gallons; 1,276 Imp gallons).
ACCOMMODATION: Crew of two or three on flight deck. Accommodation in main cabin for up to 27 fully armed troops, or 39 without equipment; up to 15 stretchers and a medical attendant in ambulance configuration; a BJ-212 Jeep-type vehicle and its crew; or other configurations according to mission. Entire accommodation heated, ventilated, soundproofed and vibration-proofed. Forward-opening crew door on each side of flight deck. Rearward-sliding door at front of cabin on starboard side. Hydraulically actuated rear-loading ramp/door.
EQUIPMENT: Equipment for SAR role can include 275 kg (606 lb) capacity hydraulic rescue hoist and two five-person liferafts. Can also be equipped with sonar, sonobuoys, search radar, or equipment for oceanography, geological survey and forest firefighting.
ARMAMENT: Can be equipped with Yu-7 torpedoes, anti-shipping missiles, or gear for minelaying (eight 250 kg mines) or minesweeping.
DIMENSIONS, EXTERNAL:
Main rotor diameter	18.90 m (62 ft 0 in)
Tail rotor diameter	4.00 m (13 ft 1½ in)

Length overall, rotors turning	23.035 m (75 ft 7 in)
Height overall, rotors turning	6.66 m (21 ft 10¼ in)
Width over main gear sponsons	5.20 m (17 ft 0¾ in)

AREAS:
Main rotor blades (each)	5.10 m² (54.90 sq ft)
Tail rotor blades (each)	0.56 m² (6.03 sq ft)
Main rotor disc	280.48 m² (3,019.1 sq ft)
Tail rotor disc	12.57 m² (135.3 sq ft)

WEIGHTS AND LOADINGS:
Manufacturer's weight empty	6,980 kg (15,388 lb)
Weight empty, equipped	7,550 kg (16,645 lb)
Max cargo payload: internal	4,000 kg (8,818 lb)
on external sling	5,000 kg (11,023 lb)
Max T-O weight: standard fuel	10,592 kg (23,351 lb)
with auxiliary fuel	13,000 kg (28,660 lb)
Max disc loading:	
standard fuel	37.8 kg/m² (7.73 lb/sq ft)
auxiliary fuel	46.35 kg/m² (9.49 lb/sq ft)
Transmission loading at max T-O weight and power:	
standard fuel	3.45 kg/kW (5.67 lb/shp)
auxiliary fuel	4.23 kg/kW (6.96 lb/shp)

PERFORMANCE (A: at T-O weight of 9,000 kg; 19,841 lb, B: at 11,000 kg; 24,251 lb, C: at 13,000 kg; 28,660 lb):
Never-exceed speed (VNE):	
A	170 kt (315 km/h; 195 mph)
B	159 kt (296 km/h; 183 mph)
C	148 kt (275 km/h; 170 mph)
Max cruising speed: A	143 kt (266 km/h; 165 mph)
B	140 kt (260 km/h; 161 mph)
C	134 kt (248 km/h; 154 mph)
Econ cruising speed: A	137 kt (255 km/h; 158 mph)
B	132 kt (246 km/h; 153 mph)
C	125 kt (232 km/h; 144 mph)
Rate of climb at S/L (15° 30′ collective pitch, OEI):	
A	690 m (2,263 ft)/min
B	552 m (1,811 ft)/min
C	396 m (1,299 ft)/min
Service ceiling: A	6,000 m (19,680 ft)
B	4,900 m (16,080 ft)
C	3,050 m (10,000 ft)
Hovering ceiling IGE: A	5,500 m (18,040 ft)
B	3,600 m (11,800 ft)
C	1,900 m (6,240 ft)
Hovering ceiling OGE: A	4,400 m (14,440 ft)
B	2,300 m (7,540 ft)
Range with max standard or auxiliary fuel, OEI, no reserves:	
A	232 n miles (430 km; 267 miles)
B	442 n miles (820 km; 509 miles)
C	431 n miles (800 km; 497 miles)
Ferry range with auxiliary fuel tanks, OEI, no reserves:	
C	755 n miles (1,400 km; 870 miles)
Endurance with max standard or auxiliary fuel, OEI, no reserves:	
A	2 h 31 min
B	4 h 43 min
C	4 h 10 min

UPDATED

CHAIG Z-11
Chinese name: Zhishengji-11 (Vertical take-off aircraft 11)
TYPE: Light utility helicopter.
PROGRAMME: First details officially released at China Air Show in Zhuhai November 1996, together with

Changhe Z-8 in service with the PLA Naval Air Force   NEW/0525833

Eight Changhe Z-11s apparently used for training by a PLA Aviation flying school   0116934

photographs showing one or two Z-11s in flight. Development by Chinese Helicopter R&D Institute began in 1991; quoted first flight date of 22 December 1994 thought to refer to re-engined modification of two second-hand ex-US AStars (Ecureuils) acquired earlier that year. In early 1997, a Chinese government agency announced that the Z-11 had flown for the first time on 26 December 1996.

The Z-11 appears identical externally to Eurocopter AS 350B Ecureuil (see International section) except for nose contours, but CHAIG publicity in mid-2001 stated that ''China owns its independent intellectual property rights''. Eurocopter (which sold eight Ecureuils to China in 1996) declines to comment on its provenance. Project reportedly launched in 1989, with development work beginning in 1992. Technical appraisal was completed in

1996, and small batch production began in 1997, first customer deliveries being made in September 1998. Design was finalised in December 2000, and the Z-11 received CAAC certification in April 2001.

A twin-engined version is planned, for which the Rolls-Royce 250 and Turbomeca Arrius 1A were in contention in early 2002. Latter also offering its Arriel 2B1 for higher-powered single-engined version.

CURRENT VERSIONS: Suitable for pilot training, police/security patrol, reconnaissance, coastguard, geological survey, rescue and forestry protection.

CUSTOMERS: Eight said to have been completed by late 1996, of which four had been delivered. All photographs released before late 1998 showed Z-11 in military camouflage with PLA insignia. One report from Aviation Expo China in October 1997 stated that the PLA had ordered 20 military

Z-11s, these apparently going to Army Aviation's training school. Production amounted to 'several dozen' by mid-2001, at which time some 10,000 hours flown and 50,000 take-offs and landings. Chongqing Three Gorges General Aviation Airlines (one ordered) reported as first civil customer in mid-2001; China Central Television of Jiangxi Province purchasing one for delivery in late 2002.

POWER PLANT: One 510 kW (684 shp) WZ8D turboshaft (licence-built Turbomeca Arriel 1D), reportedly produced by Liming engine factory. Higher-powered Arriel 2C being installed in one Z-11 for flight trials in second half of 2002.

DIMENSIONS, EXTERNAL:

Main rotor diameter	10.69 m (35 ft 0¾ in)
Tail rotor diameter	1.86 m (6 ft 1¼ in)
Length overall, main rotor turning	13.01 m (42 ft 8¼ in)
Fuselage: Length	11.24 m (36 ft 10½ in)
Max width	1.80 m (5 ft 10¾ in)
Height: over tailfin	3.02 m (9 ft 11 in)
to top of rotor hub	3.14 m (10 ft 3½ in)
Tailplane span	2.53 m (8 ft 3½ in)
Skid track	2.09 m (6 ft 10¼ in)
Fuselage/ground clearance beneath cabin	
	0.39 m (1 ft 3¼ in)

WEIGHTS AND LOADINGS:

Weight empty	1,120 kg (2,469 lb)
Max T-O weight	2,200 kg (4,850 lb)

PERFORMANCE:

Max level speed	150 kt (278 km/h; 172 mph)
Cruising speed	130 kt (240 km/h; 149 mph)
Hovering ceiling IGE	5,240 m (17,200 ft)
Max range	322 n miles (598 km; 371 miles)
Endurance	3 h 54 min

UPDATED

CHRDI

CHINESE HELICOPTER RESEARCH AND DEVELOPMENT INSTITUTE (Zhongguo Zhishengji Sheji Yangjiuso) (Subsidiary of AVIC II)

Jingdezhen, Jiangxi

This institute (workforce approximately 2,000) is the overall design authority for Chinese indigenous helicopter programmes. Few details yet available of Z-10; other projects reported to include 8 to 10 tonne heavy-lift helicopter.

UPDATED

CHRDI Z-10

Chinese name: Zhishengji-10 (Vertical take-off aircraft 10)

TYPE: Multirole helicopter.

PROGRAMME: Thought to have been initiated in about 1994; Nos. 602 and 608 Institutes reported to be taking part in airframe design. Eurocopter France became partner in programme 15 May 1997 with US$70 million to US$80 million, nine-year contract to assist in developing rotor system; joined 22 March 1999 by Agusta with contract (approximately Lit50 billion; US$30 million) to be responsible for transmission system and vibration analysis. Nomenclature of Chinese Medium Helicopter (CMH) introduced by Agusta in July 2000; officially linked with Z-10 designation by Chinese sources by mid-2002. Reports suggest that three transmission sets will be built in Italy and four, collaboratively, in China.

In size and weight class of Bell 412 and Sikorsky S-76 (MTOW variously reported as 5, 5.5 or 6 tonnes; 11,023 to 13,227 lb), with capacity for up to 14 passengers or troops. First flight thought to be targeted for 2006. Eventual production reportedly to be entrusted to Changhe.

DESIGN FEATURES: Intended to fulfil tactical military transport, attack helicopter and commercial transport roles. No design features yet officially confirmed, but intention of both military and civil use suggests that such aspects as composites construction, crashworthy airframe and seats, rotor system ballistic tolerance, engine run-dry capacity and ability to carry slung loads are likely, depending upon version.

POWER PLANT: Reported requirement is for two 969 kW (1,300 shp) class turboshafts. Talks under way in late 2000 with Turbomeca (Makila) and Pratt & Whitney Canada; latter's PT6C-67C selected (10 ordered) in mid-2001 to power prototype, and first set of these delivered by end of that year.

UPDATED

GAIC

GUIZHOU AVIATION INDUSTRY GROUP (Guizhou Hangkong Gongye Gongsi) (Subsidiary of AVIC I)

110 Jinjiang Road, Xiaohe District, Guiyang, Guizhou 550009
Tel: (+86 851) 380 80 94
Fax: (+86 851) 380 80 84
e-mail: gaic@public2.gy.gz.cn
Web: http://www.gaic.com.cn
CHAIRMAN: Zhou Wancheng
PRESIDENT: Zhang Jun

The Guizhou Aviation Industry Group incorporates many enterprises, factories and institutes engaged in various aerospace and non-aerospace activities; aerospace workforce is about 6,000. Main aircraft manufacturing plants are named Longyan, Shuangyang and Yunma; Liyang is aero-engine producer. Aviation programmes include JJ-7/FT-7 fighter trainer, two series of turbojets, air-to-air missiles and rocket launchers, plus participation in Chengdu (CAC, which see) production of single-seat J-7/F-7. GAIC is partner with EV-AT of Czech Republic in Guizhou Evektor Aircraft Corporation (see under GEAC in this section); also manufactures maintenance jigs and tools for the Airbus airliner family.

UPDATED

GAIC JJ-7

Chinese name: Jianjiji Jiaolianji-7 (Fighter training aircraft 7)
Westernised designation: FT-7

TYPE: Advanced jet trainer.

PROGRAMME: Launched October 1982; first metal cut April 1985; first flight 5 July 1985; series production began February 1986; production FT-7 first flight December 1987. Improved FT-7P made first flight 9 November 1990 and received MAS production approval 13 May 1992; in production 1996.

Unmarked GAIC FT-7P or JJ-7A, stretched version of the JJ-7/FT-7   *NEW*/0137977

CURRENT VERSIONS: **JJ-7:** Basic domestic version, based on single-seat J-7 II and MiG-21US.

JJ-7A: PLAAF equivalent of FT-7P.

FT-7: Export version of JJ-7. Two **FT-7Bs** supplied to Zimbabwe in 1991 are known locally as the FT-7BZ.

Detailed description applies to FT-7 except where indicated.

FT-7P: Stretched version of FT-7 for Pakistan: lengthened by insertion of 0.60 m (1 ft 11½ in) fuselage plug behind rear cockpit, allowing increased fuel load (for 25 per cent improvement in operational range); additional avionics and one internal 30 mm cannon; AoA sensor on port side of nose. Imported fire-control system; HUD and air data computer; improved instrument layout; two underwing pylons each side.

FT-7PG: Trainer equivalent of F-7PG for Pakistan Air Force. First flight March 2002.

CUSTOMERS: PLA Air Force (JJ-7). Export deliveries to Bangladesh Air Force (seven FT-7B); Myanmar AF (six

FT-7); Pakistan AF (15 FT-7P and six FT-7PG); Sri Lanka AF (one FT-7); Zimbabwe AF (two FT-7B). One more FT-7B was reportedly due for delivery to Bangladesh in 2000.

DESIGN FEATURES: Generally as J-7/F-7 except for twin canopies opening sideways to starboard (rear one with retractable periscope), twin ventral strakes of modified shape, and removable saddleback fuel tank aft of second cockpit. Can provide full training syllabus for all J-7/F-7 versions, plus much of that necessary for Shenyang J-8.

FLYING CONTROLS: As J-7/F-7.

STRUCTURE: As J-7/F-7.

LANDING GEAR: As J-7/F-7.

POWER PLANT: As F-7M. Internal fuel tank arrangement as for F-7M, but fuselage and wing total capacities are 1,880 and 560 litres respectively (496.5 and 148 US gallons; 413.5 and 123 Imp gallons), giving total capacity of 2,440 litres (644.5 US gallons; 536.5 Imp gallons). Internal capacity

Production line-up of standard JJ-7s   0106475

increased to 2,800 litres (740 US gallons; 616 Imp gallons) in FT-7P. External drop tanks as for F-7M.

ACCOMMODATION: Second cockpit in tandem, with dual controls. Canopies open sideways to starboard.

SYSTEMS: Cockpits pressurised and air conditioned (maximum differential 0.30 bar; 4.4 lb/sq in). Hydraulic system pressure 207 bar (3,000 lb/sq in), flow rates 36 litres (9.51 US gallons; 7.92 Imp gallons)/min in main system, 4 litres (1.06 US gallons; 0.88 Imp gallon)/min in standby system. Pneumatic system pressure 49 bar (711 lb/ sq in). Electrical power provided by QF-12C 12 kW engine-driven starter/generator; 400 A static inverters for 28.5 V DC power. YX-1 oxygen system for crew. De-icing/anti-icing standard.

AVIONICS: *Comms:* Include CT-3M VHF com transceiver. Other (unidentified) Chinese nav/com avionics designated Type 222, Type 262, JT-2A and J7L.

Flight: WL-7 radio compass and XS-6A marker beacon receiver. FJ-1 flight data recorder.

ARMAMENT: Single underwing pylon each side (two on FT-7P) for such stores as a PL-2B air-to-air missile, an HF-5A 18-round launcher for 57 mm rockets, or a bomb of up to 250 kg size; can also be fitted with Type 23-3 twin-barrel 23 mm gun in underbelly pack and HK-03E optical gunsight. Gun mounted internally in FT-7P.

DIMENSIONS, EXTERNAL: As F-7M except:

Length overall, incl probe: FT-7	14.87 m (48 ft 9½ in)
FT-7P	15.47 m (50 ft 9 in)

WEIGHTS AND LOADINGS:

Weight empty, equipped: FT-7	5,519 kg (12,167 lb)
FT-7P	5,330 kg (11,750 lb)
Max fuel weight: internal: FT-7	1,891 kg (4,169 lb)
FT-7P	approx 2,170 kg (4,784 lb)
external	558 kg (1,230 lb)
Max external stores load: FT-7	1,187 kg (2,617 lb)
Normal T-O weight with two PL-2 air-to-air missiles:	
FT-7P	7,590 kg (16,733 lb)

Max T-O weight: FT-7	8,555 kg (18,860 lb)
FT-7P	9,550 kg (21,054 lb)
Max landing weight: FT-7	6,096 kg (13,439 lb)
Max zero-fuel weight: FT-7	7,300 kg (16,094 lb)
Max wing loading: FT-7	372.0 kg/m² (76.18 lb/sq ft)
FT-7P	415.2 kg/m² (85.04 lb/sq ft)
Max power loading: FT-7	143 kg/kN (1.40 lb/lb st)
FT-7P	160 kg/kN (1.56 lb/lb st)

PERFORMANCE (FT-7):

Never-exceed speed (VNE) above 12,500 m (41,000 ft)	
	M2.35 (1,346 kt; 2,495 km/h; 1,550 mph)
Max level speed above 12,500 m (41,000 ft)	
	M2.05 (1,175 kt; 2,175 km/h; 1,350 mph)
Max cruising speed at 11,000 m (36,080 ft)	
	545 kt (1,010 km/h; 627 mph)
Econ cruising speed at 11,000 m (36,080 ft)	
	516 kt (956 km/h; 594 mph)
Unstick speed 170-181 kt (315-335 km/h; 196-208 mph)	
Touchdown speed	
	165-175 kt (305-325 km/h; 189-202 mph)

Stalling speed, flaps down	135 kt (250 km/h; 156 mph)
Max rate of climb at S/L	9,300 m (30,510 ft)/min
Service ceiling	17,300 m (56,760 ft)
Absolute ceiling	17,700 m (58,080 ft)
T-O run	827 m (2,715 ft)
T-O to 15 m (50 ft)	931 m (3,055 ft)
Landing from 15 m (50 ft)	1,368 m (4,490 ft)
Landing run	1,060 m (3,480 ft)
Range at 11,000 m (36,080 ft):	
max internal fuel (clean)	
	545 n miles (1,010 km; 627 miles)
max internal and external fuel	
	788 n miles (1,459 km; 906 miles)
max external stores 708 n miles (1,313 km; 816 miles)	
g limit with two PL-2B missiles	+7

UPDATED

GAIC FTC-2000

TYPE: Advanced jet trainer.

PROGRAMME: Revealed, in model form, at Airshow China, November 2000. Proposed as lead-in fighter trainer for CAC FC-1 (which see) and J-8 IV/F-8D. Wind tunnel testing completed by late 2000; no further reports since then, but prototype could fly by 2003.

COSTS: Predicted at US$2.4 million, excluding radar.

DESIGN FEATURES: Upgraded derivative of JJ-7/FT-7, having J-7E wings, lateral intakes and 'solid' nose to permit installation of fire-control radar. Airframe otherwise as JJ-7/FT-7.

POWER PLANT: One 59.8 kN (13,448 lb st) LMC WP7B (BM) or 64.7 kN (14,550 lb st) WP13B turbojet. In-flight refuelling probe on nose, common with that of J-8 IV/F-8D. Underfuselage and inboard underwing stations 'wet' for carriage of drop-tanks.

ARMAMENT: Stores attachments under fuselage (one) and wings (two each side).

WEIGHTS AND LOADINGS:

Normal max T-O weight	7,800 kg (17,196 lb)

PERFORMANCE (estimated):

Max level speed	M1.8
Unstick speed	135 kt (250 km/h; 155 mph)
Touchdown speed	124 kt (230 km/h; 143 mph)
Min flying speed	114 kt (210 km/h; 131 mph)
T-O and landing distance	430 m (1,410 ft)
Operational endurance	2 h

UPDATED

Model of the GAIC FTC-2000 (*Photo Link*)   0100783

GEAC

GUIZHOU EVEKTOR AIRCRAFT CORPORATION (Guizhou Evektor Feiji Gongsi)

Guizhou

This company was created in June 2000, as a joint venture between GAIC of China and EV-AT of the Czech Republic

(which see), to co-manufacture, market and support the lightplanes and ultralights of the latter concern. Launch activity is the assembly of the EV-97 Eurostar from CKD kits, the first of which was imported in August 2000 and made its maiden flight on 24 October that year. Co-production of the Wolfsberg-Letov Raven 257 is also planned.

UPDATED

HAI

HAFEI AVIATION INDUSTRY COMPANY (Hafei Hangkong Gongye Gongsi) (Subsidiary of AVIC II)

15 Youxie Street, Pingfang District, Harbin, Heilongjiang 150066

Tel: (+86 451) 650 11 22 and 653 07 36
Fax: (+86 451) 652 01 44
e-mail: hai@hafei.com
Web: http://www.hafei.com
PRESIDENT: Cui Xuewen
GENERAL MANAGER: Wang Bin
EXECUTIVE DEPUTY GENERAL MANAGER: Xu Zhanbin
PUBLIC RELATIONS OFFICER: Li Changjun

Established as Harbin Aircraft Manufacturing Corporation (Harbin Feiji Zhizao Gongsi: HAMC) 1952, subsequently producing H-5 light bomber (Soviet-designed Il-28) and Z-5

helicopter (Soviet-designed Mi-4) in large numbers, as well as smaller quantities of Chinese-designed SH-5 flying-boat and Y-11 agricultural light twin (see earlier editions of *Jane's* for details). Now core company of Harbin Aircraft Industry Group (HAIG, which see) under AVIC II. Occupies 514 ha (1,270 acre) site, including 350,000 m² (3,767,400 sq ft) of workshop space. Workforce in 1998 (latest figure provided) numbered approximately 18,000.

Currently producing own-design Y-12 utility light twin and licence-manufacturing Eurocopter Dauphin 2 as Z-9; also developing new 'Z-X' helicopter. Subcontract work includes Dauphin doors for Eurocopter.

HAI is partnered by Eurocopter and Singapore Technologies Aerospace in the Eurocopter EC 120 Colibri programme, as described in the International section; more than 200 EC-120 fuselages delivered to Eurocopter by late 2000; production rate 10 fuselages per month. In mid-2001, it became a participant in the Alliance StarLiner 100-35, as described in the HAIG (Harbin) entry in this section.

UPDATED

HAI (EUROCOPTER) Z-9 HAITUN

Chinese name: Zhishengji-9 (Vertical take-off aircraft 9)
English name: Dolphin

TYPE: Light utility helicopter.

PROGRAMME: Licence-built Eurocopter AS 365N Dauphin 2 (which see in International section). Licence agreement (Aerospatiale/CATIC) signed 2 July 1980; first (French-built) example made initial acceptance flight in China 6 February 1982; Chinese parts manufacture began 1986; initial agreed batch of 50, last of which delivered January 1992. Production continuing under May 1988 domestic contract. Plans to introduce Arriel 2C announced mid-2001, following delivery of two of these engines in April; first flight achieved in September 2001.

CURRENT VERSIONS: **Z-9:** Initial Chinese version, equivalent to French AS 365 N1 and assembled from French-built kits; 28 completed. Details in 1994-95 and earlier *Jane's*.

Z-9A: Follow-on kit-built version, to AS 365N2 standard. Final 22 of first 50 were of this version.

Red and white Z-9A used for Arctic scientific survey   0106478

Torpedo-armed Z-9C of PLA Naval Aviation   0106480

Chinese Army Aviation WZ-9s armed with rocket pods and roof-mounted sight   0116935

Z-9A-100: Effectively, prototypes for domestic licence-built version, with WZ8A engines and much increased local manufacture (72.2 per cent of airframe and 91 per cent of engine). Two built; first flight 16 January 1992; flight test programme completed 20 November 1992 after almost 200 flight hours (408 flights); Chinese type approval received 30 December 1992.

Z-9B: First indigenous production version, based on Z-9A-100. Modified Fenestron with 11 wider-chord, all-composites blades instead of 13 metal blades as in AS 365N1. Principal unarmed PLA version for SAR, artillery direction, EW, troop transport (accommodates eight), communications and other utility duties. Certified also for domestic commercial operations 19 April 2001.

Data apply to Z-9A except where indicated.

Z-9C: Version for PLA Naval Air Force, for deployment aboard certain classes of destroyers and frigates; in service by late 2000. Believed to be equivalent to Arriel 2-engined Eurocopter AS 565 Panther, but equipped with Thales HS-12 dipping sonar and possibly KLC-1 Chinese version of Agrion 15 surface search radar; armament includes two Yu-7 torpedoes or TV-guided C-701 anti-surface vessel missiles.

WZ-9: Armed (*wuzhuang*) version (also reported as Z-9W); eight Norinco HJ-8 (*Hongjian:* Red Arrow) wire-guided anti-tank missiles, twin 12.7 mm machine gun or 23 mm cannon pods, or twin 57 or 90 mm rocket pods, and gyrostabilised, roof-mounted optical sight. First flight thought to have been in early 1989; in production. Export designation **Z-9G**, available with or without roof-mounted sight.

H410A: Version with 635 kW (851 shp) Arriel 2C (WZ8C) turboshafts for improved 'hot-and-high' performance. First flight September 2001; CAAC certification 10 July 2002. Initial orders for eight.

H425: Projected next stage of H410A development: improvements to rotor system, fuel system, structure (crashworthiness), avionics and interior layout.

H450: Projected further development of H425.

CUSTOMERS: Total 110 (all versions) reportedly built by mid-2001; eight H410As ordered mid-2002 by Far Eastern Leasing Company, China State Oceanic Administration and Zhoushan Civil Aviation Development Company.

Most early production for Chinese armed services (People's Navy Aviation and Army Aviation each at least

25) and People's Armed Police (four delivered 12 September 2001). Entered service with two PLA army groups January and February 1988 (Beijing and Shenyang Military Regions respectively). People's Navy Aviation believed to use Z-9A for commando transport as well as shipboard communications. Ten based at former Royal Air Force airfield at Sek Kong in third quarter of 1997, following 1 July handover of Hong Kong by UK. First export made in late 2000 (two to Mali Air Force).

Civil models used for various duties including offshore oil rig support and air ambulance (four stretchers/two seats or two stretchers/five seats). In 1992, Flying Dragon Aviation received two (late production Z-9s, augmenting an Aerospatiale Dauphin) which are operated on behalf of the Ministry of Forestry. Five civil Z-9s ordered by Shenzhen Financial Leasing Company (SFLC) in 2001. One Z-9A deployed to Arctic regions in mid-1999, on scientific survey ship.

STRUCTURE: Transmission manufactured by Dongan Engine Manufacturing Company at Harbin, hubs and tail rotor blades by Baoding Propeller Factory.

POWER PLANT: Two 547 kW (734 shp) Turbomeca Arriel 1C1 turboshafts, produced by SAEC at Zhuzhou as WZ8A. From 2001, H410A powered by Arriel 2Cs as WZ8C. Fuel capacity 1,140 litres (301 US gallons; 251 Imp gallons).

Option for 180 litre (47.5 US gallon; 39.6 Imp gallon) auxiliary tank.

ARMAMENT (WZ-9): Up to eight HJ-8 or HJ-8E ATMs (range 1.6 n miles; 3 km; 1.9 miles); twin 12.7 mm or 23 mm gun pods; or two pods of 57 or 90 mm rockets. Possible other weapons include TY-90 IR-guided AAM (already test-flown on Z-9 in 1998; range of 3.2 n miles; 6 km; 3.7 miles) and C-701 TV-guided anti-ship missile (8.1 n miles; 15 km; 9.3 miles).

WEIGHTS AND LOADINGS:
Weight empty, equipped	2,050 kg (4,519 lb)
Max payload	2,038 kg (4,493 lb)
Max load on cargo sling	1,600 kg (3,527 lb)
Max T-O weight, internal or external load	4,100 kg (9,039 lb)

PERFORMANCE:
Max level speed	170 kt (315 km/h; 195 mph)
Max cruising speed at S/L	151 kt (280 km/h; 174 mph)

Max vertical rate of climb at S/L:
Z-9A	252 m (827 ft)/min
WZ-9	210 m (689 ft)/min

Max forward rate of climb at S/L:
Z-9A	396 m (1,299 ft)/min
WZ-9	468 m (1,535 ft)/min
Service ceiling: Z-9A	6,000 m (19,680 ft)
WZ-9	4,940 m (16,200 ft)
Hovering ceiling: IGE	2,150 m (7,050 ft)
OGE	1,150 m (3,770 ft)

Max range at 135 kt (250 km/h; 155 mph) normal cruising speed, no reserves:
standard tanks:
Z-9A	464 n miles (860 km; 534 miles)
WZ-9	358 n miles (664 km; 412 miles)

with auxiliary tank:
Z-9A	539 n miles (1,000 km; 621 miles)

UPDATED

HAI 'Z-X'

TYPE: Light helicopter.

PROGRAMME: Hafei is understood to be developing a new light helicopter. No other details could be confirmed at the time of closing for press.

UPDATED

HAI Y-12

Chinese name: Yunshuji-12 (Transport aircraft 12)
Export marketing name: Twin Panda

TYPE: Twin-turboprop light transport.

PROGRAMME: Initiated as uprated version of Y-12 (I) (1987-88 and earlier *Jane's*); first flight 16 August 1984. Y-12 (IV) (B-569L, first flight 30 August 1993) received production approval September 1995.

CURRENT VERSIONS: **Y-12 (I):** Initial version (first flight 14 July 1982), with PT6A-11 engines; three prototypes and approximately 30 production examples built; described in 1987-88 *Jane's*.

First known illustration of H410A version of Z-9   *NEW*/0524646

Y-12 flight deck   0085019

Y-12 (II): Current production version; higher rated engines, no leading-edge slats and smaller ventral fin. Certified by CAAC 25 December 1985 and UK CAA (BCAR Section K) 20 June 1990.

Detailed description applies to Y-12 (II) except where indicated.

Y-12 (IV): Improved version with sweptback wingtips; modifications to control surface actuation, main gear and brakes; redesigned seating for 18 to 19 passengers; starboard side rear baggage door; maximum payload and maximum T-O weight increased. Further changes include Rockwell Collins or Honeywell com/nav for both VFR and IFR, plus optional colour weather radar, GPS, Omega navigation, wing and tail de-icing, and oxygen system. Domestic certification received 3 July 1994 and FAR Pt 23 approval on 26 March 1995; Indonesian certification 16 November 2000. Sichuan Airlines order for 20 (7 November 2000) assumed to be of this version.

Twin Panda: Version of Y-12 (IV) intended to be completed and marketed by Canadian Aerospace Group and its Panda Aircraft Company subsidiary, primarily as a replacement for the DHC-6 Twin Otter. More powerful PT6A-34 engines, strengthened landing gear, upgraded avionics. Harbin/CAG agreement was signed on 31 March 1998, covering a 10-year period and potentially 200 Y-12s (initial 50, plus 150 options) for the North American market. However, the venture was abandoned in April 1999.

Y-12E: Next-generation version for Chinese market, with 559 kW (750 shp) PT6A-135A engines (derated to 462 kW; 620 shp) with Hartzell lightweight propellers, retractable landing gear, strengthened structure, reduced noise levels and improved avionics. Announced at Airshow China, November 2000; passed CAAC preliminary design review 16 November 2000. First flight August 2001; award of CAAC type certificate announced 26 February 2002; launch customer Sichuan Airlines (option for 10).

Future versions: Plans under consideration include stretched version and one with pressurised cabin.

CUSTOMERS: Over 110 built by Harbin by mid-2001: see table. Venezuela considering purchase of 10 in late 2001.

COSTS: Batch of 15 PT6-engined aircraft valued at US$58 million (1999).

DESIGN FEATURES: Designed to standards of FAR Pts 23 and 135 Appendix A and developed to improve upon modest payload/range of piston-engined Y-11. Constant-chord high braced wings, with small stub-wings at cabin floor level supporting mainwheel units; basically rectangular-section fuselage, upswept at rear; non-swept tail surfaces; large dorsal fin; ventral fin under tailcone.

Wing section LS(1)-0417; thickness/chord ratio 17 per cent; dihedral 1° 41'; incidence 4°.

FLYING CONTROLS: Conventional and manual. Drooping ailerons, horn-balanced elevators and rudder; trim tab in starboard aileron, rudder and each elevator; electrically actuated two-segment double-slotted flaps on each wing trailing-edge.

STRUCTURE: Conventional all-metal structure with two-spar fail-safe wings and stressed skin fuselage; Ziqiang-2 resin bonding on 70 per cent of wing structure and 40 per cent of fuselage; integral fuel tankage in wing spar box; bracing strut from each stub-wing out to approximately one-third span.

LANDING GEAR: Non-retractable tricycle type, with oleo-pneumatic shock-absorber in each unit. Single-wheel main units, attached to underside of stub-wings. Single non-steerable nosewheel. Mainwheel tyres size 640×230, pressure 5.50 bar (80 lb/sq in); nosewheel tyre size 480×200, pressure 3.50 bar (51 lb/sq in). Hydraulic brakes. Minimum ground turning radius 16.75 m (54 ft 11½ in).

POWER PLANT: Two Pratt & Whitney Canada PT6A-27 turboprops, each derated to 462 kW (620 shp) and driving a Hartzell HC-B3TN-3B/T10173B-3 three-blade constant-speed reversible-pitch propeller. All fuel in tanks in wing spar box, total capacity 1,616 litres (427 US gallons; 355.5 Imp gallons), with overwing gravity filling point each side. No recent news of domestic WJ9 turboprop (442 to 520 kW; 593 to 697 shp), said to have been under development for Y-12 and other aircraft.

ACCOMMODATION: Crew of two on flight deck, with access via forward-opening door on port side. Four-way adjustable crew seats. Dual controls. Main cabin can accommodate up to 17 passengers in commuter configuration, in three-

abreast layout (with aisle), at seat pitch of 75 cm (30 in). Alternative layouts for up to 15 parachutists, or all-cargo configuration with 11 tiedown rings. Passenger/cargo double door on port side at rear, rear half of which opens outward and forward half inward; foldout steps in passenger entrance. Emergency exit on each side at front of cabin and opposite passenger door on starboard side at rear. Baggage compartment in nose and at rear of passenger cabin, for 100 kg (220 lb) and 260 kg (573 lb) respectively.

SYSTEMS: Hamilton Sundstrand R70-3WG environmental control system. Hydraulic system (operating pressure 118 to 147 bar; 1,711 to 2,132 lb/sq in) for mainwheel brakes. Two 6 kW DC starter/generators, two 600 VA 400 Hz single-phase static inverters and one 43 Ah Ni/Cd battery for electrical power. Goodrich Type 29S-7D 5178 anti-icing system optional for wing, tailplane and fin leading-edges. Oxygen system optional.

AVIONICS: *Comms:* VHF-251 and HF-230 radio, AUD-251A intercom and TDR-950 transponder.

Radar: Honeywell 1400C, RDS-81 or RDS-82 weather radar optional.

Flight: ALT-50 radio altimeter, DME-451, ADF-650A, MKR-350 marker beacon receiver, KLN 90 GPS receiver, VIR-351 VOR, GLS-350 glideslope receiver and PN-101 pictorial navigation display. Doppler navigation with satellite responder (Y-12 (IV), Omega nav) optional (for example, in mineral detection role).

Instrumentation: 20025-11324 airspeed indicator, 510-8D10 horizon, 101420-11934 encoding altimeter, 30230-11101 vertical speed indicator, 9551-BN541 bank indicator, LC-2 magnetic compass, ZWH-1 outside air temperature indicator, and ZEY-1 flap position indicator; dual engine torquemeters, interturbine temperature indicators, gas generator tachometers, oil temperature and pressure indicators, fuel pressure and quantity indicators; 309W clock; and XDH-10B warning light box.

EQUIPMENT: Hopper for 1,200 litres (317 US gallons; 264 Imp gallons) of dry or liquid chemical in agricultural version. Appropriate specialised equipment for firefighting, geophysical survey (for example, long, kinked sensor tailboom) and other missions.

DIMENSIONS, EXTERNAL:

Wing span: Y-12 (II)	17.235 m (56 ft 6½ in)
Y-12 (IV)	19.20 m (63 ft 0 in)
Wing chord, constant	2.00 m (6 ft 6¾ in)
Wing aspect ratio: Y-12 (II)	8.7
Length overall	14.86 m (48 ft 9 in)
Height overall	5.675 m (18 ft 7½ in)
Elevator span	5.365 m (17 ft 7¼ in)
Wheel track	3.61 m (11 ft 10¼ in)
Wheelbase	4.70 m (15 ft 5 in)
Propeller diameter	2.49 m (8 ft 2 in)
Propeller ground clearance	1.325 m (4 ft 4¼ in)
Distance between propeller centres	4.94 m (16 ft 2½ in)
Fuselage ground clearance	0.65 m (2 ft 1½ in)
Crew door: Height	1.12 m (3 ft 8 in)
Width	0.65 m (2 ft 1½ in)
Passenger/cargo door: Height	1.38 m (4 ft 6¼ in)
Width (passenger door only)	0.65 m (2 ft 1½ in)
Width (double door)	1.45 m (4 ft 9 in)
Emergency exits (three, each):	
Height	0.68 m (2 ft 2¾ in)
Width	0.68 m (2 ft 2¾ in)

Y12 (I) and Y-12 (II) DELIVERIES
(at July 2001)

Country	Operator	Quantity	First Aircraft
Domestic (mostly Y-12 (I))			
China	AVIC (No. 630 Institute)	1	B-3820
	China General Aviation	5	B-3810
	China Southwest Airlines	4	B-3811
	Flying Dragon Aviation	9	B-3802
	Guizhou Aviation Corporation	1	B-3809
	State Oceanic Administration	1	?
	Subtotal	**21**	
Export*			
Australia	Aviex	1	VH-LLK
Bangladesh	Aero Bengal Airlines	2	?
Cambodia	Air Force	2	XU-016
Egypt	Air Force	2	?
Eritrea	Air Force	3	ER 800
	Red Sea Air†	1	E3-AAI
Fiji	Fiji Air	3	DQ-FHF
Iran	Air Force	9	15-2245
Kenya[1]	Air Force	12	128?
Kiribati	Air Kiribati	1	T3-ATI
Laos	Lao Aviation	7	RDPL-34115
Malaysia	Berjaya Air Charter	2	9M-TAB
Mauritania	Air Force	2	5T-MAE
Mongolia	Mongolian Airlines	6	D-0064
Namibia	Defence Force	2	?
Nepal	Nepal Airways	5	9N-ACD
Pakistan[2]	Air Force	2	96-035
	Army	2	?
Peru	Air Force	6	333
	National Police	3	224
Philippines	Philippine Eagle Airlines	3	RP-C1518
Sri Lanka	Air Force	9	CR-851
Tanzania	Air Force	2	JW9029
Zambia	Air Force	4	AF214
	Subtotal	**91**	
Total		**112**	

* Had risen to 94 by November 2000; Guyana Defence Force MoU for one in 2001
† Unclear if aircraft is ex-military
[1] Second six delivered 22 June 2000
[2] Two more ordered in July 2000

Y-12 (IV) (two P&WC PT6A-27 turboprops)   *NEW*/0106481

Baggage door (nose, port):	
Max height	0.56 m (1 ft 10 in)
Width	0.75 m (2 ft 5½ in)
DIMENSIONS, INTERNAL:	
Cabin, excl flight deck and rear baggage compartment:	
Length	4.82 m (15 ft 9¾ in)
Max width	1.46 m (4 ft 9½ in)
Max height	1.70 m (5 ft 7 in)
Floor area	7.0 m² (75.8 sq ft)
Volume	12.9 m³ (455 cu ft)
Baggage compartment volume:	
nose	0.77 m³ (27.2 cu ft)
rear	1.9 m³ (67 cu ft)
AREAS:	
Wings, gross: Y-12 (II)	34.27 m² (368.9 sq ft)
Ailerons (total, incl tab)	2.88 m² (31.00 sq ft)
Trailing-edge flaps (total)	6.00 m² (64.58 sq ft)
Fin, incl dorsal fin	2.24 m² (24.11 sq ft)
Rudder, incl tab	3.34 m² (35.95 sq ft)
Tailplane	3.10 m² (33.37 sq ft)
Elevators (total, incl tabs)	4.06 m² (43.70 sq ft)
WEIGHTS AND LOADINGS (A: Y-12 (II), B: Y-12 (IV)):	
Max usable fuel weight: A, B	1,230 kg (2,712 lb)
Max payload: A	1,700 kg (3,748 lb)
B	1,984 kg (4,374 lb)
Max T-O weight: A	5,300 kg (11,684 lb)
B	5,670 kg (12,500 lb)
Max ramp weight: A	5,330 kg (11,750 lb)
B	5,700 kg (12,566 lb)
Max landing weight: A	5,300 kg (11,684 lb)
B	5,400 kg (11,905 lb)
Max zero-fuel weight: A	4,900 kg (10,803 lb)
B	5,188 kg (11,438 lb)
Max cabin floor loading (cargo):	
A, B	750 kg/m² (154 lb/sq ft)
Max wing loading: A	145.9 kg/m² (29.90 lb/sq ft)
Max power loading: A	5.74 kg/kW (9.42 lb/shp)
B	6.14 kg/kW (10.08 lb/shp)
PERFORMANCE (A and B as above):	
Max operating speed (VMO) at 3,000 m (9,840 ft):	
A	177 kt (328 km/h; 204 mph)
B	175 kt (325 km/h; 201 mph)
Max cruising speed at 3,000 m (9,840 ft):	
A	157 kt (292 km/h; 181 mph)
Econ cruising speed at 3,000 m (9,840 ft):	
A	135 kt (250 km/h; 155 mph)
B	140 kt (260 km/h; 162 mph)

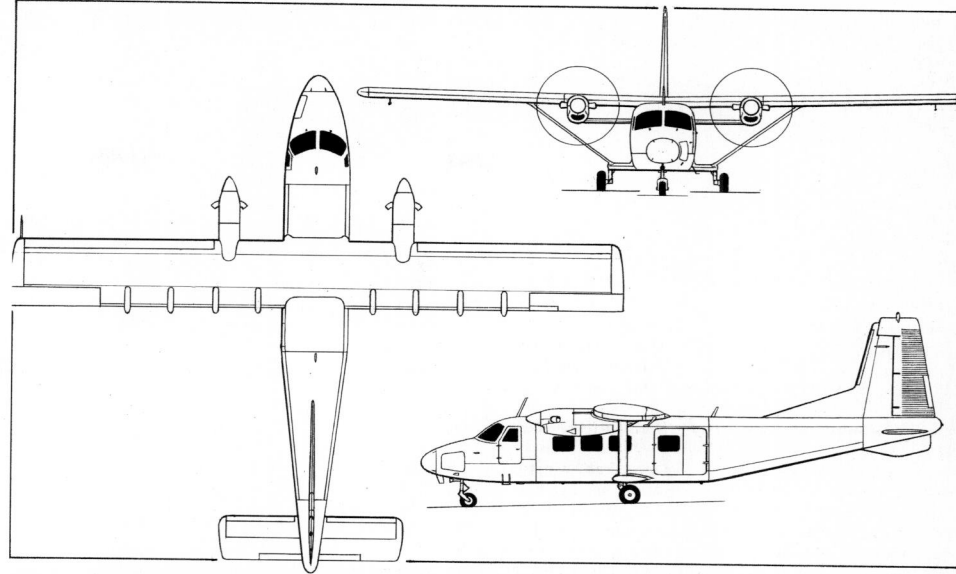

HAI Y-12 (II) twin-turboprop STOL general purpose transport (*Jane's/Dennis Punnett*)

Max rate of climb at S/L: A	486 m (1,594 ft)/min
B	468 m (1,535 ft)/min
Rate of climb at S/L, OEI: A	84 m (275 ft)/min
B	90 m (295 ft)/min
Service ceiling: A, B	7,000 m (22,960 ft)
Service ceiling, OEI, 15 m (50 ft)/min rate of climb, max	
continuous power: A	3,000 m (9,840 ft)
T-O run, 15° flap: normal: A	340 m (1,115 ft)
B	370 m (1,215 ft)
STOL: A	230 m (755 ft)
T-O to 15 m (50 ft), 15° flap:	
normal: A	420 m (1,380 ft)
B	490 m (1,610 ft)
STOL: A	370 m (1,215 ft)
Landing from 15 m (50 ft), 20° flap: with braking and	
propeller reversal:	

normal: A	480 m (1,575 ft)
STOL: A	370 m (1,215 ft)
with brakes only:	
normal: A	620 m (2,035 ft)
B	630 m (2,070 ft)
STOL: A	200 m (656 ft)
Landing run, 20° flap:	
with braking and propeller reversal: A	200 m (660 ft)
with brakes only: A, B	340 m (1,120 ft)
Range at econ cruising speed at 3,000 m (9,840 ft) with	
max fuel, 45 min reserves:	
A	723 n miles (1,340 km; 832 miles)
B	701 n miles (1,300 km; 807 miles)
Endurance, conditions as above: A	5 h 12 min
B	5 h 15 min
	UPDATED

HAIG (Harbin)

HARBIN AIRCRAFT INDUSTRY GROUP
(Harbin Feiji Gongye Gongsi) (Subsidiary of AVIC II)

15 Youxie Street, Pingfang District, PO Box 201-29, Harbin, Heilongjiang 150066
Tel: (+86 451) 650 11 22
Fax: (+86 451) 650 22 73
PRESIDENT AND GENERAL MANAGER: Cui Xuewen
VICE-PRESIDENT: Xu Zhanbin
PUBLIC RELATIONS: Chen Xiaoyi

HAIG at Harbin is the parent organisation of Hafei Aviation Industry Company (HAI, which see). It has overall responsibility for HAI's participation in the Eurocopter EC 120, for which the 200th shipset was handed over on 5 September 2000.

In June 2001, HAIG signed an agreement to become a risk-sharing partner in the now-defunct Alliance Aircraft StarLiner 100 (see US section of the 2003-04 edition), for which it would have assembled the 35-passenger version, as well as purchasing up to 50 for resale to domestic regional airlines. Also in 2001, agreement signed with Fairchild Dornier to manufacture wing/fuselage fairings for 328JET airliner and to produce the 528JET in China. However, in

early 2002 AVIC II indicated that a final choice of partner in a 30- to 50-seat regional jet programme had not yet been made, and eventually neither Alliance nor Fairchild Dornier was selected, the choice instead falling on the Embraer ERJ-145. Contract, signed 2 December 2002, will create new company (Harbin Embraer Aircraft Industry Co Ltd), located in Harbin; Embraer to invest US$25 million in new 26,477 m² (285,000 sq ft) production facility employing 220 people on ERJ-145 programme; ERJ-135/140 could also be built. (For information on the AVIC I programme for a 72- to 99-passenger regional jet, see the ACAC entry elsewhere in this section).

UPDATED

HAIG (Hongdu)

HONGDU AVIATION INDUSTRY GROUP
(Hongdu Hangkong Gongye Gongsi) (Subsidiary of AVIC II)

Xinxiqiao, PO Box 5001, Nanchang, Jiangxi 330024
Tel: (+86 791) 846 84 01 and 846 84 02
Fax: (+86 791) 845 14 91

Following the division of AVIC's responsibilities in 1999, Hongdu Aviation Industry Group became the parent organisation of Nanchang Aircraft Manufacturing Company (NAMC, which see).

UPDATED

HAIG L-15

TYPE: Advanced jet trainer/light attack jet.
PROGRAMME: Revealed as feasibility study at Aviation Expo 2001, Beijing, September 2001.
DESIGN FEATURES: Shoulder-mounted wings with curved, leading-edge wingroot strakes; tandem cockpits; twin engines; single vertical fin well forward of slab tailplane.
Following details highly provisional:

Model of HAIG L-15 shown in September 2001 *NEW*/0137971

For details of the latest updates to *Jane's All the World's Aircraft* online and to discover the additional information available exclusively to online subscribers please visit

jawa.janes.com

FLYING CONTROLS: Assumed fly-by-wire.
POWER PLANT: Twin turbofans; possibly NMKB Progress AI-25s or domestic WS11s.
ACCOMMODATION: Two zero/zero ejection seats in tandem.
AVIONICS: HOTAS controls, HUD(s) and MFDs; stores management system.
DIMENSIONS, EXTERNAL:

Wing span	11 m (36 ft)
Length overall	15 m (49 ft)

WEIGHTS AND LOADINGS:

Normal T-O weight, clean	6,500 kg (14,330 lb)
Max T-O weight	9,500 kg (20,950 lb)

PERFORMANCE:

Max level speed	M1.4
Max rate of climb at S/L	9,000 m (29,530 ft)/min
Service ceiling	16,000 m (52,490 ft)
g limits	+8/−3

UPDATED

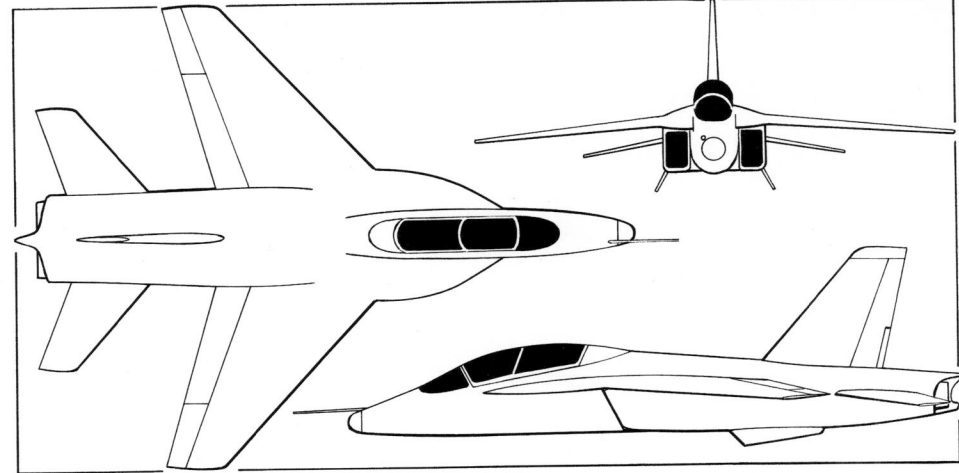

Provisional general arrangement of HAIG L-15
(*Jane's/James Goulding*)
NEW/0130858

NAI

NAI
Name reverted to NUAA (which see).

UPDATED

NAMC

NANCHANG AIRCRAFT MANUFACTURING COMPANY
(Nanchang Feiji Zhizao Gongsi)
(Subsidiary of AVIC II)
PO Box 5001-506, Nanchang, Jiangxi 330024
Tel: (+86 791) 846 84 01 and 846 84 02
Fax: (+86 791) 845 14 91
PRESIDENT: Jiang Liang
INFORMATION: Feng Jinghua

Nanchang Aircraft Manufacturing Company (NAMC), which became a core unit of the Hongdu Aviation Industry Group (HAIG) in 1998, was created in 1951; it built 379 CJ-5s (licence Soviet Yak-18s) between 1954 and 1958, and in 1960s shared in large production programme for J-6 fighter (Chinese development of MiG-19); also built (1957-68) 727 Y-5 (Chinese An-2) biplanes (see under SAIC in this section and under NAMC in 1991-92 *Jane's*). Current programmes are K-8 jet trainer and N-5A agricultural aircraft. Agreements with and via Canadian Aerospace Group in 1998 provided for North American assembly and marketing of N-5A in return for option to build CAG Windeagle and Monitor Jet in China; however, no progress had been made by mid-2002. Also built prototype of NLA AC-500 Aircar (which see).

The company occupies a 500 ha (1,235 acre) site, with 10,000 m² (107,600 sq ft) of covered space, and has a workforce of about 20,000; it delivered its 4,000th aircraft in 1993. About 80 per cent of its activities are non-aerospace.

UPDATED

NAMC K-8 KARAKORUM 8
TYPE: Basic jet trainer/light attack jet.
PROGRAMME: Launched publicly (as L-8) at 1987 Paris Air Show as proposed export aircraft to be developed jointly with international partner. Subsequently proposed to be co-

NAMC K-8 International demonstrator (*Jane's/Paul Jackson*)   *NEW*/0121039

developed with Pakistan as partner (25 per cent share); aircraft then redesignated K-8 and named after mountain range forming part of China/Pakistan border. Pakistan decided 1994 against own assembly line, but agreed to reconsider if subcontracting share (12 per cent initially, but being increased to 25 per cent in mid-2001 by addition of front fuselage) should later increase to 45 per cent.

Manufacture of four prototypes started January 1989; three flying prototypes: 001 (first flight 21 November 1990), 003 (first flight 18 October 1991) and 004; 002 is static and fatigue test aircraft.

First preproduction aircraft (1001/L8 320101) used as demonstrator. Preproduction batch of six for Pakistan Air Force evaluation (ordered 9 April 1994) handed over in China on 21 September 1994 and delivered to PAF on 10 November that year; PAF evaluation (approximately 1,200

hours) completed in August 1995; aircraft reported subsequently in use from Air Academy, Risalpur. China importing Progress AI-25 turbofans for use in domestic version (see under Power Plant), designated K-8J. One aircraft in use from 1997 by China Flight Test Establishment as variable stability testbed (K-8VSA) with fly-by-wire flight control system.
CURRENT VERSIONS: **K-8:** Initial version with TFE731 engine; *as described.*

K-8E: TFE731-powered version for Egyptian Air Force, 80 of which ordered under contract signed on 27 December 1999. First 10 Chinese-built; Arab Organisation for Industrialisation (AOI) then assembling batches of 15 and 10 from CKD kits of medium-sized and smaller components respectively before progressing to 90 per cent

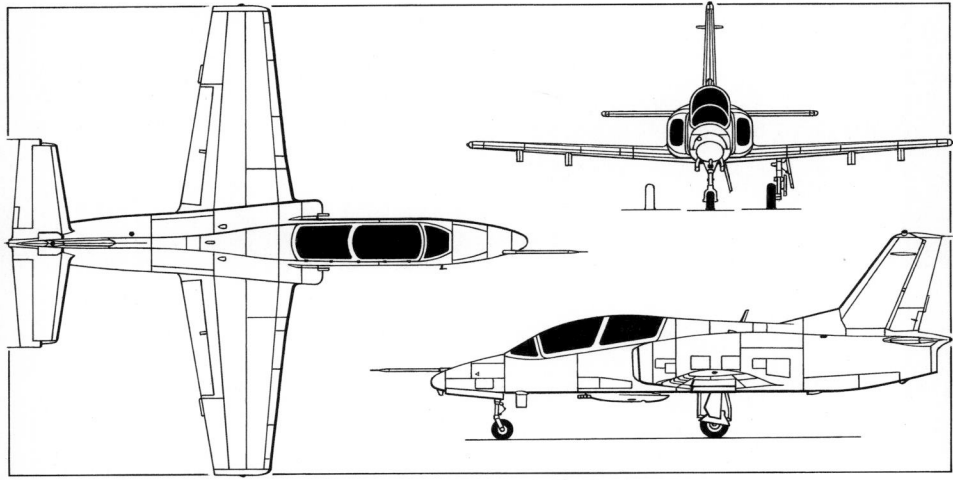

K-8 Karakorum 8 jet trainer and light attack aircraft (*Jane's/Mike Keep*)

K-8 front cockpit

local manufacture of final 45; first kit-assembled K-8E completed mid-2001. Total of 33 items newly selected, developed or upgraded for this version include instrument panels and consoles; com/nav systems; fire-control system; fuel system; environmental control system; hydraulic system; and landing gear. K-8E flown for first time on 5 July 2000.

K-8J: Reported designation of Chinese domestic version when fitted with AI-25 engine.

K-8VSA: In-flight variable stability testbed (serial number 320203), in use from 1997 by China Flight Test Establishment (CFTE). Equipped with digital fly-by-wire AFCS, sidestick controller and data acquisition system.

CUSTOMERS: Total of 12 preproduction K-8s (six each to PLA Air Force and Pakistan Air Force) delivered by end of 1996. Original joint venture agreement reportedly involved up to 75 for Pakistan Air Force, but Pakistan Secretary for Defence Production quoted in early 1996 as saying up to 100 needed eventually to replace Cessna T-37; however, SLEP for T-37 has postponed main K-8 requirement to about 2005. Chinese PLAAF originally had requirement for several hundred K-8Js, of which 25 to 30 delivered by late 1999, but now widely reported to have abandoned plans for full adoption. Myanmar Air Force last three of 12 delivered in September 1999. Zambian Air Force eight, and Namibian Air Force four, also delivered in 1999. Three of Sri Lanka six (No. 14 Squadron) destroyed in Tamil Tiger attack on Katunayake AB, 24 July 2001. Egyptian Air Force has begun to receive 80 of K-8E version to replace Aero L-29. Venezuela considering purchase of 24 in late 2001. Interest also reported from Bangladesh, Cambodia, Eritrea, Laos and Thailand. Production had totalled 50 by June 1999, according to test pilot of K-8 displayed at Paris Air Show.

K-8 EXPORT ORDERS
(at June 2001)

Country	Qty
Colombia	6
Egypt	80
Morocco	?3
Myanmar	12[1]
Namibia	4
Pakistan	6
Sri Lanka	6
Zambia	8[2]
Total	**122+**

[1] Reported total requirement for 30
[2] Reported option for further 8
[3] To replace approximately 14 Cessna T-37Bs

COSTS: US$3 million to US$3.5 million flyaway (1996) with TFE731 and Western avionics. Egyptian Air Force (80 aircraft) contract quoted as US$345 million (1999).

DESIGN FEATURES: Intended for full basic flying training plus parts of primary and advanced syllabi, but capable also of light ground attack missions. Sweptback vertical/non-swept horizontal tail surfaces.

Tapered low wings, with NACA 64A-114 (mod) root and NACA 64A-412 tip sections; sweepback 2° 13′ 8″ at quarter-chord, 1° 30′ incidence at root, 3° dihedral from roots; −2° twist.

FLYING CONTROLS: Conventional and power-assisted; ailerons have hydraulic boost and artificial feel; variable incidence tailplane; electrically operated trim tab in rudder and port elevator. Aileron travel ±18°, elevators 16° up/28° down, rudder ±28°. Two-position Fowler flaps (23° for T-O, 35° for landing), and split airbrake under fuselage just aft of mainwheel doors, are hydraulically actuated.

STRUCTURE: All-metal damage-tolerant main structure; ailerons of honeycomb, fin and rudder of composites. PAC

share (initially only tailplane and elevators) increased to include fin, rudder, rear fuselage and engine cowling/access panels. First Pakistan-built subassemblies delivered to China in mid-1997; planned output by PAC of 12 tail units and three front fuselages in 2001; 24 and 12 respectively in 2002; 36 tails and undecided number of front fuselages in 2003.

LANDING GEAR: Hydraulically retractable tricycle type, with single wheel and oleo-pneumatic shock-absorber on each unit. Main units retract inward into underside of fuselage; nosewheel, which has hydraulic steering, retracts forward. Mainwheel tyres size 561×169, pressure 6.90 bar (100 lb/sq in). Chinese hydraulic disc brakes. Anti-skid units. Minimum ground turning radius 6.69 m (21 ft 11½ in).

POWER PLANT: Including prototypes, all except K-8J have one 16.01 kN (3,600 lb st) Honeywell TFE731-2A-2A turbofan with Lucas Aerospace FADEC, mounted in rear fuselage. Production K-8J for China powered by 16.87 kN (3,792 lb st) ZMKB Progress AI-25TLK turbofan, ordered in 1997; initial batch of 58 imported.

Fuel in two flexible tanks and one inverted flight tank in fuselage and one integral tank in wing centre-section, combined capacity 1,000 litres (264 US gallons; 220 Imp gallons); single refuelling point in fuselage. Provision for carrying one 250 litre (66.0 US gallon; 55.0 Imp gallon) drop tank on outboard pylon under each wing. Oil capacity 4 kg (8.8 lb).

ACCOMMODATION: Instructor and pupil in tandem, on Jianghan TY-7 zero/zero ejection seats; rear seat elevated 28 cm (11 in). One-piece wraparound windscreen; two-piece canopy opens sideways to starboard. Cockpits pressurised and air conditioned.

SYSTEMS: Honeywell ECS 51833 air conditioning and pressurisation system, with maximum differential of 0.27 bar (3.9 lb/sq in). Hydraulic system, pressure 207 bar (3,000 lb/sq in), for operation of landing gear extension/retraction, wing flaps, airbrake, aileron boost, nosewheel steering and wheel brakes. Flow rate 15 litres (3.96 US gallons; 3.30 Imp gallons)/min, with air-pressurised reservoir, plus emergency back-up hydraulic system. Abex AP09V-8-01 pump. Electrical systems powered by 12 kW,

28.5 V DC generator (primary) and 24 V DC (auxiliary), with 115/26 V single-phase AC and 36 V three-phase AC available from 40 Ah Ni/Cd battery and two static inverters, both at 400 Hz. Liquid oxygen system for occupants. Demisting of cockpit transparencies.

AVIONICS: *Comms:* Bendix/King KTR 908 VHF or KTR 909 UHF; intercom; FJ-20 flight data recorder.

Flight: Bendix/King KNR 634A VOR/glideslope/MKR with KA 26 beacon receiver, ADF 462, Type 265 radio altimeter, HZX-4A AHRS, SS/SC-5 air data computer and KTU 709 Tacan or WL-7A radio compass.

Instrumentation: Rockwell Collins EFIS-86T in first 100 aircraft, incorporating CRT primary flight and navigation displays for each crew member plus dual display processing units and selector panels for tandem operation. Blind-flying instrumentation standard. Standby flight instruments include ASI, rate of climb indicator, barometric altimeter, emergency horizon and standby compass. HUD under development.

ARMAMENT (optional): One 23 mm gun pod under centre-fuselage; self-computing optical gunsight in cockpit, plus gun camera. Two external stores points under each wing. Twin ejector racks on inboard stations can carry total of four 6, 11.5 or 50 kg practice bombs; single-store outboard stations can each carry a PL-5E air-to-air missile, a 12-round pod of 57 mm rockets, a 200 kg, 250 kg or BL755 bomb, or a drop fuel tank.

DIMENSIONS, EXTERNAL:
Wing span	9.63 m (31 ft 7¼ in)
Wing mean aerodynamic chord	1.85 m (6 ft 0¾ in)
Wing aspect ratio	5.4
Length overall: incl nose pitot	11.60 m (38 ft 0¾ in)
excl nose pitot	10.40 m (34 ft 1½ in)
Height overall	4.21 m (13 ft 9¾ in)
Elevator span	4.20 m (13 ft 9½ in)
Wheel track	2.54 m (8 ft 4 in)
Wheelbase	4.44 m (14 ft 6¾ in)

AREAS:
Wings, gross	17.02 m² (183.2 sq ft)
Ailerons (total, incl tab)	1.095 m² (11.80 sq ft)
Trailing-edge flaps (total)	2.65 m² (28.55 sq ft)
Fin	1.92 m² (20.67 sq ft)
Rudder, incl tab	1.115 m² (12.02 sq ft)
Tailplane	2.84 m² (30.57 sq ft)
Elevators (total, incl tab)	1.32 m² (14.21 sq ft)

WEIGHTS AND LOADINGS:
Weight empty, equipped	2,757 kg (6,078 lb)
Max fuel: internal	780 kg (1,720 lb)
external (two drop tanks)	388 kg (855 lb)
Max external stores load	943 kg (2,080 lb)
T-O weight: clean	3,700 kg (8,157 lb)
with two 250 litre drop tanks	4,204 kg (9,268 lb)
Max T-O weight with external stores	4,332 kg (9,550 lb)
Max wing loading	254.5 kg/m² (52.13 lb/sq ft)
Max power loading (TFE731)	271 kg/kN (2.65 lb/lb st)

PERFORMANCE (at clean T-O weight):
Never-exceed speed (VNE)	512 kt (950 km/h; 590 mph) IAS
Max level speed at S/L	432 kt (800 km/h; 497 mph)
Unstick speed	100 kt (185 km/h; 115 mph)
Approach speed	108 kt (200 km/h; 124 mph)
Touchdown speed, 35° flap	86 kt (160 km/h; 99 mph)
Stalling speed, 35° flap	81 kt (150 km/h; 94 mph)
Max rate of climb at S/L	1,800 m (5,905 ft)/min
Service ceiling	13,600 m (44,620 ft)
T-O run	440 m (1,445 ft)
T-O to 15 m (50 ft)	600 m (1,970 ft)

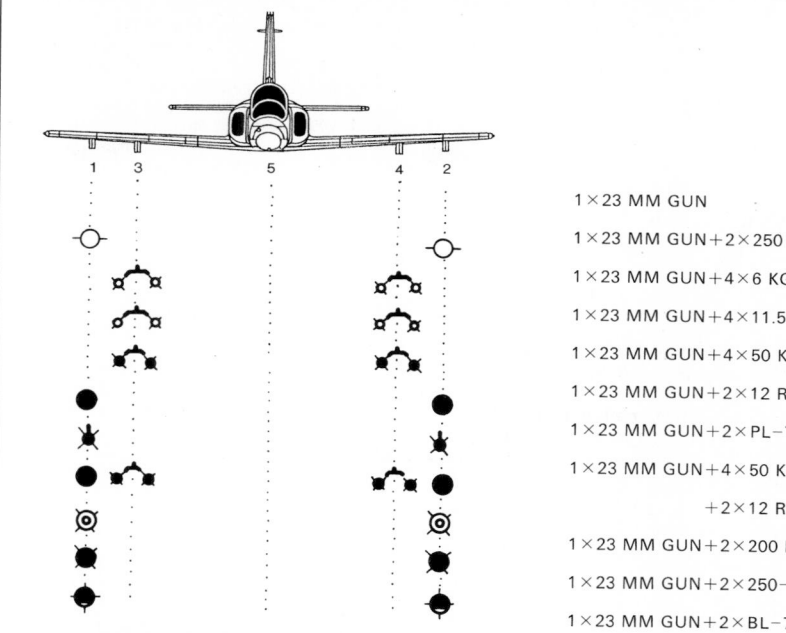

1×23 MM GUN

1×23 MM GUN+2×250 LTR DROPS

1×23 MM GUN+4×6 KG TRG BOMBS

1×23 MM GUN+4×11.5 KG TRG BOMBS

1×23 MM GUN+4×50 KG TRG BOMBS

1×23 MM GUN+2×12 ROCKETS(57-1)

1×23 MM GUN+2×PL-7 MISSILES

1×23 MM GUN+4×50 KG TRG BOMBS
+2×12 ROCKETS(57-1)

1×23 MM GUN+2×200 KG BOMBS

1×23 MM GUN+2×250-3 BOMBS

1×23 MM GUN+2×BL-755 BOMBS

Karakorum 8 weapon options

K-8s in PLAAF school service

NEW/0524647

Landing from 15 m (50 ft)	518 m (1,700 ft)
Landing run	530 m (1,740 ft)
Range:	
max internal fuel	842 n miles (1,560 km; 969 miles)
max internal/external fuel	
	1,155 n miles (2,140 km; 1,329 miles)
Endurance: max internal fuel	3 h 12 min
max internal/external fuel	4 h 12 min
g limits (clean)	+7.33/–3.0

UPDATED

NAMC N-5A
Chinese name: Nongye Feiji 5 (Agricultural aircraft 5)

PROGRAMME: Design of this agricultural sprayer began November 1987; first details revealed at Farnborough Air Show September 1988; first of three prototypes (B-501L) made first flight 26 December 1989; CAAC type certificate awarded July 1992 and production certificate in June 1995; production rate in 1997 (latest information received) reported as 10 per year, but according to CAG (see Canadian section) production suspended after 14th aircraft. Expected to resume in second half of 2001, using alternative engine, but not reported by mid-2002. Operators in 2001 included Changjiang General Aviation (four), Longken General Aviation (three) and Shenyang Beiyan General Aviation (two). A full description last appeared in the 2002-03 *Jane's All the World's Aircraft*.

UPDATED

NGA

NANJING GROEN AVIATION INDUSTRY LTD (Nanjing Groen Hangkong Youxian Gongsi)

140 Guangzhou Road, Suite 22D, Nanjing, Jiangsu 210024
Tel: (+86 25) 360 50 73
Fax: (+86 25) 360 49 53
Web (Groen): http://www.gbagyros.com

On 22 July 1996, Shanghai Energy and Chemicals Corporation (SECC) signed a letter of intent with Groen Brothers Aviation Inc of the USA (which see), committing to buy 200 of the latter company's H2X Hawk 3 three-seat autogyros with which to form an air taxi service in the Shanghai district. Parts and components would be supplied by the US company for final assembly in "an existing aircraft manufacturing plant" in Shanghai. In 1998, Groen Brothers announced new commitments from SECC for a further 200 Hawks, comprising 100 each of the five-seat Hawk 5 and eight-seat Hawk 8.

Nanjing Groen Aviation Industry Ltd (NGA) was registered as a Chinese subsidiary of Groen Brothers in 1996; it was planned that this plant, too, would import Hawk components from the USA for final assembly and sale in China, once the aircraft had received FAA certification. No further developments had been announced by mid-2002.

UPDATED

NLA

NANJING LIGHT AIRCRAFT COMPANY (Nanjing Qing Feiji Gongsi)

Room 102, Building D, 29 Yudao Street, Nanjing, Jiangsu 210016
Tel: (+86 25) 489 16 36 and 489 36 55
Fax: (+86 25) 489 16 37
e-mail: njqf@pub.jlonline.com
Web: http://www.pub.jlonline.com
PRESIDENT: Wo Dingzhu

NLA was created in 1998 by Nanjing University of Aeronautics and Astronautics (NUAA), the municipality of Nanjing, the provincial government of Jiangsu and other business interests. It has its headquarters on the NUAA campus and is responsible for the five-seat AC-500.

UPDATED

Prototype of the five-seat NLA AC-500 *(Robert Hewson)*   0105847

NLA AC-500 AIRCAR

TYPE: Five-seat lightplane.
PROGRAMME: Revealed at Airshow China in November 1998. Prototype originally planned to fly in 2000, but now rescheduled for late 2002; built and rolled out at NAMC plant, Nanchang, in third quarter of 2000 and exhibited at Airshow China in November 2000. No reports of first flight by mid-2002; FAR Pt 23 certification is intended.
CUSTOMERS: Initial order from Yuanda Air Conditioning Company.
COSTS: Estimated US$350,000 (2001).
DESIGN FEATURES: General appearance similar to that of Piper and Socata types in same weight/performance category. Constant-chord, low-mounted wings with upturned tips; sweptback vertical tail and small underfin.
FLYING CONTROLS: Conventional and manual; flaps fitted. One-piece horn-balanced elevator; ground-adjustable tab on rudder.

STRUCTURE: All-metal main structure; some non-load-bearing components manufactured from composites.
LANDING GEAR: Non-retractable tricycle type; single wheel on each unit.
POWER PLANT: One 194 kW (260 hp) Textron Lycoming IO-540 flat-six engine, driving a three-blade propeller.
ACCOMMODATION: Pilot and co-pilot or passenger in front, with two rearward-facing seats behind and a forward-facing single seat aft of these. Crew door each side of cockpit, plus passenger door aft of wing on port side.

DIMENSIONS, EXTERNAL:

Wing span	10.20 m (33 ft 5½ in)
Length overall	8.14 m (26 ft 8½ in)
Fuselage max width	1.20 m (3 ft 11¼ in)
Height overall	3.05 m (10 ft 0 in)
Wheel track	2.80 m (9 ft 2¼ in)
Wheelbase	2.18 m (7 ft 1¾ in)
Propeller diameter	1.96 m (6 ft 5 in)

WEIGHTS AND LOADINGS:

Max baggage weight	65 kg (143 lb)
Max T-O weight	1,540 kg (3,395 lb)
Max power loading	7.95 kg/kW (13.06 lb/hp)

PERFORMANCE (estimated):

Max level speed	135 kt (250 km/h; 155 mph)
Stalling speed	60 kt (110 km/h; 69 mph)
Service ceiling	3,000 m (9,840 ft)
Range	432 n miles (800 km; 497 miles)
Endurance	4 h

UPDATED

Instrument panel of the AC-500 *(Photo Link)* 0100800

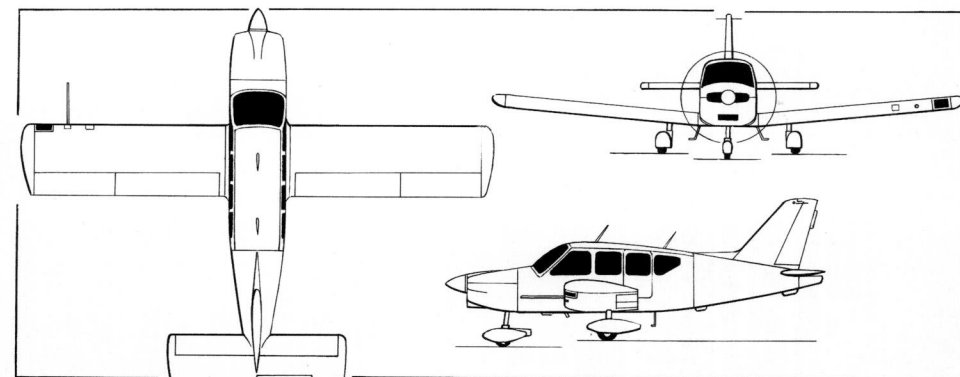

NLA AC-500 light aircraft *(Jane's/James Goulding)*   0126689

NRJ

NEW REGIONAL JET PROGRAMME MANAGEMENT COMPANY (Subsidiary of AVIC I)

Restructured in October 2002 as ACAC (which see).

UPDATED

NUAA

NANJING UNIVERSITY OF AERONAUTICS AND ASTRONAUTICS (Nanjing Hangkong Hangtian Daxue}

29 Yudao Street, Nanjing, Jiangsu 210016
Tel: (+86 25) 489 24 24
Fax: (+86 25) 489 15 12
e-mail: icedao@nuaa.edu.cn
PRESIDENT: Hu Haiyan

NUAA was responsible for design and part-production of the AD-100 and FT-300 light aircraft described in the 2001-02 *Jane's*; design of the tandem-seat AD-200 (see under BKLAIC in that edition); and production of the Chang Kong pilotless aerial target (see *Jane's Unmanned Aerial Vehicles and Targets*).

NEW ENTRY

SAC (Shaanxi)

SHAANXI AIRCRAFT COMPANY (Shaanxi Feiji Gongsi) (Subsidiary of AVIC II)

PO Box 35, Chenggu, Shaanxi 723213
Tel: (+86 916) 220 21 56
Fax: (+86 916) 220 21 58
e-mail: sac@public.hanzhong.sn.cn
PRESIDENT: Wang Wenfeng
MARKETING MANAGER: Li Yousen

Founded early 1970s; occupies a 204 ha (504 acre) site and had 1994 workforce (latest received information) of about 10,000; covered workspace includes largest final assembly building in China. Main aircraft programme, recently afforded increased priority, is Y-8 transport; non-aerospace products include coaches and small trucks.

UPDATED

Shaanxi Y-8F200, wearing test registration B-576 L (*Robert Hewson*)   NEW/0105843

SAC Y-8

Chinese name: Yunshuji-8 (Transport aircraft 8)
TYPE: Medium transport/multirole.
PROGRAMME: Redesign, as Chinese development of Antonov An-12B, started at Xian March 1969; first flight of first (Xian-built) prototype 25 December 1974, followed by second (c/n 001802, first built by SAC) 29 December 1975; production go-ahead given January 1980. Pressurised Y-8C made first flight 17 December 1990. Production rate was approximately five per year during first half of 1990s; new (mainly freighter) designations introduced with Y-8F100 and F200 from 1997; further F-prefixed variants revealed November 2000, presumably reflecting current manufacturing standard; three-person flight crews resulting from use of modernised avionics; apparently now seen to have more future in cargo and military roles, following reported difficulties with pressurisation system.
CURRENT VERSIONS: **Y-8:** Prototype and baseline military transport.
Y-8A: Helicopter carrier. Main cabin height increased by 120 mm (4.72 in) by deleting internal gantry; downward-opening rear ramp/door, as in Y-8C. Deliveries began 1987. In service.
Y-8B: Mainly unpressurised civil transport. First deliveries 1986; CAAC certification 1993. Military equipment deleted; empty weight reduced by 1,720 kg (3,792 lb); some avionics differ. In service.
Y-8C: First fully pressurised version, developed with Lockheed collaboration. Changes included redesigned (downward-opening ramp type) cargo loading door and main landing gear; handling system for standard freight containers and pallets; improved com/nav/ATC equipment, air conditioning and oxygen systems; additional emergency exits. First flight 17 December 1990 (SAC 182, converted from first Shaanxi prototype); CAAC certification 1993. Five delivered by January 1994 (latest figure received); apparently superseded by Y-8F200/F400.
Y-8D: Export baseline version, with main avionics by Rockwell Collins and Litton. Later versions designated **Y-8D II**. Deliveries began 1987 (Y-8D) and 1992 (Y-8D II); eight delivered by early 1997; none reported since then.
Y-8E: Drone carrier version of baseline Y-8 for Chang Hong (Long Rainbow) UAV. Forward pressure cabin accommodates drone controller's console; carrier/launch trapeze for one UAV under each outer wing panel. First

flight 1989; first deliveries later that year, replacing obsolete Tu-4s. In service.
Y-8F: Livestock carrier version of baseline Y-8, with cages to hold up to 500 sheep or goats. First flight early 1990; first deliveries later that year; CAAC type approval January 1994. In service.
Y-8F100: Cargo version for China Postal Airlines (three); delivered 1997 with WJ6A engines. Apparently now the baseline part-unpressurised version.
Y-8F200: Pressurised version of F100, certified by CAAC in late 1997. Increased internal passenger/cargo volume.
Detailed description applies to F-100/200 unless otherwise indicated.

Y-8F300: Freighter. Generally as Y-8F100, but with 'solid' nose and reduced (three-person) flight crew.
Y-8F400: Pressurised freighter. Generally as Y-8F200, but with 'solid' nose and reduced (three-person) flight crew. Said to have flown by September 2001 and to be scheduled for refit with PW150A engines; China Postal Airlines suggested as possible launch customer.
Y-8F600: Further Westernised, 'solid'-nosed freighter version, with Pratt & Whitney Canada PW150 turboprops and Dowty six-blade composites propellers, Honeywell Primus Epic avionics suite, three-person flight crew; otherwise generally similar to Y-8F400. Two under construction late 2000; first flight was expected by end of 2001, but not reported by mid-2002.

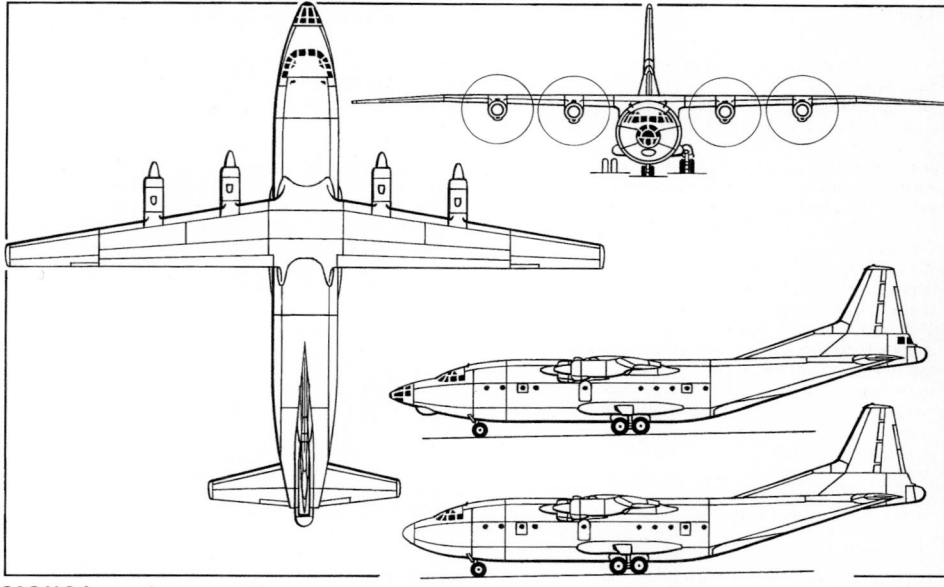

SAC Y-8 four-turboprop multipurpose transport, with additional side view (bottom) of F300/400 freighter (*Jane's/Mike Keep*)   NEW/0137951

Model of the solid-nosed Y-8F400 freighter (*Photo Link*)   0100789

Canadian turboprops and six-blade propellers characterise the Y-8F600 (*Robert Hewson*)   NEW/0105848

Poster depicting the radar-nosed 'Y-8AEW' *(Robert Hewson)*   0105849

Y-8H: Aerial survey version. No details yet known.

Y-8X: Maritime patrol prototype (B-4101), with Western com/nav, radar, surveillance and search equipment; larger chin radome. Received type approval in September 1984; no further examples confirmed, although some reports suggest PLA Navy has eight. Details in earlier editions.

'Y-8 AEW': At least one aircraft (CFTE serial number 079) fitted with BAE Systems (formerly Racal) Skymaster AEW radar, six of which ordered under US$66 million contract in 1996. Reportedly first flew November 1998. Radar, installed in bulbous nose fairing, has 360° scan and more than 200 n mile (370 km; 230 mile) range. Observed in PLA Navy exercises, reportedly transmitting information via datalink to ship-based Z-9 helicopters. Chinese official designation not yet known. Further details in *Jane's Electronic Mission Aircraft*.

CUSTOMERS: Total of about 75, including exports, reportedly delivered (of which 50 can be confirmed) by January 2001. In service in China with commercial operators, China Postal service (three) and PLA Air Force; eight military exports (Y-8D) to air forces of Myanmar (four), Sri Lanka (three) and Sudan (one). First two Sri Lankan aircraft modified locally for use as bombers (both since lost). Two (including at least one Y-8F100) leased to BonAir of Iran from May 1998.

DESIGN FEATURES: High-mounted wing; circular-section fuselage (forward section and tail turret pressurised), upswept at rear; angular tail surfaces with large dorsal fin. More pointed nose transparencies than An-12, probably from Chinese H-6 (Tu-16) production; shorter, 'solid' nose replaces this glazing in F300/F400.

Wing sections C-5-18 at root, C-3-16 at rib 15 and C-3-14 at tip, final two digits indicating thickness/chord ratio; incidence 4°; 1° dihedral on intermediate panels, 4° anhedral on outboard panels; 6° 50′ sweepback at quarter-chord; fixed-incidence tailplane.

FLYING CONTROLS: Conventional and manual. Aerodynamically balanced differential ailerons, elevators and rudder, each of which has inset trim tab; two-segment, hydraulically actuated double-slotted Fowler flaps on each wing trailing-edge; comb-shaped spoilers forward of flaps.

STRUCTURE: All-metal (aluminium alloy) conventional semi-monocoque/stressed skin; wings, tailplane and fin are all two-spar box structures; landing gear and all hydraulic components manufactured by Shaanxi Aero-Hydraulic Component Factory (SAHCF).

LANDING GEAR: Hydraulically retractable tricycle type, with Shaanxi (SAHCF) nitrogen/oil shock-struts on all units. Four-wheel main bogie on each side retracts inward and upward into blister on side of fuselage. Twin-wheel nose unit, hydraulically steerable to ±35°, retracts rearward. Mainwheel tyres size 1,050×300, pressure 28.40 bar (412 lb/sq in); nosewheel tyres size 900×300, pressure 16.70 bar (242 lb/sq in). Hydraulic disc brakes and Xingping inertial anti-skid sensor. Minimum ground turning radius 13.75 m (45 ft 1½ in).

POWER PLANT (except F600): Four 3,126 kW (4,192 shp) SAEC (Zhuzhou) WJ6A turboprops, each driving a Baoding four-blade J17-G13 constant-speed fully feathering propeller.

All fuel (F100/F300) in two integral tanks and 29 bag-type tanks in wings (20,102 litres; 5,310.5 US gallons;

4,422 Imp gallons) and fuselage (10,075 litres; 2,661.5 US gallons; 2,216 Imp gallons), giving total capacity of 30,177 litres (7,971.5 US gallons; 6,638 Imp gallons). Reduced fuel load in F200/F400 (see under Weights and Loadings). Refuelling points in starboard side of fuselage (between frames 14 and 15), mainwheel fairing, and in wing upper surface.

ACCOMMODATION: Flight crew of five (pilot, co-pilot, navigator, engineer and radio operator) in early versions and F100/F200; three crew only in F300/F400/F600. Forward portion of fuselage in F100/F300 is pressurised, and can accommodate up to 14 passengers in addition to crew. Cargo compartment (between frames 13 and 43) is unpressurised in these versions. Pressurised volume increased in F200/F400; effective length of cargo hold extended internally by 2.00 m (6 ft 6¾ in). Maximum accommodation for up to 96 troops; or 80 paratroops; or up to 92 casualties with three medical attendants; or two 'Liberation' army trucks, plus Jeep-sized vehicle on loading ramp.

Short-hold freighter versions (Y-8F100 and Y-8F300) can accommodate optimum eight 2.24 × 1.37 m (88 × 54 in) plus four 2.24 × 2.74 m (88 × 108 in) standard cargo pallets or 17 LD3 containers. Larger (Y-8F200 and Y-8F400) hold can accept optimum four 2.24 × 3.18 m (88 × 125 in) plus four 2.44 × 3.18 m (96 × 125 in) pallets or 19 LD3s. Individual cargo items of up to 7,400 kg (16,315 lb) can be airdropped.

Crew door and two emergency exits in forward fuselage. Three additional emergency exits in cargo compartment, access to which is via a large rear-loading ramp/door in underside of rear fuselage.

SYSTEMS: Forward fuselage of F100/F200 pressurised to maintain a differential of 0.20 bar (2.8 lb/sq in) at altitudes above 4,300 m (14,100 ft). (F300/F400 fully pressurised.) Two independent hydraulic systems, with operating pressures of 152 bar (2,200 lb/sq in) (port) and 147 bar (2,130 lb/sq in) (starboard), plus hand and electrical standby pumps, for actuation of landing gear extension/retraction, nosewheel steering, flaps, brakes and rear ramp/door. Electrical DC power (28.5 V) supplied by eight 12 kW generators, an 18 kW (24 hp) Xian Aero Engine Company APU (mainly for engine starting) and four 28 Ah batteries. Four 12 kVA alternators provide 115 V AC power at 400 Hz. Gaseous oxygen system for crew. Electric de-icing of windscreen, propellers and fin/tailplane leading-edges; hot air de-icing for wing leading-edges. WDZ-1 APU.

AVIONICS (Y-8F300/400): *Comms:* Rockwell Collins VHF-42B and HF-9000 com radios and TDR-94 ATC transponder.

Radar: Rockwell Collins TWR-850 colour weather radar.

Flight: Universal Avionics UNS-1K flight management system; Rockwell Collins VOR-432, DME-442, AHS-85E AHRS and EFIS-86E; Honeywell ED-55 flight data recorder.

Instrumentation: Honeywell Mk V nav display.

EQUIPMENT: Electric winch, tow, and 2,300 kg (5,070 lb) capacity hoist for cargo loading and unloading. Roller system for containerised or palletised cargo handling.

DIMENSIONS, EXTERNAL (all versions, except where indicated):

Wing span	38.00 m (124 ft 8 in)
Wing chord: at root	4.73 m (15 ft 6¼ in)
at tip	1.69 m (5 ft 6½ in)
mean aerodynamic	3.45 m (11 ft 3¾ in)
Wing aspect ratio	11.9
Length overall: F100, F200	34.02 m (111 ft 7¼ in)
F300, F400	32.93 m (108 ft 0½ in)
Fuselage:	
Max diameter of circular section	4.10 m (13 ft 5½ in)
Height overall	11.16 m (36 ft 7½ in)
Tailplane span	12.195 m (40 ft 0¼ in)
Wheel track (c/l of shock-struts)	4.92 m (16 ft 1¾ in)
Wheelbase (c/l of main bogies)	9.575 m (31 ft 5 in)
Propeller diameter	4.50 m (14 ft 9¼ in)
Propeller ground clearance	1.89 m (6 ft 2½ in)
Crew door: Height	1.455 m (4 ft 9¼ in)
Width	0.80 m (2 ft 7½ in)
Rear-loading hatch: Length	7.67 m (25 ft 2 in)
Width: min	2.65 m (8 ft 8¼ in)
max	3.10 m (10 ft 2 in)
Emergency exits (each): Height	0.55 m (1 ft 9¾ in)
Width	0.60 m (1 ft 11½ in)

DIMENSIONS, INTERNAL:

Cabin (incl flight deck, galley and toilet):	
Length: F100, F300	13.70 m (44 ft 11¼ in)
F200, F400	15.70 m (51 ft 6 in)
Width: min	3.00 m (9 ft 10 in)
max	3.50 m (11 ft 5¾ in)
Height: min	2.20 m (7 ft 2½ in)
max	2.60 m (8 ft 6½ in)
Floor area: F100, F300	55.0 m² (592 sq ft)
Volume: F100, F300	123.3 m³ (4,354 cu ft)
F200, F400	137.6 m³ (4,859 cu ft)

AREAS:

Wings, gross	121.86 m² (1,311.7 sq ft)
Ailerons (total)	7.84 m² (84.39 sq ft)
Trailing-edge flaps (total)	26.91 m² (289.66 sq ft)
Rudder	6.535 m² (70.34 sq ft)
Tailplane	27.05 m² (291.16 sq ft)
Elevators (total)	7.10 m² (76.42 sq ft)

WEIGHTS AND LOADINGS:

Weight empty, equipped: F100	34,500 kg (76,060 lb)
F200	34,760 kg (76,635 lb)
Max fuel load: F100, F300	22,909 kg (50,505 lb)
F200, F400	14,566 kg (32,115 lb)
Max payload:	
containerised: all versions	15,000 kg (33,069 lb)
bulk cargo: F100, F200	20,000 kg (44,090 lb)
Max airdroppable cargo: total	13,200 kg (29,100 lb)
single piece	7,400 kg (16,315 lb)
Max T-O weight	61,000 kg (134,480 lb)
Max ramp weight	61,500 kg (135,585 lb)
Max landing weight	58,000 kg (127,870 lb)
Max zero-fuel weight:	
F100, F300	55,278 kg (121,865 lb)
F200, F400	55,538 kg (122,440 lb)
Max wing loading	500.6 kg/m² (102.53 lb/sq ft)
Max power loading	4.88 kg/kW (8.02 lb/shp)

PERFORMANCE:

Max level speed at 7,000 m (22,960 ft):	
F100, F300	357 kt (662 km/h; 411 mph)
F200, F400	345 kt (640 km/h; 397 mph)
Max cruising speed at 8,000 m (26,250 ft):	
all versions	297 kt (550 km/h; 342 mph)
Econ cruising speed at 8,000 m (26,250 ft)	
all versions	286 kt (530 km/h; 329 mph)
Unstick speed	129 kt (238 km/h; 148 mph)
Touchdown speed at MLW	130 kt (240 km/h; 150 mph)
Max rate of climb at S/L	473 m (1,552 ft)/min
Rate of climb at S/L, OEI	231 m (758 ft)/min
Service ceiling, AUW of 51,000 kg (112,435 lb):	
F100, F300	10,400 m (34,120 ft)
F200, F400	10,050 m (32,970 ft)
Service ceiling, OEI, AUW of 51,000 kg (112,435 lb)	8,100 m (26,580 ft)
Runway ACN	15
T-O run: all versions	1,270 m (4,170 ft)
FAR T-O field length	1,900 m (6,235 ft)
T-O to 15 m (50 ft)	3,007 m (9,870 ft)
Landing from 15 m (50 ft) at MLW	2,174 m (7,135 ft)
FAR landing field length	1,650 m (5,415 ft)
Landing run at MLW: all versions	1,050 m (3,445 ft)
Range with max payload:	
F100	687 n miles (1,273 km; 791 miles)
Range with max fuel:	
F100	3,032 n miles (5,615 km; 3,489 miles)
F200, F400	1,857 n miles (3,440 km; 2,137 miles)
F300	2,861 n miles (5,300 km; 3,293 miles)

UPDATED

SAC (Shenyang)

SHENYANG AIRCRAFT CORPORATION
(Shenyang Feiji Gongsi) (Subsidiary of AVIC I)

1 Lingbei Street, Huanggu District, Shenyang, Liaoning 110034
Tel: (+86 24) 86 89 66 80 and 86 59 92 05
Fax: (+86 24) 86 89 66 89

Web: http://www.sac.com.cn
PRESIDENT: Li Fangyong

Pioneer Chinese fighter design centre; built 767 examples of J-5 (licence MiG-17F) from 1956-59 and from 1963 was major producer of several thousand of J-6 (reverse-engineered MiG-19), including 634 JJ-6 tandem-seat fighter trainers; initiated development and early production of J-7 (see under CAC). On 29 June 1994,

SAC became core enterprise in newly formed Shenyang Aircraft Industry Group (SAIG). Now occupies site area of more than 800 ha (1,976 acres) and has workforce of 30,000; only some 30 per cent of current activities are in aerospace.

Principal programme is J-8 II fighter, currently the subject of various upgrade programmes. Now also engaged in major assembly/co-production programme of Sukhoi Su-27 variants for PLA Air Force.

Aerospace subcontract manufacture includes cargo doors for Boeing 757 and baggage/service/emergency exit doors for de Havilland Dash 8Q (100th Dash 8 door delivered 12 June 1992), wing ribs and emergency exits for Airbus A319 and A320, tailcone/landing gear door/pylon components for Lockheed Martin C-130, rear fuselage and tail components for Boeing 737, and other machined parts for BAE, Boeing and EADS Deutschland. Collaboration with Hellenic Aerospace Industry Ltd announced in February 1997 on formation of Shenyang Hellenic Aircraft Repair Company.

UPDATED

SAC J-8 II

Chinese name: Jianjiji-8 (Fighter aircraft 8)
Westernised designation: F-8
NATO reporting name: Finback-B

TYPE: Multirole fighter.

PROGRAMME: Development of original J-8 started 1964; first flight 5 July 1969; initial production authorised July 1979, began December 1979. All-weather J-8 I made maiden flight 24 April 1981 and production approved July 1985; ended 1987 after between 100 and 150 J-8s and J-8 Is.

Present configuration is improved J-8 II, on which design work began 1980; maiden flight of first of four prototypes took place 12 June 1984. Peace Pearl programme (1990-91 *Jane's*), to upgrade J-8 II with Western avionics, embargoed by US government mid-1989 and cancelled by China 1990; alternative (F-8 IIM) upgrade programme now in progress (see separate entry). Trials with some parts of structure covered in Xikai SF18 radar-absorbent material were reported in early 1999.

CURRENT VERSIONS: **J-8** (Finback-A): Initial clear-weather day fighter. Single-piece forward-opening canopy; simple ranging radar in nose centrebody; single-barrelled gun on each side of nosewheel bay, with PL-2 AAMs on inboard wing pylons and provision for external tanks outboard; other details in earlier editions of *Jane's*. At least 20 built; now probably used only by 2nd Regiment of PLA Air Force training and test centre (Kunming ?) and 70th Fighter Regiment of 24th Fighter Division at Yangchun.

J-8 I (Finback-A): Improved (all-weather) version of J-8. Same power plant, but fitted from outset with Sichuan SR-4 fire-control radar in intake centrebody; single twin-barrelled 23 mm cannon on each side of lower front fuselage. Production completed; further details in 1985-86 *Jane's*. Approximately 54 in service in mid-2001, including some converted to reconnaissance role (see JZ-8 paragraph below). Units have included 1st and 3rd Regiments of 1st Fighter Division at Anshan; some (including reconnaissance versions) remain in use with 70th Fighter Regiment at Yangchun.

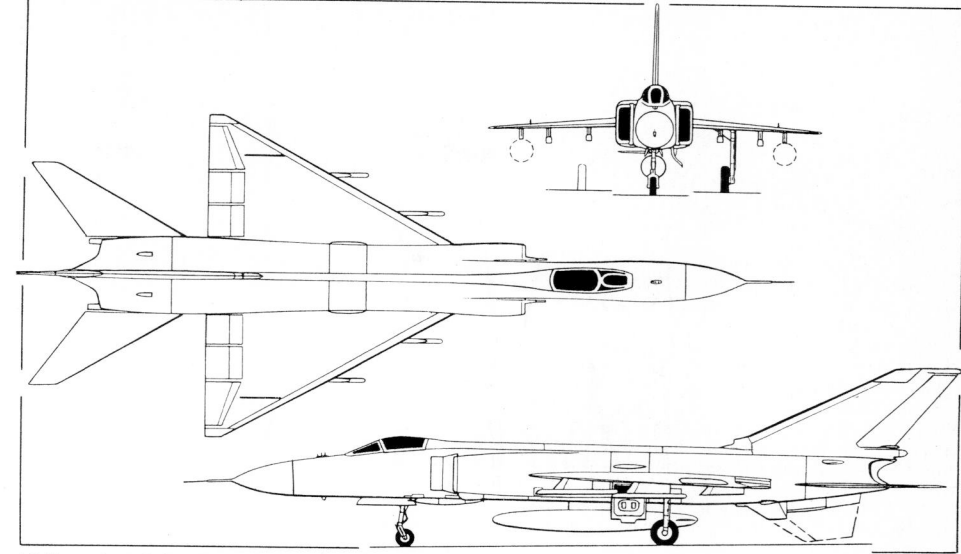

J-8 II version of the 'Finback' twin-jet fighter (*Jane's/Dennis Punnett*)

J-8 II ('Finback-B'): All-weather version, some 70 per cent redesigned compared with J-8 I. Main configuration change is to 'solid' nose and twin lateral air intakes, providing more nose space for fire-control radar and other avionics, plus increased airflow for more powerful WP13A II turbojets. In production and service, but manufactured in small economic batches rather than continuous production. *Detailed description applies to J-8 II.*

J-8 IIB ('Finback-B'): Late-production J-8 IIs have an upgraded (KLJ-1 ?) fire-control radar with a lookdown/shootdown mode compatible with PL-8 (Python 3) IR-guided and PL-11 semi-active radar-guided AAMs, plus a new KJ-8602 RWR antenna on the fin-tip. New avionics, some possibly of Israeli design or origin, include an HK-13E HUD, 563B INS, JD-3II Tacan and an RKL 800A integrated ECM suite. In service with PLA Air Force and Navy.

J-8 II ACT: Active Control Technology (fly-by-wire) testbed, which first flew 29 December 1996 and completed its 49th and last sortie on 21 September 1999; shown in model form at Airshow China in November 2000. Full-authority quad-redundant, three-axis digital AFCS; small canards mounted high on air intakes to induce instability;

two 1553B-standard flight computers with databus interface; integrated servo actuators for all moving control surfaces.

J-8 III: New production variant (Westernised designation **J-8C**), apparently based on development work with J-8 II ACT. Features include fly-by-wire flight controls, canards, WP14 turbojets (73.6 kN; 16,535 lb st with afterburning), in-flight refuelling probe and new IAI Elta EL/M-2034 or similar fire-control radar. Prototype (8301) reportedly flew in 1993, but flight test not completed until late 2001; may now be in limited use.

J-8 IV ('Finback-B Mod'): Designation (**J-8D** in Westernised form) of 12 or more J-8 IIBs built or modified for in-flight refuelling; fixed, non-retractable probe on starboard side of cockpit; combat radius increased to 648 n miles (1,200 km; 745 miles). In service with PLA Air Force and Navy.

JZ-8: Reconnaissance version (also reported as **J-8E** and **J-8 V**), believed to be converted from J-8 I; at least six known. Retains gun armament; undernose sensor package similar to that fitted to some Su-17/20/22 variants; centreline pod similar in appearance to that carried by MiG-21R, incorporating large rectangular camera window or SLAR antenna. In service with PLA Air Force.

PLA Air Force squadron of J-8 IV fighters   0105278

Model of J-8 II ACT testbed with underfuselage targeting pod and wing-mounted AAMs (*Photo Link*)   0100786

Line-up of J-8 I and J-8 II fighters at a Chinese air base   0116936

Twelfth production J-8 II of the PLA Navy's 22nd Fighter Regiment at Lingshui   0116938

F-8 II: Originally proposed export version: WP13B engines (uprated by 4 per cent to 68.7 kN; 15,430 lb st with afterburning), pulse Doppler lookdown radar, digital avionics (including HUD and two HDDs) with 1553B databus, leading-edge flaps, in-flight refuelling, seven stations for 4,500 kg (9,921 lb) stores load, maximum speed M2.2. Programme modified to become present F-8 IIM.

F-8 IIM: Upgraded version of F-8 II; *described separately.*

CUSTOMERS: PLA Air Force and Navy (approximately 150 and 40 respectively, of all versions).

DESIGN FEATURES: Extension of late 1950s Soviet heavy fighter theory. Thin-section, mid-mounted delta wings and all-sweptback tail surfaces; fuselage has area rule 'waisting', detachable rear portion for engine access, and dorsal spine fairing. Large ventral fin under rear fuselage, main portion of which folds sideways to starboard during take-off and landing, provides additional directional stability; small fence on each wing upper surface near tip; small airscoops at foot of fin leading-edge and at top of fuselage each side, above tailplane. Sweepback 60° on wing and tailplane leading-edges; wings have slight anhedral.

FLYING CONTROLS: Hydraulically boosted ailerons, rudder and low-set all-moving tailplane; two-segment single-slotted flaps on each wing trailing-edge, inboard of aileron; four door-type underfuselage airbrakes, one under each engine air intake trunk and one immediately aft of each mainwheel well.

STRUCTURE: Conventional aluminium alloy semi-monocoque/ stressed skin construction, with high-tensile steel for high load-bearing areas of wings and fuselage and titanium in high-temperature fuselage areas; ailerons, rudder and rear portion of tailplane are of aluminium honeycomb with sheet aluminium skin; dielectric skins on nosecone, tip of main fin, and on non-folding portion of ventral fin leading-edge.

LANDING GEAR: Hydraulically retractable tricycle type, with single wheel and oleo-pneumatic shock-absorber on each unit. Steerable nose unit retracts forward, main units inward into centre-fuselage; mainwheels turn to stow vertically inside fuselage, resulting in slight overwing bulge. Brake-chute in bullet fairing at base of rudder.

POWER PLANT: Two LMC (Liyang) WP13A II turbojets, each rated at 42.7 kN (9,590 lb st) dry and 65.9 kN (14,815 lb st) with afterburning, mounted side by side in rear fuselage with pen-nib fairing above and between exhaust nozzles. Lateral, non-swept air intakes, with automatically regulated ramp angle and large splitter plates similar in shape to those of MiG-23. Internal fuel capacity (four integral wing tanks plus fuselage tanks) approximately 5,400 litres (1,426 US gallons; 1,188 Imp gallons). Single-point pressure refuelling. Provision for auxiliary fuel tanks on fuselage centreline and each outboard underwing pylon. J-8 IID retrofitted with probe for in-flight refuelling from Xian H-6 (Tu-16) bombers converted as aerial tankers.

ACCOMMODATION: Pilot only, on zero/zero ejection seat under one-piece canopy hinged at rear and opening upward.

Cockpit pressurised, heated and air conditioned. Heated windscreen.

SYSTEMS: Two simple air-cycle environmental control systems, one for cockpit heating and air conditioning and one for radar cooling; cooling air bled from engine compressor. Two 207 bar (3,000 lb/sq in) independent hydraulic systems (main utility system plus one for flight control surfaces boost), powered by engine-driven pumps. Primary electrical power (28.5 V DC) from two 12 kW engine-driven starter/generators, with two 6 kVA alternators for 115/200 V three-phase AC at 400 Hz. Pneumatic bottles for emergency landing gear extension. Pop-out ram air emergency turbine under fuselage.

AVIONICS: *Comms:* VHF/UHF and HF/SSB radios; 'Odd Rods'-type IFF.

Radar: Obsolete Type 208 monopulse radar in nose. To be retrofitted with Phazotron Zhuk-8 II multimode fire-control radar (100 ordered in June 2001).

Flight: ILS, Tacan, marker beacon receiver, radio compass, radar altimeter, autopilot.

Mission: Gyro gunsight and gun camera.

Self-Defence: RWR (antenna in fin-tip); chaff/flare dispensers in tailcone.

Enlarged avionics bays in nose and fuselage provide room for modernised fire-control system and other upgraded avionics of later versions.

ARMAMENT: One 23 mm Type 23-3 twin-barrel cannon, with 200 rounds, in underfuselage pack immediately aft of nosewheel doors. Seven external stations (one under fuselage and three under each wing) for a variety of stores which can include PL-2B IR air-to-air missiles, PL-7 medium-range semi-active radar homing air-to-air missiles, Qingan HF-16B 12-round pods of 57 mm Type 57-2 unguided air-to-air rockets, launchers for 90 mm air-to-surface rockets, bombs, or (centreline and outboard underwing stations only) auxiliary fuel tanks.

DIMENSIONS, EXTERNAL:
Wing span	9.345 m (30 ft 8 in)
Wing aspect ratio	2.1
Length overall, incl nose probe	20.53 m (67 ft 4¼ in)
Height overall	6.01 m (19 ft 8½ in)
Wheel track	3.74 m (12 ft 3¼ in)
Wheelbase	7.335 m (24 ft 0¾ in)

AREAS:
Wings, gross	42.20 m² (454.2 sq ft)

WEIGHTS AND LOADINGS:
Weight empty	9,820 kg (21,649 lb)
Normal T-O weight	14,300 kg (31,526 lb)
Max T-O weight	15,288 kg (33,704 lb)
Wing loading:	
at normal T-O weight	338.9 kg/m² (69.40 lb/sq ft)
at max T-O weight	421.8 kg/m² (86.39 lb/sq ft)
Power loading:	
at normal T-O weight	108 kg/kN (1.06 lb/lb st)
at max T-O weight	135 kg/kN (1.32 lb/lb st)

PERFORMANCE:
Design max operating Mach No. (MMO)	2.2
Design max level speed	701 kt (1,300 km/h; 808 mph) IAS

Fin-tip RWR antenna and refuelling probe identify this 'Finback-B' as a J-8 IV   *NEW*/0137978

Unstick speed	175 kt (325 km/h; 202 mph)
Touchdown speed	156 kt (290 km/h; 180 mph)
Max rate of climb at S/L	12,000 m (39,370 ft)/min
Acceleration from M0.6 to M1.25 at 5,000 m (16,400 ft)	54 s
Service ceiling	20,200 m (66,275 ft)
T-O run with afterburning	670 m (2,200 ft)
Landing run, brake-chute deployed	1,000 m (3,280 ft)
Combat radius	432 n miles (800 km; 497 miles)
Ferry range	1,188 n miles (2,200 km; 1,367 miles)
g limit in sustained turn at M0.9 at 5,000 m (16,400 ft)	+4.83

UPDATED

SAC F-8 IIM

TYPE: Multirole fighter.

PROGRAMME: Upgraded J-8 II; outgrowth of earlier proposals for F-8 II export version. Developed jointly by SAC and Shenyang Aircraft Research Institute (SARI); first flight 31 March 1996, two years after delivery of drawings; flight testing of aircraft and WP13B engine completed 19 January 1998. Second F-8 IIM completed by late 1998. Additional proposed upgrades revealed by then included No. 607 Institute Blue Sky low-altitude navigation pod; Southwest China Electronic Equipment Research Institute KG 300G airborne self-protection jammer pod; No. 613 Institute FLIR/laser targeting pod; and a triple redundant digital fly-by-wire flight control system. These would presumably be similar to the systems quoted for the J-8 II ACT.

CUSTOMERS: Developed primarily for export, but none yet ordered; however, improvements now being adopted as upgrade for in-service PLAAF J-8 IIs.

DESIGN FEATURES: Main differences from J-8 II are more powerful engine, improved avionics and modernised cockpit with HOTAS controls.

POWER PLANT: Two LMC (Liyang) WP13B turbojets, each rated at 47.1 kN (10,582 lb st) dry and 68.7 kN (15,432 lb st) with afterburning.

SYSTEMS: Electrical system includes two 15 kVA starter/ generators.

AVIONICS: *Comms:* Advanced com/nav radios; IFF.

Radar: Phazotron Zhuk-8 II multifunction, look-up/ look-down pulse Doppler radar, with 38 n mile (70 km; 43.5 mile) detection range for approaching targets and 21.5 n mile (40 km; 25 mile) range for receding targets (both targets assumed to have 3 m²; 32.3 sq ft radar cross-section). N010 Zhuk-27 being flight tested by late 1998, presumably in second aircraft.

Flight: Include Tacan and datalink; locally developed HUD and INS/GPS navigation; MFDs; HOTAS controls; ARINC 429 databus.

Self-defence: Omnidirectional RWR; rear hemisphere noise jammer against threat (including pulse Doppler) radars; chaff/flare dispenser.

ARMAMENT: Up to six Chinese PL-5 or PL-9 short-range AAMs on underwing stations, or two Russian R-27R1 (AA-10 'Alamo') medium-range AAMs; up to four seven-round pods of 90 mm Type 90-1 rockets; up to 10 anti-runway bombs or 10 Type 250-III or 250-IV low-drag bombs (four under wings and six under fuselage); or eight anti-tank bombs; or five 500 kg low-drag bombs. Internal cannon as for J-8 II.

DIMENSIONS, EXTERNAL, AND AREAS: As for J-8 II

WEIGHTS AND LOADINGS:
Weight empty	10,371 kg (22,864 lb)
Normal fuel load	4,200 kg (9,259 lb)
Max external stores load	4,500 kg (9,921 lb)
Normal T-O weight	15,288 kg (33,704 lb)
Max T-O weight	18,879 kg (41,621 lb)
Max wing loading	447.4 kg/m² (91.63 lb/sq ft)
Max power loading	138 kg/kN (1.35 lb/lb st)

PERFORMANCE:
Max operating Mach No. (MMO)	2.2
Unstick speed	179 kt (330 km/h; 206 mph)
Touchdown speed	162 kt (300 km/h; 186 mph)
Level acceleration:	
M0.7 to M1.0 at 1,000 m (3,280 ft)	21 s
M0.6 to M1.25 at 5,000 m (16,400 ft)	55 s
Max rate of climb at M0.9:	
at 1,000 m (3,280 ft)	13,440 m (44,094 ft)/min
at 5,000 m (16,400 ft)	9,600 m (31,496 ft)/min

The modernised cockpit of the F-8 IIM

Service ceiling	18,000 m (59,060 ft)
T-O run with afterburning	630 m (2,070 ft)
Landing run with brake-chute	900 m (2,955 ft)

Typical mission radius:
air-to-air interception, out and back at M0.8:
at 500 m (1,640 ft) with 3 min combat

    189 n miles (350 km; 217 miles)

at 11,000 m (36,080 ft) with 5 min combat

    540 n miles (1,000 km; 621 miles)

combat air patrol, incl 10 min patrol and 5 min combat, out at M0.8 at 5,000 m (16,400 ft), back at M0.8 at 11,000 m (36,080 ft)

    324 n miles (600 km; 372 miles)

air-to-ground attack, out and back at M0.8 at 10,000 m (32,800 ft), incl 5 min combat

    486 n miles (900 km; 559 miles)

Ferry range   1,026 n miles (1,900 km; 1,180 miles)

g limits during sustained turn at M0.9:

at 1,000 m (3,280 ft)	+6.9
at 5,000 m (16,400 ft)	+4.7

VERIFIED

SAC (SUKHOI Su-27) J-11

TYPE: Air superiority fighter.

PROGRAMME: In February 1996, the Russian military sales organisation Rosvooruzheniye (now Rosoboronexport) announced a contract under which China would be licensed to manufacture the Sukhoi Su-27 'Flanker' (up to 50 per year; total of 200 in all) at Shenyang.

An initial batch of 26 Russian-built Su-27s, delivered from 1992, comprised 20 Su-27SKs ('Flanker-B') and six two-seat Su-27UBK combat trainers ('Flanker-C'). These equip the PLAAF's 3rd Division, based at Wuhu, Anhui Province, in the Nanjing Military Region. They were followed in 1996 by a further 24 (18 and six, respectively), delivered to the 2nd Division in Shuixi, Guangdong Province, Guangzhou Military Region. Agreement for an additional 50 to 60, for 1997-98 delivery, was reached in August 1997. This was preceded in February 1997 by Russian licence for Chinese manufacture at Shenyang, initially in the form of CKD kit assembly; in 1998 KnAAPO delivered the first two kits of a reported batch of 15; both of these made their first flight in December 1998, and about 20 reportedly completed by mid-2002. However, it was reported in mid-2000 that substandard work had caused Russian technicians to rebuild first two aircraft, necessitating import, from 14 December 2000 onwards, of 28 additional Russian-built two-seat Su-27UBKs to offset shortfall in Chinese production and maintain pilot training schedule. Eight of these delivered by 31 December 2000, with 10 each following in 2001 and 2002. China also reported in mid-2002 to have ordered up to 80 Russian-built Su-30MK2 naval strike variants.

F-8 IIM in attack mode, with bombs, rocket pods and air-to-air missiles   0010256

As regards Chinese production, six or seven Su-27s were planned to be assembled annually during 1999-2001, increasing to 15 to 20 per year from 2002. Chinese-built Su-27s are designated **J-11** (single-seat) and **JJ-11** (two-seat). Later production will be of upgraded Su-30MKK variant (Chinese designation possibly **J-13**, but not confirmed). Descriptions in Russian section.

UPDATED

SAC (?) 'XXJ'

The embryonic advanced combat aircraft to which the US Office of Naval Intelligence (ONI) allocated the provisional designation XXJ was last described in the 1998-99 *Jane's*, at which time it was thought likely to be a Chengdu (CAC) design. Little was then heard of it for some years, but in the early months of 2001 further US-sourced information indicated that the design had been considerably modified, now revealing a configuration suggesting some Sukhoi influence. This would appear to make Shenyang the more likely developer, although both Chinese companies are thought to have (possibly competing) advanced combat aircraft projects in hand.

According to this recent information, the 'XXJ' is of canard delta configuration, combined with clipped-delta horizontal tail surfaces and twin outward-canted fins and rudders. Conformal underfuselage intakes feed a pair of thrust-vectoring turbojets or turbofans, possibly the new 116 kN (26,000 lb) class WP15. Empty weight has been estimated at 20,000 kg (44,100 lb); stealth characteristics and FBW flight controls are assumed.

VERIFIED

SAIC (Shanghai)

SHANGHAI AVIATION INDUSTRY CORPORATION (Shanghai Hangkong Gongye Gongsi) (Subsidiary of AVIC I)

2668 Zhongshan North Road, Shanghai 200063
Tel: (+86 21) 62 57 33 51
Fax: (+86 21) 62 57 33 50

e-mail: saiccn@sh163a.sta.net.cn
Web: http://www.saic-china.com
CHAIRMAN: Li Wanxin

SAIC is the core company of the Shanghai Aviation Industry Group. It has more than 20 subordinate enterprises, of which the principal ones are Shanghai Aircraft Manufacturing

Factory (SAMF, which see); Shanghai Aircraft Research Institute (SARI); Shanghai Aero-Engine Manufacturing Factory; and Shanghai International Aero Technology. SAIC and SARI are taking part in studies for a new Chinese regional jet, as described under the ACAC heading in this section.

UPDATED

SAIC (Shijiazhuang)

SHIJIAZHUANG AIRCRAFT INDUSTRY CORPORATION (Shijiazhuang Feiji Gongye Gongsi) (Subsidiary of AVIC II)

PO Box 164, 25 Beihuanxilu, Shijiazhuang, Hebei 050062
Tel: (+86 311) 777 52 83
Fax: (+86 311) 775 29 93
e-mail: samc@yeah-net
Web: http://www.samc.com.cn
GENERAL MANAGER: Cheng Binyou
CHIEF ENGINEER: Lang Zhidong

SAIC (formerly SAMC) has produced Chinese Y-5 versions of the An-2 general purpose biplane since 1970, concentrating since 1989 on Y-5B and C customised agricultural, forestry, tourist and parachutist versions; other products have included W-5 and W-6 ultralight series (see 1992-93 and earlier *Jane's*). With the effective cessation of Polish An-2 production, SAIC is the only company now regularly building new examples of this long-serving biplane. Became part of Xian Aircraft Industrial Group July 1992, but relocated within AVIC II in 1999 reorganisation.

Occupies 46 ha (113.7 acre) site, including over 100,000 m² (1,076,400 sq ft) of covered space. Workforce of 2,803 in April 2001 included 461 engineers and technicians.

UPDATED

SAIC Y-5

Chinese name: Yunshuji-5 (Transport aircraft 5)

TYPE: Utility biplane.

PROGRAMME: Antonov An-2 has been built under licence in China since 1957, chiefly at Nanchang (727 produced up to 1968) and latterly by SAMC; production of Y-5N civil transport and general purpose version ended in 1986, at which time 221 had been completed. Y-5B dedicated agricultural and forestry version made first flight 2 June 1989; now in batch production; first nine Y-5Bs produced in 1990; total of 112 Y-5B/Cs produced by end of 2000 (latest total received). SAIC now only source of continuing production of this aircraft.

CURRENT VERSIONS: **Y-5B:** Dedicated agricultural and forestry version; certified to CCAR 23 Chinese equivalent of FAR Pt 23.

Following description applies to Y-5B, except where indicated.

Y5B-100: Designation of version displayed at Zhuhai in November 1998, fitted with wing 'tipsails' of the Y-5C and registered B-8448.

Y-5B(K): Tourist version, first flown 1993; certified to CCAR 23.

Y-5B(D): Multipurpose (agri-forest or tourist) version, first flown 1995; certified to CCAR 23.

Y-5C: Parachutist version, with wingtip vanes ('tipsails'); first flown 1996. Ordered by PLA Air Force.

CUSTOMERS: See table.

Y-5B/C PRODUCTION
(at 1 January 2001)

Customer	Qty
Military (Y-5C)	
PLA Air Force	48
Airlines (Y-5B)	
Changchun Airlines	4
Hebei Jihua Airlines	3
Xinjiang Airlines	15
Other local operators	17
General (Y-5B)	
Anhui General Aviation	3
Chaoyang General Aviation	8
Guanghan General Aviation Flying School	5
Guangzhou General Aviation	3
Hebei General Aviation	3
Jiangnan General Aviation	3
Total	**112**

DESIGN FEATURES: Unequal span, single-bay biplane; strut-braced wings (single I strut each side) and tail; fuselage circular section forward, rectangular in cabin section, oval in tail section; fin integral with rear fuselage. RPS wing section, thickness/chord ratio 14 per

Y-5B(K) civil transport (*Sebastian Zacharias*)   0010257

The 'tipsailed' Y-5C for the PLA Air Force

cent (constant); dihedral, both wings, approximately 2° 48′.

FLYING CONTROLS: Conventional and manual. Differential ailerons, plus elevators and rudder, actuated by cables and push/pull rods; electric trim tab in port aileron, rudder and port elevator; full-span automatic leading-edge slots on upper wings; electrically actuated slotted trailing-edge flaps on both wings. Wingtip vanes on some variants (see Current Versions) for additional control at low speeds.

STRUCTURE: All-metal, with fabric covering on wings aft of main spar and on tailplane. Y-5B specially treated to resist corrosion; cabin doors sealed against chemical ingress; empty weight reduced.

LANDING GEAR: Non-retractable split axle type, with long-stroke oleo-pneumatic shock-absorbers. Mainwheel tyres size 800×260, pressure 2.25 bar (33 lb/sq in). Pneumatic shoe brakes on main units. Fully castoring and self-centring tailwheel, size 470×210, with electropneumatic lock. For rough field operation, shock-absorbers can be charged from compressed air cylinder in rear fuselage. Interchangeable ski landing gear optional.

POWER PLANT: One 735 kW (985 hp) SAEC (Zhuzhou) HS5 nine-cylinder radial engine, driving a Baoding J12B-G15 four-blade variable-pitch propeller. Fuel capacity 1,240 litres (328 US gallons; 273 Imp gallons).

ACCOMMODATION: Flight crew of one or two; dual controls; seats for 12 tourist class passengers (Y-5B(K)) or 10 parachutists (Y-5C). Emergency exit on starboard side at rear. Cabin heating and ventilation improved by new ECS.

AVIONICS: *Comms:* Honeywell KHF 950 HF and KY 196 VHF radios and KMA 24 audio control panel; some electrical and other instrument installations also improved.
    Flight: GPS 155.

EQUIPMENT: Large hopper/tank with emergency jettison of contents; high flow rate, wind-driven pump; sprayers with various nozzle sizes, depending upon spray volume required.

DIMENSIONS, EXTERNAL:
Wing span: upper	18.18 m (59 ft 7¾ in)
lower	14.24 m (46 ft 8½ in)
Wing aspect ratio: upper	7.6
lower	7.3
Length overall, tail down	12.40 m (40 ft 8¼ in)
Height overall, tail down	4.01 m (13 ft 2 in)
Wheel track	3.36 m (11 ft 0¼ in)
Wheelbase	8.19 m (26 ft 10½ in)
Propeller diameter	3.40 m (11 ft 2 in)
Cargo door (port): Mean height	1.55 m (5 ft 1 in)
Mean width	1.39 m (4 ft 6¾ in)

DIMENSIONS, INTERNAL:
Cargo compartment: Length	4.10 m (13 ft 5½ in)
Max width	1.60 m (5 ft 3 in)
Max height	1.80 m (5 ft 10¾ in)

AREAS:
Wings, gross: upper	43.54 m² (468.7 sq ft)
lower	27.98 m² (301.2 sq ft)

WEIGHTS AND LOADINGS (A: Y-5B with dry chemical spreader, B: Y-5B with liquid spray system, C: Y-5C):
Max payload: A, B, C	1,500 kg (3,307 lb)
Max T-O weight: A, B, C	5,250 kg (11,574 lb)
Max wing loading: A, B, C	73.4 kg/m² (15.03 lb/sq ft)
Max power loading: A, B, C	7.15 kg/kW (11.75 lb/hp)

PERFORMANCE (A, B and C as for Weights):
Max level speed at S/L:	
A	110 kt (205 km/h; 127 mph)
B	108 kt (200 km/h; 124 mph)
C	129 kt (239 km/h; 148 mph)

Max level speed at 1,700 m (5,575 ft):	
A	119 kt (220 km/h; 137 mph)
B	116 kt (215 km/h; 133 mph)
C	138 kt (256 km/h; 159 mph)
Operating speed: A, B	86 kt (160 km/h; 99 mph)
Stalling speed: A, B, C	46 kt (85 km/h; 53 mph)
Max rate of climb at S/L: A	120 m (394 ft)/min
B	114 m (374 ft)/min
C	213 m (699 ft)/min
Rate of climb at 1,600 m (5,250 ft):	
A	133 m (436 ft)/min
B	123 m (404 ft)/min
C	225 m (738 ft)/min
Service ceiling: A, B	4,500 m (14,760 ft)
C	5,000 m (16,400 ft)
Air turning radius: A, B, C	350 m (1,150 ft)
T-O run: A	170 m (560 ft)
B	180 m (595 ft)
C	150 m (495 ft)
Landing run: A	160 m (525 ft)
B	157 m (515 ft)
C	170 m (560 ft)
Range at S/L with 670 litres (177 US gallons; 147 Imp gallons) fuel:	
A, B	456 n miles (845 km; 525 miles)
C	475 n miles (880 km; 546 miles)
Endurance, conditions as above: A, B	5 h 36 min
Swath width (A, B): HV, LV	40-50 m (130-165 ft)
ULV	60 m (200 ft)

UPDATED

SAMF

SHANGHAI AIRCRAFT MANUFACTURING FACTORY (Shanghai Feiji Zhizao Gongchang) (Subsidiary of SAIC Shanghai)

3115 Chang Zhong Road, Shanghai 200436
Tel: (+86 21) 56 68 11 22

Fax: (+86 21) 56 68 43 36
PRESIDENT: Liu Qianyou

Shanghai Aircraft Manufacturing Factory (SAMF) created 1951; now part of Shanghai Aviation Industry Corporation (SAIC, which see); occupies site area of 135.5 ha (334.8 acres); SAMF workforce 4,630 in late 2000. Built small batch of McDonnell Douglas MD-82/83 airliners in mid-1990s (see

1994-95 and earlier *Jane's*), and produced main and landing gear doors for MD-80 series from 1979; also produced cargo and service doors, avionics access doors and tailplanes for these aircraft. Current subcontract work includes tailplanes for Boeing 737 NG. Also built Q-2 ultralight.

UPDATED

SSHC

SHANGHAI SIKORSKY HELICOPTER COMPANY (Shanghai Sikorsky Zhishengji Gongsi)

PRESIDENT OF SLEC: Teng Wei

Agreement to form this private joint venture company announced in February 2002 by Sikorsky Aircraft (see US section) and Shanghai Little Eagle Science and Technology Company (SLEC). Objective is Chinese kit assembly and/or local manufacture of Schweizer 300C and 300CB (which

see), with certification and marketing assistance from Sikorsky; planned shareholding SLEC 51 per cent, Sikorsky 49 per cent.

NEW ENTRY

WHIC

WUHAN HELICOPTER INDUSTRY COMPANY (Wuhan Zhishengji Gongye Gongsi)

9th Floor, 871 Jiefang Avenue, Wuhan 430013
Tel: (+86 27) 85 79 31 13
Fax: (+86 27) 85 78 38 28
e-mail: whic@public.wh.hb.cn
Web: http://www.whic.com.cn
PRESIDENT: Dinghe Zhang

Established August 1993, under auspices of Wuhan local government, and with Y64.5 million capital, to co-produce

Enstrom F28F, 280FX, TH-28 and 480 helicopters (see US section); assembled first two (280FX and TH-28) from CKD kits October 1993; delivered to Wuhan Public Security Bureau and PLA respectively; at least one more US-built TH-28 supplied to China in May 1997. Two Enstrom 480s delivered for reassembly in June 1998. WHIC also operates an Enstrom 480 as demonstrator and for charter work. New factory completed on a 150 ha (370.7 acre) site and reportedly began operating in 1998; expected to have eventual capability for completing up to 100 helicopters per year; however, in late 2001, WHIC was still inviting potential investors to join the programme and had not confirmed the start of local production.

UPDATED

Wuhan Enstrom TH-28 in Chinese markings

0044499

XAC

XIAN AIRCRAFT INDUSTRY COMPANY
(Xian Feiji Gongye Gongsi) (Subsidiary of AVIC I)

PO Box 140-84, Xian, Shaanxi 710089
Tel: (+86 29) 684 56 65
Fax: (+86 29) 620 37 07
e-mail: zh-yang@xac.com.cn
Web: http://www.xac.com.cn
CHAIRMAN: Yang Zhong
PRESIDENT OF XIAN AIRCRAFT INDUSTRIAL GROUP, AND GENERAL
    MANAGER OF XAC: Gao Dacheng
VICE-PRESIDENT, XAC: Nie Zhongliang
AIRCRAFT MARKETING MANAGER: Wang Zhigang

Aircraft factory established at Xian 1958; current company is core enterprise of Xian Aircraft Industry Group (XAIG); it has covered area of some 300 ha (741.5 acres); of 20,000 workforce, about 90 per cent are engaged in aircraft production. Earlier programmes included licence production of Tu-16 twin-jet bomber (as H-6: see 1991-92 *Jane's All the World's Aircraft*); now being converted to aerial tanker role for J-8 II fighter and other aircraft (see current *Jane's Aircraft Upgrades*).

Major current programmes concern JH-7 attack aircraft and MA60 transport (Y-7/An-24 derivative). XAIG and Xian Aircraft Design and Research Institute (XADRI) are participants in the regional jet programme described under the ACAC heading in this section.

Subcontract work includes glass fibre header tanks, water float pylons, ailerons and various doors for Bombardier 415 amphibian; fins and tailplanes for Boeing 737/757; panel assemblies for Raytheon Beech 1900D (from late 1997); and various components for Airbus Industrie. In 1997, subcontracting for ATR (which began with ATR 42 wingtips in 1986) was extended to include ATR 72 rear fuselage sections. The long-term possibility of licensed production of the latter aircraft in China continued to be discussed in 2001-02.

UPDATED

XAC H-6

Tanker conversions of H-6 (Chinese Tu-16) bombers (see accompanying photograph) briefly described in 2002-03 edition; see *Jane's Aircraft Upgrades* for latest information.

UPDATED

XAC JH-7

Chinese name: Jianjiji Hongzhaji-7 (Fighter-Bomber aircraft 7)
Westernised designation: FBC-1 (formerly B-7)
Export name: Flying Leopard

XAC H-6U tanker accompanied by a pair of J-8s   *NEW*/0137972

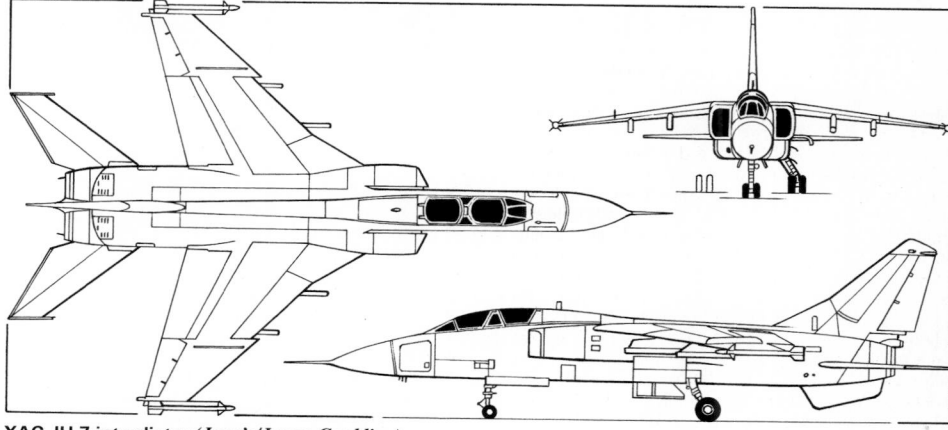
XAC JH-7 interdictor (*Jane's/James Goulding*)   0105034

A preproduction JH-7 attack fighter   0116937

Photograph believed to show a full-scale mockup of the JH-7A (*Jane's/via Yihong Chang*)   *NEW*/0096991

TYPE: Attack fighter.

PROGRAMME: Revealed publicly September 1988 as model at Farnborough International Air Show; first of four or five prototypes said to have been rolled out during previous month; first flight 14 December 1988 and first supersonic flight 17 November 1989; service entry originally scheduled for 1992-93, but delayed; first seen openly in TV broadcast of October 1995 naval exercise.

First public appearance (by third prototype '083', wearing Flight Test Establishment colours) was made in November 1998 flypast at Airshow China in Zhuhai. This coincided with announcement of new export designation and name, and statement that aircraft was "being redesigned" for export market.

CURRENT VERSIONS: **JH-7:** Domestic version.
Following description applies to JH-7 except where indicated.

JH-7A: Various China-, Russia- and US-based sources have reported that No. 603 Institute still undertaking improvements that include No. 613 Institute FLIR/laser targeting pod and No. 607 Institute Blue Sky low-altitude navigation pod; modified LETRI JL-10A Shen Ying radar; digital (and possibly quadruple) fly-by-wire controls; a health monitoring system; INS/GPS; new databus; two additional hardpoints each side; one-piece wraparound windscreen; greater use of composites; and integration of additional Russian weapons such as Kh-31P (AS-7 'Krypton') standoff missiles and KAB-500 laser-guided bombs.

Development completed in 2001 and first flight expected in 2002. Possibility of 20 to 30 being produced for service entry in 2004.

FBC-1 Flying Leopard: Designation (from 1998) of proposed export version; would have customer-defined radar, avionics and armament. No orders yet announced.

CUSTOMERS: Earlier (1997) Chinese reports, of up to 24 in service with PLA Naval Air Force in operational evaluation role in mid-1997, were apparently exaggerated (stated at 1998 Airshow China that only seven prototype/preproduction aircraft then completed). Latest best estimate (early 2001) was of 20 aircraft, including prototypes, of which 18 confirmed by known serial numbers; this is compatible with known number (50) of Spey engines initially purchased, but confirmation reported that a further 80 to 90 Speys were being delivered in mid-2001; these possibly for JH-7A. First identified JH-7 unit is PLA Naval Air Force 16th Bomber Regiment, 6th Naval Air Division, at Dachang, Shanghai.

DESIGN FEATURES: In same role and configuration class as Russian Sukhoi Su-24 'Fencer'. High-mounted wings with compound sweepback, dog-tooth leading-edges and 7° anhedral; twin turbofans, with lateral air intakes; all-swept tail surfaces, comprising large main fin, single small ventral fin and low-set all-moving tailplane with anti-flutter weights at tips; small overwing fence at approximately two-thirds span. Wings have 47° 30' sweepback on leading-edges. Quarter-chord sweep angles approximately 45° on fin, 55° on tailplane.

STRUCTURE: Conventional all-metal, except for dielectric panels. Plans to fit SF18 radar-absorbent material revealed in early 1999.

LANDING GEAR: Retractable tricycle type, with twin wheels on each unit. Trailing-link main units retract inward, nose unit rearward.

POWER PLANT: All aircraft so far built are powered by two licence-built (Xian WS9) or UK-supplied Rolls-Royce Spey Mk 202 turbofans: each 91.2 kN; 20,515 lb st with afterburning. Centreline and outboard underwing stations plumbed for auxiliary fuel tanks.

Original intention to power production JH-7 with LM (Liming) WS6 turbofans of 71.1 kN (15,990 lb st) dry rating (122.1 kN; 27,445 lb st with afterburning) abandoned, as was licensed production of Russian alternative such as Saturn/Lyulka AL-31F or SNECMA M53-P2.

ACCOMMODATION: Crew of two in tandem (rear seat elevated). HTY-4 ejection seats, operable at speeds from zero to 540 kt (1,000 km/h; 621 mph) and altitudes from S/L to 20,000 m (65,600 ft). Individual canopies, hinged at rear and opening upward. One-piece wraparound windscreen on JH-7A.

AVIONICS (FBC-1): *Comms:* Short wave and ultra short wave transceivers of Italian origin.

Radar: LETRI JL-10A Shen Ying J-band pulse Doppler fire-control radar (search range 43 n miles; 80 km; 50 miles and tracking range 22 n miles; 40 km; 25 miles in look-up mode, or 29 n miles; 54 km; 34 miles and 17 n miles; 32 km; 20 miles, respectively, in look-down mode). Scans ±60° in azimuth and can track four targets simultaneously.

Flight: Automatic flight control system; GPS/INS navigation.

Instrumentation: Includes No. 603 Institute HUD and two other MFDs.

Mission: AFCS linked to fire-control system; No. 613 Institute helmet-mounted sight.

Self-defence: Chaff/flare dispenser at base of vertical tail; omnidirectional RWR; active and passive ECM.

ARMAMENT: Twin-barrel 23 mm gun, with 200 rounds, in starboard side of lower fuselage, just forward of mainwheel bay. JH-7 has fuselage centreline stores station, plus two under each wing and rail for PL-5B, PL-7 or similar close-range air-to-air missile at each wingtip; JH-7A has additional mid-point station under each wing, plus one on lower side of each engine intake trunk. Typical underwing load for maritime attack, two C-801K or C-802K sea-skimming anti-ship missiles (inboard) and two drop tanks (outboard). Other potential weapons may include C-701 TV-guided anti-ship missiles and 500 kg LGBs.

DIMENSIONS, EXTERNAL:

Wing span	12.705 m (41 ft 8¼ in)
Wing aspect ratio	3.1
Length overall: excl probe	21.025 m (68 ft 11¾ in)
incl probe	22.325 m (73 ft 3 in)
Fuselage length (excl probe)	19.105 m (62 ft 8¼ in)
Height overall	6.575 m (21 ft 6¾ in)
Tailplane span	7.39 m (24 ft 3 in)
Wheel track	3.06 m (10 ft 0½ in)
Wheelbase	7.805 m (25 ft 7¼ in)

AREAS:

Wings, gross	52.30 m² (563.0 sq ft)

WEIGHTS AND LOADINGS:

Max fuel weight	10,050 kg (22,156 lb)
Max external stores load	6,500 kg (14,330 lb)
Max T-O weight	28,475 kg (62,776 lb)
Max landing weight	21,130 kg (46,583 lb)
Max wing loading	544.5 kg/m² (111.51 lb/sq ft)
Max power loading (WS9 engines)	156 kg/kN (1.53 lb/lb st)

PERFORMANCE:

Max level speed at 11,000 m (36,080 ft) (clean)	M1.7 (975 kt; 808 km/h; 1,122 mph)
Max operating speed	653 kt (1,210 km/h; 751 mph) IAS
Cruising speed	M0.80-0.85 (459-487 kt; 850-903 km/h; 528-561 mph)
Service ceiling (clean)	15,600 m (51,180 ft)
T-O run	920 m (3,020 ft)
Landing run	1,050 m (3,445 ft)
Combat radius	891 n miles (1,650 km; 1,025 miles)
Ferry range	1,970 n miles (3,650 km; 2,268 miles)
g limit	+7

UPDATED

Model of the projected 'Fearless Albatross' maritime patrol variant of the MA60 *(Robert Hewson)*   0105850

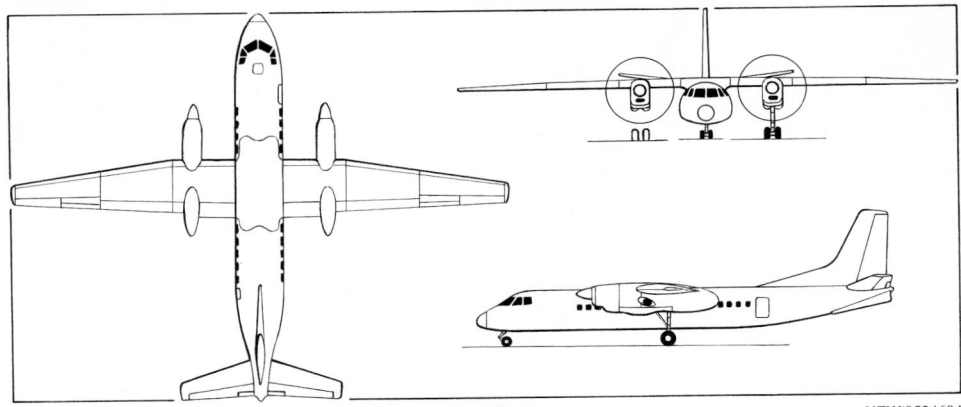

XAC MA60 twin-turboprop airliner *(Jane's/James Goulding)*   **NEW**/0524604

XAC MA60

TYPE: Twin-turboprop airliner.

PROGRAMME: Model of MA60 as proposed Y7-200A (PW127 turboprop) variant first shown at Asian Aerospace, Singapore, February 2000. First flight (prototype B-995L) reportedly 12 March 2000; CAAC type certificate issued 22 June 2000; revenue service began August 2000; public debut (Airshow China) November 2000; production certificate issued 12 December 2000.

Designation indicates Modern Ark, 60 (high density) seats. To replace earlier Y-7 variants, which from 2001 no longer meet amended Chinese airworthiness requirements. XAC production capacity 12 to 15 a year; domestic operators will lease from central leasing agency SFLC, which will be responsible for sole-source marketing (first arrangement of this kind in China).

CURRENT VERSIONS: **MA60:** Initial production passenger version; *as described.* Current life-cycle of 25,000 landings and 30,000 flight hours.

Improved MA60: Improvements planned for introduction in 2002 included reduction in empty weight by about 400 kg (882 lb); increase in 'hot-and-high' MTOW of about 900 kg (1,984 lb); increase in OEI ceiling; 162 n mile (300 km; 186 mile) increase in range; airframe drag reduction; 'glass cockpit' avionics upgrade.

MA60 Cargo: Updated MA60 equivalent of former Y7H (2000-01 *Jane's*), with rear-loading ramp/door inherited from An-26. Fully pressurised cargo compartment, with electric winch and hydraulic conveyor; cargo door can be retracted under belly to permit direct loading from truck or ground. Demonstrator bears CFTE serial number 073.

MA60-MPA (Fearless Albatross): Projected maritime patrol version (previously Y7-200BF); nose-mounted search radar; raked wingtips; auxiliary fuel tanks scabbed to fuselage sides to extend range to estimated 1,350 n miles (2,500 km; 1,553 miles); four underwing stations for air-to-air and anti-ship missiles. Shown as model at Airshow China, November 2000.

MA40: Shortened (40-passenger) version; under consideration in 2002.

CUSTOMERS: Sole-source domestic marketing by newly formed Shenzhen Finance Leasing Corporation which, on 7 November 2000, signed contract for delivery of 60 MA60s over five-year period. Launch operator Sichuan Airlines (five ordered, plus five on option) introduced first leased example (B-3426) into revenue service August

MA60 demonstrator at Singapore in February 2002 *(Jane's/Kenneth Munson)*   **NEW**/0137944

Flight deck of the MA60

NEW/0137946

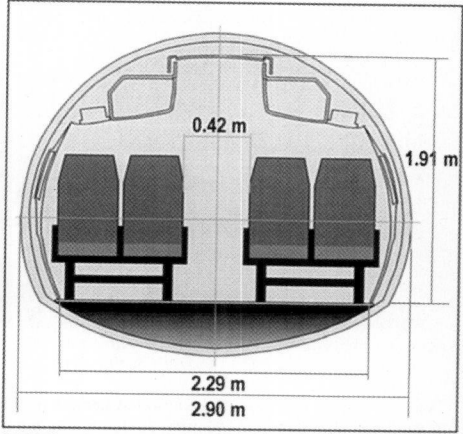

MA60 cabin cross-section   NEW/0137983

2000. Other reported orders by March 2002 from China Northern (five), Wuhan Airlines (three) and Lao Aviation (three). Of these 16, 12 said to have been completed at that time. Most of first 60 expected to be of Improved MA60 standard.

COSTS: Approximately US$11 million (2002).

DESIGN FEATURES: Further upgrade of Y-7 series, mainly through Western engines and avionics and improved passenger comfort; incorporates best features of earlier Y7-200A and B programmes. Airframe life 40,000 hours, 25,000 cycles or 25 years.

Non-swept high-mounted wings, with 2° 12′ 12″ anhedral on tapered outer panels; basically circular-section fuselage; sweptback fin and rudder; 9° dihedral on tailplane; twin ventral strakes under tailcone. Wing sweepback 6° 50′ at quarter-chord of outer panels; incidence 3°.

FLYING CONTROLS: Conventional mechanical. Mass-balanced, servo-compensated ailerons with electrically actuated trim tabs; tab in each elevator; electric trim/servo tab in rudder. Hydraulically actuated single-slotted (inboard) and double-slotted (outboard) trailing-edge flaps; landing deflection 30°.

STRUCTURE: Conventional light alloy; two-spar wing, with spot-welded skins; bonded/welded semi-monocoque fuselage.

LANDING GEAR: Retractable tricycle type with twin wheels on all units. Hydraulic actuation, with emergency gravity extension. Multidisc carbon brakes on mainwheels; hydraulically steerable (±45°) and castoring nosewheel unit.

POWER PLANT: Two 2,051 kW (2,750 shp) Pratt & Whitney Canada PW127J (PW127G in cargo version) turboprops, each driving a Hamilton Sundstrand 247F-3 four-blade, composites, fully-reversible 'scimitar' propeller. Fuel in integral wing tanks immediately outboard of nacelles, and bag-type tanks in centre-section, total usable capacity 5,200 litres (1,374 US gallons; 1,144 Imp gallons). Provision for additional tanks in centre-section. Optional pressure refuelling point in starboard engine nacelle; gravity fuelling point above each tank.

ACCOMMODATION: Crew of two on flight deck, plus one or two cabin attendants. Standard layout has four-abreast seating, with centre aisle, for 56 or 60 (with forward wardrobe removed) passengers at 72 cm (28 in) pitch in air conditioned, soundproofed and pressurised cabin. Galley and lavatory at rear on starboard side. Wardrobes forward and aft of passenger cabin, plus overhead stowage bins in cabin. Passenger airstair door on port side at rear of cabin; emergency exit opposite. Freight door forward, starboard. All doors open inward.

SYSTEMS: Hamilton Sundstrand GTCP36-150CY APU.

AVIONICS: *Comms:* Bendix/King dual KTR 908 VHF, single KHF 950 HF and CVR; Becker 3100 audio system; Chinese FJ-30C FDR.

Radar: Rockwell Collins WXR-350 weather radar; Honeywell CAS-67A collision avoidance system.

Flight: Dual VIR-32 VOR/ILS, dual ADF-60A, DME-42 and ALT-55B low-altitude radio altimeter (all Rockwell Collins); Universal UNS-1M nav system; Sextant H321AKM1 standby horizon.

Instrumentation: EFIS-85 (B14), ADS-85 air data system, APS-85 autopilot and dual AHS-85 attitude and heading systems (all Rockwell Collins); Honeywell Mk VII GPWS.

ARMAMENT (MA60-MPA): Six external stores stations: two on fuselage sides, each for 1,000 kg (2,205 lb) load; two inboard (each 1,500 kg; 3,307 lb) and two outboard (each 500 kg; 1,102 lb) under wings. Fuselage stations can each support two torpedoes, a 1,000 kg bomb, sonobuoy container or gun pod. Inboard wing stations suitable for drop fuel tank, anti-shipping missile or AAM. Outer wing stations able to carry single torpedo, 500 kg bomb, searchlight, gun pod or additional AAM.

DIMENSIONS, EXTERNAL:

Wing span	29.20 m (95 ft 9½ in)
Wing chord: at root	3.50 m (11 ft 5¾ in)
at tip	1.095 m (3 ft 7 in)
Wing aspect ratio	11.4
Length overall	24.71 m (81 ft 0¾ in)
Fuselage max width	2.90 m (9 ft 6¼ in)
Height overall	8.855 m (29 ft 0½ in)
Wheel track (c/l of shock-struts)	7.90 m (25 ft 11 in)
Wheelbase	9.565 m (31 ft 4½ in)
Propeller diameter	3.96 m (13 ft 0 in)
Passenger door: Height	1.405 m (4 ft 7¼ in)
Width	0.75 m (2 ft 5½ in)
Forward freight door: Height	1.22 m (4 ft 0 in)
Width	1.19 m (3 ft 10¾ in)
Rear baggage door: Height	1.41 m (4 ft 7½ in)
Width	0.75 m (2 ft 5½ in)

Emergency exits (two, each):	
Height	0.93 m (3 ft 0½ in)
Width	0.51 m (1 ft 8 in)

DIMENSIONS, INTERNAL:

Passenger cabin: Length	10.795 m (35 ft 5 in)
Max width	2.70 m (8 ft 10¼ in)
Max height	1.91 m (6 ft 3¼ in)
Aisle width	0.42 m (1 ft 4½ in)
Width at floor	2.29 m (7 ft 6¼ in)
Baggage volume	9.5 m³ (335 cu ft)

AREAS:

Wings, gross	74.98 m² (807.1 sq ft)

WEIGHTS AND LOADINGS (A: passenger, B: cargo, C: MPA):

Operating weight empty: A	13,700 kg (30,203 lb)
Max fuel: A, B	4,030 kg (8,885 lb)
C	7,500 kg (16,535 lb)
Baggage capacity	840 kg (1,852 lb)
Max payload: A, B	5,500 kg (12,125 lb)
Max T-O weight: A, B	21,800 kg (48,060 lb)
C	24,000 kg (52,910 lb)
Max ramp weight: A, B	21,900 kg (48,281 lb)
Max landing weight: A, B	21,200 kg (46,738 lb)
Max zero-fuel weight: A, B	19,200 kg (42,329 lb)
Max wing loading: A, B	290.7 kg/m² (59.55 lb/sq ft)
C	320.1 kg/m² (65.56 lb/sq ft)
Max power loading: A, B	5.32 kg/kW (8.74 lb/shp)
C	5.86 kg/kW (9.62 lb/shp)

PERFORMANCE (A, B and C as above):

Max cruising speed: A	278 kt (514 km/h; 319 mph)
Econ cruising speed: A	232 kt (430 km/h; 267 mph)
Max operating altitude: A	7,620 m (25,000 ft)
C	8,500 m (27,880 ft)
Service ceiling, OEI: B	4,000 m (13,120 ft)
T-O run: A	1,420 m (4,660 ft)
C	969 m (3,180 ft)
T-O field length: A	1,425 m (4,675 ft)
B	1,400 m (4,595 ft)
Landing field length: A	1,460 m (4,790 ft)
Landing run: A	1,470 m (4,825 ft)
C	650 m (2,135 ft)
Range: with 56 passengers, 100 n miles +45 min reserves:	
A	864 n miles (1,600 km; 994 miles)
B	567 n miles (1,050 km; 652 miles)
with max fuel:	
A	1,322 n miles (2,450 km; 1,522 miles)
B	1,187 n miles (2,200 km; 1,367 miles)
C	2,591 n miles (4,800 km; 2,982 miles)
Max endurance: C	12 h

UPDATED

XAC Y-7

Production completed. Full description in 2001-02 *Jane's*; shortened version in 2002-03 edition.

UPDATED

COLOMBIA

GAVILÁN

EL GAVILÁN SA

Carrera 3 No 56-19, Apartado Aéreo 180112, Santa Fé de Bogotá DC
Tel: (+57 1) 676 50 01
Fax: (+57 1) 676 06 50
e-mail: sales@elgavilan.com
Web: http://www.elgavilan.com
GENERAL MANAGER: James S Leaver
TECHNICAL DIRECTOR: Eric C Leaver
OPERATIONS AND PRODUCTION DIRECTOR: Jaime E Silva
MARKETING MANAGER: Omar Porras Pineda
ENGINEERING MANAGER: David Fernando Muñoz Galeano

Formerly Aero Mercantil (see 1991-92 and earlier *Jane's*), associated with Piper Aircraft Corporation as dealer and then distributor since 1952; sold its shares in AICSA and in 1991 formed new Gavilán (Sparrowhawk) company to pursue development and manufacture of own-design EL-1 utility aircraft.

Advanced Wing Technologies (AWT) of Canada had planned to build Gavilán 358 under late 1999 agreement, but this had been abandoned by mid-2000, at which time alternative North American production facilities being sought; by April 2002, no evidence that this had been accomplished.

UPDATED

EL GAVILÁN EL-1 GAVILÁN 358 and 508T

English name: Sparrowhawk

TYPE: Light utility transport.

PROGRAMME: Launched March 1986; sponsored by Leaver Group, of which El Gavilán SA is a member; construction began May 1987; first flight (HK-3500-Z) 27 April 1990; after 50 hours' flying, fuselage stretched 0.305 m (1 ft 0 in) immediately forward of main spar and gross weight increased by 136 kg (300 lb); new first flight 7 November 1990; wing incidence increased 1991 to improve cruising speed, and passenger windows reshaped; damage from emergency landing in USA on 1 February 1992 delayed progress and company decided to obtain FAA (FAR Pt 23,

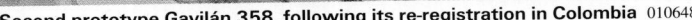

Second prototype Gavilán 358, following its re-registration in Colombia 0106488

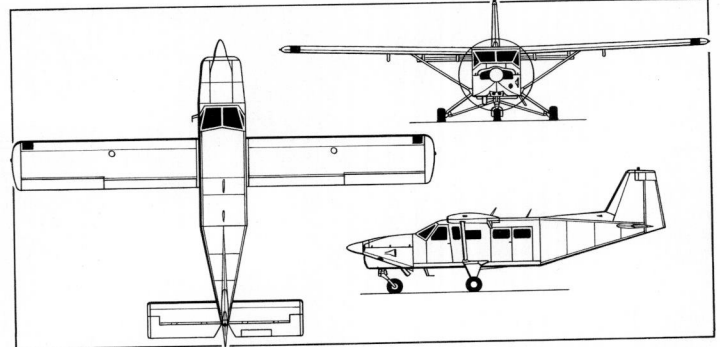

Gavilán 358 seven-passenger utility transport (*Jane's/Mike Keep*) 0044532

Amendment 45) certification from outset; this gained by second prototype, built by General Aviation Technical Services in USA; funding arranged December 1994 for FAA certification. Second aircraft (N358EL/HK-4120-Z) flown (in USA) 29 May 1996 and gained FAR Pt 23 (to Amendment 46) certification 26 May 1998. First production aircraft flew on 7 August 1997; used as North American demonstrator. Potential North American assembly site still being sought.

CURRENT VERSIONS: **Gavilán 358:** Piston-engined standard passenger, parachutist or military version.
Following description applies to Gavilán 358.
   Gavilán 508T: Turboprop version, announced May 1998; under development; choice of P&WC PT6A-34 or Walter M 601E-11 engine. Programme scheduled to start in mid-2001, but had not done so by April 2002.

CUSTOMERS: Nine confirmed orders by November 1998, including four (delivered 29 July to 30 September 1998) for Colombian Air Force (FAC), which has option on eight more. Eleven Gavilán 358s delivered by September 2000, including FAC (four), Colombian Navy (one military, delivered 23 December 1999), Colombian commercial operators (two) and Tikal Jets of Guatemala (one). Apparently no further production by April 2002.

COSTS: Standard Gavilán 358, VFR equipped US$379,000; fully equipped US$393,000 (2000).

DESIGN FEATURES: Rugged utility transport with large side door and wide track landing gear, optimised for operation from short and unprepared airstrips.
   Constant chord, unswept braced wing of NACA 4412 section; thickness/chord ratio 12 per cent; dihedral 1° 30′; incidence 5° 30′; washout 2° 30′.

FLYING CONTROLS: Conventional and manual (pushrods and cables). Mass-balanced ailerons; elevators and rudder horn balanced; spring bias on rudder; trim tab actuator in starboard elevator; manual single-slotted flaps (settings 15, 30 and 40°) on offset hinges actuated by handle on central console. Autopilot planned.

STRUCTURE: All-metal, mainly of 2024-T3 aluminium alloy sheet; fuselage frame of 4130N steel tube skinned with 2024-T3; two-spar wing with single strut each side; two-spar fin and tailplane. Entire airframe extensively corrosion-proofed.

LANDING GEAR: Non-retractable tricycle type, with elastomeric shock-absorption and single wheel on each unit. Steel tube mainwheel legs; trailing-link nose unit, with fully castoring nosewheel. McCreary tyres, sizes 7.00×6 (8 ply) main and 7.00×6 (6 ply) nose; pressures 3.59 bar (52 lb/sq in) and 2.41 bar (35 lb/sq in) respectively. Cleveland hydraulic mainwheel brakes.

POWER PLANT: One Textron Lycoming TIO-540-W2A flat-six engine (261 kW; 350 hp at 2,600 rpm), driving a three-blade constant-speed Hartzell propeller. Three rubber fuel cells in each wing, combined capacity 405 litres (107 US gallons; 89.2 Imp gallons), of which 394 litres (104 US gallons; 86.7 Imp gallons) usable. Refuelling point in top of each tank. Gravity feed system with auxiliary electric pump and header tanks. Oil capacity 11.4 litres (3.0 US gallons; 2.5 Imp gallons).

ACCOMMODATION: Pilot and co-pilot or one passenger at front; dual-control wheels. All transparencies of flat Plexiglas. Two quickly removable rows of three seats to rear of pilots, facing each other. Door at front on each side, plus larger double door at rear on port side. Accommodation for four stretchers and one attendant in air ambulance version; up to eight jumpers in parachutist version.

SYSTEMS: Engine-mounted vacuum pump for gyro instruments, driven at 0.31 bar (4.5 lb/sq in). Electrical power from 28 V 70 A engine-driven alternator and 24 V 18 Ah battery. Hydraulic system for brakes only. Gaseous oxygen system optional.

AVIONICS: *Comms:* Bendix/King KT 76A transponder and AK-350 encoder.
   Radar: Colombian Air Force aircraft planned to be equipped with Telephonics RDR 2000 weather radar in wing leading-edge radome.
   Flight: Bendix/King KLX 135A GPS/com.
   Instrumentation: Standard VFR.

EQUIPMENT: Standard cargo net and cargo tiedowns in cargo version. Optional cargo kit and air ambulance kit for passenger versions.

ARMAMENT: Colombian Navy aircraft equipped with two 7.62 mm SS-77 machine guns.

DIMENSIONS, EXTERNAL:

Wing span	12.80 m (42 ft 0 in)
Wing chord, constant	1.55 m (5 ft 1 in)
Wing aspect ratio	7.8
Length overall	9.53 m (31 ft 3 in)
Fuselage max width	1.42 m (4 ft 8 in)
Height overall	3.73 m (12 ft 3 in)
Tailplane span	4.68 m (15 ft 4¼ in)
Wheel track	3.35 m (11 ft 0 in)
Propeller diameter	2.18 m (7 ft 2 in)
Propeller ground clearance	0.17 m (6¾ in)
Crew doors (two, each):	
Height	1.17 m (3 ft 10 in)
Max width	0.86 m (2 ft 10 in)
Height to sill	1.02 m (3 ft 4 in)
Passenger/cargo door: Height	1.17 m (3 ft 10 in)
Width	1.24 m (4 ft 1 in)
Height to sill	1.02 m (3 ft 4 in)

DIMENSIONS, INTERNAL:

Cabin: Length	3.40 m (11 ft 2 in)
Max width	1.37 m (4 ft 6 in)
Max height	1.37 m (4 ft 6 in)
Floor area: excl flight deck	4.09 m² (44.0 sq ft)
incl flight deck	4.70 m² (50.6 sq ft)
Volume: excl flight deck	4.0 m³ (140 cu ft)
incl flight deck	6.4 m³ (226 cu ft)
Baggage compartment volume	0.48 m³ (17.0 cu ft)

AREAS:

Wings, gross	18.95 m² (204.0 sq ft)
Ailerons (total)	0.97 m² (10.40 sq ft)
Trailing-edge flaps (total)	1.12 m² (12.00 sq ft)
Fin	1.48 m² (15.91 sq ft)
Rudder	0.91 m² (9.81 sq ft)
Tailplane	1.51 m² (16.30 sq ft)
Elevators (total, incl tab)	0.90 m² (9.70 sq ft)

WEIGHTS AND LOADINGS:

Weight empty, standard equipped	1,270 kg (2,800 lb)
Max fuel	291 kg (642 lb)
Max payload with 30 min fuel at 75% power	488 kg (1,076 lb)
Max T-O and landing weight	2,041 kg (4,500 lb)
Max wing loading	107.7 kg/m² (22.06 lb/sq ft)
Max power loading	7.82 kg/kW (12.86 lb/hp)

PERFORMANCE:

Never-exceed speed (VNE)	155 kt (287 km/h; 178 mph) IAS
Max level speed at 4,575 m (15,000 ft)	140 kt (259 km/h; 161 mph)
Cruising speed at 3,050 m (10,000 ft):	
at 75% power	135 kt (250 km/h; 155 mph)
at 65% power	130 kt (241 km/h; 150 mph)
at 55% power	126 kt (233 km/h; 145 mph)
Stalling speed, power off:	
flaps up	65 kt (121 km/h; 75 mph) IAS
40° flap	58 kt (108 km/h; 67 mph) IAS
Max rate of climb at S/L	244 m (800 ft)/min
Max certified altitude	3,180 m (12,500 ft)
Service ceiling	6,860 m (22,500 ft)
T-O run, 15° flap	375 m (1,230 ft)
T-O to 15 m (50 ft)	595 m (1,955 ft)
Landing from 15 m (50 ft)	500 m (1,640 ft)
Landing run	327 m (1,075 ft)
Range, 45 min reserves:	
at 75% power	770 n miles (1,426 km; 886 miles)
at 65% power	820 n miles (1,518 km; 943 miles)
at 55% power	945 n miles (1,750 km; 1,087 miles)

UPDATED

CZECH REPUBLIC

AERO

AERO HOLDING AS

Beranových 130, CZ-199 04 Praha 9-Letňany
Tel: (+420 2) 88 40 65 and 88 27 47
Fax: (+420 2) 88 65 81, 88 40 65 and 88 27 47
DIRECTOR GENERAL: Ing František Petrašek
EXECUTIVE DIRECTOR, TECHNICAL AND COMMERCIAL STRATEGY:
   Ing Jan Bartoň

This joint stockholding management organisation replaced the state-owned Aero Concern of Czechoslovak Aerospace Industry on 1 December 1990; airframe, engine and equipment factories and research centres became limited companies on 1 January 1991. The joint stock company was partly privatised in 1992, basically changing Aero Holding from an operational holding to a financial one (100 per cent ownership of 10 subsidiaries); full privatisation was approved by the Czech government in June 1993. Activities now comprise organisation, co-ordination and financing of research, development, production and sale of aircraft and other aviation products.

   Shareholdings in wholly owned subsidiaries were decreased from September 1993 by sales to various foreign and domestic partners, as detailed in 2002-03 and earlier editions. Last entity to be privatised was the Aeronautical Research and Test Institute (VZLU), shares in which were contracted to PAL AS, a wholly owned subsidiary of the Czech National Property Fund (FNM) on 18 December 2000. Aero Holding is expected to be wound up by the end of 2005.
UPDATED

AERO

AERO VODOCHODY AS

CZ-250 70 Odolena Voda
Tel: (+420 2) 86 03 11 11 and 86 03 31 10
Fax: (+420 2) 83 97 00 38 and 687 25 05
e-mail: mr@aero.cz

Web: http://www.aero.cz
PRESIDENT AND CHAIRMAN: Antonín Jakubše
CEO: Vladimir Nývlt
VICE-PRESIDENTS:
   Vladimír Jaroš (Finance)
   Jiří Fidranský (Research and Development)
   Viktor Kučera (Military Aircraft Programmes)

Miloš Vališ (Civil Aircraft Programmes)
František Bílek (Aero Structures)
Martin Paloda (Marketing)
Václav Pavlíček (Services)
CHIEF DESIGNERS:
   Zdeněk Stucklík (Military Programmes)
   Josef Jironč (Civil Programmes)

Established on 1 July 1953, Aero produced (with Let) 3,665 L-29 Delfin jet trainers between 1961 and 1974.

The company underwent a change of ownership as part of a restructuring programme approved on 31 July 1996, when it had yet to recover fully from debts occurring in 1990-91. It is now a joint-stock company in which the majority shareholder is the Czech government, through shares held by Letka AS (35.66 per cent) and the Konsolidační Banka (29.00 per cent). From 17 August 1998, a further 35.29 per cent was acquired by Boeing Česká sro, a joint venture comprising Boeing and CSA Czech Airlines, owned 90 per cent by Boeing and 10 per cent by CSA. The remaining 0.05 per cent of Vodochody is owned by other shareholders. Workforce at end of 2000 was 2,067.

In May 1998, Aero Vodochody became 100 per cent owner of Technometra Radotin, producer of landing gears and other components for the aircraft of Aero and Let Kunovice. The Prague company Letov (which see), produces wings, rear fuselages and tail assemblies for Aero's jet trainers. Aero owns 10 per cent of Aero Media and 5.3 per cent of import/export agency Aero Trade.

With effect from 1 October 1999, the company was reorganised into three main business units: Military Aircraft, Commercial Aircraft and Aero Structures. Current aircraft programmes are, respectively, the L-39/139/59/159 family of jet trainers described below; the Ae 270 Ibis civil transport, in a 50/50 venture with AIDC of Taiwan (see Ibis entry in International section); and airframe components for Bombardier de Havilland (Dash 8Q tailplanes), Boeing Military (gun bay doors for F/A-18E/F) and Boeing Commercial Airplane Group (747/757). Delivery of L 159As to the Czech Air Force enabled the company to return the operating profit of Kcs470 million in 2000, compared with a Kcs132 million loss in 1999.

A US$230 million deal with Sikorsky was announced in June 2000, under which Aero Vodochody is undertaking fabrication and subassembly of S-76+ helicopter airframes (but not their dynamic systems), assisted by Fischer Advanced Composite Components of Austria as subcontractor. The agreement calls for delivery of 15 airframes per year over a seven-year period; first set, assembled from US-built kits, was redelivered to Sikorsky at Stratford, Connecticut, in January 2001 and eight by 2 August 2001. In-country manufacture by Aero Vodochody had begun by the end of that year.

BAE Systems has indicated its interest in acquiring a stake in Aero Vodochody if the Czech government should approve the purchase of JAS 39 Gripens for the Czech Air Force.

UPDATED

Many L-39s are now in private hands, such as this US-owned example (*Jane's/Paul Jackson*)   NEW/0131708

L-39 and L-59 CUSTOMERS

Country	L-39 C	L-39 V	L-39 Z0	L-39 ZA	L-39 MS	L-59	Cum Total	First Delivery
Afghanistan	26						26	1977
Algeria	7			32[10]			39	1987
Bangladesh				8			8	1995
Bulgaria				36			36	1986
Cuba	30						30	1982
Czechoslovakia	33	8[2]		30	5		76[1]	1972
Egypt						48	48	1993
Ethiopia	28						28	1983
Germany (East)			52[3]				52	1977
Iraq	22		59				81	1975
Libya			181[4]				181	1978
Lithuania				2[9]			2	1998
Nigeria				24[5]			24	1986
Romania				32			32	1981
Syria			55	44			99	1980
Thailand				40			40	1993
Tunisia						12	12	1995
USA	5						5[7]	1996
USSR	2,080						2,080[8]	1974
Vietnam	24						24	1980
Yemen	12						12	1999
Totals	**2,267**	**8**	**347**	**248**	**5**	**60**	**2,935[6]**	

Notes:
[1] Aircraft delivered to former Czechoslovakia divided between Czech and Slovak (nine Cs, two Vs and nine ZAs) air forces in 1992
[2] Two transferred to East Germany
[3] 20 transferred to Hungary
[4] 10 transferred to Egypt
[5] 1992 order for further 27 not delivered but at least 16 completed and stored; Nigerian government approved additional purchase in November 2001 (quantity not stated)
[6] Excluding five prototypes
[7] Private ownership; several others acquired second-hand
[8] Final delivery 25 January 1991; subsequently redistributed to Azerbaijan, Kazakhstan, Kyrgizia, Lithuania, Russia and Ukraine
[9] From stock (built 1992)
[10] Order for further 17 announced 5 December 2001

AERO L-39 and L-139 ALBATROS

TYPE: Basic jet trainer/light attack jet.
PROGRAMME: First flight 4 November 1968; 10 preproduction aircraft from 1971; selected 1972 to succeed L-29 as standard trainer for USSR, Czechoslovakia and East Germany and production started same year; service trials in USSR and Czechoslovakia 1973; entered service with Czechoslovak Air Force 1974. Worldwide fleet has accumulated more than 4 million flying hours. Modular upgrades being offered for all L-39 variants.
CURRENT VERSIONS: **L-39 C:** Initial pilot trainer, with two underwing stations; details in 1994-95 and earlier *Jane's*. Most recent export customer (Yemen) ordered 12 in February 1999; delivered September to December same year. Eight Czech Air Force aircraft recently refurbished with new wings, front and rear fuselages and other changes under contracts placed in June and November 1999, for one and seven respectively; first aircraft (0113) redelivered 15 December 1999, remainder by May 2000.
L-39 V: Target towing version for Czech and former East German use; eight only.
L-39 Z0 (Z = *Zbrojni*: armed): Reinforced wings with four underwing stations; first flight (X-09) 25 August 1975. Production completed; weight, performance and other data in 1991-92 and earlier *Jane's*.
L-39 ZA: Ground attack and reconnaissance version of Z0; four underwing stations and centreline gun pod; reinforced wings and landing gear; prototypes (X-10 and X-11) flown 29 September 1976 and 16 May 1977; production continuing in 2000 for possible future orders. Customer versions completed with Western avionics (HUD, mission computer, Honeywell avionics and navigation equipment) as **L-39 ZA/MP** (for multipurpose). Version for Thailand has Elbit avionics and is designated **ZA/ART;** deliveries to RTAF began in 1993. Repeat orders placed by Algeria and Nigeria in 2001.

Following shortened description applies to ZA version except where indicated; further details in 2001-02 and earlier editions.

L-39 MS: Developed version, similar to L-59 and described in 2001-02 *Jane's*.
L-59: New advanced training version with more powerful engine and improved avionics; described in 2001-02 *Jane's*.
L-139 Albatros 2000: Trainer, powered by 18.15 kN (4,080 lb st) TFE731-4-1T turbofan under preliminary Aero/AlliedSignal agreement signed June 1991; Flight Visions HUD and Bendix/King avionics; VS-2A zero/zero seats. First flight (5501) 8 May 1993. Demonstrated to air forces of Colombia and Venezuela in 2000. No orders by

April 2002; possible upgrade with some L 159 avionics being considered.
CUSTOMERS: Total of 2,935 L-39/59s built (excluding five prototypes) by December 1999; none further by 2002; see table. Eight former Czech Air Force L-39ZAs refurbished by Israel Aircraft Industries for Cambodian Air Force; deliveries by IAI began in late 1996/early 1997. Late 2001 orders by Algeria (17) and Nigeria (quantity unknown), which probably to be taken from undelivered 27 ordered in 1992 by latter country.
DESIGN FEATURES: Tandem two-seater with ejection seats and pressurisation; fixed tip tanks also contain navigation/landing lights; rear fuselage and tail, attached by five bolts, allow easy removal for access to engine; tapered wing has NACA 64A012 Mod.5 section; quarter-chord sweepback 1° 45'; dihedral 2° 30' from roots; incidence 2°.
FLYING CONTROLS: Conventional and manual, by pushrods. Electrically actuated trim tab in each elevator; balance tab in rudder; mass-balanced ailerons with balance tabs (port tab actuated electrically for trimming); airbrake panels under fuselage just below wing leading-edge, operated by single hydraulic jack, extend automatically as speed approaches M0.8; double-slotted flaps extended by rods from single hydraulic jack, retracting automatically as airspeed reaches 167 kt (310 km/h; 193 mph).
LANDING GEAR: Retractable tricycle type, with single wheel and oleo-pneumatic shock-absorber on each unit; designed for touchdown sink rate of 3.4 m (11.15 ft)/s at AUW of 4,600 kg (10,141 lb). Retraction and extension operated hydraulically, with electrical control. All wheel well doors close automatically after wheels are lowered, to prevent ingress of dirt and debris. Mainwheels retract inward into wings (with automatic braking during retraction), nosewheel forward into fuselage. K28 mainwheels with 610×215 tyres and K27 nosewheel with 450×165 tyre. Hydraulic disc brakes and anti-skid units on mainwheels; shimmy damper on nosewheel leg. Minimum ground turning radius (about nosewheel) 2.50 m (8 ft 2 in). L-39

ZA can operate from grass strips with bearing strength of 6 kg/cm² (85 lb/sq in).
POWER PLANT: One 16.87 kN (3,792 lb st) ZMKB Progress AI-25 TL turbofan in rear fuselage, with semi-circular lateral air intake, with splitter plate, on each side of fuselage above wing centre-section.

Fuel in five rubber main bag tanks aft of cockpits, with combined capacity of 1,055 litres (279 US gallons; 232 Imp gallons), and two 100 litre (26.5 US gallon; 22.0 Imp gallon) non-jettisonable wingtip tanks. Total internal fuel capacity 1,255 litres (332 US gallons; 276 Imp gallons). Gravity refuelling points on top of fuselage and on each tip tank. Provision for two 150 or 350 litre (39.6 or 92.5 US gallon; 33.0 or 77.0 Imp gallon) drop tanks on inboard underwing pylons, increasing total overall fuel capacity to a maximum of 1,955 litres (517 US gallons; 430 Imp gallons). Fuel system permits up to 20 seconds of inverted flight.
ACCOMMODATION: Crew of two in tandem, on Czech VS-1-BRI rocket-assisted ejection seats, operable at zero height and at speeds down to 81 kt (150 km/h; 94 mph); individual canopies hinge sideways to starboard and are jettisonable. Rear seat elevated. One-piece windscreen hinges forward for access to front instrument panel. Internal transparency between cockpits. Dual controls standard.
AVIONICS (L-39 ZA/ART): *Comms:* ARC-186 VHF (30 to 78 MHz/FM and 108 to 151.75 MHz/AM) and ARC-164 VHF (220 to 399.975 MHz) radios; KXP 756 transponder; IFF. Orbit crew intercom.
Flight: Bendix/King KNR 634A VOR/glideslope/MKR; KTU 709 Tacan; RV-5M radar altimeter.
Instrumentation: Elbit WDNS (weapon delivery and navigation system) with HUD and video camera in front cockpit and monitor in rear cockpit.
ARMAMENT (L-39 ZA/ART): Underfuselage pod below front cockpit, housing a single 23 mm GSh-23 two-barrel gun; ammunition (maximum 150 rounds) housed in fuselage

above gun pod. Gun/rocket/missile firing and weapon release controls in front cockpit only. Four underwing hardpoints, inboard pair each stressed for up to 500 kg (1,102 lb) and outer pair for up to 250 kg (551 lb) each; maximum underwing stores load 1,000 kg (2,205 lb). Non-jettisonable pylons, each comprising an MD3-57D stores rack. Typical underwing stores can include various combinations of bombs (two of up to 500 kg or four of up to 250 kg); four rocket launchers for 2.75 in FFAR or CRV7 rockets; AIM-9 air-to-air missiles (outboard stations only); two 150 or 350 litre drop tanks (see under Power Plant) (inboard stations only); or two training dispensers. See 1993-94 and earlier editions for L-39 ZA armament; 2001-02 and earlier for L-139 weapons diagram.

DIMENSIONS, EXTERNAL:

Wing span, incl tip tanks	9.46 m (31 ft 0½ in)
Wing chord (mean)	2.15 m (7 ft 0½ in)
Wing aspect ratio, incl tip tanks	4.8
Length overall	12.13 m (39 ft 9½ in)
Height overall	4.77 m (15 ft 7¾ in)
Tailplane span	4.40 m (14 ft 5 in)
Wheel track	2.44 m (8 ft 0 in)
Wheelbase	4.39 m (14 ft 4¾ in)

AREAS:

Wings, gross	18.80 m² (202.4 sq ft)

WEIGHTS AND LOADINGS (L-39 ZA):

Weight empty, equipped	3,565 kg (7,859 lb)
Fuel load: fuselage tanks	824 kg (1,816 lb)
wingtip tanks	156 kg (344 lb)
Max external stores load	1,290 kg (2,844 lb)
T-O weight clean	4,635 kg (10,218 lb)
Max T-O weight	5,600 kg (12,346 lb)
Max wing loading	297.9 kg/m² (61.01 lb/sq ft)
Max power loading	332 kg/kN (3.25 lb/lb st)

WEIGHTS AND LOADINGS (L-139):

Basic weight empty	3,460 kg (7,628 lb)
Fuel weights (fuselage and wingtip tanks) as for L-39 ZA	
Max external fuel (four 350 litre tanks)	
	1,088 kg (2,398 lb)
Max underwing stores	1,500 kg (3,307 lb)
Max ramp weight: clean	4,600 kg (10,141 lb)
with external stores	6,000 kg (13,227 lb)
T-O weight clean	4,550 kg (10,031 lb)
Max landing weight	4,800 kg (10,582 lb)
Wing loading at clean T-O weight	
	241.9 kg/m² (49.55 lb/sq ft)
Power loading at clean T-O weight	
	251 kg/kN (2.46 lb/lb st)

PERFORMANCE (L-39 ZA at max T-O weight, except where indicated):

Max limiting Mach No.	0.80
Max level speed: at S/L	329 kt (610 km/h; 379 mph)
at 5,000 m (16,400 ft)	340 kt (630 km/h; 391 mph)
Stalling speed	103 kt (190 km/h; 118 mph)
Max rate of climb at S/L	810 m (2,657 ft)/min
Time to 5,000 m (16,400 ft)	10 min
Service ceiling	7,500 m (24,600 ft)
T-O run (concrete)	970 m (3,185 ft)
Landing run (concrete)	800 m (2,625 ft)

g limits:

operational: at 4,200 kg (9,259 lb) AUW	+8/−4
at 5,500 kg (12,125 lb) AUW	+5.2/−2.6
ultimate, at 4,200 kg (9,259 lb) AUW	+12

PERFORMANCE (L-139 at clean T-O weight, except where indicated):

Never-exceed speed (VNE)	491 kt (910 km/h; 565 mph)
Max level speed at 6,100 m (20,000 ft)	
	410 kt (760 km/h; 472 mph)
Touchdown speed	99 kt (183 km/h; 114 mph)
Stalling speed, flaps down	87 kt (160 km/h; 100 mph)
Max rate of climb at S/L	1,344 m (4,409 ft)/min
Service ceiling	12,000 m (39,360 ft)
T-O run	540 m (1,775 ft)
Landing run	560 m (1,840 ft)
Radius of action with gun, two drop tanks and two Mk 83 (1,000 lb) bombs:	
lo-lo-lo	291 n miles (539 km; 335 miles)
hi-lo-hi	401 n miles (743 km; 461 miles)

Second (single-seat) L 159 prototype (*Jane's/Paul Jackson*)   NEW/0131709

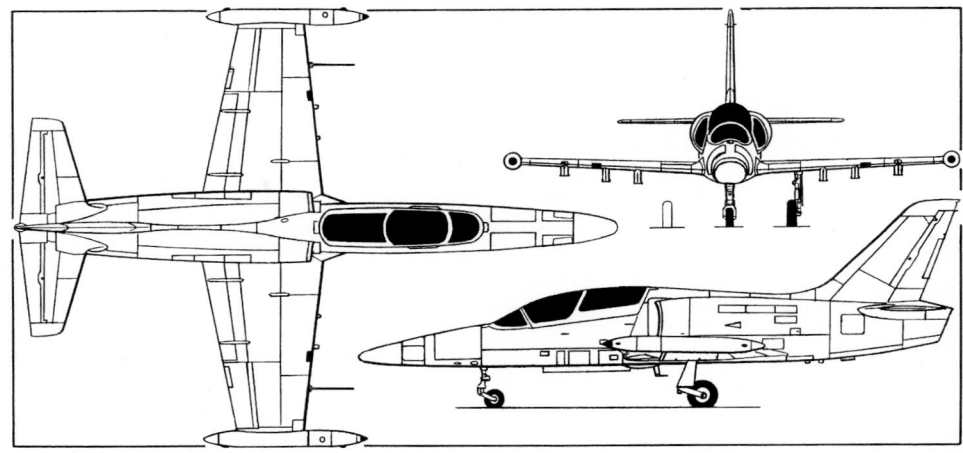

Aero L 159 light attack aircraft (*Jane's/Paul Jackson*)

Max range on internal fuel, 10% reserves	730 n miles (1,352 km; 840 miles)
Max endurance, conditions as above	3 h 19 min
g limits	+8/−4

UPDATED

AERO L-39 MS and L-59
Tunisian Air Force name: Fennec

TYPE: Advanced jet trainer/light attack jet.

PROGRAMME: Early development under designation L-39 MS; versions of L-39 with more powerful engine and improved avionics. First flight in definitive configuration (X-22 prototype OK-184) 30 September 1986; two more prototypes (X-24, X-25) flown 26 June and 6 October 1987; first flight of first L-59 E made on 22 April 1992; deliveries of L-59 E began 29 January 1993. Tunisian (L-59T) deliveries completed early 1996. No orders since then; see 2001-02 *Jane's* for detailed description.

CURRENT VERSIONS: **L-39 MS:** First version of L-59; five delivered to Czechoslovak Air Force (three in 1991, two in 1992); sixth (0001) retained for trials.

L-59 E: Production two-seat version for Egyptian Air Force; 48 delivered 1993-94.

L-59 T: Version for Tunisian Air Force; deliveries (12) completed in early 1996.

L 159: Light ground attack/lead-in fighter trainer version of L-59; described separately.

UPDATED

AERO L 159

TYPE: Light fighter/advanced jet trainer.

PROGRAMME: Design started late 1992; F124 turbofan selected as power plant mid-1994; Rockwell North American (now Boeing) awarded avionics contract late 1994. Development approved by Czech government 7 April 1995, followed by production commitment later the same month and 72-aircraft production contract 4 July 1997. Two prototypes (plus two airframes for static and dynamic test) built. First (two-seat) prototype/engine demonstrator (5831) rolled out 12 June and made first flight 2 August 1997.

Second (single-seat) prototype (5832), in Czech Air Force configuration (first flight 18 August 1998) named ALCA (Advanced Light Combat Aircraft); is first with full avionics and is avionics/weapons trials aircraft. Official first flight of 5832 was 20 August; public debut at Hradec Králové 29 August 1998. Gun firing tests in Czech Republic followed by air-launched weapon trials in Norway in April and May 1999; nav/attack system trials August/September 1999. The two prototypes had flown 1,120 hours (506 by single-seater, 614 by two-seater) by 25 August 2000.

First flight of production L 159A (6001) 20 October 1999. Deliveries to CzAF due to have begun late 1999 (five aircraft, followed by 16 in 2000, 26 in 2001 and final 25 in 2002); however, first two (6006 and 6007) not delivered (to 4 Tactical Fighter Base at Caslav) until 29 December 2000, with first pilot conversion training beginning in mid-March 2001. Contract criticised in late 2000 due to deteriorating exchange rate between Czech crown and US dollar, previous government having priced contract in latter currency. On 1 February 2001, Czech MoD demanded more than Kcs100 million compensation for non-delivery of 14 aircraft during 2000; deliveries then rescheduled as 37 by end of 2001, 24 during 2002 and final 11 during first half of 2003. However, in May 2001, during which month aircraft evaluated by US Navy Test Pilots' School, Czech defence minister stated that CzAF will operate only 37 aircraft, remaining 35 being instead offered for export.

The L 159A fleet (then 25 aircraft) was grounded on 14 November 2001 following accidents to two aircraft apparently due to avionics malfunctions. Weapons qualification tests, at Cazeaux in France, had begun in the preceding month.

CURRENT VERSIONS: **L 159A ALCA:** Single-seater.
Detailed description applies to L 159A except where indicated.

L 159B: Tandem-seat trainer version; also available for export as lead-in fighter trainer or attack aircraft. Further US$15 million invested by Boeing in July 2001 to develop software for this version. Evaluated by Indian Air Force in

A production L 159A undergoing manufacturer's testing (*R J Malachowski*)   NEW/0131692

NEW/0100854

Aero Vodochody L 159A, cutaway drawing key

1 Glass fibre radome
2 Planar radar scanner
3 Scanner tracking mechanism
4 Mounting bulkhead
5 Lower VHF/UHF antenna
6 Grifo L multimode, pulse Doppler radar
7 Nosewheel housing
8 Nose compartment door, port and starboard
9 Nose avionics equipment bay
10 Battery
11 Ice detector, temperature probe to starboard
12 Electroluminescent formation lighting strip
13 Nosewheel leg strut
14 Taxying light
15 Trailing axle suspension
16 Forward-retracting nosewheel
17 Hydraulic steering jack
18 Cockpit front pressure bulkhead
19 Rudder pedals
20 Cockpit section main longeron
21 Fresh air ram intake
22 Instrument panel shroud
23 Frameless windscreen panel
24 Flight Visions FV-3000 head-up display
25 Starboard side console panel
26 Instrument panel, dual Honeywell multifunction full-colour LCD head-down displays
27 Control column
28 Incidence transmitting vane
29 Side-mounted engine throttle lever, full HOTAS controls
30 Port side console panel
31 Front cockpit floor level
32 Kick-in step
33 Fold-out boarding step
34 Port side equipment bay, air conditioning pack to starboard
35 Canopy external latch
36 Canopy assist handle, manual opening
37 Pilot's VS-2 zero/zero ejection seat
38 Cockpit canopy, hinged to starboard
39 Avionics equipment housing
40 Rear cockpit avionics bay shelf
41 Rear canopy latch
42 Cockpit section framing
43 Forward fuel cell, tanks with inert gas protection system
44 Rear cockpit floor level, L 159B lead-in trainer variant
45 Maintenance platform
46 Pressure refuelling
47 Port ventral airbrake

48 Extended chord wingroot leading-edge
49 Central airbrake hydraulic jack
50 Centre-fuselage bag-type fuel cells (two)
51 Cockpit sloping rear pressure bulkhead
52 Junction boxes
53 Rear cockpit avionics
54 Control rod linkages
55 Cockpit pressurisation valve
56 Rear canopy, hinged to starboard
57 Starboard wing pylon hardpoints
58 Leading-edge aileron control rod installation
59 Starboard wing missile
60 Pitot head
61 Starboard landing light
62 Forward oblique RWR
63 Starboard navigation light
64 Permanent wingtip tank

65 Rear oblique RWR
66 Starboard aileron
67 Aileron hydraulic actuator
68 Starboard double-slotted flap
69 Flap operating linkage, hydraulically from central jack
70 Starboard air intake
71 Composites fairing
72 Tailplane control rods
73 Fuselage fuel tank gravity filler
74 Boundary layer spill duct
75 Boundary layer diverter
76 Tank venting ram air intake
77 Intake lip bleed air de-iced
78 Fuselage lower main longeron
79 Wing main spar bolted attachment joint

80 Port intake duct
81 Intake flank fuel cell
82 Generator cooling air intake
83 Fuel system recuperator, port and starboard
84 Engine intake compressor face
85 Upper anti-collision beacon
86 GPS antenna
87 Starboard Vicon countermeasures dispenser
88 IFF antenna
89 Engine bay venting flush air intakes
90 Engine bay access hatches
91 Honeywell/ITEC F124-GA-100 turbofan engine
92 Port countermeasures dispenser
93 Starboard side engine bleed air precooler
94 Fin root fairing
95 Two-spar and rib fin torsion box structure
96 Rudder actuator and yaw damper
97 Rudder control rod
98 Starboard tailplane

99 Remote compass transmitter
100 Formation lighting strip
101 VOR localiser antenna
102 Fin-tip antenna fairing
103 Upper VHF/UHF antenna
104 Tail navigation light
105 Rudder rib structure
106 Two-segment rudder tab
107 Elevator hydraulic actuator
108 Elevator hinge link and inboard mass balance
109 Exhaust nozzle shroud
110 Port elevator
111 Elevator rib structure
112 Static dischargers
113 Elevator outboard mass balance
114 Port tailplane
115 Tailplane two-spar and rib torsion box structure
116 Tailplane spar attachment joint
117 Fin spar root fairing
118 Tailplane root fairing
119 Tail bumper
120 Rear equipment compartment, port and starboard

121 Jetpipe
122 Detachable tail section joint frame (engine removal)
123 Rear engine mounting
124 Lower anti-collision beacon
125 Rear fuselage formation lighting strip
126 Main engine mounting
127 Engine accessory equipment including Safir APU to starboard
128 Engine oil tank
129 Accessory compartment doors
130 Wingroot trailing-edge fillet
131 Mainwheel housing
132 Main landing gear hydraulic jack
133 Wing main spar
134 Port wing rib structure 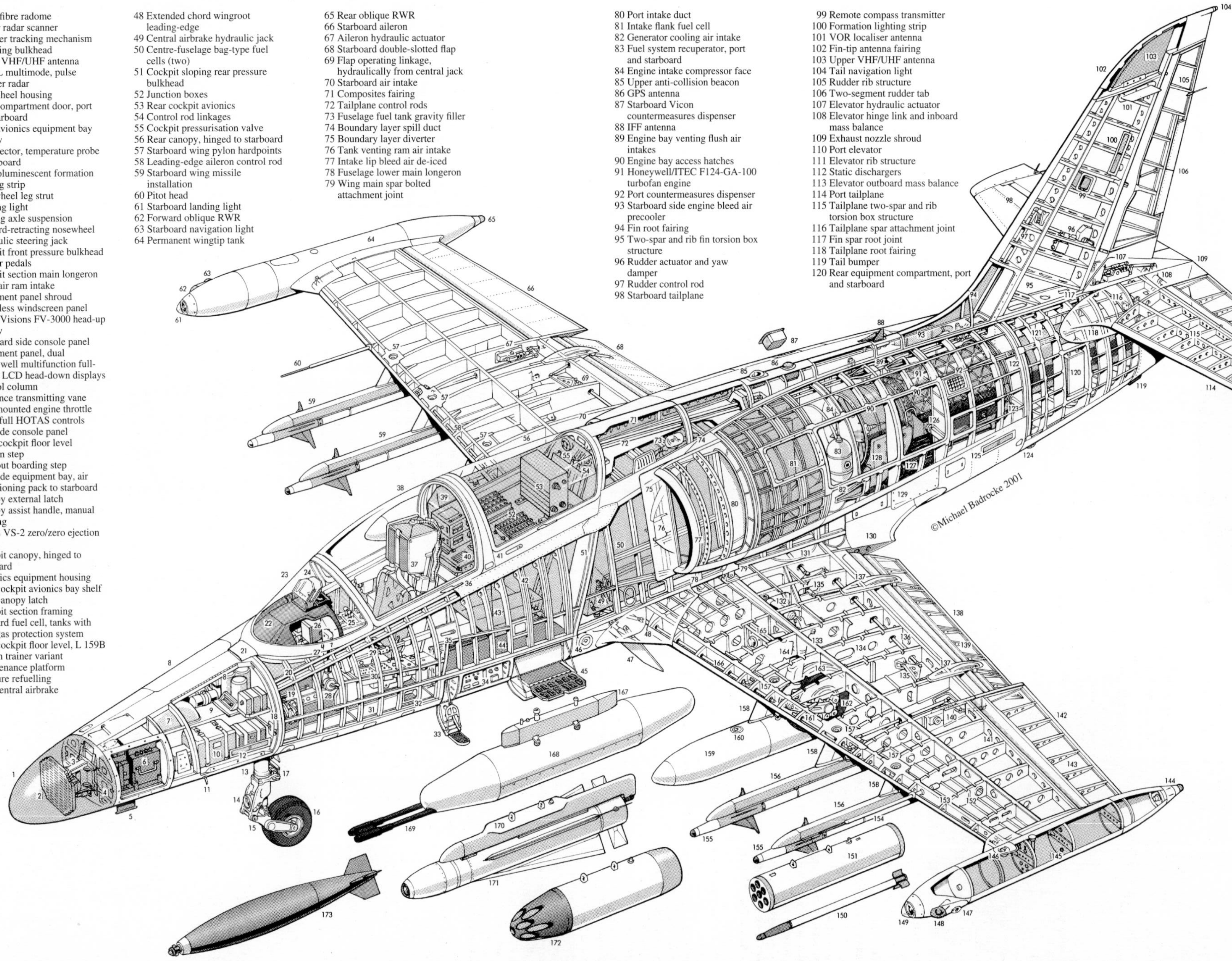
135 Flap operating linkage
136 False rear spar
137 Flap guide rails and fairings
138 Port double-slotted flap
139 Flap rib structure
140 Port aileron hydraulic actuator
141 Aileron spar
142 Port aileron
143 Aileron rib structure
144 Port rear/oblique RWR
145 Permanent wingtip fuel tank
146 Tank gravity filler
147 Port navigation light
148 Forward/oblique RWR
149 Port landing light
150 70 mm rocket projectile
151 CRV-7 seven-round rocket launcher
152 Outer wing panel rib structure
153 Front spar
154 Port pitot head
155 AIM-9L Sidewinder air-to-air missiles
156 Missile launch rails
157 Port wing pylon hardpoints
158 Port stores pylons
159 External fuel tank
160 Tank filler
161 Trailing axle mainwheel
162 Hydraulic wheel brake
163 Port mainwheel
164 Main landing gear leg strut
165 Mainwheel leg pivot mounting
166 Leading-edge aileron control
167 Fuselage centreline pylon
168 Plenam gun pod
169 20 mm twin-barrel cannon
170 Maverick launch rail adaptor
171 AGM-65 Maverick air-to-surface missile
172 18-round rocket pack
173 Mk 82 500 lb HE bomb

© Mike Badrocke 2001

Instrument panel of the L 159 (*Jane's/Paul Jackson*)
0062963

First production L 159B, which made its maiden flight 1 June 2002   *NEW*/0143350

May 2002 as possible (and cheaper) replacement for stalled order for BAE Systems Hawks. First production aircraft (6073) conducted maiden flight 1 June 2002; 45 hours flown by 24 July 2002, when initial CzAF order placed for two aircraft, for delivery before end of 2003.

CUSTOMERS: Czech Air Force order for 72 L 159As valued at approximately US$1 billion. Production lots 1 and 2 comprised five and 10 aircraft respectively, of which 12 delivered by 31 March 2001. CzAF took delivery of L 159As Nos. 38 to 42 on 27 July 2002. Primary roles will be close air support, counter-insurgency, anti-ship missions, tactical reconnaissance, point air defence, air defence against slow/low-flying targets and border patrol. Prospect of exports to other East European and Baltic countries, including Poland, to whom a Maverick-armed version was offered in late 2000. Company said to be preparing offsets for sale to Slovak Air Force in July 2001, and to be negotiating with other 'very potential' customers. India expressing close interest in L 159B in mid-2002; Greece also being targeted.

COSTS: L 159B US$8.5 million (2002).

DESIGN FEATURES: Configuration generally as L-39/L-59 except for larger, redesigned nose to accommodate radar; longer fuselage; single-seat armoured cockpit; HOTAS controls; Western avionics; 297 kg (655 lb) additional fuel tank in lieu of second seat; permanent wingtip fuel tanks; seven external stores stations.

Wing section NACA 64A-012; sweepback 6° 30′ on leading-edges; dihedral 2° 30′; incidence 2°; twist 0°.

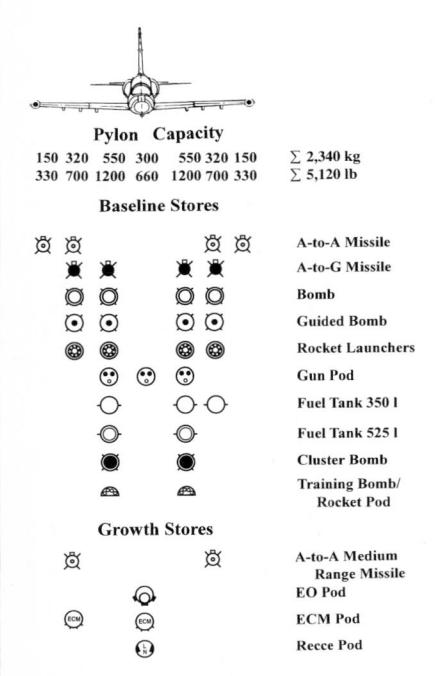

Pylon Capacity

| 150 | 320 | 550 | 300 | 550 | 320 | 150 | Σ **2,340 kg** |
| 330 | 700 | 1200 | 660 | 1200 | 700 | 330 | Σ **5,120 lb** |

Baseline Stores

A-to-A Missile
A-to-G Missile
Bomb
Guided Bomb
Rocket Launchers
Gun Pod
Fuel Tank 350 l
Fuel Tank 525 l
Cluster Bomb
Training Bomb/ Rocket Pod

Growth Stores

A-to-A Medium Range Missile
EO Pod
ECM Pod
Recce Pod

L 159 weapon options   0085020

FLYING CONTROLS: Conventional primary surfaces; ailerons and elevators actuated hydraulically, rudder mechanically with electrically operated trim tab. Artificial feel standard. Twin underfuselage airbrakes and double-slotted trailing-edge flaps, also actuated hydraulically. Maximum flap deflection 44°.

STRUCTURE: Based on that of L-59, but substantially modified. Approximately 200 Czech subcontractors involved in programme.

LANDING GEAR: Hydraulically retractable tricycle type, with single wheel and oleo-pneumatic shock-absorber on each unit. Main units retract inward into wings, nosewheel forward. Moravan K36 wheels and 610×215 tyres (pressure 6.00 to 8.00 bar; 87 to 116 lb/sq in) on main units; nose unit has Moravan K37 wheel and 460×180 tyre (pressure 3.52 to 4.48 bar; 51 to 65 lb/sq in). Moravan K52 air-cooled hydraulic brakes with anti-skid system. Nose unit hydraulically steerable ±59°. Minimum ground turning radius 5.70 m (18 ft 8½ in).

POWER PLANT: One 28.2 kN (6,330 lb st) Honeywell/ITEC F124-GA-100 turbofan with dual-redundant FADEC. Internal fuel of L 159A (total capacity 1,980 litres; 523 US gallons; 436 Imp gallons) is contained in six fuselage and two wingtip tanks and protected by OBIGGS (onboard inert gas generation system); L 159B omits one 350 litre (92.5 US gallon; 77.0 Imp gallon) fuselage tank, internal capacity thus 1,630 litres (431 US gallons; 359 Imp gallons) and does not have OBIGGS. Up to four external tanks can be carried underwing: two 500 litre (132 US gallon; 110 Imp gallon) tanks on inboard stations, two 350 litre (92.5 US gallon; 77.0 Imp gallon) tanks on inboard or centre stations. Pressure fuelling point in port side of lower front fuselage; gravity points in upper fuselage aft of canopy and in each wingtip tank. In-flight refuelling capability began flight testing on L 159B in August 2002.

ACCOMMODATION: Single VS-2B ejection seat in armoured cockpit. Canopy opens sideways manually in L 159A, upwards hydraulically in L 159B.

SYSTEMS: PBS simple cycle environmental control system, using engine bleed air; Honeywell cockpit pressure control system (maximum differential 0.28 bar; 4.1 lb/sq in). Dual hydraulic systems, each at pressure of 210 bar (3,045 lb/sq in), actuating ailerons, elevators, flaps, airbrakes, landing gear, wheel brakes and nosewheel steering. System maximum flow rate 35 litres (9.25 US gallons; 7.7 Imp gallons)/min. Hydraulic tank pressurised by nitrogen at 2.50 bar (36.3 lb/sq in). Pneumatic system (1.80 bar; 26.1 lb/sq in) for canopy sealing.

Hamilton Sundstrand 40 kVA integrated drive generator provides 200/115 V three-phase main AC power at 400 Hz; 28 V DC power from 6 kW VUES LUN-2134 back-up generator and Varta/Aero 25 Ah battery. Intertechnique OBOGS oxygen system. Windscreen and air intakes de-iced by engine bleed air. PBS Safir 5F APU, operable up to 9,000 m (29,520 ft) altitude, for engine starting and emergency hydraulic or electrical power.

AVIONICS: Integration of complete avionics suite by Boeing.
Comms: Two Rockwell Collins ARC-210 UHF/VHF radios; Honeywell AN/APX-100 IFF.
Radar: Galileo Grifo L multimode pulse Doppler radar (L 159A only).
Flight: Bendix/King KNR 634A VOR/glideslope/MKR, KDM 706A DME and air data computer; Honeywell H-746G ring laser gyro INS with embedded GPS.
Instrumentation: Flight Visions FV-3000 HUD; dual Honeywell 102 × 102 mm (4 × 4 in) liquid crystal colour head-down MFDs; Astronautics mechanical ADI and HSI. Optional helmet-mounted display/sight and NVG compatible lighting.
Mission: Dual Flight Visions 6400-100 mission computers with HUD symbology generator; provision for reconnaissance and jamming pod. Hamilton Sundstrand weapon delivery system. Dynamic Control Corporation stores management system.
Self-defence: BAE Systems Sky Guardian 200 RWR; Thales (Vinten) Vicon 78 Series 455 countermeasures dispenser.

ARMAMENT: No built-in armament. Seven external stores stations: one under fuselage and three under each wing. Fuselage station stressed for 300 kg (661 lb) load; inboard,

centre and outboard wing pylons stressed for 550, 320 and 150 kg (1,213, 705 and 331 lb) each respectively. Primary armament comprises AIM-9 Sidewinder AAMs, AGM-65 Maverick ASMs, and SUU-20 or CRV7 unguided rocket pods. Has also been trialled with Brimstone missile and TIALD targeting pod. For typical weapon loads, see accompanying illustration.

DIMENSIONS, EXTERNAL:

Wing span: excl tip tanks	8.70 m (28 ft 6½ in)
incl tip tanks	9.54 m (31 ft 3½ in)
Wing chord: at root	2.80 m (9 ft 2¼ in)
at tip	1.40 m (4 ft 7 in)
Wing aspect ratio, incl tip tanks	4.8
Length: overall	12.72 m (41 ft 8¾ in)
fuselage	12.59 m (41 ft 3¾ in)
Height overall	4.87 m (15 ft 11¾ in)
Tailplane span	4.70 m (15 ft 5 in)
Wheel track	2.42 m (7 ft 11¼ in)
Wheelbase	4.56 m (14 ft 11½ in)

AREAS:

Wings, gross	18.80 m² (202.4 sq ft)
Ailerons (total)	1.69 m² (18.19 sq ft)
Trailing-edge flaps (total)	2.68 m² (28.85 sq ft)
Fin	2.59 m² (27.88 sq ft)
Rudder, incl tab	0.91 m² (9.80 sq ft)
Tailplane	3.90 m² (41.98 sq ft)
Elevators (total)	1.40 m² (15.07 sq ft)

WEIGHTS AND LOADINGS:

Weight empty, equipped	4,350 kg (9,590 lb)
Max fuel weight:	
internal, L 159A	1,547 kg (3,411 lb)
internal, L 159B	1,320 kg (2,910 lb)
external, both (two 500 litre and two 350 litre)	1,620 kg (3,571 lb)
Max external stores load	2,700 kg (5,952 lb)
Max T-O and ramp weight	8,000 kg (17,637 lb)
Max landing weight	6,000 kg (13,228 lb)
Max zero-fuel weight	6,810 kg (15,013 lb)
Max wing loading	425.5 kg/m² (87.15 lb/sq ft)
Power loading, L 159A clean with 50% internal fuel	284 kg/kN (2.79 lb/lb st)

PERFORMANCE (C: clean, S: with 1,470 kg; 3,241 lb of external stores, except where indicated):

Never-exceed speed (VNE)	518 kt (960 km/h; 596 mph)
Max level speed at S/L: C	505 kt (936 km/h; 581 mph)
S	468 kt (867 km/h; 539 mph)
Stalling speed at 5,500 kg (12,125 lb) AUW, engine idling, flaps and gear down 100 kt (185 km/h; 539 mph)	
Max rate of climb at S/L: C	3,726 m (12,220 ft)/min
S	2,280 m (7,480 ft)/min
Sustained turning radius at 5,000 m (16,400 ft):	
C	612 m (2,008 ft)
S	882 m (2,894 ft)
Sustained turn rate at 5,000 m (16,400 ft):	
C	11.1°/s
S	7.7°/s
Sustained load factor:	
at S/L: C	5.7 g
S	3.8 g
at 5,000 m (16,400 ft): C	3.5 g
S	2.4 g
Service ceiling	13,200 m (43,300 ft)
T-O run at S/L:	
C at 5,925 kg (13,062 lb) AUW	470 m (1,545 ft)
S at 7,190 kg (15,851 lb) AUW	853 m (2,800 ft)
T-O to 15 m (50 ft) at S/L:	
C at 5,925 kg (13,062 lb) AUW	720 m (2,365 ft)
S at 7,190 kg (15,851 lb) AUW	1,276 m (4,185 ft)
Landing from 15 m (50 ft) at S/L:	
C at 4,489 kg (9,897 lb) AUW	992 m (3,255 ft)
S at 5,257 kg (11,590 lb) AUW	1,254 m (4,115 ft)
Landing run at S/L:	
C at 4,489 kg (9,897 lb) AUW	628 m (2,060 ft)
S at 5,257 kg (11,590 lb) AUW	816 m (2,675 ft)

Typical mission radius (A: L 159A, B: L 159B):
with gun pod, two Mk 82 bombs, two AIM-9 missiles and two 500 litre drop tanks:

A, lo-lo-lo	305 n miles (565 km; 351 miles)
A, hi-lo-hi	426 n miles (790 km; 490 miles)
B, lo-lo-lo	264 n miles (490 km; 304 miles)
B, hi-lo-hi	364 n miles (675 km; 419 miles)

with gun pod, two Mk 82 bombs and two 500 litre
drop tanks:

A, lo-lo-lo	310 n miles (575 km; 357 miles)
A, hi-lo-hi	442 n miles (820 km; 509 miles)
B, lo-lo-lo	270 n miles (500 km; 310 miles)
B, hi-lo-hi	378 n miles (700 km; 435 miles)

with two 500 litre drop tanks only:

| A, lo-lo-lo | 345 n miles (640 km; 397 miles) |
| A, hi-lo-hi | 526 n miles (975 km; 605 miles) |

| B, lo-lo-lo | 299 n miles (555 km; 344 miles) |
| B, hi-lo-hi | 450 n miles (835 km; 518 miles) |

Range, reserves of 10% of internal fuel:
L 159A, max internal fuel only
848 n miles (1,570 km: 975 miles)
L 159B, max internal fuel only
659 n miles (1,220 km; 758 miles)
L 159A, max internal and external fuel
1,366 n miles (2,530 km; 1,572 miles)

L 159B, max internal and external fuel
1,242 n miles (2,300 km; 1,429 miles)
g limits (ultimate)   +8/–4
UPDATED

Ae 270 IBIS

This aircraft is described under the Ibis Aerospace heading in the International section.

AEROS

AEROS s.r.o.
Nádražní 211, CZ-284 01 Kutná Hora
Tel: (+420 327) 56 22 44
Fax: (+420 327) 56 25 58
e-mail: flamingo@quick.cz or aeros@iol.cz
MANAGING DIRECTOR: Miroslav Staněk
INFORMATION OFFICER: Miss Ilona Černá

Known until 2002 as Lusus s.r.o., Aeros has 10 employees at its 700 m² (7,550 sq ft) plant.

NEW ENTRY

AEROS UL-2000 FLAMINGO

TYPE: Side-by-side ultralight.
PROGRAMME: Revealed at Friedrichshafen Air Show, April 2001.
CURRENT VERSIONS: **UL-2000 S Flamingo Sport:** Baseline version, with 48.5 kW (65 hp) Hirth 2706 engine and two-blade fixed-pitch propeller.
UL-2000 T Flamingo Trainer: As Sport, but with 74.6 kW (100 hp) Hirth F30 and two-blade ground-adjustable pitch propeller; GPS navigation; dual controls.
UL-2000 A Flamingo Aerobatic: As Trainer except for 82.0 kW (110 hp) Hirth and three-blade ground-adjustable propeller; or 73.5 kW (98.6 hp) Rotax 912 ULS or 84.6 kW (113.4 hp) Rotax 914. Export-only version.
UL-2000 B Business Flamingo: Generally similar to Trainer, but without dual controls.
UL-2000 D Flamingo De Luxe: Customised version, with modified 74.6 kW (100 hp) engine.
DESIGN FEATURES: Unbraced high-wing cabin monoplane with sweptback vertical tail. Wings detach for transportation and storage.
FLYING CONTROLS: Manual. Non-balanced rudder; large one-piece elevator with inset, in-flight-adjustable trim tab. Flaps.
STRUCTURE: All-composites.
LANDING GEAR: Tricycle type; fixed. Composites cantilever self-sprung main legs with 400×100 tyres and cable brakes; nosewheel tyre 12×4.

UL-2000 Flamingo side-by-side ultralight *(Jane's/Paul Jackson)*   *NEW*/0131725

POWER PLANT: See Current Versions above. Fuel capacity 45 litres (11.9 US gallons; 9.9 Imp gallons) in UL-2000 S, T, A and D; 60 litres (15.9 US gallons; 13.2 Imp gallons) in B.
ACCOMMODATION: Upward-opening window/door each side; hatshelf behind seats.
SYSTEMS: 12 V electrical system.
DIMENSIONS, EXTERNAL:

Wing span	8.60 m (28 ft 2½ in)
Length overall	6.36 m (20 ft 10½ in)
Height overall	2.50 m (8 ft 2½ in)

DIMENSIONS, INTERNAL:

| Cabin max width | 1.10 m (3 ft 7¼ in) |

AREAS:

| Wings, gross | 10.50 m² (113.0 sq ft) |

WEIGHTS AND LOADINGS:

| Max T-O weight (all versions) | 450 kg (992 lb) |

PERFORMANCE (S: Sport, T: Trainer, A: Aerobatic, B: Business, D: De Luxe):

Never-exceed speed (VNE):

| all | 145 kt (270 km/h; 167 mph) |
| Max level speed: all | 129 kt (240 km/h; 149 mph) |

Cruising speed:

S, T, B, D	97 kt (180 km/h; 112 mph)
A	103 kt (190 km/h; 118 mph)
Stalling speed: all	36 kt (65 km/h; 41 mph)

Max rate of climb at S/L: S   270 m (886 ft)/min

| T, B | 390 m (1,280 ft)/min |
| A | 420 m (1,378 ft)/min |

T-O to 15 m (50 ft): S   180 m (590 ft)

| T, A, B | 150 m (495 ft) |

Landing from 15 m (50 ft): all   250 m (820 ft)
Range with max fuel:

S, T, A	432 n miles (800 km; 497 miles)
B	540 n miles (1,000 km; 621 miles)
g limits: all	±6

UPDATED

AIROX

AIROX s.r.o
Dolní Radová, CZ-463 03 Stráž nad Nisou
Tel/Fax: (+420 48) 515 92 24
e-mail: airox@airox.cz
Web: http://www.airox.cz
INFORMATION: Tomáše Valouška

VERIFIED

AIROX F-31

TYPE: Side-by-side ultralight twin.
PROGRAMME: Public debut (F-31A) at aviation exhibition in Prague, September 2000.
CURRENT VERSIONS: **F-31A:** Tailwheel version; *as described.* Prototype is of this version.

F-31B: Tricycle landing gear; under development.
F-31C: Modified lifting and control surface geometry; under development.
F-31D: Multiseat, non-ultralight version; under development.
DESIGN FEATURES: Low-wing monoplane with dihedral from roots and upturned wingtips; cruciform tailplane.
FLYING CONTROLS: Conventional and manual; flaps fitted.
STRUCTURE: All-composites (CFRP sandwich shell); single-spar wing.
LANDING GEAR: Tailwheel type; fixed. Cantilever self-sprung main units.
POWER PLANT: Two 33.6 kW (45 hp) Hirth F23 flat-twin engines, each driving a SportProp 1360W two-blade, glass fibre propeller. Fuel capacity 75 litres (19.8 US gallons; 16.5 Imp gallons). Other suitable engines include Rotax 503 UL-2V, 582 DCDI-2V or 618 DCDI-2V.

AVIONICS: VFR instrumentation standard; IFR to customer's requirements, including Bendix/King KY 97 transceiver Garmin GPS.
DIMENSIONS, EXTERNAL:

| Wing span | 9.60 m (31 ft 6 in) |
| Length overall | 6.60 m (21 ft 7¾ in) |

AREAS:

| Wings, gross | 12.00 m² (129.2 sq ft) |

WEIGHTS AND LOADINGS:

| Weight empty | 278 kg (613 lb) |
| Max T-O weight | 450 kg (992 lb) |

PERFORMANCE:

Never-exceed speed (VNE)	152 kt (282 km/h; 175 mph)
Max level speed	135 kt (251 km/h; 156 mph)
Cruising speed	112 kt (208 km/h; 129 mph)
Stalling speed, flaps down	36 kt (65 km/h; 41 mph)
Max rate of climb at S/L	480 m (1,575 ft)/min

UPDATED

Prototype Airox F-31A twin-engined ultralight   0110980

Airox F-31A (two Hirth F23 flat-twins)   0110981

ATEC

ATEC v.o.s
Opolanská 171, CZ-289 07 Libice nad Cidlinou
Tel: (+420 324) 67 73 71
Fax: (+420 324) 61 30 02
e-mail: sales@atecvos.cz
Web: http://www.atecvos.cz
PRODUCTION DIRECTOR: Petr Voledjnik
SALES DIRECTOR: Jan Franěk

ATEC was formed in 1992 to develop, manufacture and sell ultralight aircraft.

UPDATED

ATEC Zephyr 2000 two-seat ultralight *(Jane's/Paul Jackson)*   *NEW*/0131746

ATEC ZEPHYR

TYPE: Side-by-side ultralight.
PROGRAMME: Prototype (OK-CUA 01) built 1997.
CURRENT VERSIONS: **Zephyr Standard:** Original version, with 10.60 m (34 ft 9¼ in) wing span and 59.6 kW (79.9 hp) Rotax 912 UL engine. Other details in 2000-01 *Jane's*.
  Zephyr 2000: Shorter wing span and Rotax 912 ULS engine; otherwise generally as Zephyr Standard. Details in 2001-02 *Jane's*.
  Zephyr 2000C: Introduced April 2001; C suffix indicates change to all-composites fuselage construction; also reverts to original 10.60 m wing span
Description below applies to Zephyr 2000C.
  Zephyr Solo: Single-seat version. Prototype (OK-FUH-01) made initial flight 29 October 2000.
CUSTOMERS: Over 70 (all versions) built and sold by December 2001. Exported to Australia, Canada, France, Germany, Netherlands, Portugal and Sweden. Marketed in Germany by Fläming Air (see German section).
COSTS: Zephyr 2000C: US$40,000 (2001).
DESIGN FEATURES: Low-wing monoplane with T tail.
FLYING CONTROLS: Conventional and manual; spring trim. Electric flaps and elevator trim optional. Dual controls.
STRUCTURE: Fuselage, wing leading-edge, wingtips, elevator, rudder, mainwheel legs and engine cowling of carbon fibre composites. Wing main spar, wing trailing-edge, ailerons and flaps made of wood.
LANDING GEAR: Tricycle type, fixed. Cantilever mainwheel legs. Aerodynamic composites nosewheel leg and integral wheel fairing. Castoring nosewheel. Tyres sizes 400×100 (main), 300×100 (nose). Hydraulic disc brakes. Speed fairings on all three wheels.
POWER PLANT: One 59.6 kW (79.9 hp) Rotax 912 UL or 73.5 kW (98.6 hp) Rotax 912 ULS flat-four engine; FITI two- or three-blade ground-adjustable pitch propeller. German distributor also offers 59.7 kW (80 hp) Verner engine as option. Fuel capacity 60 litres (15.9 US gallons; 13.2 Imp gallons) standard, 80 litres (21.1 US gallons; 17.6 Imp gallons) optional.
DIMENSIONS, EXTERNAL:
Wing span   10.60 m (34 ft 9¼ in)
Length overall   6.20 m (20 ft 4 in)
Height overall   1.90 m (6 ft 2¾ in)

AREAS:
Wings, gross   10.10 m² (108.7 sq ft)
WEIGHTS AND LOADINGS:
Weight empty   275 kg (606 lb)
Max T-O weight   450 kg (992 lb)
PERFORMANCE (912 ULS engine):
Max level speed   132 kt (245 km/h; 152 mph)
Econ cruising speed   97 kt (180 km/h; 112 mph)
Stalling speed, flaps down   36 kt (65 km/h; 41 mph)
Max rate of climb at S/L   390 m (1,280 ft)/min
T-O run   120 m (395 ft)
Landing run   200 m (660 ft)
Range: standard fuel   432 n miles (800 km; 497 miles)
  optional fuel   594 n miles (1,100 km; 683 miles)
g limits   +4/−2

UPDATED

CZAW

CZECH AIRCRAFT WORKS s.r.o.
Luční 1824, CZ-686 02 Staré Město
Tel: (+420 632) 54 34 56
Fax: (+420 632) 54 36 92
e-mail: aircraft@czaw.cz
Web: http://www.airplane.cz
CEO: Chip W Erwin

In 1997 Zenair Ltd (see Canadian section) formed a joint venture with CZAW, established earlier that year, for manufacture in the Czech Republic of its STOL CH 701 and Zodiac CH 601 series of light aircraft (described under the AMD and Zenith headings in the US section). First Czech components were manufactured in 1998. Factory output totalled 164 by 30 June 2000, including 30 in 1997, 44 in 1998 and 54 in 1999. The STOL CH 801 and CH 2000 are now also available from CZAW. Workforce in early 2001 was approximately 80.

CZAW offers the CH 601, 701 and 801 in three different stages of completion: as a factory-jigged, fast-build kit; and in full flyaway form with Rotax 912 engine, avionics, instruments and two-tone paint finish. A total of 85 CH 701s, with options for 48 more, were ordered by the Indian National Cadet Corps in late 2000 for the training of Indian Air Force pilots; 25 of these had been delivered by 21 February 2001. Development was completed in late 1998 of a modified CH 701 for ULV agricultural applications. A Zodiac floatplane, flown by CZAW's CEO, won the 1999 Schneider Cup seaplane race at Desenzano del Garda, Italy, on 19 September 1999. Additionally, CZAW is European agent for the AMD CH 2000 (which see in US section).

CZAW produces a complete line of wheeled and non-wheeled aircraft floats, with displacements ranging from 250 to 1,000 kg (551 to 2,205 lb), suitable for such aircraft as the GlaStar, Kitfox, Zenair and Van's types. 'Jump-start' wings, ailerons, flaps and tail surfaces for the GlaStar kitplane, as well as wings for its German OMF-160 Symphony counterpart, are also manufactured.

In addition, CZAW refurbished the prototype Luscombe 8F under an agreement with Renaissance Aircraft LLC and Zenair Ltd.

UPDATED

CZAW-built Zodiac CH 601 XL floatplane, winner of the 2001 Schneider Cup, and first XL built by the company   *NEW*/0131719

CZAW (ZENAIR) CH 701-AG STOL

TYPE: Agricultural sprayer.
PROGRAMME: Developed by CZAW following cessation of Zenith production of a dedicated agricultural version.
COSTS: US$12,360 (kit); US$38,000 ready-to-fly (both 2002).
DESIGN FEATURES: STOL performance, plus ULV crop treatment system optimised for low cost and precise application.
Differences from basic CH 701 (see US section) as follows:

CH 701-AG STOL with CZAW amphibious floats *(Jane's/Paul Jackson)*   0106490

Indian Air Force Zenair CH 701 produced by CZAW   0109647

EQUIPMENT: Spray Miser ULV crop treatment system, comprising 113.6 litre (30.0 US gallon; 25.0 Imp gallon) belly tank and 12 wind-driven rotary atomisers with controlled droplet application. Hydraulic nozzles produce droplet sizes varying from 1 to 500 microns in size. Electrically powered pump (12 V DC), with remotely controllable on/off switch, can deliver liquid at up to 30.3 litres (8.0 US gallons; 6.7 Imp gallons)/min. Tank has emergency dump facility.

WEIGHTS AND LOADINGS:
Weight empty, equipped	270 kg (595 lb)
Max T-O weight	450 kg (992 lb)

PERFORMANCE:
Application speed range	39-65 kt (72-121 km/h; 45-75 mph)
Max rate of climb at S/L	305 m (1,000 ft)/min
Swath width	9.1-30.5 m (30-100 ft)
Range	217 n miles (402 km; 250 miles)

UPDATED

DELTA

DELTA SYSTEM-AIR AS

Bratři Štefanů 101, CZ-500 03 Hradec Králové
Tel: (+420 49) 541 14 06
Fax: (+420 49) 563 00 10
e-mail: delta.system.air@telecom.cz
Web: http://www.dsa.cz
MANAGER, FLIGHT OPERATIONS: Radek Dlouhy

As well as producing the Pegass, Delta offers maintenance for a wide range of aircraft including single-engined Cessnas and most of the Piper range.

UPDATED

DELTA PEGASS

TYPE: Side-by-side ultralight.
PROGRAMME: First flight 1995.
CUSTOMERS: Total 12 built by end of 1999; none known in 2000, although promotion is continuing.
DESIGN FEATURES: Strut-braced high-wing cabin monoplane.
FLYING CONTROLS: Conventional and manual.
LANDING GEAR: Non-retractable tricycle type.
POWER PLANT: One 59.6 kW (79.9 hp) Rotax 912 UL or 47.8 kW (64.1 hp) Rotax 582 UL engine; three-blade propeller. Fuel capacity 35 litres (9.25 US gallons; 7.7 Imp gallons).

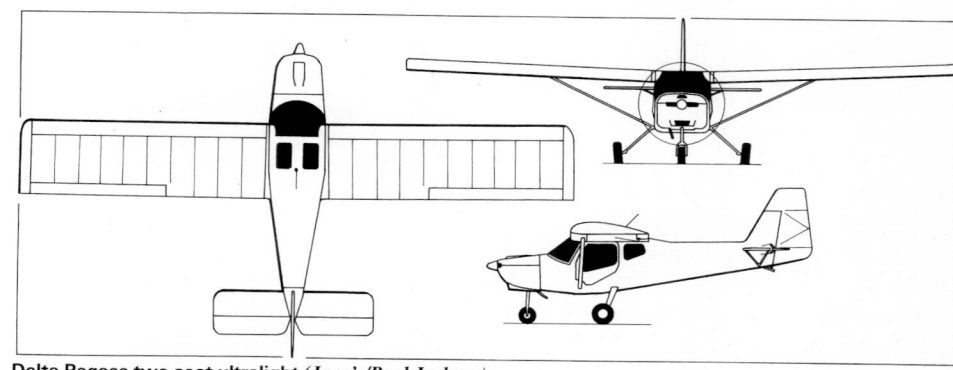

Delta Pegass two-seat ultralight (*Jane's/Paul Jackson*)   0010601

ACCOMMODATION: Door with large window each side.
DIMENSIONS, EXTERNAL:
Wing span	10.19 m (33 ft 5¼ in)
Length overall	5.70 m (18 ft 8½ in)
Height overall	2.37 m (7 ft 9¼ in)

AREAS:
Wings, gross	14.26 m² (153.5 sq ft)

WEIGHTS AND LOADINGS:
Weight empty	225 kg (496 lb)
Max T-O weight	450 kg (992 lb)

PERFORMANCE (Rotax 912 UL):
Max level speed	92 kt (170 km/h; 105 mph)
Stalling speed	36 kt (65 km/h; 41 mph)
Max rate of climb at S/L: solo	210 m (689 ft)/min
two persons	150 m (492 ft)/min
Endurance	4 h

UPDATED

EV-AT

EVEKTOR-AEROTECHNIK AS

Kunovice-Letiště, CZ-686 04 Kunovice
Tel: (+420 632) 53 73 17
Fax: (+420 632) 53 79 10
e-mail: marketing@evektor.cz
Web: http://www.evektor.cz
CHAIRMAN: Jaroslav Ružička
MANAGING DIRECTOR: Jiří Habarta
COMMERCIAL DIRECTOR: Milan Břístěla
MARKETING MANAGER: Jan Fridrich

The Evektor aircraft design office (workforce 150 in 1999) was formed in 1991, a major part of its initial work consisting of contribution to the designs of the L 159 ALCA and Ae 270 Ibis under contract to Aero Vodochody. On 2 February 1996 it acquired a 100 per cent shareholding in Aerotechnik CZ, and in 1996-97 the two companies (now a single joint-stock company) developed and delivered preseries and test front fuselage components of the L 159 to Aero.

Aerotechnik was formed in 1970, originally as an enterprise of the former Czechoslovak defence organisation Svazarm. Its first aircraft programmes involved single- and two-seat rotorcraft. Two facilities, at Kunovice and Moravská Třebová, have a combined floor space of 7,340 m² (79,000 sq ft). The combined Evektor-Aerotechnik workforce totalled 350 in mid-2000.

In recent years Aerotechnik produced kits for the Pottier P 220 S Koala all-metal lightplane (see 1999-2000 *Jane's*), powered by its own Mikron III engine. This design was further developed via the P 220 UL into the EV-97 Eurostar (which see) with a Rotax 912 UL engine, and became the first aeroplane to bear the Evektor name. A VLA version of the Eurostar, the Harmony, followed in 2001.

Evektor-Aerotechnik assisted in early stages of the Wolfsberg Raven 257 programme, including assembly of the prototype; this aircraft is now described under the Letov Air heading in this section. A separate joint venture was formed in October 1999 with GAIC of China, leading to establishment in June 2000 of Guizhou Evektor Aircraft Corporation (which see) for local production, sales and support of EV-AT aircraft in that country. The first EV-97 CKD kit for Chinese assembly was delivered in August 2000. This arrangement superseded that previously agreed for Guizhou to become a licensed production centre for the Raven 257. On 26 July 2000, Evektor-Aerotechnik signed an agreement with Seabird Aviation Australia to manufacture and market the latter's Seeker SB7L-360 light observation aircraft (which see). By early 2002, nothing further had been heard of the venture, however.

Other activities include development and production of Mikron III AE and III B piston engines; hot-air balloons and other commercial inflatables; spares and components for light and ultralight aircraft; and the overhaul and re-engining of L-13SV Vivat and L-60 Brigadýr aircraft. The company is the first in the Czech Republic to achieve JAR 21 certification standards.

UPDATED

EV-AT EV-97 TEAMEUROSTAR

TYPE: Side-by-side ultralight/kitbuilt.
PROGRAMME: Developed as P 220 UL from Pottier P 220 S Koala; extensive redesign started in September 1996; prototype (OK-CUR 97), then known as Eurostar, made first flight May 1997; first customer delivery October 1997. Certified as ultralight in Czech Republic, Germany and Slovak Republic. First UK-built example (G-NIDG, by distributor Cosmik Aviation) flown in April 2000; local assembly in China by Guizhou Evektor Aircraft Corporation (GEAC, which see); first flight by Chinese-assembled EV-97, 24 October 2000.

British-assembled EV-AT EV-97 teamEurostar built by Mr E O Otun (*Jane's/Paul Jackson*)   NEW/0137968

CURRENT VERSIONS: **Eurostar 2000:** Introduced late 1999 and described in 2000-01 *Jane's*. Increased fuselage width (4 cm; 1½ in); modified wingtips and increased span; larger ailerons; steerable nosewheel; flush, oval nose air intake. Available factory-built or as kit.

teamEurostar 2001: Further-improved version, introduced in early 2001; choice of Rotax 912 UL or more powerful 912 ULS engine. Name changed to avoid confliction with Eurostar France-UK rail link. US marketing name is **SportStar.**
CUSTOMERS: More than 150 (all versions) ordered, including 36 Model 2000, by April 2001; production of 65 was planned during 2001. Customers in Belgium, Czech Republic, Finland, France, Germany, Italy, Netherlands, Portugal, Slovak Republic, Spain, Sweden, UK and USA.
COSTS: DM120,257 flyaway, including 16 per cent tax (2001); Chinese version approximately RMB700,000 Yuan (2000).
Following description applies to teamEurostar 2001.
DESIGN FEATURES: Low-wing monoplane; modified NACA 4415 aerofoil section; dihedral 3°. Wing folding optional.
FLYING CONTROLS: Conventional and manual. Four-position (0/15/30/50°) split flaps. Elevators, starboard aileron and rudder have trim tabs.
STRUCTURE: Mostly of riveted metal; wingtips and mainwheel legs in GFRP.

EV-AT EV-97 teamEurostar 2001 two-seat ultralight (*Jane's/James Goulding*)   0059970

Cockpit of EV-AT Harmony (*Jane's/Paul Jackson*)
NEW/0131749

LANDING GEAR: Fixed tricycle with hydraulic mainwheel disc brakes and (in current version) steerable nosewheel. Wheel sizes 350×140 (main), 300×100 (nose). Optional wheel fairings or mudguards.

POWER PLANT: One 59.6 kW (79.9 hp) Rotax 912 UL flat-four engine standard, with VZLU V230C two-blade, fixed-pitch wooden propeller; 73.5 kW (98.6 hp) Rotax 912 ULS and V534BD-UL three-blade, constant-speed (hydraulic), wooden propeller optional. Fuel capacity 70 litres (18.5 US gallons; 15.4 Imp gallons) standard, optionally 50 litres (13.2 US gallons; 11.0 Imp gallons).

ACCOMMODATION: Front-hinged, upward-opening canopy, with additional transparency aft of rear frame. Space for 15 kg (33 lb) of baggage aft of seats. Early versions had canopy/rear transparency break line further forward. Optional replacement of rear transparency by metal fairing.

SYSTEMS: 12 V (battery) electrical system.

AVIONICS: Bendix/King KY 97A transceiver and VFR instrumentation standard; ELT and other avionics at customer's option.

EQUIPMENT: Optional ballistic parachute, wing-folding mechanism and aero-tow hook.

DIMENSIONS, EXTERNAL:
Wing span	8.10 m (26 ft 7 in)
Length overall	5.98 m (19 ft 7½ in)
Fuselage max width	1.04 m (3 ft 5 in)
Height overall	2.34 m (7 ft 8¼ in)
Wheel track	1.60 m (5 ft 3 in)
Wheelbase	1.40 m (4 ft 7 in)

DIMENSIONS, INTERNAL:
Cabin max width	1.00 m (3 ft 3¼ in)

AREAS:
Wings, gross	9.84 m² (105.9 sq ft)

WEIGHTS AND LOADINGS:
Weight empty	262 kg (578 lb)
Max T-O weight	450 kg (992 lb)*

480 kg (1,058 lb) permitted in Slovak Republic, Germany and USA

PERFORMANCE (at 450 kg; 992 lb MTOW. A: Rotax 912 UL engine, B: Rotax 912 ULS):
Never-exceed speed (VNE)	
A, B	145 kt (270 km/h; 167 mph) IAS
Max level speed: A	121 kt (225 km/h; 139 mph) IAS
B	132 kt (245 km/h; 152 mph) IAS
Max cruising speed: A	97 kt (180 km/h; 112 mph) IAS
B	108 kt (200 km/h; 124 mph) IAS
Stalling speed, flaps down:	
A, B	36 kt (65 km/h; 41 mph) CAS
Max rate of climb at S/L: A	330 m (1,083 ft)/min
B	480 m (1,574 ft)/min
Service ceiling: A, B	5,000 m (16,400 ft)
T-O run: A	145 m (475 ft)
B	100 m (330 ft)
Landing run: A, B	200 m (660 ft)
Max range, standard fuel:	
A, B	432 n miles (800 km; 497 miles)
g limits: A, B	+6/−3

OPERATIONAL NOISE LEVELS:
To German LSL Chapter 6	60 dBA

UPDATED

EV-AT HARMONY

TYPE: Two-seat lightplane.

PROGRAMME: Revealed (then called Eurostar VLA) 1999; renamed Harmony and made first flight (OK-VLA) 6 April 2001; certification late 2001 and first deliveries early 2002. Initial batch of six, to be followed by further 12.

COSTS: €79,136 to €95,474 (excluding VAT), depending upon avionics fit (2001).

DESIGN FEATURES: Heavier/more powerful version of teamEurostar, conforming to JAR-VLA and BCAR Section S. Reinforced fuselage aft of cockpit; slightly wider wheel track. Flaps deflect 0/15/40°.

Description as for teamEurostar 2001 except as follows:

POWER PLANT: One 73.5 kW (98.6 hp) Rotax 912 ULS flat-four engine; VZLU V230E two-blade, fixed-pitch, wooden propeller. Standard fuel capacity 67 litres (17.7 US gallons; 14.7 Imp gallons).

ACCOMMODATION: Two seats side by side; dual controls. Front-hinged canopy, with larger size of rear transparency (refer EV-97). Baggage space aft of seats.

AVIONICS: VFR instrumentation only in basic aircraft. 'Effect' version adds VSI, turn co-ordinator, VHF com antenna, PM 1000 intercom, Bendix/King KT 76A transponder, ACK encoder, KY 97A VHF and DME/transponder antenna. Top-of-range 'VFR night' suite omits KY 97A and ACK encoder, but adds KLN 35A GPS with KA 92 antenna, KX 125 nav/com, VOR/glideslope antenna, artificial horizon and directional gyro.

DIMENSIONS: As for teamEurostar

AREAS:
Wings, gross	10.13 m² (109.0 sq ft)

WEIGHTS AND LOADINGS:
Weight empty	330 kg (728 lb)
Max baggage weight	15 kg (33 lb)
Max T-O weight	550 kg (1,212 lb)
Max wing loading	54.3 kg/m² (11.12 lb/sq ft)
Max power loading	7.48 kg/kW (12.29 lb/hp)

PERFORMANCE (preliminary):
Never-exceed speed (VNE)	145 kt (270 km/h; 167 mph) IAS
Max level speed	129 kt (240 km/h; 149 mph) IAS
Max cruising speed	111 kt (205 km/h; 127 mph) IAS
Stalling speed, flaps down	44 kt (80 km/h; 50 mph) CAS
Max rate of climb at S/L	300 m (984 ft)/min
Service ceiling	5,000 m (16,400 ft)
T-O run	130 m (425 ft)
T-O to 15 m (50 ft)	260 m (855 ft)
Landing from 15 m (50 ft)	480 m (1,575 ft)
Landing run	240 m (790 ft)
Range	388 n miles (720 km; 447 miles)
g limits	+5.7/−2.25

UPDATED

Third prototype EV-97, built 1997, showing original canopy dimensions (*Jane's/Paul Jackson*)   **NEW**/0131747

Prototype EV-AT Harmony VLA-standard two-seat lightplane (*Jane's/Paul Jackson*)   **NEW**/0131748

FANTASY AIR

FANTASY AIR GROUP
ulica Krbu 35, CZ-100 00 Praha 10
Tel: (+420 2) 74 77 25 57
Fax: (+420 2) 74 77 33 43
e-mail: sales@fantasyair.com
Web: http://www.fantasyair.com
MANAGING DIRECTOR: Miloš Dermíšek

WORKS: Kollàrova 511, CZ-397 01 Pisek

Fantasy Air established at Písek 1995; initial product (see 1999-2000 *Jane's*) launched as Cora, in Legato and Allegro versions. By mid-2000 was standardising on variants of Allegro, as described in following entry.

UPDATED

FANTASY AIR ALLEGRO and ARIUS

TYPE: Side-by-side ultralight/kitbuilt.

PROGRAMME: Certified in Czech Republic, Finland, France and Germany; public debut at Friedrichshafen Air Show, May 1997. All are subvariants of Fantasy **Cora**. Legato subvariant no longer marketed. Production rate three per month in 2001.

A new wing profile was first tested in May 2002. Revealed in the same month were plans to fit a Diesel

Air D280 twin-piston engine, derated to 60.0 kW (80 hp).

CURRENT VERSIONS: **Allegro ST**: Baseline version; 47.8 kW (64.1 hp) Rotax 582 UL. Details in 2001-02 *Jane's*; remains available.

Allegro 200: As ST, but with 55.0 kW (73.8 hp) Rotax 618 UL. Details in 2001-02 *Jane's*; remains available.

Allegro 2000: Introduced from 1 March 2000 (kit version introduced September 2000); 59.6 kW (79.9 hp) Rotax 912 UL or (Allegro S) 73.5 kW (98.6 hp) 912 ULS. SportProp glass fibre or VZLU V331-3NC wooden three-blade propeller. Quoted build time 510 hours. Certified in Czech Republic, Denmark, Finland and Germany.

Allegro SW: Short-span (Speed Wing) version, without taper on outer panels. New SM 701 aerofoil section wing and 47.8 kW (64.1 hp) Rotax 582 UL or 59.6 kW (79.9 hp) Rotax 912 UL. German marketing only with Rotax 912 UL and propeller options as for Allegro 2000. Four to France in September 1999, where certification subsequently awarded; also certified in Czech Republic and Spain. Kit version available from 1 March 2000.

Arius F: Standard-span wing (SM 701 section) and Rotax 912 UL or 912 ULS. Two to Finland in August 1999, where this version certified at 520 kg (1,146 lb) MTOW to permit flying with twin-float gear. First flight (converted Allegro, OK-DUU 03) 5 November 1999.

CUSTOMERS: Total 70 (51 kits and 19 factory-built, including a few Cora Legatos) registered in Czech Republic by November 2000. Exports to Finland and Hungary in 2000. Total sales by April 2001 exceeded 100, including 70 in Czech Republic, 18 in France and 12 in Germany; 32 built in 2000, 35 planned for 2001.

COSTS: Allegro 2000 €50,363; Allegro 2000 S €52,172; Allegro SW €47,933 (all 2000, incl tax). DM97,800 with 912 ULS (April 2001).

DESIGN FEATURES: Braced high-wing monoplane; T tail; tapered outer panels on 2000 and Arius, constant chord throughout on SW. UA-2 aerofoil section on ST and 200, improved SM 701 on 2000, SW and Arius; dihedral 1° 10′ on Allegro 2000; 1° 30′ on Allegro SW.

FLYING CONTROLS: Conventional and manual. Electrically actuated flaperons (settings 0/15/48°). In-flight-adjustable tab in port half of elevator; ground-adjustable rudder tab.

STRUCTURE: Fabric- and metal-covered metal wings with welded steel tube centre-section and full-span metal flaperons; laminated glass fibre fuselage and integral fin; all-metal rudder, tailplane and one-piece elevator.

LANDING GEAR: Fixed tricycle with Kevlar cantilever mainwheel legs and 350 mm wheels, rubber-block shock-absorbers, hydraulic mainwheel brakes and steerable nosewheel; wheel spats. Full Lotus twin-float gear to be certified for Arius F; testing being undertaken on local reservoir (with OK-DUU 03) in August 2001.

POWER PLANT: One Rotax engine (details under Current Versions); Sportprop three-blade, glass fibre or Junkers Profly two-blade, ground-adjustable pitch, carbon fibre propeller. Fuel capacity 55 litres (14.5 US gallons; 12.1 Imp gallons).

EQUIPMENT: Junkers recovery parachute optional.

DIMENSIONS, EXTERNAL (A: 2000, B: Arius (wheels), C: SW):

Wing span: A, B	10.81 m (35 ft 5½ in)
C	9.60 m (31 ft 6 in)
Length overall: A, B	6.36 m (20 ft 10½ in)
C	6.10 m (20 ft 0¼ in)

Fantasy Air Allegro 2000 (*Jane's/Paul Jackson*)   NEW/0131750

Height overall: A, B	2.05 m (6 ft 8¾ in)
C	2.00 m (6 ft 6¾ in)
DIMENSIONS, INTERNAL:	
Cabin max width: A, B, C	1.08 m (3 ft 6½ in)
AREAS:	
Wings, gross: A, B	11.37 m² (122.4 sq ft)
C	11.50 m² (123.8 sq ft)
WEIGHTS AND LOADINGS:	
Weight empty: A, B, C	285-295 kg (628-650 lb)
Max T-O weight: A, B, C	450 kg (992 lb)
Arius (floats)	520 kg (1,146 lb)
PERFORMANCE, POWERED (Rotax 912 UL):	
Never-exceed speed (VNE):	
A	118 kt (220 km/h; 136 mph)
B	108 kt (200 km/h; 124 mph)
C	105 kt (195 km/h; 121 mph)

Max cruising speed at 75% power:	
A	89 kt (165 km/h; 103 mph)
Stalling speed: A, B, C	36 kt (65 km/h; 41 mph)
Max rate of climb at S/L: A	360 m (1,181 ft)/min
B	240 m (787 ft)/min
C	300 m (984 ft)/min
T-O run: A, C	150 m (495 ft)
B	160 m (525 ft)
Landing run: A, C	100 m (330 ft)
B	120 m (395 ft)
Max range: A, C	270 n miles (500 km; 310 miles)
B	324 n miles (600 km; 372 miles)
PERFORMANCE, UNPOWERED:	
Best glide ratio: A, B	12
C	10

UPDATED

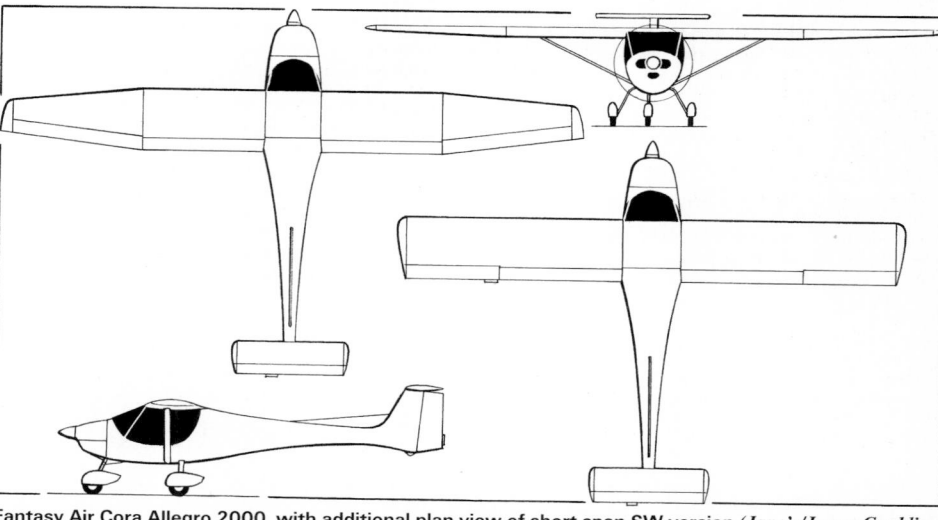

Fantasy Air Cora Allegro 2000, with additional plan view of short-span SW version (*Jane's/James Goulding*)
NEW/0137210

INTERPLANE

INTERPLANE spol s.r.o.
Letiště Zbraslavice, CZ-285 21 Zbraslavice
Tel/Fax: (+420 327) 59 13 81
e-mail: InPlane@mira.cz
Web: InterplaneAircraft.com
SALES MANAGER AND EXECUTIVE: Andrea Greplová
DIRECTOR OF MANUFACTURING: Radek Hurt
CHIEF DESIGNER: Jaroslav Dostal

NORTH AND SOUTH AMERICAN SALES:
Interplane USA
39440 South Avenue, Zephyrhills, Florida 33540, USA
Tel: (+1 813) 782 79 00
Fax: (+1 813) 788 66 00
e-mail: Inplane@gte.net
PRESIDENT: Ben Dawson

Interplane was created as an ultralight manufacturer in 1992; design and product development is by Gryf Development company at Hodonin. Earlier variations between versions led the company to rationalise its Skyboy and Griffon product line in 1999. There are now two basic versions of the Skyboy: the EX (standard version for European customers and US and Australian kitbuilders) and the UL (to US FAR Pt 103 ultralight standards). Earlier Griffons are now replaced by the UL 103 version for the US market. Griffons and Skyboys now flying in Australia, Belgium, France, Germany, Luxembourg, South Africa, Spain, Sweden, Turkey and USA, as well as in Czech and Slovak Republics.

Long-term projects include **P6 Diblík** single-seat biplane (wing span 6.00 m; 19 ft 8¼ in, area 13.00 m² ; 139.9 sq ft;

Interplane Skyboy UL ultralight (*Jane's/Paul Jackson*)   NEW/0137214

MTOW 270 kg; 595 lb), **P15 Classic** two-seat ultralight (wing span 9.50 m; 31 ft 2 in, 73.5 kW; 98.6 hp Rotax 912 ULS, 81 kt; 150 km/h; 93 mph cruising speed) and **P16 Nemesis** improved version of Skyboy.

UPDATED

INTERPLANE SKYBOY

TYPE: Side-by-side ultralight/kitbuilt.

PROGRAMME: Prototype designed and flown 1993; entered production 1994; conforms to Czech UL-2 and German BFU-DULV standards. Certified in Czech Republic (1994), France (1996), Slovak Republic and Germany (excluding noise tests, 1999). Current EX and UL versions

introduced in 1999. New canopy design permits operation without doors for hot-weather flying.

CURRENT VERSIONS: **Skyboy**: Initial version, with Rotax 503 or 582 engine as standard; no longer produced. Details in 1998-99 *Jane's*.

Skyboy S: Reinforced version for higher MTOW and speeds; Rotax 582 engine and choice of wing spans; no longer produced. Details in 1999-2000 *Jane's*.

Skyboy EX: Standard version from 1999; for European customers; also for customers in Australia and USA, where it is classified in the FAA Experimental category.

Skyboy UL: Specially lightened (empty weight 225 kg; 496 lb) to meet US FAR Pt 103 ultralight trainer regulations. Wing span 9.08 m (29 ft 9½ in), area 12.80 m²

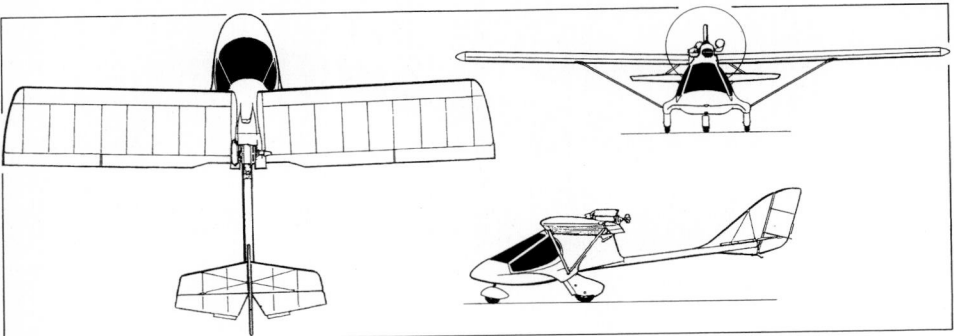

Interplane Skyboy, showing optional flaps (*Jane's/Paul Jackson*)   0110846

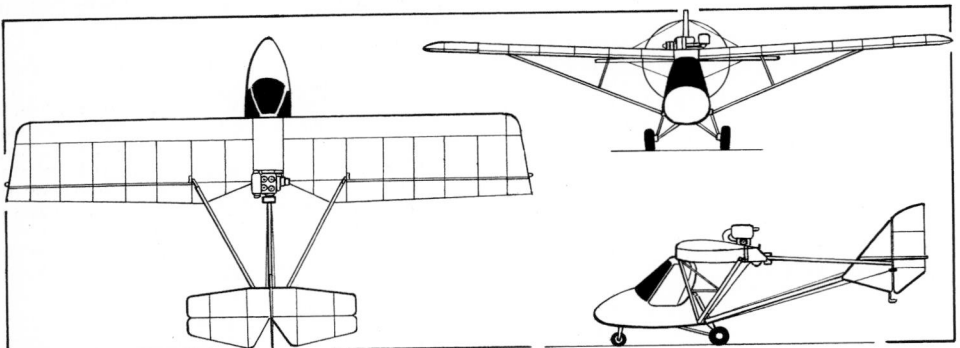

Current version of Interplane Griffon, fitted with 9.00 m (29 ft 6¼ in) wings (*Jane's/James Goulding*) 0092146

(137.8 sq ft); stalling speed, flaps down 23 kt (42 km/h; 27 mph); Rotax 582 engine; fuel capacity 38 litres (10.0 US gallons; 8.3 Imp gallons); range 130 n miles (241 km; 150 miles).

CUSTOMERS: Completion of 44th aircraft (all versions, including prototypes) effected in March 2002.

COSTS: *EX:* Kit US$14,900, flyaway US$24,900; *UL:* flyaway US$19,900 (all 2002).

Description applies to Skyboy EX.

DESIGN FEATURES: High, constant-chord wing, boom-mounted empennage and pusher propeller. Available factory-built or (with or without power plant and instruments) as kit. Wing struts and tail surfaces can be folded for transportation and storage.

Wing aerofoil section NACA 4412; dihedral 1° 30′; incidence 4° 30′. Leading-edges swept forward 2°.

FLYING CONTROLS: Conventional and manual. Tabs on rudder and each elevator; flaps optional.

STRUCTURE: Aluminium tube and glass fibre; fabric covering.

LANDING GEAR: Non-retractable tricycle type; coil spring/hydraulic (motorcar-type) shock-absorbers on trailing-link mainwheels and castoring nosewheel; mechanical or hydraulic brakes on nosewheel only. Ski and float gear optional.

POWER PLANT: One 47.8 kW (64.1 hp) Rotax 582 UL two-cylinder in-line two-stroke engine standard; 59.6 kW (79.9 hp) Rotax 912 UL or 73.5 kW (98.6 hp) Rotax 912 ULS flat-four; Rotax 462, 503 or 618; Hirth 2706; or Verner SVS 1400 optional. Wide choice of two- to six-blade wooden or laminate pusher propellers (three-blade, carbon fibre Junkers Profly standard). Standard fuel capacity 42 litres (11.1 US gallons; 9.2 Imp gallons); optional fuel capacity 57 litres (15.0 US gallons; 12.5 Imp gallons).

ACCOMMODATION: Optional cabin doors.

EQUIPMENT: Optional Magnum 450 ballistic recovery parachute.

DIMENSIONS, EXTERNAL:
Wing span	9.50 m (31 ft 2 in)
Length overall	6.37 m (20 ft 10¾ in)
Height overall	2.13 m (6 ft 11¾ in)

DIMENSIONS, INTERNAL:
Cabin max width	1.20 m (3 ft 11¼ in)

AREAS:
Wings, gross	13.50 m² (145.3 sq ft)

WEIGHTS AND LOADINGS:
Weight empty, equipped	265 kg (584 lb)
Max T-O weight	450 kg (992 lb)

PERFORMANCE (Rotax 582):
Max level speed	76 kt (140 km/h; 87 mph)
Econ cruising speed	49 kt (90 km/h; 56 mph)
Stalling speed	34 kt (63 km/h; 40 mph)
Max rate of climb at S/L	244m (800 ft)/min
T-O run	150 m (495 ft)
Landing run	100 m (330 ft)
Range with standard fuel	162 n miles (300 km; 186 miles)

UPDATED

INTERPLANE GRIFFON 103

TYPE: Single-seat ultralight/kitbuilt.

PROGRAMME: First flight by Gryf ULM-1 prototype (OK-WUC 01) made on 17 March 1989. Early versions detailed in 1999-2000 *Jane's*. Available factory-built or (with or without power plant and instruments) as kit. Conforms to FAR Pt 103; Griffon 103 introduced 1999.

CUSTOMERS: Approximately 15 to 20 (including kitbuilt examples) completed by late 1999 (latest figure known).

DESIGN FEATURES: Generally as for Skyboy (which see).

FLYING CONTROLS: Conventional and manual. Differential ailerons.

STRUCTURE: Generally similar to that of Skyboy.

LANDING GEAR: Non-retractable tricycle type; castoring nosewheel, with optional mechanical brakes.

POWER PLANT: One 37.0 kW (49.6 hp) Rotax 503 UL-2V or 31.0 kW (41.6 hp) Rotax 447 UL-2V two-cylinder two-stroke in-line engine; Junkers Profly three-blade, carbon fibre, pusher propeller. Fuel capacity 18.9 litres (5.0 US gallons; 4.2 Imp gallons).

DIMENSIONS, EXTERNAL:
Wing span	9.00 m (29 ft 6¼ in)
Length overall	5.73 m (18 ft 9½ in)
Height overall	2.20 m (7 ft 2½ in)

AREAS:
Wings, gross	13.96 m² (150.3 sq ft)

WEIGHTS AND LOADINGS (Rotax 447):
Weight empty	115 kg (254 lb)
Max T-O weight	235 kg (518 lb)

PERFORMANCE (Rotax 447):
Max level speed	55 kt (102 km/h; 63 mph)
Econ cruising speed	49 kt (90 km/h; 56 mph)
Stalling speed	25 kt (45 km/h; 28 mph)
Max rate of climb at S/L	210 m (689 ft)/min

UPDATED

Interplane Griffon single-seat ultralight   *NEW*/0131720

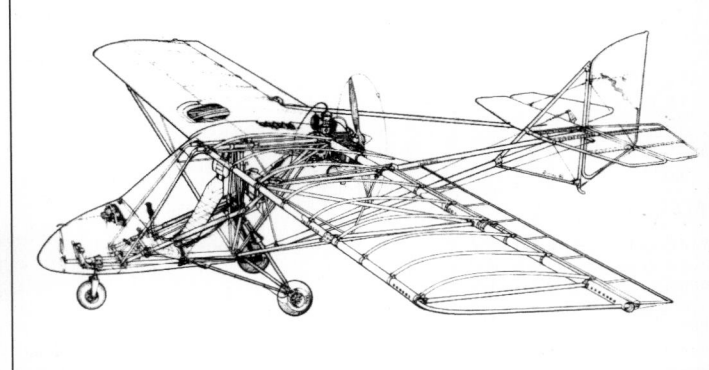

Structural details of Interplane Griffon   0106491

KAPPA

KAPPA 77 a.s.

Brtnická 21, CZ-586 01 Jihlava
Tel: (+420 66) 731 00 18
Fax: (+420 66) 730 81 22
e-mail: info@kappa77.cz
Web (1): http://www.kappa77.cz
Web (2): http://www.ultralight.cz
MANAGING DIRECTOR: Ing Miroslav Navrátil
CHIEF DESIGNER: Doc Ing Antonin Pištěk
MARKETING DIRECTOR: Vit Tausch
PUBLIC RELATIONS MANAGER: Ing Tomáš Navrátil

Kappa established 1991, initially for machine manufacture; founders previously employed by Aero. Design of KP-2U began 1994; Kappa 77 a.s. created as subsidiary in June 1997; Sova production started 1998.

VERIFIED

KP-2U Sova ultralight (*Jane's/Paul Jackson*)   *NEW*/0131724

KAPPA KP-2U SOVA

English name: Owl

TYPE: Side-by-side ultralight/kitbuilt.

PROGRAMME: Designed by Kappa company and Institute of Aerospace Engineering at Brno Technical University. First of two flying prototypes (OK-BUU 23) rolled out and first

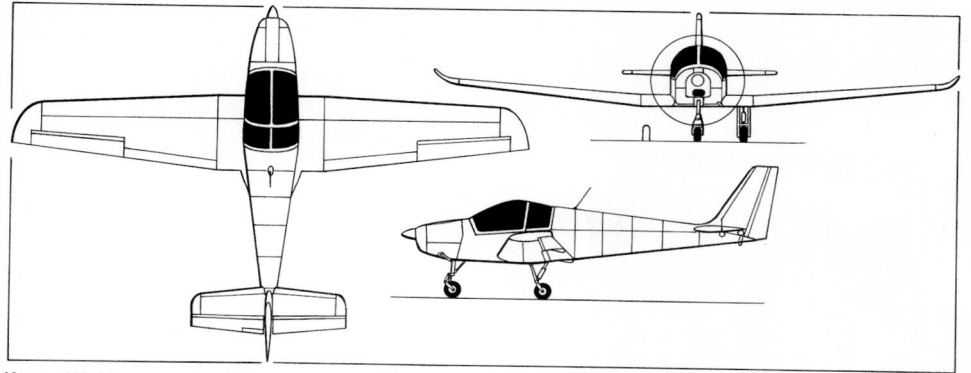

Kappa KP-2U Sova two-seat ultralight (*Jane's/James Goulding*)　　　　0062594

flown 26 May 1996; first public demonstration the following month at Jihlava Aero '96. Czech certification 23 September 1997; German certification, and of kit by FAA, followed in June 1999.

CUSTOMERS: Total of 46 sold by 1 November 2000; customers in Chile, Czech Republic, Ecuador, Germany (known as KPD-2U), Italy, Netherlands, Slovak Republic, Spain and USA. Marketed in USA by Interplane (which see).

COSTS: US$53,900 flyaway with Rotax 912 UL (2000); DM77,060 (€39,400) excluding engine (2001).

DESIGN FEATURES: Low-wing monoplane conforming to BFU 95 regulations; JAR-VLA version being developed.
　Wing sections NASA GA(W)-1 at root and GA(W)-2 at tip; upturned wingtips. Sweptback vertical tail.

FLYING CONTROLS: Conventional and manual. Electrically operated trim tab in port elevator; Fowler flaps (deflection 10 and 35°).

STRUCTURE: All-metal, except for CFRP cockpit frame.

LANDING GEAR: Electrically retractable tricycle type (all units rearward); rubber shock-absorption; mainwheels semi-

exposed when retracted. Steerable nosewheel; mechanically operated brakes. Fixed gear optional.

POWER PLANT: One 59.6 kW (79.9 hp) Rotax 912 UL engine, with 4:1 reduction drive to a two-blade, flight-adjustable or three-blade ground-adjustable pitch propeller; can alternatively accommodate Rotax 582 UL, 618 UL, 912 ULS or 914. Fuel capacity (two wing tanks) 64 litres (16.9 US gallons; 14.1 Imp gallons).

ACCOMMODATION: Forward-hinged canopy. Dual controls. Space behind seats for 14 kg (31 lb) of baggage.

DIMENSIONS, EXTERNAL:
Wing span	9.90 m (32 ft 5¾ in)
Length overall	7.17 m (23 ft 6¼ in)
Height overall	2.60 m (8 ft 6¼ in)
Propeller diameter	1.73 m (5 ft 8 in)

AREAS:
Wings, gross	11.85 m² (127.6 sq ft)

WEIGHTS AND LOADINGS:
Weight empty	285 kg (628 lb)
Max T-O weight	450 kg (992 lb)

PERFORMANCE (Rotax 912 UL engine):
Max level speed	129 kt (240 km/h; 149 mph)
Max cruising speed at 75% power	
	108 kt (200 km/h; 124 mph)
Stalling speed: flaps up	39 kt (71 km/h; 45 mph)
flaps down	26 kt (48 km/h; 30 mph)
Max rate of climb at S/L	390 m (1,279 ft)/min
T-O run	100 m (330 ft)
Landing run	140 m (460 ft)
Range with max fuel	518 n miles (960 km; 596 miles)
g limits	+4/−2

UPDATED

LET

LET AS, KUNOVICE
ADMINISTRATOR: Zlatava Davidova

Company acquired 16 July 2001 by Moravan Aeroplanes Inc and renamed Letecké Závody (LZ, which see). Purchase excluded L 610 G programme, which subsequently (early 2002) also acquired by LZ.

UPDATED

LET-MONT

LET-MONT s.r.o.
Vikýřovice 226, CZ-788 13 Šumperk
Tel: (+420 649) 21 40 85 and 21 40 81
Fax: (+420 649) 21 40 85 and 21 38 77
CONSTRUCTOR: Petr Podešva
DIRECTORS:
　O Klier
　H Podešva
GERMAN DISTRIBUTOR:
Let-Mont Deutschland GmbH Gotlandweg 105, D-59494 Soest
Tel: (+49 2921) 94 38 44
Fax: (+49 2921) 94 38 45
Web: http://www.delta-mike.pair.com/letmont

VERIFIED

LET-MONT UL TULÁK and UL PIPER
English name: Rambler

TYPE: Side-by-side ultralight/kitbuilt; tandem-seat ultralight.

PROGRAMME: Sixth factory-built aircraft delivered to Poland in 1999. Further 38 kitbuilt aircraft registered in Czech Republic by late 2000.

CURRENT VERSIONS: **UL Tulák:** Side-by-side seats; factory-built or kit.
　UL Piper: Tandem seats; otherwise as Tulák. No kit version. German name **Tandem Tulák.** Span 9.82 m (32 ft 2½ in); length 5.96 m (19 ft 6½ in); maximum cruising speed 70 kt (130 km/h; 81 mph). Engine options include 58.8 kW (78.9 hp) or 73.5 kW (98.6 hp) BMW; 58.8 kW (78.9 hp) Verner; 62.5 kW (83.8 hp) Tatra; and 51.5 kW (69.0 hp) Limbach.

CUSTOMERS: Known sales in Czech Republic, France, Germany and Poland.

COSTS: Tandem Tulák DM71,600 including tax, factory-built, with 58.8 kW (78.9 hp) BMW (2001).

Data apply to both Tulák and Piper, except where indicated.

DESIGN FEATURES: High, constant-chord wing braced by V struts; optional tapered or downturned wingtips. Wings fold back for storage and transportation. Resemble Piper designs of 1940s/1950s.

Let-Mont Tulák two-seat ultralight (*Jane's/Paul Jackson*)　　　*NEW*/0106512

Clark Y wing section, thickness/chord ratio 12.5 per cent.

FLYING CONTROLS: Conventional and manual. Flight-adjustable trim tab in each elevator. Plain flaps.

STRUCTURE: Fabric-covered metal tube fuselage; two-spar wooden wing; engine cowling metal; some composites.

LANDING GEAR: Tailwheel type; fixed. Steerable tailwheel; 420×150 mainwheels with bungee suspension and cable-operated brakes. Mainwheel spats optional.

POWER PLANT: One 37.0 kW (49.6 hp) Rotax 503 UL-2V engine; SportProp three-blade, ground-adjustable pitch, glass fibre propeller. Fuel capacity 50 litres (13.2 US gallons; 11.0 Imp gallons).

DIMENSIONS, EXTERNAL:
Wing span: tapered tips	9.60 m (31 ft 6 in)
downturned tips	9.40 m (30 ft 10 in)
Width, wings folded	3.10 m (10 ft 2 in)
Length overall	5.60 m (8 ft 4½ in)
Height overall	2.00 m (6 ft 6¾ in)

DIMENSIONS, INTERNAL:
Cabin max width: Tulák	1.14 m (3 ft 9 in)
Piper	0.70 m (2 ft 3½ in)

AREAS:
Wings, gross	13.00 m² (139.9 sq ft)

WEIGHTS AND LOADINGS:
Weight empty: Tulák	255 kg (562 lb)
Piper	265 kg (584 lb)
Max T-O weight: both	450 kg (992 lb)

PERFORMANCE (both):
Never-exceed speed (VNE)	89 kt (165 km/h; 102 mph)
Econ cruising speed	59 kt (110 km/h; 68 mph)
Stalling speed, flaps down	33 kt (60 km/h; 38 mph)
Max rate of climb at S/L	300 m (984 ft)/min
T-O run	65 m (215 ft)
Landing run	60 m (200 ft)
g limits	+6/−4

UPDATED

LETOV

LETECKÁ VÝROBA s.r.o
Beranových 65, CZ-199 02 Praha 9-Letňany

Tel: (+420 2) 859 01 40
Fax: (+420 2) 858 71 75
e-mail: kalivoda@llv.cz
Web: http://www.llv.cz

DIRECTOR GENERAL: Jan Šimúnek

Following 1998 bankruptcy, control of former Letov Air was acquired by Latécoère of France in late 2000, becoming

Wolfsberg-Letov Raven 257 prototype after modification to production configuration   *NEW*/0132901

Letecká Výroba. Company now concentrates on aerostructures manufacture for EADS and other major companies. Production of the LK-2, LK-3 and ST-4 lightplanes appears to have ended. Data on these last appeared in the 2002-03 *Jane's*.

Letov produces the rear fuselage and tail surfaces for the Aero L 159 ALCA, in addition to the ultralights described below, but gained a major programme when it was assigned design and production responsibility for the Raven 257 cargo aircraft in an announcement made on 21 November 2000.

Letov's production facilities cover 18,000 m² (193,750 sq ft). Latécoère planned to invest some €12.2 million in Letov by 2003 and increase its workforce to about 400, compared with 220 at the time of take-over. Production capacity is planned to double by 2005.

UPDATED

WOLFSBERG-LETOV RAVEN 257

TYPE: Light utility twin-prop transport.

PROGRAMME: Market study commenced June 1996; general design began June 1997, and design of major structures in January 1998; manufacture of prototype by Evektor-Aerotechnik began at Kunovice, Czech Republic, in June 1998, first flight (OK-RAV) 28 July 2000. Six aircraft (prototype, structural test and four pre-production airframes) were to have been used in the flight test and certification programme; original target date for FAR Pt 23 approval was April 2002; only one flying prototype built

by early 2002, however. Production at Letňany was scheduled to have begun in June 2001, but was rescheduled after design changes including increase of wing span to 15.40 m (50 ft 6L in); MTOW increase to 3,100 kg (6,834 lb); shorter tailbooms; tailplane of reduced span; redesign of cockpit area; repositioning of forward door to provide direct access to the flight deck, and substantial increase in cabin height. Flight testing of modified prototype resumed in February 2002. FAA FAR Pt 23 certification scheduled for December 2002. Completions of 'green' airframes will be carried out at centres in the Czech Republic, France, Netherlands and the UK.

CURRENT VERSIONS: **Baseline:** 224 kW (300 hp) engines. *As described.*

Turbo: Turbocharged TSIO-550 engines, each 261 kW (350 hp); MTOW 3,000 kg (6,613 lb); maximum level speed 157 kt (291 km/h; 181 mph).

COSTS: Approximately US$695,000 (February 2002).

DESIGN FEATURES: High-wing, twin-boom configuration with square section fuselage and high-set tailplane. Design goals include ability to operate from short unprepared strips, simple systems and easy field maintenance for operation in less-developed areas, combined with low acquisition and operating costs; seen as modern replacement for aircraft in the B-N Islander and Piper Aztec class, and as a 'step up' for operators of single-engined utility aircraft such as the Cessna 206, with applications in passenger, cargo, medevac and airdrop roles.

NACA 23018 section wings, with 2° 30′ dihedral; 3° incidence; NACA 0012 section horizontal and vertical tail surfaces.

FLYING CONTROLS: Conventional and manual, via push-pull rods. Control surfaces statically balanced. Two-section elevator with independent controls and trim tab on port side. Maximum elevator deflection +27/–22°, aileron deflection +25/–15°; rudder deflection ±25°. Trim tab on starboard rudder. Four-section electrically actuated flaps, T-O setting 20°, landing 48°.

STRUCTURE: Conventional, mostly riveted metal, with glass fibre wingtips, nosecone, cargo hatch and engine cowlings.

LANDING GEAR: Non-retractable tricycle type, based on, and mostly interchangeable with, that of B-N Islander, with twin wheels on each main unit and single nosewheel; oleo strut suspension; mainwheel legs braced to bottom of fuselage. Mainwheel tyre size 7.00-6; nosewheel 6.00-6; Cleveland 40-90F brakes on mainwheels; 40-76D on nosewheel.

POWER PLANT: Two 224 kW (300 hp) Teledyne Continental IO-550-N8B flat-six engines, driving two-blade Hartzell HC-C2YF-2CUF/FC8475K-6/SM8 composites propellers. Fuel contained in two integral tanks in wing centre-section, combined capacity 536 litres (141.5 US gallons; 118 Imp gallons), with filler port in upper surface of each wing.

ACCOMMODATION: Two-crew cockpit (although aircraft will be certified for single-pilot operation); up to eight passengers. Crew door on port side; sliding semi-bulkhead separates flight deck from cabin/cargo area, which has flat floor and quick-change features, with passenger door at rear on port side and upward-hinged rear cargo door. In cargo configuration, fuselage can accommodate ATA D and IATA 8, 12 and 17-type containers, 304 × 139 cm pallets, 1.2 × 1.0 m Europallets and bulk cargo under netting. Baggage door on port side, emergency exit on starboard side.

SYSTEMS: 24 V 74 A electrical system includes two engine-driven alternators and 24 V 19 Ah accumulator.

AVIONICS: Dual VHF nav/com, ADF, transponder, GPS.

Data refer to production version.

DIMENSIONS, EXTERNAL:	
Wing span	15.40 m (50 ft 6¼ in)
Wing chord, constant	1.85 m (6 ft 0¾ in)
Wing aspect ratio	8.4
Length overall	11.28 m (37 ft 0 in)
Height overall	4.00 m (13 ft 1½ in)
Tailplane span	4.80 m (15 ft 9 in)
Wheel track	4.00 m (13 ft 1½ in)
Wheelbase	3.86 m (12 ft 8 in)
Propeller diameter	1.98 m (6 ft 6 in)

DIMENSIONS, INTERNAL:	
Cabin, excl flight deck: Length	3.91 m (12 ft 10 in)
Max width	1.47 m (4 ft 10 in)
Max height	1.44 m (4 ft 8¾ in)
Volume	8.30 m³ (293 cu ft)

AREAS:	
Wings, gross	28.40 m² (305.7 sq ft)
Ailerons (total)	1.48 m² (15.93 sq ft)
Trailing-edge flaps (total): inner	1.06 m² (11.41 sq ft)
outer	1.60 m² (17.22 sq ft)
Tailplane	4.54 m² (48.87 sq ft)
Elevators, incl tabs	3.36 m² (36.17 sq ft)
Fin (total)	3.84 m² (41.33 sq ft)
Rudders (total)	2.36 m² (25.40 sq ft)

WEIGHTS AND LOADINGS:	
Weight empty	1,800 kg (3,968 lb)
Max T-O weight	3,100 kg (6,834 lb)
Max zero-fuel weight	3,000 kg (6,614 lb)
Max wing loading	109.2 kg/m² (22.36 lb/sq ft)
Max power loading	6.93 kg/kW (11.39 lb/hp)

PERFORMANCE (estimated):	
Max level speed	145 kt (268 km/h; 167 mph)
Cruising speed, 75% power	138 kt (255 km/h; 158 mph)
Econ cruising speed	130 kt (240 km/h; 149 mph)
Stalling speed	53 kt (98 km/h; 61 mph)
Max rate of climb at S/L	444 m (1,457 ft)/min
Rate of climb at S/L, OEI	108 m (354 ft)/min
Range with max fuel	715 n miles (1,324 km; 822 miles)

UPDATED

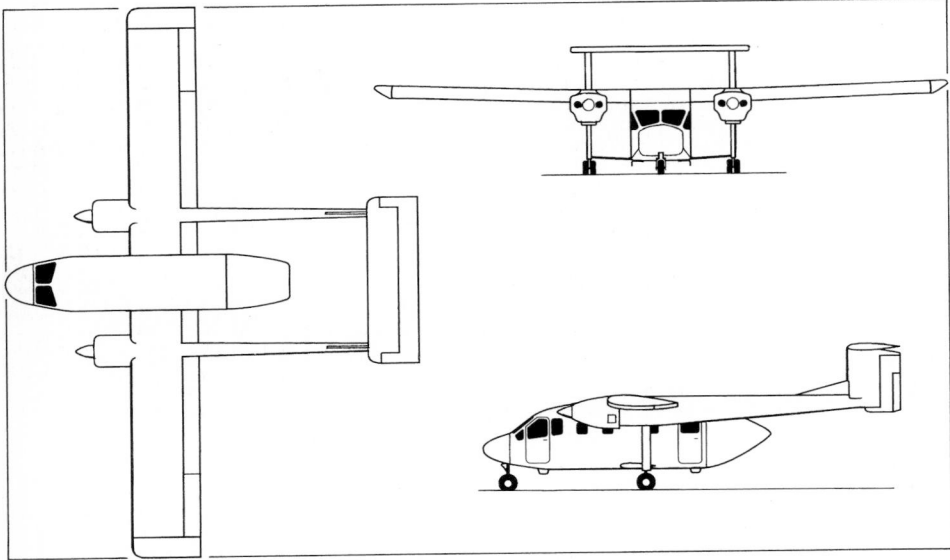

Wolfsberg-Letov Raven 257 utility aircraft (*Jane's/James Goulding*)   *NEW*/0132944

LUSUS

LUSUS s.r.o.
Renamed Aeros (which see).

UPDATED

LZ

LETECKÉ ZÁVODY a.s. (LZ Aeronautical Industries Inc)
PO Box 1177, Uherské Hradiště, CZ-686 04 Kunovice
Tel: (+420 632) 81 61 61
Fax: (+420 632) 81 61 63

e-mail: let@let.cz
Web: http://www.moravan.cz
CEO: Libor Soska
TECHNICAL DIRECTOR: Zdenek Hradsky
COMMERCIAL DIRECTOR: Lubor Směřička
MANUFACTURING DIRECTOR: Pavel Zabrana
CHIEF DESIGNER: Miroslav Pěsák

Letecké Závody (formerly Let AS, Kunovice) was established in 1936 as a repair shop for Ba 33, B-534, Junkers W 34 and Arado Ar 96B aircraft. Construction of new plant started in 1950, producing Yak-11 trainer as C-11; subsequently involved in programmes for Aero Ae 45/145, L-200 Morava, L-29 Delfin, L-13 Blanik sailplane, and Zlin Z-37. Ayres Aviation Holdings Inc, an affiliate of Ayres

Corporation of the USA (which see), acquired a 93.3 per cent holding in Let AS in August 1998. However, this lapsed when, in mid-2000, Ayres funding was suspended, all 1,400 Let employees were sent on leave and production was halted due to expiry of credit arrangement with largest creditor, Konsolidacni Banka, which filed a bankruptcy suit against Let in September 2000; Let officially declared bankrupt on 24 October 2000.

Let had focused recently on upgrading the L-410/420; repairing, modernising and selling used L-410s, especially low-time aircraft from the former USSR; continuing the L-610G certification programme; production of L-13/23/33 series of Blanik sailplanes; and attracting subcontract work. Let was to have been responsible for all wings and tail units for the Ayres LM-200 Loadmaster, as well as fuselages for Loadmasters assembled in the Czech Republic, and had delivered some components to USA before mid-2000 stoppage. First Loadmaster deliveries by Let, for European operations by Federal Express and others, were originally due in March 2001, at a rate of one per month. By then, funding only to maintain Let's contribution to Ayres Loadmaster programme had apparently been obtained, and L 610 G programme had been put up for sale.

Search for new owner resolved on 16 July 2001 when Moravan Aeroplanes (see under Zlin heading in this section) acquired company for Kcs200 million (US$5 million). Major investor is Don Jewitt, CEO of former Alberta Aerospace, Canada. Moravan plans to continue marketing of L-410/420 and Let's sailplane range, and to place Z-400 Rhino and Phoenix SigmaJet work with the factory as well, but future of Czech participation in Loadmaster programme still uncertain by early 2002, at which time it was announced that LZ had acquired the L 610 G programme for continued development.

NEW ENTRY

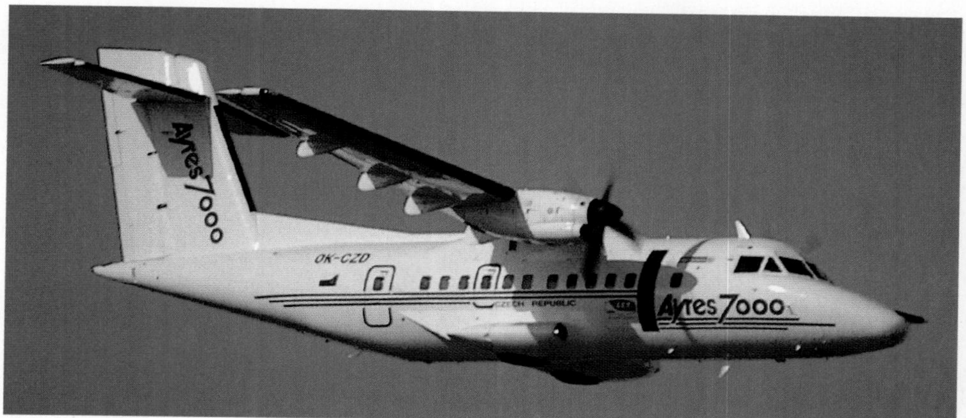

Second prototype L 610 G, marked as an Ayres 7000 *(Jane's/Paul Jackson)* ***NEW***/0131710

LZ L 410UVP-E and L 420

Full-scale manufacture of this twin-turboprop transport ended in 1992, although deliveries have proceeded at a low rate from stock, hampered by ready availability of ex-military aircraft. A spokesman for LZ reported orders for 20 L 410/420s in November 2001; however, existing stock is believed to be sufficient to cover these. Nevertheless, LZ chairman quoted in March 2002 as saying production of up to 12 a year to be resumed. One L 410UVP-E, configured for forest fire monitoring, delivered to South Korea at that time, with four more under negotiation. A description last appeared in the 2002-03 *Jane's*

UPDATED

LZ L 610 G

TYPE: Twin-turboprop transport.

PROGRAMME: First flight (X01/OK-130) 28 December 1988; five development aircraft (X02/OK-WZA, X03/OK-024, X05/OK-134, OK-XZA and OK-136), plus one for static test (X04) and one demonstrator (OK-CZD). Two versions originally planned: L 610 M with 1,358 kW (1,822 shp) Walter M 602 turboprops and L 610 G with General Electric CT7s; contract 18 January 1991 for General Electric to provide CT7-9Ds for L 610 G (first two delivered shortly afterward); first flight of this version (OK-XZA) 18 December 1992. Programme for L 610 M (see 1996-97 *Jane's*) abandoned 1996. In November 1999, Swiss marketing agency Airlines Partners SA sold two aircraft to Burundi, using marketing name Ayres 7000. Certification to FAR Pt 25 was then expected about nine months after contract signature by first US customer, but not achieved by April 2002. Twelve production aircraft were in various stages of completion by June 2000, but Let and Ayres bankruptcies later that year followed by entire L 610 programme (including putative L 610 M) being offered for sale in March 2001. Aircraft was not included in July 2001 sale of Let to Moravan Aeroplanes, but L 610 G added to LZ portfolio in early 2002.

CURRENT VERSIONS: **L 610 G**: General Electric CT7-9D turboprops, four-blade propellers and Rockwell Collins digital avionics including EFIS, weather radar and autopilot. First prototype used for Czech and US certification testing; second prototype (OK-CZD) completed and was exhibited at Farnborough Air Show in September 1998; third (first series production) aircraft under construction at that time.

Ayres 7000: Described by Ayres as 'face-lift' upgrade of L 610 G; also projected in freighter (7000F) and military (7000M) versions: see 2002-03 *Jane's* for further details.

CUSTOMERS: Seven, including two production L 610 Gs, flown by mid-2000. Options for 16 from CSA and other (unidentified) operators. Pan Pacific Airways (USA) announced in June 1999 that it planned to acquire two. Two Ayres 7000 ordered in November 1999 by City Connexion of Burundi; no evidence of delivery by April 2002.

COSTS: US$8.75 million (2000).

DESIGN FEATURES: Intended to meet FAR Pt 25 requirements; high wing for propeller ground clearance and pannier-mounted landing gear to reduce mainwheel leg length and simplify loading/unloading; high-mounted tailplane for optimum control authority.

Wing sections MS(1)-0318D at root, MS(1)-0312 at tip; thickness/chord ratios 18.29 (root) and 12 per cent (tip); dihedral 2°; incidence 3° 8′ 38″ at root, 0° at tip; quarter-chord sweepback 1°.

FLYING CONTROLS: Conventional and manual. Ailerons are horn balanced. Rudder trim tab and both aileron trim tabs actuated by electromechanical strut. Elevator trim tabs actuated mechanically by screw-nut mechanism. Automatic spring tab in rudder. Electrohydraulically actuated single-slotted Fowler flaps. Ground spoilers. Lateral control spoilers deflected proportionately to aileron deflection. Electrically actuated gust lock.

STRUCTURE: All-metal, fail-safe stressed skin structure; circular-section fuselage between flight deck and tail; wing contains high-grade aluminium and high-strength steel; honeycomb spoiler panels.

LANDING GEAR: Retractable tricycle type, with single wheel on each unit. Hydraulic actuation, mainwheels retracting inward to lie flat, without doors, in fairing each side of fuselage; nosewheel retracts forward. Oleo-pneumatic shock-absorber in each unit. Mainwheels are type K 34-3100-7, with 15.00-16 (16 ply) or 1050×390-480 tubeless tyres; type K 35-1100-7 nosewheel has a 720×310-10 or 29×11.00-10 (10 ply) tubeless tyre. Hydraulic disc brakes and electronically controlled anti-skid units. Minimum ground turning radius 18.33 m (60 ft 1¾ in) about wingtip, 7.83 m (25 ft 8¼ in) about nosewheel.

POWER PLANT: Two 1,305 kW (1,750 shp) General Electric CT7-9D turboprops, each driving a Hamilton Sundstrand HS-14RF-23 four-blade fully feathering metal propeller with reversible pitch. Fuel in two integral wing tanks, combined usable capacity 3,420 litres (903.5 US gallons; 752 Imp gallons). Pressure refuelling point in fuselage, gravity points in wings. Oil capacity 30 litres (7.9 US gallons; 6.6 Imp gallons).

ACCOMMODATION: Crew of two on flight deck, plus one cabin attendant. Standard accommodation for 40 passengers, four-abreast at seat pitch of 76 cm (30 in). Galley, two wardrobes, lavatory, freight and baggage compartment, all located at rear of cabin. Alternative mixed (passenger/cargo) and all-cargo layouts available; latter can accommodate five 1.63 × 2.08 × 1.57 m (64 × 82 × 62 in) containers or pallets in main cabin plus a further 200 kg (441 lb) in aft bulk cargo area at rear of main cargo hold. Military version has capacity for 30 fully equipped troops, 30 paratroops plus jumpmaster, 19 stretchers or five critically ill patients plus medical attendants, or up to 5,000 kg (11,023 lb) of cargo plus a loadmaster.

Passenger door at rear of fuselage, cargo door at front, both opening outward on port side. Outward-opening service door on starboard side, opposite passenger door, serving also as emergency exit; outward-opening emergency exit beneath wing on each side. Paratroop version has inward-opening/sliding door aft of wing on port side. Entire accommodation pressurised and air conditioned.

SYSTEMS: Dual Hamilton Sundstrand R 79-3W engine bleed air air conditioning systems. Nord Micro digital, fully automatic pressurisation system gives 0.36 bar (5.2 lb/sq in) differential at flight level of 7,200 m (23,625 ft) and a cabin altitude of 2,400 m (7,875 ft). Duplicated hydraulic systems (one main and one standby), operating at pressure of 210 bar (3,045 lb/sq in). Optional 20 kW Safir 5 K/G APU in tailcone, for engine starting and auxiliary on-ground and in-flight power.

Electrical system powered by two 115/200 V 25 kVA variable frequency AC generators, plus a third 8 kVA 115/200 V three-phase AC generator driven by APU. System also includes two 115 V 400 Hz inverters (each 1.5 kVA), two 27 V DC transformer-rectifiers (each 4.5 kW), and a 25 Ah Ni/Cd battery for APU starting and auxiliary power supply. Gumotex Břeclav or Goodrich pneumatic de-icing boots on wing and tail unit leading-edges; ACT electric anti-icing system for engine inlets; electric de-icing of propeller blade roots, windscreen, pitot static system and horn balances. Oxygen system for crew and four passengers.

AVIONICS: Rockwell Collins Pro Line II suite and EFIS-86 standard.

Comms: Dual 760-channel VHF; single HF (optional); ATC transponder; intercom/PA system; cockpit voice recorder.

Radar: WXR-350 weather radar.

Flight: APS-65 autopilot; AHS-85 AHRS; dual ILS with two LOC/glideslope receivers and two marker beacon receivers; single or dual ADF; dual compasses; single or dual radio altimeters; navigation computer; flight data recorder; Cat. II approach aids.

Instrumentation: Five-tube EFIS-86 with EADI and HSI for each crew member and central MFD; weather radar data can be displayed on HSI and/or MFD.

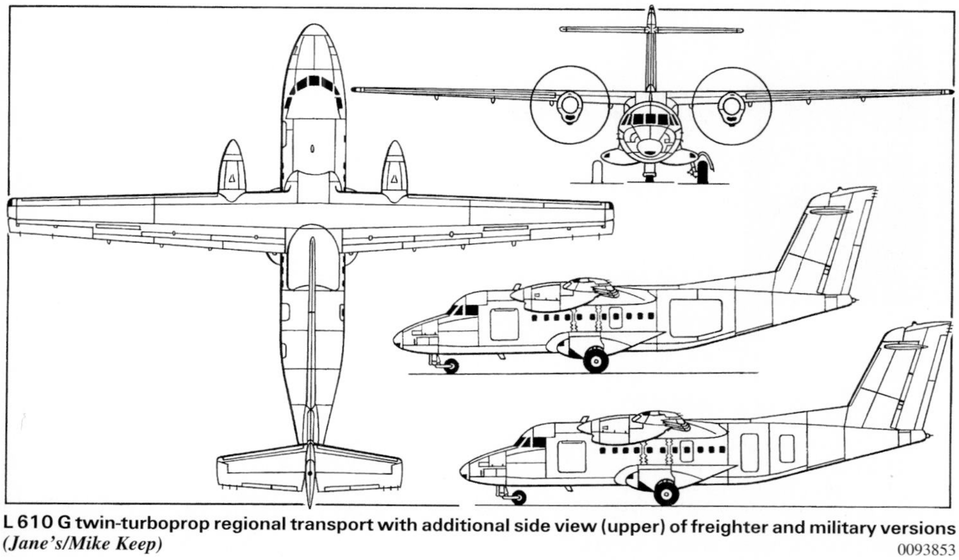

L 610 G twin-turboprop regional transport with additional side view (upper) of freighter and military versions *(Jane's/Mike Keep)*
0093853

DIMENSIONS, EXTERNAL:

Wing span	25.60 m (84 ft 0 in)
Wing chord: at root	2.92 m (9 ft 7 in)
at tip	1.46 m (4 ft 9½ in)
Wing aspect ratio	11.7
Length overall	21.72 m (71 ft 3 in)
Fuselage: Length	20.53 m (67 ft 4¼ in)
Max diameter	2.70 m (8 ft 10¼ in)
Height overall	8.19 m (26 ft 10½ in)
Tailplane span	7.91 m (25 ft 11½ in)
Wheel track	4.59 m (15 ft 0¾ in)
Wheelbase	6.61 m (21 ft 8¼ in)
Propeller diameter	3.35 m (11 ft 0 in)

Propeller fuselage clearance	0.66 m (2 ft 2 in)
Propeller ground clearance	1.71 m (5 ft 7¼ in)
Distance between propeller centres	7.00 m (22 ft 11½ in)
Passenger door: Height	1.625 m (5 ft 4 in)
Width	0.76 m (2 ft 6 in)
Height to sill	1.45 m (4 ft 9 in)
Freight door: Height	1.30 m (4 ft 3¼ in)
Width	1.25 m (4 ft 1¼ in)
Height to sill	1.45 m (4 ft 9 in)
Service door: Height	1.29 m (4 ft 2¾ in)
Width	0.61 m (2 ft 0 in)
Emergency exits (underwing, each):	
Height	0.915 m (3 ft 0 in)
Width	0.515 m (1 ft 8¼ in)

DIMENSIONS, INTERNAL:

Cabin (excl flight deck): Total length	11.10 m (36 ft 5 in)
Passenger cabin: Max length	7.84 m (25 ft 8¾ in)
Max width	2.54 m (8 ft 4 in)
Width at floor	2.02 m (6 ft 7½ in)
Aisle width	0.51 m (1 ft 8 in)
Max height	1.835 m (6 ft 0¼ in)
Floor area	22.4 m² (241 sq ft)
Volume	34.2 m³ (1,208 cu ft)
Wardrobe volume (total)	1.0 m³ (35 cu ft)
Overhead bins volume (total)	1.975 m³ (69.75 cu ft)
Baggage/freight hold volume (total)	6.4 m³ (226 cu ft)
Available volume for freight/cargo	44.7 m³ (1,579 cu ft)

AREAS:

Wings, gross	56.0 m² (602.8 sq ft)
Ailerons (total)	3.27 m² (35.20 sq ft)
Trailing-edge flaps (total)	11.29 m² (121.52 sq ft)
Spoilers (total)	3.54 m² (38.10 sq ft)
Fin	8.46 m² (91.06 sq ft)
Rudder, incl tabs	6.84 m² (73.63 sq ft)
Tailplane	8.07 m² (86.86 sq ft)
Elevators (total, incl tabs)	5.43 m² (58.45 sq ft)

WEIGHTS AND LOADINGS:

Operating weight empty	9,860 kg (21,738 lb)
Max fuel	2,700 kg (5,952 lb)
Max payload: civil and freighter	4,536 kg (10,000 lb)
military	5,500 kg (12,125 lb)
Max ramp weight	15,140 kg (33,378 lb)
Max T-O weight	15,100 kg (33,289 lb)
Max landing weight	14,800 kg (32,628 lb)
Max zero-fuel weight	13,800 kg (30,424 lb)
Max wing loading	269.6 kg/m² (55.23 lb/sq ft)
Max power loading	5.79 kg/kW (9.51 lb/shp)

PERFORMANCE:

Max operating Mach No. (M_{MO}) above 4,875 m	
(16,000 ft)	0.443
Max operating speed (V_{MO}) up to 4,875 m (16,000 ft)	
	213 kt (396 km/h; 246 mph) IAS
Max cruising speed at 6,100 m (20,000 ft)	
	238 kt (440 km/h; 273 mph)

Long-range cruising speed at 7,315 m (24,000 ft)	
	205 kt (380 km/h; 236 mph)
Min paratroop deployment speed	
	95 kt (176 km/h; 110 mph) IAS
Approach speed	99 kt (184 km/h; 114 mph) IAS
Stalling speed: flaps up	98 kt (182 km/h; 113 mph) EAS
flaps down	76 kt (141 km/h; 88 mph) EAS
Max rate of climb at S/L	498 m (1,634 ft)/min
Rate of climb at S/L, OEI	132 m (433 ft)/min
Max operating altitude	7,315 m (24,000 ft)
Service ceiling, OEI	4,500 m (14,760 ft)
FAR/JAR 25 required T-O field length at S/L:	
ISA	1,110 m (3,640 ft)
ISA + 15°C	1,241 m (4,070 ft)
FAR/JAR 25 required landing field length at MLW, at S/L:	
with propeller reversal	1,030 m (3,380 ft)
without propeller reversal	1,171 m (3,840 ft)
Range, reserves for 100 n mile (185 km; 115 mile) and 45 min hold diversion:	
with 40 passengers	680 n miles (1,259 km; 782 miles)
with max fuel	1,305 n miles (2,416 km; 1,501 miles)

OPERATIONAL NOISE LEVELS:
Comply with FAR Pt 36, Annexe 16, Section 5

UPDATED

NA DESIGN

NA DESIGN COMPANY INC

Brněnská 415, CZ-664 34 Kuřim
Tel: (+420 603) 45 44 94
Fax: (+420 5) 41 23 06 94 and 850 77 41
e-mail: nadc@centrum.cz
Web: http://www.bongo.cz and http://www.nadc.cz
CHAIRMAN AND CHIEF DESIGNER: Jan Námisňák
CUSTOMER SERVICE DIRECTOR: Jaroslav Najman

NA Design was created in third quarter of 1999, in co-operation with Moravan Aeroplanes Inc (see under Zlin heading in this section). Development of the (formerly UNIS) Bongo was continuing in early 2002, at which time NA Design was seeking a Western partner to assist with certification, production and marketing. Some US$2.5 million said to be required to bring aircraft to production readiness. Designer Jan Námisňák had previously built the Caprice 21 tandem-seat, high-wing pusher lightplane.

UPDATED

NA DESIGN NA 40 BONGO

TYPE: Two-seat ultralight helicopter.
PROGRAMME: Design work 1992 followed by model tests 1993; construction of Bongo technology demonstrator started 1996; public debut at Brno International Machinery Fair 1997 (then called UNIS NA40, and with two-blade main rotor) by first prototype OK-CIU. Began 70 hours of tethered hover flights March 1998. Two further prototypes completed by mid-1998. NA Design plans Normal category certification to FAR Pts 27, 33, 34 and 36 by late 2003, subject to finding joint venture Western partner to

share certification, manufacturing and marketing costs. However, by early 2002, financial support of US$2.5 million was still being solicited, and company was expecting production to begin by end 2002.
CURRENT VERSIONS: **NA 40 Bongo:** Basic civil version; *as described.*
NA 42 Barracuda: Proposed military/police version. Power plant as NA 40 or single Rolls-Royce 250 engine; stub-wings for weapons carriage; undernose sensor turret; intercom. Otherwise generally as NA 40.
NA 44 Bion: Projected UAV version of NA 42 for various military and civil applications.
COSTS: Approximately US$750,000 (2001).
DESIGN FEATURES: Three-blade teetering rotor; pod and boom fuselage with inverted Y-tail unit. Cocomo patented anti-torque system eliminates need for tail rotor, reduces transmission complexity and operating noise level.
FLYING CONTROLS: Conventional cyclic and collective controls. Ducted air anti-torque system.
STRUCTURE: Mainly composites, including rotor blades; some aluminium in fuselage, otherwise double-curvature monocoque sandwich; laminated, vacuum-formed elastomeric rotor head.
LANDING GEAR: Tubular twin-skid gear with ground handling wheels. Inflatable permanent or emergency floats optional.
POWER PLANT: Two 86 kW (115 shp) První Brnenská Strojirna PBS Velká Bíteš TE 50B turboshafts, with FADEC and dual ignition. Engines mounted side by side behind cockpit; transmission via combining gearbox. Single self-sealing fuel tank beneath engines, capacity 210 litres (55.5 US gallons; 46.2 Imp gallons).
ACCOMMODATION: Adjustable, foldable and anatomically shaped seats for pilot and passenger. Baggage space aft of seats; provision for additional, aerodynamic baggage

container under fuselage. Gull-wing window/door each side, hinged on centreline and opening upward.
SYSTEMS: Dual 27 V DC electrical systems; external power receptacle.
AVIONICS: *Instrumentation:* VFR standard; IFR to be offered later.
EQUIPMENT: Rocket-deployed emergency recovery parachute optional.

DIMENSIONS, EXTERNAL:

Rotor diameter	7.48 m (24 ft 6½ in)
Fuselage: Length	6.15 m (20 ft 2¼ in)
Max width	1.30 m (4 ft 3¼ in)
Height to top of rotor head	2.35 m (7 ft 8½ in)
Skid track	1.76 m (5 ft 9¼ in)

DIMENSIONS, INTERNAL:

Baggage space	0.23 m³ (8.1 cu ft)

AREAS:

Rotor disc	43.94 m² (473.0 sq ft)

WEIGHTS AND LOADINGS:

Weight empty, equipped: NA 40	480 kg (1,058 lb)
NA 42	495 kg (1,091 lb)
Max load on external sling	250 kg (551 lb)
Max T-O weight	950 kg (2,094 lb)
Max disc loading	21.62 kg/m² (4.43 lb/sq ft)

PERFORMANCE:

Never-exceed speed (V_{NE})	151 kt (280 km/h; 174 mph)
Max cruising speed	135 kt (250 km/h; 155 mph)
Econ cruising speed	124 kt (230 km/h; 143 mph)
Max rate of climb at S/L	480 m (1,575 ft)/min
Hovering ceiling IGE	4,000 m (13,120 ft)
Max range	270 n miles (500 km; 310 miles)
Endurance: NA 42	2 h 30 min

UPDATED

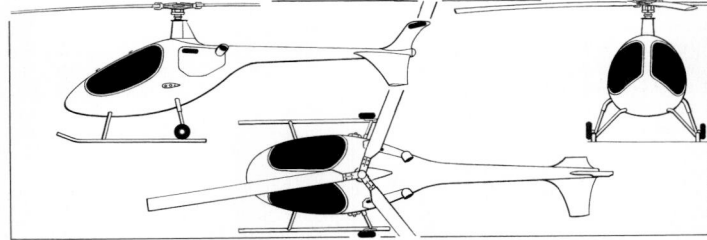

General arrangement of the NA 40 Bongo light helicopter
(Jane's/James Goulding)   0110505

NA 40 Bongo two-seat light utility helicopter
0044014

PACKO

ING LUDOVIT PACKO

Květinková 348/6, CZ-130 00 Praha 3
Tel: (+420 2) 793 36 43
Fax: (+420 5) 44 22 19 21
e-mail: Ludovit.packo@volny.cz

VERIFIED

PACKO LARUS

TYPE: Side-by-side ultralight.
PROGRAMME: Design work started 1991; prototype (OK-EUC 01) made first flight 22 October 1997.
DESIGN FEATURES: Typical small lightplane low-wing configuration. Wing sections NACA 2305 at root, NACA 2312 at tip.
FLYING CONTROLS: Conventional and manual. Horn-balanced

rudder and one-piece elevator; central trim tab in elevator. Flaps fitted.
STRUCTURE: Mainly composites.
LANDING GEAR: Tricycle type; fixed. Cantilever self-sprung main units; castoring nosewheel.
POWER PLANT: One 70.8 kW (95 hp) Aero Max 2400 piston engine; two-blade propeller. Fuel capacity 40 litres (10.6 US gallons; 8.8 Imp gallons).

DIMENSIONS, EXTERNAL:
Wing span 9.80 m (32 ft 1¾ in)
Length overall 6.30 m (20 ft 8 in)
Height overall 2.80 m (9 ft 2¼ in)
AREAS:
Wings, gross 12.86 m² (138.4 sq ft)
WEIGHTS AND LOADINGS:
Weight empty 305 kg (672 lb)
Max T-O weight 450 kg (992 lb)
PERFORMANCE:
Max level speed 126 kt (235 km/h; 146 mph)
Cruising speed 100 kt (185 km/h; 115 mph)
Stalling speed, flaps down 44 kt (80 km/h; 50 mph)
Service ceiling 3,000 m (9,840 ft)

UPDATED

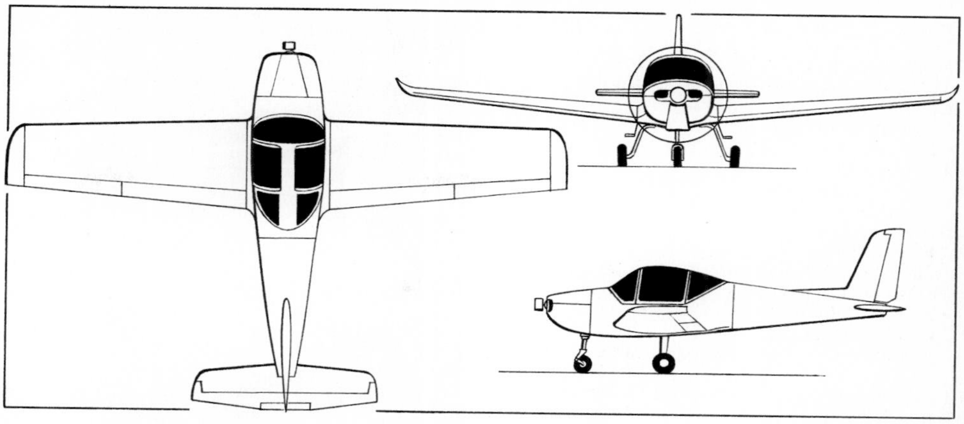

Packo Larus two-seat ultralight
(Jane's/James Goulding)
0121274

TeST

TeST s.r.o.
Dobrovskeho 78, CZ-612 00 Brno
Tel/Fax: (+420 5) 49 24 90 73
e-mail: test@infoline.cz
Web: http://www.test.infoline.cz
PRODUCTION DIRECTOR: Zdeněk Teply
MARKETING DIRECTOR: Zbynek Jaros

Company was established 1992 to design, develop and produce ultralight sailplanes, powered sailplanes and aircraft; first product was TST-1 Alpin sailplane. Capacity increased to 10 aircraft per year in 1995 and to 40 per year from September 1998. Developed versions of Alpin (TST-3 and TST-8), available in both sailplane and motor glider variants) still form core of production, along with TST-5, TST-6 and TST-9 described in the following entries; all are available in plans-only, various kits or full flyaway form. Workforce was 15 in 2001.

UPDATED

TeST TST-5 VARIANT

TYPE: Side-by-side ultralight/kitbuilt.
PROGRAMME: Developed in mid-1990s.
DESIGN FEATURES: Braced high-wing monoplane; forward-swept, constant chord wings (wing section NACA 4415). Basically a short-wing version of TST-6 (see next entry).
FLYING CONTROLS: Conventional and manual (ailerons and elevators rod-actuated, rudder by cables).
STRUCTURE: Two-spar wing, with plywood and fabric covering; glass fibre wingtips; V bracing struts. All-wood fuselage except for glass fibre panel behind cockpit.
LANDING GEAR: Tricycle; fixed. Self-sprung main legs with 400×100 wheels; 300 mm diameter steerable nosewheel. Mainwheel brakes.
POWER PLANT: One 47.8 kW (64.1 hp) Rotax 582 UL engine and two-blade propeller standard; or suitable equivalent. Fuel capacity 36 litres (9.5 US gallons; 7.9 Imp gallons).
DIMENSIONS, EXTERNAL:
Wing span 10.50 m (34 ft 5½ in)
Wing aspect ratio 8.0
Length overall 6.40 m (21 ft 0 in)
Height overall 2.30 m (7 ft 6½ in)
DIMENSIONS, INTERNAL:
Cabin max width 1.02 m (3 ft 4¼ in)
AREAS:
Wings, gross 13.70 m² (147.5 sq ft)

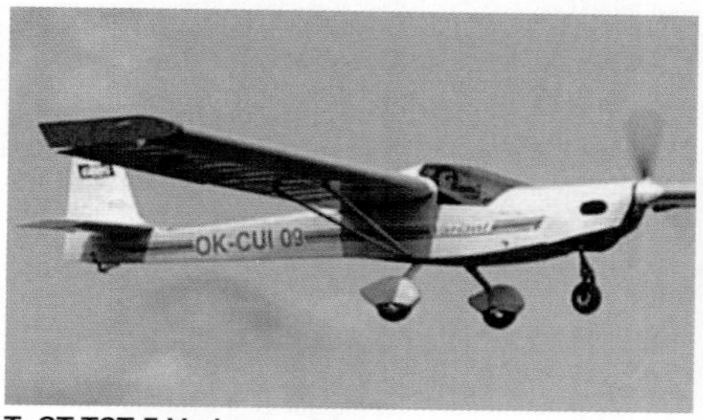

TeST TST-5 Variant 0114394

WEIGHTS AND LOADINGS:
Weight empty 260 kg (573 lb)
Max T-O weight 450 kg (992 lb)
PERFORMANCE:
Never-exceed speed (VNE) 91 kt (170 km/h; 105 mph)
Max cruising speed 81 kt (150 km/h; 93 mph)
Econ cruising speed 49-65 kt (90-120 km/h; 56-75 mph)
Stalling speed 36 kt (65 km/h; 41 mph)
Max rate of climb at S/L 210 m (689 ft)/min
Service ceiling 4,000 m (13,120 ft)
T-O and landing run 150 m (495 ft)
g limits +4/−2

UPDATED

TeST TST-6 DUO

TYPE: Side-by-side ultralight/kitbuilt.
PROGRAMME: Developed in mid-1990s.
DESIGN FEATURES: Forward-swept shoulder-wing monoplane.
FLYING CONTROLS: Conventional and manual; primary surfaces as for TST-5. Two-position trailing-edge flaps; upper-surface spoilers.
STRUCTURE: As described for TST-5 except wing has Wortmann FX61-184 section, has one main and two auxiliary spars is entirely plywood-covered and has glass fibre wingtips.
LANDING GEAR: Tricycle; fixed. Self-sprung main legs with 400×100 wheels; 300 mm diameter steerable nosewheel. Mainwheel brakes.
POWER PLANT: One 47.8 kW (64.1 hp) Rotax 582 UL engine and three-blade propeller standard; or suitable equivalent. Fuel capacity 36 litres (9.5 US gallons; 7.9 Imp gallons).
DIMENSIONS, EXTERNAL (As TST-5 except):
Wing span 14.60 m (47 ft 10¾ in)
Wing aspect ratio 16.5

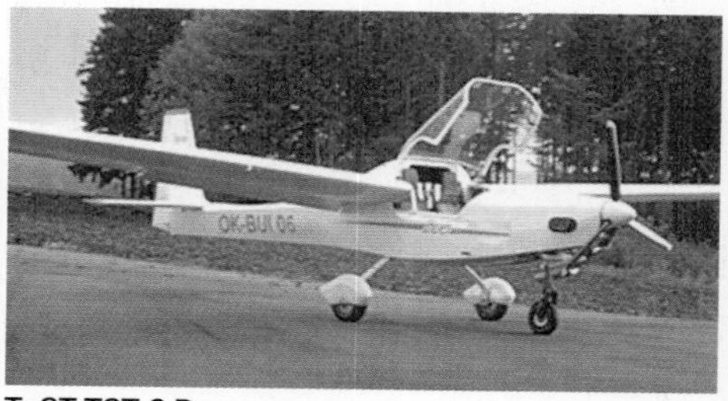

TeST TST-6 Duo 0114395

AREAS:
Wings, gross 12.90 m² (138.9 sq ft)
WEIGHTS AND LOADINGS:
Weight empty 265 kg (584 lb)
Max T-O weight 450 kg (992 lb)
PERFORMANCE, POWERED (As TST-5 except):
Max speed with spoilers extended 86 kt (160 km/h; 99 mph)
Econ cruising speed 49-59 kt (90-110 km/h; 56-68 mph)
Service ceiling 5,000 m (16,400 ft)
PERFORMANCE, UNPOWERED:
Best glide ratio at 54 kt (100 km/h; 62 mph) 16
Min rate of sink 2.00 m (6.56 ft)/s

UPDATED

TeST TST-9 JUNIOR 2000

TYPE: Single-seat motor glider/kitbuilt.
PROGRAMME: Developed in late 1990s.
DESIGN FEATURES: Forward-swept, high-wing monoplane. Improved version of earlier TST-7 Junior, from which it differs mainly in having circular section fuselage and T tail.
FLYING CONTROLS: Conventional and manual; similar to TST-6 except for one-piece elevator and absence of flaps.
STRUCTURE: Wings and fuselage as for TST-6; all-wood tail unit.
LANDING GEAR: Tricycle; fixed. Self-sprung main legs with 300×100 wheels; 250 mm diameter steerable nosewheel.
POWER PLANT: One 31.0 kW (41.6 hp) Rotax 447 engine and two-blade SportProp 1600 propeller standard; or suitable equivalent. Fuel capacity 25 litres (6.6 US gallons; 5.5 Imp gallons).
DIMENSIONS, EXTERNAL:
Wing span 12.40 m (40 ft 8¼ in)
Wing aspect ratio 15.4
Length overall 5.90 m (19 ft 4¼ in)
Height overall 2.10 m (6 ft 10¾ in)
DIMENSIONS, INTERNAL:
Cabin max width 0.59 m (1 ft 11¼ in)
AREAS:
Wings, gross 10.00 m² (107.6 sq ft)
WEIGHTS AND LOADINGS:
Weight empty 175 kg (386 lb)
Max T-O weight 300 kg (661 lb)
PERFORMANCE, POWERED:
Never-exceed speed (VNE), and max speed with spoilers extended 97 kt (180 km/h; 111 mph)
Max cruising speed 89 kt (165 km/h; 103 mph)
Econ cruising speed 49-65 kt (90-120 km/h; 56-75 mph)
Stalling speed 35 kt (63 km/h; 40 mph)
Max rate of climb at S/L 270 m (886 ft)/min
Service ceiling 5,000 m (16,400 ft)
T-O run 100 m (330 ft)
Landing run 120 m (395 ft)
g limits +4/−2
PERFORMANCE, UNPOWERED:
Best glide ratio at 49 kt (90 km/h; 56 mph) 20
Min rate of sink 1.50 m (4.92 ft)/s

UPDATED

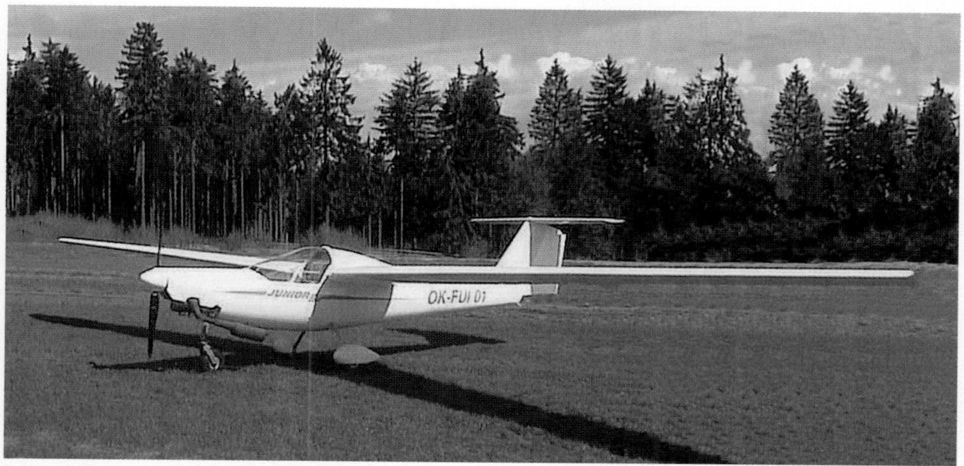

TeST TST-9 Junior single-seat kitbuilt 0114396

TL

TL ULTRALIGHT
Dobrovského 734, CZ-500 02 Hradec Králové
Tel: (+420 49) 61 33 78 and 61 17 53
Fax: (+420 49) 61 33 78
e-mail: info@tl-ultralight.cz
Web: http://www.tl-ultralight.cz
DIRECTOR: Ing Ivan Lelák

GERMAN DISTRIBUTOR:
Martin Wezel Flugzeugtechnik
Erlenbachstrasse 38, D-72768 Reutlingen
Tel: (+49 7121) 684 08
Fax: (+49 7121) 67 72 38
e-mail: wezel.martin@t-online.de
Web: http://www.Wezel-Flugzeugtechnik.de

TL also markets the TL-22 Duo/Eso Rogallo-wing trike. Formed in 1990, the company previously produced the TL-1, TL-2 and TL-32 ultralights, progressing then to the TL-132, predecessor of the current TL-232.

UPDATED

TL-96 STAR and TL-2000 STING CARBON
TYPE: Side-by-side ultralight.
PROGRAMME: Prototype Star first flew November 1997; has German certification.
CURRENT VERSIONS: **TL-96 Star:** Baseline version; *as described.*
  TL-2000 Sting Carbon: All-CFRP version; prototype OK-GUU 10; announced 2000.
CUSTOMERS: Total 44 registered in Czech Republic by late 2000; exports include several registered in Netherlands; one to Hungary in 2001. More than 90 flying in Europe by April 2001.
COSTS: DM117,000, including tax (basic aircraft, 2001).
DESIGN FEATURES: Conventional, low-wing monoplane with constant-chord wings and tailplane, plus sweptback fin. MS 313 wing aerofoil section.
FLYING CONTROLS: Manual. All-moving tailplane with anti-balance tab. Plain, two-position flaps.
STRUCTURE: All-composites (GFRP and CFRP).
LANDING GEAR: Fixed tricycle type; GFRP cantilever legs and wheel speed fairings. Steerable nosewheel. Tyre sizes 4.00-8/400×100 (main), 4.00-4 (nose). Hydraulic mainwheel brakes.
POWER PLANT: One 59.6 kW (79.9 hp) Rotax 912 UL-DCDI or 84.6 kW (113.4 hp) Rotax 914 flat-four engine; three-blade Albastar propeller. Optional 67.1 kW (90 hp) Aero Prag AP-45 flat-four and two-blade wooden propeller. Fuel capacity (RON 95) 50 litres (13.2 US gallons; 11.0 Imp gallons).
ACCOMMODATION: Heated as standard.
DIMENSIONS, EXTERNAL (A: TL-96, B: TL-2000):

Wing span: A	9.20 m (30 ft 2¼ in)
B	8.44 m (27 ft 8¼ in)
Length overall: A	6.50 m (21 ft 4 in)
B	5.93 m (19 ft 5½ in)
Height overall: A	2.15 m (7 ft 0¾ in)
B	2.30 m (7 ft 6½ in)

DIMENSIONS, INTERNAL:

Cabin max width: A	1.15 m (3 ft 9¼ in)

AREAS:

Wings, gross: A	12.20 m² (131.3 sq ft)
B	9.80 m² (105.5 sq ft)

WEIGHTS AND LOADINGS:

Weight empty: A	290 kg (639 lb)
B	275 kg (606 lb)
Max T-O weight: A, B	450 kg (992 lb)

PERFORMANCE, POWERED (Rotax 912):

Never-exceed speed (VNE):	
A	148 kt (275 km/h; 170 mph)
B	167 kt (310 km/h; 192 mph)
Max level speed: A	138 kt (255 km/h; 158 mph)
B	145 kt (270 km/h; 167 mph)
Max cruising speed: A	119 kt (220 km/h; 137 mph)
B	140 kt (260 km/h; 162 mph)
Stalling speed: A	34 kt (63 km/h; 40 mph)
B	36 kt (65 km/h; 41 mph)
Max rate of climb at S/L: A	360 m (1,181 ft)/min
B	420 m (1,378 ft)/min

PERFORMANCE, UNPOWERED:

Best glide ratio: A	16

UPDATED

TL-232 CONDOR
TYPE: Tandem-seat ultralight/kitbuilt.
PROGRAMME: Side-by-side seat TL-132 (Rotax 503 engine) first flown 1993; entered production, and TL-232 introduced, in 1994. TL-132 subsequently discontinued.
CURRENT VERSIONS: **TL-232 Condor Plus:** Compared to TL-132, has modified wing profile; option of more powerful engine; cut-down rear fuselage and additional glazing for rear cockpit.

TL-96 Star side-by-side ultralight (*Jane's/Paul Jackson*)   NEW/0131726

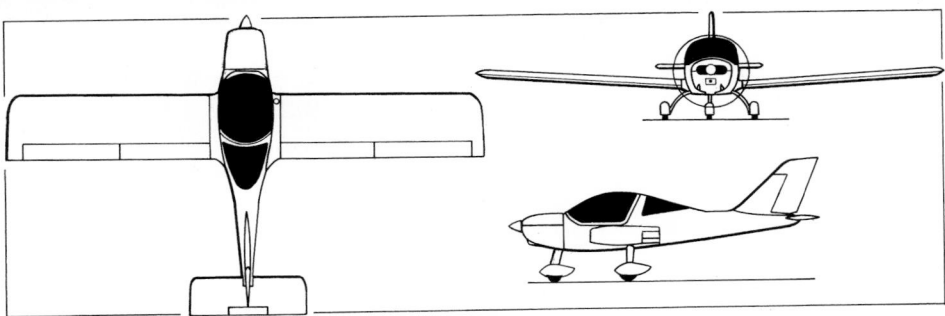

The two-seat TL-96 Star (*Jane's/Paul Jackson*)   0092120

Power Condor: Version with 73.5 kW (98.6 hp) Rotax 912 ULS and glider-towing equipment; able to aero-tow sailplanes of up to 400 kg (882 lb) AUW.
CUSTOMERS: Operators in Czech Republic, Germany, Netherlands, Poland and Sweden. Total of 51 (including nine kits) registered in Czech Republic by late 2000, at which time more than 100 flying in Europe.
COSTS: Condor Plus, DM71,400 to DM88,800, depending upon engine (March 2001).
DESIGN FEATURES: Constant-chord, high wing with V-strut bracing; mid-mounted tailplane; sweptback fin with long fillet. Wings and horizontal tail surfaces foldable for transportation and storage.
FLYING CONTROLS: Conventional and manual. In-flight-adjustable tab in starboard elevator. Two-position flaps. Dual controls.
STRUCTURE: Fabric-covered metal. Composites rear top-decking.
LANDING GEAR: Non-retractable. Choice of tricycle or tailwheel configuration; tyre sizes 16×4 (main), 400×100 (nose). Cable brakes. Leg and wheel speed fairings optional.
POWER PLANT: One 47.8 kW (64.1 hp) Rotax 582 UL-DCDI two-cylinder two-stroke in-line engine in basic TL-232,

driving a three-blade Albastar ground-adjustable pitch propeller. Optional alternatives include 59.6 kW (79.9 hp) Rotax 912 UL-DCDI or 73.5 kW (98.6 hp) Rotax 912 ULS flat-four with Ivoprop three-blade composites propeller. Fuel capacity 50 litres (13.2 US gallons; 11.0 Imp gallons) standard, 70 litres (18.5 US gallons; 15.4 Imp gallons) optional.
DIMENSIONS, EXTERNAL:

Wing span	10.60 m (34 ft 9¼ in)
Length overall: tailwheel	approx 6.50 m (21 ft 4 in)
tricycle	6.08 m (19 ft 11¼ in)
Height overall: tailwheel	approx 1.85 m (6 ft 1 in)
tricycle	2.30 m (7 ft 6½ in)

AREAS:

Wings, gross	14.84 m² (159.7 sq ft)

WEIGHTS AND LOADINGS:

Weight empty	265-280 kg (584-617 lb)
Max T-O weight	450 kg (992 lb)

PERFORMANCE (Rotax 912 UL-DCDI engine):

Max level speed	97 kt (180 km/h; 111 mph)
Cruising speed	81 kt (150 km/h; 93 mph)
Stalling speed	33 kt (60 km/h; 38 mph)
Max rate of climb at S/L	276 m (906 ft)/min

UPDATED

TL Ultralight TL-232 Condor Plus   NEW/0097996

CAP

CAP AVIATION

9 rue de l'Aviation, F-21121 Dijon-Darois
Tel: (+33 3) 80 35 65 10
Fax: (+33 3) 80 35 65 15
e-mail: info@capaviation.com
Web: http://www.capaviation.com
PRESIDENT, CEO AND HEAD OF DESIGN: Dominique Roland
SALES MANAGER, EUROPE: Alain Ruelloux

CAP Aviation (known as Akrotech Europe until January 1999) is owned by holding company Apex International (formerly known as Aéronautique Service – AES) (which see), also responsible for the Robin and BUL concerns. It formed in 1997, initially to market the Giles G-202 two-seat aerobatic aircraft, now also marketed in factory-completed form as the CAP 222.

Apex took over production of the Mudry series of aerobatic aircraft on 12 May 1997 following that company's bankruptcy. All aerobatic aircraft produced by Apex are marketed by CAP Aviation: the CAP 232 built at Bernay and CAP 10 and CAP 222 at Darois.

UPDATED

Prototype CAP 10 C two-seat aerobatic light aircraft (*Jane's/Paul Jackson*)   NEW/0137729

CAP 10

TYPE: Aerobatic two-seat sportplane.
PROGRAMME: Derived from Piel CP-30/CP-301 Emeraude, first flown 19 June 1954 and built by amateur constructors as well as by 11 commercial plants, under various names. Super Emeraude modified with taller fin and 119 kW (160 hp) engine as CP-100 prototype, flown 13 August 1966; with broader chord rudder and 134 kW (180 hp) engine, became prototype CAP 10 B, first flight (F-WOPX) 22 August 1968; further three preseries aircraft; certified 4 September 1970; FAA certification for day and night VFR 1974. Produced by Mudry until 1996.

Production transferred to Dijon, beginning with CAP 10 C in 2001.

CURRENT VERSIONS: **CAP 10 B:** Previous version; now out of production, but can be upgraded to CAP 10 C in one day's work.

CAP 10 C: Announced in early 1999; features carbon fibre reinforced wing spar for reduced weight and increased speed and roll rate.

Improvements effected in CAP 10 C include substitution of pushrods for aileron control, replacing cables; larger (25 per cent) ailerons of increased chord with repositioned axis for faster roll rate, compensated by two spade-type servos per side; electric actuation of flaps; electric fuel gauges and flush filler caps; optional constant-speed propeller (in prospect); upgraded instrumentation with four standard options; redesigned landing gear to improve ground handling; local airframe reinforcement and landing gear attachments; and revised canopy design to reduce cockpit noise levels.

Static test of wing undertaken 27 April 2000; prototype first flight (CAP 10 B HB-SAX retrofitted with new wing) 5 March 2001; first 'true' CAP 10 C (F-WWNN, c/n 300) first flown 15 May 2001 and following initial flight testing was delivered to CEV during July. DGAC certification achieved in early 2002. Roll rate improved by 30 per cent. New wing available for retrofitting to existing CAP 10 Bs.
Description refers to CAP 10 C unless otherwise stated.
CUSTOMERS: Total 284 of earlier versions (including prototypes) built to mid-1999, of which 279 completed by Mudry before 1996 bankruptcy. Most recent five produced in temporary factory at Bernay and include exports to UK (three) and USA (one); last, in 1999, was G-LORN (c/n 282) to UK. Production then paused, pending launch of CAP 10 C and transfer of production to Dijon. Previous major users, French Air Force (56) and Mexican Air Force (20) began disposals in 1994-95. Two more recent aircraft operated by 208 Squadron of South Korean Air Force; French Navy has eight; at end 2000, around 60 were still in military service and a further 180 in civil use.

By mid-2001, 10 CAP 10 Cs had been sold. First production aircraft (c/n 301) G-CPXC registered in December 2001 to Cole Aviation in UK; third is US demonstrator.

COSTS: FFr950,000 plus tax (2001). US price from US$148,966 to US$181,013 depending on equipment level (2001).
DESIGN FEATURES: Simple, sporting lightplane with low, basically elliptical wing, mid-mounted tailplane and curved rudder.

Wing section NACA 23012; dihedral 5°; incidence 0°.
FLYING CONTROLS: Conventional and manual. Electric trim tabs on port elevator; balance tab on rudder; tailplane incidence adjustable on ground; electric three-position (0, 15, 40°) flaps.
STRUCTURE: Main spar of carbon fibre caps and birch ply webs; wooden ribs; birch ply skin; carbon fibre wingtips; birch ply ailerons, flaps, fin/rudder and tailplane/elevator. Wooden fuselage, with okoumé ply upper decking and polyester fabric covering elsewhere.
LANDING GEAR: Tailwheel type; fixed. Mainwheel legs of light alloy, with ERAM type 9 270 C oleo-pneumatic shock-absorbers. Single wheel on each main unit, tyre size 380×150/15×6.00-5. Solid tailwheel tyre, size 6×2.00. Tailwheel is steerable by rudder linkage but can be disengaged for ground manoeuvring. Hydraulically actuated mainwheel disc brakes (controllable from port seat) and parking brake. Streamline fairings on mainwheels and legs.
POWER PLANT: One 134 kW (180 hp) Textron Lycoming AEIO-360-B2F flat-four engine, driving a Hoffmann HO29HM-180-170 two-blade, fixed-pitch wooden propeller; optional Evra 3.180-170-H5.F. Christen fuel system for unlimited inverted flight. Standard fuel tank aft of engine fireproof bulkhead, capacity 72 litres (19.0 US gallons; 15.8 Imp gallons). Auxiliary tank, capacity 78 litres (20.6 US gallons; 17.2 Imp gallons), beneath baggage compartment. Inverted fuel and oil (Aviat/Christen) systems permit continuous inverted flight.
ACCOMMODATION: Side-by-side adjustable seats for two persons, with provision for back parachutes, under rearward-sliding and jettisonable moulded transparent canopy. Special aerobatic shoulder harness standard. Space for 20 kg (44 lb) of baggage aft of seats in training and touring models.
SYSTEMS: Electrical system includes Delco-Rémy 40 A engine-driven alternator and STECO ET24 Ni/Cd battery.
AVIONICS: Honeywell and Garmin suites recommended.
Comms: Optional GMA 340 audio panel and Sigtronics SPA400 intercom; GTX 320 or GTX 327 transponder; ACK 30 encoder.
Flight: Garmin GNC 250 XL, GNS 420 or GNS 430 GPS; optional GI 106 VOR/ILS.

DIMENSIONS, EXTERNAL:
Wing span	8.06 m (26 ft 5¼ in)
Wing aspect ratio	6.0
Length overall	7.00 m (22 ft 11½ in)
Height overall	2.55 m (8 ft 4½ in)
Tailplane span	2.90 m (9 ft 6 in)
Wheel track	2.06 m (6 ft 9 in)

DIMENSIONS, INTERNAL:
Cabin: Length		1.05 m (3 ft 5½ in)
Max width		1.05 m (3 ft 5½ in)
Max height		0.98 m (3 ft 2½ in)

AREAS:
Wings, gross	10.85 m² (116.8 sq ft)
Vertical tail surfaces (total)	1.32 m² (14.25 sq ft)
Horizontal tail surfaces (total)	1.86 m² (20.02 sq ft)

WEIGHTS AND LOADINGS (A: Aerobatic, U: Utility):
Weight empty, equipped: A, U	530 kg (1,168 lb)
Fuel weight: A	54 kg (119 lb)
U	108 kg (238 lb)
Max T-O weight: A	780 kg (1,719 lb)
U	830 kg (1,830 lb)
Max wing loading: A	71.9 kg/m² (14.72 lb/sq ft)
U	76.5 kg/m² (15.67 lb/sq ft)
Max power loading: A	5.82 kg/kW (9.56 lb/hp)
U	6.19 kg/kW (10.16 lb/hp)

PERFORMANCE:
Never-exceed speed (VNE)	183 kt (340 km/h; 211 mph)
Max level speed at S/L	148 kt (274 km/h; 170 mph)
Max cruising speed at 75% power	140 kt (260 km/h; 162 mph)
Stalling speed: flaps down	46 kt (85 km/h; 53 mph) IAS
Max rate of climb at S/L	488 m (1,600 ft)/min
Service ceiling	5,000 m (16,400 ft)
T-O run	350 m (1,150 ft)
T-O to 15 m (50 ft)	395 m (1,300 ft)
Landing from 15 m (50 ft)	600 m (1,970 ft)
Landing run	360 m (1,185 ft)
Range with max fuel	540 n miles (1,000 km; 621 miles)
g limits	+6/−4.5

UPDATED

CAP 222

TYPE: Aerobatic two-seat sportplane.
PROGRAMME: Conceived as European, factory-built version of US Experimental category Akrotech Giles G-202 with JAR 23 certification.

Three G-202 kits imported for development, initially designated G222. Pending certification, CAP Aviation was marketing G-202 kits and had sold seven by mid-2001 when apparent structural failure of No. 5 (F-PQUX) on 21 July 2001 resulted in grounding order.

The first European-built CAP 222 (F-WWMY) flew on 12 June 1997; made its static debut at Le Bourget two days later; and performed its first public air display on 18 June. Second and third followed by January 1999. Failure to meet JAR 23 certification criteria for wing structural strength demanded redesign which was completed in April 1999.

First production CAP 222 used for static tests; second made first flight on 15 May 2000 (F-WWMZ) and initial public appearance at Aerofair, North Weald, UK on 2 June 2000. To CEV for certification trials, which were completed by May 2001.

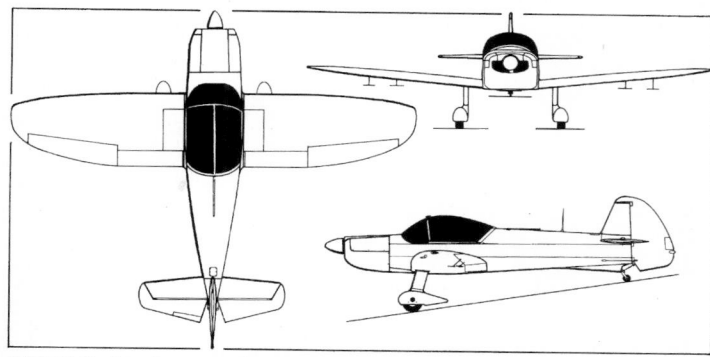

CAP 10 C, showing new ailerons (*Jane's/Paul Jackson*)   NEW/0137207

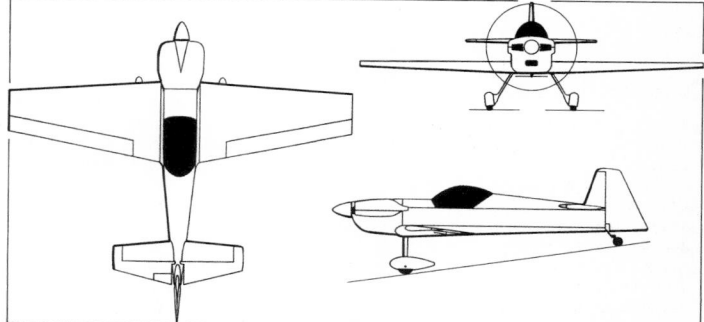

CAP Aviation CAP 232 (*Jane's/Paul Jackson*)

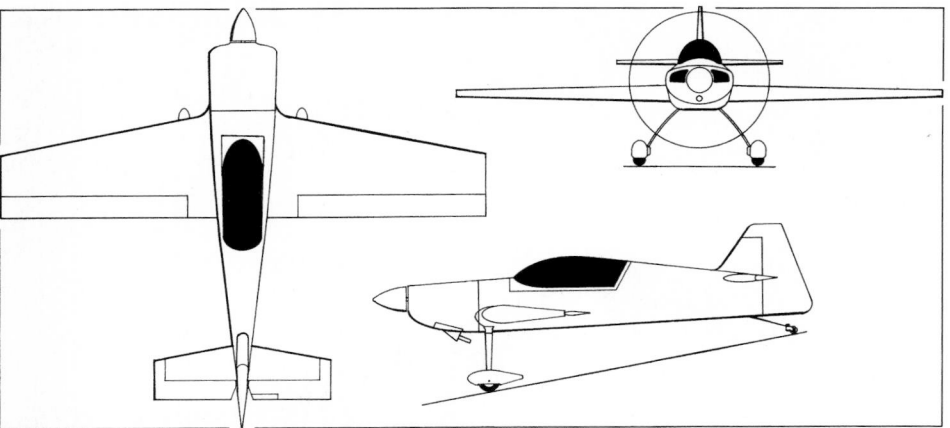

CAP 222 two-seat sportplane (*Jane's/Paul Jackson*)   0100426

First production CAP 222 (*Jane's/Paul Jackson*)   0110945

COSTS: US$220,000 (2000).
Data generally as for Akrotech Giles G-202 (US section, 2001-02 and previous Jane's) *except that below.*
POWER PLANT: One 149 kW (200 hp) Textron Lycoming AEIO-360-A1E driving an MT-Propeller three-blade, variable-pitch propeller. Fuel in main 68 litre (18.0 US gallon; 15.0 Imp gallon) fuselage tank and optional 155 litre (41.0 US gallon; 34.2 Imp gallon) wing tank.
WEIGHTS AND LOADINGS:

Weight empty	517 kg (1,139 lb)

PERFORMANCE:

Manoeuvring speed	180 kt (333 km/h; 207 mph)
Stalling speed	57 kt (106 km/h; 66 mph)
Rate of roll	500°/s
g limits	±10

UPDATED

CAP 231 EX
PROGRAMME: This predecessor of the CAP 232 was last described in detail in the 1998-99 *Jane's All the World's Aircraft*, manufacture having ended in 1993. However, in 2002, three further aircraft were reported to be on the production line, apparently being built to special order for the Moroccan Air Force.

NEW ENTRY

CAP 232
TYPE: Aerobatic single-seat sportplane.
PROGRAMME: Prototype (F-WZCH) first flew 7 July 1994; French certification and first sales March 1995. First place in 2000 World Aerobatic Championship in France was taken by Eric Vazeille in CAP 232 F-GKDX; Catherine Maunoury was fifth (women's first) in F-GXCM.
CUSTOMERS: Total 33 built by June 2001. Mainly French operators (including two to French Air Force), but others

sold to Australia (one, 1996), Switzerland (one), UK (three, 1998 and 2001) and USA (two, 1998 and four, 2000). Moroccan Air Force placed order for nine in mid-2000 to re-equip Marche Verte aerobatic team; deliveries began early 2001.
COSTS: FFr1.25 million plus tax (1999); US price US$250,000 (2001).
DESIGN FEATURES: Optimised for world-class aerobatic competition. Similar to, and improvement on, CAP 231 EX. Strong, carbon fibre wing of broad root chord, sharply tapered for high roll rates; tapered tail surfaces.
FLYING CONTROLS: Conventional and manual. Electrically actuated elevator tab; elevator servo tab to reduce stick forces. Almost full-span ailerons for high roll rates.
STRUCTURE: Two-spar carbon fibre wings; 10 ribs.

LANDING GEAR: Fixed tailwheel type; mainwheel tyres 5.00-5.
POWER PLANT: One 224 kW (300 hp) Textron Lycoming AEIO-540-L1B5D flat-six engine, driving a variable-pitch (hydraulic) MT-Propeller MTV-14-BC-C190-17 four-blade propeller. Total fuel capacity 189.3 litres (50.0 US gallons; 41.6 Imp gallons) in one fuselage tank and two wing tanks. Fuel and oil system designed for prolonged inverted flight.
ACCOMMODATION: Single seat under rear-hinged canopy. Baggage space behind pilot.
SYSTEMS: Electrical system includes engine-driven alternator and 12 V 60 Ah battery.
AVIONICS: Customer specified.
   Comms: Radio and transponder optional.
   Flight: GPS optional.
EQUIPMENT: Sighting frame can be attached to wingtip for judging exact verticals in competition aerobatics.
DIMENSIONS, EXTERNAL:

Wing span	7.39 m (24 ft 3 in)
Wing chord: at root	1.83 m (6 ft 0 in)
at tip	0.91 m (3 ft 0 in)
Wing aspect ratio	5.4
Length overall	6.76 m (22 ft 2 in)
Tailplane span	2.74 m (9 ft 0 in)
Wheel track	1.75 m (5 ft 9 in)
Propeller diameter	1.90 m (6 ft 2¾ in)
Propeller ground clearance	0.30 m (1 ft 0 in)

AREAS:

Wings, gross	10.15 m² (109.5 sq ft)
Ailerons (total)	1.00 m² (10.76 sq ft)
Fin	0.55 m² (5.92 sq ft)
Rudder	0.77 m² (8.29 sq ft)
Tailplane	1.02 m² (10.99 sq ft)
Elevators	1.11 m² (11.93 sq ft)

WEIGHTS AND LOADINGS (A: Aerobatic, N: Normal category):

Weight empty	585 kg (1,290 lb)
Max payload	227 kg (500 lb)
Fuel weight	123 kg (270 lb)
Max T-O and landing weight: N	730 kg (1,610 lb)*
A	821 kg (1,810 lb)*
Max wing loading	80.9 kg/m² (16.56 lb/sq ft)
Max power loading	3.67 kg/kW (6.03 lb/hp)

PERFORMANCE:

Never-exceed speed (VNE)	219 kt (405 km/h; 252 mph)
Max level speed	189 kt (349 km/h; 217 mph)
Max cruising speed at 75% power	
	180 kt (333 km/h; 207 mph)
Manoeuvre speed (VA)	170 kt (314 km/h; 195 mph)
Econ cruising speed	145 kt (269 km/h; 167 mph)
Stalling speed, power off	56 kt (104 km/h; 65 mph)
Max rate of climb at S/L	1,003 m (3,290 ft)/min
Rate of roll at manoeuvre speed	420°/s
Service ceiling	4,575 m (15,000 ft)
T-O run	150 m (495 ft)
T-O to 15 m (50 ft)	180 m (595 ft)
Landing from 15 m (50 ft)	450 m (1,480 ft)
Range with max fuel, 45% power	
	650 n miles (1,203 km; 748 miles)
g limits	±10

* *As quoted by manufacturer*

UPDATED

CAP 232 flown in international competitions by Tom Caselles (*Jane's/Paul Jackson*)   *NEW*/0131733

DASSAULT

DASSAULT AVIATION
9 Rond-Point Champs Elysées-Marcel Dassault, F-75008 Paris
Tel: (+33 1) 53 76 93 00
Fax: (+33 1) 53 76 93 20
ADMINISTRATION AND COMMUNICATIONS OFFICE: 78 quai Marcel Dassault, Cedex 300, F-92552 St Cloud Cedex
Tel: (+33 1) 47 11 86 90
Fax: (+33 1) 47 11 87 40
Web: http://www.dassault-aviation.fr
WORKS: F-92214 St Cloud, F-95100 Argenteuil, F-33127 Martignas,

F-33701 Bordeaux-Mérignac,
F-33260 Cazaux, F-64205 Biarritz-Parme,
F-13801 Istres, F-74370 Argonay,
F-59472 Lille-Seclin, F-86580 Poitiers
HONORARY CHAIRMAN: Serge Dassault
CHAIRMAN AND CEO: Charles Edelstenne
VICE-CHAIRMAN: Bruno Revellin-Falcoz
SENIOR EXECUTIVE VICE-PRESIDENT, OPERATIONS:
   Christian Decaix
SENIOR EXECUTIVE VICE-PRESIDENT, INTERNATIONAL:
   Pierre Chouzenoux
VICE-PRESIDENT, COMMUNICATIONS: Gérard David
INFORMATION OFFICER: Luc Berger

Former Avions Marcel Dassault-Breguet Aviation formed from merger of Dassault and Breguet aircraft companies in December 1971; French government acquired 20 per cent of stock in January 1979, raised to 45.76 per cent in November 1981; present company name adopted June 1990. Of government shareholding, 10.75 per cent held double voting rights so that French government had majority (55 per cent) control. On 17 September 1992, Dassault joined with Aerospatiale in state-owned joint holding company, SOGEPA (which see), which has 35 per cent holding in Dassault Aviation; Dassault and Aerospatiale then pooled resources and co-ordinated R&D strategy, although each remained separate entity.

French government intention to merge Aerospatiale and Dassault Aviation implemented after State relinquished control of Aerospatiale; agreement, ratified by shareholders on 23 December 1998, allocated government's shares in Dassault Aviation to Aerospatiale, but without double voting rights. Aerospatiale later became Aerospatiale Matra and then incorporated into EADS. Last-mentioned affirmed continued stake in Dassault on 27 June 2000 and this endorsed by shareholders on 4 September 2000.

Related measures include separation from Dassault Aviation of Dassault Systèmes, this becoming Dassault Participations, owned 45.76 per cent by French government, 49.93 per cent by Groupe Industriel Marcel Dassault and 4.31 per cent privately. Dassault intention to separate civil and military businesses, with Falcon business jet activities concentrated at Biarritz, Martignas and Seclin, plus Mérignac for flight test, vetoed by Aerospatiale Matra on 30 June 1999. However, *de facto* partition implemented on 1 January 2000, on ''internal and informal basis''.

Employees total some 8,600 at 10 industrial sites: St Cloud (2,800), Argenteuil (1,350), Argonay (550), Biarritz (1,200), Mérignac (1,150), Martignas (400), Istres (800), Cazaux (50), Poitiers (150) and Seclin (150). Sales in 2000 totalled €4.13 billion, of which 71 per cent civil, 25 per cent home military and 4 per cent military exports.

In 2001, Dassault flew the prototype of a stealthy UAV (the Aéronef de Validation Expérimentae: see *Jane's Unmanned Aerial Vehicles and Targets*) and announced a derivative business jet, the Falcon 7X.

Dassault assembles and tests its civil and military aircraft in its own factories, but operates wide network of subcontractors. Other products include flight control system components, maintenance and support equipment. Related companies include Dassault Falcon Jet Corporation, Dassault Falcon Service and SOGITEC. All are subordinate to Groupe Industriel Marcel Dassault, previously including Dassault Electronique, which became part of Thomson-CSF Detexis (now Thales) on 1 January 1999.

Dassault Aviation shared in Atlantique programme with Belgium, Germany, Italy, Netherlands and UK; and Alpha Jet with Germany. Offset manufacture of Dassault aircraft components arranged with Belgium, Egypt, Greece, Spain and UAE. European Aerosystems Ltd established with BAE Systems as a joint venture military aircraft company.

Dassault has produced more than 1,500 executive jets for service in 66 countries and received orders for over 2,700 Mirages of all types. The company has delivered more than 7,500 aircraft to 75 countries and these have accumulated 20 million flying hours.

UPDATED

DASSAULT MIRAGE 2000
Indian Air Force name: Vajra (Divine Thunder)
TYPE: Multirole fighter.
PROGRAMME: Selected as main French Air Force combat aircraft 18 December 1975; first of four single-seat prototypes flew 10 March 1979, followed by two-seat

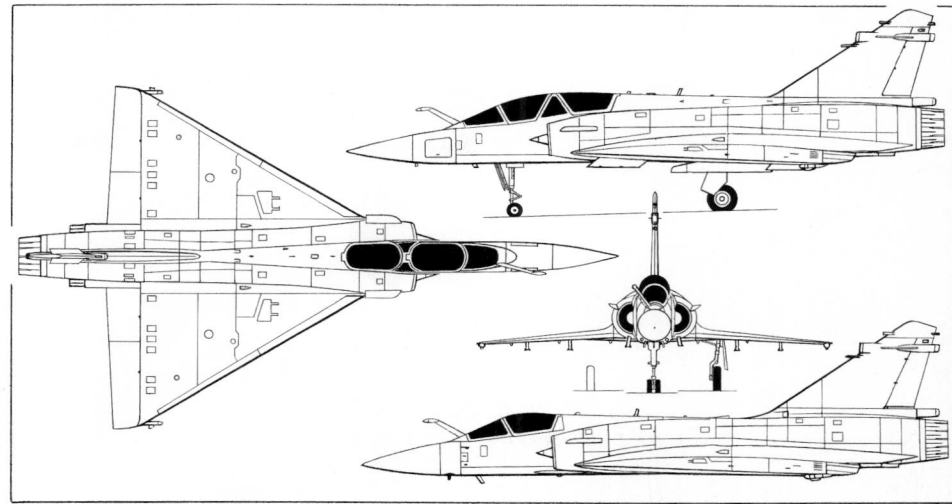

Dassault Mirage 2000-5DDA, with added side view (bottom) of Mirage 2000-5Ei *(Jane's/Dennis Punnett)*
0062696

version on 11 October 1980; initially developed as interceptor with SNECMA M53 power plant and Thomson-CSF (now Thales) RDM multimode pulse Doppler radar; M53-5 in early production aircraft succeeded by M53-P2; fitted with RDI radar from 38th French Air Force 2000C onwards; first flight of production 2000C 20 November 1982; first flight of production two-seat 2000B, 7 October 1983; first unit, EC 1/2 'Cigognes', formed at Dijon 2 July 1984. Subsequently developed for strike/attack as 2000N/D.

Second-generation Mirage 2000-5 first flown as prototype 24 October 1990; initial production aircraft flown in October 1995 and type qualification granted (for SF1 French production standard) by DGA procurement agency on 13 June 1997; initial export delivery in May 1997, followed by first to French Air Force in December 1997. By 2001, Mirage 2000s of eight air forces had accumulated 900,000 flying hours.

Potential engine upgrade revealed in early 2000; M53 PX3 under study to provide additional 8 to 10 per cent thrust.

CURRENT VERSIONS: **2000AT:** Advanced Trainer; announced November 2001. Based on 2000-5 Mk 2, including 'glass cockpit', but with operational equipment, such as radar and EW, removed. JTIDS-type (Link 16) datalink to provide realistic sensor imputs for training, obviating need for real air combat targets. Cost 20 to 25 per cent less than operational Mirage 2000; savings in training costs to be achieved by extending turboprop basic stage, then transferring student directly to 2000AT, thus eliminating intermediate jet trainer.

2000B: *(Biplace:* two-seat). Trainer counterpart of 2000C; Nos. 501 to 514 (Series) S3 with RDM radar and M53-5 power plant; Nos. 515 to 520, S4 with RDI J1-1 radar and M53-5, but Nos. 516 and 517 retrofitted with RDM; No. 521 S4-2 with RDI J2-4 and M53-5; No. 522 also S4-2, but M53-P2; Nos. 523 to 530 S5 with RDI J3-13 and M53-P2. Final delivery in December 1994. See also Mirage 2000DA below. One or two Mirage 2000Bs distributed to each 2000C operating squadron, but most operated by OCU, EC 2/5 at Orange, which assumed task from EC 2/2 at Dijon on 1 July 1998. S3 aircraft have been upgraded with RDI radar.

One aircraft, No. 504, converted to **2000BOB** *(Banc Optronique Biplace:* two-seat electro-optics testbed); flown 28 June 1989 after modification by CEV at Brétigny-sur-Orge; trials of Rubis FLIR, VEH-3020 holographic HUD, night vision goggles, helmet-mounted display and other electro-optical systems. Trials mainly in connection with the Dassault Rafale programme, but flew with Shehab/Nahar designator/FLIR system for Mirage 2000-9 programme. Fitted with OBOGS.

2000BR: Version of 2000-5 proposed for Brazil, 2001. Development and promotion agreement signed by Dassault, Embraer, Snecma and Thales 3 April 2002.

2000C: *(Chasse:* fighter). Standard interceptor; Nos. 1 to 37 built as series S1, S2 and S3 with RDM radar and M53-5 power plants; since upgraded to S3 radar standard. Loosely called Mirage **2000RDM**; Mirage 2000B and C collectively known as Mirage **2000DA** *(Défense Aérienne).* Later aircraft (loosely **2000RDI**) have RDI

French Air Force Mirage 2000-5F carrying four MICA and two Magic AAMs, plus three fuel tanks *(Aviaplans/F Robineau)*
0110849

Internal :
Two DEFA 554 30 mm (1) guns
EM and IR decoys

	300	1830	400	400	1400	400	400	1830	300	
			●	●		●	●			Mica active air-to-air missile
	●								●	Magic IR missile
		●						●		Underwing extra fuel tank
					●					Fuselage extra fuel tank
		●●	●	●	●●	●	●	●●		500 lb/250 kg bomb
		●			●					2.000 lb/1.000 kg laser guided bomb
			●	●	●●	●	●			Durandal bomb
					18					BAP 100 penetration bomb
		●	●	●	●	●	●			Beluga grenade dispenser
					●					F 2 practice bomb launcher
	●	●						●	●	F 4 (18) 68 mm rocket launcher
					●					E.O. or I.R. targeting pod
		●						●		AS 30 laser guided missile
					●					Apache missile
		●						●		Anti-radiation missile
		●						●		AM 39 anti-ship missile
					●					Flir pod navigation
					●					Twin gun pod
					●					Recce pod
					●					Offensive or Intelligence ECM pod
					●					Buddy refuelling pod

Mirage 2000-5 potential weapons options (excluding two internal 30 mm cannon and chaff/flare system). Hardpoint maximum loads are given in kilogrammes; (1) internal guns in single-seat versions only
0010578

Dassault Mirage 2000 carrying an APACHE standoff missile on the centreline *(Aviaplans/F Robineau)*   NEW/0113491

radar and M53-P2 power plants: Nos. 38 to 48 Series S4, delivered from July 1987 and later upgraded to S4-1; Nos. 49 to 63 S4-1; Nos. 64 to 74 S4-2; Nos. 75 to 124 Series S5, delivered between late 1990 and June 1995. Equipment standards of Mirage 2000B/Cs are **S3**, incapable of launching Matra BAe 530D (530F only); **S4** RDI J1-1 radar; **S4-1** retrofit of all S4s with improved J1-2; **S4-2** further radar upgrade to J2-4; **S4-2A** retrofit of all S4-1 and S4-2 aircraft (Nos. 38 to 74, 515 and 518 to 522) with HOTAS-type throttle and improved J2-5 radar; **S5** definitive standard with J3-13 radar and, from No. 93 onwards, Spirale chaff/flare dispenser; and **S5-2C** retrofit introduced 1995 (aircraft of EC 1/12) providing better anti-jamming protection for RDI radar. Conversion completed of 37 2000B/Cs to Mirage 2000-5F (which see). Withdrawal of S3 began in January 1998 for upgrading with RDI radar and final aircraft returned to service (at Orange) in March 1999; EC 2/2 gained late production (RDI) aircraft, with which it became operational in the air defence role on 1 August 1998.

Detailed description applies to 2000C except where indicated.

2000D: Two-seat conventional attack version of 2000N, lacking ASMP missile interface and nose pitot but with HOTAS controls, additional display screens for both crew members, Antilope 5-3D terrain-following/terrain-reference radar, GPS and improved (ICMS Mk 2) ECM; functions of pilot and navigator more clearly demarcated. First flight (D01, ex-N01) 19 February 1991; second prototype, D02 (ex-N02), flown 24 February 1992 ; first 2000D (No. 601) delivered CEAM Mont-de-Marsan for trials 9 April 1993; first squadron, EC 1/3 'Navarre', achieved limited IOC (six aircraft only) 29 July 1993 at CEAM. Combat debut was 5 September 1995, launching AS 30L ASMs in Bosnia. Deliveries in 1999 totalled 11, followed by 10 in 2000 and final seven in 2001.

Initial six aircraft built to 'interim baseline' configuration known as **R1N1L**, with ability to launch laser-guided weapons and Magic missiles only. Further few, designated **R1N1**, had slightly expanded weapon options. Later **R1** aircraft have full range of current French Air Force armament (BAP 100, BAT 120, Belouga, 68 mm rockets, AS 30L, GBU-12 (Mk 82), BGL 1000, Magic 2 and PDLCT designator). PDLCT-S (*Synergie*), first employed over Yugoslavia in March 1999, is modified version with improved imagery contrast. All interim 2000Ds were modified to full R1 standard by June 1995. **R2** standard, operational July 2001 (including retrofit of all earlier aircraft), introduces APACHE standoff weapons

dispenser (IOC delayed to 2002), Eclair fully integrated self-defence suite and Atlis II laser designator pods (from Jaguar force). First two R2 conversions completed early 2000 at Nancy; further 68 followed at three/four per month. At least 24 aircraft to receive Thales IMEWS improved electronic warfare system. Planned **R3** version, with SCALP SOM and a reconnaissance pod, was abandoned in June 1996, but had been reinstated by 1999, envisaging carriage of a new, modular AASM SOM, JTIDS and MIDS datalinks for full NATO interoperability and service entry in 2004-06. By 2002, Mirage 2000D earmarked as defence-suppression aircraft, possibly with avionics to be developed in conjunction with Germany.

2000E: Multirole fighter for export; M53-P2 power plant throughout. Details in Customers subsection. Baseline version for India, Egypt and Peru, differences from 2000C including RDM radar with CW illumination for Super 530D AAM; two main computers, with expanded memory; ULISS 52 INS; improved ECM (integrated system with VCM-65 display or, alternatively, Remora and Caiman pods); VE-130 HUD; VMC-180 head-down; and expanded weapon options. Abu Dhabi and Greek 2000Es have extra computing power, further armament options and improved self-defence (SAMET system for Abu Dhabi; ICMS Mk 1 for Greece).

2000ED: Two-seat trainer counterpart of 2000E.

2000N: Two-seat low-altitude penetration version to deliver ASMP nuclear standoff missile; two prototypes; first flight 3 February 1983; one preseries aircraft (No. 301) built at Istres and first flown 3 March 1986; first 24 production aircraft (Nos. 302-325) were 2000N-K1 with ASMP capability only; from July 1988 remaining aircraft, designated 2000N-K2, have full conventional and ASMP capability; production ended 1993; all K1s (initially of squadrons 3/4 and part of 2/4) retrofitted to K2 for conventional attack, programme completed by late 1998. Equipment includes Antilope 5 terrain-following radar, two SAGEM inertial platforms, two improved AHV-12 radio altimeters, colour head-down CRT, two Magic self-defence missiles, and ICMS (integrated countermeasures system) comprising Sabre jamming system, Serval RWR and Spirale automatic chaff/flare dispenser system.

Further upgrade, 2000N K3, expected to be launched in early 2003 for incorporation in 45 aircraft of EC 1/4, 2/4 and 3/4, gaining IOC in 2006-07. Dedicated WSO position in rear cockpit; Thales Reco-NG reconnaissance pod; new self-protection avionics and upgraded ASMP-Ameliore, authorised late 2000 for 2011 service entry.

2000N' (N Prime): Initial designation of 2000D.

2000R: Single-seat day/night reconnaissance export version of 2000E but with normal radar nose; various sensor pods possible (see Avionics paragraph).

2000-3: Private venture upgrade, begun 1986, with Rafale-type multifunction (five-CRT) cockpit displays known as APSI (advanced pilot system interface); prototype BY1/F-ZJTB (ex-No. B01) flew 10 March 1988. Later received RDY radar.

2000-4: Private venture integration of Matra BAe MICA AAM. First guided flight against target drone, 9 January 1992.

2000-5: Multirole upgrade incorporating -3 and -4 improvements, plus Thales RDY radar and new central processing unit, Thales VEH 3020 HUD and ICMS Mk 2 countermeasures; laser-guided bombs and ASMs, or APACHE standoff dispenser in air-to-ground role. First Mirage flight of RDY radar in BY1 (later numbered BY2) May 1988; first flight of full 2000-5 (two-seat) 24 October 1990 (same aircraft; initially no serial number; later reverted to B01); first single-seater, 01 (conversion of trials aircraft CY1) flown 27 April 1991.

Export orders from Taiwan and Qatar; air-to-air firing trials with MICA completed July 1996; air-to-ground trials completed May 1997; operational use (export; Taiwan) from June 1997.

FFr4,600 million (US$830 million) conversion programme announced November 1992 for upgrading 37 French Air Force 2000Cs (34 S4-2As, Nos. 38 to 49, 51 to 59, 61 to 63 and 65 to 74 and three S5s, Nos. 77 to 79) mainly from EC 2/5 and 3/5 at Orange to **2000-5F** for continued service; contract awarded 25 November 1993; only one reworked aircraft funded in 1994 defence budget; 10 more in 1995 funding; and 23 in 1996. Delivery schedule (not achieved) was one in 1997; 11 in 1998; 22 in 1999; and three in 2000; one lost on 16 March 1999. Rework (at Argenteuil, with reassembly at Bordeaux) required six months per aircraft and involved complete dismantling and return of fuselages to Argenteuil. RDI radars from these upgraded aircraft retrofitted in early production (RDM) Mirage 2000Cs. Prototypes were conversions of Nos. 51 and 77 (latter an S5); both modified at Istres; initial aircraft reflown 26 February 1996; second followed two months later.

First 'production' conversion, No. 38 handed over 30 December 1997 at Istres test centre before delivery in April 1998 to CEAM, Mont-de-Marsan, to begin conversion of pilots of EC 1/2 from Dijon. IOC at Dijon (12 aircraft) 31 March 1999; FOC 4 February 2000; re-equipment of EC 2/2 began late 1999; deliveries complete by 2001.

Identification features are additional (LAM) 'bullet' antenna on fin leading-edge, absence of nose pitot and four horizontal antenna bands on radome. Normal external configuration is two MICA and two Magic AAMs, plus two 2,000 litre (528 US gallon; 440 Imp gallon) and one 1,300 litre (343 US gallon; 286 Imp gallon) external tanks. Optional configuration (Standard Kilo) is four MICAs and two Magics. MICA delivered (first 25 rounds) in radar mode from August 1999 followed by IR version in 2003. Early 2000-5F deliveries were to **SF1** standard; SF1C upgrade due to follow, improving radar range by 15 per cent and adding non-co-operative target mode which analyses returns from target's engines. SF2 projected with GPS, provision for helmet-mounted sight and ability to carry unspecified long-range identification aid.

2000-5 Mk 2: Announced early 1999; purchased initially by Greece. Features in common with 2000-9 include modular avionics, laser gyro INS, upgraded ECM, expanded aircraft-missile datalink, Damoclès laser-designation pod (known as Shehab on 2000-9), Nahar

Two-seat trainer Mirage 2000B of French Air Force trials unit CEAM *(Jane's/Paul Jackson)*   0110916

Prototype Dassault Mirage 2000B posing as a 2000-5 Mk 2, with four MICA AAMs under the fuselage, two Magic 2s outboard underwing and two LGBs on the centreline, complete with laser designation pod *(Aviaplans/F Robineau)* 0110848

FRENCH MIRAGE 2000 SQUADRONS

Squadron	Base	Type	Commissioned	Remarks
EC 1/2 'Cicognes'	Dijon	2000C (RDM)	2 Jul 84	Temp stood down Jan 98[4]
		2000-5F	31 Mar 99	Operational 4 Feb 00
EC 2/2 'Cote d'Or'	Dijon	2000B/C (RDM)	27 Jul 86	OCU until Jan 98
		2000C	1 Aug 98	
		2000-5F	2000	
EC 3/2 'Alsace'	Dijon	2000C (RDM)	Mar 86	Disbanded 31 Jul 93
EC 1/3 'Navarre'	Nancy	2000D	1 Apr 94[1]	
EC 2/3 'Champagne'	Nancy	2000N	30 Aug 91	Conventional attack only
		2000D	24 Apr 97[2]	Converted 1996-98
EC 3/3 'Ardennes'	Nancy	2000D	21 Aug 95	
EC 1/4 'Dauphine'	Luxeuil	2000N	1 Jul 88	
EC 2/4 'La Fayette'	Luxeuil	2000N	1 Jul 89	
EC 3/4 'Limousin'	Istres	2000N	1 Jul 90	
EC 1/5 'Vendée'	Orange	2000C	1 Apr 89	
EC 2/5 'Ile de France'	Orange	2000C	1 Apr 90	
		2000B/C	1 Jul 98	OCU
EC 3/5 'Comtat Venaissin'	Orange	2000C	3 Sep 90[3]	Disbanded 31 Aug 97
EC 1/12 'Cambrésis'	Cambrai	2000C	1 Jul 92	
EC 2/12 'Picardie'	Cambrai	2000C	1 Sep 93	

[1] Limited IOC 29 Jul 93
[2] When only partly equipped with 2000D
[3] First delivery
[4] And equipped with Alpha Jets to maintain pilot currency

navigational FLIR in Damoclès pylon, upgraded version of RDY radar (RDY2, with multitarget air-to-sea search and track, high-resolution DBS mapping mode and search and track of mobile land targets), increased MTOW of 17,500 kg (38,580 lb), new multichannel recording system, new rear cockpit colour display repeater and, possibly, helmet-mounted sight. Can carry six MICA AAMs in addition to air-to-surface weapons.

2000-8: Mirage 2000EAD/RAD/DAD supplied to Abu Dhabi/UAE from 1989. Standard AD8.

2000-9: Version of 2000-5 for United Arab Emirates incorporating long-range air-to-ground capability with weapons including Black Shahine and Hakim. Configuration includes M53-P2 engines, RDY-2 radar with synthetic aperture and beam-sharpening modes, Thales Totem 3000 laser INS, upgraded air conditioning system, Elettronica IMEWS (integrated modular electronic warfare system) and digital terrain system. First 2000-9s undertaking flight testing and proving (three aircraft) at Istres from early 2001; first flight (DAD10; two-seat version) 14 December 2000; sole single-seat 2000-9RAD (RAD19) flew 25 January 2001. Further features as for 2000-5 Mk 2. Deliveries due in 2004.

2000S: Export attack version of 2000D; promotion discontinued.

CUSTOMERS: See table for rapid reference. **France** required seven prototypes and 372 production aircraft; reduced in late 1991 to 318 by abandonment of final 24 2000Cs and 30 2000Ds and transfer of some single-seat aircraft to trainer contract; no Mirages funded in 1992 or 1993, but one 2000B and 14 2000Cs cancelled, then re-ordered in 1994 defence budget as 2000Ds. French orders subsequently amended to 30 2000Bs, 124 2000Cs, 75 2000Ns and 86 2000Ds (total 315); last in 2001. Mirage 2000C equipped three squadrons of EC 2 at Dijon (1984-86), three of EC 5 at Orange (1988-90) and two of EC 12 at Cambrai.

Mirage 2000N deliveries began at Luxeuil 30 March 1988; EC 1/4 'Dauphine' formed 1 July 1988, now with

full dual-role K2 series aircraft, followed by two more squadrons. Pending 2000D, EC 2/3 'Champagne' operational at Nancy 1 September 1991 with 2000N in conventional role. EC 1/3 'Navarre' fully operational with 2000D from 31 March 1994; EC 3/3 'Ardennes' began re-equipment, July 1994; EC 2/3 began converting to 2000D (ex-2000N) in mid-1996, completing conversion in 1999. Mirage 2000N depot servicing required every 900 hours or 36 months, whichever sooner.

Abu Dhabi (United Arab Emirates) ordered 18 aircraft on 16 May 1983 and took up 18 options in 1985 for a total of 22 2000EADs, eight 2000RADs and six 2000DADs; these to Standard AD8; deliveries delayed from 1986 to 1989 by provision for US weapons such as Sidewinder; deliveries to Maqatra/Al Dhafra completed November 1990 for Nos. 1 and 2 Shaheen (Warrior) Squadrons. Abu Dhabi 2000RADs carry COR2 multicamera pod, but alternatives include Raphaël-type SLAR 2000 or Harold pods; second 18 have Elettronica ELT/158 threat warning receivers and ELT/558 self-protection jammers; all with Spirale chaff/flare system. Self-defence suite code-named SAMET. Deliveries 7 November 1989 to November 1990. Weapons include BAE Systems PGM Hakim ASM.

Follow-on batch authorised in late 1996; reported 16 December 1997 as 30 new 2000-9s and upgrade to this standard of 33 earlier aircraft; contract signed 18 November 1998; new element later alleged to be 32 aircraft, while upgrades reduced by attrition to approximately 30; total value estimated as US$3.4 billion for aircraft, or US$5.5 billion, including weapons and support. Weapons include Black Shahine version of Storm Shadow/SCALP SOM, IR and active radar MICA AAMs; ICMS Mk 3, incorporating Spirale and Eclair ECM systems; DDM missile approach warning. Shehab target designator pod (export version of Damoclès/PDLCT-S) and Nahar navigation FLIR mounted in starboard shoulder pylon.

Initial deliveries, from mid-2002 at one per month, are upgrades of existing fleet to Standard AD91 with MICA

AAM capability. New aircraft, delivered from 2004, will be fully capable Standard AD92s, with additional air-to-ground radar modes and weapons.

Egypt ordered 16 2000EMs and four BMs in December 1981; deliveries 30 June 1986 to January 1988; based at Beni Suef with 82 Squadron in interceptor role.

Greece ordered 36 2000EGs and four 2000BGs on 20 July 1985; handed over from 21 March 1988 and delivered from 27 April 1988 for 331 'Aegeas' and 332 'Geraki' Mire Pandos Kairou within 114 Pterix Mahis at Tanagra; deliveries suspended October 1989 at 28th aircraft; resumed 1992 and completed 18 November 1992. RDM3 radar; Spirale chaff/flare system installed, as part of ICMS self-defence suite; full ICMS system operational from August 1995. Some upgraded to 2000EG-SG3 standard with provision for AM 39 Exocet anti-ship missile. Further 15 Mirage 2000-5 Mk 2s confirmed 30 April 1999 and formally ordered on 21 August 2000, together with upgrade to this standard of 10 earlier aircraft by HAI. Deliveries from 2004, with equipment including Thales Damoclès laser designation pods and ICMS Mk 3. New weapons include MICA AAM (Standard EG 51) and SCALP EG ASM (EG 52). Option held on further three new aircraft and five upgrades.

India first ordered 36 2000Hs and four THs in October 1982; 26 Hs and four THs temporarily powered by M53-5; final 10 by M53-P2 from outset; first flight of 2000H (KF-101) 21 September 1984; first flight of 2000TH (KT-201) early 1985; No. 7 IAF Squadron 'Battle Axe' formed at Gwalior AB 29 June 1985, coincident with first arrivals in India. Named *Vajra* (Divine Thunder); second Indian order for six Hs and three THs signed March 1986 and delivered April 1987 to October 1988 to complete No. 1 'Tigers' Squadron. Third order placed 19 September 2000 for four 2000Hs and six 2000THs to be delivered in 2003/04. Despite grey-and-blue air defence camouflage, unconfirmed report claimed Indian Mirages optimised for attack, with Antilope 5 radar and twin INS. Role (at least) confirmed by 1996 selection (subject to operational trials) of the Rafael Litening laser-designator pod for IAF Mirage 2000s and Jaguars. By 1993 IAF aircraft appearing in experimental brown-and-green low-level colours, with Spirale chaff dispensers. By 1998, 38 remaining 2000Hs had been upgraded by HAL at Bangalore with local flare dispensers and were preparing for receipt of LGBs.

Peru ordered 24 2000Ps and two 2000DPs in December 1982, but reduced this to 10 2000Ps and two 2000DPs; first 2000DP handed over 7 June 1985; deliveries to Peru from December 1986; Escuadron de Caza-Bombardeo 412 of Grupo Aéreo de Caza 4 at La Joya inaugurated 14 August 1987.

Qatar ordered 12 Mirage 2000-5s under contract 'Falcon' on 31 July 1994, together with MICA and Magic 2 AAMs. Versions are 2000-5EDA (nine; single-seat) and 2000-5DDA (three; two-seat). Equipment includes GPS, ICMS Mk 2 and provision for Spirale. First flight late 1995; first three handed over at Bordeaux on 8 September 1997; four delivered to Qatar 18 December 1997; four more on 1 April 1998.

Taiwan ordered 60 Mirage 2000-5s with M53-P2 power plants on 18 November 1992. First export sale for -5; first flight late 1995; first handed over on 9 May 1996 for training in France; initial five arrived in Taiwan as sea freight on 5 May 1997; equips 41, 42 and 48 Squadrons of 2nd Tactical Fighter Wing at Hsinchu; IOC of initial squadron, November 1997. Final eight delivered late October 1998, at which time second squadron declared operational. Versions are 2000-5Ei (48 aircraft; single-seat) and 2000-5Di (12 aircraft; two-seat). Requirement revealed in 1997 for an additional 60. Armament includes MICA and Magic AAMs.

Additionally, Jordan ordered 10 2000EJs and two 2000DJs on 22 April 1988; all were cancelled in August 1991.

Total 601 firm orders (excluding seven prototypes and six company-owned trials and demonstrator aircraft) by 1 January 2001 (315 French; 286 exports); by late 2001, more than 550 had been delivered, including 229 exports. France had received all 124 Cs, 30 Bs, and 75 Ns by 31 December 1995, only 2000Ds being delivered after that date (79 by end of 2000 and last seven in 2001). Provisional agreement signed with Pakistan in January 1992 for 44 Mirage 2000Es; under further discussion in 1994-95, quantity having been revised to 36, including nine two-seat aircraft, all to 2000-5 standard. No further progress made.

Demonstration and evaluation sorties by a Mirage 2000-5 two-seat version given to Czech, Hungarian and Polish air forces in August 1996. Possible Polish final assembly by PZL Mielec is covered by an MoU announced in June 1997.

MIRAGE 2000 ORDERS

Customer	Qty	Version	First aircraft
Abu Dhabi/UAE[1]	6	2000DAD	701
	8	2000RAD	711
	22	2000EAD	731
	20	2000-9EAD/ RAD	—
	12	2000-9DAD	—
Egypt	16	2000EM	101
	4	2000BM	201
France	30	2000B	501
	124[2]	2000C	1
	86	2000D	601
	75	2000N	301
Greece	36[3]	2000EG	210
	4	2000BG	201
	15	2000-5 Mk 2	—
India	46	2000H	KF101
	13	2000TH	KF201
Peru	10	2000P	050
	2	2000DP	193
Qatar	9	2000-5EDA	QA90
	3	2000-5DDA	QA86
Taiwan	48	2000-5Ei	2001
	12	2000-5Di	2051
Subtotal	**601**		
Prototypes	7		
Development	6		
Total	**614**		

[1] Some 30 being upgraded to 2000-9DAD/RAD/EAD
[2] 37 upgraded to 2000-5F
[3] 10 to be upgraded to 2000-5 Mk 2

COSTS: Programme unit cost: Taiwan FFr 333 million (1997).

DESIGN FEATURES: Multirole combat aircraft; low-set, thin delta wing for high internal volume and low wave drag (as in Mirage III/5), but with delta's disadvantages in manoeuvrability and landing/take-off requirements offset by relaxed stability and leading-edge slats. Area-ruled fuselage.

Wing has cambered section, 58° leading-edge sweep and moderately blended root employing Karman fairings. Cleared for 9 *g* and 270°/s roll at sub- and supersonic speed carrying four air-to-air missiles.

FLYING CONTROLS: Full fly-by-wire control with SFENA autopilot; two-section elevons on wing move up 16° and down 25°; inner leading-edge slat sections droop up to 17° 30' and outer sections up to 30°; fixed strakes on intake ducts create vortices at high angles of attack that help to correct yaw excursions; small airbrakes above and below wings.

STRUCTURE: Multispar metal wing; elevons have carbon fibre skins with AG5 light alloy honeycomb cores; carbon fibre/light alloy honeycomb panel covers avionics bay; most of fin and all rudder skinned with boron/epoxy/carbon; rudder has light alloy honeycomb core.

LANDING GEAR: Retractable tricycle type by Messier-Bugatti, with twin nosewheels; single wheel on each main unit. Hydraulic retraction, nosewheels rearward, main units inward. Oleo-pneumatic shock-absorbers. Electro-hydraulic nosewheel steering (±45°). Manual disconnect permits nosewheel unit to castor through 360° for ground towing. Light alloy wheels and tubeless tyres, size 360×135-6 or 360×135R6, pressure 8.00 bar (116 lb/sq in) on nosewheels; size 750×230R15, pressure 15.00 bar (217 lb/sq in) on mainwheels. Messier-Bugatti hydraulically actuated polycrystalline graphite disc brakes on mainwheels, with anti-skid units. Compartment in lower rear fuselage for brake parachute, arrester hook or chaff/flare dispenser.

POWER PLANT: One SNECMA M53-P2 turbofan, rated at 64.3 kN (14,462 lb st) dry and 95.1 kN (21,385 lb st) with afterburning. Alternative M53-P20, rated at 98.1 kN (22,046 lb st), is no longer offered. Movable half-cone centrebody in each air intake.

Internal wing fuel tank capacity 1,480 litres (391 US gallons; 326 Imp gallons); fuselage tank capacity 2,498 litres (660 US gallons; 549 Imp gallons) in single-seat aircraft, 2,424 litres (640 US gallons; 533 Imp gallons) in two-seat aircraft. Total internal fuel capacity 3,978 litres (1,050 US gallons; 875 Imp gallons) in 2000C and E, 3,904 litres (1,030 US gallons; 859 Imp gallons) in 2000B, N, D and S. Provision for one jettisonable 1,300 litre (343 US gallon; 286 Imp gallon) RPL-522 96 kg (212 lb) fuel tank under centre of fuselage, and a 1,700 litre (449 US gallon; 374 Imp gallon) RPL-501/502 210 kg (463 lb) drop tank under each wing. Total internal/external fuel capacity 8,678 litres (2,291 US gallons; 1,909 Imp gallons) in 2000C and E, 8,604 litres (2,271 US gallons; 1,892 Imp gallons) in 2000B.

Detachable flight refuelling probe forward of cockpit on starboard side. (Availability of in-flight refuelling on export aircraft not disclosed, although probes fitted to Abu Dhabi's 2000RADs.) Dassault type 541/542 tanks of 2,000 litres (528 US gallons; 440 Imp gallons) are available for the 2000-5, 2000-9, 2000N, D and S wing attachments

Dassault Mirage 2000D tactical attack fighter No. 625 of EC 2/3 'Champagne' carrying 14 LGB and two AS 30 missile mission markings *(Jean-Louis Gaynecoetche)* ***NEW***/0131693

(and optional on 2000B/C), empty weight 240 kg (529 lb) each, and may be complemented by an RPL-522 on centreline.

ACCOMMODATION: One or two occupants (see Current Versions) on SEMMB licence-built Martin-Baker Mk 10Q zero/zero ejection seat(s), in air conditioned and pressurised cockpit. Pilot-initiated automatic ejection in two-seat aircraft; 500 microseconds delay between departures. Canopy/ies hinged at rear to open upward and, on Mirage 2000D, covered in gold film to reduce radar signature.

SYSTEMS: ABG-Semca air conditioning and pressurisation system. Two independent hydraulic systems, pressure 280 bar (4,000 lb/sq in) each, to actuate flying control servo units, landing gear and brakes. Hydraulic flow rate 110 litres (29.1 US gallons; 24.2 Imp gallons)/min. Electrical system includes two Auxilec 20110 air-cooled 20 kVA 400 Hz constant frequency alternators (25 kVA in Mirage 2000D and 2000-5), two Bronzavia DC transformers, a SAFT 40 Ah battery and ATEI static inverter. Eros oxygen system.

AVIONICS: *Comms:* Thales ERA-7000 V/UHF com transceiver, ERA-7200 UHF (with optional Have Quick II) or SCP 5000 secure voice com; Thales NRAI-7A/NRAI-11 IFF transponder/interrogator (SC10/IDEE 1 on 2000-5).

Radar: Thales RDM multimode radar or RDI pulse Doppler radar, each with operating range of 54 n miles (100 km; 62 miles). (Mirage 2000N/D have Thales Antilope terrain-following radar for automatic flight down to 61 m (200 ft) at speeds not exceeding 600 kt (1,112 km/h; 691 mph); Antilope 5TC in 2000N includes altitude-contrast updating of navigation system; Antilope 5-3C in 2000D has full terrain-reference navigation facility.) Thales RDY multimode, multitarget radar in 2000-5 has the ability to detect 24 targets while tracking eight.

Flight: SOCRAT 8900 solid-state VOR/ILS and IO-300-A marker beacon receiver, Thales radio altimeter (AHV-6 in 2000B and C, AHV-9 in export aircraft, two AHV-12 in 2000N and AHV-17 in 2000-5). Thales NC12 or Deltac Tacan. SAGEM Uliss 52 inertial platform (52D in 2000C and B; 52E for export; and two 52P in 2000N/D; 52ES in 2000-5; plus integrated GPS in 2000D and 2000-5). Thales Totem 3000 RLG in 2000-9. Thales Type 2084 central digital computer and Digibus digital databus (2084-XR in 2000D; XRI3 in 2000-5). Thales AP 605 autopilot (606 in 2000N, 607 in 2000D, 608 in 2000-5). Thales MMR (multimode receiver) in 2000-9, combining ILS, microwave landing and differential GPS). Undisclosed operator ordered BASE (BAE Systems) Terprom GPWS in 1996 for retrofit by Orbital Sciences Corporation.

Instrumentation: Thales TMV 980 data display system (VE 130 head-up and VMC 180 head-down) (two ICC 55 head-down in 2000N/D). Mirage 2000-5 has Thales Comète multidisplay system and VEM 130 HUD with Thales recording camera. Provision optional in 2000-5 and -9 for Thales Topsight E helmet-mounted sight/display system.

Mission: Mirage 2000-5 has LAM (*liaison avion-missile*) 'bullet' antenna on fin leading-edge for mid-course guidance of MICA AAMs. Sensors of strike/attack and export versions include 570 kg (1,257 lb) Thales Raphaël SLAR 2000 pod, 400 kg (882 lb) Thales COR2 multicamera pod or 680 kg (1,499 lb) Dassault AA-3-38 Harold long-range oblique photographic (Lorop) pod; 110 kg (243 lb) Thales/Intertechnique Rubis FLIR pod; Thales Atlis laser designator and marked target seeker (in pod on forward starboard underfuselage station); 340 kg (750 lb) Thales PDLCT day/night (TV/thermal imaging) laser designator pod on Mirage 2000D (or CLDP/Atlis II on export aircraft)(2000D squadrons also being issued with Atlis II designation pods – being refurbished Atlis systems from Jaguar force), while PDLCT-S available from mid-1999; one 550 kg (1,213 lb) Thales TMV 004 (CT51J) Caiman offensive or intelligence ECM pod and one 400 kg (882 lb) Thales Astac elint (interferometer) pod. Export PDLCT-S known as Damoclès (and Shehab in UAE).

Self-defence: Systems in 2000C and 2000N include Thales Serval radar warning receiver (antennas at each wingtip and on trailing-edge of fin, near tip, plus VCM-65 cockpit display); Thales Caméléon (2000N), Caméléon C2 (2000D) or Sabre (2000C) jammer at base of fin (detector on fin leading-edge); and MBDA Spirale, comprising chaff dispensers in Karman fairings at wing trailing-edge/fuselage intersection and flares in lower rear fuselage. French Air Force DDM (*Détecteur Départ Missile*) missile plume detector requirement satisfied by 1994 purchase of SAGEM SAMIR system for 1995 fitment in rear of Magic launch rails (2000D/N first, but also to 2000Cs patrolling Bosnia). Spirale fitted to 2000N-K2; retrofitted to 2000N-K1 and installed on 2000Cs from No. 93; earlier 2000Cs have Eclair system (Alkan LL5062 chaff and flare launcher) in place of braking parachute, lacking automatic operation. Spirale on some export 2000Es.

Upgrade planned of 2000E with ICMS Mk 1 (integrated countermeasures system; as in 2000EG) comprising RWR and SHR, Thales high-band jammer (leading-edge of fin and bullet fairing at base of rudder) and Spirale; automated ICMS Mk 2 of Mirage 2000-5 adds receiver/processor in nose to detect missile command links; extra pair of antennas near top of fin and additional DF antennas scabbed to existing wingtip pods (fin and secondary wingtip antennas also on Greek Mirage 2000s); ABD2000 export version of Sabre in some Mirage 2000Es. Mirage 2000-5 Mk 2 has ICMS Mk 3. External equipment can include two 182 kg (401 lb) Thales DB 3141/3163 Remora self-defence ECM pods.

EQUIPMENT: Optional 250 kg (551 lb) Intertechnique 231-300 buddy-type in-flight refuelling pod.

ARMAMENT: Two 30 mm DEFA 554 guns in 2000C, 2000E and single-seat 2000-5 (not fitted in B, D or N), with 125 rds/gun. Nine attachments for external stores, five under fuselage and two under each wing. On 2000-5, fuselage centreline stressed for 1,400 kg (3,086 lb) loads; other four fuselage points for 400 kg (882 lb) each; inner wing pylons for 1,830 kg (4,034 lb) each; and outboard wing points for 300 kg (661 lb) each.

Typical interception weapons comprise two 275 kg (606 lb) Super 530D or (if RDM radar not modified with target illuminator) 250 kg (551 lb) 530F missiles (inboard) and two 90 kg (198 lb) 550 Magic or Magic 2 missiles (outboard) under wings. Alternatively, each of four underwing hardpoints can carry a Magic. MICA AAM (110 kg; 243 lb) optional on Mirage 2000-5. Primary weapon for 2000N is 900 kg (1,984 lb) ASMP tactical nuclear missile mounted on LM-770 centreline pylon.

In air-to-surface role, the Mirage 2000 can carry up to 6,300 kg (13,890 lb) of external stores, including MBDA 250 kg retarded bombs or 32.5 kg (72 lb) TDA BAP 100 anti-runway bombs; 16 MBDA Durandal 219 kg (483 lb) penetration bombs; one or two 990 kg (2,183 lb) MBDA BGL 1000 laser-guided bombs; five or six 305 kg (672 lb) MBDA Belouga cluster bombs or 400 kg (882 lb) TDA BM 400 modular bombs; one Rafaut F2 practice bomb launcher; US Mk 20, Mk 82, GBU-10 and GBU-12 bombs; two 520 kg (1,146 lb) AS 30L, Armat anti-radar, or 655 kg (1,444 lb) AM 39 Exocet anti-ship, air-to-surface missiles; four 185 kg (408 lb) MBDA LR F4 rocket launchers, each with eighteen 68 mm rockets; two packs of 100 mm rockets; or a 765 kg (1,687 lb) Dassault CC 630 gun pod, containing two 30 mm guns and total 600 rounds of ammunition.

Mirage 2000D (and export 2000-5) to receive 1,200 kg (2,646 lb) APACHE-AP standoff (216 n mile; 400 km; 249 mile) weapons dispenser. APACHE-SCALP (*Système de Croisière Autonome à Longue Portée*) selected December 1994 to satisfy APTGD (*Arme de Précision Tirée à Grande Distance*) requirement for 216 to 324 n mile (400 to 600 km; 249 to 374 mile) range stealthy cruise missile; service entry in 2002. For air defence weapon training, a Cubic Corporation AIS (airborne instrumentation subsystem) pod, externally resembling a Magic missile, can replace Magic on launch rail,

enabling pilot to simulate a firing without carrying actual missile.

DIMENSIONS, EXTERNAL:

Wing span	9.13 m (29 ft 11½ in)
Wing aspect ratio	2.0
Length overall: 2000C, E	14.36 m (47 ft 1¼ in)
-5	14.65 m (48 ft 0¾ in)*
2000B, N	14.55 m (47 ft 9 in)*
Height overall: 2000C, E, -5	5.20 m (17 ft 0¾ in)
2000B, N, D	5.15 m (16 ft 10¾ in)
Wheel track	3.40 m (11 ft 1¾ in)
Wheelbase	5.00 m (16 ft 4¾ in)

*2000D, S and -5 versions lack nose pitot

AREAS:

Wings, gross	41.0 m² (441.3 sq ft)

WEIGHTS AND LOADINGS:

Weight empty: 2000C, E, -5	7,500 kg (16,534 lb)
2000B, N, D	7,600 kg (16,755 lb)
Max internal fuel: 2000C, -5	3,160 kg (6,967 lb)
2000B, N, D	3,100 kg (6,834 lb)
Max external fuel	4,100 kg (9,039 lb)
Max external stores load	6,300 kg (13,890 lb)
Combat weight: 2000-5	9,500 kg (20,944 lb)
T-O weight clean: 2000C, E, -5	10,860 kg (23,940 lb)
2000B, N, D	10,960 kg (24,165 lb)
Max T-O weight: except -5, -9:	
normal	14,000 kg (30,864 lb)
overload	17,000 kg (37,480 lb)
-5, -9	17,500 kg (38,580 lb)
Max landing weight	16,700 kg (36,817 lb)
Max wing loading	414.6 kg/m² (84.92 lb/sq ft)
Max power loading (M53-P2)	179 kg/kN (1.75 lb/lb st)

PERFORMANCE (M53-P2 power plant):

Max level speed: at height	M2.2
at S/L	M1.2
Max continuous speed: 2000C, E	M2.2
2000N, D	M1.4
Min speed in stable flight	100 kt (185 km/h; 115 mph)
Authorised min flight speed	zero
Approach speed	140 kt (259 km/h; 161 mph)
Landing speed	125 kt (232 km/h; 144 mph)
Max rate of climb at S/L:	
M53-P2	17,060 m (56,000 ft)/min
M53-P20	18,290 m (60,000 ft)/min
Time to 11,000 m (36,080 ft) and M1.8:	
2000-5	approx 5 min
Time from brake release to intercept target flying at M3.0	
at 24,400 m (80,000 ft)	less than 5 min
Service ceiling: 2000C	16,460 m (54,000 ft)
2000-5	18,290 m (60,000 ft)
Range: hi-hi-hi	1,000 n miles (1,852 km; 1,151 miles)
interdiction, hi-lo-hi	800 n miles (1,480 km; 920 miles)
attack, hi-lo-hi	650 n miles (1,205 km; 748 miles)
attack, lo-lo-lo	500 n miles (925 km; 575 miles)

with one 1,300 litre and two 1,700 litre drop tanks 1,800 n miles (3,333 km; 2,071 miles)

Operational loiter (2000-5) at M0.8 at 7,620 m (25,000 ft) with three external tanks, four MICA and two Magic AAMs 2 h 30 min

Operational range (2000-5) for 5 min combat at M0.8 at 9,145 m (30,000 ft) with four MICA and two Magic AAMs, tanks jettisoned 780 n miles (1,445 km; 898 miles)

g limits:	+9.0/–3.2 normal
	+11.0/–4.5 ultimate

UPDATED

DASSAULT RAFALE

English name: Squall

TYPE: Multirole fighter.

PROGRAMME: Ordered (as *Avion de Combat Tactique; ACT*) to replace French Air Force Jaguars and (as *Avion de Combat Marine; ACM*) Navy Crusaders and Super Etendards; for early development history, see 1990-91 and earlier *Jane's*; first flight of Rafale A prototype (F-ZJRE) 4 July 1986; first flight with SNECMA M88 replacing one GE F404, 27 February 1990 (was 461st flight overall); 867th and final sortie, 24 January 1994. ACE International (*Avion de Combat Européen*) GIE set up in 1987 by Dassault Aviation, SNECMA, Thomson-CSF (now Thales) and Dassault Electronique, partly to attract international partners; none found. Four preproduction aircraft, as described under Current Versions (specific). Production launch officially authorised, 23 December 1992 (and 31 December 1992 for M88-2 power plant). First Rafale B and Rafale M ordered 26 March 1993; first production aircraft (Rafale B No. 301) flew 24 November 1998 and made 'inaugural' (official) first flight 4 December; to CEV at Istres, early 1999, for development of F2 production standard. First production Rafale M (No. 1) flew at Bordeaux 7 July 1999. Rafale had by then accumulated over 4,000 sorties.

Test airframe, in Rafale M configuration, delivered to CEAT at Toulouse for ground trials 10 December 1991. Between 17 December 1991 and 2 March 1993, completed 10,000 simulated flights, including 3,000 catapult take-offs and 3,000 deck landings. Rafale structural validation achieved 15 December 1993.

French Air Force preference switched to operational two-seat (pilot and WSO) derivative of Rafale C in 1991;

Dassault Rafale B01, the first two-seat aircraft, during trials with conformal fuel tanks (*Aviaplans/F Robineau*)
NEW/0113497

announced 1992 that 60 per cent of procurement to be two-seat, although 16 aircraft deleted from requirements at this time. Procurement target further reduced, as detailed in Customers paragraph. Two-seat version of naval Rafale announced September 2000.

Funding constraints and French government demands for cost reductions resulted in suspension of Rafale programme in November 1995 and temporary blocking of most 1996 funds. Plans were simultaneously abandoned for three progressively more sophisticated service standards of Rafale (*Standard Utilisateur* 0, 1 and 2) and replaced by a common French military standard and an export parallel, although three basic software standards (F1 to F3) will phase-in operational capabilities, as described under Current Versions (general). Work on production Rafales temporarily halted in April 1996. On 22 January 1997, Dassault and French defence ministry agreed on 48-aircraft multiyear procurement (1997-2002) in return for a 10 per cent cost reduction, effectively relaunching the programme. This initiative lapsed with the change of government after June 1997, but reinstated in January 1999, with firm orders for 28, plus 20 options, covering deliveries between 2002 and 2007. Contract awarded January 2000 for development of F2 Standard Rafale, representing first capability upgrade.

First production aircraft completed late 1998; second and third (No. 302 and M1) delivered to CEV trials unit late 1999/early 2000; M2 and M3 to Landivisiau naval air station, December 2000; first naval squadron, 12 F, formed 18 May 2001 with aim of completion (10 Rafales) by June 2002; first operational carrier deployment by four aircraft (M2 to M5) aboard FS *Charles de Gaulle* for exercise Trident d'Or in Mediterranean, 21 to 29 May 2001. Rafale B No. 303 is first for air force, in late 2002; first air force squadron equipped with 20 aircraft during 2006; 80th aircraft due in 2009, 140th in 2016; final (294th) French delivery due in 2019. By January 2002, 10 production Rafales (eight Ms and two French Air Force Bs) had been delivered; further 13 deliveries planned for 2002.

Associated programmes include Thales electronic scanning RBE2 (*Radar à Bayalage Electronique deux plans*) multimode radar, ordered November 1989; test flights begun in Falcon 20 No. 104, 10 July 1992; first

RBE2 flight in Rafale 7 July 1993 (B01); first production RBE2 flew on 16 October 1997 in a Falcon 20 before being refitted in Rafale B01 from November 1997. Development authorised in early 1999 of upgraded RBE2 version (full air-to-ground weapons capability) for 2003 delivery and installation in F2 standard Rafales.

Thales/MBDA defensive aids package named Spectra (*Système pour la Protection Electronique Contre Tous les Rayonnements Adverses*); wholly internal IR detection, laser warning, electromagnetic detection, missile approach warning, jamming and chaff/flare launching; nine prototypes ordered; total weight 250 kg (551 lb); Spectra trials begun on Mirage 2000 in 1992, while full suite installed in Falcon 20 No. 252 by Dassault at Istres between December 1992 and September 1994 before flight trials; Spectra flown in Rafale M02 on 20 September 1996 at launch of integration programme at CEV Istres. Development contract awarded 1991 for Thales OSF (*Optronique Secteur Frontal*) with IRST, FLIR and laser range-finder in two modules ahead of windscreen; surveillance, tracking and lock-on by port module; target identification, analysis and optical identification by starboard module; combined output in pilot's head-level display; initially tested in Mirage 2000BOB (which see).

M88 engine, which has flown only on Rafales, achieved type certification on 22 March 1996. First of initial production batch of 42 M88-2 engines delivered by SNECMA 30 December 1996. Conformal fuel tanks flight-tested in April 2001. Integration of Matra BAe MICA AAM completed 5 July 2000, after 27 launches.

Rafale offered for export; Greek evaluation in January 2000; model displayed February 2000 carried two conformal fuel tanks, each of 1,250 litres (330 US gallons; 275 Imp gallons), increasing range with two SCALP missiles to 1,000 n miles (1,852 km; 1,150 miles), hi-lo-hi. Rafale B No. 302 equipped with supplementary software to enable demonstration of LGB capability; test separation of GBU-12 bombs in April 2000, followed by live drop in October 2000.

CURRENT VERSIONS (general): **Rafale B:** Originally planned two-seat, dual-control version for French Air Force; weight initially envisaged as 350 kg (772 lb) more than Rafale C; 3 to 5 per cent higher cost than Rafale C. Being

Rafale M carrier landing (*Aviaplans/F Robineau*)
NEW/0113499

No. 302, the second French Air Force trials Rafale B (*Jean-Louis/Gaynecoetche*)
NEW/0110852

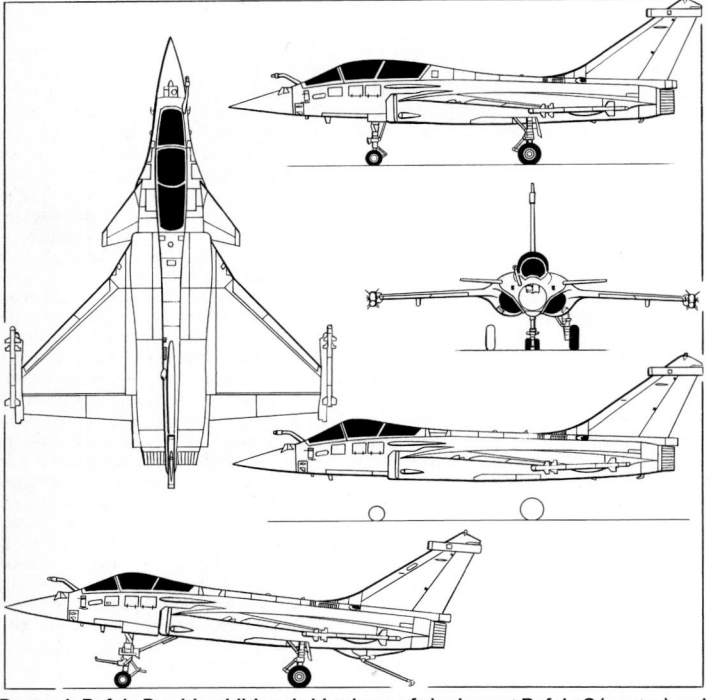

Dassault Rafale B, with additional side views of single-seat Rafale C (centre) and navalised Rafale M (bottom) (*Jane's/James Goulding*)

Rafale weapon options   0010585

		TIP R	EXT R	MED R	INT R	AFT LAT R	FWD LAT R	FUS CEN	FWD LAT L	AFT LAT L	INT L	MED L	EXT L	TIP L
MISSILES														
AIR/AIR	MAGIC	×	×	×								×	×	×
AIR/AIR	MICA	×	×	×		×		⊠		×		×	×	×
AIR/AIR	SIDEWINDER	×	×	×								×	×	×
AIR/AIR	ASRAAM	×	×	×								×	×	×
AIR/AIR	AMRAAM				×		×	×	×		×			
AIR/SEA	EXOCET/AM 39				⊠	⊠				⊠	⊠			
AIR/SEA	PENGUIN 3				⊠	⊠				⊠	⊠			
AIR/SEA	HARPOON				⊠	⊠				⊠	⊠			
AIR/GROUND	AS 30 L				⊠	⊠				⊠	⊠			
AIR/GROUND	APACHE						●	●	●					
AIR/GROUND	ALARM				×	×		×		×	×			
AIR/GROUND	HARM				⊠	⊠		×		⊠	⊠			
AIR/GROUND	MAVERICK				¤	¤				¤	¤			
BOMBS														
LG	1000 kg LG				⊠			⊠			⊠			
LG	400 kg LG			×	×			×			×	×		
LG	GBU 12			×	×			×			×	×		
LG	GBU 10				⊠			⊠			⊠			
CONVENT.	250 kg/Mk 82			××	×	×		××		×	×	××		
CONVENT.	400 kg/Mk 83			×	××			××			××	×		
CONVENT.	BELOUGA			¤	×××			×××			×××	¤		
CONVENT.	Bap 100/Bat 120				≣			≣			≣			
CONVENT.	Durandal			¤¤	≣≣≣			≣≣≣			≣≣≣	¤¤		
CONVENT.	Rockets			◉	◉						◉	◉		
TANKS														
FUEL	RL 2000 l Underwing/fus						⊕	⊕	⊕					
FUEL	RL 3000 l Underfus							⊕						
FUEL	RL 1250 l Underwing			⊕	⊕							⊕	⊕	
FUEL	RL 1250 l Underfus							⊕						
PODS														
NAV	PDLCT - TV							⊙						
NAV	FLIR								⌀					
ECM	offensive jammer		○					○					○	
RECCE	IR/opt.RECCE						⊕	⊞						
RECCE	SLAR							○						
RECCE	HAROLD							◈						
MISC	buddy-buddy refueling							⊕						
MISC	twin gun pod (600 rounds)							⌀						

developed into fully operational variant for either pilot/WSO or single-pilot combat capability. Serial numbers begin at 301. First two assigned to CEV.

Rafale C: Single-seat combat version for French Air Force. Serial numbers begin at 101. Deliveries beginning 2002.

Rafale D: Original configuration from which production versions derived; now '*Rafale Discret*' (stealthy) generic name for French Air Force versions.

Rafale M: Single-seat carrierborne fighter; serial numbers begin at 1. Navalisation weight penalty, 610 kg (1,345 lb); has 80 per cent structural and equipment commonality with Rafale C, 95 per cent systems commonality. Navy's financial share of French programme cut in 1991 from 25 to 20 per cent.

Initial operational software standard for Rafale M, designated F1, will permit air defence missions against multiple targets using Magic and radar-homing version of MICA; self-defence provided by Spectra system. F2 will apply to both naval and air force Rafales delivered from late 2004 (15 M/Ns, 22 Bs and 11 Cs) and combine F1 with air-to-ground radar modes and the ability to launch IR-guided MICA, SCALP and AASM weapons as well as OSF electro-optics suite and MIDS datalink; operational in 2006. Funding for F2 development granted 31 December 1998; first stage of F2, known as E12, is OSF, SCALP and standard air-to-surface ordnance. Finally, in 2008 (operational 2009), F3 standard (see also below) provides full capabilities to naval and air force Rafales, including air-to-sea attack, AM39 and ASMP-A weapons, refuelling and reconnaissance pods, and helmet-mounted display. Unspecified and unfunded F4 envisaged for 2010, but early naval aircraft will all have been upgraded to F2 and F3 by 2008; Meteor AAM and associated electronically scanned radar antenna are key elements of F4.

Rafale N: Two-seat naval version; also known briefly as Rafale BM; announced September 2000; prototype, No. 15, for trials in 2005; requirement for 25 aircraft within overall Navy purchase; first production aircraft to be No. 21, in 2006; service entry 2005 with squadron 11 F. Cost 5 per cent more than single-seat M; 250 kg (551 lb) heavier, but 260 kg (573 lb) less internal fuel; 85 per cent commonality with three previous versions; development

cost FFr1,500 billion (€228 million); deletion of internal canon replaces 200 litres (52.8 US gallons; 44.0 Imp gallons) of fuel lost with addition of second seat.

'Rafale R': Initial studies launched by procurement agency, DGA, late 1997, into stealthy sensor pod which would allow Rafales to replace Mirage F-1CRs and naval Super Etendards in reconnaissance role.

'Rafale Mk 2': Export version, under active consideration by 2000, featuring active antenna radar, M88-3B engines of 88.3 kN (19,850 lb st) each. Available from 2006; conformal tanks and Damoclès laser target designator. Development cost estimated in 2001 as €1.3 billion; joint venture agreed in January 2001 by Dassault, Thales and SNECMA; offered to South Korea.

CURRENT VERSIONS (specific): **C01:** Single-seat Rafale C prototype, C01/F-ZWVR, ordered 21 April 1988; flown 19 May 1991; officially flight tested at CEV in October 1991, two months ahead of schedule; 100th sortie 12 May 1992. First Rafale gun firing, 5 March 1993; first Magic 2 AAM launch, 26 March 1993. Continued high AoA trials in 2000, having exceeded 30°. M88-2 Stage 4 engine trials in late 2000.

Second Rafale C order (C02) not placed; abandoned 1991.

M01: First navalised prototype, F-ZWVM, ordered 6 December 1988; flown 12 December 1991. Assigned to structural qualification, FCS and aerodynamic trials. Catapult trials ashore at US Naval Air Warfare Center, Patuxent River, and Lakehurst, 13 July/23 August 1992; second series of US trials 15 January/18 February 1993 followed by deck trials on *Foch*; first deck landing 19 April 1993; first deck launch 20 April 1993, although first take-offs with 'jump strut' nosewheel leg began in following month; third US trials series 18 November/16 December 1993, carrying external loads; fourth series, October/December 1995, including dummy deck launch at maximum weight of 22,300 kg (49,163 lb). On 8 June 1995, M01 made the first launch of a MICA AAM against an aerial target acquired by RBE2 radar fitted in a Rafale. Employed on development trials for F2 production standard. Total 438 hours in 723 sorties by October 1997. In storage 1999; F1 standard avionics trials in 2000.

B01: Two-seat dual-control trainer Rafale B prototype, B01/F-ZWVS, ordered 19 July 1989 as first with RBE2 radar and Spectra defensive systems; first flight 30 April 1993; first flight with RBE2 7 July 1993. Longest Rafale sortie, Istres to Dubai, November 1995: approximately 3,020 n miles (5,600 km; 3,475 miles) in 6 hours 30 minutes with three aerial refuellings (including one precautionary). Heavy configuration trials (23,400 kg; 51,588 lb, including two APACHEs) completed in February 1997. Total 990 sorties by mid-1999. First Rafale to fly with conformal tanks, 18 April 2001.

M02: Second naval prototype, ordered 4 July 1990; first flight 8 November 1993; assigned to operational and maintenance testing aboard ship and navigation/weapons trials. Joint carrier trials with M01 aboard *Foch* (second series) 27 January/4 February 1994. Third series of deck trials (M02 only) aboard *Foch* begun 17 October 1994 for three weeks (total 28 launches including two at night); included maintenance, electromagnetic compatibility, RBE2 radar and Spectra ECM tests. Fitted with model of OSF in late 1994, for vibration tests. Flew Singapore to Istres (6,300 n miles; 11,668 km; 7,250 miles) in under 15 hours, February 1996. First launch of Magic 2 AAM at moving target, 4 April 1996. Total 297 hours in 348 sorties by October 1997. With B01, employed by 1998 on development of F2 and F3 avionics standards. On 6 July 1999 became only second jet fighter (following Super Etendard) to land on new carrier FS *Charles de Gaulle*. Continued Spectra trials in 2000. Table of Rafale deck trials appeared in 2001-02 and previous editions.

CUSTOMERS: Anticipated worldwide market for 500 aircraft in addition to originally planned 250 for French Air Force (225 Cs and 25 Bs) and 86 for French Navy; former service announced revised requirement for 234, comprising 95 Rafale Cs and 139 two-seat (pilot and WSO) combat versions, in 1992. Defence economies in 1996 included reduction of requirements to 60 Ms.

Naval deliveries began with No. 2 to CEPA trials unit on 19 July 2000, followed by No.3 in September 2000; service familiarisation began 4 December 2000 when Nos. 2 and 3 delivered to Landivisiau naval air base; Landivisiau's 12 Flottille re-formed 18 May 2001 with four aircraft and planned to achieve IOC with 10 aircraft in 2002 for

Two aspects of the Dassault Rafale in B (left) and M configuration *(Aviaplans/F Robineau)*

NEW/0131694

interceptor duties aboard *Charles de Gaulle* (commissioned 1999); balance replaces Super Etendards, 11 Flottille being first recipient, in 2005; third squadron to follow. Navy to receive 10 aircraft in F1 configuration, 15 F2s and 35 F3s, last in 2012, all to be based at Landivisiau.

Air Force deliveries originally planned for 1996-2009, including first 20 in interim configuration; two year postponement announced 1992; further slips in development funding delayed first receipts to 2002 and IOC to 2006, when first squadron to be established with 20 Rafale Bs. Only Air Force's first three aircraft 301, 302 and 101 (two Bs and one C) to F1 standard.

Official authorisation to launch production given 23 December 1992. Initial production contract in 1993 defence budget (formally awarded 26 March 1993), comprising one aircraft each for Air Force and Navy. Total 13 by 1996, while 1997-2002 plan envisaged 33 B/Cs and 15 Ms (total 48) to be ordered (and two Bs and 12 Ms delivered), followed by orders for 15 B/Cs per year from 2003. In early 1997, French defence procurement agency, DGA, agreed with Dassault a multiyear procurement of the 48 aircraft in return for a 10 per cent cost reduction, but this later suspended. Authorisation to order the first 13 Rafales was only granted in May 1997, however. At the same time, separate plans were being formulated for acceleration into service of the first 10 air force Rafales to equip an export-promoting and operational trials (half) squadron, but those plans also soon abandoned. Eventually, go-ahead given on 14 January 1999 for 48 aircraft, including 20 options, these confirmed on 21 December 2001. Deliveries in 2003-08 five-year plan will be 57 to Air Force and 19 to Navy. Export versions of naval variant were available to potential customers from 1999 onwards. China expressed interest in 1996-97. By 2015 French Air Force combat arm is expected to comprise 140 Rafales in front-line units.

RAFALE PROCUREMENT

Year	Rafale C	Rafale B	Rafale M/N	Total
1993		1	1	2
1994		1	2	3
1995			5	5
1996		1	2	3
1997				
1998				
1999†	7	14	7	28
2001	5	7	8	20
Totals	**12**	**24**	**25**	**61**

† start of F2 production version; earlier aircraft are F1; 62nd and subsequent will be F3
Note: Orders as announced. Third Rafale B was subsequently exchanged with first production Rafale C.

COSTS: Programme estimated at FFr155 billion (1991), including FFr40 billion for R&D; revised to FFr178 billion in 1993, FFr198.4 billion in 1995 and FFr202.37 billion in 1996. Last-mentioned total comprises FFr48.62 billion (of which 25 per cent paid by industry) for development, FFr17.583 billion for industrialisation, FFr76.25 billion for 234 Rafale B/Cs, FFr20.89 billion for 60 Rafale Ms, FFr37.812 billion for spares and FFr1.215 billion for simulators. Early 1997 agreement on 10 per cent cost reduction resulted in flyaway price falling to FFr282 million for a Rafale C, FFr299 million for Rafale B and

FFr315 million for Rafale M. Total of FFr30 billion spent by 1995. Second production order for eight aircraft (1994/1995 authorisations) estimated at FFr1.5 billion, excluding engines, radar and weapons system. In 1998, however, new cost estimate for 294 aircraft was FFr320 billion as a consequence of programme delays. Cost (1999) of 48 aircraft given as FFr17.2 billion.

DESIGN FEATURES: Multirole combat aircraft, rivalling Eurofighter Typhoon. Minimum weight and volume structure to hold costs to minimum; thin, mid-mounted delta wing with moving canard; individual fixed, kidney-shaped intakes without shock cones. HOTAS controls, with sidestick controller on starboard console and small-travel throttle lever.

Wing leading-edge sweepback approximately 48°.

FLYING CONTROLS: Fully fly-by-wire controls with fully modulated two-section leading-edge slats and two elevons per wing; canard incidence automatically increased to 20° when landing gear lowered; airbrake panels in top of fuselage beside leading-edge of fin. Specification includes 30° AoA in stable flight.

STRUCTURE: Most of wing components made of carbon fibre including elevons; slats in titanium; wingroot and tip fairings Kevlar; canard made mainly by superplastic forming and diffusion bonding of titanium; fuselage 50 per cent carbon fibre; fuselage side skins of aluminium-lithium alloy; wheel and engine doors carbon fibre; fin made primarily of carbon fibre with aluminium honeycomb core in rudder. Composites account for 25 per cent by weight of structure and 20 per cent of surface area; weight saving directly attributable to composites is 300 kg (661 lb) – equivalent to a 1 tonne reduction in empty weight.

LANDING GEAR: Hydraulically retractable tricycle type supplied by Messier-Dowty, with single 790×275-15 (20 ply) or 790×275R15 mainwheels and twin, hydraulically steerable, 360×135-6 or 360×135R6 nosewheels. All wheels retract forward. Designed for impact at vertical speed of 3 m (10 ft)/s, or 6.5 m (21 ft)/s in naval version, without flare-out. Rafale M has same mainwheels; but 520×140R10.5 nosewheels. Messier-Bugatti carbon brakes on all three units, controlled by fly-by-wire system.

Rafale M has 'jump strut' nosewheel leg which releases energy stored in shock-absorber at end of deck take-off run, changing aircraft's attitude for climb-out without need for ski-jump ramp. 'Jump strut' advantage equivalent to 9 kt (16 km/h; 10 mph) or 900 kg (1,984 lb) extra weapon load; not to be used aboard carrier *Foch*, which to have 1° 30′ ramp giving 20 kt (37 km/h; 23 mph) or 2,000 kg (4,409 lb) advantage. Dowty Aerospace Yakima holdback fitting. Naval nosewheel steerable ±70°; or almost 360° under tow. Hydraulic (Rafale M) or tension-stored (Rafale B/C) arrester hook. Landing gear management (braking and steering) by Thales computer.

POWER PLANT: Two SNECMA M88-2 augmented turbofans, each rated at 48.7 kN (10,950 lb st) dry and 72.9 kN (16,400 lb st) with afterburning. Stage 1 standard engines in first 13 (F1) production Rafales; these limited to 300 hour hot-section TBO, but Stage 4, flight tested from late 2000, will achieve initial 600 hours and eventual 800 hours, being installed in F2 Rafales (15 Ms, 22 Bs and 11 Cs). M88-3 of 88.3 kN (19,840 lb st) maximum rating offered as follow-on, and being developed for export Rafale.

Internal tanks in single-seat versions for approximately 5,700 litres (1,506 US gallons; 1,254 Imp gallons) of fuel. Rafale B internal fuel confirmed as 5,300 litres (1,400 US gallons; 1,166 Imp gallons). Fuel system by Lucas Air Equipement, Lebozec and Zenith Aviation; equipment by

Intertechnique. Five 'wet' hardpoints: centreline, two inboard wing and two centre wing; all able to accommodate a 1,250 litre (330 US gallon; 275 Imp gallon) external tank; alternative 2,000 litre (528 US gallon; 440 Imp gallon) centreline and inboard; alternative 3,000 litre (793 US gallon; 660 Imp gallon) tank on centreline. Max external fuel capacity 9,500 litres (2,510 US gallons; 2,090 Imp gallons). Total internal and external fuel (single-seat versions) approximately 15,200 litres (4,015 US gallons; 3,344 Imp gallons); slightly less in two-seat version. Pressure refuelling in 7 minutes, or 4 minutes for internal tanks only. Fixed (detachable) in-flight refuelling probe on all versions.

Conformal (spine) tanks, length 7.50 m (24 ft 7¼ in), under development; total capacity 2,300 litres (608 US gallons; 506 Imp gallons), and able to accept in-flight replenishment.

ACCOMMODATION: Pilot only, on SEMMB (Martin-Baker) Mk 16 zero/zero ejection seat, reclined at angle of 29°. One-piece Sully Produits Spéciaux blister windscreen/canopy, hinged to open sideways to starboard. Canopy gold-coated to reduce radar reflection.

SYSTEMS: Technofan cockpit air conditioning system; Cryotechnologies avionics cooling system. Dual hydraulic circuits, pressure 350 bar (5,075 lb/sq in), each with two Messier-Bugatti pumps and Bronzavia ancillaries. Auxilec electrical system, with two 30/40 kVA Auxilec variable frequency alternators. Triplex digital plus one dual analogue fly-by-wire flight control system, integrated with engine controls and linked with weapons system. Air Liquide OBOGS; EROS oxygen system; L'Hotellier fire detection system; Microturbo APU.

AVIONICS: Provision for more than 780 kg (1,720 lb) of avionics equipment and racks.

Comms: EAS V/UHF and Thales TRA 6032 SATURN UHF radios; Thales TSB 2500 transponder/interrogator. TEAM intercom; Thales voice-activated radio controls and voice alarm warning system. BAE Systems aerials.

Radar: GIE Radar (Thales) RBE2 look-down/shoot-down radar, able to track up to eight targets simultaneously, with automatic threat assessment and allocation of priority.

Flight: Thales TLS-2020 integrated ILS/MLS, VOR/DME; SAGEM Sigma 95N (RL-90) RLG INS (SAGEM Telemir interface with carrier's navigation on Rafale M); Thales NC 12E Tacan interrogator; Thales AHV-17 radio altimeter and SFIM/Thales ESPAR static memory flight recorder. Thales GPS.

Instrumentation: Digital display of fuel, engine, hydraulic, electrical, oxygen and other systems information on two 127 × 127 mm (5 × 5 in) lateral multifunction touch-sensitive colour LCD displays by Thales. Third cockpit screen is 20 × 20° head-level tactical navigation/sensor display. Thales CTH3022 wide-angle, holographic HUD (30 × 22° field of view) incorporating Signaal USFA OTA-1320 CCD camera and recorder. Thales/Intertechnique Topsight E helmet-mounted sight.

Mission: Thales OSF electro-optical sensors. MIDS (Multifunctional Information Distribution System) datalink (equivalent to JTIDS/Link 16). Various reconnaissance, ECM, FLIR and laser designation pods, including Thales Damoclès target designator. Terrain-following system cleared (1999) to 152 m (500 ft); eventual 30 m (100 ft) over land and 15 m (50 ft) over water is envisaged.

Self-defence: Spectra radar warning and ECM suite by Thales and MBDA. Thales DAL (*Détecteur d'Alerte Laser*) system.

EQUIPMENT: Integral, electrically operated, folding ladder in Rafale M.

ARMAMENT: One 30 mm Giat DEFA 791B cannon in side of starboard engine duct (except naval two-seat). Fourteen external stores attachments: two on fuselage centreline, two beneath engine intakes, two astride rear fuselage, six under wings and two at wingtips; of these, five stressed for heavy stores anad fuel tanks. Forward centreline position deleted on Rafale M. Normal external load 6,000 kg (13,228 lb); maximum permissible, 9,500 kg (20,944 lb); see weapon options table. In strike role, one ASMP standoff nuclear weapon. In interception role, up to eight MICA AAMs (with IR or active homing) and two underwing fuel tanks; or six MICAs and three external fuel tanks. In air-to-ground role, typically sixteen 227 kg (500 lb) bombs, two MICAs and two 1,250 litre (330 US gallon; 275 Imp gallon) tanks; or two APACHE standoff weapon dispensers, two MICAs and three tanks; or FLIR pod, Atlis laser designator pod, two 1,000 kg (2,205 lb) laser-guided bombs, two AS.30L laser ASMs, four MICAs and single tank. In anti-ship role, two Exocet sea-skimming missiles, four MICAs and two external fuel tanks. Future weapons will include AASM (*Armement Air-Sol Modulaire*) powered LGB.

DIMENSIONS, EXTERNAL:
Wing span, incl wingtip missiles	10.80 m (35 ft 5¼ in)
Wing aspect ratio	2.6
Length overall	15.27 m (50 ft 1¼ in)
Height overall (Rafale D)	5.34 m (17 ft 6¼ in)

AREAS:
Wings, gross	45.70 m² (491.9 sq ft)

WEIGHTS AND LOADINGS (estimated):
Basic weight empty, equipped:	
Rafale C	9,850 kg (21,716 lb)
Rafale B	10,450 kg (23,038 lb)
Rafale M	10,460 kg (23,060 lb)
Rafale N	10,710 kg (23,611 lb)
External load (incl fuel): normal	6,000 kg (13,228 lb)
max	9,500 kg (20,944 lb)
Max fuel weight: internal:	
single seat	4,500 kg (9,921 lb)
two seat: Rafale B	4,350 kg (9,590 lb)
Rafale N	4,240 kg (9,348 lb)
underwing	7,500 kg (16,535 lb)
conformal	1,850 kg (4,079 lb)
Max T-O weight: early production	19,500 kg (42,990 lb)
subsequent production	22,500 kg (49,604 lb)
developed version	24,500 kg (54,013 lb)
Max landing weight	22,500 kg (49,605 lb)
Max wing loading:	
initial version	426.7 kg/m² (87.39 lb/sq ft)
developed version	536.1 kg/m² (109.80 lb/sq ft)
Max power loading:	
initial version	134 kg/kN (1.31 lb/lb st)
developed version (M88-3)	141 kg/kN (1.38 lb/lb st)

PERFORMANCE (estimated):
Max level speed: at altitude	M1.8
at low level	750 kt (1,390 km/h; 864 mph)
Approach speed	120 kt (223 km/h; 139 mph)
Max rate of climb at S/L	approx 18,290 m (60,000 ft)/min
Roll rate	270°/s
Max instantaneous turn rate	up to 30°/s
Service ceiling	16,765 m (55,000 ft)
T-O distance: air defence	400 m (1,315 ft)
attack	600 m (1,970 ft)
Landing distance	450 m (1,480 ft)

Radius of action: low-level penetration with 12 × 250 kg bombs, four MICA AAMs and 4,000 litres (1,056 US gallons; 880 Imp gallons) of external fuel in three tanks   570 n miles (1,055 km; 655 miles)
air-to-air, long-range with eight MICA AAMs and 6,000 litres (1,585 US gallons; 1,320 Imp gallons) of external fuel in four tanks, 12,200 m (40,000 ft)
transit   950 n miles (1,759 km; 1,093 miles)
Operational loiter	up to 3 h
g limits	+9.0/−3.2

UPDATED

DASSAULT ATLANTIC 3

In late 2001, Dassault finally abandoned promotion of the ATL3 maritime patrol twin-turboprop, having been unable to make an economic case for relaunching production to meet a potential 20-24 aircraft requirement under consideration by Germany and Italy.

UPDATED

DASSAULT FALCON 7X

TYPE: Long-range business tri-jet.

PROGRAMME: Announced, under temporary name/designation Falcon Next/NXT, at the Paris Air Show in June 2001; then became Falcon FNX; formal designation Falcon 7X announced 29 October 2001. First flight scheduled for early 2005; certification scheduled for mid-2006; first customer deliveries anticipated in late-2006.

CUSTOMERS: Market estimated at 400 aircraft over unspecified period. Letters of intent for 41 aircraft held by October 2001, split approximately evenly between US and other customers.

Model of Falcon 7X long-range business tri-jet   *NEW*/0113486

COSTS: Development cost estimated at US$600 million to US$700 million. Unit cost US$35 million to US$36 million (all 2001).

DESIGN FEATURES: Similar in configuration to Falcon 900C/900EX, but with cabin some 20 per cent longer, redesigned nose, double-curved windscreen panels and an entirely new high-subsonic section wing with 20 per cent fewer parts, 5° more sweepback, (34° on inboard section, 30° on outboard section), 40 per cent more area and featuring full-span, two-section leading-edge slats. The wing design will be employed on future developments within the Falcon range, expected to be designated Falcon 5X and 9X.

FLYING CONTROLS: Fly-by-wire controls, possibly with sidestick controllers.

STRUCTURE: Programme partner suppliers include: CASA (horizontal stabiliser); Fokker (trailing-edge control surfaces); Hurel Hispano/Aermacchi (engine nacelles and thrust reversers); Latécoère (T5 fuselage section); Socata (T34 upper fuselage section and body fairing); Sonaca (wing leading-edges); and Sully Saint Gobain (windscreen and cabin windows). Final assembly in new 25,000 m² (269,097 sq ft) facility at Bordeaux.

POWER PLANT: Three Pratt & Whitney Canada PW307A turbofans, each flat-rated to 27.1 kN (6,100 lb st) at ISA +18°C.

LANDING GEAR: Retractable tricycle type by Messier-Dowty. Wheels and brakes by ABSC.

ACCOMMODATION: Typical configuration will provide three lounge areas, lavatories, galleys, crew rest area and a large flight-accessible baggage compartment. Pressurisation system will provide a 1,830 m (6,000 ft) cabin environment at high cruise altitudes.

SYSTEMS: Honeywell air management system. Pressurisation system maintains 1,830 m (6,000 ft) cabin environment at maximum operating altitude. Honeywell 35-150(FN) APU. Honeywell/Parker hydraulic power generation system. Parker hydraulic system; Intertechnique oxygen system; L'Hotelier fire detection and extinguishing; TRW electrical generation and distribution.

AVIONICS: Dassault EASy flight deck as core system, incorporating Honeywell Primus Epic platform.

DIMENSIONS, EXTERNAL:
Wing span	25.13 m (82 ft 5½ in)
Length overall	23.19 m (76 ft 1 in)
Height overall	7.77 m (25 ft 6 in)

WEIGHTS AND LOADINGS (estimated):
Max T-O weight	28,803 kg (63,500 lb)

PERFORMANCE (estimated):
Max operating Mach No. (M_{MO})	0.90
Max operating speed (V_{MO})	370 kt (685 km/h; 425 mph) IAS
Landing speed	104 kt (193 km/h; 120 mph)

Range: max   5,700 n miles (10,556 km; 6,559 miles)
with eight passengers, NBAA IFR reserves:
at M0.87	4,600 n miles (8,519 km; 5.293 miles)
at M0.85	5,100 n miles (9,445 km; 5,869 miles)

NEW ENTRY

DASSAULT FALCON 50

Spanish Air Force designation: T.16

TYPE: Business tri-jet.

PROGRAMME: First flight of prototype Falcon 50 (F-WAMD) 7 November 1976; second prototype 18 February 1978; first (and only) preproduction 13 June 1978. French certification 27 February 1979; FAA certification 7 March

Typical Dassault Falcon 50EX interior
(Bud Shannon/Dassault)   0110857

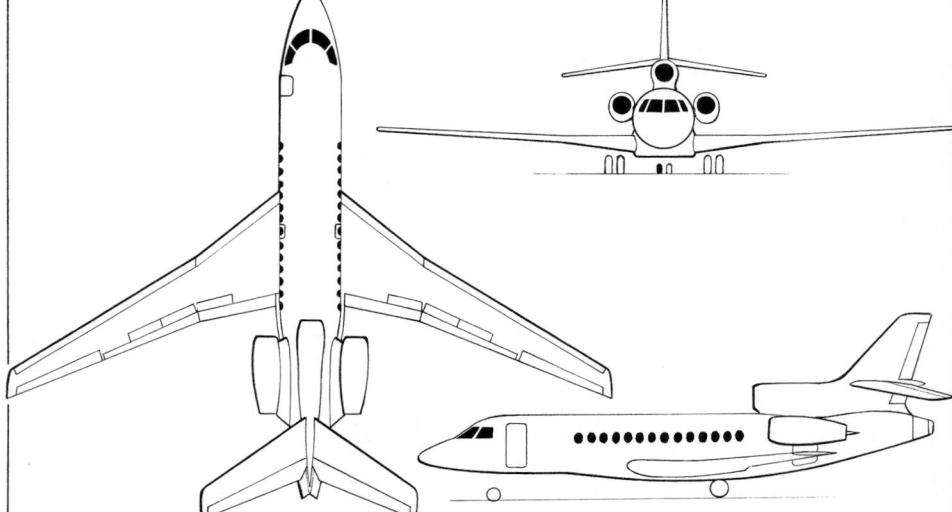

Provisional configuration of Dassault Falcon 7X *(Jane's/James Goulding)*   0121273

Dassault Falcon 50EX long-range business jet   0110855

1979; deliveries began July 1979. Available in sigint version from 1994.

Falcon 50EX announced 26 April 1995. Long-range variant; uprated turbofans provide a 7 per cent improvement in fuel consumption and improved initial cruising altitude of 12,500 m (41,000 ft). Production of initial batch of 40 Falcon 50EX airframes began 1995. First flight (F-WOND, c/n 251) 10 April 1996; new avionics suite installed in Falcon 50 c/n 252 for integration flight testing leading to avionics certification by FAA on 15 July 1996; DGAC certification achieved 15 November 1996, followed by FAA approval 20 December; initial production aircraft (F-WWHA, c/n 253) first flight October 1996; flown in 'green' condition Le Bourget, Paris, to Teterboro, New Jersey, 16 November, thence to Orlando, Florida, for US debut at National Business Aircraft Association Convention; first customer delivery (to Volkswagen of Germany) February 1997. Estimated market for 150 to 200 of EX version over unspecified period.

CURRENT VERSIONS: **Falcon 50:** Previous version, now superseded by Falcon 50EX.

Falcon 50-40: Retrofit of earlier version with Honeywell TFE731-40 engines, each 18.91 kN (4,250 lb). See *Jane's Aircraft Upgrades.*

Falcon 50EX: *As described.*

Falcon 50M Surmar: Maritime surveillance version. Order for French Navy announced 12 November 1996 covering four aircraft (compared with five in original June 1995 statement) at estimated total cost of FFr750 million. Sensors include a Thomson-CSF Ocean Master 100 search radar and Thomson-TTD Chlio FLIR. Three mission specialist/console stations, two at front of cabin, one at rear; airdrop door; provision for carriage of up to eight 25-person airdroppable liferafts; two observation windows. Capabilities include 4 hours on station at 400 n miles (740 km; 460 miles) from base, cruising at 200 kt (370 km/h; 230 mph) at 915 m (3,000 ft).

First flight (No. 36/F-ZWTA) November 1998; first delivery in January 2000; operational with Flottille 24 at Lann-Bihoué from September 2000. Fourth delivery in early 2002.

Falcon 50 Sigint: Model of a potential signals intelligence Falcon 50 displayed at Dubai Air Show in November 1997.

CUSTOMERS: Total 320 Falcon 50s and 50EXs delivered for outfitting by 1 January 2002, including 18 in 2000 and 13 (all 50EXs) in 2001. Adopted by governments of Burundi, Djibouti, France, Iraq (later Iran), Italy, Jordan, Libya, Morocco, Portugal, Rwanda, South Africa, Spain, Sudan and Yugoslavia; three of Italian Air Force convertible for medevac.

COSTS: US$18.23 million (2001).

DESIGN FEATURES: Three-engine layout permits overflight of oceans and desert areas within public transport regulations. Sharply waisted rear fuselage and engine pod designed by computational fluid dynamics; wing has compound leading-edge sweep (24° 50′ to 29° at quarter-chord) and optimised section.

FLYING CONTROLS: Fully powered controls with pushrods, dual-barrel hydraulic actuators and artificial feel; variable incidence anhedral tailplane with dual electrical actuation by screwjack; drooped leading-edge inboard and slats outboard; double-slotted flaps; three two-position airbrake/spoiler panels on each wing.

STRUCTURE: All-metal, circular-section fuselage with rear baggage compartment inside pressure cabin; wing boxes are integral fuel tanks bolted to carry-through box. Carbon fibre horizontal tail surfaces introduced during 2001. Latécoère, Potez, Reims Aviation, SEFCA and Socata are subcontractors on the programme; other suppliers include Dassault Equipements (flight controls, flaps, slats and airbrakes), Liebherr-Aerospace Toulouse (engine bleed air

system); SARMA (flight control actuating rods) and Sully Produits Spéciaux (windscreen panels).

LANDING GEAR: Retractable tricycle type by Messier-Dowty, with twin wheels on each unit. Hydraulic retraction, main units inward, nosewheels forward. Nosewheels steerable ±60° for taxying, ±180° for towing. ABS wheels, brakes and braking system. Mainwheel tyres size 26×6.6 (14 ply) or 26×6.6R14 tubeless, pressure 14.34 bar (208 lb/sq in). Nosewheel tyres size 14.5×5.5-6 (14 ply) tubeless, pressure 8.96 bar (130 lb/sq in). Four-disc brakes designed for 400 landings with normal energy braking. Minimum ground turning radius (about nosewheels) 13.54 m (44 ft 5 in).

POWER PLANT: Three Honeywell TFE731-40 turbofans, each rated at 16.46 kN (3,700 lb st) at ISA + 17°C in Falcon 50EX. Two engines pod-mounted on sides of rear fuselage, third attached by two top mounts. Thrust reverser on centre engine. Fuel in integral tanks, with capacity of 5,787 litres (1,529 US gallons; 1,273 Imp gallons) in wings and 2,976 litres (786 US gallons; 655 Imp gallons) in fuselage tanks. Total fuel capacity 8,763 litres (2,315 US gallons; 1,928 Imp gallons). Single-point pressure fuelling. Intertechnique fuel distribution and gauges.

ACCOMMODATION: Standard accommodation for two crew and nine passengers. In typical arrangement, cabin is divided into a forward section with four armchairs and two fold-out tables, and a rear section with three-seat sofa (convertible into a single bed) and two armchairs separated by a fold-out table; two galleys at forward end of cabin, toilet at rear. Alternative layouts to customer choice, with accommodation for a maximum of 19 passengers. Cabin and rear baggage compartment are pressurised and air conditioned; Barber-Colman temperature controls. Access is by separate door on port side.

SYSTEMS: Air conditioning system utilises bleed air from all three engines or APU. Maximum pressure differential 0.61

bar (8.8 lb/sq in). Pressurisation maintains a maximum cabin altitude of 2,440 m (8,000 ft) to a flight altitude of 14,935 m (49,000 ft). Two independent Messier-Bugatti and Vickers-Sterer hydraulic systems, pressure 207 bar (3,000 lb/sq in), with three engine-driven pumps and one emergency electric pump, actuate primary flying controls, flaps, slats, landing gear, wheel brakes, airbrakes and nosewheel steering. Plain reservoir, pressurised by bleed air at 1.47 bar (21 lb/sq in). 28 V DC electrical system, with a 9 kW 28 V DC Auxilec starter/generator on each engine and two 23 Ah batteries. Labinal electrical harnesses. Wing leading-edge, centre engine S-duct and engine nacelles have engine bleed air anti-icing. Automatic emergency oxygen system. Honeywell 36-150 APU standard.

AVIONICS: *Comms:* Dual Rockwell Collins VHF and mode S transponders; dual Rockwell Collins HF-9000 HF transceivers with Selcal; Teledyne Controls MagnaStar or Honeywell Flitefone 800 radiotelephones; cockpit voice recorder and ELT standard.

Radar: Rockwell Collins TWR 850 Doppler turbulence detection weather radar.

Flight: Dual Rockwell Collins VIR-432 VOR, ADF-462 and DME-442. Rockwell Collins APS-4000 autopilot, ADC-850C air data systems, dual Honeywell FHS-6100 with integrated GPS receiver, Honeywell EGPWS and dual Honeywell Laseref III laser gyro inertial reference systems standard. Dual Universal UNS-1C FMS optional, replacing GNS-XES.

Instrumentation: Rockwell Collins Pro Line 4 (EFIS-4000) four-tube EFIS; Thales three-tube LCD engine indicating electronic display (EIED).

DIMENSIONS, EXTERNAL:
Wing span	18.86 m (61 ft 10½ in)
Wing chord (mean)	2.84 m (9 ft 3¾ in)
Wing aspect ratio	7.6
Length: overall	18.52 m (60 ft 9¼ in)
fuselage	17.66 m (57 ft 11 in)
Height overall	6.98 m (22 ft 10¾ in)
Tailplane span	7.74 m (25 ft 4¾ in)
Wheel track	3.98 m (13 ft 0¾ in)
Wheelbase	7.24 m (23 ft 9 in)
Passenger door: Height	1.52 m (4 ft 11¾ in)
Width	0.80 m (2 ft 7½ in)
Height to sill	1.30 m (4 ft 3¼ in)
Emergency exits (each side, over wing):	
Height	0.92 m (3 ft 0¼ in)
Width	0.51 m (1 ft 8 in)
Baggage door: Height	0.73 m (2 ft 4¾ in)
Width	0.99 m (3 ft 3 in)

DIMENSIONS, INTERNAL:
Cabin, incl forward baggage space and rear toilet:	
Length	7.16 m (23 ft 6 in)
Max width	1.86 m (6 ft 1¼ in)
Max height	1.80 m (5 ft 10¾ in)
Volume	20.2 m³ (712 cu ft)
Baggage space	0.71 m³ (25.0 cu ft)
Baggage compartment (rear)	3.26 m³ (115 cu ft)

AREAS:
Wings, gross	46.83 m² (504.1 sq ft)
Horizontal tail surfaces (total)	13.35 m² (143.69 sq ft)
Vertical tail surfaces (total)	9.82 m² (105.70 sq ft)

WEIGHTS AND LOADINGS:
Weight empty, equipped	9,603 kg (21,170 lb)

Falcon 50EX flight deck *(Bud Shannon/Dassault)*   0110856

Basic operating weight	9,888 kg (21,800 lb)
Baggage capacity (rear)	1,000 kg (2,205 lb)
Max payload: normal	1,710 kg (3,770 lb)
with max fuel	1,170 kg (2,579 lb)
Max fuel	7,040 kg (15,520 lb)
Max T-O weight: standard	18,007 kg (39,700 lb)
optional	18,497 kg (40,780 lb)
Max ramp weight: standard	18,098 kg (39,900 lb)
optional	18,497 kg (40,780 lb)
Max zero-fuel weight	11,600 kg (25,570 lb)
Max landing weight	16,200 kg (35,715 lb)
Max wing loading	384.5 kg/m² (78.75 lb/sq ft)
Max power loading	365 kg/kN (3.58 lb/lb st)

PERFORMANCE:

Max operating Mach No. (M$_{MO}$)	0.86
Max operating speed (V$_{MO}$):	
at S/L	350 kt (648 km/h; 402 mph) IAS
at 7,225 m (23,700 ft)	
	370 kt (685 km/h; 425 mph) IAS
Max cruising speed	
	M0.85 or 487 kt (902 km/h; 560 mph)
Normal cruising speed	
	M0.80 or 459 kt (850 km/h; 528 mph)
Long-range cruising speed at 10,670 m (35,000 ft)	
	M0.75 (430 kt; 797 km/h; 495 mph)
Approach speed, eight passengers and NBAA IFR	
reserves	106 kt (196 km/h; 122 mph)
Initial cruising altitude	12,497 m (41,000 ft)
Max certified altitude	14,935 m (49,000 ft)
FAR Pt 25 balanced T-O field length, S/L, ISA with eight	
passengers and max fuel	1,490 m (4,890 ft)
FAR Pt 91 landing distance at MLW with eight	
passengers and NBAA IFR reserves:	666 m (2,185 ft)
Range with eight passengers and NBAA IFR reserves:	
at M0.80	3,075 n miles (5,695 km; 3,538 miles)
at M0.75	3,285 n miles (6,083 km; 3,780 miles)

OPERATIONAL NOISE LEVELS:

T-O	83.8 EPNdB
Approach	95.2 EPNdB
Sideline	91.9 EPNdB

UPDATED

DASSAULT FALCON 900
Spanish Air Force designation: T.18

TYPE: Long-range business tri-jet.

PROGRAMME: Falcon 900 announced 27 May 1983; first flight of prototype (F-GIDE *Spirit of Lafayette*) 21 September 1984, second aircraft (F-GFJC) 30 August 1985; flew non-stop 4,305 n miles (7,973 km; 4,954 miles) Paris to Little Rock, Arkansas, September 1985; returned Teterboro, New Jersey, to Istres, France, at M0.84; French and US certification March 1986, including status close to FAR Pts 25 and 55 for damage tolerance of entire airframe.

Prototype Falcon 900 in use as testbed (first flight 12 April 1994) for new laminar flow wing section intended to provide significant reductions in drag. New section installed as sleeve on inner wing and designed to demonstrate hybrid laminar flow with boundary layer suction via seven channels in laser-drilled titanium skin over some 10 per cent chord of upper wing surface. Following test programme, modified aircraft has returned to Dassault Falcon Service to validate laminar flow capability under normal commercial operating conditions. The 250th Falcon 900 series aircraft was delivered in mid-2000.

CURRENT VERSIONS: **Falcon 900B:** French and UK certification received end 1991; complies with FAR Pt 36 Stage III and ICAO 16 noise requirements; approved for Cat. II approaches, for operations from unpaved fields; re-engined with TFE731-5BR-1C turbofans, to give 5.5 per cent power increase; initial cruising altitude 11,855 m

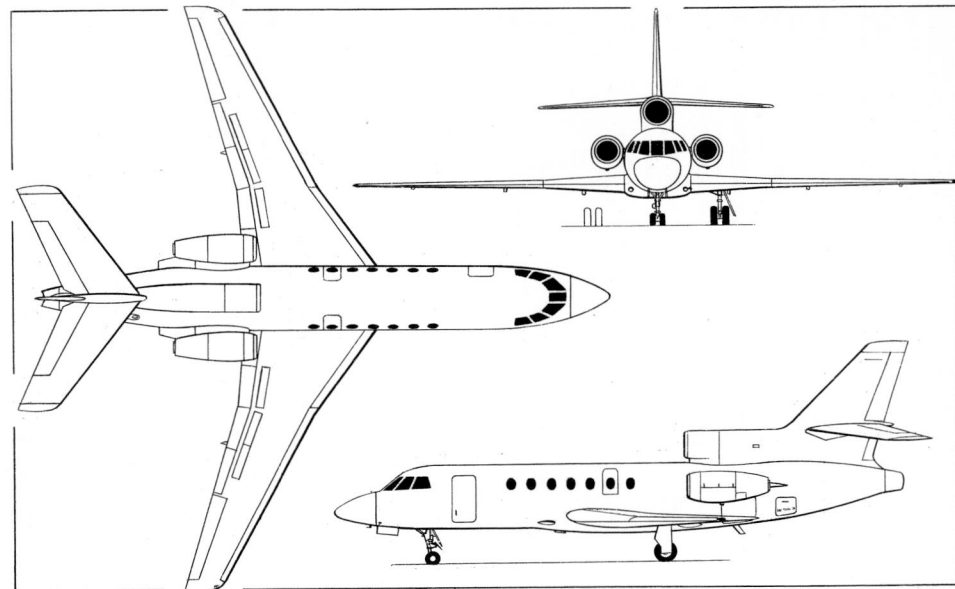

Dassault Falcon 50 long-range three-turbofan business transport (*Jane's/Dennis Punnett*)

(39,000 ft) and NBAA IFR range increased by 100 n miles (185 km; 115 miles); retrofit offered to existing operators. Supplanted by Falcon 900C.

Falcon 900C: Announced 26 June 1998. Combines airframe, engines and cabin of 900B with Honeywell Primus 2000 avionics of 900EX, but without autothrottles; 900B F-WWFP/F-GRDP (c/n 169) served as prototype for certification; first flown 17 December 1998; deliveries began in December 1999 (c/n 180 to Sony Aviation Corp in USA), replacing 900B on production line. French DGAC certification achieved 15 June 1999, with FAA certification following on 26 August 1999.

Detailed description applies to Falcon 900C, except where indicated.

Falcon 900EX: Long-range development of 900B, announced October 1994. Re-engined with 22.24 kN (5,000 lb st) (ISA+17°C) Honeywell TFE731-60 turbofans, to give 5.8 per cent increase in retained thrust at 12,200 m (40,000 ft) and more than 8 per cent improvement in cruise specific fuel consumption. Engine nacelles, pylons, thrust reversers and portions of centre engine S-duct redesigned; maximum fuel capacity increased to 11,865 litres (3,134 US gallons; 2,610 Imp gallons) by increasing capacity of centre-fuselage tank to 591 kg (1,303 lb) and addition of tank in rear fuselage, capacity 240 kg (530 lb).

Upgraded standard avionics comprise fully integrated Honeywell Primus 2000 suite with five-tube 20 × 17.75 cm (8 × 7 in) colour EFIS; one engine instrument display; three IC-800 integrated avionics computers; dual FMZ-2000 flight management systems with a third optional; dual fail-operational autopilots; T-O to landing autothrottle; Honeywell EGPWS; dual Laseref III inertial reference systems with third optional; Primus colour weather radar; optional single or dual 12-channel GPS, multichannel satcom, communications management unit (CMU) and Flight Dynamics HGS-2850 head-up display. Dassault EASy integrated flight deck with four 330 × 254 mm (13 × 10 in) active matrix liquid crystal displays (AMLCDs) to be introduced on 900EX from 2003. First flight with EASy cockpit was made by F-WNCD (c/n 97) on 22 February 2002.

Risk-sharing partners, representing 20 per cent of total development investment, are Honeywell (engines and primary avionics), SABCA (centre engine intake cowlings), Hellenic Aircraft Industries (rear fuselage fuel tank), Latécoère (T5 fuselage section and engine pylons), and Alenia (nacelles and centre engine thrust reverser).

Prototype (F-WREX) rolled out 13 March 1995; first flight 1 June 1995; flew Luton, England, to Las Vegas, Nevada, non-stop on 24 September 1995, completing the 4,700 n mile (8,704 km; 5,409 mile) flight in 11 hours 40 minutes including 30 minutes hold for air traffic delays; DGAC certification 31 May 1996; FAA approval granted 19 July after 350-hour flight test programme; first customer delivery to Anheuser-Busch Companies Inc (N200L) 1 November 1996; production aircraft delivered to Little Rock, Arkansas, for outfitting.

Japan MSA: Two Falcon 900s for long-range maritime surveillance entered service with the Japan Maritime Safety Agency September 1989; US search radar, special communications radio, operations control station, U-125A-style search windows and drop hatch for sonobuoys, markers and flares.

CUSTOMERS: Total 194 Falcon 900s and 103 Falcon 900EXs delivered to completion centres by 1 January 2002. Government/VIP versions operated by Algeria, Australia, Belgium, Equatorial Guinea, France, Gabon, Italy, Malaysia, Nigeria, Russia, Saudi Arabia, Spain, Syria and United Arab Emirates. Production totalled 20 in 1998, 24 in 1999, 29 (six 900C and 23 900EX) in 2000, and 27 (six 900C and 21 900EC) in 2001.

COSTS: Standard equipped 900C US$27.81 million (2001); 900EX US$31.19 million (2001).

DESIGN FEATURES: Larger cross-section and cabin length than Falcon 50; added economy and further power increase of engines achieved by mixer compound nozzle tailpipe, mixing cold and hot flows.

Wing adapted from Falcon 50 but increased span and area, and optimised for M0.84 cruise; compound leading-edge sweep (24° 50′ to 29° at quarter-chord); dihedral 0° 30′.

FLYING CONTROLS: Fully powered flying controls with artificial feel and variable-incidence tailplane as for Falcon

Typical cabin layout of a Falcon 900EX (*Bud Shannon/Dassault*)   0110859

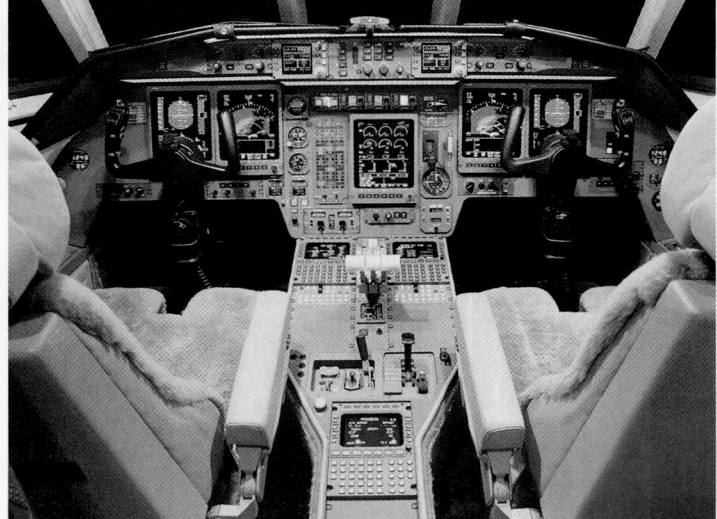

Flight deck of Falcon 900EX (*Bud Shannon/Dassault*)   0110860

Dassault Falcon 900C business jet *(F Robineau/Dassault)*

NEW/0113478

50; full-span slats and double-slotted Fowler flaps; three-position airbrakes.

STRUCTURE: Design and manufacture computer assisted; damage-tolerant structure; extensive use of carbon fibre and aramid (Kevlar); Kevlar radome, wingroot fairings and tailcone; secondary rear cabin pressure bulkhead allows access to baggage in flight and additional protection against pressure loss. Nosewheel doors of Kevlar; mainwheel doors of carbon fibre. Kevlar air intake trunk for centre engine, and rear cowling for side engines. Carbon fibre central cowling around all three engines.

New horizontal tail surface featuring cast titanium central box with resin transfer moulding composite spars and carbon fibre skin panels, resulting in a 13.6 kg (30 lb) reduction in weight, certified by the DGAC in December 1999 and introduced as standard on Falcon 900C and 900EX from December 2000 deliveries.

LANDING GEAR: Retractable tricycle type by Messier-Bugatti, with twin wheels on each unit. Hydraulic retraction, main units inward, nosewheels forward. Oleo-pneumatic shock-absorbers. Mainwheels fitted with Michelin radial tyres size 29×7.7-15, pressure 13.30 bar (193 lb/sq in). Nosewheel tyres size 17.5×5.75R8, pressure 9.80 bar (142 lb/sq in). Hydraulic nosewheel steering (±60° for taxiing, ±180° for towing). Messier-Bugatti triple-disc carbon brakes and anti-skid system. Minimum ground turning radius (about nosewheels) 13.54 m (44 ft 5 in).

POWER PLANT: Three Honeywell TFE731-5BR-1C turbofans, each rated at 21.13 kN (4,750 lb st) at ISA + 10°C. Thrust reverser on centre engine. Fuel in two integral tanks in wings, centre-section tank, and two tanks under floor of forward and rear fuselage. Total fuel capacity 10,825 litres (2,860 US gallons; 2,381 Imp gallons).

ACCOMMODATION: Type III emergency exit on starboard side of cabin permits wide range of layouts for up to 19 passengers. Flight deck for two pilots, with central jump-seat. Flight deck separated from cabin by door, with crew wardrobe and baggage locker on either side. Galley at front of main cabin, on starboard side opposite main cabin door. Passenger area is divided into three lounges. Forward zone has four 'sleeping' swivel chairs in facing pairs with tables. Centre zone is dining area, with two double seats facing a transverse table. On starboard side, storage cabinet contains foldaway bench, allowing five to six persons to be seated around table, while leaving emergency exit clear. In rear zone, inward-facing three-seat settee on starboard side converts into a bed. On port side, two armchairs are separated by a table. At rear of cabin, a door leads to toilet compartment, on starboard side, and a second structural plug door to large rear baggage area. Baggage door is electrically actuated.

Other interior configurations available. Alternative eight-passenger configuration has bedroom at rear and three personnel seats in forward zone. A 15-passenger layout divides a VIP area at rear from six (three-abreast) chairs forward; full fuel can still be carried with 15 passengers. The 18-passenger scheme has four rows of three-abreast airline-type seats forward, and VIP lounge with two chairs and settee aft. Many optional items, including additional windows, front toilet unit, video system with one or more monitors, 'Airshow 200' navigation display system, compact disc deck, aft cabin partition, one or two couches in aft cabin convertible to bed (s), storage cabinet in baggage hold, aft longitudinal table, individual listening devices for passengers, lifejackets and rafts.

SYSTEMS: Air conditioning system uses engine bleed air or air from Honeywell GTCP36-150 APU installed in rear fuselage. Softair pressurisation system, with maximum differential of 0.64 bar (9.3 lb/sq in), maintains sea level cabin environment to height of 7,620 m (25,000 ft), and cabin equivalent of 2,440 m (8,000 ft) at 15,550 m (51,000 ft). Cold air supply by single oversize air cycle unit. Two independent hydraulic systems, pressure 207 bar (3,000 lb/sq in), with three engine-driven pumps and one emergency electric pump, actuate primary flying controls, flaps, slats, landing gear retraction, wheel brakes, airbrakes, nosewheel steering and thrust reverser. Bootstrap hydraulic reservoirs. DC electrical system supplied by three 9 kW 28 V Auxilec starter/generators and two 23 Ah batteries. Heated bleed air anti-icing of wing leading-edges, intakes and centre engine duct; electrically heated windscreens. Eros (SFIM/Intertechnique) oxygen system.

AVIONICS: *Comms:* Honeywell Primus 2000 as core system.
Radar: Honeywell Primus 870 colour weather radar.
Flight: Dual autopilot. Honeywell FMZ-2000 FMS with two AZ-840 micro air data systems and two Laseref III LINS; Rockwell Collins dual VIR-432 VOR/ILS marker receiver, dual ADF-462 and DME-442; Honeywell AA-300 radar altimeter and IC-800 autopilot; and Honeywell EGPWS.
Instrumentation: Honeywell Primus 2000 five-tube EFIS, comprising two MFDs, two PFDS and one EIED, each measuring 203 × 178 mm (8 × 7 in). Flight Dynamics HGS-2850 head-up guidance system optional.

DIMENSIONS, EXTERNAL:
Wing span	19.33 m (63 ft 5 in)
Wing chord: at root	4.08 m (13 ft 4¾ in)
at tip	1.12 m (3 ft 8 in)
Wing aspect ratio	7.6
Length overall	20.21 m (66 ft 3¾ in)
Fuselage: Max diameter	2.50 m (8 ft 2½ in)
Height overall	7.55 m (24 ft 9¼ in)
Tailplane span	7.74 m (25 ft 4¾ in)
Wheel track	4.45 m (14 ft 7¼ in)
Wheelbase	7.90 m (25 ft 11 in)
Passenger door: Height	1.72 m (5 ft 7¾ in)
Width	0.80 m (2 ft 7½ in)
Height to sill	1.64 m (5 ft 4½ in)
Emergency exit (overwing, stbd):	
Height	0.92 m (3 ft 0¼ in)
Width	0.53 m (1 ft 8¾ in)
Baggage door:	
Height	0.75 m (2 ft 5½ in)
Width	0.95 m (3 ft 1½ in)

DIMENSIONS, INTERNAL:
Cabin, excl flight deck, incl toilet and baggage compartments: Length	10.11 m (33 ft 2 in)
Max width	2.34 m (7 ft 8¼ in)
Width at floor	1.91 m (6 ft 3¼ in)
Max height	1.88 m (6 ft 2 in)
Volume	35.8 m³ (1,264 cu ft)
Rear baggage compartment volume	3.6 m³ (127 cu ft)
Flight deck volume	3.8 m³ (132 cu ft)

AREAS:
Wings, gross	49.00 m² (527.4 sq ft)
Vertical tail surfaces (total)	9.82 m² (105.7 sq ft)
Horizontal tail surfaces (total)	13.35 m² (143.7 sq ft)

WEIGHTS AND LOADINGS:
Weight empty, equipped (typical):	
900EX	10,829 kg (23,875 lb)
Basic operating weight:	
900C	10,977 kg (24,200 lb)
900EX	11,204 kg (24,700 lb)
Max payload:	
900C	1,969 kg (4,341 lb)
900EX	2,796 kg (6,164 lb)
Payload with max fuel:	
900C	1,204 kg (2,654 lb)
900EX	1,270 kg (2,800 lb)
Max fuel:	
900C	8,693 kg (19,165 lb)
900EX	9,525 kg (21,000 lb)

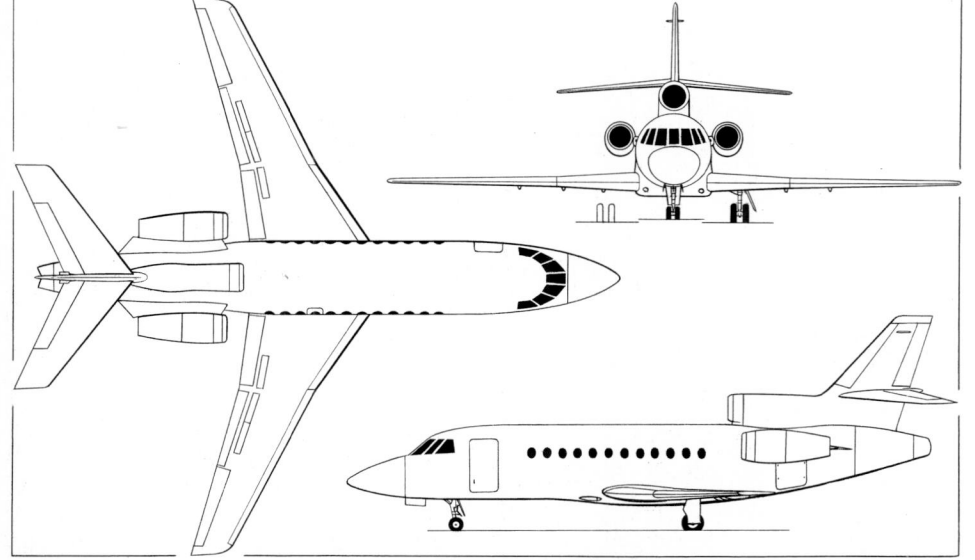

Dassault Falcon 900C (three Honeywell TFE731 turbofans) *(Jane's/Dennis Punnett)*

Current instrument panel of Falcon 2000   0110862

Dassault EASy flight deck for Falcon 2000EX   0110863

Max ramp weight:

900C standard	20,729 kg (45,700 lb)
900C optional	21,183 kg (46,700 lb)
900EX standard	22,000 kg (48,500 lb)
900EX optional	22,317 kg (49,200 lb)

Max T-O weight:

900C standard	20,640 kg (45,500 lb)
900C optional	21,092 kg (46,500 lb)
900EX standard	21,909 kg (48,300 lb)
900EX optional	22,226 kg (49,000 lb)

Max landing weight:

900C, 900EX standard	19,050 kg (42,000 lb)
900EX optional	20,185 kg (44,500 lb)

Normal landing weight, eight passengers and fuel reserves:

900C	12,494 kg (27,545 lb)
900EX	12,846 kg (28,321 lb)

Max zero-fuel weight: 900C 12,800 kg (28,220 lb)
900EX 14,000 kg (30,865 lb)

Max wing loading:

900C standard	421.2 kg/m² (86.27 lb/sq ft)
900C optional	430.5 kg/m² (88.17 lb/sq ft)
900EX standard	447.1 kg/m² (91.58 lb/sq ft)
900EX optional	453.6 kg/m² (92.91 lb/sq ft)

Max power loading:

900C standard	326 kg/kN (3.19 lb/lb st)
900C optional	333 kg/kN (3.26 lb/lb st)
900EX standard	328 kg/kN (3.22 lb/lb st)
900EX optional	333 kg/kN (3.27 lb/lb st)

PERFORMANCE (at AUW of 12,250 kg; 27,000 lb, except where indicated):

Max operating speed (V_{MO}): 900C, 900EX:
 at S/L   M0.87 (350 kt; 648 km/h; 403 mph IAS)
 between 3,050-7,620 m (10,000-25,000 ft)
          M0.84 (370 kt; 685 km/h; 425 mph IAS)

Max cruising speed:
 900C, 900EX   M0.84 or 481 kt (891 km/h; 554 mph)

Normal cruising speed:
 900C, 900EX   M0.80 or 459 kt (850 km/h; 528 mph)

Long-range cruising speed:
 900C, 900EX   M0.75 or 430 kt (796 km/h; 495 mph)

Approach speed, eight passengers and NBAA IFR fuel reserves:

900C	107 kt (199 km/h; 124 mph)
900EX	109 kt (202 km/h; 126 mph)

Stalling speed:
 900C, 900EX, clean   106 kt (196 km/h; 122 mph)
 900C, 900EX, landing configuration
                85 kt (158 km/h; 98 mph)

Initial cruising altitude:
 900C, 900EX   11,885 m (39,000 ft)

Max certified altitude:
 900C, 900EX   15,550 m (51,000 ft)

FAR Pt 25 balanced T-O field length at standard max T-O weight:

900C	1,504 m (4,935 ft)
900EX	1,590 m (5,215 ft)

FAR Pt 91 landing distance:
 with NBAA IFR reserves:

900C with five passengers	707 m (2,320 ft)
900EX with eight passengers	724 m (2,375 ft)

Range with NBAA IFR reserves:
 900C with five passengers at M0.75
                4,000 n miles (7,408 km; 4,603 miles)
 900EX with eight passengers:

at M0.84	3,810 n miles (7,056 km; 4,384 miles)
at M0.80	4,335 n miles (8,028 km; 4,988 miles)
at M0.75	4,500 n miles (8,334 km; 5,178 miles)

OPERATIONAL NOISE LEVELS:

T-O: 900C, 900EX	79.8 EPNdB
Approach: 900C	91.7 EPNdB
900EX	92.3 EPNdB

Sideline: 900C	91.2 EPNdB
900EX	90.5 EPNdB

UPDATED

DASSAULT FALCON 2000

TYPE: Business jet.

PROGRAMME: Announced Paris Air Show 1989 as Falcon X; follow-on to Falcon 20/200; launched as Falcon 2000 on 4 October 1990, following first orders; Alenia joined as 25 per cent risk-sharing partner February 1991, with responsibility for rear fuselage section and engine nacelles; selection of CFE738 engine announced 2 April 1990; first flight (F-WNAV) 4 March 1993; one prototype only; third airframe (F-WWFA) was second to fly, on 10 May 1994; ferried 'green' to US subsidiary Dassault Falcon Jet Corporation at Little Rock, Arkansas, 13 July 1994 for completion as US demonstrator, and set two world speed records 31 October to 1 November 1994 flying Los Angeles (Chino) to Bangor, Maine, in 4 hours 36 minutes 27 seconds and Bangor to Paris in 5 hours 26 minutes 12 seconds; Dassault demonstrator, second airframe (F-WNEW), flew 11 July 1994.

JAA certification to JAR 25 obtained 30 November 1994, at which time five aircraft (prototype, two demonstrators and two customer aircraft) had accumulated 1,055.5 flight hours; FAR Pt 25 and FAR Pt 36 Stage 3 noise levels certification 2 February 1995; first delivery (F-WNEW/ZS-NNF to South African customer) 16 February 1995. Approved to operate into London City Airport during 1996. Commonwealth of Independent States certification 14 April 1997. HGS-2850 HUD first flown in prototype 8 December 1995; initial JAA certification granted 30 August 1996, followed by approval for low-visibility take-offs (down to 91 m; 300 ft RVR) and hand-flown Cat. II and Cat. IIIa instrument approach approved by FAA 30 July 1997. Full JAA certification 16 December 1997; FAA certification 18 May 1998.

CURRENT VERSIONS: **Falcon 2000:** Initial production version. *As described.*

 Falcon 2000EX: Extended-range version. Development began October 1999; announced at the NBAA Convention in New Orleans, 10 October 2000.

Prototype (c/n 01/F-WMEX) rolled out 19 July 2001 and first flown 25 October 2001, with certification scheduled for the third quarter of 2002 and customer deliveries of outfitted aircraft in the second quarter of 2003 (ProLine avionics). EASy flight deck available from second quarter of 2004. Falcon 2000EX will be manufactured in parallel with Falcon 2000.

New features include two Pratt & Whitney Canada PW308C turbofans, with FADEC, each rated at 31.1 kN (7,000 lb st) for take-off at ISA +15°C; Nordam advanced single pivot (ASP) thrust reversers; revised fuel system with 1,730 kg (3,814 lb) increase in fuel capacity; strengthened main landing gear with heavy-duty braking system and Falcon 900EX nose landing gear; and new bleed air system. The first 40 Falcon 2000EXs will retain the current Collins Pro Line 4 avionics suite, but aircraft delivered from March 2004 will be equipped with Dassault EASy flight deck with four-tube flat panel displays based on those of the Honeywell Primus Epic modular avionics system.

Preliminary specifications include: Basic operating weight 10,505 kg (23,160 lb); max payload 2,966 kg (6,539 lb); max fuel 9,024 litres (2,384 US gallons; 1,985 Imp gallons), weight 7,244 kg (15,970 lb); max ramp weight 18,552 kg (40,900 lb); max T-O weight 18,461 kg (40,700 lb); max landing weight 17,373 kg (38,300 lb); max zero-fuel weight 13,472 kg (29,700 lb); max operating Mach No. (M_{MO}) 0.85-0.86; normal cruising speed Mach 0.80; time to climb to 12,495 m (41,000 ft) 21 min; T-O balanced field length 1,756 m (5,760 ft); landing run 802 m (2,635 ft); range with six passengers, NBAA IFR reserves at Mach 0.80 3,800 n miles (7,037 km; 4,373 miles).

CUSTOMERS: Production totalled 20 in 1997, 15 in 1998, 34 in 1999, 26 in 2000, and 35 in 2001. Falcon 2000 No. 159/N259QS, delivered to Executive Jet on 11 September 2001, was also 1,500th Falcon/Mystère business jet produced by Dassault. Total 200 ordered by July 2001, of which 176 had been delivered to completion centres by I January 2002. Total of 72 Falcon 2000s, and 25 Falcon 2000EXs, plus 25 Falcon 2000EXs on option, ordered by Executive Jet International for its NetJets and NetJets Middle East fractional ownership programmes in Europe,

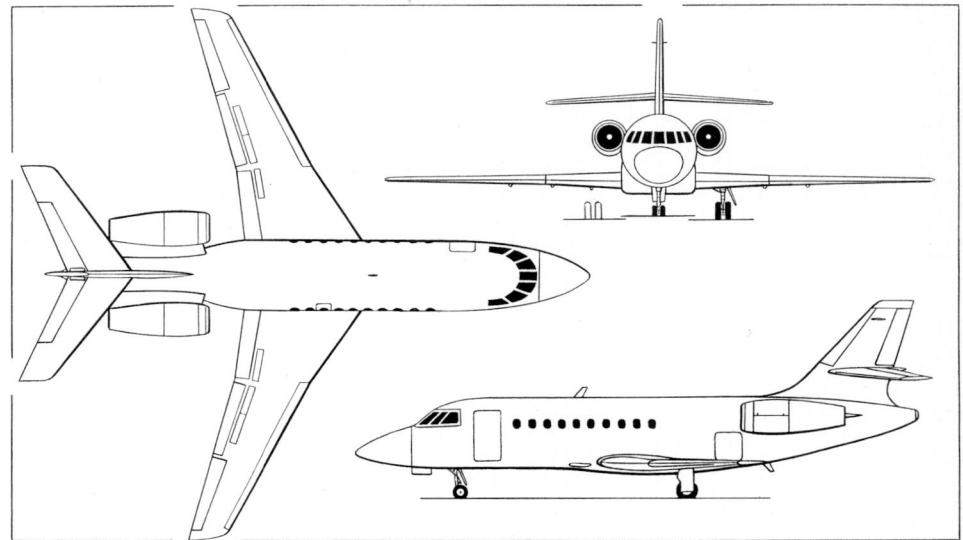

Dassault Falcon 2000 twin-engined business transport (*Jane's/Dennis Punnett*)

Typical cabin of a Falcon 2000
(Bud Shannon/Dassault)   NEW/0110861

Middle East and USA, of which 20 were in operation by April 2001. Royal Australian Air Force ordered three in 2000 to replace five leased Falcon 900s. Recent customers include Avolar (formerly United Bizjets Holdings), which ordered 30 2000s and 2000EXs, plus 50 2000EXs on option, in June 2001, for delivery beginning in late 2002. Total market estimated at more than 300 in 10 years.

COSTS: Basic price 2000 US$20.6 million (2001); 2000EX US$24 million (2000).

DESIGN FEATURES: Same fuselage cross-section as Falcon 900, but 1.98 m (6 ft 6 in) shorter. Falcon 900 wing with modified leading-edge and no inboard slats; sweepback at quarter-chord 24° 50' to 29°.

STRUCTURE: Largely as for Falcon 900. Rear fuselage, including engine pods and pylons, by Alenia and Piaggio; thrust reversers by Dee Howard. Agreement reached in early 2002 for production of fuselage components of 2000EX by Chengdu Aircraft of China. New horizontal tail surface featuring cast titanium central box with resin transfer moulding composite spars and carbon fibre skin panels, resulting in a 13.6 kg (30 lb) reduction in weight, certified by the DGAC in December 1999 and introduced as standard from December 2000 deliveries.

LANDING GEAR: Retractable tricycle type; mainwheels retract inwards, nosewheel forwards. Main tyres 26×6.6R14 (14 ply) tubeless; nose 14.5×5.5R6 tubeless.

POWER PLANT: Two GE/Honeywell CFE738-1-1B turbofans, each rated at 26.3 kN (5,918 lb st) at ISA + 15°C. Usable fuel capacity 6,865 litres (1,814 US gallons; 1,510 Imp gallons). Dee Howard clamshell-type thrust reversers certified by FAA and JAA during 1995.

ACCOMMODATION: Up to 19 passengers and two flight crew; standard passenger accommodation is four seats in forward lounge and four seats and a two-person sofa in aft lounge. Pressurised, flight accessible baggage compartment at rear of cabin.

AVIONICS: Rockwell Collins Pro Line 4.
Comms: Dual com/nav; dual TRD-94D Mode-S transponders; dual RTU-4420 radio tuning units; Rockwell Collins HF-9000 transceiver.
Radar: Rockwell Collins WXR-840 colour weather radar standard; TWR-850 Doppler weather radar optional.
Flight: Rockwell Collins FMS-6100 flight management system and Cat. II autopilot standard. Rockwell Collins dual DME-442, dual ADF-462; dual AHRS and dual digital air data computers linked to dual-channel integrated avionics processor system (IAPS). Dual Honeywell Laseref III, Honeywell EGPWS, IRS, second FMS (with GPS), flight data recorder, traffic alert and collision avoidance system (TCAS) and GPWS all optional.
Instrumentation: Rockwell Collins Pro Line 4 four-tube EFIS-4000. Sextant Avionique three-tube engine indicating electronic display (EIED); Flight Dynamics HGS-2850 HUD.
Mission: Optional satcom antenna in fintip pod.

DIMENSIONS, EXTERNAL:
Wing span	19.33 m (63 ft 5 in)
Wing chord (mean)	2.89 m (9 ft 5¾ in)
Wing aspect ratio	7.6
Length overall	20.22 m (66 ft 4 in)
Height overall	7.06 m (23 ft 2 in)
Wheel track	4.45 m (14 ft 7¼ in)
Wheelbase	7.39 m (24 ft 3 in)
Passenger door: Height	1.72 m (5 ft 7¾ in)
Width	0.80 m (2 ft 7½ in)
Emergency exits (Type III): Height	0.92 m (3 ft 0¼ in)
Width	0.53 m (1 ft 8¾ in)
Baggage door: Height	0.775 m (2 ft 6½ in)
Width	0.75 m (2 ft 5½ in)

DIMENSIONS, INTERNAL:
Cabin: Length	7.98 m (26 ft 2¼ in)
Max width	2.34 m (7 ft 8¼ in)
Max height	1.88 m (6 ft 2 in)
Volume	29.0 m³ (1,024 cu ft)
Baggage volume: standard	3.8 m³ (134 cu ft)
optional	5.4 m³ (191 cu ft)

AREAS:
Wings, gross	49.02 m² (527.6 sq ft)

WEIGHTS AND LOADINGS:
Weight empty, equipped	9,405 kg (20,735 lb)
Basic operating weight	9,798 kg (21,600 lb)
Max payload	3,270 kg (7,210 lb)
Max fuel weight	5,513 kg (12,154 lb)
Max ramp weight: standard	16,329 kg (36,000 lb)
optional	16,647 kg (36,700 lb)
Max T-O weight: standard	16,238 kg (35,800 lb)
optional	16,556 kg (36,500 lb)
Max landing weight	14,970 kg (33,000 lb)
Max zero/fuel weight	13,000 kg (28,660 lb)
Max wing loading: standard	331.3 kg/m² (67.85 lb/sq ft)
optional	337.8 kg/m² (69.18 lb/sq ft)
Max power loading: standard	308 kg/kN (3.02 lb/lb st)
optional	314 kg/kN (3.08 lb/lb st)

PERFORMANCE:
Max operating Mach No. (MMO)	0.85-0.86
Max operating speed (VMO)	350-370 kt (648-685 km/h; 403-425 mph) IAS
Max cruising speed	M0.84 or 481 kt (891 km/h; 554 mph)
Normal cruising speed	M0.80 or 459 kt (850 km/h; 528 mph)
Long-range cruising speed	M0.75 or 430 kt (796 km/h; 495 mph)
Approach speed	109 kt (202 km/h; 125 mph)
Initial cruising altitude: at M0.80	12,497 m (41,000 ft)
at M0.75	13,106 m (43,000 ft)
Max certified altitude	14,330 m (47,000 ft)
FAR Pt 25 balanced T-O field length, S/L, ISA:	
with eight passengers and max fuel	1,597 m (5,240 ft)
at max T-O weight	1,658 m (5,440 ft)
Landing field length:	
FAR Pt 91 with eight passengers and NBAA IFR reserves	780 m (2,560 ft)
at max landing weight	1,591 m (5,220 ft)
Range with eight passengers and NBAA IFR reserves:	
at M0.75	3,120 n miles (5,778 km; 3,590 miles)
at M0.80	3,000 n miles (5,556 km; 3,452 miles)
g limits	+2.64/−1

OPERATIONAL NOISE LEVELS:
T-O	79.4 EPNdB
Approach	93.1 EPNdB
Sideline	86.4 EPNdB

UPDATED

DASSAULT FALCON FNX

Redesignated Falcon 7X (which see).

UPDATED

Dassault Falcon 2000EX was added to the range in 2001 *(François Robineau/Aviaplans)*   NEW/0137209

DYN'AERO

SOCIETE DYN'AERO

19 rue de l'Aviation, F-21121 Darois
Tel: (+33 3) 80 35 60 62
Fax: (+33 3) 80 35 60 63
e-mail: dynaero@aol.com
Web: http://www.dynaero.com
CHAIRMAN: Christophe Robin
DESIGN ASSOCIATE: Michel Colomban

NORTH AMERICAN AGENTS:
American Ghiles Aircraft
522 East Washington Street, Orlando, Florida 32801, USA
Tel: (+1 386) 740 71 40
Fax: (+1 386) 740 86 21
e-mail: contact@aircraftkit.com
Web: http://www.aircraftkit.com

Dyn'Aéro formed September 1992 by Christophe Robin, son of the founder of Avions Pierre Robin; his initials (CR) appear in aircraft designations. Initial product was CR100, flown two weeks before formation of company. Two new designs derived from CR100 announced in March 1995: CR110 and CR120; simultaneously, MCR01 also announced. The new MCR01 Club flew in June 1998 and a four-seat version in June 2000. Production facilities at Darois aerodrome comprise 2,450 m² (26,375 sq ft) of workshops, storage and offices. Plans announced in February 2002 for establishment of a subsidiary, Dyn'Aéro Iberia at a new factory in Portugal, permitting an increase in annual production from 50 to 100 aircraft (initially of MCR ULC), with first airframe from new facility expected within 18 months.

By April 2002, 250 Dyn'Aéro kits had been sold of which over 70 per cent had been delivered and more than 150 aircraft were flying.

UPDATED

DYN'AERO CR100, CR110 and CR120

TYPE: Aerobatic, two-seat sportplane/kitbuilt.

PROGRAMME: Prototype CR100 (F-PDYN) first flew 27 August 1992; third aircraft (F-PAFC) was designated CR100D; first five CR100s flew total of 600 hours of aerobatics in first year of operations. Available factory-built or as a kit.

CURRENT VERSIONS: **CR100**: Baseline version.

CR100T: Tricycle landing gear version, first flown December 2000.

Description applies to CR100, except where otherwise stated.

CR110: Uprated version with 149 kW (200 hp) Textron Lycoming engine.

CR120: High-agility version. First flight (F-PPCR) September 1996; second aircraft (F-PTCR) flying by 1998. Engine as CR110, but airframe all carbon fibre; longer span ailerons (2.70 m; 8 ft 10¼ in), no flaps, shorter wing span (7.80 m; 25 ft 7 in), 270°/s rate of roll at 140 kt (260 km/h; 162 mph) compared with CR100's 195°/s, MTV-12-B-C/C-183-17e two-blade propeller, 85 litres (22.4 US gallons; 18.7 Imp gallons) aerobatic fuel in starboard wingroot and 105 litres (27.7 US gallons; 23.1 Imp gallons) in port wingroot, tyre pressure 2.00 bar (29.0 lb/sq in). Aircraft are registered under CR100 designation.

CUSTOMERS: Two CR100s ordered March 1995 by l'Equipe de Voltige de l'Armée de l'Air at Salon de Provence, of which first (No. 22 'CJ') delivered 21 July 1995; however, both withdrawn by 1997 and sold in 2000, one to UK owner Cole Aviation. Total of at least 37 CR100/110/120 kits sold; of these, 16 had been registered by early 2002.

COSTS: FFr500,000 as a kit without engine; about FFr900,000 assembled, with engine and instruments.

DESIGN FEATURES: Single-engined low-wing monoplane of conventional layout.

Aerofoil section NACA 21015/12 modified; dihedral 0°.

FLYING CONTROLS: Conventional and manual. Elevator tab for pitch trim. Ground-adjustable tab on ailerons on rudder. Plain flaps.

STRUCTURE: Wood and fabric, but including carbon fibre main wing spar.

LANDING GEAR: Tailwheel type; fixed. Oleo-pneumatic shock-absorption. Speed fairings on mainwheel legs. Disc brakes. Mainwheel tyre size 380×150; pressure 2.30 bar (33.0 lb/sq in).

POWER PLANT: One 134 kW (180 hp) Textron Lycoming AEIO-360-B2F flat-four engine, driving Arplast Dyn'Aéro CR100 two-blade propeller. Compatible with alternative engines of 119 kW (160 hp) and above. Fixed-pitch or constant-speed propellers can be fitted. Fuel capacity for aerobatics 85 litres (22.4 US gallons; 18.7 Imp gallons), of which 82 litres (21.7 US gallons; 18.0 Imp gallons) are usable; max fuel capacity 160 litres (42.3 US gallons; 35.2 Imp gallons). Oil capacity 7.6 litres (2.0 US gallons; 1.7 Imp gallons).

ACCOMMODATION: Two seats, side by side, with dual controls. Fixed windscreen and rearward-sliding canopy.

SYSTEMS: Electrical system: 12 V 30 Ah battery.

AVIONICS: Customer specified.

DIMENSIONS, EXTERNAL:

Wing span	8.50 m (27 ft 10½ in)
Wing aspect ratio	6.8
Length overall	7.10 m (23 ft 3½ in)
Tailplane span	2.80 m (9 ft 2¼ in)
Wheel track	2.37 m (7 ft 9¼ in)
Propeller diameter	1.90 m (6 ft 3 in)

AREAS:

Wings, gross	10.63 m² (114.4 sq ft)
Ailerons (total)	1.215 m² (13.08 sq ft)
Trailing-edge flaps (total)	1.21 m² (13.02 sq ft)
Vertical tail surfaces (total)	1.71 m² (18.40 sq ft)
Horizontal tail surfaces (total)	2.20 m² (23.68 sq ft)

WEIGHTS AND LOADINGS:

Weight empty	550 kg (1,213 lb)
Max T-O weight: Aerobatic	760 kg (1,676 lb)
Normal	850 kg (1,873 lb)
Max wing loading: Aerobatic	71.5 kg/m² (14.64 lb/sq ft)
Normal	80.0 kg/m² (16.38 lb/sq ft)
Max power loading: Aerobatic	5.66 kg/kW (9.31 lb/hp)
Normal	6.34 kg/kW (10.41 lb/hp)

PERFORMANCE:

Never-exceed speed (V_NE)	205 kt (379 km/h; 235 mph)
Max level speed	171 kt (317 km/h; 197 mph)
Max cruising speed	154 kt (285 km/h; 177 mph)
Manoeuvring speed (V_A)	140 kt (260 km/h; 161 mph)
Max rate of climb at S/L	480 kt (1,575 ft)/min
T-O run	153 m (500 ft)
T-O to 15 m (50 ft)	351 m (1,150 ft)
g limits	+8/−6

UPDATED

DYN'AERO MCR01 BAN BI

North American marketing names: Lafayette I Sportster (VLA); II Touring (Club); III Bushplane (ULM)

TYPE: Side-by-side kitbuilt; side-by-side ultralight kitbuilt.

PROGRAMME: Composites version of the metal MC100 Ban-Bi designed by Michel Colomban. MCR identifies Michel Colomban/Christophe Robin. Prototype (F-PECH) flew 1995.

CURRENT VERSIONS: **MCR01 VLA**: Certified version meeting JAR-VLA requirements; **VLA 912** or **VLA 914**, according to engine. JAR-VLA certification awarded 7 December 2001 in Peru.

MCR01 Club: Version of MCR01 VLA 912 intended for flight training. First flight (F-PCLB) June 1998. Certification to JAR-VLA due in 2001.

MCR01 ULM: Ultralight version meeting FAI ULM requirements; 59.7 kW (80 hp) engine. Canadian **MCR01 ULC** has 490 kg (1,080 lb) MTOW and is known as the **MCR01 SLA** in the UK, where the first example (G-BZXG) made its first flight on 30 November 2001.

Motor Glider: Under development. 'Winggrid' wingtips developed in conjunction with La Roche Consulting of Switzerland; 59.7 kW (80 hp) engine, best glide ratio 30:1.

MCR M: Tailwheel version.

Electra-Plane: Being developed by Advanced Technology Products; stated to be world's first fuel cell-powered aircraft. Prototype is based on MCR01; second prototype will utilise a Hoffmann H-36 Dimona as base aircraft. Three stages planned: first, with lithium-ion batteries, will have 86 n mile (160 km; 100 mile) range; second, with combination of batteries and 10 to 15 kW (13 to 20 hp) fuel cell, will have 217 n mile (402 km; 250 mile) range. Final version will be powered solely with fuel cells and have 434 n mile (804 km; 500 mile) range.

Description applies to all versions except where indicated.

CUSTOMERS: Total of 190 kits sold, and 100 flying by January 2001, in Austria, Germany, Italy, Japan, Netherlands, New Zealand, Spain, Switzerland, UK and USA.

COSTS: Kits, less engine and instruments: VLA FFr141,000, ULC/ULM FFr163,000, Club FFr155,000 and M FFr 181,000 (all 2001).

DESIGN FEATURES: Single-engined low-wing monoplane with T tail; design goals included extremely low structural weight, fast kit production and high performance in comfort and safety.

FLYING CONTROLS: Manual. All-moving tailplane with trim tab, which is full span on ULM version; electrically actuated flaps and pitch trim; ULM version has full-span double-slotted flaperons. VLA has flaperons. Club has separate flaps and ailerons.

STRUCTURE: Wing and control surfaces built on carbon fibre spars and cellular ribs, to which preformed aluminium skins are glued in partial vacuum; monocoque carbon fibre fuselage; quoted build time of 500 to 1,000 hours from three modular kits without special tools. Airframe disassembles for storage or road-towing, with recommended disassembly/reassembly time of 4 minutes, single-handed.

US-built Ban Bi carrying its American name, Sportster *(Jane's/Paul Jackson)*   NEW/0137953

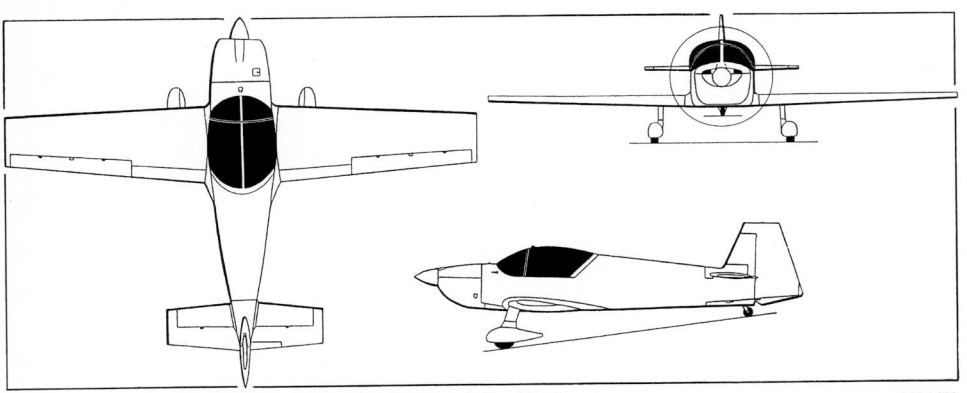
General arrangement of the Dyn'Aéro CR100 *(Jane's/Paul Jackson)*   0044672

Dyn'Aéro CR120 (marked as CR100) *(Jane's/Paul Jackson)*   0085615

LANDING GEAR: Fixed tricycle type with speed fairings on all three units; nosewheel steerable. Large wheels optional for rough field operation. Tailwheel configuration optional on MCR01 club and MCR01 ULC models.

POWER PLANT: One 59.6 kW (79.9 hp) Rotax 912 UL; 63.4 kW (85 hp) JPX 4T75; or 73.5 kW (98.6 hp) normally aspirated Rotax 912 ULS or turbocharged Rotax 914 flat-four driving a two-blade fixed-pitch propeller optimised for cruise; three-blade constant-speed propeller optional. Jabiru-powered version first flown 19 April 2002. Standard fuel capacity 80 litres (21.1 US gallons; 17.6 Imp gallons) in all versions; optional additional 90 litre (23.75 US gallon; 19.8 Imp gallon) tank in Club and ULM or 264 litre (69.7 US gallon; 58.1 Imp gallon) tank in VLA.

AVIONICS: To customer's choice.

EQUIPMENT: Optional road-towing trailer/storage hangar. BRS ballistic parachute recovery system optional on ULM.

DIMENSIONS, EXTERNAL:

Wing span: VLA	6.63 m (21 ft 9 in)
Club	6.92 m (22 ft 8½ in)
ULM	8.28 m (27 ft 2 in)
Motor Glider	9.80 m (32 ft 1¾ in)
Wing aspect ratio: VLA	8.5
Club	7.0
ULM	8.6
Length overall: all	5.66 m (18 ft 6¾ in)

DIMENSIONS, INTERNAL:

Cabin max width: all	1.12 m (3 ft 8 in)

AREAS:

Wings, gross: VLA	5.20 m² (56.0 sq ft)
Club	6.45 m² (69.4 sq ft)
ULM	7.95 m² (85.6 sq ft)
Motor Glider	9.35 m² (100.6 sq ft)

WEIGHTS AND LOADINGS:

Weight empty, equipped: VLA	235 kg (518 lb)
Club	245 kg (540 lb)
ULM	255 kg (562 lb)
Motor Glider	262 kg (578 lb)
Max T-O weight: VLA, ULM	450 kg (992 lb)
Club	490 kg (1,080 lb)

PERFORMANCE (59.7 kW; 80 hp engine):

Never-exceed speed (VNE): normal:	
VLA	172 kt (320 km/h; 198 mph)
Club, ULM	162 kt (300 km/h; 186 mph)
with BRS: all	145 kt (270 km/h; 167 mph)
Max level speed: VLA 912	163 kt (302 km/h; 188 mph)
VLA 914	197 kt (364 km/h; 226 mph)
Club	151 kt (281 km/h; 174 mph)
ULM	150 kt (278 km/h; 172 mph)
Motor Glider	135 kt (250 km/h; 155 mph)
Max cruising speed at 75% power:	
at 2,440 m (8,000 ft):	
VLA 912	160 kt (296 km/h; 184 mph)
Club	150 kt (278 km/h; 173 mph)
ULM	145 kt (269 km/h; 169 mph)
Motor Glider	130 kt (240 km/h; 149 mph)
at 3,350 m (11,000 ft):	
VLA 914	178 kt (329 km/h; 204 mph)
Stalling speed, flaps down:	
VLA	47 kt (87 km/h; 54 mph)
Club	43 kt (79 km/h; 49 mph)
ULM	35 kt (64 km/h; 40 mph)
Motor Glider	30 kt (55 km/h; 35 mph)
Max rate of climb at S/L: VLA, Club	
	609 m (2,000 ft)/min
ULM	426 m (1,400 ft)/min
T-O run: VLA	285 m (935 ft)
Club	300 m (985 ft)
ULM	195 m (640 ft)
T-O to 15 m (50 ft): VLA, Club	424 m (1,395 ft)
ULM	298 m (980 ft)
Range at econ cruising speed and altitude:	
with normal fuel:	
VLA	894 n miles (1,657 km; 1,029 miles)
Club	829 n miles (1,536 km; 954 miles)
ULM	834 n miles (1,545 km; 960 miles)
with max optional fuel:	
VLA	3,452 n miles; (6,394 km; 3,973 miles)
Club	1,623 n miles (3,006 km; 1,867 miles)
ULM	1,611 n miles (2,985 km; 1,854 miles)
g limits: all	+4/−2

UPDATED

DYN'AERO MCR4S

North American marketing name:
Lafayette Revolution

TYPE: Four-seat kitbuilt.

PROGRAMME: Mockup first shown at Aero '99, Friedrichshafen, April 1999. Kits available from late 1999. Prototype (FWWUZ) first flew 14 June 2000 and displayed at PFA International Air Rally, Cranfield, UK, 23-25 June 2000. First customer aircraft, built by noted French light aircraft designer/manufacturer Pierre Robin, flew in mid-2001. PFA approval expected by early 2003.

CURRENT VERSIONS: **Three-seat:** With 59.6 kW (79.9 hp) Rotax 912 UL or JPX 4TX75 engine; 640 kg (1,411 lb) MTOW.

Four-seat: With 73.5 kW (98.6 hp) normally aspirated engine and fixed-pitch propeller; 750 kg (1,653 lb) MTOW.

Four-seat Performance: 84.6 kW (113.4 hp) Rotax 914 UL turbocharged engine and constant-speed propeller; optimised for cruise at 3,050 m (10,000 ft).

CITEC: On 29 July 2001, Dyn'Aero and Wilksch Airmotive of the UK announced a joint programme to develop the MCR4S CITEC, powered by the Wilksch WAM-120 diesel engine. Launch customer for this version was M. Serge Blanchard.

COSTS: Kit FFr250,000, less engine; FFr400,000, FFr445,000 or FFr485,000 with Rotax 912 ULS plus fixed propeller, 912 ULS plus c/s MT propeller or 914 plus c/s propeller (2001).

DESIGN FEATURES: Three/four-seat version of MCR01, having 75 per cent commonality. Dihedral 3°. Winglets introduced in 2001.

Description generally as for MCR01, except that below.

FLYING CONTROLS: Ailerons, plus two-segment flaps. Aileron max deflection +20°/−10°. Tailplane (all moving) max deflection +5°/−10°. Double-slotted flaps, deflections 0, 17 and 45°.

POWER PLANT: See Current Versions. Fuel in two wing tanks; total capacity 120 litres (31.7 US gallons; 26.4 Imp gallons).

ACCOMMODATION: Four persons, in two side-by-side pairs. Forward-hinged, two-piece canopy/windscreen, plus two fixed side windows.

EQUIPMENT: Ballistic parachute optional.

DIMENSIONS, EXTERNAL:

Wing span	8.72 m (28 ft 7¼ in)
Wing chord (mean)	0.96 m (3 ft 1¾ in)
Wing aspect ratio	9.2
Length overall	6.72 m (22 ft 0½ in)
Height overall	1.95 m (6 ft 4¾ in)
Tailplane span	2.50 m (8 ft 2½ in)

DIMENSIONS, INTERNAL:

Cabin max width	1.20 m (3 ft 11¼ in)

AREAS:

Wings, gross	8.30 m² (89.3 sq ft)

WEIGHTS AND LOADINGS:

Weight: empty	325 kg (717 lb)
equipped	350 kg (772 lb)
Max T-O weight	640-750 kg (1,411-1,653 lb)

PERFORMANCE (Rotax 912, 912 ULS and 914 UL engines):

Never-exceed speed (VNE)	172 kt (320 km/h; 198 mph)
Max level speed: 912	143 kt (265 km/h; 165 mph)
912 ULS	156 kt (288 km/h; 179 mph)
914 UL	173 kt (320 km/h; 199 mph)
Manoeuvring speed	117 kt (217 km/h; 134 mph)
Max cruising speed: 912	136 kt (252 km/h; 157 mph)
912 ULS	150 kt (277 km/h; 172 mph)
914 UL	155 kt (287 km/h; 178 mph)
Econ cruising speed: 912	131 kt (242 km/h; 150 mph)
912 ULS, 914 UL	146 kt (270 km/h; 168 mph)
Stalling speed: flaps up	54 kt (100 km/h; 63 mph)
flaps down	45 kt (82 km/h; 51 mph)
T-O run with c/s prop: 912	317 m (1,040 ft)
912 ULS	203 m (670 ft)
914 UL	153 m (505 ft)
T-O to 15 m (50 ft) with c/s prop: 912	455 m (1,495 ft)
912 ULS	295 m (970 ft)
914 UL	221 m (725 ft)
Landing run with c/s prop: 912 ULS	201 m (660 ft)
Range with max fuel:	
912	1,034 n miles (1,916 km; 1,190 miles)
912 ULS, 914 UL	921 n miles (1,707 km; 1,060 miles)
g limits	+3.8/−1.5

UPDATED

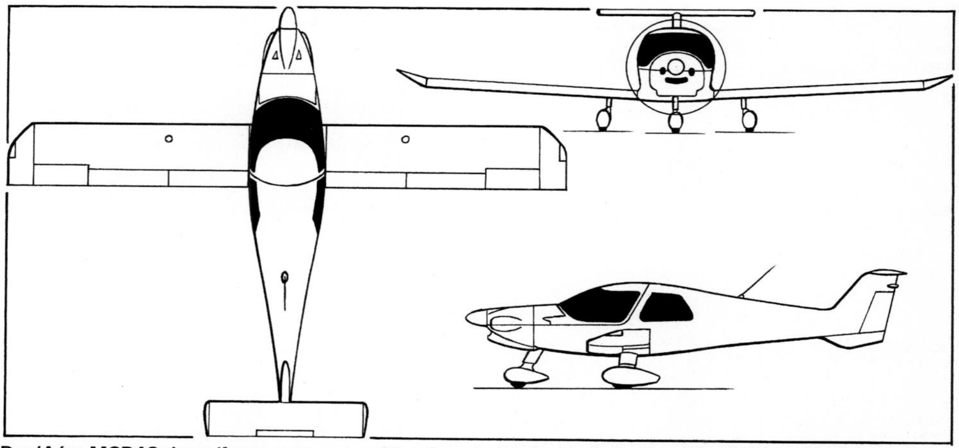

Dyn'Aero MCR01 ULM, with additional side view (lower) and half-plan view of VLA, plus scrap side view of Club (*Jane's/Paul Jackson*) 0092148

Dyn'Aéro MCR4S three/four-seat tourer with original wingtip configuration (*Jane's/James Goulding*) 0092126

Dyn'Aero MCR4S four-seater (*Jane's/Paul Jackson*) *NEW*/0137954

EADS

EADS FRANCE
37 boulevard de Montmorency, F-75781 Paris Cedex 16
Tel: (+33 1) 42 24 24 24
Fax: (+33 1) 42 24 26 19

The former Aerospatiale Matra has been incorporated into EADS, as described in the International section.
UPDATED

HUMBERT

HUMBERT AVIATION
rue du Menil, F-88160 Ramonchamp
Tel: (+33 3) 29 25 05 75
Fax: (+33 3) 29 25 98 97
DESIGNER: Jean-Jacques Humbert

Humbert's most recent design is the Tétras, although the earlier Moto du Ciel remains available.
VERIFIED

HUMBERT TÉTRAS

TYPE: Side-by-side ultralight/kitbuilt.
PROGRAMME: Prototype unveiled at 1992 RSA Rally at Moulins. Named for a fish of the genus *Characidae*.
CUSTOMERS: At least 10 on French ultralight register by December 2001.
COSTS: Kit, with Rotax 912 UL engine, €15,000; flyaway €31,500 (2001).
DESIGN FEATURES: High wing with single bracing strut on each side; strut-braced tailplane. Wing section NACA 23012. Quoted build time 150 to 250 hours.
FLYING CONTROLS: Conventional and manual. Three-position plain flaps, deflections 0, 15 and 45°.
STRUCTURE: Steel tube fuselage with Dacron covering; composites wing; light alloy ailerons and flaps.
LANDING GEAR: Non-retractable tailwheel type with bungee cord suspension on main units.

Humbert Tétras two-seat ultralight (*Jane's/Paul Jackson*)   0110868

POWER PLANT: One 53.7 kW (72 hp) Humbert-Volkswagen HW 2000 or 59.6 kW (79.9 hp) Rotax 912 UL flat-four, driving a two-blade composites propeller. Other options available, including JPX engine. Fuel in two wing tanks, combined capacity 60 litres (15.9 US gallons; 13.2 Imp gallons).

DIMENSIONS, EXTERNAL:
Wing span	10.10 m (33 ft 1¾ in)
Length overall	6.50 m (21 ft 4 in)
Height overall	2.06 m (6 ft 9 in)
Propeller diameter	1.71 m (5 ft 7¼ in)

AREAS:
Wings, gross	15.70 m² (169.0 sq ft)

WEIGHTS AND LOADINGS:
Weight empty	260 kg (573 lb)
Max T-O weight	450 kg (992 lb)

PERFORMANCE (Rotax 912 UL):
Max level speed	105 kt (194 km/h; 121 mph)
Cruising speed at 75% power at 3,050 m (10,000 ft)	101 kt (187 km/h; 116 mph)
Stalling speed, flaps down	27 kt (50 km/h; 32 mph)
Max rate of climb at S/L	302 m (990 ft)/min
T-O run	50 to 80 m (164 to 262 ft)
g limits	+6/−3

UPDATED

ISSOIRE

ISSOIRE AVIATION
ZA la Béchade, BP1, Aérodrome Issoire-Le Broc, F-63501 Issoire
Tel: (+33 4) 73 89 01 54 and 73 89 78 11
Fax: (+33 4) 73 89 54 59
e-mail: iav@issoire-aviation.com
Web (1): http://www.issoire-aviation.com
Web (2): http://www.apm20lionceau.com
PRESIDENT AND CEO: Philippe Moniot
DIRECTOR: Laurent Bourdier
SALES MANAGER: Yves Moyne-Bressand

Issoire Aviation was established in 1978 as a successor to Wassmer Aviation. Its subsidiary, Rex Composites, is responsible for carbon fibre airframes. Current product is the APM-20/21 lightplane.
UPDATED

ISSOIRE APM-20 LIONCEAU and APM-21 LION

English name: Lion Cub

TYPE: Two-seat lightplane.
PROGRAMME: Design (as Moniot APM-20) started 1992 by Les Industries de Composites d'Auvergne Réunités (ICAR); prototype (F-WWMP) exhibited at Paris Air

Show 1995 before first flight on 21 November 1995; third (second flying) aircraft (F-WWXX) exhibited statically at Paris Air Show, June 1997, fitted with JPX flat-four engine, which is not offered on production aircraft. No. 4 (also F-WWXX) exhibited at Paris in June 1999 and No. 5 (F-GRRE) in 2001. Certified to JAR-VLA 17 May 1999 and to JAR-21 in 2000. First all-carbon fibre, single-engine aircraft to gain JAR-VLA certification. Further four aircraft (Nos. 6 to 9) registered in late 2001. APM-21 will follow at unspecified date.

CURRENT VERSIONS: **APM-20:** Primary trainer *As described.*

APM-21 Lion: Aerial work version powered by 73.5 kW (98.6 hp) Rotax 912 ULS engine driving a two-blade, constant-speed Hoffmann propeller. Prototype (F-WWMP, modified from prototype Lionceau) demonstrated at Paris Air Show in June 1999.

Four-seat: Under development; no major structural changes.

CUSTOMERS: By mid-2001 Issoire Aviation had received 15 options. Initial production rate planned of two per month, possibly rising to 30 per year. Break-even point estimated at 200 aircraft.

COSTS: Development cost around FFr10 million. Unit cost FFr620,000, minimally equipped or FFr700,000 with standard equipment (2001). Operating cost FFr350 per hour (1998).

DESIGN FEATURES: Low wing, NACA 63618 aerofoil, thickness/chord ratio 18 per cent, dihedral 3°, incidence 2°, twist 1°.

FLYING CONTROLS: Conventional and manual. Spring elevator tab for pitch trim. Electrically operated slotted flaps to about two-thirds span.

STRUCTURE: Carbon fibre/epoxy.
LANDING GEAR: Fixed tricycle type with spats; oleo-sprung steerable nose leg, composites main legs. Mainwheels and nosewheel diameter 330 mm; maximum pressure 2.35 bar (34 lb/sq in).
POWER PLANT: One 59.6 kW (79.9 hp) Rotax 912A four-cylinder four-stroke driving an Evra two-blade, fixed-pitch propeller. Fuel capacity 68 litres (18.0 US gallons; 15.0 Imp gallons), of which 65 litres (17.2 US gallons; 14.3 Imp gallons) are usable. Refuelling point on port side of fuselage. Oil capacity 4 litres (1.1 US gallons; 0.9 Imp gallon).
ACCOMMODATION: Two, side by side under rearward-sliding tinted canopy. Dual controls standard. Baggage compartment at rear of seats.
AVIONICS: *Instrumentation:* VFR panel standard.

DIMENSIONS, EXTERNAL:
Wing span	8.66 m (28 ft 5 in)
Wing chord: at root	1.10 m (3 ft 7¼ in)
at tip	0.84 m (2 ft 9 in)
Wing aspect ratio	7.9
Length overall	6.60 m (21 ft 7¾ in)
Max width of fuselage	1.15 m (3 ft 9¼ in)
Height overall	2.40 m (7 ft 10½ in)
Tailplane span	2.50 m (8 ft 2½ in)
Wheel track	1.92 m (6 ft 3½ in)
Propeller diameter	1.64 m (5 ft 4½ in)
Propeller ground clearance	0.31 m (1 ft 0¼ in)

AREAS:
Wings, gross	9.50 m² (102.3 sq ft)
Ailerons (total)	0.50 m² (5.38 sq ft)
Trailing-edge flaps (total)	1.40 m² (15.07 sq ft)

Fifth Issoire APM-20 Lionceau (*Jane's/Paul Jackson*)
***NEW*/**0131734

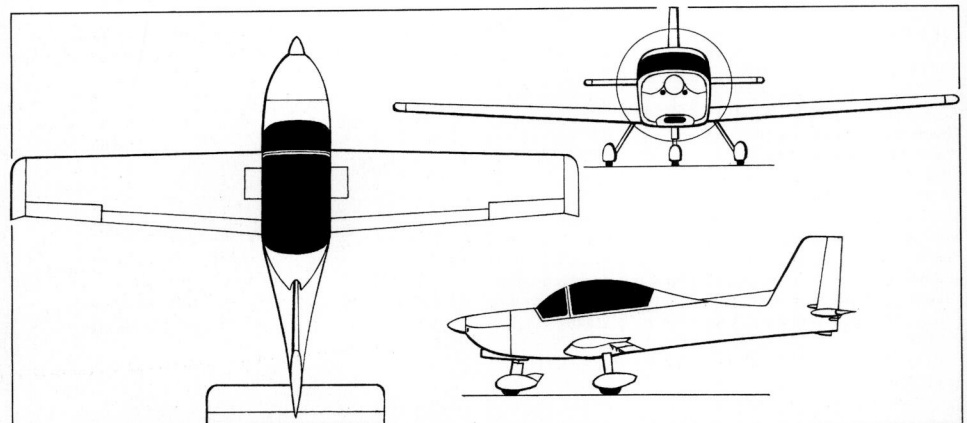

APM-20 Lionceau (Rotax 912A engine) (*Jane's/James Goulding*)   0064443

Fin	1.80 m² (19.37 sq ft)	Max wing loading	65.3 kg/m² (13.37 lb/sq ft)	Stalling speed: flaps up	54 kt (100 km/h; 63 mph)	
Tailplane	2.00 m² (21.53 sq ft)	Max power loading	10.39 kg/kW (17.06 lb/hp)	25° flap	44 kt (80 km/h; 50 mph)	
Elevator, incl tab	0.50 m² (5.38 sq ft)	PERFORMANCE:		Max rate of climb at S/L	204 m (669 ft)/min	
WEIGHTS AND LOADINGS:		Never-exceed speed (VNE)	135 kt (250 km/h; 155 mph)	T-O run	240 m (787 ft)	
Weight empty	380 kg (838 lb)	Cruising speed: at S/L	113 kt (210 km/h; 130 mph)	T-O to 15 m (50 ft)	460 m (1,509 ft)	
Baggage capacity	20 kg (44 lb)	at 1,676 m (5,500 ft)	124 kt (230 km/h; 143 mph)	Landing run	150 m (492 ft)	
Max fuel	49 kg (108 lb)	Econ cruising speed	94 kt (175 km/h; 109 mph)	Endurance	4 to 5 h	
Max T-O weight	620 kg (1,366 lb)	Manoeuvring speed	108 kt (200 km/h; 124 mph)		*UPDATED*	

JURCA

MARCEL JURCA

3 Allées des Bordes, F-94430 Chennevières-sur-Marne
Tel/Fax: (+33 1) 45 94 01 38
e-mail: divers@marcel-jurca.com
Web: http://www.marcel-jurca.com

M Marcel Jurca has designed a series of high-performance light aircraft and replicas of Second World War fighters. Most have wood or metal tube fuselages; stress analysis is verified by the ESTACA university (see Association Zéphyr entry in this section).

Jurca aircraft (all available to amateur builders) are MJ2 Tempête; MJ3 Dart; MJ4 Shadow; MJ5 Sirocco; MJ23 Orage; MJ51 Spérocco; MJ53 Autan; and MJ55 Biso (Provence wind) two-seat, fully aerobatic trainer which employs NACA 21012 section wing of MJ7/MJ77, 134 kW (180 hp) Textron Lycoming engine and first flew (F-PJJR) in October 1997.

Replicas comprise MJ7 Gnatsum two-third size Mustang Replica; MJ77 Gnatsum three-quarter size, two-seat Mustang (early- and late-style canopies); MJ8 three-quarter size Fw 190 Replica; MJ9 three-quarter size Messerschmitt Bf 109 Replica (one under construction in Canada); MJ10 three-quarter size Spitfire Replica; MJ12 three-quarter size P-40 Replica; MJ100 full-size Spitfire Replica (prototype first flown 14 October 1994; two others under construction at Prescott, Arizona, with 1,044 kW; 1,480 hp Allison piston engines; one with 1,044 kW (1,400 hp) derated Rolls-Royce Griffon Mk 58 at Annemasse, France); MJ90 full-size Messerschmitt Bf 109 Replica (410 kW; 550 hp Double Ranger V12-770C-1 engine); MJ80 full-size Focke-Wulf Fw 190 Replica (first under construction in Germany with 895 kW; 1,200 hp Pratt & Whitney R-1830 radial engine); MJ70 full-size P-51D Mustang Replica (under development by January 1996, will be powered by 1,044 kW; 1,400 hp Allison engine).

Under development in 2000 were the MJ11 single-seat, 119 kW (160 hp) biplane based on the Bücker Bü 133 Jungmeister; and MJ16 racer, with sportplane derivative.

Some 150 MJ aircraft have been registered and flown and a further 20 are under construction. Leading types are MJ2 (46

MJ12 replica at three-quarter scale of Curtiss P-40, built by Russel Moynagh at Santa Ana, California 0100412

built) and MJ5 (70). Other Jurca projects include the MJ15 Delta, MJ22 Twin Tempête, MJ50 Bise, MJ52 Zéphir, MJ54 Silas, MJ56 Sirocco S, MJ6 Crivats and MJ14 Fourtouna.

UPDATED

JURCA MJ5 SIROCCO

TYPE: Two-seat lightplane.
PROGRAMME: Prototype (F-PJSX) first flight 3 August 1962, powered by 78.5 kW (105 hp) Potez 4 E-20 engine; factory-built model subsequently awarded type certificate in Utility category. Version for amateur construction (70 built by January 2000) is generally similar to factory-built version.
CUSTOMERS: Over 60 completed by early 2002.
DESIGN FEATURES: Tandem two-seat development of MJ2 Tempête. Versions available with fixed or retractable gear; retractable tailwheel also available. Aerobatic version with 149 kW (200 hp) engine has strengthened wing spar. Quoted build time 2,000 hours.
LANDING GEAR: Choice of fixed or retractable.
POWER PLANT: Engines available include the Textron Lycoming series of the following powers: 85.75 kW

(115 hp), 112 kW (150 hp), 119 kW (160 hp), 134 kW (180 hp) and 149 kW (200 hp).
Details which follow refer to a Sirocco with a 149 kW (200 hp) Textron Lycoming IO-360 engine.
DIMENSIONS, EXTERNAL:

Wing span	7.00 m (23 ft 0 in)
Wing aspect ratio	4.9
Length overall	6.15 m (20 ft 2 in)
Height overall, tail up:	
modified rudder	2.60 m (8 ft 6¼ in)
standard rudder	2.80 m (9 ft 2¼ in)
AREAS:	
Wings, gross	10.00 m² (107.6 sq ft)
WEIGHTS AND LOADINGS:	
Weight empty	565 kg (1,246 lb)
Max T-O weight	850 kg (1,874 lb)
Max wing loading	85.5 kg/m² (17.51 lb/sq ft)
Max power loading	5.74 kg/kW (9.43 lb/hp)
PERFORMANCE:	
Max level speed	162 kt (300 km/h; 186 mph)
Cruising speed	140 kt (260 km/h; 162 mph)
Stalling speed	59 kt (110 km/h; 68 mph)
Max rate of climb at S/L	838 m (2,750 ft)/min
Service ceiling	6,000 m (19,680 ft)
T-O run	280 m (920 ft)
Landing run	500 m (1,640 ft)
Endurance	3 h 30 min
g limits	+6/−4

UPDATED

Jurca MJ100 full size, wooden Supermarine Spitfire, powered by a 522 kW (700 hp) Hispano V-12 engine
0110695

Jurca MJ5 Sirocco, with 149 kW (200 hp) Lycoming, built by Daniel Lioult 0084372

NOIN

NOIN AERONAUTIQUE ALPAERO

R N 85, Chateauvieux, F-05130 Tallard
Tel/Fax: (+33 4) 92 54 15 04
e-mail: info@alpaero.com
Web: http://www.alpaero.com
DIRECTOR: Claude Noin

The Exel is the fourth design produced by Claude Noin, whose father was a noted soaring pilot of the 1950s and was first to cross the Alps by glider. The company's earlier Sirius motor glider is no longer available, although it will return to active status if interest in design is increased.

UPDATED

NOIN CHOUCAS
English name: Jackdaw
TYPE: Side-by-side ultralight kitbuilt.
PROGRAMME: Design started 1994 by Claude Noin; construction of prototype started 1995; first flight February

Prototype Noin Choucas (Rotax 503 piston engine)

1996; construction of production prototype started 1996, not been notified to *Jane's* by April 2001. Marketed as basic kit for advanced builders, quoted build time 1,200 hours. A special version for physically handicapped pilots is to be launched, if there is sufficient demand.

CUSTOMERS: One flying and one under construction by October 2001.
COSTS: Kit (glass fibre) €7,738, (carbon fibre) €8,756 (2002).
DESIGN FEATURES: Tail-less design, stated to be stallproof and spinproof; design aimed at meeting ultralight category

certification requirements while matching performance of modern gliders and conventional three-axis ultralight aircraft. Dihedral 4°; zero twist.

FLYING CONTROLS: Unconventional flying wing; pitch control by elevators in inboard wing trailing-edge; ailerons outboard; large rudder. Trim tabs on elevators; spoilers/airbrakes on upper surface of wing.

STRUCTURE: Standard fuselage moulded in half-shells of carbon fibre/epoxy; glass fibre/epoxy fuselage available at lower cost for advanced builders, but with weight penalty; composites wing structure with carbon fibre main spar, glass fibre/epoxy-stiffened Styrofoam ribs and birch plywood D-section leading-edge, covered with heat-shrink Dacron; moulded carbon fibre/epoxy ailerons; Lexan/polycarbonate canopy with carbon fibre frame.

LANDING GEAR: Tailwheel type, fixed; streamline fairings on main legs and wheels; drum brakes.

POWER PLANT: One 37.0 kW (49.6 hp) Rotax 503 UL-2V with 2.58:1 reduction gear driving a two-blade ground-adjustable DUC 20 carbon fibre propeller. Fuel capacity 36 litres (9.5 US gallons; 7.9 Imp gallons), of which 35 litres (9.25 US gallons; 7.7 Imp gallons) are usable.

ACCOMMODATION: Two, side by side beneath one-piece upward-hinging canopy; baggage shelf to rear of seats.

SYSTEMS: 14 V DC alternator provides electrical power.

AVIONICS (factory-built aircraft): *Comms:* VHF comm and intercom.

  Instrumentation: ASI, altimeter, variometer, slip indicator, compass, tachometer, fuel gauge, chronometer, CHT and EGT gauges and voltmeter.

EQUIPMENT: BRS 5 ballistic parachute recovery system optional.

DIMENSIONS, EXTERNAL:
Wing span	14.35 m (47 ft 1 in)
Length overall	5.35 m (17 ft 6½ in)
Height overall	1.95 m (6 ft 4¾ in)
Propeller diameter	1.60 m (5 ft 3 in)

DIMENSIONS, INTERNAL:
Cabin max width	0.97 m (3 ft 2 in)

AREAS:
Wings, gross	21.30 m² (229.3 sq ft)

WEIGHTS AND LOADINGS:
Weight empty	260-290 kg (573-639 lb)
Max T-O weight:	
FAI ultralight certification	450 kg (992 lb)
motor glider certification	510 kg (1,124 lb)

PERFORMANCE, POWERED:
Max level speed	97 kt (180 km/h; 111 mph)
Cruising speed at 75% power	70 kt (130 km/h; 81 mph)
Stalling speed	33 kt (60 km/h; 38 mph)
Max rate of climb at S/L	180 m (590 ft)/min
T-O run	100 m (330 ft)
Landing run	150 m (495 ft)
Range at 75% power, no reserves	
	175 n miles (325 km; 201 miles)

PERFORMANCE, UNPOWERED:
Best glide ratio	23
Min rate of sink	less than 1.00 m (3.28 ft)/s

UPDATED

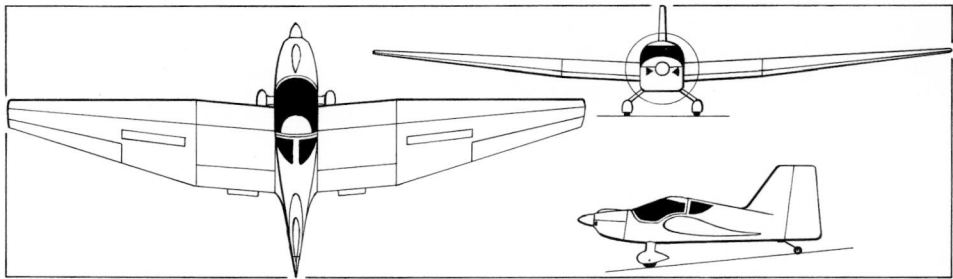

Noin Choucas tail-less light aircraft *(Jane's/James Goulding)*   0044680

Noin Exel single-seat ultralight motor glider   0098690

NOIN EXEL

TYPE: Single-seat ultralight/motor glider.

PROGRAMME: Design started 1997; first flight of prototype September 1998; construction of first production aircraft began 1999. Marketed as fast-build kit or ready to fly.

CUSTOMERS: Total of seven ordered and five built by December 2001, including one export to Chile.

COSTS: Kit less engine, propeller and instruments, €22,800, plus tax (2002).

DESIGN FEATURES: Pod-and-boom configuration, with T tail. Aimed at meeting FAI ultralight and JAR 22 certification criteria.

FLYING CONTROLS: Conventional and manual; cable-operated elevators, flaperons and rudder; three-position flaps, deflections −5, 0 and +5°. Airbrakes on upper wing surfaces.

STRUCTURE: Fuselage moulded in half-shells of glass fibre/epoxy with plywood bulkheads. Wings have carbon fibre spar caps and glass fibre skins. Fabric-covered rudder. Optional winglets. Quoted build time 300 hours.

LANDING GEAR: Non-retractable monowheel type; tailwheel incorporated in base of rudder for directional control on the ground; small outrigger wheel at each wingtip. Drum brake on mainwheel.

POWER PLANT: One 13.4 kW (18 hp) JPX D-320 two-stroke flat twin with 2.38:1 reduction drive to a carbon fibre two-blade folding pusher propeller (three-blade propeller being tested October 2001). Fuel capacity 19 litres (5.0 US gallons; 4.2 Imp gallons).

EQUIPMENT: Optional ballistic parachute recovery system.

DIMENSIONS, EXTERNAL:
Wing span	13.74 m (45 ft 1 in)
Length overall	5.90 m (19 ft 4¼ in)
Propeller diameter	1.28 m (4 ft 2½ in)

AREAS:
Wings, gross	11.62 m² (125.1 sq ft)

WEIGHTS AND LOADINGS:
Weight empty	196 kg (432 lb)
Max T-O weight	310 kg (683 lb)

PERFORMANCE, POWERED:
Never-exceed speed (VNE)	97 kt (179 km/h; 111 mph)
Cruising speed at 75% power	65 kt (120 km/h; 75 mph)
Stalling speed	34 kt (63 km/h; 40 mph)
Max rate of climb at S/L	120 m (394 ft)/min
T-O run	150 m (492 ft)
Landing run	140 m (459 ft)
Endurance	3 h
g limits	+4.4/−2.2

PERFORMANCE, UNPOWERED:
Best glide ratio	30
Min rate of sink	0.75 m (2.5 ft)/s

UPDATED

PJB

PJB AEROCOMPOSITE SARL

Zone Industrielle Aérodrome, Batiment 43, F-10150 St Leger sous Brienne
Tel: (+33 3) 25 27 57 11
Fax: (+33 3) 25 27 75 62
e-mail: contact@pjbaerocomposite.com
Web (1): http://www.pjb-aviation.com
Web (2): http://www.pjb-aerocomposite.com
SALES MANAGER: Pierre-Jean Back

PJB Aerocomposite, a division of spares supplier PJB Aviation Support, took over construction and development of the Aviakit Hermes, renaming it Vega. New 40,000 m² (430,550 sq ft) (total area) assembly plant at Brienne-le-Chateau inaugurated in 2002.

UPDATED

PJB Vega 2000 two-seat lightplane *(Jane's/Paul Jackson)*   NEW/0131735

PJB SPECTRA

TYPE: Side-by-side ultralight.

PROGRAMME: Announced 2001. High-wing version of Vega (see following entry). Formal launch in 2002, when further data were to be released.

UPDATED

PJB VEGA 2000

TYPE: Side-by-side ultralight.

DEVELOPMENT: Originally marketed as a kitbuilt by Aviakit as Hermes and around 20 built before project taken over by PJB Aerocomposite on 1 September 2001.

CURRENT VERSIONS: Four options of engine and two options of landing gear. **Vega 2000 TJ, TR5, TR9** and **TR10** with tricycle gear and Jabiru, Rotax 582, 912 UL and 912 ULS engines; **Vega 2000 CJ, CR5, CR9** and **CR10** with *Classique* (classic; tailwheel) gear and same engines.

CUSTOMERS: Total of 16 orders and 24 options by mid-2001.

COSTS: Kit: Without engine €25,875, with Rotax 912 €38,325. Completed aircraft €45,582 with Jabiru, €48,745 with Rotax 912 (2001).

DESIGN FEATURES: Low-wing tourer with high-mounted canopy and centre fuselage of sharply reduced cross-section aft of cockpit; strut-braced, mid-mounted tailplane; small underfin. Changes to design made by PJB include cutaway base of rudder and redesigned wing of shorter span.

NACA 23015 wing section with 5° dihedral. Quoted build time is 300 hours.

FLYING CONTROLS: Conventional and manual. Large ground-adjustable tab added to rudder by 1998. Plain flaps.

STRUCTURE: Fuselage is carbon-epoxy construction. Wings are wood/composites with carbon spar; strut-supported mid-mounted tailplane.

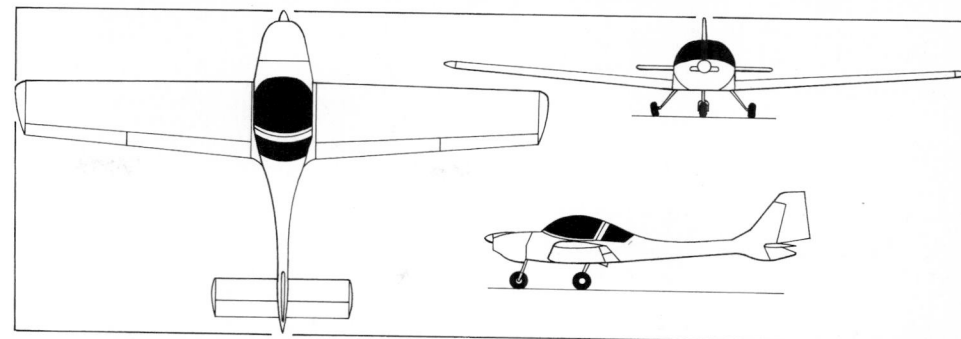

General arrangement of PJB Vega 2000 (*Jane's/Paul Jackson*)   *NEW*/0131722

LANDING GEAR: Choice of fixed tricycle or tailwheel layout; former has speed fairings.

POWER PLANT: Choice of 73.5 kW (98.6 hp) Rotax 912 ULS, 59.6 kW (79.9 hp) Rotax 912 UL or 59.7 kW (80 hp) Jabiru 2200 engines; 47.8 kW (64.1 hp) Rotax 582 UL is offered as option. Arplast two-blade, fixed-pitch, carbon fibre propeller. Fuel capacity 80 litres (21.1 US gallons; 17.6 Imp gallons) in two wing tanks.

DIMENSIONS, EXTERNAL:
Wing span	9.35 m (30 ft 8 in)
Length overall	6.50 m (21 ft 4 in)
Height overall	1.80 m (5 ft 10¾ in)

DIMENSIONS, INTERNAL:
Cabin max width	1.10 m (3 ft 7¼ in)

AREAS:
Wings, gross	11.50 m² (123.8 sq ft)

WEIGHTS AND LOADINGS:
Weight empty: Jabiru		255 kg (562 lb)
Rotax 912 UL		261 kg (575 lb)
Max T-O weight	450 kg (992 lb) or 472 kg (1,040 lb)*	

*According to local regulations

PERFORMANCE (with Rotax 912 UL engine):
Never-exceed speed (VNE)	135 kt (250 km/h; 155 mph)
Max operating speed (VMO)	124 kt (230 km/h; 143 mph)
Cruising speed	108 kt (200 km/h; 124 mph)
Stalling speed	34 kt (63 km/h; 40 mph)
Max rate of climb at S/L	420 m (1,380 ft)/min
T-O run	150 m (495 ft)
Landing run	100 m (330 ft)
Range	432 n miles (800 km; 497 miles)

UPDATED

POTTIER

AVIONS POTTIER

4 rue de Poissy, F-78130 Les Mureaux
Tel: (+33 1) 30 99 13 85
Fax: (+33 1) 34 92 97 26
e-mail: 113015.217@compuserve.com
DIRECTOR: Jean Pottier

The prolific Jean Pottier has designed numerous light aircraft for amateur construction or assembly. The P 200 series has been built commercially by several manufacturers in addition to being available to amateur builders. Most recent design is the **P 320UL**, the prototype of which (79-BC) was shown at the RSA Rally at Chambley in 2002.

UPDATED

POTTIER P 200 SERIES

TYPE: Four-seat kitbuilt.

PROGRAMME: First aircraft in P 200 series was single-seat P 210 S Coati with tailwheel-type landing gear. This does not appear to have entered production, but several two- to four-seat kitbuilt versions have appeared subsequently.

CURRENT VERSIONS: **P 220 S Koala:** Side-by-side two-seat version; 55.9 kW (75 hp) VW/Limbach engine standard. At least 20 kits delivered by early 1996; six flying and 20 building reported by early 1997. Modified version with Walter Mikron engine formerly produced in Czech Republic by Evektor-Aerotechnik (which see) and now marketed, with Limbach or Rotax engine, as the AT-3 by Aero Sp z.o.o of Poland. (see appropriate entry). Rotax-, Lycoming- and Continental-engined versions built in Czech Republic by EV-AT and in Italy by SG Aviation (both which see).

P 230 S Panda: Three-seater; 74.6 kW (100 hp) Continental engine standard. Prototype first flew 1990. Reportedly also available as **P230i** ultralight with 63.4 kW (85 hp) Rotax 912 engine.

P 240 S Saiga: Four-seat version (2 + 2); 134 kW (180 hp) Textron Lycoming engine. No specification data received for this version, and may no longer be current.

Pottier P 230 Panda built by a French amateur constructor (*Jane's/Paul Jackson*)   0109147

P 250 S Xerus: Two-seater; essentially tandem-seat version of Koala, but some dimensions differ.

P 270 S Amster: Four-seater (2 + 2); similar to P 240 S but with 112 kW (150 hp) Textron Lycoming engine.

CUSTOMERS: Sold to customers in Czech Republic, Finland, France, Germany, Netherlands, Norway, Poland and elsewhere.

DESIGN FEATURES: Typical low-wing monoplane with sweptback vertical tail; wing aerofoil section NACA 4415 on all versions.

FLYING CONTROLS: Conventional and manual. Aileron deflection +20/−15° on all versions; all-moving tailplane with automatic balance/trim tab. All versions have flaps (settings 0, 10 and 30°).

STRUCTURE: Mainly metal; some components in composites.

LANDING GEAR: All versions have non-retractable tricycle gear. Mainwheel/nosewheel sizes 330 and 270 mm respectively on two-seaters, 350 and 300 mm on three- and four-seaters. Hydraulic mainwheel brakes. Wheel speed fairings optional.

POWER PLANT: See under Current Versions above.

ACCOMMODATION: See under Current Versions above. Dual controls standard. One-piece canopy on P 250 S, two-piece on other versions.

DIMENSIONS, EXTERNAL:
Wing span: 220	6.50 m (21 ft 4 in)
230, 270	8.10 m (26 ft 7 in)
250	6.40 m (21 ft 0 in)
Wing chord, constant: all	1.25 m (4 ft 1¼ in)
Wing aspect ratio: 220	5.3
230, 270	6.6
250	5.2
Length overall: 220	5.54 m (18 ft 2 in)
230	6.35 m (20 ft 10 in)
250	5.82 m (19 ft 1¼ in)
270	6.45 m (21 ft 2 in)
Height overall: 220, 250	1.95 m (6 ft 4¾ in)
230	2.05 m (6 ft 8¾ in)
270	2.15 m (7 ft 0¾ in)
Tailplane span: 220, 230, 250	2.60 m (8 ft 6½ in)
270	3.00 m (9 ft 10 in)
Wheel track: 220	1.70 m (5 ft 7 in)
230, 270	1.75 m (5 ft 9 in)
250	1.40 m (4 ft 7 in)
Wheelbase: 220	1.32 m (4 ft 4 in)
230, 270	1.45 m (4 ft 9 in)
250	1.55 m (5 ft 1 in)
Propeller diameter: 220	1.44 m (4 ft 8¾ in)
230	1.72 m (5 ft 7¾ in)
250	1.50 m (4 ft 11 in)

DIMENSIONS, INTERNAL:
Cabin: Length: 220	1.64 m (5 ft 4½ in)
230, 250	2.00 m (6 ft 6¾ in)
270	2.10 m (6 ft 10¾ in)
Max width: 220	1.04 m (3 ft 5 in)
230, 270	1.10 m (3 ft 7¼ in)
250	0.70 m (2 ft 3½ in)
Max height: 220	1.05 m (3 ft 5¼ in)
230, 270	1.10 m (3 ft 7¼ in)
250	1.00 m (3 ft 3¼ in)

AREAS:
Wings, gross: 220	8.00 m² (86.1 sq ft)
230, 250	10.00 m² (107.6 sq ft)
250	7.90 m² (85.0 sq ft)
Ailerons (total): 220, 250	0.40 m² (4.31 sq ft)
230, 270	0.60 m² (6.46 sq ft)
Flaps (total): 220, 250	0.90 m² (9.69 sq ft)
230, 270	1.10 m² (11.84 sq ft)

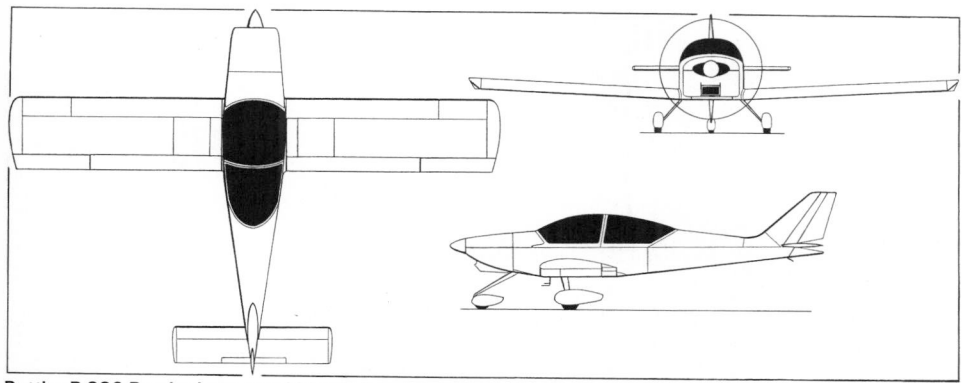

Pottier P 230 Panda three-seat kitbuilt (*Jane's/Paul Jackson*)   0044683

Vertical tail surfaces (total):

220, 230, 250	0.80 m² (8.61 sq ft)
270	1.00 m² (10.76 sq ft)

Horizontal tail surfaces (total):

220, 230, 250	1.80 m² (19.38 sq ft)
270	2.10 m² (22.60 sq ft)

WEIGHTS AND LOADINGS:

Weight empty: 220	275 kg (606 lb)
230	380 kg (838 lb)
250	240 kg (529 lb)
270	450 kg (992 lb)
Max T-O weight: 220	500 kg (1,102 lb)
230	700 kg (1,543 lb)
250	470 kg (1,036 lb)
270	870 kg (1,918 lb)
Max wing loading: 220	62.5 kg/m² (12.80 lb/sq ft)
230	70.0 kg/m² (14.34 lb/sq ft)
250	59.5 kg/m² (12.19 lb/sq ft)
270	87.0 kg/m² (17.82 lb/sq ft)
Max power loading: 220	8.95 kg/kW (14.70 lb/hp)
230	9.39 kg/kW (15.43 lb/hp)
250	9.70 kg/kW (15.94 lb/hp)
270	7.78 kg/kW (12.78 lb/hp)

PERFORMANCE:
Never-exceed speed (V$_{NE}$):

220, 230, 250	135 kt (250 km/h; 155 mph)
270	156 kt (290 km/h; 180 mph)
Max level speed: 220	113 kt (210 km/h; 130 mph)
230	121 kt (225 km/h; 140 mph)
250	119 kt (220 km/h; 137 mph)
270	146 kt (270 km/h; 168 mph)
Max cruising speed: 220	103 kt (190 km/h; 118 mph)
230	113 kt (210 km/h; 130 mph)
250	108 kt (200 km/h; 124 mph)
270	130 kt (240 km/h; 149 mph)

Stalling speed, flaps up:

220, 230	49 kt (90 km/h; 56 mph)
250	48 kt (88 km/h; 55 mph)
270	54 kt (100 km/h; 63 mph)

Stalling speed, flaps down:

220, 250	44 kt (80 km/h; 50 mph)
270	49 kt (90 km/h; 56 mph)

Rate of climb at S/L: 220	210 m (689 ft)/min
230	240 m (787 ft)/min
250	222 m (728 ft)/min
270	396 m (1,299 ft)/min
T-O run: 220	200 m (660 ft)
230	210 m (690 ft)
250	180 m (590 ft)
270	330 m (1,085 ft)
Landing run: 220	220 m (725 ft)
230	240 m (790 ft)
250	200 m (660 ft)
270	370 m (1,215 ft)

Range with max fuel:

220, 270	378 n miles (700 km; 435 miles)
230	405 n miles (750 km; 466 miles)
250	351 n miles (650 km; 403 miles)
g limit: all	+5.7

UPDATED

REIMS AVIATION

REIMS AVIATION SA

Aérodrome de Reims-Prunay, BP 2745, F-51062 Reims Cedex
Tel: (+33 3) 26 48 46 46
Fax: (+33 3) 26 49 13 60
*e-mail:*reims.aviation@reims-aviation.fr
Web: http://www.reims-aviation.fr
CEO: Jean-Paul Chaufour
VICE-PRESIDENT, MARKETING AND SALES: Gildas Illien
AREA SALES MANAGERS: Laurent Mesmin, Mathieu Quoi
EXTERNAL RELATIONS: Max Boirame

Originally Avions Max Holste, founded 1933; Cessna acquired 49 per cent share in February 1960; name then changed to Reims Aviation, which licensed to manufacture Cessna aircraft for sale in Europe, Africa and Asia, but stopped making small piston-engined types when Cessna did (see below); Reims had built 6,351 aircraft of all types by late 2001. Reims developed twin-turboprop F406 Caravan II (see below); but Cessna sold its share in Reims Aviation to Compagnie Française Chaufour Investissement (CFCI) in early 1989. Remains French distributor for Cessna 172R Skyhawk, 182S Skylane and 206 Stationair.

Reims makes components for Dassault Falcons and Mirage 2000, ATR 42/72, Airbus A300/A310/A320 series/A330/A340, Fairchild Dornier 728JET, Embraer ERJ-145 and ERJ-170/190. Other activity is maintenance of general aviation aircraft via its Reims Aviation Maintenance et Service (RAMS) subsidiary. At the Paris Air Show in June 1999, Reims announced plans to build a new maintenance facility for aircraft in the ATR and Boeing 737 class at the former NATO reserve air base at Vatry. The company offers a total rebuild of Cessna F 152 light aircraft, effectively to 'zero life' standard. Interest was shown by the company in 1999 in becoming European agent for the Antonov An-140 twin-prop airliner (which see).

Office and factory floor space covers 28,300 m² (304,500 sq ft).

UPDATED

REIMS F406 CARAVAN II

TYPE: Utility turboprop twin.
PROGRAMME: Announced mid-1982; first flight (F-WZLT) 22 September 1983; French certification 21 December 1984, FAA later; first flight production F406 (F-ZBEO) 20 April 1985.
CURRENT VERSIONS: **F406 Caravan II:** Initial production version, available in passenger, freight (with underbelly cargo pod), medevac, skydiving, aerial survey, training, navaid calibration and target towing variants.

F406 Mark II: Upgraded version announced (as F406 NG New Generation) in October 2000. First delivery due late 2003 to Daihyaku Shoji, Reims' Japanese agent. Features include 473.5 kW (635 shp) PT6A-135A engines driving four-blade propellers; reduced structural weight; 227 kg (500 lb) increase in maximum take-off weight; quick-release access panels for easier maintenance; new lightweight instrument panel with liquid crystal displays; redesigned cabin interior by Air Esthetic featuring composite panels that can be painted or upholstered, improved acoustic and thermal insulation, new ergonomically designed seats with integral harnesses for improved comfort on long missions and integral window blinds; and standard Bendix/King Silver Crown avionics with Silver Crown Plus comms and two-tube EFIS display. Three standard cabin interiors offered: commuter, special missions and VIP. Take-off, climb and OEI rate of climb performance are improved, and endurance increased by up to one hour.

Additionally, six versions of the basic **Special Missions** configuration, each with a ventral 360° radar:
Vigilant: Surveillance version (land or sea) with mission equipment to suit customer requirement, including (in latest configuration) Thales Airborne Maritime Situation Control System (AMASCOS) which comprises belly-mounted Telephonics AN/APS-143, Texas Instruments 1022 or Thales/DASA Ocean Master 100/200 360° surveillance radars, chin-mounted FLIR turret for FSI Ultra, Inframetrics Mk III, SAGEM Hesis, Star Safire, Thales Chlio or Wescam FLIR, mission management and communications systems and single operator's console in Srs 100 form; or with additional console for electronic surveillance operator in Srs 200. BAE Systems Seaspray 2000 radar for Scottish Fisheries; Texas Instruments AN/APS-134 radar for Australian Customs.

Vigilant Frontier: Border patrol and anti-drug surveillance version with mission equipment including surveillance radar in anti-drug configuration, FLIR and datalink.

Vigilant Polmar II: Pollution surveillance (maritime police) version with SAGEM Cyclope 2000 IR linescanner and Ericsson or Terra SLAR.

Vigilant Polmar III: As Polmar II, but based on F406 Mark II airframe. One ordered by French Customs in 2001 for delivery in third quarter of 2002.

Vigilant Surmar: Surveillance (maritime) version with 360° radar, such as Texas Instruments AN/APS-134 or Telephonics AN/APS-143 and optional equipment including Litton night vision system. In 1996 Reims developed underwing hardpoints for the F406 which enable it to carry light weapons such as machine gun pods or rocket launchers, or camera pods or SAR liferafts.

Vigilant Surpolmar: Maritime surveillance version for Hellenic Coast Guard, combining Surmar and Polmar missions. Equipment specific to Greek aircraft comprises Surmar suite of Honeywell 1500B radar and FLIR. Systems Star Safire turret aided by Caledonian Airborne Systems track-while-scan system for 20-target tracking; and Polmar suite of Ericsson SLAR, Swedish Space Corporation MSS 5000 and SAGEM Cyclope 2000 IR linescan

Vigilant Comint/Imint: Dedicated communications and imaging intelligence version of the Caravan II, developed jointly with Thomson-CSF Communications, first shown at Paris '97. Equipped with Thomson TCC SAS airborne Elint system, an operator's console with MS Windows NT-based workstation and an array of direction-finding antennas housed in a ventral radome. Is also first Vigilant with a pair of underwing pylons for carriage of stores which can include gun or rocket pods and airdroppable SAR equipment.

CUSTOMERS: One prototype and 89 production aircraft built by mid-2001. F406 is operated by 45 customers in 25 countries including ALH (Netherlands), Acrop Mad (Salvador), Aerovia (Spain), Aerotuy (Venezuela), Air Atlantic (Netherlands), Air Guyane (French Guyana), Air St Martin (Guadeloupe), Aircraft Ing (Kenya), Arcus Air Logistic (Germany), Aviazur (New Caledonia), Blue Bird Aviation (Kenya), Comair (South Africa), DDF (Zimbabwe), Direct Flight (UK), Excel Air (Malta), Fehlhaber Flugdienst (Germany), French Army (two used for target towing), French Customs Service (see below), Grupo America (Salvador), Hellenic Ministry of Merchant Marine (see below), Hellras (Kenya), Kensoma Ltd (UK), Luchtvaartmij (Netherlands), Namibian Fisheries, National Jet Systems (Australia, see below), NDG Aviation (USA), Overflight Support (South Africa), Polarwing (Finland), Republic of Korea Navy (see below), SAL (Angola), Scan Equipment (Norway), Sembawang (Singapore), Seychelles Coast Guard, Somacvram (Madagascar), Span Aviation (India), TCS (Pakistan), TAAG (Angola), Tanzanian Air, Tawakal Airlines, TCS (Saudi Arabia), Trackmark (Kenya), US Customs Service, US Navy and Wing Airline (Belgium).

French Customs Service received 10 in Surmar and two in Polmar I/II configuration. One in Polmar III configuration (No. 90), ordered in late 2000 for delivery by 2002. First Surmar (No. 74, although registered as a Vigilant) handed over 11 January 1995, followed by Nos. 75 and 77 on 12 September 1995, the latter two replacing Cessna 404s in Martinique, French Antilles; National Jet Systems Australia ordered three Vigilants for delivery early 1996 for Surveillance Australia programme (aircraft Nos. 76, 78 and 79) and Scottish Fisheries has four Vigilants for fisheries patrol, the latest delivered in April 1998.

Recent customers for the F406 include the Republic of Korea Navy, which ordered five in target tug configuration, the first delivered on 23 November 1998 and the last in mid-June 1999, total contract value being US$24 million (1997); and the Hellenic Ministry of Merchant Marine, which, in June 1999, ordered two for delivery in late 2000 as Surpolmar variants and additional one in November 1999; the first Surpolmar (F-WWSS/AC-21) for this contract made its maiden flight on 21 August 2000 and was delivered on 27 March 2001; remaining two received by December 2001, at which time a fourth was under negotiation. Total 90 sold by June 2001 (excluding prototype). An agreement with the Brazilian government signed in April 1998 provides for the purchase of three aircraft in multisensor configurations. Daihyaku Shoji of Japan ordered one F406 Mark II during 2000, for delivery

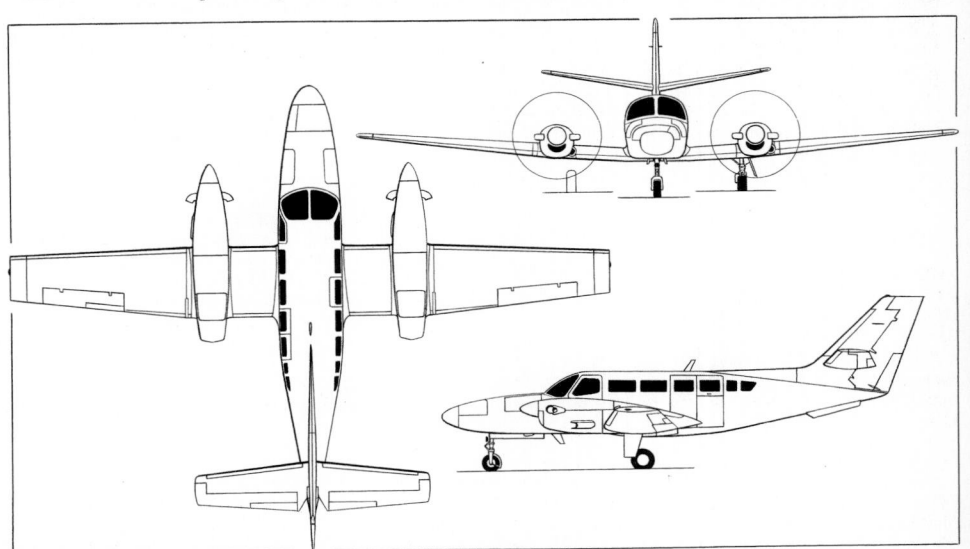

Reims F406 Caravan II light business and utility transport (*Jane's/Dennis Punnett*)

in mid-2003 and operation by an undisclosed local customer.

COSTS: Standard commuter aircraft US$2.1 million (2000). Maritime patrol versions typically US$3 million to US$4 million with mission equipment. Korean Navy contract for five aircraft valued at US$24 million, including technician training and one-year technical support (1997). Greek contract for two Surmars valued at FFr90 million (1999); single Polmar FFr33 million (1999). Direct operating cost US$338 per hour (1997).

DESIGN FEATURES: Extrapolated from Cessna Conquest airframe; wing section NACA 23018 at root and 23012 at tip; dihedral 3° 30′ on centre-section, 4° 55′ on outer panels; twist –3°; incidence 2° at root; fin offset 1° to port; tailplane dihedral 9°; cabin not pressurised.

FLYING CONTROLS: Conventional and manual; Trim tabs in elevators, port aileron and rudder; hydraulically operated Fowler flaps.

STRUCTURE: Conventional light metal with three-spar fail-safe wing centre-section to SFAR 41C; two-spar outer wings.

LANDING GEAR: Hydraulically retractable tricycle type with single wheel on each unit. Mainwheel tyre size 22×7.75-10, nosewheel 6.00-6. Main units retract inward into wing, nosewheel rearward. Emergency extension by means of a 138 bar (2,000 lb/sq in) rechargeable nitrogen bottle. Cessna oleo-pneumatic shock-absorbers. Main units of articulated (trailing link) type. Single-disc hydraulic brakes. Parking brake.

POWER PLANT: Two Pratt & Whitney Canada PT6A-112 turboprops (each 373 kW; 500 shp), each driving a McCauley 9910535-2 three-blade, reversible-pitch, and automatically feathering metal propeller. Fuel capacity 1,823 litres (481 US gallons; 401 Imp gallons) of which 1,798 litres (475 US gallons; 395.5 Imp gallons) usable. Oil capacity 17.2 litres (4.5 US gallons; 3.8 Imp gallons).

ACCOMMODATION: Crew of two and up to 12 passengers, in pairs facing forward, with centre aisle, except at rear of cabin in 12/14-seat versions. Alternative basic configurations for six VIP passengers in reclining seats in business version, and for operation in mixed passenger/freight role. Business version has partition between cabin and flight deck, and toilet on starboard side at rear. Split main door immediately aft of wing, on port side, with built-in airstair in downward-hinged lower portion. Optional cargo door forward of this door to provide single large opening. Overwing emergency exit on each side. Passenger seats removable for cargo carrying, or for conversion to ambulance, air photography, maritime surveillance and other specialised roles. Baggage compartments in nose, with three doors, at rear of cabin and in rear of each engine nacelle. Ventral cargo pod optional.

SYSTEMS: Freon air conditioning system of 17,500 BTU capacity, plus engine bleed air and electric boost heating. Electrical system includes 28 V 250 A starter/generator on each engine and 39 Ah Ni/Cd battery. Hydraulic system, pressure 120 bar (1,750 lb/sq in), for operation of landing gear. Separate hydraulic system for brakes. Optional Goodrich pneumatic de-icing of wings and tail unit, and electric windscreen de-icing.

AVIONICS: Standard Honeywell Silver Crown; Gold Crown optional.

Comms: Dual Honeywell transceivers.

Radar: Honeywell RDR 2000 weather radar optional. Maritime surveillance radar as described under Current Versions.

Flight: Dual ADF and marker beacon receiver. Autopilot optional.

Instrumentation: Provision for equipment to FAR Pt 135A standards, including dual controls and instrumentation for co-pilot.

EQUIPMENT: Optional cargo interior includes heavy-duty sidewalls, utility floorboards, cabin floodlighting and cargo restraint nets. Optional pylons under each wing for carriage of stores including gun or rocket pods and airdroppable SAR equipment.

DIMENSIONS, EXTERNAL:
Wing span	15.08 m (49 ft 5½ in)
Wing aspect ratio	9.7
Length overall	11.89 m (39 ft 0¼ in)
Height overall	4.01 m (13 ft 2 in)
Tailplane span	5.87 m (19 ft 3 in)
Wheel track	4.28 m (14 ft 0½ in)
Wheelbase	3.81 m (12 ft 6 in)
Propeller diameter	2.36 m (7 ft 9 in)
Cabin door: Height	1.27 m (4 ft 2 in)
Width	0.58 m (1 ft 10¾ in)
Cargo double door (optional):	
Total width	1.24 m (4 ft 1 in)

DIMENSIONS, INTERNAL:
Cabin (incl flight deck): Length	5.71 m (18 ft 8¾ in)
Max width	1.42 m (4 ft 8 in)
Max height	1.31 m (4 ft 3¼ in)
Min height (at rear)	1.21 m (3 ft 11½ in)
Width of aisle	0.29 m (11½ in)
Volume	8.6 m³ (305 cu ft)
Baggage compartment (nose):	
Length	2.00 m (6 ft 6¾ in)
Volume	0.74 m³ (26.0 cu ft)
Nacelle lockers: Length	1.55 m (5 ft 1 in)
Width	0.73 m (2 ft 4¾ in)
Baggage volume: total, internal	2.2 m³ (79 cu ft)
incl cargo pod	3.5 m³ (124 cu ft)

AREAS:
Wings, gross	23.48 m² (252.8 sq ft)
Ailerons (total)	1.36 m² (14.64 sq ft)
Trailing-edge flaps	3.98 m² (42.84 sq ft)
Fin	4.05 m² (43.59 sq ft)
Rudder, incl tab	1.50 m² (16.15 sq ft)
Tailplane	5.81 m² (62.54 sq ft)
Elevators, incl tabs	1.66 m² (17.87 sq ft)

WEIGHTS AND LOADINGS:
Standard empty weight	2,283 kg (5,033 lb)
Max payload	2,219 kg (4,892 lb)
Max fuel	1,444 kg (3,183 lb)
Max ramp weight	4,502 kg (9,925 lb)
Max T-O and landing weight	4,468 kg (9,850 lb)
Max zero-fuel weight	3,856 kg (8,500 lb)
Max wing loading	190.3 kg/m² (38.97 lb/sq ft)
Max power loading	5.99 kg/kW (9.85 lb/shp)

PERFORMANCE:
Max operating Mach No. (MMO)	0.52
Max operating speed (VMO)	229 kt (424 km/h; 263 mph) IAS
Max cruising speed	246 kt (455 km/h; 283 mph)
Econ cruising speed	200 kt (370 km/h; 230 mph)
Stalling speed:	
wheels and flaps up	94 kt (174 km/h; 108 mph) IAS
wheels and flaps down	81 kt (150 km/h; 93 mph) IAS
Max rate of climb at S/L	564 m (1,850 ft)/min
Rate of climb at S/L, OEI	121 m (397 ft)/min
Service ceiling	9,145 m (30,000 ft)
Service ceiling, OEI	4,935 m (16,200 ft)
T-O run	526 m (1,725 ft)
T-O to 15 m (50 ft)	803 m (2,635 ft)
Landing from 15 m (50 ft), without reverse pitch	674 m (2,215 ft)
Range with max fuel, at max cruising speed, 45 min reserves	1,153 n miles (2,135 km; 1,327 miles)
g limits	+3.6/–1.44

OPERATIONAL NOISE LEVELS:
Flyover	72.0 EPNdB

UPDATED

REIMS FUTURE CARGO AIRCRAFT
At the Paris Air Show in June 1999, Reims CEO Jean-Paul Chaufour revealed plans to develop, or produce under licence, at Vatry, a new cargo aircraft with a 50 to 60 m³ (1,766 to 2,119 cu ft), 4,000 to 5,000 kg (8,818 to 11,023 lb) cargo hold and range of 378 to 1,080 n miles (700 to 2,000 km; 435 to 1,243 miles). By late 2001, nothing further had been heard of the proposal.

UPDATED

Reims F406 Surpolmar of Hellenic Coast Guard   NEW/0102907

ROBIN
ROBIN AVIATION
1 route de Troyes, F-21121 Darois
Tel: (+33 3) 80 44 20 50
Fax: (+33 3) 80 35 60 80
Web: http://www.robinaviation.com
NON-EXECUTIVE CHAIRMAN: Philippe Corne
CEO AND GENERAL MANAGER: Guy Pellissier
SALES MANAGER, EUROPE: Pierre Pelletier

Formed October 1957 as Centre Est Aéronautique; name changed to Avions Pierre Robin 1969; acquired July 1988 by Compagnie Française Chaufour Investissement (CFCI) and incorporated into Aéronautique Service group (now Apex International) with Robin SA (after-sales support company of Dijon Val-Suzon); Pierre Robin left company 1990. Total of 3,700 aircraft produced by early 2002 (excluding 35 Sperwer UAVs). Factory area (Constructions Aéronautiques de Bourgogne) 9,000 m² (96,875 sq ft); workforce 100.

Former Robin subdivisions, Constructions Aéronautiques de Bourgogne and Aéronautique Finance et Services, are now separate companies. The first-mentioned is responsible for the construction of Robin aircraft, as explained in the entry for Aéronautique Service.

Latest addition to Robin range is R 2120 Alpha trainer, introduced 2001.

UPDATED

ROBIN HR 200
TYPE: Primary prop trainer/sportplane.
PROGRAMME: New production version, since 1993, of original Robin HR 200/120B, which had first flown in 1971 and was built through much of 1970s (see 1977-78 *Jane's*); incorporates minor modifications.

CURRENT VERSIONS: **Robin HR 200/120B:** *As described.*

Robin HR 200/160: Power plant as Robin 2160, but with prominent exhaust below cabin and retaining Robin 200's rudder. Available from 1999.

CUSTOMERS: Total of 180 built between 1971 and late 2000, including 109 in first series. Combined 330 HR 200/R 2120/R 2160s built by early 2002.

COSTS: FFr420,550 with Package 1 avionics; FFr497,785 with Package 2.

DESIGN FEATURES: Typical Robin low-wing, tricycle landing gear tourer/trainer. Constant-chord, unswept wings and tailplane, latter all-moving. Relaunched version has new instrument panel, adjustable seats, new engine cowlings and propeller spinner, and improved anti-corrosion treatment.

Robin HR 200/120B two-seat trainer, with additional scrap views of HR 200/160 forward fuselage and R 2160 empennage (*Jane's/Paul Jackson*)   NEW/0137208

Robin HR 200/160 (one Textron Lycoming flat-four) (*Jane's/Paul Jackson*)   0093691

Wing section NACA 64A515 (mod); dihedral 6° 18' from roots; incidence 6°; no sweepback.

FLYING CONTROLS: Manual. Conventional rudder and cable-actuated Frise-type ailerons. One-piece all-moving tailplane with trim and anti-balance tabs; electrically actuated trailing-edge slotted flaps.

STRUCTURE: All-metal; aluminium alloy stressed skin and ribs.

LANDING GEAR: Non-retractable tricycle type, with single wheel on each unit; nosewheel leg offset to starboard, steered by rudder bar; streamline leg and wheel fairings; damped tailskid; hydraulic disc brakes and parking brake. All three wheels and tyres are size 380×150.

POWER PLANT: One 88.0 kW (118 hp) Textron Lycoming O-235-L2A flat-four engine, driving a Sensenich two-blade, fixed-pitch aluminium propeller. Fuel capacity 120 litres (31.7 US gallons; 26.4 Imp gallons). Auxiliary tanks optional.

ACCOMMODATION: Pilot and passenger side by side under forward-sliding jettisonable canopy with anti-glare tint; dual stick controls; dual left-hand throttles; dual toe brakes.

SYSTEMS: Cabin ventilated and heated, with windscreen defrosting standard. Electrical system includes 12 V 32 Ah battery, 12 V 50 A alternator and starter.

AVIONICS: Customer selection; following are available.
  Comms: Bendix/King KY 97 VHF or KX 155-38 nav/com/glideslope; SPA 400 intercom; dual PTT buttons; KT 76A transponder with ACK 30 height encoder.
  Flight: KI 208 VOR/LOC. Flight hour recorder.
  Instrumentation: Vacuum-driven gyro horizon and direction indicator (engine-driven pump), electric turn and slip indicator, magnetic compass, rate of climb indicator and rest of blind-flying instruments.

EQUIPMENT: Navigation lights, strobe lights, landing light.

DIMENSIONS, EXTERNAL:
Wing span	8.33 m (27 ft 4 in)
Wing chord, constant	1.50 m (4 ft 11 in)
Wing aspect ratio	5.6
Length overall	6.35 m (20 ft 10 in)
Height overall	1.94 m (6 ft 4½ in)
Tailplane span	2.64 m (8 ft 8 in)
Wheel track	2.88 m (9 ft 5½ in)
Wheelbase	1.465 m (4 ft 9½ in)

DIMENSIONS, INTERNAL:
Cabin max width	1.07 m (3 ft 6 in)

AREAS:
Wings, gross	12.50 m² (134.5 sq ft)
Ailerons, total	1.06 m² (11.41 sq ft)
Trailing-edge flaps, total	1.34 m² (14.42 sq ft)
Elevators, incl tabs	2.03 m² (21.85 sq ft)

WEIGHTS AND LOADINGS:
Weight empty	525 kg (1,157 lb)
Baggage capacity	35 kg (77 lb)
Max T-O weight	780 kg (1,719 lb)
Max wing loading	62.4 kg/m² (12.78 lb/sq ft)
Max power loading	8.86 kg/kW (14.57 lb/hp)

PERFORMANCE:
Cruising speed:	
at 75% power	115 kt (213 km/h; 132 mph)
at 65% power	105 kt (194 km/h; 121 mph)
Stalling speed	52 kt (96 km/h; 60 mph)
Max rate of climb at S/L	235 m (770 ft)/min
Service ceiling	3,900 m (12,800 ft)
T-O and landing run	230 m (755 ft)
T-O to 15 m (50 ft)	510 m (1,675 ft)
Landing from 15 m (50 ft)	445 m (1,460 ft)
Range	566 n miles (1,050 km; 652 miles)
Endurance	4 h 35 min

UPDATED

ROBIN R 2120 ALPHA

TYPE: Primary prop trainer/sportplane.

PROGRAMME: Announced in early 2001. Prototype (F-WZZX) completed mid-2000; demonstrator (F-WZZY) shown at Aero '01, Friedrichshafen, April 2001. First production aircraft (EC-HMG) to Spain by early 2001.

CURRENT VERSIONS: **R 2120U:** Baseline (utility) version.
  Alpha 120T: Marketing designation.

CUSTOMERS: Deliveries in 2001 included six to New Zealand dealer Izard Pacific Aviation. Others to UK.

DESIGN FEATURES: Combines fuselage, vertical tail surfaces, fuel capacity and power plant of Robin HR 200/120B with wings and horizontal tail surfaces of R 2160.

Differences from HR 200/120B listed below:

LANDING GEAR: Wheel fairings optional. Tyres 300×150.

EQUIPMENT: Landing/taxi lights relocated to port wingtip.

DIMENSIONS, EXTERNAL:
Wing span	8.33 m (27 ft 4 in)
Wing chord, constant	1.555 m (5 ft 1¼ in)
Length overall	6.64 m (21 ft 9½ in)
Tailplane span	3.04 m (9 ft 11¾ in)
Tailplane chord	0.725 m (2 ft 4½ in)
Wheelbase	1.43 m (4 ft 8¼ in)
Propeller diameter	1.83 m (6 ft 0 in)

WEIGHTS AND LOADINGS:
Weight empty	544 kg (1,199 lb)
Max T-O weight	800 kg (1,764 lb)
Max wing loading	64.0 kg/m² (13.11 lb/sq ft)
Max power loading	9.09 kg/kW (14.94 lb/hp)

PERFORMANCE:
Cruising speed at 75% power at 2,591 m (8,500 ft):	104 kt (193 km/h; 120 mph)
Max rate of climb at S/L	192 m (630 ft)/min
T-O to 15 m (50 ft)	490 m (1,610 ft)
Landing from 15 m (50 ft)	480 m (1,575 ft)
Range with max fuel	510 n miles (944 km; 586 miles)
g limits	+6/−3.5

UPDATED

ROBIN R 2160

TYPE: Two-seat lightplane.

PROGRAMME: Certified in France in mid-1978 and in USA (FAR Pt 23 Aerobatic and Utility category) 15 November 1982. Some aircraft assembled in Canada (1983-85). Production then ceased, but restarted in France January 1994.

CURRENT VERSIONS: **R 2160 D:** Baseline version, *as described below.*
  R 2160i: Certified 8 July 1998, with 119 kW (160 hp) Textron Lycoming AEIO-320-D2B fuel-injected flat-four. Three delivered by end of 1998, including two to CATC training college in Thailand.
  Alpha 160A: Aerobatic version.
  Alpha 160Ai: Aerobatic version of R 2160i.
  Alpha 160T: Basic trainer; introduced 2002.

Demonstrator Robin R 2120 Alpha on show at Friedrichshafen in April 2001 (*Jane's/Paul Jackson*)   0110869

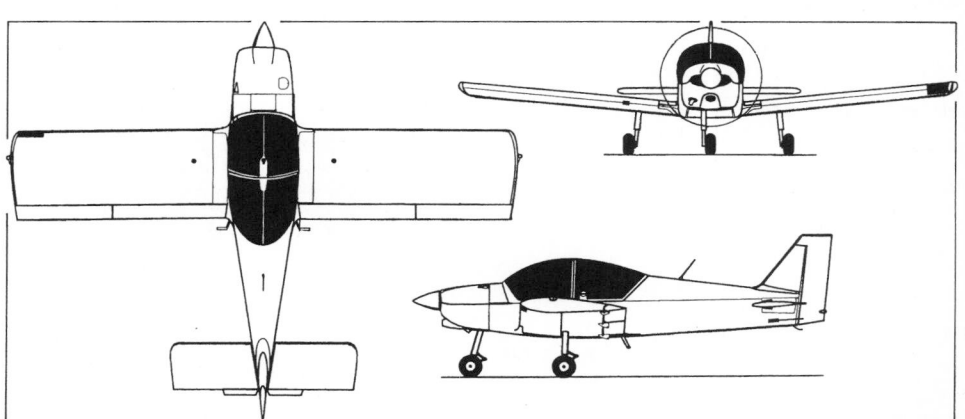

Robin R 2120 Alpha   0110870

CUSTOMERS: Total of over 130 sold in Europe, Australia, New Zealand, Canada and USA by early 2001, including 97 in first series.

DESIGN FEATURES: Generally as Robin HR 200, but with extended rudder chord, underfin and other detail changes. Wing section NACA 23015; dihedral 6° 20′; incidence 3° at root.

FLYING CONTROLS: Manual. Conventional fully balanced slotted ailerons and horn-balanced rudder without tabs. All-moving tailplane with anti-balance tabs each side. Slotted flaps.

STRUCTURE: Aluminium alloy wing spars and skinning; semi-monocoque fuselage.

LANDING GEAR: Non-retractable tricycle type with fairing, oleo-pneumatic shock-absorber and tyre (380×150) on each leg. Cleveland disc brakes on mainwheels. Nosewheel steering through rudder bar.

POWER PLANT: One 119 kW (160 hp) Textron Lycoming O-320-D2A flat-four engine, driving a Sensenich 74DM6S5-2-64 two-blade, fixed-pitch aluminium propeller. Provisional design studies completed for installation of 134 to 149 kW (180 to 200 hp) engine. Christen inverted oil system standard. Fuselage fuel tank, capacity 120 litres (31.7 US gallons; 26.4 Imp gallons). If operated in Utility category, optional fuel tank of 160 litres (42.3 US gallons; 35.2 Imp gallons).

ACCOMMODATION: Two seats side by side.

SYSTEMS: 12 V 24 Ah electrical system.

AVIONICS: *Flight:* VOR, ADF, ILS, GPS and other items, at customer's choice.
Instrumentation: Blind-flying panel optional.

DIMENSIONS, EXTERNAL:
Wing span	8.33 m (27 ft 4 in)
Wing chord, constant	1.56 m (5 ft 1¼ in)
Wing aspect ratio	5.3
Length overall	7.10 m (23 ft 3½ in)
Height overall	2.13 m (7 ft 0 in)
Tailplane span	3.03 m (9 ft 11¼ in)
Wheel track	2.91 m (9 ft 6½ in)
Wheelbase	1.44 m (4 ft 8½ in)
Propeller diameter	1.88 m (6 ft 2 in)

DIMENSIONS, INTERNAL:
Cabin: Max width	1.07 m (3 ft 6 in)
Height (seat cushion to canopy):	
R 2160	0.92 m (3 ft 0¼ in)

AREAS:
Wings, gross	13.01 m² (140.0 sq ft)

WEIGHTS AND LOADINGS:
Weight empty	555 kg (1,224 lb)
Max T-O weight: Aerobatic	800 kg (1,764 lb)
Utility	900 kg (1,984 lb)
Max baggage weight	40 kg (88 lb)
Max wing loading: Aerobatic	61.5 kg/m² (12.59 lb/sq ft)
Utility	69.2 kg/m² (14.17 lb/sq ft)
Max power loading: Aerobatic	6.71 kg/kW (11.02 lb/hp)
Utility	7.55 kg/kW (12.40 lb/hp)

PERFORMANCE (O-320 engine at Aerobatic max T-O weight):
Never-exceed speed (V_NE)	180 kt (333 km/h; 207 mph)
Max level speed	138 kt (256 km/h; 159 mph)
Cruising speed:	
at 75% power	131 kt (241 km/h; 151 mph)
at 65% power	120 kt (222 km/h; 138 mph)
Stalling speed: flaps up	52 kt (96 km/h; 60 mph)
flaps down	46 kt (86 km/h; 53 mph)
Max rate of climb at S/L	314 m (1,030 ft)/min
Service ceiling (30.5 m; 100 ft/min rate of climb)	4,575 m (15,000 ft)
T-O and landing run	230 m (755 ft)
T-O to 15 m (50 ft)	410 m (1,345 ft)
Landing from 15 m (50 ft)	425 m (1,395 ft)
Range at 75% power:	
standard fuel	363 n miles (672 km; 418 miles)
optional fuel	484 n miles (896 km; 557 miles)
g limits	+6/−3

UPDATED

ROBIN DR 400 DAUPHIN
English name: Dolphin

TYPE: Four-seat lightplane.

PROGRAMME: First flight original DR 400 Petit Prince 15 May 1972; French and UK certification 1977; DR 400 Dauphin introduced 1979; improvements introduced 1988 and 1993.

CURRENT VERSIONS: **DR 400/120 Dauphin 2+2:** Production version with 88.0 kW (118 hp) engine, to carry two adults and two children.

DR 400/125i: Version of Dauphin 2+2 with fuel-injected 93.2 kW (125 hp) Teledyne Continental IO-240 driving three-bladed Mühlbauer constant-speed propeller, diameter 1.70 m (5 ft 7 in); prototype (F-WNNK) first

Robin R 2160 (Textron Lycoming O-320 flat-four) (*Jane's/Paul Jackson*)   NEW/0131736

flown 1995 and made its public debut at the Paris Air Show in June 1995; engine/propeller combination improves take-off, climb and cruising performance and reduces noise levels; one delivered in 1996. Weights and performance data in 2000-01 and earlier editions. No evidence of production.

DR 400/140B Dauphin 4: Full four-seater with 119 kW (160 hp) engine. Previously known as Major 80.

DR 400/160 and **DR 400/180:** Described separately.

CUSTOMERS: Some 1,815 of DR 400 series built by early 2002. Recent exports mainly to Germany, Switzerland and UK.

DESIGN FEATURES: Generally as for HR 200.
Wing section NACA 23013.5 modified with leading-edge droop; centre panels parallel chord, slight twist; outer panels tapered with dihedral 14°; twist −6°.

FLYING CONTROLS: Manual. Conventional ailerons and rudder. All-moving tailplane with trimmable anti-balance tab on each side; plain flaps.

STRUCTURE: All-wood; single box spar with ribs threaded over box; plywood-covered leading-edge box; fabric covering elsewhere; fuselage plywood-covered; flaps all-metal and interchangeable; ailerons interchangeable.

LANDING GEAR: Non-retractable tricycle type, with oleo-pneumatic shock-absorbers and hydraulically actuated disc brakes. All three wheels and tyres are size 380×150, pressure 1.57 bar (23 lb/sq in) on nose unit, 1.77 bar (26 lb/sq in) on main units. Nosewheel steerable via rudder bar. Fairings over all three legs and wheels. Tailskid with damper. Toe brakes and parking brake.

POWER PLANT: *Dauphin 2+2:* One 88.0 kW (118 hp) Textron Lycoming O-235-L2A flat-four engine, driving a Sensenich 72CKS6-0-56 two-blade, fixed-pitch aluminium propeller, or Hoffmann two-blade wooden propeller.
Dauphin 4: One Textron Lycoming O-320-D2A flat-four engine developing 104 kW (140 hp) at 2,300 rpm and 119 kW (160 hp) at 2,700 rpm. Sensenich 74DM6S5-2-64 propeller.
Both versions have fuel tank in fuselage, capacity 110 litres (29.1 US gallons; 24.2 Imp gallons) with filler port on left side; Dauphin 4 has optional 51 litre (13.5 US gallon; 11.2 Imp gallon) auxiliary tank, with filler port on right side. Oil capacity 5.7 litres (1.5 US gallons; 1.25 Imp gallons).

ACCOMMODATION: Enclosed cabin, with seats for three or four persons. Maximum weight of 154 kg (340 lb) on front pair and 136 kg (300 lb), including baggage, at rear in Dauphin

2+2. Additional 55 kg (121 lb) of disposable load in Dauphin 4. Access via forward-sliding jettisonable transparent canopy. Dual controls standard. Cabin heated and ventilated. Baggage compartment with internal access.

SYSTEMS: Standard equipment includes a 12 V 50 A alternator, 12 V 32 Ah battery, electric starter, audible stall warning and windscreen de-icing.

AVIONICS: Radio, blind-flying equipment, and navigation, landing and anti-collision lights, to customer's requirements.

DIMENSIONS, EXTERNAL:
Wing span	8.72 m (28 ft 7¼ in)
Wing chord:	
centre-section, constant	1.71 m (5 ft 7½ in)
at tip	0.90 m (2 ft 11½ in)
Wing aspect ratio	5.6
Length overall	6.96 m (22 ft 10 in)
Height overall	2.23 m (7 ft 3¾ in)
Tailplane span	3.20 m (10 ft 6 in)
Wheel track	2.60 m (8 ft 6¼ in)
Wheelbase	5.20 m (17 ft 0¾ in)
Propeller diameter: Dauphin 2+2	1.83 m (6 ft 0 in)
Dauphin 4	1.88 m (6 ft 2 in)

DIMENSIONS, INTERNAL:
Cabin: Length	1.62 m (5 ft 3¾ in)
Max width	1.10 m (3 ft 7¼ in)
Max height	1.23 m (4 ft 0½ in)
Baggage volume	0.39 m³ (13.8 cu ft)

AREAS:
Wings, gross	13.60 m² (146.4 sq ft)
Ailerons, total	1.15 m² (12.38 sq ft)
Flaps, total	0.70 m² (7.53 sq ft)
Fin	0.61 m² (6.57 sq ft)
Rudder	0.63 m² (6.78 sq ft)
Horizontal tail surfaces, total	2.88 m² (31.00 sq ft)

WEIGHTS AND LOADINGS:
Weight empty, equipped: 2+2	550 kg (1,212 lb)
4	580 kg (1,279 lb)
Max baggage: 2+2, 4	40 kg (88 lb)
Max T-O and landing weight: 2+2	900 kg (1,984 lb)
4	1,000 kg (2,205 lb)
Max wing loading: 2+2	66.2 kg/m² (13.56 lb/sq ft)
4	73.5 kg/m² (15.05 lb/sq ft)
Max power loading: 2+2	10.23 kg/kW (16.81 lb/hp)
4	8.38 kg/kW (13.78 lb/hp)

PERFORMANCE:
Never-exceed speed (V_NE):	
2+2, 4	166 kt (308 km/h; 191 mph)

Older Robin DR 400/140B Dauphin 4 wearing its original name 'Major' (*Jane's/Paul Jackson*)   NEW/0131737

Max level speed at S/L:
2+2	130 kt (241 km/h; 150 mph)
4	143 kt (265 km/h; 165 mph)

Cruising speed at 75% power:
2 + 2	116 kt (215 km/h; 133 mph)
4	128 kt (237 km/h; 147 mph)

Stalling speed, flaps down:
2+2	45 kt (82 km/h; 51 mph)
4	47 kt (87 km/h; 54 mph)

Max rate of climb at S/L: 2+2	183 m (600 ft)/min
4	264 m (865 ft)/min
Service ceiling: 2+2	3,660 m (12,000 ft)
4	4,265 m (14,000 ft)
T-O run: 2+2	235 m (775 ft)
4	245 m (805 ft)
T-O to 15 m (50 ft): 2+2	535 m (1,755 ft)
4	485 m (1,595 ft)
Landing from 15 m (50 ft): 2+2	460 m (1,510 ft)
4	470 m (1,545 ft)
Landing run: 2+2	200 m (660 ft)
4	220 m (725 ft)

Range with standard fuel at 65%, no reserves:
2+2	494 n miles (915 km; 568 miles)
4	447 n miles (828 km; 514 miles)

Range, Dauphin 4 with optional fuel at max cruising
speed, no reserves   645 n miles (1,195 km; 742 miles)
UPDATED

Robin DR 400/180 Régent (Textron Lycoming O-360-A engine) *(Jane's/Paul Jackson)*   0110927

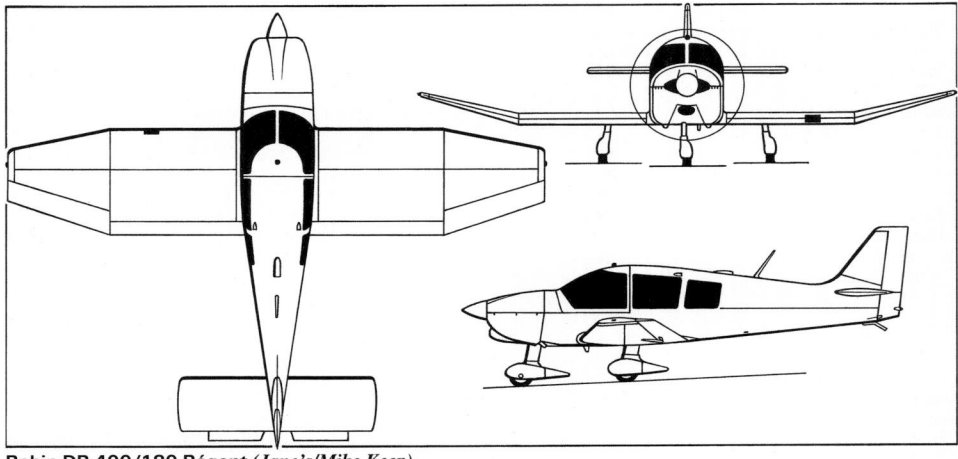

Robin DR 400/180 Régent *(Jane's/Mike Keep)*

ROBIN DR 400/160 MAJOR

TYPE: Four-seat lightplane.

PROGRAMME: First flight of original DR 400 Chevalier 29 June 1972; certified France and UK same year; Major introduced 1980.

CUSTOMERS: See main DR 400 entry; total 140 delivered by December 1998. Six built in 2000 and at least one in 2001.

DESIGN FEATURES: Main differences from Dauphin (see preceding entry) are additional rear cabin window on each side, external baggage compartment door on port side and extended wingroot leading-edges to house additional fuel tanks.

Differences from Dauphin listed below:

POWER PLANT: One 119 kW (160 hp) Textron Lycoming O-320-D flat-four engine, driving a Sensenich 74DM6S5-2-64 two-blade, fixed-pitch aluminium propeller. Fuel tank in fuselage, capacity 110 litres (29.1 US gallons; 24.2 Imp gallons), and two tanks in wingroot leading-edges, giving total capacity of 190 litres (50.2 US gallons; 41.8 Imp gallons), of which 182 litres (48.1 US gallons; 40.0 Imp gallons) are usable. Provision for auxiliary tank, raising total capacity to 240 litres (63.4 US gallons; 52.8 Imp gallons). Oil capacity 7.6 litres (2.0 US gallons; 1.7 Imp gallons).

ACCOMMODATION: Seating for four persons, on adjustable front seats (maximum load 154 kg; 340 lb total) and rear bench seat (maximum load 154 kg; 340 lb total). Forward-sliding transparent canopy, but port rear window/door provides outside access to baggage area. Up to 40 kg (88 lb) of baggage can be stowed aft of rear seats when four occupants are carried.

DIMENSIONS, EXTERNAL:
Propeller diameter	1.83 m (6 ft 0 in)
Baggage door: Height	0.47 m (1 ft 6½ in)
Width	0.55 m (1 ft 9½ in)

AREAS:
Wings, gross	14.20 m² (152.8 sq ft)

WEIGHTS AND LOADINGS:
Weight empty, equipped	598 kg (1,318 lb)
Max T-O and landing weight	1,050 kg (2,315 lb)
Max wing loading	74.0 kg/m² (15.15 lb/sq ft)
Max power loading	8.81 kg/kW (14.47 lb/hp)

PERFORMANCE:
Never-exceed speed (VNE)	166 kt (308 km/h; 191 mph)
Max level speed at S/L	146 kt (271 km/h; 168 mph)
Max cruising speed at 75% power at 2,440 m (8,000 ft)	132 kt (245 km/h; 152 mph)
Econ cruising speed at 65% power at 3,200 m (10,500 ft)	130 kt (241 km/h; 150 mph)

Stalling speed: flaps up	56 kt (103 km/h; 64 mph)
flaps down	50 kt (93 km/h; 58 mph)
Max rate of climb at S/L	255 m (836 ft)/min
Service ceiling	4,115 m (13,500 ft)
T-O run	295 m (970 ft)
T-O to 15 m (50 ft)	590 m (1,940 ft)
Landing from 15 m (50 ft)	545 m (1,790 ft)
Landing run	250 m (820 ft)

Range at econ cruising speed, no reserves:
standard fuel	812 n miles (1,505 km; 935 miles)
optional fuel	1,026 n miles (1,900 km; 1,180 miles)
UPDATED

ROBIN DR 400/180 REGENT

TYPE: Four-seat lightplane.

PROGRAMME: First flight 27 March 1972; certified 10 May 1972.

CUSTOMERS: See main DR 400 entry; total 363 delivered by early 1999. Some nine built in 2000 and at least 11 in 2001.

DESIGN FEATURES: Generally as for DR 400 series; two-seat rear bench.

Differences from DR 400/160 listed below:

POWER PLANT: One 134 kW (180 hp) Textron Lycoming O-360-A flat-four engine driving a Sensenich 76EM8S5-0-64 two-blade, fixed-pitch aluminium propeller. Fuel tankage unchanged.

ACCOMMODATION: Basically as for DR 400/160. Baggage capacity 60 kg (132 lb).

DIMENSIONS, EXTERNAL:
Propeller diameter	1.93 m (6 ft 4 in)

WEIGHTS AND LOADINGS:
Weight empty, equipped	610 kg (1,345 lb)

Max T-O and landing weight	1,100 kg (2,425 lb)
Max wing loading	77.48 kg/m² (15.87 lb/sq ft)
Max power loading	8.21 kg/kW (13.47 lb/hp)

PERFORMANCE (at max T-O weight):
Max level speed at S/L	150 kt (278 km/h; 173 mph)
Max cruising speed at 75% power at 2,285 m (7,500 ft)	140 kt (260 km/h; 162 mph)
Econ cruising speed at 60% power at 3,660 m (12,000 ft)	132 kt (245 km/h; 152 mph)
Stalling speed: flaps up	57 kt (105 km/h; 65 mph)
flaps down	52 kt (95 km/h; 59 mph)
Max rate of climb at S/L	252 m (825 ft)/min
Service ceiling	4,720 m (15,475 ft)
T-O run	315 m (1,035 ft)
T-O to 15 m (50 ft)	515 m (1,690 ft)
Landing from 15 m (50 ft)	530 m (1,740 ft)
Landing run	249 m (820 ft)

Range, no reserves:
standard fuel	753 n miles (1,395 km; 866 miles)
optional fuel	950 n miles (1,760 km; 1,094 miles)
UPDATED

ROBIN DR 400/180R REMO 180

TYPE: Glider tug.

PROGRAMME: First flight and certification 1972 as DR 400/180R (*Remorqueur*, abbreviated Remo); flown 1985 with Porsche PFM 3200 engine as DR 400RP or Remo 212; became first Porsche-powered aircraft to be certified; Remo 212 production ceased 1990 after 29 had been built (details in 1991-92 *Jane's*).

CUSTOMERS: See main DR 400 entry; total 311 Remos delivered by December 1998. Eight 180Rs produced in 2000; none known in 2001.

DESIGN FEATURES: Same as Régent, except no external baggage door, and baggage compartment covered in transparent Plexiglas to maximise rearward view; towing hook under tail; Dauphin wing (13.60 m²; 146.4 sq ft) without extended wingroot leading-edges.

Differences from DR 400/180 listed below:

POWER PLANT: One 134 kW (180 hp) Textron Lycoming O-360-A flat-four engine, driving (for glider towing) a Sensenich 76EM8S5-0-58 or Hoffmann HO27HM-180/138 two-blade, fixed-pitch metal propeller. For touring, a coarser pitch Sensenich 76EM8S5-0-64 propeller can be fitted. Fuel 110 litres (29.1 US gallons; 24.2 Imp gallons) normal; additional 50 litre (13.2 US gallon; 11.0 Imp gallon) tank optional.

WEIGHTS AND LOADINGS:
Weight empty, equipped	592 kg (1,305 lb)
Max T-O and landing weight	1,000 kg (2,205 lb)
Max wing loading	73.5 kg/m² (15.05 lb/sq ft)
Max power loading	7.46 kg/kW (12.25 lb/hp)

PERFORMANCE (A: with 300 kg; 661 lb glider, B: with 600 kg; 1,323 lb glider):
Max level speed	146 kt (270 km/h; 168 mph)

Robin DR 400/160, showing additional rear transparencies and port rear door to baggage area
(Jane's/Paul Jackson)   0062942

Robin Remo 180 glider tug (*Jane's/Paul Jackson*)   **NEW**/0131738

Cruising speed at 70% power at 2,440 m (8,000 ft)	
	124 kt (230 km/h; 143 mph)
Stalling speed, flaps down: A	47 kt (87 km/h; 54 mph)
B	45 kt (84 km/h; 52 mph)
Max rate of climb at S/L: A	282 m (925 ft)/min
B	210 m (690 ft)/min
Service ceiling	6,100 m (20,000 ft)
T-O run: A	205 m (675 ft)
B	300 m (985 ft)
T-O to 15 m (50 ft): A	400 m (1,315 ft)
B	470 m (1,542 ft)
Landing from 15 m (50 ft)	470 m (1,545 ft)
Range, no reserves:	
max normal fuel	494 n miles (915 km; 568 miles)
with supplementary fuel	
	585 n miles (1,085 km; 674 miles)
	UPDATED

ROBIN DR 400/200R REMO 200

TYPE: Glider tug.

PROGRAMME: First aircraft, HB-KFH, delivered to Segel-u. MFC at Grenchen, Switzerland, mid-2000; second built in 2001 for Charles G Kalko of Honesdale, Pennsylvania.

Differences from DR 400/180 listed below:

POWER PLANT: One 149 kW (200 hp) Textron Lycoming IO-360 flat-four engine, driving a constant-speed propeller.

WEIGHTS AND LOADINGS:	
Weight empty, equipped	650 kg (1,433 lb)
Max T-O and landing weight	1,100 kg (2,425 lb)
Max wing loading	77.5 kg/m² (15.87 lb/sq ft)
Max power loading	7.38 kg/kW (12.12 lb/hp)
PERFORMANCE (A: with 300 kg; 661 lb glider, B: with 600 kg; 1,323 lb glider):	
Cruising speed at 75% power	
	135 kt (250 km/h; 155 mph)
Max rate of climb at S/L: A	312 m (1,024 ft)/min
B	258 m (846 ft)/min
T-O to 15 m (50 ft): A	400 m (1,312 ft)
B	415 m (1,362 ft)
Range, no reserves:	
max normal fuel	447 n miles (828 km; 514 miles)
with supplementary fuel	
	610 n miles (1,130 km; 702 miles)
	UPDATED

ROBIN DR 500 PRESIDENT

TYPE: Four-seat lightplane.

PROGRAMME: Prototype (F-WZZY), then known as DR 400/200i, first flew 5 June 1997; revealed at Paris Air Show, 13 June. Renamed as above but certified as **DR 400/500**.

CURRENT VERSIONS: **DR 500/200i Président:** *As described.*

DR 500 Super Régent: 180 hp engine, lacks fuel injection. None built.

CUSTOMERS: Deliveries began in 1998 with eight aircraft. Total 36 built by early 2002 for owners in France, Germany, New Zealand, Spain, Switzerland and UK.

DESIGN FEATURES: Uprated version of DR 400/180 Régent, including that version's 14.20 m² (152.8 sq ft) wing; 149 kW (200 hp) Textron Lycoming IO-360-A1B6 engine driving a Hartzell F7 666A-2 constant-speed propeller; width of cabin increased by 10 cm (4 in) and headroom by same amount; IFR instrumentation as standard; Honeywell avionics, including KAP 140 autopilot. Fuel capacity increased. Flaps electrically operated.

Differences from DR 400/180 listed below:

LANDING GEAR: Nosewheel size 15 × 6.00-5, maximum pressure 1.8 bar (26 lb/sq in); mainwheel size 330×150, pressure 2.0 bar (29 lb/sq in).

POWER PLANT: One 149 kW (200 hp) Textron Lycoming IO-360-A1B6 driving a Hartzell two-blade, constant-speed metal propeller. Fuel contained in four tanks: two, each of capacity 40 litres (10.6 US gallons; 8.8 Imp gallons), in wingroots; main tank, capacity 105 litres (27.7 US gallons; 23.1 Imp gallons), below cabin floor; and fourth tank, capacity 90 litres (23.8 US gallons; 19.8 Imp gallons), below baggage compartment. Total capacity of 275 litres (72.6 US gallons; 60.5 Imp gallons).

ACCOMMODATION: Four adults, or two plus three children. Baggage shelf at rear of seats; upward-opening rear window on port side doubles as baggage door.

DIMENSIONS, EXTERNAL:	
Length overall	7.22 m (23 ft 8¼ in)
Propeller diameter	1.93 m (6 ft 4 in)
DIMENSIONS, INTERNAL:	
Cabin max width	1.20 m (3 ft 11¼ in)
WEIGHTS AND LOADINGS:	
Weight empty	650 kg (1,433 lb)
Baggage capacity	60 kg (132 lb)
Max T-O weight	1,150 kg (2,535 lb)
Max wing loading	81.0 kg/m² (16.59 lb/sq ft)
Max power loading	7.72 kg/kW (12.68 lb/hp)
PERFORMANCE:	
Max level speed	147 kt (272 km/h; 169 mph)
Cruising speed at 75% power	
	140 kt (260 km/h; 162 mph)
Max rate of climb at S/L	305 m (1,000 ft)/min
T-O to 15 m (50 ft)	460 m (1,509 ft)
Landing from 15 m (50 ft)	530 m (1,739 ft)
Range, standard fuel, no reserves:	
at 75% power	995 n miles (1,842 km; 1,145 miles)
at 65% power	1,001 n miles (1,855 km; 1,152 miles)
	UPDATED

Robin DR 500/200i Président (*Jane's/Paul Jackson*)   **NEW**/0131739

Robin DR 500/200i Président standard IFR instrument panel (*Jane's/Paul Jackson*)

SAUPER

SAUPER AVIATION SA

33 rue Fortuny, F-75017 Paris
WORKS: BP 1035, F-41010 Blois Cedex
Tel: (+33 2) 54 42 94 88
Fax: (+33 2) 54 42 00 11
e-mail: contact@sauper-aviation.com
Web: http://site.voila.fr/sauper_aviation
MANAGING DIRECTOR: Didier Juery

Sauper formed in 1992 to build the Chéreau J.300 Joker ultralight, which is now marketed under the name of the parent company. Personnel total 13 in a new 1,000 m² (10,750 sq ft) plant at Bois le Breuil.

UPDATED

SAUPER J.300 SERIES 3 JOKER

TYPE: Tandem-seat ultralight.

PROGRAMME: Designed by Tony Minguet in 1987. Series 3, introduced in 2000, has modified empennage and new instrument panel; JAR-VLA certification is planned.

CURRENT VERSIONS: **Civilian version** is used for private flying, crop-spraying and aerial photography. **Military version** features larger tyres and a full instrument panel including artificial horizon, UHF radio, GPS and transponder. Float-equipped aircraft are offered as an option.

Sauper J.300 Joker ultralight (*Geoffrey P Jones*)   0044658

CUSTOMERS: Total 85 built by 2001, including 15 in Africa. Two evaluated by the French Army in 1996.

COSTS: €38,200 (2001).

DESIGN FEATURES: Braced, high-wing monoplane with mid-mounted, braced tailplane; constant chord wing with tapered tips; 2.2° anhedral.

FLYING CONTROLS: Conventional and manual. Horn-balanced rudder.

STRUCTURE: Alloy tube airframe; wood and fabric covering.

LANDING GEAR: Tailwheel type; fixed. Mainwheels 8.00-6; speed fairings optional; hydraulic Grimeca brakes. Nosewheel version under development.

POWER PLANT: One 59.6 kW (79.9 hp) Rotax 912 or 73.5 kW (98.6 hp) Rotax 912 ULS flat-four driving a DUC Hélices three-blade propeller is standard; options include engines from BMW, JPX and Limbach. Two fuel tanks beneath seats, each 48.0 litres (12.7 US gallons; 10.6 Imp gallons) capacity. Oil capacity 3.0 litres (0.8 US gallon; 0.7 Imp gallon).

SYSTEMS: 12 V 17 Ah battery.
DIMENSIONS, EXTERNAL:
Wing span	9.80 m (32 ft 1¾ in)
Length overall	5.00 m (16 ft 4¾ in)
Height overall	2.00 m (6 ft 6¾ in)

AREAS:
Wings, gross	16.50 m² (177.6 sq ft)

WEIGHTS AND LOADINGS:
Weight empty	280 kg (617 lb)
Max T-O weight	450 kg (992 lb)

PERFORMANCE:
Never-exceed speed (V$_{NE}$)	108 kt (200 km/h; 124 mph)
Max operating speed (V$_{MO}$)	102 kt (190 km/h; 118 mph)
Max cruising speed	97 kt (180 km/h; 112 mph)
Econ cruising speed	76 kt (140 km/h; 87 mph)
Stalling speed	32 kt (58 km/h; 37 mph)
Max rate of climb at S/L	457 m (1,500 ft)/min
T-O run	60 m (197 ft)
Landing run	70 m (230 ft)
Range	875 n miles (1,620 km; 1,007 miles)
Endurance	7 h

UPDATED

SKYDESIGN

GECI INTERNATIONAL

105 bis boulevard Malesherbes, F-75008 Paris
Tel: (+33 1) 53 53 00 53
Fax: (+33 1) 53 53 00 50
e-mail: marc.demontessus@geci.net
Web: http://www.geci.net
CHAIRMAN AND CEO: Serge Bitboul
GENERAL MANAGER: Marc de Montessus

ASIA BRANCH:
PT GECI Nusantara
Gedung BBU, 8th Floor, Jl Asia Africa No 141-149,
Bandung 40112, Indonesia
Tel: (+62 22) 424 19 06
Fax: (+62 22) 420 42 90
e-mail: alexander.supelli@geci.co.id

Formed in 1979, GECI International is a transport-orientated engineering and consulting company with a staff of 400. It has contributed to the Fairchild-Dornier 728 and various business jet and Airbus airliner programmes. In 2001, it formed Skydesign SAS to develop the Skylander transport.

NEW ENTRY

SKYDESIGN SB100 SKYLANDER

TYPE: Twin-turboprop light transport.
PROGRAMME: Announced 17 October 2001 at Seoul Air Show. Design supervision by Desmond Norman, co-designer of the BNG Islander. FAA/JAA certification intended in September 2004.
Programme structured for joint funding (initially US$120 million) by three risk-sharing partners. Korean Aerospace Industries (which see) being first, on 17 October 2001, to pledge US$30 million to secure responsibility for wing production. Assembly of complete aircraft will be by a French company under contract to GECI. ˈ
CURRENT VERSIONS: **Skylander:** *As described.* Freighter, passenger and maritime patrol versions.
Skylander L: Lightened version with 559 kW (750 shp) PT6A engines, 2,000 kg (4,400 lb) payload and reduced price.
COSTS: Estimated US$3.5 million unit cost (2001). Sales goal over 20 years is some 1,300, being equivalent to 30 per cent of estimated market for 4,500 aircraft in this class.
DESIGN FEATURES: Intended for operation from rough airfields with minimal maintenance facilities. Fixed landing gear and conventional alloy structure; unpressurised. Certifiable to FAR Pt 23.
Fuselage of square cross-section, with upswept rear; tailplane at base of fin. High wing with single bracing strut each side.
FLYING CONTROLS: Conventional and manual. Horn-balanced rudder.

STRUCTURE: Principally of aluminium alloy.
LANDING GEAR: Tricycle type; fixed. Steeerable nosewheel. Tyre sizes 12.50-16 (main) and 8.50-10 (nose).
POWER PLANT: Two Pratt & Whitney Canada PT6A-65B turboprops, each 820 kW (1,100 shp), driving five-blade propellers. Total fuel 3,031 litres (800 US gallons; 667 Imp gallons).
ACCOMMODATION: Two pilots; 18 passengers (two plus one abreast) and one cabin attendant. Alternative combi (nine to 12 passengers ahead of LD3 container and separated by 9 *g* bulkhead) and all-freight configurations. Latter include three LD3s plus 4.0 m³ (141 cu ft) to rear; and four 1.83 × 1.40 m (72 × 55 in) pallets plus 5.4 m³ (191 cu ft). Baggage hold beneath flight deck.
Crew door forward, port; double freight door (divided vertically, with forward- and rearward-opening elements) rear, port.

DIMENSIONS, EXTERNAL:
Wing span	21.74 m (71 ft 4 in)
Wing aspect ratio	11.2
Length overall	13.57 m (44 ft 6¼ in)
Height overall	5.58 m (18 ft 3¾ in)
Wheel track	4.11 m (13 ft 5¾ in)
Wheelbase	4.675 m (15 ft 4 in)
Propeller diameter	2.82 m (9 ft 3 in)

DIMENSIONS, INTERNAL:
Hold: Length: parallel section	5.64 m (18 ft 6 in)
total flat floor	6.17 m (20 ft 3 in)
Width: max	1.94 m (6 ft 4½ in)
mean	1.88 m (6 ft 2 in)
at floor	1.83 m (6 ft 0 in)
Max height	1.83 m (6 ft 0 in)
Volume: parallel section	21.7 m³ (766 cu ft)
rear fuselage	5.3 m³ (187 cu ft)
Baggage hold volume	0.85 m³ (30.0 cu ft)

AREAS:
Wings, gross	42.25 m² (454.8 sq ft)

WEIGHTS AND LOADINGS:
Weight empty, equipped: cargo	4,315 kg (9,513 lb)
passenger	4,711 kg (10,386 lb)
Useful load: cargo	4,085 kg (9,005 lb)
passenger	3,689 kg (8,132 lb)
Max payload	3,000 kg (6,613 lb)
Max T-O weight	8,400 kg (18,518 lb)
Max wing loading	198.8 kg/m² (40.72 lb/sq ft)
Max power loading	5.12 kg/kW (8.42 lb/hp)

PERFORMANCE:
Cruising speed	220 kt (407 km/h; 253 mph)
Max rate of climb at S/L	579 m (1,900 ft)/min
Rate of climb at S/L, OEI	183 m (600 ft)/min
T-O to, and landing from, 11 m (35 ft)	600 m (1,970 ft)
Range: with max fuel	1,200 n miles (2,222 km; 1,380 miles)
with max payload	150 n miles (277 km; 172 miles)

NEW ENTRY

Model of Skydesign Skylander (*Jane's/Paul Jackson*)　　　　*NEW*/0132843

SOCATA

SOCATA Société de Constructions d'Avions de Tourisme et d'Affaires (Subsidiary of EADS)

Le Terminal Bât. 413, Zone d'Aviation d'Affaires, F-93352
Le Bourget Cedex
Tel: (+33 1) 49 34 69 69
Fax: (+33 1) 49 34 69 71
Web: http://www.socata.com

WORKS AND AFTER-SALES SERVICE:
Aérodrome de Tarbes-Ossun-Lourdes, BP 930, F-65009
Tarbes Cedex
Tel: (+33 5) 62 41 76 00
Fax: (+33 5) 62 41 76 54
PRESIDENT: Stéphane Bernard
CHAIRMAN AND CEO: Philippe Debrun
MANAGING DIRECTOR: Jean-François Trassard
SALES DIRECTOR: Christophe van den Broek
TECHNICAL DIRECTOR: Dominique Deschamps
DIRECTOR, PUBLIC RELATIONS: Philippe de Segovia

US OPERATING AND SERVICE FACILITY:
Socata Aircraft, North Perry Airport, 7501 Pembroke
Road, Pembroke Pines, Florida 33023
Tel: (+1 954) 893 14 00
Fax: (+1 954) 893 14 02

Socata TB 10 Tobago GT　　　　*NEW*/0131755

Formed 1966 as a subsidiary of Aerospatiale responsible for light aircraft. By mid-2000, Socata had produced over 2,000 TB series aircraft, excluding the TBM 700. In 1999, sold and delivered 40 TBs; sold 16 and delivered 21 TBM 700s. Reduced deliveries of 45 TBs and 33 TBMs in 2000; target for 2001 was 100+ and 33, respectively, but 63 and 33 actually achieved. Revenue for 2000 was expected to total US$160 million.

Also makes components for Airbus A300/320/330/340, Lockheed Martin C-130, ATR 42/72, Dassault Falcons, and Eurocopter Super Puma, Dauphin, Ecureuil and satellite structures. Covered floor area 57,000 m² (613,450 sq ft); workforce 1,200.

Following formation of EADS on 10 July 2000, Socata is 100 per cent owned by that company.

VERIFIED

SOCATA TB 9 TAMPICO, TB 10 and TB 200 TOBAGO

TYPE: Four-seat lightplane.

PROGRAMME: Design of original TB series launched 1975; first flight of original TB 10 (F-WZJP), powered by 119 kW (160 hp) Textron Lycoming O-320, 23 February 1977; second prototype powered by 134 kW (180 hp) Textron Lycoming.

In June 1999, Socata announced its Nouvelle Génération (New Generation) series of single piston-engined models. The title Generation Two (GT) was subsequently adopted. These feature aerodynamic and other improvements that may subsequently be incorporated in the Socata MS diesel-engined light aircraft described elsewhere in this entry. Modifications include a curved dorsal fairing for the fin, upturned wingtips, raised cabin roofline with new single-piece carbon fibre/honeycomb roof, revised window pillar design, flush-mounted windows, larger baggage door, redesigned interior and new fuel filler door.

Descriptions of the earlier TB9C Tampico Club, TB 9 Sprint, TB 10 Tobago, TB 10 Tobago Privilège and TB 200 Tobago XL will be found in the 2001-02 and earlier editions of *Jane's*. In February 2000, those models were succeeded in production by the GT series, described below.

CURRENT VERSIONS: **TB 9 Tampico Sprint GT:** Baseline model. 119 kW (160 hp) Textron Lycoming O-320-D2A flat-four engine driving a Sensenich two-blade fixed-pitch metal propeller. Fuel capacity 158 litres (41.7 US gallons; 34.8 Imp gallons), of which 152 litres (40.2 US gallons; 33.4 Imp gallons) are usable. Oil capacity 8 litres (2.1 US gallons; 1.8 Imp gallons).

TB 10 Tobago GT: 134 kW (180 hp) Textron Lycoming O-360-A1AD flat-four engine. *Detailed description applies to this version.*

TB 200 Tobago GT: Similar to TB 10 Tobago GT, but with 134 kW (180 hp) fuel-injected Textron Lycoming IO-360-A1B6 engine

CUSTOMERS: Total of 468 Tampicos and Tampico Clubs in service by June 1999 with customers in Africa and Middle East, Asia-Pacific, Australasia, Europe, and North America. Total of 637 Tobagos and 87 Tobago XLs delivered by June 1999, to customers in Africa and Middle East, Australasia, Europe, North America and elsewhere. Production of first-generation TBs (9, 10, 200, 20 and 21) totalled 1,928, including prototypes.

Deliveries in 2000 totalled 45 and in 2001 totalled 22, comprising two TB 9s, eight TB 10s and 12 TB 200s. Launch order for six TB 200 GTs, plus six options, placed by Aeronautical Academy of Europe, based in Portugal.

DESIGN FEATURES: Conventional light tourer and trainer.

Wing section RA 16.3C3; thickness/chord ratio 16 per cent; dihedral 4° 30′.

FLYING CONTROLS: Manual, with pushrod actuation for ailerons, rudder and all-moving tailplane; cable-actuated anti-balance tab on tailplane, plus ground-adjustable tabs on ailerons and rudder; strakes on lower edges of fuselage just aft of wing control turbulence under rear fuselage; electrically actuated flaps.

STRUCTURE: Conventional light alloy; single-spar wing; GFRP tips, engine cowlings, cabin roof structure and dorsal strake. Triple anti-corrosion protection.

LANDING GEAR: Non-retractable tricycle type, with steerable nosewheel. Oleo-pneumatic shock-absorber in all three units; trailing-link suspension main landing gear legs. Mainwheel tyres size 6.00-6 (6 ply), pressure 2.30 bar (33 lb/sq in) on TB 10 GT/200 GT; 15×6.00-6 (6 ply) on TB 20 GT/21 GT. Nosewheel 5.00-5 on TB 10 GT/200 GT and 5.00-4 on TB 20 GT/21 GT. Glass fibre wheel fairings on all three units. Hydraulic disc brakes. Parking brake.

POWER PLANT: One 134 kW (180 hp) Textron Lycoming O-360-A1AD flat-four engine, driving a Hartzell two-blade constant-speed propeller. Two integral fuel tanks in wing leading-edges; total capacity 210 litres (55.5 US gallons; 46.2 Imp gallons), of which 204 litres (53.9 US gallons; 44.9 Imp gallons) are usable. Oil capacity 8 litres (2.1 US gallons; 1.8 Imp gallons).

ACCOMMODATION: Four or five seats in enclosed cabin, with dual controls. Separate, adjustable front seats with inertia reel seat belts. Two separate rear seats or removable three-place bench seat with safety belts. Sharply inclined low-drag windscreen. Access via upward-hinged window/

SOCATA AIRCRAFT AT CIVIL FLIGHT CENTRES
(2001)

Country	School	TB9	TB 10	TB 200	TB 20
Angola	Sonalgol		1		
Australia	AFTS Sydney	3			1
	British Aerospace FC		8		
			50		6
	WAFC Perth		2		5
Canada	White Aviation				3
China	CAAC FC			38	
Denmark	CAT Flying	5			33
			4		
France	AIF				3
	CIPRA		3		4
	ESMA	2			3
	IAAG	5			
	SEFA		48		10
	TAF				55
Gabon	ENAC		6		
Germany	VHM		23		25
India	IGRUA				13+1*
Indonesia	EAT		39		
Italy	AeCI	82			
Jordan	Royal Academy				2
Morocco	Royal Air Maroc				10
Mozambique	ENA		2		
Netherlands	Flight Academy		4		
	Martinair				5+2*
	NLS				2
New Zealand	Massey University		2		
Nigeria	NCAT	25			
Philippines	PSCA	2	2		
Poland	OKL	3			
Portugal	AAE			12*	
	Vega	4			4*
Singapore	RSFC	2		1	1
Spain	Panavia	5	5		
	SENASA		20		5
Thailand	CATC	5			
Tunisia	EACM	4			7
Turkey	ESAC	3			2
USA	Airlease	2			6
	Embry-Riddle	25			1
	Everything Flyable	5	1		
	Frederick Aviation	4			2
	Northeast Aviation	2	2		2
	Parks College St Louis	18	1		
	Sunquest Aviation	1			
	University of N Dakota	3	1		
	Westair	5			
Uruguay	IAC		3		
Totals (717)		**215**	**228**	**50**	**224**

* GT version

Note: Military users comprise French Air Force (150 TB 30 Epsilons), Portuguese Air Force (18 TB 30 Epsilons) and Turkish navy (six TB 20s). Additionally, EAT has two TBM 700s
Orders placed in February 2002 for 10 TB 20 GTs for SEFA and two TB 9 GTs for Airways Formation France)

Socata TB 200 Tobago GT and TB 20 Trinidad GT operated by Air Academy of Europe   *NEW*/0098730

Socata TB 9 Tampico Sprint GT accompanied by TB 20 Trinidad GT   *NEW*/0132426

doors of glass fibre. Baggage compartment aft of cabin, with external door on port side. Cabin carpeted, soundproofed, heated and ventilated. Windscreen defrosting standard.

SYSTEMS: Electrical system includes 28 V 70 A alternator and 24 V 10 A battery. Hydraulic system for brakes only.

AVIONICS: Honeywell Silver Crown avionics suite to customer choice, including Bendix/King KLN 90 GPS.

Instrumentation: Typical IFR package includes dual altimeters; dual true airspeed indicators; dual artificial horizons; vertical speed indicator, electric turn co-ordinator; directional gyro, tachometer; oil temperature, oil pressure, fuel pressure, fuel quantity, manifold pressure, CHT/EGT and OAT gauges; ammeter, voltmeter and compass.

EQUIPMENT: Includes armrests for all seats, map pockets, anti-glare visors, stall warning indicator, tiedown fittings and towbar, landing and navigation lights, four individual cabin lights and instrument panel lighting.

DIMENSIONS, EXTERNAL (TB 9 GT, TB 10 GT and TB 200 GT):
Wing span	10.01 m (32 ft 10 in)
Wing chord, constant	1.22 m (4 ft 0 in)
Wing aspect ratio	8.0
Length overall: TB 9 GT	7.72 m (25 ft 4 in)
TB 10 GT/200 GT	7.75 m (25 ft 5 in)
Height overall	3.02 m (9 ft 11 in)
Tailplane span	3.20 m (10 ft 6 in)
Wheel track	2.30 m (7 ft 6½ in)
Wheelbase	1.96 m (6 ft 5 in)
Propeller diameter	1.88 m (6 ft 2 in)
Propeller ground clearance	0.10 m (4 in)
Cabin doors (each): Width	0.90 m (2 ft 11½ in)
Height	0.76 m (2 ft 6 in)
Baggage door: Width	0.64 m (2 ft 1¼ in)
Max height	0.53 m (1 ft 9 in)

DIMENSIONS, INTERNAL (TB 9, TB 9C, TB 10 and TB 200):
Cabin: Length:	
firewall to rear bulkhead	2.53 m (8 ft 3½ in)
panel to rear bulkhead	2.00 m (6 ft 6¾ in)
Max width: at rear seats	1.28 m (4 ft 2¼ in)
at front seats	1.15 m (3 ft 9¼ in)
Max height	1.20 m (3 ft 11¼ in)

AREAS (TB 9 GT, TB 10 GT and TB 200 GT):
Wings, gross	11.90 m² (128.1 sq ft)
Ailerons (total)	0.91 m² (9.80 sq ft)
Trailing-edge flaps (total)	3.72 m² (40.04 sq ft)
Fin	0.88 m² (9.47 sq ft)
Rudder	0.63 m² (6.78 sq ft)
Horizontal tail surfaces (total)	2.56 m² (27.56 sq ft)

WEIGHTS AND LOADINGS:
Weight empty, with unusable fuel and oil:	
TB 9 GT	684 kg (1,508 lb)
TB 10 GT	730 kg (1,609 lb)
TB 200 GT	745 kg (1,642 lb)
Baggage: all	65 kg (143 lb)
Max T-O weight: TB 9 GT	1,060 kg (2,336 lb)
TB 10 GT, TB 200 GT	1,150 kg (2,535 lb)
Max wing loading: TB 9 GT	89.1 kg/m² (18.25 lb/sq ft)
TB 10 GT, TB 200 GT	96.6 kg/m² (19.79 lb/sq ft)
Max power loading: TB 9 GT	8.91 kg/kW (14.64 lb/hp)
TB 10 GT	8.57 kg/kW (14.08 lb/hp)
TB 200 GT	7.72 kg/kW (12.68 lb/hp)

PERFORMANCE:
Max cruising speed:	
TB 9 GT at 70% power at 1,830 m (6,000 ft)	
	115 kt (213 km/h; 132 mph)
TB 10 GT at 75% power at 1,830 m (6,000 ft)	
	127 kt (235 km/h; 146 mph)
TB 200 GT at 75% power at 2,590 m (8,500 ft)	
	130 kt (240 km/h; 149 mph)
Econ cruising speed:	
TB 9 GT at 65% power at 2,440 m (8,000 ft)	
	105 kt (194 km/h; 121 mph)

TB 10 GT at 60% power at 2,440 m (8,000 ft)
109 kt (202 km/h; 125 mph)
TB 200 GT at 60% power at 2,590 m (8,500 ft)
115 kt (213 km/h; 132 mph)

Stalling speed:
flaps up: TB 9 GT	58 kt (107 km/h; 67 mph)
TB 10 GT	61 kt (112 km/h; 70 mph)
flaps down: TB 9 GT	48 kt (89 km/h; 56 mph)
TB 10 GT	52 kt (97 km/h; 60 mph)
TB 200 GT	53 kt (98 km/h; 61 mph)
Max rate of climb at S/L: TB 9 GT	203 m (665 ft)/min
TB 10 GT	240 m (787 ft)/min
TB 200 GT	286 m (937 ft)/min
Max certified altitude: TB 9 GT	3,350 m (11,000 ft)
TB 10 GT	3,965 m (13,000 ft)
TB 200 GT	4,875 m (16,000 ft)

T-O to 15 m (50 ft):
TB 9 GT	570 m (1,870 ft)
TB 10 GT	505 m (1,657 ft)
TB 200 GT	475 m (1,558 ft)
Landing from 15 m (50 ft): TB 9 GT	420 m (1,378 ft)
TB 10 GT	460 m (1,509 ft)
TB 200 GT	449 m (1,473 ft)
Max range: TB 9 GT	564 n miles (1,046 km; 650 miles)
TB 10 GT	697 n miles (1,290 km; 802 miles)
TB 200 GT	637 n miles (1,179 km; 733 miles)

UPDATED

SOCATA TB 20 and TB 21 TRINIDAD
Israel Defence Force name: Pashosh (Lark)

TYPE: Four-seat lightplane.

PROGRAMME: First flight TB 20 (F-WDBA) 14 November 1980; French certification 18 December 1981; FAA certification 27 January 1984; first delivery (F-WDBB) 23 March 1982; first flight TB 21, 24 August 1984; French certification 23 May 1985; FAA certification 5 March 1986.

CURRENT VERSIONS: **TB 20 Trinidad GT:** At the Paris Air Show in June 1999, Socata announced its Nouvelle Génération (New Generation) series of single piston-engined aircraft, the prototype of which, TB 20 NG F-WWRG, had first flown on 21 April 1999. The title Generation Two (GT) was subsequently adopted. The TB 20 GT received DGAC certification on 31 January 2000. The GT series was formally launched at Tarbes on 2 February 2000 with roll-out of the first production TB 20 GT, which became US demonstrator (N163GT c/n 2000) and was exhibited at Sun 'n' Fun, April 2000.

GT variants feature aerodynamic and other improvements that will be incorporated in the Socata Morane diesel-engined series described below. Identifying features include upturned wingtip, similar to those on the TBM 700 turboprop; a curved dorsal fillet; raised cabin roofline with new single-piece carbon fibre/honeycomb roof; revised cabin window/pillar design with flush-mounted windows; plus redesigned interior, larger baggage door, new fuel filler door, retractable footstep and optional three-blade Hartzell propeller.

Details of earlier TB 20 Trinidad, TB 20 Trinidad Excellence, TB 20 C and TB 21 Trinidad TC variants (of

Socata TB 21 Trinidad GT Turbo   0110986

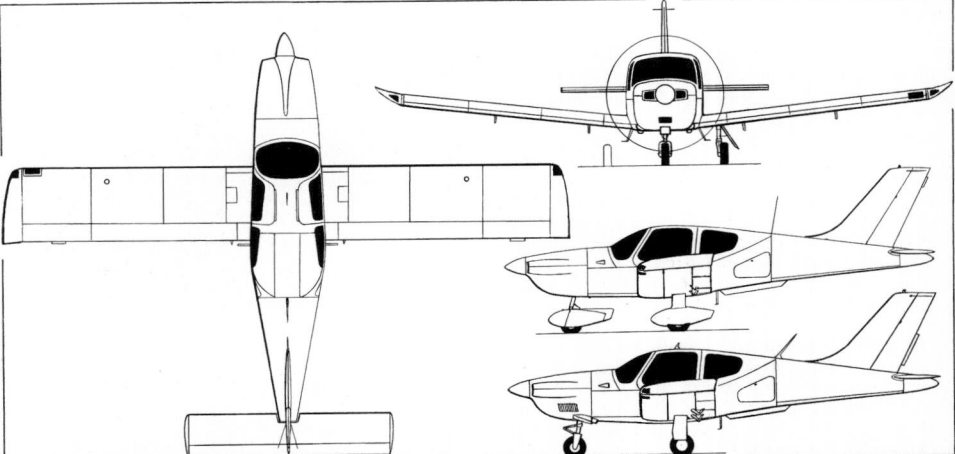

Socata TB 21 Trinidad GT, with additional side view (upper) of TB 10 Tobago GT *(Jane's/Dennis Punnett)*
0110866

SR305 engine testbed Trinidad   0110988

Artist's impression of the Socata MS 200 RG
0062691

which some 700 built) may be found in the 2000-01 and previous editions of *Jane's*.

TB 21 Trinidad GT Turbo: As above, but with computer-controlled turbocharger.

CUSTOMERS: At time of launch of GT series, the Aeronautical Academy of Europe in Evora, Portugal, took options on four TB 20 GTs. By mid-2000, TB 20 GTs had been delivered to customers in France, UK and USA from planned first year production of 72. Orders in 2000 included 69 from New Avex, distributor for US West Coast; recent customers include the Nigerian College of Aviation Technology (NCAT) of Zaria, Kaduna State, which ordered 10 Trinidad GTs for delivery beginning mid-2001. Total of 33 TB 20 GTs and eight TB 21 GTs delivered in 2001.

DESIGN FEATURES: Mainly as for Tobago; dihedral 6° 30′.

FLYING CONTROLS: As Tobago, but rudder trim and flap preselector added.

STRUCTURE: Largely as Tobago.

LANDING GEAR: Hydraulically retractable tricycle type, with single wheel on each unit. Free-fall emergency extension. Steerable nosewheel retracts rearward. Main units, with trailing-link suspension, retract inward into fuselage. Hydraulic disc brakes. Parking brake.

POWER PLANT: *TB 20 GT:* One 186 kW (250 hp) Textron Lycoming IO-540-C4D5D flat-six, driving a two-blade Hartzell constant-speed propeller.

TB 21 GT: One 186 kW (250 hp) Textron Lycoming TIO-540-AB1AD turbocharged flat-six. Two-blade Hartzell constant-speed propeller; three-blade propeller optional.

Fuel in two integral wing tanks, total capacity 336 litres (88.8 US gallons; 73.9 Imp gallons), of which 326 litres (86.1 US gallons; 71.7 Imp gallons) usable. Oil capacity 12.6 litres (3.3 US gallons; 2.8 Imp gallons).

ACCOMMODATION: Generally as for TB 9 Sprint GT, TB 10 Tobago GT and TB 200 XL GT; new headrests for all seats introduced on GT variants; rear seat can be removed for carrying 250 kg (551 lb) of cargo.

SYSTEMS: Electrical system comprises 28 V 70 A alternator and 24 V 10 Ah battery.

AVIONICS: TB 20/21 GT basic navigation package comprises single nav/com and VOR receiver. IFR packages and EFIS optional.

DIMENSIONS, EXTERNAL: As for TB 10 GT, except:
Wing span	9.97 m (32 ft 8½ in)
Wing aspect ratio	8.2
Length overall	7.75 m (25 ft 5 in)
Height overall	2.85 m (9 ft 4¼ in)
Tailplane span	3.68 m (12 ft 0¾ in)
Wheel track	2.17 m (7 ft 1½ in)
Wheelbase	1.91 m (6 ft 3¼ in)
Propeller diameter	2.03 m (6 ft 8 in)

DIMENSIONS, INTERNAL (TB 20 GT):
Cabin: Width	1.28 m (4 ft 2½ in)
Max height	1.20 m (3 ft 11¼ in)

AREAS: As for TB 10 GT, except:
Horizontal tail surfaces (total)	3.06 m² (32.94 sq ft)

WEIGHTS AND LOADINGS:
Operating weight empty: TB 20 GT	800 kg (1,764 lb)
TB 21 GT	867 kg (1,911 lb)
Max baggage: TB 20 GT, TB 21 GT	65 kg (143 lb)
Max T-O and landing weight:	
TB 20 GT, TB 21 GT	1,400 kg (3,086 lb)
Max wing loading:	
TB 20 GT, TB 21 GT	117.6 kg/m² (24.10 lb/sq ft)
Max power loading:	
TB 20 GT, TB 21 GT	7.51 kg/kW (12.35 lb/hp)

PERFORMANCE:
Cruising speed at 75% power:	
TB 20 GT at 1,981 m (6,500 ft)	
	163 kt (302 km/h; 188 mph)
TB 21 GT at 7,620 m (25,000 ft)	
	190 kt (352 km/h; 219 mph)
Econ cruising speed at 65% power:	
TB 20 GT at 2,591 m (8,500 ft)	
	150 kt (278 km/h; 173 mph)
TB 21 GT at 7,620 m (25,000 ft)	
	169 kt (313 km/h; 194 mph)
Rate of climb at S/L: TB 20 GT	366 m (1,200 ft)/min
TB 21 GT	344 m (1,130 ft)/min
TB 21 GT at 5,180 m (17,000 ft)	244 m (800 ft)/min

Certified ceiling: TB 20 GT	6,100 m (20,000 ft)
TB 21 GT	7,620 m (25,000 ft)
T-O to 15 m (50 ft): TB 20 GT	595 m (1,955 ft)
TB 21 GT	595 m (1,955 ft)
Landing from 15 m (50 ft): TB 20 GT, TB 21 GT	
	540 m (1,775 ft)
Max range:	
TB 20 GT	1,100 n miles (2,037 km; 1,265 miles)
TB 21 GT	1,035 n miles (1,916 km; 1,191 miles)

UPDATED

SOCATA MS LIGHT AIRCRAFT

In January 1997 Aerospatiale and Renault Sport formed a jointly owned subsidiary, Société de Motorisations Aéronautiques (SMA), to develop a new generation of light aircraft piston engines, under the marketing name Morane Renault. Their first application is a new range of Socata Morane aircraft, based on the current piston-engined TB series. Two engines are planned, sharing common core components; 169 kW (227 hp) SR305, and 221 kW (296 hp) SR 460. The SR 305 achieved French DGAC certification in late April 2001. The SR 460 engine will also be certified for aerobatic use.

The engines are four-cylinder, horizontally opposed, turbocharged diesels with 19:1 compression ratio, designed to run on jet fuel. They feature single-lever power control, electronic management and integral recording and checking systems, and are significantly quieter and more fuel-efficient than existing aero-engines, with estimated fuel consumption 30 to 40 per cent less than for comparable light aircraft power plants. The engines will be capable of maintaining 75 per cent of rated power up to 7,620 m (25,000 ft); TBO will be 3,000 hours.

Began ground running trials in mid-1997 and, by early 2001, had logged 3,000 hours in test rigs. A single TB 20 Trinidad served as flying testbed for both engines; this aircraft was rolled out on 24 January 1998 and first flew (F-WWRS) with an MR 200 (now SR305) on 3 March 1998.

At the Paris Air Show in June 1997, Socata announced a range of light aircraft to be powered by the Morane Renault engines. The aircraft also revive the Morane title of the former Morane-Saulnier company. Three types will be available initially; brief details of the Morane range are given below, where these differ from corresponding TB-series models.

MS 200 FG: Previously MS 180. Socata TB 20 Trinidad-derived fixed landing gear, four/five-seat tourer fitted with 169 kW (227 hp) SR305 engine with direct drive to three-blade constant-speed Hartzell propeller, diameter 1.88 m (6 ft 2 in). First delivery expected in 2003.

MS 200 RG: Previously MS 250. Socata TB 20 Trinidad-derived retractable landing gear tourer with SR305 engine with geared drive to three-blade, constant-speed Hartzell propeller, diameter 2.03 m (6 ft 8 in). Mockup displayed at Paris Air Show in June 1997.

MS 300 Epsilon II: Initially named Sabre; Socata TB 30 Epsilon-derived two-seat fully aerobatic trainer with 221 kW (296 hp) MR 300 engine with geared drive to three-blade constant-speed Hartzell propeller, diameter 1.98 m (6 ft 6 in). Empty weight 928 kg (2,046 lb). 'Prototype' (actually Epsilon demonstrator, No. 3, in non-flying condition as 'F-EMK2' displayed at Paris Air Show in June 1999. Available for delivery 24 months from launch order. The Epsilon was last described in the 1990-91 edition and appears in the current *Jane's Aircraft Upgrades*.

PERFORMANCE (estimated):
Cruising speed:	
200 FG at 3,050 m (10,000 ft)	
	134 kt (248 km/h; 154 mph)
200 RG at 4,575 m (15,000 ft)	
	189 kt (350 km/h; 217 mph)
300 at S/L	206 kt (382 km/h; 237 mph)
300 at 3,050 m (10,000 ft)	223 kt (413 km/h; 267 mph)
Stalling speed, landing configuration	
300	63 kt (117 km/h; 73 mph)
Max rate of climb at S/L:	
200 FG	more than 229 m (750 ft)/min
200 RG	more than 335 m (1,100 ft)/min
Rate of climb at 3,050 m (10,000 ft):	
300	567 m (1,860 ft)/min
Service ceiling: 200 FG	5,180 m (17,000 ft)
200 RG	7,010 m (23,000 ft)
T-O to 15 m (50 ft): 200 FG	505 m (1,660 ft)
200 RG	655 m (2,150 ft)
300	590 m (1,940 ft)
Landing from 15 m (50 ft):	
200 FG	460 m (1,510 ft)
200 RG	533 m (1,750 ft)
300	660 m (2,165 ft)
Max range: 200 FG	996 n miles (1,844 km; 1,146 miles)
200 RG	1,433 n miles (2,654 km; 1,649 miles)
300 at 222 kt (411 km/h; 255 mph) at 4,575 m	
(15,000 ft)	830 n miles (1,537 km; 955 miles)
Endurance at 6,100 m (20,000 ft) at max cruising speed:	
200 RG	8 h 10 min

UPDATED

SOCATA TBM 700

TYPE: Business turboprop.

PROGRAMME: Three prototypes built: first flights 14 July 1988 (F-WTBM), 3 August 1989 (F-WKPG) and 11 October 1989 (F-WKDL); French certification received 31 January 1990; FAR Pt 23 type approval awarded 28 August 1990; first delivery 21 December 1990; Canadian public transport certification 1993. Wider, single-piece, upward opening door replaced horizontally split door, port side, rear, from late 1998 and is standard. Aircraft used for freighting have optional crew door, port, forward. Increased weight version (TBM 700C2) introduced late 2002.

TBM 700 business turboprop in landing configuration (*Jane's/Paul Jackson*)   *NEW*/0131740

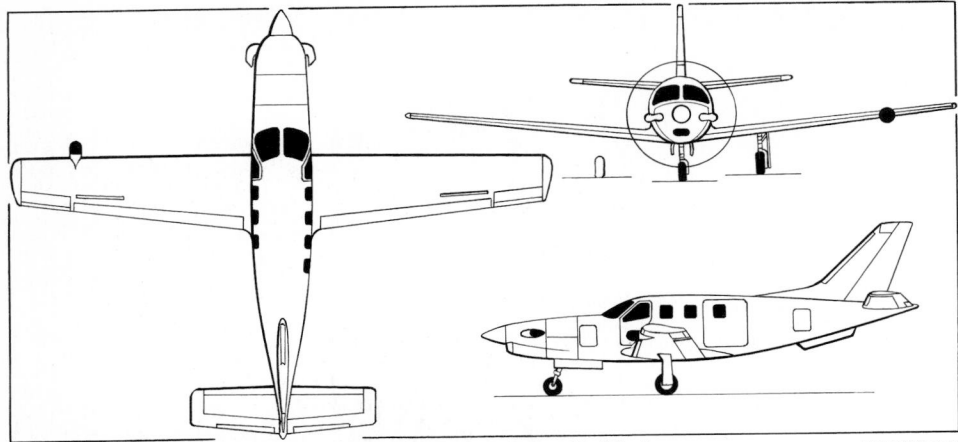

Socata TBM 700 business and multirole aircraft *(Jane's/James Goulding)*   *NEW*/0526457

CURRENT VERSIONS: **TBM 700:** In addition to basic transport, Socata offers multimission versions, including military medevac, target towing, ECM, freight, maritime patrol, law enforcement, navaid calibration and vertical photography versions. Prototype of last-mentioned (F-GLBF) exhibited at the 1996 Farnborough Air Show.

TBM 700B Freighter: Cargo version with port side cargo door, 1.19 m (3 ft 10¾ in) × 1.07 m (3 ft 6 in), and separate port side cockpit door; reinforced metal cargo floor (188 kg/m²; 38.5 lb/sq ft) with tiedown points in rails; optional port side cockpit door; maximum cargo capacity 825 kg (1,819 lb); volume 3.5 m³ (124 cu ft); target maximum T-O weight 3,300 kg (7,275 lb); quick conversion to six/seven-seat passenger configuration; announced June 1995; initially known as TBM 700C; French certification 23 November 1998. Launch customer Air Open Sky of France, which began TBM 700 freight operations in November 1999. Three handed over to French Army on 28 June 2000. Quest Diagnostics of the USA has ordered six, of which first three (N700QD, N701QD and N702QD) delivered 26 November 2001.

TBM 700C: First of two prototypes (F-WWRL/N220MA) flew February 2002; officially announced 10 September 2002 at NBAA Convention, Orlando, Florida; sole production variant from October 2002. Changes include option of increased (**TBM 700C2**) or original (**TBM 700C1**) MTOW, according to local certification rules; strengthened wing structure; stronger wheels and 8-ply tyres; additional (unpressurised) baggage compartment to rear of cabin; new interior with 20 g seats; improved air conditioning.

Description and specification refer to TBM 700C.

CUSTOMERS: Total 212 delivered by September 2002 (see table). Deliveries to French Air Force began 27 May 1992 with first of initial six for liaison duties with Groupe Aérien d'Entrainement et de Liaison (GAEL) and ETE 43; further six supplied 1993-94 (also for ETE 41 and ETE 44, plus CEAM); total Air Force requirement is 22 (of which 17 funded by mid-1998); two officially handed over to French Army 13 January 1995 for 3 GHL at Rennes, followed by six more, last three of which delivered (in TBM 700B configuration) 28 June 2000. Two delivered to Indonesian Civil Aviation Academy in September 1996. Total of 14 delivered in 2000, 33 in 2001, and 15 in the first six months of 2002.

FIRST 700 DELIVERIES

Country	Total
Australia	3
Austria	3
Canada	2
France	
Civil	3 (prototypes)
	24
Air Force	19
Army	8
CEV	1
Germany	8
Indonesia	3
Ireland	1
Italy	1
Japan	3
Luxembourg	3
Netherlands	1
Switzerland	1
Thailand	1
USA	150
Unknown	5
Total	**240**

Note: Based on initial permanent registration; transfers have since been effected

COSTS: US$2.4 million, typically equipped (2000); direct operating cost US$230 per hour (2000).

DESIGN FEATURES: High-speed, long-range, single-turboprop business transport. Low wing; twin strakes under rear fuselage; sweptback fin (with dorsal fin) and mass balanced rudder; non-swept tailplane with mass balanced elevators. Airframe life 12,000 cycles/16,000 hours from TBM 700C.

Wing of Aerospatiale RA 16-43 root section with 6° 30′ dihedral from roots.

FLYING CONTROLS: Conventional and manual. Pushrod and cable actuated with electric trim tabs in port aileron, rudder and each elevator; 'scaled-down ATR' single-slotted Fowler flaps, also electrically actuated, along 71 per cent of each wing trailing-edge; slotted spoiler forward of each flap at outer end, linked mechanically to aileron; yaw damper.

STRUCTURE: Mainly of light alloy and steel except for control surfaces, flaps, most of tailplane and fin of Nomex honeycomb bonded to metal sheet; wing leading-edges and landing gear doors GFRP/CFRP; tailcone and wingtips GFRP; two-spar torsion box forms integral fuel tank in each wing.

LANDING GEAR: Hydraulically retractable tricycle type, with emergency manual operation. Inward-retracting main units of trailing-link type; rearward-retracting steerable nosewheel (±28°). Main tyres 18×5.5 (8 ply) tubeless; nose 5.00-5 (10 ply) tubeless. Parker hydraulic disc brakes. Minimum ground turning radius (based on nosewheel) 23.98 m (78 ft 8 in).

POWER PLANT: One 1,178 kW (1,580 shp) Pratt & Whitney Canada PT6A-64 turboprop, flat rated at 522 kW (700 shp), driving a Hartzell HC-E4N-3/E9083S(K) four-blade constant-speed, fully feathering, reversible-pitch, metal propeller. Fuel in integral tank in each wing, combined capacity 1,100 litres (290.5 US gallons; 242 Imp gallons), of which 1,066 litres (282 US gallons; 234 Imp gallons) usable. Gravity filling point in top of each tank. Oil capacity 12 litres (3.2 US gallons; 2.6 Imp gallons).

ACCOMMODATION: Adjustable seats for one or two pilots at front. Dual controls standard. Four seats in club layout aft of these, with centre aisle, or five seats in high-density layout. Large upward-opening door on port side aft of wing; overwing emergency exit on starboard side. Oxygen system, comprising three under-seat bottles with individual emergency oxygen mask for each passenger and masks with integral microphones for crew. Pressurised baggage compartment at rear of cabin, with internal access only; additional unpressurised compartments in nose, between engine and firewall, and behind cabin, both with external access via doors on port side. Nose compartment of TBM 700C reduced in size by two-thirds, compared with 700B, due to installation of environmental control system. Optional crew door, port side, front. Reinforced metal floor with tie-down points.

Cabin interior of Socata TBM 700   0110867

Prototype TBM 700C2 taking shape at Tarbes   *NEW*/0526450

TBM 700C2 port side, showing new baggage door in addition to freight door   *NEW*/0526448

TBM 700C with Honeywell avionics   *NEW*/0526451

SYSTEMS: Engine bleed air pressurisation (to 0.43 bar; 6.2 lb/sq in) and Honeywell environmental control system/vapour control system. Hydraulic system for landing gear only. Electrical system powered by two 28 V 200 A engine-driven starter/generators (one main, one standby) and a 28 V 40 Ah lead/acid (optionally Ni/Cd) battery. Pneumatic rubber-boot de-icing of wing/tailplane/fin leading-edges. Propeller blades anti-iced electrically, engine inlets by exhaust air. Electric anti-icing and hot air demisting of windscreen. Gaseous oxygen system.

AVIONICS: Honeywell Silver Crown digital IFR package.

Comms: Bendix/King KY 196 VHF, KT 76 transponder; KMA 24H audio control/MKR. ELT.

Radar: Honeywell RDR-2000 weather radar optional.

Flight: KX 165 nav/com/glideslope; KNS 80 R/Nav; KR 21 MKR; KR 87 ADF; KFC 275 autopilot with KAS 297C altitude preselect/alerter; KDR 510 weather datalink dual Garmin GNS 530 GPS and/or Honeywell IHAS 8000 (Goodrich Skywatch in 2002 production only).

Instrumentation: Three-screen EFIS with KMD 850 MFDs. Shadin ETM 700 engine trend monitoring equipment standard.

DIMENSIONS, EXTERNAL:

Wing span	12.68 m (41 ft 7¼ in)
Wing chord, mean aerodynamic	1.51 m (4 ft 11½ in)
Wing aspect ratio	8.9
Length overall	10.645 m (34 ft 11 in)
Height overall	4.35 m (14 ft 3¼ in)
Tailplane span	4.99 m (16 ft 4½ in)
Wheel track	3.87 m (12 ft 8¼ in)
Wheelbase	2.91 m (9 ft 6½ in)
Propeller diameter	2.31 m (7 ft 7 in)
Crew door (optional): Height	1.02 m (3 ft 4 in)
Width	0.78 m (2 ft 6¾ in)
Cabin door: Height	1.19 m (3 ft 10¾ in)
Width	1.08 m (3 ft 6½ in)
Rear baggage door: Height	0.39 m (1 ft 3¼ in)
Width	0.25 m (9¾ in)

DIMENSIONS, INTERNAL:

Cabin: Length	4.05 m (13 ft 3½ in)
Max width	1.21 m (3 ft 11¾ in)
Max height	1.22 m (4 ft 0 in)
Volume	3.5 m³ (124 cu ft)
Baggage compartment volume:	
front	0.08 m³ (2.8 cu ft)
rear	0.10 m³ (3.5 cu ft)

AREAS:

Wings, gross	18.00 m² (193.75 sq ft)
Vertical tail surfaces (total)	2.56 m² (27.55 sq ft)
Horizontal tail surfaces (total)	4.76 m² (51.24 sq ft)

WEIGHTS AND LOADINGS:

Weight empty, basic	2,110 kg (4,652 lb)
Baggage capacity (all areas)	110 kg (243 lb)
Payload: max	626 kg (1,380 lb)
with max fuel	407 kg (897 lb)
Fuel weight (usable)	866 kg (1,910 lb)
Baggage: front	50 kg (110 lb)
rear	100 kg (220 lb)

TBM 700C with mixture of Garmin and Honeywell avionics   *NEW*/0526452

Max T-O weight: TBM 700C1	2,984 kg (6,578 lb)
TBM 700C2	3,354 kg (7,394 lb)
Max ramp weight	3,000 kg (6,613 lb)
Max zero fuel weight	2,736 kg (6,032 lb)
Max landing weight	2,835 kg (6,250 lb)
Max wing loading: TBM 700C1	
	165.8 kg/m² (33.95 lb/sq ft)
TBM 700C2	186.3 kg/m² (38.16 lb/sq ft)
Max power loading: TBM 700C1	
	5.72 kg/kW (9.48 lb/shp)
TBM 700C2	6.43 kg/kW (10.56 lb/shp)

PERFORMANCE:

Max cruising speed at FL260	300 kt (555 km/h; 345 mph)
Econ cruising speed at FL310	
	255 kt (472 km/h; 293 mph)

Stalling speed: clean	82 kt (152 km/h; 95 mph) IAS
flaps and landing gear down	
	65 kt (121 km/h; 75 mph) IAS
Max rate of climb at S/L	725 m (2,380 ft)/min
Time to FL260	24 min
Max certified altitude	9,450 m (31,000 ft)
T-O to 15 m (50 ft)	865 m (2,840 ft)
Landing from 15 m (50 ft)	742 m (2,435 ft)
Range at econ cruising speed, NBAA reserves:	
with max payload:	
	1,184 n miles (2,192 km; 1,362 miles)
with max fuel:	1,678 n miles (3,107 km; 1,931 miles)
g limits	+3.8/−1.5

UPDATED

GEORGIA

TAM

TBILISI AEROSPACE MANUFACTURING

ulitsa Khmelnitska 181B, 380036 Tbilisi
Tel: (+995 32) 98 53 90 and 70 88 38
Fax: (+995 32) 96 43 69 and 98 25 51
GENERAL DIRECTOR: Pantiko Tordia

TAM (known until privatisation in 2000 as Tbilisi Aviation State Association) began aircraft production in 1941, with the LaGG-3 fighter, and progressed through La-5, Yak-3, Yak-15, Yak-17/17UTI, Yak-23, MiG-15 and MiG-17 to building 1,677 MiG-21U/UMs (for which it offers an upgrade programme) between 1957 and 1984. The plant also began Su-25/25U production in 1978 and has built 875 basic single-seat versions, including 50 Su-25BM target-tugs. In total, it has built over 8,070 manned aircraft, 2,555 UAVs and over 36,000 AAMs (30,000 R-60/AA-8s and 6,000 R-73/AA-11s). Current manufacture, at low rate, is of the Su-25UB two-seat trainer; also produced small batch of its Su-25T (Su-39) derivative. Production is also stated to be under way of a batch of 20 Yak-58 light transports, although there is no evidence of recent sales. Upgrade activities include a 'Frogfoot' modernisation, the Su-25KM Scorpion, a prototype of which flew in 2001.

The 250,000 m² (2,691,000 sq ft) TAM plant employed 15,000 personnel at its peak, but currently has a workforce of about 4,000. A 1999 alliance with Kelowna Flightcraft of Canada (see *Jane's Aircraft Upgrades*) provides for TAM to assist with production of materials for the Convair 5800 conversion programme and eventually undertake the entire programme, if demand warrants. TAM also plans to fabricate components for the ViperJet two-seat kitbuilt, described in the US section.

UPDATED

GERMANY

AEROSTYLE

AEROSTYLE ULTRALEICHT FLUGZEUGE GmbH
Norderreeg 2, D-25852 Bordelum
Tel: (+49 4671) 93 13 93
Fax: (49 4671) 93 13 94
e-mail: info@Aerostyle-GmbH.de
Web: http://www.Aerostyle-GmbH.de
DIRECTOR: Ralf Magnussen

The company's first product, the Breezer, has undergone slight modification prior to entering production.

UPDATED

AEROSTYLE BREEZER
TYPE: Side-by-side ultralight.
PROGRAMME: Prototype (D-MOOV) exhibited unflown at Aero '99 at Friedrichshafen in April 1999; first flight was made in December 1999. Noise tests under way in mid-2000; German certification was then anticipated in August or September 2000, but not yet reported. Prototype was exhibited at Friedrichshafen in April 2001 with modifications including steerable nosewheel replacing original trailing-link type.
CUSTOMERS: Two complete aircraft and three kits sold by mid-2000, representing capacity for that year.
COSTS: DM85,905 excluding taxes (2001).
DESIGN FEATURES: Low, constant-chord wing, with wingtips upturned at rear. Sweptback fin. Wing aerofoil section NACA 4414.
FLYING CONTROLS: Conventional and manual. Horn-balanced elevators and rudder; trim tab in port elevator; half-span Fowler flaps.
STRUCTURE: Mainly riveted aluminium, with glass fibre for some non-structural fairings.
LANDING GEAR: Non-retractable tricycle type with composites cantilever main gear legs and composites steerable nose leg; mainwheel size 14×4, nosewheel 4.00×4.
POWER PLANT: One 67.1 to 74.6 kW (90 to 100 hp) fuel-injected Take Off-BMW engine driving a three-blade ground-adjustable Neuform T3 propeller. Optionally a 73.5 kW (98.6 hp) Rotax 912 ULS can be fitted. Fuel capacity 55 litres (14.5 US gallons; 12.1 Imp gallons).
DIMENSIONS, EXTERNAL:

Wing span	6.40 m (21 ft 0 in)
Length overall	8.71 m (28 ft 7 in)
Height overall	2.12 m (6 ft 11½ in)

AREAS:

Wings, gross	11.85 m² (127.6 sq ft)

WEIGHTS AND LOADINGS (estimated):

Weight empty	285-295 kg (628-650 lb)
Max T-O weight	450 kg (992 lb)

PERFORMANCE (estimated, A: 67.1 kW/90 hp engine; B: 73.5 kW/98.6 hp engine):

Max level speed: A	119 kt (220 km/h; 137 mph)
B	124 kt (230 km/h; 143 mph)

Prototype Aerostyle Breezer   0099161

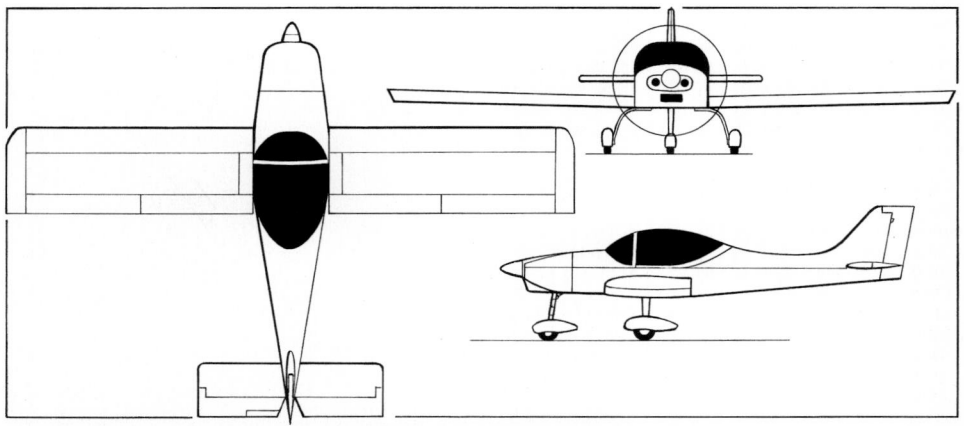

Aerostyle Breezer all-metal two-seat ultralight aircraft (*Jane's/James Goulding*)   0110592

Cruising speed: A	105 kt (195 km/h; 121 mph)	T-O and landing run	120 m (395 ft)
B	113 kt (210 km/h; 130 mph)	*g* limits	+4/−2
Stalling speed: A, B	34 kt (63 km/h; 40 mph)		*UPDATED*
Max rate of climb at S/L	270 m (886 ft)/min		

AIR LIGHT

AIR LIGHT
This company has been renamed Ultraleichtbau-International; see under ULBI heading in this section.

UPDATED

AKAFLIEG MÜNCHEN

AKAFLIEG MÜNCHEN eV
Arcisstrasse 21, D-80333 München
Tel: (+49 89) 28 91 59 78
Fax: (+49 89) 28 91 59 79
e-mail: publia@akaflieg.vo.tu-muenchen.de
Web: http://www.akaflieg.vo.tu-muenchen.de

Akademische Fliegergruppe München (Munich Academy Flying Group) formed in 1924 to provide students with theoretical and practical knowlege of aircraft design and construction; now affiliated to Munich Technical University. Over 20 types have been built; most have been sailplanes, but the latest venture is the Mü30 Schlacro powered aerobatic design, first flown on 16 June 2000.

On 8 February 2002, the Academy announced it had started work on the Mü31, a self-launching sailplane with a 15 m (49 ft 2½ in) wing span.

UPDATED

AQUILA

AQUILA TECHNISCHE ENTWICKLUNGEN GmbH
Flugplatz, D-14959 Schönhagen
Tel: (+49 33) 731 70 70
Fax: (+49 33) 73 17 07 11
e-mail: info@aquila-aero.com
Web: http://www.aquila-aero.com
DIRECTORS:
Peter Grundhoff

Alfred Schmiderer
Markus Wagner
MARKETING MANAGER: Siegfried Dörfler
SALES MANAGER: Gregor Bremer

Company formed August 1996 to develop A210; employed 12 staff at Schönhagen aerodrome by April 2000; increased to 20 during 2001 and expected to reach 35 by December 2002, following mid-2000 commissioning of new production hangar; capacity 50 aircraft per year.

UPDATED

AQUILA A210
TYPE: Two-seat lightplane.
PROGRAMME: Launched 1995; design, using CAD/CAM techniques, began 1997, assisted by Berlin Technical University; prototype (D-EQUI) displayed at Aero '99, Friedrichshafen, in April 1999; first flight 5 March 2000; initially flew without winglets; certification to JAR-VLA received September 2001; deliveries began May 2002 with first aircraft (D-ETHG) to Flugschule Hans Grade at Schönhagen.

First Aquila A210 delivery, May 2002 (*Jane's/Paul Jackson*)   *NEW*/0137223

DIMENSIONS, EXTERNAL:	
Wing span, excl winglets	10.33 m (33 ft 10¼ in)
Wing aspect ratio	10.1
Length overall	7.33 m (24 ft 0½ in)
Height overall	2.40 m (7 ft 10½ in)
Fuselage max width	1.20 m (3 ft 11¼ in)
Tailplane span	3.00 m (9 ft 10 in)
Wheel track	1.94 m (6 ft 4½ in)
Wheelbase	1.68 m (5 ft 6¼ in)
Propeller diameter	1.75 m (5 ft 9 in)
Baggage door: Max width	0.51 m (1 ft 8 in)
DIMENSIONS, INTERNAL:	
Cabin max width	1.15 m (3 ft 9¼ in)
Baggage volume	0.50 m³ (17.66 cu ft)
AREAS:	
Wings, gross	10.50 m² (113.0 sq ft)
WEIGHTS AND LOADINGS:	
Weight empty	490 kg (1,080 lb)
Baggage capacity	40 kg (97 lb)
Max T-O weight	750 kg (1,653 lb)
Max wing loading	71.4 kg/m² (14.63 lb/sq ft)
Max power loading	10.20 kg/kW (16.76 lb/hp)
PERFORMANCE:	
Max level speed	164 kt (305 km/h; 189 mph)
Cruising speed at 75% power	
	130 kt (240 km/h; 149 mph)
Stalling speed: flaps up	56 kt (102 km/h; 64 mph)
flaps down	44 kt (81 km/h; 51 mph)
Max rate of climb at S/L	229 m (750 ft)/min
Service ceiling	4,420 m (14,500 ft)
T-O run	250 m (820 ft)
T-O to 15 m (50 ft)	470 m (1,542 ft)
Landing from 15 m (50 ft)	500 m (1,640 ft)
Landing run	200 m (656 ft)
Range with max fuel at 55% power at FL500, 45 min reserves	
	620 n miles (1,148 km; 713 miles)
	UPDATED

CUSTOMERS: Total 20 firm orders by May 2002, including pilot trainers for Lufthansa, and club aircraft for DaimlerChrysler and airline Swiss.

COSTS: €112,000 excluding tax (2002).

DESIGN FEATURES: Design goals included crashworthy cabin structure, forgiving handling characteristics and low operating costs. Low wing with sharply tapered fuselage aft of cabin and sweptback fin; high aspect ratio wing with upturned tips.

Laminar flow wing section Horstmann/Quast HQ-42 modified. Dihedral 4° 30′.

FLYING CONTROLS: Conventional and manual. Actuation via pushrods for elevators and ailerons; cables for rudder. Externally hinged, single-slotted Fowler flaps, deflections 15 and 35°.

STRUCTURE: Entirely CFRP/GFRP. Fuselage is a GFRP monocoque with CFRP stringers and GFRP/CFRP frames; wing shell structure and tail unit and control surfaces are GFRP/polyurethane foam sandwich with CFRP wingspar, load-bearing structures and local reinforcements.

LANDING GEAR: Non-retractable tricycle type; spring steel main legs; steerable nose leg with rubber-in-compression suspension; all wheels size 5.00-5; speed fairings on all wheels; hydraulic brakes; combined ventral fillet/tailskid.

POWER PLANT: One 73.5 kW (98.6 hp) Rotax 912S flat-four engine driving an MT-Propeller MTV-21-A/175-05 two-blade, variable-pitch (hydraulic) propeller. Fuel in two wing tanks, total capacity 120 litres (31.7 US gallons; 26.4 Imp gallons). Fuel filler in upper surface of each wing.

ACCOMMODATION: Two persons side by side under forward- and upward-opening windscreen/canopy; baggage door on port side of aft wing. Four-point safety harnesses for both occupants. Accommodation heated and ventilated.

SYSTEMS: 14 V electrical system, including 40 Ah alternator and 12 V battery.

AVIONICS: *Comms:* Honeywell KX 125 com/nav; KT 76C transponder; encoding altimeter, PS Engineering PM 501 intercom, and ELT, all standard. Becker AR 4201 second com, Honeywell KT 76C digital transponder with encoder, Garmin GTX 327 transponder, KMA 28 audio panel, GMA 340 audio panel, and 406 MHz ELT, all optional.

Flight: Honeywell KX 155 com/nav with GI 106A VOR/LOC/GS indicator and KMD 150 GPS with colour MFD, Garmin GNS 430 com/nav/GPS with GI 106A VOR/LOC/GS indicator, optional.

Instrumentation: Three-point altimeter, ASI, VSI, magnetic compass, artificial horizon, directional gyro, turn co-ordinator, CHT gauge, oil temperature/pressure gauges, fuel gauge, tachometer, manifold pressure gauge, OAT gauge, hour meter, Davtron M800 chronometer, all standard.

EQUIPMENT: Navigation and anti-collision lights and tiedown fittings, standard. Optional equipment includes auxiliary/external power receptacles, tinted canopy, fire extinguisher and aero-tow system.

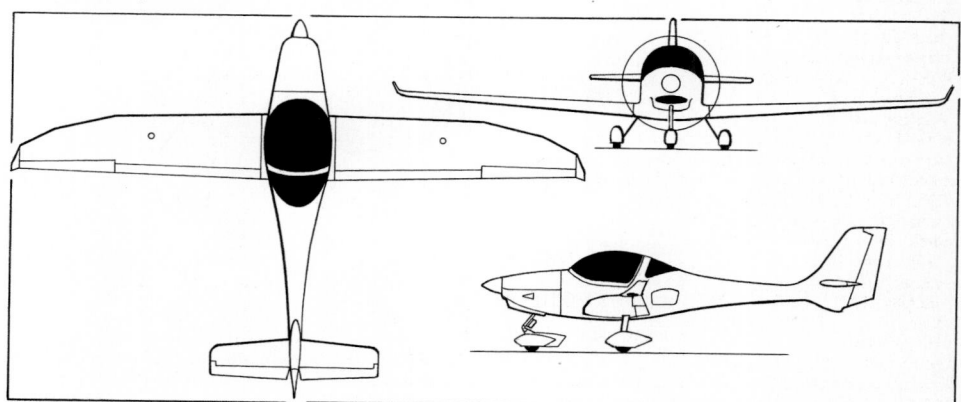

Aquila A210 all-composites light aircraft (Rotax 912 ULS flat-four) (*Jane's/James Goulding*)   *NEW*/0137230

B&F

B&F TECHNIK VERTRIEBS GmbH

Am neuen Rheinhafen 10, D-67346 Speyer
Tel: (+49 62) 327 20 76
Fax: (+49 62) 327 20 78

The FK series is designed by Dipl Ing Otto Funk and Peter Funk. B&F continues to market the original FK 9 ultralight in parallel to a new version in composites. Latest products are the nostalgic FK 12 and FK 14 Polaris composites ultralights and a version of FK 9 powered by a Smart motorcar engine.

Work on B&F's new 900 m² (9,700 sq ft) factory on the same airfield began in January 2002.

UPDATED

B&F FK 9 MARK 3

TYPE: Side-by-side ultralight.

PROGRAMME: The earlier versions of B&F's FK 9 (see 1990-91 *Jane's*) were built at Krosno, Poland, employing metal tube fuselages and composites wings, both fabric-covered. The Mark 3 all-composites version was announced in 1997; 85 per cent of airframe is manufactured in Poland, for completion in Germany. Launched at Friedrichshafen Air Show, April 1997; UK sales promotion from July 1998. Following trials in earlier FK 9 (D-MCFK), 40.5 kW (54.2 hp) Suprex three-cylinder engine, developed from that of DaimlerChrysler Smart motorcar, offered as alternative power plant from 2000 in version designated **FK 9 Smart**.

CUSTOMERS: More than 170 FK 9s of all series built by April 2001.

COSTS: DM87,500 including tax (1998).

STRUCTURE: Composites wing and fuselage; metal tailplane; fabric-covered ailerons, slotted flaps and elevators.

LANDING GEAR: Fixed tricycle type with speed fairings.

B&F FK 9 Mk 3 two-seat ultralight (*Jane's/Paul Jackson*)   *NEW*/0132405

POWER PLANT: One 59.6 kW (79.9 hp) Rotax 912 UL or 40.5 kW (54.2 hp) Suprex driving a Junkers three-blade, ground-adjustable pitch, carbon fibre propeller. Max fuel capacity 65 litres (17.2 US gallons; 14.3 Imp gallons).

DIMENSIONS, EXTERNAL:	
Wing span	9.85 m (32 ft 3¾ in)
Length overall	5.85 m (19 ft 2¼ in)
AREAS:	
Wings, gross	11.60 m² (124.9 sq ft)
WEIGHTS AND LOADINGS:	
Weight empty	275 kg (606 lb)
Max T-O weight	450 kg (992 lb)
PERFORMANCE:	
Cruising speed:	
at 75% power	103 kt (190 km/h; 118 mph)
at 65% power	97 kt (180 km/h; 112 mph)
Stalling speed	36 kt (65 km/h; 41 mph)
Max rate of climb at S/L	330 m (1,080 ft)/min
Range	540 n miles (1,000 km; 621 miles)
	UPDATED

B&F FK 12 COMET

TYPE: Tandem-seat ultralight biplane.

PROGRAMME: Design, by Peter Funk, started late 1994; construction of prototype began 1995; first flight (D-MPLI) March 1997; first flight of production aircraft May 1997; certification was proceeding in 1999 but not notified by early 2001.

CUSTOMERS: At least 47 built by January 2002, of which 40 were then flying.

COSTS: Kit DM43,000, ready-to-fly DM89,000, including taxes (1998).

DESIGN FEATURES: Wings, of laminar aerofoil section, have sweepback, and fold back for storage.

FLYING CONTROLS: Conventional and manual rudder and elevator; full-span flaperons on upper and lower wings.

STRUCTURE: Welded steel tube forward fuselage with riveted aluminium tube rear section; composites and GFRP laminate skinning; wings have carbon fibre main spar and leading-edge with aluminium secondary spar and flaperons, rear section fabric-covered.

LANDING GEAR: Tailwheel type; non-retractable.

POWER PLANT: One 59.6 kW (79.9 hp) Rotax 912 UL, 73.5 kW (98.6 hp) Rotax 912 ULS or turbocharged 84.6 kW (113.4 hp) Rotax 914 driving an MT-Propeller two-blade wooden propeller; Junkers three-blade, carbon fibre propeller optional. Fuel capacity 42 litres (11.1 US gallons; 9.2 Imp gallons).

DIMENSIONS, EXTERNAL:

Wing span	6.70 m (21 ft 11¾ in)
Length overall	5.30 m (17 ft 4¾ in)
Height overall	1.98 m (6 ft 6 in)

AREAS:

Wings, gross	13.40 m² (144.2 sq ft)

WEIGHTS AND LOADINGS:

Weight empty	265 kg (584 lb)
Max T-O weight	450 kg (992 lb)

PERFORMANCE:

Max level speed	100 kt (185 km/h; 115 mph)
Cruising speed	89 kt (165 km/h; 103 mph)
Stalling speed, flaperons down	36 kt (65 km/h; 41 mph)
Max rate of climb at S/L	305 m (1,000 ft)/min
T-O run	120 m (394 ft)
Landing run	183 m (600 ft)
Range with max fuel	243 n miles (450 km; 279 miles)

UPDATED

B&F FK 12 Comet ultralight aerobatic biplane (*Jane's/Paul Jackson*) NEW/0132396

B&F FK 14 Polaris (*Jane's/Paul Jackson*) NEW/0132395

B&F FK 14 POLARIS

TYPE: Side-by-side ultralight.

PROGRAMME: Development began early 1998; prototype (D-MVFK), then unflown, exhibited at Aero '99 at Friedrichshafen in April 1999. In production by 2000.

CUSTOMERS: At least nine built by April 2001.

COSTS: With Rotax 912, DM119,480; with Rotax 912 ULS, DM123,540; both including tax (1999).

DESIGN FEATURES: Low wing with upturned wingtips incorporating navigation lights; wing section designed with assistance of Stuttgart University wind tunnel. Quoted build time 650 hours.

FLYING CONTROLS: Conventional and manual; two-section flaps extend on rails; trim tab in each elevator; horn-balanced rudder.

STRUCTURE: Mainly composites, with metal ailerons, tailplane and first section of flaps; elevators, rudder and second section of flaps fabric-covered.

LANDING GEAR: Non-retractable tricycle type; composites cantilever main legs; steerable nosewheel; brakes; speed fairing on all wheels.

POWER PLANT: One 59.6 kW (79.9 hp) Rotax 912 UL or 73.5 kW (98.6 hp) Rotax 912 ULS driving a Junkers three-blade, ground-adjustable pitch, carbon fibre propeller. Fuel contained in single tank behind cockpit, capacity 42 litres (11.1 US gallons; 9.2 Imp gallons).

EQUIPMENT: BRS 5 emergency parachute recovery system optional.

DIMENSIONS, EXTERNAL:

Wing span	9.10 m (29 ft 10¼ in)
Length overall	5.35 m (17 ft 6½ in)
Height overall	1.88 m (6 ft 2 in)

AREAS:

Wings, gross	9.40 m² (101.2 sq ft)

WEIGHTS AND LOADINGS:

Weight empty	272 kg (600 lb)
Max T-O weight	450 kg (992 lb)

PERFORMANCE (estimated):

Cruising speed	140 kt (260 km/h; 162 mph)
Min flying speed	35 kt (65 km/h; 41 mph)
Max rate of climb at S/L	390 m (1,280 ft)/min
T-O run	122 m (400 ft)
Landing run	152 m (500 ft)
Range	350 n miles (648 km; 403 miles)

UPDATED

BUSSARD

BUSSARD DESIGN GmbH

Marie-Curiestrasse 6, D-85055 Ingolstadt
Tel: (+49 841) 901 44 60
Fax: (+49 841) 901 44 62
e-mail (1): info@bussard-design.de
e-mail (2): info@raptor-online.de
Web (1): http://www.bussard-design.de
Web (2): http://www.raptor-online.de
DEVELOPMENT DIRECTOR: Georg Heinrich

A design services subcontractor to the aerospace and automotive industries, Bussard plans to enter lightplane manufacturing with its own concept, the Raptor.

UPDATED

BUSSARD RAPTOR

TYPE: Side-by-side ultralight.

PROGRAMME: Design started June 1999; prototype construction began September 1999; scale model shown at ILA, Berlin, June 2000; assembly of first production aircraft commenced November 2000; first flight of prototype expected in September 2001 but not yet notified to *Jane's* by May 2002; first customer delivery March 2002. Available factory-built only.

CUSTOMERS: Two ordered by May 2001.

DESIGN FEATURES: Pod-and-boom configuration with pusher propeller; high wing with constant chord on inner three-quarters and two-stage sweepback on outer panels. Mid-mounted tailplane and sweptback fin.

FLYING CONTROLS: Conventional and manual. Electric trim.

STRUCTURE: Aluminium fuselage structure with carbon fibre sandwich covering, except Kevlar reinforcement around cockpit. Carbon fibre wings.

LANDING GEAR: Tricycle type; fixed. Steerable nosewheel. Speed fairings on all wheels.

POWER PLANT: One digitally controlled 84.6 kW (113.4 hp) Rotax 914 F flat-four driving a pusher propeller. 59.6 kW (79.9 hp) Rotax 912 UL optional. Fuel capacity 70 litres (18.5 US gallons; 15.4 Imp gallons).

EQUIPMENT: BRS 05 UL softback parachute recovery system.

DIMENSIONS, EXTERNAL:

Wing span	8.50 m (27 ft 10¾ in)
Length overall	6.20 m (20 ft 4 in)

DIMENSIONS, INTERNAL:

Cockpit max width	1.15 m (3 ft 9¼ in)

Model of Bussard Raptor shown at Berlin, June 2000 (*Jane's/Paul Jackson*) 0092053

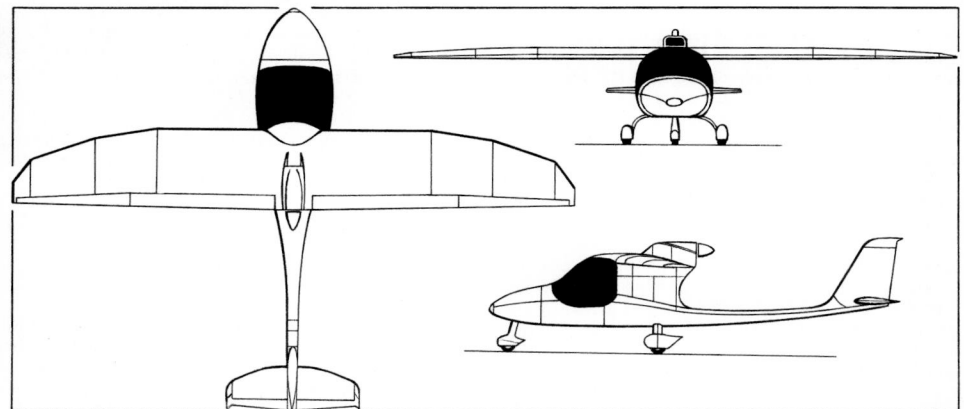

Bussard Raptor two-seat ultralight (*Jane's/James Goulding*) 0121272

AREAS:

Wings, gross	10.50 m² (113.0 sq ft)

WEIGHTS AND LOADINGS:

Weight empty	200 kg (441 lb)
Baggage capacity	20 kg (44 lb)
Max T-O weight	450 kg (992 lb)

PERFORMANCE (estimated):

Max level speed	135 kt (250 km/h; 155 mph)
Normal cruising speed	124 kt (230 km/h; 143 mph)
Stalling speed	36 kt (65 km/h; 41 mph
Max rate of climb at S/L	360 m (1,181 ft)/min
T-O and landing run	190 m (623 ft)

Range	Up to 540 n miles (1,000 km; 621 miles)
g limits	+6/−3

UPDATED

EADS DEUTSCHLAND

EADS DEUTSCHLAND GmbH

PO Box 801109, D-81663 München
Tel: (+49 89) 60 70
Fax: (+49 89) 60 72 64 81
Web: http://www.eads-nv.net

PRESIDENT AND CEO: Rainer Hertrich
VICE-PRESIDENT, PRESS AND INFORMATION: Rainer Ohler

On 14 October 1999, former Daimler-Chrysler Aerospace and France's Aerospatiale Matra signed agreement for merger as basis of EADS (see International section), CASA of Spain following suit on 2 December 1999. EADS officially formed 10 July 2000.

See International section for further details and 2002-03 and previous editions for history of German component.
Among components of EADS located in Germany is Military Aircraft.

UPDATED

EADS MILITARY AIRCRAFT

D-81663 München
Tel: (+49 89) 60 70
Fax: (+49 89) 60 72 64 81
PRESIDENT: Aloysius Rauen
EXECUTIVE VICE-PRESIDENT, SERIES PRODUCTION PROGRAMMES:
  Erwin Obermeier
EXECUTIVE VICE-PRESIDENT, SUPPORT PROGRAMMES:
  Rüdiger Sanitz
HEAD OF COMMUNICATION: Wolfdietrich Hoeveler

Major activities of Group include light fighter/trainer aircraft, airborne reconnaissance systems, aircraft armament,

disarmament verification systems (the former PRISMA system and BICES), simulation and training systems, and research into advanced aircraft systems, materials and manufacturing technologies. Division also makes Airbus subassemblies. Augsburg, Manching and Ottobrunn plants employed 5,750 personnel in 2000; EADS Military Aircraft employees then totalled 7,400 in all countries.
Refer to International section for Military Aircraft Business Unit participation in the **Mako** and **Eurofighter Typhoon** programmes, plus *Jane's Aircraft Upgrades* for the DASA **F-4F Phantom ICE** (improved combat effectiveness) programme and **Panavia Tornado**. Unit

offers upgrades for MiG-29 under agreement signed with RSK 'MiG' in June 1999; completed radar upgrade of NATO E-3A AWACS fleet in 2000.
Spanish inputs include support of EF-18 Hornet, AV-8B Harrier, Mirage F1, P-3 Orion, C-101 Aviojet, F-5B Freedom Fighter and E.26 Tamiz.

UPDATED

EXTRA

EXTRA-FLUGZEUGBAU GmbH

Flugplatz Dinslaken, Schwarze Heide 21, D-46569 Hünxe
Tel: (+49 2858) 913 70
Fax: (+49 2858) 91 37 30
MANAGING DIRECTOR: Walter Extra
INFORMATION CONTACT: Bruno Van Waeyenberghe

EXTRA 400 EUROPEAN DISTRIBUTION:
  Diamond Aircraft Service GmbH
  Flughafen Siegerland, Werfthalle 1, D-57299 Burbach
  Tel: (+49 2736) 442 80
  Fax: (+49 2736) 44 28 50
  e-mail: info@diamond-aircraft.de
  Web: http://www.diamond-aircraft.de

Extra produces aerobatic sportplanes and two variants of a business aircraft. Workforce is approximately 210.

UPDATED

EXTRA 200

TYPE: Aerobatic two-seat sportplane.
PROGRAMME: Prototype (D-ETEL) first flown 2 April 1996 and almost immediately shipped to USA for demonstrations and public debut at EAA Sun 'n' Fun in Lakeland, Florida, later that month; second prototype and principal flight test aircraft (D-EAZG) exhibited at ILA, Berlin, May 1996; German certification scheduled for 15 June 1996, followed by FAA certification (initially in Experimental/Exhibition Category) on 20 December 1996. Engineering designation **Extra 300/200**.
CUSTOMERS: Prototype and five production aircraft delivered in 1996; 15 in 1997; five in 1998; further two in mid-1999; and one in mid-2000; further two completed in late 2001 for customers in USA and Ireland; total 31. Exports to USA account for 88 per cent of production.
COSTS: €205,000 (2002).
Data generally as for Extra 300, except that below:
POWER PLANT: One 149 kW (200 hp) Textron Lycoming AEIO-360-A1E flat-four engine with Gomolzig exhaust, driving an MT-Propeller MTV-12-B-C/C183-17e three-blade, constant-speed propeller. Standard fuel capacity 116 litres (30.6 US gallons; 25.5 Imp gallons), of which 114 litres (30.0 US gallons; 25.0 Imp gallons) are usable; optional long-range tanks increase total capacity to 154 litres (40.6 US gallons; 33.8 Imp gallons), of which 151 litres (40.0 US gallons; 33.3 Imp gallons) are usable.

DIMENSIONS, EXTERNAL:

Wing span	7.50 m (24 ft 7¼ in)
Wing aspect ratio	5.4
Length overall	6.81 m (22 ft 4 in)
Height overall	2.56 m (8 ft 4¾ in)

AREAS:

Wings, gross	10.40 m² (111.9 sq ft)

WEIGHTS AND LOADINGS:

Weight empty	544 kg (1,199 lb)
Max T-O weight: Aerobatic, solo	608 kg (1,340 lb)
Aerobatic, two-seat	838 kg (1,848 lb)
Normal	868 kg (1,914 lb)

Max wing loading:

Aerobatic, solo	57.75 kg/m² (11.97 lb/sq ft)

Extra 200 aerobatic/training aircraft *(Jane's/Paul Jackson)*   **NEW**/0132414

Aerobatic, two-seat	80.6 kg/m² (16.50 lb/sq ft)
Normal	83.5 kg/m² (17.09 lb/sq ft)

Max power loading:

Aerobatic, solo	4.08 kg/kW (6.70 lb/hp)
Aerobatic, two-seat	5.62 kg/kW (9.24 lb/hp)
Normal	5.82 kg/kW (9.57 lb/hp)

PERFORMANCE (N: at Normal category T-O weight, SA: at solo Aerobatic weight):
Never-exceed speed (V_{NE}):

N	220 kt (407 km/h; 253 mph)

Manoeuvring speed	158 kt (293 km/h; 182 mph)

Cruising speed at 75% power:

N	150 kt (278 km/h; 173 mph)

Stalling speed: N 56 kt (104 km/h; 64 mph)
SA 52 kt (96 km/h; 60 mph)
Max rate of climb at S/L: N 488 m (1,600 ft)/min
SA 686 m (2,250 ft)/min
Service ceiling: N 4,575 m (15,000 ft)
Range with single pilot at 75% power at 2,440 m (8,000 ft), 45 min reserves:
  N: standard fuel 450 n miles (833 km; 517 miles)
  optional fuel 600 n miles (1,111 km; 690 miles)
g limits: SA ±10

UPDATED

EXTRA 300 and 330

TYPE: Aerobatic two-seat sportplane.
PROGRAMME: Design began January 1987; first flight (D-EAEW) 6 May 1988; LBA certification 16 May 1990; certified FAR Pt 23 Amendment 33 Normal and Aerobatic categories for USA 26 February 1993; production started October 1988.
CURRENT VERSIONS: **Extra 300:** Baseline, mid-wing, two-seat aircraft.
Details apply to Extra 300 except where indicated.

Extra 300L: Low-wing, two-seat version with other structural changes; low wing enhances visibility during take-off and landing and eases installation of full IFR instrumentation; new ailerons give slightly higher roll rate (approximately 400°/s) and improved snap-rolling capability, both erect and inverted; shorter fuselage with all-composites side and belly access panels; cockpit interior modified for more comfort, with new reclined carbon fibre seats with optional leather trim; first aircraft (D-EUWH) delivered November 1994. FAA certification 31 March 1995. Current production version.

Extra 300S: Single-seat version of 300 with same power plant; wing shortened by 50 cm (19½ in) and more powerful ailerons; first flight (D-ESEW) 4 March 1992; certified March 1992; FAA certification 28 October 1993.

Extra 330L: Strengthened version of 300L with CFRP fuselage panels, fin and rudder; wider-chord rudder and elevators, all with horn balances; 246 kW (330 hp) Textron Lycoming AEIO-580 engine. Prototype (D-ESEW, modified from prototype 300S) first flown January 1998 and since converted to 330L. First production aircraft (D-EPET) flew 11 May 1998 and delivered to Hungarian aerobatic pilot Peter Besenyei; normally flown with single-seat cockpit canopy; first public performance at North Weald, UK, 29 May 1998.

Extra 330LX: Two-seat version of 330; prototype D-EDGE first flown early 1999.

Extra 330XS: Designation applied to the former 300S (including D-EJKS of Aerobatics unlimited and N76PK of display pilot Gene Soucy) retrofitted with wide-chord rudder.

CUSTOMERS: Total 66 (including prototype) two-seat original ('mid-wing') EA 300s delivered by April 1997; additional one to USA in November 1998. Further 28 EA 300Ss built by August 1995, plus one in 1997 and one in 1999. EA

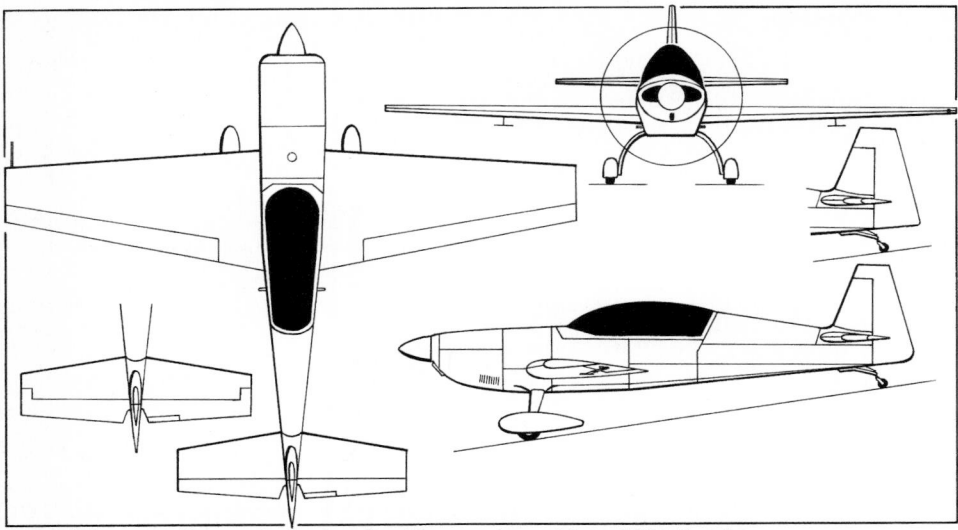

Extra 300L competition aerobatic aircraft, with scrap views of Extra 330 tail surfaces *(Jane's/James Goulding)* 0064457

300L production totalled 144 by December 2001. EA 330L/LX production three (including prototype). Almost all for customers in USA, although recent sales included one to Israel, six to Romania, one to South Africa and two to the United Kingdom for Brian Lecomber's Firebird Aerobatics.

COSTS: Basic price DM340,000 (January 1996); 330, DM350,000 (1998).

DESIGN FEATURES: Designed for unlimited competition aerobatics; tapered, square-tipped mid-mounted wing with 4° leading-edge sweepback; aerofoil symmetrical MA-15S at root, MA-12S at tip; no twist or dihedral.

FLYING CONTROLS: Conventional and manual. Rod and cable operated; nearly full-span ailerons, assisted by spade-type suspended tabs; trim tab in starboard elevator.

STRUCTURE: Fuselage (excluding tail surfaces) of steel tube frame with part aluminium, part fabric covering; wing spars of carbon composites and shells of carbon composite sandwich; tail surfaces of carbon spars and glass fibre shells.

LANDING GEAR: Fixed cantilever composites arch main gear with single polyamid-faired wheels; Cleveland brakes; leaf-sprung steerable tailwheel.

POWER PLANT: One 224 kW (300 hp) Textron Lycoming AEIO-540-L1B5 flat-six engine, with Gomolzig exhaust, driving an MT-Propeller MTV-9-B-C/C200-15 three-blade, constant-speed propeller (four-blade MTV-14-B-C/C190-17 optional). With either standard or optional propeller and Gomolzig System 3 silencer, Extra 300s meet German and US noise limits. Christen Industries inverted oil system. Fuel capacity: 300L standard capacity of 171 litres (45.1 US gallons; 37.6 Imp gallons), of which 169 litres (44.6 US gallons; 37.1 Imp gallons) usable; optional long-range tanks raise total fuel capacity to 208 litres (55.1 US gallons; 45.9 Imp gallons), of which 206 litres (54.5 US gallons; 45.4 Imp gallons) usable. 300S standard fuel capacity 160 litres (42.3 US gallons; 35.2 Imp gallons).

ACCOMMODATION: Pilot and co-pilot/passenger in tandem under single-piece canopy opening to starboard; additional transparencies in lower sides of cockpit.

SYSTEMS: 12 V generator and battery.

AVIONICS: To customer's requirements.

DIMENSIONS, EXTERNAL:

Wing span: 300	8.00 m (26 ft 3 in)
300L	7.70 m (25 ft 3 in)
300S	7.50 m (24 ft 7¼ in)
Wing chord:	
at root: 300, 300S	1.85 m (6 ft 0¾ in)
at tip: 300	0.83 m (2 ft 8¾ in)
300S	0.93 m (3 ft 0½ in)
Wing aspect ratio: 300	6.0
300S	5.4
Length overall: 300	7.12 m (23 ft 4¼ in)
300L	6.94 m (22 ft 9¼ in)
300S	6.65 m (21 ft 9¾ in)
Height overall: 300, 300S	2.62 m (8 ft 7¼ in)
Tailplane span: 300	3.20 m (10 ft 6 in)
Wheel track: 300	1.80 m (5 ft 10¾ in)
Propeller diameter: MTV-9	2.00 m (6 ft 6¾ in)
MTV-14	1.90 m (6 ft 2¾ in)

AREAS:

Wings, gross: 300, 300L	10.70 m² (115.2 sq ft)
300S	10.44 m² (112.4 sq ft)
Ailerons, total: 300, 300L	1.71 m² (18.41 sq ft)

WEIGHTS AND LOADINGS:

Weight empty: 300	630 kg (1,389 lb)
300L	653 kg (1,440 lb)
300S	609 kg (1,343 lb)
Max T-O weight:	
300, Aerobatic, solo and 300S	821 kg (1,810 lb)
300, Aerobatic, two-seat	870 kg (1,918 lb)
300, 300L, 300S, Normal	950 kg (2,095 lb)

Max wing loading:

300, Aerobatic, solo	76.5 kg/m² (15.67 lb/sq ft)
300, 300L, 300S, Normal	88.8 kg/m² (18.19 lb/sq ft)

Max power loading:

300, Aerobatic, solo	3.72 kg/kW (6.12 lb/hp)
300, Normal	4.24 kg/kW (6.98 lb/hp)
300S	3.66 kg/kW (6.02 lb/hp)

PERFORMANCE:

Never-exceed speed (V_{NE}):	
300L, 300S	220 kt (407 km/h; 253 mph)
Max level speed: 300S	185 kt (343 km/h; 213 mph)
Max manoeuvring speed: 300L, 300S	
	158 kt (293 km/h; 182 mph)
Cruising speed, 45% power at 2,440 m (8,000 ft):	
300L	170 kt (315 km/h; 196 mph)
Stalling speed:	
300L, 300S at solo Aerobatic max T-O weight	
	55 kt (102 km/h; 64 mph)
300L, 300S at Normal max T-O weight	
	60 kt (111 km/h; 69 mph)
Max rate of climb at S/L:	
300L, 300S	975 m (3,200 ft)/min
Service ceiling: 300L, 300S	4,875 m (16,000 ft)
T-O run	199 m (655 ft)
T-O to 15 m (50 ft)	545 m (1,790 ft)
Range at 65% power cruising speed at 2,440 m (8,000 ft), 45 min reserves:	
300S	440 n miles (814 km; 506 miles)
Range at 45% power cruising speed, at 2,440 m (8,000 ft) 45 min reserves:	
300L, standard fuel	415 n miles (768 km; 477 miles)
300L, long-range fuel	510 n miles (944 km; 586 miles)

Endurance: 300S — 2h 30 min
g limits: Aerobatic, solo — ±10
Two-seat Aerobatic — ±8
UPDATED

EXTRA 400

TYPE: Business prop.

PROGRAMME: Announced February 1993; designed in collaboration with Delft University; first flight (D-EBEW) 4 April 1996; initial LBA certification to FAR Pt 23 received 23 April 1997. Second prototype (D-EGBU) flew April 1998, featuring downturned wingtips and ventral stake, both of which had been added to first aircraft before its retirement in 1997. This aircraft lost on delivery to first customer, 21 August 1998. FAA certification 15 April 1998.

CURRENT VERSIONS: **Extra 400**: As described.
Extra 500: Turboprop. Described separately.

CUSTOMERS: Two prototypes and 18 production aircraft registered by late 2001, including four to USA and one to Switzerland. Total 24 completed by May 2002.

COSTS: Typically equipped, US$998,500 (2000).

DESIGN FEATURES: Cantilever high-wing, T tail layout; spacious cabin; optimised for low internal and external noise.

FLYING CONTROLS: Conventional and manual. Frise ailerons; two-section Fowler flaps approximately two-thirds of span, deflection 0, 15 and 30°; horn-balanced elevators; trim tab in starboard elevator; ground-adjustable tab on rudder. Interconnected aileron and rudder controls, with override.

STRUCTURE: Composites throughout. Airframe consists of skin with integral longerons and frames. Wing built on double front spar and rear spar, interconnected by ribs; tailplane has front and rear spar. Skins generally of carbon fibre facings and honeycomb; longerons, frames, spars and ribs of carbon fibre with foam core; wing leading-edge of glass fibre with honeycomb and galass fibre ribs. Service life 25 years/20,000 hours/30,000 cycles/22,500 pressurisation cycles.

LANDING GEAR: Hydraulically retractable tricycle type; designed and manufactured by Gomolzig; single wheels throughout; main units retract forward into fuselage sides; nosewheel retracts rearwards. Mainwheel size 15×6.0-6, nosewheel size 5.00-5. Minimum turning radius 20.4 m (67 ft). Independent hydraulic brakes on mainwheels. Nosewheel steerable ±30°.

POWER PLANT: One 261 kW (350 hp) Teledyne Continental Voyager TSIOL-550-C six-cylinder liquid-cooled engine, driving an MT-Propeller MTV-14-D/195-30a four-blade, constant-speed (hydraulic) propeller; three-blade propeller optional. Fuel in two integral wing tanks, combined capacity 468 litres (123.6 US gallons; 102.9 Imp gallons), of which 404 litres (107 US gallons; 88.9 Imp gallons) are usable. Oil capacity 12.3 litres (3.2 US gallons; 2.7 Imp gallons).

ACCOMMODATION: Pilot and five passengers in pressurised cabin; rear seats in club configuration; single door on port side is horizontally split, with airstair in bottom half; top half hinges upwards. Emergency exit starboard side, opposite door.

One of two Extra 300Ls used by Firebird Aerobatics *(Jane's/Paul Jackson)* *NEW*/0132416

Extra 330XS retrospective upgrade *(Jane's/Paul Jackson)* *NEW*/0132415

SYSTEMS: Electrical system 24 V, with direct-driven 100 A alternator, belt-driven 85 A alternator and 24 V battery in tailcone. External power receptacle in tailcone. Pressurisation system, maximum differential 0.38 bar (5.5 lb/sq in). Wing and empennage leading-edges de-iced by Teflon inflatable boots.

AVIONICS: Bendix/King package as standard, including two-tube EFIS with EFS 40 electronic EHSI. Dual Garmin GNS 430 package with Litef LCR 92 laser gyro AHRS and Meggitt System 55 autopilot optional.

Radar: Optional Telephonics RDR 2000 colour weather radar, housed in streamlined pod on port wingtip.

Instrumentation: Becker LCD engine monitoring system standard.

DIMENSIONS, EXTERNAL:

Wing span	11.50 m (37 ft 8¾ in)
Wing aspect ratio	9.3
Length overall	9.56 m (31 ft 4½ in)
Max width of fuselage	1.46 m (4 ft 9½ in)
Height overall	3.09 m (10 ft 1¾ in)
Tailplane span	3.80 m (12 ft 5½ in)
Wheel track	2.20 m (7 ft 2½ in)
Wheelbase	2.56 m (8 ft 4¾ in)
Propeller diameter	1.95 m (6 ft 4¾ in)
Cabin door:	
Width	0.68 m (2 ft 2¾ in)
Height	1.15 m (3 ft 9¼ in)

DIMENSIONS, INTERNAL:
Cabin (between forward and aft bulkheads):

Length	4.13 m (13 ft 6½ in)
Max width	1.39 m (4 ft 6¾ in)
Max height	1.24 m (4 ft 0¾ in)

AREAS:

Wings, gross	14.26 m² (153.5 sq ft)

WEIGHTS AND LOADINGS:

Basic operating weight empty	1,389 kg (3,062 lb)
Baggage capacity	90 kg (198 lb)

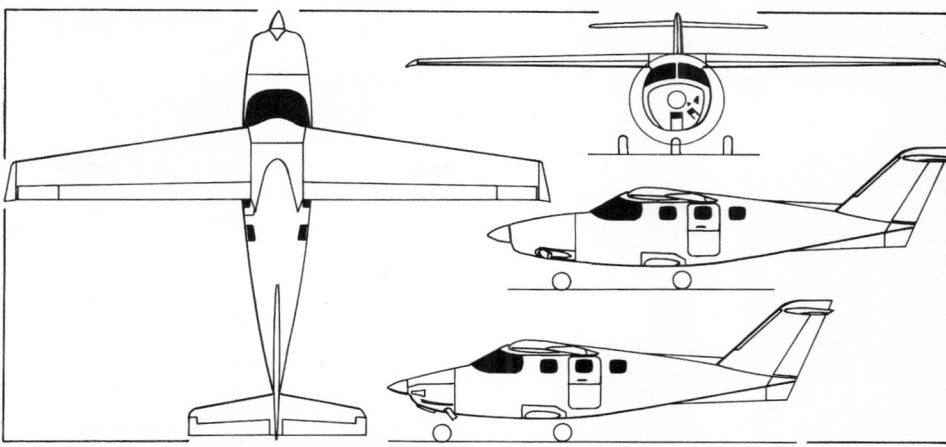
Extra 400 with additional side view (upper) of Extra 500 (*Jane's/Paul Jackson*)   NEW/0137336

Useful load	611 kg (1,347 lb)
Max T-O and landing weight	1,999 kg (4,407 lb)
Max zero-fuel weight	1,772 kg (3,907 lb)
Max wing loading	140.2 kg/m² (28.71 lb/sq ft)
Max power loading	7.66 kg/kW (12.58 lb/hp)

PERFORMANCE:

Never-exceed speed (VNE)	219 kt (405 km/h; 252 mph) IAS
Manoeuvring speed	156 kt (289 km/h; 179 mph) IAS
Max cruising speed at 75% power at FL200	225 kt (417 km/h; 259 mph)
Econ cruising speed at 65% power at FL200	210 kt (389 km/h; 242 mph)

Extra 400 (Teledyne Voyager flat-six) (*Jane's/Paul Jackson*)   NEW/0132417

Latest standard flight deck of the Extra 400   NEW/0137232

Stalling speed in landing configuration	59 kt (110 km/h; 68 mph)
Max rate of climb at S/L	326 m (1,070 ft)/min
Max certified altitude	7,620 m (25,000 ft)
T-O run	500 m (1,640 ft)
T-O to 15 m (50 ft)	860 m (2,820 ft)
Landing from 15 m (50 ft)	625 m (2,050 ft)
Landing run	275 m (900 ft)
Range at 55% power	1,160 n miles (2,148 km; 1,335 miles)
g limits	+4.0/−1.6

UPDATED

EXTRA 500

TYPE: Business turboprop.

PROGRAMME: Turboprop version of Extra 400; prototype (D-EKEW), employing previously reserved airframe, first flew 26 April 2002; damaged in forced landing during third sortie, 30 April 2002; announced at, but unable to participate in, Berlin Air Show 6 to 12 May 2002.
Data as for Extra 400, except that below.

CURRENT VERSIONS: **500 PA:** Personal Airliner; standard fuel capacity.
500 PA LR: Long-Range version; additional fuel.

POWER PLANT: One 336 kW (450 shp) Rolls-Royce 250-B17F/2 turboprop, driving a five-blade propeller. Fuel capacity, standard, 468 litres (124 US gallons; 103 Imp gallons), of which 404 litres (107 US gallons; 88.9 Imp gallons) are usable; Long-Range 687 litres (181 US gallons; 151 Imp gallons, of which 656 litres (173 US gallons; 144 Imp gallons) are usable.

DIMENSIONS, EXTERNAL:

Length overall	9.95 m (32 ft 7¾ in)
Propeller diameter	2.00 m (6 ft 6¾ in)

WEIGHTS AND LOADINGS:

Weight empty	1,360 kg (2,998 lb)
Max T-O weight: PA	2,000 kg (4,409 lb)
PA LR	2,130 kg (4,695 lb)
Max landing weight: PA, PA LR	2,000 kg (4,409 lb)
Max wing loading: PA	140.3 kg/m² (28.73 lb/sq ft)
PA LR	149.4 kg/m² (30.59 lb/sq ft)
Max power loading: PA	5.96 kg/kW (9.80 lb/shp)
PA LR	6.35 kg/kW (10.43 lb/shp)

PERFORMANCE:

Manoeuvring speed	158 kt (293 km/h; 182 mph)
Cruising speed at FL200:	
at 90% power	225 kt (417 km/h; 259 mph)
at 75% power	210 kt (389 km/h; 242 mph)
Stalling speed	61 kt (113 km/h; 71 mph)
Max rate of climb at S/L	548 m (1,800 ft)/min

Range at FL200, 45 min reserves:

at 90% power:	
PA	900 n miles (1,666 km; 1,035 miles)
PA LR	1,540 n miles (2,852 km; 1,772 miles)
at 75% power:	
PA	980 n miles (1,815 km; 1,127 miles)
PA LR	1,700 n miles (3,148 km; 1,956 miles)

Endurance, as above:

at 90% power: PA	4 h
PA LR	7 h
at 75% power: PA	4 h 35 min
PA LR	8 h

NEW ENTRY

FAIRCHILD DORNIER

FAIRCHILD DORNIER

Flugplatz Oberpfaffenhofen, PO Box 1103, D-82230 Wessling
Tel: (+49 8153) 300
Fax: (+49 8153) 30 20 07
Web: http://www.fairchilddornier.com
CEO: Thomas Brandt

Dornier GmbH, formerly Dornier-Metallbauten, formed 1922 by late Professor Claude Dornier; Daimler-Benz AG acquired majority holding in 1985, later assumed by Daimler-Benz Aerospace AG and subsequently by DaimlerChrysler Aerospace AG.

In June 1996, Fairchild Aerospace of San Antonio, Texas, acquired an 80 per cent shareholding and took over most of Dornier's operations, including manufacture of the 328 regional aircraft, Airbus subcontract work, life extension programme for German armed forces' Bell UH-1D helicopters, business and commuter aircraft service and support operations and international logistics activities. Support of NATO AWACS aircraft, previously undertaken by Dornier Luftfahrt GmbH, was not included in the acquisition. Company acquired by an investment fund led by Clayton, Dubilier & Rice and Allianz Capital Partners in April 2000 and renamed Fairchild Dornier on 13 April 2000; acquisition resulted in US$1.2 billion recapitalisation. Former Fairchild Aerospace (see US section of 2000-01 *Jane's*) transferred to Germany, apart from US sales element.

Recent activities at Fairchild Dornier Oberpfaffenhofen facility have included manufacture, sales and support of the Dornier 328JET; development of the Fairchild Aerospace 528JET, 728JET and 928JET jets; support of the Dornier 328

Fairchild Dornier 328JET twin-turbofan regional transport in service with China's Hainan Airlines   *NEW*/0132392

and Fairchild Metro range of twin-turboprop aircraft; support of German military Bell UH-1Ds; and manufacture of large components and subassemblies for Airbus A318, A319, A320, A321, A330 and A340; 10,000th Airbus component delivered in December 2001. Employees in early 2002 totalled over 4,400. Construction began on 9 October 2001 of 17,640 m² (189,875 sq ft) assembly facility for 728 family; completion then due March 2003. German production of Dornier 228 ended in 1999; HAL in India now sole source; description of aircraft can be found under that heading.

Fairchild Dornier filed for insolvency protection under German law in early April 2002, and received a US$90 million German government guaranteed loan over the following three months to keep the company operating while the administrator sought a buyer. With no immediate prospects of a resumption of work on the 728/928 airliners, 1,650 staff were laid off in September 2002, leaving a similar number to continue in the still profitable aerostructures sector of the company. Efforts to find a new owner continued up to the end of 2002 when creditors approved plans to sell 328JET production line and customer support to Avcraft of USA and investment group Dimeling Schreiber & Park; 728/928 programme and Airbus aerostructures to Aviacor and Basic Element of Russia; and maintenance facilities to RUAG of Switzerland.

UPDATED

DORNIER 228

See under HAL in the Indian section for a description.

FAIRCHILD DORNIER 328JET

TYPE: Regional jet airliner.
PROGRAMME: Development of Dornier 328 turboprop twin begun in mid-1980s, but then suspended; relaunched 3

August 1988; rolled out 13 October 1991; first flight (D-CHIC) 6 December 1991; first flight of first production 328-100 (D-CITI) 23 January 1993. JAA 25 certification 15 October 1993; FAA certification 10 November 1993; first delivery, to Air Engiadina, 21 October 1993. Production (including three prototypes) totalled 111; final aircraft (OE-LKC) delivered to Air Alps Aviation, Austria, 13 October 1999.

Twin turbofan version announced 5 February 1997 and formally launched at Paris Air Show in June 1997; marketed as 328JET, engineering designation being 328-300; high-speed wind tunnel tests, confirming cruise performance, and ground vibration tests completed by November 1997; prototype D-BJET, converted from second prototype 328 turboprop (D-CATI), rolled out 6 December 1997; first flight 20 January 1998; public debut 4 February 1998; three further prototypes built on the production line; second (D-BWAL) dedicated to performance certification testing, flew 20 May 1998; third (D-BEJR), used for avionics certification, flew 10 July 1998; fourth (D-BALL), used for function and reliability testing and simulated airline operation trials, flew on 15 October 1998; JAA certification was achieved on 8 July 1999 after completion of 1,560 flight hours in 950 sorties; FAA certification 15 July 1999; high gross weight certification achieved April 2000.

First delivery in July 1999 to Skyway Airlines (N351SK); total of 15 delivered in 1999 and 33 in 2000. By June 2001 the in-service fleet of aircraft had logged 100,000 hours, with a 99.23 per cent flight completion rate; high-time aircraft, operated by Hainan Airlines, had then logged nearly 4,000 hours. Production suspended in late 2002, with 20 almost complete aircraft awaiting delivery.

CURRENT VERSIONS: **328JET:** Regional airliner, *as described*; 32 passengers at 79 cm (31 in) seat pitch. Engineering designations **328-300** and **328-310**.

328JET SMA: Proposed multirole special missions aircraft for troop transport, VIP missions, surveillance, SAR, AEW and medevac. Under study in 2002.

Freighter: All-cargo version with large door in aft fuselage for loading of containers or palletised freight. No longer being promoted.

Envoy 3: Business jet and corporate shuttle version, launched at the 1997 National Business Aviation Association Convention in Dallas, Texas; typical executive accommodation for 12 to 14 passengers with forward service galley and wardrobe and aft galley and lavatory; optional long-range fuel tank extends maximum range to 2,000 n miles (3,704 km; 2,302 miles). Corporate Shuttle seating for 19 to 32 passengers. Option of increased cabin height from mid-2002.

Total of 10 orders by December 2001. Announced customers include Aircraft Jetcharter, which has ordered a Corporate Shuttle variant for operation on behalf of an Ohio-based customer, and Grupo Protexa of Monterey, Mexico, which has ordered one aircraft in executive configuration. Aircraft Completions of Tyler, Texas, is US completions centre.

CUSTOMERS: Total of 141 orders and 91 options quoted by early 2002; regular deliveries suspended after February 2002, with only one further aircraft supplied up to 31 December 2002.

FAIRCHILD DORNIER 328JET PRODUCTION
(at 1 January 2003)

Customer	Variant	First delivery	Qty
Air Vallee	300	18 Apr 00	1
	310	12 Jun 01	1
Atlantic Coast Airlines	310	27 Apr 00	34
Gandalf Airlines	300	17 Nov 99	8
Grossman Air Service	Envoy	8 Feb 01	1
Hainan Airlines	300	3 Nov 99	20
Johnson Controls	Envoy	8 Oct 99	1
Philip Morris Management Corp	Envoy	26 Apr 01	1
Skyway Airlines	300	30 Jul 99	9
	310	20 Feb 02	1
Great Plains (ex-Ozark)	300	27 Oct 99	2
Shell Nigeria (Bristow)	Envoy	3 Aug 99	3
Tyrolean Jet Service	Envoy	10 Sep 99	2
Ultimate Air Charters	Envoy	21 May 01	1
Wanair Tahiti	300	15 Dec 99	1
Welcome Air	300	24 Oct 02	1
Subtotal			**87**
Fairchild Dornier	300	-	6
unfinished	300/310		12
Total			**105**

Note: Excludes prototype converted from turboprop.

COSTS: Regional airliner US$13 million (2002); Corporate US$13.65 million, outfitted (2000). Airliner direct operating costs estimated at US$3.16 per n mile, based on 32 passengers and 300 n mile sectors.

DESIGN FEATURES: Combines basic TNT supercritical wing of Dornier 228 with new pressurised fuselage from NRT (Neue Rumpf Technologien) programme; internal volume designed to give passengers more seat width than in a Boeing 727 or 737 and stand-up headroom in aisle. Jet retains high commonality with turboprop model; major changes include new pylon/nacelle and centre wing attachments, strengthened landing gear and brakes, APU

Cabin of Envoy 3 Corporate Shuttle   *NEW*/0132388

0100856

Fairchild Dornier 328JET, cutaway drawing key

1 Glass fibre radome
2 Weather radar scanner
3 Radome hinge, opens to starboard
4 Radar mounting bulkhead
5 ILS glideslope antenna
6 Nosewheel housing
7 Front pressure bulkhead
8 Twin nosewheels, forward retracting
9 Nosewheel leg door
10 Taxying light
11 Rudder pedals
12 Five-tube EFIS instrument panel
13 Instrument panel shroud
14 Windscreen wipers
15 Windscreen panels, electrically heated
16 Overhead systems switch panel
17 First officer's seat
18 Observer's folding seat (provision)
19 Captain's seat
20 Curved side window panel
21 Control column handwheel
22 Nosewheel steering tiller
23 Document hatch
24 Incidence vane
25 Dual pitot heads
26 Underfloor control linkages
27 Ground power socket
28 Combined entry door/airstairs
29 Folding handrail
30 Entry lobby
31 Wing inspection light
32 Cabin attendant's folding seat
33 Avionics racks port and starboard
34 Cockpit bulkhead with sliding door
35 Cockpit roof escape hatch
36 Starboard side emergency exit hatch
37 Composites forward fuselage dorsal fairings
38 ADF antenna
39 ATC Nos. 1 and 2 antennas
40 Fuselage skin panelling
41 Overhead baggage lockers
42 Conditioned air delivery ducts from starboard pack to underfloor mixer unit
43 Wardrobe

44 Door surround structure
45 Exterior light
46 Cabin sidewall seat mounting rail
47 Three-abreast passenger seating; standard 31-seat layout shown
48 Cabin wall insulation blankets
49 Conditioned air delivery duct from port pack
50 Underfloor air mixing unit
51 Underfloor air distribution ducting
52 Cabin window panels
53 Floor beams and seat rails
54 Machined wing spar and landing gear mounting main frames
55 Centre-cabin passenger seating
56 Conditioning system heat exchanger ram-air intake
57 Ground running fan
58 Dual air conditioning packs
59 GPS antenna
60 Heat exchanger exhaust duct

61 Wing centre-section rib structure
62 Inboard integral fuel tank
63 Wingroot leading-edge chord extension
64 Overwing fuel filler
65 Engine bleed air ducting to conditioning pack
66 Starboard engine installation
67 Hinged cowling panels
68 Main engine mounting yoke
69 Starboard nacelle pylon
70 Fuel system vent surge tank
71 Outer wing panel integral fuel tank
72 Fuel venting intake and ducting
73 Pressure refuelling manifold and control panel, ventral access
74 Leading-edge de-icing boot
75 Outboard vent tank
76 Outer wing panel dry bay
77 Composites wingtip fairing
78 Starboard navigation and strobe lights
79 Static dischargers
80 Starboard aileron
81 Aileron hinge control
82 Geared tab
83 Roll control spoiler, linked to aileron
84 Ventral flap hinges
85 Two-segment ground spoilers/lift dumpers
86 Spoiler hydraulic jacks

87 Flap hydraulic jack
88 Nacelle pylon tail fairing
89 Starboard single-slotted Fowler flap, extended
90 Control system access panel
91 Wing control surface linkages
92 Flap interconnecting torque shaft
93 VHF communication antenna
94 Composites trailing-edge fairing panels
95 Starboard service/emergency exit door
96 Auxiliary power unit (APU) air supply duct
97 Extended fin root fillet, composites structure
98 ELT antenna
99 Fin root attachment joints
100 Carbon fibre-reinforced plastic (CFRP) fin structure
101 VOR localiser antenna
102 Elevator control rods
103 Fin leading-edge de-icing
104 Fin/tailplane attachment spar joints
105 Composites fin-tip fairing
106 Starboard fixed tailplane
107 Starboard elevator
108 Elevator trim tab
109 Rudder horn balance
110 Elevator hinge control links
111 Trim tab actuator
112 Port elevator trim tab
113 Elevator composites structure
114 Static dischargers

115 Elevator horn balance
116 Port tailplane CFRP structure
117 Tailplane leading-edge de-icing boot
118 Rudder trim tab
119 Composites rudder structure
120 Rudder hinge control
121 Geared tab
122 APU exhaust
123 Tail navigation light
124 Honeywell 36-150 (DD) APU
125 APU access door
126 Ventral fin, port and starboard
127 APU fire suppression bottle
128 Control cable quadrants
129 Composites fin spar mounting bulkheads
130 CFRP tailcone structure
131 Tailcone joint frame
132 Rear pressure bulkhead
133 Cabin pressure relief outflow valve
134 Baggage compartment
135 Baggage door, open
136 Door surround structure

137 Rear avionics rack with flight- and cockpit voice recorders
138 Baggage restraint net
139 Fuselage aluminium alloy frame and stringer structure
140 Cabin rear bulkhead
141 Overhead passenger service units and cabin roof lighting
142 Lavatory
143 Rear cabin seating
144 Galley unit
145 Port emergency exit window hatch
146 Composites shroud structure
147 Wing spar attachment link
148 Diagonal drag strut
149 Aileron hinge control linkage
150 Machined wing spars
151 Engine pylon mounting support structure
152 Pylon mounting support ribs
153 Outer wing panel joint rib
154 Engine pylon rear attachment joint
155 Port ground spoiler/lift dumper panels
156 Flap composites structure
157 Port single-slotted Fowler flap
158 Port roll control spoiler
159 Aileron hinge control linkage
160 Port aileron tab
161 Aileron composites structure
162 Static dischargers
163 Port navigation and strobe lights
164 Glass fibre honeycomb wingtip fairing
165 Diagonal tip rib
166 Outer wing panel rib structure
167 Wing bottom skin/stringer panel
168 Leading-edge de-icing boot
169 Port wing integral fuel tank
170 Diagonally braced aluminium alloy wing ribs
171 Composites leading-edge structure
172 Port engine nacelle
173 Port mainwheel sponson hydraulic equipment bays, oxygen bottle to starboard
174 Pylon attachment joints
175 Engine pylon structure
176 Mainwheel bay door
177 Twin mainwheels, inward retracting
178 Trailing axle suspension
179 Shock-absorber strut
180 Mainwheel leg pivot mounting
181 Battery bay
182 Sponson-mounted landing light
183 Acoustically lined engine intake
184 Pratt & Whitney Canada PW 306 high-bypass ratio turbofan engine
185 FADEC engine controller
186 Accessory equipment gearbox
187 Multilobe exhaust mixer

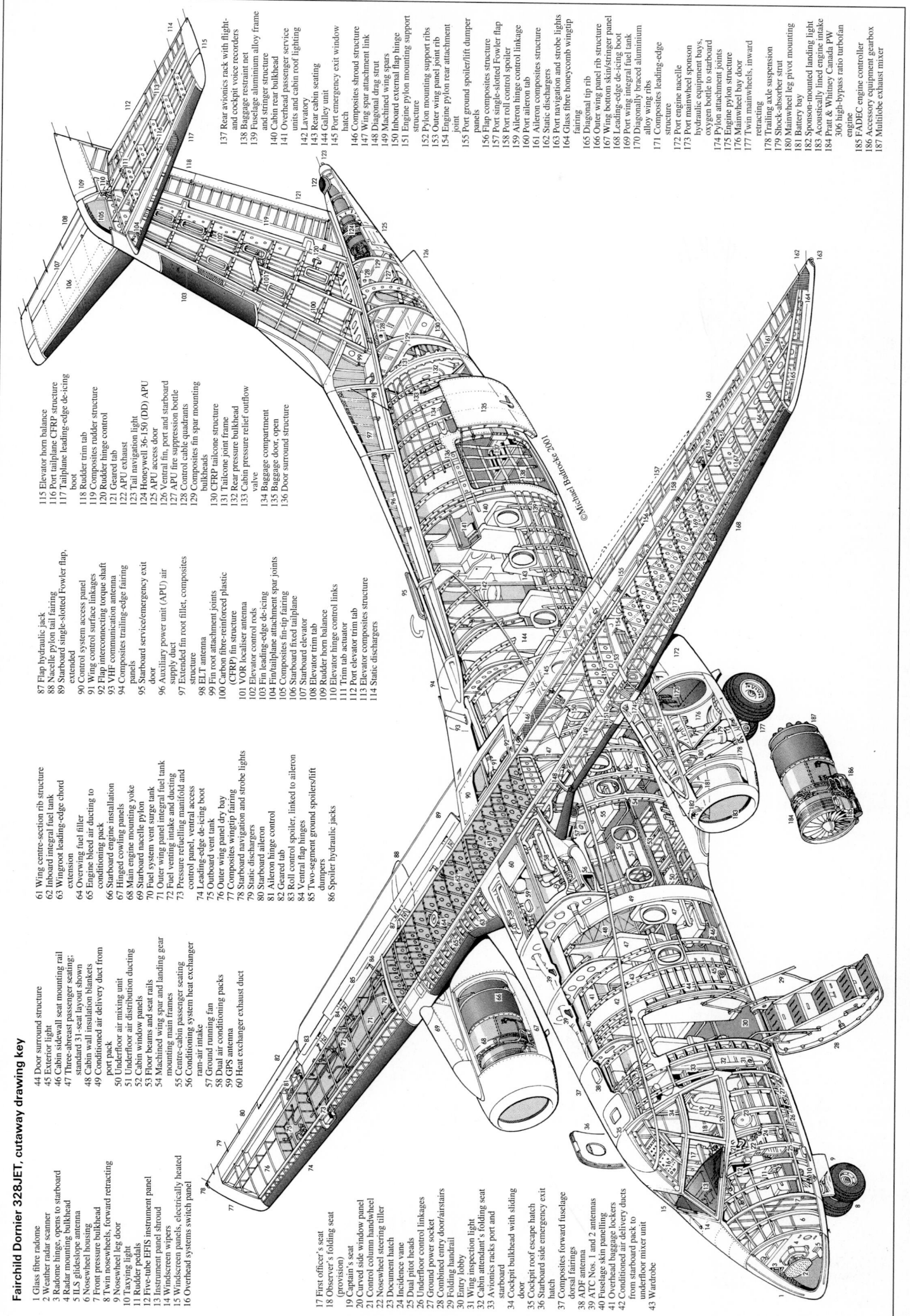

©Michael Badrocke 2001

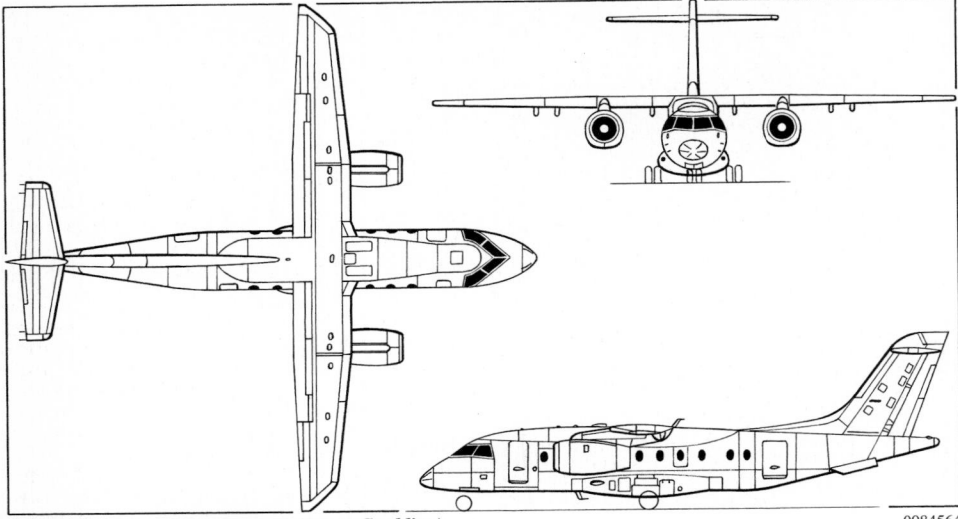

Fairchild Dornier 328JET *(Jane's/James Goulding)*   0084564

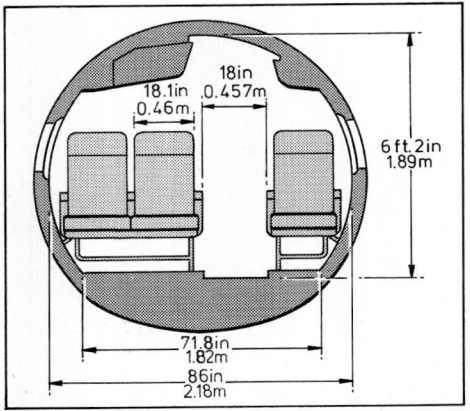

Cabin cross-section of Fairchild Dornier 328JET *(Jane's/Mike Keep)*

as standard, and modifications to Honeywell Primus 2000 avionics software, FADEC and environmental control systems.

FLYING CONTROLS: Conventional and manual. Optional lateral control (one) and ground (two) spoilers ahead of each aileron; horn-balanced elevators; trim tab in each elevator and rudder; single-slotted Fowler flaps.

STRUCTURE: Wing mainly light alloy structure; entire rear fuselage and tail surfaces of CFRP, except dorsal fin made of Kevlar/CFRP sandwich and aluminium alloy tailplane leading-edge; Kevlar/CFRP sandwich also used for wing trailing-edge structure, nosecone, tailcone and for long wing/fuselage fairing housing system components outside pressure hull; cabin doors of superplastic formed aluminium alloy; engine nacelles of superplastic formed titanium and carbon composites.

OGMA of Portugal manufactures fuselage shells that are assembled into complete fuselages by risk-sharing partner Aermacchi in Italy, which also manufactures flight deck structure; engine nacelles and doors by Westland Aerostructures; wing fairings by HAIG (Harbin) of China; rear fuselage and tail surfaces assembled by Fairchild Dornier from carbon fibre components; wing produced by Fairchild Dornier in San Antonio, USA; final assembly at Oberpfaffenhofen.

LANDING GEAR: ERAM (with SHL of Israel) retractable tricycle type, with twin Honeywell wheels on each unit; nose unit retracts forward, main units into Kevlar/CFRP sandwich unpressurised fairings on fuselage sides. Main tyres size 24×7.7 (14 ply), nose tyre 19.5×6.75-8 (10 ply); tyre pressures 4.40 bar (64 lb/sq in) on nose unit, 8.00 bar (116 lb/sq in) on main units; alternative 25.5×8.75-10 (14 ply) flotation main tyres; Honeywell brakes.

POWER PLANT: Two 26.9 kN (6,050 lb st) Pratt & Whitney Canada PW 306/9 turbofans with FADEC, pod-mounted under wings. Fuel capacity 4,268 litres (1,128 US gallons; 939 Imp gallons).

ACCOMMODATION: Flight crew of two and cabin attendant. Main cabin seats 32 to 34 passengers, three-abreast at

79 cm (31 in) or 76 cm (30 in) pitch, with single aisle; galley to rear of passenger seats; lavatory at rear of cabin, baggage compartment between passenger cabin and rear pressure bulkhead, with external access via baggage door in port side; crew/passenger airstair door at front on port side, with Type III emergency exit opposite; Type III emergency exit on port side at rear of cabin, with service door Type II exit at rear on starboard side.

SYSTEMS: Air conditioning and pressurisation systems standard (maximum differential 0.47 bar; 6.75 lb/sq in), with GKN Westland ECS. Hydraulic and two independent AC/DC electrical systems housed in main landing gear fairings. Tailcone-mounted Honeywell 36-150 (DD) APU as standard.

AVIONICS: Honeywell Primus 2000 suite.

Comms: Dual Primus II integrated radio system and Mode S transponder standard. HF comms optional.

Radar: Primus 650 weather radar standard; Primus 870 weather radar optional.

Flight: AFCS, AHRS, dual integrated avionics computer, dual digital air data reference unit, TCAS, GPWS with windshear detection system standard. Optional GPS with GPS approach and head-up guidance system allowing Cat.II landings.

Instrumentation: Five-tube EFIS (203 × 178 mm; 8 × 7 in screens) with EICAS.

DIMENSIONS, EXTERNAL:

Wing span	20.98 m (68 ft 10 in)
Wing aspect ratio	11.0
Length overall	21.23 m (69 ft 7¾ in)
Fuselage: Length	20.92 m (68 ft 7¾ in)
Max width	2.415 m (7 ft 11 in)
Max depth	2.425 m (7 ft 11½ in)
Height overall	7.05 m (23 ft 1½ in)
Elevator span	6.70 m (21 ft 11¾ in)
Wheel track (c/l of shock-struts)	3.22 m (10 ft 6¾ in)
Wheelbase	7.42 m (24 ft 4¼ in)
Passenger door (fwd, port): Height	1.70 m (5 ft 7 in)
Width	0.70 m (2 ft 3½ in)
Service door (rear, stbd): Height	1.25 m (4 ft 1¼ in)
Width	0.51 m (1 ft 8 in)
Baggage door (rear, port): Height	1.40 m (4 ft 7 in)
Width	0.92 m (3 ft 0¼ in)

DIMENSIONS, INTERNAL:

Cabin (excl flight deck): Length	10.33 m (33 ft 10¾ in)
Max width	2.18 m (7 ft 2 in)
Width at floor	1.83 m (6 ft 0 in)
Max height in aisle	1.89 m (6 ft 2½ in)
Baggage hold volume	6.4 m³ (226 cu ft)

AREAS:

Wings, gross	40.00 m² (430.6 sq ft)
Ailerons (total)	2.42 m² (26.00 sq ft)
Trailing-edge flaps (total)	7.61 m² (81.90 sq ft)
Fin	11.06 m² (119.00 sq ft)
Rudder	3.92 m² (42.20 sq ft)
Tailplane	9.03 m² (97.20 sq ft)
Elevators	3.08 m² (33.20 sq ft)

WEIGHTS AND LOADINGS (A: standard, B: high gross weight):

Operating weight empty: A	9,420 kg (20,600 lb)
B	9,394 kg (20,710 lb)
Max fuel: A, B	3,646 kg (8,039 lb)
Max payload: A	3,650 kg (7,200 lb)
B	3,676 kg (8,104 lb)

Max T-O weight: A	15,200 kg (33,510 lb)
B	15,660 kg (34,524 lb)
Max landing weight: A, B	14,390 kg (31,724 lb)
Max ramp weight: A	15,350 kg (33,841 lb)
B	15,780 kg (34,789 lb)
Max zero-fuel weight: A	12,610 kg (27,800 lb)
B	13,070 kg (28,814 lb)
Max wing loading: A	380.0 kg/m² (77.83 lb/sq ft)
B	391.5 kg/m² (80.18 lb/sq ft)
Max power loading: A	282.0 kg/kN (2.77 lb/lb st)
B	291.0 kg/kN (2.85 lb/lb st)

PERFORMANCE:

Max cruising speed, M$_{MO}$ at 7,620 m (25,000 ft), ISA +10°:	
A, B	400 kt (741 km/h; 460 mph)
Service ceiling: A, B	10,670 m (35,000 ft)
Service ceiling, OEI: A, B	7,528 m (24,700 ft)
T-O required field length: A	1,382 m (4,535 ft)
B	1,422 m (4,665 ft)
Landing required field length: A	1,306 m (4,285 ft)
B	1,387 m (4,550 ft)

Range: 328JET with 32 passengers, JAA reserves, and allowance for 100 n mile (185 km ; 115 mile) diversion:

B	1,130 n miles (2,092 km; 1,300 miles)

Envoy with 6 passengers, optional fuel increase and NBAA reserves 2,000 n miles (3,704 km; 2,301 miles)

UPDATED

FAIRCHILD DORNIER 528, 728 and 928

TYPE: Regional jet airliner.

PROGRAMME: Baseline 728 initially designated X28JET, then 728JET; JET suffix discontinued in late 2001. Design study, to Lufthansa requirements, announced at the Dubai Air Show in October 1997; launched at ILA Berlin Aerospace Show 19 May 1998 in five-abreast configuration; General Electric CF34 engine selection announced 3 August 1998; baseline configuration frozen December 1999; first metal cut by SABCA (Belgium) on 18 September 2000; design freeze December 2000; assembly of first three fuselages began in February 2001; final assembly began October 2001; prototype 728-100 (TAC 1/D-AEVA) rolled out 21 March 2002; five aircraft (four test aircraft, TACs, and the first production machine) to fly 1,800-hour test programme, mostly at Granada, Spain, but also in Germany and USA, complementing 'iron bird' engineering rig, which commissioned in November 2000 and completed 210,000 cycles in the initial fatigue test phase by June 2001; initial 10-flight test programme for fly-by-wire FCS completed in first quarter of 2001 using Veridian Flight Research Group's Learjet 25 configured to reproduce stability and control characteristics of 728. Company's financial problems forced suspension of test programme, and 728 remained unflown in February 2003.

Service entry of 728-100 with launch customer Lufthansa originally planned for July 2003. Fairchild Dornier stopped marketing the 728-100 in March 2002 and intended from 2004 to concentrate manufacturing on the 728-200. Initial production target (all versions) 100 per year, with potential to increase to 120 per year by 2005.

CURRENT VERSIONS: **728-100:** Formerly 728JET. *As described.*

728-200: Announced 28 January 2002; formerly 728JET-XP; increased range and weight. Strengthened outboard wing (as 928) and horizontal tail surfaces; reinforced landing gear and fuselage. Range 1,800 n miles (3,333 km; 2,071 miles).

728 Special Mission Aircraft (SMA): Proposed airborne early warning variant featuring dorsal radar dish housing Northrop Grumman radar. Target capability includes 100 n mile (185 km; 115 mile) transit to and from search area and on-station endurance of more than eight hours at 10,670 m (35,000 ft).

528: Also known as 'X28'. Projected shortened version with accommodation for 55 to 63 passengers in single or mixed class layouts; derated CF34-8D3 engines. Service entry in 2007.

928-100: Growth version with accommodation for 110 passengers in single- or 95 in mixed-class layouts; fuselage

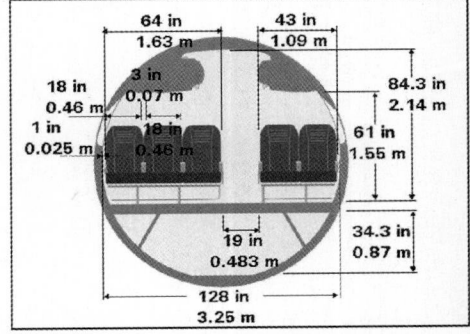

Cabin cross-section of Fairchild Dornier 728 series

NEW/0114830

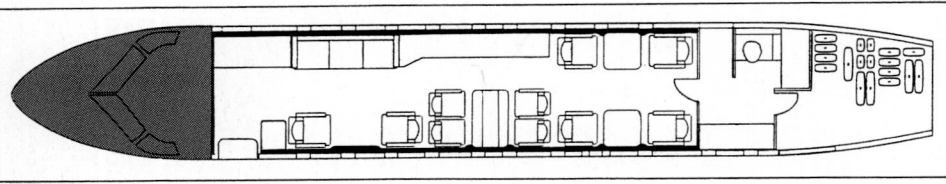

Typical Fairchild Dornier Envoy 3 internal arrangement   0044702

Roll-out of the Fairchild Dornier 728 prototype

NEW/0530191

stretch by means of two plugs (more correctly, extensions of existing assemblies) fore and aft of centre-section; new centre wing box increasing wing span and area. No overwing emergency exits.

GE CF34-10D5 engines; strengthened main landing gear; uprated environmental control system. Initial 100-hour high-speed wind tunnel testing phase using 1/12 scale half-model completed in April 2001; first flight scheduled for November 2003, followed by three-aircraft, 27-month flight test programme leading to service entry in second quarter of 2005. Total orders stood at six in June 2001, comprising four firm and two options for launch customer Bavaria International Aircraft Leasing GmbH.

928-200: Extended-range version; GE CF34-10D6 engines.

928 SMA: Proposed special missions aircraft; optional cargo door facilitates loading of up to seven 463 lb military pallets and/or containerised cargo.

Envoy 7: Corporate version of 728, with increased range, CF34-8D6 engines, (each rated at 58.1 kN; 13,055 lb st); additional underfloor fuel tanks raising max fuel weight to 15,200 kg (33,510 lb), and hybrid 'super shark' winglets; launched at NBAA Convention in Las Vegas, October 1998; launch customer Flight Options Inc announced order for 25 for its fractional ownership programme at the Paris Air Show on 14 June 1999; full-scale cabin mockup exhibited at NBAA Convention at Atlanta, Georgia, in October 1999. First flight scheduled for July 2003, followed by three-aircraft, 19-month flight test and certification programme. Lufthansa Technik AG of Hamburg, Germany, appointed European completion centre on 10 October 2000. Garrett Aviation Services is US completion centre; total orders stood at 29 by June 2001.

CUSTOMERS: Total of 118 orders and 162 options for 728 and four orders and two options for 928 by February 2002; launch (728) customer Lufthansa CityLine ordered 60, with 60 options, in April 1999, for delivery between 2003 and 2006. Other announced customers include CSA Czech Airliners, which announced orders for eight (four on lease from GECAS) at the Paris Air Show in June 2001, for delivery at the rate of three each in 2003 and 2004 and one each in 2005 and 2006. GE Capital Aviation Services Inc (GECAS), which ordered 50, with up to 100 options, in June 2000, for delivery from 2003, but cancelled the order in April 2002; Bavaria International Aircraft Leasing GmbH, which ordered two, with two options, in June 2000; and Sol Air of Italy, which has ordered two.

COSTS: Programme US$1.1 billion for all versions. Unit cost: 528 US$26.50 million; 728-100 US$29.50 million; 728-200 US$31 million; 928-100 US$35 million; 928-200 US$36.5 million (all 2002).

DESIGN FEATURES: Design objectives included seamless transition of passengers to large aircraft of large airlines; superior seat-mile economics; good payload/range performance; flight deck design for reduced pilot workload; and high dispatch reliability. Conventional twin-jet airliner with low-mounted wings carrying podded engines; sweptback fin and mid-mounted tailplane.

FLYING CONTROLS: Digital fly-by-wire. Ailerons, rudder and elevator; incidence-trimmable tailplane. Four-section slats on each wing leading-edge outboard of engines; Krüger flap inboard. Three-section spoilers on outboard section of each wing upper surface (Nos. 2, 3 and 4); single spoiler (No. 1) inboard. Two-section flaps on each wing.

STRUCTURE: Airframe participants include: Honeywell (avionics, environmental control system and APU); Goodrich (wheels, tyres, brakes, nosewheel steering system and fuel system); SABCA (flight deck section and rear fuselage); CASA/EADS (wing box and tail surfaces); General Electric (power plant); Hurel-Dubois (engine nacelles); Parker Aerospace (hydraulic system and primary FCS); Reims Aviation (728 lower cockpit, under subcontract from SABCA); and Hamilton Sundstrand (electrical system secondary FCS and ram air turbine).

Airframe design life: **728-100:** 80,000 cycles; **728-200, 928-100** and **928-200:** 65,000 cycles; **Envoy 7:** 25,000 cycles.

LANDING GEAR: Retractable tricycle type with twin wheels on each unit.

POWER PLANT: *728:* Two General Electric CF34-8D1 turbofans, each rated at 60.37 kN (13,572 lb st) with APR, pod-mounted below wings. Standard fuel capacity 12,000 litres (3,170 US gallons; 2,640 Imp gallons).
928: Two CF34-10D5s, each 82.3 kN (18,500 lb st) with APR. ER version has 85.2 kN (19,150 lb st) -10D6s.

ACCOMMODATION: Flight deck crew of two with two flight attendants and 70 to 80 passengers; in typical single-class configuration, cabin accommodates 75 passengers, five-abreast, at 84 cm (33 in) seat pitch, or in high-density layout, 85 passengers, five-abreast at 76 cm (30 in) seat pitch with single 48 cm (19 in) aisle; alternative two-class configurations accommodate 60 passengers in five-abreast seating at 84 cm (33 in) seat pitch and eight or ten in four-abreast seating at 96 cm (38 in) pitch at forward end of cabin; galley, lavatory and wardrobe forward, galley and lavatory aft. Baggage bins offer 0.06 m³ (2.13 cu ft) and 12.7 kg (28 lb) per passenger. Front crew/passenger airstair Type C door and aft Type C emergency exit on port side; forward Type C emergency exit and aft Type C service door/emergency exit and aft baggage door on starboard side. Two underfloor baggage compartments with doors, port side, forward and aft of wing; upper baggage compartment at rear of cabin. Cabin pressurised, maximum differential 0.57 bar (8.3 lb/sq in), maintaining sea level pressure to 6,431 m (21,000 ft)..

SYSTEMS: Honeywell RE220FD APU.

AVIONICS: Honeywell Epic avionics suite.
Comms: Dual Primus II integrated radio system with VHF com and Mode S transponder standard.
Radar: Primus 660 weather radar.
Flight: VOR, DME, ADF, ILS standard, with integrated sensor suite containing inertial measurement unit, dual air data modules, AFCS and GPS sensor modules; TCAS 2000; EGPWS; windshear detection system.
Instrumentation: Six-tube EFIS with LCD PFDs, MFDs and EICAS.

DIMENSIONS, EXTERNAL:

Wing span: 728	27.13 m (89 ft 0 in)
928	28.81 m (94 ft 6¼ in)
Length overall: 728	27.04 m (88 ft 8½ in)
928	30.96 m (101 ft 6¾ in)
Height overall: 728	9.045 m (29 ft 8 in)
928	9.97 m (32 ft 8½ in)
Wheel track: 728	4.85 m (15 ft 11 in)

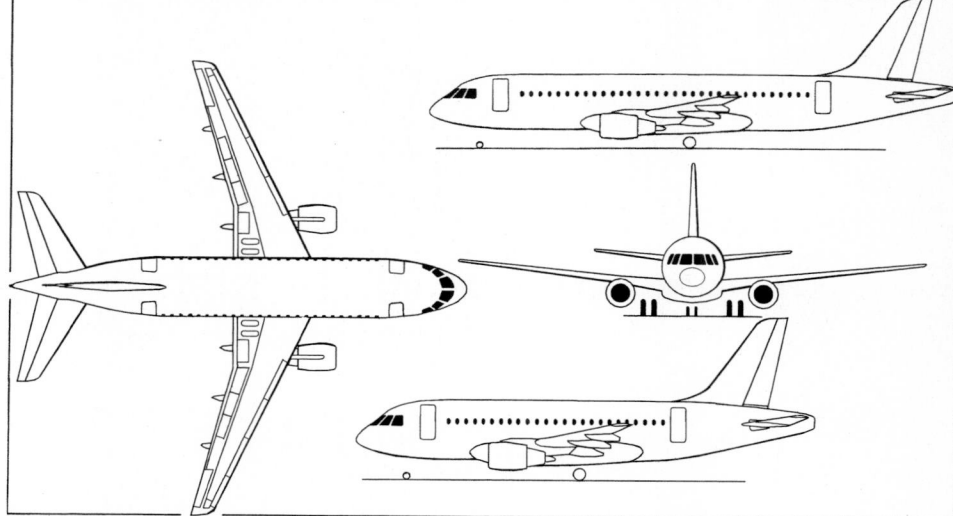

Fairchild Dornier 728 with additional side view (upper) of 928 *(Jane's/Paul Jackson)*

0110844

Computer-generated image of Fairchild Dornier 528JET twin-turbofan airliner

0084573

Artist's impression of Fairchild Dornier 728 SMA

NEW/0132390

Cabin doors (four): Height	1.85 m (6 ft 1 in)
Width	0.77 m (2 ft 6¼ in)
Height to sill	2.22 m (16 ft 2 in)
Baggage doors: Height	0.82 m (2 ft 8¼ in)
Width	1.20 m (3 ft 11¼ in)
Height to sill: 728	1.21 m (3 ft 11½ in)

DIMENSIONS, INTERNAL (728):

Cabin (excl rear upper baggage area):	
Length, incl galleys	18.02 m (59 ft 1½ in)
Max width	3.25 m (10 ft 8 in)
Max height	2.14 m (7 ft 0¼ in)
Rear upper baggage area: Length	1.27 m (4 ft 2 in)
Volume: Forward under floor	6.68 m³ (236 cu ft)
Rear under floor	7.47 m³ (264 cu ft)
Overhead bins (total)	4.70 m³ (166 cu ft)

AREAS:

Wings, gross: 728	75.00 m² (807.3 sq ft)
928	84.40 m² (908.5 sq ft)

WEIGHTS AND LOADINGS (estimated):

Weight empty: 728-100	19,380 kg (42,725 lb)
728-200	19,635 kg (43,288 lb)
928-100, 928-200	24,440 kg (53,881 lb)
Operating weight empty: 728-100	21,950 kg (48,391 lb)
728-200	22,210 kg (48,965 lb)
928-100, 928-200	27,510 kg (60,649 lb)
Envoy 7	24,117 kg (53,169 lb)
Interior completion allowance (Envoy 7)	
	3,447 kg (7,600 lb)
Max payload: 728-100	8,810 kg (19,423 lb)
928-100, 928-200	13,060 kg (28,792 lb)

Max usable fuel weight: 728-100	10,100 kg (22,267 lb)
928-100, 928-200	12,456 kg (27,461 lb)
Max ramp weight: 728-100	35,400 kg (78,043 lb)
728-200	38,100 kg (83,996 lb)
928-100	48,100 kg (106.040 lb)
928-200	49,990 kg (110,205 lb)
Max T-O weight: ISA: 728-100	35,200 kg (77,602 lb)
728-200	37,990 kg (83,753 lb)

928-100	47,870 kg (105,535 lb)
928-200	49,700 kg (109,570 lb)
Envoy 7	39,500 kg (87,082 lb)
ISA +20°C at 1,830 m (6,000 ft):	
728-100, 728-200	as ISA
928-100	46,085 kg (101,600 lb)
928-200	47,390 kg (104,475 lb)
Max landing weight: 728-100	32,420 kg (71, 474 lb)
728-200	35,250 kg (77,713 lb)
928-100, 928-200	46,220 kg (101,900 lb)
Max zero-fuel weight: 728-100	30,300 kg (66, 800 lb)
728-200	31,450 kg (69,335 lb)
928-100, 928-200	40,570 kg (89,441 lb)
Max wing loading: 728-100	496.3 kg/m² (96.13 lb/sq ft)
728-200	505.3 kg/m² (103.50 lb/sq ft)
928-100	567.2 kg/m² (116.17 lb/sq ft)
928-200	588.9 kg/m² (120.61 lb/sq ft)
Envoy 7	526.7 kg/m² (107.87 lb/sq ft)
Max power loading: 728-100	292 kg/kN (2.86 lb/lb st)
928-100	291 kg/kN (2.85 lb/lb st)
928-200	292 kg/kN (2.86 lb/lb st)

PERFORMANCE:

Max operating speed: 728, 928	
	M0.82 (335 kt; 620 km/h; 386 mph CAS)
Max cruising speed at 95% MTOW:	
728 at FL350	M0.805 (464 kt; 859 km/h; 534 mph)
928 at FL330	M0.8 (467 kt; 864 km/h; 537 mph)
Max certified altitude	12,495 m (41,000 ft)
Service ceiling, OEI, 95% MTOW:	
728-100	6,200 m (20,350 ft)
928-100	6,125 m (20,100 ft)
928-200	5,730 m (18,800 ft)
Envoy 7	6,685 m (21,930 ft)
T-O required field length: 728-100	1,465 m (4,800 ft)
928-100	1,675 m (5,500 ft)
928-200, Envoy 7	1,740 m (5,700 ft)
Landing required field length: 728-100	1,420 m (4,660 ft)
928-100, 928-200	1,705 m (5,600 ft)

Range at FL370, M0.78, with IFR reserves:

728-100: 70 passengers	
	1,430 n miles (2,648 km; 1,645 miles)
80 passengers	1,120 n miles (2,074 km; 1,288 miles)
728-200: 70 passengers	
	1,880 n miles (3,481 km; 2,163 miles)
80 passengers	1,590 n miles (2,944 km; 1,829 miles)
928-100: 95 passengers	
	1,890 n miles (3,500 km; 2,175 miles)
110 passengers	
	1,520 n miles (4,986 km; 3,098 miles)
928-200: 95 passengers	
	2,200 n miles (4,074 km; 2,531 miles)
110 passengers	
	1,900 n miles (3,518 km; 2,186 miles)
Range, Envoy 7 with eight passengers and NBAA	
reserves	4,000 n miles (7,408 km; 4,603 miles)

UPDATED

Cabin mockup of Fairchild Dornier Envoy 7 business jet   *NEW*/0132389

Computer-generated image of Fairchild Dornier 928 110-passenger airliner   *NEW*/0132391

Flight deck of Fairchild Dornier 728JET   *NEW*/0114831

Alternative seating arrangements for Fairchild Dornier 728JET   0104359

75 Seats at 33" Seat Pitch

59 ft 2 in (18.02 m)

68-seat with 8 First Class seats

FLÄMING AIR

FLÄMING AIR GmbH

Flugplatz Oehna-Zellendorf, D-14913 Zellendorf
Tel/Fax: (+49 33742) 61 70
e-mail: flugplatzoehna@t-online.de
Web: http://www.flaemingair.de

In addition to operating a flying school and air charter business, Fläming Air imports and sells Eastern European aircraft to the German market. These include the ATEC Zephyr, Urban Lambáda/Samba, FSU Saphir and the Podeswa Trener Baby, which is marketed under the Fläming Air name.

UPDATED

FLÄMING AIR FSU SAPHIR

English name: Sapphire
TYPE: Side-by-side ultralight/kitbuilt.
PROGRAMME: Incomplete prototype exhibited at Aero 2001 at Friedrichshafen in April 2001; first flight then scheduled for August or September 2001; D-MATR exhibited at Berlin, May 2002.
CURRENT VERSIONS: **UL:** Ultralight version, *as described.*
  Experimental: Kitbuilt lightplane version; maximum T-O weight 570 kg (1,256 lb).
COSTS: (Rotax 912) €70,000 (2002).
DESIGN FEATURES: Low-wing monoplane with slightly sweptback wings and upturned wingtips. Combines wing and tail unit of Urban UFM 10 Samba with cockpit section of ATEC Zephyr, both of which are described in the Czech Republic section.
  Wing section SM 701.
FLYING CONTROLS: Conventional and manual. One-piece elevator; plain flaps. Electrically actuated elevator trim tab and flaps.
STRUCTURE: All-composites.
LANDING GEAR: Tricycle type; fixed. Mainwheel diameter 400 mm, nosewheel 300 mm. Speed fairings on all three wheels.
POWER PLANT: One 59.6 kW (79.9 hp) Rotax 912 UL or 73.5 kW (98.6 hp) Rotax 912 ULS, driving a Kremen SR

Trener Baby scale replica of Zlin 126 *(Jane's/Paul Jackson)*   NEW/0137225

200 three-blade, ground-adjustable pitch wooden propeller; optional Kremen SR 2000 variable-pitch (electric), wooden propeller. To be offered with 59.7 kW (80 hp) Diesel Air D-280 power plant. Fuel in two tanks, combined capacity 100 litres (26.4 US gallons; 22.0 Imp gallons).

DIMENSIONS, EXTERNAL:
Wing span	10.30 m (32 ft 9¾ in)
Length overall	5.90 m (19 ft 4¼ in)
Height overall	2.13 m (6 ft 11¾ in)
Propeller diameter	1.70 m (5 ft 7 in)

DIMENSIONS, INTERNAL:
Cabin: Max height	1.06 m (3 ft 5¾ in)

AREAS:
Wings, gross	9.25 m² (99.6 sq ft)

WEIGHTS AND LOADINGS:
Weight empty	270 to 290 kg (595 to 639 lb)
Max T-O weight	450 kg (992 lb)

PERFORMANCE, POWERED:
Max level speed	135 kt (250 km/h; 155 mph)
Cruising speed	108 kt (200 km/h; 124 mph)

PERFORMANCE, UNPOWERED:
Stalling speed	36 kt (65 km/h; 41 mph)
Best glide ratio	19
Min rate of sink	90 m (295 ft)/min

UPDATED

FLÄMING AIR TRENER BABY

TYPE: Tandem-seat ultralight/kitbuilt.
PROGRAMME: Designed and built as ultralight prototype by Peter Podeswa; scale replica of Czechoslovak Zlin 126 Trener (member of Zlin 26 family, which was produced between late 1940s and mid-1970s). Production version built by son, Tomas Podeswa at Hranice; certified and distributed by Fläming Air; second aircraft (OK-FUD 01) exhibited at ILA, Berlin, June 2000. Certification in mid-2001.
CURRENT VERSIONS: **Ultralight:** Wooden wings; MTOW limited to 450 kg (992 lb). Also available as kit, meeting '51 per cent rule'.
  Experimental: Metal wings; 520 kg (1,146 lb) MTOW; kitbuilt.
COSTS: DM90,000, factory-built (2000).
DESIGN FEATURES: Low wing with sweptback leading-edge and unswept trailing-edge; mid-mounted, braced tailplane. Replica of Zlin 126 aerobatic trainer to 80 per cent scale, except full-scale cockpit.
FLYING CONTROLS: Conventional and manual. Split flaps. Flight-adjustable tabs on both elevators; ground-adjustable tab on rudder and port aileron.
STRUCTURE: Steel tube fuselage with fabric covering, except for sheet metal upper surface and engine cowling; fabric-covered empennage. Wings wood or metal, according to version, with aluminium leading-edge and fabric covering. Composites nose fairing.
LANDING GEAR: Tailwheel type; fixed. Mainwheels 5.00-5; tailwheel 200×50. Hydraulic brakes.
POWER PLANT: One 55.1 kW (73.9 hp) Walter Mikron III four-cylinder, in-line piston engine driving a Kremen two-blade, variable-pitch (electric), wooden propeller or Woodcomp fixed-pitch propeller. (VZLU two-blade wooden propeller on prototype.) Fuel capacity 35 litres (9.2 US gallons; 7.7 Imp gallons).
AVIONICS: To customer's specification.

DIMENSIONS, EXTERNAL:
Wing span	9.20 m (30 ft 2¼ in)
Length overall	6.20 m (20 ft 4 in)
Height overall	1.80 m (5 ft 10¾ in)

WEIGHTS AND LOADINGS (UL: Ultralight, E: Experimental):
Weight empty: UL	285 kg (628 lb)
E	320 kg (705 lb)
Max T-O weight: UL	450 kg (992 lb)
E	520 kg (1,146 lb)

PERFORMANCE (UL, E, as above):
Never-exceed speed (VNE):	
UL	145 kt (270 km/h; 167 mph)
E	162 kt (300 km/h; 186 mph)
Cruising speed at 75 % power	
	108 kt (200 km/h; 124 mph)
Stalling speed: UL	36 kt (65 km/h; 41 mph)
E	38 kt (70 km/h; 44 mph)
Max rate of climb at S/L	270 m (886 ft)/min
Max range	432 n miles (800 km; 497 miles)
g limits	+6/−3

UPDATED

Early production FSU Saphir *(Jane's/Paul Jackson)*   NEW/0137224

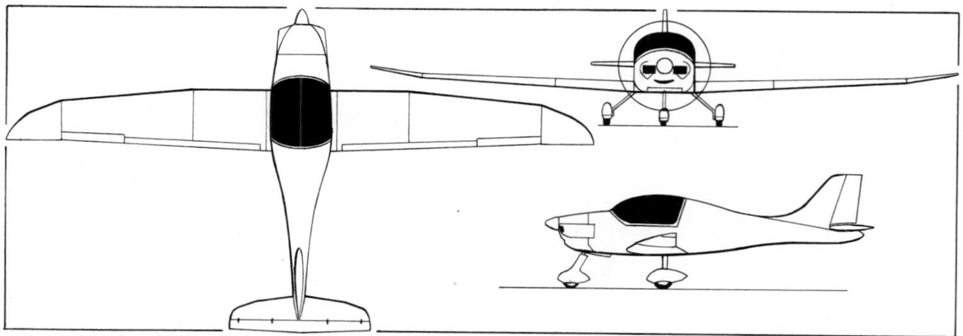

Flaming Air FSU Saphir (Rotax 912 UL four-stroke) *(Jane's/James Goulding)*   NEW/0137231

FLIGHT DESIGN

FLIGHT DESIGN GmbH

Sielminger Strasse 65, D-70771 Leinfelden-Echterdingen
Tel: (+49 711) 90 28 70
Fax: (+49 711) 902 87 99
e-mail: info@flightdesign.com
Web: http://www.flightdesign.com

PRODUCTION CENTRE:
Ost-Vest Konsalting SP
ulítsa Neftyanikov 15-A, 325000 Kherson, Ukraine

Tel: (+38 55) 223 51 35
Fax: (+38 55) 229 90 32
e-mail: owc@tlc.kherson.ua
DIRECTORS: Andrei Ya Prtygin, Aleksandr A Bondar

Flight Design collaborated with ASO Flugsport of Bottrop in producing the CT. Fabrication of parts is at Flight Design's Ukrainian factory producing composites to German specifications, with ASO responsible for certification.

UPDATED

FLIGHT DESIGN CT

TYPE: Two-seat lightplane; side-by-side ultralight.
PROGRAMME: Original CT version certified in Germany on 6 June 1997, having flown some 1,000 hours. CT and kitbuilt CTX version now discontinued; descriptions in 2001-02 and earlier editions.
CURRENT VERSIONS: **CT2K:** Standard version, *as described.* Lighter airframe; downturned wingtips.
  CTU: Longer span version to achieve 38 kg/m² (7.78 lb/sq ft) wing loading required in some countries; extra

Flight Design CT ultralight *(Jane's/Paul Jackson)*   NEW/0132407

Tailplane chord	0.70 m (2 ft 3½ in)
Wheel track	1.70 m (5 ft 7 in)
Wheelbase	1.45 m (4 ft 9 in)
Propeller diameter	1.73 m (5 ft 8 in)
Propeller ground clearance	0.28 m (11 in)

DIMENSIONS, INTERNAL:
Cabin max width	1.24 m (4 ft 0¾ in)

AREAS:
Wings, gross	10.80 m² (116.3 sq ft)
Ailerons (total)	0.98 m² (10.55 sq ft)
Fin and rudder (total)	1.14 m² (12.27 sq ft)
Tailplane (total including elevators and tab)	
	1.65 m² (17.76 sq ft)
Flaps (total)	1.51 m² (16.25 sq ft)

WEIGHTS AND LOADINGS:
Weight empty	262 kg (578 lb)
Max T-O weight	600 kg (1,322 lb)*
Max baggage weight	60 kg (132 lb)
Max wing loading	55.6 kg/m² (11.38 lb/sq ft)
Max power loading: 912 UL	10.07 kg/kW (16.54 lb/hp)
912 ULS	8.16 kg/kW (13.41 lb/hp)

* or 450 kg (992 lb) to meet national limits

PERFORMANCE (downturned wingtips, A: 912 UL; B: 912 ULS):
Never-exceed speed (V$_{NE}$):	
A, B	167 kt (310 km/h; 192 mph) IAS
Max level speed: A, B	146 kt (270 km/h; 168 mph) IAS
Cruising speed at 75% power:	
A	121 kt (225 km/h; 140 mph) IAS
B	127 kt (235 km/h; 146 mph) IAS
Stalling speed at 450 kg (992 lb) MTOW:	
A, B	34 kt (62 km/h; 39 mph) IAS
Service ceiling: A, B	4,270 m 14,000 ft)
Max rate of climb at S/L: A	210 m (689 ft)/min
B	300 m (984 ft)/min
T-O run	90 m (295 ft)
T-O to 15 m (50 ft)	160 m (525 ft)
Range with max fuel, 30 min reserves	
	1,079 n miles (2,000 km; 1,242 miles)
g limits: A, B: at 450 kg (992 lb) MTOW	+6
at 600 kg (1,322 lb) MTOW	+4.3

UPDATED

1.00 m (3 ft 3¼ in) span and 1.20 m² (12.9 sq ft) increase in area; available to order only.

CUSTOMERS: Total 180 of all versions built by April 2001, when production was running at six per month.

COSTS: Complete, with Rotax 912 UL-2, propeller and Junkers ballistic recovery system €53,987 excluding tax (2002). Kits no longer available.

DESIGN FEATURES: Optimised for safe handling, speed and range. High-mounted, cantilever, constant chord wing, sharply waisted fuselage and sweptback fin with large underfin. CT2K features new trim tab and optional downturned wingtips.
Wing dihedral 1° 40′.

FLYING CONTROLS: Conventional ailerons and horn-balanced rudder, manually actuated; all-moving tailplane with anti-balance tab. Electrically actuated plain flaps, deflection −12° and +40°; internal (external on early aircraft) mass balances for ailerons and tailplane.

STRUCTURE: Principally of composites, including carbon/glass fibre and carbon/aramid sandwich construction.

LANDING GEAR: Fixed tricycle type with swept forward aluminium tube main and nose legs and steerable nosewheel. Optional speed fairings. Hydraulic mainwheel brakes.

POWER PLANT: One 59.6 kW (79.9 hp) Rotax 912 UL-2 or 73.5 kW (98.6 hp) Rotax 912 ULS flat-four, driving a Neuform two-blade C2-V variable-pitch or C2 fixed-pitch propeller. Fuel in two wing leading-edge tanks combined capacity 130 litres (34.3 US gallons; 28.6 Imp gallons).

ACCOMMODATION: Two persons, side by side; dual controls; upward-hinged door for each occupant. Baggage space behind seats, with external door on each side.

AVIONICS: Optional avionics include Becker Radio AR 4201 com, Honeywell KX 125 nav/com, KT 76A Mode A/C transponder, Garmin GNC 300 XL nav/com/GPS with

Jeppesen database, Garmin Pilot III GPS, Flightcom headsets and ELT.
Instrumentation: Winter variometer and Rotax FlyDat engine management system optional.

EQUIPMENT: Optional equipment includes cabin comfort pack; anti-collision and position lights; and Junkers High Speed or BRS UL 1050 ballistic parachute recovery systems.

DIMENSIONS, EXTERNAL:
Wing span	9.31 m (30 ft 6½ in)
Wing chord, constant	1.17 m (3 ft 10 in)
Wing aspect ratio	8.0
Length overall	6.22 m (20 ft 5 in)
Height overall	2.165 m (7 ft 1¼ in)
Tailplane span	2.38 m (7 ft 9¾ in)

General arrangement of the Flight Design CT2K *(Jane's/James Goulding)*   0110593

FLIGHT TEAM

FLIGHT TEAM ULTRALEICHTFLUGZEUGE und FLUGSCHULE

Hauptstrasse 63, D-97258 Ippesheim
POSTAL ADDRESS: Lessingstrasse 6, D-97072 Würzburg

Tel: (+49 9339) 12 97 or (+49 171) 770 54 92
Fax: (+49 9339) 998 51
e-mail: flight.team@t-online.de
Web: http://www.home.t-online.de/home/flight.team
DIRECTOR: Peter Götzner

As well as its own three-axis, flexwing ultralights, Flight Team markets the Sinus ultralight and motor glider developed by Pipistrel in Slovenia (which see). Promotion was continuing in 2001, but there have been no known sales to German customers.

UPDATED

FLUG WERK

FLUG WERK GmbH

Kothingried 4, D-85408 Gammelsdorf
Tel: (+49 87) 668 44
Fax: (+49 87) 66 13 51
Web: http://www.flugwerk.com
DIRECTORS: Claus Colling
Hans Günther Wildmoder

Formed on 13 June 1996 and specialising in the reproduction of classic German aircraft, Flug Werk plans to augment its Fw 190 project with resumed manufacture of the Messerschmitt Bf 109E and of kits for the Arado Ar 96b (with Chevrolet V-12 engine). A Heinkel He 112 replica is also planned. New 600 m² (6,450 sq ft) hangar inaugurated 31 March 2000. Recent activities include repair of a genuine Bf 109G after a taxying accident and overhaul of a North American T-6.

UPDATED

FLUG WERK Fw 190

TYPE: Single-seat sportplane kitbuilt.
PROGRAMME: Flug Werk GmbH is assembling a batch of 12 Focke-Wulf Fw 190 Second World War fighters, in Fw

Flug Werk/Focke-Wulf Fw 190A-8/N under construction *(Jane's/Paul Jackson)*   0110622

190A-8/N and Fw 190D-9/N configurations, their /N designation indicating *nachbau:* replica. Components are being manufactured in seven countries (principally in Romania by Aerostar) from original drawings, resulting

in airframes claimed to be 98 per cent faithful to the original.
Power plant of the 10 A-8s is a 1,400 kW (1,877 hp) ASh-82FN 14-cylinder twin-row radial engine, a Russian

version of the original BMW 801D, driving a three-blade, constant-speed MT composites propeller. The first aircraft was completed in early 2000 and sold to a museum at Hanover for static display; second (D-FWKC) shown statically, April 2000, but due to integration problems with the power plant, first engine runs were postponed to the

third quarter of 2002. Three Fw 190A-8/Ns will be sold in Europe, the remaining seven going to the USA, South Africa and Australia; all had found firm buyers by early 2002. Price for kits (including engine, but not avionics, instruments, electrical wiring looms or paint) is US$525,000 ex-works (2002).

The remaining two aircraft, for which buyers were being sought in 2002, are the long-nose Fw 190D 'Dora', the original having a 1,402 kW (1,880 hp) Jumo 213.

UPDATED

FLUGTECHNIK

FLUGTECHNIK DAMME
Am Flugplatz 20, D-49401 Damme
Tel: (+49 54) 91 14 04
Fax: (+49 54) 91 44 19

e-mail: info@flugtechnik.de
Web: http://www.pioneer300.com

Flugtechnik Damme markets kits and factory-built models of the Italian Alpi Asso Vs in Germany as the **Pioneer 300**; this name is also used for US marketing, which began at Sun 'n' Fun 2001, Oshkosh. A description will be found under Alpi Aero in the Italian section.

UPDATED

GROB

BURKHART GROB LUFT- UND RAUMFAHRT e K (Associated with Grob-Werke e K)
Lettenbachstrasse 9, D-86874 Tussenhausen-Mattsies
Tel: (+49 8268) 99 80
Fax: (+49 8268) 99 81 14
e-mail: sales@grob-aerospace.de
Web: http://www.grob-aerospace.de
OWNER: Dr hc Dipl Ing Burkhart Grob
CHAIRMAN: Dipl Ing Christian Grob
CEO: Dr Andreas Plesske
SALES MANAGER: Dipl Ing Hans Doll

Company founded 1928; began aviation activities in 1971, since when has built more than 3,500 aircraft; name changed in 1988 from Burkhart Grob Flugzeugbau (formed 1974) to Burkhart Grob Luft- und Raumfahrt GmbH with light and heavy sections; light section produces sailplanes and the G 115, G 120 and G 140 powered aircraft; heavy section (see 1997-98 *Jane's*) supports Egrett and Strato 2C and deals with space activities in support of Weltraum-Institut Berlin. During 1999-2002, Grob delivered 99 G 115s for RAF University Air Squadrons and 74 to Egyptian Air Force Academy, plus seven G 120s to Lufthansa.

Factory covered area 28,000 m² (301,400 sq ft); workforce 100 in late 2002.

UPDATED

GROB G 115

Royal Air Force name: Tutor

TYPE: Aerobatic two-seat lightplane.

PROGRAMME: Developed from G 110 (first flown 6 February 1982) via G 112 (4 May 1984). First flight of Grob G 115 (D-EBGF) 15 November 1985; first flight of second prototype second quarter of 1986 with taller fin and rudder and relocated tailplane; LBA certification to FAR Pt 23 on 31 March 1987; British certification February 1988; FAA certification 21 December 1988; later gained full public transport certification and German spinning clearance; production terminated August 1990 after total of 103 (including prototypes) G 115/115As. Power plants and equipment updated and designations changed late 1992 for 1993 product line; total of 203 built by December 1999, comprising 103 early versions, one G115T, 14 G 115TAs, single G 115B and 84 G 115C/Ds. Five for Royal Navy pilot training designated **Heron**.

Descriptions, specifications and photographs of the Grob G 115B, G115C1/C2/C2 IFR Trainer, G115D and G115D2 may be found in the 1999-2000 and earlier editions of *Jane's;* Grob G 115TA in 2000-01 edition. Current versions are G 115E and G 115EG.

CURRENT VERSIONS: **G 115E:** Selected in June 1998 for UK Ministry of Defence requirement to replace BAe Bulldogs in RAF University Air Squadrons (UAS) and Air Cadet Air Experience Flights, plus a squadron of the Central Flying School (CFS). Total of 99 ordered for operation at 13 sites in civil markings under contract managed by Vosper

Egyptian Grob G 115EG   *NEW*/0073532

Thornycroft Aerospace (formerly Bombardier Defence Services). Development aircraft (D-ERAF) first flown early 1999; first production aircraft (D-EUKB/G-BYUA) and three others delivered to UK 15 July 1999; official handover of first five aircraft (G-BYUB/C/D/E/F) at RAF Cranwell, Lincolnshire, 13 September 1999, for CFS; deliveries to Cambridge UAS began 14 September 1999; final aircraft (G-BYYB) registered in August 2001.
Description applies to G 115E RAF version.

G 115EG: Version based on G115E for the Egyptian Air Force, which ordered 74 in May 2000 for service at the Egyptian Air Force Academy at Bilbays and Bilbays 2 Air Bases. AEIO-360-B1B engine and MTV-12-B-C/C-183-17e propeller. First aircraft (D-EGYB/EAF 6801) formally handed over in Germany 9 October 2000; first batch of eight handed over by 1 December; deliveries continued at the rate of four per month until mid-2001, then eight per month until contract completion in February 2002. FAA certification 6 February 2001.

CUSTOMERS: By early 2002, production totalled 203 of early versions and 174 G 115E/EGs, or 377 in all, including

prototypes and demonstrators, but excluding two G 110s and two G 112s.

DESIGN FEATURES: Low-wing monoplane of composites construction optimised for flight training. Wing trailing edge moderately swept forward; slightly tapered horizontal tail surfaces; sweptback fin. G 115E based on earlier G 115s, but with fuselage entirely of carbon fibre, saving more than 40 kg (88 lb) and thus allowing aerobatics to be performed at MTOW.

Wing section Eppler E696; dihedral 5°; incidence +2° at root, −2° at tip; tailplane section NACA 64010.

FLYING CONTROLS: Conventional and manual, via pushrods. Horn-balanced elevators with mechanically actuated trim tab in left unit; elevator deflection +34/−20°; mass-balanced ailerons with ground-adjustable trim tabs, aileron deflection +20° 25′/−18° 25′; horn-balanced rudder with ground adjustable tab, rudder deflection ±30°. Electrically actuated plain flaps, deflection 0, 15 and 60°; intermediate 45° setting optional.

STRUCTURE: Entirely of CFRP; semi-monocoque fuselage has rigid CFRP shell formed in two vertically split halves with frames and web members and integral fin; wing of CFRP/honeycomb sandwich with I-beam main spar and carbon fibre roving spar caps; CFRP/honeycomb horizontal tail surfaces of two-spar construction; rudder, ailerons and flaps are GFRP/rigid foam sandwich. Airframe is protected from lightning strike damage by means of aluminium fibres embedded in the outer layer of carbon fabric and copper mesh in the wing surfaces above and below the fuel tanks, all bonded to metal ground strips. Structural service life (aerobatic) 24,000 flight hours.

LANDING GEAR: Non-retractable tricycle type with GFRP wheel fairings; cantilever spring steel suspension on main units, gas damping on nose unit; mainwheel tyre size 6.00-6, nosewheel tyre size 5.00-5; nosewheel steering via rudder pedals, maximum deflection ±9° or ±47° with use of differential braking; hydraulic disc brakes.

POWER PLANT: One 134 kW (180 hp) Textron Lycoming AEIO-360-B1F/B flat-four driving Hoffmann HO-V343K-V/183GY a three-blade, constant-speed (hydraulic), wood/composites propeller. Christen inverted oil system. Fuel in two integral wing tanks, combined

General arrangement of Grob G 115E Tutor *(Jane's/James Goulding)*   0085599

Grob Tutor instrument panel (*Jane's/Paul Jackson*)
0085668

capacity 150 litres (39.6 US gallons; 33.0 Imp gallons) of which 143 litres (37.8 US gallons; 31.5 Imp gallons) are usable, with collector tank, capacity 5.4 litres (1.4 US gallons; 1.2 Imp gallons) for up to three minutes' inverted flight; fuel filler port in top of each wing. Oil capacity 7.6 litres (2.0 US gallons; 1.7 Imp gallons).

ACCOMMODATION: Two seats side by side, under one-piece, rearward-sliding framed canopy; dual controls (sticks) standard; seats can accommodate backpack parachutes and have five-piece harnesses; baggage space behind seats. Cockpit is heated and ventilated.

SYSTEMS: Electrical system comprises 28 V DC 35 Ah engine-driven generator and 24 V high-capacity battery for engine starting and to provide 30 minutes' emergency power in the event of generator or main bus failure.

AVIONICS: Basic aircraft supplied without avionics. Optional minimum avionics pack includes Bendix/King KX 155A nav/com/glideslope, KI 203 VOR/LOC and KT 76A transponder. Typical customer-specified Bendix/King avionics fit for military trainer detailed below.

Comms: KTR 909 UHF selector; with KFS 599A; dual KX 155A with KI 204 VOR/LOC/GS; KT 76A; KMA 28 audio panel; ELT; PTT switches on control columns and NATO-standard quick-release sockets for headsets.

Flight: KCS 55A compass with KG 102A slaved gyro unit; KI 525A HSI; KA 51B slaving control unit with KMT 112 magnetic azimuth transmitter; KN 63 DME with KDI 572 indicator; KG 102A slaved directional gyro; Filser LX500TR differential GPS.

Instrumentation: Standard ASI, VSI, turn and slip indicator, attitude gyro, directional gyro, tachometer, magnetic compass, CHT/fuel pressure gauge, manifold pressure/fuel flow gauge, oil pressure/temperature gauge, OAT/EGT gauge, fuel quantity gauge, voltage indicator, *g* meter and clock.

EQUIPMENT: Standard equipment includes navigation, anti-collision, landing and instrument panel lights; map light on flexible arm between seats optional.

DIMENSIONS, EXTERNAL:
Wing span	10.00 m (32 ft 9¾ in)
Wing chord: at root	1.43 m (4 ft 8¼ in)
at tip	0.94 m (3 ft 1 in)
Wing aspect ratio	8.2
Length overall	7.79 m (25 ft 6¾ in)
Height overall	2.82 m (9 ft 3 in)
Tailplane span	3.50 m (11 ft 5¾ in)
Wheel track	2.56 m (8 ft 4¾ in)
Wheelbase	1.51 m (4 ft 11½ in)
Propeller diameter	1.83 m (6 ft 0 in)
Propeller ground clearance	0.19 m (0 ft 7½ in)

DIMENSIONS, INTERNAL:
Cabin (incl baggage area):	
Length	2.14 m (7 ft 0¼ in)
Max width	1.03 m (3 ft 4½ in)
Max height	1.13 m (3 ft 8½ in)
Baggage volume	0.22 m³ (7.8 cu ft)

AREAS:
Wings, gross	12.21 m² (131.4 sq ft)
Ailerons (total)	1.12 m² (12.06 sq ft)
Fin	1.05 m² (11.30 sq ft)
Rudder	0.64 m² (6.89 sq ft)
Tailplane	1.86 m² (20.02 sq ft)
Elevators (total)	0.86 m² (9.26 sq ft)

WEIGHTS AND LOADINGS:
Weight empty, basic	670 kg (1,477 lb)
Baggage capacity	55 kg (121 lb)
Max fuel	103 kg (227 lb)
Max T-O weight	990 kg (2,182 lb)
Max wing loading	81.08 kg/m² (16.61 lb/sq ft)
Max power loading	7.38 kg/kW (12.12 lb/hp)

PERFORMANCE:
Never-exceed speed (V$_{NE}$)	184 kt (341 km/h; 212 mph)
Max level speed	135 kt (250 km/h; 155 mph)
Cruising speed at 75% power at 1,525 m (5,000 ft) ISA	
	124 kt (230 km/h; 143 mph)
Long-range cruising speed	96 kt (178 km/h; 110 mph)
Stalling speed: flaps up	52 kt (99 km/h; 61 mph)
60° flap	49 kt (91 km/h; 57 mph)
Max rate of climb at S/L	320 m (1,050 ft)/min
Max operating altitude	6,095 m (20,000 ft)
T-O run	248 m (815 ft)
T-O to 15 m (50 ft)	461 m (1,512 ft)
Landing from 15 m (50 ft)	457 m (1,500 ft)
Landing run	171 m (560 ft)

Grob G 120A-I Snunit of Israel Defence Force/Air Force (*Elbit Systems*)
NEW/0531201

Range, with 45 min reserves:
75% power at 1,220 m (4,000 ft)	
	446 n miles (826 km; 513 miles)
55% power at 2,440 m (8,000 ft)	
	569 n miles (1,053 km; 654 miles)
45% power at 2,440 m (8,000 ft)	
	620 n miles (1,148 km; 713 miles)
g limits	+6/−3

UPDATED

GROB G 120A

Israel Defence Force name: Snunit (cyclone).

TYPE: Aerobatic two-seat lightplane.

PROGRAMME: First flight (D-ELHU; unannounced) 1999; development of G 115 series to meet modern airline pilot and military training requirements, having advanced features including EFIS. With retractable landing gear and 194 kW (260 hp) Textron Lycoming AEIO-540 flat-six engine.

Announced January 2001, when Lufthansa ordered three (plus four options, of which three taken up in June 2001) for its US Airline Training Center Arizona (ATCA) operation at Goodyear, near Phoenix. German LBA provisional certification achieved 22 November 2001, followed by FAA FAR Pt 23 approval on 29 January 2002 and full LBA certification on 27 February 2002. Deliveries began (with N861AF) in February 2002.

CURRENT VERSIONS: **G 120A:** Baseline version; aerobatic MTOW 1,350 kg (2,976 lb).

G 120A-I Snunit: Israel Defence Force version. Certified by LBA on 27 September 2002 and by Israeli CAA on 14 October 2002. *As described.*

G 120A Observer: Long-range/endurance version; usable fuel capacity 378 litres (99.8 US gallons; 83.2 Imp gallons).

CUSTOMERS: Lufthansa, seven. On 19 February 2002, Cyclone Aviation, a subsidiary of Elbit Systems, contracted by Israel Defence Force/Air Force to supply pilot screening and primary training services over 10 years with company-owned and -operated fleet of G 120As; first three handed over at Hatzerim AFB on 27 October 2002.

Description for G 120A generally as for G 115, except that below:

DESIGN FEATURES: Aerofoil Eppler E884; dihedral 2°; incidence 0°. Tailplane aerofoil NACA 64-010; fin 64-009. Airframe life 15,000 hours.

FLYING CONTROLS: Conventional and manual. Pushrod actuation for all primary surfaces. Slotted flaps electrically actuated; max deflection 60°. Servo tab and mass balance

on each aileron; elevator trim tab starboard side with electric and manual actuation; rudder and elevator have mass balances, plus horn balance on rudder.

STRUCTURE: Semi-monocoque fuselage comprises a self-supporting carbon fibre-reinforced plastic shell with frame and web members. Cantilever wing of single-trapezoidal cross-section has an I-beam main spar with caps of carbon fibre rovings. Wing shell of honeycomb sandwich, except tank section of PVC foam sandwich. Interconnection of wings is via the spar stubs, bolted together with two large bolts which connect the stub spars. Each wing is attached to the fuselage by three bolts. The main spar and strong webs hold mountings for the main gear leg. An auxiliary spar closes the wing trailing-edge and carries the flaps and ailerons.

Flaps with upper and lower CRP skins in rigid PVC foam core. Aileron structure same as flaps. Fin, integrated with fuselage, comprises main and end spar of honeycomb sandwich design and a carbon-reinforced full laminate shell. Structural configuration of tailplane as for wings. Tailplane attached to the fuselage by four fittings. Both elevators have top and bottom CRP shells; structure as for ailerons. Airframe is protected from moisture and ultra-violet radiation by UP gel-coat. A white filler on the outer surface of the aircraft conducts electricity, being part of lightning protection system. White two-component polyurethane lacquer covers the filler and completes the finish.

LANDING GEAR: Tricycle type; retractable. Mainwheels 6.00-6; nosewheel 5.00-5. Nosewheel steerable ±10°.

POWER PLANT: One 194 kW (260 hp) Textron Lycoming AEIO-540-D4D5 flat-six, driving a Hartzell HC-C3YR-1RF/F7663R three-blade, constant-speed propeller. Fuel capacity 262 litres (69.2 US gallons; 57.6 Imp gallons), of which 256 litres (67.6 US gallons; 56.3 Imp gallons) are usable. Max oil capacity 11.4 litres (3.0 US gallons; 2.5 Imp gallons).

SYSTEMS: Grob 120A-5700 landing gear hydraulic system, pressure 103 bar (1,500 lb/sq in); Grob ECS 115TA-7301 air conditioning system. Electrical system 28 V DC with 80 A alternator and Concorde RG-24-20 24 V 19 Ah battery. External power socket, port side, to rear of cockpit.

AVIONICS: *Comms:* Bendix/King KMA 24H-70 audio control/intercom, KY 196A VHF, KT 76C transponder.

Flight: Bendix/King KX 165A nav/com/glideslope, KN 62 DME, KLN. 94 GPS, KCS 55A compass, KR 87 ADF with KI 227 indicator.

DIMENSIONS, EXTERNAL:
Wing span	10.19 m (33 ft 5¼ in)
Length overall	8.065 m (26 ft 5½ in)

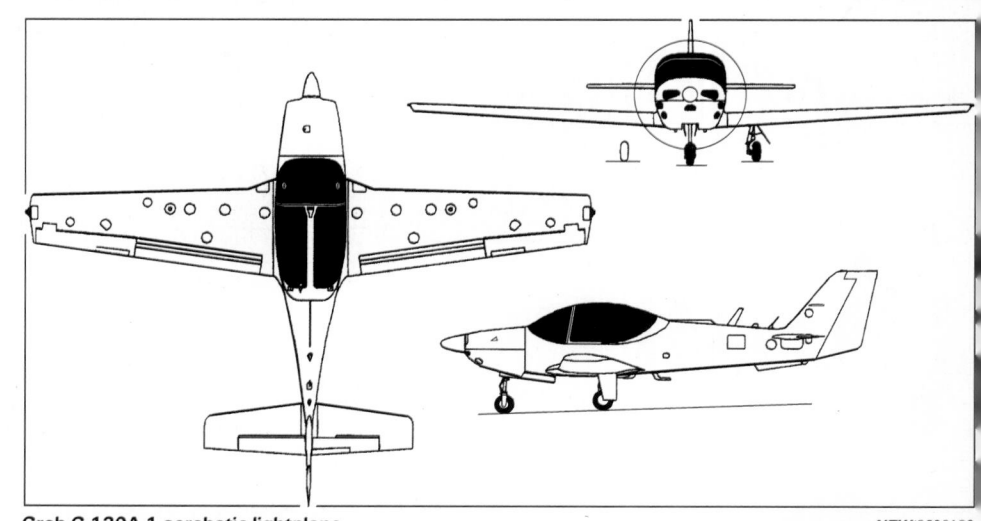

Grob G 120A-1 aerobatic lightplane
NEW/0530193

Height overall	2.57 m (8 ft 5¼ in)
Tailplane span	3.80 m (12 ft 5½ in)
Wheel track	2.42 m (7 ft 11¼ in)
Wheelbase	1.87 m (6 ft 1½ in)
Propeller diameter	1.98 m (6 ft 6 in)
AREAS:	
Wings, gross	13.29 m² (143.1 sq ft)
Ailerons (total)	0.915 m² (9.85 sq ft)
Trailing-edge flaps (total)	1.61 m² (17.33 sq ft)
Fin	1.71 m² (18.41 sq ft)
Rudder	0.77 m² (8.33 sq ft)
Tailplane	3.04 m² (32.72 sq ft)
Elevator	0.70 m² (7.53 sq ft)
WEIGHTS AND LOADINGS:	
Weight empty	960 kg (2,116 lb)
Baggage capacity	50 kg (110 lb)
Max T-O and landing weight	1,440 kg (3,174 lb)
Max zero-fuel weight	1,315 kg (2,899 lb)
Max wing loading	108.3 kg/m² (22.18 lb/sq ft)
Max power loading	7.42 kg/kW (12.21 lb/hp)
PERFORMANCE:	
Never-exceed speed (V_{NE})	235 kt (435 km/h; 270 mph)
Max level speed	172 kt (319 km/h; 198 mph)
Manoeuvre speed	165 kt (306 km/h; 190 mph)
Cruising speed at 75% power at 1,525 m (5,000 ft)	
	166 kt (307 km/h; 191 mph)
Stalling speed	55 kt (102 km/h; 64 mph)
Service ceiling, all	5,490 m (18,000 ft)
Max rate of climb at S/L	390 m (1,280 ft)/min
T-O to 15 m (50 ft)	656 m (2,150 ft)
Landing from 15 m (50 ft)	563 m (1,845 ft)
Range at 45% power, 30 min reserves:	
G 120A at 2,440 m (8,000 ft)	
	830 n miles (1,537 km; 955 miles)
Observer at 1,830 m (6,000 ft)	
	1,360 n miles (2,518 km; 1,565 miles)
Endurance at 45% power at 610 m (2,000 ft), 30 min reserves:	
Observer	11 h 40 min
g limits: G 120A Aerobatic	+6/−4
G 120A Utility, Observer	+4.4/−1.76
	UPDATED

Prototype Grob G 140TP on display at Paris in June 2001 (*Jane's/Paul Jackson*)   NEW/0132418

GROB G 140TP

TYPE: Aerobatic business turboprop.

PROGRAMME: Development started early 2001. Prototype (D-ETPG), then unflown and previously unannounced, exhibited at Paris Air Show from 16 to 24 June 2001.

Design freeze scheduled for June 2002 following incorporation of changes based on feedback from potential customers. First flight expected in mid-2002, with JAR-23 certification a year later.

CUSTOMERS: Potential market for 20 to 40 per year, for military and civilian markets. Interest reported from several air forces by early 2002.

COSTS: Unpressurised version US$1 million (2002).

DESIGN FEATURES: Generally similar to G 115/120 series.

FLYING CONTROLS: Conventional and manual. Trim tab on port elevator and each aileron; horn-balanced rudder.

STRUCTURE: Entirely GFRP. Structural service life (aerobatic) 15,000 flight hours.

LANDING GEAR: Retractable tricycle type with trailing-link suspension on main units. Main legs retract inwards into wing root area, nosewheel rearwards.

Artist's impression of Grob G 350 twin-turboprop business aircraft   0110974

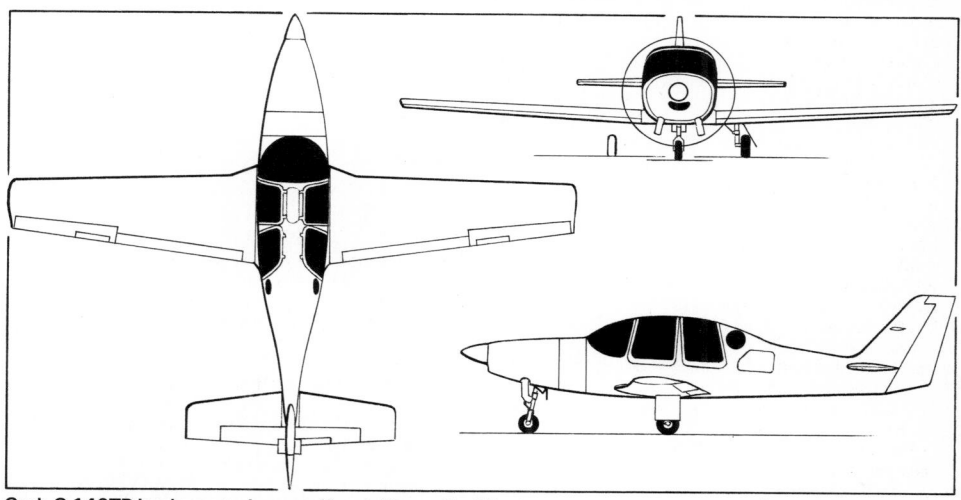

Grob G 140TP business turboprop (*Jane's/James Goulding*)   NEW/0110975

POWER PLANT: One 336 kW (450 shp) Rolls-Royce 250-B17F turboprop driving an MT-Propeller MTV-5 five-blade, constant-speed (hydraulic) propeller.

ACCOMMODATION: Four persons in individual seats. Separate upward-opening gullwing-type door for each seat, with circular 'porthole' window at rear of cabin on each side. Baggage door at rear of cabin on port side. Cabin pressurisation will be offered as an option.

DIMENSIONS, EXTERNAL:

Wing span	10.30 m (33 ft 9½ in)
Wing aspect ratio	8.0
Length overall	8.90 m (29 ft 2½ in)
Height overall	2.80 m (9 ft 2¼ in)

AREAS:

Wings, gross	13.30 m² (143.2 sq ft)

WEIGHTS AND LOADINGS: (A: Aerobatic; B: Utility)

Weight empty	1,100 kg (2,425 lb)
Max T-O and landing weight: A	1,500 kg (3,307 lb)
B	1,650 kg (3,637 lb)
Max wing loading	124.1 kg/m² (25.41 lb/sq ft)
Max power loading	4.92 kg/kW (8.08 lb/shp)

PERFORMANCE (estimated):

Cruising speed: max at 3,050 m (10,000 ft)	
	213 kt (394 km/h; 245 mph)
long range at 3,050 m (10,000 ft)	
	170 kt (315 km/h; 196 mph)
Stalling speed: A	55 kt (102 km/h; 64 mph)
B	57 kt (106 km/h; 66 mph)
Max rate of climb at S/L: A	823 m (2,700 ft)/min
B	701 m (2,300 ft)/min
T-O to 15 m (50 ft): A	374 m (1,225 ft)
B	452 m (1,480 ft)
Landing from 15 m (50 ft): A	587 m (1,925 ft)
B	691 m (2,265 ft)
Range: with max payload	
	575 n miles (1,064 km; 661 miles)
with max fuel	1,150 n miles (2,129 km; 1,323 miles)
g limits: U	+4.4/−1.76
A	+6/−4
	UPDATED

GROB G 350

In mid-2001 Grob was continuing development of this all-composites twin-turboprop, pusher configuration business aircraft, last described in the 1999-2000 edition. Projected performance includes a cruising speed of 300 kt (556 km/h; 332 mph) at 9,150 m (30,000 ft) and range of more than 2,000 n miles (3,704 km; 2,216 miles).

VERIFIED

HIGH PERFORMANCE

HIGH PERFORMANCE AIRCRAFT
Bayerwaldstrasse 28, D-94350 Falkenfels
Tel: (+49 9961) 91 01 60
Fax: (+49 9961) 91 01 61
e-mail: heiko.teegen@pilotundflugzeug.de
CEO: Heiko Teegen

Company founded 10 April 2002 as joint venture by professional aviation magazine *Pilot und Flugzeug* and local government of Mecklenburg-Vorpommern. Intention is to produce TT62 light twin-prop, for which full funding was in place by time of public launch at Berlin Air Show, 6 to 12 May 2002. Secondary purpose is restoration of Rostock-Laage airfield as centre of aviation development with initial creation of 41 jobs at High Performance plant.

NEW ENTRY

HIGH PERFORMANCE TT62

TYPE: Business twin-prop.

PROGRAMME: Promotion begun at Berlin Air Show, 6 May 2002. Deliveries targeted for 2005, following maiden flight in previous year.

COSTS: €450,000 plus tax (2002).

DESIGN FEATURES: Mid-mounted wing with moderate leading-edge sweepback and winglets. T tail. Two internal turbo-diesel engines with propellers mounted on pylons at fuselage shoulders, immediately behind wing trailing-edge.

Design aims include speed and economy, achieved in part by internal mounting of engines (reduced frontal area) and use of diesel power. Consumption 79 litres (20.9 US gallons; 17.4 Imp gallons)/h at normal cruise; 58 litres (15.3 US gallons; 12.8 Imp gallons)/h at econ cruise. Specific range, respectively, 3.84 n miles/kg and 4.80 n miles/kg.

FLYING CONTROLS: Conventional and manual. Fowler flaps.
POWER PLANT: Two 228 kW (306 hp) Thielert TAE 310 V-8 turbocharged diesel engines, each driving a four-blade, constant-speed propeller. Fuel capacity 617 litres (163 US gallons; 136 Imp gallons).
ACCOMMODATION: Pilot and four passengers in pressurised cabin; differential 0.37 bar (5.5 lb/sq in); door port side, ahead of wing.
SYSTEMS: De-icing standard.
AVIONICS: IFR standard.

DIMENSIONS, EXTERNAL:
Wing span	10.50 m (34 ft 5½ in)
Wing aspect ratio	6.6
Length overall	10.80 m (35 ft 5¼ in)
Fuselage max diameter	1.50 m (4 ft 11 in)
Height overall	3.75 m (12 ft 3¾ in)
Propeller diameter	1.70 m (5 ft 7 in)

DIMENSIONS, INTERNAL:
Cabin: Length	4.20 m (13 ft 9¼ in)
Max width	1.40 m (4 ft 7 in)

AREAS:
Wings, gross	16.70 m² (179.8 sq ft)

Model of High Performance TT62 *(Jane's/Paul Jackson)* **NEW**/0137340

WEIGHTS AND LOADINGS:
Weight empty	1,895 kg (4,178 lb)
Max T-O weight	2,550 kg (5,621 lb)
Max wing loading	152.7 kg/m² (31.27 lb/sq ft)
Max power loading	5.59 kg/kW (9.18 lb/hp)

PERFORMANCE:
Max cruising speed at FL200	
	240 kt (444 km/h; 276 mph)*
Econ cruising speed at FL250	
	180 kt (333 km/h; 207 mph)*
Max rate of climb at S/L	549 m (1,800 ft)/min
Time to FL250	18 min

Range, 45 min reserves:
with max payload:
normal cruise	695 n miles (1,287 km; 799 miles)
econ cruise	866 n miles (1,603 km; 996 miles)

with max fuel:
normal cruise	1,743 n miles (3,228 km; 2,005 miles)
econ cruise	2,178 n miles (4,033 km; 2,506 miles)

* at 2,300 kg (5,071 lb)

NEW ENTRY

HK

HK AIRCRAFT TECHNOLOGY AG

Betriebsstätte Flughafen Obermehler, Otto Lilienthalstrasse 2, D-99996 Obermehler/Thüringen
Tel: (+49 9521) 61 85 03 and 61 85 05
Fax: (+49 9521) 61 85 04
e-mail: info@hk-aircraft.de
Web: http://www.hk-aircraft.de

UPDATED

HK WEGA

English name: Vega
TYPE: Two-seat lightplane.
PROGRAMME: Prototype (D-ETHK) first flew on 9 April 1998; designation Wega Ex (experimental) applied to this aircraft; second prototype (D-EPHK) displayed at Aero '01, Friedrichshafen, April 2001. JAR-VLA certification scheduled for August 2001, but not reported by early 2002.
COSTS: DM222,720 (2001) flyaway, including tax, with standard instrumentation.
DESIGN FEATURES: German production version of the Pulsar Aircraft Corporation (Aero Designs) Turbo Pulsar kitbuilt, with modifications to enable it to be certified to JAR-VLA. Production version has cockpit width increased by 10 cm (4 in).
FLYING CONTROLS: Conventional and manual with mass-balanced rudder and elevator. Electrically actuated pitch trim via tab on port elevator; ground-adjustable tab on starboard aileron; electrically actuated flaps.
STRUCTURE: Principally GFRP with carbon fibre or aluminium reinforcement in high stress areas and foam cores. Wing skins of composites.
LANDING GEAR: Fixed tricycle type with speed fairing on each unit.
POWER PLANT: One modified 84.6 kW (113.4 hp) Rotax 914 turbocharged flat-four with intercooler, driving a Hoffmann HO-V383F three-blade, constant-speed

Second prototype HK Wega, German series production version of the Pulsar Aircraft Turbo Pulsar *(Jane's/Paul Jackson)* 0113415

propeller. Fuel in two tanks, total capacity 140 litres (37.0 US gallons; 30.8 Imp gallons), of which 135 litres (35.7 US gallons; 29.7 Imp gallons) usable.
ACCOMMODATION: Two, side by side in reclined bucket seats; forward-sliding, one-piece canopy. Central control column. Air-conditioning standard.
AVIONICS: *Comms:* Becker nav/com and transponder standard.

DIMENSIONS, EXTERNAL:
Wing span	7.62 m (25 ft 0 in)
Wing aspect ratio	7.8
Length overall	6.10 m (20 ft 0 in)
Height overall	2.07 m (6 ft 9½ in)

DIMENSIONS, INTERNAL:
Cabin max width	1.06 m (3 ft 5¾ in)
Baggage compartment volume	0.65 m³ (22.9 cu ft)

AREAS:
Wings, gross	7.43 m² (80.0 sq ft)

WEIGHTS AND LOADINGS:
Weight empty	400 kg (882 lb)
Baggage capacity	30 kg (66 lb)
Max T-O weight	700 kg (1,543 lb)
Max wing loading	94.2 kg/m² (19.30 lb/sq ft)
Max power loading	8.27 kg/kW (13.59 lb/hp)

PERFORMANCE:
Never-exceed speed (VNE)	180 kt (333 km/h; 207 mph)
Max cruising speed	170 kt (315 km/h; 196 mph)
Stalling speed	45 kt (84 km/h; 52 mph)
Max rate of climb at S/L	610 m (2,000 ft)/min
Service ceiling	6,100 m (20,000 ft)
T-O run	350 m (1,150 ft)
Landing run	380 m (1,250 ft)
Range with max fuel	
	1,133 n miles (2,100 km; 1,304 miles)
g limits	+3.8/–2

UPDATED

IKARUS

IKARUS DEUTSCHLAND (Comco Ikarus Gerätebau GmbH)

Am Flugplatz 11, Flugplatz Mengen, D-88367 Hohentengen
Tel: (+49 7572) 600 80
Fax: (+49 7572) 33 09
e-mail: post@comco-ikarus.de
Web: http://www.comco-ikarus.de

Undertakes sales and distribution of the C 42 ultralight in Europe. Ikarus also produces the C 22 open cockpit ultralight, described in the 1992-93 *Jane's*, and several hang glider designs. Over 1,500 Ikarus aircraft have been sold.

UPDATED

IKARUS C 42

TYPE: Side-by-side ultralight/kitbuilt.
PROGRAMME: Simultaneously launched in Europe (Friedrichshafen Air Show) and North America (Sun 'n' Fun) in second quarter of 1997. European deliveries began immediately. On 10 May 1998 an Ikarus C 42 set a new ultralight world speed record over a closed circuit of 54 n miles (100 km; 62 miles) at 90.4 kt (167.5 km/h; 104.1 mph), and on 13 May 1998 established another world record over a 270 n mile (500 km; 311 mile) closed circuit at 71.78 kt (132.93 km/h; 82.60 mph). Upgraded version, introduced April 1999, offers optional winglets of GFRP and further option of GFRP ailerons, flaps, elevators and rudder.

CURRENT VERSIONS: **Ikarus C 42:** A*s described.*
C 42 Competition I/II: Kitbuilt versions with additional refinements.
CUSTOMERS: First production C 42 to Netherlands owner (PH-2Y5) mid-1997. Others sold in Austria, Estonia, Finland, Germany, Norway and Sweden. United Kingdom PFA type clearance, and first deliveries, early 2002.
COSTS: €35,284 plus tax (2001).
DESIGN FEATURES: Strut-braced high-wing monoplane; strut-braced tail unit. Wings fold for storage.
NACA 2412 wing section.
FLYING CONTROLS: Conventional and manual. Ailerons and elevators pushrod-operated, rudder cable-operated. Plain flaps standard; electrically actuated flaps optional; flight adjustable trim tab on port elevator.

For details of the latest updates to *Jane's All the World's Aircraft* online and to discover the additional information available exclusively to online subscribers please visit
jawa.janes.com

Floatplane version of Ikarus C-42
(Jane's/Paul Jackson)   0110624

Standard version of Ikarus C 42 *(Jane's/Paul Jackson)*   *NEW*/0132408

STRUCTURE: Glass fibre fuselage assembled around full-length aluminium boom; tubular aluminium wings and tail surfaces, covered with Mylar/polyester laminate (GT-foil). Aluminium wings optional. Glass fibre-reinforced epoxy resin fairings; stainless steel fittings; Makrolon transparencies. Optional carbon fibre winglets.

LANDING GEAR: Fixed tricycle type with oleo-pneumatic suspension and speed fairings on all three units; hydraulic brakes on main units. Full Lotus inflatable floats optional.

POWER PLANT: One 59.6 kW (79.9 hp) Rotax 912 UL flat-four driving a Warp Drive two-blade, carbon fibre propeller via 1:2.27 reduction gearing or 73.5 kW (98.6 hp) Rotax 912 ULS driving a GSC three-blade, ground-adjustable pitch propeller via 1:2.43 reduction gearing. Standard fuel capacity 50 litres (13.2 US gallons; 11.0 Imp gallons). 100 litres (26.4 US gallons; 22.0 Imp gallons) optional.

EQUIPMENT: Optional BRS 50L4 or Magnum Speed S ballistic parachute, and ELT.

DIMENSIONS, EXTERNAL:

Wing span: standard	9.45 m (31 ft 0 in)
winglet option	9.50 m (31 ft 2 in)
Length overall	6.25 m (20 ft 6 in)
Height overall	2.34 m (7 ft 8¼ in)

DIMENSIONS, INTERNAL:

Cabin max width	1.22 m (4 ft 0 in)

AREAS:

Wings, gross	12.50 m² (134.5 sq ft)

WEIGHTS AND LOADINGS:

Weight empty: Standard	268 kg (591 lb)
Competition	259 kg (571 lb)
Max T-O weight, all	450 kg (992 lb)

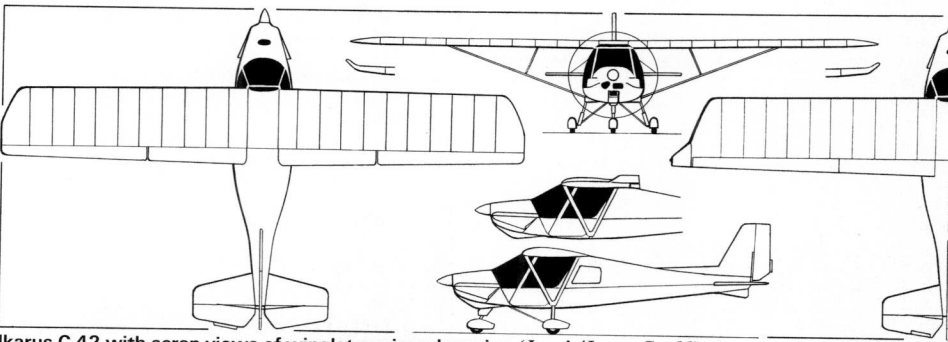

Ikarus C-42 with scrap views of winglet-equipped version *(Jane's/James Goulding)*   0110597

PERFORMANCE (Rotax 912 UL):

Cruising speed at 75% power	94 kt (175 km/h; 109 mph)
Stalling speed	34 kt (63 km/h; 40 mph)
Max rate of climb	300 m (984 ft)/min
T-O run	90 to 100 m (295 to 330 ft)
Range with standard fuel	324 n miles (600 km; 372 miles)

UPDATED

IMPULSE

IMPULSE AIRCRAFT GmbH
Otto-Lilienthalstrasse 1, D-06188 Oppin/Halle
Tel: (+49 345) 678 77 74
Fax: (+49 345) 678 69 13
e-mail: info@impulse-aircraft.de
Web: http://www.impulse-aircraft.de

Impulse was established to produce the lightplane of the same name.

UPDATED

IMPULSE IMPULSE

TYPE: Two-seat lightplane; side-by-side ultralight.

PROGRAMME: First flight 28 July 2000. Tailwheel version first shown at Berlin in May 2002, having achieved certification on 25 April 2002.

CURRENT VERSIONS: **100:** Restricted weight and power to meet 450 kg (992 lb) limit.

130: Uprated version within US Experimental category limit.

130TD: Tailwheel version introduced in 2002.

360: Higher-powered version; under development in 2002; to be certified under VLA requirements.

360TD: Tailwheel version of 360; under development.

CUSTOMERS: Total 21 sold and 18 built by May 2002.

COSTS: 130FG kit only, DM79,500 (2002).

DESIGN FEATURES: Streamlined, low-wing configuration, with fuselage sharply waisted to rear of cockpit; laminar flow wing; sweptback fin with mid-mounted tailplane. Wide speed range. Available in Experimental and Ultralight weight categories. Quoted build time 450 hours.

FLYING CONTROLS: Conventional and manual. Horn-balanced rudder; flight-adjustable trim tab in port elevator. Three-quarter-span Fowler flaps.

STRUCTURE: Composites airframe; aluminium landing gear legs; Plexiglas hood.

LANDING GEAR: Fixed tricycle or tailwheel (TD suffix) type. Hydraulic brakes.

POWER PLANT: *100:* One 73.5 kW (98.6 hp) Rotax 912 S flat-four, driving an MT three-blade, variable-pitch propeller. Fuel in wing tanks, capacity 110 litres (29.1 US gallons; 24.2 Imp gallons).

130 and 130TD: Options of 95 kW (128 hp) Limbach L2400ET flat-four and 118 kW (158 hp) LOM M332 in-line four. Fuel capacity 160 litres (42.3 US gallons; 35.2 Imp gallons).

360 and 360TD: One 134 kW (180 hp) Textron Lycoming IO-360-M1A. Fuel capacity 300 litres (79.3 US gallons; 66.0 Imp gallons).

Optional power plants include BMW motorcycle engine and Thielert TAE 125 turbo-diesel.

ACCOMMODATION: Two persons, side by side, beneath single-piece hood; baggage space behind seats.

DIMENSIONS, EXTERNAL (UL: Ultralight, E: Experimental):

Wing span	8.70 m (28 ft 6½ in)
Wing aspect ratio	8.2
Length: overall: 100	6.30 m (20 ft 8 in)
130,130TD, 360, 360TD	6.10 m (20 ft 0¼ in)

DIMENSIONS, INTERNAL:

Cockpit max width	1.18 m (3 ft 10½ in)

AREAS:

Wings, gross	9.20 m² (99.0 sq ft)

WEIGHTS AND LOADINGS (100 with Rotax, 130 with LOM engine):

Weight empty: 100	260 kg (573 lb)
130	320 kg (705 lb)
130TD	350 kg (772 lb)
360, 360TD	400 kg (882 lb)
Baggage capacity: 100	20 kg (44 lb)
130	50 kg (110 lb)
Max T-O weight: 100	450 kg (992 lb)
130, 130TD	720 kg (1,587 lb)
360, 360TD	750 kg (1,653 lb)
Max wing loading: 130, 130TD	78.3 kg/m² (16.02 lb/sq ft)
360, 360TD	81.5 kg/m² (16.70 lb/sq ft)
Max power loading: 130	6.11 kg/kW (10.04 lb/hp)
360, 360TD	5.59 kg/kW (9.18 lb/hp)

PERFORMANCE (100, 130 as above):

Never-exceed speed (VNE):	
100	145 kt (270 km/h; 167 mph)
130, 130TD	226 kt (420 km/h; 261 mph)
Max cruising speed at 75% power:	
100	130 kt (240 km/h; 149 mph)
130	162 kt (300 km/h; 186 mph)
130TD	170 kt (315 km/h; 196 mph)
360	190 kt (352 km/h; 219 mph)
360TD	200 kt (370 km/h; 230 mph)
Stalling speed, flaps up: 100	34 kt (63 km/h; 40 mph)
130	41 kt (75 km/h; 47 mph)
130TD, 360, 360TD	44 kt (82 km/h; 51 mph)
T-O run: 100	60 m (197 ft)
130, 130TD	100 m (330 ft)
360	120 m (395 ft)
T-O to 15 m (50 ft): all	200 m (656 ft)
Range: 100	647 n miles (1,200 km; 745 miles)
130, with auxiliary tank	3,509 n miles (6,500 km; 4,038 miles)

UPDATED

BMW-engined, tailwheel Impulse Impulse 100 on display at Berlin in May 2002 *(Jane's/Paul Jackson)*
NEW/0137959

KAISER

KAISER FLUGZEUGBAU GmbH

Flugplatz Schönhagen, D-14959 Schönhagen
Tel: (+49 3045) 02 28 58 or (+49 3329) 629 40
Fax: (+49 3329) 629 41
e-mail: service@kaiser-flugzeugbau.de
Web: http://www.Kaiser-Flugzeugbau.de
DIRECTOR: Dipl Ing Jörg Kaiser

WORKS:

Wojskowe Zakłady Lotnicze Nr. 3
PL-085-21 Deblin-3, Poland
Tel: (+48 81) 883 01 22
Fax: (+48 81) 883 02 48

Kaiser's first product, the Magic, had its debut as a static
exhibit at the Berlin Air Show in May 1998. A second design,
the Whisper, has been added to the company's portfolio.
VERIFIED

KAISER MAGIC

Polish designation: DEKO-9

TYPE: Aerobatic two-seat biplane.
PROGRAMME: Design begun in 1995; prototype almost
complete when exhibited at Berlin, May 1998, but maiden
flight delayed pending static testing of new structural
methods. Registered SP-PCM in early 2000; first flight
September 2000. First four aircraft, all Franklin-engined,
had been expected to be completed in 1999, but no
evidence of these by early 2002. Initial certification was to
be in Poland during 2001, followed by Germany.
CURRENT VERSIONS: **Four-cylinder:** See Power Plant.
    Six-cylinder: As prototype; see Power Plant.
COSTS: DM159,500 (four-cylinder) or DM198,000 (six-
cylinder) 2001.
DESIGN FEATURES: Fully aerobatic, classic, single-bay biplane
with N-type interplane struts and wire-braced empennage.
Extra airframe strength achieved by reinforcement of
strategic metal tube joints with composites cord binding.
German design. Short span optimises roll rate, which
approximately 360°/s.
FLYING CONTROLS: Conventional and manual. Ailerons on
both wings, interconnected by rods. No flaps.
STRUCTURE: Aluminium tube fuselage with Ceconite
covering, except for composites engine cowling, cockpit
rim and wing/fuselage fairings. Aluminium-covered
ailerons and wing leading-edges; Ceconite elevator and
rudder. Glass fibre cantilever mainwheel legs and tailwheel
arm. Upper wing on aluminium tube cabane ahead of
cockpit; N interplane bracing strut each side, plus bracing
wires inboard; fin and tailplane mutually wire-braced.
Polish construction by Military Aircraft Works No. 3 at
Deblin; available factory-built only.
LANDING GEAR: Tailwheel type; fixed, with speed fairings on
mainwheel legs; main tyres 15×6.00-6; solid, steerable
tailwheel. Hydraulic mainwheel brakes.
POWER PLANT: One Polish (PZL)-built Franklin horizontally
opposed piston engine: 93.2 kW (125 hp) four-cylinder
4A-235B4 or 153 kW (205 hp) six-cylinder 6A-350C1R.
MT-Propeller MTV-9-D-C/C188-18a three-blade,
variable-pitch (hydraulic) propeller for six-cylinder
version; type for four-cylinder version not yet selected;
suitable for engines up to 283 kW (380 hp). Fuel tank
ahead of cockpit; capacity 100 litres (26.4 US gallons; 22.0

Kaiser Magic aerobatic biplane wearing Polish registration *(Jane's/Paul Jackson)*   *NEW*/0132409

Imp gallons) with six-cylinder engine or approx 160 litres
(42.2 US gallons; 35.2 Imp gallons) with four-cylinder
engine.
ACCOMMODATION: Two persons in tandem, with dual controls,
under single-piece acrylic canopy in six-cylinder version;
in open cockpit in four-cylinder version.
AVIONICS: *Flight:* GPS standard.
    Instrumentation: Front cockpit has basic instruments
only.
EQUIPMENT: Navigation lights standard.
DIMENSIONS, EXTERNAL:
Wing span	6.80 m (22 ft 3¾ in)
Length overall	6.50 m (21 ft 4 in)
Height overall	2.20 m (7 ft 2½ in)

AREAS:
Wings, gross	15.40 m² (165.8 sq ft)

WEIGHTS AND LOADINGS (A: four-cylinder, B: six-cylinder):
Weight empty: A	450 kg (992 lb)
B	500 kg (1,102 lb)
Max T-O weight: A, B	750 kg (1,653 lb)
Max wing loading: A, B	48.7 kg/m² (9.97 lb/sq ft)
Max power loading: A	8.05 kg/kW (13.22 lb/hp)
B	4.91 kg/kW (8.06 lb/hp)

PERFORMANCE (estimated; A, B as above):
Never-exceed speed (VNE): A, B	
	189 kt (350 km/h; 217 mph)
Max rate of climb at S/L: A	600 m (1,969 ft)/min
B	900 m (2,953 ft)/min
Range: A	432 n miles (800 km; 497 miles)
g limits: B	+6/−3

UPDATED

KAISER WHISPER

TYPE: Side-by-side ultralight kitbuilt.
PROGRAMME: Exhibited incomplete at Aero '99 (April 1999),
ILA, Berlin, June 2000, and Aero '01. First flight planned
for late 2001.

COSTS: Ready to fly, DM68,500; kit, less engine, instruments
and parachute recovery system, DM21,500 (both 2001,
including tax).
DESIGN FEATURES: Pod-and-boom fuselage. Wings braced by
V-type struts; mid-mounted tailplane braced to lower
fuselage by parallel struts.
FLYING CONTROLS: Conventional and manual. Two-section
Junkers-type flaps, plus separate ailerons. Dual controls.
STRUCTURE: Principally riveted aluminium; Ceconite
covering and glass fibre pod and control surfaces.
LANDING GEAR: Tailwheel type; fixed. Carbon fibre spring
main gear; speed fairings on mainwheels; floats optional.
POWER PLANT: One 59 kW (79 hp) Hirth F30 flat-four in
prototype; production version with BMW horizontally
opposed, two-cylinder piston engine; both driving a
Junkers three-blade, carbon fibre, ground-adjustable pitch
pusher propeller turning at 2,200 rpm, or a ducted pusher
propeller. Suitable for other engines in 60 to 89 kW (80 to
120 hp) class.
SYSTEMS: Ballistic parachute recovery system.
DIMENSIONS, EXTERNAL:
Wing span	9.00 m (29 ft 6¼ in)
Length overall	6.60 m (21 ft 7¾ in)
Height overall	2.60 m (8 ft 6¼ in)

AREAS:
Wings, gross	12.90 m² (138.9 sq ft)

WEIGHTS AND LOADINGS:
Weight empty	230 kg (507 lb)

PERFORMANCE (estimated):
Cruising speed	76 kt (140 km/h; 87 mph)
Max rate of climb at S/L	360 m (1,180 ft)/min

UPDATED

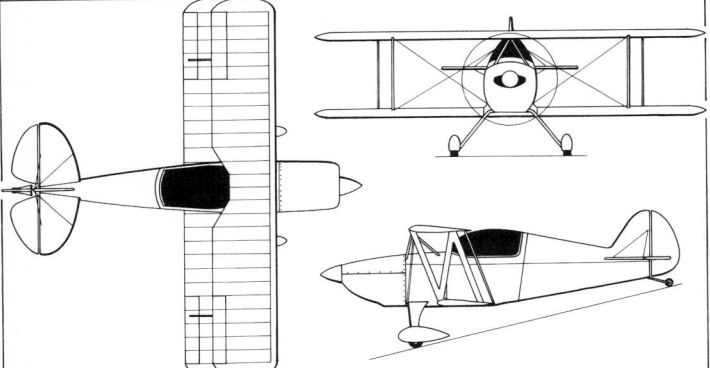

General arrangement of the Kaiser Magic *(Jane's/Paul Jackson)*   0022156

Partly complete prototype Kaiser Whisper in April 2001 *(Jane's/Paul Jackson)*   0110625

MK

MK HELICOPTER GmbH

Hans-Böcklerstrasse 1, D-65468 Trebur-Astheim
Tel: (+49 6147) 91 91 28
Fax: (+49 6147) 91 91 29
e-mail: MK-helicopter-Engineering@t-online.de
TECHNICAL MANAGER: Uwe Mathes

The MK II light helicopter, last described in the 2001-02
Jane's, has been developed into the MK III.
VERIFIED

MK HELICOPTER MK III

TYPE: Three-seat helicopter.
PROGRAMME: Design (initially as Ultralight Mk 2 and later
MK Helicopter Mk II) started 1995. Construction of
prototype scheduled to begin in December 2001; first flight
expected June 2002; construction of first production
helicopter anticipated in January 2003; JAR 27
certification process began in March 2001 and completion
expected in June 2003.
COSTS: Target US$180,000 (2001).
DESIGN FEATURES: Two-blade semi-articulated main rotor;
blades comprise glass/carbon-reinforced plastic (GRP) skin

over Rohacell foam core with lead inlay. Two-blade tail
rotor on port side with aluminium blades. Main rotor blade
section Eppler E361; tail rotor blade section NACA 81012.
Engine drives single-stage bevel-gear 1:2.5 reduction
transmission with belt drive for main rotor; 90° bevel gear
drive for tail rotor. Sweptback dorsal and ventral fins.
STRUCTURE: Tubular main frame with aluminium sheet box
structure; Nomex/carbon fibre sandwich fuselage.
LANDING GEAR: Skid type.
POWER PLANT: One 164 kW (220 hp) engine of undisclosed
type. Fuel contained in two tanks behind cabin, combined
capacity 180 litres (47.6 US gallons; 39.6 Imp gallons).

ACCOMMODATION: Pilot and two passengers, or pilot and one student with dual controls. Cabin heated and ventilated.
SYSTEMS: 24 V battery for electrical system.
DIMENSIONS, EXTERNAL:
Main rotor diameter	9.00 m (29 ft 6¼ in)
Tail rotor diameter	1.50 m (4 ft 11 in)
Main rotor blade chord	0.24 m (0 ft 9½ in)
Distance between rotor centres	5.30 m (17 ft 4¾ in)
Length: overall, rotors turning	9.90 m (32 ft 5¾ in)
fuselage	7.60 m (24 ft 11¼ in)
Height overall	2.62 m (8 ft 7¼ in)
Skid track	2.10 m (6 ft 10¾ in)

AREAS:
Main rotor disc	63.60 m² (684.6 sq ft)
Tail rotor disc	1.77 m² (19.05 sq ft)

WEIGHTS AND LOADINGS:
Weight empty, equipped	540 kg (1,190 lb)
Baggage capacity	55 kg (121 lb)
Max T-O weight	1,000 kg (2,204 lb)
Max disc loading	15.72 kg/m² (3.22 lb/sq ft)
Transmission loading at max T-O weight and power	6.10 kg/kW (10.02 lb/hp)

PERFORMANCE (estimated):
Never-exceed speed (V$_{NE}$)	102 kt (190 km/h; 118 mph)
Max cruising speed at S/L	86 kt (160 km/h; 99 mph)
Econ cruising speed	65 kt (120 km/h; 75 mph)
Max rate of climb at S/L	240 m (787 ft)/min
Hovering: IGE	2,000 m (6,560 ft)
OGE	1,600 m (5,249 ft)
Range with max fuel at S/L	421 n miles (780 km; 484 miles)

UPDATED

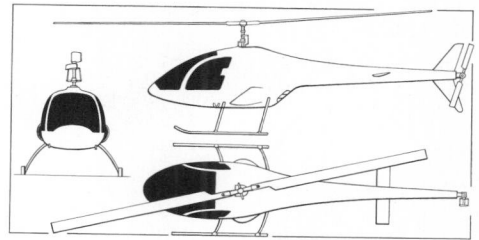

MK II helicopter, on which MK III is based
(Jane's/James Goulding) NEW/0010532

MYLIUS

MYLIUS FLUGZEUGWERK GmbH & Co KG
Am Tower 10, D-54634 Bitburg
Tel: (+49 6561) 950 50
Fax: (+49 6561) 95 05 50
e-mail: verwaltung@mylius-flugzeugwerke.de
Web: http://www.mylius-flugzeugwerke.de
MANAGING DIRECTOR: Albert Mylius
FINANCIAL DIRECTOR: Sabine Palzer
PRODUCTION DIRECTOR: Ludwig Pflüger
SYSTEMS ENGINEER AND INFORMATION OFFICER: Jim Byrum

Mylius Flugzeugwerk was founded in 1996 by Albert Mylius, son of Hermann Mylius whose MHK-101 design was the proof-of-concept prototype for the MBB (Bölkow) BO 209 Monsun light aircraft, last described in the 1973-74 *Jane's*. The company is developing a new range of light aircraft, based on Hermann Mylius' designs and similar in appearance to the Monsun, which employs modular, interchangeable structures for manufacturing efficiency, minimal spares stockage and low maintenance down-time. All versions will be certified initially to JAA standards and later to FAR Pt 23, but development efforts are currently concentrated on the two-seat MY-103.

VERIFIED

MYLIUS MY-103
TYPE: Aerobatic two-seat sportplane.
PROGRAMME: Two-place development of MY-102 (see 1999-2000 *Jane's*); design began 1996. Prototype (MY-103/200 D-ETMY) exhibited unflown at Aero '97 at Friedrichshafen in April 1997; first flight 23 May 1998. Construction of first of five preproduction aircraft began 1998; first (D-ENMY) exhibited unflown at Aero '01, Friedrichshafen, April 2001, showing detail changes to engine cowling and canopy; first flight 10 July 2001.
CURRENT VERSIONS: **MY-103/180 Standard:** Lycoming IO-360 engine.

MY-103/200 Basic Trainer: Commercial airline or military pilot trainer, also suitable for glider/banner towing; Lycoming AEIO-360 engine; enlarged canopy for helmet clearance in accordance with MIL-STD 1333B.
CUSTOMERS: Four on order by mid-2001; five built and further five under assembly by that time. However, only first production aircraft had been registered by early 2002.
COSTS: DM250,000, plus tax (2001).
DESIGN FEATURES: Intended as an affordable aerobatic aircraft, trainer and glider/banner tug for private or club use. Design goals for this and MY-104 include high commonality of parts, responsive controls, stable and predictable handling, high payload/empty weight ratio and low noise emissions. Low-wing monoplane. Tapered wing with upturned tips; swept fin with dorsal strake.
FLYING CONTROLS: Conventional ailerons and rudder, manually actuated; all-moving tailplane. Ailerons and elevator are pushrod-operated; rudder cable-operated; electrically actuated flaps. Ailerons and flaps are each of

First preproduction Mylius MY-103/200 *(Jane's/Paul Jackson)* NEW/0132410

approximately half-span; electrically actuated anti-servo tab on tailplane; external mass balance on each side of rudder.
STRUCTURE: All-metal (sheet aluminium and steel), apart from some carbon fibre/glass fibre in non-structural components. Wing has single-piece main spar with D-section nose skin forming part of the load-bearing structure; flaps and ailerons hinged to rear auxiliary spar. Wing attached to fuselage by two main and one auxiliary bolts. Rectangular cross-section fuselage of stringer/aluminium skin construction with stainless steel firewall.
LANDING GEAR: Tricycle type; spring steel main legs are non-retractable; pneumatically damped nosewheel steerable and retracts rearwards with electrical actuation. Streamlined fairings on mainwheels. Parker/Cleveland wheel assemblies and 30-239 disc brakes. Mainwheel tyre size 6.00-5; nosewheel 5.00-5; tyre pressures (all) 2.76 to 3.10 bar (40 to 45 lb/sq in).
POWER PLANT: *MY-103/180 Standard* has one 134.2 kW (180 hp) Textron Lycoming IO-360 flat-four driving an MTV four-blade constant-speed propeller.
  MY-103/200 Basic Trainer has one 149 kW (200 hp) Textron Lycoming AEIO-360 flat-four driving an MTV-22-B/C174-12 four-blade constant-speed propeller. Two- and three-blade propellers optional.
  Fuel in two wing leading-edge tanks, total capacity 210 litres (55.5 US gallons; 46.2 Imp gallons), of which 205 litres (54.2 US gallons; 45.1 Imp gallons) are usable. Gravity refuelling. Oil capacity 7.6 litres (2.0 US gallons; 1.7 Imp gallons) optional. Inverted engine lubricating system optional.
ACCOMMODATION: Two, side by side under large rearwards-sliding, Plexiglas bubble canopy with fixed windscreen. Baggage space behind seats. Cockpit ventilated and heated. Dual controls standard.
SYSTEMS: Electrical system includes 24 V 13.6 Ah battery. Hydraulic system, 69 bar (1,000 lb/sq in), for brakes.

AVIONICS: VFR or IFR avionics to customer's choice, installed in modular instrument panel facilitating maintenance and replacement.
DIMENSIONS, EXTERNAL:
Wing span	8.65 m (28 ft 4½ in)
Wing aspect ratio	7.2
Length overall	6.50 m (21 ft 4 in)
Height overall	2.34 m (7 ft 8¼ in)

AREAS:
Wings, gross	10.45 m² (112.5 sq ft)

WEIGHTS AND LOADINGS (A: MY-103/180; B: MY-103/200):
Weight empty: A	583 kg (1,285 lb)
B	620 kg (1,367 lb)
Max fuel weight: A, B	151 kg (333 lb)
Max T-O and landing weight (Normal category):	
A, B	950 kg (2,095 lb)
Max wing loading: A, B	90.91 kg/m² (18.62 lb/sq ft)
Max power loading: A	7.08 kg/kW (11.63 lb/hp)
B	6.38 kg/kW (10.48 lb/hp)

PERFORMANCE (A, B as above):
Never-exceed speed (V$_{NE}$):	
A, B	180 kt (333 km/h; 207 mph)
Max cruising speed at 2,440 m (8,000 ft):	
A, B	150 kt (278 km/h; 173 mph)
Econ cruising speed at 3,660 m (12,000 ft):	
	140 kt (259 km/h; 161 mph)
Stalling speed, power off, flaps down:	
A	61 kt (113 km/h; 71 mph)
Manoeuvring speed: A, B	144 kt (267 km/h; 166 mph)
Max rate of climb at S/L: A	360 m (1,181 ft)/min
B	366 m (1,200 ft)/min
Service ceiling: B	6,100 m (20,000 ft)
T-O run: B	300 m (985 ft)
T-O to 15 m (50 ft): B	500 m (1,640 ft)
Landing from 15 m (50 ft): B	500 m (1,640 ft)
Landing run: B	250 m (820 ft)
Range with max fuel, VFR reserves:	
B	850 n miles (1,574 km; 978 miles)
g limits: A, B	+6/−4.5

UPDATED

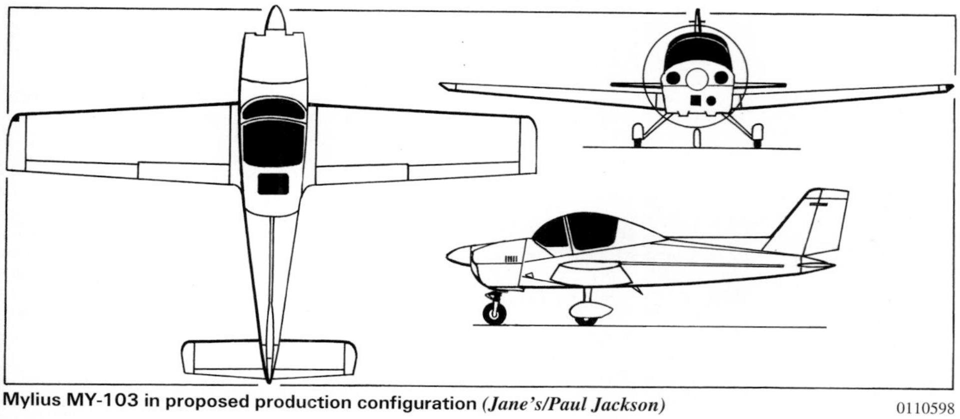

MYLIUS MY-104/200
TYPE: Four-seat lightplane.
PROGRAMME: Registration D-ETMJ reserved in 1998. First flight was planned in 2000, but has been postponed to 2003/04.
Description as for MY-103 except as follows:
DESIGN FEATURES: Wing span increased by means of new centre-section with dihedral only on outer panels; lengthened fuselage and fully retractable landing gear.
FLYING CONTROLS: Flaps have increased span.
LANDING GEAR: Retractable tricycle type; electric actuation; main legs retract inwards, nosewheel rearwards.
POWER PLANT: One 149 kW (200 hp) Textron Lycoming IO-360 flat-four driving an MTV three-blade constant-speed propeller. Fuel in two wing tanks, total capacity 276 litres (72.9 US gallons; 60.7 Imp gallons).

Mylius MY-103 in proposed production configuration *(Jane's/Paul Jackson)* 0110598

Model of the Mylius MY-104/200
(Jane's/Paul Jackson)   0064475

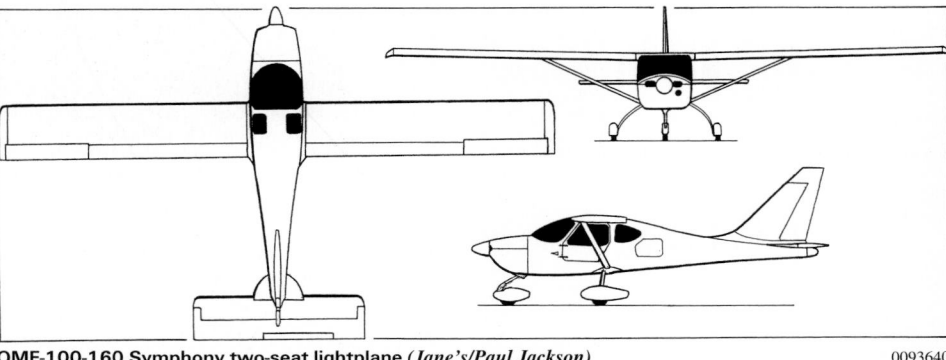

Mylius MY-104/200 four-seat tourer *(Jane's/James Goulding)*   0064481

ACCOMMODATION: Four in two pairs, under large rearward-sliding bubble canopy with fixed windscreen. Cockpit is heated and ventilated.

DIMENSIONS, EXTERNAL:

Wing span	10.40 m (34 ft 1¼ in)
Wing aspect ratio	8.0
Length overall	7.50 m (24 ft 7¼ in)
Height overall	2.38 m (7 ft 9¾ in)

AREAS:

Wings, gross	13.47 m² (145.0 sq ft)

WEIGHTS AND LOADINGS:

Weight empty	679 kg (1,498 lb)
Max fuel	200 kg (441 lb)
Payload with max fuel	318 kg (700 lb)
Max T-O weight (Normal category)	1,258 kg (2,774 lb)
Max wing loading	93.4 kg/m² (19.13 lb/sq ft)
Max power loading	8.44 kg/kW (13.87 lb/hp)

PERFORMANCE (estimated):

Never-exceed speed (VNE)	216 kt (400 km/h; 248 mph)
Normal cruising speed	148 kt (274 km/h; 170 mph)
Manoeuvring speed	135 kt (250 km/h; 155 mph)
Landing speed	57 kt (107 km/h; 66 mph)
Max landing crosswind limit	22 kt (41 km/h; 25 mph)
Max rate of climb at S/L	348 m (1,142 ft)/min
Service ceiling	5,490 m (18,000 ft)
Range with max fuel, 15 min reserves	986 n miles (1,826 km; 1,135 miles)
Endurance at 65% power	7 h 0 min
g limits, Utility	+4.4/−2.4

VERIFIED

OMF

OSTMECHLENBURGISCHE FLUGZEUGBAU GmbH

Flughafenstrasse, D-17039 Trollenhagen
Tel: (+49 395) 42 56 00
Fax: (+49 395) 425 60 20
e-mail: info@omf-aircraft.com
Web: http://www.omf-aircraft.com

DIRECTORS:
   Derek Stinnes
   Mathias Stinnes

OMF acquired aerodynamic technology for the Stoddard-Hamilton GlaStar (which see in US section under GlaStar LLC) from Arlington Aircraft Developments in March 1998. Company's 1,500 m² (16,146 sq ft) factory on Trollenhagen airfield at Neubrandenburg was opened on 10 May 2000. In September 2001, the company had 54 employees; it hopes to expand to 300 by 2004.

During 2001, OMF entered into partnership with AMD of Eastman, Georgia, to increase production and marketing in the USA. Collaborative agreements with several specialist firms in Europe and US allow OMF to accelerate production to meet demand.

UPDATED

OMF-100-160 SYMPHONY

TYPE: Two-seat lightplane.
PROGRAMME: Redesign of GlaStar to JAR 23 certification standards began in May 1998. First prototype (D-ETCW, c/n P1) was essentially unchanged and first flew May 1999; second prototype D-EMVP (c/n 0001), with modifications to meet JAR 23, first flown 7 October 1999; third aircraft (D-ENVG) had not flown by June 2000, when it made type's public debut at Berlin Air Show. LBA evaluation began mid-July 2000 and certification and production certificate both awarded 29 August 2000; FAA certificate A43CE awarded 9 April 2001. Customer deliveries began November 2001.

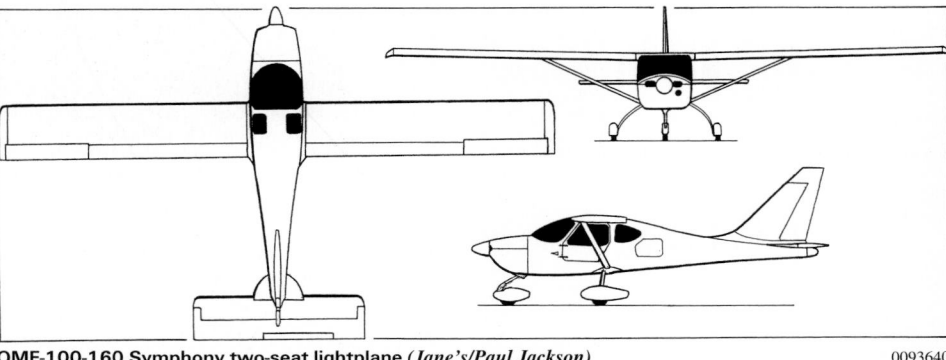

OMF-100-160 Symphony two-seat lightplane *(Jane's/Paul Jackson)*   0093640

CURRENT VERSIONS: Base model is two-seater; three- and four-seat versions are planned, plus diesel-powered and IFR-capable variants.
CUSTOMERS: By October 2000, deposits held for 37 aircraft and production positions reserved for further 52, mainly by North American customers; some 22 production aircraft delivered by March 2002, many in USA. Order for 10 from Chile, where certification received February 2002. Company has a target production rate of 300 aircraft per year.
COSTS: US$120,000 equipped (2001).
DESIGN FEATURES: Intended to be an inexpensive trainer and lightplane. Completely redesigned landing gear and wing; retains only some 10 per cent commonality with GlaStar. Crashworthy cockpit and seats. Around 60 per cent of aircraft built by OMF; remainder sourced from abroad, including Czech Republic (aluminium flying surfaces), Poland, France, UK and USA. Service life of 18,000 hours. Braced, unswept high wing of constant chord; foldable for storage. Tailplane has curved, leading-edge root strakes; tall, sweptback fin. Centre-section of top wing thickened in late 2001, increasing maximum T-O weight.

FLYING CONTROLS: Conventional and manual. Mass-balanced Frise ailerons with three hinges each (compared to GlaStar's two), travel up 23°/down 17°; electrically operated, metal, three-position Fowler flaps (0, 20 and 40°). Horn-balanced, single-piece elevator with anti-balance tab, travel up 21°/down 20°; horn-balanced rudder, travels 21° in each direction. Vortex generators added to wings in front of ailerons and at wing roots.
STRUCTURE: Fuselage of GFRP with 4130 steel tube frame strengthening in cockpit area. Metal wings, struts, flaps, ailerons, tailplane, elevator and rudder; composites wingtips, fin fillet, tailplane strakes and wheel fairings.
LANDING GEAR: Tricycle type; fixed. Castoring nosewheel. Sprung steel mainwheel legs; Cleveland hydraulic brakes. All wheels 5.00-5 and have speed fairings. Nosewheel turning circle 6.7 m (22 ft).
POWER PLANT: One 119 kW (160 hp) Textron Lycoming O-320-D2A flat-four, driving a two-blade fixed-pitch MT-186R-140-3D P-244-3 propeller. Fuel in two main tanks, each 58.5 litres (15.45 US gallons; 12.9 Imp gallons), and one 6 litre (1.6 US gallon; 1.3 Imp gallon) feeder tank, total capacity 123 litres (32.5 US gallons; 27.1 Imp gallons) of which 110 litres (29.0 US gallons; 24.2 Imp gallons) usable. Oil capacity 7.6 litres (2.0 US gallons; 1.7 Imp gallons).
ACCOMMODATION: Two persons side by side on leather-covered glass fibre-reinforced composites material seats capable of withstanding 26 g forward and 19 g vertical crash; front-hinged door each side; baggage compartment behind crew with separate door on port side as alternative access.
SYSTEMS: Electric system 24 V 70 A DC. Vision Microsystems VM1000 engine monitoring system.
AVIONICS: Bendix/King KX 125 nav/com, KLX 135 GPS/com and KT 76 transponder as standard; Skyforce colour or black and white GPS optional.

DIMENSIONS, EXTERNAL:

Wing span	10.67 m (35 ft 0 in)
Wing chord, constant	1.12 m (3 ft 8 in)
Wing aspect ratio	9.5
Width, wings folded	2.43 m (7 ft 11¾ in)
Length overall	6.96 m (22 ft 10 in)
Max width of fuselage	1.22 m (4 ft 0 in)
Height overall	2.82 m (9 ft 3 in)
Tailplane span	3.28 m (10 ft 9 in)
Wheelbase	1.73 m (5 ft 8 in)
Propeller diameter	1.85 m (6 ft 1 in)

Production OMF-100-160 Symphony with appropriate US registration *(Jane's/Paul Jackson)*   ***NEW*/0137955**

Propeller ground clearance	0.18 m (7 in)		
Passenger door: Height	0.80 m (2 ft 7½ in)		
Max width	0.81 m (2 ft 7¾ in)		
Height to sill	0.84 m (2 ft 9 in)		
Baggage door: Height	0.45 m (1 ft 5¾ in)		
Max width	0.44 m (1 ft 5¼ in)		
Height to sill	0.84 m (2 ft 9 in)		

DIMENSIONS, INTERNAL:
Cabin: Length	1.22 m (4 ft 0 in)
Max width	1.10 m (3 ft 7¼ in)
Max height	1.14 m (3 ft 9 in)
Floor area	1.34 m² (14.4 sq ft)
Volume	1.53 m³ (54 cu ft)
Baggage compartment: volume	0.06 m³ (2.1 cu ft)

AREAS:
Wings, gross	11.93 m² (128.4 sq ft)
Tailplane	2.81 m² (30.25 sq ft)

WEIGHTS AND LOADINGS:
Weight empty	658 kg (1,450 lb)
Baggage capacity	75 kg (165 lb)
Max T-O weight	975 kg (2,150 lb)
Max wing loading	81.8 kg/m² (16.74 lb/sq ft)
Max power loading	8.18 kg/kW (13.44 lb/hp)

PERFORMANCE:
Never-exceed speed (V~NE~)	162 kt (300 km/h; 186 mph)
Max level speed	105 kt (194 km/h; 121 mph)
Max cruising speed at 75% power	
	131 kt (243 km/h; 151 mph)

Econ cruising speed at 2,440 m (8,000 ft)	
	130 kt (240 km/h; 149 mph)
Stalling speed, power off:	
flaps up	57 kt (106 km/h; 66 mph)
flaps down	48 kt (89 km/h; 56 mph)
Max rate of climb at S/L	259 m (850 ft)/min
Service ceiling	5,000 m (16,400 ft)
T-O run	280 m (919 ft)
T-O to 15 m (50 ft)	350 m (1,150 ft)
Landing from 15 m (50 ft)	450 m (1,480 ft)
Landing run	230 m (755 ft)
Range with max fuel	522 n miles (966 km; 600 miles)
g limits	+3.8/−1.5

UPDATED

PC-FLIGHT

PC-FLIGHT FLUGZEUGBAU GmbH

Dr Herrmannstrasse 4, D-87719 Mindelheim
Tel/Fax: (+49 8261) 204 55
e-mail: calin.gologan@pc-aero.de
Web: http://home.t-online.de/home/calin.gologan/pretty
PRESIDENT AND DESIGNER: Dipl Ing Calin Gologan

PC-Flight derived its name from the initials of its founders, designer Calin Gologan and test pilot Prof Dr Peter Maderitch. The company's first product, the Pretty Flight, has been built in Romania from 1996 and assembled in Germany by Nitsche Flugzeugbau GmbH (now Aircraft Philipp GmbH).

UPDATED

PC-FLIGHT PRETTY FLIGHT

TYPE: Side-by-side ultralight/kitbuilt.
PROGRAMME: Prototype (D-MNPF) first flew in November 1996; first flight of production aircraft 10 November 1997; German certification achieved 25 September 1998; four prototypes then completed; manufacture of first batch of 10 series production aircraft began in November 1998, but no further information received, although promotion was continuing in 2002.
CURRENT VERSIONS: **Ultralight:** MTOW 450 kg (992 lb); *as described.*
   Experimental: Projected version in Experimental category with 550 kg (1,212 lb) MTOW.
COSTS: DM88,888 (1999).
DESIGN FEATURES: Metal structure for high utilisation and reliability; available only in complete form. Designed using company's own CAD programme. High-wing braced monoplane with upturned wingtips, sweptback fin, and tailplane tip fences. Quick-build kit version also

Third prototype PC-Flight Pretty Flight (*Jane's/Paul Jackson*)   0064483

available, with quoted build time of 300 hours. Wing section GAW PC-1.
FLYING CONTROLS: Manual. Conventional horn-balanced rudder; 75 per cent span flaperons; and all-moving tailplane with anti-balance tabs.
STRUCTURE: Predominantly metal, with fabric-covered rear wing panels and control surfaces. Production in Romania at Star Tech Impex SRL, with final assembly by Nitsche Flugzeugbau GmbH (which see) in Germany.

LANDING GEAR: Tricycle type, with cantilever spring main legs and optional speed fairings.
POWER PLANT: One 59.6 kW (79.9 hp) Rotax 912 UL flat-four driving a Fiti Speed three-blade fixed-pitch propeller. Fuel in two tanks, combined capacity 100 litres (26.4 US gallons; 22.0 Imp gallons).
EQUIPMENT: BRS 1050 or USH 520 parachute recovery system.

DIMENSIONS, EXTERNAL:
Wing span	10.00 m (32 ft 9¾ in)
Length overall	6.25 m (20 ft 6 in)
Height overall	2.57 m (8 ft 5¼ in)

DIMENSIONS, INTERNAL:
Cabin max width	1.16 m (3 ft 9¾ in)

AREAS:
Wings, gross	11.66 m² (125.5 sq ft)

WEIGHTS AND LOADINGS:
Weight empty, including parachute	290 kg (639 lb)
Max T-O weight	450 kg (992 lb)

PERFORMANCE:
Max level speed	113 kt (210 km/h; 130 mph)
Cruising speed at 75% power	
	100 kt (185 km/h; 115 mph)
Stalling speed	34 kt (62 km/h; 39 mph)
Max rate of climb at S/L	300 m (984 ft)/min
T-O run	60 m (197 ft)
Range	702 n miles (1,300 km; 807 miles)

UPDATED

PC-Flight Pretty Flight (Rotax 912 engine) (*Jane's/James Goulding*)   0064484

PHILIPP

AIRCRAFT PHILIPP GmbH

Windseestrasse 40, D-83246 Unterwössen
Tel: (+49 8641) 69 00 26
Fax: (+49 8641) 69 00 27
e-mail: info@aircraft-philipp.de
Web: http://www.aircraft-philipp.de

PRODUCTION UNIT: Streichenweg 21, D-83246 Unterwössen

In mid-2001, Philipp terminated production of the Samburo motor glider in order to concentrate on manufacture of aerostructures for Airbus, Eurocopter and others. Current workforce is eight and factory area 1,200 m² (12,900 sq ft).

UPDATED

PHILIPP AVo 68-R SAMBURO

Details of this motor glider last appeared in the 2002-03 *Jane's All the World's Aircraft.*

UPDATED

REMOS

REMOS AIRCRAFT GmbH

Waldweg 1, D-85283 Eschelbach
Tel: (+49 8442) 96 77 77
Fax: (+49 8442) 96 77 96
e-mail: aircraft@remos.com
Web: http://www.remos.com
PRESIDENT: Lorenz Kreitmayr

Remos company formed in 1983, and Remos Aircraft in 1990; currently markets the G-3 Mirage and announced the

G-4 at Friedrichshafen in April 2001. Non-aviation activities of the parent company involve production of automotive components, including an enclosed road trailer for the Mirage.

UPDATED

REMOS G-3 MIRAGE

TYPE: Side-by-side ultralight.
PROGRAMME: Prototype (D-MRAE) exhibited at Friedrichshafen Aero '97, April 1997. First flight 20 September 1997. First prize in 1998 German two-seat

ultralight championships. Production version, shown in April 1999, featured rudder horn balance and minor changes to ailerons and elevator. **Mirage S**, certified 16 August 1999 with 74.6 kW (100 hp) Rotax 912 ULS, has 113 kt (210 km/h; 130 mph) cruising speed. In June 2000, D-MPCJ shown (then unflown) at ILA, Berlin, with 53.7 kW (72 hp) Swiss Auto SAB 430 turbocharged, two-cylinder, four-stroke motorcar engine, which installation weighs 30 kg (66 lb) less than Rotax. **Mirage RS**, with redesigned rudder and landing gear, plus addition of roof window and more extensive instrumentation, introduced at Aero '01, April 2001.

Wing folding on a Remos G-3 Mirage
(Jane's/Paul Jackson)   NEW/0132413

Remos Mirage prototype, converted to latest G-3 RS standard *(Jane's/Paul Jackson)*   NEW/0132412

CUSTOMERS: Over 100 sold by 31 December 2001, when 73 on German ultralight register; had increased to 112 by May 2002. Exports include Argentine police and undisclosed military agency in Romania; two to Netherlands in 1998-99.

COSTS: €59,500, excluding tax (2002).

DESIGN FEATURES: Meets FAR Pt 23. Intended for recreational and instructional use. Waisted rear fuselage, sweptback fin and braced wing with tapered outer panels. Wings fold for transportation and storage.

FLYING CONTROLS: Conventional and manual. Single-piece elevator with flight-adjustable inset tab. Horn-balanced rudder, ailerons and elevator. No rudder tab. Horn-balanced control surfaces on production version. Electrically operated flaps and trim.

STRUCTURE: Composites fuselage; wings of sandwich composites; leading edges and tips composites, remainder fabric-covered; all tail surfaces composites. Tail bumper. GFRP ailerons and flaps. Composites parts produced in Poland; metal parts in Germany.

LANDING GEAR: Tricycle type; fixed. Cantilever main gear with hydraulic drum brakes and integral speed fairings; rubber-in-compression nosewheel leg with speed fairing. Mainwheels 4.00-6; nosewheel 4.00-6.

POWER PLANT: One 59.6 kW (79.9 hp) Rotax 912 UL four-stroke engine, driving a GT-2/169.5 two-blade wooden propeller. Optionally, a 73.5 kW (98.6 hp) Rotax 912 ULS can be fitted. Fuel capacity: normal 57 litres (15.0 US gallons; 12.5 Imp gallons); optional 70 litres (18.5 US gallons; 15.4 Imp gallons).

AVIONICS: To customer's specification.

EQUIPMENT: Junkers or BRS ballistic parachute.

DIMENSIONS, EXTERNAL:
Wing span	9.80 m (32 ft 1¾ in)
Length	6.47 m (21 ft 2¾ in)
Height overall	1.70 m (5 ft 7 in)

AREAS:
Wings, gross	12.04 m² (129.6 sq ft)

WEIGHTS AND LOADINGS:
Weight empty	284 kg (626 lb)
Max T-O weight	450 kg (992 lb)

Remos G-3 Mirage *(Jane's/James Goulding)*   0089515

PERFORMANCE, POWERED (A: Rotax 912UL, B: Rotax 912 ULS):
Max level speed	105 kt (195 km/h; 121 mph)
Normal cruising speed: A	97 kt (180 km/h; 112 mph)
B	105 kt (195 km/h; 121 mph)
Stalling speed	34 kt (63 km/h; 40 mph)
Max rate of climb at S/L: A	390 m (1,280 ft)/min
B	348 m (1,141 ft)/min
T-O run	60 m (197 ft)
T-O to 15 m (50 ft)	180 m (590 ft)
Landing from 15 m (50 ft)	200 m (660 ft)
Range more than	540 n miles (1,000 km; 621 miles)
g limits	+4/−2

PERFORMANCE, UNPOWERED:
Best glide ratio	17
Min rate of sink	1.80 m (5.91 ft)/s

UPDATED

REMOS G-4 VELOCITY

TYPE: Side-by-side ultralight.

PROGRAMME: Remos revealed artist's impression of its new G-4 high-performance, mid-wing ultralight at Aero '01, Friedrichshafen, in April 2001. The tricycle-gear aircraft, which is generally based on the existing G-3 Mirage, but with repositioned wings and tailplane, will have a 9.00 m (29 ft 6¼ in) wing span, be 6.50 m (21 ft 4 in) long and have an empty weight of 290 kg (639 lb). Fitted with a 73.5 kW (98.6 hp) Rotax 912 ULS, planned cruising speed is 126 kt (233 km/h; 145 mph) with a never-exceed speed (VNE) of 154 kt (285 km/h; 177 mph) and maximum sea level rate of climb of 395 m (1,300 ft)/min. By early 2002, Remos was predicting a maiden flight in November of that year. Projected cost is DM160,000 per aircraft, including tax, avionics and GPS, with the first five being discounted to DM129,000 (2001 prices).

UPDATED

Model of the Remos G-4 *(Jane's/Paul Jackson)*   0110632

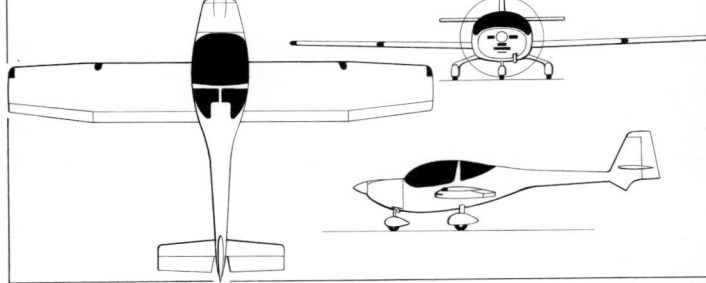

Remos G-4 general arrangement *(Jane's/James Goulding)*   NEW/0137228

RMT

RMT AVIATION GmbH & Co KG

Dorfstrasse 16 A, D-15913 Siegadel
e-mail: rmtaviation@aol.com or rmtaviation@hotmail.com
Web: http://www.rmtaviation.com
MANAGING DIRECTOR: André von Schoenebeck

RMT Aviation markets the Bateleur kit, designed by its managing director, having moved to Germany from South Africa in 2001. In 2002 RMT was building a 2,500 m² (26,900 sq ft) plant in South Africa. A European facility will be located at Bremgarten aerodrome, Germany, while plans for a North American plant were then under negotiation. RMT envisages production of over 200 Bateleurs per year by 2006.

NEW ENTRY

RMT 03 BATELEUR
English name: Impostor

TYPE: Tandem-seat ultralight kitbuilt.

PROGRAMME: Derivative of Richter Delta Dart (last described in 1998-99 *Jane's*). Construction of prototype began in August 1993; this (ZU-AND; initially registered as Delta Dart) first flew on 11 September 1994; work on first production aircraft began in May 1996. Design refinement work undertaken by Pretoria University. South African ultralight certification 23 June 1997. By late 2002, ZU-AND had returned to South Africa, been withdrawn from use and placed on static display at Cato Ridge.

CURRENT VERSIONS: **Bateleur 100:** Baseline kit version, powered by 73.5 kW (98.6 hp) Rotax 912 ULS.
Bateleur 115T: Higher-powered version, *as described*.
VLA: Higher-weight version for VLA category. Engine 97 to 112 kW (130 to 150 hp).

Sport: Intended for US Sport class; fixed landing gear; 560 kg (1,234 lb) MTOW.

COSTS: Bateleur 100 kit €78,800 (2002).

DESIGN FEATURES: Deisgned for ease of handling, manoeuvrability and quiet approach. Propeller and engine shielded for damage through positioning above the wing. Airframe life 25 years. Pusher layout with large, low-positioned delta wing, wingtip fins and close-coupled midships canard.

FLYING CONTROLS: Hydraulic. Ailerons; twin rudders (operating outwards only); elevators on wing. Electrically actuated flaps on canards also used for lateral trim.

STRUCTURE: Glass fibre, carbon fibre and Aromite.

LANDING GEAR: Retractable tricycle type; mainwheels retract inwards and nosewheel rearwards. Pneumatic actuation; emergency gravity deployment. Mainwheel tyres size

41 cm (16 in), nosewheel tyres size 33 cm (13 in). Hydraulic disc brakes.

POWER PLANT: One 84.6 kW (113.4 hp) Rotax 914S turbocharged engine, driving Rospeller three-blade variable-pitch propeller. Fuel capacity 76 litres (20.1 US gallons; 16.7 Imp gallons) of which 74 litres (19.5 US gallons; 16.3 Imp gallons) are usable. Additional, optional 35 litre (9.2 US gallon; 7.7 Imp gallon) long-range tank between wheel wells. Oil capacity 4.0 litres (1.1 US gallons; 0.9 Imp gallon).

ACCOMMODATION: Pilot and passenger in separate tandem cockpits with dual controls. Pilot's canopy opens forward; that of passenger sideways. Baggage compartment behind passenger seat.

SYSTEMS: 8 bar (116 lb/sq in) pneumatic system operates landing gear and canopy. 12 V 25 Ah battery.

EQUIPMENT: Optional ballistic parachute.

DIMENSIONS, EXTERNAL:

Wing span	6.25 m (20 ft 6 in)
Wing chord: at root	2.50 m (8 ft 2½ in)
at tip	0.93 m (3 ft 0½ in)
Canard span	4.765 m (15 ft 7½ in)
Length: fuselage	4.335 m (14 ft 2½ in)
except noseprobe	5.90 m (19 ft 4¼ in)
overall	6.02 m (19 ft 9 in)
Height overall	1.50 m (4 ft 11 in)
Wheelbase	1.70 m (5 ft 7 in)

AREAS:

Wings, gross	11.25 m² (121.1 sq ft)
Canard	3.75 m² (40.36 sq ft)

WEIGHTS AND LOADINGS:

Weight empty	250 kg (512 lb)
Max T-O weight: ultralight	450 kg (992 lb)
VLA	600 kg (1,322 lb)

PERFORMANCE (Rotax 914 engine):

Never-exceed speed (Vne):	172 kt (320 km/h; 198 mph)
Max level speed	151 kt (280 km/h; 174 mph)
Cruising speed: at 75% power	
	140 kt (260 km/h; 162 mph)
econ	124 kt (230 km/h; 143 mph)
Stalling speed, flaps down	36 kt (65 km/h; 41 mph)*
Max rate of climb at S/L (pilot plus half fuel)	
	518 m (1,700 ft)/min
Service ceiling: Rotax 912	3,660 m (12,000 ft)
Rotax 914	5,485 (18,000 ft)
T-O to 15 m (50 ft)	160 m (525 ft)
Landing from 15 m (50 ft)	100 m (330 ft)
Range with max fuel/max payload	
	755 n miles (1,400 km; 869 miles)

* No conventional stall; enters controllable sink.

NEW ENTRY

RMT 03 Bateleur making its first public appearance at Friedrichshafen in 2001 *(Jane's/Paul Jackson)*
NEW/0137956

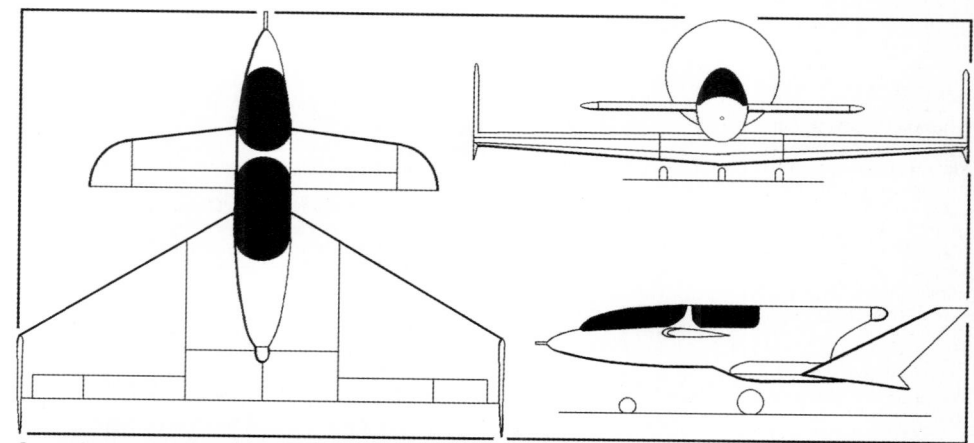

General arrangement of RMT 03 Bateleur *(Jane's/Paul Jackson)*   NEW/0530189

SCHEIBE

SCHEIBE FLUGZEUGBAU GmbH

August-Pfaltz-Strasse 23, PO Box 1829, D-85221 Dachau
Tel: (+49 8131) 720 83 and 720 84
Fax: (+49 8131) 73 69 85
e-mail: SFFlugzeug@t-online.de
Web: http://www.scheibe-flugzeugbau.de
CO-CHAIRMAN: Matthias Nährlich
CO-CHAIRMAN AND MARKETING DIRECTOR: Werner Hoffman

Scheibe has produced hundreds of gliders since it was founded in 1951 by the late Dipl Ing Egon Scheibe. Current motor glider production comprises the steel tube and wood SF 25C Falke and Rotax Falke, although the SF 36 R (1997-98 *Jane's*) was still being advertised in late 2001; manufacture of the SF 40C ultralight (2001-02 *Jane's*) has been suspended.

UPDATED

SCHEIBE SF 25C FALKE

TYPE: Motor glider.

PROGRAMME: Bergfalke sailplane first flew 5 August 1951; motorised as SF-25A Motor Falke; D-KEDO first flew May 1963. Built in several versions, including under licence by Spartavia-Pützer, Slingsby (T.61) and Loravia (France).

Current SF 25C version first flew (D-KBIK) in March 1971 and gained its type certificate in September 1972. Falke 1700, with 48.5 kW (65 hp) Limbach 1700 engine, has been discontinued. Falke 2100, with Sauer SS2100 engine, tested 1988 (D-KIAC), but only two others built.

CURRENT VERSIONS: **Falke 2000:** Limbach L 2000 EA flat-four engine; monowheel landing gear or conventional tailwheel with two mainwheels available.

Rotax Falke: Prototype D-KIAJ flown 1989. Has essentially the same performance as Falke 2000, but is powered by a water-cooled Rotax 912 A or, from 1999, Rotax 912 S flat-four engine; nosewheel landing gear optional. Most production now of this version; some earlier Falkes have been upgraded with Rotax engines. Tricycle landing gear first tested on Falke 2000 D-KBUG in 1987; deliveries from 1990.

Superschlepper: Glider tug version; has max T-O weight of 600 kg (1,323 lb), in this mode. Rotax 912 A or 912 S. Towing trials by D-KIEK in 1998-99.

Limbach-engined Scheibe SF 25C Falke with monowheel landing gear *(Jane's/Paul Jackson)*   NEW/0131700

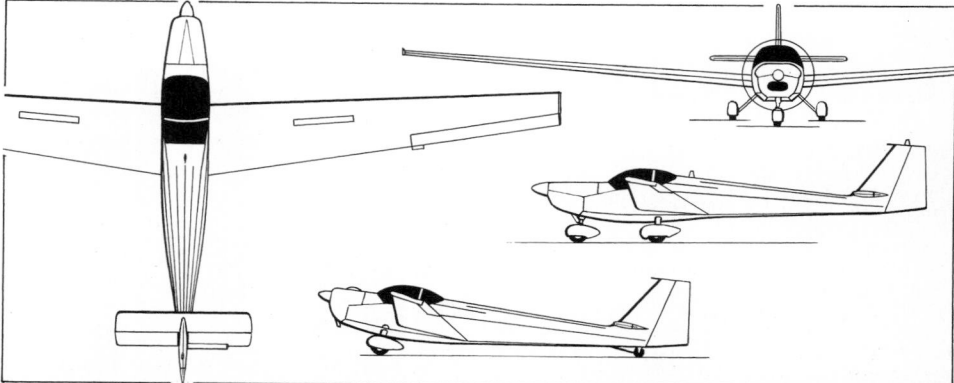

Scheibe SF 25C Falke powered by Rotax 912 A engine, with second side view (lower) of Falke 2000 with tailwheel landing gear *(Jane's/Mike Keep)*   0092142

CUSTOMERS: Over 1,234 Falkes of all types delivered by October 2001, including 673 SF-25Cs by the parent firm. More than 20 Rotax Falkes used as glider tugs.

COSTS: Falke 2000 with Limbach engine and monowheel landing gear, DM154,500; nosewheel landing gear DM5,400 extra; with Rotax 912 A2 engine DM7,900 extra (2001).

DESIGN FEATURES: Docile aerofoil with moderate aspect ratio, designed for safe and simple handling; aircraft can be easily dismantled and transported by trailer. Mid-mounted wing with taper and slight forward sweep; constant chord tailplane; sweptback fin. Various landing gear and engine options.

FLYING CONTROLS: Conventional and manual. Door-type wing spoilers at one-third span; upper wing only. Trim tab on starboard aileron; 13 mm (0.5 in) end plates on wing.

STRUCTURE: Steel tube and fabric fuselage; wooden wing and empennage; wing fabric-covered rearwards from one-third chord; fabric-covered elevators and fin.

LANDING GEAR: Standard landing gear is central monowheel, outriggers under wings, and tailwheel; two-wheel main landing gear with steerable tailwheel, and tricycle landing gear (only with Rotax engine), are optional; mainwheels and nosewheel (where fitted) 5.00-4.

POWER PLANT: Choice of three engines: a 59.6 kW (79.9 hp) Rotax 912 A with water-cooled cylinder heads, running at 5,800 rpm and turning the propeller at 2,500 rpm; a 73.5 kW (98.6 hp) Rotax 912 S running at 5,800 rpm and turning the propeller at 2,380 rpm; or a 59.7 kW (80 hp) Limbach L 2000 EA driving the propeller at 3,450 rpm. All engines have electric starter and alternator; propeller feathering optional. Glider tug has MTV-1-A/175-05 constant-speed or MT 175R130-2A fixed-pitch propeller. Standard fuel tankage is 55 litres (14.5 US gallons; 12.0 Imp gallons); optional fuel tankage 80 litres (21.1 US gallons; 17.6 Imp gallons).

ACCOMMODATION: Side-by-side seating under forward-hinged canopy; optional large cockpit opening; instrument panel space for radio.

SYSTEMS: 12 V electrical system.

AVIONICS: Choice of optional nav/com radios.

DIMENSIONS, EXTERNAL:

Wing span	15.30 m (50 ft 2¼ in)
Wing aspect ratio	12.9
Length overall	7.60 m (24 ft 11¼ in)
Height overall: tailwheel	1.68 m (5 ft 6¼ in)
nosewheel	2.50 m (8 ft 2½ in)

AREAS:

Wings, gross	18.20 m² (195.9 sq ft)

WEIGHTS AND LOADINGS (Rotax 912 A):

Manufacturer's weight empty	approx 438 kg (966 lb)
Max T-O weight	650 kg (1,433 lb)
Max wing loading	35.7 kg/m² (7.31 lb/sq ft)
Max power loading	10.90 kg/kW (17.91 lb/hp)

PERFORMANCE, POWERED (L: Limbach, A: Rotax 912 A, S: Rotax 912 S, fixed-pitch (FP) or constant-speed (CS) propeller):

Max level speed (all)	102 kt (190 km/h; 118 mph)
Cruising speed: L	81 kt (150 km/h; 93 mph)
A	92 kt (170 km/h; 106 mph)
S	97 kt (180 km/h; 111 mph)
Stalling speed (all)	35 kt (65 km/h; 41 mph)
Max rate of climb at S/L: L	192 m (630 ft)/min
A: FP	240 m (787 ft)/min
CS	288 m (945 ft)/min
S: FP	240 m (787 ft)/min
CS	330 m (1,082 ft)/min
T-O run: L	140 m (460 ft)
A	100 m (330 ft)
S	90 m (295 ft)
Range, standard fuel:	
L, A	378 n miles (700 km; 435 miles)
S	324 n miles (600 km; 372 miles)
Endurance, standard fuel:	
L, A	4-5 h
S	4 h

PERFORMANCE, UNPOWERED:

Best glide ratio	24
Min rate of sink	1.10 m (3.61 ft)/s

UPDATED

SILENCE

SILENCE AIRCRAFT

Kapellenweg 54a, D-33415 Verl
Tel: (+49 5246) 70 28 45
Fax: (+49 5246) 70 37 46
e-mail: thomasstrieker@aol.com
Web: http://www.silence-aircraft.de
PARTNERS: Matthias Strieker
  Thomas Strieker
  Michael Rudbach
  Herbert Funke

Silence's first project is an ultralight kitbuilt with flying surfaces similar in shape to those of the Supermarine Spitfire. The partners are seeking a manufacturer willing to launch production.

UPDATED

SILENCE SILENCE

TYPE: Single-seat ultralight/kitbuilt.

PROGRAMME: Developed over a period of four years in a garden gazebo. Prototype (D-MTMH, incorporating partners' initials) made first flight 30 September 2000; LBA certification received May 2002. Partners at present making kits themselves, but are looking for a subcontractor.

CUSTOMERS: Two prototypes flying by May 2002, with third aircraft then under construction.

COSTS: Kit €28,000; complete aircraft €60,000 including engine (2001). Engine cost €8,000.

DESIGN FEATURES: Compact, low-wing configuration, with elliptical wings and tailplane. Extensive use of carbon and glass fibre honeycomb construction throughout to maintain strength and minimise weight. Wings detachable in 10 minutes for transportation and storage.

FLYING CONTROLS: Conventional and manual. Plain flaps. Horn-balanced rudder.

STRUCTURE: Monocoque single-piece fuselage. Single-piece wing spar.

LANDING GEAR: Tailwheel type. Electric retraction; carbon fibre main legs retract rearwards. Tailwheel non-retractable. Mainwheels 4.00-6. Hydraulic brakes.

POWER PLANT: Engines in the region of 70 kW (93.9 hp) can be fitted, to drive a three-blade, ground-adjustable pitch Helix carbon fibre propeller; prototype has 40 kW (53.6 hp) Wankel engine; second prototype (D-MTMO) has 37.3 kW (50 hp) Mid-West AE 50 Harrier engine. Fuel capacity 80 litres (21.1 US gallons; 17.6 Imp gallons).

EQUIPMENT: Junkers ProFly ballistic parachute system.

DIMENSIONS, EXTERNAL:

Wing span	7.50 m (24 ft 7¼ in)
Length overall	6.18 m (20 ft 3¼ in)

AREAS:

Wings, gross	8.72 m² (93.9 sq ft)

WEIGHTS AND LOADINGS:

Weight empty	190 kg (419 lb)
Max T-O weight	340 kg (749 lb)

PERFORMANCE:

Never-exceed speed (VNE)	162 kt (300 km/h; 186 mph)
Cruising speed	108 kt (200 km/h; 124 mph)
Stalling speed	35 kt (65 km/h; 41 mph)
Max rate of climb at S/L	335 m (1,100 ft)/min
Range with max fuel	540 n miles (1,000 km; 621 miles)
g limits	+6/−4

UPDATED

Prototype Silence single-seat ultralight (*Jane's/Paul Jackson*)   NEW/0137730

STEMME

STEMME GmbH & Co KG

Strasse F2, Flugplatz, D-15344 Strausberg
Tel: (+49 3341) 361 20
Fax: (+49 3341) 36 12 30
e-mail: S10info@stemme.de
Web: http://www.stemme.de
PRESIDENT: Dr Reiner Stemme
TECHNICAL DIRECTOR: Dipl Ing Gottfried Freudenberger
SALES DIRECTOR: Dipl Ing Sebastian Loewer
MARKETING: Astrid Gaeckler

NORTH AMERICAN REPRESENTATIVE:
  Stemme USA Inc, 1401 S Brentwood Blvd, Suite 760, Saint Louis, Missouri 63144

Tel: (+1 314) 721 59 04
Fax: (+1 314) 726 51 14
e-mail: info@stemme.com
Web: http://www.stemme.com
PRESIDENT: Barbara Pfifferling

Founded in 1985, Stemme produces components for other manufacturers as well as its own high-performance motor gliders. Flight test and operations moved to Strausberg in 1991; factory area 3,000 m² (32,292 sq ft); 75 employees. 100th aircraft delivered September 1998. Turnover in 1999 was DM7.6 million; expected turnover for 2000 was DM8.6 million, expected to rise to euro equivalent of DM30 million in 2004; 75 per cent of products are exported. In May 2000 the company announced two new derivatives of the S 10: the all-purpose S 8 motor glider and S 7 sailplane; latter designation was revised in April 2001 to S 2, when the S 6 motor glider was added to the range. Production rate of S 10 series in 2000 was 20 per year.

UPDATED

STEMME S 10 and S 15

US marketing name: Chrysalis
US Air Force designation: TG-11A
TYPE: Motor glider.

PROGRAMME: Prototype (D-KKST), with Glaser-Dirks DG500 wing, flown 6 July 1986; first flight of second prototype (D-KCHS) with definitive wing 2 June 1987; S 10-VC surveillance and observation platform introduced in 1989; designed to JAR 22. Certified in Germany 31 December 1990, in UK 29 October 1991 and in USA, to

Stemme S 10 two-seat motor glider with propeller in operation (*Jane's/Paul Jackson*)   NEW/0131699

Solar panels on the spine of an S 10 (*Jane's/Paul Jackson*)   NEW/0143331

Proof-of-concept Stemme S 15 (*Jane's/Paul Jackson*)   *NEW*/0143325

FAR Pt 21, 8 July 1992. Production of S 10 suspended March 1994; resumed late 1994 with S 10-V.

CURRENT VERSIONS: **S 10:** Original version; 54 produced by March 1994, including prototypes.

Following description applies to S 10, except where indicated.

S 10-VC: Surveillance/observation conversion with underwing and small wingtip sensor pods for pollution control and resources investigation. One former S 10 (D-KGCM/N600PL) used by Greenpeace since May 1994; one former S 10-V (PH-1055) operated by Aerial Surveyance den Hollander, Netherlands, since 1995.

S 10-V: Variable-pitch propeller version; demonstrator (D-KGCX) completed mid-1994 as conversion of S 10 but written off in Tanzania 10 March 1995; one other converted; production version from 1994; certified September 1994; 29 built by end of 1998; none subsequently, but at least six further conversions from S 10. (S 10-V has engineering designation S 14.)

S 10-VT: Turbocharged version with Rotax 914 engine. Prototype (D-KGCR) built 1997; LBA certification August 1997, FAA certification September 1997 and CAA certification May 1998; total of at least 68 produced by January 2002; of these, 38 to USA. (S 10-VT has engineering designation S 11.) Set world soaring record of 1,328 n miles (2,459 km; 1,528 miles) on 27 November 2000.

S 15: Launched at the 1996 Berlin Air Show; has a Rotax 914 engine and a 20.00 m (65 ft 7½ in) wing span; two underwing hardpoints for sensor pods in law-enforcement or scientific research roles suitable for loads up to 65 kg (143 lb) each, including FLIR, night sun searchlights, video cameras, and environmental monitoring equipment being developed by Berlin Technical University's Geography Institute. Prototype (D-ESTE) registered 1996, but no record of completion. S 15B, light surveillance version, based on S 10-VT, was then due to enter service in 2002, followed by S 15A advanced surveillance model in 2004; both sponsored by Deutsche Bundesstiftung Umwelt. First S 10-VT (D-KGCR) shown at Berlin, June 2000 (and again in May 2002) as POC S 15.

S-UAV: Unpiloted surveillance version. One S 10-V (N600V) currently in use as testbed with Platforms International Corporation, Mojave, California.

CUSTOMERS: Owners in Australia, Austria, Belgium, Brazil, Canada, France, Germany, Mexico, Netherlands, South Africa, Spain, Switzerland, UK and USA. Two delivered in 1995 to 94th Air Training Squadron, USAF Academy, Colorado Springs, Colorado. By mid-2001, 150 S 10 variants (including prototypes) built.

COSTS: S 10-VT €181,886 (2002). S 10 and S 10-V available to special order.

DESIGN FEATURES: Shoulder-wing, T-tail sailplane with mid-mounted engine and completely retractable propeller. Transition time between soaring and powered flight is 5 seconds. Winglets option introduced from 1997. Behaviour tuned for both low and high airspeeds and docility during tight turns; wing not affected by bug accretion or water droplets. Outer wings can be folded by one person for taxying and hangarage; wing sections carried in gantries inside trailer so that one person can fit centre and outer sections straight from trailer to fuselage. Wing section Horstmann & Quast HQ41/14.35; dihedral 1°.

FLYING CONTROLS: Conventional and manual. Flap/aileron linkage for manoeuvrability and docility at low airspeeds; Schempp-Hirth-type airbrakes in outer ends of centre wing. Six-position flaps.

STRUCTURE: CFRP wings in three sections detachable from fuselage; CFRP rear fuselage and tail mounted to steel tube centre frame carrying wing, engine and landing gear; nose section, of CFRP structure with Kevlar safety lining, mounted at front of centre frame. Engine fully accessible and horizontal firewall separates engine from wing and flying controls. Winglets optional.

LANDING GEAR: Tailwheel type; inward-retracting, narrow track mainwheels and non-retractable tailwheel. Mainwheel tyres 355×122, with electrically actuated, mechanical standby disc brakes. Tailwheel: 210×65 tyre, steerable with rudder, fairing optional on S 10 and standard on S 10-VT. Larger 369×136 main tyres are optional. Tyre pressure 2.90 bar (42 lb/sq in).

POWER PLANT: *S 10* and *S 10-V:* One air-cooled 69.4 kW (93 hp) Limbach L 2400 EB1.AD flat-four engine mounted in the central fuselage just aft of cockpit; cooling by adjustable ram air intake; 1.90 m (6 ft 2¾ in) CFRP extension shaft in Kevlar tunnel drives folding, nose-mounted, two-blade, fixed-pitch (S 10) or variable-pitch (S 10-V) propeller through flexible coupling, sliding spline joint and five-belt 1.18:1 reduction gear; engine starting in flight takes 5 seconds; nosecone is moved forward and spring-folded blades emerge through peripheral slot under centrifugal force; centrifugal clutch protects against overspeed, damps starting shocks and allows propeller to slow down independently of engine on shutdown.

S 10-VT: One 84.6 kW (113.4 hp) Rotax 914 F2/S1 turbocharged liquid/air-cooled flat-four; two-blade, variable-pitch propeller, driven via 0.9:1 reduction gearbox.

Fuel (all versions) in two 45 litre (11.9 US gallon; 9.9 Imp gallon) fuel tanks in outer ends of centre wing; 60 litre (15.8 US gallon; 13.2 Imp gallon) tanks optional. Total capacities 90 litres (23.8 US gallons; 19.8 Imp gallons) or 120 litres (31.7 US gallons; 26.4 Imp gallons). US market examples have modified fuel system.

ACCOMMODATION: Two pilots side by side; dual controls standard; seats adjustable for position and rake; canopy hinged at forward end and held open by gas struts; heating provided by engine cooling air.

SYSTEMS: 12 V 26 Ah battery and generator; 35 Ah battery optional; full night lighting and landing light optional. Solar cells can be fitted behind cockpit to provide 12 V 37.5 W on S 10; is standard fitting on S 10-VT.

AVIONICS: *Comms:* Intercom, VHF radio and transponder optional, as is special cut-down panel for mountain soaring.

DIMENSIONS, EXTERNAL:

Wing span, excl winglets	23.00 m (75 ft 5½ in)
Wing aspect ratio	28.3
Width, wings folded	11.40 m (37 ft 4¾ in)
Length overall	8.42 m (27 ft 7½ in)
Fuselage max width	1.18 m (3 ft 10½ in)
Height over tailplane	1.80 m (5 ft 10¾ in)
Wheel track	1.15 m (3 ft 9¼ in)
Wheelbase	5.42 m (17 ft 9½ in)
Propeller diameter	1.63 m (5 ft 4¼ in)
Fuselage ground clearance at mainwheels	
	0.72 m (2 ft 4¼ in)

DIMENSIONS, INTERNAL:

Cockpit: Width	1.16 m (3 ft 9½ in)
Height	0.93 m (3 ft 0½ in)

AREAS:

Wings, gross	18.70 m² (201.3 sq ft)

WEIGHTS AND LOADINGS:

Weight empty: except S 10-VT	640 kg (1,411 lb)
S 10-VT	645 kg (1,422 lb)
Max T-O weight: all	850 kg (1,874 lb)
Max wing loading: all	45.5 kg/m² (9.31 lb/sq ft)
Max power loading:	
except S 10-VT	12.25 kg/kW (20.13 lb/hp)
S 10-VT	10.06 kg/kW (16.53 lb/hp)

PERFORMANCE, POWERED:

Never-exceed speed (VNE), smooth air:	
all	146 kt (270 km/h; 168 mph)
Manoeuvring speed: all	97 kt (180 km/h; 112 mph)
Cruising speed:	
S 10	89 kt (165 km/h; 102 mph)
S 10-V	121 kt (225 km/h; 139 mph)
S 10-V at FL100	140 kt (259 km/h; 161 mph)
Stalling speed: all	42 kt (78 km/h; 48 mph)
Max rate of climb at S/L:	
S 10	150 m (492 ft)/min
S 10-V	210 m (689 ft)/min
S 10-VT	240 m (787 ft)/min
Service ceiling: S 10-VT	9,140 m (30,000 ft)
T-O run: on grass: S 10	300 m (984 ft)
S 10-V	195 m (640 ft)
on concrete: S 10-VT	184 m (605 ft)
Max range with standard fuel:	
S 10	648 n miles (1,200 km; 745 miles)
S 10-V	863 n miles (1,600 km; 994 miles)
S 10-VT	696 n miles (1,290 km; 801 miles)
Max range with max optional fuel:	
S 10	863 n miles (1,600 km; 994 miles)
S 10-V	1,187 n miles (2,200 km; 1,367 miles)
S 10-VT	934 n miles (1,730 km; 1,075 miles)

PERFORMANCE, UNPOWERED (all):

Best glide ratio at 57 kt (106 km/h; 66 mph)	50
Min rate of sink at 45 kt (83 km/h; 52 mph)	
	0.57 m (1.87 ft)/s
g limits: all	+5.3/–2.65

OPERATIONAL NOISE LEVELS:

To German light aircraft rules	57.3 dBA
	UPDATED

STEMME S 6 and S 8

TYPE: Motor glider.

PROGRAMME: S 8 announced at ILA, Berlin, May 2000; S 6 at Aero '01, Friedrichshafen, April 2001. According to timetable as first announced, initial deliveries due first half of 2002 (S 6) and early 2003 (S 8). However, in May 2002, maiden flight of S 6 was scheduled for late 2002, deliveries following in 2003. Second version to be marketed is S 2, with S 8 becoming available in 2003.

CURRENT VERSIONS: **S 2:** Two-seat, high-performance training (20 m) and aerobatic (18 m) sailplane (originally designated **S 7** when announced in May 2000).

S 8: Two-seat, fixed landing gear, touring motor glider. Rotax 912S engine.

S 8T: As above, but Rotax 914F.

S 8R: Two-seat, retractable landing gear, touring motor glider. Rotax 912S engine.

S 8RT: As above, but Rotax 914F.

S 6: Two-seat, fixed landing gear, sports motor glider. Rotax 912S engine.

S 6T: As above, but Rotax 914F. Suitable for glider towing (max 750 kg; 1,653 lb).

S 6R: Two-seat, retractable landing gear, sports motor glider. Rotax 912S.

S 6RT: Retractable gear and Rotax 914F.

All powered versions have same basic dimensions.

DESIGN FEATURES: Based on the Stemme S 10 (which see), including mid-mounted engine, but with new wing profile. Propeller does not retract in flight. Can be taxied with wings folded to aid ground handling.

FLYING CONTROLS: As S 10.

STRUCTURE: Generally as S 10.

LANDING GEAR: Tricycle type. Fixed-gear versions have speed fairings on all wheels.

POWER PLANT: One 73.5 kW (98.6 hp) Rotax 912S or 84.5 kW (113.3 hp) Rotax 914F flat-four engine and two- or (914F) three-blade propeller. Standard fuel capacity of all versions 70 litres (18.5 US gallons; 15.4 Imp gallons); optional capacity 140 litres (37.0 US gallons; 30.8 Imp gallons).

ACCOMMODATION: Pilot and passenger/student side by side, under single-piece, forward-hinged canopy. Baggage compartments behind seats and in tailboom of S 8.

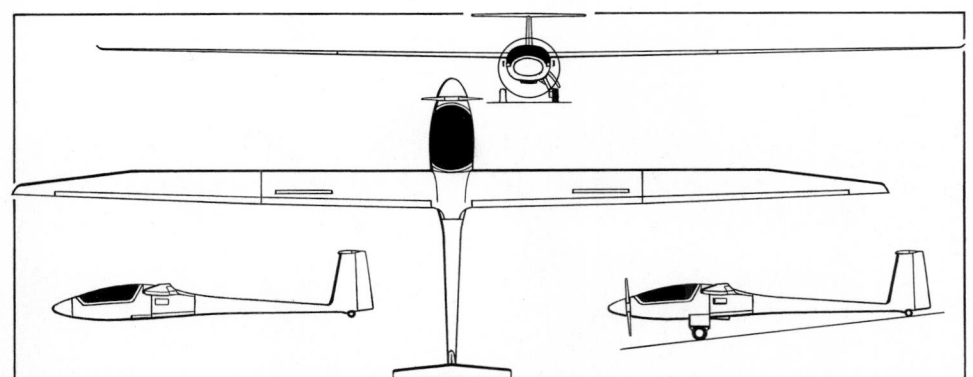

Stemme S 10 showing additional side view (left) with landing gear and propeller retracted (*Jane's/Mike Keep*)

DIMENSIONS, EXTERNAL:

Wing span	18.00 m (59 ft 0¾ in)
Wing aspect ratio	18.1
Width, wings folded	7.20 m (23 ft 7½ in)
Length overall	8.50 m (27 ft 10¾ in)
Height to top of tail	2.45 m (8 ft 0½ in)
Wheel track	2.00 m (6 ft 6¾ in)
Wheelbase	2.00 m (6 ft 6¾ in)

AREAS:

Wings, gross	17.42 m² (187.5 sq ft)

WEIGHTS AND LOADINGS:

Weight empty: S 6	577 kg (1,272 lb)
S 6R	585 kg (1,290 lb)
S 8	587 kg (1,294 lb)
S 8R	592 kg (1,305 lb)
Max T-O weight: all	850 kg (1,873 lb)
Max wing loading: all	48.8 kg/m² (9.99 lb/sq ft)
Max power loading:	
Rotax 914F	10.06 kg/kW (16.53 lb/hp)
Rotax 912S	11.56 kg/kW (19.0 lb/hp)

PERFORMANCE, POWERED (Rotax 912S):

Normal cruising speed at mid-flight weight: at 75% power:	
all	151 kt (280 km/h; 174 mph)
at 55% power: S 6R	130 kt (240 km/h; 149 mph)
S 8R	127 kt (236 km/h; 147 mph)
Max rate of climb at S/L:	
S 6R at mid-flight weight	300 m (984 ft)/min
S 8R at MTOW	268 m (879 ft)/min

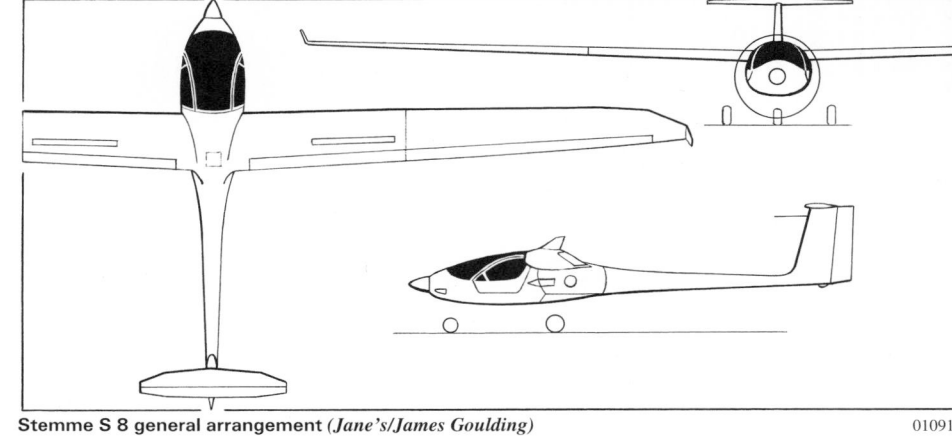

Stemme S 8 general arrangement (*Jane's/James Goulding*)   0109132

Service ceiling	3,960 m (13,000 ft)	S 6R	39
Range with max optional fuel at 55% power:		S 8	32
S 6R	1,072 n miles (1,986 km; 1,234 miles)	S 8R	38
S 8R	1,054 n miles (1,953 km; 1,213 miles)	Min rate of sink: S 6R	0.73 m (2.40 ft)/s
PERFORMANCE, UNPOWERED:		S 8R	0.74 m (2.43 ft)/s
Best glide ratio: S 6	33		**UPDATED**

Artist's impression of S 6   *NEW*/0143330

Projected Stemme S 8   *NEW*/0143328

TECHNOFLUG

TECHNOFLUG LEICHTFLUGZEUGBAU GmbH

Dr Kurt Steimstrasse 6, D-78713 Schramberg
Tel: (+49 74) 22 84 23
Fax: (+49 74) 22 87 44
e-mail: technoflug@t-online.de
Web: http://www.technoflug.de
CEO: Rolf Schmid
DESIGN ENGINEER: Berthold Karrais
INTERNATIONAL SALES MANAGER: Tom Dietrich

Technoflug is known for its sailplanes and motor gliders; for details of the company's Piccolo ultralight glider (over 120 sold) see 1989-90 *Jane's*. Its newest product, the Carat, was formally announced in 1998 and was certified on 15 July 2001; it is to be built by AMS Flight in Slovenia, which also handles production of some Glaser-Dirks sailplanes.

UPDATED

TECHNOFLUG TFK-2 CARAT

TYPE: Motor glider.
PROGRAMME: First flown 16 December 1997; flight testing carried out during 1998. Initial production aircraft displayed at Aero '99, Friedrichshafen, in April 1999. Folding propeller certified 7 November 2000; certified by LBA 15 July 2001.

CUSTOMERS: Expected total sales of 350, at around 30 per year, production to start at one or two per month, but none reported built between 1999 and 2001.
COSTS: US$69,635 (2001) without avionics.
DESIGN FEATURES: Low wing of 21.3 aspect ratio; optional winglets; T tail. Uses wings and horizontal tail (but not water tank) of Schempp-Hirth Discus sailplane, with main spar passing under pilot's knees; wings and tailplane can be detached for storage and aircraft will fit into sailplane trailer. Optional winglets.
FLYING CONTROLS: Conventional and manual. Schempp-Hirth airbrakes on upper wing surfaces and 'turbulators' below.
STRUCTURE: Generally of glass fibre and carbon fibre composites. Wings have PVC foam cores.
LANDING GEAR: Mainwheels, size 4.00-4 in, retract inwards and· forwards electrohydraulically with manual override;

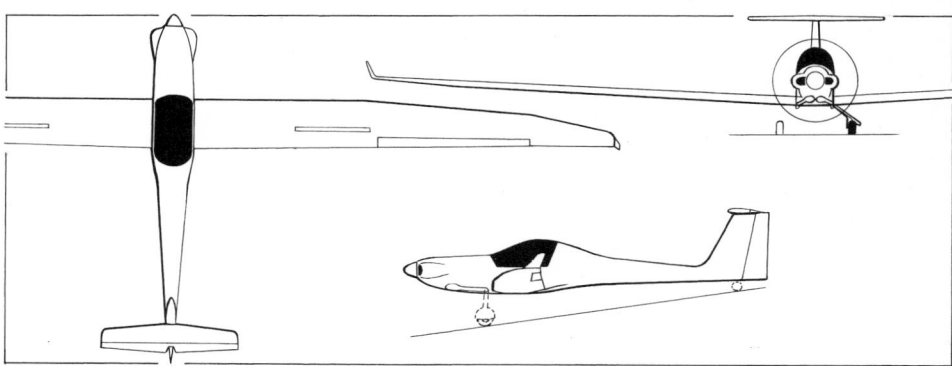

Technoflug Carat (Volkswagen motorcar engine) (*Jane's/James Goulding*)   0092145

Instrument panel of the Carat   0062971

Prototype Technoflug Carat being assembled by one person   *NEW*/0132832

fixed steerable 210×65 mm tailwheel. Hydraulic disc brakes. Parking brake.

POWER PLANT: One 1,800 cc (109.8 cu in) Sauer E1S conversion of Volkswagen four-stroke, air-cooled motorcar engine, producing 40 kW (53.6 hp) at max continuous power, driving a 1.40 m (4 ft 7 in) Technoflug KS-F3-1A/140-1 fixed-pitch propeller, blades of which are held forward by gas damping springs located under the spinner when not turning. Other planned engines include 59.6 kW (79.9 hp) Rotax 912. Fuel capacity 45 litres (12.0 US gallons; 10.0 Imp gallons) in single fuel tank situated between cockpit and engine.

SYSTEMS: Electric starter and generator; 12 V 16 Ah battery. Electric fuel pump.

AVIONICS: Suggested package includes ATR720A radio, Honeywell transponder and Filser DX50FAI GPS.

DIMENSIONS, EXTERNAL:	
Wing span	15.00 m (49 ft 2½ in)
Length overall	6.21 m (20 ft 4½ in)
DIMENSIONS, INTERNAL:	
Baggage compartment volume	0.1 m³ (3.53 cu ft)
Baggage capacity	145 kg (320 lb)
AREAS:	
Wings, gross	10.58 m² (113.9 sq ft)
WEIGHTS AND LOADINGS:	
Weight empty	325 kg (717 lb)
Max T-O weight	470 kg (1,036 lb)
PERFORMANCE, POWERED:	
Never-exceed speed (VNE)	135 kt (250 km/h; 155 mph)
Normal cruising speed at 75% power	116 kt (214 km/h; 133 mph)

Stalling speed	44 kt (80 km/h; 50 mph)
Max rate of climb at S/L	210 m (689 ft)/min
T-O run	226 m (741 ft)
Landing run	137 m (450 ft)
Range with max fuel, 30 min reserves	486 n miles (900 km; 559 miles)
g limits	+5.3/–2.65
PERFORMANCE, UNPOWERED:	
Stalling speed	44 kt (80 km/h; 50 mph)
Best glide ratio	35
Min rate of sink	0.75 m (2.46 ft)/s

UPDATED

ULBI

ULTRALEICHTBAU INTERNATIONAL

Flugplatzstrasse 18, D-97437 Hassfurt
Tel: (+49 9521) 61 83 93
Fax: (+49 9521) 61 83 95
e-mail: info@ulbigmbh.de
Web: http://www.ulbi-750.de
CEO: Angelika Kaiser

The Romanian-developed Wild Thing lightplane is marketed by Ultraleichtbau, known until 2001 as Air Light. A nosewheel version was introduced in 1999.

NEW ENTRY

ULBI WILD THING

TYPE: Side-by-side ultralight/kitbuilt.
PROGRAMME: Developed by Aerostar (which see) in Romania; prototype registered YR-6107. Launched at Friedrichshafen Air Show, April 1997; certification August 1997. Initial production aircraft had Jabiru 2200 engines; alternatives available from 1998. By early 2001, examples were operating in Austria, Czech Republic, France, Germany, Hungary and Netherlands.
CURRENT VERSIONS: **WT01**: Baseline version; tailwheel.
WT02: Nosewheel version introduced early 1999 (D-MWTA, 50th production aircraft, as demonstrator).
CUSTOMERS: Total of 65 built by April 2001.
COSTS: Quick-build kit, excluding engine, DM39,900 (2000). Power plant DM17,000 to DM26,000 extra. Factory-built DM84,500 (Jabiru 2200); DM95,000 (Rotax 912 UL); DM98,500 (Jabiru 3300), DM96,500 (Rotax 912 ULS); DM105,000 (Mid West Hawk). Nosewheel version DM5,500 extra (2001).
DESIGN FEATURES: Bears strong outward resemblance to Murphy Rebel (which see in Canadian section). Built by Aerostar in Romania for distribution out of Germany. Wings fold alongside fuselage for storage or transportation on aluminium trailer.
FLYING CONTROLS: Conventional and manual elevators and ailerons pushrod-operated, rudder cable-operated. Horn-balanced tail surfaces; trim tab in port elevator; ground-adjustable tab on rudder. Flaps.

WT02 nosewheel version of ULBI Wild Thing (*Jane's/Paul Jackson*)   *NEW*/0137226

STRUCTURE: Main structure is metal monocoque, with fabric-covered ailerons and flaps.
LANDING GEAR: Non-retractable tailwheel type with steerable tailwheel via spring links to rudder horn. Mainwheel size 8.00-6, tailwheel size 200×50. Oversize Tundra wheels and tyres optional. Cable-operated disc brakes. Nosewheel (rubber-in-compression, with repositioned mainwheel legs) optional; nosewheel size 140×6.
POWER PLANT: One 59.7 kW (80 hp) Jabiru 2200, 59.6 kW (79.9 hp) Rotax 912 UL, 78.3 kW (105 hp) Mid West Hawk, 89.5 kW (120 hp) Jabiru 3300 or 73.5 kW (98.6 hp) Rotax 912 ULS driving a Junkers ground-adjustable pitch, carbon fibre or Kremen SR 2000 XC three-blade, variable-pitch (electric) wooden propeller. Fuel in two wing tanks, combined capacity 80 litres (21.1 US gallons; 17.6 Imp gallons).
ACCOMMODATION: Prototype had third seat in rear of cabin, hence a spacious cockpit area.

AVIONICS: Becker, Filser and Icom avionics to customer's choice.
EQUIPMENT: Junkers Magnum speed parachute emergency rescue system and gliding towing equipment both optional.

DIMENSIONS, EXTERNAL:	
Wing span	9.20 m (30 ft 2¼ in)
Length overall	6.50 m (21 ft 4 in)
Height overall	1.92 m (6 ft 3½ in)
AREAS:	
Wings, gross	13.94 m² (150.0 sq ft)
WEIGHTS AND LOADINGS:	
Weight empty (typical)	300 kg (661 lb)
Max T-O weight	450 kg (992 lb)
PERFORMANCE (Jabiru 2200 engine):	
Never-exceed speed (VNE)	108 kt (200 km/h; 124 mph)
Max level speed	86 kt (160 km/h; 99 mph)
Cruising speed	76 kt (140 km/h; 87 mph)
Stalling speed	32 kt (58 km/h; 36 mph)
Max rate of climb at S/L	152 m (500 ft)/min
T-O run	98 m (320 ft)
Landing run	44 m (145 ft)
Range with max fuel	432 n miles (800 km; 497 miles)

UPDATED

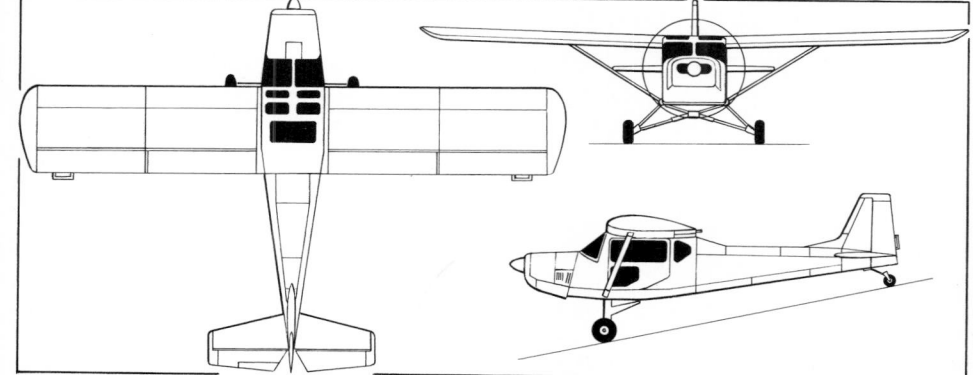
ULBI WT01 Wild Thing (*Jane's/James Goulding*)   *NEW*/0137227

ULBI WT01 Wild Thing fitted with Junkers carbon fibre propeller (*Jane's/Paul Jackson*)   0110610

USA

ULTRALEICHTFLUGZEUGWERKE SACHSEN-ANHALT

Dorfstrasse 25, D-39291 Wörmelitz
Tel: (+49 3922) 49 76 77
Fax: (+49 3921) 98 42 90
e-mail: ul-flugzeugwerke.Spirit@t-online.de
CHIEF DESIGNER: Fritz Schussler

At the Friedrichshafen Aero '99 show in April 1999, USA launched the SF 45 SA Spirit, a locally modified version of the 3Xtrim lightplane described in the Polish section. Its

designation reflects the initials of the designer, the maximum take-off weight and the region in which the company is based. The project failed to develop as rapidly as anticipated, although it remains active in 2002.

UPDATED

USA SF 45 SA SPIRIT

TYPE: Side-by-side ultralight.
PROGRAMME: Upgraded version of Aerotrade Racek 2, of which around 20 built in the Czech Republic, and parallel to certified, Polish-built 3Xtrim (which see). Prototype D-MSBI was due to fly mid-1999, with certification

intended soon thereafter. Currently factory built; kit version to be launched at later date. Prototype had reverted to Czech registration OK-FUU 02 by April 2001.
CUSTOMERS: Four under construction by mid-1999, three of which had been sold; no further sales by April 2001.
COSTS: DM89,900 (2001) including tax.
DESIGN FEATURES: Braced, high, constant-chord wing and tapered rear fuselage. Extensive use of composites throughout. Wing aerofoil TsAGI R-3.
FLYING CONTROLS: Conventional and manual. Electric elevator trim. Single-piece, horn-balanced elevator, with trim tab on port side. Three-position (0, 15, 30°) flaps. Frise ailerons.

STRUCTURE: Differs from Racek 2 and 3Xtrim by having all-composites sandwich wing.

LANDING GEAR: Tricycle type; fixed. Rubber-in-compression; nose gear with steerable wheel. Carbon fibre cantilever sprung main gear; hydraulic disc brakes on mainwheels.

POWER PLANT: One 59.6 kW (79.9 hp) Rotax 912 UL flat-four driving a ground-adjustable pitch, two-blade Kevlar propeller; 73.5 kW (98.6 hp) Rotax 912 ULS optional. Fuel tank, capacity 70 litres (18.5 US gallons; 15.4 Imp gallons), behind seats. Baggage capacity 18 kg (40 lb).

DIMENSIONS, EXTERNAL:
Wing span	10.03 m (32 ft 11 in)
Length overall	6.87 m (22 ft 6½ in)
Height overall	2.40 m (7 ft 10½ in)

AREAS:
Wings, gross	12.40 m² (133.5 sq ft)

WEIGHTS AND LOADINGS:
Weight empty	296 kg (653 lb)
Max T-O weight	450 kg (992 lb)

PERFORMANCE:
Max level speed	113 kt (210 km/h; 130 mph)
Cruising speed	100 kt (185 km/h; 115 mph)
Stalling speed	36 kt (65 km/h; 41 mph)
Max rate of climb at S/L	204 m (670 ft)/min
Range with max fuel	432 n miles (800 km; 497 miles)

VERIFIED

Prototype USA Spirit at Aero '01, Friedrichshafen, April 2001 (*Jane's/Paul Jackson*)   0110635

WD

WD FLUGZEUGLEICHTBAU GmbH

Sudetenstrasse 57/2, D-73540 Heubach
Tel: (+49 7173) 92 99 90
Fax: (+49 7173) 92 99 99
e-mail: info@dallach.de
Web: http://www.dallach.de

WD Flugzeugleichtbau markets the Fascination D4 BK and D5 Evolution, designed by Wolfgang Dallach. The earlier D3 Sunwheel is still available; WD is at present working on the **Revolution**, which will be powered by a 265 kW (355 hp) VOKBM M-14P.

UPDATED

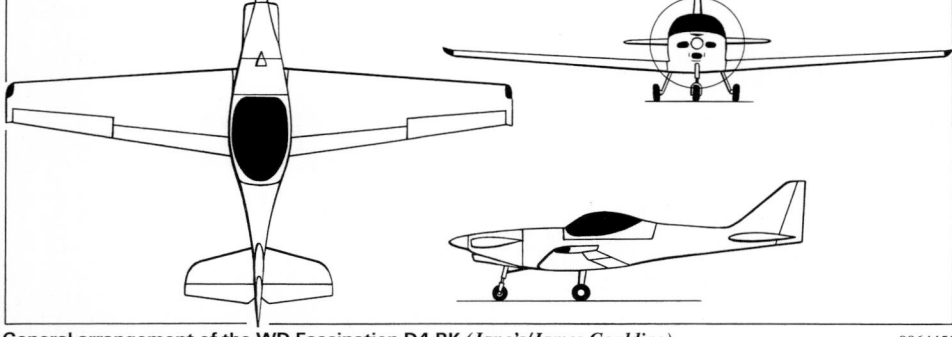

WD D5 Evolution prototype flying at Berlin in May 2002 (*Jane's/Paul Jackson*)   NEW/0137731

WD D4 BK FASCINATION

TYPE: Side-by-side ultralight/kitbuilt.

PROGRAMME: Original D4, of steel, wood, fabric and composites construction, first flew in early 1996; public debut (D-MOVE and D-MBWH) at the Réseau du Sport de l'Air rally at Epinal, France, in July 1996. Prototype D4 BK (D-MPMM, carrying spurious registration D-MMMM) exhibited at Aero '99, Friedrichshafen, in April 1999, shortly before maiden flight; US debut at EAA Air Venture at Oshkosh, Wisconsin, in July 2000.

CURRENT VERSIONS: **D4:** Initial version, last described in 1999-2000 edition.

D4 BK: All-composites version, *as described*. BK indicates *bugrad, kraftstoff:* nosewheel, composites.

D5: High-wing version; described separately.

CUSTOMERS: Total of 90 D4s sold (including 40 as kits), of which 50 were flying by April 1999, including 15 in Brazil. D4 BK sales totalled five by April 1999.

COSTS: Fully equipped €79,992 including tax (2002). Kit version 15 per cent less.

DESIGN FEATURES: Wings and tail surfaces detach for storage/transport. Quoted kit build time 800 hours.

FLYING CONTROLS: Conventional and manual. Fowler flaps.

STRUCTURE: All composites.

LANDING GEAR: Standard fixed tricycle type, with nosewheel power steering. Retractable optional; main units retract outwards, nosewheel rearwards. Cable-operated disc brakes.

POWER PLANT: One 59.6 kW (79.9 hp) Rotax 912 UL (standard), 73.5 kW (98.6 hp) Rotax 912 ULS (optional) or (on completion of development) 86.8 kW (116.3 hp) DZ 100 flat-four engine driving a Rospeller two-blade electric constant-speed propeller. Fuel capacity 80 litres (21.1 US gallons; 17.6 Imp gallons).

AVIONICS: GPS and transponder optional.

EQUIPMENT: Ballistic recovery parachute optional.

DIMENSIONS, EXTERNAL:
Wing span	9.00 m (29 ft 6¼ in)
Length overall	6.98 m (22 ft 10¾ in)
Height overall	1.85 m (6 ft 0¾ in)

AREAS:
Wings, gross	10.40 m² (111.9 sq ft)

WEIGHTS AND LOADINGS:
Weight empty	290 kg (639 lb)
Max T-O weight	450 kg (992 lb)

PERFORMANCE (Rotax 912 ULS):
Max level speed	157 kt (290 km/h; 180 mph)
Cruising speed	151 kt (280 km/h; 174 mph)
Stalling speed	35 kt (64 km/h; 40 mph)
Max rate of climb at S/L	390 m (1,280 ft)/min
T-O and landing run	110 m (360 ft)
Range with max fuel	594 n miles (1,100 km; 683 miles)

UPDATED

General arrangement of the WD Fascination D4 BK (*Jane's/James Goulding*)   0064455

WD D5 EVOLUTION

TYPE: Side-by-side, ultralight/kitbuilt.

PROGRAMME: Announced at Aero '01 at Friedrichshafen April 2001; first flight of prototype (D-MPMM) then expected within two months but not achieved until 16 February 2002.

COSTS: Factory-built €87,227 including tax (2002).

DESIGN FEATURES: Utilises wings, Fowler flaps and tail surfaces of D4 Fascination. 'Glass cockpit'. Wings and tail surfaces detach for storage and transportation.

FLYING CONTROLS: Conventional and manual. Dual controls standard. Fowler flaps.

STRUCTURE: Primarily composites, with metal landing gear doors.

LANDING GEAR: Retractable tricycle type. All three units retract rearwards, with main legs housed within lower fuselage.

POWER PLANT: One 59.6 kW (79.9 hp) Rotax 912 UL driving Rospeller three-blade constant-speed propeller via 1:2.27 reduction gearing, or 73.5 kW (98.6 hp) Rotax 912 ULS driving same propeller via 1:1.43 reduction gearing. Fuel capacity 120 litres (31.7 US gallons; 26.4 Imp gallons).

EQUIPMENT: Ballistic parachute recovery system optional.

DIMENSIONS, EXTERNAL:
Wing span	9.10 m (29 ft 10¼ in)

WD Fascination D4 BK composites kitbuilt in the USA (*Jane's/Paul Jackson*)   NEW/0132844

		WEIGHTS AND LOADINGS:				
Length overall	6.62 m (21 ft 8½ in)	Weight empty	270 kg (595 lb)	Stalling speed	35 kt (64 km/h; 40 mph)	
Height overall	2.04 m (6 ft 8¼ in)	Max T-O weight	450 kg (992 lb)	Max rate of climb at S/L	390 m (1,280 ft)/min	
DIMENSIONS, INTERNAL:		PERFORMANCE (Rotax 912 ULS):		T-O run: solo	75 m (246 ft)	
Cabin: max width	1.25 m (4 ft 1¼ in)	Max level speed	157 kt (290 km/h; 180 mph)	dual	100 m (330 ft)	
AREAS:		Cruising speed	146 kt (270 km/h; 168 mph)	Range	918 n miles (1,700 km; 1,056 miles)	
Wings, gross	10.40 m² (111.9 sq ft)				*UPDATED*	

GREECE

HAT

HELLENIC AERONAUTICAL TECHNOLOGIES
40 Moraiti Street, GR-11525 Athens
Tel/Fax: (+30 1) 677 91 52
GENERAL MANAGER: Anastasios Makrikostas

HAT is a specialist composites materials company. Its initial aircraft, the LS2, is the first Greek design intended for the homebuilt market. By 2002, however, no reports had been received regarding production.

UPDATED

HAT LS2
All known details of this two-seat lightplane last appeared in the 2002-03 *Jane's*.

UPDATED

INDIA

ADA

AERONAUTICAL DEVELOPMENT AGENCY
PO Box 1718, Vimanapura Post Office, Bangalore 560 017
Tel: (+91 80) 523 30 60 and 523 20 83
Fax: (+91 80) 523 84 93 and 523 44 93
e-mail: pai@ada.ernet.in
Web: http://www.ada.gov.in
LCA PROGRAMME DIRECTOR: Dr Kota Harinarayana
PROJECT DIRECTOR, TECHNOLOGY DEVELOPMENT AND LCA NAVY: Dr T G Pai

The ADA, an autonomous organisation under the Indian Ministry of Defence, was created in 1984 to be responsible for design and development of the next-generation Light Combat Aircraft (LCA), which made its maiden flight in early 2001; its manufacturing partner is Hindustan Aeronautics Ltd (HAL, which see). Several laboratories of the Defence Research and Development Organisation (DRDO) have also contributed to the LCA programme.

UPDATED

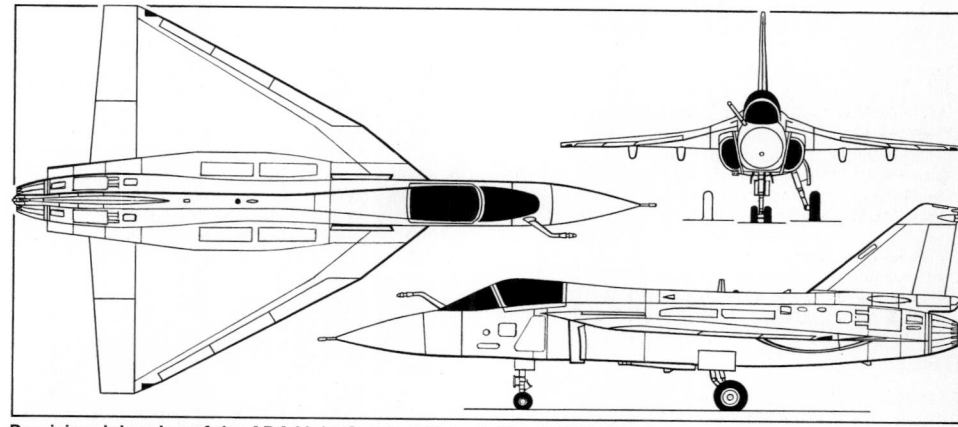

Provisional drawing of the ADA Light Combat Aircraft (*Jane's/Mike Keep*)   0110517

ADA LIGHT COMBAT AIRCRAFT (LCA)
TYPE: Multirole fighter.
PROGRAMME: Development approved by Indian government in 1983 as MiG-21 replacement; project definition begun second quarter 1987, completed late 1988, with first flight then predicted for April 1990 and service entry in about 1995. However, due to early slippages, basic design not finalised until 1990; construction by HAL started mid-1991. First aircraft (TD1: technology demonstrator; serial number KH2001) was nine months behind revised schedule at roll-out on 17 November 1995; several subsequent postponements of new (June 1996) target date for first flight, which eventually took place on 4 January 2001; completed first phase of flight tests with 12th flight on 2 June 2001, reaching 8,000 m (26,250 ft) altitude, 18° AoA and M0.71 speed. First supersonic flight targeted for January 2002.

TD2 (KH2002) rolled out on 14 August 1998 and scheduled to fly in September 2001, but did not do so until 6 June 2002. This has marginally lighter empty weight, increased usable fuel and Indian-developed HUD compared with TD1. Both aircraft powered by F404-GE-F2J3 afterburning turbofan, ground runs of which began on 9 April 1998; indigenous Kaveri engine being developed by Gas Turbine Research Establishment (GTRE) Bangalore and will undertake flight trials in Russia in Tu-16 testbed. Bench trials of this engine had reportedly totalled nearly 1,000 hours by early 2001, but flight trials still not expected to start before early 2003. SNECMA and Eurojet reported to have offered technological assistance in developing Kaveri engine, which otherwise considered

unlikely to be ready until at least 2004. Meanwhile, to offset delays in Kaveri programme, prototypes PV1 to PV5 and at least first eight production LCAs will be fitted with F404 engine, further 40 of which received US export approval in 2002. PV1 planned to fly in October 2002. Four of the PVs to be single-seaters (including one in naval configuration); the fifth will be a two-seat operational trainer having commonality with the naval variant of LCA.

Programme delays further exacerbated by May 1998 US embargo on supplies and assistance following India's refusal to abandon nuclear weapons testing, necessitating all-Indian completion of development of integration of digital AFCS and some other systems (LCA is inherently unstable and thus unable to fly without AFCS). However, sanctions lifted on 22 September 2001. Development of the indigenously developed radar is said to be on schedule; this is to be flight tested initially in a HAL HS 748 and first installed in the PV1 LCA. Work on PV1 and PV2 was under way in late 1998; additional US$125 million released in May 1999 to accelerate remaining development.

Latest (mid-2002) estimate indicated third (PV1) prototype due to fly in 2002; fourth and fifth (PV2 and 3) in 2003; two-seater (PV5) in 2004; IOC now expected in 2006 and final operational clearance by 2008.
CURRENT VERSIONS: **LCA:** Single-seat version for Indian Air Force. *As described.*

LCA Trainer: Tandem two-seat operational trainer; under development.

LCA Navy: Single-seat carrierborne version; drooped nose, arrester hook, long-stroke landing gear, retractable canards and movable vortex controls at the wingroot.

Design approved early 1999, but still in planning stage. First flight (PV4) targeted for 2004.
CUSTOMERS: Indian Air Force (requirement for 200 single-seat and 20 trainers); Indian Navy may order up to 40 single-seaters. First eight for IAF approved in MoU March 2002; production by HAL at Bangalore.
COSTS: Phase 1 (TD1 and TD2) costs, including engine, estimated at approximately Rs30 billion (US$675 million) by late 2000. Estimated unit cost, based on 220 production total, US$17 million to US$20 million (2001).
DESIGN FEATURES: Tail-less delta planform with relaxed static stability; shoulder-mounted delta wings with compound sweep on leading-edges; large twist from inboard to outboard leading-edges; wing-shielded, side-mounted, fixed-geometry Y-duct air intakes. Advanced materials for minimum structural weight; fly-by-wire and HOTAS controls; high agility; supersonic at all altitudes; wide range of external stores.
FLYING CONTROLS: Two-segment trailing-edge elevons and three-section leading-edge slats; vortex-shedding inboard leading-edges with inboard slats to form vortices over wingroot and fin; airbrake in top of fuselage each side of vertical fin.
STRUCTURE: Advanced materials to include aluminium-lithium alloy, carbon composites and titanium alloys; CFRP wings (including elevons), fin and rudder; Kevlar radome. Wings manufactured with one-piece CFRP top and bottom skins, bolted on to wing box; majority of spars and ribs in composites. Fin, rudder, elevons, airbrakes and landing gear doors embody co-cured, co-bonded techniques. Carbon fibre composites to be increased from 30 per cent of airframe weight in TD1/2 to 45 per cent in PVs, with corresponding reduction of aluminium alloys from 57 to 43 per cent.

Hindustan Aeronautics (HAL) responsible for fuselage, communications system, electrical system, mechanical system LRUs and utility systems management; Aeronautical Development Establishment (ADE) flight control system, cockpit displays and display processors; National Aerospace Laboratories (NAL) responsible for development of fin and fabrication of rudder.
LANDING GEAR: Hydraulically retractable tricycle type. Single mainwheels; twin-wheel nose unit. DRDL carbon disc brakes; computer-controlled brake management system. ADRDE brake-chute in fairing at base of rudder.
POWER PLANT: One 80.1 kN (18,000 lb st) General Electric F404-GE-F2J3 afterburning turbofan in TD1 and TD2, prototypes PV1 to PV5 and initial production batch; Indian GTRE GTX-35VS Kaveri turbofan (52.0 kN; 11,700 lb st dry, 80.5 kN; 18,100 lb st with afterburning), with digital engine control unit (KADECU), under development for

TD1 first prototype of the LCA during its maiden flight on 4 January 2001   *NEW*/0132902

later production aircraft. Multi-axis thrust vectoring nozzle planned.

Internal fuel in wing and fuselage tanks. Fixed in-flight refuelling probe on starboard side of front fuselage. Provision for up to three 1,200 or five 800 litre (317 or 211 US gallon; 264 or 176 Imp gallon) external fuel tanks on wing inboard, mid-board and underfuselage stations.

ACCOMMODATION: Pilot only, on Martin-Baker zero/zero ejection seat. Canopy opens sideways to starboard. Development will include two-seat training version.

SYSTEMS: Hydraulic system (by HAL) for powered flying controls, brakes and landing gear; electrical system for fly-by-wire and avionics power supply; Spectrum Infotech environmental control system for cockpit, radar and fuel tanks; lox system; HAL GTSU-110 gas turbine starter unit; aircraft-mounted accessory gearbox; USMS for aircraft utilities systems management.

AVIONICS (production version): *Comms:* HAL V/UHF (with built-in ECCM) and standby UHF radios and audio management unit; HAL air-to-air/air-to-ground secure datalink; HAL IFF transponder/interrogator.

Radar: Electronics Research and Development Establishment (ERDE)/HAL multimode radar with multitarget search and track-while-scan and ground mapping functions; coherent pulse Doppler system, with look-up/look-down modes, Doppler beam sharpening and moving target indication.

Flight: Quadruplex digital fly-by-wire AFCS. RLG-based INS; provision for GPS/INS; HAL RAM 1701A radio altimeter.

Instrumentation: NVG-compatible 'glass cockpit'; two Bharat Electronics active matrix, reconfigurable colour LCD MFDs; CSIO (India) collimated HUD; dedicated LCD 'get-you-home' panel; LCD multifunction keyboard.

Mission: Three multiplexed MIL-STD-1553B digital databusses; 32-bit mission computer operating in Ada, backed up by a second, equally powerful computer; IRST; laser range-finder/designator pod. Provision for reconnaissance, EW or other sensor pods. Helmet-mounted display/sight.

LCA instrument panel   0052370

Self-defence: EW suite, by Advanced Systems Integration and Evaluation Organisation (ASIEO), includes radar warning receiver, self-protection jammer, laser warning system, missile approach warning system, countermeasures dispensing system and chaff/flare dispenser.

ARMAMENT: GSh-23 twin-barrel 23 mm gun with 220 rounds in blister beneath starboard engine intake trunk. Eight external stores stations (three under each wing, one on fuselage centreline and one beneath port intake trunk) for wide range of short/medium-range air-to-air missiles, PGMs, air-to-surface (including anti-ship) missiles, targeting or ECM pod, unguided rockets, conventional and retarded bombs, and cluster bomb dispensers. Indigenous Astra active radar-guided ASM under development by DRDL. All except outboard underwing stations are wet for carriage of drop tanks.

DIMENSIONS, EXTERNAL:

Wing span	8.20 m (26 ft 10¾ in)
Wing aspect ratio	1.8
Length overall	13.20 m (43 ft 3¾ in)

Height overall	4.40 m (14 ft 5¼ in)
Wheel track	2.20 m (7 ft 2½ in)
Wheelbase	4.34 m (14 ft 2¾ in)
AREAS:	
Wings, gross	approx 38.40 m² (413.3 sq ft)
WEIGHTS AND LOADINGS:	
Weight empty	approx 5,500 kg (12,125 lb)
Max external stores load	more than 4,000 kg (8,818 lb)
T-O weight (clean)	approx 8,500 kg (18,740 lb)
Wing loading (clean)	approx 221.4 kg/m² (45.35 lb/sq ft)
Power loading (TD1 and 2, clean)	
	106 kg/kN (1.04 lb/lb st)
PERFORMANCE (estimated):	
Max level speed at altitude	M1.8
Service ceiling	above 15,240 m (50,000 ft)
g limits	+9/−3.5
	UPDATED

ADA MEDIUM COMBAT AIRCRAFT (MCA)

The ADA is studying an advanced version of the LCA under the above designation, as a potential replacement for Indian Air Force Jaguars from about 2010. The twin-engined MCA is intended to embody stealth technology, be powered by a non-afterburning variant of the GTRE Kaveri turbofan with thrust vectoring, and have a greater range than the LCA. Other design features are planned to include the adoption of radar-absorbent materials; twin outward-canted tailfins; and overwing external fuel tanks. Project definition had not begun by late 2001; design and development costs have been provisionally estimated at US$2.3 billion in 1996 dollars.

WEIGHTS AND LOADINGS (approx):

T-O weight: clean	12,000 kg (26,455 lb)
max	18,000 kg (39,685 lb)
	UPDATED

ADA MULTIROLE TRANSPORT AIRCRAFT (MTA)

See Ilyushin/IAPO/HAL entry for Il-214 in the International section.

BHEL

BHARAT HEAVY ELECTRICALS LTD

BHEL House, Siri Fort, New Delhi 110049
Tel: (+91 11) 649 34 37
Web: http://www.bhel.com
CHAIRMAN AND MANAGING DIRECTOR: K G Ramachandran
WORKS: Heavy Electrical Equipment Plant, Ranipur, Hardwar PIN 249 403, Uttar Pradesh
Tel: (+91 133) 42 64 59
Fax: (+91 133) 42 64 62 and 42 62 54
EXECUTIVE DIRECTOR: H W Bhatnagar
GENERAL MANAGER, AVIATION: P K Awasthi

BHEL was approved for the manufacture of light aircraft by the Indian Ministry of Industry in March 1991 and by the Indian Directorate General of Civil Aviation in May 1991. Its most recent known product was the Swati light training aircraft.

UPDATED

BHEL LT-IIM SWATI

TYPE: Primary prop trainer/sportplane.
PROGRAMME: Designed and developed by Technical Centre of Directorate General of Civil Aviation (DGCA), India; initial design was with tailwheel; first flight (VT-XIV, as LT-I) 17 November 1990; technical certificate issued 1991 by Indian Institute of Technology, Kanpur; manufacturing agreement between DGCA and BHEL, April 1991, for 10 aircraft, all to be delivered by March 1994; type certification awarded January 1992; follow-on order for 30 placed September 1992; first four aircraft delivered March 1993. See 1994-95 *Jane's* for description of this version.

Design modified by DGCA Technical Centre in 1993 under new designation LT-IIM; prototype was conversion of second production aircraft, VT-STB. Nos. 5 to 7 also converted to LT-IIM; No. 8 and onwards built as such.

Description applies to this version.

CUSTOMERS: By 2000, 20 production aircraft (including LT-Is) had been registered, of which 17 then delivered. Production aircraft owned by Directorate General of Civil Aviation and assigned to various schools and flying clubs (FC), including Bihar Flying Institute, Bombay FC, Kerala Aviation Training Centre, Madhya Pradesh FC, North India FC, Pinjore FC and Rajasthan State Flying School.

COSTS: Approximately Rs3.8 million (1998).
DESIGN FEATURES: Designed to FAR Pt 23 (Utility and Normal categories) but currently limited to Normal category only at 730 kg (1,609 lb) all-up weight; applications include flying training, sport flying, surveillance, photography and short hauls; capable of stall turns, lazy eights and chandelles.

Wing section NASA GA(W)-1 (constant); dihedral 4°; incidence 2°.

FLYING CONTROLS: Conventional and manual. Actuation by pushrods and cables. Plain flaps; horn-balanced rudder and elevators.

BHEL LT-IIM Swati two-seat club trainer

STRUCTURE: Wings constant chord with single box spar of Himalayan spruce, ply and fabric covered; ailerons and flaps similar; fuselage of chromoly steel tube with metal/GFRP engine cowling and skin on forward fuselage; fabric covering aft; GFRP fairings; tail surfaces light metal. Design of metal wings was undertaken in 1998.

LANDING GEAR: Non-retractable bungee-sprung mainwheels, on side Vs and half-axles, with dual hydraulic brakes. Non-retractable steerable nosewheel. Mainwheel tyres 15×6.00-6, pressure 2.00 bar (29 lb/sq in); nosewheel tyre 5.00-5, pressure 2.07 bar (30 lb/sq in).

POWER PLANT: One 80.5 kW (108 hp) Textron Lycoming O-235-N2C flat-four engine, driving a Hoffmann HO-14-178115 two-blade fixed-pitch composites propeller without spinner. Fuel capacity 90 litres (23.8 US gallons; 19.8 Imp gallons).

ACCOMMODATION: Two seats, with safety harnesses, side by side under one-piece rearward-sliding canopy. Gull-wing canopy demonstrated in mockup form, 1998.

AVIONICS: VFR instrumentation and VHF radio standard.

DIMENSIONS, EXTERNAL:

Wing span	9.20 m (30 ft 2¼ in)
Wing chord, constant	1.30 m (4 ft 3¼ in)
Wing aspect ratio	7.1
Length overall (flying attitude)	7.21 m (23 ft 7¾ in)
Height overall (flying attitude)	2.78 m (9 ft 1½ in)
Tailplane span	2.60 m (8 ft 6½ in)
Wheel track	2.00 m (6 ft 6¾ in)

Wheelbase	1.725 m (5 ft 8 in)
Propeller diameter	1.78 m (5 ft 10 in)
AREAS:	
Wings, gross	11.96 m² (128.7 sq ft)
Ailerons (total)	1.10 m² (11.84 sq ft)
Trailing-edge flaps (total)	1.04 m² (11.19 sq ft)
Fin	0.75 m² (8.07 sq ft)
Rudder	0.62 m² (6.67 sq ft)
Tailplane	0.93 m² (10.01 sq ft)
Elevators (total, incl tab)	0.94 m² (10.12 sq ft)
WEIGHTS AND LOADINGS (Normal category):	
Weight empty, equipped (typical)	540 kg (1,190 lb)
Max T-O weight	730 kg (1,609 lb)
Max wing loading	61.0 kg/m² (12.50 lb/sq ft)
Max power loading	9.07 kg/kW (14.90 lb/hp)
PERFORMANCE (at Normal category max T-O weight):	
Never-exceed speed (VNE)	143 kt (265 km/h; 164 mph)
Max level speed	104 kt (193 km/h; 120 mph)
Cruising speed at 75% power	
	100 kt (185 km/h; 115 mph)
Stalling speed	44 kt (82 km/h; 51 mph)
Max rate of climb at S/L	183 m (600 ft)/min
T-O run	259 m (850 ft)
Landing run	205 m (675 ft)
Range with max fuel	272 n miles (505 km; 313 miles)
Max endurance	3 h 30 min
g limits	+3.8/−1.52
	UPDATED

CSIR
COUNCIL OF SCIENTIFIC AND INDUSTRIAL RESEARCH

Centre for Civil Aircraft Design and Development, C-MMACS Campus, Belur, Bangalore 560 037
Tel: (+91 80) 526 32 19
Fax: (+91 80) 526 77 81 and 529 82 92
e-mail: narayan@ccadd.cmmacs.ernet.in
Web: http://www.cmmacs.ernet.in-/nal/pages/saraspg/sarashm.htm
HEAD OF CCADD AND SARAS PROJECT MANAGER:
Dr K Yegna Narayan

Design and development of the Saras twin-turboprop business and commuter transport is nearing fruition as an Indian programme, having previously appeared in the International section under the Myasishchev/NAL heading and latterly under that of the Indian National Aerospace Laboratories (NAL). It has received funding from various Indian government agencies and is currently under the leadership of CSIR, which has created the Centre of Civil Aircraft Design and Development (CCADD) to act as the nodal agency to monitor and manage the programme. Many Indian public and private sector industries are participating in the venture, with manufacture of production components being carried out in 35 work centres across the country. Manufacture of aircraft will be by HAL.

UPDATED

CSIR SARAS
English name: Crane

TYPE: Twin-turboprop transport.
PROGRAMME: Originated as six/nine-passenger light transport shown in model form at Moscow Aerospace '90 exhibition; revised design announced by Myasishchev 1993, with name Delphin; India had comparable project; general agreement concluded 1993 to combine Myasishchev and Indian programmes, Russian version being known as M-102 Duet and Indian version as Saras (species of Indian crane). Full-scale mockup shown at 1993 and 1995 Moscow Air Shows. Prototype construction (two in Russia, one in India originally) planned to begin in September 1994, but Indian share of private venture capital not secured until early 1996. Detailed engineering began in April 1996 following finalisation of design late in previous year; however, in 1997 Myasishchev was proceeding with unilateral development of the similar M-202, as described in the Russian Federation section of 2001-02 and earlier *Jane's*; by 2001, was again promoting M-102 version. In early November 2001, Russia and India signed a protocol on construction of the M-102/Saras.

India elected to continue with the Saras as a national programme, although some collaboration with the former Russian partner was not ruled out. Government secured private backing of Taneja Aerospace and Kumaran Industries, plus financial support from the Indian Technology Development Board. Hindustan Aeronautics Ltd also became a major partner in the development programme and to build production aircraft at Nasik. Release of further government funding in mid-1999 enabled work to begin on manufacture (by HAL) of two flying prototypes and a structural test airframe. First prototype reported in February 2001 to be 80 per cent complete, with first flight then expected by end of 2001, but in following month, programme management board called for ''rigorous weight control exercise''. First flight, planned late 2001, not achieved, and in March 2002 was stated to have been rescheduled for December 2002. Eventually rolled out 4 February 2003; first flight (VT-XSD) planned mid-2003; second prototype (VT-XRM) to follow.

Prototype CSIR Saras at roll-out, 4 February 2003   *NEW*/0530195

CUSTOMERS: At least 200 Indian orders forecast over 10 years.
COSTS: Design and development cost quoted in February 1996 as US$40 million. Late in 1996, cost re-estimated as US$60 million, including flight test phase. Indian government approved Rs1.3 billion (US$30 million) in June 1999 to enable prototype manufacture to begin. Estimated unit price US$3.75 million (2001).
DESIGN FEATURES: Suitable for operations in hot-and-high conditions (typically, airfield at 2,000 m; 6,560 ft, at up to 45°C), and from semi-prepared runways. Unconventional twin pusher propeller configuration. High aspect ratio, low-mounted wings, with straight taper; pressurised circular-section fuselage; T tailplane on sweptback fin; engines pylon-mounted each side of rear fuselage.

Wing section GA(W)-2 (modified); leading-edge sweep 5°; thickness/chord ratio 15 per cent; 4° dihedral; 2° incidence; 2° twist.
FLYING CONTROLS: Conventional and manual. Tabs in all control surfaces; electrically operated pitch, roll and yaw trim; yaw damper; spoilers and 25 per cent chord, single-slotted Fowler flaps; dual-channel three-axis autopilot. Highly swept ventral fins provide nose-down pitching moment at high AoA.
STRUCTURE: Mixed construction of aluminium alloy (fuselage, engine pylons and fixed portion of wing) and composites (nosecone, wing/body fairing, wing moving surfaces and tips and engine cowlings); designed for 30,000 hour life; fail-safe philosophy for all primary structures and major attachments; incorporates damage tolerance features. Two-spar wing with integrally milled skin riveted to stringers; two-spar flaps with honeycomb sandwich filler; single-spar ailerons with honeycomb filler; single-piece, two-spar metal tailplane and three-spar metal fin.
LANDING GEAR: Hydraulically retractable tricycle type, with single mainwheels and single steerable (±53°) nose unit. Mainwheels retract inward into wingroot fairings, nosewheel forward. Oleo-pneumatic shock-absorbers. Mainwheels 26×8.75, pressure 5.52 bar (80 lb/sq in); nosewheel 17.5×6.3, pressure 3.79 bar (55 lb/sq in); air-cooled carbon brakes. Minimum turning radius 8.10 m (26 ft 7 in) at nosewheel.
POWER PLANT: Two 634 kW (850 shp) Pratt & Whitney Canada PT6A-66 turboprops, pylon-mounted on sides of rear fuselage; Hartzell five-blade, constant-speed pusher propellers, rotating in opposite directions. Integral fuel tank in each wing, combined capacity 1,608 litres (425 US gallons; 354 Imp gallons), of which 1,595 litres (421 US gallons; 351 Imp gallons) are usable. Gravity refuelling point in port wing upper surface; optional single-point pressure refuelling point at starboard wingroot.

ACCOMMODATION: Two-person flight deck with dual controls, but to be certified also for one-pilot operation. Alternative layouts for 14 economy class passengers in single seats at 76 cm (30 in) pitch, with centre aisle, rear lavatory and total of 1.1 m³ (39 cu ft) for baggage in forward and rear compartments; or 18 passengers in commuter layout. Typical executive interior could have eight seats, tables, wardrobe, baggage compartment, galley and lavatory. Ambulance version capable of carrying six stretchers, with seats for two medical attendants, medical supplies storage, baggage compartment and lavatory. Various passenger/cargo arrangements or special mission interiors optional in combi version. Door at front on port side, hinged downward, with integral airstairs; overwing Type III emergency exit on each side. Baggage compartment door on port side of tailcone.
SYSTEMS: Bootstrap environmental control system; cabin pressure differential 0.45 bar (6.5 lb/sq in). Hydraulic system (pressure 207 bar; 3,000 lb/sq in) for landing gear actuation, brakes and nosewheel steering; 4 litres (1.06 US gallons; 1.27 Imp gallons)/min electrically powered hydraulic pump; hand pump for emergency lowering of landing gear.

Primary electrical power is 28 V DC, supplied by two 400 A starter/generators; 43 Ah Ni/Cd battery provides emergency power for essential loads for approximately 15 minutes, including three engine restarts. Two solid-state inverters supply 115/26/5 V AC power at 400 Hz for instrumentation; AC power for anti-icing and galley/windscreen heating obtained via alternators. Emergency oxygen system for crew and passengers (two 1,400 litre; 49.4 cu ft bottles), plus two 122 litre (4.3 cu ft) portable bottles and masks for first aid. Engine fire and cabin smoke detection systems. Halon fire extinguishing system. De-icing boots on leading-edges.
AVIONICS: Integrated digital system (ARINC 429 compatible).
Comms: VHF (two), optional HF, intercom/PA system, TCAS II transponder and CVR.
Radar: Weather radar.
Flight: Autopilot (dual-channel, three-axis), ADF, VOR/ILS, marker beacon receiver, DME, radar altimeter, air data sensor and computer, AHRS, flight control computer and flight director standard; GPS, satcom, Selcal and flight management system optional.
Instrumentation: Four-tube EFIS; 'return home' standby instrumentation.

DIMENSIONS, EXTERNAL:

Wing span	14.70 m (48 ft 2¾ in)
Wing chord: at c/l	2.65 m (8 ft 8¼ in)
at tip	0.85 m (2 ft 9½ in)
Wing aspect ratio	8.4
Length overall	15.02 m (49 ft 3¼ in)
Fuselage: Length	13.90 m (45 ft 7¼ in)
Max diameter	1.95 m (6 ft 4¾ in)
Height overall	5.20 m (17 ft 0¾ in)
Tailplane span	6.09 m (19 ft 11¾ in)
Wheel track	3.20 m (10 ft 6 in)
Wheelbase	6.465 m (21 ft 2½ in)
Propeller diameter	2.16 m (7 ft 1 in)
Propeller ground clearance	1.30 m (4 ft 3¼ in)
Distance between propeller centres	3.94 m (12 ft 11 in)
Passenger door (fwd, port): Height	1.48 m (4 ft 10¼ in)
Width	0.75 m (2 ft 5½ in)

DIMENSIONS, INTERNAL:

Cabin: Length	7.74 m (25 ft 4¾ in)
Max width	1.80 m (5 ft 10¾ in)
Aisle width at floor	0.54 m (1 ft 9¼ in)
Max height	1.72 m (5 ft 7¾ in)
Floor area	7.08 m² (76.2 sq ft)
Volume	21.3 m³ (752 cu ft)
Baggage hold volume: fwd, stbd	0.64 m³ (22.6 cu ft)
rear, port	0.46 m³ (16.2 cu ft)

AREAS:

Wings, gross	25.70 m² (276.6 sq ft)
Ailerons (total)	1.335 m² (14.37 sq ft)
Trailing-edge flaps (total)	4.07 m² (43.81 sq ft)
Spoilers (total)	0.65 m² (7.00 sq ft)
Fin	6.38 m² (68.67 sq ft)
Rudder, incl tab	1.815 m² (19.54 sq ft)

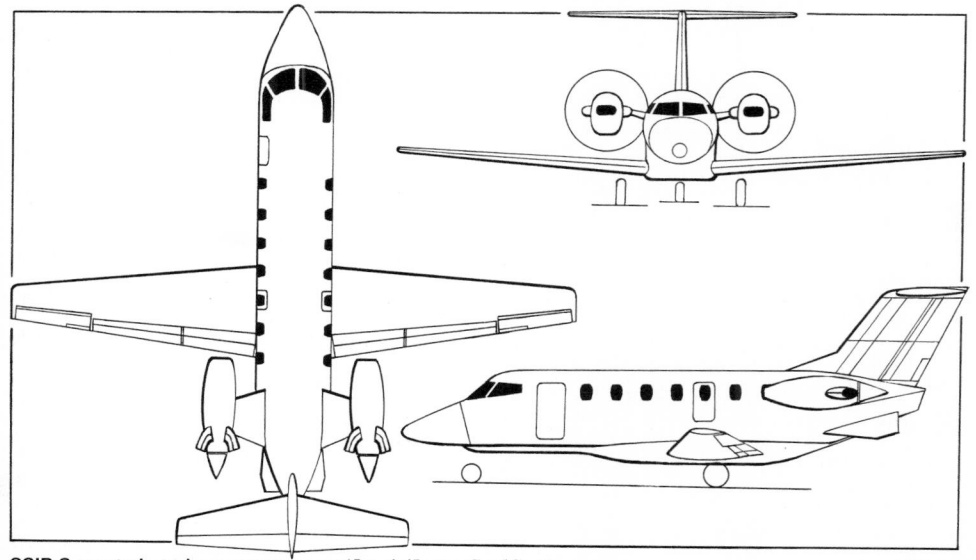
CSIR Saras twin-turboprop transport (*Jane's/James Goulding*)   0130859

Tailplane	7.00 m² (75.35 sq ft)		
Elevators, incl tabs	2.05 m² (22.07 sq ft)		

WEIGHTS AND LOADINGS:
Weight empty, equipped	4,116 kg (9,074 lb)
Max fuel weight	1,326 kg (2,923 lb)
Max payload	1,232 kg (2,716 lb)
Max T-O and landing weight	6,100 kg (13,448 lb)
Max zero-fuel weight	5,348 kg (11,790 lb)
Max wing loading	237.4 kg/m² (48.61 lb/sq ft)
Max power loading	4.81 kg/kW (7.91 lb/shp)

PERFORMANCE (estimated):
Never-exceed speed (V_{NE})	372 kt (690 km/h; 428 mph)

Max level speed at 9,000 m (29,520 ft)	335 kt (620 km/h; 385 mph)
Econ cruising speed at 9,000 m (29,520 ft)	270 kt (500 km/h; 311 mph)

Stalling speed at S/L, power off:
flaps up	88 kt (163 km/h; 101 mph)
flaps down	72 kt (132 km/h; 82 mph)
Max rate of climb at S/L	672 m (2,205 ft)/min
Rate of climb at S/L, OEI	230 m (754 ft)/min
Max certified altitude	9,000 m (29,520 ft)
Service ceiling	11,500 m (37,720 ft)
Service ceiling, OEI	7,500 m (24,600 ft)

T-O run	430 m (1,410 ft)
T-O field length	605 m (1,985 ft)
Landing field length	671 m (2,200 ft)
Landing run	342 m (1,125 ft)
Runway LCN	10

Range, IFR reserves:
with 8 passengers	976 n miles (1,809 km; 1,124 miles)
with 14 passengers	462 n miles (856 km; 531 miles)
self-ferry	1,048 n miles (1,942 km; 1,206 miles)
Endurance	6 h

UPDATED

HAL
HINDUSTAN AERONAUTICS LIMITED

CORPORATE OFFICE: PO Box 5150, 15/1 Cubbon Road, Bangalore 560 001
Tel: (+91 80) 286 46 36/7 and 286 46 39/43
Fax: (+91 80) 286 71 40 and 286 87 58
e-mail: marketing@hal-india.com
Web: http://www.hal-india.com
CHAIRMAN: N R Mohanty
EXECUTIVE DIRECTOR, PLANNING: B K Banerjee
DIRECTOR, CORPORATE PLANNING AND MARKETING: S C Das

GENERAL MANAGER, MARKETING: D Chowdhury
MANAGER, PUBLIC RELATIONS: A Nageswara Rao

Formed 1 October 1964; has 14 manufacturing divisions at eight locations (seven at Bangalore, one each at Nasik, Koraput, Hyderabad, Barrackpore, Kanpur, Lucknow and Korwa), plus Design and Development Complex; nine R&D centres located with manufacturing divisions; total workforce 31,650 in mid-2002. Additional production centres at Nasik and Koraput due to open during 2002. Hyderabad Division manufactures avionics for all aircraft produced by HAL, plus air route surveillance and precision approach radars;

Lucknow Division produces instruments and other accessories under licence from manufacturers in France, Russia and the UK; Korwa manufactures inertial navigation and nav/attack systems.

Former 100-seat transport project now a joint venture with Ilyushin and IAPO of Russian Federation and thus now described in International section.

Sales during 2000-01 totalled Rs2,447 crores (US$510 million); pre-tax profit during this period was Rs265 crores (US$55 million).

UPDATED

DESIGN AND DEVELOPMENT COMPLEX
PO Box 1789, Bangalore 560 017
Tel: (+91 80) 526 34 57
Fax: (+91 80) 526 30 96
Telex: 845 8083
DIRECTOR: Ashok K Baweja
GENERAL MANAGERS:
   Yogesh Kumar (ARDC)
   K S Sudheendra (Executive Director, RWRDC)

HAL currently has nine Research and Design (R&D) Centres, including one each for fixed-wing aircraft (ARDC) and rotary-wing (RWRDC) programmes. Former's main programmes are LCA (see ADA entry) and HJT-36; latter concerned mainly with Dhruv ALH. Transport Aircraft R&D Centre was responsible for developing oversize cabin door, integration of Super Marec radar, gun pod, IR/UV scanner and flight data recorder for HAL-built Dornier 228 programme. Earlier designs have included HT-2, Pushpak, Krishak, Basant, Marut, Kiran, Deepak and HTT-34. Jet trainer successor to Kiran was revealed as HJT-36 in 1998; prototypes are now under construction.

UPDATED

HAL DHRUV
English name: Polaris
TYPE: Light utility helicopter.
PROGRAMME: Agreement signed with MBB (Germany) July 1984 to support design, development and production of Advanced Light Helicopter (ALH); design started November 1984; ground test vehicle runs began April 1991; five flying prototypes (two basic, one air force/army, one naval and one civil); PT1 first prototype (Z3182) rolled

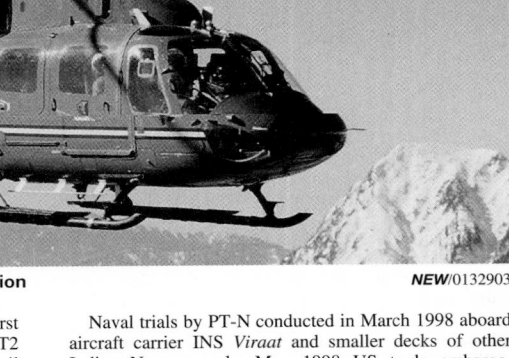

Fourth prototype in Indian Navy (PT-N) configuration   *NEW*/0132903

out 29 June 1992; first flight 20 August and 'official' first flight 30 August 1992; not then fitted with ARIS; PT2 second prototype (Z3183) made its first flight 18 April 1993; PT-A (army/air force prototype Z3268) on 28 May 1994; PT-N (naval prototype), with CTS 800 engines, flew for first time (IN901) on 23 December 1995. Total hours flown, including 'hot-and-high' trials in environments of 50°C (122°F) and more than 6,000 m (19,680 ft), were about 1,480 by May 2002. Military certification of air force/army version completed by end of February 2002; clearance of naval version expected shortly afterwards.

Naval trials by PT-N conducted in March 1998 aboard aircraft carrier INS *Viraat* and smaller decks of other Indian Navy vessels. May 1998 US trade embargo, imposed following India's refusal to sign nuclear test ban treaty, blocked import of CTS 800 engines (30 ordered) and delayed planned first flight of PTC-2 civil fifth prototype (VT-XLH) with this engine. Instead, all variants now to be powered by TM 333, including retrofit of PT-N prototype. Weight reduction programme initiated in mid-1998; RFPs issued later same year for cockpit display system. By the end of 1998, manufacture was well advanced of three preproduction aircraft (PPN-1, PPA-2 and PPA-3: one for each of the three armed services).

Deliveries (four each to Indian Air Force and Army, two each to Navy and Coast Guard) were due to begin in late 1999 but programme has slipped; new schedule provided for first two (to Army) instead becoming due for delivery by end of 2001, to be followed by two each to Coast Guard, Navy and Air Force by the end of March 2002. Seven deliveries actually achieved by this date: Army two (IA-1101 and -1103) starting 4 January 2002, Air Force two (J-401/402 on 20 March), Navy two (IN-701/702 on 28 March) and Coast Guard (CG-851 on 18 March) one. Initial batch of 30 TM 333-2B2 engines ordered in mid-1999 to power first 12 (including two civil) production Dhruvs; all for delivery by 2002. Further 52 engines ordered mid-2000 to power next 20.

Development and marketing agreement between HAL and Israel Aircraft Industries announced in late 2002; involves both Dhruv and LAH derivative; IAI to concentrate on avionics and other internal systems.
CURRENT VERSIONS: **Air force/army:** Skid gear, crashworthy fuel tanks, bulletproof supply tanks, IR and flame suppression; night attack capability; roles to include attack and SAR.
   Naval: Retractable tricycle gear, harpoon decklock, foldable tailboom, pressure refuelling; fairings on fuselage sides to house mainwheels, flotation gear and batteries.
   Civil: Roles to include passenger and utility transport, commuter/offshore executive, rescue/emergency medical service and law enforcement. Wheel landing gear. Prototype targeted to fly in 2001, but this not achieved until late March 2002; DGCA certification to be followed by FAA/JAA type approval. Ambulance version planned to enter production in 2003.

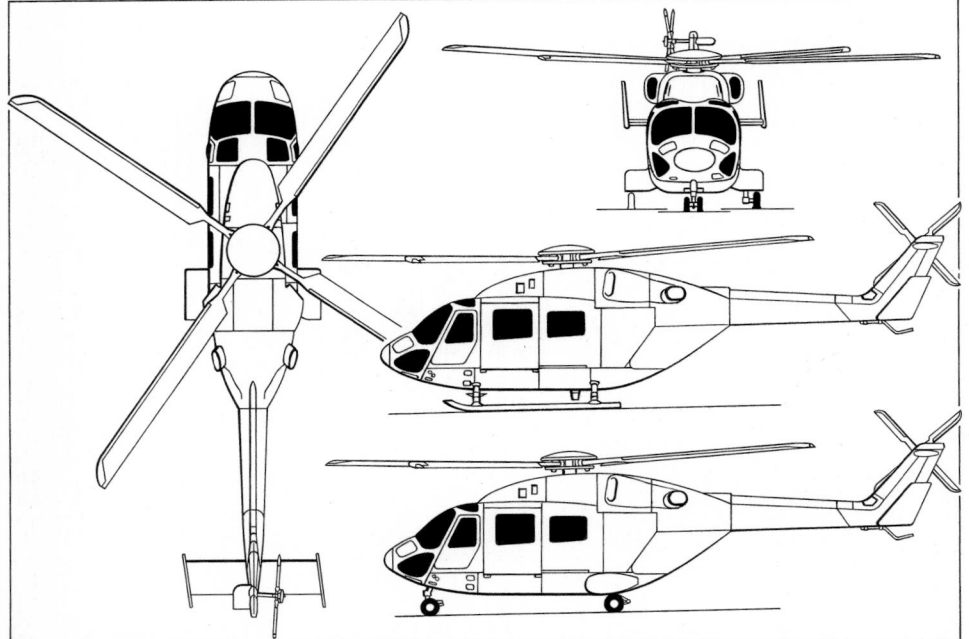

Naval version of the Dhruv Advanced Light Helicopter, with additional side view (centre) of air force/army variant (*Jane's/Mike Keep*)

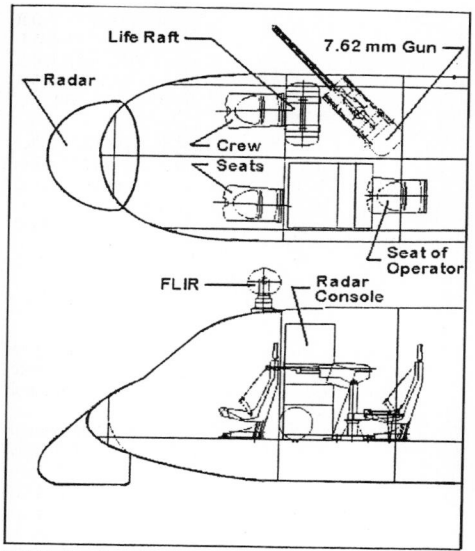

Cabin detail of Dhruv Coast Guard version   0099548

Air ambulance (EMS) configuration demonstrated by ALH second prototype   0110515

Coast Guard: High commonality with naval version; nose-mounted surveillance radar; roof-mounted FLIR; starboard side, cabin-mounted 7.62 mm machine gun; radar console and operator's seat; liferaft, loudhailer.

LAH: Light attack development: described separately.

LOH: Light observation development: described separately.

CUSTOMERS: Indian government requirement for armed forces and Coast Guard, to replace Chetaks/Cheetahs; letter of intent for 300 (Army 110, Air Force 150, Navy/Coast Guard 40) followed by contract for 100 in late 1996, but allocation revised by 2001 as Army 120, Navy 120, Air Force 60 and Coast Guard seven. Second production lot contains 20. HAL predicts total military/civil domestic orders for about 650.

COSTS: Unit price of basic aircraft approximately Rs250 million (US$5.1 million) (2002). Total programme costs US$170 million by 1997 (latest information provided).

DESIGN FEATURES: First modern helicopter of local design and construction. Conventional layout, including high-mounted tailboom to accommodate rear-loading doors. Four-blade hingeless main rotor with advanced aerofoils and sweptback tips; Eurocopter FEL (fibre elastomer) rotor head, with blades held between pair of cruciform CFRP starplates; manual blade folding and rotor brake standard; integrated drive system transmission; four-blade bearingless crossbeam tail rotor on starboard side of fin; fixed tailplane; sweptback endplate fins offset to port; vibration damping by ARIS (anti-resonance isolation system), comprising four isolator elements between main gearbox and fuselage. Folding tailboom on naval variant.

Main rotor blade section DMH 4 (DMH 3 outboard); tail rotor blade section S 102C (S 102E at tip). Rotor speeds 314 rpm (main), 1,564 rpm (tail).

FLYING CONTROLS: Integrated dynamic management by four-axis AFCS (actuators have manual as well as AFCS input); constant-speed rpm control, assisted by collective anticipator (part of FADEC and stability augmentation system acting through AFCS).

STRUCTURE: Main and tail rotor blades and rotor hub glass fibre/carbon fibre; Kevlar nosecone, crew/passenger doors, cowling, upper rear tailboom and most of tail unit; carbon fibre lower rear tailboom and fin centre panels; Kevlar/

carbon fibre cockpit section; aluminium alloy sandwich centre cabin and remainder of tailboom.

LANDING GEAR: Non-retractable metal skid gear standard for air force/army version. Hydraulically retractable tricycle gear on naval and civil versions, with twin nosewheels and single mainwheels, latter retracting into fairings on fuselage sides which also (on naval version) house flotation gear and batteries; rearward-retracting nose unit; naval version has harpoon decklock system. Spring skid under rear of tailboom on all versions, to protect tail rotor. FPT Industries (UK) Kevlar inflatable flotation bags for prototypes, usable with both skid and wheel gear.

POWER PLANT: First three, and fifth, prototypes each powered by two Turbomeca TM 333-2B2 or -2C turboshafts, with FADEC, rated at 746 kW (1,000 shp) for T-O, 788 kW (1,057 shp) maximum contingency and 663 kW (889 shp) maximum continuous. LHTEC CTS 800-4N (969 kW; 1,300 shp) selected late 1994 and test-flown in fourth prototype, but subsequently embargoed; all now to have TM 333-2B2 until availability of 895 kW (1,200 shp) class Ardiden (Shakti) in about 2006.

Transmission ratings (two engines) 1,240 kW (1,663 shp) for 5 minutes for T-O and 1,070 kW (1,435 shp) maximum continuous; OEI ratings 800 kW (1,073 shp) for 30 seconds (super contingency), 700 kW (939 shp) for 2½ minutes, 620 kW (831 shp) for 30 minutes and 535 kW (717 shp) maximum continuous. Transmission input from both engines combined through spiral bevel gears to collector gear on stub-shaft. ARIS system (see Design Features) gives 6° of freedom damping. Power take-off from main and auxiliary gearboxes for transmission-driven accessories.

Total usable fuel, in self-sealing crashworthy underfloor tanks (three main and two supply), 1,400 litres (370 US gallons; 308 Imp gallons). Pressure refuelling in naval version.

ACCOMMODATION: Flight crew of two, on crashworthy seats in military/naval versions. Main cabin seats 12 persons as standard, 14 in high-density configuration. EMS interior (first flown by PT2/Z3183 in January 2001) can accommodate two stretchers and four medical attendants, or four stretchers and two medical personnel. Crew door and rearward-sliding passenger door on each side; clamshell cargo doors at rear of passenger cabin.

SYSTEMS: DC electrical power from two independent subsystems, each with a 6 kW starter/generator, with

battery back-up for 15 minutes of emergency operation; AC power, also from two independent subsystems, each with a 5/10 kVA alternator. Three hydraulic systems (pressure 207 bar; 3,000 lb/sq in, maximum flow rate 25 litres; 6.6 US gallons; 5.5 Imp gallons)/min: systems 1 and 2 for main and tail rotor flight control actuators, system 3 for landing gear, wheel brakes, decklock harpoon (naval variant) and optional equipment. Oxygen system.

AVIONICS: *Comms:* V/UHF, HF/SSB and standby UHF com radio, IFF and intercom.

Radar: Weather radar optional. Surveillance radar in Coast Guard version.

Flight: SFIM four-axis AFCS, Doppler navigation system, TAS system, ADF, radio altimeter, heading reference standard; Omega nav system optional.

Mission: Roof-mounted FLIR in Coast Guard version. EMS version equipped with navaids, patient monitoring, data recording systems, and datalink to transmit medical information to ground-based hospitals.

EQUIPMENT: Depending on mission, can include two to four stretchers, external rescue hoist, liferaft and 1,500 kg (3,307 lb) capacity cargo sling.

ARMAMENT: Cabin-side pylons for two torpedoes/depth charges or four anti-ship missiles on naval variant; on army/air force variant, outrigger booms which can be fitted with eight anti-tank guided missiles, four pods of 68 mm or 70 mm rockets or two pairs of air-to-air missiles. Army/air force variant can also be equipped with ventral 20 mm gun turret or sling for carriage of land mines. Cabin-mounted 7.62 mm machine gun in Coast Guard version, firing from starboard side doorway.

DIMENSIONS, EXTERNAL:

Main rotor diameter	13.20 m (43 ft 3¾ in)
Main rotor blade chord: inboard	0.50 m (1 ft 7¾ in)
at tip	0.165 m (6½ in)
Tail rotor diameter	2.55 m (8 ft 4½ in)
Length:	
overall, both rotors turning	15.87 m (52 ft 0¾ in)
fuselage (except Coast Guard version)	
	13.43 m (44 ft 0¾ in)
Height: overall, tail rotor turning:	
army/air force version	4.98 m (16 ft 4 in)
naval version	4.91 m (16 ft 1¼ in)
to top of main rotor head:	
army/air force version	3.93 m (12 ft 10¾ in)
naval version	3.76 m (12 ft 4 in)

Dhruv instrument panel
0010517

ALH civil commuter interior   0099549

Fuselage max width	2.00 m (6 ft 6¾ in)
Width over mainwheel sponsons (naval version)	
	3.15 m (10 ft 4 in)
Tail unit span (over fins)	3.19 m (10 ft 5½ in)
Wheel track (naval version)	2.80 m (9 ft 2¼ in)
Wheelbase (naval version)	4.37 m (14 ft 4 in)
Skid track (army/air force version)	2.60 m (8 ft 6¼ in)
Tail rotor ground clearance	2.34 m (7 ft 8¼ in)

DIMENSIONS, INTERNAL:

Cabin, excl flight deck: Max width	1.97 m (6 ft 5½ in)
Max height	1.42 m (4 ft 8 in)
Volume	7.33 m³ (259 cu ft)
Cargo compartment volume	2.16 m³ (76.3 cu ft)

AREAS:

Main rotor disc	136.85 m² (1,473.0 sq ft)
Tail rotor disc	5.11 m² (55.0 sq ft)
Main fin	2.126 m² (22.88 sq ft)
Endplate fins (total)	1.45 m² (15.61 sq ft)
Tailplane	2.40 m² (25.83 sq ft)

WEIGHTS AND LOADINGS (A: army/air force standard, B: army/ air force alternative gross weight version, C: naval version):

Basic weight empty: A, B	2,550 kg (5,622 lb)
C	2,685 kg (5,919 lb)
Max fuel weight: A, B	1,075 kg (2,370 lb)
C	1,125 kg (2,480 lb)
Max sling load: A, B	1,000 kg (2,205 lb)
C	1,500 kg (3,307 lb)
Max T-O weight: A	4,500 kg (9,920 lb)
B	5,500 kg (12,125 lb)
C	5,555 kg (12,246 lb)
Max disc loading: A	32.9 kg/m² (6.74 lb/sq ft)
B	40.2 kg/m² (8.23 lb/sq ft)
C	40.6 kg/m² (8.31 lb/sq ft)

Transmission loading at max T-O weight and power:

A	3.63 kg/kW (5.97 lb/shp)
B	4.44 kg/kW (7.29 lb/shp)
C	4.48 kg/kW (7.36 lb/shp)

PERFORMANCE (A: at 4,500 kg; 9,920 lb AUW, B: at 5,500 kg; 12,125 lb AUW, C at 5,555 kg; 12,246 lb AUW, all at S/L, ISA + 15°C):

Never-exceed speed (V$_{NE}$):	
A	178 kt (330 km/h; 205 mph)
B, C	162 kt (300 km/h; 186 mph)
Max cruising speed: A	143 kt (265 km/h; 165 mph)
B	135 kt (250 km/h; 155 mph)
C	129 kt (239 km/h; 148 mph)
Econ cruising speed: A	119 kt (220 km/h; 137 mph)
B	115 kt (213 km/h; 132 mph)
C	110 kt (204 km/h; 127 mph)
Max rate of climb at S/L: A	660 m (2,165 ft)/min
B	550 m (1,804 ft)/min
C	560 m (1,837 ft)/min
Service ceiling: A	6,500 m (21,320 ft)
B	4,580 m (15,020 ft)
C	4,400 m (14,440 ft)
Hovering ceiling:	
IGE: A	4,300 m (14,100 ft)
B	2,200 m (7,220 ft)
C	2,000 m (6,560 ft)
OGE: A	3,700 m (12,140 ft)
B	1,550 m (5,080 ft)
C	1,370 m (4,500 ft)

Range with max fuel, 20 min reserves:

A	378 n miles (700 km; 435 miles)
B	351 n miles (650 km; 403 miles)
C	364 n miles (675 km; 419 miles)

Endurance, conditions as above:

A	4 h 20 min
B, C	4 h 0 min

UPDATED

HAL LAH

TYPE: Attack helicopter.

PROGRAMME: Derivative of Dhruv (which see); displayed in model form at Paris Air Show, June 2001. Formal launch expected in 2002; engine under joint development for planned certification in 2005.

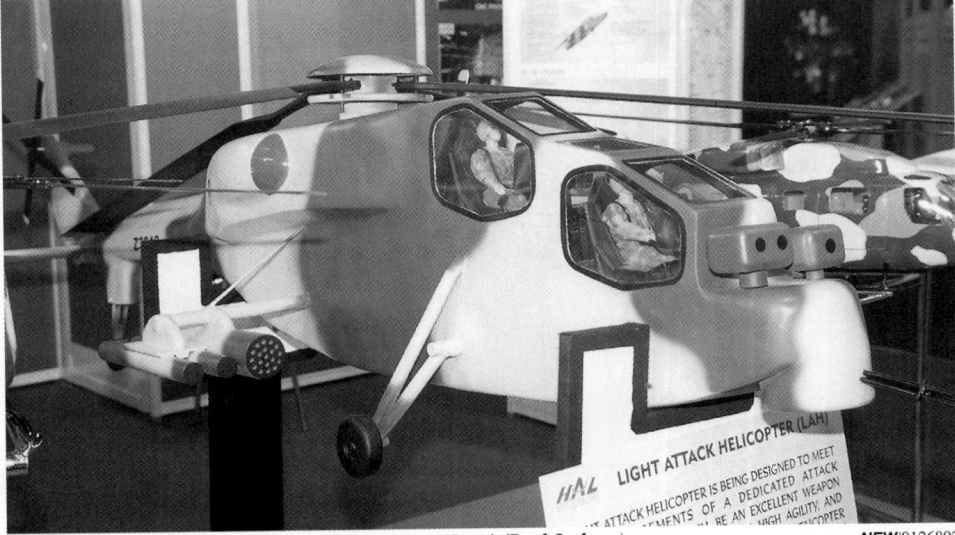

Model of HAL's projected light attack helicopter (*Jane's/Paul Jackson*)   *NEW*/0126892

DESIGN FEATURES: Slimmed-down 'gunship' fuselage with stealth characteristics and tandem crew seating. Four-blade hingeless main rotor with swept blade tips. Intended for anti-tank, close air support, air-to-air combat and scout roles.

STRUCTURE: Extensive use of composites to reduce radar signature.

LANDING GEAR: Non-retractable and crashworthy.

POWER PLANT: Two 895 kW (1,200 shp) Turbomeca/HAL TM333-2C2 Ardiden turboshafts (Indian name Shakti) with FADEC.

ACCOMMODATION: Crew of two in tandem; ergonomic cockpit.

SYSTEMS: Four-axis autostabilisation system; anti-resonance isolation system (ARIS).

ARMAMENT: Undernose 20 mm cannon or 30 mm Chain Gun. Stub-wing hardpoints for ATMs, AAMs or rocket launchers.

DIMENSIONS, EXTERNAL:

Length overall	15.87 m (52 ft 0¾ in)
Width: over stub-wings	3.19 m (10 ft 5½ in)
over armament booms	4.54 m (14 ft 10¾ in)
Height overall	4.71 m (15 ft 5½ in)

WEIGHTS AND LOADINGS (approx):

Max T-O weight	5,000 kg (11,023 lb)

PERFORMANCE (estimated):

Max cruising speed	135 kt (250 km/h; 155 mph)

UPDATED

HAL LIGHT OBSERVATION HELICOPTER (LOH)

TYPE: Light observation helicopter.

PROGRAMME: Announced at Air India Show in December 1998; design study stage at that time. Following brief details known by early 2002. Manufacture of first eight due to begin in March 2002.

DESIGN FEATURES: Envisaged as replacement for Cheetah and Chetak; configuration similar to ALH, but smaller and lighter with shorter tailboom and ducted tail rotor; bearingless main rotor; construction mainly of composites; 'glass cockpit'; health and usage monitoring system. Major roles to be day or night reconnaissance and surveillance.

LANDING GEAR: Twin skids, but provision for wheel type.

POWER PLANT: One 746 kW (1,000 shp) Turbomeca TM 333-2B2 turboshaft.

ACCOMMODATION: Up to seven persons, including pilot(s).

AVIONICS: *Radar:* Weather radar standard.

DIMENSIONS, EXTERNAL:

Main rotor diameter	approx 11.40 m (37 ft 4¾ in)

Fuselage: Length	10.355 m (33 ft 11¾ in)
Max width	1.60 m (5 ft 3 in)
Tailplane span	2.50 m (8 ft 2½ in)
Height over tailfin	4.185 m (13 ft 8¾ in)
Skid track	2.10 m (6 ft 10¾ in)

WEIGHTS AND LOADINGS (approx):

Max payload	600 kg (1,323 lb)
Max T-O weight	2,600 kg (5,732 lb)

PERFORMANCE (estimated):

Max cruising speed	130 kt (240 km/h; 149 mph)
Service ceiling	7,000 m (22,960 ft)
Hovering ceiling IGE	3,000 m (9,840 ft)
Range with max fuel	351 n miles (650 km; 403 miles)
Endurance	4 h

UPDATED

HAL HJT-36

TYPE: Basic jet trainer/light attack jet.

PROGRAMME: Revealed at Singapore Air Show, February 1998; design started 1997. Conceived as successor to IAF HJT-16 Kiran. In design development stage; mockup, with Thales (Sextant) avionics suite, exhibited at Air India Show in December 1998, differed considerably in appearance from general arrangement released at Singapore. Indian government Rs1.8 billion (US$42 million) contract awarded in July 1999, covering completion, flight test and certification of two prototypes. By early 2000, metal was being cut, although selection of risk-sharing local partners had not been announced by February 2002. Stated at that time that fuselage was ready to receive Larzac engine (due in July 2002) and first flight expected to take place December 2002; second prototype to fly six months later.

CUSTOMERS: Indian Air Force and Navy as replacement for existing fleet of approximately 170 Kirans from about

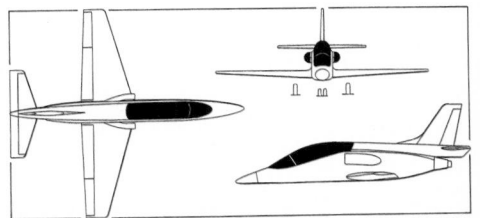

Provisional general arrangement of the HAL HJT-36 jet trainer (*Jane's/Paul Jackson*)   0121076

Model of the HAL LOH (*Jane's/Paul Jackson*)
0110518

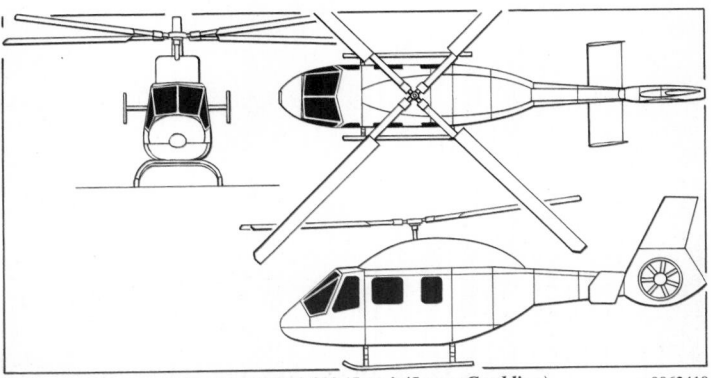

General arrangement of the HAL LOH (*Jane's/James Goulding*)   0062410

2004. Government approval for 211 awarded in January 2001, including 24 for Indian Navy.

DESIGN FEATURES: Capable of high-speed training, but with simple handling at low speeds; cockpit layout compatible with combat aircraft flying. Conventional, low-wing CAD/CAM design with moderate (18°) wing leading-edge sweepback and slight dihedral.

FLYING CONTROLS: Conventional and manual, with three-axis trim; horn-balanced elevators.

STRUCTURE: Prototypes generally of light alloy, with composites as option for production aircraft. Intended fatigue life of more than 7,500 hours, extendable to 10,000 hours.

LANDING GEAR: Retractable tricycle type. Inward-retracting main units, rearward-retracting twin nosewheels.

POWER PLANT: Prototypes each powered by one 14.03 kN (3,153 lb st) SNECMA Larzac 04-H20 non-afterburning turbofan in rear fuselage, fed by bifurcated air intake.

ACCOMMODATION: Crew of two in tandem, on lightweight zero/zero ejection seats; rear (instructor's) seat raised. One-piece canopy opens sideways to starboard.

AVIONICS: Integrated system by Smiths Aerospace, comprising dual V/UHF comms, open systems architecture mission computer, HUD, HUD repeater, GPS, AHRS, LCDs (Thales), air data computers and rear cockpit data entry panel.

ARMAMENT: One underfuselage and four underwing weapons pylons for bombs, rocket pods and gun pods.

DIMENSIONS, EXTERNAL:
Wing span   10.00 m (32 ft 9¾ in)

Length overall	11.00 m (36 ft 1 in)
Height overall	4.40 m (14 ft 5¼ in)
WEIGHTS AND LOADINGS (approx):	
Max external stores load	1,000 kg (2,204 lb)
T-O weight: clean	3,500 kg (7,716 lb)
max	4,500 kg (9,920 lb)
PERFORMANCE (estimated):	
Max operating Mach No.	0.75
Max permissible diving speed	
	405 kt (750 km/h; 466 mph)
Service ceiling	9,000 m (29,520 ft)
g limits (ultimate)	+7/−2.5
	UPDATED

BANGALORE COMPLEX

Post Bag 1785, Bangalore 560 017
Tel: (+91 80) 526 82 30
Fax: (+91 80) 527 95 64
Telex: 845 2234
MANAGING DIRECTOR: Behari Lal

Aircraft Division

Post Bag 1788, Bangalore 560 017
Tel: (+91 80) 526 89 69
Fax: (+91 80) 526 51 88
Telex: 845 2234 HALM IN
GENERAL MANAGER: K Umamaheswar

Main programmes Jaguar International; composites and metal drop tanks; Dornier 228 landing gears; sheet metal items; forward passenger doors for Airbus A319/A320/A321 (600 shipsets); overwing emergency exit doors for Boeing 757 and 777.

Helicopter Division

Post Bag 1790, Bangalore 560 017
Tel: (+91 80) 523 86 02
Fax: (+91 80) 523 47 17
Telex: 845 2764
GENERAL MANAGER: S M Kapoor

Manufacture and overhaul of Cheetah, Chetak, Lancer and ALH light helicopters.

Bangalore Complex comprises Aircraft Division, Helicopter Division, Aerospace Division, Engine Division, Overhaul Division, Foundry and Forge Division, Industrial and Marine Gas Turbines Division, and Airport Services Centre. Programmes include subcontract work for leading aerospace companies such as BAE Systems, EADS, Boeing, Fairchild Dornier and Latécoère. Aircraft Division will complete 50 of the 66 BAE Hawks which may be ordered for the Indian Air Force (eight from CKD kits, 42 by local manufacture, after delivery of 16 UK-built aircraft from BAE).

Aerospace Division manufactures light alloy structures and assemblies for satellites and launch vehicles. Engine Division manufactures, overhauls and repairs Adour Mk 811, Artouste IIIB and TPE331-5 engines; it also overhauls and repairs Adour Mk 804E, Dart, Gnome, Orpheus and Avon engines. Overhaul of Jaguar, Kiran, Mirage and An-32 aircraft, Cheetah helicopters and Pratt & Whitney and Textron Lycoming piston engines is undertaken by Overhaul Division.

UPDATED

HAL (SEPECAT) JAGUAR INTERNATIONAL
Indian Air Force name: Shamsher (Assault Sword)

TYPE: Attack fighter.
Comprehensive details in Jane's Aircraft Upgrades; *following refers generally to Jaguar S, and particularly to Indian version.*

PROGRAMME: Forty (including five two-seat) Jaguar Internationals with Adour Mk 804 engines delivered from UK, beginning March 1981; 45 more with Adour Mk 811s and new DARIN nav/attack system assembled in India, making first flight (JS136) 31 March 1982; further 31

INDIAN AIR FORCE JAGUARS

Mark/Variant	Serial Nos.	Assembled[1]	Remarks	Qty
GR. Mk 1	JI003 to 018	BAe	Loaned by/returned to RAF[2]	(16)
S(I)	JS101 to 135	BAe	Phase 2; Adour 804, NAVWASS	35
	JS136 to 170	HAL	Phase 3; Adour 811, DARIN	35
	JS171 to 194	HAL	Phase 4; Adour 811, DARIN	24[3]
	JS195 to 205	HAL	Phase 5; Adour 811, DARIN	11
Maritime	JM251 to 258	HAL	Phase 4; Adour 811, DARIN, Agave	8[3]
	JM259 to 262	HAL	Phase 5; Adour 811, DARIN, Agave	4
	JM263 et seq	HAL	Phase 6	2?
T. Mk 2	JI001 to 002	BAe	Loaned by/returned to RAF[2]	(2)
B(I)	JT051 to 055	BAe	Phase 2; Adour 804, NAVWASS	5
	JT056 to 065	HAL	Phase 3; Adour 811, DARIN	10
	JT066 *et seq*	HAL	Phase 6; night attack, Adour 811, DARIN	15
Total				**148**
				(18)

.Notes
[1] All co-built by (SEPECAT) BAe and Dassault
[2] Adour 804 engines and NAVWASS nav/attack avionics; two single-seat lost in Indian service
[3] Phase 4 originally planned as 56 aircraft of entirely local manufacture; amended to 32 kits

manufactured under licence in India (first delivery early 1988); deliveries of further 15, ordered 1993, were completed by March 1999 with delivery of last three of this batch. Indian government ordered additional batch of 17 two-seaters in March 1999. These will be equipped with upgraded DARIN II nav/attack system and employed in a night attack role with laser-guided weapons. Deliveries to Indian Air Force due to begin in March 2002. Letter of intent for further 20 single-seaters has been received, which would extend Jaguar production until 2007; these will have upgraded DARIN II nav/attack system and new LRMTS.

CURRENT VERSIONS: **Jaguar B:** Two-seat trainer version; 15 delivered; further 17 on order with DARIN II nav/attack system, laser designator pod and modified fin.

Jaguar S: Standard single-seat attack version; 106 delivered; further 20 ordered in 2000.

Maritime Jaguar: Jaguars of IAF No. 6 Squadron, assigned to anti-shipping role, have Agave radar, interfaced with DARIN nav/attack system and Sea Eagle anti-shipping missiles; first modified aircraft delivered January 1986; 12 ordered, of which all delivered by end of 1999. To be upgraded by substitution of Elta EL/M-2032 for Agave, with which flight trials were under way in 2002. In 2001, deliveries began of a further Jaguar M batch, of which two aircraft confirmed.

CUSTOMERS: Indian Air Force has received 131, comprising 116 single-seat (including 12 Maritime) and 15 two-seat combat-capable trainers. Further 17 Jaguar Bs/Ms (ordered late 1998) in production for delivery from 2001; additional 20 single-seaters ordered in 2000. Basic strike version equips Nos. 5 and 14 Squadrons at Ambala and Nos. 16 and 27 Squadrons at Gorakhpur; anti-shipping version equips No. 6 Squadron at Pune.

DESIGN FEATURES: Purpose-designed attack aircraft. Shoulder-wing monoplane. Anhedral 3°. Sweepback 40° at quarter-chord. Outer leading-edges fitted with slat which also gives effect of extended chord. Tail unit sweepback at quarter-

chord 40° on horizontal, 43° on vertical surfaces; tailplane anhedral 10°. Ventral fins beneath rear fuselage.

FLYING CONTROLS: Fairey Hydraulics powered flying controls. No ailerons: control by two-section spoilers, forward of outer flap on each wing, in association (at low speeds) with differential tailplane. Hydraulically operated (by screwjack) full-span double-slotted trailing-edge flaps. Leading-edge slats can be used in combat. All-moving slab-type tailplane, the two halves of which can operate differentially to supplement the spoilers.

STRUCTURE: Wing is an all-metal two-spar torsion box structure; skin machined from solid aluminium alloy, with integral stiffeners. Main portion built as single unit, with three-point attachment to each side of fuselage. Fuselage all-metal, mainly aluminium, built in three main units and making use of panels and, around the cockpit(s), honeycomb panels. Local use of titanium alloy in engine bay area. Two door-type airbrakes under rear fuselage, immediately aft of each mainwheel well. Structure and systems aft of cockpit(s), identical for single-seat and two-seat versions. The tail unit is a cantilever all-metal structure, covered with aluminium alloy sandwich panels. Rudder and outer panels and trailing-edge of tailplane have honeycomb core.

LANDING GEAR: Messier-Hispano-Bugatti retractable tricycle type, all units having Dunlop wheels and low-pressure tyres for rough field operation. Hydraulic retraction, with oleo-pneumatic shock-absorbers. Forward-retracting main units each have twin wheels, tyre size 615×225-10, pressure 5.8 bar (84 lb/sq in). Wheels pivot during retraction to stow horizontally in bottom of fuselage. Single rearward-retracting nosewheel, with tyre size 550×250-6 and pressure 3.9 bar (57 lb/sq in). Twin landing/taxying lights in nosewheel door. Dunlop hydraulic brakes. Anti-skid units and arrester hook standard. Irvin brake parachute of 5.5 m (18 ft 0 in) diameter in fuselage tailcone.

POWER PLANT: Two HAL-built Rolls-Royce Turbomeca Adour Mk 811 turbofans (Phase 3 aircraft onwards), each rated at 25.0 kN (5,620 lb st) dry and 37.4 kN (8,400 lb st) with afterburning. Fixed-geometry air intake on each side of fuselage aft of cockpit. Fuel in six tanks, one in each wing and four in fuselage. Total internal fuel capacity 4,200 litres (1,110 US gallons; 924 Imp gallons). Armour protection for critical fuel system components. Provision for carrying three auxiliary drop tanks, each of 1,200 litres (317 US gallons; 264 Imp gallons) capacity, on fuselage and inboard wing pylons. Provision for in-flight refuelling, with retractable probe forward of cockpit on starboard side.

ACCOMMODATION: Jaguar S has enclosed cockpit for pilot, with rearward-hinged canopy and Martin-Baker Mk 9B zero/zero ejection seat. Jaguar B crew of two in tandem on Martin-Baker Mk 9B Srs II zero/zero ejection seats. Individual rearward-hinged canopies. Rear seat 38 cm (15 in) higher than front seat. Windscreen bulletproof against 7.5 mm rifle fire.

HAL Maritime Jaguar of No. 6 Squadron, Indian Air Force (*Simon Watson*)   NEW/0132398

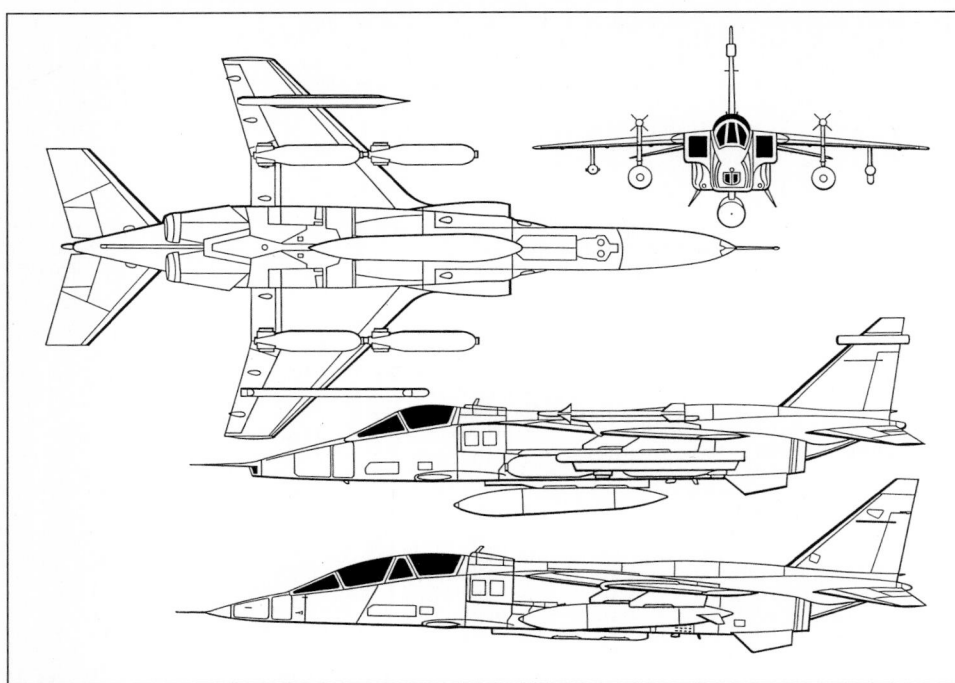

SEPECAT Jaguar single-seat tactical support aircraft and two-seat operational trainer *(Jane's/Mike Keep)*

SYSTEMS: Air conditioning and pressurisation systems for cockpit(s) also control temperature in certain equipment bays. Two independent hydraulic systems, powered by two engine-driven pumps. Hydraulic pressure 207 bar (3,000 lb/sq in). First system (port engine) supplies one channel of each actuator for flying controls, hydraulic motors which actuate flaps and slats, landing gear retraction and extension, brakes and anti-skid units. Second system supplies other half of each flying control actuator, two further hydraulic motors actuating slats and flaps, airbrake and landing gear emergency extension jacks, nosewheel steering and wheel brakes. In addition, there is a hydraulic power transfer unit and an emergency electrohydraulic pump.

Electrical power provided by two 15 kVA AC generators, either of which can sustain functional and operational equipment without load shedding. DC power provided by two 4 kW transformer-rectifiers. Emergency power for essential instruments provided by 40 Ah battery and static inverter. De-icing, rain clearance and demisting standard. Liquid oxygen system, which also pressurises pilot's anti-*g* trousers.

Jaguar is fully power-controlled in all three axes and is automatically stabilised as a weapons platform by gyros which sense disturbances and feed appropriate correcting data through a computer to the power control assemblies, in addition to human pilot manoeuvre demands. Power controls are all of duplex tandem arrangement, with mechanical and electrical servo-valves of Fairey platen design. Air-to-air combat capability can be enhanced by inclusion of roll/yaw dampers, to increase lateral stability, and by increasing slat and flap angles.

AVIONICS: HAL-manufactured DARIN (display attack and ranging inertial navigation) nav/attack system initially, incorporating INS, HUDWAC, COMED, interconnected MIL-STD-1553B dual-redundant databus and interfaced with LRMTS. The 17 new two-seaters and additional 20 strike versions will be fitted with new RLG/INS with GPS, new HUD and smart MFDs (see below); HOTAS upgrade postponed.

Comms: Main (20-channel) V/UHF; standby UHF and HAL COM 326A HF transceivers (being replaced by Incom integrated com system); IFF-400 AM transponder (single-seaters only).

Radar: Thales Avionics Agave originally in Maritime version, interfaced with DARIN system; planned to be replaced by Elta EL/M-2032.

Flight: SAGEM ULISS 82 INS (to be replaced by RLG/INGPS); HAL ADF and radar altimeter.

Instrumentation: Smiths HUDWAC (head-up display and weapon aiming computer) (to be replaced by Elop HUD with Smiths/HAL co-developed mission computer); BAE Systems COMED 2045 (combined map and electronic display) to be replaced by Thales (Sextant) digital autopilot and 152 × 152 mm (6 × 6 in) smart multifunction display (SMFD), ordered November 1999.

Mission: Laser ranger and marked target seeker (LRMTS). Provisional selection of Rafael Litening laser designation pod announced February 1996.

Self-defence: RWR; active and passive ECM.

ARMAMENT: Two 30 mm Aden guns in lower fuselage aft of cockpit in single-seater, with 150 rds/gun; single Adenon port side in two-seater (to be removed to accommodate self-protection jammer). One stores attachment on fuselage centreline and two under each wing. Centreline and inboard wing points can each carry up to 1,134 kg (2,500 lb) of weapons, outboard underwing points up to 567 kg (1,250 lb) each. Typical alternative loads include one air-to-surface missile and two 1,200 litre (317 US gallon; 264 Imp gallon) drop tanks; eight 1,000 lb bombs; various combinations of free-fall, laser-guided, retarded or cluster bombs; overwing R.550 Magic missiles; air-to-surface rockets; or a reconnaissance camera pack. Maritime Jaguars equipped with one or two Sea Eagle anti-shipping missiles.

DIMENSIONS, EXTERNAL:
Wing span	8.72 m (28 ft 7¼ in)
Wing aspect ratio	3.1
Length overall, incl probe:	
single-seat	16.955 m (55 ft 7½ in)
two-seat	17.52 m (57 ft 6 in)
Height overall	4.89 m (16 ft 0½ in)
Wheel track	2.40 m (7 ft 10½ in)
Wheelbase	5.69 m (18 ft 8 in)

AREAS:
Wings, gross	24.03 m² (258.7 sq ft)

WEIGHTS AND LOADINGS:
Typical weight empty	7,000 kg (15,432 lb)
Max external stores load (incl overwing)	4,763 kg (10,500 lb)
Normal T-O weight (single-seater, with full internal fuel and ammunition for built-in cannon)	10,954 kg (24,149 lb)
Max T-O weight with external stores	15,700 kg (34,612 lb)
Max wing loading	653.3 kg/m² (133.79 lb/sq ft)
Max power loading	210 kg/kN (2.06 lb/lb st)

PERFORMANCE:
Max level speed at S/L	M0.98 (648 kt; 1,200 km/h; 745 mph)
Max level speed above 6,000 m (19,680 ft)	M1.5 (907 kt; 1,680 km/h; 1,044 mph)
Touchdown speed	115 kt (213 km/h; 132 mph)
Service ceiling	13,715 m (45,000 ft)
T-O run, clean:	565 m (1,855 ft)
with four 1,000 lb bombs	880 m (2,890 ft)
with eight 1,000 lb bombs	1,250 m (4,100 ft)
T-O to 15 m (50 ft) with typical tactical load	940 m (3,085 ft)
Landing from 15 m (50 ft) with typical tactical load	785 m (2,575 ft)
Landing run:	
normal weight, with brake-chute	470 m (1,540 ft)
normal weight, without brake-chute	680 m (2,230 ft)
overload weight, with brake-chute	670 m (2,200 ft)
Typical attack radius, internal fuel only:	
hi-lo-hi	460 n miles (852 km; 530 miles)
lo-lo-lo	290 n miles (537 km; 334 miles)
Typical attack radius with external fuel:	
hi-lo-hi	760 n miles (1,408 km; 875 miles)
lo-lo-lo	495 n miles (917 km; 570 miles)
Range with max external fuel	1,400 n miles (2,593 km; 1,611 miles)
g limits	+8.6 (+12 ultimate)/−3

UPDATED

HAL CHEETAH

TYPE: Light utility helicopter.

PROGRAMME: Licence-built Aerospatiale SA 315B Lama. Production by HAL for Indian armed forces started 1972. In early 2000, HAL received further orders for production for several years and has taken up possible employment of Turbomeca TM 333-2B2 turboshaft to enhance performance. See 1996-97 and earlier *Jane's* for full description; shortened version in 1997-98 edition.

CUSTOMERS: HAL had delivered 260 by mid-2002.

UPDATED

HAL LANCER

TYPE: Armed observation helicopter.

PROGRAMME: Upgraded, counter-insurgency version of Cheetah. Prototype, rebuilt from late production aircraft Z2867, unveiled late 1998: cabin has lightweight composites armour protection for pilot, control linkages and fuel tank; bullet-proof flat-plate transparencies; anti-tank missiles or twin gun/rocket pods. Can be rebuild or new production. See *Jane's Aircraft Upgrades* for further details.

CUSTOMERS: Indian Army order for 12 (converted from Cheetah), of which three delivered by end of 2001; remainder due to follow during 2002. Ten ordered in November 2001 by Nepal for police use.

ACCOMMODATION: Crew of two, side by side; cleared for single-pilot operation.

ARMAMENT: Tubular metal outrigger each side of cabin, outboard of landing skids, each supporting jettisonable pod

HAL Lancer prototype, an armed and armoured version of the Cheetah   0099550

containing one 12.7 mm machine gun and three 70 mm unguided air-to-surface rockets. Pilot's gunsight.

WEIGHTS AND LOADINGS:
Basic weight empty: Cheetah	1,090 kg (2,403 lb)
Lancer, incl weapon pods	1,350 kg (2,976 lb)
Fuel weight	393 kg (866 lb)
Ammunition (500 rds) and six rockets	130 kg (287 lb)
Max T-O weight	1,950 kg (4,299 lb)

PERFORMANCE:
Never-exceed speed (VNE)	113 kt (210 km/h; 130 mph)
Cruising speed at S/L, ISA + 20°C	95 kt (176 km/h; 109 mph)
Max rate of climb at S/L	330 m (1,083 ft)/min

Service ceiling	5,400 m (17,720 ft)
Operational radius, incl 30 min over target and 20 min fuel reserves, ISA + 20°C	78 n miles (145 km; 90 miles)
Endurance	2 h 30 min

UPDATED

HAL CHETAK

TYPE: Light utility helicopter.
PROGRAMME: Licence-built Aerospatiale SA 316B Alouette III; remains in production only in India. HAL output had

totalled at least 359 (of which 35 built from French kits) by mid-2002. One was due for delivery to Sri Lanka Navy in late 2000. Further orders, to prolong production for several years, received in early 2001. Turbomeca TM333-2B2 under consideration as alternative power plant. Shortened description in 1997-98 edition; see 1996-97 and earlier *Jane's* for more detailed version.

UPDATED

TRANSPORT AIRCRAFT DIVISION

PO Box 225, Chakeri, Kanpur 208 008
Tel: (+91 512) 45 03 61 and 40 27 74
Fax: (+91 512) 45 05 05
e-mail: halknp@sancharnet.in
GENERAL MANAGER: B K Banerjee

This Division, established in 1960, is responsible for manufacture of the Dornier 228, and for overhaul and repair of the Dornier 228, Hawker Siddeley 748, HAL HPT-32 and civil aircraft. Co-operation agreement with Israel Aircraft Industries 21 February 2002 involves HAL in IAI (Bedek) Boeing 737 freighter conversion programme. Division also assisting in design and development, and manufacturing prototypes, of the CSIR Saras (which see); series production at Kanpur is planned.

Division holds ISO 9001 and ISO 14001 accreditation and is equipped with Transport Aircraft Research and Design Centre (TARDC) specialising in role modifications, sensor integration, aircraft upgrades and repair technology.

UPDATED

GERMAN/INDIAN DORNIER 228 DELIVERIES
(to March 2001)

Country	Operator	Series	Dornier	HAL/CKD	HAL	Total
India	Air Force	201		5	20	25[1]
	Airports Authority	201	1	1		2
	Coast Guard	101	3	4	11	18
		201			3	3[3]
	Navy	201	1		14	15[2]
	Oil and Natural Gas Commission	101	1			1
	UB Air	101		1		1
	Vayudoot	201	5	5		10
Mauritius	Coast Guard	101			1	1
Totals			**11**	**16**	**49**	**76**

Notes:
[1] Further order for 18 expected
[2] Further four in production
[3] Four more on order

HAL (DORNIER) 228

TYPE: Twin-turboprop transport.
PROGRAMME: First flight of Dornier 228-100 prototype (D-IFNS) in Germany 28 March 1981; first flight 228-200 prototype (D-ICDO) 9 May 1981; British CAA certification 17 April 1984, FAR Pt 23 and Appendix A Pt 135 11 May 1984, Australian 11 October 1985. For European production (including 11 for India), see table below and German section of 1996-97 and previous *Jane's*.

Contract for licensed manufacture of up to 150 Dornier 228s in India signed 29 November 1983; one pattern aircraft supplied complete from Germany; production planned in four phases (six, 10, 10 and 39 aircraft respectively), fourth phase representing full local manufacture after production from locally assembled kits; first Phase 1 Kanpur-assembled aircraft (VT-EJN, c/n 1002) handed over to Vayudoot on 1 November 1985 and made maiden flight on 31 January 1986. Full local manufacture from 18th onwards (CG758, first flight 7 March 1991). Indian production, programme dormant from first quarter of 1997, but now continuing. Following discontinuation of German production, worldwide sales and marketing rights held by HAL since November 1998; HAL pursuing export market and anticipates future orders during remainder of current decade.
CURRENT VERSIONS: **Regional Airliner:** Ten **228-201**s (five each by Dornier and HAL) delivered to Vayudoot.

Maritime Surveillance: Thirty-six ordered for Indian Coast Guard (including three from Germany, delivered from July 1986); CG754 first flight 18 February 1988; first

18 from Indian production are **228-101**s and were delivered by March 1997. Go-ahead for next batch (seven **228-201**s) received in September 1999; these ordered under US$72 million contract in early 2000, of which three delivered by March 2001. One also to Mauritius Coast Guard in 1990.

In service by early 2002 with Coast Guard Air Squadrons 700 (Dum Dum), 744 (Chennai), 745 (Andaman and Nicobar Islands) and 750 (Daman) Squadrons at Daman and Chennai; for coastal, environmental and anti-smuggling patrol; 360° Marec 2 search radar under fuselage (replaced in 15 aircraft by Super Marec); Matra infra-red/ultraviolet linescanner for pollution detection. Search and rescue liferafts, searchlight, hand-held aerial camera, bubble windows, side-mounted loudhailer, marine markers, and provision for two Micronair underwing spraypods to combat oil spills; in-flight-operable roller door for parajumping/paradropping. Normal crew two pilots, radar operator and observer. Optional armament includes two underwing 7.62 mm twin-gun pods.

Anti-ship: Indian Navy has acquired 15 specially equipped Dornier **228-201**s (ordered in batches of five and 10) with anti-ship missiles and Super Marec radar. Deliveries began (IN221) on 24 August 1991 for service with INAS 310 ('Cobras') at Dabolim; last of second batch completed in 2000. Integration of Elta EL/M-2022A search radar on these 10 aircraft was scheduled for completion by March 2002. Apart from radar (360° scan, near-SAR/MTI and track-while-scan) under forward fuselage, current aircraft have IAI Tamam Airborne Multimission Optronic Stabilised Payload (AMOSP) (TV/

FLIR/laser range-finder) ball turret in starboard main gear fairing; air/ground data downlink; and cabin consoles for four systems operators with MFDs and moving map display. Further four of this version in production.
Utility transport: Indian Air Force received 25 **228-201**s; deliveries started (HM667 and HM668) 15 December 1987 to Nos. 41 'Otters' and 59 'Hornbills' Squadrons; can carry 21 field-equipped troops on inward-facing composites folding seats; large HAL-developed cargo double door, to meet IAF requirement, at rear, port side. Expectation of IAF follow-on order for further 18.
Executive/Air taxi: Various configurations including six- or 10-seat executive or 15-passenger air taxi, with customised interior, cabin attendant, work table and galley/wardrobe/lavatory; built-in APU for air conditioning and lighting in flight or on ground. One (VT-EIX) for Indian Oil and Natural Gas Commission.
CUSTOMERS: Total 244 German aircraft built (including prototypes and Indian kits) by late 1998; further three available. HAL deliveries are given in the accompanying table.
DESIGN FEATURES: Optimised for efficient cruising flight. High wing for propeller clearance, despite minimised fuselage ground clearance and pannier-mounted landing gear. Variable sweep wing leading-edge; unswept tailplane; sweptback fin with dorsal fillet.
Special Dornier wing with Do A-5 supercritical aerofoil; 8° leading-edge sweep on outer wing panels; raked tips; no dihedral or anhedral.
FLYING CONTROLS: Conventional and manual. Variable incidence tailplane with actuator switch on aileron wheel; horn-balanced elevators; rudder trim tab; single-slotted Fowler flaps augmented by drooping ailerons; two strakes under rear fuselage for low-speed stability.
STRUCTURE: Two-spar wing box; mainly light alloy structure, but with CFRP wingtips and tips of tailplane and elevators; GFRP nosecone, tips of rudder and fin; Kevlar landing gear fairings and in part of wing ribs; hybrid composites in fin leading-edge; fuselage unpressurised, built in five sections.
LANDING GEAR: Retractable tricycle type, with single mainwheels and twin-wheel nose unit; main units retract inward into fuselage fairings; hydraulically steerable nosewheels retract forward; Goodyear wheels and tyres, 25.5×8.75-10 (650×220-10) (10/12 ply) or 8.50-10 (10 ply) on mainwheels, 6.00-6 (8 ply), on nosewheels.
POWER PLANT: HAL-built Dornier 228-201s have two Honeywell TPE331-5-252D turboprops, each flat rated at 533 kW (715 hp) to ISA +18°C and driving a Hartzell four-blade fully feathering propeller with reverse pitch. Post-2000 production aircraft to have HAL-built TPE331-10 series engines; performance improvements of -5 engines through turbine blade enhancements also planned.
Primary wing box forms integral fuel tank with standard usable capacity 2,386 litres (630 US gallons; 525 Imp gallons); can be increased to 2,850 litres (753 US gallons; 627 Imp gallons). Oil capacity per engine 5.9 litres (1.56 US gallons; 1.30 Imp gallons).
ACCOMMODATION: Crew of one or two; pilots' seats adjustable fore and aft; two-abreast seating with central aisle; maximum capacity 19 (more in military versions); flight deck door on port side; combined two-section passenger

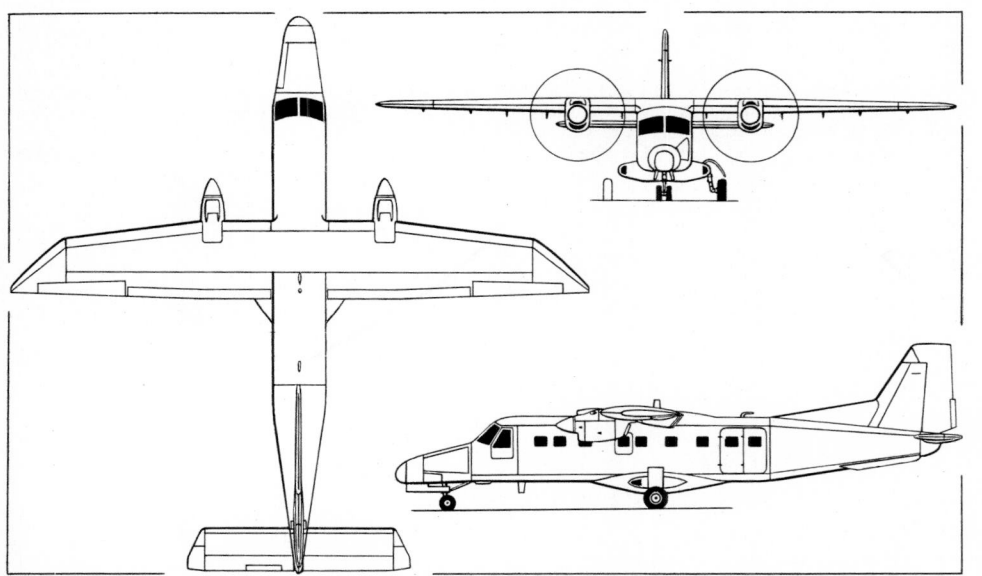

HAL (Dornier) 228-201 (two TPE331 turboprops) (*Jane's/Dennis Punnett*)

Indian Navy HAL Dornier 228-201 *(Simon Watson)*   *NEW*/0132400

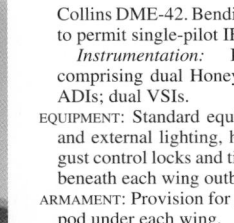

Pollution control spraypods on Indian Coast Guard HAL (Dornier) 228   0110513

and freight door, with integral steps, on port side of cabin at rear; one emergency exit on port side of cabin, two on starboard side; baggage compartment at rear of cabin, accessible externally and from cabin; capacity 210 kg (463 lb). Enlarged baggage door optional; additional baggage space in fuselage nose, with separate access; capacity 120 kg (265 lb); modular units using seat rails for rapid changes of role.

SYSTEMS: Entire accommodation heated (by engine bleed air) and ventilated. Hydraulic system, pressure 207 bar (3,000 lb/sq in), for landing gear, brakes and nosewheel steering; hand pump for emergency landing gear extension. Primary 28 V DC electrical system, supplied by two 28 V 250 A engine-driven starter/generators and two 24 V 25 Ah Ni/Cd batteries; two 350 VA inverters supply 115/26 V 400 Hz AC system. Air intake anti-icing standard; de-icing system optional for wing and tail unit leading-edges, fuselage, windscreen and propellers.

AVIONICS: *Comms:* Standard avionics include HAL UHS 190A UHF and COM 328A HF radios; Becker audio selector and intercom; Motorola Selcal; H R Smith ELT-503.

Radar: HAL-built Primus 500 weather radar; Elta or Super Marec 360° search radars.

Flight: Rockwell Collins VIR-32A VOR/ILS; dual or single Rockwell Collins ADF-62A; dual or single Bendix/ King KNI 582 RMI indicators; dual or single Rockwell

Collins DME-42. Bendix/King KFC 250 autopilot optional to permit single-pilot IFR operation.

Instrumentation: IFR instrumentation standard, comprising dual Honeywell GH14B gyro horizons; dual ADIs; dual VSIs.

EQUIPMENT: Standard equipment includes complete internal and external lighting, hand fire extinguisher, first aid kit, gust control locks and tiedown kit. Stores attachment point beneath each wing outboard of engine nacelle.

ARMAMENT: Provision for gun pod, other weapons or mission pod under each wing.

DIMENSIONS, EXTERNAL:

Wing span	16.97 m (55 ft 8 in)
Wing aspect ratio	9.0
Length overall	16.56 m (54 ft 4 in)
Height overall	4.86 m (15 ft 11½ in)
Tailplane span	6.45 m (21 ft 2 in)
Wheel track	3.30 m (10 ft 10 in)
Wheelbase	6.29 m (20 ft 7½ in)
Propeller diameter	2.69 m (8 ft 10 in)
Propeller ground clearance	1.08 m (3 ft 6½ in)
Passenger door (port, rear): Height	1.35 m (4 ft 5 in)
Width	0.64 m (2 ft 1¼ in)
Height to sill	0.91 m (2 ft 11¾ in)
Freight door (port, rear): Height	1.35 m (4 ft 5 in)
Width, incl passenger door	1.28 m (4 ft 2½ in)
Emergency exits (each): Height	0.66 m (2 ft 2 in)
Width	0.48 m (1 ft 7 in)
Baggage door (nose): Height	0.50 m (1 ft 7½ in)
Width	1.20 m (3 ft 11¼ in)
Standard baggage door (rear):	
Height	0.90 m (2 ft 11½ in)
Width	0.53 m (1 ft 9 in)

DIMENSIONS, INTERNAL:

Cabin, excl flight deck and rear baggage compartment:	
Length	7.08 m (23 ft 2¾ in)
Max width	1.346 m (4 ft 5 in)
Max height	1.55 m (5 ft 1 in)
Floor area	9.6 m² (103 sq ft)
Volume	14.7 m³ (519.1 cu ft)
Rear baggage compartment volume	2.6 m³ (92 cu ft)
Nose baggage compartment volume	0.89 m³ (31.4 cu ft)

AREAS:

Wings, gross	32.00 m² (344.3 sq ft)
Ailerons (total)	2.71 m² (29.17 sq ft)
Trailing-edge flaps (total)	5.87 m² (63.18 sq ft)
Fin, incl dorsal fin	4.50 m² (48.44 sq ft)
Rudder, incl tab	1.50 m² (16.15 sq ft)
Horizontal tail surfaces (total)	8.33 m² (89.66 sq ft)

WEIGHTS AND LOADINGS (Indian-built 228-201):

Operating weight empty	3,687 kg (8,128 lb)
Max T-O weight: civil	6,200 kg (13,668 lb)
military	6,400 kg (14,110 lb)
Max ramp weight: civil	6,230 kg (13,734 lb)
military	6,430 kg (14,175 lb)
Max landing weight	5,900 kg (13,007 lb)
Max zero-fuel weight	5,590 kg (12,324 lb)
Max wing loading: civil	193.8 kg/m² (39.68 lb/sq ft)
military	200.0 kg/m² (40.96 lb/sq ft)
Max power loading: civil	5.82 kg/kW (9.56 lb/shp)
military	6.00 kg/kW (9.87 lb/shp)

PERFORMANCE (228-201):

Never-exceed speed (V_NE)	
	255 kt (472 km/h; 293 mph) IAS
Max operating speed (V_MO)	
	223 kt (413 km/h; 256 mph) IAS
Max cruising speed: at S/L	222 kt (411 km/h; 255 mph)
at 3,050 m (10,000 ft)	231 kt (428 km/h; 266 mph)
Stalling speed: flaps up	79 kt (147 km/h; 91 mph) IAS
flaps down	63 kt (117 km/h; 73 mph) IAS
Max rate of climb at S/L	582 m (1,909 ft)/min
Service ceiling, 30.5 m (100 ft)/min rate of climb	
	8,535 m (28,000 ft)
Service ceiling, OEI, 15 m (50 ft)/min rate of climb	
	4,300 m (14,100 ft)
T-O run	442 m (1,450 ft)
T-O to 15 m (50 ft)	655 m (2,150 ft)
Accelerate/stop distance, with anti-skid	762 m (2,500 ft)
Landing from 15 m (50 ft) at MLW	536 m (1,760 ft)
Range at 3,050 m (10,000 ft) with 19 passengers, reserves for 50 n mile (93 km; 57 mile) diversion, 45 min hold and 5% fuel remaining:	
at max cruising speed	
	560 n miles (1,038 km; 645 miles)
at max range speed	630 n miles (1,167 km; 725 miles)
Range with 775 kg (1,708 lb) payload, conditions as above: at max cruising speed	
	1,160 n miles (2,148 km; 1,335 miles)
at max range speed	
	1,320 n miles (2,445 km; 1,519 miles)

UPDATED

HAL (Dornier) 228 7.62 mm underwing gun pod
NEW/0110514

MiG COMPLEX

Ojhar Township Post Office, Nasik, Maharashtra 422 207
Tel: (+91 2550) 751 00
Fax: (+91 2550) 758 21
MANAGING DIRECTOR: K P Puri
GENERAL MANAGERS:
M S Nadgir (Nasik Aircraft Division)
J R Mehta (Koraput Engine Division)

This complex comprises the Nasik Aircraft and Koraput Engine Divisions of HAL.

Licensed manufacture of Sukhoi Su-30MKI is undertaken by this complex. Agreement signed 28 December 2000 for local manufacture of up to 140 (due to start in late 2004) following earlier order for initial batch of 40 Russian-built Su-30MKIs (of which 28 delivered by late 2002); contract

valued at US$3.3 billion; HAL investment in programme reportedly valued at US$650 million. HAL manufacture starting in 2003 and targeted for completion in 2017; first deliveries to Indian Air Force due 2004. Other present activities are MiG-21/27 upgrade, overhaul and repair; see *Jane's Aircraft Upgrades*.

UPDATED

ACCESSORIES COMPLEX

PO Box 215, Lucknow 226 016
Tel: (+91 522) 34 03 27
Fax: (+91 522) 34 03 42
MANAGING DIRECTOR: L M Bhardwaj

Comprises Lucknow (Accessories), Hyderabad (Avionics) and Korwa (Avionics) Divisions.

UPDATED

RAJ HAMSA

RAJ HAMSA ULTRALIGHTS PVT LTD

40 Goshala Road, Mahadevapura, Bangalore 560 048
Tel: (+91 80) 851 69 37
Fax: (+91 80) 560 80 25
e-mail: contact@rajhamsa.com
Web: http://www.rajhamsa.com

MANAGING DIRECTOR: Joël Koechlin

PARTNER AND DISTRIBUTOR:
Randkar SA
Canal de la Martinière, F-44320 Frossay, France
Tel: (+33 2) 40 64 21 66
Fax: (+33 2) 40 64 15 22

e-mail: contact@randkar.fr
Web: http://www.randkar.fr

Company was founded in 1980, initially building hang gliders; powered hang gliders introduced 1983 and X-Air range 10 years later.

VERIFIED

RAJ HAMSA X-AIR

TYPE: Side by side ultralight kitbuilt.

PROGRAMME: Launched 1993 as development of US Weedhopper; has BCAR Section S certification.

CURRENT VERSIONS: **X-Air:** Baseline version.

X-Air F Gumnam: Has more efficient wing, of higher aspect ratio (6.4 instead of 6.0) and fitted with flaps; deeper rear fuselage, providing space for 20 kg (44 lb) of baggage aft of seats; optional auxiliary fuel tank in wing; wire-braced tailplane. Known simply as **X-Air F** for European and US marketing; or in UK as **Falcon**, with Rotax 912 flat-four.

CUSTOMERS: Over 500 in operation worldwide, including more than 40 in India and more than 80 in UK.

Following description applies to both versions, except where indicated.

DESIGN FEATURES: High-wing monoplane with sweptback leading-edges and wing-mounted engine. Quick-build kit (quoted build time only 40 hours). Potential applications include basic training, crop-spraying, aerial observation and surveillance.

NACA 4412 section wing on Gumnam.

FLYING CONTROLS: Conventional and manual. Differential ailerons; fixed tab on port elevator. All control surfaces non-balanced.

STRUCTURE: Aluminium alloy subframe, with composites nose module; rear fuselage, wings and tail surfaces fabric-covered. Braced wings (V struts); tailplane strut-braced on X-Air, wire-braced on X-Air F/Gumnam.

LANDING GEAR: Tricycle type; fixed; shock absorbers on all three units. Mainwheel drum brakes; nosewheel steerable by rudder pedals. Optional mainwheel speed fairings (standard on Gumnam) and nosewheel guard. Small tailskid at base of fin.

POWER PLANT: *X-Air:* Typically, one 37.0 kW (49.6 hp) Rotax 503 UL-2V, 47.8 kW (64.1 hp) Rotax 582 UL or 48.5 kW (65 hp) Hirth 2706 engine, all with dual electronic ignition, electric self-starter and reduction gearbox (2.58:1, 3:1 and 3:1 respectively); two-blade (optionally three-blade) wooden propeller standard. Zanzottera MZ202, HKS 700 and Jabiru 2200 installations have also been made.

Gumnam: Rotax 582 or Hirth 2706 only.

Raj Hamsa X-Air two-seat ultralight *(Jane's/Paul Jackson)*   NEW/0137211

Fuel (Mogas/oil mixture) in two tanks aft of seats, combined capacity 60 litres (15.9 US gallons; 13.2 Imp gallons); auxiliary wing tank optional for Gumnam.

ACCOMMODATION: Open-sided cockpit behind large windscreen standard on X-Air; option to enclose fully with Lexan door each side; doors standard on Gumnam. Dual controls standard.

AVIONICS: *Comms:* Com radio and intercom optional.
Flight: Compass and GPS nav receiver optional.
Instrumentation: Tachometer, ASI, altimeter, turn and slip indicator, battery charge indicator and CHT or water temperature gauge standard.

EQUIPMENT: Ballistic recovery parachute optional.

DIMENSIONS, EXTERNAL:
Wing span   9.80 m (32 ft 1¾ in)

Length overall	5.65 m (18 ft 6½ in)
Height overall	2.55 m (8 ft 4½ in)
Wheel track	1.60 m (5 ft 3 in)
Wheelbase	1.45 m (4 ft 9 in)
AREAS:	
Wings, gross: X-Air	16.00 m² (172.2 sq ft)
Gumnam	15.00 m² (161.5 sq ft)
WEIGHTS AND LOADINGS:	
Weight empty (both): Rotax 503	230 kg (507 lb)
Rotax 582, Hirth 2706	240 kg (529 lb)
Max T-O weight (both)	410 kg (903 lb)
PERFORMANCE, POWERED (both):	
Max level speed at S/L	65 kt (120 km/h; 75 mph)
Econ cruising speed at S/L	
	43-54 kt (80-100 km/h; 50-62 mph)
Min practical flying speed	35 kt (65 km/h; 40 mph)
Stalling speed	25 kt (45 km/h; 28 mph)
Max rate of climb at S/L	180 m (591 ft)/min
Practical ceiling	3,660 m (12,000 ft)
T-O run	100 m (330 ft)
T-O to 15 m (50 ft)	220 m (725 ft)
Landing from 15 m (50 ft)	230 m (755 ft)
Endurance at cruising speed, standard fuel	more than 3 h
g limits: X-Air	+6/−3
Gumnam	+6/−4
PERFORMANCE, UNPOWERED (both):	
Best glide ratio	9

UPDATED

Deeper rear fuselage characterises the Gumnam, or Falcon, version of X-Air F *(Jane's/Paul Jackson)*   NEW/0137212

RAJ HAMSA HANUMAN

TYPE: Two-seat ultralight kitbuilt.

PROGRAMME: Under development; prototype nearing completion in mid-2001.

DESIGN FEATURES: Further development of X-Air/Gumnam, redesigned for even faster kit assembly. Wings have full-span flaperons; wings and horizontal tail foldable for storage and transportation.

POWER PLANT: One 82.0 kW (110 hp) Hirth F30 in prototype. Designed for Rotax 912, Jabiru or similar engines in 59.7 to 89.5 kW (80 to 120 hp) range.

VERIFIED

TAAL

TANEJA AEROSPACE AND AVIATION LTD

1010 Tenth Floor, Prestige Meridian-1, 29 MG Road, Bangalore 560 001
Tel: (+91 80) 555 06 09 and 555 06 10
Fax: (+91 80) 555 09 55
e-mail: taal@vsnl.com
Web: http://www.tanejaaerospace.com
CHAIRMAN: Khushroo Rustumji
MANAGING DIRECTOR: Salil Taneja
JOINT MANAGING DIRECTOR: Arvind Nanda
DEPUTY GENERAL MANAGER, AIRCRAFT SALES:
Santosh Despande

TAAL is part of Indian Seamless Metal Tubes Group; has modern plant at Hosur, near Bangalore, with 1,300 m (4,265 ft) captive runway, hangars, laboratories, paint shops and other facilities for aircraft manufacture and overhaul. Entered into technical agreement second quarter 1992 with Partenavia of Italy to produce P.68C/TC, P.68 Observer and AP.68TP-600 Viator light twins in India. Installed capacity of plant is 24 aircraft per year.

TAAL's other major aircraft programme is the Hansa-3 light trainer. Also active in air taxi operations and manufacture of UAVs for Aeronautical Development Establishment and structural parts for space launchers.

UPDATED

TAAL-assembled P.68C in Taneja house colours   0079821

TAAL (PARTENAVIA) P.68 SERIES

TYPE: Light utility twin-prop transport

PROGRAMME: Production of P.68C by Partenavia in Italy started 1978, followed by P.68TC in 1980. Production agreement made with TAAL in 1992; European manufacture ended in 1994, but subsequently resumed by VulcanAir, which disassociated itself from Indian production.

TAAL programme aimed at meeting growing demand for this category of aircraft within India. Initial plans covered two demonstrators (assembled from Italian kits). First

TAAL-assembled aircraft, a P.68 Observer 2 (VT-TAA), made its first flight 17 March 1994; second was Viator. Three more P.68Cs assembled from Italian parts and sold to Indian operators; two for Mardia Chemicals and Triveni Sheet Glass; and one P.68TC demonstrator. Last of this kit-built batch completed mid-1996.

Local manufacture of the aircraft began as planned on 1 January 1997. Roll-out of first TAAL-built aircraft (VT-TAH, a P.68C) took place 20 January 1998; six (including one Observer 2 for Sanmar Alloy Casting) delivered by

February 2001. Fractional ownership scheme introduced in 1999.

For detailed description of P.68/Observer series, see under VulcanAir in Italian section.

CUSTOMERS: Six locally manufactured aircraft delivered to Indian operators and corporate owners by February 2001; further 18 P.68Cs on order at that time, all scheduled for 2001 delivery, but only known delivery by year-end was of VT-TLD to Lafarge India Ltd on 25 July.

UPDATED

NAL/TAAL HANSA-3
English name: Swan

TYPE: Primary prop trainer/sportplane.

PROGRAMME: Developed under agreement with Indian National Aerospace Laboratories and originally known as NALLA: NAL Light Aircraft. Design started May 1989; construction of Hansa-2 prototype began December 1991, and this aircraft (VT-XIW) made first flight 23 November 1993. Prototype received Indian DGCA Experimental category type certificate; re-engined 1995 as Hansa-2RE with more powerful IO-240 replacing original 74.6 kW (100 hp) O-200 flat-four. Retired to HAL's museum after 128 hours of flying. Deliveries of production Hansa-3s began on 20 March 2001.

In early 2002, NAL was in the early stages of designing a 'stretched' Hansa with between four and six seats.

CURRENT VERSIONS: **Hansa-2:** Prototype in original configuration; described in 1995-96 *Jane's.*

Hansa-2RE: Prototype (VT-XIW) re-engined with 93 kW (125 hp) IO-240B; span increased and flaps added; first flight 26 January 1996.

Hansa-3: Production version of Hansa-2RE, first flown (VT-XAL; designated Hansa-3I) on 25 November 1996. First flight with Rotax 914 F3 engine, 11 May 1998, made by second (and lighter) prototype VT-XBL (designated Hansa-3II); provisional JAR-VLA certification awarded to 'XBL in December 1998 in day VFR category. Current version (prototype VT-HNS) has lightning protection for night operations; made first flight 14 May 1999 and received DGCA certification on 1 February 2000 under FAR Pt 23, Amendment 23-42 (JAR-VLA night operation requirements).

Following description applies to this version.

CUSTOMERS: Four prototypes, including Hansa-2, of which VT-HNS handed over on 22 March 2002 for use by Department of Aerospace Engineering, IIT Kanpur

Three ordered by DGCA for Indian flying clubs (VH-HNT, 'HNU and 'HNV), of which first handed over to DGCA on 20 March 2001, then presented to Andhra Pradesh Aviation Academy on 12 April 2001 and delivered to Hyderabad on 8 May 2001. Second and third to DGCA

on 22 March 2002, of which 'NU to Kerala Aviation Training Centre.

COSTS: Programme development US$40 million (1997); standard aircraft Rs4.5 million (2001).

DESIGN FEATURES: Docile handling qualities for *ab initio* training; robust construction and low acquisition/operating cost. Low-wing monoplane with circular-section waisted fuselage; outer wings tapered on leading-edges, with unswept trailing-edge; sweptback fin and rudder; shallow ventral strake; conventional unswept, straight-tapered horizontal tail surfaces.

Laminar-flow wing, with NASA LS(1)-0415 aerofoil section on constant-chord inboard portion, linearly tapered to LS(1)-0413 from there to tip; sweepback 12° 18′ 36″ on outer leading-edges; dihedral 4° from roots; incidence 0°; twist 0° between root and kink; washout −2° between kink and tip.

FLYING CONTROLS: Conventional and manual. Frise ailerons (100 per cent internal mass balance); horn-balanced plain elevators and large rudder, all actuated by pushrods; pitch trim by electrically operated tab in port elevator. Single-slotted Fowler flaps, deflection 20°.

STRUCTURE: Built entirely of composites (CFRP/GFRP reinforced epoxy) with hand-laid, vacuum-bagged, room temperature-cured sandwich shells; components are post-cured before assembly. Conventional rib and three-spar wings, with moulded sandwich shell and PVC foam core; two-spar tail surfaces similar. Fuselage is also a moulded sandwich shell with PVC foam core.

LANDING GEAR: Non-retractable tricycle type, with cantilever steel spring mainwheel legs and steerable (±33°) nosewheel. Cleveland wheels and McCreary tyres (main 6.00-6, nose 5.00-5), pressure 2.07 bar (30 lb/sq in) on all units. Cleveland hydraulic disc brakes. Minimum ground turning radius 2.64 m (8 ft 8 in).

POWER PLANT: One 84.6 kW (113.4 hp at 5,800 rpm) Rotax 914 F3 turbocharged flat-four engine; Hoffmann NOV 352FQ+8 two-blade constant-speed propeller. Single composites fuel tank aft of cockpit, usable capacity 85 litres (22.5 US gallons; 18.7 Imp gallons). Single gravity refuelling point on top of fuselage, port side. Oil capacity 3 litres (0.8 US gallon; 0.7 Imp gallon).

ACCOMMODATION: Two seats side by side; dual controls standard. Upward-opening gullwing doors. Baggage compartment aft of pilot's seat.

SYSTEMS: Manual (toe-operated) hydraulic mainwheel brakes. Electrical (DC) system powered by 14 V, 40 A generator and 12 V, 18 Ah Gill G25M lead-acid battery.

AVIONICS: *Comms:* Honeywell KLX 125 combined VHF/VOR com/nav unit with concealed foil antennas; ACK Technologies E-01 ELT; Clark intercom.

Flight: Optional GPS.

Instrumentation: Conventional VFR (IFR version

planned). For day operations: ASI, VSI, altimeter, magnetic compass, slip/skid indicator and OAT gauge. For night operations: artificial horizon, directional gyro, turn co-ordinator, navigation lights, landing lights, anti-collision lights, panel and map lights.

DIMENSIONS, EXTERNAL:

Wing span	10.47 m (34 ft 4¼ in)
Wing chord: at root	1.30 m (4 ft 3¼ in)
at tip	0.80 m (2 ft 7½ in)
Wing aspect ratio	8.8
Length overall	7.66 m (25 ft 1½ in)
Fuselage max width	1.13 m (3 ft 8½ in)
Height overall	2.615 m (8 ft 7 in)
Tailplane span	3.60 m (11 ft 9¾ in)
Wheel track	2.20 m (7 ft 2½ in)
Wheelbase	1.95 m (6 ft 4¾ in)
Propeller diameter	1.73 m (5 ft 8 in)
Propeller ground clearance	0.33 m (1 ft 1 in)

DIMENSIONS, INTERNAL:

Cabin: Length	2.15 m (7 ft 0¾ in)
Max width	1.07 m (3 ft 6¼ in)
Max height	1.00 m (3 ft 3¼ in)
Floor area	0.96 m² (10.3 sq ft)
Volume	1.4 m³ (49.9 cu ft)
Baggage compartment volume	0.24 m³ (8.5 cu ft)

AREAS:

Wings, gross	12.47 m² (134.2 sq ft)
Ailerons (total)	0.97 m² (10.44 sq ft)
Flaps (total)	1.515 m² (16.31 sq ft)
Fin	0.75 m² (8.07 sq ft)
Rudder	0.70 m² (7.53 sq ft)
Tailplane	1.19 m² (12.81 sq ft)
Elevators (total, incl tab)	0.85 m² (9.15 sq ft)

WEIGHTS AND LOADINGS:

Weight empty: normal	485 kg (1,069 lb)
night operations	545 kg (1,202 lb)
Max fuel weight	65 kg (143 lb)
Max T-O and landing weight	750 kg (1,653 lb)
Max wing loading	60.1 kg/m² (12.32 lb/sq ft)
Max power loading	8.87 kg/kW (14.57 lb/hp)

PERFORMANCE:

Max cruising speed at 3,050 m (10,000 ft)	115 kt (213 km/h; 132 mph)
Stalling speed, flaps down	48 kt (89 km/h; 56 mph) IAS
Max rate of climb at S/L	198 m (650 ft)/min
T-O run	413 m (1,355 ft)
T-O to 15 m (50 ft)	450 m (1,480 ft)
Landing from 15 m (50 ft)	600 m (1,970 ft)
Landing run	540 m (1,775 ft)
Range with max fuel	455 n miles (842 km; 523 miles)
Endurance	4 h

UPDATED

Night-operational version of the NAL/TAAL Hansa-3 civil trainer *(Jane's/Craig Hoyle)* 0130622

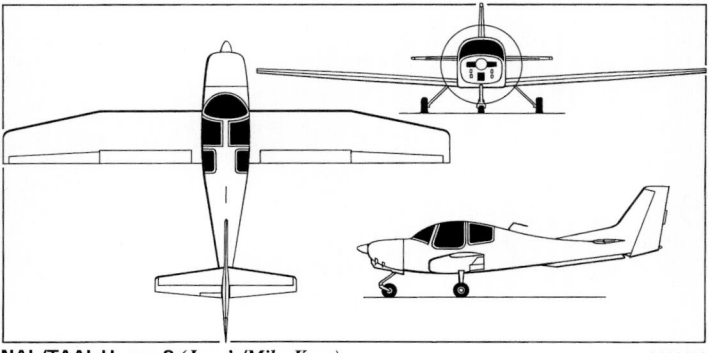

NAL/TAAL Hansa-3 *(Jane's/Mike Keep)* 0121113

INDONESIA

DIRGANTARA

PT DIRGANTARA INDONESIA (Indonesian Aerospace)

HEAD OFFICE AND WORKS: PO Box 1562 BD, GPM 4th Floor, Jalan Pajajaran 154, Bandung 40174
Tel: (+62 22) 603 45 26, 601 07 54 and 601 07 59
Fax: (+62 22) 601 95 38, 607 56 71 and 603 16 96
e-mail: pub-rel@indonesian-aerospace.com
Web: http://www.indonesian-aerospace.com
PRESIDENT DIRECTOR AND CEO: Jusman Syafeii Djamal
EXECUTIVE VICE-PRESIDENTS:
Salomo Pandaitan (Operations)
Agung Nugroho (Technology)
Ilham Akbar Habibie (Commerce)
Ending Ridwan (Finance)
STRATEGIC BUSINESS UNIT MANAGERS:
Tjiptu Santoso (Manufacturing Services)
Bagas Prasetyowibowo (Interiors)
Mulya Tirtosudiro (Technology & Engineering Services)
Bambang Soeriawan (Aircraft Services)
M Mochajan (Helicopters)
F X Sudharmono (Aerospace Systems)

Sindu Otomo (Weapon Systems)
Ina Juniarti (Information Technology Centre)
Lundi Farida (Advanced Technology Education Centre)
CORPORATE PUBLIC RELATIONS OFFICER: Soleh Affandi
PRESS OFFICER: Alex Lim

SALES OFFICE:,
PO Box 3752 JKT, 14th and 15th Floor, BBD Plaza Building, Jalan Imam Bonjol 61, Jakarta 10310
Tel: (+62 21) 32 22 47
Fax: (+62 21) 310 00 81 and 32 53 19

SUBSIDIARIES:
PT Nusantara Turbin and Propulsi, Jalan Pajajaran 154, Kawasan Pabrik IV, Bandung 40174
Tel: (+62 22) 603 19 85 and 603 18 51
e-mail: umc@bdg.centrin.net.id

PT GE Nusantara Turbine Services, Jalan Pajajaran 154, Kawasan Pabrik IV, Bandung 40174
Tel: (+62 22) 603 57 20
Fax: (+62 22) 603 64 41

PT GE Technology Indonesia, Menara Batavia, 5th Floor,
Jalan KH Mas Mansyur Kav 126, Jakarta 10220
Tel: (+62 21) 574 71 23
Fax: (+62 21) 574 71 17

PT Nusantara Systems International, Wisma Bisnis Indonesia, 14th Floor, Jalan S Parman Kav 12, Jakarta 11480
Tel: (+62 21) 530 72 22
Fax: (+62 21) 530 72 23
e-mail: info@nsi.co.id
Web: http://www.nsi.co.id

IPTN North America Inc, 1035 Andover Park West, Suite B, Tukwila, Seattle, Washington 98188-7681, USA
Tel: (+1 206) 575 65 07
Fax: (+1 206) 575 03 18
e-mail: iptn250@aol.com
PRESIDENT: M S Noer

IEU GmbH, Papenstrasse 23, D-22089 Hamburg, Germany
Tel: (+49 40) 357 68 40
Fax: (+49 40) 35 76 84 44
e-mail: SJAM-IEU@t-online-de

Originally formed by Indonesian government as PT Industri Pesawat Terbang Nurtanio (Nurtanio Aircraft Industry Ltd) 23 August 1976 to centralise all aerospace facilities in one company; renamed PT Industri Pesawat Terbang Nusantara (IPTN) in late 1985. Company restructured in January 2000, with PT Bahana Pengelola Industri Strategis as major shareholder, becoming limited liability company with nine strategic business units, six (now four) profit centres, four resource centres and six subsidiaries/corporate functions. New name PT Dirgantara Indonesia (internationally, Indonesian Aerospace, or IAe) inaugurated 24 August 2000 (start of company's 25th year of existence). Workforce in May 2001 was about 10,000, including 2,500 engineers and 5,000 technicians and operators, compared with some 15,500 in 1998; had further reduced to 9,500 by early 2002. Site area is 79.3 ha (196 acres) including 600,000 m² (6,458,350 sq ft) covered area.

International Monetary Fund (IMF) bail-out of Indonesia's collapsed economy, beginning in 1997, demanded immediate suspension of further state funding for N-250 and N-2130 programmes. Despite this, N-250 was then expected to survive; alternative support for N-2130 being sought, but no new partners for either venture reported by early 2002. Further estimated investment of US$90 million said to be needed for N-250; N-2130 programme (2001-2002 *Jane's*) now moribund.

Manufacturing SBU co-manufactures CN-235 Series 200/220 (see under Airtech in International section), produces C-212 Series 200 under licence from CASA of Spain, and was responsible for indigenous N-250 and N-2130 regional airliners. Two CN-235s delivered to South Korea in December 2001 and six more due to follow by October 2002; Pakistan Air Force has ordered four. Agreement 19 June 1991 with BAe (now BAE Systems) to collaborate on production and assembly of Hawks for Indonesian Air Force, although all were assembled in UK with Indonesia receiving offset work. Co-development and manufacturing agreement with AeroCourier Group (see US section) signed 27 February 2002, apparently as renewed attempt to gain FAA certification of N-250.

Helicopter SBU is responsible for licensed production of Eurocopter BO 105 and Super Puma (as NBO-105 and NAS-332 respectively), and Bell 412 (as NBell-412). Deliveries resumed in late 1999. Agreement with Bell, June 1996, to include Bell 407 and 430, has lapsed.

Weapon Systems SBU in Menang Tasikmalaya, West Java (plus smaller plant at Batu Poron, Madura), develops and produces weaponry for military aircraft.

Aerospace Systems SBU main business comprises satellite ground systems and components, application technology, offset and satellite engineering services.

Subcontract work includes production of components for Boeing 737 and 767 and Lockheed Martin F-16. Company's 1,400 m² (15,070 sq ft) gas-turbine maintenance centre can maintain, overhaul and repair Rolls-Royce 250, P&W JT8D, P&WC PT6T-3/3B, R-R Dart RDa.7, Honeywell LTS101 and TPE331 and General Electric CT7 engines, plus some components of Allison T56 and R-R Tay.

Indonesian government plans to begin privatising Dirgantara by the end of 2003, possibly by sell-off of individual subsidiaries.

UPDATED

DIRGANTARA N-250

TYPE: Twin-turboprop airliner.
PROGRAMME: Project (Indonesia's first fully indigenously designed transport) announced at Paris Air Show 15 June 1989; engine selected July 1990; first metal cut August 1992; decided mid-1993 to launch with N-250-100 as 64/68-seat sized model instead of originally planned 50/54-seat N-250 (though first prototype remaining configured as 50-seater); this prototype (PA1/PK-XNG *Gatotkoco*) rolled out 10 November 1994; first flight 10 August 1995. Was to be followed by three N-250-100 development aircraft (PA2 *Koconegoro*, PA3 *Krincingwesi* and PA4 *Putut Guritno*) plus two static/fatigue test airframes, with PA3 (systems certification) and PA4 (flight deck certification) to be brought up to production standard later.

Further change of plan in late 1995 made 50-seater an alternative production version after all. A fifth development aircraft (PA5) to this (modified) configuration (see Current Versions) was planned but subsequently cancelled.

PA2 (PK-XNK) originally planned to fly in May 1996, but delayed until 19 December; PA1 had flown 650 hours and PA2 40 hours by December 1997; PA3 and PA4 then expected to fly in second quarter 1998, but suspension of further state funding caused cancellation of PA4 and PA5, as well as effective termination of development programme. 'Maintenance' flying continued with PA1 and PA2 after 1998 (some 800 hours flown by late 1999) but this not counted as development time. By April 2001, certification still awaited and PA3 still unable to fly until replacement funding obtained; if secured, part will be used to complete PA3 to undertake also the development tasks intended for PA4.

Dirgantara N-250-50 (short fuselage) prototype (*Jane's/Paul Jackson*)
0109148

Boeing pledged assistance (non-financial) towards gaining FAA certification, but plans for second assembly line (for N-270 stretched version) in Mobile, Alabama, USA, under the aegis of American Regional Aircraft Industries (AMRAI), have been frozen. No further progress reported by May 2002. See 2002-03 and earlier editions for description.

UPDATED

DIRGANTARA (CASA) NC-212-200

TYPE: Twin-turboprop transport.
PROGRAMME: Original Spanish production of Series 200 ended 1987 (see 1987-88 *Jane's* for details). C-212 manufactured under licence in Indonesia as NC-212 since 1976, IPTN producing 28 NC-212-100s before switching to NC-212-200. Description in 1998-99 *Jane's*.
CUSTOMERS: Original intention was production of 126, this later reduced to 108. Total of 95 (28 Srs 100 and 67 Srs 200) produced by December 1997: see 1998-99 *Jane's* for customer details. At least five more completed and five others on production line by October 1998. Outstanding orders at 1 August 1998 totalled 16 Srs 200 (six MPA for Indonesian Navy; six, including one VIP, transports for Indonesian Army; and four for Indonesian Police). Subsequently reported that further production had ceased, and deliveries withheld due to customers' inability to pay for these last aircraft; however, two standard transports were delivered to Indonesian Navy in September 1999. Company was then scheduled to deliver six MPAs (apparently from stock, within overall total of 108 completed) to same customer in 2000-2001, but this order said to be under review in early 2000 and no further deliveries reported by May 2002.

UPDATED

DIRGANTARA (EUROCOPTER) NBO-105

TYPE: Light utility helicopter.
PROGRAMME: Manufactured under licence from MBB (now Eurocopter) as NBO-105 since 1976; only rotors and transmission now supplied from Germany; originally **NBO-105 CB**, but stretched **NBO-105 CBS** available from 101st aircraft onwards; **NBO-105 MPDS** (multipurpose delivery system) can carry 50 mm to 81 mm unguided rockets, single or twin 0.30 in or 0.50 in machine gun pods; and reconnaissance or FLIR pods. Also available in **FAC** (forward air control) version.
CUSTOMERS: At least 121 completed by mid-1996 (partial customer list in previous editions). Three ordered by Indonesian Navy, for over-horizon targeting, in June 1996, equipped with Ocean Master radar and Gemini navigation computer; these largely complete by 1999, and delivery due in 2000. Other completed NBO-105s reportedly remained unsold in mid-1999, but deliveries had restarted (including two to Indonesian Navy) by end of that year; a third NBO-105 was handed over to the Indonesian Navy on 4 May 2001, completing current orders from that service. Royal Malaysian Navy reportedly considering purchase of six in 2001 to replace 12 ageing Westland Wasps withdrawn in 2000.

UPDATED

DIRGANTARA (EUROCOPTER) NAS-332 SUPER PUMA

TYPE: Multirole medium helicopter.
PROGRAMME: Assembly of 11 SA 330J Pumas agreed 1977 and began 1981; switched to AS 332C and L Super Puma in early 1983; available configurations include **ASW/MA** (anti-submarine warfare/maritime attack). First NAS-332 for Pelita Air Service rolled out 22 April 1983.
CUSTOMERS: Deliveries include four as commando and general purpose transports for Indonesian Navy and two VVIP; see 1997-98 edition for partial customer list. Eighteen Super Pumas completed or under construction by mid-1996. Sales of seven civil NAS-332s to Iran, reportedly for oil-rig duties, was approved in October 1996 but has been in abeyance (interest reportedly renewed in late 2000); 16 on order for Indonesian Air Force as S-58T replacements. Latter, nearly all completed or in production by April 1998, comprise two VVIP, nine tactical transports and five for combat SAR. Other completed NAS-332s reportedly remained unsold in mid-2001, but first two of Indonesian Air Force 16 delivered in May and August 2001, with three more due to follow by end of that year. Next seven due to be delivered in 2002 and final four in 2003, although all 16 had reportedly been completed by August 2001.

UPDATED

DIRGANTARA (BELL) NBELL-412

TYPE: Multirole medium helicopter.
PROGRAMME: Licence agreement for Bell 412 (see Canadian section) signed 12 November 1982; covers 100 helicopters; production started 1984; first flight April 1986; NBell-412HP is 40 per cent Indonesian manufactured and employs PT6-3BE engines. FN Herstal **EMA** (external mounting assembly) for 7.62 mm and 12.7 mm gun pods and 70 mm rocket pods, already certified for Canadian and Italian Agusta-built Bell 412s, qualified for NBell-412 and fitted to several helicopters; **CAS** (close air support) configuration also available.
CUSTOMERS: Production temporarily halted at 27th helicopter, September 1995, restarted three months later (Indonesian Army ordered one in 1996), but suspended again as condition of IMF bail-out of Indonesian economy in 1997. None delivered in 1999 or 2000, but one 412EP delivered to Pelita Air Service in late 2001 or early 2002. See 1997-98 edition for partial customer list.

UPDATED

OTHER AIRCRAFT

See under Airtech in International section for details of **CN-235** twin-turboprop military and civil transport. Two CN-235-220s delivered to Republic of Korea Air Force in December 2001; six more from Indonesian production line following in February, April and November (two each), completing order placed in October 1997. Final pair are in VIP and VVIP configuration. The Pakistan Air Force ordered four CN-235-220s in June 2001.

In 2001, Dirgantara was contracted by the AeroCourier Group of Minneapolis, USA (which see), to undertake wing and fuselage design and manufacture of the latter's single-turboprop utility aircraft.

UPDATED

Dirgantara NBO-105 search and rescue helicopter of the Indonesian Air Force (*Jane's/Paul Jackson*)
NEW/0131721

INTERNATIONAL

AGUSTAWESTLAND

AGUSTAWESTLAND

Web: http://www.agustawestland.com
CHAIRMAN: Kevin Smith
CEO: Amedeo Caporaletti
MANAGING DIRECTOR: Richard Case
COO: A Gustapane (Italy)
   A Johnson (UK)
MARKETING MANAGER: G Orsi
PUBLIC RELATIONS MANAGER: Gianluca Grimaldi (Italy)
   David Bath (UK)
PARTICIPATING COMPANIES
   Finmeccanica: see Agusta SpA under Italy
   GKN: see Westland Helicopters Ltd under UK

MoU covering joint merger of helicopter divisions signed by
Finmeccanica and GKN on 16 April 1998; heads of
agreement 18 March 1999; details finalised (subject to
regulatory approval) and name AgustaWestland announced

26 July 2000; company declared fully operational 12
February 2000.
   Westland also contributed its 50 per cent share in EH 101,
GKN's aerospace transmissions business and 50 per cent
share in Aviation Training International (joint venture with
Boeing to support British Army WAH-64 Apaches). Agusta
also contributed half of EH 101, transmissions and
aerostructures business and own shares in NH90 (32 per cent)
and Bell/Agusta (45 per cent). Joint revenue in 2001 was
US$2,100 million, and order backlog at January 2002 was
US$6,400 million. Employees totalled 10,546 in mid-2001.
   Together, the two companies have built over 7,100
helicopters for customers in more than 80 countries. Sales in
2001 were 173, all but six twin-engined.
   Wholly and part-owned subsidiaries are detailed below.

Finmeccanica:
   Agusta Aerospace Corporation (USA)
   Agusta Aerospace Services BV (Belgium)

EHI Ltd (50 per cent)
Bell/Agusta Aerospace Corporation (USA) (45 per cent)
NHI sarl (27 per cent)
other subsidiaries and holdings

GKN:
   Westland Industrial Products Ltd
   Westland Transmissions Ltd
   EHI Ltd (50 per cent)
   Aviation Training International Ltd (50 per cent)
   other subsidiaries and holdings.

On 31 October 2001, AgustaWestland initiated a joint
marketing effort with Lockheed Martin, under which the
latter will promote a US market version of the EHI EH101
(which see), known as the US101.

UPDATED

AIRBUS

AIRBUS SAS

1 Rond Point Maurice Bellonte, F-31707 Blagnac Cedex,
France
Tel: (+33 5) 61 93 33 33
Fax: (+33 5) 61 93 37 92
Web: http://www.airbus.com
CHAIRMAN OF SUPERVISORY BOARD: Manfred Bischoff
PRESIDENT AND CEO: Noël Forgeard
COO: Gustav Humbert
EXECUTIVE VICE-PRESIDENTS:
   Alain Garcia (Engineering)
   John Leahy (Commercial)
   Patrick Gavin (Customer Services)
   Gérard Blanc (Programmes)
   Charles Champion (Large Aircraft Division)
VICE-PRESIDENT, CORPORATE COMMUNICATIONS:
   Michel Guérard
REGIONAL MANAGER, MEDIA RELATIONS: David Velupillai
AIRFRAME PRIME CONTRACTORS:
   EADS: see this section
   BAE Systems Airbus: see under UK

Airbus Industrie set up 18 December 1970 as Groupement
d'Intérét Economique (GIE: a company which makes no
profits or losses in its own right) and has been profitable since
1990, producing an operating surplus that is shared among its
partners. Its first task was to manage development,
manufacture, marketing and support of A300; this
management now extends to A300-600, A310, A318, A319/
ACJ, A320, A321, A330, A340 and A380. Large Airliner
Division created in March 1996 to oversee A3XX (later
A380) project. Airbus delivered its 3,000th aircraft, an A320
for JetBlue Airways, on 18 July 2002. It is currently studying
concepts for a short-range, 200-passenger and long-range
250-passenger airliner as potential replacements for the A300
and A310 from about 2010, once development of the A380 is
completed.
   A new single-aisle internal flight deck door, complying
with pre-existing and new FAA anti-intrusion regulations,
received JAA design approval on 21 May 2002 and is in
production; it was certified for the A330/A340 family on 26
June 2002, and subsequently for all Airbus aircraft.
   Plans for establishment of a single corporate entity (SCE)
overtaken and modified by formation of European
Aeronautic, Defense and Space Company (which see).
Planned restructuring of the consortium into Airbus
Integrated Company (AIC) was revealed in March 2000 and
approved by the EC in late 2000; official starting date was 1
January 2001, although completion of formalities was
delayed. EADS holds 80 per cent and BAE Systems 20 per
cent of AIC, which incorporated in France on 11 July 2001 as
an SAS (Société par Actions Simplifiées) following a formal
decision on 23 June 2000.

Airbus responsible for all work by partner companies and
has some 45,000 employees, including workers at its spare
parts centre in Hamburg and its US and Chinese subsidiaries.
Approximate European breakdown is Germany 16,000, France
14,000, UK 8,300 and Spain 2,400. Stork (formerly Fokker) is
an associate in A300 and A310 and Belairbus (Belgian
consortium) in A310, A320, A330 and A340. Alenia
manufactures front fuselage plug for A321. An engineering
centre is being established in Moscow in a joint venture
(registered in December 2002) with the Kaskol group, which
has shares in Sokol, Hydromash, Rostvertol and other Russian
aerospace companies; this was scheduled to become
operational in the second quarter of 2003, and will develop and
co-produce components for the A380 and other Airbus types.
   Subsidiaries include Airbus Industrie of North America,
Airbus Finance Company (AFC) formed in 1994, Airbus
Training Centre (Miami) and Airbus Industrie China. Airbus
Industrie's training centre in Toulouse, previously a
subsidiary known as Aeroformation, and the main spares
centre, Airbus Industrie Materiel Support, formerly Airspares
Hamburg, have been integrated within the consortium's
Customer Services Directorate; a new, purpose-built mockup
centre was added to the Toulouse facilities during 1999 and
was fully operational by end 2000. A training centre was
opened in Miami, Florida, in October 1999.
   In July 1996, Airbus Training and Support Centre in
Beijing was opened, becoming operational in October 1997.

UPDATED

Airbus A300-600 twin-turbofan airliner in the insignia of Japan Air System   *NEW*/0527075

AIRBUS A300-600

TYPE: Wide-bodied airliner.
PROGRAMME: Launched 29 May 1969; initial variants were
A300B1 (first flight 28 October 1972, service entry
November 1974: see 1971-72 *Jane's* for details), A300B2
(first flight June 1973, service entry May 1974) and
A300B4 (first flight December 1974, service entry June
1975; 248 built: see 1984-85 and previous editions). A300-
600 go-ahead 16 December 1980; first flight (F-WZLR) 8
July 1983; certified (with JT9D-7R4H1 engines) 9 March
1984; first delivery (to Saudia) 26 March 1984.
   Improved version with CF6-80C2 engines and other
changes (see Current Versions) made first flight 20 March
1985; French certification for Cat. IIIb take-offs and
landings 26 March 1985; first delivery of improved version
(to Thai Airways) 26 September 1985. Extended-range
A300-600R (then known as -600ER) made first flight 9
December 1987, receiving European and FAA certification
10 and 28 March 1988 respectively, deliveries (to
American Airlines) beginning 20 April 1988; A300-600
powered by GE CF6-80C2A5 with FADEC granted 180-
minute ETOPS April 1994. CIS certification granted May
1996. Bulk of production backlog are freighters for United
Parcel Service.
CURRENT VERSIONS: **A300-600:** Advanced version of
A300B4-200; major A300 version since early 1984.
Passenger and freight capacity increased by fitting rear
fuselage of A310 with pressure bulkhead moved aft; wings

TOTALS OF AIRBUS AIRLINERS
(at 1 January 2003)

	A300	A310	A318	A319	A320	A321	A330	A340	A380	Totals
Firm orders	583	260	84	856	1,597	421	419	317	95	4,632
Delivered	517	255	0	492	1,128	256	251	228	0	3,127
Operating	424	246	0	492	1,118	256	248	226	0	3,010

AIRBUS ORDERS and DELIVERIES 2002

	A300	A310	A318	A319	A320	A321	A330	A340	A380	Totals
Net orders	0	0	−30	162	52	6	12	21	10	233
Delivered	9	0	0	85	116	35	42	16	0	303
Backlog	66	5	84	364	469	165	168	89	95	1,505

have simple Fowler flaps and increased trailing-edge camber; forward-facing two-person flight deck with EFIS; new digital avionics; new braking control system; new APU; simplified systems; weight saving by use of composites for some secondary structural components; payload/range performance and fuel economy improved by comprehensive drag clean-up. Further improvements introduced in 1985 included CF6-80C2 or PW4000 as engine options, carbon brakes, wingtip fences and 'New World' flight deck; basic equipment of aircraft delivered from late 1991 further improved by incorporating standard options.

Cargo conversions of A300-600 and earlier A300B4 are offered; see *Jane's Aircraft Upgrades* for details.
Detailed description applies to current production A300-600/600R except where indicated.

A300-600R: Extended-range version of A300-600, differing mainly in having fuel trim tank in tailplane and higher maximum T-O weight.

A300-600 Convertible: Convertible passenger/cargo version, described separately.

A300-600 Freighter: Non-passenger version, described separately.

Airbus Super Transporter: A300-600R conversion as Super Guppy replacement; see under SATIC in 2001-02 edition.

CUSTOMERS: Total of 583 of all A300 versions ordered, of which 517 delivered, by 31 January 2003.

AIRBUS A300 ORDERS
(at 1 January 2003)

Customer	Qty
Air Afrique	3
Air France	23
Air India	3
Air Inter Europe	8
Alitalia	8
American Airlines	35
Amiri Flight	2
Ansett	8
Australian Airlines	5
China Airlines	15
China Eastern Airlines	7
China Northern Airlines	6
China Northwest Airlines	3
CityBird	2
Continental Airlines	3
Cruzeiro	2
Eastern Airlines	34
Egyptair	17
Emirates	5
Federal Express	36
Finnair	2
Garuda	9
Hapag Lloyd	7
Iberia	6
Indian Airlines	10
ILFC	9
Iran Air	8
Japan Air System	32
Japan Fleet Service	2
Korean Air	32
Kuwait Airways	8
Laker	3
LaTur	2
Lufthansa	23
Malaysia Airlines	4
Monarch	4
Olympic	10
Pakistan International Airlines	4
Pan Am	12
Philippine Airlines	5
Polaris Aircraft Leasing	5
Saudi Arabian Airlines	11
SAS	4
Singapore Airlines	8
SOGERMA SOCEA	1
South African Airways	7
Thai Airways	33
Trans European Airways	1
Tunis Air	1
United Parcel Service	90
Varig	2
VASP	3
Total	**583**

COSTS: US$109.9 million (2001).
DESIGN FEATURES: Mid-mounted wings with 10.5 per cent thickness/chord ratio, 28° sweepback at quarter-chord, and (since 1985) tip fences; circular-section pressurised fuselage; all-swept tail unit.
FLYING CONTROLS: Power-assisted. Each wing has three-segment, two-position (T-O/landing) leading-edge slats (no cutout over engine pylon), small Krueger flap at leading-edge wingroot, three cambered tabless flaps on trailing-edge, all-speed aileron between inboard flap and outer pair, and seven spoilers forward of flaps on each

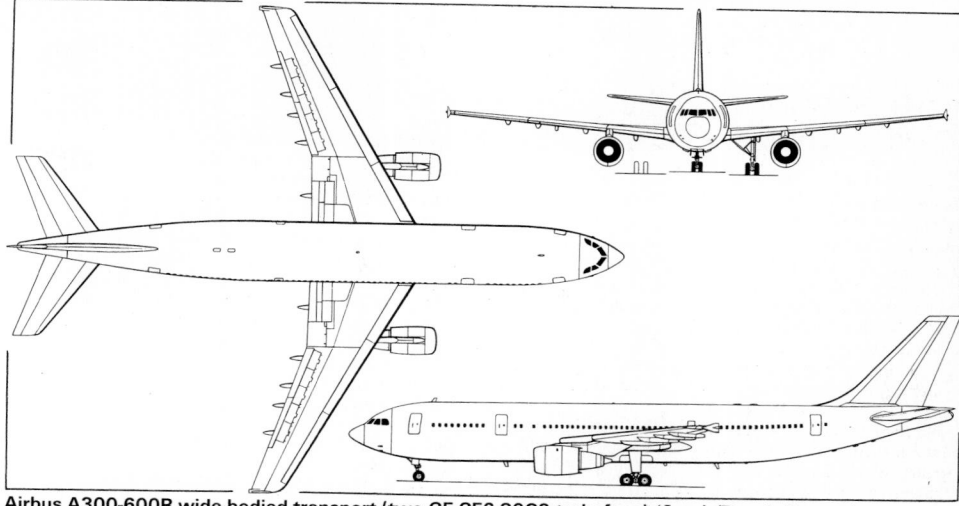

Airbus A300-600R wide-bodied transport (two GE CF6-80C2 turbofans) *(Jane's/Dennis Punnett)*

wing; flaps occupy 84 per cent of trailing-edge, increasing wing chord by 25 per cent when fully extended; ailerons deflect 9° 2′ downward automatically when flaps are deployed; all 14 spoilers used as lift dumpers: outboard 10 for roll control and inboard 10 as airbrakes; variable incidence tailplane. Ailerons/elevators/rudder fully powered by hydraulic servos (three per surface), controlled mechanically; secondary surfaces (spoilers/flaps/slats) fully powered hydraulically with electrical control, tailplane by two independent hydraulic motors electrically controlled with additional mechanical input; preselection of spoiler/lift dump lever permits automatic extension of lift dumpers on touchdown; flaps and slats have similar drive mechanisms, each powered by twin motors driving ball screwjacks on each surface with built-in protection against asymmetric operation.
STRUCTURE: Two-spar main wing box, integral with fuselage and incorporating fail-safe principles; third spar across inboard sections; semi-monocoque fuselage (frames and open Z-section stringers), with integrally machined skin panels in high-stress areas; primary structure is of high-strength, damage-tolerant aluminium alloy, with steel or titanium for some critical fuselage components, honeycomb panels or selected glass fibre laminates for secondary structures; metal slats, flaps and ailerons. CFRP fins replaced aluminium alloy unit from 1988; secondary structure composites include AFRP for flap track fairings, rear wing/body fairings, cooling air inlet fairings and radome; GFRP for wing upper surface panels above mainwheel bays, fin leading/trailing-edges, fin-tip, fin/fuselage fairings, tailplane trailing-edges, elevator leading-edges, tailplane and elevator tips and elevator actuator access panel; carbon-reinforced GFRP for elevators and rudder; CFRP for spoilers, outer flap deflector doors and fin box; all CFRP moving surfaces have aluminium or titanium trailing-edges. Nosewheel doors and mainwheel leg fairing doors also of CFRP. Nose gear is structurally identical to that of B2/B4/A310; main gear is generally reinforced, with a new hinge arm and a new pitch damper hydraulic and electrical installation. Nacelles have CFRP cowling panels and are subcontracted to Rohr (California); pylon fairings are of AFRP.
Aerospatiale Matra builds nose (including flight deck), lower centre-fuselage, four inboard spoilers, wing/body fairings and engine pylons; Airbus Deutschland builds forward fuselage (flight deck to wing box), upper centre-fuselage, rear fuselage (including tailcone), vertical tail, 10 outboard spoilers and some cabin doors; it also equips wings and installs interiors and seats; BAE Systems (formerly BAe) designed wings and builds wing box; Airbus España manufactures horizontal tail, port and starboard forward passenger doors and mainwheel/nosewheel doors; Fokker produces wingtips, ailerons, flaps, slats and main gear leg fairings. Large, fully equipped and inspected airframe sections airlifted by Beluga to EADS France at Toulouse for assembly and painting, aircraft then being flown to Hamburg for outfitting and returned to Toulouse for customer acceptance.
LANDING GEAR: Hydraulically retractable tricycle type, of Messier-Bugatti design, with Messier-Bugatti/Liebherr/Dowty shock-absorbers and wheels standard; twin-wheel nose unit retracts forward, main units inward into fuselage; free-fall extension; has four-wheel main bogies interchangeable left with right. Standard bogie size is 927 × 1,397 mm (36½ × 55 in); wider bogie of 978 × 1,524 mm (38½ × 60 in) is optional. Mainwheel tyres size 49×17-20 or 49×17.0R20 (30 ply) (standard) or 49×19-20 (30 ply) (wide bogie), with respective pressures of 12.41 and 11.10 bar (180 and 161 lb/sq in). Nosewheel tyres size 40×14 or 40×14.0R16 (22 ply), pressure 9.38 bar (136 lb/sq in). Steering angles 65°/95°. Messier-Bugatti/Liebherr/Dowty hydraulic disc brakes standard on all mainwheels. Normal braking powered by 'green'

hydraulic system, controlled electrically through two master valves and monitored by a brake system control box to provide anti-skid protection. Standby braking (powered automatically by 'yellow' hydraulic system if normal 'green' system supply fails) controlled through a dual metering valve; anti-skid protection is ensured through same box as normal system, with emergency pressure supplied to brakes by accumulators charged from 'yellow' system. Automatic braking system optional. Bendix or Goodrich wheels and brakes available optionally. Minimum ground turning radius (effective, aft CG) 22.00 m (72 ft 2¼ in) about nosewheel, 34.75 m (114 ft 0 in) about wingtips.
POWER PLANT: Two turbofans in underwing pods. A300-600 was launched with 249 kN (56,000 lb st) Pratt & Whitney JT9D-7R4H1 and currently available with 249 kN (56,000 lb st) Pratt & Whitney PW4156 or 262 kN (59,000 lb st) General Electric CF6-80C2A1. A300-600R is offered with 274 kN (61,500 lb st) CF6-80C2A5 or 258 kN (58,000 lb st) PW4158. CF6-80C2A5 and PW4158 also available as options on A300-600.
Fuel in two integral tanks in each wing, and fifth integral tank in wing centre-section, giving standard usable capacity of 62,000 litres (16,379 US gallons; 13,638 Imp gallons). Additional 6,150 litre (1,625 US gallon; 1,353 Imp gallon) fuel/trim tank in tailplane (-600R only) increases this total to 68,150 litres (18,004 US gallons; 14,991 Imp gallons). Optional extra fuel cell in aft cargo hold can increase total to 73,000 litres (19,285 US gallons; 16,058 Imp gallons) in -600R. Two standard refuelling points beneath starboard wing; similar pair optional under port wing.
ACCOMMODATION: Crew of two on flight deck, plus two observers' seats. Passenger seating in main cabin in six-, seven-, eight- or nine-abreast layout with two aisles; typical mixed class layout has 266 seats (26 first class and 240 economy), six/eight-abreast at 96/86 cm (40/32 in) seat pitch with two galleys and one lavatory forward, one galley and two lavatories at Door 2 position, and one galley and four lavatories at rear; typical economy class layout for 285 passengers eight-abreast at 86 cm (34 in) pitch. Maximum capacity (subject to certification) 361 passengers. Closed overhead baggage lockers on each side (total capacity 10.5 m³; 370 cu ft) and in double-sided central 'super-bin' installation (total capacity 14.5 m³; 512 cu ft), giving 0.03 to 0.09 m³ (1.2 to 3.2 cu ft) per passenger in typical economy layout.
Two outward parallel-opening Type A plug-type passenger doors ahead of wing on each side, and one on each side at rear. Type I emergency exit on each side aft of wing. Underfloor baggage/cargo holds fore and aft of wings, with doors on starboard side; forward hold can accommodate 12 LD3 containers, or four 2.24 × 3.17 m (88 × 125 in) pallets or, optionally, 2.43 × 3.17 m (96 × 125 in) pallets, or engine modules; rear hold can accommodate 10 LD3 containers; additional bulk loading of freight provided for in an extreme rear compartment with usable volume of 17.3 m³ (611 cu ft); alternatively, rear hold can carry 11 LD3 containers, with bulk cargo capacity reduced to 9.0 m³ (318 cu ft); bulk cargo compartment can be used to transport livestock. Entire accommodation is pressurised, including freight, baggage and avionics compartments.
SYSTEMS: Air supply for air conditioning system taken from engine bleed and/or APU via two high-pressure points; conditioned air can also be supplied direct to cabin by two low-pressure ground connections; ram air inlet for fresh air ventilation when packs not in use. Pressure control system (maximum differential 0.574 bar; 8.32 lb/sq in) consists of two identical, independent, automatic systems (one active, one standby); automatic switchover from one to other after each flight and in case of active system failure; in each system, pressure controlled by two electric outflow valves, function depending on preprogrammed cabin pressure

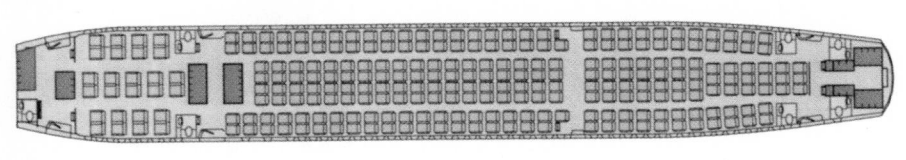

Two-class A300-600 interior for 26 first and 240 economy class passengers   0044877

altitude and rate of change of cabin pressure, aircraft altitude, and preselected landing airfield elevation. Automatic prepressurisation of cabin before take-off, to prevent noticeable pressure fluctuation during take-off. Modular box system provides passenger oxygen to all installation areas.

Hydraulic system comprises three fully independent circuits, operating simultaneously; each system includes reservoir of direct air/fluid contact type, pressurised at 3.52 bar (51 lb/sq in); fire-resistant phosphate ester-type fluid; nominal output flow 136 litres (35.9 US gallons; 30 Imp gallons)/min delivered at 207 bar (3,000 lb/sq in) pressure; 'blue' and 'yellow' systems have one pump each, 'green' system has two pumps. The three circuits provide triplex power for primary flying controls; if any circuit fails, full control of aircraft is retained without any necessity for action by crew. All three circuits supply ailerons, rudder and elevators; 'blue' circuit additionally supplies spoiler 7, spoiler/airbrake 4, airbrake 1, yaw damper and slats; 'green' circuit additionally supplies spoiler 6, flaps, Krueger flaps, slats, landing gear, wheel brakes, steering, tailplane trim, artificial feel, and roll/pitch/yaw autopilot; 'yellow' circuit additionally supplies spoiler 5, spoiler/airbrake 3, airbrake 2, flaps, wheel brakes, cargo doors, artificial feel, yaw damper, tailplane trim, and roll/pitch/yaw autopilot. Ram air turbine pump provides standby hydraulic power should both engines become inoperative.

Main electrical power supplied under normal flight conditions by two integrated drive generators, one on each engine; third (auxiliary) generator, driven by APU, can replace either of main generators, having same electromagnetic components but not constant-speed drive; each generator rated at 90 kVA, with overload ratings of 112.5 kVA for 5 minutes and 150 kVA for 5 seconds; APU generator driven at constant speed through gearbox. Three unregulated transformer-rectifier units (TRUs) supply 28 V DC power. Three 25 Ah Ni/Cd batteries used for emergency supply and APU starting; emergency electrical power taken from main aircraft batteries and emergency static inverter, providing single-phase 115 V 400 Hz output for flight instruments, navigation, communications and lighting when power not available from normal sources.

Hot air anti-icing of engines, engine air intakes, and outer segments of leading-edge slats; electrical heating for anti-icing flight deck front windscreens, demisting flight deck side windows, and for sensors, pitot probes and static ports, and waste water drain masts.

Honeywell 331-250F APU in tailcone, exhausting upward; installation incorporates APU noise attenuation. Self-contained fire protection system, and firewall panels protect main structure from an APU fire. APU provides bleed air to pneumatic system, and drives auxiliary AC generator during ground and in-flight operation; APU drives 90 kVA oilspray-cooled generator, and supplies bleed air for main engine start or air conditioning system. For current deliveries of A300-600, APU has improved relight capability, and can be started throughout flight envelope.

For new A300-600s and -600Rs, two optional modifications offered for compliance with full extended-range twin-engined operations (ETOPS) requirements: hydraulically driven fourth generator and increased cargo hold fire suppression capability. ETOPS kit qualified for aircraft with CF6-80C2 and JT9D-7R series engines and, since mid-1988, for those with PW4000 series.

AVIONICS: *Comms:* Standard communications radios include two VHF, with provision for a third, two HF, two transponders, one Selcal, interphone and passenger address systems, ground crew call system and cockpit voice recorder. Provision for Mode S transponders.

Radar: Weather radar standard, with provision for second.

Flight: Radio navigation avionics include two VOR, two ILS, two DME, one ADF, two marker beacon receivers and two radio altimeters; TCAS and GPWS. Most other avionics are to customer requirements, only those relating to the instrument landing system (Honeywell or Rockwell Collins ILS and Rockwell Collins or TRT radio altimeter) being selected and supplied by the manufacturer. Two Honeywell digital air data computers standard; basic digital AFCS has dual flight control computers (FCCs) for flight director and autopilot functions (for Cat. III automatic landings), single thrust control computer (TCC) for speed and thrust control, and two flight augmentation computers (FACs) to provide yaw damping, electric pitch trim, and flight envelope monitoring and protection. Options include second FCC (for Cat. III automatic landing); second TCC; two flight

management computers (FMCs) and two control display units for full flight management system. Basic aircraft also fitted with ARINC 717 data recording system with digital flight data acquisition unit, digital flight data recorder and three-axis linear accelerometer; optional additional level of windshear protection is available. Honeywell Enhanced GPWS available from March 1997.

Instrumentation: Six identical and interchangeable CRT electronic displays (four electronic flight instrument system and two electronic centralised aircraft monitor), plus digitised electromechanical instruments with liquid crystal displays.

DIMENSIONS, EXTERNAL:

Wing span	44.84 m (147 ft 1 in)
Wing chord at root	9.40 m (30 ft 10 in)
Wing aspect ratio	7.7
Length overall	54.08 m (177 ft 5 in)
Fuselage: Length	53.30 m (174 ft 10½ in)
Max diameter	5.64 m (18 ft 6 in)
Height overall	16.51 m (54 ft 2 in)
Tailplane span	16.26 m (53 ft 4 in)
Wheel track	9.60 m (31 ft 6 in)
Wheelbase (c/l of shock-absorbers)	18.62 m (61 ft 1 in)
Passenger doors (each): Height	1.93 m (6 ft 4 in)
Width	1.07 m (3 ft 6 in)
Height to sill: forward	4.60 m (15 ft 1 in)
centre	4.80 m (15 ft 9 in)
rear	5.50 m (18 ft 0½ in)
Emergency exits (each): Height	1.60 m (5 ft 3 in)
Width	0.61 m (2 ft 0 in)
Height to sill	4.87 m (15 ft 10 in)
Underfloor cargo door (forward):	
Height	1.71 m (5 ft 7¼ in)
Width	2.69 m (8 ft 10 in)
Height to sill	3.07 m (10 ft 1 in)
Underfloor cargo door (rear):	
Height	1.71 m (5 ft 7¼ in)
Width	1.81 m (5 ft 11¼ in)
Height to sill	3.41 m (11 ft 2¼ in)
Underfloor cargo door (extreme rear):	
Height (projected)	0.95 m (3 ft 1 in)
Width	0.95 m (3 ft 1 in)
Height to sill	3.56 m (11 ft 8 in)

DIMENSIONS, INTERNAL:

Cabin, excl flight deck: Length	40.70 m (133 ft 6¼ in)
Max width	5.28 m (17 ft 4 in)
Max height	2.54 m (8 ft 4 in)
Underfloor cargo hold:	
Length: forward	10.60 m (34 ft 9¼ in)
rear	7.95 m (26 ft 1 in)
extreme rear	3.40 m (11 ft 2 in)
Max height	1.76 m (5 ft 9 in)
Max width	4.20 m (13 ft 9¼ in)
Underfloor cargo hold volume:	
forward	75.1 m³ (2,652 cu ft)
rear	55.0 m³ (1,942 cu ft)
extreme rear	17.3 m³ (611 cu ft)

AREAS:

Wings, gross	260.00 m² (2,798.6 sq ft)
All-speed ailerons (total)	7.06 m² (75.99 sq ft)
Trailing-edge flaps (total)	47.30 m² (509.13 sq ft)
Leading-edge slats (total)	30.30 m² (326.15 sq ft)
Krueger flaps (total)	1.115 m² (12.00 sq ft)
Spoilers (total)	5.40 m² (58.13 sq ft)
Airbrakes (total)	12.59 m² (135.52 sq ft)
Fin	45.20 m² (486.53 sq ft)
Rudder	13.57 m² (146.07 sq ft)
Tailplane	44.80 m² (482.22 sq ft)
Elevators (total)	19.20 m² (206.67 sq ft)

WEIGHTS AND LOADINGS (A: CF6-80C2A1/A5 engines, B: PW4156/4158 engines, both in 266-seat configuration)*:

Manufacturer's weight empty:	
A (600)	79,210 kg (174,630 lb)
A (600R)	80,070 kg (176,525 lb)
B (600)	79,151 kg (174,500 lb)
B (600R)	79,320 kg (174,870 lb)
Operating weight empty:	
A (600)	90,115 kg (198,665 lb)
A (600R)	91,040 kg (200,700 lb)
B (600)	90,065 kg (198,565 lb)
B (600R)	90,965 kg (200,550 lb)
Max payload (structural): A (600)	39,885 kg (87,931 lb)
A (600R)	38,962 kg (85,896 lb)
B (600)	39,993 kg (88,169 lb)
B (600R)	39,037 kg (86,061 lb)
Max usable fuel:	
600: standard	49,786 kg (109,760 lb)
600R: standard	54,721 kg (120,640 lb)
with optional cargo hold tank	
	58,618 kg (129,230 lb)

Max T-O weight (A and B):		
600		165,000 kg (363,765 lb)
600R (standard)		170,500 kg (375,885 lb)
600R (option)		171,400 kg (377,870 lb)
Max ramp weight (A and B):		
600		165,900 kg (365,745 lb)
600R (standard)		171,400 kg (377,870 lb)
600R (option)		172,600 kg (380,520 lb)
Max landing weight (A and B):		
600		138,000 kg (304,240 lb)
600R (standard)		140,000 kg (308,645 lb)
Max zero-fuel weight (A and B):		
600, 600R (standard)		130,000 kg (286,600 lb)
Max wing loading: 600	634.6 kg/m² (129.98 lb/sq ft)	
600R (standard)	655.8 kg/m² (134.32 lb/sq ft)	

Production aircraft from late 1996 onward. See 1989-90 and previous editions for original versions; 1996-97 and six previous editions for intermediate versions

PERFORMANCE (A and B as for Weights and Loadings):

Max operating speed (VMO) from S/L to FL267		
		335 kt (621 km/h; 386 mph) CAS
Max operating Mach No. (MMO) above FL267		0.82
Max cruising speed at FL250	480 kt (890 km/h; 553 mph)	
Max cruising speed at FL300		
	M0.82 (484 kt; 897 km/h; 557 mph)	
Typical long-range cruising speed at FL310		
	M0.80 (472 kt; 875 km/h; 543 mph)	
Approach speed: 600	135 kt (249 km/h; 155 mph)	
600R	136 kt (251 km/h; 156 mph)	
Max operating altitude	12,200 m (40,000 ft)	
Runway ACN for flexible runway, category B:		
standard bogie and tyres: 600		56
600R		59
600R (option)		60
optional bogie and tyres: 600		52
600R		55
600R (option)		56
T-O field length at S/L, ISA + 15°C:		
600: A		2,378 m (7,800 ft)
B		2,270 m (7,450 ft)
600R: A (C2A5 engines)		2,408 m (7,900 ft)
B (PW4158 engines)		2,362 m (7,750 ft)
Landing field length: 600		1,536 m (5,040 ft)
600R		1,555 m (5,100 ft)

Range (1996 and subsequent deliveries) at typical airline OWE with 266 passengers and baggage, reserves for 200 n miles (370 km; 230 miles):

600, GE/PW engines	
	3,700 n miles (6,852 km; 4,257 miles)
600R, GE/PW engines, standard fuel	
	4,050 n miles (7,500 km; 4,660 miles)
600R, GE/PW engines, optional fuel	
	4,157 n miles (7,700 km; 4,784 miles)

OPERATIONAL NOISE LEVELS (A300-600R, ICAO Annex 16, Chapter 3):

T-O: A	91.1 EPNdB (96.3 limit)
B	92.2 EPNdB (96.3 limit)
Sideline: A	98.6 EPNdB (99.9 limit)
B	97.7 EPNdB (99.9 limit)
Approach: A	99.8 EPNdB (103.3 limit)
B	101.7 EPNdB (103.3 limit)

UPDATED

AIRBUS A300-600 CONVERTIBLE and A300-600 FREIGHTER

TYPE: Twin-jet freighter.

PROGRAMME: Specialised versions of A300-600. First flight of A300-600F 2 December 1993; certified April 1994 and entered service with Federal Express in same month; A300-600F powered by GE CF6-80C2A5 with FADEC was first A300-600 version to operate with 180-minute ETOPS in May 1994.

CURRENT VERSIONS: **Convertible:** For all-passenger or all-cargo configuration. Typical options include accommodation (in mainly eight-abreast seating) for maximum 375 passengers (subject to certification) on the main deck; or up to twenty 2.24 × 3.17 m (88 × 125 in) pallets; or five 88 × 125 in plus nine 2.44 × 3.17 m (96 × 125 in) pallets.

Freighter: For freighting only; no passenger systems provided; various systems options give airlines ability to adapt basic aircraft to specific freight requirements; Airbus offers conversion with port-side forward freight door. Freighter conversions, offered by BAE Systems and EFW, are detailed in *Jane's Aircraft Upgrades;* over 65 under conversion or completed by June 2001; maiden flight 13 December 2001 (D-ASAE, c/n 477); certified by LBA and FAA eight days later.

General Freighter: Similar to A300-600F, but with side door and cargo loading system able to handle all sizes of freight from small packets to large containers. Launch customer Air Hong Kong, which announced order for six in January 2003 (deliveries from 2004), with further four on option.

CUSTOMERS: Federal Express became A300-600 Freighter launch customer July 1991 with order for 25 and commitments for 50 more, of which 11 confirmed as orders in September 1996; all now delivered. UPS placed order on 9 September 1998 for 30 PW4158 powered A300F4-600R Freighters plus options on a further 45 (later confirmed and a further 15 firm orders, to total 90, of which 24 delivered and in operation by mid-2002); 30 more on option. Deliveries began mid-2000; UPS has an option to convert some of the order to A380s if it wishes. First (of two) for CityBird of Belgium delivered 23 July 1999; production set to continue to at least 2009.

STRUCTURE: Generally similar to A300-600. Main differences are large port-side main deck cargo door, reinforced cabin floor, smoke detection system in main cabin; main deck cargo door is on opposite side to door of forward underfloor hold, allowing simultaneous loading or unloading at all positions.

POWER PLANT: Options as for A300-600R; first example was first Airbus aircraft powered by GE CF6-80C2A5 with FADEC.

DIMENSIONS, EXTERNAL: As A300-600R, plus:
Main deck cargo door (fwd, port):
Height (projected)	2.57 m (8 ft 5¼ in)
Width	3.58 m (11 ft 9 in)
Height to sill	4.91 m (16 ft 1 in)

DIMENSIONS, INTERNAL:
Cabin main deck usable for cargo:
Length	33.45 m (109 ft 9 in)
Min height	2.01 m (6 ft 7 in)
Max height:	
ceiling trim panels in place	2.22 m (7 ft 3½ in)
without ceiling trim panels	2.44 m (8 ft 0 in)
Volume	192.0-203.0 m³ (6,780-7,169 cu ft)

WEIGHTS AND LOADINGS (basic Convertible. A: with CF6-80C2A5 engines, B: with PW4158 engines):
Manufacturer's weight empty:
A, passenger mode	82,555 kg (182,000 lb)
B, passenger mode	82,470 kg (181,815 lb)
A, freight mode	80,345 kg (177,130 lb)
B, freight mode	80,260 kg (176,945 lb)
Operating weight empty:
A, passenger mode	93,550 kg (206,240 lb)
B, passenger mode	93,475 kg (206,075 lb)
A, freight mode	81,600 kg (179,895 lb)
B, freight mode	81,525 kg (179,730 lb)
Max payload (structural):
A, passenger mode	36,448 kg (80,354 lb)
B, passenger mode	36,523 kg (80,519 lb)
A, freight mode	48,400 kg (106,705 lb)
B, freight mode	48,475 kg (106,870 lb)
Max T-O weight: A, B	170,500 kg (375,900 lb)
Max landing weight: A, B	140,000 kg (308,650 lb)
Max zero-fuel weight: A, B	130,000 kg (286,600 lb)

WEIGHTS AND LOADINGS (basic Freighter variant of -600R):
Manufacturer's weight empty:
A	78,335 kg (172,700 lb)
B	78,250 kg (172,510 lb)
Operating weight empty: A	79,050 kg (174,275 lb)
B	78,980 kg (174,120 lb)
Max payload (structural):
A, range mode	50,950 kg (112,325 lb)
B, range mode	51,020 kg (112,480 lb)
A, payload mode	54,750 kg (120,705 lb)
B, payload mode	54,820 kg (120,855 lb)

Max T-O weight: A, B:
range mode	170,500 kg (375,900 lb)
payload mode	165,100 kg (363,980 lb)
Max landing weight: A, B:
range mode	140,000 kg (308,650 lb)
payload mode	140,600 kg (309,970 lb)
Max zero-fuel weight: A, B:
range mode	130,000 kg (286,600 lb)
payload mode	133,800 kg (294,980 lb)

PERFORMANCE:
Range with max (structural) payload, allowances for 30 min hold at 460 m (1,500 ft) and 200 n mile (370 km; 230 mile) diversion:
A, B, range mode
2,650 n miles (4,908 km; 3,050 miles)
A, B, payload mode
1,900 n miles (3,519 km; 2,186 miles)

UPDATED

AIRBUS A310
Canadian Forces designation: CC-150 Polaris

TYPE: Wide-bodied airliner.

PROGRAMME: Launched July 1978; first flight (F-WZLH) 3 April 1982; initial French/German certification 11 March 1983; first deliveries (Lufthansa and Swissair) 29 March 1983, entering service 12 and 21 April respectively; JAA Cat. IIIa certification (France/Germany) September 1983; UK certification January 1984; JAA Cat. IIIb November 1984; FAA type approval early 1985. First flight of extended-range A310-300 8 July 1985 (certified with JT9D-7R4E engines 5 December 1985, delivered to launch customer Swissair 17 December); wingtip fences introduced as standard on A310-200 from early 1986 (first delivery: Thai Airways, 7 May); certification/delivery of A310-300 with CF6-80C2 engines April 1986, with PW4152s June 1987. Russian State Aviation Register certification October 1991 (first Western-built aircraft to achieve this status). A310s powered by PW4000 series and CF6-80C2 approved for 180-minute ETOPS. Enhanced GPWS installed March 1997, following certification.

CURRENT VERSIONS: **A310-200:** Basic passenger version.
Detailed description mainly applies to A310-200.

A310-200C: Convertible version of A310-200; first delivery (Martinair) 29 November 1984.

A310-200F: Conversion of A310 marketed by Elbe Flugzeugwerke GmbH (EFW); max payload 40,600 kg (89,508 lb) and max T-O weight 142,000 kg (313,056 lb). Main cargo deck accommodates sixteen 2.24 × 3.17 m (88 × 125 in) pallets plus further three pallets of this size in underfloor hold, in addition to six LD3s; alternatively 14 LD3s can be carried in underfloor hold. See *Jane's Aircraft Upgrades* for further details.

A310-300: Extended-range passenger version, launched March 1983; second member of Airbus family to introduce delta-shaped wingtip fences as standard. Extra range provided by increased basic maximum T-O weight (150,000 kg; 330,695 lb) and greater fuel capacity (higher maximum T-O weights optional); standard extra fuel capacity is in tailplane, allowing in-flight CG control for

improved fuel efficiency. For extra long range, one or two ACTs (additional centre tanks) can be installed in part of cargo hold; modification certified November 1987 (first customer Wardair of Canada).

A310-300 also available as conversion by EADS/EFW; initial example (for Federal Express) first flown 30 January 2001. See *Jane's Aircraft Upgrades* for further details.

CUSTOMERS: Total of 260 sold, of which 255 delivered, by 1 January 2003.

Belgian Air Force obtained two second-hand A310-200s in 1997; Canadian Forces operate five A310-300s, German Luftwaffe seven and French Air Force three; last-named have ETOPS for 180 minutes; Spanish intention to buy two for VVIP use, following modification by EADS CASA, announced December 2000. Details of projected military versions appear under Airbus Military Company at the end of this entry.

AIRBUS A310 ORDERS
(at 1 January 2003)

Customer	Qty
Aeroflot	5
Air Afrique	4
Air Algérie	2
Air France	11
Air India	8
Air Niugini	2
Austrian Airlines	4
Balair	4
Biman Bangladesh	2
British Caledonian Airways	2
China Eastern Airlines	5
Condor Flugdienst	5
Cyprus Airways	4
Czech Airlines	2
Delta	9
Ecuatoriana	2
Emirates	8
Hapag Lloyd Flug	7
Interflug	3
ILFC	7
Iraqi Airways	5 (not delivered)
Kenya Airways	2
KLM	10
Kuwait Airways	11
Lufthansa	20
Martinair	2
Nigeria Airways	4
Oasis Group	2
Pakistan International Airlines	6
Pan Am	18
Royal Jordanian	6
Royal Thai Air Force	1
Sabena	3
Singapore Airlines	23
Somali Airlines	1
Swissair	9
TAP-Air Portugal	5
Tarom	2
Thai Airways	2
THY	14
Trans European Airways	1
Uzbekistan Airways	1
Wardair	12
Yemenia	2
Undisclosed	2
Total	**260**

COSTS: US$86.9 million (2001).

DESIGN FEATURES: Retains same fuselage cross-section as A300, but with cabin 11 frames shorter and overall fuselage 13 frames shorter than A300B2/B4-100 and -200; new advanced-technology wings of reduced span and area; new

Air Niugini Airbus A310-300   *NEW*/0527077

Airbus A300-600F freighter of United Parcel Service   *NEW*/0527076

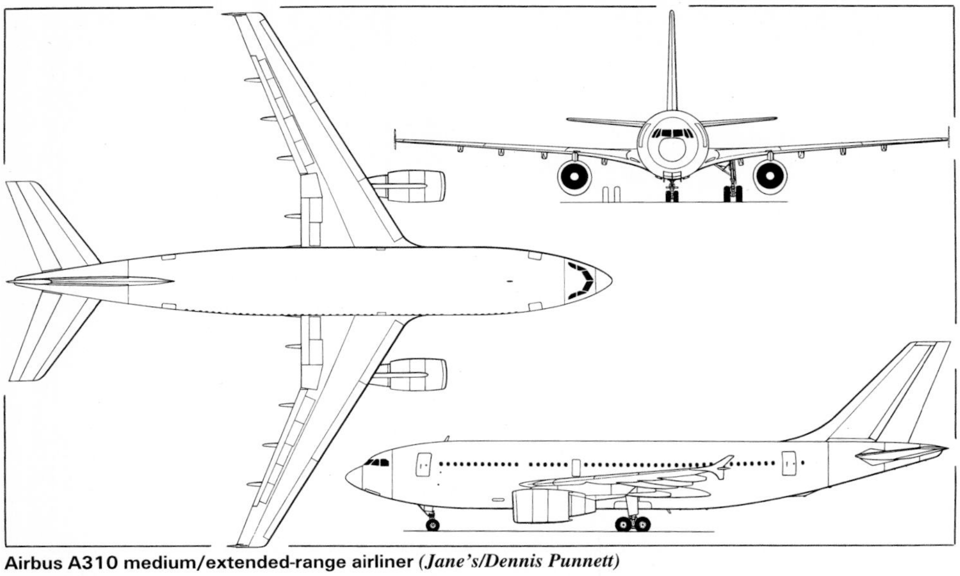

Airbus A310 medium/extended-range airliner *(Jane's/Dennis Punnett)*

and smaller horizontal tail surfaces; common pylons able to support all types of GE and PW engines offered; advanced digital two-man cockpit; landing gear modified to cater for size and weight changes. Wings have 28° sweepback at quarter-chord, root incidence 5° 3′, dihedral 11° 8′ (inboard) and 4° 3′ (outboard) at trailing-edge, and thickness/chord ratios of 15.2 (root), 11.8 (at trailing-edge kink) and 10.8 per cent (tip).

FLYING CONTROLS: Wing leading-edge movable surfaces as for A300-600; trailing-edges each have single Fowler flap outboard, vaned Fowler flap inboard, lateral control by inboard all-speed aileron and fly-by-wire outboard spoilers, without outboard ailerons; all 14 spoilers used as lift dumpers, inner eight also as airbrakes; fly-by-wire spoiler panels controlled by two independent computer systems with different software to ensure redundancy and operational safety. Tail control surfaces as for A300-600.

STRUCTURE: Mainly of high-strength aluminium alloy except for outer shrouds (structure in place of low-speed ailerons), spoilers, wing leading-edge lower access panels and outer deflector doors, nosewheel doors, mainwheel leg fairing doors, engine cowling panels, elevators and fin box, which are all of CFRP; A310 was first production airliner to have carbon fin box, starting with A310-300 for Swissair in December 1985; flap track fairings, flap access doors, rear wing/body fairings, pylon fairings, nose radome, cooling air inlet fairings and tailplane trailing-edges made of AFRP; wing leading-edge top panels, panel aft of rear spar, upper surface skin panels above mainwheel bays, forward wing/body fairings, glideslope antenna cover, fin leading/trailing-edges, fin and tailplane tips (GFRP); and rudder (CFRP/GFRP). Wing box is two-spar multirib metal structure, with top and bottom load-carrying skins. Undertail bumper beneath rear fuselage, to protect structure against excessive nose-up attitude during T-O and landing.

Manufacturing breakdown differs in detail from that of A300-600: EADS France builds nose section (including flight deck), lower centre-fuselage and wing box, rear wing/body fairings, engine pylons and airbrakes, and is responsible for final assembly; EADS Deutschland Airbus builds forward fuselage, upper centre-fuselage, rear fuselage and associated doors, tailcone, fin and rudder, flaps and spoilers, and fits control surfaces and equipment to main wing structure produced by BAE Systems; EADS CASA's contribution includes horizontal tail surfaces, nose-gear and mainwheel doors, and forward passenger doors; Fokker manufactures main landing gear leg doors, wingtips, all-speed ailerons and flap track fairings; wing leading-edge slats and forward wing/fuselage fairings produced by Belgian Belairbus consortium. An A310 of German Luftwaffe is currently testing a fuselage panel constructed from a new lightweight composites material, Glare, in preparation for its use in A3XX; this is first employment of Glare in a primary structure.

LANDING GEAR: Hydraulically retractable tricycle type. Twin-wheel steerable nose unit (steering angle 65°/95°) as for A300. Main gear by Messier-Bugatti, each bogie having two tandem-mounted twin-wheel units. Retraction as for A300-600. Standard tyre sizes: main, 46×16-20 (28/30 ply), pressure 11.24 bar (163 lb/sq in); nose, 40×14 or 40×14.0R16 (22/24 ply), pressure 9.03 bar (131 lb/sq in). Two options for low-pressure tyres on main units: (1) size 49×17.0R20 (30/32 ply), pressure 9.86 bar (143 lb/sq in); (2) size 49×19-20, pressure 8.89 bar (129 lb/sq in). Messier-Bugatti brakes and anti-skid units standard; Bendix type optional on A310-200. Carbon brakes standard since 1986. Minimum ground turning radius (effective, aft CG) 18.75 m (61 ft 6 in) about nosewheel, 33.00 m (108 ft 3¼ in) about wingtips.

POWER PLANT: Launched with two 213.5 kN (48,000 lb st) Pratt & Whitney JT9D-7R4D1 or 222.4 kN (50,000 lb st) General Electric CF6-80A3 turbofans; currently available with 238 kN (53,500 lb st) CF6-80C2A2, or 231 kN (52,000 lb st) Pratt & Whitney PW4152. Available from late 1991 with 262 kN (59,000 lb st) CF6-80C2A8 or 249 kN (56,000 lb st) PW4156A.

Total usable fuel capacity 54,920 litres (14,509 US gallons; 12,081 Imp gallons) in A310-200. Increased to 61,070 litres (16,133 US gallons; 13,434 Imp gallons) in A310-300 by additional fuel in tailplane trim tank. Further 7,200 litres (1,902 US gallons; 1,584 Imp gallons) can be carried in each, of up to two, additional centre tanks (ACT) in forward part of aft cargo hold. Two refuelling points, one beneath each wing outboard of engine.

ACCOMMODATION: Crew of two on flight deck; provision for third and fourth crew seats. Cabin, with six-, seven-, eight- or nine-abreast seating, normally for 210 to 250 passengers, although certified for up to 280; typical two-class layout for 220 passengers (20 first class, six-abreast at 102 cm; 40 in seat pitch, plus 200 economy class mainly eight-abreast at 81 cm; 32 in pitch); maximum capacity for 280 passengers nine-abreast in high-density configuration at pitch of 76 cm (30 in). Standard layout has two galleys and lavatory at forward end of cabin, plus two galleys and four lavatories at rear; depending on customer requirements, second lavatory can be added forward, and lavatories and galleys can be located at forward end at class divider position. Overhead baggage stowage as for A300-600, rising to 0.09 m³ (3.2 cu ft) per passenger in typical economy layout. Four passenger doors, one forward and one aft on each side; oversize Type I emergency exit over wing on each side. Underfloor baggage/cargo holds fore and aft of wings, each with door on starboard side; forward hold accommodates eight LD3 containers or three 2.24 × 3.17 m (88 × 125 in) standard or three 2.44 × 3.17 m (96 × 125 in) optional pallets; rear hold accommodates six LD3 containers, with optional seventh LD3 or LD1 position; LD3 containers can be carried two-abreast, and/or standard pallets installed crosswise.

SYSTEMS: Honeywell 331-250 APU. Air conditioning system, powered by compressed air from engines, APU or a ground supply unit; two separate packs; air is distributed to flight deck, three separate cabin zones, electrical and electronic equipment, avionics bay and bulk cargo compartment; ventilation of forward cargo compartments optional. Pressurisation system has maximum normal differential of 0.57 bar (8.25 lb/sq in). Air supply for wing ice protection, engine starting and thrust reverser system bled from various stages of engine compressors, or supplied by APU or ground supply unit.

Hydraulic system (three fully independent circuits operating at 207 bar (3,000 lb/sq in) details as described for A300-600).

Electrical system, similar to that of A300-600, consists of a three-phase 115/200 V 400 Hz constant frequency AC system and a 28 V DC system; two 90 kVA engine-driven brushless generators for normal single-channel operation with automatic transfer of busbar in the event of a generator failure; each has overload rating of 135 kVA fo 5 minutes and 180 kVA for 5 seconds; third (identical) AC generator, directly driven at constant speed by APU, can be used during ground operations, and also in flight to compensate for loss of one or both engine-driven generators; current production A310s have APU with improved relight capability, which can be started and operated throughout the flight envelope. Any one generator can provide sufficient power to operate all equipment and systems necessary for indefinite period of safe flight; DC power is generated via three 150 A transformer-rectifiers three Ni/Cd batteries.

Flight crew oxygen system fed from rechargeable pressure bottle of 2,166 litres (76.5 cu ft) capacity standard options are second 76.5 cu ft bottle, a 3,256 litre (115 cu ft) bottle, and an external filling connection emergency oxygen sets for passengers and cabin attendants. Anti-icing of outer wing leading-edge slats and engine air intakes by hot air bled from engines; and of pitot probes, static ports and plates, and sensors, by electric heating.

For current production A310s, an ETOPS modification kit, as for the A300-600, is available.

AVIONICS: As described for A300-600. A310 was first airline to introduce CRTs with its ECAM system.

DIMENSIONS, EXTERNAL:	
Wing span	43.90 m (144 ft 0 in)
Wing chord: at root	8.38 m (27 ft 6 in)
at tip	2.18 m (7 ft 1¾ in)
Wing aspect ratio	8.
Length overall	46.66 m (153 ft 1 in)
Fuselage: Length	45.13 m (148 ft 0¾ in)
Max diameter	5.64 m (18 ft 6 in)
Height overall	15.80 m (51 ft 10 in)
Tailplane span	16.26 m (53 ft 4¼ in)
Wheel track	9.60 m (31 ft 6 in)
Wheelbase (c/l of shock-absorbers)	15.21 m (49 ft 10¾ in)
Passenger door (forward, port): Height	1.93 m (6 ft 4 in)
Width	1.07 m (3 ft 6 in)
Height to sill at OWE	4.54 m (14 ft 10¾ in)
Passenger door (rear, port): Height	1.93 m (6 ft 4 in)
Width	1.07 m (3 ft 6 in)
Height to sill at OWE	4.85 m (15 ft 11 in)
Servicing doors (forward and rear, stbd)	as corresponding passenger door
Upper deck cargo door (A310C/F)	as A300-600
Emergency exits (overwing, port and stbd, each):	
Height	1.39 m (4 ft 6¾ in)
Width	0.67 m (2 ft 2½ in)
Underfloor cargo door (forward):	
Height	1.71 m (5 ft 7¼ in)
Width	2.69 m (8 ft 10 in)
Height to sill at OWE	2.61 m (8 ft 6¾ in)
Underfloor cargo door (rear):	
Height	1.71 m (5 ft 7¼ in)
Width	1.81 m (5 ft 11¼ in)
Height to sill at OWE	2.72 m (8 ft 11 in)
Underfloor cargo door (aft bulk hold):	
Height	0.95 m (3 ft 1½ in)
Width	0.95 m (3 ft 1½ in)
Height to sill at OWE	2.75 m (9 ft 0¼ in)
DIMENSIONS, INTERNAL:	
Cabin, excl flight deck: Length	33.25 m (109 ft 1 in)
Max width	5.28 m (17 ft 4 in)
Max height	2.33 m (7 ft 7¾ in)
Volume	210.0 m³ (7,416 cu ft)
Forward cargo hold: Length	7.63 m (25 ft 0½ in)
Max width	4.18 m (13 ft 8½ in)
Height	1.71 m (5 ft 7¼ in)
Volume	50.3 m³ (1,776 cu ft)
Rear cargo hold: Length	5.03 m (16 ft 6¼ in)
Max width	4.17 m (13 ft 8¼ in)
Height	1.67 m (5 ft 5¾ in)
Volume	34.5 m³ (1,218 cu ft)
Aft bulk hold: Volume	17.3 m³ (611 cu ft)
Total overall cargo volume	102.1 m³ (3,605 cu ft)
AREAS:	
Wings, gross	219.00 m² (2,357.3 sq ft)
Ailerons (total)	6.86 m² (73.84 sq ft)
Trailing-edge flaps (total)	36.68 m² (394.82 sq ft)
Leading-edge slats (total)	28.54 m² (307.20 sq ft)

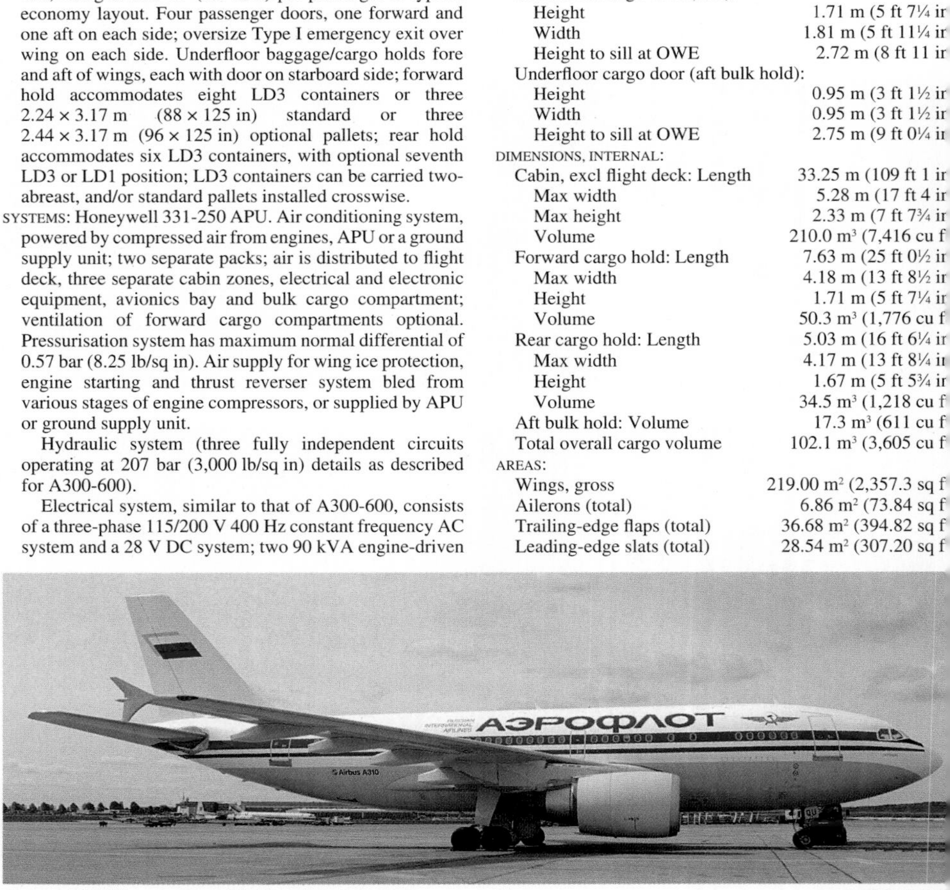

Airbus A310 in service with Russia's Aeroflot ***NEW***/052707

Spoilers (total)	7.36 m² (79.22 sq ft)	
Airbrakes (total)	6.16 m² (66.31 sq ft)	
Vertical and horizontal tail surfaces		as A300-600

WEIGHTS AND LOADINGS (220-seat configuration. C2: CF6-80C2A2 engines, P2: PW4152s, C8: CF6-80C2A8s, P6: PW4156As):

Manufacturer's weight empty:

200: C2	71,660 kg (157,975 lb)
P2	71,600 kg (157,850 lb)
300: C2	72,140 kg (159,040 lb)
P2	72,080 kg (158,910 lb)
C8	72,525 kg (159,890 lb)
P6	72,455 kg (159,735 lb)

Operating weight empty:

200: C2	80,140 kg (176,685 lb)
P2	80,125 kg (176,645 lb)
300: C2	81,205 kg (179,025 lb)
P2	81,165 kg (178,940 lb)
C8	81,610 kg (179,920 lb)
P6	81,545 kg (179,775 lb)
Max payload: 200: C2	32,858 kg (72,439 lb)
P2	32,875 kg (72,476 lb)
300: C8	32,388 kg (71,403 lb)
P6	32,456 kg (71,553 lb)
Max usable fuel: 200	44,100 kg (97,224 lb)*
300	49,039 kg (108,110 lb)
Max T-O weight: 200	142,000 kg (313,050 lb)
300	150,000 kg (330,675 lb)
options (300)	153,000 kg (337,300 lb)
	or 157,000 kg (346,125 lb)
	or 164,000 kg (361,550 lb)
Max landing weight: 200, 300	123,000 kg (271,150 lb)
options (200 and 300)	124,000 kg (273,375 lb)
Max zero-fuel weight: 200, 300	113,000 kg (249,120 lb)
options (200 and 300)	114,000 kg (251,325 lb)

*optional additional tank in aft cargo hold adds 5,779 kg (12,740 lb) of fuel and increases OWE/reduces max payload by 726 kg (1,600 lb). Two additional tanks add 11,560 kg (25,485 lb) of fuel and increase OWE/reduce max payload by 1,536 kg (3,386 lb)

PERFORMANCE (at basic max T-O weight except where indicated; engines as under Weights and Loadings):

Typical long-range cruising speed at FL310-410:

C2, P2, C8, P6	M0.80
Max operating Mach No. (MMO)	0.84

Approach speed at max landing weight:

C2, P2, C8, P6	135 kt (250 km/h; 155 mph)

T-O field length at S/L, ISA + 15°C:

200: C2	1,960 m (6,430 ft)
P2	1,890 m (6,200 ft)
300: C2 (at 150 tonne MTOW)	2,410 m (7,910 ft)
P2 (at 150 tonne MTOW)	2,180 m (7,155 ft)
C8 (at 164 tonne MTOW)	2,485 m (8,155 ft)
P6 (at 164 tonne MTOW)	2,360 m (7,745 ft)

Landing field length at S/L, at max landing weight (200 and 300): C2

	1,479 m (4,850 ft)
P2	1,555 m (5,100 ft)

Runway ACN for flexible runway, category B:

standard tyres: 200	43
300	49
optional tyres: 200	41
300	47

Range (1991 and subsequent deliveries) at typical airline OWE with 220 passengers and baggage, international reserves for 200 n mile (370 km; 230 mile) diversion:

200, GE engines	3,600 n miles (6,667 km; 4,142 miles)
200, PW engines	3,650 n miles (6,759 km; 4,200 miles)
300, GE engines	4,300 n miles (7,963 km; 4,948 miles)
300, PW engines	4,350 n miles (8,056 km; 5,005 miles)

300, option, at T-O weight 157,000 kg (346,125 lb):

GE engines	4,750 n miles (8,797 km; 5,466 miles)
PW engines	4,800 n miles (8,889 km; 5,523 miles)

300, option, at T-O weight 164,000 kg (361,560 lb):

GE engines	5,150 n miles (9,537 km; 5,926 miles)
PW engines	5,200 n miles (9,630 km; 5,984 miles)

OPERATIONAL NOISE LEVELS (ICAO Annex 16, Chapter 3):

T-O: 200: C2	89.6 EPNdB (95.3 limit)
300: C2	91.2 EPNdB (95.6 limit)
Sideline: 200: C2	96.4 EPNdB (99.2 limit)
300: C2	96.3 EPNdB (99.4 limit)
Approach: 200, 300: C2	98.6 EPNdB (102.9 limit)

UPDATED

AIRBUS A320

TYPE: Twin-jet airliner.

PROGRAMME: Launched 23 March 1984; four-aircraft development programme (first flight 22 February 1987 by F-WWAI); JAA (UK/French/German/Dutch) certification of A320-100 with CFM56-5 engines, for two-crew operation, awarded 26 February 1988; first deliveries (Air France and British Airways) 28 and 31 March 1988 respectively; JAA certification of A320-200 with CFM56-5s received 8 November 1988, followed by FAA type approval for both models 15 December 1988; certification with V2500 engines (first flown 28 July 1988) received 20 April (JAA) and 6 July 1989 (FAA), deliveries with this power plant (to Adria Airways) beginning 18 May 1989; FAA approved common type rating on A320 and A321 without further training in early 1994; 500th A320 delivered, 20 January 1995, to United Airlines; 1,000th member of family (an A319) and 1,001st (A320 for United Airlines) were delivered 15 April 1999. Chosen by Thales as platform for its competitor in the South Korean Air Force's E-X AEW&C requirement. In 2000, Airbus agreed to transfer A320 assembly from Toulouse to Hamburg, where A319 and A321 already produced.

CURRENT VERSIONS: **A320-100:** Initial version (21 ordered); details in 1987-88 Jane's. Superseded by A320-200.

A320-200: Now called simply **A320**. Standard version from third quarter 1988; differs from initial A320 in having wingtip fences, wing centre-section fuel tank and higher maximum T-O weights.

Detailed description applies to A320-200.

MPA 320: Proposed maritime patrol version as replacement for Dassault-Breguet Atlantic 1; see AMC entry in this section.

A320 research: An A320 was used for riblet research 1989-91. In mid-1998 Airbus Industrie began testing an experimental laminar-flow fin on an A320; air is sucked through small holes in the leading-edge to reduce drag and save fuel.

A320 Freighter, Convertible and **Quick-Convertible:** Freight variants under consideration by Airbus and Hindustan Aeronautics; cargo capacities would be 20,000 kg (44,090 lb), 22,000 kg (48,500 lb) and 30,000 kg (66,140 lb), respectively. EADS EFW began feasibility studies in 2001 into producing freight versions of the A320 and A321; considering the start of a conversion line in 2008.

A318: Shortened version of A320; described separately.

A319: Shortened version of A320; described separately.

A321-100 and -200: Stretched versions of A320; described separately.

CUSTOMERS: Total 1,597 sold by 1 January 2003, of which 1,128 then delivered. Airbus announced in June 2000 that production of single-aisle aircraft would increase to 30 per month by 2003.

COSTS: Average cost of A320-200 is US$53.7 million, depending on choice of engines, customisation and design weights (2001).

DESIGN FEATURES: First subsonic commercial aircraft to have composites for major primary structures, and centralised maintenance system; advanced-technology wings have 25° sweepback at quarter-chord, 5° 6' 36" dihedral plus experience from A310 and significant commonality with other Airbus aircraft where cost-effective; 6° tailplane dihedral.

FLYING CONTROLS: A320 is first subsonic commercial aircraft equipped with fly-by-wire (FBW) control throughout entire normal flight regime, and first to have sidestick controller (one for each pilot) instead of control column and aileron wheel. Thales/SFENA digital FBW system features five main computers and operates, via electrical signalling and hydraulic jacks, all primary and secondary flight controls; pilot's pitch and roll commands are applied through sidestick controller via two different types of computer; these have redundant architecture to provide safety levels

at least as high as those of mechanical systems they replace; system incorporates flight envelope protection features to a degree that cannot be achieved with conventional mechanical control systems and its computers will not allow aircraft's structural and aerodynamic limitations to be exceeded; even if pilot holds sidestick fully forward, it is impossible to go beyond aircraft's maximum operating speed (VMO) for more than a

AIRBUS A320 ORDERS
(at 1 January 2003)

Customer	Qty
ACES	8
Adria Airways	5
Aer Lingus	6
Aero Lloyd	4
Aerostar	12
Air 2000	4
Air Canada	28
Air France	32
Air Inter Europe	22
Air Jamaica	4
Air Malta	2
Air New Zealand	5
Alitalia	11
All Nippon Airways	28
America West Airlines	21
Ansett Australia	19
Austrian Airlines	8
Boullioun Aviation Services	16
British Airways	32
British Midland Airways	6
Canadian Airlines International	2
China Eastern Airlines	30
China Northwest Airlines	13
China Southern Airlines	20
CIT Group	39
Condor Flugdienst	12
Croatia Airlines	2
Cyprus Airways	8
Debis AirFinance	14
Dragonair	5
Edelweiss Air AG	3
Egyptair	12
Finnair	9
Flightlease	9
GATX/CL AIR	35
GATX/Flightlease	2
GB Airways	7
GECAS	74
GPA	51
Gulf Air	14
Iberia	59
Iberworld	2
Indian Airlines	31
Intl Lease Finance Corp	196
JetBlue Airways	76
Kawasaki Leasing Intl	8
Kuwait Airways	4
Kuwait Finance House	4
LAN Chile	25
Lombard	5
Lotus Airline	1
LTU	4
Lufthansa	37
Mexicana	18
Monarch Airlines	2
MyTravel Airways	6
Northwest Airlines	84
Nouvelair	2
ORIX Corporation	24
Philippine Airlines	4
Qatar Airways (incl Amiri Flight)	11
Royal Jordanian	3
SALE	42
Sabena	3
Shorouk Air	2
Sichuan Airlines	2
Silkair	10
South African Airways	22
Spanair	11
SriLankan Airlines	2
Sudan Airways	1
Swissair	17
Syrian Arab Airlines	6
TACA	40
TAM	25
TAP-Air Portugal	5
Thomas Cook	2
TransAsia Airways	5
Tunis Air	12
United Airlines	117
US Airways	45
Zhejiang Airlines	3
undisclosed cancellations	(−15)
Total	**1,597**

Airbus A320 operated by JetBlue Airways   NEW/0527080

NEW/0527079

Alitalia Airbus A320

few seconds; if pilot holds sidestick fully back, aircraft is controlled to an 'alpha floor' angle of attack, a safe airspeed above stall and throttles opened automatically to ensure positive climb. Nor is it possible to exceed *g* limits while manoeuvring. If a bank angle of more than 30° is commanded with the sidestick, the bank angle is automatically returned to 30° when pressure is released.

Fly-by-wire system controls ailerons, elevators, spoilers, flaps and leading-edge slats; rudder movement and tailplane trim connected to FBW system, but also signalled mechanically when used to provide final back-up pitch and yaw control, which suffices for basic instrument flying. Each wing has five-segment leading-edge slats (one inboard, four outboard of engine pylon), two-segment Fowler trailing-edge flaps, and five-segment spoilers forward of flaps; all 10 spoilers used as lift dumpers, inner six as airbrakes, outer eight and ailerons for roll control and outer four and ailerons for gust alleviation; slat and flap controls by Liebherr and Lucas.

STRUCTURE: Generally similar to A310, but with AFRP for fuselage belly fairing skins; GFRP for fin leading-edge and fin/fuselage fairing; CFRP for wing fixed leading/ trailing-edge bottom access panels and deflectors, trailing-edge flaps and flap track fairings, spoilers, ailerons, fin (except leading-edge), rudder, tailplane, elevators, nosewheel/ mainwheel doors, and main gear leg fairing doors. A320 was first airliner to go into production with CFRP tailplane.

EADS France builds entire front fuselage (forward of wing leading-edge), cabin rear doors, nosewheel doors, centre wing box and engine pylons, and is responsible for final assembly; centre and rear fuselage, tailcone, wing flaps, fin, rudder and commercial furnishing undertaken by

EADS Deutschland Airbus; BAE Systems builds main wings, including ailerons, spoilers and wingtips, and main landing gear leg fairings; Belgian consortium Belairbus produces leading-edge slats; EADS CASA responsible for tailplane, elevators, mainwheel doors, and sheet metal work for parts of rear fuselage; Mitsubishi builds wingroot shroud box under BAE Systems subcontract; AVIC I of China provides wing components and signed an MoU in November 2000 to increase its participation, possibly leading to complete wing production by 2007; GKN Aerospace providing cargo door actuators from January 2002. Final assembly undertaken at Toulouse until 2000; now at Hamburg.

LANDING GEAR: Hydraulically retractable tricycle type, with twin wheels and oleo-pneumatic shock-absorber on each unit (four-wheel main-gear bogies, for low-strength runways, optional); Dowty main units retract inward into wing/body fairing; steerable Messier-Bugatti nose unit retracts forward; nosewheel steering angle ±75° (effective turning angle ±70°). Tyre size 46×16 or 46×17.0R20 (30 ply) on main gear and 30×8.8 or 30×8.8-R15 (16 ply) on nose gear; optional tyres for main gear are 49×17 or 49×17R20 or 49×19R20 or 46×16-20 or 49×19-20. Tyres for main-gear bogie option are 915×300R16 or 36×11 or 46×17.0R20. Carbon brakes standard. Minimum width of pavement for 180° turn 23.1 m (75 ft 9½ in).

POWER PLANT: Two 111.2 to 120.1 kN (25,000 to 27,000 lb st) class CFM International CFM56-5A1 turbofans for first aircraft delivery in 1988, with 113.4 kN (25,500 lb st) IAE V2500-A1 engines available for aircraft delivered from May 1989 and 117.9 kN (26,500 lb st) CFM56-5A3 from November 1990; 120.1 kN (27,000 lb st) CFM56-5B4 and 117.9 kN (26,500 lb st) IAE V2527-A5 available from

1994. New aircraft now (1998 onwards) delivered with either 120.1 kN (27,000 lb st) CFM56-5B4/P or 117.9 kN (26,500 lb st) IAE V2527E-A5. Nacelles by Rohr Industries; thrust reversers by Hispano-Suiza (pivoting door type) for CFM56 engines, by IAE (cascade type) for V2500s. Dual-channel FADEC system standard on each engine.

Standard fuel capacity in wing and wing centre-section tanks is 23,860 litres (6,303 US gallons; 5,249 Imp gallons); one or two additional centreline tanks (ACTs), each holding 2,900 litres (766 US gallons; 638 Imp gallons); can be fitted in rear underfloor baggage/cargo hold.

ACCOMMODATION: Standard crew of two on flight deck, with one (optionally two) forward-facing folding seats for additional crew members; seats for four cabin attendants. Single-aisle main cabin has seating for up to 179 (FAR) or 180 (JAR) passengers, depending upon layout, with locations at front and rear of cabin for galley(s) and lavatory(ies). Multiple customer choice of four-, five- and six-abreast layouts and standard or double-width aisle. Typical two-class layout has 12 seats four-abreast at 91.5 cm (36 in) pitch in 'super first' and 138 six-abreast at 81 cm (32 in) pitch economy class; alternative 152 six-abreast seats (84 business + 68 economy) at 86 and 78 cm (34 and 31 in) pitch respectively; single-class economy layout could offer 164 seats at 81 cm (32 in) pitch, or up to 179 in high-density configuration. Compared with existing single-aisle aircraft, fuselage cross-section is significantly increased, permitting use of wider triple seats to provide higher standards of passenger comfort; five-abreast business class seating provides standard equal to that offered as first class on major competitive aircraft. In addition, wider aisle permits quicker turnrounds. Overhead stowage space superior to that available on existing aircraft of similar capacity, and provides ample carry-on baggage space; best use of underseat space for baggage is provided by improved seat design and optimised positioning of seat rails.

Passenger doors at front and rear of cabin on port side, forward one having optional integral airstairs; service door opposite each of these on starboard side. Two overwing emergency exits each side. Fuselage double-bubble cross-section provides increased baggage/cargo hold volume and working height, and ability to carry seven containers derived from standard interline LD3 type. As base is same as that of LD3, all existing wide-body aircraft and ground handling equipment can accept these containers without modification. Forward and rear underfloor baggage/cargo holds, plus overhead lockers; with 164 seats, overhead stowage space per seat is 0.06 m³ (2.0 cu ft). Mechanised cargo loading system will allow up to seven LD3-based containers to be carried in freight holds (three forward and four aft).

SYSTEMS: Liebherr/ABG-Semca air conditioning, Hamilton Sundstrand/Nord-Micro pressurisation. Honeywell 36-300 APU standard; Honeywell 131-9(A) or APIC APS 3200 available as standard options. All are interchangeable on A318/319/320/321 family. Primary electrical system powered by two Hamilton Sundstrand 90 kVA constant frequency generators, providing 115/200 V three-phase

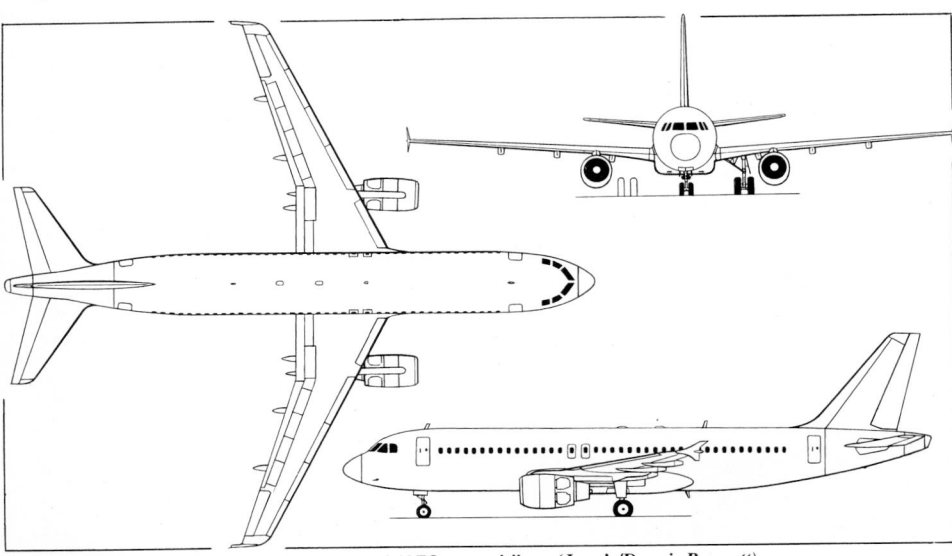

Airbus A320 twin-turbofan single-aisle 150/179-seat airliner (*Jane's/Dennis Punnett*)

AC at 400 Hz; third generator of same type, directly driven at constant speed by APU, can be used during ground operations and, if required, during flight.

AVIONICS: *Flight:* Fully equipped ARINC 700 digital avionics including advanced digital automatic flight control and flight management systems; AFCS integrates functions of SFENA autopilot and Honeywell FMS; Honeywell air data and inertial reference system.

Instrumentation: Each pilot has two Thales/VDO electronic flight instrumentation system (EFIS) displays (primary flight display and navigation display); PFD was first on an airliner to incorporate speed, altitude and heading. Between these two pairs of displays are two Thales/VDO electronic centralised aircraft monitor (ECAM) displays developed from the ECAM systems on A310 and A300-600; upper display incorporates engine performance and warning, lower display carries warning and system synoptic diagrams.

DIMENSIONS, EXTERNAL:

Wing span	34.09 m (111 ft 10 in)
Wing chord at root	6.10 m (20 ft 0 in)
Wing aspect ratio	9.5
Length overall	37.57 m (123 ft 3 in)
Fuselage: Max width	3.94 m (12 ft 11 in)
Max depth	4.14 m (13 ft 7 in)
Height overall	11.76 m (38 ft 7 in)
Tailplane span	12.45 m (40 ft 10 in)
Wheel track (c/l of shock-struts)	7.59 m (24 ft 11 in)
Wheelbase	12.65 m (41 ft 6 in)
Passenger doors (port, forward and rear), each:	
Height	1.85 m (6 ft 1 in)
Width	0.81 m (2 ft 8 in)
Height to sill	3.415 m (11 ft 2½ in)
Service doors (stbd, forward and rear), each	
	as corresponding passenger doors
Overwing emergency exits (two port and two stbd), each:	
Height	1.02 m (3 ft 4¼ in)
Width	0.51 m (1 ft 8 in)
Underfloor baggage/cargo hold doors (stbd, forward and rear), each: Height	1.25 m (4 ft 1¼ in)
Width	1.82 m (5 ft 11½ in)

DIMENSIONS, INTERNAL:

Cabin, excl flight deck: Length	27.50 m (90 ft 2¾ in)
Max width	3.68 m (12 ft 1 in)
Max height	2.16 m (7 ft 1 in)
Aisle width: standard	0.48 m (1 ft 7 in)
option 1	0.58 m (1 ft 11 in)
option 2	0.635 m (2 ft 1 in)
option 3	0.69 m (2 ft 3 in)
Underfloor baggage/cargo holds:	
Length: forward	4.85 m (15 ft 11 in)
rear	9.80 m (32 ft 2 in)
Max width	2.63 m (8 ft 7½ in)
Max height	1.24 m (4 ft 1 in)
Volume: forward	13.30 m³ (469 cu ft)
rear	24.15 m³ (853 cu ft)

AREAS:

Wings, gross	122.40 m² (1,317.5 sq ft)
Ailerons (total)	2.74 m² (29.49 sq ft)
Trailing-edge flaps (total)	21.10 m² (227.12 sq ft)
Leading-edge slats (total)	12.64 m² (136.06 sq ft)
Spoilers (total)	8.64 m² (93.00 sq ft)
Airbrakes (total)	2.35 m² (25.30 sq ft)
Vertical tail surfaces (total)	21.50 m² (231.4 sq ft)
Horizontal tail surfaces (total)	31.00 m² (333.7 sq ft)

WEIGHTS AND LOADINGS (typical 150-passenger configuration. A: CFM56-5B4/P engines, B: V2527-A5s):

Operating weight empty: A	42,100 kg (92,815 lb)*
B	42,482 kg (93,657 lb)
Max payload: A	18,633 kg (41,079 lb)
B	18,518 kg (40,825 lb)
Max fuel	19,159 kg (42,238 lb)
Max T-O weight: standard	73,500 kg (162,040 lb)
1st option	75,500 kg (166,445 lb)
2nd option	77,000 kg (169,755 lb)
Max ramp weight: standard	73,900 kg (162,920 lb)

Airbus A320 flight deck with sidestick controllers outboard and six-screen EFIS   *NEW*/0527082

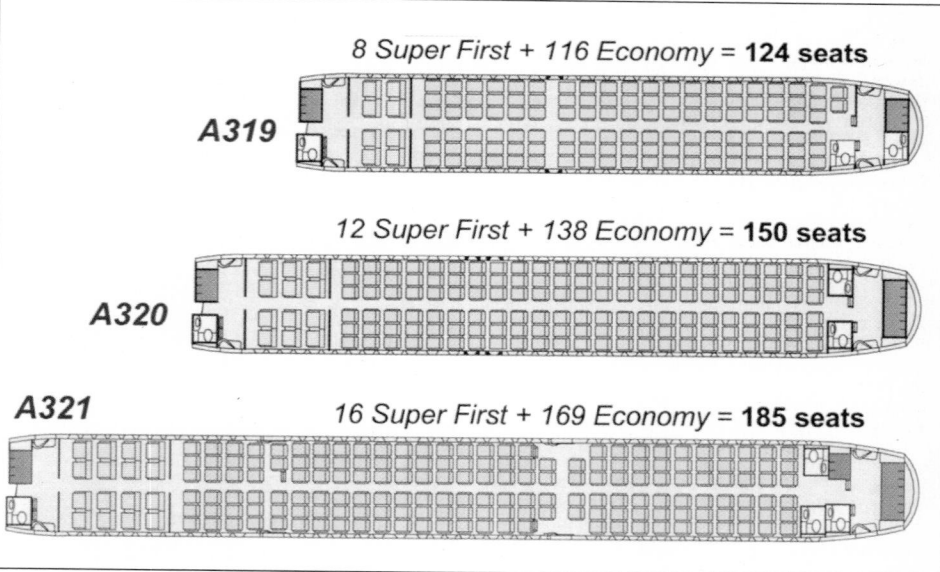

8 Super First + 116 Economy = **124 seats**

A319

12 Super First + 138 Economy = **150 seats**

A320

A321   16 Super First + 169 Economy = **185 seats**

Typical Airbus interiors for the A320 family   0044884

Tunisair has bought 12 Airbus A320s   *NEW*/0527081

Airbus A320 leased to Premiair (*Jane's/Paul Jackson*)   NEW/0526952

1st option	75,900 kg (167,330 lb)
2nd option	77,400 kg (170,635 lb)
Max landing weight: standard	64,500 kg (142,195 lb)
options	66,000 kg (145,505 lb)
Max zero-fuel weight: standard	61,000 kg (134,480 lb)
options	62,500 kg (137,789 lb)
Max wing loading:	
standard	599.5 kg/m² (122.79 lb/sq ft)
1st option	615.8 kg/m² (126.13 lb/sq ft)
2nd option	628.1 kg/m² (128.64 lb/sq ft)
Max power loading (V2527 engines):	
standard	312 kg/kN (3.06 lb/lb st)
1st option	320 kg/kN (3.14 lb/lb st)
2nd option	327 kg/kN (3.20 lb/lb st)

Options raise OWE to a maximum of 43,000 kg (94,799 lb)

PERFORMANCE (engines A and B as for Weights and Loadings, C: CFM56-5A3/5B4, D: 77,000 kg; 169,756 lb max T-O weight):

Max operating Mach No. (Mмo)	0.82
Optimum cruising speed	M0.78
Max rate of climb at S/L	152 m (500 ft)/min
Initial cruise altitude	11,280 m (37,000 ft)
Service ceiling	11,890 m (39,000 ft)
Service ceiling, OEI	5,945 m (19,500 ft)
T-O distance at S/L, ISA + 15°C: A	1,960 m (6,430 ft)
B	1,950 m (6,400 ft)
C	2,180 m (7,155 ft)
D	2,250 m (7,385 ft)
Landing distance at max landing weight:	
A, B, C, D	1,490 m (4,890 ft)

Runway ACN (flexible runway, category B):
twin-wheel, standard 45×16R20 tyres	41
four-wheel bogie option, 36×11-16 Type VII or 900×300-R16	22

Range with 150 passengers and baggage in two-class layout, FAR domestic reserves and 200 n mile (370 km; 230 mile) diversion:

A, standard	2,592 n miles (4,800 km; 2,982 miles)
B, standard	2,596 n miles (4,807 km; 2,987 miles)
A, optional	2,800 n miles (5,185 km; 3,222 miles)
B, optional	2,871 n miles (5,317 km; 3,303 miles)
2nd option:	
A, C	3,045 n miles (5,639 km; 3,504 miles)
B	3,065 n miles (5,676 km; 3,527 miles)

OPERATIONAL NOISE LEVELS (ICAO Annex 16, Chapter 3; A, B and C as for performance):

T-O: A	88.0 EPNdB (91.5 limit)
B	84.8 EPNdB (91.5 limit)
C	85.4 EPNdB (91.5 limit)
Sideline: A	94.4 EPNdB (96.8 limit)
B	92.8 EPNdB (96.8 limit)
C	93.8 EPNdB (96.8 limit)
Approach: A	96.4 EPNdB (100.5 limit)
B	95.9 EPNdB (100.5 limit)
C	96.0 EPNdB (100.5 limit)

UPDATED

AIRBUS A318

TYPE: Twin-jet airliner.

PROGRAMME: Short-bodied version of A319 (itself a truncated A320). Formally announced at Farnborough Air Show, September 1998, although known to be under consideration (as A319M5) since late 1997 when AVIC/AIA/STPL AE316/AE317 venture became uncertain. Smallest aircraft in Airbus family; programme launched 26 April 1999 with orders, commitments and options for 109 aircraft; launch customer Air France (via ILFC). Final assembly at Hamburg, resulting in some A319 production being moved to Toulouse. First metal cut November 2000; final assembly began 9 August 2001; three scheduled for completion in 2002 with production increasing to four per month by mid-2003.

Prototype (F-WWIA; 1,599th of A320 family) made maiden flight 15 January 2002, powered by PW6000 engines. Public debut at ILA, Berlin, 6 to 12 May 2002. Second prototype, F-WWIB, flew 3 June 2002; shown at Farnborough July 2002. Certification originally planned for October and service entry December 2002; however, delays in PW6000 programme (not now expected to be available until mid-2005) caused revised schedule; as a result, F-WWIA re-engined with CFM56-5B/Ps, making maiden flight in this form on 29 August 2002. Approximately 140 more hours of certification flying remained at that time, and this version said to be on schedule for service entry in third quarter 2003. A second prototype will be allocated to development of the PW6000-powered version.

CURRENT VERSIONS: **A318-100:** Baseline version.

CUSTOMERS: First customer, International Lease Finance Corporation (ILFC), signed MoU for up to 30 aircraft on 17 November 1998, subject to launch of project, followed by firm order 30 April 1999. First conditional airline customer, TWA, signed LoI on 9 December 1998 for 50, confirming 25 by firm order 14 December 1999 but cancelled when TWA taken over by American Airlines; other early customers include Egyptair (three on 17 July 1999 – first firm airline order, followed by two further examples, but subsequently all cancelled). Total orders 84 at 1 January 2003.

AIRBUS A318 ORDERS
(at 1 January 2003)

Customer	Engines	Qty
Air France	CFM	15
America West Airlines	PW	15
CIT Group	CFM	4
Frontier Airlines	CFM	5
GECAS	CFM	30
ILFC	CFM	15
Total		**84**

COSTS: Programme cost estimated as US$300 million. Unit price US$41.7 million (2001).

DESIGN FEATURES: Has 95 per cent commonality with other A320 family members, including A319 wing, pylon and interface. Laser welding (rather than riveting) used on fuselage to reduce costs and weight; first use of this technology on airliner. Common pilot type rating with A319, A320 and A321.

FLYING CONTROLS: As A320.

STRUCTURE: As A320, but fuselage 4½ frames (2.39 m; 7 ft 10 in) shorter than A319; 0.79 m (2 ft 7 in) removed forward and 1.60 m (5 ft 3 in) aft of wing. Redesigned cargo doors. Fin has tip extension. Assembled in Germany by EADS Deutschland Airbus. AVIC of China will have share of production work to offset AE31X project cancellation.

LANDING GEAR: As A320.

POWER PLANT: Two 97.9 kN (22,000 lb st) Pratt & Whitney PW6122 or two 106.8 kN (24,000 lb st) PW6124 turbofans with clamshell-type thrust reversers; A318 is launch aircraft for PW6000 family. Alternative power plant is 97.9 kN (22,000 lb st) or 103.6 kN (23,300 lb st) CFM International CFM56-5B, officially announced 4 August 1999. First CFM-powered A318 will enter service mid-2003. Thrust reversing and deflection, fuel tank capacities and locations as A320.

ACCOMMODATION: Two flight crew plus three cabin crew. Typical passenger capacity eight first class and 99 second class with 96/81 cm (38/32 in) seat pitch, or 117 in single-class seating; high-density seating for 129 passengers. Front and rear passenger doors on port side; service door opposite each on starboard side. Overwing emergency exit each side. A318 will not carry containerised freight due to smaller baggage doors (of which size reduced to maintain same engine nacelle clearance for loading vehicles as A319).

SYSTEMS: As A320.

AVIONICS: As A320.

DIMENSIONS, EXTERNAL: As A320 except:
Length overall	31.45 m (103 ft 2 in)
Height overall	12.55 m (41 ft 2 in)
Wheelbase	10.25 m (33 ft 7½ in)
Baggage doors (each): Height	1.24 m (4 ft 1 in)
Width	1.28 m (4 ft 2½ in)

DIMENSIONS, INTERNAL: As A320 except:
Cabin: Length	21.38 m (70 ft 1¾ in)
Underfloor baggage/cargo holds:	
Length: forward	2.57 m (8 ft 5in)
rear	6.07 m (19 ft 11 in)
Volume: forward	6.51 m³ (230 cu ft)
rear	14.70 m³ (519 cu ft)

AREAS: As A320 except:
Fin and rudder (total, incl tab)	21.50 m² (231.4 sq ft)

WEIGHTS AND LOADINGS:
Operating weight empty	39,035 kg (86,057 lb)
Max payload	13,965 kg (30,788 lb)

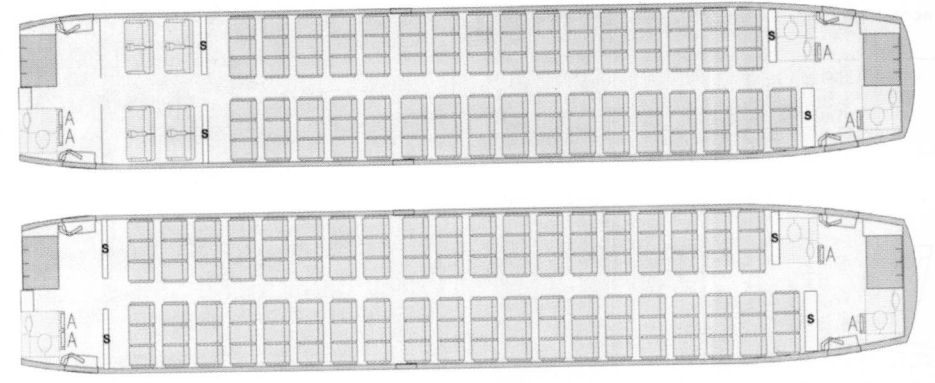

Airbus A318 representative cabin layouts for 107 seats (top) and 117 seats
A: attendant's seat, S: screen   0044887

Airbus A318, flown for the first time on 15 January 2002   NEW/052708

Max T-O weight: normal	59,000 kg (130,075 lb)
option 1	61,500 kg (135,585 lb)
option 2	66,000 kg (145,505 lb)
option 3	68,000 kg (149,915 lb)
Max ramp weight: normal	59,400 kg (130,955 lb)
options	61,900 kg (136,465 lb)
Max landing weight: normal	56,000 kg (123,460 lb)
options	57,500 kg (126,765 lb)
Max zero-fuel weight: normal	53,000 kg (116,845 lb)
options	54,500 kg (120,150 lb)
Max wing loading: normal	481.2 kg/m² (98.57 lb/sq ft)
option 1	501.6 kg/m² (102.74 lb/sq ft)
option 2	538.3 kg/m² (110.26 lb/sq ft)

Max power loading, PW6122 engines:
normal	301 kg/kN (2.96 lb/lb st)
option 1	313 kg/kN (3.07 lb/lb st)
option 2	337 kg/kN (3.31 lb/lb st)

PERFORMANCE (estimated):
Runway ACN at 59,000 kg (130,075 lb): flexible Cat. B
runway   29
T-O run at normal MTOW 500 n mile (925 km;
  575 mile) mission, elevation 610 m (2,000 ft), ISA
  +15°C: PW6122   1,670 m (5,479 ft)
  CFM56   1,630 m (5,348 ft)
Landing run at MLW, S/L, ISA:
  PW6122   1,332 m (4,370 ft)
  CFM56   1,355 m (4,446 ft)
Range with 107 passengers, FAR domestic reserves,
  200 n miles (370 km; 230 miles) diversion, max
  payload:

Airbus A319 delivered to Air Mauritius NEW/0527106

normal MTOW:
| PW6122 | 1,462 n miles (2,707 km; 1,682 miles) |
| CFM56 | 1,455 n miles (2,694 km; 1,674 miles) |
option 1 MTOW:
| PW6122 | 1,960 n miles (3,630 km; 2,255 miles) |
| CFM56 | 1,980 n miles (3,667 km; 2,278 miles) |

option 2 MTOW:
PW6122	2,820 n miles (5,222 km; 3,245 miles)
CFM56	2,880 n miles (5,333 km; 3,314 miles)
option 3 MTOW	3,250 n miles (6,019 km; 3,740 miles)
OPERATIONAL NOISE LEVELS (estimated):
T-O with PW6122 at MTOW	79.7 EPNdB
Sideline	90.4 EPNdB
Approach	89.7 EPNdB
UPDATED

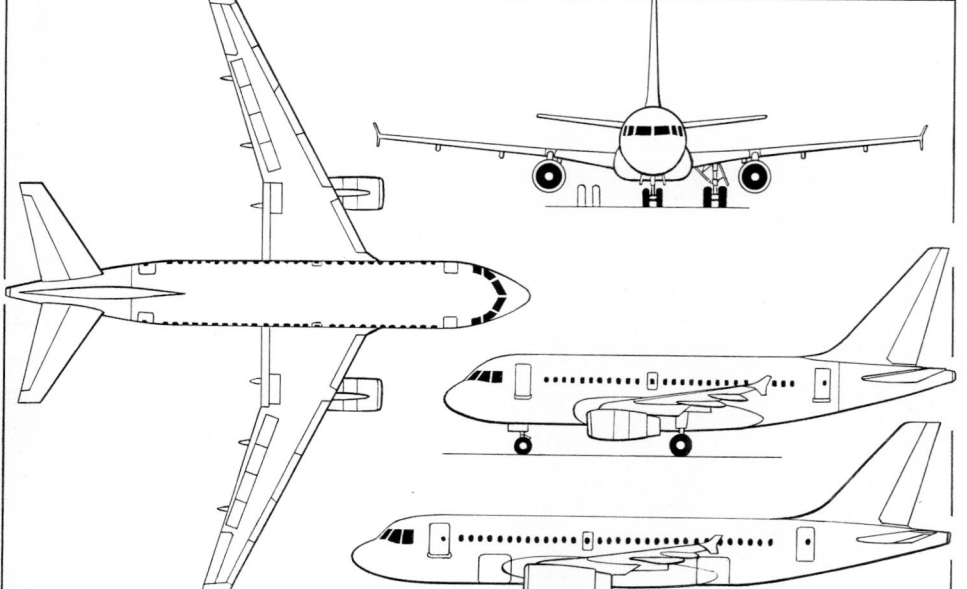
Second prototype A318 displaying at Farnborough in July 2002 (*Jane's/Paul Jackson*) NEW/0526951

AIRBUS A319

TYPE: Twin-jet airliner.

PROGRAMME: Short-fuselage A320. Airbus Board officially authorised start of sales 22 May 1992; programme launched June 1993. Final assembly of first aircraft (F-WWDB, the 546th of the A319/320/321 family) began 23 March 1995; rolled out at Hamburg 24 August; first flight (with CFM56-5A engines) on 29 August; second aircraft (F-WWAS, No. 572) flew (with CFM56-5A engines and commercial interior) 31 October 1995; 650 hour flight test programme resulted in initial certification (CFM56-5B) on 10 April 1996 (CFM56-5A and V2500-A5 followed); first delivery, HB-IPV to ILFC on 25 April 1996 and immediately to Swissair on 30 April, flying first service on 8 May; F-WWAS re-engined with IAE V2524 engines, first flight 22 May 1996 and certified on 18 December 1996; other 1996 first receipts included Air Inter (F-GPMA 21 June); Lufthansa (D-AILA 19 July); and Air Canada (C-FYIY 12 December). JAA 120 minute ETOPS approval granted 14 February 1997 and extended to ACJ in December 2000. 1,000th member of family (an A319 for Air France via ILFC) delivered 15 April 1999.

CURRENT VERSIONS: **A319-100:** Baseline version.
Description applies to A319-100.

ACJ: Airbus Corporate Jetliner. Announced at 1997 Paris Air Show. Standard aircraft will carry up to 40 passengers over a range of 6,300 n miles (11,667 km; 7,250 miles), cruising at 12,500 m (41,000 ft) at speeds of up to M0.82. Certified as a commercial airliner to Cat. IIIb landing criteria and 120-minute ETOPS, and will convert easily to airliner configuration. First customer, announced on 18 December 1997, is Mohamed Abdulmohsin Al Kharafi of Kuwait. First aircraft (G-OMAK, 913th of A320 family), type A319-132, first flew 12 November 1998 and delivered to Jet Aviation, Switzerland, for outfitting on 31 December 1998; customer receipt 8 November 1999, operated by Twinjet Aircraft for Al Kharafi.

Other customers include DaimlerChrysler, Italian Air Force (first of two aircraft delivered 7 March 2000; second in August 2000), Finnair, French Air Force (two; first delivered February 2002) and Aero Service Executive (one). Orders and commitments totalled more than 30 by mid-2002, of which 14 then delivered. FAA (FAR Pt 121) certification for both scheduled service and private operation in USA received October 2002.

Set world record non-stop 15 hour 13 minute flight of 6,918 n miles (12,812 km; 7,961 miles) from Santiago to Le Bourget on 16 June 1999. JAR certification (as amendment of A319 certificate) received August 1999.

Airbus announced in 2000 that production of the ACJ would be restricted to four per year until 2003; this was increased to six, due to demand. By mid-2002, six companies were recommended for outfitting: Lufthansa Technik, Hamburg; Jet Aviation, Basle; Air France Industries, Toulouse; Ozark Aircraft Systems, Bentonville, Arkansas; and EADS Sogerma, Toulouse.

CUSTOMERS: Total of 856 sold, of which 492 delivered, by 1 January 2003, including 14 ACJs.

General arrangement of the Airbus A318 with additional side view (lower) of A319 (*Jane's/James Goulding*) 0121271

For details of the latest updates to *Jane's All the World's Aircraft* online and to discover the additional information available exclusively to online subscribers please visit
jawa.janes.com

AIRBUS A319 and ACJ ORDERS
(at 1 January 2003)

Customer	Qty
Aer Lingus	4
Aero Services Executive	1*
Air Bosna	2
Air Canada	37
Air China	8
Air France	17
Air Inter Europe	9
Air Mauritius	2
Al Kharafi Group	1*
Alitalia	12
America West Airlines	30
Austrian Airlines	7
Boullioun Aviation Services	10
British Airways	36
CIT Group	16
Croatia Airlines	4
Cyprus Airways	2
DaimlerChrysler Aviation	1*
Debis AirFinance	9
Easyjet	120
Eurowings	6
Finnair	5
French Air Force	2*
Frontier Airlines	12
GATX/CL AIR	1
GECAS	31
ILFC	130
Italian Air Force	2*
JAT	8
Lufthansa	20
Northwest Airlines	78
Qatar Airways	1*
SALE	2
Sabena	15
Silkair	6
South African Airways	11
Swissair	6
TACA	9
TAM	18
TAP-Air Portugal	13
Tunis Air	3
United Airlines	78
US Airways	69
Venezuelan Air Force	1*
Undisclosed	4
Undisclosed cancellations	(−3)
Total	**856**

* ACJ

COSTS: Total development cost estimated at US$275 million, entirely financed by Airbus Industrie. A319-100 cost approximately US$48.7 million depending on choice of engines and level of customisation (2001); ACJ cost US$36 million (2001), excluding outfitting of between US$4 million and US$10 million.

DESIGN FEATURES: Seven fuselage frames shorter than A320 (1.60 m; 5 ft 3 in forward of wing, 2.13 m; 7 ft 0 in aft); modified rear cargo hold; bulk hold door and forward overwing emergency exit deleted; derated engines; otherwise little changed. Seats 124 passengers in typical two-class layout, compared with 150 in A320 and 185 in

German carrier Eurowings is an operator of the Airbus A319 (*Jane's/Paul Jackson*)   *NEW*/0526950

Interior of ACJ operated by Aero Services Executive   *NEW*/0527104

A321; range of 3,550 n miles (6,574 km; 4,085 miles) said to be the longest in this category of airliner; common pilot type rating with A320 and A321.

FLYING CONTROLS: Same flight deck and flying control system as A320.

STRUCTURE: As A320. Assembled in Germany by EADS Deutschland Airbus, alongside the stretched A321; partner workshares rearranged to maintain overall workshare balance between France and Germany.

LANDING GEAR: As A320, but 46×16 or 46×17.0R20 main tyre options only.

POWER PLANT: *Standard:* two CFMI CFM56-5A4 or IAE V2522-A5 turbofans, each giving 97.9 kN (22,000 lb st).
    Options: two CFMI CFM56-5A5, CFM56-5B/P or IAE V2524-A5, each giving 104.5 kN (23,500 lb st).
    Fuel capacities in A319, including ACTs, as for A320; ACJ can accommodate six additional tanks in underfloor

hold, increasing total capacity to 37,470 litres (9,899 US gallons; 8,242 Imp gallons).

ACCOMMODATION: Typically 124 passengers (eight in 'super first' class plus 116 economy); maximum 145 passengers in all-economy configuration. Built-in airstairs in ACJ.

DIMENSIONS, EXTERNAL: As A320 except:

Length overall	33.83 m (111 ft 0 in)
Wheelbase	11.05 m (36 ft 3 in)

DIMENSIONS, INTERNAL: As A320 except:

Cabin length (excl flight deck)	23.77 m (77 ft 11¾ in)

Underfloor baggage/cargo holds:

Length: forward	3.35 m (11 ft 0 in)
rear	7.67 m (25 ft 2 in)
Volume: bulk total	27.64 m³ (976 cu ft)

WEIGHTS AND LOADINGS:

Typical operating weight empty:

standard	40,160 kg (88,537 lb)
option	41,203 kg (90,837 lb)
Max payload	16,653 kg (36,714 lb)
Max T-O weight: standard	64,000 kg (141,095 lb)
option	75,500 kg (166,450 lb)
Max landing weight: standard	61,000 kg (134,480 lb)
option	62,500 kg (137,790 lb)
Max zero-fuel weight: standard	57,000 kg (125,665 lb)
option	58,500 kg (128,970 lb)

Max wing loading:

standard	522.9 kg/m² (101.09 lb/sq ft)
option	616.8 kg/m² (126.34 lb/sq ft)
Max power loading: standard	327 kg/kN (3.21 lb/lb st)
option	361 kg/kN (3.54 lb/lb st)

PERFORMANCE (A: with CFM56-5A4, CFM56-5B5 or V2522-A5; B: with CFM56-5A5, CFM56-5B6 or V2524-A5):

Max operating Mach No. (MMO)	0.82
Typical operating Mach No.	0.78
Max rate of climb	152 m (500 ft)/min
Service ceiling: A319	11,890 m (39,000 ft)
ACJ	12,500 m (41,000 ft)
Service ceiling, OEI	6,400 m (21,000 ft)

T-O distance, S/L ISA+15°C:

standard: A	1,720 m (5,643 ft)
option: B	2,640 m (8,661 ft)
Landing distance, S/L ISA: A, B	1,430 m (4,692 ft)

Range on normal fuel tankage with 124 passengers and baggage, FAR domestic reserves and 200 n mile (370 km; 230 mile) diversion:

A, at T-O weight 64,000 kg (141,095 lb)
    1,813 n miles (3,357 km; 2,086 miles)
B, at T-O weight 75,500 kg (166,450 lb)
    3,697 n miles (6,846 km; 4,254 miles)

Airbus Corporate Jetliner of the French Air Force   *NEW*/0527105

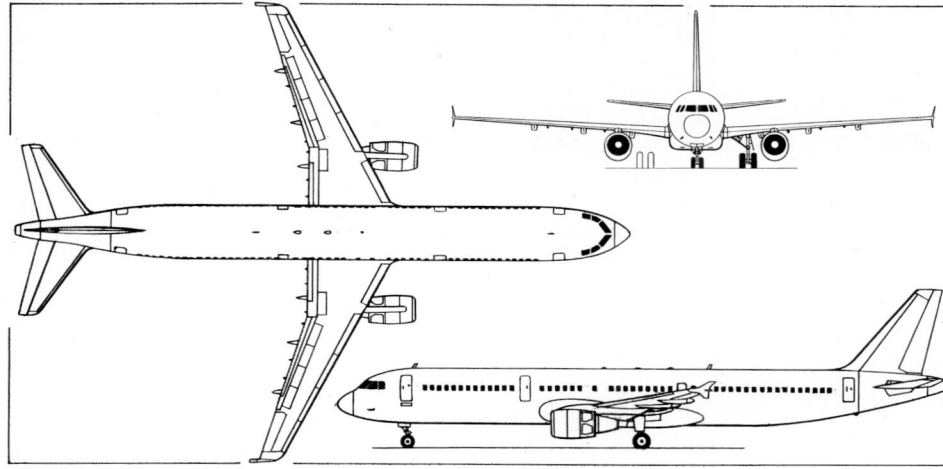

A321 stretched development of the Airbus A320 (*Jane's/Dennis Punnett*)

Airbus A321 operated by Spanair of Spain   NEW/0527103

OPERATIONAL NOSE LEVELS (A, B as above):

T-O: A	82.5 EPNdB
B	82.4 EPNdB
Sideline: A	92.6 EPNdB
B	92.2 EPNdB
Approach: A	93.9 EPNdB
B	94.2 EPNdB
	UPDATED

AIRBUS A321

TYPE: Twin-jet airliner.

PROGRAMME: Stretched version of A320. Announced 22 May and launched 24 November 1989; four development aircraft; rolled out 3 March 1993, first flight with V2530 lead engine 11 March 1993 (F-WWIA), second aircraft with alternative CFM56-5B engine in May 1993; V2530-powered version received European JAA certification 17 December 1993; CFM56-5B2-powered version certified by JAA 15 February 1994; CFM56-5B1 certified by JAA 27 May 1994; first delivery (D-AIRA to Lufthansa) 27 January 1994; first service with Lufthansa 18 March 1994. A321 powered by CFM56-5Bs passed cold-weather trials in Kiruna, January 1994; first with alternative engine handed over to Alitalia 18 March 1994. JAA approval for Cat. III automatic landings achieved in December 1994. 120-minute ETOPS granted 29 May 1996. The 100th A321 (Alitalia's I-BIXZ) flew 1 July 1998. A321 also recommended by Northrop Grumman as platform for Joint STARS, competing for NATO's Airborne Ground Surveillance requirement (later dropped) and by Raytheon for the South Korean Air Force's E-X AEWUC competition.

CURRENT VERSIONS: **A321-100:** Initial version.
Description applies to A321-100.

A321-200: Extended-range version, launched April 1995; features reinforced structure, higher-thrust versions of existing engines and optional additional centre tank (ACT), capacity 2,900 litres (766 US gallons; 638 Imp gallons) which increases maximum T-O weight to 89,000 kg (196,210 lb) and range by 350 n miles (648 km; 402 miles). A321-200 expected to have increased market appeal on North American domestic routes; charter routes between northern and southern Europe; and on scheduled routes between Europe and Middle East. First aircraft flew 12 December 1996 at Hamburg; became G-OZBC of Monarch Airlines and delivered 24 April 1997. Higher-MTOW version, with option of second ACT, launched January 1999; first customer Spanair, for delivery September 2000.

A321CJ: Airbus Industrie reported to be considering a corporate version with additional fuel tanks; no formal plans to launch known by late 2000.

A321 Freighter: EADS EFW conducting feasibility studies into an A321 Freighter with capacity for 14 standard pallets.

CUSTOMERS: Total of 421 sold, of which 256 delivered by 1 January 2003.

COSTS: A321-2001 unit cost approximately US$65.6 million, depending on engine choice and customisation level (2001).

DESIGN FEATURES: Compared with A320, A321 has 4.27 m (14 ft 0 in) fuselage plug immediately forward of wing and 2.67 m (8 ft 9 in) plug immediately aft; pairs of wing fuel tanks unified and system simplified; other changes include local structural reinforcement of existing assemblies, slightly extended wing trailing-edge with double-slotted flaps, uprated landing gear and higher T-O weights.

STRUCTURE: As for A320 except for airframe changes noted under Design Features; front fuselage plug by Alenia and rear one by BAE Systems; final assembly and outfitting by EADS Deutschland Airbus at Hamburg.

LANDING GEAR: Uprated, with 22 in wheel rims, 49×18.0-22 or 46×17.0R20 (30 ply) mainwheel tyres and increased energy brakes; wheels and brakes by Aircraft Braking Systems.

POWER PLANT: *Standard:* two CFM56-5B/P or IAE V2530-A5 turbofans, each rated at 133.4 kN (30,000 lb st). Option: two CFM56-5B2 engines of 137.9 kN (31,000 lb st).

Options: on A321-200 are two CFM56-5B3 of 142.3 kN (32,000 lb st) or two V2533-A5 of 146.8 kN (33,000 lb st).

Normal fuel capacity 23,700 litres (6,261 US gallons; 5,213 Imp gallons); one or two optional 2,900 litre (766 US gallon; 638 Imp gallon) additional centre tanks.

ACCOMMODATION: Typically offers 24 per cent more seats and 40 per cent more hold volume than A320; examples are 185 passengers in two-class layout (16 'super first' class at 91 cm; 36 in seat pitch and 169 economy class at 81 cm; 32 in), or 220 passengers in all-economy high-density configuration. Each fuselage plug incorporates one pair of emergency exits, replacing single overwing pair of A320.

SYSTEMS: Choice of Honeywell 36-300 or APIC APS3200 APU; Honeywell 131-9(A) available from 1998; full commonality with A320 installation.

DIMENSIONS, EXTERNAL: As for A320 except:

Length overall	44.51 m (146 ft 0 in)
Height overall	11.81 m (38 ft 9 in)
Wheelbase	16.92 m (55 ft 6¼ in)
Emergency exits (forward stbd and rear port/stbd, each):	
Height	1.52 m (5 ft 0 in)
Width	0.76 m (2 ft 6 in)
Emergency exit (forward port, usable also as passenger	
door): Height	1.85 m (6 ft 1 in)
Width	0.76 m (2 ft 6 in)

DIMENSIONS, INTERNAL: As A320 except:

Cabin, excl flight deck: Length	34.44 m (113 ft 0 in)
Underfloor baggage/cargo holds:	
Length: forward	8.15 m (26 ft 9 in)
rear	11.40 m (37 ft 5 in)
Volume: forward	22.81 m³ (806 cu ft)
rear	28.93 m³ (1,022 cu ft)

Aero Lloyd Airbus A321 in special colours (*Jane's/Paul Jackson*)   NEW/0526949

WEIGHTS AND LOADINGS (A321-100, except where indicated; typical 185-passenger layout. C1: CFM56-5B1, V: V2530-A5, C2: CFM56-5B2, C3: CFM56-5B3, V2: V2533-A5):

Operating weight empty: C1, C2	48,085 kg (106,010 lb)
V	48,200 kg (106,265 lb)
C3	48,084 kg (106,005 lb)
V2	48,200 kg (106,265 lb)
Max payload: C1, C2	21,416 kg (47,214 lb)
V	21,301 kg (46,960 lb)
Max fuel: C1, V, C2	19,031 kg (41,956 lb)

Max T-O weight:

A321-100: standard	83,000 kg (182,980 lb)
option	85,000 kg (187,390 lb)
A321-200: standard	89,000 kg (196,210 lb)
option	93,000 kg (205,030 lb)

Max landing weight:

A321-100: standard	73,500 kg (162,035 lb)
option	74,500 kg (164,245 lb)
A321-200: standard	75,500 kg (166,450 ft)
option	77,800 kg (171,520 lb)

Max zero-fuel weight:

A321-100: standard	69,500 kg (153,220 lb)
option	70,500 kg (155,430 lb)
A321-200: standard	71,500 kg (157,630 lb)
option	73,800 kg (162,705 lb)

Max wing loading:

standard	678.1 kg/m² (138.89 lb/sq ft)
option	694.4 kg/m² (142.23 lb/sq ft)

Max power loading:

standard	311 kg/kN (3.05 lb/lb st)
option	308 kg/kN (3.02 lb/lb st)

PERFORMANCE:

Max operating Mach No. (MMO)	0.82
Typical operating Mach No.	0.78
Max rate of climb at S/L	152 m (500 ft)/min
Service ceiling	11,890 m (39,000 ft)
Service ceiling, OEI: A321-100	5,180 m (17,000 ft)
A321-200	5,000 m (16,400 ft)

T-O distance at max T-O weight, S/L, ISA +15°C:

C1	2,220 m (7,285 ft)
V	2,065 m (6,775 ft)
C2	2,270 m (7,450 ft)
C3	2,330 m (7,645 ft)
V2	2,341 m (7,680 ft)

Landing distance at max landing weight:

C1, V	1,540 m (5,055 ft)
C2	1,530 m (5,020 ft)
C3, V2	1,577 m (5,175 ft)

Runway ACN (flexible runway, category B):

standard	48

Range on normal fuel tankage with 185 passengers and baggage at typical airline OWE, FAR domestic reserves and 200 n mile (370 km; 230 mile) diversion:

A321-100:
C1, C2:

standard	2,138 n miles (3,959 km; 2,460 miles)
optional	2,250 n miles (4,167 km; 2,589 miles)
V: standard	2,253 n miles (4,172 km; 2,592 miles)
optional	2,386 n miles (4,418 km; 2,745 miles)

A321-200:
with one ACT

	2,700 n miles (5,000 km; 3,107 miles)

with two ACT

	3,000 n miles (5,556 km; 3,452 miles)
V2: standard	2,384 n miles (4,415 km; 2,743 miles)

with one ACT

	2,714 n miles (5,026 km; 3,123 miles)

OPERATIONAL NOISE LEVELS (ICAO Annex 16, Chapter 3, estimated):

T-O: C1	87.0 EPNdB (92.1 limit)
V	85.4 EPNdB (92.1 limit)
Sideline: C1	96.2 EPNdB (97.2 limit)
V	94.5 EPNdB (97.2 limit)
Approach: C1	96.1 EPNdB (100.9 limit)
V	95.4 EPNdB (100.9 limit)

UPDATED

Air Jamaica's colourful scheme applied to an Airbus A321 *NEW*/0527102

AIRBUS A330

Follows entry for the A340, to which it is closely related.

AIRBUS A340

TYPE: Wide-bodied airliner.

PROGRAMME: Launched 5 June 1987 as combined programme with A330, differing mainly in number of engines and in engine-related systems; A330 and A340 were first airliners created with 100 per cent computer-aided design (CAD); first flight 25 October 1991 (A340-300); A340-200 and -300 certified simultaneously by 18 European joint airworthiness authorities (JAA) on 22 December 1992; A340-300 and -200 entered service with Lufthansa January 1993; both received FAA certification 27 May 1993. A340/A330 certified by JAA for GPS satellite navigation January 1994; successful trials conducted at Toulouse in October 1994 using differential global positioning (DGPS) for fully automatic landings, including roll-out. The 200th aircraft from A330/340 assembly line flew in November 1997; 100th A340 delivered to

Singapore Airlines in February 1997; 180-minute ETOPS for Rolls-Royce Trent 700 engines awarded May 1996. A340-500 and -600 were launched on 8 December 1997, at an estimated investment cost of US$2.9 billion. 2,000th Airbus was A340-300 for Lufthansa, delivered 18 May 1999. 400th A330/A340 family member first flown 23 March 2001.

Cathay Pacific took delivery of an A340-300 covered with 700 m² (7,535 sq ft) of plastic film riblets in October 1996 in a trial intended to reduce fuel consumption (average ¾ tonne per medium- or long-range flight) by reducing drag (see illustrations in 1998-99 edition).

CURRENT VERSIONS: **Initial A340-300:** Four-engined higher-capacity version, carrying up to 375 passengers (standard) or 440 (optional) and powered initially by CFM56-5C2 turbofans. Able to carry typical load of 295 passengers over distances of 7,300 n miles (13,519 km; 8,400 miles).

Current A340-300 (engineering designation -300X) powered by 151.2 kN (34,000 lb st) CFM56-5C4 turbofans; maximum T-O weight 271,000 kg (597,450 lb) with optional MTOW of 275,000 kg (606,275 lb); stronger landing gear, and aerodynamic and engine refinements, plus optional additional centre tank (ACT) for increased

fuel capacity compared with basic A340-300. First flight (F-WWJH; CFM56-5C4 engines) 25 August 1995; first delivery, to Singapore Airlines, 17 April 1996.

A340-300 Enhanced: Upgraded CFM56-5C/P engines; trial engine first flew on an A340 on 19 November 2002; advantages expected to include 1 per cent fuel burn reduction. Initial customer, South African Airways, ordered six in 2002.

Initial A340-200: Short-fuselage, longer-range version of A340-300, with same initial power plant; seating capacity 420 passengers, but more typically 303 in a two-class or 263 in three-class configuration; first flight (F-WWBA) 1 April 1992; entered service January 1993 with Lufthansa.

Current A340-200: Advanced version (previously referred to as A340-8000) with additional fuel in two tanks in rear cargo hold, strengthened fuselage and wings, CFM56-5C4 engines and 275,000 kg (606,275 lb) maximum T-O weight; first order placed March 1996 by Prince Jefri of Brunei for one VVIP aircraft, which first flew 19 December 1997 and was retained by Airbus during 1998; stored at Schönefeld-Berlin in 'green' condition, in1999. Sold to Jordan for Royal Flight.

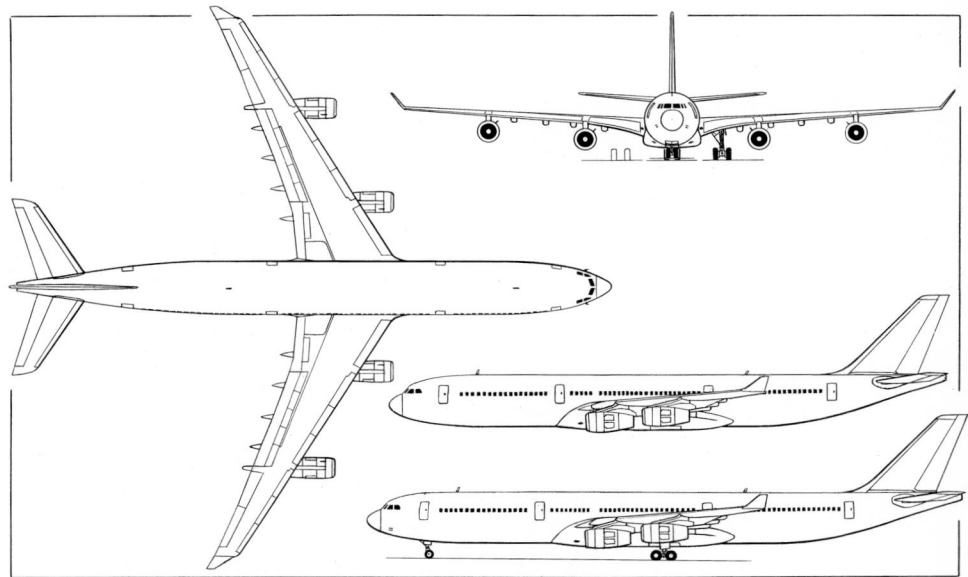

Airbus A340-300 four-turbofan long-range airliner, with additional side view (upper) of A340-200 *(Jane's/Dennis Punnett)*

A340-200 *12 First + 36 Business + 213 Economy* = **261 seats**

A340-300 *12 First + 42 Business + 241 Economy* = **295 seats**

A340-500 *12 First + 42 Business + 259 Economy* = **313 seats**

A340-600 *12 First + 54 Business + 314 Economy* = **380 seats**

Seat pitches: First 62in, Business 40in, Economy 32in

Potential interior arrangements of the Airbus A340   0044886

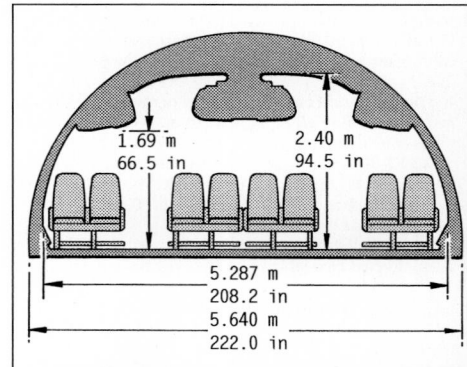

Airbus A330/340 passenger cabin cross-section
(*Jane's/Mike Keep*)   0106843

March 1999. A340-500 initial launch commitments received from Air Canada, Emirates, SIA and ILFC. Maiden flight (F-WWTE, c/n 394) took place 11 February 2002; certified 3 December and first delivery (to Air Canada) due by end of 2002.

A340-600: Derivative of A340-300 with fuselage stretch, 267 kN (60,000 lb st) class engines, improved aerodynamics and fly-by-wire flight-control system, additional fuel capacity, four-wheel central landing gear and 365,000 kg (804,690 lb) standard maximum T-O weight; can carry 380 three-class passengers or 485 in all-economy class up to 7,500 n miles (13,890 km; 8,630 miles) and designed as Boeing 747 replacement with significantly lower costs and full commonality with A330/A340 family. Compared with A340-300, has same wing chord/fuselage centre-section, wing span, tailplane and fin modifications as -500, but with further 5.87 m (19 ft 3 in) plug ahead of wing and 3.20 m (10 ft 6 in) to the rear. Airbus and GE Aircraft Engines agreed in April 1996 that latter should be sole power plant source for -600, but this accord dissolved in February 1997 and Rolls-Royce Trent 556 of 249 kN (56,000 lb st) chosen on non-exclusive basis. Ram-air turbine fitted in underwing pod between engines on starboard wing. Launch commitments from Aerolineas Argentinas, Air Canada, Egyptair, Lufthansa, Swissair and Virgin Atlantic; initial firm order placed by Virgin on 15 December 1997 for eight, plus options. First metal cut 27 July 1998. Final assembly began June 2000; airframe completed September 2000 and engines fitted November 2000. Prototype (c/n 360) rolled out 23 March 2001 and first flew (F-WWCA) 23 April 2001; public debut at Paris Air Show, June 2001; second aircraft (c/n 371) handed to test department 8 June 2001 and first flown (F-WWCB) 18 June 2001; third test example, (c/n 376) flew 24 September 2001 (F-WWCC/G-VATL); three-aircraft 1,600 hour test programme planned; 250 hours completed by August 2001. Programme of weight reduction and engine improvements implemented as a result of test data; weight savings to be achieved gradually by structural modifications, optimum configuration to be

A340-400: Series discontinued in favour of A340-600; basic details in 1998-99 *Jane's*.

A340-500: Ultra-long-range variant of A340-600 able to carry (typically) 313 passengers in three classes across 8,650 n miles (16,020 km; 9,954 miles) and available at the same time as the A340-600. Fuselage 3.20 m (10 ft 6 in) longer than A340-300. Launched at 1997 Paris Air Show and received go-ahead December 1997. Commonality with existing A330/A340 variants allows cross-crewing and employment of common cargo containers and interiors to minimise training, staffing, provisioning and maintenance costs. Stated to be the world's longest-range airliner. Development time of A340-500/-600 series reduced by 25 per cent by use of Airbus Concurrent Engineering (ACE) shared CADD-CAD/CAM systems.

Compared with -300, has 0.53 m (1 ft 9 in) fuselage plug ahead of wings and 1.07 m (3 ft 6 in) plug to the rear; wings of increased chord (incorporating a third fuselage plug of 1.60 m; 5 ft 3 in) with span stretched to 63.45 m (208 ft 2 in); 31.1° sweepback at quarter-chord; taller vertical stabiliser is married to a new horizontal stabiliser to give increased chord fin with 0.50 m (1 ft 7¾ in) height extension. Max T-O weight 368,000 kg (811,300 lb). Landing gear adaptation involves replacement of central twin-wheel unit with forward-retracting brakeable double-bogie unit. Rolls-Royce Trent 553 of 236 kN (53,000 lb st) chosen (non-exclusively) as power plant; this combines fan diameter of Trent 700 with scaled IP and HP compressors and turbines of Trent 800, plus new high-lift LP turbine; test flying of engine began 20 June 2000. Honeywell 331-600 APU chosen for A340-500/-600 in

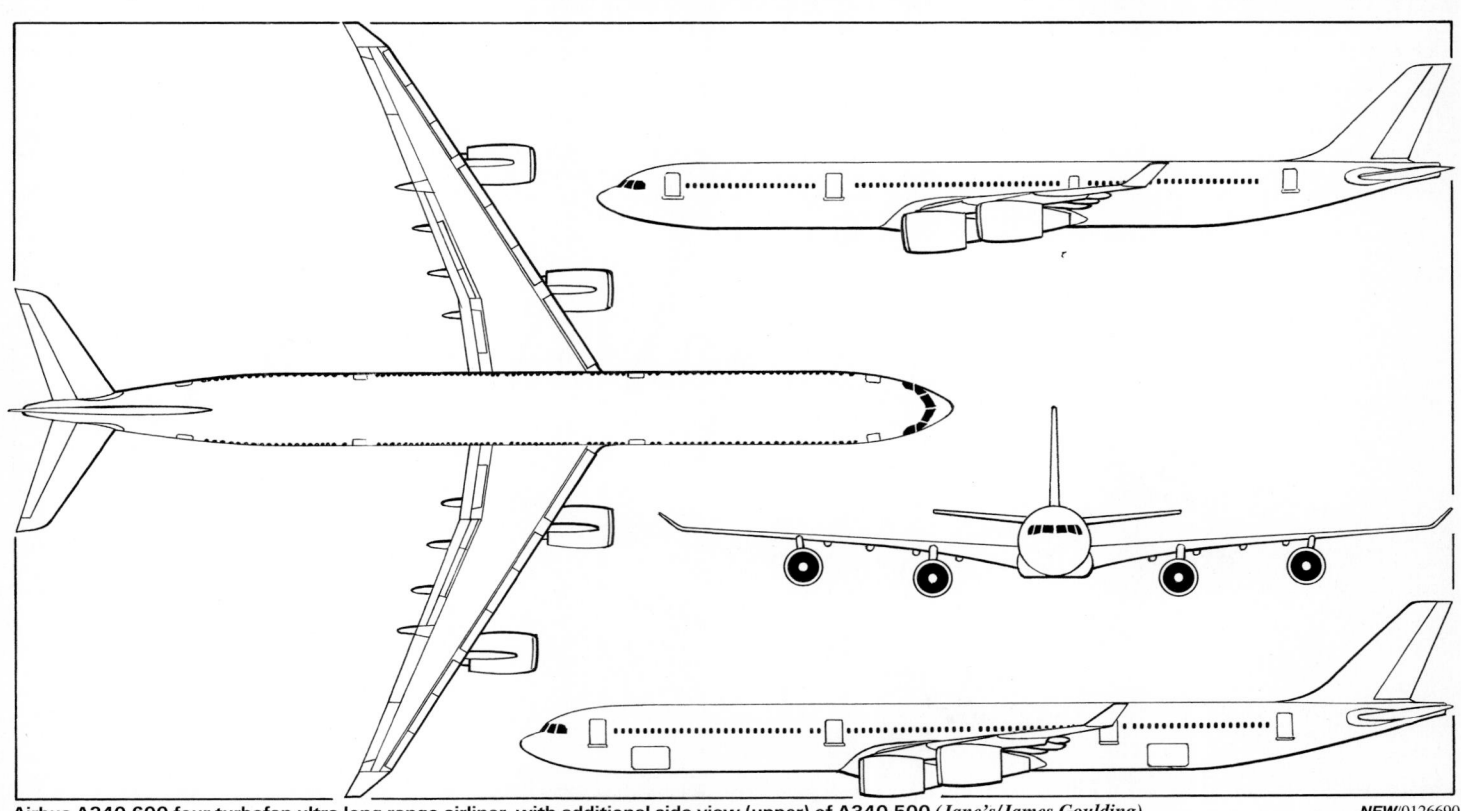

Airbus A340-600 four-turbofan ultra-long-range airliner, with additional side view (upper) of A340-500 (*Jane's/James Goulding*)   NEW/0126690

reached in second quarter of 2003 on aircraft No. 24. Certification by JAA awarded 29 May 2002; first delivery (fourth aircraft, G-VSHY, to Virgin Atlantic) 22 July 2002.

Total of 68 firm orders from 10 customers for A340-500 and -600 by July 2002, at which time five -500s and 13 -600s were in production.

A340-600F: Proposed freighter version; considered by UPS for order eventually placed for MD-11 freighters; airline has option to convert some of its 60 A300-600Fs to A340-600Fs if required.

CUSTOMERS: Sales at 1 January 2003 totalled 317, of which 228 delivered. See table.

AIRBUS A340 ORDERS
(at 1 January 2003)

Customer	Qty
Aerolineas Argentinas	6
Air Canada	13
Air China	3
Air France	14
Air Mauritius	3
Air Tahiti Nui	3
Austrian Airlines	4
Cathay Pacific	11
China Airlines	7
China Eastern Airlines	10
China Southwest Airlines	3
Egyptair	5
Emirates	6
Flightlease	2
Gulf Air	6
Iberia	21
ILFC	29
Kuwait Airways	4
LAN Chile	7
Lufthansa	45
Olympic Airways	4
Philippine Airlines	8
Qatar Airways	1
Sabena	5
SAS	7
Singapore Airlines	22
South African Airways	12
SriLankan Airlines	3
Swiss International Air Lines	12
TAP-Air Portugal	4
Turk Hava Yollari	7
UTA	7
Virgin Atlantic Airways	17
Undisclosed	6
Total	**317**

COSTS: Development cost of A340-500/-600 set at US$2.9 billion, 20 per cent lower than basic A340. Unit cost of A340-300 US$153.5 million; A340-500 US$177 million; A340-600 US$177.9 million (2001 prices).

DESIGN FEATURES: A340 capitalises on commonality with A330 (identical wing/cockpit/tail unit and same basic fuselage) to create aircraft for different markets, and also has much in common (for example, existing Airbus wide-body fuselage cross-sections, A310/A300-600 fin (except on -500/-600), advanced versions of A320 cockpit and systems) with rest of Airbus range; FAA has approved cross-crew qualification for A320 series, A330 and A340.

New design wing (by BAE Systems), approximately 40 per cent larger than that of A300-600, has 30° sweepback at quarter-chord and winglets raked at 29° 42′; thickness/chord ratios 15.25 per cent at root, 11.27 per cent at inner kink, 9.86 per cent at outer kink and 10.60 per cent at tip. A340-500/600 wing 20 per cent larger than basic A340, has increased sweepback of 30° 6′ and 1.60 m (5 ft 3 in) (removable) extension to each wingtip, rake angle 31° 30′.

FLYING CONTROLS: In A330/A340 electronic flight control system (EFCS), roll axis is controlled by two individual outboard ailerons and five outboard spoiler panels on each wing; pitch axis control is by trimmable tailplane and separate left and right elevators; tailplane can also be mechanically controlled from flight deck, but fly-by-wire computer inputs are superimposed; single rudder is

First class accommodation aboard A340-600   NEW/0527109

directly linked to rudder pedals, with dual yaw damping inputs superimposed. High-lift devices consist of full-span slats, flaps and aileron droop; speed braking and lift dumping by raising all six spoilers on each wing and raising all ailerons. Slats and flaps controlled outside main fly-by-wire complex by duplicated slat and flap control computers (SFCC). See diagram.

Control surface actuation by three hydraulic systems (green, yellow, blue); two powered control units (PCU) at each aileron and elevator are controlled either by primary or secondary flight control computers; single actuators at spoiler panels controlled by primary or secondary flight control computers; dual PCUs for fly-by-wire tailplane trimming, and for centrally located flap and slat actuators; three PCUs at rudder.

Fly-by-wire computers include three flight control primary computers (FCPC) and two flight control secondary computers (FCSC); each computer has two processors with different software; primary and secondary computers have different architecture and hardware; power supplies and signalling lanes are segregated; system provides stall protection, overspeed protection and manoeuvre protection as in the A320, but the A330/340 computer arrangement maintains the protections for longer in the face of failures of sensors and inputs; FCPC and FCSC all operate continuously and provide comparator function to active channels, but only one in active control at any one control surface; reconfiguration logic can provide alternative control after failures; different normal and alternative control laws apply fly-by-wire basically as a g demand in pitch and rate demand in roll, plus complex manoeuvre limitations; if all three inertial systems fail (removing attitude information), system reverts to direct mode in which control surface angle is directly related to sidestick position; ultimate control mode is direct control of rudder and tailplane angle from rudder pedals and manual trim wheel, which is sufficient for accurate basic instrument flight.

Pilots have sidestick controllers and normal rudder pedals; EFIS instrumentation consists of duplicated primary flight displays (PFD), navigation displays (ND) and electronic centralised aircraft monitors (ECAM); three display management computers, with separate EFIS and ECAM channels, can each control all six displays in their four possible formats; flight path control by duplicated flight management and guidance and envelope computers

(FMGEC); they control every phase of flight including course, attitude, engine thrust and flight planning using information from GPS and inertial systems; point of no return calculations made automatically for long-range flights in A340/A330. Control system data are collected for maintenance purposes by two flight control data concentrator (FCDC) computers. Honeywell Pegasus flight management system evaluated 1999, before January 2000 certification.

In normal flight, bank angle limited to 33° hands-off (autopilot control) and 67° with full stick displacement; airspeeds limited to 305 kt (565 km/h; 350 mph) and M0.82 under automatic flight control; if stick is held fully forward, the nose is automatically raised when airspeed reaches VMO + 15 kt (28 km/h; 17 mph); if nose is raised, equivalent protections apply; 'alpha max' (13° clean and 19° with flap), slightly below maximum lift coefficient, is the greatest achievable with sidestick; 'alpha floor' is the angle of attack beyond which throttles progressively open to go-around power and airspeed is finally held steady just above stall, even if stick is continuously held back; alpha protection applied at 'normal' and 'hard' modes according to alpha rate; below protection speed (VPROT) of 142 kt (263 km/h; 163 mph) automatic and manual trimming stops; outer ailerons remain centred at over 200 kt (371 km/h; 230 mph); if a spoiler panel fails, the symmetrical opposite panel stops operating; if rudder yaw damping fails, pairs of spoiler panels are used instead; minimum-speed marker on PFD adjusts to changing aircraft configuration, airbrake and pitch rate; fuel automatically transferred between wing and tailplane tanks to minimise trim drag when cruising above 7,620 m (25,000 ft).

Engines controlled by setting throttle levers to marks on quadrant, such as climb (CLB), maximum continuous and flexible take-off (Flex T-O); digital engine control makes detailed settings appropriate to altitude and temperature.

During landing and take-off, nosewheel steering, by rudder pedals, automatically disengages above 100 kt (185 km/h; 115 mph); demands for more than 4°/s nose-up pitch restrained near ground; maximum airspeeds for flaps and slats signalled on airspeed scale; trim automatically cancels effect of flaps, landing gear and airspeed; ailerons droop with full flap selection and deflect 25° up with spoilers in lift dump after touchdown; thrust reverser failure automatically countered by cancelling symmetrical opposite reverser; voice warning demands throttle closure at 6 m (20 ft) during landing flare; autothrottle disengaged when throttles closed at touchdown; on touch-and-go landing, trim automatically reset for take-off when Flex T-O power selected; engine failure in flight compensated automatically, with wings held level, slight heading drift and spoiler panels sucked down to avoid unnecessary drag.

STRUCTURE: A330 and A340 wings almost identical except latter strengthened in area of outboard engine pylon with appropriate modification of leading-edge slats 4 and 5; main three-spar wing box and leading/trailing-edge ribs and fittings of aluminium alloy, with Al-Li for some secondary structures; steel or titanium slat supports; approximately 13 per cent (by weight) of wings is of CFRP, GFRP or QFRP, including outer flaps and flap track fairings, ailerons, spoilers, leading/trailing-edge fixed surface panels and winglets; common fuselage for all

Airbus A340-300 in the colours of Turkish Airlines   NEW/0527112

Airbus A340-500, which made its maiden flight on 11 February 2002

initial versions, except in overall length (A340-300 same size as A330-300; A340-200 and A330-200 respectively eight and ten frames shorter; A340-500 and A340-600 described above; see also dimensions, below); construction generally similar to that of A310 and A300-600 except centre-section to accept new wing; tail unit (common to all versions except A330-200 and A340-500/600, which have larger tailplane, fin and rudder) utilises same CFRP fin as A300-600 and A310; new tailplane incorporates trim fuel tank and has CFRP outer main boxes bridged by aluminium alloy centre-section. A340-600 has Aircelle nacelles constructed from carbon composites.

Work-sharing along lines similar to those for A310 and A300-600, with percentages similar to those held in consortium. Airbus France thus responsible for flight deck, engine pylons, part of centre-fuselage, centre-section cabin doors wing-to-body fairings and final assembly and outfitting at Toulouse; BAE Systems (with Textron Aerostructures, USA, as subcontractor) for wings and landing gear; EADS Airbus Deutschland for most of fuselage, tailcone, fin, wing moving surfaces, cabin doors and interior; EADS Airbus España for tailplane and forward starboard cabin doors; Belgian consortium Belairbus for leading-edge slats and slat tracks. Aircelle for engine nacelles. Before final assembly at Toulouse, Saint Nazaire (Airbus France) pre-assembles flight deck and three forward fuselage sections; and two (of three) rear fuselage sections.

LANDING GEAR: Main (four-wheel bogie) and twin-wheel nose units identical on all A330/340 versions. Main tyres size 54×21.0-23 or 46×17.0R20 or 1400×530R23 (32/36 ply); nose tyres 40.5×15.5-6 or 30×8.8R15 or 1050×395R16 (28 ply). Early A340s have additional twin-wheel auxiliary unit on fuselage centreline amidships, retracting rearward; A340-500 and -600 have four-wheel centre bogie, retracting forward. Goodyear tyres on all units. Landing gear and surrounding structure reinforced for extended-range A340-300.

POWER PLANT: Four 138.8 kN (31,200 lb st) CFM56-5C2 turbofans initially; 144.6 kN (32,500 lb st) CFM56-5C3 and 151.2 kN (34,000 lb st) CFM56-5C4, Trent 553 and Trent 556 engines for A340-500/-600; CFM56 upgrade announced in July 2000, combining CFM56-5C with core of CFM56-5B/P.

Maximum fuel capacity (-200 and -300 until 1996) 138,600 litres (36,614 US gallons; 30,488 Imp gallons); (-200 1996 and subsequent deliveries) 140,640 litres (37,154 US gallons; 30,937 Imp gallons) (-300, 1996 and subsequent deliveries) 141,500 litres (37,381 US gallons; 31,126 Imp gallons); (-213X additional centre tanks) 155,040 litres (40,958 US gallons; 34,105 Imp gallons); 147,800 litres (39,045 US gallons; 32,512 Imp gallons) (A340-300E with one ACT); (-500) 214,810 litres (56,745 US gallons; 47,251 Imp gallons); (-600) 194,880 litres (51,483 US gallons; 42,869 Imp gallons). Totals include 6,230 litres (1,646 US gallons; 1,370 Imp gallons) in tailplane trim tank.

ACCOMMODATION: Crew of two on flight deck (all versions); flight deck can be supplied with humidifier system. Passenger seating typically six-abreast in first class, six-abreast in business class and eight-abreast in economy (nine-abreast optional), all with twin aisles; two-class configurations seat 335 passengers in A340-300, and 303 passengers in A340-200; more typically, a three-class layout seats 295 in A340-300 and 239 in the A340-200. Single-class capacity of A340-300 is 440 and of A340-200 is 420. Optional lower deck facilities in rear cargo hold of -600, with staircase to main deck, include an area with stand-up headroom and up to six lavatories plus, typically, eight crew rest bunks. A similar facility is offered as an

option in the -500 and -600, with up to 10 full-length beds for passengers. A340-600 (except prototype) has two overwing Type III emergency exits in addition to standard eight Type A doors. Underfloor cargo holds house up to 32 LD3 containers or 11 standard 2.24 × 3.17 m (88 × 125 in) pallets in A340-300, 26 LD3s or nine pallets in A340-200, 30 LD3s or 10 pallets in A340-500, and 42 LD3s or 14 pallets in A340-600; front and rear cargo holds have doors wide enough to accept 2.44 × 3.17 m (96 × 125 in) pallets; all versions have a 19.7 m³ (695 cu ft) bulk cargo hold aft of the rear cargo hold.

AVIONICS: Airbus Future Air Navigation System (FANS-A), comprising Smith's Industries digital control and display system married to Honeywell FMS, underwent testing end 1999 and was certified July 2000; can be retrofitted to all A330/A340s. Tenzing Communications in-flight e-mail/Internet access system being installed into A340-600 during late 2001; if successful, will be available for retrofitting into entire Airbus product line.

DIMENSIONS, EXTERNAL:
Wing span: A340-200/300   60.30 m (197 ft 10 in)
   A340-500/600   63.45 m (208 ft 2 in)

Wing chord at root: A340-200/300	10.60 m (34 ft 9¼ in)
A340-500/600	12.20 m (40 ft 0¼ in)
Wing aspect ratio: A340-200/300	10.1
A340-500/600	9.3
Length overall: A340-200	59.42 m (194 ft 11¼ in)
A340-300	63.68 m (208 ft 11 in)
A340-500	67.51 m (221 ft 6 in)
A340-600	74.96 m (245 ft 11 in)
Fuselage: Max diameter (all versions)	5.64 m (18 ft 6 in)

Height overall:
A340-200/-300	16.84 m (55 ft 3 in)
A340-500/-600	17.75 m (58 ft 3 in)
Tailplane span: A340-300	19.41 m (63 ft 8 in)
A340-500/-600	22.96 m (75 ft 4 in)
Wheel track (all versions)	10.69 m (34 ft 5 in)
Wheelbase: A340-200	23.47 m (77 ft 0 in)
A340-300	25.40 m (83 ft 4 in)
A340-500	27.58 m (90 ft 6 in)
A340-600	32.89 m (107 ft 11 in)

DIMENSIONS, INTERNAL:
Cabin: Length (excl flight deck):
   A340-200   46.08 m (151 ft 2 in)

Flight deck of the A340-600

Fourth A340-600 F-WWCD/G-VSHY landing at Farnborough on 22 July 2002 shortly after hand-over to Virgin Atlantic, as the first -600 delivery (*Jane's/Paul Jackson*)
NEW/0526948

A340-300	50.34 m (165 ft 2 in)
A340-500	53.54 m (175 ft 8 in)
A340-600	60.99 m (200 ft 1 in)
Max width	5.19 m (17 ft 0½ in)
Max height	2.40 m (7 ft 10½ in)

Underfloor baggage/cargo holds:

Max height	1.70 m (5 ft 7 in)
Width at floor	3.175 m (10 ft 5 in)

Volume: forward hold:

A340-300/-500 standard	69.1 m³ (2,442 cu ft)
A340-300/-500 option	80.5 m³ (2,844 cu ft)
A340-600 standard	92.2 m³ (3,256 cu ft)
A340-600 option	107.4 m³ (3,792 cu ft)

Volume: rear hold:

A340-300 standard	55.6 m³ (1,965 cu ft)
A340-300 option	62.6 m³ (2,212 cu ft)
A340-500 standard	46.1 m³ (1,628 cu ft)
A340-500 option	53.7 m³ (1,896 cu ft)
A340-600 standard	69.1 m³ (2,442 cu ft)
A340-600 option	80.5 m³ (2,844 cu ft)
Volume: bulk hold (all versions)	19.7 m³ (695 cu ft)

AREAS:

Wings, gross:

A340-200/-300	361.60 m² (3,892.2 sq ft)
A340-500/-600	437.00 m² (4,703.8 sq ft)

WEIGHTS AND LOADINGS:

Operating weight empty:

A340-200 current	129,500 kg (285,500 lb)
A340-300 standard	130,000 kg (286,600 lb)
A340-500	170,300 kg (375,445 lb)
A340-600	177,100 kg (390,440 lb)
Max payload: A340-200 initial	44,000 kg (97,005 lb)
A340-200 current	45,530 kg (100,375 lb)
A340-300 standard	48,100 kg (106,040 lb)
A340-500	54,700 kg (120,595 lb)
A340-600	64,900 kg (143,080 lb)

Max T-O weight:

A340-200, -300	275,000 kg (606,275 lb)
A340-300 option	276,500 kg (609,575 lb)
A340-500	368,000 kg (811,300 lb)
A340-600 standard	365,000 kg (804,675 lb)
A340-600 option	368,000 kg (811,300 lb)

Max landing weight:

A340-200 initial	185,000 kg (407,850 lb)
A340-200 current	188,000 kg (414,475 lb)
A340-300 standard	190,000 kg (418,875 lb)
A340-300 option	192,000 kg (423,275 lb)
A340-500	240,000 kg (529,105 lb)
A340-600 standard	256,000 kg (564,380 lb)
A340-600 option	259,000 kg (570,995 lb)

Max zero-fuel weight:

A340-200	173,000 kg (381,400 lb)
A340-300 standard	178,000 kg (392,425 lb)
A340-300 option	181,000 kg (399,025 lb)
A340-500	225,000 kg (496,040 lb)
A340-600 standard	242,000 kg (533,520 lb)
A340-600 option	245,000 kg (540,130 lb)

Max wing loading:

A340-200, -300	760.5 kg/m² (155.76 lb/sq ft)
A340-500, -600	835.2 kg/m² (171.07 lb/sq ft)

PERFORMANCE (provisional; definitions as for Weights and Loadings):

Max operating Mach No. (Mмo)	0.86
Typical operating Mach No.: A340-200/300	0.82
A340-500/600	0.83

Stalling speed:

A340-300, at 267,000 kg (588,625 lb), wheels up:

flaps up	161 kt (299 km/h; 186 mph)
flaps down	133 kt (247 km/h; 153 mph)

T-O field length at S/L, ISA + 15°C:

A340-200	3,017 m (9,900 ft)
A340-300	3,125 m (10,250 ft)
A340-300 option	3,170 m (10,400 ft)
A340-500 (estimated)	3,280 m (10,750 ft)
A340-600	3,140 m (10,300 ft)
A340-600 option	3,185 m (10,450 ft)

Range at typical OWE, international allowances and 200 n mile (370 km; 230 mile) diversion:

A340-200 with 239 passengers
    8,000 n miles (14,816 km; 9,206 miles)

A340-300 with 295 passengers:

standard	7,300 n miles (13,519 km; 8,400 miles)
optional	7,500 n miles (13,890 km; 8,630 miles)

A340-500 with 313 passengers
    8,650 n miles (16,019 km; 9,954 miles)

A340-600 with 380 passengers:

standard	7,500 n miles (13,890 km; 8,630 miles)
optional	7,650 n miles (14,167 km; 8,803 miles)

OPERATIONAL NOISE LEVELS (A340-300 standard):

T-O, fly-over	95.6 EPNdB
Sideline	96.1 EPNdB
Approach	96.9 EPNdB

UPDATED

AIRBUS A330

TYPE: Wide-bodied airliner.

PROGRAMME: Developed simultaneously with four-engined A340; launched 5 June 1987; first flight (F-WWKA) with GE engines 2 November 1992; first R-R Trent 700-powered A330 flew 31 January 1994; simultaneous European and US certification with initial GE CF6-80E1 engines received 21 October 1993; first delivery (Air Inter) December 1993, entered service January 1994; certification with PW4164/4168 obtained 2 June 1994; A330 powered by GE CF6-80E1 with FADEC granted 120-minute ETOPS approval May 1994; Aer Lingus flew first ETOPS services across the Atlantic in May 1994; A330 powered by PW4164/4168 granted 90-minute ETOPS approval November 1994; certification with R-R Trent achieved 22 December 1994; 90-minute ETOPS approval granted before first delivery, to Cathay Pacific Airways, 24 February 1995. Further extensions of ETOPS to 180 minutes granted on 6 February 1995 (GE engines), 4 August 1995 (P&W) and 17 June 1996 (R-R); 180-minute ETOPS for PW4168A-powered A330-300 granted July 1999.

All main structural and systems information common to both A330 and A340 is listed in A340 entry. Differences in A330 given here.

CURRENT VERSIONS: **A330-300:** Baseline version; seating capacity as for A340-300; 335 passengers in standard two-class, 440 passengers maximum. Payload increase of 7,000 kg (15,432 lb) offered for standard A330-300 in October 1993. Maximum T-O weight increased to 217,000 kg (478,400 lb) from November 1995 to allow typical 335 passengers to be carried 4,850 n miles (8,982 km; 5,581 miles), further increase to 230,000 kg (507,050 lb) with range increased to 5,600 n miles (10,370 km, 6,444 miles) offered in 1997; first aircraft to this standard ordered by Air Canada in 1997, but first in service was a Korean Air example in May 1999. Optional higher maximum T-O weight of 233,000 kg (513,675 lb) available from early 2001; earlier 230,000 kg (507,050 lb) versions can be retrofitted. During August 2000, Airbus was said to be looking at a 240,000 kg (529,100 lb) high gross weight model with increased range.

A330-200: Extended-range version; launched 24 November 1995; 10-frame reduction in fuselage length to 59.00 m (193 ft 6¾ in); maximum T-O weight 230,000 kg (507,050 lb); higher MTOW available from early 2001, as per A330-300. 253 passengers in three classes or 293 passengers in two classes; range with 253 passengers 6,650 n miles (12,315 km; 7,652 miles); engine choice as for A330-300; initial order, for 13, placed in March 1996 by ILFC. First flight (c/n 181, F-WWKA, with CF6-80E1 engines) 13 August 1997 was followed by public debut at Dubai Air Show in November 1997. Canada 3000 first user (on lease from ILFC) with first flight of production aircraft (C-GGWA) 20 January 1998 and delivery on 29 May 1998 following FAA/JAA/Transport Canada certification on 31 March 1998. First direct airline order from Korean Air, received second built in June 1998 (c/n 195, with PW4000 engines) following 4 December 1997 first flight and May 1998 certification. Version with Rolls-Royce Trents (re-engined first prototype) flew 24 June 1998 and following JAR and Transport Canada

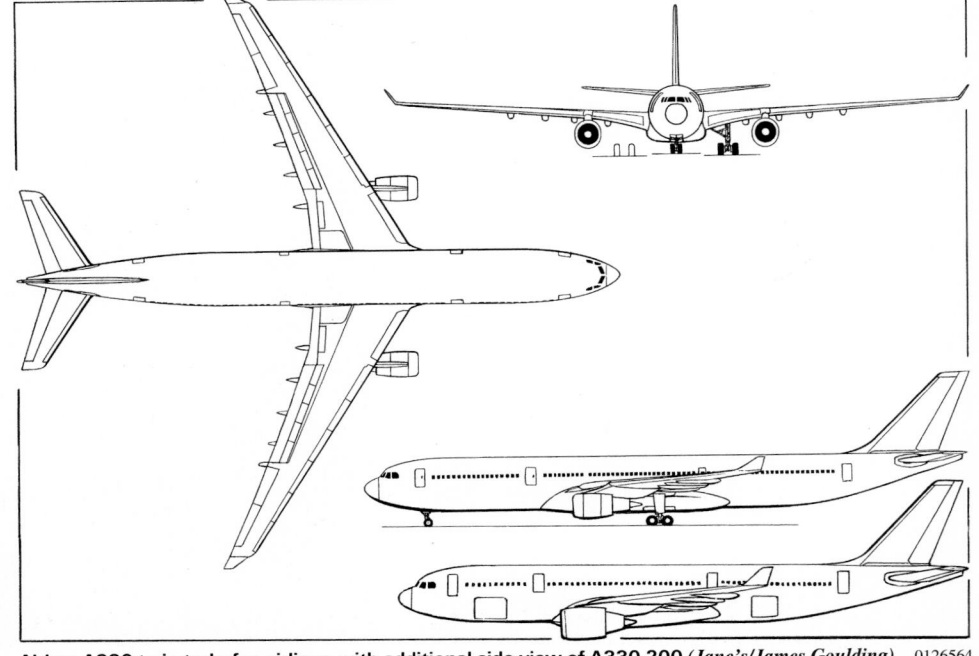

Airbus A330 twin-turbofan airliner, with additional side-view of A330-200 (*Jane's/James Goulding*)   0126564

certification in January 1999 was delivered to Air Transat in February 1999. Estimated development cost US$450 million (1995). Version with CF6-80E1A3s delivered to Air France 17 December 2001.

AirTanker consortium bidding for UK Strategic Tanker Aircraft contract is offering A330-200 tankers, using new airframes converted by Cobham, which would enter service in 2007.

A330-100: Proposed nine-frame reduction of A330-200; abandoned in favour of A330-500; see *Jane's* 2000-01 for details.

A330-200F: Proposed freighter version; possible DC-10/MD-11 replacement; planned launch at Paris Air Show in June 2001 was deferred to unspecified date. Maximum payload 63,050 kg (139,000 lb) over 4,250 n miles (7,871 km; 4,890 miles). Capacity for 20 2.24 × 3.17 × 2.44 m (88 × 125 × 96 in) pallets or containers on main deck and a combination of up to 14 LD3 plus up to 6 LD6 containers on the lower deck. Estimated market for 200 aircraft over next 10 years.

A330-500: Proposed eight-frame reduction of A330-200, announced July 2000 and initially internally designated A330-M18; details in 2002-03 *Jane's All the World's Aircraft*. Was due to be officially launched at end of 2000 with in-service date of early 2004, but suspended in favour of A380 programme.

CUSTOMERS: Total of 419 sold, of which 251 delivered, by 1 January 2003. See table.

AIRBUS A330 ORDERS
(at 1 January 2003)

Customer	Qty
Aer Lingus	3
Air Caledonie International	2
Air Canada	8
Air France	8
Air Inter Europe	7
Asiana Airlines	6
Austrian Airlines	3
British Midland Airways	1
Cathay Pacific Airways	23
CIT Group	19
Corsair	2
Dragonair	5
Emirates	28
EVA Air	2
Flightlease	9
Garuda Indonesia	9
GECAS	18
Gulf Air	6
ILFC	93
KLM	6
Korean Air	19
LTU	5
Lufthansa	10
Malaysia Airlines	10
Monarch Airlines	2
MyTravel Airways	7
Northwest Airlines	32
Philippine Airlines	8
Qantas	13
Qatar Airways	7
Sabena	3
SAS	4
SriLankan Airlines	6
Swissair	4
TAM	5
Thai Airways International	12
US Airways	10
Undisclosed	4
Total	**419**

Airbus A330-200 in Airtours International Airways colours   *NEW*/0527108

COSTS: A330-200 US$135.5 million; A330-300 US$150.6 million (2001).

POWER PLANT: First deliveries with two 300 kN (67,500 lb st) GE CF6-80E1A4 turbofans; alternative engines, using common pylon and mount, GE CF6-80E1A3 (each 320 kN; 72,000 lb st), P&W PW4168A (303 kN; 68,000 lb st) or R-R Trent 772B (316 kN, 71,100 lb st).

Fuel capacity 97,530 litres (25,765 US gallons; 21,454 Imp gallons) in A330-300; 139,100 litres (36,746 US gallons; 30,599 Imp gallons) in A330-200. Both totals include 6,230 litre (1,646 US gallon; 1,370 Imp gallon) tailplane trim tank.

DIMENSIONS, EXTERNAL: As A340 except:
Length overall: A330-200   59.00 (193 ft 7 in)
        A330-300   63.58 m (208 ft 7 in)
Height overall: A330-200   17.88 m (58 ft 8 in)
        A330-300   16.84 m (55 ft 3 in)

DIMENSIONS, INTERNAL: As A340 except:
Underfloor baggage/cargo holds:
Volume: forward hold:
        A330-200 standard   55.0 m³ (1,944 cu ft)
        A330-200 option   62.6 m³ (2,212 cu ft)

A330-300 standard   69.1 m³ (2,442 cu ft)
A330-300 option   80.5 m³ (2,844 cu ft)
Volume: rear hold:
        A330-200 standard   46.1 m³ (1,628 cu ft)
        A330-200 option   53.7 m³ (1,896 cu ft)
        A330-300 standard   55.6 m³ (1,965 cu ft)
        A330-300 option   62.6 m³ (2,212 cu ft)
Volume: bulk hold (all versions)   19.7 m³ (695 cu ft)

WEIGHTS AND LOADINGS (C: with CF6-80E1A3/A4, P: with PW4168A, T: with Trent 772B):
Typical airline OWE: A330-200:
        C   122,140 kg (269,270 lb)
        P   124,410 kg (274,275 lb)
        T   124,200 kg (273,815 lb)
A330-300:
        C   124,520 kg (274,520 lb)
        P   124,820 kg (275,180 lb)
        T   124,620 kg (274,740 lb)
Max payload: A330-300:
        C   48,480 kg (106,880 lb)
        P   48,180 kg (106,220 lb)
        T   48,380 kg (106,660 lb)

Common flight deck of the A330 (illustrated) and A340 permits crew cross-qualification   *NEW*/0116958

Airbus A330-300 leased to Airtours' Danish subsidiary Premiair (*Jane's/Paul Jackson*)   *NEW*/0526947

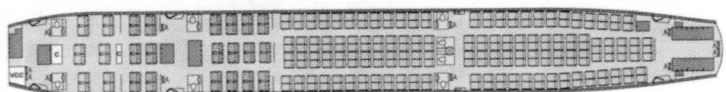

A330-200 12 First + 36 business + 205 Economy = **253 seats**

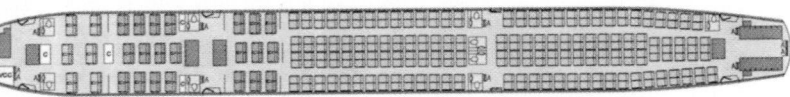

A330-300 12 First + 42 Business + 241 Economy = **295 seats**

Seat pitches: First 62in, Business 40in, Economy 32in

Airbus A330 alternative interiors   0044885

Max T-O weight:
A330-200/-300 standard	230,000 kg (507,060 lb)
A330-200/-300 option	233,000 kg (513,675 lb)

Max landing weight:
A330-200 standard	180,000 kg (396,825 lb)
A330-200 option	182,000 kg (401,250 lb)
A330-300 standard	185,000 kg (407,855 lb)
A330-300 option	187,000 kg (412,265 lb)

Max zero-fuel weight:
A330-200 standard	168,000 kg (370,375 lb)
A330-200 option	170,000 kg (374,775 lb)
A330-300 standard	173,000 kg (381,400 lb)
A330-300 option	175,000 kg (385,810 lb)

Max wing loading:
A330-200/-300 standard	633.4 kg/m² (129.74 lb/sq ft)
A330-200/-300 option	644.4 kg/m² (131.97 lb/sq ft)

Max power loading:
A330-200/-300 standard: C	359 kg/kN (3.52 lb/lb st)
P	380 kg/kN (3.73 lb/lb st)
T	364 kg/kN (3.57 lb/lb st)
A330-200/-300 option: C	364 kg/kN (3.57 lb/lb st)
P	385 kg/kN (3.78 lb/lb st)
T	368 kg/kN (3.61 lb/lb st)

PERFORMANCE:
Max operating Mach No. (Mмo)	0.86
Typical operating Mach No.	0.82

T-O field length at S/L, ISA + 15°C, Trent 772B engines:
A330-200, standard MTOW	2,545 m (8,350 ft)
A330-200, option MTOW	2,652 m (8,700 ft)
A330-300, standard MTOW	2,530 m (8,300 ft)
A330-300, option MTOW	2,606 m (8,550 ft)

Landing field length at S/L, ISA + 15°C, Trent 772B engines:
A330-200, option MLW	1,722 m (5,650 ft)
A330-300, option MLW	1,753 m (5,750 ft)

Range, typical OWE plus max passengers, international allowances and 200 n mile (370 km; 230 mile) diversion:
A330-200 at standard MTOW	6,650 n miles (12,315 km; 7,652 miles)
A330-300 at standard MTOW	5,600 n miles (10,371 km; 6,444 miles)

OPERATIONAL NOISE LEVELS (A330-300: A: 217,000 kg (478,400 lb) with Trent 768; B: 230,000 kg (507,050 lb) with Trent 772):
T-O, flyover: A	89.8 EPNdB
B	98.0 EPNdB
Sideline: A	96.5 EPNdB
B	101.0 EPNdB
Approach: A	96.8 EPNdB
B	104.0 EPNdB

UPDATED

AIRBUS A380

TYPE: High-capacity airliner.

PROGRAMME: Engineering work began in early June 1994; known as A3XX until December 2000; separate A3XX directorate (Large Aircraft Division) formed within Airbus in March 1996; concept definition began April 1996; Airbus will allocate approximately 40 per cent of programme to new partners. Full-size mockup of fuselage cross-section shown at Paris in 1997; completed, including partial concept interior, in 2000.

Airbus Industry Supervisory Board authorised programme go-ahead 8 December 1999; commercial launch authorised 23 June 2000; industrial launch and A380 designation confirmed 19 December 2000, on receipt of required 50th launch order. First metal and first carbon fibre lap were cut at Nantes on 23 January 2002. Test programme calls for 2,200 flying hours over 15 months

using four prototypes; the first to explore the flight envelope, second to verify performance, third for technical and commercial adaptations and fourth for route proving. Static testing to be carried out at Toulouse and fatigue testing at Dresden. First flight due early 2005, with certification and first deliveries to Emirates and Singapore Airlines by end of first quarter of 2006. Production intended to reach four per month by 2008.

CURRENT VERSIONS: **A380-700:** Potential short-fuselage version, more closely aligned to Boeing 747 replacement market.

A380-800: Baseline version; nominal 555 passengers in three-class layout and range of 8,000 n miles (14,816 km; 9,206 miles). Engine thrust 311 kN (70,000 lb), MTOW 560,000 kg (1,234,585 lb), fuel capacity 325,000 litres (85,856 US gallons; 71,492 Imp gallons). Lower deck capacity 13 pallets or 38 LD3 containers.

Following description applies to A380-800 except where indicated.

A380-800F: All-cargo version, carrying 150,000 kg (330,700 lb) of payload over 5,600 n miles (10,371 km; 6,444 miles). Standard layout is 17 pallets on upper deck, 28 pallets on main deck and 13 pallets in cargo hold/lower deck. Optional high-volume configuration offers more than 1,133 m³ (40,000 cu ft) of capacity.

A380-900: Potential stretched version, initially known as A3XX-200; 656 seats in three-class layout and up to 990 at high density. Increased MTOW and fuel volume.

CUSTOMERS: Total of 95 firm orders by January 2003 (see table). Potential market for 1,200 and over 300 cargo aircraft in A380 class up to 2020. Initial launch customer, on 24 July 2000, was Emirates, which committed to five A380s and two A380-800Fs, plus five options, subsequently increased to 20 passenger examples plus 10 options and 2 freighters upon confirmation of the order. Federal Express became A380-800F launch customer in January 2001.

COSTS: Estimated development cost US$10.7 billion for whole A380 family given during June 2001; of which US$3.1 billion was sought from risk-sharing partners. Projected unit cost at launch was US$217 million for passenger version and US$233 million for freighter; revised to average unit price for A380-800 of US$265 million by mid-2002.

DESIGN FEATURES: First version, designated A380-800, is conventional in external appearance except for two rows of windows, but incorporates new developments in structures, materials, systems, landing gear design and aerodynamics. Dassault CATIA and IBM computer-aided design. Flight deck commonality with in-service Airbuses permits crew cross-qualification.

Fuselage is vertically orientated oval three-deck arrangement; this 'vertical ovoid' accommodates 10 passengers abreast on main deck and eight-abreast on upper deck, offering greater space per passenger than Boeing 747. Seating ranges from a nominal 555 (22 first class and 328 economy on main deck and 102 business and 103 economy on upper deck) in three-class layout to more than 800 in high-density, shorter-range applications. Typical launch customer layouts are for about 525 passengers. Dual-lane boarding stair allows four-aisle boarding and deplaning through main deck.

Lower deck can accommodate shop, bar, restaurant and/or normal range of 36 cargo containers or 13 pallets and 18.4 m³ (650 cu ft) of bulk freight; main deck is large enough to accommodate two 2.44 × 2.44 m (8 ft 0 in × 8 ft 0 in) containers side by side in the freighter version.

Modifications to engines, nacelles and aerodynamics at customer request late in the launch phase have resulted in major reductions in noise levels.

Wing sweep 33° 30' at 25 per cent chord.

FLYING CONTROLS: Single-slotted flaps incorporate droop-nose device to improve climb performance. Two ailerons and two actuators on each wing, plus eight spoilers with individual actuators. Elevators have two panels and actuators on each side; rudder also has two panels and actuators. Flaps, ailerons and engines have all been specifically positioned to keep wake vortex at a minimum.

STRUCTURE: Extensive use of composites for all flaps and spoilers, rear pressure bulkhead, centre wing box, all tail

AIRBUS A380 ORDERS
(at 1 January 2003)

Airline	Passenger	Freighter	Options	Engines	Order confirmed	First delivery
Air France	10	–	4	GP7200	18 Jun 01	Nov 2006
Emirates	20	2	10	GP7200	4 Nov 01	Mar 2006
Federal Express	–	10	20	GP7200	Jul 02	2008
ILFC	5	5	–	Trent 900	17 Jun 01	2006
Lufthansa	15	–	–	Trent 900	20 Dec 01	2007
Qantas Airways	12	–	12	Trent 900	6 Mar 01	2006
Singapore Airlines	10	–	15	Trent 900	16 Jul 01	Mar 2006
Virgin Atlantic Airways	6	–	6	Trent 900	25 Apr 01	–
Total	**78***	**17**	**67**			

* Plus commitments by Qatar Airways 27 February 2001 for two of passenger version and Malaysia Airlines (January 2003) for six.

Computer-generated image of the twin-deck Airbus A380-800 in the livery of launch customer Emirates
NEW/0527128

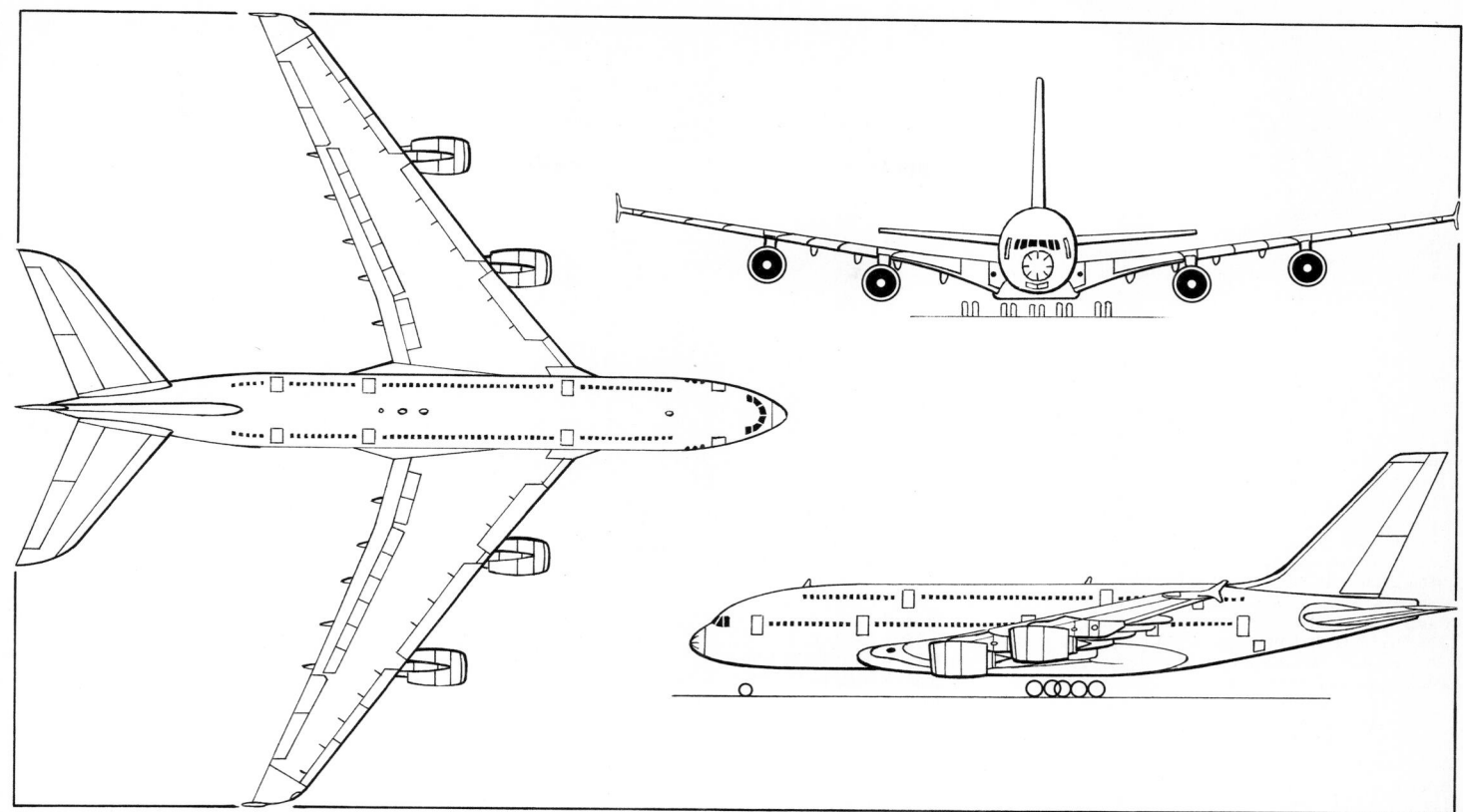

Airbus A380-800 555-seat airliner (*Jane's/James Goulding*)   0103547

surfaces, tailcone aft of fin leading-edge intersection with fuselage and engine cowlings. New 'Glare' material, consisting of alternate layers of aluminium and glass fibre-reinforced adhesive which offers significant weight reduction and fatigue/damage resistance, has been tested on a German Air Force A310 since October 1999 and will be used for upper fuselage panels. Laser beam welding, which reduces cost and weight, used to attach stringers to lower fuselage panels. Wing leading-edge constructed of thermoplastics. Outer wings will be metal bonded.

Upper floor beams on A380-800 will be constructed of composites; those in A380-800F will be aluminium; throughout the structure, lighter 2524 aluminium alloys will be used in place of more traditional 2024.

Work allocation is Airbus France (St Nazaire): flight deck and centre fuselage; Airbus Deutschland (Hamburg): forward centre fuselage, rear fuselage; Airbus Deutschland (Stade): fin and rudder; Airbus UK (Broughton): wing main panels; Airbus España: wing/fuselage fairings, belly fairing and fixed horizontal tail; Airbus France (Toulouse): engine pylons and final assembly; Saab: mid and outer wing leading-edges; Latécoère: lower part of nose section. Flaps, ailerons, winglets, spoilers, elevators and tailcone not allocated by end of 2001.

Transport of major components undertaken by purpose-built ship from Hamburg via UK and St Nazaire (forward centre fuselage disembarked, joined to flight deck and re-embarked, accompanied by centre fuselage), the aircraft set then transferred to barge at Bordeaux and joined by Spanish-built elements for river transport to within 80 km (50 miles) of Toulouse, completing journey by road. Work on the new component assembly hall at Airbus Deutschland's Hamburg site began 6 December 2001 and will be completed in 2003; Airbus France's final assembly hall at Toulouse will be completed in 2004 and the first convoy is due to be undertaken at end 2003. However, both ship- and truck-borne movements are being opposed by local residents.

Once completed, the 'green' aircraft will be flown to Hamburg for internal fitting out. European and Middle Eastern aircraft will be delivered from Hamburg; those for the rest of the world will return to Toulouse before despatch to customer.

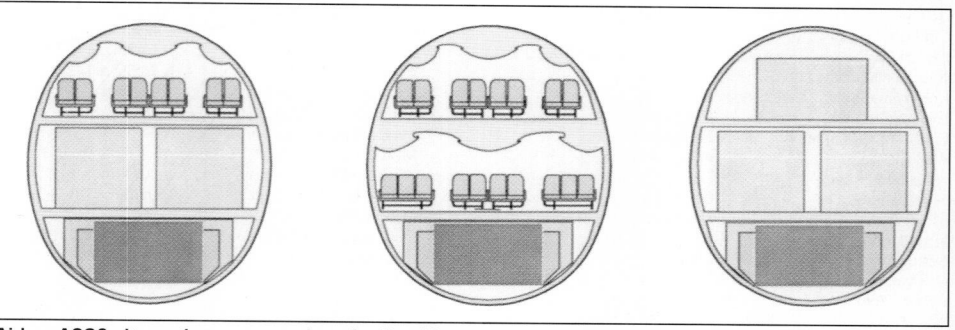

Airbus A380 alternative cross-sections for Combi, passenger and freighter service   0044883

A380 MAJOR INDUSTRIAL PARTNERS
(at July 2002)

Country	Company
Australia	Hawker de Havilland
Austria	Fischer Advanced Composites
Belgium	Asco; Belairbus*; Sabca*; Sonaca*
Europe	Eurocopter*
Finland	Patria (Finavitec)*
France	EADS Socata*; EADS Sogerma*; Labinal*; Latécoère*; L'hotellier; Mecachrome
Italy	Alenia*
Japan	Fuji*; Mitsubishi; Nippi*
Korea, South	KAI
Malaysia	CTRM*
Netherlands	EADS*; Fokker/Stork*
Spain	Aries; Gamesa*; MASA; Sacesa
Sweden	Saab*
Switzerland	RUAG*
UK	BAE Systems*; GKN*
USA	Goodrich

* Risk-sharing partner

LANDING GEAR: Goodrich main landing gear; each four-wheel wing-mounted unit weighs 2,310 kg (5,093 lb) and each six-wheel underfuselage unit weighs 4,080 kg (8,995 lb). Messier-Dowty twin-wheel nose landing gear. Underfuselage main gear set slightly aft of underwing units. Normal steering on nose gear and rear axle of body gear; back-up steering of nose gear in event of hydraulic power loss. Carbon brakes on all main group wheels. Manoeuvring compatible with 45 m (148 ft) wide runways and 23 m (75.5 ft) wide taxiways; U-turn possible on 60 m (200 ft) wide runways, by differential braking or asymmetrical thrust.

POWER PLANT: A380 is offered with a choice of Alliance (GE/P&W) GP7200 or Rolls-Royce Trent 900; Airbus Industrie and Rolls-Royce signed an MoU specifying the Trent as favoured power plant in November 1996; was due for first test run in December 2002, first flight in late 2004 and certification in 2005; ILFC, Lufthansa, Singapore Airlines, Virgin and Qantas have selected R-R power plant; Air France, Emirates and FedEx have selected GP7200, which will make its first flight in 2005. Trent 900 will be initially certified at 356 kN (80,000 lb st) but derated to 302 kN (68,000 lb st), with eventual growth to 374 kN (84,000 lb st).

Alliance GP7200 uses same core as power plants for Boeing 747X and long-range 767. Detailed design work was due to have started in August 2001 but this has slipped to early 2003, with a first run in April 2004, certification at 363 kN (81,500 lb st) in July 2005 and first flight in early 2006.

Standard fuel capacity of both models is 325,000 litres (85,856 US gallons; 71,492 Imp gallons); extra fuel tanks would be fitted in wing centre box for longer-range models.

Revised bids for thrust reversers (inner engines only) and nacelles submitted by Goodrich and Aircelle in April 2001 (latter selected) following decision to increase engine fan diameter from 2.79 m (110 in) to 2.95 m (116 in) due to concerns regarding future noise limits, and to extend nacelle length by 43 cm (17 in) at inlet and 13 cm (5 in) at exit.

Computer-aided impression of A380-800F in the colours of Federal Express   *NEW*/0527165

ACCOMMODATION: Flight deck crew of two; rest areas are provided for crew in flight deck area; see Design Features for details of passenger cabins. Five main deck and three upper deck emergency exits on each side of fuselage; Goodrich evacuation slides, those for upper decks stored within the airframe rather than the door. Seat pitches: first class 173 cm (68 in), business class 122 cm (48 in), economy class 81 or 84 cm (32 or 33 in).

SYSTEMS: Integrated modular avionics (IMA) system provided by Thales in conjunction with Diehl Avionik Systeme using computing modules slotted into cabinets throughout aircraft. Rockwell Collins supplying Ethernet avionics communications infrastructure at 100 Mbit/s speed with full duplex networking. Variable frequency AC electrical generating systems to be incorporated. Cameras fitted to top of fin and under fuselage for taxying assistance. Fuel management systems provided by Parker Aerospace including in-tank sensors and wiring, avionics and software to fit into IMA system.

Two 241 bar (3,500 lb/sq in) hydraulic systems (yellow and green) and two electrical systems (red and orange) are used for flight controls, each of the latter using at least two different systems in case of failure of any one; each system fully independent. Hydraulic systems use lighter, more compact pipework compared with earlier Airbus products. Wing-mounted landing gear powered by green system; underfuselage main gear runs on yellow system. Power generation systems with 180 kVA generators for each engine to be provided by Hamilton Sundstrand, which has also been selected (with P&WC) to provide APU. Eaton 345 bar (5,000 lb/sq in) hydraulic generators using eight Vickers engine-driven pumps and four electric pumps. Onboard oxygen generator (OBOGS) to be fitted.

Impression of the A380's flight deck
NEW/0527164

A380 MAJOR SYSTEMS SUPPLIERS
(at July 2002)

System	Supplier
Air conditioning	Hamilton Sundstrand
Air data system	Goodrich
APU	Pratt & Whitney Canada
Cockpit displays	Thales
Electrical power generation	Aerolec (Thales/TRW)
Engine nacelles	Aircelle
Environmental control system	Nord-Micro
Evacuation system	Goodrich
Flight management system	Honeywell
Fuel management system	Parker Aerospace
Hydraulic system	Eaton (Vickers)
Main landing gear	Goodrich
Mainwheels and brakes	Honeywell
Nose landing gear	Messier-Dowty
Nosewheels and brakes	Dunlop
Oxygen system	Dräger
Power plant	Engine Alliance; Rolls-Royce
Pneumatic system	Liebherr/Honeywell
Primary flight controls	Goodrich (ailerons, elevators and rudder)
	Liebherr/Parker Aerospace (spoilers)
Ram-air turbine	Hamilton Sundstrand

AVIONICS: Cockpit layout, by Airbus Toulouse division, will be compatible with other Airbus family members. Eight new-generation 15 × 20 cm (6 × 8 in) LCDs form centre of package. Onboard information system (OIS) will integrate databases with operator's in-house software packages and enable flight planning and documentation updates while en route, plus ancillary operations including passenger credit card validation, Internet access and entertainment. Onboard maintenance system (OMS) provides real-time information to ground crew. Honeywell terrain guidance and on-ground navigation will be integrated into FMS.

DIMENSIONS, EXTERNAL:

Wing span	79.65 m (261 ft 4 in)
Wing chord at root	16.60 m (54 ft 5½ in)
Wing aspect ratio	7.5
Length overall	72.75 m (238 ft 8 in)
Fuselage max width	7.14 m (23 ft 5 in)
Height overall	24.08 m (79 ft 0 in)
Tailplane span	30.38 m (99 ft 8 in)
Wheel track: A380-800	14.33 m (47 ft 0¼ in)
A380-800F	14.39 m (47 ft 2½ in)
Wheelbase: A380-800	30.40 m (99 ft 8¾ in)

Rear cargo door (A380-800F):

Max width	3.43 m (11 ft 3 in)
Max height	2.54 m (8 ft 4 in)

DIMENSIONS, INTERNAL:

Cabin, A380-800: Length	50.68 m (166 ft 3¼ in)
Max width:	
main deck	6.58 m (21 ft 7 in)
upper deck	5.92 m (19 ft 5 in)
Width at floor:	
main deck	6.20 m (20 ft 4 in)
upper deck	5.33 m (17 ft 6 in)
Bulk hold volume: A380-800	18.4 m³ (650 cu ft)
A380-800F	20.0 m³ (706 cu ft)
Total hold volume: A380-800	171.0 m³ (6,039 cu ft)
A380-800F	1,134 m³ (40,047 cu ft)

AREAS:

Wings, gross	845.0 m² (9,095.5 sq ft)

WEIGHTS AND LOADINGS:

Operating weight empty: A380-800	276,800 kg (610,240 lb)
A380-800F	252,200 kg (556,000 lb)
Max payload: A380-800	83,000 kg (182,985 lb)
A380-800F	150,000 kg (330,700 lb)
Max T-O weight: A380-800	560,000 kg (1,234,580 lb)
A380-800F	590,000 kg (1,300,720 lb)
Max ramp weight: A380-800	562,000 kg (1,239,000 lb)
A380-800F	592,000 kg (1,305,140 lb)
Max landing weight:	
A380-800	386,000 kg (850,980 lb)
A380-800F	427,000 kg (941,370 lb)

Max zero-fuel weight:	
A380-800	361,000 kg (795,860 lb)
A380-800F	402,000 kg (886,260 lb)
Max wing loading:	
A380-800	662.7 kg/m² (135.74 lb/sq ft)
A380-800F	698.2 kg/m² (143.01 lb/sq ft)
Max power loading:	
A380-800	463 kg/kN (4.54 lb/lb st)
A380-800F	433 kg/kN (4.25 lb/lb st)

PERFORMANCE (estimated, Trent 900 engines):

Max operating speed M0.89 (340 kt; 630 km/h; 391 mph)	
Econ cruising Mach No.	0.85
Initial cruising altitude	10,670 m (35,000 ft)
Service ceiling	13,100 m (42,980 ft)
T-O field length, ISA + 15°C:	
A380-800	2,987 m (9,800 ft)
A380-800F	3,009 m (9.870 ft)
Runway ACN	approx 68
Time from brake release to FL350:	
A380-800	36 min
A380-800F	35 min

Range:

A380-800	8,000 n miles (14,816 km; 9,206 miles)
A380-800F	5,600 n miles (10,371 km; 6,444 miles)

OPERATIONAL NOISE LEVELS: Complies with London airports' 'Quota Count 2' category for departures (3 dB quieter than, or half the noise energy level of, most in-service Boeing 747s.

UPDATED

AIRBUS ACJ

The Airbus Corporate Jetliner is described under the A319 heading.

AIRBUS MULTIROLE TANKER-TRANSPORT (MRTT)

TYPE: Tanker-transport.

PROGRAMME: Belgian, Canadian, French, German and Thai air forces already operate A310s variously fitted for VIP, troop and/or freight transport. Airbus has delegated

Interior impression of the A380, showing passengers' bar *NEW*/0527127

Rear central area of the A380's main cabin *NEW*/0527166

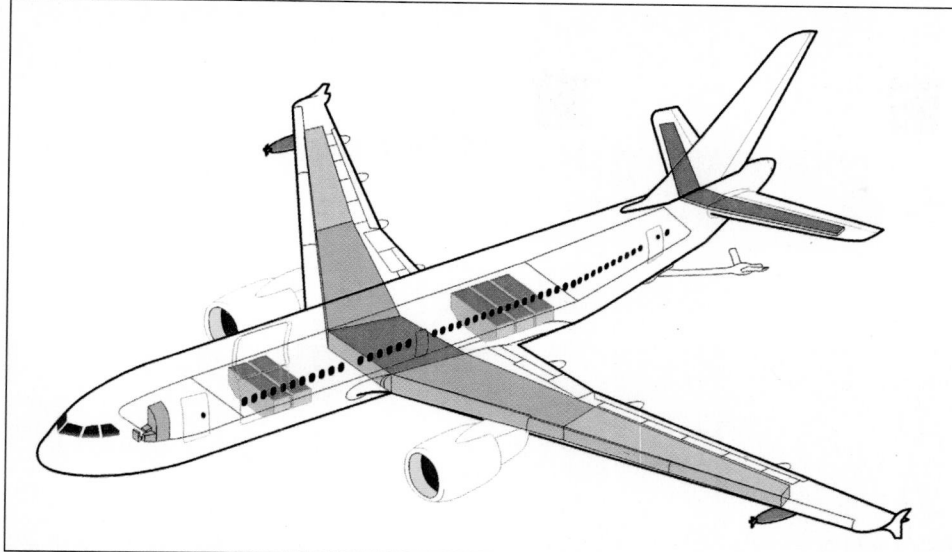

Cutaway drawing of Airbus MRTT   *NEW*/0126935

Computer-generated image of an A310 MRTT refuelling a Boeing E-3 Sentry   *NEW*/0527098

Cabin height: both	2.28 m (7 ft 5¾ in)
Max cabin width: both	5.29 m (17 ft 4¼ in)
Underfloor freight hold volume:	
310	80.0 m³ (2,825 cu ft)
330	136.0 m³ (4,803 cu ft)
WEIGHTS AND LOADINGS:	
Operating weight empty:	
310	80,830 kg (178,200 lb)
330	120,500 kg (265,650 lb)
Max non-fuel payload: 310	37,000 kg (81,571 lb)
330	61,300 kg (135,140 lb)
Max normal fuel capacity:	
310	47,940 kg (105,690 lb)
330	111,270 kg (245,300 lb)
Additional fuel: 310	28,240 kg (62,258 lb)
Max T-O weight:	
normal, 310	157,000 kg (346,125 lb)
330	230,000 kg (507,075 lb)
optional, 310	164,000 kg (361,550 lb)
330	233,000 kg (513,675 lb)
Max ramp weight: 310	164,900 kg (363,550 lb)
330	233,900 kg (515,650 lb)
Max landing weight: 310	124,000 kg (273,375 lb)
330	182,000 kg (401,250 lb)
Max zero-fuel weight: 310	114,000 kg (251,325 lb)
330	170,000 kg (374,775 lb)
PERFORMANCE:	
Refuelling speed:	
both, boom	240-320 kt (444-592 km/h; 276-368 mph)
both, hose and drogue: S/L to FL275	
	200-325 kt (370-602 km/h; 230-374 mph)
FL275 to FL350	M0.82
T-O run: 310	2,347 m (7,700 ft)
Max range, standard fuel:	
310	4,800 n miles (8,889 km; 5,523 miles)
Max range, using transferable fuel:	
310	7,200 n miles (13,334 km; 8,285 miles)
330	9,000 n miles (16,668 km; 10,357 miles)
	UPDATED

development and marketing of flight refuelling versions to its major partners, using either pre-owned or new aircraft. Demonstrator MRTT produced by conversion of former airline A310-324 N816PA; undertook compatibility trials with RAF aircraft, July 1995. Marketing efforts originally centred on the A310 version, which offered for the RAF's FSTA future tanker aircraft requirement. This planned to entail start of replacement of VC10s in 2005, although more than one type of tanker is expected to be obtained, probably by lease rather than purchase. Exclusive marketing rights for MRTT assigned to Raytheon (which will also become Design Authority), June 1999. Four Luftwaffe A310-300 MRTs will be converted into MRTTs for delivery from 2006; Canada has also announced intent to modify two Polaris MRTs into MRTTs in same programme.

A330-200 is being developed as strategic tanker/ transport and has been chosen by the Air Tanker consortium as most suitable platform for UK FSTA programme. Consortium members EADS, Rolls-Royce, Thales, FRA and Haliburton KBR are advocating a private finance initiative (PFI) solution with service entry in 2007 and programme life of 27 years. Also offered to Australia for AIR 5402 requirement.

CURRENT VERSIONS: Interim MRT (MultiRole Transport) version of A310, without tanker capability, converted by Elbe Flugzeugwerke, Dresden and Lufthansa Technik, Hamburg, for Luftwaffe. Structural strengthening and 3.50 m × 2.50 m (11 ft 5¾ in × 8 ft 2½ in) cargo door added in port forward fuselage. Capacity of up to 214 passengers; or 36 tonnes of cargo/passengers; or 56 stretchers and six intensive care patients in the casevac role. First redelivered from Dresden in June 1999; four of seven Luftwaffe A310s redelivered as MRTs by early 2002. Similarly modified A300 freighters offered in partial satisfaction of RAF's Short Term Strategic Airlifter requirement, before its abandonment in August 1999. In service, Luftwaffe aircraft proved partially incompatible with military cargo handling equipment and upper deck therefore rarely used for freight; full conversion of two MRTs to MRTT, plus conversion of further pair of standard aircraft directly to MRTT, being undertaken between 2002 and 2005. Remaining Luftwaffe A310s comprise two in VIP configuration and one passenger transport.

DESIGN FEATURES: Conversions offer greater refuelling and transport capability than earlier airliners in combination with modern aircraft with better lifetime costs and longer life expectancy. Possible roles include tanker with underwing HDUs and fuselage-mounted boom and/or hose transfer systems and carrying in excess of 111,270 kg (245,300 lb) of fuel (139,060 litres; 36,737 US gallons; 30,590 Imp gallons); and cargo and personnel transports

which can be combined with refuelling, medevac, airborne command post and reconnaissance/airborne warning.

Airbus conversions offer payloads from 35,000 to 50,000 kg (77,161 to 110,231 lb), full payload transatlantic range, long on-station time, combined boom and hosereel transfer capability, standard Airbus forward port-side freight door (projected height 2.57 m; 8 ft 5¼ in, width 3.58 m; 11 ft 9 in); quick-change main deck layout, probe or receptacle fuel receiver capability; commonality with existing airliners and same worldwide support resources, predictable spares requirements and longer remaining airframe life.

About 100 civil operators on all five continents are flying A300/310, and first-generation Airbus airliners are now becoming available on market; military rendezvous and self-protection systems can be fitted; main deck can be converted with palletised seating for up to 270 passengers in under 24 hours; up to 28,000 kg (61,729 lb) of additional fuel can be carried in tanks in underfloor cargo compartments.

DIMENSIONS, EXTERNAL (310: A310-300 MRTT, 330: A330-200 MRTT):

Length overall, including probe:

310	47.36 m (155 ft 4½ in)
330	59.69 m (195 ft 10 in)

DIMENSIONS, INTERNAL:

Usable cabin length:

310	43.90 m (144 ft 0 in)
330	45.00 m (147 ft 7¾ in)

AIRBUS A310 AEW&C

TYPE: Airborne early warning and control system.

PROGRAMME: A310 selected by Raytheon Systems as a platform for its bid for the Australian AEW&C requirement (Project Wedgetail), competing against Boeing 737 and Lockheed Martin C-130J. Incorporates Elta Phalcon 360° phased-array radar in a fixed dorsal dome; formally announced in January 1997. Australia selected Boeing 737 in July 1999, but A310 AEW&C also offered to Turkey (again unsuccessfully) and South Korea. No further developments by late 2002. Operating weight empty 80,235 kg (176,890 lb); max T-O weight 164,000 kg (361,550 lb); patrol for over 10 hours on station at 300 n miles (555 km; 345 miles) from base at M0.72 at 8,990 m (29,500 ft).

UPDATED

AIRBUS A321 AGS

TYPE: Airborne ground surveillance system.

PROGRAMME: Northrop Grumman recommended the A321 as its preferred Joint STARS platform (see E-8 entry in US section) to compete for NATO's AGS (airborne ground surveillance) requirement. Considered in 1996 and rejected on cost grounds, the A321 was reconsidered in the light of NATO's unwillingness to accept a Boeing 707-based solution. Plans called for selection in 2002 and IOC, with six aircraft, by 2007, but implementation delayed by US refusal to release all technology. Northrop Grumman's AGS proposal features the US Army's AN/APY-X RTIP (Radar Technology Insertion Program) with an underfuselage electronically scanned antenna array, giving spot/swath SAR, wide area MTI, ultra-high resolution SAR, inverse SAR and long-range high-resolution MTI modes. It would have between 10 and 12 operator stations.

Airbus A310 AEW&C proposal with non-rotating Elta radar above rear fuselage; configuration may include a large dorsal air scoop (*Jane's/Paul Jackson*)   0062385

By 2002, Northrop Grumman, EADS and Galileo Avionica promoting TIPS (Transatlantic Industrial Proposed Solution) based on A319/A320/A321 airframe and with participation of wide range of subcontractors throughout NATO countries; target IOC of 2010, given 2003 political decision, would involve estimated four aircraft and 18 ground stations to provide minimum single-orbit coverage; full requirement may be 12 and 48, respectively, although six and 24 would ensure two simultaneous missions.

UPDATED

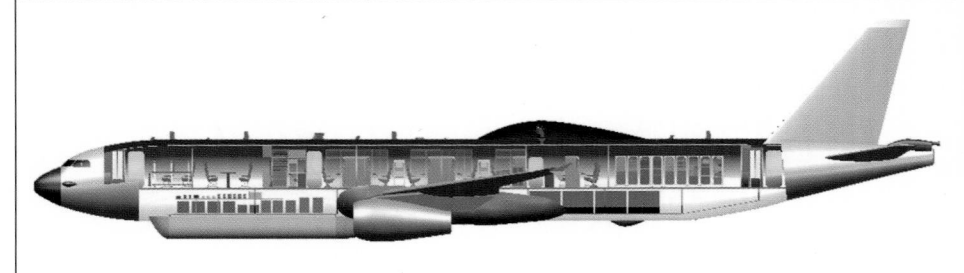

Cutaway diagram of Airbus A321 AGS *(Northrop Grumman)*   0073146

AIRBUS MPA320

TYPE: Maritime surveillance twin-jet.

PROGRAMME: Alenia Aeronautica/Finmeccanica and EADS Military Aircraft revealed at Farnborough Air Show in July 2002 that they had joined forces to submit proposal based on Airbus A320 to satisfy joint German/Italian MPA-R (Maritime Patrol Aircraft Replacement) requirement. Joint concept by EADS Deutschland (Ottobrunn) and EADS France (Toulouse). Ventral search radar; undernose EO/IR sensor; capacity for internal torpedo carriage in wing/body box; seven operator stations in cabin. Total development time five years; could fly in 2005. Request for proposals issued in November 2002 by German-Italian management team, with Airbus offer anticipating production of 24 MPA320 aircraft (10 German and 14 Italian), as well as provision of training simulators and a complete package of logistic and customer support services. Contract award is expected to occur in mid-2003 to permit service entry in 2007/08. Competing bids received (deadline 26 July 2002) from Boeing, L-3 Communications and Lockheed Martin.

Description generally as for Airbus A320, except that below:

WEIGHTS AND LOADINGS:

Weight empty: manufacturer's	35,400 kg (78,044 lb)
operational	44,600 kg (98,326 lb)
Max weapon load	5,300 kg (11,684 lb)
Max fuel weight	27,200 kg (59,966 lb)
Max T-O weight	77,000 kg (169,755 lb)
Max landing weight	66,000 kg (145,505 lb)

PERFORMANCE (estimated):

Turn radius, clean	less than 1,850 m (6,070 ft)
Ferry range	4,200 n miles (7,778 km; 4,833 miles)

NEW ENTRY

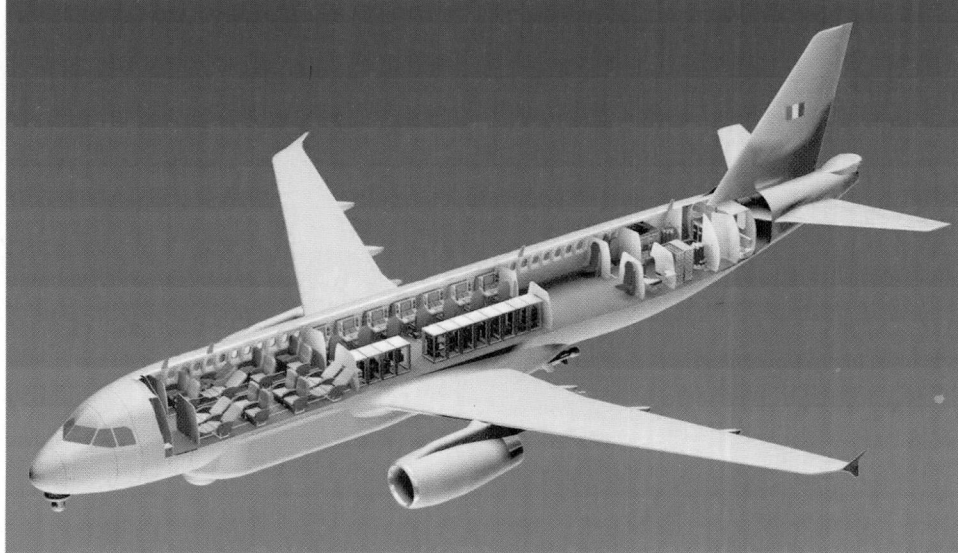

Computer-generated image of proposed Airbus MPA320   0527122

AIRTECH

AIRCRAFT TECHNOLOGY INDUSTRIES

Web: http://www.casa.es/eng_31125_cn235-vistas

PARTICIPATING COMPANIES:

CASA: see under Spain
Dirgantara: see under Indonesia

Airtech was formed by CASA (now part of EADS) and IPTN (now known as Dirgantara) to develop the CN-235 twin-turboprop transport; design and production was shared 50-50. The partnership applied only to the Series 10 and Series 100/110, later versions being exclusively CASA products, according to a statement by that company. CASA unilaterally developed the stretched C-295.

VERIFIED

AIRTECH CN-235

Spanish Air Force designations: T.19A and T.19B

TYPE: Twin-turboprop transport.

PROGRAMME: Launched as joint venture between CASA and Indonesian manufacturer IPTN (now Dirgantara, which see), which formed Airtech company to manage programme. Series 10 and Series 100/110 versions covered by this agreement; subsequent versions, notwithstanding Indonesian Series 220 and 330 equivalents, stated by CASA to be wholly Spanish.

Preliminary design began January 1980, prototype construction May 1981; one prototype completed in each country, with simultaneous roll-outs 10 September 1983; first flights 11 November 1983 (by CASA's ECT-100) and 30 December 1983 (IPTN's PK-XNC); Spanish and Indonesian certification 20 June 1986; first flight of production aircraft 19 August 1986; FAA type approval (FAR Pts 25 and 121) 3 December 1986; deliveries began 15 December 1986 from IPTN line and 4 February 1987 from CASA; entered service (with Merpati Nusantara Airlines) 1 March 1988; JAR 25 type approval October 1993.

Licence agreement with TAI (see Turkish section) announced January 1990, initially to assemble and later to manufacture locally 50 of 52 ordered; first flight of Turkish-assembled aircraft 24 September 1992; first delivery 13 November 1992; final air force delivery 10 August 1998, but TAI currently producing follow-on batch of nine maritime patrol variants.

In 1995, CASA unilaterally launched development of a stretched CN-235, as C-295; this is described under CASA heading in Spanish section.

CURRENT VERSIONS: **CN-235 Series 10:** Initial production version (15 built by each company), with CT7-7A engines; described in 1986-87 and earlier *Jane's*.

CN-235 Series 100/110: Generally as Series 10, but CT7-9C engines in new composites nacelles; replaced Series 10 in 1988 from 31st production aircraft. Series 100 is Spanish-built and, following JAA certification, was certified by FAA in February 1992. Series 110 is Indonesian-built, with improved electrical, warning and environmental systems to comply with JAR 25; certification of this version achieved in Europe (JAA), July 1995.

Detailed description applies to the above version except where indicated.

CN-235 Series 200/220: Structural reinforcements to cater for higher operating weights, aerodynamic improvements to wing leading-edges and rudder, reduced field length requirements and much-increased range with maximum payload; Series 200 is Spanish-built and was certified by FAA March 1992. Series 220 is Indonesian-built, with improvements similar to Srs 110; prototype, flown early 1996, is converted from a company development aircraft (PK-XNV, the 20th production aircraft from the Indonesian line); orders include six for Malaysian Air Force, all of which completed to Srs 220 standard (including three in maritime patrol configuration) by early 1998 (34th to 39th Indonesian-built). Revised leading-edge shape led to requirement to requalify pneumatic de-icer boots, delaying initial deliveries.

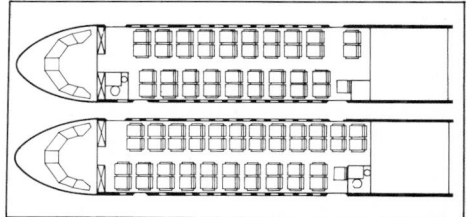

CN-235 in typical configurations for 38 (top) and 44 passengers

Further orders for Series 220 from South Korea (eight) and Pakistan (four).

CN-235 Series 300/330: IPTN originally offered Series 330 **Phoenix** (with new Honeywell avionics, ARL-2002 EW system and 16,800 kg; 37,037 lb MTOW) to Royal Australian Air Force to meet Project Air 5190 tactical airlift requirement, but was forced by financial constraints to withdraw in 1998. Separately, CASA offered its own Series 300 to meet the same specification.

CASA Series 300 under certification in 2000 with an open-systems avionics architecture, based on MIL-STD-1553B and ARINC 429 digital databusses. Full

CASA CN-235 twin-turboprop multipurpose transport, with additional side view (centre) of a representative CN-235 MPA *(Jane's/Dennis Punnett)*

NVG-compatible cockpit; four-dimensional navigation system with avionics suite, including Thales (Sextant) Topdeck colour weather radar, radios, solid-state flight data and cockpit voice recorders, enhanced TCAS, enhanced GPWS and four 152 × 203 mm (6 × 8 in) LCDs; twin HUDs and Totem 3000 ring laser gyro INS optional. Other features include in-flight refuelling capability, improved pressurisation (2,440 m; 8,000 ft cabin environment at 7,620 m; 25,000 ft) and provision for optional twin nosewheel installation to provide better soft-field taxiing capability.

CN-235 AEW: Proposals were revealed in December 1995 for fitment of an Ericsson Erieye electronically scanned phased-array radar above the fuselage of a CN-235. Initial interest was from the Indonesian Air Force, but primarily in the ocean surveillance role; retrofit of three existing aircraft was considered, but has not been undertaken. Radar, three surveillance operators' positions and associated equipment increase aircraft weight by approximately 2,000 kg (4,409 lb).

CN-235ER: Extended-range version selected by US Coast Guard in 2002 as fixed-wing element of Project Deepwater re-equipment programme. At time of announcement, in June 2002, it was also revealed that total of 35 aircraft would be purchased.

CN-235 M: Other military transport versions.

CN-235 MP Persuader and CN-235 MPA: Maritime patrol versions; described separately.

CN-235 QC: Quick-change cargo/passenger version; certified by Spanish DGAC May 1992.

CN-245: Indonesian stretched version; not built. See 1998-99 *Jane's*.

C-295: Spanish stretched version; described separately under CASA heading.

N2XXM: See CN-245 entry in 1998-99 *Jane's*. Project abandoned.

CUSTOMERS: See table; by October 1999, total orders reported by CASA as 233 (to 34 operators in 24 countries); this reflected suspension of April 1994 commitment by Merpati to buy 16 further Series 220s. Total further reduced in March 2001 by UAE decision to order four C-295s in place of CN-235s.

One (s/n 66049) acquired (presumably second-hand) by USAF in 1998. Turkey signed a lease agreement on 16 April 1999 to allow a one-year renewable lease of two Turkish Air Force CN-235s to Jordan. Switzerland briefly leased a Spanish Air Force CN-235 in 1999 to support peacekeeping operations in the former Yugoslavia. Three Merpati aircraft leased to Air Venezuela from May 1999; three leased to Asian Spirit Airlines, Philippines, from March 2000, including two on lease-purchase. Intention to buy a further two announced by Papua New Guinea in mid-1998. National Jet Systems of Australia interested in two coastal patrol variants; signed MoU for possible acquisition of two, plus five options, February 1998. One CN-235-300 (first of this subvariant in service) leased by Austrian Ministry of Defence for six months from April 2000. CN-235 is contender in Taiwanese requirement for 18 to 22 light transports, and for the US Army Airborne Common Sensor platform requirement. CN-235ER selected as winner of US Coast Guard Project Deepwater competition, with total of 35 aircraft to be acquired according to announcement of June 2002.

COSTS: US$16.9 million (1995) programme unit cost, Malaysia.

DESIGN FEATURES: Optimised for short-haul operations, enabling it to fly four 860 n mile (1,593 km; 990 mile) stage lengths (with reserves) before refuelling and to operate from paved runways or unprepared strips; high-mounted wing; pressurised fuselage (including baggage compartment) of flattened circular cross-section, with upswept rear end incorporating cargo ramp/door; sweptback fin (with dorsal fin) and rudder; low-set non-swept fixed incidence tailplane and elevators; two small ventral fins; vortex generators on rudder and elevator leading-edges; optional extended nose radome.

NACA 65_3-218 aerofoil with no-dihedral/constant chord centre-section; tapered outer panels have 3° dihedral and 3° 51′ 36″ sweepback at quarter-chord.

CN-235 PRODUCTION
(at late 2002)

Customer	Qty	First order	First aircraft	First delivery	Delivered	Mfr
Civil version:						
Austral (Argentina)	2[15]	19 Dec 1989	LV-VHM	1993	2	CASA
Binter Canarias (Spain)	4[1, 16]	10 Jun 1988	EC-EMO	22 Dec 1988	4	CASA
Binter Mediterraneo (Spain)	4[2]	19 Dec 1989	EC-FAD	4 Sep 1990	3	CASA
Mandala Airlines (Indonesia)	3	-	-	-	0	Dirgantara
Merpati Nusantara (Indonesia)	15[1]		PK-MNA	15 Dec 1986	15	Dirgantara
Military version:						
Abu Dhabi Air Force	7		810	31 Aug 1993	7	Dirgantara
Botswana Defence Force	2[1]	10 Jun 1986	OG-1	21 Dec 1987	2	CASA
Brunei Air Wing	3[3]		-	-	0	Dirgantara
	1		ATU-501	1997	1	Dirgantara
Chilean Army	4	12 Feb 1989	E-216	31 Aug 1989	4	CASA
Colombian Air Force	3[2]	Jul 1997	1260	28 Jan 1998	3	CASA
Croatian Air Force	5	Oct 1996	-	-	0	TAI
Devon Holding & Leasing Inc	1	-	N168D	2002	1	CASA
Ecuadorean Army	1	6 Jun 1989	AEE-502	6 Jun 1989	1	CASA
Ecuadorean Navy	1	27 Jul 1988	ANE-204	13 Jun 1989	1	CASA
French Air Force	15[8]	11 Apr 1990	043	28 Feb 1991	15	CASA
	5[8]	2002		2002		CASA
Gabon Air Forces	1	26 Feb 1990	TR-KJE	19 Mar 1991	1	CASA
Indonesian armed forces	24[11]		A-2301	12 Jan 1993	7	Dirgantara
Irish Air Corps	1[13]	3 Apr 1991	250	10 Apr 1991	1	CASA
	2[3]	3 Apr 1991	252	8 Dec 1994	2	CASA
Malaysian Air Force	6[7, 9]		M44-01	26 Aug 1999	6	Dirgantara
Moroccan Air Force	7[10]	19 Sep 1989	CNA-MA	27 Sep 1990	7	CASA
Oman Police	2	15 Feb 1992	A40-CU	14 Jan 1993	2	CASA
Pakistan Air Force	4[9]	29 Jun 01		2004		Dirgantara
Panama National Guard	1[1, 12]	19 Mar 1987	SAN-265	13 Sep 1988	1	CASA
Papua New Guinea Defence Force	2	26 Oct 1991	P2-0501	15 Nov 1991	2	CASA
Saudi Air Force	4[1]	5 Feb 1984	118	9 Feb 1987	4	CASA
South African Air Force (ex-Bophuthatswana)	1[1]	29 May 1990	8026	6 Jan 1991	1	CASA
South Korean Air Force	12	19 Aug 1992	078	13 Nov 1993	12	CASA
	8[9]	21 Oct 1997	-	18 Dec 2001	8	Dirgantara
Spanish Air Force	2[4]	16 Nov 1988	T.19-01	7 Dec 1988	2	CASA
	18	28 Dec 1990	T.19-03	1 Feb 1991	18	CASA
	4[3]		-	-	0	CASA
Thai Ministry of Agriculture and Co-operatives	2[9]	Oct 1996	2221	Apr 1999	2	Dirgantara
Thai Police	0[2, 14]	Apr 1995	28053	4 Mar 1996	0	CASA
Turkish Air Force	52[6]	11 Dec 1990	51	25 Jan 1992	52	TAI/CASA
Turkish Navy	6[3]	23 Sep 1998	TCB-651	23 Dec 2001	6	TAI/CASA
Turkish Coast Guard	3[3]	23 Sep 1998	TCSG-551	23 Dec 01	3	TAI/CASA
Subtotals	**234**				**196**	
Demo/trials	3[5]		EC-016		3	CASA
	5		PK-XNC		5	Dirgantara
Totals	**242**				**204**	

[1] Series 10
[2] Series 200
[3] Maritime patrol
[4] VIP version
[5] Includes one -100QC; plus one -200QC sold in 1996 to East Texas Aircraft Services Corporation, then Turbo Flight Aviation, March 1998
[6] 50 built in Turkey by TAI
[7] Option on further 12
[8] Including option on seven taken up in February 1996; first eight as Srs 100, but upgraded to Srs 200 from 1999
[9] Series 220
[10] Includes one VIP version
[11] Includes six maritime patrol of which three on firm order
[12] To Flight International (USA) 1995
[13] Withdrawn at end of lease, 1995
[14] Former demonstrator
[15] Converted from -100 to Series 200
[16] Withdrawn 1998; three to Luftmeister, South Africa; one to Turkish Army (lost 16 May 2001)

Note: All are Series 100/110 unless indicated otherwise. Croatian order not counted by CASA, which reported 189 military orders by mid-2000 and implies reduction in Indonesian requirement.

FLYING CONTROLS: Conventional and manual. Ailerons, elevators and rudder statically and dynamically balanced (duplicated actuation for ailerons); mechanical servo tab and electric trim tab in each aileron, rudder and starboard elevator, trim tab only in port elevator; single-slotted inboard and outboard trailing-edge flaps (each pair interchangeable port/starboard), actuated hydraulically by Dowty irreversible jacks.

STRUCTURE: Conventional semi-monocoque, mainly of aluminium alloys with chemically milled skins; composites (mainly glass fibre or glass fibre/Nomex honeycomb sandwich, with some carbon fibre and Kevlar) for leading/trailing-edges of wing/tail moving surfaces, wing/fuselage and main landing gear fairings, wing/fin/tailplane tips, engine nacelles, ventral fins and nose radome. Propeller blades are of glass fibre, with metal spar and urethane foam core.

CASA builds wing centre-section, inboard flaps, forward and centre fuselage, engine nacelles; Dirgantara builds outer wings, outboard flaps, ailerons, rear fuselage and tail unit; both manufacturers use numerical control machinery extensively. Final assembly line in each country. Part of tail unit built by ENAER Chile under subcontract from CASA. TAI (Turkey) initially assembled under licence before progressing gradually to local manufacture of balance of 50 aircraft for Turkish Air Force.

LANDING GEAR: Messier-Bugatti retractable tricycle type with levered suspension, suitable for operation from semi-prepared runways. Electrically controlled hydraulic extension/retraction, with mechanical back-up for emergency extension. Oleo-pneumatic shock-absorber in each unit. Each main unit comprises two wheels in tandem, retracting rearward into fairing on side of fuselage. Mainwheels semi-exposed when retracted. Single steerable nosewheel (±48°) retracts forward into unpressurised bay under flight deck. Dunlop 28×9.00-12 (12 ply) tubeless mainwheel tyres standard, pressure 5.17

Indonesian-built CN-235, in service with United Arab Emirates Air Force *(Jane's/Paul Jackson)*   0085234

CN-235 Srs 300 demonstrator, before its lease to Austria ***NEW***/0106524

bar (75 lb/sq in) on civil version, 5.58 bar (81 lb/sq in) on military version; low-pressure mainwheel tyres optional, size 11.00-12 (10 ply), pressure 3.45 bar (50 lb/sq in). Dunlop 24×7.7 (10/12 ply) tubeless nosewheel tyre, pressure 5.65 bar (82 lb/sq in) on civil version, 6.07 bar (88 lb/sq in) on military version; optional 8.50×10 (12 ply). Dunlop hydraulic differential disc brakes; Dunlop anti-skid units on main gear. Chilean Army aircraft used in Antarctic have wheel/ski gear. Minimum ground turning radius 9.50 m (31 ft 2 in) about nosewheel, 18.98 m (62 ft 3¼ in) about wingtip.

POWER PLANT: Two General Electric CT7-9C turboprops (CT7-9C3 in Srs300), each flat rated at 1,305 kW (1,750 shp) (S/L, to 41°C) for take-off and 1,394.5 kW (1,870 shp) up to 31°C with automatic power reserve. Hamilton Sundstrand 14RF-21 (14RF-37 in Srs 300) four-blade constant-speed propellers, with full feathering and reverse-pitch capability. Fuel in two 1,042 litre (275 US gallon; 229 Imp gallon) integral main tanks in wing centre-section and two 1,592 litre (421 US gallon; 350 Imp gallon) integral outer-wing auxiliary tanks; total fuel capacity 5,264 litres (1,392 US gallons; 1,158 Imp gallons), of which 5,128 litres (1,355 US gallons; 1,128 Imp gallons) are usable. Single pressure refuelling point in starboard main landing gear fairing; gravity filling point in top of each tank. Propeller braking permits No. 2 engine to be used as on-ground APU. Oil capacity 14 litres (3.7 US gallons; 3.1 Imp gallons).

ACCOMMODATION: Crew of two on flight deck, plus cabin attendant (civil version) or third crew member (military version). Accommodation in commuter version for up to 44 passengers in four-abreast seating, at 76 cm (30 in) pitch, with 22 seats each side of central aisle. Lavatory, galley and overhead luggage bins standard. Pressurised baggage compartment at rear of cabin, aft of movable bulkhead; additional stowage in rear ramp area and in overhead lockers. Can also be equipped as mixed passenger/cargo combi (for example, 19 passengers and two LD3 containers), or for all-cargo operation, with roller loading system, carrying four standard LD3 containers, five LD2s, or two 2.24 × 3.18 m (88 × 125 in) and one 2.24 × 2.03 m (88 × 80 in) pallets; or for military duties, carrying up to 57 fully equipped troops or 46 paratroops (51 troops or paratroops on Srs 300). Other options include layouts for aeromedical (18 stretchers and two medical attendants on Srs 300), electronic warfare, geophysical survey or aerial photographic duties.

Main passenger door, outward- and forward-opening with integral stairs, aft of wing on port side, serving also as a Type I emergency exit. Type III emergency exit facing this door on starboard side. Crew/service downward-opening door (forward, starboard) has built-in stairs, and serves also as a Type I emergency exit, or as passenger door in combi version; second Type III exit opposite this door on port side. Wide ventral door/cargo ramp in underside of upswept rear fuselage, for loading of bulky cargo. Accommodation fully air conditioned and pressurised.

SYSTEMS: Hamilton Sundstrand air conditioning system, using engine compressor bleed air. Honeywell electropneumatic pressurisation system (maximum differential 0.25 bar; 3.6 lb/sq in) giving cabin environment of 2,440 m (8,000 ft) up to operating altitude of 5,480 m (18,000 ft) on Srs 200; Srs 300 cabin pressurisation increased to 0.38 bar (5.5 lb/sq in), giving cabin environment of 2,350 m (7,700 ft) at altitude of 7,620 m (25,000 ft). Hydraulic system, operating at nominal pressure of 207 bar (3,000 lb/sq in), comprises two engine-driven, variable displacement axial electric pumps, a self-pressurising standby mechanical pump, and a modular unit incorporating connectors, filters and valves; system is employed for actuation of wing flaps, landing gear extension/retraction, wheel brakes, emergency and parking brakes, nosewheel steering, cargo ramp and door, and propeller braking. Accumulator for back-up braking system.

28 V DC primary electrical system powered by two 400 A Auxilec engine-driven starter/generators, with two 24 V 37 Ah Ni/Cd batteries for engine starting and 30 minutes' (minimum) emergency power for essential services. Constant frequency single-phase AC power (115/26 V) provided at 400 Hz by three 600 VA static inverters (two for normal operation plus one standby); two three-phase engine-driven alternators for 115/200 V variable frequency AC power. Fixed oxygen installation for crew of three (single cylinder at 124 bar; 1,800 lb/sq in pressure); three portable units and individual masks for passengers.

Pneumatic boot anti-icing of wing (outboard of engine nacelles), fin and tailplane leading-edges. Electric anti-icing of propellers, engine air intakes, flight deck windscreen, pitot tubes and angle of attack indicators. No APU: starboard engine, with propeller braking, can be used to fulfil this function. Engine fire detection and extinguishing system.

AVIONICS (civil): *Comms:* Two Rockwell Collins VHF-22B com radios, one Avtech DADS crew interphone, Rockwell Collins TDR-90 ATC transponder. Fairchild A-100A cockpit voice recorder, Avtech PACIS PA system. Dorne & Margolin ELT 8-1 emergency transmitter. Optional second TDR-90; optional HF-230 radio.

Radar: Rockwell Collins WXR-300 weather radar.

Flight: Two VIR-32 VOR/ILS/marker beacon receivers; DME-42; ADF-60A; two 332D-11T vertical gyros; two MCS-65 directional gyros; two ADI-85A; two HSI-85; two RMI-36; APS-65 autopilot/flight director; ALT-55B radio altimeter; two 345A-7 rate of turn sensors (all by Rockwell Collins); SFENA H-301 APM standby attitude director indicator; Hamilton Sundstrand Mk II GPWS; and Fairchild/Teledyne flight data recorder. Options include second DME-42 and ADF-60A, Rockwell Collins RNS-325 radar nav, Litton LTN-72R inertial nav or Global GNS-500A Omega navigation system.

Instrumentation: Rockwell Collins EFIS-85B five-tube CRT system standard.

AVIONICS (military) (Indonesian aircraft): *Comms:* Rockwell Collins AN/ARC-182 VHF/UHF; Rockwell Collins HF 9000 HF; IFF.

Flight: Rockwell Collins VIR-32 VHF nav; Litton LTN92 GPS-aided INS; Rockwell Collins DF-206A ADF; Rockwell Collins AN/APS-65F autopilot; GPWS.

Instrumentation: Rockwell Collins EFIS-85B(14) EFIS (four or five screens). IPTN developing cockpit lighting system compatible with night vision goggles.

AVIONICS (military): *Series 300: Thales Avionics Topdeck suite (see Current Versions) as core system.

Flight: Twin ADU 3000 air data units, GPSs and AHRSs; radar altimeter; TCAS; GPWS; weather radar; optional Totem 3000 LINS, Cat. II landing capability, MLS and satcom.

Instrumentation: Four 152 × 203 mm (6 × 8 in) LCDs; optional HUDs. Optional electro-optical sensors display imagery on LCDs. NVG compatibility.

Mission: Four-dimensional navigation FMS calculates high-altitude and computed air release points for load-dropping.

EQUIPMENT: Navigation lights, anti-collision strobe lights, 600 W landing light in front end of each main landing gear fairing, taxying lights, ice inspection lights, emergency door lights, flight deck and flight deck emergency lights, cabin and baggage compartment lights, individual passenger reading lights, and instrument panel white lighting, all standard. Hand-type fire extinguishers on flight deck (one) and in passenger cabin (two); smoke detector in baggage compartment.

ARMAMENT (military version): Three attachment points under each wing. Weapons can include Harpoon anti-ship missiles; Indonesian MPA version (which see) can be fitted with two Mk 46 torpedoes or AM 39 Exocet anti-shipping missiles.

Data follow for CASA-built Srs 300.

DIMENSIONS, EXTERNAL:

Wing span	25.81 m (84 ft 8 in)
Wing chord: at root	3.00 m (9 ft 10 in)
at tip	1.20 m (3 ft 11¼ in)
Wing aspect ratio	10.2
Length overall, standard nose	21.40 m (70 ft 2½ in)
Fuselage: Max width	2.90 m (9 ft 6 in)
Max depth	2.615 m (8 ft 7 in)
Height overall	8.18 m (26 ft 10 in)
Tailplane span	10.60 m (34 ft 9¼ in)
Wheel track (c/l of mainwheels)	3.90 m (12 ft 9½ in)
Wheelbase	6.92 m (22 ft 8½ in)
Propeller diameter: Srs 200	3.35 m (11 ft 0 in)
Srs 300	3.66 m (12 ft 0 in)
Propeller ground clearance	1.66 m (5 ft 5¼ in)
Distance between propeller centres	7.00 m (22 ft 11½ in)
Passenger door (port, rear) and service door (stbd, fwd):	
Height	1.70 m (5 ft 7 in)
Width	0.73 m (2 ft 4¾ in)
Height to sill	1.22 m (4 ft 0 in)
Paratroop doors (port and stbd, rear, each):	
Height	1.75 m (5 ft 9 in)
Width	0.90 m (2 ft 11½ in)
Height to sill	1.22 m (4 ft 0 in)
Ventral upper door (rear): Length	2.365 m (7 ft 9 in)
Width	2.35 m (7 ft 8½ in)
Height to sill	1.22 m (4 ft 0 in)
Ventral ramp/door (rear): Length	3.04 m (9 ft 11¾ in)
Width	2.35 m (7 ft 8½ in)
Height to sill	1.22 m (4 ft 0 in)
Type III emergency exits (port, fwd, and stbd, rear):	
Height	0.92 m (3 ft 0¼ in)
Width	0.51 m (1 ft 8 in)

DIMENSIONS, INTERNAL:

Cabin, excl flight deck: Length	9.65 m (31 ft 8 in)
Max width	2.70 m (8 ft 10½ in)
Width at floor	2.365 m (7 ft 9 in)
Max height	1.88 m (6 ft 2 in)
Floor area	22.8 m² (246 sq ft)
Volume	43.2 m³ (1,527 cu ft)

AREAS:

Wings, gross	59.10 m² (636.1 sq ft)
Ailerons (total, incl tabs)	3.14 m² (33.80 sq ft)
Trailing-edge flaps (total)	10.87 m² (117.00 sq ft)
Fin, incl dorsal fin	11.11 m² (119.59 sq ft)
Rudder, incl tabs	4.20 m² (45.21 sq ft)
Tailplane	21.20 m² (228.2 sq ft)
Elevators (total, incl tabs)	6.17 m² (66.41 sq ft)

WEIGHTS AND LOADINGS (Srs 300):

Operating weight empty	9,909 kg (21,846 lb)
Max payload	6,000 kg (13,228 lb)
Max fuel weight	4,230 kg (9,326 lb)
Max T-O weight	16,500 kg (36,376 lb)
Max ramp weight	16,550 kg (36,486 lb)
Max landing weight	16,500 kg (36,376 lb)
Max zero-fuel weight	15,400 kg (33,951 lb)
Max wing loading	279.2 kg/m² (57.19 lb/sq ft)
Max power loading (without APR)	6.33 kg/kW (10.39 lb/shp)

PERFORMANCE (Srs 300):

Max cruising speed	246 kt (455 km/h; 283 mph)
Max rate of climb at S/L	183 m (600 ft)/min
Service ceiling	9,145 m (30,000 ft)
Service ceiling, OEI	4,275 m (14,020 ft)
T-O run	398 m (1,305 ft)
T-O to 15 m (50 ft)	754 m (2,474 ft)
Landing from 15 m (50 ft)	603 m (1,978 ft)
Range: with max fuel	2,701 n miles (5,003 km; 3,108 miles)
with 4,000 kg (8,818 lb) payload	1,549 n miles (2,870 km; 1,783 miles)
with max payload	393 n miles (727 km; 452 miles)
g limits: at MTOW	+2.5/−1
below 14,100 kg (31,085 lb)	+3/−1

UPDATED

AIRTECH CN-235 MP PERSUADER and CN-235 MPA

TYPE: Maritime surveillance twin-turboprop.

CURRENT VERSIONS: **CN-235 MP Persuader:** CASA version; different avionics from Indonesian MPA. In service with Irish Air Corps and ordered by Spain (four) and Turkey (nine: six for Navy, three for Coast Guard, assembled by TAI at Ankara). In mid-1999, Turkey sought proposals from at least seven potential integrators of surveillance systems to provide radar, FLIR and an acoustics suite for naval CN-235s.

CN-235 MPA: Indonesian-developed version; available either with lengthened nose housing radar and IFF; or with normal CN-235 nose, plus belly radar; CN-235 prototype PK-XNC served as testbed. Maximum T-O weight 15,400 kg (33,951 lb), endurance more than 8 hours. Provision for quick-change configuration for general transport, communications or other duties. Required by Indonesian Navy (six included in national order for 24), Indonesian Air Force (three) and Brunei (three). UAE reported to have ordered four in March 1998, but these subsequently believed to have been cancelled.

Indonesia confirmed initial three firm orders in May 2000, when Thomson-CSF (now Thales) selected to supply AMASCOS airborne maritime situation control system, comprising Elettronica ALR-733 RWR, T-CSF Ghlio thermal imager and Sextant Gemini navigation computer. Brunei chose Boeing as Argo Systems integrator for its three aircraft in late 1995, specifying individual sensors in October 1996 as AN/AAQ-21 FLIR,

BAE Sky Guardian ESM, Cossor 3500 IFF and AN/APS-134 radar, plus two operators' consoles. BAE Systems Australia marketing CN-235 MPA in Asia-Pacific region under September 1997 agreement; BAE also to provide advanced systems development for proposed configurations.

AVIONICS (Persuader): *Radar:* Litton APS-504(V)5.

Mission: FLIR-2000HP undernose-mounted night vision system and Litton AN/ALR-85(V) ESM system, fully integrated via a central tactical processor with reconfigurable consoles.

AVIONICS (CN-235 MPA): *Radar:* BAE Systems Seaspray 4000, or Raytheon AN/APS-134 (LW) or Thales Ocean Master 100.

Flight: Litton LN92 ring laser gyro INS; Trimble TNL 7900 Omega/GPS.

Mission: Argo data processing and display system with multifunction consoles. BAE Systems Sky Guardian SG-300, or Argo Systems AR-700 or Litton AN/ALR-93(V)4 ESM. FLIR Systems AN/AAQ-21 Safire or BAE Systems MRT FLIR. Cossor 3500 IFF interrogator. (Trials aircraft originally equipped with APS-504 and Ocean Master; SG-300. Reconfigured by 1994 with AN/APS-134, MRT, AR-700, LN92 and TNL 7900. Further alternatives available at customer's option.)

UPDATED

Maritime surveillance CN-235 MP Persuader of the Irish Air Corps

0098155

AMC

AIRBUS MILITARY COMPANY SAS

17 avenue Didier Daurat, F-31707 Blagnac, France
Tel: (+33 5) 62 11 07 82
Fax: (+33 5) 62 11 06 11
Web: http://www.airbusmilitary.com
CEO: Francisco Fernandez Sainz
PRESIDENT: Alain Flourens
INDUSTRIAL DIRECTOR: Michele Priolo
COMMERCIAL DIRECTOR: Richard Thompson
PRODUCTION AND QUALITY CONTROL DIRECTOR: Angel Hurtado
FINANCE AND PROCUREMENT DIRECTOR: Michael Haidinger
HEAD OF MARKETING: David R Jennings
PARTICIPATING COMPANIES:

EADS France
BAE Systems Airbus (UK)
EADS CASA (Spain)
EADS Deutschland (Germany)
Alenia (Italy)
FLABEL (Belgium)
TAI (Turkey)

Airbus Military was legally established, with the above-mentioned companies, in January 1999 as a 'Société aux Actions Simplifiées' as the prospective manufacturer of the Airbus A400M, formerly known as the Future Large Aircraft (FLA). Airbus is the major (56 per cent) shareholder in AMC; TAI and FLABEL are full risk-sharing partners. EADS has assigned programme management to its Military Transport Aircraft Division located in Spain with CASA. It is also anticipated that AMC will promote military versions (including air refuelling tankers) of the Airbus series of airliners, some of which are already in air force service as transports.

Conceptual work was undertaken by the European FLA Group (Euroflag). Euroflag srl originally formed 17 June 1991, with headquarters in Alenia head office in Rome, to manage European FLA development. Aerospatiale, Alenia, British Aerospace, CASA and Daimler-Benz Aerospace Airbus (DaimlerChrysler from 1998) had equal shares in Euroflag srl; MoUs established 1992 with FLABEL (SABCA, SONACA, ASCO and BARCO) of Belgium, OGMA of Portugal and Turkish Aerospace Industries (TAI) of Turkey to allow integrated participation in FLA programme; BAE and FLABEL were industrial, not national, partners contributing their own funds, although the UK government announced in December 1994 that membership was to be upgraded to national participation.

The partners agreed in September 1994 to industrialise the programme by transferring it to their existing airliner production company; formal announcement was made on 14 June 1995 that Airbus Military Company would be established, replacing Euroflag, which then disbanded. Programme makes use of Airbus Industrie procedures and industrial infrastructure and takes advantage of technologies developed for Airbus airliners. The projected percentage shares for R&D financing were Germany 25.7, France 17.2, UK 15.5, Italy 15.1, Spain 12.4, Turkey 6.9, Belgium 4.1 and Portugal 3.1.

By early 1997, the FLA programme had lost development sponsorship by the principal participating governments, although military commitments remained, subject to the aircraft being produced with commercial funding. Programme was weakened during 1997 by unilateral German negotiations with Ukraine over Antonov An-7X (Westernised An-70). German MoD attempts to involve Russia and Ukraine continued into 1999, but these were not supported by Airbus Industrie, which declined to become the prime contractor and assume the commercial risk of a programme based on the An-70; a study commissioned by the

German MoD and carried out by DaimlerChrysler Aerospace reached a similar conclusion in September 1998. By mid-2000, senior German government sources were stressing the need for a European solution, prompting Airtruck to request assurances that its An-7X was still under consideration.

AMC submitted responses to the seven-nation FLA RFP (request for proposals, dated September 1997) on 29 January 1999 and to the competitive Future Transport Aircraft (FTA) RFP issued to Boeing, Airbus and Lockheed Martin by Belgium, France, Spain and the UK on 31 July 1998. Acceptance of A400M was formally announced by all seven members on 27 July 2000; subsequently, on 19 June 2001, MoU signed by seven of nine participating nations (Belgium, France, Germany, Luxembourg, Spain, Turkey and the UK) concerning joint procurement through OCCAR, with Italy and Portugal expected to follow suit in the near future, although Italy announced intention to withdraw from project on 25 October.

Programme encountered further difficulties in 2002, with official launch still not having occurred by year's end. Although OCCAR signed contract with AMC on 18 December 2001 for 196 aircraft for eight countries, Germany failed to secure parliamentary approval for funding before agreement expired on 31 January 2002.

UPDATED

AMC A400M

TYPE: Strategic transport.

PROGRAMME: Original FIMA programme (see 1989-90 *Jane's*) replaced April 1989 by five-nation industry MoU to develop four-turbofan, new technology transport to replace C-130 Hercules and C.160 Transall; Independent European Programme Group (IEPG) defined Outline European Staff Target (OEST) during 1991; initial studies undertaken by Euroflag organisation, which name reflected working designation Future Large Aircraft (FLA). Western European Union report in third quarter of 1991 concluded Euroflag FLA should form core of future European military transport capability to support Rapid

Reaction Corps; national armament directors of Belgium, France, Germany, Italy, Portugal, Spain and Turkey affirmed support for 12 month prefeasibility study completed by Euroflag in late 1992; UK MoD declined involvement, but retained observer status; UK participation privately maintained by BAe and Shorts (10 per cent of BAe work); European Staff Target and intergovernmental MoU signed by seven nations in 1993; full feasibility programme officially started October 1993, by which time cargo hold width and height increased from original 3.66 m (12 ft 0 in) and 3.55 m (11 ft 7¼ in), respectively; study finished May 1995 and submitted to European defence ministries. Meanwhile, FLA underwent profound change in April 1994 when turbofans deemed incapable of providing desired performance; aircraft recast with four turboprops of new design. Discussion of a 'close association' between Euroflag and Airbus Industrie began third quarter of 1993 and formalised in June 1995.

Launch of the predevelopment phase (PDP) was postponed at least six months from early 1996 as a result of funding uncertainties. Original intention was for PDP to run from 1996 to 1998 and define a comprehensive specification for the aircraft and contractual forms and conditions against which the partner nations would make commitments. Full development and production phase (DPP) scheduled to follow directly on from PDP and terminate with first flight in 2002. Customer deliveries were then planned to begin in 2004.

France announced funding withdrawal from FLA development on 13 May 1996 and UK failed to rejoin the programme later that year, despite intention announced in December 1994 (when the RAF purchased Lockheed Martin C-130J Hercules). However, Germany became first to sign a European Staff Requirement, on 24 July 1996, although having terminated official funding for FLA development in previous month. AMC accordingly announced a 'single-phase' programme in May 1996.

The new programme schedule started in mid-1998 with a set of formal prelaunch activities (PLA), largely funded by industry and lasting 12 months, leading to a fully documented proposal for the Airbus A400M. This

Artist's impression of the turboprop-powered AMC A400M

NEW/0526910

contained the technical proposal, including the aircraft specification and performance guarantees, and the commercial proposal with firm and fixed prices; a full set of contractual terms and conditions; and detailed planning of the single-phase programme. The proposal was made in the name of Airbus Military SAS (AM), formed in January 1999 and generally known as AMC. Strategic workshares (detailed enough to allow industry to provide the necessary resources to complete the proposal) were agreed at the start of PLA.

February 1999 delivery of the A400M proposal, initiated a 12-month period of negotiation of individual national requirements before planned official launch during early 2000 to meet an ESR now supported by Belgium, France, Germany, Italy, Spain, Turkey and the UK, but not Portugal. The 'PLA + single phase' programme provides industry with an uninterrupted development schedule and strong commitments from governments, while it also meets the customers' requirement that industry carries as much of the development risk as possible.

A meeting in March 2000 saw Belgium, France, Italy, Spain and Turkey identify a requirement for 131 A400Ms (37 fewer than expected) while, on 16 May, UK announced its intention to buy 25 aircraft (becoming first to fully commit). France and Germany followed on 9 June 2000. Seven participating nations announced selection of A400M on 27 July, committing to 225, including one for Luxembourg, although Turkey had reduced planned procurement to 10 at time of MoU signature in June 2001. Portugal announced requirement for four shortly thereafter and subsequently rejoined programme as a risk-sharing and industrial partner in early 2001. By June 2001, however, number of aircraft had fallen to three. MoU of 19 June 2001 on development and acquisition of A400M covered 196 aircraft, omitting 16 for Italy which in late October 2001 revealed it would not proceed with purchase. Formal launch nevertheless was expected in early 2002, but delayed for more than a year as consequence of German failure to obtain parliamentary funding approval. However, Germany made commitment in December 2002 to 60 aircraft and was then expecting official approval in second quarter of 2003.

Flight testing is expected to be at EADS CASA's Seville plant and AMC's Toulouse facility, with certification by a single authority; six prototypes to be built, of which five will be refurbished and sold on completion of test duties. EADS CASA will have sole production line, assembling components from the UK (wings), France (cockpit management and flight control systems), Germany (main fuselage), Spain (horizontal stabiliser), Belgium (wing leading edges and flaps), Italy (aft fuselage and other subsystems), Turkey (structural elements) and Portugal (wing/fuselage and undercarriage fairings). Germany committed US$4.4 billion to the programme in November 2000, though this represented 60 per cent of the amount required to guarantee Germany's 73 aircraft and planned 33 per cent workshare. France committed US$2.6 billion at the same time.

A timetable of 56 months between contract and first flight has been agreed, with first delivery (to France) 77 months after contract signature. Peak annual production rate expected to be 29.

CURRENT VERSIONS: Primarily for personnel/cargo transport and parachuting of men and equipment; basic aircraft will be capable of rapid conversion to two-point tanker with 22,000 kg (48,500 lb) of transferable fuel after a 3 hour loiter 450 n miles (833 km; 518 miles) from base. Derivatives will include a role-convertible three-point air-refuelling tanker with HDU on rear ramp. Three-point tanker could transfer 35,000 kg (77,162 lb) of fuel (in normal tanks plus roll-on/roll-off tanks in cargo bay) after a 3 hour loiter 450 n miles (833 km; 518 miles) from base

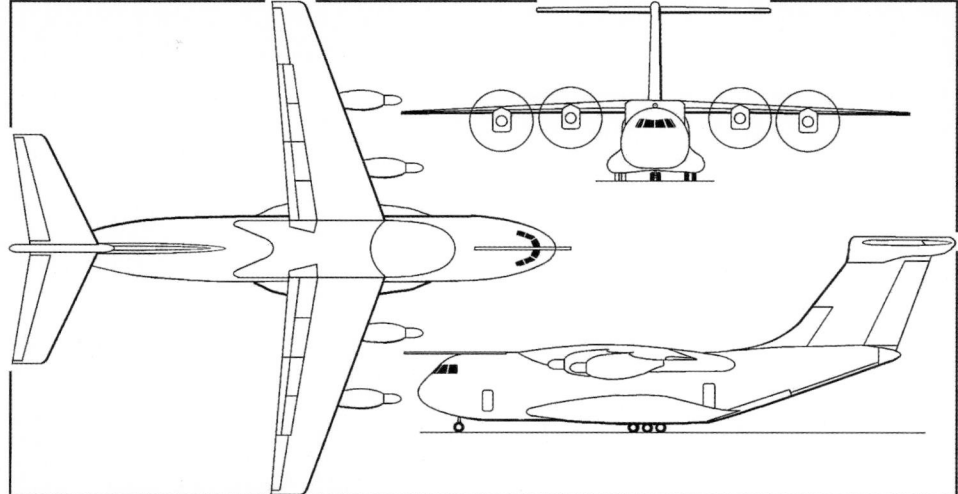

Provisional drawing of the AMC A400M military transport (*Jane's/Paul Jackson*)   NEW/0527107

but is regarded in some circles as being too slow for routine support of fast jets. Proposed maritime patrol version now abandoned and no recent news of surveillance/reconnaissance and AEW concepts.

CUSTOMERS: Procurement agency is Organisme Conjointe de Co-operation en matière d'Armement (OCCAR) in Bonn, acting for all prospective NATO purchasers. First deliveries reserved for France, followed by Turkey, then UK. Exports expected, and attempts made in early 1995 to interest Japan; Australia briefed in February 2001 and may acquire A400M as eventual replacement for existing fleet of C-130H Hercules; export market estimated at 400 aircraft over 25 years, with A400M to secure 50 per cent share.

AMC A400M REQUIREMENTS
(at December 2002)

Country	Original	Current
Belgium	12	7
France	50	50
Germany	75	60
Italy	44	0
Luxembourg	0	1
Portugal	0	3
Spain	36	27
Turkey	26	10
UK	45	25
Total	**288**	**183**

COSTS: Unit cost estimated in 1999 as US$80 million over 300 aircraft production run including amortisation of all R&D costs (estimated as some US$6 billion). In October 1999, AMC offered to 'stake-hold' by providing 20 per cent of development costs. Unit price estimated in early 2001 as US$82 million over production run of 229 aircraft and in early 2003 as €85 million.

DESIGN FEATURES: High-wing, T-tailed aircraft with rough-field landing gear and much larger cabin/hold floor area and cross-section than C-130/C.160, permitting high payload factors with low-density cargo, vehicles or mixed passenger/cargo loads. Use of propellers felt to be essential for adequate thrust-reverse performance for taxying and

short landing; for maximising power response; and for minimising FOD vulnerability. Long-range cruising speed of M0.68 up to 11,280 m (37,000 ft). Tactical mission parameters of 150 m (500 ft) AGL in IMC on predetermined route with civil standard of safety. Airbus has noted that its extensive use of new technology gives twice the volume and payload of the C-130J at the same life-cycle cost. Compared with the C-17, the A400M is said to be less than half the price and to have one third of the life-cycle cost. Minimum service life 30,000 hours, including allowance for low-level flight and short-field performance. Optimised for autonomous deployment; AMC offering 15-day away-from-base serviceability guarantee, all necessary support being within flight crew's capabilities.

Wing sweep 15° at 25 per cent chord; anhedral 2°; taper ratio 0.334; mean aerodynamic chord 5.690 m.

FLYING CONTROLS: Fly-by-wire, hydraulically powered; manually actuated electrohydrostatic back-up for ailerons, elevator and rudder. Four spoilers and two-section flaps on each wing; tailplane trimmable by screw-jack. No slats. Spoilers used for roll control, lift dumping and as speed brakes.

STRUCTURE: Aluminium alloy, with titanium alloy in highly loaded areas (around windscreen, wing/fuselage joint and landing gear anchorage) and glass fibre or carbon fibre for lightly loaded components (landing gear doors and various fairings). Tailplane has aluminium alloy central structural box and two outer composites box structures; elevator primary structure of carbon fibre. Fin has three-spar main box, trailing-edge shroud and single-piece rudder, all primarily of composites, plus metal/composites removable leading-edge. Rudder of carbon fibre, with aluminium, hinge-connecting ribs. Extensive use of composites in wing for skins, stringers, spars and (carbon fibre) moving surfaces; metal for ribs, engine mountings and fuselage pick-ups. Front spar at 15 per cent chord; composites rear spar at 67 per cent chord. Modern design and manufacturing techniques expected to afford major reductions in maintenance man-hour requirements and increases in aircraft availability/survivability.

In 2000, workshares allocated as follows. EADS France: nose/flight deck (with TAI), wing centre box and (with EADS CASA) engine nacelles. Alenia: aft fuselage (with TAI), inboard and outboard flaps and cargo handling system. BAE Systems: overall wing leadership, assembly and equipping; fuel system, ice/rain protection and landing gear. EADS CASA: tailplane (with TAI), engine nacelles (with EADS France) and final assembly. EADS Deutschland: fuselage team leaders, systems integration, centre fuselage, fin, ailerons, spoilers and wing skins. FLABEL: wing leading-edge and flap tracks. TAI: nosewheel doors and bay; forward centre fuselage; cargo doors and ramp; elevators.

LANDING GEAR: Retractable tricycle type with sufficient 'flotation' for semi-prepared and/or unsurfaced runways. Each main unit has six wheels in tandem pairs, retracting rearwards into fairings on fuselage sides. Each pair of mainwheels has independent, lever-type shock-absorbers. Twin nosewheels retract forwards. Emergency gravity extension of all units. Multidisc carbon brakes on mainwheels can operate differentially to assist steerable nosewheel in ground manoeuvring. Turning radius: landing gear 15 m (50 ft); wingtip 28.6 m (94 ft). Mainwheels can 'kneel' for unloading of large cargoes. Hydraulic strut at rear of each sponson supports and stabilises aircraft during loading and unloading.

POWER PLANT: Initial candidate engines rated at approximately 6,898 kW (9,250 shp): M138 turboprop offered by Turboprop International SNECMA (33 per cent), MTU (33 per cent), Fiat Avio (22 per cent) and ITP (12 per cent) and based on SNECMA M88-2 core; Rolls-Royce Deutschland proposed a turboprop development of the BR715 turbofan, 8,949 kW (12,000 shp) BR700-TP;

AIRDROP SYSTEM
❑ paratroops
❑ gravity, extraction, ramp supported dropping
❑ personnel and cargo simultaneous dropping

FLEXIBILITY
❑ autonomous loading/off-loading
❑ combat off-loading
❑ cross-loading
❑ aeromedical configuration
❑ troops and cargo

PARATROOP DOORS
❑ wide opening

POWERED CRANE
❑ 5 tonne capacity
❑ loading from ground
❑ cross loading

SINGLE LOADMASTER
❑ in flight reconfiguration
❑ multipurpose workstation

LOAD CARRYING RAMP
❑ up to 6000kg

REAR OPENING
❑ full cross-section for axial load movement
❑ roll-on/roll-off loading
❑ airdrop of large loads

CARGO FLOOR
❑ retractable rail and rollers
❑ accomodates civil and military pallets
❑ power drive units

Notable features of the AMC A400M   0101428

and Pratt & Whitney Canada offered a 'Twinpac' version of the existing PW150. Required engine power, as defined by Airbus, is up to 7,457 kW (10,000 shp).

Choice initially settled on three-shaft 7,457 to 9,694 kW (10,000 to 13,000 shp) turboprop TP400 developed by Fiat Avio, ITP, MTU, Rolls-Royce, SNECMA and Techspace Aero, although this also rejected in February 2002 on basis of being too costly and too heavy as well as insufficiently powerful. Engine competition subsequently re-opened, with European and US manufacturers invited to submit proposals. Airbus announced intent to choose engine by September 2002, but decision still awaited in early 2003, when AMC stated that engine choice would be customer's (OCCAR) decision. New competitors are EuroProp International (ITP, MTU, Rolls-Royce and Snecma) and Pratt & Whitney Canada.

Fuel capacity 64,030 litres (16,915 US gallons; 14,085 Imp gallons) in five tanks (no transfer tank) inside wing box; electric pumps and valves all mounted outside tanks. Detachable in-flight refuelling probe. Provision for inert gas system; wing-mounted refuelling pods; HDU in cargo hold; and additional fuel tanks, totalling 16,250 litres (4,293 US gallons; 3,575 Imp gallons), in fuselage. Pressure refuelling, with gravity back-up.

ACCOMMODATION: Two-man, NVG-compatible flight deck with dual sidestick controllers and additional forward-facing workstation for third 'mission crew member' to assist with tactical and special tasks, when required. View from flight deck exceeds JAR 25 and MIL-STD-850B. Provision for bulletproof flight deck windows, 68 mm (2¾ in) thick, and armour protection around crew's seats. Loadmaster station forward of and overlooking cargo area. Two fixed, screened urinals and fixed hand-basin, starboard, rear. Astrodrome for formation surveillance expected. Flight crew rest area with two foldaway bunks.

Two passenger doors forward; two rear. Forward, port, for normal access; forward, starboard, for emergency exit; rear doors for paratroop dropping. Three emergency exits in roof for flight crew and passengers. Cargo door, hinged at aft end, raised hydraulically to hold roof for loading via rear ramp. Closed-circuit TV surveillance of cargo hold, with imagery selectable on flight deck displays.

Cargo floor with 250 tiedown rings stressed to 4,536 kg (10,000 lb) and 60 to 11,340 kg (25,000 lb). Typical loads include Warrior or MRAV armoured transport vehicles; Super Puma or two Tiger helicopters; nine pallets (88 × 108 in military or 88 × 125 in civil); plus 57 troops and second loadmaster on permanent (tip-up) sidewall seats; two 20 ft ISO containers; Patriot SAM system; six Land Rovers, plus trailers; 66 stretchers and 10 medical attendants; or 120 armed troops on sidewall and 62 removable centreline seats. Ramp stressed for 6,000 kg (13,228 lb) loads and has three hydraulically powered toes and 90 tiedowns of 4,536 kg (10,000 lb).

SYSTEMS: FBW FCS derived from Airbus airliners, including sidestick controllers (left hand for captain, right hand for

co-pilot, with conventional central power-lever throttle quadrant).

Electrical power provided by four engine-driven generators, each of 75 kVA. Additional power from three-phase generator on APU (90 kVA) in landing gear sponson; three-phase generator on RAT; emergency battery; and external power receptacle. DC power from four 200 A transformer/rectifier units: two feed separated, main DC busbar; one to 'flight essential' busbar and emergency busbar and battery; and one to APU starting system. Two Ni/Cd batteries.

Two hydraulic systems, Blue and Yellow, each operating at 207 bar (3,000 lb/sq in). Blue (driven by Nos. 1 and 4 engines) powers port aileron and elevator, inboard (No. 4) and No. 2 spoiler on each wing, back-up brakes, cargo ramp and ground stabiliser struts. Yellow (Nos. 2 and 3 engines) responsible for starboard aileron and elevator, Nos. 2 and 4 spoilers on each side, landing gear kneeling, brakes and landing gear actuation. Both systems power flaps, rudder and tailplane trim. Each system has 140 litre (37.0 US gallon; 30.8 Imp gallon)/min engine-driven pump and 40 litre (10.6 US gallon; 8.8 Imp gallon)/min alternate current motor pump; power transfer unit between systems for emergency use.

Pneumatic system for air conditioning pressurisation; wing and engine air intake anti-icing; engine starting; and pressurisation of other onboard systems. Two computers control four engine air bleed units. Interior divided into three air conditioning zones; flight deck and two in cargo hold. Cabin pressure altitude 2,440 m (8,000 ft) when flying at 11,280 m (37,000 ft).

AVIONICS: Comms: HF, V/UHF, Selcal and optional satcom. Audio management system, cockpit voice recorder and passenger address system. Transponder, ELT and IFF.

Radar: Northrop Grumman AN/APN-241 weather radar with additional ground mapping mode.

Flight: Three inertial platforms with embedded air data systems. VOR, DME, Tacan, multimode receiver (ILS, MLS and GPS), two radar altimeters and GPWS. Optional terrain-reference navigation system.

Instrumentation: Five full-colour MFDs and two HUDs; two multifunction control and display units on centre pedestal, plus one for optional third crew member.

Mission: Secure tactical radios and optional MIDS datalink.

Self-defence: Modular DASS with optional elements including central computer, RWR, MAWS, LWR, chaff/flare dispensers, IR jammers, electronic jammer and towed radar decoy.

EQUIPMENT: Winch at forward end of cargo hold. Optional 5-tonne crane in roof above cargo ramp. Provision for one 1,500 litre (396 US gallon; 330 Imp gallon)/min refuelling pod under each wing and/or 2,250 litre (594 US gallon; 495 Imp gallon)/min HDU in rear of cargo hold.

DIMENSIONS, EXTERNAL:
Wing span   42.40 m (139 ft 1 in)

Wing aspect ratio   8.1
Length overall   42.20 m (138 ft 5½ in)
Height overall   14.73 m (48 ft 4 in)
Wheel track   6.20 m (20 ft 4 in)
Wheelbase   13.60 m (44 ft 7½ in)
Propeller diameter   5.18 m (17 ft 0 in)
Ramp: Length   5.40 m (17ft 8½ in)
   Width   4.00 m (13 ft 1½ in)
Cargo door: Length   8.10 m (26 ft 7 in)
AREAS:
Wing area, gross   221.50 m² (2,384.2 sq ft)
DIMENSIONS, INTERNAL:
Hold: Length excl ramp   17.71 m (58 ft 1¼ in)
   Width at floor, continuous   4.00 m (13 ft 1½ in)
   Height: forward of wing box   3.85 m (12 ft 7½ in)
      aft of wing box   4.00 m (13 ft 1½ in)
Floor area: incl ramp   92 m² (990 sq ft)
Volume: incl ramp (approx)   356 m³ (12,570 cu ft)
   excl ramp (approx)   274 m³ (9,680 cu ft)
WEIGHTS AND LOADINGS: (A: logistic operation at max 2.5 g, B: logistic at 2.25 g, C: tactical operation at 2.5 g):
Operating weight, empty: 66,500 kg (146,605 lb)*
Max payload: A   31,500 kg (69,445 lb)
   B   37,000 kg (81,570 lb)
   C   29,500 kg (65,036 lb)
Max T-O weight: A   126,500 kg (278,885 lb)
   B   130,000 kg (286,600 lb)
   C   116,500 kg (256,835 lb)
Max landing weight: A, B   114,000 kg (251,325 llb)
   C   106,500 kg (234,790 lb)
Max zero-fuel weight: A   98,000 kg (216,055 lb)
   B   103,500 kg (228,180 lb)
   C   96,000 kg (211,645 lb)
* including 1,000 kg (2,205 lb) allowance for optional equipment
PERFORMANCE (estimated):
Max operating speed and Mach No. (VMO/MMO).
                          M0.72 (300 kt; 555 km/h; 345 mph)
Normal cruising Mach No.   0.68
Airdrop speed 130-200 kt (241-370 km/h; 150-230 mph)
Max rate of climb at S/L   1,524 m (5,000 ft)/min
Max certified altitude   11,280 m (37,000 ft)
T-O run at 116,500 kg (256,835 lb)   1,402 m (4,600 ft)
Landing run at 93,500 kg (206,131 lb), max propeller reversal, 152 m (500 ft) roll-out   625 m (2,050 ft)
Range with 5% reserves, missed approach, 200 n mile (370 km; 230 mile) diversion and 30 min hold at 445 m (1,500 ft), B:
   with 30,000 kg (66,139 lb) payload
                        2,450 n miles (4,537 km; 2,819 miles)
   with 20,000 kg (44,092 lb) payload
                        3,550 n miles (6,574 km; 4,085 miles)
Ferry range   4,900 n miles (9,074 km; 5,638 miles)
                                              UPDATED

Computer simulation of A400M being refuelled in flight   NEW/0526929

Flight deck of the AMC A400M, created by computer imagery   0044888

AMX

AMX INTERNATIONAL

c/o Alenia, Via Giulio Vincenzo Bona 85, I-00156 Rome, Italy
Tel: (+39 06) 41 72 38 35
Fax: (+39 06) 41 72 38 75
PRESIDENT: Dott Ing Giovanni Gazzaniga (Alenia)
VICE-PRESIDENTS:
F Grandi (Aermacchi)
R Pesce (Embraer)
PARTICIPATING COMPANIES:
Alenia: see under Italy
Aermacchi: see under Italy
Embraer: see under Brazil

AMX production for the Italian and Brazilian air forces is now complete, but an export order is outstanding. A mid-life update programme involving Brazilian and Italian aircraft began in 2001.
                                              UPDATED

AMX AMX
Brazilian Air Force designations: A-1 and A-1B
Italian Air Force name: Ghibli (Desert Wind)
TYPE: Attack fighter.
PROGRAMME: Resulted from June 1977 Italian Air Force specification for small tactical fighter-bomber (see 1987-88 and previous *Jane's* for early background);

original Aeritalia/Aermacchi partnership joined by Embraer July 1980; seven single-seat prototypes built (first flight 15 May 1984: further details in 1987-88 and earlier editions); production of first 30 (Italy 21, Brazil nine), and design of two-seater, began mid-1986; first production aircraft rolled out at Turin 29 March 1988, making first flight 11 May; second contract (Italy 59, Brazil 25, including six and three two-seaters respectively) placed 1988.

Deliveries to Italian Air Force (six for Reparto Sperimentale di Volo at Pratica di Mare) began April 1989; production A-1 for Brazilian Air Force (s/n 5500) made first flight 12 August 1989, deliveries (two to Nu A-1 training nucleus at Santa Cruz) following from 17 October

1989; in-flight refuelling test programme completed (by Embraer) August/September 1989; first flight by first (of three) two-seat AMX-T prototypes 14 March 1990 (MM55024), followed by second on 16 July; first flight of Embraer two-seater (serial number 5650), 14 August 1991; third production batch authorised early 1992 (one year late); first two-seater for Brazilian Air Force (5650) delivered 7 May 1992.

Italian single-seater production temporarily halted following delivery on 1 February 1993 of 72nd aircraft (MM7160); resumed late 1994, with both AMX and first production batch of AMX-T. Final Italian single-seater delivered in 1997; total 110, comprising 74 built by Alenia and 36 by Aermacchi. Batch 4 (35 AMX and 16 AMX-T) and Batch 5 (42 AMX and 9 AMX-T) cancelled. Final Italian two-seater followed in 1998; 26 built: 17 Alenia, nine Aermacchi. Production continued in Brazil, where 50th was delivered on 1 December 1998 and last in 1999.

Istrana- and Amendola-based AMX squadrons flew 252 combat sorties (667 flying hours) during Operation Allied Force against Yugoslavia in 1999, dropping 39 Opher LGBs.

CURRENT VERSIONS: **AMX:** Replaced G91R/Y and some F-104G/S in Italian Air Force (eight squadrons originally planned) and some EMB-326GB Xavante in Brazilian Air Force for close support/interdiction/reconnaissance, sharing counter-air duties with IDS Tornado (Italy) and F-5E/Mirage 50 (Brazil); in service with five Italian Stormi (see table) and 10° and 16° Grupos (Brazil); Brazilian Air Force aircraft (designated **A-1**) differ primarily in avionics and weapon delivery systems, have two 30 mm guns instead of Italian version's single multibarrel 20 mm weapon and are usually fitted with in-flight refuelling probes.

AMX-MLU: Mid-life upgrade configuration for Italian and Brazilian air forces, with new avionics, improved pilot interface, NVG-compatible 'glass cockpit' with three MFDs, new navigation system with embedded GPS, MIL-STD-1760 stores management system, radar altimeter, helmet-mounted display system, FLIR, multimode radar, digital FCS and tactical datalink. Will be cleared with a full range of standoff weapons. First phase has been approved (for Italian AMXs) adding IN/GPS, MIL-STD-1760 SMS, JTIDS datalink and new HDD to the 18 surviving Batch 3 aircraft, at a cost of Lit170 billion (US$84.4 million). Work was due to begin in 2001. Phase 2 will add the NVG 'glass cockpit', new self-protection suite and Finmeccanica/Galileo Avionica SCP-01 Scipio radar. Italy intends to bring all 94 in active inventory to this standard.

Super AMX: Obsolete designation for two-seater which later offered to South Africa the basis of Italian upgrade. Wide-angle HUD, 'glass cockpit', improved HOTAS, GPS, HMD, integrated defensive aids, and new weapons. Also new radar – possibly Scipio already fitted to Brazilian AMX. Studies under way into replacement of Spey engine by non-afterburning 60.0 kN (13,500 lb st) EJ200, derived from Eurofighter Typhoon power plant, giving 20 per cent extra static thrust, 40 per cent extra thrust at Mach 0.8, 40 per cent increase in T-O performance and 20 per cent increase in sustained turn rate. Development of engine flight demonstrator expected, perhaps with first flight during 2000-2001. RFP issued for new digital FCS computer.

Detailed description applies to single-seater except where indicated.

AMX-T: Second cockpit accommodated by removing forward fuselage fuel tank and relocating environmental control system; dual controls, canopy, integration of rear cockpit GEC-Marconi HUD monitor, and oxygen systems, designed/redesigned by Embraer; intended both as operational trainer and, suitably equipped, for such roles as EW, reconnaissance and maritime attack; most Italian AMX-Ts assigned to operational squadrons of 32° Stormo; others assigned, one per squadron, as trainers. Brazilian designation **A-1B.**

AMX-ATA: Eight two-seaters ordered by Venezuela in September 1999 for delivery in the fourth quarter of 2001. Delivery failed to take place at the appointed time, although the order was reaffirmed by Embraer in March 2001. The Venezuelan aircraft will have Elbit avionics and may be followed by more AMX-ATA 1s.

AMX-ATA 1: Total of 24 modernised and upgraded trainers still expected to be ordered by Venezuela, based on FAB Block 3 configuration, but with digital cockpit; some ALX-based avionics in package to be supplied and integrated by Elbit. New radar, probably Alenia SCP-01 Scipio.

AMX-ATA 2: Light combat aircraft and advanced trainer for export with AMX-MLU avionics and non-afterburning EJ200 engine.

AMX-E: Proposed two-seat EW version intended as escort jammer and SEAD platform using AGM-88 HARM missiles. Flight controls removed from rear cockpit. MM55027 used for development. Feasibility study complete, further development cancelled.

CUSTOMERS: Total of 192 (136 Italy/56 Brazil, including 26 and 11 two-seaters) delivered by March 2000 (see table) to original partners, plus eight two-seaters selected by Venezuela on 9 September 1999 at undisclosed date. Italy plans to place all 19 remaining Batch 1 aircraft in reserve for possible sale. By 1998, 23

AMX Ghibli CSX7158, an upgrade trials aircraft of the Italian Air Force's Reparto Sperimentale di Volo trials unit (*Jane's/Paul Jackson*)   NEW/0131741

AMX REQUIREMENTS

| Batch | Italy | | | | Brazil | | | | Total |
	AMX	First aircraft	AMX-T	First aircraft	AMX	First aircraft	AMX-T	First aircraft	
1	19	MM7089	2	MM55024	8	5500	1	5650	30
2	53	MM7108	6	MM55026	22	5508	3	5651	84
3	38	MM7161	18	MM55034	15	5530	7	5654	78
Subtotal	**110**		**26**		**45**		**11**		**192**
4					15		4		19
5					19		0		19
Total	**110**		**26**		**79**		**15**		**230**

Notes: Brazil retains option on 23 aircraft from Batches 4 and 5. Excludes seven single-seat prototypes and eight AMX-ATAs for Venezuela.

ITALIAN AMX UNITS

Squadron	Wing	Base	Type	First aircraft and acceptance	
13° Gruppo	32° Stormo	Amendola	AMX	MM7112	Nov 94
			AMX-T[1]	MM55029	by Apr 95
14° Gruppo	2° Stormo	Rivolto[2]	AMX	MM7130	10 Jul 91
			AMX-T[1]	MM55028	24 Nov 94
28° Gruppo[5]	3° Stormo	Villafranca	AMX	MM7129	Jun 93
			AMX-T[1]	MM55029	by Dec 94
101° Gruppo[3]	32° Stormo	Amendola	AMX	MM7091	31 Jul 95
			AMX-T	MM55030	21 Nov 94
103° Gruppo	51° Stormo	Istrana	AMX	MM7098	30 Sep 89
			AMX-T[1]	MM55027	by Apr 95
132° Gruppo	51° Stormo	Istrana[6]	AMX	MM7110	15 Nov 90[4]
			AMX-T[1]	MM55037	4 Dec 95

[1] One aircraft only
[2] At Istrana until February 94; may return
[3] Unit initially re-equipped as 201° Gruppo (until 31 Jul 95); AMX OCU
[4] Initial working up at Istrana
[5] Disbanded 29 Sep 97
[6] Villafranca until March 1999

BRAZILIAN AMX UNITS

Squadron	Wing	Base	First delivery
1° Esquadrão 'Adelfi'	16 Grupo de Aviação	Santa Cruz	Formed 7 November 1990[1]
1° Esquadrão 'Poker'	10 Grupo de Aviação	Santa Maria	1996
3° Esquadrão 'Centauro'	10 Grupo de Aviação	Santa Maria	1998

[1] Out of Nucleo A-1, originally established 7 November 1988

Batch 2 aircraft had been upgraded to 'pre-FOC' (full operational capability) standard and plans were in hand to modify remaining 15 to FOC. These 38 plus all 56 remaining from Batch 3 (FOC), formed baseline fleet of 94, though AMI active inventory quoted as 104 aircraft by May 1999. MLU for AMI AMX described above. Upgrades for remaining aircraft will include Thomson CDLP laser designator and new electro-optical reconnaissance pod. Italian long-term plan is for four squadrons, including OCU.

COSTS: US$18.75 million (1999) AMX-T Venezuelan programme unit cost.

DESIGN FEATURES: Intended for high-subsonic/very low-altitude day/night missions, in poor visibility and, if necessary, from poorly equipped or partially damaged runways. Required fatigue life of 16,000 hours being extended to 24,000 hours.

Wing sweepback 31° on leading-edges, 27° 30′ at quarter-chord; thickness/chord ratio 12 per cent.

FLYING CONTROLS: Hydraulically actuated ailerons and elevators; leading-edge slats and Fowler double-slotted trailing-edge flaps (each two-segment on each wing, positions 0, 30 and 41°) actuated electrohydraulically; pair of hydraulically actuated spoilers forward of each flap pair, deployable separately in inboard and outboard pairs; fly-by-wire control of spoilers, rudder and variable incidence tailplane by Alenia/BAE Systems flight control computer; ailerons, elevators, rudder have manual reversion for fly-home capability, even with both hydraulic systems inoperative; spoilers serve also as airbrakes/lift dumpers.

STRUCTURE: Mainly aluminium alloy except for carbon fibre fin and elevators; shoulder-mounted wings, each with three-point attachment to fuselage, have three-spar torsion box with integrally stiffened skins; oval-section semi-monocoque fuselage, with rear portion (including tailplane) detachable for engine access.

Work split gives programme leader Alenia 46.7 per cent (centre-fuselage, nose radome, tail surfaces, ailerons and spoilers); Aermacchi has 23.6 per cent (forward fuselage including gun and avionics integration, canopy, tailcone) and Embraer 29.7 per cent (air intakes, wings, leading-edge slats, flaps, wing pylons, external fuel tanks and

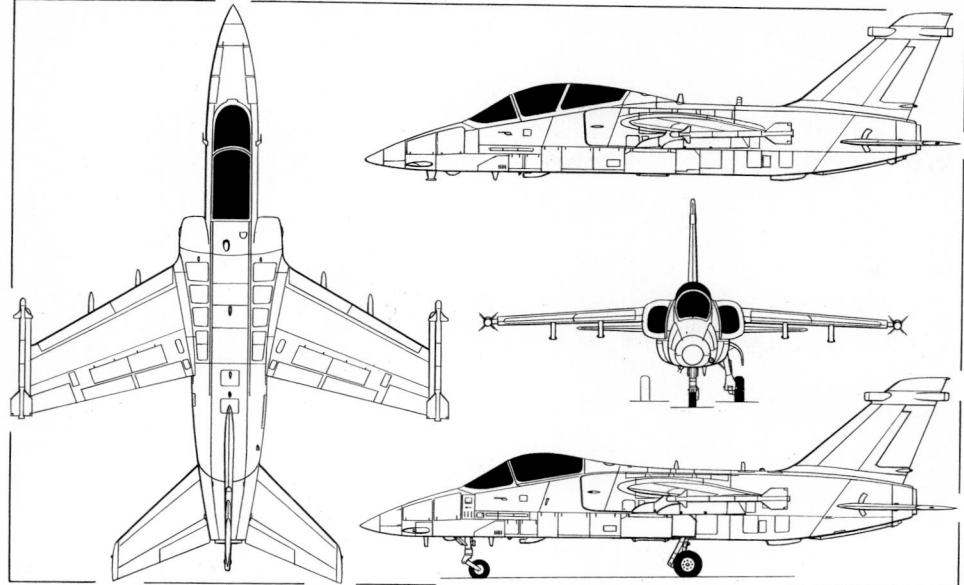

Alenia/Aermacchi/Embraer AMX; upper side view shows two-seater *(Jane's/Dennis Punnett)*

Self-defence: Elettronica active and passive ECM, including fin-mounted radar warning receiver.

ARMAMENT: One M61A1 multibarrel 20 mm cannon, with 350 rounds, in port side of lower forward fuselage of aircraft for Italian Air Force (one 30 mm DEFA 554 cannon on each side in aircraft for Brazilian Air Force). Single stores attachment point on fuselage centreline, plus two attachments under each wing, and wingtip rails for two AIM-9L Sidewinder or similar IR air-to-air missiles (MAA-1 Piranha on Brazilian aircraft). Fuselage and inboard underwing points each stressed for loads of up to 907 kg (2,000 lb); outboard underwing points stressed for 454 kg (1,000 lb) each. Twin carriers can be fitted to all five stations. Total external stores load 3,800 kg (8,377 lb). Attack weapons can include free-fall or retarded Mk 82/83/84 bombs, laser-guided bombs, cluster bombs, air-to-surface missiles (including area denial, anti-radiation and anti-shipping weapons), electro-optical precision-guided munitions and rocket launchers.

Exocet firing trials conducted 1991; Marte trials 1994; carriage trials of GBU-16 Paveway II LGB on Italian AMX in 1995 and aircraft used Elbit Opher LGB system during Operation Allied Force over Kosovo, May-July 1999.

DIMENSIONS, EXTERNAL:

Wing span:	
excl wingtip missiles and rails	8.875 m (29 ft 1½ in)
over missiles	9.97 m (32 ft 8½ in)
Wing aspect ratio	3.8
Wing taper ratio	0.5
Length: overall	13.23 m (43 ft 5 in)
fuselage	12.55 m (41 ft 2 in)
Height overall	4.55 m (14 ft 11¼ in)
Tailplane span	5.20 m (17 ft 0¾ in)
Wheel track	2.15 m (7 ft 0¾ in)
Wheelbase	4.70 m (15 ft 5 in)
AREAS:	
Wings, gross	21.00 m² (226.0 sq ft)
Ailerons (total)	0.88 m² (9.47 sq ft)
Trailing-edge flaps (total)	3.86 m² (41.55 sq ft)
Leading-edge slats (total)	2.07 m² (22.28 sq ft)
Spoilers (total)	1.30 m² (13.99 sq ft)
Fin (exposed)	4.265 m² (45.91 sq ft)
Rudder	0.83 m² (8.93 sq ft)
Tailplane (total exposed)	5.10 m² (54.90 sq ft)
Elevators (total)	1.00 m² (10.76 sq ft)

WEIGHTS AND LOADINGS (all versions):

Operational weight empty	6,730 kg (14,837 lb)
Max fuel weight: internal	2,790 kg (6,151 lb)
external	1,726 kg (3,805 lb)
Max external stores load	3,800 kg (8,377 lb)
T-O weight (clean)	9,694 kg (21,371 lb)
Typical mission T-O weight	10,750 kg (23,700 lb)
Max T-O weight	13,000 kg (28,660 lb)
Normal landing weight	7,000 kg (15,432 lb)
Combat wing loading (clean)	457.1 kg/m² (93.62 lb/sq ft)
Max wing loading	619.1 kg/m² (126.79 lb/sq ft)
Max power loading	265 kg/kN (2.60 lb/lb st)

PERFORMANCE (A at typical mission weight of 10,750 kg; 23,700 lb with 907 kg; 2,000 lb of external stores, B at max T-O weight with 2,721 kg; 6,000 lb of external stores, ISA in both cases):

Max level speed: at S/L	M0.84
at 9,140 m (30,000 ft)	M0.86
Max rate of climb at S/L	3,124 m (10,250 ft)/min
Service ceiling	13,000 m (42,650 ft)
T-O run at S/L: A	631 m (2,070 ft)
B	982 m (3,220 ft)
T-O to 15 m (50 ft) at S/L: B	1,442 m (4,730 ft)
Landing from 15 m (50 ft) at S/L: B	753 m (2,470 ft)
Landing run at S/L	464 m (1,520 ft)
Attack radius, allowances for 5 min combat over target and 10% fuel reserves	
lo-lo-lo: A	300 n miles (556 km; 345 miles)
B	285 n miles (528 km; 328 miles)
hi-lo-hi: A	480 n miles (889 km; 553 miles)
B	500 n miles (926 km; 576 miles)
Ferry range with two 1,000 litre (264 US gallon; 220 Imp gallon) drop tanks, 10% reserves	
	1,800 n miles (3,333 km; 2,071 miles)
g limits	+7.33/−3

UPDATED

reconnaissance pallets); single-sourced production, with final assembly lines in Italy and Brazil.

LANDING GEAR: Hydraulically retractable tricycle type, of Messier-Bugatti levered suspension design, built in Italy by Magnaghi (nose unit) and in France by ERAM (main units). Single wheel and oleo-pneumatic shock-absorber on each unit. Nose unit retracts forward; main units retract forward and inward, turning through approximately 90° to lie almost flat in underside of engine air intake trunks. Nosewheel hydraulically steerable (±6° normal; ±45° with full pedal movement), self-centring, and fitted with anti-shimmy device. Towing travel ±90°. Mainwheel tyres size 670×210-12 (18 ply), pressure 9.65 bar (140 lb/sq in); nosewheel tyre size 18×5.5-8 (10 ply), pressure 10.70 bar (155 lb/sq in). Hydraulic brakes and fully modulated anti-skid system. No brake-chute. Runway arrester hook. Minimum ground turning radius 7.53 m (24 ft 8½ in).

POWER PLANT: One 49.1 kN (11,030 lb st) Rolls-Royce Spey Mk 807 non-afterburning turbofan, built under licence in Italy by Fiat, Piaggio and Alfa Romeo Avio, in association with Companhia Eletro-Mecânica (CELMA) in Brazil. Self-sealing, compartmented, rubber fuselage bag tanks and two integral wing tanks with combined capacity of 3,500 litres (924.6 US gallons; 770 Imp gallons). Brazilian AMX carry 200 litres (52.8 US gallons; 44.0 Imp gallons) more internal fuel in additional saddle tank behind cockpit. Auxiliary fuel tanks of up to 1,100 litres (290 US gallons; 242 Imp gallons) capacity can be carried by Brazilian aircraft on each inboard underwing pylon, and up to 580 litres (153 US gallons; 128 Imp gallons) on each outboard pylon (Italian or Brazilian). Single-point pressure or gravity refuelling of internal and external tanks. Optional in-flight refuelling capability (probe and drogue system) is standard in Brazil.

ACCOMMODATION: Pilot only, on Martin-Baker Mk 10L zero/zero ejection seat; 18° downward view over nose. One-piece wraparound windscreen (reinforced on Brazilian aircraft); one-piece hinged canopy, opening sideways to starboard. Cockpit pressurised and air conditioned. Tandem two-seat combat trainer/special missions version also produced, with Mk 10LY-2 (front) and Mk 10LY-3 (rear) seats.

SYSTEMS: Microtecnica environmental control system (ECS) provides air conditioning of cockpit, avionics and reconnaissance pallets, cockpit pressurisation, air intake and inlet guide vane anti-icing, windscreen demisting and anti-g systems. Duplicated redundant hydraulic systems, driven by engine gearbox, operate at pressure of 207 bar (3,000 lb/sq in); both actuate primary flight control system (aileron, elevators and rudder) and secondary (flap and slat

system); No. 1 circuit additionally supplies outboard spoilers, nosewheel steering and gun; No. 2 also supplies inboard spoilers and landing gear actuation. Primary electrical system AC power (115/200 V at fixed frequency of 400 Hz) supplied by two 30 kVA IDG generators, with two transformer-rectifier units for conversion to 28 V DC; 36 Ah Ni/Cd battery for emergency use, to provide power for essential systems in the event of primary and secondary electrical system failure. Aeroeletrônica (Brazil) external power control unit. Fiat FA 150 Argo APU for engine starting. APU-driven electrical generator for ground operation. Liquid oxygen system.

AVIONICS: All avionics/equipment packages pallet-mounted and positioned for rapid access. Modular design and space provisions within aircraft permit retrofitting of alternative avionics.

Comms: UHF and VHF com, and IFF.

Radar: Pointer ranging radar in Italian AMXs is I-band set modified from Elta (Israel) EL/M-2001B and built in Italy by FIAR; AMX-MLU version will have Finmeccanica/Galileo Avionica SCP-01 Scipio radar, which is to be incorporated in second phase of upgrade project. Brazilian aircraft were originally intended to have Tecnasa/SMA-built (Alenia) SCP-01 Scipio radar, but this was not installed at time of production due to local bankruptcies of two Brazilian companies selected to produce the radar. It will now be fitted as part of the mid-life upgrade.

Flight: Litton Italia INS, with standby AHRS and Tacan, for Italian Air Force; VOR/ILS for Brazil. Data processing, with Microtecnica air data computer. BAE Systems MED 2067 video monitor display in rear cockpit of two-seater, for use by instructor/navigator as HUD monitor.

Instrumentation: Alenia computer-based weapon aiming and delivery, incorporating radar and Alenia stores management system; digital data displays (OMI/Alenia head-up, Alenia multifunction head-down, and weapons/nav selector). Provision for night vision goggles.

Mission: Italian aircraft of 3° Stormo equipped with Oude Delft Orpheus reconnaissance pods, and was an internal sensor suite being sought for deployment in 2001. This is believed to have been cancelled, but was to have comprised any one of three interchangeable Aeroeletrônica (Brazil) pallet-mounted photographic systems installed internally in forward fuselage, complementing external IR/EO pod on centreline pylon. Each system fully compatible with aircraft, and not affecting operational capability. Camera bay in lower starboard side of fuselage, forward of mainwheel bay.

ATR

AVIONS DE TRANSPORT REGIONAL INTEGRATED
1 allée Pierre Nadot, F-31712 Blagnac Cedex, France
Tel: (+33 5) 61 21 62 21
Fax: (+33 5) 61 21 63 18

Web: http://www.ataircraft.com
CEO: Jean-Michel Léonard
SENIOR VICE-PRESIDENTS:
 Paolo Revelli-Beaumont (Commercial)
 Luigi Lombardi (Operations)
 Roberto Bellino (Customer Services)
 Serge Queille (Finance)

COMMUNICATIONS MANAGER: Giancarlo Fre
MEDIA RELATIONS MANAGER: Frédéric Lahache
PARTICIPATING COMPANIES:
 Alenia Aeronautica: see under Italy
 EADS: see International section

For details of the latest updates to *Jane's All the World's Aircraft* online and to discover the additional information available exclusively to online subscribers please visit
jawa.janes.com

ATR 42-500 operated by Aerogaviota of Cuba
NEW/0527099

First Aerospatiale/Aeritalia (later Alenia) agreement July 1980; ATR programme started 4 November 1981; Groupement d'Intéret Economique (50:50 joint management company) formally established 5 February 1982 to develop ATR series of transport aircraft. Assembly or licensed production by Xian Aircraft in a new factory at Shenzhen, near Hong Kong, was discussed, but not pursued; Xian already produces components for ATR. Fuselage and wing production remain subcontracted to Alenia Aeronautica and EADS France respectively.

ATR marketing and support office opened in Washington 15 July 1986; ATR airline support centre in Singapore opened 18 November 1988; ATR Training Centre opened 1 July 1989; these functions taken over by AI(R) from 1996, but returned when this joint venture with BAe was dissolved on 1 July 1998 (see 1998-99 edition for final details of AI(R)). Phased production lines introduced in 1999 have reduced delivery times to three months from one year. The 600th ATR was delivered to Air Dolomiti on 28 April 2000.

The ATR family logged its 10 millionth flight on 10 October 2000. Sales by 1 January 2003 totalled 370 ATR 42s and 306 ATR 72s, or 676 in all, of which 369 ATR 42s and 283 ATR 72s had been delivered, representing 2002 deliveries of five and 14 respectively. The company is planning to reach proposed ICAO Stage 4 noise levels which become effective on 1 January 2006. In 2002, ATR developed a modified internal door designed to resist unauthorised flight deck incursions, in anticipation of the FAA's 9 April 2003 deadline for all airliners to be retrofitted to this standard. By mid-2002, ATR had received 100 firm orders for the modified door from American Eagle (first customer), Finnair, Jet Airways and Trans States Airlines. The new door is resistant to forcible impacts of up to 300J and penetration of 9 mm and 44 Magnum ammunition.

ATR's Asset Management arm is responsible for second-hand sales.

ATR expected to have become a single corporate entity by early 1999 but this did not materialise until 1 June 2001 with the formal establishment of ATR Integrated combining the activities of the two partners, still with GIE status and around 500 employees. Holding is 50 per cent each. Workforce had increased to 590 by mid-2002.

UPDATED

ATR 42

TYPE: Twin-turboprop airliner.
PROGRAMME: Joint launch by Aerospatiale (now in EADS) and Aeritalia (now Alenia) in November 1981, following June 1981 selection of P&WC PW120 turboprop as basic power plant; first flights of two prototypes 16 August 1984 (F-WEGA) and 31 October 1984 (F-WEGB); first flight production aircraft 30 April 1985; simultaneous certification to JAR 25 by France and Italy 24 September 1985, followed by USA (FAR Pt 25) 25 October 1985, Germany 12 February 1988, UK 31 October 1989; deliveries began 3 December 1985 to Air Littoral.

Series 500, with 'new look' interior, announced at Paris Air Show 14 June 1993; first flight F-WWEZ (c/n 443) 16 September 1994; French and UK certification 28 July 1995; first delivery (F-OHFF) to Air Dolomiti 31 October 1995. FAA certification 13 May 1996.
CURRENT VERSIONS: **ATR 42-300:** Initial version; phased out of production in 1996. Two Pratt & Whitney Canada PW120 turboprops, each flat rated at 1,342 kW (1,800 shp) for normal operation and 1,492 kW (2,000 shp) OEI; Hamilton Sundstrand 14SF four-blade constant-speed fully feathering and reversible-pitch propellers. Additional data in 1996-97 *Jane's*.
ATR 42-320: Identical to 42-300 except for optional PW121 engines for improved hot/high performance; OWE increased/payload decreased by 5 kg (11 lb). Phased out in 1996.
ATR 42-400: P&WC PW121A engines with six-blade Hamilton Sundstrand 568F propellers; maximum cruising

speed 266 kt (493 km/h; 306 mph); maximum range 825 n miles (1,527 km; 949 miles) with full payload. First flight 12 July 1995 (F-WWEF/OK-AFE); two Srs 420s ordered for CSA, and both delivered 14 March 1996, having received DGAC certification on 27 February. No further civil aircraft.
ATR 42-500: Principal ATR 42 version from 1996. Compared with Series 300, has more powerful engines, reinforced wings to allow greatly increased cruising speed and higher weights; all systems improvements of ATR 72, including flight management computers; cockpit, elevators and fin from ATR 72-210; strengthened landing gear; electrically operated main doors; reinforced fuselage and wing centre-section.
Description applies to ATR 42-500, except where indicated.
ATR 42 Tube (formerly Cargo QC): Quick-change (1 hour) interior to hold nine containers. Available as new-build or retrofit. Conversion programme includes installation of 2.95 × 1.80 m (9 ft 8 in × 5 ft 11 in) cargo door and modification of cabin into an E-class cargo compartment with strengthening of floor to 400 kg/m² (82 lb/sq ft); total volume for cargo transport 56 m³ (1,978 cu ft); maximum payload 5,883 kg (12,970 lb); parallel cabin section 10.25 m (33 ft 7½ in) long, tapered section 4.47 m (14 ft 8 in) long, with positions for up to four spider nets. MTOW in this configuration 16,900 kg (37,258 lb), MLW 16,400 kg (36,156 lb), MZFW 15,540 kg (34,260 lb). Launch customer, DHL Aviation of South Africa, received first of two converted -300s in October 2000.
ATR 42 Large Cargo Door: As ATR 42 Tube, but with 2.79 × 2.95 m (9 ft 2 in × 9 ft 8 in) upward-opening cargo door in port front fuselage aft of cockpit to permit loading of five LD3 containers. Space for 5.8 m³ (205 cu ft) of bulk cargo in rear, tapered section. Weights as ATR 42 Tube. Conversions by Aeronavali in Italy; first retrofit (for Farnair Europe) was due late 2001.
ATR 42 F: Military/paramilitary freighter with modified interior, reinforced cabin floor, port-side cargo/airdrop door can be opened in flight; can carry 3,800 kg (8,377 lb) of cargo or 42 passengers over 1,250 n miles (2,315 km; 1,438 miles). One delivered to Gabon 1989.
ATR Calibration: Projected navaid calibration version.
ATR 42L: Projected freighter with lateral cargo door; available as ATR 42 Large Cargo Door (see above).
ATR 42 Surveyor: Maritime and rescue version; *described separately.*
CUSTOMERS: Total 370 firm orders, of which 369 delivered by 1 January 2003. Four ordered and 10 delivered in 1998; 14 ordered and 12 delivered in 1999; six (plus four options) sold in 2000; five (all -500s) delivered in 2001 and five in 2002.
COSTS: ATR 42-500 development cost US$50 million; unit price US$13.8 million (2000).
DESIGN FEATURES: Designed to JAR 25/FAR Pt 25; high wing of medium aspect ratio, with constant-chord centre section and tapered outer panels; T-type tail with tapered tailplane and sweptback fin and fillet; pannier-mounted main landing gear.

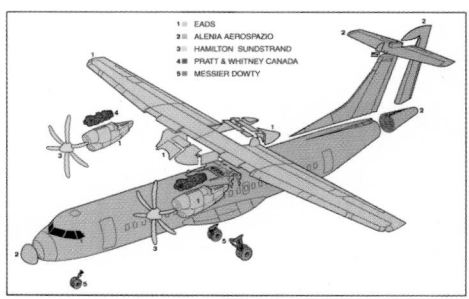

Manufacturers of the ATR airframe
0099556

Wing section Aerospatiale RA-XXX-43 (NACA 43 series derivative); thickness/chord ratio 18 per cent at root, 13 per cent at tip; constant-chord, no-dihedral centre-section with 2° incidence at root; outer panels 3° 6' sweepback at quarter-chord and 2° 30' dihedral.
FLYING CONTROLS: Conventional and manual. Lateral control assisted by single spoiler surface ahead of each outer flap; each aileron has electrically actuated trim tab; fixed incidence tailplane; horn-balanced rudder and elevators, each with electrically actuated trim tab; two-segment double-slotted flaps on offset hinges with Ratier-Figeac hydraulic actuators.
STRUCTURE: Two-spar fail-safe wings, mainly of aluminium alloy, with leading-edges of Kevlar/Nomex sandwich; wing top skin panels aft of rear spar are of Kevlar/Nomex with carbon reinforcement; flaps and ailerons have aluminium ribs and spars, with skins of carbon fibre/Nomex and carbon/epoxy respectively; fuselage is fail-safe stressed skin, mainly of light alloy except for Kevlar/Nomex sandwich nosecone, tailcone, wing/body fairings, nosewheel doors and main landing gear fairings; fin (attached to rearmost fuselage frame) and tailplane carbon structure; CFRP/Nomex sandwich rudder and elevators; dorsal fin of Kevlar/Nomex and GFRP/Nomex sandwich; engine cowlings of CFRP/Nomex and Kevlar/Nomex sandwich, reinforced with CFRP in nose and underside; propeller blades have metal spars and GFRP/polyurethane skins.

EADS France originally responsible for design and construction of wings and engine nacelles, flight deck and cabin layout, installation of power plant, flying controls, electrical and de-icing systems, and final assembly and flight testing of civil passenger versions; wing manufacture and testing reallocated to EADS Sogerma at Bordeaux from late 2001. Alenia Aeronautica builds fuselage and tail unit, installs landing gear, hydraulic system, air conditioning and pressurisation systems. ATR 42/72 manufactured at St Nazaire and Nantes (France), Pomigliano d'Arco and Capodichino (Italy), and assembled in Toulouse.
LANDING GEAR: Hydraulically retractable tricycle type, of Messier-Dowty trailing-arm design, with twin wheels and oleo-pneumatic shock-absorber on each unit. Nose unit retracts forward, main units inward into fuselage and large underfuselage fairing. Goodrich wheels and multiple-disc mainwheel brakes and Hydro-Aire anti-skid units. Mainwheel tubeless tyres, size 32×8.8R16 (10/12 ply), pressure 8.69 bar (126 lb/sq in) or H34×10.0R16 (14 ply). Nosewheel tubeless tyres, size 450×190-5 (10 ply), pressure 4.34 bar (63 lb/sq in) or 450×190R5 (10 ply). Minimum ground turning radius 17.37 m (57 ft 0 in).
POWER PLANT: Two 1,790 kW (2,400 shp) Pratt & Whitney Canada PW127E turboprops; ATR 72-210 nacelles; six-blade Ratier-Figeac/Hamilton Sundstrand 568F propellers with new electronic control giving faster response and better synchrophasing (as for ATR 72-500 from 1996). Propeller brake on starboard engine to enable engine to be used as auxiliary power unit for internal air conditioning.

Fuel in two integral tanks in spar box, total capacity 5,736 litres (1,515 US gallons; 1,262 Imp gallons). Single pressure refuelling point in starboard wing leading-edge. Gravity refuelling points in wing upper surface. Oil capacity 40 litres (10.6 US gallons; 8.8 Imp gallons).
ACCOMMODATION: Crew of two on flight deck; folding seat for observer. Seating for 42 passengers at 84 cm (33 in) pitch; or 46, 48 (standard) or 50 passengers at 76 cm (30 in) pitch; compared with Series 300, ATR 42-500 has completely new interior with new ceiling and sidewalls, indirect lighting, more sound damping; call buttons and reading lights relocated; overhead bins lengthened to 2 m (6 ft 6¼ in) to accommodate skis, golf clubs and fishing equipment carried as hand baggage. Baggage volume increased by 40 per cent. Active noise control system, previously offered as an option, is no longer available. ATR 42-500s have structural acoustic treatment

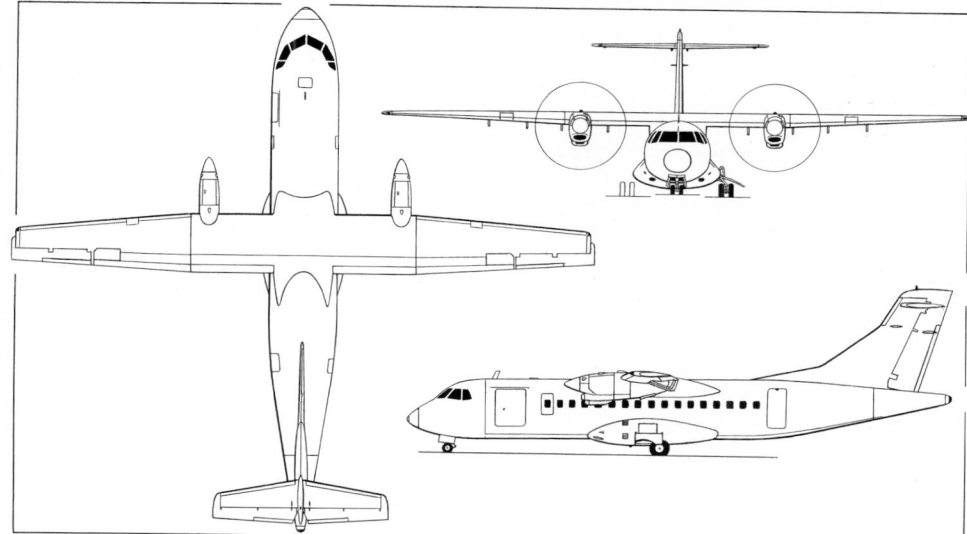

ATR 42 twin-turboprop regional transport *(Jane's/Dennis Punnett)*

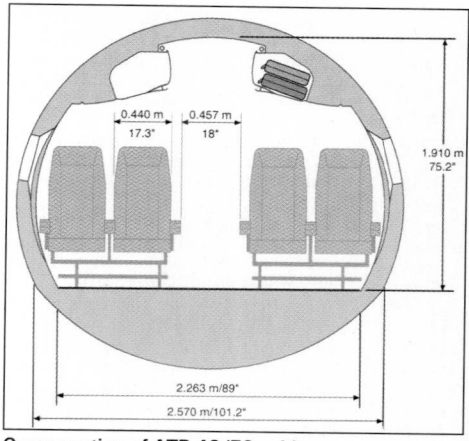

Cross-section of ATR-42/72 cabin   0099555

comprising reinforcement of seven fuselage frames adjacent propeller plane; dynamic vibration absorbers in this area; and internal aluminium skin damping material forward and aft of wing.

Passenger door, with integral steps, at rear of cabin on port side. Main baggage/cargo compartment between flight deck and passenger cabin, with access from inside cabin and separate loading door on port side; lavatory, galley, wardrobe and seat for cabin attendant at rear of passenger cabin, with service door on starboard side; rear baggage/cargo compartment aft of passenger cabin; additional baggage space provided by overhead bins and underseat stowage. Entire accommodation, including flight deck and baggage/cargo compartments, pressurised and air conditioned. Emergency exit via rear passenger and service doors, and by window exits on each side at front of cabin.

SYSTEMS: Improved in parallel with development of ATR 72; following refers to baseline aircraft. Honeywell air conditioning and Softair pressurisation systems, utilising engine bleed air. Pressurisation system (nominal differential 0.41 bar; 6.0 lb/sq in) provides cabin altitude of 2,040 m (7,000 ft) at flight altitudes of up to 7,620 m (25,000 ft).

Two independent hydraulic systems, each at pressure of 207 bar (3,000 lb/sq in), driven by electrically operated Abex pump and separated by interconnecting valve controlled from flight deck; system flow rate 7.9 litres (2.09 US gallons; 1.74 Imp gallons)/min; one system actuates wing flaps, spoilers, propeller brake, emergency wheel braking and nosewheel steering; second system for landing gear and normal braking. Kléber-Colombes pneumatic system for de-icing of wing leading-edges, tailplane leading-edges and engine air intakes; noses of aileron and elevator horns have full-time electric anti-icing.

Main electrical system is 28 V DC, supplied by two Auxilec 12 kW engine-driven starter/generators and two Ni/Cd batteries (43 Ah and 15 Ah), with two solid-state static inverters for 115/26 V single-phase AC supply; 115/200 V three-phase supply from two 20 kVA frequency-wild engine-driven alternators for anti-icing of windscreen, flight deck side windows, stall warning and airspeed indicator pitots, propeller blades and control surface horns. Eros/Puritan oxygen system. Instead of APU, starboard propeller braked and engine run to give DC and 400 Hz power, air conditioning and hydraulic pressure.

AVIONICS: Rockwell Collins com/nav equipment. Improved in parallel with development of ATR 72; following refers to baseline aircraft.
Comms: CVR, PA system.
Radar: Honeywell P-660 weather radar.
Flight: Honeywell DFZ 600 AFCS; AZ-800 ADCs; AH 600 AHRS with avionics standard communication bus;

Hamilton Sundstrand GPWS; L-3 digital FDR; Rockwell Collins DME; Honeywell FMZ-800 flight management system and dual GPS receivers installed in four Continental Airlines ATR 42s to allow autonomous approaches.
Instrumentation: EZ-820 electronic flight instrument system.

DIMENSIONS, EXTERNAL:
Wing span	24.57 m (80 ft 7½ in)
Wing chord: at root	2.57 m (8 ft 5¼ in)
at tip	1.41 m (4 ft 7½ in)
Wing aspect ratio	11.1
Length overall	22.67 m (74 ft 4½ in)
Fuselage max width	2.865 m (9 ft 4½ in)
Height overall	7.59 m (24 ft 10¾ in)
Elevator span	7.31 m (23 ft 11¾ in)
Wheel track (c/l of shock-struts)	4.10 m (13 ft 5½ in)
Wheelbase	8.78 m (28 ft 9¾ in)
Propeller diameter	3.94 m (12 ft 11 in)
Distance between propeller centres	8.10 m (26 ft 7 in)
Propeller fuselage clearance	0.835 m (2 ft 8¾ in)
Propeller ground clearance	1.10 m (3 ft 7¼ in)
Passenger door (rear, port): Height	1.73 m (5 ft 8 in)
Width	0.64 m (2 ft 1¼ in)
Height to sill (at OWE)	1.375 m (4 ft 6¼ in)
Service door (rear, stbd): Height	1.22 m (4 ft 0 in)
Width	0.61 m (2 ft 0 in)
Height to sill	1.375 m (4 ft 6¼ in)
Cargo/baggage door (fwd, port):	
Height	1.575 m (5 ft 2 in)
Width	1.295 m (4 ft 3in)
Height to sill (at OWE)	1.15 m (3 ft 9¼ in)
Emergency exits (fwd, each): Height	0.91 m (3 ft 0 in)
Width	0.51 m (1 ft 8 in)
Crew emergency hatch (flight deck roof):	
Length	0.51 m (1 ft 8 in)
Width	0.48 m (1 ft 7 in)

DIMENSIONS, INTERNAL:
Cabin: Length (excl flight deck, incl toilet and baggage compartments)	14.72 m (48 ft 3½ in)
Max width	2.57 m (8 ft 5¼ in)
Max width at floor	2.26 m (7 ft 5 in)
Max height	1.91 m (6 ft 3¼ in)
Floor area	31.0 m² (334 sq ft)
Volume	58.0 m³ (2,048 cu ft)
Baggage/cargo compartment volume:	
front (42-46 passengers)	6.0 m³ (212 cu ft)
front (48 passengers)	4.8 m³ (170 cu ft)
front (50 passengers)	3.6 m³ (127 cu ft)
rear	4.8 m³ (170 cu ft)

AREAS:
Wings, gross	54.50 m² (586.6 sq ft)
Ailerons (total)	3.12 m² (33.58 sq ft)
Flaps (total)	11.00 m² (118.40 sq ft)
Spoilers (total)	1.12 m² (12.06 sq ft)
Fin, excl dorsal fin	12.48 m² (134.33 sq ft)
Rudder, incl tab	4.00 m² (43.05 sq ft)
Tailplane	11.73 m² (126.26 sq ft)
Elevators (total, incl tabs)	3.92 m² (42.19 sq ft)

WEIGHTS AND LOADINGS:
Operating weight empty	11,250 kg (24,802 lb)
Max fuel weight	4,500 kg (9,921 lb)
Max payload	5,450 kg (12,015 lb)*
Max ramp weight	18,770 (41,380 lb)
Max T-O weight	18,600 kg (41,005 lb)
Max landing weight	18,300 kg (40,345 lb)
Max zero-fuel weight	16,700 kg (36,817 lb)*
Max wing loading	341.3 kg/m² (69.90 lb/sq ft)
Max power loading	5.20 kg/kW (8.54 lb/shp)

* Optional increase of 300 kg (661 lb)

PERFORMANCE:
Max cruising speed at FL170 at 97% MTOW	300 kt (556 km/h; 345 mph)
Time to climb to FL170	9.9 min
Service ceiling OEI, ISA +10°C, 97% MTOW	5,485 m (18,000 ft)
T-O distance: ISA, S/L	1,165 m (3,825 ft)

ISA +10°C at 915 m (3,000 ft) S/L for 300 n mile (556 km; 345 mile) stage with 48 passengers
1,163 m (3,815 ft)
Landing field length:
S/L at landing weight with max passengers
1,040 m (3,415 ft)
S/L at MLW   1,126 m (3,695 ft)
Runway ACN for flexible runway, category B   10
Range with max fuel
1,600 n miles (2,963 km; 1,841 miles)
Max range with 48 passengers
840 n miles (1,555 km; 966 miles)

OPERATIONAL NOISE LEVELS:
Flyover	76.6 EPNdB
Sideline	80.7 EPNdB
Approach	92.4 EPNdB

UPDATED

ATR 42 MP SURVEYOR

TYPE: Maritime surveillance twin-turboprop.

PROGRAMME: Variant of ATR 42 airliner developed by Alenia Aeronautica. Exhibited in model form at Dubai Air Show, November 1995. Initially designated **SAR 42** and **ATR 42MP**. First airframe modified by Officine Aeronavali's Capodichino plant; maiden flight (CMX62166) in Surveyor configuration, but without equipment installed, 1 February 1999, was also delivery to Alenia at Caselle for systems integration. Italian civil certification received 24 October 1999; first delivery 14 December 1999.

CURRENT VERSIONS: Other missions include exclusive economic zone protection, environmental protection, law enforcement, medical evacuation and VIP/troop/cargo/corporate/humanitarian transport.

ATR 42 MP Surveyor: Basic model, *as described.*

Detail of optional large cargo door for container loading   *NEW*/0526925

First of two ATR 42 Surveyors for the Italian Coast Guard *(Jane's/Paul Jackson)*   *NEW*/0526946

CUSTOMERS: Two ordered by Guardia di Finanza (Italian customs service) 1996 and first aircraft (MM62165) delivered that November but in normal transport configuration for training; converted to operational version in 2000; is model ATR 42-400 for first two aircraft; future aircraft based on 42-500.

Italian Guardia Costiera (Coast Guard) ordered one plus another option for unarmed version; first delivery (MM62170 '10-01') 30 May 2001; second aircraft due for delivery 2003.

Data as ATR 42, except particulars below.

POWER PLANT: Two 1,611 kW (2,160 shp) Pratt & Whitney Canada PW127E turboprops driving Ratier-Figeac/Hamilton Sundstrand 568F six-blade propellers.

ACCOMMODATION: Flight crew of two, plus three mission operators in modified cabin. Rest/debrief area towards rear of cabin; galley/lavatory facilities at rear. Observers' stations with bubble windows one each side of fuselage aft of wing. Rear door modified for in-flight opening; civil-style freight door in forward port side. SAR package behind rest area.

AVIONICS: *Comms:* VHF/UHF transceiver, AM-FM, VHF/FM/HF com transceiver, transponder/IFF, interphone and optional secure datalink.

Radar: Raytheon SV 2022 360° search radar; weather radar from ATR 42 retained.

Flight: ADS, VOR, DME, ADF, optional Tacan, FMS, IRS, INS/GPS, radio altimeter, direction-finder.

Mission: Galileo Avionica Airborne Tactical Observation and Surveillance system (ATOS) comprises mission management system; communications subsystem; Wescam turret beneath starboard landing gear pannier containing Galileo FLIR; daylight TV sensor; Spectrolab SX16E searchlight and Thiokol LUU-2B/B flare launcher all mounted in pod on starboard side of forward fuselage; Elettronica ALR-733 ESM. MIL-STD-1553B, RS-422 and ARINC 429 databus system. Two multifunction operator consoles with 48 cm (19 in) display, 25 cm (10 in) sensor display, keyboard, trackball, joystick and colour printer on starboard side; communications console. Provision for future growth of sensor suite.

EQUIPMENT: SAR equipment includes searchlight, loudspeakers and flare launcher. IR/UV scanner and optional SLAR and MVR for pollution detection.

ARMAMENT: Optional FN Herstal HPM twin machine gun pod on port side of forward fuselage.

WEIGHTS AND LOADINGS:
Typical operating weight empty	13,275 kg (29,266 lb)
Max payload	3,425 kg (7,551 lb)
Optional auxiliary fuel	650 kg (1,433 lb)
Max T-O weight	18,600 kg (41,005 lb)
Max landing weight	18,300 kg (40,345 lb)
Zero-fuel weight: typical	13,200 kg (29,101 lb)
max	16,700 kg (36,817 lb)

PERFORMANCE (at max T-O weight except where indicated):
Max cruising speed at FL160 at 97% MTOW
281 kt (520 km/h; 323 mph)
Typical patrol speed at FL20 at 90% MTOW:
for max range 189 kt (350 km/h; 217 mph)
for max endurance 135 kt (250 km/h; 155 mph)
Service ceiling, OEI, at 97% MTOW 3,960 m (13,000 ft)
T-O distance at S/L, ISA 1,050 m (3,445 ft)
Landing distance at MLW, S/L, ISA 1,150 m (3,775 ft)
Ferry range 2,019 n miles (3,740 km; 2,323 miles)
Endurance on station: at 200 n mile (370 km; 230 mile)
radius 8 h
at 600 n mile (1,111 km; 690 mile) radius 3 h 30 min
UPDATED

ATR 72

TYPE: Twin-turboprop airliner.

PROGRAMME: Stretched version of ATR 42; announced at 1985 Paris Air Show; launched 15 January 1986; three development aircraft built: first flights 27 October 1988 (F-WWEY), 20 December 1988 (F-WWEZ, c/n 108) and 18 April 1989 (OH-KRA, c/n 126); French and US certification 25 September and 15 November 1989 respectively; deliveries, to Kar Air of Finland, began 27 October 1989 (OH-KRB); UK certification 30 July 1993.

CURRENT VERSIONS: **ATR 72-200:** Initial production version; two Pratt & Whitney Canada PW124B turboprops, each rated at 1,611 kW (2,160 shp) for normal take-off and 1,790 kW (2,400 shp) with ATPCS; Hamilton Sundstrand 14SF-11 four-blade propellers. Also cargo version, capable of carrying 13 small containers. Now discontinued.

ATR 72-210: Improved hot/high performance version with PW127 engines rated at 1,849 kW (2,480 shp) and Hamilton Sundstrand 247F propellers with composites blades on steel hubs; ATPCS power 2,059 kW (2,760 shp); carries 17 to 19 more passengers than standard ATR 72 in WAT-limited conditions; French and US certification 15 and 18 December 1992, German on 24 February 1993; first delivery December 1992.

ATR 72-500: Launched as ATR 72-210A. Improved hot/high performance version with PW127 engines, six-blade propellers and redesigned interior. First flight 19 January 1996; DGAC certification achieved 14 January 1997; first delivery (to American Eagle) 31 July 1997.

Description applies to ATR 72-500, except where indicated.

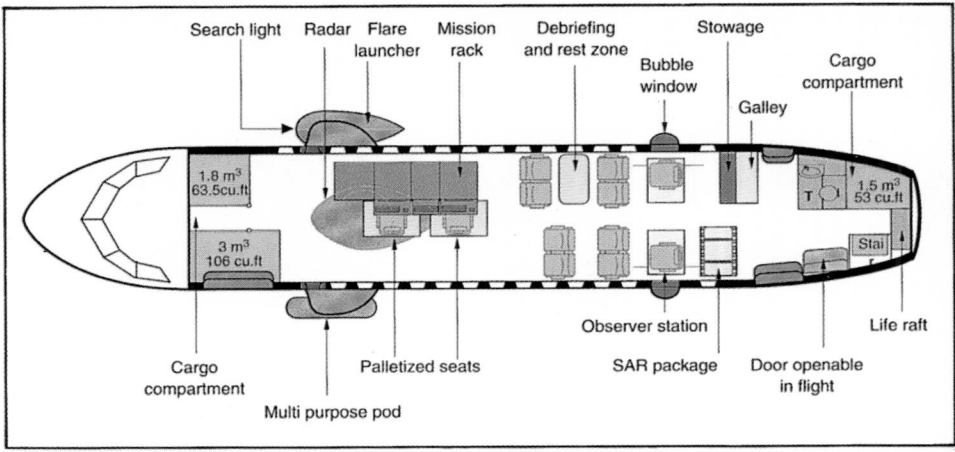

Search light · Radar · Flare launcher · Mission rack · Debriefing and rest zone · Bubble window · Stowage · Galley · Cargo compartment
1.8 m³ 63.5cu.ft · 3 m³ 106 cu.ft · 1.5 m² 53 cu.ft · Stai
Cargo compartment · Multi purpose pod · Palletized seats · Observer station · SAR package · Door openable in flight · Life raft

Typical internal and external features of the ATR 42 Surveyor 0085584

ATR 42 MP starboard landing gear pannier, showing emergency locator beacon ejection system and sensor turret (*Jane's/Paul Jackson*) NEW/0526945

ATR 72 Tube: Planned cargo version (formerly Cargo QC) similar to ATR 42 Tube; maximum payload will be 9,180 kg (20,238 lb) and total cargo volume 75.5 m³ (2,666 cu ft). Door size as ATR 42 Tube, six spider nets standard. Parallel section length 14.75 m (48 ft 4¾ in). MTOW in this configuration 22,500 kg (49,604 lb), MLW 22,350 kg (49,273 lb), MZFW 20,500 kg (45,195 lb).

ATR 72 Large Cargo Door: As ATR 72 Tube; 1.80 × 2.95 m (5 ft 11 in × 9 ft 8 in) upward-opening cargo door in port front fuselage behind cockpit to permit loading of seven LD3 containers or five ULD pallets. Space for 5.8 m³ (205 cu ft) cargo in tapered rear fuselage section. Weights as ATR 72 Tube. First retrofit (to c/n 108) completed for Farnair by Aeronavali (subsidiary of Alenia Aeronautica) in June 2002.

ATR 72 ASW: Projected anti-submarine warfare version, based on ATR 42 but also offering 1,270 kg (2,800 lb) payload including torpedoes, depth charges and anti-ship missiles. None yet built.

CUSTOMERS: Total 306 ATR 72s ordered up to 1 January 2003, at which time 283 delivered. Deliveries in 1998 totalled 21 and in 1999 numbered 23. Orders in 2000 totalled 18 plus six options; 15 (all -500s) delivered in 2001 and 14 in 2002.

DESIGN FEATURES: As ATR 42 (which see), but with more power, more fuel, greater wing span/area, and longer fuselage for up to 74 passengers.

FLYING CONTROLS: As for ATR 42 but vortex generators ahead of ailerons and aileron horn balances shielded by wingtip extensions; vortex generators under leading-edge of elevators.

STRUCTURE: Generally as for ATR 42, but new wings outboard of engine nacelles have CFRP front and rear spars, self-stiffening CFRP skin panels and light alloy ribs, resulting in weight saving of 120 kg (265 lb); sweepback

on outer panels 2° 18′ at quarter-chord. Trials of an all-composites tail assembly were conducted in 1997 and the structure incorporated in all production aircraft from 1998. The major airframe inspection period for the ATR 72 was increased on 2 October 1997 from 24,000 to 36,000 cycles, with a corresponding reduction in maintenance cost, bringing it in line with the ATR 42 family.

LANDING GEAR: Messier-Hispano-Bugatti units with Dunlop wheels (tyres size H34×10.0R16 (14 ply), pressure 7.86 bar; 114 lb/sq in) and structural carbon brakes; nosewheel tyre as ATR 42. Minimum ground turning radius 19.76 m (64 ft 10 in).

POWER PLANT: *ATR 72-500:* Two Pratt & Whitney Canada PW127F turboprops, each rated at 1,864 kW (2,500 shp) for normal flight and 2,051 kW (2,750 shp) for take-off, driving Ratier-Figeac/Hamilton Sundstrand 568F six-blade, all-composites propellers. Fuel capacity 6,337 litres (1,674 US gallons; 1,394 Imp gallons), comprising ATR 42 tanks, plus additional 637 litres (168 US gallons; 140 Imp gallons) in outer wings; pressure refuelling point in starboard main landing gear fairing.

ACCOMMODATION: Basic 68 passengers at 79 cm (31 in) seat pitch; other seating configurations range from 64 seats at 81 cm (32 in) to 72 seats at 76 cm (30 in); plus second cabin attendant's seat. Single baggage compartment at rear of cabin; two at front. Forward door, with a service door opposite on starboard side. Service door on each side at rear, that on port side replaced by a passenger door when cargo door is fitted at front. Two additional emergency exits (one each side); both rear doors also serve as emergency exits. Increased-capacity air conditioning system.

AVIONICS: More advanced avionics of ATR 72-500 have been transferred into ATR 42-500.

DIMENSIONS, EXTERNAL: As ATR 42 except:
Wing span	27.05 m (88 ft 9 in)
Wing chord at tip	1.59 m (5 ft 2½ in)
Wing aspect ratio	12.0
Length overall	27.17 m (89 ft 1¾ in)
Height overall	7.65 m (25 ft 1¼ in)
Wheelbase	10.77 m (35 ft 4 in)
Passenger door (fwd, port): Height	1.575 m (5 ft 2 in)
Width	1.295 m (4 ft 3 in)
Height to sill	1.12 m (3 ft 8 in)

DIMENSIONS, INTERNAL:
Cabin: Length (excl flight deck, incl toilet and baggage compartments) 19.21 m (63 ft 0¼ in)
Cross-section as for ATR 42
Floor area 41.7 m² (449 sq ft)
Volume 76.0 m³ (2,684 cu ft)

ATR72-500 in the livery of Denmark's Cimber Air NEW/0527100

Baggage/cargo compartment volume:
front (68 passengers with front cargo door)

	5.8 m³ (205 cu ft)
rear	4.8 m³ (170 cu ft)

AREAS: As ATR 42 except:

Wings, gross	61.0 m² (656.6 sq ft)
Ailerons (total)	3.75 m² (40.36 sq ft)
Flaps (total)	12.28 m² (132.18 sq ft)
Spoilers (total)	1.34 m² (14.42 sq ft)

WEIGHTS AND LOADINGS:

Operating weight empty	12,950 kg (28,550 lb)
Max fuel weight	5,000 kg (11,023 lb)
Max payload: standard	7,050 kg (15,543 lb)
optional	7,350 kg (16,204 lb)
Max T-O weight: standard	22,000 kg (48,501 lb)
optional	22,500 kg (49,604 lb)
Max ramp weight: standard	22,170 kg (48,876 lb)
optional	22,670 kg (49,979 lb)
Max landing weight: standard	21,850 kg (48,171 lb)
optional	22,350 kg (49,273 lb)
Max zero-fuel weight: standard	20,000 kg (44,092 lb)
optional	20,500 kg (45,195 lb)
Max wing loading: standard	360.7 kg/m² (73.87 lb/sq ft)
optional	368.9 kg/m² (75.55 lb/sq ft)
Max power loading: standard	5.36 kg/kW (8.81 lb/shp)
optional	5.49 kg/kW (9.01 lb/shp)

PERFORMANCE:
Max cruising speed at FL160, at 97% MTOW:

standard	276 kt (511 km/h; 318 mph)
optional	275 kt (509 km/h; 316 mph)

Econ cruising speed at FL230, at 95% MTOW

	248 kt (459 km/h; 285 mph)
Service ceiling	7,620 m (25,000 ft)

Service ceiling, OEI, ISA + 10°C, 97% MTOW
4,330 m (14,200 ft)
T-O balanced field length:

at S/L, ISA: basic	1,223 m (4,015 ft)
optional	1,290 m (4,235 ft)

at FL30, ISA +10°C, at T-O weight, 68 passengers,
both   1,300 m (4,265 ft)

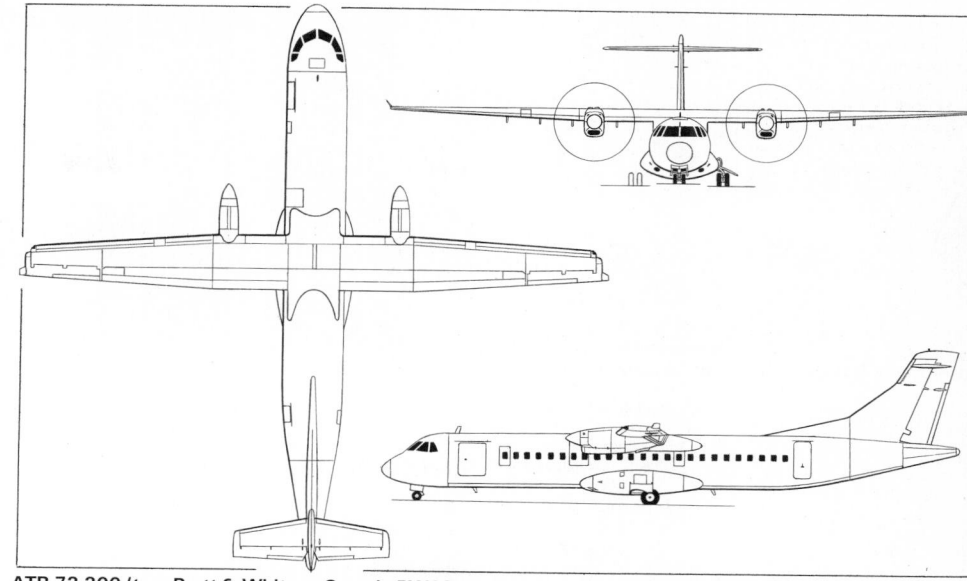

ATR 72-200 (two Pratt & Whitney Canada PW124B turboprops) *(Jane's/Dennis Punnett)*

Landing field length at S/L, at MLW:

basic	1,048 m (3,438 ft)
optional	1,067 m (3,500 ft)
Runway ACN for flexible runway, category B	13

Still air range, reserves for 87 n mile (161 km; 100 mile)
diversion and 45 min:

max payload: basic	715 n miles (1,324 km; 822 miles)
optional	890 n miles (1,648 km; 1,024 miles)

max fuel and zero payload
1,956 n miles (3,622 km; 2,251 miles)

OPERATIONAL NOISE LEVELS:

Flyover	79.7 EPNdB
Sideline	83.2 EPNdB
Approach	92.2 EPNdB

UPDATED

BELL/AGUSTA

BELL/AGUSTA AEROSPACE COMPANY (BAAC)

PO Box 901073, Fort Worth, Texas 76101, USA
Tel: (+1 817) 278 96 00
Fax: (+1 817) 278 97 26
Web: http://www.bellagustaaerospace.com
MANAGING DIRECTOR: Jim Rogers (Bell)
EXECUTIVE MANAGING DIRECTORS:
Don Barbour (BA609 Programme)
Antonio Giovannini (AB139 Programme)
PARTICIPATING COMPANIES:
Bell Helicopter Textron: see under USA
Agusta: see under Italy

Bell and Agusta announced on 8 September 1998 that they had agreed to establish a joint venture to manage development of two new aircraft: the BA609 tiltrotor, previously a Bell and Boeing programme, and the AB139, a new helicopter announced on the same day. Following approval of both boards, a definitive agreement was signed on 6 November 1998. Bell is the majority shareholder and will undertake final assembly for AB139s delivered to North America. Agusta, which has built Bell helicopters under licence since 1952, is investing and participating in development of the BA609, manufacturing some components and assembling those sold in Europe and certain other parts of the world. Additionally, Agusta is responsible for the AB139's development and certification, with participation by Bell. A military version was revealed in July 2000. Flight testing began in February 2001 and three had flown by October 2001.

UPDATED

BELL/AGUSTA AB139

TYPE: Light utility helicopter.
PROGRAMME: Announced at Farnborough Air Show, 8 September 1998, as joint venture between Agusta and Bell; to complement, rather than replace, Bell 412. Full-scale mockup unveiled at Paris Air Show 12 June 1999. Agusta responsible for development, certification to JAR/FAR 29 and transition to production, with participation by Bell on a 75:25 per cent work-share basis; final assembly by Agusta at Vergiate, and by Bell (possibly at Mirabel, Canada) for American and Pacific Rim customers. No designated "prototype"; first pre-production aircraft undertook maiden flight on 3 February 2001 followed by second aircraft on 4 June 2001 and third on 22 October 2001. Assembly of first production aircraft began in late November 2001. Three preproduction aircraft and one tie-down helicopter (TDH) undertaking flight test programme leading to certification to JAR/FAR Pt 29 and first deliveries in late 2002, but first pre-production aircraft lost in crash on 22 April 2002. Full-scale mockup of AB139 Military unveiled at Farnborough International 2000 in July 2000; development will follow certification of civilian version.

First and second Bell/Agusta AB139 preproduction airframes
0110891

Risk-sharing collaborators include GKN Westland (tail rotor drive train), Honeywell (avionics), Kawasaki (transmission input module), Liebherr Germany (landing gear and air conditioning system), Pratt & Whitney Canada (power plant) and PZL Świdnik (airframe components).
CURRENT VERSIONS: **AB139:** Commercial version, *as described.*
AB139 Military: Multirole military helicopter with provision for armoured crew seats, electronic warfare protection, IR suppressors, two internal pintle-mounted machine guns and easily removable stub-wing weapons supports for gun pods, rocket launchers and AAMs.
CUSTOMERS: Launch customer Bristow Helicopters of UK announced order for two on 26 September 2000; for delivery in 2002. Hawker Pacific ordered four on 12 February 2001 for corporate, utility and offshore operations in the Arabian Gulf. Anticipated market for 900 over 20 years, some 55 per cent for military use; 34 per cent of sales projected in Europe, 23 per cent in Middle East, 18 per cent in Far East, 13 per cent in South America and 12 per cent in North America.
COSTS: Commercial version US$6.985 million (2001).
DESIGN FEATURES: Design goals include high manoeuvrability and agility, low pilot workload, night/all-weather operation, low acoustic and infra-red emissions and mission flexibility for commercial and military operators. Intended for offshore support, medevac, corporate/VIP transport, SAR and military operations. Able to operate at

maximum T-O weight from Class A helipads at 945 m (3,100 ft) at ISA + 20°C. Five-blade, fully articulated, ballistic tolerant main rotor and four-blade tail rotor. Some transmission and rotor elements based on Agusta A129 Mangusta.
FLYING CONTROLS: Four-axis, digital AFCS.
LANDING GEAR: Heavy-duty, retractable tricycle type with twin wheels on nose unit; single wheels on main units, which retract into side sponsons.
POWER PLANT: Two Pratt & Whitney Canada PT6C-67C turboshafts, each rated at 1,252 kW (1,679 shp) for T-O and 1,142 kW (1,531 shp) maximum continuous; OEI ratings 1,289 kW (1,729 shp) for two minutes and 1,252 kW (1,679 shp) maximum continuous. Fuel tanks behind main cabin. Main transmission can run for up to 30 minutes without oil.
ACCOMMODATION: Up to 15 passengers on crashworthy seats in three rows of five, two forward facing, one rearward facing, in unobstructed cabin with flat floor; flight-accessible baggage compartment at rear of cabin. Alternatively, six stretchers and four attendants in medevac configuration. Plug-type sliding door on each side of cabin, with separate crew doors.
SYSTEMS: Systems duplicated and separated. Main and tail rotor ice protection optional. Smiths Aerospace HUMS; Eaton Corporation Smart Zapper chip detection system for main transmission, intermediate and tail rotor gearboxes.

Mockup of Bell/Agusta AB139 Military   0098868

Instrument panel of the AB 139 Military   0093644

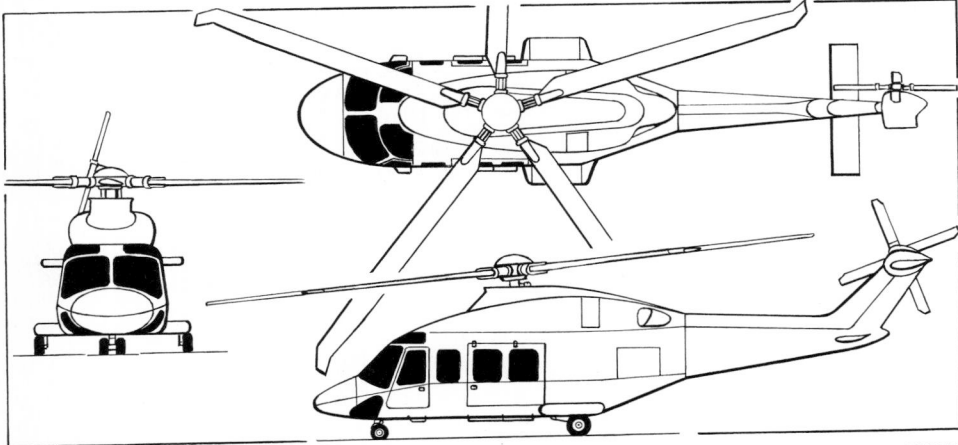

General arrangement of the Bell/Agusta AB139 (*Jane's/James Goulding*)   0079280

AVIONICS: Honeywell Primus Epic as core system. Provision for up to four 203 × 254 mm (8 × 10 in) high-definition colour active matrix liquid crystal displays for MFD, PFD and FLIR/video functions, and four-axis modular digital autopilot with flight director for hands-off operation and SAR modes.

DIMENSIONS, EXTERNAL:

Main rotor diameter	13.80 m (45 ft 3¼ in)
Tail rotor diameter	2.65 m (8 ft 8¼ in)
Length overall, rotors turning	16.65 m (54 ft 7½ in)
Fuselage: Length	13.53 m (44 ft 4¾ in)
Max width: across cabin	2.26 m (7 ft 5 in)
across sponsons	3.20 m (10 ft 6 in)
Height, rotors turning	4.95 m (16 ft 3 in)
Tail rotor ground clearance	2.30 m (7 ft 6½ in)
Cabin doors (two, each): Height	1.35 m (4 ft 5 in)
Width	1.68 m (5 ft 6 in)

DIMENSIONS, INTERNAL:

Cabin: Length	2.70 m (8 ft 10¼ in)
Max width	2.00 m (6 ft 6¾ in)
Max height	1.42 m (4 ft 8 in)
Volume	8.0 m³ (283 cu ft)
Baggage compartment volume	3.4 m³ (120 cu ft)

AREAS:

Main rotor disc	149.57 m² (1,610.0 sq ft)
Tail rotor disc	5.52 m² (59.37 sq ft)

WEIGHTS AND LOADINGS:

Useful load	2,500 kg (5,512 lb)
Max external load	2,750 kg (6,063 lb)
Max T-O weight	6,000 kg (13,227 lb)
Max disc loading	40.11 kg/m² (8.22 lb/sq ft)

PERFORMANCE (estimated):

Never-exceed speed (VNE)	167 kt (309 km/h; 192 mph)
Max cruising speed	155 kt (287 km/h; 178 mph)
Max rate of climb at S/L	610 m (2,000 ft)/min
Service ceiling OEI	2,926 m (9,600 ft)
Hovering ceiling OGE	3,660 m (12,000 ft)
Max range, no reserves	400 n miles (740 km; 460 miles)
Endurance	3 h 54 min

UPDATED

BELL/AGUSTA BA609

TYPE: Light utility tiltrotor.

PROGRAMME: Having earlier joined in partnership to develop the military V-22 tiltrotor, Bell Boeing revealed in February 1996 that studies were in progress for a nine-passenger civil tiltrotor aircraft in the 6,350 kg (14,000 lb) weight class, with the preliminary designation D-600. Subsequently, on 18 November 1996, the two companies announced that a joint venture was being established to design, develop, certify and market a six- to nine-passenger civil tiltrotor as the Bell Boeing 609.

Boeing withdrew as a partner on 1 March 1998 and Bell subsequently teamed with Agusta to develop, produce and market the tiltrotor as the BA609; this arrangement was formally announced at the Farnborough Air Show in September 1998. Agusta is investing and participating in

BA609 development and will be responsible for assembly of BA609s sold in Europe and elsewhere.

Preliminary design review completed May 1997. Manufacture of parts for prototypes began at Philadelphia, August 1997. Full-size mockup first exhibited at Paris Air Show, June 1997.

First flight of the BA609 rescheduled for late 2002, following decision in April 2002 to slow development and certification as consequence of delays in Bell Boeing V-22 Osprey programme (which see), the latter being leader in technology development. However, engine running did not begin on prototype N609TR until 6 December 2002. Four prototypes are being produced for 18-month flight test

programme leading to certification under FAR Pt 25 (fixed-wing aircraft) and Pt 29 (helicopters), plus Pt 21.17(b) Special Conditions for unique components. To be capable of single-pilot IFR operation.

CURRENT VERSIONS: **BA609**: Initial version *as described*.

HV-609: Multimission version proposed by Bell and Lockheed Martin to satisfy US Coast Guard 'Deepwater' re-equipment programme as potential replacement for Dassault HU-25 Guardian, Eurocopter HH-65 Dolphin and Sikorsky HH-60J Jayhawk. Missions could include drug interdiction and SAR, with 30 to 50 HV-609s possibly being acquired.

UV-609: Utility version conceived by Bell for US Army combat role, including casualty evacuation, command and control, logistic support and light utility/troop transport tasks. Manufacturer also promoting UV-609 to US Marine Corps as potential training system for V-22 Osprey.

619: Projected 19-seat version aimed at commuter market.

620: Proposed 22-seat version conceived by Bell Boeing team; no recent news received and concept may have lapsed.

CUSTOMERS: Order book opened 2 February 1997 at Heli Expo, with first order placed soon after by unspecified customer. Total of 80 ordered by 42 customers in 18 countries by December 2001; purchasers identified to date include Helitech Pty of Australia; Lider of Brazil; Canadian Helicopter Corporation and Northern Mountain Helicopters Inc of Canada; Petroleum Tiltrotors International of Dubai; Aero-Dienst GmbH of Germany; Mitsui of Japan; United Industries of South Korea; Helikopter Services of Norway; Lloyd's Investments of Poland; Bristow of the UK; Massachusetts Mutual Life Insurance, Austin Jet and Petroleum Helicopters of the USA; and Air Center Helicopters Inc of the US Virgin Islands. Briefing given to US Coast Guard, late 1997, followed by demonstration by XV-15 tiltrotor concept

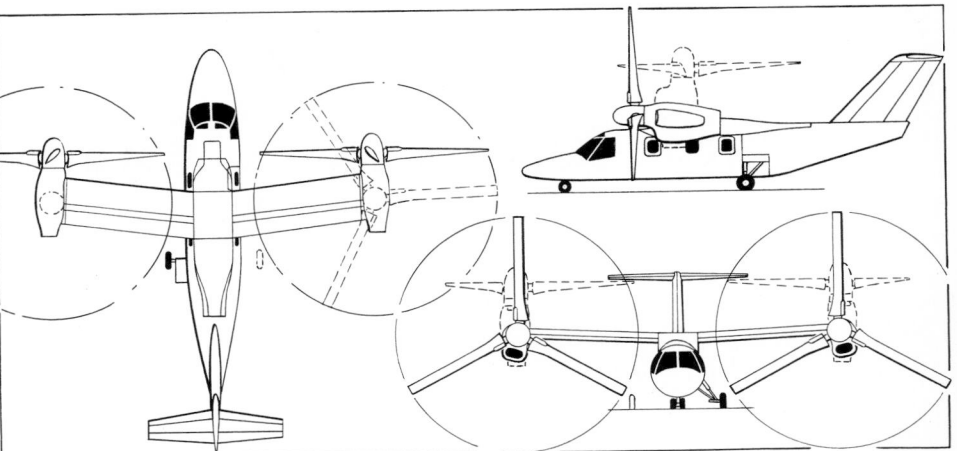

Bell/Agusta BA609 tiltrotor aircraft (*Jane's/James Goulding*)   0044896

Artist's impression of the Bell/Agusta BA609   0114828

Cabin interior of Bell/Agusta BA609 mockup 0079278

demonstrator aboard Coast Guard cutter *Mohawk* off Key West, Florida, in May 1999.

COSTS: US$8 million to US$10 million, depending on configuration.

DESIGN FEATURES: Combines the most favourable aspects of helicopter and aeroplane performance in passenger-carrying role. T tail configuration instead of endplate fin layout used by earlier tiltrotor designs (XV-15 and V-22 Osprey); this raises horizontal tailplane above rotor wake to minimise fore and aft pitching moment at transition phase of flight. Size of wing determined by requirement for it to hold all fuel for CG and safety considerations. Composites cross-shafts keep both proprotors turning in event of engine failure. Manual screwjack facility exists whereby the proprotors can be tilted into helicopter mode if cross-shafts fail. Designed using three-dimensional CATIA digital computer design system.

Refer also to Bell Boeing V-22 entry in US section for explanation of tiltrotor concept and for its extension to four-proprotor configuration.

FLYING CONTROLS: BAE Systems triplex digital fly-by-wire flight control system, with Dowty Aerospace actuators. T tail with conventional elevators; no rudder. Two-segment trailing-edge flaperons.

STRUCTURE: Aluminium fuselage structure with composites skinning; composites wing. Production fuselages, including cockpit, cabin and systems installation, will be built by risk-sharing partner Fuji Heavy Industries of Japan, which may also establish a third production line if

substantial orders are won from the Japanese government; cabin doors and fuselage tailcone supplied by Kawasaki; wing and nacelles by Bell at Fort Worth, with final assembly line at new Bell tiltrotor manufacturing facility in Amarillo, Texas. Fuselage in three major sections; nose, centre and tail, with fuselage skin incorporating graphite stringers with Japanese Toray composites material. Same used for wing and nacelles, with upper and lower wing surfaces produced as single pieces.

LANDING GEAR: Retractable tricycle type, with twin nosewheels and single wheel on each main unit. Messier-Dowty overseeing design, development and manufacture of integrated landing gear system, including legs, wheels, tyres, brakes, brake control and landing gear control systems.

POWER PLANT: Two 1,447 kW (1,940 shp) Pratt & Whitney Canada PT6C-67A turboshaft engines, installed in tilting nacelles at wingtips, each driving a three-blade proprotor. Nacelle interface units by AMETEK Aerospace Systems. Rockwell Collins EICAS; modified oil system, with dual pumps, to generate sufficient oil pressure when operating in vertical mode; several planetary gears also removed to achieve direct drive 30,000 rpm output. Fuel in integral wing tanks; usable capacity 1,401 litres (370 US gallons; 308 Imp gallons). Provision for auxiliary fuel tanks.

ACCOMMODATION: Crew of two, side by side on flight deck, with dual controls. Maximum of nine passengers in standard aircraft. Crew and passenger door on starboard side, forward of wing. Accommodation pressurised and air conditioned; pressurisation differential 0.38 bar (5.50 lb/sq in). Transparencies by Sully Produits Speciaux of France.

SYSTEMS: Equipped for flight into known icing. Lucas Aerospace DC electrical power systems. Intertechnique brushless electric pumps and motor-operated shut-off valves.

AVIONICS: Rockwell Collins Pro Line 21 package as standard.
Comms: Dual VHF radios; Mode S transponder. Cockpit voice recorder included as standard.
Radar: Optional Rockwell Collins WXR-800 solid-state weather radar.
Flight: Dual VOR/ILS, DME and ADF, with integrated control of sensors by dual Rockwell Collins RTU-4200 radio tuning units. Rockwell Collins ALT-4000 radar altimeter optional. GPS included as standard, along with TCAS and FDR.
Instrumentation: Three 250×200 mm $(10 \times 8$ in) active matrix colour LCD adaptive flight displays, including two primary flight displays and one multifunction display. Standby instrument system by Goodrich Aerospace.

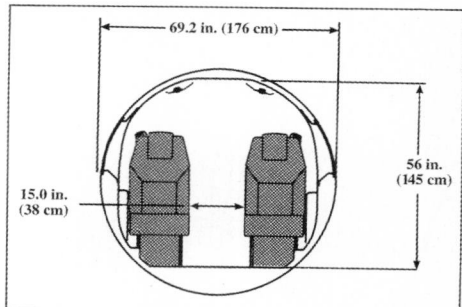

Cabin cross-section of BA609 tiltrotor 0079276

Following data are provisional.

DIMENSIONS, EXTERNAL:
Proprotor diameter	7.9 m (26 ft)
Width, rotors turning	18.3 m (60 ft)
Length overall	13.4 m (44 ft)
Max diameter of fuselage	1.75 m (5 ft 9 in)
Height to top of fin	4.6 m (15 ft)
Wheel track	3.0 m (10 ft)
Wheelbase	5.8 m (19 ft)
Distance between proprotor centres	10.0 m (33 ft)
Passenger door: Width	0.76 m (2 ft 6 in)

DIMENSIONS, INTERNAL:
Cabin: Length	5.33 m (17 ft 6 in)
Max width	1.52 m (5 ft 0 in)
Max height	1.42 m (4 ft 8 in)
Baggage hold, volume	1.4 m³ (50 cu ft)

AREAS:
Rotor discs, each	49 m² (530 sq ft)

WEIGHTS AND LOADINGS:
Weight empty	4,765 kg (10,500 lb)
Max T-O weight	7,265 kg (16,000 lb)
Max disc loading	147 kg/m² (30 lb/sq ft)

PERFORMANCE:
Max cruising speed	275 kt (509 km/h; 316 mph)
Max certified altitude	7,620 m (25,000 ft)

Range: normal fuel capacity with 2,500 kg (5,500 lb) payload at 250 kt (463 km/h; 288 mph)
750 n miles (1,389 km; 863 miles)
with auxiliary fuel tanks
1,000 n miles (1,852 km; 1,151 miles)
UPDATED

Prototype BA609 during initial ground running on 6 December 2002   *NEW*/0530176

BOEING/BAE SYSTEMS

BOEING/BAE SYSTEMS

PARTICIPATING COMPANIES:
Boeing (Military Aircraft and Missile Systems): see under USA
BAE Systems: see under UK

Original McDonnell Douglas Corporation (MDC) and Hawker Siddeley companies initially associated in 1969 through US procurement of BAe Harrier; relationship developed with US Navy selection of BAE Systems Hawk; both types built at St Louis, although new production of the Harrier ended in 1997, leaving only remanufacture and upgrade work on the AV-8B, as described in *Jane's Aircraft*

Upgrades. Joint work undertaken on advanced STOVL combat aircraft (later known as JSF), with additional participation of Northrop Grumman, until next-stage contracts were awarded to competing designs in November 1996. Boeing took over MDC in August 1997. Production of T-45 Goshawk continues, albeit at less than minimum economic rate.

UPDATED

BOEING/BAE SYSTEMS T-45 GOSHAWK

TYPE: Advanced jet trainer.
PROGRAMME: Carrier-capable version of BAE Systems Hawk selected 18 November 1981 (from five other candidates) as

winner of US Navy VTXTS (now T45TS) competition for undergraduate jet pilot trainer to replace T-2C Buckeye and TA-4J Skyhawk; original plan was for initial 54 'dry' (land-based) T-45Bs followed by 253 carrier-capable 'wet' T-45As; B model eliminated in FY84 in favour of 300 (and then 302) T-45As; FSD phase began October 1984; construction of two prototypes by Douglas began February 1986; funding approved 16 May 1986 for first three production lots (including 60 T-45As and 15 flight simulators during FY88-90); Lot 1 production contract (12 aircraft) awarded 26 January 1988; FSD prototypes made first flights 16 April (162787) and November 1988 (162788); original planned date for first deliveries (October 1989) delayed by further airframe and power plant changes requested by US Navy.

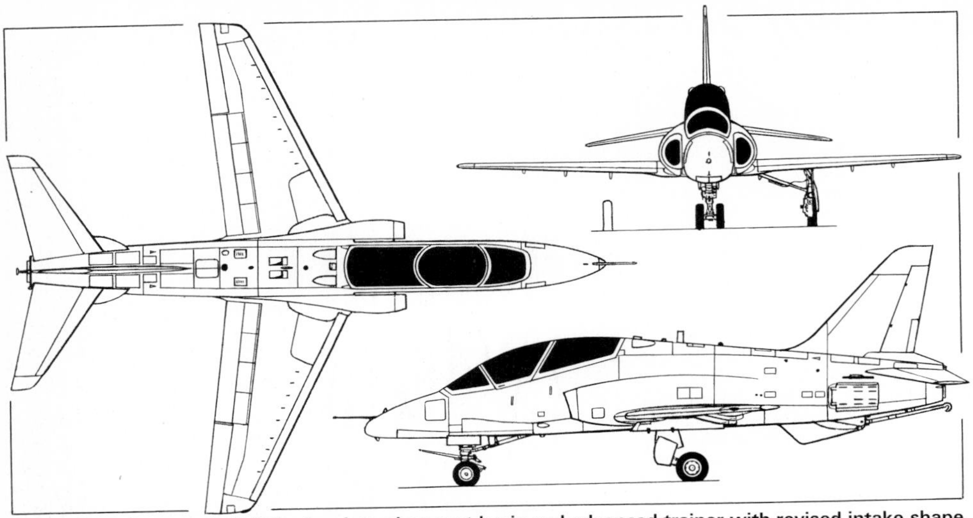

Boeing/BAE Systems T-45C Goshawk tandem-seat basic and advanced trainer with revised intake shape 0099603
(Jane's/Dennis Punnett)

Announced 19 December 1989 that entire T45TS programme to be transferred to McDonnell Aircraft Co at St Louis; modified FSD prototypes made first flights September and October 1990; two Douglas production aircraft (163599 and '600) delivered to NATC Patuxent River, Maryland, on 10 October and 15 November 1990; first carrier landing (162787 on USS *John F Kennedy*) 4 December 1991; first McAir production aircraft (163601) flew at St Louis 16 December 1991 and handed over to USN 23 January 1992. Production one per month in 1995 following successful passing of US DoD Milestone III review on 17 January 1995 authorising full-rate production. Final delivery due in 2005. The planned 187 aircraft will sustain the fleet at its designated level until 2020, but 40 more aircraft would be needed to reach an out-of-service date of 2035 at the end of the Goshawk's airframe life.

Digital/'glass cockpit' developed as 'Cockpit 21'; first to this standard (37th aircraft, 163635) made first flight 19 March 1994; planned production line introduction at 73rd aircraft, to be delivered October 1996, was delayed; flight trials of prototype aboard USS *John C Stennis* conducted April to May 1996 and successfully completed with 30 May approval for 'Cockpit 21' installation from 84th aircraft (165080), which first flew on 21 October 1997 and formally rolled-out on 31 October. Earlier aircraft to be retrofitted with 'Cockpit 21' from FY03, following a validation conversion of 163651 in 1998.

Beginning in 2000, all T-45s fitted with modified engine air intakes during routine servicing at NAS Kingsville; new intake has the top 'lip' extended forward, giving a raked profile, improving engine airflow at high angles of attack and reducing number of surges and compressor stalls.

CURRENT VERSIONS: **T-45A**: Baseline version, with analogue instruments.

T-45C: Digital ('Cockpit 21') avionics version with new computer, digital databus, data transfer cartridge and Litton GINA (GPS/INS assembly); first aircraft 165080; otherwise as T-45A; sole external difference is T-45C's GPS antenna on spine, immediately behind canopy.

CUSTOMERS: US Navy: two FSD prototypes and 187 production aircraft currently envisaged, of which 173 funded up to FY02. Final delivery of FY00 was 127th production aircraft; 139th was delivered to US Navy in October 2001. Long-term requirement is for 234 aircraft to sustain training up to 2030. Complete T45TS programme also involves 19 (originally 32) flight simulators (built by Hughes Training Inc); 48 (originally 49) computer-aided instructional devices, one training integration system mainframe, six electronic classrooms, 155 terminals, plus academic materials and contractor-operated logistic support. USN T45TS requirement stipulated 42 per cent size reduction in (TA-4 and T-2) training fleet, 25 per cent fewer flight hours, and 46 per cent fewer personnel.

Four training squadrons (VT) to equip: VT-21 and VT-22 of Training Wing 2 at Kingsville, Texas, in 1992-96; VT-9 and VT-23 (which redesignated VT-7 on 1 October 1999) of TW-1 at Meridian, Mississippi, 1997-2003. TW-2 assigned early aircraft; TW-1 receiving 90 'Cockpit 21' Goshawks (first official handover to VT-23 15 December 1997, aircraft having been received on 10 December). TW-2 will convert to the T-45C as its aircraft are retrofitted. Small numbers of T-45s at both Kingsville and Meridian wear 'Marines' titles to reflect the proportion (about 13 per cent) of USMC student pilots being trained on the type.

VT-21 operational 27 June 1992 for instructor training; four-aircraft operational evaluation begun by VT-21 on 18 October 1993 for one month first phase; student training begun 4 January 1994; first student flight 11 February 1994; first solo 23 March 1994; deployed to Miramar for deck landing course on aircraft carrier, September 1994; course graduated 5 October 1994. Second phase of operational evaluation (advanced tactics/weapons and carrier qualification) on USS *Dwight D Eisenhower* ended with clearance for fleet introduction being recommended on 5 July 1994. Primary sea platform is training carrier USS *John F Kennedy*. VT-23 (now VT-7) of TW-1 began student training on 8 July 1998; VT-9 equips in 2003. T-45 syllabus is 119 sorties (156 hours) plus 100 simulator sessions (95.4 hours).

In April 2000, 1,000th student graduated from T45TS course; fleet passed 200,000 hours in March 1999. By May 2001, T-45s had flown 342,000 hours, made 22,000 deck landings and trained 1,328 pilots; first aircraft exceeded 5,000 hours in April 2001. Fleet passed 450,000 hours and 28,500 carrier landings in February 2003, by which time 155 T-45s in service and 1,800 students graduated. Individual utilisation is 630 hours per year.

US NAVY PROCUREMENT

FY	Lot	Qty	First aircraft
88	1	12	163599
89	2	24	163611
92	3	12	163635
93	4	12	163647
94	5	12	165057
95	6	12	165069*
96	7	12	165081
97	8	12	165457
98	9	15	165469
99	10	15	165484
00	11	15	165499
01	12	14	165607
02	13	6	
03	14	8	
04	15	15	
05	16	8	
Total		**204**	

Notes: Procurement rate slowed by 1994 defence review; was to have involved 24 in FY95 and 30 per year in FY96-00. Figures for 2004 and 2005 not yet approved.
* 'Cockpit 21' installed from 12th of this batch (165080).

COSTS: Original US$316 million for 12 aircraft in FY96 procurement augmented by US$6,830,421 for addition of 'Cockpit 21' modifications. FY01 batch of 14 allocated US$306.5 million. Official unit cost US$17.2 million (1999).

DESIGN FEATURES: Carrier-capable adaptation of existing trainer design for cost and risk reduction. Generally as for two-seat BAE Systems Hawk Srs 60 (see UK section), but redesigned (including deeper and longer forward fuselage) and strengthened to accommodate new landing gear and withstand carrier operation (incidentally increasing fatigue life to 14,400 flying hours, although 28,800 hours achieved on static testing airframe); twin airbrakes of composites material; fin height increased by 15.2 cm (6 in) and single ventral fin added; rudder modified; tailplane span increased by 10.2 cm (4 in); wingtips squared off; nose tow launch bar added; underfuselage arrester hook, deployable 20° to each side of longitudinal axis. No provision for gun or outboard underwing hardpoints.

FLYING CONTROLS: Differences from two-seat BAE Systems Hawk include electrically actuated/hydraulically operated full-span wing leading-edge slats (operation limited to landing configuration); aileron/rudder interconnect; two fuselage-side airbrakes instead of one under fuselage and associated autotrim system for horizontal stabiliser when brakes deployed; BAE Systems yaw damper computer and addition of 'smurf' (side-mounted unit root fin), a small curved surface forward of each tailplane leading-edge root, to eliminate pitch-down during low-speed manoeuvres; Dowty actuators for slats and airbrakes.

STRUCTURE: BAE Systems (principal subcontractor) builds wings, centre and rear fuselage, fin, tailplane, windscreen, canopy and flying controls. Intended fatigue life of 14,000 hours. Batch of 54 composites tailplanes being built by Boeing for retrofit to extant T-45s.

LANDING GEAR: Wide-track hydraulically retractable tricycle type, stressed for vertical velocities of 7.28 m (23.9 ft)/s. Single wheel and long-stroke oleo (increased from 33 cm; 13 in of standard Hawk to 63.5 cm; 25 in) on each main unit; twin-wheel steerable nose unit with 40.6 cm (16 in) stroke. Articulated main gear, by AP Precision Hydraulics, is of levered suspension (trailing arm) type with a folding side-stay. Cleveland Pneumatic nose gear, with Sterer digital dual-gain steering system (high gain for carrier deck operations). Nose gear has catapult launch bar and holdback devices. Main units retract inward into wing, forward of front spar; nose unit retracts forward. All wheel doors sequenced to close after gear lowering; inboard mainwheel doors bulged to accommodate larger trailing arm and tyres. Gear emergency lowering by free-fall. Goodrich wheels, tyres and brakes. Mainwheel tyres size 24×7.7-10 (20 ply) tubeless; nosewheels have size 19×5.25-10 (12 ply) tubeless tyres. Tyre pressure (all units) 22.40 bar (325 lb/sq in) for carrier operation; reduced for land operation. Hydraulic multidisc mainwheel brakes with Dunlop adaptive anti-skid system.

POWER PLANT: One 26.00 kN (5,845 lb st) nominal rating Rolls-Royce Turbomeca F405-RR-401 (navalised Adour Mk 871) non-afterburning turbofan; installed rating 24.59 kN (5,527 lb st). (FSD aircraft powered by a 24.24 kN; 5,450 lb st F405-RR-400L, equivalent to Adour Mk 861-49.) Honeywell F124-GA-400 (28.02 kN, 6,300 lb st) assessed as alternative power plant under US Congressional directive; power plant change abandoned January 1994. Plans cancelled for assembly of F405 by Allison (due to have started with the 112th engine in October 1997). Air intakes and engine starting as described for BAE Systems Hawk. Fuel system similar to BAE Systems Hawk, but with revision for carrier operation. Total internal capacity of 1,635 litres (432 US gallons; 360 Imp gallons). Retrofit kits ordered in October 1993 to add 133 litres (35 US gallons; 29 Imp gallons) in engine air intake tanks, but these cancelled in 1994. Provision for carrying one 591 litre (156 US gallon; 130 Imp gallon) drop tank on each underwing pylon.

ACCOMMODATION: Similar to BAE Systems Hawk, except that ejection seats are of Martin-Baker Mk 14 NACES (Navy aircrew common ejection seat) zero/zero rocket-assisted type.

SYSTEMS: Air conditioning and pressurisation systems, using engine bleed air. Duplicated hydraulic systems, each 207 bar (3,000 lb/sq in), for actuation of control jacks, slats, flaps, airbrakes, landing gear, arrester hook and anti-skid wheel brakes. No. 1 system has flow rate of 36.4 litres (9.6 US gallons; 8.0 Imp gallons)/min, No. 2 system a rate of 22.7 litres (6.0 US gallons; 5.0 Imp gallons)/min. Reservoirs nitrogen pressurised at 2.75 to 5.50 bar (40 to 80 lb/sq in). Hydraulic accumulator for emergency operation of wheel brakes. Pop-up Dowty Aerospace ram air turbine in upper rear fuselage provides emergency hydraulic power for flying controls in event of engine or No. 2 pump failure. No pneumatic system. DC electrical power from single brushless generator, with two static inverters to provide AC power and two batteries for standby power. Onboard oxygen generating system (OBOGS).

AVIONICS: Avionics and cockpit displays optimised for carrier-compatible operations.

Comms: Rockwell Collins AN/ARN-182 UHF/VHF, Honeywell APX-100 IFF.

Flight: AN/ARN-144 VOR/ILS by Rockwell Collins, Honeywell AN/APN-194 radio altimeter, Sierra AN/ARN-136A Tacan, BAE Systems yaw damper computer. Digital avionics aircraft (No. 84 onwards) have revised navigation package comprising mission data input device (MDID); Litton LN-100G ring laser gyro and Rockwell Collins five-channel GPS linked by 12-state Kalman filter.

Instrumentation: US Navy AN/USN-2 standard attitude and heading reference system (SAHRS), Smiths Industries Mini-HUD (front cockpit), Racal Acoustics avionics management system, and Teledyne caution/warning system. Digital avionics from 84th production aircraft onwards: two 127 × 127 mm (5 × 5 in) Elbit monochrome multifunction screens in both cockpits, MIL-STD-1553B databus and Smiths HUD.

Mission: Electrodynamics airborne data recorder.

ARMAMENT: No built-in armament, but weapons delivery capability for advanced training is incorporated. Single pylon under each wing for carriage of practice multiple bomb rack, rocket pods or auxiliary fuel tank. Provision also for carrying single stores pod on fuselage centreline. CAI Industries weapon-aiming sight in rear cockpit.

DIMENSIONS, EXTERNAL:

Wing span	9.39 m (30 ft 9¾ in)
Wing chord: at root	2.87 m (9 ft 5 in)
at tip	0.89 m (2 ft 11 in)
Wing aspect ratio	5.0
Length, overall, incl nose probe	11.98 m (39 ft 4 in)
Height overall	4.11 m (13 ft 6 in)
Tailplane span	4.59 m (15 ft 0¾ in)
Wheel track (c/l of shock-struts)	3.90 m (12 ft 9½ in)
Wheelbase	4.31 m (14 ft 1¾ in)

AREAS:		PERFORMANCE:			
Wings, gross	17.66 m² (190.1 sq ft)	Normal T-O weight	5,783 kg (12,750 lb)	Approach speed (typical)	125 kt (232 km/h; 144 mph)
Ailerons (total)	1.05 m² (11.30 sq ft)	Max T-O weight	6,123 kg (13,500 lb)	Max rate of climb at S/L	2,440 m (8,000 ft)/min
Trailing-edge flaps (total)	2.50 m² (26.91 sq ft)	Max wing loading	346.7 kg/m² (71.02 lb/sq ft)	Time to 9,150 m (30,000 ft), clean	7 min 40 s
Airbrakes (total)	0.88 m² (9.47 sq ft)	Max installed power loading	249 kg/kN (2.44 lb/lb st)	Service ceiling	12,950 m (42,500 ft)
Fin	2.61 m² (28.10 sq ft)			T-O to 15 m (50 ft)	1,100 m (3,610 ft)
Rudder, incl tab	0.58 m² (6.24 sq ft)	Design limit diving speed at 1,000 m (3,280 ft)		Landing from 15 m (50 ft)	1,009 m (3,310 ft)
Tailplane	4.43 m² (47.64 sq ft)		575 kt (1,065 km/h; 662 mph)	Ferry range	700 n miles (1,296 km; 805 miles)
WEIGHTS AND LOADINGS:		Max true Mach No. in dive	1.04	g limits	+7.33/−3
Weight empty	4,261 kg (9,394 lb)	Max level speed at 2,440 m (8,000 ft)			*UPDATED*
Internal fuel: early production	1,312 kg (2,893 lb)		543 kt (1,006 km/h; 625 mph)		
enhanced capacity	1,433 kg (3,159 lb)	Max level Mach No. at 9,150 m (30,000 ft)	0.84		
		Carrier launch speed	121 kt (224 km/h; 139 mph)		

Boeing/BAE Systems Goshawk of Training Wing Two *(BAE Systems)*   NEW/0131756

EADS

EUROPEAN AERONAUTIC, DEFENCE AND SPACE COMPANY NV

Le Carré, Beechavenue 130-132, NL-1119 PR Schiphol-Rijk, Netherlands
Tel: (+31 20) 655 48 00
Fax: (+31 20) 655 48 01
Web: http://www.eads.net
PARTICIPATING COMPANIES:
Aerospatiale Matra: see EADS France in French section
CASA: see CASA in Spanish section
DaimlerChrysler Aerospace (DASA): see EADS Deutschland in German section
CHAIRMEN: Jean-Luc Lagardière
Manfred Bischoff
CHIEF EXECUTIVES: Phillipe Camus
Rainer Hertrich

CORPORATE COMMUNICATIONS:
D-81663 München
Tel: (+49 89) 60 70
e-mail: press@eads.net
SENIOR VICE-PRESIDENT, COMMUNICATIONS: Christian Poppe

Decision to merge Aerospatiale Matra of France and DASA of Germany was announced on 14 October 1999; previously mooted amalgamation of Spain's CASA with DASA was reconfirmed on 2 December 1999, increasing size of prospective company to 90 sites. Officially formed 10 July 2000, immediately becoming world's third-largest aerospace company. Asset distribution comprises 30.80 per cent stock flotation; 30.29 per cent directly, plus 2.75 per cent indirectly DaimlerChrysler; 5.53 per cent SEPI; 30.29 per cent SOGEADE (Lagardière and French government jointly); and 0.34 per cent directly by French government.

Five major operating divisions of EADS are Airbus (CEO Noël Forgeard), Aeronautics (Dietrich Russell), Military Transport Aircraft (Fransisco Fernández Sainz), Space (François Auque) and Systems and Defence Electronics (Thomas Enders), backed by Strategic Co-ordination (Jean-Louis Gergorin), Marketing (Jean-Paul Gut) and Finance (Axel Arendt) divisions. Aeronautics division comprises Eurocopter and ATR (both in International section of this edition); Socata (French section); EADS Military Aircraft (within EADS Deutschland); and EADS Sogerma and EADS EFW for overhaul and maintenance.

Company controls Socata and Eurocopter (100 per cent); also holds 80 per cent of Airbus and 56 per cent of Airbus Military Company; and has significant shares in Dassault (45.8 per cent), Eurofighter (43 per cent), ATR (50 per cent) and missile and space programmes. Aircraft produced by EADS are described in their appropriate national sections with the exception of the Mako, which is first to bear the EADS name.

Corporate headquarters established in Netherlands and EADS is subject to Dutch company law. Turnover in 2001 totalled €30,798 million; orders €60,208 million; backlog at 1 January 2002 €183,256 million; workforce 88,879, of which 40,322 (45.4 per cent) in France; 35,892 (40.4 per cent) in Germany; 7,430 (8.4 per cent) in Spain; 2,806 (3.1 per cent) in UK; and 2,429 (2.7 per cent) in rest of the world. Workforce by division in 2000 was 38 per cent Airbus, 26 per cent Aeronautics, 20 per cent Systems and Defence Electronics, 11 per cent Space and 4 per cent Military Transport Aircraft; remaining 1 per cent are HQ staff. Attempted to form joint venture with Finmeccanica of Italy under title of EMAC, as described later in this section. By 2012, EADS work distribution planned to be 60 per cent commercial, 30 per cent military and 10 per cent space.
UPDATED

EADS MAKO
English name: Shark
TYPE: Light fighter/advanced jet trainer.
PROGRAMME: Joint Aermacchi/DASA AT-2000 programme launched in April 1989, Aermacchi withdrew in 1994; DASA maintained the programme in hope of a joint development venture with South Korea. Following the latter's expressed preference in 1996 for a partnership with Lockheed Martin to satisfy its KTX-2 requirement, DASA began seeking an alternative collaborator in addition to Denel in South Africa, which was to provide avionics. Hyundai of South Korea expressed interest in the venture in early 1997, intending to build non-composites components of the centre and rear fuselage and tail. Design revealed in October 1996, after completion of wind tunnel tests and when radar cross-section trials in hand. Feasibility study completed by DASA in late 1997. Named

Artist's impression of Mako cockpit   0113984

Mako in 1998, by which time was foreseen as a 'modular design concept' with different versions for air-to-air, air-to-ground and training missions.

Predefinition phase began in early 1998. Full definition began by mid-1999, following a decision by the DASA board; full-scale mockup shown at 1999 Paris Air Show; cockpit demonstrator (conventional and VR simulator) at Paris 2001. Potential suppliers represented on demonstrator comprised BAE Systems North America (throttle and stick), Goodrich (front ejection seat), Martin-Baker (rear ejection seat) and CDC (displays and control

Model of Mako displayed in UAE markings at Dubai in November 2001 *(Jane's/Carn Wright)*   0121966

system). MoUs signed at Dubai in February 2001 and at Paris in June 2001 with SNECMA (M88-2 engines), BAE Systems (FCS and control systems), BGT/Diehl (computer and weapons systems and training aids), APPH Precision Hydraulics (landing gear and hydraulics) and Fairey Hydraulics (flight control actuation). Original target partners (South Africa and South Korea) have not joined the programme and there is no specific home-market (Luftwaffe) interest.

Launch of 18-month definition phase had been expected at Dubai Air Show in November 2001, but was delayed; however, MoUs signed at that time with Autoflug (fuel systems); Flight Visions (displays and controls); Rockwell Collins (navigation and communications equipment, displays and controls). MoUs are for down-selection only, and do not confer subcontractor status. Financing based on risk and revenue sharing with major suppliers, national partners and industrial partners. Joint programme discussed with UAE. In addition, Mako AT is candidate for AEJPT advanced European jet pilot training programme being considered by 12 air forces, including Luftwaffe. Separate interest in Mako LCA from Greece, UAE and Brazil. First of two prototypes to fly in 2005; production start in 2006; deliveries from 2009.

CURRENT VERSIONS: **Mako-AT:** Advanced trainer. Original baseline advanced two-seat trainer, available with or without radar and internal gun. Capable of in-flight weapons system simulation, with generation of synthetic threats and targets. Also suitable as companion trainer in squadrons operating advanced fighters.

Mako-LCA: Light combat aircraft. Multirole fighter with **F1** single-seat (air-to-air) and **F2** two-seat (air-to-ground) versions. Modular systems concept allows aircraft to be tailored to customer's requirements. EADS perceives a world market for 1,600 light combat aircraft between 2009 and 2030.

CUSTOMERS: Predicted 400 sales by 2030, of which 40 per cent would be trainer version. Entered (1999) in Brazilian F-X BR fighter replacement competition (to replace Mirage III and F-5E). MoUs signed with UAE November 1999 and March 2001, but these do not involve payment. Candidate for 'Eurotraining' (AEJPT) initiative. Initial wave of international orders anticipated in 2008-2012, with second period of high interest foreseen in 2018.

COSTS: Estimated development cost US$1,300 million; unit cost US$15 million to US$22 million.

DESIGN FEATURES: Modular design for ease of meeting different roles; simple structure permits low production cost. Conventional configuration described as 'transonic layout with supersonic performance'. Advanced trainer version is capable of combat missions, employing LO technology and optional thrust vectoring. LO features include chined forward section, wing/forebody blending and 'caret' engine air intakes.

Wing leading-edge sweep 45°.

FLYING CONTROLS: Reprogrammable quadruplex digital FBW FCS with 'carefree handling'. Large, single-section flaperons; all-moving tailplane (tailerons); inset rudder; and full-span wing leading-edge slats. Four door-type airbrakes adjacent to jetpipe.

STRUCTURE: Centre-fuselage of aluminium; wing-, taileron- and fin skins of carbon fibre.

LANDING GEAR: Retractable tricycle type; single wheel on each leg.

POWER PLANT: One Eurojet EJ200 turbofan; installed derated to 75.0 kN (16,860 lb st) with afterburning; optionally with thrust vectoring (provision for which is made in the FCS and airframe structure); or with full afterburning (approximately 90 kN; 20,250 lb st), according to customer's requirement. General Electric F414 and SNECMA M88-2 or -3 under consideration as alternative power plants. FCS incorporates provision for vectoring nozzle. Single-seater carries 3,300 kg (7,275 lb) of internal fuel; trainer has 3,000 kg (6,614 lb).

ACCOMMODATION: One or two pilots, according to version; layout and instrumentation generally as in Eurofighter Typhoon. Twin-seater has stepped cockpits, with 15° view over nose and 40° each side; seats inclined rearwards at 18°. Integral windscreen/front cockpit canopy slides forward; rear canopy slides aft.

SYSTEMS: OBOGS. APU. Optional OBIGGS.

AVIONICS: *Radar:* Optional 'AN/APG-67 class' multimode radar; other contenders include FIAR Grifo, Thales Avionics RD-400 and BAE Systems Bluehawk.

Instrumentation: Three MFDs in each cockpit, reflecting modern combat aircraft (such as Eurofighter) environment. Modular concept to allow easy upgrade or tailoring to individual customer requirements. Expected to include provision for FLIR, HMS and similar equipment.

Mission: Embedded in-flight simulation for combat training.

ARMAMENT: Optional Mauser BK 27 mm internal gun and seven low RCS external stores hardpoints: one at each wingtip, two under each wing and one below centreline. Centreline and inboard wing pylons stressed to 1,350 kg (2,976 lb); outboard to 675 kg (1,488 lb); wingtip for AAM only.

All data are provisional.

DIMENSIONS, EXTERNAL:

Wing span over missile rails	8.6 m (28 ft)
Wing aspect ratio	2.6
Length overall	13.75 m (45 ft)
Height overall	4.5 m (15 ft)
Wheel track	2.4 m (8 ft)
Wheelbase	4.8 m (16 ft)

AREAS:

Wings, gross	26.70 m² (287.4 sq ft)

WEIGHTS AND LOADINGS:

Weight empty: LCA: F1	6,020 kg (13,272 lb)
F2	6,370 kg (14,043 kg)
AT	5,800 kg (12,787 lb)
Max weapon load	4,500 kg (9,220 lb)
Max internal fuel weight: LCA F1	3,300 kg (7,275 lb)
LCA F2, AT	3,000 kg (6,614 lb)
Max T-O weight	13,000 kg (28,660 lb)
Max wing loading	486.9 kg/m² (99.72 lb/sq ft)
Max power loading (full afterburning)	149 kg/kN (1.46 lb/lb st)

PERFORMANCE (estimated):

Max level speed	M1.5
Service ceiling	15,240 m (50,000 ft)
T-O run	450 m (1,480 ft)
Landing run	750 m (2,460 ft)
Radius of action: interception, 54 min CAP plus dogfight, plus reserves	100 n miles (185 km; 115 miles)
ASM launch, hi-hi-hi	1,200 n miles (2,222 km; 1,381 miles)
Ferry range	2,000 n miles (3,704 km; 2,301 miles)
g limits	+9/−3

UPDATED

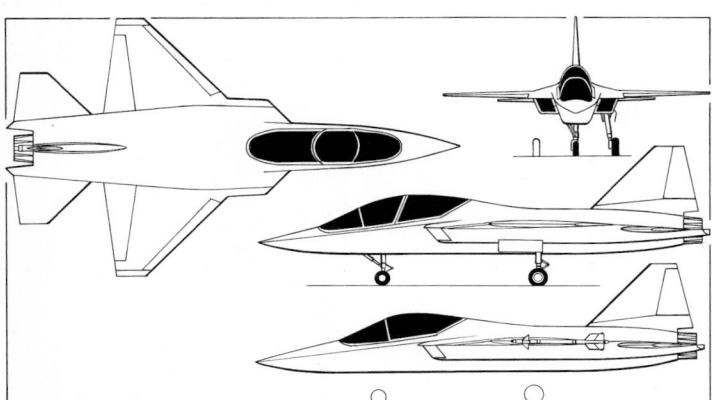

Provisional general arrangement of the Mako-AT advanced trainer, with additional side view (bottom) of single-seat Mako-LCA attack aircraft (*Jane's/James Goulding*)   NEW/0137725

Air-to-Air										
SRAAM	✕	✕	✕						✕	✕
MRAAM		✕	✕					✕	✕	
Air-to-Surface										
500 lbs Class	✕✕	✕✕	✕✕		✕✕	✕✕		✕✕	✕✕	
2,000 lbs Class		◉	◉					◉	◉	
A/S Missiles	✕	✕	✕		✕			✕	✕	
Dispensers		◉	◉					◉	◉	
Anti-Ship Missiles		✕						✕		
Fuel Tanks										
330 gal		○			○			○		
1,500 l		○						○		

Mako weapon options
0113985

EAeS

EUROPEAN AEROSYSTEMS LIMITED

W121, Warton Aerodrome, Preston, Lancashire PR4 1AX, UK
Tel: (+44 1772) 63 33 33
Fax: (+44 1772) 63 47 24
CHAIRMAN: Bruno Revellin-Falcoz (Dassault Aviation)
MANAGING DIRECTOR: Dr William E Martin (BAE Systems)

Following from a technology-sharing accord of December 1992, Dassault Aviation and British Aerospace Defence signed an agreement on 31 October 1995, providing for the formation of a joint venture military aircraft company. Intention to establish EAeS was announced on 4 September 1998, each participant to hold 50 per cent and contribute three members to directors' board. EAeS remains outside EADS.

Initial tasks of EAeS were to define and undertake research and development projects using integrated teams, formulating technologies and processes applicable both to existing products and to possible technology demonstrators for future fighter aircraft programmes. In May 2000, EAeS was awarded two technology contracts by the UK and French ministries of defence for development of weapons integration and vehicle system concepts.

Expectation of involvement in Royal Air Force's Future Offensive Air System diminished in 2001 when investigation begun into adaptations of later versions of Eurofighter Typhoon and Lockheed Martin F-35 to meet FOAS requirements.

UPDATED

EAeS could be involved in a future generation of combat aircraft

EHI

EH INDUSTRIES LIMITED (Subsidiary of Agusta Westland)

Pyramid House, Solartron Road, Farnborough, Hampshire GU14 7PL, UK
Tel: (+44 1252) 37 21 21
Fax: (+44 1252) 38 64 80

EH Industries formed June 1980 by Westland Helicopters and Agusta (50 per cent each), to undertake joint development of new anti-submarine warfare helicopter for Royal Navy and Italian Navy within an integrated programme under which naval, army and civil variants now being produced. Programme handled on behalf of both governments by UK Ministry of Defence; Westland has design leadership for commercial version, Agusta for rear-loading military/utility version; naval version developed jointly for UK and Italian navies and export. IBM Federal Systems (now Lockheed Martin Aerospace Systems Integration Corporation) selected to manage Royal Navy programme in 1991, overseeing and taking responsibility for RN-specific development activity, systems integration and aircraft production and delivery. Canadian programme launched October 1992 and terminated 12 months later, but EH101 reselected in December 1997 for purchase as new CH-149 Cormorant SAR helicopter.

Kawasaki of Japan, in conjunction with trading company Okura, signed an agreement in June 1995 for joint marketing and support of the EH101, this to include local manufacture, if warranted by orders. Further export contracts gained in 2001.

Partnership between AgustaWestland, Bell Helicopter Textron and Boeing Canada is promoting EH101 in North American market, for the Canadian maritime helicopter programme and for US requirements to replace HH-60G, MH-53E and VH-3D.

UPDATED

EH INDUSTRIES EH101

Royal Navy designation: Merlin HM. Mks 1 and 2
RAF designation: Merlin HC. Mk 3
Canadian Forces designation: CH-149 Cormorant
TYPE: Multirole medium helicopter.
PROGRAMME: Stems from Westland WG 34 (see UK section of 1979-80 *Jane's*), selected in mid-1978 by UK MoD to meet SR(S) 6646 for a Sea King replacement; broadly similar requirement by Italian Navy led to 1980 joint venture with Agusta; subsequent market research confirmed compatibility of basic EH 101 (later standardised as EH101, without space) design with commercial payload/range and tactical transport/logistics requirements, resulting in decision to develop naval, civil and military variants based on common airframe.

Nine month project definition phase approved by UK/Italian governments 12 June 1981; full programme go-ahead announced 25 January 1984; design and development contract signed 7 March 1984; selected by Canadian government August 1987; Italian-built iron bird ground test airframe followed by nine preproduction aircraft (PP1-9: see Current Versions); RTM 322 engines selected for RN Merlin June 1990; T700-GE-T6A engines selected for Italian Navy version September 1990; fourth UK/Italian government MoU signed 30 September 1991 (starting industrialisation phase); UK MoD commitment to 44 Merlins 9 October 1991; Canadian order for 50 (35 CH-148 and 15 CH-149) announced 24 July 1992; UK and Italian civil certification of Srs 300 and 500 planned for late 1993, but eventually achieved on 24 November 1994; US FAA approval gained on following day.

Flight testing halted following January 1993 loss of PP2; resumed 24 June 1993. US$1 million study completed for US Marine Corps, 1993, assessing EH101 as 30-troop or cargo transport as fallback in event of Bell/

Merlin HM. Mk 1 undertaking trials aboard HMS *Lancaster* in August 2000 (*Jane's/H M Steele*) 0073569

Boeing MV-22 cancellation. First flight with RTM 322 engines, 6 July 1993; two aircraft to fly total of 260 hours for development. Canadian requirement cut to 43 by deletion of seven CH-148s in August 1993 as cost-saving measure, but incoming government cancelled entire programme on 4 November 1993, despite award of substantial offsets to Canadian firms. UK government confirmed EH101 as next RAF tactical transport helicopter on 1 December 1993; formal announcement of order for 22 made 9 March 1995; Italian order confirmed October 1995 for 16, plus eight options. First production aircraft, RN01/ZH821 for Royal Navy, first flew 6 December 1995 and officially rolled out on 6 March 1996, although lacking mission avionics. Military Aircraft Release trials began in May 1998, using RN02/ZH822 (first flight 14 January 1997) and the first iteration of the Release was issued in the last quarter of 1998, allowing the start of Intensive Flight Trials with the commissioning of No. 700M Squadron (the Merlin IFTU) at RNAS Culdrose on 1 December 1998, and of RN aircrew training. First Merlin transferred to RN charge was RN05/ZH825 on 12 November 1998; first to civil customer, March 1999; first production EH101 for Italian Navy flew 6 December 1999; first squadron (824 NAS, Royal Navy) commissioned 2 June 2000 at RNAS Culdrose.

Total EH101 hours exceeded 15,000 by June 2001. Intensive flight operations programme (IFOP) by PP7 and PP9 concluded early, in June 2000, after 2,600 flights and 1,200 sorties (2,340 and 2,230 flying hours for each aircraft). During the previous year, the helicopter had demonstrated an MTOW of 15,500 kg (34,170 lb), or 900 kg (1,984 lb) over current limits.

On 31 October 2001, joint agreement announced by AgustaWestland and Lockheed Martin under which latter will promote – and, if necessary, build – EH101 in USA under new designation, US101. Targets include USAF's combat SAR requirement and other government agencies requiring EH101's range and payload combination.
CURRENT VERSIONS (general): **Srs 100/Naval:** Primary roles ASW, ASV, anti-ship surveillance/tracking, amphibious operations and SAR; other roles may include AEW, vertrep and ECM (deception, jamming and missile seduction); designed for fully autonomous all-weather operation from land bases, large and small vessels (including merchant ships) and oil rigs, and specifically from a 3,500 tonne frigate, with dimensions tailored to frigate hangar size. Capabilities include Type 23 frigate

launch and recovery in conditions up to Sea State 6 with ship on any heading and windspeed (from any direction) up to 50 kt (93 km/h; 57 mph); endurance and carrying capacity needed to meet expanding maritime tactical requirements of 21st century, including ability to operate distantly for up to 5 hours with state-of-the-art equipment and weapons. See under Italian Navy and Merlin headings below.
Srs 300 Heliliner: Commercial passenger version.
Utility (Srs 400) and civil **(Srs 500)** versions with rear ramp; naval **(Srs 200)** version without ramp.
US101: Versions optimised for US market, promoted by Lockheed Martin.
CURRENT VERSIONS (specific): **PP1:** Westland-built first preproduction aircraft; first flight (ZF641) 9 October 1987; completed official test centre review 1991; over 375 hours flown by September 1992, at which time being refitted with T700-GE-T6A engines and 3,878 kW (5,200 shp) main gearbox. To Italy, May 1995 for 21 hour trials programme, following which it was retired from flying after some 700 hours. Used for severe weather operating trials in February 1997, operating from Type 23 frigate HMS *Grafton*. Trials included 500 blade folding cycles in up to Sea State 6, and winds of up to 68 kt (126 km/h; 78 mph).
PP2: Agusta-built second preproduction aircraft; first flight (02) 26 November 1987; deck trials aboard Italian Navy *N Grecale* and *Maestrale* July 1990; was to lead work on achieving new maximum T-O weight of 14,288 kg (31,500 lb), and had been fitted with ACSR (see Design Features), but lost in crash in Italy, 21 January 1993, following a rotor brake fire during noise measurement trials.
PP3: Westland-built; first flight (G-EHIL) 30 September 1988; first civil-configured aircraft; rotor vibration trials and flights with 'soft link' engine mounting struts; icing trials at CFB Shearwater, Canada, for five months from November 1993 (total 23 sorties); continuing programme also includes optimisation of ACSR, and weapon and development trials. Serialled ZH647 for weapons carriage in 1993. Subsequently flew the high-altitude trials programme at up to 4,570 m (15,000 ft). Used for trials to extend MTOW to 15,500 kg (34,170 lb) in support of Cormorant programme. Withdrawn from use for preservation 18 February 1999
PP4: Westland-built; first flight (ZF644) 15 June 1989; general naval variant; overwater navigation equipment trials and AFCS development; flown with RTM 322 engines 6 July 1993, initiating two-year programme of ground running and 160 flight hours. This to have been followed by 50 hours training RN aircrew for operational trials including hot-and-high (military) at Mesa, Colorado, 1995 but aircraft lost after drive train control rod failure on 7 April 1995 after 463 hours in 385 sorties. Tasks assumed by RN01/ZH821.
PP5: Westland-built; dedicated Merlin development aircraft; first flight (ZF649) 24 October 1989; Type 23 frigate interface trials in HMS *Norfolk* (including decklock, recovery handling, weapons handling, tail and blade folding) August 1991; seagoing trials in HMS *Iron Duke* (including overwater sonobuoy release) completed December 1992. Second RTM 322 testbed. Full Merlin avionics by early 1996 for operational trials of datalink, sonar, digital map, colour displays and GPS; also ship-to-helicopter operations limits trials. Used for torpedo trials during mid-1997, making 15 drops of the Sting Ray in Cardigan Bay, operating from DTEO Aberporth. Royal Navy trials, Bahamas, 1999. Storage by 2000.
PP6: Agusta-built; dedicated Italian Navy development aircraft; first flight (I-RAIA) 26 April 1989; seagoing trials in *Giuseppe Garibaldi* and *Andrea Doria* completed mid-October 1991; rotor downwash trials on behalf of Canadian Forces to assess acceptability for SAR role. Trials of prototype (Mod 1) HELRAS dipping sonar, reserialled MMX605. In storage by mid-2000; 620 hours.

Merlin HC. Mk 3 showing open rear ramp 0113416

Canada's first CH-149 Cormorant during pre-delivery flight trials in Italy   NEW/0113428

PP7: Agusta-built; military utility development aircraft with rear-loading ramp and horizontal tail surfaces to port as well as starboard; first flight 18 December 1989; low-speed handling trials and tail rotor performance assessment 1992; fitted with ACSR 1992. Registered I-HIOI in November 1993. Severely damaged in attempted run-on landing after tail rotor pitch control failure on 20 August 1996; 450 hours of flight time. Flew again after repair 21 January 1999; re-serialled ZK101 and modified with representative in-flight refuelling probe and FLIR turrets of RAF HC. Mk 3; departed by sea on 12 June 1999 to undertake major demonstration tour along US East Coast, including evaluations by USN, USAF and Coast Guard.

PP8: Westland-built; civil variant; first flight (G-OIOI) 24 April 1990; evaluations of ADF and DME, area nav, electronic instrumentation, civil AFCS and communications equipment; fitted with ACSR. Allocated ZJ116 for trials, 1995. To Brindisi 27 March 1996 for intensive flying trials, with PP9; to Aberdeen, 14 September 1998. Attached to 700M Squadron for one week in May 1999, during which it flew demonstration flights to the Scilly Isles for BIH. Unrefuelled flight of 8 hours 15 minutes, March 2000, simulating 945 n mile (1,750 km; 1,087 mile) SAR mission with 5,500 kg (12,125 lb) of internal fuel. PP8 is fleet leader, with 3,500 flying hours. Withdrawn from use in late 2001.

PP9: Agusta-built; civil variant; final development aircraft (second with rear ramp); first flight (I-LIOI) 16 January 1991; extensive flight controls survey involving both ground and in-flight trials. Hot-and-high (civil) trials at Mesa, USA, 1995. To Brindisi May 1996, with PP8; to Aberdeen, 14 September 1998. Accommodated 55 people in a simulated SAR/evacuation mission, in support of the Cormorant programme. Had accumulated 2,800 flying hours by July 2000.

Civil Utility: For commercial operators requiring rear-loading facility; represented by PP9. First production **Mk 510** (I-AGWH) built for certification as a 'white tail'; rolled out in Italy 28 May 1997, and flew 17 June. Subsequently used for cold weather trials, operating from Fairbanks, Alaska, at temperatures from −5 to −32°C (23 to −26°F) until April 2000, and then for hot-and-high trials in USA. First order in November 1996 for a Mk 510 for Tokyo Metropolitan Police Agency; first flew September 1997; delivered from Italian production line in January 1998 as first to civil customer; fitting-out at Kawasaki Heavy Industries was delayed by bankruptcy of original financiers; eventually handed over at Gifu on 25 March 1999. Named *Ozora Ichigo* (Number One in the Big Sky) in service.

Heliliner: Commercial variant; main certification programme being flown by PP3, with PP8 as demonstrator; intended to offer 360 n mile (666 km; 414 mile) range, with full IFR reserves, carrying 30 passengers and baggage; flight crew of two, provision for cabin attendant, stand-up headroom, airline style seating, overhead baggage stowage, full environmental control, passenger entertainment, and provision for lavatory and galley. Category A VTO performance, capable of offshore/oil rig operations or scheduled flights into city centres at high all-up weights under more rigorous future civil operating rules; rear-loading ramp optional. An executive version is also proposed.

Italian Navy ASW/ASVW/Mk 110: Development aircraft PP4 (basic) and PP6; will operate from both shore bases and aircraft/helicopter carriers against surface and underwater targets with HELRAS dipping sonar and armed with two Marte Mk. 2/S ASMs. First of eight (MMX 84180 '2-01') flew on 6 December 1999; first two

delivered in December 2000 for service trials; four more handed over by end of 2001.

Italian Navy AEW/Mk 112: Requirement revealed in 1994 for AEW helicopter; system unspecified. Four ordered in 1995, with first due for delivery in 2003-04. Also known as **ASVW/E** (Anti-Submarine and Vessel Warfare, Enhanced). MM/APS-784-based Eliradar HEW-784 air and surface surveillance radar; radome almost doubled in size (from 1.80 m; 5 ft 10¾ in) to accommodate 3.00 m (9 ft 10 in) antenna, but other avionics similar to Italian Navy Srs 100s.

Italian Navy Utility/Mk 410: Based on rear-ramp Srs 400, but with Officine Galileo GaliFlir FLIR (also specified for other Italian versions), cargo hook, basic avionics and Elettronica ESM/ECM and self-defence suite similar to Italian ASW version. Weather radar. Four ordered in 1995. Will serve in the commando support role with Nucleo Elicotteristico per la Lotta Anfibia at Grottaglie, which due to take delivery in 2002-03. Italian

Interior of RAF Merlin HC. Mk 3   0099600

Winching exercise by a Royal Navy Merlin   NEW/0113420

air force also has a requirement for EH101s to replace two AS-61A-4 VIP helicopters.

Italian Navy Amphibious Support: Four ordered at beginning of 2002. Will have upgraded avionics for night missions (including NVG compatible cockpit) as well as weather radar, self-protection systems, personnel locator systems and armour and machine guns; plumbing for in-flight refuelling equipment will be installed, although it is not intended to fit refuelling probes at present.

Merlin HM. Mk 1: Royal Navy ASW version; EH101 **Mk 111**; PP4 (basic) and PP5 were development aircraft; in satisfaction of Staff Requirement (Sea) 6646; Loral ASIC is prime contractor for systems integration; will operate from Type 23 frigates, 'Invincible' class carriers, RFAs and other ships, and land bases. Initial production aircraft, RN01/ZH821, first flew 6 December 1995; second, RN02/ZH822, first with mission avionics, flew 14 January 1997. Of first seven production Merlins, RN01 and RN02 initially assigned to operational performance acceptance procedure (OPAP) trials at the fully instrumented Atlantic Underwater Test and Evaluation Center (AUTEC) in Bahamas, 1998-99, and to assistance in sea trials; RN03 to DTEO, Boscombe Down, in September 1997 for a six-phase military aircraft release trials programme; joined by RN02 for first, pre-IFTU stage of release, which was achieved in November 1998. The fully instrumented RN04 followed in late 1997; RN05-RN08 formed Intensive Flight Trials Unit (No. 700M Squadron) on 1 December 1998; PP5, RN02 and RN03 underwent some operational tests at AUTEC's range in the Bahamas from February 1999 (Operation Pearly King). Subsequent deep water trials were undertaken from Benbecula in 1999. In April 2000, RN12 from Boscombe Down and RN17 from Culdrose undertook extended Ship Helicopter Operational Limit trials aboard the RFA *Argus*, experiencing winds of up to 50 kt (93 km/h; 58 mph) and up to Sea State 10. Snow and icing trials accomplished by RN aircraft in Canada in first half of 2001.

RN02, RN03 and RN14 undertook ASuW trials from Benbecula from late June 2000, following a further 30-day series of ASW operational trials off Bahamas, during Operation Trial Wizard in March-April 2000. These trials included the dropping of eight Sting Ray torpedoes and 800 sonobuoys.

By the end of 1998 the Royal Navy had seven Merlins including two with No. 700M Squadron. No. 700M received its third aircraft on 21 April 1999. 700M evolved from IFTU to OEU on 1 September 2001. Training role passed to No. 824 Squadron at Culdrose on 2 June 2000, with No. 814 Squadron forming as the first front-line squadron on 5 October 2001 for deployment aboard HMS *Ark Royal*. Embarked operational capability was expected in May 2002 with four helicopters although complement will eventually increase to six. The aircraft will then equip a small-ships squadron (No. 829, with six Merlin flights) and a second carrier squadron, No. 820. The Royal Navy received its 25th Merlin in August 2000 and total of 35 had been delivered by September 2001. An upgrade to the sonar signal and data processing system with COTS hardware and VME architecture is planned for the HM. Mk 1 from 2001.

Merlin HM. Mk 2: Possible upgrade of Mk 1 to embrace Anti-Surface Warfare (ASuW) role with advanced, next-generation anti-ship missile (FASGW – Future Air-to-Surface Guided Weapon) to give autonomous ASuW capability; to meet SR (Sea Air) 903 under study by 1993, with improved sensors and processing equipment, plus uprated transmission for later versions of RTM 322. For retrofit from 2009; projected

Heliliner version of EH101 (*Jane's/Paul Jackson*)   NEW/0131743

follow-on purchase of up to 22 RN Merlins must now be considered unlikely.

Military Utility: Tactical or logistic transport variant (represented by PP7) with rear-loading ramp; able to airlift 6 tons or up to 30 combat-equipped troops. Tail- and rotor-folding Utility version under consideration, with role options including mine countermeasures, towing EDO Mk 106 sled. Version will probably form the basis of the EHI bid to meet the RN Future Amphibious Support Helicopter/Support Amphibious Battlefield Rotorcraft (FASH/SABR) requirement to replace Sea King HC. Mk 4 RAF Pumas, and SAR Sea King HAR. Mk. 3s.

Merlin HC. Mk 3: Bid for RAF contract entered May 1994; revised cockpit layout for low-level operations; provision for pintle-mounted machine guns in side doors. Order for 22 announced 9 March 1995, in satisfaction of Staff Requirement (Air) 440; manufacturer's designation: EH101 **Mk 411**. Each has provision for rapid installation of chin-mounted FLIR turret and (non-telescopic) refuelling probe beneath the nose, offset to starboard (though neither routinely fitted). Uprated engines. Other features include integrated defensive aids system (with Nemesis directional IR countermeasures, AN/AVR-2A(V) laser warning and Sky Guardian 200 RWR), NVG compatibility, crash-attenuating seats for all occupants (two pilots; loadmaster; optional fourth crewman; and 24 troops), active noise-reduction headsets for all passengers, non-folding main rotor blades and tailboom, improved navigation suite (compared with RN version), variable-speed cargo winch and roller conveyor for cargo handling, SAR hoist to starboard and cargo hook for external loads (installation in floor resulting in slightly reduced fuel capacity). Will be capable of carrying long wheelbase Land Rover; an overload of 30 troops can board and strap-in within 2 minutes, and de-plane within 40 seconds. Critical design review completed July 1997. Assembly of first (RAF01/ZJ117) began November 1997; rolled out 25 November 1998; first flew 24 December 1998. First aircraft delivered to DERA (re-named QinetiQ) Boscombe Down 19 January 2000. Sixth aircraft handed over to Defence Procurement Agency on 27 June 2000, this date thereby becoming official acceptance into RAF inventory; formal release to service signed on 11 January 2001. Fifth and sixth aircraft used for conversion of first RAF instructors at Yeovil in June 2000.

Deliveries due between September 1999 (first to Boscombe Down for DTEO trials) and October 2001, but latter date was not met (13 received by November 2001). Initial units, from April 2000, were Rotary Wing Operational Evaluation and Training Unit (initially at Boscombe Down) and the Operational Conversion Flight (actually an element of No. 28 Squadron at Benson) with a limited Release-to-Service for day training. Delivery of the first aircraft, due on 6 November 2000, was delayed by a worldwide EH101 grounding, following the loss of an HM. Mk 1. ZJ122 to Benson for technical familiarisation 11 December 2000; formal handover of five aircraft to RAF on 7 March 2001; by mid-June 2001, RAF had six Merlins at Benson, while further seven engaged in flight trials with QinetiQ. No.28 Squadron officially re-formed 17 July 2001 and this undertook first field deployment during 17-21 September 2001, when participated in exercise 'Pegasus Trial', which included deployed operations from RAF Honington and first tactical troop-lift in an exercise scenario. A new Medium Support Helicopter Aircrew Training Facility with two Merlin simulators opened at Benson on 17 July 2000. Both No. 28 and No. 72 (at Aldergrove) Squadrons were to form operational flights in the first quarter of 2002. Allocation is to be 12 to Benson (No. 28 and, eventually, Rotary Wing OEU); six to Aldergrove; and four in-use reserves will be used in support of 16 Air Assault Brigade.

Merlin AEW: UK MoD funded study, submitted August 1998, of possible Sea King AEW. Mk 7 replacement to meet Royal Navy's FOAEW requirement. Ventral radome and stub-wings; latter would permit speeds up to 250 kt (463 km/h; 288 mph).

'CH-X': Offered to USMC; chin gun turret; not adopted.

CH-148 Petrel and **CH-149 Chimo:** Intended Canadian ASW and SAR versions; cancelled after

Flight deck of Royal Navy Merlin HM. Mk 1   0099601

payment of US$353 million in compensation; see 1994-95 and previous *Jane's*.

CH-149 Cormorant (AW320): Adaptation of civil EH101 (with ramp) offered to Canada in early 1995 as a CH-113 replacement (in place of the CH-149 Chimo) for SAR missions; selected late 1997. Deck landing capability; reduced cost achieved by reliance on mainly commercial avionics and by original proposed use of twin hoists in single door, rather than one per side. Since then, the hoist arrangement has been revised, with a primary winch in the starboard door and a secondary in the forward port door. Features Honeywell RDR-1400 radar, Litton LN-100G embedded laser INS/GPS, FLIR, Spectrolab SX-16 searchlight, crashworthy fuel tanks, Breeze-Eastern primary and secondary hoists and an underslung load hook. Planned to include provision for internal extended range tanks (ERTs) and hover in-flight refuelling from ships under way and wire-strike protection. Selected in preference to Chinook, Cougar and Sikorsky S-70 'Maplehawk'.

Total of 15 ordered from Italian production to replace CH-113 Labradors, with deliveries scheduled to begin in October 2000 and be complete by end of 2002; however, delays in certification and qualification resulted in deliveries also being deferred. First (149901) flew at Vergiate on 7 March 2000, with an official first flight on 31 May 2000; initial two handed over on 8 October 2001. To equip 442 Squadron (19 Wing) at Comox, followed by 413 Squadron (14 Wing) at Greenwood, 424 Squadron (8 Wing) at Trenton and 103 Squadron (9 Wing) at Gander; first two helicopters (149904 and '905) delivered 29 September 2001 and arrived Comox on 8 October 2001, with subsequent deliveries at rate of one every six weeks.

Estimated to cost 40 per cent less than original CH-149 with 108 per cent offsets ('industrial regional benefits') totalling C$629 million. Associated industrial team includes Bombardier, Bristol Aerospace, Spar Aerospace and CAE, with Canadian Helicopters to provide a leasing and follow-on maintenance option. EHI has demonstrated a 15,500 kg MTOW (900 kg overweight) which would give a 1,080 n mile (2,000 km; 1,242 mile) SAR radius.

Cormorant MHP: Tentative designation for EH101 subvariant offered by EHI-led AW320 'Team Cormorant' to meet remaining Canadian requirement (Maritime Helicopter Program) for a shipborne ASW Sea King replacement with multifunction radar, sonobuoy processor, dipping sonar IFF, an EO sensor and a datalink. To have a four-man crew and be capable of carrying a stretcher and two passengers or a six-man boarding party; endurance of 2 hours 50 minutes with 30-minute reserves; able to carry at least two Mk 46 torpedoes. Total 28 required for delivery from 2005 under a programme budgeted at C$2.9 billion (US$1.96 billion). To deploy aboard 12 'Halifax' class frigates, four modernised 'Iroquois' class destroyers and two fleet replenishment ships. Boeing joined the team in mid-1999 to supply and integrate the maritime patrol mission system, based on that of the Nimrod MRA. Mk 4; other members include Bell and CAE, with latter responsible for provision of simulators and mission training systems in event of EH101 being selected.

CUSTOMERS: Orders totalled 128 by January 2002; see table. EH101 selected in 2001 by Portugal for combat SAR and fisheries protection and by Denmark for SAR/Utility tasks; latter expects to receive first five helicopters in 2004.

Total 66 for UK, comprising 44 Royal Navy Merlins and 22 RAF Merlins. Italian Navy ordered 16 (reduced from 36, and then from 24), comprising eight ASW (and ASV) versions, four AEW versions and four marines tactical transports with blade- and tail-folding; option on four of further eight converted to firm order at start of 2002; to appear as amphibious support helicopters. Westland produced one Merlin in 1995, three in 1997, six in 1998, 13 in 1999 and 12 in 2000; Agusta built two civil EH101s in 1997 and delivered first in 1998, the other being retained for trials and demonstration. Long-range utility version for logistic and tactical support role also under consideration by Italy and proposed to US Marine Corps, with RAF Merlin HC. Mk 3 used for month-long demonstration tour of USA in May 2001. VVIP version offered to Saudi Arabia in 1995. Singapore requires eight naval helicopters for its newly ordered 'La Fayette' class frigates and up to 13 utility helicopters; EH101 in contention, with other candidates including NH90, AS532SC Cougar, S-70B Seahawk and SH-2G Super Seasprite.

Prototypes assigned a 3,750 hour flight development programme, but had flown 5,000 hours by mid-1999. Additional 5,600 hours flown by PP8 and PP9 from Brindisi (Italy) and, from September 1998, Aberdeen (Scotland) in trial between March 1996 and June 2000, to improve reliability and prove extended (2,000 hour) overhaul intervals.

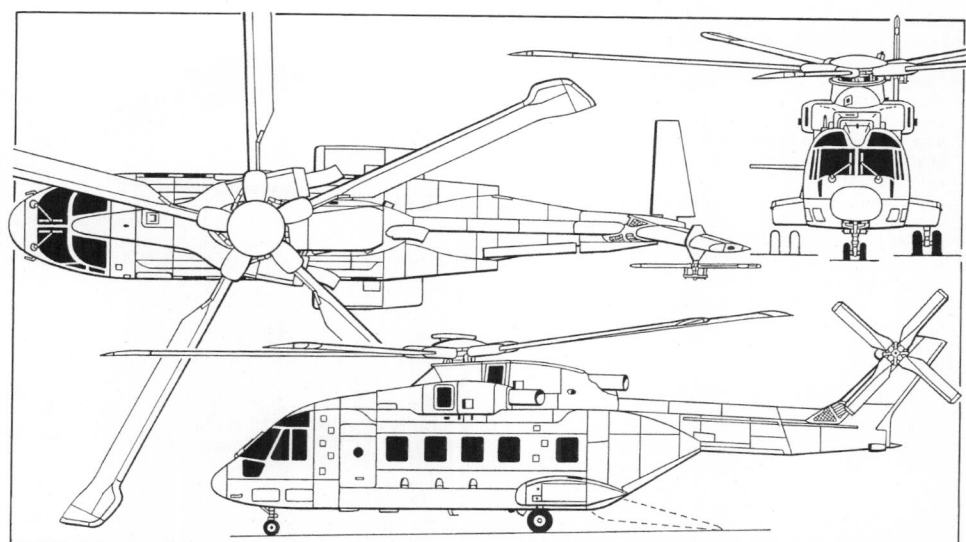

Utility version of the EH101, showing modified rear fuselage with rear-loading ramp/door (*Jane's/Mike Keep*)

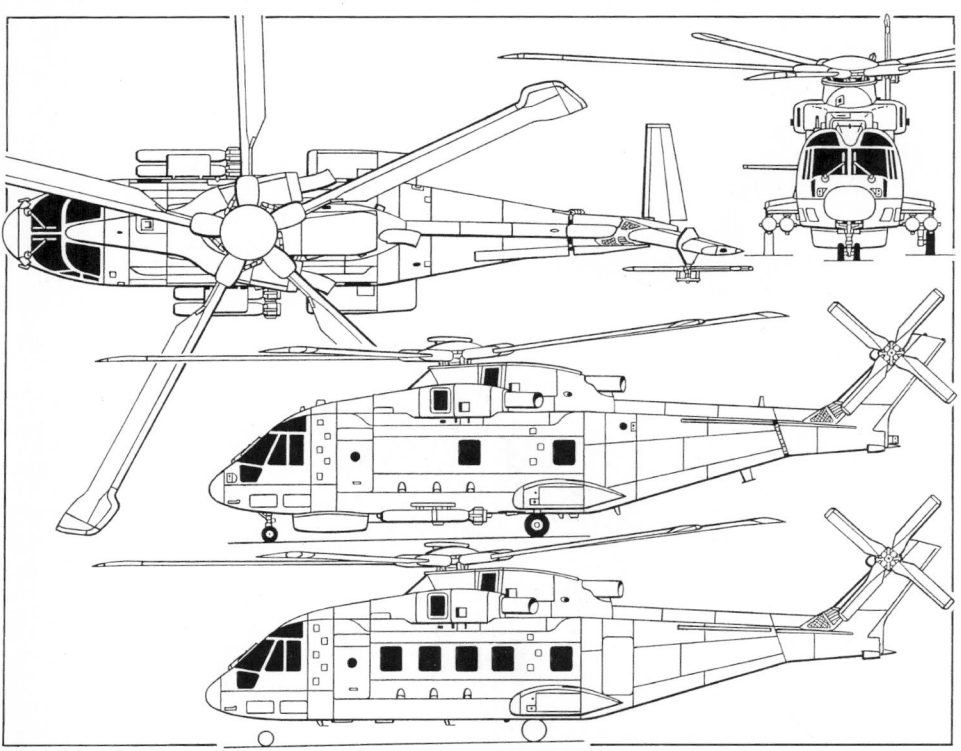

Three-view of the naval EH101, with additional side view (bottom) of the Heliliner (*Jane's/Mike Keep*)

EH101 ORDERS

Country	Service	Quantity	Commitment	Mk/Version	First aircraft
Canada	Armed Forces	15	5 Jan 98	AW320 (SAR)	149901
Denmark	Air Force	14	Dec 01*	(SAR/Utility)	
Italy	Navy	8	Oct 95	110 (ASW/ASVW)	MMX84180
	Navy	4	Oct 95	112 (AEW)	
	Navy	4	Oct 95	410 (Util)	
	Navy	4	Jan 02	(Amphib Support)	
Japan	Tokyo Police	1	Nov 96	510 (SAR)	JA01MP
Portugal	Air Force	12	Dec 01	(SAR)	
UK	Navy	44	9 Oct 91	111/HM. Mk 1	ZH821
	RAF	22	9 Mar 95	411/HC. Mk 3	ZJ117

*Contract signature

COSTS: Royal Navy £1.5 billion (1991) for 44 Merlin; RAF £500 million for 22 Merlin HC. Mk 3s; Canada C$4.4 billion (1992) for 50 CH-148/149, reduced to C$579 million for 15 CH-149 (AW320). Production investment phase (design, tooling, maturity and product support) valued at £200 million (1993). First Italian batch (16) valued at Lit1,250 billion (US$775 million), 1995, of which Westland's share is £150 million. UK official audit of early 1996 reported Merlin HM. Mk 1 as £351 million (36.1 per cent) over budget and 60 months late. Danish order for 14 reportedly worth US$343.3 million.

DESIGN FEATURES: Three-engine power margin with long-endurance 120 kt (222 km/h; 138 mph) cruise possible on two engines and good twin-engined hover performance. Fail-safe/ damage-tolerant airframe and rotating components, high system redundancy, onboard monitoring of engines/transmission/avionics/utility systems; airframe/ power plant/rotor and transmission systems/flight controls/ utility systems common to all variants; five-blade main rotor with multiple load path hub and elastomeric bearings; blades of advanced aerofoil section with BERP-derived high-speed tips; four-blade teetering tail rotor; transmission has minimum 30 minutes (60 minutes demonstrated) run-dry capacity; fuselage in four main modules (front and centre ones common to all variants; modified rear fuselage and slimmer tailboom on military utility variant to accommodate rear-loading ramp); automatic power folding of main rotor blades and tail rotor pylon on naval variant, with emergency manual back-up (tail section folds forward/downward, stowing starboard half of tailplane beneath rear fuselage). Folding version of utility (rear ramp) EH101 has been designed. New active vibration cancelling system ACSR (active control of structural response) reduces vibration by 80 per cent at blade passing frequency.

Airframe overhaul interval 1,000 hours on service entry; eventual target 3,000 hours. Intended service life of 40,000 hours.

FLYING CONTROLS: Dual-redundant digital AFCS.

STRUCTURE: Rotor head of composites surrounding a metal core; composites blades; fuselage mainly aluminium alloy, with bonded honeycomb main panels; composites for such complex shapes as forward fuselage, upper cowling panels, tailfin, tailplane and windscreen. Engine air intakes of Kevlar reinforced with aero-web honeycomb. Tail unit of carbon epoxy and Kevlar epoxy skinned sandwich panels over central skeleton of metal- or foam-cored composites ribs and longerons; Kevlar-Nomex-Kevlar sandwich for leading-edge of tailfin. Single-sourced series production, with final assembly lines in Italy and UK.

LANDING GEAR: Hydraulically retractable tricycle type, with single mainwheels and steerable twin-wheel nose unit, designed and manufactured by AP Precision Hydraulics in association with Officine Meccaniche Aeronautiche. Main units retract into fairings on sides of fuselage. Goodrich wheels, tyres and brakes: main units have size 8.50-10 wheels with 24×7.7 tyres, unladen pressure 6.96 bar (101 lb/sq in); nosewheels have size 19.5×6.75 tyres, unladen pressure 8.83 bar (128 lb/sq in). Twin-mainwheel gear optional for all variants; adopted for all Italian Navy helicopters and RAF Merlin. FPT Industries emergency flotation bags.

POWER PLANT: Three Rolls-Royce Turbomeca RTM 322-01/8 turboshafts in Royal Navy Merlin (maximum contingency rating 1,724 kW; 2,312 shp, T-O rating 1,566 kW; 2,100 shp and maximum continuous rating 1,394 kW; 1,870 shp); RTM 322-02/8 in RAF version, T-O rating 1,670 kW (2,240 shp); General Electric T700-GE-T6A turboshafts in Italian naval variant, and T6A1 in Cormorant, rated at 1,521 kW (2,040 shp) for T-O and 1,327 kW (1,780 shp) maximum continuous. Engines for Italian naval variant will be assembled by Alfa Romeo Avio and Fiat; Cormorant engines assembled in Canada. Commercial and utility variants powered by three General Electric CT7-6A turboshafts (CT7-6A in PP3) with ratings of 1,491 kW (2,000 shp) for take-off and OEI and 1,282 kW (1,718 shp) maximum continuous.

Transmission rated at 4,161 kW (5,580 shp) for T-O, 3,715 kW (4,982 shp) maximum continuous and 2,769 kW (3,713 shp) OEI maximum continuous.

Standard fuel in three tanks, each of 1,074 litres (276 US gallons; 230 Imp gallons) capacity; total 3,222 litres (851 US gallons; 709 Imp gallons). Each tank feeds separate engine, except on selection of emergency cross-feed; self-sealing optional. Additional fourth or fifth tanks (all same size) optional; maximum fuel capacity 5,370 litres (1,417 US gallons; 1,181 Imp gallons). Merlin HC. Mk 3 capacity approximately 4,075 litres (1,077 US gallons; 896 Imp gallons), augmented by optional tank in cargo hold. Computerised fuel management system. Pressure refuelling point on starboard side; maximum transfer rate 682 litres (180 US gallons; 150 Imp gallons)/min; individual gravity refuelling positions on port side. Provision for detachable refuelling probe on RAF Merlins.

ACCOMMODATION: One or two pilots on flight deck (naval version will be capable of single-pilot operation, if required, and RN will operate with pilot, observer and crewman; commercial variant will be certified for two-pilot operation). ASW version will normally also carry observer and acoustic systems operator. Italian Navy ASW crew members are designated pilot/mission commander, co-pilot/tactical co-ordinator, and two sensor operators. CH-149 Cormorant operates with pilot, co-pilot/ navigator, flight engineer and two crewmen. Martin-Baker crew seats in naval version, able to withstand 10.7 m (35 ft)/s impact. Socea or Ipeco crew seats in commercial variant. Commercial version able to accommodate 30 passengers four-abreast at approximate seat pitch of 76 cm (30 in), plus cabin attendant, with lavatory, galley and baggage facilities (including overhead bins). Offshore variant offers 'club 4' grouped seating to facilitate rapid egress through windows in event of ditching. Military variant can accommodate up to 30 (seated) or 45 (non-seated) combat-equipped troops, 16 stretchers plus a medical team, palleted internal loads, or can carry externally slung loads of up to 5,443 kg (12,000 lb). SAR version can seat 28 survivors; or four seated patients, four stretchers and six medics; or 20 fully equipped Arctic rescuers with skis. It has also demonstrated an emergency evacuation capability with 55 in the cabin.

Main passenger door/emergency exit at front on port side with additional emergency exits on starboard side and on each side of cabin at rear, above main landing gear sponson. Large sliding door at mid-cabin position on starboard side, with inset emergency exit. Commercial variant has baggage bay aft of cabin, with external access via door on port side. Cargo loading ramp/door at rear of cabin on military and utility versions. Cabin floor loading 976 kg/m² (200 lb/sq ft) on PP1.

SYSTEMS: Hamilton Sundstrand/Microtecnica environmental control system. Dual-redundant integrated hydraulic system, pressurised by three Vickers pumps each supplying fluid at 207 bar (3,000 lb/sq in) nominal working pressure, with flow rates of 55, 59 and 60 litres

Fifth production Italian military EH101, a naval ASW variant, on display at Paris in June 2001 (*Jane's/Paul Jackson*)

0131742

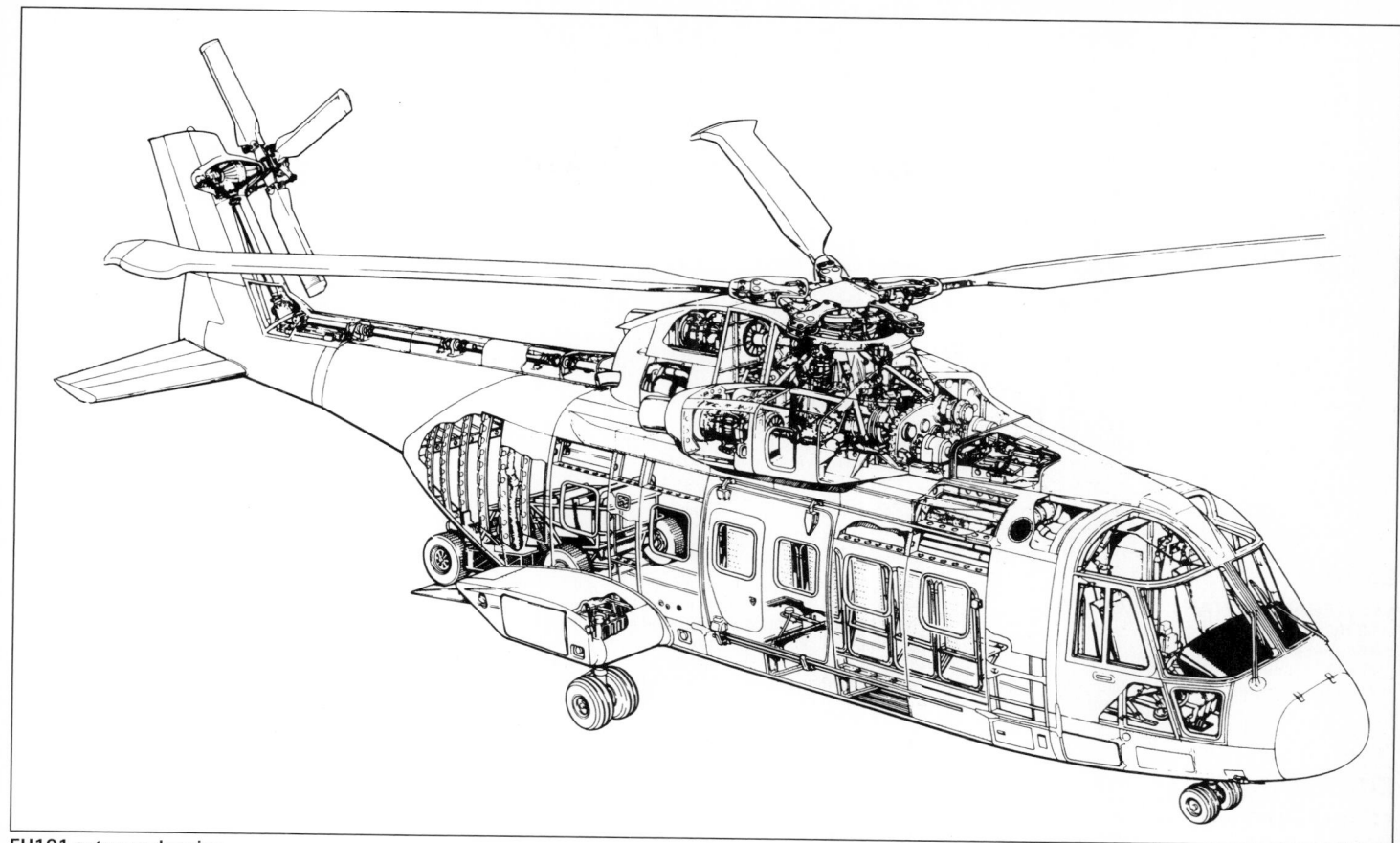

EH101 cutaway drawing

(14.5, 15.6 and 15.9 US gallons; 12.1, 13.0 and 13.2 Imp gallons)/min respectively. Hydraulic system reservoirs are of the piston load pressurised type, with a nominal pressure of 0.97 bar (14 lb/sq in).

Primary electrical system is 115/200 V three-phase AC, powered by two Lucas brushless, oilspray-cooled 45 kVA generators (90 kVA if Lucas Spraymat blade ice protection system fitted), with one driven by main gearbox and the other by accessory gearbox, plus a third, separately driven standby alternator. APU for main engine air-starting, and to provide electrical power, plus air for ECS, without running main engines or using external power supplies. Lucas Spraymat electric de-icing of main/tail blades standard on naval variant, optional on others; Dunlop electric anti-icing of engine air intakes. Fire detection and suppression systems by Graviner and Walter Kidde respectively.

AVIONICS (military): Integrated systems based on two MIL-STD-1553B multiplex databusses that link basic aircraft management, avionics and mission systems.

Comms: Royal Navy version has BAE Systems communications subsystem including internal voice intercommunication to six positions, plus secure voice transmission via two AD3400 V/UHF radios, UHF and HF, plus M/A-COM Ltd ARI 5983 I-band transponder and Link 11. Racal RA 800 Light secure communications control system for RAF Merlin. Italian Navy equipment, by Elmer, includes three SRT-651 V/UHF, two SRT-170L HF, TD8503 Link 11, SP-1450 intercom and Italtel Mk 12 IFF. RAF Merlins have ADELT (automatically deployed emergency loader transmitter) containing FDR and CVR and locator beacon, scabbed on to starboard rear fuselage.

Radar: ASW version fitted with 360° search radar (pulse-compressed, frequency-agile BAE Systems Blue Kestrel 5000 in UK's Merlin; Eliradar MM/APS-784 in Italian helicopters). AEW version has Eliradar HEW-784 coherent pulse Doppler, 360° scan radar. Italian Navy utility helicopters have second-hand (from Agusta-Bell 212ASW) Officine Galileo MM/APS-705B search and weather radar which is removable, if demanded by certain missions.

Flight: Smiths Industries OMI SEP 20 dual-redundant digital AFCS is standard, providing fail-operational autostabilisation and four-axis autopilot modes (auto hover, auto transitions to/from hover standard on naval variants, optional on commercial and military variants). AFCS sensors on naval variant include BAE Systems LINS 300 ring laser gyro inertial reference unit (IRU) and Litton Italia LISA-4000 strapdown AHRS; IRU also provides self-contained navigation, with Racal Doppler 91E velocity sensor (Elmer system for Italy); Cossor Electronics GPS receiver selected for Royal Navy variant, Elmer GPS for Italian Navy aircraft. Other avionics on naval variants include Thomson-CSF AHV 16 radar altimeters (two) (Elettronica J-band units for Italian Navy), BAE Systems low-airspeed sensing and air data system, and Alenia/BAE Systems aircraft management computer.

Instrumentation: Litton EFIS with six Smiths multifunction screens.

Mission: On naval variant, main processing element of management system is a dual-redundant aircraft management computer which carries out navigation, control and display management, performance computation and health and usage monitoring of principal systems (engines, drive systems, avionics and utilities); it also controls basic bus. Alenia/Racal cabin mission display unit. Surveillance radar (see above). Underwater detection in Royal Navy Merlin is by active/passive sonobuoys and Thomson Marconi Sonar TMS 118 ADS (active dipping sonar) system, incorporating same company's AQS-903 acoustic processor for LOFAR, DIFAR, VLAD, Barra, DICAS and CAMBS buoys and Thomson Sintra folding lightweight acoustic system for helicopters (FLASH) expandable array, which has 750 m (2,460 ft) of cable and is manoeuvred by a high-speed winch. Merlin has Racal Lightweight Common Control Unit as interface between crew and tactical navigation and communications system. The HM. Mk 1 has four units (one each for pilot, co-pilot and two operators); HC. Mk 3 has one in the cabin primarily for maintenance management. Royal Navy Merlin also has Racal Orange Reaper ESM, Normalair-Garrett mission recorder and sonobuoy/flare dispenser, Chelton sonobuoy homing system and Ultra processor for Link 11 datalink. The TMS 118 sonar is to be upgraded from 2001 with new, commercial, off-the-shelf processors, adding to speed and capacity. Honeywell HELRAS Mod 2 dipping sonar, Alenia AYK-204 processors (four) and Alenia SL/ALR-735 ESM in Italian ASW version, which also includes two operators' consoles with 350 mm (14 in) displays. ASST (anti-ship surveillance and tracking) version will carry equipment for tactical surveillance and OTH (over the horizon) targeting, to locate and relay to a co-operating frigate the position of a target vessel, and for mid-course guidance of frigate's missiles. On missions involving patrol of an exclusive economic zone it can also, with suitable radar, monitor every hour all surface contacts within area of 77,700 km² (30,000 sq miles) and can patrol an EEZ 400 × 200 n miles (740 × 370 km; 460 × 230 miles) twice in one sortie; and can effect boarding and inspection of surface vessels during fishing protection and anti-smuggling missions. FLIR specified for all Italian Navy versions; BAE Systems MST-S FLIR turret, including magnifying mode, can be rapidly installed on RAF Merlins.

Self-defence: RAF Merlins have integrated defensive aids including Raytheon laser detection, BAE Systems Sky Guardian 2000 RWR, Doppler-based MAWS, Northrop Grumman AN/AAQ-24 Nemesis DIRCM and BAE Systems North America AN/ALE-47 chaff/flare dispensers. Chaff and flare dispensers on all Italian versions, co-ordinated by Elettronica ELT/156X(V2) RWR and BAE Systems RALM/1 laser warner.

AVIONICS (civil): Integrated avionics system of commercial variant based on ARINC 429 data transfer bus.

Comms: Racal intercom system; Rockwell Collins or Honeywell communications system.

Radar: Honeywell weather radar.

Flight: BAE Systems Canada CMA-900 flight management system for fuel flow, fuel quantity and specific range computations; tuning of nav/com radios; interfaces with electronic instrument systems; two-dimensional multisensor navigation; and built-in navigational database with update service. AFCS sensors on commercial variant include two Litton Italia LISA-4000 strapdown AHRS. Standard avionics include Penny and Giles air data system.

Instrumentation: Smiths Industries/OMI electronic instrument system (EIS) providing colour flight instrument, navigation and power systems displays. CMA-900 includes colour CRT display with graphics and alphanumeric capability.

EQUIPMENT: ASW variants have two sonobuoy dispensers, external rescue hoist and Fairey Hydraulics (Merlin) decklock. BAJ Ltd four-float emergency flotation gear.

ARMAMENT (naval and military utility versions): Naval version able to carry up to four homing torpedoes (BAE Systems Sting Ray on Merlin; Mk 46 or Eurotorp MU90 on Italian version) or other weapons, including Mk 11 Mod 3 depth charges. ASV version designed to carry two air-to-surface missiles (Marte Mk 2/S for Italian Navy) and other weapons, for use as appropriate, from strikes against major units using sea-skimming anti-ship missiles to small arms deterrence of smugglers. Armament optional on military utility versions; options include pintle-mounted machine guns in doorway/rear ramp, chin turret for 12.7 mm machine gun and stub-wings for rocket pods.

DIMENSIONS, EXTERNAL (A: naval variant, B: Heliliner, C: military/utility variant):

Main rotor diameter	18.59 m (61 ft 0 in)
Tail rotor diameter	4.01 m (13 ft 2 in)
Length:	
overall, both rotors turning	22.80 m (74 ft 9¾ in)
fuselage	19.53 m (64 ft 1 in)
main rotor and tail pylon folded:	
A	15.75 m (51 ft 8 in)
C	16.00 m (52 ft 6 in)
Width: cabin	2.80 m (9 ft 2¼ in)
fuselage overall (port sponson to starboard tailplane)	
	5.09 m (16 ft 8¼ in)
over sponsons	4.61 m (15 ft 1½ in)
main rotor and tail pylon folded:	
A	5.20 m (17 ft 0¾ in)
C	5.60 m (18 ft 4½ in)
crew door open	5.32 m (17 ft 5½ in)
Height: overall, both rotors turning	6.62 m (21 ft 8¾ in)
main rotor and tail pylon folded:	
A	5.20 m (17 ft 0¾ in)
C	5.30 m (17 ft 4¾ in)
Tailplane half-span	2.78 m (9 ft 1½ in)
Wheel track	4.55 m (14 ft 11¼ in)
Wheelbase	6.98 m (22 ft 11 in)

Passenger door (fwd, port): Height	1.70 m (5 ft 7 in)	
Width	0.91 m (3 ft 0 in)	
Sliding cargo door (mid-cabin, stbd):		
Height	1.55 m (5 ft 1 in)	
Width	1.83 m (6 ft 0 in)	
Baggage compartment door (rear, port, B):		
Height	1.38 m (4 ft 6 in)	
Width	0.55 m (1 ft 10 in)	
Rear-loading ramp/door (rear, military/utility variant):		
Height	1.95 m (6 ft 4¾ in)	
Width	2.26 m (7 ft 5 in)	
Main rotor ground clearance (turning)	4.70 m (15 ft 5 in)	

DIMENSIONS, INTERNAL:
Cabin:
Length: A	7.09 m (23 ft 3 in)
C	6.50 m (21 ft 4 in)
Max width	2.49 m (8 ft 2 in)
Width at floor	2.26 m (7 ft 5 in)
Max height: B	1.90 m (6 ft 2¾ in)
Volume: A	29.0 m³ (1,024 cu ft)
B	27.5 m³ (970 cu ft)
Baggage compartment volume (B)	3.8 m³ (135 cu ft)
Rear ramp (C): Length	2.10 m (6 ft 10¾ in)
Width	1.80 m (5 ft 10¾ in)

AREAS:
Main rotor disc	271.51 m² (2,922.5 sq ft)
Tail rotor disc	12.65 m² (136.2 sq ft)

WEIGHTS AND LOADINGS (A, B, C, as above):
Operating weight empty (estimated):
A	10,500 kg (23,149 lb)
B (IFR, offshore equipped)	9,300 kg (20,503 lb)
C	9,350 kg (20,613 lb)
Merlin HC. Mk 3	10,250 kg (22,597 lb)

Max fuel weight (four internal tanks, total):
A (JP-1)	3,406 kg (7,509 lb)
B, C (JP-4)	3,360 kg (7,408 lb)
Merlin HC. Mk 3	3,200 kg (7,055 lb)

Max fuel weight (five internal tanks, total):
B, C (JP-4)	4,200 kg (9,259 lb)

Disposable load/payload:
A (four torpedoes)	960 kg (2,116 lb)
B (30 passengers plus baggage)	2,850 kg (6,283 lb)
C (24 combat-equipped troops)	3,120 kg (6,878 lb)
Max underslung load	5,443 kg (12,000 lb)
Max T-O weight: A, B, C	14,600 kg (32,188 lb)
Max disc loading: A, B, C	53.8 kg/m² (11.01 lb/sq ft)

Transmission loading at max T-O weight and power
A, B, C	3.51 kg/kW (5.76 lb/shp)

PERFORMANCE:
Never-exceed speed (V_{NE}) at S/L, ISA	
	167 kt (309 km/h; 192 mph) IAS
Average cruising speed	150 kt (278 km/h; 173 mph)
Best range cruising speed	125 kt (232 km/h; 144 mph)
Best endurance speed	80 kt (148 km/h; 92 mph)
Service ceiling	4,575 m (15,000 ft)
Hovering ceiling: IGE	2,225 m (7,300 ft)
OGE	1,128 m (3,700 ft)

Range (B):
four tanks, offshore IFR equipped, with reserves
610 n miles (1,129 km; 702 miles)
five tanks, offshore IFR equipped, with reserves
750 n miles (1,389 km; 863 miles)
Ferry range: C (four tanks plus internal auxiliary tank)
1,130 n miles (2,093 km; 1,300 miles)
SAR radius for 26 survivors, with two minute hover per
survivor	350 n miles (648 km, 403 miles)
Endurance	5 h
g limit	+3

UPDATED

EMAC

EUROPEAN MILITARY AIRCRAFT COMPANY

PARTICIPATING COMPANIES
EADS: see this section.
Finmeccanica: see Alenia under Italy.

On 14 April 2000, Finmeccanica and EADS agreed to fund a joint (50:50) venture company with the working title of EMAC and an official starting date of 1 January 2001. Subsequently, this was delayed slightly, to first quarter of 2001, then to mid-2001, then 1 January 2002; meanwhile EADS and Finmeccanica (and BAE Systems) established

missile company MBDA with agreement signed 27 April 2001. By mid-2002, there had been no further progress.
UPDATED

EUROCOPTER

EUROCOPTER SAS (An EADS company)

Aéroport International Marseille-Provence, F-13725 Marignane Cedex, France
Tel: (+33 4) 42 85 85 85
Fax: (+33 4) 42 85 85 00
Web: http://www.eurocopter.com
CHAIRMAN: Jean-François Bigay
DIRECTOR OF COMMUNICATIONS: Xavier Poupardin
CHIEF, PRESS & INFORMATION: Jean-Louis Espes

Eurocopter SA formed 16 January 1992 by merger of Aerospatiale and MBB (DASA) helicopter divisions. Share capital then held on two levels: Eurocopter Holding owned 60 per cent by Aerospatiale and 40 per cent by DASA; capital of Eurocopter SA held 75 per cent by Eurocopter Holding and 25 per cent by Aerospatiale. However, on 30 May 1997, the separate elements of Eurocopter were merged into a single management structure. Former Eurocopter France, Eurocopter International and Eurocopter Participations became a single entity (Eurocopter France, trading as Eurocopter), with Eurocopter Deutschland (in Germany) as a wholly owned subsidiary. Following formation of EADS, Eurocopter reconstituted on 18 September 2000 as simplified stock company (SAS) within EADS Aeronautics group.

Eurocopter products cover 75 to 80 per cent of the range of helicopters in terms of size and capacity, but company intends to expand this to 95 per cent, particularly with co-operative development of Russian Mil Mi-38 (see Euromil entry in this section). Company claimed 57 per cent of civil helicopter market in 2001 and has averaged 15 per cent of military market in recent years.

Employees total some 10,000. By January 2002, Eurocopter and its predecessors had produced some 11,600 aircraft (excluding Sikorsky licence) for over 1,970 customers in 133 countries. In early 2001, some 8,700 Eurocopter helicopters were active.

Orders received in 2001 for 375 helicopters, comprising 81 Colibris, 103 single-engine and 17 twin-engine Ecureuil/ Fennecs, 27 EC 130s, 38 EC 135s, two BK 117s, 10 Dauphin/ Panthers, eight EC 145s, 13 EC 155s, 26 Super Puma/Cougars (including EC 225/725s), two BO 105s and 48 NH 90s. Deliveries for the same period totalled 280. Order book in early 2002 was €2,825 million.

Eurocopter subsidiaries employ some 1,250 persons and deliver about 150 aircraft per year (2001); managed by Eurocopter Subsidiaries and Participations division (ownership percentage in parentheses): American Eurocopter (100), Eurocopter Canada (100), Hélibras (Brazil; which see; 76.5), Eurocopter España (Spain; see below; 60), Eurocopter Romania (see below; 51) and McAlpine (UK; 10) are all first-tier subsidiaries responsible for distribution, service and industrial work, including assembly and component manufacture (Canada and Hélibras), customisation (American, Canada, Hélibras and McAlpine) and repair and overhaul. Second-tier subsidiaries (distribution and service) comprise Eurocopter South East Asia (Singapore; 65), Eurocopter Mexico (100), Eurocopter Chile (100), Eurocopter South Africa (100), Eurocopter International Pacific (Australia; 100), and Euroheli (Japan; 10). Service subsidiaries are Eurocopter Philippines (70), Eurocopter Service Japan (51) and EuroAircraft Services (Malaysia; 34).

Eurocopter
Address as above

Company headquarters and main production centre at Marignane (Marseille); works at La Courneuve (Paris). In 2002, these employed 5,057 and 771 personnel, respectively.

Eurocopter Deutschland
Industriestrasse 4, Postfach 1353, D-86609 Donauwörth, Germany
Tel: (+49 906) 710
Fax: (+49 906) 71 45 75
CHAIRMAN: Friedrich Dörhöfer
MEDIA RELATIONS EXECUTIVE: Christina Gotzhein

Industrial concern in charge of production and support of products originating in German part of Eurocopter. Wholly owned subsidiary with plants at Donauwörth (2,674 personnel in 2002) and Ottobrunn (634).

Current programmes are EC 135, BK 117C-1, EC 145 and Tiger; NH90 (industrialisation); also maintenance/upgrade work for CH-53G, Lynx and Sea King. Research and development programmes concentrate on noise abatement, fly-by-light technologies, all-weather capability of helicopters and optimisation of in-flight comfort and safety. Eurocopter Deutschland closely co-operates with other industrial partners in the aerospace industry as well as the German and French aerospace institutions DLR and ONERA, respectively.

Eurocopter Deutschland, the Federal Ministry of Defence and German Aerospace Research Establishment are jointly providing DM40 million to 50 million funding for development of **Helicopter Simulator for Technology, Operations and Research (HeSTOR)**, based on EC 135, to replace BO 105-based Advanced Technology Testing Helicopter System (ATTHeS) which was lost in an accident on 19 May 1995.

Eurocopter España SA
EADS CASA, Avenida de Aragón 404, E-28022 Madrid
PUBLIC RELATIONS EXECUTIVE: Francisco Salido

Established 28 September 2000, replacing former subsidiary Helicopteros España (HESA). Works at Quatro Vientos responsible for engineering, production and customer support, on par with French and German divisions. Owned 60 per cent Eurocopter and 40 per cent EADS CASA.

Eurocopter Romania SA
SC IAR SA, R-2200 Brasov

Agreement to form Eurocopter Romania signed with IAR (which see) on 22 December 2000. Joint venture (51 per cent Eurocopter-owned) markets Eurocopter products in Romania; performs subcontract work on support of Puma and Alouette for Eurocopter; and maintains, supports and upgrades Romanian and export Pumas.

Additionally, Eurocopter has entered into two alliances with manufacturers in Asia in order to develop specific products. These are regarded by the company as wholly Eurocopter aircraft.

Eurocopter/CATIC/ ST Aero
PARTICIPATING COMPANIES:
Eurocopter: see this entry
CATIC: see under HAMC in Chinese section
Singapore Technologies Aerospace: see under Singapore

This partnership formed to develop the EC 120 Colibri, production of which is now fully under way. Shares are Eurocopter 61 per cent (and programme leader), CATIC/HAMC 24 per cent and ST Aero 15 per cent.

Eurocopter/Kawasaki:
PARTICIPATING COMPANIES:
Eurocopter: see this entry
Kawasaki: see under Japan

Companies formed a partnership on 25 February 1977 to develop the BK 117 multipurpose helicopter. Low-rate production continues in both Germany and Japan. A development, the EC 145, was certified in December 2000.
UPDATED

EUROCOPTER 665 TIGER/TIGRE

TYPE: Attack helicopter.
PROGRAMME: France and Germany agreed in 1984 to develop a common combat helicopter; Eurocopter Tiger GmbH formed 18 September 1985 to manage development and manufacture for French and German armies; was not a full member of Eurocopter because it was working on a single government contract; executive authority for programme is DFHB (Deutsch Französisches Hubschrauberbüro) in Koblenz; procurement agency is German government BWB (Bundesamt für Wehrtechnik und Beschaffung).

Original 1984 MoU amended 13 November 1987; FSD approved 8 December 1987; main development contract awarded 30 November 1989, when name Tiger (Germany)/Tigre (France) adopted; five development aircraft built, including three unarmed aerodynamic prototypes, used also for core avionics testing (PT1, 2 and 3), one (PT4) in HAP (initially called Gerfaut) configuration and one (PT5) as UHT prototype; PT1 rolled out 4 February 1991; first flight 27 April 1991; fifth prototype flew on 21 February 1996, at which time the first four aircraft had accumulated 1,090 flying hours; further details below; total of 2,869 hours flown by five prototypes up to June 2001. Germany confirmed purchase of full 212 required, 1994, having considered cut to 138.

Industrialisation phase brought forward by two years to strengthen export prospects and Franco-German MoU signed 30 June 1995; timetable then was first deliveries in 1999 to France (approximately 10) and for export, but France announced spending moratorium in November 1995, postponing authorisation of further funding commitments until signature of a FFr2.5 billion (DM733.6 million) production investment contract on 20 June 1997. Deliveries then expected in 2001, but further delayed to July 2003 (for HAP; 2011 for HAC) by May 1996 defence plan, which envisaged procurement of only 25 Tigres in 2000-2002 budgets.

Tiger pilots occupy the front cockpit (left), the gunner located behind   0105082

In October 1996, Germany announced a 12-month delay in launching Tiger production because of funding constraints. However, the government planned to recoup lost time by accelerating production when eventually begun, maintaining in-service date (ISD) of 2001 and having 50 delivered by 2006, after which the manufacturing tempo would be reduced. However, by 1999, first UHT delivery planned in December 2002. France indicated in early 1997 that it would be prepared to see a single Tiger production line located at Donauwörth in Germany which, combined with other economies, would reduce French expenditure by FFr13.5 billion, but a second assembly line at Marignane was subsequently confirmed. Production investment agreed June 1997.

On 20 May 1998, France and Germany signed a commitment to order an initial joint batch of 160 Tigers. However, planned late 1998 placing of contracts was delayed by requirement of new German government to conduct a defence review; options included delaying ISD; or reducing numbers; or even cancelling UHT and procuring French HAP version. Production contract was finally signed on 18 June 1999 for the full 160 aircraft; first deliveries in 2002. HAP deliveries to include two in 2003, eight in 2004 and 10 per year in 2005-10; first production aircraft rolled out in Germany, 22 March 2002. Production of the first batch of 320 engines (plus 12 spares) began during 2000, and will continue through 2011. June 1999 contract also formalised German contract change from PAH2 to UHT and French change from HAP to HAP-F (*Finalisé*).

Joint team at Marignane is flight testing basic helicopter, updating avionics during trials, and testing HAP variant; similar team at Ottobrunn is qualifying basic avionics, Euromep mission equipment package, and integrating weapons system. Rotor downwash problems resulted in trial forward positioning of horizontal stabiliser; by mid-1994 definitive solution adopted of reversion to original position, but halving area. By January 1998, the design had been frozen, and the development programme was more than 90 per cent complete.

First export order for Tiger confirmed 14 August 2001, when Australia announced selection to meet AIR 87 requirement.

CURRENT VERSIONS (general): Original partners require three versions in two basic configurations with about 80 per cent commonality: **U-Tiger** is basis of the UHT and HAC, both with mast-mounted sight and Trigat missiles; **HCP** (*Hélicoptère de Combat Polyvalent*) is basis of the HAP (roof sight and turreted gun). Other variants proposed to meet export requirements.

Tigre HAP: *Hélicoptère d'Appui et de Protection*; name Gerfaut dropped late 1993; escort and fire support version for French Army; armed with 30 mm Giat AM-30781 automatic cannon in undernose turret, with 150 to 450 rounds of ammunition; four Mistral air-to-air missiles and two pods each with twenty-two 68 mm unguided TDA rockets delivering armour-piercing darts, mounted on stubwings, or 12-round rocket pod instead of each pair of Mistrals, making total of 68 rockets; roof-mounted sight, with TV, FLIR, laser range-finder and direct view optics sensors; image intensifiers integrated in helmets; and extended self-defence system. HAP configuration was approved by late 1998, permitting type qualification in December 2002. Deliveries to begin from Marignane in 2003; final aircraft due in 2010.

UHT: *Unterstützungshubschrauber Tiger* (previously designated UHU); German Army multirole 'utility' or multirole anti-tank and fire-support helicopter for delivery from 2002; replaces dedicated anti-tank PAH-2 Tiger; type qualification due in December 2002; final assembly at Donauwörth. Underwing pylons for HOT 3 or (from 2006) Trigat missiles, Stinger self-defence missiles, unguided rockets, gun pod and extended self-defence system; mast-mounted TV/FLIR/laser ranger sight for gunner; nose-mounted FLIR for piloting. A mid-life upgrade for the UHT may integrate the Mauser 30 mm gun in a chin turret which traverses ±140° in azimuth and from +20 to −45° in elevation.

Tigre HAC: *Hélicoptère Anti-Char*; anti-tank variant for French Army; final assembly at Donauwörth. Type qualification due in third quarter of 2011; same weapon options (except Mistral AAM in place of Stinger), mast-mounted sight and pilot FLIR system as UHT. A mid-life upgrade to the HCP could see addition of a mast-mounted automatic air surveillance and warning system, in the form of DAV pulse Doppler radar, together with HUMS and an IR jammer.

Export: A basic export version combining features of French and German versions; offered (unsuccessfully) to UK and Netherlands.

Aussie Tiger (HCP): Hybrid Tiger variant to meet Australian Army Air 87 Requirement. Based on French HAP, with undernose Giat 30-781 30 mm cannon, roof-mounted sight and provision for underwing rocket pods, but with added anti-tank capability, initially with HOT missile then (from 2006) with Trigat AC3G. Australia also requires integration of the AGM-114 Hellfire ATM. Maximum mission weight of 6,100 to 6,300 kg (13,448 to 13,889 lb). A\$1,300 million (US\$674 million) contract for 22 signed 21 December 2001.

HCP Tiger (HCP): Export version based on French Army HAP with the same undernose gun turret and roof-mounted sight, HOT 3 and Trigat missiles; Hellfire optional. Strix roof sight (direct view optics optional) and additional laser designator plus video signal interfacing for Trigat operation. Either Mistral or Stinger air-to-air missiles. A mid-life upgrade to the HCP could see addition of a DAV mast-mounted air surveillance radar (pulse Doppler type) or a mast-mounted MMW radar for automatic ground and air surveillance. No gun pod option.

Tigre HAD: *Hélicoptère d'Appui-Destruction*. Multirole version; development cost estimated as €152 million (2002). Offered to Spain. Uprated engines, roof sight for Trigat ATMs.

Tiger T800: LHTEC T800 or CTS800 engines proposed as an engine option for the Tiger. Turkey is sales prospect, as Army has reservations about growth potential of MTR390.

CURRENT VERSIONS (specific): **PT1/F-ZWWW:** Aerodynamic prototype; basic avionics; first flight 27 April 1991. Successively fitted with aerodynamic mockups of mast-mounted and roof-mounted sights, nose-mounted gun and weapon containers. Relegated to ground fatigue testing and static display in early 1996 on completion of flight programme. Flown 502 hours.

PT2/F-ZWWY: HAP aerodynamic configuration; full core avionics; rolled out 9 November 1992; first flight 22 April 1993. Used for radar cross-section and detectability

tests. Retrofit with HAP systems completed in November 1996; redesignated PT2R. Mistral launch trials at Landes ranges 14/15 December 1998; technical assessment by French Army at Valence between 17 May and 3 June 1999; rocket qualification, June 1999. Used for HAP version qualification (redesignated PT2R2) at Landes test centre between 4 April and 12 May 2000. Redesignated PT2X in 2001 to serve as multimission demonstrator, adding LFK/SAGEM sighting system for HOT 3 anti-tank missiles in addition to original Mistral missiles and rockets.

PT3/9823: Full core avionics (including navigation and autopilot); first flight (as F-ZWWT) 19 November 1993. Retrofit with UHT systems began in February 1997; redesignated PT3R; Euromep C (see Avionics) from late 1997. HOT launches with mast sight at extreme range in night and smoke conditions, June 1999; hot weather trials at Bateen AB in Abu Dhabi September 1999. Moved back to France for HAC development.

PT4/F-ZWWU: HAP aerodynamic configuration and avionics (including roof sight, HUD and Topowl helmet sight; first Tiger with live weapons system); first flight 15 December 1994. Sighting system trials early 1995; Giat cannon trials (15 ground-based tests) completed at Toulon, April 1995; full testing began at CEV Cazaux, 21 September 1995 and, by late November, had demonstrated airborne cannon firing and launch of Mistral AAM (without seeker); by 1 January 1997 had fired eight Mistrals, 3,000 cannon rounds and 50 rockets; 1997 trials included two more Mistrals, rockets and tests of gun controls. Painted in three-tone disruptive camouflage. Winter trials in Sweden, early 1997 with skid/skis landing gear. Flown 296 hours to 1 December 1997; crashed during night low-level evaluation by Australian Army 17 February 1998.

PT5/9825: Full UHT avionics; first flight 21 February 1996. Undertook German Army weapon trials (Stinger, HOT 2 and 12.7 mm podded gun) in 1997 including the firing of six HOT 2s using Euromep Osiris mast-mounted sight. Retrofitted as PT5R with production-standard weapon system; first flight 8 October 1999.

PT6 and **PT7:** Static test airframes for fatigue and crash-resistance trials.

PS1/F-ZVLJ: Pre-series HAP built at Marignane on production tooling; laid down in third quarter of 1998; first flight 21 December 2000. Tasks include validation of production methods and planned production configuration.

UHT S01: First true production aircraft; planned to fly on 1 March 2002, but not rolled out until 22 March. To be followed by S02 late 2002. Deliveries December 2002 and April 2003, respectively. S01 will be used for six-month techeval/opeval trials, replacing PT5R.

HAP S01: First production French Tiger, planned to fly during early 2003; delivery to French Army in July 2003.

CUSTOMERS: Original requirement was for 427 (France 75 HAP and 140 HAC, Germany 212 PAH-2); UHU (later UHT) version substituted for PAH-2s in 1993; French order amended by 1994 to 115 HAP and 100 HAC but may be reduced to overall total of 180; in mid-2001 French Army expressed preference for multirole Hélicoptère d'Appui-Destruction (HAD) version in place of two subvariants now on order. Germany committed to 212, of which 112 to be funded between 2001 and 2009; initial commitment of 20 May 1998 confirmed 80 each by France (70 HAP, 10 HAC) and Germany (80 UHT), as agreed by Franco-German Security Council on 9 December 1996. Germany's Tigers are required to equip four 48-aircraft regiments, each supporting an Army division. A joint training school at le Luc is being established as the Ecole Franco-Allemand, or EFA with a Thomson Training and Simulation/STN ATLAS aircrew training system, including six-axis motion simulators and wide-angle visual systems. The planned training course will last 28 weeks for a crew chief, or 19 weeks for a pilot, courses will begin in 2001-2002. Fleet of 14 German and 13 French Tigers will be assigned by 2006, with eleven simulators. In October

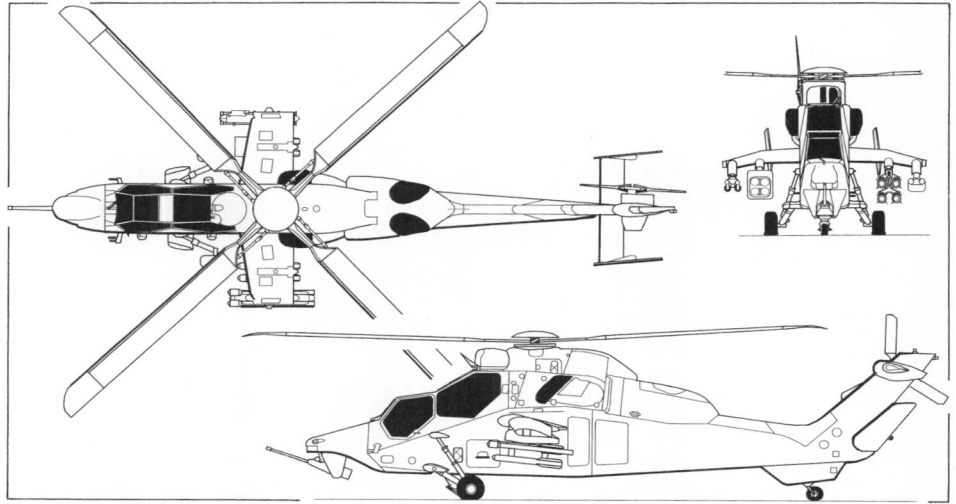

Eurocopter Tigre HCP combat escort and fire support helicopter (*Jane's/Paul Jackson*)

2000 there were reports that the German MoD was considering reducing its Tiger buy to 100 helicopters.

Australia selected Tiger for AIR 87 requirement on 14 August 2001; contract signed 21 December 2001 for 22 helicopters. Interest shown by Spain in 36 Tigers to be delivered from 2010; in early 2002, contract discussions reached impasse over R&D funding for specific Spanish multirole HAD variant. Exports of 200 Tigers thought possible.

COSTS: Tiger current development cost, shared equally by France and Germany, reported DM2.2 billion. Production tooling cost FFr2.6 billion (US$500 million) (1996). Unit cost (1996) for UH estimated as US$11 million, including launchers and all government-furnished equipment. Initial batch of 160 assigned FFr21.5 billion, of which FFr13 billion for 80 German helicopters and FFr8.5 billion for 80 French (1998). Australian programme unit cost US$30.6 million (2001).

DESIGN FEATURES: Robust, tandem-seat design with pylon-mounted armament, representing current combat helicopter style. FEL (fibre elastomer) main rotor designed for simplicity, manoeuvrability and damage tolerance; has infinite life except for inspection of elastomeric elements at more than 2,500-hour intervals; hub consists of titanium centrepiece (including duct for mast-mounted sight) with composites starplates bolted above and below; flap and lead/lag motions of blades allowed by elastic bending of neck region and pitch change by elastic part of elastomeric bearings; lead/lag damping by solid-state viscoelastic damper struts faired into trailing-edge of each blade root; equivalent flapping hinge offset of 10.5 per cent gives high control power; SARIB passive vibration damping system between transmission and airframe; three-blade Spheriflex tail rotor has composite blades with fork roots; built-in ram air engine exhaust suppressors.

FLYING CONTROLS: Fully powered hydraulic flying controls by SAMM/Liebherr; Labinal/Electrométal servo trim; horizontal tail mounted beneath tail rotor; autopilot is part of basic avionics system (see under Avionics heading).

STRUCTURE: 80 per cent CFRP, block and sandwich and Kevlar sandwich; 6 per cent titanium and 11 per cent aluminium; airframe structure protected against lightning and EMP by embedded copper/bronze grid and copper bonding foil; stub-wings of aluminium spars with CFRP ribs and skins; titanium engine deck may be replaced by GFRP; airframe tolerates crash impacts at 10.5 m/s (34.4 ft/s) and meets MIL-STD-1290 crashworthiness standards; titanium main rotor hub centrepiece and tail rotor Spheriflex integral hub/mast; blade spars filament-wound; GFRP, CFRP skins and subsidiary spars and foam filling.

French plants building transmission, tail rotor, centre-fuselage (including engine installation), aerodynamics, fuel and electrical systems, weight control, maintainability, reliability and survivability; Eurocopter Deutschland responsible for main rotor, flight control and hydraulic systems, front and rear fuselage (including cockpits), prototype assembly, flight characteristics and performance, stress and vibration testing and simulation.

Pre-series Eurocopter Tigre HAP PS1 (*Jane's/Paul Jackson*)   NEW/0137356

LANDING GEAR: Tailwheel type, non-retractable, with single wheel on each unit. Designed to absorb impacts of up to 6 m (20 ft)/s. Main gear by Messier-Bugatti, tail gear by Liebherr Aerotechnik.

POWER PLANT: Two MTU/Rolls-Royce/Turbomeca MTR 390 modular turboshaft engines mounted side by side above centre-fuselage, divided by armour plate 'keel' (engine first flown in Panther testbed 14 February 1991); power ratings are maximum T-O 958 kW (1,285 shp), super emergency 1,160 kW (1,556 shp), maximum continuous 873 kW (1,171 shp). LHTEC has proposed the T800-801 as a potential alternative power plant for export variants of the Tiger.

Self-sealing crashworthy fuel tanks, with explosion suppression and with non-return valves which minimise leakage in a crash; total capacity 1,360 litres (359 US gallons; 299 Imp gallons). Provision for two external tanks, one on each inboard pylon, each of approximately 350 litres (92 US gallons; 77 Imp gallons) capacity. Gearbox has specified 30 minutes' dry running capability (demonstrated 65 minutes, November 1994).

ACCOMMODATION: Crew of two in tandem, with pilot in front and weapons system operator at rear; full dual controls; both crew members can perform all tasks and weapon operation except that anti-tank missile firing only available to gunner. Armoured, impact-absorbing seats; stepped cockpits, with flat-plate windscreens and slightly curved non-glint transparencies.

SYSTEMS: Redundant hydraulic, electrical and fuel systems. Primary power generation by two 20 kVA alternators; DC power generation by two 300 A 28 V transformer/rectifiers and two 23 Ah Ni/Cd batteries.

Dual redundant AFCS provides four-axis command and stability augmentation. Basic AFCS modes: attitude hold, heading hold. Higher AFCS modes: Heading/acquire/hold, barometric altitude capture/hold, altitude acquire, airspeed

hold, vertical speed acquire/hold, nav coupling, radar height hold, Doppler hover hold, line of sight acquisition/hold. Other AFCS functions: gun recoil force compensation, axis decoupling and tactical mode (follow-up trim on override of break-out forces).

AVIONICS: *Basic or core avionics* common to all three versions include bus/display system, com radio (French and German systems vary), autonomous nav system and radio/Doppler navaids, Thales TSC 2000 IFF Mk 12, NH 90-based ECM suite (including laser warning) and AFCS, all connected to and controlled through redundant MIL-STD-1553B data highway.

Flight: Navigation system, by Thales, Teldix and EADS, is fully redundant; system contains two Thales PIXYZ three-axis ring laser gyro units, two air data computers, two magnetic sensors, one Teldix/BAE Canada CMA 2012 Doppler radar, a radio altimeter and GPS; these sensors also provide signals for flight control, information display and guidance; integrated duplex AFCS by Thales and Nord Micro; AFCS computers produced by Thales, VDO-Luft and Litef.

Instrumentation: Colour liquid crystal flight displays showing symbology and imagery (two per cockpit for flight and weapon/systems information) by Thales and VDO-Luft; each crewman has central control/display unit for inputting all radio, electronic systems and navigation selections; digital map display system by Dornier and VDO-Luft (incorporating NH 90's Eurogrid map generation system); engine and systems data are fed into the databus for in-flight indication and subsequent maintenance analysis. BAE Systems Knighthelm fully integrated day and night helmet ordered for German Tigers; French Tigres have similar Thales Topowl helmet-mounted sights, with integrated night vision (image intensifiers), FLIR, video and synthetic raster symbology.

Mission: Euromep (European mission equipment package) includes SATEL Condor 2 pilot vision subsystem (PVS), air-to-air subsystem (Stinger or Mistral), mast-mounted sight and missile subsystem and Euromep management system all connected to separate MIL-STD-1553B data highway. Euromep Standard B avionics first flew February 1995 (PT5); Standard C testing began in late 1997 (PT3R). PVS has 40 × 30° instantaneous field of view (with ±110 × 35° total field of view) thermal imaging sensor steered by helmet position detector giving both crewmen day/night/bad weather vision, flight symbology and air-to-air aiming in helmet-mounted display; mast-mounted sight, gunner sight electronics and gunner's head-in target acquisition display and ATGW 3 subsystem connected by separate data highway; HOT 3 missile system also available. Thales armament control panel and fire-control computer.

HAP combat support mission equipment package includes SFIM STRIX gyrostabilised roof-mounted sight (with IRCCD IR channel) above rear cockpit; includes direct view optics with folding sight tube, television and IR channels and laser ranger/designator.

Self-defence: EADS C-model EW suite (as in NH 90) is one element in HAP's fully integrated avionics suite. This has an EADS laser warning receiver, EADS missile launch warning device (*Lenkflugkörpersysteme*) and Thales EW processor and radar warning receiver. Also chaff and flare dispensers (with up to 144 cartridges, sequenced by a Saphir M system). UH is similar, but with option of fitting IR jammer.

ARMAMENT: Tiger has four outboard weapon stations for typical four HOT, four Trigat, two Mistral or Stinger launchers or pods of 12 or 22 rockets; optional 12.7 mm gun pod with 250 rounds or auxiliary fuel tank. HCP (HAP) options all include one 30 mm Giat AM-30781 automatic cannon with up to 450 rounds (traversing from +33 to −30° in elevation and through ±90° in azimuth). Possible weapon configurations include (a) HAP: 44 rockets on outer stations, or 24 rockets and four Mistrals, or mix of launchers, (b) HAC: eight maximum HOT or eight Trigat or HOT/Trigat mix on inner stations, plus four Mistrals on outer stations, (c) UHT: 44 rockets, eight HOT, eight Trigat, or two gun pods on inner stations; four Stingers on outer stations. Combination of auxiliary fuel tank and weapon launcher (rocket,

Fifth Tiger prototype in German UHT configuration (*Eurocopter/Patrick Penna*)   NEW/0137439

HOT, Trigat, gun pod) on inner stations for extended-range missions is possible, (d) HCP (Export): nose-mounted 30 mm gun with 450 rounds, plus all stores combinations of UHT and HAP (except 12.7 mm gun pods).

DIMENSIONS, EXTERNAL (UT: UHT and HAC, HCP: HAP, where different):

Main rotor diameter	13.00 m (42 ft 7¾ in)
Tail rotor diameter	2.70 m (8 ft 10¼ in)
Length overall, rotors turning	15.80 m (51 ft 10 in)
Length of fuselage: UT	14.08 m (46 ft 2¼ in)
HCP: incl cannon	15.00 m (49 ft 2½ in)
excl cannon	14.77 m (48 ft 5½ in)
Height: to top of rotor head	3.84 m (12 ft 7¼ in)
to top of tail rotor disc	4.32 m (14 ft 2 in)
to top of mast sight (UT only)	5.20 m (17 ft 0¾ in)
Width over weapon pylons	4.53 m (14 ft 10¼ in)
Wheel track	2.38 m (7 ft 9¾ in)
Wheelbase	7.65 m (25 ft 1 in)

AREAS:

Main rotor disc	132.70 m² (1,428.7 sq ft)
Tail rotor disc	5.72 m² (61.63 sq ft)

WEIGHTS AND LOADINGS:

Weight empty (HCP)	4,200 kg (9,259 lb)
Fuel weight: internal	1,080 kg (2,381 lb)
external (two tanks)	555 kg (1,224 lb)
Mission T-O weight	5,300-6,100 kg (11,685-13,448 lb)
Max T-O weight	6,100 kg (13,448 lb)
Main rotor disc loading (max mission T-O weight)	45.2 kg/m² (9.26 lb/sq ft)

PERFORMANCE (UT and HCP, as above):

Never-exceed speed (V$_{NE}$):	
UHT, HAC	171 kt (298 km/h; 185 mph)
HCP, HAP	175 kt (322 km/h; 200 mph)
Max level speed: UT	140 kt (259 km/h; 161 mph)
HCP	150 kt (278 km/h; 173 mph)
Cruising speed	124 kt (230 km/h; 143 mph)
Max rate of climb at S/L: UHT	642 m (2,106 ft)/min
HAP	690 m (2,263 ft)/min
Vertical rate of climb at S/L: UHT	312 m (1,023 ft)/min
HAP	384 m (1,259 ft)/min
Hovering ceiling OGE: UHT	3,200 m (10,500 ft)
HAP	3,500 m (11,480 ft)
Range: on internal fuel	432 n miles (800 km; 497 miles)
with ferry tanks	691 n miles (1,280 km; 795 miles)
Endurance: operational mission	2 h 50 min
max internal fuel	3 h 25 min

UPDATED

EUROCOPTER SUPER PUMA Mk I and COUGAR Mk I

Spanish military designations: HD.21 and HT.21
Swedish Air Force designation: Hkp 10

TYPE: Multirole medium helicopter.

PROGRAMME: Early history and versions listed in 1985-86 and earlier *Jane's*; first flight AS 332 Super Puma (F-WZJA) 13 September 1978; six prototypes; deliveries began mid-1981; present version powered by Turbomeca Makila 1A1 introduced 1986; military versions renamed Cougar in 1990; first AS 332L (stretched fuselage) certified to French IFR Cat. II 7 July 1983 and delivered to Lufttransport of Norway; certified for flight into known icing 29 June 1983; FAA Cat. II certification with SFIM CDV 85 P4 four-axis AFCS and FAR Pt 25 Appendix C known icing clearance. Mk I continues in production as long as orders (mainly making good attrition) continue.

Some AS 332/532s built under licence by IPTN in Indonesia and TAI in Turkey; some assembled by CASA in Spain; and some equipped by F+W (now SF) in Switzerland. Super Puma/Cougar worldwide fleet had accumulated 2,134,000 hours by January 2001.

CURRENT VERSIONS: Designation suffixes: U: military unarmed utility, A: armed, S: anti-ship/submarine, C: military court (short) fuselage, L: long fuselage, military or civil.

AS 332L1 Super Puma: Standard Mk I civil version with long fuselage and airline interior for 20 passengers; UK CAA IFR certification 21 April 1992.

AS 532 Cougar 100: Simplified 9 tonne tactical transport; fixed landing gear reduces cruising speed by 5 kt (9 km/h; 6 mph); price reduced by 15 to 20 per cent. External tanks cannot be fitted. Two versions are proposed:
AS 532UB Short fuselage version and **AS 532UE** stretched fuselage version (first flight due 2000, but not confirmed); only AS 532UE currently marketed. Brazilian Army announced as first customer (for eight) on 16 June 2001; deliveries will take place during January-June 2002.

AS 532UC Cougar: Military Mk I short fuselage unarmed utility; seats up to 21 troops and two crew; cabin floor reinforced for 1,500 kg/m² (307 lb/sq ft). Crashworthy self-sealing fuel tanks standard. Equipment includes IR suppressors, RWR, chaff dispensers, flare launcher and armoured protection for crew and troops.

AS 532UL Cougar: Military Mk I unarmed long fuselage version; cabin lengthened by 0.76 m (2 ft 6 in); extra fuel and two large windows in forward cabin plug; carries up to 25 troops and two crew. Crashworthy self-sealing fuel tanks standard. Equipment as per AS 532UC; sliding forward cabin doors can be fitted to allow carriage of forward-firing 7.62 mm machine guns.

AS 532AC Cougar: Armed AS 532UC (Mk I).

Eurocopter AS 332L1 Super Puma of Aero Asahi, Japan (*H Seo/Eurocopter*)   *NEW*/0132427

AS 532AL Cougar: Armed AS 532UL (Mk I).

AS 532SC Cougar: Naval short Mk I version with ASW/ASV equipment; folding tail rotor pylon and main rotor blades; deck harpoon landing aid. Crew of two plus up to two operators.

AS 532UL Horizon: Battlefield surveillance radar helicopter; flight trials of small Orphée radar under an AS 330B Puma started 1986; French Army wanted 20 AS 532 Cougar Mk II fitted with radar, dedicated ECM and datalink (Orchidée programme). First flight of full-scale AS 532/Orchidée June 1990; programme cancelled for budgetary reasons August 1990; prototype without datalink flew 24 missions in Gulf War February 1991 in Operation Horus.

Horizon programme (*Hélicoptère d'Observation Radar et d'Investigation sur Zone*) with more effective operational concept and reduced development costs replaced Orchidée; development contract awarded to Eurocopter October 1992 for two (later increased to four) AS 532UL Horizons with same radar capabilities and ECM as Orchidée, but in AS 532UL Cougar Mk I, with standard ECM and longer mission endurance. Antenna rotates at either 2° 24' or 8°/s; radar range 108 n miles (200 km; 124 miles) with helicopter cruising at 97 kt (180 km/h; 112 mph) at 4,000 m (13,120 ft); normal operational range is 81 n miles (150 km; 93 miles); target speed resolution is 2 m (6.6 ft)/s. Military features include jet efflux deflector/diluters, anti-icing systems, NVG-compatible cockpit lighting, weather radar, encrypted communications, INS/GPS and active/passive countermeasures. First flight Horizon with full radar 8 December 1992; first delivery 24 June 1996; second (plus a ground station) in December 1996; third delivered in 1997 and fourth (plus second ground station) in 1998.

AS 332L2 Super Puma Mk II: Stretched version with Spheriflex rotor heads, new cockpit, enlarged rotors and more powerful Makila 1A2 engines. See separate entry.

AS 532U2/A2 Cougar Mk II: Stretched version with Spheriflex rotor heads, new cockpit, enlarged rotors and more powerful Makila 1A2 engines. See separate entry.

CUSTOMERS: Total 575 AS 332/532s ordered, of which 520 delivered to nearly 100 customers in 50 countries (two-thirds of them military Cougars with 26 air forces, eight armies and five navies in 33 countries) by 1 June 2001; orders for Super Pumas and Cougars of all marks in 2001 totalled 26, compared with seven in 1998, 22 in 1999 and five in 2000. 23 were delivered in 2000, compared to 17 in 1999. 500th helicopter in family (an AS 332L2) was delivered to Bristow Helicopters on 17 June 1999. Some 110 AS 332L1/L2 Super Pumas in civil offshore oil industry support operations; French military orders include six for French Air Force (three for nuclear test facilities in Pacific and three for government VIP flying); French Army (ALAT) has supplemented SA 330 Pumas with AS 532 Cougars; 22 delivered to *Force d'Action Rapide* between late 1988 and end 1991.

Export customers include Abu Dhabi (eight including two VIP of which five to be upgraded with Exocet, sonar and torpedoes under 1995 contract), Brazil (25 including six AS 532SCs, one AS 332M1 and eight ordered January 2001), Cameroon (one), Chile (army, two; navy, four AS 532SCs), China (six), Ecuador (eight including six army), Finland (three for border police), Gabon (two, Presidential Guard), Germany (three, border police), Greece (Ministry of Merchant Marine, four ordered August 1998 and first two delivered 21 December 1999; remaining pair March 2000), Greek Air Force (four plus two options in 2001 for delivery between 2001 and 2003); Indonesia (built under licence: see IPTN entry); Japan (three, army/VIP), Jordan (eight), South Korea (three army/VIP, one AS 332L1), Mexico (two VIP), Nepal (two, Royal Flight), Nigeria (two), Oman (two, Royal Flight), Panama (one VIP), Saudi Arabia (12 AS 532SCs), Singapore (36), Slovenian Army (two ULs), Spain (10 SAR HD.21s, two VIP HT.21s, 18 army tactical transport HT.21s plus 15 AS 532ULs ordered February 1996 and delivered from 1998), Sweden (12 SAR), Switzerland (15 AS 332Ms; plus 12 AS 532ULs ordered December 1998 for delivery from 2000; two to be built by Eurocopter and remainder subcontracted to RUAG Aerospace, which see), Togo (one), Turkey (30 AS 532ULs and 20 AS 532ALs in SAR configuration, including 28 assembled by TAI; first Turkish-built AS 532AL handed over 1 May 2000; deliveries to be completed by February 2003; first Turkish-built AS 532UL handed over 31 May 2000), Venezuela (six), Zaïre (one VIP). Competing for Taiwanese UH-1 and Greek UH-1/AB 205 replacement programmes. Bristow Helicopters acquired 31 examples of 19-passenger AS 332L (see 1993-94 *Jane's*) for offshore oil support. See also Super Puma/Cougar Mk II.

Eurocopter AS 332C1 operated by Greek Air Force on behalf of Coast Guard (*Patrick Penna/Eurocopter*)
NEW/0132428

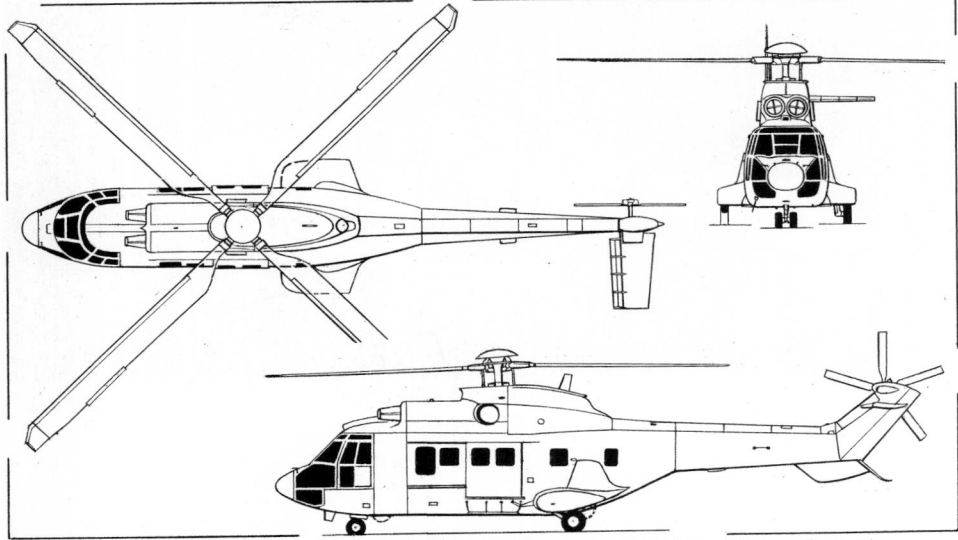

Eurocopter AS 532UL Cougar (long-fuselage military Mk I unarmed utility) *(Jane's/Dennis Punnett)*

COSTS: AS 332L1 US$14.25 million, IFR; direct operating cost US$1,542 per hour (both 1997).

DESIGN FEATURES: Four-blade fully articulated main rotor turning clockwise seen from above; five-blade tail rotor on starboard side of tailboom; engines mounted above cabin have rear drive into main transmission at 23,840 rpm; main rotor turns at 265 rpm, tail rotor at 1,278 rpm; various lateral sponsons available housing partly retracted main landing gear and combinations of additional fuel or pop-out floats; optional air conditioning system housed in casing on port forward flank of cabin; all civil versions certified for IFR category A and B to FAR Pt 29.

FLYING CONTROLS: Dual fully powered hydraulic controls with full-time autostabilisation and yaw damping; machine remains flyable with autostabiliser switched off; cyclic trimming by stick friction adjustment; inverted slot on tailplane to maintain attitude-holding effect at low climb speeds; large ventral fin; saucer fairing on rotor head to smooth wake of hub; four-axis SFIM 155 autopilot standard.

STRUCTURE: Conventional light alloy airframe with some titanium; crashworthy fuel system, impact-absorbing landing gear and other features; main rotor blades of GFRP with CFRP stiffening and Moltoprene filler; elastomeric drag dampers.

LANDING GEAR: Retractable tricycle high energy absorbing design by Messier-Bugatti; all units retract rearward hydraulically, mainwheels into sponsons on sides of fuselage; dual-chamber oleo-pneumatic shock-absorbers; twin-wheel self-centring nose unit, tyre size 7.00-6 (8 ply) tubeless, pressure 7.00 bar (102 lb/sq in); single wheel on each main unit with tyre size 615×225-10 (12 ply) tubeless or 640×230-10, pressure 9.00 bar (130 lb/sq in); hydraulic differential disc brakes, controlled by foot pedals; lever-operated parking brake; emergency pop-out flotation units can be mounted on main landing gear fairings and forward fuselage.

POWER PLANT: Two Turbomeca Makila 1A1 turboshafts, each with maximum contingency rating of 1,400 kW (1,877 shp) and take-off rating of 1,357 kW (1,820 shp). Maximum continuous rating of 1,184 kW (1,588 shp). Air intakes protected by a grille against ingestion of ice, snow and foreign objects; Centrisep multipurpose intake optional for flight into sandy areas. Transmission rated at 2,238 kW (3,000 shp) for T-O.

AS 532UC/AC have five flexible fuel tanks under cabin floor, with total usable capacity of 1,497 litres (395 US gallons; 329 Imp gallons); AS 532SC has total basic capacity of 2,141 litres (565 US gallons; 471 Imp gallons); AS 332L1/532UE/532UL/532AL have a basic fuel system of six flexible tanks with total capacity of 2,020 litres (533 US gallons; 444 Imp gallons) in the 332 and 2,003 litres (529 US gallons; 440 Imp gallons) in the 532; provision for additional 1,900 litres (502 US gallons; 418 Imp gallons) in four auxiliary ferry tanks installed in cabin; two external auxiliary tanks with total capacity of 650 litres (172 US gallons; 143 Imp gallons) standard on AS 532SC, optional on other versions; AS 532UB Cougar 100 has capacity of 1,533 litres (405 US gallons; 337 Imp gallons). For long-range missions (mainly offshore) in AS 332L1, a special internal auxiliary tank can be fitted in cargo sling well, in addition to the two external tanks, to raise total usable fuel capacity to 2,994 litres (791 US gallons; 658 Imp gallons). Refuelling point on starboard side of cabin; fuel system designed to avoid leakage following a crash; self-sealing tanks standard on military versions, optional on other versions; other options include a fuel dumping system and pressure refuelling.

ACCOMMODATION: One pilot (VFR) or two pilots side by side (IFR) on flight deck, with jump seat for third crew member or paratroop dispatcher; provision for composite light alloy/Kevlar armour for crew protection on military models; door on each side of flight deck and internal doorway connecting flight deck to cabin; dual controls, co-pilot instrumentation and crashworthy flight deck and cabin floors; maximum accommodation for 21 troops in AS 532UC, 20 passengers in AS 332L1, 24 in AS 332L2, and 25 troops in AS 532UL, 20 troops in AS 532UE; interiors available for VIP, air ambulance with six stretchers and 11 seated casualties/attendants; strengthened floor for cargo carrying, with lashing points; jettisonable sliding door on each side of main cabin or port side door with built-in steps and starboard side double door in VIP configuration; removable panel on underside of fuselage, at rear of main cabin, for longer loads; removable door with integral steps for access to baggage racks optional; hatch in floor below main rotor contains hook for slung loads up to 4,500 kg (9,920 lb) on internally mounted cargo sling; cabin and flight deck heated, ventilated and soundproofed; demisting, de-icing, washers and wipers for pilots' windscreens.

SYSTEMS: Two independent hydraulic systems, supplied by self-regulating pumps driven by main gearbox. Each system supplies one set of powered flying controls; left-hand system also supplies autopilot, landing gear, rotor brake and wheel brakes; hydraulically actuated systems can be operated on ground from main gearbox (when a special disconnect system is installed to permit running of port engine with rotors stationary), or by external power through ground power receptacle. Emergency landing gear lowering by standby pump.

Three-phase 200 V AC electrical power supplied by two 20 kVA 400 Hz alternators, driven by port side intermediate shaft from main gearbox and available on ground under same conditions as hydraulic ancillary systems; two 28.5 V DC transformer-rectifiers; main battery used for self-starting and emergency power in flight.

AVIONICS: *Comms:* Optional equipment includes VHF, UHF, tactical HF and HF/SSB radio and intercom.

Radar: Offshore models have nose-mounted radar; search and rescue version has nose-mounted RDR 1400 or chin-mounted 1500 search radar (as on Swedish Hkp 10s); naval ASW and ASV versions can have nose-mounted Thales Ocean Master radar, linked to a tactical table in the cabin.

Flight: ADF, VOR/ILS, radio altimeter, GPS, VLF Omega, Decca navigator and flight log, Doppler, SFIM 155 autopilot, with provision for coupling to self-contained navigation and landing systems; search and rescue version has Doppler, and Thales Nadir or Decca self-contained navigation system (Nadir Mk 2 in French Army and Greek versions), including navigation computer with SAR patterns, polar indicator, roller map display, hover indicator, route mileage indicator and groundspeed and drift indicator; SFIM CDV 155 autopilot coupler contains automatic nav track including search patterns, transitions and hover; multifunction video display shows radar and route images, SAR patterns and hover indication; Swedish Hkp 10s have Thales RAMS flight management system including BAE Systems AHRS, Decca Doppler and GPS. Optional Honeywell Mk XXII EGPWS.

Instrumentation: Full IFR instrumentation optional.

Mission: Naval ASW versions have Thomson Marconi HS 312 acoustics suite; upgraded FLASH version currently available.

EQUIPMENT: Optional fixed rescue hoist (capacity 275 kg; 606 lb) starboard side and electrical back-up hoist; equipment for naval missions can include sonar, MAD and sonobuoys.

ARMAMENT (optional): Typical alternatives for army/air force missions are two 20 mm guns or two 7.62 mm machine guns or two 68 mm rocket launchers. Armament for naval missions includes two AM 39 Exocet missiles or two lightweight torpedoes.

DIMENSIONS, EXTERNAL:	
Main rotor diameter	15.60 m (51 ft 2¼ in)
Tail rotor diameter	3.05 m (10 ft 0 in)
Main rotor blade chord	0.60 m (1 ft 11½ in)
Length: overall, rotors turning	18.70 m (61 ft 4¼ in)
fuselage, incl tail rotor:	
AS 532UC/AC/SC/UB	15.53 m (50 ft 11½ in)
AS 332L1/532AL/UL/UE	16.29 m (53 ft 5½ in)
tail and rotors folded (AS 532SC)	12.95 m (42 ft 5¾ in)
Width, blades folded:	
AS 532UB/UC/AC/AL/UL/332L1	3.79 m (12 ft 5¼ in)
AS 532SC	3.86 m (13 ft 3 in)
Width over sponsons:	
AS 332L1/AS 532AL/UL/UC/AC	3.38 m (11 ft 1 in)
AS 532SC	3.56 m (11 ft 8¼ in)
Fuselage max width	2.00 m (6 ft 6 in)
Height: overall	4.92 m (16 ft 1¾ in)
blades and tail pylon folded:	
AS 532UE/UC/AC/SC	4.80 m (15 ft 9 in)
to top of rotor head	4.60 m (15 ft 1 in)
Tailplane half-span	2.11 m (6 ft 11 in)
Wheel track	3.00 m (9 ft 10 in)
Wheelbase: AS 532UC/AC/SC	4.49 m (14 ft 8¾ in)
AS 332L1/532AL/UL/UE	5.25 m (17 ft 2¾ in)
Passenger cabin doors, each: Height	1.35 m (4 ft 5 in)
Width	1.30 m (4 ft 3¼ in)
Floor hatch, rear of cabin: Length	0.98 m (3 ft 2¾ in)
Width	0.70 m (2 ft 3½ in)

DIMENSIONS, INTERNAL:	
Cabin: Length: AS 532UC/AC/SC	6.05 m (19 ft 10½ in)
AS 332L1/532AL/UL/UE	6.81 m (22 ft 4 in)
Max width	1.80 m (5 ft 10¾ in)
Max height	1.55 m (5 ft 1 in)
Floor area: AS 532UC/AC/SC	7.80 m² (84.0 sq ft)
AS 332L1/532AL/UL/UE	9.18 m² (98.8 sq ft)
Usable volume:	
AS 532UC/AC/SC	11.4 m³ (403 cu ft)
AS 332L1/532AL/UL/UE	13.3 m³ (470 cu ft)

AREAS:	
Main rotor disc	191.13 m² (2,057.3 sq ft)
Tail rotor disc	7.31 m² (78.64 sq ft)

WEIGHTS AND LOADINGS	
Weight empty (standard aircraft):	
AS 532UC/AC	4,330 kg (9,546 lb)
AS 532SC	4,500 kg (9,920 lb)
AS 332L1/532AL/UL	4,460 kg (9,832 lb)
AS 332UE	4,485 kg (9,888 lb)
Max external payload	4,500 kg (9,920 lb)
Max T-O weight: internal load:	
AS 532UC/AC/AL/UB/UL/SC/UB/UE	9,000 kg (19,841 lb)
AS 332L1	8,600 kg (18,960 lb)
external slung load	9,350 kg (20,615 lb)
Max disc loading, external load	48.9 kg/m² (10.02 lb/sq ft)
Transmission loading at max T-O weight and power:	
internal load:	
AS 532UC/AC/AL/UB/UL/SC	4.03 kg/kW (6.61 lb/shp)
AS 332L1	3.85 kg/kW (6.32 lb/shp)
external load	4.18 kg/kW (6.87 lb/shp)

PERFORMANCE:	
Never-exceed speed (V NE)	150 kt (278 km/h; 172 mph)
Cruising speed at S/L:	
AS 532UC/AC/AL/UL/SC	139 kt (257 km/h; 160 mph)
AS 532UB/UE	134 kt (249 km/h; 155 mph)
AS 332L1	141 kt (261 km/h; 162 mph)
Max rate of climb at S/L:	
AS 532UC/AC/AL/UL	420 m (1,378 ft)/min
AS 532UB/UE	432 m (1,417 ft)/min
AS 532SC	372 m (1,220 ft)/min
AS 332L1	486 m (1,594 ft)/min
Service ceiling: AS 332L1	4,600 m (15,080 ft)
AS 532UC/AC/SC/AL/UL/UB/UE	4,100 m (13,450 ft)
Hovering ceiling IGE:	
AS 532AL/UL/AC/UC/SC/UB/UE	2,800 m (9,180 ft)
AS 332L1	3,250 m (10,660 ft)
Hovering ceiling OGE:	
AS 532AL/UL/AC/UC/SC/UB/UE	1,650 m (5,420 ft)
AS 332L1	2,300 m (7,540 ft)
Range at S/L, standard tanks, no reserves:	
AS 332L1	449 n miles (831 km; 516 miles)
AS 532UC/AC	334 n miles (618 km; 384 miles)
AS532UB	309 n miles (573 km; 356 miles)
AS 532SC	492 n miles (911 km; 566 miles)
AS 532AL/UL	455 n miles (842 km; 523 miles)
AS 332L1	470 n miles (870 km; 540 miles)
AS 532UE	416 n miles (770 km; 478 miles)
Range with auxiliary internal tank:	
AS 532UE	486 n miles (900 km; 559 miles)
AS 532UB	380 n miles (703 km; 437 miles)
Range at S/L with external (two 338 litre) and auxiliary (one 320 litre) tanks, no reserves:	
AS 332L1	659 n miles (1,220 km; 758 miles)
AS 532UC/AC	528 n miles (977 km; 607 miles)
AS 532AL/UL	637 n miles (1,179 km; 733 miles)

UPDATED

EUROCOPTER SUPER PUMA Mk II and COUGAR Mk II
Spanish Army designation: HU.21L
TYPE: Multirole medium helicopter.
PROGRAMME: Derived from Super Puma/Cougar Mk I (which see); first flight of development vehicle 6 February 1987; French certification 2 April 1992; UK BCAR 29 certification 16 November 1992; first delivery August 1993. Earlier AS 332L1 and 532U/A Cougar Mk I remain on production line. Further development is under consideration.
CURRENT VERSIONS: **AS 332L2 Super Puma Mk II:** Current production civil transport; innovations include Spheriflex rotor heads, super-emergency engine rating, EFIS flight deck and built-in health and usage monitoring system, duplex four-axis AFCS, and hydraulically powered standby electrics for 2 hours' operation after complete main generator failure. First delivery 28 August 1993.
AS 332L2 Super Puma Mk II VIP: Eight- to 15-passenger arrangement with attendant and fully equipped galley and lavatory; two four-seat lounges; fine materials and fittings throughout; range with eight passengers and attendant, 635 n miles (1,176 km; 730 miles). Used in 28 countries.
AS 532U2 Cougar Mk II: Unarmed military tactical transport; longest member of Cougar family; carries 29 troops and two-man crew.
AS 532A2 Cougar Mk II: Armed version. First order (for four) by French Air Force for combat SAR, 1995; known as Cougar RESCO (REcherche et Sauvetage en COmbat); prototype (F-ZVMC) undertook maker's trials in 1998; was delivered to CEAM military test unit on 9 September 1999 for one year of evaluation; second aircraft, to determine operational fit, was due to be delivered in 2000. Potential equipment includes RWR, missile approach warner, chaff, flares, rescue hoist, two outboard 20 mm cannon, two door-mounted 12.7 mm machine guns, encrypted Personnel Locator System, third-generation NVG-compatible lighting, air-to-air refuelling capability. Nadir Mk 2 navigation computer, SFIM PA-165 four-axis digital autopilot, VOR, Tacan, DME, GPS, searchlight and FLIR. Alternative maximum T-O weight of 11,200 kg (24,692 lb) permits an 800 n mile (1,482 km; 920 mile), plus 30 min hover and 30 min reserve, unrefuelled radius of action to rescue two persons.
EC 225 Super Puma Mk II+: Growth version, announced June 1998, as competitor to Sikorsky S-92 in offshore support role; has military counterpart in **EC 725 Cougar Mk II+**. Cabin volume increased by 25 per cent to 20.0 m³ (706 cu ft) through increases of 35 cm (13¾ in) in height, 70 cm (2 ft 3½ in) in length and 25 cm (9¾ in) in width. Reinforced gearbox and 1,800 kW (2,414 shp) Makila 1A4 engines with FADEC produce 14 per cent power increase over earlier versions. Five-blade main rotor blades have composites spars and parabolic tips, diameter unchanged. MTOW increased to 10,400 kg (21,560 lb) for civil and 11,000 kg (24,251 lb) for military version (11,200 kg; 24,691 lb with external load). New integrated flight and display system (IFDS) using four 152 × 203 mm (6 × 8 in) multifunction LCDs plus four-axis autopilot. FAR/JAR-29 certification expected late 2002 and first deliveries to French Air Force, April 2003; sales of 200 anticipated. First prototype (No. 2338, F-ZVLR, converted from standard Mk II) made maiden flight 27 November 2000, at which time Eurocopter revealed an order for four

from the French Air Force, which requires an eventual 14, including one earlier Mk 2 raised to this standard. JAR 29 certification expected by end-2002. Launch customer for EC 225 is CHC Helikopter, which ordered one (plus one AS332L2) in early 2001. French Air Force took delivery of initial aircraft 15 January 2001 for trials. First pre-production EC 725 (with Makila 1A4 engines) was due to fly during mid-2001, with optional de-icing system to be fitted late 2001. First fully equipped production example due for delivery September 2003 to EH1/67 'Pyrénées' at Cazaux.
CUSTOMERS: Helikopter Service ordered four AS 332L2 Mk II in May 1992 and a further four in June 1997, latter delivered from 27 May 1998 onwards; RNethAF ordered 17 AS 532U2s October 1993 and deliveries began 1 April 1996 (with 400th AS 332/532) and last was received in October 1997; first of follow-on order for two for Bond Helicopters delivered 27 August 1998. Eventual 14 AS 532A2s required for combat SAR by French Air Force, first of four ordered delivered 9 September 1999; remainder to be delivered in three batches by 2015; these orders converted to EC 725 version late 2000. Saudi Arabia ordered 12 AS 532A2s in July 1996 under 'Al Fahd' contract for combat SAR, deliveries beginning in 1998; in-flight refuelling tests successfully completed during second half of 2000. VIP orders include Samsung (one AS 332L2) and the German government (three AS 532U2s delivered from 4 November 1997 and operated by the Luftwaffe). Three ordered for SAR by Hong Kong Government Flying Service for delivery mid-2001, the first being handed over 18 June 2001. Total of 46 of both versions in service by June 2000.
COSTS: 17 Netherlands AS 532U2 cost FFr1.4 billion, plus 120 per cent offset and technology offset over 12 years. Spanish order for 15 valued at US$207 million in 1996.
DESIGN FEATURES: Generally as for Mk I, but gross weight increased by 150 kg (331 lb) in 1993; composites plug in rear cabin area accommodates one extra row of seats and moves tail rotor rearwards; Spheriflex main and tail rotor heads with elastomeric spherical bearings and Kevlar retention bands; main rotor with longer blades having parabolic tips; enlarged composites sponsons can contain fuel, liferafts, air conditioning and pop-out floats.

Thai Air Force AS 332L2 Super Puma Mk 2 *(Gérard Deulin/Eurocopter)*   NEW/0132429

FLYING CONTROLS: Fully powered hydraulically actuated with SFIM 165 four-axis digital AFCS and coupler.
STRUCTURE: Rear fuselage 55 cm (1 ft 9½ in) plug made of composites giving extra row of seats; large windows at rear of cabin; increased use of composites compared with Mk I; fuselage frames strengthened to retain same degree of crashworthiness.
LANDING GEAR: As Super Puma Mk I.
POWER PLANT: Two Turbomeca Makila 1A2 rear drive free turbines; maximum emergency power each (OEI 30 seconds) 1,573 kW (2,109 shp), 12 per cent higher than AS 332 Mk I; intermediate emergency power (OEI 2 minutes) 1,467 kW (1,967 shp); take-off power 1,376 kW (1,845 shp); maximum continuous power 1,236 kW (1,657 shp); protective intake grilles standard; optional Centrisep multipurpose intakes for dusty conditions. Transmission ratings: twin-engined maximum 2,410 kW (3,229 shp), maximum from single engine 1,666 kW (2,232 shp), maximum transient 20 seconds 2,651 kW (3,552 shp), maximum continuous 1,555 kW (2,084 shp); transmission has extended run-dry capability.
Standard fuel tankage under cabin floor 2,020 litres (535 US gallons; 444 Imp gallons); auxiliary tankage includes 324 litres (85.6 US gallons; 71.3 Imp gallons) in cargo hook well; sponson tanks each holding 325 litres (85.9 US gallons; 71.5 Imp gallons); 600 litres (159 US gallons; 132 Imp gallons) in cabin fuel tank; one to five internal ferry tanks each holding 475 litres (126 US gallons; 104.5 Imp gallons). Tankage (self-sealing tanks standard from 1997) AS 532A2 1,896 litres (500 US gallons; 417 Imp gallons); hook well tank 320 litres (84.5 US gallons; 70.4 Imp gallons); 325 litres (85.9 US gallons; 71.5 Imp gallons) in each sponson tank; crashproofing is standard in AS 532 and optional in 332; pressure refuelling optional. Military versions can be fitted with probe for in-flight refuelling; under consideration for EC 725.
ACCOMMODATION: Single pilot in DGAC Category B civil operation; single pilot plus licensed crewman in Category A VFR; two pilots in IFR; military operation one pilot in VFR, two in IFR; civil transport 24 passengers and attendant in airline interior for 220 n miles (408 km; 253 miles) or 19 passengers at 81 cm (32 in) seat pitch for 350 n miles (648 km; 403 miles); military capacity, squad chief plus 28 troops; SAR version four crew plus 15 survivors over radius of 285 n miles (527 km; 328 miles). VIP version for eight to 15 passengers plus attendant; ambulance version holds 12 stretchers and four seated casualties plus attendant.
Flight: Optional Bendix/King Mark XXII EGPWS.
Radar: RNethAF AS 532U2s have been retrofitted with Honeywell weather radar.
AVIONICS: *Instrumentation:* Civil and military cockpits have Sextant Avionique four-tube EFIS integrated flight and display system (IFDS) with four 15 × 15 cm (6 × 6 in) screens; subsystems include smart multimode display (SMDS), automatic flight control system (AFCS) and flight data system (FDS); military cockpit compatible with night vision goggles; civil and military IFR system offered, plus military communications radio; SAR system includes radar and FLIR coupled to display system as well as navigation computer with Doppler and GPS sensors; can also be fitted with inertial nav sensor for combat SAR configuration at customer's request; SAR system can include automatic search pattern, transition and hover hold.
Equipment: Pop-out floats, rescue winch (272 kg; 600 lb), external sling. Military equipment (U2/A2) includes IR suppressors, flare/chaff dispensers, RWR, MAWS and armoured protection for crew and troops.

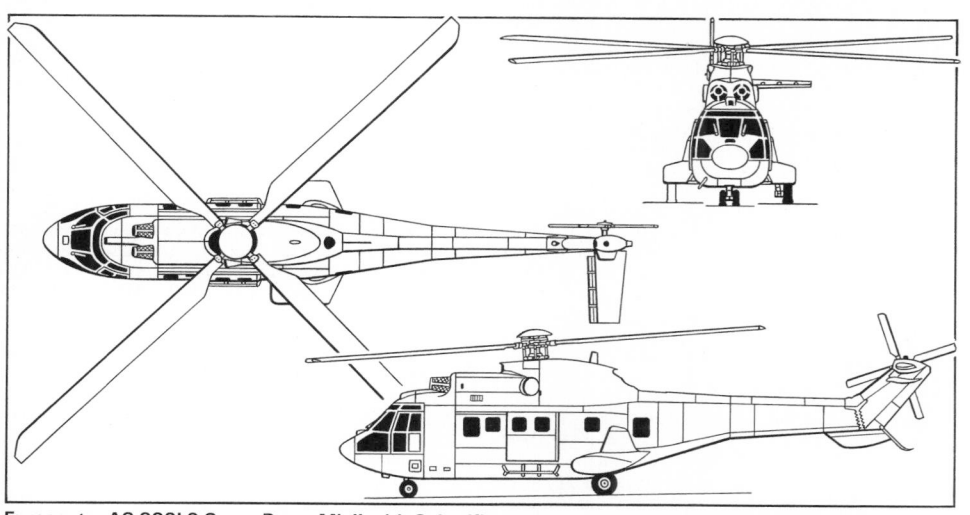
Eurocopter AS 332L2 Super Puma Mk II with Spheriflex main and tail rotor heads *(Jane's/Mike Keep)*

French Air Force Cougar RESCO during service trials (*Jean-Louis Gaynecoetche*) NEW/0132399

Prototype EC 725 Cougar posing as an EC 225 civil variant (*Gérard Deulin/Eurocopter*) NEW/0132430

ARMAMENT: Armament choice as for AS 532A Cougar Mk I except for Exocet. Special sliding doors on forward cabin windows allow forward firing 7.62 mm machine guns to be fitted.

DIMENSIONS, EXTERNAL:
Main rotor diameter	16.20 m (53 ft 1½ in)
Tail rotor diameter	3.15 m (10 ft 4 in)
Length overall: rotors turning	19.50 m (63 ft 11 in)
main rotor folded	16.79 m (55 ft 0½ in)
Width: fuselage	2.00 m (6 ft 6 in)
over sponsons	3.38 m (11 ft 1 in)
overall, main rotor folded	3.86 m (12 ft 8 in)
Height: overall, tail rotor turning	4.97 m (16 ft 4 in)
to top of rotor head	4.60 m (15 ft 1 in)
Tailplane span	2.12 m (6 ft 11½ in)
Wheel track	3.00 m (9 ft 10 in)
Wheelbase	5.25 m (17 ft 2¾ in)
Sliding cabin doors, each: Height	1.35 m (4 ft 5 in)
Width	1.30 m (4 ft 3¼ in)
Floor hatch: Length	0.98 m (3 ft 2¾ in)
Width	0.70 m (2 ft 3½ in)

DIMENSIONS, INTERNAL:
Cabin: Max length	7.87 m (25 ft 10 in)
Floor length	6.15 m (20 ft 2¼ in)
Max width	1.80 m (5 ft 10¾ in)
Max height	1.45 m (4 ft 9 in)
Volume	15.0 m³ (530 cu ft)

AREAS:
Main rotor disc	206.12 m² (2,218.7 sq ft)
Tail rotor disc	7.79 m² (83.88 sq ft)

WEIGHTS AND LOADINGS:
Manufacturer's weight empty: L2	4,686 kg (10,331 lb)
U2	4,945 kg (10,902 lb)
A2	5,012 kg (11,050 lb)
Useful load: L2	4,595 kg (10,130 lb)
U2	4,805 kg (10,593 lb)
A2	4,738 kg (10,445 lb)
EC 225	5,380 kg (11,861 lb)
EC 725	5,670 kg (12,500 lb)
Standard fuel weight: L2	1,596 kg (3,519 lb)
U2, crashworthy tanks	1,548 kg (3,412 lb)
Max normal T-O weight: L2	9,300 kg (20,502 lb)
U2	9,750 kg (21,495 lb)
EC 225	10,400 kg (22,928 lb)
EC 725	11,000 kg (24,250 lb)
A2 Combat SAR	11,200 kg (24,692 lb)
Max slung load: A2/U2, EC 725	5,000 kg (11,023 lb)
L2	4,500 kg (9,920 lb)

Max flight weight with slung load:
U2	10,500 kg (23,148 lb)
EC 225, EC 725	11,200 kg (24,692 lb)

Max disc loading, normal T-O weight:
L2	45.1 kg/m² (9.24 lb/sq ft)
U2	47.3 kg/m² (9.69 lb/sq ft)
U2, alternate	54.3 kg/m² (11.13 lb/sq ft)

Transmission loading at max T-O weight and power:
L2	3.85 kg/kW (6.34 lb/shp)
U2 with slung load	4.35 kg/kW (7.16 lb/shp)
U2 alternate	4.65 kg/kW (7.64 lb/shp)

PERFORMANCE:
Never-exceed speed (VNE):
L2/U2/EC 725	170 kt (315 km/h; 195 mph)
Fast cruising speed: L2	150 kt (277 km/h; 172 mph)
U2	147 kt (273 km/h; 170 mph)
EC 225	156 kt (289 km/h; 180 mph)
EC 725	160 kt (296 km/h; 184 mph)
Econ cruising speed: L2	136 kt (252 km/h; 157 mph)
U2	132 kt (244 km/h; 152 mph)

Rate of climb at 70 kt (130 km/h; 81 mph) at S/L:
L2	441 m (1,447 ft)/min
U2	384 m (1,260 ft)/min

Service ceiling (45.7 m; 150 ft/min climb), ISA:
L2	5,180 m (17,000 ft)
U2	4,100 m (13,450 ft)

Hovering ceiling IGE, normal T-O weight, ISA, T-O power:
L2	3,120 m (10,240 ft)
U2	2,690 m (8,820 ft)

Hovering ceiling OGE, normal T-O weight, ISA, T-O power:
L2	2,250 m (7,380 ft)
U2	1,900 m (6,240 ft)

Hovering ceiling OGE, slung load weight, ISA, T-O power: L2/U2 960 m (3,150 ft)

Range, no reserves, standard fuel, econ cruise:
L2	447 n miles (828 km; 514 miles)
U2	427 n miles (791 km; 491 miles)

Range, no reserves, max fuel, econ cruise:
L2	656 n miles (1,215 km; 755 miles)
U2	636 n miles (1,178 km; 732 miles)
EC 225, EC 725	783 n miles (1,450 km; 901 miles)

Ferry range: EC 725 1,000 n miles (1,852 km; 1,150 miles)

Endurance, standard fuel, at 70 kt (130 km/h; 81 mph):
L2	4 h 54 min
U2	4 h 20 min
EC 725	6 h 30 min

UPDATED

EUROCOPTER AS 350 ECUREUIL/ASTAR and AS 550 FENNEC

Brazilian Air Force designations: CH-50 and TH-5 Esquilo
Brazilian Army designation: HA-1 Esquilo
Brazilian Navy designation: UH-12 Esquilo
UK forces designation: Squirrel HT. Mk 1 and HT. Mk 2

TYPE: Light utility helicopter.

PROGRAMME: First flight (F-WVKH) powered by Lycoming LTS101 turboshaft 27 June 1974; first flight second prototype (F-WVKI) powered by Turbomeca Arriel 1A February 1975; first production version was AS 350B powered by 478 kW (641 shp) Arriel 1B and certified 27 October 1977; LTS101 powered AStar sold only in USA. AS 350BA Ecureuil was powered by 478 kW (641 shp) Turbomeca Arriel 1B and fitted with large main rotor blades of AS 350B2 (see below); maximum T-O weight increased by 150 kg (331 lb); AS 350B upgraded to AS 350BA in field; replaced AS 350B during 1992; French VFR certification 1991; UK and US certifications 1992, Japanese 1993. Production of AS 350BA ended in 1998; see *Jane's* 1999-2000 for details.

Delivery of the 2,000th AS 350/550 (F-WWPZ, later ZJ270 of UK Defence Helicopter Flying School) effected in July 1997. In February 2001, delivery effected of 3,000th of Ecureuil family (including AS 355/555), which also was first EC 130 upgraded version. Total 10,986,000 Ecureuil flying hours by January 2001.

Built at Marignane, France, and under licence by Helibrás in Brazil (which see). See also the CHAIC Z-11 entry in the Chinese section.

CURRENT VERSIONS: **AS 350B2 Ecureuil:** Uprated engine and transmission; wide-chord new section main and tail rotor blades originally developed for AS 355 twin; certified 26 April 1989; known as **SuperStar** in North America; supplied to UK Ministry of Defence as **AS 350BB**, designated **Squirrel HT. Mk 1** with normal instrumentation and **Squirrel HT. Mk 2** with provision for pilot's NVGs.

AS 350B1, 350BA and 350B versions can be upgraded to 350B2.

AS 350B3 Ecureuil: Current variant; 632 kW (847 shp) Arriel 2B and wide-chord tail rotor; optimised for high-altitude operation; uprated transmission and digital engine control. First flight (F-WWPB) 4 March 1997; French VFR certification December 1997 after 150 hour test programme; first delivery, to Osterman Helicopter, January 1998; Australian and New Zealand certification achieved late 1998. By 1 June 2001, 180 AS 350B3s were in service with 120 customers, in 25 countries; 100th AS 350B3 delivered 14 December 1999 to Japanese customer.

AS 350D AStar: Version of 350B for North American market; improved LTS101 engine, with 6 per cent more power, offered early 1999.

AS 350 Firefighter: Conair system able to pick up water load in 30 seconds while hovering over water; demonstrated 1986; Isolair system certified September 1995.

AS 550C3 Fennec: Light anti-tank version armed with four TOW missiles; has same power plant as AS 550A3 and 2 hour 45 minute endurance.

AS 550A3 Fennec: Light attack version of AS 350B2 with 3-hour endurance; powered by 632 kW (847 shp), Arriel 2B; standard features include taller landing gear, sliding doors; NVG-compatible cockpit; reinforced airframe; provision for armoured seats and engine cowlings; can be armed with single 20 mm cannon or two rocket launchers.

AS 550U3 Fennec: Military utility version with Arriel 2B power plant; capable of performing observation, commando and transport duties.

EC 130: *Described separately.*

CUSTOMERS: Total 3,028 AS 350s, AS 355s and EC 130s ordered by at least 70 countries by January 2001, of which 2,675 then in service; 1,948 single-engined Ecureuils in service at May 2000.

Military customers include Singapore armed forces (10 AS 550B2 and 10 AS 550C2); Australia (RAAF 18 for training; RAN six utility); Danish Army (12 AS 550C2 with ESCO HeliTOW system ordered 1987, delivered 1990); French Army ALAT (originally expressed need for up to 100 to replace Alouette IIs); French CEV (flight test centre), and FBS Ltd, which ordered 38 AS 350BBs, with deliveries beginning November 1996 and service entry in April 1997. Of these, 26 are Squirrel HT. Mk 1s with Defence Helicopter Flying School at Shawbury and 12 are NVG-compatible HT. Mk 2s with 670 Squadron, Army Air Corps, at Middle Wallop; by June 2001 the fleet had amassed 100,000 hours. Other orders include 11 for Brazilian Navy for liaison and patrol work; 30 for Brazilian Air Force for training; 36 for Brazilian Army (16 for liaison and 20 for training and firefighting); eight for China (see also CHAIC Z-11) and 40 AS 350B2 options for US Customs Service (augmenting five B2s delivered from 21 October 1998 to supplement 19 earlier AStars); six AS 550C3 and 13 AS 350B3 ordered by United Arab Emirates in 2000. Orders in 1997-2001 were 102, 73, 95, 137 and 103; deliveries in 1999 and 2000 were 95 and 105.

DESIGN FEATURES: Conventional pod-and-boom design with power plant above cabin; fin, underfin and twin horizontal stabilising surfaces. Starflex bearingless glass fibre main rotor head; all versions now have lifting section composites main rotor blades. Main rotor turns at 394 rpm, tail rotor at 2,086 rpm.

FLYING CONTROLS: Single fully powered controls with accumulators to delay manual reversion following hydraulic failure until airspeed can be reduced; cyclic trim by adjustable stick friction; inverted aerofoil tailplane to adjust pitch attitude in climb, cruise and descent; saucer fairing on rotor head to smooth wake; swept fins above and below tail. Variety of autostabilisers and autopilots available (see Avionics).

STRUCTURE: Main rotor head and much of airframe of glass fibre and aramids; main rotor blades automatically manufactured in composites; self-sealing composites fuel tank in military versions.

LANDING GEAR: Steel tube skid type. Taller version standard on military aircraft. Emergency flotation gear optional.

POWER PLANT: AS 350B2 powered by one 546 kW (732 shp) Turbomeca Arriel 1D1 with transmission rated at 440 kW (590 shp) for T-O; AS 350B3/AS 550C3 powered by one 632 kW (847 shp) Turbomeca Arriel 2B turbine engine controlled by FADEC, with transmission rated at 500 kW (671 shp) for T-O; AS 350D powered by AlliedSignal LTS101 sold only in USA. Plastics fuel tank (self-sealing on AS 550) with capacity of 540 litres (143 US gallons; 119 Imp gallons). Optional 475 litre (125.5 US gallon; 104.5 Imp gallon) auxiliary tank in cabin.

Soloy offers conversion using Rolls-Royce 250-C30M turboshaft; see *Jane's Aircraft Upgrades*.

ACCOMMODATION: Two individual bucket seats at front of cabin and four-place rear bench standard, optional two-place front bench seat; optional ambulance layout; large forward-hinged door on each side of versions for civil use; optional sliding door at rear of cabin on port side; (sliding doors standard on military version); baggage compartment aft of cabin, with full-width upward-hinged door on starboard side; top of baggage compartment reinforced to provide work platform on each side.

SYSTEMS: Hydraulic system includes four single-body servo units, operating at 40 bar (570 lb/sq in) pressure, and accumulators to delay reversion to manual control; electrical system includes a 4.5 kW engine-driven starter/generator, a 24 V 16 Ah Ni/Cd battery and a ground power receptacle; cabin air conditioning system optional.

AVIONICS: *Comms:* VHF/AM radios, HF/SSB transponder and ICS.

Flight: VOR/ILS, ADF, marker beacon receiver and DME. SFIM PA 85T31 (IFR), GPS; full-colour LCD VEMD (vehicle and engine multifunction display) on AS 350B3 and AS 550C3.

EQUIPMENT: Options include 907 kg (2,000 lb) capacity cargo sling (1,160 kg; 2,557 lb for AS 350B2/550, 1,400 kg; 3,086 lb on AS 350B3), 135 kg (297 lb) capacity electric hoist (optional 204 kg; 450 lb on AS 350B3), a TV camera for aerial filming, SX-16 searchlight, Wescam and FLIR observation systems, and a 735 litre (194 US gallon; 161 Imp gallon) Simplex agricultural spraytank and boom system.

ARMAMENT (AS 550C3): Provision for wide range of weapons, including 20 mm Giat M621 gun, FN Herstal TMP twin 7.62 mm and 12.7 mm machine gun pods, Thomson Brandt 68.12 launchers for twelve 68 mm rockets, Forges de Zeebrugge launchers for seven 2.75 in rockets, and ESCO HeliTOW anti-tank missile system.

DIMENSIONS, EXTERNAL:

Main rotor diameter	10.69 m (35 ft 0¾ in)
Main rotor blade chord	0.35 m (13¾ in)
Tail rotor diameter	1.86 m (6 ft 1¼ in)
Tail rotor blade chord: AS 350D	0.185 m (7¼ in)
AS 350B2/B3/550C3	0.205 m (8 in)
Length: overall, rotors turning	12.94 m (42 ft 5½ in)
fuselage	10.93 m (35 ft 10½ in)
Width: fuselage (max)	1.80 m (5 ft 10¾ in)
overall, blades folded (horizontal stabiliser span)	
	2.53 m (8 ft 3¾ in)
Height overall: AS 350/B2/B3/D	3.14 m (10 ft 3½ in)
AS 550C3	3.34 m (10 ft 11½ in)
Tailplane span	2.53 m (8 ft 3½ in)
Skid track: AS 350/B2/B3/D	2.17 m (7 ft 1½ in)
AS 550C3	2.28 m (7 ft 5¾ in)
Cabin doors (civil versions, standard, each):	
Height	1.15 m (3 ft 9¼ in)
Width	1.10 m (3 ft 7¼ in)

DIMENSIONS, INTERNAL:

Cabin: Length	2.42 m (7 ft 11¼ in)
Width at rear	1.65 m (5 ft 5 in)
Height	1.35 m (4 ft 5 in)
Floor area	2.60 m² (28.0 sq ft)
Baggage compartment volume	1.0 m³ (35 cu ft)

AREAS:

Main rotor disc	89.75 m² (966.1 sq ft)
Tail rotor disc	2.72 m² (29.25 sq ft)

WEIGHTS AND LOADINGS:

Weight empty: AS 350B2	1,171 kg (2,582 lb)
AS 350B3	1,175 kg (2,590 lb)
AS 350D	1,123 kg (2,475 lb)
AS 550C3	1,220 kg (2,689 lb)
Max T-O weight:	
AS 350B2/B3/550C3	2,250 kg (4,960 lb)
AS 350D	1,950 kg (4,300 lb)

Eurocopter AS 350B3 on firefighting duties in Chile *(Gérard Deulin/Eurocopter)*   NEW/0132424

Max weight with slung load:

AS 350B2	2,500 kg (5,511 lb)
AS 350B3/AS 550C3	2,800 kg (6,173 lb)
AS 350D	2,100 kg (4,630 lb)

Max disc loading:

AS 350D with slung load	23.4 kg/m² (4.79 lb/sq ft)
AS 350B2/550 clean	25.1 kg/m² (5.13 lb/sq ft)
AS 350B2 with slung load	27.9 kg/m² (5.71 lb/sq ft)
AS 350B3/550C3 with slung load	
	30.6 kg/m² (6.28 lb/sq ft)

Transmission loading at max T-O weight and power:

AS 350B2 clean	5.11 kg/kW (8.40 lb/shp)
AS 350B2 with slung load	5.68 kg/kW (9.34 lb/shp)
AS 350B3/550C3 clean	4.50 kg/kW (7.39 lb/shp)
AS 350B3/550C3 with slung load	
	5.50 kg/kW (9.04 lb/shp)

PERFORMANCE (AS 350D at normal T-O weight, AS 350B2/B3/550 at 2,200 kg; 4,850 lb):

Never-exceed speed (V_{NE}) at S/L:

AS 350/B2/B3/550	155 kt (287 km/h; 178 mph)
AS 350D	147 kt (272 km/h; 169 mph)

Max cruising speed at S/L:

AS 350B2	134 kt (248 km/h; 154 mph)
AS 350B3	140 kt (259 km/h; 161 mph)
AS 350D	124 kt (230 km/h; 143 mph)
AS 550C3	133 kt (246 km/h; 153 mph)

Max rate of climb at S/L:

AS 350B2	534 m (1,752 ft)/min
AS 350B3/550C3	618 m (2,028 ft)/min
Service ceiling: AS 350B2	4,800 m (15,750 ft)
AS 350B3/550C3	5,280 m (17,323 ft)
Hovering ceiling: IGE: AS 350B2	3,200 m (10,500 ft)
AS 350B3/550C3	4,260 m (13,976 ft)
OGE: AS 350B2	2,550 m (8,360 ft)
AS 350B3/550C3	3,630 m (11,909 ft)

Range with max fuel at recommended cruising speed, no reserves:

AS 350B2	360 n miles (666 km; 414 miles)
AS 350B3	357 n miles (661 km; 410 miles)
AS 550C3	350 n miles (648 km; 402 miles)

UPDATED

EUROCOPTER EC 130B

TYPE: Light utility helicopter.

PROGRAMME: Derived from AS 350 Ecureuil; PT-1 prototype first flew (unannounced) June 1999 and registered F-WQEV in May 2000; second prototype (F-WQES) assisted in 200 hour test programme, including high-

altitude testing in Albuquerque, USA, September 2000; JAA certification 14 December 2000; FAA certification 21 December 2000; revealed only in February 2001, when initial production aircraft (also 3,000th of Ecureuil family) handed over to Blue Hawaiian Helicopters at HeliExpo. Four preproduction helicopters delivered to launch customers in first half of 2001.

CURRENT VERSIONS: **EC 130B4:** Initial production version.

CUSTOMERS: Announced orders include Blue Hawaiian (10 in 2001), Rocky Mountain Helicopters (10 in 2001), Mont Blanc Hélicoptères (two) and five to UK customers; deliveries started in June 2001 and were scheduled to total 15 by year end; 2002 production set at 45. Orders in 2001 totalled 27, in which year more than 80 were delivered.

COSTS: US\$1.45 million (2001).

DESIGN FEATURES: As for AS 355B3, but considerably modified external appearance, with windscreen side panels, doors and landing gear from EC 120, plus Fenestron based on EC 135, although symmetrical, carried by redesigned tailboom. Dual hydraulic system from AS 355N. Cabin width increased by 25 cm (10 in) to give additional 23 per cent internal space, accommodating seven or eight seats and enlarged (10 per cent) baggage compartment; optional air conditioning. Retains AS 355B3 engine and transmission, but automatic system (in conjunction with dual-channel FADEC, plus third, digital, back-up channel) matches rotor speed to flight conditions for noise reduction; 84.3 EPNdB flyover rating is 7 dB below ICAO limit and 0.5 dB under special Grand Canyon National Park restrictions achieved by use of parabolic rotor tips.

LANDING GEAR: Emergency flotation system available. Tailskid. Handling wheels.

POWER PLANT: One 543 kW (728 shp) Turbomeca Arriel 2B1 turboshaft with transmission rated at 632 kW (847 shp) for T-O, with FADEC. Fuel capacity 540 litres (143 US gallons; 119 Imp gallons) in single tank.

ACCOMMODATION: Pilot and up to seven passengers in transport role; or pilot and one or two stretcher patients plus two medical attendants in casualty evacuation configuration. Dual controls.

SYSTEMS: 4.5 kW 28 V DC starter-generator. 15 Ah battery; 28 V DC cabin outlet.

AVIONICS: *Comms:* Shadin 8800 T altitude encoder; Bendix/King KT 76C Mode C transponder and NAT AA95 ICS.

Flight: Two Bendix/King KX 165A nav/com/glidescope; Garmin GNS 430 GPS; Artex 100 HM ELT; AIM 1110 HSI; Bendix/King KI 525 HSI; full-colour VEMD (Vehicle and Engine Multifunction Display).

Second prototype Eurocopter EC 130B, derived from AS 350B Ecureuil *(Jane's/Paul Jackson)*   NEW/0132419

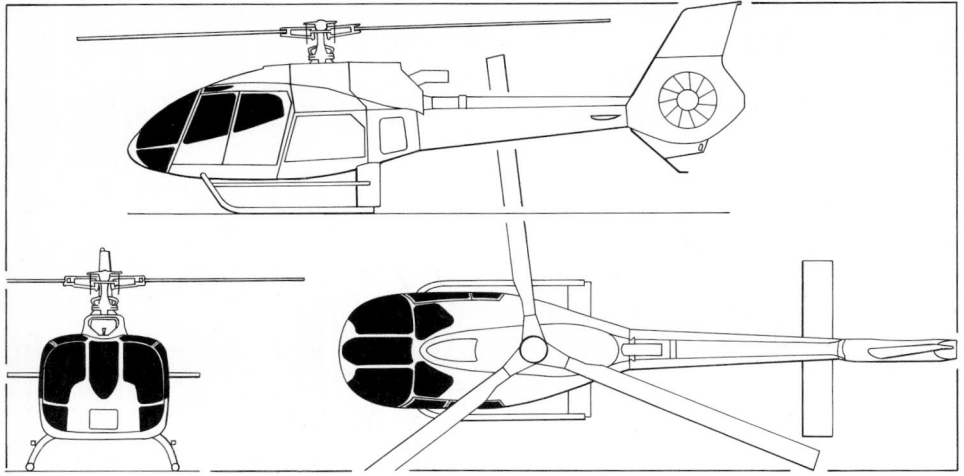

Eurocopter EC 130B (Turbomeca Arriel turboshaft) *(Jane's/Paul Jackson)* 0062408

DIMENSIONS, EXTERNAL:
Main rotor diameter	10.69 m (35 ft 0¾ in)
Length: overall, rotors turning	12.64 m (41 ft 5½ in)
fuselage	10.68 m (35 ft 0½ in)
Fuselage max width	2.03 m (6 ft 8 in)
Height: to top of fin	3.61 m (11 ft 10¼ in)
to top of rotor head	3.34 m (10 ft 11½ in)
Tailplane span	2.73 m (8 ft 11½ in)
Skid track	2.31 m (7 ft 7 in)
Cabin door: Height	1.18 m (3 ft 10½ in)
Max width: front	0.84 m (2 ft 9 in)
rear	0.92 m (3 ft 0¼ in)

DIMENSIONS, INTERNAL:
Cabin: Length	2.19 m (7 ft 2¼ in)
Max width	1.865 m (6 ft 1½ in)
Max height	1.28 m (4 ft 2½ in)
Volume	3.7 m³ (131 cu ft)
Baggage hold, volume:	
front, port	0.29 m³ (10.2 cu ft)
front, starboard	0.25 m³ (8.8 cu ft)
rear	0.57 m³ (20.1 cu ft)

AREAS:
Main rotor disc	89.75 m² (966.1 sq ft)

WEIGHTS AND LOADINGS:
Weight empty	1,360 kg (3,000 lb)
Max T-O weight:	
internal load	2,400 kg (5,291 lb)
external load	2,800 kg (6,173 lb)
Max disc loading:	
internal load	26.74 kg/m² (5.48 lb/sq ft)
external load	31.20 kg/m² (6.39 lb/sq ft)
Max power loading: internal load	
	4.42 kg/kW (7.27 lb/shp)
external load	5.16 kg/kW (8.48 lb/shp)

PERFORMANCE (two occupants):
Never-exceed speed (V$_{NE}$)	155 kt (287 km/h; 178 mph)
Max operating speed	127 kt (235 km/h; 146 mph)
Max rate of climb at S/L	698 m (2,290 ft)/min
Service ceiling	7,010 m (23,000 ft)
Hovering ceiling: IGE	5,875 m (19,280 ft)
OGE	5,320 m (17,460 ft)

OPERATIONAL NOISE LEVELS (ICAO Annex 16 Ch 8):
Average	86.8 EPNdB
Overflight	84.3 EPNdB
	UPDATED

EUROCOPTER AS 355 ECUREUIL 2/ TWINSTAR and AS 555 FENNEC

Brazilian Air Force designations: CH-55 and VH-55 Esquilo
Brazilian Navy designation: UH-12B Esquilo
Royal Air Force name: Twin Squirrel

TYPE: Light utility helicopter.

PROGRAMME: Twin-engined version of AS 350. First flight of first prototype (F-WZLA) 28 September 1979; details of early production AS 355E/F versions in 1984-85 *Jane's*; AS 355F superseded by AS 355F1 in January 1984; AS 355F2 certified 10 December 1985; AS 355N powered by two Turbomeca TM 319 Arrius 1A certified in France in 1989, UK and USA in 1992; deliveries of this version began early 1992.

CURRENT VERSIONS: Current production AS 355 Ecureuil 2 and AS 555 Fennec are powered by two Turbomeca Arrius1A turboshafts.

AS 355N Ecureuil 2: Current civil production version, adaptable for passengers, cargo, police, ambulance, slung loads, carrying harbour pilots, working on high-tension cables and other missions; Category A OEI performance; known as **TwinStar** in USA.

Details refer to AS 355N, except where stated.

AS 555UN Fennec: Military utility version and French Army ALAT IFR pilot trainer.

AS 555AN Fennec: Armed version; later machines adapted for centreline-mounted 20 mm gun and pylon-mounted rockets.

AS 555MN Fennec: Naval unarmed version. Can carry a 360° chin-mounted radar.

AS 555SN Fennec: Armed naval version for ASW and over-the-horizon targeting, operating from ships of 600 tonnes upwards; armament includes one lightweight homing torpedo or the cannon and rockets of land-based versions; avionics include Honeywell RDR-1500B 360° chin-mounted radar, Thales Nadir 10 navigation system, Thales RDN 85 Doppler and SFIM 85 T31 three-axis autopilot.

CUSTOMERS: Total 592 twin-engined Ecureuils in service at May 2000. French Air Force ordered 52 Fennecs, first eight as AS 355F1s powered by Rolls-Royce 250; flown by 67e Escadre d'Hélicoptères at Villacoublay and other units for communications and, with side-mounted Giat M61 20 mm gun pod, by ETOM 68 in Guyana; delivery of remaining 44 (AS 555AN powered by Arrius) began 19 January 1990; from 24th onwards, provision for centrally mounted 20 mm cannon and T-100 sight, plus Mistral missiles; French Army ordered 18 AS 555UNs for IFR training (delivery from February 1992); Brazilian Air Force has 13 AS 555s: 11 with armament, designated CH-55, and two VIP transports, designated VH-55; Brazilian Navy acquired 11 UH-12Bs; Brazilian AS 555s assembled by Helibrás in Brazil (which see); four more navies ordered eight AS 555s in 1992, two in 1993, four in 1994 and one in 1995. Two AS 355F1s, leased from OSS of Hayes, UK, supplied to the RAF's No. 32 (The Royal) Squadron at

Northolt on 1 April 1996 for VIP transport; third subsequently added. Total of six AS 3555Ns ordered by Malaysian Navy on 13 October 2001 to replace Westland Wasps with deliveries at end 2003. Orders in 1997-2001 totalled nine, 20, 10, 13 and 17.

DESIGN FEATURES: Starflex main rotor; two engine shafts drive into combiner gearbox containing freewheels; main rotor turns at 394 rpm and tail rotor at 2,086 rpm; otherwise substantially as AS 350.

FLYING CONTROLS: Powered, full dual flying controls without manual reversion; trim by adjustable stick friction; inverted aerofoil tailplane; swept fins above and below tailboom.

STRUCTURE: Light metal tailboom and central fuselage structure; thermoformed plastics for cabin structure.

LANDING GEAR: As for AS 350B2/550.

POWER PLANT: Two Turbomeca Arrius 1A turboshafts, each rated at 357 kW (479 shp) for take-off and 302 kW (406 shp) maximum continuous; 388 kW (520 shp) 30 minutes OEI emergency; 407 kW (547 shp) 2.5 minutes OEI emergency; full authority digital engine control (FADEC) allows automatic sequenced starting of both engines, automatic top temperature and torque limiting and preselection of lower limits for practice OEI operation. Transmission rating 511 kW (686 shp) for T-O. Two integral fuel tanks, with total usable capacity of 730 litres (193 US gallons; 160 Imp gallons), in body structure. Optional 475 litre (126 US gallon; 105 Imp gallon) auxiliary tank in cabin.

ACCOMMODATION: As for single-engined Ecureuil and Fennec.

SYSTEMS: As for AS 350B2/550, except two hydraulic pumps, reservoirs and tandem powered flying control units and electric generators.

AVIONICS: Options include a second VHF/AM and radio altimeter; provision for IFR system, SFIM 85 T31 three-axis autopilot and CDV 85 T3 nav coupler.

EQUIPMENT: Casualty installations, TV, FLIR, searchlight and 204 kg (500 lb) capacity winch available. Cargo hook optional; capacity 1,134 kg (2,500 lb). See also Current Version descriptions.

ARMAMENT (AS 555): Optional alternative weapons include TDA or Forges de Zeebrugge rocket packs, Matra or FN Herstal machine gun pods or a Giat M621 20 mm gun. Naval version carries one homing torpedo in ASW role, or SAR winch.

DIMENSIONS, EXTERNAL: As for AS 350B2/550C3, except:
Width: fuselage	1.87 m (6 ft 1½ in)
overall, blades folded	3.05 m (10 ft 0 in)
Tailplane span	3.05 m (10 ft 0 in)

DIMENSIONS, INTERNAL: As for AS 350B2/550C3

WEIGHTS AND LOADINGS:
Weight empty	1,436 kg (3,166 lb)
Max sling load	1,134 kg (2,500 lb)

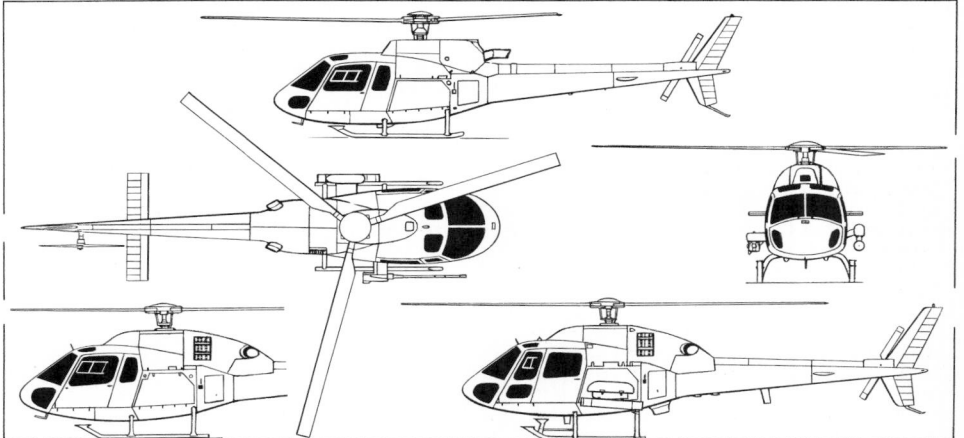

Eurocopter AS 555AN Fennec twin-turbine armed helicopter with side view (top) of single-engined AS 350B3 Ecureuil and scrap view (bottom) of AS 355N civil twin *(Jane's/Dennis Punnett)* 0085583

Eurocopter AS 355N light helicopter *(G Deulin/Eurocopter)* 0110874

Max T-O weight:
internal load	2,600 kg (5,732 lb)
max sling load	2,600 kg (5,732 lb)

Max disc loading:
internal load	29.0 kg/m² (5.93 lb/sq ft)
external load	29.0 kg/m² (5.93 lb/sq ft)

Transmission loading at max T-O weight and power:
internal load	5.09 kg/kW (8.36 lb/shp)
external load	5.09 kg/kW (8.36 lb/shp)

PERFORMANCE (ISA):
Never-exceed speed (VNE)	150 kt (278 km/h; 172 mph)

Max cruising speed at S/L:
AS 355N	120 kt (222 km/h; 138 mph)
AS 355UN/AN/MN/SN	118 kt (219 km/h; 136 mph)
Econ cruising speed	117 kt (217 km/h; 135 mph)
Max rate of climb at S/L	384 m (1,260 ft)/min
Rate of climb at S/L, OEI	156 m (512 ft)/min
Service ceiling	3,800 m (12,460 ft)
Service ceiling, OEI	1,450 m (4,760 ft)
Hovering ceiling: IGE	2,000 m (6,560 ft)
OGE	750 m (2,460 ft)

Radius, SAR, two survivors
	70 n miles (129 km; 80.5 miles)

Range with max fuel at S/L, no reserves:
AS 355N	385 n miles (713 km; 443 miles)
AS 355UN/AN	386 n miles (714 km; 444 miles)
AS 355MN/SN	350 n miles (648 km; 402 miles)

Endurance, no reserves:
one torpedo, or cannon or rocket pods	2 h 20 min
cannon plus rockets	1 h 50 min

UPDATED

EUROCOPTER AS 365N DAUPHIN 2 and AS 565 PANTHER

US Coast Guard designation: HH-65A Dolphin
Israel Defence Force name: Dolpheen

TYPE: Light utility helicopter.

PROGRAMME: Certified for VFR in France November 1989; 500th Dauphin (all versions) delivered in November 1991 was 83rd AS 365N2. Dauphin family had flown 2.86 million flying hours by June 2001.

CURRENT VERSIONS: **AS 365N2 Dauphin 2:** Recent baseline version; now produced in parallel with AS 365N3.
Details below refer to N2 version.

AS 365N3 Dauphin: Improved hot-and-high version; technology for N3 and N4 versions developed by experimental DGV (Dauphin *Grande Vitesse*) F-WDFK; DGAC certification in October 1997 and first delivery to Noordzee Helikopter Vlaanderen of Belgium in early 1998. Turbomeca Arriel 2C engines offer maximum T-O weight up to 48°C (118°F), rather than N2's 36°C (97°F). FADEC as standard. Orders had reached 40 by mid-2000. Upgrade of AS 365N1 and 365N2 is offered by Eurocopter.

AS 365N4 Dauphin: Redesignated EC 155; see following entry.

AS 366G1 (HH-65A Dolphin): US Coast Guard version (99 built, plus two ex-trials aircraft bought for evaluation by Israel as Dolpheen); trial installation of 895 kW (1,200 shp) LHTEC T800 turboshaft ordered February 1990, but abandoned November 1991 when LTS101 performance promised to improve; re-engining with French Arriels also declined. For description see 1991-92 and earlier *Jane's*.

AS 565UB Panther: Army version; last described in 2001-02 edition.

AS 565SB Panther: Naval version; last described in 2001-02 edition.

Ambulance/EMS: Flight crew of two plus two or four stretchers along cabin sides loaded through rear doors; up to four seats can replace stretchers on one side; doctor sits in middle with equipment; rear doors open through 180° instead of sliding.

Offshore support: Pilot plus up to 12 passengers, autopilot and navigation systems, pop-out floats.

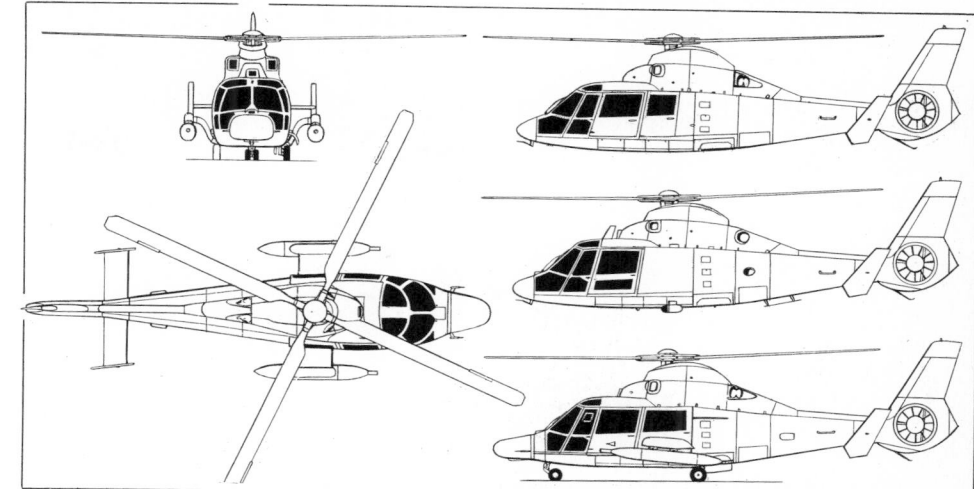

Military Panthers: Eurocopter AS 565AA with side views (centre) of AS 366 HH-65A Dolphin and (top) AS 365N2 Dauphin 2 *(Jane's/Dennis Punnett)*

DTV5 *(Démonstrateur Technologie Véhicule):* Demonstrator of new Dauphin variant; new rotor of 13.50 m (44 ft 3½ in) diameter; Tigre transmission; and two 809 kW (1,085 shp) Turbomeca TM 333-2E1 turboshafts. Conversion of Dauphin N4 prototype, subsequently renamed Dolphin VHS (very high speed).

Dolphin FWB: Testbed for fly-by-wire technology used in NH 90.

CUSTOMERS: Total 684 AS 365/366/565 ordered for civil and military use by over 187 customers from 53 countries by January 2001; 599 then in service; totals include 50 produced in Harbin Z-9/9A in China and 101 in HH-65A Dolphin US Coast Guard versions; Chinese production by HAMC (which see) continuing. Dauphin and Panther orders totalled 10 in 2001, compared with 18 in both 2000 and 1999, 28 in 1998 and 16 in 1997. Eurocopter has entered offset agreement with LOT of Slovak Republic to allow modification of AS 365s for East European customers; first sale announced 1998.

DESIGN FEATURES: Progressively tapered rear fuselage with stabilising surfaces comprising horizontal tail with endplate fins, plus central fin; Fenestron anti-torque rotor; engines above cabin.

AS 365N2 and N3 have Starflex main rotor hub with four blades; 11-blade Fenestron; main rotor blades have quick disconnect pins for manual folding; ONERA OA 212 (thickness/chord ratio 12 per cent) at root to OA 207 (thickness/chord ratio 7 per cent) at tip; adjustment tab near tip; leading-edge of tip swept at 45°; main rotor rpm 350; Fenestron rpm 3,665; rotor brake standard. AS 365N3 can be retrospectively fitted with EC 155's five-blade Spheriflex main rotor and 10-blade Fenestron rotor with unevenly spaced blades to reduce vibration.

FLYING CONTROLS: Hydraulic dual fully powered; cyclic trim by adjustable stick friction damper; fixed tailplane, with endplate fins offset 10° to port; IFR systems available.

STRUCTURE: Mainly light alloy; machined frames fore and aft of transmission support; main rotor blades have CFRP spar and skins with Nomex honeycomb filling; Fenestron duct and fin of CFRP and Nomex/Rohacell sandwich; nose and power plant fairings of GFRP/Nomex sandwich; centre and rear fuselage assemblies, flight deck floor, roof, walls and bottom skin of fuel tank bays of light alloy/Nomex sandwich.

LANDING GEAR: Hydraulically retractable tricycle type; twin-wheel self-centring nose unit retracts rearward; single wheel on each main unit; main legs retract into troughs in fuselage without cover doors; all three units have oleo-pneumatic shock-absorber; mainwheel tyres 15×6.00-6 or 380×150-6 (6 ply) tubeless, pressure 8.60 bar (125 lb/sq in); nosewheel tyres size 13×5.00-4, pressure 5.50 bar (80 lb/sq in); hydraulic disc brakes.

POWER PLANT: *AS 365N2:* Two Turbomeca Arriel 1C2 turboshafts, each rated at 551 kW (739 shp) for T-O and 471 kW (631 shp) maximum continuous, mounted side by side aft of transmission with stainless steel firewall between them. Transmission rating 965 kW (1,294 shp) for T-O and maximum continuous.

AS 365N3: Two FADEC-equipped Turbomeca Arriel 2C turboshafts, each rated at 635 kW (851 shp) for T-O, 597 kW (800 shp) maximum continuous and 729 kW (977 shp) for 30 seconds.

Standard fuel in four tanks under cabin floor and fifth tank in bottom of centre-fuselage; total capacity 1,135 litres (300 US gallons; 249.5 Imp gallons); provision for auxiliary tank in baggage compartment, with capacity of 180 litres (47.5 US gallons; 39.5 Imp gallons); or ferry tank in place of rear seats in cabin, capacity 475 litres (125.5 US gallons; 104.5 Imp gallons); refuelling point above landing gear door on port side. Oil capacity 14 litres (3.7 US gallons; 3.1 Imp gallons).

ACCOMMODATION: Standard accommodation for pilot and co-pilot or passenger in front, and two rows of four seats to rear; high-density seating for one pilot and 12 passengers; VIP configurations for four to six persons in addition to pilot; three forward-opening doors on each side; freight hold aft of cabin rear bulkhead, with door on starboard side; cabin heated and ventilated.

SYSTEMS: Air conditioning system optional. Duplicated hydraulic system, pressure 60 bar (870 lb/sq in). Electrical system includes two 4.8 kW starter/generators, one 24 V 27 Ah battery and two 250 VA 115 V 400 Hz inverters.

AVIONICS: *Comms:* Optional transponder.
Radar: Optional weather radar.
Flight: SFIM 155 duplex autopilot; optional VOR, ILS, ADF, DME, GPS and SFIM CDV 85 autopilot coupler. Optional Bendix/King Mark XXII EGPWS.
Instrumentation: Two-pilot IFR instrument panel; optional EFIS.

EQUIPMENT: Includes 1,600 kg (3,525 lb) capacity cargo hook, and 275 kg (606 lb) capacity hoist with 90 m (295 ft) cable.

DIMENSIONS, EXTERNAL (N2 and N3):
Main rotor diameter:	11.94 m (39 ft 2 in)
Fenestron diameter	1.10 m (3 ft 7¼ in)
Main rotor blade chord: basic	0.385 m (1 ft 3¼ in)
outboard of tab	0.405 m (1 ft 4 in)
Length: overall, rotor turning	13.73 m (45 ft 0½ in)
fuselage	11.63 m (38 ft 2 in)
Width, rotor blades folded	3.25 m (10 ft 8 in)
Height: to top of rotor head	3.47 m (11 ft 4½ in)
overall (tip of fin)	4.06 m (13 ft 4 in)
Wheel track	1.90 m (6 ft 2¾ in)
Wheelbase	3.64 m (11 ft 11¼ in)

Main cabin door (fwd, each side):
Height	1.16 m (3 ft 9½ in)
Width	1.14 m (3 ft 9 in)

Main cabin door (rear, each side):
Height	1.16 m (3 ft 9½ in)
Width	0.87 m (2 ft 10¼ in)

Baggage compartment door (stbd):
Height	0.51 m (1 ft 8 in)
Width	0.73 m (2 ft 4¾ in)

DIMENSIONS, INTERNAL:
Cabin: Length	2.30 m (7 ft 6½ in)
Max width	1.92 m (6 ft 3½ in)
Max height	1.16 m (3 ft 9¾ in)
Floor area	4.20 m² (45.2 sq ft)
Volume	5.0 m³ (176 cu ft)
Baggage compartment volume	1.6 m³ (56 cu ft)

AREAS:
Main rotor disc	111.97 m² (1,205.2 sq ft)
Fenestron disc	0.95 m² (10.23 sq ft)

Eurocopter SA 365N3 Dauphin of Swiss operator, Heli Link *(Gérard Deulin/Eurocopter)*   NEW/0132425

WEIGHTS AND LOADINGS:

Weight empty: N2	2,281 kg (5,028 lb)
N3	2,302 kg (5,075 lb)
Max T-O weight: N2	4,250 kg (9,369 lb)
N3	4,300 kg (9,480 lb)
Max disc loading: N2	37.9 kg/m² (7.78 lb/sq ft)
N3	38.4 kg/m² (7.87 lb/sq ft)

Transmission loading at max T-O weight and power:

N2	4.40 kg/kW (7.24 lb/shp)

PERFORMANCE:

Never-exceed speed (VNE):

N2 and N3	155 kt (287 km/h; 178 mph)
Max cruising speed: N2 and N3	
	150 kt (278 km/h; 173 mph)
Econ cruising speed: N2	137 kt (254 km/h; 158 mph)
N3	148 kt (274 km/h; 170 mph)
Max rate of climb at S/L: N2	408 m (1,339 ft)/min
Rate of climb at S/L, OEI: N2	78 m (256 ft)/min
N3	183 m (602 ft)/min
Service ceiling: N2	3,700 m (12,140 ft)
N3	4,700 m (15,420 ft)
Service ceiling, OEI: N2	1,200 m (3,940 ft)
N3	1,870 m (6,140 ft)
Hovering ceiling: IGE: N2	2,000 m (6,560 ft)
N3	2,620 m (8,600 ft)
OGE: N2	1,200 m (3,940 ft)
N3	1,150 m (3,780 ft)

Max range with standard fuel:

N2	464 n miles (859 km; 534 miles)
N3	440 n miles (815 km; 506 miles)
Endurance with standard fuel: N2	4 h 18 min
N3	4 h 6 min

UPDATED

EUROCOPTER EC 155B

TYPE: Light utility helicopter.

PROGRAMME: Programme launched September 1996, as further development of Dauphin 2 (see previous entry). Announced at 1997 Paris Air Show, when known as AS 365N4 Dauphin 2. First flight 17 June 1997 as conversion of DGV testbed F-WDFK (see AS 365N Dauphin 2 entry); 1,000 flying hours achieved by February 1998. New, EC 155 designation revealed at HAI convention, February 1998. First production EC 155 (F-WWOZ) flew 11 March 1998. JAR certification received 9 December 1998, FAR certification was due late 1999. Certification for single pilot in IFR issued 25 January 2000.

CURRENT VERSIONS: **EC 155B:** Baseline version; last described in 2002-03 *Jane's*; MTOW 4,800 kg (10,582 lb) and Arriel 2C1 engines.

EC 155B1: Upgraded version from 2002; *as described*. Features include new engine cowlings, new hydraulic cooling system, chip detectors, cargo fire protection, jettisonable cockpit doors; fixed cockpit foot steps and Thales AHV 16 radar altimeter display. Available with standard Corporate, Offshore and Parapublic equipment packages.

CUSTOMERS: First customer, German Border Guard (Bundesgrenzschutz; BGS), ordered 13 for delivery between 16 March 1999 and 2000 and further two in February 2002 for delivery in 2003. German Interior Ministry ordered two for Baden-Württemberg regional government in February 2000 for delivery in March 2001; Hong Kong Government Flying Service ordered five with deliveries to be completed by late 2002; firm orders stood at 33 by June 2000, of which eight were placed in 1999. Orders in 2000 totalled seven, six of which were placed by Shell Nigeria, whose first aircraft was delivered on 26 September 2001. Swedish Helicopter Service ordered three in June 2001 for delivery between October and December 2002. By October 2001, 44 had been ordered of which 18 were in service. First Nigerian example handed over 26 September 2001. Sales in 2001 totalled 13, including three for COHC (China), two for SFC (Vietnam) and three for SHS (Sweden). Further 13 purchased by BGS in February 2002.

COSTS: US$5.5 million (1998).

DESIGN FEATURES: Compared to earlier Dauphin models, has bulged sliding cabin doors, redesigned cabin windows and 40 per cent larger cabin area. Five-blade Spheriflex main rotor and 10-blade Fenestron rotor with unevenly spaced blades to reduce noise.

FLYING CONTROLS: As AS 365N.

STRUCTURE: As AS 365N.

LANDING GEAR: As AS 365N.

POWER PLANT: Two FADEC-equipped Turbomeca Arriel 2C2 turboshafts, each rated at 697 kW (935 shp) for T-O, 645 kW (865 hp) max continuous power and 780 kW (1,046 shp) for 30 seconds. Standard fuel in six tanks, total capacity 1,257 litres (332 US gallons; 277 Imp gallons); provision for auxiliary tank in baggage compartment, with capacity for 180 litres (47.6 US gallons; 39.6 Imp gallons); or ferry tank in place of rear seats in cabin, capacity 460 litres (121.5 US gallons; 101 Imp gallons); refuelling point above landing gear door on port side. Oil capacity 14 litres (3.7 US gallons; 3.1 Imp gallons).

ACCOMMODATION: Standard accommodation for pilot and co-pilot or passenger in front, and three rows of four seats to rear; high-density seating for one pilot and up to 14 passengers; VIP configurations for between four and eight

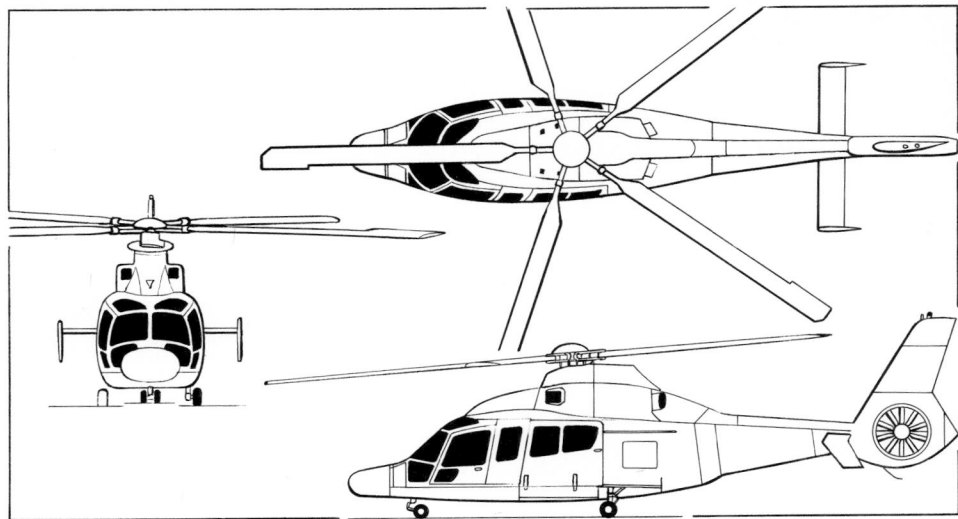

Eurocopter EC 155B *(Jane's/James Goulding)*   0130860

persons in addition to pilot; 12 cabin seats in offshore version; up to six stretchers in casevac role; one crew door and one large sliding door on each side; freight hold aft of cabin rear bulkhead, with door on both sides. Option of hinged cabin door on VIP versions.

SYSTEMS: Electrical system includes two starter/generators, each 160 A, 28 V DC and 43 Ah Ni/Cd battery. Duplicated hydraulic system. Optional 10 kVA alternator.

AVIONICS: *Comms: Corporate:* Twin Collins 422A VHF/AM; Collins TDR 94D transponder; NAT AA20-431 passenger address system; Team BA 1920 passenger interphone. Kannad 406AF ELT. Optional Maritime NPX 138, Collins HF 9100 HF/SSB and Collins TDR 94D transponder; Racal CPT 609 ELT. *Offshore:* Twin Collins 422A; TRD 94D; NPX 138 standard; AA20-431; BA 1920; Kannad 406. Optional second TDR 94D and CPT 609. *Parapublic:* Twin Collins 422A; TDR 94D; AA20-431; BA 1920; Kannad 406. Optional NPX 138, HF 9100, CPT 609.

Radar: Corporate: Telephonics RDR 2000 with VRU. *Offshore:* Telephonics RDR 1400C standard; RDR 2000 optional replacement. *Parapublic:* RDR 1400C.

Flight: Corporate: Universal UNS-1D flight management system. Twin collins VIR 432 VOR/ILS; Collins DME 442; Collins ADF 462. Optional Honeywell CAS 66A TCAS, Euroavionics Euronav 3 moving map, Racal V 694 voice alerting system. *Offshore:* UNS-1D; Twin Collins VIR 432; DME 442; ADF 462. Optional Trimble TNL 2101 (replacing UNS-1D), Bendix CAS 66A TCAS, Euronav 3, V 694. *Parapublic:* UNS-1D; Twin VIR 432; DME 442; ADF 462. Optional CAS 66A, Euronav 3, V 694.

Instrumentation: As AS 365N, but optional eighth MFD available (152 × 203 mm; 6 × 8 in) for mission equipment display.

Mission: German Border Guard aircraft to have Hellas obstacle warning system.

EQUIPMENT: Standard role packages, included in equipped weight, comprise:

Corporate: Air conditioning, hinged cabin doors, fin and belly strobe lights, retractable passenger steps both sides, improved soundproofing and carpets.

Offshore: Windscreen washer, high-intensity lights, emergency flotation gear, strobe lights, liferaft stowage and improved soundproofing.

Parapublic: Air Equipment SAR hoist (272 kg; 600 lb, 90 m; 295 ft), Spectrolab SX-16 searchlight to port, strobe lights, loudhailer and siren, rappelling rope stays, improved soundproofing and 12 cabin seats.

DIMENSIONS, EXTERNAL:

Main rotor diameter	12.60 m (41 ft 4 in)

Fenestron diameter	1.10 m (3 ft 7¼ in)
Main rotor blade chord: basic	0.385 m (1 ft 3¼ in)
outboard of tab	0.405 m (1 ft 4 in)
Length: overall, rotor turning	14.30 m (46 ft 11 in)
fuselage	12.71 m (41 ft 8½ in)
Tailplane span and width, rotor blades folded	
	3.48 m (11 ft 5 in)
Height: to top of rotor head	3.64 m (11 ft 11¼ in)
overall (tip of fin)	4.35 m (14 ft 3¼ in)
Wheel track	1.90 m (6 ft 2¾ in)
Wheelbase	3.91 m (12 ft 10 in)
Cockpit door (each side): Height	1.34 m (4 ft 4¾ in)
Width	1.19 m (3 ft 10¾ in)
Baggage compartment door (stbd):	
Height	0.69 m (2 ft 3¼ in)
Width	0.73 m (2 ft 4¾ in)

DIMENSIONS, INTERNAL:

Cabin: Length	2.55 m (8 ft 4½ in)
Max width	2.10 m (6 ft 10¾ in)
Max height	1.34 m (4 ft 4¾ in)
Floor area	5.09 m² (54.8 sq ft)
Volume	6.7 m³ (237 cu ft)
Baggage compartment: Floor area	2.95 m² (31.8 sq ft)
Volume	2.5 m³ (88 cu ft)

AREAS:

Main rotor disc	124.69 m² (1,342.1 sq ft)
Fenestron disc	0.95 m² (10.23 sq ft)

WEIGHTS AND LOADINGS:

Weight: empty	2,615 kg (5,765 lb)
equipped: Corporate	3,153 kg (6,951 lb)
Offshore	3,027 kg (6,673 lb)
Parapublic	2,876 kg (6,340 lb)
Max underslung load	1,600 kg (3,527 lb)
Max T-O weight	4,850 kg (10,692 lb)
Max disc loading	38.9 kg/m² (7.97 lb/sq ft)

PERFORMANCE:

Never-exceed speed (VNE)	175 kt (324 km/h; 201 mph)
Max cruising speed at S/L	144 kt (267 km/h; 166 mph)
Econ cruising speed at FL60	146 kt (270 km/h; 168 mph)
Max rate of climb at S/L	354 m (1,161 ft)/min
Service ceiling	4,570 m (15,000 ft)
Hovering ceiling: IGE	2,290 m (7,520 ft)
OGE at 4,400 kg (9.700 lb)	2,365 m (7,760 ft)

Max range with standard fuel, no reserves:

standard tanks	424 n miles (785 km; 487 miles)
one auxiliary tank	488 n miles (903 km; 561 miles)
Endurance: standard tanks	4 h 0 min
one auxiliary tank	4 h 35 min

UPDATED

Eurocopter EC 155B alighting on an oil platform *(D Bauchere/Eurocopter)*   *NEW*/0132422

EUROCOPTER BO 105

Swedish designation: Hkp 9

TYPE: Light utility helicopter.

PROGRAMME: First flight of prototype 16 February 1967; LBA certification 15 October 1970; production of 100 BO 105 M (VBH) and 212 BO 105 P (PAH-1) (see 1985-86 *Jane's* and current *Jane's Aircraft Upgrades*) completed 1984. Also assembled by Eurocopter Canada (see Canadian section) and Dirgantara (Indonesian section). By January 2001, the type had accumulated 5.66 million flight hours; German Army fleet celebrated one million hours on 22 May 2000. European production to order only.

CURRENT VERSIONS: **BO 105 CBS:** Basic model with cabin stretched 0.25 m (10 in) in rear seat area; additional window aft of rear door; FAA certification to SFAR Pt 29-4 early 1983; superseded by CBS-5.

BO 105 CBS-5: Main production model since 1993, formerly known as EC Super Five. Upgraded version of BO 105 CBS-4, derived from German Army PAH-1 programme; announced at Heli-Expo February 1993 and certified late 1993.

Has new main rotor blades with parallel-chord DM-H4 aerofoil to 0.8 radius and tapered DM-H3 to tip; rotor lift increased 150 kg (330 lb); airframe vibration reduced to less than 0.1 *g*; improved stability. Improvements over CBS-4 also include Cat. A T-O weight in ISA + 20°C increased by 130 kg (287 lb); OGE hover ceiling at maximum T-O weight in ISA + 20°C increased by 500 m (1,640 ft); power required to hover OGE at S/L in ISA + 20°C reduced by 26 kW (35 shp); maximum weight in OGE hover, OEI with 2.5 minutes' power in ISA + 20°C, increased by 240 kg (529 lb); service ceiling OEI at maximum weight increased by 900 m (2,950 ft).

Details here apply to BO 105 CBS-5, except where indicated.

BO 105 CBS-3: Hot-and-high variant powered by two Rolls-Royce 250-C28C each rated at 410 kW (550 shp) for 30 seconds; produced exclusively by Eurocopter Canada (which see).

BO 105 LSA-3 Super Lifter: Developed specifically for lifting external loads; maximum T-O weight 2,850 kg (6,283 lb). Transmission rating 620 kW (832 shp) for T-O; 588 kW (788 shp) maximum continuous.

CUSTOMERS: Total 1,394 BO 105s of all models delivered to 42 countries by January 2001; further two ordered in 2001; some 1,154 remained in service in 2001. Apart from the German Army, major operators include Mexican Navy (12), Spanish Army (70 including 18 armed reconnaissance, 14 observation and 28 anti-tank), German Interior Ministry (BGS) (22), Swedish Army (20), Swedish Air Force (four BO 105 CBS for IFR search and rescue), Netherlands Army (30).

Purchases of CBS-5 include two by Bahrain Navy, 1994, with rescue hoist and nose-mounted 360° radar; two, equipped with rescue hoist, searchlights and auxiliary fuel tanks, to TsENTROSPAS (ministry of the Russian Federation for Civil Defence, Emergencies and Elimination of Consequences of Natural Disasters) delivered May 1995 for operation by MChS (State Unitary Aviation Enterprise). 17 ordered in January 1996 for emergency rescue duties by German Ministry of the Interior, deliveries to BGS Katastrophenschutz beginning immediately and ending early 1997.

In November 1997, South Korean Army selected the CBS-5 for reconnaissance and liaison; initial two delivered end 1999 (first German-built BO 105s produced since 1997); further 10 being assembled by Daewoo from 2000. Deliveries in 1999 and 2000 were three and 10. No orders received in 1999 or 2000.

DESIGN FEATURES: Traditional pod-and-boom configuration; tail rotor mounted on top of fin; auxiliary horizontal and vertical stabilising surfaces; rear doors for large cargo.

Four-blade main rotor with rigid titanium hub; only articulation is roller bearings for blade pitch change (max 3°); all-composites blades of NACA 23012 section with drooped leading-edge and reflexed trailing-edge; 8° linear twist; pendulous vibration damper near each blade root;

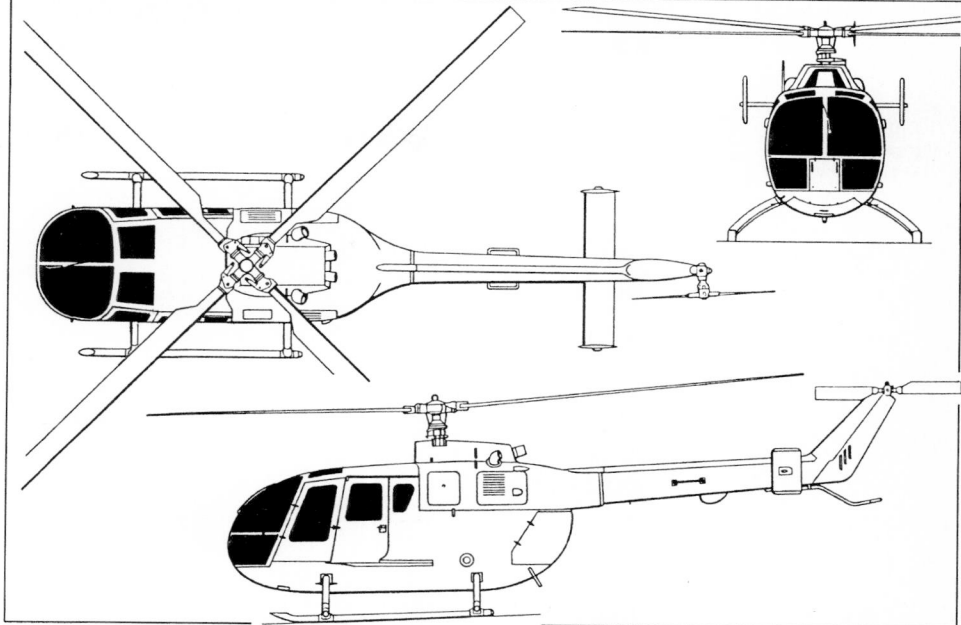

Eurocopter BO 105 CBS-5 multirole light helicopter (*Jane's/Paul Jackson*)

main rotor brake standard; two blades can be folded; two-blade semi-rigid tail rotor; main rotor rpm 424, tail rotor rpm 2,220.

FLYING CONTROLS: Fully powered controls through hydraulic actuation pack mounted on transmission casing; very high control power and aerobatic capability, including claimed sustained inverted 1 *g* flight; IFR certification without autostabiliser.

STRUCTURE: Light alloy structure with GFRP main and tail rotor blades and top fairing. Main production source is Eurocopter Deutschland factory at Donauwörth, but Spanish models assembled by EADS CASA, Indonesian models manufactured and assembled by Dirgantara and BO 105 LS produced exclusively by Eurocopter Canada in Ontario (which see).

LANDING GEAR: Skid type, with aluminium alloy cross-tubes designed for energy absorption by plastic deformation in event of heavy landing. Inflatable emergency floats can be attached to skids.

POWER PLANT: Two 313 kW (420 shp) Rolls-Royce 250-C20B turboshafts, each with a maximum continuous rating of 298 kW (400 shp). Transmission rating 566 kW (759 shp) for T-O and maximum continuous. Bladder fuel tanks under cabin floor, usable capacity 570 litres (150.6 US gallons; 125.3 Imp gallons); fuelling point on port side of cabin; optional auxiliary tanks in freight compartment. Oil capacity: engine 12 litres (3.2 US gallons; 2.6 Imp gallons), gearbox 11.6 litres (3.06 US gallons; 2.55 Imp gallons).

ACCOMMODATION: Pilot and co-pilot or passenger on individual longitudinally adjustable front seats with four-point harnesses; optional dual controls. Bench seat at rear for three or four persons, removable for cargo and stretcher carrying; EMS versions available; entire rear fuselage aft of seats and under power plant available as freight and baggage space, with straight-in access through two clamshell doors at rear; two standard stretchers side by side in ambulance role; one forward-opening hinged and jettisonable door and one sliding door on each side of cabin; ram air and electrical ventilation system; heating system optional.

SYSTEMS: Dual hydraulic systems, pressure 103.5 bar (1,500 lb/sq in), for powered main rotor controls; system

flow rate 6.2 litres (1.64 US gallons; 1.36 Imp gallons)/min; bootstrap/fluid reservoir, pressurised at 1.70 bar (25 lb/sq in); electrical system powered by two 150 A 28 V DC starter/generators and a 24 V 25 Ah Ni/Cd battery; external power socket.

AVIONICS: Wide variety of avionics available including weather radar, Doppler and GPS navigation, Honeywell RDR 1500B 360° search radar, FLIR, TV broadcast and microwave datalink.

EQUIPMENT: Standard equipment includes pilot's windscreen wiper, 250 W retractable landing light in nose, single-hand starting device, scavenge oil filter, heated pitot, tiedown rings in cargo compartment, cabin and cargo compartment dome lights, position lights and collision warning lights; options include heating system, rescue winch, searchlight, externally mounted loudspeaker, fuel dump valve, external load hook, settling protectors, snow skids, manual main rotor blade folding, emergency floats, sand filters, and a wide range of EMS and VIP arrangements. NVG-compatible cockpit, first used operationally on 5 February 1999 by Rocky Mountain Helicopters.

DIMENSIONS, EXTERNAL:

Main rotor diameter	9.84 m (32 ft 3½ in)
Tail rotor diameter	1.90 m (6 ft 2¾ in)
Main rotor blade chord	0.27 m (10¾ in)
Tail rotor blade chord	0.18 m (7 in)
Distance between rotor centres	5.95 m (19 ft 6¼ in)
Length: incl main and tail rotors	11.86 m (38 ft 11 in)
excl rotors	8.81 m (28 ft 11 in)
fuselage pod	4.55 m (14 ft 11 in)
Fuselage max width	1.58 m (5 ft 2¼ in)
Height to top of main rotor head	3.00 m (9 ft 10 in)
Skid track: unladen	2.53 m (8 ft 3½ in)
laden	2.58 m (8 ft 5½ in)
Rear-loading doors: Height	0.64 m (2 ft 1 in)
Width	1.40 m (4 ft 7 in)

DIMENSIONS, INTERNAL:

Cabin, incl cargo compartment:	
Max width	1.40 m (4 ft 7 in)
Max height	1.25 m (4 ft 1 in)
Volume	4.8 m³ (169 cu ft)
Cargo compartment: Length	1.85 m (6 ft 0¾ in)
Max width	1.20 m (3 ft 11¼ in)
Max height	0.57 m (1 ft 10½ in)
Floor area	2.25 m² (24.2 sq ft)
Volume	1.3 m³ (46 cu ft)

AREAS:

Main rotor disc	76.05 m² (818.6 sq ft)
Tail rotor disc	2.835 m² (30.5 sq ft)

WEIGHTS AND LOADINGS:

Weight empty, basic: CBS-5	1,301 kg (2,868 lb)
LSA-3 Super Lifter	1,374 kg (3,029 lb)
Max external load	1,200 kg (2,646 lb)
Standard fuel (usable)	456 kg (1,005 lb)
Max fuel, incl auxiliary tanks	776 kg (1,710 lb)
Max T-O weight: CBS-5:	
internal load	2,500 kg (5,511 lb)
external load	2,600 kg (5,732 lb)
LSA-3 Super Lifter:	
internal load	2,600 kg (5,732 lb)
external load	2,850 kg (6,283 lb)
Max disc loading:	
CBS-4, internal load	32.9 kg/m² (6.74 lb/sq ft)
LSA-3 Super Lifter	34.2 kg/m² (7.00 lb/sq ft)
Transmission loading at max T-O weight and power, internal load:	
CBS-5	4.86 kg/kW (7.98 lb/shp)
LSA-3 Super Lifter	5.05 kg/kW (8.29 lb/shp)

Eurocopter BO 105 CBS-5 (*Patrick Penna/Eurocopter*)   NEW/0132431

PERFORMANCE:
Never-exceed speed (VNE) at S/L:
  CBS-4   131 kt (242 km/h; 150 mph)
  LSA-3 Super Lifter with underslung load
                           80 kt (148 km/h; 92 mph)
Max cruising speed at S/L:
  CBS-5   132 kt (245 km/h; 152 mph)
  Best range speed at S/L   110 kt (204 km/h; 127 mph)
Max rate of climb at S/L, max continuous power
                           570 m (1,870 ft)/min
Vertical rate of climb at S/L, T-O power
                           90 m (295 ft)/min
Service ceiling, OEI, 30 min power   1,700 m (5,580 ft)
Max operating altitude:
  at 2,500 kg (5,511 lb)   3,050 m (10,000 ft)
  at 2,300 kg (5,070 lb)   5,180 m (17,000 ft)
Hovering ceiling, T-O power: IGE 3,200 m (10,500 ft)
  OGE   2,430 m (7,960 ft)
Range with standard fuel and max payload, no reserves:
  at S/L   310 n miles (574 km; 357 miles)
  at 1,525 m (5,000 ft)   321 n miles (596 km; 370 miles)
Ferry range with auxiliary tanks, no reserves:
  at S/L   519 n miles (961 km; 597 miles)
  at 1,525 m (5,000 ft)
                           550 n miles (1,020 km; 634 miles)
Endurance with standard fuel and max payload, no
  reserves: at S/L   3 h 24 min
                           UPDATED

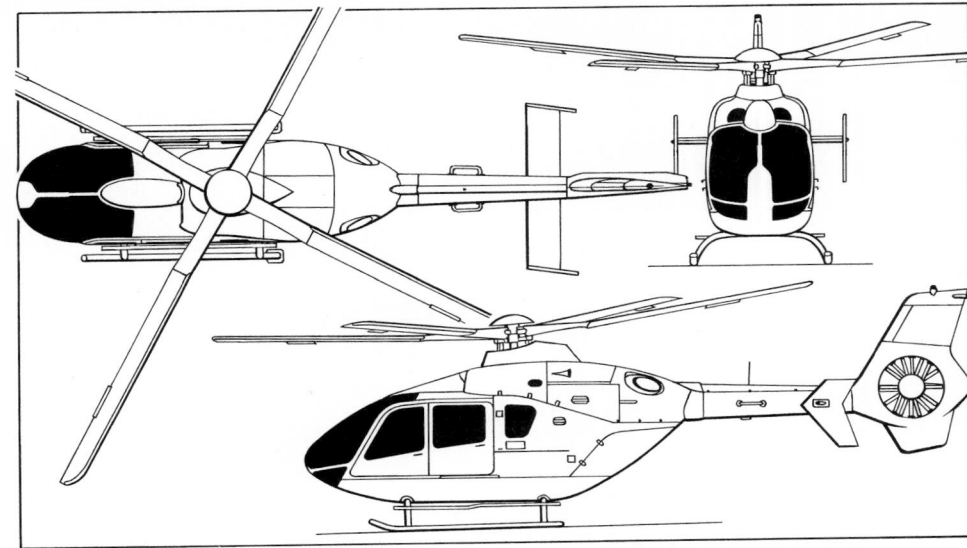

Eurocopter EC 135 seven-seat, twin-engined helicopter (*Jane's/Mike Keep*)

EUROCOPTER EC 135 and EC 635

TYPE: Light utility helicopter.
PROGRAMME: First flight on 15 October 1988 of technology prototype, D-HBOX, previously known as BO 108, powered by two Rolls-Royce 250-C20R turboshafts with conventional tail rotor; new all-composites bearingless tail rotor tested during 1990; Eurocopter announced in January 1991 that BO 108 was to succeed BO 105; first flight of second prototype (D-HBEC) powered by two Turbomeca TM 319-1B Arrius, 5 June 1991; production main and tail rotors flight tested during 1992 in preparation for certification programme; design revised late 1992 to increase maximum seating to seven; advanced Fenestron adopted.

Two preproduction prototypes D-HECX and D-HECY made first flights respectively 15 February and 16 April 1994 powered by Turbomeca Arrius 2B and P&WC PW206B intended as production alternatives; third preproduction prototype (D-HECZ) made first flight 28 November 1994, powered by Arrius 2B, and subsequently made type's US debut at HeliExpo '95 in Las Vegas in January 1995; total flight time of first three preproduction EC 135s was nearly 1,600 hours by end 1996, by which time all three preproduction prototypes had been retired. VFR certification to JAR 27 16 June 1996 and to FAR Pt 27 with Category A provisions and for both engine options, 31 July 1996; IFR certification awarded jointly by DGAC (France) and LBA (Germany) on 9 December 1998, while that for CAA attained late 2000; LBA (JAA) single pilot IFR certification achieved 2 December 1999; certified in 17 countries by June 2000; first two production aircraft delivered to Deutsche Rettungsflugwacht on 31 July 1996.
CURRENT VERSIONS: **EC 135P1:** Pratt & Whitney engine version (PW206B). First two (D-HQQQ and D-HYYY; c/ns 0005 and 0006) delivered on 31 July 1996 to Deutsche Rettungsflugwacht.
  EC 135P2: Introduced August 2001. PW206B2 engine with improved contingency ratings.
  EC 135T1: Turbomeca engine version. First (0010/N4037A) delivered to USA in November 1996. Early helicopters had 435 kW (583 shp) Arrius 2B engines, later replaced by 500 kW (670 shp) Arrius 2B1. Uprated Arrius 2B1A engine certified April 2001.
  EC 135T2: Deliveries from September 2002. Arrius 2B2 engines.
  EC 135 ACT/FHS: Active control technology and flying helicopter simulator; German fly-by-light trials

programme; first flight of modified production EC 135 (D-HECV) from Eurocopter's Ottobrunn facility 28 January 2002.
  EC 135 APH: Advanced Police Helicopter. Unified mission fit offered by McAlpine Helicopters of UK, 1997; allows simple outfitting with sensors and equipment, according to tasking, using under-fuselage pod; typical equipment, including loudspeakers, searchlights, microwave downlink and multisensor turret, can be fitted externally; TV and video equipment carried internally. Pod reduces maximum speed by 5 kt (9 km/h; 6 mph).
  EC 635: Military version; mockup was conversion of first preproduction EC 135 (D-HECX). Offered (unsuccessfully) to South Africa and unveiled at Aerospace Africa Air Show on 28 April 1998. First customer is Portuguese Army, which ordered nine EC 635T1s on 22 October 1999 for delivery from June 2001.
CUSTOMERS: Announced customers include Boeing Executive Flight Operations (two in mid-2002) plus one option, Bond, UK (15), Deutsche Rettungsflugwacht (German Air Rescue; two); EuropAvia, Switzerland; MFL, France; Helicap of Paris, France (13 EMS versions); Basque Police, Spain; Bavarian Police, Germany (9 with PW206 engines); Sultan of Pahang, Malaysia; Heliair/OAMTC, Austria (five, delivered from 10 August 1997 followed by order for 11 on 27 September 2000 for delivery between December 2000 and early 2002); DAO, Argentina; Laidlaw, UK; Hood, UK; Manichi Press, Japan; Elifriulia, Italy; Proeteus, France; Air Service 51, France; Center for Emergency Medicine of Pittsburgh, Pennsylvania (one); Norsk Luftambulanse (six); Petroleum Helicopters Inc (three); Temsco of Ketchikan, Alaska (one), TexAir of Houston, Texas (one), ADAC, Germany (12 for EMS), Osterman, Swedish Police (seven; first aircraft handed over 10 August 2001) Haier, China (one in VIP configuration).
  Law enforcement users include Mecklenburg-West Pomerania, Germany (two), Rheinland-Palatinate, Germany (one for delivery April 2002), Saxony Police (one), Abu Dhabi Police (one), Chile Police (one), Greek Police (two), Travis County Police Department, USA (one), plus UK police forces (seven). German Army ordered 15 at cost of DM95 million in August 1997 for delivery in 2000 to replace Alouette II in training role of which the first was delivered on 13 September 2000. German Border Guard (Bundesgrenzschutz; BGS) ordered nine (plus two options) in December 1997; first three delivered 21 September 2000, with eight in service by end of 2001.

Eurocopter expects to win 700 sales out of a world market for 1,350 during the period 1998 to 2007. Total of 260 on order by 50 customers in September 2001. 100th delivered (to Bavarian Police) on 16 June 1999, some two-thirds with Turbomeca engines; 30 ordered in 1999, 40 in 2000 and 38 in 2001; deliveries totalled 37 in 1999 and 31 in 2000.
COSTS: Operating cost 25 per cent lower than BO 105; development programme funded by Eurocopter Deutschland and Eurocopter Canada, suppliers, and German Ministries of Economics and Research and Technology; flyaway cost US$2.39 million (1996); programme unit cost DM6.67 million (US$4.4 million); Bavarian police (1996). Portuguese contract for nine valued at €35 million (1999).
DESIGN FEATURES: Designed to FAR Pt 27 including Category A and European JAR 27; pod-and-boom configuration, with Fenestron; forward flight stability by two horizontal and four vertical (fin, underfin and two endplates) surfaces; four-blade FVW bearingless main rotor, single-piece rotor head/mast; rotor rpm are variable; composites blades mounted on controlled flexibility composites arms giving flap, lag and pitch-change freedom; control demands transmitted from rods to root of blade by rigid CFRP pitch cuffs; main rotor blades have DM-H3 and -4 aerofoils with non-linear twist and tapered transonic tips; main rotor axis tilted forward 5°.
  Airframe drag 30 per cent lower than BO 105 by clean and compact external shape; cabin height retained by shallow two-stage transmission; vibration reduced by ARIS mounting between transmission and fuselage; all dynamically loaded components to have 3,500 hours MTBR or be maintained on-condition. Fenestron has 10 asymmetrically spaced blades.
  Second BO 108 prototype had EFIS-based IFR system; fuselage stretched 15 cm (5.9 in) and interior cabin width extended by 10 cm (3.9 in); main rotor diameter extended to 10.20 m (33 ft 5½ in); for EC 135, tail rotor replaced in 1992 by New Generation Fenestron with 11 fixed flow-straightening vanes in fan efflux designed to avoid momentum losses and improve fan figure of merit; vanes are swept relative to radius and fan has different number of blades to avoid shocks and reduce noise; fan blade tip speed is only 185 m (607 ft)/s; maximum T-O weight increased to 2,720 kg (5,997 lb).
FLYING CONTROLS: Conventional hydraulic fully powered controls with integrated electrical SAS servos; objective is single-pilot IFR with cost-effective stability augmentation. Electric cyclic trim system.
STRUCTURE: Airframe mainly Kevlar/CFRP sandwich composites, except aluminium alloy sidewalls, pod lower module and cabin floor, tailboom and around cargo area; some titanium in engine bay; composites tailplane.
LANDING GEAR: Skid type; ground handling wheels can be fitted.
POWER PLANT: Choice of turboshaft engines. Turbomeca-engined aircraft have two Arrius 2B2s, each giving 452 kW (606 shp) at T-O, 426 kW (571 shp) maximum continuous, 528 kW (708 shp) OEI continuous, 580 kW (777 shp) for 2 minutes with OEI and 609 kW (816 shp) for 30 s. Alternative power plant is two Pratt & Whitney Canada PW206B2s, each giving 463 kW (621 shp) at T-O, 419 kW (562 shp) maximum continuous, 528 kW (708 shp) OEI continuous, 580 kW (777 shp) for 2 minutes with OEI and 609 kW (816 shp) for 30 s. Both types of engine have FADEC. Transmission rating 616 kW (826 shp) maximum T-O, 567 kW (760 shp) maximum continuous, 353 kW (473 shp) OEI continuous, 513 kW (687 shp) for 2 minutes with OEI and 526 kW (705 shp) for 30 s.
  Fuel capacity of first 249 aircraft 673 litres (178 US gallons; 148 Imp gallons) of which 663 litres (175 US gallons; 146 Imp gallons) are usable. Capacity 700 litres (185 US gallons; 154 Imp gallons) from No. 250.

Eurocopter EC 135 light utility helicopter (*Jane's/Paul Jackson*)   NEW/0132421

Additional long-range tank optional, usable capacity 198.5 litres (52.4 US gallons; 43.7 Imp gallons). Optional self-sealing fuel tanks. Oil capacity 8 litres (2.0 US gallons; 1.75 Imp gallons).

ACCOMMODATION: Seven persons, including one or two pilots, in standard version, or six persons in VIP version; optional max capacity of eight. Four-point harnesses for front seats; three-point harnesses for remaining seats. Forward-hinged doors for two front occupants; sliding doors for five persons in cabin. Rear of pod has clamshell doors for bulky items/cargo; flights permissible with clamshell doors removed; optional window in each rear door. Unobstructed cabin interior. EMS variant can accommodate one pilot with two stretcher cases and two seated medical staff/attendants; alternative layouts for one, one, three, or two, one, three, or two, two, two.

SYSTEMS: Redundant 28 V DC electrical supply systems to JAR/FAR 27 standards; two 160 A 28 V starter/generators and 24 V 17 Ah Ni/Cd batteries in Arrius 2B variant, two 200 A 28 V starter/generators and 24 V 25 Ah Ni/Cd batteries in PW206B variant. Fully redundant dual hydraulic systems. NATO standard external power connector.

AVIONICS: Bendix/King Gold Crown and Thales equipment. Total of 10 standard avionics and equipment packages available.

Radar: Provision for integrated weather radar.

Flight: Air data computer; SFIM automatic flight control system (AFCS); GPS. Honeywell combined solid-state flight data and cockpit voice recorder tested on EC 135 late 1998. BGS aircraft have EADS-Dornier Hellas obstacle warning system. Optional Bendix/King Mark XXII EGPWS available for EC 135 APH.

Instrumentation: Liquid crystal dual-screen (Thales SMD45) vehicle and engine management displays with AN equipment.

Mission: Optional FLIR and NVGs for military; police and ambulance roles. BGS EC 135s are first to be equipped with Hellas laser obstacle warning system.

EQUIPMENT: Options include cargo hook, external loudspeakers and searchlights, rescue winch (230 kg; 507 lb) with 50 m (164 ft) cable, emergency floats, sand filter, wire-strike protection system and light armour protection.

ARMAMENT: EC 635 undergoing testing with FN Herstal HMP 400 12.7 mm machine gun, Giat NC621 20 mm gun and 12-round 70 mm rocket launcher.

DIMENSIONS, EXTERNAL:

Main rotor diameter	10.20 m (33 ft 5½ in)
Fenestron diameter	1.00 m (3 ft 3¼ in)
Length: overall, rotor turning	12.16 m (39 ft 10¾ in)
fuselage: incl boom	10.20 m (33 ft 6½ in)
excl boom	5.87 m (19 ft 3 in)
Height: overall	3.51 m (11 ft 6 in)
to top of rotor head	3.35 m (11 ft 0 in)
Width: without rotor blades (tailplane span)	2.65 m (8 ft 8¼ in)
fuselage (max)	1.56 m (5 ft 1½ in)
Skid length	3.20 m (10 ft 6 in)
Skid track	2.00 m (6 ft 6¾ in)
Ground clearance: fuselage	0.40 m (1 ft 3¾ in)
tailboom underfin	0.66 m (2 ft 2 in)

DIMENSIONS, INTERNAL:

Cabin and cockpit: Length: normal	3.06 m (10 ft 0½ in)
with EMS floor extension	4.11 m (13 ft 5¾ in)
Max width	1.50 m (4 ft 11 in)
Max height	1.26 m (4 ft 1½ in)
Floor area	4.3 m² (46 sq ft)
Volume: EC 135	4.8 m³ (170 cu ft)
EC 635	4.6 m³ (162 cu ft)
Baggage compartment:	
Length	1.05 m (3 ft 5¼ in)
Max width	1.23 m (4 ft 0½ in)
Max height	0.70 m (2 ft 3½ in)
Floor area	1.20 m² (12.9 sq ft)
Volume	1.1 m³ (39 cu ft)

AREAS:

Main rotor disc	81.71 m² (879.5 sq ft)
Fenestron disc	2.84 m² (30.52 sq ft)

WEIGHTS AND LOADINGS:

Weight empty: EC 135	1,490 kg (3,284 lb)
EC 635	1,530 kg (3,373 lb)
Max fuel: standard (No. 250 on)	560 kg (1,235 lb)
with long-range tank	720 kg (1,587 lb)
Max external load	1,360 kg (2,998 lb)
Max T-O weight: normal	2,835 kg (6,250 lb)
with external load	2,900 kg (6,393 lb)
Max disc loading: normal	34.7 kg/m² (7.11 lb/sq ft)
with external load	35.5 kg/m² (7.27 lb/sq ft)
Transmission loading at max T-O weight and power:	
Normal T-O	4.61 kg/kW (7.57 lb/shp)
T-O with external load	4.71 kg/kW (7.73 lb/shp)

PERFORMANCE (A: Arrius 2B2, B: PW206B2, at normal MTOW):

Never-exceed speed (VNE):	
A, B	140 kt (259 km/h; 161 mph)
Max cruising speed: A, B	138 kt (256 km/h; 159 mph)
Econ cruising speed: A	129 kt (239 km/h; 148 mph)
B	124 kt (230 km/h; 143 mph)
Max rate of climb at S/L: A, B	457 m (1,500 ft)/min
Rate of climb at S/L, OEI: A, B	65 m (215 ft)/min
Max certified altitude: A, B	3,050 m (10,000 ft)
Service ceiling, OEI: A	3,050 m (10,000 ft)
B	2,925 m (9,600 ft)
Max certified hovering ceiling, IGE: A, B	3,050 m (10,000 ft)
Hovering ceiling, OGE: A, B	2,195 m (7,200 ft)
Range at S/L, standard fuel (700 litres):	
A	335 n miles (620 km; 385 miles)
B	349 n miles (646 km; 401 miles)
Ferry range with long-range tank:	
A	431 n miles (798 km; 460 miles)
B	450 n miles (833 km; 517 miles)
Endurance at S/L, standard fuel (700 litres):	
A	3 h 24 min
B	3 h 47 min

UPDATED

Eurocopter EC 135 employed in the UK as an air ambulance ***(Jane's/Paul Jackson)*** ***NEW***/0132420

EUROCOPTER EC 145

The Eurocopter EC 145 is a development of the Eurocopter/Kawasaki BK 117 and is described following that entry.

EUROCOPTER EC 155

This Dauphin development is described immediately following the Eurocopter AS 565 Panther entry.

EUROCOPTER EC 165

Studies are continuing for a new helicopter in the 6,000 kg (13,228 lb) class, to replace the AS 365N Dauphin 2. The DTV5 programme (see Dauphin 2 entry) will contribute to technology which may be used in the EC 165.

UPDATED

OTHER AIRCRAFT

Refer elsewhere in this section for **Euromil (Eurocopter/Mil/Kazan/Klimov) Mi-38**; Eurocopter also participates in **NH Industries NH 90** international programme. Refer to *Jane's Aircraft Upgrades* for BO 105M PAH-1A1 upgrades.

VERIFIED

EUROCOPTER EC 120 B COLIBRI

English name: Hummingbird

Spanish Air Force designation: HE.25

TYPE: Light utility helicopter.

PROGRAMME: Definition phase of original P120L launched 15 February 1990; subsequently redesigned with 500 kg (1,102 lb) lower gross weight and new engine and rotor; development contract signed October 1992; redesignated EC 120, January 1993; design definition completed mid-1993; assembly of first of two prototypes began at Eurocopter France at Marignane in early 1995; first flight (F-WWPA) 9 June 1995; second prototype (F-WWPD) flown 17 July 1996; certification to JAR 27 was achieved on 16 June 1997 (following Arrius 2F engine approval by DGAC on 22 January 1997); FAR Pt 27 certification on 22 January 1998; operations in cold weather certified late 1998; first production EC 120 (c/n 1005) flew 5 December 1997 before delivery to Japanese distributor Nosaki on 23 January 1998 for eventual use (as JA120B) by Nosaki Sangyo, Osaka.

Eurocopter has 61 per cent share and responsible for instrument panel, landing gear, seats, rotor system, transmission, final assembly, flight test and certification; CATIC (China National Aero-Technology Import & Export Corporation in the form of Hafei Aviation Industry Company; 24 per cent) for cabin structure and doors, engine cowlings, pod central and intermediate structure and fuel system; and Singapore Technologies Aerospace (15 per cent) for tailboom, fin, horizontal stabiliser, Fenestron, general doors and instrument pedestal. Joint design team working in Eurocopter France at Marignane. Global support system, with 13 product centres, being initiated. Australian assembly centre established as part of offsets for 2001 purchase of Tiger attack helicopters by Australian Army.

CUSTOMERS: Recent customers include the Spanish Air Force, which ordered 15 in 1999 for basic training; deliveries began 26 July 2000 and were completed in July 2001; and Air Logistics, which took delivery of six in 2001 for offshore operations in the Gulf of Mexico. Three ordered for Indonesian Navy and 12 for Indonesian Air Force; deliveries began June 2001. 100th aircraft handed over 19 April 2000 to a German customer; 200th to a Swedish customer on 29 March 2001. Deliveries in 2000 totalled 91, compared to 52 in 1999; orders for the same periods were 61 and 36 respectively, followed by 81 in 2001.

COSTS: US$840,000 (1999). Spanish contract for 15 valued at €15 million (1999), including training and spares.

DESIGN FEATURES: Pod-and-boom layout; horizontal stabilisers, fin and underfin for directional stability; anti-torque Fenestron. Three-blade main rotor on Spheriflex hub rotating clockwise (nominal rotor speed 406 rpm) integrated with main shaft and transmission; two-stage reduction gear; rotor brake; eight asymmetrically positioned bladed New Generation Fenestron (nominal tail rotor speed 4,567 rpm). External noise almost 7 dB below ICAO limits. Single level cabin floor. Engine mounted to left of main rotor mast to improve balance by counteracting main rotor downwash.

FLYING CONTROLS: Control forces on collective and cyclic reduced by three electrically actuated hydraulic servos operating at 37 bar (537 lb/sq in). Tab on each main rotor blade trailing-edge at three-quarter span.

STRUCTURE: Main rotor blades have carbon fibre rib with foam filler and glass fibre skin, plus stainless steel attachment bushes and leading-edge. Titanium alloy Spheriflex head and shaft made as single composites assembly; metal centre-fuselage; light alloy skid landing gear; crashworthy seats and fuel system. Light alloy tail rotor shaft.

LANDING GEAR: 'Moustache' configuration of fixed skids, having sweptback forward supports and full-width boarding step. Optional 1.50 m (4 ft 11 in) skis for snow operations.

POWER PLANT: One Turbomeca TM 319 Arrius 2F engine selected for first 300 EC 120s; rating 376 kW (504 shp) for T-O, 335 kW (449 shp) max continuous. Transmission rating 330 kW (442 shp). Fuel capacity 416 litres (109.9 US gallons; 91.5 Imp gallons) in two tanks (one located beneath cabin floor and one above baggage compartment); usable fuel 406 litres (107 US gallons; 89.3 Imp gallons).

ACCOMMODATION: Pilot and four passengers (or pilot, patient and paramedic in HEMS configuration); two front seats and one rear bench seat. Large door, front, starboard; hinged front and rear sliding door, port. Compartment below engine on same level as cabin floor, accessible from cabin, by external side door and rear door. Seating conforms to new FAR Pt 27 requirements for 30 *g* vertical and 18 *g* horizontal deceleration. Dual controls optional. Accommodation ventilated and heated; air conditioning optional.

SYSTEMS: VEMD (vehicle and engine multifunction display), fitted as standard, is a fully duplex three-module processing system using two glass screens to monitor performance and maintenance requirements. Electrical system includes 4.5 kW 28 V DC starter/generator and Ni/Cd battery.

AVIONICS: To customer's requirements. Six standard packages available for passenger transport (two), law enforcement, training (two) and utility.

Comms: Bendix/King: KX 165 nav/com/glideslope and secondary KY 196A VHF, Jolet JE2 NG ELT, Bendix/King KT 76A transponder and Team SIB 120 interphone in typical passenger configuration.

Flight: Trimble TNL 2101 GPS with moving map in typical passenger configuration.

EC 120 B Colibri five-seat light helicopter (*Gérard Deulin/Eurocopter*)   *NEW*/0132961

Instrumentation: Standard fit of VEMD, ASI, NR/NF dual indicator and altimeter.

EQUIPMENT: Optional Aerazur emergency flotation bags, and cable cutters. Optional pilot-controllable swivelling landing light. Optional windscreen wipers.

DIMENSIONS, EXTERNAL:

Main rotor diameter	10.00 m (32 ft 9¾ in)
Main rotor blade chord	0.26 m (10¼ in)
Fenestron diameter	0.75 m (2 ft 5½ in)
Fenestron blade chord	0.06 m (2¼ in)
Length overall, rotors turning	11.52 m (37 ft 9½ in)
Fuselage: Length	9.60 m (31 ft 6 in)
Max width of cabin	1.50 m (4 ft 11 in)
Max height, excl skids:	
to engine cowling	1.67 m (5 ft 5¾ in)
to rotor head	2.52 m (8 ft 3¼ in)
Tailplane span	2.60 m (8 ft 6¼ in)
Height: overall	3.40 m (11 ft 1¾ in)
to top of rotor head	3.08 m (10 ft 1¼ in)
to top of engine cowling	2.23 m (7 ft 3¾ in)
Skid track	2.07 m (6 ft 9½ in)
Height of skids	0.56 m (1 ft 10 in)
Height of boarding step	0.50 m (1 ft 7¾ in)
Large door, starboard: Width	1.15 m (3 ft 9¼ in)
Height	1.16 m (3 ft 9¾ in)
Front door, port: Width	0.81 m (2 ft 8 in)
Height	1.16 m (3 ft 9¾ in)
Sliding door, port: Width	0.90 m (2 ft 11½ in)
Height	1.16 m (3 ft 9¾ in)

DIMENSIONS, INTERNAL:

Cabin: Length	2.30 m (7 ft 6½ in)
Max width	1.35 m (4 ft 5¼ in)
Max height	1.25 m (4 ft 1¼ in)
Floor area	3.0 m² (32.29 sq ft)
Volume	2.14 m² (75.6 sq ft)
Baggage compartment: Length	1.31 m (4 ft 3½ in)
Max width	0.72 m (2 ft 4¼ in)
Max height	1.02 m (3 ft 4¼ in)
Volume	0.8 m³ (28.25 cu ft)

AREAS:

Main rotor disc	78.54 m² (845.4 sq ft)
Fenestron disc	0.44 m² (4.73 sq ft)

WEIGHTS AND LOADINGS:

Weight empty: standard	960 kg (2,116 lb)
typical passenger fit	1,007 kg (2,220 lb)
typical law enforcement fit	985 kg (2,172 lb)
advanced training fit	1,021 kg (2,251 lb)
Max sling load	700 kg (1,543 lb)
Max useful load	755 kg (1,664 lb)
Max T-O weight: internal load	1,715 kg (3,781 lb)
external load	1,800 kg (3,968 lb)
Max disc loading:	
internal load	21.8 kg/m² (4.47 lb/sq ft)
external load	22.9 kg/m² (4.69 lb/sq ft)
Transmission loading at max T-O weight and power:	
internal load	5.20 kg/kW (8.54 lb/shp)
external load	5.45 kg/kW (8.96 lb/shp)

PERFORMANCE (at 1,715 kg; 3,781 lb):

Never-exceed speed (VNE)	150 kt (278 km/h; 173 mph)
Cruising speed: max	122 kt (226 km/h; 140 mph)
econ	103 kt (191 km/h; 119 mph)
Max rate of climb at S/L	366 m (1,200 ft)/min
Service ceiling	5,180 m (17,000 ft)
Hovering ceiling: IGE	2,820 m (9,260 ft)
OGE	2,320 m (7,620 ft)
Range, no reserves	393 n miles (727 km; 452 miles)
Endurance, no reserves	4 h 32 min

UPDATED

EUROCOPTER BK 117

TYPE: Light utility helicopter.

PROGRAMME: Developed jointly by partners; four prototypes, first flight 13 June 1979; one preproduction aircraft, first flight 6 March 1981; first flights of production aircraft 24 December 1981 (JQ1001 in Japan) and 23 April 1982 (in Germany); certified in Germany and Japan 9 and 17 December 1982 respectively, followed by US FAA 29 March 1983 (FAR Pt 29, Categories A and B, including Amendments 29-1 to 29-16); deliveries began early 1983. See 1991-92 and previous editions for earlier A series and B-1. BK 117 fleet reached 1,274,000 flying hours by January 2001, at which time 416 delivered and 379 remaining in service.

In October 2001, Eurocopter announced plans for transfer of production from Donauwörth to Trento, Italy, where seven to be produced annually by Avionline (jointly owned by Aerosud Elicotteri and Helicopters Italia); marketing and delivery remain unchanged.

CURRENT VERSIONS: **BK 117 B-2:** Production completed in 1998; performance figures in 1999-2000 *Jane's.*

BK 117 C-1: Turbomeca Arriel 1E2 engines first flown in converted BK 117A/B (F-WMBB) 6 April 1990; German LBA certification 17 January 1992; US FAA certification 7 December 1992; first delivery (to USA)

December 1992. French DGAC certification 15 July 1993. Performance similar to that of BK 117 B-2; better hot-and-high performance.

Production BK 117 C-1 additionally features higher transmission and engine OEI ratings, new tail rotor blades and variable rotor speed to improve tail rotor thrust and reduce external noise, and torque matching system to reduce pilot workload.

Prototype C-1 (D-HECD) built 1992; exports began, to USA, December 1992. German LBA certification achieved 28 April 1994, followed by US FAA certification 29 September 1994, Italian RAI certification 24 November 1994, and UK CAA certification 28 July 1995.

Japanese (Kawasaki-built) version which includes both re-engine and above improvements was approved 8 June 1995 by Japanese CAB.

Details below apply to BK 117 C-1 version.

BK-117 C-2: See separate entry for EC 145.

NBK-117: Designation of aircraft built by IPTN (now Dirgantara) (see Indonesian section, 1991-92 *Jane's*) under November 1982 agreement with MBB. Three only; no longer produced.

BK 117 P5: Advanced technology demonstrator in Japan; fly-by-wire control system; described in 1999-2000 and earlier *Jane's.*

EC-Futura BK117: Technology demonstrator, first flown July 1999 at Ottobrunn, used as part of All Weather Rescue Helicopter programme.

EC 145: See separate entry.

CUSTOMERS: Total of 291 (including prototypes) built by MBB/Eurocopter up to early 2002, comprising 71 BK 117 A-1s, 50 A-3s, 18 A-4s, 104 B-1s, nine B-2s and 40 C-1s; Kawasaki total was 131 by early 2002, mostly B-1/B-2s, but including some 20 C-1s. Total production 422 by early

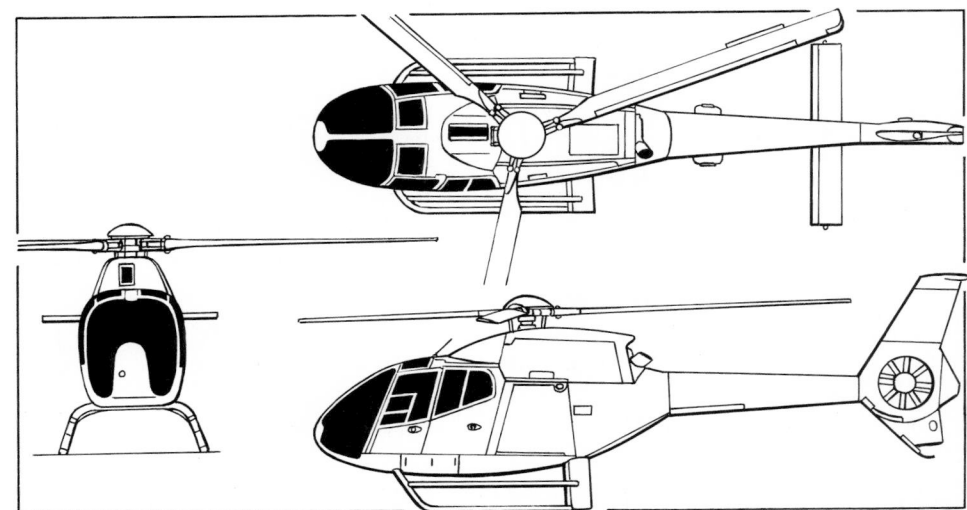

Eurocopter EC 120 B Colibri single-engined light helicopter (*Jane's/James Goulding*)   0099566

Eurocopter BK 117 used for medical evacuation in USA   *NEW*/0132960

2002. Two BK117 family ordered in 2001, compared with seven in 2000, three in 1999, five in 1998 and seven in 1997. Customers include TsENTROSPAS of Russia (two, delivered 1998 and December 2000), Greek Fire Brigade (two, delivered June 2001), Telmex of Mexico (2001), First Security Bank of USA (October 2002), Virginia State Police (2002) and Italian EMS operators Aeroveneta, (2000) and Alidaunia (2001).

COSTS: BK 117 B-2 US$2.815 million (1996); BK 117 C-1 US$3.1 million (1996).

DESIGN FEATURES: Pod-and-boom configuration, latter mounted high for cabin access via rear freight doors; tail rotor mounted at fin-tip; large auxiliary fins at tip of each horizontal stabiliser; engines above cabin.

System Bölkow four-blade main rotor head, almost identical to that of BO 105; main rotor blades similar to but larger than those of BO 105, with NACA 23012/23010 (modified) section; optional two-blade folding. Two-blade semi-rigid tail rotor with MBB-S102E performance/noise optimised blade section; rotor rpm 383 (main), 2,169 (tail).

FLYING CONTROLS: Equipped as standard for single-pilot VFR operation; dual controls and dual VFR instrumentation optional; rotor brake and yaw CSAS standard on German-built versions, optional on Kawasaki aircraft; options include IFR instrumentation, two-axis (pitch/roll) CSAS and Honeywell SPZ-7100 dual digital AFCS. Mast moment indicator discourages excessive cyclic control inputs.

STRUCTURE: Main rotor has one-piece titanium hub with pitch-change bearings; fail-safe GFRP blades with stainless steel anti-erosion strip. Tail rotor, mounted on port side of central fin, has GFRP blades of high-impact resistance. Main fuselage pod and tailboom are aluminium alloy with single-curvature sheets and (on fuselage) bonded aluminium sandwich panels; secondary fuselage components are compound curvature shells of Kevlar sandwich. Engine deck, to which tailboom is integrally attached, forms cargo compartment roof and is of titanium adjacent engine bays. Detachable tailcone carries main fin/tail rotor support, and horizontal stabiliser with offset endplate fins.

Eurocopter responsible for rotor systems, tailboom, tail unit, skid landing gear, hydraulic system, engine firewall and cowlings, powered controls and systems integration; Kawasaki for fuselage, transmission, fuel and electrical systems, and standard equipment. Components single-sourced and exchanged for separate assembly lines at Donauwörth and Gifu; some components and accessories interchangeable with those of BO 105 (which see), from which hydraulic-powered control system is also adapted.

LANDING GEAR: Non-retractable tubular skid type, of aluminium construction. Skids are detachable from cross-tubes. Ground handling wheels standard. Emergency flotation gear, settling protectors and snow skids optional.

POWER PLANT: BK 117 B-2 has two Honeywell LTS 101-750B-1 turboshafts, each rated at 410 kW (550 shp) for take-off and maximum continuous power, 441 kW (592 shp) for 30 minutes with OEI. BK 117 C-1 has two Turbomeca Arriel 1E2 turboshafts each rated at 550 kW (738 shp) for take-off, 516 kW (692 shp) maximum continuous and 574 kW (770 shp) for 2½ minutes OEI.

Kawasaki KB 03 main transmission rated at 736 kW (986 shp) for twin-engine T-O, 632 kW (848 shp) maximum continuous; for single-engine operation, 550 kW (738 shp) allowed for 2½ minutes, 404 kW (542 shp) for maximum continuous (see also BK 117 C-1 improvements).

Fuel in four flexible bladder tanks (forward and aft main tanks, with two supply tanks between), in compartments under cabin floor. Two independent fuel feed systems for engines and common main fuel tank. Total standard fuel capacity 697 litres (184 US gallons; 153 Imp gallons), of which 685 litres (181 US gallons; 151 Imp gallons) are usable; single or twin internal auxiliary fuel tanks, each of 200 litres (52.8 US gallons; 44.0 Imp gallons) capacity, and two external auxiliary fuel tanks, each of 150 litres (39.6 US gallons; 33.0 Imp gallons), optional.

ACCOMMODATION: Pilot and up to six (executive version), seven (Eurocopter standard version) or nine passengers (Kawasaki version). High-density layouts available for up to 10 passengers in addition to pilot. Level floor throughout cockpit, cabin and cargo compartment. Jettisonable forward-hinged door on each side of cockpit, pilot's door having an openable window. Jettisonable rearward-sliding passenger door on each side of cabin, lockable in open position. Fixed steps on each side. Two hinged, clamshell doors at rear of cabin, providing straight-in access to cargo compartment. Rear cabin window on each side. Aircraft can be equipped for offshore support, medical evacuation (one or two stretchers side by side and up to six attendants), firefighting, search and rescue, law enforcement, cargo transport or other operations.

SYSTEMS: Ram air and electrical ventilation system. Fully redundant tandem hydraulic boost system (one operating and one standby), pressure 103.5 bar (1,500 lb/sq in), for flight controls. System flow rate 8.1 litres (2.14 US gallons; 1.78 Imp gallons)/min. Bootstrap/oil reservoir, pressure 1.70 bar (25 lb/sq in). Main DC electrical power from two 150 A 28 V (200 A 28 V for C-1) starter/generators (one on each engine) and a 24 V 25 Ah Ni/Cd battery. AC power provided by inverter; second AC inverter optional; emergency busbar provides direct battery power to essential services, external DC power receptacle.

AVIONICS: *Comms:* VHF-AM/FM, UHF and HF radios to customer's requirements.

Radar: Multirole radar optional.

Flight: Long-range navaids optional.

Instrumentation: Basic instrumentation for single-pilot VFR operation includes airspeed indicator with two electrically heated pitot tubes and static ports, two hydraulic pressure indicators, encoding altimeter, instantaneous vertical speed indicator, 4 in artificial horizon, 3 in standby artificial horizon, HSI, (4 in and 3 in artificial horizons and HSI optional on Kawasaki-built aircraft), gyro magnetic heading system, magnetic compass, ambient air temperature thermometer and clock.

EQUIPMENT: Standard basic equipment includes rotor brake, annunciator panel, two master caution lights, rotor rpm/engine fail warning control unit, fuel quantity indicator and low-level sensor, outside air temperature indicator, engine and transmission oil pressure and temperature indicators, two exhaust temperature indicators, dual torque indicator, triple tachometer, two N1 tachometers, full internal and external lighting, ground handling wheels, pilot's windscreen wiper, floor covering, interior panelling and sound insulation, ashtrays, map/document case, tiedown rings in cabin and cargo compartment, engine compartment fire warning indicator, engine fire extinguishing system, portable fire extinguisher, first aid kit, and single-colour exterior paint scheme.

Optional equipment includes long-range fuel tanks, high-density seating arrangement, bleed air heating system, crashworthy seats, emergency flotation gear, settling protectors, snow skids, main rotor blade folding kit, dual pilot operation kit, stretcher installation, external cargo hook for 1,500 kg (3,307 lb) maximum load, rescue hoist (90 m; 295 ft cable, maximum load 270 kg; 595 lb, variable winch speed), Spectrolab SX 16 remotely controlled searchlight, external loudspeaker, sand filter and kits for rescue, law enforcement, firefighting and VIP transport.

Active anti-vibration system certified in Japan in mid-1997; available as option or for retrofit.

DIMENSIONS, EXTERNAL:

Main rotor diameter	11.00 m (36 ft 1 in)
Tail rotor diameter	1.96 m (6 ft 5 in)
Tail rotor blade chord	0.22 m (8¾ in)
Length: overall, both rotors turning	13.01 m (42 ft 8¼ in)
fuselage, tail rotor blades vertical	9.93 m (32 ft 7 in)
Fuselage: Max width	1.60 m (5 ft 3 in)
Height: overall, both rotors turning	3.85 m (12 ft 7½ in)
to top of main rotor head	3.36 m (11 ft 0¼ in)
Tailplane span (over endplate fins)	2.71 m (8 ft 10¾ in)
Ground clearance: tail rotor	1.89 m (6 ft 2½ in)
fuselage	0.35 m (1 ft 1¾ in)
Skid track	2.50 m (8 ft 2½ in)

DIMENSIONS, INTERNAL:

Combined cabin and cargo compartment:

Max length	3.02 m (9 ft 11 in)
Width: max	1.59 m (5 ft 2½ in)
min	1.27 m (4 ft 2 in)
Height: max	1.28 m (4 ft 2½ in)
min	0.99 m (3 ft 3 in)
Useful floor area	3.70 m² (39.8 sq ft)
Volume	5.0 m³ (177 cu ft)
Cabin only: Length	2.56 m (8 ft 4¾ in)

AREAS:

Main rotor blades (each)	1.76 m² (18.94 sq ft)
Tail rotor blades (each): B-2	0.098 m² (1.05 sq ft)
C-1	0.108 m² (1.16 sq ft)
Main rotor disc	95.03 m² (1,022.9 sq ft)
Tail rotor disc	3.00 m² (32.24 sq ft)

WEIGHTS AND LOADINGS:

Basic weight empty: B-2	1,745 kg (3,846 lb)
C-1	1,764 kg (3,890 lb)
Fuel: standard usable	558 kg (1,230 lb)
incl first auxiliary tank	718 kg (1,583 lb)

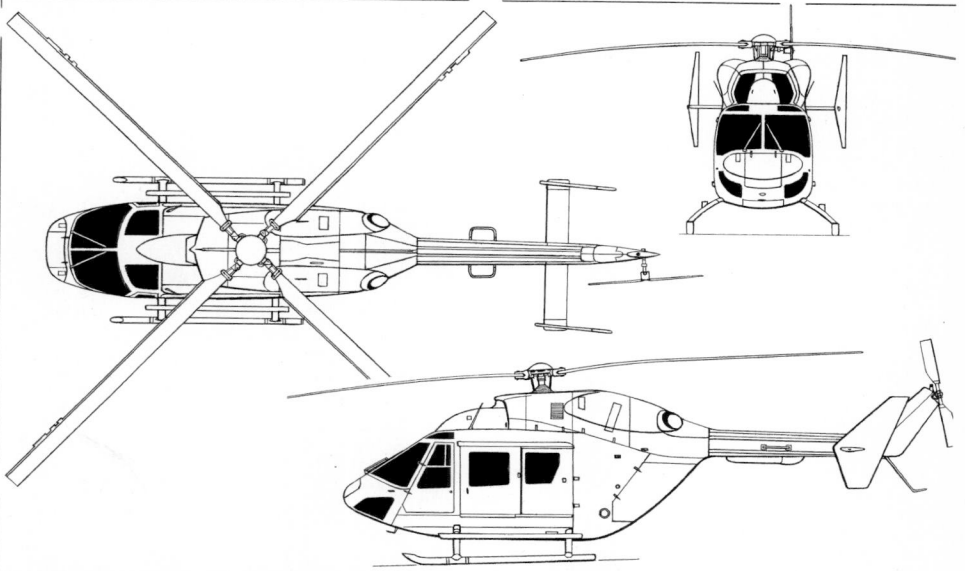

Eurocopter BK 117 B-2 twin-turboshaft multipurpose helicopter (*Jane's/Dennis Punnett*)

Max T-O weight: internal payload   3,350 kg (7,385 lb)
  external payload   3,500 kg (7,716 lb)
Max disc loading:
  internal payload   35.25 kg/m² (7.22 lb/sq ft)
  external payload   36.8 kg/m² (7.54 lb/sq ft)
Transmission loading at max T-O weight and power:
  internal payload   4.55 kg/kW (7.48 lb/shp)
  external payload   4.76 kg/kW (7.83 lb/shp)
PERFORMANCE (main values for BK 117 C-1; A: at gross weight of 3,000 kg; 6,614 lb, B: at 3,200 kg; 7,055 lb, C: at 3,350 kg; 7,385 lb):
Never-exceed speed (VNE) at S/L:
  A   150 kt (277 km/h; 172 mph)
  B, C   140 kt (259 km/h; 161 mph)
Max cruising speed: A   135 kt (250 km/h; 155 mph)
  B   134 kt (248 km/h; 154 mph)
  C   133 kt (246 km/h; 153 mph)
Econ cruising speed: A   127 kt (235 km/h; 146 mph)
  B   126 kt (233 km/h; 145 mph)
  C   125 kt (231 km/h; 144 mph)
Max rate of climb at S/L: A   655 m (2,150 ft)/min
  B   587 m (1,925 ft)/min
  C   538 m (1,765 ft)/min
Max certified altitude: A, B, C   4,575 m (15,000 ft)
Service ceiling: A, B   5,480 m (18,000 ft)
  C   5,090 m (16,700 ft)
Hovering ceiling IGE (zero wind):
  A   3,690 m (12,100 ft)
  B   3,050 m (10,000 ft)
  C   2,530 m (8,300 ft)
Hovering ceiling IGE (20 kt; 37 km/h; 23 mph)
  crosswind: A   3,200 m (10,500 ft)
  B   2,530 m (8,300 ft)
  C   1,920 m (6,300 ft)
Hovering ceiling OGE: A   3,520 m (11,550 ft)
  B   3,000 m (9,840 ft)
  C   1,480 m (4,860 ft)
Range: C   292 n miles (540 km; 336 miles)
Endurance: B, C   2 h 50 min
UPDATED

EUROCOPTER EC 145
Japanese designation: BK 117C-2
TYPE: Light utility helicopter.
PROGRAMME: Incorporation of EC 135 technology into BK 117 (see previous entry) began in 1997; named EC 145 in late 1999, but retains engineering designation **BK 117C-2** and is marketed in Japan as such. First flight of German aircraft (D-HMBK) (unannounced) 12 June 1999; first flight of Japanese prototype 15 March 2000; third prototype (D-HMBL) joined programme 14 April 2000; fourth (D-HMBM) on 27 October 2000. Kawasaki builds tail section; Eurocopter responsible for forward section. Certification by LBA received 12 December 2000; commercial launch at Paris Air Show, June 2001; FAA certification awarded 14 February 2002, coincident with type's formal 'introduction' at HeliExpo, Orlando, Florida.
CUSTOMERS: Launch customer was French Sécurité Civile, which ordered 32 in December 1997 for delivery between 2001 and 2006 to replace Alouette III; two preproduction examples delivered May 2001 for familiarisation; first production example (F-ZBPA) formally handed over at Nimes-Garons 24 April 2002. Second customer is French Gendarmerie, with firm order for eight placed in 1999. ADAC (German Automobile Club) ordered two in June 2001 to become civilian launch customer; delivery in 2002. Eight ordered in 2001, including four for Rega HEMS in Switzerland.
COSTS: Sécurité Civile contract valued at US$170 million. Flyaway cost reported as US$4.9 million (2000).
DESIGN FEATURES: Fuselage redesigned forward of engines; new nose, based on EC 135, provides improved visibility. Main rotor blades are same diameter as BK 117, but have EC 135 profile. Redesigned tail rotor. Optimised for SAR and emergency roles.
STRUCTURE: As BK 117.
LANDING GEAR: As BK 117.
POWER PLANT: Two Turbomeca Arriel 1E2 turboshafts, each rated at 550 kW (738 shp) for take-off, 516 kW (692 shp) maximum continuous and 574 kW (770 shp) for 2½ minutes' OEI. Main transmission rated at 735 kW (986 shp) for twin-engine T-O, 632 kW (848 shp) maximum continuous; for single-engine operation 551 kW (739 shp) allowed for 2½ minutes, 404 kW (542 shp) for maximum continuous. Standard fuel contained in main tank, usable capacity 741.5 litres (195.7 US gallons; 163.1 Imp gallons) and left and right supply tanks, usable capacities 59 litres (15.6 US gallons; 13.0 Imp gallons) and 67 litres (17.7 US gallons; 14.7 Imp gallons) respectively, for total capacity of 867.5 litres (229.0 US gallons; 190.8 Imp gallons). Optional long-range tanks increase usable capacity to 1,086 litres (287 US gallons; 239 Imp gallons).
ACCOMMODATION: Compared with BK 117, cabin is more spacious through removal of centre post and door supports.
SYSTEMS: As BK 117.
AVIONICS: As BK 117.
EQUIPMENT: As BK 117.
DIMENSIONS, EXTERNAL: As BK 117C-1, except:
  Main rotor diameter   11.0 m (36 ft 1 in)
  Tail rotor diameter   1.96 m (6 ft 5¼ in)

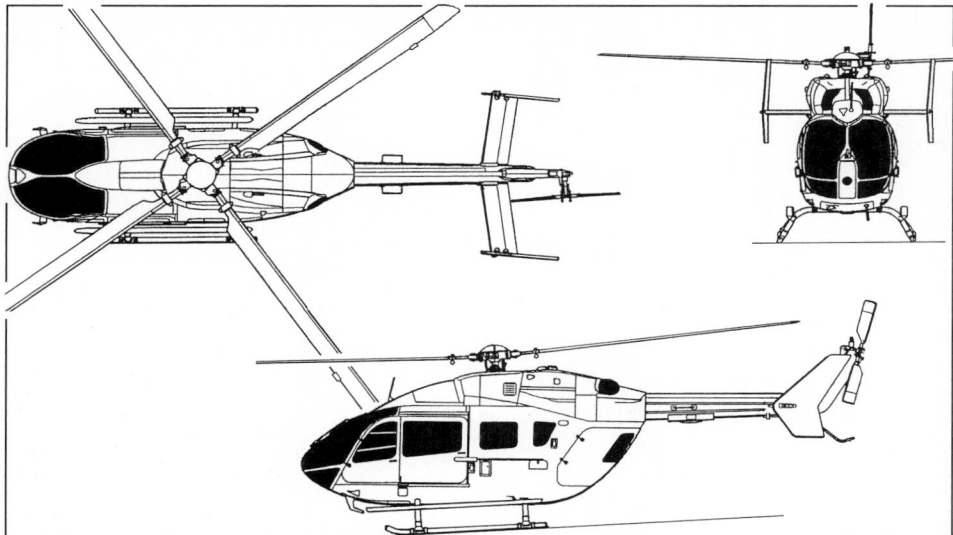

Eurocopter EC 145/Kawasaki BK 117C-2 (*Jane's/Paul Jackson*)   *NEW*/0526898

Length: overall, both rotors turning   13.03 m (42 ft 9 in)
  fuselage, tail rotor blades vertical   10.19 m (33 ft 5¼ in)
Fuselage max width   1.84 m (6 ft 0½ in)
Height: overall, both rotors turning   3.96 m (13 ft 0 in)
  to top of main rotor head   3.45 m (11 ft 3¾ in)
Tailplane span (over endplate fins)   3.12 m (10 ft 2¾ in)
Ground clearance: tail rotor   2.00 m (6 ft 6¾ in)
  fuselage   0.45 m (1 ft 5¾ in)
Skid track   2.40 m (7 ft 10½ in)
DIMENSIONS, INTERNAL:
Cabin: Length: excl cockpit   3.42 m (11 ft 2½ in)
  incl cockpit   4.55 m (14 ft 11¼ in)
  Max width   1.40 m (4 ft 7 in)
  Max height   1.22 m (4 ft 0 in)
  Floor area: excl cockpit   4.72 m² (50.8 sq ft)
  total   5.43 m² (58.5 sq ft)
  Volume (total)   6.8 m³ (240 cu ft)
AREAS:
  Main rotor disc   95.03 m² (1,022.9 sq ft)
  Tail rotor disc   3.02 m² (32.48 sq ft)
WEIGHTS AND LOADINGS:
  Weight empty   1,804 kg (3,977 lb)
  Max usable fuel: standard   694 kg (1,530 lb)
  optional   869 kg (1,916 lb)
  Max T-O weight, internal or external payload   3,585 kg (7,903 lb)
  Max underslung load   1,500 kg (3,307 lb)
  Max disc loading   37.7 kg/m² (7.73 lb/sq ft)
  Transmission loading at max T-O weight and power   4.88 kg/kW (8.02 lb/shp)
PERFORMANCE (A: at gross weight of 2,200 kg; 4,850 lb, B: at 2,700 kg; 5,952 lb, C: at 3,000 kg; 6,614 lb; D: at 3,585 kg; 7,903 lb):
Never-exceed speed (VNE):
  A, B, C   150 kt (278 km/h; 173 mph)
  D   145 kt (268 km/h; 166 mph)
Max cruising speed: A   138 kt (256 km/h; 159 mph)

  B   137 kt (254 km/h; 158 mph)
  C   136 kt (252 km/h; 156 mph)
  D   133 kt (246 km/h; 153 mph)
Normal cruising speed: A   127 kt (235 km/h; 146 mph)
  B   129 kt (239 km/h; 148 mph)
  C   130 kt (241 km/h; 150 mph)
  D   131 kt (243 km/h; 151 mph)
Max rate of climb: A   978 m (3,210 ft)/min
  B   780 m (2,560 ft)/min
  C   674 m (2,210 ft)/min
  D   488 m (1,660 ft)/min
Service ceiling: A, B, C   5,485 m (18,000 ft)
  D   5,240 m (17,200 ft)
Hovering ceiling IGE: A, B   5,485 m (18,000 ft)
  C   4,695 m (15,400 ft)
  D   2,925 m (9,600 ft)
Hovering ceiling OGE: A   5,485 m (18,000 ft)
  B   5,120 m (16,800 ft)
  C   4,345 m (14,250 ft)
  D   370 m (1,215 ft)
Range at normal cruising speed, no reserves:
  standard fuel:
  B   380 n miles (705 km; 438 miles)
  C   378 n miles (700 km; 435 miles)
  D   367 n miles (680 km; 422 miles)
  optional fuel:
  C   472 n miles (875 km; 543 miles)
  D   461 n miles (855 km; 531 miles)
Endurance at 65 kt (120 km/h; 75 mph), no reserves:
  standard fuel:
  B   3 h 55 min
  C   3 h 50 min
  D   3 h 35 min
  optional fuel:
  C   4 h 50 min
  D   4 h 30 min
UPDATED

Sécurité Civile Eurocopter EC 145 (*Eurocopter/Wolfgang Obrusnik*)   *NEW*/0527169

EUROFIGHTER

EUROFIGHTER JAGDFLUGZEUG GmbH
Am Söldermoos 17, D-85399 Hallbergmoos, München,
Germany
Tel: (+49 811) 80 15 55
Fax: (+49 811) 80 15 57
e-mail: eurofighter.pr@ibm.net
Web: http://www.eurofighter.com
CHAIRMAN: Pablo de Bergia
MANAGING DIRECTOR: Bob Haslam
CEO: Filippo Bagnato
DIRECTOR, MARKETING SALES SUPPORT: Andrew Lewis
VICE-PRESIDENT, COMMUNICATIONS: Ian Bustin

Eurofighter GmbH formed to manage EFA (European
Fighter Aircraft) programme June 1986, followed shortly
after by Eurojet Turbo GmbH to manage engine programme.
Eurofighter GmbH is owned by Alenia (Italy), BAE Systems
(UK) and EADS (formerly CASA; Spain and DASA;
Germany); development workshares are 21, 33 and 46 per
cent, respectively. Eurojet Turbo participants are Fiat
Aviazione (Italy), ITP (Spain), MTU-München (Germany)
and Rolls-Royce (UK). Radar is provided by the Euroradar
consortium of BAE Systems (UK), FIAR (Italy), EADS
(Germany) and ENOSA (Spain). NETMA (NATO
Eurofighter and Tornado Management Agency) supervises
the programme on behalf of the customer air forces.

All four participating countries agreed on 22 December
1997 to proceed with production and signed the appropriate
authorisation on 30 January 1998. The aircraft was formally
named Typhoon on 2 September 1998 although, initially, that
title was only used for marketing outside Europe. However,
on 23 July 2002, Typhoon name was formally adopted by
RAF.

On 4 November 1999, (then) four partners announced
impending formation of Eurofighter International (EFI) as
dedicated sales organisation with target of securing half of
available market for 800 combat aircraft over following 30
years. Relations with NETMA remain unchanged. First
export commitment issued by Greece on 7 March 2000.

Former subsidiary marketing organisation, Eurofighter
International, was disbanded in 2002.

UPDATED

EUROFIGHTER TYPHOON
**Spanish Air Force designations: C.16 (single-seat) and
CE.16 (two-seat)**
TYPE: Multirole fighter.
PROGRAMME: **Politico-industrial** history began with outline
staff target for common combat aircraft issued December
1983 by air chiefs of staff of France, Germany, Italy, Spain
and UK; initial feasibility study launched July 1984;
France withdrew July 1985, shareholdings then being
readjusted to 33 per cent each to UK and Germany, 21 per
cent Italy and 13 per cent Spain; project definition phase
completed September 1986; definitive ESR-D (European
Staff Requirement – Development) issued September
1987, giving military requirements in greater detail;
definition refinement and risk reduction stage completed
December 1987; main engine and weapons system
development contracts signed 23 November 1988.

Programme re-examined in 1992 following German
demands for substantial cost reduction and studies of
alternative proposals, which submitted in October 1992,
although none was adopted; Italy and Spain froze EFA
work mid-October. Defence ministers' conference of 10
December 1992 relaunched aircraft as Eurofighter 2000,
delaying service entry by three years, to 2000, and

allowing Germany to incorporate off-shelf avionics (AN/
APG-65 radar suggested), lower standard of defensive aids
and other deletions to effect 30 per cent price cut. By 1996,
however, these downgrades had been abandoned and
Eurofighter GmbH was planning for German production
almost identical to the common standard.

Additional difficulties resulted from German under-
funding and demands for further cost cuts. Political re-
apportionment of production work-shares negotiated in
1995, following reduction of German requirement.
Revised European Staff Requirement – Development
signed by four air forces, 21 January 1994 and re-
orientation of programme agreed in MoU 4, July 1995.
MoU 5, covering work-shares, delayed to 1996 by German
claims for 30 per cent, despite 140 aircraft requirement
representing only 23 per cent of production. Compromise,
agreed January 1996, involves addition of at least 40 (and
possibly up to 60) ground attack aircraft to German
requirement after 2012 and reduction of UK needs to 232.
Work-shares for production phase finally agreed as 30 per
cent to Germany, 37 per cent to UK, 19 per cent to Italy and
14 per cent to Spain. However, for first 148 aircraft, agreed
1998, average shares are 30, 36.33, 20 and 13.67 per cent,
respectively. Initial production details contained in
Quotation 4, submitted in March 1996; this envisaged start
of manufacturing in January 1998 and (after minor
restructuring) production of three in 2001, 12 in 2002, 37 in
2003, 46 in 2004 and 52 per year thereafter.

Eurofighter GmbH held open the terms of Quotation 4
throughout 1996 as Germany repeatedly postponed a
production decision as a consequence of financial
constraints. UK was first to declare a firm production
commitment, on 2 September 1996, and Spain followed on

21 October 1996, when terms for start-up funding were
agreed with industry. All four governments declared
support for a production launch on 5 December 1996,
although Germany and Italy did not grant funds until 26
November and 9 December 1997, respectively, allowing
defence ministers' conference on 22 December 1997 to
launch production phase. First metal for a production
aircraft was cut at DASA's Augsburg plant in May 1998.
Each partner nation is assembling its own aircraft on lines
at Manching (Germany), operational from December
2000; Caselle (Italy), operational November 2000; Getafe
(Spain), operational June 2001 (officially 26 July 2001);
and Warton (UK), operational September 2000. Locations
for assembly of export aircraft yet to be announced.
Assembly of major components for first production aircraft
began at Augsburg (Germany) in February 1999 and
Caselle in March. Each nation final assembly of this aircraft began at
Warton on 8 September 2000 following arrival of centre
fuselage from Germany on 31 August. First production
aircraft flew (in Italy) on 5 April 2002; deliveries due from
December 2002 onwards.

Typhoon is participating in several fighter competitions,
as detailed in the table below.

Engineering and flight-testing programme originally
to be based on eight development aircraft (no prototypes
apart from BAe EAP – see 1991-92 and earlier *Jane's*);
reduced to seven (DA1-7) early 1991, coincident with 11
per cent cut in intended flight test programme to 4,500
hours. Total to be achieved in 2,990 sorties by DA series
(635, 575, 430, 420, 385, 315 and 290 for DA1 to DA7,
respectively, although this considerably changed in
practice) plus 1,700 sorties by first five production aircraft
(IPA), which are fitted with test instrumentation. Total

EUROFIGHTER TYPHOON EXPORT ORDERS AND PROSPECTS

Nation	Requirement	Timeframe	Status
Requests for proposals (RFP):			
Austria	24	Deliveries from 2004	Selection announced 2 July 2002. Negotiations under way
Greece	60 plus 30 option	2004	Selection confirmed April 1999 EF Office Athens opened July 1999. Intentions confirmed March 2000. Order delayed by financial constraints, April 2001
South Korea	40	Decision late 2001 Deliveries 2004-2007	EF Office Seoul opened April 1998 RFP received June 1999 Proposal submitted 8 September 1999 Flight evaluation expected early 2000 Final RFP submitted 28 June 2000 Flight Evaluation Team December 2000
Norway	48-60	Deliveries 2010	EF Office Oslo opened March 1999 RFP received February 1999 Proposal submitted 1 June 1999 Offsets proposal submitted 1 December 1999 Competition suspended 2000
Requests for information (RFI):			
Australia	60	Decision after 2000	Discussions under way in 1999
Czech Republic	Possible 36	Long-term requirement	RFI received June 1999 RFI response September 1999
Netherlands	60-70	Decision 2005 Deliveries 2010	RFI received June 1999 RFI response October 1999
Poland	60	Long-term requirement	RFI received June 1999 RFI response July 1999
Saudi Arabia	possible 50-70	Decision after 2000	Early discussions under way in 1999
Singapore	up to 60		RFI response submitted late 1999 Evaluation late 2001

First production Eurofighter for the UK, ZJ699

NEW/0145046

First take-off by German production Eurofighter IPA3, 8 April 2002   NEW/0127061

1,614 hours in 1,935 sorties accumulated by April 2002; at that time, further 2,691 sorties (including 1,516 by IPAs) envisaged before full operational capability.

No formal roll-out; DA1 and DA2 remained unflown for some 18 months after completion for exhaustive cross-checking of flight control system (FCS). First flight eventually achieved on 27 March 1994, but FCS development resulted in later aircraft flying out of sequence, DA6 being fourth to fly (31 August 1996), by which time earlier aircraft had completed 241 sorties. DA5 flew Eurofighter's 500th sortie (almost 450 hours total airborne time) on 21 October 1997; the 750th sortie (over 630 hours) was flown in June 1998 and the 1,000th in May 1999. Total 1,135 sorties and 931 hours by fourth quarter of 1999; 1,000th hour flown February 2000. Some 90 per cent of flight envelope had been explored by April 1999. Further details of individual aircraft appear below.

Defensive aids subsystem development contract awarded to Euro-DASS 13 March 1992, but Germany and Spain initially declined to participate; production contract allocated June 2001. 'A' version of ECR 90 radar first flew in nose of modified BAe One-Eleven testbed (ZE433) at Bedford, 8 January 1993; 'C' version is first ECR 90 packaged to fit Eurofighter; flown in One-Eleven from July 1996. First development standard radar delivered to DASA in June 1996; flight testing (in DA5) began 24 February 1997. Radar named Captor in 2000; first production unit delivered in February 2001. Electronically scanned version to fly in One-Eleven by 2003, for possible service in 2010.

Major airframe fatigue test (AFT) fuselage at Ottobrunn achieved 6,000 hours in May 1995 and target of 18,000 hours (equivalent to 6,000 hours of service use) on 4 September 1998.

In 1998, BAE undertook a UK Ministry of Defence funded study of a possible short take-off but arrested recovery (STOBAR) Eurofighter which could operate from aircraft carriers. Officially confirmed in early 2000 that navalised Eurofighter was one option under consideration for Royal Navy's Future Carrier-Borne Aircraft (FCBA) requirement although Lockheed Martin F-35 is selected type.

Production version of EJ200 engine certified 8 March 2001. Thrust vectoring nozzles for EJ200 had undergone 78 hours of bench testing by mid-2000. Study launched by ITP of Spain in 1994, using private funding. Trial system demonstrated 23° 30′ vector angle and 110°/s slew rate; advantages of installation for regular use would include 3 per cent reduction in cruise drag, 7 per cent improvement in sustained turn rate, 7 per cent increase in installed thrust, 25 per cent reduction of take-off run and 3 per cent reduction of mission fuel burn.

Three production Tranches subdivided into eight Blocks, each of increased capability: Tranche 1 (148 aircraft) begins with PSP 1 (Block 1) software standard, suitable only for basic air defence training; PSP 2 (Block 2)

configuration (2003-04) fully air defence capable (AMRAAM, ASRAAM, AIM-9) through addition of direct voice input (DVI), digital datalink (MIDS) and MLA, plus limited ground attack capability and interim defensive aids subsystem (DASS); PSP 3 (Block 5), representing 2005 production, gives 'swing role' capability with unguided air-to-surface weaponry (augmented by Paveway II and ALARM modifications for only UK; GBU-16 for Italy), enhanced situational awareness and improved survivability through incorporation of PIRATE passive tracking system, helmet-mounted display, towed radar decoy and full sensor fusion. Following five instrumented production aircraft, PSP 1 was to have been achieved in June 2002 on delivery of second two-seater for RAF (production number BT002), which is seventh series production (SPA7). PSP 2 follows in December 2003 with SPA44 (fourth RAF single-seat; BS004); and PSP 3 in April 2005 with SPA115 (20th German single-seat; GS020). See adjacent table

TRANCHE 1 BREAKDOWN

Batch	Software	Hardware	Block	Qty
IPA				5
1	PSP1	CA1A	1	38
2	PSP2	CA1	2	70
2	PSP3	CA1	5	35
Total				**148**

Notes: Batch 1 (including 13 RAF) to be refurbished to Batch 2 standard after initial training use. Batch 2 includes 39 RAF.

Tranche 2 aircraft, totalling 236, begin with Block 8 (2006 production), having new hardware standard, including updated mission computer; Block 10 (2007-08 production) will be to EOC 1 (enhanced operational capability) standard (adding Paveway III, IRIS-T, AIM-120C-5, plus full-standard DASS and other interoperability features) and EOC 2 (Block 15, built 2009 to late 2010) will add Meteor AAM, Storm Shadow, Taurus and GPS/IN guided weapons and reconnaissance capability; increased MTOW of 24,500 kg (54,013 lb) under consideration.

Tranche 3, comprising 236 aircraft, envisaged as Block 20 and 25, built late 2010 to early 2015, has yet to be defined.

In April 2002, industrial partners completed a 12-month study into further development of Eurofighter to maintain operational effectiveness beyond 2040. Options addressed (some previously mooted unofficially) included electronically scanned radar, two-dimensional thrust vectoring, extended range and signature reduction. These to be applicable to Tranche 2 and 3 aircraft. Germany and

UK seeking to add long-range attack capability in Tranche 3; BAE Systems announced completion in May 2002 of wind tunnel trials of conformal fuel tanks offering 25 per cent extension of range and mockup shown at Farnborough in July 2002.

CURRENT VERSIONS (general): **Single-seater:** Standard version.

Two-seater: Combat-capable conversion trainer. Slightly reduced internal fuel capacity.

CURRENT VERSIONS (specific): **DA1/9829:** (DASA-built at Ottobrunn; airframe No. 01; Luftwaffe serial number 9829.) By road to Manching 11 May 1992; first flight, 27 March 1994, using Phase 0 software; planned transfer to Warton for handling and envelope expansion trials (wearing UK serial number ZH586) cancelled; remained at Manching; nine sorties to June 1994, when grounded for FCS upgrade to Phase 2; reflown 18 September 1995; first sortie by a German military pilot (Lt Col Heinz Spolgen), March 1996 at start of initial military evaluation phase which completed on 24 April 1996. Total 123 flights by late 1997, when stood down for retrofit with EJ200 series 03Z engines, plus parallel avionics upgrade to 3910 standard and installation of Martin-Baker Mk 16 ejection seats. Upgrade completed in November 1998; returned to flying in third quarter of 1999. Laid up 11 September 2000 with 184 hours 24 minutes in 232 sorties; flight control system upgrade; remained unflown in February 2001. Made first Eurofighter visit to JBG 38 at Jever on 3 July 2001 for two weeks of flights over North Sea air combat range. Refuelled from Panavia Tornado, August 2001. Total 313 flights and 254 hours by April 2002.

DA2/ZH588: (BAE at Warton; airframe No. 02.) First engine run 30 August 1992, first flight 6 April 1994; assigned to envelope expansion, 'carefree' handling and load trials; nine sorties to June 1994, when stood-down for FCS upgrade; reflown 17 May 1995 with Phase 2 version of flight software and made Eurofighter's world public debut (Paris, 11 June 1995, static) and UK debut (Fairford 22 July 1995, flying). On 9 November 1995, 57th sortie was also first with an RAF pilot (Sqdn Ldr Simon Dyde). Demonstrated 25° AoA in May 1997, followed by radar decoy trials and demonstration of full 'carefree' handling. First M2.0 Eurofighter sortie 23 December 1997; first aerial refuelling (RAF VC10) 14 January 1998. Retrofitted with EJ200 engines, upgraded avionics and Mk 16 ejection seat; reflown late August 1998. Achieved 15,240 m (50,000 ft), April 1999. Prepared for load testing during mid-1999. First to fly with 2B2 software, 7 July 2000, by which time painted black overall, to cover installation of more than 500 pressure transducers for airflow measurement. Laid up 18 December 2000 with 303 hours 3 minutes in 345 sorties; fuel system upgrade; engine relight trials, 2001; ASRAAM compatibility trials, early 2002. Total 404 flights and 362 hours by April 2002. 'Carefree handling' trials completed by mid-2002; DASS decoy trials.

DA3/MMX602: (Alenia at Turin/Caselle; airframe No. 04.) First with EJ200 power plants (series -01A) for engine trials (originally scheduled from March 1993 but postponed) and gun/weapon release trials. Initial flight 4 June 1995 on Phase 1 software (20° AoA and +6 g limits). Total 53 sorties by 1 September 1996 (all with Phase 1 software). Refitted with EJ200-01C engines in 1996; in-flight engine relight demonstrated December 1996; first sortie with two 1,000 litre underwing tanks 5 December 1997; EJ200-03A engine by early 1998; 1,500 litre tanks, February 1999; M1.6 with two 1,000 litre tanks, March 1999; 200th sortie, November 1999. Initiated ground dropping trials of air-to-surface weapons, 1999. Laid up 31 March 2000 with 191 hours 51 minutes in 246 sorties; gun and ejection seat upgrade; remained unflown for most of 2001. First Eurofighter gun firing, 13 March 2002. Total 250 flights and 195 hours by April 2002.

DA4/ZH590: (BAE; airframe No. 03.) First two-seat and first with full avionics (including ECR 90) for radar development, and 'carefree' handling trials. Rolled out 4 May 1994; first flight 14 March 1997; 14 sorties up to end of first phase of trials on 8 September 1997. First scheduled demonstration of supercruise 20 February 1998; lightning strike trials in test rig at Warton May-June 1998; rear cockpit activated (as on other two-seaters) 19 April 1999; autopilot autothrottle activated 28 April 1999; first flight of helmet-mounted sight 17 June 1999; Eurofighter 1,000th hour flown February 2000 on aircraft's 75th sortie. First to fly with active missile approach warning. First two-seat Eurofighter night flight. Laid up 13 April 2000 with 98 hours 57 minutes in 97 sorties; DASS ground trials 2001; upgrades to avionics and power generation system; returned to flying in November 2001; weapons integration trials, including first guided AMRAAM firing against drone target 9 April 2002. Total 127 flights and 138 hours by April 2002; 128th sortie was first two-seat air refuelling, first refuelling with external tanks, first night refuelling and longest Eurofighter sortie (4 hours 22 minutes). DASS ESM trials 2002.

DA5/9830: (DASA.) Construction begun 2 November 1992; maiden flight 24 February 1997; first with ECR 90 radar; autopilot and weapons trials. DS-X radar software standard upgraded to DS-C1 in June 1997, at which time first EJ200-03A engines fitted. Reportedly is testbed for latest standard of radar-absorbent materials. Deployed to

Eurofighter Typhoon with scrap view of two-seat version *(Jane's/Paul Jackson)*   NEW/0143737

Rygge, Norway, June 1998 (first visit to prospective customer outside original partner nations), for evaluation. Flown by Norwegian test pilot, 15 December 1998. New standard software (Phase 2B1; also in DA4) flown 1 April 1999, permitting autopilot and autothrottle operation. Radar trials against four simultaneous targets, mid-1999; flew Typhoon's 1,000th sortie 18 May 1999, representing 830 hours. In-flight icing trials, February 2000. Completed radar trials, 15 to 29 March 2001, with 20-ship tests in various modes. Total 243 flights and 218 hours by April 2002.

DA6/XCE.16-01: (EADS CASA at Seville.) Second two-seat; performance (including 'carefree' handling), environmental systems, MIDS integration and helmet integration trials. Scheduled (by 1995 reprioritisation) to be fourth to fly; early 1996 date set back by six months for re-checking of Phase 2A software. First flight 31 August 1996; high-temperature trials at Morón, Spain, from 20 July 1998; trials of LCSS (see Systems) cooled aircrew clothing, June 1999; environmental trials (with DA1), including ground tests at Boscombe Down, completed May 2000. Direct voice input trials begun 2001. Total 271 flights and 251 hours by April 2002.

DA7: (Alenia.) Nav/com, performance and weapons integration trials. First flight 27 January 1997. First firing of AIM-9L Sidewinder 15 December 1997; first AIM-120 AMRAAM jettison two days later; first (1,000 litre) external wing tank jettison 17 June 1988. Total 94 hours 21 minutes in 177 sorties to 2 February 2001. AMRAAM and AIM-9L launch trials, Sardinia, from April 2001; ASRAAM launch, June 2001. PIRATE sensor trials 2001. Total 327 flights and 196 hours by April 2002.

Further eight ground testing part-airframes and five instrumented production aircraft.

IPA1/ZJ699: (UK.) Two-seat; first production Eurofighter Typhoon. Production numbers PT001/BTT1. Final assembly began 8 September 2000; delivery was due in late 2001; however, first flight postponed to 14 April 2002; short lay-up after six sorties for addition of refuelling probe, flight test instrumentation and paint. Defensive aids trials. Planned 414 sorties to FOC.

IPA2/MMX614: (Italy.) Two-seat; PT002; air-to-surface weapon and sensor fusion trials. Final assembly began 3 November 2000. First flight 5 April 2002. Planned 301 sorties to FOC.

IPA3/9803: (Germany.) Two-seat; PT003; air-to-surface weapon integration. Rolled out 2 March 2002. First flight 8 April 2002. Planned 408 sorties to FOC.

IPA4: (Spain.) Single-seat; PS001; AAM and gun trials; environmental system trials. Planned 170 sorties to FOC.

IPA5: (UK.) Single-seat; PT001; air-to-air and -surface weapons trials. Serial number ZJ700. Planned 223 sorties to FOC.

CUSTOMERS: Originally declared requirements for 765 (UK and Germany 250 each; Italy 165 and Spain 100). In January 1994, Spain announced firm requirement for 87; Germany revised needs to 180 (including at least 40

First production Eurofighter to take to the air, on 5 April 2002, was Italy's IPA2   NEW/0145050

fighter-bombers post-2012) under January 1996 work-share agreement. Final total is 620 aircraft, including 1,382 engines; plus options on 90 and 183, respectively (see accompanying table). MoUs 6 and 7 (production and logistical support) of 3 December 1997 and Production Investment contract of 10 December followed on 22 December 1997 by political agreement of all four nations to fund production phase.

Production contract (Supplement 1) for all 620 aircraft (plus 90 options) signed 30 January 1998; Supplement 2 agreement of 18 September 1998 authorised first 148 of these on fixed-price terms, together with 363 engines. UK parliament told in early 2002 that all four participating countries were reviewing aircraft delivery schedule.

Deliveries to air forces to have begun in June 2002 with RAF and Italy; rescheduled acceptance of BT001/ZJ800 by RAF December 2002; first RAF unit, formed September 2002 under Wg Cdr David Chan, is Operational Evaluation Unit (OEU, also known as No. 17 (Reserve) Squadron) at BAE Warton, where first 16 pilots to be converted using initial 13 (Batch 1) aircraft in 1,300 flying hours; in January 2004, Coningsby will receive OEU and form No. 29 (Reserve) Squadron as an OCU, followed by two operational squadrons in 2005 and 2006; two squadrons form at Leeming in 2006 and 2007; three at Leuchars in 2008-2010. First of seven combat squadrons to be air defence tasked; remainder to comprise three air defence, two multirole and one offensive support.

RAF announced in May 2000 that no ammunition would be procured for Eurofighter's cannon, despite installation of Mauser in Tranche 1 aircraft. Decision had been reversed by 2002.

First German wing will be JG 73 at Laage, followed by JG 74 at Neuberg and JG71 at Wittmund in air defence role; additionally JG 31 (between 2007 and 2010) and JG 33 (2012-2015) in attack role; final German delivery in 2014, but further two squadrons (single- or two-seat to be decided) may be obtained for air-to-ground operations. Italian re-equipment begins with 4° Stormo at Grosseto (9° Gruppo from 2002 and 20° Gruppo from 2004), other bases being Cameri (53° Stormo) and Trapani (37° Stormo). First Spanish squadron to be Escuadron 113 of Ala 11 at Morón de la Frontera. Other details in accompanying table.

Export orders also being solicited: Norway entered exploratory discussions in 1997 with a view to acquiring up to 30 aircraft to replace Northrop F-5A/Bs; appointed a liaison officer to NETMA in 1998 and issued formal request for proposal (RFP) on 15 February 1999; suspended programme in early 2000. Greece announced intention to obtain Typhoon on 12 February 1999, government confirmation following in March 2000; local final assembly under consideration. Austria announced selection of Eurofighter on 2 July 2002. South Korean RFP received June 1999; in 2001, South Korea offered non-NATO partnership in Eurofighter, including final assembly. Eurofighter predicts export of 400 Typhoons between 2005 and 2030, worth in excess of £35 billion (2000).

Contract for Eurofighter ground support system signed 24 May 2000. UK has four training rigs, numbered ZJ695 to ZJ698. Aircrew Synthetic Training Aids programme contract signed 27 April 2001, including 18 full mission simulators and nine cockpit trainers, deliverable from March 2004 to all four participants.

COSTS: Estimated £25 to 26.5 million, UK, 1992 unit cost; DM127 million, Germany early 1992, 10 year system price; reduced to DM89 million by late 1992 economies. UK National Audit Office report of early 1996 estimated British share of the development phase to be £1,253 million (43.7 per cent), of which £407 million resulted from restructuring to keep Germany in the programme. Same source reported programme 46 per cent over budget and 36 months late by mid-1997. Total development cost estimated as US$21 billion by 1998. Combined

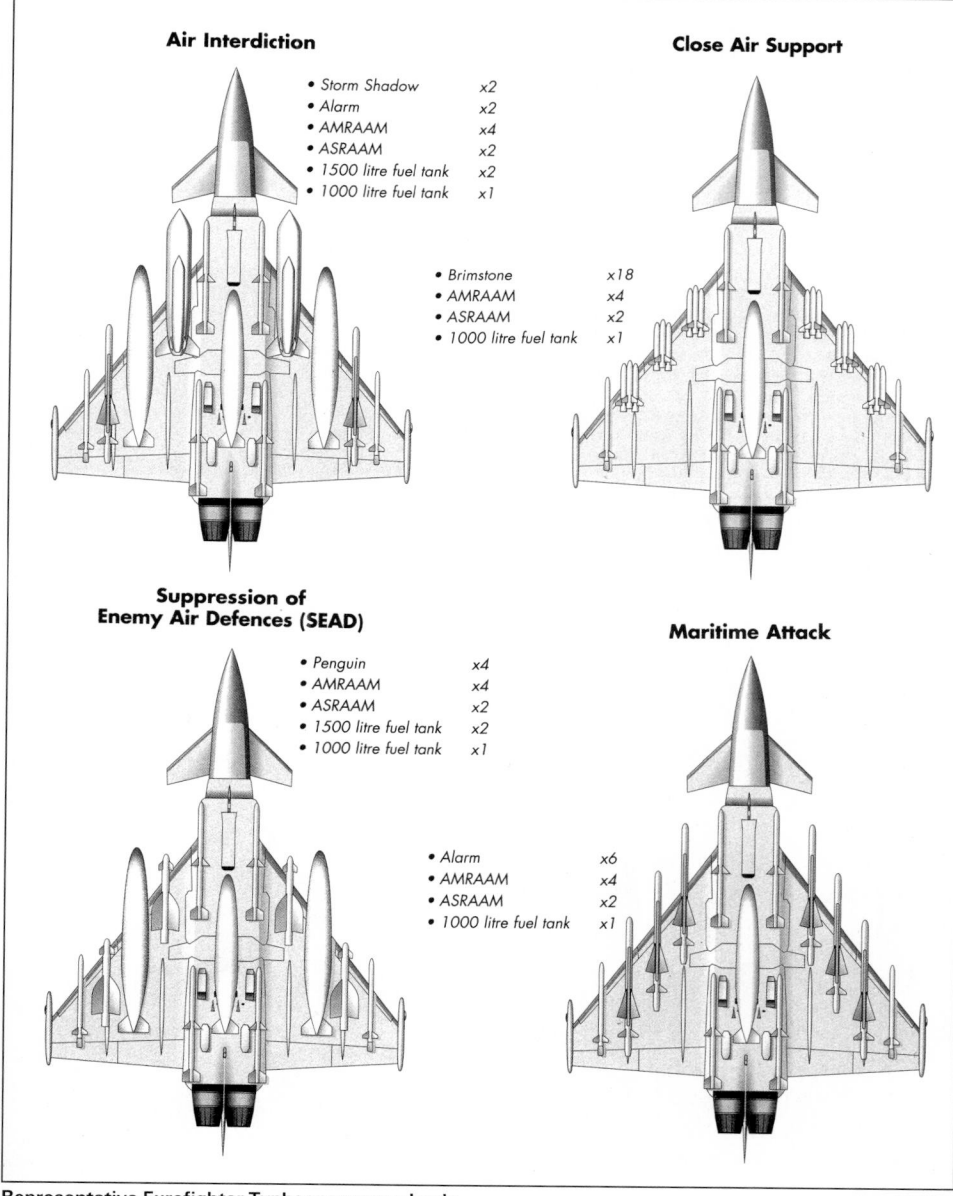

Air Interdiction
- Storm Shadow x2
- Alarm x2
- AMRAAM x4
- ASRAAM x2
- 1500 litre fuel tank x2
- 1000 litre fuel tank x1

Close Air Support
- Brimstone x18
- AMRAAM x4
- ASRAAM x2
- 1000 litre fuel tank x1

Suppression of Enemy Air Defences (SEAD)
- Penguin x4
- AMRAAM x4
- ASRAAM x2
- 1500 litre fuel tank x2
- 1000 litre fuel tank x1

Maritime Attack
- Alarm x6
- AMRAAM x4
- ASRAAM x2
- 1000 litre fuel tank x1

Representative Eurofighter Typhoon weapon loads   0110879

EUROFIGHTER TYPHOON PRODUCTION PLAN

Year	Combined total	Cum total
2001	3	3
2002	12	15
2003	40	55
2004	45	100
2005	52	152
2006	52	204
2007	52	256
2008	52	308
2009	52	360
2010	52	412
2011	52	464
2012	52	516
2013	52	568
2014	50	618
2015	35	653
2016	27	680
2017	21	701
2018	9	710

Note: Presumes all options taken up

EUROFIGHTER TYPHOON REQUIREMENTS BY PARTNER NATIONS

	Delivery	Germany		Italy		Spain		UK		Totals		
Tranche		S	T	S	T	S	T	S	T	S	T	Comb
1	2001-05	28	16	19	10	12	8	37	18	96	52	148[1,2]
2	2005-10	58	10	43	3	27	6	83	6	211	25	236[3]
3	2010-14	61	7	44	2	33	1	75	13	213	23	236[4]
Subtotals		**147**	**33**	**106**	**15**	**72**	**15**	**195**	**37**	**520**	**100**	**620**
		180		**121**		**87**		**232**				
Options		nil		9		16		65				90
Totals		**180**		**130**		**103**		**97**				**710**

Notes: S is single-seat; T is two-seat
German aircraft numbered from 3001 onwards (except IPA3); RAF Tranche 1 are ZJ700 and ZJ910 to ZJ945 for single-seat, and ZJ699 and ZJ800 to ZJ815 for 17 of two-seat aircraft
[1] Including five instrumented trials aircraft, one (20th production) for static testing and 37 'IOC' standard aircraft assigned to training
[2] Plus 363 engines and 147 radars
[3] Plus 519 engines and 236 radars
[4] Plus 500 engines and 236 radars

development investment estimated (1996) as DM18 billion (US$12.25 billion); non-recurring production investment estimated (1996) as DM12 billion (US$8.15 billion).

Revised data give flyaway price of DM75 million to DM85 million (US$51 million to US$58 million) and system price of DM150 million to DM170 million (US$102 million to US$116 million) at 1996 levels. UK NAO estimated (1996) £15.4 billion for 250 RAF aircraft, including £9.5 billion for production (unit cost £38 million) and this reaffirmed (US$58 million/£37 million) in mid-1998. UK programme cost officially estimated at £16.1 billion (mid-1999); of this £5,444 million had been committed by 31 March 2001.

In January 1997, German government agreed a weapons system price (including logistics support) of DM125.4 million (US$79 million), equivalent to DM23 billion for 180 aircraft. Supplement 2 (148 aircraft) valued at DM14 billion.

DESIGN FEATURES: Agile fighter; subsonic instability exceeds 35 per cent (as achieved by Grumman X-29 research aircraft). Collaborative design by BAE, DASA, Alenia and EADS CASA, incorporating some design and technology (including low detectability) from BAe EAP programme. No official requirement for thrust vectoring (TV), supercruise or high order of 'stealthiness'; however, TV nozzle for Eurofighter is under private development; supercruise was 'inadvertently' demonstrated at high altitude in 1997; and RAF has confirmed that aircraft meets low-observables specification.

Low-wing, low-aspect ratio tailless delta with 53° leading-edge sweepback; underfuselage box with side-by-side engine air intakes, each with fixed upper wedge/ramp and vari-cowl (variable position lower cowl lip) with Dowty actuators.

Intended service life, 6,000 hours or 25 years. Integrated structural health-and-usage monitoring system (first in any combat aircraft) calculates structural fatigue at 20 positions on the airframe 16 times per second during flight. Maintainability features include 10 mmh/fh and single engine change by four engineers in 45 minutes. Operational turn-round by six ground crew in 25 minutes. Germany, and possibly UK, will contract-out maintenance on 'power by the hour' arrangement with private industry.

FLYING CONTROLS: Two-segment automatic slats on wing leading-edges, inboard and outboard flaperons on trailing-edges; all-moving foreplanes below windscreen; rudder; hydraulically actuated airbrake aft of canopy, forming part of dorsal spine; Liebherr primary flight control actuators. Full-authority quadruplex ACT (active control technology) digital fly-by-wire flight control system (team leader was DASA; Bodenseewerk and ENOSA flight control computer) combines with mission adaptive configuring and aircraft's instability in pitch to provide required 'carefree' handling, gust alleviation and high sustained manoeuvrability throughout flight envelope; pitch and roll control via foreplane/flaperon ACT to provide artificial longitudinal stability; yaw control via rudder; no manual reversion.

STRUCTURE: Fuselage, wings (including inboard flaperons), wing-fuselage fairings, fin and rudder mainly of CFC (carbon fibre composites) except for foreplanes, outboard flaperons and exhaust nozzle fairings (titanium); nose radome and fin-tip (GFRP); leading-edge slats, wingtip pods, fin base, fin leading-edge, rudder trailing-edge, cockpit side strake and canopy-to-airbrake fairing (aluminium-lithium alloy); and canopy surround (aluminium). CFC constitutes 70 per cent of surface area, with metal 15 per cent, GFRP 12 per cent and other materials 3 per cent. Manufacture includes such advanced techniques as superplastic forming and diffusion bonding; EADS CASA-led joint structures team. On development aircraft (only) UK responsible for front fuselage, foreplanes, starboard leading-edge slats and flaperons; Germany the centre-fuselage and fin, Italy the port wing, port leading-edge and flaperons, and stages 2 and 3 of rear fuselage; Spain the rear fuselage stage 1; and Spain and UK the starboard wing; no duplication of tooling.

Work-share on production aircraft involves UK for front fuselage, canards, windscreen, canopy, dorsal fairing inboard flaperons, fin and rear fuselage stage 1; Germany for centre-fuselage; Italy for port wing, outboard flaperons and rear fuselage stages 2 and 3; and Spain for starboard wing and leading-edge slats. Assembly at Caselle (Italy), Getafe (Spain), Manching (Germany) and Warton (UK).

LANDING GEAR: Dowty Aerospace retractable tricycle type with SICAMB mainwheels and Magnaghi/OMA nose gear; Dunlop Aviation wheels, brakes and braking system; Ultra Electronics landing gear computer. Single-wheel main units retract inward into fuselage; nosewheel unit forward. Nosewheel steering is subfunction of DFCS. Tyre sizes 28×9.5R15 main; 18×7.75R6 nose. Elektro Metall braking parachute at base of fin.

POWER PLANT: Two Eurojet EJ200 advanced technology turbofans (each of approximately 60 kN; 13,490 lb st dry and 90 kN; 20,250 lb nominal thrust with afterburning), mounted side by side in rear fuselage with ventral intakes. EJ200-01A initially; -1C for early flight tests; -03A first flown June 1997 (DA5); -03B followed in late 1998; and 03Z in December 1999. Staged EJ200 improvements available (but not funded) to 103 kN (23,155 lb st) (designated EJ 230) and then 117 kN (26,300 lb st). MTU FADEC. Lucas Aerospace fuel management system; Aeroquip GmbH fuel ducts; Autoflug sensors; Teldix flowmeters; VDO computer and gauges; Smiths fuel measurement system. First two development aircraft originally powered by two Turbo-Union RB199-122 (Mk 104E) afterburning turbofans (each more than 71.2 kN; 16,000 lb st). Both were retrofitted with EJ200s in 1998.

Internal fuel capacity classified, but believed to total approximately 5,700 litres (1,506 US gallons; 1,254 Imp gallons) in two fuselage tanks and two integral wing tanks. Two-seat trainer lacks forward transfer tank, but partly offsets loss of capacity with auxiliary tank in the enlarged spine. Pressure refuelling point below fuselage, immediately behind air intake. Provision for in-flight refuelling and up to three suspended, external fuel tanks: two 1,000 litre (264 US gallon; 220 Imp gallon) or 1,500 litre (396 US gallon; 330 Imp gallon) underwing, plus one 1,000 litre (264 US gallon; 220 Imp gallon) centreline tank. Only the smaller tanks are rated for supersonic flight.

In early 1998, UK was reported to be designing upper fuselage conformal tanks to increase combat radius to 1,500 n miles (2,778 km; 1,726 miles); work subcontracted to GKN Engage in Australia. Wind tunnel trials of conformal tanks completed May 2002, size 1,500 litres (396 US gallons; 330 Imp gallons) each, providing for additional total of 2,400 kg (5,291 lb) of fuel, for 25 per cent range increase. Conformal tanks can be installed or removed in 75 minutes. Tranche 2 aircraft have structural and piping modifications to accept conformal tanks.

ACCOMMODATION: Pilot(s) on Martin-Baker Mk 16A zero/zero ejection seat(s). Single-piece Aerospace Composite

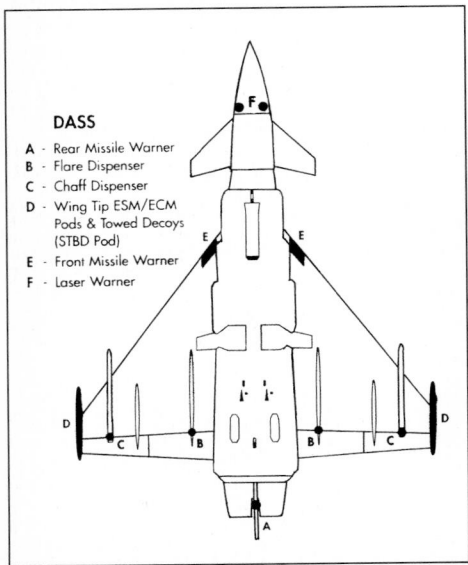

DASS
A - Rear Missile Warner
B - Flare Dispenser
C - Chaff Dispenser
D - Wing Tip ESM/ECM Pods & Towed Decoys (STBD Pod)
E - Front Missile Warner
F - Laser Warner

Configuration of the full Eurofighter defensive aids subsystem   0044917

Eurofighter DA5, the second German prototype   *NEW*/0143727

windscreen and single-piece, rear-hinged canopy on both versions. Optional liquid-cooled vest for pilot. Helmet-mounted display. Anti-*g* trousers augmented by pressure breathing system. Two equipment/baggage stowage bays, each 0.01 m³ (0.35 cu ft), above air intakes, port and starboard.

SYSTEMS: Responsibility for systems delegated to Eurofighter GmbH; participants are UK for electrics, Spain for environmental control and Germany for hydraulics.

Electrical system has Lucas Aerospace as leading supplier and is designed to minimise risk of total power loss by using high level of redundancy and by system positioning. Engine start, systems test, avionics activation and alignment, need no external power. Elements include ZF accessories drive gearbox, two AC and two DC engine-driven Hamilton Sundstrand generators, Honeywell APU in forward port wingroot, Varta battery in corresponding starboard position, two Ferranti Technologies transformer/rectifier units and power converter from same company.

Environmental control system (ECS), with Normalair Garrett as leading supplier, provides conditioned and partly conditioned air to cockpit canopy seal, anti- and de-misting, pilot's anti-*g* clothing, radar, FLIR, avionics and general equipment. Main oxygen provided by molecular sieve generation system (MSOGS). Precooler (at base of fin), heat exchanger, cold air unit and MSOG all located in aircraft's spine. Liquid cooling subsystem (LCSS) connects aircrew vest to ECS.

Hydraulic system (Magnaghi as leading supplier) comprises two independent circuits supplying power to flight control system (Dowty Boulton Paul actuators), landing gear (including nosewheel steering and brakes), port and starboard utilities, gun, canopy, airbrake and refuelling probe.

Utilities control system is integrated within overall system architecture and provides for continuous monitoring and fault detection, comprising front computer, fuel computer, secondary power system computer, landing gear computer and maintenance data panel. Integrated monitoring and recording system constantly checks status of all other systems, airframe and engine, to provide rapid, onboard fault diagnosis; functions include crash-survivable memory, bulk storage device, video-voice recording, mission data loading and portable data store, maintenance data panel, portable maintenance data store and air-to-ground relay of data.

AVIONICS: BAE Systems has overall team leadership for avionics development and integration. All avionics, flight control and utilities control systems integrated through STANAG 3910 databus highways with appropriate redundancy levels, using fibre optics and microprocessors.

Some functions activated by direct voice input, with 100 word vocabulary.

Comms: Rohde & Schwarz or Elmer VHF/UHF communications, both secure and non-secure. BAE video and voice recorder. EADS and Cossor IFF; BAE antennas.

Radar: Euroradar ECR 90 Captor multimode pulse Doppler radar.

Flight: Smiths mission data loader and radar altimeter. Litton laser INS; Marconi SpA microwave landing system; BAE Systems TERPROM and ground proximity warning; Tacan; provision for terrain reference navigation; Elmer SpA crash survival memory unit. FCS software updated to Phase 2 in late 1995; Phase 2A 'carefree' handling throughout the subsonic flight envelope, including 25° AoA and over +6 *g* (cleared by DA2 in January 1998); Phase 2B allows carriage of heavy stores and comprises 2B1 (flown April 1999) for 28° AoA and +7.25 *g*, and 2B2 for 30-35° and 9 *g* (first flown 7 July 2000). Phase 3 is IOC standard in 2001 (9 *g* envelope), DA2, DA4 and DA6 receiving Phase 3R1 in that year; Phase 4 covers air-to-surface weapons; and full combat capability Phase 5. DFCS incorporates auto-recovery mode ('panic button') for immediate return to straight-and-level flight in emergency. Ada language, apart from time-critical subroutines in Assembler.

Instrumentation: Special attention given to reducing pilot workload. New cockpit techniques simplify safe and effective operation to limits of flight envelope while monitoring and managing aircraft and its operational systems, and detecting/identifying/attacking desired targets while remaining safe from enemy defences. This achieved through high level of system integration and automation, including HOTAS controls; BAE wide-angle (30° azimuth; 25° elevation) HUD able to display, in addition to other symbology, FLIR pictures from PIRATE sensor mounted externally to port side of cockpit; helmet-mounted sight (HMS), with helmet tracking system and direct voice input (DVI) for appropriate functions; and three Smiths Industries multifunction head-down colour CRT displays (MHDD) and Smiths glareshield standby displays. EADS, EDS Defence and CANAVA digital map generator; Teldix cockpit interface unit.

Mission: Alenia and Computing Devices nav/attack computers. Eurofirst (FIAR consortium) PIRATE (Passive Infra-Red Airborne Tracking Equipment) port side of windscreen. Secure datalink. RAF aircraft to have optional reconnaissance capability with a long-range electro-optical pod, for which SR(A) 1368 was issued in 1995.

Self-defence: Advanced integrated defensive aids subsystem (DASS), contracted to Euro-DASS consortium, led by BAE Systems and including Indra (Spain) and

Elettronica (Italy); includes RWR and active jamming pod at each wingtip, laser warning receiver (each side, adjacent to windscreen), missile approach warning (wing leading-edges, inboard, and at rear base of fin), Elettronica Aster/GAMESA/CelsiusTech expendables (flares in flap actuator fairings; chaff in aileron actuator fairings) and towed radar decoys. (Germany initially did not join Euro-DASS, but is now to adopt standard equipment, with contract awarded October 2001). Two BAE Systems radar decoys in the starboard wingtip pod, each on a 100 m (320 ft) fibre optic cable. Italy considering Cross Eye ECM system as alternative to towed decoy. Spain also declined to join Euro-DASS, but is now participating; UK and Spain are only nations to have LWR. Initial contract for 103 systems placed June 2001; to be augmented by German requirement and addition of EADS to consortium.

EQUIPMENT: Hella lighting; Logic anti-collision beacons.

ARMAMENT: Total of 13 external stores stations: five (including one wet) under fuselage and four (including one wet) under each wing. Internally mounted 27 mm Mauser gun on starboard side.

Planned weapons (with maximum load in parentheses) include:

Air-to-air: Metcor (six), AIM-120 AMRAAM (six), ERAAM (six), FMRAAM (six), AIM-9L Sidewinder (six), ASRAAM (six) and IRIS-T (six).

Air-to-air surface: ALARM (six), Penguin (four), Harpoon (four), Brimstone (18), Taurus, Storm Shadow, GBU-10 (four), GBU-16 (four), Paveway III (three), CRV-7 (four pods of 19 rockets each), 500 lb bombs (12) and reduced quantity of larger weapons up to 2,000 lb Mk 84 (four).

DIMENSIONS, EXTERNAL:

Wing span over ECM pods	10.95 m (35 ft 11 in)
Wing aspect ratio	2.4
Length overall	15.96 m (52 ft 4¼ in)
Height overall	5.28 m (17 ft 4 in)

AREAS:

Wings, gross	50.0 m² (538.2 sq ft)
Foreplanes (total)	2.40 m² (25.83 sq ft)

WEIGHTS AND LOADINGS (approx):

Weight empty	11,150 kg (24,582 lb)
Internal fuel load	4,500 kg (9,920 lb)
External stores load (weapons and/or fuel):	
normal	6,500 kg (14,330 lb)
overload	7,500 kg (16,535 lb)
Max T-O weight: normal	21,000 kg (46,297 lb)
overload	23,500 kg (51,809 lb)
Max wing loading	470.0 kg/m² (96.26 lb/sq ft)
Max power loading	130 kg/kN (1.28 lb/lb st)

PERFORMANCE (design):

Max level speed	M2.0
Time to 10,670 m (35,000 ft)/M1.5	up to 2 min 30 s
Service ceiling	16,765 m (55,000 ft)
Runway requirement	700 m (2,300 ft)
T-O run, air combat mission	300 m (985 ft)
Combat radius, single-seat: ground attack, lo-lo-lo	
	325 n miles (601 km; 374 miles)
ground attack, hi-lo-hi with three LGBs, designator pod and seven AAMs	
	750 n miles (1,389 km; 863 miles)
air defence with 3 hour CAP	
	100 n miles (185 km; 115 miles)
air defence with 10 minute loiter	
	750 n miles (1,389 km; 863 miles)
g limits with full internal fuel and four AIM-120s	+9/–3

UPDATED

Plan view of Italian prototype DA7 (*Geoffrey Lee/Eurofighter*)   *NEW*/0143728

Eurofighter Typhoon cockpit   0080295

EUROMIL

PARTICIPATING COMPANIES:
Eurocopter: see this section
Mil: see under Russian Federation
Kazan: see under Russian Federation
GENERAL DIRECTOR: Vladimir Yablokov

Agreement signed in Moscow 29 September 1994 between Eurocopter, Mil Moscow Helicopter Plant, Kazan Helicopter Plant (KVZ) and Klimov Corporation to form equal share joint venture company, operating under Russian law and covering development and production of 30-passenger Mi-38 helicopter. Eurocopter plans to use Mi-38 to extend its range of products upwards; expects to invest about US$100 million in programme, from total cost (1999) of US$500 million. Each of above-named had one-third share, but risk-sharing partnership expanded to include Sextant Avionique and Pratt & Whitney Canada.

Mil, Kazan and Eurocopter took a board decision in December 1998 to proceed with Mi-38 on new timetable and signed contract in Moscow on 18 August 1999 for completion of demonstrator, as preliminary to manufacture of representative prototype. Final months of 1999 used to establish operational organisation of Euromil, before obtaining design and certification approvals, plus exclusive

programme rights from Russian authorities. Russian airworthiness authorities inspected Mil facilities in November 2000 as prelude to assembly of prototype.
                                                              VERIFIED

EUROMIL Mi-38
TYPE: Medium transport helicopter.
PROGRAMME: Fuselage of demonstrator (plus two static test airframes) had been built by mid-1997, but engine and transmission suppliers had not then been selected. Programme then delayed by Russian financial collapse. P&WC now to supply 2,461 kW (3,300 shp) PW127 turboshafts for demonstrator and production-standard PW127T/S variant for Western exports; Klimov TVA-3000 for CIS operators, subject to completion of development. Revised timetable called for Mi-38 demonstrator (PT-1) first flight in 2001; four prototypes to follow by 2003; and series production from 2006, following FAR/JAR 29 certification; this apparently further delayed, maiden flight target having moved to 2002, according to report in September 2001. Sales predicted of 200 in CIS, plus 100 exports. Reports of impending French withdrawal were denied on 16 July 2000.

Mil (see Russian Federation section for Mi-38 description) responsible for development, including

Model of the projected Mi-38 medium transport helicopter *(Jane's/Paul Jackson)*   0105078

general design, drawings, component testing and flight testing; Eurocopter leading in flight deck, avionics system and passenger accommodation and responsible for preparation for export from Russia; Kazan manufactures fuselage and rotor blades and undertakes final assembly for domestic market. Major Russian subcontractors include Krasny Oktyabr for transmission and Stupino for rotor head.
                                                              UPDATED

EUROPATROL

EUROPATROL

Details of the European maritime patrol aircraft project last appeared in the 2002-03 *Jane's*. Contenders include the Dassault Atlantic 3 (which see in French section).
                                                              UPDATED

EUROTILT

EUROPEAN TILTROTOR PROGRAMME
PARTICIPATING COMPANIES:
AgustaWestland: see Italian/UK sections
Eurocopter: see this section
TYPE: Multimission tiltrotor.
PROGRAMME: Beginning in 1987, the European Commission sponsored feasibility and definition studies for a tiltrotor transport, as last described under EuroFAR in the 1998-99 *Jane's*. It was hoped by Eurocopter that this ground-work could be incorporated in a forthcoming venture, announced in October 1998, for which other European partners were sought. Eurocopter envisaged a 19/20 seat aircraft with an MTOW of 25,000 kg (55,116 lb) and a service-entry date of around 2015. In early 1999 it undertook a study, sponsored by the French defence ministry, into the military potential of tiltrotors.

By 1999, 33 companies from nine European countries had joined with Eurocopter to apply for EC funding, initially to build a test rig for a proposed 10,000 kg (22,046 lb), 12/19-seat tiltrotor, provisionally known as Eurotilt. Envisaged shares in the project, which could have led to a prototype flying in 2004-05 and entry into service in 2008, included France 36 per cent, Spain 25.7 per cent, Germany 22.3 per cent and Italy 11.6 per cent. Work-shares included Fiat Avio for transmission, CASA for wings and Rolls-Royce Turbomeca for RTM 332 engines.

Performance parameters were 300 kt (556 km/h; 345 mph) cruising speed over ranges between 200 and 800 n miles (370 and 1,481 km; 230 and 920 miles) and a service ceiling of 7,620 m (25,000 ft). Development cost (1999) estimated as US$1.05 billion, of which US$87.5 million for ground test rig.

Meanwhile, in July 1999, Agusta (which see) announced the 10-tonne, 20-seat Erica Tiltrotor and was seeking €90 million of funding from EU's 5th Framework research programme to build a ground test vehicle (GTV). Testing was envisaged five years after GTV go-ahead; prototype first flight in further two years. Backing received from 16 companies, including GKN Westland (UK), ZF Luftfahrttechnik (Germany), IAI (Israel), Aermacchi (Italy), NLR (Netherlands), Gamesa (Spain) and Saab (Sweden).

Erica was second-generation tiltrotor, incorporating improvements over BA609 and V-22. Engines mounted inboard, underwing, driving connecting shafts to proprotors at wingtips. Outboard half of wing was to tilt with proprotors, obviating blocking effect of fixed wing and increasing vertical lift by 12 per cent. This improvement would permit smaller proprotor diameter, increasing cruising speed and allowing almost conventional rolling landings, with outboard wing tilted by only 5 to 7° to prevent ground contact.

In October 1999, European Commission rejected separate funding of competing European tiltrotors and urged merger of Erica with Eurotilt.

Common research project, known as 2Gether (second-generation European tilting highly efficient rotorcraft), submitted to European Commission on 31 March 2000 by Eurocopter, Westland and Agusta (since merged as AgustaWestland). Timetable envisaged initial €95 million study between 2000 and 2004, followed by demonstrator first flight in 2005 (additional €250 million) and production programme (€1,000 million) leading to series manufacture beginning in 2010. Launch version would have had 10-tonne MTOW and carried 20 passengers.

European Commission rejected 2Gether as too costly and insufficiently innovative. New development proposal submitted in March 2001; cost €40 million to €60 million, with industry providing half; features tilting outer wing, as proposed by Agusta; early research devoted to overcoming additional technical difficulties of this configuration; flight demonstration could begin in 2007-08. Study team also to include Mecaer and Televio of Italy; Gamesa and Sener of Spain; UK's FHL; ZF of Germany; ONERA, DLR, NLR and CIRA research institutes; Israel Aircraft Industries; and Pratt & Whitney Canada. These joined by Flight Science and Technology Laboratory at Liverpool University, UK, which inaugurated in June 2001 a flight simulator to be used in developing a control system for Eurocopter's Eurotilt submission.
                                                              UPDATED

Eurocopter's concept of a European tiltrotor has been rejected   0093643

Agusta Erica; current studies are based on the tilting outboard wing envisaged by this design
(Jane's/James Goulding)
0081740

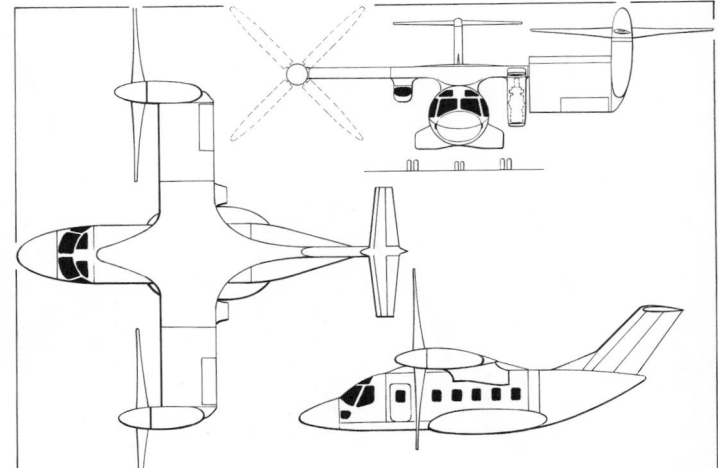

IBIS

IBIS AEROSPACE LTD

PARTICIPATING COMPANIES:
Aero Vodochody: see under Czech Republic
AIDC: see under Taiwan

EUROPEAN SALES OFFICE:
250 70 Odolena Voda, Czech Republic
Tel: (+420 2) 86 03 31 19
Fax: (+420 2) 83 97 00 55

US SALES OFFICE:
804 Water Street, Kerrville, Texas 78028
Tel: (+1 830) 257 82 00
Fax: (+1 830) 257 82 01
e-mail: ibisaerospace@ktc.com
Web: http://www.ibisaerospace.com
MARKETING DIRECTOR: Jeffrey V Conrad

FAR EAST SALES OFFICE:
1141-4, Lane 68, Fu-Hsing North Road, Taichung,
Taiwan
Tel: (+886 4) 22 56 28 61
Fax: (+886 4) 22 56 23 25

Ibis Aerospace Ltd is a 50-50 joint venture company created
by an agreement of 15 March 1997 under which Aero
Vodochody and AIDC are co-developing and producing the
Ae 270 Ibis single-turboprop utility aircraft. Final assembly
is only in the Czech Republic, at least initially.
UPDATED

First prototype Ae 270 P Ibis in flight   *NEW*/0132833

IBIS Ae 270 IBIS

TYPE: Light utility turboprop.
PROGRAMME: Announced by Aero Vodochody of Czech
Republic in early 1990, originally as L-270; configuration
modified 1991; originally planned in two versions (Ae 270
U and Ae 270 MP: see 1993-94 *Jane's*), but revised late
1993 and name Ibis introduced; design frozen 1995. Chief
designer Jan Mikula. Three flying prototypes (Nos. 1, 3
and 5), plus one each for static and fatigue testing (Nos. 2
and 4, respectively); to be certified under FAR Pt 23
(Normal category) and be suitable for FAR Pt 135 single-
pilot IFR. Wings for first prototype (Ae 270 P) received
from Taiwan August 1999. Roll-out delayed by late arrival
of some components, but took place on 10 December 1999;
first flight (OK-EMA) 25 July 2000 and to VZLU test
centre 23 October 2000. Third prototype (second flying,
OK-SAR) rolled out 2 November 2001; first flight 23
December 2001.

Certification of Ae 270 P originally expected mid-2002,
followed by Ae 270 HP in fourth quarter; first deliveries
(Ae 270 P) anticipated in October 2002, with Ae 270 HP
deliveries commencing in early 2003. This not achieved,
and at NBAA Convention at Orlando, Florida, in
September 2002, Ibis announced suspension of Ae 270 P in
order to concentrate resources on Ae 270 HP, with
intention of JAR-23 certification in late 2003 and
deliveries from early 2004, coincident with FAR Pt 23
(single pilot) approval. Eligible for FAR Pt 135.
CURRENT VERSIONS: **Ae 270 P:** Basic pressurised version,
with 634 kW (850 shp) flat rated P&WC PT6A-42A
engine, Honeywell avionics and retractable gear.
Development suspended in mid-2002.

Ae 270 HP: High-performance version, announced 7
October 2000 at NBAA Convention in New Orleans;

PT6A-66A engine offering improved speed, climb and
altitude performance and greater fuel efficiency. Aircraft
05 is prototype for this version. *As described.*

Ae 270 W: Non-pressurised model (previously called
Ae 270 U), with fixed landing gear, Walter M 601F engine
and Czech avionics. Plans for this version in abeyance in
2001. Details in 2001-02 *Jane's*.

Ae 270 FP and **FW:** Wheeled-float versions of P and
W. Not currently promoted.
CUSTOMERS: Initial orders announced at Paris Air Show, June
2001, involved 51 aircraft for six regional distributors
(four in USA, one each in Australia and South Africa).
Firm orders for 60, plus nine options, by September 2002.

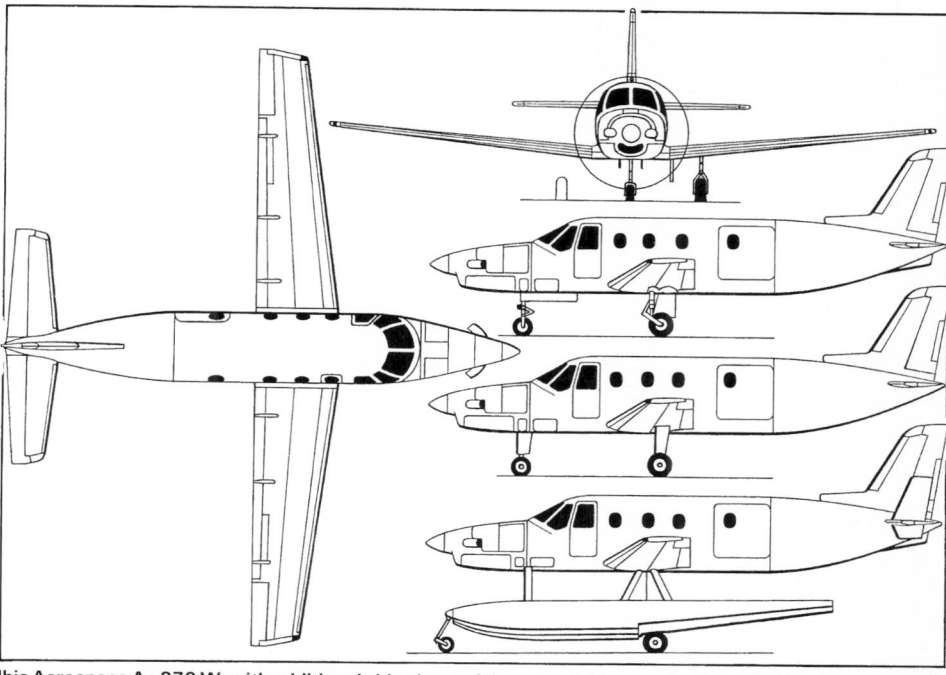
Ibis Aerospace Ae 270 W, with additional side views of the retractable-gear Ae 270 P/HP (top) and wheeled-
float Ae 270 FW/FP (*Jane's/Mike Keep*)   0109168

COSTS: US$1.995 million (Ae 270 P) and US$2.195 million
(Ae 270 HP) (2002).
DESIGN FEATURES: Sized between Socata TBM 700 and Pilatus
PC-12. High-aspect ratio low wing, circular cabin
windows, sweptback vertical tail.

Medium-speed aerofoil section (thickness/chord ratio
17 per cent at root, 12 per cent at tip); leading-edge
sweepback 7° 42'; dihedral 6°; incidence 3°; twist 3°.
FLYING CONTROLS: Conventional and manual, augmented by
upper-wing roll control spoilers interconnected to ailerons.
Elevators trimmed mechanically, ailerons and rudder
electromechanically; trim tab in rudder. Wide-span single-
slotted Fowler flaps (70 per cent of trailing-edge; positions
0, 20 and 40°) actuated hydraulically. Autopilot optional.
STRUCTURE: All-metal semi-monocoque, with fail-safe
structural elements in fuselage and two-spar wings.
Fuselage of conventionally formed sheet metal bulkheads,
stringers and skin; wings (built by AIDC) have formed
sheet metal ribs, stringers and chemically milled skin.
Production in Czech Republic (including engine
compartment by Moravan, cabin floor panels by LZ and
rear fuselage, fin and rudder by EV-AT) except for wings
and landing gear, produced in Taiwan by AIDC; final
assembly by Aero Vodochody. Other suppliers include:
Eros (oxygen system), Fischer (crew and passenger seats),
Gumotex Breclav (de-icer boots), Honeywell (air
conditioning and pressurisation components),
Intertechnique (fuel gauges), Jihlavan (hydraulic system),
Mikrotechna Modrany (instrumentation), Moravan
(wheels and brakes), and PPG Industries (windows).
LANDING GEAR: Retractable tricycle type by Technometra
Radotin, with Moravan wheels and brakes; steerable
nosewheel (60° by brakes, 15° by rudder pedals). Inward
retraction for mainwheels, rearward for nosewheel. Oleo-
pneumatic shock-absorbers in all units; hydraulic disc
brakes on mainwheels. Dunlop tubeless tyres: size 6.50-10,
pressure 6.90 bar (100 lb/sq in) on mainwheels, size
6.00-6, pressure 3.80 bar (55 lb/sq in) on nosewheel.
Minimum ground turning radius (based on nosewheel)
4.00 m (13 ft 1½ in).
POWER PLANT: One 634 kW (850 shp) (flat rated) Pratt &
Whitney Canada PT6A-66A turboprop, driving a Hartzell
HC-D4N-3/D9511FK four-blade, constant-speed,
feathering and reversing propeller.

Integral fuel tank in each wing centre-section, combined
usable capacity 1,170 litres (309 US gallons; 257 Imp
gallons). Gravity refuelling point in top of each wing. Oil

Ae 270 state-of-the-art flight deck with flat panel displays   0100521

Maiden sortie of the second flying Ae 270 Ibis   NEW/0123951

capacity 5.7 litres (1.5 US gallons; 1.25 Imp gallons) in Ae 270 P.

ACCOMMODATION: Flight crew of two standard, but to be certified for single-pilot operation. Main cabin suitable for up to eight passengers or 1,200 kg (2,645 lb) of cargo, or combinations of both. Six/seven-seat business or four-seat club layouts permit inclusion of lavatory. Medevac configuration provides accommodation for two stretcher cases and two medical attendants. Forward-opening crew door at front on port side; upward-opening passenger/cargo door on port side aft of wing with optional airstair door inset; overwing emergency exit on starboard side. Baggage door on starboard side at rear. Cockpit and cabin air conditioning, pressurisation and windscreen heating standard.

SYSTEMS: Electrical power provided by 28 V 250 A DC engine-driven starter/generator and 24 V 42 Ah lead-acid battery; 28 V DC external power connector. Standby generator optional. Hydraulic system, pressure 150 bar (2,175 lb/sq in), for actuation of flaps, mainwheel brakes and landing gear extension/retraction; flow rate 11 litres (2.9 US gallons; 2.4 Imp gallons)/min. Landing gear and mainwheel brakes also controllable by separate emergency hand-operated valves and parking brake.

Honeywell air conditioning and pressurisation system maintains differential of 0.30 bar (4.4 lb/sq in) up to altitude 7,500 m (24,600 ft). Gumotex pneumatic (engine bleed air) de-icing of wing and tailplane leading-edges; electric de-icing of windscreens, propeller blades; hot air de-icing of engine air intake; stall warning sensor and pitot tube standard on both models. Eros emergency oxygen system for crew and passengers.

AVIONICS: Standard flight, navigation and engine instrumentation (VFR or IFR) to comply with FAR Pt 23.
Radar: Weather radar optional.
Flight: IFR package standard for business versions, including VOR/ILS, ADF, Trimble GPS and S-Tec S-55 autopilot.
Instrumentation: EFIS optional.

DIMENSIONS, EXTERNAL:
Wing span	13.82 m (45 ft 4 in)
Wing chord: at root	1.88 m (6 ft 2 in)
at tip	1.04 m (3 ft 5 in)
Wing aspect ratio	9.1
Length overall	12.23 m (40 ft 1½ in)
Fuselage: Length	12.19 m (40 ft 0 in)
Max width	1.60 m (5 ft 3 in)
Max depth	1.75 m (5 ft 9 in)
Height overall	4.78 m (15 ft 8¼ in)
Elevator span	5.40 m (17 ft 8½ in)
Wheel track	2.83 m (9 ft 3½ in)
Wheelbase	3.53 m (11 ft 7 in)
Propeller diameter: Hartzell	2.44 m (8 ft 0 in)
McCauley	2.39 m (7 ft 10 in)

Propeller ground clearance:
Hartzell	0.41 m (1 ft 4¼ in)
McCauley	0.44 m (1 ft 5¼ in)

Passenger/cargo door (port, rear):
Height	1.30 m (4 ft 3¼ in)
Width	1.25 m (4 ft 1¼ in)
Height to sill	1.30 m (4 ft 3¼ in)

Crew door (port, fwd): Height 1.20 m (3 ft 11¼ in)
Width	0.70 m (2 ft 3½ in)
Height to sill	1.30 m (4 ft 3¼ in)

Emergency exit (stbd, overwing):
Height	0.71 m (2 ft 4 in)
Width	0.50 m (1 ft 7¾ in)

DIMENSIONS, INTERNAL:
Cabin: Length: excl flight deck	4.98 m (16 ft 4 in)
incl flight deck	5.50 m (18 ft 0½ in)
Max width	1.45 m (4 ft 9 in)
Max height	1.36 m (4 ft 5½ in)
Volume	7.5 m³ (265 cu ft)
Baggage compartment volume	0.73 m³ (25.8 cu ft)

AREAS:
Wings, gross	21.00 m² (226.0 sq ft)
Ailerons (total)	1.00 m² (10.76 sq ft)
Trailing-edge flaps (total)	4.22 m² (45.42 sq ft)
Spoilers (total)	0.325 m² (3.50 sq ft)
Fin, incl dorsal fin	1.96 m² (21.10 sq ft)
Rudder, incl tab	1.11 m² (11.95 sq ft)
Tailplane	2.975 m² (32.02 sq ft)
Elevators (total)	1.885 m² (20.29 sq ft)

WEIGHTS AND LOADINGS (Ae 270 HP):
Weight empty, equipped	2,300 kg (5,071 lb)
Max fuel weight	1,034 kg (2,280 lb)
Max T-O weight	3,700 kg (8,157 lb)*
Max landing weight	3,700 kg (8,157 lb)
Max zero-fuel weight	3,500 kg (7,716 lb)
Max wing loading	176.2 kg/m² (36.09 lb/sq ft)
Max power loading	5.84 kg/kW (9.60 lb/shp)

* To be increased retrospectively to 3,800 kg (8,377 lb) after initial certification

PERFORMANCE (Ae 270 HP):
Max cruising speed: at S/L	220 kt (407 km/h; 253 mph)
at FL 200	270 kt (500 km/h; 311 mph)
Stalling speed, engine idling:	
flaps up	85 kt (158 km/h; 98 mph)
flaps down	66 kt (123 km/h; 76 mph)
Max rate of climb at S/L	521 m (1,710 ft)/min
Max certified altitude	9,140 m (30,000 ft)
T-O run	270 m (890 ft)
T-O to 15 m (50 ft)	550 m (1,805 ft)
Landing from 15 m (50 ft)	500 m (1,640 ft)
Landing run	167 m (1,545 ft)
Range with max fuel at FL 300: 30 min reserves	1,610 n miles (2,981 km; 1,852 miles)
NBAA IFR reserves	1,420 n miles (2,629 km; 1,634 miles)

UPDATED

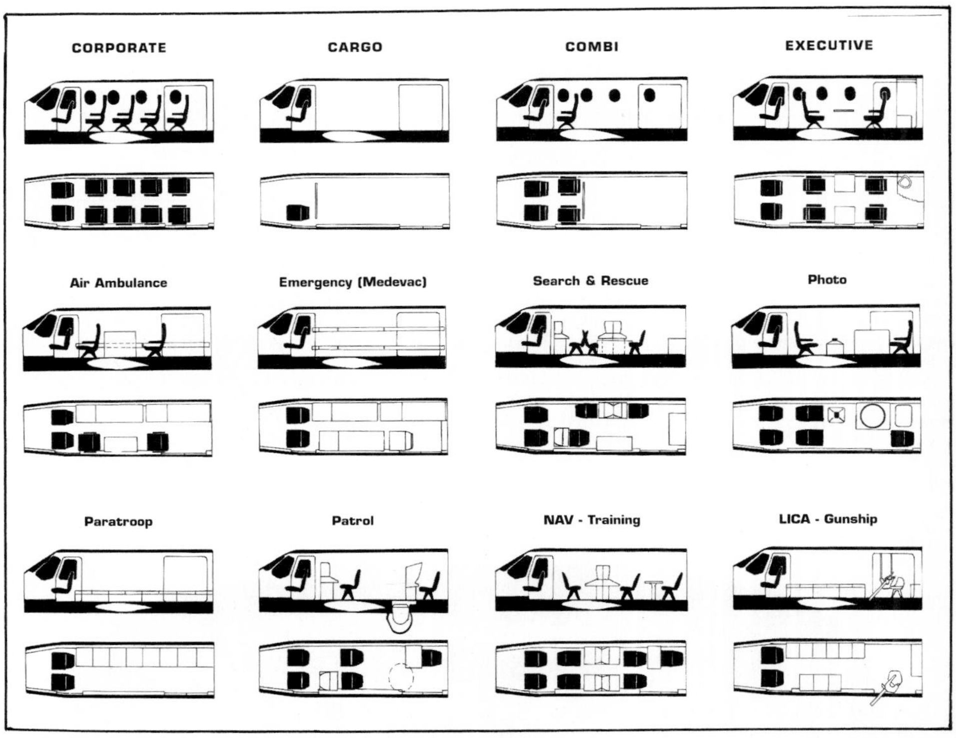

CORPORATE	CARGO	COMBI	EXECUTIVE
Air Ambulance	Emergency (Medevac)	Search & Rescue	Photo
Paratroop	Patrol	NAV - Training	LICA - Gunship

Potential interior configurations of the Ibis utility transport   0093627

ILYUSHIN/IAPO/HAL

ILYUSHIN/IAPO/HAL

PARTICIPATING COMPANIES:
Ilyushin: see under Russian Federation
IAPO: see under Russian Federation
HAL: see under India

Agreement announced between these three companies in June 2001 for joint design, development and co-production of a multirole military and civil transport, based on Ilyushin Il-214 and funded mainly by IAPO. Production lines in both countries are planned. Under agreement of January 2002, Russian partners will transfer conceptual draft for series production and metallurgical information to HAL; India will create infrastructure, manufacturing programme and commercial production.

UPDATED

ILYUSHIN/IAPO/HAL MULTIROLE TRANSPORT AIRCRAFT (MTA)

Russian designation: Il-214

TYPE: Medium transport/multirole.
PROGRAMME: This Indian requirement became the subject of a joint venture between HAL and Ilyushin (Russian designation Il-214) following the opening of negotiations in March 1999; joint product signed later in the same year; MoU in late 2000 followed in June 2001 by joint design, development and co-production agreement involving these two companies plus Irkutsk Aviation Production Organisation.

Configuration now differs considerably from the original Indian design of 1996, details of which last appeared in the 2000-01 *Jane's.* Programme is included in Russian government's aeronautical development plan for 2001-15. Full-scale development due to be launched in second quarter of 2002, leading to flight testing during 2005-06, and production start in 2007 for first deliveries in 2008.

By early 2002, India reportedly seeking Western avionics and engines for incorporation in MTA; revised engine specification anticipated in May 2002.
CURRENT VERSIONS: **Airliner:** With cabin windows and seating for 98 to 108. Russian designation **Il-214-100.**
Tactical transport: Military version with rear-loading ramp; provision for 82 paratroops. Russian designation **Il-214T.**
Commercial freighter: With rear loading ramp.
CUSTOMERS: Required by Indian and Russian air forces; manufacture by HAL and IAPO. Initial commitment (50 firm and 100 options), announced in October 2001, believed to be from Indian Air Force. Estimated market for 200 to 300.
COSTS: Development estimated as US$300 million. Unit cost US$12 million to US$15 million with PS-12 engines. Target price is 20 per cent less than for comparable aircraft.
POWER PLANT: Two turbofans, each of some 108 to 118 kN (24,000 to 26,500 lb st). Options include CFM56, Perm PS-12 (under development) or Rolls-Royce BR715.

Model of projected MTA/Il-214 in civil transport guise 0110516

Military MTA/Il-214 in model form (*Jane's/Paul Jackson*) NEW/0137213

Following details all provisional:

DIMENSIONS, EXTERNAL:
Wing span	29.20 m (95 ft 9½ in)
Length overall	34.50 m (113 ft 2¼ in)
Height overall	8.75 m (28 ft 8½ in)

DIMENSIONS, INTERNAL:
Cargo hold: Length	13.50 m (44 ft 3½ in)
Max width at floor	3.15 m (10 ft 4 in)
Max height	3.00 m (9 ft 10 in)

WEIGHTS AND LOADINGS (design):
Max payload	20,000 kg (44,090 lb)
Max fuel weight	13.500 kg (29,762 lb)
Max T-O weight	55,000 kg (121,250 lb)
Max landing weight	49,500 kg (109,125 lb)

PERFORMANCE (estimated):
Max cruising speed	459 kt (850 km/h; 528 mph)
Normal cruising speed	432 kt (800 km/h; 497 mph)
Airdropping speed	135 kt (250 km/h; 155 mph)
Max operating altitude	11,000 m (36,080 ft)
Service ceiling	12,000 m (39,370 ft)
T-O to 15 m (50 ft)	1,300 m (4,265 ft)
Landing from 15 m (50 ft)	1,200 m (3,940 ft)

Range: airliner with max payload
1,889 n miles (3,500 km; 2,174 miles)
tactical transport with 4,500 kg (9,920 lb) payload and
auxiliary fuel 3,239 n miles (6,000 km; 3,728 miles)

UPDATED

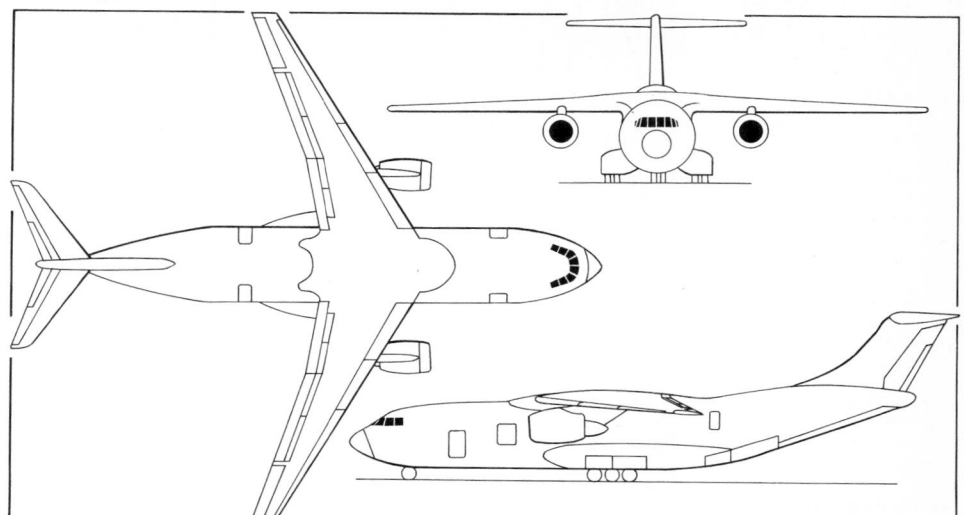

Ilyushin Il-214, which is India's new Multirole Transport Aircraft. Note detail differences from Indian conception in model form (*Jane's/Paul Jackson*) 0062411

INTRACOM

INTRACOM GENERAL MACHINERY SA

Chemin de Faguillon 1, CH-1223 Geneva, Switzerland
Tel: (+41 22) 786 27 57
Fax: (+41 22) 786 27 62
e-mail: intracom@bluewin.ch
DIRECTOR: Nik Schmidt

Intracom, formed in 1994, is the aircraft design and manufacturing arm of General Machinery SA. It joined with the Khrunichev company of Russia to offer the GM-17 light transport in Western markets, but by 2002 had formed a new alliance with Technoavia, and its associated SmAZ production plant, of the same country (which see).

At a later stage, it is planned to produce a diverse family of aircraft, brief details of which last appeared in the 2001-02 *Jane's*.

UPDATED

INTRACOM GM-17 VIPER

TYPE: Light utility turboprop.
PROGRAMME: Designed within Khrunichev bureau in Moscow by Evgeny P Grunin. Considerably modified Piper PA-31P Pressurised Navajo. First flight (GM-17-000, registered RA-01559; the former N38RG) 6 December 2000; 10 sorties and 6 hours flown by February 2001, when laid-up for static testing. Further two prototypes were to have followed in 2001, using Navajos G-OIEA and SE-IGB as basis; programme delayed by change of partner company to Technoavia (chief designer Vyacheslav Kondratiev) and renaming of aircraft from Feniks (Phoenix) to Viper; three PA-31P airframes (GM-17-001, 002 and 003) delivered to Technoavia in November 2001 and entered modification at SmAZ, Smolensk.

Intracom had previously proposed a collaborative project with Piper, to be known as PA-31TXL, but this was rejected by the US company; projected new production, to begin in 2003, will not use Piper-sourced airframes, therefore. Intended production of 32 aircraft by 2005; manufacture at Smolensk and flight testing in Switzerland.

First Smolensk-built Intracom GM-17 Viper NEW/0533370

COSTS: US$860,000 (2000).
DESIGN FEATURES: Single-engined turboprop for passenger and light freight transport; suitable for executive ferry, medical evacuation, patrol, geological survey and training. Revised structure includes fireproof steel bulkhead (Frame 3) behind engines; new engine cowlings forward of Frame 4; replacement wing panels in former engine positions; tailplane root fillets; new window frames; ventral fin; revised tailplane incidence (+2°, replacing −2°); composites fintip; replaced fuel tanks, oil system, electrical system (some modified components), electrical system (retaining some PA-31 parts), fire warning system and hydraulic system, de-icing system and oxygen system (reduced capacity).

Low-mounted, tapered wing with wingtip tanks or winglets; mid-positioned, tapered tailplane and sweptback fin.

Wing section NACA 63_2415 at root; 63_1212 at tip; leading-edge sweepback 5°; thickness chord ratio 14 per cent at root, 12 per cent at tip; dihedral 5°; incidence 1°. Tailplane incidence 2°; wingroot nib sweepback 28°.
FLYING CONTROLS: Conventional and manual. Trim tab in starboard aileron, rudder and both elevators. Flaps.
LANDING GEAR: Tricycle type; hydraulically actuated; mainwheels retract inward; nosewheel rearward; single wheel on each unit. Nosewheel steerable ±28°; free castoring angle ±80°. Goodyear mainwheels and 6.50-10 (8-ply) tyres; Cleveland nosewheel with 6.00-6 (8-ply) tyres. Hydraulic brakes.
POWER PLANT: One 560 kW (757 shp) Walter M 601E turboprop, driving a five-blade V510 constant-speed, fully reversing propeller. Three prototypes have M 601D engines and three-blade V508E/99 propellers; Pratt & Whitney PT6A-42A on fourth aircraft (GM-17-004) and

---

Intracom GM-17-001 prototype, remanufactured from a Piper Navajo   *NEW*/0533369

available thereafter as option; TVD-100 turboprop to be offered as alternative on production aircraft. Total fuel capacity 916 litres (242 US gallons; 202 Imp gallons in six wing tanks of 212 litres (56.0 US gallons; 46.7 Imp gallons) (inner pair), 151.5 litres (40.0 US gallons; 33.3 Imp gallons) and 95 litres (25.0 US gallons; 20.8 Imp gallons) (outboard pair). Optional wingtip tanks of Piper design.

ACCOMMODATION: Two pilots and six passengers; forward four passengers in club arrangement; lavatory and baggage area at rear. Externally accessed baggage compartment in nose. Single door, port, rear. Emergency exit, starboard. Cabin protected by up to 25 mm (1 in) of soundproofing material. Accommodation pressurised and air-conditioned.

SYSTEMS: Hydraulic system, pressure 150 bar (2,175 lb/sq in); 27 V electrical system; emergency oxygen. De-icing system on all leading-edges.

AVIONICS: *Comms:* Garmin GNS 430 and GNS 530 VHF/NAV. Garmin GTX 27 transponder. ELT.
Radar: Optional Honeywell RDS 81 four-colour weather radar in pod on starboard wing.
Flight: Bendix/King KR 87 radio compass and KI 228 ADF. S-Tec 55 autopilot.

DIMENSIONS, EXTERNAL:
Wing span: over winglets	12.73 m (41 ft 9¼ in)
over optional tanks	13.04 m (42 ft 9½ in)
Wing chord: at root	2.28 m (7 ft 5¾ in)
at tip	0.985 m (3 ft 2¾ in)
Wing aspect ratio	7.5
Length: overall	10.60 m (34 ft 9¼ in)
fuselage	10.28 m (33 ft 8¾ in)
Fuselage max width	1.55 m (5 ft 1in)
Height overall	4.08 m (13 ft 4¾ in)

Tailplane span	6.05 m (19 ft 10¼ in)
Wheel track	4.19 m (13 ft 9 in)
Wheelbase	2.69 m (8 ft 10 in)
Propeller diameter	2.50 m (8 ft 2½ in)
Propeller ground clearance	0.49 m (1 ft 7¼ in)
Passenger door: Height	1.32 m (4 ft 4 in)
Width	0.78 m (2 ft 6¾ in)
Emergency exit: Height	0.54 m (1 ft 9¼ in)
Width	0.60 m (1 ft 11¾ in)
Nose baggage door: Height	0.53 m (1 ft 9 in)
Width	0.66 m (2 ft 2 in)

DIMENSIONS, INTERNAL:
Cabin: Length	4.92 m (16 ft 1¾ in)
Max width	1.30 m (4 ft 3¼ in)
Max height	1.32 m (4 ft 4 in)
Baggage hold volume: cabin	0.62 m³ (22 cu ft)
nose	0.57 m³ (20 cu ft)

AREAS:
Wings, gross	21.60 m² (232.5 sq ft)
Ailerons (total)	1.245 m² (13.40 sq ft)
Trailing-edge flaps (total)	3.135 m² (33.74 sq ft)
Winglets (total)	2.86 m² (30.78 sq ft)
Rudder (incl tab)	1.18 m² (12.70 sq ft)
Tailplane	6.08 m² (65.44 sq ft)
Elevators (total)	2.16 m² (23.25 sq ft)

WEIGHTS AND LOADINGS:
Weight empty	1,900 kg (4,189 lb)
Max T-O and landing weight	3,300 kg (7,275 lb)
Max wing loading	152.8 kg/m² (31.29 lb/sq ft)
Max power loading	5.90 kg/kW (9.70 lb/shp)

PERFORMANCE (at max T-O weight except where indicated):
Max cruising speed at FL600	235 kt (435 km/h; 270 mph)
Econ cruising speed at FL600	225 kt (417 km/h; 259 mph)
Stalling speed, flaps down	67 kt (125 km/h; 78 mph)*
Max rate of climb at S/L	402 m (1,320 ft)/min*
Service ceiling	7,165 m (23,500 ft)
T-O run	510 m (1,675 ft)
T-O to 15 m (50 ft)	800 m (2,625 ft)
Landing from 15 m (50 ft)	508 m (1,670 ft)
Landing run with propeller reversal	350 m (1,150 ft)
Range with max fuel, plus reserves	1,560 n miles (2,889 km; 1,795 miles)

* At 2,990 kg (6,592 lb)

UPDATED

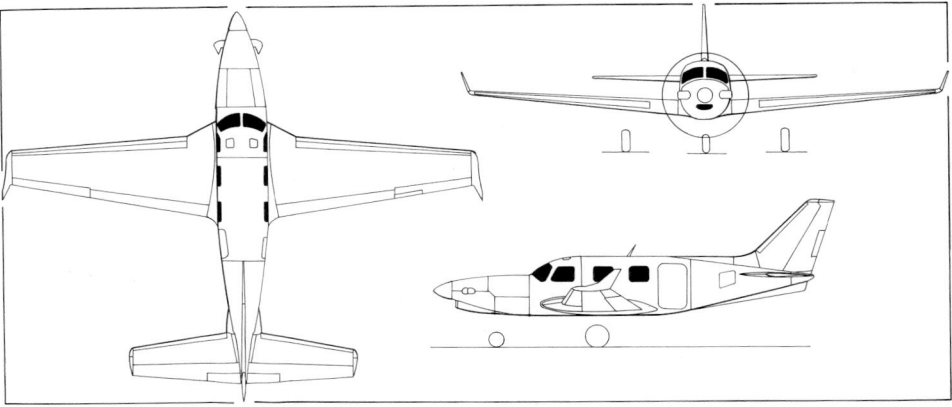

Intracom GM-17 Viper (Walter M 601 turboprop) *(Jane's/Paul Jackson)*   *NEW*/0533393

NH INDUSTRIES

NH INDUSTRIES sarl
Le Quatuor, Batiment C, 42 route de Galice, F-13090 Aix-en-Provence, France
Tel: (+33 4) 42 95 97 00
Fax: (+33 4) 42 95 97 48/49
e-mail: nhi90@nhindustries.com
Web: http://www.nhindustries.com
GENERAL MANAGER: Philippe Stuckelberger
PARTICIPATING COMPANIES:
Agusta: see under Italy
EADS Eurocopter: see this section
EADS Eurocopter Deutschland: see this section
Stork Fokker: see *Jane's Aircraft Upgrades*

NH Industries established August 1992 to manage design and development of NH 90 and is also responsible for production, logistic support, marketing and sales; in October 1995 workshares were slightly amended to Agusta 28.2 per cent, Eurocopter Deutschland 23.7 per cent, Eurocopter France 41.6 per cent, Fokker Aircraft 6.5 per cent. NFT (Norway) joined in 1994 as risk-sharing partner of Eurocopter France. Further amendment occurred on 21 June 2001, when Portugal joined programme and allocated 1.2 per cent share to OGMA. Netherlands activities transferred to Fokker Aerostructures BV following 1996 bankruptcy of the former Fokker Aircraft. Industrial ownership (and production share) is now EADS 61.7 per cent, Finmeccanica (Agusta) 31.6 per cent, Stork Fokker 5.5 per cent and Portuguese industry (as Eurocopter and Agusta subcontractor) 1.2 per cent.

Joint agency NAHEMA (NATO Helicopter Management Agency) formed February 1992 by the four governments, within NATO framework, to manage programme; now additionally represents Portuguese government; NAHEMA located alongside NH Industries in Aix.

UPDATED

Fifth NH 90 prototype during trials aboard *Etna* in May 2001   0113506

NH INDUSTRIES NH 90
TYPE: Multirole medium helicopter.
PROGRAMME: Initial studies by NIAG (NATO Industrial Advisory Group) SG14 in 1983-84; September 1985 MoU between defence ministers of France, Germany, Italy, Netherlands and UK (USA and Canada having already withdrawn) preceded 14 month feasibility/predefinition study for new naval/army NH 90 (NATO helicopter for the 1990s); initial design phase approved December 1986; second MoU in September 1987 led to predefinition phase and completion of weapons system definition in 1988; UK withdrew from programme April 1987; German workshare reduced early 1990, Italian participation renegotiated later 1990; French and German launch decision 26 April 1990;

two ministerial MoUs in December 1990 and June 1991 cover design and development responsibilities; intercompany agreement signed March 1992; design and development contract signed 1 September 1992; five prototypes and GTV (ground test vehicle) planned, with three prototypes produced in France, TTH in Germany and NFH in Italy; assembly of first prototype began at Marignane, October 1993; first flight 18 December 1995. All five prototypes flying by December 1999.

Development suspended May 1994; restarted July 1994, but all new-development items subject to rigorous cost examination; preference for off-the-shelf components. Programme doubts dispelled by inclusion of NH 90 in French defence plan adopted in June 1996, which requires

NH 90, MMX613, carrying full NFH sensor suite   0110878

countermeasures, FLIR and four FIM-92 Stinger self-defence AAMs.

'Kits on Option': To accommodate differing budgetary and operational requirements, a baseline configuration has been agreed, to which specific packages could be added. French shipborne utility version combining troop seats and ramp could replace Super Frelon. This version could also form basis of dedicated VIP variant, and even of a civil variant. Eurocopter France is responsible for development of the basic helicopter.

Netherlands NFH: All 20 equipped with common core avionics, FCS, 360° radar, FLIR, ESM, ECM and sonics, but only 14 with full mission system; rest with provision for, but without, LRUs. Proposal to equip six for night special operations/insertion duties. To serve aboard eight M-type frigates, four LCF frigates, two LPDs and two AORs.

CURRENT VERSIONS (specific): **PT1/F-ZWTH:** Prototype; common basic configuration; assembled at Marignane; rolled out 29 September 1995; first flight 18 December 1995 with RTM 322 engines; formal debut (official first flight) 15 February 1996; public debut at Berlin, May 1996; 75 hours airborne by 1 January 1997; tactical (green) colours; with 160 hours logged, PT1 flew to Agusta's Cascina Costa plant on 29 July 1997 for installation of Alfa Romeo Avia/GE T700-T6E engines before an 18-month test programme funded by Italy; first flew as such on 13 March 1998; had amassed 205 hours with these engines by August 2000, including sea trials with French 'La Fayette' class frigate *Courbet*, making 62 deck landings in two days (22/23 July 1998). Flown with Italian serial MMX612. Total 365 hours by 31 August 2000, when scheduled flight activities were completed.

PT2/F-ZWTI: Common basic configuration; rolled out November 1996 at Marignane; first flight 19 March 1997; naval (light grey) colours; first with fly-by-wire controls. Initially flew for 5 hours with mechanical back-up flight controls, then first flew with analogue FBW on 2 July 1997; total 20 hours by December 1997; digital FBW installed early 1998 and first flown on 15 May. PT2 also first with automatic electric blade folding intended for NFH, and due to receive IR exhaust suppressors. Total 305 hours (including 106 in FBW mode) by 25 July 2001.

PT3/F-ZWTJ: Common basic configuration; first with core avionics system with AFCS, navigation, communications, IFF and full 'glass' cockpit; retains mechanical control back-up; first flight at Marignane 27 November 1998. Core avionics trials with production-standard LCDs; replaced PT2 in extending flight envelope. De-icing system tests began 13 April 2001, lasting 20 hours. Total 195 hours (including 109 FBW) to 25 July 2001.

PT4/9890: TTH configuration; first flight 31 May 1999, at Ottobrunn. First NH 90 to be fitted with C-model prototype of DaimlerChrysler Aerospace electronic warfare suite as one element in a fully integrated avionics suite, with LCD cockpit displays, FLIR, helmet sight, tactical control system, secure communications, and radar (with digital mapping). Cleared operation of rear loading ramp and doors open in flight up to 140 kt (259 km/h; 161 mph). Night flights with FLIR and rescue hoist demonstration. Total 117 hours (51 in FBW mode) by 25 July 2001.

PT5/MMX613: NFH mission system; first flight at Cascina Costa 22 December 1999. Performed flotation system test. Deck operating campaign aboard replenishment tanker *Etna*, May 2001, with 77 landings and 42 deck lock engagements, all in FBW mode. Total 81 hours (54 in FBW) by 25 July 2001.

GTV: Ground test vehicle at Cascina Costa. Instrumented to monitor 300 parameters. First run 28 September 1995; 539 hours accumulated by June 2000.

first NFH delivery to Aéronavale in 2004 and first TTH to ALAT in 2011, although delivery of the last ALAT aircraft has been delayed to 2025. Bid for production investment and Lot 1 manufacture submitted 5 October 1997; four nations signed political authorisation for first production batch of 154 (amended later to 151) helicopters on 19 May 1998; pricing proposals submitted by NHI to NAHEMA on 15 July 1998; industrial contract was due in mid-1999 to permit first deliveries in 2003 for Netherlands Navy, but was delayed and latter date rescheduled to 2007.

In September 1999, Germany revealed a cut of 24 in its total order and now plans to receive first TTH in 2004 and NFH in 2007, instead of withdrawing current Westland Lynx in 2003. Italian deliveries (both versions) to begin in April 2004, continuing until 2017-18. NH 90 to be built on parallel production lines in France, Germany and Italy. Netherlands confirmed in November 1998 that its helicopters would be built in Italy; German NFHs will be built in France. Further, in 1998, a basic NH 90 variant was proposed to which 'Kits on Option' could be added to produce specific variants at lower cost than purpose-build.

Rescheduled production contract signature, at Berlin Air Show, 8 June 2000, replaced by MoU committing to 243, plus 55 options and minimum buy of 366 from stated requirements for 595. Production investment and production contract finally signed in Paris on 30 June 2000 on these terms and quoting cost for 298 as €6.6 billion, with industrial participants providing 25 per cent of production investment. Portuguese contract signed at Paris Air Show on 21 June 2001 added fifth member to consortium and 10 aircraft to previous figures, for new totals of 253 firm, 376 minimum purchase and 605 stated requirement. This further increased by 52 on 13 September 2001 on selection of NH 90 for Nordic Standard Helicopter Project.

On 6 October 1995, NAHEMA contracted NHI to undertake additional work and national customisation, involving integration of T700 engine to meet Italian requirement; design of rear ramp installation for TTH version; and reinforcement for carriage of stores up to 700 kg (1,543 lb). Associated national requirements include study of command post version; cannon pod; sand filter; radiometer; second VHF/FM radio for TTH; and

sonobuoy data relay, Tacan and rear ramp for NFH. Total value of €58.23 million slightly amends national workshares. Italy formally confirmed T700 engine selection in June 2000; Germany and Netherlands simultaneously confirmed RTM322, which earlier selected by France.

First five aircraft had flown more than 1,065 hours by July 2001. Flight envelope expanded by that time to 190 kt (352 km/h; 219 mph), 6,096 m (20,000 ft) altitude and T-O weight of 10,600 kg (23,369 lb). The aircraft has also demonstrated fast rolling landings at 50 kt (93 km/h; 58 mph) and landings on slopes of up to 12°. Flight test activity in 2001 mainly dedicated to mission system qualification.

CURRENT VERSIONS (general): **NFH** (NATO Frigate Helicopter): Naval version, primarily for autonomous ASW and ASVW; additional applications include OTH targeting, vertrep, SAR, transport and anti-air warfare support; designed for all-weather/severe ship motion environment; fully integrated mission system for crew of three (optionally four); ECM, anti-radar and IR protection systems. Italy considering a unique sensor fit from FIAR/Honeywell consortium. Agusta, with its EH 101 and S-61 experience, is responsible for NFH development and integration.

TTH (Tactical Transport Helicopter): Land-based army/air force version, primarily for tactical transport, airmobile operations and SAR with up to 20 armed troops or 2.5 tonnes of supplies; additional applications include tactical support, special EW, airborne command post, VIP transport and training; defensive weapons suite; rear-loading ramp/door, loading winch and rolling strips to be provided for French, Italian and German armies to accommodate light armoured anti-tank missile vehicle; 4-tonne external underslung load hook; high manoeuvrability and survivability for NOE operation near front line. Eurocopter Deutschland is responsible for development of the TTH.

C-SAR: Combat search and rescue version, yet to be fully defined, but probably equipped with refuelling probe, 'bolt-on' armour on lower fuselage, two external fuel tanks, flotation gear, secure datalink, integrated

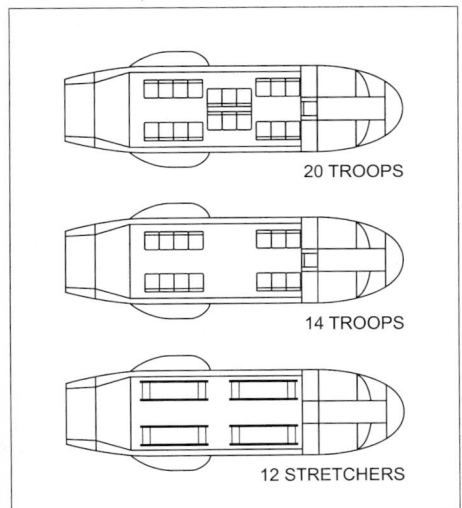

20 TROOPS

14 TROOPS

12 STRETCHERS

NH 90 TTH seating/carrying options   0113515

Fourth NH 90, 9890, in TTH guise with rear ramp (*Jane's/Paul Jackson*)   0110952

CUSTOMERS: Requirements originally estimated as 726 (France 220, Germany 272, Italy 214 and Netherlands 20); reduced to 647 in July 1996, then to 642 in 1998 and 595 in late 1999. These comprise Germany 181 TTH and 38 NFH (= 219); France 133 TTH and 27 NFH (= 160); Italy 150 TTH and 46 NFH (= 196); and Netherlands 20 NFH. In mid-2001, total of 308 on order or option; see table. France scheduled eventually to buy 133 TTH for the Armée de l'Air from 2011. Additionally, over 600 export orders are anticipated; with Portugal joining the programme in June 2001, when it contracted for 10 examples of the TTH to be delivered to the army in 2006-07. NHI has received an RFP for 30 TTH from Oman, and is bidding to meet Singapore requirement for 12 NFHs. Selected for Nordic Standard Helicopter Project 13 September 2001; Norway requires 14, plus 10 options; Finland 20 and Sweden 18.

Customer	Firm		Option	
	TTH	NFH	TTH	NFH
France				
Navy		27		
Germany				
Air Force	30		24	
Army	50		30	
Italy				
Air Force			1	
Navy	10	46		
Army	60			
Netherlands				
Navy		20		
Portugal				
Army	10			
Totals	**160**	**93**		
		253	**55**	
		308		

Notes: Second contract will be for 68 French TTH, completing guaranteed purchase of 376 by original four partners. Further 229 required to complete stated needs of five partners. German Air Force total includes 23 for combat SAR

COSTS: Development cost €1,376.15 million (FFr 9.6 billion) (January 1988 values) agreed September 1992, provided by governments and industry: €421.83 million French government plus €161.24 million Eurocopter France; €256.41 million Germany plus €74.59 million Eurocopter Deutschland; €89.44 million Netherlands plus €2.64 million Fokker; and €370 million by Italian government. Additional development, agreed October 1995, adds €58.23 million. By January 1997, an estimated 75 per cent of R&D costs had been disbursed. Projected flyaway prices (1995) are FFr90 million for TTH and FFr145 million for NFH, but further 15 per cent cut demanded by French military procurement office in February 1996. Initial batch of French NFHs have 2000 unit cost of FFr200 million, including spares and tax; NHI quotes FFr115 million, pre-tax. Netherlands paying NLG1.7 billion (US\$510 million) giving a unit price of US\$25.5 million per aircraft. Cost of 10 TTH for Portugal quoted as €200 million.

DESIGN FEATURES: Compact external dimensions with large cabin (larger and slightly higher than Sikorsky Black Hawk). Low vulnerability/detectability, with aerodynamic

NH 90 flight deck   0113512

and low-observables design, reduced maintenance requirements, and day/night operability within temperature range of −40 to +50°C. Specification requires fewer than 250 failures per 1,000 flight hours; 97.5 per cent mission reliability rate; 87 per cent availability; MTBF better than 4 hours; and under 2.5 mmh/fh (excluding engines). Service life of 10,000 hours over 30 years.

Titanium Spheriflex main rotor hub with elastomeric spherical thrust bearings; four blades with advanced aerofoils and curved tips; main rotor turns anticlockwise at 256.6 rpm with 219 m (719 ft)/s tip speed; bearingless tail rotor with four blades of similar design and construction; tail rotor with cross-beam hub turns at 1,235.4 rpm with 207 m (679 ft)/s tip speed; automatic electronic folding of main rotor blades and tail pylon in NFH; manual blade folding (automatic optional) in TTH. Large unobstructed internal volume.

FLYING CONTROLS: Quadruplex fly-by-wire controls eliminate cross-coupling between control axes. NH 90 is the world's first FBW transport helicopter. Flight control system developed by Eurocopter France; optimised for NOE flight and minimal vulnerability to small arms damage.

STRUCTURE: All-composites fuselage; fail-safe design of structure, rotating parts and systems for high safety levels. First set of main and tail rotor blades completed for ground testing May 1994; blades have multibox structure and glass and carbon fibre skin; leading-edge box and roved parts of blade prefabricated to reduce time in main mould; resin used reduces material ageing in damp and hot regions; blades designed and manufactured using CATIA CAD/CAM.

Eurocopter France builds rotor, rear gearbox, auxiliary gearbox, fairings and produces wiring, electrical system and air conditioning. Eurocopter Deutschland builds front and centre fuselage and fuel system. Agusta responsible for rear fuselage, main gearbox, hydraulic system and rear ramp; Fokker for tailboom, sliding doors, landing gear and fairings and intermediate gearbox. Assembly lines in

France, Germany (TTH only) and Italy. Portuguese industries supply Agusta and Eurocopter with cabin steps, some electrical wiring and engine supports.

LANDING GEAR: Retractable crashworthy tricycle gear by DAF Special Products with twin-wheel nose unit (tyre size 6.00-6; 10 ply) and single-wheel main units (tyre size 615×225-10; 12 ply); all units retract rearwards; four-balloon emergency flotation gear effective up to Sea State 3. NFH has harpoon-type deck haul-down gear.

POWER PLANT: Two engines; required power from each engine during normal operation at S/L is 1,790 kW (2,400 shp) for 30 minutes, 1,663 kW (2,230 shp) maximum continuous, OEI maximum contingency (2½ minutes) 1,955 kW (2,622 shp), emergency (30 sec) 2,158 kW (2,895 shp) and continuous (1 h) 1,802 kW (2,417 shp). Engine RFP issued April 1992. Decided on 25 January 1994 that NFH and TTH will be developed and qualified with 1,566 kW (2,100 shp) RTM 322-01/9 made by European consortium of Turbomeca, Rolls-Royce, MTU and Piaggio; during development, the 1,521 kW (2,040 shp) GE T700-T6E1 is being qualified by Alfa Romeo and GE to meet Italian military requirements. In June 2001, US\$1 billion contract signed by Eurocopter for supply of production RTM 322-01/9 engines, including initial tranche of 254 power plants for installation in aircraft for France, Germany and Netherlands. Engine control by FADEC.

Transmission rating 2,925 kW (3,922 shp) with both engines, 2,170 kW (2,910 shp) for 30 seconds OEI. Modular main gearbox features integrated lubrication system with innovative integral monitoring and diagnostic system and, like the two accessory gearboxes, can run dry for 30 minutes.

Fuel system has eight Uniroyal crash-resistant self-sealing cells and AFG management system. Total internal capacity approx 2,500 litres (660 US gallons; 550 Imp gallons). Provision for pressure refuelling and hover refuelling. Provision for two external tanks (modified from Alpha Jet) each of 750 litres (198 US gallons; 165 Imp gallons) plus four ferry tanks in cabin.

ACCOMMODATION: Minimum crew one pilot VFR and IFR, NFH crew, pilot, co-pilot/TACCO on flight deck and one system operator in cabin; optionally, two pilots on flight deck and one TACCO, one SENSO in cabin; TTH, two pilots or one pilot and one crewman on flight deck and 14 to 20 equipped troops or one 2 tonne tactical vehicle in cabin. Sogerma crew seats and Autoflug composites passenger seats withstand 22 g or 11 m (35 ft)/s vertical descent crash loading.

SYSTEMS: Full redundancy for all vital systems; hydraulic system has two Vickers mechanically driven and one electrohydraulic and one electric pumps; redundant flight control hydraulic system has two main separated and independent circuits; independent utilities system provides hydraulic power for landing gear and ancillary operations; electrically driven pump as back-up in event of one circuit failing; other circuit permanently supplied in parallel by a utility system pump. Electrical system has two Saft batteries and three 40 kVA Auxilec AC generators (two driven by main gearbox, one by remote accessory gearbox) feeding DC busses through transformer-rectifiers; 70 kW Microturbo Saphyr 100 APU for electrical engine starting, environmental control on ground and emergency use in flight; Kollmorgen-Artus emergency DC generator (driven by remote accessory gearbox); fire detection and suppression in engine bays, APU and cabin. Microtecnica air conditioning system (with two identical independent units) operating in three zones: cockpit, cabin (both vapour cycle) and main avionics bays (filtered external air). Anti-icing of main blades (standard on TTH, optional on NFH) by heated leading-edge mat (heating carbon coating and not metal parts). Anti-icing system also on tail rotor blades

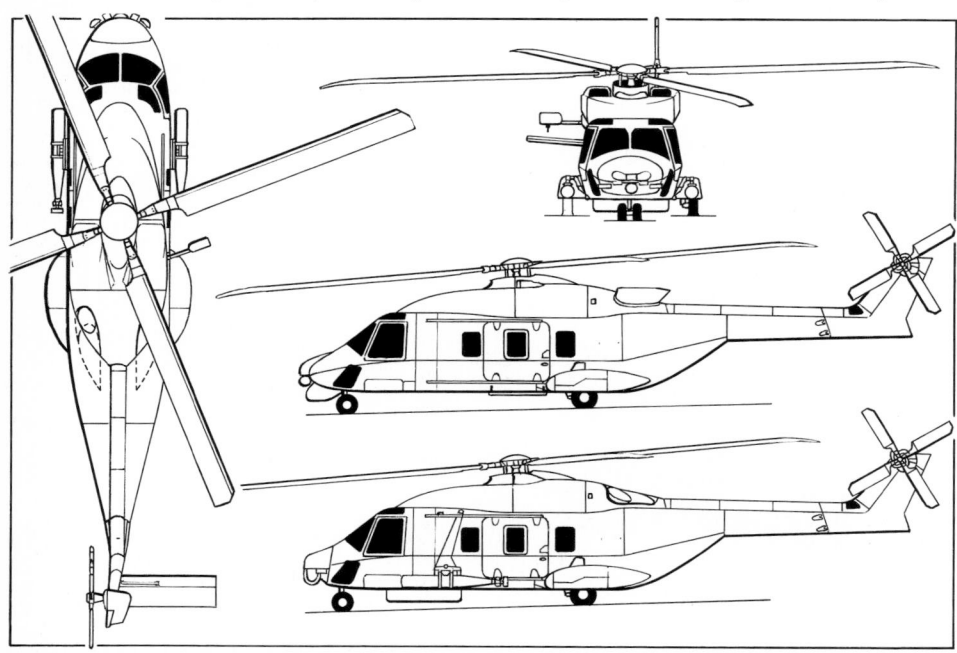

NH Industries NH 90 NFH, with additional side view (upper) of TTH (*Jane's/Mike Keep*)   0121430

and horizontal stabiliser, and on air intakes. Ice protection system effective in maximum conditions defined by DEF-STAN 000-970.

AVIONICS: Core avionic system based on dual MIL-STD-1553B data highways (one for the core avionics system, one for the mission system) allows integration of aircraft and mission equipment through several computers. Eurocopter France responsible for basic avionics; Eurocopter Deutschland for TTH avionics; Agusta-Westland for NFH mission system.

Comms: Integrated management of communications and identification systems. Two V/UHF; V/UHF DF/homer; HF SSB and Thales TSC 2000 IFF. Elmer intercom.

Radar: NFH has Thales ENR 360° track-while-scan surveillance radar with target recognition capability; TTH has Honeywell weather radar.

Flight: Twin INS with embedded GPS and Doppler ground speed sensor; twin air data computers; provision for radio altimeter and MLS. SFIM/Alenia flight control computer and Thales/BAE Systems microwave landing system.

Instrumentation: NVG compatible. Five 200 mm (8 in) square Thales LCDs in cockpit of NFH; up to three more in cabin; four (fifth optional) in TTH. TTH has digital map. Central warning system, two remote frequency indicators and conventional analogue back-up instruments. Provision for helmet-mounted display and sight.

Mission: NFH: Alenia or Thales dipping sonar, high focal-length tactical FLIR, IFF interrogator, Link 11 datalink including ship-to-air umbilical, ESM and telebrief; electronic warfare subsystem. Dedicated video/audio network distributes tactical information to all crew members. TTH: FLIR.

Self-defence: Provision for MAWS, RWR, LWR, Matra chaff/flares and IR jammer. TTH has self-protection suite with passive threat warning system known as EWS (electronic warfare suite) with EADS laser warning receiver, LFK AN/AAR-60 MILDS missile launch warning system and Thales EW processor and radar warning receiver. Active elements of self-

protection suite include chaff/flare dispensers and IR jammer.

EQUIPMENT: Rosemount windscreen washers/wipers; Siren 5 tonne cargo hook and rescue hoist. NFH has Magnaghi 5 ton deck restraint harpoon and automatic main rotor blade and tail folding.

ARMAMENT: TTH can have area suppression and self-defence armament; NFH to carry air-to-surface missiles weighing up to 700 kg (1,543 lb) and anti-submarine torpedoes (including Murène 90); air-to-air missiles optional. Italian NFH will carry Marte Mk 2/S.

DIMENSIONS, EXTERNAL:

Main rotor diameter	16.30 m (53 ft 5½ in)
Tail rotor diameter	3.20 m (10 ft 6 in)
Main rotor blade chord	0.65 m (2 ft 1½ in)
Tail rotor blade chord	0.32 m (1 ft 0½ in)
Length: overall, rotors turning	19.56 m (64 ft 2 in)
fuselage	16.14 m (52 ft 11½ in)
folded (NFH)	13.50 m (44 ft 3½ in)
Height: folded (NFH)	4.10 m (13 ft 5½ in)
overall, tail rotor turning	5.31 m (17 ft 5 in)
Width: max	4.52 m (14 ft 10 in)
over mainwheel fairings	3.63 m (11 ft 11 in)
fuselage max	2.60 m (8 ft 6¼ in)
folded (NFH)	3.80 m (12 ft 5½ in)
Tailplane half-span (stbd)	2.79 m (9 ft 1¾ in)
Wheel track	3.20 m (10 ft 6 in)
Wheelbase	6.08 m (19 ft 11½ in)
Rear-loading ramp: Length	1.58 m (5 ft 2¼ in)
Width	1.78 m (5 ft 10 in)
Cabin door: Width	1.60 m (5 ft 3 in)
Height	1.50 m (4 ft 11 in)

DIMENSIONS, INTERNAL:

Cabin: Length: excl rear ramp	4.80 m (15 ft 9 in)
incl rear ramp	6.17 m (20 ft 3 in)
Min usable width	2.00 m (6 ft 6¾ in)
Min usable height	1.58 m (5 ft 2¼ in)
Volume	18.0 m³ (636 cu ft)

AREAS:

Main rotor disc	208.67 m² (2,246.1 sq ft)
Tail rotor disc	8.04 m² (86.57 sq ft)

WEIGHTS AND LOADINGS:

Weight empty: basic: NFH	6,287 kg (13,860 lb)
TTH	5,945 kg (13,106 lb)
equipped: TTH	6,200 kg (13,669 lb)
Standard fuel: NFH (usable)	2,036 kg (4,489 lb)
Max payload: TTH	2,500 kg (5,511 lb)

T-O weight, TTH, NFH:

typical mission	10,000 kg (22,046 lb)
max	10,600 kg (23,369 lb)
Max disc loading	50.8 kg/m² (10.40 lb/sq ft)

Transmission loading at max T-O weight and power

3.62 kg/kW (5.96 lb/shp)

PERFORMANCE (TTH at 1,000 m; 3,280 ft ISA +15°C, NFH at S/L ISA +10°C, except where stated):

Max cruising speed:	
NFH	157 kt (291 km/h; 181 mph)
TTH	160 kt (298 km/h; 185 mph)
Econ cruising speed: NFH	132 kt (244 km/h; 152 mph)
TTH	140 kt (259 km/h; 161 mph)
Max rate of climb at S/L: NFH	528 m (1,732 ft)/min
TTH	522 m (1,713 ft)/min
Rate of climb at S/L, OEI: NFH	61 m (200 ft)/min
TTH	45 m (148 ft)/min
Hovering ceiling: IGE: NFH	3,140 m (10,300 ft)
TTH	2,960 m (9,720 ft)
OGE: NFH	2,515 m (8,260 ft)
TTH	2,355 m (7,720 ft)

Mission loiter: NFH, 50 n miles (93 km; 58 miles) from base, 20 min reserves   3 h 20 min

Range: max operational:

NFH	490 n miles (907 km; 563 miles)
TTH	430 n miles (796 km; 494 miles)
ferry: NFH	655 n miles (1,213 km; 753 miles)
TTH	650 n miles (1,203 km; 748 miles)

Radius of action:

TTH with 2,000 kg; 4,409 lb payload, 30 min reserves

160 n miles (296 km; 184 miles)

Max endurance: NFH	4 h 45 min
TTH	4 h 35 min

UPDATED

SEPECAT

SOCIETE EUROPEENNE DE PRODUCTION DE L'AVION E. C. A. T.

PARTICIPATING COMPANIES:
BAE Systems: see under UK
Dassault Aviation: see under France
PRESIDENT: J P Weston (BAE)
VICE-PRESIDENT: R Dubost (Dassault)

Anglo-French company formed May 1966 by Breguet Aviation and British Aircraft Corporation to design and

produce Jaguar strike fighter/trainer; production now in India only (see HAL entry in Indian section); RAF upgrades described in previous editions; further data in *Jane's Aircraft Upgrades*.

UPDATED

SEPECAT JAGUAR

Details of the Jaguar International export version appear in the Indian section (which see); data particular to RAF GR. Mk 1A last appeared under SEPECAT heading in the 1993-94 *Jane's*. Further upgrades, including a current

re-engining programme, have taken place to produce GR. Mk 3/3A and T. Mk 4/4A versions, as last described in 2001-02 edition. French Jaguars have been withdrawn from service.

UPDATED

IRAN

AII

AVIATION INDUSTRIES OF IRAN

Km 6, Karaj Old Road (PO Box 14195-111), next to Kan River, Tehran
Tel: (+98 21) 680 92 70 to 680 82 72
Fax: (+98 21) 680 92 68
Web: http://www.aii-co.com
MANAGING DIRECTOR: Dr S H Pourtakdoust
COMMERCIAL DIRECTOR: A Sardari

AII was formed in April 1993. It is affiliated to the Industrial Development and Renovation Organisation (IDRO) of the Iranian Ministry of Industries to produce, repair and maintain various types of aircraft. Following the AVA-101 glider, it flew its first powered aircraft, the AVA-202 two-seat trainer, in mid-1997.

In 1998 the company began construction of its own 75 ha (185.3 acre) airport facility at Azadi, some 65 n miles (120 km; 75 miles) from Tehran; this was completed in 2000, and will be used to support future programmes. AII plans to establish an aeroclub at this location for pilot training and customer demonstrations, and it may eventually become a base for future private owner/operators of the company's products.

Under the designation AVA-303, AII is undertaking licensed assembly of the PZL-18B Dromader agricultural aircraft to supplement those already purchased directly from PZL. By the end of 2002 the first set of Polish-supplied components had been received and assembled. The programme is expected to yield 12 to 15 aircraft for the Iranian Ministry of Agriculture, built at a new production line to be located at Azadi. AII plans some design changes, including a modified cockpit canopy.

Other programmes stated to be current in late 2002 included a new 19-seat transport (thought to be the

Fourth example of the AII AVA-202 trainer *(Robert Hewson)*   NEW/0530167

AVA-404); a small helicopter (AVA-505) similar in appearance to the HESA Shahed 278; a new six-seat aircraft; and studies on a joint-venture project for a 150-passenger transport.

UPDATED

AII AVA-202

TYPE: Primary prop trainer/sportplane.
PROGRAMME: Intended for domestic market to avoid dependence on foreign imports; prototype (originally known as IR-02: see 1997-98 *Jane's All the World's Aircraft*) made first flight on 3 June 1997; second had been built by 3 January 1999 when one was lost in a fatal accident while testing an emergency parachute. Series production launched in 2000. Made its public debut (fourth aircraft shown) at the Iran International Air Show in

October/ November 2002, at which time certification still awaited.
CUSTOMERS: Four completed by late 2002.
DESIGN FEATURES: Similar configuration to Van's RV-6A (US section), but with greater wing span; meets JAR 22 and JAR-VLA standards. Intended for pilot training, general aviation, surveillance and recreational flying.

NASA aerofoil sections: 63₂-215 (wing), 63₂-015 (horizontal tail) and 63-010 (vertical tail).
FLYING CONTROLS: Conventional and manual. Trim tab in port elevator. Electrically actuated single-slotted flaps.
STRUCTURE: Mainly aluminium alloy; steel alloys and glass fibre in some areas.
LANDING GEAR: Non-retractable tricycle type. Self-sprung cantilever legs; castoring nosewheel with shimmy damper. Hydraulic mainwheel disc brakes. Streamlined composites fairings over all three wheels.

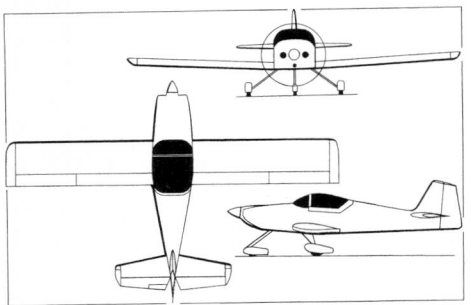

Aviation Industries of Iran AVA-202 trainer
(Jane's/Paul Jackson)   0062613

POWER PLANT: One Textron Lycoming O-320-B2B flat-four engine (119 kW; 160 hp at 2,700 rpm), driving a Sensenich M 74 DM two-blade, fixed-pitch metal propeller. Optional 112 kW (150 hp) Lycoming O-320-E2D.
ACCOMMODATION: Two seats side by side. Dual controls. Wraparound windscreen; one-piece rearward-sliding canopy. Cabin heated and ventilated. Rear baggage compartment with internal access.
SYSTEMS: 12 V electrical system
AVIONICS: To customer requirements; can include Bendix/King KX 155 nav/com radio, ELT, and ADF/VOR/GPS
DIMENSIONS, EXTERNAL:

Wing span	8.74 m (28 ft 8 in)
Wing chord, constant	1.24 m (4 ft 0¾ in)
Wing aspect ratio	7.0
Length overall	6.02 m (19 ft 9 in)
Height overall	2.04 m (6 ft 8¼ in)
Propeller diameter	1.88 m (6 ft 2 in)

AREAS:

Wings, gross	10.87 m² (117.0 sq ft)
Vertical tail surfaces (total)	1.04 m² (11.19 sq ft)
Horizontal tail surfaces (total)	2.40 m² (25.83 sq ft)

WEIGHTS AND LOADINGS:

Weight empty	500 kg (1,102 lb)
Max T-O and landing weight	750 kg (1,653 lb)
Max wing loading	69.0 kg/m² (14.13 lb/sq ft)
Max power loading	6.29 kg/kW (10.33 lb/hp)

PERFORMANCE (estimated):

Max level speed at S/L	140 kt (259 km/h; 161 mph)
Max cruising speed, 75% power	135 kt (250 km/h; 155 mph)
Stalling speed, flaps down	45 kt (84 km/h; 52 mph)
Max rate of climb at S/L	457 m (1,500 ft)/min
Service ceiling	6,400 m (21,000 ft)
T-O run	250 m (820 ft)
Range with max fuel	540 n miles (1,000 km; 621 miles)

UPDATED

AII AVA-404

TYPE: Twin-turboprop light transport.
PROGRAMME: In design stage 1996; no manufacturing timetable announced by early 2002. See 2002-03 and earlier editions for last known details.

UPDATED

AII AVA-505 THUNDER

TYPE: Light utility helicopter.
PROGRAMME: In design stage by 1996, under former designation of IR-H5; first flight then planned in 1999, but no evidence of manufacture of first flight reported by early 2002. See 2002-03 and earlier Jane's for description..

UPDATED

AIO

AVIATION INDUSTRIES ORGANISATION OF THE ISLAMIC REPUBLIC OF IRAN
107 Gharani Avenue, Tehran
Tel: (+98 21) 881 04 81 to 881 04 87
Fax: (+98 21) 882 50 46
e-mail: aviation-id@isiran-net.com

AIO, a subdivision of the Iranian Ministry of Defence, is the policy maker, co-ordinator and planner for the manufacturing, overhaul, repair and support of aircraft by Iran's aviation industry. Created to counter problems arising from the Gulf wars and subsequent economic sanctions, its primary concern has been in supplying spares for aircraft in service and obtaining the technologies required for manufacturing various types of aircraft for the Iranian armed forces. However, it has now assumed a much greater role in aircraft development and production, including civil programmes such as the Ir.an-140.

According to Jane's sources in December 2001, the Iranian government was to consider a bill calling for the establishment of an Aerospace Organisation of the Defence and Armed Forces Logistics Ministry, into which all aerospace manufacturing and research units, as well as Iranian MoD factories, would be merged.

Nothing has been heard for some years of earlier proposals for a co-production programme involving the Ilyushin Il-114, but talks between Iranian officials and the Tupolev JSC have continued concerning possible Iranian participation (including local assembly) in the Tu-204 and manufacture of up to 100 Tu-334s over a 15-year period.

SUBSIDIARIES:
Iran Aircraft Manufacturing Industries (IAMI)
HEAD OFFICE: 5th floor, 107 Gharani avenue (PO Box 14155-5568), Tehran
Tel: (+98 21) 882 73 75 and 882 83 55
Fax: (+98 21) 882 83 55
WORKS: PO Box 83145-311, Shahin Shahr, Esfahan
Tel: (+98 311) 221 42 18
Fax: (+98 311) 221 42 19
DEPUTY MANAGING DIRECTOR: Dr M A Vaziri
E-MAIL: info@hesaco.com

Otherwise known as **HESA**, this state-owned company is controlled jointly by the Iranian Ministry of Defence and Ministry of Industries, and is the main aircraft production centre of the AIO, with divisions responsible for trainers, helicopters, the Ir.An-140 transport and UAVs. Major programmes include the HT-80, a replacement for (but based on) the Pilatus PC-7 Turbo-Trainer.

Esfahan factory construction initiated 1975 by Bell Helicopter Textron for production of Bell 214, but suspended 1978 after Islamic revolution when only 11 per cent completed. Reactivated 1983; began piston engine manufacture 1988 and Northrop F-5B overhaul 1989. Helicopter (Bell AH-1J and other) and hovercraft overhaul started 1993; first production piston engine completed in 1996.

Following a February 1996 agreement, a factory was set up, with Ukrainian government assistance, for licensed assembly of the Antonov An-140 twin-turboprop transport. A similar agreement with MAPO organisation, for manufacture of 60 Klimov TV7-117VMA turboprop engines, was announced at the 1997 Paris Air Show and valued at US$180 million. These engines were said to be for installation on 'Ukrainian- and Russian-built' airliners.

Qods Aviation Industries (QAI)
PO Box 13445-873, 4 Km Karaj Road, Tehran
Tel: (+98 21) 450 33 06
Fax: (+98 21) 450 33 07
e-mail: mohajer@isiran.com

Established 1984 to design, develop and manufacture surveillance/reconnaissance and attack RPV/UAVs and aerial targets (see *Jane's Unmanned Aerial Vehicles and Targets*). Subsidiary of IAMI/HESA. Also manufactures parachutes for various applications. Factory area 40,000 m² (430,550 sq ft); 1998 workforce approximately 400.

Iran Aircraft Industries (IACI)
PO Box 14155-1449, Tehran
Tel: (+98 21) 603 56 06
Fax: (+98 21) 600 00 75
e-mail: karhava@isiran.com
MANAGING DIRECTOR: Nasser Marzebani
AERO-ENGINE PROGRAMMES MANAGER: Eng Amiri

Established 1970; first Iranian aviation company to obtain ISO 9002 certification. Known locally as **SAHA**, factory has overhauled, maintained and repaired approximately 4,000 aircraft and aero-engines. Like IAMI/HESA, it is controlled by the Ministry of Defence. It occupies a 45,000 ha (111,195 acre) site, and had a 1998 workforce of more than 1,000. It is designated to produce the Tu-204, and the Tu-334 and its D-436 engines, if these programmes come to fruition. Earlier aircraft projects included the Fajr (Dawn) piston-engined trainer described briefly in the 1990-91 *Jane's All the World's Aircraft*.

Turbine Engine Manufacturing (TEM)
Manufactures engine components at a 100,000 m² (1.08 million sq ft) site with shop floor area of 33,000 m² (355,200 sq ft). Also has well-equipped research laboratories and active design team. May be producer of Tolloue 4 small jet engine, unveiled in late 1999 as power plant for UAVs and aerial targets.

Iran Helicopter Support and Renewal Company (IHSRC)
PO Box 13185-1688, Mehrabad Airport Road, Meradj Avenue, Tehran
Tel: (+98 21) 600 34 14
Fax: (+98 21) 601 58 62
e-mail: panha@isiran.com
MANAGING DIRECTOR: Nasser Akhavan
DEPUTY MANAGING DIRECTOR: Abdollah Kheirollahi

Established 1969 with assistance of Agusta (Italy) and Bell Helicopter (USA); overhaul and maintenance of Agusta-Bell 205, 212, 214A/B/C; Bell 206/206L JetRanger/LongRanger and AH-1J/JT Cobra series; Boeing CH-47C Chinook; and Sikorsky SH-3D Sea King and RH-53D Sea Stallion. In 1999, it was stated that experience gained with extensive rebuild and refurbishment of AH-1J SeaCobras had given IHSRC (known locally as **PANHA**) the potential for full manufacture, if so required.

UPDATED

IACI (SAHA) SHAFAGH
English name: The light before dawn
Russian name: Rassvet
TYPE: Advanced jet trainer/light attack jet.
PROGRAMME: Existence revealed at Aero 2000 air show in Tehran; name also transliterated as Safak or Shafak. Programme controlled by Tehran's Aviation University Complex (part of the Malek Ashtar University of Technology). Design based on earlier (twin-engined) proposals from Russian OKB headed by Fatidin Mukhamedov, with characteristically unorthodox configuration. Wind tunnel testing of one-seventh scale model in Tehran followed by completion of full-size mockup. Integration of landing gear, hydraulic system and

HESA model of the HT-80 Iranian development of the PC-7, targeted for roll-out in about 2005
(Robert Hewson)   NEW/0530184

Underside view of Shafagh model showing external weapon stations (*Robert Hewson*)
NEW/0531277

View emphasising the unique circular centre-section of the Shafagh's wing (*Robert Hewson*)
NEW/0531276

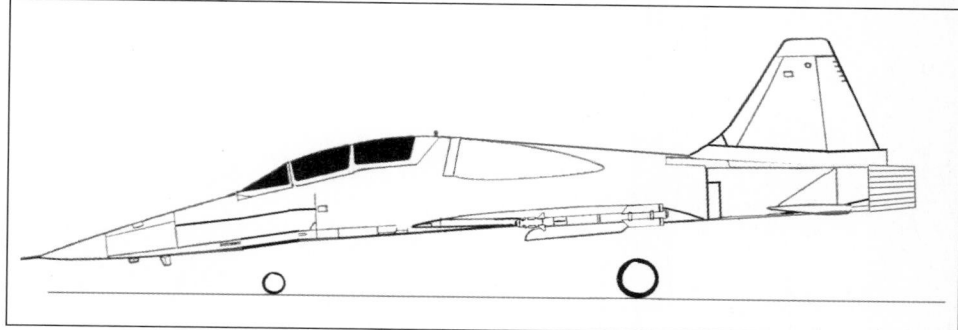

Possible appearance of Azarakhsh fighter, based on some reports (*Jane's Defence Weekly*)
0067681

avionics under way in late 2002. Displayed in model form at Iran International Air Show in late 2002, at which time prototype roll-out targeted for 2008.

CUSTOMERS: Potentially, Russian Air Forces (to replace Aero L-39 C) and Iranian Air Force.

DESIGN FEATURES: Close-coupled configuration. Circular wing centre-section with pronounced LERX and tapered outer panels; twin outward-canted fins and rudders and (probably all-moving) tailplane halves.

POWER PLANT: One RD-33 turbofan (General Electric J79 turbojet in initial design studies).

ACCOMMODATION: Crew of two in tandem.

AVIONICS: Row of three MFDs; HUD; HOTAS controls on centrally mounted column.

ARMAMENT: Seven external stores stations (one on centreline and three beneath each wing).

DIMENSIONS, EXTERNAL (approx):
Wing span   12 m (39.4 ft)
Length overall   14 m (45.9 ft)
WEIGHTS AND LOADINGS (approx):
Weight empty   5,000 kg (11,023 lb)
PERFORMANCE (estimated):
Max level speed   M1.3
UPDATED

IAMI (HESA) AZARAKHSH
English name: Lightning
TYPE: Attack fighter.

PROGRAMME: Development said to have begun in 1986; programme name *Oaj* (Zenith); stated to have passed acceptance tests by April 1997 and completed flight testing in June. Designers' names given as Morteza Sanai and Morteza Satari. Announced by Islamic Republic News Agency (IRNA) in late September 1997, quoting Army Chief of Staff Maj Gen Ali Shahbazi, that the aircraft had taken part, earlier that month, in domestic military exercises, during the course of which it had dropped two 113 kg (250 lb) napalm bombs. Announced at that time that production was to begin 'soon', but this now known to be incorrect.

In June 1999, Air Force Gen Habibollah Baghai told Iranian journalists that series production had begun, but stated by Lt Gen Iraj Ossareh (IRIAF Deputy Commander) in May 2000 that four aircraft completed at that time, with a fifth under construction. Current plans then to build up to 30 over next three years. Taxi trials reported completed by May 2001, with first flight believed imminent; at that time, fifth prototype said to be completed and sixth under construction. An Iranian news report of 15 August 2001, stated that production had been deferred pending resolution of 'serious structural problems'. At the same time, a parallel fighter programme known as **Owaz** was also reported.

Following details highly provisional and unconfirmed:

DESIGN FEATURES: In early 2000, Azarakhsh was reported to be a 10 to 15 per cent scaled-up, reverse-engineered Northrop F-5F, re-engined with two Klimov RD-33 turbofans, having shoulder-mounted air intakes. (However, earlier reports had claimed it to be based on the McDonnell Douglas F-4 Phantom.)

STRUCTURE: Extensive use of composites reported.

ACCOMMODATION: Two seats in tandem.

AVIONICS: *Radar:* Reported to be locally modified Phazotron N019ME Topaz with BVR capability and enhanced air-to-ground performance.

Self-defence: Said to have chaff/flare dispenser(s) in fairing beneath rear fuselage.

WEIGHTS AND LOADINGS (approx):
Weight empty   8,000 kg (17,637 lb)
Max external stores load 4,000-4,400 kg (8,818-9,700 lb)
PERFORMANCE (estimated):
Max level speed   M1.6 class
UPDATED

IAMI (HESA) SHAHBAZ
English name: Aurora
TYPE: Basic jet trainer.

PROGRAMME: During a military exercise in May 1998, Brig Gen Reza Pardis of the Islamic Republic of Iran Air Force (IRIAF) was quoted by the *Iran Daily* as saying that prototypes of a jet trainer had then completed 120 test flights and that the aircraft was to enter production shortly. IRIAF Gen Habibollah Baghai told Iranian journalists in June 1999 that the trainer was then in production, whereas Lt Gen Ossareh stated in May 2000 that production would start shortly. The aircraft is sometimes referred to by the name Dorna, the programme title under which it was developed. It reportedly resembles the Chinese K-8, but no further details had emerged at the time of going to press.

VERIFIED

IAMI (HESA) IR.AN-140 FARAZ
TYPE: Twin-turboprop transport.

PROGRAMME: Licence-built Antonov An-140 (see Ukrainian section). Selected by Iran in 1995, against competition from 12 other contenders; initial agreement leading to Iranian manufacture signed February 1996; first stage is assembly from imported KhAPO CKD kits at rate of up to 12 per year (first two kits received by early 2000); long-term objective is gradual progress to local parts manufacture; Iranian engineers installed at Antonov from 1997. Contract announced late 1998 for initial quantity of 80 aircraft, for Iranian domestic (civil and military) use only, but possibility of additional production for export to CIS states has been suggested. In May 1999, Iran Electronics and Leninets (St Petersburg) signed contract for Iran 140 avionics development. Assembly of first aircraft (c/n 90-01) began at Esfahan March 1999 and completed by September 2000; maiden flight achieved on 7 February 2001. Certification and service entry targeted for 2002. Second aircraft rolled out 26 December 2002; third almost complete at that time.

CURRENT VERSIONS: Total of 14 scheduled for completion from CKD kits by 2004. Offered for passenger (including VIP) civil transport, aeromedical (ambulance or flying hospital) use, cargo/passenger or parachutist carrying, photogrammetry, geosurvey, environmental research, fisheries protection and pollution control; military versions could include staff or personnel transport (**Ir.-140TC**), coastal and maritime patrol (**Ir.an-140MP**), radar patrol/ missile guidance, elint and crew training. Possibility of future 68-passenger stretched version.

COSTS: Reportedly US$9 million unit price (2001); set-up of production line US$273 million.

Model of the projected Ir.an-140MP maritime patrol version (*Robert Hewson*)
NEW/0531417

IAMI (HESA) 90-01, the first Iranian-assembled Ir.an-140 *(Robert Hewson)*   NEW/0530169

Display model of the Ir.an-140TC transport version *(Robert Hewson)*   NEW/0530183

CUSTOMERS: Launch customers initially reported to be Iran Air and Iran Asseman Airlines, of which latter signed order for 20 in October 1999, but Iran Air has since declined to operate the type. Kish Air now said to be first recipient, with delivery of first two aircraft expected in late 2002 or early 2003. Likely also to be procured, if funding is forthcoming, by Iranian Air Force and Navy (potential roles tactical transport, maritime patrol and EW). Stated in October 2000 that orders had been received for 127.

POWER PLANT: Iranian company quotes Klimov/ZMKB Progress TV3-117VMA-SBM1 engines, as in Antonov-built version, but Antonov spokesman indicated in 1998 that Iranian-built aircraft would have an uprated (1,864 kW; 2,500 shp) version of this engine for improved hot/high performance.

AVIONICS: Being negotiated with various European and Russian suppliers.
UPDATED

IAMI (HESA) PARASTU
English name: Swallow
TYPE: Basic prop trainer.
PROGRAMME: First revealed at Mehrabad Air Show in February 1996, when six or seven examples were reported to have been completed; reverse-engineered Raytheon (Beech) F33A Bonanza (see US section of 1997-98 and earlier *Jane's*), with modifications that include performance-improving winglets. Programme initiated September 1987; first flight 5 April 1988. According to an IRNA announcement in September 1997, series production was then due to begin in the near future. However, according to a senior IRIAF spokesman in May 2000, the 11th aircraft was then in manufacture, with full-scale production imminent. Fourteen or more were reported to be in service by April 2001.
VERIFIED

IAMI (HESA) SHAHED 278
TYPE: Light utility helicopter.
PROGRAMME: Three prototypes, first of which made maiden flight late 1997 or early 1998; most-flown prototype had accumulated 200 hours by March 2001. Public debut late 2002 as HESA 278, having previously been reported as Panha 287.
DESIGN FEATURES: Four-seater; intended for dual military/civil use, though designed primarily to meet military requirements. Designed entirely in Iran, according to manufacturer, although some sources report use of locally made airframe and dynamic components of Bell 206 JetRanger.
POWER PLANT: Single turboshaft. Prototypes have evaluated Rolls-Royce 250 and equivalents from Turbomeca and a Russian supplier, but final selection apparently not yet made. Fuel capacity 287 litres (75.8 US gallons; 63.1 Imp gallons).

PERFORMANCE:
Max level speed: at S/L	129 kt (240 km/h; 149 mph)
at FL50	124 kt (230 km/h; 142 mph)
Max rate of climb at S/L	762 m (2,500 ft)/min
Service ceiling	6,400 m (21,000 ft)
Hovering ceiling: IGE	6,218 m (20,400 ft)
OGE	4,865 m (15,960 ft)
Range	183 n miles (340 km; 211 miles)

UPDATED

IHSRC (PANHA) 2061
TYPE: Light utility helicopter.
PROGRAMME: Announced, but not seen, at launch of Shabaviz 2-75 in December 1998; five-seat, reverse-engineered or locally manufactured version of Bell 206 JetRanger, also with non-US power plant. Said to be intended for reconnaissance and training, and also officially began series production on 4 May 1999; apparently only four ordered by April 2001. Armed version, with AT-2 'Swatter' ATM launchers, made debut in 'Zolfaqar 3' defence exercise, September 2000.
UPDATED

IHSRC (PANHA) 2091
TYPE: Attack helicopter.
PROGRAMME: Almost certainly a reverse-engineered Bell (Model 209) AH-1J, possibly with indigenously developed avionics and weaponry improvements. One delivered to Iranian Army by August 2001, according to Ministry of Defence announcement.
UPDATED

IHSRC (PANHA) SHABAVIZ 2-75
English name: Owl
TYPE: Multirole medium helicopter.
PROGRAMME: First prototype built 1993 and production authorised 1996, but existence not revealed until presentation and flying display in Tehran on 10 December 1998; claimed as "first domestically produced" helicopter, but visibly a Bell 205 except for imported (non-US) engines which were not identified. Is probably an upgraded or reverse-engineered Agusta-Bell 205A, of which over 20 supplied to Iran before 1979 Islamic revolution; this appeared to be first example completed locally.

WEIGHTS AND LOADINGS:
Weight empty	682 kg (1,504 lb)
Normal T-O weight	1,451 kg (3,198 lb)

HESA Shahed 278 light utility helicopter *(Robert Hewson)*   NEW/0530166

First official photograph of the Panha 2061 (left) and Shabaviz 2-75 *(IHSRC)*   NEW/0110512

Details disclosed at unveiling included payload of 14 passengers or 2,500 kg (5,511 lb), ceiling of 3,800 m (12,460 ft) and unrefuelled range of about 270 n miles (500 km; 310 miles). Intended for both military and civil duties, including an armed version. Significance of 2-75 in designation not apparent. Series production officially launched on 4 May 1999 with opening of a military production plant to build this and the 2061 (which see), but reported in April 2001 that only four had then been ordered, of which two delivered by August of that year.

UPDATED

IHSRC (PANHA) SHAHED 274

Now thought to be product of IRGC and described under that organisation's heading.

VERIFIED

IHSRC (PANHA) SANJAQAK

English name: Dragonfly
TYPE: Two-seat helicopter.
PROGRAMME: Announced January 1997 and reported in mid-1997 to be nearing flight test at a state research centre

in Esfahan; max T-O weight 450 kg (992 lb). According to a Jane's source in March 2001, the Sanjaqak bears a close resemblance to the Bell 47 and is now in service, possibly for pilot training. However, circumstantial evidence would suggest a relationship with the Dragon Fly 333, described in the Italian section.

UPDATED

DORNA

H F DORNA COMPANY

20 6th Sarvestan Street (PO Box 16315-345), Pasdaran Avenue, Tehran 16619
Tel: (+98 21) 285 48 27
Fax: (+98 21) 284 18 31
e-mail: info@hfdorna.com
Web: http://www.hfdorna.com
MANAGING DIRECTOR: Yaghoob Antesary
CHIEF DESIGNER: Farid Najmabadi

Company established March 1989 (see 1991-92 *Jane's*) to specialise in aircraft design and development, and in composites materials technology. It has 3,000 m² (32,300 sq ft) of factory space and had a workforce of 79 in July 2001. Current product is two-seat Blue Bird.

UPDATED

DORNA D139-PT1 BLUE BIRD

TYPE: Two-seat lightplane.
PROGRAMME: Launched 1994; prototype manufacture started mid-1994; first flight was made on 27 July 1998; prototype had accumulated 32 hours' flying in 45 flights by end of November that year. Certification to JAR 21 standards by the Iranian CAA was awarded on 2 December 2001, and initial production is under way.
CUSTOMERS: First six production aircraft sold to Iranian civil aeroclubs for pilot training; manufacture of these was in process in mid-2001. Further 12 orders expected during 2002.

COSTS: Programme cost by November 1998 estimated at US$1.5 million excluding R&D; basic aircraft quoted at US$108,000 in December 2001.
DESIGN FEATURES: Conventional low-wing, fixed-gear lightplane, designed for low-cost, easy-to-fly/maintain operation.
  Wings have constant chord, NACA 63-215 aerofoil section, 2° 30′ dihedral from roots, 2° incidence and −2° twist.
FLYING CONTROLS: Conventional and manual. Ailerons and elevators actuated by pushrods; rudder by cables; elevators have electric trim. Plain flaps.
STRUCTURE: All-composites. Two-spar wing, single-spar horizontal tail, monocoque fuselage.
LANDING GEAR: Non-retractable tricycle type, with Goodyear 5.00-5 tyres (6 ply on mainwheels, 4 ply on nosewheel). Cantilever self-sprung steel leg on each unit. Cleveland mainwheel brakes. Composites speed fairings on all three wheels.
POWER PLANT: One 84.6 kW (113.4 hp) Rotax 914 F3 flat-four engine driving an MT Propeller MTV-21-A/175-05 two-blade variable-pitch propeller at 5,800 rpm; three-blade MTV-6-A/172-08 variable-pitch propeller (as on prototype) optional. Fuel tank in each wing, combined capacity 128.7 litres (34.0 US gallons; 28.3 Imp gallons); gravity filling point on each tank. Oil capacity 3.0 litres (0.8 US gallon; 0.7 Imp gallon).
ACCOMMODATION: Two seats side by side, with baggage compartment aft of seats. Narrow hardtop roof to cockpit; access via upward-opening window each side, each with smaller non-opening window to rear.

SYSTEMS: Electrical system: 12 V 20 A integral AC generator, 12 V 40 A external alternator and 35 Ah battery.
AVIONICS: *Comms:* Radio.
  Instrumentation: Conventional VFR.

DIMENSIONS, EXTERNAL:
Wing span	9.45 m (31 ft 0 in)
Wing chord, constant	1.17 m (3 ft 10 in)
Wing aspect ratio	8.1
Length overall	6.17 m (20 ft 3 in)
Fuselage max width	1.12 m (3 ft 8 in)
Height overall	2.17 m (7 ft 1½ in)
Wheel track	1.94 m (6 ft 4½ in)
Wheelbase	1.51 m (4 ft 11½ in)
Propeller diameter	1.72 m (5 ft 7¾ in)
Propeller ground clearance	0.18 m (7 in)

DIMENSIONS, INTERNAL:
Cockpit: Length	1.09 m (3 ft 7 in)
Max width	1.02 m (3 ft 4¼ in)

AREAS:
Wings, gross	11.06 m² (119.0 sq ft)
Ailerons (total)	0.95 m² (10.23 sq ft)
Trailing-edge flaps (total)	1.54 m² (16.58 sq ft)
Fin	0.42 m² (4.52 sq ft)
Rudder	0.37 m² (3.98 sq ft)
Tailplane	0.84 m² (9.04 sq ft)
Elevators (total)	0.515 m² (5.54 sq ft)

WEIGHTS AND LOADINGS:
Weight empty, equipped	490 kg (1,080 lb)
Max fuel weight	90 kg (198 lb)
Max T-O and landing weight	750 kg (1,653 lb)
Max wing loading	67.8 kg/m² (13.89 lb/sq ft)
Max power loading	8.87 kg/kW (14.58 lb/hp)

PERFORMANCE (estimated):
Never-exceed speed (VNE)	173 kt (321 km/h; 200 mph)
Max level speed at 3,660 m (12,000 ft)	122 kt (225 km/h; 140 mph)
Max cruising speed at 3,660 m (12,000 ft)	113 kt (209 km/h; 130 mph)
Econ cruising speed at 3,660 m (12,000 ft)	87 kt (161 km/h; 100 mph)
Stalling speed at S/L, flaps down	46 kt (84 km/h; 52 mph)
Max rate of climb at S/L	229 m (750 ft)/min
Service ceiling	4,265 m (14,000 ft)
T-O run	275 m (900 ft)
Landing run	244 m (800 ft)
Range with max fuel	695 n miles (1,287 km; 800 miles)

UPDATED

Easy cockpit access is one of the Blue Bird's features
0110999

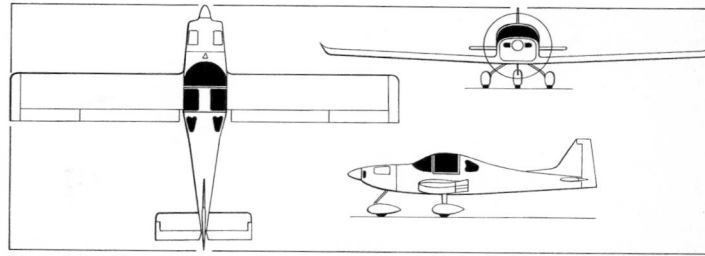
Dorna Blue Bird two-seat light aircraft (*Jane's/James Goulding*)
0062614

Three-quarter front view of the Rotax-engined Blue Bird
0111000

FACI

FAJR AVIATION AND COMPOSITES INDUSTRY

Km 5, Karaj Road, PO Box 13445-885, Tehran
Tel: (+98 21) 465 94 61
Fax: (+98 21) 465 94 60
e-mail: faci@fajr-ind.com
Web: http://www.fajr-ind.com
GENERAL DIRECTOR: Mohammad Bani Najarian

FACI established 1991, with objective of developing all-composites aircraft in Iran. Plant has 30,000 m² (323,000 sq ft) of covered space at Mehrabad International Airport; total workforce is 150. Its first design, the F-3, entered

production in 2001 and made its public debut in October 2002.

UPDATED

FACI FAJR F-3

English name: Dawn
TYPE: Four-seat lightplane.
PROGRAMME: Design began 1992; construction of prototype started 1993; first flight 26 January 1995; further airframe for static testing; JAR 23-standard domestic type certificate awarded September 2000; construction of first production aircraft began 2001; first deliveries due June 2002. Development plans so far announced include introduction of retractable landing gear and some weight

reduction measures to improve cruise performance; flight tests of modified aircraft were scheduled for June 2002.
CUSTOMERS: Domestic orders for 20 by January 2003. Five reportedly completed by that time, of which two delivered.
DESIGN FEATURES: Intended for pilot training and utility roles. Designed for low cost and easy maintenance.
  Horstmann-Quast HQ-42E advanced laminar aerofoil section, with thickness/chord ratio of 16 per cent and 2° 17′ sweepback. Dihedral 6° 30′ from roots; twist −2° 30′; incidence 2° 18′. Raked wingtips.
FLYING CONTROLS: Conventional and manual. Tab on each aileron. Electrically actuated trim tab on rudder and each elevator. Fixed-incidence tailplane. Electrically actuated Fowler flaps.

STRUCTURE: All-composites; monocoque fuselage; high-stress parts of glass fibre airframe are reinforced with carbon fibre.

LANDING GEAR: Retractable tricycle-type. Goodyear tyres: size 15×6.00-6 (main) and 5.00×5 (6 ply) (nose); pressures 4.41 bar (64 lb/sq in) and 3.38 bar (49 lb/sq in) respectively. Nose gear is steerable (±18°) through a mechanical linkage to the rudder pedals. Parker Hannifin brakes are direct-operation, hydraulically actuated disc type. Minimum ground turning radius 2.75 m (9 ft 0¼ in).

POWER PLANT: One Textron Lycoming AEIO-540-L1B5 flat-six engine, flat rated at 201 kW (270 hp), driving a three-blade Hoffmann HOV 123K-V-K200-AH constant-speed propeller. Fuel capacity 212 litres (56.0 US gallons; 46.6 Imp gallons) standard; 292 litres (77.1 US gallons; 64.2 Imp gallons) optional. Oil capacity 15.1 litres (4.0 US gallons; 3.3 Imp gallons).

ACCOMMODATION: Four persons, in pairs. Dual controls. Two upward-hinged window/doors of glass fibre. Cabin heated and ventilated.

SYSTEMS: 24 to 28 V DC electrical system powered by Gill 70 A 10 Ah alternator. Oxygen system.

AVIONICS: To customer's requirements. Options include Honeywell VHF com, ADF and GPS.

DIMENSIONS, EXTERNAL:

Wing span	10.50 m (34 ft 5½ in)
Wing chord: at root	1.76 m (5 ft 9¼ in)
at tip	0.94 m (3 ft 1 in)
Wing aspect ratio	7.9
Length overall	8.07 m (26 ft 5¾ in)
Height overall	3.05 m (10 ft 0 in)
Tailplane span	3.87 m (12 ft 8¼ in)
Wheel track	2.08 m (6 ft 10 in)
Wheelbase	1.78 m (5 ft 10 in)
Propeller diameter	2.10 m (6 ft 10¾ in)
Propeller ground clearance	0.175 m (7 in)

DIMENSIONS, INTERNAL:

Cabin: Length	2.18 m (7 ft 1¾ in)
Max width	1.25 m (4 ft 1¼ in)
Max height	1.53 m (5 ft 0¼ in)

AREAS:

Wings, gross	14.02 m² (150.9 sq ft)
Ailerons (total, incl tabs)	0.46 m² (4.95 sq ft)
Flaps (total)	0.95 m² (10.23 sq ft)
Fin	2.33 m² (25.08 sq ft)
Rudder (incl tab)	0.91 m² (9.80 sq ft)
Tailplane	2.95 m² (31.75 sq ft)
Elevators (total, incl tabs)	1.16 m² (12.49 sq ft)

WEIGHTS AND LOADINGS:

Weight empty	1,100 kg (2,425 lb)
Max fuel weight: standard	141 kg (311 lb)
optional	204 kg (450 lb)
Max T-O and landing weight	1,580 kg (3,483 lb)
Max wing loading	112.7 kg/m² (23.08 lb/sq ft)
Max power loading	7.85 kg/kW (12.90 lb/hp)

PERFORMANCE:

Never-exceed speed (V_{NE})	200 kt (370 km/h; 230 mph)
Max level speed at FL40	160 kt (296 km/h; 184 mph)

The FACI Fajr F-3 four-seat low-wing lightplane *(Robert Hewson)*   *NEW*/0530168

Cruising speed at 65% power at FL90		T-O to 15 m (50 ft)	370 m (1,215 ft)
	144 kt (267 km/h; 166 mph)	Landing from 15 m (50 ft)	720 m (2,365 ft)
Stalling speed, power off:		Landing run	360 m (1,180 ft)
flaps up	67 kt (125 km/h; 78 mph)	Range, 65% power at FL100, standard fuel	
flaps down	56 kt (104 km/h; 65 mph)		610 n miles (1,129 km; 702 miles)
Max rate of climb at S/L	314 m (1,030 ft)/min	Endurance	6 h 15 min
Service ceiling	5,480 m (18,000 ft)	*g* limits	+4.4/−1.76
T-O run	310 m (1,020 ft)		*UPDATED*

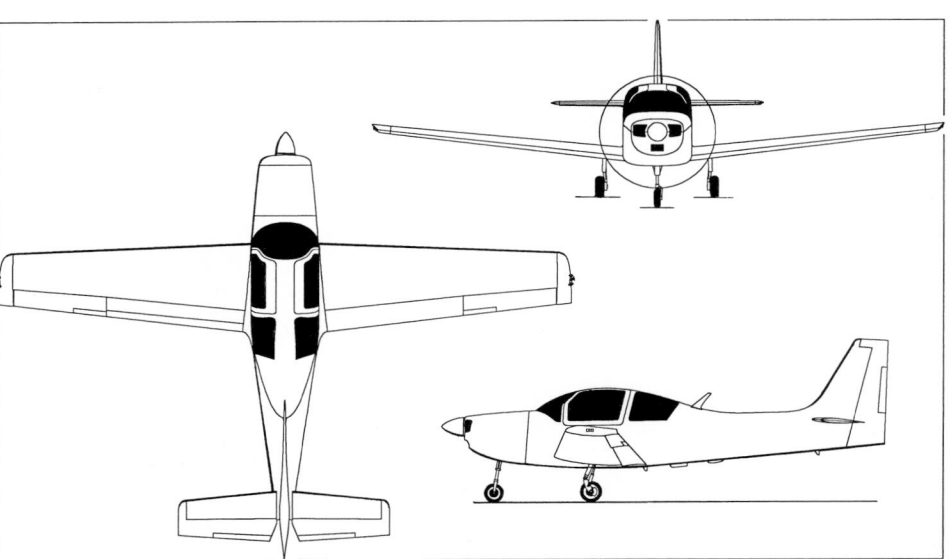

General arrangement of the FACI F-3   0110841

IRGC

IRAN REVOLUTIONARY GUARD CORPS

Development of the Shahed 274 helicopter is reported to have been undertaken by the Pasdaran, or Revolutionary Guards. The organisation may also be responsible for a new transport helicopter.

VERIFIED

IRGC SHAHED 274

TYPE: Light utility helicopter.

PROGRAMME: Sponsored (as X-5) by the Institute of Industrial Research and Development of the IRGC; reportedly due to

have made its first flight in 1997. First aircraft (71-832) handed over to IRGC 16 September 1999. International public debut, in Tehran, 30 December 2000. At least two more (74-001 and '002) in service by end of 2001. Further public appearance in air show on Kish Island, October/November 2002.

CUSTOMERS: Total of 20 (other sources suggest 30) reportedly to be built by end of 2004.

DESIGN FEATURES: Two-blade main and tail rotors; fully enclosed cabin and tailboom; upper and lower vertical fins. Intended applications include observation, rescue and light cargo-carrying.

LANDING GEAR: Twin-skid type.

POWER PLANT: One 313 kW (420 shp) Rolls-Royce 250-C20B turboshaft.

ACCOMMODATION: Seats for five persons including pilot. Forward-opening crew door and passenger door each side; baggage door aft of latter on port side.

Following data highly provisional and unconfirmed.

DIMENSIONS, EXTERNAL:

Main rotor diameter	10 m (33 ft)
Fuselage length	9 m (30 ft)
Height overall	3 m (10 ft)

WEIGHTS AND LOADINGS:

Weight empty	1,000 kg (2,200 lb)
Max T-O weight	1,500 kg (3,300 lb)

PERFORMANCE:

Max level speed	97 kt (180 km/h; 112 mph)
Ceiling	5,200 m (17,060 ft)
Max range	324 n miles (600 km; 372 miles)

NEW ENTRY

IRGC TRANSPORT HELICOPTER

TYPE: Multirole medium helicopter.

PROGRAMME: According to a government announcement, a new Iranian-built transport helicopter made its first 'official' flight on 11 August 2001. No name or designation, or other details, were given, the announcement adding only that the aircraft was intended for airlifting military personnel and for civilian relief operations. This event may be connected with reports from industry sources at about the same time that Iran was considering either the remanufacture or new production of the Bell 214, large numbers of which were supplied to Iran before 1979. Plans for in-country licensed manufacture of the Bell 214 were thwarted following that year's revolution.

NEW ENTRY

First photograph released of the Shahed 274 five-seat helicopter *(Aviation Industries Journal of Iran)* 0095152

IRIAF

ISLAMIC REPUBLIC OF IRAN AIR FORCE (Ya Hossein Project/Owj Industrial Complex)

PO Box 13885-193, Mehrabad Airport, Tehran
Tel: (+98 21) 452 08 10
Fax: (+98 21) 452 08 53

The Ya Hossein, also known as the Owj Industrial Complex, is an agency of the Iranian MoD and IRIAF, established in the early 1990s. Its activities have included design and development of the Tazarve jet trainer, which made its first flight in 2001.

UPDATED

IRIAF TAZARVE

English name: Eagle
TYPE: Basic jet trainer.
PROGRAMME: First design studies early 1990s, leading to first prototype (named Dorna), flown in or about 1995; substantially redesigned second prototype (Tondar) flown 1998; Tazarve is third (and reportedly final) configuration, embodying further wing, fuselage and systems improvements; first flight said to have taken place in mid-2001; public debut October/November 2002 at airshow on Kish Island. Fourth prototype under construction at that time and due to fly during 2003.
CUSTOMERS: Under development for IRIAF, which has ordered five pre-series aircraft for delivery by 2005 and plans acquisition of 25 production examples thereafter.
DESIGN FEATURES: Intended to provide basic, advanced and lead-in fighter training, with secondary capability for light attack. Low/mid-mounted wings, lateral air intakes, conventional sweptback tail unit. Designed for 6,000 hour operational life.
STRUCTURE: Mainly GFRP/nickel sandwich and carbon fibre.
LANDING GEAR: Retractable tricycle type, nosewheel rearward, mainwheels inward into engine intake trunks.
POWER PLANT: One General Electric J85-GE-17 non-afterburning turbojet. Fuel capacity 704 kg (1,552 lb) in wing tanks and 300 litres (79.3 US gallons; 66.0 Imp gallons) in fuselage tanks.
ACCOMMODATION: Crew of two in tandem; rear seat elevated.
AVIONICS: *Comms:* Locally produced UHF/VHF radio.
DIMENSIONS, EXTERNAL:

Wing span	8.04 m (26 ft 4½ in)
Length overall	10.70 m (35 ft 1¼ in)
Height overall	3.63 m (11 ft 11 in)

WEIGHTS AND LOADINGS:

Weight empty	2,550 kg (5,622 lb)
Max T-O weight	4,000 kg (8,818 lb)

PERFORMANCE:

Max level speed at FL180	350 kt (648 km/h; 402 mph)
Stalling speed	85 kt (158 km/h; 98 mph)
Service ceiling	11,582 m (38,000 ft)
T-O run	600 m (1,970 ft)
Landing run	650 m (2,135 ft)
Max range	405 n miles (750 km; 466 miles)
g limits	+8.0/−3.8

UPDATED

Iran's Tazarve all-new jet trainer prototype *(Robert Hewson)*   *NEW*/0531273

Tazarve front cockpit *(Robert Hewson)*   *NEW*/0530170

ISRAEL

AD & D

AERO-DESIGN & DEVELOPMENT LTD

PO Box 565, Gad Finestein Road, Rehovot Industrial Park, IL-76102 Rehovot
Tel: (+972 8) 931 55 33
Fax: (+972 8) 931 55 32
e-mail: ad_d@netvision.net.il
Web: http://www.ad-d.co.il
CEO: Hagai Zalomons
DIRECTORS:
Israel Gottesman
Janina Frankel-Yoeli

In addition to the Hummingbird described below, AD & D has developed a UAV version known as the Hornet, which began tethered flight testing in January 1999. See *Jane's Unmanned Aerial Vehicles and Targets* for details. Development of the CityHawk is now being continued by new company Romeo Yankee Ltd (which see), whose President, Dr Rafi Yoeli, continues to hold a 50 per cent share in AD & D.

VERIFIED

AD & D HUMMINGBIRD

TYPE: Lifting vehicle.
PROGRAMME: Launched as private venture, reviving concept explored by US Hiller VZ-1 in the 1950s but incorporating new features. First (tethered) hovering flight 28 August 1997; first free (untethered) flight (4X-BEB) 4 October 1998. Availability in kit form originally planned for 2000, but has been delayed; no new date reported by June 2002.
COSTS: Estimated US$30,000 (1998).
DESIGN FEATURES: Naturally stable VTOL platform which can be flown by persons with limited piloting experience. Has full engine redundancy, and can continue flying safely throughout full flight envelope after a single engine failure (initial hovering flights used approximately half of total available power). Air duct has outwards-curved lip on upper surface and diameter:length ratio of approximately 2.75:1. Quoted build time approximately 250 hours.
FLYING CONTROLS: Vertical speed (rise/descent) is controlled by pilot increasing or reducing rotor rpm via a twist-grip throttle with the left hand. Lateral (sideways) and longitudinal (fore/aft) motion is achieved simply by the pilot tilting his body slightly in desired direction. Maximum allowable speed is attained when pilot's body reaches limit imposed by circular railing at waist height. 'Nose' of vehicle can be turned by a small lever at pilot's right hand.
STRUCTURE: Engines are mounted on four sides of a square gearbox which also serves as pedestal supporting occupant. Rotors extend from lower end of gearbox. Engines and propellers are basically off-the-shelf ultralight components. Circular outer duct is of graphite/epoxy construction.
LANDING GEAR: Four non-retractable, free-castoring single wheels, each with shock-absorbing oleo. Can withstand vertical impacts of up to 3.7 m (12 ft)/s without injuring pilot or damaging airframe.
POWER PLANT: Four 16.4 kW (22 hp) Hirth F33-15A single-cylinder piston engines, each with dual ignition, own carburettor and own fuel line. Engines are coupled through torsional dampers and one-way clutches. These drive, via an AD & D gearbox equipped with a chip detector, two three-blade contrarotating propellers/rotors which are synchronised to obtain practically zero yawing moment at

Hummingbird, complete and in kit form   0100132

Hummingbird hovering in free flight   0080486

all throttle settings. Alternative engines under consideration. Standard fuel capacity 19 litres (5.0 US gallons; 4.2 Imp gallons).
ACCOMMODATION: Pilot only, standing inside tubular metal frame.
AVIONICS: *Instrumentation:* Battery voltage meter; rotor tachometer; four CHT gauges with an ignition kill switch for each engine; four warning lights (engine out, transmission chips, engine temperature and low battery);

master switch; emergency parachute activation handle (optional); and mode selector button for yaw stabilisation augmentation system.

EQUIPMENT: Ballistic parachute recovery system optional.

DIMENSIONS, EXTERNAL:

Diameter	2.20 m (7 ft 2½ in)
Height overall	1.90 m (6 ft 2¾ in)

WEIGHTS AND LOADINGS:

Weight empty	145 kg (320 lb)
Max T-O weight	260 kg (573 lb)

PERFORMANCE:

Never-exceed speed (VNE)	40 kt (74 km/h; 46 mph)
Hovering ceiling: IGE	2,440 m (8,000 ft)
OGE	1,525 m (5,000 ft)

Range, standard fuel	17 n miles (31 km; 19 miles)
Endurance, standard fuel	30 min

UPDATED

IAI

ISRAEL AIRCRAFT INDUSTRIES LTD

Ben-Gurion International Airport, IL-70100 Tel-Aviv
Tel: (+972 3) 935 31 11
Fax: (+972 3) 935 85 16
Web: http://www.iai.co.il
CHAIRMAN: Ori Orr
PRESIDENT AND CEO: Moshe Keret
EXECUTIVE VICE-PRESIDENT AND COO: Ovadia Harari
CORPORATE VICE-PRESIDENT: Menham Shmul (General Manager, Military Aircraft Group)
DEPUTY CORPORATE VICE-PRESIDENT: Dr David Harari (Director, Research and Development)
DEPUTY CORPORATE VICE-PRESIDENT, MARKETING: Shimon Eckhaus
DIRECTOR OF CORPORATE INDUSTRIAL SERVICES: Menham Tadmor
DEPUTY CORPORATE VICE-PRESIDENT, COMMUNICATIONS: Doron Suslik

Tel: (+972 3) 935 85 09
Fax: (+972 3) 935 85 12
e-mail: hpaz@hdq.iai.co.il

Founded 1953 as Bedek Aviation; ceased being unit of Ministry of Defence and became government-owned corporation 1967; name changed to Israel Aircraft Industries 1 April 1967; number of divisions reduced from five to four February 1988 (Aircraft, Electronics, Technologies and Bedek Aviation). Further restructuring 1994 and now comprises four independent operating Groups: Commercial Aircraft, Military Aircraft, Bedek Aviation and Electronics. Covered floor space 680,000 m² (7.32 million sq ft); total workforce 14,500 in August 2001, of whom one-third are engineers or academics. IAI has ISO 9001 certification, and its products and technologies are fully backed by integrated logistic support, training and a complete after-sales service.

Approved as repair station and maintenance organisation by Israel Civil Aviation Administration, US Federal Aviation Administration, UK Civil Aviation Authority and Israel Air Force. Products include aircraft of own design; in-house produced airframe systems and avionics; service, upgrading and retrofit packages for civil and military aircraft and helicopters; other activities include space technology, missile and ordnance development, and seaborne and ground equipment.

New orders worth US$2.6 billion were taken in 2000, followed by a further US$1.502 billion in the first nine months of 2001. Net profit at 30 September 2001 was US$92.2 million, including a US$34.2 million capital gain from the sale of Galaxy Aerospace to General Dynamics on 1 May 2001; excluding the latter, this represented a net profit decrease of 5 per cent compared with the corresponding period of 2000. Order backlog at 1 July 2001 was more than US$3.6 billion.

UPDATED

COMMERCIAL AIRCRAFT GROUP

Ben-Gurion International Airport, IL-70100 Tel-Aviv
Tel: (+972 3) 935 33 37 and 935 71 49
Fax: (+972 3) 935 42 26
GENERAL MANAGER: M Boness

Established 1994. Design, development, certification and manufacture of civil aircraft; hydraulic, electromechanical and pneumatic components; dynamic systems; and composites material structures (all backed by full computer-aided design, engineering and manufacturing processes).

VERIFIED

IAI 1125 ASTRA SPX

Now renamed Gulfstream 100 (which see under Gulfstream Aerospace entry in US section).

UPDATED

IAI 1126 GALAXY

Now renamed Gulfstream 200 (which see under Gulfstream Aerospace entry in US section).

UPDATED

IAI C-5WA AIRTRUCK

TYPE: Twin-turboprop freighter.
PROGRAMME: IAI submission chosen in principle by Federal Express of USA against competition from Saab (Sweden) and Ayres (USA) in mid-1998 as replacement for Fokker Friendship parcel transports. FedEx purchase would be 100, but IAI is seeking further 100 sales to reduce unit price to FedEx-imposed ceiling of US$10 million. Risk-sharing partners also needed to ensure programme launch, but not obtained by early 2002.
CUSTOMERS: Unconfirmed reports by late 1999 that FedEx requirement may have been reduced; Global Air (Leasing) Group ordered 50, with further 50 on option, in July 1999.
DESIGN FEATURES: High-wing twin-turboprop with T tail; optimised for night transport of up to five standard-size cargo containers.

Model of the projected IAI Airtruck *(Photo Link)*   0100796

LANDING GEAR: Retractable tricycle type. Main units retract into fairings on fuselage sides.
POWER PLANT: Two 4,098 kW (5,495 shp) class turboprops (to be selected); four-blade propellers.
ACCOMMODATION: Pressurised flight deck, with forward airstair door access on port side. Unpressurised, but temperature-controlled, main cargo compartment has large port-side cargo door aft of wing. Tailcone compartment, also unpressurised, configured for carriage of hazardous materials. Main compartment can accommodate five M1 (2.44 × 2.44 m; 8 × 8 ft) cargo pallets.
SYSTEMS: Redundant fire detection and extinguishing system in tailcone compartment.
DIMENSIONS, EXTERNAL:

Wing span	31.725 m (104 ft 1 in)
Wing aspect ratio	10.8
Length overall	29.985 m (98 ft 4½ in)
Cargo door: Height	2.59 m (8 ft 6 in)
Width	2.74 m (9 ft 0 in)

DIMENSIONS, INTERNAL:

Main cargo compartment: Max height	2.44 m (8 ft 0 in)
Max width	3.18 m (10 ft 5¼ in)
Volume	126 m³ (4,450 cu ft)

AREAS:

Wings, gross	93.00 m² (1,001.0 sq ft)

WEIGHTS AND LOADINGS (design):

Operating weight empty	17,240 kg (38,008 lb)
Max fuel weight	8,300 kg (18,298 lb)
Max payload	12,470 kg (27,492 lb)
Max T-O weight	35,380 kg (78,000 lb)
Max wing loading	380.6 kg/m² (77.96 lb/sq ft)

PERFORMANCE (estimated):

Max cruising speed	300 kt (556 km/h; 345 mph)

Range:
with max payload 950 n miles (1,759 km; 1,093 miles)
with max fuel and 10,000 kg (22,045 lb) payload
1,500 n miles (2,778 km; 1,726 miles)

UPDATED

BEDEK AVIATION GROUP

Ben-Gurion International Airport, IL-70100 Tel-Aviv
Tel: (+972 3) 935 89 64
Fax: (+972 3) 971 22 98
GENERAL MANAGER: David Arzi

Group provides total aircraft, engine and component repair, overhaul, retrofit, modification, conversion and support services to airlines and leasing companies.

VERIFIED

MILITARY AIRCRAFT GROUP

Ben-Gurion International Airport, IL-70100 Tel-Aviv
Tel: (+972 3) 935 41 36 and 971 14 71
Fax: (+972 3) 935 34 53
GENERAL MANAGER: Menham Shmul

Develops and manufactures unmanned air vehicles (see *Jane's Unmanned Aerial Vehicles and Targets* for current activities); integrates and upgrades military fixed-wing aircraft and helicopters; designs and manufactures helicopter structures, electrical harnesses and crash-attenuating aircraft seats.

VERIFIED

ELECTRONICS GROUP

PO Box 105, Industrial Zone, IL-56000 Yehud
Tel: (+972 3) 531 55 55 and 531 40 21
Fax: (+972 3) 536 52 05 and 536 39 75
GENERAL MANAGER: S Alkon

Specialises in electronics, electro-optical and inertial systems, and components for military and commercial applications.

VERIFIED

ROMEO YANKEE

ROMEO YANKEE LTD
Renamed Urban Aeronautics Ltd.

UPDATED

URBANAERO

URBAN AERONAUTICS LTD
PO Box 190, Airport City, Ben-Gurion Airport, IL-70151 Tel-Aviv
Tel: (+972 3) 683 57 98
Fax: (+972 3) 518 76 45
e-mail: info@turbohawk.com
Web: http://www.urbanaero.com
PRESIDENT: Dr Rafi Yoeli
VICE-PRESIDENT, OPERATIONS: Omni Knoller

Company formed in 2000 as Romeo Yankee Ltd (see 2002-03 *Jane's*) specifically to continue development of (former AD & D) CityHawk 'flying car' and its CityHawk II and TurboHawk derivatives. Name changed to UrbanAero in 2002.

UPDATED

URBANAERO CITYHAWK, X-HAWK and TURBOHAWK

TYPE: Lifting vehicle.
PROGRAMME: Announced April 2000; CityHawk prototype/demonstrator completed and undergoing tethered testing since August 2002. Engines had been run at 80 per cent power by September 2002, achieving lift-off of rear duct.
CURRENT VERSIONS: **CityHawk:** POC demonstrator; *as described.*

X-Hawk: Planned, FAA-certified commercial three-seater as urban-area utility transport, with twin turboshaft engines (P&WC PW207D or Turbomeca Arrius 2B2). Projected weights of 1,150 kg (2,535 lb) empty, 430 kg (948 lb) payload and 1,920 kg (4,232 lb) maximum take-off. Funding of up to US$8 million being sought to construct two prototypes.

TurboHawk: Military derivative. Subject of joint venture with LoPresti Gordon VTOL Inc (LoGo) of USA, which see, formed in early 2001 by Roy LoPresti and Larry Gordon to develop vehicles for offer to US and other Special Forces groups.
DESIGN FEATURES: Two-seat extrapolation of Hummingbird concept; T-O/landing area compatible with typical urban parking spaces and garages. Foreseen applications include personal urban transportation, police and traffic patrol, ambulance and urban evacuation, air taxi, and agricultural/environmental monitoring.
FLYING CONTROLS: Manual, from left-hand seat. Differential thrust between fans for pitch and forward speed control; combined thrust of both fans controls vertical speed (rise/descent). Roll and yaw control by cascades of vanes on upper and lower surfaces of each fan.
STRUCTURE: Fans, engines and gearboxes identical with those of Hummingbird.
LANDING GEAR: Tricycle type.
POWER PLANT: Two Hummingbird-type fans, each driven by four 16.4 kW (22 hp) Hirth F33-15A single-cylinder piston engines. Engine redundancy allows vehicle to fly down to a landing even if each fan loses one engine. Fuel capacity 60 litres (15.9 US gallons; 13.2 Imp gallons).
ACCOMMODATION: Side-by-side cockpits for two persons under separate canopies.
EQUIPMENT: Ballistic parachute for emergency recovery.
DIMENSIONS, EXTERNAL:

Length	4.70 m (15 ft 5in)
Max width	2.20 m (7 ft 2½ in)
Max height	1.80 m (5 ft 10¾ in)

WEIGHTS AND LOADINGS:

Max T-O weight	570 kg (1,256 lb)

PERFORMANCE (estimated):

Max level speed	80-90 kt (148-166 km/h; 92-103 mph)
Max operating altitude	2,440 m (8,000 ft)
T-O/landing footprint	2.5 × 5.5 m (8.2 × 18 ft)
Endurance	almost 1 h

UPDATED

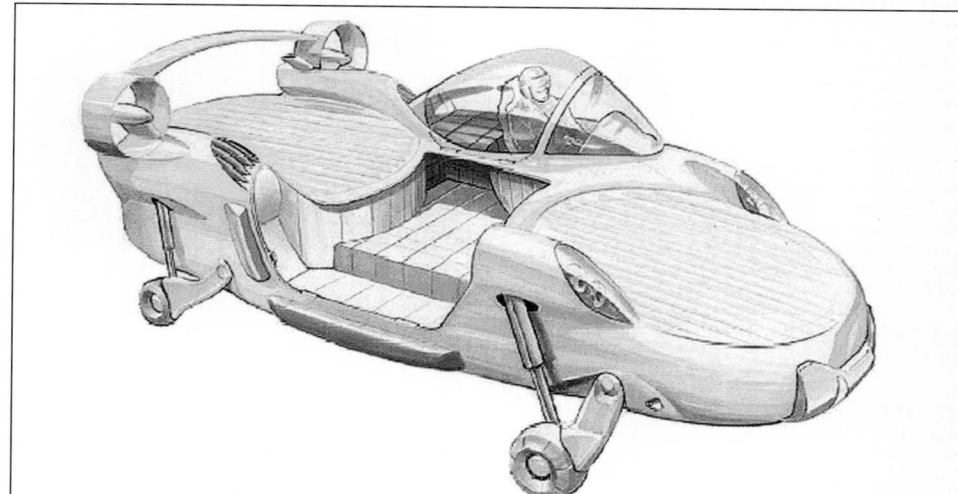
Artist's impression of projected X-Hawk   NEW/0536617

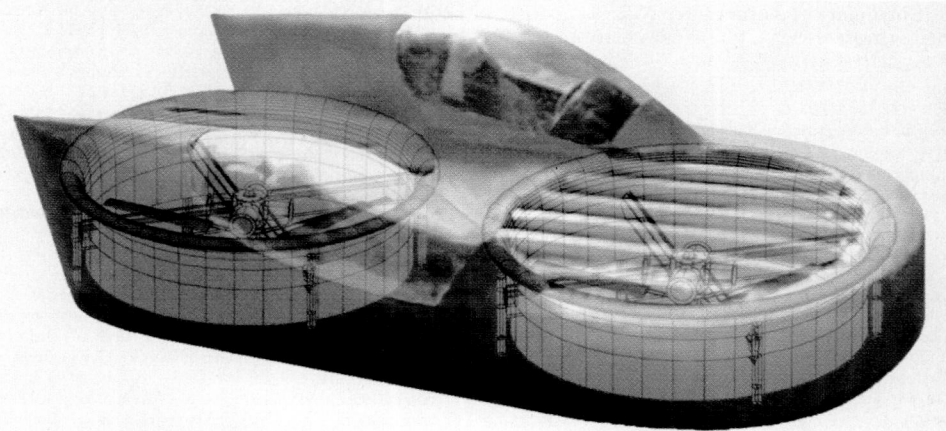
Artist's impression of twin-fan CityHawk   0100133

CityHawk demonstrator under construction
0101482

ITALY

AERONAUTICA MACCHI

AERONAUTICA MACCHI SpA

Via Ing Paolo Foresio 1, I-21040 Venegono Superiore (VA)
Tel: (+39 0331) 86 59 12
Fax: (+39 0331) 86 59 10
CHAIRMAN: Dott Fabrizio Foresio

Original Macchi company, founded 1913 in Varese, produced famous line of high-speed flying-boats and seaplanes. Aeronautica Macchi became holding company for Aermacchi SpA (see following entry); SICAMB (Società Italiana Costruzioni Aeronautiche Martin-Baker; aerostructures and equipment manufacturing, including licensed production of Martin-Baker ejection seats) at Latina; and Logic (aircraft equipment and controls). A 25 per cent holding in Aeronautica Macchi was acquired by Aeritalia, now Alenia, in 1983, the balance being held by the Foresio family. In January 1997, Aermacchi acquired the SIAI-Marchetti company from Alenia-Finmeccanica.

Also obtained in the same agreement was Finmeccanica's nacelle manufacturing activities for Airbus airliners and Dassault Falcon business jets, previously the responsibility of Alenia at Turin. Related activities formerly undertaken by SIAI included Airbus A300 and 310 tailcones; overhaul and repair of various types of aircraft (notably C-130 Hercules, DHC-5 Buffalo and Cessna Citation II); participation in national or multinational programmes, producing parts for Eurofighter Typhoon (carbon fibre structures, titanium engine cowlings, ECM pods and weapon pylons), Alenia G222/C-27J (wing and tailerons), Panavia Tornado (wing furniture) and AMX. Macchi makes engine nacelles for Airbus A 318, 319, 320, 321 and 330, Dassault Falcon 900 and Falcon 2000; 1,000th Airbus nacelle delivered in

Aeronautica Macchi's aerostructures business has included the Fairchild Dornier 328JET   *NEW*/0137401

December 2000. Was contracted to build nacelles (in collaboration with Hurel Dubois) for Fairchild 728JET and bid to do same for Embraer 170/190 families. Was risk-sharing partner in Fairchild Dornier 328JET, having supplied 100th fuselage (including earlier turboprop version) in 1998. To design and build engine nacelles for Trent 900 and GP7200 versions of Airbus A380. Space activities include elements of Ariane 5.

Workforce in 2001 totalled over 2,200, including 1,880 in Aermacchi. Activities in civil field accounted for over 50 per cent of turnover by 2000.

UPDATED

AERMACCHI

AERMACCHI SpA
(Subsidiary of Aeronautica Macchi SpA)

Via Ing Paolo Foresio 1, PO Box 101, I-21040 Venegono Superiore (VA)
Tel: (+39 0331) 81 31 11
Fax: (+39 0331) 82 75 95
e-mail: info@aermacchi.it
Web: http://www.aermacchi.it
CHAIRMAN: Dott Fabrizio Foresio
MANAGING DIRECTOR: Dott Ing Giorgio Brazzelli
GENERAL MANAGER: Ing Roberto Assereto
TECHNICAL DIRECTOR: Dott Ing Massimo Battini
COMMERCIAL MANAGER: Dott Cesare Cozzi

Aermacchi is aircraft manufacturing company of Aeronautica Macchi group; plants at Venegono airfield occupy total area of 274,000 m² (2,949,300 sq ft), including 113,000 m² (1,216,300 sq ft) covered space; flight test centre has covered space of 5,100 m² (54,900 sq ft) in total area of 28,000 m² (301,400 sq ft). Subsidiary company is Logic avionics.

Aermacchi was one of two Italian assembly centres for AMX (which see in International section), having produced

Cockpit of MB-339CD/FD   *NEW*/0132394

36, plus nine AMX-Ts; also has important roles in Eurofighter Typhoon and Yakovlev Yak-130 programmes, offering Westernised version of latter as M-346. In January 1996, Aermacchi obtained production rights to the former Valmet L-90TP Redigo turboprop trainer, which was redesignated M-290TP Redigo. On 1 January 1997, Aermacchi acquired all training activities and programmes of SIAI-Marchetti, including the SF.260 and S.211 (but not SF.600 Canguro; see VulcanAir in this section) and support of earlier S.205s and S.208s remaining in service. Production was then transferred from Sesto Calende to the Aermacchi plant at Venegono. Since 1960, 2,000 Aermacchi trainers have been bought by 40 countries.

UPDATED

AERMACCHI MB-339

TYPE: Basic jet trainer/light attack jet.
PROGRAMME: MB-339A selected by Italian Air Force on 11 February 1975 in preference to MB-338; one static test airframe and two prototypes: first flights (MM588) 12 August 1976 and (MM589) 20 May 1977; first flight of production MB-339A (MM54438) 20 July 1978; initial two deliveries (MM54439 and 54440) on 8 August 1979; first batch of 51 included three used as radio calibration aircraft at Pratica di Mare and 14 PANs for Frecce Tricolori aerobatic team.

MB-339C developed as an uprated MB-339 with a Viper 680 power plant; experimental installation in I-MABX, first flown on 9 June 1983; initial MB-339C first flight (I-AMDA) 17 December 1985; production aircraft (I-TRON) 8 November 1989. The 200th MB-339 (a 339CE for Eritrea) was delivered on 14 April 1997. MB-339s had flown some 500,000 hours by late 2002.
CURRENT VERSIONS: **MB-339A:** Initial military basic/advanced trainer (full details in 1990-91 *Jane's*). Final deliveries (small attrition replacement batch for Italy) in

1995. Main operator is 61° Stormo Scuola Volo Basico Iniziale su Aviogetti at Lecce-Galatina.

Lit110 billion MLU initiated in mid-1999 includes structural modifications on some 70 aircraft to extend service lives from 10,000 to 15,000 hours (from 20 to 30 years). Features of MLU include Litton Italia LISA-FG GPS/INS; new integrated air data transducer; provision for AN/ARC-150(V) Have Quick secure radio; Elmer crash recorder; ELT; instrument panel modifications; radio switches on control column; formation flying lights; higher capacity batteries connected in parallel with starter/generator; provision for target towing; anti-skid brakes; nosewheel splash-guard; repositioned IFF antenna; airborne stress gauge; and improved access, by means of additional removable panels, to wing/fuselage joint for inspection and engine maintenance. Depot maintenance interval will increase to 1,500 hours after MLU. Prototype (CSX54453) 20 December 1999; 'production' will take between eight and 10 years for fleet, beginning with MB-339PANs.

MB-339PAN (*Pattuglia Aerobatica Nazionale:* national aerobatic team): Special version of MB-339A for Italian Air Force; smoke generator added; wingtip tanks deleted. Official debut 27 April 1982. Total 25 built as, or converted to, PAN. Included in MLU.

MB-339AM: Special anti-ship version armed with AOSM Marte Mk 2A missile; avionics, equivalent to MB-339C, include new inertial navigator, Doppler radar, navigation and attack computers, head-up display and multifunction display. Prototype converted from MB-339A; qualification completed January 1995. No known orders.

MB-339B: Powered by 19.57 kN (4,400 lb st) Viper Mk 680-43; larger tip tanks. Prototype (demonstrator I-GROW) modified in 1993 with two LCD EFIS displays and air-to-air refuelling (AAR).

Aermacchi MB-339A MLU prototype (*Jane's/Paul Jackson*)   *NEW*/0132402

Night mission launch by an Aermacchi MB-339CD   *NEW*/0132423

MB-339RM (*Radiomisure:* radio calibration): Three produced for Italian Air Force, 1981; withdrawn and transferred to training as MB-339As.

T-Bird II: One demonstrator modified for US JPATS competition; delivered to Lockheed factory at Marietta, Georgia, on 20 May 1992. Power plant, one Rolls-Royce Viper 680-582 of 17.79 kN (4,000 lb st) with noise reduction kit. Rejected in favour of Pilatus PC-9 (Raytheon T-6A Texan II), June 1995. Continues in use as chase aircraft.

MB-339CB: For Royal New Zealand Air Force; Viper 680 engine. Served with 14 Squadron at Ohakea in training and light attack roles until 13 December 2001, when unit disbanded and aircraft offered for sale.

MB-339CD (C Digital): Developed for Italian Air Force advanced/fighter lead-in training. Rolled out 12 April 1996; first flight (conversion of MB-339A MM54544; renumbered MMX606) 24 April 1996. Rolls-Royce Viper Mk 632-43 engine; removable in-flight refuelling probe (first phase of tanker trials completed March 1996 by development MB-339 I-GROW); new avionic architecture based on a single central mission computer, MIL-STD-1553B digital databus, one ring laser gyro platform with embedded GPS, EFIS cockpits with HUD in each, three identical liquid crystal colour MFDs and HOTAS controls. Other details generally as MB-339CB, but (compared with MB-339A) nose lengthened; elevators modified to give handling characteristics closer to the fast jets to which students will progress; some structural parts and systems modified; and pilots' escape system updated. Achieved full operational capability with IAF in October 1998 after 148 sorties by prototype.

Contract signed in August 2001 for further 15 MB-339CDs as conversion of MMX606, first flown in new guise on 6 November 2001; new features of Lot 2 include 8.33 kHz radios, onboard simulation of electronic warfare (RWR, MAWS, chaff/flare), IFF Mode S, crash data recorder, ELT and underwater acoustic beacon, multifunctional information distribution system, full NVG capability, digital map system and autonomous air combat manoeuvring instrumentation. All new features to be retrofitted to first batch. Initial aircraft of Lot 2 (MM55077) delivered September 2002.

Detailed description applies to the above version.

MB-339CE: Variant of MB-339CD; Viper 680 engine. Six ordered by Eritrea on 7 November 1995; deliveries began April 1997.

MB-339FD (Full Digital): Export equivalent of CD, with Viper 680 engine. Offered to Australia in 1994; unsuccessful in competition against the BAE Systems' Hawk. Also offered to Brazil and South Africa as Xavante/Impala replacement. Purchase by Venezuela of an initial eight announced 7 July 1998, but decision subsequently reversed and competition reopened; no further announcement by early 2002.

CUSTOMERS: See table. Fifteen CDs ordered by Italian Air Force; initial delivery of two aircraft on 18 December 1996, followed by further 10 in 1997 and three in 1998 (last in November); first four to Reparto Sperimentale Volo at Pratica di Mare for trials before issue to 61° Stormo; some assigned to base flights at Gioia del Colle and Novara-Cameri for continuation training, including in-flight refuelling.

DESIGN FEATURES: Conventional subsonic jet trainer with tandem, stepped cockpits and low, moderately sweptback wing. Designed to MIL-A-8860A for 10,000 hours service life. Intended maintenance requirement of 5.95 mmh/fh; scheduled maintenance at 150 and 300 hour intervals; IRAN at 1,500 hours (extendable to 1,800 hours).

Wing section NACA 64A-114 (mod) at centreline, 64A-212 (mod) at tip; quarter-chord sweepback 8° 29′.

FLYING CONTROLS: Conventional. Power-assisted ailerons with artificial feel, plus servo tabs to assist emergency manual reversion; elevators and rudder actuated manually by pushrods; electrically controlled servo tab for control assistance and trimming on elevator; hydraulically actuated single-slotted flaps; two ventral strakes under tail; electrohydraulically actuated airbrake panel under forward fuselage; wing fence ahead of aileron inboard edge; both pilots have HUD; rear pilot elevated sufficiently to be able to aim guns and air-to-surface weapons and fly visual approaches (see nav/attack system under Avionics heading).

STRUCTURE: All-metal; stressed skin wings with main and auxiliary spars and spanwise stringers; bolted to fuselage; tip tanks permanently attached; rear fuselage detachable by four bolts for engine access.

LANDING GEAR: Hydraulically retractable tricycle type with oleo-pneumatic shock-absorbers, suitable for operation from semi-prepared runways. Hydraulically steerable nosewheel retracts forward; main units retract outward into wings. Low-pressure mainwheel tubeless tyres size 545×175-10 (14 ply); nosewheel tubeless tyre size 380×150-4 (6 ply). Emergency extension system. Hydraulic disc brakes with anti-skid system. Minimum ground turning radius 8.63 m (28 ft 3¾ in).

POWER PLANT: One Rolls-Royce Viper Mk 680-43 turbojet, installed rating 19.31 kN (4,340 lb st). MB-339Cs for Italy retain the MB-339A's 17.79 kN (4,000 lb st) Viper Mk 632-43. Fuel in two-cell rubber fuselage tank, capacity 781 litres (206 US gallons; 172 Imp gallons), and two wingtip tanks with combined capacity of 1,000 litres (264 US gallons; 220 Imp gallons). Total internal usable capacity 1,781 litres (470.5 US gallons; 392 Imp gallons). Self-sealing tanks optional. Single-point pressure refuelling point in port side of fuselage, below wing trailing-edge. Gravity refuelling points on top of fuselage and each tip tank. Provision for two drop tanks, each of 325 litres (86.0 US gallons; 71.5 Imp gallons) usable capacity, on centre underwing stations. Optional refuelling probe on starboard side of cockpits.

ACCOMMODATION: Crew of two in tandem, on Martin-Baker Mk 10LK zero/zero ejection seats in pressurised cockpit. Independent or rear-seat command ejection. Design in accordance with MIL-STD-203F. Rear seat elevated 32.5 cm (1 ft 1 in). Rearview mirror for each occupant. Two-piece moulded transparent canopy, opening sideways to starboard.

SYSTEMS: Pressurisation system maximum differential 0.24 bar (3.5 lb/sq in); cockpit designed for 40,000 pressurisation cycles. Bootstrap-type air conditioning system, also providing air for windscreen and canopy demisting. Hydraulic system, pressure 172.5 bar (2,500 lb/sq in), for actuation of flaps, aileron servos, airbrake, landing gear, wheel brakes and nosewheel steering. Back-up system for wheel brakes and emergency extension of landing gear. Main electrical DC power from one 28 V 9 kW engine-driven starter/generator and one 28 V 6 kW secondary generator; power distribution via five 28 V buses. Two 24 V 22 Ah Ni/Cd batteries for engine starting. Fixed-frequency 115/26 V AC power from two 600 VA single-phase static inverters; provision for additional inverter for three-phase AC. External power receptacle. Low-pressure demand oxygen system, operating at 28 bar (400 lb/sq in). Anti-icing system for engine air intakes.

AVIONICS: Representative fit includes following:

Comms: Honeywell AN/APX-100 IFF.

Flight: BAE 620 kbit navigation computer; RT-1159/A or Rockwell Collins AN/ARN-118(V) Tacan, or Honeywell KDM 706A DME; Rockwell Collins 51RV-4B VOR/ILS and MKI-3 marker beacon receiver; Rockwell Collins ADF-60A ADF/L (or, optionally, DF-301E V/UHF ADF); BAE AD-660 Doppler velocity sensor integrated with Litton LR-80 inertial platform; HOTAS controls. Flight recorder and airborne stress gauge.

Instrumentation: Alenia CRT multifunction display; Alenia/Honeywell HG7505 radar altimeter; Astronautics AN/ARU-50/A attitude director indicator and AN/AQU-13 HSI.

Mission: Kaiser Sabre head-up display and weapon aiming computer; Logic stores management system; provision for FIAR P 0702 laser range-finder; Lockheed Martin Fairchild video camera; photographic pod with four Vinten 70 mm cameras. Optional laser range-finder.

Self-defence: Options include ELT-156 radar warning system; single Elettronica ELT-555 ECM pod combined with AN/ALE-40 chaff/flare dispenser.

ARMAMENT: Up to 1,814 kg (4,000 lb) of external stores on six underwing hardpoints. Four inner hardpoints each stressed for up to 454 kg (1,000 lb) load, and two outer hardpoints each for up to 340 kg (750 lb) load. RNZAF aircraft fitted for AIM-9 Sidewinder and AGM-65 Maverick. Provision on two inner stations for installation of two Macchi gun pods, each containing either a 30 mm DEFA 553 cannon with 120 rounds or a 12.7 mm AN/M-3 machine gun with 350 rounds.

Other typical loads can include two Magic or AIM-9 Sidewinder air-to-air missiles on two outer stations; six general purpose or cluster bombs of appropriate weights; six AN/SUU-11A/A 7.62 mm Minigun pods, each with 1,500 rounds; six Matra 155 launchers, each for eighteen 68 mm rockets; six AN/LAU-68/A or AN/LAU-32G launchers, each for seven 2.75 in rockets; six Aerea AL-25-50 or AL-18-50 launchers, each with twenty-five or eighteen 50 mm rockets respectively; six Aerea AL-18-80 launchers, each with twelve 81 mm rockets; four AN/LAU-10/A launchers, each with four 5 in Zuni rockets; four TDA 100-4 launchers, each with four 100 mm TDA rockets; six Bristol Aerospace LAU-5002 launchers for CRV-7 high-velocity rockets; six Aerea BRD bomb/rocket dispensers; six Aermacchi 11B29-003 bomb/flare dispensers; six TDA 14-3-M2 adaptors, each with six BAP 100 anti-runway bombs or BAT 120 tactical support bombs. Marte 2A anti-ship missile completed MB-339 qualification trials in February 1995.

Following data are for MB-339FD, but generally applicable to MB-339CD.

DIMENSIONS, EXTERNAL:

Wing span over tip tanks	11.22 m (36 ft 9¾ in)
Wing aspect ratio	6.5
Length overall	11.24 m (36 ft 10½ in)
Height overall	3.945 m (12 ft 11¼ in)
Elevator span	4.16 m (13 ft 7¾ in)

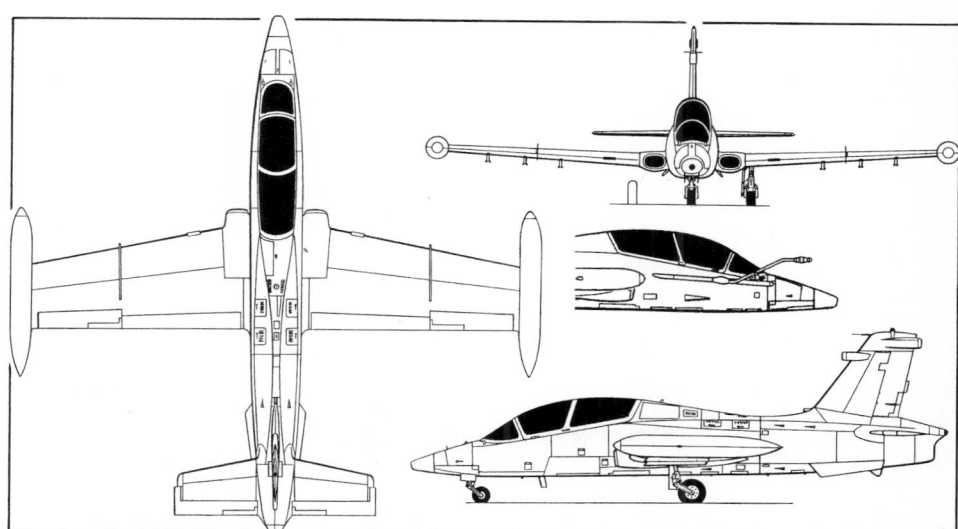

Aermacchi MB-339CD advanced trainer and attack aircraft, with additional starboard side view of optional refuelling probe (*Jane's/Dennis Punnett*)

Wheel track	2.48 m (8 ft 1¾ in)
Wheelbase	4.37 m (14 ft 4 in)

AREAS:

Wings, gross	19.30 m² (207.7 sq ft)
Ailerons (total)	1.33 m² (14.32 sq ft)
Trailing-edge flaps (total)	2.55 m² (27.45 sq ft)
Airbrake	0.52 m² (5.60 sq ft)
Fin	2.21 m² (23.78 sq ft)
Rudder, incl tab	0.68 m² (7.32 sq ft)
Tailplane	3.38 m² (36.38 sq ft)
Elevators (total incl tabs)	0.98 m² (10.55 sq ft)

WEIGHTS AND LOADINGS:

Weight empty	3,334 kg (7,350 lb)
Max weapon load	1,814 kg (4,000 lb)
Max fuel weight: internal	1,430 kg (3,153 lb)
external	522 kg (1,151 lb)
T-O weight, clean	4,950 kg (10,913 lb)
Max T-O weight with external stores	6,350 kg (14,000 lb)
Max wing loading	329.0 kg/m² (67.39 lb/sq ft)
Max power loading	329 kg/kN (3.23 lb/lb st)

PERFORMANCE (at trainer clean T-O weight, except where indicated):

Never-exceed speed (VNE)	M0.82 (500 kt; 926 km/h; 575 mph)
Max level speed at S/L:	
clean	490 kt (907 km/h; 564 mph)
with four 500 lb bombs	440 kt (815 km/h; 506 mph)
Max level speed at 9,140 m (30,000 ft)	M0.77 (441 kt; 815 km/h; 508 mph)
Max speed for landing gear extension	175 kt (324 km/h; 202 mph)
T-O speed	100 kt (185 km/h; 115 mph)
Approach speed over 15 m (50 ft) obstacle	102 kt (189 km/h; 117 mph)
Stalling speed, 20% fuel	86 kt (160 km/h; 99 mph)
Max rate of climb at S/L	2,011 m (6,600 ft)/min
Service ceiling	13,715 m (45,000 ft)
T-O run at S/L	540 m (1,775 ft)
Landing run at S/L, 20% fuel	470 m (1,545 ft)

Radius of action, 10% reserves:
with four 500 lb bombs:

lo-lo-lo	145 n miles (268 km; 166 miles)
hi-lo-hi	255 n miles (472 km; 293 miles)

with Marte ASM and two underwing tanks:

lo-lo-lo	230 n miles (426 km; 264 miles)
hi-lo-hi	440 n miles (814 km; 506 miles)

with two 500 lb bombs, two rocket launchers and two gun pods, 20 min loiter and 6 min combat, hi-lo-hi
125 n miles (231 km; 143 miles)

Ferry range: clean	950 n miles (1,759 km; 1,039 miles)
with two underwing drop tanks, 10% reserves	1,050 n miles (1,944 km; 1,208 miles)
Max endurance with drop tanks	3 h 50 min
g limit	+8.0/−4.0

UPDATED

First Lot 2 MB-339CD for Italian Air Force, delivered September 2002   *NEW*/0528618

MB-339 PRODUCTION

Customer	Variant	Qty	First Aircraft	First Delivered
Argentine Navy	AA	10[2]	0761	1980
Eritrean Air Force	CE	6	409[3]	Apr 1997
Ghana Air Force	A	2	G800	1987
		2	G802	1994
Italian Air Force	A[1]	80	MM54438	19 Feb 1979
		21	MM54533	1986
		6	MM55052	1994
	CD	30	MM55062	18 Dec 1996
Malaysian Air Force	AM	13	M34-01	1983
New Zealand Air Force	CB	18	NZ6460	13 Mar 1991
Nigerian Air Force	AN	12	301	Jun 1984
Peruvian Air Force	AP	16	452	Nov 1981
UAE (Dubai) Air Force	A	2	431	24 Mar 1984
		3	433	1987
Subtotal		**221**		
Prototypes		2	MM588	1976
Demonstrator	A	1	I-NEUN	1980
	B	1	I-GROW	1984
	C	1	I-AMDA	1985
Total		**226**		

[1] Also 339PAN and 339RM
[2] One converted to JPATS demonstrator N339L; others lost or withdrawn from service; including at least two believed transferred to UAE
[3] Eritrean aircraft serial numbered in reverse order

AERMACCHI SF-260
Uruguayan Air Force designation: T-260

TYPE: Basic prop trainer/attack lightplane.

PROGRAMME: Originated as F.250 designed by Stelio Frati and made by Aviamilano (see 1965-66 *Jane's*); civil SIAI-Marchetti SF.260A and B detailed in 1980-81 *Jane's*; SF.260C detailed in 1985-86 *Jane's*; SF.260D was later civil variant; SF.260M military trainer detailed in 1984-85 and earlier *Jane's*; current versions developed (unsuccessfully) for USAF's Enhanced Flight Screener requirement; still in low-volume production as a military trainer following transfer of design rights to Aermacchi in 1997 as SF-260 (with hyphen). By 1999, SF-260s had flown more than 1,700,000 hours.

CURRENT VERSIONS: **SF-260E:** Direct-injection engine, strengthened airframe and 100 kg (220 lb) higher aerobatic weight than superseded SF.260D. Wing aerodynamically refined to preserve original stalling speed, despite increased MTOW; automatic fuel management. Canopy height and width increased. Certified by RAI on 21 January 1992 and by FAA on 17 August 1994.

Main description applies to SF-260E and Warrior.

SF-260F: As SF-260E, but carburetted engine. Promotion discontinued.

SF-260E Warrior: Trainer/tactical support version. Two underwing pylons, for up to total 300 kg (661 lb) of external stores, and cockpit stores selection panel. Able to undertake a wide variety of roles, including low-level strike; forward air control; forward air support; armed reconnaissance; and liaison.

SF-260TP: Turboprop-powered version. Described in 2002-03 and previous editions.

CUSTOMERS: Total of some 790 produced (including turboprop version) by SIAI-Marchetti at Sesto Calende for customers in 27 countries; further 64 sold by Aermacchi. Civil operators include Sabena, Royal Air Maroc and Alitalia. Six SF-260Fs ordered by Air Force of Zimbabwe, early 1997, of which last delivered in October 1998; these were first of type built at Venegono. In June 1998, 12 SF-260Es bought by Venezuelan Air Force, and delivered to 14 Grupo at Maracay from late 2000. Delivery of 13 SF-260Es to Uruguayan Air Force began in September 1999, following first flight on 23 July, and completed in December 2000. Mexico placed order in early 1998,

initially for 13, but subsequently 30, of which first six delivered in early 2000. Five Warrior variants of SF-260EU to Mauritania in 2000. Italian Air Force is

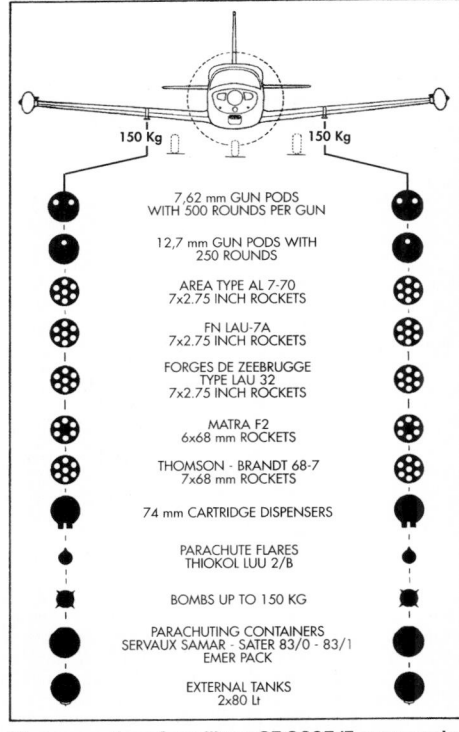

Weapon options for military SF-260E/F or two-pylon configuration for SF-260TP; latter may also carry second pairs of gun pods, parachute flares, bombs or parachuting containers on outboard pylons   0110887

150 Kg	150 Kg
7,62 mm GUN PODS WITH 500 ROUNDS PER GUN	
12,7 mm GUN PODS WITH 250 ROUNDS	
AREA TYPE AL 7-70 7x2.75 INCH ROCKETS	
FN LAU-7A 7x2.75 INCH ROCKETS	
FORGES DE ZEEBRUGGE TYPE LAU 32 7x2.75 INCH ROCKETS	
MATRA F2 6x68 mm ROCKETS	
THOMSON - BRANDT 68-7 7x68 mm ROCKETS	
74 mm CARTRIDGE DISPENSERS	
PARACHUTE FLARES THIOKOL LUU 2/B	
BOMBS UP TO 150 KG	
PARACHUTING CONTAINERS SERVAUX SAMAR - SATER 83/0 - 83/1 EMER PACK	
EXTERNAL TANKS 2x80 Lt	

Aermacchi SF-260EU of Venezuelan Air Force   *NEW*/0120121

potential SF-260E customer, requiring up to 40 for replacement of its SF-260AM fleet.

AERMACCHI SF-260 MILITARY CUSTOMERS

Operator	Variant	Qty	First Aircraft
Belgium	MB	36	ST-01
	D	9	ST-40
Bolivia	CB	6	FAB-180
Brunei	W	2	129
Burkina Faso		4[1]	
Burundi	C	3	9U-RUA
Congo Republic	MC	21	AT-101
Dubai	TP	6	401
	W	1	H
Ethiopia	TP	10	156?
Haiti	TP	6	1271
Ireland	WE	11	222
Italy	260	3	MM91004
	AM	45	MM54418
Libya	W	240	
Mauritania	EU	4†[4]	5T-MAK
Mexico	E	30†	–
Myanmar	M	20	2001
Philippines	MP	30	601
	WP	16	631
	TP	18	701
Singapore	W	12	151
	MS	16	120
Somalia	C	6	6O-SBC
	W	10	
Sri Lanka	TP	9[2]	CT121
Thailand	MT	19	15-01
Tunisia	CT	6	W41501
	WT	12	W41401
Turkey	D	41	773
Uganda		3[3]	AF504
Uruguay	EU	13†	610
Venezuela	EU	12†	0027
Zambia	MZ	9	AF-507
Zimbabwe	C	17	R9900
	W	14	R3260
	F	6†	
Subtotal		**726**	
various civil		125	
Total		**851**	

† Built by Aermacchi
[1] Plus others second-hand
[2] Plus three second-hand
[3] Plus two or three from Libya
[4] Warrior
Note: Nicaragua received some from Libya

DESIGN FEATURES: Side-by-side, piston-engined aerobatic trainer. Low-mounted wings with forward-swept trailing-edges and tip tanks; sweptback fin.
Wing section NACA 64₁-212 (modified) at root, 64₁-210 (modified) at tip; dihedral 6.5° 20′ from roots.

FLYING CONTROLS: Conventional and manual. Frise ailerons each with servo tab; trim tabs for all three axes; four-way trim button on each stick top; electrically operated slotted flaps; controls operated by dual cables.

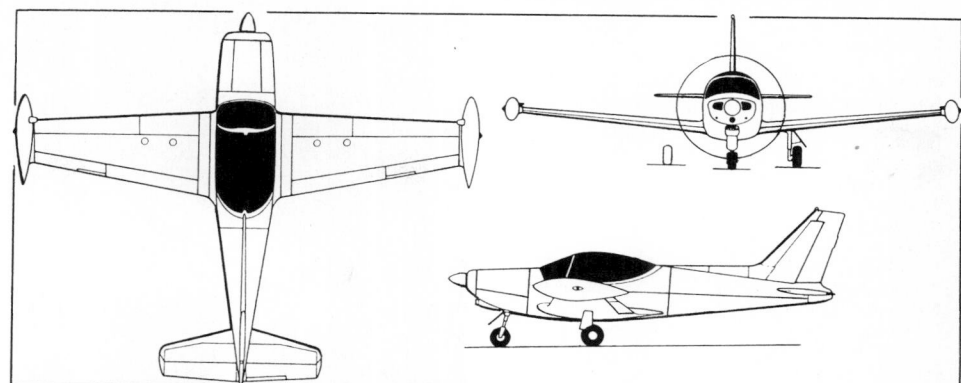

Aermacchi SF-260E/F basic trainer (*Jane's/Paul Jackson*)   0011076

STRUCTURE: All-metal stressed skin structure; wing skin formed of butt-jointed panels flush riveted; single main spar and auxiliary rear spar; press-formed ribs; wings bolted together on centreline and attached to fuselage by six bolts. Fatigue life 15,000 hours.

LANDING GEAR: Electrically retractable tricycle type, with manual emergency actuation. Inward-retracting main gear, of trailing arm type, and rearward-retracting nose unit, each embodying Magnaghi oleo-pneumatic shock-absorber (Type 2/22028 in main units). Nose unit is of leg and fork type, with coaxial shock-absorber and torque strut. Cleveland P/N 3080A mainwheels, with size 6.00-8 tube and tyre (6 ply rating), pressure 2.45 bar (36 lb/sq in). Cleveland P/N 40-77A nosewheel, with size 5.00-5 tube and tyre (8 ply rating), pressure 1.96 bar (28 lb/sq in). Cleveland P/N 3000-500 independent hydraulic single-disc brake on each mainwheel; parking brake. Nosewheel steering (±20°) operated directly by rudder pedals.

POWER PLANT: One 194 kW (260 hp) Textron Lycoming AEIO-540-D4A5 flat-six driving a Hartzell HC-C2YK-1BF/8477-8R two-blade, constant-speed, metal propeller.
Fuel in two light alloy tanks in wings each holding 49.5 litres (13.1 US gallons; 10.9 Imp gallons); and two permanent wingtip tanks each holding 72 litres (19.0 US gallons; 15.85 Imp gallons). Total internal fuel capacity 243 litres (64.2 US gallons; 53.3 Imp gallons), of which 235 litres (62.1 US gallons; 51.7 Imp gallons) are usable. Individual refuelling point on top of each tank. In addition, SF-260W may be fitted with two 80 litre (21.1 US gallon; 17.5 Imp gallon) auxiliary tanks on underwing pylons. Oil capacity (all models) 11.4 litres (3.0 US gallons; 2.5 Imp gallons).

ACCOMMODATION: Two pilots side by side in adjustable seats, with full blind-flying panel on right and reduced panel and radio on left. Third seat in rear. All three seats equipped with lap belts and shoulder harnesses. Baggage compartment aft of rear seat. Upper portion of sliding canopy tinted. Emergency canopy release handle for front seat occupant. Steel tube windscreen frame for protection in the event of an overturn. Air conditioning, oxygen and enlarged canopy optional.

SYSTEMS: Foot-powered hydraulics for mainwheel brakes only. No pneumatic system. 24 V DC electrical system of single-conductor negative earth type, including 70 A

Prestolite engine-mounted alternator/rectifier and 24 V 24 Ah battery, for engine starting, flap and landing gear actuation, fuel booster pumps, electronics and lighting. Sealed battery compartment in rear of fuselage on port side. Connection of an external power source automatically disconnects the battery. Heating system for carburettor air intake. Emergency electrical system for extending landing gear if normal electrical actuation fails; provision for mechanical extension in the event of total electrical failure. Cabin heating, and windscreen de-icing and demisting, by heat exchanger using engine exhaust air. Additional manually controlled warm air outlets for general cabin heating. Optional R134a gas air conditioning system and autopilot both available from 2000.

AVIONICS: *Comms:* Bendix/King or Rockwell Collins airways radio with dual com and nav; in SF-260E, avionics mounted aft of cabin and battery moved forward.

EQUIPMENT: Military equipment to customer's requirements. External stores can include one or two reconnaissance pods with two 70 mm automatic cameras, or two supply containers or two external tanks. Landing light in nose, below spinner. Two Servaux Samar-Sater 83/0 or 83/1 emergency packs.

ARMAMENT: Warrior has two underwing hardpoints, able to carry external stores on NATO 356 mm (14 in) shackles up to a maximum of 300 kg (661 lb) when flown as a single-seater. Typical alternative loads can include one or two SIAI gun pods, each with one or two 7.62 mm FN machine guns and 500 rounds; two 0.50 Browning machine gun pods; two Aerea AL-8-70 launchers each with eight 2.75 in rockets; two Forges de Zeebrugge LAU-32 launchers each with seven 2.75 in rockets; two Aerea AL-18-50 launchers each with eighteen 2 in rockets; two Aerea AL-8-68 launchers each with eight 68 mm rockets; two Thomson-Brandt 68-7 launchers each with seven 68 mm rockets; two Aerea AL-6-80 launchers each with six 81 mm rockets; two Thiokol LUU-2/B parachute flares; two SAMP EU 32 125 kg general purpose bombs or EU 13 120 kg fragmentation bombs; two SAMP EU 70 50 kg general purpose bombs; Mk 76 11 kg practice bombs; two cartridge throwers for 70 mm multipurpose cartridges, F 725 flares or F 130 smoke cartridges.

DIMENSIONS, EXTERNAL:
Wing span over tip tanks	8.35 m (27 ft 4¾ in)
Wing chord: at root	1.60 m (5 ft 3 in)
mean aerodynamic	1.325 m (4 ft 4¼ in)
at tip	0.785 m (2 ft 7 in)
Wing aspect ratio (excl tip tanks)	6.3
Wing taper ratio	2.2
Length overall	7.10 m (23 ft 3½ in)
Fuselage: Max width	1.10 m (3 ft 7¼ in)
Max depth	1.04 m (3 ft 5 in)
Height overall	2.41 m (7 ft 11 in)
Elevator span	3.01 m (9 ft 10½ in)
Wheel track	2.275 m (7 ft 5½ in)
Wheelbase	1.66 m (5 ft 5¼ in)
Propeller diameter	1.93 m (6 ft 4 in)
Propeller ground clearance	0.32 m (1 ft 0½ in)

DIMENSIONS, INTERNAL:
Cabin: Length	1.66 m (5 ft 5¼ in)
Max width	1.00 m (3 ft 3¼ in)
Height: from seat cushion:	
to canopy	0.98 m (3 ft 2½ in)
to enlarged canopy	1.07 m (3 ft 6 in)

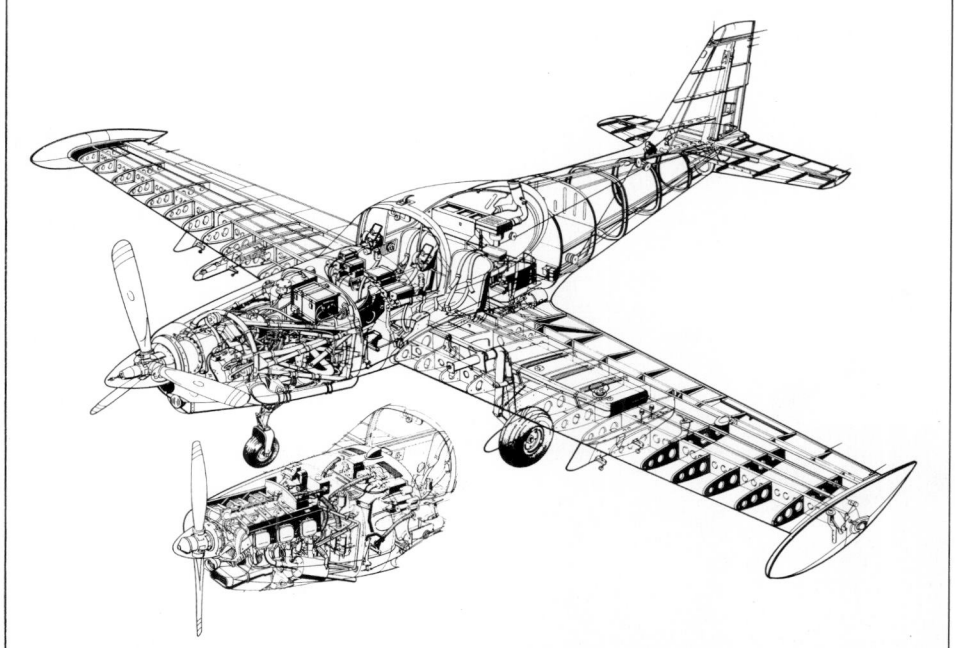

Structural details of the SF-260TP, with scrap view of the piston-engined SF-260E/F   0051082

Current SF-260E instrument panel   0110886

Volume	1.5 m³ (53 cu ft)
Baggage compartment volume	0.18 m³ (6.4 cu ft)

AREAS:

Wings, gross	10.10 m² (108.7 sq ft)
Ailerons (total, incl tabs)	0.76 m² (8.18 sq ft)
Trailing-edge flaps (total)	1.18 m² (12.70 sq ft)
Fin	0.76 m² (8.18 sq ft)
Dorsal fin	0.16 m² (1.72 sq ft)
Rudder, incl tab	0.60 m² (6.46 sq ft)
Tailplane	1.46 m² (15.70 sq ft)
Elevator, incl tab	0.96 m² (10.30 sq ft)

WEIGHTS AND LOADINGS:

Manufacturer's basic weight empty	779 kg (1,717 lb)
Baggage capacity	20 kg (44 lb)
Max fuel weight: wing internal and tip tanks (all	
versions)	169 kg (372.5 lb)
underwing tanks	115 kg (254 lb)
Max T-O weight: E	1,200 kg (2,645 lb)
Warrior	1,350 kg (2,976 lb)
Max wing loading: E	118.8 kg/m² (24.33 lb/sq ft)
Warrior	128.7 kg/m² (26.36 lb/sq ft)
Max power loading: E, M	6.18 kg/kW (10.17 lb/hp)
Warrior	6.70 kg/kW (11.02 lb/hp)

PERFORMANCE (at 1,100 kg; 2,425 lb):

Never-exceed speed (VNE)	236 kt (437 km/h; 271 mph)
Max level speed at S/L	182 kt (337 km/h; 209 mph)
Cruising speed at 75% power at 1,830 m (6,000 ft)	
	175 kt (324 km/h; 201 mph)
Stalling speed, flaps down	59 kt (110 km/h; 68 mph)
Max rate of climb at S/L	548 m (1,800 ft)/min
Service ceiling	6,100 m (20,000 ft)
T-O run	275 m (905 ft)
Landing run	270 m (885 ft)
Range with max internal fuel	
	710 n miles (1,314 km; 817 miles)
Max range, internal and external fuel	
	1,090 n miles (2,018 km; 1,254 miles)
g limits	+6/−3

UPDATED

AERMACCHI SF-260TP

There have been no recent sales of this variant of SF-260. Refer to 2002-03 *Jane's* for detailed description.

UPDATED

AERMACCHI M-346

TYPE: Advanced jet trainer/light attack jet.

PROGRAMME: Announced 25 July 2000. Westernised version of Yakovlev Yak-130 (which see in Russian section), conforming to requirements of Eurotrainer group. Three prototypes as follow-on to 300 trial flights with prototype Yak-130; mockup shown at Paris, June 2001; roll-out was due June 2002, and first flight 12 months later.

Aermacchi shares design rights and has production and modification rights for the Yak-130. Development of the aircraft was transferred to Italy in mid-1998 to prevent further delays resulting from the economic situation in Russia. Some funding provided by Russian Ministry of Industry. Dott Massimo Luchesini is Aeromacchi's AEM/Yak-130 programme director. Italian Air Force expected to certify aircraft and issue service release, although no firm local requirement yet exists.

Features generally as for Yakovlev Yak-130 (which see in Russian section). Specific aspects detailed below.

DESIGN FEATURES: Cockpit based on Eurofighter Typhoon. Chine deleted from forward fuselage. Autonomous operation made possible by self-starting, OBOGS, BITE, APU and use of standard support equipment. Design to MIL-STD-1350A and MIL-A-8660A; maintenance to MIL-STD-470B; technical publications in accordance with applicable MIL-STD. Service life 10,000 hours; extension to 15,000 hours planned.

Aermacchi M-346 cockpit mockup demonstrated by Galileo Aerospazio (*Jane's/Paul Jackson*)   NEW/0132401

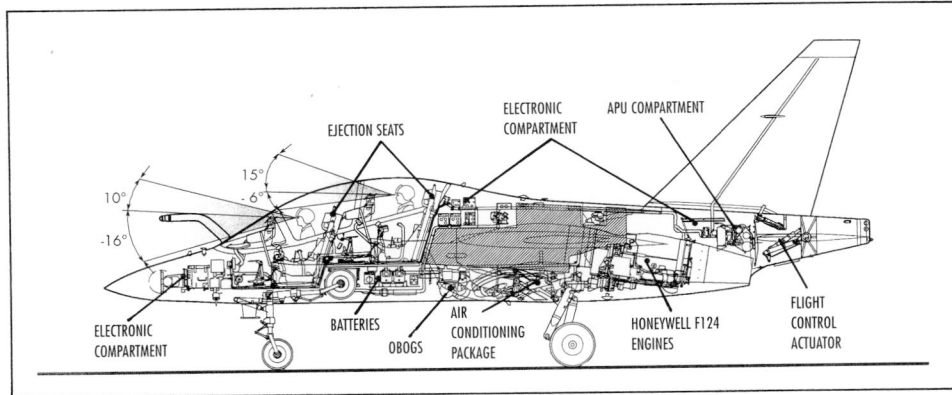

Internal profile of Aermacchi M-346   0110889

LANDING GEAR: Nosewheel leg by Liebherr. Forward-retracting mainwheel legs from AMX fighter-bomber.

POWER PLANT: Two 28.0 kN (6,300 lb st) Honeywell F124-GA-200 twin-shaft turbofans with dual-channel FADEC. Single-point pressure refuelling. Inboard wing pylons 'wet', each with provision for 591 litre (156 US gallon; 130 Imp gallon) tank. Goodrich/Secondo Mona fuel system. Optional (30 min attachment time) refuelling probe.

ACCOMMODATION: Two Martin-Baker Mk 16D zero/zero ejection seats. Canopy reduced in depth and cross-section, compared with Yak-130.

SYSTEMS: Teleavio/BAE Systems quadruple-redundant, full authority digital FCS. Environmental control by bleed air or APU; nominal cockpit pressure differential 0.25 bar (3.6 lb/sq in); OBOGS. Electrical system includes ASE 20 kVA generator on each engine (100 per cent redundancy), two 9 kW transformer-rectifiers, one 5 kW 28 V DC emergency generator (Honeywell GTCP 36-150 APU) and two Ni/Cd batteries. Two independent hydraulic systems by Microtecnica, both operating at 207 bar (3,000 lb/sq in); Dowty/Microtecnica control surface actuators.

AVIONICS: *Comms:* Two VHF/UHF transceivers; IFF transponder.

Flight: Honeywell laser gyro INS with embedded GPS; Tacan and VOR/ILS/marker beacon receiver.

Instrumentation: Alenia Difesa raster/stroke HUD; HOTAS controls; three Alenia Difesa 127 × 127 mm (5 × 5 in) colour LCDs per cockpit; provision for HMD and NVGs. Optional two-screen cockpit prepared for customers who wish to simulate Lockheed Martin F-35 environment.

Mission: Centreline attachment for Vinten Vicon 19 reconnaissance pod, Litening FLIR, or ATLIS targeting pod.

Self-defence: Provision for RWR, chaff/flares and underwing Elettronica ELT-555 ECM pod.

ARMAMENT: Nine hardpoints: one on centreline, rated at 600 kg (1,323 lb); four under each wing, rated at 1,050 kg (2,315 lb) (inboard), 550 kg (1,213 lb), 300 kg (661 lb) and 150 kg (331 lb) (wingtip). Optional 30 mm DEFA cannon pod on centreline. Typical weapon loads include six Mk 82 (500 lb), four Mk 83 (1,000 lb) or two Mk 84 (2,000 lb) bombs; four GBU-12 (500 lb), six GBU-16 (1,000 lb) or Opher Mk 82 guided bombs; six Rockeye or four BL 755 CBUs; six Durandal anti-runway dispensers; four BRD-4 or LAU-7/LAU-5002/LAU-32 rocket pods; four AIM-9 Sidewinder AAMs; four AGM-65 Maverick ASMs; or four Brimstone anti-armour missiles.

DIMENSIONS, EXTERNAL:

As for Yak-130

AREAS:

As for Yak-130

WEIGHTS AND LOADINGS:

Weight empty	4,625 kg (10,196 lb)
Max weapon load	3,000 kg (6,614 lb)
Max internal fuel weight	1,950 kg (4,299 lb)
Max T-O weight: trainer	6,700 kg (14,771 lb)
attack	9,500 kg (20,943 lb)
Max wing loading: trainer	284.9 kg/m² (58.34 lb/sq ft)
attack	403.9 kg/m² (82.73 lb/sq ft)
Max power loading: trainer	120 kg/kN (1.17 lb/lb st)
attack	169 kg/kN (1.66 lb/lb st)

PERFORMANCE:

Limiting Mach No	0.95
Max level speed at 1,525 m (5,000 ft)	
	585 kt (1,083 km/h; 673 mph)
Stalling speed, landing with 20% fuel,	
flaps down	90 kt (167 km/h; 104 mph)
Max rate of climb at S/L	6,100 m (20,000 ft)/min

Mockup of Aermacchi M-346   NEW/0096682

Service ceiling	13,715 m (45,000 ft)
T-O run	280 m (920 ft)
Landing run	590 m (1,935 ft)

Radius of action:
air combat training with gun pod and two Sidewinders, 20 min manoeuvring
100 n miles (185 km; 115 miles)
combat air patrol, as above plus two external fuel tanks, 2 h loiter and 3 min combat
160 n miles (296 km; 184 miles)

close air support, hi-lo-lo-hi, two Mk 83 and four Mk 82 bombs, two Sidewinders, gun pod, 30 min loiter
110 n miles (203 km; 126 miles)
interdiction, hi-lo-lo-hi, two Mk 83 bombs, two Sidewinders, gun pod, two external fuel tanks
330 n miles (611 km; 379 miles)
Range, 10% reserves: with max internal fuel
1,020 n miles (1,889 km; 1,173 miles)
with two external tanks
1,370 n miles (2,537 km; 1,576 miles)
UPDATED

OTHER AIRCRAFT

Production rights to the (former SIAI-Marchetti) S.211 jet trainer and light attack aircraft were transferred to Aermacchi in 1997. Promotion continued during 1999, but no production has taken place since before the sale. An illustrated description last appeared in the 1997-98 *Jane's*. The M-290TP Redigo turboprop trainer was last described in the 1999-2000 *Jane's*.

See International section for Aermacchi/Alenia/Embraer **AMX** attack aircraft.

UPDATED

AGUSTA

AGUSTA
(An AgustaWestland Company)

Via Giovanni Agusta 520, I-21017 Cascina Costa di Samarate (VA)
Tel: (+39 0331) 22 97 69
Fax: (+39 0331) 22 99 84
Web: http://www.agusta.com
CHAIRMAN AND CEO: Amedeo Caporaletti
VICE-PRESIDENT, EXTERNAL RELATIONS AND COMMUNICATIONS: Gian Luigi Ghezzi
CO-GENERAL MANAGER, MARKETING AND SALES: Giuseppe Orsi
PRESS OFFICER: Gianluca Grimaldi

VERGIATE WORKS:

Via Roma 51, I-21019 Vergiate (VA)
Tel: (+39 0331) 94 01 11
Fax: (+39 0331) 94 79 09

Formed in 1977; the Agusta group (see 1980-81 *Jane's*) completely reorganised from 1 January 1981 under new holding company Agusta; became part of Italian public holding company EFIM, employing nearly 10,000 people in 12 factories in various parts of Italy. Agusta workforce in Italy and abroad was 5,014 in 2000.

As part of recovery of liquidated state-owned EFIM, Agusta group integrated into Finmeccanica on 20 December 1996.

On 16 April 1998, Finmeccanica and the United Kingdom's GKN, owners of Agusta and Westland, respectively, announced signature of an MoU which was intended to lead to an 'alliance of equals' of their helicopter divisions. This was effected on 26 July 2000, as described under AgustaWestland in the International section. The pooling of design, manufacturing and marketing activities will not require refinancing nor affect previous programme-based agreements with other partners.

In November 1998 Agusta and Bell Helicopter Textron (which see in Canada and United States sections) signed an agreement to form Bell/Agusta Aerospace Company, to share development, manufacture and marketing of the Bell/Agusta BA609 civil tiltrotor and AB139 utility helicopter, described in the International section. See also the Eurotilt entry in that section, in which AgustaWestland also involved.

On 14 January 2002, Agusta and Denel Group of South Africa signed an agreement for licensed production of the A 109 and A 119 in South Africa. Under the terms of the agreement Denel will manufacture the helicopters at its Kempton Park facility for marketing in Africa, the Middle East, South America and Southeast Asia. Denel will manufacture 25 of the 30 A 109s ordered by the South African Air Force; negotiations were under way in early 2002 for Denel to manufacture the airframes of A 109Ms ordered

Italian Army Agusta A 129 from Lot 2 *(Jane's/Paul Jackson)*   NEW/0137357

by the Swedish armed forces.

By mid-2000, Agusta had built more than 3,770 helicopters under licence from Bell, Boeing, MD helicopters and Sikorsky.

INTERNATIONAL AFFILIATES:

Agusta Aerospace Corporation
2655 Interplex Drive, Travose, Philadelphia, Pennsylvania 19047, USA
Tel: (+1 215) 281 14 00
Fax: (+1 215) 281 04 40
Telex: 6851181

Agusta Aerospace Service SA
Keiberg Park, 23 Excelsiorlaan, B-1930 Zaventem, Belgium
Tel: (+32 2) 648 55 15
Telex: 63349

Bell/Agusta Aerospace Company
See International section.

EH Industries Ltd
See International section

NH Industries sarl
See International section

CASCINA COSTA WORKS:

Original Agusta company established 1907 by Giovanni Agusta; acquired licence for Bell 47 in 1952; first flight of first Agusta example 22 May 1954; later produced Bell 204, 205, 206 and 212; Bell AB412 still in low-volume production; also produced various versions of Sikorsky S-61 under licence and is partner with Westland in EH 101 (see under EHI in International section); participates in NH 90 programme (see International). Own designs include A 109 multirole helicopter, A 119 Koala, A 129 anti-tank helicopter and AB139 joint venture with Bell Helicopter Textron.

UPDATED

AGUSTA A 129 MANGUSTA
English name: Mongoose

TYPE: Attack helicopter.
PROGRAMME: Italian Army specification issued 1972; A 129 given go-ahead March 1978; final form settled 1980; detail design completed 30 November 1982; first flights of five development aircraft 11 September 1983, 1 July and 5 October 1984, 27 May 1985 and 1 March 1986; first delivery October 1990. In 1999, Italian government approved A 129 upgrade, which will raise first 45 army helicopters to multirole standard. Raytheon Stinger, in AAM version, selected in March 2000 to arm Italian A 129s; integration due to be complete by 2003.
CURRENT VERSIONS: **Anti-tank:** Initial Italian Army version; also demonstrated for export (*described below*). Lot 2 aircraft (from 16th production onward – MM81391) equipped with secure communications, cockpit lighting compatible with third-generation NVGs, IR suppressors, IR jammers, laser warning system, improved main computer software, auxiliary fuel tanks and folding main rotor blades.

Multirole: Italian Army is receiving aircraft Nos. 46 to 60 (MM81421 to 81435) in this configuration; originally designated **A 129 CBT** (*combattimento:* combat), but renamed **A 129C** or (*elicottero da esplorazione e scorta:* reconnaissance and escort helicopter) by 2002. Main features are five-blade rotor; transmission uprated to 1,268 kW (1,700 shp); tail rotor speed increased by 7 per cent; maximum take-off weight increased to 4,600 kg (10,141 lb); Lockheed Martin/Otobreda TM 197B 20 mm cannon in nose turret; and Stinger AAMs. Fifth A 129 prototype (MMX598) converted to this configuration by 1997. Other features are G13 software standard, small strake along port side of tailboom; long, tray-shaped ammunition tank on port side of fuselage; revised cockpit layout, including HOCAC controls; updated Galileo Avionica 'HIRNS Plus' navigation/sighting system; integrated GPS; higher landing gear; improved LO paint scheme and AN/ALQ-144 IR jammer.

Similar upgrade of first 45 Mangustas due for completion by 2007, but all will additionally receive G15 software in associated programme, permitting upgrade of defensive aids suite (*sistema integrato di autoprotezione:* integrated automatic protection system) with Marconi

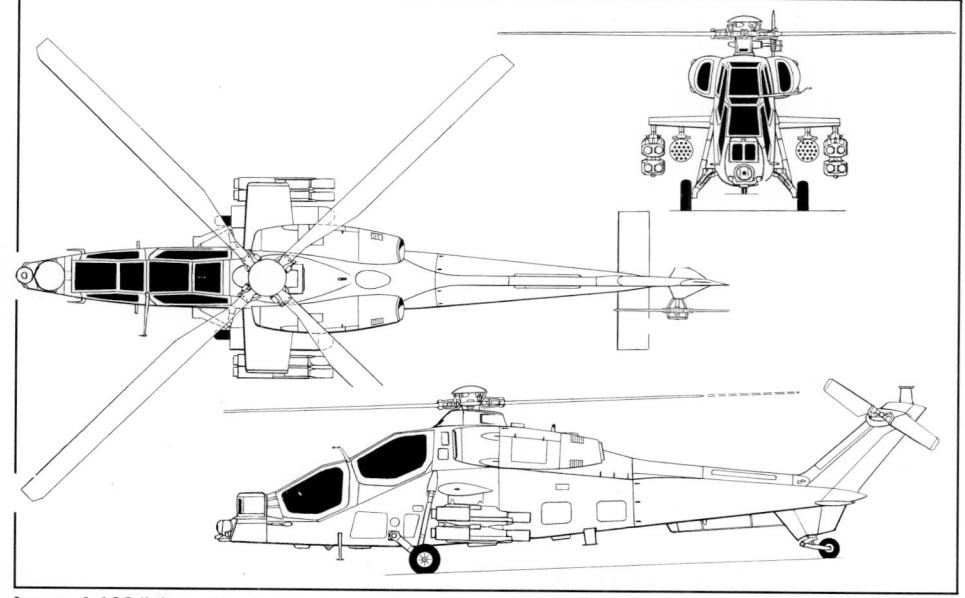

Agusta A 129 light anti-tank, attack and advanced scout helicopter *(Jane's/Dennis Punnett)*

Initial A 129C/A 129 EES undergoing trials with Italian Army in 2002   *NEW*/0526453

MILDS II missile approach warner and RALM-01(V)2 laser warner, Elettronica ELT-156X(V)4 RWR and MES ECDS-2 chaff/flare dispensers integrated with current AN/ALQ-144 IR jammer. First 45 helicopters receiving G15 first (2002 to 2007), followed by new-build A 129Cs.

Shipborne: Proposed maritime anti-ship version; no orders received by early 2000.

A 129 International: Described separately.

CUSTOMERS: First five of planned 60 for Italian Army anti-tank squadrons (15 Lot 1 and 45 Lot 2) delivered October 1990 after delay of more than a year to allow fitting of Saab/ESCO HeliTOW system with nose-mounted sight; currently operated by 493° Squadrone Elicotteri da Attacco of 49° Gruppo Squadroni at Rimini/Miramare. First Lot 2 A 129, with Honeywell/OMI helicopter IR navigation system (HIRNS), entered service in August 1993; operated by 7th Attack Helicopter Regiment 'Vega' at Casarsa and Army Aviation Centre at Viterbo; 45 in service by end of 1996; last 15 aircraft being delivered as a A 129 EES version from 2002 onwards. Upgrade contract for first 45 awarded December 2001.

DESIGN FEATURES: Fully articulated four-blade main rotor with blades retained by single elastomeric bearing and restrained by hydraulic drag damper and mechanical droop stop; main rotor blade folding on Lot 2 aircraft.

Main transmission has independent oil cooling system; intermediate and tail rotor gearboxes grease lubricated; all designed for at least 30 minutes' run dry; accessory gearbox can be run independently on ground without rotor engagement by No. 1 engine engaged by pilot-operated clutch.

FLYING CONTROLS: Full-time dual electronic flight controls, with full manual reversion, provide automatic heading hold, autohover, autopilot modes and autostabiliser modes, all selectable by pilot; gunner in front seat has cyclic side-arm controller, normal collective lever and pedals and has full access to AFCS; electrical inputs from AFCS integrated with hydraulic-powered control units.

STRUCTURE: Composites materials account for 45 per cent of fuselage weight (less engines) and 16.1 per cent of total empty weight; material used for fuselage panels, nosecone, tailboom, tail rotor pylon, engine nacelles, canopy frame and maintenance panels; each blade has CFRP and Nomex main spar, Nomex honeycomb leading- and trailing-edges, composites skins, stainless steel leading-edge abrasion strip and frangible tip; control linkage runs inside driveshaft to reduce radar signature, avoid icing and improve ballistic tolerance. Blades tolerant to 12.7 mm hits, possibly also 23 mm; delta-hinged two-blade tail rotor with broad-chord blades for ballistic tolerance. Total 70 per cent of airframe surface is composites. Bulkhead in nose and A frame running up through fuselage to rotor pylon protect crew against roll-over; overall IR suppressing paint; airframe meets MIL-STD-1290 crashworthiness covering vertical velocity changes of 11.2 m (36 ft 9 in)/s and longitudinal changes of 13.1 m (43 ft 0 in)/s.

LANDING GEAR: Non-retractable tailwheel type, with single wheel on each unit. Two-stage hydraulic shock-strut in each main unit designed to withstand normal loads and hard landings at descent rates in excess of 10 m (32.8 ft)/s.

POWER PLANT: Two Rolls-Royce Gem 1004 turboshafts, each with a T-O rating of 657 kW (881 shp) for 30 minutes; maximum continuous rating of 615 kW (825 shp) for normal twin-engined operation; intermediate contingency rating of 657 kW (881 shp) for 30 minutes; maximum contingency rating of 704 kW (944 shp) for 2½ minutes; and emergency rating (S/L, ISA) of 759 kW (1,018 shp) for 20 seconds. Transmission rating (Lot 2) is 969 kW (1,300 shp) (two engines), 704 kW (944 shp) for single-engined operation, with emergency rating of 759 kW (1,018 shp); power input into transmission is at 27,000 rpm from the RR 1004. Standard transmission rating for Lot 3 (Multirole) aircraft is 1,268 kW (1,700 shp); 1,566 kW (2,100 shp) in emergency. Production engines licence-built in Italy by Piaggio. Fireproof engine compartment, with engines widely spaced to improve survivability from enemy fire.

Two separate fuel systems, with cross-feed capability; interchangeable self-sealing and crash-resistant tanks, self-sealing lines, and digital fuel feed control. Tanks can be foam-filled for fire protection. Single-point pressure refuelling. IR exhaust suppression system (from Lot 2) and low engine noise levels. Separate independent lubrication oil-cooling system for each engine. Provision (Lot 2 aircraft) for auxiliary (self-ferry) fuel tanks on outboard underwing stations.

ACCOMMODATION: Pilot and co-pilot/gunner in separate cockpits in tandem. Elevated rear (pilot's) cockpit. External crew field of view exceeds MIL-STD-850B. Each cockpit has a flat plate, low-glint canopy with upward-hinged door panels on starboard side, blow-out port side panel for exit in emergency, and Martin-Baker crashworthy seat with sliding side panels of composites armour. Landing gear design and crashworthy seats reduce impact from 50 *g* to 20 *g* in crash.

SYSTEMS: Hydraulic system includes two main circuits dedicated to flight controls and two independent circuits for rotor and wheel braking. Main system operates at pressure of 207 bar (3,000 lb/sq in) and is fed by two independent power groups integrated and driven mechanically by the main transmission. Tandem actuators are provided for main rotor flight controls. Hydraulic system flow rate 23.6 litres (6.2 US gallons; 5.2 Imp gallons)/min in each main group. Spring-type reservoirs, pressurised at 0.39 bar (5.6 lb/sq in).

AVIONICS: *Comms:* Dual Marconi SRT-651/A U/VHF and single Marconi SRT-170/EB4 HF/SSB radios; Italtel IFF (with Mode 4 encryption unit in Lot 2 aircraft). A 129C suite additionally comprises Have Quick V secure radios and Marconi SRT-651/S SINCGARS for communication with ground forces.

Flight: Fully integrated digital multiplex system (IMS) controls navigation, flight management, weapon control, autopilot, monitoring of transmission and engine condition, fuel/hydraulic/electrical systems, caution and warning systems; IMS managed by two Agusta Sistemi/Harris central computers, each capable of operating

independently, backed by two interface units which pick up outputs from sensors and avionic equipment and transfer them, via redundant MIL-STD-1553B databusses, to main computers for real-time processing; processed information is presented to pilot and co-pilot/gunner on separate graphic/alphanumeric head-down multifunction displays (MFDs) with standard multifunction keyboards for easy access to information, including area navigation using up to 100 waypoints, weapons status and selection, radio tuning and mode selection, caution and warning, and display of aircraft performance; conventional instruments and dials are provided as back-up; IMS computer can store up to 100 preset frequencies for HF, VHF and UHF radio management; navigation is controlled by navigation computer of IMS coupled to Doppler radar and radar altimeter with low airspeed indicator, normally used for rocket aiming, providing back-up velocity data when the Doppler is beyond limits; synthetic map presentation of waypoints, target areas and dangerous areas is shown on pilot's or co-pilot's MFD; Litton strapdown inertial reference for both flight control and navigation is integrated into the IMS; AFCS provides either three-axis stabilisation or full attitude and heading hold, automatic hover, downward transition to hover or holds for altitude, heading and airspeed or groundspeed and automatic track following. Litton LISA 4000 AHRS is integrated on A 129C with GPS; later variant also has Marconi ANV-351 Doppler velocity sensor.

Instrumentation: Full day/night operational capability, with equipment designed to give both crew members a view outside helicopter irrespective of light conditions; cockpit lighting compatible with night vision goggles.

Mission: Pilot's night vision system (HIRNS: helicopter IR night system) allows nap-of-earth (NOE) flight by night with outside view generated by BAE North America FLIR sensor mounted on a GEC/OMI steerable platform at nose of aircraft and presented to both crewmen through the monocle of the Honeywell integrated helmet and display sighting system (IHADSS), to which it is slaved by helmet position sensors; flight information symbology superimposed on to image, giving true head-up reference. HeliTOW sight gives co-pilot/gunner direct view optics and FLIR, plus laser for ranging; provision for mast-mounted sight (MMS). A 129C has Galileo Avionica HIRNS Plus and TEAC video recording system for helmet and display sighting, plus voice, allowing post-mission analysis.

Self-defence: Active and passive self-protection systems (ECCM and ECM) standard on Italian Army A 129; Have Quick II frequency-hopping radio in Lot 2; onboard nav/weapon system can connect directly or by datalink with Italian CATRIN C³I combat information system; active and passive electronic warfare systems include Elettronica ELT-554 radar jammer and ELT-156 RWR (ELT-156-05 in Lot 2) radar warning receiver; BAE Italia RALM-101 laser warning receiver; Sanders AN/ALQ-144A IR jammer; provisions for chaff/flare dispenser. Upgraded defensive aids under development for Lot 3 and retrofit, including ELT-156X(V)4, Marconi MILDS II missile approach warner, MES ECDS-2 chaff/flare dispenser and BAE RALM-01(V)2 laser warner.

ARMAMENT: Four attachments beneath stub-wing stressed for loads of 200 kg (441 lb) outboard and 300 kg (661 lb) inboard or 300 kg (661 lb) outboard and 100 kg (220 lb) inboard; outboard stations incorporate articulation which allows pylon to be elevated 2° and depressed 10° to increase missile launch envelope; they are aligned with aircraft automatically, with no need for boresighting.

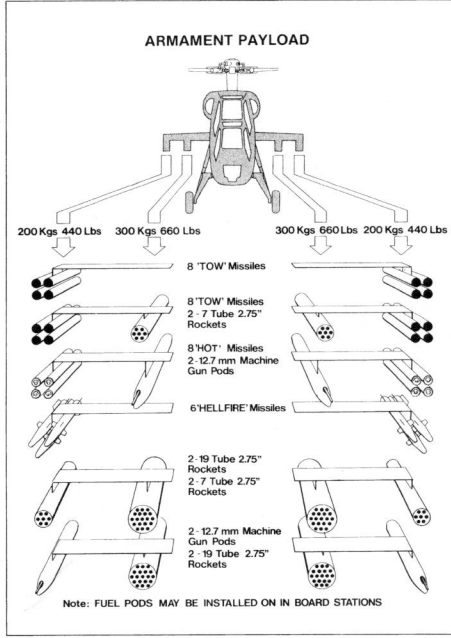

ARMAMENT PAYLOAD

200 Kgs 440 Lbs   300 Kgs 660 Lbs   300 Kgs 660 Lbs   200 Kgs 440 Lbs

8 'TOW' Missiles

8 'TOW' Missiles
2 - 7 Tube 2.75"
Rockets

8 'HOT' Missiles
2 - 12.7 mm Machine
Gun Pods

6 'HELLFIRE' Missiles

2 - 19 Tube 2.75"
Rockets
2 - 7 Tube 2.75"
Rockets

2 - 12.7 mm Machine
Gun Pods
2 - 19 Tube 2.75"
Rockets

Note: FUEL PODS MAY BE INSTALLED ON IN BOARD STATIONS

A 129 weapon options

Initial armament of up to eight TOW, ITOW or TOW 2/2A wire-guided anti-tank missiles (two, three or four in carriers suspended from each wingtip station), with Saab/ ESCO HeliTOW aiming system; with these can be carried, on inboard stations, either two 7.62, 12.7 or 20 mm gun pods, or two launchers each for seven or 19 air-to-surface rockets. For general attack missions, rocket launchers can be carried on all four stations; Italian Army has specified SNIA-BPD 81 and 70 mm rockets. Multirole and International versions can carry, in addition to the above, a Lockheed Martin/Otobreda TM 197B (standard 350 or optional 500 rounds) or Giat M621 20 mm cannon mounted in a nose turret; and Stinger AAMs. Optional upgrades offered on export versions include an autotracking sight; provision for up to eight Hellfire anti-tank missiles with autonomous laser designation capability; or a mix of four Hellfires and four TOWs. Other armament options include up to eight Hot missiles; AIM-9L Sidewinder, Mistral or Javelin AAMs; or grenade launchers. Lucas 0.50 in self-contained gun turret qualified, but not used by Italian Army.

DIMENSIONS, EXTERNAL:

Main rotor diameter	11.90 m (39 ft 0½ in)
Tail rotor diameter	2.32 m (7 ft 7 in)
Wing span	3.20 m (10 ft 6 in)
Width over TOW pods	3.60 m (11 ft 9¾ in)
Length overall, both rotors turning	14.29 m (46 ft 10½ in)
Fuselage: Length	12.275 m (40 ft 3¼ in)
Max width	0.95 m (3 ft 1½ in)
Height:	
over tailfin, tail rotor horizontal	2.75 m (9 ft 0¼ in)
tail rotor turning	3.315 m (10 ft 10½ in)
to top of rotor head	3.35 m (11 ft 0 in)
Tailplane span	2.50 m (8 ft 2¼ in)
Wheel track	2.23 m (7 ft 3¾ in)
Wheelbase	6.955 m (22 ft 9¾ in)

AREAS:

Main rotor disc	111.20 m² (1,197.0 sq ft)
Tail rotor disc	4.23 m² (45.53 sq ft)

WEIGHTS AND LOADINGS (Italian Army):

Weight empty, equipped	2,529 kg (5,575 lb)
Max internal fuel load	750 kg (1,653 lb)
Max external weapons load	1,200 kg (2,645 lb)
Max T-O weight: standard	4,100 kg (9,039 lb)
multirole version	4,600 kg (10,141 lb)
T-O weight, normal mission	3,950 kg (8,708 lb)
Max disc loading: standard	36.9 kg/m² (7.55 lb/sq ft)
multirole version	41.4 kg/m² (8.47 lb/sq ft)
Transmission loading at max T-O weight and power:	
standard	4.23 kg/kW (6.95 lb/shp)
multirole version	3.63 kg/kW (5.97 lb/shp)

PERFORMANCE (Italian Army, with eight TOW, at mission T-O weight of 3,950 kg (8,708 lb), at 2,000 m (6,560 ft), ISA+20°C, except where indicated):

Dash speed	159 kt (294 km/h; 183 mph)
Max level speed at S/L	135 kt (250 km/h; 155 mph)
Max rate of climb at S/L	612 m (2,008 ft)/min
Service ceiling	4,725 m (15,500 ft)
Hovering ceiling: IGE	3,140 m (10,300 ft)
OGE	1,890 m (6,200 ft)

Basic 2 h 30 min mission profile with eight TOW and 20 min fuel reserves:
Fly 54 n miles (100 km; 62 miles) to battle area, mainly in NOE mode, 90 min loiter (incl 45 min hovering), and return to base

Max endurance, no reserves	3 h 5 min
g limits	+3.5/−0.5

UPDATED

AGUSTA A 129 INTERNATIONAL

TYPE: Attack helicopter.

PROGRAMME: Upgraded export version of A 129 (see previous entry) with increased power and military load, plus improved avionics. First flight of prototype powered by two T800 turboshafts October 1988; demonstrated in Arabian Gulf during 1990; first flight with production standard five-blade rotor (converted prototype MMX592, 29002) 9 January 1995; tail rotor diameter slightly increased; T800 gives between 20 and 40 per cent more power than Gem 1004. First firing of AGM-114 Hellfire ATMs, November 1999; Rafael NT-D Dandy ATM integrated and launched in same test series.

CURRENT VERSIONS: **International:** *As described.*

Scorpion: Offered to Australia in 1998 to meet Project Air 87 for 25 to 30 armed reconnaissance helicopters. Based on Multirole version, but some A 129 International features. Competition won in 2001 by Eurocopter Tiger.

A 129DE: Trials aircraft registered (I-INTR) to Agusta in February 1998.

A 129 International showing its five-blade main rotor   0102392

Data generally as for Mangusta, except as below.

DESIGN FEATURES: Five-blade main rotor; pannier on each side of forward fuselage.

POWER PLANT: Two LHTEC T800-LHT-800 turboshafts, each 996 kW (1,335 shp) for twin-engined operation or 1,047 kW (1,404 shp) maximum contingency OEI; transmission rating 1,268 kW (1,700 shp). Optional 33 per cent additional internal fuel.

AVIONICS: *Flight:* INS/GPS as main navigation sensor.

Instrumentation: Two 152 × 203 mm (6 × 8 in) MFDs in each crew position.

Mission: Upgraded observation and targeting sensors (CCD TV and laser range-finder, replacing original TOW telescopic sight), permitting autonomous laser designation and maximum range launch of Hellfire ATMs.

ARMAMENT: Hellfire and/or TOW anti-tank missile, 70 and 81 mm rockets; Lockheed Martin/Otobreda TM 197B 20 mm cannon in nose turret; Stinger or Mistral AAMs.

WEIGHTS AND LOADINGS:

Primary mission T-O weight	4,800 kg (10,582 lb)
Max T-O weight	5,100 kg (11,243 lb)

PERFORMANCE (mixed weapons at mission T-O weight):

Cruising speed	150 kt (278 km/h; 173 mph)
Max rate of climb at S/L	677 m (2,220 ft)/min
Max rate of climb, OEI	274 m (900 ft)/min
Vertical rate of climb at S/L	326 m (1,070 ft)/min
Hovering ceiling: IGE	4,206 m (13,800 ft)
OGE	3,292 m (10,800 ft)

Max range, internal fuel, no reserves
303 n miles (561 km; 348 miles)

VERIFIED

AGUSTA A 109

US Coast Guard designation: MH-68A
Swedish armed forces designation: Hkp 15

TYPE: Light utility helicopter.

PROGRAMME: First flight of original A 109 4 August 1971; deliveries of A 109A started early 1976; single-pilot IFR certification 20 January 1977; deliveries of uprated A 109 Mk II began September 1981; A 109C certified 1989; first deliveries February 1989; A 109 Max was medevac version; A 109CM and A 109EOA for military use.

First flight of A 109K April 1983; first flight of production representative second aircraft March 1984. A 109K2s of REGA air ambulance fitted with Sextant AFDS 95-1 AFCS, granted FAA single-pilot IFR certification late in 1996.

A 109 Power programme begun in 1993, intending to unify A 109C and A 109K2; initially known as A 109 Unified; launched November 1993; unique A 109D development aircraft with FADEC-equipped Allison 250-C22R9s flew 1 October 1994; prototype A 109E (I-EPWS) flew 8 February 1995. A 109D became second A 109E prototype.

CURRENT VERSIONS: **A 109KM:** Military version of A 109K2; roles include anti-tank/scout, escort, command and control, utility, ECM and SAR/medevac; fixed landing gear; sliding side doors.

A 109KN: Shipboard version with equivalent roles to A 109KM, including anti-ship, over-the-horizon surveillance and targeting and vertical replenishment.

A 109K2: Special civil rescue version first sold to Swiss REGA non-profit rescue service; REGA equipment

Agusta A 109 Special Edition Power Elite   *NEW*/0527101

For details of the latest updates to *Jane's All the World's Aircraft* online and to discover the additional information available exclusively to online subscribers please visit

jawa.janes.com

First Agusta A 109LUH for South Africa *NEW*/0527097

includes Spectrolab SX-16 searchlight, 1,000 kg (2,204 lb) cargo hook, GPS, Elbit moving map display and single-pilot IFR instrumentation; NVG compatible. Equipped with Thales AFDS 95-1 AFCS from 1996.

A 109K2 Law Enforcement: Dedicated police version; optional equipment includes 907 kg (2,000 lb) cargo hook, 204 kg (450 lb) capacity variable speed rescue hoist with 50 m (164 ft) of cable, rappeling kit, wire-strike protection, SX-16 searchlight, MA3 retractable light, external loudspeakers, emergency floats, GPS, FM tactical communications, weather radar, LLTV and FLIR.

A 109 Power: Revealed at 1995 Paris Air Show, by which time prototype had accumulated more than 60 hours of flight testing. Engineering designation is **A 109E**. Based on A 109K2 airframe with new, A 129-derived lightweight, low-maintenance titanium main rotor head connected to composites material grips via single elastomeric bearing on each blade; Pratt & Whitney Canada PW206C or Turbomeca Arrius 2K1 engines; new heavy-duty high-clearance landing gear in revised position below fuselage; cabin as for A 109K2, retaining quick-change facility from passenger to EMS operation. First production aircraft (I-PWER) completed late 1995; RAI IFR certification 31 May 1996; FAA IFR certification 26 August 1996. Aircraft with Rogerson IIDS (integrated instrument display system) cockpit (I-EAPW; first production) displayed at 1997 Paris Air Show. First 120 or so Powers had P&W engines; Turbomeca variant phased in with delivery of N7YL to Erie Lifestar in early September 2001. PZL-Świdnik of Poland contracted to build fuselages for A 109E between 1996 and 2002 as alternative production source. By 2001, production stood at some 45 per year.

US Coast Guard MH-68A (see Customers, below) is equipped with night vision device (NVD) cockpit, FLIR, weather radar and machine gun. Following trials aboard the USCG *Gallatin* (WHEC 721) in early 2001, the MH-68A has been cleared for deployment aboard all helicopter-capable cutters in the USCG fleet.

A 109 Power Elite: Special edition of A 109E; improved interior and soundproofing; limited to 50 aircraft, c/ns 11151 to 11200, of which first (I-RAIB) exhibited at Paris in June 2001; initial export, fourth Elite,

was delivered to Air Harrods in UK, May 2002; some 12 built by late 2002, including first for Italian Carabinieri.

A 109M: Military version of A 109E with PW207C or Arrius 2K2 engine. Swedish armed forces ordered 20, under designation Hkp 15, on 20 June 2001; first two handed over 10 September 2002. Hkp 15s will be used for crew training, utility, ASW, SAR and Medevac missions, from shore and shipboard bases.

MH-68A: US Coast Guard version of A 109E (P & W version). Provision for MG240 machine gun in port doorway and rescue hoist over starboard door. Night vision compatible cockpit; encrypted radios.

A 109LUH: Requirement for 40 A 109s announced by South African Air Force in November 1998 of which 30 ordered, with 10 options; these likely to be equipped with IST Dynamics turret mounting a Vektor multiple-calibre gun; first five manufactured in Italy, remainder by the Denel Group in South Africa. Initial aircraft rolled out 6 September 2002; Turbomeca engines.

A 109S: Stretched (25 cm; 10 in) cabin version of A 109E Power; PW207 engines allow approximately 200 kg (441 lb) increase in MTOW; new tail rotor with scimitar-shape, composites blades for sound reduction; first flight (unannounced) 2002.

CUSTOMERS: Recent customers include the US Coast Guard, which ordered four A 109Es (with four options, all taken up and delivered by late 2001) in interdiction configuration in April 2000 to equip Helicopter Interdiction Tactical Squadron 10 (HITRON TEN) based at Jacksonville, Florida, for armed anti-drug missions; Italy's Carabinieri, which took delivery of two A 109Es in 2000; Duke Life Flight of Durham, North Carolina, which took delivery of two A 109Es in EMS configuration in December 2001 and January 2002; the Italian government, which has ordered one A 109E in VIP configuration for the Prime Minister's use; car manufacturer and Formula One racing team Ferrari, which has ordered one A 109E; Air Harrods of the UK, which has ordered one A 109E; and Careflite, of Dallas-Fort Worth, Texas, which has ordered four A 109Es in EMS configuration, plus four options, for delivery in 2002 and 2003 respectively. Greek government ordered five A 109Es in EMS configuration in 1999, and Dyfed/Powys Police Authority in the UK took delivery of an A

109E in late 1999. Three A 109K2s delivered to Dubai Police in 1995 and 1996. Switzerland ordered one A 109 Power, which was delivered in May 1998 to the Federal Office for Civil Aviation. Nigerian Navy ordered three Turbomeca-engined A 109s for anti-smuggling duties in the Niger delta, for delivery from June 2002. Deer Jet of China ordered two A 109E Powers for harbour pilot shuttle services. Total of more than 636 A 109s built by late 2002, including 336 A 109A series, 79 A 109C series, 41 A 109K2s and 180 A 109Es.

COSTS: A 109E US$3.2 million to US$3.5 million (2002). A 109K2 US$3.5 million (2001). A 109E direct operating cost US$380 per hour (2000). US Coast Guard contract for four A 109Es valued at US$18.6 million (2000). Swedish armed forces order for 20 A 109Ms valued at €130 million (2001).

DESIGN FEATURES: Fully articulated four-blade metal main rotor hub with tension/torsion blade attachment and elastomeric bearings; delta-hinged two-blade stainless steel tail rotor; manual blade folding and rotor brake optional. Main blade section NACA 23011 with drooped leading-edge; thickness/chord ratios 11.3 per cent at root, 6 per cent at tip. Tail rotor with Wortmann aerofoil and stainless steel skins; optional rotor brake. Compared with earlier models, A 109K has lengthened cabin to hold two stretchers fore and aft; modified fuel system; and smaller instrument panel.

FLYING CONTROLS: Fully powered hydraulic; IFR system with autopilot available. A 109KM has three-axis stability augmentation/attitude hold system; dual redundant IFR system and four-axis AFCS with flight path computer optional.

LANDING GEAR: A 109K series has non-retractable tricycle type, giving increased clearance between fuselage and ground. A 109E Power has tricycle type, with oleo-pneumatic shock-absorber in each unit. Single mainwheels and self-centring nosewheel castoring ±45°. Hydraulic retraction, nosewheel forward, mainwheels upward into fuselage. Hydraulic emergency extension and locking. Magnaghi disc brakes on mainwheels. All tyres are tubeless, of same size (360×135-6 or 380×135-6 or 14.5×5.5-6, 12/14 ply tubeless) and pressure (5.90 bar; 85 lb/sq in). Tailskid under ventral fin. Emergency pop-out flotation gear and fixed snow skis optional on all models.

POWER PLANT: *A 109K:* Two Turbomeca Arriel 1K1 turboshafts, each rated at 575 kW (771 shp) for 2½ minutes, 550 kW (737 shp) for take-off (30 minutes) and 471 kW (632 shp) maximum continuous power. Engine particle separator optional. Main transmission uprated to 671 kW (900 shp) for take-off and maximum continuous twin-engined operation; single-engine rating is 477 kW (640 shp) for 2½ minutes and 418 kW (560 shp) maximum continuous. Main rotor rpm 384, tail rotor 2,085. Standard usable fuel capacity 750 litres (198 US gallons; 165 Imp gallons), with optional 150 litre (39.6 US gallon; 33.0 Imp gallon) auxiliary tank for EMS operations, or 200 litre (52.8 US gallon; 44.0 Imp gallon) auxiliary tank in the A 109KM. Self-sealing fuel tanks optional. Independent fuel and oil system for each engine.

A 109E: Two Pratt & Whitney Canada PW206C engines, each rated at 477 kW (640 shp) for T-O, 423 kW (567 shp) for twin-engine operation, 546 kW (732 shp) for 2½ minutes' OEI contingency rating and 500 kW (670 shp) maximum continuous OEI; or two Turbomeca Arrius 2K1, each rated at 500 kW (670 shp) for T-O, 426 kW (571 shp) for twin-engine operation, 559 kW (750 shp) for 2½ minutes' OEI contingencies and 500 kW (670 shp) maximum continuous OEI; transmission rating 671 kW (900 shp) maximum continuous for twin-engine operation, 418 kW (560 shp) maximum continuous and 477 kW (640 shp) for 2½ minutes' OEI; FADEC and liquid crystal multifunction displays for engine management. Standard fuel capacity (three cells) 605 litres (160 US gallons; 133 Imp gallons); optional capacity 710 litres (187 US gallons; 156 Imp gallons) with one extra cell or 870 litres (230 US gallons: 191 Imp gallons) with two extra cells.

SYSTEMS: Military versions have electrical system supplied by two 160 A 28 V DC self-cooled starter/generators and 27 Ah 28 V Ni/Cd battery. Optional AC electrical system comprises two 250 VA or two high-load 600 VA 115/26 V AC 400 Hz static inverters. Optional high-load AC system comprises one 6 kVA alternator and one standby 250 VA solid-state inverter. Dual independent hydraulic systems for flight controls, each capable of operating main actuators in the event of failure of the other system; utility hydraulic system with normal and emergency accumulators for operation of rotor brake, wheel brakes and nosewheel centring.

AVIONICS: A 109E has Rockwell Collins Pro Line II or Bendix/King Silver Crown as core system. Military versions have ergonomic, NVG-compatible flight deck with provision for IFR instrumentation and role/mission dedicated displays.

Comms: VHF/AM, VHF/FM, UHF, HF, three-station intercom, IFF and ELT.

Radar: Colour weather radar optional.

Flight: ADF, VOR/GS/ILS, DME, GPS and VLF-Omega. Optional Rockwell Collins CMS80 cockpit management system with one or more centrally mounted CDUs and automatic target hand-off system compatibility.

Mission: FLIR.

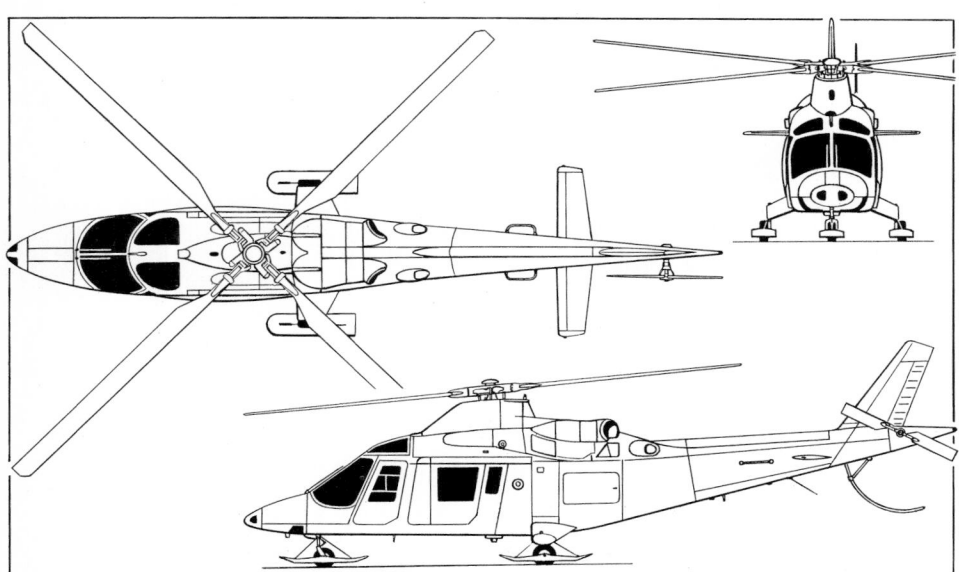

Agusta A 109K2 civil rescue and utility helicopter (two Turbomeca Arriel 1K1 turboshafts)
(Jane's/Dennis Punnett)

Self-defence: Radar/laser warning receivers and chaff/flare/smoke dispensers.

EQUIPMENT: Optional equipment for military use includes windscreen wipers, rear view mirror, bleed air heater, particle separator, engine fire extinguisher, oxygen system, environmental control unit, one- or two-stretcher installation, air ambulance kit, external loudspeaker system, high-intensity searchlight, cargo platform, external cargo hook with maximum capacity 1,000 kg (2,205 lb), rescue hoist maximum capacity 270 kg (595 lb), snow skis and emergency floats.

ARMAMENT (optional): Internal armament comprises pintle-mounted 7.62 mm machine gun and 12.7 mm machine gun in doorway. Provision for carriage of four or eight TOW, TOW 1, TOW 2 or TOW 2A missiles on external lateral pylons, each of 300 kg (661 lb) maximum capacity, with roof-mounted HeliTOW sight or APX M 334 or Helios gyrostabilised sights. Alternatively, pylons can accommodate seven- or 12-tube pods for 2.75 in or 81 mm rockets; rocket/machine gun (RPM) pods each with three 70 mm rockets and a 12.7 mm machine gun with 200 rounds; or machine gun (MG) pods with 12.7 mm gun (and 250 rounds) or 7.62 mm gun. A 109M has 12.7 mm or 20 mm nose-mounted turret gun.

Agusta A 119 Koala light utility helicopter in Australian ownership (*Jane's/Paul Jackson*)   NEW/0137358

DIMENSIONS, EXTERNAL:

Main rotor diameter	11.00 m (36 ft 1 in)
Tail rotor diameter	2.00 m (6 ft 6¾ in)
Length overall, rotors turning	13.04 m (42 ft 9 in)
Fuselage: Length	11.44 m (37 ft 6 in)
Height over tailfin	3.50 m (11 ft 5¾ in)
Tailplane span	2.88 m (9 ft 5½ in)
Width over mainwheels	2.45 m (8 ft 0½ in)
Wheelbase	3.535 m (11 ft 7¼ in)
Passenger doors (each): Height	1.06 m (3 ft 5¾ in)
Width	1.15 m (3 ft 9¼ in)
Height to sill	0.65 m (2 ft 1½ in)
Baggage door (port, rear): Height	0.51 m (1 ft 8 in)
Width	1.00 m (3 ft 3¼ in)

DIMENSIONS, INTERNAL:

Cabin: Length	2.10 m (6 ft 10¾ in)
Max width: KM, K2	1.59 m (5 ft 2½ in)
E	1.61 m (5 ft 3½ in)
Max height	1.28 m (4 ft 2½ in)
Volume, incl flight deck: K2	4.9 m³ (173 cu ft)
E	5.1 m³ (180 cu ft)
Baggage compartment volume: K2	0.85 m³ (30.0 cu ft)
E	0.95 m³ (33.6 cu ft)

AREAS:

Main rotor disc	95.03 m² (1,022.9 sq ft)
Tail rotor disc	3.14 m² (33.82 sq ft)

WEIGHTS AND LOADINGS:

Weight empty: KM	1,660 kg (3,660 lb)
K2	1,650 kg (3,638 lb)
E	1,570 kg (3,461 lb)
M	1,639 kg (3,614 lb)
Max slung load	1,000 kg (2,204 lb)
Max T-O weight: all	2,850 kg (6,283 lb)
Max T-O weight with slung load: all	3,000 kg (6,613 lb)
Max disc loading:	
KM, K2, E	30.0 kg/m² (6.14 lb/sq ft)
Transmission loading at max T-O weight and power:	
all	4.24 kg/kW (6.97 lb/shp)

PERFORMANCE:

Never-exceed speed (V$_{NE}$):	
KM, K2	152 kt (281 km/h; 174 mph)
E, M	168 kt (311 km/h; 193 mph)
Max cruising speed at S/L, clean:	
KM, K2	143 kt (264 km/h; 164 mph)
E	156 kt (289 km/h; 180 mph)
M	151 kt (280 km/h; 174 mph)
Max rate of climb at S/L: KM, K2	594 m (1,950 ft)/min
E, M	588 m (1,930 ft)/min
Rate of climb at S/L, OEI:	
KM, K2, M	274 m (900 ft)/min

Service ceiling: KM, K2	6,100 m (20,000 ft)
E	5,974 m (19,680 ft)
M	5,029 m (16,500 ft)
Service ceiling OEI: KM, K2	3,660 m (12,000 ft)
E, M	3,990 m (13,100 ft)
Hovering ceiling IGE: KM, K2	5,305 m (17,400 ft)
E	5,060 m (16,600 ft)
M	4,328 m (14,200 ft)
Hovering ceiling OGE: KM, K2	3,900 m (12,800 ft)
E	3,597 m (11,800 ft)
M	2,957 m (9,700 ft)
Max range, best height and speed, max optional (five cells) fuel: KM, K2	434 n miles (805 km; 500 miles)
E	521 n miles (964 km; 599 miles)
M	447 n miles (827 km; 514 miles)
Endurance: KM, K2	4 h 0 min
E	5 h 4 min
M	4 h 29 min

UPDATED

AGUSTA A 119 KOALA

TYPE: Light utility helicopter.

PROGRAMME: First flight (I-KOAL) early 1995; public debut at Paris Air Show June 1995; second prototype flew later in 1995; both initially with 597 kW (800 shp) Turbomeca Arriel 1 turboshaft; re-engined with PT6B turboshafts early in 1997; production target 20 to 25 per year. As part of South African purchase of 40 military A 109s, Denel (which see) will assemble and market the Koala in Africa and provide components for the Italian production line.

CUSTOMERS: Six ordered by Omniflight Helicopters in February 1996; orders and options totalled more than 20 by January 2001 from customers in Australia, Austria, Brazil, UK, USA and Venezuela. Southeast Mississippi Air Ambulance District of Hattiesburg, Mississippi, took delivery of one in EMS configuration in February 2002. Interest shown by Carabinieri (Italian military police) as replacement for AB 206 and Guardia di Finanza (customs) as NH 500 replacement. Total of 17 production helicopters registered by late 2001; none further known in first three months of 2002.

COSTS: US$1.85 million (2000).

DESIGN FEATURES: Fully articulated four-blade composites main rotor with titanium hub, composites material grips and elastomeric bearings; two-blade tail rotor; large underfin.

STRUCTURE: Aluminium alloy fuselage.

LANDING GEAR: Fixed skids.

POWER PLANT: One Pratt & Whitney Canada PT6B-37A turboshaft rated at 747 kW (1,002 shp) for T-O, and 650 kW (872 shp) maximum continuous. Transmission rating 671 kW (900 shp) for take-off and maximum continuous. Standard three-cell fuel system, combined capacity 606 litres (160 US gallons; 133 Imp gallons); optional four- and five-cell fuel tanks, combined capacities respectively 721 litres (188 US gallons; 156.5 Imp gallons) and 871 litres (230 US gallons; 191.5 Imp gallons).

ACCOMMODATION: Pilot and passenger in front; six passengers in main cabin in three-abreast club configuration claimed to be 30 per cent larger than that of any current single-turbine light helicopter; flight-accessible baggage compartment in cabin; main baggage compartment in rear fuselage with optional extensions; in EMS configuration can accommodate two stretchers and three medical attendants in main cabin without intrusion into cockpit area. Large sliding doors on each side of main cabin; forward-hinged door to cockpit on both sides. Cabin is heated, air conditioned and soundproofed.

AVIONICS: Bendix/King Silver Crown suite standard; integrated instrument display system (IIDS) optional.
  Comms: Silver Crown com/nav and ELT standard.
  Flight: Three-axis autopilot, flight director, autotrim, radar altimeter, GPS, moving map display.
  Instrumentation: NVG compatible.

EQUIPMENT: Optional equipment includes 500 kg (1,102 lb) or 1,000 kg (2,205 lb) fixed cargo hook; 200 kg (441 lb) rescue hoist; snow skids, external emergency floats, particle separator, Spectrolab SX-5 searchlight and FLIR/LLTV camera.

DIMENSIONS, EXTERNAL:

Main rotor diameter	10.83 m (35 ft 6½ in)
Tail rotor diameter	2.00 m (6 ft 6¾ in)
Length overall, rotors turning	13.01 m (42 ft 8¼ in)
Height overall	3.50 m (11 ft 5¾ in)
Fuselage: max width	1.67 m (5 ft 5¾ in)
Width overall, rotor in 'X'	7.66 m (25 ft 1½ in)
Main rotor tip ground clearance	2.52 m (8 ft 3¼ in)
Tail rotor tip ground clearance	1.29 m (4 ft 2¾ in)
Fuselage ground clearance	0.54 m (1 ft 9¼ in)

DIMENSIONS, INTERNAL:

Cabin: Length	2.10 m (6 ft 10¾ in)
Max width	1.67 m (5 ft 5¾ in)
Max height	1.28 m (4 ft 2½ in)
Floor area	2.60 m² (28.0 sq ft)
Volume, incl flight deck	4.96 m³ (175.2 cu ft)
Baggage compartment: Length	2.30 m (7 ft 6½ in)
Volume	0.95 m³ (33.5 cu ft)

AREAS:

Main rotor disc	92.12 m² (991.6 sq ft)
Tail rotor disc	3.14 m² (33.82 sq ft)

WEIGHTS AND LOADINGS:

Basic weight empty	1,430 kg (3,153 lb)
Max T-O weight: internal load	2,720 kg (5,996 lb)
external load	3,000 kg (6,613 lb)
Max disc loading:	
internal load	29.53 kg/m² (6.05 lb/sq ft)
external load	32.57 kg/m² (6.67 lb/sq ft)
Transmission loading at max T-O weight and power:	
internal load	4.05 kg/kW (6.65 lb/shp)
external load	4.47 kg/kW (7.35 lb/shp)

PERFORMANCE (max internal load):

Never-exceed speed (V$_{NE}$)	152 kt (281 km/h; 174 mph)
Max level speed	144 kt (267 km/h; 166 mph)
Hovering ceiling: IGE	3,353 m (11,000 ft)
OGE	2,682 m (8,800 ft)
Service ceiling	5,700 m (18,700 ft)
Range with max auxiliary fuel, at 1,525 m (5,000 ft), no reserves	552 n miles (1,023 km; 635 miles)
Endurance	5 h 42 min

UPDATED

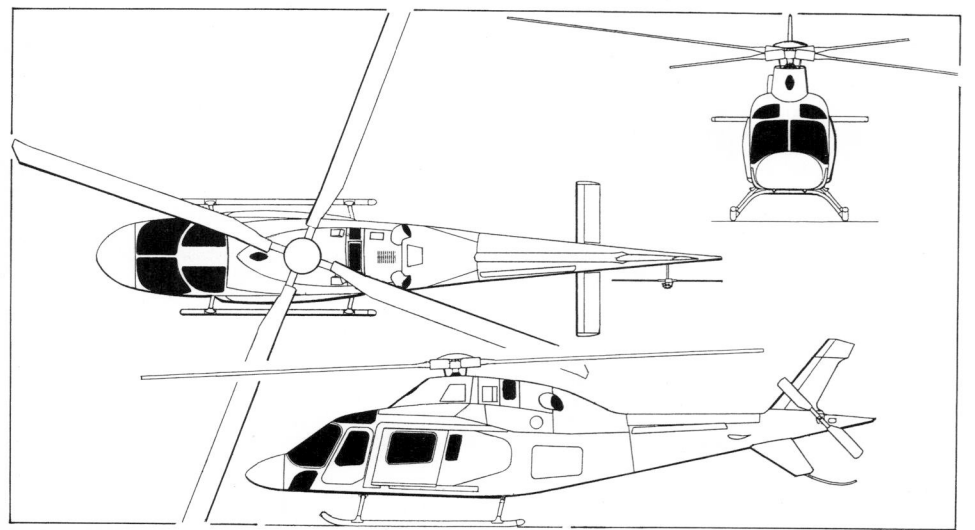

Agusta A 119 Koala (one P&WC PT6B turboshaft) (*Jane's/Paul Jackson*)   NEW/0137339

AGUSTA-BELL 412 and GRIFFON

TYPE: Multirole medium helicopter.

PROGRAMME: First flight of Bell 412SP (in USA) August 1979; deliveries started January 1981; Agusta licensed production of civil version started 1981; first flight of military Griffon August 1982; deliveries began January 1983; Bell 412HP certified 29 June 1990; Bell 412EP (see Canadian section) is a civilian version currently manufactured under licence by Agusta.

CURRENT VERSIONS: **Griffon**: Military derivative developed for direct fire support, scouting, assault transport, equipment transport, SAR and maritime surveillance.

Creso: Battlefield surveillance version; trial installation for Italian Army in AB 412 MM81196 as an element (with Mirach 26 UAVs and other sensors) of the CATRIN C^{3}I system. FIAR Creso E-band MTI radar in a circular radome below the nose, plus Elettronica emitter locators at the nose and on the tailboom. Galileo FLIR turret above pilot's seat. Maximum operational altitude of 1,500 m (4,920 ft) gives radar range of 32 to 38 n miles (60 to 70 km; 37 to 43 miles). First flight of operational system mid-1996; Creso is Italy's contender in the NATO Alliance Ground Surveillance programme.

CUSTOMERS: Those in Italy include Army (24), Carabinieri (34), Civil Protection Service (six), Coast Guard (eight; ultimate requirement for 24), national fire service (14), national forest service (nine) and Guardia di Finanza (16); others include Zimbabwe Air Force (10), Ugandan Army (two), Finnish Coast Guard (four); Royal Netherlands Air Force (three); Dubai Air Force (nine); Ghana Air Force (two, in EMS configuration); Swedish Army (five); and Dubai Police (two); Italian production totalled 240 to mid-2000. Five ordered for SAR by Turkish Coast Guard in late 1999.

COSTS: Five Turkish Coast Guard, total cost US$52 million (1999).

DESIGN FEATURES: Griffon has reinforced impact-absorbing landing gear, selective armour protection and differences noted below.

POWER PLANT: One Pratt & Whitney Canada PT6T-3D Twin-Pac rated at 1,342 kW (1,800 shp) for T-O and 1,194 kW (1,600 shp) maximum continuous (single-engine ratings

Agusta-Bell 412SP of Italy's Coast Guard   0102393

850 kW; 1,140 shp for 2½ minutes and 723 kW; 970 shp continuous). Transmission rating 1,181 kW (1,584 shp) for 5 minutes, 846 kW (1,135 shp) maximum continuous and 850 kW (1,140 shp) single engine. IR emission reduction devices optional. Fuel capacity 1,249 litres (330 US gallons; 275 Imp gallons). Two 75.7 or 341 litre (20.0 or 90.0 US gallon; 16.7 or 74.9 Imp gallon) auxiliary fuel tanks optional; single-point refuelling.

ACCOMMODATION: One or two pilots on flight deck, on energy-absorbing, armour-protected seats. Fourteen crash-attenuating troop seats in main cabin in personnel transport roles, six patients and two medical attendants in ambulance version, or up to 1,814 kg (4,000 lb) of cargo or other equipment. Space for 181 kg (400 lb) of baggage in tailboom. Total of 51 fittings in cabin floor for attachment of seats, stretchers, internal hoist or other special equipment.

SYSTEMS: Generally as for Bell 212/412.

EQUIPMENT: SAR and coastal surveillance versions equipped with 360° panoramic search radar integrated with FLIR

and TV; video and still camera systems; datalink; dual digital four-axis AFCS with auto approach to hover, mark on target and search pattern; dual GPS; EFIS-LCD IFR cockpit instrumentation; operator's console in passenger compartment; gyrostabilised binoculars; searchlight and rescue light; electric rescue hoist; liferafts and lightweight emergency floats. BAE Systems MST-S multisensor turret system supplied for undisclosed AB 412EP customer in 1996, to operate in conjunction with Honeywell RDR-1500 maritime surveillance radar.

ARMAMENT: Wide variety of external weapon options for Griffon include swivelling turret for 12.7 mm gun, two 25 mm Oerlikon cannon, four or eight TOW anti-tank missiles, two launchers each with nineteen 2.75 in SNORA or twelve 81 mm rockets, 12.7 mm machine guns (in pods or door-mounted), four air-to-air or air defence suppression missiles, or, for attacking surface vessels, four Sea Skua or similar air-to-surface missiles.

DIMENSIONS: As for Bell 212/412

WEIGHTS AND LOADINGS:

Weight empty, equipped (standard configuration)
2,914 kg (6,425 lb)
Max T-O weight   5,398 kg (11,900 lb)
Transmission loading at max T-O weight and power
4.57 kg/kW (7.51 lb/shp)

PERFORMANCE:

Never-exceed speed (V$_{NE}$) at S/L
140 kt (259 km/h; 161 mph)
Cruising speed: at S/L   122 kt (226 km/h; 140 mph)
at 1,500 m (4,920 ft)   125 kt (232 km/h; 144 mph)
at 3,000 m (9,840 ft)   123 kt (228 km/h; 142 mph)
Max rate of climb at S/L   542 m (1,780 ft)/min
Rate of climb at S/L, OEI   168 m (551 ft)/min
Service ceiling, 30.5 m (100 ft)/min climb rate
5,395 m (17,700 ft)
Service ceiling, OEI, 30.5 m (100 ft)/min climb rate
2,320 m (7,620 ft)
Hovering ceiling: IGE   3,110 m (10,200 ft)
OGE   1,585 m (5,200 ft)
Range with max standard fuel at appropriate cruising speed (see above), no reserves:
at S/L   354 n miles (656 km; 407 miles)
at 1,500 m (4,920 ft)   402 n miles (745 km; 463 miles)
at 3,000 m (9,840 ft)   434 n miles (804 km; 500 miles)
Max endurance: at S/L   3 h 36 min
at 1,500 m (4,920 ft)   4 h 12 min
VERIFIED

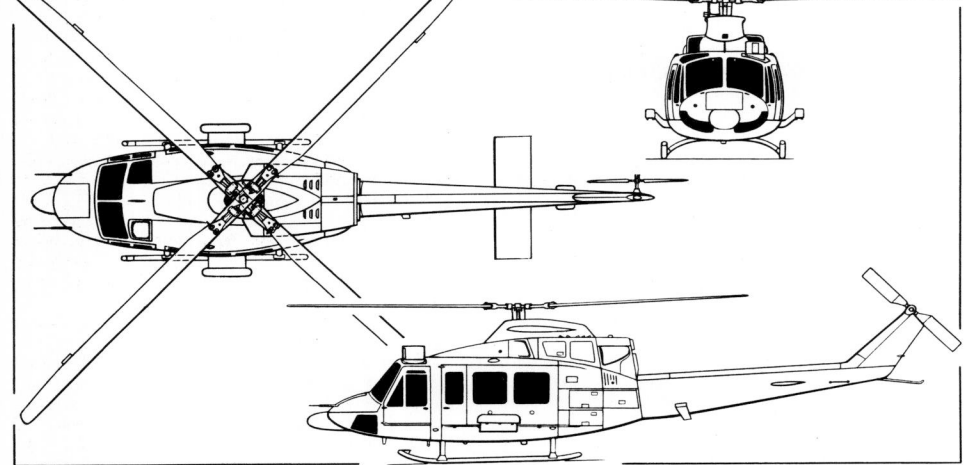

Agusta-Bell 412 Griffon military helicopter (*Jane's/Dennis Punnett*)   0052775

ALENIA

ALENIA AEROSPAZIO (A Finmeccanica company)

Via Giulio Vincenzo Bona 85, I-00156 Roma
Tel: (+39 06) 41 72 31
Fax: (+39 06) 411 44 39
Web: http://www.finmeccanica.net
PRESIDENT: Giorgio Zappa
SENIOR VICE-PRESIDENT, HEAD OF STRATEGY AND MILITARY PROGRAMMES: Carmelo Cosentino
SENIOR VICE-PRESIDENT, OPERATIONS: Gianni Cantini
HEAD OF AERONAUTICS DIVISION: Filippo Bagnato
SENIOR VICE-PRESIDENT, TECHNICAL AND PRODUCTION FACILITIES, AERONAUTICS DIVISION: Giuseppe Ragni
MARKETING COMMUNICATIONS OFFICER: Riccardo Rovere

Alenia Aerospazio, a Finmeccanica company, is the leading Italian aerospace designer and manufacturer. Employees total 12,000, engaged on projects and programmes for civil and military applications. Alenia Aerospazio is organised into two Divisions: Aeronautics and Space (Alenia Spazio SpA). On 14 April 2000, it signed a joint venture agreement with EADS to form the European Military Aircraft Company (EMAC), but early establishment was not achieved and intermittent discussions continued into 2002 before being abandoned.

Aeronautics Division dedicated to full range of activities, from design and production to modification and product support for both military and civil aircraft; most entail collaboration with other aerospace companies; in 2001, division had 9,415 employees and conducted 48 per cent of its activities in the military sphere.

Division's activities fall into categories of Military Aircraft, Regional Aircraft and Aerostructures. It also operates Officine Aeronavali Venezia, specialising in maintenance, overhaul and modification of commercial and military aircraft.

In military sector, company designs and produces directly, or through international collaborations, combat and transport aircraft such as Tornado, Eurofighter Typhoon, C-27J, A400M and ATR 42 MP Surveyor; was also responsible for updating of the F-104S/ASA-M and TF-104G-M in service with Aeronautica Militare Italiana and has assembled AV-8B Harrier II Plus for the Italian Marina Militare.

In regional aircraft sector, activities include production of the ATR turboprop family developed with Aerospatiale Matra.

In the aerostructures sector, Alenia co-operates with other major aeronautical companies manufacturing structural parts for commercial aircraft such as B767, B777, B717, A321, A300/310, A330/A340, Falcon 900EX and Falcon 2000. Is 4 per cent risk-sharing partner in Airbus A380 programme, responsible for a fuselage section.

In the modification and maintenance field, Alenia Aerospazio has a full capability of design, production, installation and tests of complex parts and systems.
UPDATED

ALENIA/LOCKHEED MARTIN C-27J SPARTAN

TYPE: Twin-turboprop transport.

PROGRAMME: Conceived in 1960s as jet-powered, V/STOL transport to NATO requirement (NBMR-4), but this version not developed. Original Fiat G222 designed by Giuseppe Gabrielli; two prototypes (lacking the pressurisation standard on later aircraft) flew on 18 July 1970 (MM582) and 22 July 1971 (MM583), both at Turin; MM582 handed over to Italian Air Force for operational evaluation on 21 December 1971. First production G222 (MM62101) flew 23 December 1975; deliveries began 21 November 1976 with single aircraft to Dubai, and to Italy on 21 April 1978. Tenth production aircraft was first to be built at Naples; 27th (March 1979) was 22nd and last built at Turin. Main users in Italy are 2° and 98° Gruppi of the 46° Brigata Aerea at Pisa. Civil category R (equivalent to FAR Pt 25) certification granted to G222SAA by Italian airworthiness authority on 1 April 1997. Several subvariants for specific roles; last detailed in 2000-01 *Jane's*.

Improved version of G222 conceived during 1995 negotiations between Lockheed Martin and Alenia on potential offsets for proposed Italian purchase of C-130J Hercules; initially designated G222J, by reason of having C-130J flight deck features and improved (T64G) versions of the G222's engines with new four-blade propellers. Formally announced as a joint project in February 1996, when commonality with the C-130J was further increased by adoption of the Allison (now Rolls-Royce) AE 2100 as power plant, allied to six-blade propellers. Accordingly redesignated C-27J, to reflect the C-27A version of G222 delivered to the US Air Force. Feasibility phase February to September 1996; definition phase September 1996 to May 1997.

Development and certification costs being shared equally between Alenia Aerospazio and Lockheed Martin; latter responsible for propulsion systems, avionics, worldwide marketing and product support; Alenia for production, flight test and certification; promotion by Lockheed Martin Alenia Tactical Transport Systems. Programme formally launched on 17 June 1997; 'propulsion test' prototype, a converted G222 demonstrator, rolled out at Turin/Caselle on 14 June 1999 and first flew (c/n 4043; I-CERX) 24 September 1999; initial testing completed second quarter of 2000; modified for certification trials and resumed flying on propulsion system, performance and handling evaluation; total 115 hours/61 sorties by July 2000.

Second C-27J and initial new-build aircraft, 4115/I-FBAX, first flew 12 May 2000; first with advanced flight deck and full avionics, new APU and new landing gear; achieved 54 hours/23 sorties in initial two months; will be sold on completion of civil certification trials. Third prototype, 4033/MMCSX62127, converted from Italian Air Force G222TCM, first flew 8 September 2000 returned to IAF after DGAA civil certification, which achieved on 20 June 2001. Military type certificate awarded 20 December 2001, following 445 sortie, 793 hour programme. First two batches, totalling 10, under construction (long lead items) by 1999. Final assembly remains in Italy only.

CURRENT VERSIONS: **Transport:** *As described.*

AEW: Airborne early warning version proposed in 1998, employing Ericsson Erieye system, as fitted to Saab S 100B Argus.

Firefighter: Proposed with 2,200 kg (4,850 lb) mission system and 6,800 kg (14,991 lb) of retardant liquid.

Aerial Sprayer: Proposed with 2,200 kg (4,850 lb) mission system; capable of covering 50 hectares (124 acres) at 150 litres/hectare (16.0 US gallons; 13.0 Imp gallons/acre).

CUSTOMERS: See table. Italian Air Force announced launch order for 12 on 11 November 1999, although this had not been signed by early 2002; deliveries tentatively planned to begin in 2004. Up to 500 sales anticipated over a 20 year period, mostly to existing Hercules operators. Lockheed Martin co-markets the aircraft. Greece expressed interest, mid-1998, in 15, plus five options, this maintained in 2002. Sales targets include Argentina, Brazil (12) and Taiwan (18 to 22); also promoted for US Army's 40-aircraft Aerial Common Sensor programme and to US Army National Guard (44) as replacement for Shorts Sherpas. Production target of 18 per year.

COSTS: US$28 million (2000). In 2001, Alenia announced intention to reduce cost by 30 per cent to increase competitiveness with CASA C-295.

C-27J's five-screen EFIS   0081742

DESIGN FEATURES: Conventional tactical transport configuration of high wing, pannier-mounted main landing gear and upswept rear fuselage with integral loading ramp.

Intended to complement the Lockheed Martin Hercules. Upgraded G222 with new, two-crew flight deck and increased performance. Propulsion system, cargo loading system and many of the avionics and flight controls are adapted from the C-130J Hercules. Compared with the G222, the C-27J is intended to provide increases of 35 per cent in range, 30 per cent cruise ceiling, 15 per cent high-speed cruise and over 200 per cent payload/range/speed; maintainability and reliability are scheduled to increase by 100 per cent for the engine, 275 per cent for propeller, 150 per cent for other systems and 30 per cent for avionics, resulting in a saving of 30 per cent in operating costs (including 5 per cent off fuel).

Wing has max thickness/chord ratio of 15 per cent. Dihedral 2° 30′ on outer panels.

FLYING CONTROLS: Conventional; manually actuated ailerons and elevators; powered rudder. Ailerons each have inset servo tab. Two-section hydraulically actuated spoilers ahead of each outboard flap segment, used also as lift dumpers on landing. Double-slotted flaps extend over 60 per cent of trailing-edge. Spoilers and flaps fully powered by tandem hydraulic actuators. Rudder fully powered by tandem hydraulic actuators. Two tabs in each elevator; no rudder tabs.

STRUCTURE: Wing of aluminium alloy three-spar fail-safe box structure, built in three portions. One-piece constant chord centre-section fits into recess in top of fuselage and is secured by bolts at six main points. Outer panels tapered on leading- and trailing-edges. Upper surface skins are of 7075-T6 alloy, lower surface skins of 2024-T3 alloy. All control surfaces have bonded metal skins with metal honeycomb core.

Pressurised fail-safe fuselage of aluminium alloy stressed skin construction and circular cross-section. Easily removable stiffened floor panels. Cantilever safe-life tail surfaces of aluminium alloy, with sweptback three-spar fin and slightly swept two-spar variable incidence tailplane.

Subcontractors include Aermacchi (outer wings), Piaggio (wing centre-section), Agusta (tail unit), Magnaghi (landing gear) and Aeronavali Venezia (airframe components).

LANDING GEAR: Hydraulically retractable tricycle type, suitable for use from prepared runways, semi-prepared strips or grass fields. Main gear built by APPH; nose gear by Magnaghi. Steerable twin-wheel nose unit retracts forward. Main units, each consisting of two single wheels in tandem, retract into fairings on sides of fuselage. Oleo-pneumatic shock-absorbers. Gear can be lowered by gravity in emergency, the nose unit being aided by aerodynamic action and the main units by the shock-absorbers, which remain compressed in the retracted position. Oleo pressure in shock-absorbers is adjustable to permit variation in height and attitude of cabin floor from ground. Low-pressure tubeless tyres on all units, size 39×13 (14/16 ply) on mainwheels, 29×11.00-12 or 29×11.00-10 (10 ply) on nosewheels. Tyre pressures 4.41 bar (64 lb/sq in) on main units, 3.92 bar (57 lb/sq in) on nose unit. Hydraulic multidisc brakes.

POWER PLANT: Two Rolls-Royce AE 2100D2 turboprops, each rated at 3,460 kW (4,640 shp), driving Dowty R391 six-blade composites propellers. Fuel in integral tanks; two in outer wings, combined capacity 6,800 litres (1,796 US gallons; 1,495 Imp gallons); two centre-section tanks, combined capacity 5,200 litres (1,374 US gallons; 1,143 Imp gallons); crossfeed provision to either engine. Total overall fuel capacity 12,000 litres (3,170 US gallons; 2,638 Imp gallons).

ACCOMMODATION: Two-pilot crew on flight deck with third seat; provision for loadmaster or jumpmaster when required. Crew door, port, forward; Type III emergency door, starboard, forward; paratroop door each side, immediately rear of sponsons; rear loading ramp; emergency hatches (three) in roof, above flight deck and level with leading- and trailing-edges.

Standard troop transport version has 34 foldaway sidewall seats and 12 stowable seats for 46 fully equipped troops (62 in high density). Paratroop transport can carry between 34 (normal) and 46 (maximum) fully equipped paratroops, and is fitted with 32 sidewall seats, plus eight stowable seats, door jump platforms and static lines. Cargo transport version can accept standard pallets of up to 2.24 m (88 in) wide, and can carry up to 9,000 kg (19,841 lb) of freight. Hydraulically operated rear-loading ramp and upward-opening door in underside of upswept rear fuselage, which can be opened in flight for airdrop operations. Five pallets of up to 1,000 kg (2,205 lb) each can be airdropped from rear opening, or single pallet of up to 5,000 kg (11,023 lb). Paratroop jumps can be made either from this opening or from rear side doors. Medical evacuation accommodation for 36 stretchers and six attendants. Entire accommodation pressurised.

SYSTEMS: Pressurisation system maintains a cabin differential of 0.41 bar (5.97 lb/sq in), giving a 1,200 m (3,940 ft) environment at altitudes up to 6,000 m (19,680 ft). Air conditioning system uses engine bleed air during flight; on ground, it is fed by compressor bleed air from APU to provide cabin heating to a minimum of 18°C. Honeywell 113 kW (152 hp) APU, installed in starboard main landing gear fairing, provides power for engine starting, hydraulic pump and alternator actuation, air conditioning on ground, and all hydraulic and electrical systems necessary for loading and unloading on ground.

Two independent hydraulic systems, each of 207 bar (3,000 lb/sq in) pressure. No. 1 system actuates flaps, spoilers, rudder, wheel brakes and (in emergency only) landing gear extension; No. 2 system actuates flaps, spoilers, rudder, wheel brakes, nosewheel steering, landing gear extension and retraction, rear ramp/door and windscreen wipers. Auxiliary hydraulic system, fed by APU-powered pump, can take over from No. 2 system in flight, if both main systems fail, to operate essential services. In addition, a standby hand pump is provided

G222 and C-27 CUSTOMERS

Customer	Version	Qty	First aircraft	Delivered
Argentine Army	G222	3	AE260	29 Mar 1977
Congo Air Force	G222	3		on order[4]
Dubai Air Force	G222	1	321	21 Nov 1976
Italy: prototypes	G222	2	MM582	21 Dec 1971
Air Force	G222TCM	40[3]	MM62101	Apr 1978
	C-27J	12		2001
Air Force	G222RM	4	MM62139	Jan 1983
Air Force	G222VS	2	MM62107	1978
SNPC	G222PROCIV	5[6]	MM62145	1987
Libyan Air Force	G222T	20	221	1981
Nigerian Air Force	G222	5	950	Sep 1984
Somali Air Force	G222	2	AM-94	1980
Venezuelan: Army	G222	1[1]	EV-8327	1983
Air Force	G222	6	1258	1984
Thai Air Force	G222	6	60307	2 May 1995
US Air Force	C-27A	10[5]	90-0170	17 Apr 1991
Alenia (demonstrator)	G222/C-27J	1	I-CERX	1983
	C-27J	1	I-FBAX	2000
Total		**124[2]**		

Notes:
[1] Transferred to air force
[2] Prototypes and 22 early aircraft built at Turin; remainder at Naples
[3] Including 10 with provision for rapid conversion to 222SAA; two to Tunisian Air Force 19 May 2001
[4] Built, but awaiting completion of contractual details
[5] Transferred to US government, with civil registrations, 1999
[6] All reverted to transports in 2001

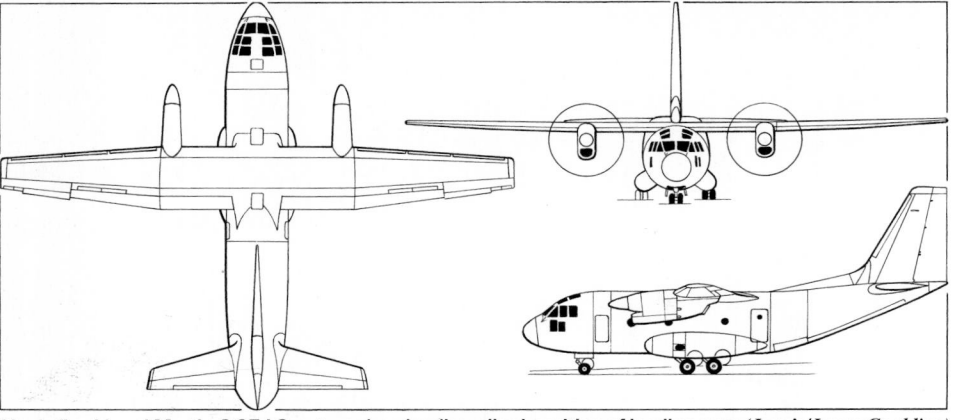

Alenia/Lockheed Martin C-27J Spartan, showing 'kneeling' position of landing gear (*Jane's/James Goulding*)
0093635

for emergency use to lower the landing gear and, on the ground, to operate the ramp/door and parking brakes.

Three 45 kVA alternators, one driven by each engine through constant-speed drive units and one by the APU, provide 115/200 V three-phase AC electrical power at 400 Hz. 28 V DC power is supplied from the main AC buses via two transformer-rectifiers, with 24 V 34 Ah Ni/Cd battery and static inverter for standby and emergency power. External AC power socket. Engine intakes anti-iced by electrical/hot air system. Pneumatically inflated de-icing boots on outer wing leading-edges, and fin and tailplane leading-edges, using engine bleed air. Liquid oxygen system for crew and passengers (with cabin wall outlets); this system can be replaced by a gaseous oxygen system if required. Emergency oxygen system available for all occupants in the event of a pressurisation failure.

AVIONICS: *Radar:* Northrop Grumman AN/APN-241.
Instrumentation: Five-screen EFIS based on C-130J flight deck. Radar integrated with moving map display. NVG-compatible.

DIMENSIONS, EXTERNAL:

Wing span	28.70 m (94 ft 2 in)
Wing aspect ratio	10.0
Length overall	22.70 m (74 ft 5½ in)
Height overall: unladen	10.57 m (34 ft 8¼ in)
fully laden	9.70 m (31 ft 10 in)
Fuselage: Max diameter	3.55 m (11 ft 7¾ in)
Tailplane span	12.40 m (40 ft 8¼ in)
Wheel track	3.67 m (12 ft 0½ in)
Wheelbase (to c/l of main units):	
unladen	6.23 m (20 ft 5¼ in)
fully laden	6.40 m (21 ft 0 in)
Propeller diameter	4.11 m (13 ft 6 in)
Distance between propeller centres	9.50 m (31 ft 2 in)
Propeller/fuselage clearance	1.04 m (3 ft 5 in)
Rear-loading ramp/door: Width	2.45 m (8 ft 0½ in)
Height	2.25 m (7 ft 4½ in)
Crew door: Width	0.70 m (2 ft 3½ in)
Height	1.52 m (4 ft 11¾ in)
Emergency door: Width	0.53 m (1 ft 8¾ in)
Height	1.01 m (3 ft 3¾ in)
Paratroop doors: Width	0.91 m (2 ft 11¾ in)
Height	1.92 m (6 ft 3½ in)
Emergency hatch: Flight deck:	
Width	0.70 m (2 ft 3½ in)
Length	0.50 m (1 ft 7¾ in)
Cabin (both): Width	0.63 m (2 ft 0¾ in)
Length	0.90 m (2 ft 11½ in)

DIMENSIONS, INTERNAL:

Main cabin: Length	8.58 m (28 ft 1¾ in)
Width	2.45 m (8 ft 0½ in)
Height	2.25 m (7 ft 4½ in)
Floor area: excl ramp	21.0 m² (226 sq ft)
incl ramp	25.7 m² (276 sq ft)
Volume	58.0 m³ (2,048 cu ft)

AREAS:

Wings, gross	82.00 m² (882.6 sq ft)
Ailerons (total)	3.65 m² (39.29 sq ft)
Trailing-edge flaps (total)	18.40 m² (198.06 sq ft)
Spoilers (total)	1.65 m² (17.76 sq ft)
Fin (incl dorsal fin)	12.19 m² (131.21 sq ft)
Rudder	7.02 m² (75.56 sq ft)
Tailplane	19.09 m² (205.48 sq ft)
Elevators (total)	4.61 m² (49.62 sq ft)

WEIGHTS AND LOADINGS:

Operating weight empty	17,000 kg (37,479 lb)
Max payload: at 2.5 *g*	9,000 kg (19,840 lb)
at 2.25 *g*	10,225 kg (22,542 lb)
airdrop	5,000 kg (11,023 lb)
Max fuel load	9,400 kg (20,725 lb)
Max T-O weight	31,800 kg (70,106 lb)
Max landing weight	30,000 kg (66,138 lb)
Max cargo floor loading	1,500 kg/m² (307.2 lb/sq ft)
Max wing loading	387.8 kg/m² (79.43 lb/sq ft)
Max power loading	4.60 kg/kW (7.55 lb/shp)

PERFORMANCE:

Max level speed	325 kt (602 km/h; 374 mph)
Time to 4,570 m (15,000 ft)	7 min
Initial cruising altitude	8,380 m (27,500 ft)
Service ceiling	9,145 m (30,000 ft)
Service ceiling, OEI	3,355 m (11,000 ft)
T-O run	410 m (1,345 ft)
T-O to 15 m (50 ft)	640 m (2,100 ft)
Landing from 15 m (50 ft)	690 m (2,265 ft)
Landing run	390 m (1,280 ft)
Radius of action: with 46 paratroops	
	1,100 n miles (2,037 km; 1,265 miles)
with 5,000 kg (11,023 lb) airdrop load	
	1,215 n miles (2,250 km; 1,398 miles)
Range: with max payload	
	1,160 n miles (2,148 km; 1,334 miles)
with 6,000 kg (13,228 lb) payload	
	2,500 n miles (4,630 km; 2,877 miles)
ferry	3,200 n miles (5,926 km; 3,682 miles)
g limit	+3.0

UPDATED

First production C-27J Spartan, on display at Paris in June 2001 (*Jane's/Paul Jackson*)
NEW/0132403

ALPI

ALPI AVIATION srl
Via Brigata Osoppo 180, I-33070 Vigonovo di Fontanafredda
Tel: (+39 0434) 37 04 96
Fax: (+39 0434) 25 39 38
e-mail: info@alpiaviation.com
Web: http://www.alpiaviation.com

CHIEF DESIGNER: Corrado Rusalen

FACTORY:
Via dei Templari 24, I-33080 San Quirino, Pordenone
US SALES AGENT:
Orlando Sanford Aircraft Sales
1920 E Airport Boulevard, Sanford, Florida 32773
Tel: (+1 800) 276 77 61 and (+1 407) 322 36 62
Fax: (+1 407) 322 47 72

The Alpi Pioneer 300S is a derivation of the Asso V designed by Vidor Guiseppe and formerly listed in *Jane's* Italian section under Rusalen & Rusalen. German marketing is by Flugtechnik at Damme. Alpi also produces the wooden Pioneer 200 and has a 5,000 m² (53,820 sq ft) factory at Pordenone; wooden components are sourced from Croatia.

NEW ENTRY

ALPI PIONEER 200

TYPE: Side-by-side ultralight kitbuilt.

PROGRAMME: Draws on experience gained from Pioneer 300 (which see). Four aircraft flying by early 2002.

DESIGN FEATURES: Similar to Pioneer 300, but with untapered wing and fixed landing gear.

FLYING CONTROLS: As Pioneer 300, but without flaps.

STRUCTURE: Broadly as Pioneer 300, but with less swept fin.

LANDING GEAR: Fixed, tricycle layout; otherwise as Pioneer 300.

POWER PLANT: One 59.6 kW (79.9 hp) Rotax 912 UL flat-four driving fixed-pitch propeller. Fuel capacity 54 litres (14.3 US gallons; 11.9 Imp gallons), of which 50 litres (13.2 US gallons; 11.0 Imp gallons) are usable, in single centrally-positioned fuel tank.

DIMENSIONS, EXTERNAL:
Wing span	7.30 m (23 ft 11½ in)
Length overall	6.20 m (20 ft 4 in)

DIMENSIONS, INTERNAL:
Cockpit: Max width	1.05 m (3 ft 5¼ in)

AREAS:
Wings, gross	10.22 m² (110.0 sq ft)

WEIGHTS AND LOADINGS:
Weight empty	275 kg (606 lb)
Max T-O weight	450 kg (992 lb)

PERFORMANCE:
Never-exceed speed (VNE)	129 kt (240 km/h; 149 mph)
Max operating speed	111 kt (205 km/h; 127 mph)
Normal cruising speed at 75% power	100 kt (185 km/h; 115 mph)
Stalling speed, flaps down	34 kt (62 km/h; 39 mph)
Max rate of climb at S/L	305 m (1,000 ft)/min
T-O and landing run	100 m (330 ft)
Range with max fuel	351 n miles (650 km; 403 miles)

NEW ENTRY

ALPI PIONEER 300

TYPE: Side-by-side ultralight-kitbuilt.

PROGRAMME: Derivation of Aerei Asso Vs, last described in *Jane's* 2002-03. First flight February 1999 (Italian ultralight category; Mid-West engine); public debut 2 April 1999 and exhibited at Aero '99, Friedrichshafen,

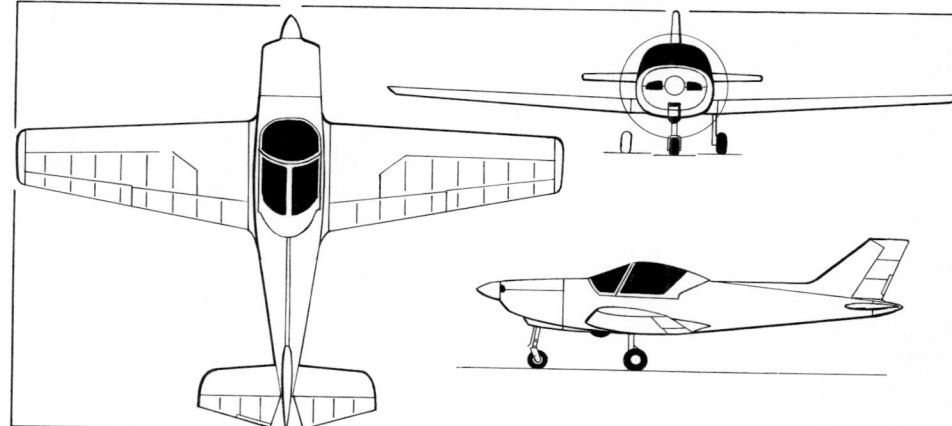

Alpi Pioneer 300 *(Jane's/James Goulding)*   NEW/0137728

later in the same month, at which time total of 11 hours flown.

CURRENT VERSIONS: **300L:** Long-span version for European market; no longer produced by Alpi.

300S: Short-span version, predominantly for US market, *as described.*

CUSTOMERS: At least 50 kits sold, of which over 30 300Ls flying in France, Germany, Italy and Spain by end 2001; four 300Ss to US in 2001.

COSTS: Quick-build kit US$26,500 (2002), excluding engine, propeller, instruments and upholstery.

DESIGN FEATURES: Streamlined, retractable gear, low-wing monoplane, adaptable for Ultralight or Experimental category operation. Tapered wings and horizontal tail surfaces; sweptback fin; upturned wingtips; NACA 2315 aerofoil. Asso Vs has biconvex asymmetric wing section. Rearward-sliding canopy; fixed windscreen. Dual controls. Baggage shelf behind seats holds 27 kg (60 lb). Maximum roll rate 120°/sec. Quoted build time 700 hours for standard kit; 350 hours for fast-build kit.

FLYING CONTROLS: Conventional and manual. Actuation by steel cables. Horn-balanced rudder. Electrically actuated flaps deflect to 30°. Trim tabs on rudder and port elevator.

STRUCTURE: Wooden airframe, including wing with single box spar; preformed composites fuselage/fin covering in left and right halves; composites engine cowling and wingtips; Dacron-covered elevators and rudder; plywood-covered ailerons, flaps and tailplane.

LANDING GEAR: Retractable, tricycle type; steerable nosewheel with helical spring suspension. Mainwheels have lever/rubber-in-compression suspension. Mainwheels retract outwards; nosewheel rearwards; electric actuation. No doors, apart from fairing fixed ahead of nosewheel leg. Ingegno wheels with disk brakes; mainwheels size 4.00-6, nosewheel 4.00-4.

POWER PLANT: One 73.5 kW (98.6 hp) Rotax 912 ULS four-stroke, driving a two-blade GT-2/173/155 fixed-pitch propeller. Three-blade propeller optional. Alternative Jabiru 3300, Mid-West AE110 and Sauer engines. Fuel tank in each wing, combined capacity 80 litres (21.1 US gallons; 17.6 Imp gallons).

AVIONICS: To customer's choice.

DIMENSIONS, EXTERNAL:
Wing span	8.10 m (26 ft 7 in)
Length overall	6.25 m (20 ft 6 in)
Height overall	2.00 m (6 ft 6¾ in)

DIMENSIONS, INTERNAL:
Cockpit max width	1.06 m (3 ft 5¾ in)

AREAS:
Wings, gross	11.00 m² (118.4 sq ft)

WEIGHTS AND LOADINGS:
Weight empty	280 kg (617 lb)
Max T-O weight (ultralight)	450 kg (992 lb)
Max T-O weight (experimental)	520 kg (1,146 lb)

PERFORMANCE (AE 110 engine):
Never-exceed speed (VNE) and max level speed	153 kt (285 km/h; 177 mph)
Cruising speed at 75% power	135 kt (250 km/h; 155 mph)
Stalling speed	33 kt (60 km/h; 38 mph)
Max rate of climb at S/L	500 m (1,640 ft)/min
T-O and landing run	120 m (395 ft)
Range	702 n miles (1,300 km; 807 miles)
g limits	+4.4/-2.2

NEW ENTRY

Alpi Pioneer 300 on display at Sun 'n' Fun, April 2002 *(Jane's/Paul Jackson)*   NEW/0137732

DRAGON FLY

DRAGON FLY SRL

e-mail: a.frommer@dragonfly-helicopter.com
Web: http://www.dragonfly-helicopter.com

Original company founded in 1993 by twin brothers Angelo and Alfredo Castiglioni specifically to produce the Dragon Fly light helicopter, rights to which were taken over from general engineering company CRAE Elettromeccanica SpA. An uprated version is under development. The company was being restructured in 2001-02.

UPDATED

DRAGON FLY 333

TYPE: Two-seat helicopter.

PROGRAMME: Developed originally by CRAE; design studies and manufacture of single-seat prototype 1985-88; ground and flight testing of this aircraft, 1989-90; two-seat prototype built and tested, 1991-93. Total of two single-seat and three two-seat prototypes, followed by four pre-series aircraft; production transferred to new factory from October 1994. Developed and tested by manufacturer to standards approaching FAR Pt 27; initial Italian certification is in ultralight class, but domestic VLR (very light rotorcraft) certification obtained 16 June 1996; JAA and FAA certification was under way by March 1998, since when no further information has been received.

Dragon Fly 333 light helicopter *(Jane's/Paul Jackson)*   NEW/0132845

In 2001, a Dragon Fly was flown at the Cielo di Volo airshow, powered by a 112 kW (150 hp) APU of undisclosed type.

CURRENT VERSIONS: **Dragon Fly 333:** *As described.*

Dragon Fly 333AC: Certified version to Italian VLR rules.

Héliot: RPV version developed in association with French companies Etudes et Développement Techniques (EDT) and CAC Systèmes; launched in June 1996 when prototype displayed at Eurosatory 1996 show at Le Bourget, Paris. Described in Issue 18 and earlier editions of *Jane's Unmanned Aerial Vehicles and Targets.*

CUSTOMERS: First delivery in May 1994 to Chinese Civil Protection Volunteers. Total of 80 ordered, of which some 70 delivered, by late 1998, the latest date for which details have been received; customers in Abu Dhabi, Australia, Belgium, Czech Republic, France, Germany, Italy (37), New Zealand, Portugal and Turkey. First delivery of Italian certified version was in 1997; initial operator was Venice Aero Club.

COSTS: US$105,000 (1997).

DESIGN FEATURES: Two-blade, semi-rigid main rotor and two-blade tail rotor; all blades of NACA 0012 aerofoil section; main rotor nominal speed 520 rpm. Can be road-towed on trailer with main blades folded. Optionally available in kit form.

STRUCTURE: Cabin is welded titanium frame with composites outer shell; aluminium alloy tailboom, rotor blades and landing skids. Full corrosion protection.

LANDING GEAR: Conventional twin-skid type. Emergency floats and skis under development.

POWER PLANT: One Dragon Fly/Hirth F30A26AK four-cylinder two-stroke, rated at 82 kW (110 hp) for T-O, 70.8 kW (95 hp) maximum continuous. Transmission driven through centrifugal clutch and two V-belts. Fuel capacity 64 litres (16.9 US gallons; 14.1 Imp gallons) of which 57 litres (15.0 US gallons; 12.5 Imp gallons) are usable.

ACCOMMODATION: Side-by-side seats for two persons. Dual controls standard. Small baggage compartment below seats.

SYSTEMS: 12 V electrical system with engine-driven generator and 12 V 24 Ah battery..

AVIONICS: *Comms:* Provision for transceiver and intercom.
Flight: Optional electric lateral trim.
Instrumentation: Standard VFR.

EQUIPMENT: Medevac and firefighting kits and external load hook under development in 1997.

DIMENSIONS, EXTERNAL:
Main rotor diameter	6.70 m (21 ft 11¾ in)
Tail rotor diameter	1.12 m (3 ft 8 in)
Length: overall, rotors turning	7.86 m (25 ft 9½ in)
fuselage	5.56 m (18 ft 3 in)
Height to top of rotor head	2.36 m (7 ft 9 in)
Skid track	1.55 m (5 ft 1 in)

DIMENSIONS, INTERNAL:
Cabin: Length	1.11 m (3 ft 7¾ in)
Max width	1.15 m (3 ft 9¼ in)
Max height	1.13 m (3 ft 8½ in)

AREAS:
Main rotor disc	35.21 m² (379.0 sq ft)

WEIGHTS AND LOADINGS:
Weight empty, standard	260 kg (573 lb)
T-O weight: normal	450 kg (992 lb)
max	500 kg (1,102 lb)

PERFORMANCE (two crew, half fuel):
Max level speed at S/L	80 kt (148 km/h; 92 mph)
Cruising speed at S/L	65 kt (120 km/h; 75 mph)
Max rate of climb at S/L	390 m (1,280 ft)/min
Service ceiling	3,050 m (10,000 ft)
Hovering ceiling: IGE	2,050 m (6,720 ft)
OGE	1,450 m (4,760 ft)
Range with 20 min reserves	162 n miles (300 km; 186 miles)
Endurance	2 h

UPDATED

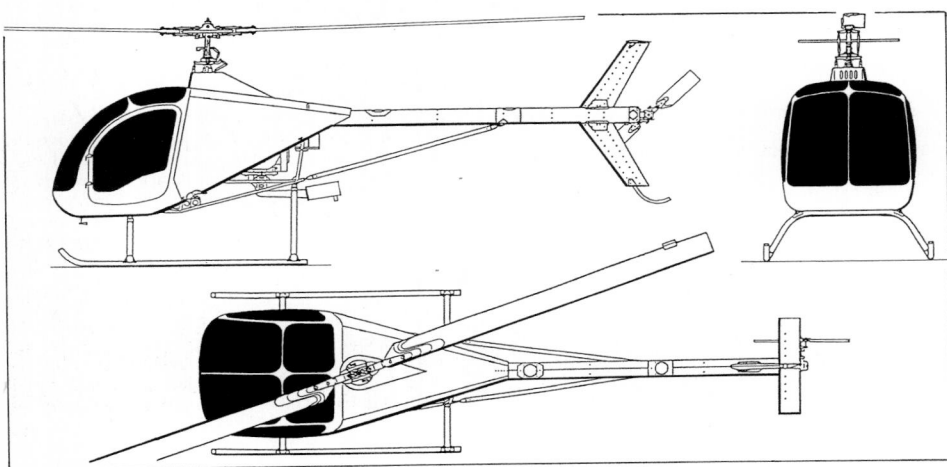

Dragon Fly 333 very light helicopter *(Jane's/Paul Jackson)*

EURO ALA

EURO ALA

C da Vibrata 136, I-64013 Corropoli (TE)
Tel: (+39 0861) 80 80 26
Fax: (+39 0861) 80 83 42
e-mail: euroala@itol.it
Web: http://www.itol.it/euroala
PRESIDENT: Alfredo Di Cesare

The Advanced Light Aircraft (ALA) company was established in 1985 and changed name to Euro ALA in 1995.
UPDATED

EURO ALA JET FOX 97

TYPE: Side-by-side ultralight.

PROGRAMME: Earlier Jet Fox 91 first flown in 1991; 140 sold by 1993. Considerably improved Jet Fox 97 first flown 10 March 1997 and exhibited at Aero '97 at Friedrichshafen the following month. Jet Fox series certified in Germany 1993; France and Belgium in 1995 and Israel in 2001.

CUSTOMERS: Total of 60 built by early 2001, including remanufactured Jet Fox 91s. Two to USA in 2001.

COSTS: £47,950 with Rotax 582; £55,650 with Rotax 912 (both 1999).

DESIGN FEATURES: Considerably refined version of earlier design, but bears structural resemblance to Fox-C22 (1988-89 *Jane's*); fully faired fuselage and engine; flaps. Wings rapidly detachable for transport.

FLYING CONTROLS: Conventional and manual. Flight-adjustable elevator trim.

STRUCTURE: Alloy tube and steel fuselage frame with composites shell; Dacron-covered aluminium wings and tail surfaces; glass fibre and carbon fibre for landing gear.

LANDING GEAR: Fixed tricycle type. Mainwheels 4.00-6. Cable-operated disc brakes. Steerable nosewheel, size 4.00-4. Speed fairings optional.

POWER PLANT: One 47.8 kW (64.1 hp) Rotax 582 UL two-stroke or 59.6 kW (79.9 hp) Rotax 912 UL four-stroke engine, driving a two-blade GT wooden propeller. Single fuel tank behind seats, capacity 59 litres (15.6 US gallons; 13.0 Imp gallons).

EQUIPMENT: Optional ballistic parachute.

DIMENSIONS, EXTERNAL:
Wing span	9.78 m (32 ft 1 in)
Length overall	5.78 m (18 ft 11½ in)
Height overall	2.80 m (9 ft 2¼ in)

AREAS:
Wings, gross	14.62 m² (157.4 sq ft)

WEIGHTS AND LOADINGS (Rotax 912):
Weight empty, including parachute	290 kg (639 lb)
Max T-O weight	450 kg (992 lb)

PERFORMANCE (Rotax 912):
Max level speed	94 kt (175 km/h; 109 mph)

One of two Euro ALA Jet Fox 97s registered in the USA during 2001 *(Jane's/Paul Jackson)*   NEW/0132846

General arrangement of the Euro ALA Jet Fox 97 *(Jane's/James Goulding)*   0054609

Cruising speed	81 kt (150 km/h; 93 mph)
Stalling speed	33 kt (60 km/h; 38 mph)
Max rate of climb at S/L	360 m (1,181 ft)/min
T-O run	100 m (330 ft)

Landing run	120 m (395 ft)
Endurance	4 h 30 min
g limits	+6/−3.5

UPDATED

FLY SYNTHESIS

FLY SYNTHESIS SRL

Via Gorizia 63, I-33050 Gonars/Udine
Tel/Fax: (+39 0432) 99 24 82 and 99 35 57
e-mail: flysynth@tin.it
Web: http://www.flysynthesis.com

The Storch ultralight produced by this company also provides the basis of the tailwheel/T tail Flight Team Sinus motor glider and ultralight, described in the Slovenian section. US marketing undertaken by American Ghiles Aircraft. Two new designs, the Texan, and Wallaby, have recently been added to the company's range.

UPDATED

FLY SYNTHESIS STORCH

US Name: Lafayette Stork
TYPE: Side-by-side ultralight/kitbuilt.
PROGRAMME: Initially marketed as Rodaro Storch.
CURRENT VERSIONS: **CL De Luxe:** Standard version, *as described.*

SS: As CL De Luxe, but with shorter overall length (5.95 m; 19 ft 6¼ in) and wing span (9.32 m; 30 ft 7 in) and Rotax 912 UL or Jabiru 2200 engine. Flaps.

HS De Luxe: As SS De Luxe, but with shorter overall length (5.75 m; 18 ft 10¾ in) and (8.71 m; 28 ft 7 in) wing span.
CUSTOMERS: More than 300 built by end of 2001.
COSTS: CL De Luxe, HS De Luxe: (Rotax 582) €32,640, (Jabiru) €36,763; SS: (Rotax 912) €41,707, Jabiru €42,743 (all 2001).
DESIGN FEATURES: Laminar flow aerofoil section. Strut-braced wings fold alongside fuselage for storage.
FLYING CONTROLS: Manual. Full-span, Junkers-type ailerons and all-moving tailplane with anti-balance tab.
STRUCTURE: Composites main fuselage shell with welded steel tube frame in cabin area and tubular alloy tailboom; light alloy constant chord wing has 6082 light alloy spar and composites skin; all-composites wing optional.
LANDING GEAR: Fixed tricycle type. Drum brakes on CL De Luxe, disc brakes on SS and HS De Luxe. Oil-damped, steerable nosewheel with maximum deflection ±40°.
POWER PLANT: CL De Luxe and HS De Luxe have one 47.8 kW (64.1 hp) Rotax 582 UL or 59.7 kW (80 hp) Jabiru 2200; SS has one 59.6 kW (79.9 hp) Rotax 912 UL or 59.7 kW (80 hp) Jabiru 2200, each driving a two- or three-blade propeller. Standard fuel capacity 60 litres (15.9 US gallons; 13.2 Imp gallons).
ACCOMMODATION: Cabin doors open upwards.
EQUIPMENT: BRS ballistic recovery parachute optional.
Data below refer to CL De Luxe with Jabiru 2200 engine.

DIMENSIONS, EXTERNAL:
Wing span	10.14 m (33 ft 3¼ in)
Length overall	6.25 m (20 ft 6 in)
Height overall	2.45 m (8ft 0½ in)

AREAS:
Wings, gross	13.02 m² (140.15 sq ft)

WEIGHTS AND LOADINGS:
Weight empty	258 kg (569 lb)
Max T-O weight	450 kg (992 lb)

PERFORMANCE:
Never-exceed speed (VNE)	97 kt (180 km/h; 112 mph)
Max level speed	86 kt (159 km/h; 99 mph)
Cruising speed at 75% power	83 kt (154 km/h; 96 mph)
Stalling speed	31 kt (57 km/h; 35 mph)
Max rate of climb at S/L	270 m (886 ft)/min
T-O run	105 m (345 ft)
Landing run	120 m (394 ft)
Range	356 n miles (659 km; 410 miles)
Endurance	3 h 18 min
g limits	+6/−3

UPDATED

FLY SYNTHESIS TEXAN

TYPE: Side-by-side ultralight.
PROGRAMME: First flown 1997. Available factory-built only.
CUSTOMERS: 10 built by August 2001, including five to Israel.
COSTS: (Jabiru 2200) €48,562; (Rotax 912 ULS) €48,572 (2001).
DESIGN FEATURES: Low-wing monoplane. Constant-chord wing of laminar flow aerofoil section; sweptback fin and rudder; rearward-sliding cockpit canopy with fixed windscreen.
FLYING CONTROLS: Manual. Ground-adjustable trim tab on starboard aileron; all-moving tailplane with anti-balance tab and mass balance weights; horn-balanced rudder. Four-position slotted flaps.
STRUCTURE: Composites throughout.
LANDING GEAR: Non-retractable tricycle type; streamlined fairings on all three wheels. Hydraulic disc brakes.
POWER PLANT: One 59.7 kW (80 hp) Jabiru 2200 or 73.5 kW (98.6 hp) Rotax 912 ULS driving a two-blade GT-2/173/155 wooden propeller. Fuel in wingroot tanks, total capacity 70 litres (18.5 US gallons; 15.4 Imp gallons).
EQUIPMENT: BRS ballistic recovery parachute system optional.

DIMENSIONS, EXTERNAL:
Wing span	8.60 m (28 ft 2½ in)
Length overall	6.90 m (22 ft 7¾ in)
Height overall	2.61 m (8 ft 6¾ in)

AREAS:
Wings, gross	12.00 m² (129.2 sq ft)

WEIGHTS AND LOADINGS:
Weight empty: Jabiru	265 kg (584 lb)
Rotax	268 kg (591 lb)
Max T-O weight	450 kg (992 lb)

PERFORMANCE (Rotax 912 ULS):
Never-exceed speed (VNE)	135 kt (250 km/h; 155 mph)
Cruising speed at 75% power	121 kt (224 km/h; 139 mph)
Stalling speed	35 kt (64 km/h; 40 mph)
Max rate of climb at S/L	420 m (1,378 ft)/min
Landing run	140 m (460 ft)

UPDATED

FLY SYNTHESIS WALLABY

TYPE: Single-seat ultralight kitbuilt.
PROGRAMME: First flight reported to have taken place in 1998. Prototype, one preproduction and one production aircraft completed by April 2001.

Fly Synthesis Storch two-seat ultralight (*Jane's/Paul Jackson*)   NEW/0132847

New-built (2001) Fly Synthesis Texan (*Jane's/Paul Jackson*)   0110639

Wingless fuselage of Fly Synthesis Wallaby (*Jane's/Paul Jackson*)   0110640

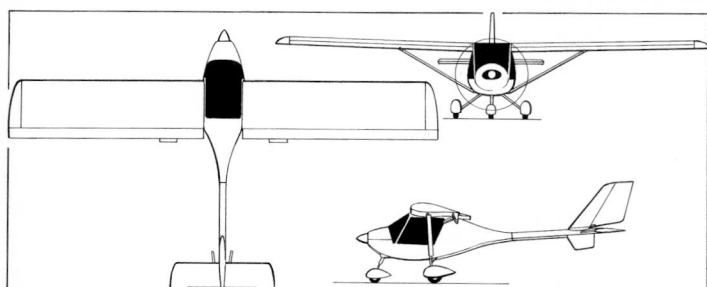
General arrangement of the Fly Synthesis Storch (*Jane's/James Goulding*)
0051090

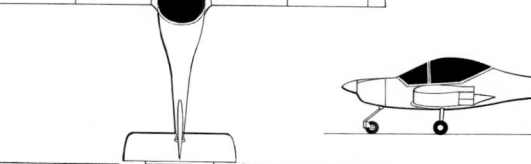
Fly Synthesis Texan two-seat ultralight (*Jane's/James Goulding*)   0109130

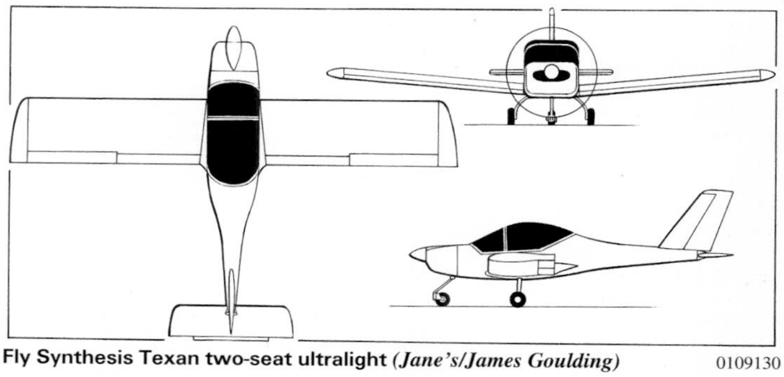

COSTS: €17,353 plus tax (2001).
DESIGN FEATURES: Strut-braced high-wing monoplane. Pod-and-boom fuselage. Open cockpit with windscreen. Sweptback fin.
FLYING CONTROLS: Manual. All-moving tailplane with anti-balance tab and mass balance weights.
STRUCTURE: Utilises Storch wing and tail unit, but in composites.
LANDING GEAR: Non-retractable tricycle type. Mainwheel size 14×4.

POWER PLANT: One 37.0 kW (49.6 hp) Rotax 503 UL driving a two-blade GT-2/170/90 propeller. Fuel capacity 40 litres (10.6 US gallons; 8.8 Imp gallons).

DIMENSIONS, EXTERNAL:
Wing span	9.42 m (30 ft 10¾ in)
Length overall	6.05 m (19 ft 10¼ in)
Height overall	2.41 m (7 ft 10¾ in)

AREAS:
Wings, gross	12.70 m² (136.7 sq ft)

WEIGHTS AND LOADINGS:
Weight empty	198 kg (437 lb)
Max T-O weight	425 kg (937 lb)

PERFORMANCE:
Never-exceed speed (V_NE)	86 kt (160 km/h; 99 mph)
Cruising speed at 75% power	62 kt (115 km/h; 71 mph)
Stalling speed	30 kt (55 km/h; 35 mph)
Max rate of climb at S/L	180 m (591 ft)/min
Landing run	90 m (295 ft)
g limits	+6/−3

UPDATED

FRATI

STELIO FRATI
Via Noe 1, I-20133 Milano
Tel/Fax: (+39 02) 204 78 31
CHIEF DESIGNER: Dott Ing Stelio Frati

In 1998, Dott Ing Frati left General Avia to resume the activities of a freelance designer – a career successfully practised since 1950. His past designs include the F.8 Falco, F.15 Picchio, F.250 (now manufactured by Aermacchi as the SF-260), SF.600A Canguro (for SIAI-Marchetti, and now VulcanAir) and F.20 Pegaso (see 1981-82 *Jane's*). The F.8 Falco is marketed as a kit aircraft by Sequoia Aircraft Corporation (which see in US section).

Frati has designed a new four-seat, all-composites tourer with a 194 kW (260 hp) engine described under the III entry in this section; a 'Falco Jet' employing an unspecified Williams turbofan; and a twin-jet commuter airliner with a fuselage possibly based on the Canguro.

UPDATED

HELISPORT

CH-7 HELISPORT srl
Strada Traforo del Pino 102, I-10132 Torino
Tel: (+39 011) 899 95 65, 899 96 18 and 899 67 30
Fax: (+39 011) 899 96 18, 898 11 44 and 899 55 50
e-mail: kompress@tin.it
Web: http://www.ch-7helicopter.com

Previously known as EliSport, the company markets the Kompress, a development of the CH-6 produced in 1987 by the Argentine designer Augusto Cicaré. US agent is Lancair (which see).

UPDATED

HELISPORT CH-7 KOMPRESS
TYPE: Two-seat ultralight helicopter kitbuilt.
PROGRAMME: CH-6 project acquired in 1989 by HeliSport; added new cockpit by sports car designer Marcello Gandini; renamed CH-7 Angel, two-seat version became Kompress.

CURRENT VERSIONS: **CH-7 Kompress:** Standard version, *as described.*
CH-7 Mariner: Float-equipped version. Multi-tube inflatable floats add 15 kg (33 lb) to empty weight.
CH-7 Angel: Single-seat version, powered by either 47.8 kW (64.1 hp) Rotax 582 UL (as Angel 582) or 59.6 kW (79.9 hp) Rotax 912 UL (as Angel 912). Max T-O weight of both versions is 300 kg (661 lb).
CUSTOMERS: Some 140 flying by 2002. Owners in Australia, Germany, Italy, South Africa, Taiwan, USA and elsewhere.
COSTS: Basic kit, less engine and tax €59,885 (2001).
DESIGN FEATURES: Small 'penny-farthing' helicopter with two-blade, semi-rigid, teetering main and two-blade rotors; pod-and-boom fuselage; sweptback upper and lower fins, latter with tailskid. Main rotor transmission by single 10 cm (4 in) five-groove Goodyear belt. Main rotor has 10,000-hour life, on condition. Quoted build time 200 hours; quick-build kit 85 hours.
Rotor blade aerofoil sections NACA 8-H-12 (main) and NACA 63-014 (tail); 1.5 kg (3.3 lb) main rotor tip weights; max 520 rpm main, 2,808 rpm tail rotor.
FLYING CONTROLS: Conventional and manual, by pushrods and bell cranks.
STRUCTURE: Welded 4130 steel tube cabin frame, with nitrogen filling; glass fibre cockpit shell and Plexiglas or Lexan canopy; 2024T3 anodised aluminium alloy tailboom and landing skids. Rotor blades of composites (main) and aluminium alloy (tail).
LANDING GEAR: Fixed; twin-skid type.
POWER PLANT: One 84.6 kW (113.4 hp) Rotax 914 piston engine. Fuel capacity 40 litres (10.6 US gallons; 8.8 Imp

gallons) standard in all versions; additional 19 litre (5.0 US gallon; 4.2 Imp gallon) tank optional.

DIMENSIONS, EXTERNAL:
Main rotor diameter	6.27 m (20 ft 6¾ in)
Tail rotor diameter	1.03 m (3 ft 4½ in)
Length: fuselage	5.31 m (17 ft 5 in)
overall, rotors turning	7.41 m (24 ft 4 in)
Fuselage max width	0.82 m (2 ft 8¼ in)
Height to top of rotor head	2.31 m (7 ft 7 in)
Skid track	1.50 m (4 ft 11 in)

DIMENSIONS, INTERNAL:
Cabin max width	0.76 m (2 ft 6 in)

AREAS:
Main rotor disc	29.90 m² (321.8 sq ft)
Tail rotor disc	0.83 m² (8.93 sq ft)

WEIGHTS AND LOADINGS:
Weight empty: Europe	245 kg (540 lb)
USA	281 kg (620 lb)
Max T-O weight: Europe	450 kg (992 lb)
USA	499 kg (1,100 lb)

PERFORMANCE:
Never-exceed speed (V_NE)	112 kt (209 km/h; 129 mph)
Max cruising speed:	86 kt (160 km/h; 99 mph)
Max rate of climb at S/L: Europe	450 m (1,476 ft)/min
USA	347 m (1,140 ft)/min
Hovering ceiling: IGE	3,500 m (11,480 ft)
OGE	2,500 m (8,200 ft)
Service ceiling	5,000 m (16,400 ft)
Range: with normal fuel	298 n miles (552 km; 343 miles)
with optional fuel	388 n miles (719 km; 477 miles)
Endurance:	3 h 40 min

UPDATED

HeliSport CH-7 Kompress two-seat light helicopter
NEW/0132834

Augusto Cicaré and his CH-6 prototype
NEW/0132959

ICP

ICP SRL
Via Torino 12, Piovà Massaia (At)
Tel: (+39 0141) 99 65 03
Fax: (+39 0141) 99 65 06
e-mail: info@icp.it
Web: http://www.icp.it

The company markets modified versions of Zenair designs (which see in US section), and is also active in the automotive industry.

UPDATED

ICP AMIGO
TYPE: Side-by-side ultralight.
PROGRAMME: Adaptation of Zenith Zodiac, most notably with conventional fin and rudder plus reinforced cabin roof with side-opening doors. Rotax 912 engine and 120 litre (31.7 US gallon; 26.4 Imp gallon) fuel capacity as standard; flight-adjustable pitch propeller. With Rotax 912S is designated **Amigo S.**
COSTS: €34,577 flyaway (2001).
Data generally as for Zodiac (which see in US section) except that which follows.

DIMENSIONS, EXTERNAL:
Wing span	8.25 m (27 ft 0¾ in)
Wing chord	1.58 m (5 ft 2¼ in)
Length overall	6.15 m (20 ft 2¼ in)
Height overall	2.15 m (7 ft 0½ in)

AREAS:
Wings, gross	13.00 m² (139.9 sq ft)

WEIGHTS AND LOADINGS:
Weight empty	286 kg (631 lb)
Max T-O weight	450 kg (992 lb)

PERFORMANCE:
Never-exceed speed (V_NE)	151 kt (280 km/h; 174 mph)

ICP Amigo   *NEW*/0137436

ICP Savannah   *NEW*/0137435

Max cruising speed:	
Rotax 912	97 kt (180 km/h; 112 mph)
ICP/Rotax 912 Turbo	116 kt (215 km/h; 134 mph)
Stalling speed	35 kt (64 km/h; 40 mph)
T-O run	70 m (230 ft)

UPDATED

ICP SAVANNAH

TYPE: Side-by-side ultralight.

PROGRAMME: Zenith CH 701 adaptation, employing conventional rudder and elevators; see Ekoflug MXP-740 in Slovak Republic section for description. Version with Rotax 503 engine is designated **Bingo 503**; with Simoni Victor 2 engine (first flown 19 October 2001) is **Super Bingo**.

COSTS: €30,919 flyaway (2001).

UPDATED

III

INIZIATIVE INDUSTRIALI ITALIANE SpA

Corso Trieste n 150, I-00198 Rome
Tel: (+39 06) 854 63 41/85 30 14 61/85 30 17 94/855 94 85
Fax: (+39 06) 855 71 62
e-mail: info@skyarrow.com
Web: http://www.skyarrow.com
PRESIDENT AND MANAGING DIRECTOR: Avv Furio Lauri

US AGENT:
Pacific Aerosystem Inc, 5760 Chesapeake Court, San Diego, California 92123, USA
Tel: (+1 619) 571 14 41
Fax: (+1 619) 571 08 03

Company formed April 1947 in Trieste, as Meteor SpA. Main R&D and production facility then in Monfalcone, where two- and four-seat aircraft produced for training and sport flying, plus target aircraft and electronic systems for military market. Present company created in 1985; relocated to new facility in Monterotondo 1989; now produces pleasure boats and general aviation aircraft and is developing the Rondine unmanned surveillance system which is based on the Sky Arrow.

VERIFIED

Computer-generated image of new Stelio Frati-designed D-130   *NEW*/0137434

III D-130, D-200, F.300 and F.300 SPORT

TYPE: Primary prop trainer/sportplane; four-seat lightplane.

PROGRAMME: Design (by Dott Ing Stelio Frati) started in 2000; manufacture of mockup, and wind tunnel testing, under way mid-2001; production scheduled to begin at end of 2002. Certification of all three variants will be to JAR/FAR Pt 23 IFR requirements.

CURRENT VERSIONS: **D-130:** Primary trainer.
D-200: Aerobatic trainer.
F.300: Tourer and business aircraft.
F.300 Sport: Aerobatic trainer and tourer; bubble canopy.

DESIGN FEATURES: Typical Stelio Frati design with low wing of tapered planform, and swept fin with long dorsal fillet.

FLYING CONTROLS: Conventional and manual; cable actuated. Horn-balanced elevators and rudder; trim tabs on each aileron and elevator; tab on rudder.

STRUCTURE: All composites; mostly carbon fibre, Kevlar and epoxy resin.

LANDING GEAR: Retractable tricycle type; main units retract inwards, nosewheel rearwards.

POWER PLANT: *D-130:* One 93.2 kW (125 hp) Teledyne Continental IO-240 flat-four driving a two-blade fixed-pitch propeller.

D-200: One 149 kW (200 hp) Textron Lycoming IO-360 flat-four driving a three-blade constant-speed propeller.

F.300 and F.300 Sport: One 224 kW (300 hp) Textron Lycoming IO-540 flat-six and three-blade constant-speed propeller.

ACCOMMODATION: D-130, D-200 and F.300 Sport: two persons, side by side; F.300, four persons. F.300 has conventional cabin accessed via doors, F.300 Sport has single-piece rearward-sliding canopy with fixed windscreen. Control sticks standard. Baggage compartment at rear of cabin on F.300, with external baggage door on port side aft of wing.

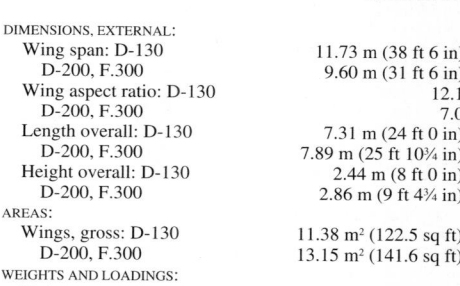

DIMENSIONS, EXTERNAL:	
Wing span: D-130	11.73 m (38 ft 6 in)
D-200, F.300	9.60 m (31 ft 6 in)
Wing aspect ratio: D-130	12.1
D-200, F.300	7.0
Length overall: D-130	7.31 m (24 ft 0 in)
D-200, F.300	7.89 m (25 ft 10¾ in)
Height overall: D-130	2.44 m (8 ft 0 in)
D-200, F.300	2.86 m (9 ft 4¾ in)
AREAS:	
Wings, gross: D-130	11.38 m² (122.5 sq ft)
D-200, F.300	13.15 m² (141.6 sq ft)
WEIGHTS AND LOADINGS:	
Weight empty: D-130	399 kg (880 lb)
D-200	726 kg (1,600 lb)
F.300	860 kg (1,897 lb)
Max T-O weight: D-130	649 kg (1,430 lb)
D-200	1,098 kg (2,420 lb)
F.300	1,402 kg (3,090 lb)
Max wing loading: D-130	56.98 kg/m² (11.67 lb/sq ft)
D-200	83.44 kg/m² (17.09 lb/sq ft)
F.300	106.53 kg/m² (21.82 lb/sq ft)
Max power loading: D-130	6.96 kg/kW (11.44 lb/hp)
D-200	7.36 kg/kW (12.10 lb/hp)
F.300	6.27 kg/kW (10.30 lb/hp)
PERFORMANCE:	
Never-exceed speed (V$_{NE}$):	
D-130	160 kt (296 km/h; 184 mph)
D-200, F.300	230 kt (426 km/h; 265 mph)
Max level speed: at S/L:	
D-130	145 kt (268 km/h; 167 mph)
D-200	180 kt (333 km/h; 207 mph)
F.300	205 kt (380 km/h; 236 mph)
Cruising speed at 75% power at S/L:	
D-130	120 kt (222 km/h; 138 mph)
D-200	160 kt (296 km/h; 184 mph)
F.300	180 kt (333 km/h; 207 mph)
Cruising speed at 75% power at 4,575 m (15,000 ft):	
F.300	174 kt (322 km/h; 200 mph)
Stalling speed, flaps down, power off:	
D-130	43 kt (80 km/h; 49 mph)
D-200	50 kt (93 km/h; 57 mph)
F.300	59 kt (109 km/h; 68 mph)
Max rate of climb at S/L:	
D-130, D-200	366 m (1,200 ft)/min
F.300	482 m (1,580 ft)/min

Computer-generated image of new Stelio Frati-designed F.300 and F.300 Sport   *NEW*/0137433

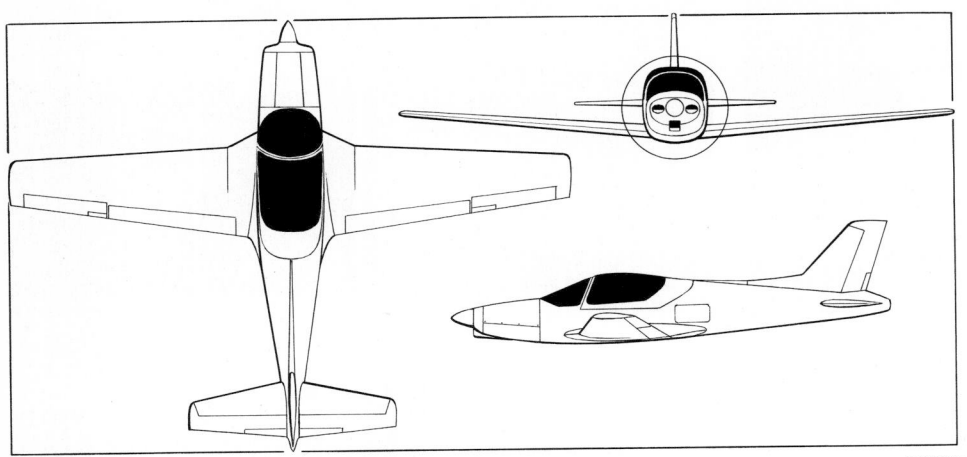

Projected III F.300 Sport aerobatic tourer (*Jane's/James Goulding*)   0126938

Service ceiling: D-130, D-200	4,575 m (15,000 ft)
F.300	6,400 m (21,000 ft)
T-O run: D-130	91 m (300 ft)
D-200	305 m (1,000 ft)
F.300	280 m (920 ft)
Landing run: D-130	91 m (300 ft)
D-200	274 m (900 ft)
F.300	326 m (1,070 ft)

Range, 75% power, 30 min reserves:
D-130	480 n miles (889 km; 552 miles)
D-200	640 n miles (1,185 km; 736 miles)
F.300	700 to 1,190 n miles
	(1,296 to 2,204 km; 805 to 1,369 miles)

Endurance, 75% power, 30 min reserves:
D-130, D-200	2 h 0 min
F.300	4 to 6 h
g limits: D-130	+7/–3.5
D-200	+8.3/–4
F.300	+6.5/–3.2

UPDATED

III SKY ARROW

TYPE: Two-seat lightplane; tandem-seat ultralight.
PROGRAMME: First flight March 1993; certified by FAA in primary aircraft sport category.
CURRENT VERSIONS: Designations refer to maximum T-O weights in kilograms.
450T, 450TS and 450TGS: For markets where MTOW for ultralights is 450 kg.
480T and 480TS: For markets where MTOW for ultralights is 480 kg.
500TF: For markets where MTOW for ultralights is 500 kg.
560TS: For markets where MTOW for ultralights is 560 kg.
650TC and 650TCS: JAR 22 and FAR Pt 33 certified models. MTOW 650 kg. First production 650TC (I-TREI) built 1996; 13 produced by mid-2001.
650TCN and 650TCNS: JAR-VLA and FAR Pt 33 certified for day or night VFR operation. MTOW 650 kg. Total 20 built by late 2001.
Description below applies to these versions, except where noted.
650 TCNS ERA: Environmental Research Aircraft variant equipped with surface surveillance sensors provided by the US National Oceanic Atmospheric Administration. Features fully retractable nosewheel and folding main leading gear.
650 TCNS RAWAS: Remotely Assisted Working Aerial System. Mission equipment can include gyrostabilised and traversable video cameras, and FLIR, IR linescan, vertical multispectral sensor, crop growth monitoring sensors, magnetometer or vertical panoramic still camera; operates as airborne relay station for RAWAS surveillance system, receiving and transmitting data at range up to 52 n miles (96 km; 60 miles) from mobile ground station. Fully retractable nosewheel and folding main landing gear.
RFG: 'Retractable Fixed Gear'. One aircraft (I-RAWE), demonstrated 2001; proof-of-concept for ERA and RAWAS versions, providing clear field of view for ventral sensors. Nosewheel retracts conventionally; regular Sky Arrow mainwheel legs splay outwards and upwards.
CUSTOMERS: Customers in Germany, Ghana, Italy, UK, USA and elsewhere.
COSTS: Sky Arrow 650TCNS US$99,500; 650TCS US$94,000 (2002). USA, Sky Arrow 650TCN US$98,500 (1998).
DESIGN FEATURES: Designed with objectives of good view from cockpit, lightness, ease of maintenance, ease of disassembly for transport and storage. Capable of being converted to an amphibian, with additional endplate fins on tailplane. High-mounted engine behind wing; pusher propeller; low-mounted pod fuselage below propeller carries T tail.

Wing aerofoil Gottingen 398 modified; constant wing chord; dihedral 1° 30'; twist 1°; strut-braced wings easily detachable at centreline joint.
FLYING CONTROLS: Conventional and manual. Ailerons and elevator are actuated by aluminium rods, rudder by cables. Electrically actuated half-span flaps and tabs; flap deflections 10, 20 and 30°; centrally mounted elevator tab for trimming. Full dual controls (sidestick and rudder) and two throttles.
STRUCTURE: Fuselage, wings, tail unit and landing gear manufactured from carbon fibre sandwich with Kevlar for strengthening in some areas, such as the cabin. Sky Arrow 450 and 480 series have aluminium wings.
LANDING GEAR: Fixed tricycle; hydraulic disc brakes; composite leaf spring main gear; steel/carbon fibre nose leg with rubber and spring shock-absorber; twin amphibious floats and aerodynamic fairings optional.
POWER PLANT: *450T, 480T and 500TF:* One 59.6 kW (79.9 hp) Rotax 912 UL flat-four four-stroke, driving a two-blade fixed-pitch Hoffmann propeller via 1:2.27 reduction gear.
450TS, 480TS and 560TS: One 73.5 kW (98.6 hp) Rotax 912 ULS driving a two-blade fixed-pitch propeller via 1:2.43 reduction gear; three-blade propeller optional on 450T and 450TS.
650TC and 650TCN: One 59.6 kW (79.9 hp) Rotax 912 A or 912 F, driving a two-blade fixed-pitch propeller.
650TCS and 650TCNS: One 73.5 kW (98.6 hp) Rotax 912 S2 driving a two-blade fixed-pitch propeller; Hoffmann propeller on 650 models.
Fuel capacity (all) 68 litres (18.0 US gallons; 15.0 Imp gallons).

ACCOMMODATION: Two seats in tandem; sideways-opening tinted canopy; baggage compartments below and behind rear seat.
SYSTEMS: Dual hydraulic brakes. Electrical system includes 12 V 14 Ah battery; alternator.
AVIONICS: *Comms:* Single Bendix/King KX-125 nav com, dual optional; KT-76A transponder; altitude encoder; audio panel with voice-activated intercom; two Sigtronics headsets; ELT.
Flight: Bendix/King KLX-1325A com/GPS optional.
Instrumentation: ASI; altimeter; rate of climb indicator; attitude indicator; directional gyro; turn-and-bank indicator; magnetic compass; pitch trim indicator; tachometer; fuel quantity, fuel pressure, oil temperature, oil pressure, water temperature and vacuum gauges; ammeter/voltmeter; and flight hours meter, all standard. Manifold pressure indicator kit optional.
EQUIPMENT: Optional handicapped pilot kit comprises sidestick controller on port cockpit console for rudder and throttle inputs, enabling the aircraft to be flown by pilots unable to use their legs; baggage compartment accommodates folding wheelchair. Fitted trailer for road transportation, with hydraulic loading system, optional. Banner/glider towing hook, incorporating LCD micro video camera for monitoring towing operation, optional.

DIMENSIONS, EXTERNAL:
Wing span: all	9.70 m (31 ft 9¾ in)
Wing chord, constant	1.40 m (4 ft 7 in)
Wing aspect ratio	7.0
Length overall: all	7.60 m (24 ft 11¼ in)
Height overall: all	2.60 m (8 ft 6¼ in)
Tailplane span: all	2.80 m (9 ft 2¼ in)
Tailplane chord: all	0.75 m (2 ft 5½ in)
Wheel track	1.75 m (5 ft 9 in)

AREAS:
Wings, gross	13.50 m² (145.3 sq ft)

WEIGHTS AND LOADINGS:
Weight empty: 450T	288 kg (635 lb)
450TS	286 kg (631 lb)
480T	302 kg (666 lb)
480TS	298 kg (657 lb)
650TC, 650TCS, 650TCN, 650TCNS	400 kg (882 lb)
Max T-O weight: 450T, 450TS, 450TGS	450 kg (992 lb)
480T, 480TS	480 kg (1,058 lb)
500TF	500 kg (1,102 lb)
560TS	560 kg (1,234 lb)
650TC, 650TCS, 650TCN, 650TCNS	650 kg (1,433 lb)

Max wing loading:
450T, 450TS, 450TGS	33.3 kg/m² (6.83 lb/sq ft)
480T, 480TS	35.6 kg/m² (7.28 lb/sq ft)
500TF	37.0 kg/m² (7.59 lb/sq ft)
560TS	41.5 kg/m² (8.50 lb/sq ft)
650TC, 650TCS, 650TCN, 650TCNS	48.2 kg/m² (9.86 lb/sq ft)

III Sky Arrow 1450L kitbuilt US version, equivalent to 650 (*Jane's/Paul Jackson*)   *NEW*/0137359

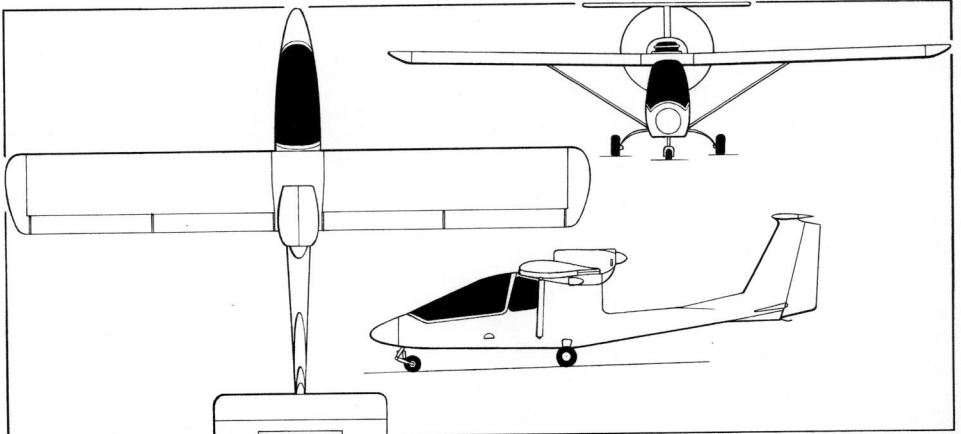

III Sky Arrow 650 very light aircraft (*Jane's/James Goulding*)   0011089

'Retractable Fixed Gear' Sky Arrow RFG
(Jane's/Paul Jackson)

0126914

Max power loading:	
450T, 450TGS	7.55 kg/kW (12.41 lb/hp)
450TS	6.12 kg/kW (10.06 lb/hp)
480T	8.05 kg/kW (13.23 lb/hp)
480TS	6.53 kg/kW (10.73 lb/hp)
500TF	8.39 kg/kW (13.78 lb/hp)
560TS	7.62 kg/kW (12.52 lb/hp)
.650TC, 650TCN	10.91 kg/kW (17.92 lb/hp)
650TCS, 650TCNS	8.84 kg/kW (15.53 lb/hp)

PERFORMANCE:
Never-exceed speed (VNE):
450T, 450TS, 480T, 480TS
121 kt (224 km/h; 139 mph)
650TC, 650TCS, 650TCN, 650TCNS
132 kt (244 km/h; 151 mph)
Max level speed at S/L:
450T, 480T, 650TC, 650TCN
104 kt (193 km/h; 120 mph)
450TS, 480TS   108 kt (200 km/h; 124 mph)
650TCS, 650TNS   107 kt (198 km/h; 123 mph)
Cruising speed at 75% power at S/L:
450T, 480T, 650TC, 650TCN
90 kt (167 km/h; 104 mph)
450TS, 480TS   92 kt (170 km/h; 106 mph)
650TCS, 650TCNS   97 kt (180 km/h; 112 mph)
Stalling speed, power off, flaps down:
450T, 450TS   33 kt (62 km/h; 38 mph)
480T, 480TS   34 kt (63 km/h; 40 mph)
650TC, 650TCS, 650TCN, 650TCNS
40 kt (75 km/h; 47 mph)
Max rate of climb at S/L: 450T   306 m (1,004 ft)/min
450TS   384 m (1,260 ft)/min
480T   294 m (965 ft)/min
480TS   372 m (1,220 ft)/min
650TC, 650TCN   216 m (709 ft)/min
650TCS, 650TCNS   258 m (846 ft)/min
T-O run: 450T   120 m (394 ft)
450TS   100 m (328 ft)
480T   138 m (453 ft)
480TS   115 m (377 ft)
650TC, 650TCN   240 m (787 ft)
650TCS, 650TCNS   177 m (581 ft)
T-O to 15 m (50 ft): 450T   215 m (705 ft)
450TS   160 m (525 ft)
480T   239 m (784 ft)

III Speed Arrow 750TCI sport and surveillance aircraft

0126937

480TS	180 m (591 ft)
650TC, 650TCN	410 m (1,345 ft)
650TCS, 650TCNS	380 m (1,247 ft)
Landing run: 450T, 450TS	75 m (246 ft)
480T, 480TS	86 m (282 ft)
650TC, 650TCS, 650TCN, 650TCNS	135 m (443 ft)
Landing from 15 m (50 ft): 450T, 450TS	170 m (558 ft)
480T, 480TS	188 m (617 ft)
650TC, 650TCS, 650TCN, 650TCNS	215 m (705 ft)

Max range, 75% power, no reserves:
450T, 480T, 650TC, 650TCN
378 n miles (700 km; 435 miles)
450TS, 480TS   340 n miles (630 km; 391 miles)
650TCS, 650TCNS   329 n miles (610 km; 379 miles)
Endurance, 75% power, no reserves:
450T, 480T   4 h 10 min
450TS, 480TS   3 h 40 min
650TC, 650TCN   4 h 15 min
650TCS, 650TCNS   3 h 25 min
g limits: all   +8/–4

UPDATED

III SPEED ARROW 750TCI

TYPE: Two-seat lightplane.
PROGRAMME: Prototype first flown mid-2000. Certification to JAR-VLA and some specific FAR Pt 23 IFR requirements in progress during 2001.
DESIGN FEATURES: Generally as for Sky Arrow 450/480/658/680 series, but with tapered cantilever wing and revised aerofoil.

LANDING GEAR: Retractable tricycle type; main units retract into streamlined sponsons.
POWER PLANT: One 84.6 kW (113.4 hp) Rotax 914 four-stroke driving a two-blade, constant-speed pusher propeller.

DIMENSIONS, EXTERNAL:	
Wing span	11.89 m (39 ft 0 in)
Wing aspect ratio	10.9
Length overall	7.59 m (24 ft 10¾ in)
Height overall	2.56 m (8 ft 5 in)
AREAS:	
Wings, gross	13.01 m² (140.0 sq ft)
WEIGHTS AND LOADINGS:	
Weight empty	499 kg (1,100 lb)
Max T-O weight	748 kg (1,650 lb)
Max wing loading	57.54 kg/m² (11.79 lb/sq ft)
Max power loading	8.86 kg/kW (14.55 lb/hp)

PERFORMANCE:
Never-exceed speed (VNE)   160 kt (296 km/h; 184 mph)
Max level speed: at S/L   115 kt (213 km/h; 132 mph)
at 4,575 m (15,000 ft)   130 kt (241 km/h; 150 mph)
Cruising speed at 75% power at S/L
100 kt (185 km/h; 115 mph)
Stalling speed, flaps down   43 kt (80 km/h; 49 mph)
Max rate of climb at S/L   305 m (1,000 ft)/min
Service ceiling   6,100 m (20,000 ft)
T-O and landing run   91 m (300 ft)
Max range, 75% power cruising speed, 30 min reserves
460 n miles (852 km; 529 miles)
Endurance, 75% power, 30 min reserves   4 h 0 min
g limits   +7/–3.5

VERIFIED

MAGNI

MAGNI GYRO

Via Puccini 10, I-21010 Besnate (VA)
Tel: (+39 0331) 27 48 16
Fax: (+39 0331) 27 48 17
e-mail: magnigyro@logic.it
Web: http://www.magnigyro.com
FOUNDER: Vittorio Magni

Founded as VPM in 1976 by Vittorio Magni, this company developed the single-seat MT5 and two-seat MT7 gyrocopters. It was renamed Magni Gyro in 1996 and currently produces a range of ultralight autogyros, including the open-cockpit single-seat M 18 Spartan and two-seat M 14 Scout and M 16 Tandem Trainer; and the enclosed-cockpit M 19 Shark and M 20 Talon, described below. In total, around 300 examples were active at the beginning of 2002.

UPDATED

MAGNI M 19 SHARK

TYPE: Two-seat autogyro ultralight.
PROGRAMME: Prototype flying; production was expected to start at end 2002.
DESIGN FEATURES: Generally as for M 20 Talon.
STRUCTURE: Generally as for M 20 Talon.
LANDING GEAR: Fixed tricycle type.
POWER PLANT: One turbocharged 84.6 kW (113.4 hp) Rotax 914 UL flat-four with electric starter, driving a ground-adjustable pitch, three-blade, carbon fibre pusher propeller. Fuel in integral cockpit seat/tank of epoxy/glass fibre, capacity 74 litres (19.5 US gallons; 16.3 Imp gallons).

Two-seat Magni M 19 Shark *(Jane's/Paul Jackson)*
NEW/0132848

DIMENSIONS, EXTERNAL:	
Rotor diameter	8.23 m (27 ft 0 in)
Length overall	4.80 m (15 ft 9 in)
Height overall	2.60 m (8 ft 6¼ in)
Width overall	1.85 m (6 ft 0¾ in)
Propeller diameter	1.70 m (5 ft 7 in)
WEIGHTS AND LOADINGS:	
Weight empty	280 kg (617 lb)
Max T-O weight	550 kg (1,212 lb)
PERFORMANCE:	
Max level speed	100 kt (185 km/h; 115 mph)
Cruising speed	78 kt (145 km/h; 90 mph)
Max rate of climb at S/L	300 m (984 ft)/min
Service ceiling	3,500 m (11,480 ft)
T-O run	90 m (295 ft)

Magni M 20 Talon single-seat ultralight autogyro
0054612

Landing run	up to 30 m (100 ft)
Range, no reserves	260 n miles (481 km; 299 miles)
Endurance	4 h 30 min

UPDATED

MAGNI M 20 TALON

TYPE: Single-seat autogyro ultralight.
PROGRAMME: Prototypes flying; production was expected to start at end 2002.
DESIGN FEATURES: Pod and boom fuselage with sweptback tailplane and triple fins; rudder on central fin only.

STRUCTURE: Welded 4130 chromoly main structure with glass fibre fuselage and wheel fairings. Two-blade main rotor of composites construction with mechanical prerotator.

LANDING GEAR: Fixed tricycle type.

POWER PLANT: One 59.6 kW (79.9 hp) Rotax 912 UL flat-four driving a ground-adjustable pitch, three-blade, carbon fibre pusher propeller. Fuel in integral cockpit seat/tank of epoxy/glass fibre, capacity 45 litres (11.9 US gallons; 9.9 Imp gallons).

DIMENSIONS, EXTERNAL:

Rotor diameter	7.62 m (25 ft 0 in)
Length overall	3.92 m (12 ft 10¼ in)
Height overall	2.37 m (7 ft 9¼ in)
Width overall	1.73 m (5 ft 8 in)
Propeller diameter	1.62 m (5 ft 3¾ in)

WEIGHTS AND LOADINGS:

Weight empty	205 kg (452 lb)
Max T-O weight	350 kg (771 lb)

PERFORMANCE:

Max level speed	96 kt (177 km/h; 110 mph)
Cruising speed	78 kt (145 km/h; 90 mph)
Max rate of climb at S/L	300 m (984 ft)/min
Service ceiling	3,500 m (11,480 ft)
T-O run	45 m (150 ft)
Landing run	up to 30 m (100 ft)
Range, no reserves	156 n miles (289 km; 179 miles)
Endurance	3 h

UPDATED

PIAGGIO

PIAGGIO AERO INDUSTRIES SpA

Via Cibrario 4, I-16154 Genova Sestri, Genova
Tel: (+39 010) 648 11
Fax: (+39 010) 648 13 09 (Marketing and Commercial Department)
e-mail: marketing@piaggioaero.it
Web: http://www.piaggioaero.com
CEO: José Di Mase
CHAIRMAN: Piero Ferrari
GENERAL MANAGER: Pier Cesare Guenzi
DEPUTY DIRECTOR, MARKETING AND COMMERCIAL: Paolo Barbaris

WORKS:
Genova Sestri and Finale Ligure (SV)

BRANCH OFFICE:
Via A Gramsci 34, I-00197 Rome

SUBSIDIARIES:
Piaggio Aero France SAS
Le Centaure-Nice Leader, 64 rue du Grenoble, F-06200 Nice, France
Tel: (+33 4) 493 71 02.69
Fax: (+33 4) 493 71 41 96
DIRECTOR: Alfonso Ruggiero

Piaggio America Inc
2 Exchange Street, Greenville, South Carolina 29605, USA
Tel: (+1 864) 277 39 79
Fax: (+1 864) 277 43 78
CEO AND PRESIDENT: Steve Harvey

Piaggio began aircraft production at Genoa Sestri in 1916 and later extended to Finale Ligure; Rinaldo Piaggio SpA formed 29 February 1964; covered floor area Sestri and Finale Ligure 120,000 m² (1,291,700 sq ft). Early in 1993, Rinaldo Piaggio restructured with Piaggio family retaining 19 per cent of stock; Alenia acquired 31 per cent holding in Piaggio in 1988 (adjusted to 24.5 per cent in 1992, but raised to 30.9 per cent in early 1993 following further restructuring and reorganisation). Industrie Aeronautiche and Meccaniche Rinaldo Piaggio SpA placed in administration in November 1994; following takeover in November 1998 by Turkish state holding company Tushav, which acquired 51 per cent of stock, was renamed Piaggio Aero Industries SpA, with balance of stock held by Italian investment group Aero Trust via Royal Bank of Canada (44 per cent), CSC of Italy (3 per cent) and the Buitoni family (2 per cent). However, Aero Trust share increased to 60 per cent in June 1999, returning company to Italian control. Workforce 1,327 in 2001. Turnover in 2000 was Lit182.5 billion and revenue Lit1.2 billion.

Two new subsidiaries were established in 2000: Piaggio Aero France SAS at Nice, France, which is a branch of Piaggio's Finale Ligure engineering department; and Piaggio America Inc, based in Greenville, South Carolina, which is the US support organisation for the P.180 Avanti.

Piaggio also licence-produces engines, including the RRTI RTM 322 (see *Jane's Aero-Engines*) and mobile shelters; manufactures subassemblies of Alenia (Lockheed Martin) G222/C-27J and Dassault Falcon 2000 and overhauls and supports Viper, T53 and T55 power plants.

VERIFIED

PIAGGIO P.166

TYPE: Utility turboprop twin.

PROGRAMME: Prototype first flew 26 November 1957; P.166DL2 described in 1978-79 *Jane's* and earlier versions in previous editions; total 113 P.166s of DL2 and earlier versions were built; first flight of DL3 version 3 July 1976 (I-PIAC); Italian and US certification 1978; production continues on demand.

CURRENT VERSIONS: **P.166-DL3SEM:** Most recent production reconnaissance version able to carry chin-mounted radar and underwing FLIR sensor; training, transport, medical and special patrol/observation versions offered.

P.166-DP1: Updated version with Pratt & Whitney Canada PT6A-121 turboprops, Collins avionics suite and updated mission systems; first flight 11 May 1999.

Abbreviated description below refers to -DP1; full details in 1990-91 Jane's.

Prototype Piaggio P.166-DP1 (PT6A turboprops)   0132282

CUSTOMERS: One DL3 to Transavio; two DL3MP to Somali Air Force; two DL3-Cargo to Somali government; six DL3APH for Italian Air Force communications (303° Gruppo at Rome/Guidonia); (two Alitalia aircraft transferred to Guardia di Finanza customs service); total 13 non-surveillance DL3s, including prototype; further 12 DL3SEM to Capitanerie di Porto coastguard service (four each for 1° Nucleo Volo Capitanerie at Catania; 2° NVC at Luni; and 3° NVC at Pescara) delivered by mid-1990; 12 DL3SEM to Guardia di Finanza between 23 January 1992 and 1994. Total 22 SEMs built by early 1995 and no more by 1999; 35 of all DL3 versions built by 1999. Total of eight DP1s ordered by Italian customs and coastguard; further 28 existing aircraft may be upgraded.

POWER PLANT: Two Pratt & Whitney Canada PT6A-121 turboprops, each rated at 458.6 kW (615 shp) maximum continuous power and driving a Hartzell HC-B3DL/LT10282-9.5 three-blade constant-speed fully feathering metal pusher propeller. Fuel in two 238.5 litre (63.0 US gallon; 52.5 Imp gallon) outer-wing main tanks, two 329 litre (86.8 US gallon; 72.35 Imp gallon) wingtip tanks, and a 185.5 litre (49.0 US gallon; 40.8 Imp gallon) fuselage collector tank; total standard internal fuel capacity 1,320.5 litres (348.6 US gallons; 290.5 Imp gallons). Auxiliary fuel system available optionally, comprising a 124.9 litre (33.0 US gallon; 27.5 Imp gallon) fuselage tank, transfer pump and controls; with this installed, total usable fuel capacity is increased to 1,445.4 litres (381.6 US gallons; 318.0 Imp gallons). Gravity refuelling points in each main tank and tip tank. Single-point pressure refuelling and fuel jettisoning system, permitting partial discharge, optional. Air intakes and propeller blades de-iced by engine exhaust.

ACCOMMODATION: Crew of one or two on raised flight deck, with dual controls. Aft of flight deck, accommodation consists of a passenger cabin, utility compartment and baggage compartment. Access to flight deck via passenger/cargo double door on port side, forward of wing, or via individual crew door on each side of flight deck. External access to baggage compartment via port side door aft of wing. Passenger cabin extends from rear of flight deck to bulkhead at wing main spar; fitting of passenger carrying, cargo or other interiors facilitated by two continuous rails on cabin floor, permitting considerable flexibility in standard or customised interior layouts. Standard seating for eight passengers, with individual lighting, ventilation and oxygen controls. Flight deck can be separated from passenger cabin by a screen. Door in bulkhead at rear of cabin provides access to utility compartment, in which can be fitted a lavatory, bar, or mission equipment for various roles. Entire accommodation heated, ventilated and soundproofed. Emergency exit forward of wing on starboard side. Windscreen hot-air demisting standard. Windscreen wipers, washers and methanol spray de-icing optional.

DIMENSIONS, EXTERNAL:

Wing span over tip tanks	14.69 m (48 ft 2½ in)
Length overall	12.15 m (39 ft 10¼ in)
Height overall	5.00 m (16 ft 5 in)
Cabin door: Height	1.38 m (4 ft 6 in)
Width	1.28 m (4 ft 2 in)

AREAS:

Wings, gross	26.56 m² (285.9 sq ft)

WEIGHTS AND LOADINGS:

Weight empty, equipped	2,853 kg (6,290 lb)
Max fuel	1,130 kg (2,491 lb)
Max T-O weight	4,500 kg (9,920 lb)
Max wing loading	169.4 kg/m² (34.70 lb/sq ft)
Max power loading	4.91 kg/kW (8.07 lb/shp)

PERFORMANCE:

Never-exceed speed (V_NE) at 3,050 m (10,000 ft)	220 kt (407 km/h; 253 mph) CAS
Max level and max cruising speed at 3,660 m (12,000 ft)	225 kt (417 km/h; 259 mph)
Max range speed at 3,660 m (12,000 ft)	165 kt (306 km/h; 190 mph)
Stalling speed	75 kt (139 km/h; 87 mph)

UPDATED

PIAGGIO P.180 AVANTI

English name: Forward

TYPE: Business twin-turboprop.

PROGRAMME: Design studies began 1979; launched 1982; Gates Learjet became partner in 1983, but withdrew for economic reasons on 13 January 1986; all existing Learjet P.180 tooling and first three forward fuselages transferred to Piaggio; first flights of two prototypes I-PJAV 23 September 1986 and I-PJAR 14 May 1987; two static test

Piaggio P.180 Avanti, twin-turboprop business aircraft   0126939

fuselages; first Italian certification 7 March 1990; first flight full production P.180 (I-RAIH/N180BP), 30 May 1990; Italian and US certification 2 October 1990; first customer delivery (N180BP to Robert Pond) 30 September 1990; French certification March 1993. P.180 is certified to Italian RAI 223 and FAR Pt 23 including single pilot, night and day, VFR/IFR and flight into known icing.

CURRENT VERSIONS: Increased gross weight giving higher payload/range decided 1991 and early aircraft retrofitted with minor modifications to allow new weights; weights increased again in 1992. Modified version relaunched in 1997 incorporates changes including fin, rudder and foreplane of aluminium alloy construction and increased fuel capacity. Economy and deluxe interior versions in development in mid-1999.

CUSTOMERS: Total 30 (including prototypes) built by early 1995. 31st followed in early 1999; eight delivered in 2000, and 12 in 2001; production plans are 26 in 2002 and 36 in 2003. Fiftieth aircraft (including prototypes) registered (in USA as N129PA) December 2001. Italian Air Force ordered six for communications; first delivery MM62159 14 May 1993; further five (three for Italian Army; two for SNPC civil protection service) ordered 1994, of which first, MM62167, handed over on 28 July 1997 and delivered 29 August 1997, and second and third delivered March-April 1998, with two SNPC aircraft following in late 1998 and mid-1999.

First new aircraft delivery since company reorganisation took place on 2 February 2000 to Aviation Services for Business Aircraft Inc, Portland, Oregon (N680JP). Recent customers include Italian car manufacturer Ferrari, which took delivery of the first of two aircraft (I-FXRB) on 18 May 2000 (these operated by Foxair); the Greek Ministry of Health, which has ordered two in EMS configuration; the Italian Air Force, which has ordered further nine (first of which completed in late 2001); Locafit (Italy); CIT Group Equipment Financing (Arizona, USA); and two undisclosed customers in the UK.

COSTS: US$4.695 million (2001).

DESIGN FEATURES: Intended to provide jet-type speeds with turboprop economy. Three-surface control with foreplane and T tail to allow unobstructed cabin with maximum headroom to be placed forward of mid-mounted wing carry-through structure; pusher turboprops aft of cabin and wing reduce cabin noise and propeller vortices on wing, assisting in achievement of 50 per cent laminar flow; mid-wing avoids root bulges of low-set wings and spar does not pass through cabin; lift from foreplane allows horizontal tail to act as lifting surface and thereby reduce required wing area by 34 per cent.

Laminar flow wing section Piaggio PE 1491 G (mod) at root, PE 1332 G at tip; thickness/chord ratio 13 per cent at tip, 14.5 per cent at root; dihedral 2°; sweepback 1° 11′ 24″; taper ratio 0.34; foreplane aerofoil Piaggio PE 1300 GN4 unswept; 5° anhedral on foreplane and tailplane; latter sweptback 29° 48′ at 25 per cent chord. Tailplane swept 40° at 25 per cent chord.

FLYING CONTROLS: Manual. Aerodynamically and mass-balanced elevators; horn- and mass-balanced rudder. Variable incidence swept tailplane for trim; electrically actuated trim tab in starboard aileron and in rudder; two strakes under tail; electrically actuated outboard Fowler and inboard single-slotted flaps on wing synchronised with single-slotted flaps in foreplanes; three flap positions; dual control circuits; foreplane stalls before wing, providing pitch-down moment. All control surfaces have sealed gaps and are aerodynamically balanced. No dampers or stick-pusher.

STRUCTURE: Airframe 90 per cent aluminium alloy and 10 per cent composites. Fuselage precision stretch-formed in large seamless sections and inner structure matched to precise outer contour in innovative 'outside-in' construction technique; CFRP in high-stress areas; 'Working skin' wing main box, with machined spars and panels, plus integral stiffeners; two main spars; third spar

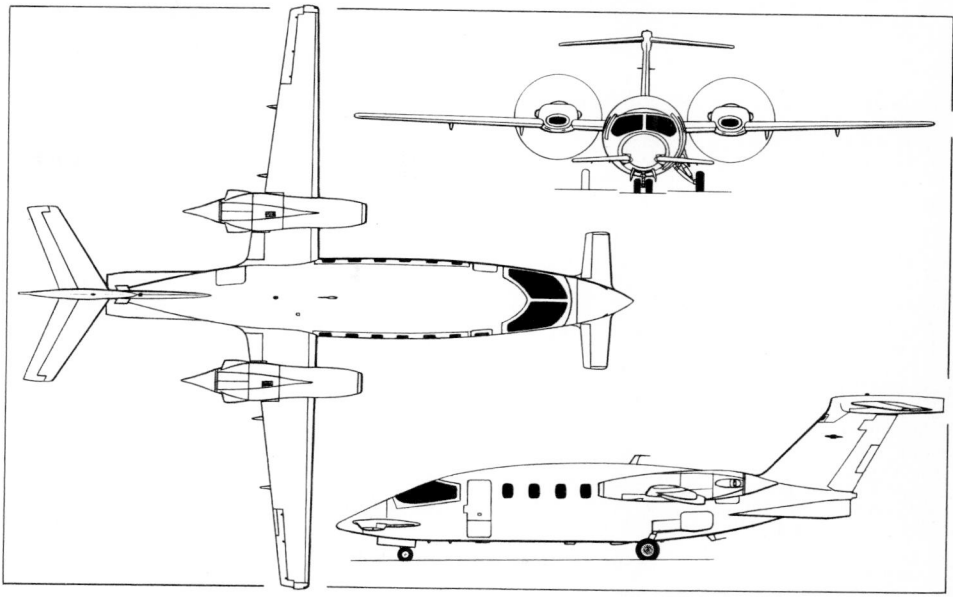

Piaggio P.180 Avanti corporate transport (*Jane's/Dennis Punnett*)

runs from nacelle to fuselage centreline; aluminium leading-edges and aluminium and composites trailing-edges, both connected to main box. Similar construction of forward wing, which connected to lower fuselage at four points and has aluminium alloy leading-edges with electric de-icing blanket and full-depth honeycomb aluminium flaps. Tailplane has two-spar sandwich construction of graphite fabric on Nomex honeycomb core; elevators are single-spar and full-depth aluminium honeycomb with aluminium skins. Fin attached to tailcone bulkheads by four vertical aluminium machined spars; chemically milled aluminium sheet skins. Two-spar rudder, with aluminium alloy skin and carbon fibre and foam core.

Composites parts manufactured by Moreggio (nacelles) and Salver (horizontal stabiliser), wings and tail section and final assembly by Piaggio in Genoa.

LANDING GEAR: Dowty Aerospace hydraulically retractable tricycle type, with single-wheel main units and twin-wheel nose unit. Main units retract rearward into sides of fuselage; nose unit retracts forward. Emergency hand pump. Dowty hydraulic shock-absorbers. Electro-hydraulic nosewheel steering available through ±20° on take-off and ±50° for taxiing. Tyre sizes 6.50-10 (main) and 5.00-5 (nose). Goodrich hydraulic, multidisc carbon brakes.

POWER PLANT: Two 1,107 kW (1,485 shp) Pratt & Whitney Canada PT6A-66 turboprops, flat rated at 634 kW (850 shp), each mounted above wing in all-composites nacelle and driving a Hartzell five-blade constant-speed fully feathering reversible-pitch pusher propeller; propellers handed to counterrotate. Fuel in two fuselage tanks and two wing tanks; total fuel capacity 1,597 litres (422 US gallons; 351 Imp gallons), of which 1,583 litres (418 US gallons; 348 Imp gallons) are usable. Tanks divided into left and right groups (plus primary and secondary collector tanks) which are independent, except during optional pressure refuelling via single point on starboard centre-fuselage. Gravity refuelling point in upper part of fuselage.

ACCOMMODATION: Crew of one or two on flight deck; certified for single-pilot operation. Seating in main cabin for up to nine passengers, with galley, fully enclosed lavatory and coat storage area; choice of nine-passenger high-density or five-seat VIP cabins. Club passenger seats are armchair type, which can be reclined, tracked and swivelled, and

locked at any angle. Foldaway tables can be extended between facing club seats. Two-piece wraparound electrically heated windscreen. Rectangular cabin windows, including one emergency exit at front on starboard side. Indirect lighting behind each window ring, plus individual overhead lights. Airstair door at front on port side is horizontally split, upper part forward-opening. Baggage compartment aft of rear pressure bulkhead, with upward-opening door immediately aft of wing on port side. Entire cabin area pressurised and air conditioned. New interior by Sergio Pininfarina under development in mid-1999; total of 15 interior options available.

SYSTEMS: Hamilton Sundstrand R70-3WG three-wheel air-cycle bleed air ECS, with maximum pressure differential of 0.62 bar (9.0 lb/sq in) maintaining sea level cabin altitude to 7,315 m (24,000 ft) and a 1,950 m (6,400 ft) cabin altitude at 12,500 m (41,000 ft). Single hydraulic system driven by electric motor, with hand pump for emergency back-up, for landing gear (up to 207 bar; 3,000 lb/sq in), brakes and steering (up to 83 bar; 1,200 lb/sq in). Electrical system powered by two 400 A 28 V starter/Lear Siegler generators and 25 V 38 Ah Ni/Cd battery, with triple-redundant essential bus. Two 250 VA static inverters for AC. External power receptacle above port mainwheel well. Scott 1.13 m³ (40 cu ft) oxygen system. Hot air anti-icing of main wing outer and inner leading-edges; electric anti-icing for foreplane and windscreen; rubber boot for engine intakes, with dynamic particle separator; propeller blades de-iced by engine exhaust. Rockwell Collins APS-65A three-axis autopilot.

AVIONICS: Comms: Dual Rockwell Collins VHF-22C transceivers; dual Rockwell Collins TDR-90 transponders. Dorne & Margolin DM ELT8 ELT.

Radar: Rockwell Collins WXR-840 weather radar standard, TWR-850 turbulence-detecting weather radar optional.

Flight: ADF-462, DME-42, ALT-55B radar altimeter, MCS-65 compass and dual VIR-32 VOR/LOC/MCR (all Rockwell Collins); Goodrich Stormscope WX-1000E; dual Aeronetics RMI-3337 radio compasses. Rockwell Collins ADS-85 air data system. Goodrich SkyWatch HP traffic advisory system. Optional Universal UNS-1K GPS and second ADF-462, and Goodrich LandMark terrain awareness and warning system.

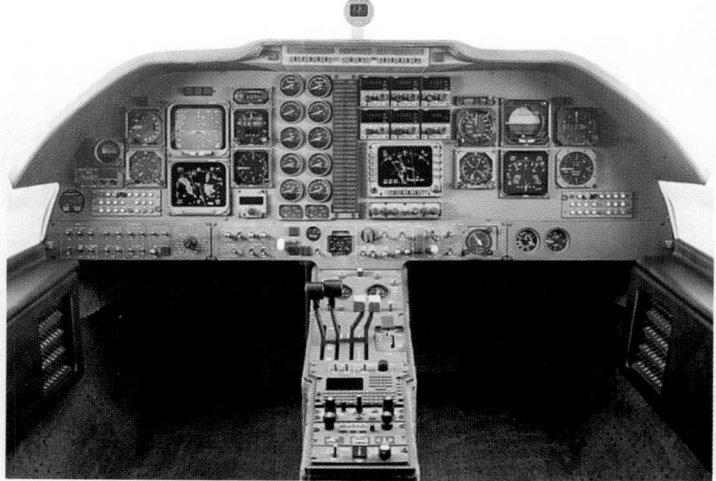

Avanti flight deck   0064740

Typical Piaggio Avanti cabin interior   0126940

Instrumentation: Rockwell Collins EFIS-85B system, with two EFD-85 dual colour CRT MFDs for captain and central MFD-85B radar display; EHSI-74 colour display for co-pilot.

Mission: Optional Global/Wulfsberg Flitefone VI in-flight telephone.

DIMENSIONS, EXTERNAL:
Wing span	14.035 m (46 ft 0½ in)
Foreplane span	3.355 m (11 ft 0 in)
Wing chord: at root	1.82 m (5 ft 11¾ in)
at tip	0.62 m (2 ft 0½ in)
Foreplane chord: at root	0.79 m (2 ft 7 in)
at tip	0.55 m (1 ft 9¾ in)
Wing aspect ratio	12.3
Foreplane aspect ratio	5.1
Length overall	14.41 m (47 ft 3½ in)
Fuselage: Length	12.53 m (41 ft 1¼ in)
Max width	1.95 m (6 ft 4¾ in)
Height overall	3.98 m (13 ft 0¾ in)
Tailplane span	4.26 m (13 ft 11¾ in)
Wheel track	2.84 m (9 ft 4 in)
Wheelbase	5.79 m (19 ft 0 in)
Propeller diameter	2.16 m (7 ft 1 in)
Propeller ground clearance	0.795 m (2 ft 7¼ in)
Distance between propeller centres	4.125 m (13 ft 6½ in)
Passenger door (fwd, port): Height	1.345 m (4 ft 5 in)
Width	0.61 m (2 ft 0 in)
Height to sill	0.58 m (1 ft 10¾ in)
Baggage door (rear, port): Height	0.60 m (1 ft 11¾ in)
Width	0.70 m (2 ft 3½ in)
Height to sill	1.38 m (4 ft 6½ in)
Emergency exit (stbd): Height	0.67 m (2 ft 2¼ in)
Width	0.48 m (1 ft 7 in)

DIMENSIONS, INTERNAL:
Passenger cabin (excl flight deck):
Length	4.55 m (14 ft 11¼ in)
Max width	1.85 m (6 ft 0¾ in)
Max height	1.75 m (5 ft 9 in)
Volume	10.6 m³ (375 cu ft)
Flight deck: Length	1.45 m (4 ft 9 in)
Volume	2.3 m³ (80 cu ft)

Baggage compartment: Floor length	1.70 m (5 ft 7 in)
Max length	1.95 m (6 ft 4¾ in)
Volume	1.25 m³ (44 cu ft)

AREAS:
Wings, gross	16.00 m² (172.2 sq ft)
Ailerons (total, incl tab)	0.66 m² (7.10 sq ft)
Trailing-edge flaps (total)	1.60 m² (17.23 sq ft)
Foreplane	2.19 m² (23.57 sq ft)
Foreplane flaps (total)	0.58 m² (6.30 sq ft)
Fin	4.73 m² (50.91 sq ft)
Rudder, incl tab	1.05 m² (11.30 sq ft)
Tailplane	3.83 m² (41.23 sq ft)
Elevators (total, incl tabs)	1.24 m² (13.35 sq ft)

WEIGHTS AND LOADINGS:
Weight empty, equipped	3,402 kg (7,500 lb)
Operating weight empty, one pilot	3,479 kg (7,670 lb)
Max usable fuel weight	1,271 kg (2,802 lb)
Max payload	907 kg (2,000 lb)
Payload with max fuel	589 kg (1,299 lb)
Max T-O weight	5,239 kg (11,550 lb)
Max ramp weight	5,262 kg (11,600 lb)
Max landing weight	4,965 kg (10,945 lb)
Max zero-fuel weight	4,309 kg (9,500 lb)
Max wing loading	327.4 kg/m² (67.07 lb/sq ft)
Max power loading	4.13 kg/kW (6.79 lb/shp)

PERFORMANCE:
Max operating Mach No (M$_{MO}$): current	0.67
future certification	0.70

Max operating speed (V$_{MO}$)
260 kt (482 km/h; 299 mph) IAS
Max level speed at 8,535 m (28,000 ft)
395 kt (732 km/h; 455 mph)
Manoeuvring speed   199 kt (368 km/h; 229 mph)
Max cruising speed with four passengers at mid-cruise weight:
at 8,535 m (28,000 ft)	391 kt (724 km/h; 450 mph)
at 11,885 m (39,000 ft)	341 kt (632 km/h; 393 mph)

Stalling speed at max landing weight:
flaps up	109 kt (202 km/h; 125 mph)
flaps down	93 kt (172 km/h; 107 mph)
Max rate of climb at S/L	899 m (2,950 ft)/min

Rate of climb at S/L, OEI	230 m (755 ft)/min
Max certified altitude	12,500 m (41,000 ft)
Service ceiling	11,885 m (39,000 ft)
Service ceiling, OEI	7,590 m (24,900 ft)

T-O to 15 m (50 ft) ISA, S/L at max T-O weight
869 m (2,850 ft)
Landing from 15 m (50 ft) at max landing weight, no propeller reversal   872 m (2,860 ft)
Range: at max cruise power, IFR reserves:
at 10,670 m (35,000 ft)
1,260 n miles (2,333 km; 1,450 miles)
at 11,885 m (39,000 ft)
1,456 n miles (2,696 km; 1,675 miles)
at max range power:
VFR reserves at 11,885 m (39,000 ft)
1,793 n miles (3,320 km; 2,063 miles)
IFR reserves:
at 9,150 m (30,000 ft)
1,304 n miles (2,415 km; 1,500 miles)
at 10,670 m (35,000 ft)
1,420 n miles (2,629 km; 1,634 miles)
at 11,885 m (39,000 ft)
1,509 n miles (2,794 km; 1,736 miles)

OPERATIONAL NOISE LEVELS (FAR Pt 36):
Appendix F	76.0 dB(A)
Appendix G	81.8 dB(A)

UPDATED

PIAGGIO AVANTI DEVELOPMENTS

At the 1999 Paris Air Show Piaggio announced that it was undertaking conceptual design work on a family of aircraft based on the Avanti. These include a stretched, 19-passenger regional airliner, and a turbofan-powered business aircraft. The turbofan version would feature a new wing and rear fuselage structure. By April 2002 there had been no further announcements.

UPDATED

RUSALEN & RUSALEN

RUSALEN & RUSALEN SdF
Via Nanetti 18, I-33077 Sacile (PN)
Tel: (+39 0335) 816 88 30
Fax: (+39 0421) 427 69
e-mail: steni@dacos.it

For details on the Rusalen & Rusalen/Aerostyle Asso Vs, see Alpi Aviation.
UPDATED

SAI

SOCIETA AERONAUTICA ITALIANA
(Italian Aeronautical Society)
Via F Bottazzi 38, I-80128 Napoli
Tel: (+39 08) 18 49 63 74
Fax: (+39 08) 18 82 97 82
e-mail: info@saiaeronautica.com
Web: http://www.saiaeronautica.com
FOUNDERS: Vincenzo Giordano
  Gaetano Iervoleno
  Gastone Iervolino
INFORMATION OFFICER: Federica Tevvolino

SAI was established in 1997 to design, market and support ultralight aircraft; conduct aerodynamic research; and manage static and flight certification test programmes. Works at Vitulazio (CE) are being augmented by a new plant at San Salvatore Telesino (BN).

VERIFIED

SAI G97 SPOTTER

TYPE: Multisensor surveillance lightplane; side-by-side ultralight.
PROGRAMME: Design began June 1997; construction of prototype G97 ultralight started November 1997; first flight June 1998, powered by 47.8 kW (64.1 hp) Rotax 582; first G97V lightplane (I-OPIS) had public debut at Aero '01, Friedrichshafen, in April 2001 and first flew 15 May 2001. Construction of first production G97 began September 2000; initial flight March 2001. Certification of G97V was due late 2001.
CURRENT VERSIONS: **G97V Spotter:** Certified light aircraft to Italian ENAC V.EL (elementary aircraft) standard. Differs from G97 prototype in having stretched forward fuselage.
  G97 Spotter: As G97V, but in Ultralight category.
CUSTOMERS: Six aircraft of all versions (including prototypes) flying by September 2001.
COSTS: G97 €46,000; G97V €60,000 (2001).
DESIGN FEATURES: Advanced ultralight able to transition to certified aircraft category; intended for police/civil protection agency use, with helicopter-like view for crew assured by pusher propeller, side-by-side seating and absence of wing external bracing. Also suitable for private

Prototype SAI G97V Spotter at Friedrichshafen in April 2001 *(Jane's/Paul Jackson)*   0110893

flying, training and light transport. Prime design objectives included sacrifice of speed to ensure utility; configuration overcomes related problems of ground cooling for engine, aerodynamic balance with large CG range, comfort in turbulent air and cabin noise. Computer modelling used in design, particularly with pod/boom interface and aerofoil tested in Department of Aircraft Design, Naples University.

Pod-and-boom fuselage; cantilever, constant chord wings, detachable for storage. Sweptback fin; tailplane below propeller slipstream.

Purpose-designed Giordano VG1-13H high-lift aerofoil; no sweep; no twist; thickness-chord ratio 13.1 per cent; dihedral 1°; incidence 2° 30'.
FLYING CONTROLS: Manual. Frise differential ailerons, rudder and all-moving tailplane with anti-balance tab and electrically actuated trim. Electrically actuated slotted flaps on 71 per cent of wing trailing-edge.

STRUCTURE: Mostly aluminium alloy structure and skin, except rudder and part of tailplane Dacron covered; wing/fuselage fairing, wingtips and tailcone of glass fibre.
LANDING GEAR: Tricycle type; fixed. Cantilever-sprung mainwheel legs of single carbon fibre/glass fibre piece, attached to fuselage by four bolts. Nosewheel with rubber shock-absorber. Hydraulic disc brakes on mainwheels operated by lever in cockpit. Steerable nosewheel. Main tyres size 6.00-5; nose 11×4.00-5.
POWER PLANT: One 59.6 kW (79.9 hp) Rotax UL flat-four driving a Tonini two-blade, wooden propeller of 1.44 m (4 ft 8½ in) pitch. Two wing leading-edge fuel tanks, combined capacity 70 litres (18.5 US gallons; 15.4 Imp gallons), of which 66 litres (17.4 US gallons; 14.5 Imp gallons) are usable. Option of 73.5 kW (98.6 hp) Rotax 912S in later versions.
ACCOMMODATION: Two persons, side by side, with four-point harnesses. Upward-opening door each side. Dual controls,

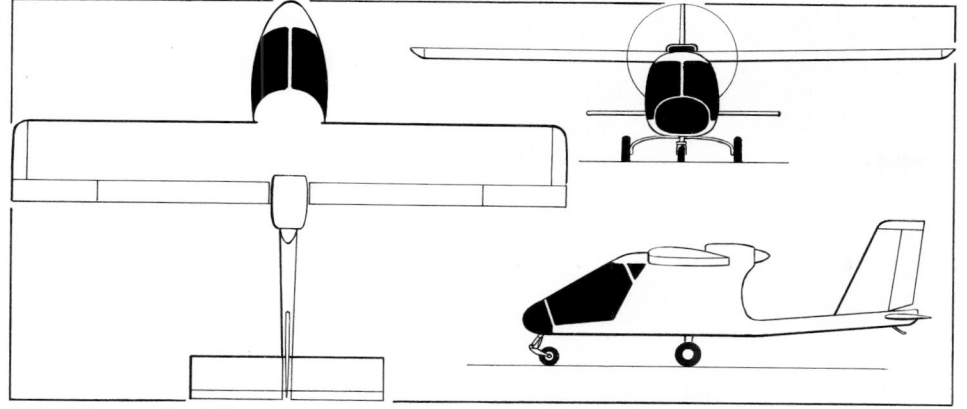

SAI G97 Spotter (*Jane's/James Goulding*)

0121431

AREAS:		
Wings, gross		10.30 m² (110.87 sq ft)
Tailplane		1.97 m² (21.20 sq ft)
WEIGHTS AND LOADINGS:		
Weight empty: G97		283 kg (624 lb)
G97V		298 kg (657 lb)
Max T-O weight: G97		450 kg (992 lb)*
G97 alternative		530 kg (1,168 lb)*
G97V		530 kg (1,168 lb)
Max wing loading: G97		43.8 kg/m² (8.95 lb/sq ft)
G97 alternative, G97V		51.5 kg/m² (10.54 lb/sq ft)
Max power loading: G97		7.56 kg/kW (12.4 lb/hp)
G97 alternative, G97V		8.90 kg/kW (14.62 lb/hp)

* According to local regulations

PERFORMANCE:	
Never-exceed speed (V_{NE})	118 kt (220 km/h; 136 mph)
Max level speed at S/L	108 kt (200 km/h; 124 mph)
Cruising speed at 75% power	94 kt (175 km/h; 108 mph)
Econ cruising speed	86 kt (160 km/h; 99 mph)
Stalling speed, power off:	
flaps up	38 kt (69 km/h; 43 mph)
flaps down	32 kt (58 km/h; 36 mph)
Max rate of climb at S/L	280 m (919 ft)/min
Service ceiling	3,505 m (11,500 ft)
T-O and landing run	85 m (280 ft)
Range with 30 min reserves	
	297 n miles (550 km; 341 miles)
g limits	+4/−2

VERIFIED

including central throttle; ventilation. Baggage compartment behind seats.

SYSTEMS: Electrical system has 12 V battery and 100 W alternator.

AVIONICS: To customer's requirements.

DIMENSIONS, EXTERNAL:

Wing span	8.25 m (27 ft 0¾ in)
Wing chord, constant	1.25 m (4 ft 1¼ in)
Wing aspect ratio	6.7
Length overall	6.24 m (20 ft 5½ in)

Height overall	2.30 m (7 ft 6½ in)
Tailplane span	2.90 m (9 ft 6¼ in)
Wheel track	2.14 m (7 ft 0¼ in)
Wheelbase	1.60 m (5 ft 3 in)
Propeller diameter: normal	1.66 m (5 ft 5¼ in)
max	1.72 m (5 ft 7¾ in)

DIMENSIONS, INTERNAL:

Cabin: Length	1.16 m (3 ft 9¾ in)
Max width	1.08 m (3 ft 6½ in)
Baggage volume	0.40 m³ (14.1 cu ft)

SG

SG AVIATION

Via Biancamano, Zona Industriale, I-04016 Sabaudia, Latina
Tel: (+39 0773) 51 52 16
Fax: (+39 0773) 51 14 07
e-mail: sg@storm-sg.it
Web: http://www.storm-sg.it
PRESIDENT: Giovanni Salsedo

NORTH AMERICAN AGENT:
Eden Technologies International LLC
Skylark Airpark, Broad Brook, Connecticut 06016
Tel/Fax: (+1 860) 668 54 74
e-mail: edentech@worldnet.att.net
DIRECTOR: Eric Dixon

In addition to the Storm series of light aircraft (based on an original Pottier design; see French section) and Sea Storm, SG Aviation manufactures spare parts and subassemblies for the Italian Air Force. The workforce numbered 70 in March 2002.

In April 2002, SG launched the high-wing Rally.

UPDATED

SG RALLY

TYPE: Side-by-side ultralight.

PROGRAMME: Aerodynamic studies began 1991; prototype flew in 1993, but development was suspended in favour of Sea Storm and Storm 300, both of which employed the Rally's wing profile. Work resumed in 2000; marketing launched in April 2002 to meet new, more liberal European sport pilot and US Experimental category legislation.

CURRENT VERSIONS: **Rally 105 S:** Experimental category or Ultralight version. *As described.*

Rally 105 UL: Ultralight version only. Empty weight 265 kg (584 lb); MTOW 435 kg (959 lb); reduced fuel and power.

DESIGN FEATURES: High-mounted, strut-braced, constant-chord wings. Low tailplane and sweptback fin mounted on steeply waisted rear fuselage.

FLYING CONTROLS: Conventional and manual. Horn-balanced tailplane and rudder. Slotted flaps.

Cabin and upward-hinged door of the Rally 105
NEW/0137407

Rally 105 S Experimental category lightplane
NEW/0137408

STRUCTURE: Generally of composites.
LANDING GEAR: Fixed tricycle type.
POWER PLANT: One 59.6 kW (79.9 hp) Rotax 912 UL-2 flat-four in Rally UL; 73.5 kW (98.6 hp) Rotax 912 ULS-2 in Rally S. Other suitable engines up to 93.2 kW (125 hp). Fuel capacity 85 litres (22.5 US gallons; 18.7 Imp gallons) in Rally UL; 100 litres (26.4 US gallons; 22.0 Imp gallons) in Rally S.

DIMENSIONS, EXTERNAL:

Wing span	9.18 m (30 ft 1½ in)
Length overall	7.08 m (23 ft 2¾ in)

DIMENSIONS, INTERNAL:

Cabin max width	1.16 m (3 ft 9¾ in)

AREAS:

Wings, gross	12.48 m² (155.9 sq ft)

WEIGHTS AND LOADINGS:

Weight empty	290 kg (639 lb)*
Max T-O weight	450 kg (992 lb)*

* if flown as Ultralight

PERFORMANCE:

Max level speed	130 kt (240 km/h; 149 mph)
Cruising speed at 75% power	
	116 kt (215 km/h; 134 mph)
Stalling speed: flaps up	36 kt (65 km/h; 41 mph)
flaps down	30 kt (55 km/h; 35 mph)

Max rate of climb at S/L	390 m (1,280 ft)/min
T-O run	70 m (230 ft)
Landing run	110 m (360 ft)
Max range	675 n miles (1,250 km; 776 miles)
g limits	+4.5/−2.2

NEW ENTRY

SG STORM

TYPE: Side-by-side lightplane/kitbuilt.

PROGRAMME: Design started 11 February 1989 (Storm 280 SI); construction of prototype started 24 November 1990; first flight March 1991; certified to Aero Club Italia ULM category. Production aircraft manufacture began 29 March 1993, first flight July 1993; first delivery August 1993.

CURRENT VERSIONS: Total of 11 versions currently available, both above and below ultralight weight limit: **Storm 280 G, 280 SI, 300** and **400 TI** have tricycle landing gear; **Storm 300 Special**, introduced in 1997, has tricycle gear; **Storm 300 Focus** is identical airframe with additional extra equipment; including speed fairings, lighting and electric trims. Available only as complete aircraft; adds 10 kg (22 lb) to empty weight. The **Storm Century** development of the Storm 300 Special was introduced in 2000 and is available in tricycle and

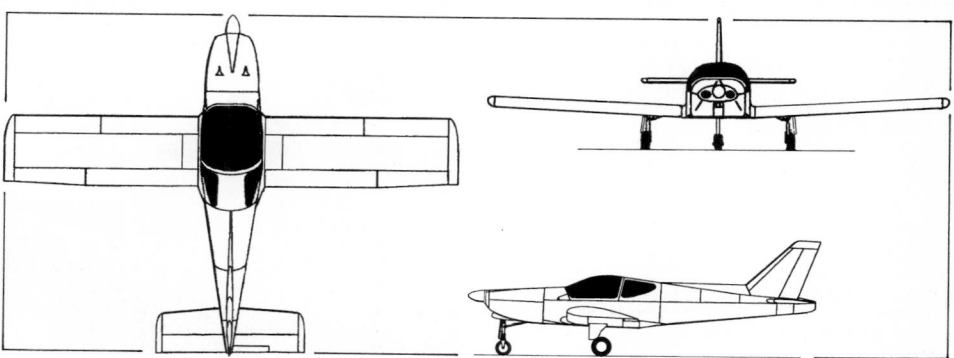

SG Storm RG retractable landing gear version
NEW/0137727

Tailwheel version of SG Storm (*Jane's/Paul Jackson*)   0110600

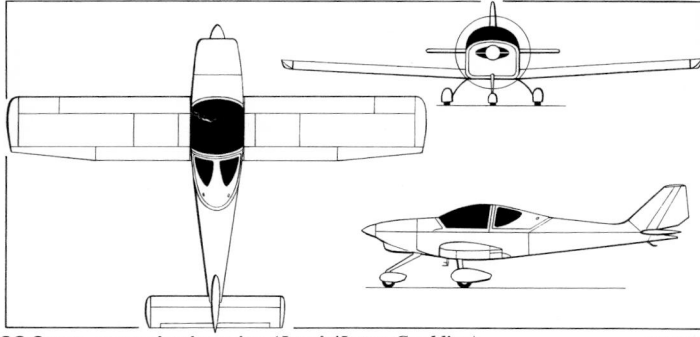

SG Storm, nosewheel version (*Jane's/James Goulding*)   0011094

First US-built Storm 400SP, owned by Raymond Taunton of Brandon, Florida (*Jane's/Paul Jackson*)
NEW/0137733

tailwheel versions. Other modifications include a lengthened engine cowling, increased fuel capacity using wing tanks, and a baggage compartment behind rear cabin. Additionally, the 119 kW (160 hp) **Storm 400 Special** is a stretched version of the 300 with a stronger spar which has been developed for the US market; **Storm 400 Focus** equipped as per 300 Focus. See Power Plant paragraph for details of engines. The Storm 280 E and 320 E are no longer available.

Storm 500: Introduced in 2001 in two versions (basic CLK and retractable-gear SLK); is full-four-seat version, available only in tricycle form.

Storm RG: First flown 2001; development of 400 with conventional tailplane and modified fin with dorsal fillet. First deliveries in 2002; 12 delivered by April 2002 on production aircraft. Prototype being tested with winglets; not standard.

Cyclone: Military version of Storm RG; hardpoint under each wing. Cockpit glazing comprises fixed windscreen; rearwood sliding canopy; and fixed rear windows. No known production.

Sea Storm: Unrelated design; see separate entry.

CUSTOMERS: By early 2002, nearly 500 Storms ordered, of which more than 250 were flying. Around 80 per cent are tricycle landing gear versions.

COSTS: Kits: Storm 280 €19,212; Storm 300 Special €18,954; Storm 400 Special €19,884; 500 CLK €21,898, 500 SLK €25,513. Complete: Storm 280 €61,045; Storm 300 Special €62,026; Storm 300 Focus €72,872; 400 Special €72,511; 400 Focus €83,356; 500 CLK €73,905, 500 SLK €79,483; RG US$82,000 (all 2002).

DESIGN FEATURES: Extensive use of modern materials and technology. All-aluminium fuselage and wings on 280 SI, 300 Special, 320 E and 400 TI. Quoted build time for kits 600 hours.

Wing sections NACA 4415 (Storm 280 SI); GA 3OU-6135 Mod (Storm 300 Special). Thickness/chord ratio 15 per cent; 3° dihedral; 3° incidence.

FLYING CONTROLS: Manual. All-moving tailplane with anti-balance tab on most versions; no rudder tab. Storm RG has conventional horizontal tail. Balanced Frise ailerons on 70 per cent of span. Electric trim controls. Flap settings: 280, 10/20/32°; 12/25/38° on 300 and 400 series; all models have Fowler flaps.

LANDING GEAR: Main legs are cantilever sprung. Tricycle versions have tubular steel sprung nosewheel leg steerable ±40°. Tailwheel leg of Storm 300 is metal sprung with a solid tyre. Hydraulic disc brakes on mainwheels. Speed fairings can be fitted. Matco tyres; mainwheels

15×6.00, pressure 2.00 bar (29 lb/sq in); tailwheel (mainwheel in Storm 300 Special) 15 cm (6 in), pressure 1.80 bar (26 lb/sq in). Storm RG has 5.00-5 wheels on all legs; main wheels retract inwards, nosewheel rearwards. No doors.

POWER PLANT: *Storm 280 SI and 280 G:* One 59.6 kW (79.9 hp) Rotax 912 UL or 84.6 kW (113.4 hp) Rotax 914 flat-four four-stroke driving either a two-blade GT166-42 fixed-pitch propeller or a variable-pitch IVO-Prop or AR-Plast propeller; or one 59.7 kW (80 hp) Moto Guzzi driving a two-blade GM fixed-pitch propeller. Fuel capacity for 280 and 300 series 85 litres (22.5 US gallons; 18.7 Imp gallons). Oil capacity 3.5 litres (0.92 US gallon; 0.77 Imp gallon).

Storm 300, 300 Special, 300 Focus and Century: One 84.6 kW (113.4 hp) Rotax 914 turbocharged four-cylinder four-stroke engine driving a two-blade Rospeller constant-speed metal propeller. Mid-West AE 110 Hawk 78.3 kW (105 hp) rotary engine optional. Fuel capacity of Storm 300 and 300 Special as Storm 280 G; Storm Century 100 litres (26.4 US gallons; 22.0 Imp gallons).

Storm 400 TI: One 86.5 kW (116 hp) Textron Lycoming O-235-N2C four-cylinder engine driving a two-blade Hartzell HC-C2YL-1BF constant-speed propeller. Fuel

capacity 140 litres (37.0 US gallons; 30.8 Imp gallons); optional tanks increase this to 200 litres (52.8 US gallons; 44.0 Imp gallons).

Storm 400 Special and 400 Focus: One 119 kW (160 hp) Textron Lycoming O-320 flat-four; engines in the range 82.0 to 119 kW (110 to 160 hp) can be fitted. Fuel capacity as Storm 400 TI.

Storm 500 CLK and 500 SLK: One 157 kW (210 hp) Textron Lycoming O-360. Fuel capacity 180 litres (47.6 US gallons; 39.6 Imp gallons); engines in the range 104 to 134 kW (140 to 180 hp) can be fitted.

Storm RG: Duc three-blade propeller.

ACCOMMODATION: Two persons, side by side, with dual controls; single-piece canopy hinged upwards and forwards. Storm 400 TI can accommodate one large or two small additional passengers. Large baggage area behind cockpit.

SYSTEMS: 12 V battery.

AVIONICS: *Instrumentation:* Option of full IFR fit.

DIMENSIONS, EXTERNAL (A: 280 G, B: 300 Special, C: 400 TI, D: 400 Special, E: Century, F: 500 CLK; G: 500 SLK; H: RG):

Wing span:	
A, B, E	7.80 m (25 ft 7 in)
C, D	7.98 m (26 ft 2¼ in)
F, G	8.60 m (28 ft 2½ in)
H	8.02 m (26 ft 4 in)
Wing chord, constant: B, C	1.30 m (4 ft 3 in)
Length overall:	
A, B	6.70 m (21 ft 11¾ in)
C, D	6.82 m (22 ft 4½ in)
E	6.80 m (22 ft 3¾ in)
F, G	7.22 m (23 ft 8¼ in)
H	6.48 m (21 ft 3 in)
Height overall: A, C, D	2.15 m (7 ft 0¾ in)
B	2.12 m (6 ft 11½ in)
H	2.39 m (7 ft 10 in)

DIMENSIONS, INTERNAL:

Cabin max width: A, B, D, E, F, G	1.10 m (3 ft 7¼ in)

AREAS:

Wings, gross:	
A, B, E	9.98 m² (107.4 sq ft)
C, D	10.45 m² (112.5 sq ft)
F, G	11.35 m² (122.2 sq ft)

WEIGHTS AND LOADINGS:

Weight empty: A	280 kg (617 lb)
B	293 kg (646 lb)
C	400 kg (882 lb)
D	480 kg (1,058 lb)
E	320 kg (705 lb)
F	510 kg (1,124 lb)
G	530 kg (1,168 lb)
Max T-O weight: A, B	450 kg (992 lb)
C	720 kg (1,587 lb)
D	850 kg (1,874 lb)
E	560 kg (1,234 lb)
F	975 kg (2,149 lb)
G	995 kg (2,193 lb)

Mockup of Storm RG shown at Sun 'n' Fun, April 2002 (*Jane's/Paul Jackson*)   *NEW*/0137734

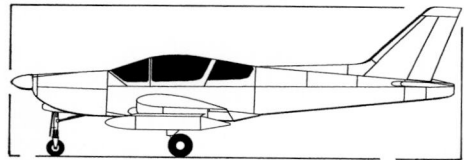

Proposed SG Cyclone military version of Storm RG
NEW/0137726

Max wing loading: A	47.0 kg/m² (9.63 lb/sq ft)
B	45.1 kg/m² (9.24 lb/sq ft)
C	68.9 kg/m² (14.11 lb/sq ft)
D	81.3 kg/m² (16.66 lb/sq ft)
E	56.1 kg/m² (11.49 lb/sq ft)
F	85.9 kg/m² (17.59 lb/sq ft)
G	87.7 kg/m² (17.96 lb/sq ft)
Max power loading: A	8.72 kg/kW (14.34 lb/hp)
B	5.32 kg/kW (8.74 lb/hp)
C	8.33 kg/kW (13.68 lb/hp)
D	7.14 kg/kW (11.74 lb/hp)
E	6.62 kg/kW (10.88 lb/hp)
F	6.21 kg/kW (10.20 lb/hp)
G	6.34 kg/kW (10.41 lb/hp)

PERFORMANCE:
Never-exceed speed (V_NE):

A, C	136 kt (252 km/h; 156 mph)
B, D, E	167 kt (310 km/h; 192 mph)
F, G	172 kt (320 km/h; 198 mph)

Max operating speed (V_MO):

A	121 kt (225 km/h; 140 mph)
B	152 kt (282 km/h; 175 mph)
C	150 kt (278 km/h; 172 mph)
E	155 kt (287 km/h; 178 mph)
D, F	157 kt (290 km/h; 180 mph)
G	165 kt (305 km/h; 190 mph)

Cruising speed at 75% power:

A	108 kt (200 km/h; 124 mph)
B	143 kt (265 km/h; 165 mph)
D	148 kt (274 km/h; 170 mph)
E	150 kt (278 km/h; 173 mph)
F	150 kt (277 km/h; 172 mph)
G	158 kt (292 km/h; 181 mph)

Stalling speed, power off, flaps down:

A	30 kt (55 km/h; 35 mph)
B	29 kt (52 km/h; 33 mph)
C, D	38 kt (70 km/h; 44 mph)
E	30 kt (55 km/h; 35 mph)
F, G	43 kt (78 km/h; 49 mph)
Max rate of climb at S/L: A	330 m (1,083 ft)/min
B, E	420 m (1,378 ft)/min
C, D, F, G	450 m (1,476 ft)/min
Service ceiling: A	4,000 m (13,120 ft)
B, D, E	6,096 m (20,000 ft)
T-O run: A	90 m (295 ft)
B, E	120 m (395 ft)
C	210 m (690 ft)
D, F, G	220 m (720 ft)
Landing run: A	150 m (495 ft)
B	180 m (590 ft)
C	240 m (790 ft)
D	200 m (660 ft)
E	120 m (305 ft)
F, G	250 m (820 ft)

Range with max fuel:

A	648 n miles (1,200 km; 745 miles)
B	702 n miles (1,300 km; 807 miles)
C	453 n miles (839 km; 521 miles)
D	594 n miles (1,100 km; 683 miles)
E	864 n miles (1,600 km; 994 miles)
F	621 n miles (1,150 km; 714 miles)
G	648 n miles (1,200 km; 745 miles)

g limits: A +5.7/−3
B +6/−3
C +3.8/−1.9
D, F +3.8/−2.0
E +4.2/−2
G +3.7/−1.9
UPDATED

SG SEA STORM

TYPE: Two/four-seat amphibian kitbuilt.
PROGRAMME: Derived from earlier ultralight version, with 59.7 kW (80 hp) engine and metal strut-braced wings, of which some 20 flying in Italy by 2002. First all-composites

SG Sea Storm amphibian nearing completion in April 2001 (*Jane's/Susan Bushell*) 0110584

Sea Storm under construction in 2000/01 by SG's North American agent, Eric Dixon and first flew October 2001.
CURRENT VERSIONS: **Z2**: Two-seat; baseline version.
Z4: Four-seat; increased wing span and power.
CUSTOMERS: Three sold in USA by April 2002.
COSTS: Z2 US$32,200 (kit), US$69,700 (complete). Z4 US$37,400 (kit), US$83,400 (complete) (2000).
DESIGN FEATURES: High, constant-chord wing, moderately sweptback tailplane and highly sweptback fin; pusher propeller immediately aft of wing trailing-edge; mid-positioned stub-wing with stabilising floats. Quoted kit build time 700 hours.
FLYING CONTROLS: Conventional and manual. Plain flaps.
LANDING GEAR: Optional retractable wheel landing gear attached immediately outboard of wingroots.
POWER PLANT: One pusher piston/rotary engine.
Z2: Option of 73.5 kW (98.6 hp) Rotax 912 ULS or 84.6 kW (113.4 hp) Rotax 914 flat-four or 70.8 kW (95 hp) Mid-West AE 100 rotary driving an adjustable-pitch propeller. Textron Lycoming engines in the 82.0 to 119 kW (110 to 160 hp) range are also suggested. Standard fuel capacity 140 litres (37.0 US gallons; 30.8 Imp gallons), with optional extra 79 litre (20.9 US gallon; 17.4 Imp gallon) tanks available.
Z4: One 157 kW (210 hp) Textron Lycoming O-360; engines recommended in the range 149 to 224 kW (200 to 300 hp). Fuel capacity 200 litres (52.8 US gallons; 44.0 Imp gallons) as standard; optional 61 litre (16.1 US gallon; 13.4 Imp gallon) tanks available.
Idrovario hydraulic, flight-adjustable pitch, two- or three-blade propeller.
ACCOMMODATION: Two (side by side) or four persons in enclosed cabin with centreline-hinged door each side.
All data following (including dimensions) are approximate.
DIMENSIONS, EXTERNAL (A: Z2; B: Z4):

Wing span: A	9.60 m (31 ft 6 in)
B	10.10 m (33 ft 1¼ in)
Length overall: A, B	7.44 m (24 ft 5 in)
Height overall: A, B	1.83 m (6 ft 0 in)

DIMENSIONS, INTERNAL:
Cabin max width: A, B 1.05 m (3 ft 5¼ in)
AREAS:
Wings, gross: A 12.96 m² (139.5 sq ft)
B 13.65 m² (146.9 sq ft)

WEIGHTS AND LOADINGS (A, B, as above):

Weight empty: A	510 kg (1,124 lb)
B	690 kg (1,521 lb)
Baggage capacity: A	30 kg (66 lb)
B	55 kg (121 lb)
Max T-O weight: A	810 kg (1,785 lb)
B	1,200 kg (2,645 lb)
Max wing loading: A	62.5 kg/m² (12.80 lb/sq ft)
B	87.9 kg/m² (18.01 lb/sq ft)
Max power loading:	
A (Rotax 912)	11.02 kg/kW (18.11 lb/hp)
B (157 kW; 210 hp)	7.64 kg/kW (12.56 lb/hp)

PERFORMANCE (A, B, as above):
Never-exceed speed (V_NE):

A	135 kt (250 km/h; 155 mph)
B	162 kt (300 km/h; 186 mph)
Max level speed: A	121 kt (225 km/h; 140 mph)
B	143 kt (265 km/h; 165 mph)

Cruising speed at 75% power at 2,440 m (8,000 ft):

A	111 kt (205 km/h; 127 mph)
B	125 kt (232 km/h; 144 mph)
Stalling speed: flaps up: A	35 kt (64 km/h; 40 mph)
B	46 kt (85 km/h; 53 mph)
flaps down: A	33 kt (61 km/h; 38 mph)
B	40 kt (74 km/h; 46 mph)
Max rate of climb at S/L: A	240 m (787 ft)/min
B	330 m (1,082 ft)/min
Service ceiling: A	3,960 m (13,000 ft)
B	4,570 m (15,000 ft)
T-O run: on land: A	180 m (591 ft)
B	220 m (722 ft)
on water: A	195 m (640 ft)
B	350 m (1,150 ft)
Landing run: on land: A	200 m (656 ft)
B	250 m (820 ft)
on water: A	146 m (480 ft)
B	210 m (690 ft)

Range, no reserves:
A 917 n miles (1,700 km; 1,056 miles)
B 540 n miles (1,000 km; 621 miles)
g limits: A +4.2/−2.0
B +3.2/−1.8
UPDATED

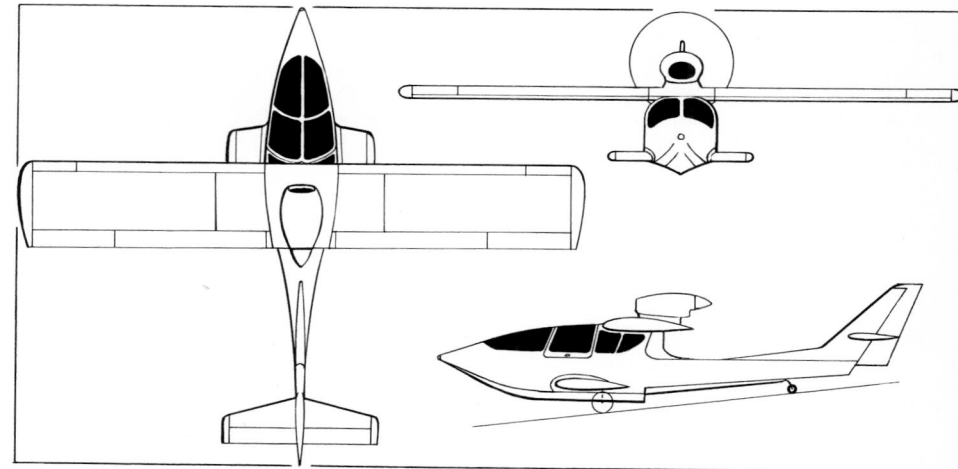

SG Sea Storm (*Jane's/James Goulding*) 0093636

TECNAM

COSTRUZIONI AERONAUTICHE TECNAM SRL

1a Traversa Via G Pascoli, I-80026 Casoria (Naples)
Tel: (+39 081) 758 32 10 and 758 87 51
Fax: (+39 081) 758 45 28
e-mail: tecnamca@tin.it
Web: http://www.tecnam.com

MANAGING DIRECTOR: Prof Luigi Pascale Langer
MARKETING AND SALES DIRECTOR: Paolo Pascale Langer

Company founded 1986, after Pascale brothers were released from the original Pascale company, Partenavia (see now VulcanAir), which had been placed under control of Alenia in 1981. Tecnam manufactures tailplane and other components of ATR 42/72 and parts of A 109, C-27J, SF-260, P.68, plus fuselage panels for Boeing. Workshops qualified to NATO AQAP4 and to Italian Aeronautical Register, JAR 21/F. Manufactures P92 Echo ultralight aircraft and P92-J very light aircraft and in 1997 launched production of the P96 and P96-J; P92 SeaSky amphibian added to range in 1998, P92-S in 1999, and 2000 RG in 2000. Pacific Aerosystem of San Diego appointed US agents in 2001.

In April 2001, Tecnam announced that it was planning to add a four-seater to its range.
UPDATED

TECNAM P92 ECHO

TYPE: Side-by-side lightplane/kitbuilt.

PROGRAMME: First flight 14 March 1993. Prototype P92-J (I- TECN) first flown second quarter 1995 and received type certificate 10 November 1995. Production rate 10 per month by end 2001.

CURRENT VERSIONS: **P92:** Standard ultralight version (Italian ULM rules).

P92-J: Detail differences in airframe and equipment to comply with JAR-VLA airworthiness requirements. First delivery (002), to US distributor, second quarter 1996; 12 built by 1998; manufacture suspended until 2001, when replaced by upgraded P92-JS.

P92-JS: Upgraded version of P-92J with higher-powered Rotax 912 ULS and slightly higher mac T-O weight.

P92-S: Added to range in 1999; redesigned rear cabin with additional window; redesigned wing profile and wing-tips, revised engine cowling and windscreen; improvements in performance, *as described*.

P92 Echo 2000 RG: Retractable landing gear version introduced in 2000. Similar to P92-S, but with shorter (8.70 m; 28 ft 6½ in) wing span and reprofiled flaps and fin tips. Fuselage lower surfaces redesigned (rounded and deeper) to accommodate retracted gear; bulged doors provide extra 4 cm (1½ in) of shoulder room.

P92 SeaSky: Amphibious floatplane version first flown 1 July 1997; strake below rear fuselage. Full Lotus floats with optional wheels for land operations. Rotax 912 ULS engine and increased empty and maximum T-O weights.

Echo 80: Either P92 or P92-S available with standard Rotax 912 UL engine.

Echo 100S: Either P92 or P92-S available with optional Rotax 912 ULS engine. Latter variant designated P92 Echo S 100S.

CUSTOMERS: Some 590 delivered by mid-2001; six to Cambodian Air Force (first two September 1994); overseas orders currently account for 51 per cent of production, including many to Israel. P92-J deliveries totalled 15 by February 2002. P92 2000 RG deliveries totalled 23 by early 2002. Production rate 10 per month during 2001.

COSTS: P92 2000 RG kit US$32,100 less engine, propeller, fuel tanks and electrical wiring (2001).

DESIGN FEATURES: Objectives were lightness, simplicity and accessibility for inspection and servicing. Braced single-spar high wing; aerofoil chosen for good performance at low Reynolds number; untapered with 1° 30′ dihedral; underside of fuselage mostly flat for ease of kit assembly; large one-piece windscreen, inward tapered inboard wing leading-edges, large windows in doors; rear view window (s).

FLYING CONTROLS: Manual. Frise ailerons; all-moving tailplane with anti-balance tab; electrically actuated flaps cover half trailing-edge.

STRUCTURE: Fuselage mainly metal except GFRP lower shell of engine cowling and wing leading-edge; ailerons and part of tailplane fabric covered; GFRP rudder tip; tailplane halves can be unpinned and removed quickly for transport; engine cowling removable by undoing four quick latches to reveal whole power plant.

LANDING GEAR: Non-retractable tricycle type with steel leaf mainlegs attached to bottom of fuselage for easy maintenance; hand-powered hydraulic disc brakes operated together from single lever in cockpit; nosewheel steered from rudder pedals; designed for grass field operation. Wheels 5.00-5 on P92-S and P92-JS; 400-6 on SeaSky. P92 2000 RG nosewheel is 11×4.00, has oleo-pneumatic shock-absorber and retracts rearwards; 14×4 mainwheels retract inward. Both tube legs are mounted on a single cantilever arm.

POWER PLANT: One 59.6 kW (79.9 hp) Rotax 912 UL (P92) or 912A (P92-J) or 73.5 kW (98.6 hp) Rotax 912 ULS flat-four engine; integrated reduction gear driving two-blade Bipala wooden propeller: GT-166/146 for Rotax 912 UL;

GT-172/164 for Rotax 912 ULS landplane; GT-180/155 for SeaSky.

Limbach 2000 (59.7 kW; 80 hp) and Rotax 582 (47.8 kW; 64.1 hp) two-stroke offered as options for P92; P92 prototype had the latter; 59.7 kW (80 hp) Jabiru 2200 flat-four under trial in 2000; added to options following successful completion of tests; 13 delivered by mid-2001. Fuel capacity 70 litres (18.5 US gallons; 15.4 Imp gallons) in two wing tanks, of which 60 litres (15.9 US gallons; 13.2 Imp gallons) are usable.

ACCOMMODATION: Side-by-side seats with three-point harness; baggage space behind seats. Full dual controls and two throttles. Bulged doors on current versions give increased cabin width.

SYSTEMS: 14 V 55 A battery; 100 W alternator. Optional BRS-5 parachute in P92 2000 RG.

AVIONICS: Customer specified; optional IFR nav/com in P92-J.

DIMENSIONS, EXTERNAL (all versions, except where indicated):

Wing span: P92 2000 RG	8.70 m (28 ft 6½ in)
P92, P92-S, P92-JS	9.30 m (30 ft 6 in)
P92-J, SeaSky	9.60 m (31 ft 6 in)
Wing chord	1.40 m (4 ft 7 in)

P92 Echo S 100S (large cabin/Rotax 912 ULS) version (*Jane's/Paul Jackson*)   NEW/0132849

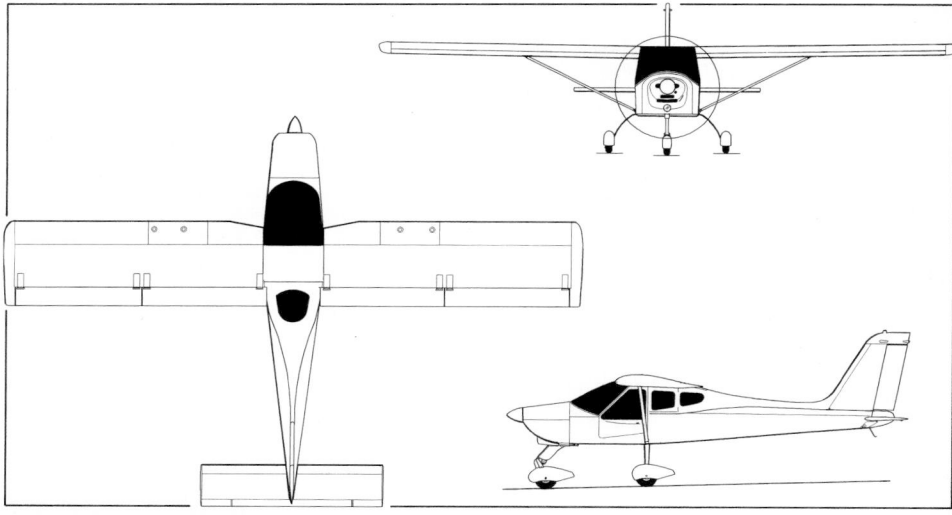

Tecnam P92-S Echo very light aircraft   0110602

Length overall: except P92 2000 RG	6.30 m (20 ft 8 in)
P92 2000 RG	6.40 m (21 ft 0 in)
Height overall: except SeaSky	2.49 m (8 ft 2 in)
SeaSky	2.97 m (9 ft 9 in)
Tailplane span	2.90 m (9 ft 6 in)
Wheel track:	
except SeaSky and P92 2000 RG	1.78 m (5 ft 10 in)
P92 2000 RG	1.73 m (5 ft 8 in)
Float track (SeaSky)	2.79 m (9 ft 2 in)
Propeller diameter:	
59.7 kW engine	1.66 m (5 ft 5¼ in)
73.5 kW engine	1.72 m (5 ft 7¾ in)

DIMENSIONS, INTERNAL:

Cabin max width	1.10 m (3 ft 7¼ in)

AREAS:

Wings, gross:	
except SeaSky and P92 2000 RG	13.20 m² (142.1 sq ft)
P92 2000 RG	11.98 m² (129.0 sq ft)
SeaSky	13.40 m² (144.2 sq ft)
Tailplane	1.97 m² (21.20 sq ft)

WEIGHTS AND LOADINGS:

Basic weight empty:	
P92, P92-S, P92 2000 RG	281 kg (619 lb)
P92-J, P92-JS	310 kg (683 lb)
SeaSky	330 kg (728 lb)
Max T-O weight:	
ultralight: P92, P92-S, P92 2000 RG	450 kg (992 lb)
SeaSky	500 kg (1,102 lb)
certified: P-92, P92-S, P92-JS	550 kg (1,212 lb)
P92-J	535 kg (1,179 lb)
SeaSky	600 kg (1,322 lb)

PERFORMANCE (P92 in Echo 80 form, P92-S in Echo 100 form):

Never-exceed speed (VNE):	
except SeaSky	135 kt (250 km/h; 155 mph)
SeaSky	108 kt (200 km/h; 124 mph)
Max level speed at S/L:	
P92, P92-J	119 kt (220 km/h; 137 mph)
P92-S, P92 2000 RG	124 kt (230 km/h; 143 mph)
P92-JS	110 kt (204 km/h; 127 mph)
SeaSky	92 kt (170 km/h; 106 mph)
Cruising speed at 75% power at 2,200 propeller rpm:	
P92, P92-J, P92-JS	100 kt (185 km/h; 115 mph)
P92-S	111 kt (206 km/h; 128 mph)
SeaSky	84 kt (156 km/h; 97 mph)
Stalling speed: flaps up: P92	36 kt (68 km/h; 42 mph)
P92-J	40 kt (74 km/h; 46 mph)
flaps down:	
P92, P92-S, P92 2000 RG	33 kt (61 km/h; 38 mph)

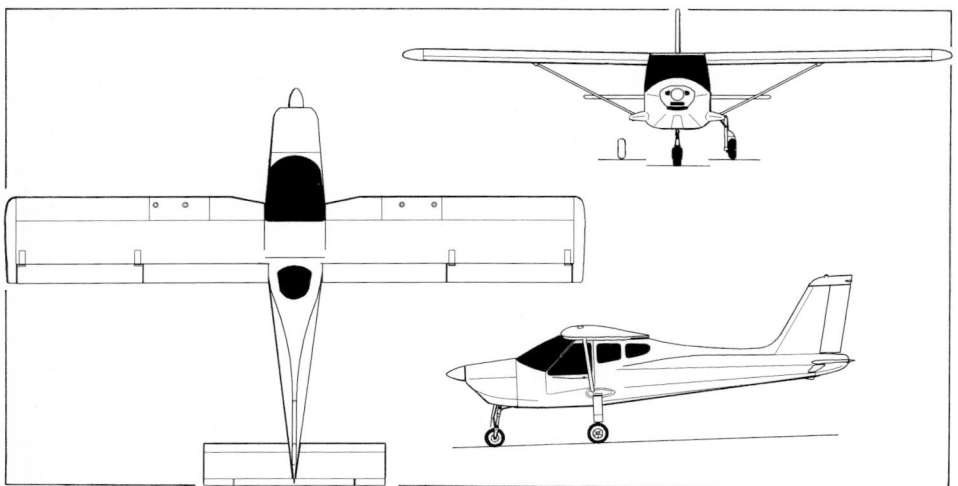

Tecnam Echo 2000 RG   NEW/0132835

US demonstrator for Echo 2000 RG retractable landing gear version *(Jane's/Paul Jackson)* **NEW**/0132850

P92-J, SeaSky	35 kt (64 km/h; 40 mph)
P92-JS	36 kt (67 km/h; 42 mph)
Max rate of climb at S/L: P92	330 m (1,082 ft)/min
P92-J	305 m (1,000 ft)/min
P92-JS	300 m (985 ft)/min
P92-S, P92 2000 RG	384 m (1,260 ft)/min
SeaSky	244 m (800 ft)/min
Service ceiling: P92, P92-J	4,265 m (14,000 ft)
P92 2000 RG, P92-S	4,510 m (14,800 ft)
P92-JS	4,300 m (14,100 ft)
SeaSky	3,505 m (11,500 ft)
T-O run: on land: P92	110 m (360 ft)
P92-J, P92-JS	143 m (470 ft)
P92-S	100 m (330 ft)
P92 2000 RG	140 m (460 ft)
SeaSky	122 m (400 ft)
on water: SeaSky	350 m (1,150 ft)
Landing run: on land:	
except P92-J, P92 2000 RG, SeaSky	100 m (328 ft)
P92 2000 RG	110 m (360 ft)
P92-J	96 m (315 ft)
SeaSky	122 m (400 ft)
on water (SeaSky)	350 m (1,150 ft)
Max range, no reserves:	
except SeaSky	400 n miles (740 km; 460 miles)
SeaSky	300 n miles (555 km; 345 miles)
Endurance: P92	4 h 30 min
P92-J	5 h
Best glide ratio: P92, P92-J	13
g limits: P92-J, P-92-S	+6/−3
P92-JS	+3.8/−1.5
SeaSky	+5.3/−2

UPDATED

TECNAM P96 GOLF

TYPE: Side-by-side ultralight.

PROGRAMME: Prototype construction began July 1996; first flight March 1997. Uprated Golf 100 added in 1999.

CURRENT VERSIONS: **P96 Golf 80:** Ultralight version.
P96 Golf 100: Higher-powered ultralight version.

CUSTOMERS: The 200th Golf registered in Canada (C-ITEC) in February 2002.

DESIGN FEATURES: Emphasis on reduction of aerodynamic drag without compromising cost. Fuselage cross-sectional area allows easy access for taller occupants by reducing the height of structural carry-through elements. Wing lower surface and fuselage bottom are flush, obviating interference drag and complex fairings. Wing-to-fuselage point optimised by strakes at wingroots. Many parts common to P92 Echo.

FLYING CONTROLS: Manual. Frise ailerons; all-moving tailplane with anti-balance tab; electrically actuated flaps cover half of trailing-edge. Ground-adjustable tab on port aileron and rudder.

STRUCTURE: Fuselage includes a composites spine that follows through to the tailfin and keeps wetted area to a minimum while allowing for good pressure recovery aft of canopy. All-moving horizontal tail reduces drag to a minimum and provides optimum longitudinal stability for stick-free operation. Wing has 5° dihedral. Fabric-covered flaps, ailerons and tailplane; metal rudder.

on guiderails and can be opened with engine on and during flight. Full dual controls and twin throttles.

SYSTEMS: 100 W 12 V alternator and battery.

AVIONICS: To customer's requirements; provision for nav/com radio.

EQUIPMENT: Junkers ballistic parachute.

DIMENSIONS, EXTERNAL:	
Wing span	8.41 m (27 ft 7 in)
Length overall	6.40 m (21 ft 0 in)
Height overall	2.29 m (7 ft 6 in)
Tailplane span	2.90 m (9 ft 6 in)
Wheel track	1.78 m (5 ft 10 in)
Wheelbase	1.60 m (5 ft 3 in)
Propeller diameter	1.65 m (5 ft 5 in)
DIMENSIONS, INTERNAL:	
Cabin max width	1.13 m (3 ft 8½ in)
AREAS:	
Wings, gross	12.20 m² (131.3 sq ft)
WEIGHTS AND LOADINGS:	
Weight empty: Golf 80	281 kg (619 lb)
Max T-O weight: both	450 kg (992 lb)
PERFORMANCE:	
Never-exceed speed (V$_{NE}$)	140 kt (259 km/h; 161 mph)
Max level speed: Golf 80	121 kt (224 km/h; 139 mph)
Golf 100	130 kt (241 km/h; 150 mph)
Cruising speed at 75% power:	
Golf 80	105 kt (195 km/h; 121 mph)
Golf 100	116 kt (215 km/h; 133 mph)
Stalling speed, power off, flaps down:	
both	36 kt (67 km/h; 42 mph)
Max rate of climb at S/L: Golf 80	271 m (890 ft)/min
Golf 100	360 m (1,180 ft)/min
Service ceiling: Golf 80	4,000 m (13,120 ft)
Golf 100	4,500 m (14,760 ft)
T-O run: Golf 80	110 m (361 ft)
Golf 100	100 m (328 ft)
Landing run: both	100 m (328 ft)
Max range: Golf 80	400 n miles (740 km; 460 miles)
Endurance: Golf 80	4 h 30 min
Best glide ratio: Golf 80	12
Ultimate g limits: both	+6/−3

UPDATED

LANDING GEAR: Fixed tricycle type; hydraulic disc brakes operated from single hand lever in cockpit. Levered suspension nosewheel leg; nosewheel steered by rudder pedals. Mainwheels 5.00-5, nosewheel 4.00-6. Speed fairings.

POWER PLANT: *Golf-80*: One 59.6 kW (79.9 hp) Rotax 912 UL flat-four engine with integrated reduction gear (1:2.27), driving a GT-2/173/155 single-piece two-blade wooden propeller.

Golf 100: 73.5 kW (98.6 hp) Rotax 912 ULS. Propeller as Golf-80.

Fuel capacity (both) 70 litres (18.5 US gallons; 15.4 Imp gallons).

ACCOMMODATION: Two seats side by side; baggage space behind seats. Fixed windscreen; canopy slides backwards

Tecnam P96 Golf 100 *(Jane's/Paul Jackson)* **NEW**/0132851

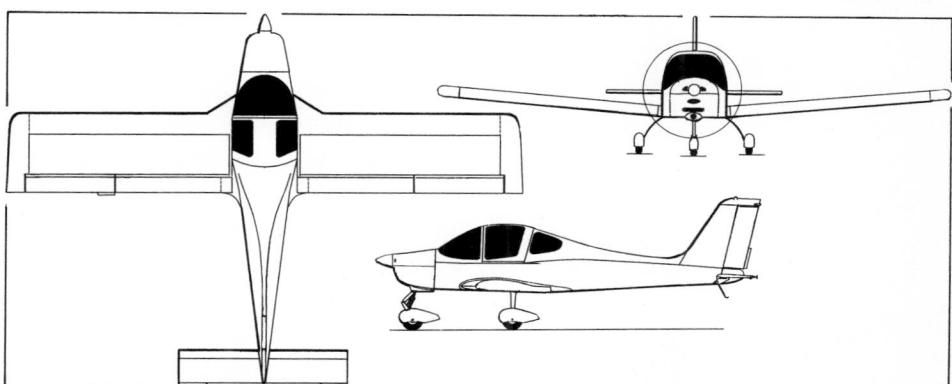

Tecnam P96 Golf two-seat low-wing ultralight aircraft (Rotax 912 piston engine) *(Jane's/Paul Jackson)*
0062982

For details of the latest updates to *Jane's All the World's Aircraft* online and to discover the additional information available exclusively to online subscribers please visit
jawa.janes.com

VULCANAIR

VULCANAIR SpA
Via G Pascoli 7, I-80026 Casoria (NA)
Tel: (+39 081) 591 81 11
Fax: (+39 081) 591 81 72
e-mail: infomarketing@vulcanair.com
Web: http://www.vulcanair.com
PRESIDENT: Carlo de Feo
MARKETING MANAGER: Gianni de Stefano

Samanta aircraft service company purchased by Carlo de Feo in 1997; Samanta renamed as VulcanAir in April 1998 when it purchased bankrupt Partenavia company. VulcanAir returned P.68 series to production in 1999 and continues to supply spares for Partenavia aircraft. Acquired rights to SF.600 Canguro from SIAI-Marchetti on latter's incorporation into Aermacchi in 1997. In June 1999, VulcanAir announced development of the VA 300 twin-diesel-engined light transport derivative of the P.68 and single-engined VA 600W, the latter a new aircraft with some commonality with the SF.600A and P.68 series.

UPDATED

VulcanAir P.68C built by the current manufacturer in 2000 and used as a demonstrator in the USA *(Jane's/Paul Jackson)*

NEW/0137360

VULCANAIR P.68 and OBSERVER

TYPE: Light utility twin-prop transport; multisensor surveillance twin-turboprop.
PROGRAMME: Prototype (I-TWIN) first flew 25 May 1970. Production of P.68 Victor and P.68B by Partenavia in Italy started 1978, followed by P.68C in 1980. Partenavia manufacture ended in 1994, but assets subsequently acquired by VulcanAir; 10 assembled in India up to 1999 by TAAL. Improvements announced in 1999 include Garmin avionics. In 2002, diesel-powered version under consideration.
CURRENT VERSIONS: **P.68C:** Basic version.
　　P.68C-TC: As P.68C, but with turbocharged engines for better hot-and-high performance.
　　P.68 Observer 2: For use by government and specialised services for patrol, surveillance and search; largely transparent nose section with lowered, compact instrument panel; 63 × 46 cm (2 ft 1 in × 1 ft 6 in) underfuselage hatch can carry variety of electro-optical sensors; slightly different equipment from other versions.
　　P.68 TC Observer TC: Turbocharged version of Observer 2 for hot/high performance.

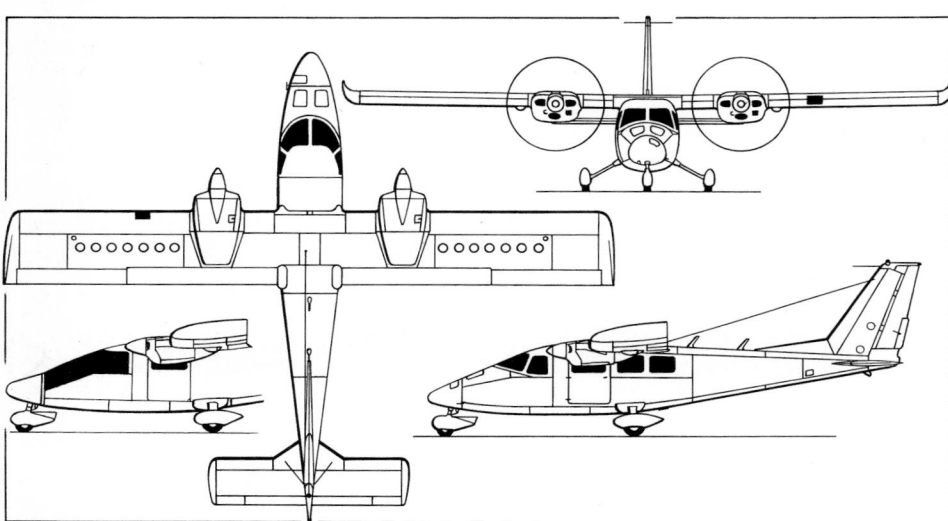

VulcanAir P.68C, with scrap view of Observer 2 *(Jane's/Mike Keep)*   0093638

VulcanAir P.68 Observer 2 *(Jane's/Paul Jackson)*   0110956

CUSTOMERS: Total 401 P.68s of all versions built in Italy by end 1998; these comprised 12 P.68s, 196 P.68Bs, 91 P.68Cs, 47 P.68C-TCs, 39 Observers, seven Observer 2s, three Observer TCs and six untraced. VulcanAir began production batch of 20 (mainly Observer 2s) in 1998; eight Observer 2s sold to Italian police; first delivery (PS-B07) 11 November 1999; further three (including two Observer 2s) registered in USA 2000-01, excluding demonstrator.
COSTS: US$475,000 (P.68C); US$485,000 (P.68 Observer 2); US$505,000 (P.68C-TC); US$515,000 (P.68 TC Observer), all IFR equipped (2001).
DESIGN FEATURES: High wing with NACA 63-3515 aerofoil section and Hoerner tips; dihedral 1°, incidence 1° 3'.
FLYING CONTROLS: Manual. Pushrod and cable actuated, with all-moving tailplane and anti-balance tab; trim tab in rudder; electrically operated single-slotted flaps.
STRUCTURE: Light alloy stressed skin fuselage with frames and longerons; stressed skin two-spar torsion box wing; metal stressed skin tailplane and fin; fuselage/wing fairings mainly GFRP.
LANDING GEAR: Non-retractable, with spring steel main legs; oleo suspension for nosewheel, steered from rudder pedals; mainwheels Cleveland 40-142 with Pirelli 6.00-6 or 7.00-6 (8 ply) tyres; nosewheel Cleveland 40-77B with Goodyear 5.00-5 or 6.00-6 (6 ply) tyre; Cleveland Type 30-61 foot-powered hydraulic disc brakes; streamlined wheel fairings optional. P.68 Observer 2 has larger mainwheel tyres as standard. Minimum ground turning radius 5.70 m (18 ft 8 in).
POWER PLANT: *P.68C:* Two 149 kW (200 hp) Textron Lycoming IO-360-A1B6 flat-four engines, each driving a Hartzell HC-C2YK-2CUF two-blade constant-speed fully feathering propeller.
　　P.68C-TC: Two 157 kW (210 hp) Textron Lycoming TIO-360-C1A6D; same propellers as P.68C.
　　Fuel capacity 696 litres (184 US gallons; 153 Imp gallons) in integral wing tank, of which 670 litres (177 US gallons; 147 Imp gallons) usable; overwing gravity refuelling. Oil capacity 7.5 litres (2.0 US gallons; 1.7 Imp gallons) for each engine.
ACCOMMODATION: One or two pilots and five or six passengers; cabin has two forward-facing seats in middle and three-seat rear bench; club seating optional; baggage door at rear and pilot door at front on starboard side; passenger door to port in centre cabin; baggage compartment accessible from inside cabin. P.68 Observer 2 has no front starboard door for pilots.
SYSTEMS: Two 24 V 130 Ah alternators and one 24 V 17 Ah battery; Goodrich pneumatic de-icing boots optional; air conditioning optional.
AVIONICS: Basic package includes IFR-equipped aircraft with two Garmin GNS 430 GPS and Bendix/King autopilot.
　　Radar: Weather radar optional in Observer 2.
　　Mission: Observer 2 can carry FLIR, ATAL video surveillance pod with data downlink and SLAR. Aerial cameras, thermal imager and video cameras can be operated through floor hatch.

DIMENSIONS, EXTERNAL (C: P.68C, TC: P.68C-TC, O: P.68 Observer 2):

Wing span	12.00 m (39 ft 4½ in)
Wing chord, constant	1.55 m (5 ft 1 in)
Wing aspect ratio	7.7
Length overall: C, TC	9.55 m (31 ft 4 in)
O	9.43 m (30 ft 11¼ in)
Height overall	3.40 m (11 ft 1¾ in)
Tailplane span	3.90 m (12 ft 9½ in)
Wheel track	2.40 m (7 ft 10½ in)
Wheelbase	3.65 m (11 ft 11¾ in)
Propeller diameter: all versions	1.83 m (6 ft 0 in)
Propeller ground clearance	0.77 m (2 ft 6¼ in)

DIMENSIONS, INTERNAL:

Cabin: Length	4.05 m (13 ft 3½ in)
Max width	1.16 m (3 ft 9½ in)
Max height	1.20 m (3 ft 11¼ in)
Baggage compartment volume	0.56 m³ (19.8 cu ft)

AREAS:

Wings, gross	18.60 m² (200.2 sq ft)
Ailerons (total)	1.79 m² (19.27 sq ft)
Trailing-edge flaps (total)	2.37 m² (25.51 sq ft)
Fin	1.59 m² (17.11 sq ft)
Rudder, incl tab	0.44 m² (4.74 sq ft)
Tailplane, incl tab	4.41 m² (47.47 sq ft)

WEIGHTS AND LOADINGS:

Weight empty equipped: C, O	1,320 kg (2,910 lb)
TC	1,350 kg (2,976 lb)
Baggage capacity	181 kg (400 lb)
Max T-O weight:	2,084 kg (4,594 lb)
Max ramp weight:	2,100 kg (4,630 lb)
Max landing weight:	1,980 kg (4,365 lb)
Max zero-fuel weight:	1,890 kg (4,167 lb)
Max wing loading	112.0 kg/m² (22.94 lb/sq ft)
Max power loading	6.99 kg/kW (11.49 lb/hp)

PERFORMANCE:

Never-exceed speed (VNE):	
C, TC	193 kt (358 km/h; 222 mph)
O	194 kt (359 km/h; 223 mph)
Max level speed:	
C, O at S/L	173 kt (320 km/h; 199 mph)
TC at S/L	174 kt (322 km/h; 200 mph)
Max cruising speed at 75% power:	
C, O at 2,285 m (7,500 ft)	165 kt (306 km/h; 190 mph)
TC at 3,660 m (12,000 ft)	171 kt (317 km/h; 197 mph)
Stalling speed, power off:	
flaps up	68 kt (126 km/h; 79 mph)
flaps down: TC	57 kt (106 km/h; 66 mph)

Retouched image, showing projected VulcanAir VA 300   0085593

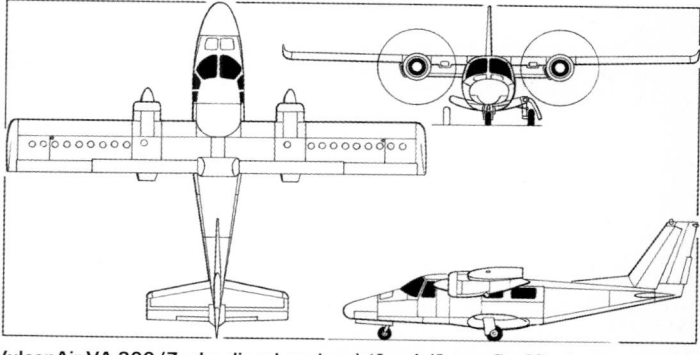

VulcanAir VA 300 (Zoche diesel engines) (*Jane's/James Goulding*)   0093611

Max rate of climb at S/L: C, O	378 m (1,240 ft)/min
TC	438 m (1,437 ft)/min
Max rate of climb, OEI: C, O	64 m (210 ft)/min
TC	78 m (256 ft)/min
Service ceiling: C	5,563 m (18,250 ft)
O	6,000 m (19,685 ft)
TC	6,100 m (20,000 ft)
Service ceiling, OEI: C	1,341 m (4,400 ft)
O	1,750 m (5,740 ft)
TC	2,957 m (9,700 ft)
T-O run: C, O	240 m (790 ft)
TC	230 m (755 ft)
T-O to 15 m (50 ft): C	400 m (1,312 ft)
TC	385 m (1,263 ft)
Landing from 15 m (50 ft): all	600 m (1,970 ft)
Landing run: all	240 m (790 ft)
Range with max payload:	
TC	300 n miles (556 km; 345 miles)
O with auxiliary fuel	
	590 n miles (1,093 km; 679 miles)
Range with max fuel:	
C	1,525 n miles (2,825 km; 1,755 n miles)
O	1,601 n miles (2,965 km; 1,842 n miles)
TC	1,403 n miles (2,600 km; 1,615 miles)
Endurance with max fuel	11 h
g limits	+3.74/−1.50

UPDATED

VULCANAIR VA 300

TYPE: Light utility twin-prop transport.

PROGRAMME: Announced at Paris, June 1999, as a concept based on the VulcanAir AP.68TP Viator. Diesel engine test-flown on a tri-motor-configured P.68C in March 2001 and Phase 1 of the certification process has been completed; flying with the diesel engines running was due to start in late 2001, but certification target date has been put back to 2005.

DESIGN FEATURES: Version of the P.68 series incorporating two diesel engines. Cantilever, untapered high wing of same span and form as P.68C but with upturned wingtips; aerofoil section NACA 63-3515, with Hoerner wingtips; dihedral 1°, incidence 1° 3′.

FLYING CONTROLS: Conventional and manual. Actuation by pushrods and cables; trim tabs on elevators, rudder and ailerons; elevator has vortex generators under leading-edge; down spring in elevator circuit; stall strips on wing leading-edges; electrically actuated single-slotted flaps.

STRUCTURE: All-metal stressed skin fuselage; two-spar torsion box wing; metal control surfaces.

LANDING GEAR: Retractable tricycle type; main gear retracts hydraulically inward into fuselage fairings, nosewheel forward; Cleveland mainwheels size 40-163EA; Cleveland 40-778 nosewheel; McCreary 6.50-8 mainwheel tyres and 6.00-6 on nosewheel; Cleveland powered hydraulic disc brakes. Minimum ground turning raduis 5.45 m (17 ft 11 in).

POWER PLANT: Two 224 kW (300 hp) Zoche ZO 02A eight-cylinder, turbocharged diesel engines, each driving a three-blade, constant-speed, fully feathering, reversible-pitch propeller. Fuel capacity 848 litres (224 US gallons; 187 Imp gallons); overwing gravity refuelling.

ACCOMMODATION: Two plus up to eight passengers; two doors to port, one for pilot and one for passengers; two doors to starboard, one for co-pilot and one for baggage; baggage compartment variable by using part of cabin; baggage accessible in flight.

SYSTEMS: Two 28 V DC 150 Ah starter/generators; one 24 V 29 Ah battery; hydraulics for brakes and landing gear actuation, pressurised by electric pump; electric anti-icing for engine intake and propellers, and pneumatic de-icing boots, standard.

AVIONICS: VFR radio and instruments standard; full IFR with weather radar and Bendix/King Silver Crown or Collins radios optional; observation equipment optional.

DIMENSIONS, EXTERNAL:

Wing span	12.00 m (39 ft 4½ in)
Wing chord, constant	1.55 m (5 ft 1 in)
Wing aspect ratio	7.7
Length overall	10.18 m (33 ft 4¾ in)
Height overall	3.63 m (11 ft 11 in)

Tailplane span	4.01 m (13 ft 1¾ in)
Wheel track	3.17 m (10 ft 4¾ in)
Wheelbase	2.37 m (7 ft 9¼ in)
Propeller diameter	1.90 m (6 ft 2¾ in)
DIMENSIONS, INTERNAL:	
Cabin (excl flight deck): Length	2.92 m (9 ft 7 in)
Max width	1.32 m (4 ft 4 in)
Max height	1.27 m (4 ft 2 in)
Volume	4.9 m³ (172 cu ft)
AREAS:	
Wings, gross	18.60 m² (200.2 sq ft)
WEIGHTS AND LOADINGS (provisional):	
Weight empty	1,730 kg (3,814 lb)
Max T-O and landing weight	2,850 kg (6,283 lb)
Max ramp weight	2,875 kg (6,338 lb)
Max zero-fuel weight	2,550 kg (5,622 lb)
Max wing loading	153.2 kg/m² (31.38 lb/sq ft)
Max power loading	6.37 kg/kW (10.47 lb/hp)
PERFORMANCE (estimated):	
Max cruising speed at S/L	190 kt (352 km/h; 219 mph)
Normal cruising speed	177 kt (328 km/h; 204 mph)
Stalling speed, flaps down	67 kt (124 km/h; 77 mph)
Max rate of climb at S/L	462 m (1,516 ft)/min
Rate of climb at S/L, OEI	102 m (335 ft)/min
Service ceiling	7,620 m (25,000 ft)
T-O run	394 m (1,295 ft)
T-O to 15 m (50 ft)	600 m (1,970 ft)
Landing from 15 m (50 ft)	650 m (2,135 ft)
Range:	
with max fuel	1,663 n miles (3,080 km; 1,913 miles)
with max payload	718 n miles (1,330 km; 826 miles)

UPDATED

VULCANAIR VF 600W MISSION

TYPE: Light utility turboprop.

PROGRAMME: Launched (as WF 600W) 1999; construction of first prototype began May 2001, at which time first flight planned for following September, but not achieved until 30 January 2003 (following first 'hop' on 4 December 2002); manufacture of first production aircraft scheduled for 2003.

COSTS: "Less than US$1 million".

DESIGN FEATURES: Single-engined derivative of Canguro. Nose-mounted engine; longer-span wings with new (modified NACA 63A3.515) aerofoil section; modified landing gear.
Following details summarise main differences from SF.600A (see 2002-03 All the World's Aircraft).

LANDING GEAR: Single mainwheels, with size 8.50-10 tyres, pressure 3.79 bar (55 lb/sq in); nosewheel tyre 6.50-8, pressure 5.17 bar (75 lb/sq in). Cleveland mainwheel disc brakes.

POWER PLANT: One 580 kW (778 shp) Walter M 601F-11 turboprop, driving an Avia V 510 five-blade constant-speed propeller. Optional Pratt & Whitney Canada PT6A turboprop installation. Fuel capacity (two wing tanks plus reservoir) 1,338 litres (353 US gallons; 294 Imp gallons), of which 1,300 litres (343 US gallons; 286 Imp gallons) are usable.

ACCOMMODATION: Pilot's door on port side; cargo/passenger sliding door on port side. Seating for between 10 and 16 passengers and crew.

DIMENSIONS, EXTERNAL:

Wing span	15.50 m (50 ft 10¼ in)
Length overall	13.12 m (43 ft 0½ in)

VulcanAir VF 600W prototype during its maiden flight   *NEW*/0530192

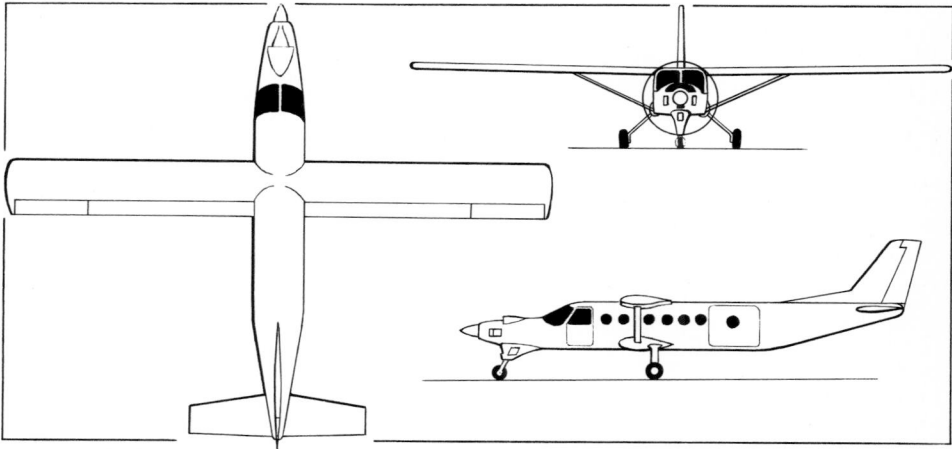

VulcanAir VF 600W single-engined utility turboprop (*Jane's/James Goulding*)   0121270

Height overall	4.55 m (14 ft 11¼ in)	PERFORMANCE (estimated):		T-O run	335 m (1,100 ft)	
Tailplane span	5.00 m (16 ft 4¾ in)	Max level speed	190 kt (352 km/h; 219 mph)	Landing run	497 m (1,630 ft)	
Wheel track	3.50 m (11 ft 5¾ in)	Cruising speed	183 kt (339 km/h; 211 mph)	Range	1,079 n miles (2,000 km; 1,242 miles)	
WEIGHTS AND LOADINGS (provisional):		Stalling speed, flaps down	61 kt (113 km/h; 71 mph)		*UPDATED*	
Max T-O weight	3,900 kg (8,598 lb)	Max rate of climb at S/L	274 m (900 ft)/min			
Max power loading	6.73 kg/kW (11.05 lb/shp)	Max certified altitude	6,095 m (20,000 ft)			

JAPAN

FUJI

FUJI JUKOGYO KABUSHIKI KAISHA (Fuji Heavy Industries Ltd)

Subaru Building, 7-2 1-chome, Nishi-Shinjuku, Shinjuku-ku, Tokyo 160-8316
Tel: (+81 3) 33 47 25 25
Fax: (+81 3) 33 47 25 88
e-mail: okak@sb.hq.subaru-fhi.co.jp
Web: http://www.fhi.co.jp
PRESIDENT: Takeshi Tanaka

Utsunomiya Manufacturing Division

1-11 Yonan 1-chome, Utsunomiya, Tochigi 320-8564
Tel: (+81 28) 684 70 55
Fax: (+81 28) 684 70 71
e-mail: nakagawan@ho.subaru-fhi.co.jp
GENERAL MANAGER: Youichi Sugimura
DEPUTY GENERAL MANAGER: Tsunenori Hoshi

Aerospace Division

SENIOR VICE-PRESIDENT AND GENERAL MANAGER:
  Hiroyuki Nakatsubo
VICE-PRESIDENT AND DEPUTY GENERAL MANAGER:
  Kisaburo Wani
DEPARTMENT GENERAL MANAGERS:
  Kenichiro Usuki (Commercial Marketing and Sales)
  Norihusa Matsuo (Defence Marketing and Sales)
  Shunji Notake (Defence Market Development)
  Takehiko Sarukawa (Administration)

Established 15 July 1953 as successor to Nakajima. Utsunomiya Manufacturing Division occupies 47.7 ha (117.9 acre) site, including 153,000 m² (1,646,880 sq ft) floor area; workforce 2,745 in April 2000.
Fuji producing Bell UH-1J (see *Jane's Aircraft Upgrades*; 76 completed by 1 October 2000); delivered last of 89 AH-1S attack helicopters (serial No. 73492) to JGSDF on 14 December 2000; has been selected to manufacture new T-7 trainer and will be prime contractor for JGSDF AH-64D; also manufactures wings, tailplanes and canopies for Kawasaki T-4 (which see). Commercial aircraft components produced are spoilers, inboard and outboard ailerons for Boeing 747; outboard flaps for Boeing 757; wing/body fairings and main landing gear doors for Boeing 767 and 777, plus front centre wing box for 777; and complete wing sets for the Raytheon Hawker Horizon business jet (see US section). Company was selected in May 2000 as risk-sharing subcontractor to produce composites fuselages for Bell/Agusta BA609 tiltrotor, with deliveries to start in FY03. Other products include BQM-34AJKai (modified Firebee) and J/AQM-1 target drones and RPH-2 unmanned helicopter (see *Jane's Unmanned Aerial Vehicles and Targets*).
Fuji has participated in such projects as design of the HOPE-X space shuttle and development of an NAL aerospaceplane. Research continues towards an SST/HST

(supersonic/hypersonic transport), including a thermal protection system, heat-resistant structures and composites materials.

UPDATED

FUJI T-3Kai/T-7

TYPE: Basic turboprop trainer.
PROGRAMME: Modification (Fuji designation KM-2D) of JASDF piston-engined T-3, which it is intended to replace on a one-for-one basis. Selected by JDA (in preference to Pilatus PC-7 Mk IIM) in third quarter 1998 to meet future T-7 requirement; two prototypes included in FY99 budget request, but in late 1998 procurement deferred for one year and Fuji suspended from all contracts until 31 December 1999. JDA reopened competition in August 1999, with bids in third quarter 2000; T-3Kai selected as basis for T-7, September 2000; first 11 T-7s approved in FY01 budget. A T-3Kai prototype has been test flown and awarded JCAB type certificate. First two production aircraft (5901 and 5902) delivered to Air Development & Test Wing at Gifu in September 2002.
CUSTOMERS: JASDF requirement for 47 T-7s over a 10-year period; 11 funded in FY01 defence budget; further 10 in FY02.
Following details refer to T-3Kai.
DESIGN FEATURES: Modifications to engine cowling, wings and tail unit compared with T-3.
  Wing section NACA 23016.5 at root, NACA 23012 at tip; dihedral 6° from roots; sweepback 0° at quarter-chord.
FLYING CONTROLS: Conventional and manual. Plain ailerons, each with balance tab (port tab controllable for trim); controllable tab in each elevator; anti-balance tab in rudder. Single-slotted flaps.
STRUCTURE: Two-spar all-metal wing; all-metal semi-monocoque fuselage.
LANDING GEAR: Electrically retractable tricycle type.
POWER PLANT: One 336 kW (450 shp) Rolls-Royce 250-B17F turboprop; three-blade propeller. Two bladder-type fuel tanks in each wing, combined capacity 375 litres (99.0 US gallons; 82.4 Imp gallons).
ACCOMMODATION: Crew of two in tandem. Dual controls standard.
SYSTEMS: Vapour cycle air conditioning system.
AVIONICS: Include VHF, UHF, ATC transponder, ICS and Tacan.
DIMENSIONS, EXTERNAL:

Wing span	10.04 m (32 ft 11¼ in)
Wing aspect ratio	6.1
Length overall	8.59 m (28 ft 2¼ in)
Height overall	2.96 m (9 ft 8½ in)
Tailplane span	3.71 m (12 ft 2 in)
Propeller diameter	2.12 m (6 ft 11½ in)

AREAS:

Wings, gross	16.50 m² (177.6 sq ft)
Fin	1.28 m² (13.78 sq ft)

First production Fuji T-7   *NEW*/0530185

Rudder, incl tab	0.66 m² (7.10 sq ft)
Tailplane	2.07 m² (22.28 sq ft)
Elevators (total, incl tabs)	1.39 m² (14.96 sq ft)

WEIGHTS AND LOADINGS:

Max T-O weight	1,585 kg (3,494 lb)
Max wing loading	96.1 kg/m² (19.67 lb/sq ft)
Max power loading	4.73 kg/kW (7.76 lb/shp)

PERFORMANCE:

Econ cruising speed at 915 m (3,000 ft)	161 kt (298 km/h; 185 mph)
Stalling speed, flaps and gear down	56 kt (104 km/h; 65 mph)
T-O to 15 m (50 ft) at S/L	608 m (1,995 ft)
Landing from 15 m (50 ft) at S/L	566 m (1,860 ft)

UPDATED

FUJI (BOEING) AH-64D

TYPE: Attack helicopter.
PROGRAMME: To replace existing JGSDF fleet of 80-plus AH-1Ss; original AH-X requirement said to be for up to 100, and government approval to acquire first 10 of selected design announced on 15 December 2000; these to be delivered by March 2006. RFP, issued 27 March 2001, resulted in submissions from Fuji/Boeing (AH-64D) and Mitsubishi/Bell (AH-1Z Viper), selection of former being announced on 27 August 2001. Selection of initially more expensive AH-64D said to be justified on grounds of better 20-year life-cycle costs, enabling smaller total number of aircraft to be procured; requirement accordingly now expected to be fulfilled by between 50 and 60 aircraft, of which only a proportion (possibly no more than one-third) to have Longbow radar.
  Initial batch (possibly all of first 10) to be US-built, with Fuji as prime contractor for remainder; funding for first two included in FY02 budget request.
POWER PLANT: General Electric T700-GE-701C turboshafts confirmed.
AVIONICS: To be decided.
ARMAMENT: To be decided.

NEW ENTRY

JADC

JAPAN AIRCRAFT DEVELOPMENT CORPORATION

Toranomon Daiichi Building, 2-3 Toranomon 1-chome, Minato-ku, Tokyo 105-0001
Tel: (+81 3) 35 03 32 25
Fax: (+81 3) 35 04 03 68
Web: http://www.iijnet.or.jp/jadc/jadc_home.htm
CHAIRMAN: Takashi Nishioka
VICE-CHAIRMAN: Teiichi Nishikawa
SENIOR MANAGING DIRECTOR: Toshinori Nishi
MANAGING DIRECTOR: Masaomi Kadoya

Known as CTDC (Civil Transport Development Corporation) from 1972 until 1982, JADC is a consortium established by airframe manufacturers Mitsubishi, Kawasaki and Fuji to promote commercial aircraft business with the support of the Japanese government. Airframe production

share (15 per cent of the Boeing 767 and 21 per cent of the 777) is managed by JADC's sister organisation, Commercial Aircraft Company (CAC). By 1 July 2001, CAC members had supplied 856 shipsets of parts for the 767 and 364 for 777.
  Current research programmes include supersonic commercial transport studies, advanced systems, innovative structures, next-generation avionics, advanced composites design and manufacturing and market research.

UPDATED

JADC YSX

TYPE: Twin-jet airliner.
PROGRAMME: Feasibility studies, continuing in 2002, are part of programme conducted over several years by Boeing under title of New Small Aircraft (NSA), intended for market slot between current regional jets and Next Generation 737 family. Japan hopes to participate in eventual aircraft, possibly as prime contractor, and

substantial portion of JADC's ¥3.6 billion FY01 budget devoted to this work, which has generic designation YSX. Boeing/JADC MoU continues to be renewed periodically; feasibility studies in 2001 included AFCS and other avionics, flight deck configuration and structural considerations.
CURRENT VERSIONS: Possibility of up to four basic versions (80, 90, 100 and 110 passengers), plus 122-passenger stretched version.
Following details are provisional.
DESIGN FEATURES: Low-wing monoplane with wing-mounted engines.
POWER PLANT: Two Rolls-Royce BR715 or Pratt & Whitney PW6000 turbofans.
PERFORMANCE:

Range:	
80 passengers	2,969 n miles (5,500 km; 3,417 miles)
122 passengers	1,997 n miles (3,700 km; 2,299 miles)

UPDATED

JASDF

NIHON KOKU JIEITAI (Japan Air Self-Defence Force)

7-45 Akasaka 9-chome, Minato-ku, Tokyo 107
Tel/Fax: (+81 3) 34 08 52 11

New aircraft programmes currently being formulated by the JASDF include the following:

C-X: Twin-turbofan transport. Development now combined with that of the MP-X maritime patroller under the prime contractor leadership of Kawasaki, within whose entry a description can now be found.

FI-X: Programme for next-generation fighter, to succeed F-15J early in the 21st century. Launched with FY95 allocation of ¥1 billion (US$10.2 million) to IHI to develop new 50 kN (11,240 lb st) class turbofan (XF-7) as power plant, to be test flown in TD-X technology demonstrator. Preliminary TRDI (Japan Defence Agency's Technology Research and Development Institute) design proposal for FI-X showed twin-engined configuration with canards, low aspect ratio tapered wings, twin fins and rudders and thrust-vectoring exhaust nozzles. Construction of up to four prototypes was originally expected to begin in FY99 and to include co-cured composites, radar-absorbent materials and digital fly-by-light and engine control systems. Wing span and length provisionally 9.15 m (30 ft) and 13.40 m (44 ft) respectively. Avionics to include conformal radar and IR seeker.

First XF-7 engine was delivered for static testing in June 1998, but FI-X programme has been stretched, and TD-X demonstrator not now expected to fly until 2007. No funding for the FI-X was requested in the FY02 budget.

KC-X: Aerial refuelling tanker. JDA down-selected the Airbus A310 and Boeing 767 as contenders in late 2001, the latter's commonality with the JASDF's AWACS platform giving it a probable advantage. Funds to purchase one example of the eventual choice were requested in the FY02 budget proposals.

PT-X: Primary trainer. To be acquired (as T-7); see entry for Fuji T-3Kai.

UPDATED

JGSDF

NIHON RIKUJYO JIEITAI (Japan Ground Self-Defence Force)

Address and tel/fax number as for JASDF

Two helicopter programmes, for combat and support, are in prospect by Japan's army:

AH-X: Attack helicopter. Boeing AH-64D selected; see Fuji entry for further details.

LH-X: Light transport and reconnaissance helicopter. Fuji and Mitsubishi competing for contract; no details of either design released by early 2002.

UPDATED

JMSDF

NIHON KAIJYO JIEITAI (Japan Maritime Self-Defence Force)

Address and tel/fax number as for JASDF

Naval aviation requires a new maritime patrol aircraft and an MH-53EJ replacement, as follows:

MP-X: Requirement for approximately 80 maritime patrol/surveillance aircraft to replace Kawasaki (Lockheed Martin) P-3C from about 2010 is to be met by a Kawasaki-led programme for a design that will also address the JASDF's C-X transport needs. See Kawasaki entry for further details.

MCH-X: Minesweeping and transport rotorcraft to replace existing Sikorsky S-80/MH-53EJ Sea Dragons. Approval for two, to be acquired during the period FY01 to FY06, was given in December 2000, but no funding had been allocated or requested by FY02. Possible contenders include the EH101, Sikorsky Seahawk and Sikorsky S-92.

UPDATED

KAC

KANEMATSU AEROSPACE CORPORATION

2-1 Shibaura 1-chome, Minato-ku, Tokyo 105-8005
Tel: (+81 3) 54 40 80 00
Fax: (+81 3) 54 40 65 03
e-mail: thashimoto@kanematsu.co.jp
Web: http://www.kanematsu.co.jp
PRESIDENT AND CEO: Tadashi Kurachi
SALES DIVISION MANAGER: Akiro Ohdoi

Kanematsu (workforce 785 at 31 March 2000) is prime contractor for outfitting Raytheon Hawker 800s (formerly BAe 125 Corporate 800) to JASDF specifications. Full description of standard aircraft under Raytheon heading in the US section. Fuji is systems integrator and responsible for U-125/U-125A maintenance. Kanematsu is also Japanese distributor for Embraer ERJ-145.

UPDATED

Raytheon Hawker U-125 navaid calibration aircraft (*Peter R Foster*)   0093618

KAC (RAYTHEON) HAWKER 800
JASDF designations: U-125 and U-125A

TYPE: Multirole twin-jet.

PROGRAMME: Raytheon Hawker 800 (see US section) selected under JASDF H-X programme to replace Mitsubishi MU-2J and MU-2E in navaid calibration and SAR roles respectively; first U-125 delivered to JASDF 18 December 1992 and first U-125A (52-3003) delivered on 11 December 1994; three U-125As delivered by 31 January 1995 in preparation for formal handover. First six U-125As assembled in UK; seventh aircraft (from Wichita production line) delivered to Kanematsu November 1997; 19th arrived in Japan on 10 December 2001 for fitting out.

CURRENT VERSIONS: **U-125:** For navaid flight check role, replacing MU-2J; operated by Flight Check Group at Iruma. Three ordered; all delivered from UK production.

U-125A: Search and rescue version, replacing MU-2E; 360° search radar, FLIR, airdroppable marker flares, liferaft and rescue equipment. Operated by detachments of the Air Rescue Wing at Chitose, Hyakuri, Komatsu and Naha, initially going to its Training Squadron at Komaki.

CUSTOMERS: JASDF (three U-125 and 22 U-125A ordered by FY01; none requested for FY02; options for total of 27 U-125As.

COSTS: ¥3.85 billion (2000).

DESIGN FEATURES: U-125A has deep observation 'patio' window each side of fuselage immediately ahead of wing,

U-125/A PROCUREMENT AND DELIVERY (JASDF)
(at 1 January 2001)

Ordered FY	U-125	U-125A	Delivered CY	First Acft	Qty
90	1		92	29-3041	1
91	1		93	39-3042	1
92	1		94	49-3043	1
92		3	95	52-3001	3
93		1	96	62-3004	1
94		1	97	72-3005	2
95		2	98	82-3007	3
96		3	99	92-3010	3
97		4	00	02-3013	3
98		3	01	12-3016	3
99		2	02	22-3019	—
00		2	—	—	—
01		1			
Totals	**3**	**22**			**21**

and dinghy/rescue pack dropping system via pressure door built into lower fuselage which is exposed for operation when landing gear is deployed.

AVIONICS (U-125A): *Radar:* Toshiba-built Raytheon 360° search radar.

Mission: Mitsubishi Electric IR imager in retractable underfuselage turret.

EQUIPMENT (U-125A): Flare and marker buoy dispenser; liferaft.

UPDATED

KAWASAKI

KAWASAKI JUKOGYO KABUSHIKI KAISHA (Kawasaki Heavy Industries Ltd)

Kobe Crystal Tower, 1-3 Higashi-Kawasaki-cho 1-chome, Chuo-ku, Kobe 650-8680

Aerospace Group/Gas Turbine & Machinery Group
World Trade Center Building, 4-1 Hamamatsu-cho 2-chome, Minato-ku, Tokyo 105-6116
Tel: (+81 3) 34 35 21 11
Fax: (+81 3) 34 36 30 37
Web: http://www.khi.co.jp/aero/airplane.html
PRESIDENT: Masamoto Tazaki

MANAGING DIRECTOR AND PRESIDENT OF AEROSPACE COMPANY: Takashi Sugoh
MANAGING DIRECTOR AND GAS TURBINE & MACHINERY GROUP
SENIOR GENERAL MANAGER: Tadashi Nishimura
INFORMATION: Masayuki Hirata
WORKS: Gifu, Nagoya 1 and 2 (Aerospace Group); Akashi and Seishin (Gas Turbine & Machinery Group)

Kawasaki Aircraft Company has built many US aircraft under licence since 1955; amalgamated with Kawasaki Dockyard Company and Kawasaki Rolling Stock Manufacturing Company to form Kawasaki Heavy Industries Ltd 1 April 1969; Aerospace Group employs some 3,200 people; Gas Turbine & Machinery Group has some 800 aerospace-related employees. Kawasaki has had 25 per cent holding in Nippi (which see in 1995-96 and earlier editions) since 1970; this scheduled to increase to 100 per cent by April 2003.

Kawasaki is currently prime contractor on T-4 programme and for OH-1 observation helicopter; now also to lead development of C-X/MP-X transport/maritime patroller programme; co-developer and co-producer, with Eurocopter, of BK 117/EC 145 helicopter (see entry in International section); is prime contractor for Japanese licensed production of CH-47 Chinooks for JGSDF and JASDF; was prime contractor for P-3C variants for JMSDF (now completed); also builds MD Helicopters MD 500 under licence.

Subcontract work includes centre fuselage for Mitsubishi F-2; currently producing forward and centre-fuselage barrel sections and wing ribs for Boeing 767 and 777, plus rear centre wing box and rear pressure bulkhead for 777. Kawasaki also responsible for design and sole-source manufacture of transmission for MD Helicopters Explorer; in 1999 signed agreement with Aerostructures of USA to manufacture cabin doors and tailcones for Bell/Agusta BA609. Risk-sharing partner on Embraer ERJ-170, and selected to manufacture wings for ERJ-190, as well as wing parts for ERJ-170. Nominated as prime contractor for maintenance and support of JASDF E-2C Hawkeyes, E-767 AWACS and C-130 Hercules.

Kawasaki also extensively involved in satellites and launch vehicles; is member of International Aero Engines consortium and produces T53 and T55 engines under licence (see *Jane's Aero-Engines*); overhauls engines; and builds hangars, docks, passenger bridges and other airport equipment.

UPDATED

KAWASAKI C-X and MP-X

TYPE: Medium transport/multirole (C-X); maritime reconnaissance four-jet (MP-X).

PROGRAMME: Initial development funding of ¥5.3 billion in FY01; RFP issued in May 2001. Foreign proposals for C-X included Airbus A310, Boeing C-17 and Lockheed Martin C-130J, but Kawasaki selected by JDA on 27 November 2001 to lead development of an indigenous design to meet both C-X and MP-X requirements, with optimum degree of structural commonality; Fuji and Mitsubishi will be involved in programme, which could also lead to a 100- to 150-seat civil transport version. Two-year design phase to begin in 2003, with prototype/preproduction development following from 2005; first flights targeted for 2006 (MP-X) and 2007 (C-X); engineering and operational testing until 2009 for MP-X, 2010 for C-X; service entry 2010 and 2011 respectively.

CURRENT VERSIONS: **C-X:** To fulfil long-standing JASDF requirement to replace current Kawasaki C-1A fleet with a transport of increased range and payload capacity. Desired features include a twin-turbofan power plant and a payload approximately double that of the Lockheed Martin C-130; earlier need for secondary aerial tanker capability no longer applicable. According to JDA's Technical Research and Development Institute (TRDI) in late 2000, design parameters include high-mounted wing, rear-loading ramp/door, digital AFCS and 'glass cockpit' avionics; outer wing, flight deck and tail unit to be common with those of JMSDF's MP-X.

MP-X: Outer wing, flight deck and tail unit as for C-X, but different fuselage cross-section, and wing to be low-mounted, carrying four (lower-powered) engines instead of two; AFCS to be fly-by-light as protection against electromagnetic interference (EMI).

CUSTOMERS: Approximately 40 C-Xs required by JASDF, to replace C-1As and C-130Hs; JMSDF needs about 80 MP-Xs to replace Kawasaki (Lockheed Martin) P-3C fleet.
Following details are provisional:
POWER PLANT: Two (C-X) or four (MP-X) 49 to 67 kN (11,000 to 15,000 lb st) class turbofans. TRDI and IHI developing the XF7-1 53.4 kN (12,000 lb st) class engine as possible candidate for MP-X; Rolls-Royce has offered Trent 500 for C-X.
AVIONICS: Not yet specified.
DIMENSIONS, EXTERNAL:
Wing span	40 m (131.2 ft)
Length overall	40 m (131.2 ft)
Height overall	13 m (42.7 ft)

WEIGHTS AND LOADINGS:
Max payload: C-X	26,000 kg (57,320 lb)
Max T-O weight:	
MP-X	50,000-70,000 kg (110,230-154,325 lb)

PERFORMANCE (design):
Cruising speed:
C-X at 12,980 m (42,600 ft)
480 kt (889 km/h; 552 mph)
MP-X at 9,140 m (30,000 ft)
450 kt (833 km/h; 518 mph)
Range: C-X with max payload
3,500 n miles (6,482 km; 4,027 miles)
MP-X 4,300 n miles (7,963 km; 4,948 miles)

NEW ENTRY

KAWASAKI T-4

TYPE: Advanced jet trainer.

PROGRAMME: Kawasaki named prime contractor 4 September 1981 by Japan Defence Agency; T-4 based on Kawasaki KA-851 design, by engineering team led by Kohki Isozaki; basic design studies completed October 1982; funding approved in FY83 and FY84 for four flying prototypes; prototype construction began April 1984; first flight of first XT-4 (56-5601) 29 July 1985; all four delivered between December 1985 and July 1986, preceded by static and fatigue test aircraft.

Production began FY86; first flight of production T-4, 28 June 1988; deliveries started 20 September 1988 to begin replacement of Lockheed T-33A and Fuji T-1A/B. Fuji and Mitsubishi each have 30 per cent share in production programme.

CUSTOMERS: JASDF T-4s used for pilot training, liaison and other duties; total of 212 (including prototypes) ordered by September 2000, of which more than 190 delivered (see table). None ordered in FY01 or requested for FY02. Used by Nos. 31 and 32 Flying Training Squadrons of 1st Air Wing at Hamamatsu, near Tokyo; and in small numbers by instrument rating/communications flights of most combat squadrons and regional HQ flights. First delivery to Blue Impulse aerobatic team (11 Squadron/4th Wing) 1994; nine operational for air show season.

COSTS: Unit cost ¥2,251 million (2000).

DESIGN FEATURES: High subsonic manoeuvrability; ability to carry external loads under wings and fuselage; anhedral mid-mounted wings, with extended chord outer panels giving dog-tooth leading-edges; tandem stepped cockpits with dual controls; baggage compartment in centre-fuselage for liaison role.

Supercritical wing section; thickness/chord ratio 10.3 per cent at root, 7.3 per cent at tip; anhedral 7° from roots; incidence 0°; sweepback at quarter-chord 27° 30′.

FLYING CONTROLS: Hydraulically actuated controls; plain ailerons with Teijin powered actuators; all-moving tailplane and rudder use Mitsubishi servo actuators; double-slotted trailing-edge flaps; no tabs; airbrake on each side of rear fuselage.

STRUCTURE: Aluminium alloy wings, with slow crack growth characteristics; CFRP ailerons, fin, rudder and airbrakes; aluminium alloy flaps (with AFRP trailing-edges), tailplane (CFRP trailing-edge) and fuselage with slow crack growth characteristics; titanium used sparingly in critical areas. Kawasaki builds forward fuselage and is responsible for final assembly and flight testing; Fuji builds rear fuselage, wings and tail unit; Mitsubishi builds centre-fuselage and engine air intakes.

LANDING GEAR: Hydraulically retractable tricycle type, with Sumitomo oleo-pneumatic shock-absorber in each unit. Single-wheel main units retract forward and inward; steerable nosewheel retracts forward. Kayaba (Honeywell) mainwheels, tyre size 22×5.5-13.8, pressure 19.31 bar (280 lb/sq in); Kayaba (Honeywell) nosewheel, tyre size 18×4.4-11.6, pressure 12.76 bar (185 lb/sq in). Kayaba (Honeywell) carbon brakes and Sumitomo (Hydro-Aire) anti-skid units on mainwheels. Minimum ground turning radius 9.45 m (31 ft 0 in).

POWER PLANT: Two 16.37 kN (3,680 lb st) Ishikawajima-Harima F3-IHI-30 turbofans, mounted side by side in centre-fuselage. Internal fuel in two 401.25 litre (106 US gallon; 88.3 Imp gallon) wing tanks and two Japanese-built Goodyear rubber bag tanks in fuselage, one of 776 litres (205 US gallons; 170.7 Imp gallons) and one of 662.5 litres (175 US gallons; 145.7 Imp gallons). Total internal capacity 2,241 litres (592 US gallons; 493 Imp gallons). Single pressure refuelling point in outer wall of port engine air intake. Provision to carry one 454 litre (120 US gallon; 100 Imp gallon) ShinMaywa drop tank on each underwing pylon. Oil capacity 5 litres (1.3 US gallons; 1.1 Imp gallons).

ACCOMMODATION: Crew of two in tandem in pressurised and air conditioned cockpit with wraparound windscreen and one-piece sideways (to starboard) opening canopy. Dual controls standard; rear (instructor's) seat elevated 27 cm (10.6 in). UPCO (Stencel) SIIS-3 ejection seats and Teledyne McCormick Selph canopy severance system, licence-built by Daicel Chemical Industries. Baggage compartment in centre of fuselage, with external access via door on port side.

SYSTEMS: Shimadzu bootstrap-type air conditioning and pressurisation system (maximum differential 0.28 bar; 4.0 lb/sq in). Two independent hydraulic systems (one each for flight controls and utilities), each operating at 207 bar (3,000 lb/sq in) and each with separate air/fluid reservoir pressurised at 3.45 bar (50 lb/sq in). Flow rate of each hydraulic system 45 litres (12 US gallons; 10 Imp gallons)/min. No pneumatic system. Electrical system powered by two 9 kW Shinko engine-driven starter/generators. Tokyo Aircraft Instruments onboard oxygen generating system.

AVIONICS: *Comms:* Mitsubishi Electric J/ARC-54 VHF/UHF com, and Nagano JRC J/AIC-103 intercom.

KAWASAKI T-4 PROCUREMENT AND DELIVERY (JASDF)

PROCUREMENT FY	Lot	Qty	Cum Total	DELIVERY CY	Qty	First Aircraft	Cum Total
83		3	3				
84		1	4				
85			4	85	2	56-5601	2
86	C-1	12	16	86	2	66-5603	4
87	C-2	20	36	87			4
88	C-3	20	56	88	8	86-5605	12
89	C-4	20	76	89	13	96-5613	25
90	C-5	19	95	90	29	06-5626	54
91	C-6	21	116	91	19	16-5655	73
92	C-7	19	135	92	19	26-5674	92
93	C-8	9	144	93	18	36-5693	110
94	C-9	9	153	94	21	46-5711	131
95	C-10	9	162	95	10	56-5732	141
96	C-11	9	171	96	10	66-5742	151
97	C-12	13	184	97	9	76-5752	160
98	C-13	9	193	98	9	86-5761	169
99	C-14	10	203	99	11	96-5770	180
00	C-15	9	212	00	11	06-5781	191
				01		16-5792	

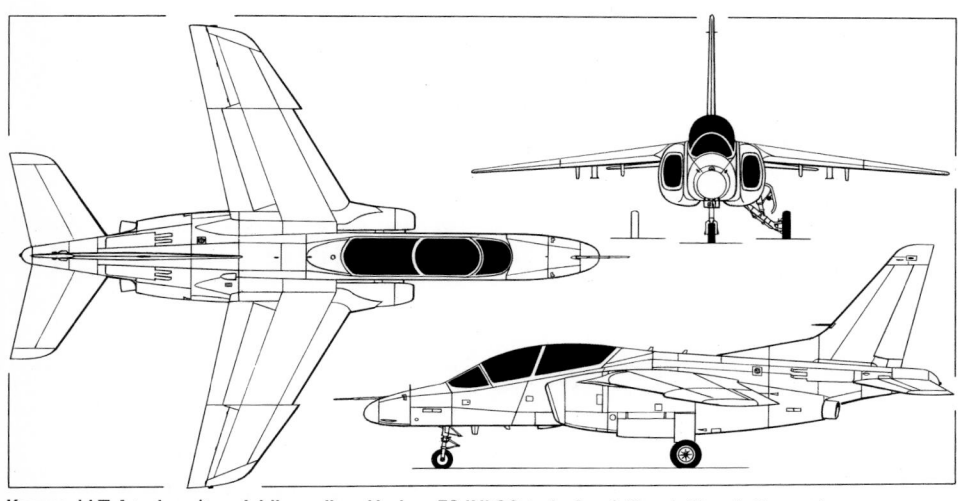

Kawasaki T-4 trainer (two Ishikawajima-Harima F3-IHI-30 turbofans) *(Jane's/Dennis Punnett)*

JASDF Kawasaki T-4 twin-turbofan trainer *(Shojiro Ootake/Kawasaki)*   *NEW*/0132397

Flight: Nippon Electric J/ARN-66 Tacan, Toshiba J/ARN-69 VOR/ILS, Toyo Communication (Teledyne Electronics) J/APX-106 SIF, Japan Aviation Electronics (Honeywell) J/ASN-3 laser gyro AHRS, Tokyo Keiki (Honeywell) J/ASK-1 air data computer, Tokyo Aircraft Instrument J/ASH-3 VGH recorder and Kanto (Smiths) J/ASH-4 FDR.
Instrumentation: Shimadzu (Kaiser) J/AVQ-1 HUD.
ARMAMENT: No built-in armament.
EQUIPMENT: Two Nippi pylons under each wing for carriage of drop tanks (see Power Plant) or travel pods; one Nippi pylon under fuselage, on which can be carried target towing equipment, ECM/chaff dispenser, travel pod or air sampling pod.

DIMENSIONS, EXTERNAL:
Wing span	9.94 m (32 ft 7½ in)
Wing chord: at root	3.11 m (10 ft 2½ in)
at tip	1.12 m (3 ft 8 in)
Wing aspect ratio	4.7
Length: overall, incl probe	13.00 m (42 ft 8 in)
fuselage	11.96 m (39 ft 3 in)
Height overall	4.60 m (15 ft 1¼ in)
Tailplane span	4.40 m (14 ft 5¼ in)
Wheel track	3.20 m (10 ft 6 in)
Wheelbase	5.10 m (16 ft 9 in)

DIMENSIONS, INTERNAL:
Cockpit: Length	3.20 m (10 ft 6 in)
Max width	0.69 m (2 ft 3 in)
Max height	1.40 m (4 ft 7¼ in)

AREAS:
Wings, gross	21.00 m² (226.1 sq ft)
Ailerons (total)	1.51 m² (16.25 sq ft)
Trailing-edge flaps (total)	2.93 m² (31.54 sq ft)
Fin	3.78 m² (40.69 sq ft)
Rudder	0.91 m² (9.80 sq ft)
Tailplane	6.04 m² (65.02 sq ft)

WEIGHTS AND LOADINGS:
Weight empty	3,840 kg (8,466 lb)
T-O weight, clean	5,730 kg (12,632 lb)
Max design T-O weight	7,500 kg (16,535 lb)
Max wing loading	357.1 kg/m² (73.15 lb/sq ft)
Max power loading	229 kg/kN (2.25 lb/lb st)

PERFORMANCE (in clean configuration. A: at weight of 4,850 kg; 10,692 lb with 50% fuel, B: at T-O weight of 5,730 kg; 12,632 lb):
Max level speed (A): at height	M0.907
at S/L	560 kt (1,038 km/h; 645 mph)
Cruising speed: B	M0.75
Stalling speed: A	90 kt (167 km/h; 104 mph)
Max rate of climb at S/L: B	3,078 m (10,100 ft)/min
Service ceiling: B	14,815 m (48,600 ft)
T-O run, 35°C: B	655 m (2,150 ft)
Landing run, 35°C: B	704 m (2,310 ft)

Range (B) at M0.75 cruising speed at 7,620 m (25,000 ft):
internal fuel only	700 n miles (1,297 km; 806 miles)
with two 120 US gallon drop tanks	900 n miles (1,668 km; 1,036 miles)
g limits	+7.33/−3

UPDATED

KAWASAKI OH-1

TYPE: Armed observation helicopter.
PROGRAMME: Developed to replace OH-6Ds of JGSDF. Japan Defence Agency (JDA) awarded ¥2.7 billion (US$22.5 million) in FY92 to cover basic design phase of helicopter then provisionally designated OH-X; RFPs issued by JDA's Technical Research & Development Institute (TRDI) 17 April 1992; Kawasaki selected as prime contractor (60 per cent of programme) 18 September 1992, with Fuji and Mitsubishi (20 per cent each) as partners; Observation Helicopter Engineering Team (OHCET), formed by these three companies, began preliminary design phase 1 October 1992. Mockup made public 2 September 1994 under Japanese name *Kogata Kansoku* (new small observation [helicopter]).

Programme includes six prototypes (four flying, two for ground test); first aircraft (32001) rolled out at Gifu on 15 March 1996 and made first flight 6 August 1996, followed by second prototype on 12 November; OH-1 designation assigned late 1996; first two XOH-1s handed over to JDA on 26 May and 6 June 1997; third flown on 9 January 1997, at which time earlier aircraft had accumulated some 30 and 20 hours, respectively; handed over 24 June 1997; fourth flown on 12 February 1997 and handed over 29 August 1997. Prototypes renumbered by 1999 from 32001-04 to 32601-04.

First three production OH-1s funded FY97 and ordered 1998; first prototype flown with more fuel-efficient TS1-10QT (replacing XTS1-10) engines, 30 March 1998. By early 1999, four prototypes had flown 450 hours and were due to complete further 450 hours by end of 1999, including operational evaluation at Akeno JGSDF base. First production OH-1 (32605) flown July 1999 and handed over to JGSDF at Gifu 24 January 2000; Hiko Jikkentai (Flight Test Squadron) formed at Akeno with first four production aircraft on 27 March 2001.
CURRENT VERSIONS: **OH-1**: Basic initial production version; *as described.*
Growth versions: Under study with more powerful engines (possibly LHTEC T800 or R-R/Turbomeca/MTU MTR 390) and uprated gearbox. **OH-1Kai** is possible candidate for AH-X requirement (see JGSDF entry) with tentative designation AH-2, armour-plated forward and centre fuselage, upgraded engines and transmission and additional weapons carriage.
CUSTOMERS: See table. Total of 18, including prototypes, ordered by FY01; further two requested for FY02. Japan Ground Self-Defence Force requirement for 150 to 200.

COSTS: Funding for development, prototypes and flight testing ¥2.5 billion in FY92, ¥10.2 billion in FY93, ¥50.1 billion in FY94 and ¥23.3 billion in FY95. Unit costs of first four production lots ¥1.924 billion (FY97), ¥2.018 billion (FY98), ¥2.229 billion (FY99) and ¥2.075 billion (FY00).
DESIGN FEATURES: Kawasaki hingeless, bearingless and 20 mm ballistic-tolerant four-blade elastomeric main rotor and transmission system; Fenestron-type tail rotor with eight unevenly angled blades (35 and 55°); stub-wings for stores carriage. Active vibration damping system.
FLYING CONTROLS: Integrated AFCS and stability control augmentation system (SCAS).
STRUCTURE: Rotor blades and hub manufactured from GFRP composites; centre-fuselage and engines by Mitsubishi, tail unit/canopy/stub-wings/cowling by Fuji, rest by Kawasaki. Some 37 per cent of airframe (by weight) in GFRP/CFRP.
LANDING GEAR: Non-retractable tailwheel type. Provision for wheel/skis on main units.
POWER PLANT: Twin 662 kW (888 shp) FADEC-equipped Mitsubishi TS1-10QT turboshafts (XTS1-10 originally in prototypes). Possibility of off-the-shelf alternative engines not ruled out. Transmission has 30-minute run-dry capability. Stub-wings can each carry a 235 litre (62.0 US gallon; 52.0 Imp gallon) auxiliary fuel tank.
ACCOMMODATION: Crew of two on tandem armoured seats (pilot in front). Flat-plate cockpit transparencies, upward-opening on starboard side for crew access.
AVIONICS: *Flight:* AFCS with stability augmentation and holding functions; dual HOCAS controls.
Instrumentation: Two Yokogawa Electric large, flat-panel, liquid crystal colour MFDs in each cockpit, linked to a MIL-STD-1553B databus; Shimadzu HUD in front cockpit.
Mission: Kawasaki electrically operated roof-mounted turret combining Fujitsu thermal imager, NEC real-time colour TV camera and NEC laser range-finder/designator; field of regard 110° in azimuth, 40° in elevation.
Self-defence: Spine-mounted IR jammer based on Sanders AN/ALQ-144.
ARMAMENT: Four Toshiba Type 91 (modified) lightweight, short-range, IR-guided AAMs on pylons under stub-wings for self-defence.
Following data all provisional:

DIMENSIONS, EXTERNAL:
Main rotor diameter	11.6 m (38 ft 0¾ in)
Main rotor blade chord, constant portion	0.38 m (1 ft 3 in)
Wing span	3.0 m (9 ft 10 in)
Fuselage: Length	12.0 m (39 ft 4½ in)
Max width	1.0 m (3 ft 3¼ in)
Height: to top of rotor head	3.4 m (11 ft 1¾ in)
over tailfin	3.8 m (12 ft 5½ in)
Tailplane span	3.0 m (9 ft 10 in)

AREAS:
Main rotor disc	105.68 m² (1,137.5 sq ft)

KAWASAKI OH-1 PROCUREMENT AND DELIVERY (JGSDF)

	PROCUREMENT			DELIVERY		
FY	Qty	Cum Total	FY	Qty	First Aircraft	Cum Total
93	2[1]	2				
95	2[1]	4				
97	3	7	97	4	32601	4
98	2	9	99	3	32605	7
99	3	12				
00	4	16				
01	2	18				

Note:
[1] XOH-1 prototypes

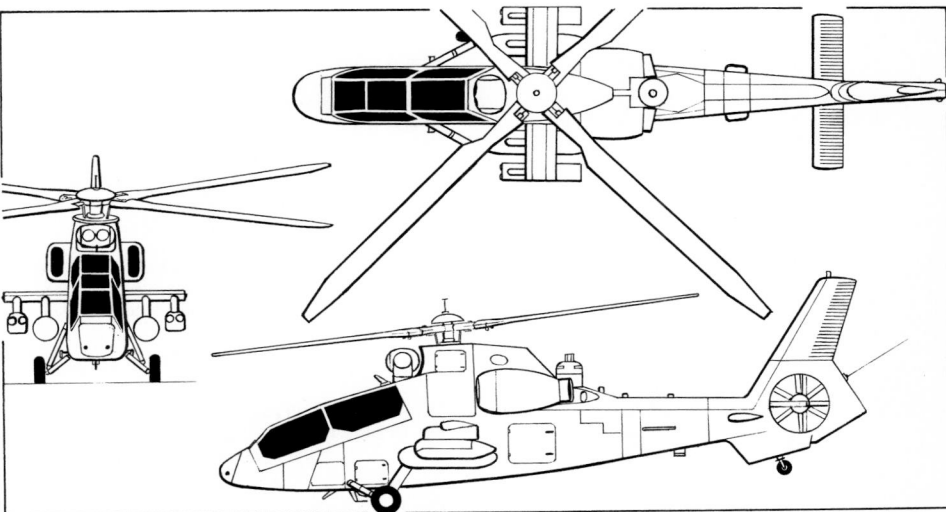

Kawasaki OH-1 observation helicopter *(Jane's/James Goulding)*   0093614

First production OH-1 in service with the JGSDF Aviation School at Akeno   0093613

A trio of Kawasaki-built Boeing CH-47J in JGSDF camouflage   *NEW*/0132393

WEIGHTS AND LOADINGS:

Weight empty	2,450 kg (5,401 lb)	
Max weapons load	132 kg (291 lb)	
T-O weight: design	3,550 kg (7,826 lb)	
max	4,000 kg (8,818 lb)	
Max disc loading	37.9 kg/m² (7.75 lb/sq ft)	

PERFORMANCE:

Max level speed	150 kt (277 km/h; 172 mph)
Combat radius	108 n miles (200 km; 124 miles)
Range	297 n miles (550 km; 342 miles)
g limits	+3.5/−1

UPDATED

KAWASAKI (BOEING) CH-47
JASDF/JGSDF designations: CH-47J and CH-47JA

TYPE: Medium lift helicopter.

PROGRAMME: FY84 defence budget approved purchase of three Boeing CH-47s, two for JGSDF and one for JASDF;

first two built in USA and delivered second quarter 1986; Nos. 3 to 7 delivered as CKD kits for assembly in Japan; Kawasaki granted manufacturing licence for Japanese services' Chinooks; first CH-47Js delivered late 1986.

CURRENT VERSIONS: **CH-47J:** Generally similar to US CH-47D. Total 34 (from all sources) delivered to JGSDF by 1996; procurement continues for JASDF.

CH-47JA: Designation introduced to describe CH-47Js delivered to army aviation from 1996 with nose-mounted weather radar; first (52951) was 35th JGSDF Chinook, 50th Japanese Chinook.

CUSTOMERS: See table. Total 58 (including US-built) delivered by 1 April 2000. JGSDF has requirement for 52, of which 42 in service by 1 April 2000. Operated by 1st and 2nd Squadrons of 1st Helicopter Brigade at Kisarazu (CH-47J), 1st Composite Brigade on Okinawa (CH-47JA) and Aviation School at Akeno. Two more JA requested for FY02.

JASDF 16 CH-47J delivered by 31 March 2000. Operated by Air Rescue Wing flights at Iruma, Kasuga, Misawa and Naha (Okinawa). Two more requested for FY02.

COSTS: Unit cost ¥4.5 billion for CH-47J, ¥5.1 billion for CH-47JA (2000).

AVIONICS: *Flight:* GPS in JGSDF aircraft from 1993; Mitsubishi Precision INS.

KAWASAKI CH-47J/JA PROCUREMENT*

FY	JASDF	First aircraft	JGSDF	First aircraft	Cum Total
84	1	67-4471	2	52901	3
85	1	77-4472	3	52903	7
86	3	87-4473	4	52906	14
87	2	97-4476	4	52910	20
88	3	07-4478	5	52914	28
89	2	27-4481	5	52919	35
90	2	37-4483	5	52924	42
91	1	47-4485	3	52929	46
92			3	52932	49
93			2†	52951	51
94			2†	52953	53
95	1	87-4486	2†	52955	56
96			2†	52957	58
97			2†	52959	60
98			1†	52961	61
99	2	07-4487	2†	52962	65
00	1		2†		68
01	1		1†		70
Totals	**20**		**50**		**70**

* First seven built in USA (two) or assembled from kits
† CH-47JA

UPDATED

KAWASAKI (MDH) MD 500D
JGSDF/JMSDF designations: OH-6D and OH-6DA

Production of this light utility helicopter has been completed. See 2002-03 *Jane's* for details.

UPDATED

MITSUBISHI

MITSUBISHI JUKOGYO KABUSHIKI KAISHA (Mitsubishi Heavy Industries Ltd)

5-1 Marunouchi 2-chome, Chiyoda-ku, Tokyo 100-8315
Tel: (+81 3) 32 12 31 11 and 32 12 56 41
Fax: (+81 3) 32 12 98 65
e-mail: QQ2100@hq.mhi.co.jp
Web: http://www.mhi.co.jp

NAGOYA AIRCRAFT SYSTEMS WORKS:
10 Oye-cho, Minato-ku, Nagoya 455
PRESIDENT: Takashi Nishioka
EXECUTIVE VICE-PRESIDENT: Kimiyuki Hanada
GENERAL MANAGER, DEFENCE AIRCRAFT AND AERO-ENGINE DEPARTMENT: Tetsuya Otabe

Present Komaki South plant built 1952, adjacent to Nagoya Airport, for assembly, outfitting and flight test. Main Nagoya facility (previously known as Komaki North) is divided into Aerospace Systems Works and Guidance & Propulsion Systems Works; former has 280,900 m² (3,023,575 sq ft) of floor space on a 58.6 ha (144.8 acre) site and had an April 1999 workforce of 3,921; latter occupies a 140.4 ha (346.9 acre) site with 84,530 m² (909,900 sq ft) of covered space and has approximately 1,700 employees. Oye plant manufactures aircraft fuselage components, spacecraft parts and other aero-related equipment; Tobishima plant responsible for aircraft fuselage subassembly, plus assembly and check-out of space systems; fuselage subassembly, final assembly, outfitting, flight test and repair undertaken by Komaki South plant.

Developed MU-2, MU-300 (now Raytheon Beechjet 400, which see in US section), T-2 supersonic trainer and the

related close support F-1 for JASDF. Built 167 HSS-2/2A/2B and 18 S-61A helicopters under Sikorsky licence (last aircraft delivered 2 March 1990), and 138 F-4EJ and 199 F-15J/DJ jet fighters under McDonnell Douglas licence (last F-15 delivered December 1999). Is currently prime contractor for F-15J/DJ modernisation programme, F-2 fighter and SH/UH-60J helicopters; has developed SH-60K and own-design MH2000 helicopter; manufactures centre-fuselage and engines of Kawasaki OH-1 and centre-fuselage and engine intakes for Kawasaki T-4.

Risk-sharing partner in Bombardier Global Express (responsible for wings and centre-fuselage), Dash 8 Q400 (centre and rear fuselage and tail unit), CRJ700 (rear fuselage) and BD 100 (wings).

Participating in Boeing 767 (rear fuselage panels and cargo doors) and 777 (rear fuselage panels, rear fuselage doors). Other subcontract work includes Boeing 737 (inboard flaps), 747 (inboard flaps) and 757-300 (stringers). Responsible for main cabin section of Sikorsky S-92 Helibus medium helicopter.

Aero-engine activities detailed in *Jane's Aero-Engines*; also produces rocket engines; participates in H-I and H-II launchers and Japanese Experimental Module for US space station.

UPDATED

MITSUBISHI F-2

TYPE: Attack fighter.

PROGRAMME: Indigenous design, plus proposals based on F-15, F-16, F/A-18 and Tornado ADV, were originally considered; modified F-16C selected as Japan's FS-X replacement for Mitsubishi F-1 on 21 October 1987; Mitsubishi appointed prime contractor November 1988;

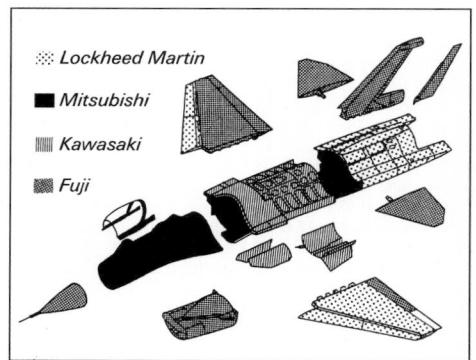

The principal F-2 contractors

initial contracts awarded for airframe design March 1989 and prototype active phased-array radar February 1990; General Electric F110-GE-129 Improved Performance Engine selected 21 December 1990. Programme delayed by questions of development sharing with General Dynamics (now Lockheed Martin Aeronautics Company (LMAC)) and technology transfer to Japan, but agreed at Japan 60 per cent and USA 40 per cent cost sharing (confirmed July 1996); first subcontracts to GD let February 1990 for design and development of rear fuselage, wing, leading-edge flaps, avionics and computer-based test equipment. Active phased-array radar, EW (ECM/ESM), mission computer and inertial reference system being developed using Japanese domestic technology.

Japan totally responsible for programme, including all funding; Japanese airframe subcontractors include Kawasaki and Fuji (see Structure). Programme involves four flying prototypes (two single-seat XF-2A first, then two tandem two-seat XF-2B) and two for static and fatigue test; construction began early 1994, final assembly mid-1994; first prototype (single-seat 63-0001) rolled out 12 January 1995; first flight 7 October 1995, followed by second prototype (63-0002) on 13 December 1995; third prototype (63-0003) flew on 18 April 1996 and fourth (63-0004; first in blue and grey colour scheme) on 24 May 1996. Japanese Cabinet approved 130-aircraft programme and allocated F-2 designation on 15 December 1995. Static test airframe (No. '991') under loading trials by 1995.

First XF-2A handed over to JDA on 22 March 1996, followed by '02 on 26 April 1997 and the two XF-2Bs on 9 August and 20 September 1997; prototypes re-serialled 63-8501/02/03/04 on 1 December 1997. Congressional approval of US participation in September 1996 followed in October by US$75 million Mitsubishi initial production contract to Lockheed Martin. First Lockheed Martin rear fuselage accepted by Mitsubishi in November 1998; first licence-built engines delivered by IHI in early 1999. Discovery in May 1998 of wing cracks, and evidence, when flying with two ASM-2 anti-ship missiles, of severe flutter, necessitated modifications to wingtips and pylon attachments, resulting in nine-month slippage of March 1999 scheduled completion date. Flight and static/fatigue testing rescheduled for completion in December 1999 and deliveries for FY99 (three) and FY00 (eight), but further wing crack problems revealed in mid-1999, causing completion of development testing to be extended to 30 June 2000 and first deliveries rescheduled to third quarter 2000. First production F-2A made maiden flight 12 October 1999 and delivered to JASDF on 25 September 2000. Total of 28 production F-2As and F-2Bs delivered by 31 March 2002.

Development of a dedicated air superiority version is under construction, to replace the F-4EJKai.

CURRENT VERSIONS: **F-2A:** Single-seat support fighter.
Description applies to F-2A except where indicated.
F-2B: Combat-capable two-seater.

CUSTOMERS: See table. Total of 57 production aircraft ordered by March 2001, with eight more requested for FY02. JASDF (sole user) originally stated requirement for approximately 72 F-2As to replace F-1s in three support fighter squadrons, plus approximately 50 F-2Bs for OCU and possibly to replace T-2/2A. Current plans are to acquire 130 single/two-seaters (83 + 47, reduced from earlier figure of 141 by cancellation of 11 two-seaters).

First recipient originally due to begin conversion in FY99 and to be completely equipped by FY00; however, first delivery (03-8503) not made until 25 September 2000 (formal acceptance 3 October). Six F-2As and eight F-2Bs delivered by year-end: one F-2A temporarily to No. 1 Technical School at Hamamatsu and seven to Rinji F-2 Hikotai (temporary F-2 Squadron) at Misawa, which became 3 Squadron of 3 Wing at same base on 27 March 2001. Remainder of first 45 aircraft will be allocated to Matsushima OCU (19 Bs). In 2006, 6 Squadron of 8 Wing at Tsuiki will convert from F-1; followed in 2007 by 8 Squadron/3 Wing at Misawa (which, pending F-2s, has converted from F-1 to F-4EJKai Phantom).

COSTS: Total JDA expenditure (1988 to 1995) US$3.27 billion, including ¥75.7 billion (US$575 million) in FY92 to include first prototype and radar development; further ¥96.5 billion (US$804 million) in FY94 provided for three more flying prototypes and two for ground test. First two Mitsubishi contracts to GD totalled US$280.5 million; follow-on contract to Lockheed (5 February 1993) valued at US$74.2 million. FY96 batch of 11 cost US$1,081.58 million (US$98.32 million per unit); FY97 batch of eight cost US$797.47 million (US$99.68 million per unit); FY00 aircraft cost ¥11.8 billion each.

DESIGN FEATURES: Configuration based on Lockheed Martin F-16. New co-cured composites wing of Japanese design, with greater span, root chord and 25 per cent more area than that of F-16; tapered trailing-edge; slightly longer radome and forward fuselage to house new radar and other mission avionics; longer mid-fuselage and shorter jetpipes;

Second Mitsubishi XF-2B two-seat operational trainer   0093615

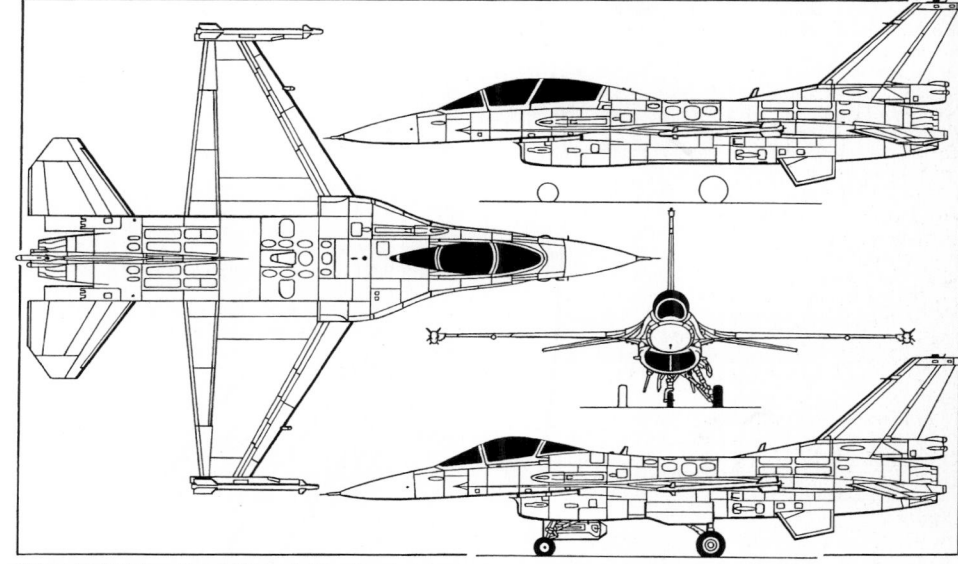

Mitsubishi F-2A, with additional side view (top) of two-seat F-2B (*Jane's/Mike Keep*)

increased-span tailplane; addition of brake-chute; adoption of increased performance engine.

Wing leading-edge sweepback 33° 12'; incidence 0°. Tailplane anhedral 8°.

FLYING CONTROLS: Full-span leading-edge flaps and trailing-edge flaperons; all-moving tailplane; rudder; twin, fixed ventral fins, canted outward 15°. Initially planned vertical canards deleted; CCV functions achieved instead by triple-redundant digital fly-by-wire system, developed jointly by Japan Aviation Electronics and Honeywell. Based on earlier Mitsubishi work with T-2 CCV testbed. Available modes include control augmentation, relaxed static stability, manoeuvre load control, decoupled yaw and manoeuvre enhancement. Single analogue back-up flight control system for roll and yaw control.

STRUCTURE: Composites structure wing (except leading-edge flaps), horizontal tail, fin (except leading-edge and base), rudder and landing gear doors; other structures also use advanced materials and structure technology, including Mitsubishi Rayon radar-absorbent material on nose, wing leading-edges and engine intakes; titanium in rear fuselage and tail unit.

Mitsubishi builds forward fuselage and wings; other Japanese airframe companies involved include Fuji (upper wing skins, wing fairings, radome, flaperons, engine air intakes and tail unit) and Kawasaki (fuselage mid-section, mainwheel doors and engine access doors). Lockheed Martin providing rear fuselage, 80 per cent of all port-side wing boxes, all leading-edge flaps, avionics systems, stores

management systems and some test equipment; first LM production rear fuselage accepted 10 November 1998; first wing box delivered March 1999.

LANDING GEAR: Retractable tricycle type, with single wheel on each unit; mainwheels retract inward, nosewheel rearward. Mainwheel tyres size 27.75×8.75R145, pressure 22.06 bar (320 lb/sq in); nosewheel tyre size 18×5.7-8, pressure 20.69 bar (300 lb/sq in). Brake-chute in fairing at base of rudder.

POWER PLANT: One General Electric F110-GE-129 turbofan (131.2 kN; 29,500 lb st with afterburning), licence-built by IHI for production aircraft. Maximum internal fuel capacity 4,637 litres (1,225 US gallons; 1,020 Imp gallons), of which 4,588 litres (1,212 US gallons; 1,009 Imp gallons) are usable; reduced to 3,948 litres (1,043 US gallons; 868.5 Imp gallons), of which 3,903 litres (1,031 US gallons; 858.5 Imp gallons) usable, in F-2B. Maximum external fuel capacity (both) 5,678 litres (1,500 US gallons; 1,249 Imp gallons) (one 1,135.5 litre; 300 US gallon; 249.8 Imp gallon and two 2,271.25 litre; 600 US gallon; 499.6 Imp gallon tanks). No provision for in-flight refuelling.

SYSTEMS: Include onboard oxygen generation system (OBOGS).

AVIONICS: *Comms:* Magnavox AN/ARC-164 UHF transceiver; NEC V/UHF transceiver; Hazeltine AIFF; Kokusai Electric HF radio.

Radar: Mitsubishi Electric active phased-array radar.

Flight: Japan Aviation Electronics/Honeywell digital AFCS and laser IRS.

MITSUBISHI F-2 PROCUREMENT AND DELIVERY (JASDF)

		PROCUREMENT					DELIVERY					
		Qty		Cum Total			Qty		First Aircraft		Cum Total	
FY	Lot	F-2A	F-2B	F-2A	F-2B	FY	F-2A	F-2B	F-2A	F-2B	F-2A	F-2B
92		1[1]	0	1								
93												
94		1[1]	2[1]	2	2							
95												
96	1	7	4	9	6	96	2	2	63-8501	63-8101	2	2
97	2	8	0	17	6	97						
98	3	2	7	19	13	98						
99	4	2	6	21	19	99						
00	5	0	9	21	28	00	5	4	03-8503	03-8103	7	6
01	6	8	4	29	32	01		1	13-8508	13-8107		7
						02				23-8108		

[1] Prototypes; originally numbered 63-0001/02 (XF-2A) and 63-8501/02 (XF-2B)

First production Mitsubishi F-2A wearing JASDF camouflage *(Lockheed Martin)*   0114393

Instrumentation: Yokogawa 127 × 127 mm (5 × 5 in) colour LCD multifunction display and two 102 × 102 mm (4 × 4 in) liquid crystal colour MFDs; Shimadzu holographic wide-angle HUD.

Mission: Mitsubishi Electric mission computer. Sanders MPS III mission planning system.

Self-defence: Mitsubishi Electric integrated EW system.
ARMAMENT: One internal M61A1 Vulcan 20 mm multibarrel gun in port wingroot, plus 13 external stores stations: Sta 6 on centreline; Sta 1 (port) and 11 at each wingtip; and five under each wing (Sta 2, 3, 4L, 4, 5, 6, 8, 8R, 9 and 10); Flight Refuelling common rail launchers, built and installed by Nippi, configured initially for AIM-7F/M Sparrow medium-range air-to-air missiles (Sta 2 and 10); other armament expected to include AIM-9L or Mitsubishi AAM-3 air-to-air missiles (Sta 1, 2, 10 and 11) and ASM-1 and ASM-2 anti-shipping missiles (Sta 3, 4, 8 and 9); 500 lb bombs (Sta 4L and 8R); 340 kg bombs (Sta 4 and 8); CBU-87/B cluster bombs (Sta 4 and 8); and JLAU-3/A or RL-4 rocket launchers (Sta 4 and 8). Centreline and inboard underwing stations (5, 6 and 7) wet for carriage of drop tanks.

DIMENSIONS, EXTERNAL:

Wing span: over missile rails	11.125 m (36 ft 6 in)
excl missile rails	10.80 m (35 ft 5¼ in)
Wing chord: at root	5.27 m (17 ft 3½ in)
at tip	1.185 m (3 ft 10¾ in)
Wing aspect ratio	3.3
Length overall	15.52 m (50 ft 11 in)
Height overall	4.96 m (16 ft 3¼ in)
Tailplane span	6.045 m (19 ft 10 in)
Wheel track	2.36 m (7 ft 9 in)
Wheelbase	4.05 m (13 ft 3½ in)

AREAS:

Wings, gross	34.84 m² (375.0 sq ft)
Leading-edge flaps (total)	4.70 m² (50.59 sq ft)
Flaperons (total)	3.96 m² (42.63 sq ft)
Fin, incl dorsal fin	4.00 m² (43.06 sq ft)
Rudder	1.08 m² (11.63 sq ft)
Ventral fins (total)	1.50 m² (16.15 sq ft)
Horizontal tail surfaces (total)	7.05 m² (75.89 sq ft)

WEIGHTS AND LOADINGS:

Weight empty: F-2A	9,527 kg (21,003 lb)
F-2B	9,633 kg (21,237 lb)
T-O weight: clean: F-2A	13,459 kg (29,672 lb)
F-2B	13,230 kg (29,167 lb)
with two short-range AAMs:	
F-2A	13,713 kg (30,232 lb)
F-2B	13,412 kg (29,568 lb)
with four short-range and four medium-range AAMs:	
F-2A	15,711 kg (34,637 lb)
F-2B	15,392 kg (33,934 lb)
with two short-range AAMs, six 500 lb bombs and two	
drop tanks: F-2A	19,888 kg (43,845 lb)
F-2B	19,569 kg (43,142 lb)
with two short-range AAMs, four anti-ship missiles	
and two drop tanks: F-2A	20,517 kg (45,232 lb)
F-2B	20,198 kg (44,529 lb)
Design max T-O weight with external stores:	
F-2A, F-2B	22,100 kg (48,722 lb)
Design max landing weight:	
F-2A, F-2B	18,300 kg (40,345 lb)
Design max wing loading	634.3 kg/m² (129.92 lb/sq ft)
Design max power loading	168 kg/kN (1.65 lb/lb st)

PERFORMANCE (F-2A):

Design max level speed: at high altitude	approx M2.0
at low altitude	approx M1.1
Typical combat radius:	
F-2A	more than 450 n miles (833 km; 518 miles)

UPDATED

MITSUBISHI (SIKORSKY) SH-60
JMSDF designations: SH-60J and SH-60K
TYPE: Naval combat helicopter.
PROGRAMME: Detail design of S-70B-3 version of Sikorsky SH-60B Seahawk (see US section), to meet JMSDF requirements, started August 1983; Japanese avionics and equipment integrated by Technical Research and Development Institute (TRDI) of JDA. First flight of first of two XSH-60J prototypes (8201), based on imported airframes, 31 August 1987; evaluation by 51st Air Development Squadron of JMSDF at Atsugi completed early 1991; first production SH-60J (8203) flown 10 May 1991, delivered 26 August.
CURRENT VERSIONS: **SH-60J:** Initial Japanese production version; *as described.*

SH-60K: Following trials begun by TRDI in 1995 on an all-composites rotor system with redesigned planforms and tips, offering 5 per cent increase in hovering efficiency and 30 per cent vibration reduction, FY97 budget requested funding for one prototype (since increased to two), upgraded with new main rotor hub and blades, increased cabin cross-section (33 cm; 13 in longer and 15 cm; 6 in more headroom), active sonar, a tactical data processing and display system, and a laser-based shipboard automatic landing system. This aircraft (8401) originally planned to enter flight test in July 2000, but production start planned for FY01 deferred, reportedly due to discovery of cracks in main rotor during fatigue tests; development costs said to have totalled some ¥40 billion

by mid-2000. Prototype eventually rolled out 8 August 2001 and made its first hovering flight on 9 September.
CUSTOMERS: See table. JMSDF requirement for about 100 SH-60Js; 103 ordered, of which more than 65 in service, by 1 April 2001; half of force land-based. Upgrade of nine earlier aircraft also funded in FY00; three new aircraft approved in FY01 to offset delays in SH-60K programme, followed by budget request for seven SH-60Ks in FY02. SH-60Js operated by Nos. 121 and 123 Squadrons (at Tateyama); Nos. 122 and 124 (at Ohmura); and part of No. 211 (training unit at Kanoya); No. 101 Squadron at Tateyama undergoing conversion in 2000.

MITSUBISHI SH-60 PROCUREMENT (JMSDF)

FY	Qty	First Aircraft	Cum Total
82	1[1]	8201	1
83	1[1]	8202	2
88	12	8203	14
89	12	8215	26
90	11	8227	37
91	5	8238	42
92	7	8243	49
93	4	8250	53
94	5	8254	58
95	6	8259	64
96	6	8265	70
97	7	8271	77
98	7	8278	84
99	9	8285	93
00	7	8294	100
01	3	8301	103
01	1[2]	8401	104
Total	**104**		**104**

Notes:
[1] Sikorsky-built XSH-60Js
[2] SH-60K prototype

COSTS: Unit cost ¥5.07 billion (2000).
POWER PLANT: T700-401C engines manufactured by IHI.
ACCOMMODATION: Crew of three plus five systems operators/observers.
AVIONICS: *Radar:* Japanese HPS-104 search radar.
Flight: Japanese automatic flight management system and ring laser gyro AHRS.

Prototype SH-60K shipboard maritime patrol helicopter *(TRDI)*   *NEW*/0528623

JMSDF No. 121 Squadron Mitsubishi (Sikorsky) SH-60J   0064742

Instrumentation: Japanese controls and displays subsystem, datalink and tactical data processor.

Mission: Japanese HQS-103 sonar; Raytheon AN/ASQ-81D2(V) MAD; Ednac AN/ARR-75 sonobuoy receiver.

Self-defence: General Instruments AN/ALR-66 (VE) RWR; Japanese HLR-108 ESM.

EQUIPMENT: RAST, sonobuoys, rescue hoist and cargo sling.

ARMAMENT (SH-60K): Anti shipping missiles and gun.

UPDATED

MITSUBISHI UH-60 PROCUREMENT

FY	UH-60J (JASDF)	First aircraft	UH-60J (JMSDF)	First aircraft	UH-60JA (JGSDF)	First aircraft	Cum total
88	3[1]	18-4551					3
89	2	28-4554	3	8961			8
90	2	28-4556					10
91	4	38-4558	3	8964			17
92	2	58-4562	2	8967			21
93	1	68-4564	2	8969			24
94	2	68-4565	1	8971			27
95	2	_8-4567	1	8972	2	43101	32
96	1	98-4569	2	8973	4	43103	39
97	3	08-4570	2	8975	4	43107	48
98	2	_8-4573	2	8977	5	43111	57
99	2				3	43116	62
00	2				3	43119	67
01	2		1	8979	2	43122	72
Totals	**30**		**19**		**23**		**72**

Notes:

[1] One Sikorsky-built, two Mitsubishi-assembled from CKD kits

Mitsubishi (Sikorsky) UH-60J rescue helicopter in JASDF markings (*Peter R Foster*)   0051107

MITSUBISHI (SIKORSKY) UH-60

JASDF/JMSDF designation: UH-60J

JGSDF designation: UH-60JA

TYPE: Medium transport helicopter.

PROGRAMME: Detail design to Japanese requirements started April 1988; one US-built S-70A-12 imported, followed by two CKD kits for licence assembly (first flight 20 December 1989 at Sikorsky; delivered to JASDF 28 February 1991; second kit aircraft first flew February 1990, delivered to JASDF 29 March 1991). Remainder being built in Japan.

CURRENT VERSIONS: **UH-60J:** Standard search and rescue version for JASDF and JMSDF; can fly one hour search at 250 n miles (463 km; 288 miles) from base.

UH-60JA: JGSDF version; navigation/weather radar and FLIR night and adverse weather vision system. First (43101) completed late 1997; initial deliveries (four) January 1998.

CUSTOMERS: See table. JASDF has total requirement for 46 UH-60Js, JMSDF for 19; total of 21 in service with JASDF and 15 with JMSDF by 1 April 2000. One more for each service requested for FY02. JASDF aircraft operated by detached flights of Air Rescue Wing (HQ: Iruma); JMSDF UH-60Js to Atsugi Rescue Squadron in 1992; subsequently Kanoya, Tokushima, Shimofusa and Hachinohe.

JGSDF plans to procure up to 80 UH-60JAs in a US$2.67 billion programme announced early 1995; procurement began that year; JGSDF will deploy 50 with five district helicopter units alongside UH-1Js, with balance assigned to VIP transport and training; six delivered by 1 April 1999, this remaining as inventory total on 1 April 2000, despite more then awaiting acceptance. First two units are Nos. 1 and 12 Squadrons. Further two JAs requested for FY02.

COSTS: Unit costs in FY00 were ¥3.37 billion for UH-60JA and ¥3.6 billion for air force UH-60J.

POWER PLANT: T700-401C turboshafts manufactured by IHI. External long-range fuel tanks.

ACCOMMODATION: Crew of four (JMSDF) or five (JASDF) plus up to 11 other persons. Bubble windows for pilot and on each side at front of main cabin.

AVIONICS: *Radar:* Nose-mounted Japanese search/weather radar.

Flight: Sikorsky self-contained navigation system.

Mission: Turret-mounted FLIR beneath nose.

EQUIPMENT: Lucas rescue hoist (JA), ETS and cargo sling.

UPDATED

MITSUBISHI MH2000

Following a product recall in August 2000, there has been no further information released concerning this light helicopter. A description last appeared in the 2002-03 *Jane's*.

UPDATED

NAL

NATIONAL AEROSPACE LABORATORY

7-44-1 Jindaijihigashi-machi, Chofu City, Tokyo 182-8522

Tel: (+81 422) 40 30 81

Fax: (+81 422) 40 30 08

e-mail: miyakawa@nal.go.jp

Web: http://www.nal.go.jp

DIRECTOR GENERAL: Susuma Toda

DIRECTOR, PLANNING OFFICE: Masaaki Murata

PUBLIC RELATIONS OFFICER: Kimio Okuhara

NAL is a governmental establishment, founded in 1958 for research and development in aeronautics and space technologies. Its major recent aeronautical activities have included developmental studies for the HOPE-X (H-II Orbiting Plane – Experimental), jointly with NASDA; a future aero-spaceplane (2000-01 *Jane's*); and a supersonic commercial transport (see 2002-03 edition).

NAL operates two research aircraft known as MuPAL (MultiPurpose Aviation Laboratory) for flight systems R&D. MuPAL-α is a modified Dornier 228-200; MuPAL-ε is a Mitsubishi MH2000. The former incorporates a fly-by-wire flight control system, direct lift control flap deflection system, second 'cockpit' in the main cabin, and a data acquisition system with telemetry downlink. Main features of the MuPAL-ε are a variable stability AFCS, cockpit graphic workstation, external view recording system (two video cameras), a hybrid DGPS/INS, main rotor sensors and a data acquisition system.

UPDATED

SHINMAYWA

SHINMAYWA KOGYO KABUSHIKI KAISHA (Shinmaywa Industries Ltd)

5-25 Kosone-cho 1-chome, Nishinomiya Hyogo 663-8122

Tel: (+81 798) 47 03 31

Fax: (+81 798) 45 57 43

e-mail: head@sa.shinmaywa.co.jp

Web: http://www.shinmaywa.co.jp

CHAIRMAN: Shiko Saikawa

PRESIDENT: Jushi Ide

Aircraft Division

1-1 Ohgi 1-chome, Higashinada-ku, Kobe 658-0027

Tel: (+81 78) 412 91 51

Fax: (+81 78) 431 36 95

BOARD DIRECTOR AND GENERAL MANAGER: Yasuo Takeuchi

BOARD DIRECTOR AND DEPUTY GENERAL MANAGER: Takao Masuda

DEPUTY GENERAL MANAGER: Shun Sonoda

WORKS: Konan and Tokushima

SALES OFFICES:

Department No. 1:

2-43 Shitte 3-chome, Tsurumi-ku, Yokohama 230-0003

Tel: (+81 45) 575 79 00

Fax: (+81 45) 575 79 01

GENERAL MANAGER: Hiroya Koda

Department No 2 (International Sales):

1-1 Ohgi 1-chome, Higashinada-ku, Kobe 658-0027

Tel: (+81 78) 412 22 10

Fax: (+81 78) 412 20 29

GENERAL MANAGER: Kazuya Makino

Former Kawanishi Aircraft Company (established 1928); became Shin Meiwa in 1949 and renamed ShinMaywa Industries in June 1992; major overhaul centre for Japanese and US military and commercial aircraft. Principal activities are production of US-1A for JMSDF, and overhaul work on amphibians. Manufactures external drop tanks for Kawasaki T-4s; tailplanes for Mitsubishi-built SH-60Js; internal cargo handling system for Kawasaki-built CH-47JA; fixed trailing-edges for Boeing 767, under subcontract to Vought; and other components for Boeing 767, under subcontract to Mitsubishi; design and manufacture of wing/body fairings for Boeing 777. Support centre for U-36A (Learjet 36A) and Fairchild aircraft; maintenance and technical support for JASDF U-4 (Gulfstream IV) since 1996.

Continues to study and look for partners to develop Amphibious Air Transport System, which is 30/50-passenger airliner powered by two wing-mounted turbofans with upper surface blowing; range would vary from 500 n miles (926 km; 575 miles) with full payload to 1,200 n miles (2,222 km; 1,381 miles) with full fuel; take-off distance 1,000 m (3,280 ft) on water and 800 m (2,624 ft) on soft ground; cruising speed between 300 and 360 kt (556 and 667 km/h; 345 and 414 mph).

UPDATED

SHINMAYWA US-1A

TYPE: Four-turboprop amphibian.

PROGRAMME: First flown 16 October 1974; first delivery (as US-1) 5 March 1975 (see 1985-86 *Jane's*); all now have T64-IHI-10J engines as US-1As.

CURRENT VERSIONS: **US-1A:** SAR amphibian, developed from PS-1 ASW flying-boat; manufacturer's designation **SS-2A.**

Data apply to US-1A.

US-1AKai: Proposed upgrade initiated in 1996: FADEC-equipped, 3,356 kW (4,500 shp) Rolls-Royce AE 2100J turboprops, six-blade Dowty R414 propellers, FBW controls, pressurised upper hull and 'glass cockpit'. Parts manufacture began in April 2000; first flight targeted for third quarter 2003 and service entry for 2007.

Firefighting amphibian: PS-1 modified in 1976 to firefighting configuration; 7,348 kg (16,200 lb) capacity

ShinMaywa US-1A four-turboprop SAR amphibian *(K Okumura/ShinMaywa)*   NEW/0120432

water tank in centre-fuselage forward of step. One US-1A modified experimentally in 1986 to 1988, with more than 13,608 kg (30,000 lb) tank capacity; details in earlier editions.

CUSTOMERS: Total of 19 US-1/1As ordered, 18 of which delivered by end of 2000; none requested for FY00, but US$283 million requested instead to resume development of US-1AKai. One US-1A requested for FY02. Orders since 1988 have been for attrition replacements and, more recently, due to phase-out of older aircraft. No. 71 SAR Squadron of the JMSDF maintains fleet structure of seven aircraft at Iwakuni and Atsugi bases.

US-1/1A PROCUREMENT (JMSDF)

FY	Qty	Cum Total
72	1	1
73	2	3
77	1	4
78	2	6
79	1	7
80	1	8
82	1	9
83	1	10
84	1	11
88	1	12
92	1	13
93	1	14
95	1	15
96	1	16
97	1	17
99	1	18
01	1	19

DESIGN FEATURES: Large turboprop-powered amphibian, suitable for maritime patrol and sea rescue missions. Boundary layer control system and extensive flaps for propeller slipstream deflection for very low landing and take-off speeds; low-speed control and stability enhanced by blowing rudder, flaps and elevators, and by use of automatic flight control system (see Flying Controls). Fuselage high length/beam ratio; V-shaped single-step planing bottom, with curved spray suppression strakes along sides of nose and spray suppressor slots in lower fuselage sides aft of inboard propeller line; double-deck interior. Large dorsal fin.

FLYING CONTROLS: Digital flight control system controlling elevators, rudder and outboard flaps. Hydraulically powered ailerons, elevators (with tabs) and rudder, all with feel trim. High-lift devices include outboard leading-edge slats over 17 per cent of wing span and large outer and inner blown trailing-edge flaps deflecting 60° and 80° respectively; outboard flaps can be linked with ailerons; inboard flaps, elevators and rudder blown by BLC system. Two spoilers in front of outer flaps on each wing. Inverted slats on tailplane leading-edge.

STRUCTURE: All-metal; two-spar wing box.

LANDING GEAR: Flying-boat hull, plus hydraulically retractable Sumitomo tricycle landing gear with twin wheels on all units. Steerable nose unit. Oleo-pneumatic shock-absorbers. Main units, which retract rearward into fairings on hull sides, have 40×14-22 tyres, pressure 7.79 bar (113 lb/sq in). Nosewheel tyres size 25×6.75-18, pressure 20.69 bar (300 lb/sq in). Three-rotor hydraulic disc brakes. No anti-skid units. Minimum ground turning radius 18.80 m (61 ft 8¼ in) towed, 21.20 m (69 ft 6¾ in) self-powered.

POWER PLANT: Four 2,535 kW (3,400 shp) Ishikawajima-built General Electric T64-IHI-10J turboprops, each driving a Sumitomo-built Hamilton Sundstrand 63E60-27 three-blade constant-speed reversible-pitch propeller. Fuel in five wing tanks, with total usable capacity of 11,640 litres (3,075 US gallons; 2,560.5 Imp gallons) and two fuselage tanks (10,849 litres; 2,866 US gallons; 2,386.5 Imp

gallons); total usable capacity 22,489 litres (5,941 US gallons; 4,947 Imp gallons). Pressure refuelling point on port side, near bow hatch. Oil capacity 152 litres (40.2 US gallons; 33.4 Imp gallons). Aircraft can be refuelled on open sea, either from surface vessel or from another US-1A with detachable at-sea refuelling equipment.

ACCOMMODATION: Crew of three on flight deck (pilot, co-pilot and flight engineer), plus navigator, radio/radar operator and medical attendant's seats in main cabin. Latter can accommodate up to 20 seated survivors or 12 stretchers, one auxiliary seat and two observers' seats. Sliding rescue door on port side of fuselage, aft of wing.

SYSTEMS: Cabin air conditioning system. Two independent hydraulic systems, each 207 bar (3,000 lb/sq in). No. 1 system actuates ailerons, outboard flaps, spoilers, elevators, rudder and control surface feel; No. 2 system actuates ailerons, inboard and outboard flaps, wing leading-edge slats, elevators, rudder, landing gear extension/retraction and lock/unlock, nosewheel steering, mainwheel brakes and windscreen wipers. Emergency system, also of 207 bar (3,000 lb/sq in), driven by 24 V DC motor, for actuation of inboard flaps, landing gear extension/retraction and lock/unlock, and mainwheel brakes.

Honeywell 85-131J APU (RE-220 in US-1AKai) provides power for starting main engines and shaft power for 40 kVA emergency AC generator. BLC system includes a C-2 compressor, driven by a 1,119 kW (1,500 shp) Ishikawajima-built General Electric T58-IHI-10-M2 gas turbine, housed in upper centre portion of fuselage, which delivers compressed air at 14 kg (30.9 lb)/s and pressure of 1.86 bar (27 lb/sq in) for ducting to inner and outer flaps, rudder and elevators (US-1AKai has variant of LHTEC CTS800).

Electrical system includes 115/200 V three-phase 400 Hz constant-frequency AC and three transformer-rectifiers to provide 28 V DC. Two 40 kVA AC generators, driven by Nos. 2 and 3 main engines. Emergency 40 kVA AC generator driven by APU. DC emergency power from two 24 V 34 Ah Ni/Cd batteries.

De-icing of wing and tailplane leading-edges. Oxygen system for all crew and stretcher stations. Fire detection and extinguishing systems standard.

AVIONICS: *Comms:* HRC-121 HF, HRC-113 VHF and HRC-115-1 U/VHF radios; HIC-3B interphone; AN/APX-68-NB IFF transponder; RRC-22 emergency transmitter.

Radar: ; Thales OM-100 Ocean Master search radar.

Flight: HRN-101 ADF; OA-8697A/ARD UHF/DF; AN/ARN-118(V)2 Tacan; HRN-115-5 GPS nav system;

HRN-107B-1 VOR/ILS receiver; AN/APN-171 (N2) radio altimeter; HPN-101B wave height meter; HSN-4 INS; N-TR-45 TAS transmitter; HRA-6-3 nav display; and N-ID-66/HRN BDHI.

EQUIPMENT: Marker launcher, 10 marine markers, six green markers, two droppable message cylinders, 10 float lights, pyrotechnic pistol, parachute flares, two flare storage boxes, binoculars, two rescue equipment kits, two droppable liferaft containers, rescue equipment launcher, lifeline pistol, lifeline, three lifebuoys, loudspeaker, hoist unit, rescue platform, lifeboat with outboard motor, camera, and 12 stretchers. Sea anchor in nose compartment. Stretchers can be replaced by troop seats.

DIMENSIONS, EXTERNAL:

Wing span	33.15 m (108 ft 9 in)
Wing chord: at root	5.00 m (16 ft 4¾ in)
at tip	2.39 m (7 ft 10 in)
Wing aspect ratio	8.1
Length overall	33.46 m (109 ft 9¼ in)
Height overall	9.95 m (32 ft 7¾ in)
Tailplane span	12.36 m (40 ft 8½ in)
Wheel track	3.56 m (11 ft 8¼ in)
Wheelbase	8.33 m (27 ft 4 in)
Propeller diameter	4.42 m (14 ft 6 in)

Rescue hatch (port side, rear fuselage):

Height	1.58 m (5 ft 2¼ in)
Width	1.46 m (4 ft 9½ in)

AREAS:

Wings, gross	135.82 m² (1,462.0 sq ft)
Ailerons (total)	6.40 m² (68.90 sq ft)
Inner flaps (total)	9.40 m² (101.18 sq ft)
Outer flaps (total)	14.20 m² (152.85 sq ft)
Leading-edge slats (total)	2.64 m² (28.42 sq ft)
Spoilers (total)	2.10 m² (22.60 sq ft)
Fin	17.56 m² (189.0 sq ft)
Dorsal fin	6.32 m² (68.03 sq ft)
Rudder	7.01 m² (75.50 sq ft)
Tailplane	23.04 m² (248.00 sq ft)
Elevators, incl tabs	8.78 m² (94.50 sq ft)

WEIGHTS AND LOADINGS (search and rescue):

Manufacturer's weight empty	23,300 kg (51,367 lb)
Weight empty, equipped	25,630 kg (56,505 lb)
Usable fuel: JP-4	17,518 kg (38,620 lb)
JP-5	18,397 kg (40,560 lb)
Max oversea operating weight	36,000 kg (79,365 lb)
Max T-O weight: from water	43,000 kg (94,800 lb)
from land	45,000 kg (99,200 lb)
Max wing loading	331.3 kg/m² (67.85 lb/sq ft)
Max power loading	4.24 kg/kW (6.97 lb/shp)

PERFORMANCE (search and rescue, land T-O. A: at 36,000 kg; 79,365 lb weight, B: at 43,000 kg; 94,800 lb, C: at max T-O weight):

Max level speed: C	276 kt (511 km/h; 318 mph)
Max level speed at 3,050 m (10,000 ft):	
A	282 kt (522 km/h; 325 mph)
Cruising speed at 3,050 m (10,000 ft):	
C	230 kt (426 km/h; 265 mph)
Max rate of climb at S/L: A	713 m (2,340 ft)/min
C	488 m (1,600 ft)/min
Service ceiling: A	8,655 m (28,400 ft)
C	7,195 m (23,600 ft)
T-O to 15 m (50 ft) on land, 30° flap, BLC on (ISA):	
C	655 m (2,150 ft)
T-O distance on water, 40° flap, BLC on (ISA):	
B	735 m (2,410 ft)
Landing from 15 m (50 ft) on land, AUW of 36,000 kg (79,365 lb), 40° flap, BLC on, with reverse pitch:	
A	810 m (2,655 ft)
Landing distance on water, AUW of 36,000 kg (79,365 lb), 60° flap, BLC on:	
A	561 m (1,840 ft)
Runway LCN requirement: B	42
Max range at 230 kt (426 km/h; 265 mph) at 3,050 m (10,000 ft)	2,060 n miles (3,815 km; 2,370 miles)

UPDATED

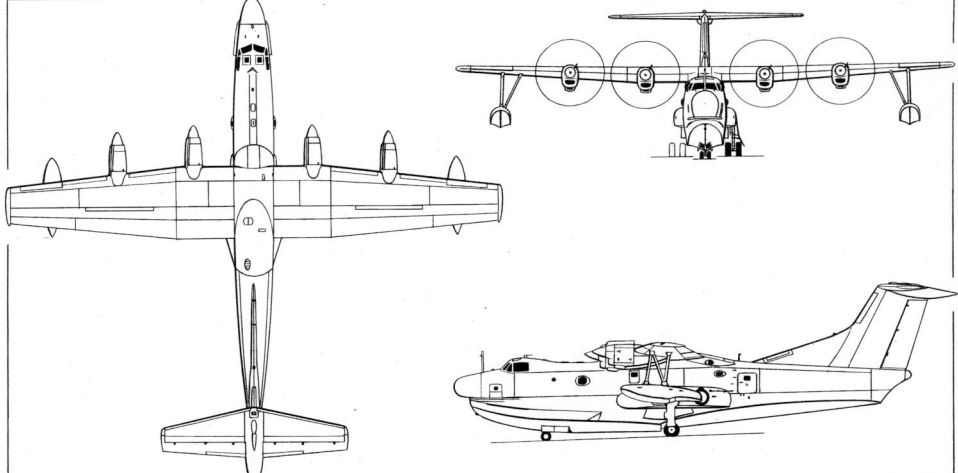

ShinMaywa US-1A ocean-going search and rescue amphibian *(Jane's/Dennis Punnett)*

YANAGISAWA

GEN YANAGISAWA ENGINEERING SYSTEM CO LTD

ESCO, Matsumoto, Nagano
Web: http://www.engineeringsystem.co.jp
PRESIDENT AND DESIGNER: Gennai Yanagisawa

US AGENT:
Ace Craft USA
Platte City, Montana
Web: http://www.acecraftusa.com

Mr Yanagisawa, whose nickname is 'Gen', developed a small, but comparatively high-powered, horizontally opposed two-cylinder engine while working for the Zenoa company. By 1985, this had been tested in a hang-glider, para-glider and model aircraft, but failed to find a market. Tests of a coaxial rotor RPV in 1987 were unsuccessful because of control problems, while a manned twin-engined design in 1992 failed to produce sufficient lift. By 1994, however, this had evolved into the BDH-1 (Boy's Dream Helicopter 1), exhibited at that year's Japan Aerospace Show. A three-engined BDH-2 followed in 1996 and was further modified as the BDH-3 which, although successful, was unable to sustain height with one engine inoperative. The BDH-4 was shown at AirVenture, Oshkosh, USA, in 1997 and continues to be refined.

NEW ENTRY

YANAGISAWA GEN H-4

TYPE: Exoskelitor flying vehicle.
PROGRAMME: Four-engined development of BDH-3; basic version shown at AirVenture, Oshkosh, 1997; yaw control by small, electrically powered, auxiliary propellers; by 1998, yaw accomplished by differentiation of rotor rpm. Marketed in kit form.
COSTS: Kit US$30,000 (2002).
DESIGN FEATURES: 'Strap-on' personal helicopter with contrarotating twin blades of tapered planform and fixed pitch. Seat, with fuel tank at rear, supported by triangular tube framework landing gear and footrest. Gimballed platform above occupant's head houses four small motors, central transmission and rotor assembly with gears. Combined hand-grip control column and instrument panel attached to front of platform. Rotor speed 800 to 900 rpm. Quoted build time 40 hours.
FLYING CONTROLS: Two-axis control by tilting of engine/rotor platform; yaw control effected by thumb switch on instrument panel, operating electric motor which differentiates speed of rotors by 1 per cent.
STRUCTURE: Framework of 5 cm (2 in) aluminium tube; glass fibre seat and backpack; rotors of carbon fibre and Kevlar, with foam core.
LANDING GEAR: Four small 'wheels' each comprise multiple rollers in circular arrangement.

Yanagisawa Gen H-4 personal helicopter (*Jane's/Paul Jackson*)   *NEW*/0137380

POWER PLANT: Four 7.5 kW (10 hp) Gen 125 two-stroke, horizontally opposed, two-cylinder piston engines with independent ignition, positioned symmetrically on pivoting platform. Fuel, of 30:1 petrol and oil premix, in seatback tank.
DIMENSIONS, EXTERNAL:
Rotor diameter (each)   4.00 m (13 ft 1½ in)
AREAS:
Rotor disc (each)   12.57 m² (135.3 sq ft)

WEIGHTS AND LOADINGS:
Weight empty   70 kg (154 lb)
Max T-O weight   220 kg (485 lb)
PERFORMANCE:
Never-exceed speed (VNE)   48 kt (90 km/h; 55 mph)
Cruising speed   43 kt (80 km/h; 50 mph)
Service ceiling   3,000 m (9,840 ft)
Endurance   up to 1 h 0 min
NEW ENTRY

KOREA, SOUTH

KA

KOREAN AIR (Aerospace Division)

KA Building, 1370 Gonghang-dong, Gangseo-gu, Seoul
Tel: (+82 2) 656 39 20
Fax: (+82 2) 656 39 17
e-mail: jypark@koreanair.co.kr
Web: http://www.koreanair.co.kr
MANAGING VICE-PRESIDENT: Sang-Mook Suh
GENERAL MANAGER, PLANNING DEPARTMENT: Yong-An Yoon

KIMHAE PLANT:
103 Daejeo 2-dong, Gangseo-gu, Pusan 618142
Tel: (+82 51) 970 56 20
Fax: (+82 51) 970 50 62
MANAGING VICE-PRESIDENT: Sang-Mook Suh
GENERAL MANAGER: Yong-An Yoon

Korean Institute of Aeronautical Technology (KIAT)
461 Jonmin-dong, Yusung-ku, Taejon 305390
Tel: (+82 42) 868 61 14
Fax: (+82 42) 868 61 28
MANAGING VICE-PRESIDENT: Si-Yoong Yoo

Aerospace Division of Korean Air Lines established 1976 to manufacture and develop aircraft; occupies 64.75 ha (160 acre) site at Kimhae, including floor area of 250,000 m² (2.69 million sq ft); 2001 workforce about 2,150. Has overhauled RoKAF aircraft since 1978; programmed depot maintenance of US military aircraft in Pacific area began 1979, including structural repair of F-4s, systems modifications for F-16s, MSIP upgrading of F-15s and overhaul of C-130s. Began assembly in 1981 of first domestically manufactured fighter (Northrop F-5E/F), completing deliveries to RoKAF 1986. Since 1988, KAL has delivered wing components for Boeing 747/777, New-Generation 737 and MD-11 (last-named

programme completed July 1999); and fuselage components for MD-80/90 and Airbus A330/A340. Deliveries of nose sections for Boeing 717 started 28 March 1997, but suspended in 2001 KAL manufactured UH-60P helicopters between 1991 and 1999, under licence from Sikorsky.

Korean Institute of Aeronautical Technology (KIAT), established as division of KAL in 1978, has grown to become a major Korean aerospace industry R&D centre. It is to add a Helicopter Development Research Centre (HDRC) as a result of October 2001 agreement with Sikorsky to collaborate on KAL's submission in KMH programme (which see).

As part of Korean industry development programme from 1988, KAL designed and developed a light aircraft (Chang-Gong 91: see 1995-96 *Jane's*); co-developed (with McDonnell Douglas) the MD 520MK military helicopter derived from the MD 500; is a major member of KFP Korean Fighter Programme (wings and rear fuselages for F-16C/D); and domestically co-developed with Daewoo the KT-1 basic trainer (see KAI entry) for the RoKAF. Incorporation into KAI still the subject of prolonged discussions in 2002.

UPDATED

KA (SIKORSKY) UH-60P

TYPE: Multirole medium helicopter.
PROGRAMME: Original licence agreement to build UH-60P in Korea (S-70A-18 variant of Sikorsky UH-60L: see US section) signed March 1990. Seven US-built aircraft imported 1991, followed by materials including kits and parts for local manufacture; first Korean-built UH-60P made first flight 15 February 1992 and delivered 26 March. Some UH-60Ps now being upgraded for search and rescue, command and control, or special land/sea missions. Initial batch of 81 followed by 57 more; deliveries of second batch began in August 1995; last delivery December 1999; details in 2002-03 and earlier *Jane's*. Follow-on

programme for Korean Navy under consideration by South Korean government in 2001, but unresolved by mid-2002.

UPDATED

KA KMH

TYPE: Multirole light helicopter (requirement).
PROGRAMME: Initiated as entrant in Korean Military Helicopter (KMH) competition for armed scout and attack roles, but M now indicating Multipurpose for appeal also to civil operators; early design assistance by Sikorsky; mockup displayed and first details released at Seoul Air Show in October 1996. Preliminary design, market survey and solicitation of customer requirements under way in late 1998. KAL signed MoU with Sikorsky Aircraft on 16 October 2001, under which the two companies will collaborate on former's submission for KMH programme.
CUSTOMERS: KAL foresees a domestic market for up to 400 military and 150 civil KMHs.
DESIGN FEATURES: Externally similar in some respects to the S-76 or UH-60 Black Hawk, except for rear fuselage (T tail and Fenestron) resembling RAH-66 Comanche (see accompanying illustration).
FLYING CONTROLS: Dual redundant fly-by-light control system.
LANDING GEAR: Non-retractable tailwheel type.
POWER PLANT: Two 671 to 820 kW (900 to 1,100 shp) class turboshafts (type not yet selected).
ACCOMMODATION: Flight crew of two; six seats in main cabin.
AVIONICS: *Mission:* Nose-mounted FLIR sighting system.
Self-defence: Jammer and RWR.
ARMAMENT: Fuselage sidebars for various combinations of Hellfire or Stinger missiles, seven-round rocket launchers, 0.50 in machine gun pods or 454 litre (120 US gallon; 100 Imp gallon) auxiliary fuel tanks.

DIMENSIONS, EXTERNAL: Not provided
WEIGHTS AND LOADINGS:
Max external sling load	approx 1,360 kg (3,000 lb)
Max T-O weight	approx 3,400 kg (7,500 lb)

PERFORMANCE (design target):
Max level speed	165 kt (305 km/h; 189 mph)
Max rate of climb at S/L	640 m (2,100 ft)/min
Service ceiling	6,100 m (20,000 ft)
Hovering ceiling IGE	4,115 m (13,500 ft)
Range	350 n miles (648 km; 402 miles)

UPDATED

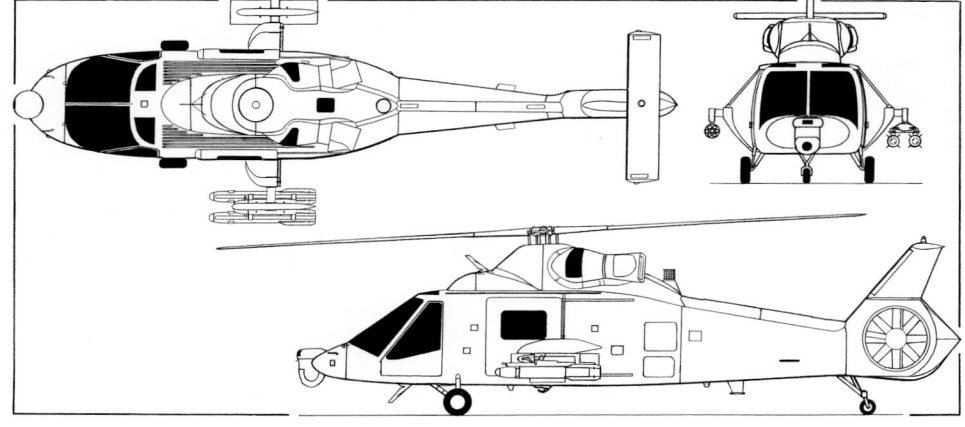

General appearance of the KIAT-designed
KAL KMH candidate
(Jane's/Paul Jackson)
0051114

KAI

KOREA AEROSPACE INDUSTRIES LTD

HEAD OFFICE: Seosomun-dong, Jung-gu, Seoul 100737
Tel: (+82 2) 20 01 31 14
Fax: (+82 2) 20 01 30 11
Web: http://www.koreaaero.com
PRESIDENT AND CEO: Kil, Hyoung-Bo
EXECUTIVE VICE-PRESIDENT: Ahn, Taek-Soon

CHANGWON PLANT: Seongju-dong, Changwon,
    Gyeongsangnam-do 641120
Tel: (+82 55) 280 65 01
Fax: (+82 55) 285 23 83

SACHEON PLANT 1: Yucheon-ri, Sanam-myeon, Sacheon,
    Gyeongsangnam-do 664942
Tel: (+82 55) 851 10 00
Fax: (+82 55) 851 10 05

SACHEON PLANT 2: Yongdang-ri, Sacheon-eup, Sacheon,
    Gyeongsangnam-do 664802
Tel: (+82 55) 851 29 99
Fax: (+82 55) 854 86 38

As detailed in earlier editions, the incoming South Korean government of 1998 ordered a restructuring of the country's aerospace industry.

Korea Aerospace Industries was established on 1 October 1999 by consolidating the aerospace business of Daewoo Heavy Industries, Samsung Aerospace and Hyundai Space & Aircraft to create a single, unified aircraft manufacturer. In 2000, the Korean government designated KAI as the sole specialised aerospace defence industry in Korea, with exclusive and preferential rights in Korean aerospace defence programmes.

Sacheon Plant 1, built on 66.4 ha (164 acres) (building area 10.6 ha; 26.1 acres) adjacent to Sacheon airport, has approximately 1,900 employees; consists of various facilities for aircraft development such as parts fabrication, assembly and R&D; has a full development capability ranging over design, manufacturing, ground and flight test. Sacheon Plant 2, built on 12.1 ha (30 acres) (building area 2.7 ha; 6.7 acres) located near Plant 1 and adjacent to Sacheon airport, has over 300 workforce for development and manufacturing; final assembly and several ground tests undertaken by its main facility. Changwon Plant, built on 24.7 ha (61 acres) (building area 7.0 ha; 17.2 acres), has manufacturing facility for fabrication and subassembly and R&D centre; also has

full-scale design, development and manufacturing capabilities with around 800 employees.

KAI has developed KT-1 and Night Intruder 300 unmanned aerial vehicle (UAV) with government Agency for Defence Development (ADD); co-developed SB 427 helicopter with Bell Helicopter; built F-16 fighters under Lockheed Martin licence for Korean Air Force and BO 105 helicopters under Eurocopter licence for Korean Army; is currently prime contractor for Korea Fighter Program (KFP) Phase II; basic trainer (KT-1) production programme; undertaking full-scale development of advanced training and light attack aircraft (T-50/A-50); will play key role in Korean Multipurpose Helicopter (KMH) as well as Maritime Patrol Aircraft (MPA).

Present and recent subcontract work includes A319/A320 wing top panels and A380 bottom panels for Airbus; wing rib assemblies for Boeing 747-400; stringers for Boeing 737/747/757; wing trailing-edges for Boeing 757/767; horizontal tail surfaces and trailing-edges for Bombardier de Havilland Dash 8; long-term contracts with Boeing, Airbus, Vought and Bombardier to produce and supply major aerostructures.

UPDATED

KAI KT-1 two-seat basic trainer of 217 TTS, RoKAF   *NEW*/0142062

KAI KT-1 WOONG-BEE
English name: Great Flight
TYPE: Basic turboprop trainer.
PROGRAMME: Started February 1988; built under KTX (Korean Trainer Experimental) programme to design by Daewoo and ADD (Agency for Defence Development); construction of first prototype began June 1991; nine prototypes (01-05 flying and 001-004 for static and fatigue test), of which 01, with 410 kW (550 shp) PT6A-25A engine, rolled out November 1991 and made first flight 12 December that year; 02, identical to 01, made first flight 5 February 1993; 03 (with PT6A-62A) made first flight on 10 August 1995; named Woong-Bee in November 1995; 04, which first flew on 10 May 1996, further modified with nose shortened and horizontal tail surfaces remounted lower and farther aft; 03 also modified to this standard. Fifth (preproduction) prototype 05 flew for the first time on 16 March 1998; development and operational testing completed 18 September 1998 after 1,474 flying hours in 1,184 sorties. Series production began 1999; first aircraft handed over to RoKAF at Sachon AB on 7 November 2000 for use by 217 Tactical Training Squadron.
CURRENT VERSIONS: **KT-1:** Standard trainer, *as described*.
    KT-1B: Export designation for Indonesian Air Force; some different avionics to customer's requirements.
    XKO-1: Forward air control (Experimental Korean Observer) and version for RoKAF. Development started in 2000 by modifying aircraft 05 to FAC configuration with weapon management system, HUD, armament management LCD MFD and four underwing hardpoints; first flight in this configuration 1 November 2001. Deliveries to begin in 2003.
    KO-1: Armed trainer (new export version); full 'glass cockpit', off-the-shelf avionics, HOTAS controls, weapon delivery system (simulated and live, up to six hardpoints and OBOGS. Under development.
CUSTOMERS: Total of 105 production aircraft approved by Republic of Korea Air Force (RoKAF). Contract 9 August 1999 for 85 KT-1 trainers, of which 36 delivered by 28 February 2002, to replace Cessna T-37s. Follow-on option for 20 XKO-1s for forward air control duties. Indonesian Air Force announced order for seven on 26 February 2001, with option on further 13; to replace Beechcraft T-34Cs of No. 102 Squadron. KT-1 is candidate for Singapore S.211 replacement.
COSTS: RoKAF order for 85 quoted as Swon600 billion (US$473 million) in 1999; Indonesian order for seven reportedly valued at US$60 million (2001).
DESIGN FEATURES: Meets requirements of FAR/JAR 23 (Aerobatic Category) and MIL-F-8785C trainer category

Prototype XKO-1 forward air control version of KT-1   *NEW*/0142068

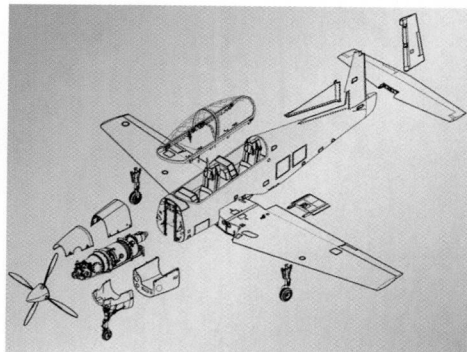
KT-1 major airframe components   0062987

Trailing-edge flaps (total)	2.22 m² (23.90 sq ft)
Fin, incl dorsal fin	1.35 m² (14.57 sq ft)
Rudder, incl tab	0.88 m² (9.43 sq ft)
Tailplane	2.32 m² (25.02 sq ft)
Elevators (total, incl tab)	1.19 m² (12.78 sq ft)

WEIGHTS AND LOADINGS:

| Weight empty | 1,910 kg (4,210 lb) |
| Max fuel weight | 408 kg (900 lb) |

Max T-O and landing weight:

Aerobatic	2,540 kg (5,600 lb)
Utility	3,205 kg (7,065 lb)
with external stores	3,311 kg (7,300 lb)

Max wing loading:

Aerobatic	158.7 kg/m² (32.50 lb/sq ft)
Utility	200.2 kg/m² (41.00 lb/sq ft)
with external stores	206.8 kg/m² (42.37 lb/sq ft)

Max power loading:

Aerobatic	3.59 kg/kW (5.89 lb/shp)
Utility	4.53 kg/kW (7.44 lb/shp)
with external stores	4.68 kg/kW (7.68 lb/shp)

PERFORMANCE:

Max level speed at 4,570 m (15,000 ft):

| KT-1 | 280 kt (518 km/h; 322 mph) |

XKO-1 with external stores

	244 kt (513 km/h; 280 mph)
Max operating speed	310 kt (574 km/h; 357 mph) IAS
Stalling speed, flaps down	71 kt (132 km/h; 82 mph) IAS
Max rate of climb at S/L	969 m (3,180 ft)/min
Absolute ceiling	11,580 m (38,000 ft)
T-O run	250 m (820 ft)
T-O to 15 m (50 ft)	494 m (1,620 ft)
Landing from 15 m (50 ft)	727 m (2,385 ft)
Landing run	397 m (1,300 ft)

Range at 7,620 m (25,000 ft), max internal fuel and 30 min reserves   720 n miles (1,333 km; 828 miles)

Ferry range at 6,100 m (20,000 ft) max internal and external fuel, 30 min reserves   1,118 n miles (2,070 km; 1,286 miles)

Endurance at 6,100 m (20,000 ft), max internal fuel and 30 min reserves   up to 4 h

| g limits: Aerobatic | +7/−3.5 |
| with external stores | +4.5/−2.25 |

UPDATED

KAI T-50 and A-50 GOLDEN EAGLE

TYPE: Advanced jet trainer/light attack jet.

PROGRAMME: Begun by Samsung Aerospace (SSA) in 1992 under designation KTX-2 (Korean Trainer, Experimental); initial design assistance by Lockheed Martin as offset in F-16 Korean Fighter Programme; early design featured shoulder-mounted wings and twin tail unit; revised later to present configuration; basic configuration established mid-1995; full-scale development originally planned to begin in 1997, subject to finding risk-sharing partner; government go-ahead given on 3 July 1997; Samsung/Lockheed Martin agreement September 1997 to continue joint development until 2005; Lockheed Martin Aeronautics at Fort Worth responsible for wings, flight control system and avionics; development phase funded 70 per cent by South Korean government, 17 per cent by Samsung/KAI and 13 per cent by Lockheed Martin.

FSD contract, signed 24 October 1997, called for two static/fatigue test airframes and four flying prototypes (two T-50 and two A-50). Work split 55 per cent in USA, 44 per cent in South Korea and 1 per cent elsewhere. Preliminary design review (PDR) completed 12 to 16 July 1999; wind tunnel testing completed (4,800 hours) and aerodynamic design frozen November 1999. Critical design review (CDR) passed in August 2000. KTX-2 redesignated T-50 and A-50 in early 2000. T-50 International Company (TFIC) established November 2000 (following July MoU) by KAI and LMAS to market aircraft outside South Korea. First prototype entered final assembly January 2001 and rolled out 31 October 2001; first flights 20 August (001) and 8 November 2002 (002); first supersonic flight 19 February 2003; third and fourth prototypes due to fly in August 2003; production aircraft to begin manufacture in August 2003. Static testing began 2 January 2002 and due to complete in August 2002; fatigue tests began 22 July 2002 and continue to June 2004. Deliveries to RoKAF planned to begin in mid-2005.

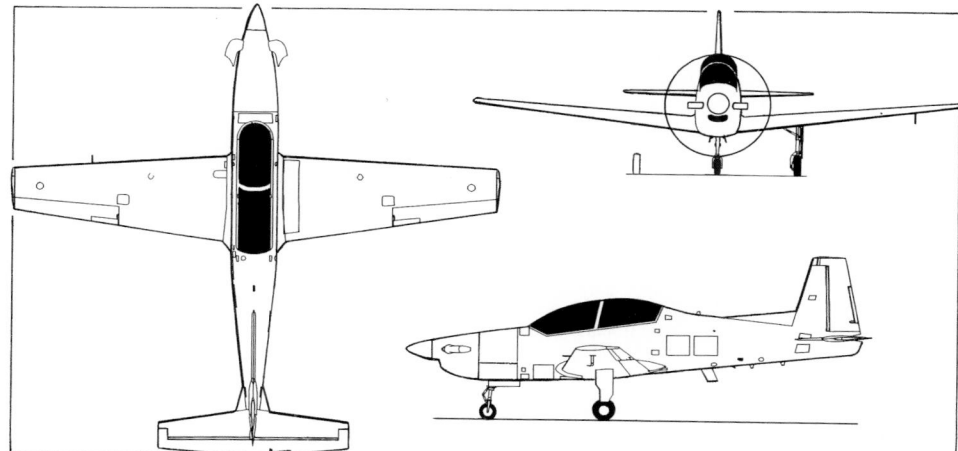
KAI KT-1 basic trainer (P&WC PT6A-62A turboprop engine) *(Jane's/Paul Jackson)*   0062986

Class IV. Able to cover entire primary and basic training syllabus and forward air control missions.

Unswept low wing with NACA 63-218 (root) and 63-212 (tip) aerofoil sections; tandem cockpits; conventional unswept vertical and horizontal tail surfaces; four weapon hardpoints; retractable tricycle landing gear.

FLYING CONTROLS: Primary control surfaces actuated mechanically, with electrically operated trim tab in starboard elevator and port aileron; automatic rudder trim compensates for propeller-induced yaw to provide jet-like handling; rudder and one-piece elevator are horn-balanced. Airbrake under centre-fuselage and split flaps on wing trailing-edge are hydraulically operated.

LANDING GEAR: Hydraulically retractable tricycle type, with single wheel and oleo-pneumatic shock-absorber on each unit. Mainwheels retract inward; nosewheel is steerable ±18° and retracts rearward. Parker Hannifin mainwheels with 18×5.5 tyres (10 ply), pressure 9.65 bar (140 lb/sq in). Dunlop nosewheel with 5.00-5 tyre (14 ply), pressure 6.90 bar (100 lb/sq in). Parker Hannifin hydraulic mainwheel brakes.

POWER PLANT: One 708 kW (950 shp) (flat rated) Pratt & Whitney Canada PT6A-62 turboprop, driving a Hartzell HC-E4N-2/E9512CB-1 four-blade constant-speed fully feathering propeller. Single-lever combined control for engine fuel and propeller pitch.

Fuel in one 275.5 litre (72.8 US gallon; 60.6 Imp gallon) integral tank in each wing, giving total internal fuel capacity of 551 litres (145.6 US gallons; 121.2 Imp gallons). Gravity fuelling point in each wing upper surface. Provision for 189 litre (50.0 US gallon; 41.6 Imp gallon) auxiliary fuel tank on each inboard underwing station. Oil capacity 5.7 litres (1.5 US gallons; 1.25 Imp gallons). Fuel system permits up to 30 seconds of inverted flight.

ACCOMMODATION: Instructor and pupil in tandem cockpits on Martin-Baker Mk 16LF zero/zero ejection seats; rear seat elevated. One-piece canopy with MDC opens sideways to starboard.

SYSTEMS: Compressed bleed air air conditioning system for cockpit ventilation, heating, cooling and demisting; heating and cooling by wheel bootstrap air cycle system driven by engine compressor. Self-pressurised main and emergency hydraulic systems, operating pressure 207 bar (3,000 lb/sq in), flow rate 17.8 litres (4.7 US gallons; 3.9 Imp gallons)/min at 7,650 rpm. Pneumatic back-up system, also 207 bar (3,000 lb/sq in), for landing gear, wheel bay doors and flaps. Electrical power (28 V DC) available from engine-driven starter/generator and built-in battery, or from external power source. Fixed-geometry inertial separator for engine intake anti-icing. Diluter demand oxygen system provides 1,367 litres (48.2 cu ft) gaseous capacity for crew; selected and controlled individually from regulator/control panel in each cockpit.

AVIONICS (KT-1): *Comms:* UHF/VHF radio, IFF, interphone and communication control system; optional ELT.

Flight: AHRS and Tacan; optional VOR/ILS.

Instrumentation: EFIS with standby ASI, electronic engine instruments, FCMS, AoA sensor and standby magnetic compass.

AVIONICS (XKO-1): *Comms:* Integrated UHF/VHF radios, IFF, interphone and control system; optional ELT.

Flight: GPS/INS and Tacan; optional VOR/ILS.

Instrumentation: As KT-1, plus HUD and MFDs. Options include mission computer, AVTR and NVG-compatible cockpit lighting system.

ARMAMENT (XKO-1): Two stores stations under each wing for two LAU-131 seven-round rocket launchers and two machine gun pods; two LAU-131s and two drop tanks; or two gun pods and two drop tanks. Option for two additional hardpoints, bombs, laser range-finder, FLIR and air-to-air training missile.

DIMENSIONS, EXTERNAL:

Wing span	10.59 m (34 ft 9 in)
Wing chord; at root	2.06 m (6 ft 9 in)
at tip	0.97 m (3 ft 2 in)
Wing aspect ratio	7.0
Length overall	10.26 m (33 ft 8 in)
Height overall: KT-1	3.68 m (12 ft 1 in)
XKO-1	3.78 m (12 ft 5 in)
Tailplane span	4.17 m (13 ft 8 in)
Wheel track	3.54 m (11 ft 7 in)
Wheelbase	2.57 m (8 ft 5 in)
Propeller diameter	2.44 m (7 ft 11 in)
Propeller ground clearance	0.38 m (1 ft 3 in)

DIMENSIONS, INTERNAL:

Cockpits: Length (total)	3.20 m (10 ft 6 in)
Max width	0.99 m (3 ft 3 in)
Max height	1.52 m (5 ft 0 in)

AREAS:

| Wings, gross | 16.01 m² (172.3 sq ft) |
| Ailerons (total, incl tabs) | 1.11 m² (11.95 sq ft) |

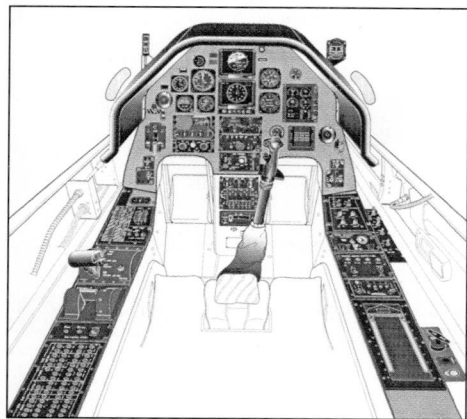
Diagram of KT-1 front cockpit   0051113

Cockpit of the T-50   *NEW*/0142067

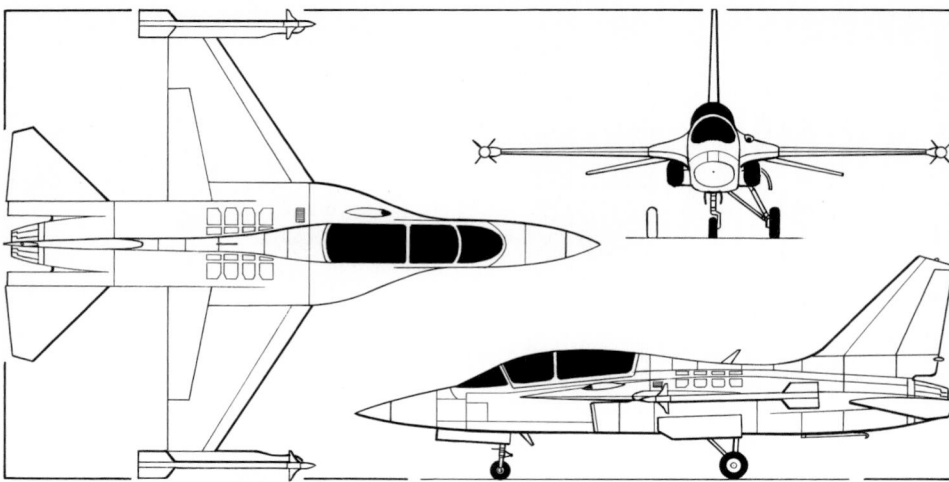

KAI T-50/A-50 Golden Eagle (*Jane's/James Goulding*)   *NEW*/0137947

CURRENT VERSIONS (planned): **T-50:** Advanced trainer; no internal gun or radar.

A-50: Lead-in fighter trainer and light combat version with radar and internal gun.

CUSTOMERS: Initial RoKAF requirement for 94 (50 T-50s and 44 A-50s), with options for up to 100 more, including further A-50s. Initially to replace RoKAF T-38s and F-5Bs, aimed also at F-5 replacement market. Exports (from 2006) estimated potentially at 600 to 800.

COSTS: Development programme cost estimated at US$2,000 million (1995); but reassessed in 1996 as US$1,500 million, and only US$1,200 million by early 1997. Initial October 1997 FSD contract valued at approximately US$1,270 million. Development phase re-estimated at US$1.8 billion to US$2.1 billion in mid-2000. Unit cost US$18 million to US$20 million for T-50, US$20 million to US$22 million for A-50 (2001).

DESIGN FEATURES: Mid-mounted, variable camber wings, swept back on leading-edges only; leading-edge root extensions (LERX); all-moving tailplane; sweptback fin leading-edge. Single turbofan engine, with twin side-mounted intakes. KAI developing fuselage and tail unit. Avionics include HUD and colour MFDs. Designed for service life of more than 8,000 hours.

FLYING CONTROLS: Digital fly-by-wire control of flaperons, flaps, tailplane and rudder. Parker and Moog hydraulic actuators. Leading-edge manoeuvring flaps; split airbrake at rear of fuselage, between exhaust nozzle and tailplane halves.

STRUCTURE: Conventional aluminium alloy except for GFRP nose radome and some fairings. Lockheed Martin, wings; KAI, fuselage, tail unit and final assembly.

LANDING GEAR: Messier-Dowty hydraulically retractable tricycle type, with single wheel and oleo-pneumatic shock-absorber on each unit. Mainwheels retract into engine intake trunks, steerable nosewheel forward. Runway emergency arrester hook.

POWER PLANT: One General Electric F404-GE-102 turbofan (78.7 kN; 17,700 lb st with afterburning), equipped with FADEC. Provision for 568 litre (150 US gallon; 125 Imp gallon) drop fuel tank on centreline and/or inboard underwing stations.

ACCOMMODATION: Crew of two in tandem; stepped cockpits; Martin-Baker ejection seats. Rear-hinged, upward-opening single canopy; one-piece wraparound windscreen.

SYSTEMS: Dual-redundant, independent hydraulic systems (A and B), each at 214 bar (3,100 lb/sq in) pressure and with flow rate of 161 litres (42.5 US gallons; 35.4 Imp gallons)/min; System A has 207 bar (3,000 lb/sq in) emergency back-up system with 86 litres (22.7 US gallons; 18.9 Imp gallons)/min flow rate. Honeywell emergency power system (EPS) operates System A emergency back-up and 1 kVA emergency generator throughout flight envelope. Hamilton Sundstrand APU provides self-contained engine starting capability on ground and up to 6,100 m (20,000 ft). Digitally controlled Hamilton Sundstrand environmental control system for avionics cooling, capacity 5.6 kW. Litton self-generating oxygen life support system.

AVIONICS: *Comms:* UHF/VHF radio; IFF.

Radar: Lockheed Martin AN/APG-67 in A-50.

Flight: Digital fly-by-wire flight controls with HOTAS; nav/attack system for fighter lead-in training; embedded GPS ring laser gyro INS Tacan; radar altimeter.

Instrumentation: BAE Systems HUD; two 127 mm (5 in) colour MFDs; Honeywell instrumentation displays (eight 76 mm; 3 in displays, including HSI, attitude indicator, electronic altimeter and Mach speed indicator).

Self-defence: A-50 provision for EW pods, chaff/flare dispenser and RWR.

ARMAMENT (A-50): Internal 20 mm three-barrel Gatling-type cannon with 200 rounds (port LERX). Seven external stations (one on centreline, two under each wing and AAM rail at each wingtip) for AAMs, ASMs, rocket pods,

bombs, munition dispensers, practice bombs or equipment, and training targets. Includes AIM-9 Sidewinder and AGM-65 Maverick missiles, CBU-58 and Mk 20 cluster munition, ACMI pod, BDU-33 and SUU-20 practice equipment, AGTS and TIX-3 training targets and Mk 82/83/84 series bombs.

DIMENSIONS, EXTERNAL:
Wing span: incl wingtip launcher	9.45 m	(31 ft 0 in)
excl wingtip launcher	9.11 m	(29 ft 10¾ in)
Length overall	13.14 m	(43 ft 1¼ in)
Height overall	4.94 m	(16 ft 2½ in)

WEIGHTS AND LOADINGS:
Weight empty	6,480 kg	(14,285 lb)
Max T-O weight: clean	9,339 kg	(20,589 lb)
with external stores	12,393 kg	(27,322 lb)

PERFORMANCE (design):
Max level speed		M1.4
Max rate of climb at S/L	10,058 m	(33,000 ft)/min
Service ceiling	14,630 m	(48,000 ft)
Max range	1,400 n miles	(2,592 km; 1,611 miles)
g limits		+8/−3

UPDATED

KAI (LOCKHEED MARTIN) F-16C/D FIGHTING FALCON

TYPE: Multirole fighter.

PROGRAMME: Korean Fighter Programme (KFP) to co-produce 108 of 120 F-16s (80 Block 52D F-16Cs and 40 F-16Ds) announced in Seoul 28 March 1991; Samsung (now KAI) was main Korean contractor, with Daewoo (also now KAI) and Hanjin as main subcontractors; see 2001-02 *Jane's* for details; 120th aircraft delivered on 19 April 2001. Decision to produce a further 20 (Block 52 standard) announced 13 May 1999; approved July 2000 as KFP 2; deliveries to start July 2003.

COSTS: US$663 million (estimated) for additional 20 (1999).

Prototype T-50 advanced trainer and light attack jet   *NEW*/0137979

NEW/0526921

KAI T-50/A-50 Golden Eagle, cutaway drawing key

1 Starboard all-moving tailplane
2 Corrugated aluminium tailplane core structure
3 Split trailing edge airbrake, upper and lower surfaces
4 Airbrake hydraulic actuator
5 Variable area exhaust nozzle
6 Nozzle seal
7 Afterburner nozzle actuator
8 Engine bay vent ejector, port and starboard
9 All-moving tailplane pivot mounting
10 Tailplane hydraulic actuator
11 Afterburner duct
12 Fin attachment to fuselage main frames
13 Fin floodlight, port and starboard
14 Engine bay upper vent ejector
15 Rudder hydraulic actuator
16 Rear RWR antenna (provision)
17 Tail navigation light
18 Rudder aluminium honeycomb core structure
19 Static dischargers
20 Anti-collision light
21 Multispar fin structure
22 Port tailplane
23 Aluminium honeycomb fin leading-edge core structure

24 Fin root attachment fittings
25 Rear engine mounting
26 Afterburner fuel manifold
27 APU exhaust
28 Chaff/flare launcher (provision), port and starboard
29 Runway emergency arrester hook
30 Fuselage side-body frame structure
31 Auxiliary Power Unit (APU)
32 APU intake
33 Main engine mounting
34 General Electric F404-102 afterburning turbofan
35 Flight control system hydraulic accumulator
36 Emergency Power Unit (EPU)
37 EPU JP-8 fuel tank
38 Engine compressor intake
39 Engine driveshaft to accessories gearbox
40 Engine oil tank
41 Starboard flaperon hydraulic actuator
42 Flaperon root fairing structure
43 Aluminium honeycomb flaperon core structure
44 Pylon mounting hardpoints
45 Starboard wing integral fuel tank
46 Multispar wing panel structure
47 Fixed portion of trailing-edge
48 Starboard formation light
49 AIM-9 Sidewinder air-to-air missile
50 Detachable wingtip missile launch rail
51 Fully programmed starboard leading-edge flap
52 Leading-edge flap rotary actuators
53 Aluminium honeycomb core flap structure
54 Outboard stores pylon
55 Mk 82 500 lb HE bomb

56 External fuel tank (may also be carried on fuselage centreline pylon)
57 Starboard mainwheel, forward retracting
58 Axle rotating link; leg lies horizontally on retraction
59 Starboard inboard (wet) stores pylon
60 Main landing gear leg strut and shock absorber
61 Hydraulic retraction jack
62 Leading-edge flap driveshaft
63 Wingroot attachment fittings
64 Machined fuselage main frames
65 Air turbine starter
66 Airframe-mounted accessory drive gearbox (AMAD)

67 Generator
68 Fuselage integral fuel tanks
69 Dorsal spine fairing
70 Upper VHF/UHF/Tacan antenna
71 Port flaperon hydraulic actuator
72 Port flaperon
73 Wing panel integral fuel tank
74 Static dischargers
75 Port formation light
76 Wingtip fairing (interchangeable with AIM-9 launcher)
77 Fully programmed port leading-edge flap
78 Leading-edge flap driveshaft and rotary actuators
79 Port wing panel root attachment fittings
80 Upper IFF antenna
81 Upper formation light
82 GPS antenna
83 Leading-edge flap hydraulic drive motor
84 Canopy hinge point
85 Transverse ammunition feed and link return chutes
86 Ammunition magazine, 208 rounds
87 Mainwheel door
88 Door hydraulic jack
89 Hydraulic system ground connectors
90 No. 2 system hydraulic reservoir (No. 1 system between mainwheel bays)
91 Ground test panel
92 Leading-edge root extension (LERX) chine member
93 Starboard navigation light
94 AGM-65 Maverick air-to-surface missile
95 Starboard engine air intake
96 Onboard oxygen generating system (OBOGS)
97 Boundary layer diverter and air system air intake
98 Conditioning system air-cycle machine
99 Air system heat exchanger
100 Air system heat exchanger
101 Heat exchanger exhaust

102 EPU air bottle
103 Canopy seal
104 Cockpit rear pressure bulkhead
105 Second pilot's Martin-Baker ejection seat
106 Internal A-50 20 mm Gatling gun, port LERX-mounted
107 Gun port
108 Rear cockpit HUD repeater
109 Cockpit port side console with engine throttle lever, full HOTAS controls
110 Rear cockpit instrument console
111 Sidestick controller, digital fly-by-wire control system
112 Rear cockpit pressure floor
113 Rear cockpit rudder pedals
114 Electrical equipment bay
115 Front cockpit pressure floor
116 Starboard side console panel
117 Sidestick controller elbow rest
118 First pilot's Martin-Baker ejection seat
119 Canopy centre arch fitting and hydraulic actuator
120 Upward hingeing one-piece cockpit canopy
121 One-piece frameless windscreen panel
122 Head-up display (HUD)
123 Front cockpit instrument console
124 Instrument panel with multifunction LCD displays
125 Front cockpit sidestick controller
126 Forward fuselage frame structure
127 Lower VHF/UHF/Tacan antenna
128 Nosewheel hydraulic steering jack
129 Forward-retracting nosewheel, single door on port side with door-mounted taxying lights
130 Nosewheel shock-absorber leg strut
131 Nosewheel leg-mounted landing lights
132 Front cockpit rudder pedals
133 Cockpit front pressure bulkhead
134 Windscreen fairing pressure frame
135 Total temperature probe
136 Avionics equipment bay, access port and starboard
137 Lateral air data sensors
138 Flight control computer
139 Ventral marker beacon antenna
140 Ventral air data sensors
141 Forward RWR antenna (provision)
142 Radar equipment bay
143 Radome hinge
144 Radar mounting bulkhead
145 Scanner tracking mechanism
146 AN/APG-67 multimode radar antenna
147 Glass fibre composites radome, hinged to starboard

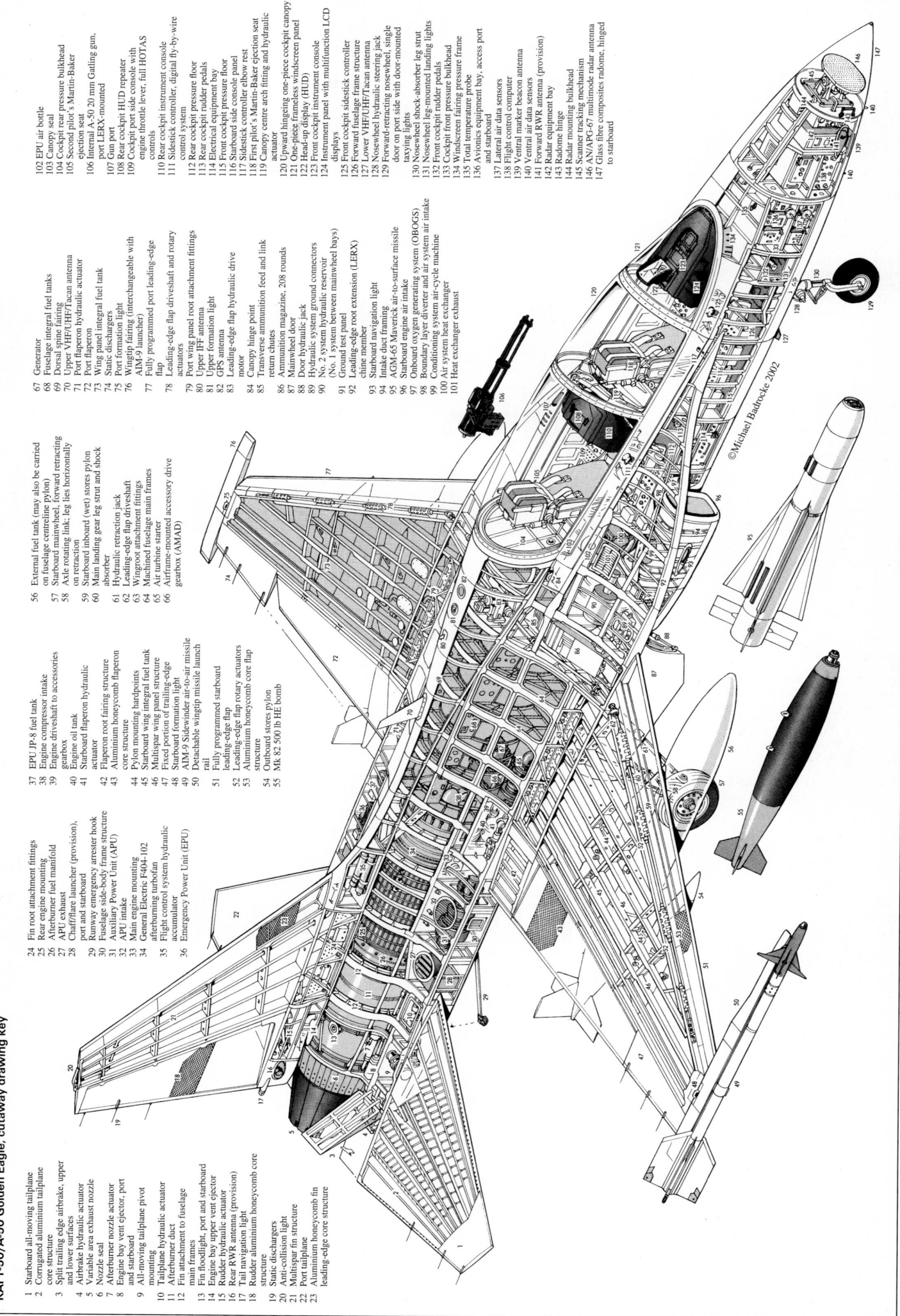

©Michael Badrocke 2002

POWER PLANT: One Pratt & Whitney F100-PW-229 turbofan.
AVIONICS: *Mission:* Lockheed Martin LANTIRN pods.
ARMAMENT: Includes AGM-84 Harpoon, AGM-88 HARM and AIM-120 AMRAAM missiles.

UPDATED

KAI KMH

TYPE: Multirole light helicopter (requirement).
PROGRAMME: Initiated as part of Republic of Korea Army (RoKA) Force Improvement Plan, with objective of replacing 500MDs, UH-1Hs and AH-1Ss with advanced type usable in combat, light attack, command and control, liaison and passenger-carrying roles. Confirmed by Korean government as acquisition programme in September 2001 and assigned FY02 budget for exploratory development; prime contractor selection decision and formal launch expected by end of 2002.
   KAI candidate displayed in model form at Asian Aerospace, February 2002, though no data offered; reportedly in 6,800 kg (15,000 lb) MTOW class, with twin engines. If selected, KAI will issue RFPs to potential domestic and foreign partners.
   Kamov Ka-60 proposed as late entrant in October 2001, in attempt to boost offset attraction of Ka-52K candidate for AH-X attack helicopter requirement. Bell reportedly considering offering a tiltrotor candidate, among other proposals.

UPDATED

KAI (BELL) SB 427

TYPE: Light utility helicopter.
PROGRAMME: KAI's Sacheon plant is local assembly centre for Bell 427s (see Canadian section) sold to customers in Southeast Asia. Canadian type certification obtained in November 1999; CAAC (China) type certificate April 2000; South Korean certificate for IFR operation in May 2000. KAI completion rate one per month until end of 2001, then scheduled for two per month thereafter. KAI is sole source of cabin and tailboom structures for Canadian production line. Licence agreement also provides for local manufacture of complete aircraft, but no evidence seen by mid-2002 of any Korean-built examples.
CURRENT VERSIONS: **SB 427:** Korean designation of standard Canadian-built Bell 427 (which see).
   SB 428: Military version proposed (then designated SB 427M) by Samsung in 1998 as candidate for Korean Multipurpose Helicopter (KMH) programme (which see) and for local area export. Fuselage-side weapon mountings, mast-mounted target acquisition and designation system (TV/FLIR), optional RWR, uprated transmission, increased rotor blade chord, strengthened skid gear, cockpit armour protection and higher operating weights.

Two-seat F-16D of Republic of Korea Air Force   *NEW*/0142059

Model of KAI projected KMH multirole helicopter, shown at Singapore in February 2002 (*Jane's/Paul Jackson*)   0131664

CUSTOMERS: First (Bell-built) 427 delivered to Broad Air Conditioning Company (China) in early 2000; second aircraft then due for delivery to Korea National Police in first quarter of 2001. Sales to China totalled 12 by December 2000, including five for Fei Ma Airlines, deliveries of which began in June 2001.
Following (provisional) details refer to SB 428:
POWER PLANT: As Bell 427, but transmission rating increased to approximately 708 kW (950 shp).

ACCOMMODATION: Armour protection for cockpit, cabin and engine cowling.
AVIONICS: Could include RWR and IRCM.
ARMAMENT: Up to four Hellfire or TOW anti-tank missiles, four air-to-air Stingers or 70 mm rocket pods on fuselage sidebars; underfuselage 12.7 mm machine gun pod.
WEIGHTS AND LOADINGS:
   Max T-O weight   3,175 kg (7,000 lb)

UPDATED

SKYGEAR

KOREAN LIGHT AIRCRAFT CORPORATION

1544 Bujeol-ri, San Dong-meon, Namwon Jeon Buk 590871
Tel: (+82 63) 626 98 88
Fax: (+82 63) 626 98 85
e-mail: master@skygear.co.kr
Web: http://www.skygear.co.kr
PRESIDENT: Hyung Jun Lee
VICE-PRESIDENT ENGINEERING: Paul Kenneth

US DISTRIBUTOR:
   Airdale Flyer Company, 4317 Aviation Way #9, Caldwell, Idaho 83607
   Tel: (+1 208) 459 62 54
   Web: http://www.airdale.com

Founded by Mr Lee, KLAC, which trades as Skygear, has a 1,300 m² (14,000 sq ft) plant, including digital design facilities, in Korea and is planning to establish a Canadian subsidiary. Its Comet is the first South Korean homebuilt to appear on the US market. The company also has a pusher design, the Skygear-2000, in the flight development stage and is also working on the Skygear-3000.

NEW ENTRY

SKYGEAR COMET

TYPE: Side-by-side ultralight kitbuilt.
PROGRAMME: Design started by Paul Kenneth in mid-1990s; made US debut at Sun 'n' Fun in 2001.
CURRENT VERSIONS: **Comet 21:** Baseline version, *as described.* Known as **Comet SP** in USA and features slight variations from Korean version.
   Comet UL: Lighter empty weight version, powered by Rotax 582 engine.
CUSTOMERS: Total of 35 sold by early 2002 to customers in Canada, Italy, Japan and USA.
DESIGN FEATURES: Braced high-wing monoplane, similar in appearance to Cessna 152. Available ready to fly or in kit form; quoted build time 300 hours.
FLYING CONTROLS: Conventional and manual. Frise ailerons; elevator electric trim. Fowler flaps.

STRUCTURE: Welded 4130 chromoly steel tube fuselage inside CFRP shell; pop-riveted, single-spar wing with drooped tips; all-CFRP tail unit. Single-strut bracing of wings and tailplane.
LANDING GEAR: Tricycle type; fixed. Rubber-sprung, steerable nosewheel; sprung steel main legs. Hydraulic brakes.
POWER PLANT: *Comet 21/Comet SP:* One 47.8 kW (64.1 hp) Rotax 582 UL or 59.6 kW (79.9 hp) Rotax 912 UL engine, driving two-blade propeller.
   Comet UL: Rotax 582 UL only.
   Suitable for other two- or four-stroke engines between 48.5 and 74.6 kW (65 and 100 hp) weighing no more than 79.5 kg (175 lb), including 59.7 kW (80 hp) Jabiru 2200.
   Standard usable fuel capacity (two tanks) 76 litres (20.0 US gallons; 16.7 Imp gallons).
ACCOMMODATION: Pilot and passenger. Forward-opening window/door each side; one-piece Plexiglas windscreen.
AVIONICS: VFR or IFR instrumentation to customer's choice.
DIMENSIONS, EXTERNAL:
   Wing span   8.69 m (28 ft 6 in)
   Length overall   5.79 m (19 ft 0 in)
   Height overall   2.44 m (8 ft 0 in)
DIMENSIONS, INTERNAL:
   Cabin max width   1.12 m (3 ft 8 in)
AREAS:
   Wings, gross   10.59 m² (114.0 sq ft)

US Comet SP demonstrator at Sun 'n' Fun, 2001 (*Jane's/Susan Bushell*)   *NEW*/0137948

WEIGHTS AND LOADINGS (A: Comet 21/SP, B: Comet UL):
   Weight empty: A   218 kg (480 lb)
      B   236 kg (520 lb)
   Max T-O weight: A, B   558 kg (1,232 lb)
   Max wing loading: A, B   52.8 kg/m² (10.81 lb/sq ft)
   Max power loading
      A and B, Rotax 582   11.70 kg/kW (19.22 lb/hp)
      A, Rotax 912   9.38 kg/kW (15.42 lb/hp)
PERFORMANCE (at 499 kg; 1,100 lb AUW, A/Rotax 912 and B as above):
   Never-exceed speed (VNE):
      A, B   117 kt (217 km/h; 135 mph)
   Max cruising speed at 215 m (700 ft) at 75% power:
      A   104 kt (193 km/h; 120 mph)
      B   96 kt (177 km/h; 110 mph)
   Stalling speed:
      flaps up: A, B   40 kt (73 km/h; 45 mph) CAS
      flaps down: A, B   34 kt (63 km/h; 39 mph) CAS
   Max rate of climb at S/L: A   366 m (1,200 ft)/min
      B   283 m (930 ft)/min
   Service ceiling: A   3,960 m (13,000 ft)
   T-O run: A   76 m (250 ft)
      B   92 m (300 ft)
   Landing run: A, B   54 m (175 ft)

NEW ENTRY

MALAYSIA

CTRM

COMPOSITES TECHNOLOGY RESEARCH MALAYSIA SDN BHD

Suite 19-14-3, Level 14, UOA Centre, 19 Jalang Pinang, 50450 Kuala Lumpur
Tel: (+60 3) 21 64 50 99
Fax: (+60 3) 21 64 50 93
e-mail: gmbd@ctrm.com.my
Web: http://www.ctrm.com.my
WORKS: Batu Berendam, Malacca
CEO: Dr Chen Lip Keong
COO: George Blower

Created in 1991, CTRM became an investor in Eagle Aircraft of Australia, via its subsidiary Eagle Aircraft (M) Sdn Bhd; component production for the Australian company's two-seat Eagle 150B began in 1997, subsequently progressing to full local manufacture. This is undertaken in a 20,903 m² (225,000 sq ft) facility at Malacca by CTRM's wholly owned subsidiary Aero-Composites Technologies Sdn Bhd (ACT). All Eagle production was transferred to Malaysia following closure of the Australian factory at the end of February 2002. ACT is currently also manufacturing composites components for the Lancair Columbia 300 (which see). It also is under contract to BAE Systems to produce composites wing leading- and trailing-edge panels for the Airbus A300 and A318/319/320/321.

CTRM-built Eagle 150 ARV *(Jane's/Kenneth Munson)* 0126944

CTRM's associate company Excelnet Sdn Bhd offers engineering solutions ranging from conceptual design to digital mockups and strength/stress analysis. It has recently been involved, with CTRM and BAE Systems, in developing the optionally piloted Eagle 150 Airborne Reconnaissance Vehicle (ARV), details of which are given in *Jane's Unmanned Aerial Vehicles and Targets*. Other current work includes wing life extension (spar modification) for the Scottish Aviation Bulldog, for which it has worldwide type certificate maintenance. **UPDATED**

FLYING POWER

FLYING POWER SDN BHD

No 7-1 Jalan 3/27B, Bandar Baru Wangsa Maju, 53300 Kuala Lumpur
Tel: (+60 3) 40 24 92 54
Fax: (+60 3) 41 41 12 61

e-mail: econwira@tn.net.my
CHAIRMAN AND CEO: Raja Sulaiman bin Raja Shariff

Flying Power announced during the LIMA Air Show in Langkawi in October 2001 that it intended to acquire the manufacturing rights to the Seawolf amphibian, plus marketing and sales rights for the aircraft in Australia, Brunei, Cambodia, China, Indonesia, Laos, Malaysia, Myanmar, New Zealand, Papua New Guinea, Philippines, Singapore, South Africa, Thailand, United Arab Emirates and Vietnam. Nothing further has been reported, and it is understood that the venture has lapsed. **UPDATED**

SME

SME AVIATION SDN BHD

Lot 14101, Pangkalan TUDM Subang, PO Box 7574, 40718 Shah Alam, Selangor Darul Ehsan
Tel: (+60 3) 746 85 77
Fax: (+60 3) 746 85 66
e-mail: smeav@po.jaring.my
Web: http://www.smeav.com.my
CEO: Lt Col (Retd) Mohamed Tarmizi Ahmed
EXECUTIVE DIRECTOR: Tommy Tay
GENERAL MANAGER: Prabhakaran Nair

US MARKETING:
SME Aero Inc
The Tallahassee Aerospace Center, 3240 Capital Circle South-West, Tallahassee, Florida 32310
Tel: (+1 850) 575 90 02
Fax: (+1 850) 575 43 54
CHAIRMAN: Richard Ledson

SME Aviation Sdn Bhd, established in September 1993, is one of four divisions of SME Technologies Sdn Bhd, a group wholly owned by the Malaysian government which provides products and services for defence, aerospace, plastics and metal-based manufacturing industries. It manufactured the SME MD3-160 AeroTiga trainer with initial technical support from the aircraft's designer, MDB Flugtechnik of Switzerland, and British Aerospace. Subcontract work includes Hawk wing pylons; fuselage components for the Alenia/Lockheed Martin C-27J; and, for Ulan-Ude Aviation Plant, parts manufacture for the Mi-171 helicopter **UPDATED**

SME MD3-160 AEROTIGA

Production completed. See 2002-03 *Jane's* for description and illustrations. **UPDATED**

NETHERLANDS

EURO-ENAER

EURO-ENAER HOLDING BV

This company was created in 1997 as a joint venture with ENAER of Chile (which see), to gain JAA and FAA certification for the latter company's Namcu all-composites two-seat trainer/tourer; design rights were acquired by Euro-ENAER in 1997 and it planned to assemble the aircraft in its 700 m² (7,525 sq ft) factory at den Helder. Production was delayed by certification problems, and although these were resolved in July 2001, the company was forced to request a €1.2 million loan in October 2001. This was not forthcoming, and bankruptcy was announced in January 2002. **UPDATED**

EURO-ENAER EE 10 EAGLET

This aerobatic two-seat lightplane was last described in the 2002-03 edition. **UPDATED**

SSVOBB

STICHTING STUDENTEN VLIEGTUIGONTWIKKELING -BOUW EN -BEHEER (Foundation for Student Aircraft Development, Manufacturing and Operation)

Kluyverweg 1, NL-2629 HS Delft
Tel: (+31 15) 278 10 57
Fax: (+31 15) 278 12 43
e-mail: ssvobb@lr.tudelft.nl
Web: http://www.delftaerospace.com/ssvobb
CHAIRMAN: Raf Segers
SECRETARY: Maaik Borst
TREASURER: Hugo Vergnes
PUBLIC RELATIONS MANAGER: Karolien Pas

Impuls project
PRODUCTION CO-ORDINATOR: Tom Dillerop
ENGINEERING CO-ORDINATOR: Alexander Vastenavond

SSVOBB was formed in 1990 by The Society of Aerospace Students 'Leonardo da Vinci' of the Delft University of Technology, Faculty of Aerospace Engineering. Its main objective is to co-ordinate the design, manufacture and maintenance of aircraft by students. A replica of the 1937 Lambach HL II aerobatic biplane was constructed between 1989 and 1995, and work is proceeding on an original lightplane design, the Impuls. The projects are financed by funds and sponsorship. **UPDATED**

SSVOBB IMPULS
English name: Impulse
TYPE: Two-seat lightplane.
PROGRAMME: Preliminary design started early 1994. Registration PH-VXM reserved March 1997. Main fuselage moulds completed early 1999. Main wings due to be completed by mid-2001. Certification to JAR 23 being undertaken under supervision of Prof dr ir Theo van Holten (flight performance); Prof dr ir J A Mulder (stability/control), ir F van Dalen (construction); Prof ir R J Zwaan (aero-elasticity); ing K van Woerkom (systems); ir van Badegom (fuel and propulsion systems).
DESIGN FEATURES: Twin-boom layout with pod-type fuselage and high wing of parallel chord; swept fins.
FLYING CONTROLS: Conventional and manual. Actuation by pushrods (ailerons and elevator), with cable and pushrod interconnected twin rudder and nosewheel system. Servo-actuated electric pitch trim and electrically operated single-slotted flaps.
STRUCTURE: All-composites wing with GFRP sandwich skin and spars; GFRP tail unit. Some GFRP parts will be produced by resin transfer moulding process. Fuselage steel tube frame with glass fibre shell will be used to finalise design of cockpit and control systems.
Wing section NACA 63₃415 with RA163CW3 flaps; twin tailbooms joined by NACA 0012 section horizontal stabiliser.

SSVOBB's first project was this replica Lambach HL II, first flown on 18 September 1995 0097640

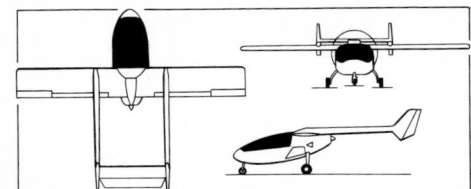

SSVOBB Impuls (MWAE rotary engine)
(Jane's/Paul Jackson) 0051122

Model of the SSVOBB Impuls two-seat light aircraft undergoing wind tunnel test 0097639

LANDING GEAR: Non-retractable tricycle type; GFRP cantilever sprung mainwheels. Non-steerable nosewheel; mainwheels have 45 cm (17¾ in) tyres pressurised to 3.14 bar (45.5 lb/sq in). Hydraulically operated brakes on mainwheels.

POWER PLANT: One 74.6 kW (100.hp) Mid-West AE110R rotary engine driving a three-blade fixed-pitch propeller. Fuel capacity 65 litres (17.2 US gallons; 14.3 Imp gallons) in single tank located between cockpit and engine. Oil capacity 3.0 litres (0.8 US gallon; 0.7 Imp gallon).

ACCOMMODATION: Side-by-side seating for pilot and passenger under one-piece, Perspex canopy. Baggage space behind seats.

DIMENSIONS, EXTERNAL:

Wing span	7.75 m (25 ft 5 in)
Wing chord, constant	1.03 m (3 ft 4½ in)
Wing aspect ratio	7.5
Length overall	6.37 m (20 ft 10¾ in)
Fuselage: Length	3.32 m (10 ft 10¾ in)
Max width	1.20 m (3 ft 11¼ in)

Height overall	2.15 m (7 ft 0¾ in)
Tailplane span	2.10 m (6 ft 10¾ in)
Wheel track	2.20 m (7 ft 2½ in)
Wheelbase	1.75 m (5 ft 9 in)
Propeller diameter	1.50 m (4 ft 11 in)

AREAS:

Wings, gross	8.00 m² (86.1 sq ft)
Ailerons (total)	0.66 m² (7.10 sq ft)
Flaps (total)	1.06 m² (11.41 sq ft)
Fins (total)	2.00 m² (21.53 sq ft)
Rudders (total)	0.24 m² (2.58 sq ft)
Tailplane	1.51 m² (16.25 sq ft)
Elevators (total)	0.62 m² (6.67 sq ft)

WEIGHTS AND LOADINGS (estimated):

Weight empty	375 kg (827 lb)
Baggage capacity	30 kg (66 lb)
Max T-O weight	575 kg (1,267 lb)
Max wing loading	71.9 kg/m² (14.72 lb/sq ft)
Max power loading	7.71 kg/kW (12.67 lb/hp)

PERFORMANCE (estimated):

Max level speed at S/L	124 kt (230 km/h; 142 mph)
Max cruising speed	113 kt (210 km/h; 130 mph)
Stalling speed	52 kt (96 km/h; 60 mph)
Max rate of climb at S/L	246 m (807 ft)/min
T-O to 15 m (50 ft)	340 m (1,115 ft)
Landing from 15 m (50 ft)	430 m (1,410 ft)
Range with max fuel	
	more than 378 n miles (700 km; 435 miles)
Endurance	5 h
g limits	+4.4/−2.2

UPDATED

STORK

STORK AEROSPACE

PO Box 75047, NL-1117 ZN Schiphol
Tel: (+31 20) 605 48 31
Fax: (+31 20) 605 48 30
e-mail: info.aerospaceindustries@stork.com
Web: http://www2.stork.com

OPERATING COMPANIES:

Fokker Services BV, PO Box 3, NL-4630 AA Hoogerheide (aircraft maintenance, repair and overhaul; holder of Fokker aircraft type certificates)

Fokker Elmo BV, PO Box 75, NL-4630 AB Hoogerheide (electrical/electronic systems and components for aviation industry)

Fokker Special Products BV, PO Box 59, NL-7900 AB Hoogeveen (industrial products for civil and military use)

OTHER BUSINESS UNITS:

Fokker Product Support, Schiphol (sale of spares; technical support of Fokker aircraft)

Fokker Aerostructures BV, PO Box 1, NL-3350 AA Papendrecht (manufacture of parts, structural and non-structural lightweight components and subassemblies for aerospace industry)

Failure of several bids to resuscitate the bankrupt aircraft manufacturing arm of Fokker resulted in the last jet airliner, Fokker 70 PH-KZK of Cityhopper, being delivered on 18 April 1997 (the final Fokker 100 preceded it as PT-MRW of TAM Brasil on 21 March 1996); and the final Fokker

aircraft of all, Fokker 50 ET-AKS, departing Schiphol for Ethiopian Airlines on 23 May 1997. Recent history of the company and descriptions of the Fokker 50, 60, 70 and 100 last appeared in the 1997-98 edition. Details of proposed production restarts by Rekkoff Restart and Forward Aircraft were detailed in the 2002-03 *Jane's*.

In late 2000, the Dutch company RDM, parent company of MD Helicopters, proposed reviving the Fokker 70 and 100 by building the aircraft in a country with low production costs, such as China; no further news has been received.

In addition to its aerospace activities, Stork is also involved in textiles, printing, poultry, food processing, industrial components and technical services.

UPDATED

NEW ZEALAND

PAC

PACIFIC AEROSPACE CORPORATION LIMITED

Private Bag HN 3027, Hamilton Airport, Hamilton
Tel: (+64 7) 843 61 44
Fax: (+64 7) 843 61 34
e-mail: pacific@aerospace.co.nz
Web: http://www.aerospace.co.nz
GENERAL MANAGER: Graeme Polley
MANAGING DIRECTOR: Brian Hare
PURCHASING MANAGER: Graeme York

Pacific Aerospace Corporation formed 1982 following acquisition of assets and undertakings of New Zealand Aerospace Industries; became wholly owned subsidiary of AeroSpace Technologies of Australia (75.1 per cent) and Lockheed Martin, USA (24.9 per cent). In October 1995, Aeromotive, a privately owned New Zealand company, purchased ASTA's 75.1 per cent share. PAC maintains production and support facilities for its own aircraft, the CT4 Airtrainer series, Fletcher FU24 series, Cresco 08-600 and 08-750, and PAC 750XL; by April 2002 some 560 aircraft had been produced. Manufacturing facility also produces items for the Boeing 747/777, Airbus A330/340 and F/A-18 Hornet; 100 employed in the company's 24,155 m² (260,000 sq ft) facility.

UPDATED

PAC CRESCO and 750XL

TYPE: Agricultural sprayer; light utility turboprop.

PROGRAMME: Design began 1977; first flight of prototype (ZK-LTP) 28 February 1979; first flight of production aircraft early 1980; entered service January 1982 as Cresco 08-600 powered by 447 kW (599 shp) LTP101-700-1A turboprop. This variant was augmented by Cresco 08-750 (PT6A-34AG version) in November 1992. First flight of PT6A-34AG version (ZK-TMN, c/n 010) 18 November 1992. Work on a wide-body development started in 2000.

Pacific Aerospace Corporation Cresco agricultural aircraft *NEW*/0137422

CURRENT VERSIONS: **Cresco 08-600:** Initial version, with 447 kW (599 shp) (flat rated) Honeywell LTP 101-700A-1A turboprop. No longer available.

Cresco 08-750: With higher-powered 559 kW (750 shp) Pratt & Whitney Canada PT6A-34AG engine; launched 1992 (first delivery 23 December) as 08-600-34AG.

PAC 750XL: Wide-body version with fuselage height overall raised 15 cm (6 in); design started January 2000 and prototype (ZK-XLA) first flown 5 September 2001. Passenger compartment 1.40 m (4 ft 7 in) wide, 3.99 m (13 ft 1 in) long, 1.42 m (4 ft 8 in) high aft of pilot; length 11.58 m (38 ft 0 in) remains same. 1.27 × 1.12 m (4 ft 2 in × 3 ft 8 in) cargo door. Optimised for parachute jumping; able to carry 17 jumpers to 4,265 m (14,000 ft) in less than 13 minutes. Other fields of use include special operations.

Total 39.2 hours in 63 sorties up to February 2002 start of certification trials. Preliminary tests showed need for leading-edge root extension and increase of maximum flap angle to 40°.

CUSTOMERS: Total 30 Crescos built by March 2002, of which five delivered in 1997, three in 1998 and three in 1999. One delivered to Kevron Pty Ltd in Australia during 1998 is used for geophysical exploration. Of 1999 deliveries, two were built for parachuting, second for NSW Skydive of Australia. Current Cresco operators include New Zealand (seven companies), Australia (two companies), Malaysia and Bangladesh. First 10 750XLs sold to US skydiving company; 750XL offered to Australian Army.

COSTS: 750XL approximately US$900,000 (2002).

DESIGN FEATURES: Turboprop development of earlier FU24 with approximately 60 per cent commonality of parts. Primary configuration as agricultural aircraft with appropriate hopper base; can also be used for firefighting.

Wing section NACA 4415 (constant); 8° dihedral on outer wing panels; incidence 2°.

FLYING CONTROLS: Conventional and manual. Horn-balanced ailerons, horn-/mass-balanced elevator and rudder; electric elevator trim with manual override; all-moving tailplane with full-span anti-servo tab; ground-adjustable tab on rudder; single-slotted flaps; electrically operated through

Prototype Pacific Aerospace Corporation 750XL prior to addition of LERX   0131831

to 30° (40° on 750XL); row of vortex generators forward of each aileron.

STRUCTURE: Conventional light alloy; two-spar wing; cockpit area stressed for 25 g impact.

LANDING GEAR: Non-retractable tricycle type; steerable nosewheel. Oleo-pneumatic shock absorption. Cleveland wheels and hydraulic disc brakes on main units. Goodyear tyres, size 8.50-10 (10 ply) on mainwheels, pressure 2.62 bar (38 lb/sq in), nose/tailwheel 8.5-6 (6 ply), pressure 2.07 bar (30 lb/sq in).

POWER PLANT: See Current Versions. Hartzell HC-B3TN-3D/T10282NS+4 three-blade constant-speed, fully feathering, reversible-pitch metal propeller. Four integral fuel tanks in wing centre-section, total capacity 500 litres (132 US gallons; 110 Imp gallons) on 08-750 and 927 litres (245 US gallons; 204 Imp gallons) on 750XL; ferry configuration capacity 2,400 litres (634 US gallons; 528 Imp gallons) using hopper as auxiliary fuel tank. Four refuelling points in upper surface of each wing. Oil capacity 5.7 litres (1.5 US gallons; 1.25 Imp gallons).

ACCOMMODATION: Cresco versions have two-seat cockpit with sliding canopy; rear compartment, with port freight door, holds six passengers, equivalent freight or chemical hopper, 750XL cabin carries nine passengers at 86 cm (34 in) pitch or 17 parachutists and has 16 floor-mounted tiedown points; cabin door on port side behind wing, plus two upward-hinged doors on each side of cockpit for crew entry.

SYSTEMS: 24 V DC electric system. Air conditioning optional on 750XL.

EQUIPMENT: 1,893 litre (500 US gallon; 416 Imp gallon) liquid or 1,860 kg (4,100 lb) dry hopper on Cresco versions.

DIMENSIONS, EXTERNAL:
Wing span	12.81 m (42 ft 0 in)
Wing chord, constant	2.13 m (7 ft 0 in)
Wing aspect ratio	6.0
Length overall: Cresco	11.07 m (36 ft 4 in)
750XL	11.46 m (37 ft 7 in)
Fuselage:	
Max width: Cresco	1.22 m (4 ft 0 in)
750XL	1.52 m (5 ft 0 in)
Max height: 750XL	1.57 m (5 ft 2 in)
Height overall	3.84 m (12 ft 7 in)
Tailplane span	4.93 m (16 ft 2 in)
Wheelbase: Cresco	2.77 m (9 ft 1¼ in)
750XL	3.15 m (10 ft 4 in)
Wheel track	3.76 m (12 ft 4 in)
Propeller diameter	2.69 m (8 ft 10 in)
Propeller ground clearance: Cresco	0.305 m (1 ft 0 in)
750XL	0.41 m (1 ft 4 in)
Cargo door (port): Height	0.94 m (3 ft 1 in)
Width	0.94 m (3 ft 1 in)
Height to sill	0.91 m (3 ft 0 in)

DIMENSIONS, INTERNAL:
Passenger/cargo compartment (aft of hopper):
Length	3.12 m (10 ft 3 in)
Max width	1.09 m (3 ft 7 in)
Max height	1.24 m (4 ft 1 in)
Floor area	2.79 m² (30.0 sq ft)
Volume: passenger version	3.4 m³ (120 cu ft)
cargo version	3.8 m³ (134 cu ft)
Hopper volume	1.86 m³ (66 cu ft)

AREAS:
Wings, gross	27.31 m² (294.0 sq ft)
Ailerons (total)	1.97 m² (21.18 sq ft)
Trailing-edge flaps (total)	2.92 m² (31.42 sq ft)
Fin	1.53 m² (16.47 sq ft)
Rudder	0.63 m² (6.77 sq ft)
Tailplane (excluding elevators)	3.13 m² (33.64 sq ft)
Elevator, incl tab	2.59 m² (27.92 sq ft)

WEIGHTS AND LOADINGS:
Weight empty, equipped: 600	1,270 kg (2,800 lb)
750	1,338 kg (2,950 lb)
750XL	1,406 kg (3,100 lb)
Max fuel: 600, 750	435 kg (960 lb)
750XL	739 kg (1,630 lb)
Typical payload: 750, Normal	1,111 kg (2,450 lb)
750, Agricultural (Restricted)	2,155 kg (4,750 lb)
750XL	1,905 kg (4,200 lb)
Max disposable load (fuel + hopper):	
600, Normal	1,578 kg (3,480 lb)
600, Agricultural (Restricted)	1,828 kg (4,030 lb)
750, Normal	1,524 kg (3,360 lb)
750, Agricultural (Restricted)	2,429 kg (5,356 lb)
Max T-O weight: 600, Normal	2,925 kg (6,450 lb)
600, Agricultural (Restricted)	3,175 kg (7,000 lb)
750, Normal	2,925 kg (6,450 lb)
750, Agricultural (Restricted)	3,742 kg (8,250 lb)
750XL	3,402 kg (7,500 lb)
Max landing weight: 600, 750	2,925 kg (6,450 lb)
750XL	3,231 kg (7,125 lb)
Wing loading at Normal max T-O weight:	
600, 750	107.1 kg/m² (21.94 lb/sq ft)
750XL	124.6 kg/m² (25.51 lb/sq ft)

Wing loading at Agricultural (Restricted) max T-O weight:
600	116.3 kg/m² (23.81 lb/sq ft)
750	137.1 kg/m² (28.08 lb/sq ft)
Power loading at Normal max T-O weight:	
600	6.55 kg/kW (10.77 lb/shp)
750	5.23 kg/kW (8.60 lb/shp)
750XL	6.09 kg/kW (10.00 lb/shp)
Power loading at Agricultural (Restricted) max T-O weight:	
600	7.11 kg/kW (11.69 lb/shp)
750	6.70 kg/kW (11.01 lb/shp)

PERFORMANCE (at max Normal T-O weight, except where indicated):
Never-exceed speed (VNE):	
600, 750	173 kt (320 km/h; 199 mph)
750XL	176 kt (326 km/h; 202 mph)
Max level speed at S/L:	
600	148 kt (274 km/h; 170 mph)
750	152 kt (282 km/h; 175 mph)
750XL	155 kt (287 km/h; 178 mph)
Max cruising speed at 75% power at 305 m (1,000 ft):	
600	133 kt (246 km/h; 153 mph)
750	140 kt (259 km/h; 161 mph)
Stalling speed, flaps down, power off:	
600 at 2,767 kg (6,100 lb) AUW	52 kt (97 km/h; 60 mph)
750	57 kt (106 km/h; 66 mph)
750XL at 3,402 kg (7,500 lb) AUW	60 kt (112 km/h; 70 mph)
Max rate of climb at S/L: 600	379 m (1,245 ft)/min
750	475 m (1,560 ft)/min
750XL	518 m (1,700 ft)/min
Absolute ceiling: 600	5,485 m (18,000 ft)
750	8,230 m (27,000 ft)
750XL	9,140 m (30,000 ft)
T-O run: 600	323 m (1,060 ft)
750	227 m (745 ft)
T-O run at 1,406 kg (3,100 lb) AUW:	
750	45 m (150 ft)
T-O to 15 m (50 ft): 600	436 m (1,430 ft)
750	325 m (1,065 ft)
Landing from 15 m (50 ft): 600	500 m (1,640 ft)
750	427 m (1,400 ft)
Landing run at 1,406 kg (3,100 lb) AUW with propeller pitch reversal: 750	86 m (285 ft)

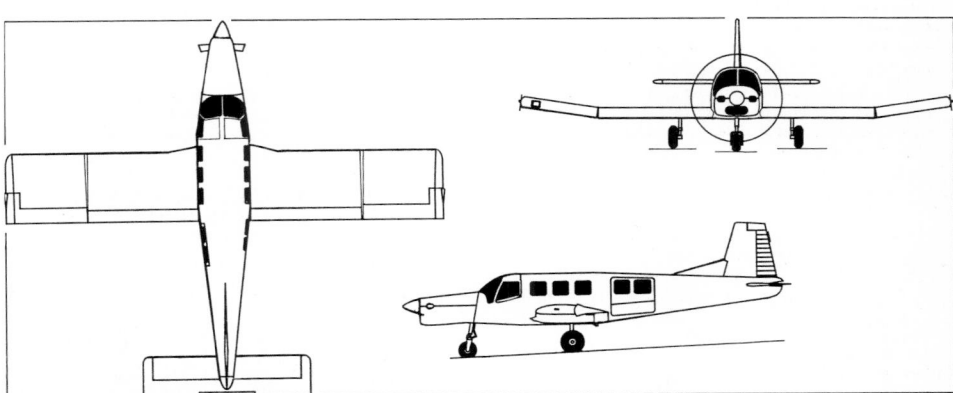

Pacific Aerospace Corporation 750XL general arrangement   0131830

Production Cresco 750 VH-MOS, operated for parachuting by Alan Moss of Mile High P/L from Barwon Heads, Victoria, Australia (*Jane's/Paul Jackson*)   *NEW*/0530186

Range at 75% power with standard fuel, no reserves:
600 at 3,175 kg (7,000 lb) MTOW
        460 n miles (852 km; 529 miles)
750   305 n miles (676 km; 420 miles)
750XL   592 n miles (1,097 km; 682 miles)
Range as above with fuel in hopper:
750   2,037 n miles (3,774 km; 2,345 miles)
Endurance with standard fuel:
750 at 75% power   2 h 40 min
750 at 60% power   3 h 0 min
750XL at 75% power   4 h 16 min
UPDATED

PAC AIRTRAINER CT4E

Royal Thai Air Force designation: BF 16

TYPE: Primary prop trainer/sportplane.
PROGRAMME: New Zealand redesign of Australian Victa Aircruiser; first flight 23 February 1972; deliveries began October 1973; 94 (plus two prototypes) built before production ended 1977 (75 CT4A and 19 special military variants); production line reopened 1991 to build 12 civil CT4Bs for Ansett Flying College and six for Thailand; last completed August 1992. Piston-engined CT4A and B last described fully in 1994-95 *Jane's*. First flight of CT4C turboprop prototype (ZK-FXM, converted RAAF CT4B) 21 January 1991.
CURRENT VERSIONS: **T350:** Turboprop version (previously known as CT4C) with 313 kW (420 shp) Rolls-Royce 250-B17D (throttle limited to 261 kW; 350 shp) in lengthened nose. See 1994-95 *Jane's* for description. Certification was expected 1998 in anticipation of Southeast Asian order for 30, but this failed to materialise. Some interest still being shown in this version.
 CT4E: Developed version of CT4B with more powerful engine; wing mounted slightly farther forward than on CT4B; first flight (ZK-EUN, converted from RAAF CT4A) 14 December 1991; NZ certification (FAR Pt 23 Amendment 36) 8 May 1992.
Detailed description applies to CT4E.
CUSTOMERS: PAC previously built 75 CT4As and 38 CT4Bs; further production of CT4E began in 1997, of which 13 supplied on 20 year lease to the Royal New Zealand Air Force as replacements for CT4Bs; contract signed 18 August 1998; first (NZ1985) delivered to Pilot Training Squadron at Ohakea in same month; last (NZ1997) on 8 June 1999; by late 2000 the fleet had accumulated 10,000 flying hours. Also used by Red Checkers aerobatic team. Thailand has a requirement for 24 CT4Es, initial order being for 12, deliveries of which began in July 1999 and were completed in June 2000 (final Thai aircraft first flew 11 May 2000). Total 26 built up to May 2000 temporary suspension of manufacture, one of which is demonstrator. Production for Thai Air Force being undertaken during 2001, four of second batch delivered by February 2002. Further production reportedly for Singapore.
DESIGN FEATURES: Metal translation of 1953-vintage wooden tourer. Conventional, low-wing lightplane. Tapered wings and constant-chord tailplane. Detachable wingtips to allow fitting of optional wingtip fuel tanks.
 Wing section NACA 23012 (modified) at root, NACA 4412 (modified) at top; dihedral 6° 45′ at chord line; incidence 3° at root; twist 3°; root chord increased by forward sweep of inboard leading-edges; small fence on outer leading-edges.

PAC CT4E Airtrainer of Royal Thai Air Force
                NEW/0137423

FLYING CONTROLS: Conventional and manual. Aerodynamically and mass-balanced bottom-hinged ailerons and statically balanced one-piece elevator actuated by pushrods; rudder by rod and cable linkage; ground-adjustable tab on rudder, electric trim control for elevator and rudder; electrically actuated single-slotted flaps.
STRUCTURE: Light alloy stressed skin except for Kevlar/GFRP wingtips and engine cowling; ailerons and flaps have fluted skins.
LANDING GEAR: Non-retractable tricycle type, with cantilever spring steel main legs; steerable (±25°) nosewheel carried on telescopic strut and oleo shock-absorber. Main units have Dunlop Australia wheels and tubeless tyres size 15×6.00, pressure 1.65 bar (24 lb/sq in); nosewheel has tubeless tyre size 14×4.5, pressure 1.24 bar (18 lb/sq in). Single-disc toe-operated hydraulic brakes, with hand-operated parking lock. Minimum ground turning radius 2.90 m (9 ft 6 in). Landing gear designed to shear before any excess impact loading is transmitted to wing, to minimise structural damage in event of crash landing.
POWER PLANT: One 224 kW (300 hp) Textron Lycoming AEIO-540-L1B5 flat-six engine with inverted oil system; three-blade Hartzell HC-C3YF-4BF/FC7663-2R constant-speed metal propeller. Standard fuel capacity of 204.4 litres (54.0 US gallons; 45.0 Imp gallons) of which 200 litres (52.8 US gallons; 44.0 Imp gallons) are usable; wingtip tanks, each of 77.6 litres (20.5 US gallons; 17.1 Imp gallons) capacity, available optionally. Gravity fuelling point in each wing. Oil capacity 11.4 litres (3.0 US gallons; 2.5 Imp gallons).
ACCOMMODATION: Two seats side by side under hinged, fully transparent jettisonable Perspex canopy. Space to rear for optional third seat or 77 kg (170 lb) of baggage or equipment. Dual controls standard.
SYSTEMS: 28 V DC electrical system; ground power receptacle.
DIMENSIONS, EXTERNAL:
Wing span   7.92 m (26 ft 0 in)
Wing chord: at root   2.18 m (7 ft 2 in)
 at tip   0.94 m (3 ft 1 in)

Wing aspect ratio   5.2
Length overall   7.16 m (23 ft 5¾ in)
Height overall   2.59 m (8 ft 6 in)
Fuselage: Max width   1.12 m (3 ft 8 in)
 Max depth   1.40 m (4 ft 7¼ in)
Tailplane span   3.61 m (11 ft 10 in)
Wheel track   2.97 m (9 ft 9 in)
Wheelbase   1.66 m (5 ft 5½ in)
DIMENSIONS, INTERNAL:
Cabin: Length   2.74 m (9 ft 0 in)
 Max width   1.08 m (3 ft 6½ in)
 Max height   1.34 m (4 ft 5 in)
AREAS:
Wings, gross   11.98 m² (129.0 sq ft)
Ailerons (total)   1.07 m² (11.56 sq ft)
Flaps (total)   2.10 m² (22.60 sq ft)
Fin   0.78 m² (8.44 sq ft)
Rudder, incl tab   0.58 m² (6.26 sq ft)
Tailplane   1.51 m² (16.24 sq ft)
Elevator   1.56 m² (16.74 sq ft)
WEIGHTS AND LOADINGS:
Weight empty, equipped   807 kg (1,780 lb)
Max fuel weight   149 kg (328 lb)
Max T-O weight   1,179 kg (2,600 lb)
Max wing loading 98.38 kg/m² (20.15 lb/sq ft)
Max power loading   5.26 kg/kW (8.67 lb/hp)
PERFORMANCE:
Never-exceed speed (V$_{NE}$) 209 kt (387 km/h; 240 mph)
Max level speed at S/L 163 kt (302 km/h; 188 mph)
Cruising speed at 2,590 m (8,500 ft) at 75% power
         152 kt (282 km/h; 175 mph)
Stalling speed at S/L: flaps up 60 kt (112 km/h; 69 mph)
 flaps down   44 kt (82 km/h; 51 mph)
Max rate of climb at S/L, ISA 558 m (1,830 ft)/min
Service ceiling   5,550 m (18,200 ft)
Time to service ceiling   13 min
T-O run at S/L, ISA   187 m (612 ft)
Landing from 15 m (50 ft)   244 m (800 ft)
Landing run   169 m (553 ft)
Range at S/L with max fuel at 75% power, ISA, no
 reserves   520 n miles (963 km; 599 miles)
UPDATED

PAKISTAN

PAC

PAKISTAN AERONAUTICAL COMPLEX

Kamra, District Attock
WORKS: F-6 Rebuild Factory; Mirage Rebuild Factory; Kamra Avionics and Radar Factory; Aircraft Manufacturing Factory (all at Kamra)
Tel: (+92 51) 927 01 11 and 927 01 17
Fax: (+92 51) 927 01 07 and 927 01 00
e-mail: info@pac.org.pk
Web: http://www.pac.org.pk
CHAIRMAN: Air Marshal Pervez A Nawaz
MANAGING DIRECTORS:
 Air Cdre S Sohail H Sadiq (AMF)
 Air Cdre Iftikhar A Gul (F-6RF)
 Air Cdre Basit Raza Abbasi (MRF)
 Air Cdre Saleem Akbar (KARF)
SALES AND MARKETING DIRECTOR:
 Air Cdre Raja Tariq Mahmood
CHIEF TEST PILOT: Wg Cdr Ahmed Faiq

Pakistan Aeronautical Complex (PAC) is the nucleus of the aeronautical industry in Pakistan. Its professional standards and reliability enable domestic customers and foreign agencies alike to undertake joint production with PAC in various fields. PAC Kamra comprises four factories: the Mirage Rebuild Factory (MRF), F-6 Rebuild Factory (F6-RF), Aircraft Manufacturing Factory (AMF) and Kamra Avionics and Radar Factory (KARF). All four are ISO 9000 certified. The main roles and capabilities of these factories are as follows:

Mirage Rebuild Factory (MRF) was established in 1978 to provide in-country maintenance for the Dassault Mirage weapon system. Main mission today is complete overhaul of Mirage III/V aircraft, Atar 09C engines, F100 engine modules (for F-16 aircraft) and all associated aircraft components and engine accessories. Equipped with modern machinery, test equipment, jigs and fixtures, the capabilities include: major structural repair of fuselage and accessories; wing refurbishment; No. 4 rib replacement; CNC Aerospace pipe manufacturing facility; overhaul of ejection seat; repair of 561 subassemblies and components; aerospace coatings on Mirage, F-16, T-37 and Falcon 20 aircraft; and repair of fuel cells. The engine overhaul facility includes general overhaul of Atar 09C engine and 65 major subassemblies; life core modification and overhaul of F100 engine (for F-16); electron beam welding, plasma spray, NDI and multi-engine test cell. Overhaul experience has been used in recovery of old Australian Mirages, avionics upgrade of Mirage III, and upgrade of F100-PW-200 engine jet fuel starter to F100-PW-200E configuration. The MRF has produced 840 engines and 400 aircraft, including more than 110 overhauls, mid-life painting and servicing, avionics upgrade and major electrical wiring replacement and repair. It has an effective quality control system to verify processes and improve product quality; ISO 9000 certification was achieved on 14 September 1995.

F-6 Rebuild Factory (F6-RF): Although originally established, in 1980, for the F-6 (Chinese version of the MiG-19), last of which was withdrawn from PAF service in 2002, the F-6 RF also currently overhauls Chinese origin F-6, FT-6, FT-5, A-5 III, F-7P (variant of MiG-21) and FT-7P aircraft. All the components and instruments of these aircraft are also rebuilt at the factory. Modern technical facilities for aircraft structural repairs and spares manufacture include casting, forging, heat treatment, surface treatment and various machining processes, both conventional and CNC (turning, milling, cutting, copying, welding and EDM die-making). F6-RF also manufactures jettisonable fuel tanks, hardware parts, compression/tension/torsion springs, aluminium and steel washers and clamps, standard harness and rubberised items. It manufactures its own aeronautical grade synthetic rubber and has a moulding and forming facility where a large variety of rubber parts of the aircraft are formed. In addition, the F6-RF has extensive and well-equipped laboratories and a testing facility to meet all aviation standard requirements, including a wide range of inspections ranging from X-rays to non-destructive inspection, geometrical standards (PME), tensile testing (probably the only such facility in the country), hardness testing, flow and pressure measuring, spectrophotometer analysis, chemical testing, rubber testing and salt spray testing.

Aircraft Manufacturing Factory (AMF): In operation since mid-1981, the AMF spearheads the aircraft manufacturing industry in Pakistan. It is the only aircraft manufacturing concern in Pakistan and is at present engaged in the manufacture of the light, robust, basic trainer-cum-surveillance aircraft Mushshak; the Super Mushshak – a variant with a more powerful engine, cockpit air conditioning, electric instruments and several other improvements; the Baaz and Ababeel low-speed target drones for anti-aircraft gunnery training; and development and manufacture, with NAMC of China, of the K-8 (Karakoram) jet trainer aircraft, a few of which are already flying with the Pakistan Air Force. In addition, the AMF has excellent facilities to rebuild and repair the Mushshak at factory level, along with after-sales spares support to Mushshak operators worldwide. AMF also has the skills and capabilities of conventional/CNC machining, sheet metal forming, hand layup, contact suction hot press moulding, stretch moulding, gravity moulding, reaction injection moulding, mould/die and pattern making in woodwork, surface treatment/painting, heat treatment, tig welding (AC and DC), electric arc welding, gas welding, bronzing, soldering, tube banding, material testing, spectroanalysis, tensile testing, precision measurement chemical analysis, CNS calibration, co-ordinate measurement, mould/dies and fixture designing in tooling facility, CNC copy milling and EDM (spark erosion) milling.

A four-seat trainer for flying club use is under development by the AMF.

Kamra Avionics Radar Factory (KARF): KARF, the fourth factory of PAC, came into operation in 1987. It is an electronics centre with proven capabilities for rebuilding of radars, control and reporting centres (CRC), generators, and manufacture and repair of avionics systems. Currently, it is producing ESM equipment and airborne radar systems. All production activities are carried out in large, spaciously built shops equipped with state-of-the-art machinery. Apart from the production of RWR and Fiar Grifo 7 radar systems and the overhauling of MPDRs, CRCs and generators, KARF offers services in many fields such as surface mount technology, cable repair and manufacturing inspection/testing of microwave components, testing of gears and synchro bridge, environmental testing, testing hydraulic components, rebuilding of three-phase air conditioning systems and general engineering and painting facilities.

UPDATED

PAC (AMF) MUSHSHAK and SUPER MUSHSHAK

English name: Proficient
TYPE: Three-seat lightplane.
PROGRAMME: Fifteen Mushshaks supplied complete from Sweden; 10 then assembled from SKD kits and 82 from CKD kits at Risalpur between 1975 and 1981; completely indigenous production followed, with 180 delivered by December 1996; 113 repaired at Kamra including 100 from 1980 to 1996; engines, instruments, electrical equipment and radios imported; complete airframe manufactured locally. Shahbaz (prototype 86-5147; first flight July 1987; English name Falcon) received US FAR Pt 23 certification 1989. Super Mushshak (see Current Versions) introduced 1997. Production rate (new aircraft and conversion kits) stated to be 24 per year in late 1999 (latest information received).
CURRENT VERSIONS: **Mushshak:** Standard production version; used for training, communications and observation; more than half a million hours flown by late 1999. AMF then upgrading 16 to Super Mushshak for PAF, of which six delivered to Air Force Academy by May 2001.
Shahbaz: As Mushshak except for 157 kW (210 hp) Teledyne Continental TSIO-360-MB turbocharged

PAC Super Mushshak prototype (*Jane's/Paul Jackson*)
0126891

engine; four completed by September 1993; no further examples since then. Details in 1997-98 and earlier *Jane's*.
Super Mushshak: Further increase in power by adopting IO-540 flat-six engine. Prototype 95-5385X made first flight 15 August 1996. Revealed at Dubai Air Show in November 1997, at which time five completed; total had risen to 16 by May 2001, then in store but delivered by November to Pakistan Army, which has reported requirement for 48.
Description applies to above version, except where indicated.
CUSTOMERS: See table. Total of 315, comprising 15 imported aircraft plus 280 Mushshaks, 16 Super Mushshaks and four Shahbaz acquired/produced by May 2001. Mushshak deliveries to Iran in 1988-91; six supplied to Syria and three to Oman in 1994. Last-named became first export customer for Super Mushshak (five) in December 2001; these delivered in mid-2002. Iranian Air Force also reported to have ordered 16 Supers. Sale of 30 to 40 Super Mushshaks to Saudi Arabia was reportedly being negotiated in late 2001.

MUSHSHAK PRODUCTION
(at November 2001)

Customer	Qty
At Risalpur:	
Pakistan Air Force/Army	92[1]
At Kamra:	
Pakistan Air Force	36[2]
Pakistan Army	117
Pakistan (Air Force and/or Army)	13
Iran	25
Oman	8[4]
Syria	6
Pakistan Aeronautical Complex	4
Total	**301**[3]

[1] Saab kits (10 SKD, 82 CKD)
[2] First aircraft 83-5116 (numbers 5200-5299 not allocated); equip Nos. 1 and 2 Primary Flying Training Squadrons and three Headquarters Flights
[3] Plus complete aircraft from Saab to PAF
[4] Includes five Super Mushshaks

DESIGN FEATURES: Based on Swedish Saab Safari/Supporter, but without latter's armament option. Optimised for

military support (originally including light attack) from semi-prepared airstrips; wing position aids ground observation.

Wing thickness/chord ratio 10 per cent; dihedral 1° 30′; incidence 2° 48′; sweepforward 5° from roots.
FLYING CONTROLS: Manual. Mass-balanced ailerons with servo tab in starboard unit; rudder with trim tab; all-moving mass-balanced tailplane with large anti-balance tab. Electrically actuated plain sealed flaps; leading-edge slots.
STRUCTURE: All-metal, except for GFRP tailcone, engine cowling panels, wing strut/landing gear attachment fairings and fin-tip.
LANDING GEAR: Non-retractable tricycle type. Cantilever composites spring main legs; Cessna oleopneumatic nosewheel shock-strut. Goodyear wheels and Flight Custom tyres, size 6.00-6 (6 ply) on mainwheels and 5.00-5 (6 ply) on steerable (±30°) nosewheel. Tyre pressure (all units) 2.07 bar (30 lb/sq in). Parker Hannifin hydraulic disc brakes on main units.
POWER PLANT: *Mushshak:* One 149 kW (200 hp) Textron Lycoming IO-360-A1B6 flat-four engine, driving a Hartzell HC-C2YK-4F/FC7666A-2 two-blade constant-speed metal propeller.
Super Mushshak: One 194 kW (260 hp) Textron Lycoming IO-540-V4A5 flat-six; Hartzell HC-C2YK-1BF two-blade constant-speed propeller.
Fuel (both versions) in two integral wing fuel tanks, total capacity 178 litres (47.0 US gallons; 39.1 Imp gallons), of which 159 litres (42.0 US gallons; 35.0 Imp gallons) are usable. Oil capacity (both) 7.5 litres (2.0 US gallons; 1.6 Imp gallons). From 10 to 20 seconds inverted flight (limited by oil system) permitted.
ACCOMMODATION: Side-by-side adjustable seats, with provision for back-type or seat-type parachutes, for two persons beneath fully transparent upward-hinged canopy. Dual controls standard. Space aft of seats for 100 kg (220 lb) of baggage (with external access on port side) or, optionally, a rearward-facing third seat. Upward-hinged door, with window, beneath wing on port side. Cabin heated and ventilated in Mushshak, air conditioned in Super Mushshak.
SYSTEMS: Hydraulic reservoir for mainwheel brakes. 28 V DC electrical system (70 A alternator and 70 Ah battery).
AVIONICS: *Comms:* Honeywell KLX 125 and KLX 135 radios; KT 76A/78A ATC transponder; KA 134 audio panel.
Flight: KLX 125 VOR; KLX 135 GPS; KR 87 ADF.
Instrumentation: Provision for full blind-flying instrumentation.

Instrument panel of the Super Mushshak
0084283

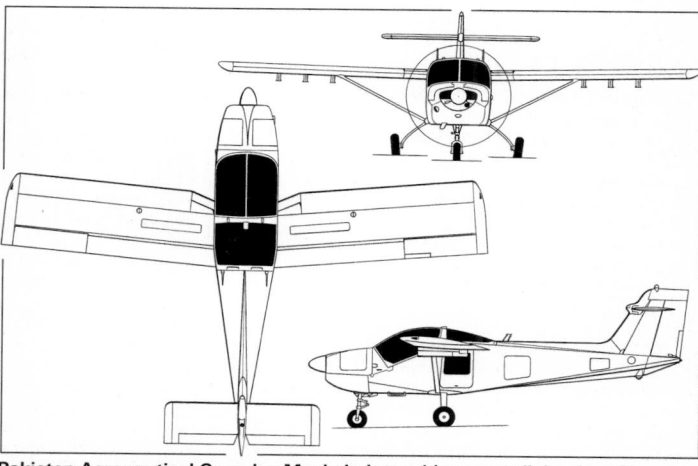
Pakistan Aeronautical Complex Mushshak two/three-seat light aircraft (*Jane's/Dennis Punnett*)

EQUIPMENT: Provision for six underwing attachment points, inner two stressed to carry up to 150 kg (330 lb) each and outer four up to 100 kg (220 lb) each. Options include ULV crop-spraying kit, target towing kit, or underwing supply/relief containers.

DIMENSIONS, EXTERNAL:

Wing span	8.85 m (29 ft 0½ in)
Wing chord: at root	1.21 m (3 ft 11¾ in)
outer panels, constant	1.36 m (4 ft 5½ in)
Wing aspect ratio	6.6
Length overall: Mushshak	7.00 m (22 ft 11½ in)
Super Mushshak	7.15 m (23 ft 5½ in)
Height overall	2.60 m (8 ft 6½ in)
Tailplane span	2.80 m (9 ft 2¼ in)
Wheel track	2.20 m (7 ft 2½ in)
Wheelbase: Mushshak	1.61 m (5 ft 3½ in)
Super Mushshak	1.58 m (5 ft 2¼ in)
Propeller diameter	1.93 m (6 ft 4 in)
Propeller ground clearance	7.62 cm (3 in)
Cabin door (port): Height	0.78 m (2 ft 6¾ in)
Width	0.52 m (1 ft 8½ in)

DIMENSIONS, INTERNAL:

Cabin: Max width	1.10 m (3 ft 7¼ in)
Max height (from seat cushion)	1.00 m (3 ft 3¼ in)

AREAS:

Wings, gross	11.90 m² (128.1 sq ft)
Ailerons (total)	0.98 m² (10.55 sq ft)
Flaps (total)	1.55 m² (16.68 sq ft)
Fin	0.77 m² (8.29 sq ft)
Rudder, incl tab	0.73 m² (7.86 sq ft)
Tailplane, incl tab	2.10 m² (22.60 sq ft)

WEIGHTS AND LOADINGS (Mushshak and Super Mushshak. A: Aerobatic, U: Utility, N: Normal category):

Weight empty, equipped: Mushshak	646 kg (1,424 lb)
Super Mushshak	760 kg (1,676 lb)
Max external stores load (both)	300 kg (661 lb)
Max T-O weight: A (both)	900 kg (1,984 lb)
U (Mushshak)	1,000 kg (2,205 lb)
U (Super Mushshak)	1,030 kg (2,270 lb)
N (Mushshak)	1,200 kg (2,645 lb)
N (Super Mushshak)	1,250 kg (2,755 lb)
Max wing loading: A (both)	75.6 kg/m² (15.48 lb/sq ft)
U (Mushshak)	84.0 kg/m² (17.20 lb/sq ft)
U (Super Mushshak)	86.6 kg/m² (17.73 lb/sq ft)
N (Mushshak)	100.8 kg/m² (20.65 lb/sq ft)
N (Super Mushshak)	105.0 kg/m² (21.51 lb/sq ft)

Max power loading (Mushshak):

A	6.04 kg/kW (9.92 lb/hp)
U	6.71 kg/kW (11.02 lb/hp)
N	8.05 kg/kW (13.23 lb/hp)

Max power loading (Super Mushshak):

A	4.64 kg/kW (7.63 lb/hp)
U	5.31 kg/kW (8.73 lb/hp)
N	6.45 kg/kW (10.60 lb/hp)

PERFORMANCE (Utility category):

Never-exceed speed (V_{NE}):	
both	196 kt (363 km/h; 226 mph)
Max level speed at S/L:	
Mushshak	130 kt (240 km/h; 149 mph)
Super Mushshak	145 kt (268 km/h; 166 mph)
Cruising speed at S/L at 75% power:	
Mushshak	115 kt (213 km/h; 132 mph)
Super Mushshak	130 kt (240 km/h; 149 mph)

Stalling speed, power off:

Mushshak, flaps up	60 kt (111 km/h; 69 mph)
Super Mushshak, flaps up	57 kt (105 km/h; 66 mph)
Mushshak, flaps down	54 kt (100 km/h; 63 mph)
Super Mushshak, flaps down	52 kt (96 km/h; 60 mph)
Max rate of climb at S/L:	
Mushshak	372 m (1,220 ft)/min
Super Mushshak	518 m (1,700 ft)/min
Time to 1,830 m (6,000 ft): Mushshak	7 min 30 s
Time to 3,050 m (10,000 ft): Super Mushshak	10 min
Service ceiling: Mushshak	4,800 m (15,740 ft)
Super Mushshak	6,705 m (22,000 ft)
T-O run: Mushshak	150 m (495 ft)
Super Mushshak	183 m (600 ft)
T-O to 15 m (50 ft): Mushshak	252 m (825 ft)
Super Mushshak	275 m (900 ft)
Landing from 15 m (50 ft): Mushshak	314 m (1,030 ft)
Super Mushshak	296 m (970 ft)
Landing run: Mushshak	140 m (460 ft)
Super Mushshak	153 m (500 ft)

Range with max fuel:

Mushshak	580 n miles (1,074 km; 667 miles)
Super Mushshak	440 n miles (814 km; 506 miles)

Max endurance at 65% power at S/L, 10% reserves:

Mushshak	5 h 10 min
Super Mushshak	4 h 15 min
g limits (both): A	+6/−3
U	+5.4/−2.7
N	+4.8/−2.4

UPDATED

PHILIPPINES

PADC

PHILIPPINE AEROSPACE DEVELOPMENT CORPORATION

PO Box 7395, Domestic Airport Post Office, Lock Box 1300, Domestic Road, Pasay City, Metro Manila
Tel: (+63 2) 832 37 57 and 23 81
Fax: (+63 2) 832 25 68
PRESIDENT: Reynato R Jose
SENIOR VICE-PRESIDENT: Josefina Laquindanum

PADC established 1973 as government arm for development of Philippine aviation industry; is now an attached agency of Department of Transportation and Communication (DOTC) and has technical workforce of about 200. Main activities are aircraft manufacturing and assembly; maintenance engineering; aircraft and spare parts sales; service centres for Britten-Norman Islander.

Past programmes include licensed assembly of 67 Islanders (including 22 for Philippine Air Force), 24 Agusta S.211 jet trainers, 16 (of 18) SF-260TP turboprop trainers and 44 BO 105s, as detailed in previous editions of *Jane's*.

Six Lancair ESs and two Lancair IVs were being assembled from PAI kits in 1999 (latest information known) for delivery to the Philippine National Police (PNP). PADC's six-seat RPX-Alpha Hummingbird, the first helicopter to be designed and built in the Philippines, was described in the 1998-99 *Jane's*.

PADC has maintenance/repair/overhaul centre for Rolls-Royce 250 series turbine engines and for Textron Lycoming and Teledyne Continental piston engines of up to 298 kW (400 hp). Its Maintenance and Engineering Department undertakes FMS work on 250-C30 engines for Sikorsky helicopters and 250-B17 turboprops and propellers for PAF Nomads.

PADC's 30/70 per cent joint venture with Eurocopter, known as Eurocopter Philippines, offers sales and after-sales service and assembly of Eurocopter helicopters from CKD kits. Two Ecureuils have been supplied to the PNP and one to the Department of Environment and Natural Resources (DENR).

PADC's latest known venture is a partnership with National Airmotive Corporation (NAC) of the USA to establish a centre in the Philippines to support the aviation needs of Southeast Asia (and especially the needs of the Philippine Air Force) regarding overhaul of Rolls-Royce T56/501 and 250 series engines, QEC overhaul and repair, propeller refurbishment and C-130 Hercules airframe components.

VERIFIED

PAI

PACIFIC AERONAUTICAL INC

25 First Avenue, Mactan Export Processing Zone, Lapu-Lapu City 6015
Tel: (+63 32) 334 02 83/84/86
Fax: (+63 32) 334 02 85
e-mail: pacaero@skyinet.net
PRESIDENT: Robert C Fair

VICE-PRESIDENTS:
Wilfredo Dela Cruz (Production)
Augusto V Dayrit (Operations)

PAI continues to manufacture kits for Lancair aircraft (see US section), and during 1999 (latest details known) shipped, on average, eight Lancair IV and two Lancair ES kits per month. Planned monthly rate for 2000 was three Lancair IV, two Lancair ES and six fast-build Legacy 2000 kits.

Responsibility for the Lancair Columbia 300 lies with Pacific Aeronautical Inc (Malaysia).

Recent orders have included two Lancair IVs, two ESs and two Legacy 2000s for the Mexican Navy. The first batch of Legacy 2000 kits was shipped to the USA at the end of June 2000.

VERIFIED

POLAND

3XTRIM

ZAKŁADY LOTNICZE 3XTRIM Sp. z o.o. (3Xtrim Aircraft Factory)

ulica Regera 109, PL-43 382, Bielsko-Biała
Tel: (+48 33) 818 92 51
Fax: (+48 33) 818 91 21
e-mail: biuro@3xtrim.pl
Web: http://www.3xtrim.pl
PRESIDENT AND CHIEF DESIGNER: Adam Kurbiel
VICE-PRESIDENT: Agata Kurbiel
WORKS: Bielsko-Biała and Szczecin; Warsaw factory planned

Company was formed in 1996 as WNKL A Kurbiel, changing to its present title in November 1999, but has its roots in the SZD-PZL-Bielsko which for some 40 years has been the major producer of Polish high-performance sailplanes and gliders. Its main current products are the two-seat lightplanes described in this entry, but glider production is under way and other (non-aerospace) products in composites are also manufactured.

UPDATED

3XTRIM 450 and 550

TYPE: Side-by-side ultralight; two-seat lightplane.
PROGRAMME: Development of EOL-2 Racek 2, of which some 20 built. Design started 1996, prototype 450-UL making maiden flight May 1998; construction of first 550-VLA began 1997 (prototype SP-PUP made first flight 5 February 2000 and 150 hours flown by May 2001); static/fatigue testing October 1999 to January 2000; JAR-VLA tests completed 16 May 2000. 450-UL has Czech ultralight certificate and Polish special certification; type certification for 550-VLA anticipated by end of 2001; French and German certification under way.
CURRENT VERSIONS: **3Xtrim 450-UL:** Ultralight version. Reduction of MTOW to required 450 kg (992 lb) achieved by wider use of CFRP instead of GFRP and omission of some equipment and furnishing (for example, no door stays). Rotax 912 UL engine standard, but can be offered with 912 ULS for higher performance.
3Xtrim 550-VLA: For aero club and flying school (PPL) training; also available with IFR instrumentation as border patrol and reconnaissance aircraft.

XXtrim: Modified version marketed in France by Aerotrophy. Reduced wing span and area (9.50 m; 31 ft 2 in and 11.90 m²; 128.1 sq ft) and additional option of Jabiru engine. Performance generally comparable with that of Polish versions.

SF 45 SA Spirit: Version marketed in Germany by USA (which see).
CUSTOMERS: Total 23 450-ULs ordered by May 2001, of which 20 delivered, starting 15 May 1998. Customer countries include France.
COSTS: 550-VLA DM111,000 including VAT; 450-UL DM86,000 including VAT (2001).
DESIGN FEATURES: Typical strut-braced high-wing cabin monoplane. Wing incidence 2° 30′; tailpane incidence −3°.
FLYING CONTROLS: Conventional and manual (pushrod ailerons and elevator, cable-operated rudder). Rudder and one-piece elevator horn-balanced; 450-UL has small, inset, electrically actuated trim tab in port half of elevator; tab in 550-VLA is twice as large and protrudes beyond trailing-edge. Slotted flaps deflect 33° up/15° down.
STRUCTURE: All-composites fuselage (glass and carbon fibre/epoxy cellular construction with plastics foam

filling) with integral fin; GFRP sandwich single-spar wing.

LANDING GEAR: Fixed tricycle type, with GFRP, self-sprung cantilever mainwheel legs and castoring (±15°) nosewheel. Mainwheels are size 5.00-5, with 350×135 tyres on 550-VLA; size 4.00-6 on 450-UL. Tyre pressure (both) 2.50 bar (36 lb/sq in); hydraulic disc brakes.

POWER PLANT: *550-VLA:* One 73.5 kW (98.6 hp) Rotax 912 ULS flat-four engine, driving an MT 170 R 135-A2 two-blade fixed-pitch propeller.

450-UL: One 59.6 kW (79.9 hp) Rotax 912 UL and two-blade 3Xtrim 170 propeller with ground-adjustable pitch standard. Rotax 912 ULS (see above) optional.

Fuel capacity (both) 70 litres (18.5 US gallons; 15.4 Imp gallons) standard (65 litres; 17.2 US gallons; 14.3 Imp gallons usable), in single tank behind pilot's seat; optionally, two 43 litre (11.4 US gallons; 9.5 Imp gallon) tanks in same location. Oil capacity 2.5 litres (0.66 US gallon; 0.55 Imp gallon).

ACCOMMODATION: Two seats side by side; dual controls standard. Fully enclosed, heated and ventilated cabin with upward-opening door on each side. Baggage space behind seats.

SYSTEMS: Hydraulic system for brakes only; 12 V 18 Ah battery for electrical power.

AVIONICS: Bendix/King KY 97A and KT 76A transponder; VFR instrumentation and stall warning system standard.

EQUIPMENT: BRS 5 ballistic recovery parachute.

DIMENSIONS, EXTERNAL:

Wing span	10.03 m (32 ft 11 in)
Wing chord: at root	1.30 m (4 ft 3¼ in)
at tip	1.00 m (3 ft 3¼ in)
Wing aspect ratio	8.1
Length overall	6.87 m (22 ft 6½ in)
Fuselage max width	1.18 m (3 ft 10½ in)
Height overall	2.405 m (7 ft 10¾ in)
Tailplane span	2.75 m (9 ft 0¼ in)
Wheel track	1.895 m (6 ft 2½ in)
Wheelbase	1.48 m (4 ft 10¼ in)
Propeller diameter (both)	1.70 m (5 ft 7 in)
Propeller ground clearance	0.195 m (7¾ in)

AREAS:

Wings, gross	12.40 m² (133.5 sq ft)
Ailerons (total)	1.00 m² (10.76 sq ft)
Trailing-edge flaps (total)	1.50 m² (16.15 sq ft)
Fin	0.44 m² (4.74 sq ft)
Rudder	0.38 m² (4.09 sq ft)
Tailplane	1.10 m² (11.84 sq ft)
Elevator, incl tab	0.64 m² (6.89 sq ft)

WEIGHTS AND LOADINGS (A: 450-UL, B: 550-VLA):

Weight empty: A	285 kg (628 lb)
B	325 kg (717 lb)
Max fuel weight: A	49 kg (108 lb)
B	61 kg (134 lb)
Max T-O weight: A	450 kg (992 lb)
B	550 kg (1,212 lb)

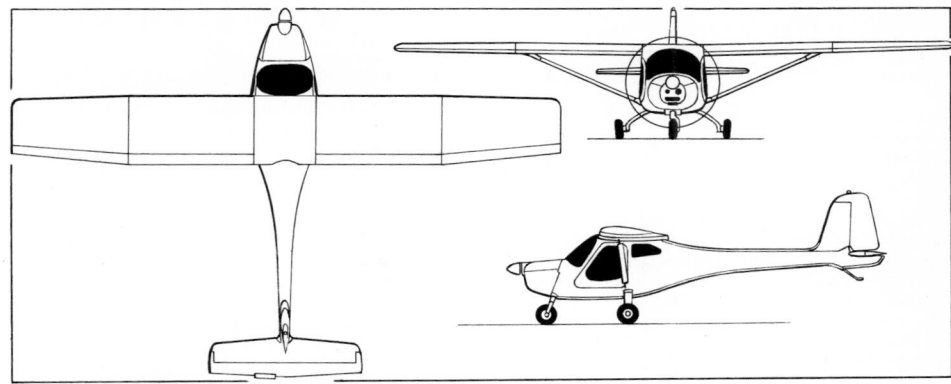

3Xtrim 550-VLA two-seat lightplane (*Jane's/James Goulding*)　　0093619

The 3Xtrim 550-VLA prototype (*Jane's/Paul Jackson*)　　0110931

Max wing loading: A	36.3·kg/m² (7.43 lb/sq ft)
B	44.4 kg/m² (9.08 lb/sq ft)
Max power loading: A	7.55 kg/kW (12.41 lb/hp)
B	7.48 kg/kW (12.29 lb/hp)

PERFORMANCE (A and B as above, Rotax 912 ULS engine):

Never-exceed speed (V_{NE}):	
A, B	118 kt (220 km/h; 136 mph)
Max cruising speed at 75% power:	
A	94 kt (175 km/h; 109 mph)
B	91 kt (170 km/h; 106 mph)
Stalling speed, flaps up: A	34 kt (63 km/h; 40 mph)
B	38 kt (70 km/h; 44 mph)
Max rate of climb at S/L: A	330 m (1,083 ft)/min
B	270 m (886 ft)/min

Service ceiling	4,000 m (13,120 ft)
T-O run	50 m (165 ft)
T-O to, and landing from, 15 m (50 ft):	
A, B	250 m (820 ft)
Range with max standard fuel:	
A, B	405 n miles (750 km; 466 miles)*
g limits: A	+4.0/−2.0
B	+3.8/−1.5

*B 675 n miles (1,250 km; 776 miles) with larger tanks

OPERATIONAL NOISE LEVEL (ICAO Chapter 10, Appendix 16):

Average	67 dBA

UPDATED

AERO

AERO Sp. z o. o.
Wał Miedzeszyński 844, PL-03-942 Warszawa
Tel: (+48 22) 616 20 87
Fax: (+48 22) 617 85 28
e-mail: aero@post.pl
Web: http://www.aero.com.pl
PRESIDENT AND CHIEF DESIGNER: Ing Tomasz Antoniewski

UK DISTRIBUTOR:
S2T Aviation, 125 High Road, North Weald, Essex CM16 6EA
Tel: (+44 1992) 52 27 97
Fax: (+44 7753) 57 59 96
e-mail: info@s2taviation.com
Web: http://www.s2taviation.com

Company formed in 1994. Three P.220 Panda kits were ordered from Aerotechnik of Czech Republic that year, being given Polish designation suffix AT-2 (AT-1 having been a single-seat homebuilt design started in 1987). These all differed dimensionally and were variously fitted with Walter Mikron or 55.9 kW (75 hp) Limbach engines (first flight 12 December 1995 by c/n 001 SP-PUL, later SP-FUL). Third aircraft subsequently developed with more powerful Limbach and other modifications for production in Poland as AT-3; now available also with similarly rated Rotax power plant.

UPDATED

AERO AT-3

TYPE: Side-by-side lightplane/kitbuilt.

PROGRAMME: Completed initially (c/n 003, SP-PUH) as P.220 S-AT2, but subsequently modified to AT-3 with 67.1 kW (90 hp) Limbach L-2400EB3AC engine, revised dimensions and higher weights; given new c/n of 001 and first flown in new configuration 18 January 1998. Received JAR-VLA certification 14 May 1999 and re-registered SP-GUH. Five production aircraft completed and five more under construction by May 2001. German certification then being sought; US planned later.

CURRENT VERSIONS: **AT-3 L100:** With 74.6 kW (100 hp) Limbach L-2400DF1 flat-four engine and MT-160L-120-2C propeller.

AT-3 R100: With 73.5 kW (98.6 hp) Rotax 912 ULS flat-four and MT-165R-152-2M propeller. Otherwise as AT-3 L100.

CUSTOMERS: Initial customer Telekomunikacja Polska, for support of Polish aero clubs; two in service by May 2001. Potential Polish market for 200 over five-year period.

Aero (Antoniewski) AT-3 two-seat lightplane (*Jane's/James Goulding*)　　0121432

AT-3 L100 instrument panel
0102502

COSTS: Flyaway US$69,000, kit US$22,000, excluding VAT (2001).

DESIGN FEATURES: Changes from AT-2/P.220 S include approximately 1 m² (10.8 sq ft) more wing area, 4 cm (1.6 in) longer cockpit, enlarged fin and rudder, and strengthened landing gear.

FLYING CONTROLS: Manual; piano-hinged ailerons, horn-balanced rudder and all-moving elevator with central anti-balance tab; two-position split flaps (maximum deflection 40°).

STRUCTURE: Primarily metal, with some skins and fairings of carbon or glass fibre.

LANDING GEAR: Non-retractable tricycle type; Cantilever self-sprung main legs with 380×150/15×6.00 wheels and Cleveland hydraulic disc brakes; levered suspension nose leg with 5.00-4 wheel and rubber-in-compression shock absorption. Wheel speed fairings optional.

POWER PLANT: One Limbach or Rotax flat-four engine (see under Current Versions) with MT-Propeller two-blade, fixed-pitch propeller. Fuel capacity (both) 70 litres (18.5 US gallons; 15.4 Imp gallons).

ACCOMMODATION: Two seats side by side; dual controls standard. Hatrack/baggage shelf behind seats, capacity 30 kg (66 lb). Upward-opening one-piece canopy, hinged at front.

AVIONICS: Bendix/King KY 97A transceiver and VFR instrumentation standard; IFR, including KY 197A, KR 87 ADF, KT 76A transponder and directional gyro, optional.

DIMENSIONS, EXTERNAL:
Wing span	7.55 m (24 ft 9¼ in)
Wing chord, constant	1.27 m (4 ft 2 in)
Wing aspect ratio	6.1
Length overall	5.88 m (19 ft 3½ in)
Height overall	2.23 m (7 ft 3¾ in)
Elevator span	2.79 m (9 ft 1¾ in)
Wheel track	2.26 m (7 ft 5 in)
Wheelbase	1.36 m (4 ft 5½ in)
Propeller diameter	1.65 m (5 ft 5 in)

AREAS:
Wings, gross	9.30 m² (100.1 sq ft)

AT-3 L100 sponsored by Telekomunikacja Polska and operated by Aeroklub Polski at Krosno (*R J Malachowski*) NEW/0137424

WEIGHTS AND LOADINGS:
Weight empty: L100	372 kg (820 lb)
R100	350 kg (772 lb)
Max T-O weight: both	582 kg (1,283 lb)
Max wing loading: both	62.6 kg/m² (12.82 lb/sq ft)
Max power loading: L100	7.81 kg/kW (12.83 lb/hp)
R100	7.92 kg/kW (13.01 lb/hp)

PERFORMANCE:
Never-exceed speed (VNE):	
both	120 kt (223 km/h; 138 mph) IAS
Max level speed: L100	110 kt (205 km/h; 127 mph) IAS
R100	116 kt (215 km/h; 133 mph) IAS
Max cruising speed:	
L100	100 kt (185 km/h; 115 mph) IAS
R100	108 kt (200 km/h; 124 mph) IAS
Max manoeuvring speed (VA):	
both	106 kt (197 km/h; 122 mph) IAS

Stalling speed: both	45 kt (83 km/h; 52 mph) IAS
Max rate of climb at S/L: L100	210 m (689 ft)/min
R100	222 m (728 ft)/min
Service ceiling: both	3,800 m (12,460 ft)
T-O run: L100	170 m (560 ft)
R100	160 m (525 ft)
T-O to 15 m (50 ft):	
L100	465 m (1,525 ft)
R100	440 m (1,445 ft)
Landing from 15 m (50 ft):	
both	445 m (1,460 ft)
Landing run: L100	160 m (525 ft)
R100	150 m (495 ft)
Max range: both	488 n miles (904 km; 561 miles)
g limits: both	+3.8/−1.5

UPDATED

AGROLOT

FUNDACJA AGROLOT (Agrolot Foundation)

aleja Krakowska 110/114, PL-00-971 Warszawa
Tel: (+48 22) 846 00 31 or 846 00 69 extn 586
Fax: (+48 22) 846 27 01
CHAIRMAN: Andrzej Słociński

This foundation (established 1990) is responsible for the PZL-126P Mrówka programme previously managed by PZL Warszawa-Okecie. The prototype made its maiden flight in October 2000, some three years later than originally planned but was not expected to enter full flight test programme until 2002.

UPDATED

AGROLOT PZL-126P MRÓWKA 2001

English name: Ant

TYPE: Agricultural sprayer.

PROGRAMME: See under PZL Warszawa-Okecie in 1993-94 and previous editions for early history. PZL-126 prototype (SP-PMA: see 1995-96 and earlier editions for details) made first flight 20 April 1990; this aircraft was used in 1994 to test a modified 44.7 kW (60 hp) PZL-F2A-120-C1 two-cylinder engine driving a two-blade fixed-pitch propeller. It was then withdrawn for rebuilding as the prototype PZL-126P planned operational version, which was designed and developed at the PZL Warszawa-Okecie factory. Project is funded by Agrolot Foundation; chief designer is Andrzej Słociński. First flight of the PZL-126P

Prototype PZL-126P Mrówka (*Mariusz Adamski/Altair*) NEW/0120256

(SP-PMB) was first envisaged for 1997, but aircraft not completed until 1998, marked 'Mrówka 2001'; first flight was eventually made on 20 October 2000. No later reports received by mid-2002, but full flight test programme

expected to begin during that year, subject to availability of funding.

CURRENT VERSIONS: **PZL-126P 2001:** Under conversion since 1995. Increased wing span and area, more powerful engine, larger fuel tanks and increased payload compared with PZL-126 prototype. Intended as economical carrier of airborne systems for forestry protection (for example, combating pests, identifying diseased vegetation, detecting/controlling forest fires, spreading non-toxic small-volume selective-action agents for plant protection such as Trichrogramma parasitic wasp eggs laid in eggs of host insect). Other potential applications include environmental monitoring, glider towing and pipeline patrol; the possibility of an unmanned version with NBC detection equipment has also been reported.

Description applies to PZL-126P 2001.

DESIGN FEATURES: Single-engined low-wing monoplane with rectangular tail surfaces; meets requirements of FAR Pt 23. Semi-monocoque fuselage with equipment bay in lower portion aft of cockpit, accessible through hatch; upswept rear fuselage facilitates access to this bay. Quick-fastening lock beneath each wingtip for attachment of integral spraypod between main wing and tip. Aircraft can be dismantled quickly for towing on own landing gear by light all-terrain vehicle.

Wing aerofoil section NASA GA(W)-1; dihedral 3°.

FLYING CONTROLS: Manual. Three-section area-increasing flaps and flaperons on each wing, actuated by pushrods and interconnected by single central system located under

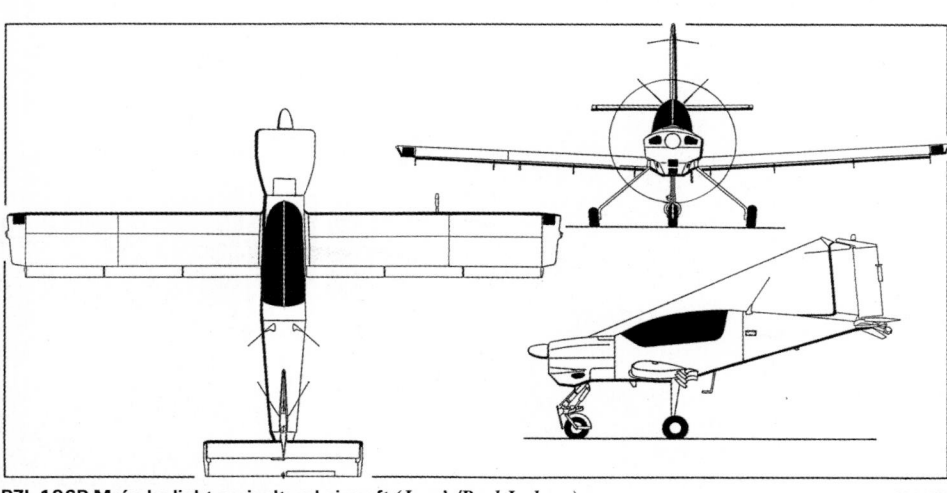

PZL-126P Mrówka light agricultural aircraft (*Jane's/Paul Jackson*) 0011923

cockpit floor; elevator actuated by pushrods, rudder by cables from adjustable rudder pedals; trim tab on elevator, ground-adjustable tab on rudder.

STRUCTURE: All-metal except for GFRP/epoxy engine cowling, laminated wingtips and fairings. Single-spar wings and fixed tail surfaces; wings mounted on fuselage mandrels and offset spars connected with single pin; front part of fuselage has two 'tusks' for engine and nosewheel mounting.

LANDING GEAR: Non-retractable tricycle type. Self-springing cantilever mainwheel legs of duralumin; castoring nosewheel with oleo-pneumatic shock-absorber. Wheel and tyre size 350×135 on all three units. Hydraulic differential mainwheel disc brakes.

POWER PLANT: One 74.6 kW (100 hp) Teledyne Continental O-200A flat-four engine, driving a McCauley 1A100MCM6950 two-blade metal propeller. Integral fuel tank in each wing, combined capacity 200 litres (52.8 US gallons; 44.0 Imp gallons).

ACCOMMODATION: Single-seat cockpit, fitted with airbag enabling in-flight seat adjustment. One-piece organic glass moulded canopy, opening sideways to starboard.

SYSTEMS: Hydraulic system for mainwheel brakes only; 24 V electrical system; hand pump to inflate seat cushion.

AVIONICS: *Comms:* Bendix/King KX 155 com/nav, KY 96A VHF transceiver and KT 76A transponder.
Flight: GPS 95XL global positioning; stall warning system.
Instrumentation: Prototype has IFR capability, but basic version fitted with VFR instrumentation. Instrument panel,

with cover, can be lifted up for access to instruments and front part of cockpit.

EQUIPMENT: Spraypod can be fitted to outer end of each main wing panel, inboard of detachable tip. Pod is an integral unit with all necessary attachments for spraying, attached by quick-fastening locks, electric multiplug socket supplying power to atomisers, and electric valve controlling outflow of liquid from tank. Optional third spray unit can be mounted under fuselage. Up to 35 ha (86.5 acres) can be sprayed in one flight with 70 litres (18.5 US gallons; 15.4 Imp gallons) of pesticide delivered at rate of 2 litres (0.53 US gallon; 0.44 Imp gallon)/ha.

Special biological spreader developed for dosing and spreading eggs of Trichrogramma wasp; eggs are carried in capsules in reel of tape covered with thin paper, spreader holding four such reels. Spreader is powered by electric motor and activated or stopped by push-button on throttle lever; 3 kg (6.6 lb) load of capsules is sufficient, at typical dispersal rate of four capsules every 50 m (164 ft) in rows 50 m apart, to seed an area of 800 ha (1,977 acres). Rotational mounting of spreader allows for full deflection and easy replacement of egg reels.

Aircraft can also carry miniaturised equipment such as photographic, video, thermal or other systems, coupled to satellite navigation system, monitoring and recording pictures or other signals from space.

DIMENSIONS, EXTERNAL:
Wing span: excl pods 7.66 m (25 ft 1½ in)
incl pods 8.46 m (27 ft 9 in)

Wing chord, constant:	
excl extended flaps/flaperons	0.75 m (2 ft 5½ in)
incl extended flaps/flaperons	0.92 m (3 ft 0¼ in)
Wing aspect ratio	8.5
Length overall	5.25 m (17 ft 2¾ in)
Height overall	2.80 m (9 ft 2¼ in)
Tailplane span	2.20 m (7 ft 2½ in)
Wheel track	2.15 m (7 ft 0¾ in)
Wheelbase	1.35 m (4 ft 5¼ in)
Propeller diameter	1.75 m (5 ft 9 in)

AREAS:
Wings, gross: excl pods	6.87 m² (73.9 sq ft)
incl pods	7.27 m² (78.3 sq ft)
Flaperons (total)	0.48 m² (5.17 sq ft)
Trailing-edge flaps (total)	0.97 m² (10.44 sq ft)
Vertical tail surfaces (total)	0.73 m² (7.86 sq ft)
Horizontal tail surfaces (total)	1.08 m² (11.63 sq ft)

WEIGHTS AND LOADINGS:
Max T-O and landing weight	575 kg (1,267 lb)
Max wing loading:	
without pods	83.7 kg/m² (17.14 lb/sq ft)
with pods	79.1 kg/m² (16.20 lb/sq ft)
Max power loading	7.72 kg/kW (12.68 lb/hp)

PERFORMANCE (estimated):
Max level speed	105 kt (195 km/h; 121 mph)
Operating speed	75 kt (140 km/h; 87 mph)
Max rate of climb at S/L	more than 360 m (1,181 ft)/min

UPDATED

EADS PZL

EADS PZL WARSZAWA-OKĘCIE SA
aleja Krakowska 110/114, PL-00-971 Warszawa
Tel: (+48 22) 846 61 52 and 846 79 96
Fax: (+48 22) 846 27 01
e-mail: wasiucionek@pzl-okecie.com.pl
Web: http://www.pzl-okecie.com.pl
PRESIDENT AND CEO: Domingo Ureña Raso
GENERAL DESIGNER: Andrzej Frydrychewicz, Msc Eng
CHIEF MARKETING AND SALES EXECUTIVE:
 Zbigniew Wasiucionek, MSc(Ec)

Okęcie factory founded in 1928; responsible for light aircraft development and production, and for design and manufacture of associated agricultural equipment for its own aircraft and those built at other Polish factories; known as PZL Warszawa-Okęcie; became public stock company, entirely owned by Ministry of Industry and Trade, on 2 January 1995; by 1 January 2000 had produced 3,777 aircraft since 1945, including 33 in 1995, eight in 1996; 11 in 1997, 20 in 1998 and eight in 1999; workforce in 2002 was over 800. ISO 9001 status was awarded on 3 December 1999.

On 17 October 2001, as part of the contract to supply eight C-295M transports to the Polish Air Force, EADS CASA (see Spanish section) acquired a 51 per cent holding in PZL Warszawa-Okęcie for some US$7 million, and is to acquire a further 34 per cent over the ensuing two years; the remaining 15 per cent will continue to be held by Okęcie employees. Integration of PZL into EADS was ahead of schedule by August 2002, with facilities being refurnished and modernised for new activities. These will include increasing importance as an aerostructures centre and as a supplier of crop-spraying/firefighting aircraft and services. Components for the CASA C-295 and CN-235 transports are already being produced by Okęcie, and modernisation of the PZL-130TC-1 Orlik is being discussed with the Polish Air Force.

Light Aircraft and Agricultural Equipment Pilot Plant
Tel: (+48 22) 846 63 50
e-mail: kubicki@pzl-okecie.com.pl
DIRECTOR: Tomasz Kubicki, MSc Eng

Main function of this plant is to develop and perform research tasks, build and flight test prototypes.

UPDATED

EADS PZL PZL-130 ORLIK
English name: Spotted Eaglet
TYPE: Basic turboprop trainer/attack lightplane.
PROGRAMME: Development of piston-engined Orlik (see 1989-90 *Jane's*) discontinued 1990. Development of turboprop derivative began 1985 and fully described in 1999-2000 and earlier editions.

First deliveries, to Polish Air Force Academy at Deblin, were two PZL-130TMs (c/n 005 and 006) in October 1992, followed by first two PZL-130TBs (c/n 012 and 013) from December 1992. Deliveries suspended in 1995; resumed in early 1998 with manufacture of what were, reportedly, last three TC-1s (043 to 045), but at least four further aircraft (046 to 049) completed since then. Fleet now being retrofitted with underwing drop tanks following Polish Air Force dissatisfaction with mission radius capability; first flight in this form in third quarter 1999.

CURRENT VERSIONS: **PZL-130TB:** For Polish Air Force. First flight 17 September 1991. Nine production TBs delivered,

PZL-130TC-1 Orlik turboprop trainer (*Jane's/Paul Jackson*) 0110898

of which 021 lost on 30 April 1994; remainder retrofitted to TC-1 standard. Details in 2000-01 and earlier editions.

PZL-130TC: Advanced version, with 708 kW (950 shp) P&WC PT6A-62 engine; Honeywell avionics, Martin-Baker Mk 11B ejection seats and Flight Visions HUD; prototype/demonstrator SP-PCE made first flight 2 June 1993; in storage by 1998. Further details in 1998-99 and earlier editions.

PZL-130TC-1: Upgraded version of PZL-130TB, first flown (converted 009/SP-PRF) 9 July 1994. Power plant as for TB, but Martin-Baker Mk 11B zero height/90 kt (167 km/h; 104 mph) seats, Honeywell GPS and multifunction flight data recorder. Modified tailplane and enlarged ventral fin trialled on 037/SP-PCH in May 1996.

Total of 23 TC-1s (15 new-build (022 to 036) and eight TB conversions) delivered to Polish Air Force by end of 1995. None delivered in 1996 or 1997; five (038 to 042) delivered and three (043 to 045) built in 1998 and four more (046 to 049) by end of 2001. Tail surface modifications accepted by Polish Air Force in September 1997. Further control surface modifications flight-tested on 037 in 1998.

Following description applies to TC-1, except where otherwise indicated; further details in 1999-2000 edition.

PZL-130TC-2: Alternative version; simplified equipment compared with TC, and 559 kW (750 shp) PT6A-25C engine. Weights, loadings and performance data in 1999-2000 *Jane's*. Initial development programme abandoned and prototype (c/n 014) relegated to static test.

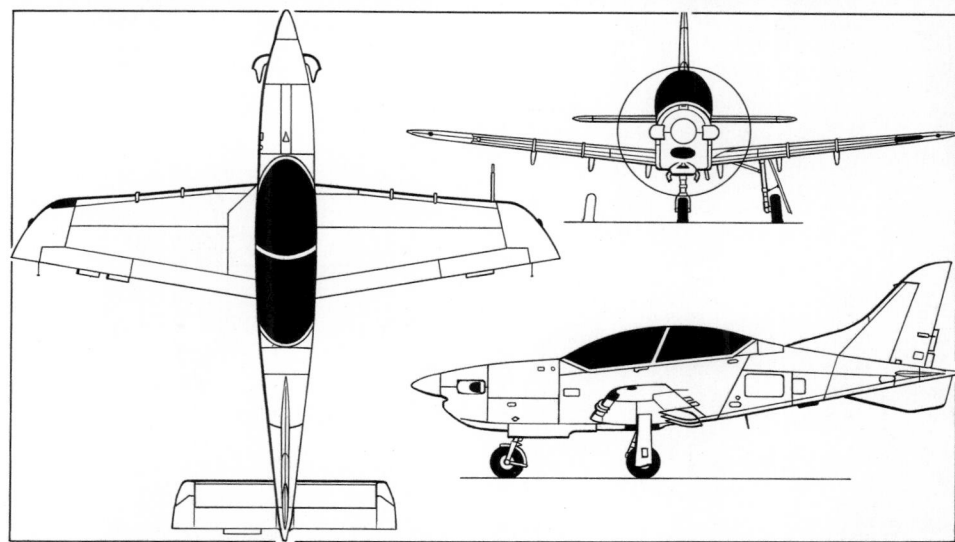

EADS PZL PZL-130TC-1 Orlik, showing current Polish Air Force in-service configuration (*Jane's/Mike Keep*)
0062999

Model of current proposal for an upgraded TC-2 version of Orlik (*R J Malachowski*) **NEW**/0143729

Wilga 2000 No. 14, the US demonstrator (*Jane's/Paul Jackson*) **NEW**/0143704

However, proposal revealed early 2000 for revised version, still with PT6A-25C engine, but revamped to offer capabilities comparable to US T-6A Texan II. Major changes would include new, extended wingtip 'salmons' of composites; strengthened main wing able to support Alkan 6170 stores carriers for light air-to-ground weapons or 7.62 mm gun pods; landing gear reinforced for higher operating weights; new nav/attack avionics; and HOTAS controls. Seen as basis for upgrade of existing Orlik fleet; Flight Visions Inc of USA announced receipt of contract in February 2002 to supply integrated weapons delivery and navigation system, including HUD, for this version. New avionics will include FV-4000 modular mission display processor and two MFDs. Work on prototype conversion due to begin in August 2002 and be ready for October 2002 maiden flight. If trials are satisfactory, a further eight Orliks could be re-engined in 2003.

PZL-140 Orlik 2000: Projected seven-seat business transport, using some PZL-130TC-1 components. Not built; described in 2000-01 *Jane's*.

CUSTOMERS: By October 2001, 34 TC-1s (including eight TB conversions) in Polish Air Force service. Four Orliks formed a Polish Air Force aerobatic team in 1998.

FLYING CONTROLS: Conventional and manual. Frise differential ailerons, elevators and rudder all aerodynamically and mass balanced; ailerons actuated by pushrods and torque tube, elevators by rods and cables, rudder by cables; electrically actuated trim tabs on port aileron (two), starboard aileron (one), port elevator (one) and rudder (two). Vortex generators ahead of ailerons on testbed 037. Three-position double-slotted trailing-edge flaps, also actuated electrically.

Alternative trim system flight tested in 1995, with Lear Astronics computer controlling a movable undertail fin.

LANDING GEAR: Hydraulically retractable tricycle type, all three units retracting into fuselage (mainwheels inward, steerable nosewheel rearward). Nosewheel protrudes slightly when retracted. PZL Warszawa-Okęcie oleo-pneumatic shock-absorber in each unit. All three wheels same size, with 500×200 tubeless tyres. Differential hydraulic multidisc brakes; parking brakes for main and nosewheels. No anti-skid system.

POWER PLANT: One 560 kW (750 shp) Walter M 601T turboprop, driving an Avia/Hamilton Sundstrand V 510T

Rear (instructor's) cockpit of the PZL-130TC-1 (*R J Malachowski*) **NEW**/0143730

five-blade propeller. Fuel in four integral wing tanks, total usable capacity 530 litres (140 US gallons; 117 Imp gallons). Overwing refuelling point for each tank. Fuel and oil systems permit up to 30 seconds of inverted flight. Retrofit with two 250 litre (66.0 US gallon; 55.0 Imp gallon) underwing drop tanks initiated in 1999.

ACCOMMODATION: Tandem seating for pupil (in front) and instructor under one-piece canopy which opens sideways to starboard. Rear seat elevated 65 mm (2.6 in). Martin-Baker Mk 11B zero height/90 kt (167 km/h; 104 mph) seats in TC/TC-1. Baggage space aft of rear seat. Full dual controls standard. Cockpit heating and ventilation; canopy demisting.

ARMAMENT: Three hardpoints under each wing, inboard and centre ones stressed for loads of 160 kg (353 lb) each and outboard stations for 80 kg (176 lb) each. Typical external loads for PZL-130TC-1 include free-fall bombs of up to 100 kg, Tejsy and Mokrzycko bomb canisters, Zeus gun pods each with two 7.62 mm machine guns, launchers for 57 mm or 80 mm rockets, or Strela IR air-to-air missiles; plus S-17 gunsight and S-13 gun camera.

DIMENSIONS, EXTERNAL:

Wing span	9.00 m (29 ft 6¼ in)
Wing aspect ratio	6.2
Length overall	9.00 m (29 ft 6¼ in)
Height overall	3.53 m (11 ft 7 in)*
Wheel track	3.10 m (10 ft 2 in)
Wheelbase	2.90 m (9 ft 6¼ in)

* Before installation of taller fin

AREAS:

Wings, gross	13.00 m² (139.9 sq ft)

WEIGHTS AND LOADINGS:

Weight empty	1,750 kg (3,858 lb)
Max external stores load	800 kg (1,764 lb)
Max T-O and landing weight:	
Aerobatic	2,150 kg (4,740 lb)
Utility	2,700 kg (5,952 lb)
Max wing loading:	
Aerobatic	165.4 kg/m² (33.87 lb/sq ft)
Utility	207.7 kg/m² (42.54 lb/sq ft)
Max power loading: Aerobatic	3.85 kg/kW (6.32 lb/shp)
Utility	4.83 kg/kW (7.94 lb/shp)

PERFORMANCE (at max Aerobatic T-O weight, clean configuration, except where indicated):

Max level speed at S/L	245 kt (454 km/h; 282 mph)
Max level speed at 6,000 m (19,680 ft)	
	270 kt (501 km/h; 311 mph)
Stalling speed, engine idling:	
flaps and gear up	84 kt (154 km/h; 96 mph) EAS
flaps and gear down	61 kt (112 km/h; 70 mph) EAS
Max rate of climb at S/L	800 m (2,625 ft)/min
Service ceiling	10,000 m (32,800 ft)
T-O run at S/L	345 m (1,135 ft)
T-O to 15 m (50 ft)	550 m (1,805 ft)
Landing from 15 m (50 ft)	570 m (1,870 ft)
Landing run at S/L	201 m (660 ft)

Range with reserves, internal fuel only:

at S/L	453 n miles (840 km; 522 miles)
at 3,000 m (9,840 ft)	
	573 n miles (1,062 km; 659 miles)

Max range with two 250 litre drop-tanks, no reserves:

at S/L	836 n miles (1,550 km; 963 miles)
at 3,000 m (9,840 ft)	
	1,028 n miles (1,905 km; 1,183 miles)
g limits: Aerobatic	+6.5/–3.0
Utility	+4.4/–1.76

UPDATED

EADS PZL PZL-104M WILGA 2000

English name: Oriole

TYPE: Four-seat lightplane.

PROGRAMME: First flight of prototype Wilga 1, 24 April 1962 (see 1968-69 *Jane's*); first flight of improved Wilga 35, 28 July 1967; production of Wilga 35 and 32 began 1968; both received Polish type certificate 31 March 1969 (Wilga 32 described in 1974-75 *Jane's*; see 1975-76 edition for Lipnur Gelatik Indonesian modified Wilga 32); first flight of Wilga 80, 30 May 1979. Both powered by one 194 kW (260 hp) PZL AI-14RA nine-cylinder supercharged radial air-cooled engine. Total of 962 of all earlier versions (except Gelatik) built by 1996, some of which stored and later supplied from stock, including one Wilga 80 in 1999. Further details of Wilga 35 and Wilga 80 variants in 1999-2000 and earlier *Jane's*.

Wilga 2000, improved version of Wilga 35, developed to appeal to Western markets, principally through use of Western engine and avionics. Wing and nose modifications tested on Wilga 35A SP-CSG. First flight of Wilga 2000 prototype (SP-PHG, later -WHG) 20 August 1996; first production (SP-AHV) in mid-1997; FAR Pt 23 certification achieved in 1997. SP-PHG won 1st World Air Games, 1997.

CURRENT VERSIONS: **PZL-104M Wilga 2000:** Standard landplane version; *as described.*

PZL-104MN Wilga 2000: Designation of 2001 and subsequent production.

PZL-104MW Wilga 2000 Hydro: Floatplane version; first flight 11 September 1999. Maximum T-O weight 1,500 kg (3,306 lb); performance generally similar to landplane except T-O run from water 180 m (590 ft) and S/L climb rate 210 m (689 ft)/min.

CUSTOMERS: Ten reported orders for Wilga 2000 by late 1996, including four for Polish National Aeroclub and three for export. First two deliveries in 1997 to Polish Aeroclub and private owner; five built in 1998 and delivered under 1997 contract to Polish Border Guard, first one being handed over 1 June 1998; one built in 1999, delivered to UK as demonstrator in May 2000. None built in 2000. At least five built in 2001, increasing production to 14 (including prototype).

COSTS: Approximately US$100,000 (1996).

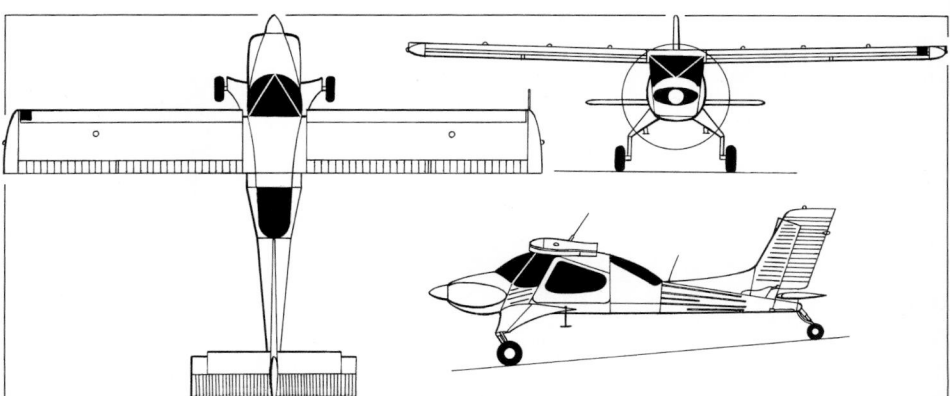

PZL-104M Wilga 2000 (224 kW; 300 hp IO-540 engine) (*Jane's/James Goulding*) 0110900

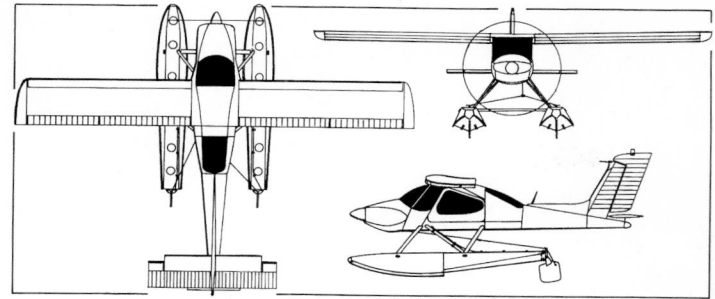

Floatplane version of Wilga 2000   0092544

Underfuselage FLIR turret on Polish Border Guard Wilga 2000
(R J Malachowski)
0110899

DESIGN FEATURES: Suitable for wide variety of military, general aviation and flying club duties; STOL performance bestowed by slats, flaps and drooping ailerons, allied to heavy-duty landing gear. High-mounted cantilever wings; braced tail unit; tall landing gear legs. In addition to Western equipment, Wilga 2000 features increased fuel, strengthened wing with integral fuel tanks and shorter mainwheel legs enclosed in fairings. Has 70 per cent structural commonality with original Wilga.
Wing section NACA 2415; dihedral 1°.

FLYING CONTROLS: Conventional and manual. Aerodynamically and mass-balanced slotted ailerons can be drooped to supplement flaps during landing; tab on starboard aileron; aerodynamically, horn- and mass-balanced one-piece elevator and rudder; trim tab in centre of elevator; manually operated slotted flaps; fixed slat on wing leading-edge along full span.

STRUCTURE: All-metal, with beaded skins; single-spar wings, with leading-edge torsion box; fuselage in two portions, forward incorporating main wing spar carry-through structure; rear section is tailcone; cabin floor of metal sandwich, with paper honeycomb core, covered with foam rubber; aluminium tailplane bracing strut.

LANDING GEAR: Non-retractable tailwheel type. Semi-cantilever main legs, of rocker type, have oleo-pneumatic shock-absorbers. Low-pressure tyres size 500×200 on mainwheels. Hydraulic brakes. Steerable tailwheel, tyre size 255×110, carried on rocker frame with oleo-pneumatic shock-absorber. Metal ski landing gear optional. Airtech Canada CAP 3000 twin-float gear on Wilga 2000 Hydro.

POWER PLANT: One 224 kW (300 hp) Textron Lycoming IO-540-K1B5 flat-six engine, driving a Hartzell HC-C3YR-1RF/F8468A-6R three-blade constant-speed propeller. Fuel in integral wing tanks, total capacity 400 litres (106 US gallons; 88.0 Imp gallons).

ACCOMMODATION: Passenger version accommodates pilot and three passengers, in pairs, with adjustable front seats. Baggage compartment aft of seats, capacity 35 kg (77 lb). Rear seats can be replaced by additional fuel tank for longer-range operation. Upward-opening door on each side of cabin, jettisonable in emergency.

AVIONICS: *Comms:* Bendix/King KX 155 main nav/com/glideslope and KY 96A standby VHF transceiver; KT 76A transponder; KMA 24 audio control/MKR; intercom.
Flight: Bendix/King KLN 89B GPS receiver and KR 87 ADF.
Instrumentation: ASI, VSI; ADF, VOR and twin indicators; artificial horizon; directional gyro; magnetic compass; altimeter; stall warning device; signalling and warning panel; clock; rpm tachometer; standard fuel/pressure/temperature gauges and indicators.
Mission: Polish Border Guard aircraft equipped with FLIR.

EQUIPMENT: Heavy-duty glider-towing hook, release handle and towing mirrors available as option.

DIMENSIONS, EXTERNAL:
Wing span	11.12 m (36 ft 5¾ in)
Wing chord, constant	1.40 m (4 ft 7 in)
Wing aspect ratio	8.2
Length overall: 2000	8.10 m (26 ft 6¾ in)
2000 Hydro	8.52 m (27 ft 11½ in)
Height overall: 2000	2.96 m (9 ft 8½ in)
2000 Hydro: on water	approx 3.00 m (9 ft 10 in)
on land	3.57 m (11 ft 8½ in)
Tailplane span	3.70 m (12 ft 1¾ in)
Wheel track	2.75 m (9 ft 0¼ in)
Wheelbase	6.70 m (21 ft 11¾ in)
Distance between float c/l:	
2000 Hydro	2.48 m (8 ft 1¾ in)
Propeller diameter	2.03 m (6 ft 8 in)
Passenger doors (each): Height	0.90 m (2 ft 11½ in)
Width	1.50 m (4 ft 11 in)

DIMENSIONS, INTERNAL:
Cabin: Length	2.20 m (7 ft 2½ in)
Max width	1.20 m (3 ft 10 in)
Max height	1.50 m (4 ft 11 in)
Floor area	2.20 m² (23.8 sq ft)
Volume	2.40 m³ (85 cu ft)
Baggage compartment	0.50 m³ (17.5 cu ft)

AREAS:
Wings, gross	15.00 m² (161.5 sq ft)
Tailplane	3.16 m² (34.01 sq ft)
Elevator, incl tab	1.92 m² (20.67 sq ft)

WEIGHTS AND LOADINGS:
Weight empty	900 kg (1,984 lb)
Max T-O weight	1,400 kg (3,086 lb)
Max wing loading	93.3 kg/m² (19.12 lb/sq ft)
Max power loading	6.26 kg/kW (10.29 lb/hp)

PERFORMANCE (landplane):
Never-exceed speed (V_{NE})	131 kt (243 km/h; 151 mph)
Max level speed	113 kt (211 km/h; 131 mph)
Cruising speed at 75% power	103 kt (190 km/h; 118 mph)
Stalling speed: flaps up	58 kt (106 km/h; 66 mph)
flaps down	49 kt (89 km/h; 56 mph)
Max rate of climb at S/L	293 m (961 ft)/min
Service ceiling	4,118 m (13,510 ft)
T-O run (concrete)	168 m (555 ft)
Landing run (concrete)	150 m (495 ft)
Max range	809 n miles (1,500 km; 932 miles)

UPDATED

EADS PZL PZL-106BT TURBO-KRUK
English name: Turbo-Raven

TYPE: Agricultural sprayer.

PROGRAMME: Original piston-engined PZL-106 designed early 1972; first flight (SP-PAS), with IO-720 engine, 17 April 1973; last piston-engined aircraft built in 1990. First flight of turboprop-powered prototype (SP-PAA) 18 September 1985; FAR Pt 23 certification early 1994; features include taller fin and increased payload. A Turbo-Kruk was flown experimentally in 1997 with a 559 kW

(750 shp) P&WC PT6A-34AG turboprop, modified cockpit and minor aerodynamic improvements.

CURRENT VERSIONS: **PZL-106BT-601:** Standard production version with Walter M 601D engine.
PZL-106BTU-34: Export version with PT6A-34AG turboprop; first flight (SP-PBW, 262nd Kruk built) 19 August 1998. Redesigned engine cowling, improved electrical system and new instrument panel. Aimed initially at Latin American market.

Following description applies to current production PZL-106BT-601 except where indicated.

CUSTOMERS: Total of 39 built by mid-2000, including one BT-34 in 1998 and two BT-601s in 1999. Export deliveries include Argentina (21), Egypt, Iran, South Africa (five) and Sudan. Overall total of 270 (all versions) built up to mid-2000, comprising eight prototypes, 143 PZL-106As (of which two converted to -106B) between 1975 and 1981, 79 PZL-106BR/BSs and 42 PZL-106BTs. No known production in 2000 and 2001, although promotion continues.

DESIGN FEATURES: Typical low-, braced-wing agricultural monoplane with hopper at CG and high cockpit with good forward view.
Wings have upward-cambered tips; braced tailplane. Sweepback 6° at quarter-chord; 6° dihedral.

FLYING CONTROLS: Conventional and manual. Slotted ailerons (each with electrically actuated trim tab); mass-balanced elevators (with port trim tab); rudder has trim tab and anti-servo tab. Full-span, four-segment slats on each wing leading-edge; trailing-edge flaps.

STRUCTURE: Duralumin wings with metal and polyester fabric covering and GFRP tips; flaps and ailerons of duralumin with polyester fabric skins; GFRP/foam core sandwich slats; duralumin V bracing struts. Welded steel tube

PZL-106BT Turbo-Kruk agricultural aircraft (Walter M 601D turboprop)   *NEW*/0143731

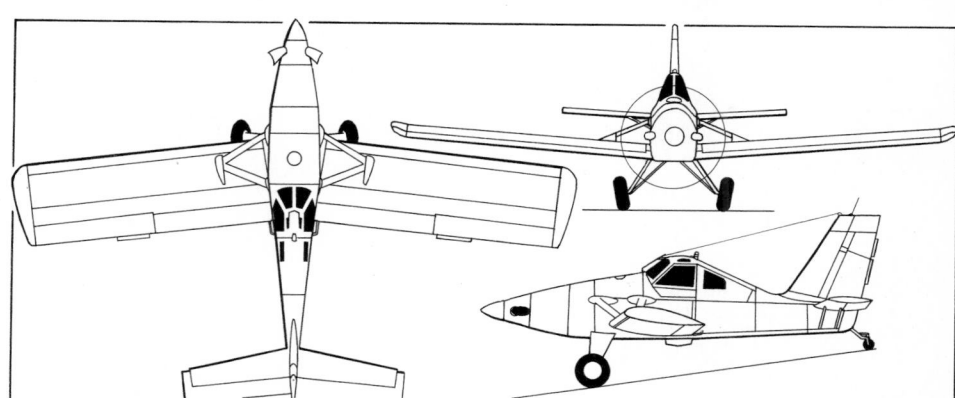

PZL-106BT Turbo-Kruk with a PT6A-34AG turboprop and aerodynamic refinements *(Jane's/Paul Jackson)*
0079002

fuselage, with polyurethane enamel coating and quickly removable light alloy and GFRP panels; structure can be pressure tested for cracks. Duralumin tail unit has metal-skinned fixed surfaces and polyester fabric-covered control surfaces. Structure corrosion resistant, and additionally protected by external finish of polyurethane enamel.

LANDING GEAR: Non-retractable tailwheel type, with oleo-pneumatic shock-absorber in each unit. Mainwheels, with 800×260 low-pressure tyres (2.00 bar; 29 lb/sq in), are each carried on a side V and half-axle. Steerable tailwheel has 350×135 tubeless tyre, pressure 2.50 bar (36 lb/sq in). Pneumatically operated mainwheel hydraulic disc brakes; parking brake.

POWER PLANT: *PZL-106BT-601:* One 548 kW (735 shp) Walter M 601D-1 turboprop; Avia V 508 D three-blade propeller.

PZL-106BTU-34: One 559 kW (750 shp) Pratt & Whitney Canada PT6A-34AG turboprop and Hartzell HC-B3TN-3D three-blade propeller.

Fuel in two integral wing tanks, total capacity 560 litres (148 US gallons; 123 Imp gallons); can be increased to total of 950 litres (251 US gallons; 209 Imp gallons) by using hopper as auxiliary fuel tank. Gravity refuelling point on each wing; semi-pressurised refuelling point on starboard side of fuselage.

ACCOMMODATION: Single vertically adjustable seat in enclosed, ventilated and heated cockpit with steel tube overturn structure. Provision for instructor's cockpit with basic dual controls, forward of main cockpit and offset to starboard, for training of pilots in agricultural duties. Optional rearward-facing second seat (for mechanic) to rear. Jettisonable window/door on each side of cabin. Pilot's seat and seat belt designed to resist 40 *g* impact. Cockpit air conditioning optional.

AVIONICS: *Comms:* VHF com transceiver standard; 720-channel UHF transceiver optional.

EQUIPMENT: Easily removable non-corroding (GFRP) hopper/tank, forward of cockpit, can carry more than 1,000 kg (2,205 lb) of dry or liquid chemical, and has maximum capacity of 1,400 litres (370 US gallons; 308 Imp gallons). Enlarged hopper (capacity 1,500 kg; 3,307 lb or 1,500 litres; 397 US gallons; 330 Imp gallons) optional.

Turnaround time, with full load of chemical, is approximately 28 seconds. Hopper quick-dump system can release 1,000 kg of chemical in 5 seconds or less. Pneumatically operated intake for loading dry chemicals optional. Distribution system for liquid chemical (jets or atomisers) powered by fan-driven centrifugal pump. Tunnel spreader, with positive on/off action for dry chemicals. For ferry purposes, hopper can be used to carry additional fuel instead of chemical.

When converted into two-seat trainer (see Accommodation), standard hopper can be replaced easily by container with reduced capacity tank for liquid chemical. Steel cable cutter on windscreen and each mainwheel leg; steel deflector cable runs from top of windscreen cable cutter to tip of fin. Windscreen washer and wiper standard. Other equipment includes clock, rearview mirror, second (mechanic's) seat (optional), landing light, anti-collision light, and night working lights (optional).

DIMENSIONS, EXTERNAL:

Wing span	15.00 m (49 ft 2½ in)
Wing chord, constant	2.16 m (7 ft 1 in)
Wing aspect ratio	7.1
Length overall	10.34 m (33 ft 11 in)
Height overall	4.34 m (14 ft 2¾ in)
Tailplane span	5.77 m (18 ft 11¼ in)
Wheel track	3.10 m (10 ft 2 in)
Wheelbase	7.41 m (24 ft 3¾ in)
Avia propeller diameter	2.50 m (8 ft 2½ in)
Avia propeller ground clearance (tail down)	
	0.63 m (2 ft 0¾ in)

Seventh production PZL-110 Koliber 160A *(R J Malachowski)*   *NEW*/0143732

Crew doors (each): Height	0.91 m (2 ft 11¾ in)
Width	1.06 m (3 ft 5¾ in)

DIMENSIONS, INTERNAL:

Cabin: Length	1.37 m (4 ft 6 in)
Max width	1.25 m (4 ft 1¼ in)
Max height	1.30 m (4 ft 3¼ in)
Floor area	1.12 m² (12.05 sq ft)

AREAS:

Wings, gross	31.69 m² (341.1 sq ft)
Vertical tail surfaces (total)	3.44 m² (37.03 sq ft)
Horizontal tail surfaces (total)	7.56 m² (81.38 sq ft)

WEIGHTS AND LOADINGS:

Weight empty, equipped: BT-601	1,680 kg (3,704 lb)
BTU-34	1,656 kg (3,651 lb)
Max chemical payload	1,500 kg (3,307 lb)
Max T-O weight	3,500 kg (7,716 lb)
Max landing weight	3,000 kg (6,614 lb)
Max wing loading	110.4 kg/m² (22.61 lb/sq ft)
Max power loading	6.39 kg/kW (10.50 lb/shp)

PERFORMANCE (at 3,000 kg; 6,614 lb AUW, S/L, ISA. A: BT-601 clean, B: BT-601 with tunnel spreader, C: BTU-34):

Never-exceed speed (VNE):	
A, B	145 kt (270 km/h; 167 mph)
Max level speed: A	130 kt (240 km/h; 149 mph)
Operating speed:	
B	86-97 kt (160-180 km/h; 99-112 mph)
Stalling speed, flaps up, power off:	
A	59 kt (108 km/h; 68 mph)
B	61 kt (112 km/h; 70 mph)
Stalling speed, flaps down: C	42 kt (76 km/h; 48 mph)
Max rate of climb: A	378 m (1,240 ft)/min
B	330 m (1,083 ft)/min
C, clean	336 m (1,102 ft)/min
Service ceiling: A	6,500 m (21,320 ft)
T-O run: A	240 m (790 ft)
B	270 m (885 ft)
C	280 m (920 ft)
Landing run: A	260 m (855 ft)
B	270 m (885 ft)
C	275 m (900 ft)
Range, wing tanks only, no reserves:	
C	594 n miles (1,100 km; 683 miles)
Swath width: B: atomisers	45 m (148 ft)
spray nozzles	30 m (98 ft)
tunnel spreader	25 m (82 ft)

UPDATED

EADS PZL PZL-110 KOLIBER 160
English name: Hummingbird

TYPE: Four-seat lightplane.

PROGRAMME: Licence-built and uprated version of Socata Rallye 100 ST; original (Morane-Saulnier) Rallye flew 10 June 1959; main production in France until 1983.

First PZL-110, modified to receive 86.5 kW (116 hp) PZL-F (Franklin) engine, first flown 18 April 1978; first flight of Koliber 150 prototype (SP-PHA) 27 September 1988; Polish type certificate for 150 awarded January 1989, FAA certificate for 150A February 1995. See 1991-92 and earlier *Jane's* for details of initial **Series I/II/III** (40 built), Koliber 150 and Koliber 150A/Koliber II. Koliber 160A introduced in 1998; production deliveries from 1999.

CURRENT VERSIONS: **Koliber 160A:** Introduced 1998 and is now the only production version. First flight (SP-WGF) 15 April 1998; second prototype (SP-WGG) followed 25 May 1998. First two production aircraft delivered to PZL International at North Weald, UK, in September and October 1999 for onward sale.

Following description applies to Koliber 160A.

Koliber 235A (PZL-111): More powerful version; described separately.

Koliber Junior (PZL-112): Two-seat, new-design primary trainer; described separately.

CUSTOMERS: Koliber 150s and 150As sold in Canada, Denmark, Germany, Netherlands, Sweden, UK and USA. Total of 79 Koliber 110/150/150As built by 1995, of which five held in stock during 1997; no production in 1996-98. Two stock 150As converted to 160A prototypes and delivered to UK as demonstrators in 1998. Eight production 160As delivered by late 2001, including one development aircraft, three supplied to UK distributor and two to USA.

COSTS: Koliber 160A, US$115,000 (1998). Avionics options US$13,200 extra (Nav I) or US$14,700 (Nav II).

DESIGN FEATURES: Winner of French competition for ideal flying club aircraft (Rallye); side-by-side seating, nosewheel landing gear, leading-edge slats for low-speed safety (minimum tendency to stall) and simple, inexpensive construction. Constant-chord wing, tailplane and fin, last-mentioned with sweepback.

Wing section NACA 63A-416 (modified); dihedral 7° 7′ 30″; incidence 4°.

FLYING CONTROLS: Conventional and manual. Pushrod-actuated, aerodynamically and mass-balanced ailerons, each with ground-adjustable tab; aerodynamically and mass-balanced elevator with central control tab, rudder with ground-adjustable tab; full-span automatic leading-edge slats; electrically actuated Fowler flaps.

STRUCTURE: All-metal; flaps, elevator and rudder have corrugated skins; wing torsion box and trailing-edge segments electrically spot-welded.

LANDING GEAR: Non-retractable tricycle type, with leg fairings and oleo-pneumatic shock-absorption; castoring nosewheel. Tyre sizes 5.00-4 (nose) and 6.00-6 (main) with pressures 1.40 and 1.80 bar (20 and 26 lb/sq in) respectively. Toe-operated hydraulic disc brakes. Parking brake.

POWER PLANT: One 119 kW (160 hp) Textron Lycoming O-320-D2A flat-four engine, driving a Sensenich 74DM6-0-58 two-blade fixed-pitch metal propeller. Fuel in two metal tanks in wings, with total usable capacity of 161 litres (42.5 US gallons; 35.4 Imp gallons). Refuelling points above wings. Oil capacity 7.6 litres (2.0 US gallons; 1.7 Imp gallons).

ACCOMMODATION: Two fore-and-aft adjustable, side-by-side seats, plus two-person bench seat at rear, under large rearward-sliding canopy. Dual controls. Heating, ventilation and soundproofing standard.

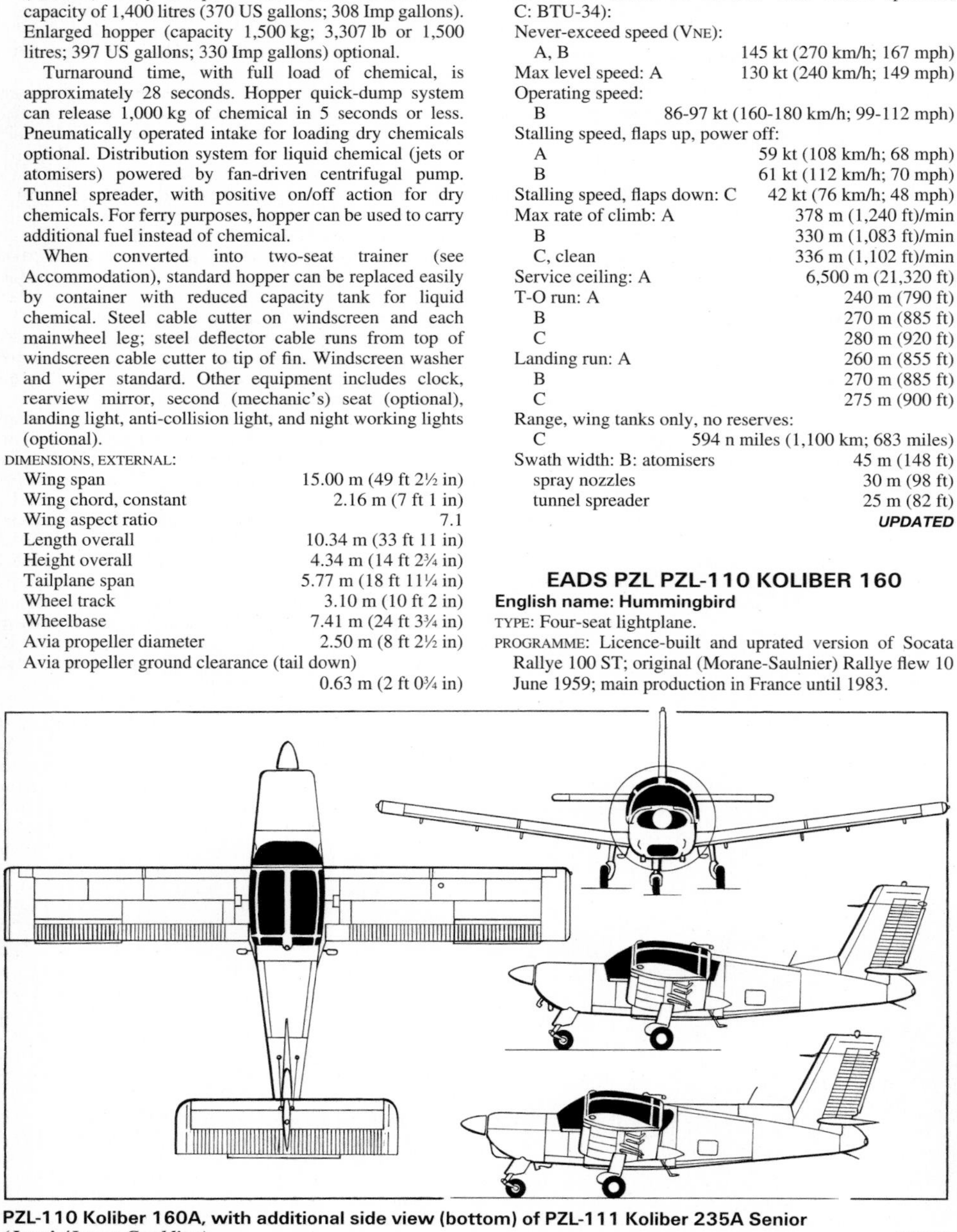

PZL-110 Koliber 160A, with additional side view (bottom) of PZL-111 Koliber 235A Senior *(Jane's/James Goulding)*   0051170

Instrument panel of a production PZL-110 Koliber 160A *(Jane's/Paul Jackson)*   0092511

SYSTEMS: 14 V electrical system, with 70 Ah alternator and 70 Ah battery.

AVIONICS: *Comms:* Bendix/King KX 155 nav/com/glideslope; KT 76A transponder; SPA 400 intercom; ELT.
   Flight: Bendix/King KI 204 VOR/LOC/GS and KMA 24 audio control/MKR. Audible stall warning system standard. Two optional Bendix/King avionics packages: *Nav I* (KLN 89 GPS, KX 155A nav/com/glideslope, KR 87 ADF and KI 209 VOR/LOC/glideslope; or *Nav II* (KLN 89B GPS/IFR, MD 41-228 GPS nav, KX 155A glideslope, KR 87 ADF and KI 209A VOR/LOC/glideslope with GPS interface).
   Instrumentation: Standard VFR; IFR applied for.

EQUIPMENT: Position, anti-collision and landing/taxying lights; instrument panel, cabin and map lights; fixed cabin entrance steps; tiedown rings; fire extinguisher; winterisation kit.

DIMENSIONS, EXTERNAL:

Wing span	9.74 m (31 ft 11½ in)
Wing chord, constant	1.30 m (4 ft 3 in)
Wing aspect ratio	7.5
Length overall	7.37 m (24 ft 2¼ in)
Height overall	2.80 m (9 ft 2¼ in)
Tailplane span	3.67 m (12 ft 0½ in)
Wheel track	2.03 m (6 ft 8 in)
Wheelbase	1.71 m (5 ft 7¼ in)
Propeller diameter	1.88 m (6 ft 2 in)

DIMENSIONS, INTERNAL:

Cabin max width	1.08 m (3 ft 6½ in)

AREAS:

Wings, gross	12.68 m² (136.5 sq ft)
Ailerons (total)	1.56 m² (16.79 sq ft)
Trailing-edge flaps (total)	2.40 m² (25.83 sq ft)
Vertical tail surfaces (total)	1.74 m² (18.73 sq ft)
Horizontal tail surfaces (total)	3.48 m² (37.50 sq ft)

WEIGHTS AND LOADINGS (Normal category):

Weight empty, equipped	607 kg (1,338 lb)
Max fuel weight	116 kg (256 lb)
Max T-O and landing weight	950 kg (2,094 lb)
Max wing loading	67.0 kg/m² (13.72 lb/sq ft)
Max power loading	7.60 kg/kW (12.49 lb/hp)

PERFORMANCE (Normal category):

Never-exceed speed (VNE)	136 kt (251 km/h; 156 mph)
Max level speed	119 kt (220 km/h; 137 mph)
Cruising speed at 75% power	105 kt (194 km/h; 121 mph)
Econ cruising speed	75 kt (140 km/h; 87 mph)
Stalling speed: flaps up	49 kt (89 km/h; 56 mph)
Max rate of climb at S/L	210 m (690 ft)/min
Service ceiling	3,500 m (11,480 ft)
T-O run at S/L	168 m (555 ft)
T-O to 15 m (50 ft) at S/L	397 m (1,305 ft)
Landing from 15 m (50 ft) at S/L	320 m (1,050 ft)
Landing run at S/L	137 m (450 ft)
Range with max fuel, no reserves:	
at 75% power	496 n miles (919 km; 571 miles)
at 65% power	518 n miles (960 km; 596 miles)
g limits	+3.8/−1.5

UPDATED

EADS PZL PZL-111 KOLIBER 235A
English name: Hummingbird

TYPE: Four-seat lightplane.

PROGRAMME: Construction of prototype of higher-powered version of now-discontinued Koliber 150A began August 1990; static testing completed March 1993. Prototype (SP-PHI; later -WHI) made its first flight on 14 September 1995. Polish certification 13 December 1996. No production aircraft had emerged by mid-2002, but continues to be offered. Proposed Koliber 180 and 200, with alternatively rated engines (see 1996-97 *Jane's*), have not yet joined the range.

Description as for Koliber 160A except as follows:

COSTS: US$158,000 (1998) with basic instrumentation. Nav I and II avionics options as for Koliber 160A; glider tow hook option US$2,250.

DESIGN FEATURES: Numerous small differences from PZL-110 include extended dorsal fin, windows to rear of canopy, modified engine cowling and (in production version)

square-topped fin. Intended for basic, navigation and general training; glider towing; sport flying; touring; executive flights; and special missions.

LANDING GEAR: Goodyear 6 ply tyres, size 350×35 (nose) and 6.00-6 Type III (main).

POWER PLANT: One 175 kW (235 hp) Textron Lycoming O-540-B4B5 flat-six engine; Hartzell HC-C2YK-1BF/F8468A4 two-blade constant-speed metal propeller; three-blade McCauley B3D32C414/G-82NDA-6 propeller also available. Fuel capacity as for Koliber 160A.

ACCOMMODATION: Baggage space aft of seats.

AVIONICS: Narco or Honeywell nav/com, including GPS, for VFR and IFR operation.

EQUIPMENT: Baggage securing straps; optional glider towing hook.

DIMENSIONS, EXTERNAL: As Koliber 160A except:

Length overall	7.52 m (24 ft 8 in)
Propeller diameter: Hartzell	2.03 m (6 ft 8 in)
McCauley	1.93 m (6 ft 4 in)

WEIGHTS AND LOADINGS (Normal category):

Weight empty	705 kg (1,554 lb)
Max baggage weight	50 kg (110 lb)
Max T-O and landing weight	1,150 kg (2,535 lb)
Max wing loading	89.2 kg/m² (18.27 lb/sq ft)
Max power loading	6.56 kg/kW (10.79 lb/hp)

PERFORMANCE (Normal category):

Max level speed	140 kt (260 km/h; 162 mph)
Max cruising speed at 75% power	134 kt (248 km/h; 154 mph)
Stalling speed, flaps up	50 kt (92 km/h; 58 mph)
Max rate of climb at S/L	312 m (1,024 ft)/min
Service ceiling	4,500 m (14,775 ft)
T-O run	135 m (445 ft)
Landing run	171 m (560 ft)
Range with max fuel, no reserves:	
at 75% power	396 n miles (734 km; 456 miles)
at 60% power	517 n miles (958 km; 595 miles)

UPDATED

PZL-112 Junior prototype *(Jane's/Paul Jackson)*   *NEW*/0143705

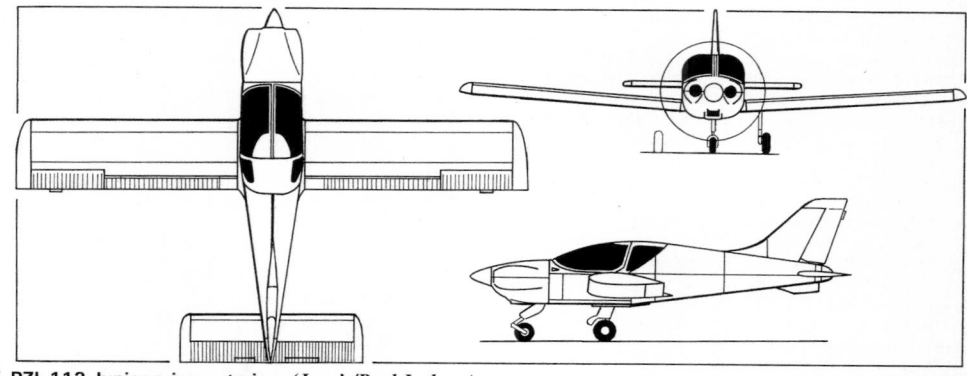

PZL-112 Junior primary trainer *(Jane's/Paul Jackson)*   *NEW*/0143735

EADS PZL PZL-112 JUNIOR

TYPE: Primary prop trainer/sportplane.

PROGRAMME: Designed in collaboration with Warsaw University of Technology; originally called Koliber Junior; prototype construction began in late 1997. Programme delayed; prototype (SP-PRG) first flew 17 June 2000, but still awaiting launch decision in 2002. To be certified to FAR/JAR 23 (VFR and IFR), including spinning and other basic aerobatics.

COSTS: Approximately US$40,000 (1997).

DESIGN FEATURES: Low-wing monoplane with sweptback vertical tail and small ventral fin; intended as inexpensive primary trainer for Polish aeroclubs; utilises some components of PZL-110 Koliber. Dihedral 4°.

FLYING CONTROLS: Conventional and manual. Two elevator trim tabs; ground-adjustable tabs on rudder and both ailerons; rudder and elevators horn-balanced. Fowler flaps.

STRUCTURE: Generally of metal. Fluted control surfaces.

LANDING GEAR: Non-retractable, trailing-link tricycle type; oleo-pneumatic shock-absorbers. Mainwheel tyres size 15×6.00-6 (4 ply rating), pressure 1.80 bar (26 lb/sq in); nosewheel tyre size 5.00×4 (6 ply rating), pressure 1.40 bar (20 lb/sq in). Hydraulic disc brakes.

POWER PLANT: One 86.5 kW (116 hp) Textron Lycoming O-235-L2C flat-four engine, driving a Hoffmann HO-14HM-175-131 two-blade, fixed-pitch propeller; Lycoming O-320-D2A, derated to 116 hp, optional. Fuel capacity 100 litres (26.4 US gallons; 22.0 Imp gallons), of which 95 litres (25.1 US gallons; 20.9 Imp gallons) are usable.

ACCOMMODATION: Two seats side by side in fully enclosed cabin. Two-piece gull-wing canopy.

DIMENSIONS, EXTERNAL:

Wing span	8.803 m (28 ft 10½ in)
Wing chord, constant	1.30 m (4 ft 3¼ in)
Wing aspect ratio	6.8
Length overall	6.561 m (21 ft 6¼ in)
Height overall	2.625 m (8 ft 7¼ in)
Tailplane span	3.15 m (10 ft 4 in)
Wheel track	1.97 m (6 ft 5½ in)
Wheelbase	1.513 m (4 ft 11½ in)
Propeller diameter	1.75 m (5 ft 9 in)
Propeller ground clearance	0.12 m (4¾ in)

AREAS:

Wings, gross	11.44 m² (123.1 sq ft)
Ailerons (total)	0.60 m² (6.46 sq ft)
Trailing-edge flaps (total)	2.40 m² (25.83 sq ft)
Fin	0.66 m² (7.10 sq ft)
Rudder	0.50 m² (5.38 sq ft)
Tailplane	1.36 m² (14.64 sq ft)
Elevators (total, incl tab)	1.48 m² (15.93 sq ft)

WEIGHTS AND LOADINGS:

Weight empty	520 kg (1,146 lb)
Max T-O weight	750 kg (1,653 lb)
Max wing loading	65.56 kg/m² (13.43 lb/sq ft)
Max power loading	8.67 kg/kW (14.25 lb/hp)

PERFORMANCE (estimated):

Never-exceed speed (VNE)	141 kt (262 km/h; 162 mph) EAS

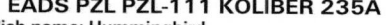

Prototype PZL-111 Koliber 235A *(R J Malachowski)*   *NEW*/0143733

Max level speed	113 kt (210 km/h; 130 mph)	Service ceiling	4,000 m (13,120 ft)	Landing run	130 m (425 ft)
Cruising speed	97 kt (180 km/h; 112 mph)	T-O run	180 m (590 ft)	Range	405 n miles (750 km; 466 miles)
Stalling speed, flaps down	44 kt (80 km/h; 50 mph)	T-O to 15 m (50 ft)	396 m (1,300 ft)	g limits	+4.4/–1.8
Max rate of climb at S/L	210 m (689 ft)/min	Landing from 15 m (50 ft)	290 m (950 ft)		*UPDATED*

E&K

E&K Sp. z o.o.
ulica Radziwiłłowska 5, PL-20-080 Lublin
Tel: (+48 81) 743 73 57
Fax: (+48 81) 743 66 12
PRESIDENT: Andrzej Izdebski

No recent reports of the SBM-03 Kos have been received,
although development appears to be continuing.

VERIFIED

E&K SBM-03 KOS

TYPE: Side-by-side ultralight.
PROGRAMME: Single-seat prototype (SP-PEG) first flew
August 1998; 55 hours by May 1999; start of certification
(in Germany), and first flight of two-seat prototype,
expected to follow shortly thereafter, but this aircraft, SP-
YAB, not produced until 2001. No reports of a first flight
received by mid-2002.
FLYING CONTROLS: Conventional and manual. Horn-balanced
rudder; flaps.
STRUCTURE: Glass fibre/epoxy construction; semi-monocoque
fuselage.
LANDING GEAR: Tailwheel type; fixed. Mainwheels on
cantilever, self-sprung leg bows.
POWER PLANT: One 59.6 kW (79.9 hp) Rotax 912 A4 (VLA
version) or 912 UL4 (ultralight) flat-four engine; KG-3-
150L three-blade propeller. Option of 47.8 kW (64.1 hp)
Rotax 582 UL in ultralight. Two fuel tanks aft of cockpit,
combined capacity 40 litres (10.6 US gallons; 8.8 Imp
gallons).
AVIONICS: *Comms:* Bendix/King KY 96A VHF radio.
   Flight: Navigation system of customer's choice.
   Instrumentation: Standard VFR.
EQUIPMENT: Ballistic parachute in front of cockpit.
DIMENSIONS, EXTERNAL:

Wing span	8.50 m (27 ft 10¾ in)
Length overall	5.95 m (19 ft 6¼ in)
Height overall	1.80 m (5 ft 10¾ in)

AREAS:

Wings, gross	10.20 m² (109.8 sq ft)

WEIGHTS AND LOADINGS:

Weight empty	265 kg (584 lb)
Max T-O weight	450 kg (992 lb)

Prototype two-seat SBM-03 Kos (*R J Malachowski*)   0110901

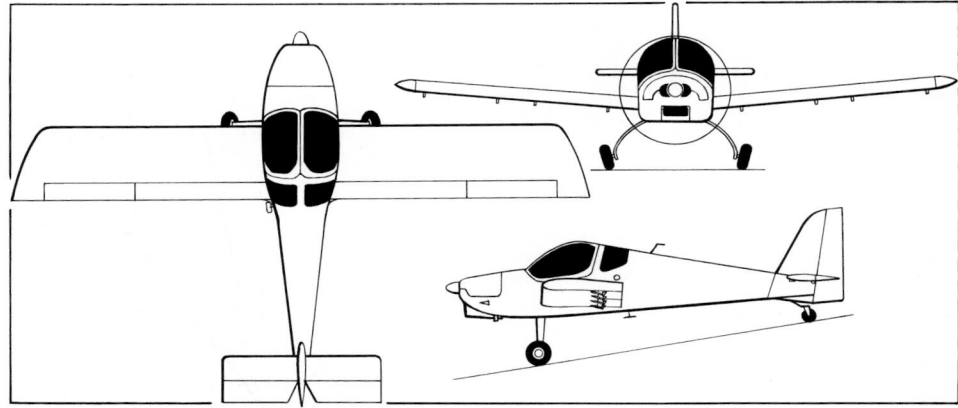

E&K SBM-03 Kos light aircraft (*Jane's/James Goulding*)   0062601

PERFORMANCE (Rotax 912 A4 engine):

Max level speed	113 kt (210 km/h; 130 mph)	Max rate of climb at S/L	540 m (1,771 ft)/min
Cruising speed at 75% power	94 kt (175 km/h; 109 mph)	Range with max fuel, 10% reserves	
Stalling speed: flaps up	45 kt (83 km/h; 52 mph)		367 n miles (680 km; 422 miles)
flaps down	39 kt (72 km/h; 45 mph)	g limits	+3.8/–2.5
			UPDATED

HE-RO'S

HE-RO'S DESIGN LUBLIN
PO Box 106, PL-20-950 Lublin 1
Tel/Fax: (+48 81) 743 47 70
MANAGER AND DESIGNER: H P Rozwadowski

UK ADDRESS:
92 Thorn Court, Manchester M6 5EL
Tel/Fax: (+44 161) 737 93 44

Design work on what was intended to become a related series
of aircraft began in 1993. In March 2000, design of the HB-2

was said to be approaching finalisation, although no further
progress had been reported by mid-2002.
   Company's first development partner is Aerokris ZPHU, a
Polish plans-based manufacturer of the Zenair CH 601;
research undertaken by Lublin Technical University.

UPDATED

IL

INSTYTUT LOTNICTWA (Aviation Institute)
aleja Krakowska 110/114, PL-02-256 Warszawa
Tel: (+48 22) 846 00 11 and 846 38 12
Fax: (+48 22) 846 44 32 and 846 42 32
e-mail: ilot@ilot.edu.pl
Web: http://www.ilot.edu.pl
GENERAL DIRECTOR: Dr Eng Witold Wisniowski

Aircraft Department
Address and fax as above
Tel: (+48 22) 846 01 71
CHIEF DESIGNER, I-23: Dr Eng Alfred Baron

Founded 1926; directly subordinate to Ministry of Heavy
and Machine Building Industry and responsible for most
research and development work in Polish aviation industry;
conducts scientific research, including investigation of
problems associated with low-speed and high-speed
aerodynamics, static and fatigue tests, development and
testing of aero-engines, flight instruments, space science
instrumentation, and other equipment, flight tests, and
materials technology; also responsible for construction of
aircraft and aero-engines. Workforce approximately 400 in
2001.
   In 1997 IL signed an agreement with Boeing covering
possible co-operation in 20 advanced technology areas
related to the latter's sales campaign for F-18 Hornets to the
Polish Air Force.

UPDATED

IL I-23 MANAGER

Responsibility for series production allocated to PZL-
Świdnik, under whose heading description can now be found.

UPDATED

IL IS-2

TYPE: Two-seat helicopter.
PROGRAMME: Initiated 1 June 1994; three prototypes planned
(first and second for static and ground resonance testing,
third for flight test); construction started 1 June 1995 but
not completed until 2002; rolled out June 2002 and first
flight expected in second half of 2002; to be certified to
JAR 27.
COSTS: Estimated programme cost US$3.5 million; target unit
price US$180,000 (2000).
DESIGN FEATURES: Designed for basic training and pipeline/
border patrol, with low cost of operation and high level of
safety. Conventional pod and boom configuration, with
three-blade main rotor and four-blade Fenestron-type
ducted tail rotor. Comparatively low rotation rates of both
rotors minimise noise.
   Main rotor blades have IL HX4A1 aerofoil section with
thickness/chord ratios of 12 per cent at root and 9 per cent
at tip.
FLYING CONTROLS: Conventional and manual.
STRUCTURE: Fuselage truss frame of steel tube and aluminium
alloy, with glass fibre skin in cockpit area; aluminium alloy
semi-monocoque tailboom; glass fibre composites tailfin;
all rotor blades of glass fibre/epoxy and carbon fibre
composites.
LANDING GEAR: Conventional metal skid type, with small
ground wheels.

POWER PLANT: One 134 kW (180 hp) Textron Lycoming
O-360 flat-four engine. Fuel in two tanks, total capacity
110 litres (29.1 US gallons; 24.2 Imp gallons).
ACCOMMODATION: Two seats side by side; forward-opening
window/door on each side. Baggage space aft of seats.
Cabin heated and ventilated.
SYSTEMS: No hydraulic or pneumatic systems; 28 V DC
electrical system. De-icing optional.
AVIONICS: *Comms:* Bendix/King KT 76A transponder.
   Flight: Bendix/King KLX 135 GPS/com and KR 87
ADF.
   Instrumentation: Conventional; includes directional
gyro and Goodrich attitude gyro.
DIMENSIONS, EXTERNAL:

Main rotor diameter	7.50 m (24 ft 7¼ in)
Main rotor blade chord: at root	0.20 m (7¾ in)
at tip	0.15 m (6 in)
Tail rotor diameter	0.865 m (2 ft 10 in)
Length overall, rotor turning	8.93 m (29 ft 3½ in)
Fuselage: Length	7.07 m (23 ft 2¼ in)
Max width	1.40 m (4 ft 7 in)
Height overall	2.83 m (9 ft 3½ in)
Tailplane span	1.18 m (3 ft 10½ in)

DIMENSIONS, INTERNAL:

Cabin: Length	1.66 m (5 ft 5¼ in)
Max width	1.35 m (4 ft 5¼ in)
Max height	1.35 m (4 ft 5¼ in)
Floor area	1.60 m² (17.2 sq ft)
Volume	2.2 m³ (78 cu ft)

AREAS:

Main rotor disc	44.18 m² (475.5 sq ft)
Tail rotor disc	0.59 m² (6.35 sq ft)

Fin, incl tail rotor cover	1.16 m² (12.49 sq ft)
Tailplane	0.47 m² (5.06 sq ft)

WEIGHTS AND LOADINGS:

Weight empty	550 kg (1,213 lb)
Max internal fuel	82 kg (181 lb)
Max payload	100 kg (220 lb)
Normal T-O weight	790 kg (1,742 lb)
Disc loading at normal T-O weight	17.9 kg/m² (3.66 lb/sq ft)
Transmission loading at normal T-O weight and power	5.89 kg/kW (9.67 lb/hp)

PERFORMANCE (estimated):

Never-exceed speed (V$_{NE}$)	118 kt (220 km/h; 136 mph)
Max level speed at 500 m (1,640 ft)	104 kt (193 km/h; 120 mph)
Max cruising speed at 500 m (1,640 ft)	92 kt (170 km/h; 106 mph)
Econ cruising speed at 500 m (1,640 ft)	84 kt (155 km/h; 96 mph)
Max rate of climb at S/L	390 m (1,280 ft)/min
Time to 2,000 m (6,560 ft)	6 min
Service ceiling	3,450 m (11,320 ft)
Hovering ceiling: IGE	1,250 m (4,100 ft)
OGE	1,000 m (3,280 ft)
Max range	216 n miles (400 km; 248 miles)
Max endurance	4 h 5 min

UPDATED

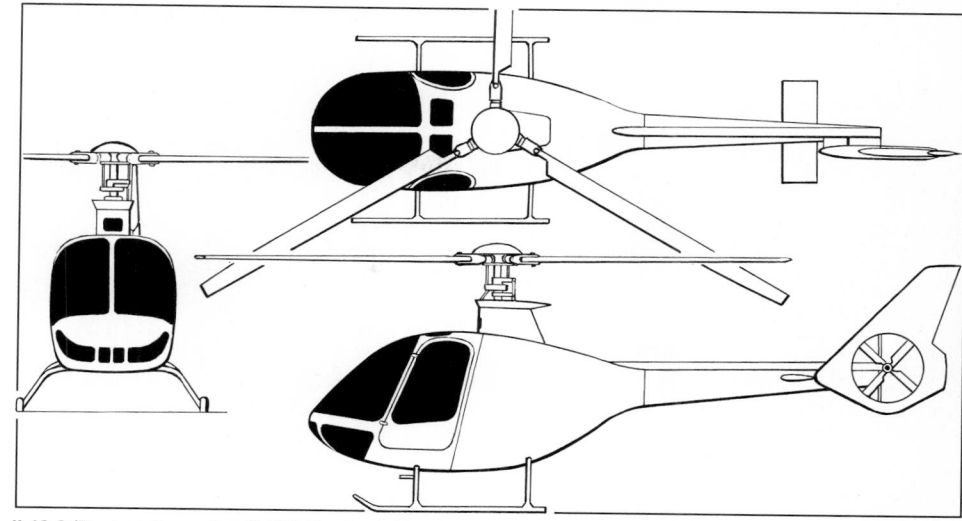

IL IS-2 (Textron Lycoming O-360 flat-four) (*Jane's/James Goulding*)   NEW/0526899

Prototype IS-2 two-seat helicopter, as seen in mid-2002 (*R J Malachowski*)   NEW/0525748

PZL

POLSKIE ZAKŁADY LOTNICZE Sp. z o.o.
(Polish Aviation Factory LLC)
ulica Wojska Polskiego 3, PL-39-300 Mielec
Tel: (+48 17) 788 79 21
Fax: (+48 17) 788 78 29
e-mail: pzl@ptc.pl
Web: http://www.pzlmielec.pl
COMMERCIAL DIRECTOR: Władysław Yasiczek
OPERATIONS DIRECTOR: Tadeusz Skubisz

As the result of a restructuring process carried out by the Polish government, the former WSK-PZL Mielec and its subsidiary PZL Mielec Aircraft Company Ltd (see 1999-2000 *Jane's* for previous history) were subjected to liquidation proceedings in 1998. To replace the parent company, a new business entity under the above title came into operation on 23 October 1998. With its more than 60 years of history, an early 2002 workforce of more than 1,600, and the production facilities, technology, design and testing capabilities inherited from its predecessor, Polish Aviation Factory LLC remains the largest and best equipped aircraft manufacturing plant in Poland. It is licensed to operate within the Europark Mielec special economic zone. In July 1999, its production facilities were certified by Poland's General Inspectorate of Civil Aviation under JAR 21 Subsection G, followed in September 1999 by design organisation approval under JAR 21 Subsection A. Quality management system (QMS) certification has been awarded by BAE Systems, and similar approval from Boeing was awarded subsequently. In mid-2001, the Factory was progressing towards ISO 9001 and AQAP-110 certification.

PZL's current indigenous aircraft programmes comprise the M-18 Dromader, M-26 Iskierka, M-28B Bryza and M-28

Skytruck. Four An-2s delivered to the Vietnamese People's Air Force in January 2002 following brief reopening of the Polish production line to complete aircraft abandoned a decade earlier; Chinese Y-5 manufacture (which see) continues. Subcontract work continues for BAE Systems (Hawk), Boeing (757), GKN Westland (for Boeing 737) and Saab (Gripen).

UPDATED

A stock of uncompleted airframes allows PZL to deliver An-2s to special order (*R J Malachowski*) NEW/0137425

PZL (ANTONOV) M-28B BRYZA
English name: Breeze
NATO reporting name: Cash
TYPE: Multirole twin-turboprop.
PROGRAMME: Developed by Antonov in former USSR as An-28; robust utility transport for service on Aeroflot's

shortest routes, particularly those operated by An-2s into places relatively inaccessible to other fixed-wing aircraft; official Soviet flight testing completed 1972; first preproduction An-28 (SSSR-19723) originally retained same engines as prototype, but re-registered SSSR-19753 April 1975 when flown with current engines; production assigned to PZL Mielec 1978; temporary Soviet NLGS-2 type certificate awarded 4 October 1978 to second Soviet-built preproduction aircraft. Polish manufacture started with initial batch of 15; first flight of Polish An-28 (SSSR-28800) 22 July 1984; version received full Soviet type certificate 7 February 1986. See 1994-95 and earlier *Jane's* for full An-28 description.

Polish production now for domestic military use only, as M-28 Bryza; current marketing effort concentrated on M-28 Skytruck derivative (which see).

Polish Navy M-28M Bryza E with boom-type SLAR antenna (one of a pair) on rear fuselage and auxiliary fuel tank above mainwheel sponson *(R J Malachowski)*   NEW/0137426

New nose shape of the MPA/ASW version of the Bryza *(R J Malachowski)*   NEW/0137405

CURRENT VERSIONS: **M-28B Bryza R:** Originally designated An-28RM Bryza 1RM (*Ratownictwa Morskiego:* maritime reconnaissance). Prototype (SP-PDC) first flown June 1992; to Polish Navy June 1993 for evaluation; later upgraded and redelivered (as 0810) March 1999. First production example (serial number 1022) delivered 25 October 1994 and since upgraded with Bendix/King Silver Crown avionics. Four more (1006, 1114, 1115 and 1116) delivered to Polish Navy during 2000; eventual requirement reportedly for 12. Upgrade continuing with introduction of Hartzell five-blade propellers; new-production aircraft also to have new (RDR-2000) weather radar and central, single-point refuelling system. Further plans include semi-retractable landing gear and provision for additional fuel tankage.

M-28 Bryza TD: (*Towaru Desantowy:* cargo airdrop): Transport/paradrop version, designated for airdrop of paratroops or cargo containers, with folding seats, static lines and roller floor as appropriate. In medevac configuration, attachment provisions for six stretchers, oxygen mask container, resuscitator and medicare containers. Clamshell rear doors replaced by single door sliding forward under fuselage, similar to that of An-26. Previously known as An-28B1T, then An-28TD.

Prototype (SP-PDE) completed as B1 September 1991; to TD configuration May 1994. First production TD (1003) to Polish Air Force October 1994, followed by prototype (as 0723) in May 1995. Two more (1008 and 1017) delivered in March 1999, followed by further pair (1117 and 1118) delivered 18 January 2002. Further 10 transports for Polish Air Force ordered 16 May 2001; deliveries began 26 March 2002 with 0203 and 0204; followed by 0205 and 0206 in May 2002. Polish Navy also operates one aircraft (1007) originally converted from an An-28 as Bryza 2RF elint aircraft (see 2001-02 and earlier *Jane's*), but delivered unequipped on 22 November 1996 and operated only in transport role.

M-28M Bryza E: Two Polish Navy An-28s (0404 and 0405, originally delivered in October 1988) equipped in 2000-01 with Swedish Space Corporation (SSC) MSS 5000 maritime surveillance system for ecological monitoring; 0405 recommissioned 28 November 2000; both redelivered to Polish Navy 18 January 2002; based at Gdynia.

M-28MPA/ASW: Provisional designation of anti-submarine version. Prototype under development. Mission subsystems yet to be selected (early 2002); include FLIR sensor in redesigned nose; MAD; EW system; sonobuoy system; and torpedoes.

CUSTOMERS: Bryza deliveries to Polish Navy totalled nine by May 2002 comprising six Rs, two Es, and one 2RF; the Rs and 2RF equip the 3rd Division at Siemirowice; the Es are based at Gdynia Baby Doły. Polish Air Force has 10 Bryza TDs, with a further four to follow; these mainly with 13th Transport Regiment at Krakow, but two operated by 36 Squadron at Warsaw-Okecie in VIP transport role. Two supplied to Venezuelan civilian operator CIACA in late 1996.

POWER PLANT: Two 716 kW (960 shp) TWD-10B/PZL-10S turboprops, each driving a 2.80 m (9 ft 2¼ in) diameter AW-24AN three-blade or a five-blade Hartzell HC-B5MP-3D propeller; first aircraft with latter was 0810, redelivered to Polish Navy on 27 August 2000. PT6A-65B engines available optionally. Fuel capacities as for M-28 Skytruck; programme to install auxiliary tanks on landing gear sponsons under development in 2002.

ACCOMMODATION (Bryza R): Pilot, co-pilot, engineer and three systems operators.

AVIONICS (Bryza R): *Comms:* LS-10C³ subsystem; Bendix/King KT 76A transponder; NATO-compatible IFF.

Radar: ARS-400 360° search and surveillance radar (86 n mile; 160 km; 99 mile range), in underfuselage radome; Honeywell RDR-2000 weather radar in nose.

Flight: Include Bendix/King KNS 81S navigation system and KLN 90B GPS receiver.

Mission: Chelton DF 707-1 emergency locator system. SSC MSS 5000.

EQUIPMENT (Bryza R): ACR/RLB-14 radio marker buoys; two 100 kg (220 lb) SAB-100NM illumination flares; five six-seat Mewa-6 dinghies, including two for crew.

DIMENSIONS, EXTERNAL:
Wing span	22.07 m (72 ft 5 in)
Length overall	13.10 m (42 ft 11¾ in)
Height overall	4.90 m (16 ft 1 in)
Cargo door: Height	2.60 m (8 ft 6¼ in)
Width	1.20 m (3 ft 11¼ in)

DIMENSIONS, INTERNAL:
Cabin: Length (excl flight deck)	5.26 m (17 ft 3 in)
Max width	1.74 m (5 ft 8½ in)
Max height	1.72 m (5 ft 7¾ in)

WEIGHTS AND LOADINGS:
Basic weight empty	4,050 kg (8,929 lb)
Weight empty, equipped	4,350 kg (9,590 lb)
Max fuel weight	1,520 kg (3,351 lb)
Max payload	1,750 kg (3,858 lb)
Max T-O weight	7,000 kg (15,432 lb)
Max landing weight	6,500 kg (14,330 lb)
Max power loading	4.89 kg/kW (8.04 lb/shp)

PERFORMANCE:
Never-exceed speed (V_{NE})	189 kt (350 km/h; 217 mph)
Max operating speed (V_{MO})	181 kt (335 km/h; 208 mph)
Stalling speed, flaps up	67 kt (123 km/h; 77 mph)
Max rate of climb at S/L	600 m (1,969 ft)/min
Time to 3,000 m (9,840 ft)	6 min 43 s
Service ceiling	6,000 m (19,680 ft)
Service ceiling, OEI	3,000 m (9,840 ft)
T-O run	265 m (870 ft)
T-O to 15 m (50 ft)	410 m (1,345 ft)
Landing from 15 m (50 ft)	560 m (1,840 ft)
Landing run	230 m (755 ft)
Max range, 45 min reserves	664 n miles (1,230 km; 764 miles)
Endurance: standard fuel	4 h
max (wing) fuel	6 h*

* 8 h with over-sponson auxiliary tanks

UPDATED

PZL M-28 SKYTRUCK

TYPE: Twin-turboprop transport.

PROGRAMME: Westernised development of the Antonov An-28, with Polish power plant and avionics replaced by P&WC PT6A-65B turboprops, Hartzell five-blade propellers and Bendix/King nav/com, weather radar and other equipment. Prototype conversion (SP-PDF) begun early 1991; first flight July 1993; Polish temporary type certificate to FAR Pt 23 Amendment 34 granted March 1994; permanent certificate March 1996, permitting MTOW increase to 7,000 kg (15,432 lb). Certified to FAR Pt 23 Amendment 42. Increased gross weight version (7,500 kg; 16,534 lb), a prototype of which was completed in 1996, offers options that include reinforced landing gear, increased fuel capacity, hydraulically operated rear door and additional emergency exit.

Bryza TD of latest Polish Air Force production batch, delivered from November 2001 *(R J Malachowski)*   NEW/0137406

First example of a Bryza R with mainwheels retracted to lie underneath the wing struts *(R J Malachowski)*   NEW/0527157

PZL M-28 Skytruck of the Venezuelan National Guard   0098661

CURRENT VERSIONS: **M-28 Skytruck:** *As described.*

M-28B Bryza: Amended designation of current production military An-28s; described separately.

M-28 Skytruck Plus: Provisional designation for stretched version; described separately.

CUSTOMERS: First customer delivery (HK-4066X to Latina de Aviación of Colombia) made in January 1996; firm orders totalled 12 at that time, including six for Venezuelan National Guard (deliveries to which began in December 1996); equip Air Detachment 5 at Caracas (four), AD 3 at Maracaibo (one) and AD 9 at Puerto Agoucho (one). One delivered to Aerogryf in Poland in mid-1996; and one to USA (subsequently to Honduras) in 1997; one to Venezuelan civil operator in 2000, but lost in July 2001. Venezuelan National Guard order subsequently (1997) increased to 12, second-batch deliveries beginning mid-1999. Also ordered (further 12) by Venezuelan Army; first deliveries April 2000; four received by July 2000. Production totalled 30 by mid-2001, comprising four civil, 24 Venezuelan military and two in stock. No further production in second half of 2001; one delivered to Royal Nepal Army on 18 September 2002 for 11th Brigade at Khatmandu.

DESIGN FEATURES: Inherited An-28's ability to operate from short, austere airfields; added Western engines and avionics to improve export sales prospects. Braced wings; will not stall because of action of automatic slats. Short stub-wing extends from lower fuselage to carry main landing gear and support wing bracing struts, curving forward and downward at front to serve as mudguards; underside of rear fuselage upswept, incorporating clamshell doors for passenger/cargo loading; twin fins and rudders, mounted on inverted-aerofoil, no-dihedral fixed incidence tailplane.

Wing section CAHI (TsAGI) R-II-14; thickness/chord ratio 14 per cent; constant chord, non-swept, no-dihedral centre-section, with 4° incidence; tapered outer panels with 2° dihedral, negative incidence and 2° sweepback at quarter-chord.

FLYING CONTROLS: Manual. Single-slotted mass and aerodynamically balanced ailerons (port aileron has trim tab), designed to droop with large, hydraulically actuated, two-segment double-slotted flaps; electrically actuated trim tabs in elevator have manual back-up; twin rudders each with electrically actuated trim tab; automatic leading-edge slats over full span of wing outer panels; slab-type spoiler forward of each aileron and each outer flap segment at 75 per cent chord; fixed slat under full span of tailplane leading-edge. If an engine fails, patented upper surface spoiler forward of aileron on opposite wing is opened automatically, resulting in wing bearing dead engine dropping only 12° in 5 seconds instead of 30° without spoiler; patented fixed tailplane slat improves handling during high angle of attack climb-out; under icing conditions, if normal anti-icing system fails, ice collects on slat rather than tailplane, to retain controllability.

STRUCTURE: Mostly metal; duralumin ailerons with fabric covering; duralumin slats with CFRP skins; CFRP spoilers and trim tabs. Air intakes lined with epoxy laminate. Two-spar wing with integral fuel tanks between spars.

LANDING GEAR: Non-retractable tricycle type, with single wheel and PZL oleo-pneumatic shock-absorber on each unit. Main units, mounted on small stub-wings, have wide tread balloon tyres, size 720×320, pressure 3.50 bar (51 lb/sq in). Steerable (±50°) and self-centring nosewheel, with size 595×185-280 Stomil (Poland) tyre, pressure 3.50 bar (51 lb/sq in). Multidisc, anti-lock hydraulic brakes on main units, and inertial anti-skid units. Reinforced landing gear and brakes optional. Minimum ground turning radius 16.00 m (52 ft 6 in). Ski gear under development.

POWER PLANT: Two 820 kW (1,100 shp) Pratt & Whitney Canada PT6A-65B turboprops, each driving a Hartzell HC-B5MP-3D10876ASK five-blade, fully feathering propeller with reverse pitch. Two main fuel tanks in outer wings, each 660 litres (174 US gallons; 145 Imp gallons); two auxiliaries in wing centre-section, each 290 litres (76.6 US gallons; 63.8 Imp gallons); two auxiliary tanks in outer

wings, each 189 litres (49.9 US gallons; 41.6 Imp gallons); total wing fuel thus 2,278 litres (602 US gallons; 501 Imp gallons). Auxiliary tanks gravity-filled; main tanks by pressure or gravity. Additional fuel can be carried in tank in main cabin. Oil capacity 16 litres (4.2 US gallons; 3.5 Imp gallons) per engine.

ACCOMMODATION: Pilot and co-pilot on flight deck, which has bulged side windows and electric anti-icing for windscreens, and is separated from main cabin by bulkhead with connecting door. Dual controls standard. Jettisonable emergency door on each side. Standard cabin layout of passenger version has seats for 19 people, with seven single seats on port side and six double seats on starboard side of aisle, at 72 cm (28 in) pitch. All seats easily foldable or removable for carriage of cargo. Aisle width 35.4 cm (13.9 in). Five passenger windows each side of cabin. Seats fold back against walls when aircraft is operated as a freighter or in mixed passenger/cargo role, seat attachments providing cargo tiedown points. Hoist of 500 kg (1,102 lb) capacity able to deposit cargo in forward part of cabin. Alternative interior configurations for 17 parachutists; or six stretchers, eight seated casualties and two medical attendants.

Entire cabin heated, ventilated, soundproofed and, optionally, air conditioned. Outward/downward-opening clamshell doors, under upswept rear fuselage, for passenger and cargo loading. Emergency exit on each side. Option for hydraulically operated door at rear and additional emergency exit aft of flight deck. Optional ventral pannier, increasing baggage/cargo capacity by 300 kg (661 lb).

SYSTEMS: No pressurisation or pneumatic systems. Hydraulic system powered by two engine-driven pumps, pressure 147 bar (2,132 lb/sq in), for flap, spoiler and (where fitted) cargo ramp actuation, mainwheel brakes and nosewheel steering, with emergency back-up system for spoiler extension and mainwheel braking. Optional air conditioning system, electrically powered with ground power supply back-up, reduces ambient temperatures of up to 40°C (104°F) by 10 to 12°C (18 to 22°F).

Primary electrical system is three-phase AC, with two 28 V DC 12 kW engine-driven starter/generators; two 200/115 V, 400 Hz, 1,500 VA inverters as secondary supply for windscreen heating. Two 26 Ah Ni/Cd batteries.

Electric anti-icing of flight deck windscreens, propellers, spinners and pitot heads, sufficient for short exposure to icing conditions. Optional pneumatic de-icing of wing, slat, tailplane and fin leading-edges. Oxygen system (for crew plus two passengers) optional. No APU.

AVIONICS: Standard Silver Crown suite by Bendix/King.

Comms: KY 196A VHF-AM and KX 165 HF radios; KMA 24-70 audio selector panel and intercom; KT 76A transponder and encoding altimeter. Optional KHF 950 short wave radio.

Radar: Optional RDR-2000 digital weather radar.

Flight: Primary navigation system based on KNS 81 VOR/DME/RNAV with HSI (KPI 552B) indicator; secondary is KX 165-based for VOR/DME. KR 22 marker beacon receiver; twin KR 87 ADF; KCS 55A gyrocompass. Optional KLN 90B GPS and flight data recorder. KFC 325 AFCS and KCP 220 computer.

DIMENSIONS, EXTERNAL:	
Wing span	22.06 m (72 ft 4½ in)
Wing chord: at root	2.20 m (7 ft 2½ in)
mean aerodynamic	1.89 m (6 ft 2½ in)
at tip	1.10 m (3 ft 7¼ in)
Wing aspect ratio	12.3
Length overall	13.10 m (42 ft 11¾ in)
Fuselage: Length	12.68 m (41 ft 7¼ in)
Max width	1.90 m (6 ft 2¾ in)
Max depth	2.14 m (7 ft 0¼ in)
Height overall	4.90 m (16 ft 1 in)
Tailplane span	5.14 m (16 ft 10¼ in)
Wheel track	3.405 m (11 ft 2 in)
Wheelbase	4.40 m (14 ft 5¼ in)
Propeller diameter	2.82 m (9 ft 3 in)
Propeller ground clearance	1.06 m (3 ft 5¼ in)
Distance between propeller centres	5.20 m (17 ft 0¾ in)
Rear clamshell doors:	
Length	2.60 m (8 ft 6¼ in)
Total width: at top	1.00 m (3 ft 3¼ in)
at sill	1.20 m (3 ft 11¼ in)
Emergency exits (each):	
Height	0.93 m (3 ft 0½ in)
Width	0.62 m (2 ft 0½ in)

DIMENSIONS, INTERNAL:	
Cabin (excl flight deck): Length	5.26 m (17 ft 3 in)
Max width	1.74 m (5 ft 8½ in)
Max height	1.72 m (5 ft 7¾ in)
Floor area	approx 7.5 m² (80.7 sq ft)
Volume	approx 14.0 m³ (494 cu ft)
Pannier (optional): Height	0.38 m (1 ft 3 in)
Width	1.53 m (5 ft 0¼ in)
Volume	1.35 m³ (47.7 cu ft)

AREAS:	
Wings, gross	39.72 m² (427.5 sq ft)
Ailerons (total)	4.33 m² (46.61 sq ft)
Trailing-edge flaps (total)	7.99 m² (86.00 sq ft)
Spoilers (total)	1.67 m² (17.98 sq ft)
Fins (total)	10.00 m² (107.64 sq ft)
Rudders (total, incl tabs)	4.00 m² (43.6 sq ft)
Tailplane	8.85 m² (95.26 sq ft)
Elevators (total, incl tabs)	2.56 m² (27.56 sq ft)

WEIGHTS AND LOADINGS:	
Weight empty, equipped	4,100 kg (9,039 lb)
Max fuel load	1,766 kg (3,893 lb)
Max payload	2,000 kg (4,409 lb)
Fuel with max payload	1,085 kg (2,392 lb)
Payload with max fuel	1,319 kg (2,908 lb)
Max T-O and landing weight	7,500 kg (16,534 lb)
Max zero-fuel weight	6,600 kg (14,550 lb)
Max wing loading	188.8 kg/m² (38.67 lb/sq ft)
Max power loading	4.57 kg/kW (7.52 lb/shp)

PERFORMANCE (at 7,000 kg; 15,432 lb MTOW):	
Max operating speed (VMO)	191 kt (355 km/h; 220 mph)
Econ cruising speed at 3,000 m (9,840 ft)	146 kt (270 km/h; 168 mph)
Max manoeuvring speed (VA)	132 kt (244 km/h; 152 mph)
Min control speed (VMCA)	83 kt (153 km/h; 96 mph)
Unstick speed	73 kt (135 km/h; 84 mph)
Approach speed	70 kt (130 km/h; 81 mph)
Stalling speed, flaps up	67 kt (123 km/h; 77 mph)
Max rate of climb at S/L	660 m (2,165 ft)/min
Time to 3,000 m (9,840 ft): de-icing off	6 min
de-icing on	9 min
Service ceiling	7,620 m (25,000 ft)
Service ceiling, OEI	3,750 m (12,300 ft)
T-O run	255 m (835 ft)
T-O to 10.7 m (35 ft)	325 m (1,065 ft)
Landing from 15 m (50 ft)	560 m (1,835 ft)
Landing run	260 m (855 ft)

M-28 Skytruck in Venezuelan Army camouflage *(Marek Dykas/PZL)*   0110895

Range at 3,000 m (9,840 ft), 45 min reserves:
    with max payload   432 n miles (800 km; 497 miles)
    with max fuel   809 n miles (1,500 km; 932 miles)
    self-ferry, with additional tanks in cabin
                    1,673 n miles (3,100 km; 1,926 miles)
Endurance   6 h 12 min
g limits   +3/−1
UPDATED

PZL M-28 03/04 SKYTRUCK PLUS

TYPE: Twin-turboprop transport.

PROGRAMME: Conceived in response to interest expressed by potential customers. Ground test prototype, completed in August 1998, used component assemblies of standard M-28. Programme continuing with a first flight targeted for the end of 2002.

CURRENT VERSIONS: **Passenger:** Commuter transport.

Passenger/Cargo: With folding seats for 19 to 27 passengers.

Cargo: In two forms. One with rear fuselage clamshell doors or single ramp/door, both operated hydraulically, plus cargo hoist; the other with starboard-side large cargo door and set of floor rollers. Former optimised for miscellaneous small cargo, latter for accepting three LD-3-size containers; both with optional 2.0 m³ (70.6 cu ft) ventral pannier and auxiliary fuel tanks. These versions also offering medevac, paradrop (up to 27 paratroops) and long-range self-ferry configurations.

DESIGN FEATURES: Generally as described for Skytruck. Operable from concrete or unprepared airstrips.

POWER PLANT: Two 875 kW (1,173 shp) turboprops. Standard fuel capacity 2,228 litres (589 US gallons; 490 Imp gallons).

DIMENSIONS, EXTERNAL:
Wing span   22.06 m (72 ft 4½ in)
Length overall   15.40 m (50 ft 6¼ in)
Height overall   5.15 m (16 ft 10¾ in)
WEIGHTS AND LOADINGS:
Fuel weight (standard)   1,726 kg (3,805 lb)
Max payload   3,000 kg (6,614 lb)
Max T-O and landing weight   8,600 kg (18,959 lb)
PERFORMANCE (estimated):
Max operating speed (V_{MO})
                199 kt (370 km/h; 229 mph) IAS
T-O to 10.7 m (35 ft)   520 m (1,705 ft)
Landing from 15 m (50 ft)   640 m (2,100 ft)
Max range at 3,000 m (9,840 ft), standard fuel, 45 min
    reserves   809 n miles (1,500 km; 932 miles)
Max endurance, conditions as above   6 h 12 min
UPDATED

PZL M-18 DROMADER

English name: Dromedary

TYPE: Agricultural sprayer.

PROGRAMME: Designed to meet requirements of FAR Pt 23; prototype (SP-PBW) first flew 27 August 1976; further two prototypes; M-18 awarded Polish type certificate 27 September 1978; eight preproduction aircraft used for operational trials; initial US certification on 23 January 1981; M-18A followed on 8 September 1987 and M-18B on 19 December 1995. Later certified in Argentina, Australia, Brazil, Canada, China, former Czechoslovakia, France, former East Germany, Italy, Lithuania, New Zealand, Spain, USA and former Yugoslavia.

Initial M-18 single-seat version (1988-89 *Jane's*) produced from 1979 to 1984; followed by M-18A two-seater (1997-98 and earlier *Jane's*; produced 1984 to 1996) and M-18AS training version. These superseded from mid-1996 by improved M-18B and BS, which remain current standard variants. In June 1996 PZL Mielec signed an agreement for the assembly of M-18Bs from Polish CKD kits by a Brazilian local authority, but this did not take place and deliveries to that country from Polish production continued in 2001. Minor airframe improvements (improved ventilation system, lengthened tailwheel leg, additional rudder hinge) introduced 2002 and available as retrofit; power plant with better hot/high performance under consideration.

CURRENT VERSIONS: **M-18B:** Standard production version from mid-1996. Improved performance development of M-18A, awarded extension of Polish type certificate 27 January 1994; FAA certification granted 19 December 1995. Elevators have smaller, centrally located trim tabs; spring interconnect between elevators and flaps, and between ailerons and rudder; flaps-down deflection increased from 15 to 30°; maximum payload and overload MTOW increased. Normal category landing run reduced; flaps-down stalling speed reduced; lower control stick force values; enhanced static and dynamic longitudinal stability. Power plant and hopper as for M-18A.

Detailed description applies to M-18B except where indicated.

M-18BS: Two-seat dual-control trainer, generally as M-18AS, but using M-18B airframe. Prototype (SP-ZUW, converted from an earlier M-18) first flew in November 1997. Polish type certificate awarded early 1998. Front (instructor's) seat reduces hopper capacity to 900 litres (238 US gallons; 198 Imp gallons) or 700 kg (1,543 lb). At least four produced for Iranian operator and delivered in

PZL M-18B Dromader agricultural aircraft (*Marek Dykas/PZL*)   *NEW*/0137344

2002; Greece ordered three more (delivered 16 July 2002) with a further three on option.

M-18C: Improved M-18B, with same payload but powered by an 895 kW (1,200 hp) PZL Kalisz K-9 engine offering enhanced performance; longer tailwheel leg to improve view over nose; modified wingtips to aid dispersal of chemical. First flown in August 1995, but subsequently abandoned.

Turbine Dromader: Turboprop versions with 761 to 875 kW (1,020 to 1,173 shp) P&WC PT6A-45A/B/R or PT6A-65R/AG/AR engines with Hartzell propeller. Developed initially by James Mills Turbines Inc in USA and continued by Turbine Conversions Ltd. Variants with Honeywell 746 kW (1,000 shp) TPE331-10 or T53-L-7 also available from respective STC holders.

CUSTOMERS: Total of 720 built (all versions) by mid-2002, of which approximately 90 per cent for export; 28th production batch under way in 2000-02; sold to operators in Argentina, Australia, Brazil, Bulgaria, Canada, Chile, China, Cuba, former Czechoslovakia and East Germany, Greece (30 for firefighting), Hungary, Iran, Italy, Morocco, Nicaragua, Poland, Portugal, Spain, Swaziland, Trinidad, Turkey, USA (more than 200), Venezuela and former Yugoslavia.

DESIGN FEATURES: Emphasis on crew safety (cockpit located behind large mass concentrations of power plant and hopper; reinforced cockpit structure and roof for higher crashworthiness in case of turnover; wire cutters forward of cockpit; windscreen bird-impact resistant; fuel tanks in outboard wings, away from cockpit; steel cable between cockpit and fin for vertical tail protection against collision with overhead power lines). No members of fuselage load-carrying truss extend through hopper. All major airframe component assemblies interchangeable, with no individual fitting required. Mainwheels not subject to lateral travel during shock-absorbers' deflection, thus reducing tyre wear-off. Expansion mandrels in centre-section-to-outboard wing attachment joints for facilitated installation and removal. Whole aircraft, in disassembled condition, with operational and ground maintenance equipment, can be carried in a single 12.2 m (40 ft) container. All parts exposed to chemical contact treated with polyurethane or epoxy enamels, or manufactured of stainless steel; detachable fuselage side panels for airframe inspection and cleaning; braced tailplane.

Wing sections NACA 4416 at root, NACA 4412 at end of centre-section and on outer panels; incidence 6°.

FLYING CONTROLS: Conventional and manual. Mass and aerodynamically balanced slotted ailerons with trim tabs, actuated by pushrods; aerodynamically and mass-balanced rudder and horn-balanced elevators with trim tabs, actuated by cables and pushrods respectively;

hydraulically actuated two-section trailing-edge slotted flaps. Control surface movements: ailerons +21/−17°; elevators +27/−17°; rudder ±23°; flaps max 30°. See also Current Versions M-18B paragraph above.

STRUCTURE: All-metal; steel-capped duralumin wing spar; fuselage mainframe of helium-arc welded chromoly steel tube, oiled internally against corrosion; duralumin fuselage side panels and stainless steel bottom covering; corrugated inboard flap and tail unit skins.

LANDING GEAR: Non-retractable tailwheel type. Oleo-pneumatic shock-absorber in each unit. Main units have 800×260 tyres and are fitted with hydraulic disc brakes, parking brake and wire cutters. Fully castoring tailwheel, lockable for take-off and landing, with tyre size 380×150.

POWER PLANT: One PZL Kalisz ASz-62IR nine-cylinder radial air-cooled supercharged engine (731 kW; 980 hp at 2,200 rpm), driving a PZL Warszawa AW-2-30 four-blade constant-speed aluminium propeller. Integral fuel tank in each outer wing panel, combined usable capacity 712 litres (188 US gallons; 157 Imp gallons). Gravity feed header tank in fuselage. For ferrying, additional fuel capacity in chemical hopper. Oil capacity 70 litres (18.5 US gallons; 15.4 Imp gallons).

ACCOMMODATION: Single adjustable seat in fully enclosed, sealed and ventilated cockpit stressed to withstand 40 *g* impact. Additional cabin located behind cockpit and separated from it by a wall. Latter equipped with rigid, rear-facing seat with protective padding and safety belt, port-side jettisonable door, windows (port and starboard), fire extinguisher and ventilation valve. Communication with pilot provided by window in dividing wall, and by intercom. In M-18BS, standard hopper is replaced by smaller one, permitting installation of bolt-on instructor's cabin. Standard hopper of M-18B and instructor's cockpit of M-18BS interchangeable. Glass fibre cockpit roof and rear fairing, latter with additional small window each side. Front cockpit of M-18BS has more extensive glazing. Adjustable shoulder-type safety harness. Adjustable rudder pedals. Quick-opening door on each side of front cockpit; port door jettisonable.

SYSTEMS: Hydraulic system, pressure 98 to 137 bar (1,421 to 1,987 lb/sq in), for flap actuation, disc brakes and dispersal system. Electrical system powered by 28.5 V 100 A generator, with 24 V 28 Ah lead-acid battery and overvoltage protection relay.

AVIONICS: *Comms:* RS-6102 (Polish-built), Bendix/King KX 155 or KY 196A com transceiver.

Instrumentation: KI 208 course indicator, VOR-OBS indicator, artificial horizon with turn and bank indicator, gyrocompass, radio compass and stall warning.

EQUIPMENT: Glass fibre epoxy hopper, with stainless steel tube bracing, forward of cockpit; capacity 2,500 litres (660

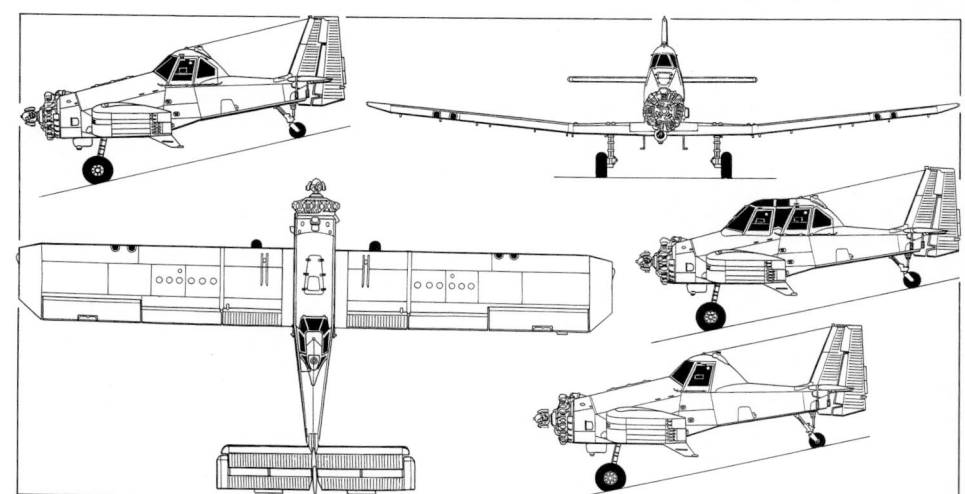

PZL M-18B Dromader, with additional side elevations (right) of M-18BS training version (upper) and original M-18 (lower)
0062994

PZL M-26 Iskierka in US ownership, wearing spurious military camouflage (*Jane's/Paul Jackson*) *NEW*/0137361

US gallons; 550 Imp gallons) of liquid or 2,200 kg (4,850 lb) of dry chemical. Smaller hopper in M-18BS. Deflector cable from cabin roof to fin.

M-18 can be fitted optionally with several different types of agricultural and firefighting systems, as follows: spray system with 54/96 nozzles on spraybooms; dusting system with standard, large or extra large spreader; atomising system with 10 atomisers; water bombing installation; fire-bombing installation with foaming agents; and multifunction firefighting system (water bombing or sequential line-drops). Aerial application roles can include seeding, fertilising, weed or pest control, defoliation, forest and bush firefighting, and patrol flights.

Special wingtip lights permit agricultural flights at night, and aircraft can operate in both temperate and tropical climates. Navigation lights, cockpit light, instrument panel lights and two rotating beacons standard. Landing lights, taxying light and night working light optional. Built-in jacking and tiedown points in wings and rear fuselage; towing lugs on main landing gear. Cockpit fire extinguisher and first aid kit.

DIMENSIONS, EXTERNAL:
Wing span	17.70 m (58 ft 0¾ in)
Wing chord, constant	2.29 m (7 ft 6¼ in)
Wing aspect ratio	7.8
Length overall	9.47 m (31 ft 1 in)
Height: over tailfin	3.70 m (12 ft 1¾ in)
overall (flying attitude)	4.60 m (15 ft 1 in)
Tailplane span	5.90 m (19 ft 4¼ in)
Wheel track	3.48 m (11 ft 5 in)
Propeller diameter	3.30 m (10 ft 10 in)
Propeller ground clearance (tail up)	0.23 m (9 in)

DIMENSIONS, INTERNAL:
Hopper volume	2.5 m³ (88 cu ft)

AREAS:
Wings, gross	40.00 m² (430.5 sq ft)
Ailerons (total)	4.33 m² (46.61 sq ft)
Trailing-edge flaps (total)	5.16 m² (55.54 sq ft)
Vertical tail surfaces (total)	3.60 m² (38.75 sq ft)
Horizontal tail surfaces (total)	7.58 m² (81.59 sq ft)

WEIGHTS AND LOADINGS (M-18B. A: Normal category, B: CAM 8):
Max payload	2,200 kg (4,850 lb)
Max fuel weight	510 kg (1,124 lb)
Max T-O weight: A	4,200 kg (9,259 lb)
B	5,300 kg (11,684 lb)
Max wing loading: A	105.0 kg/m² (21.51 lb/sq ft)
B	132.5 kg/m² (27.14 lb/sq ft)

Max power loading: A		5.75 kg/kW (9.45 lb/hp)
B		7.26 kg/kW (11.92 lb/hp)

PERFORMANCE (M-18B; A and B as above):
Never-exceed speed (VNE):		
A		151 kt (280 km/h; 174 mph)
B		124 kt (230 km/h; 142 mph)
Normal operating speed: A		124 kt (230 km/h; 143 mph)
B		108 kt (200 km/h; 124 mph)
Max airspeed with agricultural equipment:		
A, B		108 kt (200 km/h; 124 mph)
Stalling speed, flaps down: A		59 kt (108 km/h; 68 mph)
B		66 kt (121 km/h; 76 mph)
Max rate of climb at S/L: A		390 m (1,280 ft)/min
B		264 m (866 ft)/min
Service ceiling: A		6,500 m (21,320 ft)
T-O run: A		240 m (790 ft)
B		350 m (1,150 ft)
T-O to 15 m (50 ft): A		530 m (1,740 ft)
B		920 m (3,020 ft)
Landing from 15 m (50 ft): A		360 m (1,180 ft)
B		570 m (1,870 ft)
Max range, no reserves:		
A, normal tankage		523 n miles (970 km; 602 miles)
B, normal tankage		432 n miles (800 km; 497 miles)
A, with hopper fuel		
		1,079 n miles (2,000 km; 1,242 miles)
g limits: A		+3.4/–1.4
B		+2.8/–1.1

UPDATED

PZL M-26 ISKIERKA
English name: Little Spark
Export marketing name: Air Wolf

TYPE: Primary prop trainer/sportplane.

PROGRAMME: Two versions developed; first flight of first prototype (SP-PIA) with PZL-F engine 15 July 1986; first flight of Textron Lycoming-engined prototype (SP-PIB) 24 June 1987. Second prototype used as US demonstrator until destroyed on 17 August 1995. Polish certification obtained 26 October 1991; production of M-26 01 launched second quarter 1994 with initial 20 aircraft order from US distributor Melex; deliveries to USA began July 1996 with first production aircraft (N2601M). US FAR Pt 23 certification received 16 April 1998; now also has Australian certification (both for Lycoming version).

CURRENT VERSIONS: **M-26 00:** PZL-F engine and Polish avionics. No production examples by mid-2002.

M-26 01: Textron Lycoming AEIO-540-L1B5 engine and Bendix/King avionics. Most US aircraft are registered with type designation 'M2601'.

Detailed description applies to this version except where indicated.

CUSTOMERS: Initial order for M-26 01 by Melex USA Inc (20), of which two (fourth and fifth aircraft) delivered to Venezuelan National Guard Training Centre at Porlamar, Isla Margarita, in 1998. Short-listed by Israel Defence Force as Piper Super Cub replacement, but not selected. Sixth production aircraft to USA early 1999. Seventh production aircraft completed in August 2000 and retained in Poland; deliveries up to mid-2002 comprised four (plus second prototype and refurbished first prototype) to USA; two to Venezuela; and current demonstrator. No further production known up to early 2002, although three further airframes are partly complete.

DESIGN FEATURES: Optimised for civil pilot training and military pilot selection; also suitable as high-performance private aircraft. Selected parts and assemblies of M-20 Mewa used in design of wings, tail unit, landing gear, power plant, and electrical and power systems. Fixed incidence tailplane.

Wing section NACA 65₂-415 (constant chord); 7° dihedral from roots; 2° incidence; leading-edges swept forward at root. Horizontal tail surfaces NACA 0009 aerofoil section; NACA 0010 for vertical surfaces.

FLYING CONTROLS: Conventional and manual. Frise ailerons, horn-balanced rudder, and elevators with starboard trim tab; single-slotted trailing-edge flaps. Control surface movements: ailerons +24/–14°, elevators +30/–28°, rudder ±35°, flaps 25° (take-off) and 40° (landing).

STRUCTURE: Safe-life aluminium alloy throughout except for tips of wings and all tail surfaces, which are of composites.

LANDING GEAR: Retractable tricycle type, actuated hydraulically, with single wheel and oleo strut on each unit. Mainwheels retract inward into wings, nosewheel rearward. Nosewheel travel ±27°. Size 6.00-6 wheels on all three units; tyre pressures 3.43 bar (50 lb/sq in) on main units, 2.16 bar (31 lb/sq in) on nose unit. PZL Lucas hydraulic disc brakes on mainwheels. Parking brake.

POWER PLANT: One 224 kW (300 hp) Textron Lycoming AEIO-540-L1B5 flat-six engine, driving a Hoffmann HO-V123K-V/200AH-10 or Hartzell HC-C3YR-4BF three-blade constant-speed propeller. Second tank in each wing. Total fuel capacity 369 litres (97.5 US gallons; 81.2 Imp gallons); restricted to maximum 180 litres (47.6 US gallons; 39.6 Imp gallons) for aerobatics. Gravity fuelling point in top of each outer wing tank. Oil capacity 18 litres (4.75 US gallons; 4.0 Imp gallons).

ACCOMMODATION: Tandem seats for pupil (in front) and instructor, under framed canopy which opens sideways to starboard. Rear seat elevated 15 cm (5.9 in). Baggage compartment aft of rear seat. Both cockpits heated and ventilated.

SYSTEMS: Two independent hydraulic systems, one operating at 154 bar (2,233 lb/sq in) for landing gear extension/retraction and one at 103 bar (1,494 lb/sq in) for wheel braking. DC electrical power supplied by 24 V alternator (50 A in M-26 00, 100 A in M-26 01) and 25 Ah battery.

AVIONICS: *Comms:* Bendix/King KX 165 radio, KT 76A transponder, KMA 24 selector panel and intercom.
Flight: Bendix/King KR 22 marker beacon receiver, KR 87 ADF and KCS 55A gyrocompass.

EQUIPMENT: Landing light in port wing leading-edge. Melex offers simulated air combat package of smoke, video guns and laser guns.

ARMAMENT: Up to 120 kg (265 lb) of ordnance on each of two underwing hardpoints; total 240 kg (529 lb).

DIMENSIONS, EXTERNAL:
Wing span	8.60 m (28 ft 2½ in)
Wing chord: at root	1.88 m (6 ft 2 in)
at tip	1.60 m (5 ft 3 in)
Wing aspect ratio	5.3
Length overall	8.295 m (27 ft 2½ in)
Height overall	2.96 m (9 ft 8½ in)
Tailplane span	3.80 m (12 ft 5½ in)
Wheel track	2.93 m (9 ft 7¼ in)
Wheelbase	1.93 m (6 ft 4 in)
Propeller diameter	1.90 m (6 ft 2¾ in)

DIMENSIONS, INTERNAL:
Cockpits: Total length	2.91 m (9 ft 6½ in)
Max width	0.88 m (2 ft 10½ in)
Max height	1.30 m (4 ft 3¼ in)

AREAS:
Wings, gross	14.00 m² (150.7 sq ft)
Ailerons (total)	1.17 m² (12.59 sq ft)
Trailing-edge flaps (total)	1.06 m² (11.41 sq ft)
Fin	1.96 m² (21.10 sq ft)
Rudder	0.89 m² (9.58 sq ft)
Tailplane	3.30 m² (35.52 sq ft)
Elevators (total, incl tab)	1.15 m² (12.38 sq ft)

WEIGHTS AND LOADINGS (A: Aerobatic, U: Utility category):
Weight empty: A, U	1,040 kg (2,292 lb)
Max fuel weight	271 kg (597 lb)
Max payload: A	275 kg (606 lb)
U	350 kg (772 lb)
Max T-O weight: A	1,315 kg (2,899 lb)
U	1,400 kg (3,086 lb)
Max wing loading: A	93.9 kg/m² (19.24 lb/sq ft)
U	100.0 kg/m² (20.48 lb/sq ft)

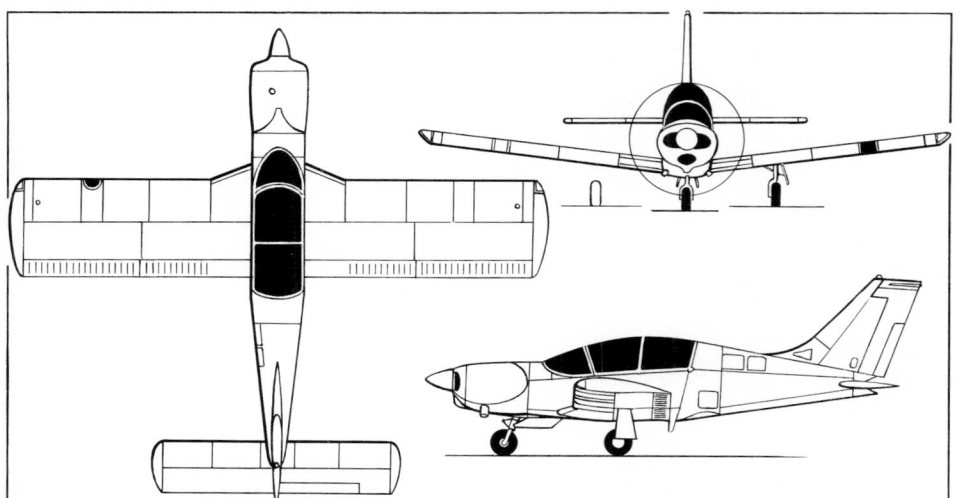

PZL M-26 Iskierka tandem two-seat primary trainer (*Jane's/James Goulding*)   0011932

Max power loading: A	5.88 kg/kW (9.66 lb/hp)	Max level speed at S/L:		Landing from 15 m (50 ft): A	636 m (2,090 ft)	
U	6.26 kg/kW (10.29 lb/hp)	A	178 kt (330 km/h; 205 mph)	U	685 m (2,250 ft)	
PERFORMANCE (M-26 01, A and U as above):		U	173 kt (321 km/h; 199 mph)	Max range, 30 min reserves:		
Never-exceed speed (VNE):		Max rate of climb at S/L: A	450 m (1,476 ft)/min	A, U	761 n miles (1,410 km; 876 miles)	
A	213 kt (395 km/h; 245 mph)	U	360 m (1,181 ft)/min	g limits: A	+6/−3	
U	200 kt (371 km/h; 230 mph)	T-O to 15 m (50 ft): A	520 m (1,710 ft)	at max T-O weight (U)	+4.4/−1.76	
		U	625 m (2,055 ft)		*UPDATED*	

PZL-ŚWIDNIK SA

WYTWÓRNIA SPRZĘTU KOMUNIKACYJNEGO (Transport Equipment Corporation) PZL-ŚWIDNIK SA

aleja Lotników Polskich 1, PL-21-045 Świdnik
Tel: (+48 81) 751 35 05 and 468 76 61
Fax: (+48 81) 751 21 73, 468 09 19 and 468 09 18
e-mail: hem@pzl.swidnik.pl
Web: http://www.pzl.swidnik.pl
PRESIDENT: Mieczysław Majewski
MANAGING DIRECTOR: Jan Miroński
COMMERCIAL DIRECTOR: Ryszard Cukierman
DIRECTOR OF ENGINEERING AND NEW PROGRAMMES CENTRE:
    Artur Wasilak
HEAD OF MARKETING: Andrzej Dyzma
HEAD OF SALES: Marzena Siwek
MARKETING MANAGER: Andrzej Stachyra

Świdnik factory recently completed 50 years of aircraft manufacture, having been established 1 January 1951; engaged initially in manufacturing wings and control surfaces for LiM-1 (MiG-15) jet fighter; began licensed production of Soviet Mi-1 helicopter in 1954 (first flight March 1956), building some 1,594 as SM-1s, followed by 86 Świdnik-developed SM-2s; design office formed at factory to work on variants/developments of SM-1 and original projects such as SM-4 Łątka. More than 7,300 helicopters of all types built to date, including four pre-series and 5,418 production Mi-2s from 1965; in addition, further six Mi-2 static test airframes produced,. Eight more incomplete Mi-2s remained in storage in mid-2000; United Arab Emirates announced intention to buy an unspecified number of Mi-2s in October 2001.

Świdnik works named after famous pre-war PZL designer, Zygmunt Pulawski; became a joint stock company on 4 January 1991, with Polish Treasury the major shareholder (now 74.39 per cent); workforce of 2,422 in December 2000. Buyer(s) for 36.7 per cent share in company still being sought in early 2002, following failure to agree terms with AgustaWestland. Production of own designs concentrates on W-3 Sokół variants and SW-4; fixed-wing I-23 lightplane added to product range in 2002. ISO 9001 status was achieved in 1994, and on 15 June 2000 Świdnik was granted AQAP-110 certification, having fully met NATO quality assurance requirements for design, development and production.

Świdnik manufactures PW-5 Smyk (from 1994) and PW-6 composites sailplanes (both recently sold to Egypt) and (also since 1994) central wing box for ATR 72; manufacturing fuselages for AgustaWestland A 109 Power and A 119 Koala between 1996 and 2002, fuselage components for Bell/ Agusta AB139, front fuselage modules for Dassault Mirage 2000-5 (25 were to be delivered in 2002) and passenger doors for Airbus A319/A320/A321; collaborating in development of Bell/Agusta AB139, for which Świdnik builds the fuselage; MoU with Saab announced in February 2002, under which Świdnik would manufacture carbon fibre components

Polish Navy Anakondas are now fitted with folding main rotor blades *(Piotr Butowski)*   0034664

(including nosewheel and mainwheel doors) for the Gripen, if ordered by the Polish Air Force.

PZL-Świdnik made its first net profit (zł7.5 million) in 2000, based on a sales total of zł140 million. Projected sales of some zł165 million were forecast for 2001. Subsidiaries include Heliseco, which operates over 100 helicopters on agricultural and similar tasks at home and in Egypt, Greece, Oman, Portugal, Spain and elsewhere.

UPDATED

PZL-ŚWIDNIK W-3 SOKÓŁ

English name: Falcon

TYPE: Multirole medium helicopter.

PROGRAMME: Developed under 1970 joint Polish/Soviet agreement; preliminary design by engineers from both countries at Mil OKB in Moscow 1972, where preliminary mockup completed 1975. Detail design and definitive mockup by Świdnik 1976; static/fatigue ground test airframe of 1978 followed by five flying prototypes, first of which (SP-PSA) made first flight 16 November 1979, and used in subsequent tiedown tests; remaining prototypes embodied changes resulting from tests; Soviet participation ended 1980. Manufacturer's flight trials resumed 6 May 1982 with second prototype (SP-PSB); third, fourth and fifth prototypes all made first flights in 1984, on 24 July (SP-PSC), 4 June (SP-PSD) and 26 November (SP-PSE) respectively; certification trials carried out in wide range of operating conditions, including heavy icing and extreme temperatures of −60 and +50°C. Provisional Polish certification 26 September

1988, followed by full certification in Poland 10 April 1990, and to Russian NLGW-1 and -2 regulations 17 December 1992; production started in 1985 and 20 early aircraft supplied to Aeroflot from 10 August 1988, but subsequently returned to Świdnik, most being reallocated to Heliseco for firefighting in Spain.

Polish (26 March), Spanish, US (31 May) and German (6 December) FAR Pt 29 (VFR) certification of W-3A received 1993; Polish ICAO-standard noise certification 1995; 100th example completed 1996. MoU 13 April 1996 with Daewoo, South Korea (now part of KAI), for purchase of 35 Sokółs with marketing rights in Asia, but payments (and hence deliveries) delayed due to regional economic difficulties.

CURRENT VERSIONS: **W-3 Sokół:** Initial standard civil and military version; production (52) completed. Two W-3s (0501 and 0502) upgraded by WZL 1 as **W-3RL** for combat SAR role in 1999 as part of Polish qualification for NATO membership; based at Bydgoszcz. New equipment includes Rockwell Collins search radar, Bendix/King com radios and intercom, Tacan, GPS, radio compass, IFF, searchlight, rescue hoist, stretchers (two) and auxiliary fuel tanks; standard crew of five. One conversion to **S-1RR** prototype (which see below); one other to **W-3U Salamandra** armed prototype (see 1995-96 *Jane's*) reverted to W-3 and delivered to Myanmar.

W-3W: Armed W-3 (W for *Wielozadaniowy:* multipurpose) with starboard-mounted 23 mm GSz-23 twin-barrel gun; Mars-2 launchers for sixteen 57 mm S-5 or 80 mm S-8 unguided rockets, ZR-8 bomblet dispensers, Platan minelaying packs, and six cabin window-mounted AK 47, 5.45 mm Tantal or PKM machine guns. Twenty-two delivered to Polish Ministry of National Defence; entered service with Polish Air Force 47 Szkolny Pulk Smiglowcow (Helicopter School Regiment) at Nowe Miasto (five) and Polish Army (17).

W-3A: Improved version for Western certification; redesign started 1989; first flight 30 July 1992; FAA type approval to FAR Pt 29 received 31 May 1993, German LBA certification 6 December 1993. Dual hydraulic systems, new de-icing system, Western instrumentation. First delivery, to Saxony Police Department, Germany, 20 December 1993. One W-3A (c/n 370508, SP-PSL) fitted in early 1998 with Sextant Avionique AFDS 95-1 AFCS (four-axis digital autopilot) with a view to obtaining Polish (GILC) and US (FAA) single-pilot IFR certification.

Detailed description applies to W-3A, except where indicated.

W-3AM: Version of W-3A with six inflatable flotation bags. Thirteen, mostly for South Korea, produced by January 2002.

W-3WA: Combat support version of W-3A; armament includes GSz-23L gun, Strzala-2 AAMs and Polish Gad fire-control system. Total of 23 delivered to Polish armed forces by January 2002: Air Force five, Army 18. Three in combat SAR configuration (presumably based on W-3RL) delivered in second quarter 2000 to 7th Cavalry Regiment at Mazowiecki: new comms, NVG-compatible instrumentation, new IFF, classified ESM and armoured crew seats. Four adapted in 2001 for Polish Army smoke-laying role, equipped with PWD Pylia smoke generation system developed by WSK-PZL Rzeszow and the Polish

PZL-Świdnik W-3 Sokół operated by Polish police air wing *(R J Malachowski)*   0110897

Armed Forces' Chemistry and Radiometry Military Institute; equip 66th Air Squadron.

W-3WB (*wsparcia bojowego*: combat support): Armed prototype (c/n 360318), equipped 1993-94 by Kentron (hence alternative designation **W-3K**) with South African weapon systems; details in 1995-96 and earlier *Jane's*. Returned to Świdnik 1994 and became HOT/Viviane testbed (see W-3H paragraph below).

W-3PPD Gipsowka: Airborne command post (*powietrzny punkt dowodzenia*) version of W-3A. Four to be procured by Polish Army as Mi-2PPD replacements by 2006, first of which was delivered to 66th Air Squadron in first half of 2001. Specialised command, control and communications (C³) avionics include Thales RRC-9500 UHF and RRC-3500 HF radios, Totem 300 INS and a MIL-STD-1553B databus. Crew includes four operator workstations, each with a 40.6 cm (16 in) flat-panel colour display; flight deck is NVG-compatible. Also to have EW suite (as in W-3), including Thales SPS-H RWR and chaff/ flare dispenser. Defensive armament is carried on external outriggers.

W-3RM Anakonda: Offshore search and rescue (*ratowinczy morski*: sea rescue) version of W-3; watertight cabin, six inflatable flotation bags, additional window in lower part of each flight deck door. In service with Polish Navy (five delivered, of which two since lost; retrofit with folding main rotor blades from late 2000); Ministry of Interior (one) and for Świdnik trials (one). Latest examples, with US FSI Ultra 4000 FLIR, are designated **W-3WARM** (first aircraft, 360813, delivered 21 May 1998; second, 360815, 15 March 1999; third, 360906, 18 January 2002), the A indicating American (FAA) certification standard and W indicating armament. Upgrade under test from 2000 includes a more advanced engine control system and a deck lock for shipboard deployment.

S-1RR Procjon: Electronic combat reconnaissance (*rozpoznanie radioelektronicznego*) version of W-3; prototype (c/n 310203) converted from W-3, first flown 1996 and delivered to Polish Air Force by 1 January 1997. Subsequent aircraft designated **SRR-10**; first of these (c/n 370720) first flown in 1998 and fully equipped in 1999. ECR suite includes two-place console and chaff/flare dispensers; large external antenna housings include one on nose, one on cabin roof and one on starboard side of fuselage; last-named can be rotated downwards for full 360° scan capability.

W-3H: Unofficial designation for armed support helicopter, derived from W-3WA; pursued as upgrade under now-abandoned **Huzar** programme. Local requirement for 96. Avionics and weapons fit decided in favour of Israeli (Elbit avionics and Rafael NT-D ATM) equipment in October 1997, but selection immediately overturned by new Polish government; rival avionics offered by Boeing and Sextant, partnered by Hellfire II and HOT 3, respectively. Programme reinstated October 1998 but foundered due to inability to demonstrate NT-D in Poland within deadline of 30 November, and Israeli deal cancelled by Polish government on 8 December 1998.

September 1998 co-operation agreement between Świdnik and Euromissile led to integration of HOT 3 ATM on Sokół, followed by successful firing tests during demonstration of HOT/Viviane system at Polish firing range in Nowa Deba on 4 March 1999, using Świdnik trials aircraft SP-SUW (c/n 360318). This aircraft utilised in 2000 as prototype to test new main rotor blades with modified leading-edges and increased damage tolerance. Świdnik continues to offer W-3/HOT 3 variant as potential export version, but domestic requirement for W-3H now shelved in favour of a 50-aircraft support role upgrade of existing Sokóls, attack requirement instead being met by upgrading Mi-24 fleet.

W-3 Sep: Naval combat derivative; described separately.

SW-5: Provisional designation, allocated to an improved version of the W-3 Sokół under study since at least 1997. Principal changes envisaged include simplified rotor head, wider use of composites, upgraded avionics by Thales (Sextant) (including four-axis autopilot with AFDS 95-1 flight director) and 746 kW (1,000 shp), FADEC-equipped Pratt & Whitney Canada PT6B-67B engines. No recent news of programme status received.

Instrument panel of the combat SAR W-3RL
(R J Malachowski)   0093622

CUSTOMERS: See tables. Total of 146 (excluding five flying prototypes) completed by early 2002. Three reportedly purchased in early 2002 by Ukraine Border Guard.

COSTS: US$3.156 million for basic VFR, single-pilot W-3A (1999).

DESIGN FEATURES: Conventional utility helicopter of pod-and-boom layout and with engines above cabin. Four-blade fully articulated main rotor and three-blade tail rotor; main rotor has pendular Salomon-type vibration absorber for smooth flight and low vibration levels. Transmission driven via main, intermediate and tail rotor gearboxes. Tailfin integral with tailboom; fixed incidence horizontal stabiliser, not interconnected with main rotor control system.

Main rotor blades have NACA 23012M aerofoil section and (on Polish Navy W-3RM) manual folding. Rotor brake standard. Rotor rpm 268.5 (main) and 1,342 (tail); main rotor blade tip speed 220.7 m/s (724 ft/s).

FLYING CONTROLS: Three hydraulic boosters for longitudinal, lateral and collective pitch control of main rotor; one booster for tail rotor control. Constant-speed rpm control for continuous operation (manual rpm control also available). Two-axis stability augmentation system with pitch and roll hold. Three- and four-axis AFCS available from late 1994.

STRUCTURE: Rotor blades (main and tail) and single-spar horizontal stabiliser of laminated GFRP impregnated with epoxy resin; tail rotor driveshaft of duralumin tube with splined couplings; duralumin fuselage; GFRP fin trailing-edge.

LANDING GEAR: Non-retractable tricycle type, plus tailskid beneath tailboom. Twin-wheel castoring and self-centring nose unit; single wheel on each main unit. Oleo-pneumatic shock-absorber in each unit. Mainwheel Stomil Poznań tyres size 700×250; nosewheel tyres size 400×140. Tyre pressures 4.90 and 4.40 bar (71 and 64 lb/sq in) respectively. Pneumatic disc brakes on mainwheels. Metal ski landing gear optional. Six inflatable flotation bags on Anakonda and W-3AM.

POWER PLANT: Two WSK PZL-Rzeszów PZL-10W turbo-shafts, each with rating of 671 kW (900 shp) for T-O and

W-3 SOKÓŁ PRODUCTION
(to 1 January 2002)

Version	Lot 1 1986-89	Lot 2 1989-93	Lot 3 1990-93	Lot 4 1992-98	Lot 5 1994-95	Lot 6 1996-98	Lot 7 1997-98	Lot 8 1999-	Total
W-3	10[1]	22[2]	17	3[3]					52
W-3A			1	4		12		2	19
W-3PPD							1		1
W-3AM				2		5	6		13
W-3RM			1	6					7
W-3W		1		5	17				23
W-3WA					3	2	10	8	23
W-3WARM							2	1	3
SRR-10						1			1
Unidentified							1		1
Static test		2	1						3
Totals	**10**	**25**	**20**	**20**	**20**	**20**	**20**	**11**	**146**

Notes:
[1] One converted as S-1RR Procjon prototype
[2] One to W-3U prototype; reverted to W-3 before delivery
[3] Two upgraded to W-3RL by WZL 1

W-3 SOKÓŁ CUSTOMERS
(at 1 January 2002)

Country	Customer	Qty	Version	First aircraft	First delivery
Czech Republic	Air Force	11	W-3A	0709	Sep 96
Germany	Saxony Police	2	W-3A	D-HSNA	Mar 94
Korea, South	Choong Nam Fire Dept	1	W-3AM	119	Dec 99
	CitiAir	4	W-3AM	HL-9220	Dec 95
	Daegu Fire Dept	1	W-3AM	?	Nov 01
	Daewoo	1	W-3A	SP-FSU	not delivered
		1	W-3AM	HL-9257	Oct 96
	Heli Korea	4	W-3AM	HL-9259	Jul 97
Myanmar	Air Force	13	W-3	6501	Nov 90
Nigeria	Okada Air	1	W-3	5N-UYI	Dec 91
Poland	Air Force	7	W-3	0415	Jul 93
		5	W-3W	0516	Apr 94
		5	W-3WA	0618	?
	Army	17	W-3W	0601	Sep 94
		18	W-3WA	0806	97
	Heliseco	2	W-3	SP-SUB	Feb 97
		1	W-3AM	SP-SYI	Jul 96
	Lublin Medical Aviation Centre	1	W-3	SP-SXU	Feb 94
	Ministry of Defence	1	W-3/S-1RR	0203	96
		1	SRR-10	0720	98
		1	W-3PPD	0816	Nov 98
	Ministry of Interior	2	W-3	0315	Apr 90
		1	W-3A	0420	Jun 94
		1	W-3RM	0510	Apr 93
	Navy	2	W-3	0209	Jul 89
		5	W-3RM	0505	Jul 92
		3	W-3WARM	0813	Aug 98
	Polish Telecom	1	W-3	SP-SUI	Apr 92
	PZL Świdnik	1[1]	W-3	SP-SUA	Jan 89 †
		1	W-3W prot	SP-SUW	Mar 93 †
		1	W-3RM prot	0411/SP-SYG	Sep 95 †
		1[2]	W-3A	SP-PSL	May 98 †
	Zakopane Mountain Rescue	1	W-3	SP-SXT	Feb 93
		1	W-3A	SP-SXZ	Jan 96
Russia	Aeroflot	20[3]	W-3	SSSR-04101	Aug 88
UAE	Ministry of Internal Affairs	2	W-3A	?	02
Vietnam	SFC Vietnam	1	W-3AM	VN-417	Jan 98
unidentified		1			
Total		**143**			

Notes:
† Date of registration
[1] Firefighting demonstrator
[2] Sextant AFDS avionics testbed
[3] 18 transferred to Heliseco from January 1996

emergency ratings of 746 kW (1,000 shp) and 858 kW (1,150 shp) for 30 and 2½ minutes OEI respectively.

Particle separators on engine intakes, and inlet de-icing, standard. Power plant equipped with advanced electronic fuel control system for maintaining rotor speed at pilot-selected value amounting to ±5 per cent of normal rpm, and also for torque sharing as well as for supervising engine limits during start-up and normal or OEI operation. Engines and main rotor gearbox mounted on bed frame, eliminating drive misalignment due to deformations of fuselage. Transmission rating 1,342 kW (1,800 shp) maximum for T-O, 1,163 kW (1,560 shp) maximum continuous and 857 kW (1,150 shp) OEI. Engine input rpm 23,615.

Four bladder fuel tanks beneath cabin floor, with combined capacity of 1,720 litres (454 US gallons; 378 Imp gallons). Auxiliary tank, capacity 1,100 litres (291 US gallons; 242 Imp gallons), optional (not FAA approved). Oil capacity 14 litres (3.7 US gallons; 3.1 Imp gallons) per engine.

ACCOMMODATION: Pilot (port side), and co-pilot or flight engineer, side by side on W-3 flight deck, on adjustable seats with safety belts. W-3A can be flown by single pilot in VFR, with extra passenger in co-pilot seat. Dual controls and dual flight instrumentation optional. Accommodation for 12 passengers in main cabin or up to eight survivors plus two-person rescue crew and doctor in Anakonda SAR version. Seats removable for carriage of internal cargo. Medevac version can carry four stretcher cases and medical attendant (EMS version, one stretcher, three medical personnel and intensive care suite). Baggage space, capacity 180 kg (397 lb), at rear of cabin.

Door with bulged window on each side of flight deck; large sliding door for passenger and/or cargo loading on port side at forward end of cabin; second sliding door at rear of cabin on starboard side. Optically flat windscreens, improving view and enabling wipers to sweep a large area. Accommodation soundproofed, heated (by engine bleed air) and ventilated.

SYSTEMS: Two independent hydraulic systems, working pressure 90 bar (1,300 lb/sq in), for controlling main and tail rotors, unlocking collective pitch control lever, and feeding damper of directional steering system; automatic power changeover if one system fails. Flow rate 11 litres (2.9 US gallons; 2.4 Imp gallons)/min in each system. Vented gravity feed reservoir, at atmospheric pressure. Pneumatic system for actuating hydraulic mainwheel brakes. Electrical system providing 200/115 V three-phase AC power at 400 Hz and 28 V DC power. Electric anti-icing of all rotor blades. Fire detection/extinguishing system. Air conditioning and oxygen systems optional. Neutral gas system optional, for inhibiting fuel vapour explosion.

AVIONICS: Standard VFR and IFR nav/com avionics permit adverse weather operation by day or night. Bendix/King avionics standard; alternatives at customer's option.

Comms: Chrom (NATO 'Pin Head') IFF transponder in military versions.

Radar: Bendix/King RDS-82 weather radar in W-3A; 5A-813 radar in W-3RM.

Flight: Stability augmentation system standard. AP Decca navigator in W-3RM.

Mission: SPOR search and detection system in W-3RM.

Self-defence: Modified Syrena RWR in military versions initially; to be replaced by Thales EWR-99 (SPS-H in W-3PPD).

EQUIPMENT: Cargo version equipped with 2,100 kg (4,630 lb) capacity external hook and 150 kg (331 lb) capacity rescue hoist. W-3RM has 272 kg (600 lb) capacity electric hoist; stretchers, two-person rescue basket, rescue belts, two six-person liferafts, rope ladder, portable oxygen equipment, electric blankets and vacuum flasks, various types of buoy (light, smoke and radio) and marker, binoculars, flare pistol and searchlights. Firefighting version: one 1,590 litre (420 US gallon; 350 Imp gallon) 'Bambi bucket' on cargo sling or 1,500 litre (396 US gallon; 330 Imp gallon) expandable underbelly tank.

ARMAMENT: As described under Current Versions.

DIMENSIONS, EXTERNAL:
Main rotor diameter	15.70 m (51 ft 6 in)
Tail rotor diameter	3.03 m (9 ft 11¼ in)
Main rotor blade chord	0.44 m (1 ft 5¼ in)
Distance between rotor centres	9.50 m (31 ft 2 in)
Length: overall, rotors turning	18.79 m (61 ft 7¾ in)
fuselage	14.21 m (46 ft 7½ in)
Fuselage max width	1.75 m (5 ft 9 in)
Height: to top of rotor head	4.20 m (13 ft 9½ in)
overall, rotors turning	5.135 m (16 ft 10¼ in)
Stabiliser span	2.385 m (7 ft 10 in)
Wheel track	3.15 m (10 ft 4 in)
Wheelbase	3.55 m (11 ft 7¾ in)
Passenger/cargo doors:	
Height (each)	1.18 m (3 ft 10½ in)
Width: port	0.94 m (3 ft 1 in)
starboard	1.25 m (4 ft 1¼ in)
Height to sill	0.86 m (2 ft 10 in)

DIMENSIONS, INTERNAL:
Cabin: Length	3.21 m (10 ft 6½ in)
Max width	1.55 m (5 ft 1 in)
Max height	1.38 m (4 ft 6¼ in)
Floor area	4.80 m² (51.7 sq ft)
Volume	6.9 m³ (243 cu ft)

Polish Army W-3WA (left), with W-3PPD command post version beyond (*R J Malachowski*) NEW/0137427

AREAS:
Main rotor disc	193.6 m² (2,083.8 sq ft)
Tail rotor disc	7.21 m² (77.6 sq ft)
Main rotor blades (each)	2.90 m² (31.22 sq ft)
Tail rotor blades (each)	0.28 m² (3.01 sq ft)
Fin	1.00 m² (10.76 sq ft)
Elevator	2.16 m² (23.25 sq ft)

WEIGHTS AND LOADINGS:
Min basic weight empty	3,300 kg (7,275 lb)
Basic operating weight empty (multipurpose versions)	3,850 kg (8,488 lb)
Max fuel weight	1,326 kg (2,923 lb)
Max cargo payload, internal or external	2,100 kg (4,630 lb)
Normal T-O weight	6,100 kg (13,448 lb)
Max T-O weight	6,400 kg (14,110 lb)
Max disc loading	33.1 kg/m² (6.78 lb/sq ft)
Transmission loading at max T-O weight and power	4.77 kg/kW (7.84 lb/shp)

PERFORMANCE (A: at normal T-O weight, B: at max T-O weight, ISA, except where indicated):
Never-exceed speed (VNE):		
A, B	140 kt (260 km/h; 161 mph)	
Max cruising speed: A	129 kt (238 km/h; 148 mph)	
B	127 kt (235 km/h; 146 mph)	
Econ cruising speed: A	121 kt (225 km/h; 140 mph)	
B	120 kt (222 km/h; 138 mph)	
Max rate of climb at S/L: A	558 m (1,831 ft)/min	
B	510 m (1,673 ft)/min	
Service ceiling: A	4,910 m (16,100 ft)	
B	4,520 m (14,820 ft)	
Hovering ceiling IGE:		
A at ISA	3,020 m (9,900 ft)	
B at ISA	2,550 m (8,360 ft)	
A at ISA + 20°C	2,150 m (7,050 ft)	
B at ISA + 20°C	1,690 m (5,540 ft)	
Hovering ceiling OGE:		
A at ISA	1,900 m (6,220 ft)	
B at ISA	660 m (2,160 ft)	
A at ISA + 20°C	1,020 m (3,340 ft)	
B at ISA + 20°C	400 m (1,310 ft)	

Max range at econ cruising speed, no reserves:
A (standard tanks)	402 n miles (745 km; 463 miles)
B (standard tanks)	396 n miles (734 km; 456 miles)
A (with ferry tank)	622 n miles (1,152 km; 715 miles)
B (with ferry tank)	671 n miles (1,244 km; 773 miles)

Max endurance at 67 kt (125 km/h; 78 mph), no reserves:
A (standard tanks)	4 h 18 min
B (standard tanks)	4 h 12 min
A (with ferry tank)	6 h 42 min
B (with ferry tank)	6 h 24 min

PERFORMANCE (W-3RM at typical mission T-O weight of 5,500 kg; 12,125 lb):
Never-exceed speed (VNE)	140 kt (260 km/h; 161 mph)
Max cruising speed	131 kt (243 km/h; 151 mph)
Max rate of climb at S/L	672 m (2,205 ft)/min
Service ceiling	5,830 m (19,120 ft)
Hovering ceiling: IGE	4,000 m (13,120 ft)
OGE	2,970 m (9,740 ft)
Max range, no reserves	411 n miles (761 km; 473 miles)
Max endurance, no reserves	4 h 30 min

UPDATED

PZL-ŚWIDNIK W-3 SEP

English name: Vulture

TYPE: Naval combat helicopter.

PROGRAMME: Reported in 2000 as new design derived from W-3RM Anakonda.

CURRENT VERSIONS: Projected in three versions: anti-ship, ASW (**W-3ZOP**; *zwalczania okrętów podwodnych*) and reconnaissance/target acquisition (**W-3RWC**: *rozpoznania i wskazywania celów*).

DESIGN FEATURES: Shore-based but retaining manual blade-folding to improve shipboard compatibility; uprated (746 kW; 1,000 shp) engines; wider-track main landing gear.

AVIONICS: Planned to include ARS-410 undernose radar and BAE Systems FLIR.

UPDATED

PZL-ŚWIDNIK SW-4

TYPE: Light utility helicopter.

PROGRAMME: Development began 1985; full-scale mockup completed 1987 (see description and illustrations in 1991-92 and earlier *Jane's*); major redesign undertaken 1989-90, using Allison (now Rolls-Royce) 250 engine in more streamlined fuselage with modified tail unit. Prototype (c/n 600102), rolled out December 1994, is non-flying testbed for ground and equipment tests; '101 is static test airframe, '103 and '104 are flying prototypes. First flight made by 600103 (red overall; later registered SP-PSW) on 26 October 1996 ('official' flight three days later).

Trials in 1997 demonstrated requirement for a new rotor head design, enlarged horizontal stabiliser and more robust hydraulic system. Following 70 hours of test-flying, SP-

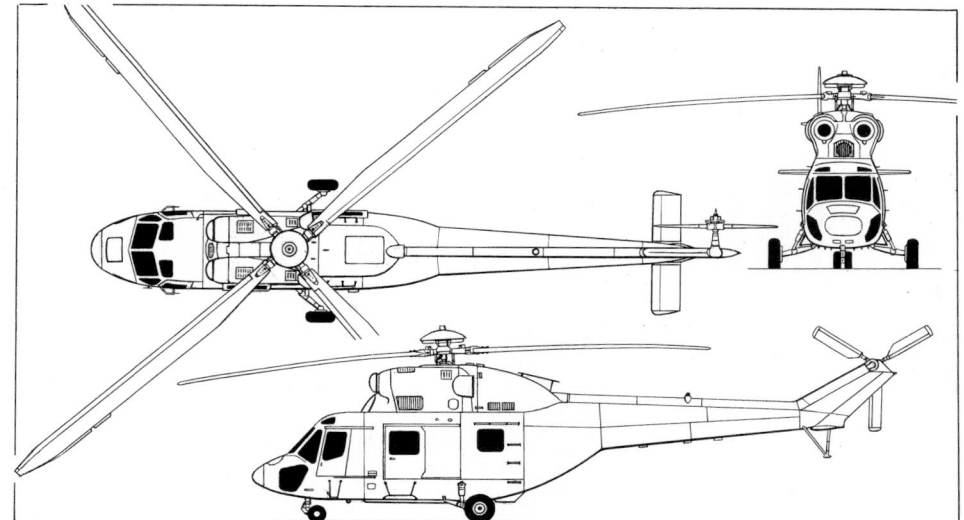

PZL-Świdnik W-3A Sokół twin-turboshaft helicopter (*Jane's/Dennis Punnett*)

Second flying prototype of the SW-4 four/five-seat light helicopter *(Jane's/Paul Jackson)*   NEW/0137362

AVIONICS: *Instrumentation*: Bendix/King VFR/IFR in first prototype; IFR optional in production aircraft.

DIMENSIONS, EXTERNAL:

Main rotor diameter	9.00 m (29 ft 6¼ in)
Tail rotor diameter	1.50 m (4 ft 11 in)
Distance between rotor centres	5.25 m (17 ft 2¾ in)
Length:	
overall, both rotors turning	10.55 m (34 ft 7½ in)
fuselage: incl tailskid	9.075 m (29 ft 9¼ in)
excl tailskid	8.25 m (27 ft 0¾ in)
Max width: cabin	1.515 m (4 ft 11¾ in)
over tail unit endplates	1.83 m (6 ft 0 in)
over skids	2.28 m (7 ft 5¾ in)
Height to top of rotor head	2.97 m (9 ft 9 in)
Skid track	2.20 m (7 ft 2½ in)
Fuselage ground clearance	0.55 m (1 ft 9¾ in)

DIMENSIONS, INTERNAL:

Cabin: Length	2.14 m (7 ft 0¼ in)
Max width	1.40 m (4 ft 7 in)
Max height	1.25 m (4 ft 1¼ in)
Volume	3.85 m³ (136 cu ft)
Baggage compartment volume	0.85 m³ (30.0 cu ft)

AREAS:

Main rotor disc	63.62 m² (684.8 sq ft)
Tail rotor disc	1.77 m² (19.05 sq ft)

WEIGHTS AND LOADINGS:

Weight empty	950 kg (2,094 lb)
Max payload: internal	550 kg (1,213 lb)*
on external sling	700 kg (1,543 lb)
Normal T-O weight (internal payload)	
	1,600 kg (3,527 lb)
Max T-O weight with sling load	1,800 kg (3,968 lb)
Max disc loading	26.7 kg/m² (5.47 lb/sq ft)
Transmission loading at max T-O weight and power	
	5.07 kg/kW (8.33 lb/shp)

* 400 kg (882 lb) in main cabin, 150 kg (331 lb) in baggage compartment

PERFORMANCE (at normal T-O weight, ISA):

Never-exceed speed (V$_{NE}$)	140 kt (260 km/h; 161 mph)
Max cruising speed	122 kt (226 km/h; 140 mph)
Normal cruising speed	108 kt (200 km/h; 124 mph)
Max rate of climb at S/L	480 m (1,575 ft)/min
Service ceiling	5,000 m (16,400 ft)
Hovering ceiling, IGE	1,650 m (5,420 ft)
Max range, standard fuel, no reserves	
	413 n miles (766 km; 476 miles)
Endurance, as above	4 h 51 min

UPDATED

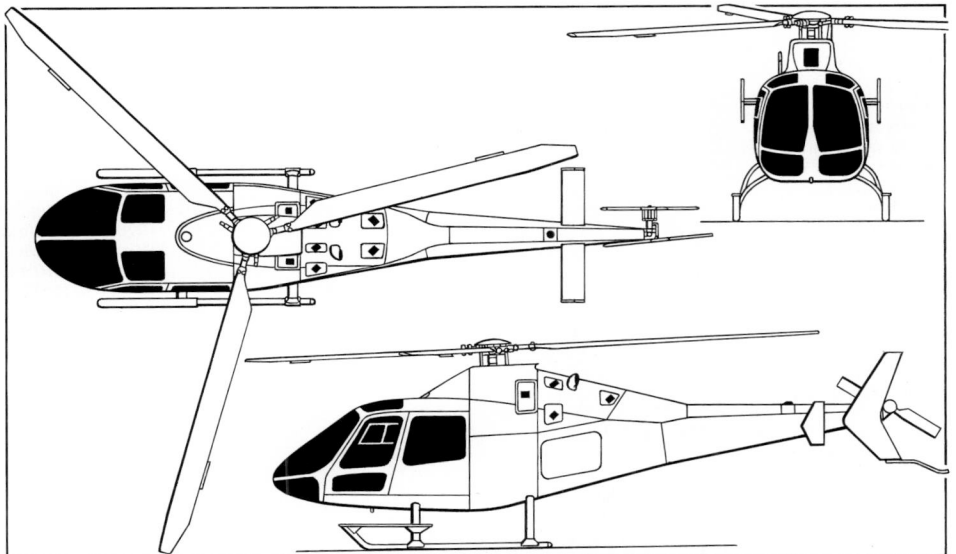

PZL-Świdnik SW-4 (Rolls-Royce 250-C20R/2 turboshaft) *(Jane's/Mike Keep)*   0100525

PSW was grounded in late 1997 for installation of SAMM-designed hydraulic flight control system, with which it was then due to return to flying in 1998. Second flying prototype (yellow overall), with improved skids, exhibited statically at Paris Air Show, June 1997; registered as SP-PSZ in October 1998, but not flown until early 2001. Some 640 hours (total) flown by July 2002, when certification programme said to be almost complete, with domestic certificate to FAR/JAR 27 expected in October 2002. First six production aircraft started during first quarter 1999; first deliveries expected in 2003. Second prototype shown at Paris in June 2001; first at Berlin in May 2002.

CURRENT VERSIONS: Will be marketed in two versions:

Economy: With one 336 kW (450 shp) Rolls-Royce 250-C20R engine and 1,600 kg (3,527 lb) normal T-O weight.

High Performance: With one 459 kW (615 shp) Pratt & Whitney Canada PW200/9 engine and 1,800 kg (3,968 lb) normal T-O weight.

Provision has been made for later installation of two engines to fulfil ICAO requirements for OEI operation (including hover).

CUSTOMERS: Polish Air Force requirement confirmed in mid-2002 for 47 by 2010, of which seven planned to enter service in 2005 and 14 in 2006; for use in training role. Two of first six production aircraft scheduled for Polish Police and Polish MoD (one each). Other orders reported from three unidentified German customers.

COSTS: US$700,000 (2002).

Following description applies to first flying prototype:

DESIGN FEATURES: Compact light helicopter, originally envisaged as Eastern Bloc parallel to Eurocopter, Agusta and similar types, but reorientated with Western engines and avionics as rival design. Intended applications include passenger and cargo transport, medevac, border patrol, armed scout and training use.

Three-blade main rotor; arrowhead tailfin on port side, with two-blade tail rotor to starboard; narrow tailplane with small endplate fins; skid landing gear.

STRUCTURE: GFRP for approximately 20 per cent of airframe, including all rotor blades; remainder mainly of aluminium alloy.

LANDING GEAR: Skid type; able to accommodate heavy landing sink rate of 3.1 m (10 ft)/s by elastic deflection of cross-tubes.

POWER PLANT: One 336 kW (450 shp) Rolls-Royce 250-C20R/2 turboshaft. Transmission rating 336 kW (450 shp) for T-O, 283 kW (380 shp) maximum continuous; 30 minute run-dry capability. Standard fuel capacity 500 litres (132 US gallons; 110 Imp gallons) in tank below main gearbox.

ACCOMMODATION: Pilot and up to four passengers or one stretcher patient and a medical attendant. One front-hinged and one rearward-sliding door on each side of cabin.

PZL-ŚWIDNIK I-23 MANAGER

TYPE: Four-seat lightplane.

PROGRAMME: Launched as Instytut Lotnictwa development programme June 1992; funded by grant from State Committee for Scientific Research; construction started January 1994 of one static test (001) and two flying prototypes; first flight, by 002/SP-PIL, 20 February 1999; original name Ibis changed to Manager 1999; Polish type certificate BB-215 awarded 3 October 2001; second prototype had not been produced by early 2002, at which time first aircraft was re-registered SP-GIL; series production by PZL-Świdnik planned to begin 2002, followed by certification to Normal category FAR Pt 23 Amendment 42.

CURRENT VERSIONS: **I-23N:** Standard Normal category, retractable gear IFR version with dual controls.

Description applies to this version except where indicated.

I-23D: Normal category IFR executive or patrol version of I-23N with single controls and uprated, fuel-injection engine.

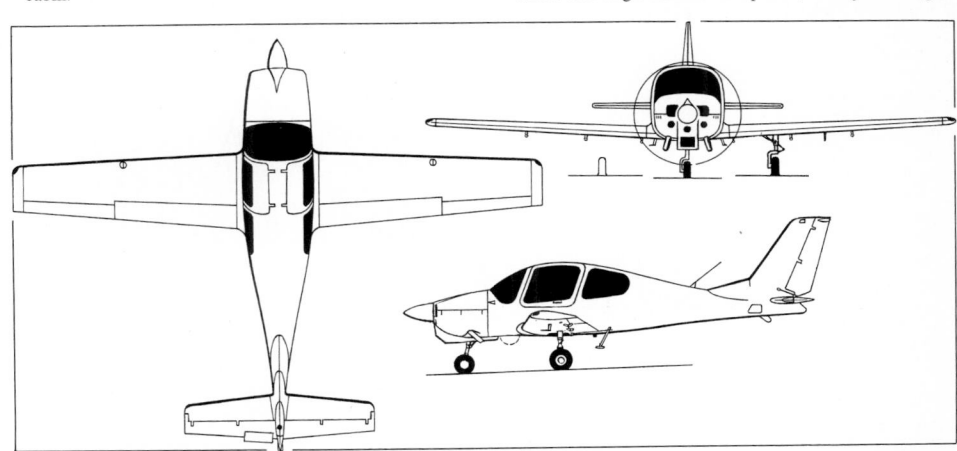

PZL-Świdnik I-23 Manager four-seat light aircraft *(Jane's/Mike Keep)*   NEW/0137337

For details of the latest updates to *Jane's All the World's Aircraft* online and to discover the additional information available exclusively to online subscribers please visit

jawa.janes.com

I-23S: Utility category basic trainer. Lower-powered engine, dual controls, VFR instrumentation, fixed landing gear and reinforced structure.

CUSTOMERS: Two ordered by Polish Ministry of Defence.

DESIGN FEATURES: Intended as safe, low-cost, low fuel consumption private owner aircraft. Conventional low-wing, tricycle-gear design with fully enclosed cabin.

Wing has ILL-217 (root) and ILL-213 (tip) aerofoil sections, 2° 30′ leading-edge sweepback, 4° dihedral, 2° 30′ incidence and -1° twist. Tailplane NACA 015 profile; fin NACA 0018.

Design permits use of any Textron Lycoming flat-four engine from 119 to 149 kW (160 to 200 hp), option for fixed or retractable landing gear, and equipment to customer's specifications.

FLYING CONTROLS: Conventional and manual. Ailerons deflect 22° up/17° down, elevators 30° up/24° down, rudder 21° left/right. Trim tab in port elevator. Flap settings 0/20/30°.

STRUCTURE: Mainly GFRP/CFRP/epoxy, with sandwich skins; CFRP for main spar and in region of landing gear attachments. Wing ribs and landing gear support plate of duralumin; steel fittings. Skin panels attached by adhesives, reinforced by bolts and rivets. First Polish aircraft with prepreg composites.

LANDING GEAR: Electrohydraulically retractable tricycle type on I-23N and D, with oleo-pneumatic shock-absorbers; steerable (±25°) nosewheel; mainwheels retract inward, nosewheel rearward, latter remaining half exposed when retracted; Parker wheels and brakes; Goodrich 5.00-5 tyres (main and nose), pressure 3.80 bar (55 lb/sq in) on mainwheels and 3.10 bar (45 lb/sq in) on nosewheel. Non-retractable gear on I-23S.

POWER PLANT: *I-23N:* One Textron Lycoming O-360-A1A flat-four (134 kW; 180 hp at 2,700 rpm), driving an MT-built Hartzell HC-C2YR-1BF/F766A-4 two-blade constant-speed propeller.

I-23D: 147 kW (200 hp) Lycoming IO-360-A1A.

I-23S: 118 kW (160 hp) Lycoming O-320-D2A and Sensenich fixed-pitch propeller.

Integral fuel tank in each wing, combined capacity 190 litres (50.2 US gallons; 41.8 Imp gallons) in I-23N/I-23D, 110 litres (29.1 US gallons; 24.2 Imp gallons) in I-23S. Gravity fuelling points on each wing. Oil capacity (all versions) 7.6 litres (2.0 US gallons; 1.7 Imp gallons).

ACCOMMODATION: Four persons (pilot and three passengers) on two rows of seats. Upward-opening door each side. Cabin heated and ventilated.

SYSTEMS: Electrohydraulic system for landing gear actuation; hydraulic system for mainwheel brakes only. Electrical power from 14 V 70 A alternator and 12 V 35 Ah battery. Heated pitot head.

AVIONICS: *Comms:* Bendix/King com transceiver and ATC transponder standard.

Flight: Bendix/King nav transceiver, ADF, GPS and directional gyro standard; optional full IFR system with KNS 81 integrated nav/RNav, DME and KCS 55A HSI.

DIMENSIONS, EXTERNAL:
Wing span	8.95 m (29 ft 4¼ in)
Wing chord: at root	1.31 m (4 ft 3½ in)
mean	1.14 m (3 ft 8¾ in)
at tip	0.84 m (2 ft 9 in)
Wing aspect ratio	8.0
Length: overall	7.10 m (23 ft 3½ in)
fuselage	5.70 m (18 ft 8½ in)
Height overall	2.49 m (8 ft 2 in)
Tailplane span	3.20 m (10 ft 6 in)
Wheel track	2.89 m (9 ft 5¾ in)
Wheelbase	1.65 m (5 ft 5 in)
Propeller diameter	1.83 m (6 ft 0 in)
Propeller ground clearance	0.20 m (7¾ in)

DIMENSIONS, INTERNAL:
Cabin: Length	2.35 m (7 ft 8½ in)
Max width	1.13 m (3 ft 8½ in)
Max height	1.20 m (3 ft 11¼ in)
Baggage compartment:	
Length	0.80 m (2 ft 7½ in)
Max width	0.68 m (2 ft 2¾ in)
Max height	0.35 m (1 ft 1¾ in)

AREAS:
Wings, gross	10.00 m² (107.6 sq ft)
Ailerons (total)	0.71 m² (7.64 sq ft)

First prototype I-23 Manager (*Jane's/Paul Jackson*)   NEW/0143327

Trailing-edge flaps (total)	1.48 m² (15.93 sq ft)		
Fin	1.33 m² (14.32 sq ft)		
Rudder	0.95 m² (10.23 sq ft)		
Tailplane	2.08 m² (22.39 sq ft)		
Elevators (total)	0.83 m² (8.93 sq ft)		

WEIGHTS AND LOADINGS (A: I-23N, B: I-23D, C: I-23S):
Weight empty: A, B	690 kg (1,521 lb)
C	650 kg (1,433 lb)
Fuel weight: A, B	130 kg (287 lb)
C	80 kg (176 lb)
Max payload: A	330 kg (728 lb)
B	320 kg (705 lb)
C	300 kg (661 lb)
Max T-O weight: A, B	1,150 kg (2,535 lb)
C	1,050 kg (2,315 lb)
Max wing loading: A, B	115.0 kg/m² (23.55 lb/sq ft)
C	105.0 kg/m² (21.51 lb/sq ft)
Max power loading: A	8.57 kg/kW (14.08 lb/hp)
B	7.71 kg/kW (12.68 lb/hp)
C	8.81 kg/kW (14.47 lb/hp)

PERFORMANCE (A, B and C as above):
Never-exceed speed (VNE):	
A	162 kt (300 km/h; 186 mph)
B	170 kt (315 km/h; 195 mph)
C	153 kt (285 km/h; 177 mph)
Max cruising speed at S/L:	
A	154 kt (285 km/h; 177 mph)
B	162 kt (300 km/h; 186 mph)
C	140 kt (260 km/h; 162 mph)
Max manoeuvring speed: A, B, C	121 kt (225 km/h; 140 mph)
Stalling speed: flaps up:	
A, B	68 kt (125 km/h; 78 mph)
flaps down: A, B	62 kt (113 km/h; 71 mph)
Max rate of climb at S/L: A	270 m (886 ft)/min
B	330 m (1,083 ft)/min
C	264 m (866 ft)/min
T-O to 15 m (50 ft): A, B	480 m (1,575 ft)
Landing from 15 m (50 ft): A, B	300 m (985 ft)
T-O run: A	250 m (820 ft)
B	220 m (725 ft)
C	255 m (840 ft)
T-O to 15 m (50 ft): A, B	480 m (1,575 ft)
Landing from 15 m (50 ft): A, B	300 m (985 ft)
C	280 m (920 ft)
Range with max fuel, 45 min reserves:	
A	755 n miles (1,400 km; 869 miles)
B	782 n miles (1,450 km; 901 miles)
C	432 n miles (800 km; 497 miles)
g limit	+3.8

UPDATED

I-23 instrument panel (*Jane's/Paul Jackson*)   NEW/0143329

SSH

SERWIS SAMOLOTOW HISTORYCZNICH-JANUSZ KARASIEWICZ (Janusz Karasiewicz Historic Aircraft Service)

PL-43 385 Jasienica 829
Tel: (+48 33) 815 34 91
Fax: (+48 33) 815 34 92
e-mail: jungmann@pro.onet.pl
MANAGING DIRECTOR: Janusz Karasiewicz
CUSTOMER SERVICE MANAGER: Julius Zulauf

GERMAN DISTRIBUTOR:
Bücker Flugzeugbau GmbH, Langwiederstrasse 1, D-85757 Karlsfeld-Gröbenried
Tel: (+49 8131) 789 26
Fax: (+49 8131) 803 14
e-mail: rangsdorf@bueckerflug.de
Web: http://www.bueckerflug.de

SSH is certified by the Polish Aviation Authority as an approved production organisation under JAR 21, Subpart G. In addition to producing the Bü 131, it was building a prototype of a related aircraft, the Bü-133P Jungmeister, in late 2001. It was hoped to fly this in 2002, using the original Siemens Sh-14A radial engine.

UPDATED

SSH (BÜCKER) T-131 JUNGMANN

TYPE: Aerobatic two-seat biplane.

PROGRAMME: Original German design first flew 27 April 1934; civil certification 3 October 1938; name refers to naval cadet; in addition to some 3,000 by parent firm (ending March 1941), licensed manufacture undertaken in Japan (339 K9W floatplanes and 1,037 Ki-86As), Spain (555 CASA 1.131s), Czechoslovakia (10 Tatra 131s and 300 Bu 131s, plus 260 Aero C.104s after 1945) and Switzerland (130 by Dornier). Last of some 5,700 were those produced in Spain, 1960.

SSH T-131 launched in March 1990, based on engineering drawings of Czech Bu 131D version (T and P in designations indicate Tatra and Poland). First of four pre-series aircraft flew 8 July 1994; first production aircraft exported to Austria in kit form December 1994 and made maiden flight 7 June 1997.

CURRENT VERSIONS: **T-131P:** Early version, with 78.3 kW (105 hp) Walter Minor 4-III four-cylinder inverted in-line engine and Hoffmann HO-40AHM-180 two-blade fixed-pitch metal propeller. Production complete.

T-131PA: More powerful version, with 103 kW (138 hp) LOM M 332AK four-cylinder, in-line engine. Prototype (SP-FPY) first flew 9 August 1995. Current baseline version; Polish certification under way in 2001, as step towards main objective of full standard certification in Germany.

Description applies to T-131PA.

CUSTOMERS: Three (including prototype) T-131Ps built 1994-95, of which two exported to Sweden. Total of 15 T-131PAs registered by early 2002, including two in Austria; at least seven to German owners, but retaining Polish registrations, and one each to Poland and Switzerland.

COSTS: Development Zl 15 million (2000); standard T-131PA €100,000 (2001).

DESIGN FEATURES: Single-bay biplane with interchangeable upper and lower wings, braced by pair of I-struts each side. This version based on Bü 131D, with ball bearings in control pivots and tailwheel, although Hirth 504A-2 engine replaced.

Wings have Göttingen aerofoil section (thickness/chord ratio 10.5 per cent) and 11° leading-edge sweepback; dihedral 3° 30′ (upper), 1° 30′ (lower); incidence −1° 30′; twist −1° 30′.

FLYING CONTROLS: Conventional and manual. Mass-balanced ailerons on all four wings; non-balanced rudder; trim tab in each elevator. No flaps.

STRUCTURE: Welded 4130 chromoly steel tube fuselage and tail unit; wooden wing spars (two) and ribs. Mostly Ceconite fabric covering except for metal engine cowling and around cockpit area. Fuel tank of glass fibre and carbon fibre.

LANDING GEAR: Non-retractable, with sprung 200×50 tailwheel; mainwheels 420×150 on forward-raked legs. Tyre pressures: main 1.50 bar (53 lb/sq in), tail 2.50 bar (88 lb/sq in). Hydraulic brakes.

POWER PLANT: One 104 kW (140 hp) LOM M 332AK four-cylinder in-line engine, driving an MT32.¹¹⁰/₁₂ two-blade

SSH T-131PA replica of Bücker Jungmann *(Jane's/Paul Jackson)*   **NEW**/0143706

fixed-pitch wooden propeller. Fuselage fuel tank, capacity 85 litres (22.5 US gallons; 18.7 Imp gallons), of which 80 litres (21.1 US gallons; 17.6 Imp gallons) are usable. Oil capacity 8.5 litres (2.25 US gallons; 1.9 Imp gallons).

ACCOMMODATION: Two persons in tandem, open cockpits.

SYSTEMS: 24 V DC electrical system (generator and two batteries).

AVIONICS: Basic VFR standard; other avionics to customer's requirements.

DIMENSIONS, EXTERNAL:

Wing span	7.40 m (24 ft 3¼ in)
Wing chord (each, constant)	1.00 m (3 ft 3¼ in)
Wing aspect ratio	4.1
Length overall	6.62 m (21 ft 8½ in)
Height overall	2.25 m (7 ft 4½ in)
Tailplane span	2.50 m (8 ft 2½ in)
Wheel track	1.75 m (5 ft 9 in)
Wheelbase	4.10 m (13 ft 5½ in)
Propeller diameter	1.88 m (6 ft 2 in)
Propeller ground clearance	0.63 m (2 ft 0¾ in)

AREAS:

Wings, gross	13.50 m² (145.3 sq ft)
Ailerons (four, total)	2.00 m² (21.53 sq ft)
Fin	0.50 m² (5.38 sq ft)
Rudder	0.80 m² (8.61 sq ft)
Tailplane	1.25 m² (13.45 sq ft)
Elevators (total, incl tabs)	1.15 m² (12.38 sq ft)

WEIGHTS AND LOADINGS (approx):

Weight empty	470 kg (1,036 lb)
Fuel weight	63 kg (139 lb)
Max T-O and landing weight: Normal	720 kg (1,587 lb)
Utility	670 kg (1,477 lb)
Aerobatic	590 kg (1,300 lb)
Max wing loading	53.3 kg/m² (10.92 lb/sq ft)
Max power loading	6.90 kg/kW (11.33 lb/hp)

PERFORMANCE:

Never-exceed speed (V_{NE})	151 kt (280 km/h; 174 mph)
Max level speed	102 kt (190 km/h; 118 mph)
Max cruising speed at S/L	97 kt (180 km/h; 112 mph)
Econ cruising speed at S/L	86 kt (160 km/h; 99 mph)
Stalling speed, engine idling	46 kt (85 km/h; 53 mph)
Max rate of climb at S/L	210 m (689 ft)/min
Service ceiling	4,000 m (13,120 ft)
T-O run	180 m (590 ft)
T-O to 15 m (50 ft)	400 m (1,315 ft)
Landing from 15 m (50 ft)	300 m (985 ft)
Landing run	150 m (495 ft)
Max range, 30 min reserves	
	283 n miles (525 km; 326 miles)

UPDATED

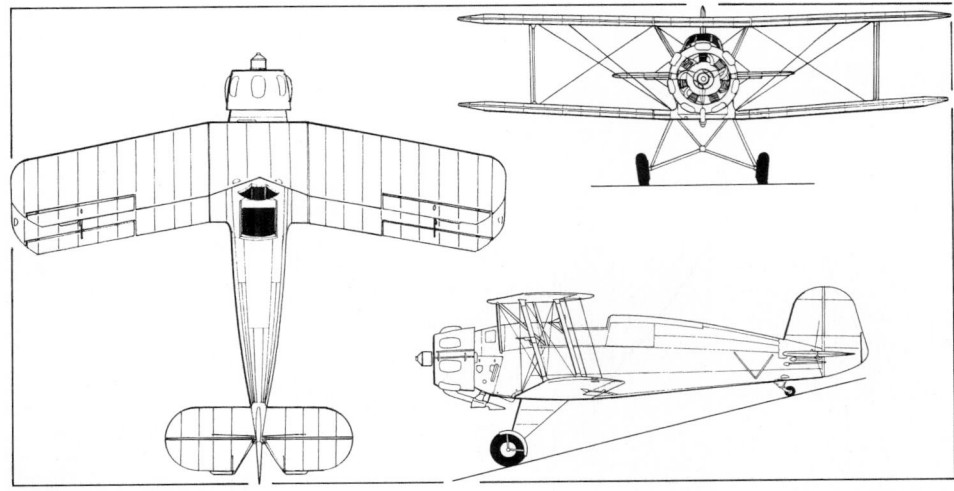

SSH (Bücker) Bü-133P Jungmeister   0110902

SSH (BÜCKER) BÜ-133P JUNGMEISTER

English name: Young Champion

TYPE: Aerobatic single-seat biplane.

PROGRAMME: Entered service in 1936; also licence-built in Spain and Switzerland. Small number of replicas built in late 1960s by Bitz at Augsburg and Hirth at Nabern. SSH project initiated in late 1990s; prototype (SP-PUV) planned to fly in 2002. Perpetuates original German Bü 133C of 1935. (Correct German designation is unhyphenated.)

DESIGN FEATURES: Generally similar to those of Jungmann.

FLYING CONTROLS: As Jungmann.

STRUCTURE: Fabric-covered steel tube fuselage and tail unit, except for wooden turtledeck; two-spar wooden wings with composites leading-edges and metal interplane struts.

LANDING GEAR: Tailwheel type; fixed. Mainwheel tyres size 6.00-6.

POWER PLANT: One 119 kW (160 hp) Siemens-Halske Sh-14a seven-cylinder radial engine.

ACCOMMODATION: Pilot only, in open cockpit.

DIMENSIONS, EXTERNAL:

Wing span	6.60 m (21 ft 7¾ in)
Length overall	6.02 m (19 ft 9 in)
Height overall	2.20 m (7 ft 2½ in)

AREAS:

Wings, gross	12.00 m² (129.2 sq ft)

WEIGHTS AND LOADINGS:

Weight empty	425 kg (937 lb)
Max T-O weight	615 kg (1,355 lb)
Max wing loading	51.25 kg/m² (10.50 lb/sq ft)
Max power loading	5.15 kg/kW (8.47 lb/hp)

PERFORMANCE (estimated):

Max level speed	118 kt (220 km/h; 136 mph)
Time to 1,000 m (3,280 ft)	2 min 48 s
Service ceiling	4,500 m (14,760 ft)
Range	270 n miles (500 km; 310 miles)

UPDATED

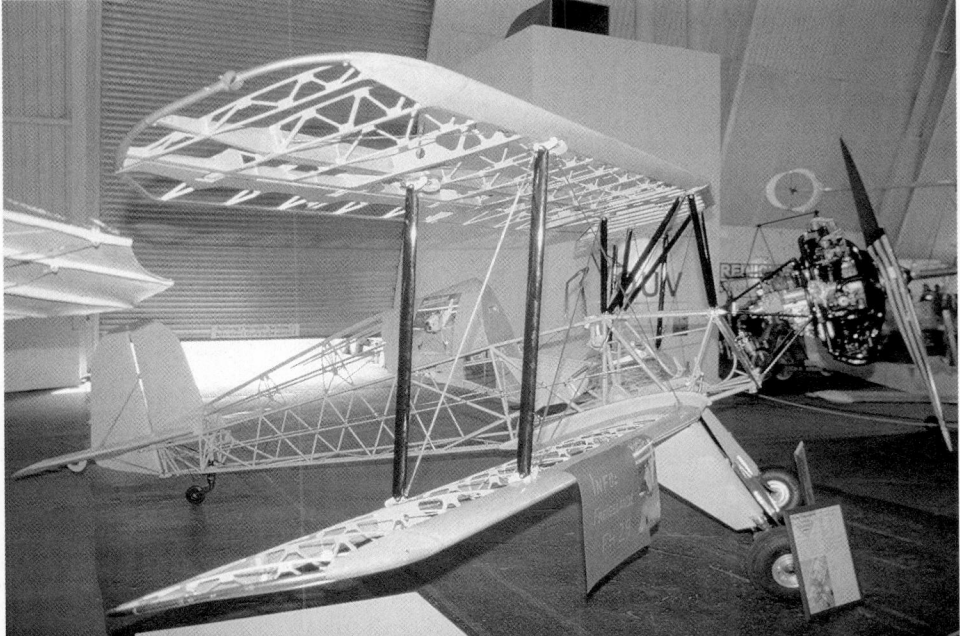

First SSH-built Jungmeister before covering *(Jane's/Paul Jackson)*   0110937

ZRPSL

ZAKŁAD REMONTÓW I PRODUKCJI SPRZĘTU LOTNICZEGO (Aviation Equipment Repair and Production Works)

ulica Cieszyńska 321, PL-43-300 Bielsko-Biała
Tel/Fax: (+48 33) 815 01 10
e-mail: e.marganski@pro.onet.pl
Web: http://www.marganski.home.pl
MANAGING DIRECTOR: Edward Marganski

Edward Marganski has been connected with the design of several Polish light aircraft. His latest design is the Bielik, envisaged as a modern successor to the TS-11 Iskra jet trainer.

UPDATED

ZRPSL BIELIK and FENIX

TYPE: Basic jet trainer.
PROGRAMME: Started as private venture under project name Iskra II, but renamed Bielik (a native Polish eagle) after press competition before public debut third quarter 2001 at exhibitions in Katowice and Radom. First flight (prototype SP-YEM) was then anticipated before end of 2001, but had not been reported by mid-2002.

CURRENT VERSIONS: **Bielik:** Military basic and advanced training version, aimed at Polish Air Force requirement following failure of Iryda.
Fenix (English name Phoenix): Civil general aviation version, including aerobatics; intended for export.
DESIGN FEATURES: Mid-wing, blended wing/body design with forward-extending curved strakes; twin outward-canted fins and rudders. Designed for high-*g* and high-AoA manoeuvres.
Supercritical wing aerofoil section, thickness/chord ratio 6 per cent.
FLYING CONTROLS: Manual, with hydraulically boosted all-moving slab tailplane, with tabs, for pitch and roll control; twin rudders deflect outwards only. Full-span leading- and trailing-edge flaps.
STRUCTURE: All-composites (carbon fibre and epoxy resin) except for light alloy and titanium rear fuselage. Single-spar wing.
LANDING GEAR: Hydraulically retractable tricycle type, with single wheel on each unit. Mainwheels retract inward, nosewheel forward.
POWER PLANT: One 12.75 to 17.65 kN (2,866 to 3,968 lb st) class turbojet in production aircraft; prototype has 13.52 kN (3,040 lb st) General Electric J85. Thrust vector control planned for production aircraft. All fuel in forward and centre fuselage tanks.
ACCOMMODATION: Crew of two in tandem, on zero/zero ejection seats, in pressurised cockpit.

SYSTEMS: Hydraulic system for landing gear, flap, airbrake and wheel brake actuation and control surface boost. Electrical system. Crew oxygen system.
AVIONICS: Bielik basic trainer and Fenix to have standard nav/com avionics. Bielik advanced trainer version planned to be equipped with combat application electronic simulations of armament operation; real or virtual, radar- and IR-based observation and sighting systems; evaluation of sighting and weapon delivery techniques; and hostile air defence facilities.

DIMENSIONS, EXTERNAL:
Wing span	6.60 m (21 ft 7¾ in)
Wing aspect ratio	3.7
Length overall	9.00 m (29 ft 6¼ in)
Height overall	2.50 m (8 ft 2½ in)

AREAS:
Wings, gross	11.90 m² (128.1 sq ft)

WEIGHTS AND LOADINGS:
Weight empty	1,700 kg (3,748 lb)
Fuel weight	850 kg (1,874 lb)
Max T-O weight	2,500 kg (5,511 lb)
Max wing loading	210.1 kg/m² (43.03 lb/sq ft)
Max power loading (prototype)	185 kg/kN (1.81 lb/lb st)

PERFORMANCE (estimated):
Max level speed	540 kt (1,000 km/h; 621 mph)
Stalling speed	84 kt (155 km/h; 97 mph)
Max rate of climb at S/L	4,500 m (14,760 ft)/min
Range with max fuel	1,349 n miles (2,500 km; 1,553 miles)

UPDATED

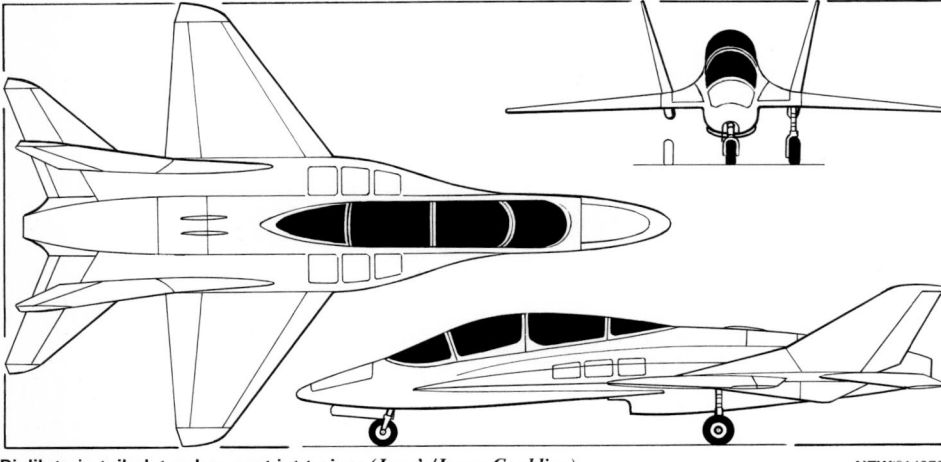

Bielik twin-tailed, tandem-seat jet trainer (*Jane's/James Goulding*)   NEW/0143736

Bielik cockpit detail (*R J Malachowski*)   0130589

Two aspects of ZRPSL (Marganski) Bielik high-subsonic jet trainer (*Wozciech Gorgolewski*)   NEW/0145049

ROMANIA

IAROM

SC IAROM SA

44A Bulevardul Ficusului, Sector 1, R-71547 Bucuresti
Tel: (+40 1) 232 63 16
Fax: (+40 1) 232 63 46
GENERAL MANAGER: Mihail Toncea
COMMERCIAL AND BUSINESS DEVELOPMENT MANAGER: Antonela Banea
RESEARCH DEPARTMENT MANAGER: Vlad Atanase

IAROM (Industria Aeronautica Romana) is holding company for Romanian aircraft industry, created 1991 to replace former CNIAR (1991-92 and earlier *Jane's*). Aircraft and aero-engine companies and R&D institutes given full autonomy from 1990, when all became joint stock companies as prelude to privatisation. Shareholding was initially divided between State Ownership Fund (70 per cent) and Private Ownership Funds (30 per cent); full privatisation was achieved in 1998. IAROM has a 65 per cent holding in Aerostar (which see).

Main function is to establish general strategy and policy for Romanian aviation industry in fields of research, design, development, production and marketing, including technical support services for Romanian and foreign companies; also acts as agent for foreign companies doing business within Romania.

UPDATED

AEROSTAR

SC AEROSTAR SA

9 Condorilor Street, R-5500 Bacău
Tel: (+40 34) 17 50 70
Fax: (+40 34) 17 20 23 and 17 22 59
e-mail: aerostar@aerostar.ro
Web: http://www.aerostar.ro
PRESIDENT AND GENERAL DIRECTOR: Dipl Eng Grigore Filip
DIRECTORS:
   Dipl Eng Ovidiu Buhai (Systems Division)
   Dipl Eng Petru Placinta (Technology Division)
   Dipl Eng Dorin Panfil (Commercial Division)
HEAD OF MARKETING DEPARTMENT: Serban Iosipescu
PUBLIC RELATIONS: Doina-Gabriela Matanie

Factory celebrates 50th anniversary in 2003, having been founded 1953 as URA (later IRAv, then IAv Bacau), originally as repair centre for Romanian Air Force Yak-17/23, MiG-15/17/19/21 and Il-28/H-5 front-line military aircraft and Aero L-29/L-39 jet trainers; built first Romanian prototype of IAR-93 between 1972 and 1974. Five other specialised work sections: (1) landing gears, hydraulic and pneumatic equipment; (2) special production; (3) light aircraft; (4) engines and reduction gears; (5) avionics. Site area 45 ha (111.2 acres), including 25 ha (61.8 acres) of covered workshop space. Workforce in March 2001 totalled 2,800.

Romanian State Ownership Fund (SOF) originally owned 69.99 per cent of Aerostar. In preparation for full privatisation, Aerostar reorganised 1997 into three semi-autonomous divisions: systems, commercial, and technology support, reporting to a strategic fourth division. Creation of a joint venture (Aerothom Electronics) with Thomson-CSF (now Thales) Communications announced June 1999; Aerostar owns 60 per cent of this company, formed to supply IFF equipment to Romanian MoD. Aerostar became fully privatised company 11 February 2000 as part of Aerostar-PAS-IAROM, a consortium of Aerostar management and employees (PAS) with IAROM (which see), acquiring Aerostar stock previously held by SOF. The new shareholdings are IAROM 65 per cent; SIF 'Moldova' (financial investment company) 11 per cent; PAS-Aerostar 5 per cent; others 19 per cent. SOF retains 'golden share' entitling it to veto matters affecting company's defence production capability. Consortium agreed to take over all Aerostar's obligations and debts; not to dissolve company for 10 years; and to keep its core business intact for at least five years. Company turnover in 2000 was Lei429.5 billion, of which exports accounted for approximately 65 per cent.

Aerostar's main product for many years has been Iak-52 trainer, joined in 2001 by Aerostar 01. Manufactures airframes for ULBI Wild Thing ultralight (see German section) and components for replicas of Focke-Wulf Fw 190 (see Flug Werk in German section) and North American P-51 Mustang (available 2003 from Fighter Factory LLC, 13545 Sycamore Avenue, San Martin, California 95046, USA; (*tel:* +1 408 683 25 31; *fax:* +1 408 683 25 33). Aerostar also responsible for Lancer and Sniper Romanian Air Force upgrade programmes for MiG-21s and MiG-29s; see *Jane's Aircraft Upgrades* for details.

Landing gears and/or hydraulic/pneumatic equipment have been produced for IAR-316B (Alouette III), IAR-330 (Puma), IAR-93, IAR-99, Romaero (BAC) One-Eleven and Iak-52. Engines include M-14P for Iak-52, M-14V26 for Ka-26 and RU-19A-300 APU for An-24; reduction gears include R-26. Avionics factory produces radio altimeters, radio compasses, marker beacon receivers, IFF and other radio/radar items.

UPDATED

Instrument panel of Aerostar 01   0059971

Aerostar 01 side-by-side ultralight in flight over Airclub Romania *(Arh Mihai Andrei/Aerostar)* *NEW*/0525750

AEROSTAR 01

TYPE: Side-by-side ultralight/kitbuilt.
PROGRAMME: Illustrated (as project) in 1999 company brochure, but no details then given. Made debut at Paris Air Show, June 2001, following first flight (YR-6138) on 31 May. Offered in factory-built or kit form.
CUSTOMERS: Initial orders received at time of debut (believed from Romanian Airclub); planned eventual worldwide sales and output of 40 to 50 per year. First two production aircraft registered in USA (N1062N and N1062M) in July 2001.
COSTS: US$35,000 to US$55,000 flyaway, depending upon engine/avionics fitted (2001).
DESIGN FEATURES: Designed to Canadian standard TP10141E; derivative of Air Light Wild Thing (see German section), utilising wings, nose, standard engine, rear fuselage and tail unit from that aircraft. Principal differences are enlarged cabin, nosewheel landing gear, and relocation of wing in low-mounted position, without strut bracing, but still foldable alongside fuselage for transportation and storage.
   Wing aerofoil section NACA 4415.
FLYING CONTROLS: Conventional and manual. Flaps and ailerons pushrod-actuated, elevators and horn-balanced rudder by cables; mechanically or electrically actuated elevator trim tab. Dual controls standard.
STRUCTURE: Aluminium alloy, except for fabric-covered flaps and ailerons.
LANDING GEAR: Tricycle type, fixed; main units have cantilever legs, hydraulic disc brakes and Matco tyres; steerable, trailing-link nosewheel with rubber shock absorption.
POWER PLANT: One 59.7 kW (80 hp) Jabiru 2200 flat-four standard, driving a two-blade wooden propeller. Fuel tank in each wing.
   Optional engines include 59.6 kW (79.9 hp) Rotax 912 UL, 63.4 kW (85 hp) Limbach L 2000, 70.8 kW (95 hp) Hirth F 30, or Textron Lycomings or Continentals of 89.5 to 112 kW (120 to 150 hp).
ACCOMMODATION: Space provision for one or two child seats behind main pair. Two upward-opening, gull-wing doors.
AVIONICS: Basic VFR instrumentation standard; other avionics to customer's choice.

DIMENSIONS, EXTERNAL:
Wing span	9.17 m (30 ft 1 in)
Length overall	6.55 m (21 ft 5¾ in)

AREAS:
Wings, gross	14.00 m² (150.7 sq ft)

WEIGHTS AND LOADINGS:
Weight empty	295 kg (650 lb)
Max T-O weight	450-480 kg (992-1,058 lb)*

* *Depending upon local regulations*

PERFORMANCE (estimated, Jabiru engine):
Max level speed	91 kt (170 km/h; 105 mph)
Stalling speed, flaps down	33 kt (60 km/h; 38 mph)
Max rate of climb at S/L	240 m (787 ft)/min
Service ceiling	3,000 m (9,840 ft)
T-O run	70 m (230 ft)
Landing run	60 m (200 ft)
Range with max fuel	405 n miles (750 km; 466 miles)
g limits	+4/−2

UPDATED

AEROSTAR (YAKOVLEV) Iak-52

TYPE: Primary prop trainer/sportplane.
PROGRAMME: Design of Yak-52 in USSR began 1975; series production assigned to Comecon programme (Council for Mutual Economic Assistance), following Romanian-USSR intergovernmental agreement of 1974; designation is transliteration of 'Yak' in Romanian alphabet, but civil aircraft are registered as Yaks.
   Construction began 1977 and first Romanian prototype (c/n 780102) made first flight May 1978; series production began 1979; first deliveries (to DOSAAF, USSR) 1975; 1,000th aircraft delivered in 1986 and 1,500th in 1990. After cessation of deliveries to USSR in 1991, production had almost come to a standstill, but interest reawakened by improved Iak-52W (first flight April 1999, debut at Paris Air Show, June 1999). Iak-52W began development 1998 and first flew (12201) March 1999; currently offered as alternative to standard version. Aerostar planned five-year programme of 30 per year; deliveries started August 1999. Tailwheel (TW) version announced early 2000; first flew ('26', c/n 0112226, later N52SD) 2 July 2001; launched at EAA AirVenture, Oshkosh, 24 to 30 July 2001.
CURRENT VERSIONS: **Iak-52:** Standard model.
Full description of standard Iak-52 follows, augmented by details of Iak-52W and TW, where different.
   Iak-52W: Upgraded ('Westernised') version; introduced 1999. Main changes are increased fuel tankage, three-blade propeller, Western avionics, and metal-skinned control surfaces. No known deliveries since 2001, but some believed held in stock.
   Iak-52TW: Tailwheel version of Iak-52W, revealed early 2000. Production aircraft available from July 2001 (four sold by end of first day of promotion at Oshkosh); US marketing by GeSoCo Industries of Swanton, Vermont, USA. Features include extended (rounded) wingtips, uprated engine with underslung oil radiator and modified front shutters; each cockpit lengthened by 10 cm (4 in).
   Yak-54: Armed military version with underwing pylons; designation duplicates that of current tailwheel-equipped aircraft described in Russian section. Three built:

Aerostar Iak-52W in spurious military markings *NEW*/0525751

one lost in service in Afghanistan; one to Monino museum; third (Yak-54B ES-FYA) currently with AS Strovest in Netherlands.

CUSTOMERS: Estimated 1,815, in 121 production batches, built by 1998, mainly for USSR and Romanian Air Force but latterly also for West European, Australian and American civil markets; Hungarian Air Force received 12 in early months of 1994. Production averaged 10 per year between 1992 and 1996 (almost all to USA or Hungary), augmenting large number of military surplus sales. Twelve delivered to Vietnam in 1997. No known deliveries in 1998.

Iak-52W/TW production began during 122nd batch (first being 26th in batch); exports began to USA in mid-1999; one also to Australia in mid-2000; total 11 registered by August 2002.

COSTS: Iak-52W approximately US$120,000 (1999); Iak-52TW US$125,000 for first 10, US$139,000 thereafter (2000).

DESIGN FEATURES: Tandem-cockpit variant of Yak-50, with unchanged span and length, but with semi-retractable landing gear to reduce damage in wheels-up landing (mainwheels fully retractable in Iak-52TW). Simple, robust, low-wing design with long landing gear legs to accommodate large propeller. Moderately tapered wings and tailplane; curved fin; 'glasshouse' canopy. Current production aircraft lack wingtip fairings. Strengthened wingtips on Iak-52W for optional fitment of tip-tanks. Increased wing span in Iak-52TW.

FLYING CONTROLS: Conventional and manual. Actuation of mass-balanced slotted ailerons by pushrods; mass-balanced elevators by pushrods/cables; and horn-balanced rudder by cables; manually operated trim tab in port elevator; ground-adjustable tab on rudder and each aileron. Pneumatically actuated trailing-edge split flaps.

STRUCTURE: All-metal except for fabric-covered primary control surfaces (these also metal-skinned on Iak-52W/TW). Single-spar wings; modified spar, post-1986 (and some retrofits), permits higher load factors. Iak-52W wingtips strengthened to carry optional tip-tanks.

LANDING GEAR: Semi-retractable tricycle type, with single wheel and oleo-pneumatic shock-absorber on each unit. Iak-52TW mainwheels (size 6.00-6) fully retractable, on Aerostar pneumatically actuated legs, retracting inward; flush doors. Cleveland wheels and Western tyres on Iak-52W/TW; Scott steerable tailwheel, with 10×3.50-4 tyre, on TW. Pneumatic actuation, nosewheel retracting rearward, main units forward. All three wheels remain fully exposed to airflow, against undersurface of fuselage and wings respectively, to offer greater safety in event of wheels-up emergency landing. Pneumatic drum (hydraulic disc on Iak-52W and TW) mainwheel brakes, operated differentially from rudder pedals. Non-retractable plastics-coated duralumin skis, with shock-struts, can be fitted in place of wheels for winter operations.

POWER PLANT: One 265 kW (355 hp) Aerostar-built VOKBM (Bakanov) M-14P nine-cylinder air-cooled radial, driving a V-530TA-D35 two-blade constant-speed wooden propeller (three-blade MT-Propeller MTV-9-B-C/CL250-27 on Iak-52W and TW). Iak-52TW has 298 kW (400 hp) M-14P-XDK. Two aluminium alloy fuel tanks, in wingroots forward of spar; collector tank in fuselage supplies engine during inverted flight.

Total internal fuel capacity (standard Iak-52) 122 litres (32.2 US gallons; 26.8 Imp gallons). Additional integral tank in each wing of Iak-52W and TW, raising capacity to 280 litres (74.0 US gallons; 61.6 Imp gallons). Oil capacity 10 litres (2.6 US gallons; 2.2 Imp gallons). Provision on Iak-52W/TW for wingtip tanks.

ACCOMMODATION: Tandem seats for pupil (at front) and instructor under long 'glasshouse' canopy, with separate rearward-sliding hood over each seat. Hooker Harness seat belts for both occupants. Dual controls standard. Seats and rudder pedals adjustable. Heating and ventilation standard. Optional 0.20 m³ (7.06 cu ft) baggage compartment in

Aerostar Iak-52TW first production aircraft in camouflage colour scheme (*Jane's/Paul Jackson*) **NEW**/0525734

standard version, accessible from rear seat; Iak-52W and TW have externally accessible baggage and battery compartments on port side, aft of rear cockpit, plus permanently mounted retractable ladder for cockpit access.

SYSTEMS: Independent main and emergency pneumatic systems, pressure 50 bar (725 lb/sq in), for flap and landing gear actuation, engine starting and brake control. Electrical system (27 V DC) supplied by 3 kW engine-driven generator and (in port wing) 12 V 23 Ah ASAM battery (two 27 V in Iak-52W); two static inverters in fuselage for 36 V AC power at 400 Hz. Oxygen system available optionally.

AVIONICS: *Comms:* Balkan 5 VHF radio and SPU-9 intercom in standard Iak-52; replaced in Iak-52W/TW by Western instruments including Garmin IC-AQ200 com radio and GTX transponder, AmeriKing ELT and altitude encoder, and NAT AA80-20 intercom.

Flight: ARK-15M automatic radio compass, eight-channel ADF and GMK-1A gyrocompass.

Other avionics available at customer's option.

EQUIPMENT: Strobe and navigation lights, recessed landing and taxying lights on Iak-52W and TW; roll bar in Iak-52TW.

DIMENSIONS, EXTERNAL:

Wing span: except 52TW	9.30 m (30 ft 6¼ in)
52TW	9.90 m (32 ft 5¾ in)
Length overall: except 52TW	7.745 m (25 ft 5 in)
52TW	7.98 m (26 ft 2¼ in)
Height overall, propeller turning:	
52	2.70 m (8 ft 10¼ in)
52W, 52TW	2.50 m (8 ft 2½ in)
Wheel track: all	2.70 m (8 ft 10¼ in)
Wheelbase: except 52TW	1.86 m (6 ft 1¼ in)
Propeller diameter: 52	2.40 m (7 ft 10½ in)
52W, 52TW	2.00 m (6 ft 6¾ in)

DIMENSIONS, INTERNAL:

Cockpit: Max width	0.74 m (2 ft 5¼ in)
Max height	1.12 m (3 ft 8 in)

AREAS:

Wings, gross: except 52TW	15.00 m² (161.5 sq ft)
52TW	15.42 m² (166.0 sq ft)
Ailerons (total)	1.98 m² (21.31 sq ft)
Flaps (total)	1.03 m² (11.09 sq ft)
Fin	0.61 m² (6.57 sq ft)
Rudder	0.87 m² (9.36 sq ft)
Tailplane	1.32 m² (14.21 sq ft)
Elevators (total)	1.53 m² (16.47 sq ft)

WEIGHTS AND LOADINGS:

Weight empty, equipped: 52	1,015 kg (2,238 lb)
52W	960 kg (2,116 lb)
52TW	980 kg (2,160 lb)
Max T-O weight: 52	1,305 kg (2,877 lb)
52W	1,315 kg (2,899 lb)
52TW	1,320 kg (2,910 lb)
Max wing loading: 52	87.0 kg/m² (17.82 lb/sq ft)
52W	87.7 kg/m² (17.96 lb/sq ft)
52TW	85.6 kg/m² (17.53 lb/sq ft)
Max power loading: 52	4.93 kg/kW (8.10 lb/hp)
52W	4.97 kg/kW (8.17 lb/hp)
52TW	4.43 kg/kW (7.27 lb/hp)

PERFORMANCE (Iak-52W):

Never-exceed speed (V_{NE})	194 kt (360 km/h; 223 mph)
Max level speed: at S/L	154 kt (285 km/h; 177 mph)
at 1,000 m (3,280 ft)	145 kt (270 km/h; 167 mph)
Econ cruising speed at 1,000 m (3,280 ft)	
	103 kt (190 km/h; 118 mph)
Landing speed	60 kt (110 km/h; 69 mph)
Stalling speed, flaps down, engine idling	
	49 kt (90 km/h; 56 mph)
Max rate of climb at S/L	420 m (1,378 ft)/min
Time to 4,000 m (13,120 ft)	15 min
Service ceiling	4,000 m (13,120 ft)
T-O run	170 m (560 ft)
T-O to 15 m (50 ft)	200 m (660 ft)
Landing from 15 m (50 ft)	350 m (1,150 ft)
Landing run	300 m (985 ft)
Range at 500 m (1,640 ft), max fuel, 20 min reserves:	
52	296 n miles (550 km; 341 miles)
52W	648 n miles (1,200 km; 745 miles)
g limits: up to c/n 867215	+5/−3
from c/n 877301	+7/−5

PERFORMANCE (Iak-52TW, estimated):

Max cruising speed	162 kt (300 km/h; 186 mph)
Stalling speed, flaps down	46 kt (85 km/h; 53 mph)
Max rate of climb at S/L	360 m (1,181 ft)/min
Service ceiling	4,000 m (13,120 ft)
T-O run	140 m (460 ft)
Landing run	270 m (885 ft)
Range at 500 m (1,640 ft) at cruising speed, 20 min reserves:	
standard fuel	297 n miles (550 km; 341 miles)
max fuel	648 n miles (1,200 km; 745 miles)
g limits	+7/−5

UPDATED

Front cockpit instrument panel of the upgraded Iak-52W
0079292

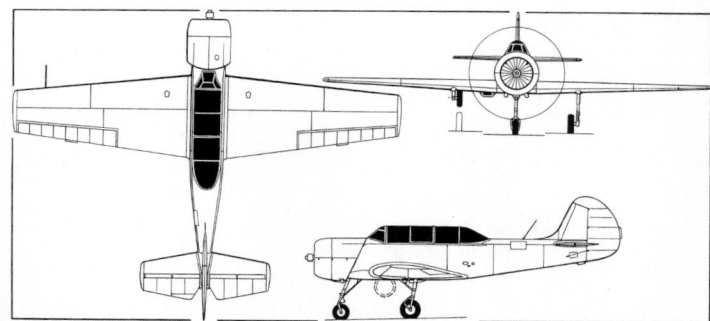

Aerostar (Yakovlev) Iak-52 tandem two-seat primary trainer (*Jane's/Dennis Punnett*)   0011952

AVIOANE

AVIOANE CRAIOVA SA

1 Aeroportului Street, PO Box 492, R-1100 Craiova
Tel: (+40 51) 40 29 01
Fax: (+40 51) 40 20 40
e-mail: deneanu@acv.ro
Web: http://www.acv.ro
GENERAL MANAGER: Nicolae Deneanu
MARKETING MANAGER: Mircea Popa

Founded 1 February 1972 as Intreprinderea de Avioane Craiova, beginning aircraft manufacture with joint Romanian/Yugoslav IAR-93/Orao attack aircraft; changed to present name 29 March 1991; range of products and services for military and civil aviation. Aircraft and equipment design and manufacture, repair and overhaul, life cycle management and integrated logistics support. Granted ISO 9001 status 1998. IAR-99 lead-in trainer and close air support aircraft is now factory's major programme. Design studies for a small, 336 kW (450 shp) turboprop trainer, provisionally designated IAR-XXX, was under way in the second half of 2001. Site area 1.70 ha (4.20 acres), including 47,500 m² (511,275 sq ft) shop floor area. Workforce 1,300 (April 2002).

UPDATED

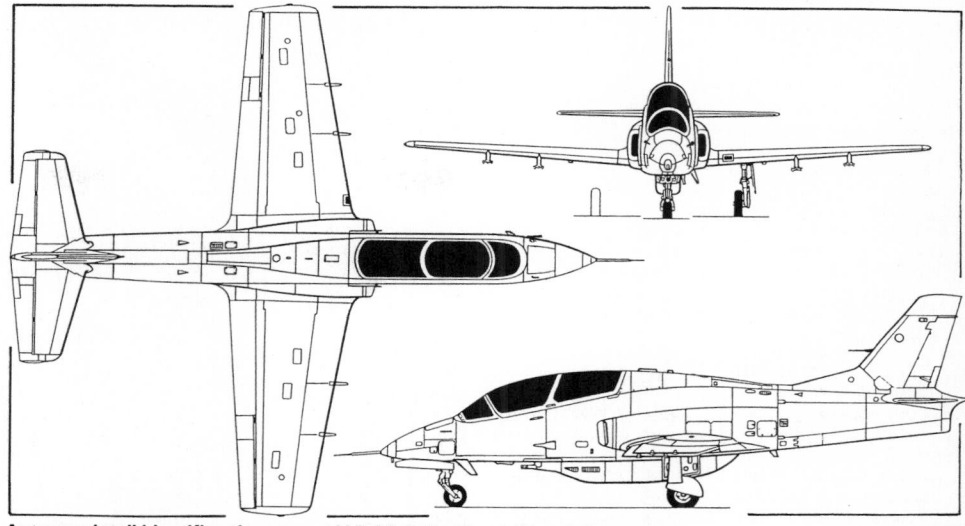

Antenna detail identifies the current IAR-99 Şoim (*Jane's/Dennis Punnett*)   0079327

AVIOANE IAR-99 ŞOIM

English name: Little Hawk

TYPE: Advanced jet trainer/light attack jet.
PROGRAMME: Design by INAv (Institutul de Aviatie) at Bucharest started 1975; three prototypes, of which S-001 made first flight 21 December 1985; S-002 was static test airframe; S-003 second flying prototype. Initial production batch of 17 (serial numbers 701-717), deliveries of which began 1988; two of these (708 and 709) completed for proposed avionics upgrade programme with Jaffe Aircraft of USA in 1991, which not pursued; one (712) equipped with Collins avionics in 1992; similar venture with IAI resulted in upgraded IAR-109 Swift prototype (7003, converted S-003), flown in November 1993 (see 1995-96 and earlier *Jane's*), but no production of this version ensued. Order for 24 new aircraft with Elbit modernised avionics announced in September 1998 and received Romanian MoD contract in 2000, following first flight of prototype (serial number 718) on 22 May 1997; first four (719 to 722) due for delivery during 2002, of which 719 rolled out on 31 July and handed over 1 August 2002.
CURRENT VERSIONS: **Original IAR-99:** Initial production version (17) for Romanian Air Force; in service with 49th and 67th Fighter-Bomber Groups at Ianca and Craiova respectively.
Current IAR-99: Elbit (Israel) teamed with Avioane to produce this version as lead-in fighter trainer and close air support aircraft for Romanian Air Force and for export. First upgraded aircraft (718; actually constructed 1994) served as development demonstrator; US$21 million contract for 24 aircraft announced in September 1998 and contains options for further 16 to follow.

Following description applies to this version.
CUSTOMERS: Total of 17 of original production version delivered to Romanian Air Force, of which about 13 believed still in service; deliveries of 24 of current version began in 2002.
COSTS: Current version approximately US$6 million (2002).
DESIGN FEATURES: Typical basic/advanced jet trainer with tandem, stepped cockpits and moderately tapered wings. Provision for armament increases versatility.

Wing section NACA 64₁A-214 (modified) at centreline; 64₁A-212 (modified) at tip; dihedral 3° from roots; quarter-chord sweepback 6° 35′; incidence 1° at root. Tail unit sweepback at quarter-chord 34° on fin, 9° 8′ on tailplane.

Main feature of Elbit avionics suite is a data transfer system that processes navigational information received via datalink from other aircraft or from a ground station equipped with the same system. This information is presented on a simulated 'virtual radar' display in the cockpit, so that the pilot can be trained in the use of radar without a need for the IAR-99 itself to be fitted with a radar. Current production aircraft have a large blade antenna beneath nose on port side.
FLYING CONTROLS: Conventional and partly assisted. Statically balanced ailerons hydraulically actuated with manual reversion; horn-balanced elevators and statically balanced rudder actuated mechanically by push/pull rods; servo tab in port aileron, trim tabs in rudder and each elevator, all operated electrically; ailerons deflect 15° up/15° down, elevators 20° up/10° down, rudder 25° to left and right. Hydraulically actuated single-slotted flaps, deflecting 20° for T-O and 40° for landing, retract gradually when airspeed reaches 162 kt (300 km/h;

186 mph); twin hydraulically actuated airbrakes under rear fuselage.
STRUCTURE: All-metal; aluminium honeycomb ailerons/elevators/rudder; semi-monocoque fuselage includes honeycomb panels for fuel tank compartments; machined wing skin panels form integral fuel tanks.
LANDING GEAR: Retractable tricycle type, with single wheel and oleo-pneumatic shock-absorber on each unit. Mainwheels retract inward, castoring nosewheel forward, all being fully enclosed by doors when retracted. Landing light in port wingroot leading-edge. Mainwheels fitted with tubeless tyres, size 552×164-10, pressure 7.5 bar (109 lb/sq in), and hydraulic disc brakes with anti-skid system. Nosewheel has tubeless tyre size 445×150-6, pressure 4.0 bar (58 lb/sq in).
POWER PLANT: One Turbomecanica Romanian-built Rolls-Royce Viper Mk 632-41M turbojet, rated at 17.79 kN (4,000 lb st). Fuel in two flexible bag tanks in centre-fuselage, capacity 900 litres (238 US gallons; 198 Imp gallons), and four integral tanks between wing spars, combined capacity 470 litres (124 US gallons; 103 Imp gallons). Total internal fuel capacity 1,370 litres (362 US gallons; 301 Imp gallons). Gravity refuelling point on top of fuselage. Provision for two drop tanks, each of 225 litres (59.5 US gallons; 49.5 Imp gallons) capacity, on inboard underwing stations. Maximum internal/external fuel capacity 1,820 litres (481 US gallons; 400 Imp gallons).
ACCOMMODATION: Crew of two in tandem, on Martin-Baker Mk 10L zero/zero ejection seats in pressurised and air conditioned cockpit. Rear seat elevated 35 cm (13.8 in). Dual controls standard. One-piece canopy with internal screen, opening sideways to starboard.

Avioane IAR-99 Şoim current demonstrator, with armament display (*R J Malachowski*)   0109170

SYSTEMS: Engine compressor bleed air for pressurisation, air conditioning, anti-*g* suit and windscreen anti-icing system, and to pressurise fuel tanks. Hydraulic system, operating at pressure of 206 bar (2,990 lb/sq in), for actuation of landing gear and doors, flaps, airbrakes, ailerons and mainwheel brakes. Emergency hydraulic system for operation of landing gear doors, flaps and wheel brakes. Main electrical system, supplied by 9 kW 28.5 V DC starter/generator, with 28.5 V 36 Ah Ni/Cd battery, ensures operation of main systems, in case of emergency, and engine starting. Two static inverters (one 3 kVA and one 750 VA) supply two secondary AC networks: 115 V/400 Hz and 26 V/400 Hz. Oxygen system for two crew for 2 hours 40 minutes.

AVIONICS: Elbit Systems integrated suite, based on MIL-STD-1553B.

Comms: Voice-activated intercom; IFF.

Flight: VOR/ILS; DME; ADF; Litton Italia INS with Trimble GPS nav.

Instrumentation: Modular multirole computer; HOTAS controls; colour CRT MFD in each cockpit; head-up display with up-front control panel (HUD/UFCP) in front cockpit, with aft station HUD monitor (ASHM) in rear (instructor's) cockpit; dual ADIs with HSIs; ADC; radio altimeter; radio marker; SAIMS flight recorder; Elta data transfer system (DTS) with pilot's virtual radar display; display and sight helmets (DASH).

Self-defence: Elta RWR and ECM pod; chaff/flare dispenser.

ARMAMENT: Five external stores stations. Centreline station can be occupied by ECM, laser designator or aerial reconnaissance pod, a drop fuel tank, or by a twin-barrel 23 mm GSh-23 gun pod with 200 rounds. All armament controlled by stores management system slaved to avionics mission computer. Four underwing hardpoints stressed for loads of 250 kg (551 lb) each. Typical underwing stores can include four 100 kg or 250 kg bombs; four triple carriers each for three 50 kg bombs (or two 100 kg and one 50 kg); laser- or IR-guided bombs; R-3, R-13, R-60 or Python 3 AAMs (outer pylons only); four L-16-57 launchers each containing sixteen or thirty-two 57 mm air-to-surface rockets; and auxiliary fuel tanks (see under Power Plant) on centreline and inboard pylons.

DIMENSIONS, EXTERNAL:	
Wing span	9.85 m (32 ft 3¾ in)
Wing chord: at root	2.305 m (7 ft 6¾ in)
mean	1.965 m (6 ft 5¼ in)
at tip	1.30 m (4 ft 3¼ in)
Wing aspect ratio	5.2
Length overall	11.01 m (36 ft 1½ in)
Height overall	3.90 m (12 ft 9½ in)
Elevator span	4.12 m (13 ft 6¼ in)
Wheel track	2.685 m (8 ft 9¾ in)
Wheelbase	4.38 m (14 ft 4½ in)
AREAS:	
Wings, gross	18.71 m² (201.4 sq ft)
Ailerons (total)	1.56 m² (16.79 sq ft)
Flaps (total)	2.54 m² (27.34 sq ft)
Fin, incl dorsal fin	1.92 m² (20.67 sq ft)
Rudder	0.63 m² (6.78 sq ft)
Tailplane	3.125 m² (33.63 sq ft)
Elevators (total)	1.25 m² (13.45 sq ft)
WEIGHTS AND LOADINGS:	
Weight empty, equipped	3,200 kg (7,055 lb)
Max fuel weight: internal	1,100 kg (2,425 lb)
external	350 kg (772 lb)
Max T-O weight: trainer	4,400 kg (9,700 lb)
ground attack: pilot only	5,480 kg (12,081 lb)
two crew	5,560 kg (12,258 lb)
Max wing loading: trainer	235.2 kg/m² (48.17 lb/sq ft)
ground attack	297.2 kg/m² (60.86 lb/sq ft)
Max power loading: trainer	247.5 kg/kN (2.42 lb/lb st)
ground attack	312.7 kg/kN (3.06 lb/lb st)
PERFORMANCE:	
Max operating Mach number (MMO)	0.76
Max level speed at S/L:	
trainer, clean	467 kt (865 km/h; 537 mph)
Max speed for landing gear actuation	135 kt (250 km/h; 155 mph)
Max rate of climb at S/L	2,100 m (6,890 ft)/min
Service ceiling	12,900 m (42,325 ft)
Min air turning radius	330 m (1,083 ft)
T-O run: trainer	450 m (1,477 ft)
ground attack	960 m (3,150 ft)
T-O to 15 m (50 ft): trainer	750 m (2,461 ft)
ground attack	1,350 m (4,430 ft)
Landing from 15 m (50 ft): trainer	740 m (2,428 ft)
ground attack	870 m (2,855 ft)
Landing run: trainer	550 m (1,805 ft)
ground attack	600 m (1,969 ft)
Typical combat radius (one pilot, ventral gun, internal fuel only):	
lo-lo-hi, four 16-round rocket pods, AUW 5,000 kg (11,023 lb)	189 n miles (350 km; 217 miles)
hi-lo-hi, two 16-round rocket pods, two 50 kg and four 100 kg bombs, AUW 5,280 kg (11,640 lb)	186 n miles (345 km; 214 miles)
hi-hi-hi, four 250 kg bombs, AUW 5,480 kg (12,081 lb)	208 n miles (385 km; 239 miles)
Max range with internal fuel:	
trainer	593 n miles (1,100 km; 683 miles)
ground attack	522 n miles (967 km; 601 miles)
Max endurance with internal fuel: trainer	2 h 40 min
ground attack	1 h 46 min
g limits	+7/−3.6

UPDATED

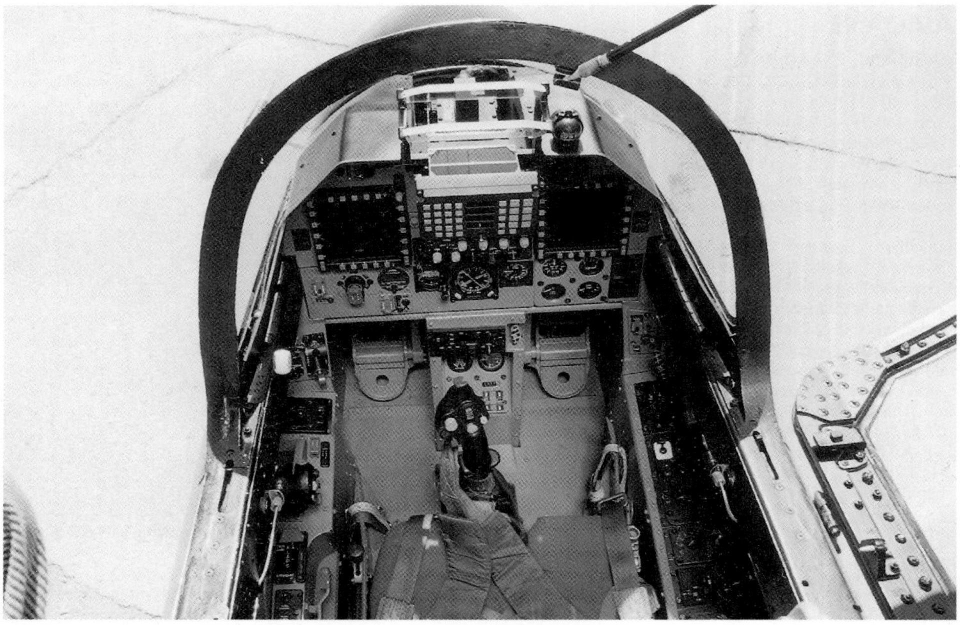

IAR-99 modernised front cockpit with up-front control panel 0051183

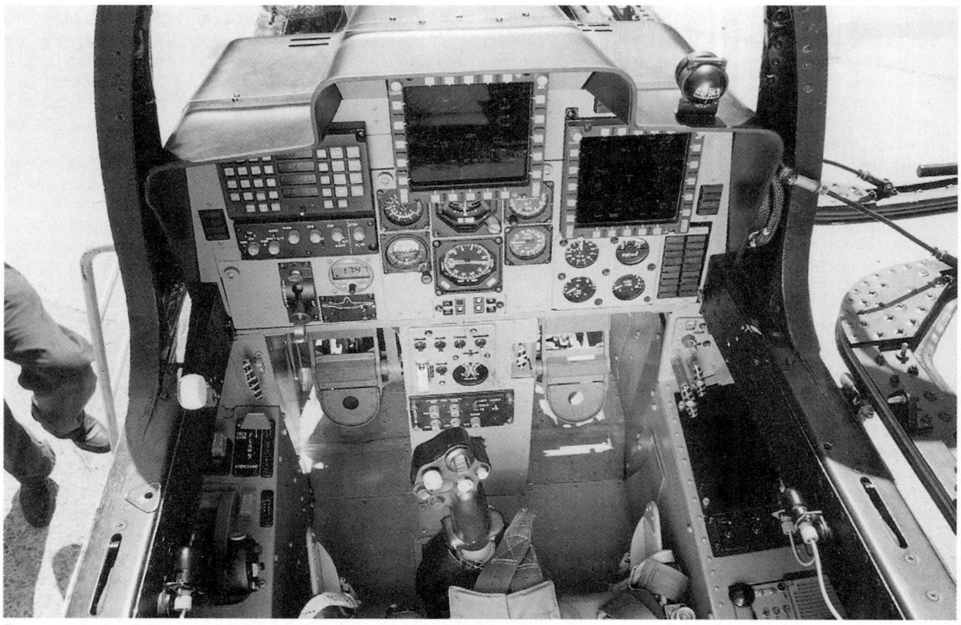

Rear cockpit of the current production standard IAR-99 0051184

IAR

SC IAR SA

1 Aeroportului Street, PO Box 198, R-2200 Brasov
Tel: (+40 268) 47 51 08
Fax: (+40 268) 47 69 81
e-mail: iar.sa@rdslink.ro
Web: http://www.iar.ro
GENERAL DIRECTOR: Dipl Eng Neculai Banea
COMMERCIAL AND FINANCIAL DIRECTOR:
Dipl Ec Ion Dumitrescu
MARKETING MANAGER: Stefan Paunescu

Factory, created 1968, continues work begun in 1925 by IAR-Brasov; occupies 136 ha (336 acre) site, including 185,400 m² (1,995,625 sq ft) factory area, and had September 2001 workforce of more than 1,700. Now the sole Romanian manufacturer of helicopters, it has produced more than 360 Alouette IIIs and Pumas under French licence. Currently has full capacity and capabilities for design and development; manufacturing aircraft structural items; general assembling; integration of modern avionics and systems; maintenance, overhaul and repair works; modernisation programme; flight test and certification; and customer support. Main product is the IAR-330 L Puma, manufactured under Eurocopter licence; currently carrying out Puma SOCAT upgrading programme to meet Romanian MoD requirements for an advanced armed helicopter (see *Jane's Aircraft Upgrades* for details). Based on Eurocopter Romania SA joint venture created 21 June 2001, IAR also provides worldwide Puma fleet logistic support, customisation work and upgrading such as Makila IAI re-engining; advanced avionics; new autopilot; and new optional equipment.

IAR also manufactures own-designed IAR-46S very light aircraft; has produced some 830 IS-28M2/GR motor gliders and IS-28/29/35 series gliders; spares for IAR-330 Puma and IAR-316 B Alouette III helicopters; aircraft components.

In July 2000, IAR Brasov was named as first partner of ATG (see UK part of Lighter than Air section) in production of latter's new SkyCat 200 airship. First deliveries of major subassemblies were due in 2001.

UPDATED

IAR IAR-46

TYPE: Two-seat lightplane.

PROGRAMME: Definition phase and marketing studies started early 1991; detail design began late 1991; development phase initiated mid-1992; first of two prototypes

(YR-1037) shown at Paris Air Show June 1993 and made first flight November that year, followed by flight test evaluation, static test and ground testing for JAR-VLA compliance. Development of 02 second prototype began at end of 1995; this aircraft (YR-BVC) made first flight in April 1997; manufacturer's preliminary flight test programme completed August 1999; JAR-VLA certification (Utility category) by Romanian CAA began mid-1998 and received November 1999; YR-BVC registered to Airclub Romania in August 2000; FAA certification expected in September 2000 but still awaited in mid-2002. Further three production aircraft built and stored by late 1998; registered April 2001.

CURRENT VERSIONS: **IAR-46:** Basic version.

IAR-46S: Entered production in 2000. First three (c/n 03 to 05) registered to Airclub of Romania in April 2001. 73.5 kW (98.6 hp) Rotax 912 ULS engine.

DESIGN FEATURES: Low-wing monoplane; GA(W)-1 aerofoil section) with high aspect ratio; raked tips; T tail. Dihedral 2° from centre-section; incidence 4° at root; no twist. Foldable horizontal tail surfaces.

FLYING CONTROLS: Conventional and manual. Actuation by pushrods and cables; trim tab in each elevator. Manually operated plain trailing-edge flaps.

STRUCTURE: All-metal except fabric covering on elevators and rudder and GFRP for non-stressed fairings. Wing has main and auxiliary spars. Fluted aluminium skins on ailerons and flaps.

LANDING GEAR: Retractable (manual/mechanical) single mainwheels with hydraulic shock-absorbers and hydraulic, toe-operated disc brakes; steerable, non-retractable tailwheel with rubber-in-compression shock-absorption. Rearward-retracting 5.00×5 Matco or Cleveland mainwheels, with 360×110 tyres, pressure 3.00 bar (43.5 lb/sq in). Tost 210×65 tailwheel and tyre, pressure 2.50 bar (36 lb/sq in).

POWER PLANT: One 59.6 kW (79.9 hp) Rotax 912 F3 (IAR-46) or 73.5 kW (98.6 hp) Rotax 912 ULS (IAR-46S) flat-four engine, with 2.43:1 reduction gearing to a Hoffmann HO-V352F/170FQ two-blade constant-speed propeller ('170FQ+6 in IAR-46S). Fuel in single tank in fuselage, capacity 70 litres (18.5 US gallons; 15.4 Imp gallons).

ACCOMMODATION: Two adjustable backrest seats side by side; dual controls standard; adjustable rudder pedals. Fixed windscreen and rearward-sliding jettisonable canopy. Baggage compartment aft of seats. Cockpit heated and ventilated.

SYSTEMS: Electrical power supplied by 250 W 14 V DC alternator and 12 V 25 Ah battery. Hydraulic system for mainwheel brakes only, with parking brake valve.

AVIONICS: *Instrumentation:* Standard VFR instrumentation to JAR 22.1303 and JAR 22.1305. Options include horizon and directional gyros, turn co-ordinator and Becker nav/com and transponder avionics package.

EQUIPMENT: Anti-collision and position lights standard; ground power receptacle optional.

DIMENSIONS, EXTERNAL:
Wing span	12.05 m (39 ft 6½ in)
Wing chord: at root	1.40 m (4 ft 7 in)
at tip	0.93 m (3 ft 0½ in)
Wing aspect ratio	9.4
Length overall	7.85 m (25 ft 9 in)
Height overall	2.15 m (7 ft 0½ in)
Tailplane span	3.48 m (11 ft 5 in)
Wheel track	1.59 m (5 ft 2½ in)
Wheelbase	4.94 m (16 ft 2½ in)
Propeller diameter	1.70 m (5 ft 7 in)
Propeller ground clearance	0.295 m (11½ in)

DIMENSIONS, INTERNAL:
Cockpit: Length	1.54 m (5 ft 0¾ in)
Max width	1.04 m (3 ft 5 in)
Max height (from seat cushion)	0.85 m (2 ft 9½ in)

AREAS:
Wings, gross	13.87 m² (149.3 sq ft)
Ailerons (total)	0.82 m² (8.83 sq ft)
Trailing-edge flaps (total)	1.36 m² (14.64 sq ft)
Fin	0.60 m² (6.46 sq ft)

Second prototype IAR-46 (*Jane's/Paul Jackson*)   0062606

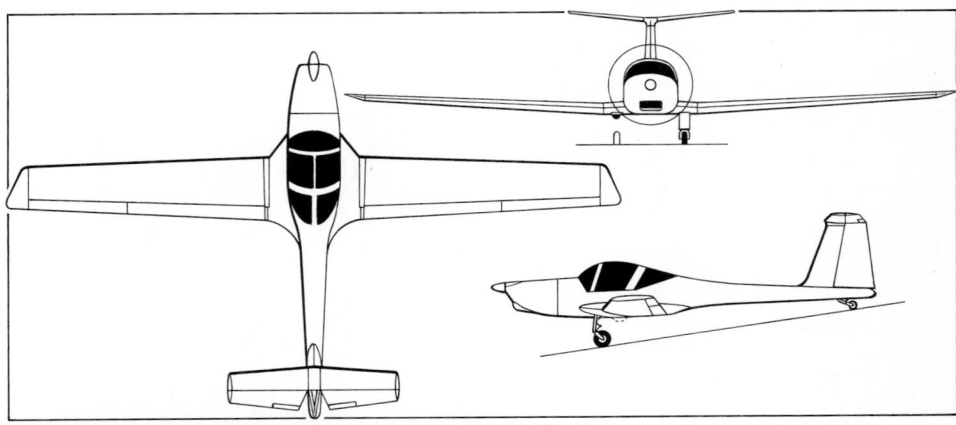
IAR-46 side-by-side two-seat very light aircraft (*Jane's/Mike Keep*)

Rudder	0.89 m² (9.58 sq ft)
Tailplane	1.64 m² (17.65 sq ft)
Elevators (total, incl tabs)	1.10 m² (11.84 sq ft)

WEIGHTS AND LOADINGS:
Weight empty	550 kg (1,213 lb)
Max fuel weight	53 kg (117 lb)
Max T-O and landing weight	750 kg (1,653 lb)
Max wing loading	54.1 kg/m² (11.08 lb/sq ft)
Max power loading	12.58 kg/kW (20.67 lb/hp)

PERFORMANCE (A: IAR-46, B: IAR-46S):
Never-exceed speed (VNE):	
A, B	145 kt (270 km/h; 167 mph) IAS
Max level speed: B	116 kt (215 km/h; 133 mph)
Max cruising speed: A	97 kt (180 km/h; 112 mph) IAS
B	103 kt (190 km/h; 118 mph) IAS
Econ cruising speed: A	81 kt (150 km/h; 93 mph) IAS
B	86 kt (160 km/h; 99 mph) IAS
Max manoeuvring speed: B	93 kt (172 km/h; 107 mph)
Max speed for flap extension:	
B	76 kt (140 km/h; 87 mph)
Optimum climbing speed: B	60 kt (112 km/h; 70 mph)
Stalling speed at 1,000 m (3,280 ft), engine idling:	
A, B, flaps up	50 kt (91 km/h; 57 mph) IAS
A, B, flaps down	47 kt (87 km/h; 55 mph) IAS
Max rate of climb at S/L: A	150 m (492 ft)/min
B	213 m (699 ft)/min
Service ceiling: A, B	4,000 m (13,120 ft)

T-O run: A	193 m (635 ft)
B	185 m (610 ft)
T-O to 15 m (50 ft): A	465 m (1,525 ft)
B	409 m (1,340 ft)
Landing from 15 m (50 ft): A, B	171 m (560 ft)
Landing run: A, B	110 m (360 ft)
Range, with reserves:	
A, B	297 n miles (550 km; 341 miles)
g limits	+4.4/–2.2

UPDATED

IAR (EUROCOPTER) AS 350BA ECUREUIL and AS 355N ECUREUIL 2

On 29 May 1996 an agreement was signed for licensed manufacture, subject to French and Romanian government approval, of up to 80 AS 350BA single-engined Ecureuil and twin-engined AS 355N Ecureuil 2 helicopters, described under Eurocopter in the International section. Purchases are anticipated from local civilian and public service operators, while export orders would be handled jointly by Eurocopter and IAR. Agreement to form Eurocopter Romania (which see) subsequently announced (December 2000), but no starting date for this programme had been revealed by mid-2002.

UPDATED

ROMAERO
SC ROMAERO SA
44 Bulevardul Ficusului, Sector 1 (PO Box 18), R-71544 Bucuresti 1
Tel: (+40 1) 232 37 35 and 232 60 60
Fax: (+40 1) 232 20 82
e-mail: management@romaero.rdsnet.ro
CHAIRMAN AND CEO: Francisc Toba
DIRECTOR, SALES AND MARKETING: Lucian Popescu
CHIEF OF PUBLIC RELATIONS: Carmen Gheorghiu

Established 1951 and operated successively under several names (see 1991-92 *Jane's* for details); became commercial company under present name 20 November 1990. Manufactures BNG Islander and Defender 4000; major subassemblies and components for Boeing 757; cabin subassemblies and floats for Bombardier (Canadair) CL-415; tail units for Learjet 45 (subcontracted from Shorts, UK); rear

fuselages for Gulfstream 200; also undertakes maintenance and repair of various types of aircraft, including Islander, BAC One-Eleven and Boeing 707/727/737. Plans to establish an FAA-approved maintenance and repair station for larger aircraft of various types. Romaero's 36.43 ha (90.0 acre) Baneasa Airport site includes a 163,670 m² (1.76 million sq ft) production area; workforce in mid-2000 was 1,280.

The sale of Romaero to Britten-Norman of the UK in January 1999 was aborted when the latter ceased trading three months later, resulting in suspension of Islander deliveries to the newly formed B-N Group until July 2000.

UPDATED

ROMAERO (BNG) ISLANDER and DEFENDER
TYPE: Light utility twin-prop transport.
PROGRAMME: First flight of Romanian Islander (assembled from UK kit by former IRMA) 4 August 1969; first

Romanian-built flew 18 October 1969. Defender 4000 production started in 1995; first delivery (G-BWPU) 27 September 1997.

CUSTOMERS: Initial commitment to build 215 completed January 1977; total of 509 standard Islanders, six Turbine Islanders and six Defender 4000s delivered to Britten-Norman by 26 September 1998. No further deliveries then until 19 July 2000 (one BN-2B-20 which had first flown on 10 July 2000). By mid-2001, production and deliveries again suspended pending new agreement on a mutually viable annual output; contract with BNG signed July 2002 for 24 aircraft over a two-year period, this including three taken over from earlier contract, 21 new-build and options on Trislander versions. First of new-build batch was expected to be delivered to BNG by early 2003.

UPDATED

RUSSIAN FEDERATION

AEROPRAKT

AEROPRAKT OOO (Aeropract JSC)
a/ya 9863, 443008 Samara
Tel: (+7 8462) 27 09 65
Fax: (+7 8462) 42 96 13
e-mail: ela@dionis.samtel.ru
PRESIDENT: Vladimir N Pavlyuk
CHIEF CONSTRUCTOR: Viktor K Kurshev

Formed in 1974 to design and build gliders, small flying-boats and amphibians, OKB Aeroprakt established a second centre at Kiev in 1986, only to be divided on dissolution of the USSR. Became LM Aeropract Samara, a joint Russo-Finnish venture, in 1991; KB Aeropract in 1993; and Aeropract JSC in 1997.

The Russian Aeropract adopts a -ct ending in transliteration to differentiate it from the Aeroprakt now in Ukraine (which see). The two companies are no longer connected. Chief designer Igor Vakhrushev was killed in the crash of an Aeropract-23M in June 2000, after which production was suspended. A modified version of Aeropract-27, the 27M, was shown at Moscow in August 2001.

For details of the -21M, -23M and -25 see *Jane's* 2001-02 edition.

UPDATED

AEROPRACT-27

TYPE: Side-by-side lightplane/kitbuilt.
PROGRAMME: Prototype (FLARF-02647) first flew (in floatplane configuration) 22 June 1998; public debut at Gelendzhik 98 seaplane show, and in landplane version at MAKS, Moscow, August 1999; evaluated during 1999 and 2000 by AvtoZavod Flying Club in Togliatti. Seventh

Aeropract-27M Advantage two-seat lightplane *(Jane's/Paul Jackson)* **NEW**/0132852

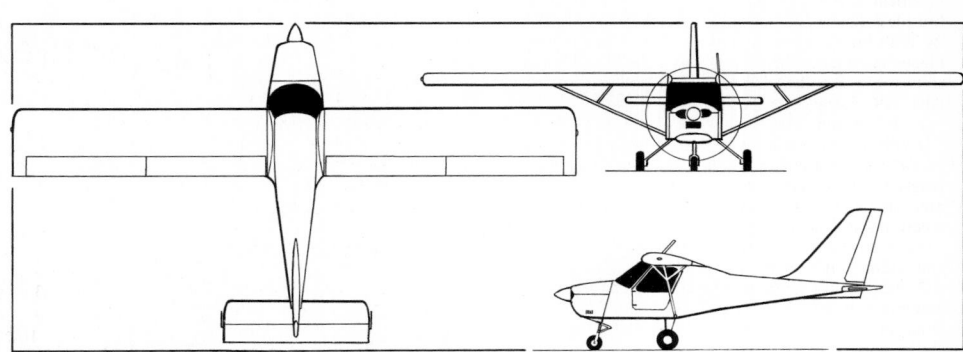

Aeropract-27M Advantage *(Jane's/James Goulding)* 0121662

production airframe, under construction in mid-2001, is first with Rotax 914 engine.
CURRENT VERSIONS: **A-27:** Baseline version with Hirth engine.
A-27M Advantage: Version with 73.5 kW (98.6 hp) Rotax 912 ULS engine driving a Kiev ground-adjustable pitch propeller or 84.6 kW (113.4 hp) Rotax 914 ULS, driving a variable-pitch propeller. Several minor design changes, including more rounded forward lower fuselage surfaces, but description generally as for A-27, except where indicated.
COSTS: Kit US$14,000, excluding instruments (US$700), landing gear and Rotax 912 engine (US$14,000). Ex-works, complete, with wheels, US$35,500 (1999). Wheel landing gear US$1,000; skis US$250 to US$350; floats US$1,200 and US$1,650.
DESIGN FEATURES: Braced, high-wing monoplane with single main and two jury struts each side. Sweptback fin and rudder; unswept wings and horizontal tail. Meets JAR-VLA certification requirements.

FLYING CONTROLS: Manual. Ailerons, rudder and all-moving tailplane with anti-balance tab. Flaps. A-27M has fixed tailplane with elevators, and ailerons with external mass balances.
STRUCTURE: Tube metal fuselage with composites skin.
LANDING GEAR: Tricycle type; fixed. Cantilever, metal mainwheel legs; telescopic, leaf-sprung nosewheel leg. Main wheel size 400×150; nose wheel size 300×125. Optional floats or skis.
POWER PLANT (A-27): One 70.8 kW (95 hp) Hirth F30 flat-four driving a four-blade, ground-adjustable pitch propeller. Fuel capacity: 80 litres (21.1 US gallons; 17.6 Imp gallons); (A-27M) 100 litres (26.4 US gallons; 22.0 Imp gallons).
ACCOMMODATION: Two persons, side by side; forward-hinged door each side. Luggage space behind seats.
AVIONICS: Standard VFR.
Data refer to both versions, except where indicated.
DIMENSIONS, EXTERNAL:

Wing span: A27	10.60 m (34 ft 9¼ in)
A-27M	10.00 m (32 ft 9¾ in)
Wing chord, constant	1.35 m (4 ft 5¼ in)
Length overall: A-27	6.30 m (20 ft 8 in)
A-27M	6.50 m (21 ft 4 in)
Height overall	2.22 m (7 ft 3½ in)

WEIGHTS AND LOADINGS:

Weight empty: A-27	335 kg (739 lb)
A-27M	325 kg (717 lb)
Max T-O weight: A-27	600 kg (1,322 lb)
A-27M	650 kg (1,433 lb)
Max power loading: A-27	8.47 kg/kW (13.92 lb/hp)
A-27M	8.84 kg/kW (14.53 lb/hp)

PERFORMANCE:

Max level speed: A-27	95 kt (175 km/h; 109 mph)
A-27M	97 kt (180 km/h; 112 mph)
Normal cruising speed: A-27	70 kt (130 km/h; 81 mph)
A-27M	70-81 kt (130-150 km/h; 81-93 mph)
T-O run	100 m (330 ft)
Range: A-27	324 n miles (600 km; 372 miles)
A-27M	540 n miles (1,000 km; 621 miles)

UPDATED

AEROPROGRESS/ROKS-AERO

AEROPROGRESS/ROKS-AERO
Gosudarstvennoye Kosmichesky Nauchno-Proizvodst-vennoye Tsentr imeni M B Khrunicheva, ulitsa Novozavodskaya 18, 121309 Moskva
Tel: (+7 095) 145 88 60
Fax: (+7 095) 145 94 77
PRESIDENT AND GENERAL DESIGNER: Evgeny P Grunin
DEPUTY GENERAL DESIGNER: Arnold I Andrianov
DEPUTY GENERAL DESIGNER, FOREIGN ECONOMIC RELATIONS: Alexander V Andreev
DESIGN BUREAU MANAGER: Sergei M Zhiganov

Known initially as ROS-Aeroprogress, this organisation was founded in 1990 to design and manufacture utility, commuter, amphibian, aerobatic, agricultural, firefighting, training and attack aircraft, WIG (wing-in-ground-effect) vehicles, replicas and other vehicles. ROKS-Aero Inc is the design bureau of Aeroprogress.

VERIFIED

AEROPROGRESS/ROKS-AERO T-101 GRACH

English name: Rook
TYPE: Light utility turboprop.
PROGRAMME: Design started by Utility Aircraft Division in September 1991, as monoplane successor to Antonov An-2/3 biplane; construction of prototype started April 1992; first flight (FLARF-01466) 7 December 1994; no evidence of reported four additional prototypes, all of which are presumed to have been used for static testing.

Series manufacture by Moscow Aviation Production Organisation (MAPO) initiated January 1993; initial production aircraft shown unpainted at Lukhovitsy factory, August 1999 (see LMZ entry in this section); further seven then substantially complete. One T-101 delivered to Chukota region in July 2000 on six-month lease, representing initial revenue-earning use; second supplied to undisclosed operator in early 2001 and third being prepared for delivery by late 2001. A Westernised version, developed via the Aeroprogress ROKS/Aero T-201, was unsuccessfully marketed as the Khrunichev T-201 (which see in 1999-2000 and previous editions). In March 2002 the company suggested it was looking for certification by 2004.
CURRENT VERSIONS: **T-101:** Basic passenger/cargo transport.
Detailed description applies specifically to T-101.
Several other variants have been proposed, as last detailed in the 2002-03 edition, but none has yet appeared.
CUSTOMERS: Two delivered by mid-2001. Third on order; Asian customer negotiating for four in late 2001. In March 2002 production of parts for 50 aircraft was under way, with 25 sets ready for assembly.
COSTS: Approximately US$700,000 (2002).
DESIGN FEATURES: Single-turboprop aircraft for Normal category passenger/cargo transportation and utility applications; suitable to replace Antonov An-2. High-mounted, unswept, constant-chord, braced wing; non-retractable landing gear and unpressurised cabin; STOL capable, with wide CG range; large passenger/freight door; sweptback fin and rudder with dorsal fin; braced constant chord tailplane and elevators.

Wing section CAHI (TsAGI) P-11-14; dihedral 3°; incidence 3° constant.

FLYING CONTROLS: Conventional and manual. Actuation by pushrods and cables. Slotted ailerons, deflection 30° up, 14° down; elevator deflection 42° up, 22.5° down; rudder deflection ±28°. Trim tabs in each aileron, each elevator and rudder. Electrically actuated single-section slotted trailing-edge flap and two-section automatic leading-edge slats on each wing; flap deflection 25° for take-off, 40° for landing.
STRUCTURE: All-metal (aluminium alloy and high-tensile steel) structure; two-spar wings, metal skinned with integral stringers, and ribs; two-spar fin and tailplane; semi-monocoque fuselage, with frames, stringers and stressed skin.
LANDING GEAR: Non-retractable tailwheel type with single wheel on each unit. Main legs of tripod type, with oleo-pneumatic shock-absorption; KT-135D mainwheels, with 720×320 tyres, pressure 3.43 bar (50 lb/sq in); K-392 tailwheel, with 380×200 tyres, pressure 3.43 bar (50 lb/sq in); hydraulic brakes and anti-skid units on mainwheels. Skis and floats optional.
POWER PLANT: One 706 kW (947 shp) Mars (Omsk) TVD-10B turboprop, driving AV-24AN three-blade constant-speed propeller with reverse pitch and full feathering. Three fuel tanks in each wing, each 200 litres (52.8 US gallons; 44.0 Imp gallons); total fuel capacity 1,200 litres (317 US gallons; 264 Imp gallons). Oil tank capacity 30 litres (7.9 US gallons; 6.6 Imp gallons).
ACCOMMODATION: Crew of one or two; up to nine passengers or equivalent freight. Forward-opening door each side of flight deck; large upward-opening freight door aft of wing on port side, with integral inward-opening passenger door; door between flight deck and cabin; starboard emergency exit. Cabin ventilated and heated by engine bleed air.

SYSTEMS: Hydraulic system for brakes; maximum flow 4 litres (1.05 US gallons; 0.88 Imp gallon)/min, at 147 bar (2,135 lb/sq in). Three-phase 120/208 V 400 Hz AC electrical system, supplied by 6 kVA 200 A BU6BK brushless alternator, with emergency DC power supply and 24 V 40 Ah Ni/Cd battery. Electric de-icing of propeller blades and spinner; engine air intake de-iced by heated oil.

AVIONICS: *Comms:* R-855A emergency locator beacon and ARB-NK emergency radio buoy.

Radar: Weather radar pod under starboard wing.

Flight: Com/nav equipment for VFR and IFR operations by day and night over all terrain; PNP-72-14 nav; A-723 long-range radio nav; ARK-M ADF; GRAN low-altitude radio altimeter; GROM satellite nav; A-611 marker beacon receiver; AP-93 autopilot.

Instrumentation: Air data system with digital airspeed and altitude indication; VBM-1PB standby altimeter; KC-MC compact compass system; AGB-96 gyro horizon.

DIMENSIONS, EXTERNAL:

Wing span	18.20 m (59 ft 8½ in)
Wing chord: at root	2.40 m (7 ft 10½ in)
at tip	2.45 m (8 ft 0½ in)
Wing aspect ratio	7.6
Length overall	15.06 m (49 ft 5 in)
Fuselage: Max width	1.80 m (5 ft 10¾ in)
Max height	2.52 m (8 ft 3¼ in)

Height overall	4.86 m (15 ft 11¼ in)
Tailplane span	5.82 m (19 ft 1¼ in)
Wheel track	3.36 m (11 ft 0¼ in)
Wheelbase	8.30 m (27 ft 3 in)
Propeller diameter	2.80 m (9 ft 2¼ in)
Propeller ground clearance	0.97 m (3 ft 2¼ in)
Flight deck doors (each): Height	1.20 m (3 ft 11¼ in)
Width: at top	0.43 m (1 ft 4¾ in)
at bottom	0.67 m (2 ft 2¼ in)
Passenger door: Height	1.42 m (4 ft 8 in)
Width	0.81 m (2 ft 7¾ in)
Freight door: Height	1.80 m (5 ft 10¾ in)
Width	1.60 m (5 ft 3 in)
Emergency exit: Height	0.54 m (1 ft 9 in)
Width	0.87 m (2 ft 10¼ in)

DIMENSIONS, INTERNAL:

Cabin: Length	4.50 m (14 ft 9¼ in)
Max width	1.60 m (5 ft 3 in)
Max height	1.85 m (6 ft 0¾ in)
Floor area	6.72 m² (72.3 sq ft)
Volume	12.1 m³ (427 cu ft)

AREAS:

Wings, gross	43.63 m² (469.6 sq ft)
Ailerons (total)	5.66 m² (60.93 sq ft)
Trailing-edge flaps (total)	4.02 m² (43.27 sq ft)
Leading-edge slats (total)	5.98 m² (64.37 sq ft)

Fin, incl dorsal fin	5.455 m² (58.72 sq ft)
Rudder, incl tab	2.955 m² (31.81 sq ft)
Tailplane	6.34 m² (68.25 sq ft)
Elevator, incl tab	4.10 m² (44.13 sq ft)

WEIGHTS AND LOADINGS:

Weight empty, equipped	3,330 kg (7,342 lb)
Max payload	1,400 kg (3,086 lb)
Max fuel	950 kg (2,095 lb)
Max T-O and landing weight	5,250 kg (11,574 lb)
Max wing loading	120.3 kg/m² (24.65 lb/sq ft)
Max power loading	7.44 kg/kW (12.22 lb/shp)

PERFORMANCE (estimated):

Max level speed at 3,000 m (9,840 ft)	162 kt (300 km/h; 186 mph)
Nominal cruising speed at 3,000 m (9,840 ft)	135 kt (250 km/h; 155 mph)
Service ceiling	4,000 m (13,120 ft)
T-O run	350 m (1,150 ft)
Landing run	200 m (660 ft)

Range:

with max payload	377 n miles (700 km; 434 miles)
with max fuel	685 n miles (1,270 km; 789 miles)

UPDATED

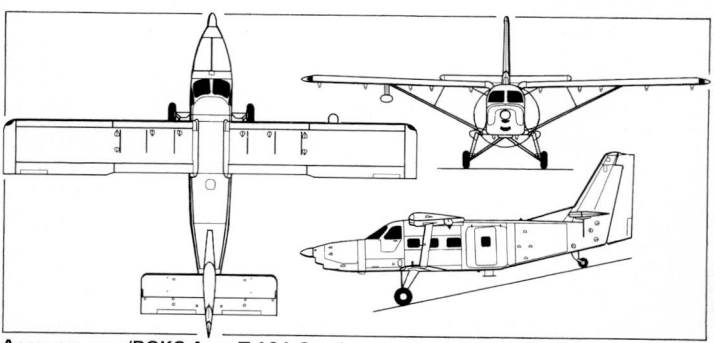

Aeroprogress/ROKS-Aero T-101 Grach turboprop utility aircraft *(Jane's/Paul Jackson)*
0079324

Aeroprogress/ROKS-Aero T-101 Grach prototype *(Jane's/Paul Jackson)* 0099617

AERO-VOLGA

AERO-VOLGA OOO NPO
ulitsa Kutyakova 8, Samara
Tel: (+7 8462) 70 20 75
e-mail: airvolga@mail.radiant.ru
Web: http://www.radiant.ru~airvolga

Two prototypes of this company's twin-engined amphibian have been flown.

UPDATED

AERO-VOLGA L-6
TYPE: Six-seat amphibian.
PROGRAMME: Prototype (01) built in six months and first flew in June 2000; public debut at Gelendzhik hydromarine exhibition in August 2000; promotion continued at MAKS, Moscow, August 2001 when L-6M first shown.
CURRENT VERSIONS: **L-6A:** Initial version. Three-step hull.
L-6M: Developed version with increased MTOW and four-step hull.
DESIGN FEATURES: High-wing amphibian. Wings strut-braced and with engines mounted on upper surface; stabilising floats at approximately three-quarter span. V-type twin fins have conventional tailplane mounted at tips. Two-step hull. Downturned wingtips.
FLYING CONTROLS: Manual, with twin fins, each having rudder. Single-piece elevator with flight-adjustable tab. Two-section slotted flaps each side. Control actuation by pushrods.
STRUCTURE: Composites, with monocoque hull.
LANDING GEAR: Tailwheel type; retractable. Mainwheels, size 400×150, retract forwards into hull; tailwheel size 255×110. Hydraulic mainwheel brakes.
POWER PLANT: Two VAZ RPD-416 piston engines, each 121 kW (163 hp), driving two-blade, fixed-pitch propellers. Two fuel tanks, combined capacity 340 litres (89.7 US gallons; 74.8 Imp gallons).
ACCOMMODATION: Six persons, including pilot(s), in three side-by-side pairs; short, sideways bench each side of rear

Aero-Volga L-6M, showing revised shape of hull and engine cowlings, compared with L-6A *(Jane's/Paul Jackson)*
NEW/0132853

of cabin. Dual controls. Single hatch, forward hinged, aft of wing trailing-edge.
Data apply to both versions, except where indicated.

DIMENSIONS, EXTERNAL:

Wing span	14.20 m (46 ft 7 in)
Wing aspect ratio	9.4
Length overall	10.00 m (32 ft 9¾ in)
Height overall: flying attitude	3.00 m (9 ft 10 in)
normal attitude	2.40 m (7 ft 10½ in)
Wheel track	1.60 m (5 ft 3 in)
Distance between propeller centres	2.80 m (9 ft 2¼ in)

AREAS:

Wings, gross	21.50 m² (231.4 sq ft)

WEIGHTS AND LOADINGS:

Weight empty	1,100 kg (2,425 lb)
T-O weight: normal	1,600 kg (3,527 lb)
max: L-6A	1,800 kg (3,968 lb)
L-6M	2,000 kg (4,409 lb)

Max wing loading: L-6M	93.0 kg/m² (19.05 lb/sq ft)
Max power loading: L-6M	8.23 kg/kW (13.52 lb/hp)

PERFORMANCE:

Max level speed: L-6A	119 kt (220 km/h; 137 mph)
L-6M	108 kt (200 km/h; 124 mph)
Cruising speed: max	92 kt (170 km/h; 106 mph)
econ	81 kt (150 km/h; 93 mph)
Stalling speed	54 kt (100 km/h; 63 mph)
Max rate of climb at S/L: L-6A	300 m (984 ft)/min
Service ceiling	4,000 m (13,120 ft)
T-O run: on land	300 m (985 ft)
on water	350 m (1,150 ft)
Range: L-6A	513 n miles (950 km; 590 miles)
L-6M	540 n miles (1,000 km; 621 miles)
g limits	+3.8/−1.0

UPDATED

ALBATROS

TSENTR NAUCHNO-TEKHNICHESKOGO TVORCHESTVA ALBATROS (Albatross Scientific-Technical Works)
ulitsa Tsentral'naya 22/126, 142092 Troitsk, Moskovskaya oblast

Tel/Fax: (+7 096) 751 66 76
CONSTRUCTOR: Sergei V Ignatev
TEST PILOT: Mikhail Y Shevyakov

Original Aviacomplex company established in 1990 to build prototypes of the AS-2 two-seat lightplane. Testing was completed in collaboration with CAHI (TsAGI) and LII

Flight Research Institute. Production by Aviacomplex and Myasishchev Engineering Bureau, with participation by other Russian companies. At 1999 Moscow Air Show (MAKS), a production AS-2 was shown under new designation of Albatros AS-3A and first details were revealed of a development, the Sigma-4, but no further details of either aircraft have been received.

However, an aircraft similar to the Sigma-4 (see 2001-02 *Jane's*) was displayed, partly shrouded and for one day only, at the 2001 Moscow show, wearing insignia of the Zhukovsky-based LII research institute and apparently with the slightly increased MTOW of 540 kg (1,190 lb). Another of the type appears to be flying with the Sebastopol Aero Club, Ukraine.

UPDATED

Unidentified aircraft, believed to be an Albatros
Sigma-4, seen at Moscow in 2001
(Jane's/Paul Jackson)
NEW/0524605

ARSENYEV

ARSENYEVSKOYE AVIATSIONNOYE PROIZVODSTVENNOYE PREDPRIYATIE IMENI N I SAZYKINA (Arsenyev Aviation Production Enterprise 'Progress' named for N I Sazykin)

prospekt Lenina 5, 692335 Arsenyev, Primovsky Kray
Tel: (+7 423) 612 48 97
Fax: (+7 423) 612 61 30
GENERAL DIRECTOR: Vladimir Pechyonkin

Arsenyev plant previously built the Mil Mi-24/25/35 series of combat helicopters, in parallel with Rostvertol, and was also manufacturer of the Yak-55 aerobatic lightplane; a modified version of the last-mentioned, the Technoavia SP-55 (which see), was returned to production in 1999, in an initial batch of five.

Arsenyev is responsible currently for the Kamov Ka-50 (including, if ordered, the Ka-50-2 export version) and Mil Mi-34 helicopters, as well as Moskit missiles. It was reportedly assigned the new Kamov Ka-60, but that programme has been reallocated to LMZ (which see).

Russian government shareholding is 51 per cent; personnel totalled some 5,000 in 2000.

A government decree of 31 December 1997 authorised the Arsenyev plant to offer the Ka-50 for export and supply spares and support for existing Mi-24/25/35 helicopters. Overseas delivery of P-15U, P-20, P-21 and P-22 cruise missiles is also covered.

Official plans announced on 12 May 2001 called for the Arsenyev plant to be incorporated in the proposed grouping to include RSK 'MiG', Tupolev and Kamov.

UPDATED

AVIA

NAUCHNO- PROIZVODSTVENNOE OBEDINENIE AVIA LTD (Avia Scientific-Production Association JSC)

Studeni proezd 5, 129282 Moskva
Tel: (+7 095) 472 97 89 and 478 55 62
Fax: (+7 095) 479 16 09
e-mail: avia@msk.sitek.net
Web: http://www.npo-avia.ru
GENERAL DIRECTOR: Aleksandr Loshkarev
SALES DIRECTOR: Vadim Salimov

OKB AKKORD (Accord Design Bureau)
ulitsa Chaadaeva 1a, 603035 Nizhny Novgorod
Tel: (+7 8312) 46 71 41
e-mail: accord@nnov.cityline.ru
DIRECTOR: Yury Lakhtachev

EUROPEAN SALES AGENT:
Ground Support Equipment srl
Viale del Vignola 44, I-00196 Roma, Italy
Tel: (+39 06) 322 28 77
Fax: (+39 06) 361 17 15
DIRECTOR: Roberto Salmoni

The Avia company was established on 20 February 1995 to design, manufacture and support light aircraft. Its associated design bureau (itself part of the Sokol plant) is responsible for the Accord range of light twin-engined aircraft. GSE of Italy was appointed as sales agent in 2000.

UPDATED

AVIA ACCORD-201

TYPE: Light utility twin-prop transport.
PROGRAMME: Original (lightweight) Accord design began May 1991; first flew 18 April 1994 and was last described in the 1997-98 *Jane's*. Production version is Accord-201; prototype flew 18 August 1997 at Nizhny Novgorod, initially with only VFR avionics and without floats. Demonstrations to prospective customers began December 1997; series of 12 flights from water, using floats,

Avia Accord-201 in landplane guise *(Jane's/Paul Jackson)*   **NEW**/0137958

undertaken mid-1998. By early 1999, four more aircraft, including static test specimen, under construction and tooling in place for maximum possible production of 10 per month. Second prototype was due to fly in December 2001. Russian AP-23 certification anticipated October 1999, but delayed, eventually beginning on 20 April 2000. Experimental category certificate, granted mid-2000 with expiry date of 19 June 2001. Hartzell propellers received Russian certification in October 2000. Full certification was still awaited in December 2001; expected mid-2002, using Becker avionics in place of original Bendix/King equipment.
CURRENT VERSIONS: Offered for a variety of roles, including SAR (**201P**), aerial survey (**201AFS**), ecological monitoring (**201EM**), sigint (**201RC**) and jamming (**201REP**).
CUSTOMERS: First firm order received in May 1998 from Almazy Rossii-Sakha diamond mining company, which ordered two to be used for geomagnetic survey and exploration. Customers in Abu Dhabi and Liechtenstein by

2000; however, because of certification delays, former contract under renegotiation and latter cancelled by late 2001.
COSTS: Standard aircraft US$298,000 on wheels; US$328,000 on floats. Operating cost US$44 per hour (2000).
DESIGN FEATURES: High-wing monoplane with single bracing strut each side; unswept, high-lift wing section and blown flaps; pod and boom fuselage; cruciform tail surfaces with sweptback fin and rudder; mainwheels at tips of short stub-wings that support bracing struts. Constant-chord main wing panels, with engines on leading-edge; tapered outer panels.

Service life 15,000 hours/15 years. Wing section CAHI (TsAGI) P-II, with 12 per cent thickness/chord ratio at root, CAHI (TsAGI) P-III at tip; tip sweepback 45°; no dihedral, incidence 1°, no twist.
FLYING CONTROLS: Conventional and manual. Ailerons and elevators 100 per cent balanced; twin horn-balanced rudders; all cable actuated. Three-axis electric trim. Two-section slotted flaps each side.
STRUCTURE: All-metal; D16 aluminium alloy and high-strength stainless steel. Fuselage based on welded rectangular-section cabin frame of steel tubing; remainder of airframe conventional light alloy construction.
LANDING GEAR: Non-retractable tricycle type; single wheel on each unit. Shock absorption by specially shaped rubber cushion between strut and wheel suspension lever. Cleveland wheels and brakes; mainwheels type 40-142 with 8.00-6 tyres; nosewheel type 40-140 with 6.00-6 tyres; type 30-127 disc brakes. Minimum turning circle 3.2 m (10 ft 6 in). Maximum nosewheel steering angle ±60°. Optional floats (additional to, not replacing, fixed landing gear) retract upward electrically, outside mainwheels.
POWER PLANT: Two 157 kW (210 hp) Teledyne Continental IO-360-ES7B flat-six engines; Hartzell PHC-H3YF-2UF/FC7453 three-blade, constant-speed, fully feathering

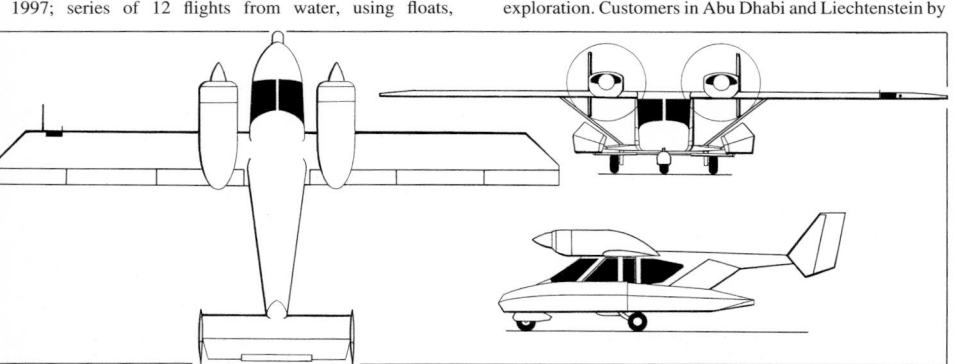

Avia Accord-201, showing position of retracted floats in nose and side views *(Jane's/Paul Jackson)*   0079325

propellers. Fuel tank in each engine nacelle; total capacity 750 litres (200 US gallons; 165 Imp gallons). Total oil capacity 15.2 litres (4.0 US gallons; 3.3 Imp gallons).

ACCOMMODATION: Pilot and up to six passengers in 2-3-2 seating, at 80 cm (31.5 in) pitch. Alternatively, pilot and three passengers and cargo, or all-cargo, or one stretcher patient, one or two attendants and medical equipment. Optional dual controls. Two large upward-hinged side doors. Baggage/cargo hold at rear of cabin, with door.

SYSTEMS: SAU-201 autopilot. Webasto Air Top 32 cabin heater. Electrical system includes two 27 V Teledyne N653344 DC generators and one 19 Ah Gill G-247 battery. Elipos-3 pneumatic de-icing system for wing, fin and tailplane leading-edges; Goodrich electric propeller de-icing. Janitrol cabin heater.

AVIONICS: Mainly Becker equipment:
Radar: Weather radar optional.

DIMENSIONS, EXTERNAL:

Wing span	13.75 m (45 ft 1¼ in)
Wing chord, constant	1.30 m (4 ft 3¼ in)
Wing aspect ratio	11.1
Length overall: landplane	8.12 m (26 ft 7¾ in)
amphibian	8.73 m (28 ft 7¾ in)
Height overall: landplane	2.94 m (9 ft 7¾ in)
amphibian	3.455 m (11 ft 4 in)
Span over tailfins	3.66 m (12 ft 0 in)
Wheel track	2.34 m (7 ft 8¼ in)
Float track	2.90 m (9 ft 6¼ in)
Wheelbase	2.845 m (9 ft 4 in)
Propeller diameter	1.93 m (6 ft 4 in)
Propeller ground clearance	1.25 m (4 ft 1¼ in)

Distance between propeller centres	2.80 m (9 ft 2¼ in)
Cabin doors (each): Height	1.10 m (3 ft 7¼ in)
Width: at top	0.75 m (2 ft 5½ in)
at bottom	1.50 m (4 ft 11 in)
Baggage/airdrop door: Height	0.90 m (2 ft 11½ in)
Width: at top	0.85 m (2 ft 9½ in)
at bottom	1.20 m (3 ft 11¼ in)

DIMENSIONS, INTERNAL:

Cabin: Length	3.50 m (11 ft 5¾ in)
Max width	1.31 m (4 ft 3½ in)
Max height	1.20 m (3 ft 11¼ in)
Floor area	3.50 m² (37.7 sq ft)
Volume	4.0 m³ (141 cu ft)
Baggage hold: Volume:	
pilot and six passengers	0.50 m³ (17.7 cu ft)
pilot and four passengers	1.5 m³ (53 cu ft)
pilot and cargo	2.5 m³ (88 cu ft)

AREAS:

Wings, gross	17.00 m² (183.0 sq ft)
Ailerons (total)	1.17 m² (12.59 sq ft)
Trailing-edge flaps (total)	3.23 m² (34.77 sq ft)
Fins (total)	1.36 m² (14.64 sq ft)
Rudders (total)	1.36 m² (14.64 sq ft)
Tailplane	2.18 m² (23.47 sq ft)
Elevators (total)	1.78 m² (19.16 sq ft)

WEIGHTS AND LOADINGS:

Operating weight empty: landplane	1,350 kg (2,976 lb)
amphibian	1,530 kg (3,373 lb)
Max T-O weight	2,200 kg (4,850 lb)
Max wing loading	129.4 kg/m² (26.51 lb/sq ft)
Max power loading	7.03 kg/kW (11.55 lb/hp)

PERFORMANCE (estimated):

Never-exceed speed (VNE)	202 kt (375 km/h; 233 mph)
Max level speed at S/L:	
landplane	161 kt (299 km/h; 186 mph)
amphibian	150 kt (277 km/h; 172 mph)
Max cruising speed at 75% power at 1,000 m (3,280 ft):	
landplane	148 kt (275 km/h; 170 mph)
amphibian	138 kt (255 km/h; 158 mph)
Econ cruising speed at 3,000 m (9,840 ft):	
landplane	119 kt (220 km/h; 137 mph)
amphibian	111 kt (205 km/h; 127 mph)
Stalling speed, flaps down, power off:	
landplane	58 kt (106 km/h; 66 mph)
Max rate of climb at S/L:	
landplane	445 m (1,460 ft)/min
amphibian	420 m (1,378 ft)/min
Service ceiling: landplane	6,100 m (20,020 ft)
amphibian	5,800 m (19,020 ft)
T-O run: on land: landplane	240 m (790 ft)
amphibian	250 m (820 ft)
on water	360 m (1,185 ft)
Landing run on land:	
landplane, amphibian	220 m (725 ft)
on water	190 m (625 ft)
Range with max fuel at 3,000 m (9,840 ft):	
landplane	1,592 n miles (2,950 km; 1,833 miles)
amphibian	1,339 n miles (2,480 km; 1,541 miles)
g limits	+3.8/−1.6

UPDATED

AVIACOR

AVIAKOR AVIATSIONNOYE ZAVOD OAO (Aviacor Aviation Depot JSC)

ulitsa Pskovskaya 32, 443052 Samara
Tel: (+7 8462) 27 03 87 and 27 04 44
Fax: (+7 8462) 27 06 91 and 27 04 77
e-mail: aviaprom@aviacor.ru
Web: http://www.aviacor.ru
GENERAL DIRECTOR: Vladimir Belogub
MARKETING DIRECTOR: Sergei Grachev

Founded in Voronezh during the 1930s and evacuated to Kuybishev (Samara) during Second World War, Aviacor

(formerly GAZ 18) is now terminating production of the Tupolev Tu-154 three-turbofan transport, the final 12 of which are being completed between 2000 and 2002. It has manufactured the Molniya-1 six-seat light aircraft and an air cushion vehicle derivative, and will produce the Tupolev Tu-354. It was also earmarked to build the Ukrainian Antonov An-70 and An-140 in Russia and to assist RSK 'MiG' with the Tu-334. The first An-140 (of six in initial batch) was rescheduled for construction in first quarter of 2002, but detailed discussions on a production timetable and contracts for delivery were not held with Antonov until 5 December 2001, implying further delay.

It was announced in September 2001 that An-70 assembly had been reallocated to Polyot because of Aviacor's inability

to provide start-up funding; some related subcontract work (possibly wings) may be allocated, however. Two new aircraft, known only as the Aist-2 and Aist-4, were reported to be under development, but nothing further has been heard. Subsidiary businesses are Aviacor-Service and Aviacor Repair. Company overhauls Tu-95MS strategic bombers and Tu-154 airliners (15 in 2000). Group members employ Westernised form of company name in place of direct transliteration, Aviakor.

Russian government shareholding is 25.5 per cent; controlling interest held by Siberian Aluminium Group since 1998. Undertakes subcontract work for Airbus and Boeing. Personnel in 2000 numbered some 4,000.

UPDATED

AVIASTAR

AVIASTAR, ULYANOVSKY AVIATSIONNYI PROMYSHLENNYI KOMPLEKS OAO (Ulyanovsk Aviation Industrial Complex 'Aviastar' JSC)

prospekt Antonova 1, 432072 Ulyanovsk
Tel: (+7 8422) 20 25 75 and 29 10 22
Fax: (+7 8422) 20 35 06 and 29 23 65
CHAIRMAN AND GENERAL DESIGNER: Igor Shevchuk
DIRECTOR GENERAL: Gennadi I Koradnev
MARKETING DIRECTOR: Aleksandr I Gladkov
EXECUTIVE DIRECTOR: Valery V Savotchenko

Founded in 1976, this 1.5 million m² (16.5 million sq ft) production facility, known as Plant 25, began manufacture of the Antonov An-124 ultra-large airlifter (which see in the Ukrainian section) in 1985, following with the Tupolev Tu-204 transport in 1987 and the related Tu-234 in 1996. In 1996, brief details were given of the Module two-seat light kitplane; no known production; last described in 1999-2000 *Jane's*.

In mid-1997, Aviastar agreed to merge with Tupolev design bureau and Aviacor (which see). Tupolev merger effected 30 June 1999, forming Tupolev OAO with inputs from Russian government (51 per cent), Aviastar (43.6 per cent) and Tupolev Design Bureau (5.4 per cent). Under Russian government plans announced on 12 May 2001, Aviastar will be incorporated in major grouping also

Tupolev Tu-204-200 airliner awaiting completion at the Aviastar plant (*Yefim Gordon*) NEW/0132836

including RSK 'MiG', Tupolev, Kamov, Aviacor and Sokol. Also in 1997, Aviastar Asia was formed in Taipei, Taiwan, as a joint venture with Asian interests to promote the Tu-204 in the Far East. Finance is provided by Aviastar Financial International, formed by Aviastar, Perm Motors and Moscow International Bank.

After five-year break, production of An-124 resumed with August 2000 delivery of one to Volga-Dnepr. State financing (Rb20 million) was received in 2000 for construction of a second assembly hall for An-124 production; however, by late 2001, Aviastar was urgently seeking new financial support to clear its debts. This reportedly achieved in early

2002 when Cato Avomatic (Egypt) and Leader Group (Russia) agreed to fund between 30 and 50 new Tu-204 airframes in 2003-05.

Tu-204 production rate remained at four per year in 2000, but entire plant operated at only 17 per cent capacity in 1999 and was targeting 30 per cent for 2001. New batch of five Tu-204s begun in 2000; no deliveries in first seven months of 2001 and, although six planned by end of that year, only one delivered and further aircraft cleared for flight testing. Other activities include manufacture of bodies for Ikarus-280 bus and plans for major overhaul of LiAZ-677 buses.

UPDATED

AVIATON

AVIATON NAUCHNO-PROIZVODSTVENNAYA AVIATSIONNAYA FIRMA (Aviaton Aviation Scientific-Production Firm)

Balashikhinsky r-on, p/o Chernaya, Mars, 143991 Moskovskaya oblast
Tel: (+7 095) 252 82 21 and 522 90 66
Fax: (+7 095) 191 68 27
GENERAL DIRECTOR: Avtandil Khachapuridze

WORKS:
Kutaisi Technical Aircraft Factory
ulitsa Shartava 2, Kutaisi, 384000 Georgia

Tel: (+995 331) 706 13 and 751 34
Fax: (+995 331) 752 54
MANAGER: Enuki Gabunia

Aviaton displayed its first product, the Merkury light twin, at the 1997 Moscow Air Show. On 27 March 1998, it purchased a controlling share (91.44 per cent) of the Kutaisi Technical Aircraft Factory and is assembling aircraft there from parts manufactured elsewhere.

VERIFIED

AVIATON MERKURY

English name: Mercury
TYPE: Four-seat utility twin.

PROGRAMME: Prototype completed, but apparently not then flown, by August 1997. Second prototype under construction at Kutaisi by late 2000; third, uprated, prototype being built by RSK 'MiG's' Voronin plant, intended for public debut at Moscow Salon in August 2001, but did not appear. No further reports by mid-2002.

CURRENT VERSIONS: **Four-seat:** *As described.*
    Six-seat: Third prototype powered by two LOM M337B six-cylinder piston engines, each 176 kW (237 hp); stretched fuselage. Estimated cost (2000) US$360,000.

CUSTOMERS: US government has reportedly financed aircraft for Georgian Frontier Troops.

DESIGN FEATURES: Robust utility design; high wing for ease of access and minimal propeller damage from semi-prepared airstrips. To meet FAR Pt 23 criteria.

FLYING CONTROLS: Conventional and manual. Slotted flaps;
horn-balanced rudder; flight-adjustable trim tabs in rudder
and starboard elevator.
STRUCTURE: Metal throughout; fin and tailplane mutually
braced.
LANDING GEAR: Non-retractable tricycle type. Mainwheels on
independently hinged, steel tube levers with vertically
mounted motorcar-type shock-absorbers; steerable
nosewheel. Mainwheel tyres 500×200; nosewheel
400×150.
POWER PLANT: Two 103 kW (138 hp) LOM M332A four-
cylinder engines. Fuel filler on port side, behind cabin; tank
capacity 360 litres (95.0 US gallons; 79.2 Imp gallons).
ACCOMMODATION: Four persons, including two pilots. Large
door each side pulls out and slides backwards on internal
bearings for access.
EQUIPMENT: Landing light in nosecone.
DIMENSIONS, EXTERNAL:

Wing span	10.12 m (33 ft 2½ in)
Length overall	6.65 m (21 ft 9¾ in)
Height overall	3.15 m (10 ft 4 in)

AREAS:

Wings, gross	12.65 m² (136.2 sq ft)

WEIGHTS AND LOADINGS:

Weight empty	900 kg (1,984 lb)
Max T-O weight	1,500 kg (3,306 lb)
Max wing loading	118.6 kg/m² (24.29 lb/sq ft)
Max power loading	7.29 kg/kW (11.98 lb/hp)

PERFORMANCE (estimated):

Max level speed	189 kt (350 km/h; 217 mph)
Normal cruising speed	140 kt (260 km/h; 162 mph)
Max rate of climb at S/L	360 m (1,181 ft)/min
Service ceiling	4,000 m (13,120 ft)
T-O run	125 m (410 ft)
Range	809 n miles (1,500 km; 932 miles)

UPDATED

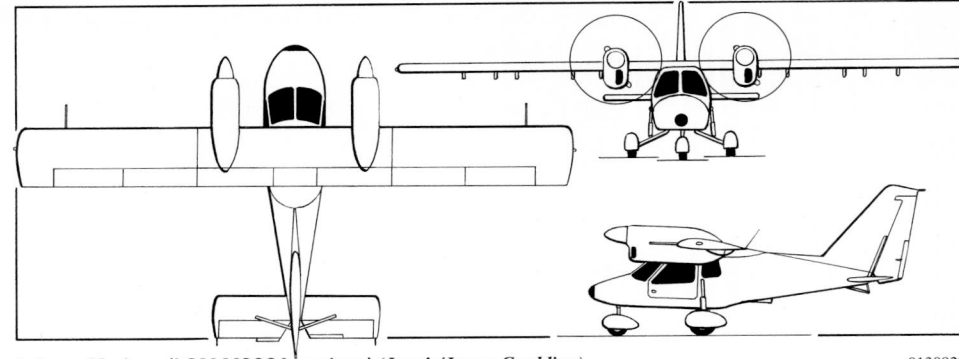

Aviaton Merkury (LOM M332A engines) *(Jane's/James Goulding)*   0130927

Prototype Aviaton Merkury light transport
(Jane's/Paul Jackson)
NEW/0137949

BERIEV

BERIEVA AVIATSIONNYI KOMPANIYA
(Beriev Aviation Company)
ploshchad Aviatorov 1, 347923 Taganrog
Tel: (+7 86344) 499 01 and 498 39
Fax: (+7 86344) 414 54
e-mail: info@beriev.com
Web: http://www.beriev.com
PRESIDENT AND GENERAL DESIGNER: Gennady S Panatov
HEAD OF MARKETING: Andrei Shishko
HEAD OF INFORMATION: Andrei I Salinkov

Original Beriev design bureau (OKB) founded in October
1934 by Georgy Mikhailovich Beriev (1902-79); except
during part of the Second World War, 1942-45, it has been
based at Taganrog, in northeast corner of Sea of Azov; since
1948 has been primary centre for Russian seaplane
development. Beriev has manufactured more than 20 types of
aircraft, most of which have entered series production. In
1990 was redesignated Taganrogsky Aviatsionnyi Nauchno-
Tekhnichesky Kompleks imeni G M Berieva (TANTK:
Taganrog Aviation Scientific-Technical Complex named for
G M Beriev); became part of AVPK Sukhoi in 1996. As a
consequence of this merger, the bureau's products are often
manufactured at factories traditionally associated with
Sukhoi, the Be-103, for example, at Komsomolsk. On 1
January 1998 adopted style, 'Beriev Aviation Company' for
international promotion, retaining TANTK in Russia. State
shareholding is 38 per cent; personnel strength in 2001 was
more than 2,000.

Beriev company now includes the experimental design
bureau, experimental production facilities, a flight test
complex, economic, financial and logistics support services,
with test bases and proving grounds at the Black Sea and Sea
of Azov. Its products are experimental prototypes of
amphibious aircraft and wing-in-ground-effect (WIG)
vehicles, together with test reports and technical
documentation for their series production. It undertakes
design and development of unconventional aircraft in
response to requests for proposals from other companies,
testing of aircraft and assemblies in maritime conditions, and
training of aircrew and ground personnel for seaplane
operation. In 2001, Beriev revealed it was developing a
multirole amphibian designated **Be-112**. A competition was
launched in early 2002 to find names for the Be-103 and
Be-200.

UPDATED

BERIEV A-42
Similar to the Beriev A-40, this aircraft project was last
described in the 1998-99 edition. Prototype 85 per cent
complete at TANTK, Taganrog, by September 2000, at
which time only Rb19.5 million assigned of Rb307 million
required to complete development. By January 2002, Beriev
was reporting that this prototype was an A-42P patrol and

**Interest, but without full funding, continues to be maintained in Beriev's A-40 (illustrated) and A-42 flying
boats** *(Yefim Gordon)*   *NEW*/0536625

Model of D-27-engined Be-42/A-42 wearing alternative designation A-40M *(Yefim Gordon)*   *NEW*/0536618

SAR variant; power plant quoted as two Progress D-27 propfans, each rated at 109.8 kN (24,690 lb st), which description applies to Beriev **A-45**, last described in 1998-99 *Jane's*. In this form, A-42P has MTOW of 96,000 kg (211,650 lb), combat load of 8,500 kg (18,739 lb) and range of 6,209 n miles (11,500 km; 7,145 miles). Main sensor suite to be Leninets Novella (Sea Dragon). Domestic needs are up to eight for military use and up to four for MChS civil protection agency; export interest (A-42PEh) from Canada, Chile, China, Finland, South Korea and Thailand. Announced in April 2002 that funding of A-42 to completion would begin in 2005, at earliest.

In mid-2002, Beriev was continuing promotion of an A-40Eh export version of the original aircraft as well as predicting a return of official interest to the A-40, which the Russian Navy abandoned in 1996, in favour of a maritime patrol version of Tupolev Tu-204 airliner.

UPDATED

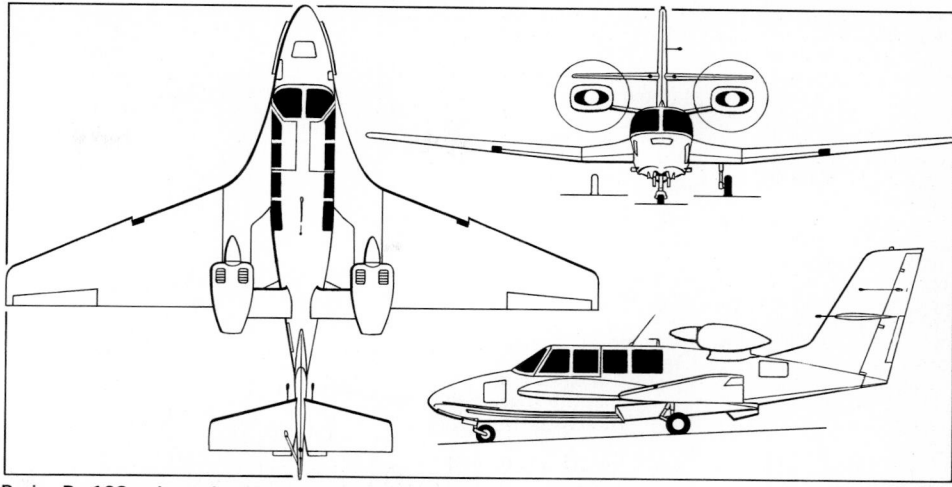

Beriev Be-103 twin-engined light multipurpose amphibian (*Jane's/Mike Keep*)   NEW/0132839

BERIEV Be-103

TYPE: Six-seat utility twin-prop amphibian.

PROGRAMME: Design started 1992; model exhibited and initial data released at Moscow Air Show '92; construction of preproduction batch of four began by KnAAPO at Komsomolsk-on-Amur 1994; prototype (RA-37019), with M-17F engines, displayed statically at Gelendzhik Hydro-aviation Show, on the Black Sea, 23 September 1996; first flew at Taganrog on 15 July 1997, but destroyed on its 28th sortie, during practice for Moscow Air Show on 18 August 1997. Second prototype (RA-03002) flew 17 November 1997; made type's first water take-off and landing on 24 April 1998 (aircraft's sixth sortie), but crashed on 29 April 1999 while returning from exhibition at Aero '99, Friedrichshafen; further two structure/static test airframes completed for use at Taganrog by early 1997.

Type certification rescheduled for second half of 1998 and first delivery for 1999, but these dates also not met; development being assisted by three preproduction aircraft, which completed by late 1997, although first of these (believed 03101) did not fly until 19 February 1999; 03103 shown at Gelendzhik in August 2000 when seen to have fixed wing leading-edge slats outboard of landing lights (as on 03102); by early 2001, 03103 had received nose radome. Further eight production aircraft reported under construction by September 1998, but none seen by late 2001, when manufacture confirmed to have involved five flying and two static test aircraft. Be-103 development included, since 1997, in Rostov regional development programme, while funding provided by government's 1999-2000 Light Aircraft Development Programme; KnAAPO paid for VFR certification, which granted by Russian Aviation Register on 26 December 2001, after 161

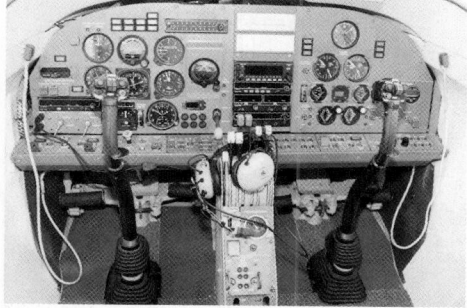

Beriev Be-103 (third preproduction) cockpit
(*Yefim Gordon*)   0121085

hours of trials, including 59 water and 204 land take-offs. FAR/JAR 23 due in 2003.

Marketed by **Takom Avia** consortium, formed by Beriev, KnAAPO and VO Mashinoexport of Moscow. In late 1999, consideration given to establishing final assembly centre in Malaysia, combining Russian airframes and US avionics and engines. Aerocorp International, appointed US dealer by 2002. Launch of US certification effort announced in September, but KnAAPO simultaneously revealed suspension of production due to lack of orders; factory will complete three to US contracts, leaving 10 unfinished.

CURRENT VERSIONS: **Be-103** *As described*.

SA-20P: Eight-seat, single-engined version for Russian Federation market; 265 kW (355 hp) VOKBM M-14P nine-cylinder radial, pylon-mounted above fuselage; conventional horizontal tail surfaces (tailplane and separate elevator); 1,600 kg (3,527 lb) empty weight; first aircraft under construction at KnAAPO by September 2001, when designation first announced. Prototype exhibited unflown at Gelendzhik Gidroavia Salon, September 2002. Reportedly flying by late October 2002. Certification to Russian AP-23 and FAR Pt 23 due 2003, for first deliveries in 2004.

CUSTOMERS: Projected sales of 600 including 230 exports. First order from Russian Border Guards for 20 (out of possible 200 required), originally for delivery from 1999; Forestry Service requires 30 to 40. First export order announced August 1999 in form of 10 for undisclosed customer; however, in January 2002 Beriev was still negotiating first firm orders, comprising possible six for forestry use and others for border patrol.

COSTS: Basic price US$600,000 to US$700,000 (2000), with IO-360 engines. SA-20P US$500,000 (2002).

DESIGN FEATURES: Low-wing monoplane, with water-displacing wings of moderate sweep with large wingroot extensions; two-step boat hull; no stabilising floats; horizontal strake each side of nose, plus vertical strake ahead of second hull step; wing centre-section increases lift during take-off and landing, partly compensating for absence of flaps. Engine pylon mounted on each side of rear fuselage, aft of wings; engines raised slightly on production version. Sweptback fin and rudder; tailplane mid-set on fin. Designed in compliance with AP-23 and FAR Pt 23 requirements; suitable for passenger and cargo transport, medical evacuation, patrol and ecological

Instrument panel of SA-20P (*Yefim Gordon*)
NEW/0536620

monitoring. Operable in wave heights up to 0.5 m (1½ ft) and on water as shallow as 1.25 m (4 ft).

Wing leading-edge sweep 22°; wing section NACA 2412M; dihedral 5° 3' on outer wings; incidence 1°. Fixed leading-edge slats.

FLYING CONTROLS: Manual. All-moving tailplane (with anti-balance tab) and ailerons actuated by pushrods; rudder by cables; two mass balances ahead of tailplane; mass balance (port side only) on rudder, which has flight-adjustable trim tab; electric trim; spring feel in tailplane control; no flaps.

STRUCTURE: All-metal semi-monocoque boat-type fuselage; all-metal single-spar wings. Extensive use of 1446 aluminium-lithium alloy.

LANDING GEAR: Pneumatically retractable tricycle type; single wheel on each unit; mainwheels retract forward into wing centre-section; nosewheel retracts forward; oleo-nitrogen shock-absorbers; brakes on mainwheels; nosewheel tyre size 400×150, mainwheel tyres 7.00-6; pressure 3.00 bar (43 lb/sq in) in all units; disc brakes; self-centring nosewheel steerable ±35°.

POWER PLANT: Two 157 kW (210 hp) Teledyne Continental IO-360-ES4 flat-four piston engines, each driving an MT-Propeller MTV-12-D-C-F-R-(M)/CFR183-17 three-blade, variable pitch (hydraulic), metal propeller. Later option will be two 221 kW (296 hp) VAZ-4263 piston engines and Avia AV-103 three-blade, variable-pitch propellers, with optional reversible pitch. Fuel tank in each wingroot, total capacity 450 litres (119 US gallons; 99.0 Imp gallons).

Third preproduction Beriev Be-103, displaying leading-edge slats and nose radome (*Yefim Gordon*)   NEW/0132838

Prototype of Komsomolsk-built SA-20P (*Yefim Gordon*)   *NEW*/0536619

ACCOMMODATION: Pilot and five passengers in pairs; dual controls optional; large upward-opening door on each side, hinged on centreline; folding seats facilitate entry/loading baggage/freight compartment aft of cabin. Optional ambulance configuration for pilot, one stretcher, three seated persons; all-cargo configuration for items up to 2.00 × 0.70 × 0.70 m (6 ft 6¾ in × 2 ft 3½ in × 2 ft 3½ in) in size; patrol; agricultural and ecological monitoring use.

SYSTEMS: Interior heated and ventilated. Pneumatic system, bottle capacity 6 litres (365 cu in) of air, pressure 49 bar (711 lb/sq in). Electrical system 27 V DC and three-phase 36 V 400 Hz AC, supplied by 3 kW DC generators, rectifiers and 25 Ah battery. Fire suppression system. Optional anti-icing.

AVIONICS: Integrated by Bendix/King. VFR standard; IFR optional. Version with Russian avionics may be offered in the future.

   Comms: KY 196A VHF com radio, KMA 24 intercom, KX 165 nav/com/glideslope, KT 76 transponder and P-855 emergency locator beacon.

   Radar: Optional RDR-2000 or RDR-1400 weather radar in extreme nose.

   Flight: KN 63 DME, KR 87 ADF, KCS 55A compass, KRA 405 radar altimeter, KLN 89B GPS. Optional KFC 150 or KAP 140 autopilot.

   Instrumentation: Conventional.

EQUIPMENT: Landing light in each wing leading-edge.

DIMENSIONS, EXTERNAL:

Wing span	12.72 m (41 ft 9 in)
Wing chord: at root	6.21 m (20 ft 4¾ in)
at tip	0.83 m (2 ft 9 in)
Wing aspect ratio	6.4
Length: overall	10.65 m (34 ft 11¼ in)
fuselage	9.96 m (32 ft 8 in)
Height overall	3.76 m (12 ft 4 in)
Tailplane span	3.90 m (12 ft 9½ in)
Wheel track	2.275 m (7 ft 5½ in)
Wheelbase	4.12 m (13 ft 6¼ in)
Propeller diameter	1.83 m (6 ft 0 in)
Distance between propeller centres	3.00 m (9 ft 10¼ in)

DIMENSIONS, INTERNAL:

Cabin: Length	3.65 m (11 ft 11¾ in)
Max width	1.25 m (4 ft 1¼ in)
Max height	1.23 m (4 ft 0½ in)
Baggage compartment: Max width	1.08 m (3 ft 6½ in)
Max height	0.88 m (2 ft 10½ in)
Volume	0.85 m³ (30.0 cu ft)

AREAS:

Wings, gross	25.10 m² (270.2 sq ft)
Ailerons, total	0.80 m² (8.61 sq ft)
Fin	2.86 m² (30.78 sq ft)
Rudder	1.54 m² (16.58 sq ft)
Tailplane, total	3.68 m² (39.61 sq ft)

WEIGHTS AND LOADINGS:

Weight empty, equipped	1,540 kg (3,395 lb)
Max payload	385 kg (849 lb)
Max T-O weight	2,270 kg (5,004 lb)
Max wing loading	90.4 kg/m² (18.52 lb/sq ft)
Max power loading	7.25 kg/kW (11.91 lb/hp)

PERFORMANCE (US version):

Max level speed	154 kt (285 km/h; 177 mph)
Max cruising speed at FL98	135 kt (250 km/h; 155 mph)
Econ cruising speed at FL98	120 kt (222 km/h; 138 mph)
Service ceiling	5,180 m (17,000 ft)
T-O run: on land	310 m (1,020 ft)
on water	550 m (1,805 ft)
Landing run: on land	390 m (1,280 ft)
on water	300 m (985 ft)
Range, 30 min reserves	
	600 n miles (1,111 km; 690 miles)
Endurance, no reserves	6 h 30 min

BERIEV Be-132

TYPE: Twin-turboprop light transport.

PROGRAMME: Original Be-30 first flown in prototype form on 3 March 1967; eight Be-30s built, but programme terminated when Aeroflot ordered Let L-410As from Czechoslovakia. Hard currency shortage revived programme 1993, with modestly upgraded version known as Be-32; one of original Be-30s (RA-67205) exhibited at 1993 Paris Air Show, as Be-32 demonstrator with original Russian TVD-10B engines; re-engined with PT6A-65B turboprops, RA-67205 first flew as Be-32K on 15 August 1995. Launch of Be-32K production continued to be discussed in 2001, although initial investment of Rb1,060 million required, plus Rb1,460 million in medium term; no announced orders.

Be-132 emerged from agreement between Motor-Sich of Ukraine, IAPO, Klimov and Beriev, signed at Paris Air Show in June 2001. Initial details released at MAKS, Moscow, August 2001. Completion of business plan and start of marketing were due for early 2002. Certification planned to AP-25.

CURRENT VERSIONS: **Be-132MK:** Initial version. Compared with Be-32, has more circular fuselage cross-section and increased length.

DESIGN FEATURES: General purpose light transport, optimised for operation from remote airfields. Conventional cantilever high-wing monoplane; three-section wings, with anhedral on outer panels; 42° sweptback vertical tail surfaces; engines at tip of centre-section each side. Ventral underfin.

FLYING CONTROLS: Conventional and manual. Trim tabs in both ailerons, both elevators and rudder. Double-slotted flaps in two sections each side, inboard and outboard of engines.

STRUCTURE: All-metal; semi-monocoque fuselage of rectangular section; spars and skin panels of wing torsion box are mechanically and chemically milled profile pressings; detachable bonded leading-edge; half of wings and most of tail unit covered with thin honeycomb panels stiffened with stringers; majority of fuselage made of adhesive-bonded panels; tips of wings and tail surfaces and wing/fuselage fillets of GFRP.

LANDING GEAR: Tricycle type; single wheel on each unit; nosewheel retracts forward, mainwheels rearward into engine nacelles; tyres size 720×320 on mainwheels, 500×150 on nosewheel; mainwheel brakes.

POWER PLANT: Two 1,103 kW (1,479 shp) Klimov VK-1500 turboprops; six-blade Aerosila AV-36 propellers; integral wing fuel tanks.

ACCOMMODATION: Basic seating for two crew and 26 passengers in two-plus-one configuration at 78 cm (30½ in) pitch. Carry-on baggage compartment on starboard side, aft of cabin seating, opposite forward-hinged door and airstairs; lavatory to rear.

DIMENSIONS, EXTERNAL:

Wing span	17.80 m (58 ft 4¾ in)
Length overall	17.925 m (58 ft 9¾ in)
Fuselage max width	1.70 m (5 ft 7 in)
Height overall	5.545 m (18 ft 2¼ in)
Tailplane span	7.40 m (24 ft 3¼ in)
Wheel track	6.00 m (19 ft 8¼ in)
Wheelbase	5.985 m (19 ft 7¾ in)
Propeller diameter	2.65 m (8 ft 8¼ in)

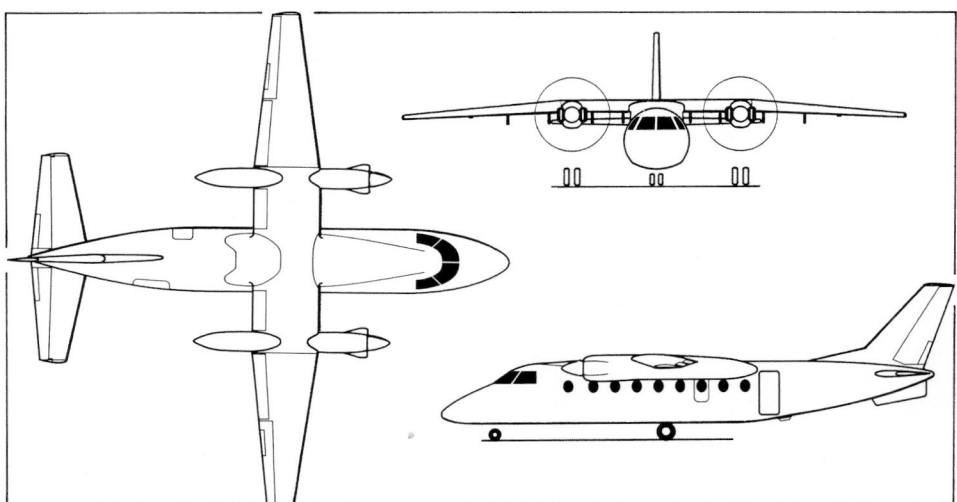

Beriev Be-132MK twin-turboprop multipurpose light transport (*Jane's/Paul Jackson*)   0121080

Model of Beriev Be-132MK   0113983

UPDATED

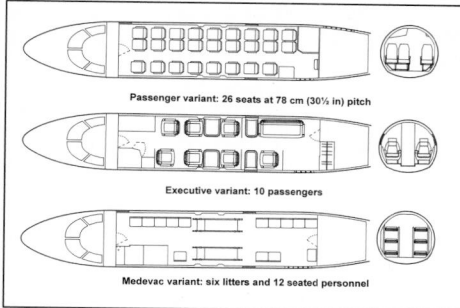

Alternative seating plans of the Beriev Be-132
0121083

Cabin door: Height	1.30 m (4 ft 3 in)
Width	0.75 m (2 ft 5½ in)
Emergency exits (each): Height	0.92 m (3 ft 0¼ in)
Width	0.60 m (1 ft 11½ in)

WEIGHTS AND LOADINGS:
Max T-O weight	10,000 kg (22,046 lb)
Max power loading	4.54 kg/kW (7.45 lb/shp)

PERFORMANCE (estimated):
Max cruising speed	248 kt (460 km/h; 286 mph)
Econ cruising speed	232 kt (430 km/h; 267 mph)
Unstick speed	113 kt (210 km/h; 130 mph)
Landing speed	108 kt (200 km/h; 124 mph)
Service ceiling	6,000 m (19,680 ft)
T-O run	700 m (2,300 ft)
Landing run	450 m (1,480 ft)
Required runway length	1,500 m (4,925 ft)
Range:	
with 26 passengers	712 n miles (1,320 km; 820 miles)
with max fuel	1,490 n miles (2,760 km; 1,715 miles)

UPDATED

BERIEV (BETAIR) Be-200

TYPE: Twin-jet amphibian.

PROGRAMME: Initiated 1989 under design leadership of Alexander Yavkin. Russian government approval for purpose-designed water bomber granted 8 December 1990. Details announced, and model displayed, at 1991 Paris Air Show; full-scale mockup constructed 1991; development by Beta Air (Betair: **Be**riev **ta**ganrog **ir**kutsk) consortium, formed 27 November 1991 from Irkutsk Aircraft Production Association (53.5 per cent), Beriev OKB (20.5 per cent), ILTA Trade Finance SA of Geneva, Switzerland (5 per cent) and the Ukrainian bank Prominvest (15 per cent) plus unidentified partner(s) (6 per cent); included in Civil Aviation Development Programme and state forestry protection programme.

Beriev responsible for development, design and documentation; systems bench-testing; static-, flight- and fatigue-testing of prototypes; certification and design support of serial production. IAPO duties comprise production preparation; manufacture of tooling; production of four prototypes and series aircraft; and spare parts manufacture. In May 2002, EADS signed joint marketing agreement with IAPO.

First prototype (001) rolled out 11 September 1996, at Irkutsk; first flight scheduled 1997, but eventually achieved (from land) 24 September 1998; 'official' first flight 17 October 1998; aircraft transferred from Irkutsk to Taganrog (by then with 19 sorties and 26½ flying hours) on 27 April 1999 to begin certification trials, including water drops; exhibited at Paris (by then registered RA-21511)

Be-200 releasing water (*Yefim Gordon*)   *NEW*/0132840

June 1999; first water landings and take-offs, 10 September 1999. Total 80 sorties, including 18 from water, by November 1999, when water operations ceased for winter. Experimental category certification achieved March 2000. Awarded limited category AP-25 certification for firefighting operations 10 August 2001, by which time had flown 337 sorties (100 from water) and achieved 202 flying hours. Second phase of testing begun October 2001 to certify patrol and passenger operations using BR715 engines, although most of that month occupied by a nine-country Far East sales tour. Prototype had accumulated 650 hours in nearly 600 sorties (216 from water) by May 2002.

Second flying prototype (003) in firefighting configuration (Be-200ChS). Two test airframes, of which first ferried from Irkutsk to Taganrog by An-124 in March 1995, followed by second in 1997; these, respectively, static test (SI : *staticheskiy ispytaniya*) and fatigue test (RI : *resursnye ispytaniya*). Four production aircraft (0101 to 0104) in hand by late 1998; fifth was taking shape in early 2002. On original schedule, second prototype was due for delivery in late 1999 to Russian Emergencies Ministry. Initial production and certification supported by Alenia of Italy under US$1.6 million programme financed by European Union. Deliveries of firefighting version scheduled to begin 2000, followed by search and rescue version. Repeated delays resulted in first flight (003/RA-21512) on 27 August 2002, following which it was exhibited at Gelendzhik Gidroavia Salon, 4 to 8 September 2002.

Avionics development and scoop trials for water bomber version being undertaken by the Beriev Be-12P-200/Be-128 testbed since August 1996. TANTK also produced four Be-12Ps (*Pozharny:* firefighting) for concept and tactics development and two Be-12NKh (*Narodno-Khozyaistvennye:* National Economy) general purpose versions, both of which lost in 1994.

CURRENT VERSIONS: **Firefighting;** TsENTROSPAS/MChS designation **Be-200ChS:** Tanks under cabin floor of centre-fuselage, capacity 12 m³ (423 cu ft) water; six tanks in cabin for 1.2 m³ (42 cu ft) liquid chemicals; two retractable water scoops forward of step, two aft; 30 fully equipped smoke jumpers can be carried on seats along sidewalls of cabin, with jump-door at rear of cabin on starboard side; 12 tonnes of water scooped from seas in 14 seconds with waves up to 1.2 m (4 ft).

Fully fuelled, Be-200 can drop total 310,000 kg (683,420 lb) of water in successive flights when airfield to reservoir distance is 108 n miles (200 km; 125 miles) and reservoir to fire zone distance is 5.4 n miles (10 km; 6.2 miles); or 140,000 kg (308,640 lb) when distances are respectively 108 n miles (200 km; 125 miles) and 27 n miles (50 km; 31 miles). Tank emptying time 0.8 to 1.0 second; minimum dropping speed 119 kt (220 km/h; 137 mph). Tanks quickly removable when aircraft carries freight. Flight deck and cargo hold sealed against smoke ingress. ARIA-200M avionics features include water source/drop zone track memory, automatic glideslope and digital flight deck/ground fire crew communications. Also outfitted for SAR, with equipment including Orion 25S inflatable boat, naval radios, 600 W loudspeaker and searchlight, storage for 20 inflatable rafts and two four-seat motorboats, observer's station, 57 seats and attachment for 30 litters.

Passenger: Two flight crew, two cabin attendants, and 72 tourist class passengers four-abreast in pairs, with centre aisle, at seat pitch of 75 cm (29.5 in).

Cargo: Payload 7,500 kg (16,534 lb) in unobstructed cabin 17.00 m (55 ft 9 in) long, 2.6 m (8 ft 6 in) wide and 1.9 m (6 ft 3 in) high.

Ambulance: Two flight crew, seven seated casualties/medical personnel, 30 stretchers in three tiers.

Patrol: Paramilitary version for possible use of Russian Frontier Guards revealed to be under development in early 2002.

Be-200PS: Search and rescue. Provisions included in Be-200ChS described above.

Be-200M: Redesignated Be-210 in 1998.

Be-200P: Projected anti-submarine version; combat radius 2,537 n miles (4,700 km; 2,920 miles); endurance 7 hours.

Be-210: Described separately.

CUSTOMERS: Seven ordered (of 20 required) by TsENTROSPAS/MChS (Russian Ministry of Emergency Situations) on 15 January 1997 for firefighting, with further orders to follow for SAR; decision in principle to acquire eighth, funded by cancellation of proposed two Antonov An74s; will equip four SAR stations on Black Sea and Baltic and two in eastern Russia; first base will be Gelendzhik. Beriev contracted in 2002 to establish training school at Taganrog for MChS pilots and technicians. Russian state forest service/firefighting agency requires up to 54, of which near-term needs total 10 to 15; 50 required by Sakhalin regional administration; five by Irkutsk regional administration. However, at time of maiden flight, only five orders were being claimed as firm. Potential market foreseen for more than 400 by 2010, of which over 60 per cent for export. South Korea evinced interest in 12 of maritime patrol version for police duties during 1998; China considering licensed production at Harbin; Italian interest in 15 reported mid-2000. Australia, France, Indonesia, Japan and Philippines expressing interest in 2001-02 and Be-200 demonstrated at Marseilles, France, 13 to 15 May 2002, and at Elefsis, Greece.

COSTS: Basic price US$25 million (2000). Break-even at 46/48th aircraft. Development cost estimated as US$250 million, of which US$190 million expended by 2001.

DESIGN FEATURES: First Russian purpose-designed water-bomber. Derived from Beriev A-40, but underwing stabilising floats moved inboard from tips, twin-wheel main landing gear units, and no booster turbojets. Conforms to FAR Pt 25 criteria. Swept wings of moderate aspect ratio, with high-lift devices; single-step hull of high length/beam ratio, which reported to provide world's first variable-rise bottom, providing a considerable improvement in stability and controllability in the water, as well as a reduction in *g* loads when landing and taking off at sea; small wedge-shape boxes ('hydrodynamic compensators') aft of step aid 'unsticking' from water in wave heights up to 2.2 m (7 ft 2½ in); all-swept T tail; high-mounted engines protected from spray by strakes on each side of nose and by wings; large underwing pod each side of hull, faired into wingroot.

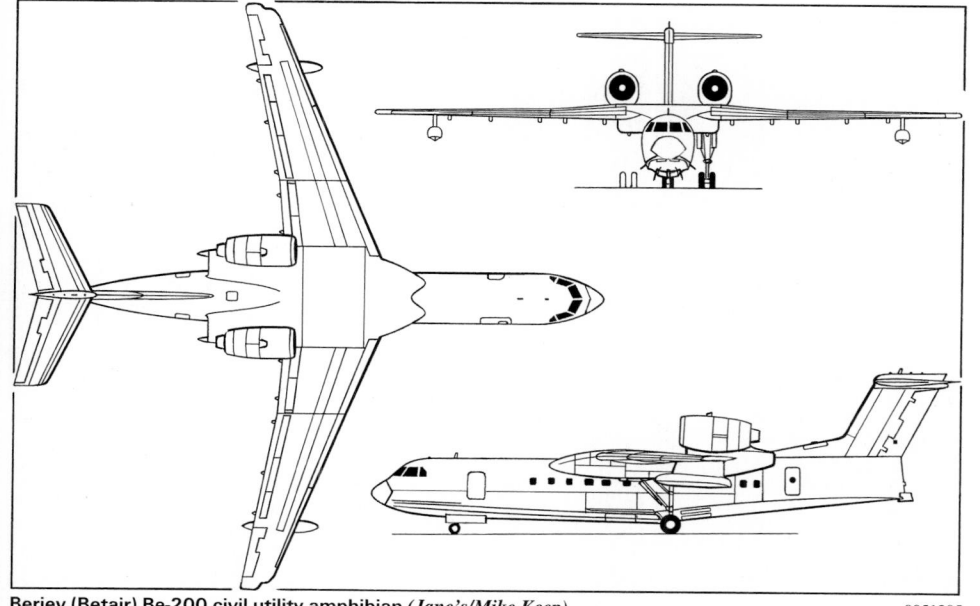

Beriev (Betair) Be-200 civil utility amphibian (*Jane's/Mike Keep*)   0051205

First production Beriev Be-200ChS displaying in public at Gelendzhik shortly after its maiden flight *(Yefim Gordon)*   *NEW*/0536623

Beriev Be-200 multirole amphibian *(Yefim Gordon)*   *NEW*/0132841

Wing leading-edge sweep 23° 13′; supercritical wing sections, thickness/chord ratio 16 per cent to 11.5 per cent.

FLYING CONTROLS: Conventional. Entire span of each wing trailing-edge occupied by aileron and two-section area-increasing single-slotted flaps; full-span leading-edge slats in three sections each side; five spoiler/lift dumper sections forward of flaps each side.

STRUCTURE: Hull made primarily of high-strength aluminium/lithium alloys; interior of composites; water tanks of ferric alloys of aluminium in firefighting version.

LANDING GEAR: Hydraulically retractable tricycle type. Twin-wheel main units, tyre size 950×300, pressure 9.80 to 10.30 bar (142 to 150 lb/sq in); twin nosewheels, tyre size 620×180, pressure 7.35 to 7.85 bar (106 to 114 lb/sq in). Mainwheels and nosewheels retract rearwards. Water rudder. Ground turning radius 17.4 m (57 ft 1 in). Nosewheel steering angle ±45°.

POWER PLANT: Two ZMKB Progress D-436TP turbofans, each 73.6 kN (16,550 lb st). Rolls-Royce BR715s and others under consideration as alternative engines. Fuel system by Pall (USA). Total oil capacity 22 litres (5.8 US gallons; 4.85 Imp gallons).

ACCOMMODATION: Two flight crew; up to 72 tourist class passengers at 75 cm (29.5 in) seat pitch and two attendants; or 10 to 32 first class and business class passengers at up to 102 cm (40 in) seat pitch, with provision for galley, lavatory and baggage stowage. Up to nine freight containers or, typically, six containers and 19 economy class seats. Cargo door, with integral passenger door, starboard side, forward, opens upwards; passenger door port, forward; emergency exits; forward and rear freight/baggage holds. Interior design by AIM Aviation (UK).

SYSTEMS: Barco Display Systems FMS system. All accommodation pressurised. Three hydraulic systems at 207 bar (3,000 lb/sq in); 150 litres (39.6 US gallons; 33.0 Imp gallons) of MGJ-5U fluid; flow rate 70 litres (18.5 US gallons; 15.4 Imp gallons)/min. Capacity of pneumatic system bottles 29 litres (1.02 cu ft). Three-phase 115/220 V 400 Hz AC electrical system; single-phase 115 V 400 Hz AC system; 27 V DC system; supplied by two engine-driven 60 kVA AC generators and three static inverters; three batteries. Gaseous oxygen bottle, pressure 147 bar (2,135 lb/sq in). Provision for de-icing tail unit, slats, engine air intakes and windscreen. TA-12 APU, operable up to 7,000 m (23,000 ft) in starboard wingroot. Propeller-driven emergency generator at base of fin.

AVIONICS: *Radar:* MN-85 weather radar in nose.

Flight: ARIA-200M digital flight and navigation system by American-Russian Integrated Avionics, a joint venture of AlliedSignal (now Honeywell) and Moscow Research Institute of Aircraft Equipment. INS standard.

Instrumentation: Honeywell EFIS displays, with six 152 × 203 mm (6 × 8 in) LCDs.

DIMENSIONS, EXTERNAL:

Wing span over winglets	32.78 m (107 ft 6½ in)
Wing chord: at root	5.58 m (18 ft 3½ in)
at tip	1.72 m (5 ft 7¾ in)
Wing aspect ratio	9.1
Length overall	31.43 m (103 ft 1½ in)
Fuselage: Length	29.18 m (95 ft 9 in)
Max diameter	2.86 m (9 ft 4½ in)
Height overall	8.90 m (29 ft 2½ in)
Tailplane span	10.115 m (33 ft 2¼ in)
Wheel track (c/l shock-absorbers)	4.30 m (14 ft 1¼ in)
Wheelbase	11.145 m (36 ft 6¾ in)
Width over stabilising floats	25.60 m (84 ft 0 in)
Passenger doors (each): Height	1.70 m (5 ft 7 in)
Width	0.90 m (2 ft 11½ in)
Cargo door: Height	1.76 m (5 ft 9¼ in)
Width	2.05 m (6 ft 8¾ in)
Emergency exits: Height	1.70 m (5 ft 7 in)
Width	0.90 m (2 ft 11½ in)

DIMENSIONS, INTERNAL:

Cabin, excl flight deck:	
Length: passenger	17.00 m (55 ft 9 in)
cargo	18.70 m (61 ft 4¼ in)
Max width: passenger	2.40 m (7 ft 10½ in)
cargo	2.50 m (8 ft 2½ in)
Max height: passenger	1.80 m (5 ft 10¾ in)
cargo	1.89 m (6 ft 2½ in)
Floor area: passenger	39.0 m² (420 sq ft)
cargo	41.0 m² (441 sq ft)

Volume: passenger:	
forward baggage hold	8.8 m³ (310 cu ft)
rear baggage hold	4.5 m³ (159 cu ft)
main cabin and freight/baggage holds, total, cargo	
configuration	84.0 m³ (2,966 cu ft)
cargo	80.8 m³ (2,853 cu ft)

AREAS:

Wings, gross	117.44 m² (1,264.2 sq ft)
Ailerons (total)	3.56 m² (38.32 sq ft)
Flaps (total)	20.43 m² (219.91 sq ft)
Slats (total)	12.61 m² (135.74 sq ft)
Spoilers (total)	4.59 m² (49.41 sq ft)
Fin	12.60 m² (135.63 sq ft)
Rudder	4.60 m² (49.52 sq ft)
Tailplane	17.96 m² (193.33 sq ft)
Elevators (total)	6.96 m² (74.92 sq ft)

WEIGHTS AND LOADINGS:

Max payload	7,500 kg (16,534 lb)
Max fuel weight	12,260 kg (27,025 lb)
Max T-O and ramp weight:	
cargo/passenger	42,000 kg (92,594 lb)
firefighting	37,200 kg (82,011 lb)
Max airborne weight (after water scooping)	
	43,000 kg (94,800 lb)
Max landing weight, land or water	35,000 kg (77,160 lb)
Max wing loading:	
cargo/passenger	357.6 kg/m² (73.25 lb/sq ft)
firefighting	316.8 kg/m² (64.88 lb/sq ft)
max airborne weight	366.1 kg/m² (74.99 lb/sq ft)
Max power loading:	
cargo/passenger	285 kg/kN (2.80 lb/lb st)
firefighting	253 kg/kN (2.48 lb/lb st)
max airborne weight	292 kg/kN (2.86 lb/lb st)

PERFORMANCE (A: passenger version; B: business; C: cargo/passenger; D: cargo; E: firefighting):

Max Mach No. in level flight: A, B, C, D	0.69
Never-exceed speed (V$_{NE}$):	
A, B, C, D	329 kt (610 km/h; 379 mph) EAS
Max level speed at 7,000 m (22,960 ft):	
A, B, C, D	388 kt (720 km/h; 447 mph)
Max cruising speed:	
A, C, D at 8,000 m (26,240 ft)	
	378 kt (700 km/h; 435 mph)

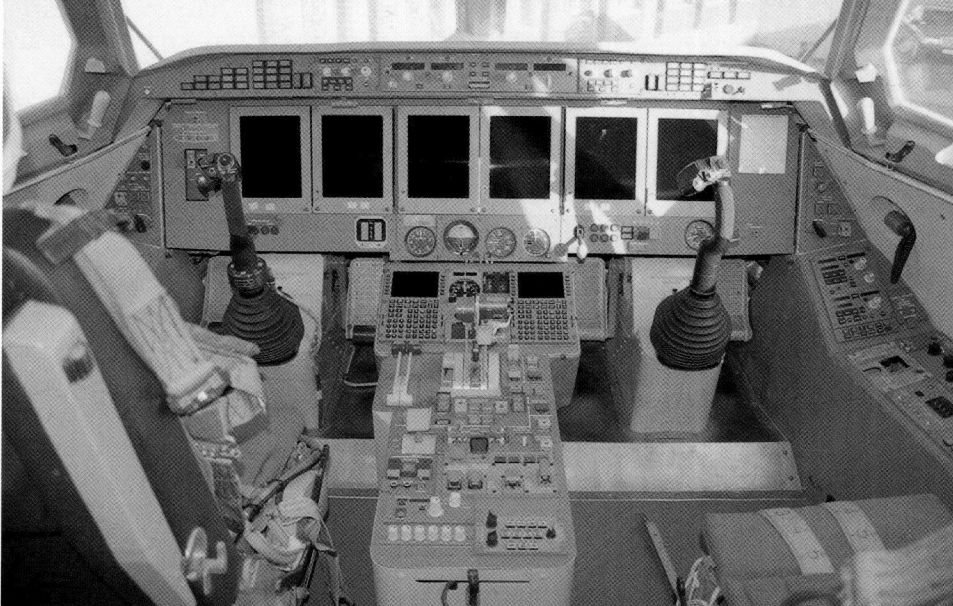

Flight deck of first production Beriev Be-200ChS *(Yefim Gordon)*   *NEW*/0536621

B at 10,000 m (32,800 ft)	378 kt (700 km/h; 435 mph)
E at 3,000 m (9,840 ft)	383 kt (710 km/h; 441 mph)

Econ cruising speed:
A, C, D at 8,000 m (26,240 ft)
297 kt (550 km/h; 342 mph)
B at 10,000 m (32,800 ft) 297 kt (550 km/h; 342 mph)
E at 3,000 m (9,840 ft) 324 kt (600 km/h; 373 mph)
Stalling speed: flaps up 116 kt (215 km/h; 134 mph)
flaps down 84 kt (155 km/h; 97 mph)
Max rate of climb at S/L 840 m (2,755 ft)/min
Rate of climb at S/L, OEI 168 m (550 ft)/min
Service ceiling: A, B, C, D 11,000 m (36,080 ft)
E 8,000 m (26,240 ft)
Service ceiling, OEI 5,500 m (18,040 ft)
T-O to 11 m (35 ft): on land:
A, B, C, D 950 m (3,120 ft)
E 700 m (2,300 ft)
on water: A, B, C, D 1,400 m (4,595 ft)
E 1,000 m (3,280 ft)

Landing from 15 m (50 ft): on land:
A, B, C, D 800 m (2,625 ft)
E 950 m (3,120 ft)
on water: A, B, C, D 950 m (3,120 ft)
E 1,300 m (4,265 ft)
Balanced runway requirement: all 1,800 m (5,905 ft)
Water scooping distance to 15 m (50 ft)
1,450 m (4,760 ft)
Range:
A with 72 passengers
998 n miles (1,850 km; 1,149 miles)
C, D with 6,500 kg (14,330 lb) freight, 1 h reserves
998 n miles (1,850 km; 1,149 miles)
E, ferry, with 1 h reserves
1,943 n miles (3,600 km; 2,236 miles)
all, with max fuel
2,078 n miles (3,850 km; 2,392 miles)
OPERATIONAL NOISE LEVELS:
T-O 96 EPNdB

Climb 90 EPNdB
Landing 98 EPNdB
UPDATED

BERIEV (BETAIR) Be-210
TYPE: Amphibious airliner.
PROGRAMME: Announced in 1998; development of Be-200 multirole amphibian for airline use in regions of minimal airport infrastructure. Seating for two pilots, two attendants and 72 passengers at 75 cm (29½ in) seat pitch. Other details may be assumed to be similar to Be-200, but airframe strengthened for extra fuel in wing centre-section (42,000 kg; 92,594 lb MTOW) and ferry range increased to 2,483 n miles (4,600 km; 2,858 miles), or 998 n miles (1,850 km; 1,149 miles) with full passenger load. In 2001, foreign investors, including Scorpion of Greece, showing interest in developing BR715-engined airliner version of Be-200; this may carry Be-210 designation.
UPDATED

CHERNOV
OPYTNYI KONSTRUKTORSKOYE BYURO CHERNOV B & M OOO (B & M Chernov Experimental Design Bureau JSC)
Tel: (+8 8462) 37 24 25
e-mail: chernovb@samaramail.ru
Web: http://www.sama.ru/~chernovb

The designs of Boris V Chernov have previously appeared under the heading of SKB-1 at Samara (which see). However, in August 1999, the Che-25 amphibian was shown at MAKS '99, credited to Chernov's own design office. The earlier Che-22 is now produced by Gidroplan (which see) and the Che-15 was last described in the 2001-02 *Jane's*. Other designs have been the Che-10 (1986), Che-20 (1989), (unbuilt) Che-23 and Che-30 (1991). Work is currently in hand to design Che-35 twin-engined flying-boat.
UPDATED

CHERNOV Che-25
TYPE: Four-seat amphibian.
PROGRAMME: Development of Che-22 Corvette described under the Gidroplan entry in this edition. Prototype (FLARF-17039) built by SKB-1 (see SGAU entry in this section) first shown at Gelendzhik Hydro-aviation Show, Russia, September 1996. Che-25M shown at MAKS '99, Moscow, had redesigned empennage with raised tailplane and deletion of rudder's horn balance.
COSTS: US$44,000 (1997).
DESIGN FEATURES: Generally as Che-22. Operable on water in 0.8 m (30 in) swell.
FLYING CONTROLS: Manual, with full-span slotted flaperons. No tabs.
STRUCTURE: Single-spar wing of riveted duralumin; fuselage, tail unit and wingtip floats of vacuum-moulded GFRP sandwich. Wings braced by struts and wires; strut-braced tailplane. Airframe life 3,000 hours.
LANDING GEAR: Optional amphibian configuration; mainwheel legs, attached to fuselage sides, turn 90° forward for retraction; mainwheel tyre size 300×125; tailskid; cable brakes. Tailwheel/water rudder immediately to rear of second hull step.
POWER PLANT: Two 47.8 kW (64.1 hp) Rotax 582 UL-2V two-cylinder piston engines driving two-blade fixed-pitch propellers. Fuel capacity 130 litres (34.3 US gallons; 28.6 Imp gallons). Version planned with Rotax 912 ULS of 73.5 kW (98.6 hp).
ACCOMMODATION: Four persons; dual controls. Gull-wing door each side of cockpit.

DIMENSIONS, EXTERNAL:
Wing span 11.80 m (38 ft 8½ in)
Wing chord, constant 1.40 m (4 ft 7¼ in)
Wing aspect ratio 8.4
Length overall 7.15 m (23 ft 5½ in)
Height overall (on wheels, fuselage horizontal)
2.76 m (9 ft 0¾ in)
Tailplane span 2.80 m (9 ft 2¼ in)
Wheel track (optional) 1.42 m (4 ft 8 in)
Propeller diameter 1.84 m (6 ft 0½ in)
Distance between propeller centres 1.84 m (6 ft 0½ in)
AREAS:
Wings, gross 16.50 m² (177.6 sq ft)

WEIGHTS AND LOADINGS:
Weight empty 510 kg (1,124 lb)
Max payload 300 kg (661 lb)
Max T-O weight 950 kg (2,094 lb)
Max wing loading 57.6 kg/m² (11.79 lb/sq ft)
Max power loading 9.94 kg/kW (16.33 lb/hp)
PERFORMANCE:
Max level speed 97 kt (180 km/h; 112 mph)
Max cruising speed 70 kt (130 km/h; 81 mph)

Unstick speed 38 kt (70 km/h; 44 mph)
Max rate of climb at S/L 300 m (984 ft)/min
Service ceiling 4,000 m (13,120 ft)
T-O run: on land 30 m (100 ft)
on water 50 m (165 ft)
Range 464 n miles (860 km; 534 miles)
g limits +4/−2
UPDATED

Che-30 prototype, a previous creation of Boris Chernov   *NEW*/0137984

Che-25 four-seat amphibian   *NEW*/0137980

The projected Chernov Che-35   *NEW*/0137975

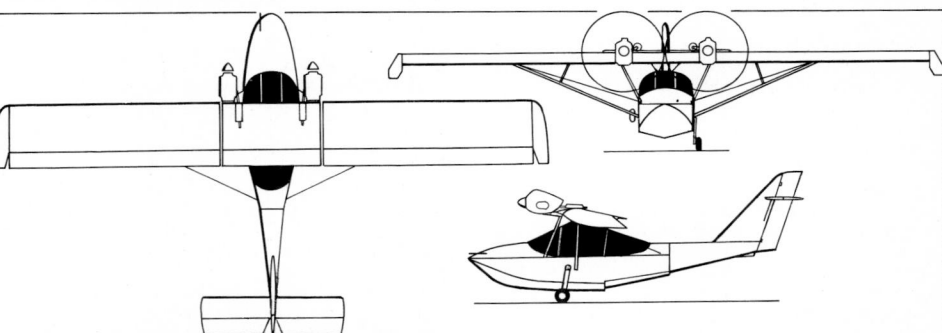

General arrangement of the Che-25 (*Jane's/Paul Jackson*)   0084428

DUBNA (DMZ)

PROIZVODSTVENNO-TEKHNICHESKY KOMPLEKS DUBNENSKOGO MASHINOSTROITELNOGO ZAVOD AO (Dubna Machine-Building Plant Production Technical Complex Co Ltd)

ulitsa Zhukovskogo 2, 141980 Dubna, Moskovskaya oblast
Tel: (+7 09621) 514 14, 516 13 and 513 79
Fax: (+7 09621) 235 24
Telex: 206132 TITAN
GENERAL DIRECTOR: Aleksey Saushkin

Founded in 1939, the Dubna Plant is an enterprise of the General Department of Aircraft Industry of the State Committee on Defence Industries of the Russian Federation. In the 1960s, it launched airframe production of the MiG-25 fighter and became a major producer of remotely piloted vehicles for Eastern bloc and other countries. As part of its restructuring under the *Konversiya* policy, since 1993, it has produced components for the Sukhoi Su-29 two-seat aerobatic aircraft and developed the Dubna-1/2 series of light aircraft. Two Su-29 fuselages were built in 2000.

The single-seat Dubna-1 Shmel was last described in the 2001-02 edition. Dubna has an agreement to build light aircraft designed by MKB Raduga, whose precision-guided weapons are also manufactured by DMZ. The company had some 1,000 employees in early 2002 and is 60 per cent owned by the State.

UPDATED

DUBNA-2 OSA

English name: Wasp

TYPE: Two-seat lightplane.
PROGRAMME: Designed by Taifun Experimental Design Bureau as DMZ-2; prototype (FLARF-01325) displayed at Moscow Air Show '95 and '97 with 59.6 kW (79.9 hp) Rotax 912 engine; promoted in China, 1996. By 1999, when definitive prototype FLARF-01268 shown at Moscow, detail changes included increased wheel track; repositioning of landing gear leg to below fuselage and of wing bracing strut to further forward; and Hirth engine, with additional fuel.

Production version of Dubna-2 Osa two-seat multipurpose light aircraft (*Jane's/Paul Jackson*)   0092528

CUSTOMERS: Total 20 delivered by late 2001. Production of 12 planned for 2002, including four to United Arab Emirates and one each to Bulgaria and Greece; others to Saratov and Orenburg regional governments of Russia in agricultural spraying configuration.
COSTS: US$32,000 (2002).
DESIGN FEATURES: As Dubna-1, suitable for passenger, freight and mail transport, cropspraying, forest surveillance, oil/gas pipeline and power line patrol, coastal fishery survey, aerial photography, *ab initio* pilot training and special missions.
FLYING CONTROLS: Conventional and manual. Flight-adjustable tab on elevator; ground-adjustable tabs on starboard aileron and rudder.
STRUCTURE: As Dubna-1.
LANDING GEAR: Non-retractable tricycle type; single wheel on each unit; arched metal leafspring mainwheel legs; trailing-link nosewheel leg. Tyres size 400×150 mm on mainwheels, 310×135 mm on nosewheel. Floats and skis optional.
POWER PLANT: One 89.5 kW (120 hp) Hirth F30/4C flat-four driving two-blade pusher propeller. Fuel tank behind cabin, capacity 80 litres (21.1 US gallons; 17.6 Imp gallons).

ACCOMMODATION: Two seats side by side, with dual controls, in fully enclosed, extensively glazed cabin; forward-hinged window/door on each side; space for baggage/freight in compartment behind cabin.
DIMENSIONS, EXTERNAL:

Wing span	9.92 m (32 ft 6½ in)
Wing aspect ratio	9.0
Length overall	5.97 m (19 ft 7in)
Width over doors	1.25 m (4 ft 1¼ in)
Height overall	2.91 m (9 ft 6½ in)
Tailplane span	3.30 m (10 ft 10 in)
Wheel track	2.40 m (7 ft 10½ in)
Wheelbase	1.90 m (6 ft 2½ in)

AREAS:

Wings, gross	10.92 m² (117.5 sq ft)

WEIGHTS AND LOADINGS:

Weight empty	296 kg (653 lb)
Baggage capacity	30 kg (66 lb)
Max T-O weight	576 kg (1,270 lb)
Max wing loading	52.7 kg/m² (10.81 lb/sq ft)
Max power loading	9.60 kg/kW (15.88 lb/hp)

PERFORMANCE:

Max level speed	89 kt (165 km/h; 102 mph)
Nominal cruising speed	62 kt (115 km/h; 71 mph)
Landing speed	41 kt (75 km/h; 47 mph)
Unstick speed	46 kt (85 km/h; 53 mph)
Max rate of climb at S/L	240 m (790 ft)/min
Service ceiling	4,000 m (13,120 ft)
T-O run	120 m (395 ft)
Max range	270 n miles (500 km; 310 miles)
g limits	+4.0/-2.5

UPDATED

DUBNA-4

TYPE: Four-seat utility twin.
PROGRAMME: Designed by MKB Raduga; based on Dubna-2; completion of prototype planned for 2002. Optimised for use by Ministry of Emergency Situations and Forestry Service.
COSTS: Estimated US$55,000 to 58,000 (2002).

NEW ENTRY

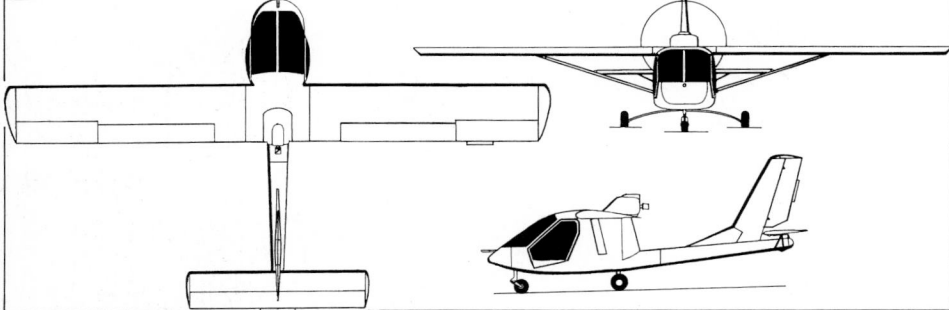

Dubna-2 Osa (Rotax 912 engine) (*Jane's/Paul Jackson*)   0092552

ELITAR

VVV-AVIA

a/ya 50, Lyubertsy, 140000 Moskovskaya oblast
Tel/Fax: (+7 095) 552 24 47
Fax: (+7 095) 552 79 69
e-mail: info@vvv-avia.ru
Web: http://www.vvv-avia.ru
DESIGN BUREAU:
Ehlitar – 101 OOO Korp 2, ulitsa B Cheremushkinskaya, 117448 Moskva

At the August 1999 MAKS air show, Ehlitar (which adopts the Western form, Elitar) displayed the newly flown IE-101 lightplane and released initial details of the Senator.

In addition to these light aircraft, described below, Elitar is also developing the E-301 and E-401 high-wing, twin-engined light transports, seating four to five and nine persons, respectively. A four-seat business 'triplane', the Premier, has been under design since 2000.

Members of the Elitar design and production team work for MiG, Molniya and Sukhoi design bureaux and have been responsible for the M-5 Oktyabr and M-9 Marafon ultralights.

UPDATED

ELITAR IE-101 ELITAR

TYPE: Side-by-side ultralight.
PROGRAMME: Prototype built at Myachkovo, Moscow; first flown 14 August 1999; public debut at MAKS '99, 17 August 1999; registered subsequently as FLARF-01699. None further built by mid-2001.
DESIGN FEATURES: Simple, robust lightplane, meeting FLARF (Russian amateur aviation) airworthiness rules. High,

unswept, constant-chord wing with single bracing strut each side. Sweptback fin augmented by underfin. Wings fold for storage.
FLYING CONTROLS: Conventional and manual. Mass-balanced ailerons with ground-adjustable tab to starboard; horn-balanced rudder with ground-adjustable tab; two-piece elevator with flight-adjustable tab to port. Fowler flaps.
STRUCTURE: Fuselage and wing of composites. Metal bracing struts and cantilever mainwheel struts.
LANDING GEAR: Tricycle type; fixed; speed fairings on each wheel. Hydraulic brakes. Mainwheels 400×150. Optional twin-float gear.

POWER PLANT: One 59.6 kW (79.9 hp) Rotax 912 UL flat-four driving a two-blade ground-adjustable pitch propeller. Fuel capacity 320 litres (84.5 US gallons; 70.4 Imp gallons). Optional engines include 85.8 kW (115 hp) Rotax 914 and 89 kW (120 hp) Hirth F30.
ACCOMMODATION: Two persons, side by side, with dual controls. Upward-hinged door each side.
EQUIPMENT: RSV K-500 ballistic parachute.
DIMENSIONS, EXTERNAL:

Wing span	8.00 m (26 ft 3 in)
Length overall	6.28 m (20 ft 7¼ in)
Width, wings folded	2.20 m (7 ft 2½ in)

Model of Elitar-401, shown at Moscow in August 2001 (*Jane's/Paul Jackson*)   0121696

Prototype Elitar IE-101 on display at Moscow, August 2001 *(Jane's/Paul Jackson)*   0121695

Model of Elitar IE-201   0121735

Height overall	2.35 m (7 ft 8½ in)
Wheel track	2.00 m (6 ft 6¾ in)
Wheelbase	1.40 m (4 ft 7 in)

DIMENSIONS, INTERNAL:
Cabin max width	1.12 m (3 ft 8 in)

AREAS:
Wings, gross	8.80 m² (94.7 sq ft)

WEIGHTS AND LOADINGS:
Weight empty	320 kg (705 lb)
Max T-O weight	700 kg (1,543 lb)

PERFORMANCE:
Never-exceed speed (VNE)	145 kt (270 km/h; 167 mph)
Max level speed	119 kt (220 km/h; 137 mph)
Max cruising speed	100 kt (185 km/h; 115 mph)
Landing speed	47 kt (86 km/h; 54 mph)
Max rate of climb at S/L	258 m (846 ft)/min
T-O run	154 m (505 ft)
Range with max fuel	1,511 n miles (2,800 km; 1,739 miles)
g limits	+4.9/−1.9

VERIFIED

ELITAR IE-201 SENATOR

TYPE: Two-seat lightplane.
PROGRAMME: Announced, and preliminary details released, August 1999. No further details published by early 2002.

DESIGN FEATURES: Low-wing cabin monoplane of composites construction, optimised for high cruising speed and long range; constant-chord, cantilever mainplanes; sweptback fin with underfin.
LANDING GEAR: Tricycle type; fixed.
POWER PLANT: One 84.6 kW (113.4 hp) Rotax 914 ULS flat-four driving a two-blade propeller. Fuel capacity 200 litres (52.8 US gallons; 44.0 Imp gallons).

DIMENSIONS, EXTERNAL:
Wing span	7.00 m (22 ft 11½ in)
Length overall	5.56 m (18 ft 3 in)

AREAS:
Wings, gross	7.70 m² (82.9 sq ft)

WEIGHTS AND LOADINGS:
Weight empty	315 kg (694 lb)
Baggage capacity	40 kg (88 lb)
Max T-O weight	650 kg (1,433 lb)

PERFORMANCE:
Max level speed	151 kt (280 km/h; 174 mph)
Max cruising speed	135 kt (250 km/h; 155 mph)
Stalling speed	46 kt (84 km/h; 53 mph)
Max rate of climb at S/L	270 m (886 ft)/min
T-O run	267 m (876 ft)
Landing run	145 m (476 ft)
Range with 60% fuel	1,080 n miles (2,000 km; 1,242 miles)

UPDATED

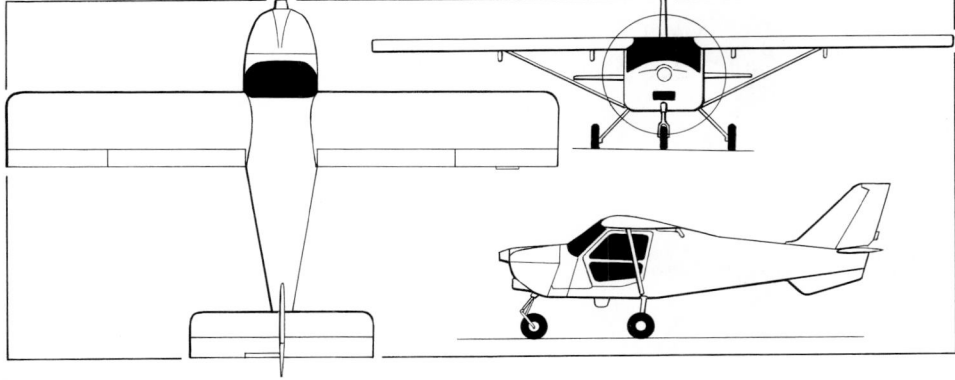

Elitar IE-101 two-seat lightplane *(Jane's/James Goulding)*   0079008

Prototype Elitar IE-201 Senator under construction   0121734

GIDROPLAN

GIDROPLAN OOO (Hydroplane Ltd)
Moskovskoye Shosse, Litera A, A', Samara 443072
Tel: (+7 8462) 53 90 19
Fax: (+7 8462) 53 39 38
e-mail: hyd@mail.radiant.ru
Web: http://www.hydroplane.ru
GENERAL DIRECTOR: Leonid Logkin
CHIEF OF DESIGN BUREAU: Rosteslav Pivovarov
GENERAL DESIGNER: Evgeny Ungerov
FINANCIAL DIRECTOR: Pavel Piotnitsa

Gidroplan Ltd was formed in 1995 and has marketed amphibian lightplanes designed by the Chernov bureau.

This venture was placed on a firmer footing in late 2000 when Gidroplan invested the equivalent of US$500,000 in acquiring 3,500 m² (37,675 sq ft) of working area from the Samara Instrument Bearing Plant and installing production equipment. Manufacturing activities include the Tsikada (Cicada) twin-engined ultralight agricultural aircraft, some six of which were manufactured before production ceased, and the Che-22 Corvette amphibian.

In recent years, designers associated with Gidroplan have produced flying prototypes of three lightplanes not promoted outside Russia. These are the **A-17** single-seat ultralight (260 kg; 573 lb MTOW, 22.4 kW; 30 hp engine, believed from SGAU); Chernov **Che-40** flying-boat (two Rotax 582 engines, briefly described in the 1999-2000 edition and now confirmed to have been built); and **Ul'tralayt** twin-tailboom ultralight (600 kg; 1,322 lb MTOW, Rotax 912 S) built in 1998 by E Yungerovym and S Seleznevym.

UPDATED

GIDROPLAN Che-22 KORVET
English name: Corvette
US marketing name: Pelican
TYPE: Three-seat amphibian/kitbuilt.
PROGRAMME: Developed on basis of four earlier amateur-built, Vikhr-engined Che-20s, winners of many light aircraft competitions; design and construction of prototype started 1988, first flight 1989; construction of production aircraft started 1992, first flight 1993; more than 5,000 hours of test and training flights, and more than 120 amateur pilots trained on prototype and production aircraft in first five years. Promotion also undertaken by Refly (which see in this section); launched as Refly Pelican at EAA AirVenture, Oshkosh, 2000.

Upgraded version under development and certification during 2001 features cantilever wing, strengthened tail unit with increased rudder area, new control system, redesigned engine support and modified instrument panel; flight testing was completed at the end of 2001 and JAR/FAR Pt 23 certification awarded in early 2002.
CURRENT VERSIONS: **Corvette 503:** With two Rotax 503 UL two-cylinder two-stroke engines.
　　Corvette 582: With two Rotax 582 UL two-cylinder two-stroke engines.
　　Corvette 912: With one Rotax 912 UL four-cylinder four-stroke engine.
　　Refly Pelican: Generally as for Corvette 582.
CUSTOMERS: First preproduction delivery to a forest control organisation 1990. One (Che-22R-2 c/n 006, RP-X1548 of OVOS WLL Co) sold in Philippines. Others to China, Spain and Vietnam. Total 60 built by February 2002, including 10 in 2000. First Pelican in US (N27NS, ex-RA-02777) registered in March 2001.
COSTS: Standard aircraft US$35,000 to US$55,000 (2002).

DESIGN FEATURES: Designed to JAR-VLA, FAR Pt 23 and AP-23 requirements, for maximum simplicity of design and construction; primarily for forestry applications, but suitable for patrol, inspection and ecological monitoring of hunting and fishing locations, border patrol, training and business operations. Quoted build time 1,500 hours, or 50 hours for fast build kit.

Braced parasol monoplane, with light central cabane and single main bracing strut and bracing wire each side; downturned wingtips serve as stabilising floats; flying boat hull with two steps; unswept strut-braced horizontal tail surfaces mounted on fin of sweptback vertical surfaces; engine(s) mounted on leading-edge of wing centre-section. US demonstrator has fin of reduced height. Unswept constant chord wing; CAHI (TsAGI) P-IIIA section, thickness/chord ratio 15 per cent; incidence 4° 30'; dihedral 1° 30'. Airframe life 3,000 flying hours.
FLYING CONTROLS: Conventional and manual, with pushrod actuation. Full-span slotted flaperons, elevators and horn-balanced rudder; fixed tab on port elevator.
STRUCTURE: Single-spar wing of riveted duralumin; fuselage, tail unit and integral wingtip floats of vacuum-moulded GFRP sandwich; flaperons fabric-covered.
LANDING GEAR: Optional tailwheel type; mainwheels, tyre size 350×120, carried on cantilever GFRP spring. GFRP skis optional.
POWER PLANT: Two 37 kW (49.6 hp) Rotax 503 UL-2V or two 47.8 kW (64.1 hp) Rotax 582 UL or 59.6 kW (79.9 hp) Rotax 912 UL. Two-blade wooden or three-blade GFRP fixed-pitch propeller(s) of Russian manufacture. Two fuel tanks in wing centre-section, 100 litres (26.4 US gallons; 22.0 Imp gallons); gravity fuelling.
ACCOMMODATION: Pilot and two passengers under large glazed blister canopy, giving exceptional field of view.

Detachable windscreen panels. Gull-wing canopy/door each side.
SYSTEMS: 12/24 V electrical system.
AVIONICS: *Comms*: VHF radios.
    Flight: Garmin GPS.
DIMENSIONS, EXTERNAL:

Wing span	11.70 m (38 ft 4¾ in)
Wing chord, constant	1.40 m (4 ft 7¼ in)
Wing aspect ratio	8.3
Length overall	6.70 m (21 ft 11¾ in)
Height overall	2.20 m (7 ft 2¾ in)
Tailplane span	2.50 m (8 ft 2½ in)
Wheel track	1.60 m (5 ft 3 in)
Wheelbase	4.55 m (14 ft 11 in)
Propeller diameter	1.50 m (4 ft 11 in)
Canopy doors (each): Height	0.70 m (2 ft 3½ in)
Width	0.80 m (2 ft 7½ in)
Height to sill	0.63 m (2 ft 0¾ in)

DIMENSIONS, INTERNAL:

Cabin: Length	1.50 m (4 ft 11 in)
Max width	1.10 m (3 ft 7¼ in)
Max height	1.20 m (3 ft 11¼ in)

AREAS:

Wings, gross	16.40 m² (176.5 sq ft)
Flaperons (total)	3.40 m² (36.60 sq ft)
Fin	0.55 m² (5.92 sq ft)
Rudder	0.35 m² (3.77 sq ft)
Tailplane	1.68 m² (18.08 sq ft)
Elevators (total)	0.80 m² (8.61 sq ft)

WEIGHTS AND LOADINGS (A: Corvette 503, B: Corvette 582, C: Corvette 912):

Weight empty: A, B, C	360 to 480 kg (794 to 1,058 lb)
Max fuel: A, B, C	58 kg (128 lb)

Gidroplan Korvet displayed at Gelendzhik in August 2002 *(Yefim Gordon)*   *NEW*/0536624

Max T-O weight: A, B, C	675 kg (1,488 lb)	T-O run: on land: A	90 m (295 ft)
Max wing loading: A	21.95 kg/m² (4.50 lb/sq ft)	B	70 m (230 ft)
C	29.27 kg/m² (5.99 lb/sq ft)	C	100 m (328 ft)
Max power loading: A	4.86 kg/kW (7.99 lb/hp)	on water: A	110 m (361 ft)
C	8.05 kg/kW (13.23 lb/hp)	B	90 m (295 ft)
PERFORMANCE:		C	120 m (394 ft)
Max level speed: A	81 kt (150 km/h; 93 mph)	Range with standard fuel:	
B, C	86 kt (160 km/h; 99 mph)	A	232 n miles (430 km; 267 miles)
Cruising speed: A	59 kt (110 km/h; 68 mph)	B	243 n miles (450 km; 279 miles)
B, C	65 kt (120 km/h; 75 mph)	C	324 n miles (600 km; 372 miles)
Max rate of climb at S/L: A	300 m (985 ft)/min	*g* limits	+3.8/−1.5
B	420 m (1,378 ft)/min	Permissible wave height: A, B, C	0.5 m (1 ft 7¾ in)
C	240 m (787 ft)/min		*UPDATED*
Service ceiling: A, B, C	3,000 m (9,840 ft)		

IAPO

IRKUTSKOYE AVIATSIONNOYE PROIZVODSTVENNOYE OBEDINENIE OAO (Irkutsk Aviation Industrial Association JSC)

ulitsa Novatorov 3, 664020 Irkutsk
Tel: (+7 3952) 32 29 09 and 32 29 42
Fax: (+7 3952) 32 29 41 and 32 29 45
e-mail: market@irkut.ru
Web: http://www.irkut.ru
PRESIDENT: Alexei I Fedorov
GENERAL DIRECTOR: Vladimir V Kovalkov
DIRECTOR, MARKETING AND SALES: Yuri Belozerov
DIRECTOR, PUBLIC RELATIONS: Vitaly P Zelenkov

Founded on 28 March 1932 and commissioned on 24 August 1934, IAPO has built some 6,500 aircraft of 16 types from Antonov, Ilyushin, MiG, Petlyakov, Sukhoi, Tupolev and Yakovlev bureaux and supplied them to 21 countries. In recent years, it has manufactured MiG-23UB trainers (1970-85); 165 kits for Indian-assembled MiG-27MLs; Su-27UB trainers (from 1986); and is currently responsible for producing the Su-30 fighter (since 1991) and Beriev Be-200 amphibian. Series manufacture of the Yak-112 lightplane has yet to begin, although IAPO has been allocated prospective manufacture of the Ilyushin/HAL Il-214 twin-jet transport (which see in International section). IAPO also undertakes Su-30 upgrades. Russian government holding is 14.7 per cent; US investor holds 15 per cent; is member of AVPK Sukhoi. Sales in 2000 totalled US$170 million.

In 2001, IAPO branched out into design and manufacture of its own products on completion of the prototype A-002 autogyro. Expansion of civil programmes is strategic objective. Under government plans announced in May 2001, IAPO is to join the aerospace group also including Sukhoi, Ilyushin, Mil, Yakovlev and their associated factories.
UPDATED

IAPO A-002

TYPE: Three-seat autogyro.
PROGRAMME: Project announced October 2000; almost complete prototype shown at Moscow, August 2001. Maiden flight had not been announced by early 2002, however.
DESIGN FEATURES: Streamlined, cabin autogyro which can be stored and serviced in a motorcar garage. Choice of engines, including those using motorcar fuel. Classic configuration, with pusher propeller; single, sweptback fin with large rudder; broad track landing gear. Engine-driven pre-rotation of rotor.
FLYING CONTROLS: Manual. All-moving tailplane mounted at one-third of fin height; horn- and mass-balanced rudder.
STRUCTURE: Generally of metal. Rudder and elevator of composites.
LANDING GEAR: Tricycle type; fixed. Cantilever-spring mainwheels; rubber-in-compression shock-absorber on nosewheel. Mainwheel tyres 400×150; nose 300×125. Hydraulic disc brakes on mainwheels. Safety tail skid on prototype; auxiliary tailwheel on production version.
POWER PLANT: Prototype has one 157 kW (210 hp) Teledyne Continental IO-320 flat-six driving a fixed-pitch, two-blade propeller. Alternative engines of 134 to 157 kW (180

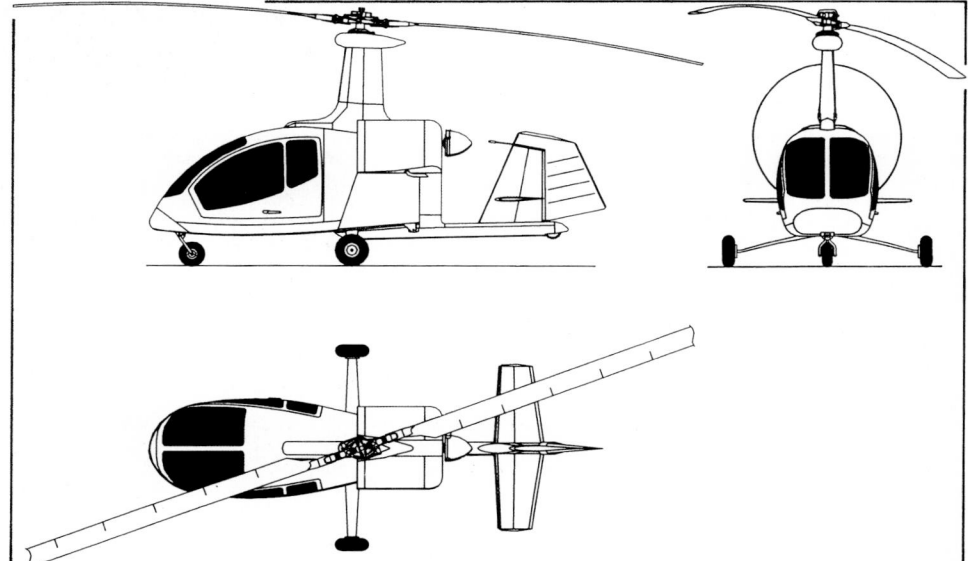

General arrangement of the IAPO A-002 autogyro   0121087

IAPO A-002 autogyro on display at Moscow in August 2001, prior to its maiden flight   *NEW*/0120308

to 210 hp) and three-blade propeller can be installed.

ACCOMMODATION: Two front seats, side by side, with dual controls; single rear seat. Upward-hinged door each side.

DIMENSIONS, EXTERNAL:

Rotor diameter	9.80 m (32 ft 1¾ in)
Length of fuselage	4.98 m (16 ft 4 in)
Height, rotor removed	3.17 m (10 ft 4¾ in)
Wheel track	2.40 m (7 ft 10½ in)
Wheelbase	2.015 m (6 ft 7¼ in)
Propeller diameter	1.90 m (6 ft 2¾ in)

AREAS:

Rotor disc	75.43 m² (811.9 sq ft)

WEIGHTS AND LOADINGS:

Weight empty	420 kg (926 lb)
Max T-O weight	750 kg (1,653 lb)
Max rotor loading	9.94 kg/m² (2.04 lb/sq ft)
Max power loading	4.78 kg/kW (7.85 lb/hp)

PERFORMANCE (157 kW; 210 hp) engine:

Max level speed	113 kt (210 km/h; 130 mph)
Min flying speed	22 kt (40 km/h; 25 mph)

Max rate of climb at S/L	150 m (492 ft)/min
Service ceiling	3,000 m (9,840 ft)
T-O run: jump start	zero
normal	40 m (135 ft)
Range with 200 kg (441 lb) payload	
	334 n miles (620 km; 385 miles)
Endurance	6 h

UPDATED

ILYUSHIN

MEZHDUNARODNYYI AVIATSIONNYI KOMPANIYA ILYUSHINA (Ilyushin International Aviation Company)

Incorporating aircraft design and production companies,

Ilyushin MAK formed April 2000 by government decree. Initial members are Ilyushin Aviation Complex (see below) and VAPO (see later in this section), although formalisation of their union was still awaited in early 2002; following negotiations with Uzbekistan, will be expanded to include TAPO (which see) as second production plant;

previous Russian and Uzbek government agreement of 1998 covered collaboration in developing and marketing the Il-76 and Il-114. These plans overtaken in May 2001 by wider national strategy of merging Ilyushin, Beriev, Mil, Sukhoi and Yakovlev.

UPDATED

ILYUSHIN – AVIATSIONNYI KOMPLEKS IMENI S V ILYUSHINA OAO (Aviation Complex named for S V Ilyushin JSC)

Leningradsky prospekt 45G, 125190 Moskva
Tel: (+7 095) 157 35 73
Fax: (+7 095) 212 02 75
e-mail: ilyushin@online.ru
CHAIRMAN OF THE BOARD OF DIRECTORS AND GENERAL DESIGNER:
  Genrikh V Novozhilov
GENERAL DIRECTOR AND CEO: Victor V Livanov
HEAD OF INTERNATIONAL RELATIONS AND CHIEF DESIGNER:
  Igor Ya Katyrev
DEPUTY GENERAL DESIGNER, MARKETING, BUSINESS DEVELOPMENT AND FOREIGN ECONOMIC RELATIONS:
  Vladimir A Belyakov

Ilyushin OKB is named after Sergei Vladimirovich Ilyushin, who died 9 February 1977, aged 82. Bureau (OKB-240) was founded 1933; has been headed by Genrikh Novozhilov since 1970. About 60,000 aircraft of Ilyushin design have been built. Personnel in 2001 totalled about 3,500; by that time, activities were 90 per cent civil orientated.

Funding problems and bank collapses have affected Ilyushin and some of its suppliers, delaying the establishment of an Il-114 leasing arm and the acquisition of Western systems for the Il-96T, which eventually abandoned. However, Bureau is jointly designing Il-214 twin-jet transport in collaboration with HAL of India (see International section). It is a partner, with Boeing and Sukhoi, in the RRJ regional jet programme announced in August 2001 and described in the Sukhoi entry. Collaboration with Boeing began in 1997 and resulted in August 2001 opening of joint design training centre in Moscow to provide engineers, mainly from Ilyushin, for Boeing's design and technical research centres in Russian capital.

UPDATED

ILYUSHIN Il-76

Production of the Il-76 is, effectively, at an end, the Tashkent plant in Uzbekistan having a stock of some 40 uncompleted airframes with which to satisfy orders in the medium-term, including any for the stretched Il-76MF placed by the Russian Air Forces or its Il-76TF civil counterpart. Situation could be altered by Chinese interest in 38 aircraft, which continued in 2001-02. Full description last appeared in the 2001-02 *Jane's*.

Il-76 is also the basis of the Beriev A-50 AEW aircraft (see 2001-02 and earlier editions, and *Jane's Aircraft Upgrades*). Recent orders for similar aircraft are expected to involve conversion of existing Il76/A-50s.

UPDATED

ILYUSHIN Il-78M

NATO reporting name: Midas
There have been no recent conversions of Il-76 to this airborne tanker configuration. However, India ordered four Il-78MKs in February 2001 and these are being converted from stocks of uncompleted Il-76s held at TAPO in Uzbekistan (which see). A full, illustrated description last appeared in the 2001-02 *Jane's*.

VERIFIED

ILYUSHIN Il-96-300

TYPE: Wide-bodied airliner.
PROGRAMME: First of two flying prototypes (SSSR-96000) flew at Khodinka 28 September 1988 and exhibited at Paris in June 1989; second (SSSR-96001) first flew 28 November 1989; further airframe used for static and fatigue testing; all three built at GAZ 30, Khodinka; areas of commonality with Il-86 permitted planned test programme to be reduced to 750 flights totalling 1,200 hours; route proving trials by second production Il-96-300 (SSSR-96005) conducted late 1991; production at VAPO, Voronezh; certification received 29 December 1992.

Ilyushin Il-96-300 four-turbofan wide-bodied passenger transport of Aeroflot Russian International Airlines *(Mark Wagner/Flight International)*

Following initially poor reliability, Aeroflot (ARIA) serviceability improved, as evidenced by TBO increase from 1,600 to 4,130 hours between 1997 and 1999. PS-90A engine in-flight shutdown had improved to one per 29,000 hours by 1999.

Westernised Il-96M/T versions developed during 1990s, but abandoned in mid-2001. Description last appeared in 2001-02 *Jane's*.

CURRENT VERSIONS: **Il-96-300:** *As described.*

  Il-96-300V: Extended-range version on order by Vnukovo Airlines, 1999. Increased fuel for additional 1,200 miles (2,222 km; 1,380 miles).

  Il-96PU: One aircraft (RA-96012) built for use of Russian president; VIP interior, additional communications facilities by OAO Relero of Omsk and medical centre. Second aircraft, designated Il-96PU(M), authorised by Russian government on 6 May 2000; this under construction by late 2001 and has different interior layout.

  Il-96-300 Freighter: Design under way by early 2000; development partnership also includes VAPO and Russian banks. To carry 70,000 kg (154,325 lb) payload over 4,859 n miles (9,000 km; 5,592 miles); uprated PS-90 engines each 176.5 kN (39,683 lb st). Estimated unit cost US$30 million (2000).

  Il-96-400T: Under development by mid-2000 (when designation Il-96-3XX briefly employed). Equivalent to Il-96M/T, but employing only Russian engines and avionics. Power plant envisaged as Aviadvigatel PS-90A modified with additional 14.67 kN (3,300 lb st) to allow full range at MTOW of 270,000 kg (595,250 lb), with 92,000 kg (202,825 lb) of cargo. Proposed to convert 10 uncompleted Il-96Ms (which see) to this standard for lease from Ilyushin Finance by Atlant Soyuz cargo airline; provisional agreement signed February 2001, envisaging 2003 service-entry; however, retrospective installation of cargo door could weaken airframe; Soyuz signed firm order for four -400Ts in August 2001, first being former Il-96T prototype, due in service early 2002; second -400 to

follow 24 months thereafter. Also in 2001, Aeroflot issued letter of intent for four -400Ts to be delivered from 2006, and was considering passenger version.

CUSTOMERS: Aeroflot Russian International Airlines (six delivered by 1996); Domodedovo Airlines (four on order, of which third was delivered 16 April 1999); Russia State Transport Company (three; one delivered 1995; two completed, but not delivered, in 1997); and Atlant Soyuz (received first production aircraft in 1999, following trials use). Total 11 production aircraft delivered: 10th in February 1996; 11th in April 1999. By early 2002, six Il-96-300 and two Il-96T airframes were incomplete at the VAPO plant.

Aeroflot signed lease agreement with Ilyushin Finance in November 1999 for further six, plus sale, refurbishment and leaseback of one of original six; contract involves uncompleted aircraft on production line, permitting deliveries between 2001 and 2003. Target was not achieved but order remained in place during 2001, despite Aeroflot commitment to Il-96-400T. China Xinjiang Airlines requires three Il-96-300s and was continuing negotiations in early 2002; Kras Air of Russia revealed requirement in late 2001.

COSTS: Il-96-300 Rb1,320 million (2002). Il-96-400T US$38 million, flyaway (2001).

DESIGN FEATURES: Superficial resemblance to Il-86, but new design, with different engines to overcome performance deficiencies of original Il-86; new structural materials and state-of-the-art technology intended to provide life of 60,000 hours and 12,000 landings; no lower deck passenger entry. Current development aiming at range of 6,475 n miles (12,000 km; 7,450 miles) with 300 passengers.

Conventional wide-bodied airliner, with low wing (and winglets) and four pod-mounted engines.

Supercritical wings, with 30° sweep at quarter-chord; sweepback at quarter-chord 37° 30' on tailplane, 45° on fin.

FLYING CONTROLS: Triplex fly-by-wire, with manual reversion; each wing trailing-edge occupied by, from root, double-slotted inboard flap, small inboard aileron, two-section single-slotted flaps, and outboard aileron used only as gust damper and to smooth out buffeting; seven-section full-span leading-edge slats on each wing; three airbrakes forward of each inboard trailing-edge flap; six spoilers forward of outer flaps; inboard pair supplement ailerons, others operate as airbrakes and supplementary ailerons; variable incidence tailplane; two-section rudder and elevators, without tabs.

STRUCTURE: Basically all-metal, including new high-purity aluminium alloy, with composites flaps, maindeck floors and underfloor holds of honeycomb and CFRP; inner wings three-spar, outer panels two-spar; each wing has seven machined skin panels, three top surface, four bottom, with integral stiffeners; circular-section semi-monocoque fuselage; leading- and trailing-edges of fin and

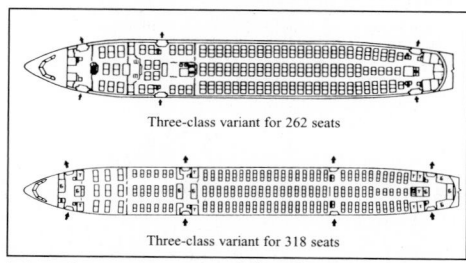

Three-class variant for 262 seats

Three-class variant for 318 seats

Internal arrangements of Ilyushin Il-96-300 with 262 passengers and (not to scale) Il-96M with 318

0011985

tailplane of composites. Some components manufactured by PZL Aircraft Factory, Poland.

LANDING GEAR: Retractable four-unit type. Forward-retracting steerable twin-wheel nose unit; three four-wheel bogie main units. Two of latter retract inward into wingroot/fuselage fairings; third is mounted centrally under fuselage, to rear of others, and retracts forward after the bogie has itself pivoted upward 20°. Oleo-pneumatic shock-absorbers. Nosewheel tubeless tyres size 1,260×460; mainwheel tubeless tyres size 1,300×480. Tyre pressure (all) 11.65 bar (169 lb/sq in).

POWER PLANT: Four Aviadvigatel PS-90A turbofans, each 156.9 kN (35,275 lb st), on pylons forward of wing leading-edges. Thrust reversal standard. Integral fuel tanks in wings and fuselage centre-section, total capacity 148,260 litres (39,166 US gallons; 32,613 Imp gallons).

ACCOMMODATION: Pilot, co-pilot and flight engineer; two seats for supplementary crew or observer. Ten or 12 cabin staff. Basic all-tourist configuration has two cabins for 66 and 234 passengers respectively, nine-abreast at 87 cm (34.25 in) seat pitch, separated by buffet counter, video stowage and two lifts from galley on lower deck. Two aisles, each 55 cm (21.65 in) wide. Two lavatories and wardrobe at front; six more lavatories, a rack for cabin staff's belongings and seats for cabin staff at rear. Seats recline, and are provided with individual tables, ventilation, earphones and attendant call button. Indirect lighting is standard. 235-seat mixed class version has front cabin for 22 first class passengers, six-abreast in pairs, at 102 cm (40 in) seat pitch and with aisles 75.5 cm (29.7 in) wide; centre cabin with 40 business class seats, eight-abreast at 90 cm (35.4 in) seat pitch and with aisles 56.5 cm (22.25 in) wide; and rear cabin for 173 tourist class passengers, basically nine-abreast at 87 cm (34.25 in) seat pitch, with aisle width of 55 cm (21.65 in).

Future Aeroflot standard, defined late 1999, is 12 first class, 21 business class and 180 economy seats, plus complete interior renewal. Refurbishment cost (including some Western avionics) US$1.5 million per aircraft.

Passenger cabin is entered through three doors on port side of upper deck, at front and rear and forward of the wings. Opposite each door, on starboard side, is emergency exit door. Lower deck houses front cargo compartment for six ABK-1.5 (LD3) containers or igloo pallets, central compartment aft of wing for 10 ABK-1.5 containers or pallets, and tapering compartment for general cargo at rear. Three doors on starboard side provide separate access to each compartment. Galley and lifts are between front cargo compartment and wing, with separate door aft of front cargo compartment door.

SYSTEMS: Four independent hydraulic systems, using fireproof and explosion-proof fluid, at pressure of 207 bar (3,000 lb/ sq in). APU in tailcone for engine starting and air conditioning on ground.

AVIONICS: *Flight:* Triplex flight control and flight management systems, together with a head-up display, permit fully automatic en route control and operations in ICAO Cat. IIIa minima. Duplex engine and systems monitoring and failure warning systems feed in-flight information to both the flight engineer's station and monitors on the ground. Autothrottle is based on IAS, without angle of attack protection.

Instrumentation: Primary flight information is presented on dual twin-screen colour CRTs, fed by triplex INS, a satellite-based and Omega navigation system and other sensors. Another electronic system provides real-time automatic weight and CG situation data.

DIMENSIONS, EXTERNAL:

Wing span: excl winglets	57.66 m (189 ft 2 in)
over winglets	60.11 m (197 ft 2½ in)
Wing aspect ratio	9.5
Length overall	55.35 m (181 ft 7¼ in)
Fuselage: Length	51.15 m (167 ft 9¾ in)
Max diameter	6.08 m (19 ft 11½ in)
Height overall	17.55 m (57 ft 7 in)
Tailplane span	20.57 m (67 ft 6 in)
Wheel track	10.40 m (34 ft 1½ in)
Wheelbase	20.07 m (65 ft 10 in)
Passenger doors (three): Height	1.83 m (6 ft 0 in)
Width	1.07 m (3 ft 6 in)
Height to sill: Nos. 1 and 2	4.54 m (14 ft 10¾ in)
No. 3	4.80 m (15 ft 9 in)
Emergency exit doors (three):	
Height	1.825 m (5 ft 11¾ in)
Width	1.07 m (3 ft 6 in)
Cargo compartment doors (front and centre):	
Height	1.825 m (5 ft 11¾ in)
Width	1.78 m (5 ft 10 in)
Height to sill: front	2.34 m (7 ft 8¼ in)
centre	2.48 m (8 ft 1¾ in)
Cargo compartment door (rear):	
Height	1.38 m (4 ft 6¼ in)
Width	0.97 m (3 ft 2¼ in)
Height to sill	2.74 m (9 ft 0 in)
Galley door: Height	1.20 m (3 ft 11¼ in)
Width	0.80 m (2 ft 7½ in)

DIMENSIONS, INTERNAL:

Cabins, excl flight deck: Height	2.60 m (8 ft 6¼ in)
Max width	Approx 5.70 m (18 ft 8½ in)
Volume	350.0 m³ (12,360 cu ft)
Cargo hold volume: front	37.1 m³ (1,310 cu ft)
centre	63.8 m³ (2,253 cu ft)
rear	15.0 m³ (530 cu ft)

AREAS:

Wings, gross	391.60 m² (4,215.1 sq ft)
Vertical tail surfaces (total)	61.00 m² (656.60 sq ft)
Horizontal tail surfaces (total)	96.50 m² (1,038.75 sq ft)

WEIGHTS AND LOADINGS:

Basic operating weight	117,000 kg (257,950 lb)
Max payload	40,000 kg (88,185 lb)
Max fuel	114,900 kg (253,310 lb)
Max T-O weight	216,000 kg (476,200 lb)
Max landing weight	175,000 kg (385,800 lb)
Max zero-fuel weight	157,000 kg (346,125 lb)
Max wing loading	551.6 kg/m² (112.97 lb/sq ft)
Max power loading	344 kg/kN (3.37 lb/lb st)

PERFORMANCE (estimated):

Normal cruising speed at 10,100-12,100 m (33,135-39,700 ft) 459-486 kt (850-900 km/h; 528-559 mph)

Approach speed 140-146 kt (260-270 km/h; 162-168 mph)

Balanced T-O runway length	2,600 m (8,530 ft)
Balanced landing runway length	1,980 m (6,500 ft)

Range, with UASA reserves: with max payload 4,050 n miles (7,500 km; 4,660 miles)

with 30,000 kg (66,140 lb) payload 4,860 n miles (9,000 km; 5,590 miles)

with 15,000 kg (33,070 lb) payload 5,940 n miles (11,000 km; 6,835 miles)

OPERATIONAL NOISE LEVELS: Il-96-300 is designed to conform with ICAO Chapter 3 Annex 16 noise requirements

UPDATED

ILYUSHIN Il-100

TYPE: Utility turboprop twin.

PROGRAMME: Announced February 2000, when in basic design stage. Draft design complete by October 2000. First flight due two years after go-ahead, which conditional upon 50 firm commitments. Manufacture of production aircraft at Lukhovitsy (LMZ). Estimated sales minimum of 200; potential 1,200 over 10 years. Following positive response from potential market, Ilyushin began search in late 2001 for launch customer to assist with costs of development.

CURRENT VERSIONS: **Il-100:** Baseline version; 12 seats.

Il-100A: Stretched version; 18-20 seats.

COSTS: Development estimated as US$20 million to US$30 million. Unit cost US$1 million (2000).

DESIGN FEATURES: Replacement for Antonov An-2 utility biplane, but with 40 per cent reduction in direct operating cost. Accordingly, stated to be "in Cessna Caravan I class". Rugged, unpressurised and simple to maintain; STOL performance. High-wing, T-tail configuration with unswept rear fuselage, but only port side freight door. Airframe life 60,000 cycles. Choice of Russian (Kuznetsov NK-123) or Western (RR 250 or PT6) turboprop engines.

WEIGHTS AND LOADINGS:

Max payload	1,500 kg (3,307 lb)

PERFORMANCE:

Cruising speed	200 kt (370 km/h; 230 mph)
T-O run	400 m (1,315 ft)
Landing run	260 m (855 ft)

UPDATED

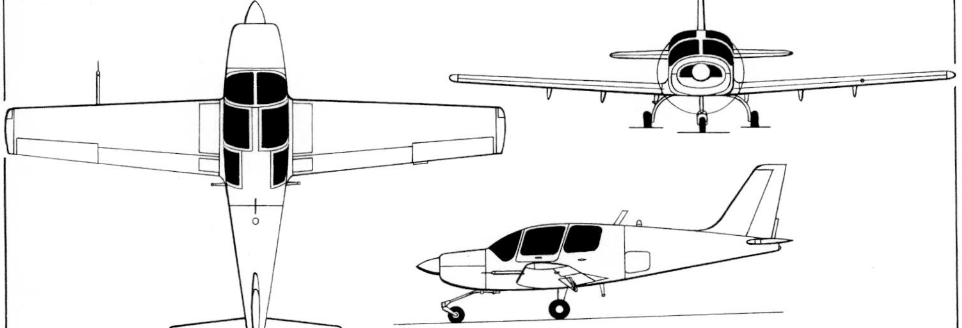

Ilyushin Il-103 two/five-seat multipurpose light aircraft ***(Jane's/James Goulding)*** 0079319

ILYUSHIN Il-103

TYPE: Four-seat lightplane.

PROGRAMME: Exhibited in model form at Moscow Aerospace '90; programme go-ahead 1990; first flight (RA-10321) 17 May 1994; second prototype (RA-10302) flew 30 January 1995 at LMZ, Lukhovitsy, following August 1993 decision to establish a production line there; preseries batch comprised three flyable and two static test aircraft, former having flown 414 sorties in two years after maiden flight; first production aircraft flew 30 January 1995.

Type certificate by Russian authorities under AP-23 received 15 February 1996; production approval certificates issued 7 March 1996 and 7 April 1998; airworthiness certificate granted 9 July 1997. Amended Russian certification granted 4 December 1997, following addition of GPS, radio compass and AHRS to avionics suite. FAR Pt 23 certification achieved 9 December 1998 (first by a Russian-built aircraft) after 110-hour/208-sortie programme. Annex of 6 September 1999 allows operation from unpaved runways.

Initial export order, six for Peruvian Air Force, agreed June 1999 and finalised in April 2000; deliveries had not begun by August 2000, due to lack of payment, but all reported in service by late 2001.

CURRENT VERSIONS: **Il-103-01:** Baseline VFR version for Russian Federation market.

Il-103-10: Export version with fully upgraded avionics suitable for international airways navigation.

Il-103-11: Export version with partly upgraded avionics suitable for local air navigation.

Initial certification is for a max T-O weight of 1,285 kg (2,832 lb), as reflected in weight and performance data below. The Il-103 will later be certified at 1,310 kg (2,880 lb) and, ultimately, 1,460 kg (3,218 lb).

Il-103SKh: Crop-sprayer (*selskokhozyaistvenny:* agricultural) version; first flown 29 March 2000.

CUSTOMERS: Initial operators (by 1999) included Il-Service at Myachkovo, Cherepovets Aviation Club, Avialesookhrana (forestry), Special Ecological Aviation Centre, Tatarstan National Flying Club, Civil Aviation Academy and St Petersburg Methodological Centre. Sales in 2000 comprised four to GP Bellesavia of Belarus in September for forest firefighting; one to Il-Service; and one for Vladimirsky Aerial Forest Protection. Despite prediction of Russian market for 1,500 and plans to produce 100 per year, sales stagnated in 2001; manufacture by LMZ was reported, in mid-2000, temporarily to have ceased.

Avialine requires fleet of 12 for air taxi service from Moscow-Domodedovo. Federal Fire Service interested in 30. Uzbekistan government announced requirement for 100 in crop-spraying role, late 1999. Russian Federal Air Transport service reportedly ordered 30 in 1999 for issue to several flying schools, but these failed to materialise. Total 23 delivered by January 2002, including 10 exports (Belarus and Peru). Further sales in 2002 included 23 to South Korean Air Force and three to Laos.

COSTS: US$164,000 to 210,000 (2000).

DESIGN FEATURES: Conventional low-wing monoplane, with non-retractable landing gear, originally to meet DOSAAF requirement for 500 military/civil pilot trainers. Designed for daytime VFR flying, in non-icing conditions and ambient air temperature of −35 to +45°C; intended service life 14,000 hours/20,000 landings/15 years.

Wing sweep 0° at quarter-chord, slight dihedral from roots, twist 0°, thickness/chord ratio 16 per cent at root, 15 per cent at tip, taper ratio 1.9.

FLYING CONTROLS: Conventional and manual. Actuation by pushrods, except cable-actuated rudder; horn-balanced rudder and single-piece elevator; single-slotted trailing-edge flaps, 0° or 10° deflection; electrically actuated elevator trim tab. Deflection angles: ailerons +25/−20°, elevator +25/−20°, rudder ±25°. Ground-adjustable trim tabs on starboard aileron and rudder.

STRUCTURE: All-metal, basically aluminium alloy, except for titanium firewall frame and wingroot attachments; bonded GFRP wingtips, elevator and rudder tips and elevator tab; and small amounts of magnesium alloys and Al-Li alloys. Single-spar wings, with front and rear false spars, integral fuel tanks and detachable leading-edge, mounted at sides of fuselage. Main spar is riveted beam with extruded caps; false spars stamped from sheet alloy; rolled sheet stringers. Semi-monocoque front/centre fuselage with inbuilt wing carry-through structure; separate rear fuselage; separate tailcone. Longitudinal fuselage members comprise two reinforced spars in cabin floor, two side-ribs of centre wing section and three beams for landing gear attachment. Two-spar, single-piece tailplane attached to fuselage by four bolts; two-spar fin. Detachable wings, fin and tailplane.

LANDING GEAR: Non-retractable tricycle type, with single wheel on each unit. Cantilever spring nose and mainwheel legs of titanium alloy; castoring nosewheel with shimmy damper; oleo-pneumatic nosewheel leg tested on RA-10300 in 1998. Mainwheel tyres size 400×150-115, pressure 3.95 bar (57 lb/sq in) on KT-214-1 wheels;

Ilyushin Il-103 operated by Il-Service (*Jane's/Paul Jackson*)   NEW/0132854

nosewheel tyre size 310×135-99, pressure 3.43 bar (50 lb/sq in) on K-290 wheels. Optional tyres 6.00-6 and 5.00-5, respectively, on Parker 40-75B and 40-77B wheels. Multidisc hydraulic anti-lock brakes on mainwheels, pedal-actuated. Turning radius (outboard wheel) 4.70 m (15 ft 6 in). Floats and skis optional.

POWER PLANT: One 157 kW (210 hp) Teledyne Continental IO-360-ES2B flat-six engine; Hartzell BHC-C2YF-1BF/F8459A-8R two-blade, constant-speed (hydraulic), metal propeller. (Manual pitch control on three development aircraft; automatic pitch selection on production aircraft.) Alternative 194 kW (260 hp) Textron Lycoming or similarly rated Teledyne Continental engine under consideration. Fuel and oil systems suitable for inverted flight; two main fuel tanks in wingroots, total capacity 200 litres (52.8 US gallons; 44.0 Imp gallons); supply tank, capacity 3 litres (0.8 US gallon; 0.66 Imp gallon), in fuselage forward of wing front spar carry-through; gravity refuelling points in wingroots.

ACCOMMODATION: Two forward-folding seats side by side at front of cabin; bench seat for two adults or three children at rear; optional control wheel in place of standard stick; optional dual controls; optional front seats for parachutes; space for 220 kg (485 lb) freight with rear bench seat removed; two gull-wing window/doors, hinged on centreline, at front of canopy. Rear windows removable for ground emergency exit. Unrestricted access to baggage hold.

SYSTEMS: Cabin ventilated and heated and windscreen demisted electrically by fan heater. Electrical system 27 V DC, with 1,800 W 60 A generator and 25 Ah battery.

AVIONICS: *Comms:* Yurok VHF radio as standard; P-855A1 optional extra. Il-103-11 equipment includes UBD transponder. Il-103-10 has Bendix/King KX 165 nav/com/glideslope (replacing Russian radios) and KT 76A transponder.

Flight: BUR-4 flight data recorder.

Il-103-11 includes MKS-1 compass and Bendix/King KR 87 ADF and KLN 89B GPS. Il-103-10 additionally equipped with VOR/ILS, KN 63 DME, KMA 24 audio control/MKR, KCS 55A compass and encoding altimeter.

DIMENSIONS, EXTERNAL:

Wing span	10.56 m (34 ft 7¾ in)
Wing aspect ratio	7.6
Wing chord: at root	1.825 m (5 ft 11¾ in)
at tip	0.96 m (3 ft 1¾ in)
Length overall	8.00 m (26 ft 3 in)
Fuselage: Length	7.81 m (25 ft 7½ in)
Max height	1.42 m (4 ft 8 in)
Max width	1.40 m (4 ft 7 in)
Height overall	3.135 m (10 ft 3½ in)
Tailplane span	3.90 m (12 ft 9½ in)
Wheel track	2.405 m (7 ft 10¾ in)
Wheelbase	2.045 m (6 ft 8½ in)
Propeller diameter	1.93 m (6 ft 4 in)
Baggage door: Width	0.70 m (2 ft 3½ in)
Height	0.34 m (1 ft 1¼ in)

DIMENSIONS, INTERNAL:

Cabin: Length	2.65 m (8 ft 8¼ in)
Max height	1.27 m (4 ft 2 in)
Max width	1.30 m (4 ft 3 in)

AREAS:

Wings, gross	14.71 m² (158.4 sq ft)
Ailerons (total)	1.137 m² (12.24 sq ft)
Flaps (total)	2.204 m² (23.72 sq ft)
Fin	0.84 m² (9.04 sq ft)
Rudder	0.56 m² (6.03 sq ft)
Tailplane	1.48 m² (15.93 sq ft)
Elevator	1.56 m² (16.79 sq ft)

WEIGHTS AND LOADINGS:

Weight empty	900 kg (1,984 lb)
Baggage capacity	60 kg (132 lb)
Max payload	270 kg (595 lb)
Max fuel	150 kg (330 lb)
Max T-O weight: Utility	1,285 kg (2,832 lb)
FAR Pt 23	1,150 kg (2,535 lb)
Max wing loading: Utility	87.4 kg/m² (17.89 lb/sq ft)
Max power loading: Utility	8.21 kg/kW (13.48 lb/hp)

PERFORMANCE:

Never-exceed speed (VNE)	183 kt (340 km/h; 211 mph)
Max level speed	119 kt (220 km/h; 137 mph)

Cruising speed	97 kt (180 km/h; 112 mph)
Stalling speed: flaps up	64 kt (117 km/h; 73 mph)
10° flap	60 kt (111 km/h; 69 mph)
Max rate of climb at S/L	190 m (623 ft)/min
Max certified altitude	3,000 m (9,840 ft)
T-O run	380 m (1,250 ft)
Landing run	320 m (1,050 ft)

Max range at cruising speed, pilot and 270 kg (595 lb) payload, 30 min reserves

432 n miles (800 km; 497 miles)

g limits: Utility	+4.4/−1.8
Aerobatic	+6/−3

OPERATIONAL NOISE LEVELS: Designed to conform with GOST 23023-85 and ICAO Annex 16

UPDATED

ILYUSHIN Il-112

TYPE: Twin-turboprop transport.

PROGRAMME: Initiated 1994, with possibility of Russian government funding as alternative to Ukrainian Antonov An-38; design since enlarged and considerably refined; planned manufacture by KAPP at Kumertau and (Il-112V) VAPO at Voronezh.

Revealed July 1999 that Il-112V selected by Russian defence ministry as potential replacement for Antonov An-24/26, Yak-40 and Let L 410; significant performance improvements to be achieved by redesign begun in same month. Presidential approval granted April 2000. First flight originally due in 2002, but revised timetable called only for final decision in that year between Il-112 and competitors.

CURRENT VERSIONS: **Il-112**: Passenger transport, *as described in detail.*

Il-112 business transport: Basically as Il-112 airline version, but cabin divided into three sections, separated by curtains. Front cabin contains four pairs of facing seats, with tables between, and centre aisle, a wardrobe on the port side between the door and flight deck bulkhead, inward-opening door to flight deck, and large baggage compartment on starboard side with outside access.

Central compartment has four armchairs on port side, with tables between, three-seat sofa on starboard side to rear of cabinets; aisle width between armchairs and sofa 86 cm (34 in); wardrobe at rear on each side. Rear section contains main passenger door with airstairs on port side, two buffets, seat for attendant, and access to rear baggage compartment and lavatory.

Il-112T: Freighter with rear-loading ramp; lavatory and seat for loadmaster at front of unobstructed main hold; ramp forms rear wall of hold when stowed. Typical payloads include four 1AK-1.5 (LD3) or five 3AK-1 containers or five PA-1.5 pallets. Airstair door at front of hold on port side; service door opposite; emergency exit on each side in centre of hold.

Il-112V (*voenny:* military): Passenger/freighter for Russian armed forces. Rear ramp; provision for 35

passengers, or 18 stretchers or 34 paratroops. Patrol/surveillance version also to be produced. Detail design under way by early 2001.

CUSTOMERS: Requirement by Russian Air Forces could total 100 to 120; estimated potential for 120 civil sales.

DESIGN FEATURES: Based on Il-114 airframe. Conventional high-wing monoplane with T tail; constant-chord wing centre-section, tapered outer panels; sweptback fin and rudder; circular-section pressurised fuselage.

FLYING CONTROLS: Entire trailing-edge of each wing made up of horn-balanced aileron and two-section, area-increasing flaps; two-section spoilers forward of flap on each side of centre-section. Horn-balanced rudder and elevators; tab in rudder.

LANDING GEAR: Retractable tricycle type; twin-wheel nose unit; single wheel on each main unit, retracting into large fairing outside fuselage pressure cell.

POWER PLANT: Two 1,839 kW (2,466 shp) Klimov TV7-117S turboprops.

ACCOMMODATION: Flight crew of two; standard seating for 40 passengers, with 11 pairs of seats on port side and nine on starboard side of 45 cm (17.7 in) wide centre aisle. Baggage/freight compartment with outside access on starboard side at front of cabin. Main airstair passenger door at rear of cabin on port side, with service door opposite; emergency exits at front of cabin on port side, in centre on starboard side. Access to rear baggage compartment and lavatory at rear of cabin; buffet and seat for attendant.

DIMENSIONS, EXTERNAL:

Wing span: except Il-112V	22.55 m (73 ft 11¾ in)
Il-112V	23.45 m (76 ft 11¼ in)
Length: except Il-112V: overall	21.78 m (71 ft 5½ in)
fuselage	20.00 m (65 ft 7½ in)
Il-112V	20.65 m (67 ft 9 in)
Height overall: except Il-112V	7.63 m (25 ft 0¼ in)
Il-112V	7.90 m (25 ft 11 in)
Tailplane span	7.14 m (23 ft 5 in)
Wheel track	3.80 m (12 ft 5½ in)
Passenger door: Height	1.70 m (5 ft 7 in)
Width	0.76 m (2 ft 6 in)
Service door: Height	1.38 m (4 ft 6¼ in)
Width	0.72 m (2 ft 4¼ in)
Forward baggage door: Height	1.30 m (4 ft 3 in)
Width	0.96 m (3 ft 1¾ in)
Emergency exits (each): Height	0.91 m (2 ft 11¾ in)
Width	0.51 m (1 ft 8 in)

WEIGHTS AND LOADINGS (A: passenger transport, B: business transport, C: Il-112V):

Weight empty, equipped: A, B, C	9,100 kg (20,062 lb)
Max payload: A, B	4,000 kg (8,818 lb)
C	6,000 kg (13,227 lb)
Max T-O weight: A	14,530 kg (32,033 lb)
B	13,490 kg (29,740 lb)
C	16,700 kg (36,817 lb)
Max power loading: A	3.95 kg/kW (6.49 lb/shp)
B	4.67 kg/kW (6.03 lb/shp)

PERFORMANCE (estimated):

Nominal cruising speed at 7,600 m (24,950 ft)

A, B	297-324 kt (550-600 km/h; 342-373 mph)
C	351 kt (650 km/h; 404 mph)
T-O run: A	520 m (1,710 ft)
B	500 m (1,640 ft)
C	650 m (2,135 ft)
T-O balanced field length: A	980 m (3,215 ft)
B	940 m (3,085 ft)
C	1,200 m (3,940 ft)
Landing balanced field length: A, B	1,000 m (3,280 lb)
Landing run: A, B	400 m (1,315 ft)
Range: A	809 n miles (1,500 km; 932 miles)
B	1,943 n miles (3,600 km; 2,236 miles)
C: with 6,000 kg (13,227 lb) payload	
	540 n miles (1,000 km; 621 miles)
with 2,000 kg (4,409 lb) payload	
	3,239 n miles (6,000 km; 3,728 miles)

UPDATED

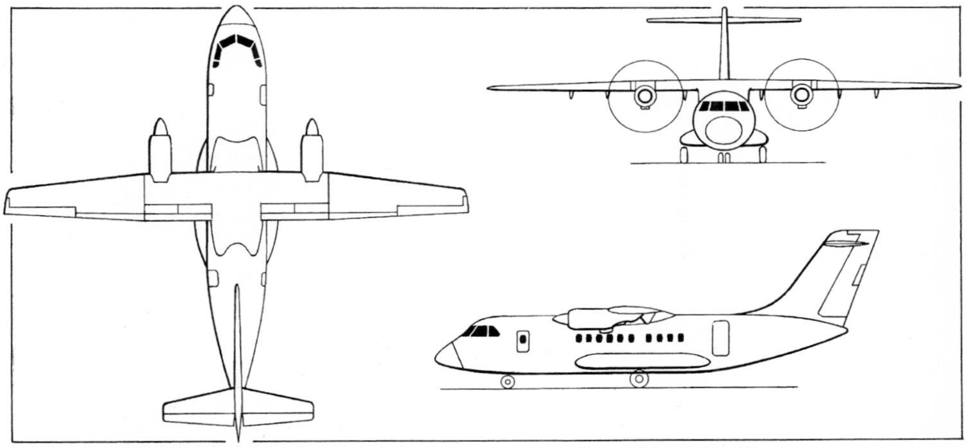

Ilyushin Il-112 twin-turboprop short-range passenger transport (*Jane's/James Goulding*)   0092139

ILYUSHIN Il-114

TYPE: Twin-turboprop airliner.

PROGRAMME: Design finalised 1986, as replacement for aircraft in An-24 class; was scheduled to enter service (with Aeroflot's Tashkent division) in 1992. Prototype (SSSR-54000) first flew at Khodinka 29 March 1990; second prototype (SSSR-54001) flew at Khodinka 24 December 1991, but lost in accident on 5 July 1993, resulting in withdrawal of government funding. Series production by TAPO at Tashkent, Uzbekistan, initial aircraft comprising Nos. 0101, 0103 and 0105 for flight development plus 0102 for static tests and 0104 for dynamic tests; first production aircraft flew 7 August 1992. Certification delayed by loss of second prototype and deferred certification of TV7 engines, but finally achieved on 26 April 1997. Negotiations reportedly under way in 1996 for production at Esfahan by Iran Aircraft Manufacturing Company; this venture lapsed upon selection of rival Antonov An-140. Recent development effort centred on -100 version.

In 1998, Russian and Uzbek governments agreed to promote Il-114; Ilyushin and Inkombank signed provisional agreement for production funding (although the bank's trading licence was revoked shortly afterwards, resulting in further delays); and Uzbekistan Airlines took delivery of production aircraft, making first commercial flight on 27 August 1998.

CURRENT VERSIONS: **Il-114:** *As described in detail.*

Il-114T: Cargo version developed for Uzbekistan Airlines; port freight door, size 3.25 × 1.715 m (10 ft 8 in × 5 ft 7½ in) in rear fuselage; removable roller floor; cargo attendants' cabin at forward end of freight hold accommodates up to two persons, is smokeproof and variable in size. MTOW and performance as for Il-114 passenger version, except where otherwise indicated. First production example (RA-91005) flew at Tashkent 14 September 1996. Second (UK-91004) converted by 1998, but lost in accident 5 December 1999. Prototype delivered to LII at Zhukovsky in early 2000 for certification trials; these were continued in Yakutia in early 2002.

Il-114-N200S: Rear loading ramp version; none yet built.

Il-114-100: With 2,051 kW (2,750 shp) P&WC PW127H turboprops, Hamilton Sundstrand 586E-7 six-blade propellers, Sextant avionics and new systems; primarily for export; English inscriptions and Imperial calibration; designated **Il-114PC** until late 1997;

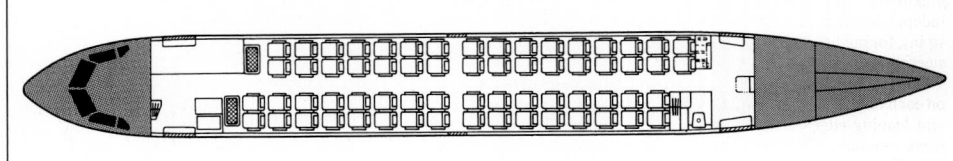

Ilyushin Il-114 interior with 64 seats at 75 cm (29½ in) pitch

passenger and cargo (**Il-114-100T**) versions envisaged. Performance generally as Il-114, but with increased range and economy. Weight empty, equipped (including two crew) 16,100 kg (35,494 lb). Joint venture agreed by Ilyushin and P&WC on 16 June 1997; first flight at Tashkent (UK-91009) 26 January 1999; CIS Interstate Aviation Committee type certificate awarded 27 December 1999.

Il-114M: With TV7M-117 turboprops, increased maximum T-O weight and 7,000 kg (15,430 lb) payload.

Il-114MA: Version of Il-114M with P&WC engines to carry 74 passengers at 324-351 kt (600-650 km/h; 373-404 mph) on 1,079 n mile (2,000 km; 1,242 mile) stages.

Il-114P and Il-114MP: Maritime patrol versions. Described in 2001-02 and previous editions.

Il-114FK: Military reconnaissance/cartographic survey version. Described in 2001-02 and previous editions.

Il-114PR: Signals intelligence and electronic warfare version; announced October 2000.

Il-140: Described separately.

CUSTOMERS: Uzbekistan Airlines purchased two pre-series aircraft in 1994; later modified with extra lavatory, new galley and accommodation reduced from 64 to 52; each flew 300 hours before engine overhaul time expiry and grounding; airline operated both Il-114Ts on service trials in 1998 and has total requirement for 10 Il-114/Ts, three of which on order in Il-114-100 configuration by April 2000 for delivery in March and April 2001, but failed to appear and only one was reported on order in early 2002, delivery date unspecified. Interest expressed by Bulgarian airlines; order for three Il-114-100s from Singaporean purchaser reported early 2000; delivery planned in 2001, although nothing further heard. Vyborg ordered three in 2001 for operations from St Petersburg, deliveries planned for 2002 in 64-seat configuration. Three (excluding prototype)

owned by Ilyushin Bureau. Total of 15 flying or substantially complete January 1998, most awaiting orders; delivery totals officially quoted as 1992 two, 1993 one, 1994 one, 1997 one and 1998 one. No further new production by end of 2001, when seven remained on assembly line, including two Il-114Ts; by early 2002, components sets for first 40 production aircraft had been manufactured wholly or in part. Transaero opened negotiations in 2001 for between five and 10.

COSTS: US$10 million for Il-114; US$8 million for Il-114T (2000).

DESIGN FEATURES: Conventional low-wing monoplane; only fin and rudder swept; slight dihedral on wing centre-section, much increased on outer panels; operation from unpaved runways practical. Service life of 30,000 cycles/30,000 hours/30 years, with overhaul at 6,000 hour intervals.

FLYING CONTROLS: Manual actuation for all except elevator; each wing trailing-edge occupied entirely by aileron, with servo and trim tabs, and hydraulically actuated double-slotted trailing-edge flaps, inboard and outboard of engine nacelle; two airbrakes (inboard) and spoiler (outboard) forward of flaps; spoilers supplement ailerons differentially in event of engine failure during take-off. Elevator control is FBW with back-up cable actuation. Trim and servo tabs in rudder, trim tab in each elevator.

STRUCTURE: Approximately 10 per cent of airframe by weight made of composites; two-spar wings; removable leading-edge on outer panels; circular-section aluminium alloy semi-monocoque fuselage built as five subassemblies; metal tail unit (CFRP tailplane and fin boxes planned for later aircraft).

LANDING GEAR: Retractable tricycle type, with twin wheels on each unit. All retract forward hydraulically; emergency extension by gravity. Oleo-pneumatic shock-absorbers. Tyres size 620×80 on nosewheels, 880×305 on mainwheels. Nosewheels steerable ±55°. Disc brakes on mainwheels. All wheel doors remain closed except during retraction or extension of landing gear.

POWER PLANT: Two 1,839 kW (2,466 shp) Klimov TV7-117S turboprops (with potential to increase to 2,088 kW; 2,800 shp), each driving a low-noise six-blade Stupino SV-34 CFRP propeller. Integral fuel tanks in wings, capacity 8,780 litres (2,319 US gallons; 1,931 Imp gallons).

ACCOMMODATION: Flight crew of two, plus stewardess. Emergency exit window each side of flight deck. Four-abreast seats for 64 passengers in main cabin, at 76 cm (30 in) seat pitch, with central aisle 45 cm (17¾ in) wide. Provision for rearrangement of interior for increased seating, removal of seats for cargo-carrying, and lengthening of fuselage for 70 to 75 passengers. Two passenger doors on port side: airstair door at front of cabin, further door at rear, both opening outward. Galley, cloakroom and lavatory at rear; emergency escape slide by service door on starboard side. Type III emergency exit over each wing. Service doors at front and rear of cabin on starboard side. Baggage compartments forward of cabin on starboard side and to rear of cabin, plus overhead baggage racks. Optional carry-on baggage shelves in lobby by main door at front.

SYSTEMS: TsNPKO-114 autopilot began trials on Il-114-100 in 2001. Dual-redundant pressurisation and air conditioning system using bleed air from both engines;

Ilyushin Il-114 short-range passenger transport, with additional side view (upper) of Il-114T freighter
(*Jane's/Mike Keep*)   0011989

Prototype Ilyushin Il-114-100 in its 2001 colours (*Yefim Gordon*)   *NEW*/0132842

maximum differential 0.44 bar (6.4 lb/sq in). Two independent hydraulic systems, pressure 207 bar (3,000 lb/sq in), for landing gear actuation, wheel brakes, nosewheel steering, airbrakes and flaps. Three-phase 115/220 V 400 Hz AC electrical system powered by 40 kW alternator on each engine. Secondary 24 V DC system. Wing and tail unit leading-edges de-iced electrically by patented pulse wave system. Electrothermal anti-icing system for propeller blades and windscreen. Engine air intakes de-iced by hot air. APU in tailcone.

AVIONICS: Digital avionics for automatic or manual control by day or night, including automatic approach and landing in limiting weather conditions (ICAO Cat. I and II).

Instrumentation: Two colour CRTs for each pilot for flight and navigation information. Centrally mounted CRT for engine and systems data.

Flight: Barco computer for FMS.

DIMENSIONS, EXTERNAL:
Wing span	30.00 m (98 ft 5¼ in)
Wing aspect ratio	11.0
Length overall	26.875 m (88 ft 2 in)
Fuselage: Length	26.20 m (85 ft 11½ in)
Max diameter	2.86 m (9 ft 4½ in)
Height overall	9.185 m (30 ft 1½ in)
Tailplane span	11.10 m (36 ft 5 in)
Wheel track	8.40 m (27 ft 6½ in)
Wheelbase	9.13 m (29 ft 11½ in)
Propeller diameter	3.60 m (11 ft 9¾ in)
Propeller ground clearance	0.50 m (1 ft 7¾ in)
Propeller fuselage clearance	0.97 m (3 ft 2¼ in)
Passenger doors (each): Height	1.70 m (5 ft 7 in)
Width	0.90 m (2 ft 11¼ in)
Service door (front): Height	1.30 m (4 ft 3¼ in)
Width	0.96 m (3 ft 1¾ in)
Service door (rear): Height	1.38 m (4 ft 6¼ in)
Width	0.72 m (2 ft 4¼ in)

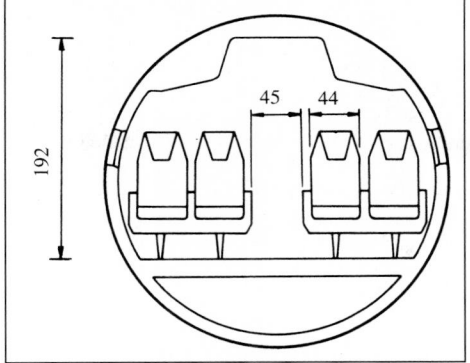

Cross-section of Il-114; dimensions in centimetres

Emergency exit (each): Height	0.91 m (3 ft 0 in)
Width	0.51 m (1 ft 8 in)

DIMENSIONS, INTERNAL:
Length between pressure bulkheads	
	22.24 m (72 ft 11½ in)
Cabin: Length	18.93 m (62 ft 1¼ in)
Width: max	2.64 m (8 ft 8 in)
at floor	2.28 m (7 ft 5¾ in)
Max height	1.92 m (6 ft 3½ in)
Cargo cabin volume (Il-114T)	76.0 m³ (2,684 cu ft)

AREAS:
Wings, gross	81.90 m² (881.6 sq ft)

WEIGHTS AND LOADINGS:
Operating weight empty	15,000 kg (33,070 lb)
Max payload	6,500 kg (14,330 lb)
Max fuel	6,500 kg (14,330 lb)
Max T-O weight	23,500 kg (51,808 lb)
Max ramp weight	23,600 kg (52,029 lb)
Max wing loading	286.9 kg/m² (58.77 lb/sq ft)
Max power loading	6.39 kg/kW (10.50 lb/shp)

PERFORMANCE:
Max level speed	270 kt (500 km/h; 310 mph)
Cruising speed	254 kt (470 km/h; 292 mph)
Approach speed	100 kt (185 km/h; 115 mph)
Landing speed	87 kt (160 km/h; 100 mph)
Optimum cruising height: Il-114	7,600 m (24,940 ft)
Il-114T	6,000 m (19,680 ft)
T-O run: paved	1,360 m (4,465 ft)
Landing run: paved or unpaved	1,260 m (4,135 ft)

Range, with reserves:
with 64 passengers	540 n miles (1,000 km; 621 miles)
with 1,500 kg (3,300 lb) payload	
	2,590 n miles (4,800 km; 2,980 miles)

UPDATED

ILYUSHIN Il-140

TYPE: Multisensor surveillance twin-turboprop.

PROGRAMME: Two derivatives of Il-114 first revealed to be in project stage in October 2000. No further announcements had been made by the end of 2001, although unofficial report at that time noted possible Russian Air Forces' interest in a surveillance version of the Il-114.

CURRENT VERSIONS: **Il-140:** Tactical air control version with airborne radar.

Il-140M: Patrol, ecological monitoring and maritime SAR version.

UPDATED

KAMOV

KAMOV OAO (Kamov JSC)

ulitsa 8 Marta 8a, Lyubertsy, 140007 Moskovskaya oblast
Tel: (+7 095) 700 32 04 and 171 37 43
Fax: (+7 095) 700 30 71 and 700 31 10
e-mail: kb@kamov.ru
Web: http://www.kamov.ru
GENERAL DESIGNER: Sergei V Mikheyev, PhD
DEPUTY GENERAL DESIGNER: Beniamin A Kasyanikov
CHIEF DESIGNERS:
   Boris Gubarev (Ka-115)
   Vyacheslav Krygin
   Evgeny Pak
   Aleksandr Piorzhnikov
   Grigory Yakemenko (Ka-50)
   Evgeny Sudarev

Formed in 1948 by Prof Dr Ing Nikolai Ilyich Kamov, this OKB originated in an experimental autogyro production plant operating in Moscow from 1940 to 1943. Kamov's initial brief was to design helicopters for naval use. Since the Ka-10 of 1949, Kamov helicopters have had coaxial contrarotating rotors; only the Ka-60/62, under development, have single main rotor, with anti-torque Fenestron. It was reported in September 2000 that a new transport helicopter, to compete against Western NH90 and S-92, is under development.

Since 1996, Kamov has been a member of the RSK 'MiG' consortium (which see). It is 49 per cent state-owned, and its design bureau and experimental construction plant employed 2,500 people in 1998. Under plans announced in May 2001 by the Russian government, Kamov and MiG will be in the major aerospace grouping which also includes Tupolev, Aviastar, Aviacor and those factories producing Kamov helicopters, namely the Arsenyev factory (Ka-50) and the Kumertau factory (Ka-32 family).

The affiliated Aero-Kamov Air Transportation Company (Tel/Fax: (+7 095) 700 31 60) operates a fleet of Ka-32s for diverse tasks, including firefighting, and has helicopters based in Russia, Canada and South Africa.

UPDATED

KAMOV Ka-29 and Ka-31

NATO reporting name: Helix-B

TYPE: Medium transport helicopter/AEW helicopter.

PROGRAMME: Developed for AV-MF, following cancellation of proposed joint-service, tandem-rotor, multirole V-50 and its replacement by what became Ka-50, meeting Army requirement only. Ka-252TB prototype (also known as Izdelie D2B or Izdelie 502) first flew 28 July 1976, possibly with Ka-25 nose or original narrow Ka-27/Ka-32 nose. Production at Kumertau (KAPP) from 1984.

Entered service with Northern and Pacific Fleets 1985; photographed on board assault ship *Ivan Rogov* in Mediterranean 1987, thought to be Ka-27B and given NATO reporting name 'Helix-B'; identified as Ka-29 combat transport at Frunze (Khodinka) Air Show, Moscow, August 1989; Ka-31 radar picket version completed initial shipboard trials on aircraft carrier *Admiral of the Fleet Kuznetsov* (then *Tbilisi*) 1990.

CURRENT VERSIONS: **Ka-29TB** ('Helix-B'): Armed derivative for day/night, VFR and IFR, transport and close support of seaborne assault troops; in-the-field conversion from one role to the other. Non-retractable landing gear and 50 cm (20 in) wider armoured flight deck. Reportedly used by Experimental Combat Group in Chechen War in 1996. No recent production known.

Detailed description generally as for Ka-32 except as under.

Ka-31 (formerly Ka-29RLD: *radiolokatsyonnogo dozora*: radar picket helicopter): Development began 1980; first flown October 1987; two examples (031 and 032) tested on *Admiral of the Fleet Kuznetsov*; state testing completed in 1996; limited production launched (for Indian Navy) at Kumertau Aircraft Plant, Bashkirskaya, 1999. Indian aircraft have 12-channel Kronshtadt GPS with Abris digital moving map and a 152 × 203 mm (6 × 8 in) AMLCD screen.

Basic airframe of Ka-27 (which see in 2001-02 and earlier *Jane's*) with broader flight deck; E-801 or E-801E (export) Oko (eye) early warning radar system by Radio Engineering Institute, Nizhny Novgorod, includes large rotating radar antenna (area 6.0 m²; 64.5 sq ft) that stows flat against underfuselage and deploys downward, turning through 90° into vertical plane before starting to rotate at 6 rpm; landing gear retracts upward to prevent interference, nosewheels into long fairings. Once system has been switched on, antenna extended and operation mode selected, data on air targets flying below helicopter's altitude, and on water surface situation, are acquired, evaluated and transmitted automatically to command centre, requiring only two crew (pilot and navigator, latter monitoring – but not operating – the system) in helicopter. Kronshtadt Kabris GPS navigation and display system. Loiter speed 54 to 65 kt (100 to 120 km/h; 62 to 75 mph) at up to 3,500 m (11,480 ft); loiter duration 2 h 30 min. Maximum surveillance radius 54 to 81 n miles (100 to 150 km; 62 to 93 miles) for fighter-size targets, 135 n miles (250 km; 155 miles) for surface vessels; up to 20 targets tracked simultaneously. Antenna can be retracted manually or explosively jettisoned in the event of a forced landing.

Two large panniers starboard side of cabin, fore and aft of main landing gear on helicopter numbered 032 (forward panniers only on 031); starboard airstair-type cabin door, aft of flight deck, divided horizontally into upward- and downward-opening sections, with box fairing in place of window; hatch window deleted above starboard rear pannier; new TA-8Ka APU positioned above rear of engine bay fairing, with slot-type air intake at front of housing, displacing usual ESM and IR jamming pods, gives radar and antenna an independent power supply. Tyre size 620×180 on mainwheels, 480×200 on nosewheels. Tailcone extended by fairing for flight recorder; no armour, gun door, stores pylons or outriggers.

Ka-33: Utility transport. Civilianised version of Ka-29TB shipborne assault transport. Designation revealed at Moscow Air Show in August 1997; no further details released and no known conversions.

CUSTOMERS: Total of 59 Ka-29s built for Russian Federation Naval Aviation (about 45) and Ukrainian Navy (about 12). Five Ka-31s ordered in August 1999 by Indian Navy for delivery in 2001 and basing aboard aircraft carriers and 'Krivak' class destroyers; further seven ordered February 2001. Additional 12 may be required. By October 2001,

Prototype Kamov Ka-31 radar picket helicopter with antenna folded and stowed *(Jane's/Paul Jackson)*

0121027

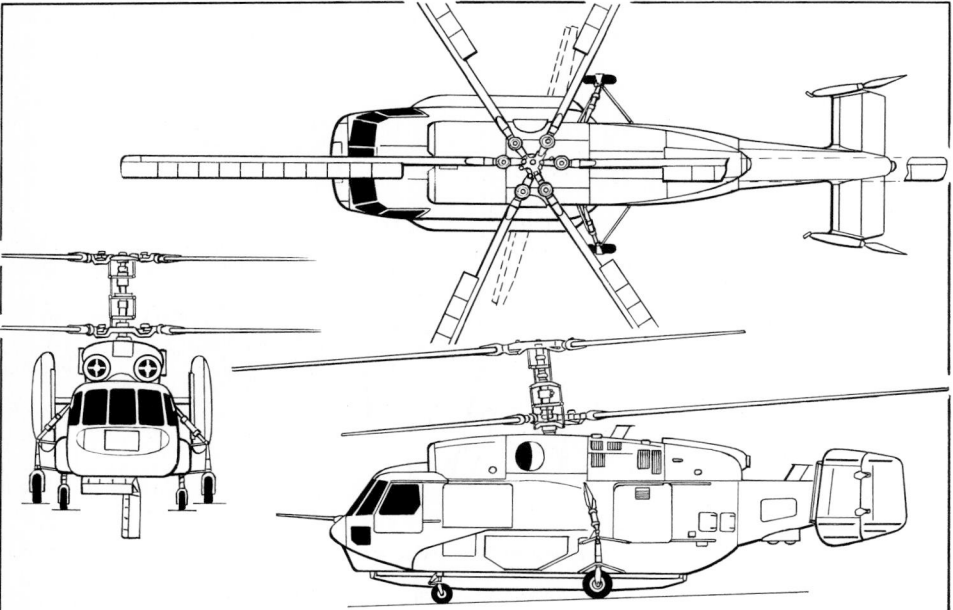

Kamov Ka-31, showing radar deployed in nose view (*Jane's/James Goulding*)   0130861

first two Indian airframes delivered from KAPP to Kamov at Moscow for avionics installation.

COSTS: Indian Navy batch of four priced at Rs4 billion (US$92 million) (2000).

POWER PLANT: Two Klimov TV3-117VMA turboshafts, each 1,633 kW (2,190 shp). Engines started by APU. Fuel tanks filled with reticulated polyurethane foam for fire suppression.

ACCOMMODATION: Wider flight deck than Ka-27 for two crew; three flat-plate windscreen glazings instead of two-piece curved transparency; 350 kg (772 lb) of armour around cockpit and engines; main cabin port-side door, aft of landing gear, divided horizontally into upward- and downward-opening sections, lower section forming step when open, to facilitate rapid exit of up to 16 assault troops; four stretcher patients, seven seated casualties and medical attendant in ambulance role; internal or slung cargo provisions.

AVIONICS: *Comms:* Two UHF and HF radios.
    Radar: Primary radar in port side of nose.
    Flight: INS; Doppler box under tailboom; IFF ('Slap Shot').
    Mission: Undernose Shturm-V missile guidance and LLTV pods; ESM 'flower pot' above rear of engine bay fairing.
    Self-defence: L-166V IR jammer ('Hot Brick'); chaff/flare dispensers.

EQUIPMENT: Station-keeping light between ESM and jammer.

ARMAMENT: Four-barrel Gatling-type GShG-7.62 7.62 mm machine gun, with 1,800 rounds, flexibly mounted behind downward-articulated door on starboard side of nose; four pylons on outriggers, for two four-round packs of 9M114 Shturm (AT-6 'Spiral') ASMs and two UV-32-57 57 or B-8V20 80 mm rocket pods. Alternative loads include four rocket packs, two pods each containing a 23 mm gun and 250 rounds, or two ZAB-500 incendiary bombs. Internal weapons bay for torpedo or bombs. Provision for 30 mm Type 2A42 gun above port outrigger, with 250-round ammunition feed from cabin.

DIMENSIONS, EXTERNAL (A: Ka-29, B: Ka-31):

Rotor diameter (each)	15.90 m (52 ft 2 in)
Blade length, aerofoil section (each)	
	5.45 m (17 ft 10½ in)
Blade chord	0.48 m (1 ft 7 in)
Vertical separation of rotors	1.40 m (4 ft 7 in)
Length overall, excl noseprobe and rotors:	
A	11.30 m (37 ft 1 in)
B	11.25 m (36 ft 11 in)
Height overall: A	5.40 m (17 ft 8½ in)
B	5.60 m (18 ft 4½ in)
Width: between centrelines of outboard pylons	
	5.65 m (18 ft 6½ in)
over tailfins and centred rudders	3.65 m (12 ft 0 in)
of flight deck	2.20 m (7 ft 2 in)
Mainwheel track	3.50 m (11 ft 6 in)
Nosewheel track: A	1.41 m (4 ft 7½ in)
B	2.41 m (7 ft 11 in)
Wheelbase: A	3.00 m (9 ft 10 in)
B	3.05 m (10 ft 0 in)

AREAS:

Rotor disc (each)	198.50 m² (2,136.6 sq ft)

WEIGHTS AND LOADINGS (A, B as above):

Weight empty: A	5,520 kg (12,170 lb)
Max load: A, internal	2,000 kg (4,409 lb)
A, external	4,000 kg (8,818 lb)
Max combat load: A	1,800 kg (3,968 lb)
Normal T-O weight: A	11,000 kg (24,250 lb)
B	12,500 kg (27,557 lb)
Max T-O weight: A, internal load	11,500 kg (25,353 lb)
B	12,500 kg (27,557 lb)
Max airborne weight:	
A, external slung load	12,600 kg (27,775 lb)

PERFORMANCE (A, B as above):

Max level speed at S/L: A	151 kt (280 km/h; 174 mph)
B	135 kt (250 km/h; 155 mph)
Nominal cruising speed: A	130 kt (240 km/h; 149 mph)
B	119 kt (220 km/h; 137 mph)
Loitering speed: B	54 kt (100 km/h; 62 mph)
Max rate of climb at S/L: A	888 m (2,910 ft)/min
Service ceiling: A	4,300 m (14,100 ft)
B	3,500 m (11,480 ft)
Loitering altitude: B	3,500 m (11,480 ft)
Hovering ceiling OGE: A	3,000 m (9,840 ft)
Combat radius, with six to eight attack runs over target:	
A	54 n miles (100 km; 62 miles)
Range:	
A, max standard fuel	248 n miles (460 km; 285 miles)
B	324 n miles (600 km; 372 miles)
A, ferry	400 n miles (740 km; 460 miles)
Loitering endurance: B	2 h 30 min

VERIFIED

KAMOV Ka-32
NATO reporting name: Helix-C

TYPE: Multirole medium helicopter.

PROGRAMME: Development of Ka-27/32 began 1969; first flight of common prototype 1973; first Ka-32 (SSSR-04173) flew 8 October 1980; prototype of utility version shown at Paris Air Show June 1985; new military versions first exhibited at Moscow Air Show '95; Ka-32S and Ka-32T versions in production by KAPP; other conversions by Kamov at Lyubertsy. Klimov VK-3000 turboshaft was to be certified in 2001 as alternative power plant, but no installations have been reported.

CURRENT VERSIONS: **Ka-32T** ('Helix-C'): Utility transport (*transportnyi*), ambulance, flying crane and agricultural sprayer; production began in 1987. Limited avionics; for carriage of internal or external freight, and passengers, along airways and over local routes, including support of offshore drilling rigs. Military 'Helix-C' similar; no undernose radome, but with dorsal ESM 'flower pot' and other military equipment. Several seen on board carriers, operating in SAR and planeguard roles. Military version understood to be designated Ka-27 or Ka-27T.

Detailed information applies to Ka-32T and Ka-32S. Ka-32A series generally similar, except as noted.

Ka-32S ('Helix-C'): Shipborne (*sudovoi*) version, intended especially for polar use; in production since 1987. More comprehensive avionics, including autonomous navigation system and Osminog (octopus) undernose radar (search radius 108 n miles; 200 km; 124 miles), for IFR operation from icebreakers in adverse weather and over terrain devoid of landmarks; 300 kg (661 lb) electric load hoist standard; additional external fuel tanks available 1994, strapped on each side at top of cabin; duties include ice patrol, guidance of ships through icefields, unloading and loading ships (up to 30 tonnes an hour, 360 tonnes a day). Simplex carbon fibre/epoxy tank, capacity 1,500 litres (396 US gallons; 330 Imp gallons) or 3,000 litres (792 US gallons; 660 Imp gallons), and 12.0 m (39 ft 4½ in) spraybar can be fitted for maritime anti-pollution work. Spraytime 6 minutes with 1,500 litre tank. In maritime search and rescue role, can loiter for 1 hour anywhere within 260 n miles (480 km; 300 miles) of base, and return carrying four crew and 5,000 kg (11,023 lb) payload. Maximum fuel capacity 2,650 litres (700 US gallons; 583 Imp gallons); weight empty 6,997 kg (15,425 lb); maximum payload 3,300 kg (7,275 lb) internally, 4,600 kg (10,141 lb) externally; maximum level and cruising speeds as Ka-32T.

Ka-32K: Flying crane (*kran*) with retractable gondola for second pilot under cabin. Prototype first flew December 1991; operational testing completed 1992. Supplied to Krasnodar Institute of Civil Aviation.

Ka-32A: Assemblies and systems of basic Ka-32 modified in 1990-93 to meet all requirements of Russian NLG-32-29 and NLG-32-33 and US FAR Pt 29/FAR Pt 33 airworthiness standards in categories A and B. First flight September 1990; Russian type certificates obtained for Ka-32A and its TV3-117VMA engines in June 1993. Production began 1996. Larger tyres. Optional pressure fuelling with reduced fuel capacity. Maximum accommodation for 13 passengers. Advanced avionics available, including Canadian Marconi dual CMA-900 flight management system, with EFIS, AFCS, CMA-2012 Doppler velocity sensor and CMA-3012 GPS sensor. Modification of helicopters to Ka-32A standard started by Kamov 1994.

Kamov Ka-32A civil utility helicopter (*Jane's/Paul Jackson*)   *NEW*/0137363

For details of the latest updates to *Jane's All the World's Aircraft* online and to discover the additional information available exclusively to online subscribers please visit

jawa.janes.com

Ka-32A1: Firefighting version of Ka-32A, first flown 12 January 1994. Equipped with Canadian or Russian variants of 'Bambi Bucket', capacity 5,000 litres (1,320 US gallons; 1,100 Imp gallons). Two operated by Moscow fire service, with doorway-mounted steerable water cannon and three types of rescue cage, able to lift two, 10 or 20 people from roofs of tall buildings. Other equipment includes searchlights and loudspeakers. Fire service aircraft and others flown by anti-riot police controlled by Aviatika Concern ISC, set up by Moscow city authorities and private investors to develop urban air transport system. Several on lease to South Korean forestry department have Simplex 10900-050 system, including 2,955 litre (780 US gallon; 650 Imp gallon) belly tank which can be refilled in 1½ minutes, plus 152 litre (40.2 US gallon; 33.4 Imp gallon) retardant tank. Ka-32 can also be fitted with 10900-055 system with two panniers totalling 5,000 litres (1,320 US gallons; 1,100 Imp gallons) of water and 250 litres (66.0 US gallons; 55.0 Imp gallons) of retardant.

One Moscow fire service helicopter (RA-31073) retrofitted with large, forward-facing nose boom for fire suppression in tall buildings; trials completed April 2001; shown statically at Moscow Salon, August 2001. System developed by Soyuz Federal Centre of Double Technologies at Dzerginsk; water supply of 2,800 litres (740 US gallons; 616 Imp gallons) carried in two underslung tanks or helicopter can be connected to fire vehicle on ground for unlimited supply; hose boom movable in vertical plane only.

Ka-32A2: Police version used by Moscow Militia, first flown 21 March 1995; seen in camouflage finish (RA-06144) at Moscow Air Show '95. Seats for 11 passengers, two of whom can operate pintle-mounted guns in port-side rear doorway and starboard rear window. Fuel tanks filled with polyurethane foam to prevent explosion after damage or catching fire. Equipped for abseiling from both sides of cabin. Hydraulic hoist; two sets of loudspeakers; L-2AG searchlight under nose. Militia reportedly has 25. Maximum T-O weight 12,700 kg (28,000 lb).

Ka-32A3: Ordered by Russian Ministry of Emergency Situations (MChS) to carry rescue and salvage equipment to disaster areas and evacuate casualties.

Ka-32A7: Armed export version (alternatively known as **Ka-327**) of Russian Border Troops' Ka-27PV developed from military Ka-27PS for frontier and maritime economic zone patrol, with Osminog (octopus) radar and pairs of Kh-25 ASMs, UPK-23-250 pods each containing a GSh-23L twin-barrel 23 mm gun with 250 rounds, or B-8V-20 pods each with twenty 80 mm S-8 rockets, on four underwing pylons. Displayed – but not yet integrated – with Kh-35 (AS-20 'Kayak') active radar-homing ASMs. Provision for 30 mm Type 2A42 gun above port outrigger. Optional twin searchlights on weapons pylons. Large oblique camera in starboard rear window. Search and rescue equipment standard, with ability to lift up to 10 survivors at a distance of 108 n miles (200 km; 124 miles) from base. Provision for 13 persons in cabin. Maximum T-O weight 11,000 kg (24,250 lb). Maximum level speed 140 kt (260 km/h; 161 mph). First flown 1995.

Ka-32-10: Announced 25 May 2001; projected 24-seat civil version with enlarged cabin; internal payload 4,000 kg (8,818 lb). Target certification date 2004.

Ka-32A11BC: Built in accordance with requirements of Transport Canada. FAR Pt 29 certification gained 11 May 1998, but full clearance achieved 26 February 1999, after installation of dual actuators in flight control system; first Russian helicopter to gain Western certification. Two development aircraft delivered to VIH Logging in May 1997; flew 4,000 hours up to February 1999; also used for firefighting; further 15 on order by 1998.

Ka-32A12: Version approved by Aviation Register of Switzerland.

Ka-32M: Under development by Kamov, to increase lifting capability to 7,000 kg (15,432 lb); retrofit with 1,839 kW (2,466 hp) TV3-117VMA-SB3 engines. Probably replaced in planning by Ka-32-10.

CUSTOMERS: Aeroflot and its successors; operators in Bulgaria (32S), Canada (32A), Laos (air force; six Ka-32T), Papua New Guinea (32A), South Africa (32A), Switzerland (32A), Yemen (32S/T). Estimated 132 Ka-32s in civilian use, of which 50 were abroad in 1998. Between December 1993 and November 2000, 36 imported by LGI of South Korea for operation by Forestry Service (23 Ka-32Ts), National Maritime Police Agency (eight Ka-32Ss), Kyonggi Provincial Fire & Disaster HQ (two Ka-32Ts) and Kyongsang Buk-do Fire Defence Aviation Corps, National Parks and Ulsan Fire Defence HQ (one Ka-32T each); further 20 expected in settlement of Russian debt, of which 10 reportedly ordered in March 2001 and three delivered by end of 2001. Following lease of two (later three) Ka-32s, Cyprus government announced intention, August 2001, to purchase three.

COSTS: US$833,300 for firefighting version (1998).

DESIGN FEATURES: Conceived as completely autonomous 'compact truck', to stow in much the same space as Ka-25 with rotors folded, despite greater power and capability, and to operate independently of ground support equipment; special attention paid to ease of handling with single pilot; overall dimensions minimised by use of coaxial rotors, requiring no tail rotor, and twin fins on short tailboom; upper rotor turns clockwise, lower rotor anti-clockwise; rotor mast tilted forward 3°; twin turbines and APU above cabin, leaving interior uncluttered; lower fuselage sealed for flotation.

FLYING CONTROLS: Dual hydraulically powered flight control systems, without manual reversion; spring stick trim; yaw control by differential collective pitch applied through rudder pedals; mix in collective system maintains constant total rotor thrust during turns, to reduce pilot workload when landing on pitching deck, and to simplify transition to hover and landing; twin rudders intended mainly to improve control in autorotation, but also effective in co-ordinating turns; flight can be maintained on one engine at maximum T-O weight.

STRUCTURE: Titanium and composites used extensively, with particular emphasis on corrosion resistance; fully articulated three-blade coaxial contrarotating rotors have all-composites blades with carbon fibre and glass fibre main spars, pockets (13 per blade) of Kevlar-type material, and filler similar to Nomex; blades have non-symmetrical aerofoil section; each has ground-adjustable tab; each lower blade carries adjustable vibration damper, comprising two dependent weights, on root section, with further vibration dampers in fuselage; tip light on each upper blade; blades fold manually outboard of all control mechanisms, to folded width within track of main landing gear; rotor hub is 50 per cent titanium/50 per cent steel; rotor brake standard; all-metal fuselage; composites tailcone; fixed incidence tailplane, elevators, fins and rudders have aluminium alloy structure, composites skins; fins toe inward approximately 25°; fixed leading-edge slat on each fin prevents airflow over fin stalling in crosswinds or at high yaw angles.

LANDING GEAR: Four-wheel type. Oleo-pneumatic shock-absorbers. Castoring nosewheels. Mainwheel tyres size 600×180 (Ka-32); 620×180, pressure 10.80 bar (156 lb/sq in) (Ka-32A). Nosewheel tyres size 400×150 (Ka-32); 480×200, pressure 5.90 bar (85 lb/sq in) (Ka-32A). Skis optional.

POWER PLANT: Two 1,633 kW (2,190 shp) Klimov TV3-117V (Ka-32) or TV3-117VMA (Ka-32A) turboshafts, with automatic synchronisation system, side by side above cabin, forward of rotor driveshaft. Main gearbox brake standard. Oil cooler fan aft of gearbox. Cowlings hinge downward as maintenance platforms. Fuel in tanks under cabin floor and inside container each side of centre-fuselage; capacity of main tanks 2,180 litres (576 US gallons; 480 Imp gallons); maximum capacity with two underfloor auxiliary tanks 3,450 litres (911 US gallons; 759 Imp gallons). Single-point pressure refuelling behind small forward-hinged door on port side, where bottom of tailboom meets rear of cabin.

ACCOMMODATION: Pilot and navigator side by side on air conditioned flight deck, in adjustable seats. Rearward-sliding jettisonable door with blister window each side. Seat behind navigator, on starboard side, for observer, loadmaster or rescue hoist operator. Alcohol windscreen anti-icing. Direct access to cabin from flight deck. Heated and ventilated main cabin of Ka-32 can accommodate freight or 16 passengers, on three folding seats at rear, six along port sidewall and seven along starboard sidewall (13 passengers in Ka-32A). Lifejackets under seats. Fittings to carry four stretchers. No provisions for lavatory or galley. Pyramid structure can be fitted on floor beneath rotor driveshaft to prevent swinging of external cargo sling loads. Rearward-sliding door aft of main landing gear on port side, with steps below. Emergency exit door opposite. Hatch to avionics compartment on port side of tailboom.

SYSTEMS: Three hydraulic systems: main system supplies servos, mainwheel brakes and hydraulic winch when fitted; standby system supplies only servos after main system failure; auxiliary system supplies brakes after main system failure and adjusts height of helicopter fuselage above ground; it can also be connected to main system for checking all functions on ground. Electrical system includes two independently operating AC generators and two batteries which cut in automatically or manually via inverters after AC generating system failure. After failure of either generator, the other is switched automatically to supply both circuits. Two rectifiers supply DC power. Electrothermal de-icing of entire profiled portion of each blade switches on automatically when helicopter enters icing conditions. Hot air engine intake anti-icing. APU in rear of engine bay fairing on starboard side, for engine

Kamov Ka-32A1 emergency helicopter of Moscow fire service (*Yefim Gordon*) *NEW*/0137345

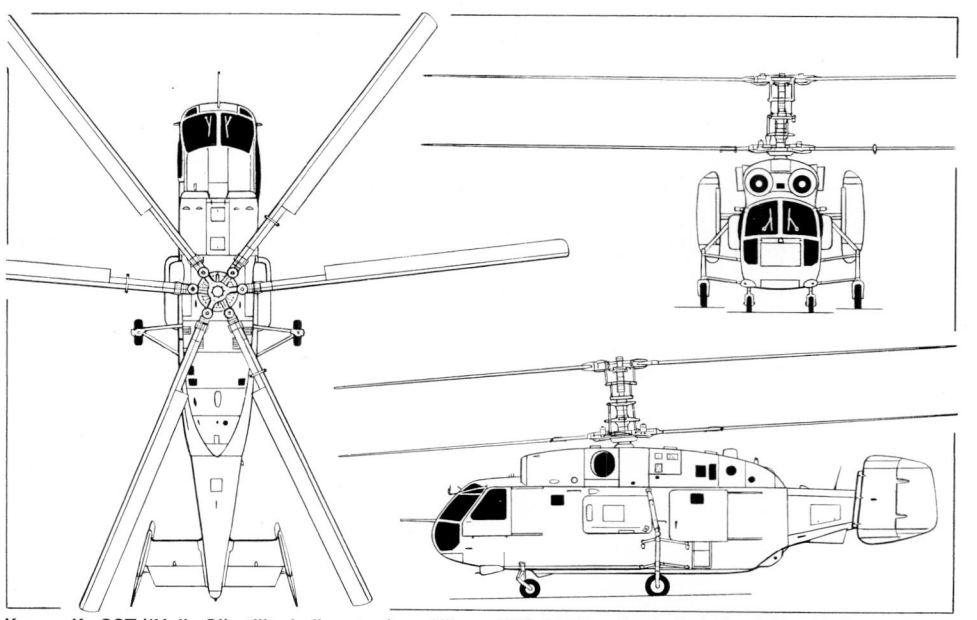

Kamov Ka-32T ('Helix-C') utility helicopter (two Klimov TV3-117V turboshafts) (*Jane's/Dennis Punnett*)

starting and to power all essential hydraulic and electrical services on ground, eliminating need for GPU.

AVIONICS: *Flight:* Include electromechanical flight director controlled from autopilot panel, Doppler hover indicator, two HSI and air data computer. Fully coupled three-axis autopilot can provide automatic approach and hover at height of 25 m (82 ft) over landing area, on predetermined course, using Doppler. Radar altimeter. Doppler box under tailboom.

EQUIPMENT: Doors at rear of fuel tank bay provide access to small compartment for auxiliary fuel, or liferafts which eject during descent in emergency, by command from flight deck. Container each side of fuselage, under external fuel containers, for emergency flotation bags, deployed by water contact. Optional rescue hoist, capacity 300 kg (661 lb), between top of door opening and landing gear. Optional external load sling, with automatic release and integral load weighing and stabilisation systems. Firefighting version of Ka-32T demonstrated in 1996.

DIMENSIONS, EXTERNAL:
Rotor diameter (each)	15.90 m (52 ft 2 in)
Length overall: excl rotors	11.27 m (36 ft 11¾ in)
rotors folded	12.25 m (40 ft 2¼ in)
Width, rotors folded	4.00 m (13 ft 1½ in)
Height to top of rotor head	5.45 m (17 ft 10½ in)
Wheel track: mainwheels	3.515 m (11 ft 6½ in)
nosewheels	1.41 m (4 ft 7½ in)
Wheelbase	3.03 m (9 ft 11¼ in)
Cabin door: Height	approx 1.20 m (3 ft 11¼ in)
Width	approx 1.20 m (3 ft 11¼ in)

DIMENSIONS, INTERNAL:
Cabin: Length	4.52 m (14 ft 10 in)
Max width	1.30 m (4 ft 3 in)
Max height	1.24 m (4 ft 0¾ in)

AREAS:
Rotor disc (each)	198.50 m² (2,136.6 sq ft)

WEIGHTS AND LOADINGS:
Weight empty	6,610 kg (14,573 lb)
Max payload: internal	3,700 kg (8,157 lb)
external	5,000 kg (11,023 lb)
Normal T-O weight	11,000 kg (24,250 lb)
Max flight weight with slung load	12,700 kg (27,998 lb)

PERFORMANCE (Ka-32):
Max level speed	140 kt (260 km/h; 162 mph)
Max cruising speed	130 kt (240 km/h; 149 mph)
Service ceiling	6,000 m (19,680 ft)
Hovering ceiling OGE	3,500 m (11,480 ft)
Hovering ceiling OGE, OEI	1,705 m (5,600 ft)
Range with max standard fuel	432 n miles (800 km; 497 miles)
Max range with auxiliary fuel	612 n miles (1,135 km; 705 miles)
Endurance with max standard fuel	4 h 30 min
Max endurance with auxiliary fuel	6 h 25 min

UPDATED

KAMOV Ka-50 CHERNAYA AKULA
English name: Black Shark
NATO reporting name: Hokum

TYPE: Attack helicopter.

PROGRAMME: Project launched in December 1977 as **V-80** (*Vertolyet* 80: Helicopter 80); first prototype (010) built by Kamov bureau and hovered at Lyubertsy 17 June 1982 and flew on 23 July, powered by TV3-117V engines; second prototype (011) flew 16 August 1983 with TV3-117VMA engines and mockup of Shkval tracking system, Merkury LLLTV, cannon and K-041 sighting system; both prototypes wore painted 'windows' to simulate fictitious rear cockpits. Initially reported in West in mid-1984, but first photograph did not appear (US Department of Defense's *Soviet Military Power*) until 1989.

First prototype lost in fatal accident on 3 April 1985; replaced by third prototype (012) with Mercury LLTV system for state comparative test programme against Mil Mi-28, which completed in August 1986. Two preproduction **V-80Sh-1**s (014 and 015) were first to be built at Arsenyev and introduced UV-26 chaff/flare dispensers; second had K-37-800 ejection system and mockup of LLLTV in articulated turret. Ordered into production in December 1987. Further three for continued development work comprised 018 (first flown at Arsenyev 22 May 1991), 020 'Werewolf' and 021 'Black Shark'. (Export marketing name was originally Werewolf, but changed to Black Shark by 1996.) State tests of Ka-50 began in mid-1991 and type was commissioned into Russian Army Aviation in August 1993 for trials at 4th Army Aviation Training Centre, Torzhok. In August 1994, the Ka-50 was included in the Russian Army inventory by Presidential decree, judged winner of the fly-off against Mi-28. The Mi-28 was nominally terminated on 5 October 1994 but the competition continued.

Further army evaluation followed when first two of four production Ka-50s were funded in 1994 and officially accepted on 28 August 1995; third and fourth received in 1996; these four numbered 20 to 23 (prompting preseries 021 to be renumbered 024 to avoid confusion). Arsenyev production was to have increased to one per month during 1997, but this did not occur. The original Ka-50 (and rival Mi-28A) were overtaken by the issue of a revised requirement which emphasised night capability –

favouring the two-seat Mi-28. The initial order for 15 Ka-50s was reportedly cancelled in September 1998, with procurement postponed until 2003. Three deployed to Mozdok during 1999 for use in Chechnya, but not used operationally. Two returned to theatre in December 2000; first firing of weapons against guerrilla forces was on 6 January 2001 (operating in conjunction with Mil Mi-24s); helicopters returned to Torzhok in March 2001.

Klimov VK-3000 turboshaft offered as alternative power plant.

CURRENT VERSIONS: **Ka-50** ('Hokum'): *As described.*

Ka-50N (*Nochnoy:* Nocturnal): Also reported as Ka-50Sh. Night-capable attack version; essentially a single-seat Ka-52. Programme began 1993; originally based on TpSPO-V and Merkury LLLTV systems, which tested on Ka-50 development aircraft. Ka-50N first reported April 1997 as conversion of prototype 018 with Thomson-CSF Victor FLIR turret above the nose and Arbalet (crossbow) mast-mounted radar, plus second TV screen in cockpit; FLIR integrated with Uralskyi Optiko-Mekhanicheskyi Zavod (UOMZ) Shamshit-50 (Laurel-50) electro-optic sighting system, incorporating French IR set. First flight variously reported as 4 March or 5 May 1997; programmed improvements included replacement of PA-4-3 paper moving map with digital equivalent; by August 1997, FLIR turret was repositioned below nose and Arbalet was removed; by mid-1998, had IT-23 CRT display replaced by TV-109, and HUD removed and replaced by Marconi helmet display. Proposed new cockpit shown in September 1998, having two Russkaya Avionika 203 × 152 mm (8 × 6 in) LCDs and central CRT for sensor imagery. Indigenous avionics intended for any local production orders; French systems as interim solution and standard for export. Republic of Korea Army evaluated both the Ka-50N and the baseline Ka-50. In 1999, preproduction aircraft 014 was exhibited with a UOMZ GOES sensor turret in place of Shkval.

Ka-50-2: Designation applies to three quite different aircraft. Basic Ka-50-2 is a variant of the Ka-50 single-seater, though the designation is also applied to two twin-seat aircraft; first of these was a version of the Ka-52 Alligator and, as such, was described in 2001-02 and earlier *Jane's*. All Ka-50-2s differ from the baseline Ka-52 in retaining attack and anti-tank role using 12 laser beam-riding AT-8 Vikhr ATGMs or 16 Rafael NT-D ATGMs; avionics to be supplied by Israel Aircraft Industries, Lahav Division; 024 used as demonstrator. The basic Ka-50-2 was proposed to China, Finland, India, South Korea, Malaysia, Myanmar, Poland, South Africa, Syria and Turkey.

Second variant of Ka-50-2 is another two-seater, intended to have conventional stepped tandem cockpits; is offered to armed forces which do not accept the single-seat or side-by-side two-seat layouts. A further subvariant of the tandem-seat Ka-50-2, the **Erdoğan** (Turkish for Born Fighter), was proposed to Turkey jointly by Kamov and Israel Aircraft Industries. This would have been fitted with longer-span wings and feature a NATO-compatible Giat 621 turret containing a single 20 mm cannon which would

Model of a projected naval Ka-50 *(Yefim Gordon)*
0121090

fold down below the belly of the helicopter in flight, for a 360° arc of fire; it would fold to starboard for landing, and could be fired directly forward, even when folded. TV3-117VMA-02 engines. Ten Turkish pilots flew Alligator '061' at Antalya, Turkey, in early 1999 as part of evaluation process; requirement was for 145. Named as second choice when Bell AH-1Z selected.

Ka-52: Two-seat version; described in 2001-02 and earlier *Jane's*.

CUSTOMERS: Four for Russian Army service trials, plus eight flying prototype and preseries helicopters; all delivered. Further 10 ordered in 1997 budget and six in 1998, of which first three were due for delivery before end of 1998; initial helicopter eventually completed in June 1999, two more being due by mid-2000. By early 2002, it was still unclear if helicopters from the first batch of 10 had been delivered to Army Aviation. Two operational Ka-50s shown at Moscow Salon in August 2001 may have been repainted trials aircraft. One army helicopter lost in accident, 17 June 1998; attributed to rotor clash.

COSTS: Unit price of Ka-50N quoted as between US$12 million and US$15 million in mid-1999.

DESIGN FEATURES: World's first single-seat close support helicopter. Coaxial, contrarotating and widely separated semi-rigid three-blade rotors, with swept blade tip, attached to hub by steel plates; small fuselage cross-section, with nose sensors; flat-screen cockpit, heavily armour protected by combined steel/aluminium armour and spaced aluminium plates, with rearview mirror above windscreen; small sweptback tailfin, with inset rudder and large tab; high-set tailplane on rear fuselage, with endplate auxiliary fins; retractable landing gear; mid-set unswept wings, carrying ECM pods at tips; four underwing weapon pylons; engines above wingroots; high agility for fast, low-flying, close-range attack role; partially dismantled can be air-ferried in Il-76 freighter. Much of fuselage skin formed by large hinged door panels, providing access to interior equipment from ground level.

FLYING CONTROLS: Kamov coaxial design; generally as Ka-32.

STRUCTURE: Fuselage built around steel torsion box beam, of 1.0 m (3 ft 3¼ in) square section. Wing centre-section passes through beam. Cockpit mounted at front of beam,

Kamov Ka-50 operated by the Russian Army in 2001 *(Yefim Gordon)*
NEW/0137346

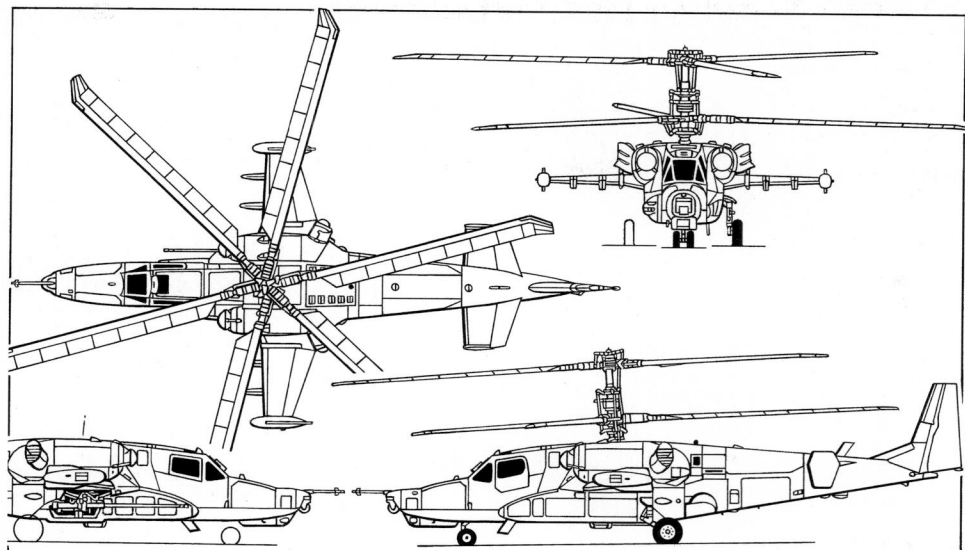

Ka-50 ('Hokum') single-seat combat helicopter with scrap view of gun installation on starboard side *(Jane's/Mike Keep)*

Height overall	4.93 m (16 ft 2 in)
Tailplane span	3.16 m (10 ft 4½ in)
Wheel track: main	2.67 m (8 ft 9 in)
nose	0.34 m (1 ft 1½ in)
Wheelbase	4.19 m (13 ft 9 in)
AREAS:	
Rotor disc (each)	165.13 m² (1,777.4 sq ft)
WEIGHTS AND LOADINGS:	
Weight empty	7,800 kg (17,196 lb)
Max external stores	3,000 kg (6,610 lb)
Normal T-O weight: Ka-50	9,800 kg (21,605 lb)
Erdoğan	9,800 kg (21,605 lb)
Max T-O weight: Ka-50	10,800 kg (23,810 lb)
Erdoğan	11,300 kg (24,912 lb)
PERFORMANCE:	
Max speed:	
in shallow dive	210 kt (390 km/h; 242 mph)
in level flight	162 kt (300 km/h; 186 mph)
in sideways flight	43 kt (80 km/h; 49 mph)
in backward flight	48 kt (90 km/h; 55 mph)
Cruising speed	146 kt (270 km/h; 168 mph)
Vertical rate of climb at 2,500 m (8,200 ft)	
	600 m (1,970 ft)/min
Service ceiling	5,500 m (18,040 ft)
Hovering ceiling OGE	4,000 m (13,120 ft)
Range: combat	243 n miles (450 km; 279 miles)
with max internal fuel	280 n miles (520 km; 323 miles)
with 4 auxiliary tanks:	
Ka-50	594 n miles (1,100 km; 683 miles)
Erdoğan	626 n miles (1,160 km; 720 miles)
Endurance: standard fuel, 10 min reserves	1 h 40 min
with 2 auxiliary tanks	2 h 50 min
g limit	+3.5

UPDATED

KAMOV Ka-52 ALLIGATOR

Only one prototype of this two-seat combat helicopter has been built. A description last appeared in the 2001-02 edition. Attempts to secure a customer continued in 2001-02 with the offer of a Ka-52K variant to South Korea.

UPDATED

KAMOV Ka-60 KASATKA

English name: Killer Whale

TYPE: Medium transport helicopter.

PROGRAMME: Original coaxial rotor, twin tail, single-engined V-60 won Soviet Army lightweight helicopter and Mi-8 replacement competition against heavier, twin-engined Mil Mi-36 in 1982; subsequently, design considerably modified to achieve greater speed through adoption of single main rotor and Fenestron-type of tail rotor with eleven blades. First flight originally due 1993, but programme slowed by funding shortages, and priority changed to promotion of civil variant (see following entry for Ka-62). Ka-60 officially revealed at Lyubertsy on 29 July 1997, when prototype close to completion; first flew (601) 10 December 1998; second sortie 21 December; first official flight 24 December; all were hovering flights; international debut at MAKS '99, Moscow, August 1999; first 'forward flight' 24 December 1999. Conflicting reports quote both Arsenyev and Ulan Ude as prospective production lines. However, LMZ (now LAPIK, part of RSK 'MiG') was reported in April 2000 to be preparing for production and in mid-2001 was building second prototype, which to be completed as a trainer in **Ka-60U** configuration. Series production LAPIK due to begin in 2003.

gearbox above and engines to sides. Carbon-based composites materials constitute 35 per cent by weight of structure, including rotors. Approximately 350 kg (770 lb) of armour protects pilot, engines, fuel system and ammunition bay; canopy and windscreen panels are 55 mm (2¼ in) thick bulletproof glass.

LANDING GEAR: Hydraulically retractable tricycle type; twin-wheel steerable nose unit and single mainwheels all semi-exposed when up; all wheels retract rearward; low-pressure tyres.

POWER PLANT: Two 1,633 kW (2,190 shp) Klimov TV3-117VMA turboshafts with VR-80 main reduction gearbox and two PVR-800 intermediate gearboxes, with air intake dust filters and exhaust heat suppressors. Later use of 1,838 kW (2,465 shp) TV3-117VMA-SB3 turboshafts intended. Two primary fuel tanks, filled with reticulated foam, inside fuselage box beam. Total internal capacity approximately 1,800 litres (485 US gallons; 404 Imp gallons). Front tank feeds port engine; rear feeds starboard and APU. Each tank protected by layers of natural rubber. Provision for four 500 litre (132 US gallon; 110 Imp gallon) underwing auxiliary fuel tanks. Transmission remains operable for 30 minutes after oil system failure.

ACCOMMODATION: Double-wall steel armoured cockpit, able to protect pilot from hits by 20 and 23 mm gunfire over ranges as close as 100 m (330 ft). Interior black-painted for use with NVGs. Specially designed Zvezda K-37-800 ejection system, ostensibly for safe ejection at any altitude (actually from 100 m; 330 ft); following explosive separation of rotor blades and opening of cockpit roof, pilot is extracted from cockpit by large rocket; alternatively, he can jettison doors and stores before rolling out of cockpit sideways. Associated equipment includes automatic radio beacon, activated during ejection, inflatable liferaft and NAZ-7M survival kit.

SYSTEMS: All systems configured for operational deployment away from base for up to 12 days without need for maintenance ground equipment; refuelling, avionics and weapon servicing performed from ground level. AI-9V APU for engine starting, and ground supply of hydraulic and electrical power, in top of centre-fuselage. Anti-icing system for engine air intakes, rotors, AoA and yaw sensors; de-icing of windscreen and canopy by liquid spray.

PrPNK Rubikon (L-041) piloting, navigation and sighting system based on five computers: four Orbita BLVM-20-751s for combat and navigation displays and target designation, plus one BCVM-80-30201 for WCS. Incorporates PNK-800 Radian navigation system, with C-061K pitch and heading data, IK-VSP-VI-2 speed and altitude and PA-4-3 automatic position plotting subsystems. Series 3 Tester U3 flight data recorder. Ekran BITE and warning system. KKO-VK-LP oxygen system with 2 litre (0.07 cu ft) supply for 90 minutes. Electrical supply from two 400 kW generators at 115 V 400 Hz three-phase AC; 500 W converter; rectifiers for 27 V DC supply.

AVIONICS: Integrated by NPO Elektro Avtomatika.

Comms: Two R800L1 and one R-868 UHF transceivers, SPU-9 intercom, P-503B headset recorder, Almaz-UP-48 voice warning system and HF com/nav; IFF ('Slap Shot').

Flight: INS; autopilot; Doppler box under tailboom; ARK-22 radio compass; A-036A radio altimeter.

Instrumentation: Conventional instruments; ILS-31 HUD; moving map display (Kronshtadt Abris on some aircraft); small IT-23MV CRT beneath HUD, with rubber hood, to display only FLIR and monochrome LLLTV imagery. Pilot has Obzor-800 helmet sight effective within ±60° azimuth and from −20 to +45° elevation; when pilot has target centred on HUD, he pushes button to lock sighting and four-channel digital autopilot into one unit. Displays compatible with OVN-1 Skosok NVGs.

Mission: To reduce pilot workload and introduce a degree of low observability, target location and designation is assigned to other aircraft; equipment behind windows in nose includes I-25IV Shkval-V daylight electro-optical search and auto-tracking system, laser marked target seeker and range-finder; FOV ±35° in azimuth +15 to −80° in elevation. FLIR turret to be added in nose for use with NVGs.

Self-defence: L150 Pastel RWR in tailcone, at rear of each wingtip EW pod and under nose; total of 512 chaff/flare cartridges (in four UV-26 dispensers) in each wingtip pod. L-140 Otklik laser detection system; L-136 Mak IR warning.

ARMAMENT: Four BD3-UV pylons on wings. Up to 80 S-8 80 mm air-to-surface rockets in four underwing B8V20A packs or 20 S-13 122 mm rockets in four B-13L pods; or up to 12 9A4172 Vikhr-M (AT-12) tube-launched laser-guided ASMs with range of 8 to 10 km (5 to 6.2 miles) capable of penetrating 900 mm of reactive armour; or mix of both; Vikhr launched from trainable UPP-800 mounts, which can be depressed to −12°; single-barrel 30 mm 2A42 gun on starboard side of fuselage, with up to 470 armour-piercing or high-explosive fragmentation rounds, can be depressed from +3° 30′ to −37° in elevation and traversed from −2° 30′ to +9° in azimuth hydraulically and is kept on target in azimuth by tracker which turns helicopter on its axis; two ammunition boxes in centre-fuselage. Front box contains 240 AP rounds, rear box 230 HE rounds. Selectable rapid (550 to 600 rds/min) or slow (350 rds/min) fire, with bursts of 10 or 20 rounds. Provision for alternative weapons, including UPK-23-250 23 mm gun pods, Igla or R-73 (AA-11 'Archer') AAMs, Kh-25MP (AS-12 'Kegler') ARMs, FAB-500 bombs or dispenser weapons.

DIMENSIONS, EXTERNAL:

Rotor diameter (each)	14.50 m (47 ft 7 in)
Length overall, rotors turning	16.00 m (52 ft 6 in)
Fuselage length, excl noseprobe	14.20 m (46 ft 7 in)
Wing span	7.34 m (24 ft 1 in)

Mockup of Kamov Ka-50-2 at Moscow, August 2001 *(Jane's/Paul Jackson)*   *NEW*/0137364

A smaller variant of Ka-60 was reported in mid-2001 to have been offered to Russian Navy.

CURRENT VERSIONS: Pilot and aircrew training (Ka-60U), utility, shipborne over-the-horizon targeting (Ka-60K) reconnaissance (Ka-60R) anti-tank and anti-helicopter versions proposed; role of Ka-60R, with Shamshit target acquisition system, transferred to Kamov Ka-52.

CUSTOMERS: None. Long-term interest maintained by Russian Army; reportedly evaluated by Iran. Russian requirements estimated as over 100 Ka-60Us between 2001 and 2010.

COSTS: US$1.7 million, Ka-60U (2000).

DESIGN FEATURES: Generally as for Ka-62 (which see), but with IR- and radar-absorbent coatings, reduced rotor speed, low-IR exhausts.

POWER PLANT: RKBM Rybinsk RD-600V turboshafts, as Ka-62. RRTM RTM322 or GE CT7 available in export versions.

ACCOMMODATION: Up to 16 infantry troops; or six stretchers and three attendants. Pilot (starboard) and co-pilot/gunner (port) side by side. Provision for dual controls; control stick top common with Ka-50/52.

SYSTEMS: All Russian; Western equivalents optional for export.

AVIONICS: As above, including Pastel RWR and Otklik laser warning system.

Radar: Arbalet MMW antenna in nose.

EQUIPMENT: Cargo hook.

ARMAMENT: One-piece transverse boom through cabin, to rear of doors, optional to provide suspension for total of two B-8V-7 seven-round 80 mm rocket pods, two 7.62 mm or 12.7 mm gun pods, or similar armament.

DIMENSIONS, EXTERNAL: As Ka-62 except:
Fuselage length   13.465 m (44 ft 2 in)

WEIGHTS AND LOADINGS: Generally as Ka-62 except:
Max payload: external   2,750 kg (6,062 lb)
UPDATED

KAMOV Ka-62

TYPE: Multirole medium helicopter.

PROGRAMME: Funded under Russian programme for development of civil aviation for 2000. Construction of prototype Ka-62 (then known as V-62) began early 1990, but apparently abandoned; one Ka-60 military version (which see) and two Ka-62s intended to undertake flight trials, although second of basic type completed as Ka-60U, delaying debut of civil version; will be certified under Russian AP and FAR Pt 29A/B standards. Republic of Buryatia officially requested Moscow for production rights for UUAP (which see) in February 2000. In April 2001, Turkish Ministry of Public Health was discussing contract for six Ka-62s, with total value of US$31.5 million. Russian government's 2002-10 aviation plan includes Rb113 million to develop Ka-62 and launch production at both UUAP and RSK 'MiG' (LAPIK).

CURRENT VERSIONS: **Ka-62:** Basic model for domestic market. *Detailed description applies to the above.*

Ka-62M: To be certified to Western standards, for sale outside Russian Federation and Associated States (CIS); two 1,212 kW (1,625 shp) General Electric T700/CT7-2D1 turboshafts; five-blade main rotor; avionics to be developed by Aviapribor. Production version is expected to be manufactured by KnAAPO at Komsomolsk.

Ka-64 Sky Horse: This development of the Ka-60/62 series is reported to be a joint venture with Agusta of Italy intended for export. Features include a conventional tail rotor, modified passenger cabin, Western avionics and option of General Electric CT7-2DL, LHTEC T800 or RRTI RTM 322 turboshaft engines. Production would be by UUAP at Ulan-Ude.

DESIGN FEATURES: Originated as military transport; all main systems and components duplicated, with main and secondaries routed on opposite sides of airframe; transmission resistant to 12.7 mm bullets; main blades to 23 mm shells; run-dry gearboxes. Advanced technology main rotors with sweptback tips. Production versions will have slower-turning five-blade rotor. Yaw control by 11-blade fan-in-fin. Reverse tricycle landing gear.

STRUCTURE: Composites account for 60 per cent, by weight, of structure, including blades of main rotor; fuselage sides, doors, floor and roof, tailboom, fin, vertical stabilisers, and fan blades of carbon-reinforced Kevlar.

LANDING GEAR: Retractable reverse tricycle type; single KT-217 mainwheels retract inward and upward into bottom of fuselage; twin rear wheels retract forward into tailboom; shock-absorber in each unit. Optional inflatable pontoons for emergency use on water.

POWER PLANT: Basic Ka-62 has two RKBM Rybinsk RD-600V turboshafts, each 956 kW (1,282 shp) max continuous; 1,140 kW (1,529 shp) emergency rating. Fuel tanks under floor, capacity 1,450 litres (383 US gallons; 319 Imp gallons). General Electric T700/CT7-2D1 offered as alternative to RD-600V.

ACCOMMODATION: Crew of one or two, side by side; optional bulkhead divider between flight deck and cabin; up to 14 passengers in four rows; forward-hinged door each side of flight deck; large forward-sliding door and small rearward-hinged door each side of cabin; baggage hold to rear of cabin. VIP configuration to be available, with five to nine seats and refreshment bar.

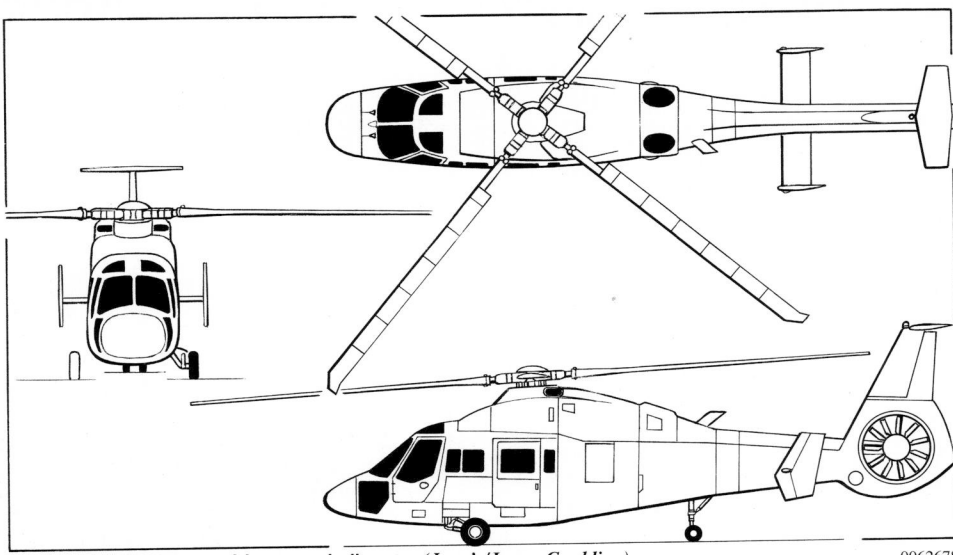

Kamov Ka-60 military multipurpose helicopter (*Jane's/James Goulding*)   0062678

Prototype Ka-60 utility helicopter (*Yefim Gordon*)   *NEW*/0137347

SYSTEMS: Interior heated and air conditioned. Thermoelectric de-icing system optional. Ivchenko AI-9V APU originally proposed, but replacement Aerosila TA-14 under development.

AVIONICS: Optional Russian or Western.

EQUIPMENT: Stretchers, hoist above port cabin door, cargo tiedowns, and other items as necessary for variety of roles, including transport of slung freight; air ambulance/operating theatre; search and rescue; patrol of highways, forests, electric power lines, gas and oil pipelines; survey of ice areas; surveillance of territorial waters, economic areas and fisheries; mineral prospecting; and servicing of offshore gas and oil rigs.

DIMENSIONS, EXTERNAL:
Main rotor diameter	13.50 m (44 ft 3½ in)
Tail rotor diameter	1.40 m (4 ft 7 in)
Length overall, rotors turning	15.60 m (51 ft 2¼ in)
Fuselage length	13.465 m (44 ft 2 in)
Height: overall	4.60 m (15 ft 1 in)
to top of rotor head	3.80 m (12 ft 5½ in)
Width over endplate fins	3.00 m (9 ft 10 in)
Wheel track	2.50 m (8 ft 2½ in)
Wheelbase	5.445 m (17 ft 10¼ in)
Cabin doors (each) Height	1.30 m (4 ft 3¼ in)
Width	1.25 m (4 ft 1¼ in)

DIMENSIONS, INTERNAL:
Cabin, excl flight deck: Length	3.40 m (11 ft 2 in)
Width	1.78 m (5 ft 10 in)
Height	1.30 m (4 ft 3¼ in)

AREAS:
Main rotor disc	143.10 m² (1,540.3 sq ft)
Tail rotor disc	1.54 m² (16.57 sq ft)

WEIGHTS AND LOADINGS:
Max payload: internal: 62, 62M	2,000 kg (4,409 lb)
external: 62, 62M	2,500 kg (5,510 lb)
T-O weight: 62, normal	6,000 kg (13,228 lb)
62, internal load, max	6,250 kg (13,779 lb)
62 external load and 62M, internal load, max	6,500 kg (14,330 lb)
62M, external load, max	6,750 kg (14,880 lb)
Max disc loading: 62	41.9 kg/m² (8.58 lb/sq ft)
62M	47.1 kg/m² (9.66 lb/sq ft)

PERFORMANCE (estimated, at max T-O weight with internal load):
Max level speed:	
62, 62M	162 kt (300 km/h; 186 mph)
Cruising speed: 62, 62M	148 kt (275 km/h; 171 mph)
Max rate of climb at S/L: 62	625 m (2,050 ft)/min
62M	690 m (2,263 ft)/min
Rate of climb at S/L, OEI: 62, 62M	123 m (403 ft)/min

Service ceiling: 62	5,150 m (16,900 ft)
62M	5,500 m (18,040 ft)
Hovering ceiling: IGE: 62	2,900 m (9,520 ft)
OGE: 62	2,100 m (6,880 ft)
62M	2,500 m (8,200 ft)
Range with max standard fuel at 2,000 m (6,560 ft), no reserves: 62	332 n miles (615 km; 382 miles)
62M	353 n miles (655 km; 407 miles)
Range with auxiliary fuel at 2,000 m (6,560 ft), no reserves: 62	566 n miles (1,050 km; 652 miles)
62M	553 n miles (1,025 km; 637 miles)
UPDATED

KAMOV Ka-115 MOSKVICH

With the exception of proposals to fit Progress (ZMKB) AI-450 or Turbomeca Arrius 2G turboshafts in place of the intended P&WC PW 206D engine, there have been no progress reports of this light utility helicopter programme in recent years. Known data last appeared in the 2001-02 *Jane's*.

UPDATED

Mockup of Kamov Ka-115   *NEW*/0137428

KAMOV Ka-215

TYPE: Light utility helicopter.

PROGRAMME: Twin-engined version of Ka-115 (which see), announced 2000, when cost estimated as more than US$1 million. No further information had been released by early 2002.

UPDATED

KAMOV Ka-226A SERGEI

TYPE: Light utility helicopter.

PROGRAMME: Turbine version of Ka-26, of which 816 built between 1968 and 1977. Announced at 1990 Helicopter Association International convention, Dallas, USA. Developed originally for Russian TsENTROSPAS disaster relief ministry, which is providing significant funding. First flight (RA-00199), at Lyubertsy, 3 September 1997; 'official' first flight on following day. Flew total of four sorties by 31 December 1997. Began AP-29 certification testing on 28 March 2001. State ground testing of Ka-226 second prototype completed at Strela's Orenburg plant on 6 March 2000. KAPP built two prototypes, of which first was rolled out at Kumertau on 29 May 1998; first production aircraft from KAPP was due to have flown in first quarter of 2002 but remained under construction in mid-2002.

Named Sergei in 1999, honouring politician Sergei Shoigu, but programme also guided by Sukhoi General Designer, Sergei Mikheev. By mid-2000, Moscow regional government had provided Rb12 million in development funding and was beginning disbursements under second programme valued at Rb4 million. Initial deliveries due in first half of 2000, but not effected. Certification now expected in third quarter of 2002.

Prototypes built jointly by Kamov, Strela, KAPP and Ufa Motors, with final assembly by KAPP and Strela for production aircraft. Strela scheduled to have delivered five preproduction helicopters to Kamov at Lyubertsy by mid-2000; these for MChS Rossii, but not now expected until 2002. KAPP designated second production plant; first batch of five under construction by 2001. Motor-Sich of Zaporozhye, Ukraine, negotiated with Kamov in June 2000 to build Ka-226s powered by indigenous ZMKB AI-450 engine; agreement on AI-450 installation signed 15 August 2001. On 19 October 2001, however, Motor Sich announced it would source all Ka-226 components with Ukrainian industry, if decision to proceed were taken. MoU on use of Turbomeca Arrius 2G signed in August 2001 with NPO Saturn and French manufacturer. Batch of Rolls-Royce engines ordered by Kamov in July 2000.

CURRENT VERSIONS: **Ka-226A**: *As described.*

Ka-226-50: Designation first revealed in September as 'improved' version.

CUSTOMERS: Orders by January 2002 totalled 66: Gazprom 50, Moscow City 10, TsENTROSPAS five and Bashkiriyan one.

Second Kamov Ka-226, showing detached cabin (*Jane's/Paul Jackson*) 0121032

Identified requirements include up to 20 for City of Moscow for patrol and medevac; some 250 for MChS Rossii/TsENTROSPAS disaster relief organisation; and up to 75 for Gazprom in gasfield support role. Firm order for 25 reportedly received from TsENTROSPAS by 1997, but quantity had reduced to 10 by 1999; by mid-2000 this quoted as five firm (to be first five production aircraft) and further 15 to be ordered by 2002; manufacture by Strela. Bashkiriyan local government ordered one Ka-226-50 in September 2001; this accepted 28 December 2001 (when still not cleared for flight) and due for trials at Zhukovsky before service entry. Funds for 22 of initial Gazprom order for 50 had been transferred by 2001, this initial batch, built by Strela, to have been received by 2005 (although formal signing of order for 50 was undertaken at Moscow Salon in

August 2001). Moscow city government signed US$1.5 million order for 10 in December 2001, delivery over two years, but later announcement indicated that funds had not been earmarked; Moscow's helicopters to be built by KAPP. Ka-226 was beaten by Kazan Ansat (which see) in competition to supply new training helicopter to Russian armed forces, announced September 2001, although small number of Ka-226s required by Russian Navy.

COSTS: US$1.5 million (2000). Development cost Rb108 million (1999).

DESIGN FEATURES: Classic Kamov utility helicopter, featuring interchangeable mission pods. Refined development of Ka-26/126; new rotor system with hingeless hubs and glass fibre/carbon fibre blades; changes to shape of nose, twin tailfins and rudders, and passenger pod; passenger cabin has much larger windows and remains interchangeable with variety of payload modules including agricultural systems with hopper capacity of 1,000 litres (264 US gallons; 220 Imp gallons); Kamov BP-226 transmission; new rotor system, interchangeable with standard coaxial system, will become available later.

FLYING CONTROLS: Assisted by irreversible hydraulic actuators. Automatic rotor constant-speed control; conventional four-channel control (longitudinal, lateral, cyclic and differential pitch). Two endplate fins and rudders, toed inward 15°; fixed horizontal stabiliser.

STRUCTURE: Primarily of aluminium alloys, steel alloys and composites sandwich panels of GFRP with honeycomb filler. Rotor blade overhaul interval 2,000 hours; total life 6,000 hours, but to be extended by increments to 18,000 hours.

LANDING GEAR: Non-retractable four-wheel type. Main units at rear, carried by stub-wings. All units embody oleo-pneumatic shock-absorber. Mainwheel tyres size 595×185, pressure 2.50 bar + 0.50 (36.25 lb/sq in + 7.25); forward tyres size 300×125, pressure 3.50 bar + 0.50 (50.75 lb/sq in + 7.25). Forward units of castoring type, without brakes. Rear wheels have pneumatic brakes.

POWER PLANT: Two 335 kW (450 shp) Rolls-Royce 250-C20R/2 turboshafts, side by side aft of rotor mast, with individual driveshafts to rotor gearbox. Two 335 kW (450 shp) Rolls-Royce 250-C20B engines in prototypes. Transmission rating 626 kW (840 shp). Alternatively, two Progress (ZMKB) AI-450 turboshafts, each 331 kN (444 shp) or two Turbomeca Arrius 2G (500 kW; 670 shp) or Klimov VK-800 turboshafts (588 kW; 789 shp).

Standard fuel capacity 770 litres (203 US gallons; 169 Imp gallons), in tanks above and forward of payload module area. Provision for two external tanks, on sides of fuselage, total capacity 320 litres (84.5 US gallons; 70.4 Imp gallons).

ACCOMMODATION: Fully enclosed and lightly pressurised flight deck, with rearward-sliding door each side; normal operation by single pilot; second seat and dual controls optional. Cabin ventilated, and warmed and demisted by air from combustion heater, which also heats passenger cabin when fitted. Space aft of cabin, between main landing gear legs and under transmission, can accommodate variety of interchangeable payloads. Cargo/passenger pod has two bench seats, each accommodating three persons; one bench faces forward, the other, rear; baggage compartment behind rear wall. Seventh passenger beside pilot on flight deck. Provision for cargo sling. Ambulance pod accommodates two stretcher patients, two seated casualties and medical attendant. For agricultural work, chemical hopper (capacity 1,000 litres; 264 US gallons; 220 Imp gallons) and dust spreader or spraybar are fitted in this position, on aircraft's CG. (Flight deck pressurisation protects crew against chemical ingress.)

Ka-226 prototype believed built by Strela NEW/0137429

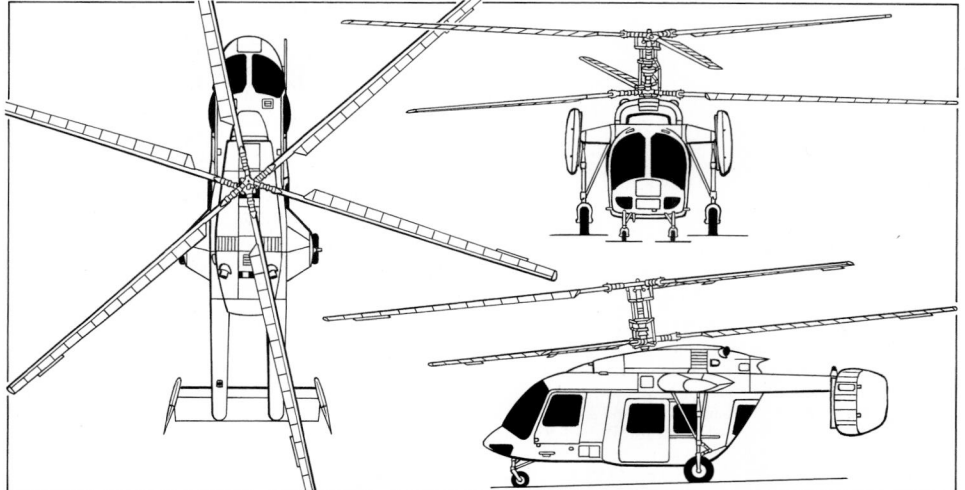

Kamov Ka-226A utility helicopter with production passenger cabin design (*Jane's/Mike Keep*) 0062676

Original Ka-226 prototype, wearing experimental probe (*Jane's/Paul Jackson*)   *NEW*/0137365

Aircraft can also be operated with either an open platform for hauling freight or hook for slinging loads at end of a cable or in a cargo net.

SYSTEMS: Single hydraulic system, with manual override, for control actuators. Main electrical system 27 V 3 kW DC, with back-up 40 Ah battery; secondary system 36/115 V AC with two static inverters; 115/200 V AC system with 16 kVA generator (6 kVA to power agricultural equipment and rotor anti-icing). Electrothermal rotor blade de-icing; hot air engine air intake anti-icing; alcohol windscreen anti-icing; electrically heated pitot. Pneumatic system for mainwheel brakes, tyre inflation, agricultural equipment

control, pressure 39 to 49 bar (570 to 710 lb/sq in). Oxygen system optional.

AVIONICS: Cockpit instrumentation and avionics to customer's choice, including Bendix/King equipment for IFR flight.

Comms: Optional Bendix/King KY 196A VHF radio; transponder.

Flight: Optional Bendix/King KN 53 ILS; KR 87 ADF; KLN 90B GPS; LCR 92 laser AHRS; and ADF.

EQUIPMENT: Specially equipped payload modules available for variety of roles, including ambulance and agricultural duties.

ARMAMENT: Optional provision for light weapons.

DIMENSIONS, EXTERNAL:

Rotor diameter (each)	13.00 m (42 ft 7¾ in)
Length of fuselage	8.10 m (26 ft 7 in)
Width over stub-wings	3.22 m (10 ft 6¾ in)
Height to top of rotor head	4.15 m (13 ft 7½ in)
Wheel track: nosewheels	0.90 m (2 ft 11½ in)
mainwheels	2.56 m (8 ft 4¾ in)
Wheelbase	3.48 m (11 ft 5 in)
Passenger pod: Length	2.35 m (7 ft 8½ in)
Width	1.40 m (4 ft 7 in)
Height	1.54 m (5 ft 0¾ in)

DIMENSIONS, INTERNAL:

Passenger pod: Length	2.04 m (6 ft 8¼ in)
Width	1.28 m (4 ft 2¼ in)
Height	1.40 m (4 ft 7 in)

AREAS:

Rotor disc (each)	132.70 m² (1,428.4 sq ft)

WEIGHTS AND LOADINGS:

Weight empty	1,952 kg (4,304 lb)
Max underslung payload	1,300 kg (2,865 lb)
Max internal fuel	600 kg (1,322 lb)
Auxiliary fuel	256 kg (564 lb)
T-O weight: normal	3,100 kg (6,835 lb)
max	3,400 kg (7,495 lb)
Transmission loading at max T-O weight and power:	
normal	4.95 kg/kW (8.14 lb/shp)
max	5.43 kg/kW (8.92 lb/shp)

PERFORMANCE (estimated):

Never-exceed speed (VNE)	115 kt (214 km/h; 133 mph)
Max level speed	111 kt (205 km/h; 127 mph)
Max cruising speed	105 kt (194 km/h; 120 mph)
Max rate of climb at S/L	610 m (2,000 ft)/min
Rate of climb, OEI	96 m (315 ft)/min
Vertical rate of climb at S/L	168 m (550 ft)/min
Service ceiling	6,200 m (20,340 ft)
Hovering ceiling OGE	2,500 m (8,200 ft)
Range: with max payload	16 n miles (30 km; 18 miles)
with max internal fuel	
	324 n miles (600 km; 372 miles)
with auxiliary fuel, no reserves	
	480 n miles (890 km; 552 miles)
Max endurance with internal fuel, no reserves	4 h 42 min

UPDATED

KAPO

KAZANSKOYE AVIATSIONNOYE PROIZVODSTVENNOYE OBEDINENIE IMENI S P GORBUNOVA (Kazan Aircraft Production Association named for S P Gorbunov)

ulitsa Dementiev 1, 420036 Kazan, Respublika Tatarstan
Tel: (+7 8432) 54 24 32
Fax: (+7 8432) 54 36 93
GENERAL DIRECTOR: Nail G Khairoullin

Since 1927, KAPO (formerly GAZ 22) has built more than 18,000 aircraft of 34 types, including the Tu-4, Tu-16, Tu-22, Tu-104 and Il-62. It currently produces the Tupolev Tu-214 and will manufacture the Tu-330 freighter, if it is ordered, and Tu-324 regional airliner. In 2000, KAPO belatedly delivered one Tu-160 strategic bomber, completion of which had been delayed by collapse of the former USSR; a second was to follow in 2002; two Tu-22Ms also in stock.

Under aerospace industry restructuring plans announced in 2001, KAPO is to be joined with RSK 'MiG', Kamov,

Tupolev and their associated plants (Sokol, Aviakor, Aviastar and KAPP). The plant also produces customer goods, including leather processing machinery.

UPDATED

KAPP

KUMERTSKOYE AVIATSIONNOYE PROIZVODSTVENNOYE PREDPRIYATIE (Kumertau Aviation Production Enterprise)

ulitsa Novozarinskaya 15A, 453350 Kumertau, Respublika Bashkortostan
Tel: (+7 34761) 422 53 and 423 00
Fax: (+7 34761) 439 13
Web: http://www.kumape.ru
GENERAL DIRECTOR: Boris S Malyshev
EXPORT DEPARTMENT: R M Rafikov

KAPP has manufactured helicopters, aeroplanes and related equipment at Kumertau since 1962. Major programmes have involved the Kamov Ka-26 and Ka-27/28/29/32 helicopter series, Myasishchev M-17/M-55, wings for the Tupolev Tu-154M transport, and unmanned air vehicles including the Tu-243 Reis-D reconnaissance system. KAPP is the only manufacturer of carbon fibre/glass fibre rotor blades in the Russian Federation and Associated States (CIS).

Current production includes main rotor blades and associated bushes for the Ka-226. KAPP built second and third prototype Ka-226s, and in mid-2002 had further helicopter under construction; will build the type for Moscow

city government, if requirement for 10 is confirmed. Ka-32 remains in low-volume production; Ka-31 being manufactured for India. KAPP has also been chosen to manufacture the Ilyushin Il-112 transport for the Russian Air Forces. Helicopter overhaul activities have included Indian Ka-28s.

The plant is to become part of the consortium led by RSK 'MiG' under aerospace restructuring plans announced by the Russian government in May 2001.

UPDATED

KAZAN

KAZANSKY VERTOLETNYI ZAVOD AO (Kazan Helicopter Plant JSC)

ulitsa Tetsevskaya 13/30, 420085 Kazan, Respublika Tatarstan
Tel: (+7 8432) 71 81 81
Fax: (+7 8432) 71 82 82
e-mail: market@kazanhelicopters.com
Web: http://www.kazanhelicopters.com
GENERAL DIRECTOR: Aleksandr P Lavrentyev
MARKETING DIRECTOR: Valery A Pashko
CHIEF ENGINEER: Igor S Bougakov
CHIEF DESIGNER: Aleksei Stepanov
ANSAT PROGRAMME DIRECTOR: Valery Kartashev

GAZ 387 founded in Leningrad in 1935 and re-established at Kazan on 15 August 1941 following evacuation. Built 11,344 Polikarpov Po-2 biplanes up to 1947, when production turned to combined harvesters. Since 1951, however, Kazan (KVZ) has marketed and built Mil helicopters comprising 30 Mi-1s,

3,257 Mi-4s, 4,066 Mi-8s, 2,620 Mi-8M/Mi-17s, 273 Mi-14s and (up to 2001) 1,094 Mi-8MTV/Mi-17-1Vs. Exports began in 1956 and now encompass over 70 countries and some 4,000 aircraft; the Mi-17 accounts for 1,200 exports to 30 countries.

Kazan became a joint stock company in 1994, owned 38 per cent by the state, 62 per cent by its employees; by 2002, ratios were 29.92 per cent state-owned, 28.82 privately owned, with Tatar regional government having deciding vote. Under plans announced by the Russian government in May 2001, it will become part of the consortium led by Sukhoi, joining the Ilyushin, Mil and Yakovlev bureaux, plus their associated manufacturing plants. At the 1995 Paris Air Show, Kazan exhibited a mockup of the first product of its own newly formed (1993) design office, the Ansat light multipurpose helicopter. Modifications and upgrades to Mi-17 have generated new business for Kazan. Russian certification for helicopter design awarded February 1997. A second light helicopter, the Aktai, was revealed at Moscow in August 1997 to be under development, although Kazan is also nominated as potential manufacturer of the Mil Mi-60.

Kazan is also a co-founder of Euromil, with Mil Helicopter Plant, Eurocopter and Klimov (see International section), and is manufacturing parts for the Mi-38 prototypes.

UPDATED

KAZAN/MIL Mi-17

TYPE: Medium transport helicopter.
Following entry refers specifically to the above variant; see Mil entry for other Mi-8M/Mi-17 versions.
PROGRAMME: Kazan's Mi-17 development intended to compete with latest Western helicopters (such as Sikorsky S-92, NHI NH 90 (see earlier entry on last year's edition) EHI EH 101 and Eurocopter AS 332) combining advanced avionics and systems with well-proven airframe and dynamic system. Prototype (RA-70937, converted from Mi-8MTV) displayed at 1995 Paris Air Show with features including widened forward, port, door; additional starboard door; and rear loading via short ramp and two clamshell doors. Further modified, with large, single-piece rear-loading ramp and other changes, for display at ILA

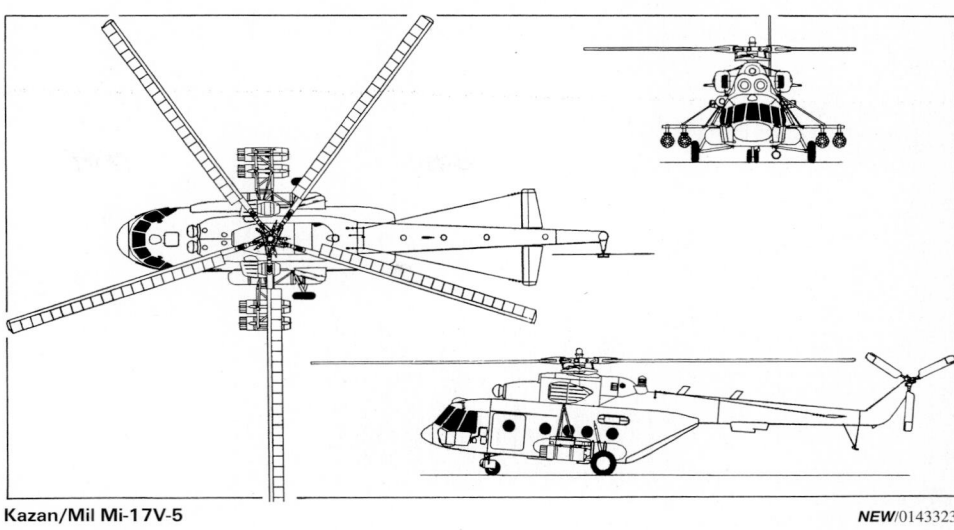

Kazan/Mil Mi-17V-5   *NEW*/0143323

Regular-size starboard passenger door of Kazan Mi-17-V5 (*Jane's/Paul Jackson*)   0121071

'96, Germany, in search and rescue form, designated Mi-17MD Night. By 1997, '70937 had gained a spine-mounted IR jammer and flight deck armour and was stated by Kazan to have the dual designation Mi-17MD/Mi-8MTV-5. At the 1999 Moscow Air Show it had lost its registration and become an 'Mi-8MTV5-1' with the alternative designations Mi-17N (*Noch:* Night) and (for export) Mi-17V-5. First flight in this guise was in January 1998; first production Mi-17V-5 flew August 1999.

Second example (RA-70877) shown at LIMA, Malaysia, November 1999, marked as Mi-17-1V, although with full Mi-17MD airframe modifications, was conversion of prototype Mi-17KF (see 2001-02 and earlier *Jane's*), destined for South Korean police force (eventually handed over, January 2000). Mi-17KF redesignated Mi-172.

Kazan demonstrator RA-70898, rebuilt in Mi-172 guise in 2000 (following accident) with partial 'glass cockpit' supplied by Elbit, GOES FLIR chin turret and additional external fuel tank (900 litres; 238 US gallons; 198 Imp gallons) each side, increasing range to 600 n miles (1,111 km; 690 miles); optimised for offshore oil support and maritime patrol; displayed at Farnborough, July 2000.

A further upgrade, designated Mi-8MTV-7, was due to have received an investment of Rb100 million during 2001.

CURRENT VERSIONS: **Mi-17V-5 (Mi-8MTV5-1):** *As described.*

Mi-17V-7: Proposed upgraded version with 1,765 kW (2,367 shp) VK-2500 engines, uprated transmission, new composites main blades, new tail rotor, 'glass cockpit', increased service ceiling and extra 500 kg (1,102 lb) payload.

CUSTOMERS: Four armed Mi-17MD exported to Rwanda, 1999, with a fifth (Mi-172) aircraft in VIP configuration. South Korea (unspecified quantity as Russian debt repayment). Indian Ministry of Defence ordered 40 Mi-17Ms from Kazan in May 2000; deliveries beginning a few months later.

DESIGN FEATURES: Generally similar to latest versions of Mi-17 (which see). Main differences include rear loading ramp and pointed 'Dolphin' nose with increased volume to allow space for radar. Conventional pod and boom configuration; five-blade main rotor, inclined forward 4° 30′ from vertical; interchangeable blades of basic NACA

230 section, twist 5°, thickness/chord ratio 13 per cent at root, 11.38 per cent at tip; spar failure warning system; drag and flapping hinges a few inches apart; flapping compensator and centrifugal blade droop stop; rotor brake standard; non-folding blades carried on machined spider; pendulum vibration damper; three-blade port tail rotor. Transmission comprises VR-14 two-stage planetary main reduction gearbox (engine input 15,000 rpm; oil capacity 49 litres; 12.9 US gallons; 10.8 Imp gallons), intermediate (8A-1515-000) and tail rotor (2Y6-1517-000) gearboxes, and drives off main gearbox for tail rotor, fan, AC generator, hydraulic pumps, compressor and tachometer generators; correct rotor speed maintained automatically by system that also synchronises output of the two engines; tail rotor pylon forms small vertical stabiliser; horizontal stabiliser near end of tailboom.

FLYING CONTROLS: Mechanical system, with irreversible hydraulic boosters; main rotor collective pitch control linked to throttles.

STRUCTURE: All-metal; semi-monocoque fuselage of riveted construction. Main rotor blades of aluminium alloy, each with extruded spar carrying root fitting, 21 honeycomb-filled trailing-edge pockets and blade tip, with titanium abrasion shielding on some of leading-edge; balance tab on each blade; each tail rotor blade made of spar and honeycomb-filled trailing-edge. Composites blades and elastomeric hub components under development.

LANDING GEAR: Non-retractable tricycle type; self-centring twin-wheel nose unit, locked in flight; single wheel on each main unit; oleo-pneumatic (gas) shock-absorbers. Mainwheel tyres 865×280 type K2-110, pressure 5.4 bar (78 lb/sq in); nosewheel tyres 595×185 type KT-97/3, pressure 4.4 bar (64 lb/sq in). Pneumatic shoe-type brakes on mainwheels. Optional mainwheel fairings. Optional Aerozur four-bag inflatable emergency floats.

POWER PLANT: Two Klimov TV3-117VM turboshafts; installed ratings of 1,618 kW (2,170 shp) in emergency; 1,471 kW (1,973 shp) for T-O; 1,250 kW (1,677 shp) normal; and 1,103 kW (1,479 shp) for cruising. Engine cowling side panels form maintenance platforms when open, with access via hatch on flight deck.

Standard fuel capacity of 2,725 litres (720 US gallons, 599 Imp gallons) in two main external tanks and internal feed tank. Up to four ferry tanks in cabin, each 915 litres (242 US gallons; 201 Imp gallons). Engine oil capacity 15 litres (4.0 US gallons; 3.3 Imp gallons).

ACCOMMODATION: Two pilots side by side on flight deck, with provision for flight engineer's station. Windscreen de-icing standard. Up to 36 troops in cabin. Hydraulically actuated rear ramp.

Sliding, jettisonable passenger doors at front of cabin; electrically operated rescue hoist can be installed at this doorway.

SYSTEMS: AP-34B autopilot. AI-9V APU, providing 3 kW DC for engine starting, mounted externally, starboard side; AC electrical supply from two 40 kW three-phase 115/220 V 400 Hz GT40P488 generators; two 25 V 20 Ah batteries. KO-50 kerosene heater in cabin. Main and auxiliary hydraulic systems, operating pressure between 44 and 64 bar (640 to 924 lb/sq in). Pneumatic system for braking; pressure 4 to 5 bar (58 to 73 lb/sq in); reservoir

US Special Forces Mil Mi-8MTV-1 built at Kazan in 1991 as c/n 95747 and currently civilian-registered in the USA as N353MA to an undisclosed owner in Bethesda, Maryland   *NEW*/0528616

Kazan-built Mi-17V-5 of Indian Air Force   *NEW*/0102982

Rear loading ramp of Kazan Mi-17-V5, showing curved sides (*Jane's/Paul Jackson*)   0121069

volume 10 litres (2.6 US gallons; 2.2 Imp gallons). Engine air intake de-icing standard. KKO oxygen system.

AVIONICS: *Comms:* Yadro-1A1 and KHF 950 VHF; P-863 VHF/UHF; ACR-500 VHF/UHF; Orlan-85ST UHF; UVD CO-70 or CO-96 transponder.

Radar: Type 8A-813, RDR 1400 or RDR 2000 weather radar optional.

Flight: Radio altimeter; radio compass; VOR/ILS; VIM-95; CD-75 VND-94 or Bendix/King KN 62 DME.

Self-defence (optional): ASO-2V chaff/flare dispensers under tailboom and L166V1AE IR jammer (NATO 'Hot Brick') at forward end of tailboom.

EQUIPMENT: Optional medical evacuation and SAR fitments. LPG-150M winch for cargo loading; optional SLG-300 300 kg (661 lb) external SAR winch.

ARMAMENT: Optional armament on two external pylons. Typical loads include UPK-23-250 23 mm cannon pods, 500 kg bombs and rocket pods.

DIMENSIONS, EXTERNAL:
Main rotor diameter	21.295 m (69 ft 10½ in)
Main rotor blade chord	0.52 m (1 ft 8½ in)
Tail rotor diameter	3.91 m (12 ft 9⅞ in)
Length: overall, rotors turning	25.31 m (83 ft 1¼ in)
fuselage	18.99 m (62 ft 3½ in)
Width over weapon pylons	7.20 m (23 ft 7½ in)
Tailplane span	3.705 m (12 ft 1¾ in)
Height: overall, rotors turning	5.545 m (18 ft 2¼ in)
to top of rotor head	4.865 m (15 ft 11½ in)
Wheel track: main	4.51 m (14 ft 9½ in)
nose	0.30 m (11¾ in)
Wheelbase	4.28 m (14 ft 0½ in)
Fwd passenger door (stbd): Height	1.405 m (4 ft 7¼ in)
Width	0.825 m (2 ft 8½ in)
Fwd passenger door (port): Height	1.405 m (4 ft 7¼ in)
Width	1.215 m (3 ft 11¾ in)
Rear cargo ramp: Height	1.56 m (5 ft 1½ in)
Width	2.30 m (7 ft 6½ in)

DIMENSIONS, INTERNAL:
Cabin: Length	5.34 m (17 ft 6¼ in)
Width	2.34 m (7 ft 8¼ in)
Height	1.80 m (5 ft 10¾ in)

AREAS:
Main rotor disc	356.16 m² (3,833.7 sq ft)
Tail rotor disc	12.01 m² (129.28 sq ft)

WEIGHTS AND LOADINGS:
Weight empty, equipped	7,468 kg (16,464 lb)
Max payload:	
internal	4,000 kg (8,820 lb)
external, on sling	5,000 kg (11,023 lb)
Max T-O weight:	
internal load	13,000 kg (28,660 lb)
external load	13,500 kg (29,762 lb)
Max disc loading:	
internal load	36.5 kg/m² (7.48 lb/sq ft)
external load	37.9 kg/m² (7.76 lb/sq ft)

PERFORMANCE:
Max level speed	135 kt (250 km/h; 155 mph)
Max cruising speed	124 kt (230 km/h; 143 mph)
Econ cruising speed	70 kt (130 km/h; 81 mph)
Max forward rate of climb at S/L	624 m (2,047 ft)/min
Service ceiling	6,000 m (19,680 ft)
Hovering ceiling, OGE	3,980 m (13,055 ft)
Range with 5% reserves:	
with 4,000 kg (8,820 lb) payload	
	189 n miles (350 km; 217 miles)
ferry: normal fuel	386 n miles (715 km; 444 miles)
with max cabin fuel	
	890 n miles (1,650 km; 1,025 miles)

UPDATED

KAZAN ANSAT

English name: Light

TYPE: Light utility helicopter.

PROGRAMME: Design begun at Kazan in 1993; design subcontracts to Kazan State Technical University for structural strength and aerodynamic calculations; Aviacon Scientific and Production Centre for rotor; and

Kazan's Mi-172 demonstrator RA-70898 *(Valery Solomakhine/KVZ)* 0121664

Ferry fuel tanks in cargo hold of an Mi-17 0102974

Aeromekhanica for transmission. Name in Tatar language variously translated as *easy, simple, ethereal* and *light*.

Fuselage mockup exhibited at 1995 Paris Air Show; considerably revised engineering mockup (001) at Paris '97; by August 1998, now marked '01', this had accumulated 10 hours of ground running with engines and rotors; total 350 hours by late 2001.

First flight scheduled for late 1997, but initial designated flight trials aircraft (02) exhibited at Farnborough in September 1998, still unflown. First flight (02) was 12 minute hover on 17 August 1999; initial forward flight on 6 October 1999. Trials halted in November 1999, after 4 hours, due to gearbox problems; resumed in second quarter

of 2000 with strengthened and redesigned main transmission, scarfed engine exhausts and new identity '902'. Total 70 hours in 260 sorties up to late 2001.

Third (second flying) prototype (03) was to have joined the programme in late 1999, but not completed until August 2001. First flown 27 December 2001, this is to pre-production standard with small pannier tanks, increased fin area, PW207 engines, additional side window and flatter windscreen combined with revised nose shape; will add 400 hours to trials programme. Certification postponed to third quarter of 2002, then further, to 2003. Production version, of which manufacture began in 2001, expected to have skids repositioned for nose-up ground angle. Trials continuing with fourth airframe for static tests (completed late 2001) and Nos. 5, 6 and 7 for flight evaluation.

On 14 September 2001, Ansat declared winner of competition to supply 100 training helicopters to Russian armed forces by 2015 (beating Mi-34 and Ka-226).

CURRENT VERSIONS: Can be optimised for transport (including underslung load), ambulance, SAR, training, patrol and other duties. Provision for attachment of sponsons or tanks to sides of cabin.

Ansat: *As described.*

Ansat-U: Optimised for training (*uchebni*). Aircraft No. 5 to this standard, including dual controls and wheel landing gear.

Ansat-AG: Version for Gazprom pipeline inspection.

Ansat Nablyudatel: (Ansat Observation) Mockup of proposed scout helicopter based on Ansat was shown at Moscow Salon in August 2001. GOES 521 turret in nose with Svishch sensors integrated by Optooil ZAO. Max useful load 1,500 kg (3,307 lb); max T-O weight 3,500 kg (7,716 lb); max level speed 162 kt (300 km/h; 186 mph); range 383 n miles (710 km; 441 miles) on internal fuel, 691 n miles (1,280 km; 795 miles) with external tanks; service ceiling 5,700 m (18,700 ft); hovering ceiling

Mockup of proposed scout helicopter based on Ansat *(Jane's/Paul Jackson)* 0121712

Prototype Ansat wearing test boom

NEW/0102977

Mockup of medical evacuation Kazan Ansat, equipped with SAR winch (*Jane's/Paul Jackson*)
0121711

Ansat engine installation *NEW*/0102979

3,800 m (12,460 ft). Powered by two PW207K turboshafts, each rated at 522 kW (700 shp).

CUSTOMERS: Armed forces of Russia to purchase 100 by 2015. Russian Federal Border Service (Federaliya Pogranichiya Sluzhba) requirement for 100 notified in 1997. Under consideration by Gazprom in 2000 for pipeline inspection (100 required). Total 10 civil sales reported by mid-2002.

COSTS: US$1.8 million to US$2.0 million for utility version (2002). Kazan's development expenditure had reached Rb200 million by mid-2000.

DESIGN FEATURES: Optimised for aerodynamic efficiency, survivability in emergency landing (1 m; 3 ft/s power-off touchdown speed) and off-base routine maintenance. Traditional metal structure for simplicity and reliability. Meets FAR Pt 29 Category A and Russian AP-29 requirements. First Russian helicopter with shock-absorbing seats for passengers. Conventional configuration, with high-mounted tailboom carrying fixed horizontal stabiliser and twin fins; power plant above cabin. Hingeless main rotor hub with glass fibre torsion bar; four main blades; two-blade tail rotor. Two-stage, VR-23 main rotor reduction gear in magnesium case ahead of engines; ratio 16.4; rotation speed 365.4 rpm; blade tip speed 220 m (722 ft)/s. Transmission rating 769 kW (1,031 shp). Tail rotor speed 2,000 rpm via single stage conical geabox. Rotor brake. Manual blade folding.

Main rotor aerofoil section NACA 23012.

STRUCTURE: Aluminium alloy fuselage; sparing use of composites; layered glass fibre main rotor blades, window frames and nosecone.

LANDING GEAR: Twin skids with Kazan transverse shock-absorbers; tail bumper to protect anti-torque rotor. Wheels optional; tricycle configuration, with Yaroslav tyres and Gidroagregat (Balashikha) brakes. Optional emergency flotation system.

POWER PLANT: Two P&W Rus XRK206S turboshafts, each rated at 477 kW (640 shp) for T-O, 418 kW (560 shp) max continuous, in prototypes. Production version with PW207Ks, rated at 470 kW (630 shp) for T-O, 410 kW (550 shp) max continuous, 529 kW (710 shp) for 30 s, 470 kW (630 shp) continuous OEI and 491 kW (659 shp) 2 minutes OEI. FADEC standard. Fuel capacity 700 litres (185 US gallons; 154 Imp gallons) in either external panniers or underfloor. Optional internal ferry fuel.

ACCOMMODATION: Up to 11 persons, including one or two pilots, on energy-absorbing seats; or two stretcher patients and three attendants; or internal or externally slung freight. Two forward-hinged doors each side of flight deck; two horizontally split doors each side of cabin, forward; baggage bay behind cabin, with rear-facing door. Baggage door also used for loading stretchers of medical variant. Accommodation ventilated and heated; optional air conditioning.

SYSTEMS: Avionika FBW controls comprise quadruplex electronic system and duplex hydraulic system. Automatic flight control is standard on all piloting functions and optional on navigation functions. Current FBW system to be replaced by KSU-A digital control system. Main transmission drives two alternators (each 200 V, 400 Hz), two generators (each 27 V), two fans and two hydraulic fuel pumps for separate systems. Electrical system 27 V, with battery; optional AC system, with second battery. Electric de-icing optional.

AVIONICS: Standard Russian avionics and instruments (apart from prototypes' BAE Systems North America engine parameters display, which to be replaced by Ulyanovsk BISK-A system); full Western avionics fit available as an option.

Radar: Provision in nosecone.

EQUIPMENT: Cabin tie-down points. Optional rescue hoist and sling.

ARMAMENT: Optional external rocket and machine gun pods.

DIMENSIONS, EXTERNAL:
Main rotor diameter	11.50 m (37 ft 8¾ in)
Main rotor blade chord	0.32 m (1 ft 0½ in)
Tail rotor diameter	2.10 m (6 ft 10½ in)
Length: overall, rotors turning	13.76 m (45 ft 1¾ in)
fuselage	11.06 m (36 ft 3½ in)
fuselage, tail rotor turning	11.54 m (37 ft 10¼ in)
pod	6.95 m (22 ft 9½ in)
Height to top of rotor head	3.40 m (11 ft 1¾ in)
Width: fuselage	1.80 m (5 ft 10¾ in)
over optional sponsons	3.61 m (11 ft 10¼ in)
Skid track	2.50 m (8 ft 2½ in)

DIMENSIONS, INTERNAL:
Cabin: Length	3.15 m (10 ft 4 in)
Max width	1.65 m (5 ft 5 in)
Max height	1.30 m (4 ft 3¼ in)
Volume	6.7 m³ (237 cu ft)

AREAS:
Main rotor disc	103.87 m² (1,118.0 sq ft)
Tail rotor disc	3.46 m² (37.28 sq ft)

WEIGHTS AND LOADINGS:
Weight empty, equipped	1,900 kg (4,189 lb)
Max fuel weight	540 kg (1,190 lb)
Max payload: internal	1,000 kg (2,204 lb)
external	1,300 kg (2,866 lb)

T-O weight, internal or external load:
normal	3,000 kg (6,613 lb)
max	3,300 kg (7,275 lb)
Max disc loading	31.8 kg/m² (6.51 lb/sq ft)

Transmission loading at max T-O weight and power
4.29 kg/kW (7.06 lb/shp)

PERFORMANCE:
Never-exceed speed (VNE)	153 kt (285 km/h; 177 mph)
Max level speed	148 kt (275 km/h; 171 mph)
Max cruising speed	135 kt (250 km/h; 155 mph)
Max rate of climb at S/L	960 m (3,150 ft)/min
Max rate of climb at S/L, OEI	240 m (787 ft)/min
Service ceiling	4,500 m (14,760 ft)
Service ceiling, OEI	2,600 m (8,540 ft)
Hovering ceiling: IGE	2,800 m (9,180 ft)
OGE	2,600 m (8,540 ft)
Range: max normal fuel	291 n miles (540 km; 335 miles)
ferry fuel	702 n miles (1,300 km; 807 miles)
Endurance	3 h 20 min

UPDATED

Profile of Kazan Ansat in proposed production configuration 0121669

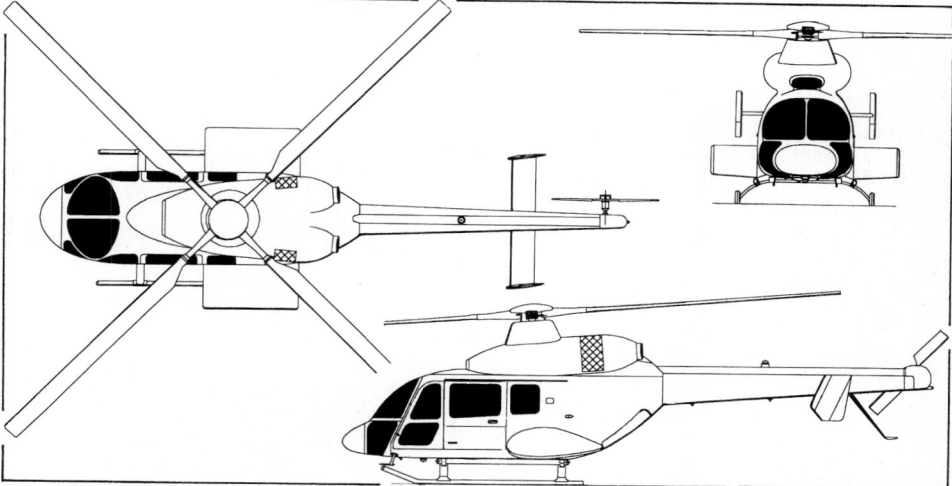

Kazan Ansat light multipurpose helicopter fitted with enlarged sponsons (*Jane's/Paul Jackson*)

KAZAN AKTAI

English name: White Colt

TYPE: Three-seat helicopter.

PROGRAMME: Work began 1996. Project revealed at Moscow Air Show, 19 August 1997, with display of mockup. Second, engineering mockup built 1998. Kazan decided in mid-2001 to begin building a prototype to fly in 2002. Production to begin in 2004.

CURRENT VERSIONS: Passenger, medical, evacuation and military configurations planned.

COSTS: US$280,000 to US$300,000, according to equipment standard (2000). Estimated (2000) market for over 1,000. Development cost (2000) US$10 million.

DESIGN FEATURES: Intended for light transport, patrol, medical evacuation and training. Certification will be to FAR Pt 27. Conventional pod and boom configuration with T tail. Semi-articulated three-blade main rotor; tapered blades on production version; max (emergency) rotation speed 2,586 rpm; rotor brake; no provision for folding; VR-10 main gearbox, transmission rating 216 kW (290 hp); engine input 6,000 rpm. XP-10 gearbox for two-blade tail rotor.

STRUCTURE: Extensive use of composites throughout, including main rotor blades.

LANDING GEAR: Conventional, non-retractable skids on arched support tubes; spring shock absorbtion.

POWER PLANT: One 201 kW (270 hp) VAZ-4265 rotary motorcar engine mounted above the cabin and behind the gearbox and operating on 92/93 grade petrol. Manual rpm control. Fuel capacity 300 litres (79.2 US gallons; 66.0 Imp gallons). Refuelling point port side. Oil capacity 20 litres (5.3 US gallons; 4.4 Imp gallons).

ACCOMMODATION: Pilot and two passengers, all side by side, in individual seats; forward-hinged door each side. Twin clamshell doors at rear of pod, below tailboom, provide access to flat freight floor; one casualty stretcher can be loaded through rear doors when passenger seat is removed. Alternative access via door on starboard side. Optional, rear-facing seat in luggage bay. Optional heating and air conditioning.

SYSTEMS: Electrical system with accumulator and M-16 generator, max rating 16 kW. Anti-icing of main rotor blades.

AVIONICS: *Comms:* Yurok radio.
   Flight: KG-1B navigation system; GPS.

EQUIPMENT: Inset toe-steps for maintenance access.

DIMENSIONS, EXTERNAL:

Main rotor diameter	10.00 m (32 ft 9¾ in)
Main rotor blade mean chord	0.24 m (9½ in)
Tail rotor diameter	1.48 m (4 ft 10¼ in)
Length overall, rotors turning	10.88 m (35 ft 8¼ in)
Fuselage: Length	8.35 m (27 ft 4¾ in)
Max width	1.70 m (5 ft 7 in)
Max height	1.48 m (4 ft 10¼ in)
Height overall	2.69 m (8 ft 10 in)
Tailplane span	1.44 m (4 ft 8¾ in)
Skid track	2.10 m (6 ft 10¾ in)
Passenger doors, both:	
Max height	1.00 m (3 ft 3¼ in)
Max width	0.80 m (2 ft 7½ in)
Baggage door, starboard:	
Max height	0.60 m (1 ft 11½ in)
Max width	0.70 m (2 ft 3½ in)

DIMENSIONS, INTERNAL:

Cabin: Length	1.20 m (3 ft 11¼ in)
Max width	1.70 m (5 ft 7 in)
Floor area	2.80 m² (30.1 sq ft)
Volume	4.0 m³ (141 cu ft)
Freight compartment: Length	1.20 m (3 ft 11¼ in)
Max width	1.00 m (3 ft 3¼ in)
Max height	1.48 m (4 ft 4 in)
Volume	2.8 m³ (99 cu ft)

AREAS:

Main rotor blades, each	0.99 m² (10.66 sq ft)
Tail rotor blades, each	0.11 m² (1.18 sq ft)
Main rotor disc	78.54 m² (845.4 sq ft)
Tail rotor disc	1.72 m² (18.52 sq ft)

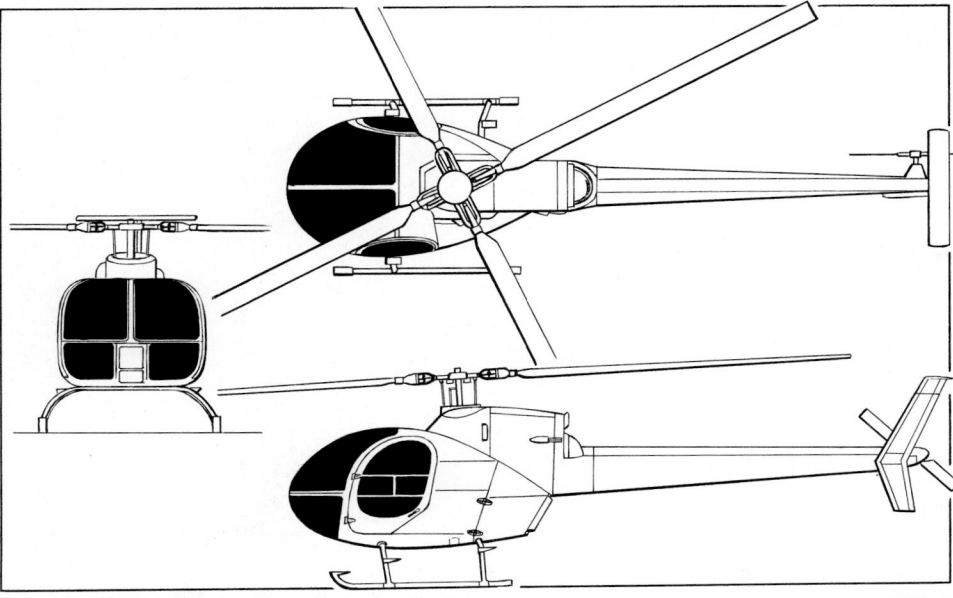

Kazan Aktai light helicopter (*Jane's/James Goulding*)   0131345

Mockup of the Kazan Aktai (*Valery Solomakhine/KVZ*)   0121670

Fins, each	0.40 m² (4.31 sq ft)	Max level speed at S/L	103 kt (190 km/h; 118 mph)
Tailplane	0.35 m (3.77 sq ft)	Econ cruising speed	84 kt (155 km/h; 96 mph)
WEIGHTS AND LOADINGS:		Max rate of climb at S/L	336 m (1,102 ft)/min
Weight empty	605 kg (1,334 lb)	Service ceiling	4,700 m (15,420 ft)
Max payload	270 kg (595 lb)	Hovering ceiling: IGE	2,300 m (7,540 ft)
Max T-O weight: normal	1,050 kg (2,314 lb)	OGE	1,300 m (4,260 ft)
max	1,150 kg (2,535 lb)	Range:	
Max disc loading	14.64 kg/m² (3.00 lb/sq ft)	with max payload	216 n miles (400 km; 248 miles)
Max power loading	5.71 kg/kW (9.39 lb/hp)	with max fuel	324 n miles (600 km; 372 miles)
PERFORMANCE (estimated):		Endurance	4 h
Never-exceed speed (V NE)	135 kt (250 km/h; 155 mph)		*UPDATED*

KHRUNICHEV

GOSUDARSTVENNYI KOSMICHESKII NAUCHNO-PROIZVODSTVENNYI TSENTR IMENI M V KHRUNICHEVA (State Research and Production Space Centre named for M V Khrunichev)

ulitsa Novozavodskaya 18, 121309 Moskva
Tel: (+7 095) 145 98 02
Fax: (+7 095) 145 92 03
e-mail: proton@online.ru
Web: http://www.khrunichev.com
GENERAL DIRECTOR: Anatoly I Kiselyov
DEPUTY GENERAL DIRECTORS:
   Alexander V Lebedev
   Anatoly A Kalinin
DIRECTOR, AVIATION PROGRAMME: Maksim Glazkov

Khrunichev Aviatekhnika
Tel: (+7 095) 145 93 33
Fax: (+7 095) 145 99 53

Khrunichev's T-208 Orel twin-turboprop utility transport (see 1998-99 *Jane's***) was again displayed in mockup form at Moscow in August 2001, but is no nearer to flight** (*Jane's/Paul Jackson*)   *NEW*/0126847

e-mail: proton@online.ru
Web: http://www.aircraft.avias.com
CHIEF DESIGNER: Evgeny P Grunin
DEPUTY CHIEF DESIGNER: Arnold I Andrianov
AVIATION DEPARTMENT MANAGER: Sergei M Zhiganov
RESEARCH AND DEVELOPMENT DIVISION MANAGER:
  Yuri I Polatsky
DESIGN DIVISION MANAGER: Mikhail M Vasyuk

The Aviation Department of this prominent rocket and space vehicle design and manufacturing centre was formed in August 1994 to develop and manufacture general purpose aircraft. It comprises a design bureau, experimental facilities and operational and maintenance services. The main task of the Aviation Department is to develop to the series production stage aircraft designed on the basis of foreign and domestic components, and then to market them worldwide. Production is being launched of the T-411 Aist, while development efforts are concentrated on the T-440 Mercury turboprop-twin and newly revealed T-419/T-517 agricultural aircraft.

UPDATED

Mockup of Khrunichev T-21 Sapsan (*Jane's/Paul Jackson*)   0126842

Prototype Khrunichev Aist (VOKBM M-14X radial) (*Jane's/Paul Jackson*)   **NEW**/0143707

KHRUNICHEV T-21 SAPSAN

English name: Hawk

TYPE: Two-seat lightplane
PROGRAMME: Announced 2000; mockup displayed statically at MAKS 2001 at Moscow. Prototype was to be built in 2002.
DESIGN FEATURES: Powered, V-strut-braced wing with boom-mounted cruciform empennage and underslung fuselage pod. Intended for training, sport flying, aerial photography/surveillance and crop-spraying.
FLYING CONTROLS: Conventional and manual; two-segment Fowler-type flaps occupy some two-thirds of each wing.
STRUCTURE: Triangular section welded steel truss fuselage with rectangular section light alloy tailboom; light alloy two-spar tailplane and fin, metal skinned, with Stits fabric covering on rudder and elevators.
LANDING GEAR: Tricycle type; fixed. Single wheel on each main unit, twin nosewheels. Mainwheel size 500×150, nosewheel size 310×135. Semi-lever type suspension on main units; castoring nosewheel.
POWER PLANT: One 104.4 kW (140 hp) Walter M-332 inline four mounted at junction of wing/tailboom. Fuel contained in a single tank aft of cockpit bulkhead.
ACCOMMODATION: Two persons, side by side in streamlined enclosed cockpit pod with large windscreen and jettisonable transparent door on each side. Cargo compartment, capacity 200 kg (441 lb) aft of cockpit, with upward-hinged access door.

DIMENSIONS, EXTERNAL:
Wing span	11.00 m (36 ft 1 in)
Length overall	6.70 m (21 ft 11¾ in)

WEIGHTS AND LOADINGS:
Weight empty	540 kg (1,190 lb)
Max fuel weight	70 kg (154 lb)
Max T-O weight	750 kg (1,653 lb)
Max power loading	7.19 kg/kW (11.81 lb/hp)

PERFORMANCE:
Cruising speed	81 kt (150 km/h; 93 mph)
T-O run	60 m (200 ft)
T-O to 15 m (50 ft)	150 m (495 ft)
Landing from 15 m (50 ft)	290 m (955 ft)
Landing run	50 m (165 ft)
Max range	486 n miles (900 km; 559 miles)

UPDATED

KHRUNICHEV T-411 AIST

English name: Stork

TYPE: Utility kitbuilt.
PROGRAMME: Programme started in 1994 when licence acquired from Aeroprogress/ROKS-Aero to produce the T-411 Aist-2 (see 1997-98 and previous editions); Aist-2 design started November 1992; construction of prototype began April 1993; first flight (marked only 'T-411') 10 November 1993 in standard form; 15 March 1994 with ski landing gear. Khrunichev prototype (RA-01585), with several detail changes, first exhibited at Moscow Air Show 1997. Second prototype reported nearing first flight in January 2002. Initially designated T-411 Wolverine, which name now transferred to T-421 (which see). North American promotion (in kit form) by Washington Aeroprogress; alternative power plant is Chevrolet eight-cylinder motorcar engine; demonstrator (M-14 engine, and possibly a grounded prototype) 'N01522' based at Boeing Field, Washington, 1998.

Manufacture intended by RSK 'MiG', whose Voronin plant contracted in March 2000 to build 15 for delivery in 2001, but delivery target slipped to first quarter 2002, with all 15 to be handed over during following six months. However, by June 2002, talks between Khrunichev and RSK 'MiG' still had not reached a satisfactory conclusion, although first aircraft then nearing completion. Manufacture of kits is also envisaged.
CURRENT VERSIONS: **T-411**: *As described.*
  T-411A: Agricultural version.
  T-411P: Floatplane version with twin floats, for operations in seas with wave height up to 0.35 m (1 ft

1¾ in); length overall 10.00 m (32 ft 9¾ in); maximum take-off weight 1,650 kg (3,637 lb); payload 440 kg (970 lb); cruising speed 113 kt (210 km/h; 130 mph); T-O run 315 m (1,033 ft); T-O to 15 m (50 ft) 475 m (1,558 ft); landing from 15 m (50 ft) 305 m (1,000 ft); landing run 195 m (640 ft); maximum range 607 n miles (1,125 km; 699 miles).
  T-411F: As T-411P, but with amphibious floats.
  T-411L: Ski-equipped version.
  T-411LY: Textron Lycoming TIO-540 powered version, see Power Plant below.
  T-411 Turbo: Walter M 601B turboprop. Incomplete prototype shown in Moscow, August 2001.
  T-421: Described in 2001-02 *Jane's*.
CUSTOMERS: Orders received from Russian forestry, fisheries and national emergency ministries, Australia, Malaysia and USA from which approximately 70 orders received, including aircraft supplied in kit form.
COSTS: US$130,000 with M-14 engine; US$20,000 more with Lycoming engine (2002).
DESIGN FEATURES: Conventional strut-braced high-wing monoplane. Constant-chord unswept wings with fixed leading-edge slat, slotted ailerons and single-slotted, three-position (0, 20, 40°) trailing-edge flaps; V bracing struts. Rectangular-section fuselage. Unswept tail unit with dorsal fin; constant-chord horizontal tail surfaces. Compared with the Aeroprogress Aist-2, the Khrunichev version has slightly stretched airframe and more roomy cabin; different landing gear and power plant.

Wing section NACA 23011; dihedral 2° from root; incidence 3° 30'.
FLYING CONTROLS: Conventional and manual. Actuation by pushrods and cables; electrically controlled trim tab in each elevator; fixed full-span leading-edge slats; slotted flaps.
STRUCTURE: Fuselage structure of 4130 steel tube covered with Dacron synthetic fabric and aluminium alloys. Two-

spar, constant-chord wings; metal skin on leading-edge and over fuel tanks at root, Dacron covered between spars. Metal tail surfaces, Dacron covered.
LANDING GEAR: Non-retractable tailwheel type; single wheel on each unit, with fairings on mainwheels. Arched cantilever tubular spring main legs. Self-centring castoring tailwheel. Mainwheel tyres 600×180; tailwheel 310×135. Brakes on mainwheels.
POWER PLANT: One 268 kW (360 hp) VOKBM M-14X nine-cylinder radial with Mühlbauer MTV-9 three-blade variable-pitch propeller, or 261 kW (350 hp) Textron Lycoming TIO-540 flat-six with Hartzell three-blade variable-pitch propeller. Two main and two auxiliary fuel tanks in wingroots, total capacity 340 litres (89.8 US gallons; 74.8 Imp gallons); gravity fuelling. Oil capacity 22.0 litres (5.8 US gallons; 4.8 Imp gallons).
ACCOMMODATION: Pilot and three or four passengers, on side-by-side front seats and rear bench seat; starboard front seat and rear bench removable, or bench seat foldable, for carrying equivalent freight; convertible into ambulance for one stretcher patient, one seated casualty and medical attendant in addition to pilot. Dual controls standard. Baggage/cargo compartment aft of rear seats, with large upward-opening door on port side, used also for loading freight or stretcher. Forward-hinged jettisonable door each side of cabin. Ventilation and heating standard.
SYSTEMS: Pneumatic system for starting engine and operating mainwheel brakes; pressure 50 bar (725 lb/sq in). Electrical system includes GSR-3000M engine-driven 27 V DC generator and 28 Ah 12CAM-28 battery.
AVIONICS: Russian or Western (Honeywell) avionics available, including GPS. Optional autopilot.

DIMENSIONS, EXTERNAL:
Wing span	13.02 m (42 ft 8½ in)
Wing aspect ratio	7.0
Length overall	9.45 m (31 ft 0 in)
Height overall	2.60 m (8 ft 6¼ in)

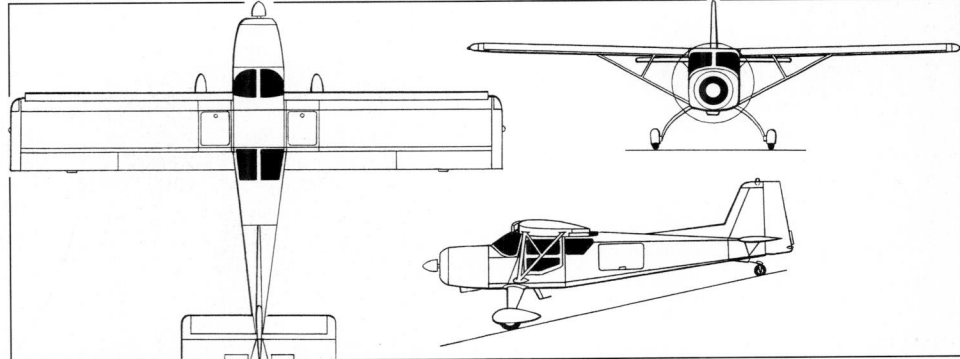
Khrunichev T-411 Aist four/five-seat light aircraft (*Jane's/James Goulding*)   0079312

Prototype Khrunichev T-411 Turbo on show at Moscow Salon, August 2001 *(Jane's/Paul Jackson)*
NEW/0143708

Wheel track	2.59 m (8 ft 6 in)
Wheelbase	6.50 m (21 ft 4 in)
Propeller diameter	2.03 m (6 ft 8 in)
Baggage door: Max height	0.65 m (2 ft 1½ in)
Width	1.22 m (4 ft 0 in)

DIMENSIONS, INTERNAL:

Cabin: Length	2.94 m (9 ft 7¾ in)
Max width	1.27 m (4 ft 2 in)
Max height	1.30 m (4 ft 3 in)

AREAS:

Wings, gross	24.30 m² (261.6 sq ft)

WEIGHTS AND LOADINGS:

Weight empty	544 kg (1,200 lb)
Max payload	361 kg (795 lb)
Max fuel weight	255 kg (562 lb)
Max T-O weight	1,650 kg (3,638 lb)
Max wing loading	67.9 kg/m² (13.91 lb/sq ft)
Max power loading	6.15 kg/kW (10.10 lb/hp)

PERFORMANCE:

Cruising speed	110 kt (204 km/h; 127 mph)
Stalling speed: flaps up	61 kt (112 km/h; 70 mph)
flaps down	51 kt (93 km/h; 58 mph)
Service ceiling	3,000 m (9,840 ft)
T-O run	85 m (279 ft)
Landing run	136 m (446 ft)
T-O to 15 m (50 ft)	210 m (690 ft)
Landing from 15 m (50 ft)	331 m (1,090 ft)
Range with max fuel	648 n miles (1,200 km; 745 miles)

UPDATED

KHRUNICHEV T-415 SNEGIR

English name: Bullfinch

TYPE: Six-seat utility transport.

PROGRAMME: Design started (by Aeroprogress/ROKS-Aero) in 1993; prototype construction began 1994; design revised 1998; first flight (RA-01522) 15 May 1999.

DESIGN FEATURES: Conventional strut-braced high-wing monoplane. Constant-chord unswept wings; V bracing struts. Rectangular section fuselage. Unswept tail unit with long dorsal fin and horn-balanced elevators and rudder.

FLYING CONTROLS: Conventional and manual, by pushrods and cables; slotted ailerons; electrically actuated single-slotted flaps; electrically actuated trim tab in each elevator. Dual controls standard.

STRUCTURE: Two-spar wing with metal skinning around wingroot fuel tanks, remainder covered with Stits Poly-Fiber synthetic fabric. Welded tube truss fuselage, covered with metal sheet and synthetic fabric. Metal tail surfaces, fabric covered.

LANDING GEAR: Tailwheel type; fixed. Single wheel on each unit, with fairings optional on mainwheels. Spring-type tailwheel unit with castoring tailwheel. Mainwheel tyre size 700×260; tailwheel 300×135. Brakes on mainwheels. Ski- and float-equipped versions under development in 2001.

POWER PLANT: One 265 kW (355 hp) VOKBM M-14P nine-cylinder radial driving an MT-Propeller MTV-9 three-blade, variable-pitch (hydraulic) propeller. *Detailed description applies to this variant.* Version with 559.3 kW (750 shp) Walter M 601 turboprop under development in 2001; future provision for versions with 224 kW (300 hp) Textron Lycoming IO-340 flat-six or 336 kW (450 shp) Rolls-Royce 250-B17F turboprop, each driving a Hartzell three-blade, constant-speed propeller. Fuel contained in four tanks in wingroots, each with filler port.

ACCOMMODATION: Pilot and up to five passengers; four passenger seats in two rows with provision for fifth passenger alongside pilot. Forward-hinged fully glazed door on each side, with large passenger/cargo door on port side of cabin at rear.

SYSTEMS: Pneumatic system for starting engine and operating mainwheel brakes; pressure 50 bar (725 lb/sq in). Electrical system includes GSR-3000M engine-driven 27 V DC generator and 12SAM-28 battery.

AVIONICS: VFR avionics standard, IFR optional.
Flight: Includes ARK-25 ADF, AGB-96B gyro horizon and GPS.

DIMENSIONS, EXTERNAL:

Wing span	12.94 m (42 ft 5½ in)
Wing chord, constant	2.00 m (6 ft 6¾ in)
Wing aspect ratio	6.5
Length overall	8.84 m (29 ft 0 in)
Height overall	2.74 m (8 ft 11¾ in)
Propeller diameter	2.49 m (8 ft 2 in)

DIMENSIONS, INTERNAL:

Cabin max width	1.40 m (4 ft 7 in)

AREAS:

Wings, gross	25.88 m² (278.6 sq ft)
Horizontal tail surfaces	5.70 m² (61.35 sq ft)
Vertical tail surfaces	2.60 m² (27.9 sq ft)

WEIGHTS AND LOADINGS:

Weight empty	1,150 kg (2,535 lb)
Max fuel	306 kg (675 lb)
Max T-O weight	1,750 kg (3,858 lb)
Max wing loading	67.62 kg/m² (13.85 lb/sq ft)
Max power loading	6.61 kg/kW (10.87 lb/hp)

PERFORMANCE:

Max level speed	124 kt (230 km/h; 143 mph)
Service ceiling	7,600 m (24,930 ft)
T-O run	220 m (725 ft)
Landing run	300 m (985 ft)
Range with max fuel	842 n miles (1,560 km; 969 miles)
g limits	+3.8/−1.5

UPDATED

KHRUNICHEV T-419 STREKOZA

English name: Dragonfly

TYPE: Agricultural sprayer.

PROGRAMME: Mockup shown at Moscow Salon, August 2001. Related to T-517 Fermer (which see in 2002-03 and previous *Jane's*), but with VOKBM M-14 radial engine. (Designation **T-411SKh** has also been used). Reported in May 2002 that prototype to fly within a few months. Cost US$100,000 (2002).

UPDATED

KHRUNICHEV T-440 MERKURY

English name: Mercury

No reports have been received of recent developments, although promotion was continuing in 2002. The aircraft was last described in the 2001-02 edition of *Jane's*.

UPDATED

Khrunichev T-419 mockup *(Jane's/Paul Jackson)*   0126846

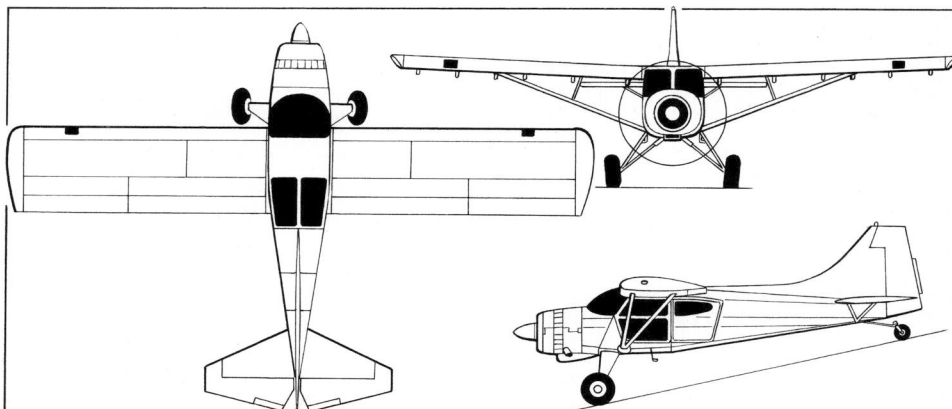

Khrunichev T-415 Snegir *(Jane's/James Goulding)*   0131845

Prototype Khrunichev T-415 Snegir   0102511

KHRUNICHEV T-507 SKVORETS
English name: Starling

TYPE: Light utility turboprop.

PROGRAMME: Design started 1994 under designation T-417; 261 kW (350 hp) Teledyne Continental TSIO-550 flat-six piston engine (see 1999-2000 edition): prototype under construction by 1996; project terminated and recast as T-507, with turboprop and tailwheel landing gear; prototype (RA-01507) displayed incomplete at MAKS 2001 at Moscow, employing Skvorets name previously assigned to T-407.

DESIGN FEATURES: Conventional strut-braced high-wing monoplane, with STOL performance and non-retractable tricycle landing gear (see illustrations).

FLYING CONTROLS: Conventional and manual. Trim tab in rudder and each elevator; single-slotted ailerons and trailing-edge flaps.

STRUCTURE: All-metal. Conventional aluminium alloy wings and tail unit; VNS-2 steel fuselage truss structure, with aluminium alloy skin and fabric covering at rear. Composites engine cowling.

LANDING GEAR: Tailwheel type; fixed. Single wheel on each unit; mainwheel size 800×260, tailwheel size 470×210.

POWER PLANT: One 560 kW (751 shp) Walter M 601E turboprop, driving a three-blade propeller; fuel tanks in wingroots, combined capacity 620 kg (1,367 lb).

ACCOMMODATION: Side-by-side seats for two persons on flight deck, with jettisonable forward-hinged door each side. Dual controls standard. Normally flown by pilot only as freighter, but provision for pilot and up to eight passengers. Large double freight door on port side of cabin, vertically split to open forward and rearward for unrestricted access. Steps beneath each door. Emergency exit on starboard side of cabin.

SYSTEMS: Interior heated and ventilated. Hydraulic system for flap actuation and wheel brakes. 27 V DC electrical system supplied by engine-driven generator and battery. Pneumatic de-icing system. Engine compartment fire warning and extinguishing system.

Following data are provisional.

DIMENSIONS, EXTERNAL:
Wing span	13.20 m (43 ft 3¾ in)
Length overall	9.00 m (29 ft 6¼ in)

WEIGHTS AND LOADINGS:
Weight empty	1,490 kg (3,285 lb)
Max fuel weight	620 kg (1,367 lb)
Max T-O weight	2,850 kg (6,283 lb)
Max power loading	5.09 kg/kW (8.36 lb/shp)

PERFORMANCE (estimated):
Cruising speed	178 kt (330 km/h; 205 mph)
T-O run	110 m (361 ft)
T-O to 15 m (50 ft)	320 m (1,050 ft)

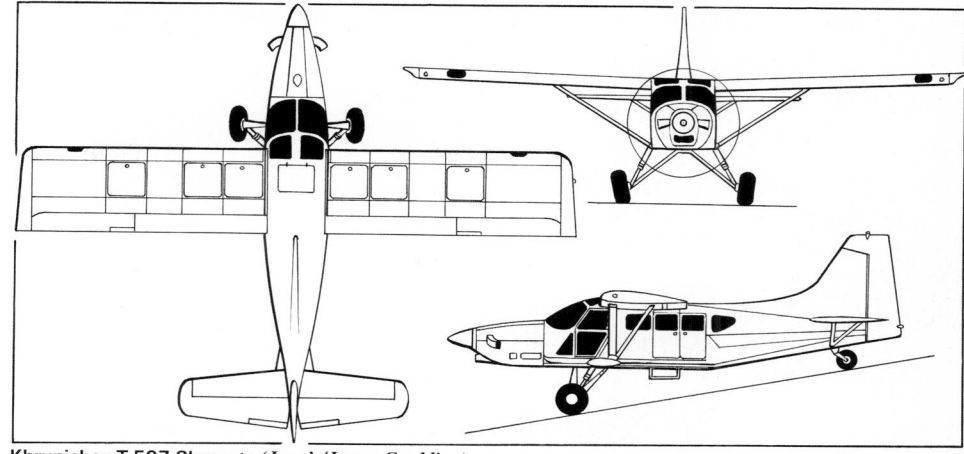

Khrunichev T-507 Skvorets *(Jane's/James Goulding)* 0126947

Khrunichev T-507 mockup at Moscow in August 2001 *(Jane's/Paul Jackson)* **NEW**/0143709

Landing from 15 m (50 ft)	350 m (1,148 ft)
Landing run	160 m (525 ft)
Range	777 n miles (1,440 km; 851 miles)

UPDATED

KHRUNICHEV T-517 FERMER

This agricultural sprayer was last described in the 2002-03 *Jane's*.

UPDATED

KnAAPO

GOSUDARSTVENNOYE UNITARNOYE PREDPRIYATIE KOMSOMOLSKOE-na-AMURE AVIATSIONNOYE PROIZVODSTVENNOYE OBEDINENIE IMENI Yu A GAGARINA (State Unitary Enterprise Komsomolsk-on-Amur Aircraft Production Association named for Y A Gagarin)

ulitsa Sovetskaya 1, 681018 Komsomolsk-na-Amure
Tel: (+7 42172) 636 09 and 635 89
Fax: (+7 42172) 634 51 and 298 51
e-mail: knaapo@kmscom.ru
Web: http://www.ceebd.co.uk/knaapo
GENERAL DIRECTOR: Viktor I Merkulov
TECHNICAL DIRECTOR: Aleksandr I Pekarsh
CHIEF, ADVANCED DEVELOPMENTS DEPARTMENT:
Sergei Drobyshev

A major production centre for Sukhoi aircraft at the present time, KnAAPO has manufactured combat aircraft of the single-seat Su-27 series, including the naval Su-33 and Su-27KUB, and the Su35/35UB/37, and is also responsible for the S-80 STOL transport, Beriev Be-103 light amphibian and, when placed in production, Kamov Ka-62M helicopter. In September 2001 announced own SA-20P programme, being modified version of Be-103. Known as GAZ 126 when established in 1934, it has built more than 10,000 aircraft for 20 countries, beginning in May 1936. Test airfield is Dzemgi. Other products include small aluminium boats, audio-video equipment, polygraphy equipment and furniture, and refurbishment of tramcars. Operates own transport airline

Sukhoi Su-30MKK produced at Komsomolsk 0120086

with four An-12s, one An-26, two An-32s, two Il-76s, two Tu-134s and three Mi-8s. The 100 per cent state holding in KnAAPO is to be reduced to 25.5 per cent, creating a joint stock company; balance will be transferred to Sukhoi Aviation Holding Company under Presidential decree of October 2001, placing Novosibirsk and Komsomolsk plants within Sukhoi. Employees in 1999 totalled 15,000.

UPDATED

For details of the latest updates to *Jane's All the World's Aircraft* online and to discover the additional information available exclusively to online subscribers please visit

jawa.janes.com

LMZ — *see RSK 'MiG'*

LVM

PROIZVODSTVENNOYE OBEDINENIE LEGKIE VERTOLETI MI OAO (Mi-Light Helicopters Production Association JSC)

a/ya 34, 103051 Moskva
Tel: (+7 095) 209 16 68
Fax: (+7 095) 200 25 34
GENERAL DIRECTOR: Grigory Bado

LVM, which trades for export as Mi-Light, was formed in April 1993 with responsibility for design, testing, series manufacture, sales, after-sales service, repair, spares provisioning, leasing and general support of the Mi-34 light helicopter, described under the Mil heading. Activities also include the Mi-52 light rotorcraft, although promotion of that product has only recently begun. A further light design, the Mi-60, was shown in mockup form at the Moscow Salon in August 2001.

Participating companies comprise Mil Moscow Helicopter Plant, NIAT – Moscow Aviation Technology Research Institute, Moscow Aviation Industry State Design and Research Institute, St Petersburg Aviation Industry State Design and Research Institute, Perm Motors, Sazykin Arsenyev Aviation Production Plant, Avitek, Aviastar (Ulyanovsk) and Aeromechanica Research and Production Association.

VERIFIED

MAI

MOSKOVSKOGO GOSUDARSTVENNOGO AVIATSIONNOGO INSTITUTA (Moscow State Aviation Institute)

Volokolamskoye shosse 4, 125871 Moskva

This academic institution includes a students' design bureau.

OTRASLEVOYE SPETSIALNOYE KONSTRUKTORSKOYE BURO EKISPERIMENTALNOGO SAMOLOTOSTROENIYA (OSKBES — Students' Special Construction Bureau for Experimental Aircraft)

Marketing address as for RSK 'MiG'
Tel: (+7 095) 158 44 68
Fax: (+7 095) 250 19 84
e-mail: oskbes@com2com.ru
Web: http://www.com2com.ru/oskbes
CHIEF DESIGNER: Nikolai P Goryunov

MAI formed students' design bureau in 1967; by government decree of 11 August 1982 became OSKBES, under leadership of the late Kazimir Zhidovetski. Products comprised Kvant aerobatic lightplane, PS-01 and Elf-D UAVs, Elf lightplane (see Dubna entry in 2001-02 and previous editions), Yunior/MAI-89/890 biplanes, Photon jet research aircraft, MAI-900 aerobatic single-seater and MAI-920 training glider with 90 per cent of parts made from standard -890 components. MAI-950 business aircraft not constructed.

Marketing and production of MAI aircraft initially by Aviatika, in which MAI was partner, with LII research institute and MAPO (now RSK 'MiG'). Aviatika applied for Aviaregister approval on 27 July 1992 and this awarded on 17 February 1993; production rights to MAI-890 awarded to MAPO on 10 April 1995. Aviatika relationship with OSKBES ended in 1996 and bureau continued to work directly with MAPO. All Aviatika rights were transferred to MAI on 19 February 1998; MAI then transferred marketing and production rights to MAPO/RSK 'MiG'. Is partner with Avitexgrup OOO in promotion of MAI-205 autogyro.

VERIFIED

AVIATIKA-MAI-890

TYPE: Single-seat ultralight biplane/side-by-side ultralight biplane.
PROGRAMME: Open cockpit Yunior (Junior) first flew July 1987 as single-seater with 5.68 m (18 ft 7½ in) span and 280 kg (617 lb) MTOW. Developed as MAI-89, 1989, span 8.11 m (26 ft 7½ in); MTOW 340 kg (750 lb). Production version, MAI-890, flew 1990 and entered series manufacture in 1991; although this halted temporarily in mid-1990s, it was announced in 2002 that production would continue. Russian certification awarded in December 1999. Manufactured by RSK 'MiG'.

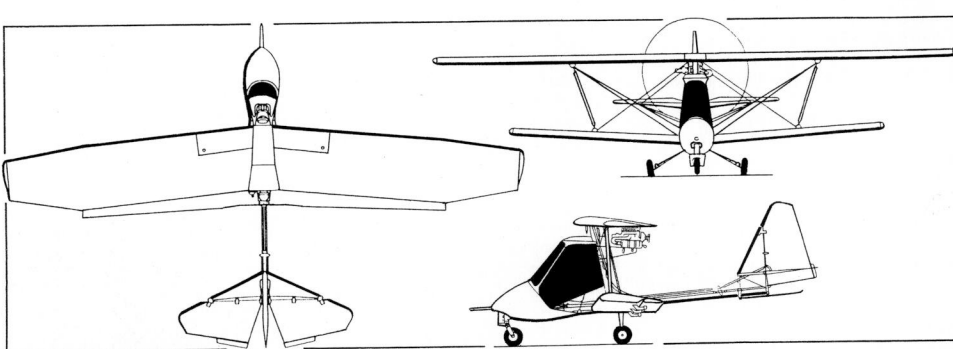

Aviatika-MAI-890 (Rotax piston engine) *(Jane's/Paul Jackson)*   0051198

CURRENT VERSIONS: **Aviatika-MAI-890:** Single-seater.
Aviatika-MAI-890U: Side-by-side two-seater (*uchebno*: training). First flown August 1991 (Rotax 582 UL). Normally with Rotax 912 UL engine (first flight 16 September 1992). Empty weight 298 kg (657 lb); MTOW 540 kg (1,190 lb); length 5.49 m (18 ft 0¼ in). Certification anticipated in early 2001 but not achieved by early 2002. Floatplane version completed trials on Lake Shartash, Ekaterinburg, September 2000.
Aviatika-MAI-890SKh: Crop-sprayer version (*selskokhozyaistvenny*: agricultural); marketing name **Farmer**. Chemical tank under engine and spraybar aft of lower wings; weight of agricultural equipment 28 kg (62 lb); chemicals 60 kg (132 lb) (105 litres; 27.7 US gallons; 23.1 Imp gallons) maximum with Rotax 582 engine. Certification was scheduled for September 2001, but rescheduled for third quarter of 2002. Local market estimated as 1,200 to 1,500.
Aviatika-MAI-890USKh: Two-seat crop-sprayer version based on MAI-890U. Tank volume 100 litres (26.4 US gallons; 22.0 Imp gallons).
Aviatika-MAI-890S: Rotax 912 ULS 73.5 kW (98.6 hp) flat-four. Prototype (ZU-BZO) converted in South Africa, 1999, by Aeroserve Pty Ltd.
CUSTOMERS: Total of "more than 300" delivered by mid-2002, of which some 60 were agricultural variants; eight sold during last six months of 2001, including six equipped for crop spraying. Approximately 30 imported by a South African dealer in 1995; these known locally as **890CSH Mai** and **890U Mai**. Total market for agricultural variant estimated at 800 to 1,000. Around 300 of all versions sold in 12 countries by early 2002.
COSTS: US$38,000 to US$40,000 (2001).
DESIGN FEATURES: Strut- and wire-braced biplane; unspinnable; conventional tail surfaces carried on tubular boom; engine mounted under trailing-edge of upper wing; slight sweepback on all wings; dihedral on lower wings only. Semi-aerobatic and capable of flying in rough air conditions. Designed to JAR-VLA and FAR Pt 23 standards. Open cockpit standard; doors optional, one jettisonable; provision for backpack parachute(s).

FLYING CONTROLS: Conventional and manual. Full-span ailerons on lower wings; large-area rudder and elevators, each with ground-adjustable tab.
STRUCTURE: Aircraft grade aluminium and titanium alloys, alloy steels; fabric covering unaffected by sun's radiation or atmospheric precipitation.
LANDING GEAR: Non-retractable tricycle type; single wheel on each unit; cantilever spring main legs. Brakes on mainwheels. Optional skis or floats. All wheels 300×125.
POWER PLANT: One 47.8 kW (64.1 hp) Rotax 582 UL two-cylinder or 59.6 kW (79.9 hp) Rotax 912 UL four-cylinder piston engine; two-blade pusher propeller. A 37.0 kW (49.6 hp) Rotax 503 UL-2V may also be fitted. Standard fuel 50 litres (13.2 US gallons; 11.0 Imp gallons); provision for 55 litre (14.5 US gallon; 12.1 Imp gallon) auxiliary tank on single-seat version.
EQUIPMENT: Provision for up to 120 kg (265 lb) payload on four attachments, under engine mountings (60 kg; 132 lb), or underbelly (100 kg; 220 lb), or at lower wingtips (each 45 kg; 99 lb), including agricultural dusting and spraygear for Aviatika-MAI-890USKh Farmer. Optional BRS ballistic parachute system.
Following details apply to single-seat 890 with Rotax 912 UL engine.

DIMENSIONS, EXTERNAL:
Wing span, upper	8.11 m (26 ft 7½ in)
Length overall, incl pitot	5.32 m (17 ft 5½ in)
Height to tip of fin	2.30 m (7 ft 6½ in)

AREAS:
Wings, gross, total	14.29 m² (153.8 sq ft)

WEIGHTS AND LOADINGS:
Weight empty	277 kg (611 lb)
Max T-O weight	450 kg (992 lb)

PERFORMANCE:
Max level speed	81 kt (150 km/h; 93 mph)
Cruising speed	54 kt (100 km/h; 62 mph)
Landing speed	35-38 kt (65-70 km/h; 41-44 mph)
Max rate of climb at S/L	360 m (1,181 ft)/min
T-O run	65 m (215 ft)
Landing run	95 m (315 ft)
Endurance	3 h 7 min
g limits	+5/−2.5

UPDATED

Aviatika-MAI-890 SKh agricultural sprayer *(Jane's/Paul Jackson)*   *NEW*/0143722

MAI-205

TYPE: Single-seat autogyro.
PROGRAMME: Prototype completed 1994; designated Aviatika-MAI-890A and powered by a 47.8 kW (64.1 hp) Rotax 582 UL. Developed MAI-205 (marked only as '205') first flew 26 January 2001.
COSTS: US$40,000 in agricultural version (2001). Development cost US$200,000.
DESIGN FEATURES: Conventional configuration, with fully enclosed cockpit, tail surfaces carried on tubular tailboom; two-blade autorotating rotor with optional pre-rotation for 'jump start'; rotor column folds for storage. Control movements emulate aeroplane, rather than autogyro, providing safer type conversion for former fixed-wing pilots. Ground-adjustable tab on rudder and each elevator. Rotor blade aerofoil NACA 23012; blade angle 4° for running take-off and flight; 10° for 'jump start'.

MAI-205 single-seat autogyro *(Jane's/Paul Jackson)*
NEW/0143723

STRUCTURE: Fuselage of aluminium tube with glass fibre fairings. Strut and wire-braced, fabric-covered tailplane.

LANDING GEAR: Non-retractable tricycle; all wheels 300×125. Brakes on mainwheels. Steerable nosewheel.

POWER PLANT: One 84.6 kW (113.4 hp) Rotax 914 ULS piston engine mounted behind top of cabin pod; two-blade pusher propeller. Fuel tank under engine, capacity 24 litres (6.3 US gallons; 5.3 Imp gallons).

EQUIPMENT: Two 50 litre (13.2 US gallon; 11.0 Imp gallon) tanks. Optional Micronair AU7000 atomisers.

DIMENSIONS, EXTERNAL:
Rotor diameter	9.20 m (30 ft 2¼ in)
Height overall	2.87 m (9 ft 5 in)
Fuselage length	5.32 m 17 ft 5½ in)
Wheel track	1.52 m (4 ft 11¾ in)
Wheelbase	1.81 m (5 ft 11¼ in)

WEIGHTS AND LOADINGS:
Weight empty	370 kg (816 lb)
Max T-O weight	530 kg (1,168 lb)

PERFORMANCE:
Max level speed	62 kt (115 km/h; 71 mph)
Max rate of climb at S/L	162 m (531 ft)/min
Service ceiling	1,750 m (5,740 ft)
T-O run	30 m (100 ft)
Landing run	15 m (50 ft)
g limits	+3.5/−0.5
	UPDATED

AVIATIKA-MAI-910 INTERFLY

TYPE: Two-seat lightplane.

PROGRAMME: Design started and construction of first prototype began 1993; first flight 22 June 1995 in MAI-910 configuration. Promotion began at MAKS '99, Moscow, August 1999; development continuing in 2002.

CURRENT VERSIONS: **Aviatika-MAI-910:** Original version with boom-mounted empennage.

Aviatika-MAI-960: Version with conventional rear fuselage, interchangeable with the above in 15 minutes and including ventral strake. This subsequently marked as MAI-910.

DESIGN FEATURES: Designed to JAR-VLA standards. Braced high-wing monoplane; single bracing strut and jury strut each side; fully enclosed cabin pod; strut- and wire-braced conventional tail unit carried on interchangeable rear fuselage of conventional thickness or on short boom. Wings fold.

Wing section CAHI (TsAGI) Sh-1-14; 1° leading-edge sweepback; thickness/chord ratio 14 per cent; constant chord except for tapered tips; no dihedral; incidence 5°.

FLYING CONTROLS: Conventional and manual. Actuation by cables; long-span flaperon on each wing; ground-adjustable tab on rudder and each elevator.

STRUCTURE: Metal structure with fabric-covered wings and tail surfaces, metal-skinned fuselage. Wings have single tubular spar, diameter 102 mm (4 in), and stamped sheet metal ribs. Materials primarily aluminium alloy, with stainless steel and titanium alloy where appropriate.

LANDING GEAR: Non-retractable tricycle type; single wheel on each unit; cantilever spring legs; mainwheel tyres 400×150 or 300×120; nosewheel tyre 300×120; tyre pressure (all) 2.53 bar (37 lb/sq in). Mechanical brakes with 300×120 tyres; hydraulic brakes with 400×150 tyres. Nosewheel steerable through ±50°. Tailwheel gear optional.

POWER PLANT: One 59.6 kW (79.9 hp) Rotax 912 UL flat-four; Russian two-blade fixed-pitch propeller. Two fuel tanks, each 25 litres (6.6 US gallons; 5.5 Imp gallons); one in each wingroot. Oil capacity 6 litres (1.6 US gallons; 1.3 Imp gallons).

ACCOMMODATION: Two seats side by side in fully enclosed cabin with wide field of view for observation, patrol, search and aerial photography. Passenger seat folds for carriage of stretcher or lengthy freight. Large rearward-sliding window/door each side. Space for baggage or freight behind seats. Cabin heated and ventilated.

SYSTEMS: 12 V AC electrical system; Teledyne battery.

AVIONICS: *Comms:* UHF radio.
Flight: GPS.

DIMENSIONS, EXTERNAL:
Wing span	10.73 m (35 ft 2½ in)
Wing chord: at root	1.10 m (3 ft 7¼ in)
at tip	0.70 m (2 ft 3½ in)

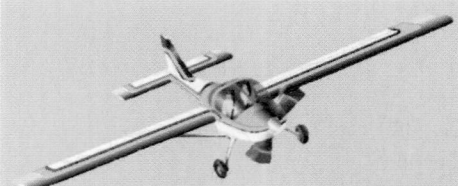

Artist's impression of MAI-217 0121736

Wing aspect ratio	10.3
Length overall	5.26 m (17 ft 3 in)
Height overall	2.50 m (8 ft 2½ in)
Wheel track	2.20 m (7 ft 2¾ in)
Wheelbase	1.60 m (5 ft 3 in)
Propeller diameter	1.90 m (6 ft 2¾ in)
Propeller ground clearance	0.30 m (11¾ in)
Doors (each): Height	0.87 m (2 ft 10¼ in)
Max width	1.10 m (3 ft 7¼ in)
Baggage door: Length	0.75 m (2 ft 5½ in)
Width	0.70 m (2 ft 3½ in)

DIMENSIONS, INTERNAL:
Cabin: Max width	1.20 m (3 ft 11¼ in)
Max height	1.08 m (3 ft 6½ in)
Floor area	1.20 m² (12.9 sq ft)
Baggage/freight space: Volume	0.70 m³ (24.7 cu ft)

AREAS:
Wings, gross	11.20 m² (120.6 sq ft)

WEIGHTS AND LOADINGS:
Weight empty	380 kg (838 lb)
Max fuel	40 kg (88 lb)
Max T-O weight	620 kg (1,366 lb)
Max wing loading	55.4 kg/m² (11.34 lb/sq in)
Max power loading	10.40 kg/kW (17.09 lb/hp)

PERFORMANCE:
Never-exceed speed (VNE)	121 kt (225 km/h; 139 mph)
Max level speed at S/L	84 kt (155 km/h; 96 mph)
Max and econ cruising speed	75 kt (140 km/h; 87 mph)
Stalling speed	41 kt (76 km/h; 48 mph)
Max rate of climb at S/L	210 m (690 ft)/min
Service ceiling	3,000 m (9,840 ft)
T-O run	150 m (495 ft)
T-O to 15 m (50 ft)	300 m (985 ft)
Landing from 15 m (50 ft)	150 m (495 ft)
Range	216 n miles (400 km; 248 miles)
g limits	+4/−2
	UPDATED

MAI-217

TYPE: Two-seat lightplane.

PROGRAMME: Development under way at OSKBES by 2000. Promotion continuing in 2002.

CURRENT VERSIONS: **MAI-217:** *As described.*

MAI-217B: (*Boyevoy:* combat) Proposed military version with cockpit armour and provision for up to 200 kg (441 lb) of light weapons.

DESIGN FEATURES: Designed to JAR-VLA criteria. Side-by-side lightplane with strut-braced, mid-positioned, constant-chord wing. Uses some Aviatika-MAI-910 components. Suitable for agricultural spraying, training and surveillance. Wings foldable for storage in 5 minutes. Convertible to tailwheel configuration by two persons in under 2 hours.

POWER PLANT: One 73.5 kW (98.6 hp) Rotax 912 ULS flat-four driving a two-blade, fixed-pitch propeller.

EQUIPMENT: Optional 200 litre (52.8 US gallon; 44.0 Imp gallon) chemical tank and spraybar.

ARMAMENT: Optional light bombs and rocket pods.

DIMENSIONS, EXTERNAL:
Wing span	11.32 m (37 ft 1¾ in)
Length overall	6.50 m (21 ft 4 in)
Height overall	1.90 m (6 ft 2¾ in)

AREAS:
Wings, gross	13.58 m² (146.2 sq ft)

WEIGHTS AND LOADINGS (estimated):
Weight empty	325 kg (717 lb)
Max T-O weight	700 kg (1,543 lb)
Max wing loading	51.5 kg/m² (10.56 lb/sq ft)
Max power loading	9.53 kg/kW (15.65 lb/hp)

PERFORMANCE (estimated):
Max level speed	108 kt (200 km/h; 124 mph)
Landing speed	39 kt (72 km/h; 45 mph)
T-O run	100 m (330 ft)
Landing run	140 m (460 ft)
Range	432 n miles (800 km; 497 miles)
	UPDATED

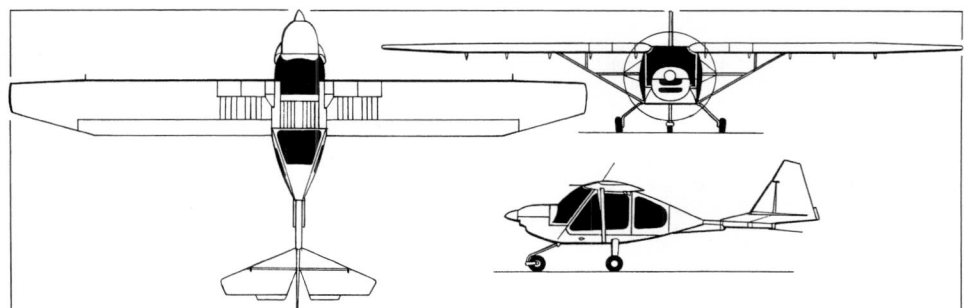

Aviatika-MAI-910 with original rear fuselage *(Jane's/James Goulding)* 0011968

Aviatika-MAI-960 light passenger/freight aircraft, produced by fitting new rear fuselage to the prototype -910 *(Jane's/Paul Jackson)* 0121697

MiG

AVIATSIONNYI NAUCHNO-PROMYSHLENNYI KOMPLEKS MiG (ANPK MiG; MiG Aviation Scientific-Industrial Complex)

a/ya 125299, Leningradskoye shosse 6, Moskva
Tel: (+7 095) 158 18 72
Fax: (+7 095) 943 00 27

ADVISER TO GENERAL DIRECTOR: Rotislav A Belyakov
CHIEF DESIGNER: Lev Shengelaya
MiG-AT PROGRAMME DIRECTOR: Vasily Shtykalo
DIRECTOR OF STRATEGIC PLANNING, INFORMATION AND MARKETING: Alexander I Ageev
CHIEF, PUBLIC RELATIONS AND ADVERTISING DEPARTMENT: Svyatoslav Yu Ribas

This bureau originated in GAZ 1 (factory No. 1) founded in 1893. The experimental design bureau, founded by Artem I Mikoyan, dissociated from the factory in 1939 and later became officially the Mikoyan Aviation Scientific-Industrial Complex 'MiG' (the final letter indicating Mikhail I Gurevich). Currently known as A Mikoyan OKB Engineering Centre. It developed 250 projects and built 120

Second MiG-29K prototype in revised colour scheme for continued trials, 2001 *(Yefim Gordon)*   *NEW*/0137348

prototypes. More than 60,000 aircraft of MiG design have been manufactured, including 11,000 for export.

In 1995, the Moscow Aircraft Production Organisation and ANPK MiG were merged into the single state company MAPO 'MiG'. The original 'MiG' acronym continued to be used, though MiG lost its separate legal identity and was prevented from selling its own aircraft. On 4 July 1997 the legal status of MAPO 'MiG' was restored. Following a change of title, the bureau is now subordinate to RSK 'MiG' (which see).

UPDATED

MiG-29

NATO reporting name: Fulcrum
Indian Air Force name: Baaz (Eagle)
TYPE: Multirole fighter.
PROGRAMME: Production of land-based MiG-29s ended in mid-1990s, leaving substantial stock of semi-complete airframes available to meet export orders. Activity currently centres on upgrades for existing aircraft; details in 2001-02 edition and in *Jane's Aircraft Upgrades.*

New production dependent upon Indian order for carrier-based MiG-29K. By 2001 − in advance of formal order − MiG constructing two pre-series MiG-29Ks, one MiG-29KUB and static test airframe in single-seat configuration. Sokol plant (which see) contracted in 2001 to build folding wing and nose section for production aircraft. Indian deliberations upon purchase of aircraft carrier continued into 2002, delaying decision on MiG-29K purchase.
CURRENT VERSIONS: **MiG-29K** (Factory index 9.31; K for *korabelnyy*; ship-based): Maritime version, used for ski-jump take-off and deck landing trials on carrier *Admiral of the Fleet Kuznetsov* (formerly *Tbilisi*), beginning 1 November 1989; two new-build prototypes, using MiG-29M structure; completely redesigned, mainly steel, wing using modified aerofoil section and of increased area (increased span, reduced leading-edge sweep) with double slotted flaps, drooping flaperons and more powerful leading-edge flaps; new spar in front of original wing box; new strengthened centre-section without overwing louvres; upward-folding outer wing panels; RD-33K turbofans with 92.2 kN (20,723 lb st) contingency rating for ski-jump take-offs. Fuel capacity reduced to 5,670 litres (1,498 US gallons; 1,247 Imp gallons). First flown (16188 '311') 23 June 1988; second prototype was 27579 '312'.

State Acceptance Trials suspended due to funding difficulties, early 1992. Further development ended initially when not selected for deployment on *Admiral of the Fleet Kuznetsov*, but resumed at Zhukovsky in September 1996, in expectation of order from India. (More details in 1993-94 *Jane's*.) First prototype currently grounded; second returned to flight status in support of MiG-29M programme. Proposed naval **MiG-29KU** (9.62) trainer derivative with two separate stepped cockpits remained unbuilt.

'MiG-29K-2002' (Factory index 9.41): The original MiG-29SMTK (factory index 9.17K; effectively a MiG-29SMT with the 9.31's folding wing and landing gear), previously offered to India along with the former helicopter carrier *Admiral Gorshkov*, is understood to have been replaced by a new, multirole, carrierborne variant based more closely on the MiG-29K/M, albeit without the expensive Al-Li alloys. Development of new MiG-29M variants (single- and two-seat) continued in 2002 with MiG-29OVT (thrust-vectoring ground testbed) and MiG-29M2 (29MRCA) which first flew on 26 September 2001 after conversion to tandem configuration.

With a MIL-STD-1553B-type bus and open systems avionics architecture, the MiG-29K-2002 is compatible with a wide range of Russian and Western weapons, and may feature the colour displays and GPS-based navigation

system of the MiG-29SMT. This variant, possibly designated MiG-29 MTK, is claimed to be able to perform 90 per cent of its missions with a 10 kt (18 km/h; 11 mph) wind-over-deck, even in tropical conditions using new autothrottle. Notable features of the new aircraft are its improved wing with enlarged trailing-edge flaps, digital FBW system, emergency fuel draining and much-reduced folded span of 5.80 m (19 ft 0¼ in), achieved by positioning the fold line much closer in to the wingroot, and by adding upward-folding tailplanes. The aircraft can also fold its radome (up and back), reducing overall length to 14.13 m (46 ft 4¼ in). Accordingly, *Admiral Gorshkov* can carry a full air wing of 24 MiG-29Ks (plus six helicopters), or (according to some sources) as many as 30.

A projected **MiG-29K-2008** upgrade configuration could add Zhuk-MF phased-array radar, RD-133 turbofans, a computer upgrade and additional electro-optic, radar, IIR and reconnaissance pods, together with take-off performance improvements. In December 1999, it was reported that India had selected the MiG-29K for use aboard *Admiral Gorshkov*, and an initial order for 50 aircraft (against a total requirement for 60 to 70 aircraft) was expected, with an unknown quantity of Kh-35 anti-ship missiles and Kh-31P ARMs. A refuelling tanker fit has been proposed. Local manufacture by HAL is expected.

MiG-29KUB (Factory index 9.47): Revised carrier-borne two-seat trainer design offered to India, based on MiG-29K-2002 with reduced-span, inboard wing fold and folding tailplanes. Assumed to feature original stepped tandem cockpits of MiG-29KU. Some reports suggest enlarged tailfins with integral fuel tanks, possibly even single centreline tailfin.
DESIGN FEATURES: Emphasis from start on high manoeuvrability, to counter US F-15, F-16 and F-18, with target destruction at distances from 200 m (660 ft) to 32 n miles (60 km; 37 miles), and with effective air-to-surface capability. All-swept mid-wing configuration, with wide ogival wing leading-edge root extensions (LERX), 40 per cent of lift provided by lift-generating centre-fuselage, twin tailfins carried on booms outboard of widely spaced engines with wedge intakes. Gap between roof of each intake and skin of wingroot extension for boundary layer bleed.

Fire-control and mission computers link radar with laser range-finder and infra-red search/track sensor, in conjunction with helmet-mounted target designator. Difficult to get into stable flat spin, reluctant to enter normal spin, recovers as soon as controls released; wing

leading-edge sweepback 73° 30' on LERX, 42° on outer panels; anhedral approximately 3°. Tailfins canted outward 6°; leading-edge sweep 47° 50' on fins, 50° on horizontal surfaces; anhedral 3° 30'. Fins canted outwards 6°. Engine replacement time 2 hours; preflight preparation 20 minutes. Design flying life 2,500 hours.
FLYING CONTROLS: Digital fly-by-wire.
STRUCTURE: Approximately 7 per cent of airframe, by weight, of composites; remainder metal; three-spar wings with three 'false spars' two ahead of, one behind, torsion box; 16 stringers and skins reinforced by stringers; trailing-edge wing flaps, ailerons and vertical tail surfaces of carbon fibre honeycomb; approximately 65 per cent of horizontal tail surfaces aluminium alloy, remainder carbon fibre; semi-monocoque all-metal fuselage built around 10 mainframes in three subassemblies, with forward (frames 1 to 3), centre (frames 4 to 7) and rear sections, the latter including the engine bays; fuselage sharply tapered and downswept aft of flat-sided cockpit area, with ogival dielectric nosecone mounting PVD-18 main pitot boom (PVD-7 auxiliary pitot mounted on side of nose); small vortex generator each side of nose helps to overcome early tendency to aileron reversal at angles of attack above 25°; tail surfaces carried on slim booms alongside engine nacelles.
LANDING GEAR: Hydromash retractable tricycle type, with oleo-pneumatic shock-absorbers. Mainwheels retract forward into wingroots, turning through 90° to lie flat above leg; nosewheels, on trailing-link oleo, retract rearward between engine air intakes. Hydraulic retraction and extension, with mechanical emergency release.
POWER PLANT: Two Klimov/Sarkisov RD-33 Series 3M 'marinised' turbofans, each 85.3 kN (19,180 lb st) with afterburning. Engines mounted 4° nose-up, and nacelles toe-in by 1° 30'. Engine ducts canted at approximately 9°, with wedge intakes, sweptback at approximately 35°, under wingroot leading-edge extensions.
ACCOMMODATION: Pressurised cockpit enclosed by frames 1 and 2. Pilot only, on 16° rearward-inclined K-36DM Series 2 zero/zero ejection seat, giving −14° view forward over the nose, and under hydraulically actuated, rearward-hinged transparent blister canopy in high-set cockpit. Sharply inclined one-piece curved windscreen of electrically de-iced triple glass. Three internal mirrors provide rearward view.
SYSTEMS: Two independent hydraulic systems powered by NP-103A variable-displacement pumps, driven by the engine accessory gearboxes, pressurised to 207 bar (3,000 lb/sq in), with 80 litres (21 US gallons; 17.5 Imp

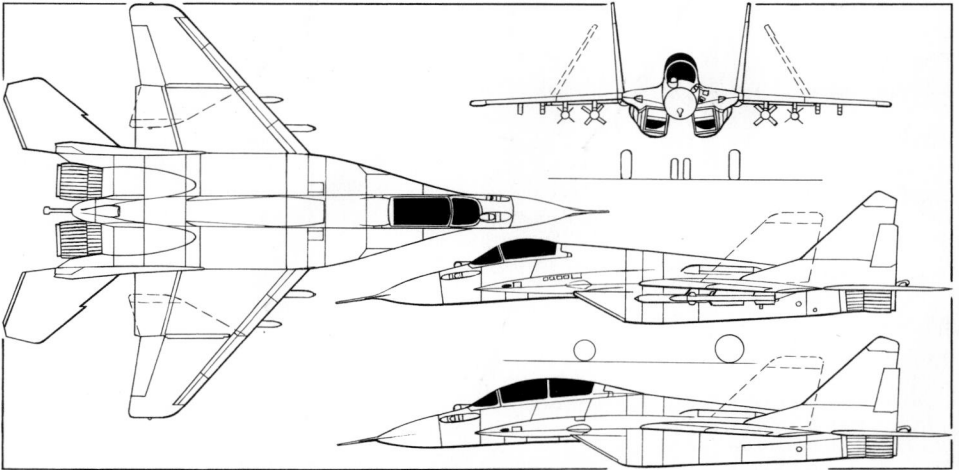

Navalised 'Fulcrums', the MiG-29K and side view (lower) of MiG-29KUB *(Jane's/James Goulding) NEW*/0131833

gallons) fluid. Main system powers one chamber of each control surface actuator, leading-edge and trailing-edge flaps, stick-pusher, artificial feel unit, landing gear extension/retraction, nosewheel steering and APU exhaust door; back-up system powers second chamber of each control surface actuator and stick-pusher, and can be powered by an emergency NS-58 pump.

Three separate pneumatic systems, with main system powering the wheel-brakes, canopy, fuel shut-offs and brake parachute actuator and jettison; and emergency system operating mainwheel brakes, and allowing emergency gear extension; final system pressurises hydraulic tanks and avionics bays.

Air conditioning uses bleed air cooled by heat exhangers and turbocooler. System also pressurises pilot's suit, demists screen and cools the gun bay.

Ozh-65 glycol-based liquid cooling system for radar.

AVIONICS: Expected to incorporate French, Russian and Indian components.

Radar: Phazotron-NIIR Zhuk-M; able to track 20 targets simultaneously and engage four. N011M Bars offered as basis for alternative radar for MiG-29K.

ARMAMENT: R-73 (AA-11 'Archer'), R-27 (AA-10A 'Alamo'), and/or R-77 (AA-12 'Adder') AAMs; Kh-35 (AS-20 'Kayak'), Kh-31 (AS-17 'Krypton') and Kh-29 (AS-14 'Kedge') ASMs; KAB-250KR TV-guided bombs; FAB-500 and FAB-250 free-fall bombs; and rocket pods. One 30 mm Gryazev/Shipunov GSh-301 (TKB-687/9A4071K) single barrel gun in port wingroot leading-edge extension, with 150 AO-18 rounds.

DIMENSIONS, EXTERNAL:

Wing span	11.99 m (39 ft 4 in)
Wing aspect ratio	3.4
Length overall: incl noseprobe	17.37 m (56 ft 11¾ in)
folded	14.13 m (46 ft 4¼ in)
Height overall	5.18 m (17 ft 0 in)

AREAS:

Wings, gross	42.00 m² (452.1 sq ft)

WEIGHTS AND LOADINGS:

Max weapon load	4,500 kg (9,921 lb)
Max fuel load	5,240 kg (11,552 lb)
T-O weight: normal	18,550 kg (40,895 lb)
max	22,400 kg (49,383 lb)
Max wing loading	533.3 kg/m² (109.24 lb/sq ft)
Max power loading	131 kg/kN (1.29 lb/lb st)

PERFORMANCE:

Max level speed:	
at height	(1,296 kt; 2,400 km/h; 1,491 mph)
at S/L	(670 kt; 1,240 km/h; 771 mph)
Max rate of climb at S/L	17,760 m (58,260 ft)/min
Service ceiling	17,500 m (57,420 ft)
Range: combat: with max internal fuel	459 n miles (850 km; 528 miles)
with external tanks	702 n miles (1,300 km; 807 miles)
ferry	1,889 n miles (3,500 km; 2,174 miles)
g limits	+8

UPDATED

MiG-AT

TYPE: Advanced jet trainer/light attack jet.

PROGRAMME: Design started late 1980s; selected in May 1992 as one of two finalists (with Yak-130, following rejection of single-engined Sukhoi S-54 and Myasishchev M-200 contenders) in Russian competition to replace Aero L-29 Delfin and L-39 Albatros; originally planned (as MiG-ATTA, a designation also applied to abortive joint venture with Promavia) with T tail and R-35 (DV-2) engines, with wing by Daewoo, landing gear by Messier-Bugatti, avionics by Sextant, Thomson-CSF (now Thales) and SFENA. Under October 1992 and subsequent agreements, prototypes and pre-series aircraft are powered by Larzac

Second MiG-AT flying at Moscow Salon in August 2001 (*Yefim Gordon*) **NEW**/0137349

04R20 engines supplied by SNECMA of France; production aircraft for domestic market could have Larzacs licence-built by Chernyshev of Moscow, though there is a new Russian engine – the 16.7 kN (3,748 lb st) Aviadvigatel (Soyuz/CIAM) RD-1700 – which is stated to be 2.5 times cheaper than the Larzac and which may remedy early shortfalls in performance, ceiling and range. Suggested that MiG-AT has bureau Type number 08, perhaps with original T-tail configuration being 8.1 and later version as 8.2, prototype configuration being 8.21.

Prototype rolled out 18 May 1995; high-speed taxi trials at Zhukovsky August 1995; first 5 minute hop 16 March 1996; first flight 21 March 1996, piloted by Roman Taskaev; three aircraft to be used for flight test programme; two static test airframes; by 27 March 2001, first prototype had flown 496 sorties and second, 47. Production of first series of 16 at RSK 'MiG' (Voronin) plant well advanced with seven in final assembly by December 1998. Subsequent progress delayed by funding difficulties; total of six airframes (including static test) completed by late 2001, although only two then flown. Fourth flying aircraft handed over to MiG (for completion) by mid-2001.

Following initial series of static trials on second (non-flying) airframe, operation at 5,600 kg (12,346 lb) cleared in 2001; subsequent increases planned to 6,500 kg (14,330 lb), then 7,800 kg (17,196 lb).

Russian requirement for 200 to 250 trainers in this category. Request for funds for first 10 aircraft made by Russian Air Forces, late 1996, at which time 10 engines ordered from Turbomeca-SNECMA (following initial five supplied for two prototypes). Further 20 engines ordered late 1997 for FFr230 million (US$39 million), while Sextant contracted to supply 30 nav/attack systems. Competition against Yak-130 continued throughout 1990s; Russian Defence Ministry committee responsible for deciding between two contenders held first meeting on 17 February 2002, having been constituted six days earlier; Yak announced (formally, on 10 April 2002) as winner, but MiG encouraged to continue development of AT for possible export.

Actively promoted in India (evaluated in Russia by IAF pilots, December 2001) and South Africa; may have been eliminated from Indian competition due to renewed IAF emphasis on timescale, though BAE Hawk's reported selection may yet be reversed. Beaten by BAE Hawk in South Africa. Greek requirement for 40 trainers was targeted in 2001. MiG-AT aircraft now being marketed as one element in AT System, incorporating simulators,

computerised classrooms, part-task simulators, and video training packages, all using unified software.

CURRENT VERSIONS (general): **MiG-AT:** Basic trainer; suitable for combat use of unguided weapons against ground and sea targets, and air-to-air missiles in conjunction with helmet-mounted target designation system. Pylons for three-weapons/stores in trainer role, seven for combat training. Version for domestic use (ATR) has Russian-built engines and avionics; export version (ATF) has Sextant Topflight avionics (French).

Production version has longer nose than prototypes, but with pitots repositioned to slightly ahead of windscreen.

MiG-UTS (MiG-ATR): Planned Russian Air Forces' version, with Russian avionics and RD-1700 engines.

MiG-ATS: Combat trainer, with increased capabilities and underwing hardpoints able to launch guided missiles, using guidance equipment pod. Second flying prototype (823 '83') to this standard and has Russian avionics.

MiG-AS: Single-seat light tactical fighter version with built-in gun and radar for all-weather use of AAMs and ASMs; intended to compete in international market with BAE Hawk 200. Single-seater, aerodynamically similar to original MiG-AT, with rear cockpit covered by metal fairing of similar outline to original canopy.

MiG-AP: Patrol coastguard fishery protection law enforcement and SAR co-ordination aircraft; based on MiG-AS; nose-mounted search radar.

All versions can be modified for deck operation on aircraft carriers at customer's request as **MiG-ATK** (French avionics), **MiG-ATSK** (armed trainer), or **MiG-ASK** (single seater). Span with optional wing folding 6.50 m (21 ft 4 in). Provision for deck arrester hook and catapult launching.

MiG-AT can undertake the following training missions, all with take-off at 50 per cent fuel (plus 150 kg; 331 lb reserves):

Basic flying training: Between 30 and 35 flight manoeuvres, three roller landings and straight-in final landing at up to 27 n miles (50 km; 31 miles) from base, altitude up to 6,000 m (19,680 ft).

Ground attack: Four diving attacks, three advanced manoeuvre attacks and circuit/landing against visual target up to 43 n miles (80 km; 50 miles) from base, transiting at up to 2,500 m (8,200 ft).

Interception/combat manoeuvring: Loiter, directed interception, attack and close combat, day VMC/IMC at up to 140 n miles (260 km; 161 miles) from base, altitudes up to 8,000 m (26,240 ft).

CURRENT VERSIONS (specific): **AT-1/821 '81':** Prototype; first flight 16 March 1996; first 'official' flight 21 March 1996. Has been modified since its initial flights, briefly gaining in overwing fence level with the leading-edge discontinuity. This since removed, but intakes have been redesigned and now project forward of the wing leading-edge; now have oval cross-section and blunter lips, rather than appearing to be sharp, inverted Us from the front. New intake configuration is understood to have first flown during May 1999; shown at MAKS '99, Moscow, August 1999. Rear cockpit now filled with test instrumentation. Tested with external fuel tanks. Conventional (non-FBW) controls. By 2001 had replaced pitot booms with shorter units not protruding beyond nosecone.

AT-3/823 '83': Second flying prototype shown (then still unflown) at MAKS '97, Moscow, August 1997; nominally representative of intended UTS in some respects, and of ATS armed trainer configuration in others. Fitted with centreline and underwing hardpoints; weapon carriage trials; Russian avionics and displays. Fitted with three-channel, digital FBW system in early 2000 (first Russian aircraft thus equipped); original 70 kg (154 lb) nose counterweight then removed, giving neutral stability. Original intake configuration with overwing fences was replaced by current design and aircraft resumed flying (with desert camouflage scheme) in June 2001, appearing at Paris later that month and first displaying wingtip missile rails.

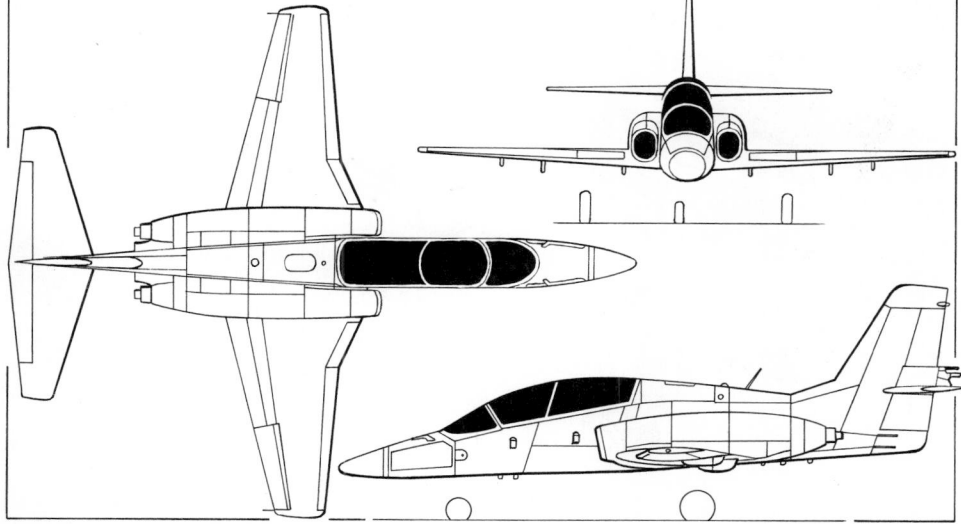

MiG-AT advanced trainer (*Jane's/James Goulding*) 0130862

Mil Mi-8MTO night attack helicopter conversion (*Yefim Gordon*)　0106504

fuselage. (Mi-17P designation used also for civil export versions.) Similar versions include **Mi-8MTP; Mi-8MTR; Mi-8MTSh; Mi-8MTPSh; Mi-8MTU; Mi-8TYa; Mi-8MTI** (Mi-17 with small horizontal array on forward part of boom and larger box-like radome on cabin side); **Mi-8MTPI; Mi-8MTTs2** and **Mi-8MTTs3** with non-rectangular ('teardrop') radome on cabin sides and less regularly shaped arrays on sides of rear cabin.

Mi-8MTV ('Hip-H'): (V=*visotnyi*: high-altitude); 1,633 kW (2,190 shp) TV3-117VM turboshafts for improved 'hot-and-high' operation. Built at Ulan-Ude (as **Mi-8AMT**) from 1991; 100 built by 1999. Civil version built at Kazan is **Mi-8MTV-1**; missile-armed, radar-equipped military version with six-hardpoint stub-wing is **Mi-8MTV-2**; export equivalent is **Mi-17-1V**, with optional armament, nose radar, flotation gear and firefighting equipment. Indian Mi-17-1V has enlarged side doors and a rear ramp to facilitate airdrops, as developed at Kazan (which see). Firefighting conversion by Airod in Malaysia adds detachable aluminium boom to starboard side, stowed facing forwards, but trainable through 90°, with hose nozzle at tip. Internal tanks, fillable via own trailing hose and onboard pump in 1½ minutes, carry 2,300 litres (608 US gallons; 506 Imp gallons) of water, which may also be dumped from belly port in 8 seconds. Addition of suffix -GA indicates civil version, often with rectangular 'Salon'-type cabin windows. **Mi-8MTV-1S** is VVIP variant for Rossiya state transport company; two believed ordered in 2001.

Mi-8MTO (or **Mi-8N**; also quoted as **Mi-8MTKO**) ('Hip-H'): (O = *ochki nochnogo videniya*: night vision goggles). Dedicated night attack conversion of Mi-8MT/ MTV to meet urgent operational requirement; tested in Chechnya ('Caucasian campaign') with aircraft equipping BEG (Boyevaya Eksperimentalnaya Gruppa: Combat Experimental Group) based at Mozdok; at least five aircraft thus upgraded by mid-2000. (Unit additionally equipped with four or six upgraded Mi-24s, and also parented trial deployment of Ka-50. Originally intended to be equipped with modified Ka-29TB and Ka-50). Upgrade produced by Mil Moscow plant, with collaboration of Russkaya Avionika, NPO Geofizika and Uralsky Optik Mechanichesky Zavod team, and tested at Russian Air Forces flight test centre at Chkalovskaya from April 1999, thereafter undergoing evaluation at Torzhok with the 696th IIVP, equipping BEG from September 1999.

Aircraft has NVG-compatible cockpit and external lights and may also feature NVG-covert, formation-keeping external lights. Crew equipped with Geofizika ONV-1 NVGs, and also use gyrostabilised GOES undernose turret containing AGEMA 1000 or SAGEM-321 FLIR, laser range-finder and LLTV. Incorporates cockpit and navigation system upgrade, with MFI-68 (or, possibly, US-supplied) colour LCD MFDs, A737-00 GPS, full-standard MTO will gain new air data system and Saratov Industrial Automatics OKB PKV-M digital autopilot. May acquire AT-X-16 Vikhr-M ATMs intended for BEG Ka-29TBs, and could operate as laser designation platform for Su-25s and other tactical fast jets. Similar export version offered to India (which has already used its existing Mi-17s in the gunship role in Kashmir, and hopes to upgrade these with an Mi-35-type weapon aiming system, improved avionics and a 'night cockpit') and others as **Mi-17N**.

Mi-17-1VA: Version of Mi-8MTV-1 produced for Ministry of Health of former USSR as flying hospital equipped to highest practicable standards for relatively small helicopter; interior, with equipment developed in Hungary; provision for three stretchers, operating table, extensive surgical and medical equipment, accommodation for doctor/surgeon and three nursing attendants.

Mi-8MTV-3/Mi-172A: As Mi-8MTV-2, also from Kazan, but with four-hardpoint stub-wing and other equipment changes and planned for certification to FAR Pt 29 standards; TV3-117VM Srs 2 engines, giving maximum cruising speed of 118 kt (218 km/h; 135 mph) and service ceiling of 6,000 m (19,680 ft); air conditioning and heating systems, main and tail rotor blade de-icing, canopy demisting and heating of engine air intakes standard; options include flotation gear, Doppler, weather radar, DME, GPS, VOR, ILS, transponder and VIP interiors for seven, nine and 11 passengers. Standard seating for up to 26 passengers.

Mi-172A export version first exhibited at 1994 Singapore Air Show; based on 'salon' (rectangular window) cabin. Seven ordered by Mesco, India, 1995; others to Vietnam.

Mi-8MTV-7: Upgraded development by Kazan; announced 2002.

Mi-172AG: Announced mid-2000; proposed version with TV3-117SB3 engines, improved composites main rotor blades and upgraded navigation avionics.

Mi-17PI: As Mi-17P but single D-band jamming system able to jam up to eight sources simultaneously over 30° sector.

Mi-17PG: As Mi-17P but with H/I-band system for jamming pulse/CW and CW interrupted equipment.

Mi-17AMT: Provisional designation for proposed Ulan-Ude-built variant offered to Malaysia to meet eventual 40-aircraft requirement, with initial order for six CSAR-configured aircraft reportedly already placed. May have rear loading ramp of Mi-8MTV-5. Mi-17 will replace 31 S-61 Nuri helicopters, and two already in use with Malaysian Fire Department. SME Aerospace appointed as partner responsible for assembly, flight test and local certification.

Mi-8AMT(Sh) 'Terminator': Current counterpart of Mi-8MTV series, built at Ulan-Ude. Crew positions protected by armour plating. Armament can include up to eight Igla-V AAMs or 9M114/9M114M Shturm-V ('Spiral') ASMs, with thimble radome on nose and chin-mounted electro-optics pod. Prototype (RA-25755) exhibited at Farnborough in September 1996. First production helicopter almost complete by late 1998, but delayed by funding shortage. State testing began 10 April 2000, and had been completed by January 2001: funding assistance provided by Interprofavia. Offered for export as

Mi-171(Sh). Night-capable variant under development by mid-2001.

Mi-171: Export version of Mi-8AMT; first displayed 1989 Paris Air Show. Mostly Western avionics, including Honeywell and Rockwell Collins radios. Built at Ulan-Ude.

In August 2001, Ulan-Ude showed upgraded Mi-171A at Moscow Salon, featuring rear cargo ramp (straight-sided, compared with Kazan version) and enlarged starboard passenger door.

Mi-17MD (Mi-8MTV-5): Described under Kazan heading in this section.

Mi-17KF: Described under Kazan heading in previous editions.

Mi-17LL: Flying testbed (*laboratoriya*: laboratory). One aircraft, RA-70880, employed by the Gromov Flight Research Institute at Zhukovsky, on trials including atmospheric sampling for contamination and searching for leakage from oil and gas pipelines. Special equipment includes a sampling probe mounted on a framework extending from the nose and a downward-viewing window on the starboard side.

Mi-18: Originally, was the designation of the prototype Mi-17 (which see). In 1979, two Mi-8MTs (93038 and 93114) were modified by Kazan to meet requirements for the campaign in Afghanistan, under the same designation. These Mi-18s were lengthened by 1.0 m (3 ft 3¼ in) and fitted with an additional sliding door on the starboard side. Payload was 5,000 kg (11,023 lb). Provision for 30 passengers. Later (1984) use of retractable landing gear raised maximum speed to 148 kt (275 km/h; 170 mph). The airframes were used eventually for static training.

Mi-19: Command relay platform; similar to earlier Mi-9. Believed to incorporate Ivolga (oriole) equipment but not used as airborne relay platform; R-111 relay antenna folds upward under fuselage on ground; rearward-inclined 'hockey stick' antennas of R-405MPB datalink project from rear of cabin and from undersurface of tailboom, aft of Doppler radar box; rearward-inclined short whip antenna above forward end of tailboom; two long plate aerials of R-826 HF radio on fuselage undersurface; R-802 VYa UHF mast antenna under forward fuselage.

Operating concept envisages aircraft flying to selected location; landing, shutting down and being linked to two masts – one 11 m (36 ft) for R-111 (deployed 26 m; 85 ft to port side of helicopter), one 16.5 m (54 ft) antenna for R-405M (at 29.5 m; 97 ft) to starboard. Mission crew of six (three controllers, seated aft, at map tables; three radio operators forward), plus pilots and flight engineer.

Mi-17Z-2: Converted from 'Hip-H' in former Czechoslovakia for electronic warfare ECM role; local designation (Z-1 is similarly tasked Antonov An-26). First seen in Czech Air Force service at Dobrany-Line airbase, near Plzen, 1991; each of two examples had a tandem pair of large cylindrical containers mounted each side of cabin; assumed that containers made of dielectric material and contain receivers to locate, analyse and jam hostile electronic emissions; each of two operators' stations in main cabin has large screens, computer-type keyboards and oscilloscope; several blade antennas project from tailboom. Currently in service with Slovak Air Force.

Long-range modification: AEFT (Auxiliary External Fuel Tanks) system by Aeroton adds a further 1,900 litres (502 US gallons; 418 Imp gallons) in two internal tanks, plus 2,850 litres (753 US gallons; 626 Imp gallons) in six tanks on the stores pylons of Mi-8MT, AMT, MTV-1, civil MTV and Mi-17 variants. Operational range with all eight auxiliary tanks is 701 n miles (1,300 km; 807 miles); ferry range 998 n miles (1,850 km; 1,149 miles). Users include Leningrad customs service.

Upgrades: Kazan offers three stages of upgrade for existing Mi-17 variants. The first can include new composites main rotor blades with a 6,000-hour service life, NVG-compatible cockpit and improved avionics; second could include more powerful 1,864 kW (2,500 shp)

Mi-8AMT(Sh) demonstrator RA-25755 (*Jane's/Paul Jackson*)　0106505

engines, uprated transmission and Delta-H 'X-type' tail rotor, latter refinements presaging increase in MTOW to 14,000 kg (30,864 lb).

Kronshtadt is offering an avionics and cockpit upgrade, with LCD screens, digital moving map and embedded GPS.

Israel Aircraft Industries announced **Peak-17** in February 2001, combining features from Kamov Ka-50-2 Erdogan and Mi-35 export 'Hind' with 'glass cockpit', improved self-defence Rafael NT-D Dandy and NT-S Spike anti-armour missiles.

Elbit Systems of Israel in 2001 was promoting 'Andor' Mi-8/17 upgrade for civil and military applications. Modular modernisation concept entails addition of various items of equipment, including electro-optical sensor turret, MFDs, HOCAS, NVGs, HUD, EW systems and choice of Eastern or Western weaponry.

Civil conversion: Helion Procopter Industries GmbH of Bitburg, Germany (joint venture by Mil, Kazan, Ulan-Ude and Wolfgang Wibbelt), demonstrated prototype of ex-military Mi-8MTV conversion with VIP interior at ILA, Berlin, June 2000; then estimating FAA/JAA certification within 18 months; Western avionics optional.

CUSTOMERS: Many operational side by side with Mi-8s in RFAS armed forces; also operated in Algeria, Angola, Armenia, Bangladesh, Bulgaria, Burkina Faso, Cambodia, China, Colombia, Costa Rica, Croatia, Cuba, Czech Republic, Ecuador, Egypt, Ethiopia, Hungary, India, Indonesia, Kenya, North Korea, Laos, Macedonia, Malaysia, Mexico, Myanmar, Nicaragua, Pakistan, Papua New Guinea, Peru, Poland, Romania, Sierra Leone, Slovak Republic, Sri Lanka, Syria, Turkey, Uganda, Ukraine, Venezuela, Vietnam, Yugoslavia. More than 810 exported by Aviaexport. Recent announced sales include 10 Mi-171 to China in 1999, 10 Mi-171V (with provision for Skosok NVGs) to Colombia, 20 Mi-17-1V to Egypt, 40 Mi-17-1V to India (contract signed 26 May 2000), eight Mi-17-1V to Indonesia, five Mi-171s to Iran, 12 Mi-17V to Laos, 19 Mi-17-1V to Turkey, two Mi-17s to Nepal in late 2001 and four civilian Mi-171s (of at least eight on order) delivered to Pakistan on 14 June 2002. The type was evaluated by the Hellenic Army during mid-2000.

COSTS: Quoted by ITAR-TASS as being from US$4 million to US$5 million; India's 40 Mi-17-1Vs cost a reported US$170 million. Kazan quoted US$3.7 million, flyaway, basically equipped, in March 2001.

DESIGN FEATURES: Conventional pod and boom configuration; clamshell rear-loading freight doors on most versions, but both Kazan and Ulan-Ude plants offer rear-loading ramp version; five-blade main rotor, inclined forward 4° 30' from vertical; interchangeable blades of basic NACA 230 section, solidity 0.0777; spar failure warning system; drag and flapping hinges a few inches apart; blades carried on machined spider; pendulum vibration damper; three-blade port tail rotor; transmission comprises VR-14 two-stage planetary main reduction gearbox, intermediate and tail rotor gearboxes, main rotor brake, and drives off main gearbox for tail rotor, fan, AC generator, hydraulic pumps and tachometer generators; correct rotor speed maintained automatically by system that also synchronises output of the two engines; tail rotor pylon forms small vertical stabiliser; horizontal stabiliser near end of tailboom. Civil versions have rectangular cabin windows.

FLYING CONTROLS: Mechanical system, with irreversible hydraulic boosters; main rotor collective pitch control linked to throttles.

STRUCTURE: All-metal; main rotor blades, mainly VD-76 aluminium alloy, each has extruded spar carrying root fitting, 21 honeycomb-filled trailing-edge pockets and blade tip, with titanium abrasion shielding on some of leading-edge; balance tab on each blade; each tail rotor blade made of spar and honeycomb-filled trailing-edge; semi-monocoque fuselage. Composites blades and elastomeric hub components under development.

LANDING GEAR: Non-retractable tricycle type; steerable twin-wheel nose unit, locked in flight; single wheel on each main unit; oleo-pneumatic (gas) shock-absorbers. Mainwheel tyres 865×280; nosewheel tyres 595×185.

Mil Mi-8MT used in Vietnam   *NEW*/0102972

Rear loading ramp of Ulan-Ude Mi-171, showing straight sides   0121070

Pneumatic brakes on mainwheels; pneumatic system can also recharge tyres in the field, using air stored in main landing gear struts. Optional mainwheel fairings.

POWER PLANT: Two 1,397 kW (1,874 shp) Klimov TV3-117MT turboshafts in basic Mi-17; should one engine stop, output of the other increased automatically to contingency rating of 1,637 kW (2,195 shp), enabling flight to continue. Alternative high-altitude TV3-117VM engine has installed ratings of 1,545 kW (2,071 shp) in emergency; 1,397 kW (1,874 shp) for T-O; 1,250 kW (1,677 shp) normal; and 1,103 kW (1,479 shp) for cruising. Newest version of engine is TV3-117VMA-SB3, which believed to be available for new-build and upgraded aircraft. Deflectors on engine air intakes prevent ingestion of sand, dust and foreign objects; engine cowling side panels form maintenance platforms when open, with access via hatch on flight deck.

Standard fuel capacity of 2,615 litres (691 US gallons, 575 Imp gallons), comprising 2,170 litres (573 US gallons; 477 Imp gallons) in main external tanks and 445 litres (118 US gallons; 98.0 Imp gallons) in feed tank. Additional 1,830 litres (483 US gallons; 402 Imp gallons) optional auxiliary fuel in two tanks.

ACCOMMODATION: Two pilots side by side on flight deck, with provision for flight engineer's station. Military versions can be fitted with external flight deck armour. Windscreen de-icing standard.

Seats and bulkheads of basic version quickly removable for cargo carrying. Standard military versions have cargo tiedown rings in floor, winch of 150 kg (330 lb) capacity and pulley block system to facilitate loading of heavy freight, an external cargo sling system (capacity 3,000 kg; 6,614 lb), and 24 tip-up seats along sidewalls of cabin; six additional centreline seats optional. Military Mi-17-1V carries up to 30 troops, up to 20 wounded or 12 stretchers in ambulance role, or weapons on six outrigger pylons; civilian Mi-17 promoted as essentially a cargo-carrying helicopter, with secondary passenger transport role. Large windows on each side of flight deck slide rearward. Sliding, jettisonable main passenger door at front of cabin on port side; electrically operated rescue hoist (capacity 150 kg; 330 lb) can be installed at this doorway. Rear of cabin made up of clamshell freight-loading doors, which are smaller on commercial versions, with downward-hinged passenger airstair door centrally at rear. Hook-on ramps used for vehicle loading.

SYSTEMS: (Mi-17-1V/171): AI-9V APU for pneumatic engine starting; AC electrical supply from two 40 kW three-phase 115/220 V 400 Hz GT40/P-48 V generators. Engine air intake de-icing standard. Provision for oxygen system for crew and, in ambulance version, for patients. Freon fire extinguishing system in power plant bays and service fuel tank compartments, actuated automatically or manually. Two portable fire extinguishers in cabin.

AVIONICS: (Mi-17-1V/171): *Comms:* Baklan-20 and Yadro-1G1 com radio.
Radar: Type 8A-813 weather radar optional.
Flight: Type A-723 long-range nav.
Instrumentation: ARK-15M radio compass, ARK-UD radio compass, DISS-32-90 Doppler; AGK-77 and AGR-74V automatic horizons; BKK-18 attitude monitor; ZPU-24 course selector; A-037 radio altimeter.
Self-defence (optional): ASO-2V chaff/flare dispensers under tailboom and L-166V IR jammer (NATO 'Hot Brick') at forward end of tailboom.

EQUIPMENT: Options on military versions include external cockpit armour, triple-lobe engine nozzle IR suppressors and a VMR-2 fit for air-dropping such stores as mines.

ARMAMENT: Various combinations of weaponry available, including 64 × 57 mm S-5 rockets in four UV-16-57 packs or 192 × 57 mm S-5 rockets in six UV-32-57 packs or four 9M17P Skorpion (AT-2 'Swatter') anti-armour missiles or six 9M14M Malyutka (AT-3 'Sagger') anti-armour missiles; can also operate with bombs and mines and able to use 23 mm GSh-23 gun packs. (AAMs and newer ASMs on Mi-8AMT(Sh).)

DIMENSIONS, EXTERNAL:
Main rotor diameter	21.25 m (69 ft 10½ in)
Tail rotor diameter	3.91 m (12 ft 9⅞ in)
Distance between rotor centres	12.66 m (41 ft 6 in)
Length: overall, rotors turning:	
Mi-17	25.35 m (83 ft 2 in)
fuselage: Mi-17	18.465 m (60 ft 7 in)
nose to tail rotor c/l: Mi-17	18.425 m (60 ft 5½ in)
Mi-172	18.855 m (61 ft 10¼ in)
Mi-17MD	18.99 m (62 ft 3¾ in)
fuselage, excl tail rotor	18.17 m (59 ft 7¼ in)

Ulan-Ude-built Mil Mi-171   0121068

Kazan's Mi-172 demonstrator RA-70898 *(Valery Solomakhine/KVZ)*   **NEW**/0121664

Width of fuselage, excl fuel tanks	2.50 m (8 ft 2½ in)
Tailplane span	3.705 m (12 ft 1¾ in)
Height: overall, rotors turning	5.545 m (18 ft 2¼ in)
to top of rotor head: Mi-8	4.73 m (15 ft 6¼ in)
Mi-17	4.745 m (15 ft 6¾ in)
Mi-17MD, Mi-171	4.755 m (15 ft 7¼ in)
Mi-172	4.865 m (15 ft 11½ in)
Wheel track: main	4.51 m (14 ft 9½ in)
nose	0.30 m (11¾ in)
Wheelbase	4.28 m (14 ft 0½ in)
Fwd passenger door: Height	1.41 m (4 ft 7½ in)
Width	0.82 m (2 ft 8¼ in)
Rear passenger door: Height	1.70 m (5 ft 7 in)
Width	0.84 m (2 ft 9 in)
Rear cargo door: Height	1.82 m (5 ft 11½ in)
Width	2.34 m (7 ft 8¼ in)
DIMENSIONS, INTERNAL:	
Passenger cabin: Length	6.36 m (20 ft 10¼ in)
Width	2.34 m (7 ft 8¼ in)
Height	1.80 m (5 ft 10¾ in)
Cargo hold (freighter, excl doors):	
Length at floor	5.34 m (17 ft 6¼ in)
Width	2.29 m (7 ft 6¼ in)
Height	1.80 m (5 ft 10¾ in)
Volume	22.5 m³ (795 cu ft)
AREAS:	
Main rotor disc	356.16 m² (3,833.7 sq ft)
Tail rotor disc	12.01 m² (129.28 sq ft)
WEIGHTS AND LOADINGS:	
Weight empty, equipped: Mi-17	7,100 kg (15,653 lb)
Mi-171	6,913 kg (15,241 lb)
Mi-17-1V	7,489 kg (16,510 lb)
Mi-17-1VA	7,586 kg (16,724 lb)
Mi-8AMT(Sh)	8,493 kg (18,724 lb)

Fuel weight, Mi-17-1V/171:	
normal	2,027 kg (4,469 lb)
plus one auxiliary tank	2,737 kg (6,034 lb)
plus two auxiliary tanks	3,447 kg (7,600 lb)
Max payload:	
internal: Mi-17, Mi-17-1V/171, Mi-8AMT(Sh)	
	4,000 kg (8,820 lb)
external, on sling:	
Mi-17, Mi-17-1V	3,000 kg (6,614 lb)
Mi-171, Mi-8AMT(Sh)	4,000 kg (8,820 lb)
Mi-17-1V	5,000 kg (11,023 lb)
Normal T-O weight:	
Mi-17, Mi-17-1V/171, Mi-8AMT(Sh)	
	11,100 kg (24,470 lb)
Max T-O weight:	
Mi-17, Mi-17-1V/171, Mi-8AMT(Sh)	
	13,000 kg (28,660 lb)
Mi-172	11,878 kg (26,186 lb)
Max disc loading:	
Mi-17, Mi-17-1V/171	36.5 kg/m² (7.48 lb/sq ft)
PERFORMANCE:	
Max level speed:	
Mi-17/172, max AUW	135 kt (250 km/h; 155 mph)
Mi-17-1V/171, normal AUW	
	135 kt (250 km/h; 155 mph)
Mi-17-1V/171, max AUW	
	124 kt (230 km/h; 143 mph)
Max cruising speed:	
Mi-17/172, max AUW	129 kt (240 km/h; 149 mph)
Mi-17-1V, normal AUW	135 kt (250 km/h; 155 mph)
Mi-171, normal AUW	124 kt (230 km/h; 143 mph)
Mi-171, max AUW	113 kt (210 km/h; 130 mph)
Service ceiling:	
Mi-17M, normal AUW	5,600 m (18,380 ft)
Mi-17M, max AUW	4,400 m (14,440 ft)
Mi-17-1V, normal AUW	6,000 m (19,680 ft)
Mi-17-1V, max AUW	4,800 m (15,740 ft)
Mi-171, normal AUW	5,700 m (18,700 ft)
Mi-171, max AUW	4,500 m (14,760 ft)
Mi-172, max AUW	5,650 m (18,535 ft)
Hovering ceiling OGE:	
Mi-17, max AUW	1,760 m (5,775 ft)
Mi-17M, normal AUW	3,900 m (12,795 ft)
Mi-17M, max AUW	1,500 m (4,920 ft)
Mi-17-1V/171, normal AUW	3,980 m (13,055 ft)
Mi-17-1V/171, max AUW	1,700 m (5,575 ft)
Mi-172, max AUW	3,300 m (10,825 ft)
Range with max standard fuel, 5% reserves:	
Mi-17, normal AUW	267 n miles (495 km; 307 miles)
Mi-17, max AUW	251 n miles (465 km; 289 miles)
Mi-172, max AUW	267 n miles (495 km; 307 miles)
Range at 500 m (1,640 ft), max AUW, 30 min reserves:	
Mi-17M/-1V, standard fuel	
	251 n miles (465 km; 289 miles)
Mi-171, standard fuel	329 n miles (610 km; 379 miles)
Mi-171, standard fuel plus one auxiliary tank	
	440 n miles (815 km; 506 miles)
Mi-17M, standard fuel plus two auxiliary tanks	
	545 n miles (1,010 km; 627 miles)
Mi-171, standard fuel plus two auxiliary tanks	
	575 n miles (1,065 km; 661 miles)
Mi-17M, standard fuel plus four auxiliary tanks	
	809 n miles (1,500 km; 932 miles)
	UPDATED

Large-size starboard passenger door of Ulan-Ude Mi-171   0121072

MIL Mi-24, Mi-25 and Mi-35

Detailed description last appeared in the 2001-02 *Jane's*. Production terminated after manufacture of more than 5,000 examples, but several organisations currently active in developing and marketing improvement packages, further details of which can be found in *Jane's Aircraft Upgrades*.

     UPDATED

MIL Mi-26

NATO reporting name: Halo

TYPE: Heavy lift helicopter.

PROGRAMME: Development started early 1970s (initially as Mi-6M); aim was payload capability 1½ to 2 times greater than that of any previous production helicopter; first prototype flew 14 December 1977; first production aircraft rolled out October 1980; one of several prototype or preproduction Mi-26s (SSSR-06141) displayed at 1981 Paris Air Show; in-field evaluation, probably with military development squadron, began early 1982; fully operational 1983; export deliveries started (to India) June 1986; production continues at low rate, with manufacture and marketing by Rostvertol (which see).

CURRENT VERSIONS: **Mi-26** (*Izdelie* 90): Basic military transport.

Detailed description applies to basic Mi-26, except where indicated.

Mi-26A: Modified military Mi-26, tested in 1985, with PNK-90 integrated flight/nav systems for automatic approach and descent to critical decision point, and other tasks. Not adopted.

Mi-26T: Basic civil transport (*Izdelie* 209), generally as military Mi-26. Production begun in 1985. Variants include **Geological Survey Mi-26** towing seismic gear, with tractive force of 10,000 kg (22,045 lb) or more, at 97 to 108 kt (180 to 200 km/h; 112 to 124 mph) at 55 to 100 m (180 to 330 ft) for up to 3 hours. The mockup of an Mi-26 two-crew flight deck was shown at the 1997 Moscow Air Show.

Mi-26TS (*sertifitsyrovannyi*: certified): Mi-26T (*Izdelie* 219), but prepared for certification and marketed (in West as **Mi-26TC**) from 1996. Preproduction version, with gondola (port, front), positioned a 16,000 kg (35,275 lb) TV tower, 30 m (98 ft) long, in Rostov-on-Don in 1996. One delivered to Samsung Aerospace Industries in South Korea on 13 September 1997; supplied with Twin Bambi Bucket fire-suppressant system and fulfils dual transport/firefighting roles. This version is subject of upgrade proposal involving installation of new avionics suite and other improvements that will reduce crew numbers from five to three and offer benefits in area of operational effectiveness; if implemented, is expected to result in improved helicopter becoming available in about 2006.

Mi-26MS: Medical evacuation version of Mi-26T, typically with intensive care section for four casualties and two medics, surgical section for one casualty and three medics, pre-operating section for two casualties and two medics, ambulance section for five stretcher patients, three seated casualties and two attendants; laboratory; and amenities section with lavatory, washing facilities, food storage and recreation unit. Civil version in use by MChS Rossii (Ministry of Emergency Situations). Alternative medical versions available, with modular box-laboratories or fully equipped medical centres that can be inserted into the hold for anything from ambulance to field hospital use. As field ambulance can accommodate up to 60 stretcher patients; or seven patients in intensive care, 32 patients on stretchers and seven attendants; or 47 patients and eight attendants in other configurations, which can include 12 bunks in four tiers forward, or patent Rostvertol box laboratory behind the first row of bunks, with 16 bunks behind.

The box includes an operating table, diagnostic equipment, anaesthetic and breathing equipment and other systems. Another configuration includes a larger theatre box by Heinkel Medizin Systeme and 12 stretchers behind, and the helicopter can be fitted with an X-ray laboratory or form the central element of a deployable air-portable field hospital.

Mi-26NEF-M: ASW version with search radar in undernose faired radome, extra cabin heat exchangers and towed MAD housing mounted on ramp.

Mi-26P: Transport for 63 passengers, basically four-abreast in airline-type seating, with centre aisle; lavatory, galley and cloakroom aft of flight deck.

Mi-26PK: Flying crane (*kran*) derivative of Mi-26P with operator's gondola on fuselage side, next to cabin door on port side. First produced in 1997.

Mi-26PP: Reported ECM version. First noted 1986; current status unknown.

Mi-26S: Hastily developed version for disaster relief tasks following explosion at Chernobyl nuclear facility; equipped with deactivating liquid tank and underbelly spraying apparatus.

Mi-26TM: Flying crane, with gondola for pilot/sling supervisor under fuselage aft of nosewheels or under rear-loading ramp. First produced in 1992.

Mi-26TP: Firefighting (*pozharnyi*) version that appeared in 1994, with internal tanks able to dispense up to 15,000 litres (3,962 US gallons; 3,300 Imp gallons) fire retardant from one or two vents, or 17,260 litres (4,560 US gallons; 3,796 Imp gallons) of water from an underslung VSU-15 bucket, or from two linked EP-8000 containers. Can fill tanks on the ground using pumps with 3,000 litres (793 US gallons; 660 Imp gallons)/min throughput. Prototype RA-06183 operated by Rostvertol. One delivered to Moscow Fire Brigade on 19 August 1999.

Mi-26TZ: Tanker version that emerged in 1998, with 14,040 litres (3,710 US gallons; 3,088 Imp gallons) of T2, TS1 or R2 aviation fuel or DL, DZ or DA diesel oil fuel and 1,040 litres (275 US gallons; 228 Imp gallons) lubricants (in 52 jerry cans), dispensed through four 60 m (197 ft) hoses for aircraft, or 10 20 m (66 ft) hoses for ground vehicles. Conversion to/from Mi-26T takes 1 hour 25 minutes for each operation.

Mi-26M: Upgrade under development; all-GFRP main rotor blades of new aerodynamic configuration, new ZMKB Progress D-127 turboshafts (each 10,700 kW; 14,350 shp), and modified integrated flight/nav system with EFIS. Transmission rating unchanged, but full payload capability maintained under 'hot and high' conditions, OEI safety improved, hovering and service ceilings increased, and greater maximum payload (22,000 kg; 48,500 lb) for crane operations.

The 1990 edition of US Department of Defense's *Soviet Military Power* stated ''New variants of 'Halo' are likely in the early 1990s to begin to replace 'Hooks' specialised for command support''. Two prototypes are reported to have been built in 1988, with designation **Mi-27**. These have new antennas along lower 'corner' of fuselage, blade and box-type and with long folded masts which are horizontal in flight, vertical when deployed on ground. Prototype wears Aeroflot colour scheme.

CUSTOMERS: Nearly 300 built by 2001. Reportedly sold to about 20 countries; operators include Belarus (15), Cambodia (two), Congo Democratic Republic (one), India (10), Kazakhstan, North Korea (two), South Korea (one), Mexico (two second-hand) in 2000, Peru (three), Russian Army (35), Russian Ministry of Emergencies, Mil-Avia and Ukraine (20). Russian Army deliveries included four in 1994 (but none subsequently). Three delivered in 2000 (two to Peru; one to Scorpion of Greece); four were scheduled for delivery in 2001 (including second for Scorpion).

COSTS: US$8 million to US$10 million (Mi-26T) (2000).

DESIGN FEATURES: Largest ever production helicopter; empty weight comparable to that of Mi-6 and, as specified, is approximately 50 per cent of maximum T-O weight; weight saved by in-house design of main gearbox providing multiple torque paths, GFRP tail rotor blades, titanium main and tail rotor heads, main rotor blades of mixed metal and GFRP, use of aluminium-lithium alloys in airframe; conventional pod and boom configuration, but first successful use of eight-blade main rotor, of smaller diameter than Mi-6 rotor; payload and cargo hold size similar to those of Lockheed C-130 Hercules; auxiliary wings not required; rear-loading ramp/doors; main rotor rpm 132; main rotor spindle inclined forwards 4°.

FLYING CONTROLS: Hydraulically powered cyclic and collective pitch controls actuated by small parallel jacks, with redundant autopilot and stability augmentation system inputs. Fly-by-wire system flight tested 1994.

STRUCTURE: Eight-blade constant-chord main rotor; flapping and drag hinges, droop stops and hydraulic drag dampers; no elastomeric bearings or hinges; each blade has one-piece tubular steel spar and 26 GFRP aerofoil shape full-chord pockets, honeycomb filled, with ribs and stiffeners and non-removable titanium leading-edge abrasion strip; blades have moderate twist, taper in thickness toward tip, and are attached to small forged titanium head of unconventional design; each has ground-adjustable trailing-edge tab; five-blade constant-chord tail rotor, starboard side, has GFRP blades, forged titanium head; conventional transmission, with tail rotor shaft inside cabin roof; all-metal riveted semi-monocoque fuselage with clamshell rear doors; flattened tailboom undersurface; engine bay of titanium for fire protection; all-metal tail surfaces; swept vertical stabiliser/tail rotor support profiled to produce sideways lift; ground-adjustable variable incidence horizontal stabiliser.

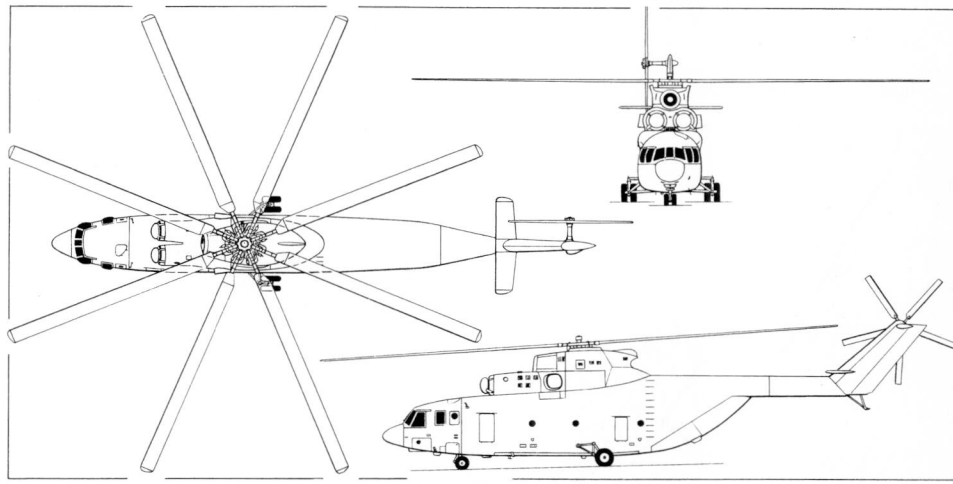

Mil Mi-26 multipurpose heavy-lift helicopter *(Jane's/Dennis Punnett)*

LANDING GEAR: Non-retractable tricycle type; twin wheels on each unit; steerable nosewheels, tyre size 900×300; mainwheel tyres size 1,120×450. Retractable tailskid at end of tailboom to permit unrestricted approach to rear cargo doors. Length of main legs adjusted hydraulically to facilitate loading through rear doors and to permit landing on varying surfaces. Device on main gear indicates take-off weight to flight engineer at lift-off, on panel on shelf to rear of his seat.

POWER PLANT: Two 8,500 kW (11,399 shp) ZMKB Progress D-136 free-turbine turboshafts, side by side above cabin, forward of main rotor driveshaft. Air intakes fitted with particle separators to prevent foreign object ingestion, and have both electrical and bleed air anti-icing systems. Above and behind is central oil cooler intake. VR-26 fan-cooled main transmission, rated at 14,914 kW (20,000 shp), with air intake above rear of engine cowlings. System for synchronising output of engines and maintaining constant rotor rpm; if one engine fails, output of other is increased to maximum power automatically. Independent fuel system for each engine; fuel in eight underfloor rubber tanks, feeding into two header tanks above engines, which permit gravity feed for a period in emergencies; maximum standard internal fuel capacity 12,000 litres (3,170 US gallons; 2,640 Imp gallons); provision for four auxiliary tanks. Mi-26TS normal capacity is 13,020 litres (3,440 US gallons; 2,864 Imp gallons). Two large panels on each side of main rotor mast fairing, aft of engine exhaust outlet, hinge downward as work platforms.

ACCOMMODATION: Crew of five on flight deck: pilot (on port side) and co-pilot side by side, tip-up seat between pilots for flight technician, and seats for flight engineer (port) and navigator (starboard) to rear; upgrade proposal revealed in early 2001 involves installation of new avionics and will result in reduction of flight deck crew to three. Four-seat passenger compartment aft of flight deck. Loads in hold include two airborne infantry combat vehicles and a standard 20,000 kg (44,090 lb) ISO container; about 20 tip-up seats along each sidewall of hold; maximum military seating for 90 combat-equipped troops; alternative provisions for 60 stretcher patients and four/five attendants. Heated windscreen, with wipers; four large blistered side windows on flight deck; forward pair swing open slightly outward and rearward. Downward-hinged doors, with integral airstairs, at front of hold on port side, and each side of hold aft of main landing gear units. Hold loaded via downward-hinged lower door, with integral folding ramp, and two clamshell upper doors forming rear wall of hold when closed; doors opened and closed hydraulically, with back-up hand pump for emergency use. Two LG-1500 electric hoists on overhead rails, each with capacity of 2,500 kg (5,511 lb), enable loads to be transported along cabin; winch for hauling loads,

capacity 500 kg (1,100 lb); roller conveyor in floor and load lashing points throughout hold. Flight deck fully air conditioned.

SYSTEMS: Two main and one emergency hydraulic systems, operating pressure 157 and 206 bar (2,276 and 2,987 lb/sq in). Electrical system three-phase 200/115 V 400 Hz; single-phase 115 V 400 Hz; three-phase, 36 V 400 Hz; single-phase 36 V 400 Hz; DC 27 V. TA-8V 119 kW (160 hp). APU under flight deck, with intake louvres (forming fuselage skin when closed) and exhaust on starboard side, for engine starting and to supply hydraulic, electrical and air conditioning systems on ground. Electrically heated leading-edge of main and tail rotor blades for anti-icing. Only flight deck pressurised.

AVIONICS: All items necessary for day and night operations in all weathers are standard.

Radar: Groza 7A813 weather radar in hinged (to starboard) nosecone.

Flight: Integrated PKV-26-1 flight/nav system and automatic flight control system, Doppler, map display, HSI, and automatic hover system. Optional GPS.

Self-defence: Military versions can have IR jammers and suppressors, IR decoy dispensers and colour-coded identification flare system.

EQUIPMENT: Hatch for load sling in bottom of fuselage, in line with main rotor shaft; sling cable attached to internal winching gear. Closed-circuit TV cameras to observe slung payloads. Specialised versions can utilise firefighting equipment.

ARMAMENT: None.

DIMENSIONS, EXTERNAL:

Main rotor diameter	32.00 m (105 ft 0 in)
Tail rotor diameter	7.61 m (24 ft 11½ in)
Length: overall, rotors turning	40.025 m (131 ft 3¾ in)
nose to turning tail rotor	35.91 m (117 ft 9¾ in)
fuselage, excl tail rotor	33.745 m (110 ft 8½ in)
Height: to top of rotor head	8.145 m (26 ft 8¾ in)
to top of fin	7.45 m (24 ft 5¼ in)
tail rotor turning	11.60 m (38 ft 0¾ in)
Width overall (outsides of mainwheels)	
	6.15 m (20 ft 2¼ in)
Tailplane span	6.02 m (19 ft 9 in)
Wheel track: c/l shock-absorbers	5.00 m (16 ft 4¾ in)
outer wheels	5.75 m (18 ft 10½ in)
Wheelbase	8.95 m (29 ft 4½ in)

DIMENSIONS, INTERNAL:

Freight hold:	
Length: excl ramp	12.08 m (39 ft 7½ in)
ramp trailed	15.00 m (49 ft 2½ in)
Max width	3.23 m (10 ft 7¼ in)
Height	2.98-3.17 m (9 ft 9¼ in-10 ft 4¾ in)
Floor area: excl ramp	39.3 m² (423 sq ft)
ramp trailed	49.2 m² (530 sq ft)
Volume: excl ramp	121.0 m³ (4,273 cu ft)
ramp trailed	135.9 m³ (4,799 cu ft)

AREAS:

Main rotor disc	804.25 m² (8,656.8 sq ft)
Tail rotor disc	45.48 m² (489.54 sq ft)

WEIGHTS AND LOADINGS:

Weight empty	28,200 kg (62,170 lb)
Max payload, internal or external	20,000 kg (44,090 lb)
Normal T-O weight:	
except Mi-26TS	49,600 kg (109,350 lb)
Mi-26TS	49,650 kg (109,455 lb)
Max T-O weight	56,000 kg (123,450 lb)
Max disc loading: Mi-26TS	69.6 kg/m² (14.26 lb/sq ft)
Transmission loading at max T-O weight and power:	
Mi-26TS	3.76 kg/kW (6.17 lb/shp)

PERFORMANCE (A: Mi-26, B: Mi-26M, C: Mi-26TS at normal T-O weight):

Max level speed: A	159 kt (295 km/h; 183 mph)
C	146 kt (270 km/h; 168 mph)
Normal cruising speed: A, C	137 kt (255 km/h; 158 mph)
Service ceiling: A	4,600 m (15,100 ft)
B	5,900 m (19,360 ft)
C	4,300 m (14,100 ft)

Mil Mi-26 before delivery to Scorpion of Greece   0097422

Hovering ceiling IGE:
  A, ISA, with 5,100 kg (11,240 lb) payload
                               1,000 m (3,280 ft)
  B, ISA + 15°C, with 12,300 kg (27,115 lb) payload
                               1,000 m (3,280 ft)
Hovering ceiling OGE, ISA: A   1,800 m (5,900 ft)
  B   2,800 m (9,180 ft)
  C   1,520 m (4,980 ft)
Range: A at 2,500 m (8,200 ft) ISA + 15°C, with
  7,700 kg (16,975 lb) payload
                270 n miles (500 km; 310 miles)
  B at 2,500 m (8,200 ft) ISA + 15°C, with 13,700 kg
    (30,200 lb) payload 270 n miles (500 km; 310 miles)
  A, S/L ISA, with max internal fuel at max T-O weight,
    5% reserves   318 n miles (590 km; 366 miles)
  A, S/L ISA, with four auxiliary tanks
         1,036 n miles (1,920 km; 1,190 miles)
                                UPDATED

Prototype Mil Mi-28A (*Yefim Gordon*)   ***NEW*/0525121**

MIL Mi-28
NATO reporting name: Havoc
TYPE: Attack helicopter.
PROGRAMME: Design started 1980 under Marat N Tishchenko; first of two flying Mi-28 prototypes (012) flew 10 November 1982; each prototype different: first and second (022) had upward-pointing exhaust diffusers and fixed undernose fairing for electro-optic equipment; first also had conventional three-blade tail rotor; second replaced this with the definitive 'Delta-H' configuration. The first Mi-28A (032) introduced the definitive downward-pointing exhaust suppressors and flew in January 1988; second Mi-28A prototype (042) demonstrated at Moscow in 1992 and represented the intended production configuration. It had the definitive moving E-O sensor turret undernose, downward-pointing exhaust diffusers and wingtip electronics/chaff dispenser pods; small-scale pre-series production planned, but not yet initiated, by Rostvertol, Rostov-on-Don, which stated in mid-2001 that it was ready to begin series production. Rival Ka-50 officially 'adopted' 5 October 1994, but competition then continued; final decision was due by early 2001, but no recent news received.
CURRENT VERSIONS: **Mi-28:** First two prototypes with 1,434 kW (1,923 shp) TV3-117BM engines and VR-28 gearbox. **Mi-28A** (Type 280): Basic version, *as described in detail*. Third and fourth aircraft built.
  Mi-28N: (*Nochnoy:* Night): Unofficial names: Night Hunter and Night Pirate. Added night/all-weather operating capability. Russian Army funding announced January 1994; demonstrator (014) modified from first Mi-28 prototype (012); first hover 14 November 1995; formal roll-out 16 August 1996; first flight 30 April 1997. Mast-mounted 360° scan millimetre wave Kinzhal V or Arbalet radar (pod soon enlarged in vertical plane); FLIR ball beneath missile-guidance nose radome and above new shuttered turret for optical/laser sensors, including Zenit low-light-level TV. EFIS cockpit. Armament of production version to include 9M114 Shturm (AT-6 'Spiral') ASMs and Igla (SA-16 'Gimlet') AAMs and R-73 AAMs. New composites rotor with sweptback blade tips added subsequently. Mi-28N introduced uprated VR-29 transmission and IKBO integrated flight/weapon aiming system, with automatic terrain-following and automatic target search, detection, identification and (in formations of Mi-28Ns) allocation; Ramenskoye Breo-28N mission control system.
  Five trials Mi-28Ns being built by Rostvertol; first flight was due early 2002, but slipped to late 2002; TV3-117VMA engines initially, but 1,839 kW (2,466 shp) Klimov VK-2500s to be installed later. Second helicopter funded jointly by Rostvertol and Southwest Sperbank
  Versions projected for naval amphibious assault support and air-to-air missions.

  Mi-28NEh: (*Nochnoy, Ehksport:* Night, Export): Version of above offered to South Korea in 2000. Evaluated by Swedish Army in 2001 against Boeing AH-64 and Eurocopter Tiger.
COSTS: Mi-28N development cost US$150 million (2000); unit cost approximately US$15 million (2000).
DESIGN FEATURES: Conventional gunship configuration, with two crew in stepped cockpits; original three-blade tail rotor superseded by low noise 'scissors' or 'Delta-H' type comprising two independent two-blade rotors set as narrow X (35°/145°) on same shaft with self-lubricating bearings; resulting flapping freedom relieves flight loads; agility enhanced by doubling hinge offset of main rotor blades compared with Mi-24; survivability emphasised; crew compartments protected by titanium and ceramic armour and armoured glass transparencies; single hit will not knock out both engines; vital units and parts are redundant, widely separated and shielded by the less vital; multiple self-sealing fuel tanks in centre-fuselage enclosed in composites second skin, outside metal fuselage skin; no explosion, fire or fuel leakage results if tanks hit by bullet or shell fragment; energy absorbing seats and landing gear protect crew in crash landing at descent rate of 12 m (40 ft)/s; crew doors are rearward-hinged, to open quickly and remain open in emergency; parachutes are mandatory for Russian Federation and Associated States (CIS) military helicopter aircrew; if Mi-28 crew had to parachute, emergency system would jettison doors, blast away stub-wings, and inflate bladder beneath each door sill; as crew jumped, they would bounce off bladders and clear main landing gear; no provision for rotor separation; port-side door, aft of wing, provides access to avionics compartment large enough to permit combat rescue of two or three persons on ground, although it lacks windows, heating and ventilation.
  Hand crank, inserted into end of each stub-wing, enables stores of up to 500 kg (1,100 lb) to be winched on to pylons without hoists or ground equipment; current 30 mm gun is identical with that of RFAS army ground vehicles and uses same ammunition; jamming averted by attaching twin ammunition boxes to sides of gun mounting, so that they turn, elevate and depress with gun; main rotor shaft has 5° forward tilt, providing tail rotor clearance; transmission capable of running without oil for 20 to 30 minutes; main rotor rpm 242; with main rotor blades and wings removed, helicopter is air-transportable in An-22 or Il-76 freighter.

FLYING CONTROLS: Hydraulically powered mechanical type; horizontal stabiliser linked to collective; controls for pilot only.
STRUCTURE: Five-blade main rotor; blades have very cambered high-lift section and sweptback tip leading-edge; full-span upswept tab on trailing-edge of each blade; structure comprises numerically controlled, spirally wound glass fibre D-spar, blade pockets of Kevlar-like material with Nomex-like honeycomb core, and titanium erosion strip on leading-edge; each blade has single elastomeric root bearing, mechanical droop stop and hydraulic drag damper; four-blade GFRP tail rotor with elastomeric bearings for flapping; rotor brake lever on starboard side of cockpit; strong and simple machined titanium main rotor head with elastomeric bearings, requiring no lubrication; power output shafts from engines drive main gearbox from each side; tail rotor gearbox, at base of tail pylon, driven by aluminium alloy shaft inside composites duct on top of tailboom; sweptback mid-mounted wings have light alloy primary box structure, leading- and trailing-edges of composites; no wing movable surfaces; provision for countermeasures pod on each wingtip, housing chaff/flare dispensers and sensors, probably RWR; light alloy semi-monocoque fuselage, with titanium armour around cockpits and vulnerable areas; composites access door aft of wing on port side; swept fin has light alloy primary box structure, composites leading- and trailing-edges; cooling air intake at base of fin leading-edge, exhaust at top of trailing-edge; two-position composites horizontal stabiliser.
LANDING GEAR: Non-retractable, tailwheel type; single wheel on each unit; mainwheel tyres size 720×320, pressure 5.40 bar (78 lb/sq in); castoring tailwheel with tyre size 480×200.
POWER PLANT: Two Klimov TV3-117VMA turboshafts, each 1,636 kW (2,194 shp), in pod above each wingroot; three jetpipes inside downward-deflected composites nozzle fairing on each side of third prototype shown in Paris 1989; upward deflecting type also tested. Deflectors for dust and foreign objects forward of air intakes, which are de-iced by engine bleed air. Internal fuel capacity 1,720 litres (454 US gallons; 378 Imp gallons). Provision for four external fuel tanks on underwing pylons.
ACCOMMODATION: Navigator/gunner in front cockpit; pilot behind, on elevated seat; transverse armoured bulkhead between; flat non-glint tinted transparencies of armoured glass; navigator/gunner's door on port side, pilot's door on starboard side.
SYSTEMS: Cockpits air conditioned and pressurised by engine bleed air. Duplicated hydraulic systems, pressure 152 bar (2,200 lb/sq in). 208 V AC electrical system supplied by two generators on accessory section of main gearbox, ensuring continued supply during autorotation. Low-airspeed system standard, giving speed and drift via main rotor blade-tip pitot tubes at −27 to +38 kt (−50 to +70 km/h; −31 to +43 mph) in forward flight, and ±38 kt (±70 km/h; ±43 mph) in sideways flight. Main and tail rotor blades electrically de-iced. Ivchenko AI-9V APU in rear of main pylon structure supplies compressed air for engine starting and to drive small turbine for preflight ground checks.
AVIONICS: *Comms:* UHF/VHF nav/com; small IFF fairing each side of nose and tail.
  Instrumentation: Conventional IFR instrumentation, with autostabilisation, autohover, and hover/heading hold lock in attack mode; pilot has HUD and centrally mounted CRT for basic TV; aircraft designed for use with night vision goggles.
  Mission: Radio for missile guidance in nose radome. Daylight optical weapons sight and laser range-finder in gyrostabilised and double-glazed nose turret above gun, with which it rotates through ±110°; wiper on outer glass protects inner optically flat panel.

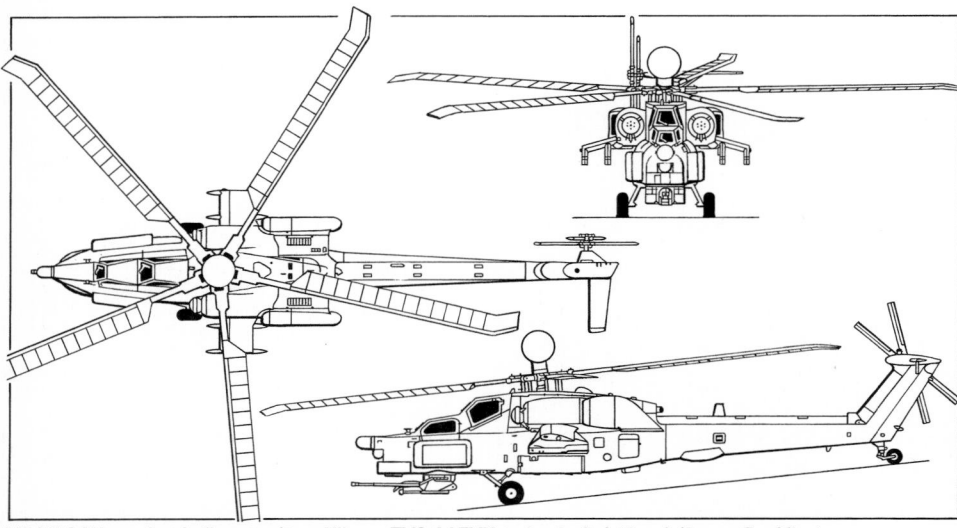

Mil Mi-28N combat helicopter (two Klimov TV3-117VK turboshafts) (*Jane's/James Goulding*)

Self-defence: Two fixed IR sensors on initial basic production Mi-28; IR suppressors, radar and laser warning receivers standard; optional countermeasures pod on each wingtip, housing chaff/flare dispensers and sensors, probably RWR. Mi-28N has integrated Vitebsk DASS with Pastel RWR, Mak IR warning system, Platan jammer and UV-26 flare dispensers.

EQUIPMENT: Two slots, one above the other on port side of tailboom, for colour-coded identification flares. Three pairs of rectangular formation-keeping lights in top of tailboom; further pair in top of main rotor pylon fairing.

ARMAMENT: One 2A42 30 mm turret-mounted gun (with 250 rounds in side-mounted boxes) in NPPU-28 mount at nose, able to rotate ±110°, elevate 13° and depress 40°; maximum rate of fire 900 rds/min air-to-air and air-to-ground. (New specially designed gun under development.) Two pylons under each stub-wing, each with capacity of 480 kg (1,058 lb), typically for total of sixteen 9M114 Shturm C (AT-6 'Spiral') radio-guided tube-launched anti-tank missiles and two UB-20 pods of eighty 80 mm S-8 or twenty 122 mm S-13 rockets or two UPK-23-250 gun pods. Alternative ATMs include 9M120/9M121F Vikhr and 9A-2200; up to eight 9M39 Igla-V AAMs in place of ATMs; in minelaying role can carry two KGMU-2 dispensers. Main 2A42 gun fired and guided weapons launched normally only from front cockpit; unguided rockets fired from both cockpits. (When fixed, gun can also be fired from rear cockpit.)

DIMENSIONS, EXTERNAL:

Main rotor diameter	17.20 m (56 ft 5 in)
Main rotor blade chord	0.67 m (2 ft 2½ in)
Tail rotor diameter	3.84 m (12 ft 7¼ in)
Tail rotor blade chord	0.24 m (9½ in)
Length overall, excl rotors, incl gun	
	17.01 m (55 ft 9¾ in)
Fuselage max width	1.85 m (6 ft 1 in)
Width over stub-wings	4.88 m (16 ft 0¼ in)
Height: overall	4.70 m (15 ft 5 in)
to top of rotor head	3.82 m (12 ft 6½ in)
Wheel track	2.29 m (7 ft 6¼ in)
Wheelbase	11.00 m (36 ft 1 in)

AREAS:

Main rotor disc	232.35 m² (2,501.0 sq ft)
Tail rotor disc	11.58 m² (124.65 sq ft)

WEIGHTS AND LOADINGS:

Weight empty, equipped: 28	7,900 kg (17,416 lb)
28A	8,095 kg (17,846 lb)
28N	8,590 kg (18,938 lb)
Fuel weight: standard internal	1,337 kg (2,947 lb)
with added tanks	1,782 kg (3,928 lb)
Normal T-O weight: 28	10,200 kg (22,487 lb)
28A	10,400 kg (22,928 lb)
28N	10,700 kg (23,589 lb)
Max T-O weight: 28	11,200 kg (24,691 lb)
28A, 28N	11,500 kg (25,353 lb)
Max disc loading: 28	48.2 kg/m² (9.87 lb/sq ft)
28A, 28N	49.5 kg/m² (10.14 lb/sq ft)

PERFORMANCE:

Max level speed: 28A	162 kt (300 km/h; 186 mph)
28N	172 kt (320 km/h; 199 mph)
Max cruising speed: 28A	143 kt (265 km/h; 164 mph)
28N	145 kt (270 km/h; 167 mph)
Max rate of climb at S/L	816 m (2,677 ft)/min
Service ceiling: 28A	5,800 m (19,020 ft)
28N (-SB3)	5,700 m (10,700 ft)
Hovering ceiling OGE: 28	3,470 m (11,380 ft)
28A, 28N	3,600 m (11,820 ft)
28N (-SB3)	4,500 m (14,760 ft)
Radius of action, standard fuel, 10 min loiter at target,	
5% reserves	108 n miles (200 km; 124 miles)
Range, max standard fuel, 10% reserves: all	
	234 n miles (435 km; 270 miles)
Ferry range, 5% reserves	
	593 n miles (1,100 km; 683 miles)
Endurance with max fuel	2 h
g limits	+3/–0.5

UPDATED

MIL Mi-34
NATO reporting name: Hermit
TYPE: Four-seat helicopter.

PROGRAMME: First flight 17 November 1986; two prototypes and structure test airframe completed by mid-1987, when exhibited for first time at Paris Air Show; first helicopter built in former USSR to perform normal loop and roll; series production began at Arsenyev plant in 1993; airframe manufactured by Carpathian Helicopter Production Association; marketed by Mi-Light Helicopters (which see in LVM entry;) Mi-34S/-34C completion was at Moscow plant of LVM, but subsequently reverted to Arsenyev; planned completion of 30 in 1994-95 hampered by lack of funding; five delivered in 1995, one in first half of 1996. State order for resumed manufacture 1996. Marketing initiative, early 1999, planned worldwide establishment of service centres in countries where more than 10 Mi-34s ordered.

Trials in 1999 by civil pilot training school at Omsk showed Mi-34 to be 2.8 times cheaper and more effective to operate than current fleet; school to acquire three Mi-34Cs and obtain up to 10 more in long term. Other schools expected to replace ageing Mi-2 with Mi-34.

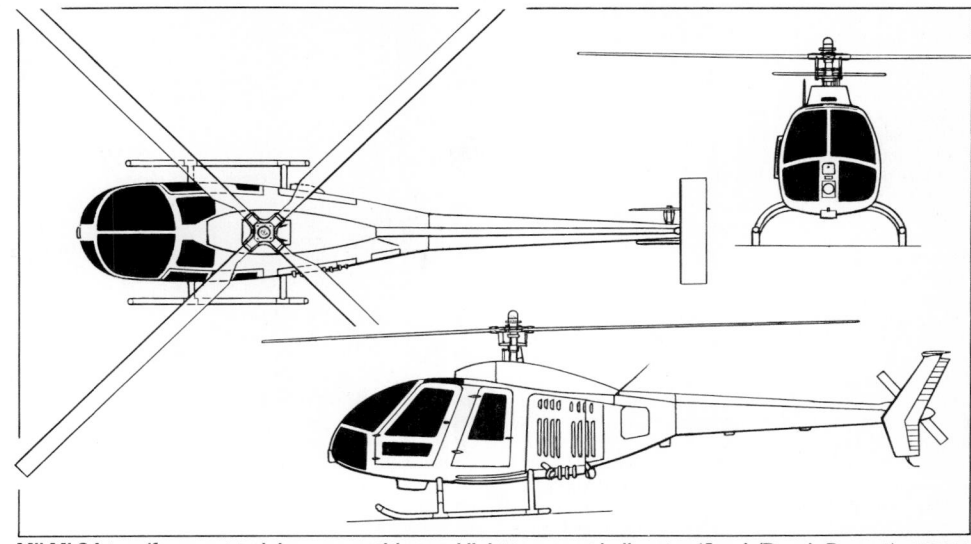

Mil Mi-34 two/four-seat training, competition and light transport helicopter (*Jane's/Dennis Punnett*)   0079310

In 2001, an upgraded variant was proposed with M-14 engine rated at 272 kW (365 hp), IFR instrumentation and auxiliary fuel tank.

CURRENT VERSIONS: **Mi-34S:** Basic version; marketed in Russia as Mi-34; certified by Interstate Aviation Committee Aviation Register (initially at 1,350 kg; 2,976 lb max T-O weight), with helicopter, engine and noise type certificates; meets FAR Pt 27 requirements.

(Note that until 1999, all marketing literature for this version used the hybrid Roman/Cyrillic '**Mi-34C**' to indicate certified status.)

Description applies to Mi-34S (Mi-34C), except where indicated.

Mi-34L: Projected version with 261 kW (350 hp) Textron Lycoming TIO-540J piston engine.

Mi-34P (*patrulnyi:* patrol): Version of Mi-34S, equipped for police duties. Renewed interest in 2001 from Gazprom for pipeline patrol.

Mi-34A: Originally with 335 kW (450 shp) Rolls-Royce 250-C20R turboshaft; mockup, with luxury interior, exhibited at Moscow Air Show '95. Promotion recommended in 1999, employing 376 kW (504 shp) Turbomeca TM 319 Arrius 2F turboshaft; MTOW 1,450 kg (3,196 lb); usable fuel increased to 340 litres (89.7 US gallons; 74.8 Imp gallons); accommodation for four passengers, plus pilot; 1.0 m³ (35 cu ft) baggage compartment. None yet built.

Mi-34M1/M2: Projected twin-turbine, six-passenger versions; MTOW 2,500 kg (5,511 lb).

Mi-34UT: Announced 2001. Dual control trainer. Unsuccessfully competed against Kazan Ansat in competition, in which decision announced September 2001, to provide 100 training helicopters for Russian armed forces.

Mi-234: Described separately.

CUSTOMERS: Total of 14 sold by 2001, including 'two or three' built in 1999. First three for Mayor's office, Moscow. Others used by Bashkir Airlines and Mi-Avia for patrol and training. One operated by Bosniac and Croat Federation Air Force. Three delivered to Nigerian Air Force in 2001, along with six Mi-35s; these were from first batch of six Mi-34s built at Arsenyev, after pause of several years; further five were due to have been delivered to Nigeria by end of 2001. LVM (which see) reportedly ordered 20 Mi-34s for construction at Arsenyev in 2001, but only delivery in that year apart from Nigerians was one to Sibneft. In June 2002, Russian sources reported foreign negotiations for ''several dozen'' Mi-34s.

COSTS: Flyaway (2001): standard, US$316,000; fully equipped, US$350,000.

DESIGN FEATURES: Aerobatic helicopter; intended initially for training and international competition flying; conventional pod and boom configuration; piston engine of same basic type as that in widely used Yakovlev fixed-wing training aircraft and Kamov Ka-26 helicopters; suitable also for light utility, mail delivery, observation and liaison duties, and border patrol; later developments concentrate on light transport role.

Aerobatic capabilities include looping; backwards flight at 70 kt (130 km/h; 81 mph); and rotation about main rotor axis at 120°/s.

FLYING CONTROLS: Manual, with no hydraulic boost.

STRUCTURE: Semi-articulated four-blade main rotor with flapping and cyclic pitch hinges, but natural flexing in lead/lag plane; blades of GFRP with CFRP reinforcement, attached by flexible steel straps to head like that of Boeing MD 500; two-blade tail rotor of similar composites construction, on starboard side; riveted light alloy fuselage; sweptback tailfin with small unswept T tailplane.

LANDING GEAR: Conventional fixed skids on arched support tubes; small tailskid to protect tail rotor.

POWER PLANT: One 239 kW (320 hp) VOKBM M-14V-26V nine-cylinder radial air-cooled engine mounted sideways in centre-fuselage. Fuel capacity 176 litres (46.5 US gallons; 38.7 Imp gallons); system for inverted flight.

ACCOMMODATION: Normally one or two pilots, side by side, in enclosed cabin, with optional dual controls. Rear of cabin contains low bench seat, available for two passengers and offering flat floor for cargo carrying. Forward-hinged door on each side of flight deck and on each side of rear cabin.

SYSTEMS: Primary electric power provided by 27 V 3 kW engine-driven generator; secondary power supplies of 115 V AC, 400 Hz, single-phase and 36 V AC, 400 Hz, three-phase; 27 V 17 Ah battery.

AVIONICS: *Comms:* Briz VHF radio.
Flight: A-037 radio altimeter; ARK-22 radio compass.
Instrumentation: Magnetically slaved compass system incorporating radio magnetic indicator.

EQUIPMENT: Gyro horizon. Version used by Moscow Police has dual controls, two rear seats and loudspeaker under rear of pod.

DIMENSIONS, EXTERNAL:

Main rotor diameter	10.00 m (32 ft 9¾ in)
Main rotor blade chord	0.22 m (8¾ in)
Tail rotor diameter	1.48 m (4 ft 10¼ in)
Tail rotor blade chord	0.16 m (6¼ in)

Mil Mi-34S light utility helicopter (*Jane's/Paul Jackson*)   0121035

Length overall, rotors turning	11.415 m (37 ft 5½ in)
Fuselage: Length	8.71 m (28 ft 7 in)
Max width	1.42 m (4 ft 8 in)
Height: overall	2.75 m (9 ft 0¼ in)
to fin-tip	2.45 m (8 ft 0½ in)
Skid track	2.175 m (7 ft 1½ in)
Fuselage ground clearance	0.36 m (1 ft 2¼ in)
Cockpit doors: Height	1.15 m (3 ft 9¼ in)
Max width	0.70 m (2 ft 3½ in)
AREAS:	
Main rotor disc	78.70 m² (847.1 sq ft)
Tail rotor disc	1.72 m² (18.52 sq ft)
WEIGHTS AND LOADINGS:	
Weight empty	950 kg (2,094 lb)
Fuel weight	128 kg (282 lb)
T-O weight: Aerobatic	1,100 kg (2,425 lb)
Normal	1,280 kg (2,822 lb)
Max	1,450 kg (3,196 lb)
Max disc loading	18.4 kg/m² (3.77 lb/sq ft)
Max power loading	6.08 kg/kW (9.99 lb/hp)
PERFORMANCE (Mi-34S/C at Normal TOW; Mi-34A at MTOW):	
Max level speed: Mi-34S/C	113 kt (210 km/h; 130 mph)
Mi-34A	121 kt (225 km/h; 140 mph)
Max cruising speed:	
Mi-34S/C	92 kt (170 km/h; 106 mph)
Mi-34A	113 kt (210 km/h; 130 mph)
Best climbing speed: Mi-34S/C	49 kt (90 km/h; 56 mph)
Service ceiling: Mi-34S/C	4,000 m (13,120 ft)
Mi-34A	5,000 m (16,400 ft)
Hovering ceiling, OGE: Mi-34S/C	900 m (2,960 ft)
Mi-34A	2,750 m (9,020 ft)
Range with max fuel at 500 m (1,640 ft), 5% reserves:	
Mi-34S/C	192 n miles (356 km; 221 miles)
Mi-34A	297 n miles (550 km; 341 miles)
g limit	+3

UPDATED

Model of Mi-38 (*Jane's/Paul Jackson*)   *NEW*/0143724

MIL Mi-38

TYPE: Medium transport helicopter.

PROGRAMME: Design begun in 1983; model shown at 1989 Paris Air Show, when aircraft at mockup stage. Modifications in evidence by 1993 included fixed landing gear with wider track and reduced base. Under December 1992 agreement, Eurocopter will integrate flight deck, avionics and passenger systems, and will adapt Mi-38 for international market; Euromil joint stock company (which see in International section) established September 1994 to advance collaboration, adding Kazan production plant (as main manufacturing and final assembly centre); funding for Euromil granted in October 1994 by European Bank for Reconstruction and Development. Sextant and Pratt & Whitney Canada added as risk-sharing parties for avionics and engine. Funding by Russian Ministry for Defence Industries 1996. Mi-38 rotor blades began test flying on an Mi-17 in early 2001.

By 1997, Euromil was anticipating first flight in 1999 and start of production two years later, following FAR Pt 29 certification. However, contracts for completion of demonstrator not signed until 18 August 1999, following unilateral decision of Euromil board in December 1998 to launch programme and fly demonstrator at Kazan in 2001, although this subsequently slipped to 2002. Demonstrator (PT-1) is third airframe, following test articles at Mil Moscow and Kazan. Four prototypes to follow by 2003; deliveries from 2006.

CUSTOMERS: Predicted sales of 200 in CIS, plus 100 exports.

COSTS: Estimated development cost (2001) US$500 million; unit cost US$12 million to US$16 million.

DESIGN FEATURES: Planned as replacement for Mi-8/17 series. Western engines optional. Conventional pod and boom configuration; power plant above cabin; six-blade main rotor with considerable non-linear twist and swept tips; main rotor has hydraulic drag dampers; single lubrication point, at driveshaft; rotor brake standard; Krasny Oktyabr transmission, comprising main, intermediate and tail gearboxes and tail rotor drive shaft; engine input 19,017 rpm; two independent two-blade tail rotors, set as narrow X on same shaft; sweptback fin/tail rotor mounting; small horizontal stabiliser; for day/night operation over temperature range −60 to +50°C.

FLYING CONTROLS: Fly-by-wire, with manual back-up.

STRUCTURE: Composites main and tail rotors by Kazan; low-profile titanium main rotor head, with elastomeric bearings, built by Stupino; fuselage, mainly composites, built by Kazan.

LANDING GEAR: Fixed tricycle type; single wheel on each main unit; oleo-pneumatic shock absorbers; twin, self-centring nosewheels; low-pressure tyres; optional pontoons for emergency use in overwater missions. Main tyres 950 mm diameter, pressure 5.88 bar (85 lb/sq in); nose tyres 600 mm diameter, pressure 4.90 bar (71 lb/sq in).

POWER PLANT: Those helicopters for CIS customers powered by two Klimov TVA-3000 (TV7-117 derivative) turboshafts, each rated at 1,838 kW (2,465 shp) for T-O; single-engine rating 2,610 kW (3,500 shp) and transmission rated for same power. Demonstrator to have two 2,461 kW (3,300 shp) P&WC PW127 turboshafts, which are also available, in PW127T/S form, as an option for Western customers. Power plant above cabin, to rear of main reduction gear; air intakes and filters in sides of cowling. Bag fuel tanks beneath floor of main cabin; provision for external auxiliary fuel tanks. Liquid petroleum gas fuel planned as alternative to aviation kerosene.

ACCOMMODATION: Crew of two on flight deck, separated from main cabin by compartment for majority of avionics; single-pilot operation possible for cargo missions. Lightweight seats for 30 passengers as alternative to unobstructed hold for 5,000 kg (11,020 lb) freight. Ambulance and air survey versions planned. Passenger door, forward, port; freight door, forward, starboard; cargo ramp at rear; hatch in cabin floor, under main rotor driveshaft, for tactical/emergency cargo airdrop and for cargo sling attachment; optional windows for survey cameras in place of hatch. Provision for hoist over port-side door, remotely controlled, hydraulically actuated rear cargo ramp, powered hoist on overhead rails in cabin, and roller conveyor system in cabin floor and ramp.

SYSTEMS: Air conditioning by compressor bleed air, or APU on ground, maintains temperature of not more than 25°C on flight deck in outside temperature of 40°C, and not less than 15°C on flight deck and in main cabin in outside temperature of −50°C. Three independent hydraulic systems; any one able to maintain control of helicopter in emergency. Electrical system has three independent generators, two at 12 kW DC and one 60 kW AC; optional fourth generator 40 or 60 kW AC; two batteries; electric rotor blade de-icing, main and tail. Independent fuel system for each engine, with automatic crossfeed; forward part of cowling houses VD-100 APU, hydraulic, air conditioning, electrical and other system components.

AVIONICS: Sextant equipment in export aircraft.

Radar: Weather/nav radar (range 54 n miles; 100 km; 62 miles).

Flight: Preset flight control system allows full autopilot, autohover and automatic landing. Avionics controlled by large central computer, linked also to automatic nav system with Doppler, ILS, satellite nav system, main radar, autostabilisation system and automatic radio compass.

Instrumentation: Six colour CRTs for use in flight and by servicing personnel on ground. Equipment monitoring, failure warning and damage control system. Closed-circuit TV for monitoring cargo loading and slung loads. Options include low-cost electromechanical instrumentation based on that of Mi-8, sensors for weighing and CG positioning of cargo in cabin, and for checking weight of slung loads.

DIMENSIONS, EXTERNAL:

Main rotor diameter	21.10 m (69 ft 2¾ in)
Tail rotor diameter	3.84 m (12 ft 7¼ in)
Distance between rotor centres	12.755 m (41 ft 10¼ in)
Length: overall, rotors turning	25.20 m (82 ft 8¼ in)
fuselage	19.95 m (65 ft 5½ in)
Width, rotors folded	3.17 m (10 ft 4¾ in)
Height: to top of rotor head	5.20 m (17 ft 0¾ in)
to top of fin	5.56 m (18 ft 3 in)
Stabiliser span	4.20 m (13 ft 9½ in)
Wheel track	4.50 m (14 ft 9¼ in)
Wheelbase	5.17 m (16 ft 11½ in)
Forward freight door: Height	1.665 m (5 ft 5½ in)
Width	1.445 m (4 ft 9 in)
Forward passenger door: Height	1.66 m (5 ft 5¼ in)
Width	0.61 m (2 ft 0 in)
Floor hatch: Length	1.15 m (3 ft 9¼ in)
Width	0.75 m (2 ft 5½ in)

DIMENSIONS, INTERNAL:

Main cabin: Length	8.70 m (28 ft 6½ in)
Max width	2.30 m (7 ft 6½ in)
Width at floor	2.20 m (7 ft 2½ in)
Height: centre	1.80 m (5 ft 10¾ in)
rear	1.85 m (6 ft 1 in)
Volume	29.5 m³ (1,042 cu ft)

AREAS:

Main rotor disc	349.67 m² (3,763.8 sq ft)
Tail rotor disc	11.58 m² (124.65 sq ft)

WEIGHTS AND LOADINGS (provisional):

Max payload: internal	5,000 kg (11,020 lb)
external	7,000 kg (15,432 lb)
Normal T-O weight	14,200 kg (31,305 lb)
Max T-O weight	15,600 kg (34,392 lb)
Max disc loading	44.6 kg/m² (9.14 lb/sq ft)
Transmission loading at max T-O weight and power	5.97 kg/kW (9.82 lb/shp)

PERFORMANCE (estimated, PW127 engines):

Max level speed	153 kt (285 km/h; 177 mph)
Max cruising speed	148 kt (275 km/h; 171 mph)
Service ceiling	6,500 m (21,325 ft)
Hovering ceiling OGE	2,500 m (8,200 ft)
Range:	
with 5,000 kg (11,020 lb) payload	256 n miles (475 km; 295 miles)
with max internal fuel	442 n miles (820 km; 509 miles)
ferry, with auxiliary fuel	728 n miles (1,350 km; 838 miles)

UPDATED

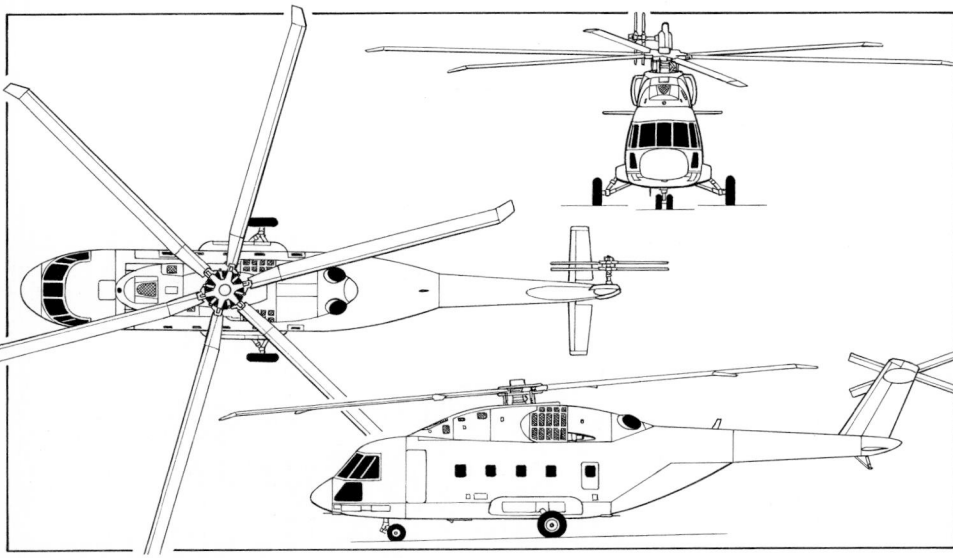

Mil Mi-38 medium transport helicopter (*Michael Badrocke*)   *NEW*/0143734

MIL Mi-52 SNEGIR

English name: Bullfinch

A description and illustration of this light helicopter project last appeared in the 1998-99 edition. The company unveiled a mockup on 1 June 2000 and announced that it was then hoping to resuscitate the venture, with a view to flying a prototype in 2001, although this failed to occur. By mid-2001, Mil was still seeking US$15 million to build, test and certify two prototypes.

Single- (Mi-52-1) and twin-engined (Mi-52-2 with 199 kW; 266 hp VAZ-4265) versions are planned with commercial load capabilities of 350 kg (772 lb) and 400 kg (882 lb), respectively. A military trainer is also envisaged. Unit cost estimated as US$600,000 to US$700,000.

UPDATED

MIL Mi-60

TYPE: Three-seat helicopter.

PROGRAMME: Announced July 2000, when mockup under construction at Kazan helicopter factory; this exhibited at Moscow Salon in August 2001, promoted by Rostvertol, which considering establishing production line. Building of prototype was to have begun in 2001; development cost US$30 million, including US$2.58 million for scientific research and US$15.77 million for prototypes and preparation for series production; estimated unit cost US$140,000 to US$150,000. Programme included in Russian federal aviation plan for 2002-10.

Development by Moscow Aviation Institute, thus designated Mi-60 MAI: financed by Ministry of Education. Maintenance-free rotor hub; airframe TBO of 2,000 hours.

LANDING GEAR: Skid type.

POWER PLANT: One or two piston engines; option of one 145 kW (195 hp) Textron Lycoming IO-360, one 177 kW (237 hp) VAZ-4261 or two 84.6 kW (113.4 hp) Rotax 914s.

DIMENSIONS, EXTERNAL:
Main rotor diameter	10.00 m (32 ft 9¾ in)
Main rotor blade chord	0.22 m (8¾ in)

WEIGHTS AND LOADINGS (estimated; L: Lycoming, R: Rotax, V: VAZ engines):
Max underslung load	90 kg (198 lb)
T-O weight: Normal: all	800 kg (1,764 lb)
Max: L	1,115 kg (2,458 lb)
R	1,260 kg (2,778 lb)
V	1,300 kg (2,866 lb)

Mockup of Mil Mi-60 MAI exhibited by Rostvertol at Moscow Salon, August 2001 *(Jane's/Paul Jackson)*
0121036

Fuel weight: L	79 kg (174 lb)	
R	70 kg (154 lb)	
V	71 kg (157 lb)	

PERFORMANCE (estimated; L, R, V as above):
Max level speed: L	108 kt (200 km/h; 124 mph)	
R	113 kt (210 km/h; 130 mph)	
V	121 kt (225 km/h; 140 mph)	
Max cruising speed: L	94 kt (175 km/h; 109 mph)	
R	103 kt (190 km/h; 118 mph)	
V	100 kt (185 km/h; 115 mph)	
Econ cruising speed: L, V	46 kt (85 km/h; 53 mph)	
R	43 kt (80 km/h; 50 mph)	
Max rate of climb at S/L: vertical: L	228 m (748 ft)/min	
R	300 m (984 ft)/min	
V	468 m (1,535 ft)/min	
inclined: L	444 m (1,457 ft)/min	
R	480 m (1,575 ft)/min	
V	630 m (2,067 ft)/min	

UPDATED

MIL Mi-234

TYPE: Four-seat helicopter.

PROGRAMME: Mockup exhibited at Moscow Air Show '92, when known as Mi-34 VAZ (military equivalent Mi-34M). New designation, and confirmation of continued development, released July 2000. No further announcements during 2001. Data last appeared in 2002-03 *Jane's*.

UPDATED

MI-LIGHT HELICOPTERS — *see LVM*

MOLNIYA

NAUCHNO-PROIZVODSTVENNOYE OBEDINENIE MOLNIYA OAO (Lightning Scientific Production Association JSC)

ulitsa Novoposelkovaya 6, 123459 Moskva
Tel: (+7 095) 492 92 35
Fax: (+7 095) 492 47 23
e-mail: molniya@dol.ru
Web: http://www.buran.ru

MANAGING DIRECTOR: Aleksandr Bashilov
DEPUTY GENERAL DIRECTOR: Valery Verobzhev
CHIEF ENGINEER: Gennady G Kryuchkov
MARKETING DIRECTOR: Prof Mikhail Y Gofin
CHIEF OF MARKETING DEPARTMENT: Dr Vladimir I Fishelovich

To compensate for reduced funding for Buran space shuttle orbiter programme, in which it was much involved, Molniya has diversified its activities to include conveyor equipment (automatic car parking and wheelchair lifts) and aeroplane design. During the early 1990s, it projected a series of civil aircraft of vastly varying size (½ to 450 tonne payloads) all of 'triplane' (canard, wing and tailplane) configuration, as described in the 1997-98 and earlier *Jane's*. The only one so far to fly has been the Molniya-1. Despite regular reports of production having begun, only the prototype has been reported. Full details last appeared in the 2001-02 *Jane's*, followed by a brief update in 2002-03.

UPDATED

MVEN

MVEN OOO (MVEN Ltd)

a/ya 282, 420141 Kazan
Tel: (+7 8432) 43 49 03 or 54 20 11
e-mail: mven@mven.kazan.ru
Web: http://www.mven.netfirms.com

Since establishment in 1990, MVEN has produced parachute ballistic recovery systems for various types of aircraft. In 1995 it was incorporated within the Scientific Research Institute of Aeronautical Technologies and began the manufacture of light aircraft fuselages and components under subcontract to Chernov and others. The company's first aircraft, the MVEN-1 agricultural sprayer was shown at the Gelendzhik Hydroaviation Salon in September 2002, three months before its maiden flight.

NEW ENTRY

MVEN-1 FERMER

English name: Farmer

TYPE: Agricultural sprayer.

PROGRAMME: Shown at Gelendzhik, 4 to 8 September 2002. First flight 9 November 2002; official evaluation flight of 11 December gave clearance for continued company testing.

DESIGN FEATURES: Classic low-wing sprayplane with chemical hopper ahead of raised cockpit giving good view over nose. Strut-braced wings and tailplane. Heavy-duty suspension and tyres.

FLYING CONTROLS: Conventional and manual. No aerodynamic balances. Ailerons and flaps occupy full extent of wing trailing-edge.

LANDING GEAR: Tailwheel type; fixed. Long-stroke, rubber-in-compression mainwheel suspension. Hydraulic brakes on mainwheels.

POWER PLANT: Not disclosed, but believed to be one 118 kW (158 hp) LOM M332B four-cylinder piston engine driving a three-blade propeller. Fuel in two tanks, combined capacity 70 litres (18.5 US gallons; 15.4 Imp gallons).

ACCOMMODATION: Two persons in tandem. Single, upward-hinged door, port side.

EQUIPMENT: Sprayer system with 13 atomisers on each of two underwing elements. MVEN ballistic parachute.

DIMENSIONS, EXTERNAL:
Wing span	10.40 m (34 ft 1½ in)
Length overall	7.10 m (23 ft 3½ in)
Height overall	2.95 m (9 ft 8¼ in)

WEIGHTS AND LOADINGS:
Max payload	300 kg (661 lb)
Normal T-O weight	965 kg (2,127 lb)

PERFORMANCE:
Max level speed	108 kt (200 km/h; 124 mph)
Normal operating speed	67-86 kt (125-160 km/h; 78-99 mph)

NEW ENTRY

MVEN-1 Fermer at its public debut, Gelendzhik, September 2002 *(Yefim Gordon)*   NEW/0527151

Cockpit of MVEN-1 *(Yefim Gordon)*   NEW/0527150

MYASISHCHEV

EKSPERIMENTALNYI MASHINOSTROITELNYI ZAVOD IMENI V M MYASISHCHEVA (Experimental Engineering Bureau named for V M Myasishchev)

140180 Zhukovsky-5, Moskovskaya oblast
Tel: (+7 095) 556 77 76 and 912 60 41
Fax: (+7 095) 556 52 98 and 728 41 30
e-mail: mdb@mastak.msk.ru
Web: http://www.corbina.ru/~kluka
GENERAL DESIGNER: Valery K Novikov
CHIEF DESIGNER, M-101: Evgeny Charsky
CHIEF DESIGNER, M-55: Leonid Sokolov
INFORMATION OFFICER: Stanislav G Smirnov

The Myasishchev OKB was founded in 1952, and led by Prof Vladimir Mikhailovich Myasishchev until his death on 14 October 1978. In the 1970s and 1980s, the bureau was engaged in development of multipurpose subsonic high-altitude aircraft, but has more recently diversified into civil aviation. In 1981, the bureau was named after Prof Myasishchev.

Attempts to forge overseas alliances proved unsuccessful (see 2002-03 *Jane's*), but in April 2001 it was announced that the KASKOL group of companies, the Nizhny Novgorod Sokol Aviation Plant and Myasishchev had formed Novi Regionalnyi Samoletnyi (NoRS, New Regional Aeroplanes), which will produce the M-101T Gzhel business aircraft at Nizhny Novgorod.

In 2001, Myasishchev was reported to be preparing a response to an anticipated official requirement for a 130/170-seat short/medium-range airliner to be developed under the 2002-10 civil aviation plan and enter service in 2015. This would be based on the GP-60 series of aerodynamic studies, last described in the 1999-2000 edition. Other aircraft in this family could include the 14-seat M-60 Svetozar, M-60 Perun (214-seat) and M-60 Kolovrat (35,000 kg; 77,161 lb payload).

UPDATED

MYASISHCHEV M-55

NATO Reporting name: Mystic

Russian defence ministry approved, in September 1999, plans to complete one of two unfinished Myasishchev M-55 high-altitude surveillance aircraft held at Smolensk production plant; funding assigned in 2001 for completion in 2002 to **M-55RM** standard, but in June 2002 it was reported that insufficient funds had been received to complete the aircraft in that year. Also in 2001, India offered unspecified export variant designated **M-55RTR**, and Myasishchev announced plans for **M-55Kh** ''space tourism'' version with 3,000 kg (6,614 lb) S-XXI suborbital module carried on spine for launch at altitudes up to 19,000 m (62,340 ft). M-55 last described in 1996-97 *Jane's*.

UPDATED

MYASISHCHEV M-70

TYPE: Regional jet airliner.
PROGRAMME: In late 2002, a project by the Myasishchev Design Bureau was revealed to be competing against the Tupolev Tu-414 and Sukhoi RRJ for official acceptance as Russia's next-generation regional jet. Designation M-70-70 is believed to indicate 70 seats.

NEW ENTRY

MYASISHCHEV M-101T GZHEL

TYPE: Light utility turboprop.
PROGRAMME: Derived from Nelli and M-70 projects; first shown in model form at Moscow Air Show '90; developed full-scale mockup exhibited at Moscow Air Show '92 with piston engine and at Moscow Air Show '93 with turboprop. Two static test airframes and three flying prototypes built at Nizhny Novgorod by Sokol; first (RA-15001) first flew 31 March 1995; this and second flying aircraft (RA-15003), shown at Moscow Air Show '95, lack ventral fin. Third prototype (RA-15004) to what then regarded as preproduction standard; first preproduction aircraft (RA-15101) displayed at Paris and Moscow Air Shows 1997. Initial batch of 15 at Sokol plant; 11 completed by September 2001; No. 6 (RA-15106) is first in full production configuration, shown at Moscow, August 2001; No. 7 for reliability tests; No. 8 is first for customer delivery. Trials aircraft crashed near Zhukovsky test aerodrome on 12 September 2001, following loss of longitudinal stability, putting programme back by six months. Certification trials were due to restart in May 2002, with completion scheduled by December 2002. In April 2002 it was reported that 12 aircraft had been built – 10 for flying trials and two for static testing.

Russian certification to AP-23 in passenger category was expected late 1998 or early 1999, following delivery of two aircraft to State Civil Aviation Research Institute (Gos NII GA) on 23 January 1998 for short-range freight services from Moscow/Sheremetyevo by Fenix Air but delayed by lack of funding; meanwhile early production aircraft used on freight services by Sokol plant, completing

Model of Myasishchev M-55Kh high-altitude aircraft carrying S-XXI suborbital module *(Jane's/Paul Jackson)*
NEW/0143725

Ventral and dorsal fin additions to production version of M-101T *(Jane's/Paul Jackson)*   0121691

over 500 missions by 2000, and one reported in use by Central African Airlines, based in Egypt. In November 2001, US$3 million was being sought to complete certification and further US$5 million to complete production tooling and retrospectively modify airframes already built.

Shortcomings in directional stability addressed by retrospective installation of additional two ventral fins and dorsal fillet on third prototype (RA-15004) in 1999, this also having extended wingtips. These additional tail surfaces standardised from RA-15106 onwards.

The aircraft is named for a type of fine porcelain made in the city of Gzhel. Some development funding was provided by Gzhel, other finance coming from Myasishchev, Sokol, Inkombank and a Czech bank. Marketing is by Gzhel-Avia.

CURRENT VERSIONS: **M-101T:** *As described.*
M-101D: Projected variant with diesel engine of Italian origin.
M-101P: Projected version with Textron-Lycoming piston engine.
M-101PW: Projected Westernised version for North American market; 1,178 kW (1,580 shp) P&WC PT6A-64 engine, Hartzell propeller and Honeywell or Becker avionics. Prototype was to have flown before end of 1998, but programme apparently abandoned.
CUSTOMERS: Undisclosed central African customer ordered one in early 2000; minimum estimated market for 300. Reported 70 to 80 orders by September. First Russian customer reported as Gazprom, which requires 10; company foresees production reaching 300 by end 2005, of which two-thirds would be exported.

COSTS: US$1.6 million (2001). Hourly operating cost estimated at US$200 (2001). By mid-1999, Sokol plant had invested Rb102.4 million in M-101, and in 2000 was planning funding of US$2.5 million to complete certification.
DESIGN FEATURES: Intended for operation in remote areas; able to serve away from base aerodrome for up to 50 flying hours and fly from unpaved runways. Conventional all-metal low-wing monoplane with pressurised cabin; sweptback vertical tail surfaces and ventral fin. Designed in accordance with Russian AP-23 and US FAR Pt 23 airworthiness requirements. Service life of 15 years or 20,000 hours.
FLYING CONTROLS: Conventional and manual. Electrically operated trim tabs in port aileron, rudder and each elevator; ground-adjustable tab on starboard aileron. Double-slotted flaps operated hydraulically.
STRUCTURE: All-metal. Wing constructed around torsion box.
LANDING GEAR: Hydraulically retractable tricycle type; single wheel on each unit; nosewheel retracts rearward, mainwheels inward into wingroots and fuselage; mainwheels uncovered by doors when retracted; tyre size 500×150-9 on mainwheels, 400×150-5 on nosewheel; levered suspension legs; castoring nosewheel with shimmy damper; designed to use paved and unpaved runways.
POWER PLANT: One 559 kW (751 shp) Walter M 601 F32 turboprop, driving Avia Hamilton Sundstrand V-510 five-blade propeller. Power plant is protected against particle ingestion. Fuel in two wing tanks and two feeder tanks, with electrical pumps.
ACCOMMODATION: One or two pilots and four passengers, in pairs in pressurised cabin; max capacity, one pilot and seven passengers. Rear-hinged door to flight deck on port side; large door for passengers and freight loading aft of wing on port side; emergency exit on starboard side above wing; provision for rapid change to cargo/passenger, freight or ambulance configuration.
SYSTEMS: Hydraulic system for landing gear and flaps. Pressurisation system maintains cabin altitude of 2,400 m (7,880 ft). Pneumatic de-icing of wing, tailplane and fin leading-edges; electric anti-icing of propeller, pitot tubes and windscreen clear-view panel. Electrical system, 27 V DC, supplied by 250SG125Q1 starter/generator and 115 A/400 Hz single-phase secondary supply with Varta 24 V emergency battery.
AVIONICS: Russian or Western equipment for single-pilot VFR or two-pilot IFR.
Comms: Bendix/King KX 165 nav/com.

Myasishchev M-101T Gzhel in its current configuration *(Jane's/Paul Jackson)*   0121692

Flight: Bendix/King KLN 89B GPS, WX-900 Stormscope and autopilot all optional.

DIMENSIONS, EXTERNAL:

Wing span	13.00 m (42 ft 8 in)
Length overall	9.975 m (32 ft 8¾ in)
Height overall	3.45 m (11 ft 3¾ in)
Tailplane span	4.32 m (14 ft 2 in)
Wheel track	3.00 m (9 ft 10 in)
Wheelbase	2.885 m (9 ft 5½ in)
Propeller diameter	2.30 m (7 ft 6½ in)
Passenger/freight door: Height	1.15 m (3 ft 9¼ in)
Width	1.23 m (4 ft 0½ in)
Flight deck door: max width	0.90 m (2 ft 11½ in)
max height	0.975 m (3 ft 2½ in)
Emergency exit: max width	0.485 m (1 ft 7 in)
max height	0.665 m (2 ft 2¼ in)

DIMENSIONS, INTERNAL:

Cabin: Length	4.56 m (14 ft 11½ in)
Max width	1.32 m (4 ft 4 in)
Max height	1.26 m (4 ft 1½ in)
Volume	7.5 m³ (265 cu ft)

AREAS:

Wings, gross	17.06 m² (183.6 sq ft)

WEIGHTS AND LOADINGS:

Weight empty	2,270 kg (5,004 lb)
Max payload	630 kg (1,389 lb)
Max fuel weight	450 kg (992 lb)
T-O weight: normal	2,900 kg (6,393 lb)
max	3,200 kg (7,054 lb)
Max wing loading	187.6 kg/m² (38.41 lb/sq ft)
Max power loading	5.18 kg/kW (8.51 lb/shp)

PERFORMANCE:

Max cruising speed	232 kt (430 km/h; 267 mph)
Stalling speed, flaps down	61 kt (113 km/h; 71 mph)
Cruising altitude	7,600 m (24,940 ft)
T-O run	380 m (1,250 ft)
Landing run	370 m (1,215 ft)

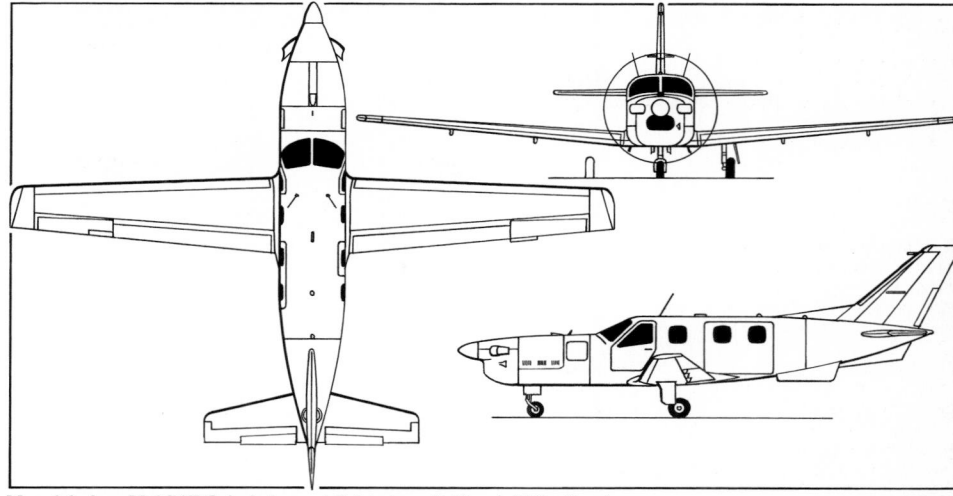

Myasishchev M-101T Gzhel six-seat light aircraft (*Jane's/Mike Keep*)   0121660

Range:

with max payload	432 n miles (800 km; 497 miles)
with max fuel, 45 min reserves	755 n miles (1,400 km; 869 miles)

UPDATED

MYASISHCHEV M-500

It was reported during 2000 that this agricultural/forestry and seven-seat transport aircraft is included in the development programme of civil aircraft up to 2015. It will be built at the Sokol factory and at Smolensk beginning 2003. The M-500 will be offered with a 316 kW (424 hp) VOKBM M-14NTK piston engine driving a three-blade MT propeller, or a 331 kW (444 hp) Ufa NPP turbocharged diesel driving a Hartzell propeller. Estimated market for 700 to 800 aircraft, with reported orders for 80 by May 2001; unit cost US$160,000 to US$170,000 (2001).

VERIFIED

NAPO

GOSUDARSTVENNOYE UNITARNOYE PREDPRIYATIE NOVOSIBIRSKOYE AVIATSIONNOYE-PROIZVODST-VENNOYE OBEDINENIE IMENI V P CHKALOVA (State Unitary Enterprise Novosibirsk Aircraft Production Association named for V P Chkalov)

ulitsa Polzunova 15, 630051 Novosibirsk
Tel: (+7 3832) 79 80 95 and 77 37 06
Fax: (+7 3832) 77 10 26
e-mail napa@mail.cis.ru
GENERAL DIRECTOR: Nikolai I Bobritsky
MARKETING DIRECTOR: Valery Sadaev

Founded in 1936 as GAZ 153, NAPO has built about 36,000 aircraft of many types, including I-16, Yak-7, Yak-9, MiG-15, MiG-17, MiG-19, Yak-28, Su-9, Su-15 and, from 1972, Su-24 multirole combat aircraft. The plant is wholly state-owned, but 74.5 per cent of shares are to be transferred to Sukhoi Aviation Holding Company. NAPO's current products include the Antonov An-38 (available with local TVD-20 engines from 2002) and Sukhoi Su-32 (Su-27IB) as well as motor boats. Is promoting Su-24 modernisation programmes for export customers. Will build Su-49 piston trainer.

NAPO also rebuilds historic aircraft from severely damaged originals, most recent being a Polikarpov I-15bis (registered FLARF-02915) from wreckage recovered in Karelian region; this exhibited at Moscow Salon in August 2001 before export to New Zealand, joining six I-16s and two I-153s similarly restored earlier.

UPDATED

Polikarpov I-15bis extensively restored by NAPO (*Yefim Gordon*)   *NEW*/0525114

POLYOT

PROIZVODSTVENNOYE OBEDINENIE POLET (Polyot Production Association)

ulitsa Bogdan Khmelnitsky, 226, 644021 Omsk
Tel: (+7 3812) 37 17 76
Fax: (+7 3812) 33 79 15
e-mail: polyot@omsknet.ru
Web: http://www.omsknet.ru/polyot
GENERAL DIRECTOR: Valentin Ziatsev

DESIGN BUREAU:
Tel: (+7 3812) 53 92 01
Fax: (+7 3812) 53 88 31
CHIEF DESIGNER: Viktor V Markelov

Established in 1941, Polyot is one of the largest aerospace enterprises in Russia, with facilities extending over more than 15 km² (5.8 sq miles) and up to 20,000 employees. During the Second World War it manufactured 3,405 Tupolev Tu-2 bombers and Yakovlev Yak-7 and Yak-9 fighters in 2½ years; post-war products included 758 Ilyushin Il-28 jet bombers and 58 Tupolev Tu-104 jet transports. It remains state-owned. The Antonov An-74 series of STOL transports is now in series production.

Polyot manufactured SS-4, SS-7 and SS-11 strategic missiles, warheads, and first-stage engines for Zenith and Energiya rocket launch vehicles. Other products include many spacecraft and a variety of consumer goods under *konversiya* programmes. Brief details of the projected VK-1 light aircraft appeared in the 1997-98 edition.

In June 1997, Polyot and Baranov Engine Manufacturing Association relaunched the Antonov An-3 programme (An-2 with 1,140 kW; 1,529 shp OMKB TVD-20 turboprop), described in *Jane's Aircraft Upgrades*. Certification of factory to build An-3 received by May 2000, but funding unavailable for production line and only conversions of An-2s being considered. An-3 work was interrupted during the first half of 2001 when Polyot repaired the prototype Antonov An-70 (which see) after it had crashed following take-off from Omsk. Revealed in September 2001 that Polyot selected to assemble An-70s ordered for Russian Air Forces; first An-70 to fly in 2003. Required investment to begin An-70 production (Polyot to build cargo hold) quoted in late 2001 as Rb3 billion, of which Russian government and Omsk regional government each donating Rb1,050 million.

UPDATED

Sketch of Polyot Yula   *NEW*/0145047

POLYOT YULA

English name: Whirlygig
TYPE: Exoskelitor flying vehicle.
PROGRAMME: Polyot announced on 22 May 2002 the successful test of this personal 'assault and attack'

helicopter, most details of which remain secret. Designed by Vyacheslav Kotel'nikov; telescopic fuselage and twin-blade main rotor; ramjet on each blade; total weight 20 kg (44 lb); max level speed 65 kt (120 km/h;

75 mph); service ceiling 1,000 m (3,280 ft); endurance 25 minutes.

NEW ENTRY

REFLY

KOMPANIYA REFLAI (Refly Company)
a/ya 21, nab Oktyabrskaya 6, 193021 Sankt Peterburg
Tel: (+7 812) 445 08 34
Fax: (+7 812) 968 51 01

e-mail: refly@rusavia.ru
Web: http://www.rusavia.ru/refly
GENERAL DIRECTOR: Pavel V Egorov

Formed in 1997, Refly sells Western aircraft kits (Revolution Mini 500, RotorWay Exec, Skystar Kitfox, American

Sportscopter Ultrasport, Aerocomp Comp Air 6) and light aviation supplies on Russian market. Chernov Che-22 (which see under Gidroplan in this section) marketed in USA as Refly Pelican. The Refly Delfin was last described in the 2001-02 edition of *Jane's*.

UPDATED

ROSTVERTOL

ROSTOVSKY VERTOLETNYI PROIZVODSTVENNYI KOMPLEKS OAO 'ROSTVERTOL' ('Rostvertol' Rostov Helicopter Production Complex JSC)
ulitsa Novatorov 5, 344038 Rostov-na-Donu
Tel: (+7 8632) 31 74 93
Fax: (+7 8632) 45 05 35
e-mail: rostvertol@gin.ru
Web: http://www.rostvertolplc.ru
DIRECTOR GENERAL: Boris N Slyusar
CHIEF DESIGNER: Sergey I Koksharov
DEPUTY DIRECTOR GENERAL, MARKETING AND SALES:
  Yuri Zaikin

DEPUTY DIRECTOR-GENERAL, MARKETING AND EXPORT SALES:
  Andrey B Shibitov
TECHNICAL DIRECTOR: Igor Semyonov

The company now known as Rostvertol was founded on 1 July 1939, and began by manufacturing wooden propellers. It progressed to aircraft production during the Second World War and to helicopter work in the mid-1950s. Past products at Rostov-on-Don included the UT-2M (1944), Po-2, Yak-14 glider, Il-10M, Il-40, Mi-1, Mi-6, Mi-10 and Mi-10K. Rostvertol is wholly privately owned.
  Government decree of 19 February 1996 authorised Rostvertol to export spares and auxiliary equipment for Mil Mi-24 and Mi-25; and to export new Mi-26, Mi-28 and Mi-35 helicopters, as described in current and previous Mil Moscow

Helicopter Plant entries. Production in 1999 included one Mi-26 delivered to the Moscow fire brigade for water-bombing and rescue duties. Mi-26 supplied to Mexican Air Force in January 2000 and second ordered in following April. First production Mi-28N being assembled in 2001, although Rostvertol expects no Russian military orders for any of its helicopters at least until 2005. Also involved in Mi-60 project.
  Rostvertol also active in helicopter overhaul and upgrading; to modify 12 to 14 Mi-35s to Mi-35M in 2002 for undisclosed foreign customer. However, independent export sales authorisation rescinded 23 July 1999, and Rostvertol exports now made through official Russian agencies. Company has exported 650 aircraft.

UPDATED

RSK 'MiG'/P A VORONIN PRODUCTION CENTRE

FEDERALNOYE GOSUDARSTVENNOYE UNITARNOYE PREDPRIYATIE, ROSSIYSKAYA SAMOLETOSTROITEL'NAYA KORPORATSIYA 'MiG' (Russian Aircraft-Building Corporation 'MiG' Federal State Unitary Enterprise)
1-ň Botkinsky proezd, Dom 7, 125040 Moskva
Postal address: a/ya No 1, 103045 Moskva
Tel: (+7 095) 252 87 47
Fax: (+7 095) 250 07 70
e-mail: mig@migavia.ru
Web: http://www.migavia.ru
GENERAL DIRECTOR: Nikolai F Nikitin
FIRST DEPUTY GENERAL DESIGNER, MIG BUREAU:
  Vladimir I Barkovsky
FIRST DEPUTY GENERAL FOR MANUFACTURING, VORONIN PRODUCTION CENTRE: Viktor M Puzanov
DEPUTY DIRECTOR GENERAL, LUKHOVITSY PLANT:
  Yuri A Tsyupko

PUBLIC RELATIONS AND MEDIA CENTRE:
Tel: (+7 095) 207 04 76
Fax: (+7 095) 207 07 57
e-mail: migpress@mail.ru

Original Moscow Aircraft Production Organisation (MAPO) renamed by official decree of 8 December 1999 as RSK 'MiG' and its Botkinsky manufacturing plant dedicated to former director (1938-1982) Pavel A Voronin. Up to 1999, MAPO (formerly MMZ No. 30 Znamaya Truda [Banner of Labour]) had built 25,000 aircraft of 40 types, from the Nieuport IV of 1913 to the MiG-29. On 5 October 1999, it was nominated as a production centre for the Tupolev Tu-334 airliner in return for contributing 50 per cent of certification costs.
  Service activities include complete maintenance support and pilot training; is also involved in Chinese Chengdu FC-1 fighter programme, China.
  Aircraft production is centred at the Moscow (Voronin)

plant alongside the Mikoyan design complex and at LAPIK, Lukhovitsy. However, in 2001, construction began at Lukhovitsy of a Tu-334 assembly hall, launching move of almost all of Voronin plant out of central Moscow, whereupon most of Botkinsky site will be sold in settlement of debts to Moscow local government, leaving only subassembly manufacture and company headquarters within city.
  Members of RSK 'MiG' are:
Full members:
  A Mikoyan OKB Engineering Centre, Moscow (design of new aircraft and upgrades; see MiG entry).
  MiG Light Civil Aircraft Moscow (formed 2001 as separate light aviation division)
  LAPIK: formerly Lukhovitsy Machine-Building Plant, Lukhovitsy (series production and flight testing; see below)
  Kalyazin Machine-Building Plant
  V Ya Klimov Plant, St Petersburg (engines, gearboxes, accessories)
  Soyuz Machine-Building Design Bureau, Tushino (engines, gearboxes, accessories)
  Ryazan State Instrument Plant, Ryazan (airborne and ground test equipment)
  Electroavtomatika Design Bureau, St Petersburg (airborne and ground test equipment)
Associate members:
  Kamov Company, Lyubertsy (helicopter design and manufacture)
  V V Chernyshov Machine-Building Enterprise, Moscow
  Krasny Oktyabr Machine-Building Enterprise, St Petersburg
  Pribor JSC, Kursk
  Aviatest, Scientific-Engineering Enterprise, Rostov-on-Don
  Instrument-Making Company, Perm
  Aviabank
  No. 121 Aircraft Repair Depot
Workforce in late 2000 was 13,000.
  On 10 November 2000, RSK 'MiG' formed MiG Light Civil Aviation company to manage assembly of Aeroprogress T-101 Grach and to design and develop new lightplanes. Four of the latter were reported to be under design in early 2001, including six- and 12-passenger types.

LUKHOVITSY AVIATSIONNYI PROIZVODSTVENNYI – ISPTATELNNYI KOMPLEKS (Lukhovitsy Aviation Production-Testing Complex)
140500 Lukhovitsy, Moskovskaya oblast
Tel: (+7 09663) 113 76
Fax: (+7 09663) 111 80
e-mail: lmz-avia@mtu-net.ru
DIRECTOR: Vladimir I Nungezer
COMMERCIAL DIRECTOR: Valery Bredikhin

Founded at Tretyakovo airfield in 1953, this state-owned subsidiary of RSK 'MiG' functioned as a flight test centre until 1968 (and continues to do so for the MiG-29). It then was assigned to Znamaya Truda production plant, replacing final assembly and test airfield at Khodinka, in the Moscow suburbs, and consequently acquired extensive equipment for manufacturing high-strength aluminium, titanium and composites components in support of RSK 'MiG' production, being renamed Lukhovitsy Mashinostroitelnyi Zavod (LMZ – Lukhovitsy Machine-Building Plant). On 7 March 1996 became first (and, currently, only) Russian factory with light aircraft production approval certificate. Assembly continues at low rate of the Aeroprogress T-101 Grach. However, manufacture of the Ilyushin Il-103 and Interavia I-1L was not being undertaken in 2001, the former due to lack of orders (since rectified by South Korea) and the latter pending launch of a marketing campaign. Also participated in manufacture of parts for Sukhoi Su-29 and Su-31 sportplanes, being responsible for complete assembly from January 2001 onwards; four built in 2001.
  Announced in June 2000 that Lukhovitsy will produce Ilyushin Il-100. By mid-2000, was preparing to build a prototype Kamov Ka-62. To build Tupolev Tu-334 regional airliners.
  Other activities include flying school for commercial or private pilot training on Il-103; freight handling; and production of jetskis and suntanning beds.
  In early 2002, 10-year plan initiated to remove remaining Moscow-based elements of RSK 'MiG' to Lukhovitsy, with majority of work to be completed by 2004. LMZ adopted current name of LAPIK by January 2002.

UPDATED

SAT 'SAVIAT'

SPETSIALNOYE AVIATSIONNOYE TEKHNOLOGY AO (Special Aviation Technologies Co Ltd)
pristan Akademika AN Tupoleva 15, 111250 Moskva
Tel: (+7 095) 265 79 12 and 263 75 14
Fax: (+7 095) 261 87 13
e-mail: ul0409@dialup.podolsk.ru
PRESIDENT: Vladimir S Eger

Trading as Saviat, SAT is developing a series of rhomboid wing light transport aircraft which will be built by RSK 'MiG'.

VERIFIED

SAVIAT E-4

TYPE: Agricultural sprayer.
PROGRAMME: Manufacture of prototype began May 1999; first flight was then due in January 2000; had not been reported by mid-2002; programme was confirmed in early 2002 as still being active and start of series manufacture by RSK 'MiG' predicted for March 2003.
CUSTOMERS: Orders for eight aircraft had been placed by late 1999. Demand estimated in 2001 as 100 over five years.
COSTS: US$75,000 (2001). Break-even total 30 to 40.
DESIGN FEATURES: Intended as small sprayer/utility aircraft with high safety margins, certifiable to JAR-VLA criteria. Specific features of this sesquiplane include newly developed aerofoil; direct side-force control (DSFC) system, which includes unusually large proportion of vertical stabilisation and control areas, for accurate positioning in

low-level flight; twin tailbooms; pusher configuration; ballistic parachute; agro-chemical tank in lower wing.
  Special aerofoil section; no leading-edge sweep; thickness/chord ratio 8.3 per cent; no dihedral; incidence 4° 30′; no twist.
FLYING CONTROLS: Manual. Conventional ailerons on upper wings; single-piece elevator with trim tab; two-element rudders on each tailfin. Additional fins and rudders between lower wingtips and upper wings, plus inboard ailerons on lower wings, all form DSFC system. Flaps on upper planes only.
STRUCTURE: Fuselage and upper wing centre-section of metal; remainder of composites sandwich.
LANDING GEAR: Tricycle type; fixed. Cantilever mainwheel legs of titanium; nosewheel shock-absorption by helical spring. Single wheel on each unit; all tyres 400×150.

POWER PLANT: One 89 kW (120 hp) Hirth F30-A36C flat-four driving a D&D, fixed-pitch, two-blade propeller. Two fuel tanks in upper wing; normal capacity 100 litres (26.4 US gallons; 22.0 Imp gallons). Provision for chemical hoppers to be used as additional fuel tanks.

ACCOMMODATION: Pilot and provision for two passengers. Single-piece windscreen; forward-hinged door, port side. Passenger seats may be removed for freight-carrying. Cabin heated, ventilated and pressurised to 0.015 bar (0.21 lb/sq in) differential to prevent chemical ingress.

SYSTEMS: Electrical system includes 12 V and 27 V DC generators and 45 Ah battery.

AVIONICS: *Flight:* GPS navigation.

EQUIPMENT: Chemical hoppers in lower wing; spraybar immediately behind, and extending beyond, lower wing. Saviat-designed ballistic parachute recovery system, glide ratio 1.5, in upper wing centre-section.

DIMENSIONS, EXTERNAL:

Wing span (upper)	10.00 m (32 ft 9¾ in)
Wing chord, constant	1.20 m (3 ft 11¼ in)
Wing aspect ratio (upper)	8.3
Length: overall	7.20 m (23 ft 7½ in)
fuselage pod	3.60 m (11 ft 9¾ in)
Fuselage pod: Max width	1.10 m (3 ft 7¼ in)
Max height	1.45 m (4 ft 9 in)
Height overall	2.65 m (8 ft 8¼ in)
Tailplane span	2.00 m (6 ft 6¾ in)
Wheel track	2.10 m (6 ft 10¾ in)
Wheelbase	2.55 m (8 ft 4½ in)
Propeller diameter	1.87 m (6 ft 1½ in)

Model of the Saviat E-4   0079011

Propeller ground clearance	0.47 m (1 ft 6½ in)
Passenger door: Max height	1.17 m (3 ft 10 in)
Max width	1.03 m (3 ft 4½ in)
Height to sill	0.40 m (1 ft 3¾ in)

DIMENSIONS, INTERNAL:

Cabin: Length	2.10 m (6 ft 1¾ in)
Max width	1.05 m (3 ft 5¼ in)
Max height	1.22 m (4 ft 0 in)
Floor area	1.35 m² (14.5 sq ft)
Volume: total	2.10 m³ (74 cu ft)
optional freight stowage	0.85 m³ (30.0 cu ft)

AREAS:

Wings, gross	17.20 m² (185.1 sq ft)
Ailerons (total): conventional	1.00 m² (10.76 sq ft)
DSFC	0.43 m² (4.63 sq ft)

Flaps	1.32 m² (14.20 sq ft)
Fins (total)	2.84 m² (30.57 sq ft)
Rudders (total)	0.74 m² (7.97 sq ft)
Tailplane	2.34 m² (25.19 sq ft)
Elevators, incl tab	0.70 m² (7.53 sq ft)

WEIGHTS AND LOADINGS:

Weight empty	420 kg (926 lb)
Max fuel weight	70 kg (154 lb)
Max T-O and landing weight	750 kg (1,653 lb)
Max wing loading	43.6 kg/m² (8.93 lb/sq ft)
Max power loading	8.38 kg/kW (13.78 lb/hp)

PERFORMANCE (estimated):

Never-exceed speed (VNE)	118 kt (220 km/h; 136 mph)
Max level speed at S/L	86 kt (160 km/h; 99 mph)
Max cruising speed at 1,000 m (3,280 ft)	
	81 kt (150 km/h; 93 mph)
Econ cruising speed at 1,000 m (3,280 ft)	
	67 kt (125 km/h; 78 mph)
Stalling speed: flaps up	44 kt (80 km/h; 50 mph)
flaps down	39 kt (72 km/h; 45 mph)
Max rate of climb at S/L	264 m (866 ft)/min
Max certified altitude	4,000 m (13,120 ft)
T-O run	120 m (395 ft)
T-O to 15 m (50 ft)	275 m (905 ft)
Landing from 15 m (50 ft)	250 m (820 ft)
Landing run	120 m (395 ft)
Range with max normal fuel	
	680 n miles (1,259 km; 782 miles)

UPDATED

SAZ

SARATOVSKY AVIATSIONNY ZAVOD ZAO (Saratov Aviation Plant JSC)

ploshchad Ordzhonikidze 1, 410015 Saratov
Tel: (+7 8452) 44 81 01
Fax: (+7 8452) 44 36 07
e-mail: saz@mail.saratov.ru
GENERAL DIRECTOR: Aleksandr V Ermishin
EXECUTIVE DIRECTOR: Nikolay I Dubrovin
DEPUTY GENERAL DIRECTOR, FOREIGN ECONOMIC AFFAIRS AND SALES: Yury N Lyalkov

During the Second World War, the Saratov plant (GAZ 292) manufactured Yakovlev Yak-1 and Yak-3 piston-engined fighters, continuing post-war with Yak-15, Yak-23, Yak-25 and Yak-27 military jet aircraft. In the 1960s and 1970s, production centred on the Yak-40 three-turbofan transport, with Yak-36 and Yak-141 V/STOL prototypes and Yak-38 V/STOL fighters also manufactured in the 1970s and 1980s. Yak-42 entered series production in 1982, although there have been no recent completions. Yak-54 trainer/sporting aircraft production began in 1994, but in early 2001, SAZ was attempting to resuscitate an American order for 48 which had previously been terminated after only six deliveries. Of five

Yak-54s to have been built in 2001, only one completed before venture was abandoned for lack of funds. Meanwhile, manufacturing is turning increasingly to agricultural tractors, buses, street cleaning equipment and catamarans. SAZ is earmarked for amalgamation within Sukhoi holding group under 2001 plans for Russian aerospace industry rationalisation.

SAZ also worked on a completely new type of lifting-body jet aircraft known as the EKIP project. It was the first company in the Russian Federation and Associated States (CIS) to receive a production certificate for commercial aircraft in compliance with international standards.

UPDATED

SGAU

SAMARSKY GOSUDARSTVENNYI AEROKOSMITSESKY UNIVERSITET (Samara State Aerospace University)

Moskovskoye shosse 34, 443086 Samara
Tel: (+7 8462) 35 18 26
Fax: (+7 8462) 35 18 36

This prominent centre of aerospace learning includes a design bureau, SKB-1, in which students can develop their skills. Although this has concentrated on the small-scale manufacture of lightplanes, it has also designed – under leadership of Albert Elatonsev, its director – an executive

transport designated Aist 92 (Stork 92). An accompanying drawing shows the general configuration, the aircraft having a 10 m (33 ft) span and contrarotating propellers driven by a 3,300 kW (4,425 shp) turbine to give a 350 kt (648 km/h; 403 mph) cruising speed with eight persons.

STUDENTSKOYE KONSTRUCTORSKOYE BURO 1 (Student Design Bureau 1)

Moskovskoye shosse 34, Korpus 3, 443086 Samara
Tel: (+7 8462) 35 72 81
e-mail: skbl@ssau.ru
Web: http://www.aero.ssau.ru/fla/skb1
DIRECTOR: Professor V M Shakhmistov

SKB-1 designs, described in 2001-02 and earlier editions, are available for commercial production.

Bureau founded 1955; has built Sverchock-1 gyroplane (1971), A-13 motor glider (1974), Shmel lightplane (1977), Strekoza lightplane (1978), Bat glider (1978), A-10 glider (1982) and A-18 Rusich motor glider (1987). Also built Che-15 and Che-25 (see Chernov and Gidroplan respectively earlier in this section) and S-202 (last described in 1998-99 *Jane's*). The A-16 aerobatic monoplane (1997), S-302 twin-engined amphibian (1995), S-400 four-seat amphibian (1998), Favorit two-seat ultralight (1998) and Crechet two-seat lightplane (1999) were all described in the 2001-02 edition.

UPDATED

SmAZ

SMOLENSKY AVIATSIONNYI ZAVOD OAO (Smolensk Aircraft Plant PLC)

ulitsa Frunze 74, 214006 Smolensk
Tel: (+7 08122) 293 13
Fax: (+7 0812) 61 97 70
e-mail: smaz@sci.smolensk.ru
GENERAL DIRECTOR: Aleksandr Miroshkin
CHIEF ENGINEER: Yury Litvinov
MARKETING MANAGER: Vladim Merzlyakov

Smolensk Aircraft Plant was established in 1926 as a repair facility. In 1934, Buro Osovikh Konstruktsii, OKB for

experimental aircraft, relocated from Moscow to Smolensk GAZ 35, which built the OKB's prototypes. Produced Il-2 attack fighter and A-2, PM and VA-3/48 gliders during Second World War; thereafter, constructed prototypes for various OKBs, Myasishchev M-17/M-55, wings for Yak-40 transports, and more than 700 Yak-18Ts from 1973. Built small number of early production Yak-42s alongside main series at Saratov (SAZ). Constructed Buran space shuttle. Government share was 25.5 per cent.

Production of Yak-18T was resumed 1993, five years after entire original fleet had been scrapped; more than 60 built up to 1996, when 15 placed in storage due to financial problems; production resumed in 1998; deliveries have included three

to Ulyanovsk Aviation School in June 2001. Built three Yak-112 lightplanes before project transferred entirely to IAPO in Irkutsk. Other work included manufacture of RPVs and cruise missiles; wings for the Yak-42, upgrading of Yak-40s for business use (including additional fuel capacity), and development and production of Technoavia SM-92 Finist seven-seat STOL aircraft. Has built wing and empennage for first three Sukhoi Su-38L agricultural aircraft and will be responsible for series production. Currently producing one further M-55. Also participated in Yak-130 programme. Overhauls some 10 to 15 Yak-18Ts per year.

VERIFIED

SOKOL

SOKOL NIZHEGORODSKY AVIATSTROITELNYI ZAVOD AOOT (Sokol Nizhny Novgorod Aircraft Manufacturing Plant JSC)

ulitsa Chaadaeva 1, 603035 Nizhny Novgorod
Tel: (+7 8312) 46 71 27 and 46 75 69
Fax: (+7 8312) 22 19 25
e-mail: wings@kis.ru
Web: http://www.sokolplant.ru
GENERAL DIRECTOR: Vasily H Pankov
CHIEF ENGINEER: Valery Drobyshevski
HEAD OF MARKETING: Aleksandr E Zaitsev

Founded (as GAZ 21) on 1 February 1932, this production plant concentrated on fighter aircraft manufacture until the start of the *konversiya* programme led to retooling for civilian production; registered as a JSC on 22 September 1994. Joined Kaskol group of companies in 2000.

Built 17,691 fighters up to end of Second World War. From 1949, in collaboration with the Mikoyan OKB, it manufactured 2,148 MiG-15s (1949-51), 2,470 MiG-17s (1951-54), 1,125 MiG-19s (1955-58), 5,278 (or 5,752) MiG-21s (1957-85) and 1,186 MiG-25s (1966-85). Russian Ministry of Property holds 38 per cent of shares, which equivalent to 50.5 per cent of voting rights; Rossiysky Credit (19.79 per cent) and Apone SA (16.99 per cent) are remaining

major (voting) shareholders. Current or recent products include the MiG-29UB (from kits supplied from Moscow), MiG-31 and Yak-130 jet trainer; 90 per cent of Yak-130 tooling in place by mid-2000. It is also responsible for supplying kits for upgrading 125 Indian Air Force MiG-21s to MiG-21-93 standard, with an initial order for 36 (see *Jane's Aircraft Upgrades* for details). Agreement reached in July 2000 on upgrading half of Russian Air Forces' 280 MiG-31s.

Civilian programmes include manufacture of the Myasishchev M-101 Gzhel six-seat turboprop light aircraft, a prototype AeroRIC Dingo amphibious aircraft with air cushion landing gear and Avia Accord twin-engined light multipurpose aircraft; plans also exist for Myasishchev M-202 production.

M-101 marketing is first task of New Regional Aircraft Company (NORS), formed 18 April 2001 by Sokol, Myasishchev and Kaskol financial group. AeroRIC Research-Industrial Company is co-located with Sokol; in 2001 its Dingo project was in suspension because of financial shortfall.

Plans to build MiG-110 twin-turboprop transport apparently not overtaken by announcement in early 2000 of its reallocation to MAPO and discussions on continued participation in programme under way with RSK MiG in mid-2000. Agreement of 2001 covers production of wings and noses for navalised MiG-29K, plus complete MiG-29KUB trainers. Production of the Eurospace F-15-F Excalibur four-seat touring aircraft was halted at an early stage when a contract for almost 100 was cancelled; some 20 airframes remained at Nizhny Novgorod in June 2000, when 10 sold to Turbine Design (which see) in USA.

Other products, in conjunction with AeroRIC, include WIGE craft, air cushion vehicles and hydrofoils. Offers 'familiarisation' flights in MiG-29UB, MiG-21U and L-29 Delfin. Employment (including part-time) was 9,000 in 2001.

UPDATED

STRELA

STRELA PROIZVODSTVENNOYE OBEDINENIE (Strela Production Association)

ulitsa Shevchenko 26, 46005 Orenburg
Tel: (+7 3532) 35 72 09 and 35 61 74
Fax: (+7 3532) 35 54 60

The wholly state-owned Strela plant manufactures UAVs and produces parts of the Kamov Ka-226 helicopter; assembled prototype Ka-226; plans for Ka-226 series manufacture included first three or four production helicopters in 2001, but production investment of Rb300 million was not announced until January 2002, first two Ka-226s being complete by late March 2002. Factory has orders for 27 Ka-226s. It has produced replica Yak-3/Yak-9s (which see) for export, the most recent in late 2001.

UPDATED

SUKHOI (AVPK SUKHOI)

GOSUDARSTVENNOYE UNITARNOYE PREDPRIYATIE AVIATSIONNYI VOYENNO-PROMYSHLENNYI KOMPLEKS SUKHOI (State Unitary Enterprise, Aviation Military-Industrial Complex Sukhoi)

a/ya 604, ulitsa Polikarpova 23A, 125284 Moskva
Tel: (+7 095) 941 76 87
Fax: (+7 095) 945 68 06
Web: http://www.sukhoi.org

GENERAL DIRECTOR: Mikhail A Pogosyan
PRESS OFFICER: Yury Chervakov

AVPK Sukhoi was created, under presidential decree, in August 1996, to bring together the Sukhoi and Beriev design bureaux, the production plants at Irkutsk (IAPO), Komsomolsk-on-Amur (KnAAPO), Novosibirsk (NAPO) and Taganrog (TANTK), and two banks, Oneximbank and Inkombank. The grouping was formally inaugurated on 30 December 1996. As originally constituted, the group comprises the wholly state-owned KnAAPO and NAPO; plus Sukhoi OKB (see following entry), IAPO and TANTK/ Beriev, APVK Sukhoi having a controlling interest in three last-mentioned.

In October 2001, decree by Russian president created Sukhoi Aviation Holding Company, which to become successor to AVPK Sukhoi. New structure initially envisages KnAAPO and NAPO as joint stock companies (held 74.5 per cent each by Sukhoi and balance by Russian state), with IAPO to follow after its flotation.

In January 2002, co-production agreement signed with Avibras for local assembly of Su-35 if chosen in Brazilian Air Force's fighter competition.

UPDATED

OPYTNYI KONSTRUKTORSKOYE BURO SUKHOGO AOOT (Sukhoi Experimental Design Bureau JSC)

ulitsa Polikarpova 23A, 125284 Moskva
Tel: (+7 095) 941 01 30
Fax: (+7 095) 945 55 70
CHAIRMAN: Yury Koptev
DEPUTY CHAIRMAN: Vladimir Chernov
GENERAL DESIGNER: Mikhail Petrovich Simonov
CHIEF DESIGNER: Vladimir Babak

OKB, which is 57 per cent owned by AVPK Sukhoi, named for Pavel Osipovich Sukhoi, who headed it from 1939 until his death in September 1975. It remains one of two primary Russian centres for development of fighter and attack aircraft, and it is notable that 60 per cent of front-line Russian Air Forces' aircraft are of Sukhoi design. Bureau has widened its activities to include civilian aircraft, under *konversiya* programme. This offsets the drastic reduction in combat aircraft production and competition in the upgrade and overhaul market.

Sukhoi Advanced Technologies
GENERAL DIRECTOR: Boris V Rakitin

This division is responsible for promotion and sales of Sukhoi light aircraft, such as the Su-29, Su-31, Su-38L and Su-49. It intends to relocate Su-29 and Su-31 production to KAPO at Kazan from 2002. Related activities include conversion of Su-26s and Su-29s with SKS-94 extraction systems.

Nauchno-Proizvodstvennoe Kontsern Shturmoviki Sukhogo (Sukhoi Stormovik Scientific-Production Concern)
GENERAL DIRECTOR: Vladimir Babak

Offers upgrades of Su-25 'Frogfoot' to Russian and export customers.

Sukhoi Civil Aircraft
GENERAL DIRECTOR: Andrey Il'in

Subdivision formed May 2000; on 5 June 2000 signed agreement with Alliance Aircraft Corporation (which see in US section) for joint development and production of StarLiner regional airliner. However, Sukhoi unilaterally cancelled agreement on 2 November 2000. In August 2001, Sukhoi revealed RRJ (Russian Regional Jet) programme in collaboration with Boeing.

UPDATED

SUKHOI Su-25 and Su-28
NATO reporting name: Frogfoot
Russian Air Forces popular name: Grach (Rook)
TYPE: Attack fighter.
PROGRAMME: Development began 1968; Novosibirsk-built prototype, known as T8-1, flew 22 February 1975, with two 25.5 kN (5,732 lb st) non-afterburning versions of Tumansky RD-9 turbojet and underbelly twin-barrel

Trio of Sukhoi-designed naval aircraft, comprising Su-25UTG (leading), Su-33 and Su-33UB
(*Jane's/Paul Jackson*)   0114595

depressable GSh-23 23 mm gun in fairing; in eventual developed form, second prototype and rebuilt first aircraft, T8-2D and T8-1D, had more powerful non-afterburning versions of Soyuz/Gavrilov R-13, designated R-95Sh, increased wing span, wingtip avionics/speed brake pods, underwing weapon pylons, and internal AO-17 30 mm gun necessitating offset nosewheel. Also Klen-PS laser range-finder replacing Fone, and new DISS-7 Doppler, new gunsight and new navigation computer; observed by satellite at Zhukovsky flight test centre 1977, given provisional US designation 'Ram-J'; production, with R-95Sh turbojets, began 1978.

T8-1D and pre-series T8-3 sent to Afghanistan April 1980, as Rhombus team, for trials and combat testing; followed in mid-1981 by 200th Independent Shturmovik Squadron, Soviet Air Force, flying 12 production Su-25s for co-ordinated low-level close support of Soviet ground forces in mountain terrain, with Mi-24 helicopter gunships; fully operational 1984; 200th OShAE and later Su-25 squadrons flew 60,000 sorties during the eight-year war in Afghanistan, losing 23 aircraft and eight pilots killed; all but two of 139 laser-guided ASMs launched in combat achieved direct hits. Attack versions built initially at Factory No 31, Tbilisi, Georgia; production at Tbilisi ended 1989 (approximately 875 built, of which 210 exported according to company statement); subsequently, restarted for small batches (four to Congo Democratic Republic in 1999; six to same customer and two – allegedly Su-25Ts, but unlikely – to Ethiopia in 2001) and factory reports sufficient parts ''in stock'' to assemble up to 75 aircraft, of which 30 were almost complete in 2001; production at Ulan-Ude completed 1991-92; see separate entry on Su-25T.
CURRENT VERSIONS: **Su-25** ('Frogfoot-A'): Single-seat close support aircraft. Cannon muzzle redesigned from ninth production series; aileron control rod fully faired and trim tab deleted from 10th series. Revised airbrakes with double fold on late production aircraft, and by retrofit. Export version **Su-25K** (*kommercheskiy*; commercial) still being promoted by Tbilisi plant in 2000.
Detailed description applies basically to late production Su-25K.

Su-25SM: Russian Air Forces decided in March 1999 to upgrade about 40 per cent of surviving Su-25s (some 220 single-seat and 65 two-seat remained operational in 2001). Single-seaters will be upgraded to Su-25SM (*Seriya, Modernizatsya*: Series, Modified) standards; two-seaters to Su-25UBM configuration; completion date is 2010. New variant uses some equipment and systems developed for Su-25TM (Su-39). Su-25SM reported initially to feature Panther fire-control system with Kopyo-25 radar in a rebuilt nose (not pod-mounted, as on Su-25TM), giving compatibility with a wider range of air-to-air and air-to-ground weapons; however, report of January 2001 specifically excluded onboard radar, which seen as option for later fitment. Klen PS laser range-finder is relocated below fuselage. Revised cockpit with a new HUD and two large colour LCD MFDs, but retains standard analogue flight instruments. Irtysh EW system, with new RWR, MAWS and automatic control of chaff and flares. Structural changes limited to life-extension modifications and provision for nose-mounted radar. Sukhoi Stormovik Concern promises improved maintainability and a service life extended to 2,500 flying hours.

Conversion of Russian Su-25s being undertaken by Air Forces' own Aircraft Repair Works No. 121, at Kubinka; programme management by Sukhoi Stormovik Concern, which agreed in mid-2000 to fund initial two aircraft as means of launching programme. First of these made debut ('33') at MAKS, Moscow, August 2001; further two, then unfunded, due before end of 2001, plus five in 2002. State Acceptance Tests were to have begun in November 2001 and be completed in following year. However, only second prototype joined programme in late 2001 and total for 2002 was reduced to two more, of which first was rolled out at Kubinka on 5 January 2002, first flew on 5 March and was due to complete State testing in September 2002 before transfer to Lipetsk combat training centre. Second, flown 20 March, was initial Su-25UBM. Russian Air Forces requirement for 100 Su-25SMs by 2008, total planned for 2003 being eight. Belarusian aircraft to be upgraded at Baranovichi; Ukrainians also being converted locally.

Su-25UBK tailcone, containing Sirena-3 RWR and ASO-2V chaff/flare dispensers ***(Jane's/Paul Jackson)***
0109121

Su-25UB ('Frogfoot-B'): Tandem two-seat operational conversion and weapons trainer; prototype T8U-1 first flew 10 August 1985; entered production (only at Ulan-Ude) 1987; first photographs early 1989; rear seat raised considerably, giving humpback appearance; separate hinged portion of continuous framed canopy over each cockpit; taller tailfin, increasing overall height to 5.20 m (17 ft 0¾ in); new IFF blade antenna forward of windscreen instead of SRO-2 ('Odd Rods'); weapons pylons and gun retained. Optional periscope over rear cockpit. Export version **Su-25UBK.**

Su-25UBM: Two-seat Su-25s are to be modernised in the same way as the Su-25SM and, in fact, accorded higher priority to allow pilot training for those destined to fly the new single-seat version. First aircraft to be upgraded in the OKB's own Moscow workshops by Sukhoi Stormovik Concern was to have been a new Su-25UTG which, in April 2000, remained at Ulan-Ude factory, awaiting payment.

Su-25UT ('Frogfoot-B'): As Su-25UB, but without weapons; intended as L-29/L-39 replacement for air forces and DOSAAF; prototype (converted from T8U-1) first flew 6 August 1985; demonstrated 1989 Paris Air Show as **Su-28**; overall length 15.36 m (50 ft 4¾ in); one or two only. Discontinued.

Su-25UTG (G for *gak*: hook) ('Frogfoot-B'): As Su-25UT, with added arrester hook under tail; first flew September 1988; used initially for deck landing training on dummy flight deck marked on runway at Saki naval airfield, Ukraine; on 1 November 1989 was third aircraft to land for trials on carrier *Admiral of the Fleet Kuznetsov*, after Su-27K and MiG-29K; 10 built in 1989-90; five in Ukrainian use at Saki; one lost; four at Severomorsk, Kola Peninsula, for service individually on *Kuznetsov*. Two on board when ship deployed to Adriatic December 1995. Russian Navy received 11th aircraft in September 1997, while 12th was due in late 1998. These reportedly being given combat role: No. 11 said to be with millimetre-wave SLAR, Pastel-K RWR and TKS coded datalink for reconnaissance and target acquisition; No. 12 reportedly with anti-ship missiles and target acquisition systems also earmarked for Kamov Ka-29 upgrade. Discussions reported in early 1998 on possible further 12 for Russian Navy, but Ulan-Ude foresaw no prospect of further production after 2000.

Su-25UBP: Further aircraft to similar (UTG) standard, produced by conversion of Su-25UBs, to compensate for losses to Ukraine and attrition.

Su-25T/TM (Su-39): See separate entry.

Su-25BM (*buksir misheney*: target towing aircraft) ('Frogfoot-A'). Standard late-series, R-195-engined Su-25 attack aircraft, with added underwing pylons for a winch, and Planyer-M miss-distance detection system and datalink, TL-70 Kometa towed target or PM-6 rocket-propelled targets released for missile training by fighter pilots; prototype (converted T8-9) first flew 1989; 50 built at Tbilisi.

Su-25KM Scorpion: Upgrade of Su-25K developed by Tbilisi Aerospace Manufacturing in Georgia (which see) in conjunction with Elbit Systems of Israel. Launched September 2000. Prototype 10629 first flew at Tbilisi on 14 April 2002 and made first 'official' flight 18 April; demonstrated at Paris, June 2001. Equipment includes two 152 × 203 mm (6 × 8 in) LCDs, HUD and Elbit weapon delivery and navigation system, improving pilot's situational awareness and all-weather capabilities. Cost US$3.5 million for upgrade; US$6 million for complete aircraft.

CUSTOMERS: Russian tactical air forces and Naval Aviation have about 200; pre-1985 aircraft now withdrawn. Exports to Angola (12, plus two two-seat), Azerbaijan, Belarus (currently 80), Bulgaria (36 plus four two-seat), Congo (10), Czech Republic (24, plus one two-seat), Ethiopia (two, plus two two-seat), Georgia (five), Iran (seven), Iraq (45), North Korea (36, plus four two-seat), Peru (10 Su-25 and 8 Su-25UB from Belarus), Slovak Republic (11, plus one two-seat), Turkmenistan (46 currently in storage) and Ukraine (currently 64). Croatia also reported to have received a 'squadron' (never uncrated, according to some reports), but denied by Croatian Air Force. Some sources suggest that Afghanistan received 12, all now out of service. Tbilisi plant quotes total of 875 single-seat Su-25s built (including 50 Su-25BMs), plus Su-25UBs. Combined production stated by Sukhoi to exceed 1,200, of which more than 200 were exported and almost 300 are two-seat. Russian estimate of 2002 reported 860 Su-25s in service with 16 countries.

Su-25 CURRENT OPERATORS

Unit	Base	Equipment
Russia (Air Force)		
16th OShAP	Taganrog	Su-25
18th GvOShAP	Galenki	Su-25
237th GvTsPAT	Kubinka	Su-25
368th OShAP	Budyennovsk	Su-25
461st OShAP	Krasnodar	Su-25
899th OShAP	Buturlinovka	Su-25
968th IISAP	Lipetsk	Su-25
Unident OShAP	Olovyannaya	Su-25
Unident OShAP	Chernigovka	Su-25
Krasnodar Higher Military Aviation School	Krasnodar	Su-25
Russia (Naval Aviation)		
279th KIAP	Severomorsk	Su-25UTG, UBP
Angola		
Unident unit	Saurimo	Su-25, UB
Armenia		
121st ShAE	Kumayari	Su-25
Azerbaijan		
Unident unit	Kyurdamir	Su-25
Belarus		
206th ShAB	Lida	Su-25
Bulgaria		
22nd Shturmova Aviobasa	Bezmer/Yambol	Su-25K, UBK
Congo DR		
n/k	Kamina	Su-25K
Czech Republic		
322 Taktica Letka	Namest	Su-25K, UBK
Ethiopia		
n/k	unknown base	Su-25K
Georgia (Republic)		
Unident unit	Kopitnari	Su-25
Georgia (Abkhazia)		
Unident unit	unknown base	Su-25
Iraq		
Unident unit[1]	unknown base	Su-25
Kazakhstan		
Unident unit	unknown base	Su-25
Korea (North)		
Unident unit	unknown base	Su-25
Peru		
Grupo 11	Vitor	Su-25UB
Slovak Republic		
33/2 Letka	Malacky-Kuchyna	Su-25K, UBK
Turkmenistan		
56 BRS	Kizyl-Arvat	Su-25[2]
Ukraine		
299th OShAP	Saki	Su-25, UTG
452nd OShAP	Chortkov	Su-25
Uzbekistan		
59th APIB	Chirchik	Su-25

Notes: OShAP is *Otdel'nyi Shturmovoy Aviatsion'nyi Polk* (Independent Stormovik Aviation Regiment); KIAP is *Korabel'nyi Istrebitel'nyi Aviatsion'nyi Polk* (Shipborne Fighter Aviation Regiment); APIB is *Aviatsion'nyi Polk Istrebetelei-Bombardirovchikov* (Aviation Fighter-Bomber Regiment); ShAB is *Shturmovoy Aviatsion'nyi Baza* (Stormovik Aviation Base); BRS is *Basis Reserv Samolety* (Base for Reserve Aircraft). Gv prefix indicates *Gvardiya* (Guards)

[1] Probably not in service. Seven survivors fled to Iran in 1991

[2] In storage

COSTS: Tbilisi factory quoting US$6 million per aircraft during 2000. Upgrade to Su-25SM costs US$2 million.

DESIGN FEATURES: Robust structure and systems redundancy increase resistance to small arms fire. Emphasis on survivability led to features accounting for 7.5 per cent of normal T-O weight, including cockpit armoured with some 600 kg (1,323 lb) of welded ABVT-20 titanium; pushrods instead of cables to actuate flying control surfaces (duplicated for elevators); damage-resistant main load-bearing members; widely separated engines in stainless steel bays; fuel tanks filled with reticulated foam for protection against explosion.

Maintenance system packaged into four pods for carriage on underwing pylons; covers onboard systems checks, environmental protection, ground electrical power supply for engine starting and other needs, and pressure refuelling from all likely sources of supply in front-line areas; engines can operate on any fuel likely to be found in combat area, including MT petrol and diesel oil.

Shoulder-mounted wings; approximately 20° sweepback; 2° 30′ anhedral from roots; thickness/chord ratio 10.5 per cent; extended chord leading-edge dogtooth on outer 50 per cent each wing; wingtip pods each split at rear to form airbrakes that project above and below pod when extended; retractable PRF-4M landing light in base of each pod, outboard of small glareshield and aft of dielectric nosecap for ECM; semi-monocoque fuselage, with 24 mm (0.94 in) welded titanium armoured cockpit; pitot on port side of nose, transducer to provide data for fire-control computer on starboard side; conventional tail unit; variable incidence tailplane, with slight dihedral.

FLYING CONTROLS: Conventional and partly assisted. Hydraulically actuated ailerons, with manual back-up; multiple tabs in each aileron; double-slotted two-section wing trailing-edge flaps; full-span leading-edge slats, two segments per wing; manually operated elevators and two-section inset rudder; upper rudder section operated through sensor vanes and transducers on nose probe and automatic electromechanical yaw damping system; tabs in lower rudder segment and each elevator.

STRUCTURE: All-metal (60 per cent duralumin, 13.5 per cent titanium alloys, 19 per cent steel, 2 per cent magnesium); three-spar wings; semi-monocoque slab-sided fuselage.

LANDING GEAR: Hydraulically retractable tricycle type; mainwheels retract to lie horizontally in bottom of engine air intake trunks. Single wheel with low-pressure tyre on each levered suspension unit, with oleo-pneumatic shock-absorber; mudguard on forward-retracting steerable nosewheel, which is offset to port; mainwheel tyres size 840×360, pressure 9.30 bar (135 lb/sq in); nosewheel tyre size 660×200, pressure 7.35 bar (106 lb/sq in); brakes on mainwheels. Twin 25 m² PTK-25 cruciform brake-chutes housed in tailcone.

POWER PLANT: Two Soyuz/Gavrilov R-95Sh turbojets in long nacelles at wingroots, each 40.21 kN (9,039 lb st), superseded in about 50 late production aircraft (from 1989) by R-195s, each 44.18 kN (9,921 lb st); these first trialled by T8-14 and T8-15; 5 mm thick armour firewall between engines; R-195 turbojets have pipe-like fitment (believed not initially cleared for export, thus not fitted to Su-25K) at end of tailcone, from which air is expelled to lower exhaust temperature and so reduce IR signature; non-waisted undersurface to rear cowlings, which have additional small airscoops (as three-view); self-sealing fuel tanks filled with reticulated foam.

Nos. 1 and 2 tanks in fuselage between cockpit and wing front spar, and between rear spar and fin leading-edge (latter acting as collector tank) contain 2,386 litres (630 US gallons; 525 Imp gallons); and in wing centre-section, capacity 1,274 litres (337 US gallons; 280 Imp gallons); total capacity 3,660 litres (967 US gallons; 805 Imp gallons); provision for four PTB-1500 external fuel tanks on underwing pylons.

ACCOMMODATION: Single Severin (Zvezda) K-36L ejection seat with zero altitude/54 kt (100 km/h; 62 mph) envelope; sideways-hinged (to starboard) canopy, with small rearview mirror on top; flat bulletproof windscreen. Folding ladder for access to cockpit built into port side of fuselage.

SYSTEMS: 28 V DC electrical system, supplied by two engine-driven generators. Dual independent hydraulic systems, pressure 206 bar (2,987 lb/sq in).

AVIONICS: *Comms:* SRO-1P ('Odd Rods') or (later) SRO-2 IFF transponder, with antennas forward of windscreen and under tail. SO-69 ATC transponder. R-862 V/UHF transceiver, R-855 emergency radio, R-828 air-to-ground radio.

Flight: RSBN Tacan, DISS Doppler, MRP-56P marker beacon receiver, RV-1S radio altimeter, ARK-15M radio compass SVS-1-72-18 air data system UUAP-72 AoA and acceleration indicator.

Ukrainian Sukhoi Su-25 ('Frogfoot-A') ***(Yefim Gordon)***
01

Sukhoi Su-25UTG navalised jet trainer *(Yefim Gordon)*   NEW/0525028

Instrumentation: ASP-17BSh-8K weapons sight incorporating AKS-750s camera gun.

Mission: KLEN-PS laser range-finder and target designator under flat sloping window in nose; strike camera in top of nosecone.

Self-defence: SPO-15LM (L-006LE) Sirena-3 radar warning system antenna above fuselage tailcone; ASO-2V-01 chaff/flare dispensers (total of 256 flares) can be carried above root of tailplane and above rear of engine ducts. Some later aircraft fitted with Gardeniya active jammer.

ARMAMENT: One twin-barrel AO-17A 30 mm gun with rate of fire of 3,000 rds/min in bottom of front fuselage on port side, with 250 rounds (sufficient for a 1 second burst during each of five attacks). Eight large pylons under wings for 4,000 kg (8,818 lb) of air-to-ground weapons (though 1,400 kg; 3,086 lb regarded as normal maximum), including UB-32A rocket pods (each 32 × 57 mm S-5), B-8M1 rocket pods (each 20 × 80 mm S-8), 240 mm S-24 and 330 mm S-25 guided rockets, Kh-23 (AS-7 'Kerry'), Kh-25ML (AS-10 'Karen') and Kh-29L (AS-14 'Kedge') ASMs, laser-guided rocket-boosted 350 kg, 490 kg and 670 kg bombs, conventional bombs, 500 kg incendiary, anti-personnel and other cluster bombs, and SPPU-22 pods each containing a 23 mm GSh-23 gun with twin barrels that can pivot downward for attacking ground targets, and 260 rounds. Some late-production R-195-engined Russian Su-25BMs may have been wired for the carriage of Kh-58U/E (AS-11) ARMs and the associated Vynga datalink pod. Some Soviet aircraft configured to carry RN-61 nuclear weapon or associated IAB-500 'shape' for role training. Two small outboard pylons for R-3S (K-13T; NATO AA-2D 'Atoll') or R-60 (AA-8 'Aphid') self-defence AAMs. SU-25BM carries TL-70 winch and Kometa towed target below port wing, with inert FAB-250 or FAB-500 to starboard.

DIMENSIONS, EXTERNAL (A, Su-25K; B, Su-25UTG):
Wing span: A, B	14.36 m (47 ft 1½ in)
Wing aspect ratio: A, B	6.1
Length overall, incl probe: A	15.53 m (50 ft 11½ in)
B	15.36 m (50 ft 4¾ in)
Height overall: A	4.80 m (15 ft 9 in)
B	5.20 m (17 ft 0¾ in)

AREAS:
Wings, gross: A, B	33.7 m² (362.75 sq ft)

WEIGHTS AND LOADINGS:
Weight empty: A	9,800 kg (21,605 lb)
T-O weight: normal:	
A	14,600 kg (32,187 lb)
B	13,005 kg (28,671 lb)
max: A	17,600 kg (38,800 lb)
B	16,245 kg (35,814 lb)
Max landing weight: A	13,300 kg (29,320 lb)
Max wing loading: A	522.2 kg/m² (106.97 lb/sq ft)
B	482.0 kg/m² (98.73 lb/sq ft)
Max power loading: A	199 kg/kN (1.95 lb/lb st)
B	184 kg/kN (1.80 lb/lb st)

PERFORMANCE:
Max level speed at S/L:	
A	M0.8 (526 kt; 975 km/h; 606 mph)
B	M0.775 (512 kt; 950 km/h; 590 mph)
Max attack speed, airbrakes open:	
A	372 kt (690 km/h; 428 mph)
Landing speed (typical): A	108 kt (200 km/h; 124 mph)
Service ceiling: clean: A, B	7,000 m (22,960 ft)
with max weapons: A	5,000 m (16,400 ft)
T-O run, typical: A	500 m (1,640 ft)
A, unpaved surface	700 m (2,300 ft)
B, on concrete	750 m (2,460 ft)
B, on carrier ramp	175 m (575 ft)
Landing run, normal: A	650 m (2,135 ft)
A, unpaved surface	750 m (2,460 ft)
B, arrested	90 m (295 ft)
B, on concrete	1,000 m (3,280 ft)

Prototype of the TAM-modified Su-25 Scorpion upgrade *(Jane's/Paul Jackson)*   NEW/0525029

First Sukhoi Su-25SM conversion, displayed at Moscow in August 2001 *(Jane's/Paul Jackson)*   0121065

Sukhoi Su-28 (Su-25UT), with added side elevation (centre) of Su-25K *(Jane's/Dennis Punnett)*

Sukhoi Su-25UBK export two-seat trainer *(Jane's/Paul Jackson)*   NEW/0525030

Range with 4,400 kg (9,700 lb) weapon load and two
  external tanks:
    at S/L: A   405 n miles (750 km; 466 miles)
    at height: A   675 n miles (1,250 km; 776 miles)
g limits: with 1,500 kg (3,307 lb) of weapons: A   +6.5
    with 4,000 kg (8,818 lb) of weapons: A   +5.2
                                                        UPDATED

SUKHOI Su-25TM (Su-39)

TYPE: Attack fighter.
PROGRAMME: Developed as an ATM-armed derivative of the
standard Su-25, with a two-crew concept similar to that
used in the Mi-24 'Hind'. Plans to put the WSO in the front
cockpit were abandoned at an early stage. Airframe is
based on that of the Su-25UB, with which it shares about
85 per cent commonality. Three **Su-25T** development
aircraft (Sukhoi T8M; numbered 08, 09 and 10) were
converted Su-25UB airframes, with humped rear cockpit
faired over and internal space used to house new avionics
and extra tonne of fuel; conversions began 1976; first flight
by converted first prototype Su-25UB 17 August 1984;
construction at GAZ 31 Tbilisi, Georgia, of initial batch of
12 pre-series aircraft for air force acceptance testing began
1989; first one flew July 1990 and deliveries to
development unit at Akhtubinsk began 1991, although
only six were completed, remainder being in long-term
storage at Tbilisi. Passed State Acceptance Tests in 1993.
Production was to be transferred to Ulan-Ude (Su-25UB/
UTG plant) where two were reportedly assembled in 1999,
with six more to follow in 2000 and 12 in 2001; however,
by mid-2002 first two aircraft still incomplete and awaiting
relaunch of Ulan-Ude production to meet Russian Air
Forces' requirement for some 50 aircraft.
    Redesignated **Su-25TM** after provision for 3 cm
waveband Kopyo-25 radar, Schkval-M targeting unit,
Kinzhal MMW radar and Khod IIR pods and other
equipment changes; T8M-3 first exhibited at Dubai '91 Air
Show as export **Su-25TK**; version (aircraft 20) with
Kopyo-25 radar pack, for improved night/bad weather
attack capability, shown at Moscow Air Show '95; at that
time, one Su-25TM had been completed at Ulan-Ude plant
and first two pre-series Su-25T aircraft had been upgraded to
same standard. Two further Su-25TMs built at Ulan-Ude
in 1997 and 1998, and followed by first batch of four
production Su-25TMs, two of which ''almost complete''
by late 2001 and still reported in same condition in May
2002. Remaining 20 production Su-25Ts to be similarly
upgraded. Known to Sukhoi and both production plants,
but not Russian Air Forces, as **Su-39 'Strike Shield'**;
aircraft 21 at Moscow Air Show, 1997, carried only this
designation and was also exhibited in 1999, as Su-25TM;
RLPK-25 radar, derived from Kopyo-25, fully installed in
pack 1996; development continues, although originally
scheduled for completion 1997. Licence for production
offered to PZL Mielec of Poland, late 1997, as debt
resettlement but not taken up.
CUSTOMERS: Russian Air Forces have eight Su-25Ts,
delivered to Akhtubinsk test centre (two) and Lipetsk
training base (six, including aircraft '83') in 1990-91. Four
Lipetsk aircraft took part in bombing of Chechnya during
2000. Initial batch of seven Su-25TMs built at Ulan-Ude;
third aircraft was first delivered to Russian Air Forces in
early 1998; fourth and fifth ordered in 1999 and scheduled
for delivery to Russia in 2001, permitting start of State
Acceptance Tests, although report of June 2002 noted
these only just begun, with total of four aircraft; due for
completion in 2004; further three planned for funding in
2001. Initial Russian requirement for 24 Su-25TMs to
serve in six regional rapid-deployment groups, each of four
Su-25TMs and 12 Su-25s, plus helicopters; Georgia has
requirement for 50 TMs, and reportedly ordered three in
1996 and more in 1998.
COSTS: US$15 million to US$19 million (2002).
DESIGN FEATURES: Basically as Su-25 (which see), but
embodies lessons learned during war in Afghanistan,
particularly survival in intense anti-aircraft defence
environment and night/bad weather capability. All sub-

variants use increased-thrust R-195 engines, allowing
operation from 1,200 m (3,940 ft) runways at elevation of
3,000 m (9,840 ft) ASL. Welded cockpit 'bathtub' of
lighter ABVT-20 titanium alloy armour, weight of
polyurethane foam fuel tank filling and protective coatings
increased. Centre-fuselage fuel pipes and control rods
strengthened; gap between fuel tanks and air ducts filled
with elastic porous material to avoid spillage into ducts
following combat damage; additional armour for fuel
system and avionics compartment. Total weight of
survivability system is 1,115 kg (2,458 lb). Reported to use
special radar absorbent paint, previously tested on at least
one Su-25.
    New nav system makes possible flights to and from
combat areas under largely automatic control; equipment
in widened nose includes TV activated some 5 n miles
(10 km; 6 miles) from target; subsequent target tracking to
an accuracy of 0.6 m (2 ft), weapon selection and release
automatic; wingtip countermeasures pods introduced; gun
transferred to underbelly position, on starboard side of
farther-offset nosewheel. New, depressible 45 mm cannon
originally planned. IR signature reduced by factor of three
to four.

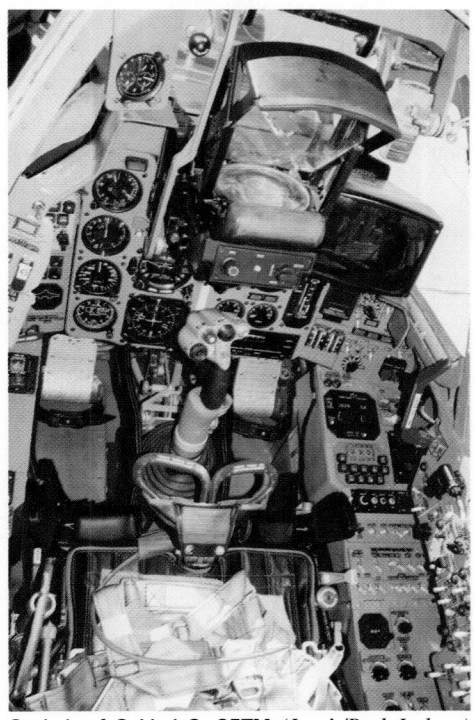

Cockpit of Sukhoi Su-25TM *(Jane's/Paul Jackson)*
                                                    0015090

Airframe total life 4,500 hours; major overhaul interval
1,500 hours. Engine total life 3,000 hours; TBO 1,000
hours.
    Wing leading-edge sweepback 19° 54′, dihedral 2° 30′,
incidence 0° 43′.
FLYING CONTROLS: Basically as Su-25. Trim only on rudder;
artificial feel in lateral and longitudinal channels.
Hydraulically actuated elevator. SAU-8 automatic control
system linked to WDNS.
LANDING GEAR: KT-163D mainwheels, with tyre size
840×360, pressure 9.30 bar (135 lb/sq in); KH-27A
nosewheel further offset to port (22 cm; 8½ in), with tyre
size 680×260, pressure 7.35 bar (106 lb/sq in); metal and
ceramic disc brakes, with anti-skid units. Two PTK-25
cruciform brake-chutes, each 13.0 m² (140 sq ft).
POWER PLANT: Two Soyuz/Gavrilov R-195Sh turbojets, each
44.1 kN (9,920 lb st). Ten internal fuel tanks, with new
Nos. 3 and 4 tanks in fuselage; total capacity 4,890 litres
(1,292 US gallons; 1,076 Imp gallons). Provision for four
800 litre (211 US gallon; 176 Imp gallon) or two 1,150 litre
(304 US gallon; 253 Imp gallon) underwing tanks. Single-
point pressure fuelling on starboard engine air intake duct;
optional gravity fuelling. Additional air intakes on upper
surface of engine nacelles to further cool jet exhaust.
ACCOMMODATION: As Su-25 but probably with zero/zero
ejection seat. Weight of titanium cockpit armour reduced
some 25 per cent to 450 kg (992 lb).
SYSTEMS: Pressurised cockpit, maximum differential 0.25 bar
(3.55 lb/sq in). Hydraulic system using AMG-10 fluid;
flow rate 76 litres (20.0 US gallons; 17.7 Imp gallons)/min;
pressure 207 bar (3,000 lb/sq in). AC and DC electrical
systems. Gaseous oxygen system.
AVIONICS: *Radar:* Kinzhal (Dagger) 8 mm MMW radar
abandoned and replaced by podded Kopye (Spear;
Westernised as Kopyo-25) system (incorporating NO27
radar) pending provision of a new Kinzhal using Russian
rather than Ukrainian parts.
    Flight: Nav system includes air data system, twin INS,
A-312 RSBN radio-nav system combining Tacan and ILS,
A-723 long-range Loran/Omega radio-nav, ShO-13A
Doppler, RV-21 radar altimeter and (from 2001) GPS.
    Instrumentation: HUD; SUV-25TM Voskhod nav/
attack system on stabilised platform, with two Orbita-20
navigation and fire-control computers, and Krasnogorsk
Shkval I-251 nose-mounted electro-optical system for
precision attacks on armour.
    Mission: Krasnogorsk I-251M Shkval-M targeting and
weapons control system, with window in nose for TV, laser
range-finder and target designator, all using same
stabilised mirror, ×23 magnification lens and cockpit CRT;
Khod (Motion) night attack IIR pack, with Merkur LLTV
and Prichal laser designator, and Phazotron Kopye radar
pod on centreline RLPK-25 pylon. Kopye detection range
13.5 n miles (25 km; 15.5 miles) for tanks, 40 n miles
(75 km; 47 miles) for small ships, 13.5 to 31 n miles (25 to
57 km; 15.5 to 35 miles) air-to-air; able to track eight
targets and engage two simultaneously.
    Self-defence: Irtysh DASS, with Pastel radar warning/
emitter location system; UV-26 chaff/flare dispensers (192
cartridges) in top of fuselage tailcone and in large
cylindrical housing at base of rudder that also contains
L-166SI Sukhogruz 6,000 W caesium lamp active IR
jammer. Optional MSP-410 Omul or Gardeniya
underwing active radar jamming pods on outboard
underwing pylons.
ARMAMENT: One NNPU-8M twin-barrel 30 mm gun (same
AO-17A as Su-25, but in new external VPU-17A
mounting), with 200 rounds, is offset 27 cm (10½ in) to
starboard; 11 external stores attachments; two eight-round
underwing APU-8 launchers for Vikhr M (AT-X-16) tube-
launched primary attack missiles able to penetrate 900 mm
of reactive armour; other weapons include laser-guided
Kh-25ML (AS-10 'Karen') and Kh-29L (AS-14 'Kedge'),
TV-guided Kh-29T (AS-14 'Kedge'), anti-ship Kh-35
(AS-20 'Kayak') and anti-radiation Kh-58U (AS-11
'Kilter') or Kh-31P (AS-17 'Krypton') ASMs,
KAB-500KR laser-guided bombs, S-25L laser-guided
rockets, conventional bombs from 50 to 500 kg each,
RBK-250 and RBK-500 cluster bombs, KGM-U
submunition dispensers, SPPU-687 gun pods (containing a
GSh-1-30 with 150 rounds) S-8, S-13 and S-25 rockets,
and R-27R/RE (AA-10 'Alamo-A/C'), RVV-AE (R-77;
AA-12 'Adder') and R-73 (AA-11 'Archer') AAMs.

Sukhoi Su-25TM (Su-39) all-weather attack aircraft *(Jane's/Paul Jackson)*
                                                    0121066

DIMENSIONS, EXTERNAL:

Wing span	14.52 m (47 ft 7¾ in)
Wing chord: at root	3.00 m (9 ft 10¼ in)
at tip	1.025 m (3 ft 4¼ in)
Wing aspect ratio	7.0
Length overall, incl probe	15.33 m (50 ft 3½ in)
Height overall	5.20 m (17 ft 0¾ in)
Tailplane span	4.58 m (15 ft 0½ in)
Wheel track	2.505 m (8 ft 2¾ in)
Wheelbase	3.58 m (11 ft 9 in)

AREAS:

Wings, gross	30.10 m² (324.0 sq ft)
Ailerons (total)	1.51 m² (16.25 sq ft)
Trailing-edge flaps (total)	4.44 m² (47.79 sq ft)
Leading-edge slats (total)	3.16 m² (34.02 sq ft)
Fin	5.28 m² (56.84 sq ft)
Rudder	0.75 m² (8.07 sq ft)
Tailplane	5.61 m² (60.39 sq ft)
Elevators (total)	1.88 m² (20.24 sq ft)

WEIGHTS AND LOADINGS:

Weight empty	10,600 kg (23,369 lb)
Max combat load	6,000 kg (13,228 lb)
Max fuel: internal	3,840 kg (8,465 lb)
external	3,070 kg (6,768 lb)
T-O weight: planned	21,500 kg (47,399 lb)
overload	19,500 kg (42,990 lb)
normal	16,950 kg (37,368 lb)
Max landing weight	13,200 kg (29,100 lb)
Max wing loading	714.3 kg/m² (146.30 lb/sq ft)
Max power loading	243 kg/kN (2.39 lb/lb st)

PERFORMANCE:

Never-exceed speed (VNE)	M0.82
Max level speed at S/L	
	M0.77 (512 kt; 950 km/h; 590 mph)
Max cruising speed at 200 m (650 ft)	
	378 kt (700 km/h; 435 mph)
Econ cruising speed	350 kt (650 km/h; 404 mph)
Stalling speed	95 kt (175 km/h; 109 mph)
Landing speed	130 kt (240 km/h; 150 mph)
Max rate of climb at S/L	3,480 m (11,415 ft)/min
Rate of climb at S/L, OEI	1,020 m (3,345 ft)/min
Service ceiling	12,000 m (39,380 ft)
Service ceiling, OEI	9,000 m (29,520 ft)
T-O run	650 m (2,135 ft)
Landing run	750 m (2,465 ft)
Combat radius with 2,000 kg (4,410 lb) weapons:	
at low altitude	216 n miles (400 km; 248 miles)
at height	340 n miles (630 km; 391 miles)
Ferry range	1,349 n miles (2,500 km; 1,553 miles)
g limit	+6.5/−2.0

UPDATED

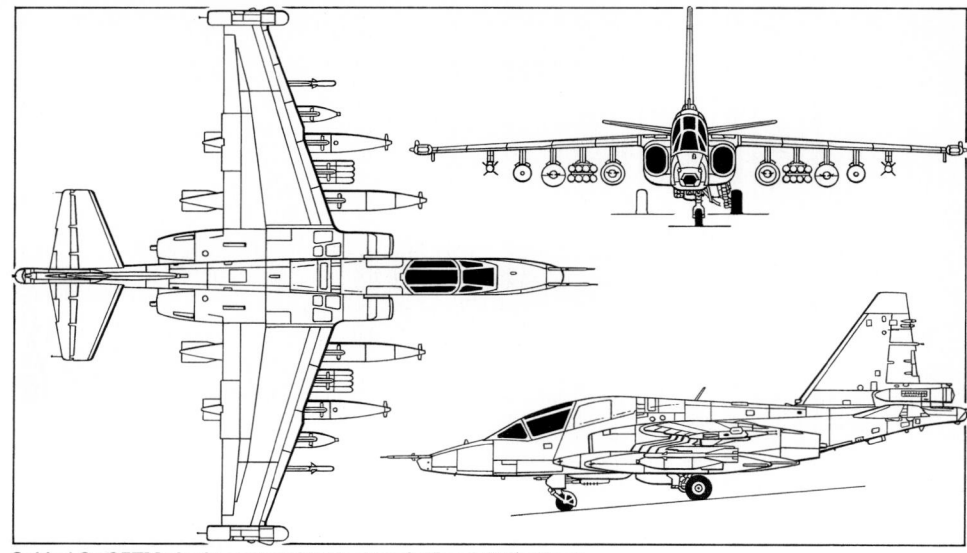

Sukhoi Su-25TM single-seat anti-tank aircraft *(Jane's/Mike Keep)*

First Su-25T, pictured during trials at Akhtubinsk *(Yefim Gordon)*   *NEW*/0525031

SUKHOI Su-27

NATO reporting name: Flanker

TYPE: Air-superiority fighter.

PROGRAMME: Development of long-range heavy fighter to meet PFI (*perspektivnyi frontovoy istrebitel:* advanced frontal fighter) requirement began 1969 under leadership of Pavel Sukhoi; augmented in high-low mix by Mikoyan LPFI (later MiG-29). T10 based on 'integral' layout of unbuilt T4MS. Construction of prototype began in 1974 under Mikhail Simonov's supervision, and it was flown 20 May 1977 by Vladimir Ilyushin. Development undertaken by nine flying **Su-27** 'Flanker-As': four prototypes (T10-1 and T10-2) produced in OKB workshops; T10-3 and T10-4 built by Komsomolsk plant and assembled by OKB; five development (T10-5, -6, -9, -10 and -11) and one static test airframe at Komsomolsk. Individual prototype designations duplicated some competing unbuilt T10 configurations. Prototypes had curved wingtips, rearward-retracting nosewheel and tailfins mounted centrally above engine housings.

Some variation between prototypes; T10-1 originally flew with 'clean' wing; four fences then added, followed by anti-flutter weights on wingtips and fins; inboard fences subsequently removed. T10-3 (first flight 23 August 1979) and T10-4 (first flight 31 October 1979) used AL-31F engines, others had AL-21F-3s. Canted tailfins from T10-3. T10-5, -6, -9, -10, and -11 had eight pylons (two below each wing) and -9 and -11 had long-chord ogival radomes; these five aircraft collectively known as T10-5 subvariant. Original configuration revealed poor controllability with inadequate stability in roll and yaw, and with poor high AoA capability; two pilots lost their lives before major airframe redesign resulted in T-10S production configuration (known internally as T10 junior).

Construction of new model began 1979, with first flight of prototype (T10-7) 20 April 1981, followed by second prototype (T10-12); T10-8 was T10S-0 static test airframe; series production by KnAAPO at Komsomolsk began with T10-15, first flown 2 June 1982; entry into service 22 June 1985 with air defence regiment co-located with Komsomolsk factory airfield; official type acceptance 23 August 1990 (Council of Ministers decree); ground attack role observed in 1991. Following purchase of complete aircraft, China signed licensed production agreement on 6 December 1996, covering further 200. Production continues for export. In late 1999, Sukhoi quoted 567 of Su-27 family in service in eight countries; foresees total of 760 sales by 2005.

Su-27 PROTOTYPES

Batch	Qty	Identity	Remarks
-	1	T10-1	'Flanker-A' AL-21 engine
-	1	T10-2	'Flanker-A' AL-21 engine; lost 7 July 1978
1	1	T10-2	'Flanker-A' AL-31 engine; first KnAAPO-built
1	1	T10-4	'Flanker-A' AL-31 engine
2	1	-	Static test airframe
2	1	T10-5	'Flanker-A' AL-21 engine
2	1	T10-6	'Flanker-A' AL-21 engine
2	1	T10-9	'Flanker-A' AL-21 engine
3	1	T10-10	'Flanker-A' AL-21 engine
3	1	T10-11	'Flanker-A' AL-21 engine
4	2	-	Static test airframes
4	1	T10-7	T10S; first 'Flanker-B'; lost 3 September 1981
4	1	T10-8	Static test airframe
4	1	T10-12	Lost 23 December 1981
4	1	T10-14	Static test airframe
1	1	-	'Flanker-C' static test airframe
1	1	T10U-1	'Flanker-C'; first KnAAPO-built two-seat
2	1	T10U-2	'Flanker-C'; became T10PU-6
2	1	T10U-3	'Flanker-C'

Su-27 PRE-SERIES

Batch	Qty	Identity	Remarks
5	1	T10-15	Later P-42
5	1	T10-17	Lost 1983
5	1	T10-21	Lost 1983
5	1	T10-16	
5	1	T10-20	Later T10KTM
5	1	T10-18	
6	1	T10-22	
6	1	T10-23	
6	1	T10-25	Hook; lost 11 November 1984
7	1	T10-24	Canards; naval trials; lost 20 January 1987
7	1	T10-26	Su-27LL-KS
7	1	T10-27	
1	1	T10U-4	Irkutsk-built; became Su-27LL-OS
1	1	T10U-5	Became Su-27PU (Su-30) T10PU-5, then Su-30MKK
1	1	-	Static test airframe
2	1	T10U-16	Became Su-27UB,-PS
2	1	T10U-6	Became T10PU-6, then Su-30MKI

Ukrainian Air Force Sukhoi Su-27 'Flanker B' (*Jane's/Paul Jackson*)   NEW/0525034

CURRENT VERSIONS: **Su-27** ('Flanker-B'): Single-seat land-based production version for air defence force (PVO); full-span leading-edge flaps, trailing-edge flaperons, 5 per cent increase in wing area, straight leading-edges, new aerofoil, square wingtips, carrying anti-flutter weights which doubled as AAM launchers; wider-spaced, uncanted tailfins outboard of engine housings; flatter canopy of reduced cross-section; extended tailcone instead of flat 'beaver-tail'; forward-retracting nosewheel; first flown (T10-7) 20 April 1981. Standard radar tracks 10 targets simultaneously, engages only one. There was early, but probably erroneous, speculation that PVO aircraft were designated Su-27P, with Su-27S designation applied to Frontal Aviation aircraft. All aircraft can carry Sorbtsya-S active ECM jammer pods on wingtips in place of wingtip launch rails. Su-27S designation little used, but differentiates production (Series) from prototype and preproduction. Four initial production aircraft (T10-15, -17, -18 and -22) used for State Acceptance Tests were part of small initial batch featuring horizontally cropped fin caps.

Detailed description applies to the above version, except where indicated.

'Su-27RV' ('Flanker-B'): Six replacement aircraft for Russian Knights aerobatic team feature GPS and Western-compatible communications equipment.

Su-27SK ('Flanker-B'): Export version of basic Su-27, using air-to-ground capabilities not exploited by Soviet/Russian Su-27s and using same weapons options and downgraded avionics. Armament, totalling up to 4,000 kg (8,818 lb), includes 250 kg and 500 kg bombs, B-8M1 packs of 20 × 80 mm rockets, B-13L packs of five 122 mm rockets, S-250FM 250 mm rockets, KMGU-2 cluster bombs, or podded SPPU-22 30 mm gun with optional downward-deflecting barrel for air-to-ground and air-to-air use. Dimensions, weights and performance generally similar to Su-27 but with reinforced landing gear giving increased (33,000 kg; 72,752 lb) MTOW. Chinese aircraft locally designated **J-11**, though this strictly applies only to locally manufactured examples. First Chinese-assembled aircraft flight tested in December 1998. Some later export aircraft may have upgraded radar and 6,200 kg (13,670 lb) weapon load, or even 8,000 kg (17,637 lb) according to some manufacturers' brochures.

Su-27SM: Mid-life update configuration in parallel to Su-27UBM and Su-30KN. First conversion undertaken in 2001.

Su-27SMK: Single-seat multirole fighter based on the basic Su-27SK, and not the Zhuk-equipped Su-27SM; revealed 1995. Compared with Su-27SK, has 12 instead of 10 hardpoints; 8,000 kg (17,637 lb) weapon load; state-of-the-art nav system, including long-range radio nav, GPS, multichannel comms and latest ECM; larger internal wing fuel tanks and provision for two 2,000 litre (528 US gallon; 440 Imp gallon) underwing tanks; flight refuelling and buddy refuelling capability; retains N001 radar. Most improvements already introduced on other Su-27 series aircraft, enabling virtually full-standard Phase One Su-27SMKs to be offered for immediate delivery. These would be configured for air-to-air roles, with the new wings containing increased internal fuel, provision for underwing tanks, flight refuelling, 12 hardpoints and RVV-AE (R-77; AA-12 'Adder') AAM capability. Su-30KI demonstrator (see below) is regarded as Phase One standard. Phase Two aircraft, with improved avionics and weapon systems for air-to-surface missions, theoretically available later; weapons options to include Kh-29 and Kh-31 ASMs, plus KB-500 LGBs although this seems unlikely without much more extensive upgrade. See accompanying charts for alternative weapon loads. Dimensions as Su-27, except wing span over wingtip R-73E missiles 14.95 m (49 ft 0½ in). Weights and performance: see tabulated data.

Su-27PD: Long-endurance test aircraft with retractable AAR probe, offset IRST and recontoured tail 'sting'; sometimes described as single-seat Su-30, originally Su-27P. Probably stripped of weapons system for research at Gromov Flight Research Institute and demonstration flying. Added satellite navigation; no radar or weapon control system; PC 486 with LCD for flight assessment of

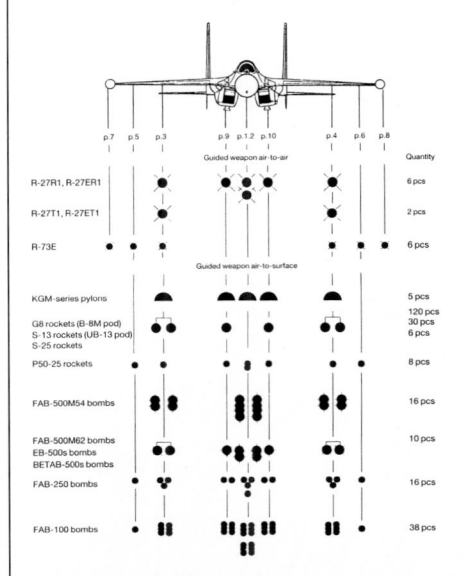

Weapon options for the export Su-27SK

indication formats. At least one aircraft (3720 '598', last known Russian military Su-27); by early 1998, this was marked 'Su-27 Upgrade' and undertaking development flying for **Su-30KI**, later Su-30KM programme (which see). Duties included two flights to North Pole in July and September 1999, testing SRNK satnav system. Another Su-27P (3711 '595') lacked AAR probe.

Su-30KI 'Flanker C': The Su-30KI prototype (4002, wearing the basic Chinese dark grey colour scheme, but with a disruptive camouflage superimposed on the wings, tailplanes and outer surfaces of the tailfins) first flew on 28 June 1998, and was then sent to Chkalov Flight Test Centre at Akhtubinsk, where it was evaluated and used for launch trials of the RVV-AE (AA-12 'Adder') AAM, and for avionics and refuelling probe verification. The Su-30KI was described as a 'single-seat Su-30', and was tailored to meet an Indonesian requirement, before the Asian economic crisis halted that programme. Extent to which the aircraft incorporated the 'extended endurance' features of the Su-30 (apart from refuelling probe, GPS and Western VOR/DME navigation equipment) remains uncertain, however. Su-30KI also used as basis of Sukhoi's latest proposal to upgrade Russia's in-service Su-27s, replacing Su-27SM. Upgrade programme still confusingly referred to by Sukhoi as the Su-30KI, although Russian Air Forces understood to use neither the Su-30 designation nor the KI suffix.

Su-27UB ('Flanker-C'): Tandem two-seat trainer version of 'Flanker-B' with full combat capability (Sukhoi designation T10U); four prototypes (first of which for static testing) built at Komsomolsk; first flown 7 March

1985 (T10U-1). Series manufacture by Irkutsk Aircraft Production Association began 1986; first production UB (T10U-4) flew on 10 September 1986. Instructor in raised rear seat with 6° view forward over the nose; taller fin; length same as 'Flanker-B'; overall height 6.36 m (20 ft 10¼ in), maximum combat load 8,000 kg (17,637 lb). 1,500 kg increase in empty weight, no reduction in internal fuel capacity. Export version is **Su-27UBK**.

Su-27UBM: Russian Air Forces upgrade. First modified Su-27UB (1201 '20') delivered from Irkutsk plant to LII at Zhukovsky 6 March 2001; further seven conversions planned before end of same year; two completed by September 2001; parallel programme to Su-30KN upgrade. Equipment includes new computer, GPS, three 152 × 203 mm (6 × 8 in) MFDs, RVV-AE (AA-12) missiles and ground mapping mode for N001 radar. Later improvements may include radar-absorbent paint and refuelling probe.

'Aircraft 02-01' ('Flanker-C'): Second Su-27UB flying prototype (Komsomolsk 0201) converted as in-flight refuelling systems testbed with retractable probe and provision for centreline 'buddy' pod; first flown 6 June 1987. Later became Su-27PU.

Su-27M: Advanced development of Su-27. Still the official designation for the aircraft known to the OKB as the Su-35. See **Su-35**

Su-27/Su-30 (Su-30KI Minor Modernisation): Proposed upgrade for export Su-27SK, bestowing same multirole standard as Su-30MKK, with inflight refuelling probe, glass cockpit, N-011M radar, GPS, VOR, DME, new mission computer, new navigation system with GPS, expanded EW and provision for new targeting pods. Development undertaken by Su-27PD 3720 '598' and, later, by Su-30KI 4002, first flown 28 June 1998. Phase I of upgrade is known in Russia as Su-30KI, despite the two-seat associations of that designation. In February 1998, Su-27PD 3720 '598' was displayed at the Singapore Air Show marked 'Su-27 Upgrade' and – notwithstanding the fact that it is a single-seat aircraft – stated to be undertaking development work for a variant to be designated Su-30KI. Latter is phase one of Su-27SMK programme, duties of trials aircraft 4002 including satellite-based navigation; upgraded computers; RVV-AE AAMs; Kh-29, Kh-31P and Kh-59M ASMs; KAB-500 and KAB-1500 TV-guided bombs; and multifunction cockpit displays.

Su-27/Su-30 Major Modernisation: Proposed advanced upgrade configuration bringing any 'first-generation' 'Flanker' to virtual Su-27M/Su-30MKI standards, with a new MIL-STD-1553B-based avionics system, N011M radar with a phased-array antenna, and with the option of adding thrust vectoring and/or canard foreplanes. To be undertaken on Russian Su-27s from 2005 onwards; designation **Su-27BM** (*bolshaya modernizatsya:* big modification).

Su-27LL-KS: Su-27 (T10-26; 0702) with axisymmetric afterburner nozzle. Also known as Su-27LLUV(KS) (*upravlyayayemy vektor tyagi; krugloye soplo:* thrust vector control; axisymmetric nozzle) or Su-27-KSI. Evaluated against two-dimensional nozzle testbed, described below. First flew 21 March 1989.

Su-27UB-PS: Su-27UB (Irkutsk 0202; '08') modified in 1990 for thrust-vectoring development, with large two-dimensional box nozzle on port tailpipe. Also known as Su-27LL-PS or Su-27LL-UV(PS) (*upravlyayemy vektor tyagi; ploskoye soplo:* thrust vector control; flat nozzle).

Su-27LMK: CCV (*lyotno-modeliruyushchy kompleks:* flight simulation complex) conversion of production aircraft 2405 with FADEC and side-stick, and fitted with spin-recovery rockets, tested subsequently with axisymmetric nozzle on starboard tailpipe. Trials at Zhukovsky test centre, under direction of TsIAM and Saturn/Lyulka OKB, began 1990. Advanced control testbed, 2001.

Su-27LL-OS: Missile testbed; 1989 conversion of first production Su-27UB 0101 (T10U-4).

Su-27K: Described separately.

Su-27IB: Described separately.

Su-27PU: Two prototypes only; described in 1993-94 *Jane's*. In production as **Su-30** (which see).

P-42: Specially prepared Su-27 (T10-15; first production Su-27; originally flown 2 June 1982); set 31 official world records between 1986 and 1988, including

The unique Su-30KI demonstrator 4002 (*Yefim Gordon*)   NEW/0525035

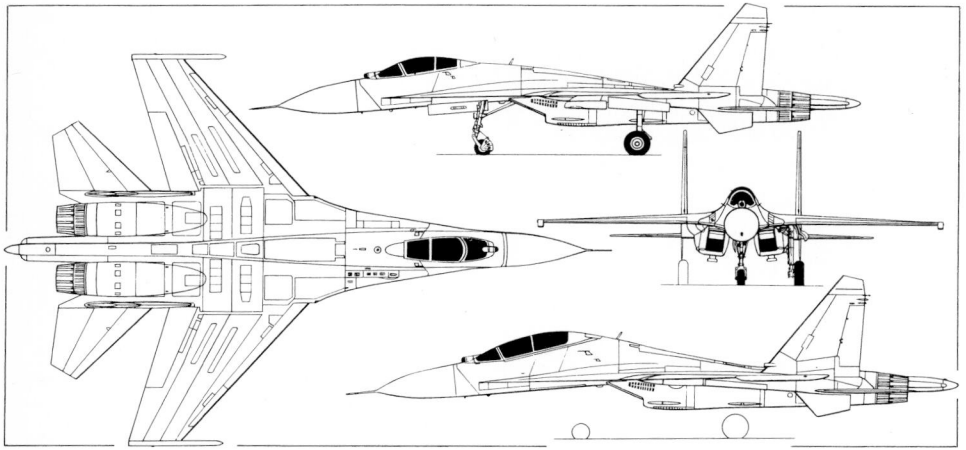

Sukhoi Su-27P, with added side elevation (bottom) of two-seat Su-27B (*Jane's/Dennis Punnett*)

climb to 12,000 m (39,370 ft) in 55.542 seconds, and to 22,250 m (73,000 ft) with 1,000 kg (2,205 lb) payload; some records are in FAI category for STOL aircraft.

CUSTOMERS: Approximately 567 in service by December 1999. About 395 in service with Russian Air Forces (or 340 plus 10 with training units by mid-2000, according to some sources); these reports refer to active inventory; evidence suggests that main production batches for former USSR contained up to 600 Su-27s and 140 Su-27UBs.

China received 26 (24 Su-27SK single-seat, two Su-27UBK two-seat) in 1991-92, followed by 24 more (14 single-seat, 10 two-seat) in 1995-96; original agreements called for licensed manufacture of 200 as J-11, at planned rate of 50 a year, in new purpose-built factory (owned by Shenyang Aircraft Company, which see) from 1998 (when first two shipsets delivered from Komsomolsk; first Chinese-assembled Su-27 rolled out December 1998; only eight delivered by late 2000, following considerable quality control problems, leading to the procurement of 28 extra Russian-built Su-27UBKs, delivered 2000-02, first four on 14 December 2000, second four in December 2000; to Nanjing/Yuxikou air base, where 10 supplied in 2001 and balance of 10 following in 2002. China's licence agreement does not allow exports. The first 50 aircraft are being assembled from Russian-supplied kits of diminishing completeness and the Chinese aircraft will always include at least 30 per cent Komsomolsk content; first 10 kits stated to have been 100 per cent complete but, by 2000, China seeking more in this state to expedite deliveries. No licence has been granted for Chinese production of the AL-31 engine. China's 50 Su-27s were initially based at Wuhan, with the 9th Fighter Regiment,

3rd Air Division though 24 subsequently moved to Liancheng, where 17 were damaged (three beyond repair) in an April 1997 hurricane. Two more were written off in accidents. Production will now cease after 80 J-11s, thereafter switching to the Su-30MKK.

Vietnam has first batch of about six (including one Su-27UB, although further two UBs delivered by airfreight on 1 December 1997 and second pair, representing balance of follow-on order, destroyed in An-124 crash six days later, but replaced by 'Su-30s'); second batch of six now

delivered, bringing in-service total to seven Su-27SKs and five Su-27UBK/Su-30Ks; 24 more on order. Upgrade expected, since Vietnam requires aircraft to use R-77 AAMs, Kh-29, Kh-31 and Kh-59 ASMs. Kazakhstan is receiving 32 from Russia, most recently six in 1997, four in January 1999 and four during 2000 (with 12 then outstanding), as compensation for return of Tu-95s and for alleged environmental and ecological damage. Belarus has 22 or 23 (though these may have been sold-on, perhaps to Angola); Ukraine 66 to 70 and Uzbekistan 30; others reportedly went to Armenia, Azerbaijan and Georgia; Syria has requested 14 (some reports suggest 17 in service others that only four were in service by mid-2000, two at Minkah AB and two at Damascus). Ethiopia received the first of eight second-hand from Russia in late 1998 and Yemen is negotiating for 'a squadron'. In late 1999, there were reports that Angola had taken delivery of eight Su-27s at Catumbela in August, with the balance of seven expected imminently. Pilots had reportedly trained in Belarus, the presumed source of the aircraft, though technical support came from Ukraine. Japan has been reported to be interested in two aircraft for evaluation/ aggressor use, but Sukhoi unwilling to sell less than six. Two Japanese pilots underwent a US$300,000 46-day training programme on the Su-27 in early 1998, however. More recently Malaysia was reportedly 'considering' an Su-27 purchase, while the aircraft was also offered to New Zealand as an alternative to its intended second-hand F-16 buy. Sukhoi offered an Su-30 10-year lease, with targeting pods, PGMs and support, for a quoted cost of NZ$124.8 million (US$68.8 million). On 26 November 1995, the first of two Su-27s was delivered to the USA inside an An-124 for an unknown purpose, believed to be support of military training exercises.

First upgraded Sukhoi Su-27UBM ('Flanker-C') two-seat operational trainer (*Jane's/Paul Jackson*) NEW/0525036

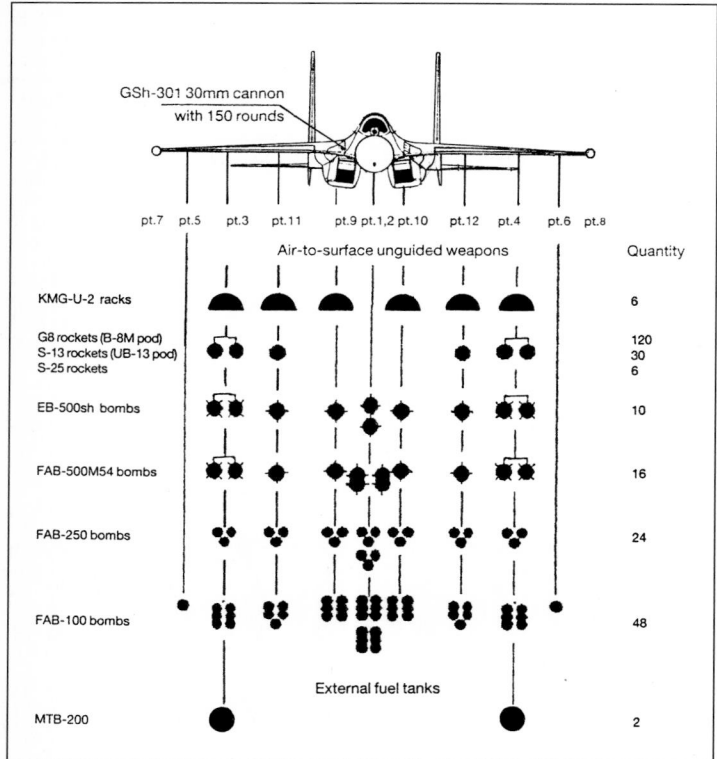

Weapons options for the Su-27SMK multirole fighter: unguided air-to-surface (above) and guided air-to-air and air-to-surface (right). X-29, X-31, X-25 and X-59 have alternative Westernised designations Kh-29, Kh-31, Kh-25 and Kh-59

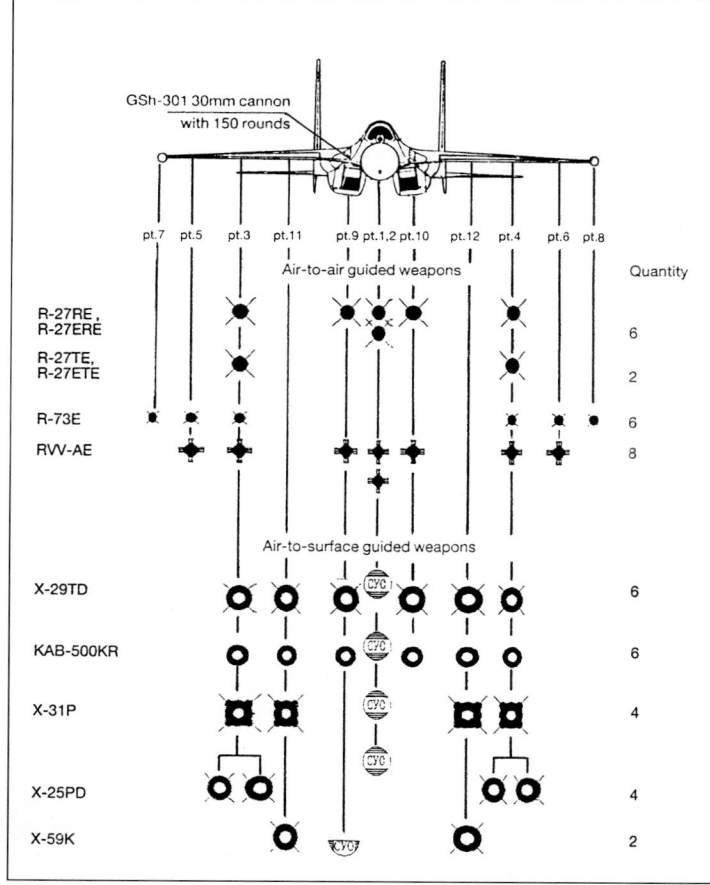

Su-27 CIS (USSR) PRODUCTION

Batch	Qty	Remarks
'Flanker-B' at Komsomolsk		
8	20	Full quantity unconfirmed
9	20	Including T10-30 (0906)
10	20	1003 with sidestick controller
11	20	
12	20	
13	20	Some to Ukraine
14	20	Some to Ukraine
15	20	
16	20	1602 to prototype Su-27M
17	20	Some to 582 IAP
18	20	
19	20	Some to 582 IAP
20	20	Some to 159 and 582 IAP
21	20	Some to 159 IAP
22	20	Some to 159 IAP
23	20	Some to 159 IAP
24	20	2404/T10-41; 2405 Su-27LMK; Some to 159 IAP
25	20	
26	20	Some to 159 and 582 IAP
27	20	Some to 159 and 582 IAP
28	20	
29	20	
30	20	Some to 61 IAP
31	20	Some to 237 GvTsPAT
32	20	
33	20	
34	20	Some to 61 IAP
35	20	
36	20	
37	20	3711 Su-27P; 3720 Su-27PD
'Flanker-C' at Irkutsk		
03	10	Some to Ukraine
04	10	
05	10	
06	10	
07	10	Some to 582 IAP
08	10	Some to 237 GvTsPAT
09	10	Some to 61 and 582 IAP
10	10	
11	10	Some to 582 IAP
12	10	Some to 582 IAP; 1201 first Su-27UBM
13	10	
14	10	
15	10	
16	10	Some to 237 GvTsPAT

Note: Data from unofficial sources. Batches 9 and 10 may have contained fewer aircraft. IAP is Interceptor Air Regiment; GvTsPAT is Guards Centre for Display of Aviation Equipment

Su-27 EXPORT PRODUCTION

Country	Type	Qty	First Aircraft	Remarks
China	Su-27SK	24	'01'	Batch 38 and 39; 1991-92
	Su-27SK	14		Batch 39; 1995-96
	Su-27SK	80		Local assembly in progress
	Su-27UBK	2	'26'	Delivered 1991-92
	Su-27UBK	10		1995-96
	Su-27UBK	28		2000-02
Vietnam	Su-27	5	6001	
		6		
	Su-27UBK	1	8521	
		4		Two lost in airfreight delivery
		2		Replacements for above

Note: China to assemble further 120 Su-30s

DESIGN FEATURES: Developed to replace Yak-28P, Su-15 and Tu-28P/128 interceptors in APVO, for dual-role ground attack/air combat and to escort Su-24 deep-penetration strike missions; basic requirement was effective engagement of F-15 and F-16 and other future aircraft and cruise missiles; exceptional range on internal fuel made flight refuelling unnecessary until Su-24s received probes when more range was required in the escort role; external fuel tanks still not considered necessary; all-swept blended fuselage/mid-wing configuration, with long curved wing leading-edge root extensions, lift-generating fuselage, twin tailfins and widely spaced engines with wedge intakes; rear-hinged doors in intakes hinge up to prevent ingestion of foreign objects during take-off and landing; integrated fire-control system with pilot's helmet-mounted target designator; exceptional high-Alpha performance; basic wing sweepback 42° on leading-edge, 37° at quarter-chord; no dihedral or incidence.

FLYING CONTROLS: Four-channel analogue SDU-27 fly-by-wire, with no mechanical back-up; artificial feel; relaxed longitudinal stability; no ailerons; full-span leading-edge flaps and plain inboard flaperons controlled manually for take-off and landing, computer-controlled in flight; differential/collective tailerons operate in conjunction with flaperons and rudders for pitch and roll control; flight control system limits *g* loading to +9 and normally limits angle of attack to 30 to 35°; angle of attack limiter can be overruled manually for certain flight manoeuvres; large door-type airbrake in top of centre-fuselage.

STRUCTURE: All-metal, with extensive use of aluminium-lithium alloys and titanium but no composites; comparatively conventional three-spar wings; basically circular section semi-monocoque fuselage with load-bearing spine, sloping down sharply aft of canopy; cockpit high-set behind drooped nose; large ogival dielectric nosecone; long rectangular steel blast panel forward of gun on starboard side, above wingroot extension; two-spar fins and horizontal tail surfaces; uncanted vertical surfaces on narrow decks outboard of engine housings; fin extensions beneath decks form parallel, widely separated ventral fins.

LANDING GEAR: Hydraulically retractable tricycle type, made by Hydromash, with single wheel on each unit; KT-156D mainwheels turn 90° while retracting forward into wingroots; hydraulically steerable non-braking KN-27 nosewheel, with mudguard, also retracts forward; mainwheel tyres 1,300×350, pressure 12.25 to 15.70 bar (178 to 227 lb/sq in); nosewheel tyre 680×260, pressure 9.30 bar (135 lb/sq in); hydraulic carbon disc brakes with two-signal anti-skid system; electric brake cooling fan in each mainwheel hub; further brake in nosewheel; brake-chute housed in fuselage tailcone.

POWER PLANT: Two Saturn/Lyulka AL-31F turbofans, each 122.6 kN (27,557 lb st) with afterburning. Large spring-loaded auxiliary air intake louvres in bottom of each three-ramp engine duct near primary wedge intake; two rows of small vertical louvres in each sidewall of wedge, and others in top face; fine grille of titanium hinges up from bottom of each duct to shield engine from foreign object ingestion during take-off and landing.

Fuel in four integral tanks: three in fuselage and one split between each outer wing. Max internal fuel capacity approximately 11,775 litres (3,110 US gallons; 2,590 Imp gallons); normal operational fuel load 6,600 litres (1,744 US gallons; 1,452 Imp gallons). Higher figure represents internal auxiliary tank for missions in which manoeuvrability not important. No provision for external fuel tanks, except in versions where specifically indicated. Pressure or gravity fuelling. Flight refuelling capability optional; Su-27UB operated as buddy tanker during development of system.

ACCOMMODATION: Pilot only, on Zvezda K-36DM Series 2 zero/zero ejection seat, under large rear-hinged transparent blister canopy, with low sill; 14° view downward over the nose.

SYSTEMS: Automatically regulated cockpit air conditioning. Two independent, duplicated, hydraulic systems, pressure 275 bar (4,000 lb/sq in), for actuation of control surfaces, airbrake, landing gear, wheel brakes, air intake ramps and FOD screens. Electrohydraulic (flight control) parts of system quadruplicated. APU in top of rear fuselage for ground and emergency in-flight power. Pneumatic system pressure 210 bar (3,045 lb/sq in) for back-up landing gear extension avionics bay pressurisation and canopy operation and sealing. Electrical supply 27 V DC, 115 V and 200 V 400 Hz AC; two type NKBN-25 Ni/Cd batteries. AC supplied by two integral engine-driven generators with three-phase and single-phase converters. Gaseous oxygen for 4 flight hours.

AVIONICS: Systems integrated by NPO Elektroavtomatika.

Communications: R-800 UHF radio, R-864 HF, intercom and cockpit voice recorder. SO-69 ATC transponder; various IFF fits, according to production batch.

Flight: PNK-10 flight instrumentation and navigation suite encompasses the usual, traditional flight instruments (the IK-VSP altitude and speed data system), SAU-10 autopilot, Ts-050 computer, ARK-19 or ARK-20 ADF, a radio altimeter and A-317 Uron SHORAN. Aircraft capable of ICAO Cat. I autoland.

Radar: RLPK-27 radar sighting system with NIIP N001 Mech (Sword, NATO 'Slot Back') track-while-scan coherent pulse Doppler look-down/shoot-down radar (long-chord twist-cassegrain antenna diameter approximately 1.0 m; 3 ft 4 in) with search range of up to 54 n miles (100 km; 62 miles), tracking range 35 n miles (65 km; 40 miles), in forward hemisphere against MiG-21 size target (ability to track 10 targets and engage two simultaneously in current Su-27SK/UB upgrade configurations) and TsVM-80 computer.

Instrumentation: Integrated fire-control system enables radar, IRST and laser range-finder to be slaved to pilot's helmet-mounted target designator and displayed on wide-angle HUD; autopilot able to restore aircraft to right-side-up level flight from any attitude when 'panic button' depressed.

Mission: Duplex SUV-27 weapons targeting complex integrates radar and OEPS-27 electro-optic sighting system with Model 36Sh OLS-27 IR search/track (IRST) sensor, range 27 n miles (50 km; 31 miles), collimated laser range-finder, range 4.3 n miles (8 km; 5 miles), functioning through common optics in transparent housing forward of windscreen and NSTs-27 Schchel-3U helmet-mounted sight. Beryuza tactical/GCI datalink. Provision for reconnaissance pod on centreline pylon.

Self-defence: SPO-15LM Beryoza 360° radar warning antennas, outboard of each bottom air intake lip and at tail. Gardeniya active ECM jamming system. Three banks of APP-50 chaff/flare dispensers (total 96 cartridges) in bottom of long tailcone extension and top of tailsting. Tailcone widened to provide extra chaff/flare dispensers from batch 18 (mid-1987).

ARMAMENT: One 30 mm Gryazev/Shipunov 9A-4071K GSh-30-1 gun in starboard wingroot extension, with 150 rounds. Up to 10 AAMs in air combat role, on tandem pylons under fuselage between engine ducts, beneath each duct, under each centre-wing and outer-wing, and at each wingtip. Typically, two short-burn semi-active radar homing R-27R (NATO AA-10A 'Alamo-A') in tandem under fuselage; two short-burn IR homing R-27T (AA-10B 'Alamo-B') on centre-wing pylons; and long-burn semi-active radar homing R-27ER (AA-10C 'Alamo-C') or IR R-27ET (AA-10D 'Alamo-D') beneath each engine duct. The four outer pylons carry either R-73A (AA-11 'Archer') or R-60 (AA-8 'Aphid') close-range IR AAMs. R-33 (AA-9 'Amos') AAMs optional in place of AA-10s. Up to eight 500 kg bombs, sixteen 250 kg bombs or four launchers for S-8, S-13 or S-25 rockets.

DIMENSIONS, EXTERNAL (Su-27):

Wing span	14.70 m (48 ft 2¾ in)
Wing aspect ratio	3.5
Length overall, excl nose probe	21.94 m (71 ft 11½ in)
Height overall	5.93 m (19 ft 5½ in)
Fuselage: Max width	1.50 m (4 ft 11 in)
Tailplane span	9.88 m (32 ft 5 in)
Distance between fin tips	4.30 m (14 ft 1¼ in)
Wheel track	4.36 m (14 ft 3¾ in)
Wheelbase	5.88 m (19 ft 3½ in)

AREAS:

Wings, gross	62.04 m² (667.8 sq ft)
Wing leading-edge flaps (total)	4.60 m² (49.51 sq ft)
Flaperons (total)	4.90 m² (52.74 sq ft)
Fins (total)	11.90 m² (128.10 sq ft)
Rudders (total)	3.50 m² (37.67 sq ft)
Horizontal tail surfaces (total)	12.30 m² (132.40 sq ft)

WEIGHTS AND LOADINGS (A: Su-27, B: Su-27UB, C: Su-27SK, D: Su-27SMK):

Operating weight empty: A	16,380 kg (36,110 lb)
B	17,500 kg (38,580 lb)
Max fuel weight: A	9,400 kg (20,723 lb)
Normal T-O weight: A	23,000 kg (50,705 lb)
B	24,140 kg (53,220 lb)
D, with two R-73 and two R-27 missiles	23,700 kg (52,250 lb)
Max T-O weight: A, C, D	33,000 kg (72,750 lb)
B	33,500 kg (73,850 lb)
Max wing loading: A, C, D	531.9 kg/m² (108.94 lb/sq ft)
B	540.0 kg/m² (110.60 lb/sq ft)
Max power loading: A, C, D	135 kg/kN (1.32 lb/lb st)
B	124 kg/kN (1.22 lb/lb st)

PERFORMANCE:

Max level speed: at height:	
A, B	M2.35 (1,350 kt; 2,500 km/h; 1,550 mph)
C	M2.3 (1,319 kt; 2,443 km/h; 1,518 mph)
D	M2.17 (1,241 kt; 2,300 km/h; 1,429 mph)
at S/L:	
A, B	M1.1 (725 kt; 1,345 km/h; 835 mph)
C, D	M1.14 (756 kt; 1,400 km/h; 870 mph)
Stalling speed: A	108 kt (200 km/h; 125 mph)
Acceleration: D:	
from 324 to 594 kt (600 to 1,100 km/h; 373 to 683 mph)	15 s
from 594 to 701 kt (1,100 to 1,300 km/h; 683 to 808 mph)	12 s
Rate of roll: A	approx 270°/s
Service ceiling: A, D	18,000 m (59,060 ft)
B	17,500 m (57,420 ft)
C	18,500 m (60,700 ft)
T-O run: A	450 m (1,475 ft)
B	550 m (1,805 ft)
C, D	650 m (2,135 ft)
Landing run: A, D	620 m (2,035 ft)
B	700 m (2,300 ft)
Combat radius: A, B	810 n miles (1,500 km; 930 miles)
D	840 n miles (1,560 km; 970 miles)
Range with max fuel:	
A, C	1,985 n miles (3,680 km; 2,285 miles)
B	1,620 n miles (3,000 km; 1,865 miles)
Range, D:	
at S/L, internal fuel	745 n miles (1,380 km; 857 miles)

at S/L, with external tanks (dropped when empty)
  894 n miles (1,656 km; 1,029 miles)
at height, internal fuel
  2,046 n miles (3,790 km; 2,355 miles)
at height, with external tanks (dropped when empty)
  2,370 n miles (4,390 km; 2,727 miles)
max, with flight refuelling
  2,807 n miles (5,200 km; 3,231 miles)
g limit (operational): A, B, C, D   +9
                                    UPDATED

SUKHOI Su-27IB (Su-32 and Su-34)

TYPE: Attack fighter.

PROGRAMME: Side-by-side two-seat long-range fighter-bomber (*istrebitel bombardiroshchik*) Su-27 variant intended as tactical strike/attack replacement for Su-24 and Su-25; project designation T10V; redesignated Su-34 by Sukhoi ("to stress father-and-son relationship" to Su-24), but Russian Air Forces retained Su-27IB. Su-32FN/Su-32MF assigned to proposed export versions but, in 2000, all Su-34s were redesignated Su-32.

Conceptual design ordered 21 January 1983, production authorised 19 June 1986. Designed under the direction of Rollan Martirosov. Prototype (T10V-1 '42') built in Sukhoi's own workshops and first flown 13 April 1990; first seen in Tass photograph showing this aircraft approaching (but not landing on) the carrier *Admiral of the Fleet Kuznetsov*; described as deck landing trainer, but no wing folding or deck arrester hook; although with foreplanes and twin nosewheels like Su-27K. Designation Su-27KU quoted, though dedicated side-by-side carrier trainer then officially known as T10KM-2 or Su-27KM-2 for which the new side-by-side cockpit had been designed. Russian unofficial name 'Platypus'; exhibited to Russian Federation and Associated States (CIS) leaders at Machulishche airfield, Minsk, February 1992, with simulated attack armament on 10 external stores pylons (under each intake duct, on each wingtip, three under each wing); Kh-31A/P (AS-17 'Krypton') ASMs under ducts, R-73A (AA-11 'Archer') AAMs on wingtips; a 500 kg laser-guided bomb inboard, TV/laser-guided Kh-29 (AS-14 'Kedge') ASM on central pylon and RVV-AE (R-77; AA-12 'Adder') AAM outboard under each wing.

Production originally planned for Irkutsk, but eventually located at Novosibirsk. First series production aircraft flew 28 December 1994. Four were in assembly at Novosibirsk by early 1997. Twelve were once scheduled for delivery by 1998; intention was to replace all Su-24s by 2005; reconnaissance and electronic warfare versions reportedly under development. In September 2000, following increase in funding, Russian Air Forces C-in-C predicted first Su-32 deliveries in 2004, although by early 2002 this had slipped further to "after 2005". In mid-2001, Sukhoi predicted completion of State testing of Su-32 by end of 2001, using five aircraft based at Akhtubinsk and Zhukovsky.

Side-by-side cockpit to form basis of proposed Su-30-2 long-range interceptor and Su-33UB (Su-27KUB) carrier trainer (which see).

CURRENT VERSIONS (specific): **T10V-1 '42'**: First flying prototype; detailed above. Converted from Su-27UB airframe by Sukhoi OKB workshops, with new nose built at Novosibirsk, and reportedly fitted with Su-33 main landing gear.

T10V-2 '43': Flown 18 December 1993; first aircraft to be built at Novosibirsk; introduced twin mainwheel bogies. Sometimes described as first production Su-34, in that it had Su-35-type four-hardpoint wing panels and larger internal fuel cells, reinforced wing centre-section, new main landing gear and fixed-geometry engine air intakes. However, Sukhoi now identifies this aircraft as Su-32 prototype, designed to meet Russian Air Forces requirement of 1998.

T10V-3: Static test airframe

T10V-4 '44': First flown late 1996. Reported at Leninets radar plant, Pushkino, early 1997; first with full avionics and weapons systems, except EW package. Exhibited at Paris Air Show, June 1997. Also became known as Su-32.

Sukhoi Su-27IB second prototype (*Jane's/Paul Jackson*)   0121003

Fourth flying Su-32 T10V-5, demonstrated at Moscow in August 2001 (*Jane's/Paul Jackson*)   0121004

T10V-5 '45': First Su-27IB with full Leninets mission avionics fit. Sometimes described as first series-produced T10V. First flew 28 December 1994.

T10V-6 '46': First flown January 1998.

T10V-7: First flown 22 December 2000.

T10V-8: By mid-2001, Novosibirsk production stated to be eight, including two static test airframes.

CURRENT VERSIONS (general): **Su-27R**: Proposed version to replace Su-24MR and MiG-25RB in tactical reconnaissance roles. BKR (*bortovoi kompleks razvedki:* onboard reconnaissance complex) suite expected to include nose-mounted Pika SLAR and ESM, electro-optical, laser and IRLS reconnaissance equipment.

Su-27IBP: Proposed tactical jammer to replace Yak-28PP and Su-24MP.

'Su-27IB Interceptor': Proposed ultra-long endurance combat air patrol variant. OKB's internal designation not known. May be confused reference to Su-33UB-based Su-30K-2 project.

Su-32: Initially export version; applied also to domestic version from 2000.

Su-32FN/MF: Preseries Su-32FN (T10V-4 '45') first flown 28 December 1994, exhibited at 1995 Paris Air Show; then stated to be in production to replace Su-24s of Russian Naval Aviation; programme reportedly suspended early 1997 before a fully equipped true prototype could fly. Su-32MF designation first appeared in 1999 to describe a 'multifunction', export version. Some French sources suggested that the type had received a new ASCC reporting name 'Fallback', by July 2000.

This version designed to attack hostile submarines and surface vessels by day and night in all weathers, although official drawing shows slightly different shape to nose compared with land attack version; was intended for parallel manufacture at Novosibirsk. Probably common to both types are Su-32MF's active artificial intelligence system to support pilot in critical situations; active gust alleviation smooth-flight system to damp turbulence in low-level flight at high speeds; liquid-crystal EFIS with seven CRTs; and Sorbtsya active ECM jamming pods on wingtips. Planned specialised equipment includes Leninets

Sea Dragon avionics suite, with 'Sea Snake' coherent maritime search radar, a ventral sonobuoy pod containing 72 buoys of various types, MAD, IIR, IRTV system and laser range-finder. By early 2002, NIIP offering Osa electronically scanned radar for rear protection and precision weapon targeting because of reported development delays with Leninets suite.

Su-32 'Escort Jammer': Proposed in 2001 by Knirti Institute; four podded jammers (including two at wingtips) and two anti-radiation missiles.

Su-34: Initially domestic version; discontinued in 2000.

Details generally as for single-/tandem-seat Su-27, except those below.

DESIGN FEATURES: One third heavier empty weight, with 50 per cent increase in MTOW, 30 per cent increase in internal capacity, 10 per cent increase in mid-section. Completely new and wider front fuselage built as titanium armoured tub, 17 mm ($^{11}/_{16}$ in) thick; armour adds 1,480 kg (3,262 lb); new EFIS cockpit containing two seats side by side; side-by-side arrangement avoids some duplication of controls and instruments, while promoting better crew co-operation. New avionics suite integrated by Ramenskoye Instrument-making Design Bureau; wing extensions taken forward as chines to blend with dielectric nose housing nav/attack and terrain-following/avoidance radar; deep fairing behind wide humped canopy; small foreplanes; louvres on engine air intake ducts reconfigured; new landing gear; broader chord and thicker tailfins, containing fuel; no ventral fins; and a longer, larger diameter tailcone. This has been raised and now extends as a spine above the rear fuselage to blend into the rear of the cockpit fairing. It houses at its tip a rearward-facing radar to detect aircraft approaching from the rear.

LANDING GEAR: Retractable tricycle type; strengthened twin nosewheel unit with KN-27 wheels, tyre size 680×260, farther forward than on Su-27 and retracting rearward; main units have small tandem KT-206 wheels with tyres size 950×400, carried on links fore and aft of oleo. New down-lock fairings. Twin cruciform brake-chutes repositioned in spine to rear of spine/fairing juncture.

POWER PLANT: Two Saturn/Lyulka AL-31F turbofans; each 74.5 kN (16,755 lb st) dry and 122.6 kN (27,557 lb st) with afterburning. Later, two AL-31FM or AL-35F turbofans, each 125.5 to 137.3 kN (28,220 to 30,865 lb st) with afterburning. Production version to use 175 kN (39,240 lb st) AL-41F with FADEC and TVC, according to some sources. Additional fuel in tailfins and increased capacity No. 1 tank raising total to 12,100 kg (26,676 lb) plus provision for three external tanks totalling 7,200 kg (15,873 lb). Retractable flight refuelling probe beneath port windscreen.

ACCOMMODATION: Two crew side by side on modified K-36DM zero/zero ejection seats with built-in massage function. Access to cockpit via built-in extending ladder to door in nosewheel bay; area protected with 17 mm (⅔ in) thick titanium armour; lavatory and galley with air-stove inside deep fuselage section aft of cockpit.

AVIONICS: *Radar:* Leninets B004 multifunction phased-array radar with high resolution; rearward-facing radar in tailcone.

Instrumentation: Colour CRT, multifunction displays and helmet-mounted sight for pilot and navigator.

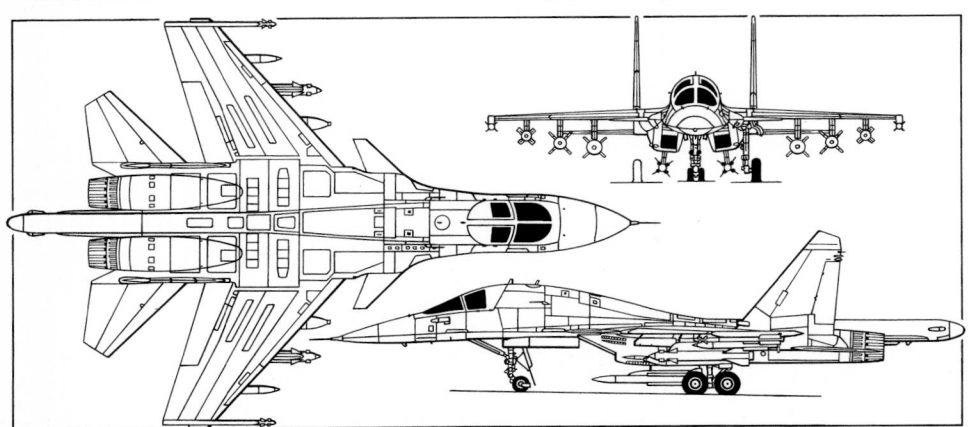
Three-view drawing of Sukhoi Su-27IB twin-turbofan theatre bomber (*Jane's/Mike Keep*)

Mission: Built-in UOMZ EO IRST sighting system with TV and laser channels, optimised for air-to-ground use. Separate Geofizika podded thermal imaging system planned. New Argon main computer. Sorbtsya active ECM jamming pods under test on Su-27IB prototype 1995.

Self-defence: TsNIRTI electronic warfare system.

ARMAMENT: One 30 mm GSh-301 gun with 150 rounds. Twelve pylons for high-precision self-homing and guided ASMs, comprising Kh-59ME Ovod, Kh-31P/Kh-31A(P) (AS-17 'Krypton'), Kh-29T/TE/L (AS-14 'Kedge') and Kh-41/3M80 Moskit; and KAB-500 and KAB-1500 laser-guided bombs with ranges of 0 to 135 n miles (250 km; 155 miles); R-27 (AA-10 'Alamo'), R-73 (AA-11 'Archer') and RVV-AE (R-77; AA-12 'Adder') AAMs.

DIMENSIONS, EXTERNAL:

Wing span	14.70 m (48 ft 2¾ in)
Wing aspect ratio	3.5
Foreplane span	6.40 m (21 ft 0 in)
Length (without probe)	23.335 m (76 ft 6¾ in)
Height overall	6.50 m (21 ft 4 in)
Wheel track	4.40 m (14 ft 5¼ in)
Wheelbase	6.60 m (21 ft 7¾ in)

AREAS:

Wings, gross	62.00 m² (667.4 sq ft)

WEIGHTS AND LOADINGS (Su-32):

Max external stores	8,000 kg (17,637 lb)
T-O weight: normal	38,240 kg (84,304 lb)
max	44,350 kg (97,774 lb)

PERFORMANCE:

Max speed:		
at height	M1.8 (1,025 kt; 1,900 km/h; 1,180 mph)	
at S/L	M1.14 (756 kt; 1,400 km/h; 870 mph)	
Service ceiling	15,000 m (49,220 ft)	
Combat radius (internal fuel):		
hi-hi-hi	594 n miles (1,100 km; 683 miles)	
lo-lo-lo	324 n miles (600 km; 372 miles)	
Range with max internal fuel		
	2,159 n miles (4,000 km; 2,485 miles)	

UPDATED

SUKHOI Su-30 (Su-27PU)

TYPE: Air superiority fighter.

PROGRAMME: Design started 1986; proof-of-concept T10PU (T10U-2 'Aircraft 01-02'; see Su-27 entry) first flew 6 June 1987; construction of two prototypes (T10PU-5 '05' and T10PU-6 '06', converted from Su-27UBs T10U-5 and 0201/T10U-6) began at Irkutsk in 1987, as Su-27PU; first flown 31 December 1989; prototype flew 7,252 n miles (13,440 km; 8,351 miles) in 15 hours 42 minutes non-stop during round trips Moscow-Novaya Zemlia-Moscow and Moscow-Komsomolsk-Moscow. First pre-series Su-27PU flew at Irkutsk on 14 April 1992; intitial two aircraft (27596 and 27597 c/ns 0101 and 0102), without military equipment, delivered to 'Test Pilots' aerobatic team at Zhukovsky flight test centre, possibly as Su-27PUDs. By 1999, 27597 was flight-testing MFI-68 152 × 203 mm (6 × 8 in) LCDs for Russian 'Flanker' upgrade programmes while 27596 had been used in support of the Su-30MK programme, renumbered '603'.

CURRENT VERSIONS: **Su-30** (Sukhoi T10PU, *Izdelie* 10-4PU): Unofficial OKB designation for basic two-seat long-range interceptor for Russian Air Forces (to which it is still the Su-27PU); deliveries under way by 1996 to 54 Interceptor Air Regiment at Savostleyka advanced training base, though production very limited, and unit relies heavily on Su-27 and Su-27UB. Apparently, five Su-30s are in frontline use (Red 50, 51, 52, 53 and 54). Designed for mission of 10 hours or more with two in-flight refuellings; systems proved for extended duration sorties, including group missions with four Su-27s; only Su-30 would operate radar, enabling it to assign targets to Su-27s by radio datalink; can carry bombs and rockets but not guided air-to-surface weapons. Su-27UB training capability retained. Canards and thrust vectoring to be optional (see Su-30MKI). Export designation **Su-30K** (T10-4PK).

Russian Air Forces upgrade for the Su-30 was developed in 'Project 302' *(Yefim Gordon)* NEW/0525041

Su-30 PROTOTYPES AND PRODUCTION

Batch	Qty	Identity	Remarks
Russia			
-	(1)	T10PU-5 '05'	Converted from T10U-5; became Su-30MKK
-	(1)	T10PU-6 '06'	Converted from T10U-2; became Su30MKI
1	1	'596'	Became '603' for Su-30MK programme
1	1	'597'	Testbed at Zhukovsky (LII)
2/3	5	'50' – '54'	54 IAP PVO, Savostleyka
3	1	'302'	Upgrade prototype (Su-30KN)
India			
PU	8	SB001-008	Delivered March-April 1997
K	10	SB009-018	Delivered October 1999

Situation confused by tendency to describe all current Irkutsk-built two-seaters as Su-30s, even standard Su-27UB trainers and by emergence of upgrade using MiG-29SMT cockpit, using the same Su-30K designation. This Su-30K was reported to have been tested at Akhtubinsk by June 2000 but may be the same as the Su-30KM described below. Reported in early 2002 that 'Russian Knights' aerobatic team to convert to Su-30s from Su-27s.

Su-30KI: Single-seat configuration for Indonesia, subsequently offered more widely as an upgrade. Described in main Su-27 entry.

Su-30I: One foreplane-equipped Su-27PU prototype only; served as Su-30MKI prototype (which see); believed offered as upgrade for Su-27 and as naval trainer, but neither taken up.

Su-30K-2: Variant of Su-33UB and described in that entry.

Su-30KN: An undelivered production Su-30 (0302 '302') first flew in early March 1999 as the testbed (also referred to as Su-30K and Su-30KM) for a Russian Air Forces upgrade of UB Su-27s and Su-30s. Prime purpose of 'Project 302' was to convert fighter into multirole attack aircraft by adding terrain-mapping and moving target indication to the N001 radar. This achieved by adding new bypass circuit (*obvodnoy kanal*, abbreviated to *Oko* – 'eye'). Cockpit initially unchanged, apart from MFI-55 127 × 127 mm (5 × 5 in) MFDs, an SUV-30K weapons control system comprising a new MVK computer added to existing SUV-27 system (permitting new types of AAM and ASM to be carried) and an A737 GPS; joint venture is undertaken by Sukhoi, Irkutsk (IAPO) and Russkaya Avionika and has high commonality with MiG-29SMT upgrade. Prototype tested at Air Forces Research Institute from mid-1999. State certification awarded 9 November 2001. Added weapons are Kh-29T short-range, or two Kh-59ME long-range TV-guided ASMs; up to six KAB-500KR bombs; four Kh-31P ARMs; and six Kh-31P/A anti-ship missiles. By late 2000, the Su-30KM

designation had been replaced by Su-30KN and potential maritime capabilities were being stressed, including possible future compatibility with Kh-59, Yakhont and Alfa ASMs. Up to November 2001, upgrades comprised prototype '302' and three air force aircraft ('51' being the first), all used for trials and modernised at IAPO's expense; order signed by Russian Air Forces in October 2001 for eight upgrades in 2002, 10 in 2003 and 12 in 2004. Su-27UBM and Su-27SM are parallel programmes. 12 Su-30Ks sought by Vietnam believed to be to this standard with Kh-29, Kh-31 and Kh-59 ASMs, KAB-500 PGMs and R-77 AAMs.

Su-30KNM: Subsequent modifications to 0302 involve larger (152 × 203 mm; 6 × 8 in), MFI-68 screens (three for pilot and four for WSO), Pero ('Feather') phased-array radar and equivalent of MIL-STD-1553B databus. Project began in 2002. Upgrade of air force aircraft could begin 2005.

Su-30M (Sukhoi T-10PM): Multirole version; described separately.

CUSTOMERS: See table.

COSTS: Chinese order for 40 Su-30MKKs valued at US$1.5 billion (2000).

DESIGN FEATURES: Development of Su-27/27UB, with latter's tandem seating and new avionics, and without Su-35's advanced radar, foreplanes (in basic version), advanced control system and new power plant. Designed for effective engagement of fighters at long distances from

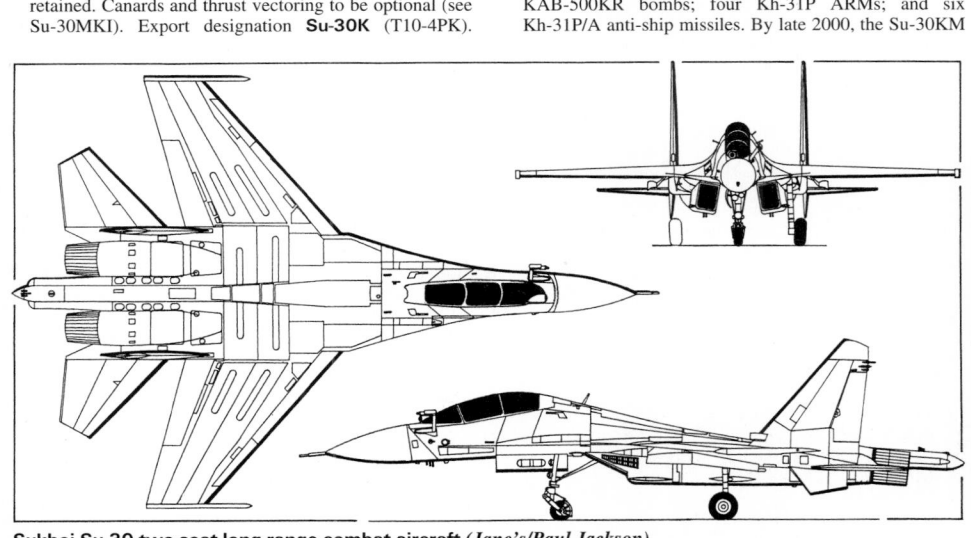

Sukhoi Su-30 two-seat long-range combat aircraft *(Jane's/Paul Jackson)*

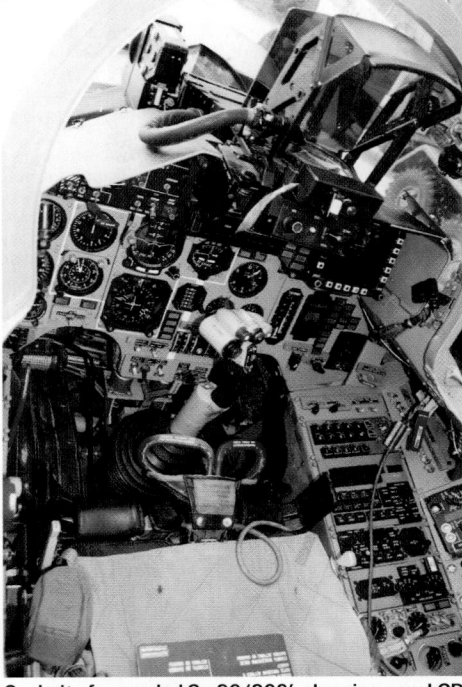

Cockpit of upgraded Su-30 '302', showing new LCD at upper right of panel *(Yefim Gordon)* 0121094

base, and to destroy bombers and intercept cruise missiles. Integral configuration similar to Su-27UB, with unstable aerodynamic characteristics. Automatic control system standard.

FLYING CONTROLS: As Su-27UB.

STRUCTURE: As Su-27UB.

LANDING GEAR: As Su-27UB.

POWER PLANT: As Su-27UB, but flight refuelling probe and buddy refuelling capability standard.

ACCOMMODATION: Two crew in tandem in identical cockpits, on K-36DM zero/zero ejection seats, with rear seat raised.

SYSTEMS: As Su-27UB, except gaseous oxygen for 10 hours' flight.

AVIONICS: *Radar:* NIIP N001 Myech ('Slot Back') coherent pulse Doppler look-down/shoot-down radar offered, detection range up to 54 n miles (100 km; 62 miles), tracking range 35 n miles (65 km; 40 miles); ability to track 10 targets and engage two simultaneously; probably not available on current in-service aircraft.

Flight: New navigation system based on GPS, Loran and Omega.

Instrumentation: Integrated fire-control system enables radar, IRST and laser range-finder to be slaved to pilot's helmet-mounted target designator and displayed on wide-angle HUD.

Mission: Provision for fitting foreign-made airborne and weapon systems at customer's request.

Self-defence: SPO-15LM Beryoza 360° radar warning system; chaff/flare dispensers.

ARMAMENT: One 30 mm GSh-301 gun, with 150 rounds; 12 hardpoints for up to six R-27R1E and R-27T1E (AA-10 'Alamo') radar homing and IR long-range AAMs, and six R-73E (AA-11 'Archer') IR close-range AAMs; alternative RVV-AE (R-77; AA-12 'Adder') AAMs; unguided bombs or rockets as Su-27; reconnaissance or EW pods.

DIMENSIONS, EXTERNAL: As Su-27UB

WEIGHTS AND LOADINGS:

Fuel weight: normal	5,090 kg (11,222 lb)
max	9,400 kg (20,723 lb)
Max combat load	8,000 kg (17,635 lb)
Max external stores	8,000 kg (17,635 lb)
Normal T-O weight	24,550 kg (54,123 lb)
Max T-O weight	33,000 kg (72,752 lb)

PERFORMANCE:

Max level speed:	
at height	M2.35 (1,350 kt; 2,500 km/h; 1,550 mph)
at S/L	M1.14 (756 kt; 1,400 km/h; 870 mph)
Service ceiling	17,500 m (57,420 ft)
T-O run	550 m (1,805 ft)
Landing run	700 m (2,300 ft)
Combat range: with max internal fuel	
	1,619 n miles (3,000 km; 1,865 miles)
with one in-flight refuelling	
	2,805 n miles (5,200 km; 3,230 miles)
g limit	+8.5

UPDATED

SUKHOI Su-30M

TYPE: Multirole fighter.

PROGRAMME: Design started 1991; demonstration and development work by Su-27UB 0806 '321' and 0403 '56'; 'Blue 56' first flew in Su-30M configuration on 14 April 1992, but it was not then canard-equipped and did not have full avionics package. Conversion of first true prototype ('06') began 1993, using 0201/T10PU-6; 0101 '603' (formerly first prototype Su-27PU-5) first demonstrated at Berlin Air Show 1994; thrust vectoring and canards under development as options 1997. Canards and AL-37PP engines first flown on '56' 1 July 1997, this regarded by Sukhoi as first flight of Su-30MK; second prototype (06) flew 23 April 1998. In production for India.

Su-30M designation initially associated with canard fitment. However, this now appears to identify multirole aircraft with upgraded airframe capable of 38,800 kg (85,539 lb) MTOW, irrespective of aerodynamic configuration. Chinese Su-30MKKs are described by Sukhoi as Su-30MK family members, despite having twin nosewheels and square-topped fins previously regarded as Su-35 features.

CURRENT VERSIONS: **Su-30M:** Basic version.

Su-30MK: Irkutsk-built. As Su-30M, for export.

Su-30MK: Designation re-used for advanced two-seater derived from Su-27SK by KnAAPO plant and combining two-seat Su-30 concept with avionics, canards and thrust vectoring of Su-37. Design, under A F Barkovsky, began 1994. Known internally (by KnAAPO) as Su-35UB (T10UBM) and Su-37UB. First and second true Su-30MKs, 01 (produced through the retrofit of canards to '56') and 06 (converted from T10PU-6 in Sukhoi OKB's own workshop) first flown on 1 July 1997 and 23 April 1998, respectively. Equipped with foreplanes and AL-37FP thrust-vectoring engines; demonstrated to Indian officials at Zhukovsky, 15 June 1998 as Su-30MK-1 and Su-30MK-6, respectively. Su-30MK-1 had the new twin-wheel nosegear, which may become a feature of the production MKI, but was lost while displaying at Paris Air Show on 12 June 1999. Additionally, 01 was exhibited at Aero India, December 1998. AL-37FP power plant, as specified for India, extends length by 40 cm (15¾ in) and

Third development Sukhoi Su-30MKI *(Yefim Gordon)*   **NEW**/0525042

incurs weight penalty of 110 kg (243 lb), engine life remaining unchanged at 5,000 hours (1,000 hours TBO). Nozzle movement is 15° up or down. Flight control system helps pilot to set power and thrust vector for each engine, according to required manoeuvre. Indian radar choice will be between improved version of Phazotron N010 (Zhuk-27) and NIIP N011M multimode, dual-frequency radar with electronically scanned antenna. Expected to be known as Su-30MKR if produced (at Irkutsk) for Russian Air Forces.

Su-30MKI: Version for India in four configurations, initially referred to as Su-30MKI, MKII, MKIII and MKIV. Indian contract signed 30 November 1996. First eight delivered March 1997 to basic Su-30PU (Su-30K or even Su-37UB) standard, with AL-31F engines; eight for 1998 delivery were expected to have French Sextant avionics including VEH 3000 HUDs, high-resolution colour LCD MFDs, a new flight data recorder, a Totem ring laser gyro dual INS with embedded GPS, Israeli EW equipment, a new UOMZ OLS-30 electro-optic targeting system and rearward-facing radar in tailcone; 12 originally to have been delivered 1999 were to add canards, as on Su-37; final 12 (originally scheduled for delivery in 2000) were to have AL-37FP engines with single-axis thrust-vectoring nozzles inclined outwards 32° from the centreline for improved yaw control, especially in single-engined case. AL-37PP claimed to offer 3-D thrust vectoring, with nozzle actuation via the fuel, and not the hydraulic, system.

Completion and delivery of balance of 32 was repeatedly delayed; decision taken that all these would be completed to final standard before delivery. (In interim, India contracted on 18 December 1998 for 10 standard Su-30Ks which had been cancelled by Indonesia; all had been delivered by October 1999.) First prototype full-specification Su-30MKI flew at Irkutsk on 26 November 2000; Irkutsk built four prototypes, last of which handed over to Sukhoi design bureau on 11 August 2001.

First production aircraft flew 28 December 2001. Under renegotiated contract, IAF to receive six full-specification aircraft by 2002 and balance of 26 in batches beginning 18 to 24 months later; thereafter eight 1997-delivery aircraft will be raised to full standard. First two departed Irkutsk inside An-124 transport 22 June 2002. However, 'full-specification' still being received in three standards, first 10 being Stage I. Stages II and III, of which quantities undecided, to introduce additional weapons and upgraded flight control system. Licensed production of 140 Su-30MKIs by HAL was agreed in September 2000, followed by contract signature in Irkutsk on 28 December 2000. Sometimes referred to by KnAAPO as Su-35UB.

Su-30MKK ('Flanker-C'): Second K stands for *Kitaya*, or China. Two-seat, multirole version, with an N001VE radar (with expanded air-to-ground capabilities, including mapping); 'glass cockpit' with two 178 × 127 mm (7 × 5 in) MFI-9 colour LCD MFDs in

front, plus single MFI-9 and 204 × 152 mm (8 × 6 in) MFI-10 in rear; ILS-31 HUD; A737 GPS; expanded EW capability; provision for various new TV- and EO-based targeting pods.

First Su-27PU (T10PU-5) was refurbished and rebuilt to serve as an Su-30MKK development aircraft, first flying in its new guise on 9 March 1999. First KnAAPO-built prototype ('501') flew 19 May 1999 (sometimes reported as 20 February 1999), with second ('502', in basic Chinese camouflage scheme but marked simply as Su-30MK, not MKK) following later in 1999. '501' and '502' representative of planned production configuration, with tall, flat-topped Su-35-type tailfins, retractable in-flight refuelling probes and (according to some sources) Su-35-type radomes. Further pair ('503' and '504') built by mid-2001.

Chinese production of the 'Flanker' will switch from the J-11 to the Su-30MKK after 80 aircraft. However, China reportedly ordered 45 KnAAPO-built Su-30MKKs and placed supplementary order (initially quoted as 24, but later stated to be 40) in June 2001. First batch of 10 Su-30MKKs left Russia on delivery to China on 20 December 2000; nine followed in March/April 2001 and 10 more on 21 August. Later Chinese aircraft, known as Series III, will switch to Zhuk-MS radars. Russian weapon deliveries began in January 2001 with Kh-59ME (AS-18 'Kazoo'), Kh-29T (AS-14 'Kedge') and Kh-31P (AS-17 'Krypton') ASMs and KAB-500Kr guided bombs. Early Su-30MKKs variously reported with 'Three Swords' Air Regiment at Yuxikou, near Nanjing or at Wuhu, Anhui; those delivered in August 2001 were supplied to Cangzhou, Hebei.

CUSTOMERS: Indian Air Force initial US$1.8 billion order for 40 signed 30 November 1996; deliveries from Irkutsk began to No. 24 'Hunting Hawks' Squadron at Pune in March 1997; declared operational 11 June 1997; 10 more ordered September 1998. Option taken up on licensed production of 140 more by HAL, India. On 29 August 1997, Indonesia ordered for eight 'single-seat Su-30s' and four two-seat, but this was cancelled on 9 January 1998. Interest reportedly renewed in 2001.

COSTS: Indian aircraft quoted as US$20 million each (flyaway, 1998), with US$8 million extra per aircraft to integrate Indian-specific systems and avionics.

Description refers to two-seat Su-30MK which generally as for Su-30, except as follows:

DESIGN FEATURES: Improvement on combat capabilities of Su-30 by compatibility with high-precision guided air-to-surface weapons with standoff launch range up to 65 n miles (120 km; 75 miles), in addition to Su-30's ability to engage two airborne targets simultaneously.

POWER PLANT: Two Saturn/Lyulka AL-35F turbofans, each 123 kN (27,558 lb st).

AVIONICS: In addition to standard Su-30 systems, Su-30M has more accurate navigation system, a TV command guidance system, a guidance system for anti-radiation missiles, a

Su-30M PROTOTYPES AND PRODUCTION

Batch	Qty	Identity	Remarks
-	(1)	'56'; later '01'	Lost 12 June 1999
-	(1)	06	Previously T10PU-6
China			
MKK	45*	501 *et seq*	Ordered 1999; first four for trials in Russia
MKK	24†		Ordered June 2001
India			
MKI	4	'02' – '05'	Development, built 2000-01
MKI	28		Production at Irkutsk 2001-03
MKI	140		Production in India 2004-17

* Variously quoted as 38, 40 and 45

† Second batch may be increased to 40. A third batch, also reportedly of 40, was being discussed in mid-2002.

Sukhoi Su-30MKK development aircraft for Chinese version *(Yefim Gordon)* NEW/0525043

larger monochrome TV display system in rear cockpit for ASM guidance, and ability to carry one or two pods, typically for laser designation or ARM guidance in association with Pastel RWR and APK-9 datalink. Western avionics, guidance pods and weapons can be fitted optionally. Sextant Avionique package for Indian aircraft includes VEH3000 or Elop HUD, Totem or Sigma 9SN/MF INS/GPS and liquid-crystal multifunction displays (six 127 × 127 mm; 5 × 5 in MFD 55 and one 152 × 152 mm; 6 × 6 in MFD 66 per aircraft).

ARMAMENT: One 30 mm GSh-301 gun, with 150 rounds; 12 external stations for 8,000 kg (17,635 lb) of stores, including FAB-250, FAB-500, OFAB-250-270, OFAB-100-120 and guided KAB-500KR and KAB-1500KR bombs; B-8M-1 (20 × 80 mm), B-13L (5 × 130 mm) and O-25 (single 266 mm) rocket packs; up to six R-27ER (AA-10C 'Alamo-C'), R-27ET (AA-10D 'Alamo-D') or RVV-AE (R-77; AA-12 'Adder') medium-range AAMs; or two R-27ETs and six R-73E (AA-11 'Archer') IR homing close-range AAMs; and a variety of air-to-surface weapons such as four ARMs, six guided bombs or short-range missiles with TV homing, six laser homing short-range missiles, or two long-range missiles with TV command guidance; these include Kh-29L/T (AS-14 'Kedge'), Kh-31A/P (AS-17 'Krypton') and Kh-59M; (AS-18 'Kazoo') with APK-9 pod.

DIMENSIONS, EXTERNAL: As Su-27 except:
Height overall   6.355 m (20 ft 10¼ in)

WEIGHTS AND LOADINGS (Su-30MK):
Weight empty   17,700 kg (39,022 lb)
Max external weapon load   8,000 kg (17,637 lb)
Max internal fuel weight   9,640 kg (21,253 lb)
T-O weight: normal
24,960 kg (55,027 lb) max   34,500 kg (76,059 lb)
overload   38,800 kg (85,539 lb)

PERFORMANCE (Su-30MK):
Max level speed:
at height   (1,144 kt; 2,120 km/h; 1,317 mph)
at S/L   M1.14 (729 kt; 1,350 km/h; 839 mph)
Max rate of climb at S/L   13,800 m (45,275 ft)/min
Service ceiling   17,300 m (56,760 ft)
T-O run, normal weight   550 m (1,805 ft)
Landing run with parachute   750 m (2,460 ft)

Combat range:
internal fuel   1,620 n miles (3,000 km; 1,865 miles)
with one in-flight refuelling
                 2,805 n miles (5,200 km; 3,230 miles)
g limit   +9
UPDATED

SUKHOI Su-32FN and MF (Su-27IB)
Described under Su-27IB entry.

SUKHOI Su-33 (Su-27K)
TYPE: Multirole fighter.
PROGRAMME: Development began 1976; based on production Su-27, but embodying folding wings, other features for shipboard operation, and movable foreplanes (last-mentioned first flown on T10-24 in May 1985; the aircraft also used as CCV/aerodynamic research vehicle and for ski-jump take-off trials. Navalised development Su-27 (T10-25), with arrester hook, flew 1984; extensively used for arrested landing trials and ski-ramp take-offs, with T10-3 and T10-24 (first ramp take-off was 22 August 1992 by T10-3; T10-25 followed on 25 September 1985, having made first arrested landing on 1 September 1985). 'Aircraft 02-01' (see Su-27 entry) was fitted with arrester hook and 'Resistor' ACLS and extensively used in support of Su-27K programme.
Full-scale development of definitive Su-27K began in 1984 as T10K, under Konstantin Marbashev. Full-scale

(non-flying) T10-20KTM mockup with crude (unpowered) wing and tailplane folding produced by conversion of grounded fifth production T10S. Prototype (T10K-1 '37') flew 17 August 1987; lacked folding wing, double-slotted flaps and strengthened landing gear; retrofitted with folding wing and reflown on 25 August 1988, but written-off on 28 September 1988; second (T10K-2 '39') made initial flight on 22 December 1987; first conventional (non-V/STOL) landing by Soviet aircraft on ship, the *Admiral of the Fleet Kuznetsov* (then *Tbilisi*), 1 November 1989 (26 minutes before MiG-29K); first landing by service pilot followed on 26 September 1991.

Production at Komsomolsk began with static test airframe, and followed by seven pre-series aircraft: T10K-3 (first flight 17 February 1990), T10K-4 '59', T10K-5 '69', T10K-6 '79', T10K-7, T10K-8 (lost 11 July 1991) and T10K-9 '109'. Between 16 and 20, probably 18 (of which 16 positively identified), delivered of 24-aircraft production batch (plus trials aircraft) and shore-based with 279th KIAP (formerly 100th KIAP) at Severomorsk on Kola Peninsula beginning with initial four in April 1993; five months of intensive flying from *Kuznetsov*, in Barents Sea, in second half 1994; carrier deployed temporarily to Adriatic late 1995, with at least 10 Su-27Ks (including eight of 'red' squadron and T10K-9 operational trials aircraft 'blue 109'). Complement of 24 aircraft in 279th Regiment believed to include at least four preproduction machines, with numbers made up by standard land-based 'Flankers'. Naval variant known as both Su-27K (preferred by Russian Air Forces) and Su-33, until latter was formally adopted on 31 August 1998, coincident with presidential decree of acceptance into service. Reported radar and ESM/ECM problems have reduced aircraft's usefulness while accidents have accounted for several Su-27Ks. Not allocated 'Flanker-D' reporting name, despite some reports to this effect.

CURRENT VERSIONS: **Su-33**: *As described.*

Su-33 Upgrade: A 1999 Sukhoi/KnAAPO/Rosvooruzheni brochure described an Su-33 Upgrade configuration, though there is no real evidence that any such programme has been funded. The cockpit is to gain two large colour LCD multifunction displays, with a modernised flight control/navigation system and enhanced air-to-ground capabilities. Upgraded Su-33 is to be compatible with the R-77 (AA-12 'Adder') AAM and with a range of air-to-ground PGMs, including TV-guided Kh-29, Kh-59N, KAB-500Kr and KAB-1500Kr and laser-guided Kh-29L, KAB-500L, and KAB-1500L. Use of the Kh-31P will give an effective SEAD capability.

Su-28: Early, stillborn AEW variant, with spine-mounted rotodome.

Su-33 PROTOTYPES AND PRODUCTION

Batch	Qty	Identity	Remarks
-	2	T10K-1, 2	Built by Sukhoi OKB
1	1	-	Static test airframe; first by Komsomolsk
2	3	T10K-3, 4, 5	T10K-4 to Su-27KUB
3	4	T10K-6, 7, 8, 9	T10K-9 for operational trials
-	(1)	T10KM-2	Su-27KUB
4-8	16	'60' et seq	Up to '88' with gaps

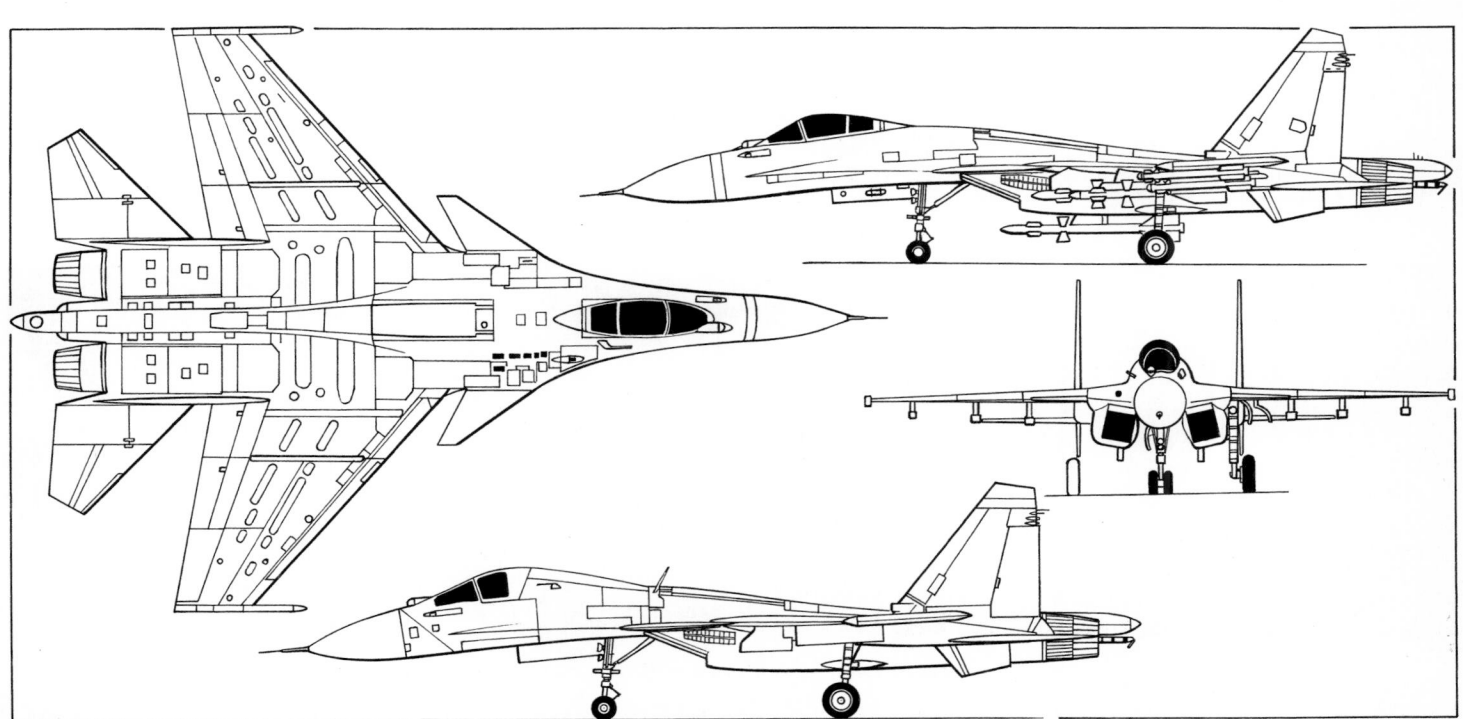

Sukhoi Su-33 single-seat carrierborne fighter, with additional side view (lower) of two-seat Su-33UB operational trainer *(Jane's/James Goulding)* 0075937

Su-27KM: Proposed definitive version, with Su- 27M FCS, Zhuk radar, full air-to-ground capability and vectoring engine nozzles; probably abandoned.

Su-27KRT (*korabelnyi razvedchik-tseleukazatel:* shipboard reconnaissance and target acquisition platform): Reportedly flying in prototype form by 1999. Equipped with Leninets MMW radar, encrypted communications and datalink equipment and radio/electronic reconnaissance suite. Production Su-27KRT may be based on Su-27KUB/Su-33UB airframe; configuration of prototype unknown.

Su-27KPP (*korabelnyi postanovshchik pomekh:* shipboard EW/command post): Reportedly under development.

Su-27KUB: Described under Su-33UB heading.

DESIGN FEATURES: Airframe generally similar to Su-27, but with folding wings and power-folding tailplane; fins allegedly 'slightly shortened' due to hangar deck constraints; added foreplanes; arrester hook and other features for carrierborne operations; strengthened landing gear with twin nosewheels; long tailcone of land-based versions shortened to prevent tailscrapes during take-off and landing on ship; brake parachute replaced by SPO-15 (L-006) Beryoza RHAWS relocated from the intake sides; IRST with wider angle of view.

FLYING CONTROLS: Basically as Su-27, but with sweptback (52°) foreplanes that operate at all speeds and only collectively, not differentially (7° up/70° down). Standard Su-27 fly-by-wire, automatic control system and central control column. Slow-speed devices comprise drooping ailerons plus two-section Fowler flaps mounted on partly drooping wing trailing-edge.

STRUCTURE: Generally as Su-27, but hydraulically folding outer wings (through 135°) and upward folding horizontal tail surfaces. Riveted and welded structure of aluminium and titanium alloys and steel.

LANDING GEAR: Generally as Su-27, but strengthened and with increased stroke; mainwheel tyres size 1,030×350, twin nosewheel tyres size 620×180; nosewheels steerable through ±60°. Shackles on main oleos for tiedown and for connection to Mustang hangar deck conveyor belt system. Arrester hook under tailcone.

POWER PLANT: Two Saturn AL-31F3 turbofans, each rated at 75.2 kN (16,909 lb st) dry and 125.5 kN (28,220 lb st) with afterburning. Internal fuel capacity 12,100 litres (3,196 US gallons; 2,662 Imp gallons). Retractable flight refuelling probe beneath windscreen on port side; maximum receipt rate 2,300 litres (608 US gallons; 506 Imp gallons)/min. Provision for UPAZ-1A centreline buddy refuelling pack.

ACCOMMODATION: As Su-27. K-36DM Srs 2 seat.

SYSTEMS: As Su-27.

AVIONICS: *Flight:* Nav systems optimised for use over sea; navalised TKS coded datalink. Resistor K42 ACLS allowing ICAO Cat. II autoland capability. Active radar beacon on nose oleo to increase gear-down RCS.

Self-defence: As Su-27, but with reduced number of chaff/flare dispensers (24 per side).

EQUIPMENT: Lights for optical/laser-based Luna-3 visual approach system.

ARMAMENT: Basically as Su-27; 12 pylons. Up to six R-27R1/ER1 (AA-10 'Alamo'), two R-27T1/ET1 and up to six R-73E (AA-11 'Archer') AAMs. Despite reports of precision-guided weapons, official sources only claim basic attack capability with eight FAB-500 500 kg bombs or RBK-500 cluster bombs; 31 FAB-250 250 kg bombs, 50 OFAB-100-120 100 kg bombs, 80 S-8 rockets in four B-8M1 pods, 20 S-13s in four B-13L pods or four S-25-OFM rockets in single O-25 launchers.

DIMENSIONS, EXTERNAL:

Wing span	14.70 m (48 ft 2¾ in)
Wing aspect ratio	3.2
Length: overall, incl noseprobe	21.185 m (69 ft 6 in)
folded	19.20 m (63 ft 0 in)
Width, wings folded	7.40 m (24 ft 3½ in)
Height overall	5.72 m (18 ft 9¼ in)
Tailplane span	9.90 m (32 ft 6 in)
Wheel track	4.44 m (14 ft 6¾ in)

Sukhoi Su-33 ship-based fighter (*Yefim Gordon*)   NEW/0525044

Wheelbase	5.87 m (19 ft 3 in)

AREAS:

Wings, gross	67.84 m² (730.2 sq ft)
Foreplanes (total)	3.00 m² (32.29 sq ft)
Ailerons (total)	2.40 m² (25.83 sq ft)
Trailing-edge flaps (total)	6.60 m² (71.04 sq ft)
Leading-edge flaps (total)	5.40 m² (58.13 sq ft)
Fins (total)	11.60 m² (124.87 sq ft)
Rudders (total)	3.50 m² (37.67 sq ft)
Horizontal tail surfaces (total)	12.30 m² (132.40 sq ft)

WEIGHTS AND LOADINGS:

Max military load	6,500 kg (14,330 lb)
Max fuel weight	9,500 kg (20,944 lb)
Normal T-O weight	25,000 kg (55,115 lb)
Max carrier T-O weight	30,000 kg (66,135 lb)
Max T-O weight	33,000 kg (72,752 lb)
Max landing weight	24,500 kg (54,013 lb)
Max wing loading	486.4 kg/m² (99.63 lb/sq ft)
Max power loading	131 kg/kN (1.29 lb/lb st)

PERFORMANCE:

Never-exceed speed (VNE) at 11,000 m (36,000 ft)	
	M2.165 (1,240 kt; 2,300 km/h; 1,430 mph)
Max speed at S/L M1.06 (702 kt; 1,300 km/h; 807 mph)	
Approach speed	130 kt (240 km/h; 150 mph)
Service ceiling	17,000 m (55,780 ft)
T-O run on carrier: normal	195 m (640 ft)
with 14° ramp	105 m (345 ft)
Landing run, arrested	90 m (295 ft)
Range with max internal fuel:	
at S/L	540 n miles (1,000 km; 621 miles)
at altitude	1,620 n miles (3,000 km; 1,865 miles)
g limits	+8/−2

UPDATED

SUKHOI Su-33UB (Su-27KUB)

TYPE: Multirole fighter.

PROGRAMME: Original Su-27IB prototype initially described by Sukhoi design bureau as 'Su-27KU' (*korabelnyi uchebno:* shipborne trainer) and described as carrier trainer, despite lack of arrester hook, folding wings, strengthened landing gear and carrier landing system. Aircraft did make dummy carrier approaches, and may have flown from dummy deck at Saki experimental airbase to evaluate suitability of new cockpit for carrier operation. Naval strike Su-32MF (which see) may once have been intended for carrier operation.

Need for dedicated trainer for Su-33 (which see) became increasingly clear, and development of T10KM-2-based **Su-27KUB** (*korabelnyi uchebno boevoi:* as T10KU shipborne fighter trainer) began in late 1990s. Formally acknowledged 21 October 1998. Probably being developed as a private venture, with no firm commitment from Russian Navy. Likely to become **Su-33UB**, following redesignation of carrierborne single-seat

Su-27K as Su-33.

Three prototypes, with noses built by KnAAPO, as T10KU-1, -2, and -3, incorporating lessons from the *Kuznetsov*'s 1996 Mediterranean cruise. Construction of T10KU-1 began in 1998 mating a new nose and new wings and tailplanes to an existing T10K prototype (T10K-4). Powered by AL-31F engines, the aircraft first flew on 29 April 1999 and made first arrested landing on dummy deck at NIUTK ('Nitka') test centre, Saki, 3 September 1999; first take-off from deck ramp followed on 6 September; first landing and take-off from carrier *Kuznetsov* on 6 October 1999. T10KU-2 and -3 reported to be under construction in 1999, perhaps using new-build airframes, but had not been seen by January 2001. T10KU-1 test flown by Indian pilots, September 1999, but Su-27 judged too large for planned carriers. Further test series at Saki begun December 2000.

Any production is likely to be by KnAAPO; baseline variant due to be extrapolated to produce trainer, reconnaissance and AEW versions, the last-mentioned with a phased-array mounted on the spine, between the composites antenna tailfins. Increased thrust, thrust-vectoring AL-31FP, AL-31FM or AL-41F engines mooted for production version.

CURRENT VERSIONS: **Su-33UB:** *As described.*

Su-30K-2: Two-seat interceptor version based on Su-33 fuselage under construction at Komsomolsk, late 1999; due to fly late 2000, but no further reports received.

DESIGN FEATURES: Has navalised features of Su-33, including folding wing (with fold further outboard and with a larger fold angle). Some sources suggest that the Su-33UB's tailplanes do not fold, since they reach only as far as the new outboard wing fold. Double slotted flaps, unslotted, 'adaptive' leading-edge, arrester hook, datalink and carrier landing system. However, compared with Su-27IB, forward fuselage is slightly narrower, with seats closer together and has much less pronounced dorsal hump. New 'glass cockpit' with five colour LCD displays (one 53 cm; 21 in diagonally; rest 38 cm; 15 in) with provision for central or sidestick and with helmet-mounted sighting system. Aircraft has OBOGS and OBIGGS and so does not need oxygen or nitrogen bottles. N014 solid-state, phased-array radar, with enhanced air-to-ground and over-water capabilities, planned eventually, but prototypes have ballast or Zhuk-MS and production aircraft may initially use NIIR N610-27 (Zhuk 27); circular-section radomes replace flattened 'platypus' nose associated with Su-27IB. Some reports suggest that the Su-33UB's new wing is 12 per cent larger in span (16 m; 52½ ft) and area (70 m²; 750 sq ft), as are the canards and tailplanes. Production Su-27KUB will feature a higher set, square-section, lengthened tail-sting, possibly mounting rear warning radar; tailcone folds (upwards) to reduce stowed length; prototypes have the standard Su-33 tailcone.

UPDATED

Sukhoi Su-27KUB in naval colour scheme (*Yefim Gordon*)   NEW/0525045

SUKHOI Su-34
See Sukhoi Su-27IB entry.

SUKHOI Su-35 and Su-37 (Su-27M)
TYPE: Multirole fighter.

PROGRAMME: Development of Su-27M authorised on 29 December 1993 by Council of Ministers. Experimental version of Su-27 with foreplanes (T10-24) flew May 1985; improved FBW system and refuelling probe tested by T10U-2. Five prototypes produced by conversion of production Su-27s, retaining single nosewheel and standard tailfins: T10M-1 '701' (ex-1602), then lacking radar and weapon control system (successively T10S-70, T10M, Su-27M, Su-35) flew 28 June 1988; T10M-2 flown 18 January 1989; T10M-5, T10M-6 '706' and T10M-7 '707'; used mainly by NII VVS at Akhtubinsk, flown by service pilots.

Production at KnAAPO, Komsomolsk, beginning with static test airframe T10M-4; first flight (T10M-3 '703') 1 April 1992; latter exhibited at 1992 Farnborough Air Show. Of further six ordered (T10M-8 to 13; '708' to '713'), final aircraft was cancelled. NIIP N011M Bars phased-array radar tested in '711' and '712'. Large-scale series production originally planned 1996-2005 as interim fighter pending availability of Mikoyan MFI.

Three production aircraft (Blue 86, 87 and 88) delivered to Akhtubinsk from Komsomolsk in 1996 or 1997.

KnAAPO revealed in 2001 that production Su-35s and Su-35UBs will have thrust vectoring AL-31FP engines, thereby negating Su-37 designation.

CURRENT VERSIONS: **Su-35:** Baseline single-seater; *as described.*

Su-35UB: Two-seat (tandem) derivative of the Su-35 with same FCS, canard foreplanes, tall square-topped tailfins, 12-pylon wing, Zhuk main AI radar and N012 rearward-facing, tailcone-mounted radar. Developed as a demonstrator and trainer for the Su-35, the construction of the prototype (Blue 801) at Komsomolsk may have been prompted by the needs of Sukhoi's campaign to sell the Su-35 to South Korea. The aircraft first flew on 7 August 2000 and was reported to be undergoing trials at Akhtubinsk in October 2000. Has 123 kN (27,558 lb st) AL-31FP (AL-31F) thrust-vectoring engines.

Su-35UB has 38,800 kg (83,775 lb) MTOW; 8,000 kg (17,637 lb) combat load; 12,400 litres (3,276 US gallons; 2,728 Imp gallons) internal fuel load; 1,090 kt (2,020 km/h; 1,255 mph) max speed; 1,619 n mile (3,000 km; 1,864 mile) unrefuelled range; wing span 14.70 m (48 ft 2¾ in); length overall 21.94 m (71 ft 11½ in) (shorter tailboom); and height overall 6.355 m (20 ft 10¼ in).

Su-37: Technology demonstrator for proposed production fighter resulting from vectored-thrust programme of the Su-27UB-PS, Su-27UBL, Su-27LMK No. 2405 and the penultimate prototype Su-35 ('711') which made its first flight, with nozzles fixed, on 2 April 1996 (or 12 April, according to some sources). This aircraft had previously been first of two N011M radar testbeds, flying in blue/grey camouflage. Designated Su-37 by Sukhoi bureau. By September 1996, when demonstrated at Farnborough Air Show (Western debut), in tropical camouflage, '711' had made 50 flights with hydraulically actuated nozzles able to move ±15° in pitching plane at rate of 30°/s under control of aircraft's flight control system. Probably toed-out 32° from the centreline, like Su-30 MKI,

Prototype Su-35UB two-seat version *(Yefim Gordon)*   *NEW*/0525046

Su-35 PROTOTYPES AND PRODUCTION

Batch	Qty	Identity	Remarks
-	(1)	T10M-1 '701'	Previously Su-27 1602
-	(1)	T10M-2 '702'	Previously Su-27
-	(1)	T10M-5 '705'	Previously Su-27
-	(1)	T10M-6 '706'	Previously Su-27
-	(1)	T10M-7 '707'	Previously Su-27
10	1	T10M-4	Static test airframe
10	1	T10M-3 '703'	First Komsomolsk-built
11	1	T10M-8 '708'	
11	1	T10M-9 '709'	
11	1	T10M-10 '710'	
11	1	-	Static test airframe
11	1	T10M-11 '711'	Converted to Su-37
12	3	'86' et seq	Production
n/k	1	T10M-12 '712'	Radar testbed
n/k	1	'801'	Su-35UB

Note: Former Su-27 aircraft may include 2209 and 2219

generating powerful yawing moment when actuated differentially. An emergency pneumatic system returns the nozzles to level flight setting in the event of system failure.

Any production Su-37 was expected to feature uprated AL-31FU engines (142.2 kN; 31,967 lb st). In 1999, Sukhoi was using designation Su-37MR for proposed production version, apparently reflecting a major avionics upgrade then under way. By 2001, '711' had been retrofitted with non-vectoring AL-31F engines, revised Ramenskoye instrumentation and a new flight control system which was stated to provide TV-type agility without recourse to vectoring; used in trials for Indian Su-30MKI Stage 3.

Second Su-37 reportedly converted from T10M-12 and first flew in mid-1998, initially powered by AL-31F; installation of thrust-vectoring AL-37FPs was expected in late 1998; status of this aircraft, if still extant, remains unclear. Development of AL-31F engine being undertaken unilaterally by MMPP Salyut, version promising 10 to 15 per cent extra thrust having first flown on 24 January 2002 (on Su-27 testbed and lacking Klivit thrust vectoring nozzle).

CUSTOMERS: Once scheduled for entry into Russian Air Forces service as Su-27M from 1995 onwards, for effective operation until 2015-2020; programme in suspension.

COSTS: Estimated US$35 million to US$40 million (2002).

DESIGN FEATURES: Advanced multirole development of Su-27 to counter latest versions of USAF F-15 Eagle and F-16 Fighting Falcon, with better dogfighting characteristics, higher AoA limits, lighter weight and new BVR armament; proposed to include 216 n mile (400 km; 248 mile) range AAM-L (one AAM-L contender was Novator KS-172). Also planned to have greater autonomy from GCI control. Airframe, power plant, avionics and armament all upgraded; quadruplex digital fly-by-wire controls under development by Avionika, though prototypes retain analogue system; longitudinal static instability; 'tandem triplane' layout, with foreplanes; double-slotted flaperons; taller, square-tip twin tailfins with integral fuel tanks; reprofiled front fuselage for larger-diameter radar antenna; enlarged tailcone for rearward-facing radar; twin-wheel nose landing gear; axisymmetric thrust-vectoring nozzles under development for use on production aircraft (see Su-37).

STRUCTURE: Higher proportion of carbon fibre and aluminium-lithium alloy in fuselage; composites used for components such as leading-edge flaps, nosewheel door and radomes.

POWER PLANT: Production Su-27M planned to use two Saturn/Lyulka AL-35FP turbofans; each 123 kN (27,558 lb st)

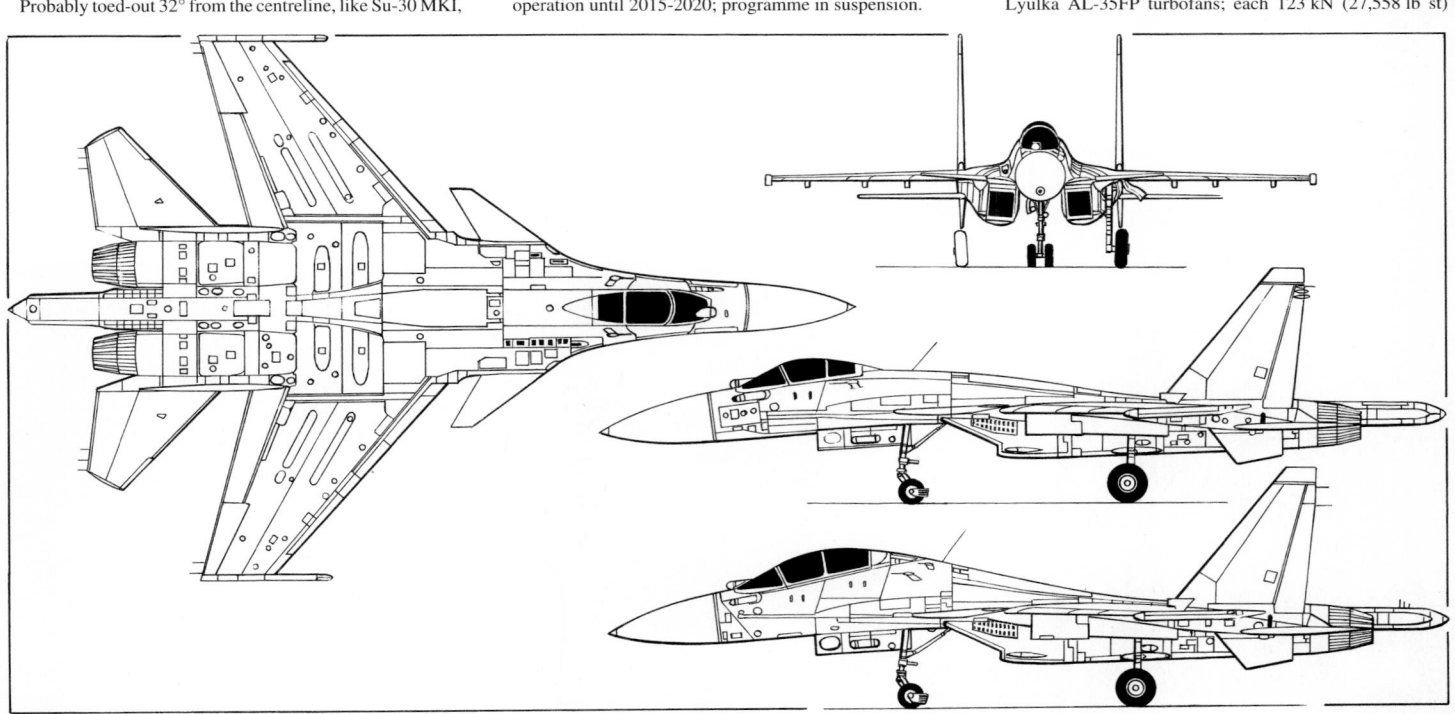
Sukhoi Su-35 single-seat counter-air and ground attack fighter, with additional side view (lower) of Su-35UB *(Jane's/James Goulding)*   *NEW*/0525932

Sukhoi Su-35 of the Akhtubinsk test centre (*Jane's/Paul Jackson*)   0114593

with afterburning; prototypes retain standard AL-31F. Increased internal tankage through use of welded aluminium-lithium tanks and new tanks in tailfins and finroots. Total fuel capacity 12,900 litres (3,408 US gallons; 2,838 Imp gallons). Retractable flight refuelling probe on port side of nose.

ACCOMMODATION: Pilot(s), on Zvezda K-36D-3.5E zero/zero ejection seat, this now angled back 30°.

AVIONICS: *Radar:* Originally planned to incorporate NIIP N011 Zhuk-27 multimode low-altitude terrain-following/avoidance radar, search range up to 54 n miles (100 km; 62 miles) against advancing target, 30 n miles (55 km; 34 miles) against retreating target; able to track 15 targets and engage four to six simultaneously. Phazotron Zhuk-Ph phased-array radar under development as alternatives or for retrofit; search range for fighter-size targets 75 n miles (140 km; 87 miles) with simultaneous tracking of 24 air targets and ripple-fire engagement of six to eight; N012 rearward-facing radar, range approximately 2 n miles (4 km; 2.5 miles), may enable firing of rearward-facing IR homing AAMs.

Flight: Fully automatic flight modes and armament control against ground, maritime and air targets, including automatic low-altitude flight and automatic target designation. RPKB nav system includes laser-gyro INS and Glonass GPS.

Instrumentation: Two-seat variant has side-by-side MFI-10-5 screens and HUD in front cockpit; vertically stacked MFI-10-5 screens in rear.

Mission: New-type IRST moved to starboard; small external TV pod; all combat flight phases computerised. Shown at Farnborough with GEC-Marconi TIALD (thermal imaging airborne laser designator) night/adverse visibility pod fitted for possible future use.

Self-defence: Entirely new integrated EW suite with ECM, including active jammer and wingtip Sorbtsa-S G- or J-band ECM/ESM pods; Pastel RWR; Mak IR-based MAWS.

ARMAMENT: One 30 mm GSh-30 gun in starboard wingroot extension, with 150 rounds. Mountings for up to 14 stores pylons, including R-27 (AA-10 'Alamo-A/B/C/D'), R-40 (AA-6 'Acrid'), R-60 (AA-8 'Aphid'), R-73E (AA-11 'Archer') and RVV-AE (R-77; AA-12 'Adder') AAMs, Kh-25ML (AS-10 'Karen'), Kh-25MP (AS-12 'Kegler'), Kh-29T (AS-14 'Kedge'), Kh-31P (AS-17 'Krypton') and Kh-59 (AS-18 'Kazoo') ASMs, S-25LD laser-guided rockets, S-25IRS IR-guided rockets, GBU-500L and GBU-1500L laser-guided bombs, GBU-500T and GBU-1500T TV-guided bombs, KMGU cluster weapons, KAB-500 bombs and rocket packs. Maximum weapon load 8,200 kg (18,077 lb).

DIMENSIONS, EXTERNAL:
Wing span over ECM pods	15.16 m (49 ft 8¾ in)
Wing aspect ratio	3.5
Length overall	22.185 m (72 ft 9½ in)
Height overall	6.36 m (20 ft 10¼ in)

AREAS:
Wings, gross	62.04 m² (667.8 sq ft)

WEIGHTS AND LOADINGS:
Weight empty	17,000 kg (37,479 lb)
Max fuel weight	10,250 kg (22,597 lb)
Max combat load	8,500 kg (18,739 lb)

T-O weight: normal max	34,000 kg (74,957 lb)
overload	38,800 kg (85,539 lb)
Max wing loading (normal)	548.4 kg/m² (112.32 lb/sq ft)
Max power loading (normal)	139 kg/kN (1.36 lb/lb st)

PERFORMANCE:
Max level speed:
at height	M2.35 (1,350 kt; 2,500 km/h; 1,555 mph)
at S/L	M1.14 (756 kt; 1,400 km/h; 870 mph)
Service ceiling	17,200 m (56,440 ft)
Balanced runway length	1,200 m (3,940 ft)

Range: with max internal fuel
1,835 n miles (3,400 km; 2,112 miles)
with flight refuelling
3,401 n miles (6,300 km; 3,914 miles)
g limit	+9

UPDATED

SUKHOI Su-39
See Sukhoi Su-25TM.

SUKHOI Su-47 BERKUT
The S-37 multirole fighter technology demonstrator was redesignated Su-47 in 2000. Although no further production is planned (and the Irkutsk-built second flying prototype will not be completed) the sole Su-47 continues to fly as a technology demonstrator for a future generation of Russian combat aircraft. It had accumulated some 150 sorties by early 2002. A description last appeared in the 2001-02 *Jane's*.

UPDATED

SUKHOI Su-49
TYPE: Basic prop trainer.

PROGRAMME: Design of military derivative of Su-26 and Su-29 aerobatic aircraft, to meet trainer requirement, started July 1992, under former designation Su-32; designation changed to Su-39 in 1995, Su-49 in 1996. Construction of prototype began 1994; first flight scheduled second half of 1996, but delayed to third quarter of 2002. Design further refined by mid-1997 to include LERX, raised rear portion of canopy and twin nosewheels. P&W Rus PK6A-25 turboprop reportedly considered as replacement for M-9 (M-14 previously intended) piston engine in at least a proportion of production aircraft; this to be tested from early 2002 in modified Su-29KS. In October 2001 was named winner of competition against Yak-152 for Russian Air Forces trainer requirements; prototype first flight scheduled for 2002. Production planned at NAPO, Novosibirsk.

CURRENT VERSIONS: Two equipment standards envisaged for different stages of training syllabus. Weapon training and light attack versions under consideration.

CUSTOMERS: Sukhoi anticipates orders (including exports) for 800 to 1,200, mainly for Russian air force flying schools and ROSTO (formerly DOSAAF); air force version has more complex equipment.

COSTS: US$300,000 (baseline) to US$700,000 (full avionics) (2000). Direct operating cost US$150 to US$200 per hour.

DESIGN FEATURES: Based on Su-26 and Su-29 aerobatic aircraft, with objectives of short take-off and landing and high manoeuvrability, plus optional carriage of guided and unguided weapons and suitability for counter-insurgency, patrol and coastal protection missions. Service life 10,000 hours or 30,000 landings.

Wing leading-edge sweepback 5° 30′, wing section NACA 23012, dihedral 1° 30′, incidence 0°.

FLYING CONTROLS: Conventional and manual. Elevators and rudder horn-balanced; pneumatically operated wing trailing-edge slotted, area-increasing flaps and ventral airbrake. Flaps extend 35° for landing, 20° for take-off. Flight-adjustable tab in port elevator.

STRUCTURE: Forward fuselage of welded steel tubes. Two-spar wings; single-spar fin and tailplane. Fuselage longerons and wing spars of CFRP; wing, fuselage and tail unit skin panels of Kevlar-type composites and GFRP.

LANDING GEAR: Pneumatically retractable tricycle type; single wheel on each main unit; twin nosewheels; main units retract inward, nosewheel rearward; oleo-pneumatic shock-absorbers; Rubin mainwheels type KT-214; tyres size 400×150-140, pressure 3.00 to 3.50 bar (43.5 to 50.75 lb/sq in); hydraulic brakes on mainwheels.

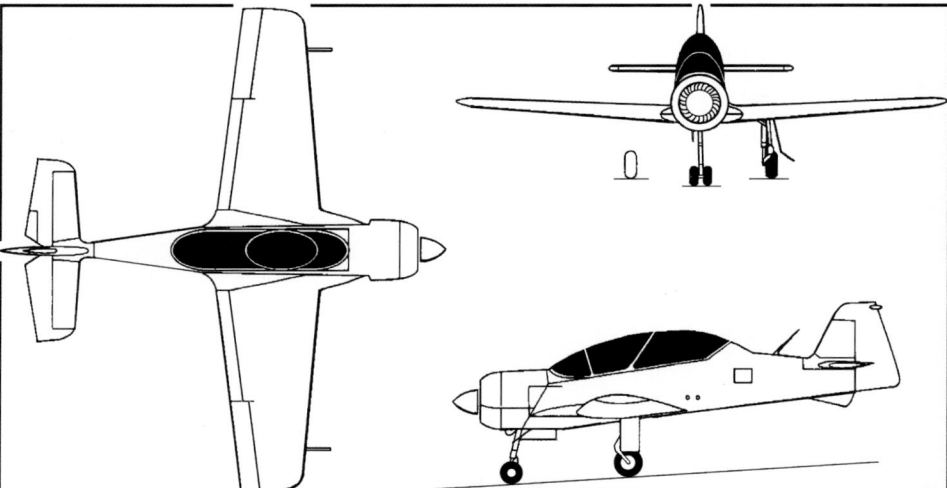

Sukhoi Su-49 primary trainer (*Jane's/Paul Jackson*)   *NEW*/0525378

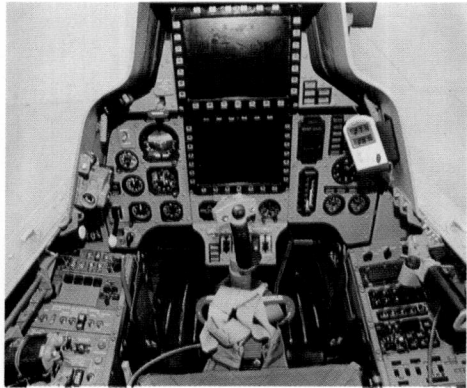

Su-35UB rear cockpit   *NEW*/0525047

Sukhoi's Su-47 Berkut continues its flight trials programme (*Yefim Gordon*)   *NEW*/0525048

POWER PLANT: One 309 kW (414 hp) VOKBM M-9F nine-cylinder radial engine; three-blade, constant-speed MTV-9 propeller. Alternative P&WC Klimov PK6A-25 turboprop for export versions. Fuel tank in each wing, capacity 150 litres (39.6 US gallons; 33.0 Imp gallons); header tank in fuselage, capacity 20 litres (5.3 US gallons; 4.4 Imp gallons); total internal fuel capacity 32.0 litres (84.5 US gallons; 70.4 Imp gallons); provision for two 100 litre (26.4 US gallon; 22.0 Imp gallon) underwing tanks; gravity fuelling; oil capacity 20 litres (5.3 US gallons; 4.4 Imp gallons).

ACCOMMODATION: Two seats in tandem; rear (instructor's) seat raised. Zvezda SKS-94M-49 crew extraction system (54 to 215 kt; 100 to 400 km/h; 62 to 248 mph/10 m; 33 ft); crew extracted through canopy on diverging trajectories in emergency, without seats. Canopy hinged to starboard. Baggage space behind cockpits, capacity 0.20 m³ (7.0 cu ft).

SYSTEMS: Cockpit air conditioning and pressurisation standard. Brake hydraulic system pressure 24.50 bar (355 lb/sq in). Pneumatic system pressure 50.60 bar (735 lb/sq in). One 3.5 kW generator for 27 V DC electrical system. Propeller anti-iced and windscreen demisted by hot air.

AVIONICS: To customer's requirements; standard export suite described below.

Comms: Bendix/King KX 165 VHF nav/com/glideslope, KT 76A transponder and KR 87 AD.

Radar: Provision for external radar pod.

Flight: Optional automatic approach and Bendix/King KLN 90 GPS. Provision for autopilot and autoland.

Self-defence: RWR and IR warning system in military versions.

ARMAMENT: Provision for integral gun, bombs, air-to-air and anti-tank missiles in combat versions.

DIMENSIONS, EXTERNAL:
Wing span	8.50 m (27 ft 10¾ in)
Wing chord: at root	1.88 m (6 ft 2 in)
at tip	0.885 m (2 ft 10¾ in)
Wing aspect ratio	5.9
Length overall	7.285 m (23 ft 10¾ in)
Height overall	2.60 m (8 ft 6½ in)
Tailplane span	2.90 m (9 ft 6¼ in)
Wheel track	2.00 m (6 ft 6¾ in)
Wheelbase	1.78 m (5 ft 10 in)
Propeller diameter	2.50 m (8 ft 2½ in)
Propeller ground clearance	0.24 m (9¼ in)

DIMENSIONS, INTERNAL:
Cockpit: Length	2.70 m (8 ft 10¼ in)
Max width	0.60 m (1 ft 11½ in)
Max height	1.10 m (3 ft 7¼ in)
Floor area	2.50 m² (26.9 sq ft)

AREAS:
Wings, gross	13.10 m² (141.0 sq ft)
Ailerons (total)	2.24 m² (24.11 sq ft)
Trailing-edge flaps (total)	3.30 m² (35.52 sq ft)
Fin	0.30 m² (3.23 sq ft)
Rudder	0.90 m² (9.69 sq ft)
Tailplane	0.98 m² (10.55 sq ft)
Elevators (total)	1.56 m² (16.79 sq ft)

WEIGHTS AND LOADINGS:
Weight empty, equipped	850 kg (1,874 lb)
Max fuel: internal	260 kg (573 lb)
external	200 kg (441 lb)
Normal T-O weight	1,400 kg (3,086 lb)
Max wing loading	106.9 kg/m² (21.89 lb/sq ft)
Max power loading	4.54 kg/kW (7.45 lb/hp)

PERFORMANCE (estimated, at normal T-O weight):
Never-exceed speed (VNE)	286 kt (530 km/h; 329 mph)
Max level speed	232 kt (430 km/h; 267 mph)
Max cruising speed	178 kt (330 km/h; 205 mph)
Landing speed	65 kt (120 km/h; 75 mph)
Stalling speed, flaps down	63 kt (115 km/h; 72 mph)
Max rate of climb at S/L	780 m (2,559 ft)/min
Service ceiling	4,000 m (13,120 ft)
T-O run	300 m (985 ft)
Landing run	240 m (790 ft)

Su-26MX converted to non-standard Sukhoi Shark, with a 560 kW (751 shp) Walter M 601 turboshaft, by Ray Vetsch of Albuquerque, New Mexico. Thrust: weight ratio is approximately 2.8:1 *(Jane's/Paul Jackson)*

NEW/0525050

Range:
with max payload and 30 min reserves	
	647 n miles (1,200 km; 745 miles)
with max internal fuel	
	809 n miles (1,500 km; 932 miles)
with max internal and external fuel	
	1,080 n miles (2,000 km; 1,242 miles)
g limits	+9/–4.5

UPDATED

SUKHOI LFS

Projected renamed PACFA (which see).

UPDATED

SUKHOI PACFA

TYPE: Multirole fighter.

PROGRAMME: Successor programme to original quest for fifth-generation fighter, for which MiG 1.42 and Sukhoi S-37/Su-47 Berkut demonstrators were built (see 2001-02 and earlier *Jane's*). Development began in 1998 to meet Russian Air Forces TTZ (tactical technical assessment) of the same year. Sukhoi-led co-operative project launched May 2001; initially termed LFS (*legkiy frontovoy samolet*: Light Frontline Aircraft), although competition also known as LFI (*legkiy frontovoy istrebitel*: Light Frontline Fighter). Air Forces' project name is PACFA (*perspektivnnyi aviatsionnyi kompleks frontovoi aviatsyi*: Prospective Aviation Complex for Frontal Aviation).

Examining committee established 10 January 2002 to assess competing bids from RSK 'MiG' and AVPK Sukhoi, each of which nominated Yakovlev as associate. On 26 April 2002, Russian Federation Ministry of Industry, Science and Technology declared Sukhoi as lead developer of new aircraft, to be assisted by RSK 'MiG' and Yakovlev as subcontractors. In May 2002, development agreement signed by AVPK Sukhoi, Sukhoi OKB, State Research Institute of Aviation Systems (Gos-NIIAS), Central Aero- and Hydrodynamic Institute (TsAGI), Research Institute of Aero Engine Technology and Production (TsIAM), Central Research Institute of Material (VIAM), National Institute of Aviation Technologies (NIAT), Lyulka/Saturn engine bureau, Ramenskoye Instrument Design Bureau (RPKB), Aviapribor holding company, Aviakosmitcheskoye Oborudvanyye and Vympel and Strela weapon companies. Several manufacturing plants expected to join at later stage. Participation of KnAAPO production plant was agreed in June 2002.

Timetable for PACFA is draft design by end of 2002; first flight in 2006; production from 2010. Official funding equivalent to US$1.5 billion promised for R & D, but this considered inadequate, representing some 20 per cent of needs, excluding production investment. Unit cost expected to be US$35 million to US$40 million (2002 prices), based on production of between 500 and 600 aircraft.

Parameters believed to include 20 tonne MTOW – between MiG-29 and Su-27 types it is due to replace – supersonic cruising speed, low observables, high manoeuvrability and short-field performance. Engine choice yet to be made, but NPO Saturn is offering a derivative of AL-41F1, which was due to begin bench testing in 2002.

NEW ENTRY

SUKHOI Su-26

Production of this single-seat aerobatic aircraft ceased in 1996. Brief details may be found in the 2001-2002 *Jane's*, and a full description in the 1996-97 and earlier editions.

UPDATED

SUKHOI Su-29

TYPE: Aerobatic two-seat sportplane.

PROGRAMME: Announced at Moscow Air Show '90; design started 1990; construction of first of three prototypes and two static test airframes began 1991; prototype first flew 1991, first production aircraft May 1992; entered service July 1992. AP-23 type certificate awarded in 1994. Assessed by Russian Air Forces experimental centre at Akhtubinsk in 1996-97, but no orders placed.

Following structural failure of Su-29 wing in 1996; remedial manufacturing practices in place and recertified by 1998. Resumed civilian export production now featured M-14PF engine and new propeller, but M-9F engine introduced in 1999. Retrospective installation of SKS-94 pilot's emergency extraction system offered by Sukhoi from 2000.

Production transferred to LAPIK division of RSK 'MiG' (formerly LMZ) at Lukhovitsy from early 2001, earlier aircraft having been built by Sukhoi Advanced Technologies using components manufactured by LMZ, Dubna and others. Only one built in 2001.

CURRENT VERSIONS: **Su-29:** Basic two-seat training/aerobatic aircraft.

Description applies to baseline Su-29.

Su-29AR: Aircraft for Argentina have German propeller, Swiss canopies and US wheel assemblies and avionics, including GPS.

Su-29KS: Development vehicle for Zvezda SKS-94 lightweight crew extraction system. Weight empty, equipped 800 kg (1,764 lb). First exhibited at 1994 Farnborough Air Show. One only (RA-01485); carries designation 'Su-29kS'.

Su-29M: Production version from 1999 (initial series of 10 under construction); M-9F engine; weights as Su-29; no extraction system.

Su-49: Military trainer (which see).

CUSTOMERS: Total 52 basic Su-29s and one Su-29KS built and sold by 1997; many exported to Pompano Air Center, Florida, USA, for reassembly and delivery worldwide (including eight in 1992; 12 in 1993; seven in 1994 and four in 1995); others to Australia (three), Italy, South Africa (two) and UK. Relaunched production from 1999 (initial batch of 10). Eight ordered by Argentine Air Force, for training, March 1997; delivery between September 1997 and August 1998 to Escuadrilla Cruz del Sur (Southern Cross Squadron) based at Mendoza AFB. By mid-2001, Sukhoi had built 166 aerobatic light aircraft of the Su-26/29/31 series, of which all but 10 had been exported, including 80 to the USA. Sukhoi presented two to ruler of unspecified Gulf state, but these not used and eventually sold. Total of three scheduled for delivery in 2002, including one each to the Czech Republic and South Korea.

COSTS: US$220,000 to US$260,000 (2002).

DESIGN FEATURES: Typical aerobatic competition aircraft; mid-wing of specially developed symmetrical section, variable along span, slightly concave in region of ailerons

Model of Sukhoi Su-49 *(Jane's/Paul Jackson)*

NEW/0525049

Sukhoi Su-29KS ejection system testbed, modified with separate windscreens *(Yefim Gordon)* **NEW**/0525051

to increase their effectiveness; leading-edge somewhat sharper than usual to improve responsiveness to control surface movement. Two-seat development of Su-26M single-seat aerobatic competition aircraft; wing span and overall length increased; improved aerodynamics and reduced stability margin for enhanced manoeuvrability. Service life 1,250 hours.

Wing leading-edge sweepback 3° 28′, symmetrical section, thickness/chord ratio 16 per cent at root, 12 per cent at tip, dihedral 0°, incidence 0°.

FLYING CONTROLS: Conventional and manual. Elevators and rudder horn-balanced; elevator trim; two suspended triangular balance tabs under each aileron. Later Su-29s have Su-31-type wing with ailerons extending to wingtips. No flaps.

STRUCTURE: Composites comprise more than 60 per cent of airframe weight; one-piece wing, covered with honeycomb composites panels; foam-filled front box spar with CFRP booms and wound glass fibre webs; channel section rear spar of CFRP; titanium truss ribs; plain ailerons have CFRP box spar, GFRP skin and foam filling; fuselage has basic welded truss structure of VNS-2 high-strength stainless steel tubing; lower nose section of truss removable for wing detachment; quickly removable honeycomb composites skin panels; light alloy engine cowlings; integral fin and tailplane construction same as wings; rudder and elevator construction same as ailerons; titanium exhaust, battery box and firewall; forged magnesium control linkages. Aircraft assembled by Sukhoi from components produced by LMZ, Dubna (DMZ) and NPO Technologia (at Omsk).

LANDING GEAR: Non-retractable tailwheel type; arched cantilever mainwheel legs of titanium alloy; mainwheels size 400×150, with hydraulic disc brakes; steerable tailwheel, on titanium spring, connected to rudder. Optional composites fairings for mainwheels.

POWER PLANT: One 265 kW (355 hp) VOKBM M-14PT or 294 kW (394 hp) M-14PF nine-cylinder radial engine; three-blade MTV-3-8-S/L250-21 or MTV-9-260 propeller. Current production employs VOKBM M-9F (M-14 derivative) rated at 309 kW (414 hp). Steel tube engine mounting. Fuel tank in fuselage forward of front spar; capacity 63 litres (16.6 US gallons; 13.8 Imp gallons); tank in each wing leading-edge; capacity 106.5 litres (28.15 US gallons; 23.4 Imp gallons); total fuel capacity 276 litres (72.9 US gallons; 60.6 Imp gallons); gravity fuelling. Oil capacity 20 litres (5.3 US gallons; 4.4 Imp gallons). Fuel and oil systems adapted for inverted flight; pneumatic engine starting system.

ACCOMMODATION: Pilot only for aerobatic competition, two persons in tandem for training. Canopy normally opens sideways to starboard; but upward and rearward in emergency to jettison. Dual controls standard. Space for 5 kg (11 lb) baggage in rear fuselage.

SYSTEMS: Electrical system 24/28 V, with 3 kW generator, batteries and external supply socket.

AVIONICS: *Comms:* Briz VHF radio; optional Becker or Bendix/King com/nav and Garmin GPS.

EQUIPMENT: Optional provision for smoke generation.

DIMENSIONS, EXTERNAL:

Wing span	8.20 m (26 ft 10¾ in)
Wing chord: at root	1.985 m (6 ft 6¼ in)
at tip	1.04 m (3 ft 4¾ in)
Wing aspect ratio	5.5
Length overall	7.285 m (23 ft 10¾ in)
Height overall	2.885 m (9 ft 5¾ in)
Tailplane span	2.90 m (9 ft 6¼ in)
Wheel track	2.40 m (7 ft 10½ in)
Wheelbase	5.08 m (16 ft 8 in)
Propeller diameter	2.50 m (8 ft 2½ in)
Propeller ground clearance	0.425 m (1 ft 4¾ in)

DIMENSIONS, INTERNAL:

Cockpit: Length	2.60 m (8 ft 6¼ in)
Max width	0.82 m (2 ft 8¼ in)
Max height	1.05 m (3 ft 5¼ in)

AREAS:

Wings, gross	12.20 m² (131.4 sq ft)
Ailerons (total)	2.32 m² (24.97 sq ft)
Fin	0.28 m² (3.01 sq ft)
Rudder	0.90 m² (9.69 sq ft)
Tailplane	0.98 m² (10.55 sq ft)
Elevators (total)	1.56 m² (16.79 sq ft)

WEIGHTS AND LOADINGS (two persons):

Weight: empty	735 kg (1,620 lb)
empty, equipped	780 kg (1,720 lb)
Max fuel	207 kg (456 lb)
Max T-O weight: pilot only	860 kg (1,896 lb)
two persons	1,200 kg (2,645 lb)
Max wing loading	98.4 kg/m² (20.15 lb/sq ft)
Max power loading: M-14PT	4.53 kg/kW (7.45 lb/hp)
M-14PF	4.09 kg/kW (6.71 lb/hp)
M-9F	3.89 kg/kW (6.39 lb/hp)

PERFORMANCE (M-14PT engine):

Never-exceed speed (VNE)	242 kt (450 km/h; 279 mph)
Max level speed	175 kt (325 km/h; 202 mph)
Landing speed	65 kt (120 km/h; 75 mph)
Stalling speed	62 kt (115 km/h; 72 mph)
Max rate of climb at S/L	960 m (3,150 ft)/min
Service ceiling	4,000 m (13,120 ft)
Max rate of roll	360°/s
T-O run	120 m (395 ft)*

Landing run	380 m (1,250 ft)*
Range with max fuel	647 n miles (1,200 km; 745 miles)
g limits	+12/–10

*at 914 kg (2,015 lb) AUW

UPDATED

SUKHOI Su-31

TYPE: Aerobatic single-seat sportplane.

PROGRAMME: Design started 1991; prototype construction began 1992, flew June 1992 as Su-29T, demonstrated at 1992 Farnborough Air Show; followed by two more prototypes and two static test airframes; first production aircraft by Sukhoi Advanced Technologies (RA-01405) flown 1994. Su-31M2 introduced in 1999. Manufacture transferred to RSK 'MiG' (which see) in early 2001; two Su-31Ms, one Su-31 built in 2001; contract for 10 Su-29s/Su-31Ms for an undisclosed buyer pending in early 2002.

In XXI World Aerobatic Championships, June 2001, Su-31s gained first place and seven of next 14 places.

CURRENT VERSIONS: **Su-31:** Basic version; non-retractable landing gear. Alternatively known as **Su-31T** (*Turnirnyi:* Competition). See Su-29 drawings for side view.

Su-31M: As Su-31T but with Zvezda SKS-94 pilot's extraction system under modified canopy with deeper frame. Empty weight 760 kg (1,676 lb); normal T-O weight 880 kg (1,940 lb). Prototype RA-01486 converted from Su-31T.

Su-31M2: Improved version of Su-31M with larger wing, airframe weight reduced by approximately 80 kg (176 lb), ergonomically redesigned cockpit, and upgraded Zvezda SKS-94M pilot's extraction system. Prototype scheduled to fly in mid-2003.

Su-31X: Export version of Su-31T.

Su-31U: As Su-31T but retractable landing gear. None yet built.

CUSTOMERS: Total 25 (including five Su-31Ms) built by late 1998 and exported to Australia, Brazil, Italy, Lithuania, South Africa, Spain, Switzerland, Ukraine, UK and USA. Su-31Ms operated in Italy, Russia and Switzerland. Production batch of five Su-31M2s, 1999-2000, of which at least two to USA; RSK 'MiG' to build initial batch of nine at its LAPIK division (three Su-29s and six Su-31s). Total of six Su-31s scheduled for delivery in 2002, including two Su-31Ms for the Czech Republic and one Su-31M for a Swiss customer. Production in 2003 includes two for UK dealership.

Sukhoi Su-29 two-seat sportplane *(Jane's/Paul Jackson)* **NEW**/0525052

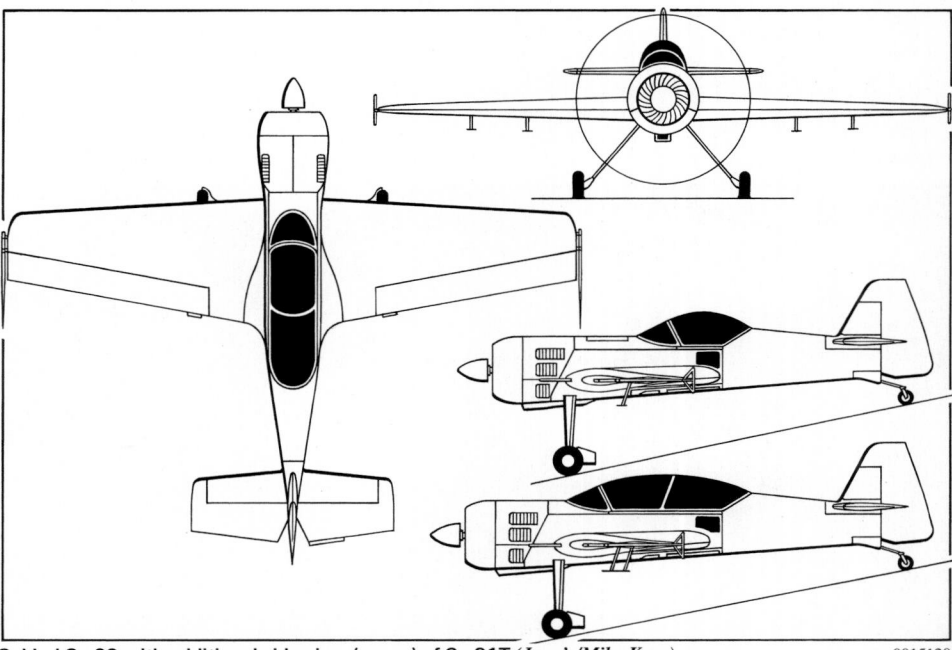

Sukhoi Su-29 with additional side view (upper) of Su-31T *(Jane's/Mike Keep)* 0015120

COSTS: US$220,000 to US$260,000 (2002).

DESIGN FEATURES: Basically single-seat version of Su-29 with uprated engine; new landing gear; 35° inclination of seat enables pilot to employ repeatedly a g load of +12/−10, giving advantages in controlling aircraft within limited flying area and to perform very complicated manoeuvres; improved field of view; two baggage compartments.

STRUCTURE: More than 70 per cent composites by weight; centre-fuselage is welded truss of high-strength stainless steel tube, with detachable skin panels of honeycomb-filled composite sandwich; rear fuselage is semi-monocoque of composites with honeycomb filler; one-piece two-spar wing with carbon fibre main spar, titanium ribs, covered with honeycomb sandwich skin; tail unit is all-composites; mainwheel legs titanium.

POWER PLANT: One 294 kW (394 hp) VOKBM M-14PF nine-cylinder radial engine in Su-31T/M; MTV-9 three-blade propeller. Basic fuel capacity 78 litres (20.6 US gallons; 17.2 Imp gallons), in fuselage tank; provision in Russian version for 210 litre (55.5 US gallon; 46.2 Imp gallon) centreline drop tank for ferrying; alternatively, in export version, two tanks in wings, total capacity 200 litres (52.8 US gallons; 44.0 Imp gallons).

ACCOMMODATION: Pilot only; windscreen and separate canopy, opening as Su-29, except single-piece windscreen/canopy and pilot extraction system in Su-31M and Su-31M2.

DIMENSIONS, EXTERNAL (Su-31T):
Wing span	7.80 m (25 ft 7 in)
Wing chord: at root	1.99 m (6 ft 6¼ in)
at tip	1.04 m (3 ft 4¾ in)
Wing aspect ratio	5.2
Length overall	6.83 m (22 ft 4¾ in)
Height overall	2.76 m (9 ft 0¾ in)
Tailplane span	2.90 m (9 ft 6¼ in)
Wheel track	2.40 m (7 ft 10½ in)
Wheelbase	4.90 m (16 ft 1 in)
Propeller diameter	2.50 m (8 ft 2½ in)
Propeller ground clearance	0.425 m (1 ft 4¾ in)

AREAS (Su-31T): As Su-29 except:
Wings, gross	11.83 m² (127.3 sq ft)

WEIGHTS AND LOADINGS (Su-31T, except where otherwise indicated):
Weight: empty	680 kg (1,499 lb)
empty, equipped: Su-31T	740 kg (1,631 lb)
Su-31M	750 kg (1,653 lb)
Max fuel: internal	53 kg (117 lb)
external	209 kg (461 lb)
T-O weight: normal	835 kg (1,841 lb)
max	968 kg (2,134 lb)
Max wing loading	82.0 kg/m² (16.80 lb/sq ft)
Max power loading: Su-31T/M	3.25 kg/kW (5.34 lb/hp)

PERFORMANCE (Su-31T):
Never-exceed speed (V_NE)	243 kt (450 km/h; 280 mph)
Max level speed	178 kt (330 km/h; 205 mph)
Stalling speed	61 kt (113 km/h; 71 mph)
T-O speed	60 kt (110 km/h; 69 mph)
Landing speed	62 kt (115 km/h; 72 mph)
Max rate of climb at S/L	1,440 m (4,725 ft)/min
Service ceiling	4,000 m (13,120 ft)
Max rate of roll	400°/s
T-O run	110 m (360 ft)
Landing run	300 m (985 ft)
Range, internal fuel	156 n miles (290 km; 180 miles)
Ferry range	432 n miles (800 km; 497 miles)
g limits	+12/−10

UPDATED

Single-piece canopy identifies the extraction system-equipped Sukhoi Su-31M *(Jane's/Paul Jackson)*
NEW/0525054

Sukhoi Su-31T single-seat aerobatic competition aircraft with separate windscreen and canopy *(Jane's/Paul Jackson)*
NEW/0525053

SUKHOI Su-38L

TYPE: Agricultural sprayer.

PROGRAMME: Design started, as Su-38, under Boris Rakitin, August 1993; originally based closely on Su-29 sportplane; construction of prototype began January 1994 but curtailed due to financial situation in Russia. Promotion resumed in 1998, under S-38L designation and with considerable change of detail design, including replacement of VOKBM M-14P radial engine. Announced June 2000 that order signed to build prototype, now known as Su-38L. Prototypes fabricated by SmAZ; target of five complete (including two static test) aircraft by second quarter of 2002; first two under assembly at Sukhoi OKB workshops by September 2000. First flight (01) 27 July 2001; public debut at Moscow Salon, 14-19 August 2001, when 01 shown with (static) and without (flying) winglets and then unflown 02 exhibited statically. Compared with earlier drawings and mockup, prototypes have increased length due to additional fuselage plug ahead of wings. By June 2002 two (of three planned) prototypes had completed 60 missions of projected 100- to 150-sortie flight programme, intended to lead to initial certification in March 2003. Initial production by SmAZ with target of 20 in 2003, rising to 40 in 2004, and potential for up to 100 aircraft per year. Second production line intended by Razdanmash AOZT of Armenia, subject to acquisition of US$23 million for flight test and tooling.

Sukhoi Su-38L prototype flying with winglets removed
NEW/0525055

Sukhoi Su-80 prototype on an early test flight near Zhukovsky

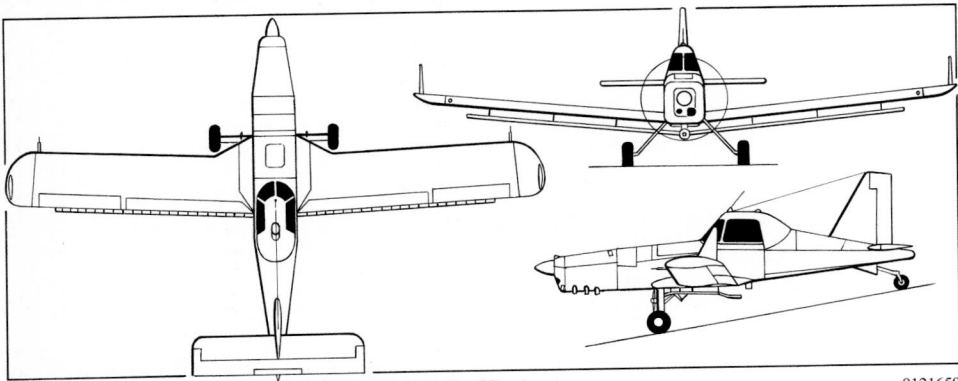

Sukhoi Su-38L agricultural aircraft *(Jane's/James Goulding)*   0121659

CUSTOMERS: Initial orders for 20 anticipated by December 2002; estimated market for 500 in Russia.

COSTS: Unit cost US$100,000 to US$150,000 (2002); direct operating cost US$300 per hour (2001).

DESIGN FEATURES: Conventional low-wing monoplane. Constant-chord wings swept slightly forward and with winglets and root nibs; sweptback fin, plus underfin doubling as tailwheel mount. Chemical hopper between engine and cockpit, capacity 500 litres (132 US gallons; 110 Imp gallons). Service life 3,000 hours/10 years.

FLYING CONTROLS: Conventional and manual. Horn-balanced single-piece elevator and rudder; trim tab in elevator; trailing-edge flaps.

LANDING GEAR: Tailwheel type; fixed. Mainwheel size 8.00-6, tailwheel size 310×135. Hydraulic brakes.

POWER PLANT: One 184 kW (247 hp) LOM M337S six-cylinder engine operating on a 1:2 mix of A-76 and A-92 mogas; LOM B-546 three-blade ground-adjustable propeller. Plans for uprated, 186 kW (250 hp) version in production aircraft. Fuel tank in each wing; total capacity 210 litres (55.5 US gallons; 46.2 Imp gallons).

ACCOMMODATION: Pilot only; baggage shelf behind seat. Cockpit is pressurised to prevent ingress of chemicals, and incoming air is filtered.

EQUIPMENT: Transland underwing spraybar. Cable cutter/deflector standard.

DIMENSIONS, EXTERNAL: None released.

WEIGHTS AND LOADINGS:
Normal T-O and landing weight   1,200 kg (2,646 lb)

PERFORMANCE (estimated):
Operating speed	81-97 kt (150-180 km/h; 93-112 mph)
Stalling speed	41 kt (76 km/h; 48 mph)
Typical operating altitude	1 to 15 m (3 to 50 ft)
Ferry range:	
with spraybars	486 n miles (900 km; 559 miles)
without spraybars	648 n miles (1,200 km; 745 miles)

UPDATED

SUKHOI Su-80

TYPE: Twin-turboprop transport.

PROGRAMME: First, largest and most advanced design by Sukhoi under *konversiya* programme of former Soviet industry; to replace L-410, An-28, Yak-40 and An-24; certification intended to FAR Pt 25, JAR 25 and AP-25. Work began 1989 on medical evacuation aircraft, then designated S-80, under order from Ministry of Public Health, by Sukhoi-Europe/Asia joint stock company, with founder members Sukhoi Design Bureau ASIC, KnAAPO (which see), Rybinsk Motor Engineering Design Bureau, Ramenskoye Instrument Engineering Design Bureau, and Instrument Engineering R&D Institute. Included in government's civil aviation plan, but minimal funding by

public money. Model displayed 1989 Paris Air Show; funding ended with collapse of USSR; project revived 1992, with priority on more marketable passenger and passenger/cargo variants, with imported engines, propellers and avionics.

Manufacture of flying prototype and test airframe under way at KnAAPO's Komsomolsk-on-Amur plant by 1993; Rybinsk TVD-500 turboprop engines originally intended; agreement with General Electric to fit CT7-9 turboprops early 1995; design further refined and military versions reintroduced 1996; first flight scheduled second half 1996 but repeatedly postponed; prototype (RA-82911, in S-80GP configuration) shown on production line on 15 May 1998 to representatives of Border Guards, MoD, and Magadan, Khabarovsk and Petropavlovsk-Kamchatski airlines; was handed over to Sukhoi for flight test preparation in January 2000 and relocated to Zhukovsky; first flight 4 September 2001; had flown 40 sorties by May 2002. Second prototype was to follow in late 2002; two static airframes (02 and 04) for structural testing, of which first was delivered to SIBNIA at Novosibirsk in early 1998 and second was to have been completed in third quarter of 2001. By early 2002, had been redesignated Su-80. Russian AP-25 certification planned for 2003, followed by FAR Pt 25 in 2004. First four production aircraft being manufactured at KnAAPO by early 2002; these have 1.40 m (4 ft 7 in) fuselage stretch, compared with prototype, increasing passenger capacity from original 26 to 30.

CURRENT VERSIONS: **S-80GP:** Basic cargo/passenger *(gruzo/passazhirski)* version.

Detailed description applies specifically to S-80GP; generally to all versions:

S-80A: Optimised for Arctic operation; otherwise as S-80GP.

S-80GR: Geological survey version.

S-80M: Medical evacuation (10 casualties) version.

S-80P: Passenger *(passazhirski)* version.

S-80PT: Patrol transport, embodying Leninets Strizh (martin) avionics suite, with undernose 360° search radar and rotating electro-optic (FLIR/LLLTV) sensor turret under centre-fuselage; one operator's console in cabin. Able to perform 6 to 9 hour patrol missions over sea and land frontiers up to 189 n miles (350 km; 217 miles) from base; personnel and cargo transportation; and airdropping up to 20 paratroops and/or cargo. Flight crew of two or three. Overall length 17.45 m (57 ft 3 in).

S-80R: Fisheries patrol version.

S-80TD: Troop, paratrooop (total 21), freight and medevac transport, generally similar to civil S-80GP, but with mechanised cargo-handling equipment.

CUSTOMERS: Market for 360 to 500 domestic sales estimated in 2001. Interest from Delta-K of Yakutia in 2000; export prospects include Argentina, China, Czech Republic, Indonesia, Malaysia, Thailand and Vietnam. Proposed for Russian Air Forces' future tactical military transport aircraft requirement. By December 2001, 13 Russian airlines had announced intentions to acquire total of 68 Su-80GPs.

COSTS: Notionally funded by Russian Federation government but, by early 1997, KnAAPO had invested over Rb20 billion in development, compared with Rb1.5 billion received from official sources. Of US$20 million expended by mid-2000, two thirds provided by KnAAPO; thereafter, Sukhoi providing funds for flight test programme, estimated as further US$20 million. Flyaway price US$5.5 million to US$6 million (2001 estimate).

DESIGN FEATURES: Utility freighter with unobstructed rear access for loading and wide track landing gear. Basically conventional high-wing, podded fuselage, twin-boom, rear-loading configuration, but with short tandem-wing surfaces between each tailboom and rear fuselage; unswept wings of high aspect ratio with no dihedral or anhedral; large-span constant chord inner panels; small sweptback winglets on tapered outer panels; sweptback vertical tail surfaces, toed slightly inward, with bridging horizontal surfaces. Systems, accessories and components of Su-25, Su-27 and Su-35 embodied in S-80. Automatic built-in systems testing.

FLYING CONTROLS: Conventional and manual. Actuation by rods and cables; flaps in three sections per wing with one section inboard of boom; split ailerons with two trim tabs

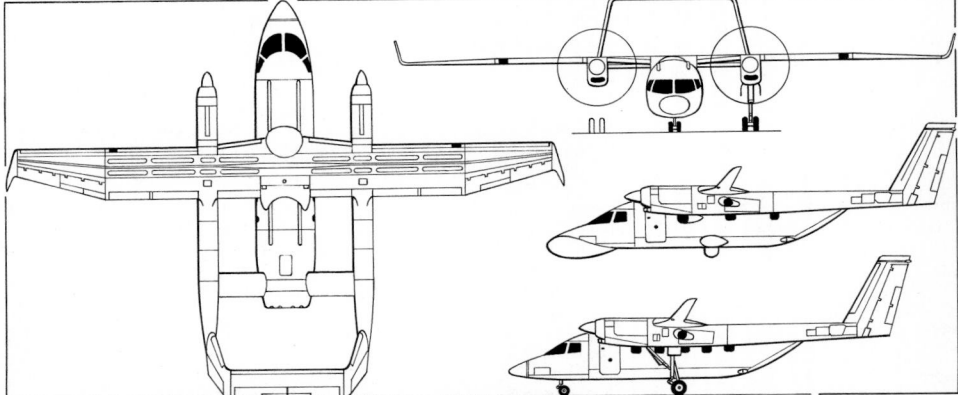

Prototype Sukhoi Su-80GP, with additional side view (upper) of S-80PT patrol transport *(Jane's/James Goulding)*   0121656

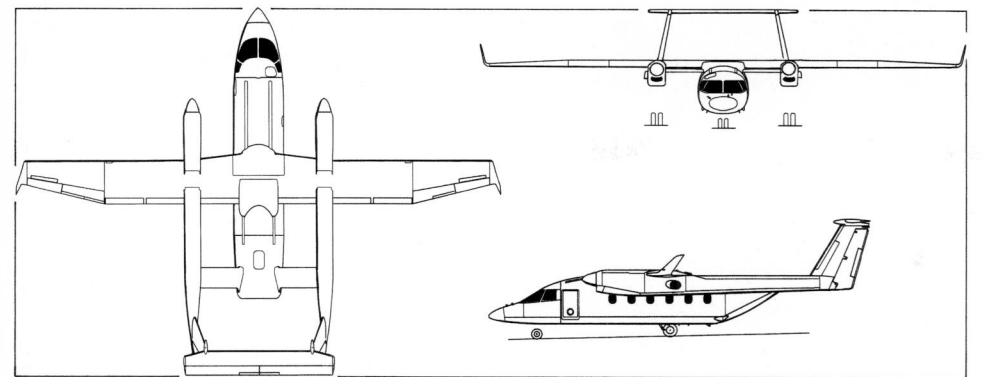

Production version of Sukhoi Su-80   0100419

on port side and one on starboard, electrically actuated; electrically actuated trim tab in each rudder.

STRUCTURE: Materials used in construction comprise 70 per cent aluminium alloy, 8 per cent composites, 6 per cent steel, 6 per cent titanium, 2.5 per cent stainless steel, 7 per cent other non-metals and 0.5 per cent other metals.

Fuselage reinforcement band in line with propellers.

LANDING GEAR: Retractable tricycle type; main units retract rearwards and are enclosed by four sequenced doors per side; nose unit retracts rearwards with two sequenced doors; twin wheels on each unit; main units retract into tailbooms; mainwheel tyre size 660×200-356; nosewheel tyres 500×170-254; nosewheels steerable ±38°.

POWER PLANT: Two 1,305 kW (1,750 shp) General Electric CT7-9B turboprops; production aircraft with engines built locally by GE-Rybinsk Aero Engines; Hamilton Sundstrand 14RF-35 or Dowty feathering and reversible-pitch four-blade propellers. Two fuel tanks, total capacity 2,350 litres (621 US gallons; 517 Imp gallons).

ACCOMMODATION: Two pilots and 29 passengers; optionally 30th passenger in place of co-pilot. All accommodation pressurised. Baggage space, lavatory, wardrobe or galley at front, as specified by customer. Available in 'Salon' configuration with nine, 12 or 16 passenger seats. Typical freighter has a row of seats behind flight deck and unobstructed main hold for a small vehicle or cargo. Door at centre of cabin on port side; hydraulically actuated; rear-loading ramp; emergency door on starboard side, opposite main door.

SYSTEMS: Electrical system developed by Lucas Aerospace and Auxilec. Anti-icing system. APU for autonomous operation at remote, unprepared sites.

AVIONICS: Integration by Elektroavtomatika.

Comms: Com/nav, identification and ATC equipment of Russian manufacture. VOR/DME/ILS for ICAO Cat. II operation by Rockwell Collins.

Flight: Elektroavtomatika PNK-80 navigation system. AFCS and autopilot by Rockwell Collins; satellite nav system with Litton computer.

Instrumentation: Rockwell Collins Pro Line 2 five-screen EFIS.

EQUIPMENT: Options include equipment for air photography.

ARMAMENT: (S-80PT): Typically 23 mm GSh-23L gun pod pylon-mounted on starboard side of cabin; four underwing pylons for electro-optical ASM, 20-round rocket pack, cluster of eight Vikhr tube-launched missiles and R-60 (AA-8 'Aphid') self-defence AAM.

DIMENSIONS, EXTERNAL:	
Wing span	23.17 m (76 ft 0¼ in)
Wing chord: at root	2.16 m (7 ft 1 in)
at tip	1.20 m (3 ft 11¼ in)
Wing aspect ratio	12.2
Length overall: prototype	18.26 m (59 ft 10¾ in)
production	19.66 m (64 ft 6 in)
Height overall	5.52 m (18 ft 1¼ in)
Tailplane span	4.88 m (16 ft 0¼ in)
Wheel track	5.60 m (18 ft 4½ in)
Wheelbase: prototype	6.50 m (21 ft 4 in)
Propeller diameter (each)	2.65 m (8 ft 8½ in)
Propeller ground clearance	1.10 m (3 ft 7¼ in)
Distance between propeller centres	5.60 m (18 ft 4½ in)
Passenger door: Height	1.75 m (5 ft 9 in)
Width	0.90 m (2 ft 11½ in)
Service door: Height	1.34 m (4 ft 4¾ in)
Width	0.64 m (2 ft 1¼ in)
Type III emergency exits (each):	
Height	0.905 m (2 ft 11½ in)
Width	0.54 m (1 ft 9¼ in)

DIMENSIONS, INTERNAL:	
Cabin: Length: prototype	7.80 m (25 ft 7 in)
production	9.20 m (30 ft 2¼ in)
Max width	2.17 m (7 ft 1½ in)
Max height	1.83 m (6 ft 0 in)
Volume, approx	20.0 m³ (706 cu ft)

AREAS:	
Wings, gross	44.00 m² (473.6 sq ft)
Ailerons (total)	3.27 m² (35.20 sq ft)
Trailing-edge flaps (total)	7.05 m² (75.90 sq ft)
Leading-edge slats (total)	4.68 m² (50.38 sq ft)
Fins (total)	6.64 m² (71.47 sq ft)

Rudders (total)	6.22 m² (66.95 sq ft)
Horizontal tail surfaces	5.86 m² (63.08 sq ft)

WEIGHTS AND LOADINGS (prototype):	
Max payload	3,300 kg (7,275 lb)
Max fuel weight	2,350 kg (5,181 lb)
Max T-O weight	13,500 kg (29,762 lb)
Max landing weight	13,350 kg (29,432 lb)
Max wing loading	306.8 kg/m² (62.84 lb/sq ft)
Max power loading	5.18 kg/kW (8.50 lb/shp)

PERFORMANCE (estimated):	
Max cruising speed	254 kt (470 km/h; 292 mph)
Max certified altitude	7,600 m (24,940 ft)
Balanced field length	850 m (2,790 ft)
T-O run	555 m (1,820 ft)
Landing run	840 m (2,756 ft)
Landing run with propeller reversal	460 m (1,509 ft)
Range: with 30 passengers	
	756 n miles (1,400 km; 870 miles)
with 1,950 kg (4,299 lb) payload	
	1,322 n miles (2,450 km; 1,522 miles)

OPERATIONAL NOISE LEVELS: Designed to conform to FAR Pt 36 standards

UPDATED

SUKHOI RRJ

TYPE: Regional jet airliner.

PROGRAMME: In early 2001, Rosaviakosmos and The Boeing Company agreed to joint development and marketing of the Russian Regional Jet (RRJ; transliterated into Cyrillic as RRZh, but also known by Sukhoi as *grazhdanskie samolet*: Civil Aircraft). At the Paris Air Show on 21 June 2001 it was announced that AVPK Sukhoi (Sukhoi Civil Aircraft) will lead the design and manufacture of the aircraft, with Ilyushin holding responsibility for certification, and Boeing handling marketing, sales, leasing and after-sales support. Three firms signed agreement 20 July 2001. Provisional configuration and specification announced 13 August 2001, on eve of Moscow Salon. Initial phase of feasibility studies between June and December 2001; second phase completed in July 2002. Full launch awaits receipt of 200 orders. First flight expected in 2004 or 2005, with Russian AP-25 and FAR/JAR 25 certification and service entry in 2007.

CURRENT VERSIONS: **RRJ-55:** Shrunk version; two fuselage plugs, each of 1.62 m (5 ft 4 in), removed fore and aft of wing; **RRJ-55ER:** Extended-range version. **RRJ-55LR:** Long-range version.

RRJ-75: Baseline version.

RRJ-75ER: Extended-range version.

RRJ-75LR: Long-range version.

RRJ-95: Stretched version; two plugs, each of 3.25 m (10 ft 8 in), fore and aft of wing.

RRJ-95ER: Extended-range version.

RRJ-95LR: Long-range version.

CUSTOMERS: Aeroflot has initial requirement for up to 30, having signed MoU in August 2001. Russian market estimated as 150; additional 500 export sales possible by 2020.

COSTS: Development costs estimated at US$600 million (2002), plus similar amount for engine.

DESIGN FEATURES: Swept wings and horizontal and vertical tail surfaces.

LANDING GEAR: Retractable tricycle type.

POWER PLANT: Two turbofans of unspecified type, pod-mounted beneath wings. Engine selection process began in 2002 with General Electric (CF34-8E), Snecma/NPO Saturn (SM146), Rolls-Royce Deutschland (BR710) and P&WC/Aviadvigatel (PW800) among announced contenders. By mid-2002 choice narrowed to SM146 and PW800. Thrust 53 to 67 kN (12,000 to 15,000 lb) each, according to version. Fuel capacity 13,000 litres (3,434 US gallons; 2,860 Imp gallons) in all versions.

ACCOMMODATION: RRJ-55 seats up to 55 passengers; RRJ-75 seats up to 75 passengers; RRJ-95 seats up to 95 passengers, all at 81 cm (32 in) seat pitch, five abreast.

All data are provisional.

DIMENSIONS, EXTERNAL:	
Wing span	26.50 m (86 ft 11¼ in)
Length overall: RRJ-55	25.60 m (83 ft 11¾ in)
RRJ-75	28.84 m (94 ft 7½ in)
RRJ-95	32.09 m (105 ft 3¼ in)

RRJ-55, the shortest Sukhoi regional jet *(Jane's/Paul Jackson)*   0121686

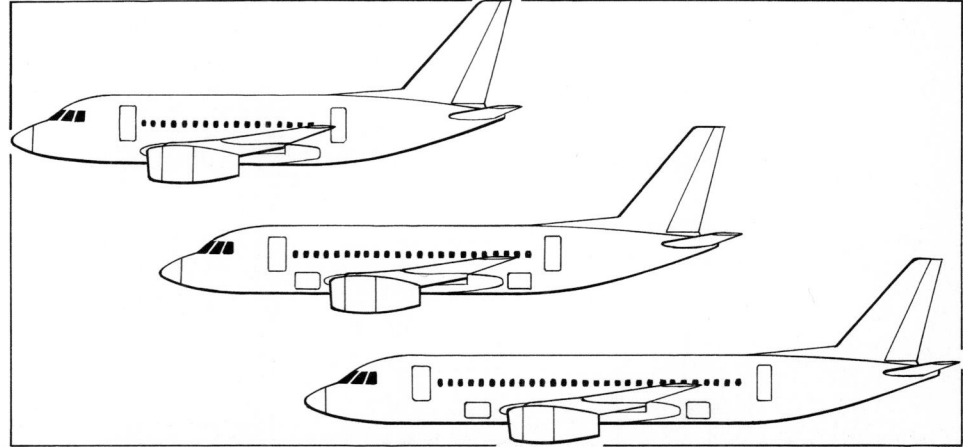

The three projected versions of Sukhoi RRJ: 55 (top), 75 and 95 *(Jane's/James Goulding)*   0130926

WEIGHTS AND LOADINGS:
Max payload: RRJ-55	6,050 kg (13,338 lb)
RRJ-75	8,250 kg (18,188 lb)
RRJ-95	10,450 kg (23,038 lb)
Max T-O weight: RRJ-55	31,640 kg (69,754 lb)
RRJ-55ER	33,245 kg (73,292 lb)
RRJ-55LR	34,915 kg (76,974 lb)
RRJ-75	35,910 kg (79,168 lb)
RRJ-75ER	37,510 kg (82,695 lb)
RRJ-75LR	39,180 kg (86,377 lb)
RRJ-95	40,525 kg (89,342 lb)
RRJ-95ER	42,130 kg (92,880 lb)
RRJ-95LR	43,500 kg (96,562 lb)

PERFORMANCE:
Range: RRJ-75	1,450 n miles (2,685 km; 1,668 miles)
RRJ-75LR	3,239 n miles (6,000 km; 3,728 miles)

UPDATED

SUKHOI Su-XX

In 2001, Sukhoi, in conjunction with its UK-based sales agent Richard Goode Aerobatics, was reported to be developing a new Unlimited category competition aerobatic aircraft under the provisional designation Su-XX. The aircraft will be of similar size to the Su-31, with an airframe making more extensive use of composites, and may feature retractable landing gear. Initially it will be a single-seater, but a tandem-seat version may follow, if demand warrants. Power plant will be a Voronezh M9F. First flight expected in 2003 with service entry in time for participation in the 2005 World Air Sports Olympics.

VERIFIED

TAGANROG (TANTK)

TAGANROGSKY AVIATSIONNYI NAUCHNO-TEKHNICHESKY KOMPLEKS (Taganrog Aviation Scientific-Technical Complex)

Address as for Beriev (which see).

This plant, which is integrated with the Beriev design bureau and now part of AVPK Sukhoi, was responsible for production of the Tupolev Tu-142 'Bear' strategic bomber (until 1994) and conversion of Ilyushin Il-76 transports to Beriev A-50 AWACS (until 1992). Current activities centre on Tu-142 overhaul. Plans for manufacture of Tu-334 airliners were halted following transfer of the programme to RSK 'MiG' by a Presidential decree of 1999; and establishment of a production line for the Beriev Be-32/Be-132 twin-turboprop transport has yet to show results. The plant is 44 per cent owned by the Russian state, and part by the Tupolev holding company.

UPDATED

TECHNOAVIA

NAUCHNO-KOMMERCHESKY FIRMA TECHNOAVIA (Technoavia Scientific and Commercial Firm)

bulvar Kronshtadsky 7A, 125212 Moskva
Tel: (+7 095) 452 56 03
Fax: (+7 095) 452 56 94
e-mail: info@tehnoavia.ru
Web: http://www.tehnoavia.ru
GENERAL DIRECTOR: Vyatcheslav Kondratiev
CHIEF AERODYNAMICIST: Yuri Kovyzhenko

Technoavia was formed in 1991 by former Sukhoi designer Vyatcheslav Kondratiev and currently produces the SM-92 and SP-55M light aircraft at the Smolensk Aircraft Plant (SmAZ).

UPDATED

Technoavia SM-92 Finist (VOKBM M-14 radial engine) *(Jane's/Paul Jackson)*   0121722

TECHNOAVIA SM-92 FINIST

TYPE: Light utility transport.
PROGRAMME: Design to FAR Pt 23 and JAR 23 standards, started July 1992; construction of first of two prototypes (RA-44482) began January 1993; first flight, as SM-92 Finist (name of a magical bird that was transformed into a prince), made 28 December 1993; second Finist (RA-44484) completed round-the-world flight through Europe, Atlantic, Canada, Alaska and Siberia in August 1995 covering 16,200 n miles (30,000 km; 18,640 miles) in 160 flying hours; components production at SmAZ, Smolensk. Sixth aircraft (RA-44493) completed as **SM-92P** armed version (described in 2001-02 and previous editions). AP-23 certification received by late 1998 and Hungarian FAR Pt 23 certification in August 2000.

In 1991, Moravan (Zlin) of Czech Republic (which see) bought plans of SM-92 for US$500,000 and launched local production as Z 400 Rhino. Russian manufacture appears to have ceased.

CURRENT VERSIONS: **SM-92:** *As described.*

SM-92P: Border guard version. Described in 2001-02 and previous editions. Prototype, RA-44493, displayed at Moscow Air Show '95; second aircraft (RA-44494) complete by 1997, but converted to SM-92 Finist. No further manufacture known.

Z 400 Rhino: See Zlin entry in Czech Republic section.
CUSTOMERS: Fifteenth Finist delivered June 2001 to customer in Ukraine. Production of 16 for Russian Federation border guard (which eventually requires 300) was disrupted by problems at Smolensk aircraft factory; alternative manufacturer being sought. Regional government of Yakutia ordered 14 in October 1997, but none reported in service by 2002. First delivery (aircraft No. 3, RA-44485) to Mike Crymble in UK, 21 January 1995; RA-44487 to Sport-Para Centrum, Antwerp, Belgium, 6 July 1995.

COSTS: US$240,000 (2001).
DESIGN FEATURES: Rugged light transport; in approximate class of out-of-production DHC-2 Beaver, but less powerful and lower in cost than most used Beavers. Sweptback fin and rudder; small dorsal fin; tailplane mounted on fin, with single bracing strut each side. Wide CG range.

Basic aircraft accommodates up to seven persons with baggage. Convertible under field conditions to transport 600 kg (1,323 lb) freight; two stretcher patients and attendant with medical equipment; to drop six trainee parachutists or four firefighters with parachutes and firefighting equipment. Can carry hopper for 600 kg (1,323 lb) agricultural chemicals in cabin, with spraybar, or cameras for forest surveillance, patrolling electric power lines, gas pipeline inspection and similar duties.

Special wing section by Technoavia and CAHI (TsAGI). No sweep; thickness/chord ratio 15 per cent; dihedral 2°; incidence 3° at root, 1° at tip.
FLYING CONTROLS: Conventional and manual. Ailerons and elevators pushrod-actuated; rudder cable-actuated. Electrically operated, three-position (0, 20, 40°), two-section single-slotted flaps on each wing; horn-balanced rudder and elevators, with fluted skin; trim tab on starboard elevator; ground-adjustable tab on rudder; three-section tab along entire trailing-edge of each aileron, centre section being adjustable on ground.
STRUCTURE: All-aluminium, stressed-skin semi-monocoque construction. Simple, reliable structure, with no expensive or exotic materials; repair possible under field conditions. Airframe life 10,000 hours or 20,000 landings, with 'on condition' extension.
LANDING GEAR: Non-retractable tailwheel type with medium-pressure mainwheel tyres, size 600×180 on KT-317 wheels. Cantilever faired tubular steel main legs; steerable, semi-castoring tailwheel with 255×110 tyre on tubular steel strut. Tyre pressure 2.45 bar (35.5 lb/sq in) on mainwheels; 2.95 bar (43 lb/sq in) on tailwheel. Wheel/skis and amphibious or plain floats optional. Pedal-operated disc brakes, with parking lock. Minimum turning radius 6.6 m (21 ft 8 in).
POWER PLANT: One VOKBM M-14Kh air-cooled nine-cylinder radial engine, rated at 265 kW (355 hp) for take-off and 213 kW (286 hp) maximum continuous, driving a Mühlbauer MTV-3-B-C/L250-21 three-blade variable-pitch propeller. Engine TBO 1,000 hours; total life 3,000 hours. Two fuel tanks in wing leading-edge, total capacity 392 litres (104 US gallons; 86.2 Imp gallons); usable capacity 380 litres (100.4 US gallons; 83.5 Imp gallons). Wingtip tanks optional, capacity 200 litres (52.8 US gallons; 44 Imp gallons) each. Oil capacity 30 litres (7.9 US gallons; 6.6 Imp gallons).
ACCOMMODATION: Pilot and six passengers, in pairs; quickly removable seats with folding armrests and back; dual controls standard for pilot training; adjustable rudder pedals. Small baggage container (or medical equipment stowage for ambulance version) on port side at rear of cabin. Forward-hinged, jettisonable door each side of flight deck; large rearward-sliding passenger/freight door on port side of cabin, openable in flight. Blister windows to flight deck and cabin. Steps on mainwheel legs and cable handholds for access to flight deck; removable tubular steel ladder beneath cabin door.

Convertible in field to transport 600 kg (1,323 lb) of freight; two stretcher patients and two attendants; six trainee parachutists or four smoke-jumpers with parachutes and firefighting equipment. Provision for carrying hopper for 600 kg (1,323 lb) of agricultural chemicals in cabin, or cameras for forest surveillance, patrolling electric power lines, gas pipeline inspection and similar duties. Clearance adequate for underbelly pannier. Cabin heated and ventilated.
SYSTEMS: Pneumatic system for engine starting, pressure 49 bar (710 lb/sq in). Electrical system provides 36/115 V AC power at 400 Hz and 28.5 V DC power, with 20NKBN-25 battery.
AVIONICS: *Comms:* Two Bendix/King KY 96A VHF transceivers, KA 134 audio control, KR 87A ADF and KT 76A transponder; Garmin GPS 150.

DIMENSIONS, EXTERNAL:
Wing span	14.60 m (47 ft 10¾ in)
Wing chord, constant	1.40 m (4 ft 7 in)
Wing aspect ratio	10.4
Length overall	9.30 m (30 ft 6¼ in)
Height: overall	3.08 m (10 ft 1¼ in)
fuselage horizontal	3.94 m (12 ft 11 in)
Tailplane span	5.53 m (18 ft 1¾ in)
Wheel track	2.95 m (9 ft 8¼ in)
Wheelbase	6.33 m (20 ft 9¼ in)
Propeller diameter	2.50 m (8 ft 2½ in)

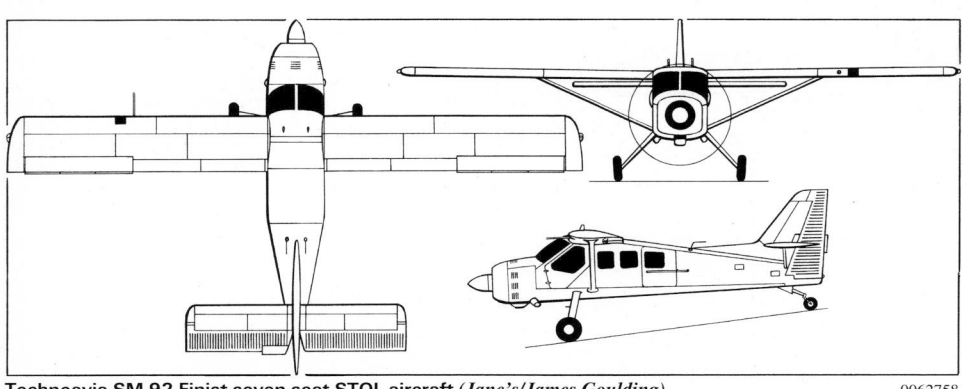

Technoavia SM-92 Finist seven-seat STOL aircraft *(Jane's/James Goulding)*   0062758

Propeller ground clearance (tail up)	0.22 m (8¾ in)
Freight door: Height	1.12 m (3 ft 8 in)
Width	1.35 m (4 ft 5¼ in)

DIMENSIONS, INTERNAL:
Cabin: Length	3.40 m (11 ft 1¾ in)
Max width	1.27 m (4 ft 2 in)
Max height	1.38 m (4 ft 6¼ in)
Volume	5.2 m³ (184 cu ft)

AREAS:
Wings, gross	20.44 m² (220.0 sq ft)
Ailerons (total)	2.38 m² (25.62 sq ft)
Flaps (total)	3.28 m² (35.31 sq ft)
Fin	2.13 m² (22.93 sq ft)
Rudder	1.63 m² (17.55 sq ft)
Tailplane	3.73 m² (40.15 sq ft)
Elevators (total)	2.91 m² (31.32 sq ft)

WEIGHTS AND LOADINGS:
Operating weight empty	1,500 kg (3,307 lb)
Payload: max	600 kg (1,323 lb)
with full fuel	550 kg (1,212 lb)
Max T-O and landing weight	2,350 kg (5,180 lb)
Max wing loading	115.0 kg/m² (23.55 lb/sq ft)
Max power loading	8.87 kg/kW (14.59 lb/hp)

PERFORMANCE:
Never-exceed speed (V$_{NE}$)	140 kt (260 km/h; 161 mph)
Max level speed	124 kt (230 km/h; 143 mph)
Max cruising speed	108 kt (200 km/h; 124 mph)
Econ cruising speed	92 kt (170 km/h; 106 mph)

Stalling speed, power off:
flaps up	63 kt (115 km/h; 72 mph)
T-O flap	57 kt (105 km/h; 66 mph)
full flap	54 kt (100 km/h; 62 mph)
Max rate of climb at S/L	300 m (985 ft)/min
Service ceiling	3,000 m (9,840 ft)
T-O and landing run on grass	250 m (820 ft)

Range, 40 min reserves: with max payload:
at max level speed	332 n miles (615 km; 382 miles)
at econ cruising speed	
	594 n miles (1,100 km; 683 miles)
with max fuel and 550 kg (1,212 lb) payload	
	647 n miles (1,200 km; 745 miles)
Endurance with max fuel	7 h 30 min

UPDATED

Technoavia-converted Turbo Finist with wingtip pods *(Yefim Gordon)*   *NEW*/0525942

Technoavia SMG-92 sport parachuting turboprop *(G Badovszky)*   0121675

TECHNOAVIA SMG-92 TURBO FINIST

TYPE: Light utility turboprop.

PROGRAMME: Developed from SM-92 Finist, with emphasis on sport parachutist dropping; commissioned and financed by G-92 Commerce Ltd, of Nagybanyai 61, H-1025 Budapest, Hungary (e-mail: budata@euroweb.hu). First flight (newly built aircraft, HA-YDF) 7 November 2000; Hungarian type certificate awarded 14 December 2000; delivery to Wingglider Ltd in UK, January 2001. Second conversion, HA-YDG, produced from RA-44485, exhibited at Aero '01, Friedrichshafen, April 2001, and stationed with Skydive Center, Bad Saulgau, Germany, from June 2001. By mid-2002 Finist prototype RA-44482 had also received a Walter M 601 engine as well as wingtip pods.

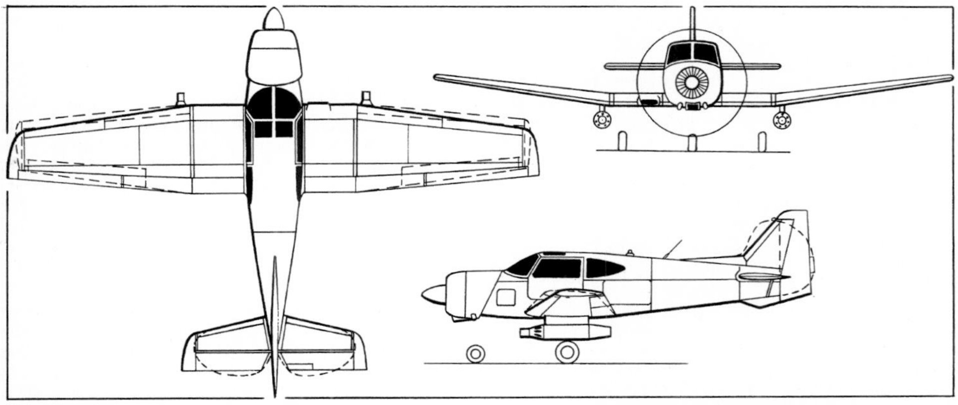

Airframe and engine mounting produced by SmAZ at Smolensk; engine installation and cowling, paradrop step and handrails in vicinity of cabin door, new electrical system, instrument panel and engine controls by Aerotech Slovakia AS at Bratislava. Marketing by PB Aircraft, St Moritz, Switzerland (Web: http://www.turbofinist.com).

POWER PLANT: One 400 kW (536 shp) Walter M 601D-2 turboshaft, driving an Avia V-508D-2 three-blade, fully feathering, reversible-pitch propeller. Optional ferry tank.

ACCOMMODATION: Pilot and 10 parachutists; alternatively, pilot and up to seven passengers on lightweight seats.

Data generally as for SM-92 Finist, except that below.

DIMENSIONS, EXTERNAL:
Length overall	9.925 m (32 ft 6¾ in)
Wheelbase	6.59 m (21 ft 7½ in)

WEIGHTS AND LOADINGS:
Weight empty	1,450 kg (3,197 lb)
Max T-O weight: paradrop	2,700 kg (5,952 lb)
passenger	2,350 kg (5,180 lb)
Max landing weight	2,350 kg (5,180 lb)

PERFORMANCE:
Never-exceed speed (V$_{NE}$)	164 kt (305 km/h; 189 mph)
Max operating speed (V$_{MO}$)	143 kt (265 km/h; 165 mph)
Max level speed at 915 m (3,000 ft)	
	159 kt (295 km/h; 183 mph)
Cruising speed	130 kt (240 km/h; 149 mph)
Max rate of climb at S/L	457 m (1,500 ft)/min

Stalling speed, engine off:
flaps up	63 kt (115 km/h; 72 mph)
flaps down	54 kt (100 km/h; 63 mph)
Time to 4,000 m (13,120 ft)	11 min
Max operating altitude	5,940 m (19,500 ft)
Descent from 4,000 m (13,120 ft)	5 min 30 s
T-O run	250 m (820 ft)
T-O to 15 m (50 ft)	457 m (1,500 ft)
Landing run with propeller reversal	145 m (475 ft)
Max range with optional fuel	
	324 n miles (600 km; 372 miles)

UPDATED

Technoavia SM-94, with Yak-18T configuration shown by broken lines *(Jane's/James Goulding)*   0015130

TECHNOAVIA SM-94

This proposed aerodynamic update of Yak-18T (which see) was last described fully in the 1999-2000 edition, with further brief additional details in 2000-01, following which nothing further was heard. However in August 2002, a development aircraft (RA-44542) was seen, apparently being the first Yak-18T fitted with the straight-edged fin and rudder intended for the SM-94, but not its modified wing.

NEW ENTRY

Yak-18T/SM-94 fitted with the latter's straight-edged fin/rudder and two-piece windscreen seen on recently-built Yak-18s *(Yefim Gordon)*   *NEW*/0525971

TECHNOAVIA SP-55M

TYPE: Aerobatic single-seat sportplane.

PROGRAMME: Initial batch of five aircraft under construction in 2000 at Progress plant at Arsenyev; Russian prototype unregistered; initial production aircraft, RA-44547, delivered in January 2001 to Richard Goode Aerobatics in UK.

COSTS: US$105,000 (2002).

DESIGN FEATURES: Mid-wing configuration with symmetrical section of original Technoavia design; no dihedral, anhedral or incidence. Development of Yak-55M, last described in 1998-99 *Jane's*. Modifications include redesigned rear fuselage turtledeck; new fin and tailplane fillets; reprofiled wingtips; redesigned flying surfaces with composites skin; steel engine firewall; new engine cowling; new instrument panel with US instruments and avionics; modified fuel and oil systems to FAR Pt 23 requirements; modified pneumatic system; pilot-controlled oil and carburettor heat doors; new entry steps and assist handle on fuselage and landing gear legs; inclined seat for increased *g* tolerance; new safety harness; baggage compartment and smoke system.

FLYING CONTROLS: Conventional and manual. Near-full-span ailerons have narrower chord than those of Yak-55M, without horn balances; elevators and rudder of greater area than Yak-55M. Trim tab on each elevator. Suspended balance tab below each aileron.

STRUCTURE: Generally similar to Yak-55M. Metal, single-spar wing with composites-covered ailerons and composites tips. Metal fuselage with steel firewall; composites-covered elevator and rudder.

LANDING GEAR: Non-retractable tailwheel type with spring steel legs on all three units; KT-98 mainwheels with 400×150 tyres; lockable, free-castoring tailwheel.

POWER PLANT: One 265 kW (355 hp) VOKBM M-14P air-cooled, nine-cylinder radial engine driving an MT-Propeller MTV-9-B-C/CL260-27 three-blade, constant-speed (hydraulic) metal propeller. Fuel in wing tanks, total capacity 120 litres (31.7 US gallons; 23.4 Imp gallons); inverted fuel and oil systems.

ACCOMMODATION: One person in inclined seat with recess for backpack parachute; fixed windscreen and one-piece sliding canopy.

SYSTEMS: Electrical system comprises lightweight 10 Ah generator and storage battery, both of US origin

AVIONICS: Avionics fit of US origin.

EQUIPMENT: Optional smoke generation system; or glider hook.

DIMENSIONS, EXTERNAL:

Wing span	8.00 m (26 ft 3 in)
Wing aspect ratio	5.3
Length overall	7.48 m (24 ft 6½ in)
Height overall	2.225 m (7 ft 3½ in)
Tailplane span	3.51 m (11 ft 6¼ in)
Wheel track	2.76 m (9 ft 0½ in)
Propeller diameter	2.50 m (8 ft 2½ in)

AREAS:

Wings, gross	12.17 m² (131.0 sq ft)
Ailerons (total)	2.66 m² (28.63 sq ft)
Fin	0.40 m² (4.31 sq ft)
Rudder	1.02 m² (10.98 sq ft)
Tailplane	1.40 m² (15.07 sq ft)
Elevators (total)	1.79 m² (19.27 sq ft)

First production Technoavia SP-55M aerobatic lightplane (*Jane's/Paul Jackson*)   NEW/0525058

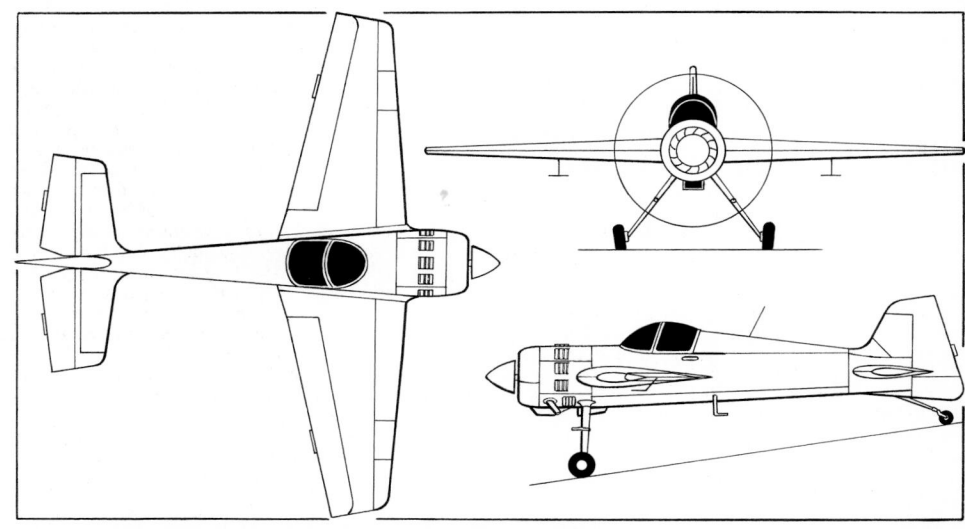

Technoavia SP-55M single-seat sportplane (*Jane's/James Goulding*)   NEW/0525057

WEIGHTS AND LOADINGS:

Weight empty	690 kg (1,521 lb)
Max T-O weight: training	855 kg (1,885 lb)
ferry	955 kg (2,105 lb)
Max wing loading: training	70.3 kg/m² (14.39 lb/sq ft)
ferry	78.5 kg/m² (16.07 lb/sq ft)
Max power loading: training	3.23 kg/kW (5.31 lb/hp)
ferry	3.61 kg/kW (5.93 lb/hp)

PERFORMANCE:

Never-exceed speed (VNE)	194 kt (360 km/h; 223 mph)
Max level speed	173 kt (320 km/h; 199 mph)
Stalling speed	57 kt (105 km/h; 66 mph)
Max rate of climb at S/L	1,080 m (3,543 ft)/min
Max roll rate	360°/s
T-O run	170 m (558 ft)
Range with max fuel at econ cruising speed at 1,000 m (3,280 ft), 7% fuel reserves	
	415 n miles (770 km; 478 miles)
g limits	+9/−6

UPDATED

TUPOLEV

TUPOLEV OAO (Tupolev JSC)

ulitsa Bakrushina 23, Korpus 1, 11250 Moskva
Tel: (+7 095) 238 34 13
Fax: (+7 095) 238 68 41
e-mail: tu@tupolev.ru
Web: http://www.tupolev.ru
PRESIDENT: Aleksandr Polikov

Under decree of 30 June 1999, Tupolev JSC formed 30 July 1999, combining Tupolev design bureau (see below) and Aviastar production plant (which see) at Ulyanovsk, with financial interest from Russian government amounting to 51 per cent, plus 43.6 per cent from Aviastar. JSC also holds intellectual rights to Tupolev designs previously owned by government. New entity was recertified by Interstate Aviation Committee Aviation Registrar in December 2000. Kazan plant will be incorporated later; proposal submitted July 2000.

Other activities include sales and leasing of jet transports, aircraft operation and after-sales support. Tupolev Aviation Leasing Company supplies Tu-204, Antonov An-124 and Yakovlev Yak-40 transports.

Government plans to rationalise Russian aerospace industry, announced May 2001, call for incorporation of Tupolev (with Aviakor, Aviastar and Kazan plants) into RSK 'MiG' grouping, also including Kamov bureau and Sokol, KAPP and Arsenyev plants.

UPDATED

AVIATSIONNY NAUCHNO-TEKHNISHESKY KOMPLEKS IMENI A N TUPOLEVA OAO (Aviation Scientific-Technical Complex named for A N Tupolev JSC)

Naberezhnaya Akademika Tupoleva 17, 111250 Moskva
Tel: (+7 095) 267 25 33
Fax: (+7 095) 267 27 33
e-mail: tu@tupolev.ru
Web: http://www.tupolev.ru
GENERAL DESIGNER: Igor Shevchuk
HEAD OF ECONOMIC RELATIONS: Evgeny Efimov

Tupolev Bureau was founded in 1922 and concentrated primarily on large military and civil aircraft until the early 1990s; has designed 300 aircraft, of which 35 placed in production; current effort is 80 per cent civil programmes, although it was suggested in 1998 that Tupolev had begun preliminary design of a new bomber to replace the Tu-160 and, probably, Tu-95/142; this would enter service some time after 2010 and may be based on earlier studies, such as the

Tu-202 bomber of the 1980s or the more recent Tu-404 700/ 850-seat airliner. The Bureau remains heavily committed to the development of a civil supersonic transport aircraft and of cryogenic fuels for jet and turboprop transports.

ANTK Tupolev has a 5.4 per cent share in Tupolev AO production group.

Tupolev's head office, main design bureau and experimental facility are in Moscow; Tomilino branch and flight research centre at Zhukovsky; and design offices at Samara, Kazan and Voronezh. Share of 2001 government funds for aviation development was 15 per cent, the largest element for certification of Tu-334 airliner.

UPDATED

TUPOLEV Tu-160

NATO reporting name: Blackjack
Unofficial name: Belyi Lebed (White Swan)
TYPE: Strategic bomber.

PROGRAMME: Designed as Aircraft 70 under leadership of V I Bliznuk; programme began 1967, but relaunched following issue of more modest specification in 1970; derived from unbuilt Tu-135 bomber and Tu-144 derivatives; some features from rival Myasishchev M-18. First of two prototypes (70-01) observed by intelligence source at Zhukovsky flight test centre 25 November 1981 (photographed from landing airliner; see 1982-83 *Jane's*); first flew 18 or 19 December 1981; first exceeded M1.0 February 1985; second (production-standard) aircraft first flew 7 October 1984. Third prototype (70-03) set world records for altitude, speed in a closed circuit and weight-to-altitude on 31 October 1989, 20 of these remaining unbroken in 2001. Further nine closed-circuit records established 22 May 1990 by aircraft '70-304'.

Second production aircraft lost pre-delivery, March 1987; US Defense Secretary Frank Carlucci invited to inspect 12th aircraft built, at Kubinka airbase, near Moscow, 2 August 1988; deliveries to 184th Guards Heavy Bomber Aviation Regiment, Priluki airbase,

Tupolev Tu-160 'Blackjack' *(Yefim Gordon)*   *NEW*/0525119

Ukraine, began April 1987; equipment of 1096th HBAR at Engels from 16 February 1992, but only six received before production at Kazan airframe plant terminated 1992; of 100 aircraft due to be built, at least 32 (including prototypes) accounted for by mid-1990s; unconfirmed reports suggest total of 40 having flown, plus three uncompleted at Kazan.

In 2001, Russia was reportedly formulating a Tu-160 upgrade, to be undertaken at Kazan, which includes provision for a conventionally armed cruise missile. Upgrade of 15 aircraft was agreed late 2001 and formally announced at Kazan on 18 January 2002; work will extend service lives to 2030; first upgrade candidate was delivered to Kazan on 5 April 2002.

CURRENT VERSIONS: **Tu-160** ('Blackjack'): Strategic bomber.

Tu-160SK: Commercialised, demilitarised version, as carrier component of Burlak aviation space launch system; Burlak-Diana two-stage rocket, carrying payload, under fuselage on centreline mount. Announced at Singapore Air Show '94; proposed by Russian partners MKB-Raduga, OKB MEI and Tupolev, with German company OHB-System. At the 1995 Paris Air Show, Tu-160 0401 was exhibited statically (the type's Western debut) with a model Burlak rocket below the fuselage. Development of the system continued even after the German government withdrew funding in 1998, and may form the basis of the Ukrainian/PIC HAAL-2000 programme. Further data in 2001-02 and previous *Jane's*.

Tu-160M: Proposed stretched variant carrying two 2,700 n mile (5,000 km; 3,107 mile) range hypersonic Kh-90 missiles.

Tu-160P: Proposed very long-range escort fighter.

Tu-160PP: Proposed escort jammer.

Tu-160R: Proposed strategic reconnaissance platform.

CUSTOMERS: In January 2001, Russia had seven operational Tu-160s; a further eight undergoing refurbishment for delivery over following months; and one under construction.

Of original deliveries, Ukraine government seized 19 at Priluki on achieving independence, pre-empting planned transfer to Engels; purchase of these by Russia was subject to protracted negotiations, and aircraft deteriorated in storage; March 1996 agreement on transfer of 10 best airframes was not implemented; attempts to purchase eight failed in March 1998 and Russia then supposedly abandoned hopes of expanding Tu-160 fleet. However, October 1999 announcement revealed eight to be returned to Russia for refurbishment (together with three Tu-95MSs, for a total of US$285 million) and these were delivered between 5 November 1999 and 21 February 2000. Six others scrapped with US assistance, last being destroyed on 4 February 2001. August 2000 report mentioned further three Tu-160s which Ukraine could return to Russia in part-payment for natural gas deliveries. One further Ukrainian Tu-160 was flown to Poltava Kondratyuk-Shagray aerospace museum on 5 April 2000.

Russia maintained force of six (declared as ALCM carriers under START) at Engels (where 1096th HBAR was redesignated as part of the 121st Guards HBAR, within 22 Air Division, in 1994), plus flying testbed at Zhukovsky; at least four more were derelict at Zhukovsky by 1995. Despite this, 1998 US estimates of Russian combat forces reported total holding of nominally serviceable aircraft as 25, though this would not seem to be possible. Eight more to re-enter service from 2001 after purchase from Ukraine, while plan announced in June 1999 to complete one unfinished Tu-160 at Kazan. This aircraft ('07' *Aleksandr Molodchyi*) was delivered on 5 May 2000. Another, brand new Tu-160 (the first of three more incomplete, unfinished aircraft at Kazan) was due to be delivered in late 2002, and the remaining two will follow, increasing the fleet to 18. At one time there were plans to refurbish some (perhaps four) of the grounded aircraft at Zhukhovsky. Ex-Ukrainian aircraft failed to enter service when planned due to poor condition and need for extensive refurbishment; requirements were still being discussed in early 2002. Tu-160 may re-enter limited production, to meet a stated requirement for 25 operational 'Blackjacks' by 2003, allowing formation of a second regiment. Five of original six at Engels are also named: '01' *Mikhail Gromok*; '02' *Vasily Retsetnikov*; '04' *Ivan Yargin*; '05' *Il'ya Muromets* (1) and '06' *Il'ya Muromets* (2).

On 3 March 1999, the Russian Commonwealth Aerospace Technology Consortium (RCATC) was authorised by the Ukrainian government to sell three

demilitarised Ukrainian Air Force Tu-160s, plus spares, to Platforms International Corporation of the USA, with which it has finalised a strategic partnership. The US$20 million deal includes a 20 per cent interest in Orbital Network Services Corporation, which plans to use the aircraft as reusable communications satellite launchers in its HAAL-2000 High-Altitude Air Launch programme. The aircraft would probably be modified to Tu-160SK standards and continue to be based at Priluki, maintained and flown by Ukrainian crews, but flown to customer countries for individual space launch missions.

DESIGN FEATURES: Intended for high-altitude standoff role carrying ALCMs and for defence suppression, using short-range attack missiles similar to US Air Force SRAMs, along path of bomber making low-altitude penetration to attack primary targets with free-fall nuclear bombs or missiles; this implies capability of subsonic cruise/supersonic dash at almost M2 at 18,300 m (60,000 ft) and transonic flight at low altitude. About 20 per cent longer than USAF B-1B, with greater unrefuelled combat radius and much higher maximum speed; low-mounted variable geometry wings, with very long and sharply swept fixed root panel; small diameter circular fuselage; horizontal tail surfaces mounted high on fin, upper portion of which is pivoted one-piece all-moving surface; large dorsal fin; engines mounted as widely separated pairs in underwing ducts, each with central horizontal V wedge intakes and jetpipes extending well beyond wing centre-section trailing-edge; manually selected outer wing sweepback 20, 35 and 65°; when wings fully swept, inboard portion of each trailing-edge flap hinges upward and extends above wing as large fence; unswept tailfin; sweptback horizontal surfaces, with conical fairing for brake-chute aft of intersection.

FLYING CONTROLS: Quadruplex fly-by-wire with mechanical reversion. Full-span leading-edge flaps, long-span double-slotted trailing-edge flap and inset drooping aileron on each wing; five-section spoilers forward of flaps; all-moving vertical and horizontal one-piece tail surfaces.

STRUCTURE: Slim and shallow fuselage blended with wing-roots and shaped for maximum hostile radar signal deflection; 20 per cent titanium, including leading-edges and wing centre-section spar box.

LANDING GEAR: Twin nosewheels retract rearward; main gear comprises two bogies, each with three pairs of wheels; retraction very like that on Tu-154 airliner; as each leg pivots rearward, bogie rotates through 90° around axis of centre pair of wheels, to lie parallel with retracted leg; gear

retracts into thickest part of wing, between fuselage and inboard engine on each side; so track relatively small. Nosewheel tyres size 1080×400; mainwheel tyres size 1260×425.

POWER PLANT: Four purpose-designed Samara NK-321 turbofans, each 137.3 kN (30,865 lb st) dry, 245 kN (55,115 lb st) with afterburning. In-flight refuelling probe retracts into top of nose. Fuel in centre-section spar box and in outer wings.

ACCOMMODATION: Four crew members in pairs, on individual Zvezda K-36LM zero/zero ejection seats, in pressurised compartment; one window each side of flight deck can be moved inward and rearward for ventilation on ground; flying controls use fighter-type sticks rather than yokes or wheels; crew enter via extending ladder in nosewheel bay. Cooking facilities and lavatory.

AVIONICS: Systems utilise around 100 digital processors and eight digital nav computers.

Radar: Obzor (NATO 'Clam Pipe') nav/attack radar in slightly upturned dielectric nosecone with separate Sopka radar providing terrain-following capability.

Flight: K-042K astro-inertial nav with map display.

Instrumentation: Analogue instruments. No HUD or CRTs.

Mission: OPB-15 strike sight fairing with flat glazed front panel, under forward fuselage, for video camera to provide visual assistance for weapon aiming.

Self-defence: Baikal self-protection system, with integrated RHAWS, chaff/flare dispensers in tailcone and active jamming.

ARMAMENT: No guns. Internal stowage for free-fall bombs, mines, short-range attack missiles or ALCMs; two tandem 12.80 m (42 ft) long weapon bays; MKU-6-5U rotary launcher for six (or up to 12) Kh-55MS (AS-15 'Kent') or RKV-500B (AS-15B 'Kent-B') ALCMs or 12 to 24 Kh-15P (AS-16 'Kickback') SRAMs in each bay. Aircraft upgraded from 2002 onwards have ability to launch Kh-555 missiles; later plans envisage carriage of up to 12 non-nuclear Kh-101 ALCMs, when available.

DIMENSIONS, EXTERNAL:

Wing span: fully spread (20°)	55.70 m (182 ft 9 in)
35° sweep	50.70 m (166 ft 4 in)
fully swept (65°)	35.60 m (116 ft 9¾ in)
Wing aspect ratio: fully spread	8.6
Length overall	54.10 m (177 ft 6 in)
Height overall	13.10 m (43 ft 0 in)
Tailplane span	13.25 m (43 ft 5¾ in)
Wheel track	5.40 m (17 ft 8½ in)
Wheelbase	17.88 m (58 ft 8 in)

DIMENSIONS, INTERNAL:

Weapons bay (each): Volume	43.0 m³ (1,518 cu ft)

AREAS:

Wings, gross: fully swept	360.00 m² (3,875.0 sq ft)
fully spread	approx 400.00 m² (4,305.6 sq ft)
moving areas, fully swept (total)	approx 180.00 m² (1,937.5 sq ft)

WEIGHTS AND LOADINGS:

Weight empty	110,000 kg (242,505 lb)
Weight empty, equipped	117,000 kg (257,940 lb)
Max fuel	171,000 kg (376,990 lb)
Max weapon load	40,000 kg (88,185 lb)
Normal T-O weight	267,600 kg (589,950 lb)
Max T-O weight	275,000 kg (606,260 lb)
Max landing weight	155,000 kg (341,710 lb)
Max power loading	280 kg/kN (2.75 lb/lb st)

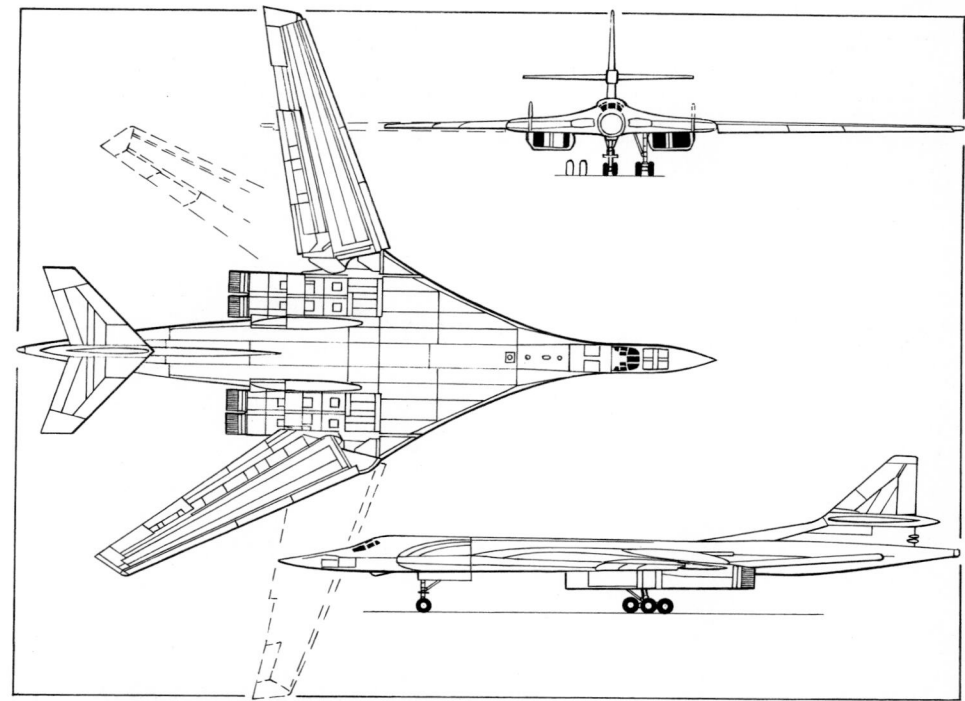

Tupolev Tu-160 strategic bomber *(Jane's/James Goulding)*

Tupolev Tu-154M of Sibir Airlines (*Jane's/Paul Jackson*)   *NEW*/0525754

PERFORMANCE:

Max level speed at 12,200 m (40,000 ft)
         M2.05 (1,200 kt; 2,220 km/h; 1,380 mph)
Cruising speed at 13,700 m (45,000 ft)
         M0.9 (518 kt; 960 km/h; 596 mph)
Max rate of climb at S/L   4,200 m (13,780 ft)/min
Service ceiling   15,000 m (49,200 ft)
T-O run at max AUW   2,200 m (7,220 ft)
Landing run at max landing weight   1,600 m (5,250 ft)
Radius of action at M1.5
         1,080 n miles (2,000 km; 1,240 miles)
Max unrefuelled range
         6,640 n miles (12,300 km; 7,640 miles)
g limit   +2

                 UPDATED

TUPOLEV Tu-154M
NATO reporting name: Careless

TYPE: Tri-jet airliner.

PROGRAMME: Basic Tu-154 announced second quarter 1966 to replace Tu-104, Il-18 and An-10 on Aeroflot medium/long stages up to 3,240 n miles (6,000 km; 3,725 miles); SSSR-85000, first of six prototype/preproduction models flew 3 October 1968; first passenger flight 9 February 1972; regular services began 25 February 1972; 606 prototype and production Tu-154s and Tu-154As, Bs and B-2s with uprated turbofans and other refinements delivered, over 500 to Aeroflot (last described 1985-86 *Jane's*); prototype Tu-154M (SSSR-85317), converted from standard Tu-154B-2 (see 1990-91 *Jane's*), first flew 1982; first two production aircraft delivered to Aeroflot from GAZ 18 Kuybyshev (now known as Samara) December 1984. Aviacor contracted in 2000 for assembly of two Tu-154Ms by GTK Russia. Aviacor plant planned to end production in 2002 (see Customers paragraph below). Flight testing completed 30 May 2002 of one aircraft (for Ural Airlines) equipped with engine 'hush kit' (*zvukopoglayushchaya zashchita*), satellite navigation system and European collision aviodance system to conform to European Union noise and navigation standards; further three to this standard planned by end of 2002.

CURRENT VERSIONS: **Tu-154M:** Basic airliner with alternative standard configurations for up to 180 passengers; executive version available; with all passenger seats removed can carry light freight.

Detailed description applies to Tu-154M.

    Tu-154M-100: Upgraded Zhasmin (Jasmine) avionics system and new Aviacor interior furnishings; 12 ordered by Iranian airlines in 1997; delivered to Iran Air Tours (nine) and Bon Air (three) in 1997-98 from surplus former

Aeroflot aircraft, despite which Aviacor was still optimistic in 1998 of up to 10 new-build orders from Iran. Zhasmin includes GPS and collision-avoidance radar, and bestows Cat. II landing capability.

    Tu-154-200: Projected version with two Samara NK-93 turbofans.

    Tu-154M-LK-1: Two aircraft for head-of-state use.

    Tu-154M2: Modernised, twin-engined version; Perm/Soloviev PS-90A turbofans, consuming 62 per cent as much fuel per passenger as Tu-154M. Intended originally to fly 1995, but apparently abandoned. Life 20,000 hours or 15,000 cycles.

    Tu-154C: Specialised freight version, announced in third quarter of 1982; offered primarily as Tu-154B conversion; unobstructed main cabin cargo volume 72 m³ (2,542 cu ft); freight door 2.80 m (9 ft 2¼ in) wide and 1.87 m (6 ft 1½ in) high in port side of cabin, forward of wing, with ball mat inside and roller tracks full length of cabin floor; typical load nine standard international pallets 2.24 × 2.74 m (88 × 108 in) plus additional freight in standard underfloor baggage holds, volume 38 m³ (1,341 cu ft); nominal range 1,565 n miles (2,900 km; 1,800 miles) with 20,000 kg (44,100 lb) cargo.

    Tu-154 Retrofit: In 1998, Rybinsk Motors was promoting the CFM56 turbofan as a re-engine option.

    Tu-154M/OS: Version for Russian 'Open Skies' Treaty observation flights, with side-looking synthetic aperture radar developed under 1995 co-operative agreement between German Ministry of Defence and Defence Ministry of the Russian Federation. Russian designation **Tu-154M-ON** (*otkrytoye nebo:* open skies). Sole aircraft is RA-85655, a former Tu-154M-LK-1. Additional duties can include environmental monitoring, ice patrol, mapping and geological survey. Radar developed by Kulon Russian research institute and Dornier GmbH of Germany, but installation plans terminated prior to this being fitted. See *Jane's Aircraft Upgrades* for more details.

    Tu-154M Elint: First four of reported 12 ex-airline aircraft entered service by 1999 with Nanjing Military Region, China, following modification by military upgrade plant at Nan Yuan, Beijing.

CUSTOMERS: Total of 318 Tu-154Ms (and 924 of all variants) built or substantially complete by late 1996. Total increased by two up to mid-2002; some recently supplied from stock, including 1998 deliveries of six: to Azerbaijan government (one), Slovak government (one), SKA Slovak Airlines (three) and Tyumen Avia (one). Sole 1996 delivery was one to the Czech government on 14 December; single aircraft in 1997 was delivered to Tyumen Avia in June. None known in 1999, leaving 15 Tu-154Ms in storage at Samara minus engines and other fitments. These include one for Ukrainian government as

VIP transport, ordered 1996, but still awaiting completion in 2002. Aviacor reported sale of four in 2000 and two in 2001; final eight being completed in 2002 before abandoning programme, but will continue to offer freighter conversions of Tu-154B and Tu-154M. RA-85833, delivered to Ural Airlines on 3 June 2002, was 926th (including prototype) and, reportedly, last of type to be built, although few deliveries from stock expected to follow. Known Tu-154M civil operators number 73, with some 280 aircraft.

COSTS: US$8.5 million (2002).

DESIGN FEATURES: Conventional all-swept low-wing configuration; two podded turbofans on sides of rear fuselage, third in extreme rear fuselage with intake at base of fin; nacelle to house retracted main landing gear on trailing-edge of each wing; wing sweep 35° at quarter-chord; anhedral on outer panels; geometric twist along span; circular section fuselage; sweepback at quarter-chord 40° on T tailplane; leading-edge sweep 45° on fin.

    In 2002, Sibir Airlines was operating four aircraft retrofitted with modified wingtips in trial against original aircraft, anticipating 5 per cent fuel economy.

FLYING CONTROLS: Conventional and power-operated. Hydraulically actuated ailerons, triple-slotted flaps, four-section spoilers forward of flaps on each wing; electrically actuated slats on outer 80 per cent each wing leading-edge; tab in each aileron; electrically actuated variable incidence tailplane; rudder and elevators hydraulically actuated by irreversible servo controls; tab in each elevator.

STRUCTURE: All-metal; riveted three-spar wings, centre spar extending to just outboard of inner edge of aileron; semi-monocoque fail-safe fuselage; rudder and elevators of honeycomb sandwich construction.

LANDING GEAR: Hydraulically retractable tricycle type; mainwheels 930×305; nosewheels 800×225. Main bogies, each three pairs of wheels in tandem, retract rearward into fairings on wing trailing-edge; rearward-retracting anti-shimmy twin-wheel nose unit, steerable through ±63°; disc brakes and anti-skid units on mainwheels.

POWER PLANT: Three Aviadvigatel D-30KU-154-II turbofans, each 104 kN (23,380 lb st), in pod each side of rear fuselage and inside extreme rear of fuselage; two lateral engines have clamshell thrust reversers. Integral fuel tanks in wings: four tanks in centre-section and two in outer wings; all fuel fed to collector tank in centre-section and thence to engines; single-point refuelling.

ACCOMMODATION: Crew of three; two pilots and flight engineer, with provisions for navigator and five cabin staff. Two passenger cabins, separated by service compartments; alternative configurations for 166 tourist class passengers, at 75 cm (29.5 in) pitch, with hot meal service, or 134 tourist plus separate first class cabin seating 12 persons; mainly six-abreast seating with centre aisle; washable non-flammable materials used for all interior furnishing. Fully enclosed baggage containers. Toilet, galley and wardrobe to customer's requirements. Executive and light cargo configurations available. Passenger doors forward of front cabin and between cabins on port side, with emergency and service doors opposite; all four doors open outward; six emergency exits: two overwing and one immediately forward of engine nacelle each side. Two pressurised baggage holds under floor of cabin, with two inward-opening doors; smaller unpressurised hold under rear of cabin.

SYSTEMS: Air conditioning pressure differential 0.58 bar (8.4 lb/sq in). Three independent hydraulic systems, working pressure 207 bar (3,000 lb/sq in), powered by engine-driven pumps; Nos. 2 and 3 systems each have additional electric back-up pump; systems actuate landing gear retraction and extension, nosewheel steering, and operation of ailerons, rudder, elevators, flaps and spoilers. Three-phase 200/115 V 400 Hz AC electrical system supplied by three 40 kVA alternators; additional 36 V 400 Hz AC and 27 V DC systems and four storage batteries; TA-92 APU in rear fuselage. Hot air anti-icing of wing, fin and tailplane leading-edges, and engine air intakes; wing slats heated electrically. Engine fire extinguishing system in each nacelle; smoke detectors in baggage holds.

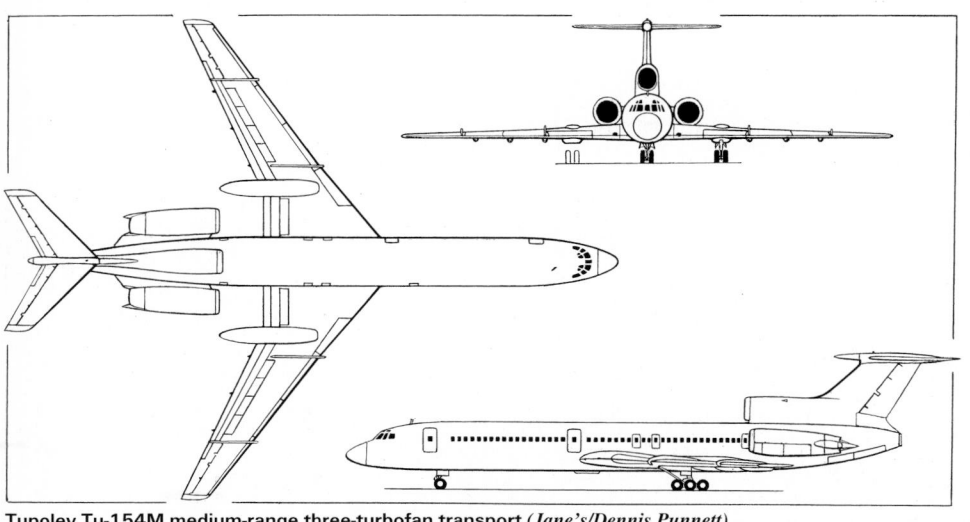

Tupolev Tu-154M medium-range three-turbofan transport (*Jane's/Dennis Punnett*)

AVIONICS: Avionics meet ICAO standards for Cat. II weather minima.

Comms: Dual HF and VHF com and emergency VHF, voice recorder, and transponder.

Radar: Weather radar.

Flight: Automatic flight control system standard until 1995 operates throughout flight except during take-off to 400 m (1,312 ft) and landing from 30 m (100 ft); automatic go-round and automatic speed control provided by autothrottle down to 10 m (33 ft) on landing; triplex INS, Doppler and GPWS. New Zhasmin system available since 1995 offers automatic flight control at all stages of flight including automatic landing to ICAO Cat. II standards; equipment can include TCAS, Orion, Omega and ABSU-154 nav systems.

DIMENSIONS, EXTERNAL:

Wing span	37.55 m (123 ft 2½ in)
Wing aspect ratio	7.0
Length overall	47.92 m (157 ft 2½ in)
Height overall	11.40 m (37 ft 4¾ in)
Diameter of fuselage	3.80 m (12 ft 5½ in)
Tailplane span	13.40 m (43 ft 11½ in)
Wheel track	11.50 m (37 ft 9 in)
Wheelbase	18.92 m (62 ft 1 in)
Passenger doors (each): Height	1.70 m (5 ft 7 in)
Width	0.80 m (2 ft 7½ in)
Height to sill	3.10 m (10 ft 2 in)
Servicing door: Height	1.28 m (4 ft 2½ in)
Width	0.61 m (2 ft 0 in)
Emergency door: Height	1.28 m (4 ft 2½ in)
Width	0.64 m (2 ft 1¼ in)
Emergency exits (each): Height	0.90 m (2 ft 11½ in)
Width	0.48 m (1 ft 7 in)
Main baggage hold doors (each):	
Height	1.20 m (3 ft 11¼ in)
Width	1.35 m (4 ft 5 in)
Height to sill	1.80 m (5 ft 10¾ in)
Rear (unpressurised) hold: Height	0.90 m (2 ft 11½ in)
Width	1.10 m (3 ft 7¼ in)
Height to sill	2.20 m (7 ft 2½ in)

DIMENSIONS, INTERNAL:

Cabin: Width	3.58 m (11 ft 9 in)
Height	2.02 m (6 ft 7½ in)
Volume	163.2 m³ (5,763 cu ft)
Baggage hold: Volume: front	21.5 m³ (759 cu ft)
rear	16.5 m³ (582 cu ft)
Rear underfloor hold: Volume	5.0 m³ (176 cu ft)

AREAS:

Wings, gross	201.45 m² (2,168.4 sq ft)
Horizontal tail surfaces (total)	42.20 m² (454.24 sq ft)

WEIGHTS AND LOADINGS:

Basic operating weight empty	55,300 kg (121,915 lb)
Max payload	18,000 kg (39,680 lb)
Max fuel	39,750 kg (87,633 lb)
Max ramp weight	100,500 kg (221,560 lb)
Max T-O weight	100,000 kg (220,460 lb)
Max landing weight	80,000 kg (176,365 lb)
Max zero-fuel weight	74,000 kg (163,140 lb)
Max wing loading	496.4 kg/m² (101.67 lb/sq ft)
Max power loading	321 kg/kN (3.14 lb/lb st)

PERFORMANCE:

Nominal cruising speed	504 kt (935 km/h; 581 mph)
Max cruising height	11,900 m (39,000 ft)
Service ceiling at 85,000 kg (187,390 lb) AUW	
	12,100 m (39,700 ft)
Balanced field length for T-O and landing	
	2,500 m (8,200 ft)
Range: with max payload	
	1,997 n miles (3,700 km; 2,299 miles)
with 12,000 kg (26,455 lb) payload	
	2,805 n miles (5,200 km; 3,230 miles)
with max fuel and 5,450 kg (12,015 lb) payload	
	3,563 n miles (6,600 km; 4,100 miles)

UPDATED

TUPOLEV Tu-156

TYPE: Tri-jet airliner.

PROGRAMME: Details of 1988-89 flight trials of Tu-155, a Tu-154 modified with a Kuznetsov NK-88 turbofan operating on liquid hydrogen and liquefied natural gas fuels, last appeared in 1990-91 *Jane's*. On 23 April 1994, Russian government allocated funding for conversion of three Tu-154s to Tu-156 standard, delivery of 12 NK-89 turbofans and six cryogenic fuel systems; an installation to supply liquefied natural gas (LNG) will be established at

Model of Tupolev Tu-156M cryogenic aircraft (*Jane's/Paul Jackson*)   0015138

Tupolev Tu-204-100 operated by Kras Air (*Jane's/Paul Jackson*)   *NEW*/0525756

Samara. Order reduced to one Tu-156; delivery scheduled for 1998, but had not taken place by late 2002; federal funding of Rb7.5 million and Rb17 million was planned for allocation in 2001 and 2002 respectively to cover construction of a ground test rig for the cryogenic fuel system; total cost of development programme estimated at Rb200 million to 2005, to be funded by Rosaviakosmos (70 per cent), with balance shared equally between Tupolev and Kuznetsov.

CURRENT VERSIONS: **Tu-156S:** Initial kerosene/LNG-powered conversion of Tu-154B, carrying 13,000 kg (28,660 lb) of LNG at rear of cabin and 3,800 kg (8,378 lb) in two underfloor tanks, plus 10,200 kg (22,487 lb) of kerosene in wing tanks; payload 14,000 kg (30,865 lb) or 130 passengers; range 1,403 n miles (2,600 km; 1,615 miles).

Tu-156M: Developed version converted from Tu-154M, carrying 135 passengers; fuel load and performance as Tu-156S.

Tu-156M2: Proposal only; converted from Tu-154M2; 20,000 kg (44,092 lb) of LNG only, carried in two tanks above centre-fuselage allowing passenger load increase to 160; two NK-94 engines; range 2,159 n miles (4,000 km; 2,485 miles).

POWER PLANT: Experimental power plant is 103.0 kN (23,150 lb st) Samara NK-89; tanks for cryogenic fuel mounted in rear of cabin and in forward baggage hold of Tu-156S/M. Aircraft will operate on mixed LNG/kerosene, as Tu-155, kerosene being used for flights out of non-LNG aerodromes and for emergencies, when in-flight switch from LNG can be made in 5 seconds. Tu-156M2 would have only two NK-94 engines, wholly LNG powered.

UPDATED

TUPOLEV Tu-204 and Tu-214

TYPE: Twin-jet airliner.

PROGRAMME: Development, to replace Tu-154 and Il-62, announced 1983; preliminary details available 1985; programme finalised 1986; first prototype (SSSR-64001), with PS-90A engines, flown 2 January 1989 by Tupolev chief test pilot A Talalakine; two more prototypes (RA-64003 and 004) followed, plus two (002 and 005) for structural and fatigue testing. Second version, with RB211-535E4-B engines, flew 14 August 1992 (fourth flying prototype, RA-64006) and was demonstrated at Farnborough Air Show in following month.

Production of basic Tu-204 series by Aviastar at Ulyanovsk began 1990, Tu-214 by KAPO at Kazan in 1994; Russian certification of basic Tu-204 received 12 January 1995; used initially only for freight services. First revenue-earning passenger flight, Moscow to Mineralnye Vody, operated by Vnukovo Airlines 23 February 1996. Restrictions on PS-90-engined Tu-204s were eventually removed in mid-1998. Deliveries outside CIS began on 2 November 1998 with handover of -120 and -120C freighter to Air Cairo on seven-year lease.

Avionics upgrade under discussion with NII in mid-2000; goal is six LCD, two-pilot flight deck; estimated cost US$5 million to US$8 million. First application on Tu-204-300.

CURRENT VERSIONS: **Tu-204:** Basic medium-haul airliner for up to 214 passengers or maximum payload of 21,000 kg (46,295 lb); 158.3 kN (35,580 lb st) Aviadvigatel PS-90A turbofans. Marketed 1989. RVSM approval granted 28 December 2000.

Tu-204C: Cargo version of basic Tu-204. (Western marketing designation is Tu-204C, but correct Cyrillic designation, used in Russian literature, is **Tu-204S**.)

Tu-204-100: Extended-range passenger version; additional fuel in wing centre-section; dimensions, payload and power plant unchanged; maximum T-O weight 103,000 kg (227,070 lb). Marketed 1993. First flight 9 August 2001.

Tu-204-100C: Freighter, as Tu-204-100. Payload increased at expense of reduced maximum range. Marketed 1994. Bulk freight volume 180.0 m³ (6,357 cu ft), not including fore and aft underfloor holds.

Tu-204-120: As Tu-204-100, but with Rolls-Royce RB211-535-E4 turbofans; maximum T-O weight 103,000 kg (227,070 lb); Russian avionics; prototype (RA-64006) flew 14 August 1992. First production aircraft (RA-64027) flew 7 March 1997; gained limited Russian certification on 16 July 1997 and full certification 12 months later. JAA certification effort began May 2001. First five Sirocco 120/120Cs have Russian avionics; later aircraft are Phase II with Honeywell items, including VIA 2000 suite.

Tu-204-120C: As -100C, but RB211 engines; two for Air Cairo (via Sirocco). First aircraft (RA-64028) flew 10 November 1997; certification completed by August 1998. JAA certification effort began May 2001. First delivery 2 November 1998. High gross weight version, with 30,000 kg (66,139 lb) payload, was due for roll-out in August 1999 and Russian certification in May 2000.

Tu-204-122: As Tu-204-120, but with Rockwell Collins avionics. None produced.

Tu-204-200: Further increase in payload and T-O weight; small increase in fuel in wing centre-section and adjacent baggage hold; dimensions and power plant as Tu-204-100; strengthened landing gear; marketed 1994. First example of Series 200 was a Tu-214 freighter, first flown in March 1996; first true Tu-204-200 (RA-64036) was nearing completion at Ulyanovsk in late 1998; PS-90 engines. By mid-2000, Tu-214 designation was being used for this aircraft.

Tu-204-200C: As Tu-204-200. Payload increased at expense of reduced maximum range. Marketed 1994. Also designated Tu-214C.

Tu-204-200C³ (Cargo, Converted, Containerised): Combi version, marketed as **Tu-214C³**.

Tu-204P: Proposed military patrol (*patrulnyi*) version of Tu-204-200. It was reported in January 1997 that the Russian Navy had approved an anti-submarine version of Tu-204 to replace its Il-38 'May' and shorter-range Be-12 'Mail' patrollers. Leninets Novella (export name Sea Dragon) surveillance suite. By 2002, reported that interest declining in favour of reinstated Beriev A-40 flying-boat.

Tu-204-220: As Tu-204-200, but 191.7 kN (43,100 lb st) Rolls Royce RB211-535E4 or RB211-535F5 turbofans; Tupolev-funded variant with 185.5 kN (41,700 lb st) Pratt & Whitney PW2240 turbofans; promotion began 1998. Is Tu-214 variant, but also designated Tu-224.

Tu-204-220C: Cargo version of Tu-204-220.

Tu-204-222: As Tu-204-220, but with Rockwell Collins avionics.

Tu-204-230: With 176.5 kN (39,683 lb st) Samara NK-93 turbofans. At initial project stage.

Tu-204 Business Jet: Scheduled to fly in 1998, but failed to do so.

Tu-204-300: See Tu-234.

Tupolev Tu-214 version of the Tu-204-200 (*Jane's/Paul Jackson*)   0092557

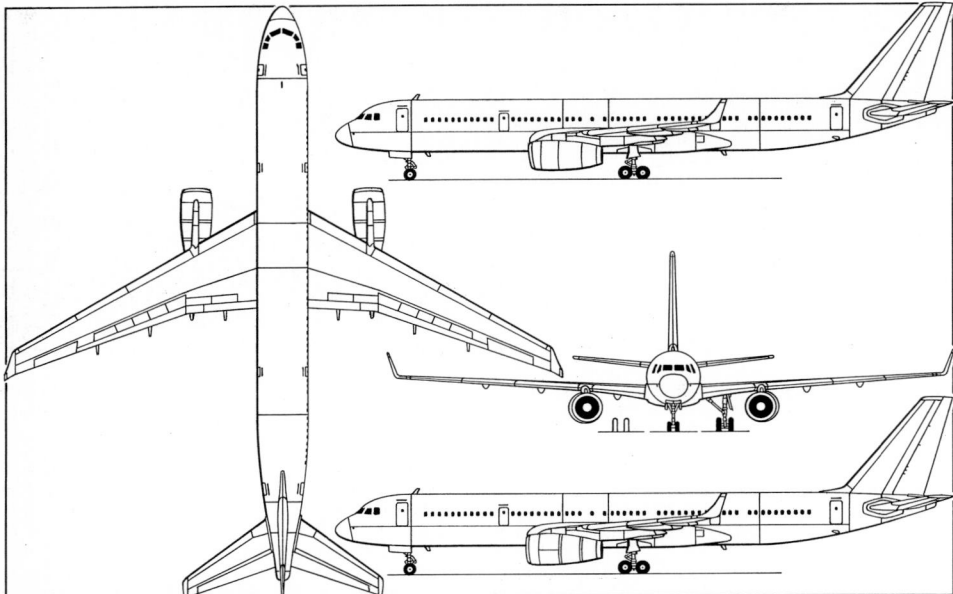

Tupolev Tu-204-200 medium-range transport (two Aviadvigatel PS-90A turbofans), with additional side view (top) of Tu-204-220 *(Jane's/Mike Keep)*

Tu-204-400: Version for 240 passengers; digital avionics; uprated PS-90A2 turbofans for 120,000 kg (264,550 lb) MTOW. Aviastar production.

Tu-204-500: PS-90A2 engines and fuselage stretch to 52.00 m (170 ft 7¼ in); possible wing redesign for higher cruising speeds; private venture outside government's aviation plan. Aviastar production. Designation first reported in early 2000.

Tu-304 and **Tu-306:** Projected 400-seat versions, latter fuelled by LNG. Described in 1999-2000 *Jane's*.

Tu-206 and **Tu-216:** Projected dual-fuel versions. Tupolev 'gasification' programme for Tu-204 now being referred to by **Tu-204K** designation.

Tu-214: Longer-range, increased-weight version of Tu-204, produced by KAPO. Prototype (RA-64501) rolled out at Kazan 5 February 1996; first flew 21 March 1996; public debut at Farnborough Air Show, September 1996. Provisional certification awarded 1998. PW2037 being assessed as an alternative to PS-90A. By early 2000, Kazan stock included seven partly complete Tu-214s from initial production batch of 18. Flight testing of prototype suspended after some 300 sorties, but resumed in May 2000 with further series of 85 flights leading to 'second class' AP-25 certification in March 2000 and full approval on 29 December 2000. First of two aircraft for Dal'avia (s/n 64502) made its first flight on 10 April 2001, and was delivered on 23 May 2001, for service on routes between Moscow and Asia; second received 20 October 2001.

Tu-214C[3]: Combi version, based on Tu-204-200; also designated -200C[3]; announced 1995. Typical loads in main cabin (additional to those in underfloor compartments) comprise 16 LD3 containers and 30 passengers; or six containers and 130 persons; or 18 LD3s only.

Tu-214C (alternatively Tu-214-200C): Dedicated cargo version; 3.405 × 2.18 m (11 ft 2 in × 7 ft 1¾ in) door. Alternative designation **Tu-214F** quoted in 2002. Atlant-Soyuz may become launch customer.

Tu-214D: Long-range version, first mentioned mid-2002. Auxiliary fuel tanks in underfloor holds, potentially extending range to 4,967 n miles (9,200 km; 5,716 miles). Prototype under construction at Kazan.

Tu-214VSSN: VVIP version (*vozdushnoe sudno specialnogo naznacheniya*) for Russian president. Fourth production Tu-214 (RA-64504) is in this configuration; first flight 22 June 2002.

Tu-224 (alternatively Tu-204-220) and **Tu-234** (alternatively Tu-204-300): Announced 1994; trunk route versions with fuselage shortened by 5.80 m (19 ft 0¼ in): 3.00 m (9 ft 10 in) forward and 2.80 m (9 ft 2¼ in) aft of wing; short-range and mid-range models for 166 passengers, long-range model for 99 to 160 passengers; Tu-224 has RB211-535E4 engines; Tu-234 has 158.3 kN (35,580 lb st) PS-90A engines. Prototype Tu-234 (rebuild of original Tu-204 prototype RA-64001) rolled out on 24 August 1995 at Moscow Air Show, but not reported to have flown by mid-2002. First production aircraft (RA-64026, built by Aviastar at Ulyanovsk) rolled out, minus engines, in mid-1996; fitted with PS-90s in early 2000 and first flew 8 July 2000 (ahead of prototype).
Tu-234C cargo version planned. First Tu-224 will be RA-64503.

CUSTOMERS: Include GTK Rossiya division of Aeroflot (two), Aeroflot Russian International (four), Vnukovo Airlines (four, three more on order). Russian government

authorised guarantee in April 1997 for 20 PS-90-powered Tu-204s for Moscow Aviation International Leasing. The Kato Group of Egypt ordered an initial batch of 13 Tu-204-120s, with 17 options (some of which now taken up), in August 1996, with Honeywell avionics; these to be leased through Sirocco Aerospace to KrasAir (Siberia) (10, subsequently cancelled because of high Russian internal taxes) and Air Cairo (five); by mid-2000, four in service with Air Cairo and two more deliveries due before year's end. Volga-Dnepr ordered two Tu-204C-120 freighters in late 1997 and Armenian Airlines was to have taken two (plus three options) in 1998. KrasAir subsequently ordered two Tu-204-100s from Aviastar with regional government backing; first delivery August 2000; further eight options. Transeuropean (Russia) ordered three, of which first (RA-64018) entered service 15 May 1999. Transaero (Russia) announced intention to order 10 Tu-204-100s from Aviastar plant, 1999; firm commitment awaited, but US$110 million loan being arranged in mid-2000. Sibir (Russia) negotiating in late 1999 for three Tu-214s from Kazan, but leased, then bought, one from Perm engine plant and switched to Ulyanovsk as source of second and third aircraft; Tatarstan Airlines planning for two, while Dal'avia of Khabarovsk ordered two Tu-214s in June 2000, both being delivered during 2001. AirRep (UK) was to receive three Tu-204-100Cs, first of which delivered to Manston on 4 April 2000, but this later to KMV. Aviastar Asia Corporation of Taipei, Taiwan, plans to order 20 Tu-204-120/220/234 transports annually for sale and lease in Eastern Europe, the Russian Federation and Associated States (CIS), Asia, the Middle East and Africa. Aeroflot Russian Airlines has announced requirement for eight to 10 Tu-204-120s; Aeroflot requires 36 Tu-214s, nine of which would be put into service from 2003. One -100 reportedly ordered by Aviaexpresscruise of Moscow, 2000. In September 2001 Sirocco brokered sale of five Tu-204-120Cs to China Northwest and China Southwest, delivery beginning in June 2003.

Many early Tu-204s have changed ownership; position in late 2001 shown in the accompanying table. In addition Tupolev has four aircraft, while two were used only for static testing. KMV (Kaminvodyavia) aircraft owned by Perm engine plant, which took two Tu-204s in lieu of cash payment for PS-90s; one Perm aircraft sold to Sibir in December 1999. Siberia Airlines also planning to acquire three Tu-214s from Kazan line.

COSTS: With PS-90 engines and Russian avionics US$27 million (2001); with RB211-535 engines US$38 million (1996).

DESIGN FEATURES: Conventional low/mid-wing configuration, with all surfaces sweptback, and winglets; wing dihedral

TUPOLEV Tu-204 OPERATORS
(at October 2001)

Airline	Variant	In service	On order
Aeroflot	214/204-200		2
Air Cairo	204-120C	1	
China Northwest	204-120C		2
China Southwest	204-120C		3
Dal'avia	214/204-200	2	4
Egyptair	204-120	3	
Inversiya (Air Rep)	204-100C	1	
Inversiya	204-120C		2
KMV	204-100	2	
Kras Air	204-100	2	
Pulkovo	214/204-200		2
Rossiya	214/204-200		2
Sibir	204-100	2	
	214/204-200		2
Transaero	204-100/200		10
Tsentr Avia	204-100	2	
Totals		**15**	**29**

Flight deck of Tupolev Tu-204-120C *(Jane's/Paul Jackson)*   0015140

Model of Tu-204-300, alias Tu-234 *(Jane's/Paul Jackson)*   NEW/0525758

from roots; semi-monocoque oval section pressurised fuselage; torsion box of fin forms integral fuel tank, used for automatic trimming of CG in flight; design life 45,000 flights, 60,000 flight hours or 20 years.

Wing section supercritial; sweepback 28°; thickness/chord ratio 14 per cent at root, 9 to 10 per cent at tip; negative twist.

FLYING CONTROLS: Triplex digital fly-by-wire, with triplex analogue back-up; conventional 'Y' control yokes selected after evaluation of alternative sidestick on Tu-154 testbed; inset aileron outboard of two-section double-slotted flap on each wing trailing-edge; two-section upper surface airbrake forward of each centre-section flap; five-section spoiler forward of each outer flap; four-section leading-edge slat over full span of each wing; conventional rudder and elevators; no tabs.

STRUCTURE: Approximately 18 per cent of airframe by weight of composites; three-piece two-spar wing, with metal structure, part composites skin; carbon fibre skin on spoilers, airbrakes and flaps; glass fibre wingroot fairings; all-metal fuselage, utilising aluminium-lithium and titanium; nose radome and some access panels of composites; extensive use of composites in tail unit, particularly for leading-edges of fixed surfaces and for rudder and elevators.

LANDING GEAR: Hydraulically retractable tricycle type; electrohydraulically steerable twin-wheel nose unit (±10° via rudder pedals; ±70° by electric steering control) retracts forward; four-wheel bogie main units retract inward into wing/fuselage fairings. Carbon disc brakes, electrically controlled. Tyre size 1,070×390 on mainwheels, 840×290 on nosewheels.

POWER PLANT: Two turbofans (see notes on versions), underwing in composites cowlings. Tu-204-200 series carries fuel in six integral tanks in wings, in centre-section and adjacent baggage hold, and in tailfin, total capacity 40,730 litres (10,760 US gallons; 8,960 Imp gallons); torsion box of fin forms integral fuel tank for automatic trimming of CG in flight, as well as for optional standard use.

ACCOMMODATION: Can be operated by pilot and co-pilot, but Aeroflot specified requirement for a flight engineer; provision for fourth seat for instructor or observer. Three basic single-aisle passenger arrangements in Tu-204-200 series: (1) 184 seats, with 30 business class seats six-abreast at pitch of 96 cm (38 in) at front, and 154 tourist class seats six-abreast at pitch of 81 cm (32 in) at rear; (2) 200 six-abreast tourist class seats at pitch of 81 cm (32 in); (3) 212 seats six-abreast at tourist class pitch of 81 cm (32 in). Tu-234 has 162 seats in short-range version, 145 seats at same pitch in mid-range and 102 seats in long-range version, all at 81 cm (32 in) pitch. All configurations have buffet/galley and lavatory immediately aft of flight deck; 184-seat and 200-seat configurations have two more lavatories in tourist cabin and a further large buffet/galley and service compartment at rear of passenger accommodation; 212-seat configuration has two lavatories and galley at rear; overhead stowage for hand baggage. Other layouts optional, with increased galley and lavatory provisions.

Passenger doors at front and rear of cabin on port side; service doors opposite. Type I emergency exit doors fore and aft of wing each side; inflatable slide for emergency use at each of eight doors. Tu-204-200 has two underfloor baggage/freight holds for total of eight LD-3-46 international class containers, three in forward hold, five in rear hold (AK-07 containers in Tu-204 and 204-100). Fully automatic container loading system; manual back-up.

SYSTEMS: Triplex fly-by-wire digital control system, with triplex analogue back-up; Barco Display Systems selected in 2000 to supply FMS computers. Three independent hydraulic systems, pressure 207 bar (3,000 lb/sq in). Ailerons, elevators, rudder, spoilers and airbrakes operated by all three systems; flaps, leading-edge slats, brakes and nosewheel steering operated by two systems; landing gear retraction and extension by all three systems. Electrical power supplied by two 200/115 V 400 Hz AC generators and a 27 V DC standby system. Type TA-12-60 APU in tailcone.

AVIONICS: Avionics of Russian design or, optionally, integrated by Sextant Avionique or Honeywell, with standard nav/com equipment by Rockwell Collins.
Comms: VHF and HF radio, intercom.
Radar: Weather radar.
Flight: KSPNO-204 triplex automatic flight control and navigation system (being upgraded in 2002), VOR, DME, automatic approach and landing system, for operation to ICAO Cat. IIIa minima; Series 100/200 have RLG INS (Litton LTN-101 or Honeywell HG115OBD02), GPS (Litton LTN-2001) and TCAS II (Honeywell).
Instrumentation: EFIS equipment comprises two colour CRTs for flight and nav information for each pilot, plus two central CRTs for engine and systems data. All six to have been replaced by LCDs by end of 2002.

DIMENSIONS, EXTERNAL (A: 204, B: 204-100 (and 204C unless otherwise stated), C: 204-120 (and 204-120C unless otherwise stated), D: 204-200 and Tu-214, E: 204-220, F: 234 short-range, G: 234 mid-range, H: 234 long-range, J: 224):

Wing span: all	41.80 m (137 ft 1¾ in)
Wing aspect ratio: all	9.6

Tupolev Tu-204-100 painted to mark 25th anniversary of the Aviastar production plant (*Jane's/Paul Jackson*)
0126917

Length overall: A, B, C, D, E	46.16 m (151 ft 5¼ in)
F, G, H, J	40.20 m (131 ft 10¾ in)
Fuselage cross-section:	
all	3.80 × 4.10 m (12 ft 5½ in × 13 ft 5½ in)
Height overall: all	13.90 m (45 ft 7¼ in)
Tailplane span: A, B, C, D, E	15.00 m (49 ft 2½ in)
Wheel track: A, B, C, D, E	7.82 m (25 ft 8 in)
Wheelbase: A, B, C, D, E	17.00 m (55 ft 9¼ in)
Passenger doors (each): Height: all	1.85 m (6 ft 0¾ in)
Width: all	0.84 m (2 ft 9 in)
Service doors (each): Height: all	1.60 m (5 ft 3 in)
Width: all	0.65 m (2 ft 1½ in)
Cargo door (appropriate versions):	
Height	2.18 m (7 ft 1¾ in)
Width	3.405 m (11 ft 2 in)
Emergency exit doors (each):	
Height: all	1.44 m (4 ft 8¾ in)
Width: all	0.61 m (2 ft 0 in)
Baggage holds: Height to sill: all	2.71 m (8 ft 10¾ in)

DIMENSIONS, INTERNAL:

Cabin, excl flight deck:	
Length: A, B, C, D, E	30.18 m (99 ft 0 in)
Max width: all	3.57 m (11 ft 8½ in)
Max height: all	2.28 m (7 ft 6 in)
Fwd cargo hold: Height: all	1.16 m (3 ft 9¾ in)
Volume: A, B, C	11.0 m³ (388 cu ft)
Rear cargo hold: Height:	
A, B, C, D, E	1.16 m (3 ft 9¾ in)
Volume: A, B, C	15.4 m³ (544 cu ft)

AREAS:

Wings, gross: all	182.40 m² (1,963.4 sq ft)

WEIGHTS AND LOADINGS:

Operational weight empty: A	58,300 kg (128,530 lb)
B	58,800 kg (129,630 lb)
C	59,300 kg (130,735 lb)
D, E	59,000 kg (130,070 lb)
Tu-214 freighter	57,000 kg (125,665 lb)
Max payload: A, B, C	21,000 kg (46,296 lb)
204C	27,000 kg (59,524 lb)
204-120C	25,000 kg (55,116 lb)
D, E (weight limited), J	25,200 kg (55,555 lb)
Tu-214 freighter	30,000 kg (66,139 lb)
F, space limited (196 seats)	19,565 kg (43,132 lb)
F, G	18,000 kg (39,682 lb)
H	12,000 kg (26,455 lb)
Payload with max fuel: 204C	13,900 kg (30,644 lb)
204-120C	13,200 kg (29,101 lb)
Max baggage/freight: fwd hold: E	3,625 kg (7,990 lb)
rear hold: E	5,075 kg (11,190 lb)
Max fuel: A	24,000 kg (52,910 lb)
B, C, D, E, J	32,700 kg (72,090 lb)
F	24,300 kg (53,572 lb)
G, H	35,000 kg (77,162 lb)
Max ramp weight: E	111,750 kg (246,360 lb)
Max T-O weight: A	94,600 kg (208,550 lb)
B, C	103,000 kg (227,075 lb)
D, E, J	110,750 kg (244,155 lb)
F	88,000 kg (194,005 lb)
G, H	103,000 kg (227,070 lb)
Max landing weight: B, C	89,500 kg (197,310 lb)
F, G, H	88,000 kg (194,005 lb)
Max zero-fuel weight: E	84,200 kg (185,625 lb)
Max wing loading: A	518.6 kg/m² (106.23 lb/sq ft)
B, C	564.7 kg/m² (115.66 lb/sq ft)
D, E, J	607.2 kg/m² (124.36 lb/sq ft)
F	482.5 kg/m² (98.81 lb/sq ft)
G, H	564.7 kg/m² (115.66 lb/sq ft)
Max power loading: A	299 kg/kN (2.93 lb/lb st)
B	325 kg/kN (3.19 lb/lb st)
C	269 kg/kN (2.63 lb/lb st)
D	350 kg/kN (3.43 lb/lb st)
E	289 kg/kN (2.83 lb/lb st)
F	278 kg/kN (2.73 lb/lb st)
G, H	325 kg/kN (3.19 lb/lb st)

PERFORMANCE:

Nominal cruising speed at 11,100-12,100 m	
(36,400-39,700 ft):	
A, B, C, D, E	437-459 kt (810-850 km/h; 503-528 mph)
F, G, H	448-459 kt (830-850 km/h; 515-528 mph)

Approach speed: F	119 kt (220 km/h; 137 mph)
G, H	122 kt (225 km/h; 140 mph)
Cruising height	12,100 m (39,700 ft)
T-O run: F	1,450 m (4,760 ft)
G, H	2,050 m (6,725 ft)
Required T-O runway at max T-O weight:	
A	2,030 m (6,660 ft)
B (except 204C), D, E	2,045 m (6,710 ft)
204C	2,160 m (7,090 ft)
C	1,830 m (6,005 ft)
J	2,630 m (8,630 ft)
Required landing runway: F, G, H	2,000 m (6,565 ft)
Range:	
with max payload:	
A	1,312 n miles (2,430 km; 1,509 miles)
B	2,321 n miles (4,300 km; 2,671 miles)
Tu-204C	2,321 n miles (4,300 km; 2,671 miles)
C	2,213 n miles (4,100 km; 2,547 miles)
Tu-204-120C	1,808 n miles (3,350 km; 2,081 miles)
D	2,591 n miles (4,800 km; 2,982 miles)
Tu-214 passenger	
	3,374 n miles (6,250 km; 3,883 miles)
E	2,483 n miles (4,600 km; 2,858 miles)
with design payload:	
F	1,403 n miles (2,600 km; 1,615 miles)
G	3,617 n miles (6,700 km; 4,163 miles)
H	4,990 n miles (9,250 km; 5,750 miles)
with max fuel:	
Tu-204C	3,671 n miles (6,800 km; 4,225 miles)
Tu-204-120C	3,542 n miles (6,560 km; 4,076 miles)
J	3,617 n miles (6,700 km; 4,163 miles)

UPDATED

TUPOLEV Tu-214

Described under Tu-204 heading.

TUPOLEV Tu-224

Described under Tu-204 heading.

TUPOLEV Tu-234

Described under Tu-204 heading.

TUPOLEV Tu-324

TYPE: Regional jet airliner.
PROGRAMME: Announced at 1995 Paris Air Show; proposed manufacture by KAPO at Kazan, Tatarstan, from 2003. Specification revised 1996 and further refined in 1997 when governments of Russia and Tatarstan agreed on 20 August to 50:50 sharing of development costs and latter placed symbolic first order for a presidential transport (since superseded by Tu-214 VSSN, which see). Mockup completed by late 1997; technical documentation issued 2000. First of five prototypes to begin construction in 2002 subject to funding, but no finance allocated up to mid-year. Is first Russian aircraft programme funded totally by commercial backers and designed entirely with use of computers.
CURRENT VERSIONS: **Tu-324A:** (*Administrativny:* executive) VIP transport, carrying up to 10 passengers.

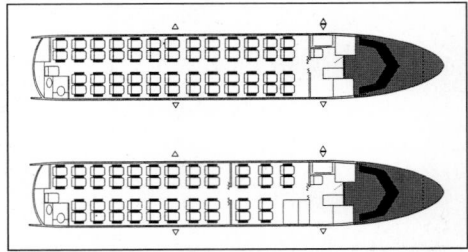

Tupolev Tu-324R configurations include 50 single-class passengers (top) and 10 business-class plus 34 or 36 tourist-class passengers (bottom) 0024039

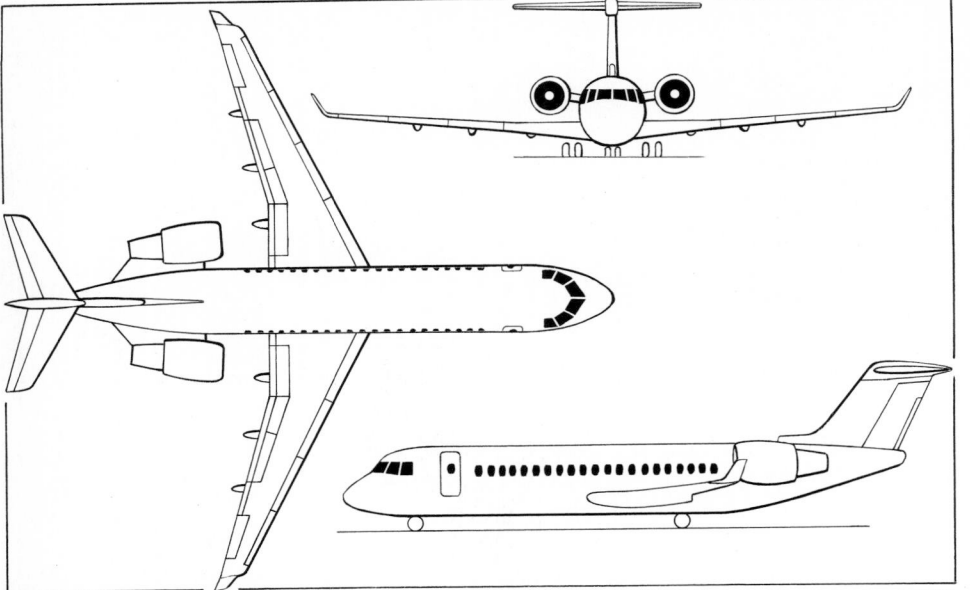

Tupolev Tu-324R twin turbofan (AI-22) regional transport (*Jane's/James Goulding*)   0079303

Tu-324R: Regional airliner: 52 seats in single-class configuration: 10 + 34 or 36 in two-class.

Tu-324 stretch: Long-term project for stretching to 70-seat airliner, possibly with Rolls-Royce Deutschland BR715 turbofans.

CUSTOMERS: Estimated market for 400 to 450, including 150 Tu-324A.

COSTS: Estimated total programme cost US$330 million; Tu-324A US$10 million, export versions US$15 million to US$20 million (2001).

DESIGN FEATURES: Wings scaled down from Tu-204; illustrations show the aircraft with and without winglets.

LANDING GEAR: Messier-Dowty main contractor.

POWER PLANT.: Two General Electric CF34-3B1 turbofans, each 41.0 kN (9,220 lb st), in pods on sides of rear fuselage. Alternatives under consideration include Progress AI-22 and Soyuz TRDD R-126-300.

AVIONICS: Honeywell main supplier.

DIMENSIONS, EXTERNAL:
Wing span	23.20 m (76 ft 1½ in)
Length overall: Tu-324A	23.50 m (77 ft 1¼ in)
Tu-324R	25.50 m (83 ft 8 in)
Height overall	7.40 m (24 ft 3¼ in)

WEIGHTS AND LOADINGS:
Max payload: Tu-324A	1,800 kg (3,968 lb)
Tu-324R	5,500 kg (12,125 lb)
Max T-O weight: Tu-324A, Tu-324R	
	23,700 kg (52,249 lb)

PERFORMANCE (estimated, AI-22 engines):
Max cruising speed	
	448-459 kt (830-850 km/h; 516-528 mph)
Balanced runway length	1,800 m (5,910 ft)
Range:	
Tu-324R: with 52 passengers	
	1,349 n miles (2,500 km; 1,553 miles)
with 46 passengers	
	1,619 n miles (3,000 km; 1,864 miles)
Tu-324A, 8 passengers	
	4,233 n miles (7,840 km; 4,871 miles)

UPDATED

TUPOLEV Tu-330

TYPE: Twin-jet freighter.

PROGRAMME: Announced early 1993 as replacement for Antonov An-12; government support announced April 1994; first 10 to be built at Kazan; first flight scheduled 1997, for service from 1998; neither target achieved. Production at both KAPO and Samara. Stated by commander of Russian Military Transport Aviation to be one of three basic types to be operated by his force in 21st century. Production drawings released by Tupolev to KAPO at Kazan in late 1997 and construction then begun. By early 2000 the venture was being offered as a potential partnership.

Tu-330 included in federal plans for civil aviation up to 2015, requiring investment of equivalent of US$340 million over seven years. Tupolev expenditure to early 2000 totalled Rb80 million. Kazan plant hoping for production launch in 2002, but funding still awaited in mid-year. However, Russian Air Forces' reluctance to accept An-70 seen to improve Tu-330's chances of go-ahead.

CURRENT VERSIONS: **Tu-330:** *As described.*

Tu-330K (previously Tu-338): Proposed cryogenic-fuelled version with Samara NK-94 engines and 20,000 kg (44,092 lb) of liquid natural gas.

COSTS: US$25 million to US$35 million (2000, estimate).

DESIGN FEATURES: See accompanying three-view drawing: said to have 75 per cent commonality with Tu-204. Wing design basically similar to Tu-204; engine pylons as Tu-204; rear-loading ramp.

FLYING CONTROLS: Each wing has four-section leading-edge flaps, three-section trailing-edge flaps, eight spoilers forward of flaps, and aileron. Elevators and rudder each in two sections.

LANDING GEAR: Retractable tricycle type; twin-wheel, rearwards-retracting nose unit; each main unit has three pairs of wheels in tandem. Able to operate from concrete or grass.

POWER PLANT: Two Aviadvigatel PS-90A turbofans. Rolls-Royce and Pratt & Whitney turbofans under consideration for export sales. Tu-330K is projected for LNG transportation, with three tanks in the cabin for 22,800 kg (50,265 lb) of LNG.

DIMENSIONS, EXTERNAL:
Wing span over winglets	43.50 m (142 ft 8½ in)
Wing aspect ratio	10.6
Length overall	42.00 m (137 ft 9½ in)
Height overall	14.00 m (45 ft 1¼ in)
Rear-loading ramp: Length	4.00 m (13 ft 1½ in)
Width	4.00 m (13 ft 1½ in)

DIMENSIONS, INTERNAL:
Cargo hold: Length	19.50 m (63 ft 11½ in)
Width	4.00 m (13 ft 1½ in)
Height: min	3.55 m (11 ft 7¾ in)
Max	4.00 m (13 ft 1½ in)
Volume	350.0 m³ (12,360 cu ft)

AREAS:
Wings, gross	177.80 m² (1,913.8 sq ft)

WEIGHTS AND LOADINGS:
Payload: in hold	30,000 kg (66,139 lb)
on ramp	5,000 kg (11,023 lb)
Max	35,000 kg (77,162 lb)
Max T-O weight	103,500 kg (228,175 lb)
Max wing loading	582.1 kg/m² (119.23 lb/sq ft)

PERFORMANCE (estimated):
Nominal cruising speed at 11,000 m (36,000 ft)	
	431-458 kt (800-850 km/h; 497-528 mph)
T-O run	2,200 m (7,220 ft)
Nominal range:	
With 15,000 kg (33,069 lb) payload	
	3,779 n miles (7,000 km; 4,349 miles)
With 20,000 kg (44,090 lb) payload	
	3,020 n miles (5,600 km; 3,480 miles)
With 30,000 kg (66,139 lb) payload	
	1,620 n miles (3,000 km; 1,865 miles)

UPDATED

TUPOLEV Tu-334, Tu-336 and Tu-354

TYPE: Twin-jet airliner.

PROGRAMME: Launched 1986, with target first flight in 1991 and service entry in 1995; first prototype Tu-334 (RA-94001) eventually rolled out at Zhukovsky during Moscow Air Show, 25 August 1995; then scheduled to fly 1996 with D-436T1 engines, but delayed by funding

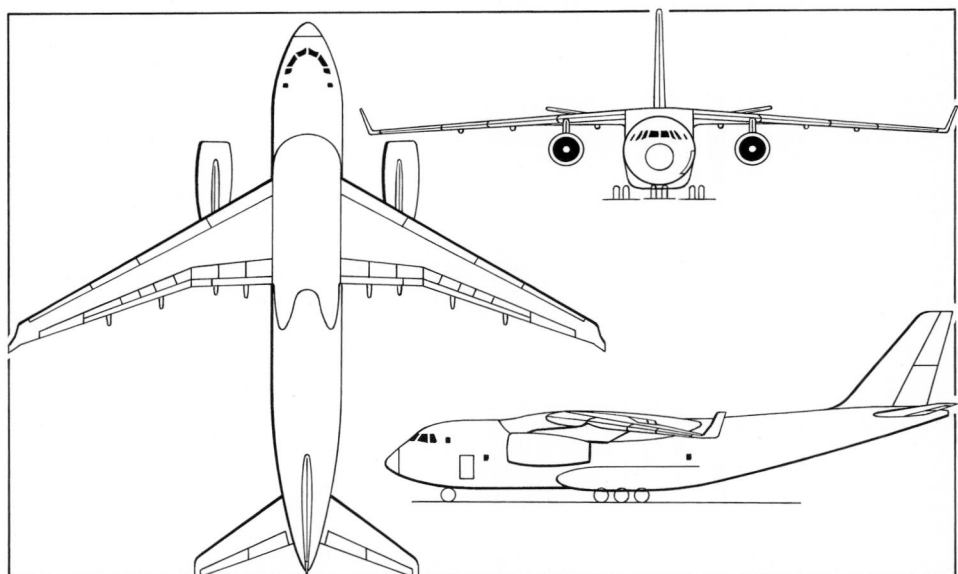

Tupolev Tu-330 twin-turbofan freight transport (*Jane's/Mike Keep*)

Model of the Tu-324A project with CF34 engines (*Jane's/Paul Jackson*)   0092533

Artist's impression of the Tu-330   *NEW*/0525832

NEW/0526922

Tupolev Tu-334, cutaway drawing key

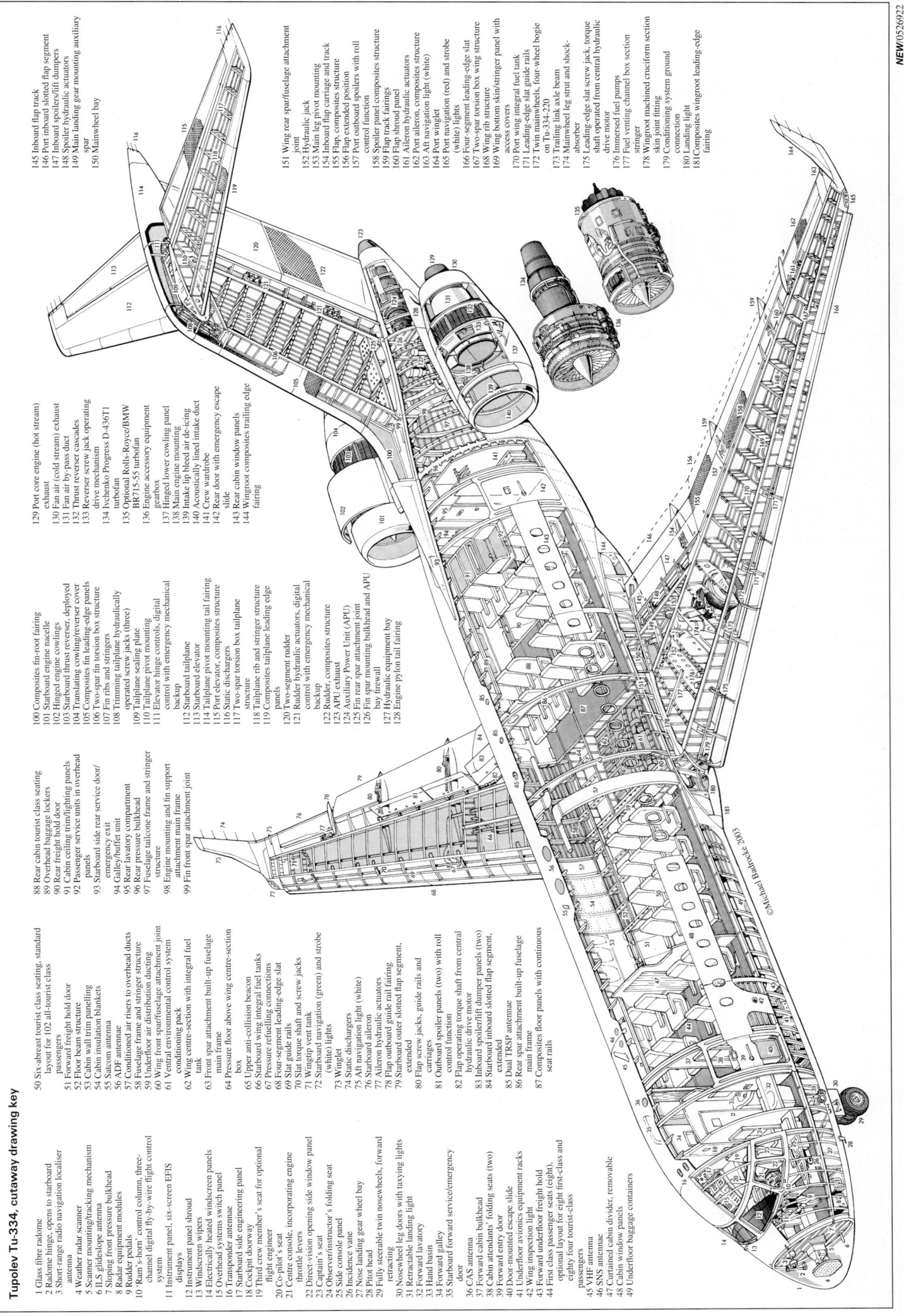

1 Glass fibre radome
2 Radome hinge, opens to starboard
3 Short-range radio navigation localiser antenna
4 Weather radar scanner
5 Scanner mounting/tracking mechanism
6 ILS glideslope antenna
7 Sloping front pressure bulkhead
8 Radar equipment modules
9 Rudder pedals
10 'Ram's-horn' control column, three-channel digital fly-by-wire flight control system
11 Instrument panel, six-screen EFIS displays
12 Instrument panel shroud
13 Windscreen wipers
14 Electrically heated windscreen panels
15 Overhead systems switch panel
16 Transponder antenna
17 Starboard side engineering panel
18 Cockpit doorway
19 Third crew member's seat for optional flight engineer
20 Co-pilot's seat
21 Centre console, incorporating engine throttle levers
22 Direct-vision opening side window panel
23 Captain's seat
24 Observer/instructor's folding seat
25 Side console panel
26 Incidence vane
27 Nose landing gear wheel bay
28 Pitot head
29 Fully steerable twin nosewheels, forward retracting
30 Nosewheel leg doors with taxying lights
31 Retractable landing light
32 Forward lavatory
33 Hand basin
34 Forward galley
35 Starboard forward service/emergency door
36 CAS antenna
37 Forward cabin bulkhead
38 Cabin attendants' folding seats (two)
39 Forward entry door
40 Door-mounted escape slide
41 Underfloor avionics equipment racks
42 Wing inspection light
43 Forward underfloor freight hold
44 First class passenger seats (eight), optional layout for eight first-class and eighty four tourist-class passengers
45 VHF antenna
46 SNS antennae
47 Curtained cabin divider, removable
48 Cabin window panels
49 Underfloor baggage containers
50 Six-abreast tourist class seating, standard layout for 102 all-tourist class passengers
51 Forward freight hold door
52 Floor beam structure
53 Cabin wall trim panelling
54 Cabin insulation blankets
55 Satcom antenna
56 ADF antenna
57 Conditioned air risers to overhead ducts
58 Fuselage frame and stringer structure
59 Underfloor air distribution ducting
60 Wing front spar/fuselage attachment joint
61 Ventral environmental control system conditioning pack
62 Wing centre-section with integral fuel tank
63 Front spar attachment built-up fuselage main frame
64 Pressure floor above wing centre-section box
65 Upper anti-collision beacon
66 Starboard wing integral fuel tanks
67 Pressure refuelling connections
68 Four-segment leading-edge slat
69 Slat guide rails
70 Slat torque shaft and screw jacks
71 Wingtip vent tank
72 Starboard navigation (green) and strobe (white) lights
73 Winglet
74 Static dischargers
75 Aft navigation light (white)
76 Starboard aileron
77 Aileron hydraulic actuators
78 Flap outboard guide rail fairing
79 Starboard outer slotted flap segment, extended
80 Flap screw jacks, guide rails and carriages
81 Outboard spoiler panels (two) with roll control function
82 Flap operating torque shaft from central hydraulic drive motor
83 Inboard spoiler/lift dumper panels (two)
84 Starboard inboard slotted flap segment, extended
85 Dual TRSP antennae
86 Rear spar attachment built-up fuselage main frame
87 Composites floor panels with continuous seat rails
88 Rear cabin tourist class seating
89 Overhead baggage lockers
90 Rear freight hold door
91 Cabin ceiling trim/lighting panels
92 Passenger service units in overhead panels
93 Starboard side rear service door/emergency exit
94 Galley/buffet unit
95 Rear lavatory compartment
96 Rear pressure bulkhead
97 Fuselage tailcone frame and stringer structure
98 Engine mounting and fin support attachment main frame
99 Fin front spar attachment joint
100 Composites fin-root fairing
101 Starboard engine nacelle
102 Hinged engine cowlings
103 Starboard thrust reverser, deployed
104 Translating cowling/reverser cover
105 Composites fin leading-edge panels
106 Two-spar fin torsion box structure
107 Fin ribs and stringers
108 Trimming tailplane hydraulically operated screw jacks (three)
109 Tailplane sealing plate
110 Tailplane pivot mounting
111 Elevator hinge controls, digital control with emergency mechanical backup
112 Starboard tailplane
113 Starboard elevator
114 Tailplane pivot mounting tail fairing
115 Port elevator, composites structure
116 Static dischargers
117 Two-spar torsion box tailplane structure
118 Tailplane rib and stringer structure
119 Composites tailplane leading edge panels
120 Two-segment rudder
121 Rudder hydraulic actuators, digital control with emergency mechanical backup
122 Rudder, composites structure
123 APU exhaust
124 Auxiliary Power Unit (APU)
125 Fin rear spar attachment joint
126 Fin spar mounting bulkhead and APU bay firewall
127 Hydraulic equipment bay
128 Engine pylon tail fairing
129 Port core engine (hot stream) exhaust
130 Fan air (cold stream) exhaust
131 Fan air by-pass duct
132 Thrust reverser cascades
133 Reverser screw jack operating drive mechanism
134 Ivchenko Progress D-436T1 turbofan
135 Optional Rolls-Royce/BMW BR715-55 turbofan
136 Engine accessory equipment gearbox
137 Hinged lower cowling panel
138 Main engine mounting
139 Intake lip bleed air de-icing
140 Acoustically lined intake duct
141 Crew wardrobe
142 Rear door with emergency escape slide
143 Rear cabin window panels
144 Wingroot composites trailing edge fairing
145 Inboard flap track
146 Port inboard slotted flap segment
147 Inboard spoilers/lift dumpers
148 Spoiler hydraulic actuators
149 Main landing gear mounting auxiliary spar
150 Mainwheel bay
151 Wing rear spar/fuselage attachment joint
152 Hydraulic jack
153 Main leg pivot mounting
154 Inboard flap carriage and track
155 Wing rib structure
156 Flap, composites structure
157 Port outboard spoilers with roll control function
158 Spoiler panel composites structure
159 Flap track fairings
160 Flap shroud panel
161 Aileron hydraulic actuators
162 Port aileron, composites structure
163 Aft navigation light (white)
164 Port winglet
165 Port navigation (red) and strobe (white) lights
166 Four-segment leading-edge slat
167 Two-spar torsion box wing structure
168 Wing rib structure
169 Wing bottom skin/stringer panel with access covers
170 Port wing integral fuel tank
171 Leading-edge slat guide rails
172 Twin mainwheels, four-wheel bogie on Tu-334-220
173 Trailing link axle beam
174 Mainwheel leg strut and shock-absorber
175 Leading-edge slat screw jack, torque shaft operated from central hydraulic drive motor
176 Immersed fuel pumps
177 Fuel venting channel box section
178 Wingroot machined cruciform section skin joint fitting
179 Conditioning system ground connection
180 Landing light
181 Composites wingroot leading-edge fairing

©Michael Badrocke 2003

Prototype Tu-334 RA-94001 displaying at Moscow salon (*Jane's/Paul Jackson*)   *NEW*/0525760

shortages. Refinancing by Ukrainian government in early 1997 resulted in new target date of May/June 1997 for first Tu-334s built by Aviant at Kiev; this also missed; second and third identical, but non-flying, prototypes under construction at Zhukovsky by 1996; second used for static tests at CAHI (TsAGI) 1996; unidentified test airframe delivered to Aviatest LNK at Riga from Kiev in mid-2000 to simulate 20 year fatigue life; further two prototypes and three production aircraft under construction at Kiev by early 1999. RA-94001 eventually flown 8 February 1999 and had achieved eight sorties by time of public debut at Moscow in August 1999. Plans for 200 flights in 2001 not achieved (114 only), and total of only some 180 hours in 144 flights logged by April 2002. Second prototype currently expected to complete final assembly by January 2003, followed by third (first by RSK 'MiG') in fourth quarter of 2003.

AP-25/FAR Pt 25 certification programme was due for completion in 2001 after nearly 900 flying hours, but victim of above-mentioned delays, created by erratic funding and delayed delivery of D-436 engines; certification and deliveries now seen as unlikely before 2004 at earliest. TANTK at Taganrog was to have been second source for -100, but Russian government decree of 5 October 1999 nominated RSK 'MiG' (which is to contribute half of certification costs, first payment being made in January 2002); first Taganrog fuselage transported to Voronin Production Centre, Moscow, arriving 7 April 2000. Completion and final assembly was to have been in second quarter of 2001, but this not achieved (see above); Taganrog horizontal tail delivered early 2002. Half of all assemblies and components for RSK 'MiG' production to be supplied by Aviant.

On 16 June 1997 Rolls-Royce Deutschland agreed loan of BR710 engines to build prototype of Tu-334-120; first flight of -120 had been expected late 1998, but delayed by funding shortages. Revised plans for BR710s to go into -100 prototype, after short test programme with D-436s, also thwarted by aircraft's continued grounding. R-R support now directed at versions powered by BR715 engines.

During mid-1997, AIO of Iran showed interest in establishing a kit assembly line at Esfahan over 15 years. Negotiations still proceeding in 2002, based on eventual fuselage detailed assembly in Iran, with complete wings from Ukraine and empennage from Russia; peak production of 12 per year envisaged.

CURRENT VERSIONS: **Tu-334-100:** Basic version, with D-436T1 engines, for 72 mixed class or 102 tourist class passengers. Prototypes are of this version.

Tu-334-100C: Combi version of -100; up to six 2.24 × 3.17 m (88 × 125 in) pallets in underfloor compartments.

Tu-334-100M: Variant of -100; MTOW 47,700 kg (105,160 lb); range 1,684 n miles (3,120 km; 1,938 miles).

Tu-334-100D: Extended-range version, announced at Moscow Air Show '95. Basically similar to Tu-334-100, but with increased wing span and uprated D-436T2 engines; increased fuel and maximum T-O weight; 102 seats.

Tu-334-120: As Tu-334-100, but two 88.9 kN (19,995 lb st) Rolls-Royce Deutschland BR715-55 turbofans. Third flying prototype to be in this configuration.

Tu-334-120D: Rolls-Royce Deutschland BR715-55 engines; increased weight and range. Announced 1998.

Tu-334-200: Basically similar to Tu-334-100D and alternatively known as **Tu-354**; D-436T2 or BR715-55 engines; fuselage lengthened by 3.90 m (12 ft 9½ in) to accommodate up to 126 passengers at 81 cm (32 in) pitch; increased wing span; four-wheel bogies on main landing gear units. Construction of four by Aviacor in Samara reportedly begun by early 1997, but no recent news from

this quarter and Aviacor no longer mentioned in connection with Tu-334 programme.

Tu-334-200C: Variant of -200; combi with same payload.

Tu-334-220: Further increased MTOW; BR715-55 engines. Announced 1998.

Tu-334C: Cargo version; at design study stage 1995.

Tu-336: Cryogenic-fuelled version of Tu-334 with 13,000 kg (28,660 lb) of liquid natural gas in two faired tanks extending full length of cabin roof; Ivchenko Progress D-436T2 engines; 102 passengers; range 1,457 n miles (2,700 km; 1,677 miles). Revealed 1998; no recent news.

CUSTOMERS: Provisional orders in 1995 for approximately 40 aircraft, for Rossiya Airlines, Bashkiri Airlines, Tyumen Airlines and Tatarstan Airlines; total increased to 160 from 13 airlines by 1997. MoU for 20 Tu-334-120s signed by Aeroflot in June 2000; other letters of intent from Atlant Soyuz (two) and Pulkovo (15 to 20) during first half of 2002. CIS market estimated as about 350 by Rosaviakosmos; by 1999, Iranian airlines committed to purchase 50 from local assembly.

COSTS: Tu-334-100 US$13 million to US$18 million; Tu-334-200 US$20 million (both 2001). Break-even at 48th aircraft, according to RSK 'MiG'.

DESIGN FEATURES: Replacement for Tu-134 and Yak-42, to meet requirements of Russian Federation and Associated States (CIS) airlines. Wings have much in common with those of Tu-204 and the fuselage is shortened version of Tu-204, with identical flight deck. Configuration is all-swept low/mid-wing, with rear-mounted engines and T tail; circular-section semi-monocoque fuselage; wings have dihedral from roots. Service life 60,000 hours.

Wings have supercritical section, 24° sweepback, with winglets.

FLYING CONTROLS: Main and standby fly-by-wire; emergency hydraulic and mechanical back-up, except for ailerons. Two-section, single-slotted, trailing-edge flaps, two-section airbrakes forward of inner flap and two-section spoilers forward of outer flap on each wing; four-section leading-edge slat over full span each wing; conventional ailerons, elevators and two-section rudder; no tabs. Variable-incidence tailplane.

STRUCTURE: Composites and other lightweight materials make up 20 per cent of structure by weight. Metal alloy wing, fin and horizontal tail, each constructed on two spars,

have composites leading-edges and covered by milled panels. Elevators, ailerons, airbrakes, spoilers, flaps, rudder and floor panels of composites honeycomb. Fuselage of metal alloy with riveted skin. Final assembly at Aviant (Ukraine) and RSK 'MiG' plants.

LANDING GEAR: Retractable tricycle type; twin wheels on each unit of Tu-334-100, four-wheel bogies on main units of Tu-334-200; main units retract inward into wing/fuselage fairings; trailing-link mainwheel legs on Tu-334-100; nosewheels retract forwards, forward elements of two-section door each side remaining closed except during cycling. Some recent Tupolev illustrations have shown all versions of Tu-334 with four-wheel main bogies. Tyre size 1,070×390-480R on mainwheels; 680 × 260-355R on nosewheels. Stout grille behind mainwheels of prototype to prevent ice and slush entering turbofans.

POWER PLANT: Two Ivchenko Progress D-436T1 turbofans, each rated at 73.6 kN (16,535 lb st) on Tu-334-100. Two D-436T2 turbofans, each rated at 80.4 kN (18,078 lb st), on Tu-334-100D and -200. Two similar D-436T2 turbofans, or Rolls-Royce Deutschland BR715-55 turbofans each rated at 88.9 kN (19,995 lb st), on Tu-334-120/120D and -220. D-436T1 production by consortium of Motor Sich at Zaporozhye (Ukraine), Salyut and UMPO. Fuel in integral wing tanks, total capacity (Tu-334-100) 9,540 kg (21,030 lb).

ACCOMMODATION (Tu-334-100 and -120): Crew of two or three on 'Tu-204 compliant' flight deck; provision for fourth seat for instructor or observer. Two basic single-aisle passenger arrangements: (1) 72 seats, with 12 seats four-abreast in first class cabin at front, at pitch of 102 cm (40 in) and with 73 cm (28.75 in) aisle, and 60 tourist class seats six-abreast at 78 cm (30.7 in) pitch with 49.5 cm (19.5 in) aisle; (2) 102 seats, all tourist, at 81 cm (32 in) pitch. Both configurations have buffet/galley, coat compartment and lavatory immediately behind flight deck, a further lavatory and service compartment at rear; 72-seater has additional galleys at front and rear; overhead stowage for hand baggage. Other arrangements at customer's option. Passenger doors at front and rear of cabin on port side; service doors opposite. Underfloor baggage/freight holds; doors on starboard side.

ACCOMMODATION (Tu-334-200/Tu-354): Flight deck unchanged. Two basic single-aisle passenger arrangements: (1) 110 seats, with eight seats four-abreast

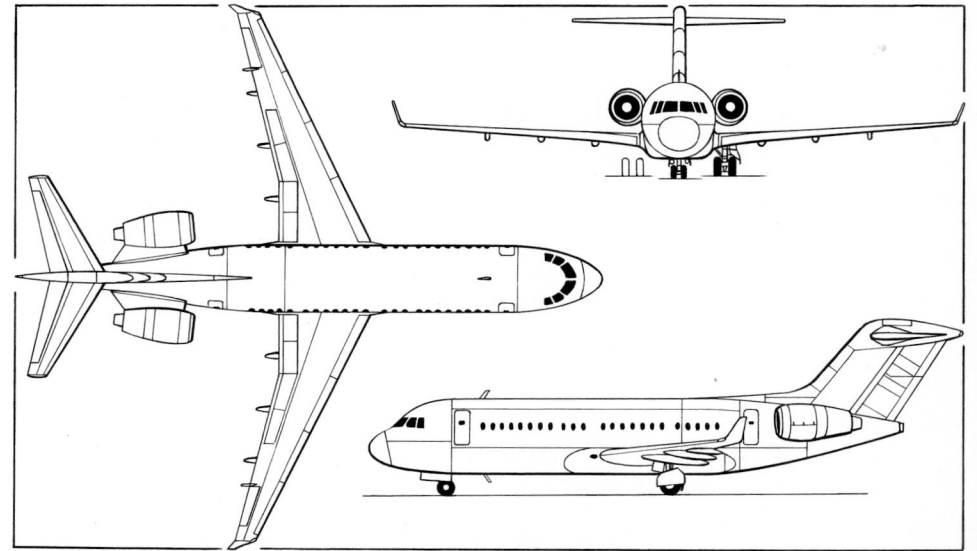

Tupolev Tu-334-100 twin-turbofan medium-range transport (*Jane's/James Goulding*)   0084429

in first class cabin at front, at pitch of 99 cm (39 in), and 102 tourist class seats six-abreast at 81 cm (32 in) pitch; (2) 126 seats, all tourist, at 81 cm (32 in) pitch. Facilities as Tu-334-100/120, plus emergency exit over wing each side.
SYSTEMS: APU in tailcone. Hydraulic system for actuation of flying controls. Air conditioning unit in wing centre-section.
AVIONICS: *Radar:* Nose-mounted weather radar.
Instrumentation: EFIS standard, with six CRTs. Satcom optional. Equipment for landings in ICAO Cat. IIIa conditions.

DIMENSIONS, EXTERNAL:
Wing span: 100, 120	29.77 m (97 ft 8 in)
100D, 200	32.61 m (107 ft 0 in)
Wing aspect ratio: 100, 120	10.2
100D, 200	10.6
Length overall: 100, 120	31.26 m (102 ft 6¾ in)
100D	31.80 m (104 ft 4 in)
200	35.16 m (115 ft 4½ in)
Fuselage cross-section:	
All	3.80 × 4.10 m (12 ft 5½ in × 13 ft 5½ in)
Height overall: all	9.38 m (30 ft 9¼ in)
Wheel track (c/l shock struts): all	4.80 m (15 ft 9 in)
Wheelbase: 100, 120	11.75 m (38 ft 6¾ in)
Baggage door:	
forward: Height	1.03 m (3 ft 4½ in)
Width	1.165 m (3 ft 9¾ in)
Height to sill	1.26 m (4 ft 1½ in)
rear: Height	0.95 m (3 ft 1½ in)
Height to sill	0.93 m (3 ft 0½ in)

DIMENSIONS, INTERNAL:
Cabin: Length: 100, 120	17.84 m (58 ft 6¼ in)
Height:	
above aisle: all	2.185 m (7 ft 2 in)
beneath hand baggage racks: all	1.70 m (5 ft 7 in)
Max width: all	3.57 m (11 ft 8½ in)
Volume: 100, 120	118.0 m³ (4,167 cu ft)
Baggage holds:	
Max length: 100, 120, forward	4.10 m (13 ft 5¼ in)
100, 120, rear	3.00 m (9 ft 10 in)
Max width: 100, 120	2.50 m (8 ft 2½ in)
Max height: 100, 120	1.20 m (3 ft 11¼ in)
Volume: 100, forward	11.7 m³ (413 cu ft)
Total volume: 100, 100D, 120	16.2 m³ (572 cu ft)
200	23.0 m³ (812 cu ft)

AREAS:
Wings, gross: 100, 120	83.23 m² (895.8 sq ft)
100D, 120D, 200, 220	100.0 m² (1,076.4 sq ft)

WEIGHTS AND LOADINGS:
Operating weight empty: 100	28,950 kg (63,824 lb)
100D	31,920 kg (70,371 lb)
120	28,350 kg (62,501 lb)
200	33,975 kg (74,902 lb)
Max payload:	
100C, 120D	9,690 kg (21,363 lb)
100, 100D	12,000 kg (26,455 lb)
120	11,000 kg (24,251 lb)
200, 200C, 220	11,970 kg (26,389 lb)
Max fuel: 100, 120	9,540 kg (21,030 lb)
100D, 200	13,790 kg (30,400 lb)
Max T-O weight: 120, 120C	46,100 kg (101,630 lb)
100	47,900 kg (105,600 lb)
100D, 120D	54,420 kg (119,975 lb)
200	55,470 kg (122,290 lb)
220: BR715	54,800 kg (120,815 lb)
D-436	55,470 kg (122,290 lb)
Max landing weight: 100	43,500 kg (95,900 lb)
Max wing loading:	
120	553.9 kg/m² (113.44 lb/sq ft)
100	575.5 kg/m² (117.87 lb/sq ft)
100D, 120D	544.2 kg/m² (111.46 lb/sq ft)
200, 220: BR715	548.0 kg/m² (112.24 lb/sq ft)
220: D-436	554.7 kg/m² (113.61 lb/sq ft)
Max power loading: 100	325 kg/kN (3.19 lb/lb st)
120	334 kg/kN (3.27 lb/lb st)
100D	338 kg/kN (3.31 lb/lb st)
120D	306 kg/kN (3.00 lb/lb st)
200	340 kg/kN (3.34 lb/lb st)
220: D-436	344 kg/kN (3.38 lb/lb st)
BR715	308 kg/kN (3.02 lb/lb st)

PERFORMANCE (estimated):
Nominal cruising speed at 10,600-11,100 m (34,775-36,400 ft): 100, 100D, 200
431-442 kt (800-820 km/h; 497-510 mph)
Balanced runway length at 30°C:
100   2,100 m (6,890 ft)
Range with max payload, 60 min reserves:
100	1,101 n miles (2,040 km; 1,267 miles)
100D	1,630 n miles (3,020 km; 1,876 miles)
200	529 n miles (980 km; 608 miles)

Model of Tupolev Tu-414 *(Jane's/Paul Jackson)*
NEW/0526930

Range with design payload:
100, 100C	1,700 n miles (3,150 km; 1,957 miles)
100D	2,213 n miles (4,100 km; 2,547 miles)
120	1,685 n miles (3,122 km; 1,939 miles)
120D	2,267 n miles (4,200 km; 2,609 miles)
200, 220	1,187 n miles (2,200 km; 1,367 miles)

OPERATIONAL NOISE LEVELS (ICAO, estimated):
T-O	85.8 EPNdB
Approach	97.4 EPNdB
Sideline	92.0 EPNdB
	UPDATED

TUPOLEV Tu-354
Refer to Tu-334/354 entry.

TUPOLEV Tu-414
TYPE: Regional jet airliner.
PROGRAMME: This project was last described in the 1997-98 edition, since which time little further news has emerged. However, in late 2002, the 50/70-seat twin-turbofan was confirmed as a competitor to the Myasishchev M-70 and Sukhoi RRJ for official funding.
NEW ENTRY

UNIKOMTRANSO

UNIKOMTRANSO AO (Unicomtranso JSC)
ulitsa Sovetskoi Armii 165-40, 443090 Samara
Tel/Fax: (+7 8462) 28 63 39 and 59 75 49
e-mail: inteltec@mail.ru
GENERAL DIRECTOR: Oleg N Voronkov

Last previously known product of this company was Model 11 amphibian, shown at 1995 Moscow Salon and illustrated in 1997-98 and earlier *Jane's*. Don was first exhibited in 2001.
UPDATED

UNIKOMTRANSO DON
TYPE: Agricultural sprayer.
PROGRAMME: Designed by S F Tsarkov, whose previous products included Leader, Frigat and Flamingo lightplanes. Production at Progress plant, Samara. Two-seat prototype (c/n 1010101; registration pending) shown at Moscow, August 2001.
CURRENT VERSIONS: **Single-seat:** With 300 litre (79.3 US gallon; 66.0 Imp gallon) chemical tank. Starboard side pilot's door.
Two-seat: With 200 litre (52.8 US gallon; 44.0 Imp gallon) chemical tank; dual controls. Additional port side door for rear occupant.
DESIGN FEATURES: Pod-and-boom configuration with pusher propeller and high, constant chord wing, having single bracing strut each side. Tailplane, of inverted aerofoil section, braced to fin by struts. Sweptback, narrow chord fin. Can be dismantled or reassembled by two persons in 1 hour. Airframe protected for 25 years of open-air operation.
FLYING CONTROLS: Conventional and manual. Pushrod and cable actuation. No aerodynamic balances. Ground-adjustable tabs on starboard aileron and port elevator. Plain flaps.

Prototype Unikomtranso Don on display at Moscow in August 2001 *(Jane's/Paul Jackson)*   NEW/0525059

STRUCTURE: Metal throughout.
LANDING GEAR: Tricycle type; fixed. Cantilever main legs with hydraulic wheel brakes. Trailing link nosewheel mounted rigidly on forward-projecting horizontal arm. All tyres size 400×150. Optional floats or skis.
POWER PLANT: One VAZ-2184-Avia adapted motorcar engine or any suitable power plant in range 75 to 90 kW (100 to 120 hp). Prototype has 73.5 kW (98.6 hp) Rotax 912 ULS flat-four driving a three-blade Kiev ground-adjustable pitch propeller.
ACCOMMODATION: One or two seats (in tandem).
EQUIPMENT: Chemical tank behind seats(s). Spraybar beneath wings; coverage rate of 1 hectare (2.5 acres)/min.

DIMENSIONS, EXTERNAL:
Wing span	10.40 m (34 ft 1½ in)
Length overall	6.40 m (21 ft 0 in)
Height overall	2.30 m (7 ft 6½ in)
Wheel track	2.00 m (6 ft 6¾ in)

WEIGHTS AND LOADINGS:
Weight empty	350 kg (772 lb)
Max T-O weight	750 kg (1,653 lb)
Max power loading (Rotax 912)	10.20 kg/kW (16.77 lb/hp)

PERFORMANCE:
Max level speed	86 kt (160 km/h; 99 mph)
Normal cruising speed	65 kt (120 km/h; 75 mph)
Min operating speed	43 kt (80 km/h; 50 mph)
Unstick speed	27 kt (50 km/h; 31 mph)
T-O and landing run	70 m (230 ft)
Range	135 n miles (250 km; 155 miles)
Endurance	2 h 0 min
	UPDATED

UUAP

ULAN-UDENSKY AVIATSIONNYI ZAVOD OAO (Ulan-Ude Aviation Plant JSC)
ulitsa Khorinskaya 1, 670009 Ulan-Ude, Buryatia
Tel: (+7 3012) 25 74 75 and 25 35 55
Fax: (+7 3012) 25 21 47
e-mail: uuaz@buriatia.ru

GENERAL DIRECTOR: Leonid Ya Belykh
TECHNICAL DIRECTOR: Sergey V Solomin
DEPUTY DIRECTOR-GENERAL, SALES AND MARKETING:
Tsydyp Ts Galdanov
MANAGER, COMMERCIAL DEPARTMENT: Andrei Polyutin

Founded on 1 January 1939 as GAZ 99, the Ulan-Ude Aviation Plant formed on 26 October 2000 and was formally incorporated as such on 23 January 2001, having previously been known as Ulan-Ude Aviation Industrial Association. It manufactures a wide range of items, from helicopters and aeroplanes to spares and domestic equipment. Is only Russian Federation plant currently producing both fixed- and rotary-wing aircraft. Employees in 2000 totalled 5,056.
The factory has built An-24 transports, Yak-25RV reconnaissance aircraft and MiG-27 fighter-bombers.

Helicopter production began in 1956, progressing through Ka-15, Ka-18 and Ka-25 to the 1970 launch of Mi-8 manufacture; total of 3,700 Mi-8 series built. Current products include modern developments of the Mil Mi-8/ Mi-17 series of helicopters (including ten Mi-171s delivered to China in 1999 and 21 to Iran in 2000-01); and the Sukhoi Su-25UBK combat trainer and its related Su-39 attack aircraft. In 2001 revealed Mi-171 variant with rear loading ramp of own design, thus directly competing with Kazan's Mi-8MTV-5. Gained Far East order for five Mi-171Sh

(Mi-8AMTSh) combat helicopters in 2001 and sold 10 Mi-171s to Malaysia. Four Mi-171s to Pakistan, June 2002, with more to follow. Also in 2001, was building two Su-25TMs with own funds. Will build Kamov Ka-62 and Ka-64. Also undertakes overhaul and upgrades, including five refurbished Mi-8AMTs supplied to Iranian Navy in 1999. Government shareholding is 38 per cent. Joint venture with SME Aviation of Malaysia, 1999, for promotion of Mi-8/Mi-17 and Su-25.

UPDATED

Mi-8 helicopters under construction at Ulan-Ude
0092558

VAPO

VORONEZHSKOYE AVIATSIONNOYE PROIZVODSTVENNOYE OBEDINENIE OAO (Voronezh Aviation Production Association JSC)

VAPO –see VASO.

UPDATED

VASO

VORONEZHSKOYE AKTSIONERNOYE SAMOLETOSTROITELNOYE OBSHCHESTVO OAO (Voronezh Shareholder Aircraft-Building Society JSC)

ulitsa Tsiolkovskoga 27, 394029 Voronezh
Tel: (+7 80732) 49 93 97
Fax: (+7 80732) 49 90 17
e-mail: admin@air.vrn.ru

Web: http://www.vaso.newmail.ru
GENERAL DIRECTOR: Albert Mikhailov
CHIEF, FOREIGN ECONOMIC RELATIONS: Vyatcheslav K Kudinov

VASO is an integral part of the Ilyushin Aviation Complex, responsible for production of the Ilyushin Il-96 transport aircraft, although this has virtually ceased in recent years due to lack of funding. It plans to form a joint holding with Ilyushin, initially by uniting the state shareholdings (30 per cent and 60 per cent, respectively) in each company; this will later be incorporated within AVPK Sukhoi as part of aviation

industry restructuring plans. Russian government holding is 20 per cent. Non-aviation manufacture includes a family of small motor boats.

Plant originated in GAZ 84, established 1932, and later was responsible for Tupolev Tu-144 supersonic airliner and Ilyushin Il-86 wide-bodied airliner. Previously known as VAPO. Plans to build Tupolev Tu-54 agricultural aircraft. Also produces boats, tractors, washing machines, food containers, motorcar silencers and other consumer items.

NEW ENTRY

VITEK

KOMPANIYA VITEK (Vitek Company)

ulitsa Generala Panfilova D20, Korpus 3, 125080 Moskva
Tel: (+7 095) 158 23 67
Fax: (+7 095) 158 65 14
e-mail: info@vitek-ltd.ru
Web: http://www.avia.vitek-ltd.ru
DIRECTOR: Yury D Bazhanov
CONSTRUCTOR: Eduard B Bazhanov

Vitek is a building materials company which also operates a light aviation division. Promotion continues of the Ovod lightplane (2001-02 and 2002-03 *Jane's*), although only two have been built; cost is US$25,000 (2002). In March 2002, the company announced that it had completed design of its Ezhik agricultural aircraft. A three-seat lightplane and four/ six-seat amphibian are also being designed.

Company has 3,000 m² (32,300 sq ft) manufacturing facility in Moscow suburb of Istrinsky, capable of producing 20 lightplanes per year. It is Russian agent for the Canadian Safari kitbuilt helicopter.

UPDATED

VITEK EZHIK

TYPE: Agricultural sprayer.
PROGRAMME: Design begun in 1999 and completed by March 2002; construction of prototype then begun; public debut was then planned for August 2003 Moscow Salon.
DESIGN FEATURES: Economic sprayer, with same payload as Sukhoi Su-38L, but at half purchase price. Raised cockpit with good forward/downward view. Constant-chord, low-mounted, cantilever wings; sweptback tail surfaces with mutual wire bracing; narrow, deep fuselage.

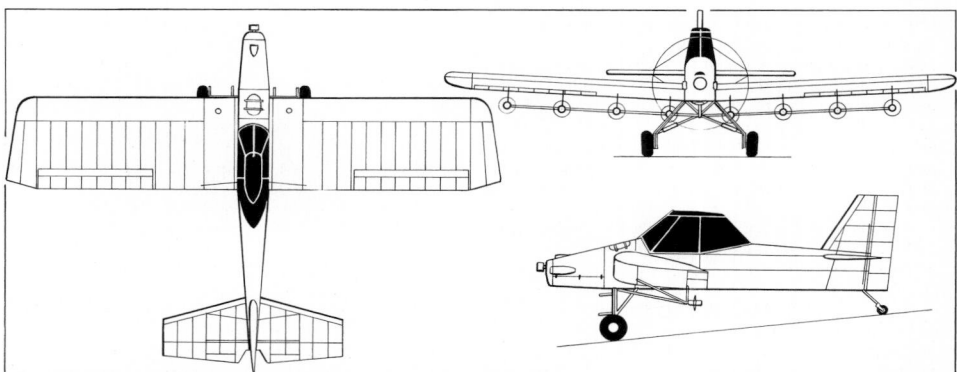

Vitek Ezhik agricultural sprayer *(Jane's/James Goulding)*   *NEW*/0525933

FLYING CONTROLS: Conventional and manual. No horn balances.
LANDING GEAR: Tailwheel type; fixed.
POWER PLANT: One 154 kW (207 hp) LOM M337A six-cylinder piston engine operating on A-76 motorcar fuel.
ACCOMMODATION: Pilot only, in enclosed cockpit.
EQUIPMENT: Chemical hopper; eight rotary atomisers beneath wings.
DIMENSIONS, EXTERNAL:

Wing span	12.00 m (39 ft 4½ in)*
Wing chord, constant	2.20 m (7 ft 2½ in)
Length overall	8.325 m (27 ft 3¾ in)
Height overall	3.35 m (11 ft 0 in)
Tailplane span	4.30 m (14 ft 1¼ in)

Company literature also gives 11.00 m (36 ft 1 in)

AREAS:

Wings, gross	24.25 m² (261.0 sq ft)
WEIGHTS AND LOADINGS:	
Weight empty	520 kg (1,146 lb)
Payload: normal	300 kg (661 lb)
max	360 kg (794 lb)
Max T-O weight	960 kg (2,116 lb)
Max wing loading	39.6 kg/m² (8.11 lb/sq ft)
Max power loading	6.22 kg/kW (10.22 lb/hp)
PERFORMANCE:	
Max level speed	97 kt (180 km/h; 112 mph)
Spraying speed	49-81 kt (90-150 km/h; 56-93 mph)
T-O run	90 m (295 ft)
Landing run	55 m (180 ft)

NEW ENTRY

YAKOVLEV

OPYTNO-KONSTRUKTORSKOYE BYURO IMENI A S YAKOVLEVA OAO (Experimental Design Bureau JSC named for A S Yakovlev)

Leningradsky prospekt 68, 125315 Moskva
Tel: (+7 095) 158 36 61
Fax: (+7 095) 787 28 44
e-mail: yakokb@cityline.ru

GENERAL DIRECTOR: Aleksandr G Efanov
CHAIRMAN AND PRESIDENT: Oleg F Demchenko
FIRST DEPUTY GENERAL DIRECTOR: Leonid L Galperin
DEPUTY GENERAL DIRECTORS: Arkady I Gurtovoy (External Economic Relations and Marketing)
   Roman P Taskaev (Flight Testing)
CHIEF DESIGNERS: Konstantin F Popovich (Yak-130)
   Aleksei G Rakhimbaev (Yak-40, Yak-42)
   Dmitry K Drach (Yak-54, Yak-152, Yak-112)
   Yuri I Yankevich (UAVs)

DEPARTMENTAL CHIEFS:
   Anatoly S Ivanov (Marketing and Sales)
   Evgeny M Tarasov (External Relations)
   Yuri V Zasypkin (Information/Press Service)

More than 200 aircraft designs and variants have been completed by Yakovlev since 1927, of which about 100 have been manufactured in series. About 70,000 aircraft of Yakovlev design have been built. Sales in 2001 were more than Rb1 billion.

By mid-2000, Yakovlev working on design of short/medium-range, 130 to 170 seat transport conforming to Russian aviation technology plan up to 2015. Yak-242 (see 1997-98 *Jane's*) is basis for this venture.

UPDATED

YAKOVLEV Yak-130

TYPE: Advanced jet trainer/light attack jet.

PROGRAMME: One of designs by five OKBs to meet official Russian requirement for 200 aircraft to replace Aero L-29 and L-39 Albatros for all aspects of flying training, basic to combat simulation, and combat; known initially as Yak-UTS; work began in 1991; preliminary down-selection (with MiG-AT) in 1995; developed in partnership with Aermacchi of Italy with definition completed in December 1999 resulting in a common basic configuration, from which all current versions are derived; originally designated Yak/AEM-130 outside Russian Federation and Associated States (CIS) but renamed Aermacchi M-346 (which see) in July 2000. Yak-130 announced as winner of Russian Air Forces' trainer competition on 10 April 2002 (decision having been reached on 16 March), in preference to MiG-AT. However, Yakovlev bureau funded 84 per cent of programme to this stage and will require export sales of between 60 and 80 aircraft to generate funds for completion of Russian Air Forces' variant, which required for service from 2010 onwards. Westernised version being funded 50 per cent by Aermacchi, and 25 per cent each by Yakovlev and Sokol aircraft factory at Nizhny-Novgorod. Aermacchi undertaking 5,000 hours of wind tunnel tests. Aermacchi has rights to establish a second assembly line in Italy and contribute tail section and wing assemblies and FBW control system and avionics to both production lines, all components to be single-sourced. Yakovlev designed, and will manufacture, wing, tail surfaces, canopy, and digital FBW FCS and will integrate avionics and weapon systems on non-RFAS aircraft.

Yak-130D development aircraft shown to press 30 November 1994; first flight 25 April 1996 at Zhukovsky (later registered RA-43130). Flew 52 sorties (36 hours) in first year, including six without winglets, pending fitment of differently shaped units. To Italy for flight trials July/August 1997, during which (18 July) the 100th sortie was flown; low-speed, high-AoA programme (30 flights) completed in Italy by prototype 31 January 1999; AoAs of up to 41° (maximum), 35° in stabilised flight and 28° in landing configuration were flown successfully. Aircraft returned to Russia for maintenance, followed by evaluation at Akhtubinsk and Armavir test centres; again in Italy by 2000. Total 305 sorties and 254 hours flown by mid-2001; 350 sorties by mid-2002; painted in tactical camouflage for display at MAKS, Moscow, August 2001.

Ten planned preseries aircraft, slightly smaller than prototype, were to comprise three for Zhukovsky test centre for evaluation originally planned to begin in 1998 and seven for air force evaluation planned for 1999 although development programme transferred to Italy, mid-1998, to escape financial instability in Russia, leading to far-reaching reappraisal. By September 1999 it was expected that funding for four preproduction prototypes would soon be obtained, together with three aircraft for evaluation; contract for initial four Yak-130-01s (including two for static tests) signed February 2001 between Yakovlev and Sokol production plant, all to be built in 2003-04; first flight in fourth quarter of 2003; certification testing to be complete by end of 2004; series production from 2005. By mid-2002, consideration being given to increasing flying preproduction fleet to three, leaving only one for ground testing. First prototype was to receive RD-2500 engines in third quarter of 2001, while 10 interim AI-222 engines were due to be delivered in 2002 from Motor-Sich for use in pre-series aircraft.

Navalised version proposed for carrier training; tender now includes Penza (Russia)/CAE Electronics (Canada) simulators and Yak-152 for screening and primary training.

CURRENT VERSIONS: **Yak-130D:** Development prototype RA-43130.

Yak-130: For Russian air force; indigenous equipment; service entry originally scheduled for 2000; delivery now planned by 2004/2005 "at best".

Yak/AEM-130: Hybrid export version; also known as AY-130; programme frozen. Items listed under Systems and Avionics refer mainly to this version. Viewed as low flyaway cost solution to Western trainer requirements by using Yak's good aerodynamic configuration and by addressing poor man-machine interface, low thrust-to-weight ratio and some other "problem areas".

Aermacchi M-346: See Italian section. Yakovlev and Aermacchi assigned separate marketing areas for rival products.

CUSTOMERS: Initial batch of two being produced, with Russian avionics, by Sokol, for evaluation. US$241 million funding provided by Italian government for development, although IAF has no immediate requirement. Interest expressed by Slovak Republic, which received demonstration in July 1997; Slovakia announced on 5 August 1998 that debts owed by Russia will be repaid in part by delivery of 12 Yak-130s, although no firm contract yet placed. The Yak-130 remains a competitor in Slovakia's search for a subsonic multirole fighter with the AMX, L 159, Hawk and MiG-AT, but has not been officially selected. Its Slovak engines give it competitive advantage. Algerian interest was reported in early 2002. Sales anticipated of up to 630, of potential market for 2,300, plus 200 or more for Russian Air Forces.

COSTS: US$12 million to US$15 million, depending on standard (2002). US$70 million reportedly already spent by Aermacchi and US$30 million by Yakovlev, with another US$100 million of R&D spending to follow.

DESIGN FEATURES: Designed from outset to conform to Western specifications for structure, flight controls and maintenance (MSG-3 and MIL-STD-470B). All-swept mid-wing monoplane, except for straight wing and tailplane trailing-edges; no dihedral or anhedral; two-piece full-span automatic leading-edge slats originally on prototype, but dogtooth leading-edge discontinuity tested on RA-43130 will be adopted on production version; all-moving low-mounted tailplane, with dogtooth leading-edge; forward-hinged, door-type airbrake on spine, ahead of fin; engines in ducts under wingroots, beneath LERX extending almost to windscreen; optional air intake screens

obviate ingestion damage when operating from unpaved airstrips. Wing leading-edge sweepback 31°. Prototype's winglets reduced in size and moved inboard to wingroot as 'LEX fence'; may be deleted in production version. Strakes ahead of windscreen by 1999 replicate characteristics of planned rounded nose. Series production versions will be 1 tonne lighter than the prototype with a more slender, shallower fuselage, lacking chines.

Service life 10,000 hours; extension planned to 15,000 hours. However, in mid-2002, 6,000 hour life was stated to be sufficient for 30-year operational life.

FLYING CONTROLS: Avionika full-authority, analogue, four-channel fly-by-wire system, with digital 'fourth channel' and 'carefree handling', slaved to laser-gyro platform, developed from that of Yak-141 V/STOL combat aircraft. Demonstrator has two selectable FCS models, one simple, one in MiG-29 class; will be replaced by digital system developed by BAE Italy and Lear Astronics. Prototype inherently stable; production aircraft intended to have 5 per cent longitudinal instability to reproduce handling characteristics of MiG-29/Su-27 series. Handling characteristics can be reprogrammed to simulate other types. Max AoA 42°.

STRUCTURE: Mainly of light alloys, but carbon fibre composites for most control surfaces. Structural design accords to MIL-STD-1530A and MIL-A-8660A.

Combat version has Kevlar armour around power plant, cockpits and avionics compartment aft of cockpits.

LANDING GEAR: Retractable tricycle type; single wheel on each unit; mainwheels retract into engine ducts; low-pressure tyres. Nosewheel steering to MIL-STD-8812 Class A. Anti-skid system.

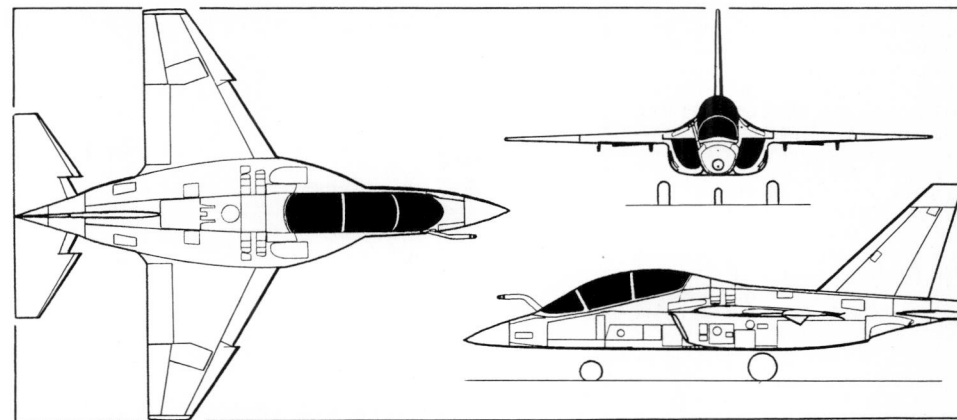

Production version of Yakovlev Yak-130 advanced trainer/combat aircraft (*Jane's/Paul Jackson*)   0051417

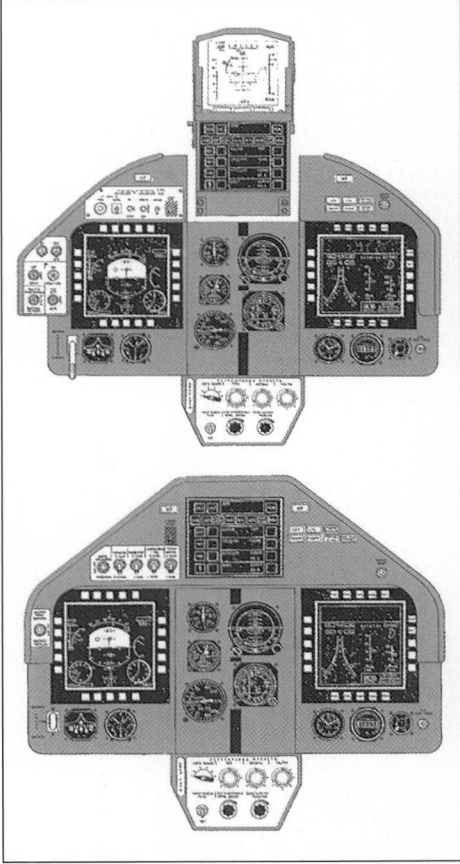

Front/student's (top) and rear/instructor's instrument panels of proposed Russian preproduction Yak-130. Western versions will have three MFDs

Some items of potential armament available to the Yakovlev Yak-130 (*Yefim Gordon*)   **NEW**/0525060

POWER PLANT: Two Povazské Strojàrne ZMK DV-2S (Klimov RD-35) turbofans each 21.58 kN (4,852 lb st). Autonomous engine starting; 20 second negative *g* fuel supply. Production version of DV-2, designated ZMKB Progress AI-222-25, considered as interim power plant until availability of Soyuz RD-2500; both potential production engines rated at 24.5 kN (5,512 lb st).

Two fuel tanks in wings; one in centre-fuselage; combined capacity approximately 2,250 litres (594 US gallons; 495 Imp gallons), but normal training fuel restricted to approximately 1,060 litres (280 US gallons; 233 Imp gallons). Provision for two external fuel tanks, each of 560 litres (148 US gallons; 123 Imp gallons) beneath wings. Single-point pressure refuelling; gravity filling optional. Detachable aerial refuelling probe. Prototype has intake FOD doors of MiG-29 type which will be option for production version. Production aircraft may receive new 2,500 kg Tushino-Soyuz RD-2500 engines, based on existing RD-1700. Other engine choices include the Klimov RD-35, Lyulka/Saturn AL-55 and Zaparozhe's AI-222.

ACCOMMODATION: Two crew in tandem under blister canopy, on Zvezda K-36LT3.5 zero/zero ejection seats with built-in computer, side-force rocket engine and extreme attitude performance; export option of Martin-Baker Mk 16 standard on M-346; explosive cord canopy penetration; rear seat raised. Accommodation air conditioned and pressurised up to differential of 0.26 bar (3.8 lb/sq in). Front pilot has maximum view over nose of −16°; rear pilot −6°.

SYSTEMS: Flight control system includes four BAE Systems computers and provides artificial stability, flying qualities in accordance with MIL-STD-1797A, controllability up to at least 35° AoA, carefree handling (*g* limitation, stall/spin prevention, AoA limitation) and adaptability to various degrees of automation and autopilot modes; includes automatic reversionary modes in the event of damage or failure; and built-in test capability. Honeywell GTCP 36-150 APU, providing normal and emergency electric, hydraulic and pneumatic power. OBOGS. Dual independent hydraulic systems, pressure 207 bar (3,000 lb/sq in); single failure leaves 75 per cent control surface movement; separate accumulators for emergency landing gear extension and brake. Dual Auxilec 20 kVA generators (one per engine); two Auxilec 6 kW transformer/rectifiers; one APU-driven 5 kW 28 V DC generator; system rating 200/115 V AC, 400 Hz. Two Ni/Cd batteries. Anti-icing system for windscreen and engine inlet guide vanes. Fire suppression system for engine and APU.

AVIONICS: Western avionics optional to replace standard Russian equipment. Primary suppliers Leninets (Russia) and GFSA/FIAR (Italy).

Comms: Dual V/UHF transceivers; intercom.

Radar: Optional. Kopyo and Osa under consideration.

Flight: Nav computer, laser gyro INS with embedded GPS, air data system, short-range radio nav Tacan and VOR/ILS, ADF and radio altimeter; IFF.

Instrumentation: HUD in front cockpit as part of collimated flight and sighting display in conjunction with pilot's helmet-mounted target designator; two Elektroavtomatika 130 mm (5.1 in) square colour multifunction liquid-crystal displays in each cockpit (three 152 × 203 mm; 6 × 8 in MFDs in export aircraft) with standby electromechanical flight/nav instruments. Optional helmet-mounted display. NVG compatibility and HOTAS controls.

Mission: Optional laser/TV weapon guidance system; Platan and Sapsan under consideration. Mission data entry system; optical weapon aiming computer and weapons control system; simulator of moving targets, guidance commands, weapons preparation and tactical situations; flight data and crew actions recorder; TV monitor of eyes and hands positions, with video recorder, in front cockpit.

Self-defence: Provision for chaff/flare dispenser, RWR and active electronic countermeasures.

Yakovlev Yak-130 prototype displaying in its camouflage colour scheme *(Yefim Gordon)*   **NEW**/0525061

Reproduction Yakovlev Yak-3M fighter imported into the USA by Shadetree Aviation, Carson City, Nevada
NEW/0525356

Front cockpit of the Yak-130 first prototype *(Yefim Gordon)*   0121099

Yak-130 prototype rear cockpit *(Yefim Gordon)*   0121100

ARMAMENT (optional): Seven (optionally nine, including wingtip AAM stations) hardpoints for up to 3,000 kg (6,614 lb) of rockets, guns, missiles, guided and unguided bombs, including eight 250 kg bombs, four 500 kg laser-guided bombs, submunitions containers, Kh-25ML (AS-10 'Karen') or AGM-65 Maverick ASMs, R-73 (AA-11 'Archer'), AIM-9L Sidewinder or Magic 2 AAMs.

DIMENSIONS, EXTERNAL (production aircraft):

Wing span	9.725 m (31 ft 10¾ in)
Wing aspect ratio	4.0
Length overall	11.495 m (37 ft 8½ in)
Height overall	4.76 m (15 ft 7½ in)
Wheel track	2.53 m (8 ft 3½ in)
Wheelbase	3.90 m (12 ft 9½ in)

AREAS:

Wings, gross	23.52 m² (253.2 sq ft)

WEIGHTS AND LOADINGS (production aircraft):

Weight empty	4,600 kg (10,141 lb)
Max weapon load	3,000 kg (6,614 lb)
Max internal fuel weight	1,750 kg (3,858 lb)
T-O weight: trainer, typical	5,700 kg (12,566 lb)
attack, normal	8,090 kg (17,835 lb)
attack, max	9,000 kg (19,841 lb)
Max wing loading: trainer	242.3 kg/m² (49.64 lb/sq ft)
attack/trainer	382.7 kg/m² (78.37 lb/sq ft)
Max power loading: trainer	116 kg/kN (1.14 lb/lb st)
attack/trainer	184 kg/kN (1.80 lb/lb st)

PERFORMANCE (clean, estimated):

Max level speed	572 kt (1,060 km/h; 659 mph)
Unstick speed	108 kt (200 km/h; 124 mph)
Landing speed	103 kt (190 km/h; 118 mph)
Stalling speed, 20% fuel	89 kt (165 km/h; 103 mph) CAS
Max rate of climb at S/L	4,500 m (14,763 ft)/min
Service ceiling	12,500 m (41,020 ft)
T-O run	340 m (1,115 ft)
Landing run	550 m (1,805 ft)
Range with max internal fuel	1,079 n miles (2,000 km; 1,242 miles)
Radius of action, 5 min combat: interdiction, two Mk 83 1,000 lb bombs, two AAMs, gun pod, two external tanks, hi-lo-hi	440 n miles (815 km; 506 miles)
close air support, two Mk 82 500 lb and two Mk 83 1,000 lb bombs, two AAMs, gun pod, hi-lo-hi, 30 min loiter	120 n miles (222 km; 138 miles)
combat air patrol, two AAMs, gun pod, two external tanks, 2 h loiter at 11,600 m (38,060 ft)	240 n miles (444 km; 276 miles)
Max rate of roll	200°/s
Sustained *g* limit at M0.8 at 4,575 m (15,000 ft) with 50% internal fuel	+4.8
g limits	+8/–3

UPDATED

YAKOVLEV Yak-3M and Yak-9U-M

TYPE: Aerobatic single-seat sportplane.

PROGRAMME: Reproduction Second World War fighter. Prototype of original wooden-skinned Yak-3 flew 1943; all-metal version flew 1945; deliveries to Soviet air forces began July 1944, totalled 4,848 (of 36,737 Yakovlev single-engined Second World War fighters); described often as lightest weight and most agile monoplane of 1939-45 period; able to turn 360° in 18.5 seconds. All-metal version now available, with Western power plant and uprated instrumentation; new-build 'prototype' (0470101, later N854DP) displayed 1993 Paris Air Show; 11 built by Strela at Orenburg to meet orders from Gunnell Museum, USA, but at least 14 eventually produced; one later to New Zealand (then Australia in 1999) and one to South Africa.

Production turned in 1996 to Yak-9U of similar appearance. Original Yak-9 entered service in 1942 and 16,769 were built; all-metal Yak-9U-M replicas include at least seven exported to USA by mid-1999, of which five then flying; one to France; distribution by Shadetree Aviation of Carson City, Nevada (Tel: +1 775 841 10 68; Web: http://www.shadetreeusa.com). Further Yak-9 completed by Strela in late 2001 for export to US.

UK importer, Richard Goode Aerobatics (see Yak-18T entry), commissioned five further Yak-9s for delivery in 2002, three having already been sold by May 2002.

COSTS: US$365,000 plus tax (2002).

Following data apply to Yak-3M.

DESIGN FEATURES: Precise reproduction in metal of Second World War airframe, except for repositioned carburettor air intake above engine cowling to suit changed engine. Conventional cantilever low-wing configuration, with tapered, round-tipped wings. Mainwheel tyres size 600×180.

POWER PLANT: Reconditioned 925 kW (1,240 hp) Allison V-1710-39 12-cylinder V liquid-cooled piston engine replaces original 925/969 kW (1,240/1,300 hp) VK-105PF-2; three-blade propeller. Fuel capacity 320 litres (84.5 US gallons; 70.4 Imp gallons).

DIMENSIONS, EXTERNAL:

Wing span	9.20 m (30 ft 2¼ in)
Wing aspect ratio	5.7
Length overall	8.49 m (27 ft 10¼ in)
Height overall	2.39 m (7 ft 10 in)

AREAS:

Wings, gross	14.85 m² (159.8 sq ft)

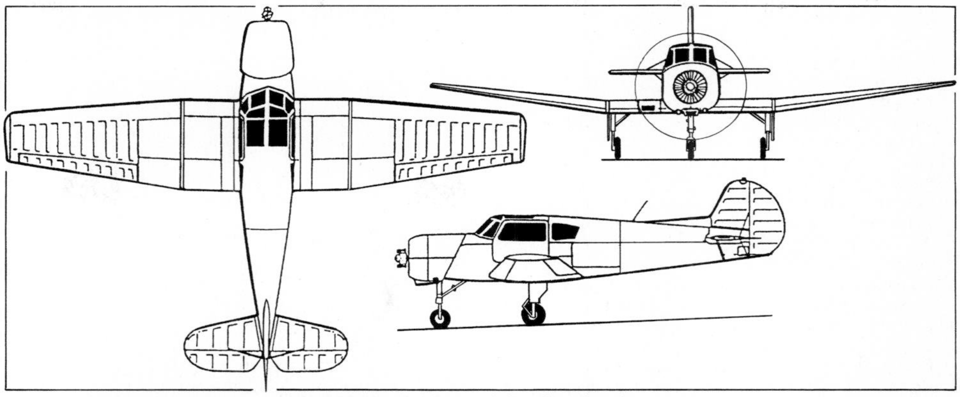

Yakovlev Yak-18T (VOKBM M-14P radial engine) ***(Jane's/Paul Jackson)*** 0015144

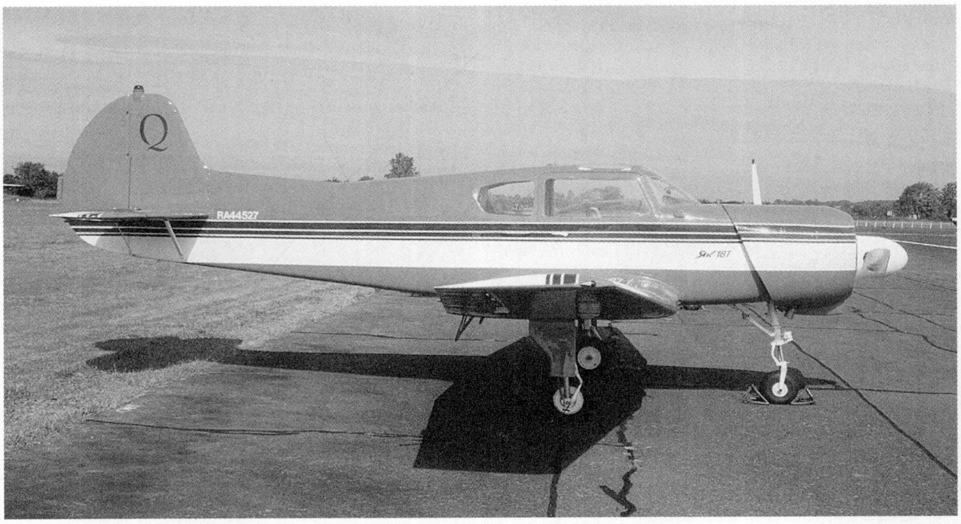

Late production Yak-18T with two-piece windscreen ***(Jane's/Paul Jackson)*** ***NEW***/0525062

WEIGHTS AND LOADINGS:

Weight empty	1,988 kg (4,383 lb)
Max T-O weight	2,600 kg (5,732 lb)
Max wing loading	175.1 kg/m² (35.86 lb/sq ft)
Max power loading	2.81 kg/kW (4.62 lb/hp)

PERFORMANCE:

Max level speed: at height	350 kt (648 km/h; 402 mph)
at S/L	307 kt (570 km/h; 354 mph)
Time to turn 360° at 1,000 m (3,280 ft)	19 s
Range	486 n miles (900 km; 559 miles)

UPDATED

YAKOVLEV Yak-18T

TYPE: Four-seat lightplane.

PROGRAMME: Prototype first flew mid-1967 as extensively redesigned cabin version of veteran Yak-18 trainer; over 700 built at Smolensk Aircraft Plant (SmAZ), including trainer, ambulance, communications and light freight versions; production resumed by Smolensk in 1993 to meet new contracts and one converted to Technoavia SM-94 prototype (see 2000-01 and previous editions). Second production run ended in 1996 due to bankruptcy of SmAZ. In 1998, work recommenced on 15 stored, part-built airframes, which being completed with two-piece windscreen, designed for Technoavia SM-94. New aircraft delivered as either SM-94 or distributed in Western Europe (eight reserved) by Richard Goode Aerobatics of White Waltham, UK (+44 1963 36 27 46; fax +44 1963 36 37 51; e-mail richard.goode@russianacros.com). Further production now appears to be in hand.

CUSTOMERS: First resumed (1993) deliveries to Luxembourg, Philippines, Russia, Switzerland, Turkey, UAE, UK and USA. Recent deliveries include three to Ulyanovsk Aviation School in June 2001. Production in 2001 reportedly 10 to 12 per year; total of 80 built between 1993 and mid-2001.

COSTS: US$95,000 (1999, Goode-sourced).

DESIGN FEATURES: Adaptation of aerobatic trainer. Conventional cantilever low-wing monoplane; three-section two-spar wings; semi-monocoque fuselage of basically square section; strut- and wire-braced tail surfaces. Airframe life 5,000 hours or 20 years.

Wing section Clark YH, thickness/chord ratio 14.5 per cent at root, 9.3 per cent at tip; dihedral 7° 20′ on outer panels only.

FLYING CONTROLS: Conventional and manual. Plain flaps and elevator trim.

STRUCTURE: All-metal construction, except for fabric covering on parts of outer wing panels, ailerons and tail surfaces.

LANDING GEAR: Retractable tricycle type; single wheel on each unit; pneumatic retraction, K-141/T-141 mainwheels inward into centre-section, Type 44-1 nosewheel rearward; Hydromash oleo-nitrogen shock-absorbers; castoring nosewheel is non-steerable; Yaroslavl tyres size 500×150 on mainwheels, 400×150 on nosewheel; tyre pressure (all) 2.95 bar (43 lb/sq in); pneumatic plate-type brakes on mainwheels. Minimum ground turning radius 2.52 m (8 ft 3¼ in).

POWER PLANT: One 265 kW (355 hp) VOKBM M-14P air-cooled radial piston engine; optional 294 kW (394 hp) M-14PF in 1999 version; V-530TA-D35 two-blade variable-pitch metal propeller; wingroot fuel tanks, combined capacity 190 litres (50.2 US gallons; 41.8 Imp gallons); gravity fuelling points on wing upper surface. Optional usable fuel capacities of 321 litres (84.7 US gallons; 70.6 Imp gallons), 360 litres (95.3 US gallons; 79.4 Imp gallons) or 614 litres (162 US gallons; 135 Imp gallons) available in 1999 version. Oil capacity 20 litres (5.3 US gallons; 4.4 Imp gallons). Inverted fuel and oil systems optional.

ACCOMMODATION: Cabin of main production version seats four persons in pairs, with forward-hinged jettisonable door each side; rear bench seat removable for freight carrying; ambulance configuration accommodates pilot, stretcher patient and medical attendant; baggage compartment aft of rear seat. Option to increase seating to six has not been pursued. Cabin heated and ventilated.

SYSTEMS: Pneumatic system for landing gear and flaps, operating pressure 49 bar (711 lb/sq in); 36/115 V AC electrical system at 400 Hz, 28.5 V DC and 20NKBN-25 battery for red panel lighting, nav and landing lights, and fin-tip anti-collision beacon.

AVIONICS: *Comms:* Baklan-5 UHF radio, intercom.

Instrumentation: ARK-15M radio compass, radio altimeter and flight recorder standard.

Customised avionics available in 1999 version.

EQUIPMENT: Optional smoke system for 1999 version.

DIMENSIONS, EXTERNAL:

Wing span	11.16 m (36 ft 7¼ in)
Wing chord: at root	2.00 m (6 ft 6¾ in)
at tip	1.06 m (3 ft 5¾ in)
Wing aspect ratio	6.6
Length overall	8.39 m (27 ft 6½ in)
Height overall	3.40 m (11 ft 1¾ in)
Tailplane span	3.54 m (11 ft 7½ in)
Wheel track	3.12 m (10 ft 2¾ in)
Wheelbase	1.955 m (6 ft 5 in)
Propeller diameter	2.40 m (7 ft 10½ in)
Propeller ground clearance	0.16 m (6¼ in)

Cabin doors (each): Height	0.88 m (2 ft 10¾ in)
Width	1.18 m (3 ft 10½ in)
Baggage door: Height	0.50 m (1 ft 7¾ in)
Width	1.00 m (3 ft 3¼ in)

DIMENSIONS, INTERNAL:

Cabin: Max width	1.28 m (4 ft 2¼ in)
Max height	1.25 m (4 ft 1¼ in)

AREAS:

Wings, gross	18.80 m² (202.4 sq ft)
Ailerons (total)	1.92 m² (20.66 sq ft)
Flaps (total)	1.60 m² (17.22 sq ft)
Fin	0.72 m² (7.73 sq ft)
Rudder	0.98 m² (10.57 sq ft)
Tailplane	1.95 m² (21.00 sq ft)
Elevators (total)	1.235 m² (13.30 sq ft)

WEIGHTS AND LOADINGS (A: instructor and one pupil, B: four persons):

Weight empty	1,217 kg (2,683 lb)
Max payload: A	306 kg (675 lb)
B	338 kg (745 lb)
Max fuel	140 kg (308 lb)
Max T-O weight: A	1,500 kg (3,307 lb)
B	1,650 kg (3,637 lb)
Max landing weight: B	1,650 kg (3,637 lb)
Max wing loading: A	79.8 kg/m² (16.34 lb/sq ft)
B	87.8 kg/m² (17.98 lb/sq ft)
Max power loading: A	5.66 kg/kW (9.32 lb/hp)
B	6.22 kg/kW (10.25 lb/hp)

PERFORMANCE:

Never-exceed speed (VNE)	162 kt (300 km/h; 186 mph)
Max level speed: A, B	159 kt (295 km/h; 183 mph)
Max cruising speed: B	135 kt (250 km/h; 155 mph)
Econ cruising speed	113 kt (210 km/h; 130 mph)
Landing speed	68 kt (125 km/h; 78 mph)
Max rate of climb at S/L: B	300 m (985 ft)/min
Service ceiling: A, B	5,520 m (18,120 ft)
T-O run: B	260-280 m (855-920 ft)
Landing run: B	345-365 m (1,135-1,200 ft)
Range with standard max fuel, with reserves:	
A, B	313 n miles (580 km; 360 miles)
g limits: A	+6.4/−3.2

UPDATED

Yakovlev Yak-54 (VOKBM M-14 radial engine) *(Yefim Gordon)*   *NEW*/0525063

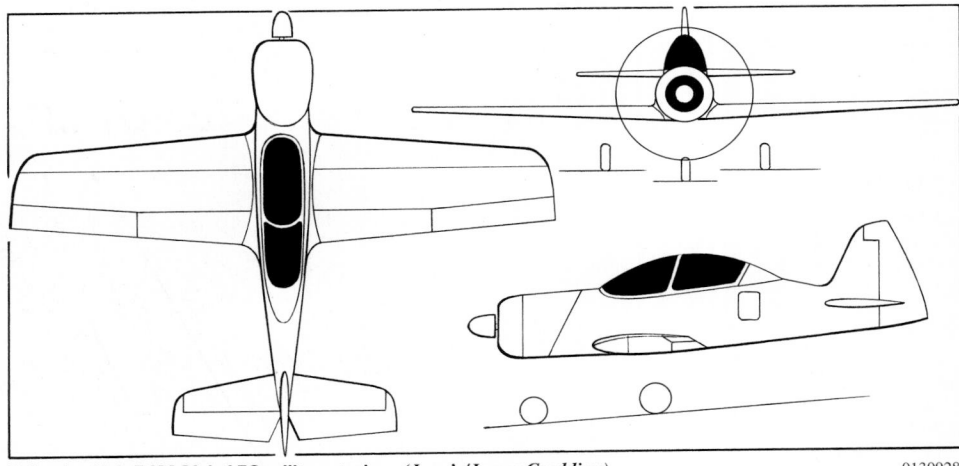

Yakovlev Yak-54M/Yak-152 military trainer *(Jane's/James Goulding)*   0130928

YAKOVLEV Yak-52

See Aerostar Iak-52 in Romanian section.

YAKOVLEV Yak-54

TYPE: Aerobatic two-seat sportplane.

PROGRAMME: Yak-54 announced 1992; prototype at 1993 Paris Air Show; first flown 23 December 1993; third prototype at Paris in 1995. In production from 1995 at Saratov Aviation Plant (SAZ), with maximum capacity of more than 100 per year. However, production suspended in 1998; to resume after certification achieved, early 2002; initial batch of five funded (and to be marketed) by ZAO Gorki Yu-2 company, including two for Promety (Prometheus) aerobatic group (representing total 2002 production) and one for export to Australia; a Slovakian customer has also been mentioned. However, in February 2002, SAZ again suspended production, having built only one aircraft (in 2001), this then due to receive Russian certification in following month.

CURRENT VERSIONS: **Yak-54:** *As described.*

Yak-56: Primary trainer derivative; also designated **Yak-54M**; redesignated **Yak-152** in late 2000 (although Yak-54M designation was still being used in 2001); 294 kW (394 hp) M-14PF, retractable tricycle landing gear and SKS-94 crew ejection system. Prototype under construction in 1999. Wing span 8.80 m (28 ft 10½ in); length overall 7.30 m (23 ft 11½ in); height overall 2.80 m (9 ft 2¼ in); empty weight 950 kg (2,094 lb); fuel weight 200 kg (441 lb); maximum take-off weight 1,300 kg (2,866 lb); manoeuvring speed 243 kt (450 km/h; 280 mph); stalling speed 54 kt (100 km/h; 63 mph); maximum range 540 n miles (1,000 km; 621 miles); g limits +9/−7.

In October 2001, Yak-152 was unsuccessful in competition against Sukhoi Su-49 to provide new equipment for ROSTO (successor to DOSAAF) paramilitary training organisation. Despite this, prototype intended to fly by early 2003 and type to enter production at Saratov (SAZ). Commitment reconfirmed by Yakovlev's chairman in April 2002, but engine reverted to 265 kW (355 hp) M-14Kh and MTOW 1,320 kg (2,910 lb).

Yak-57: Single-seat sportplane; under development by 1999.

CUSTOMERS: Total 48 Yak-54s ordered in January 1997 by Northwest Aerobatic Center, Ephrata, Washington, USA, and Dancing Bear air show team; 15 built up to 1998 suspension of production; five (of six exported) flying in USA, plus one (second-hand) delivered to Australia during 1999. In March 2002, Yak-54 fleet stated to be four in Russia, five in US, plus two destroyed in accidents. US marketing by Yakovlev Aircraft USA, Scotts Mills, Oregon; e-mail: info@yak54.com.

COSTS: Yak-54 US$160,000 (2002); Yak-152 US$200,000 (2002).

DESIGN FEATURES: Optimised for aerobatics; derived from Yak-55 (1998-99 and earlier *Jane's*). Conventional mid-wing configuration; symmetrical section; no dihedral, anhedral or incidence; almost full-span ailerons, elevators and rudder all horn-balanced; each aileron has large suspended balance tab. Designed on basis of systems and units of Yak-55M.

FLYING CONTROLS: Conventional and manual. Ailerons occupy 90 per cent of wing trailing-edge and have both horn balance and suspended tab; horn-balanced tail surfaces.

STRUCTURE: All-metal; two-spar wings; semi-monocoque fuselage; conventional tail unit.

LANDING GEAR: Non-retractable tailwheel type, with titanium spring cantilever main legs and small wheels, tyre size 400×150; tailwheel tyre size 200×80.

POWER PLANT: One 265 kW (355 hp) VOKBM M-14P nine-cylinder air-cooled radial engine; MTV-9 three-blade variable-pitch propeller.

ACCOMMODATION: Two seats in tandem under continuous transparent canopy, hinged to starboard.

DIMENSIONS, EXTERNAL:

Wing span	8.16 m (26 ft 9¼ in)
Wing aspect ratio	5.2
Length overall	6.91 m (22 ft 8 in)

AREAS:

Wings, gross	12.89 m² (138.75 sq ft)

WEIGHTS AND LOADINGS:

Max T-O weight: one pilot	850 kg (1,874 lb)
two occupants	990 kg (2,182 lb)
Max wing loading	76.8 kg/m² (15.73 lb/sq ft)
Max power loading	3.74 kg/kW (6.15 lb/hp)

PERFORMANCE:

Never-exceed speed (VNE)	243 kt (450 km/h; 280 mph)
Stalling speed	60 kt (110 km/h; 69 mph)
Rate of roll	345°/s
Max rate of climb at S/L	900 m (2,950 ft)/min
Ferry range	377 n miles (700 km; 435 miles)
g limits	+9/−7

UPDATED

YAKOVLEV Yak-SKh

TYPE: Agricultural sprayer.

PROGRAMME: Announced late 2000, when still at project stage; provisional SKh designation indicates *selskokhozyaistvenny:* agricultural; derived from Yak-54 aerobatic sportplane, retaining M-14 engine. Single-seat version to be standard; second seat optional.

VERIFIED

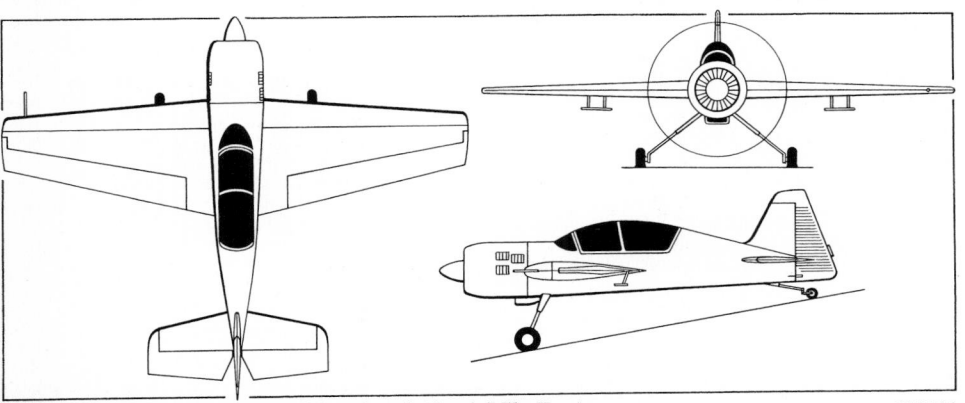

Yak-54 sporting and aerobatic training aircraft *(Jane's/Mike Keep)*   0051406

SINGAPORE

ST Aero

SINGAPORE TECHNOLOGIES AEROSPACE LTD (Division of Singapore Technologies Engineering Ltd)

540 Airport Road, Paya Lebar, Singapore 539938
Tel: (+65) 287 11 11
Fax: (+65) 280 97 13 and 280 82 13
e-mail: mktg.aero@stengg.com
Web: http://www.stengg.com
CHAIRMAN AND CEO: C B Lim
DEPUTY CHAIRMAN: S F Boon
PRESIDENT: K K Tay
DEPUTY PRESIDENT AND MILITARY MARKETING PRESIDENT: G Yeo
EXECUTIVE VICE-PRESIDENT, INTERNATIONAL MARKETING: O L Heong
EXECUTIVE VICE-PRESIDENT, MILITARY BUSINESS: Y S Ho
HEAD OF CORPORATE COMMUNICATIONS: Celina Low

Formed early 1982 as government-owned Singapore Aircraft Industries Ltd, controlled by Ministry of Defence Singapore Technology Holding Company Pte Ltd; renamed Singapore Aerospace April 1989; 15,000 m² (161,450 sq ft) new facility at Paya Lebar opened October 1983; workforce more than 4,500 in early 2002. Currently organised into Commercial Business Group, Military Business Group and Engineering and Development Centre. Provides comprehensive high-echelon commercial and military aircraft maintenance and engineering services, including structural modification and refurbishment, engine and aircraft component repair and overhaul, precision manufacturing, spares and material support.

ST Aero is partner in EC 120 helicopter programme with Eurocopter and CATIC (see International section); is manufacturing 100 shipsets of nosewheel doors worth US$12 million for Boeing 777 (option for another 100). Partners Boeing in freighter conversions of 757 (17 ordered by DHL) and MD-11 (13 for UPS); seeking participation in Airbus A380. Recently upgraded F-5Es and F-5Fs of RSAF to F-5S and F-5T standard with new avionics and WDNS (IOC January 1998); now partnering IAI and Elbit of Israel for avionics upgrade (first flight June 2001) of 48 F-5E/Fs of Turkish Air Force; and involved with Lockheed Martin, BAE Systems, FIAR, Rafael, Thales and others in promoting 'Falcon ONE' upgrade for F-16A/B Fighting Falcon.

ST Aero is the aerospace arm of Singapore Technologies Engineering Ltd. In 2001 it was restructured into the following three operating segments and subsidiaries:

Aircraft Maintenance and Modification (AMM)
ST Aviation Services Co Pte Ltd
ST Mobile Aerospace Engineering Inc*
DalFort Aerospace LP*
ST Aerospace Engineering Pte Ltd
Pacific Flight Services Pte Ltd
* Subsidiary of Singapore Technologies Engineering Ltd

Component/Engine Repair and Overhaul (CERO)
ST Aerospace Engines Pte Ltd
ST Aerospace Systems Pte Ltd
DalFort Components

Engineering and Materials Services (EMS)
Engineering and Development Centre
ST Aerospace Supplies Pte Ltd
Airline Rotables Ltd
ST Aerospace International Structures Pte Ltd
iShopAero Pte Ltd
VisionTech Engineering Pte Ltd
ST Aviation Resources Pte Ltd

UPDATED

SLOVAK REPUBLIC

AEROPRO

AEROPRO s r.o.
Kostolna 42, SK-949 01 Nitra
Tel/Fax: (+421 87) 652 63 55
e-mail: aeropro@flynet.sk

GERMAN DISTRIBUTOR:
Ikarusflug Leichtflugzeuge GbR
Mennwanger Strasse 3, D-88682 Salem
Tel: (+49 7553) 17 70
Fax: (+49 7553) 605 38

Light aviation supplies company Aeropro has taken over production of the Eurofox ultralight formerly marketed by EV-AT of Czech Republic (which see).

VERIFIED

AEROPRO EUROFOX

TYPE: Side-by-side ultralight/kitbuilt.
PROGRAMME: Adapted from SkyStar (Denney) Kitfox, primarily for sale in Germany by Ikarusflug. Certification to German Bauvorschriften für Ultraleichtflugzeuge (BFU) and Slovakian ultralight requirements was achieved in March 1996. Certification of Eurofox Pro followed. The 1999 models introduced increased cabin width (Eurofox Space) and option of Rotax 912 ULS with provision for glider towing.
Following description refers to versions marketed by Ikarusflug.
CURRENT VERSIONS: **Eurofox:** Tailwheel ultralight version; two-blade propeller. MTOW 450 kg (992 lb).
　Eurofox Pro: Nosewheel version of above; three-blade propeller.
　Eurofox Space: Introduced 1999; JAR-VLA version with 480 kg (1,058 lb) MTOW; nosewheel or tailwheel. Improvements include increased cabin width, increased leg room (partly through raised instrument panel) and modified engine cowling.
　Note that many aircraft are registered with the type name **Fox**.
CUSTOMERS: The 100th was built in 2001 for a Dutch owner; at least 125 registered by mid-2002.
COSTS: Eurofox Space DM86,700 (2000), including tax, basic.
DESIGN FEATURES: Extensive redesign of Kitfox (see SkyStar in US section). Main changes are lengthened fuselage to improve longitudinal stability, NACA 4412 modified wing section, Junkers combined flaps/ailerons, more spacious cabin dimensions and revised landing gear.
FLYING CONTROLS: Conventional rudder and elevator, manually actuated; 80 per cent span, mass-balanced Junkers-type flaperons, maximum deflection 20°.
STRUCTURE: Folding wings have two alloy tubing spars with diagonal bracing, alloy full and half ribs, glass fibre flaps and leading-edges and Ceconite covering. Fuselage of welded steel tubing.
LANDING GEAR: Non-retractable tailwheel type on Eurofox, with cantilever main legs and hydraulic disc brakes. Eurofox Pro has tricycle gear with smaller mainwheels (380×150 instead of 420×150) and steerable 300×100 nosewheel with rubber-in-compression suspension.

Aeropro Eurofox Space with nosewheel and wheel fairings (*Jane's/Paul Jackson*)　0110506

POWER PLANT: One 59.6 kW (79.9 hp) Rotax 912 UL flat-four four-stroke driving Kremen SR30 two-blade or SR200 three-blade fixed-pitch propeller via a 1:2.273 reduction gear. Alternatively, one 73.5 kW (98.6 hp) Rotax 912 ULS. Fuel capacity 58 litres (15.3 US gallons; 12.8 Imp gallons), of which 56 litres (14.8 US gallons; 12.3 Imp gallons) are usable.
SYSTEMS: Electrical system with 16 Ah 250 W DC generator.
AVIONICS: Optional avionics by Becker and Honeywell, including GPS, to customer's choice.
EQUIPMENT: USH 520 or BRS5UL4 parachute recovery system optional.
DIMENSIONS, EXTERNAL (A: Eurofox, B: Eurofox Pro, C: Eurofox Space):

Wing span: A, B, C	9.20 m (30 ft 2¼ in)
Length: overall: A	6.00 m (19 ft 8¼ in)
B, C	5.75 m (18 ft 10½ in)
wings folded: A	6.10 m (20 ft 0¼ in)
B	6.20 m (20 ft 4 in)
Height overall:	
tailwheel (top of wing)	1.80 m (5 ft 10¾ in)
nosewheel (over tail)	2.28 m (7 ft 5¾ in)
Width overall, wings folded	2.40 m (7 ft 10½ in)
Wheel track	1.94 m (6 ft 4½ in)
Wheelbase: A	4.10 m (13 ft 5½ in)
B	1.23 m (4 ft 0½ in)
Propeller diameter: two-blade	1.80 m (5 ft 10¾ in)
three-blade	1.70 m (5 ft 7 in)

DIMENSIONS, INTERNAL:

Cabin max width: A, B	1.06 m (3 ft 5¾ in)
C	1.12 m (3 ft 8 in)

AREAS:

Wings, gross: A, B	11.50 m² (123.8 sq ft)

WEIGHTS AND LOADINGS:
Weight empty (with parachute recovery system):

A	279 kg (615 lb)
B	285 kg (628 lb)
C	270 kg (595 lb)
Max T-O weight: A, B	450 kg (992 lb)
C	480 kg (1,058 lb)

PERFORMANCE (at 450 kg; 992 lb):

Never-exceed speed (VNE)	99 kt (185 km/h; 115 mph)
Cruising speed at 75% power	86 kt (160 km/h; 99 mph)
Stalling speed	35 kt (65 km/h; 41 mph)
Max rate of climb at S/L	300 m (984 ft)/min
T-O to 15 m (50 ft)	191 m (625 ft)
Range with max fuel	378 n miles (700 km; 435 miles)

UPDATED

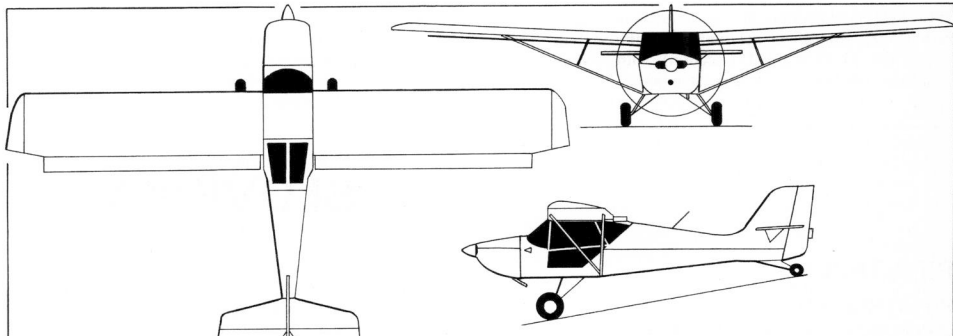

Aeropro Eurofox Space in its tailwheel version (*Jane's/Paul Jackson*)　0110508

refuelling point in starboard mainwheel fairing. Additional fuel can be carried in one 1,000 litre or two 750 litre (264 or 198 US gallon; 220 or 165 Imp gallon) optional ferry tanks inside cabin, and/or two 500 litre (132 US gallon; 110 Imp gallon) auxiliary underwing tanks. Oil capacity 4.5 litres (1.2 US gallons; 1.0 Imp gallon) per engine.

ACCOMMODATION: Crew of two on flight deck; cabin attendant in civil version. Ergonomically redesigned flight deck; improved cabin soundproofing, sidewall design, lighting and toilet; 400M, foldaway seats along each side of cabin.

For troop transport role, main cabin can be fitted with 25 inward-facing seats along cabin walls, to accommodate 24 paratroops with instructor/jumpmaster; or 25 fully equipped troops. As ambulance, cabin is normally equipped to carry 12 stretcher patients and four medical attendants. As freighter, up to 2,950 kg (6,504 lb) of cargo can be carried in main cabin, including two LD1, LD727/DC-8 or three LD3 containers, or light vehicles. Cargo system, certificated to FAR Pt 25, includes roller loading/unloading system and 9 *g* barrier net. Photographic version equipped with two Leica RC-20/30 vertical cameras and darkroom. Navigation training version has individual desks/consoles for instructor and five pupils, in two rows, with appropriate instrument installations.

Civil passenger transport version has standard seating for up to 26 in mainly three-abreast layout at 72 cm (28.5 in) pitch, with provision for quick change to all-cargo or mixed passenger/cargo interior. Lavatory, galley and 400 kg (882 lb) capacity baggage compartment standard, plus additional 150 kg (330 lb) in nose bay. VIP transport version can be furnished to customer's requirements.

Forward/outward-opening door on port side immediately aft of flight deck; forward/outward-opening passenger door on port side aft of wing; inward-opening emergency exit opposite each door on starboard side. Additional emergency exit in roof of forward main cabin. Two-section underfuselage loading ramp/door aft of main cabin can be opened in flight for discharge of paratroops or cargo, and can be fitted with optional external wheels for door protection during ground manoeuvring. Interior of rear-loading door can be used for additional baggage stowage in civil version. Entire accommodation heated and ventilated; air conditioning optional.

SYSTEMS: Freon cycle or (on special mission versions) engine bleed air air conditioning system optional. Hydraulic system, operating at service pressure of 138 bar (2,000 lb/sq in), provides power via electric pump to actuate mainwheel brakes, flaps, nosewheel steering and rear cargo ramp/door. Hand pump for standby hydraulic power in case of electrical failure or other emergency. Electrical system supplied by two 9 kW starter/generators, three batteries and three static converters. Pneumatic boot and engine bleed air de-icing of wing and tail unit leading-edges; electric de-icing of propellers and windscreens. Oxygen system for crew (including cabin attendant); two portable oxygen cylinders for passenger supply. Engine and cabin fire protection systems.

AVIONICS: *Comms:* Rockwell Collins VHF, ATC transponder, intercom and PA system standard; ELT; Rockwell Collins HF, UHF and second transponder, and Fairchild CVR, optional.

Radar: Honeywell weather radar standard. RDR-1500B search radar in ventral radome on Patrullero versions.

Flight: Rockwell Collins VOR/ILS, ADF, DME and radio altimeter, and Honeywell AFCS and directional gyro, standard; second Rockwell Collins ADF, Global Omega nav, Dorne & Margolin marker beacon receiver and Fairchild flight data recorder optional. Flight

Flight deck of the C-212 Series 400 0051420

management system (FMS) incorporates VOR, ADF, DME and GPS nav receiver.

Instrumentation: EFIS with four CRTs; IEDS (integrated engine data system) with two colour LCDs. Blind-flying instrumentation standard.

EQUIPMENT: 1,000 kg (2,205 lb) capacity cargo winch optional.

ARMAMENT (military versions, optional): Two machine gun pods or two rocket launchers, or one launcher and one gun pod, on hardpoints on fuselage sides (capacity 250 kg; 551 lb each).

DIMENSIONS, EXTERNAL:

Wing span	20.27 m (66 ft 6 in)
Wing chord: at root	2.50 m (8 ft 2½ in)
at tip	1.25 m (4 ft 1¼ in)
Wing aspect ratio	10.0
Length overall	16.15 m (52 ft 11¾ in)
Fuselage max width	2.30 m (7 ft 6½ in)
Height overall	6.60 m (21 ft 7¾ in)
Tailplane span	8.40 m (27 ft 6¾ in)
Wheel track	3.10 m (10 ft 2 in)
Wheelbase	5.46 m (17 ft 11 in)
Propeller diameter	2.79 m (9 ft 2 in)
Propeller ground clearance (min)	1.44 m (4 ft 8¾ in)
Distance between propeller centres	5.27 m (17 ft 3¾ in)
Passenger door (port, rear): Height	1.58 m (5 ft 2¼ in)
Width	0.70 m (2 ft 3½ in)
Crew and servicing door (port, fwd):	
Max height	1.10 m (3 ft 7¼ in)
Width	0.58 m (1 ft 10¾ in)
Rear-loading door: Max length	3.66 m (12 ft 0 in)
Max width	1.68 m (5 ft 6¼ in)
Max height	1.80 m (5 ft 10¾ in)
Emergency exit (stbd, fwd): Height	1.10 m (3 ft 7¼ in)
Width	0.58 m (1 ft 10¾ in)
Emergency exit (stbd, rear): Height	0.95 m (3 ft 1½ in)
Width	0.56 m (1 ft 10 in)

Two of the three CASA C-212-400Ms delivered to the Venezuelan Navy 0110903

DIMENSIONS, INTERNAL:

Cabin (excl flight deck and rear-loading door):	
Length: passenger	7.27 m (23 ft 10¼ in)
cargo/military	6.55 m (21 ft 5¾ in)
Max width	2.10 m (6 ft 10¾ in)
Max height	1.80 m (5 ft 10¾ in)
Floor area: passenger	13.5 m² (145 sq ft)
cargo/military	12.2 m² (131 sq ft)
Volume: passenger	23.8 m³ (840 cu ft)
cargo/military	22.0 m³ (777 cu ft)
Cabin: volume incl flight deck and rear-loading door	30.4 m³ (1,074 cu ft)
Baggage compartment volume	3.6 m³ (127 cu ft)

AREAS:

Wings, gross	41.00 m² (441.3 sq ft)
Ailerons (total, incl tab)	1.875 m² (20.18 sq ft)
Trailing-edge flaps (total)	2.14 m² (23.03 sq ft)
Fin, incl dorsal fin	6.58 m² (70.83 sq ft)
Rudder, incl tab	2.16 m² (23.25 sq ft)
Tailplane	12.57 m² (135.30 sq ft)
Elevators (total, incl tabs)	1.78 m² (19.16 sq ft)

WEIGHTS AND LOADINGS:

Manufacturer's weight empty	3,780 kg (8,333 lb)
Weight empty, equipped (cargo)	4,550 kg (10,031 lb)
Max cargo payload	2,950 kg (6,504 lb)
Max fuel: standard	1,600 kg (3,527 lb)
with underwing auxiliary tanks	2,400 kg (5,291 lb)
Max T-O and landing weight	8,100 kg (17,857 lb)
Max ramp weight	8,150 kg (17,967 lb)
Max zero-fuel weight	7,500 kg (16,535 lb)
Max cabin floor loading	732 kg/m² (150 lb/sq ft)
Max wing loading	197.6 kg/m² (40.46 lb/sq ft)
Max power loading	6.04 kg/kW (9.92 lb/shp)

PERFORMANCE (Series 400M):

Max cruising speed at 3,050 m (10,000 ft)	195 kt (361 km/h; 224 mph)
Econ cruising speed at 3,050 m (10,000 ft)	163 kt (302 km/h; 188 mph)
Time to 3,050 m (10,000 ft)	8 min
Service ceiling	7,925 m (26,000 ft)
Service ceiling, OEI	3,275 m (10,740 ft)
T-O to 15 m (50 ft)	587 m (1,925 ft)
Landing from 15 m (50 ft)	527 m (1,730 ft)
Required runway length for STOL operation	402 m (1,320 ft)
Range:	
with max payload	233 n miles (431 km; 268 miles)
with max standard fuel and 2,000 kg (4,409 lb) payload	800 n miles (1,481 km; 920 miles)
with max standard and auxiliary fuel and reduced payload	1,375 n miles (2,546 km; 1,582 miles)

UPDATED

CASA C-212 PATRULLERO

TYPE: Maritime surveillance twin-turboprop.

DESIGN FEATURES: Special missions versions of C-212.

CURRENT VERSIONS: **MP:** Maritime patrol version; belly-mounted 360° search radar and FLIR/TV turret. Other features include operator's console, datalink via Inmarsat, external loudspeakers, searchlight, observation bubble windows, camera, and seating for 10 passengers. Initial C-212-400MP deliveries to Spanish Ministry of Agriculture, Fisheries and Food (one in December 1998 and a second on 29 March 2001) and Suriname Air Force (one in May 1999); third for Spanish MAFF due in early 2003.

Pollution control: Equipment as for MP, plus dedicated sensor for detection, measurement and control of

harvest plagues, marine resources and pollution. This equipment has been fitted in Swedish Coast Guard and Portugese Air Force C-212s.

CUSTOMERS: See table with main C-212 entry. More than 56 Patrulleros sold by February 2002.

POWER PLANT: As relevant C-212 Series. Two 500 litre (132 US gallon; 110 Imp gallon) underwing auxiliary fuel tanks.

ACCOMMODATION: Flight crew of two; mission sensor operator and observer in MP. Two observer stations at bubble windows in rear of cabin. Operator console with search radar, FLIR/TV, moving map and communications control. Radar repeater display and searchlight control added to flight deck. Galley, lavatory and rest area equipped with commuter seats or stretchers.

AVIONICS (MP): *Comms:* Single VHF, UHF/VHF and HF radios; intercom; IFF.

Radar: 360° scan underfuselage radar.

Flight: Redundant flight management system with dual GPS nav.

Instrumentation: Four-screen EFIS; IEDS.

Mission: FLIR/TV turret, Inmarsat datalink, electronic cartography and camera in MP. Additional to, or independent of, this equipment, Pollution control version can be fitted with a SLAR, IR/UV scanner or microwave radiometer.

EQUIPMENT: Survival kit and raft launcher in rear ramp; searchlight on fuselage hardpoint.

ARMAMENT: Hardpoint (capacity 500 kg; 1,102 lb) on each side of fuselage, on which can be carried machine gun pods or rocket launchers.

PERFORMANCE:

Endurance with underwing fuel tanks   8 h 30 min
UPDATED

C-212-400MP Patrullero delivered to the Spanish Ministry of Agriculture, Fisheries and Food   0098126

VOL MEDITERRANI

VOL MEDITERRANI SL

Finca 'Les Umbertes', PO Box 83, E-08180 Moià
Tel/Fax: (+34 93) 830 12 52

Company headquarters are at the 'Les Umbertes' aerodrome, in the Barcelona region.

NEW ENTRY

VOL MEDITERRANI VM-1 ESQUAL

English name: Shark

TYPE: Side-by-side ultralight kitbuilt.

PROGRAMME: Prototype EC-YZZ. UK debut at PFA Rally, Cranfield, June 2002 (demonstrator EC-ZFF).

CURRENT VERSIONS: **Fixed gear:** Nosewheel landing gear. *As described.*

Retractable gear: Tailwheel type. Retractable mainwheels; fixed tailwheel with speed fairing. Prototype EC-ZFX had flown by mid-2002.

COSTS: Kit, excluding engine, instruments and propeller: €27,768 glass fibre spar, €36,080 carbon fibre spar (2002).

DESIGN FEATURES: High-performance ultralight. Low-mounted, tapered wing and mid-tailplane. Sweptback fin. No composites work required for assembly.

FLYING CONTROLS: Conventional and manual; ailerons and elevator pushrod actuated. Two-part (right and left), linked ailerons. Variable incidence (electric) tailplane for trim. Flaps also deflect upwards to optimise cruising speed.

STRUCTURE: Glass fibre monocoque fuselage; cockpit area reinforced. Steel tube centre-section. Wing main spar of carbon fibre or glass fibre (according to customer's choice; latter with 10 kg; 22 lb weight saving), with secondary trailing-edge spar and glass fibre sandwich skin.

LANDING GEAR: As described under Current Versions. Hydraulic brakes.

POWER PLANT: Option of 73.5 kW (98.6 hp) Rotax 912 ULS flat-four or 89 kW (120 hp) Jabiru 3300 flat-six; latter adds 8.5 kg (19 lb) to empty weight. Two-blade fixed-pitch propeller or three-blade ground-adjustable pitch propeller. Fuel in two wing tanks, combined capacity 70 litres (18.5 US gallons; 15.4 Imp gallons).

EQUIPMENT: Optional ballistic parachute.

DIMENSIONS, EXTERNAL:

Wing span	9.25 m (30 ft 4¼ in)
Length overall	5.86 m (19 ft 2¾ in)
Height overall	1.94 m (6 ft 4½ in)

Demonstrator Vol Mediterrani VM-1 Esqual on display at PFA Rally, Cranfield, June 2002 (*Jane's/Paul Jackson*)   *NEW*/0137965

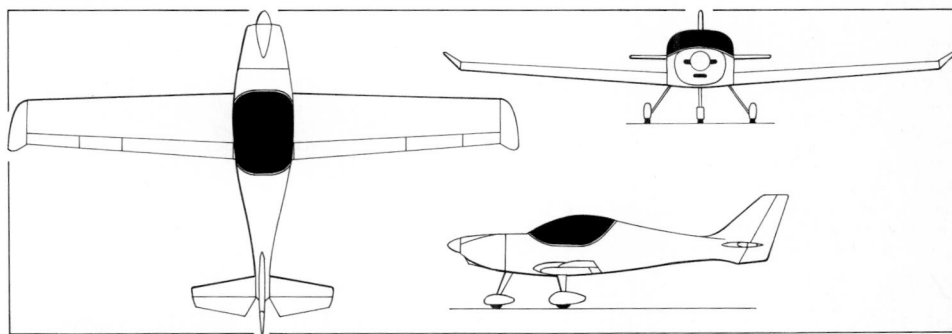

Vol Mediterrani VM-1 Esqual side-by-side kitbuilt (*Jane's/Paul Jackson*)   *NEW*/0137969

DIMENSIONS, INTERNAL:

Cabin max width	1.10 m (3 ft 7¼ in)

AREAS:

Wings, gross	9.00 m² (96.9 sq ft)

WEIGHTS AND LOADINGS (Rotax 912 ULS and glass fibre wing spar):

Weight empty	258 kg (569 lb)
Max T-O weight	450 kg (992 lb)

PERFORMANCE (Rotax 912 ULS, unless otherwise specified):

Max level speed: Rotax	140 kt (260 km/h; 161 mph)
Jabiru	178 kt (330 km/h; 205 mph)

Cruising speed at 75% power

	124 kt (230 km/h; 143 mph)
Stalling speed, flaps down	33 kt (60 km/h; 38 mph)
Max rate of climb at S/L	457 m (1,500 ft)/min
T-O and landing run	80 m (265 ft)
Range	567 n miles (1,050 km; 652 miles)
g limits	+9/−3 (ultimate)

NEW ENTRY

SRI LANKA

JABIRU

JABIRU ASIA
Koggala

Formation of Jabiru Asia announced mid-1998 as joint venture between Jabiru (see Australian section) and CDE Aviation Company of Sri Lanka (member of Lionair group) to manufacture Jabiru ST3 at new factory in Koggala. Work was due to begin before end of 1998, but start not reported by mid-2002.

UPDATED

SWEDEN

SAAB

SAAB AB

SE-581 88 Linköping
Tel: (+46 13) 18 00 00
Fax: (+46 13) 18 00 11
e-mail: infosaab@saab.se
Web: http://www.saab.se
PRESIDENT AND CEO: Bengt Halse
EXECUTIVE VICE-PRESIDENT AND CFO: Göran Sjöblom
SENIOR VICE-PRESIDENT, COMMUNICATIONS: Jan Nygren
PRESS DIRECTOR: Anders Stålhammar

Svenska Aeroplan AB founded at Trollhättan 1937 to make military aircraft; amalgamated 1939 with Aircraft Division of Svenska Järnvägsverkstäderna rolling stock factory at Linköping; renamed Saab Aktiebolag May 1965; merged with Scania-Vabis 1968 to combine automotive interests; Malmö Flygindustri acquired 1968. Bid for Celsius AB 16 November 1999 and acquisition completed 8 March 2000; enlarged company reorganised in 2000 into six main business areas and several independent operations:

Saab Aerospace: Combining military aircraft, future aerospace systems and commercial programmes.

Saab Aviation Services: Commercial aircraft maintenance and engine and component maintenance. Comprises Celsius Aviation Services, Saab Aircraft and Saab Aircraft Leasing; holds type certificates for Saab 340 and Saab 2000 regional turboprop airliners, production of which ended in 1999.

Saab Systems and Electronics: Electronic warfare, simulation and training and radar control, including Combitech Systems and Ericsson Saab Avionics.

Saab Bofors Dynamics: Missiles, anti-armour and underwater systems.

Saab Technical Support and Services: Including aircraft maintenance.

Saab Ericsson Space: Digital, microwave and mechanical products.

Saab workforce was 14,028 in January 2002. Continued subcontract work includes structural floor assemblies, pylons and landing gear doors for the Airbus A340-500/600 series. Saab also partner in A380 and A400M programmes. Builds A380 wing panels; crew doors and wingtip panels for Boeing 737; and wing components for Boeing 777. Developing new tactical system for NH 90 helicopter under contract placed in early 2002. Saab group sales include 40 per cent exports and are 70 per cent defence-related.

It was announced on 30 April 1998 that Saab's owner, Investor AB, had approved the sale to British Aerospace plc (now BAE Systems) of a 35 per cent voting shareholding. Investor remains Saab's leading owner, with 36 per cent of the votes and 20 per cent of the capital. Purchase rights to the remaining 29 per cent of votes and 45 per cent of capital continue to be offered to Investor shareholders.

More than 4,000 military and commercial aircraft delivered since 1940; has held dealership for MD Helicopters (formerly Schweizer/Hughes, McDonnell Douglas and Boeing) products in Scandinavia and Finland since 1962. In September 2000, Saab announced entry into design of UAVs, in conjunction with Swedish Defence Materiel Administration and other local industries (see *Jane's Unmanned Aerial Vehicles and Targets*).

UPDATED

SAAB AEROSPACE

SE-581 88 Linköping
Tel: (+46 13) 18 00 00
Fax: (+46 13) 18 24 11
e-mail: coms@saab.se
Web: http://www.saabaerospace.com
HEAD OF BUSINESS AREA: Ake Svensson
DEPUTY HEAD (Commercial): Hans Krüger
INFORMATION MANAGER: Peter Larsson

Gripen International

SE-581 88 Linköping
Tel: (+46 13) 18 00 00
Fax: (+46 13) 18 00 55
Web: http://www.gripen.com
MANAGING DIRECTOR: Ian McNamee
DEPUTY MANAGING DIRECTOR: Kjell Möller
MARKETING DIRECTOR: Bob Kemp
COMMUNICATIONS DIRECTOR: Owe Wagermark
Web: http://www.gripen.co.uk

Original Industrigruppen JAS AB (JAS Industry Group) formed 1981 to represent (then) Saab-Scania (65 per cent share), Ericsson (16 per cent), Volvo Flygmotor (15 per cent) and FFV Aerotech (4 per cent) in JAS 39 Gripen programme; continues to act as contractor for Försvarets Materielverk (Defence Materiel Administration, FMV) and co-ordinated JAS 39 Gripen programme within Sweden. Volvo Flygmotor became Volvo Aero in 1994 and FFV Aerotech share passed to Celsius in 1990 through merger with Bofors. Employees of Saab Aerospace totalled 4,120 in 2001. International marketing of JAS 39 Gripen is now responsibility of Saab and BAE Systems, acting jointly since November 1995; current members of JAS Industry Group are subcontractors.

Workshare on export orders would be Saab 55 per cent/ BAE 45 per cent under agreement signed 12 June 1995. BAE involvement also includes manufacture of main landing gear, the first of which was delivered from the Brough factory in April 1996 and incorporated in 49th production Gripen. From May 1999, BAE also responsible for assembly of wing attachment unit and associated component join-up for two-seat aircraft (single-seaters to follow).

UK and Swedish export guarantee departments signed an agreement in August 1996 to finance any third-party sales. A contract of 16 November 1995 provides for Danubian Aircraft Company of Hungary to manufacture Gripen components at Tököl as part of a wider Swedish-Hungarian defence information exchange agreement signed 18 December 1995, which may lead to Hungarian purchase of the JAS 39. First Hungarian parts (tailcone fittings) delivered April 1996. Further Swedish-Hungarian agreement of 22 January 1997 provided for comprehensive industrial offsets in the event of a Gripen purchase. First South African parts – Denel-built assembly set for stores pylon – handed over April 2000. Contract awarded to Denel in March 2001 for manufacture of section of centre fuselage for 40 Lot 3 Gripens, plus 28 for SAAF aircraft. Polish part manufacture (metal subassemblies) began July 2000 at PZL (Polish Aircraft Factory).

Gripen International formed 3 September 2001 to strengthen Saab-BAE Systems marketing ventures; is jointly owned and staffed, but registered in Sweden.

UPDATED

SAAB JAS 39 GRIPEN

English name: Griffin
TYPE: Multirole fighter.
PROGRAMME: Funded definition and development began June 1980; initial proposals submitted 3 June 1981; government approved programme 6 May 1982; initial FMV development contract 30 June 1982 for five prototypes and

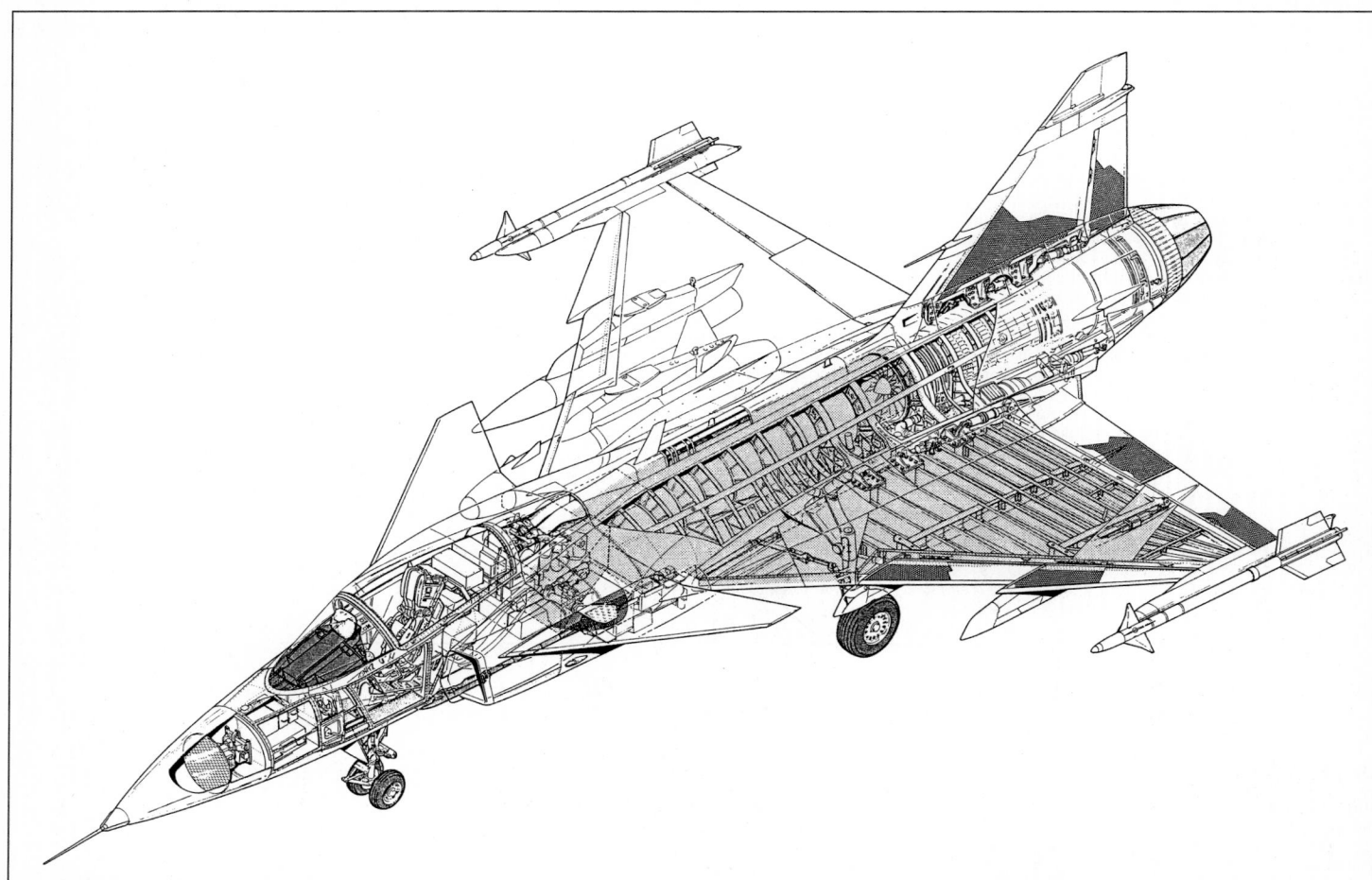

Interior details of the JAS 39A Gripen

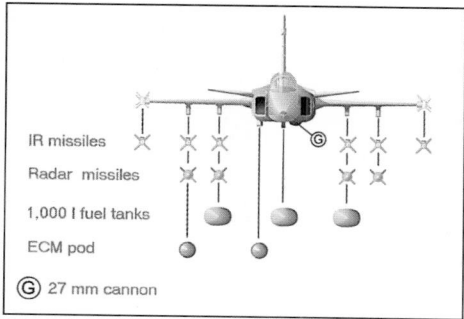

Gripen air-to-air armament 0106499

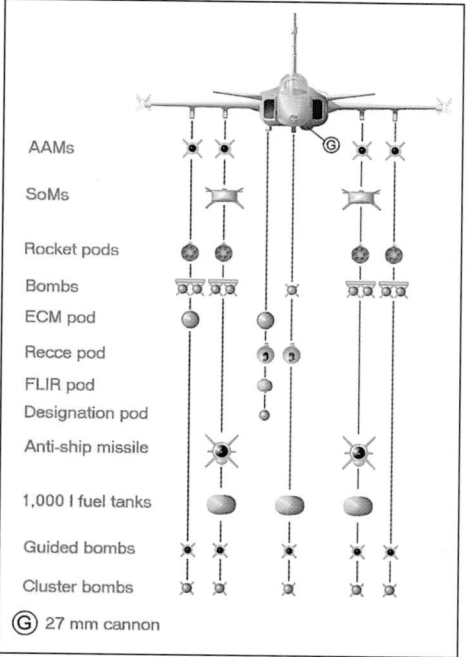

Gripen air-to-surface armament 0106500

Depiction of Lot 3 Gripen cockpit *NEW*/0536705

SwAF JAS 39 CONTRACTS

Batch	Qty	Type	Serials	Remarks
Prototypes (5)	5	JAS 39	39-1 to 39-5	Plus 39-6 (below)
Lot 1 (30)	29	JAS 39A	39101 to 39129	39130 completed as JAS 39B
	1	JAS 39B	39800	
Lot 2 (110)	76	JAS 39A	39131 to 39206	
	14	JAS 39B	39801 to 39814	
	1	JAS 39C	39207/39-6	Testbed
Lot 3 (64)	19	JAS 39C	39208 to 39226	Interim standard
	50	JAS 39C	39227 to 39276	Not fully funded
	14	JAS 39D	39815 to 39828	Not fully funded
Total	**209**			

30 production aircraft, with options for next 110; overall go-ahead confirmed second quarter 1983; first test runs of RM12 engine January 1985; Gripen HUD first flown in Viggen testbed February 1987; study for two-seat JAS 39B authorised July 1989.

First of five single-seat prototypes (39-1) rolled out 26 April 1987; made first flight 9 December 1988 but lost in landing accident after fly-by-wire problem 2 February 1989; only six sorties flown. Subsequent first flights 4 May 1990 (39-2), 20 December 1990 (39-4), 25 March 1991 (39-3) and 23 October 1991 (39-5); the 2,000th Gripen sortie was flown (by 39-4) on 22 December 1995; modified Viggen (37-51) retired at end of 1991 after assisting with avionics trials (nearly 250 flights); two single-seat fatigue test airframes (39-51 discarded 1993; 39-52 began 16,000 hour programme, August 1993 and achieved 8,000 hours in early 1996. Second production batch (Lot 2: 110 aircraft) approved 3 June 1992; first production Gripen (39101) made first flight 10 September 1992 and joined test programme in lieu of 39-1; flight test programme in 1995-96, included high-AoA (at least 28° achieved) and spin trials by 39-2 and trials of an APU (for Lot 2 production) by 39-4. All development work in the original (Lot 1) contract had been completed by late 1996; total programme was over 1,800 hours in 2,300 sorties by six aircraft. By 1996 had demonstrated M1.08 cruise without reheat. Follow-on trials with mockup aerial refuelling probe conducted by 39-4 on eight sorties between 2 and 17 November 1998 from RAF VC10 K. Mk 4. Captive flight of two KEPD 150 Taurus SOMs on inner wing pylons of 39145 of F7 conducted on 27 August 1998. Live Raytheon AIM-120 AMRAAM firings by 39-5 in April 1998.

Initial production aircraft for Swedish Air Force (39102) made first flight 4 March 1993 and was handed over to FMV 8 June 1993; flight control software modified following loss of 39102 in crash on 8 August 1993 and installed from December 1994; further software upgrade to new-generation P11 standard (introducing 11 filters to prevent pilot-induced oscillations) first flown 22 March 1995 in trials aircraft and installed in test Gripens from late 1995; in production aircraft (known then as R11) built after early 1996 (and retrofitted from 1997 as R11:9); modified control stick introduced on production aircraft 39108 (first flight 11 April 1995). R12:3 flight control software under trial by late 1999, bringing Gripen up to original design goals; installed during following year as R12:4. Next stage is R14, increasing MTOW by some 1,000 kg (2,205 lb).

PP12 display processor for Lot 2 colour displays first flown August 1995. Thrust-vectoring under consideration

for Gripen; first stage is proposed participation in further trials programme of Rockwell/DASA X-31. JAS 39B prototype rolled out 29 September 1995; first flight 29 April 1996.

Initial 30 production JAS 39As delivered 1993-96 (five in 1994; six in 1995, comprising 39108 to 112 and 120; and last of balance – 39129 – on 13 December 1996); by 2002, these all upgraded to Lot 2 standard; deliveries of second lot of 110 began 19 December 1996, and completed in 2003; Swedish Parliament authorised third batch of 64 on 13 December 1996; contract formally placed on 26 June 1997; deliveries between 2003 and 2007. Saab contracted on 24 November 1999 to supply new EWS 39 defensive aids suite, designed by Ericsson Saab Avionics and with substantial CelsiusTech content. In same month, Saab made first flight with FADEC, as destined for Lot 3 aircraft.

First unit was F7 Wing at Såtenäs; maintenance training begun May 1994 at Linköping; conversion scheduled to begin 1 October 1995 but postponed to 1996, with pilot training centre at Såtenäs officially opened 9 June 1996; Gripen IOC achieved September 1997, following three-week field exercise by 2/F7 Squadron. Last of F7's previous Viggens withdrawn in October 1998. By mid-2002 SwAF Gripens had flown 34,000 sorties and 25,000 hours. SwAF received 100th Gripen on 12 March 2001 and 113 by 31 December that year. Final JAS 39A (and first JAS 39C) delivered 6 September 2002.

Gripen being promoted in several fighter competitions, as described in Customers section. On 18 November 1998, it was announced that the aircraft had been selected for purchase by the South African Air Force and this was confirmed by contract signature on 3 December 1999. In

connection with SAAF requirement, Gripen completed initial series of trials with helmet-mounted display (Thales Optronics Guardian) in February 2001.

CURRENT VERSIONS: **JAS 39A:** Standard single-seater. Final aircraft delivered 6 September 2002.

Description applies to JAS 39A except where indicated.

JAS 39B: Two-seater with 0.655 m (2 ft 1¾ in) fuselage plug and lengthened cockpit canopy. Primary roles conversion and tactical training, but also combat-capable. Not used for instruction until January 2002, when first students from pilot training began conversion. Avionics essentially as for JAS 39A and both cockpits identical, except no HUD in rear; instead, HUD image from front seat can be presented on flight data display in rear cockpit. Boosted environmental control system (ventral air intake replaces twin scoops on fuselage sides); inflatable airbag protects rear occupant during pre-ejection canopy fracturing. Reduced internal fuel; no internal gun.

Prototype entered final assembly 1 September 1994; first flight (39800) 29 April 1996; first production two-seater (39801) completed final assembly on 29 February 1996 and flew on 22 November; production deliveries began with 39802 on 19 May 1998.

Fatigue test specimen (39-71) also built; began a simulated 16,000 hour programme in February 1996.

JAS 39C and D: New features warrant revised designations for JAS 39A and B; improvements include full-colour displays, FADEC, helmet-mounted sight, new (Modular Airborne Computer System) processor for PS-05/A radar, Saab Dynamics IR-OTIS IR search and tracking system, in-flight refuelling probe and enhanced EW systems. IR-OTIS tested on Saab Viggen in 1997.

Originally planned for introduction at start of Lot 3, but final 19 Lot 2 aircraft (plus one testbed 39207/39-6) completed to interim standards with refuelling probes and upgraded computers and designated JAS 39C. First delivery (39208) to FMV for service trials 6 September 2001.

An electronically scanned radar antenna is under development for a potential Gripen MLU in 2010. By 2000, plans in hand to deliver all 14 Lot 3 two-seat Gripens in **JAS 39D C²** configuration with entirely redesigned rear cockpit for command and control duties.

JAS 39X: Potential export version; developed by Saab and BAE Systems; likely to include improvements considered for Swedish Lot 3 aircraft such as colour cockpit displays, integrated EW suite, NATO standard radios, air-to-air refuelling probe (above port air intake trunk), OBOGS, uprated environmental control system and NATO pylons.

Enhanced Gripen: Envisaged for 2010; first details mid-2001. Could include improved EW systems with laser warning, missile approach warning linked to towed decoys and EW function incorporated in electronically scanned radar. Range enhanced by standardisation on two-seat airframe, with fuel in place of rear cockpit, and/or conformal tanks; new engine, possibly EJ200, M88 or F414, under consideration. Navalised Gripen reportedly under consideration by India in 2002.

CUSTOMERS: Swedish Air Force requirement was originally 280, to equip 16 squadrons, second eight replacing JA and AJS Viggens; this reduced when third batch authorisation (December 1996) covered only four squadrons, thus amending requirement to 204 (including 28 JAS 39B two-seat versions); however, prototype JAS 39B added to first batch as conversion of aircraft under production, further amending contracts to 175 single-seat and 29 two-seat. By 2002, funding assigned for only 160 aircraft for eight squadrons, although 204 aircraft officially remained on order.

First 30 ordered with prototypes and full-scale development 30 June 1982; next 110 (of which first – 39131 – flew on 20 August 1996) include 14 JAS 39Bs and 20 JAS 39Cs; black radomes on 39109 to 39127; low-visibility markings from 39128, augmented by light grey (previously medium grey) radomes from 39131. Third lot of 64 (also including 14 two-seaters) for delivery between 2003 and 2007 are to JAS 39C/D standard.

Some Lot 3 improvements incorporated early in final 20 Lot 2 aircraft, (which redesignated JAS 39C), comprising refuelling probe and substantially improved computer

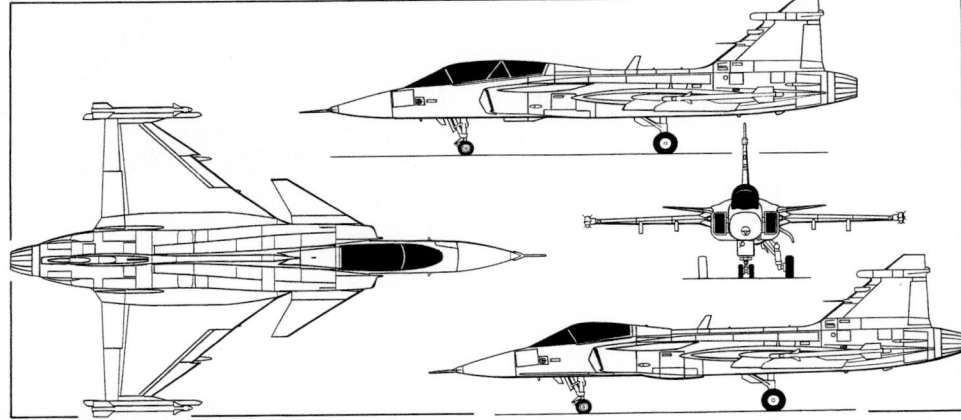

JAS 39A Gripen multirole combat aircraft for the Swedish Air Force, with additional side view (top) of two-seat JAS 39B (*Jane's/Dennis Punnett*)   0064907

(unofficially known as 'Mk 3') including D96 (MACS) processor to simplify possible later upgrade to colour displays.

Total 120 (including two-seat) delivered by July 2002. First operational squadron was 2/F7, September 1997, followed by 1/F7 12 months later. F10 at Angelholm followed in 1999-2000, conversion of pilots (by F7) having begun on 11 March 1999, followed by arrival of first two aircraft on 30 September 1999; first squadron (2/F10) operational September 2000, with second following in mid-2001. However, F10 disbanded by 31 December 2002 and transferred its equipment to F17 at Kallinge, latter being inaugurated as Gripen unit on 14 June 2002. F4 at Froson-Ostersund to convert in 2004, while F21 at Luleå received first two Gripens on 17 January 2002 (although its second squadron will not become operational until 2005). From 1 January 2004, one squadron of F21 is Swedish Air Force Rapid Reaction Unit, for worldwide deployment at 30 days' notice, equipped with Lot 3 aircraft.

Exports of some 250 anticipated over 20 years from 1996; by then, Saab and BAE engaged in 12 export campaigns; presentations made to Austria, Brazil, Czech Republic, Chile and Poland in 1996-97. Promotion in Philippines, Slovenia and South Africa, 1997. Details of competitions summarised below.

Australian Air 6000 requirement received Gripen response on 1 February 2002, but Lockheed Martin F-35 selected June 2002.

Austrian RFP received October 2001 for up to 30 new fighters; Gripen proposal delivered 22 January 2002, based on either 24 single- and four two-seat aircraft for early delivery; or 12 leased aircraft as lead-in to later delivery of 24 plus six new-build Gripens. Revised proposals, delivered to Austria on 30 April 2002, envisaged accelerated delivery between mid-2005 and mid-2007, but Eurofighter Typhoon eventually selected by parliament on 2 July 2002, although almost immediately suspended.

Brazilian RFP received August 2001 for potential 24 aircraft. Revised proposals delivered to Brazil on 3 May 2002.

Czech Republic RFP issued 27 December 2000; invitation to tender for new fighter issued 9 January 2001; Saab response delivered 31 May 2001 for 24 or 36 aircraft with 150 per cent offsets. Plan officially confirmed on 10 December 2001 for 24 Gripens valued at Kcs50 billion, including spares and training; first squadron to be operational in 2005 and second and final in 2008. Formal approval given 22 April 2002, contract signing due before end of year.

Hungary received Gripen proposal on 30 November 2000, this amended on 9 February 2001 with lease proposal; selected Gripen on 10 September 2001, initially with lease of 14 aircraft (including two tandem-seat) for 10 years at cost of Ft130 billion to Ft140 billion. MoU signed 23 November 2001 and agreement finalised 20 December 2001 requiring initial deliveries in fourth quarter of 2004 and last in June 2005. Modifications from Swedish standard include new communications, APU, general electronic control unit and FADEC, plus changes to avionics (IFF, ILS, secure radios) and instrument calibration (Imperial units). Operating unit based at Kecskemét.

Polish negotiations begun September 2001 in search for 60 new fighters. Final assembly of required 44 to 60 Polish Gripens would have been by PZL Aircraft Factory, under 1997 agreement, as covered by June 1999 proposal; however, revised offer of January 2001 restricted to five-year lease of 16 aircraft, including two trainers.

South Africa selected Gripen on 18 November 1998 for planned purchase of 28; order placed 15 September 1999

Montage of Saab JAS 39B Gripen armed with Python 4 and R-Darter missiles (*Katsuhiko Tokunaga/Saab*)   *NEW*/0536707

Saab JAS 39A Gripens over Visby, Gotland (*Peter Liander/Saab*)   *NEW*/0536709

Impression of a possible Enhanced Gripen with conformal fuel tanks and towed decoys *NEW*/0536711

First refuelling probe installation on Gripen *NEW*/0118104

for nine two-seat Gripens, with option on further 19 of single-seat version; formal signature 3 December 1999, but first deliveries postponed at that time from 2002 to mid-2006. However, only one two-seat Gripen due in that year for trials, followed by remaining eight between November 2007 and September 2009. Single-seat Gripens due between November 2009 and late 2011.

COSTS: Planned cost of SEK25.7 billion in 1982, increased to SEK48.5 billion in 1991 by inflation; FMV has reported total cost increase of SEK9.3 billion for period 1982-2001; total budget SEK60.2 billion decided by Swedish Parliament 1993. SEK22.7 billion spent by 1 July 1993, including SEK14.5 billion to IG JAS. Up to SEK300 million approved late 1991 for JAS 39B development. Costs for 300-aircraft programme (subsequently reduced to 204) estimated in 1994 as SEK15 billion for development and SEK48 billion for production. South African purchase of 28 estimated programme cost R10.9 billion (US$1.9 billion) (1998).

DESIGN FEATURES: Intended to replace AJ/SH/SF/JA/AJS versions of Saab Viggen, in that order, and remaining J 35 Drakens; to operate from 800 m (2,625 ft) Swedish V90 road strips; simplified maintenance and quick turnround with ground crew comprising one technician and five conscripts.

By 2000, Gripen demonstrating 12 mmh/fh; 7.6 hours MTBF readiness rate; and 10 minute turnround in fighter configuration, 20 minutes in attack role. Targets are 10 mmh/fh and 9.0 hours MTBF.

Mid-mounted delta wing, with squared tips for missile rails, has three-section leading-edge, of which inboard and outboard sections sweptback at 55° and centre at 52°; foreplanes, independently movable, have leading-edge sweep of approximately 58°. Sweptback fin carries various antennas. Moderately high cockpit. Near-rectangular engine air intakes, each with splitter plate.

FLYING CONTROLS: BAE Systems (Lot 1) or Lockheed Martin SA11 (Lot 2) triplex fly-by-wire system with Moog electrically signalled servo valves on powered control units; Saab Combitech aircraft motion sensors and throttle actuator; mini-stick and HOTAS controls.

Leading-edge with dog-tooth and automatic flaps (one inboard/one outboard of dog-tooth, inner one outboard of canard) on Lucas Aerospace 'geared hinge' rotary actuators; two elevon surfaces at each trailing-edge; individual all-moving foreplanes, which also 'snowplough' for aerodynamic braking after landing; airbrake each side of rear fuselage.

STRUCTURE: Airframe is 60 per cent aluminium by weight, 6 per cent titanium and 5 per cent other metals; most of remainder is carbon fibre. First 3½ carbon fibre wing sets produced by BAE; all subsequent carbon fibre parts made by Saab, including wing boxes, foreplanes, fin and all major doors and hatches. BAE produces landing gear and, from 1999, responsible for assembly of centre fuselage for two-seat version, with single-seat to follow. Denel of South Africa contracted in 2000 to design and produce stores pylons for export Gripens. PZL-Mielec of Poland began delivering tailcones in late 2000. By 2002, late production JAS 39As had received additional 200 kg (441 lb) of strengthening to landing gear and wings for MTOW increase to 14,000 kg (30,864 lb) and service life increase from 3,000 to 4,000 hours.

LANDING GEAR: AP Precision Hydraulics retractable tricycle gear, single mainwheels retracting hydraulically forward into fuselage; steerable twin-wheel nose unit retracts rearward. Goodyear wheels. Carbon disc brakes and ABS anti-skid units. Nosewheel braking. Entire gear designed for high rate of sink. Mainwheel tyres 25.5×8.0-14 (16 ply); nosewheel 14×5.5-6 (8 ply).

POWER PLANT: One General Electric/Volvo Flygmotor RM12 (F404-GE-400) turbofan, rated initially at approximately 54 kN (12,140 lb st) dry and 80.5 kN (18,100 lb st) with afterburning. RM12UP version, in Lot 3 aircraft, incorporates FADEC, improved flame holder and redesigned turbine. Fuel in integral tanks in fuselage and wings. Intertechnique fuel management system; Dowty fuel health monitoring system for emergency (leak/battle damage) conservancy management. Optional FRL telescopic, retractable, hydraulically actuated in-flight refuelling probe mounted in port engine air intake; available on export aircraft and fitted to Swedish single-seat aircraft from 39207 (106th) onwards.

ACCOMMODATION: Pilot only in JAS 39A, on Martin-Baker Mk 10L zero/zero ejection seat. Hinged canopy (opening sideways to port) and one-piece windscreen by Lucas Aerospace. Two seats in tandem in JAS 39B; command sequence in two-seat aircraft ejects rear occupant first, simultaneously inflating an airbag between the two cockpits to protect the rear pilot from Perspex splinters.

SYSTEMS: Hymatic environmental control system for cockpit air conditioning, pressurisation and avionics cooling. Hughes-Treitler heat exchanger. Two hydraulic systems, with Dowty equipment and Abex pumps. Hamilton Sundstrand main electrical power generating system (40 kVA constant speed, constant frequency at 400 Hz) comprises an integrated drive generator, generator control unit and current transformer assembly. Lucas Aerospace auxiliary and emergency power system, comprising gearbox-mounted turbine, hydraulic pump and 10 kVA AC generator, to provide auxiliary electric and hydraulic power in event of engine or main generator failure. In emergency role, the turbine is driven by engine bleed or APU air; if this is not available, the stored energy mode, using thermal batteries, is selected automatically. APU and air turbine starter for engine starting, cooling air and standby electrical power. Original APU was Microturbo TGA15-090; changed to TGA15-328 from 40th JAS 39A and retrofitted to all in early second batch; Hamilton Sundstrand APU from 106th production single-seat aircraft (39207) and will be retrofitted. Optional OBOGS on export aircraft. Lot 3 Gripens have single Ericsson Saab Avionics GECU general electronic control unit, replacing previous three controllers for air, fuel and hydraulic systems.

AVIONICS: Service entry with E11 avionics software; upgraded via E12, E12.5 and E14 to E15 in 2001 and E15.1 (full AMRAAM and additional air-to-ground modes) in 2002.

Comms: CelsiusTech dual VHF/UHF transceivers and IFF in early aircraft; Rohde & Schwarz Series 6000 in third production lot and last 20 Lot 2 aircraft; option for installation in entire fleet; Series 6000 is element of tactical radio system (TARAS), providing secure communications via Data Link 39. Export aircraft to have Avitronics (Grintek) GUS 1000 audio management system.

Radar: Ericsson/BAE PS-05/A multimode pulse Doppler target search and acquisition (lookdown/shootdown) radar (weight 156 kg; 344 lb). For fighter missions, system provides fast target acquisition at long range; search and multitarget track-while-scan; quick scanning and lock-on at short ranges; and automatic fire control for missiles and cannon. In attack and reconnaissance roles, operating functions are search against sea and ground targets; mapping, with normal and high resolution; and navigation. Upgrade with electronically scanned antenna is intended in 2010.

Flight: Ericsson SDS 80 central computing system (D80 computer, Pascal/D80 high-order language and programming support environment; upgraded D80E computer flown mid-1994 and introduced from 39108; D96 computer from 92nd single-seat aircraft, 39193); three MIL-STD-1553B databusses, one of which links flight data, navigation, flight control, engine control and main systems; Honeywell laser INS and radar altimeter; Nordmicro air data computer. BAE three-axis strapdown gyromagnetic unit provides standby attitude and heading information. Navigation data fusion in prospect via NINS (new integrated navigation system), under development by 1998; this to be followed by NILS (new integrated landing system) for autonomous Cat. I landing capability.

Instrumentation: Ericsson EP-17 electronic display system, incorporating Kaiser (Hughes in Lot 1) wide-angle HUD and using advanced diffraction optics to combine symbology and video images; display processor (replacing PP1 and PP2 in Lot 1) makes colour imagery possible; this facility is not required on Swedish Lot 2 aircraft), but colour capability introduced from 106th single-seat, 39207 (which will be regarded as sixth 'prototype' 39-6); three Ericsson CRT HDDs, each 120 × 150 mm (4¾ × 6 in), but 152 × 203 mm (6 × 8 in) MFID 68 active matrix LCDs in Lot 3. Left-hand (flight data) HDD normally replaces all conventional flight instruments; central display shows computer-generated map of area surrounding aircraft with tactical information superimposed; right-hand CRT is a multisensor display showing information on targets acquired by radar, FLIR and weapon sensors. Minimum of conventional analogue instruments for back-up only. Thales Guardian helmet-mounted display undertook compatibility trials at Linköping in early 2001.

Mission: Rafael/Zeiss Optronik Litening targeting and navigation pod under development in 2000 for carriage under starboard air intake trunk, forward of wing leading-edge, providing heat picture of target on right-hand HDD. IR-OTIS (a combined TRST and FLIR) under development by Saab Avionics for installation ahead of windscreen, slightly offset to port. CelsiusTech datalink shares information with up to four aircraft simultaneously. Contract for SEK600 million awarded to Saab in late 2001, for Modular Airborne Reconnaissance System 39 pods to be delivered from 2004 onwards and become operational in 2005.

Vinten Vicon 70 Srs 72C modular reconnaissance pod available for export aircraft.

Self-defence: CelsiusTech RWR in early aircraft replaced by second-generation (Gen 2) version in 2000. CelsiusTech countermeasures, including chaff/flare and jamming. Saab EWS 39 electronic warfare suite ordered 1999 to replace existing system in Swedish aircraft, and is similar to export equipment; includes BOL 500 (BO2D) towed radar decoy under port wing, two pylon-mounted BOP 402 (BOP B) dispensers, laser warning system and missile approach warning.

ARMAMENT: Internally mounted 27 mm Mauser BK27 automatic cannon in port side of lower front fuselage and two wingtip-mounted Rb74 (AIM-9L) Sidewinder IR AAMs standard. (No internal gun in JAS 39B.) Six other external hardpoints (two under each wing, one on centreline and one below starboard air intake trunk) for short- and medium-range air-to-air missiles such as Rb74, MICA or Rb99 (AIM-120) AMRAAM; air-to-surface missiles such as Rb75 (Maverick); anti-shipping missiles such as Saab RBS 15F; DWS 39 (BK 90) munitions dispenser; KEPD 150 and KEPD 350 SOMs; conventional or retarded bombs; air-to-surface rockets; or external fuel tanks. Swedish MoD contract of October 2001 covers integration of GBU-16 and GBU-24 LGBs. IRIS-T AAM being integrated for 2004 IOC with Swedish AF. New weapons under consideration for integration include Brimstone and Meteor. MUPSOW standoff weapon and V-3E A-Darter AAM for South Africa. Agreement of June 2001 adds Rafael weapons to potential armoury, initially Python 4 AAM and Spice guided bomb.

DIMENSIONS, EXTERNAL:

Wing span, incl missile rails	8.40 m (27 ft 6¾ in)
Length, excl pitot tube: JAS 39A	14.10 m (46 ft 3 in)
JAS 39B	14.755 m (48 ft 5 in)
Height overall	4.50 m (14 ft 9 in)
Wheel track	2.40 m (7 ft 10½ in)
Wheelbase: JAS 39A	5.20 m (17 ft 0¾ in)
JAS 39B	5.90 m (19 ft 4¼ in)

WEIGHTS AND LOADINGS:

Operating weight empty: JAS 39A	6,622 kg (14,600 lb)
JAS 39C	approx 6,820 kg (15,036 lb)
JAS 39B	8,000 kg (17,637 lb)
Internal fuel weight	2,268 kg (5,000 lb)
T-O weight, clean	approx 8,500 kg (18,740 lb)
Max T-O weight with external stores:	
R12 software	12,500 kg (27,557 lb)
R14 software	14,000 kg (30,864 lb)

PERFORMANCE:

Min flying speed	less than 100 kt (185 km/h; 115 mph)
Max level speed	supersonic at all altitudes
T-O and landing strip length	approx 800 m (2,625 ft)
Combat radius	approx 432 n miles (800 km; 497 miles)
g limit	+9

UPDATED

SWITZERLAND

ABS

ABS AIRCRAFT AG

Bühlstrasse 2, CH-8700 Küsnacht/ZH
Tel: (+41 1) 274 20 99
Fax: (+41 1) 274 20 99
e-mail: abs.kb@ggaweb.ch

VERIFIED

ABS RF-9

TYPE: Motor glider.

PROGRAMME: French Fournier RF-9 adapted for production; rights originally obtained by ABS, but programme taken over by Herbert Gomolzig in Germany in 1995; former ABS demonstrator purchased and completed as D-KHGO; first flight 30 June 1995; German production was due to be launched in late 1997, but did not take place; ABS Aircraft reacquired production rights during 1999; first aircraft sold in 2000; two more in production for delivery in late 2001 and early 2002. More powerful versions, suitable for glider-towing, were in development during 1999 (latest information received).

COSTS: €100,000 (1999).

DESIGN FEATURES: Outer wing panels fold inward for hangarage.

FLYING CONTROLS: Conventional and manual. Wing-mounted Schempp-Hirth-type spoilers.

STRUCTURE: All-wood.

LANDING GEAR: Tailwheel layout with inward-retracting wide-track mainwheels; foot-operated wheel brakes.

POWER PLANT: One 59.6 kW (79.9 hp) Rotax 912A-3 water-cooled four-cylinder four-stroke driving a Hoffmann HO-V62 hydraulic constant-speed and feathering propeller.

ACCOMMODATION: Side-by-side seating with adjustable seats and parachute recesses; heating by engine-cooling liquid and three-position fresh air intake; baggage compartment.

DIMENSIONS, EXTERNAL:

Wing span	17.30 m (56 ft 9 in)
Wing aspect ratio	16.6
Width, wings folded	10.00 m (32 ft 9¼ in)
Length overall	8.06 m (26 ft 5¼ in)
Height overall	1.93 m (6 ft 4 in)

AREAS:

Wings, gross	18.00 m² (193.7 sq ft)

WEIGHTS AND LOADINGS:

Manufacturer's weight empty	500 kg (1,102 lb)
Max T-O weight	750 kg (1,653 lb)
Max wing loading	41.4 kg/m² (8.47 lb/sq ft)
Max power loading	12.58 kg/kW (20.68 lb/hp)

PERFORMANCE, POWERED:

Never-exceed speed (V$_{NE}$)	135 kt (250 km/h; 155 mph)
Max level speed	119 kt (220 km/h; 137 mph)
Cruising speed	102 kt (190 km/h; 118 mph)
Stalling speed	38 kt (70 km/h; 44 mph)

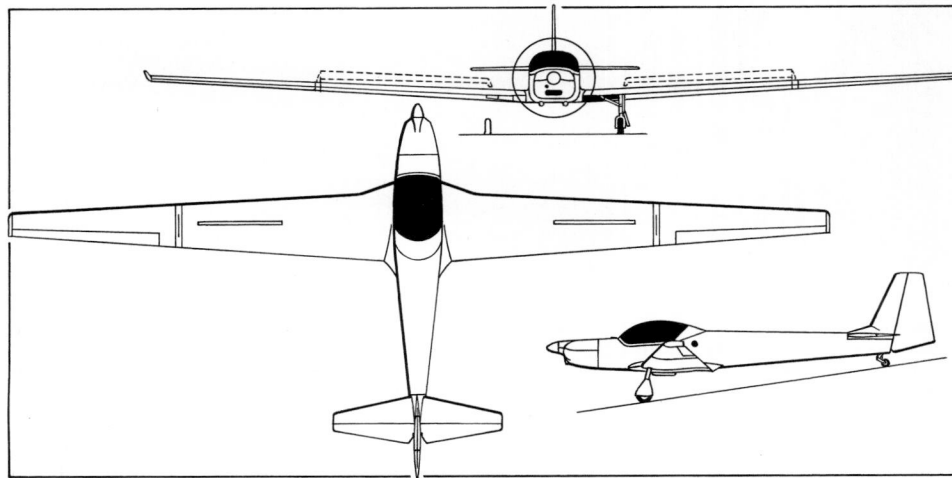

Fournier RF-9, to be produced in Switzerland by ABS Aircraft *(Jane's/Mike Keep)*

ABS-built RF-9 formerly used as a demonstrator by Gomolzig   0059985

Max rate of climb at S/L	216 m (708 ft)/min	PERFORMANCE, UNPOWERED:	
Service ceiling	6,500 m (21,320 ft)	Best glide ratio	29
T-O to 15 m (50 ft)	400 m (1,312 ft)	Min rate of sink	0.80 m (2.62 ft)/s
Landing from 15 m (50 ft)	300 m (985 ft)	NOISE LEVEL:	
Range, cruising with engine		To German LSL Chapter 6	55 dBA
	486 n miles (900 km; 559 miles)		*UPDATED*

ACEAIR

ACEAIR SA

Via Cantonale 35b, CH-6928 Manno
Tel: (+41 91) 605 55 46
Fax: (+41 91) 605 55 19
e-mail: info@aeriks.com
Web(1): http://www.aceair.ch
Web(2): http://www.aeriks.com
CHAIRMAN: Ugo Wyss
GENERAL MANAGER AND CEO: Antonio Latella
CHIEF ENGINEER AND DESIGNER: Ing Igor Medici
PRODUCT MANAGER: Ing Massimo Stoppa

Aceair's unorthodox Aeriks A-200 kitbuilt sportplane was unable to make its debut as scheduled at the EAA's AirVenture at Oshkosh in July 2001, but made a successful first flight in May 2002.

UPDATED

ACEAIR AERIKS A-200

TYPE: Tandem-seat sportplane kitbuilt.

PROGRAMME: Concept phase began June 1997; design started 1998; officially launched April 1999 after trials with two radio-controlled scale models; engineering phase, dynamic (radio-controlled) model flight tests and wind tunnel test activity completed October 1999. Full-scale engineering mockup shown at EAA AirVenture, Oshkosh, July 2000; detail design started two months later; prototype construction began mid-2001; debut planned for EAA AirVenture at Oshkosh, July 2001, but failed to occur. Rolled out 21 April 2002 and first flight eventually made (HB-YKS) on 29 May 2002. More than 10 hours flown in 14 flights by 30 June 2002. Original name of Aeris was changed to Aeriks in July 2002 due to pre-existing trade name protection in US. Public debut at EAS rally, Lodrino, 16-18 August 2002. Production of kits was planned to begin in January 2003.

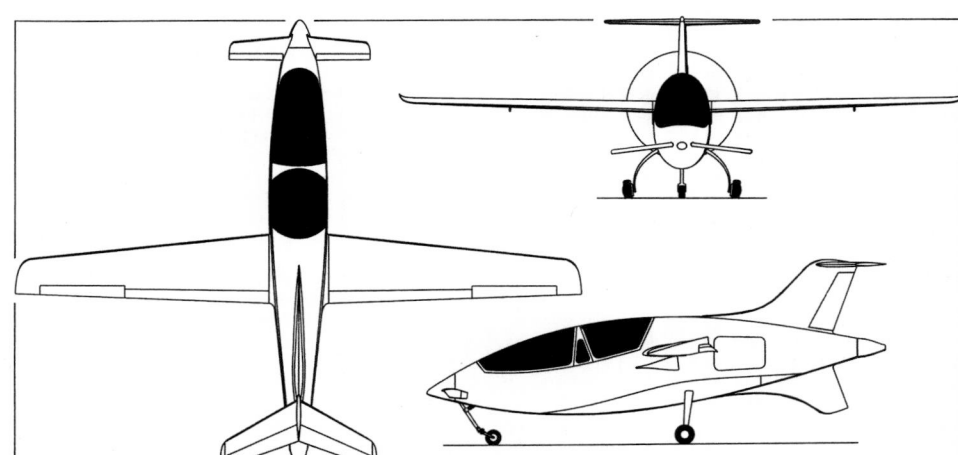

Aceair Aeriks (GIAE-110R engine) *(Jane's/Paul Jackson)*   NEW/0533394

COSTS: US$55,000 to US$65,000 (2002).

DESIGN FEATURES: Three lifting surfaces aerodynamic configuration; single pusher engine. Straight-tapered, low-swept (4° at 25 per cent chord) main wing amidships, with 2° dihedral from roots, features natural laminar flow sections based on NASA NLF(1)-0215F; main wing twist (2° 30′ washout from root chord) and planform are selected for stable low-speed characteristics. Foreplane aerofoil Wortmann FX-75-141 (4° sweepback at quarter-chord), 6° anhedral. Lifting surfaces relative setting angles optimised for low drag at normal cruising speed. Standard symmetrical sections on T-tail configured, 30° swept horizontal surface. Separate fins above and below fuselage (lower portions serve as over-rotation bumper and propeller strike protector); rudder mounted on upper vertical fin.

Quoted build time 1,000 to 1,500 hours, plus 200 to 400 hours for fully retractable landing gear option.

FLYING CONTROLS: Mechanical via push/pull rods for both ailerons and T-tail-mounted elevators; ⅛ in steel cables for rudder control. Foreplanes have electric trim. Electrically actuated single-slotted flaps on main wing, synchronised with plain flaps on foreplanes; all flaps can be partially deflected for T-O; landing flap setting is 40° on main wing, 20° on foreplane (relative to wing chord).

STRUCTURE: All-composites. Main wings detachable for storage and/or transportation; C-section carbon fibre main spars and rear auxiliary spars; 30 per cent chord single-

Aceair Aeris/Aeriks on its maiden flight, 29 May 2002   *NEW*/0525771/0525772

slotted flap and 25 per cent chord aileron in each wing; glass fibre ribs. Monocoque fuselage. Primary structures built from wet laminated glass fabric laid-up on CNC milled all-composites female moulds; all shells of honeycomb sandwich. Glass fibre flaps and control surfaces. Primary foreplanes and tail assembly of carbon fibre C-spars, composites ribs and glass fibre-based honeycomb sandwich skins.

LANDING GEAR: Tricycle type. Electrically actuated nose gear retracts rearward into fuselage; non-retractable main gear units on cantilever self-sprung glass fibre struts with Cleveland heavy-duty disc brakes and wheels. Nosewheel steerable by differential braking. Hydraulic mainwheel brakes and parking brake.

POWER PLANT: One Mid-West GIAE 110R rotary engine (78.3 kW; 105 hp), driving a three-blade constant-speed MT-Propeller MTV-7-D/RD157-106 pusher propeller via an aluminium driveshaft. Maximum propeller speed 2,500 rpm. Engine is mounted far aft of passenger cabin and has two NACA-type scoop air inlets under main wingroot. Fuel in single 110 litre (29.1 US gallon; 24.2 Imp gallon) integral fuselage tank aft of passenger seat.

ACCOMMODATION: Pilot and co-pilot or passenger in tandem in individual cockpits; dual sidestick controls. Separate one-piece flush-type canopy hinged on starboard side. Baggage compartment, capacity 20 kg (44 lb), under passenger seat. Central carbon fibre roll bar for flip-over protection.

SYSTEMS: Single voltage (14 V DC) electrical system powered by two engine-driven generators to provide power for gear and flap extension/retraction; main bus powers flaps, trim, landing gear, standby fuel pump and other flight-essential equipment.

AVIONICS: To customer's requirement. Can be equipped to full IFR standard.

EQUIPMENT: Anti-collision strobe light and navigation light in each wingtip; one white strobe light in tail.

DIMENSIONS, EXTERNAL:
Wing span	8.00 m (26 ft 3 in)
Foreplane span	2.00 m (6 ft 6¾ in)
Wing aspect ratio	11.4
Foreplane aspect ratio	6.7
Length overall	6.50 m (21 ft 4 in)
Height over tailplane	2.46 m (8 ft 0¾ in)
Wheel track	1.50 m (4 ft 11 in)
Wheelbase	2.72 m (8 ft 11 in)
Fuselage ground clearance at mainwheels	
	0.57 m (1 ft 10½ in)
Propeller diameter	1.57 m (5 ft 1¾ in)
Propeller ground clearance	0.565 m (1 ft 10¼ in)

DIMENSIONS, INTERNAL:
Cockpits: Max combined length	2.75 m (9 ft 0¼ in)
Max width	0.74 m (2 ft 5¼ in)
Max height	1.00 m (3 ft 3¼ in)

AREAS:
Wings, gross	6.20 m² (66.7 sq ft)
Foreplanes, gross	0.60 m² (6.46 sq ft)
Tailplane	1.06 m² (11.41 sq ft)
Fins (total)	1.35 m² (14.53 sq ft)
Ailerons (total, incl tabs)	0.375 m² (4.04 sq ft)
Rudder	0.24 m² (2.58 sq ft)
Elevators (total incl tabs)	0.425 m² (4.57 sq ft)

WEIGHTS AND LOADINGS:
Weight empty	400 kg (882 lb)
Fuel weight	82 kg (181 lb)
Payload with max fuel	200 kg (441 lb)
Max T-O weight	650 kg (1,433 lb)
Max wing/foreplane loading	95.59 kg/m² (19.58 lb/sq ft)
Max power loading	8.31 kg/kW (13.65 lb/hp)

PERFORMANCE (estimated):
Never-exceed speed (VNE)	200 kt (370 km/h; 230 mph)
Max level speed at S/L	155 kt (287 km/h; 178 mph)
Manoeuvring speed (VA)	130 kt (241 km/h; 150 mph)
Max cruising speed at 3,660 m (12,000 ft)	
	140 kt (259 km/h; 161 mph)
Stalling speed at S/L: flaps up	65 kt (121 km/h; 75 mph)
flaps down (two-seat)	58 kt (108 km/h; 67 mph)
flaps down (solo)	54 kt (100 km/h; 63 mph)
Max rate of climb: at S/L	426 m (1,398 ft)/min
at 3,500 m (11,480 ft)	180 m (591 ft)/min
Service ceiling	3,660 m (12,000 ft)
T-O run at S/L	350 m (1,150 ft)
T-O to 15 m (50 ft) (JAR 23.59) at S/L	520 m (1,705 ft)
Landing from 15 m (50 ft) (JAR 23.75) at S/L	
	550 m (1,805 ft)
Landing run at S/L	230 m (755 ft)
Max range at 3,660 m (12,000 ft), no reserves	
	688 n miles (1,275 km; 792 miles)
Max endurance at 3,660 m (12,000 ft), no reserves	5 h
g limits	+4.4/−1.8

UPDATED

FFA BRAVO

FFA BRAVO AG
Flughafenstrasse 11, CH-9423 Altenrhein
Tel: (+41 52) 657 35 83
Fax: (+41 52) 657 24 05
e-mail: b.widmer@ffa-aircraft.com
CHAIRMAN: Carl M Holliger
TECHNICAL DIRECTOR: Bruno Widmer
VICE-PRESIDENT, MARKETING: Rolf Boehm

Company founded in 1999 to manufacture and distribute AS 202 Bravo lightplane, previously managed by FFA Flugzeugwerke Altenrhein AG. Subsidiary, FFA Bravo (Germany) GmbH, assigned to produce the related Bravo 3000, although no news has been received of manufacture.
UPDATED

FFA BRAVO AS 202 BRAVO

Although production of this two-seat lightplane was to have restarted in 1999-2000, no new aircraft had been reported by mid-2002 and no further reports received from the manufacturer.
UPDATED

FFA BRAVO 3000

TYPE: Four-seat lightplane.
PROGRAMME: Under development by FFA Bravo; over 200 hours of flight testing accumulated by August 2000 using prototype FFT Eurotrainer 2000A, as last described in 1992-93 *Jane's*.

Originally fitted with 200 kW (268 hp) flat-rated Textron Lycoming AEIO-540-L1B5, all-composites prototype D-EJDZ first flew on 29 April 1991; FFT ceased trading on 31 December 1992 and planned purchase of eight by Swissair failed to materialise. D-EJDZ restored to German civil register on 21 December 1999. Bravo 3000 to have 224 kW (300 hp) engine and 200 kt (370 km/h; 230 mph) cruising speed. Programme launch scheduled for late October 2000 and public debut planned for mid-2001, but neither reported by mid-2002.
UPDATED

MSW

MSW AVIATION
Rigackerstrasse 24, CH-5610 Wohlen
Tel: (+41 56) 622 18 07
Fax: (+41 56) 611 00 55
e-mail: info@mswaviation.com
Web: http://www.mswaviation.com

MSW Aviation was formed in 1991. Its latest product is the Votec 322 aerobatic two-seater.
VERIFIED

MSW VOTEC 322 and 332

TYPE: Aerobatic two-seat sportplane/kitbuilt.
PROGRAMME: Inspired by, and derived from, DR 107 One Design single-seater, one of which acquired by MSW and flown as HB-YJM. Prototype Votec 322 (HB-YJY) made first flight 6 April 2001 and public debut at Friedrichshafen Air Show later that month.
CURRENT VERSIONS: **Votec 322:** Factory-built version; *as described.*
    Votec 332: Kit version (Experimental category).
DESIGN FEATURES: Low-wing monoplane with symmetrical aerofoil section and angular flying surfaces.

Prototype MSW Votec 322 on display at Aero '01, Friedrichshafen, in April 2001 (*Jane's/Paul Jackson*) 0110940

FLYING CONTROLS: Conventional and manual. Near full-span ailerons; horn-balanced rudder and elevators; in-flight-adjustable trim tab in each elevator.

STRUCTURE: Steel tube fuselage main frame; wooden wings; carbon fibre fuselage skin and tail unit.

LANDING GEAR: Tailwheel type; fixed. Cantilever self-sprung main units; steerable tailwheel. Mainwheel speed fairings.

POWER PLANT: One 246 kW (330 hp) Textron Lycoming AEIO-540 flat-six engine, driving an MT-Propeller MTV-9-B-C/C203-20 three-blade propeller. Fuel in integral 40 litre (10.6 US gallon; 8.8 Imp gallon) tank in each wingroot, plus 110 litre (29.0 US gallon; 24.2 Imp gallon) fuselage tank; total capacity thus 190 litres (50.2 US gallons; 41.8 Imp gallons).

ACCOMMODATION: Two persons in tandem under one-piece canopy which opens sideways to starboard.

AVIONICS: Described as 'glass cockpit' standard.

DIMENSIONS, EXTERNAL:
Wing span	7.30 m (23 ft 11½ in)
Length overall	6.50 m (21 ft 4 in)
Height overall	2.50 m (8 ft 2½ in)

AREAS:
Ailerons (total)	1.50 m² (16.15 sq ft)

WEIGHTS AND LOADINGS:
Weight empty	600 kg (1,323 lb)
Max T-O weight	840 kg (1,851 lb)
Max power loading	3.41 kg/kW (5.61 lb/hp)

PERFORMANCE:
Never-exceed speed (VNE)	208 kt (386 km/h; 240 mph)
Max cruising speed	156 kt (290 km/h; 180 mph)
Stalling speed	57 kt (105 km/h; 65 mph)
Max rate of climb at S/L	3,000 m (9,843 ft)/min
Max rate of roll	400°/s
T-O run	112 m (370 ft)
Landing run	350 m (1,150 ft)
Endurance	3 h
g limits	±10
	VERIFIED

PILATUS

PILATUS AIRCRAFT LTD

CH-6371 Stans
Tel: (+41 41) 619 62 96
Fax: (+41 41) 610 62 24
Web: http://www.pilatus-aircraft.com
CHAIRMAN: Peter Küpfer
PRESIDENT AND CEO: O J Schwenk
SENIOR VICE-PRESIDENT: O Bründler (Controlling, Finance and Administration)
VICE-PRESIDENTS:
Dr O Masefield (R&D)
J Roche (Marketing and Sales)
I Gretener (General Aviation)
M Kälin (Maintenance)
M Zuberbühler (Logistics)
H Niederberger (Manufacturing)
A Waldispühl (Aircraft Assembly)
SALES MANAGER: Karl Scheuber
PUBLIC RELATIONS: Urs Maienfisch

PC-12 MARKETING:
Pilatus Business Aircraft Ltd
Jefferson County Airport, Terminal Building B-7, 11755 Airport Way, Broomfield, Colorado 80021, USA
Tel: (+1 303) 465 90 99
Fax: (+1 303) 465 91 90
Web: http://www.pc12.com
PRESIDENT AND CEO: Angelo Fiataruolo

Formed December 1939; became part of Oerlikon-Bührle Group; Britten-Norman (Bembridge) Ltd of UK (which see), acquired on 24 January 1979, reverted to UK ownership in 1998, reducing Pilatus workforce to 980, subsequently increasing to 1,146 worldwide by the end of 2000. Transairco SA of Geneva was purchased in April 1997. In December 1998, as part of a move to dispose of its aerospace activities, Oerlikon-Bührle (now renamed Unaxis) revealed that it was seeking a buyer for Pilatus. New owners were named on 1 March 2001 as a consortium comprising Jörg Burkart, IHAG Holding, Hoffman-LaRoche pension fund and Karl Nicklaus. Company renamed Pilatus Aircraft Ltd. A public share offering is expected in about 2004.

Current products include PC-6 Turbo Porter, PC-7 and PC-7 Mk IIM Turbo Trainer, PC-9M Turbo Trainer, PC-12 and Eagle and PC-21. Pilatus also overhauls Swiss Air Force aircraft. Modified (Mk II) PC-9 selected to fulfil US Air Force/Navy JPATS trainer requirement as T-6A Texan II.

UPDATED

PILATUS PC-7 TURBO TRAINER

There have been no recent deliveries of baseline PC-7s. Brief details last appeared in the 2002-03 *Jane's*.

UPDATED

PILATUS PC-7 Mk II M TURBO TRAINER
SAAF name: Astra

TYPE: Basic turboprop trainer.

PROGRAMME: First production PC-7 flew 18 August 1978; baseline version last described fully in 1997-98 *Jane's*. Mk II launched 1992 and developed to bid for South African Air Force requirement; two prototypes (HB-HMR and

-HMS), first of which made maiden flight 28 September 1992; SAAF contract announced June 1993; first production aircraft started January 1994 and made first flight August 1994. Since 1996, PC-7 Mk IIs and PC-9s (which see) have been built from common fuselages, former being designated PC-7 Mk II M (for 'modular').

CUSTOMERS: South Africa (60); Brunei (four); Royal Malaysian Air Force (nine in 2001).

PILATUS PC-7 and PC-7 Mk II PRODUCTION

Customer	Qty	First aircraft	First delivery
Military:			
Angola	25	R-401	1982
Austria	16	3H-FA	Oct 1983
Bolivia	24	450	Apr 1979
Bophuthatswana	3[1]	T400	Dec 1989
Botswana	7	OD-1	Feb 1990
Brunei	4[2]	ATU-301	1997
Chile	10	N-210	May 1980
France	5	576	1991
Guatemala	12	218	Apr 1980
Iran	35	8-9901	1983
Iraq	52	5000	Jul 1980
Malaysia	46[6]	M33-01	Feb 1983
Mexico	88	501	Jul 1979
Myanmar	17	2300	1979
Netherlands	13	L-01	Jan 1989
South Africa	60[2]	2001	1994
Suriname	3[3]	111	Oct 1986
Switzerland	40	A-902	Aug 1979
UAE (Abu Dhabi)	31	901	Apr 1982
Uruguay	6	301	1992
Subtotal	**497**		
Civil:			
Pilatus	1 (P)	HB-HAO	
	2	HB-HOO	
	1	(stock)	
USA (various)	8	N701RB	
CIPRA	2[4]	F-GDME	
Martini[5]	4	HB-HMA	
Total	**517**		

[1] To South African Air Force; returned to Pilatus and resold to private owners in 1998-99
[2] Mk II
[3] No longer in service
[4] Both to Chad Air Force
[5] 1982-90; to Patrouille Ecco (1991-96); Adecco (1997-98)
[6] Including nine Mk IIs delivered in 2001

DESIGN FEATURES: Generally as for PC-9. Reduced-drag airframe based on that of PC-9M; higher-powered PT6A-25C engine with four-blade propeller; Martin-Baker Mk 11A ejection seats standard; optimised flying controls.

FLYING CONTROLS: Generally as for PC-9M, including trim aid device; stall strips and ventral fin added to improve stall characteristics and directional stability.

STRUCTURE: Generally as for PC-9M.

LANDING GEAR: Hydraulically actuated retractable tricycle type, with emergency manual extension. Mainwheels retract inward, nosewheel rearward. Oleo-pneumatic shock-absorber in each unit. Castoring nosewheel, with shimmy dampers. Tyres 6.50-8 (8 ply) all round, pressure 8.62 bar (125 lb/sq in) on mainwheels, 4.83 bar (70 lb/sq in) on nosewheel. Hydraulic disc brakes on mainwheels. Parking brake. Alternatively 6.50-8 main and 6.00-6 tailwheel tubeless tyres. Steering ±61° by differential braking; nosewheel steering (±12°) standard. Minimum ground turning radius 11.125 m (36 ft 6 in), based on nosewheel.

POWER PLANT: One 522 kW (700 shp) (flat-rated) Pratt & Whitney Canada PT6A-25C turboprop, driving a Hartzell HC-D4N-2A four-blade constant-speed fully feathering propeller. Two integral wing fuel tanks, each of 253 litres (66.8 US gallons; 55.7 Imp gallons) usable capacity, plus a 12 litre (3.2 US gallon; 2.6 Imp gallon) aerobatic tank in fuselage, giving total internal usable capacity of 518 litres (137 US gallons; 114 Imp gallons).

Provision for 155 litre (41.0 US gallon; 34.1 Imp gallon) long-range auxiliary tank or 246 litre (65.0 US gallon; 54.1 Imp gallon) ferry tank under each wing. Single overwing gravity fuelling point each side. Oil capacity 16 litres (4.2 US gallons; 3.5 Imp gallons). Capability for 30 seconds of inverted flight.

ACCOMMODATION: Instructor and pupil in tandem on stepped Martin-Baker Mk 11A ejection seats. One-piece canopy opens sideways to starboard. Cockpit unpressurised.

SYSTEMS: Vapour cycle environmental control system. Hydraulic system, pressure 207 bar (3,000 lb/sq in), for landing gear actuation, nosewheel steering and airbrake; system maximum flow rate 18.8 litres (4.97 US gallons; 4.14 Imp gallons)/min. Supplementary system as for PC-9M. Electrical system (28 V DC) powered by 200 A starter/generator and 24 V Ni/Cd battery. OBOGS standard.

AVIONICS (standard): *Comms:* Two VHF com radios, transponder, ELT and audio/intercom to customer's choice. UHF com optional.

Flight: Range of navigation systems including ADF, DME, GPS, dual VHF nav, AHRS, RMI, EADI and EHSI available at customer's option.

Instrumentation: Standard IFR installation includes Mach/airspeed indicators; VSIs; main (encoding) and standby altimeters; standby attitude indicator; standby magnetic compass; combined aileron/rudder/elevator trim indicators; accelerometer indicators; electric clocks; and engine indication system with LCD displays. Advanced IFR with Flight Visions FV-2000 HUD with video camera, rear cockpit repeater and video recorder, ADC and radar altimeter optional.

AVIONICS (Astra): See 1998-99 and earlier editions.

EQUIPMENT: As for PC-9M.

DIMENSIONS, EXTERNAL, INTERNAL AND AREAS: As for PC-9M

WEIGHTS AND LOADINGS (A: Aerobatic, U: Utility category with underwing stores)
Basic weight empty, equipped	1,670 kg (3,682 lb)
Max underwing stores load	1,040 kg (2,293 lb)
Max T-O weight: A	2,250 kg (4,960 lb)
U	2,850 kg (6,283 lb)
Max ramp weight: A	2,260 kg (4,982 lb)
U	2,860 kg (6,305 lb)
Max landing weight: A	2,250 kg (4,960 lb)
U	2,750 kg (6,062 lb)
Max zero-fuel weight: A	1,900 kg (4,189 lb)
Max wing loading: A	138.1 kg/m² (28.29 lb/sq ft)
U	175.0 kg/m² (35.83 lb/sq ft)
Max power loading: A	4.31 kg/kW (7.09 lb/shp)
U	5.46 kg/kW (8.98 lb/shp)

PERFORMANCE (A and U as above):
Max operating Mach number (MMO): A, U	0.60
Max operating speed (VMO):	
A, U	300 kt (555 km/h; 345 mph) EAS
Max cruising speed:	
A at S/L	242 kt (448 km/h; 278 mph)
A at 3,050 m (10,000 ft)	251 kt (465 km/h; 289 mph)
A at 7,620 m (25,000 ft)	225 kt (417 km/h; 259 mph)
Max manoeuvring speed:	
A	210 kt (389 km/h; 241 mph) EAS
U	200 kt (370 km/h; 230 mph) EAS
Max speed with flaps and/or landing gear down:	
A, U	150 kt (277 km/h; 172 mph) EAS

Pilatus PC-7 Mk II M Turbo Trainer built for Malaysia in 2001 (*Jane's/Paul Jackson*)   0110507

Stalling speed, engine idling:

flaps and gear up: A   75 kt (139 km/h; 87 mph) EAS
U   84 kt (156 km/h; 97 mph) EAS
flaps and gear down: A   68 kt (126 km/h; 79 mph) EAS
U   75 kt (139 km/h; 87 mph) EAS
Max rate of climb at S/L: A   866 m (2,840 ft)/min
Time to 4,575 m (15,000 ft)   6 min 55 s
Max operating altitude: A, U   7,620 m (25,000 ft)
Service ceiling: A, U   9,150 m (30,000 ft)
T-O run at S/L: A   260 m (850 ft)
T-O to 15 m (50 ft) at S/L: A   415 m (1,360 ft)
Landing from 15 m (50 ft): A   674 m (2,210 ft)
Landing run: A   336 m (1,100 ft)
Max range at long-range cruising speed, 5% and 20 min
fuel reserves: A   810 n miles (1,500 km; 932 miles)
Max endurance at 110 kt (204 km/h; 127 mph) IAS,
conditions as above: A   4 h 40 min
g limits: A   +7/−3.5
U   +4.5/−2.25
UPDATED

PILATUS PC-9M ADVANCED TURBO TRAINER

Slovene Military Aviation name: Hudournik (Swift)
Royal Thai Air Force designation: BF. 19

TYPE: Basic turboprop trainer.

PROGRAMME: Design began May 1982; aerodynamic elements tested on PC-7 1982-83; first flights by two preproduction PC-9s: HB-HPA on 7 May 1984 and HB-HPB on 20 July 1984; aerobatic certification 19 September 1985. PC-9M introduced early 1997.

CURRENT VERSIONS: **PC-9:** Standard production version until 1997. Description in 1997-98 and earlier *Jane's*.

PC-9/A: Australian version of PC-9.

PC-9B: German target towing version of PC-9, operated for Luftwaffe by Condor Flugdienst; increased fuel for a 3 hour 20 minute mission; two Meggitt RM-24 winches on inboard pylons with targets stowed aft of winch; TAS 06 acoustic scoring system.

PC-9 Mk II: Modified for US Air Force/Navy JPATS trainer programme (T-6A Texan II); described (with export variants) under Raytheon heading in US section.

PC-9M: Upgraded version introduced 1997 and now the standard production model. Enlarged dorsal fin fairing (as on PC-7 Mk II M) improves longitudinal stability and reduces stick forces; modified wingroot fairings improve low-speed characteristics; 'exciter strips' on wing leading-edges improve stall characteristics; new engine/propeller control system separates flight idle and ground idle modes, reducing acceleration time and improving handling.

Detailed description applies to PC-9M.

Hudournik (Swift): Dedicated weapons training version in service with Slovene Military Aviation, modified after delivery by Radom Aviation Systems of Israel. First example to this standard (L9-61) made first flight early May 1999. Mission systems installation includes Litton INS-100G inertial navigation with GPS; HOTAS controls; Flight Visions HUD with up-front control and rear cockpit video repeater; a central flight data recorder; chaff/flare dispenser and control panel; Honeywell EFIS and displays; new weapon selection and armament control unit and displays; MIL-STD-1553B dual databus; and RS-422/ARINC 429 datalink. Maximum external stores load (see Armament paragraph below) increased to 1,250 kg (2,756 lb).

CUSTOMERS: Total of 259 delivered by March 2000; none further by March 2001, at which time eight more were under assembly for an undisclosed customer, although these had not emerged by mid-2002. See table. Saudi aircraft prepared for delivery by British Aerospace; first pair from second batch (Nos. 31 and 32) handed over in UK on 4 December 1995.

Slovenian order for nine PC-9M placed December 1997 and completed by December 1998. Omani order placed January 1999; all delivered by March 2000, replacing BAC Strikemasters.

DESIGN FEATURES: Typical turboprop trainer; stepped cockpits; aerobatic; performance and handling ease student's transition to jets. Meets FAR Pt 23 (Amendment 52) and special Swiss federal civil conditions for both Aerobatic and Utility categories; complies with selected parts of US military training specifications.

Conventional, low-wing monoplane; structural commonality with PC-7 Mk II M; parallel chord wing centre-section; tapered outer panels; tailplane LERX.

Wing section PIL15M825 at root, IL12M850 at tip; quarter-chord sweepback 1°; dihedral 7° from centre-section; incidence 1° at root; twist −2°.

FLYING CONTROLS: Conventional and manual. Cable-operated elevator and rudder; ailerons operated by pushrods; mass-balanced, electrically actuated trim tab in each aileron and starboard half of elevator; electrically actuated trim/anti-balance tab in rudder controlled from rocker switch on power control lever. Trim aid device (TAD) automatically adjusts rudder trim tab setting in response to changes in engine torque or aircraft airspeed. This counteracts aircraft yaw induced by variations in effect of propeller slipstream, depending upon airspeed and engine power setting. TAD can be selected on or off by switch on trim control panel on left-hand side panel of front cockpit. Electrically operated split flaps; hydraulically operated 'perforated' airbrake under centre-fuselage.

STRUCTURE: All-metal with some GFRP wing/fuselage fairings; one-piece wing with auxiliary spar, ribs and stringer-reinforced skin.

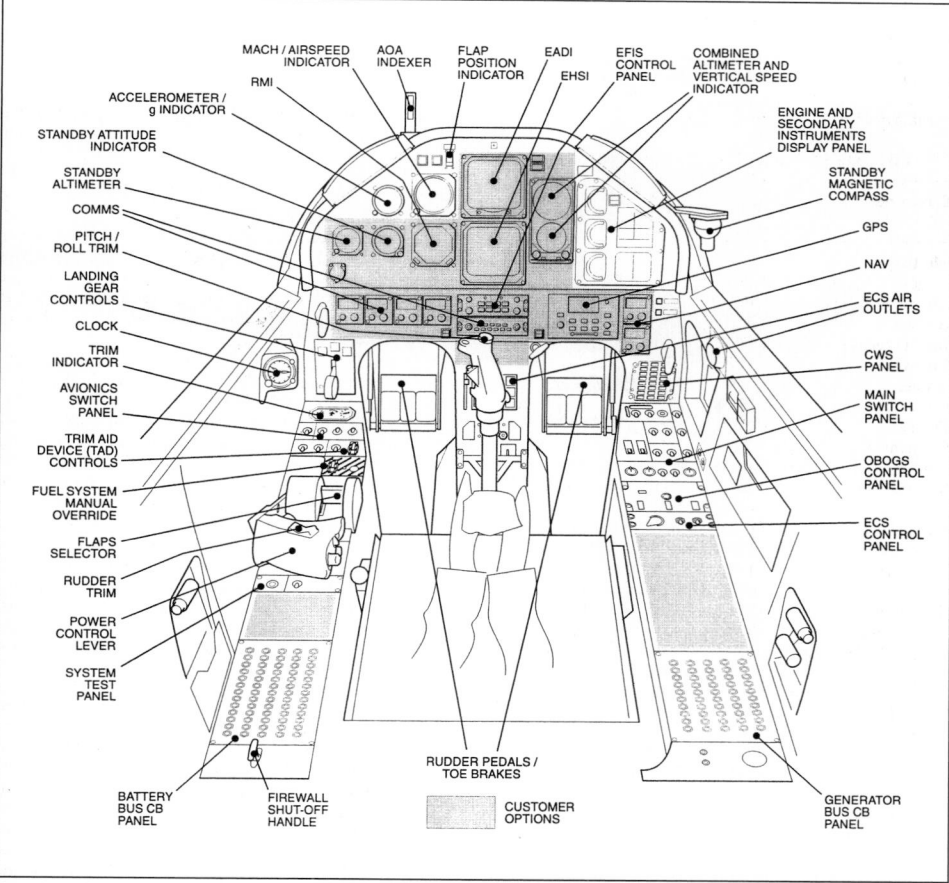

Standard front cockpit of the PC-7 Mk II M and PC-9M; rear cockpit is generally similar   0079300

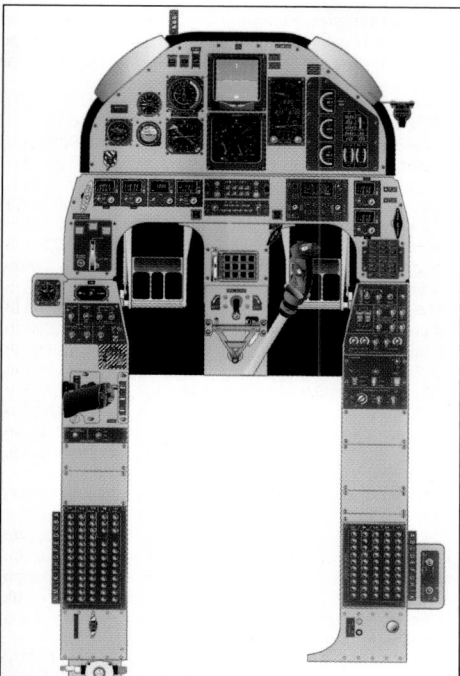

PC-7 Mk II M/PC-9M front cockpit (standard avionics)   0079299

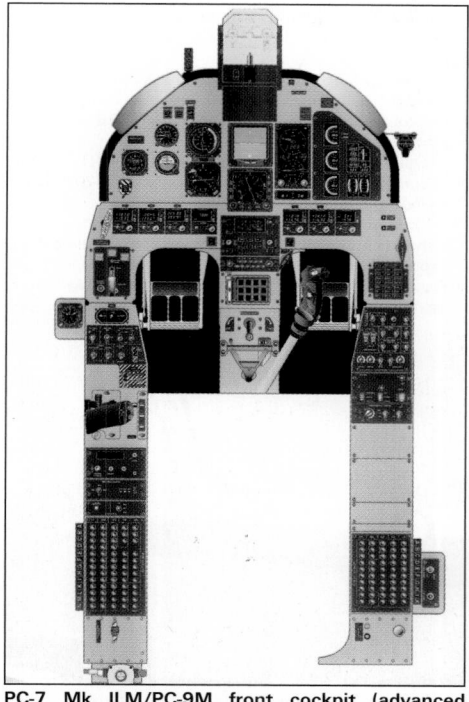

PC-7 Mk II M/PC-9M front cockpit (advanced avionics with HUD)   0079298

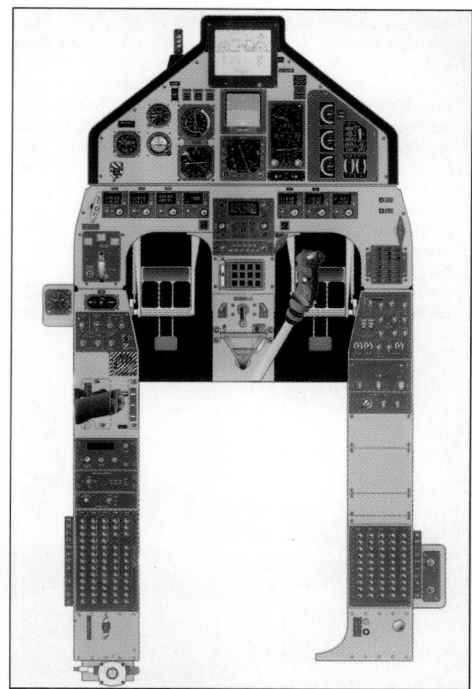

PC-7 Mk II M/PC-9M rear cockpit (advanced avionics and HUD repeater)   0079295

PILATUS PC-9 PRODUCTION
(excluding Raytheon/Beech Mk II)

Customer	Version	Qty	First aircraft	First delivery
Military:				
Angola	PC-9	4	(c/n 115)	1987
Australia	PC-9/A	67[1]	A23-001	Oct 1987
Croatia	PC-9M	17[4]	054	1997
Cyprus	PC-9	2	901	1989
Iraq	PC-9	20	5801	1987
Myanmar	PC-9	10	3601	Apr 1986
Oman	PC-9M	12	426	1999
Saudi Arabia	PC-9	50	2201	Jan 1987
Slovenia	PC-9M	9[5]	L9-61	Nov 1998
Switzerland	PC-9	14[2]	C-401	1987
Thailand	PC-9	36	1/34	1991
US Army	PC-9	3[3]	91-071	1991
Subtotal		**244**		
Civil:				
Pilatus	PC-9	2	HB-HPA	
	PC-9 Mk II	1	HB-HPC	
	PC-9M	1	HB-HPJ	
Condor	PC-9B	10[6]	D-FAMT	Sep 1990
BAE	PC-9	1	ZG969	Jul 1989
Total		**259**		

[1] First 17 (after two from Pilatus) assembled by Hawker de Havilland from Swiss kits; final 48 locally built
[2] First two for evaluation; returned to Pilatus
[3] Transferred to Slovenia in 1995; now L9-51 *et seq*
[4] Plus three second-hand (051 *et seq*)
[5] Modified to Hudournik standard in Israel after delivery
[6] Plus ex-demonstrator, August 2000

LANDING GEAR: Retractable tricycle type, with hydraulic actuation. Mainwheels retract inward to lie semi-recessed in wing centre-section covered by bulged doors; nosewheel retracts rearward; all units enclosed when retracted. Oleo-pneumatic shock-absorber in each leg. Hydraulically actuated nosewheel steering. Goodrich wheels and tubeless tyres, with Goodrich multipiston hydraulic disc brakes on mainwheels. Main tyres 20×4.4 (8 ply), nose tyre 16×4.4 (8 ply). Low-pressure tyres optional. Parking brake.

POWER PLANT: One 857 kW (1,150 shp) Pratt & Whitney Canada PT6A-62 turboprop, flat rated at 708 kW (950 shp), driving a Hartzell HC-D4N-ZA/09512A four-blade constant-speed fully feathering propeller. Single-lever engine control. Fuel in two integral tanks in wing leading-edges, total usable capacity 518 litres (137 US gallons; 114 Imp gallons). Overwing refuelling point on each side. Fuel system includes 12 litre (3.2 US gallon; 2.6 Imp gallon) aerobatics tank in fuselage, forward of front cockpit, which permits up to 60 seconds of inverted flight. Provision for two 154 or 248 litre (40.7 or 65.5 US gallon; 33.9 or 54.5 Imp gallon) drop tanks on centre underwing attachment points. Total oil capacity 16 litres (4.2 US gallons; 3.5 Imp gallons).

ACCOMMODATION: Two Martin-Baker Mk 11A adjustable ejection seats, each with integrated personal survival pack and fighter-standard pilot equipment. Stepped tandem arrangement with rear seat elevated 15 cm (5.9 in). Seats operable, through canopy, at zero height and speeds down to 60 kt (112 km/h; 70 mph) EAS. Anti-g system optional. One-piece acrylic Perspex windscreen; one-piece framed canopy, incorporating roll-over bar, opens sideways to starboard. Dual controls standard. Cockpit heating, cooling, ventilation and canopy demisting standard. Space for 25 kg (55 lb) of baggage aft of seats, with external access.

SYSTEMS: Vapour cycle air conditioning system and engine bleed air for cockpit heating/ventilation and canopy demisting. Fairey Systems hydraulic system, pressure 207 bar (3,000 lb/sq in), for actuation of landing gear, mainwheel doors, nosewheel steering and airbrake; system maximum flow rate 18.8 litres (4.97 US gallons; 4.14 Imp

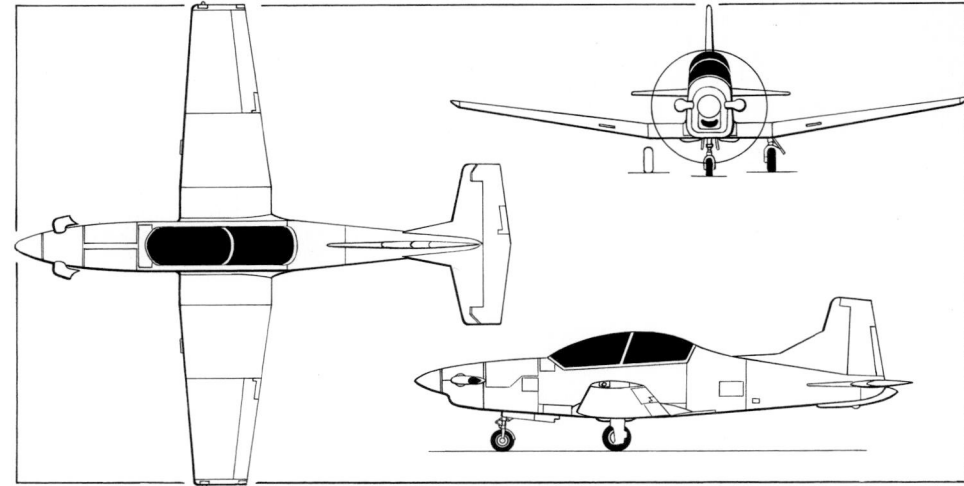

Pilatus PC-7 Mk II M and PC-9M basic/advanced trainer (*Jane's/James Goulding*)   0051488

Hudournik conversion of PC-9M of Slovene Military Aviation (*Jane's/Paul Jackson*)   **NEW**/0525774

gallons)/min. Bootstrap oil/oil reservoir, pressurised at 3.45 to 207 bar (50 to 3,000 lb/sq in). A supplementary high-pressure, one-shot nitrogen storage bottle provides power to lower the landing gear in an emergency. This bypasses the main hydraulic systems by feeding directly into shuttle valves in the down side of the main door and landing gear actuators.

Primary electrical system (28 V DC operational, 24 V nominal) powered by 30 V 300 A starter/generator and 24 V 40 Ah Ni/Cd battery; two optional static inverters can supply 115/26 V AC power at 400 Hz. Ground power receptacle. Electric anti-icing of pitot tube, static ports and AoA transmitter standard; electric de-icing of propeller blades optional. Diluter demand oxygen system, capacity 1,350 litres (47.6 cu ft) selected and controlled individually from panel in each cockpit. OBOGS standard.

AVIONICS: Both cockpits fully instrumented to customer specifications, with Logic computer-operated integrated engine and systems data display. Customer-specified equipment provides flight environmental, attitude and direction data, and ground-transmitted position determining information.

Comms: Single or dual VHF, UHF and/or HF radios to customer's requirements. Audio integrating system controls audio services from com, nav and interphone systems.

Optional avionics include Honeywell CRT displays (electronic ADI and HSI), Flight Visions FV-2000 HUD, encoding altimeter and ELT.

Instrumentation: See accompanying diagram.

EQUIPMENT: Three hardpoints under each wing: inboard and centre stations each stressed for 250 kg (551 lb), outboard stations for 110 kg (242.5 lb) each. Retractable 250 W landing/taxiing light in each main landing gear leg bay.

Optional equipment can include smoke generators for aerial displays, target towing kit, electronic jammer for EW training, and blind-flying hood.

ARMAMENT: FN Herstal package for Hudournik/Swift comprises HMP-250 12.7 mm gun pods; LAU-7A seven-round unguided rocket launchers; Alkan Type 65 adaptor and dispensers for IBDU-33 practice bombs; and a laser range-finder pod.

DIMENSIONS, EXTERNAL:

Wing span	10.125 m (33 ft 2½ in)
Wing chord: mean aerodynamic	1.65 m (5 ft 5 in)
mean geometric	1.61 m (5 ft 3½ in)
Wing aspect ratio	6.3
Length overall	10.14 m (33 ft 3¼ in)
Fuselage max width	0.97 m (3 ft 2¼ in)
Height overall	3.26 m (10 ft 8⅓ in)
Tailplane span	3.665 m (12 ft 0¼ in)
Wheel track	2.54 m (8 ft 4 in)
Wheelbase	2.31 m (7 ft 7 in)
Propeller diameter	2.44 m (8 ft 0 in)
Propeller ground clearance	0.18 m (7 in)

DIMENSIONS, INTERNAL:

Baggage compartment volume	0.14 m³ (4.9 cu ft)

AREAS:

Wings, gross	16.29 m² (175.3 sq ft)
Ailerons (total)	1.57 m² (16.90 sq ft)
Trailing-edge flaps (total)	1.77 m² (19.05 sq ft)
Airbrake	0.30 m² (3.23 sq ft)
Fin	0.86 m² (9.26 sq ft)
Rudder, incl tab	0.90 m² (9.69 sq ft)
Tailplane	1.80 m² (19.38 sq ft)
Elevator, incl tab	1.60 m² (17.22 sq ft)

WEIGHTS AND LOADINGS (A: Aerobatic, U: Utility category with underwing stores):

Basic weight empty, equipped	1,725 kg (3,803 lb)
Max underwing stores load	1,040 kg (2,293 lb)
Max T-O weight: A	2,350 kg (5,180 lb)
U	3,200 kg (7,054 lb)
Max ramp weight: A	2,360 kg (5,203 lb)
U	3,210 kg (7,076 lb)
Max landing weight: A	2,350 kg (5,180 lb)
U	3,100 kg (6,834 lb)
Max zero-fuel weight: A	2,000 kg (4,409 lb)
Max wing loading: A	144.3 kg/m² (29.55 lb/sq ft)
U	196.4 kg/m² (40.23 lb/sq ft)
Max power loading: A	3.32 kg/kW (5.45 lb/shp)
U	4.52 kg/kW (7.42 lb/shp)

PERFORMANCE (A and U as above, propeller speed 2,000 rpm):

Max operating Mach number (M_{MO}): A, U	0.65
Max operating speed (V_{MO}):	
A, U	320 kt (593 km/h; 368 mph) EAS
Max cruising speed:	
A at S/L	271 kt (501 km/h; 311 mph)
A at 3,050 m (10,000 ft)	297 kt (550 km/h; 342 mph)
A at 7,620 m (25,000 ft)	300 kt (556 km/h; 345 mph)

Pilatus PC-9/A Advanced Turbo Trainer of Australia's Roulettes aerobatic team (*Jane's/Paul Jackson*)
NEW/0525773

Max manoeuvring speed:
A 205 kt (380 km/h; 236 mph) EAS
U 200 kt (370 km/h; 230 mph) EAS
Max speed with flaps and/or landing gear down:
A, U 150 kt (278 km/h; 172 mph) EAS
Stalling speed, engine idling:
  flaps and gear up:
  A 77 kt (143 km/h; 89 mph) EAS
  U 90 kt (167 km/h; 104 mph) EAS
  flaps and gear down:
  A 69 kt (128 km/h; 80 mph) EAS
  U 80 kt (149 km/h; 93 mph) EAS
Max rate of climb at S/L: A 1,247 m (4,090 ft)/min
Time to 4,575 m (15,000 ft): A 4 min 5 s
Max operating altitude: A, U 7,620 m (25,000 ft)
Service ceiling: A, U 11,580 m (38,000 ft)
T-O run at S/L: A 243 m (795 ft)
T-O to 15 m (50 ft) at S/L: A 391 m (1,280 ft)
Landing from 15 m (50 ft) at S/L at MLW:
A 700 m (2,295 ft)
Landing run at S/L at MLW:
  A (normal braking action) 351 m (1,150 ft)
Max range at 210 kt (389 km/h; 242 mph) at 7,620 m
  (25,000 ft), 5% and 20 min fuel reserves:
  A 830 n miles (1,537 km; 955 miles)
Max endurance at 110 kt (204 km/h; 127 mph) IAS,
  conditions as above: A 4 h 30 min
g limits: A +7/−3.5
U +4.5/−2.25
UPDATED

PILATUS PC-21

TYPE: Basic turboprop trainer.
PROGRAMME: Project launched November 1998; development began January 1999; provisional designation chosen to indicate 21st Century technology, but was retained for commercial use; first brief details revealed late 1999; modified PC-7 Mk II development aircraft (HB-HMS) flying by October 2000 with a sweptback fin, spoilers, power management system and five-blade propeller; later received PC-21 avionics and mission management system.
   Prototype (HB-HZA) rolled out 30 April 2002; first flight 1 July 2002; public debut at Royal International Air

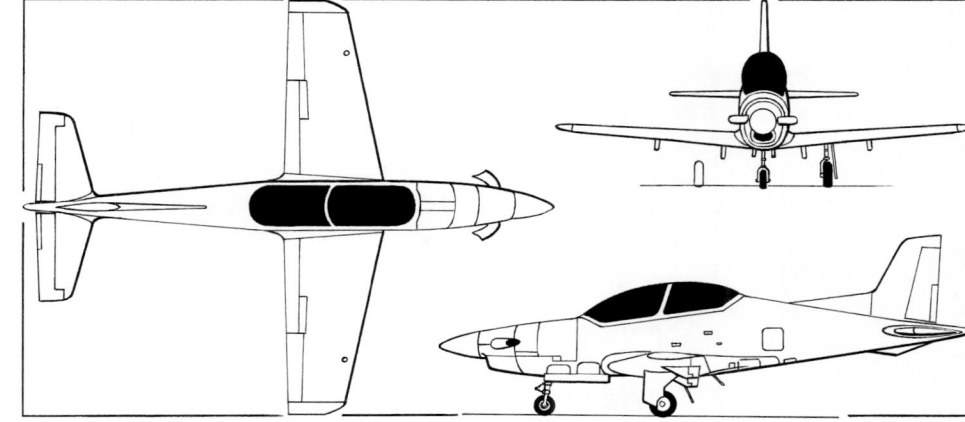

Pilatus PC-21 advanced turboprop trainer (*Jane's/James Goulding*)   *NEW*/0536697

Tattoo, Fairford, UK, 20-21 July 2002, at which time the aircraft had flown some 14 hours, including ferry time; appeared at Farnborough Air Show from 22 July. Second development aircraft scheduled to fly in late 2003; production to start in mid-2004, with certification (achieved by first and second development PC-21s) and first production delivery expected in November 2004.
CUSTOMERS: Estimated market for 1,000 aircraft in class over 20-year period, of which Pilatus hopes to capture 50 per cent; Australia, South Africa and UK seen as primary marketing targets for launch orders.
COSTS: Company-funded development cost estimated at SFr200 million; unit cost SFr5 million to SFr6 million (both 2002).
DESIGN FEATURES: Extrapolation of PC-7/PC-9 series (which see in current and previous editions); optimised for advanced flying (pilot) and weapons training (pilot and WSO), but at turboprop cost. Avionics able to simulate aiming and release of multiple weapons (in excess of aircraft's actual carrying capability) without resort to training rounds, although addition of underwing stores

pylons is possible. Jet-like response assisted by engine governor which restricts output to 671 kW (900 shp) below 80 kt (148 km/h; 92 mph) or when landing gear unloaded, increasing by graduations to full power at 230 kt (426 km/h; 265 mph) and above. Propeller and spinner angled 4° down and 4° starboard to offset full torque, while wing and fin slightly offset for similar reason; optional automatic yaw compensation. Design aims included superior aerodynamic performance; integrated and cost-effective training system; and (30-year) life-cycle support cost not exceeding current turboprop trainers.
   Wing leading-edge sweepback 12° 42′; tailplane sweepback approximately 16°; dihedral from roots.
FLYING CONTROLS: Conventional and manual. Flaps. Door-type, hydraulically actuated spoiler in each wing upper surface, ahead of outboard one-third of flaps, to enhance roll. Flight-adjustable trim tabs in rudder, each aileron and each half of tailplane.
STRUCTURE: Generally as PC-9, but redesigned wing with additional spar forward. Wing leading-edges of impact-absorbent material for birdstrike resistance.
LANDING GEAR: Retractable tricycle type, with single wheel on each unit. Retraction inwards (mains) and rearward (nose). Mainwheel tyres size 20×14.4R12, pressure 16.6 bar (240 lb/sq in); nosewheel tyre size 10×4.4R8, pressure 11.0 bar (160 lb/sq in).
POWER PLANT: One 1,193 kW (1,600 shp) Pratt & Whitney Canada PT6A-68B turboprop, driving a Hartzell E8991KX five-blade graphite/titanium propeller. Provision for two underwing fuel tanks.
ACCOMMODATION: Two in tandem on Martin-Baker Mk 16L zero zero ejection seats in pressurised cockpit. Two-section acrylic canopy with thickened front for birdstrike resistance without resort to separate windscreen. Rear seat height increased in comparison with PC-9; front occupant has −11° forward view; rear occupant −4°.
SYSTEMS: OBOGS standard.
AVIONICS: *Flight:* Optional ground proximity warning. Selected systems can be invalidated for training purposes, but immediately restored in the event of an emergency.
   Instrumentation: 'Glass cockpit' featuring three 152 × 203 mm (6 × 8 in) AMLCD PFDs and two 76 mm (3 in) Meggitt AMLCD secondary flight displays in each cockpit; Flight Visions SparrowHawk HUD with FVD-4000 HUD symbol generator standard in front cockpit; HUD repeater in rear cockpit optional; cockpit displays are NVG-compatible. Digital recording of all displays for post-flight debriefing. Optional decoupling of instructor's displays from student's in the event of latter's incorrect inputs or to access data denied to student.
ARMAMENT: Provision for external stores on one centreline and four underwing hardpoints.

Tandem cockpits of the Pilatus PC-21   *NEW*/0525776

Prototype PC-21 taking off on its maiden flight (*Daniel Buehler/Pilatus*)   *NEW*/0525777

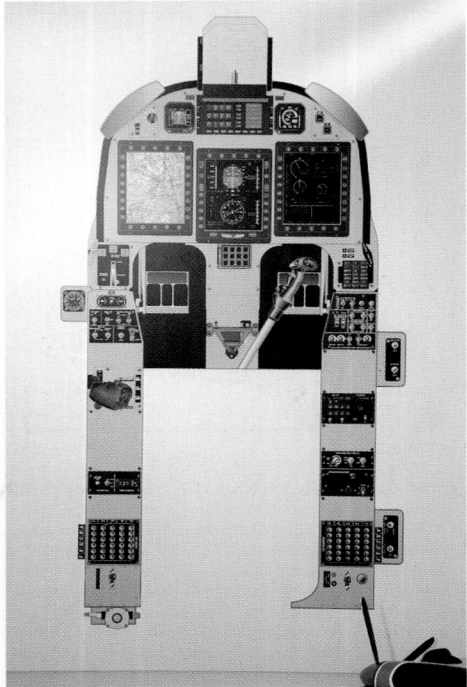

Three-screen cockpit of the PC-21   *NEW*/0525775

DIMENSIONS, EXTERNAL:
Wing span	8.765 m (28 ft 9 in)
Length overall	11.19 m (36 ft 8½ in)
Height overall	3.915 m (12 ft 10 in)
Fuselage: Max depth	2,015 m (6 ft 7¼ in)
Max width	1.00 m (3 ft 3¼ in)
Tailplane span	4.00 m (13 ft 1½ in)
Wheel track	2.735 m (8 ft 11¾ in)
Wheelbase	2.495 m (8 ft 2¼ in)

WEIGHT AND LOADINGS:
Weight empty	2,250 kg (4,960 lb)
T-O weight: Aerobatic (clean)	3,100 kg (6,834 lb)
with underwing tanks	3,600 kg (7,936 lb)
max	4,250 kg (9,370 lb)
Max external load	1,150 kg (2,535 lb)
Max power loading	3.56 kg/kW (5.86 lb/shp)

PERFORMANCE (estimated):
Max operating speed	M0.72 (370 kt; 685 km/h; 426 mph)
Design diving speed	M0.80 (420 kt; 778 km/h; 483 mph)
Max level speed at FL100	340 kt (630 km/h; 301 mph)
Stalling speed, flaps and gear down	
	80 kt (148 km/h; 92 mph)
Max certified altitude	7,620 m (25,000 ft)
Service ceiling	11,582 m (38,000 ft)
g limits	+8/−4

UPDATED

PILATUS PC-12

TYPE: Business turboprop.

PROGRAMME: Announced at NBAA Convention October 1989; first flight P.01 (HB-FOA) 31 May 1991; first flight of P.02 second prototype (HB-FOB) 28 May 1993; Swiss certification to FAR Pt 23 Amendment 42 (covering FAR Pt 135 commercial and Pt 91 general operations) received 30 March 1994, FAA type approval 15 July 1994, and FAR Pt 25 certification for flight into known icing conditions in 1995. Deliveries (N312BC to Carlston Leasing Corporation in USA) began September 1994.

Higher gross weight option (4.5 tonnes, hence **PC-12/45**) introduced in 1996, this becoming production standard by 1997. FAA FAR Pt 135 approval for commercial single-engined IFR operations announced in third quarter 1997; production increased from three to four per month in August 1997. First airline scheduled service operator (1997) was Kelner Airways of Canada. Total of 76 ordered and 51 delivered in 1998 (including 100th in April); 29 ordered in first half of 1999; 200th sale announced at NBAA Convention in October 1999; production rate increased in 1999 from 48 to 60 per year and targeted to reach 100 or more per year by 2002. Worldwide fleet hours passed 300,000 mark in April 2002. Now certified in 12 countries. 1,500th Pilatus single-engined turboprop, manufactured in mid-2001, was PC-12 N377PC.

In 1999, Pilatus was designing a new cockpit, featuring a 230 × 250 mm (9 × 10 in) IS&S flat-panel display. Other modifications and improvements, including performance gains from design studies for PC-21, were under development by 2000.

CURRENT VERSIONS: **Standard:** Nine-passenger commuter or passenger/cargo combi.

Detailed description applies to standard PC-12/45 except where indicated.

Executive: Six to eight passenger seats.

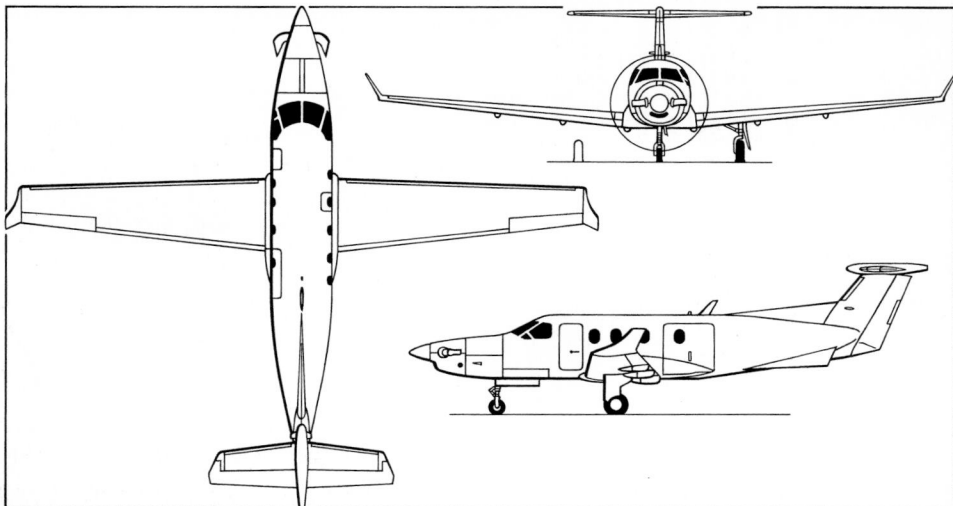

Pilatus PC-12/45 pressurised light utility and business transport *(Jane's/Mike Keep)*

Eagle: Surveillance and special missions version; described separately.

CUSTOMERS: Completion of 300th production aircraft effected in 2001. Deliveries in 2000 and 2001 totalled 70 each compared with 55 in 1999. Some 53 per cent of first 200 aircraft are registered in USA, further 18 per cent in Canada (including nine with Royal Canadian Mounted Police) and 11 per cent in South Africa (including two for Red Cross and one with Air Force). US Drug Enforcement Agency (DEA) received two in transport configuration in December 1999 and October 2000, and predicts further orders. Other customers include those in Argentina (Border Guard), Australia (Royal Flying Doctor Service with 16 aircraft, including three delivered in early 2002), Austria, Bermuda, Brazil, Denmark, France, Germany, Japan, Kenya, Switzerland and Zimbabwe.

Approximately 10 per cent of those sold are in air ambulance configuration. First single-engine turboprop fractional ownership scheme involves PC-12s of Alpha Flying's PlaneSense programme at Nashua, New Hampshire, USA. Similar scheme introduced 2001 by SkyAir TimeJet of Zurich, Switzerland.

COSTS: Basic price US$2,712,760 (2002). Direct operating cost US$334.84 per hour (2001).

DESIGN FEATURES: Able to fly three 200 n mile (370 km; 230 mile) sectors in 6 hour flight; approved for single-pilot VFR/IFR operation into known icing.

Low-wing monoplane with tapered wing and T tail; latter's mounting reduces trim changes with power; CG range 13 to 46 per cent of MAC. Modifications following early flight trials include introduction of winglets, increased wing span, paired elevators instead of single surface, sweptback tailplane/elevator tips, addition of tailplane/fin bullet fairing, and enlarged dorsal fin and ventral strakes.

Wing sections (modified NASA GA(W)-1 series), LS(1)-0417-MOD at root and LS(1)-0313 at tip.

FLYING CONTROLS: Conventional and manual. Actuation by pushrods and cables; servo tab in each aileron; electrically actuated Flettner tab in rudder; short-span mass-balanced ailerons; electrically actuated Fowler flaps cover 67 per cent of wing trailing-edge; electrically actuated (dual redundant) variable incidence tailplane.

STRUCTURE: Aluminium alloy primary skin and structure; composites for ventral strakes and dorsal fin (Kevlar/honeycomb sandwich), wingtips (glass fibre), fairings, engine cowling (glass fibre/honeycomb sandwich) and interior trim; titanium firewall; two-spar wing with integral fuel tankage; airframe proved for 20,000 hour life; complete internal and external corrosion protection.

All parts fabricated in Switzerland, but wings and fuselage assembled by OGMA in Portugal before return to Swiss production line. North American interiors installed in USA finishing centre; others outfitted at Stans.

LANDING GEAR: Pilatus hydraulically retractable tricycle type, with single wheel on each unit; nosewheel mechanically steerable ±60°. Emergency extension system; toe-operated brakes; parking brake. Suitable for operation from grass strips. Goodrich wheels and low-pressure tyres on all units: size 22×8.50-10 on main gear, 17.5×6.25-6 on nose unit; tyre pressure 4.14 bar (60 lb/sq in) on nose unit, 3.79 bar (55 lb/sq in) on main units. Propeller ground clearance maintained with nose leg compressed and nosewheel tyre flat. Trailing-link main gear retracts inward into wings, nose gear rearward under flight deck. Ground turning radius about wingtip 10.21 m (33 ft 6 in), about main gear 4.50 m (14 ft 9 in).

POWER PLANT: One 1,197 kW (1,605 shp) Pratt & Whitney Canada PT6A-67B turboprop, flat rated to 895 kW (1,200 shp) for T-O and 746 kW (1,000 shp) for climb and cruise. Hartzell HC-E4A-3D/E10477K constant-speed, fully feathering reversible-pitch four-blade aluminium propeller, turning at 1,700 rpm. Two integral fuel tanks in wings, total capacity 1,540 litres (407 US gallons; 339 Imp gallons), of which 1,522 litres (402 US gallons; 335 Imp gallons) are usable. Gravity fuelling point in top of each wing. Oil capacity 11 litres (2.9 US gallons; 2.4 Imp gallons).

ACCOMMODATION: Two-seat flight deck: approved for single pilot, with dual controls; second flight instrument panel optional. Limit of nine passengers under FAR Pt 23, or executive layout for six, with lavatory. Two or three stretcher patients, plus life support systems and medical attendants, in ambulance configuration. Downward-opening airstair crew/passenger door at front, upward-opening cargo door at rear, both on port side; Type III emergency exit above wing on starboard side.

SYSTEMS: Normalair Garrett engine bleed air ECS, maximum pressure differential 0.4 bar (5.8 lb/sq in), maintaining a 2,440 m (8,000 ft) cabin altitude at 7,620 m (25,000 ft). Vickers Systems (Germany) hydraulic system, pressure 207 bar (3,000 lb/sq in), for landing gear actuation. Electrical power system (28 V DC) supplied by 300 A engine-driven starter/generator), 130 A back-up alternator and 24 V 40 Ah Ni/Cd battery. Second Ni/Cd battery optional. Goodrich pneumatic boot de-icing of wing and tailplane leading-edges; Goodrich electric anti-icing of windscreen, propeller blades, pitot probes and stall warning sensor; exhaust air anti-icing of engine air intake. Oxygen system for crew and passengers.

AVIONICS: *Comms:* Bendix/King dual KX 155A VHF transceivers, KMA 26 audio control panel and intercom with marker beacon switch and KT 70A transponder, and Narco ELT-910 emergency locator transmitter, standard; Bendix/King KHF 950 HF radio, AMS 44 dual audio systems and second KT 70A optional.

Radar: Honeywell RDR 2000 weather radar optional.

Flight: Honeywell KFC 325 autopilot system, KN 63

This Pilatus PC-12 was the manufacturer's 1,500th production single-engine turboprop   *NEW*/0525064

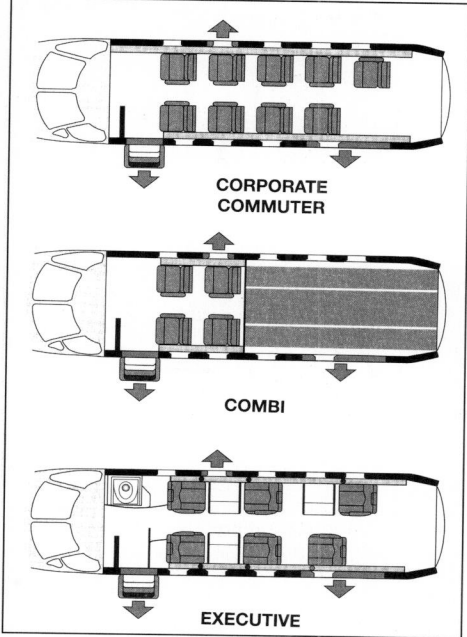

PC-12/45 typical cabin layouts   0051493

CORPORATE COMMUTER

COMBI

EXECUTIVE

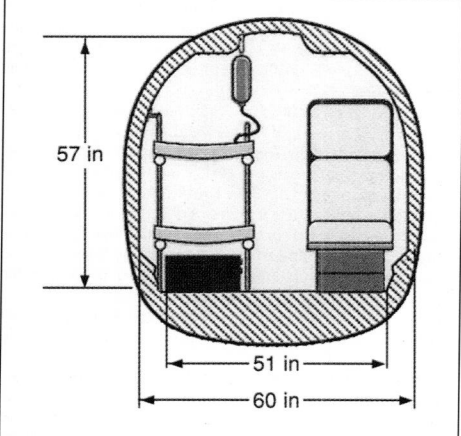

PC-12/45 cabin cross-section in medical evacuation fit   0051494

57 in / 51 in / 60 in

DME, KR 87 ADF, KNI 582 RMI and KEA 130A encoding altimeter and Litef LCR-92 AHRS standard; Bendix/King KLN 90B or KLN 900 GPS optional. Other options include KRA 405 radar altimeter, second LCR-92, WX 1000E Stormscope, TCAS 66A, and Mk VI GPWS with CIC 8800M air data computer.

Instrumentation: Bendix/King EHI 40/50 EFIS with 102 mm (4 in) display standard; MFD for EHI 40, and 127 mm (5 in) EFIS on pilot and co-pilot panels, optional.

DIMENSIONS, EXTERNAL:
Wing span	16.23 m (53 ft 3 in)
Wing aspect ratio	10.2
Length overall	14.40 m (47 ft 3 in)
Height overall	4.27 m (14 ft 0 in)
Elevator span	5.21 m (17 ft 1 in)
Wheel track	4.52 m (14 ft 10 in)
Wheelbase	3.53 m (11 ft 7 in)
Propeller diameter	2.67 m (8 ft 9 in)
Propeller ground clearance	0.32 m (1 ft 0½ in)
Passenger door: Height	1.35 m (4 ft 5 in)
Width	0.635 m (2 ft 1 in)
Cargo door: Height	1.32 m (4 ft 4 in)
Width	1.35 m (4 ft 5 in)
Emergency exit: Height	0.66 m (2 ft 2 in)
Width	0.48 m (1 ft 7 in)

DIMENSIONS, INTERNAL:
Cabin: Length, excl flight deck	5.16 m (16 ft 11 in)
Max width	1.52 m (5 ft 0 in)
Width at floor	1.295 m (4 ft 3 in)
Max height	1.45 m (4 ft 9 in)
Volume	9.34 m³ (330 cu ft)
Baggage compartment volume	1.13 m³ (40 cu ft)

AREAS:
Wings, gross	25.81 m² (277.8 sq ft)

WEIGHTS AND LOADINGS:
Weight empty: standard	2,600 kg (5,732 lb)
executive	2,903 kg (6,400 lb)
Combi	2,536 kg (5,591 lb)
Max usable fuel	1,226 kg (2,703 lb)
Max payload: standard	1,410 kg (3,108 lb)
Executive	1,243 kg (2,740 lb)
Combi	1,474 kg (3,250 lb)
Payload with max fuel: executive	388 kg (855 lb)

Current PC-12 flight deck with optional co-pilot's EFIS   0093858

Max ramp weight	4,520 kg (9,965 lb)
Max T-O and landing weight	4,500 kg (9,920 lb)
Max zero-fuel weight	4,100 kg (9,040 lb)
Max wing loading	174.3 kg/m² (35.71 lb/sq ft)
Max power loading	5.03 kg/kW (8.27 lb/shp)

PERFORMANCE:
Max operating Mach number (M_{MO})	0.48
Max operating speed (V_{MO})	240 kt (444 km/h; 276 mph) IAS
Max cruising speed at 7,620 m (25,000 ft)	270 kt (500 km/h; 311 mph)
Stalling speed:	
flaps and gear up	92 kt (171 km/h; 106 mph) IAS
flaps and gear down	64 kt (119 km/h; 74 mph) IAS
Max rate of climb at S/L	512 m (1,680 ft)/min
Max certified altitude	9,150 m (30,000 ft)
Service ceiling	10,670 m (35,000 ft)
T-O run	452 m (1,480 ft)
T-O to 15 m (50 ft)	701 m (2,300 ft)
Landing from 15 m (50 ft) at MLW	558 m (1,830 ft)
Landing run at MLW	288 m (945 ft)

Max range at 9,150 m (30,000 ft), VFR reserves:
standard	2,261 n miles (4,187 km; 2,602 miles)
executive	2,197 n miles (4,070 km; 2,529 miles)
combi	2,282 n miles (4,226 km; 2,626 miles)
g limits: flaps up	+3.4/−1.36
flaps down	+2.0

UPDATED

PILATUS PC-12M EAGLE

TYPE: Multisensor surveillance turboprop.

PROGRAMME: Announced at Dubai Air Show in November 1995; second PC-12 HB-FOB converted as demonstrator (first flight October 1995). Second demonstrator HB-FOG (c/n 134) followed in September 1996 with airframe modifications and different equipment (see Design Features and Current Versions paragraphs below). By early 2001, both prototypes had ventral sensor panniers and fintip extensions removed and HB-FOG was undergoing extensive overhaul.

CURRENT VERSIONS: **HB-FOB:** Ventral pannier carrying a variety of electro-optical sensors. A sensor management system (SMS) controls and monitors sensors and displays data on two consoles (one electro-optical, one radar, for comint and elint) inside cabin; includes three COTS

processors. Consoles have four displays: colour displays for navigation, situational awareness and ESM/IR/TV; and one 325 mm (14½ in) square LCD XGA high-resolution colour display for synthetic aperture radar (SAR). Sensor operator's map display is 'North up', featuring coastlines, rivers, roads and political boundaries.

Sensor options include: (1) WF-160DS turret with FLIR having 7.2 × 9.8° wide field of view and 2.16 × 2.95° narrow field, plus daylight camera with 20 to 280 mm zoom lens; (2) SAR installations of several types, either mechanical or electronically scanned (one or two dimensions, from 45 to 200 cm; 18 to 79 in long, giving up to 36 cm; 14¼ in resolution); and (3) RISTA (Reconnaissance, Infra-red Surveillance, Target Acquisition), derived from US Army's Airborne Minefield Detection System and combining FLIR (for battle damage assessment) and IR linescan (covering a 7 n mile; 13 km; 8 mile ground track) with real-time downlink to ground station.

HB-FOG: Sensors include WF-160DS IR/E-O system with tracker and Geotrack cueing; Raytheon Sea Vue SV 1021 radar with ISAR capability 1997 upgrade for SAR; 1998 upgrade for track-while-scan of 100 targets and MTI on SAR; communications digital datalink; 108 n mile (200 km; 124 mile) range video downlink to ground station; and sensor management system.

In late 2001, HB-FOG was delivered to Gruppe Rustung of Swiss Defence Procurement Agency and based at Emmen for trials connected with new FLORAKO air defence ground environment. Pannier removed and fintip extension not fitted; standard PC-12 winglets.

Eagle is also capable of carrying such other sensors as 2 to 18 GHz (optionally 40 GHz) elint, including automatic jammer; forward- and backward-looking sensor management systems; a passive missile warning system, including automatic chaff/flare dispensers; LOROP; and a bidirectional secure datalink. Schemes for operational employment have included airborne control of a flight of armed PC-9s.

CUSTOMERS: One delivered to Swiss Defence Procurement Agency. Interest reported from Brunei, Malaysia and Thailand in early 2002.

DESIGN FEATURES: FLIR turret at front of ventral pannier; winglets replaced by conventional wingtips on HB-FOB (weather radar in starboard tip pod); undertail strakes

Second Eagle prototype, with pannier fitted, escorted by Pilatus' PC-7 Mk II demonstrator   0106533

enlarged (to maximum depth of 53 cm; 21 in) to maintain stability. HB-FOG fitted 1997 with new 'tiplets' (wingtips with small vertical winglets), newly defined undertail strakes and additional fin area above the elevator, increasing overall height by 48 cm (19 in). Eagle can undertake various reconnaissance and surveillance missions, but retains ability to revert to passenger and other missions at short notice.

DIMENSIONS, EXTERNAL: As standard PC-12/45 except:
Wing span   16.115 m (52 ft 10½ in)
Height overall   4.74 m (15 ft 6½ in)
WEIGHTS AND LOADINGS: As standard PC-12/45 except:
Weight empty   2,688 kg (5,926 lb)
Max payload   1,402 kg (3,091 lb)
PERFORMANCE:
Max cruising speed at 6,100 m (20,000 ft)
                                                230 kt (426 km/h; 265 mph)
Stalling speed, flaps and gear down
                                                65 kt (121 km/h; 76 mph) IAS
Max rate of climb at S/L   408 m (1,340 ft)/min
Max operating altitude   9,145 m (30,000 ft)
T-O run   450 m (1,475 ft)
Landing run   280 m (920 ft)
Max range at 9,145 m (30,000 ft) at 200 kt (370 km/h;
230 mph), 45 min reserves
                            1,633 n miles (3,080 km; 1,913 miles)
                                                              UPDATED

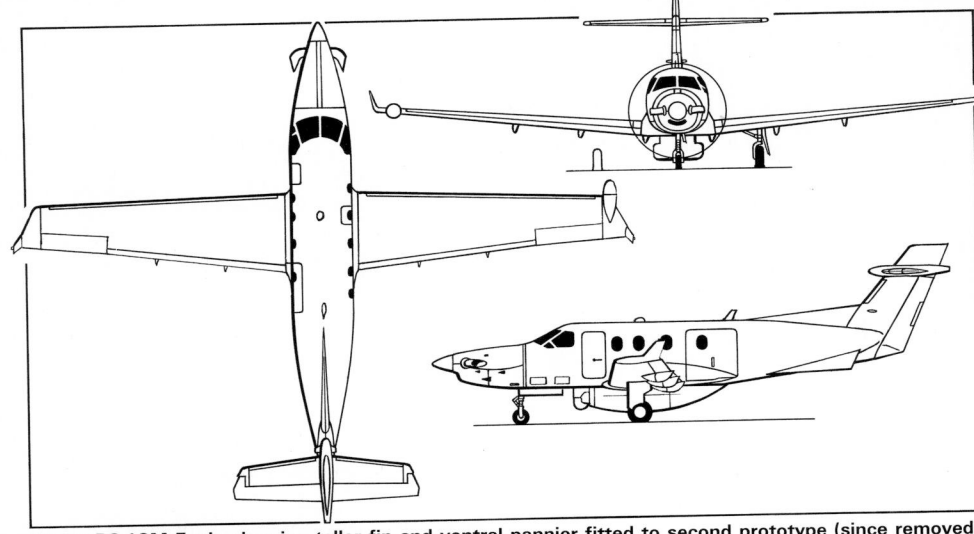

Pilatus PC-12M Eagle showing taller fin and ventral pannier fitted to second prototype (since removed) *(Jane's/James Goulding)*   0051495

RUAG

RUAG AEROSPACE (Subsidiary of RUAG Holding)

PO Box 301, CH-6032 Emmen
Tel: (+41 41) 268 41 11
Fax: (+41 41) 260 25 88
e-mail: marketing.aerospace@ruag.com
Web: http://www.ruag.com
MANAGING DIRECTOR: Peter Schneuwly
MARKETING MANAGER:: René Diehl

In 1996, the former Swiss Federal Aircraft Factory Emmen merged with the industrial division of the Swiss Air Force Logistic Command and parts of the Federal Ordnance Office. On 1 January 1999, it was converted from a federal ordnance establishment into a public diversified private company within the RUAG group and renamed Swiss Aircraft and Systems Enterprise Corporation. This title was changed to RUAG Aerospace in 2001, with sites at Emmen, Dübendorf, Stans, Zweisimmen, Interlaken, Lodrino and Alpnach. At that time, the company had a total of 79,300 m² (853,575 sq ft) factory area, office area of 22,500 m² (242,190 sq ft) and a workforce of 1,700.

Main activities are the maintenance, manufacture, modernisation and assembly under licence of military and civil aircraft and guided missiles. R&D activities involve four departments, of which Aerodynamics and Flight Mechanics department has four wind tunnels for speeds up to M4 to M5, and test cells for piston and jet engines with and without afterburners. Structural and Systems Engineering department deals with aircraft and space hardware. Electronics and Missile Systems department deals with all systems aspects of aircraft avionics and missiles. Fourth R&D department is for prototype fabrication, flight test, instrumentation and system environmental testing. The Production department can handle mechanical, sheet metal and composites and electronic, electrical, electromechanical and electro-optical subassemblies.

Recent programmes have included structural improvements to Swiss Air Force Tiger F-5E/Fs and the integration of new systems (smoke systems, target towing, missiles); manufacture and delivery of Rafale drop tanks to Dassault Aviation; assembly of 19 BAe Hawk Mk 66s; integration of AS 532UL Cougars for the Swiss Air Force and assembly of 32 single- and two-seat F-18 Hornets for the Swiss Air Force; it also established engineering support and manufactured leading-edges, rudders and its own low-drag pylons for the last-named.

Current work includes the serial production of rudders for the Boeing 717, wingtips for the Airbus A319/A320/A321, entire wing fixed trailing-edges for the Airbus A380, the tailplane for the Pilatus PC-12 and the manufacture and ground support of the Ranger UAV and its hydraulic launcher. During 2001 and 2002 it was assembling 10 of the follow-on order for 12 AS 532UL Cougars ordered for the Swiss Air Force in December 1998.

RUAG Aerospace is sole manufacturer of Dragon anti-tank missile; TOW missile programme began 1996; Stinger missile programme began 1989.

RUAG Aerospace is accredited to maintain GE J85 (Tiger), Adour (Hawk), GE F404 (F/A-18), Makila (Cougar) and PT6A (PC-6, PC-7, PC-9) engines in own workshops. JAR 145 accreditation allows the company to carry out maintenance work at all levels. This includes inspections, repairs, modifications, prototyping and fault elimination. Other specialities include non-destructive testing procedures such as ultrasonic, eddy current and X-ray as well as comprehensive service packages for systems re-engineering.

In collaboration with Contraves, RUAG Aerospace co-manufactures all payload fairings for Ariane IV and V space launchers on behalf of the ESA, ESTEC, CNES and private industry.
                                                              UPDATED

RUAG-assembled Eurocopter AS 532UL Cougar of the Swiss Air Force *(S F Fotodienst)*   NEW/0536713

SMARTFISH

SMARTFISH

Junkerngasse 43, CH-3011 Bern
Tel: (+41 31) 311 49 59
e-mail: info@smartfish.ch
Web: http://www.smartfish.ch
FOUNDER: Koni Schafroth

SmartFish was established to exploit the aerodynamic advantages of a unique configuration, patented in November 2001, which is based on its piscine namesake. To illustrate the advantages of the design in comparison to existing aircraft, SmartFish has formulated a hypothetical business jet and a sportplane.
                                                              NEW ENTRY

SMARTFISH BUSINESS JET

TYPE: Business jet.
PROGRAMME: Project initiated in 1998 by small team of designers, inventors and scientists. Computer simulations at Lausanne followed by flight trials of scale model.

Model of the SmartFish concept   NEW/0527123

DESIGN FEATURES: Characteristic aerofoil includes marked arrow-shaped leading edge for speeds in excess of 486 kt (900 km/h; 559 mph). Low-aspect ratio minimises torsional forces and obviates flutter; low wing loading permits slow speed without lift-enhancing aids. Concept combines advantages of flying wing and lifting body designs, as well as retaining stability and controllability of conventional designs. Low manufacturing costs and specific fuel consumption.
POWER PLANT: Two small turbofans.
ACCOMMODATION: Nine passengers, plus two crew.
DIMENSIONS, EXTERNAL:

Wing span	9.75 m (31 ft 11¾ in)
Wing aspect ratio	1.6
Length overall	13.90 m (45 ft 7¼ in)
Height without landing gear	4.20 m (13 ft 9¼ in)

WEIGHTS AND LOADINGS:

Max payload	900 kg (1,984 lb)
Max fuel weight	2,000 kg (4,409 lb)
Max T-O weight	5,700 kg (12,566 lb)
Max zero-fuel weight	2,800 kg (6,173 lb)

PERFORMANCE (estimated):

Never-exceed speed (VNE)	M0.9
Stalling speed, clean	68 kt (125 km/h; 78 mph)
T-O and landing run	750 m (2,460 ft)
Range	3,158 n miles (5,850 km; 3,635 miles)

NEW ENTRY

SMARTFISH SPORTPLANE
TYPE: Two-seat jet sportplane.
PROGRAMME: As for Business Jet.
POWER PLANT: One turbofan, approximately 3.56 kN (800 lb st).
ACCOMMODATION: Two pilots in tandem.

DIMENSIONS, EXTERNAL:

Wing span	4.50 m (14 ft 9¼ in)
Wing aspect ratio	1.6
Length overall	6.00 m (19 ft 8¼ in)
Height without landing gear	1.57 m (5 ft 1¾ in)

WEIGHTS AND LOADINGS:

Max payload	300 kg (661 kg)
Max fuel weight	300 kg (661 kg)
Max T-O weight	1,000 kg (2,204 lb)
Max zero-fuel weight	400 kg (882 lb)

PERFORMANCE:

Never-exceed speed (VNE)	M0.85
Stalling speed, clean	68 kt (125 km/h; 78 mph)
T-O and landing run	700 m (2,300 ft)
Range	2,159 n miles (4,000 km; 2,485 miles)

NEW ENTRY

TAIWAN

AIDC

AEROSPACE INDUSTRIAL DEVELOPMENT CORPORATION
111-6 Lane 68, Fu-Hsing North Road, Taichung 407
Tel: (+886 4) 259 00 01
Fax: (+886 4) 256 23 94
e-mail: joelchu@ms.aidc.com.tw
Web: http://www.aidc.com.tw
CHAIRMAN: Chuen-Huei Tsai
VICE-PRESIDENT, BUSINESS: Chou Yeuan-Yuan
MANAGER, PUBLIC RELATIONS: Michael H L Chang

Established 1 March 1969; became subsidiary of Chung Shan Institute of Science and Technology (CSIST, which see) under Ministry of National Defense on 1 January 1983 and known as Aero Industry Development Center until June 1996. See 2001-02 and earlier editions for aircraft types produced during that time. New 35,700 m² (384,270 sq ft) assembly facility opened in 1989. Current facilities at Taichung, Kangshan and Sha Lu; total site area 109.3 ha (270 acres); workforce was 4,400 in mid-2001.

From 1 July 1996 was officially transferred under Ministry of Economic Affairs with title as above; completed production of IDF Ching-Kuo fighter in January 2000. AIDC is a participant (with 5 per cent share) in the Sikorsky S-92 Helibus (see US section), responsible for the cockpit structure, and (from 1999) is sole-source supplier of crew and passenger doors for the S-76. Other subcontract work includes wing design of the Ibis Aerospace Ae 270 (see International section); Boeing 717-200 (tail unit; first example delivered mid-December 1997); Dassault Falcon 900 and 2000 (rudders); Eurocopter/CATIC/ST Aero EC 120 Colibri (rear fuselage components); Alenia C-27J Spartan (tail unit); and Bombardier Continental (rear fuselage and tail unit). Assembled 13 Bell OH-58Ds for Republic of China Army (first one delivered 16 November 1999; last in third quarter 2001). It also participates in such programmes as the Lockheed Martin F-16 combat aircraft, and the Honeywell TFE1042 and Rolls-Royce 572K gas-turbine engines. However, despite this range of work, AIDC revenue had reportedly shrunk from about US$750 million in 1999 to less than US$300 million in 2001, and additional subcontract work was being actively sought in 2002.

As a newly government-owned commercial entity, AIDC is dedicated to research and development for Taiwan's aerospace industry (military and commercial aircraft). Full privatisation, originally planned for fourth quarter of 1999, has been repeatedly postponed following a mid-2000 company restructuring based on three core activities: defence systems and technologies, aerostructures and engines. In 2002, Taiwan government set a 'must keep' deadline of December 2003, but by October 2002 an 'en bloc' sale appearing likely to be replaced by proposals to sell off divisions individually.

UPDATED

AIDC F-CK-1 CHING-KUO
TYPE: Air superiority fighter.
PROGRAMME: Final two aircraft (88-8134 and 88-8135) delivered to RoCAF on 14 January 2000; second wing (1st TFW) declared operational July 2000; detailed description in 2000-01 and earlier editions. An update is under development for existing aircraft and two further versions (tentative designations IDF 'C' and 'D') have been proposed to permit production to be restarted. Latter reportedly initiated by AIDC with initial US$30 million, but further US$200 million said to be required to continue programme to flight test stage; this not yet realised. Meanwhile, in 2002 company began flight testing an F-5E Tiger II, loaned by RoCAF, with an upgraded avionics suite (APG-67 radar, MFDs, HUDs and 1553 databus) having high level of commonality with that proposed for IDF.
CURRENT VERSIONS: **F-CK-1A:** Single-seat fighter. Three prototypes; 103 production. In 1999 F-CK-1A 1417 was undergoing modification by AIDC as prototype of mid-life update, believed to include upgraded GD-53 radar and GPS antennas ahead of windscreen.
F-CK-1B: Two-seat operational trainer. One prototype; 28 production.
IDF 'C': Development programme, with initial funding in mid-2000, for improved single-seater with same upgrades as D version, plus new weapons; projected service-entry date 2010.
IDF 'D': A lead-in fighter trainer version of the Ching-Kuo, based on the two-seater minus its ECM and internal gun, was proposed to the RoCAF in 1995. This engineering modification programme, known as Derivative IDF, in preliminary design stage by late 1999; other changes would include simplified avionics and additional 771 kg (1,700 lb) of internal fuel. AIDC still seeking an international partner to join a next-generation trainer programme and market this IDF derivative globally.

UPDATED

MODUS

MODUS VERTICRAFT INC
26, 41st Road, PO Box 46-155, Taichung 407
Tel: (+886 939) 92 87 19
Fax: (+886 4) 23 55 10 05
e-mail: modus@verticraft.com
Web: http://www.verticraft.com
PRESIDENT: Charles Lin

Verticraft designs still awaiting venture capital in 2002. Company also marketing kits for Dominator two-seat autogyro.

UPDATED

MODUS VERTICRAFT JET DEFENDER
TYPE: New-concept rotorcraft.
PROGRAMME: Originally Military Scout variant (see 2002-03 *Jane's*) now totally redesigned for more commonality with Verticraft and for variety of military roles. Still at project stage in 2002.
DESIGN FEATURES: See accompanying drawing and main Verticraft entry.
LANDING GEAR: Retractable tricycle type.
Following data are provisional:
WEIGHTS AND LOADINGS:

Max T-O weight	1,814-2,721 kg (4,000-6,000 lb)

PERFORMANCE (estimated):

Max cruising speed	500 kt (926 km/h; 575 mph)
Service ceiling	10,670 m (35,000 ft)
Range at cruising speed	up to 1,500 n miles (2,778 km; 1,726 miles)

UPDATED

MODUS VERTICRAFT
TYPE: New-concept rotorcraft.
PROGRAMME: Revealed at HeliExpo 99 air show in February 1999; wind-tunnel-tested at Cheng Kung University; plans then were to flight-test quarter-scale remotely controlled prototype by mid-year; prototypes said to be under construction, but no evidence of this received by October 2002. At that time, turboshaft version (see illustration in 2002-03 edition) appeared to have been shelved and name Verticraft applied instead to jet-engined version previously known as VertiJet. Conventional fin and rudder now added to original design.
CURRENT VERSIONS: **Verticraft:** Proof-of-concept version; twin turbojets pod-mounted on sides of rear fuselage for forward propulsion.
DESIGN FEATURES: Hybrid design, patented in Taiwan and USA. Instead of conventional rotors, a pair of contrarotating discs is pylon-mounted above fuselage, each with a series of rigid bladelets around its rim. Disc-type rotor is said to generate more lift than a conventional rotor of same diameter, enabling Verticraft to employ a smaller diameter than traditional rotorcraft and enhancing manoeuvrability. Rotor disc is employed in conjunction with direct propulsion to achieve speeds at which the disc provides sufficient lift to permit disengagement of rotors; bladelets are then retracted into discs, discs are stopped, and because of their close spacing they then act as a single supercritical aerofoil. In this configuration, aircraft is said to be capable of approaching supersonic speeds. Design permits autorotational landing in the event of loss of power.
FLYING CONTROLS: Collective pitch of bladelets is preset; proximity of opposing bladelets to each other minimises separation between rotor discs, enabling dynamic platform to be gimballed and eliminating need for cyclic pitch controls. Vertical movement is accomplished by varying rotational speed of discs; control for hover and horizontal flight by CG transfer and varying angle of attack.
Following data apply to POC prototype:
WEIGHTS AND LOADINGS:

Max payload	more than 45.4 kg (100 lb)

PERFORMANCE (estimated):

Max level speed	up to 500 kt (926 km/h; 575 mph)
Range	up to 869 n miles (1,609 km; 1,000 miles)

UPDATED

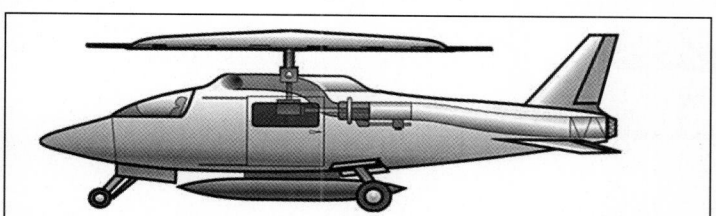
General appearance of the projected Jet Defender *NEW*/0536704

Model of the Modus Verticraft (*Jane's/James Goulding*) *NEW*/0536702

TURKEY

ALP

ALP HAVACILIK (Alp Aviation)
Organize Sanayi Bölgesi, 8. Cadde, Eskişehir
Tel: (+90 222) 236 13 00 and 236 12 61
Fax: (+90 222) 236 12 85
e-mail: alphavacilik@alp.com.tr
Web: http://www.alp.com.tr
MANAGING DIRECTOR: Ahmet Aytekin

Formed May 1999 as joint venture company by Sikorsky Aircraft Turkey Inc and The Alpata Group (50 per cent each) to manufacture high-technology, precision-machined aerospace and defence components and assemblies for global market in modern 6,500 m² (70,000 sq ft) facility. Company has 16 CNC machine tools and early 2001 workforce of 80. Is exporting helicopter components to Sikorsky USA under two contracts awarded in 1998 with total potential value of more than US$60 million. Also a supplier to TAI (which see) and TUSAŞ Engine Industries.

VERIFIED

TAI

TURKISH AEROSPACE INDUSTRIES (TUSAS Havacilik ve Uzay Sanayii A S)
PO Box 18, TR-06692 Kavaklidere, Ankara
Tel: (+90 312) 811 18 00
Fax: (+90 312) 811 14 25 and 811 14 08
e-mail: bcelik@tai.com.tr
Web: http://www.tai.com.tr
GENERAL MANAGER: Kaya Ergenç
GENERAL CO-ORDINATOR: Lt Gen (Ret'd) Vural Avar
EXECUTIVE DIRECTORS:
 Brig Gen (Ret'd) Enver Dayanir (Programmes)
 Ercan Turken (Production and Materials)
 Aydin Oniz (Business Development and Management)
 Fatih Tezok (Design and Engineering)
MARKETING AND SALES MANAGER: Yilmaz Güldoğan
PUBLIC RELATIONS CO-ORDINATOR: Nursel Köran
PUBLIC RELATIONS SPECIALIST: Bican Çelik

Turkish Aerospace Industries (TAI), established on 15 May 1984, is a majority-owned Turkish company made up of Turkish (51 per cent) and US (49 per cent) partners. Shareholders are Turkish Aircraft Industries Inc (49 per cent), Turkish Armed Forces Strengthening Foundation (1.9 per cent), Turkish Air League (0.1 per cent), Lockheed Martin of Turkey Inc (42 per cent) and General Electric International Inc (7 per cent). Workforce in mid-2001 was 2,000.

TAI's aircraft facilities, located in Ankara, occupy a 23 ha (56.8 acre) site, including more than 160,000 m² (1.72 million sq ft) of covered space. They are furnished with high-technology machinery and equipment that provide extensive capabilities ranging from parts manufacturing to aircraft assembly, flight test and delivery from aerodrome and works at Akinci. Quality control meets world standards including NATO AQAP-110, ISO-9001, MIL-Q-9858A and Boeing D6-82479.

With experience in aircraft and aerostructures manufacturing business, TAI is a supplier for Lockheed Martin; MD Helicopters (MD 902 fuselages); Boeing (737 wingtips and flight deck panels); Airbus (A319/320/321 fuselage panels); Eurocopter (rear doors and engine cowlings for EC 135); Sikorsky (S-70A tail booms, tail rotor pylons and tailplanes, and tailplanes for S-76); EADS CASA; AgustaWestland; Sonaca; Northrop Grumman; and others.

TAI's experience includes co-production of 278 F-16 fighters (see 2000-01 *Jane's*), CN-235 light transports (59), SF-260D trainers (34) and AS 532 Cougar (28), as well as design and development of UAVs, target drones and the ZIU agricultural aircraft. The company is also prime contractor for the Turkish armed forces attack helicopter programme.

As a full member of the Airbus Military Company, TAI is engaged in the development of the A400M with major European companies. In 2002 it became the seventh international participant in the Lockheed Martin JSF programme.

TAI's core business also includes modernisation, modification and systems integration programmes and after-sales support. Major programmes of this kind are structural modification and EW retrofit of Turkish Air Force F-16s; modification of S-2E Tracker maritime patrol aircraft into firefighting aircraft (new avionics, Honeywell TPE331 turboprops and a 4,500 kg; 9,921 lb capacity water tank); CN-235 modification for Open Skies; CN-235 and Black Hawk modifications for Special Forces; modification of CN-235 platforms for MPA/MSA missions for the Turkish Navy and Coast Guard; Cougar modification and modernisation; and 'glass cockpit' retrofit of S-70 helicopters.

UPDATED

TAI (AIRTECH) CN-235M
TYPE: Twin-turboprop transport.
PROGRAMME: Under a contract with EADS CASA of Spain, TAI produced 50 CN-235 transports for the Turkish Air Force between 1991 and 1998, manufacturing 92 per cent of the total airframe, including 20 per cent made of composites. See earlier *Jane's* for further details. Production is continuing with six as surveillance aircraft for the Turkish Navy and three for maritime patrol with the Turkish Coast Guard, ordered on 23 September 1998 and delivered from 23 December 2001 onwards. Airframe percentage in these nine aircraft is increased to 95. Seven deliveries (Turkish Navy five, Coast Guard two) had been made by 1 November 2002.

UPDATED

TAI (BELL) AH-1Z KING COBRA
TYPE: Attack helicopter.
PROGRAMME: Turkish Army ATAK programme. Bids from Boeing (Apache Longbow), AgustaWestland (A 129 International) and Eurocopter (Tiger); Bell AH-1Z (King Cobra) and Kamov/IAI (Ka-50-2 Erdoğan); AH-1Z (see Bell entry in US section) selected as preferred option in July 2002, with reserve choice of Ka-50-2. Contract negotiations with Bell protracted due to US insistence on retention of US (instead of proposed Turkish indigenous) mission computer; supposedly resolved in second quarter 2002 by agreement for co-production of US equipment, with contract signature expected during third quarter. However, in early September 2002, Turkish defence procurement organisation said to be re-evaluating Kamov Ka-50-2 in wake of further delays over access to source codes and performance guarantees. If AH-1Z is proceeded with, TAI to be King Cobra prime contractor, with Bell Helicopter Textron as primary subcontractor.
CUSTOMERS: Turkish Army requirement for 145, to be acquired in batches of 50, 50 and 45; delivery schedule November 2003 to January 2011.
COSTS: Initial 50-aircraft purchase quoted as US$1.5 billion (2000); estimated US$4 billion for full 145.
ARMAMENT: ATM selection (reportedly between Brimstone, Hellfire 2 and Rafael NT-D) still awaited in mid-2002.

UPDATED

TAI (EUROCOPTER) COUGAR Mk I
TYPE: Multirole medium helicopter.
PROGRAMME: Within the framework of the Eurotai consortium, TAI is co-producing 28 of 30 Cougar Mk I helicopters (10 for the Turkish Army and 20 for the Turkish Air Force), under a 1997 contract (Phenix II programme) valued at US$434 million. These comprise 24 AS 532ULs (Army six utility, four SAR; Air Force 14 SAR) and six AS 532ALs (Air Force, combat SAR). The first two helicopters were search and rescue aircraft, manufactured in France by Eurocopter and delivered in April and May 2000; the TAI programme covers fabrication, assembly, systems integration and flight test. First Turkish-assembled Cougar (an AS 532UL) was handed over to the Army on 31 May 2000, at which time 11 others were under construction; deliveries totalled 25 (Army six, Air Force 19) by 1 November 2002 and were due to be completed in February 2003.

UPDATED

TAI ZIU
TYPE: Agricultural sprayer.
PROGRAMME: Name is acronym of *Zirai Ilaçlama Uçaği:* agricultural aircraft. Turkish government preliminary design specification issued 1 April 1997; conceptual and preliminary design phases completed 1 July 1997; programme go-ahead 28 November 1997; TAI's first indigenous manned aircraft design. Public debut September/October 1999; maiden flight achieved on 26 June 2000.
 Planned second prototype apparently not completed and production not now intended. See 2002-03 edition for description and illustrations.

UPDATED

UKRAINE

AEROKOPTER

AEROKOPTER OOO (Aerocopter Ltd)
ulitsa Lishchevinkov 27, 36014 Poltava
Tel: (+38 0532) 66 98 58 and 66 87 60
Fax: (+38 0532) 66 07 10
e-mail: slcopter@kot.poltava.ua

Aerokopter was established on 14 December 1999 to develop light rotorcraft. It was briefly associated with what, on 3 May 2000, became the separate Aviaimpex design bureau, described later in this section.

NEW ENTRY

AEROKOPTER ZA-6 SAN'KA
TYPE: Two-seat helicopter.
PROGRAMME: First flown in 1999.
CURRENT VERSIONS: **ZA-6:** As described.
 AK-1: Development; described separately.
CUSTOMERS: Operators include Ukraine AeroClub.
DESIGN FEATURES: Conventional pod-and-boom helicopter with five-blade main rotor and two-blade tail rotor mounted on port side. High agility. Power plant, fuel tanks and gear box mounted externally, to rear of crew pod.

Aerokopter ZA-6 San'ka operating with the Ukraine Aero Club *(Aeroklub Ukrainy)* NEW/0527148

FLYING CONTROLS: Manual.
STRUCTURE: Metal and composites.
LANDING GEAR: Twin skids.
POWER PLANT: One 119 kW (160 hp) Subaru EJ22 flat-four. Fuel capacity 80 litres (21.1 US gallons; 17.6 Imp gallons).
ACCOMMODATION: Two persons, side by side.
DIMENSIONS, EXTERNAL:
Main rotor diameter	6.00 m (19 ft 8¼ in)
Main rotor blade chord	0.15 m (6 in)
Tail rotor diameter	1.24 m (4 ft 0¾ in)
Tail rotor blade chord	0.115 m (4½ in)

AREAS:
Main rotor disc	28.27 m² (304.3 sq ft)
Tail rotor disc	1.21 m² (13.00 sq ft)

WEIGHTS AND LOADINGS:
Weight empty	400 kg (882 lb)
Max T-O weight	750 kg (1,653 lb)
Max disc loading	26.5 kg/m² (5.43 lb/sq ft)
Max power loading	6.29 kg/kW (10.33 lb/hp)

PERFORMANCE:
Max level speed	103 kt (190 km/h; 118 mph)
Max cruising speed	76 kt (140 km/h; 87 mph)
Econ cruising speed	49 kt (90 km/h; 56 mph)
Max rate of climb at S/L:	
at optimal forward speed	642 m (2,106 ft)/min
657 rpm main rotor speed	444 m (1,457 ft)/min
vertical ascent	282 m (925 ft)/min
Service ceiling	2,100 m (6,880 ft)
Range with max fuel	270 n miles (500 km; 310 miles)

NEW ENTRY

Aerokopter AK-1-5 five-blade prototype (*Yefim Gordon*)   NEW/0527152

AEROKOPTER AK-1

TYPE: Two-seat helicopter.
PROGRAMME: Two-stage development of ZA-6, initially involving one of the latter (registered GL-0478, shown at Manufacturing & Security Exhibition, Kiev, September 2002), designated AK-1-5 (five-blade main rotor). At the same event, the uncompleted AK-1-3 prototype was shown with the definitive three-blade rotor, starboard side tail rotor and twin vertical fins mounted on horizontal tail surfaces. Marketing agreements are in place with agents in Australia, France, Italy and Spain.

NEW ENTRY

Prototype AK-1-3, showing three-blade main rotor and starboard side tail rotor (*Yefim Gordon*)   NEW/0527149

AEROPRAKT

AEROPRAKT OOO (Aeroprakt JSC)

a/ya 112, 03148 Kiev
Tel: (+38 044) 457 92 93
Fax: (+38 044) 457 91 59
e-mail: air@prakt.kiev.ua
Web: http://www.aeroprakt.kiev.ua
MANAGER: Oleg V Litovshenko
CHIEF DESIGNER: Yuri V Yakovlev

US AGENT:

Spectrum Aircraft Corporation
PO Box 1381, Sebring, Florida 33872
Tel: (+1 941) 314 97 88
Fax: (+1 941) 314 02 85
e-mail: jhunter@strato.net
Web: http://www.spectrumaircraft.com

Aeroprakt was formed in Kiev in 1986, after Yuri Yakovlev, who had designed the T-8 lightplane for the Aeropract organisation of Samara (which see in Russian section) was invited to work at Antonov Design Bureau. Yakovlev then produced the A-20 Chervonets and joined with fellow engineer Oleg Litovshenko in an initially part-time venture. Established in present, commercial form in 1991. Retaining the 'k for Kiev' in its name, the company now employs only even-numbered designations for its aircraft, the latest to fly being the A-28. Company has agencies in Australia, Germany, Italy, Latvia, Russia, UAE, UK and USA. Design and production personnel total 40; factory space 2,000 m² (21,525 sq ft), although some production subcontracted to Aviant and Antonov.

UPDATED

AEROPRAKT-20

US marketing name: Spectrum SA-20 Vista
TYPE: Tandem-seat ultralight/kitbuilt.
PROGRAMME: Design started September 1990; construction of prototype began November 1990; first flight 5 August 1991; construction of first production aircraft started

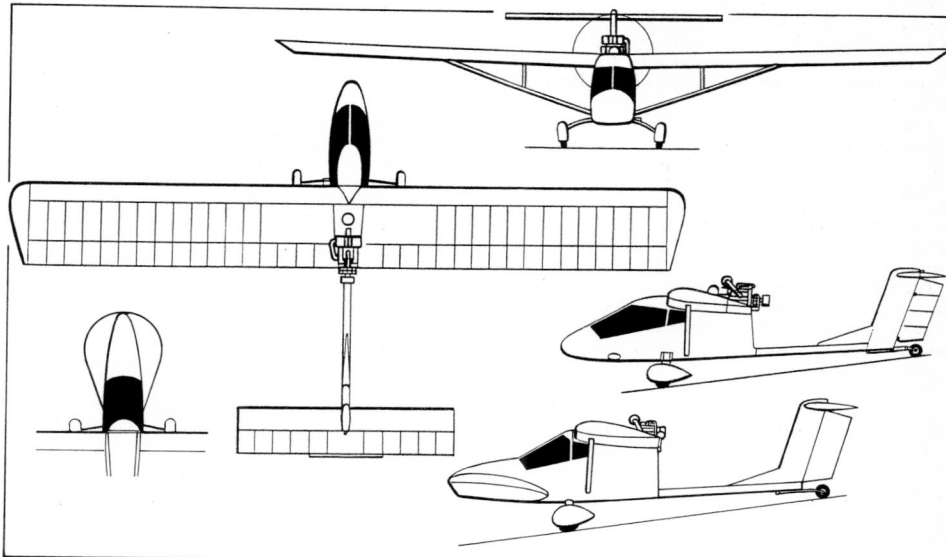

Aeroprakt-20, with additional side and scrap plan views of the radiometer-equipped A-20S (*Jane's/James Goulding*)

September 1991, first flown 15 August 1993; prototype appeared abroad for first time in Hodkovice, Czech Republic, at European Microlight Championship, August 1993, gaining ninth place; fourth production A-20 won third place at World Microlight Championship, Poznan, Poland, August 1994; seventh production A-20 gained second place at EMC '95, Little Rissington, UK.
CURRENT VERSIONS: **A-20 Vista:** Basic aircraft, *as described in detail.*
　A-20R912: Sky Cruiser: Described separately.
　A-20S: Variant under development by 1997, with large

multichannel radiometer built into its nose for a wide range of ecological survey work. No known manufacture.
　A-20SKh: Agricultural (*selbskokhozyaistvennyi:* agricultural) with 59.6 kW (79.9 hp) Rotax 912 UL, three-blade ground-adjustable pitch Ivoprop propeller, sweptback (1° 30′) wing, two chemical tanks, each 90 litres (23.8 US gallons; 19.8 Imp gallons) scabbed to fuselage sides, and streamlined spraybar with four or six atomisers. Prototype construction began July 1997; first flight September 1997; no subsequent production reported. A fuller description and three-view will be found in the 2001-02 edition.

US registered Spectrum Aircraft SA-20 Vista, otherwise known as Aeroprakt-20R912 (*Jane's/Paul Jackson*)　　*NEW*/0525065

A-20 Vista STOL: With 47.8 kW (64.1 hp) Rotax 582 UL; cruising speed 76 kt (140 km/h; 87 mph).

A-20 Vista SS: Rotax 582 and wing span of 10.20 m (33 ft 5½ in); cruising speed 82 kt (152 km/h; 94 mph).

A-30 Vista Speedster: Prototype under test at Sebring, Florida, by early 2002.

CUSTOMERS: First customer delivery August 1993; exports to Czech Republic, Germany, Hungary, Jordan, Poland, Russia, Singapore and United Arab Emirates; last-mentioned acquired seven for Umm-al Qaiwain Flying Club. Total of 29 A-20s of all types produced by mid-2000 (latest information supplied).

COSTS: US$35,000 complete (2000).

DESIGN FEATURES: High-wing, T-tail configuration, optimised for rapid disassembly for transport by trailer. Tail surfaces mounted on aluminium tube which detaches at joint with the nacelle.

Unswept wing; P-IIIa-15 wing section, thickness/chord ratio 15 per cent; chamfered tips; dihedral 1° 30′; incidence at roots 3° 30′, twist 3°.

FLYING CONTROLS: Manual. Single-piece, rod-operated flaperons extend full length of wing trailing-edge; no tabs. Single-piece elevator with protruding, in-flight adjustable tab. Full-height rudder with no tab.

STRUCTURE: Riveted aluminium wing structure; leading-edge D box closed by I-section main spar; stamped wing ribs; bent sheet rear false spar; partially fabric covered. Control surfaces similar to wings. All-composites honeycomb sandwich fuselage pod.

LANDING GEAR: Mainwheels, with speed fairings, on single leaf-spring; steerable tailwheel linked to rudder.

POWER PLANT: One 37.0 kW (49.6 hp) Rotax 503 UL-2V two-stroke piston engine driving a Junkers-Profly (Czech) three-blade ground-adjustable pitch pusher propeller via 3:1 reduction gearbox. Rotax 462, 582 (47.8 kW; 64.1 hp), 618 and 912 engines optional. Fuelling point on starboard side of nacelle. Standard capacity 38 litres (10.0 US gallons; 8.4 Imp gallons); optional 90 litres (23.8 US gallons; 19.8 Imp gallons).

EQUIPMENT: Ballistic recovery parachute.

DIMENSIONS, EXTERNAL:
Wing span	11.34 m (37 ft 2½ in)
Length overall	6.72 m (22 ft 0½ in)
Height: cabin roof	1.74 m (5 ft 8½ in)
propeller turning	2.20 m (7 ft 2½ in)

AREAS:
Wings, gross	15.70 m² (169.0 sq ft)

WEIGHTS AND LOADINGS:
Weight empty	218 kg (481 lb)
Max T-O weight	455 kg (1,003 lb)

PERFORMANCE (Rotax 582 engine):
Max level speed	86 kt (160 km/h; 99 mph)
Stalling speed, flaperons down	26 kt (48 km/h; 30 mph)
Max rate of climb at S/L	300 m (984 ft)/min
T-O and landing run	80 m (265 ft)
Max range	216 n miles (400 km; 248 miles)
Endurance	4 h 30 min
g limits	+4/−2

UPDATED

AEROPRAKT-20R912 SKY CRUISER
US marketing name: Vista Cruiser

TYPE: Tandem-seat ultralight/kitbuilt.

PROGRAMME: See Aeroprakt-20; designation indicates increased engine power provided by flat-four Rotax 912.

Alternatively designated **Aeroprakt-20M.** Construction of prototype Sky Cruiser began February 1997; first flight 11 May 1997. Gained second place in 1st World Air Games, Turkey, September 1997 and second place in 1998 World Microlight Cup, Hungary.

CUSTOMERS: Five aircraft ordered and built by early 1999 (latest data); production continues.

DESIGN FEATURES: Development of A-20 with increased power, shorter wing and tailplane spans, engine cowling, constant-speed propeller, fully balanced control surfaces, smaller wheels and more streamlined wing struts and landing gear. Sweepback 1° 30′.

Description of A-20 applies generally, except as below.

POWER PLANT: One 73.5 kW (98.6 hp) Rotax 912 ULS flat-four driving a three-blade Ivoprop constant-speed propeller. Standard fuel capacity 90 litres (23.8 US gallons; 19.8 Imp gallons); optional capacity 128 litres (33.8 US gallons; 28.8 Imp gallons).

DIMENSIONS, EXTERNAL:
Wing span	10.17 m (33 ft 4½ in)
Height: cabin roof	1.68 m (5 ft 6¼ in)
propeller turning	2.33 m (7 ft 7¾ in)

AREAS:
Wings, gross	14.00 m² (150.7 sq ft)

WEIGHTS AND LOADINGS:
Weight empty	262 kg (577 lb)
Max T-O weight	455 kg (1,003 lb)

PERFORMANCE:
Max level speed	113 kt (210 km/h; 130 mph)
Cruising speed	105 kt (195 km/h; 121 mph)
Stalling speed, power off, flaps down	29 kt (53 km/h; 33 mph)
Max rate of climb at S/L	360 m (1,181 ft)/min
T-O and landing run	80 m (265 ft)
Range with normal tankage	647 n miles (1,200 km; 745 miles)
Endurance	5 h 24 min

UPDATED

AEROPRAKT-22
French marketing name: Vision
UK marketing name: Foxbat
US marketing name: Valor

TYPE: Side-by-side ultralight/kitbuilt.

PROGRAMME: Design started February 1990, construction of prototype began September 1994; first flight 21 October 1996; certified to German BFU-95; production began in early 1999.

CURRENT VERSIONS: **Aeroprakt-22 Shark:** Original version, with 59.6 kW (70.9 hp) Rotax 912 UL engine. Data in 1999-2000 and previous *Jane's.*

Following versions have uprated engine.

FUL A22: Marketed from 1999 by FUL of Damme, Germany. Plans for floatplane and glider tug versions; alternative engine under consideration.

Foxbat: UK version, marketed from 2000 by Small Light Aeroplane Company Ltd at Otherton, Staffordshire (www.foxbat.co.uk). Prototype, G-FBAT (16th airframe) first flown after kit assembly in UK 12 August 2000. BCARS certification in early 2002; several built under auspices of PFA.

Valor: US version.

Vision: Marketed in France by Aerotrophy of Montagu.

Description applies to Valor.

CUSTOMERS: More than 20 A-22s built by early 2002, including sales to USA and Latvia, including at least one for Umm-al Qaiwain Flying Club.

COSTS: Foxbat kit £14,375, plus engine £7,300, excluding VAT (2000).

DESIGN FEATURES: Constant-chord wings and horizontal tail surfaces. Wings swept forward 2° 30′; P-IIIa-15 wing section, thickness/chord ratio 15 per cent; chamfered tips; dihedral 1° 30′; incidence at root 4°; twist 2° 30′.

FLYING CONTROLS: Manual, by pushrods and cables. Full-span slotted flaperons with trim tab to starboard; single-piece elevator with tab; and sweptback rudder with ground-adjustable tab.

Aeroprakt-22 two-seat ultralight, wearing '22F designation following undisclosed modification (*Jane's/Paul Jackson*)　　*NEW*/0525066

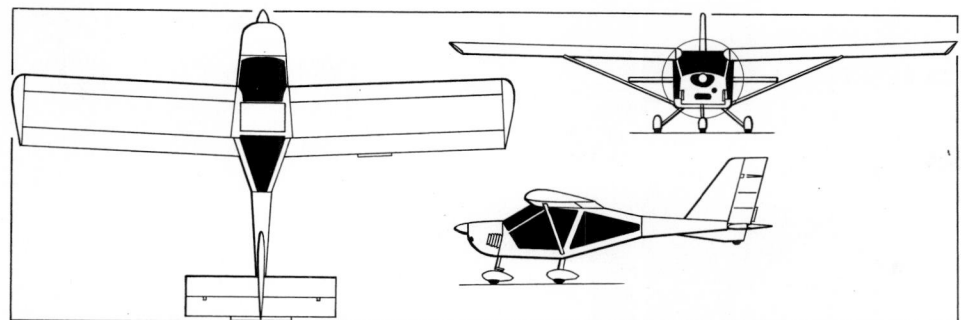

Aeroprakt-22 ultralight (*Jane's/James Goulding*)   0054615

STRUCTURE: Riveted aluminium wing structure; leading-edge D box closed by I-section main spar; stamped wing ribs; bent sheet rear false spar; Ceconite covering on wings (except metal leading-edge), rudder and elevator. Fin and tailplane similar to wings. Aluminium fuselage with profiled sheet and fluted skin, stamped bulkheads, steel and aluminium tubing. Extensive glazing. Glass fibre engine cowling, wheel spats, wing fillets and fin-tip.

LANDING GEAR: Non-retractable tricycle type. Cantilever composites spring mainwheel legs. Hydraulic mainwheel brakes. Steerable nosewheel; small tailwheel protects ventral strake from nose-high landings.

POWER PLANT: One 59.6 kW (79.9 hp) Rotax 912 UL or 73.5 kW (98.6 hp) Rotax 912 ULS flat-four driving an Aeroprakt three-blade ground-adjustable pitch propeller or (Foxbat) Newton two-blade. Fuel tank in each wingroot; combined capacity 90 litres (23.8 US gallons; 19.8 Imp gallons). Optional capacity 135 litres (35.6 US gallons; 29.7 Imp gallons).

ACCOMMODATION: Two persons, side by side. Large upward-hinged window/door on each side.

EQUIPMENT: Ballistic parachute.

Note that data vary slightly between German, UK and US variants.

DIMENSIONS, EXTERNAL:
Wing span	10.10 m (33 ft 1¾ in)
Length overall	6.30 m (20 ft 8 in)
Height overall	2.40 m (7 ft 10½ in)
Tailplane span	3.00 m (9 ft 10 in)
Propeller diameter	1.68 m (5 ft 6¼ in)

DIMENSIONS, INTERNAL:
Cabin: Length	1.60 m (5 ft 3 in)
Max width	1.30 m (4 ft 3¼ in)
Height	1.10 m (3 ft 7¼ in)

AREAS:
Wings, gross	13.70 m² (147.5 sq ft)

WEIGHTS AND LOADINGS:
Weight empty, equipped	260 kg (573 lb)
Max T-O weight	450 kg (992 lb)

PERFORMANCE (Rotax 912 UL):
Max level speed	92 kt (170 km/h; 106 mph)
Cruising speed	86 kt (160 km/h; 99 mph)
Stalling speed	30 kt (55 km/h; 35 mph)
Max rate of climb at S/L	300 m (984 ft)/min
T-O and landing run	90 m (295 ft)
Range with max fuel	594 n miles (1,100 km; 683 miles)
g limits	+4/−2

UPDATED

AEROPRAKT-24
US marketing name: Viking

TYPE: Three-seat amphibian kitbuilt.

PROGRAMME: Design under way by 1996. At least two flying by early 2002, when stated to be in series production.

CURRENT VERSIONS: **Viking:** *As described*

Twin Viking: Twin-engined version. Two Rotax 582s, each 47.8 kW (64.1 hp); empty weight 431 kg (950 lb); MTOW 726 kg (1,600 lb); max level speed 87 kt (161 km/h; 100 mph); T-O run 138 m (450 ft); stalling speed 33 kt (62 km/h; 38 mph).

DESIGN FEATURES: Braced high-wing monoplane, with several detail changes from drawing which last appeared in 2001-02 *Jane's*, and weight in Experimental category; single bracing strut and jury strut each side; flying-boat hull with single step; sweptback vertical tail with unswept horizontal surfaces at tip.

FLYING CONTROLS: Manual. Full-span flaperons.

Artist's impression of Aeroprakt Twin Viking 0110573

LANDING GEAR: Tailwheel type, with retractable, but externally stowed, mainwheels. Stabilising float under each outer wing.

POWER PLANT: One 73.5 kW (98.6 hp) Rotax 912 ULS flat four projecting from wing centre-section. Fuel capacity 87 litres (23.0 US gallons; 19.1 Imp gallons).

DIMENSIONS, EXTERNAL:
Wing span	11.12 m (36 ft 5¾ in)
Length overall	7.28 m (23 ft 10½ in)
Height: top of engine cowlings	2.41 m (7 ft 11 in)
propellers turning	3.19 m (10 ft 5½ in)

DIMENSIONS, INTERNAL:
Cabin: Length	2.10 m (6 ft 10¾ in)
Max width	1.15 m (3 ft 9¼ in)
Max height	1.07 m (3 ft 6¼ in)

AREAS:
Wings, gross	15.00 m² (161.5 sq ft)

WEIGHTS AND LOADINGS:
Weight empty	440 kg (970 lb)
Max T-O weight	750 kg (1,653 lb)
Max wing loading	50.0 kg/m² (10.24 lb/sq ft)
Max power loading	10.20 kg/kW (16.76 lb/hp)

PERFORMANCE:
Max level speed	89 kt (165 km/h; 103 mph)
Cruising speed	76 kt (140 km/h; 87 mph)
Stalling speed, flaperons down	38 kt (70 km/h; 43 mph)
Max rate of climb at S/L	240 m (787 ft)/min
T-O and landing run: on land	150 m (492 ft)
Range with max fuel	432 n miles (800 km; 497 miles)
Endurance	4 h 30 min
g limits	+4/−2

UPDATED

AEROPRAKT-26
US marketing name: Vulcan

TYPE: Tandem-seat ultralight twin/kitbuilt.

PROGRAMME: Development of A-20 with two engines; commissioned by Gulf Aviation Technologies; prototype construction began March 1996; first flight 18 November 1997.

CURRENT VERSIONS: **SA-26 Vulcan:** Lower-powered variant; Rotax 503 engines. Marketed in US by Spectrum Aircraft.
Vulcan SS: With Rotax 582 engines.

CUSTOMERS: One prototype built and flown by early 1999. One reported in Umm-al Qaiwain Flying Club, February 1999. More than eight sold in USA.

Description of A-20 applies generally, except as below.

DESIGN FEATURES: Introduces twin-engine safety margins to A-20 design; able to take off on one engine. Sweepback increased to 3°; tailfin height and rudder chord both increased.

POWER PLANT: Two 47.8 kW (64.1 hp) Rotax 582 UL two-cylinder two-stroke engines, each driving a three-blade, ground-adjustable pitch Ivoprop propeller. Alternatively, two 38.3 kW (51.6 hp) Rotax 462 or 34.0 kW (45.6 hp) Rotax 503 twin-piston engines. Fuel capacity 38 litres (10.0 US gallons; 8.4 Imp gallons) in SA-26 or 90 litres (23.8 US gallons; 19.8 Imp gallons) standard in Vulcan SS; 180 litres (47.5 US gallons; 39.6 Imp gallons) optional.

SYSTEMS: Electrical system with 12 V DC, 14 Ah battery for electric starter.

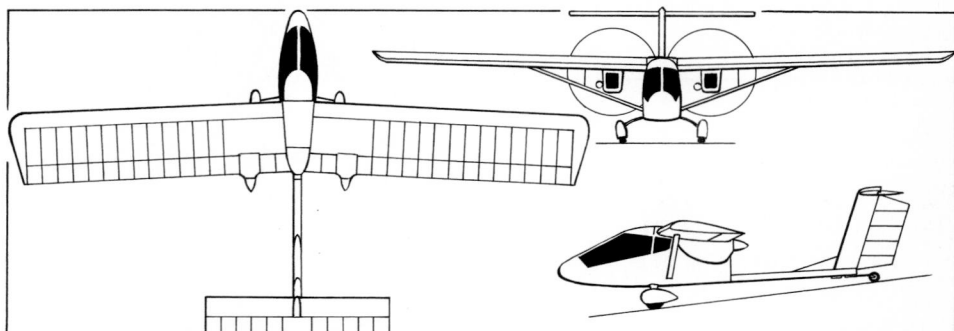

Aeroprakt-26 twin-engined ultralight (*Jane's/James Goulding*)   0051502

A-26 Vulcan SS assembled by a US builder (*Jane's/Paul Jackson*)   NEW/0525068

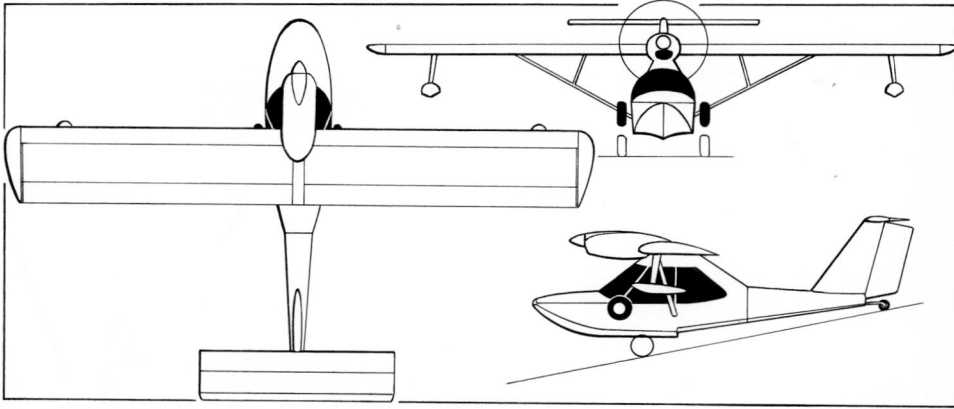

Aeroprakt-24 kitbuilt amphibian (*Jane's/James Goulding*)   NEW/0525067

DIMENSIONS, EXTERNAL:
Wing span	11.34 m (37 ft 2½ in)
Length: fuselage	5.75 m (18 ft 10½ in)
overall	6.72 m (22 ft 0½ in)
Height overall	1.90 m (6 ft 2¾ in)
Propeller diameter	1.70 m (5 ft 7 in)

DIMENSIONS, INTERNAL:
Cabin: Length	2.20 m (7 ft 2½ in)
Max width	0.64 m (2 ft 1¼ in)
Max height	1.20 m (3 ft 11¼ in)

AREAS:
Wings, gross	15.70 m² (169.0 sq ft)

WEIGHTS AND LOADINGS:
Weight empty: Vulcan	300 kg (660 lb)
Vulcan SS	318 kg (701 lb)
Max T-O weight	550 kg (1,212 lb)*
*or local ultralight limit	

PERFORMANCE (Vulcan SS):
Max level speed	97 kt (180 km/h; 112 mph)
Cruising speed: max	70 kt (130 km/h; 81 mph)
econ	49 kt (90 km/h; 56 mph)
Stalling speed, power off:	
flaps up	36 kt (65 km/h; 41 mph)
flaps down	30 kt (55 km/h; 35 mph)
Max rate of climb at S/L	600 m (1,969 ft)/min
T-O and landing run	100 m (330 ft)
Range with max optional fuel	
	334 n miles (619 km; 384 miles)
Endurance	5 h 42 min

UPDATED

Prototype Aeroprakt-28 on the US civil register *(Jane's/Paul Jackson)*   NEW/0525069

AEROPRAKT-28

US marketing name: SA-28 Victor

TYPE: Four-seat kitbuilt twin.

PROGRAMME: Project began in December 1997; first flight October 1999; after almost 100 hours, prototype shipped to Dubai for one year of field tests. Initial eight aircraft were reported to be in production by 2000. Following Dubai trials, prototype refurbished at Kiev and registered to John Hunter of Sebring, Florida, in April 2001. US Marketing by Spectrum Aircraft.

CURRENT VERSIONS: **A-28**: As described.

SA-28: US production version. Several changes as detailed below. Target max level speed of 162 kt (300 km/h; 186 mph) with constant-speed propellers; altered fuselage shape. Optional nosewheel configuration, with and without retraction option.

COSTS: US$120,000 (2002).

DESIGN FEATURES: Low-wing, twin-engined configuration with T tail; sole four-seater in current Aeroprakt range.

Unswept, constant-chord wings of P-IIIa-15 section; thickness/chord ratio 15 per cent; washout 5°; chamfered wingtips. Dihedral 5°; incidence 5°; twist 3°. US production version has NACA 633-618 laminar flow aerofoil.

FLYING CONTROLS: Manual. Conventional ailerons, slotted flaps and rudder; single-piece, mass-balanced elevator. Flight-adjustable trim tabs on fin and rudder. Flap deflections 0, 14, 33 and 40°.

STRUCTURE: Composites cabin with riveted 2024-T3 aluminium monocoque rear fuselage; tailplane, of composites ribs and skin built on metal spar, is reinforced version of A-20 unit; aluminium rudder structure with Stitts Poly-Fiber covering. Aluminium wings with fabric-covered ailerons. Composites engine cowlings and wheel fairings. Aluminium 6061-T256 mainwheel legs.

LANDING GEAR: Tailwheel type; fixed. Mainwheels size 6.00-6; tailwheel 260×85. Speed fairings on each main unit. Citroen motorcar shock-absorbers.

POWER PLANT: Two 59.6 kW (79.9 hp) Rotax 912 UL flat-fours. Fuel capacity 180 litres (47.5 US gallons; 39.6 Imp gallons) in mid-wing tanks. US versions have 73.5 kW (98.6 hp) Rotax 912 ULS engines and total of 91 litres (24.0 US gallons; 20.0 Imp gallons). Aeroprakt three-blade ground-adjustable pitch propellers.

ACCOMMODATION: Up to four persons in side-by-side pairs. Two door/windscreens open forwards on centreline hinges.

DIMENSIONS, EXTERNAL:
Wing span	12.10 m (39 ft 8½ in)
Wing chord, constant	1.40 m (4 ft 7 in)
Wing aspect ratio	8.8
Tailplane span	1.17 m (3 ft 10 in)
Length overall	7.045 m (23 ft 1¼ in)
Height overall	2.33 m (7 ft 7¾ in)
Wheel track	3.60 m (11 ft 9¾ in)
Wheelbase	4.90 m (16 ft 1 in)
Propeller diameter	1.85 m (6 ft 0¾ in)

AREAS:
Wings, gross	16.70 m² (179.8 sq ft)

WEIGHTS AND LOADINGS:
Weight empty	530 kg (1,168 lb)
Max T-O weight	950 kg (2,094 lb)
Max wing loading	56.9 kg/m² (11.65 lb/sq ft)
Max power loading	7.98 kg/kW (13.10 lb/hp)

PERFORMANCE:
Max level speed	130 kt (240 km/h; 149 mph)
Max cruising speed	81 kt (150 km/h; 93 mph)
Stalling speed	41 kt (75 km/h; 47 mph)
Max rate of climb at S/L	300 m (984 ft)/min
T-O run	95 m (315 ft)
Landing run	110 m (360 ft)
Range with max fuel	809 n miles (1,500 km; 932 miles)
Endurance	10 h

UPDATED

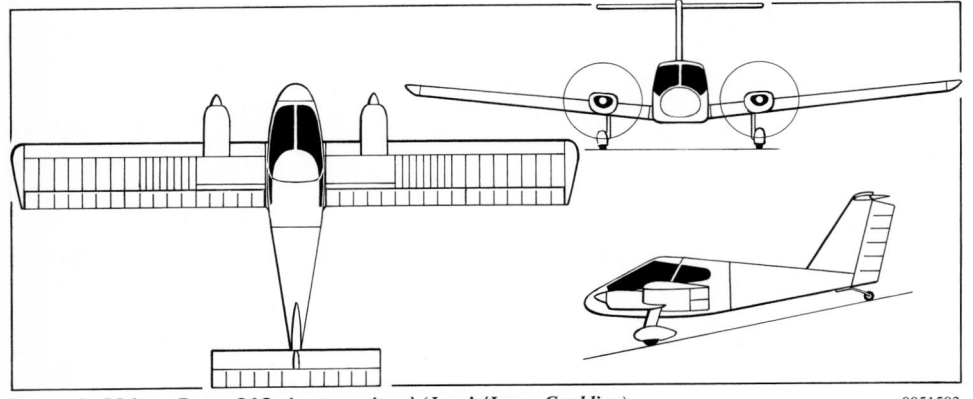

Aeroprakt-28 (two Rotax 912 piston engines) *(Jane's/James Goulding)*   0051503

ANTONOV

AVIATSIONNY NAUCHNO-TEKHNICHESKY KOMPLEKS IMENI O K ANTONOVA (Aviation Scientific-Technical Complex named for O K Antonov)

ulitsa Tupoleva 1, 03062 Kiev
Tel: (+38 044) 442 70 98
Fax: (+38 044) 443 00 05
e-mail: info@antonov.com

GENERAL DESIGNER AND PRESIDENT, MEDIUM TRANSPORT AIRCRAFT INTERNATIONAL CONSORTIUM: Pyotr Balabuyev
GENERAL DIRECTOR, MEDIUM TRANSPORT AIRCRAFT INTERNATIONAL CONSORTIUM: Leonid Terentyev
CHIEF DESIGNER: Anatoly Vovnyanko
DEPUTY CHIEF DESIGNER: Genrich G Ongirsky
PUBLIC RELATIONS OFFICER: Andrey Sovenko

Antonov OKB was founded in 1946 by Oleg Konstantinovich Antonov, who died 4 April 1984, aged 78. More than 22,000 aircraft of over 100 types and versions of Antonov design have been built; more than 1,500 have been exported, to 42 countries. Other production includes trolley-buses, trams and racing bicycles. Antonov received Aviation Register approval on 30 December 1992 to develop civil aircraft. It also operates its own cargo airline with a fleet of one An-225, eight An-124s, one An-22, three An-12s, two An-32Ps, two An-24s and one each An-26 and An-74. Plans for privatisation of the state-owned Antonov have been in preparation since 1997. Company parents the Medium Transport Aircraft International Consortium created by Russia and Ukraine in February 1996 and formally established on 18 May 1999.

Production plants associated with ANTK Antonov include Aviastar, Aviakor, AAK Progress, KiGAZ, Kharkov Air Carrier, Polyot (Omsk) and TAPO.

UPDATED

ANTONOV An-2

NATO reporting name: Colt

TYPE: Utility biplane.

PROGRAMME: Prototype flew as SKh-1 on 31 August 1947. An-2s were built at Kiev-Svyetoshino until mid-1960s. Soviet production comprised about 3,480 An-2s and 506 An-2Ms. Production then transferred to PZL Mielec, Poland, from where some 11,650 delivered from 1960 onwards. Antonov estimates over 15,500 built in former USSR, Poland and China, of which over 4,000 (2,315 in Russia, 2000) remain in service with civil operators and with 22 air arms, including 180 with Russian forces; China acquired licence and has built **Yunshuji-5 (Y-5)** versions from 1957 to date at both Nanchang and Shijiazhuang (see SAMC entry in Chinese section). However, PZL in Poland has stock of semi-complete airframes, from which occasional aircraft are completed.

UPDATED

ANTONOV An-28

See PZL (Polish Aviation Factory) in Polish section.

ANTONOV An-32

NATO reporting name: Cline
Indian Air Force name: Sutlej
TYPE: Twin-turboprop transport.

Current generation of Antonov transports: An-140 (top), An-70 and An-225 *(Jane's/Paul Jackson)*
0121010

Model of Antonov An-32B-300 0121011

Antonov An-38-100 on lease to Malaysia's Layang Layang Aerospace *(Jane's/Paul Jackson)* NEW/0525070

PROGRAMME: No new production of the An-32 has been reported since 1997, although promotion was continuing in mid-2002 and a number of incomplete airframes remains available for completion to order (Aviant quotes 361 production aircraft built, of which only 346 accounted for, implying 15 incomplete). Full description last appeared in 2001-02 *Jane's*.

A new, Westernised, version was first revealed in model form at the 2001 Moscow air show.

CURRENT VERSIONS: **An-32B-300:** Derivative powered by two Rolls-Royce AE 2100D turboprops, each 3,424 kW (4,591 shp), turning Dowty R391 six-blade propellers. Max T-O weight 27,000 kg (59,525 lb); max level speed 286 kt (530 km/h; 329 mph); max payload 7,500 kg (16,525 lb); range 1,474 n miles (2,730 km; 1,696 miles); two-crew flight deck.

UPDATED

ANTONOV An-38

TYPE: Twin-turboprop transport.

PROGRAMME: Requirement for 25-30 seat development of An-28 (see PZL M-28 in Polish section) emerged during 1989 sales tour of India. Development of all-new An-38 approved by Soviet Ministry of Aviation, late 1990. Details announced, and model displayed, at 1991 Paris Air Show; initial batch of six built at production factory, NAPO, Novosibirsk, Russia: one prototype (01001; first flight 23 June 1994, with TPE331 engines), four trials aircraft and one (01002) for static testing at Kiev; certification to AP-25 granted 22 April 1997. In December 1995, Antonov and NAPO formed joint venture company, Siberian Antonov Aircraft, to produce, market and provide after-sales service for the An-38. Indian demonstration tour undertaken in July and August 1997, followed by appearance at Aero India in December 1998 and February 2001.

CURRENT VERSIONS: **An-38-100:** With Honeywell TPE331 engines. First and second (01003; exhibited Moscow 1997) flying aircraft to this standard. Trials of international navigation avionics completed March 2000.

An-38-110: Reduced avionics fit in comparison with -100.

An-38-120: Enhanced avionics fit in comparison with -100; equipment includes VOR/DME, Opal-B voice recorder and SPPZ-2000 ground proximity warner. NAPO-Aviations' aircraft to this standard.

An-38-200: With Omsk MKB 'Mars' TVD-20-03 engines. Third and fourth prototypes were planned to this standard when engine development complete; however, minor problems with Stupino (Aerosila) AV-36M propeller delayed programme, but maiden flight achieved (at NAPO) 11 December 2001. Equipment standard as for An-38-120, but with addition of TCAS-2000 traffic collision avoidance system.

An-38K: Convertible version of An-38-100; large upward-hinging side door at rear on port side; able to carry four LD-3 (KMP-500) or five LD-3K containers (= *konteinernyi*); cargo handling equipment removable for conversion to 30-passenger transport.

Versions with RKBM TVD-1500 engines said to be under construction in 2000, but none had emerged by mid-2002. All versions can be equipped for aerial photography (An-38F: *fotografiya*), survey (An-38GF *geofizichesky*), forest patrol (An-38D: *desantnyi*), VIP transport, ambulance (An-38S: *sanitarnyi*; six stretchers, nine seated, with attendant) and fishery/ice patrol duties (An-38LR: *ledovoi razvedki*). An assault transport, also designated An-38D, capable of carrying 22 paratroops, 26 troops or 3 tonnes of cargo, was revealed to be in the design stage in August 2001; -100 or -200 engine options will be available.

CUSTOMERS: Eight produced by mid-2000: two prototypes (one at Antonov; one at NAPO), one static test airframe and five with airlines (Vostok, three; Alrosa, two); total unchanged at end of 2001. Two Vostok aircraft to Malaysia during 2001 for use by Layang Layang Aerospace for tourist flights, cargo transport and aerial photography.

First three (subsequently increased to eight) An-38-100s ordered for Vostok Airlines and received by mid-1995 for one year of intensive trials before passenger certification. Second firm customer is Chukotavia (10; although initial batch is two); letters of intent from Petropavlovsk-Kamchatsky, Merninsky, Novosibirsky, Ulyanovsky and Nikolaevsk-na-Amur. In 1999-2000, second prototype was being operated by NAPO-Aviatrans, the airline of the NAPO aircraft factory. In 1998, Siberia Airlines was considering purchase of two. Alrosa-Avia of Zhukovsky has ordered five for diamond mining support, of which first in service by early 2000; second followed in July 2000. Indian Air Force interest expressed in initial six to 10; estimated market for 40 with Indian regional airlines (2001), of which 20 covered by letters of intent. Interest in -200 from Vietnam Airlines, which in 2001 signed lease for NAPO's own -120.

NAPO anticipates sales of 170 by 2010.

COSTS: An-38-100 basic price US$4 million (2000); An-38-200 US$2.6 million to US$3.0 million (2000).

DESIGN FEATURES: Developed from PZL Mielec (Antonov) An-28 (see Polish section) to replace An-24s, Let L 410s and Yak-40s. New high-efficiency engines; lengthened passenger cabin; optional weather radar and automatic flight control system; improved sound and vibration insulation; reduced external noise; wheel or ski landing gear; rear cargo door and cargo handling system; able to operate from unpaved runways; operating temperatures from −45 to +45°C, including 'hot and high' conditions. Service life 30,000 hours. Maintenance requirement 4 man-hours/flying hour.

FLYING CONTROLS: Conventional and manual. Single-slotted mass and aerodynamically balanced ailerons (port aileron has trim tab), designed to droop with large, hydraulically actuated, two-segment double-slotted flaps; electrically actuated trim tabs in each elevator have manual back-up; twin rudders each with electrically actuated trim tab; automatic leading edge slats over full span of wing outboard of engines; slab-type spoiler forward of each aileron and each outer flap segment at 75 per cent chord.

LANDING GEAR: Tricycle type; fixed. Mainwheels 610×320-330; nosewheel 600×320-254.

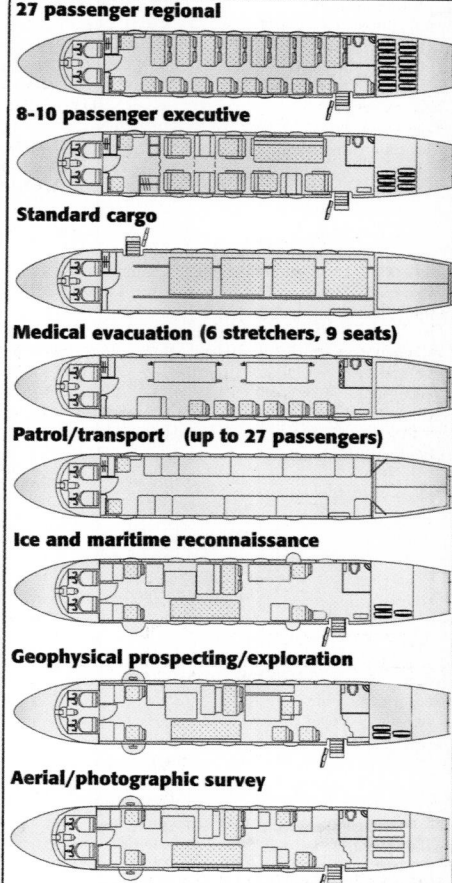

27 passenger regional

8-10 passenger executive

Standard cargo

Medical evacuation (6 stretchers, 9 seats)

Patrol/transport   (up to 27 passengers)

Ice and maritime reconnaissance

Geophysical prospecting/exploration

Aerial/photographic survey

Alternative cabin configurations for An-38 current and projected versions 0052899

POWER PLANT: Two Honeywell TPE331-14GR-801E turboprops, each 1,118 kW (1,500 shp), driving Hartzell HC-B5MA five-blade propellers rotating at 1,522 rpm; or two Omsk MKB 'Mars' TVD-20 turboprops, each 1,029 kW (1,380 shp), driving Aerosila AV-36M quiet reversible-pitch propellers rotating at 1,827 mph.

ACCOMMODATION: Two crew side by side on flight deck; passenger cabin equipped normally with 26 seats, basically three-abreast, with centre aisle; 27 seats at 75 cm (29½ in) pitch optional; ambulance version for six stretchers, eight seated casualties and medical attendant, executive versions with eight to 10 seats and forest surveillance/paradrop version for 26 smoke-jumpers or trainee paratroops (reduced to 22 with full kit) available; seats and baggage compartment can be folded quickly against cabin wall to provide clear space for 2,500 kg (5,510 lb) of freight. Maximum practical cargo dimensions in combi variant are 1.40 m (4 ft 7 in) height and 1.05 m (3 ft 5¼ in) width with seats removed or 0.95 m (3 ft 1½ in) with seats stowed against walls. Door with airstairs on port side, with service door opposite; emergency exit each side. Optional cargo door under upswept rear fuselage slides forward under cabin for direct loading/unloading of freight.

AVIONICS: Russian or Western equipment; latter comprises Bendix/King Silver Crown range; former listed below.
Comms: SO-72 transponder.
Radar: A813Ts weather radar.
Flight: BSFK-1 navigation system; VEM-72PB-3A altimeter and A-037 radar altimeter; twin US-450K airspeed indicators; SAU-28 AFCS; M3 GPS; VMD-94 DME; SPPZ-200 GPWS; TCAS-2000 TCAS; and KURS-93M autoland.

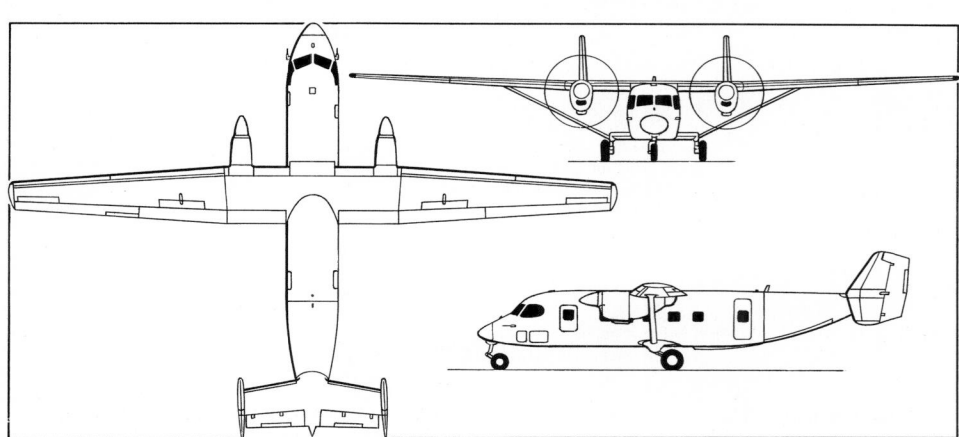

Antonov An-38-100 27-seat transport *(Jane's/Mike Keep)* 0051508

EQUIPMENT: Hand-operated travelling overhead winch in cabin: capacity 500 kg (1,102 lb).

DIMENSIONS, EXTERNAL:

Wing span	22.065 m (72 ft 4¾ in)
Length overall	15.67 m (51 ft 5 in)
Height overall	5.05 m (16 ft 6¾ in)
Span over tailfins	5.14 m (16 ft 10½ in)
Wheel track	3.515 m (11 ft 6½ in)
Wheelbase	6.345 m (20 ft 9¾ in)
Distance between propeller centres	5.58 m (18 ft 3¾ in)
Propeller diameter: Hartzell	2.85 m (9 ft 4 in)
AV-36	2.65 m (8 ft 8¼ in)
Cargo ramp:	
Length	2.40 m (7 ft 10½ in)
Width: at floor	1.40 m (4 ft 7 in)
at rear	1.00 m (3 ft 3¼ in)
Passenger door: Width	0.70 m (2 ft 3½ in)
Height	1.55 m (5 ft 1 in)

DIMENSIONS, INTERNAL:

Cargo hold: Length, excl ramp	7.83 m (25 ft 8¼ in)
Width: max	1.74 m (5 ft 8½ in)
at floor	1.55 m (5 ft 1 in)
Max height	1.70 m (5 ft 7 in)
Volume, incl ramp	24.7 m³ (872 cu ft)

WEIGHTS AND LOADINGS (A: An-38-100, B: An-38-200, K: An-38K):

Weight empty: A	5,300 kg (11,684 lb)
Max payload: A	2,500 kg (5,510 lb)
B	2,800 kg (6,173 lb)
K	3,200 kg (7,055 lb)
Max fuel: A	2,210 kg (4,872 lb)
Max T-O weight: A	9,500 kg (20,943 lb)
B	9,930 kg (21,891 lb)

PERFORMANCE (estimated, with TPE331 engines):

Max level speed	219 kt (405 km/h; 252 mph)
Nominal cruising speed: A, B	
	205 kt (380 km/h; 236 mph)
Nominal max cruising altitude: A, B	4,200 m (13,780 ft)
T-O run: A	350 m (1,150 ft)
K	480 m (1,575 ft)
Landing run: A	270 m (885 ft)
K	440 m (1,445 ft)
Balanced field length: A	895 m (2,940 ft)
B	1,050 m (3,445 ft)
Range at FL100, no reserves:	
with max fuel: A	378 n miles (700 km; 435 miles)
B	421 n miles (780 km; 484 miles)
with max fuel (1,300 kg; 2,866 lb payload)	
A	944 n miles (1,750 km; 1,087 miles)
B	961 n miles (1,780 km; 1,106 miles)

UPDATED

ANTONOV An-70

TYPE: Strategic transport.

PROGRAMME: Development began 1975 to replace some An-12s remaining in air force service from 2002-2003; announced by *Izvestia* 20 December 1988; at 1991 Paris Air Show Antonov OKB reported prototype being assembled at Kiev; funding by Russian (80 per cent) and Ukrainian governments under agreement of 24 June 1993; preliminary details released and model displayed at Moscow Aero Engine and Industry Show April 1992; prototype first flight 16 December 1994 (was also delivery flight to Gostomel test airfield); this aircraft lost during fourth sortie following in-flight collision with chase An-72 on 10 February 1995; second prototype (without nose-mounted instrumentation boom) produced by upgrading of static test airframe (for which replacement under construction in 2000); rolled out 24 December 1996; first flight (UR-NTK) 24 April 1997; international debut at Moscow Air Show, August 1997; handed over to Russian Air Forces' test centre at Akhtubinsk, August 1998. Had flown one-third of planned 780-sortie test programme by January 2000; high AoA trials mid-2000; first stage of State testing completed October 2000, confirming safety in all flight regimes. Crash-landed and fuselage broken into two parts immediately after take-off from Omsk on 27 January 2001 following double engine failure. Airframe transferred to Polet aircraft plant and repaired at cost of US$3 million; reflown 5 June 2001; certification rescheduled to first quarter of 2002; appeared at MAKS '01, Moscow, August 2001.

Bilateral agreement between Russia and Ukraine revised 18 May 1999, and underlined decision taken to order first 10 (and 50 engines) before planned production at Aviant plant, Kiev, from 1999 and Aviacor plant, Samara, from 2000; each plant building initial batch of five, which to enter service by 2003. Russian production specification issued 4 December 1999; Russian government decree of 2000 covers purchase of 164 An-70s, but, at same time, Samara regional government announced that local manufacturers were withdrawing from An-70 project. Ukrainian government resolution guaranteeing An-70 purchase passed 12 October 2000; total of 65 to be obtained by 2018. First five for Ukraine ordered from Aviant on 2 April 2001. However, in September 2001, Polyot was allocated Russian An-70 final assembly, augmented by Novosibirsk's NAPO and Voronezh's VASO. Aviacor no longer involved. Russian share of series production is 72 per cent; Ukraine 28 per cent.

Second prototype Antonov An-70 displaying at Moscow in August 2001 *(Yefim Gordon)*   *NEW*/0525780

Offered as alternative to Airbus A400M FLA; Germany and Ukraine agreed in December 1997 to explore possible industrial collaboration and German government strongly promoted the An-7X as the basis of the FLA, though this was rejected by other FLA partners and the Luftwaffe. Aircraft evaluated by DaimlerChrysler Aerospace, which determined that it could be modified to meet FLA requirement. Aim of wide co-operation between Russian, Ukrainian and West European aircraft industries (in effect bid to meet European FLA requirement) stated in February 1998 in joint declaration by Russian and Ukrainian Presidents. Certification is planned to FAR Pt 25 and equivalents. Antonov obviated effects of intermittent official funding by investing income from its own An-124 charter operations.

Medium Transport Aircraft International Consortium (MTA; also known as Medium-Size Transport Plane consortium or AirTruck) formed February 1996 to co-ordinate development, certification, sales and after-service, comprising four Ukrainian and six Russian companies: ANTK Antonov (designer and component manufacturer), ZMKB Progress (engine designer), Aviant (airframe manufacturer) and Motor Sich (engine manufacturer) from Ukraine; and Aviacor (airframe manufacturer), Ufa (engine manufacturer), Aviapribor (flight control system manufacturer), Elektroavtomatika (avionics), Leninets (airborne monitoring and diagnostic system) and Aerosila (propfan) from Russia. See 'Structure'.

Third aircraft to fly will be An-70T commercial variant, construction of which announced mid-1998; first flight anticipated 2002. In September 2000, Volga-Dnepr airline pledged to fund An-70 production if military contracts not forthcoming. Fourth batch (of seven) D-27 engines, authorised in 2000, includes those for An 70T prototype.

CURRENT VERSIONS: **An-70:** Military STOL transport; proposed production version. Stated to have double the payload of Lockheed Martin C-130J, with similar STOL and rough-field performance, yet only 40 kt (74 km/h; 46 mph) slower than the Boeing C-17A Globemaster III, which carries 15 per cent more payload.

Detailed description applies to baseline An-70.

An-70-100: Proposed military STOL transport, as An-70 but with two-crew cockpit.

An-77: Proposed military STOL transport for export customers; as An-70-100 but with cockpit for two or three crew. Runway length of 1,900 m (6,235 ft) required with 35,000 kg (77,161 lb) payload for 2,051 n mile (3,800 km; 2,361 mile) range.

An-70T: Commercial transport, generally as An-70, with improved runway capability but no requirement for STOL. Two or three crew. To carry 35,000 kg (77,161 lb)

payload 2,051 n miles (3,800 km; 2,361 miles) from 1,900 m (6,235 ft) runway, or 20,000 kg (44,092 lb) for 2,915 n miles (5,400 km; 3,555 miles) from 1,300 m (4,265 ft) runway. Stated to have load-carrying capability of Il-76, but runway requirements of An-74. Promoted as Il-76/An-12 replacement. Certification planned to FAR Pt 25 and equivalents. First fuselage delivered to Samara from Kiev November 1999.

An-70T-100: Development of An-70T with two D-27 propfans, two crew and revised landing gear, to carry 30,000 kg (66,138 lb) payload 540 n miles (1,000 km; 621 miles) from 2,500 m (8,205 ft) runway, or 10,000 kg (22,046 lb) for 1,187 n miles (2,200 km; 1,367 miles) from 1,300 m (4,265 ft) runway. Promoted as An-12 and An-74 replacement

An-70T-200: Powered by two Kuznetsov NK-93 turbofans.

An-70T-300: With two CFM56-5C4 turbofans.

An-70T-400: With four CFM56-5C4 turbofans.

An-70TK: Convertible cargo/passenger transport for 30,000 kg (66,138 lb) of freight or 150 passengers, with seats in removable modules.

An-7X: Designation for provisional variant offered to meet multinational FLA requirement at 40 per cent of anticipated A400M cost. AirTruck GmbH formed 20 May 1999 by eight supporting German companies (Aerodata, ASL Aircraft Services, Autoflug, BGT, R-R Deutschland, ESG, Liebherr and VDO) to promote An-7X in conjunction with MTA consortium.

Antonov-supplied (on 29 January 1999) data evaluated by DaimlerChrysler Aerospace, which assessed technical risk as minor and performance to be compliant; however, specific fuel consumption targets not met, concern expressed over noise, and power plant felt to need Western FADEC. Further recommendations include wiring insulation change to meet Western standards; landing gear modifications; addition of in-flight refuelling and fuel-dumping; completely revised NVG-compatible, two-crew cockpit; permanently installed cargo system; and minor flying control and manufacturing changes.

Assembly of 75 An-7Xs for Luftwaffe would be at Lemwerder, Germany. If rejected by Germany, An-7X programme to continue in readiness for alternative export prospects for non-STOL version.

An-170: Heavy transport derivative carrying 45,000-50,000 kg (99,210-110,230 lb) of cargo.

An-171: Proposed stretched version with increased wing span and more powerful engines; design work under way by 2001.

Adaptation of military An-70 for tanker, AEW, SAR (An-70PS), naval patrol, and of An-70T or An-70T-100 for

Antonov An-70 second prototype, showing spoilers, trailing-edge double-slotted flaps and slats *(Jane's/Paul Jackson)*   *NEW*/0525782

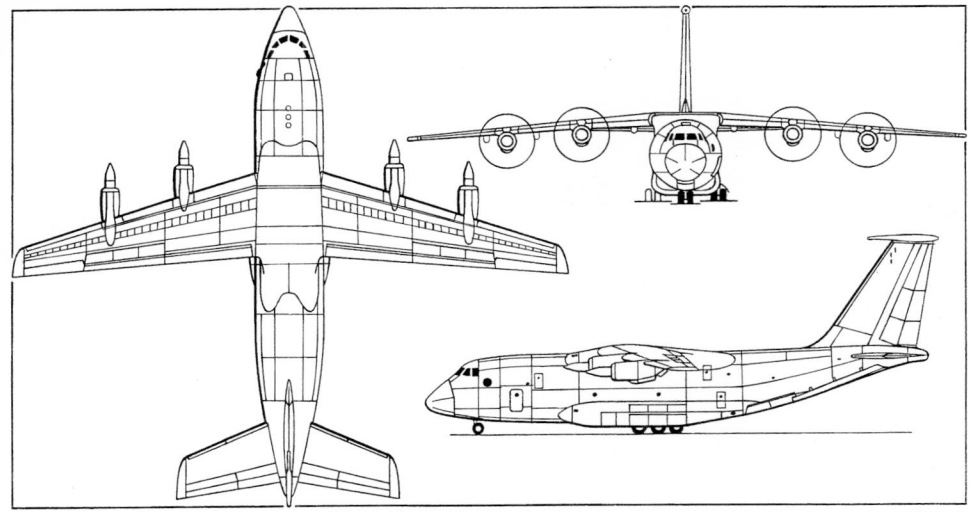

Antonov An-70 strategic transport (*Jane's/James Goulding*)   0130863

firefighting, ecological monitoring and ambulance duties being studied.

CUSTOMERS: Requirements originally expressed by Russian Air Forces for up to 500 and by Ukrainian Air Force for 100. This modified by 1999 to 164 for Russia and 65 for Ukraine, with in-service date of 2002. German requirement potentially for 75 aircraft. By mid-1999 potential civil operators had signed documents of intent for some 100 further aircraft. Chinese interest reported in 2000. Czech Republic reportedly finalising a contract for three in 2002, delivery to be from Omsk at one per year, beginning 2005.

COSTS: Military version US$60 million (2002). Civil An-70T to cost 20 to 30 per cent less than military variant. Total of US$3,500 million reportedly spent on development by 2002, of which Russian defence ministry owes Antonov US$49 million; engines cost US$75 million up to 2002, and further US$30 million allegedly required to address 30 recorded malfunctions during testing.

DESIGN FEATURES: First aircraft to fly powered only by propfans. Slightly larger than projected European FLA transport, much smaller than US Boeing C-17A; conventional high-wing configuration, with wings and tail surfaces slightly sweptback; supercritical wing section; anhedral from roots; loading ramp/doors under upswept rear fuselage with adjustable sill height and built-in cargo handling system; horizontal tail surfaces on rear fuselage; propfans mounted conventionally on wing leading-edge; propeller wash doubles wing lift during take-off and landing. Multiple-section control surfaces provide redundancy in event of battle damage or physical obstruction. Claimed features include independent operational capability at non-equipped airfields for 30 days. Design life 15,000 cycles and 45,000 flying hours in 25 years. Operable 3,500 hours per year, with eight to 10 man-hours of maintenance per flying hour. Cost-effective with only 200 flying hours per month.

FLYING CONTROLS: Prototypes have fly-by-wire system with three digital and six analogue channels; primary controls are quadruplex; back-up by unique fly-by-hydraulics system, in which pilot or autopilot inputs are relayed (via conventional 'mini-wheels') to actuators by commands in hydraulic control channels, unaffected by electromagnetic interference. Production aircraft will have a four-channel, all-digital primary FCS, rather than the hybrid digital/analogue system of the prototypes. Secondary FCS controls two independent flap systems, leading-edge slats and blown flaps. Three-section double-winged rudder. Double-slotted trailing-edge blown Fowler flaps in two sections on each wing; forward element maximum deflection 60°, rear element 80°; intermediate settings (forward element) 5, 10, 15, 20, 25, 30, 35, 40 and 50°. Three-section spoilers forward of each outer flap. Leading-edge has flaps inboard; slats centre and outboard. Two-section, double-hinged elevators forward section maximum deflections +28/−20°, rear section +50/−40°, to enhance low-speed authority; horizontal stabilisers are fitted with automatic leading-edge slats.

STRUCTURE: Approximately 28 per cent of airframe, by weight, made of composites, including complete tail unit, ailerons and flaps. Fuselage stringer/skin joints are spot-welded and hot-bonded, manually.

Single source for all major components. Russian assembly by Polyot, which also to build cargo hold; NAPO producing the centre section; VASO, the wing (which previously assigned to Chkalov plant at Tashkent, Uzbekistan). Aviant at Kiev to assemble Ukrainian aircraft and build flight deck, empennage and engine nacelles.

LANDING GEAR: Twin-wheel nose unit; each main unit has three pairs of wheels in tandem, retracting into large fairing on side of cabin; can operate from unpaved surfaces of bearing ratio 8 kg/cm² (114 lb/sq in). All tyres 1,120×450. Steel-steel brakes. Nosewheel turning angle ±55° for taxying; nosewheel turning radius 16.3 m (53½ ft);

wingtip turning radius 29.5 m (97¾ ft); required taxiway width for 180° turn 27.4 m (90 ft).

POWER PLANT: Four ZMKB Progress/Ivchenko D-27 propfans, each 10,290 kW (13,800 shp). Aerosyla Stupino SV-27 contrarotating propellers, each with eight composites blades in front and six at rear. Reversible-pitch blades of scimitar form, with electric anti-icing. Export versions proposed with CFM56-5C4 turbofans.

ACCOMMODATION: Three flight crew (two pilots and flight engineer or 'tactical pilot') plus loadmaster; navigation station on captain's left is optionally operated by fourth member of flight crew; provision for converting cockpit for two-crew operation, with co-pilot operating flight engineer's station; seats in forward fuselage for two cargo attendants; freight loaded via rear ramp using four built-in, powered hoists (each of 3 tonne capacity) reaching out 6.6 m (22 ft) from aircraft. Hoists can be combined for heavier loads. Freight can be carried on PA-5.6 rigid pallets, PA-3, PA-4 and PA-6.8 flexible pallets, in UAK-2.5, UAK-5 and UAK-10 containers; unpackaged freight, wheeled and tracked vehicles, food and perishables can be carried; seats for 300 troops, or 206 stretchers, can be installed using optional, prefabricated (10 section) upper deck or optional, easily removable seven-section upper deck (each segment holding 1.5 tonnes) in cargo hold; vehicles, freight and paratroops can be airdropped; maximum single airdrop item weight 20,000 kg (44,092 lb); crew door at front of cabin on port side; two upper deck doors each side, front and rear; cargo hold pressurised and air conditioned.

SYSTEMS: Aircraft systems automated to simplify operation and decrease probability of crew errors. Electronpribor engine control system; Leninets monitoring and information system.

AVIONICS: Integrated by Aviapribor, Leninets and Elektroavtomatika. Flight data, navigation and radio-navigation systems to ARINC 700 requirements; digital multiplex data interface equivalent to Western MIL-STD-1553B.

Comms: Integrated system by Gorkiski.

Flight: Ring laser INS; SKI-77 HUD; flight management system; designed for operation in adverse weather and for landing in ICAO Cat. II and IIIa conditions. BASK-70 onboard diagnostic system collects data from subsystems, registering and analysing 8,000 in-flight parameters.

Instrumentation: Ten-screen EFIS by Elektroavtomatika comprises six main screens, each 200 × 200 mm (7¾ × 7¾ in), facing pilots and two each at navigation and flight engineer's stations, plus smaller secondary LCD screens and roof-mounted HUDs for pilot and co-pilot on production aircraft.

EQUIPMENT: Four electric hoists in hold.

DIMENSIONS, EXTERNAL:
Wing span	44.06 m (144 ft 6¾ in)
Length overall	40.73 m (133 ft 7½ in)
Fuselage diameter	4.80 m (15 ft 9 in)
Height overall	16.38 m (53 ft 9 in)
Wheel track (bogie centres)	5.21 m (17 ft 1 in)
Wheelbase: front mainwheels	16.65 m (54 ft 7½ in)
centre mainwheels	18.47 m (60 ft 7¼ in)
rear mainwheels	20.43 m (67 ft 0¼ in)
Propeller diameter	4.50 m (14 ft 9 in)
Propeller ground clearance (outer)	3.00 m (9 ft 10 in)
Rear-loading aperture: Height	4.10 m (13 ft 5½ in)
Width	4.00 m (13 ft 1½ in)

DIMENSIONS, INTERNAL:
Cargo hold: Floor length: excl ramp	19.10 m (62 ft 8 in)
incl loadable ramp	22.40 m (73 ft 6 in)
incl unusable ramp	30.60 m (100 ft 4¾ in)
Max width	4.80 m (15 ft 9 in)
Max width at floor	4.00 m (13 ft 1½ in)
Height: max	4.10 m (13 ft 5½ in)
min	4.00 m (13 ft 1½ in)
Floor area, incl ramp	89.0 m² (958 sq ft)
Volume: pressurised	400.0 m³ (14,126 cu ft)
total	425.0 m³ (15,008 cu ft)

WEIGHTS AND LOADINGS:
Weight empty	72,800 kg (160,500 lb)
Normal payload (incl 5,000 kg; 11,025 lb on ramp)	35,000 kg (77,161 lb)
Normal payload from unpaved runway	30,000 kg (66,138 lb)
Payload: max	47,000 kg (103,615 lb)
restricted runway: option	35,000 kg (77,161 lb)
600 m runway option	20,000 kg (44,092 lb)
Max T-O weight	130,000 kg (286,600 lb)
Max power loading	3.16 kg/kW (5.19 lb/shp)

PERFORMANCE (estimated):
Cruising speed: long range	405 kt (750 km/h; 466 mph)
max short range	432 kt (800 km/h; 497 mph)
Nominal cruising height	9,100-11,000 m (29,860-36,080 ft)
Runway length required: for normal operation	
T-O	1,800 m (5,905 ft)
Landing	2,200 m (7,220 ft)

Range (runway length A: 1,800 m; 5,905 ft, B: 700 m; 2,300 ft:
with 47 tonnes:		
A	1,619 n miles (3,000 km; 1,864 miles)	
B	not an option	
with 35 tonnes:		
A	2,699 n miles (5,000 km; 3,106 miles)	
B	not an option	
with 30 tonnes:		
A	3,239 n miles (6,000 km; 3,728 miles)	
B	647 n miles (1,200 km; 745 miles)	
with 20 tonnes:		
A	3,563 n miles (6,600 km; 4,101 miles)	
B	1,619 n miles (3,000 km; 1,864 miles)	
with max fuel: all options		
	4,319 n miles (8,000 km; 4,971 miles)	

UPDATED

ANTONOV An-72 and An-74
NATO reporting name: Coaler

TYPE: Twin-jet transport.

PROGRAMME: Originated as private venture military transport based on stillborn An-60 64-73 seat civil airliner designed to meet 1967 specification. Revised design, following issue of military requirement in 1968, included relocation of engines above wings; chief designer Yuri G Orlov. Two static test airframes; first of four prototype An-72s, built at Kiev, flew (SSSR-197744) 31 August 1977. Production order placed in December 1980 for improved An-72A version; manufacture transferred to Kharkov, Ukraine, where first production An-72 flew 22 December 1985; An-74 also produced at Kharkov from December 1989. An-74 announced February 1984; An-72P maritime patrol version demonstrated 1992. Production of An-74 also

Antonov An-74 twin-turbofan STOL transport of Russian energy company, Gazprom (*Jane's/Paul Jackson*)
NEW/0525071

started by Polyot Industrial Association at Omsk, Russia, in 1993 (assisted by Progress at Arsenyev); first Polyot aircraft (RA-74050) was flown 25 December 1993. Development of An-74-200 and An-74TK-200 by Antonov started 1995; An-74TK certified by Interstate Aviation Committee in August 1995.

CURRENT VERSIONS: **An-72A** ('Coaler-C'): STOL transport for military use. Compared with An-72 ('Coaler-A') prototypes, wing span increased by 6.00 m (19 ft 8¼ in), fuselage length by 1.50 m (4 ft 11 in), fuel load by 2,500 kg (5,512 lb). An-72 prototype 003/SSSR-19795 converted to An-72A ('Arctic') and first flew (SSSR-780334) 29 September 1983. Additionally, on production version, tail unit and engine air intake de-icing improved over An-72; Buran radar in enlarged radome; advanced navigation aids, including inertial navigation system; provision for wheel/ski landing gear.

An-72P: Maritime patrol version, described in 2001-02 and previous editions.

An-72R: One (SSSR-783573), at Akhtubinsk operational research centre, appears to have a large, flat, side-looking airborne radar (SLAR) built into each side of its upswept rear fuselage, possibly for standoff battlefield surveillance. Engineering designation **An-88**. First noted 1995, but possibly by then out of service.

An-72S: Military VIP version with appropriate *'Salon'* interior.

An-72V: Export version; two crew only.

An-72-100: Civilianised An-72, certified 1997; upgraded avionics.

An-74: Designation refers, generally, to all civilian versions (as under). Airframe identical with An-72 except for two blister windows at rear of flight deck and front of cabin on port side. Convertible in field for ambulance, firefighting and other duties. Proving trials 1985-86. First preproduction An-74 (SSSR-58642) first flew at Kharkov on 26 June 1986; further five development aircraft built by 1989, these having smaller radomes. Maiden flight of production An-74 (c/n 0706) December 1989; features include new APU, Buran-74 radar in larger radome and improved navigation aids. Type certificate awarded August 1991; initially operated by Yakutsk division of Aeroflot.

An-74-200: Freight version with D-36 Series 3A engines of unchanged rating; increased payload and maximum T-O weight. Able to carry four YAK-2.5 containers. Crew of four/five.

An-74T-200A ('Coaler-B'): Transport (*transportnyi*) version with longer hold; payload 10,000 kg (22,045 lb); loading winch; roller conveyors in floor; crew of two. Range with maximum payload up to 728 n miles (1,350 km; 838 miles), with 3,500 kg (7,716 lb) payload 2,159 n miles (4,000 km; 2,485 miles).

An-74TK-200 ('Coaler-B'): Convertible transport/passenger aircraft (*transportnyi konvertiruyemnyi*) with twin seats for 52 passengers that fold against cabin walls, and with baggage racks, buffet/galley and lavatory. Alternative all-cargo or all-passenger or combi layouts. Typical combi options include 12 passengers plus 6,000 kg (13,228 lb) of freight and 20 plus 4,500 kg (9,921 lb). Built-in loading equipment. Crew of two. Range with 10,000 kg (22,045 lb) payload, 1 hour reserve, 430 n miles (800 km; 497 miles).

An-74TK-200D Salon: Business transport for 10 to 16 passengers, with increased cabin comfort. Equipment includes telephone, fax, video, bar, refrigerator, galley and separate rest area. Optional compartment for car at rear. Power plant and weights as basic An-74. Crew of four/five.

An-74T-100 ('Coaler-B'): As An-74T-200, with navigator station (crew of four).

An-74TK-100: As An-74TK-200, with navigator and flight engineer stations (crew of four). Russian type certificate issued 4 August 1995.

An-74TK-100C: One VIP/medical evacuation aircraft (RA-74005) delivered to Gazprom on 26 February 2002. Air Ambulance Technology (Austrian) equipment. Designation remains '-100C' in both Cyrillic and Roman alphabets.

An-74TK-300: Described separately.

An-74-400: Stretched version; under development by 1998; to be offered in passenger and cargo versions.

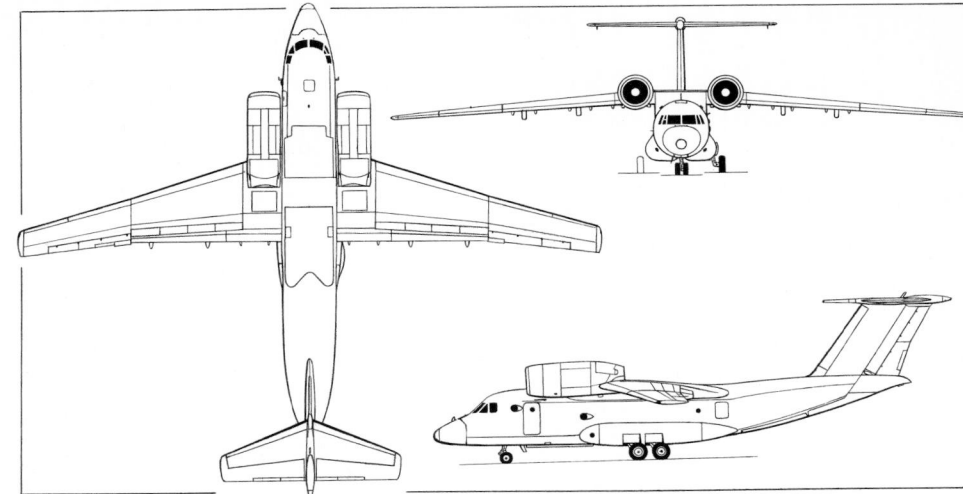

Antonov An-74 ('Coaler-B') STOL transport (two ZMKB Progress/Ivchenko D-36 turbofans) (*Jane's/Dennis Punnett*)

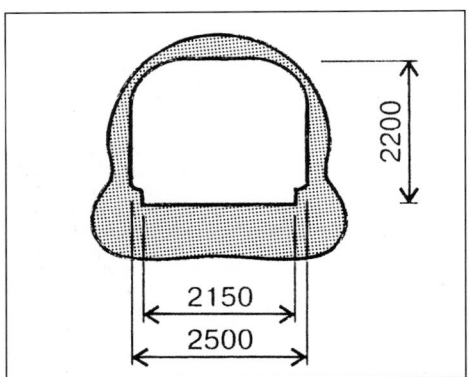

Cross-section of Antonov An-74 cabin/hold, dimensions in millimetres   0051512

An-76: Engineering designation for military An-72Ps.

An-79: US military sources use this designation for a version of An-72 transport.

An-174: Proposed stretched An-74TK-300; described separately.

An-71: AEW version; last described in 1997-98 *Jane's.*

An-71P: Radio relay version for Soviet anti-aircraft defence. Three development aircraft converted from An-72s before cancellation. Engineering designation **An-76**.

CUSTOMERS: More than 160 An-72/74s (including 96 military) built before An-74 production additionally established at Omsk; 20 in Russian Air Forces; four in Peruvian Air Force; 26 in Ukrainian Air Force; others in Kazakhstan and Moldova. Order placed by Iran in 1997 for 12 An-74TKs includes some for Presidential Guard; other recent deliveries to Laotian government, Ukraine Border Guard, MChS Rossii, Vitair and Gazpromavia. By mid-2000, civil operators included five with 12 An-72s and 17 with 37 An-74s. No new civil deliveries reported in 1998; two An-74TK-200s built at Kharkov in 1999. Indian interest being pursued in 2000; Chinese interest reported late 2001; Dominican Republic civil aviation interest in 2002. Omsk received order for one for MChS Rossii and one of required three for FPS (border guard) to be delivered in 2001; by late 2000, Omsk had delivered total of only four aircraft since 1993 and had further three in various stages of completeness.

COSTS: US$9 million (Omsk, 2000).

DESIGN FEATURES: Primary role as STOL replacement for turboprop An-26, with emphasis on freight carrying. High-wing, T-tail configuration, with upswept rear fuselage for freight access. Ejection of exhaust efflux over upper wing

surface and down over large multislotted flaps gives considerable increase in lift; high-set engines avoid foreign object ingestion; special ramp/door as An-26; low-pressure tyres and multiwheel landing gear for operation from unprepared strips, ice or snow; sweptback fin and rudder.

Wing leading-edge sweepback 17°; anhedral approximately 10° on outer wings; normal T-O flap setting 25 to 30°, maximum deflection 60°.

FLYING CONTROLS: Conventional and assisted. Power-actuated ailerons, with two tabs in port aileron, one starboard; double-hinged rudder, with tab in lower portion of two-section aft panel; during normal flight only lower rear rudder segment is used; both rear segments used in low-speed flight; forward segment is actuated automatically to offset thrust asymmetry; horn-balanced and mechanically actuated, aerodynamically balanced elevators, each with two tabs; hydraulically actuated full-span wing leading-edge flaps outboard of nacelles; trailing-edge flaps double-slotted in exhaust efflux, triple-slotted between nacelles and outer wings; four-section spoilers forward of triple-slotted flaps; two outer sections on each side raised before landing, remainder opened automatically on touchdown by sensors actuated by weight on main landing gear; inverted leading-edge slat on tailplane linked to wing flaps.

STRUCTURE: All-metal; multispar wings mounted above fuselage; wing skin, spoilers and flaps of titanium aft of engine nacelles; circular semi-monocoque fuselage, with rear ramp/door; tapered fairing forward of T tail fin/tailplane junction, blending into ogival rear fairing.

LANDING GEAR: Hydraulically retractable tricycle type, primarily of titanium. Rearward-retracting steerable twin-wheel nose unit. Each main unit comprises two trailing-arm legs in tandem, each with a single wheel, retracting inward through 90° so that wheels lie horizontally in bottom of fairings, outside fuselage pressure cell. Oleo-pneumatic shock-absorber in each unit. Low-pressure tyres, size 720×310 on nosewheels, 1,050×400 on mainwheels. Hydraulic disc brakes. Telescopic strut hinges downward, from rear of each side fairing, to support fuselage during direct loading of hold with ramp/door under fuselage.

POWER PLANT: Two ZMKB Progress/Ivchenko D-36 high-bypass ratio turbofans (Srs 2A in An-74; Srs 3A in An-74-200, T-100/200 and TK-100/200), each 63.74 kN (14,330 lb st). Integral fuel tanks between spars of outer wings. Thrust reversers standard.

ACCOMMODATION: Pilot and co-pilot/navigator side by side on flight deck of basic An-72, plus flight engineer, with provision for fourth person. Heated windows. Two windscreen wipers. Flight deck and cabin pressurised and air conditioned. Main cabin designed primarily for freight, including four YAK-2.5 containers or four PAV-2.5 pallets each weighing 2,500 kg (5,511 lb); An-72 has folding seats along sidewalls and removable central seats for 68 passengers. It can carry 57 parachutists, and has provision for 24 stretcher patients, 12 seated casualties and an attendant in ambulance configuration. An-74 can carry eight mission staff in combi role, with tables and bunks. Bulged observation windows on port side for navigator and hydrologist. Provision for wardrobe and galley. Movable bulkhead between passenger and freight compartments, with provision for 1,500 kg (3,307 lb) of freight in rear compartment. Reinforced, movable bulkhead in combi versions protects passengers from shifting cargo in the event of sudden deceleration.

Downward-hinged and forward-sliding rear ramp/door for loading trucks and tracked vehicles, and for direct loading of hold from trucks. It is openable in flight, enabling freight loads of up to 7,500 kg (16,535 lb), with a maximum of 2,500 kg (5,511 lb) per individual item, to be airdropped by parachute extraction system. In normal freight role, 1,000 kg (2,204 lb) of payload can be placed on ramp. Maximum size of containers up to

Antonov An-74TK-200 convertible passenger/cargo aircraft (*Jane's/Paul Jackson*)   0121014

1.90 × 2.44 × 1.46 m (6 ft 3 in × 8 ft × 4 ft 9½ in), pallets up to 1.90 × 2.42 × 1.46 m (6 ft 3 in × 7 ft 11 in × 4 ft 9½ in). Main crew and passenger door at front of cabin on port side. Emergency exit and servicing door at rear of cabin on starboard side.

SYSTEMS: Air conditioning system, altitude limit 10,000 m (32,810 ft), with independent temperature control in flight deck and main cabin areas; used to refrigerate main cabin when perishable goods carried. Maximum cabin pressure differential 0.49 bar (7.1 lb/sq in). Hydraulic system for landing gear, flaps and ramp. Electrical system powers auxiliary systems, flight deck equipment, lighting and mobile hoist. Thermal de-icing system for leading-edges of wings and tail unit (including tailplane slat), engine air intakes and cockpit windows. Provision for TA-12 APU in starboard landing gear fairing. This can be used to heat cabin; under cold ambient conditions, servicing personnel can gain access to major electric, hydraulic and air conditioning components without stepping outside.

AVIONICS: *Comms:* HF com, VHF com/nav. 'Odd Rods' IFF standard.
   Radar: Navigation/weather radar in nose.
   Flight: ADF. Compatible with DME, Tacan, VOR, ILS and SP systems. Doppler-based automatic navigation system, linked to onboard computer, is preprogrammed before take-off on push-button panel to right of map display.
   Instrumentation: Failure warning panels above windscreen display red lights for critical failures, yellow lights for non-critical failures, to minimise time spent on monitoring instruments and equipment.
   An-74 has enhanced avionics, including INS.

EQUIPMENT: Removable mobile winch, capacity 2,500 kg (5,511 lb), assists loading. Cargo straps and nets stowed in lockers on each side of hold when not in use. Provision for roller conveyors in floor.

DIMENSIONS, EXTERNAL:
Wing span	31.89 m (104 ft 7½ in)
Wing aspect ratio	10.3
Length overall	28.07 m (92 ft 1¼ in)
Fuselage: Max diameter	3.10 m (10 ft 2 in)
Height overall	8.65 m (28 ft 4½ in)
Wheel track	4.09 m (13 ft 5 in)
Wheelbase	8.68 m (28 ft 5¾ in)
Min loading clearance beneath rear fuselage	
	2.80 m (9 ft 2¼ in)
Distance between engine centrelines	4.15 m (13 ft 7½ in)
Crew/passenger door: Height	1.65 m (5 ft 5 in)
Width	0.90 m (2 ft 11¼ in)
Rear-loading door: Length	7.10 m (23 ft 3½ in)
Width	2.40 m (7 ft 10½ in)
Height to sill	1.54 m (5 ft 0¾ in)

DIMENSIONS, INTERNAL:
Cabin: Length: excl ramp: An-74	9.50 m (31 ft 2 in)
An-74T	10.50 m (34 ft 5¼ in)
incl ramp: An-74T	14.30 m (46 ft 11 in)
Width: at floor level	2.15 m (7 ft 0½ in)
max	2.50 m (8 ft 2½ in)
Height	2.20 m (7 ft 2½ in)
Floor area: An-74T	22.5 m² (242 sq ft)

Ramp area	8.2 m² (88.3 sq ft)
Volume: An-74T (total)	73.3 m³ (2,589 cu ft)

AREAS:
Wings, gross	98.53 m² (1,060.6 sq ft)

WEIGHTS AND LOADINGS (A: An-72, C: An-74, D: An-74 Salon, E: An-74-200, F: An-74T-200 and An-74TK-200):
Weight empty: A	19,050 kg (42,000 lb)
F	21,820 kg (48,105 lb)
Max fuel: A	12,950 kg (28,550 lb)
C, D, E, F	13,200 kg (29,100 lb)
Max payload: C, D	7,500 kg (16,535 lb)
A, E, F	10,000 kg (22,045 lb)
Max T-O weight: from 1,800 m (5,905 ft) runway:	
A	34,500 kg (76,060 lb)
from 1,500 m (4,920 ft) runway:	
A	33,000 kg (72,750 lb)
from 600-800 m (1,970-2,630 ft) runway:	
A	27,500 kg (60,625 lb)
C, D	34,800 kg (76,720 lb)
E, F	36,500 kg (80,468 lb)
Max landing weight: A	33,000 kg (72,750 lb)
F	36,500 kg (80,468 lb)
Max wing loading: A	349.8 kg/m² (71.62 lb/sq ft)
Max power loading: A	271 kg/kN (2.65 lb/lb st)

PERFORMANCE (An-72. A: at T-O weight of 33,000 kg; 72,750 lb, B: at T-O weight of 27,500 kg; 60,625 lb on 1,000 m; 3,280 ft unprepared runway. C, D, E, F, An-74 series as above):
Max level speed at 10,000 m (32,810 ft):	
A	380 kt (705 km/h; 438 mph)
Max level speed at 10,100 m (33,135 ft):	
C, D, E, F	377 kt (700 km/h; 434 mph)
Cruising speed at 10,000 m (32,810 ft):	
A, B	297-324 kt (550-600 km/h; 342-373 mph)
Approach speed: A	97 kt (180 km/h; 112 mph)
Service ceiling: A	10,700 m (35,100 ft)
B	11,800 m (38,720 ft)
Service ceiling, OEI: A	5,100 m (16,740 ft)
B	6,800 m (22,300 ft)
T-O run: A	930 m (3,055 ft)
B	620 m (2,035 ft)
T-O to 10.7 m (35 ft): A	1,170 m (3,840 ft)
B	830 m (2,725 ft)
Landing run: A	465 m (1,525 ft)
B	420 m (1,380 ft)
Max length of runway required:	
C, D	1,200-1,800 m (3,940-5,905 ft)
E, F	1,400-2,150 m (4,595-7,055 ft)
Range, 45 min reserves:	
A with max payload	430 n miles (800 km; 497 miles)
A with 7,500 kg (16,535 lb) payload	
	1,080 n miles (2,000 km; 1,240 miles)
A with max fuel	2,590 n miles (4,800 km; 2,980 miles)
B with 5,000 kg (11,020 lb) payload	
	430 n miles (800 km; 497 miles)
B with max fuel	1,760 n miles (3,250 km; 2,020 miles)
F with 10,000 kg (22,046 lb) payload	
	809 n miles (1,500 km; 932 miles)
F with 5,000 kg (11,023 lb) payload	
	1,943 n miles (3,600 km; 2,236 miles)

Range, 1 hour reserves:	
C, F with 7,500 kg (16,535 lb) payload	
	944 n miles (1,750 km; 1,087 miles)
E with 7,500 kg (16,535 lb) payload	
	1,160 n miles (2,150 km; 1,336 miles)
C, E with 5,000 kg (11,020 lb) payload	
	1,511 n miles (2,800 km; 1,739 miles)
F with 5,000 kg (11,020 lb) payload	
	1,403 n miles (2,600 km; 1,615 miles)
C with max fuel and 800 kg (1,763 lb) payload	
	2,375 n miles (4,400 km; 2,734 miles)
D with max fuel and 16 passengers	
	2,429 n miles (4,500 km; 2,796 miles)
E with max fuel and 2,500 kg (5,511 lb) payload	
	2,294 n miles (4,250 km; 2,640 miles)
F with max fuel and 5,000 kg (11,020 lb) payload	
	2,321 n miles (4,300 km; 2,671 miles)
Endurance: F	6 h 50 min

UPDATED

ANTONOV An-74-300 and An-174

TYPE: Twin-jet transport.

PROGRAMME: First variant, An-74TK-300, announced mid-1998. Model of An-174 shown at Paris, June 1999. Prototype An-74TK-300 modified by KhGAPP from An-72 c/n 1910; work began December 1999; first official flight (UR-74300) 20 April 2001; international debut at Paris Salon, June 2001; certification trials completed July 2002 after 219 sorties.

CURRENT VERSIONS: Both are An-74 derivatives in which podded, underslung engines of increased power replace the normal An-72/74 installation.
   An-74T-300: Baseline transport. *As described.*
   An-74TK-300: Combi version. Internal dimensions as for An-74T-100/200.
   An-74MP-300: Proposal announced 2002. Maritime patrol version of An-74TK-300; also capable of carrying 22 paratroops, or 44 soldiers or 16 stretchers. Provision for GSh-23L gun, rockets and 100 kg bombs.
   An-174: Stretched version with 6 m (19½ ft) fuselage extension provided by plugs fore and aft of wing. Progress D-436T1 engines, each 74.3 kN (16,700 lb st). Studies had

Engine installation of An-74-300
(Jane's/Paul Jackson)   0121016

Prototype An-74-300 at Moscow in August 2001 *(Yefim Gordon)*   *NEW*/0525072

Model of Antonov An-174 stretched version of An-74-300 *(Jane's/Paul Jackson)*   0121103

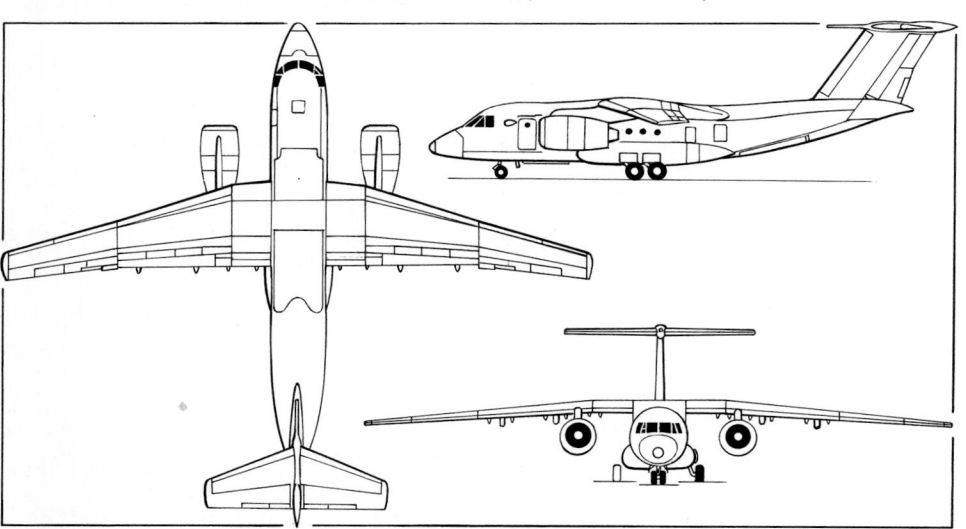

Antonov An-74T-300 STOL transport *(Jane's/James Goulding)*   0079288

begun by 1998. Potential replacement for An-12. Also known as **An-74-400.**

DESIGN FEATURES: Revised engine installation decreases T-O run, despite loss of Coanda effect; new power plants reduce maintenance requirements and fuel consumption (up to 29 per cent); new position simplifies access. Compared to An-74T-200's range with 45 min reserves, An-74T-300 offers additional 372 n miles (690 km; 428 miles) with 10,000 kg (22,046 lb) payload and 518 n miles (960 km; 596 miles) with 5,000 kg (11,023 lb). Cruising speed (at nominal 10,600 to 11,000 m; 34,780 to 36,420 ft) also noticeably improved.

Cabin enlarged by repositioning of partitions; integral airstairs; upgraded air conditioning; improved versions of Buran radar and Malva navigation system; improved instrumentation, including provision for LCDs.

POWER PLANT: Two 63.7 kN (14,330 lb st) Progress D-36-4A turbofans.

AREAS:
Wings, gross   98.62 m² (1,061.5 sq ft)
WEIGHTS AND LOADINGS:
Max payload   10,000 kg (22,045 lb)
Max T-O weight   34,500 kg (76,060 lb)
PERFORMANCE:
Max level speed   405 kt (750 km/h; 466 mph)
Max cruising speed   391 kt (725 km/h; 450 mph)
Service ceiling   10,100 m (33,140 ft)
Balanced field length   1,900 m (6,235 ft)
Range: with 10,000 kg (22,046 lb) payload
                            809 n miles (1,500 km; 932 miles)
  with 52 passengers
                            1,889 n miles (3,500 km; 2,174 miles)
  ferry   2,807 n miles (5,200 km; 3,231 miles)
                                              UPDATED

ANTONOV An-124
NATO reporting name: Condor
TYPE: Outsize freighter.
PROGRAMME: Bureau design number 305; originally designated An-40. Prototype (SSSR-680125) first flew 26 December 1982; second aircraft registered SSSR-680210 but static test airframe only. First production aircraft (SSSR-82002 *Ruslan*, named after giant hero of Russian folklore immortalised by Pushkin)

exhibited 1985 Paris Air Show; lifted payload of 171,219 kg (377,473 lb) to 10,750 m (35,269 ft) on 26 July 1985, exceeding by 53 per cent C-5A Galaxy's record for payload lifted to 2,000 m and setting 20 more records. Entered service January 1986; set closed-circuit distance record 6 to 7 May 1987 by flying 10,880.625 n miles (20,150.921 km; 12,521.201 miles) in 25 hours 30 minutes.

Deliveries to VTA (Russian Air Forces transport arm), to replace An-22, began 1987; in September 1990, during Gulf crisis, an An-124 carried 451 Bangladeshi refugees from Amman to Dacca, after being fitted with chemical

toilets, a 570 litre (150 US gallon; 125 Imp gallon) drinking water tank and foam rubber cabin lining in lieu of seats. Carried heaviest single commercial load transported by air: 135.2 tonnes in 1993; and heaviest commercial shipment moved in one flight; 146 tonnes in 1994.

Service life extension programme begun by Aviastar on first of 17 Antonov Airlines and Volga-Dnepr aircraft in 2000; includes new avionics; upgraded crew rest compartment; and cargo floor and loading equipment strengthening. First, RA-82078 of Volga-Dnepr, redelivered 14 March 2000. Continued production assured by Volga-Dnepr's requirement for expansion by one or two An-124s per year; company plans to register two future aircraft in UK, following JAR 25 certification.
CURRENT VERSIONS: **An-124:** Baseline transport.
Detailed description applies to above version.
   An-124-100: Commercial transport; civil type certificate granted by AviaRegistr of Interstate Aviation Committee of Russian Federation and Associated States (CIS) on 30 December 1992. Civil-operated An-124s are now to this standard. Maximum T-O weight restricted to 392,000 kg (864,200 lb) and maximum payload to 120,000 kg (264,550 lb).
   An-124-100M: As An-124-100, but with Western avionics, including Litton LTN-92 INS, Rockwell Collins GPS, ACARS, weather radar and TCAS-2. Series 3 versions of D-18T engine, offering 6,000 hour overhaul interval and 24,000 hour total life. Crew reduced to four by removal of radio operator and navigator. Prototype (RA-82079) completed at Ulyanovsk late 1995, but not flown until June 2000; delivered Volga-Dnepr 3 August 2000. Service life extension to 24,000 hours formally certified on 22 February 2001. Version offered to RAF in 1999 as alternative to An-124-210 (which see). Final details were agreed in November 2001 for upgrades of remainder of Volga-Dnepr, plus Antonov Airlines fleets.
   An-124-102: Flight deck EFIS equipped, with dual sets of CRTs. Crew reduced to three (two pilots and flight engineer).
   An-124-130: Under study in 1996 with General Electric CF6-80 turbofans. Prototype reportedly will be 36th Ulyanovsk aircraft.
   An-124-200: Proposed version with GE CF6-80C2 engines, each 263 kN (592,000 lb st).
   An-124-210: Joint proposal with Air Foyle to meet UK's Short Term Strategic Airlifter (STSA) requirement; 273 kN (60,600 lb st) Rolls-Royce RB211-524H-T engines and Honeywell avionics. Weight empty 184,000 kg (405,650 lb); payload and MTOW as An-124-100. Range (30 min reserves plus 5 per cent) 2,267 n miles (4,200 km; 2,609 miles) with 120,000 kg (264,550 lb) max payload; 3,855 n miles (7,140 km; 4,436 miles) with 80,000 kg (176,375 lb); or 7,424 n miles (13,750 km; 8,543 miles) with max fuel. JAR 25 runway length 2,300 m (7,545 ft). Three flight crew. STSA competition was abandoned in August 1999, then reinstated and won by Boeing C-17A.
   An-124FFR: Water-bomber project, able to drop 200 tonnes of fire retardants including 70 tonnes in centre fuel tank. Convertible to freighter.
   An-124 Turboprop: Retrofit with four Aviadvigatel NK-93 propfans considered by Volga-Dnepr in 1997.
   An-124-100VS: Russian government approval given late 1998 to modify two An-124s to carry the Vozdushny Start booster, capable of placing a 1,630 kg (3,593 lb) satellite into 200 km (124 mile) orbit. ORIL launch system has development cost of US$130 million (2000) and will

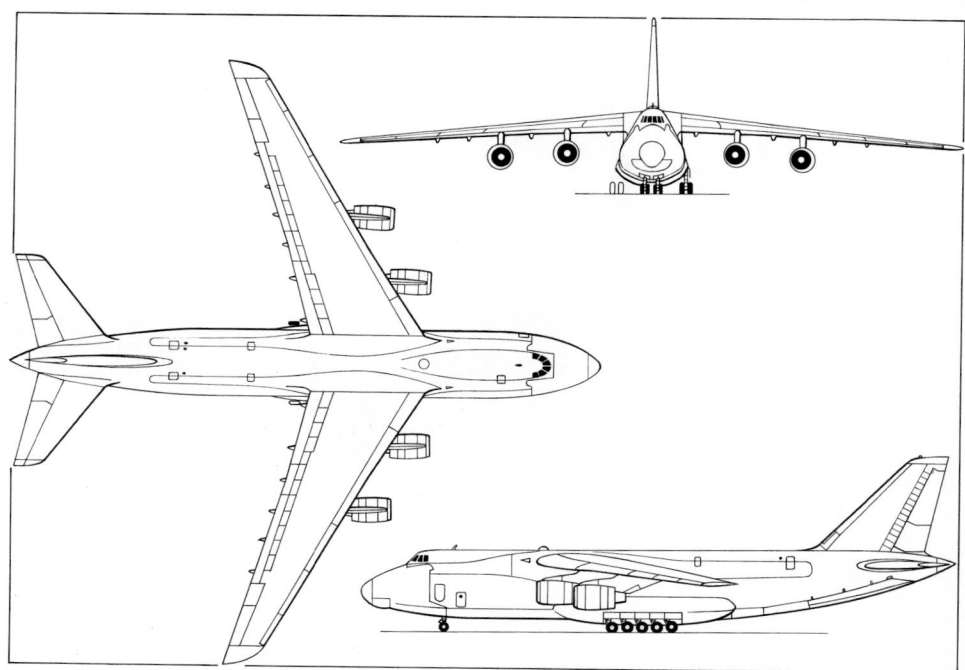

Antonov An-124 (four ZMKB Progress/Ivchenko D-18T turbofans) *(Jane's/James Goulding)*

cost up to US$20 million per launch; An-124 upgrades are US$13 million each. Booster is extracted from An-124's rear by parachute before rocket ignition. Alternative (Anglicised) designation **An-124AL**. Aircraft include RA-82010. IOC by early 2003.

Aircraft transferred from Russian Air Forces to Ulyanovsk for modification, last arriving on 8 February 2001; first aircraft's public debut was at Ulyanovsk on 24 August 2001; second was due for completion before end of 2001; operated by Polet alongside further two ex-military transfers.

CUSTOMERS: Manufactured at Kiev by Aviant and (beginning with eighth production aircraft, SSSR-82005, in late 1985) by Aviastar at Ulyanovsk; 55 completely or substantially built by late 1995, comprising 19 at Kiev (including prototype, but not test airframe) and 36 at Ulyanovsk; by June 2000, only 18 and 34, respectively, had flown. Production at Kiev temporarily halted by 1991 after 16 aircraft and briefly resumed with two more in 1993-94. Ulyanovsk has continued throughout, producing 33 by 1995; in 2001, No. 35 was scheduled for delivery to Polet in 2002. In early 2002, further two aircraft under construction at Aviastar for Volga-Dnepr, first of which formally ordered (company's 10th) on 24 July 2002.

First international commercial operator was Air Foyle of Luton, UK, which wet leases from Antonov OKB; others previously available to HeavyLift (UK) from Volga-Dnepr (Russia), for charter operations until partnership dissolved in 2000 and Air Foyle Heavylift joint venture launched in October 2001. Operators in 2001 comprised Antonov Airlines (nine), Polet (six), Volga-Dnepr (nine), Russian Air Forces (11) and Libyan Cargo Airlines (one transferred from Antonov Airlines in 2001). Further 11 Russian Air Forces aircraft stored pending possible resale; four destroyed; one (prototype) in storage, completing total of 52 aircraft flown up to 2001; the 19th and last Aviant aircraft was due for completion and delivery to Libyan Cargo in 2002.

COSTS: US$40 million (2001).

DESIGN FEATURES: World's largest production aircraft; upward-hinged visor-type nose and rear fuselage ramp/ door for simultaneous front and rear loading/unloading; titanium floor throughout constant-section main hold, which is lightly pressurised, with a fully pressurised cabin for passengers above; landing gear for operation from unprepared fields, hard packed snow and ice-covered swampland; steerable nosewheels and mainwheels permit turns on 45 m (148 ft) wide runway.

Service life initially 7,500 hours; contract for extension to 12,000 hours signed between Volga-Dnepr and Antonov, July 2000; aircraft delivered from 2000 have 24,000 hour airframe and engine life.

Supercritical wings, with anhedral; sweepback approximately 35° on inboard leading-edge, 32° outboard; all tail surfaces sweptback.

FLYING CONTROLS: All surfaces hydraulically actuated; manual control of automatic control systems, control surface actuators, control system manual linkage and trimming system. Two-section ailerons, three-section single-slotted Fowler flaps (two outer, one inner) and six-section full-span leading-edge flaps on each wing; small slot in outer part of two inner flap sections each side to optimise aerodynamics; 12 spoilers on each wing, forward of trailing-edge flaps (four lateral control spoilers outboard; four glissage spoilers; and four airbrakes inboard); no wing fences, vortex generators or tabs; hydraulic flutter dampers on ailerons; rudder and each elevator in two sections, without tabs but with hydraulic flutter dampers; fixed incidence tailplane; control runs (and other services) channelled along fuselage roof.

STRUCTURE: Basically conventional light alloy, but 5,500 kg (12,125 lb) of composites make up more than 1,500 m² (16,150 sq ft) of surface area, giving weight saving of more than 2,000 kg (4,410 lb); each wing has one-piece root-to-tip upper surface extruded skin panel, strip of carbon fibre skin

panels on undersurface forward of control surfaces, and glass fibre tip; front and rear of each flap guide fairing of glass fibre, centre portion of carbon fibre; central frames of semi-monocoque fuselage each comprise four large forgings; fairings over intersection of fuselage double-bubble lobes in line with wing, from rear of flight deck to plane of fin leading-edge, primarily of glass fibre, with central, and lower underwing, portions of carbon fibre; other glass fibre components include tailplane tips, nosecone, tailcone and most bottom skin panels forming blister underfairing between main landing gear legs; carbon fibre components include strips of skin panels forward of each tail control surface, nose and main landing gear doors, some service doors, and clamshell doors aft of rear-loading ramp.

LANDING GEAR: Hydraulically retractable nosewheel type, made by Hydromash, with 24 wheels. Two independent forward-retracting and steerable twin-wheel nose units, side by side. Each main gear comprises five independent inward-retracting twin-wheel units; front two units on each side steerable. Each mainwheel bogie enclosed by separate upper and lower doors when retracted. Nosewheel doors and lower mainwheel doors close when gear extended. All wheel doors of carbon fibre. Main gear bogies retracted individually for repair or wheel change. Mainwheel tyres size 1,270×510. Nosewheel tyres size 1,120×450.

Aircraft can 'kneel', by retracting nosewheels and settling on two extendable 'feet', giving floor of hold a 3.5° slope to assist loading and unloading. Process takes 3 minutes to lower aircraft and 6½ minutes to raise. Rear of cargo hold lowered by compressing main gear oleos. Carbon brakes normally toe-operated, via rudder pedals. For severe braking, pedals depressed by toes and heels. Turning radius (outboard wheels) 19.6 m (64 ft 4 in).

POWER PLANT: Four ZMKB Progress/Ivchenko D-18T turbofans, each 229 kN (51,590 lb st); thrust reversers standard. Optional hush kit certified to ICAO Chapter 3 in mid-1997. Engine cowlings of glass fibre; pylons have carbon fibre skin at rear end. All fuel in 10 integral tanks in wings, total capacity 348,740 litres (92,128 US gallons; 76,714 Imp gallons), not all of which is utilised in civil versions.

ACCOMMODATION: Crew and passenger accommodation on upper deck; freight and/or vehicles on lower deck. Flight crew of six, in pairs, on flight deck, with place for loadmaster in lobby area (10 to 12 cargo handlers and servicing staff carried on commercial flights). Pilot and co-pilot on fully adjustable seats, which rotate for improved access. Two flight engineers, on wall-facing seats on starboard side, have complete control of master fuel cocks, detailed systems instruments, and digital integrated data system with CRT monitor. Behind pilot are navigator and communications specialist, on wall-facing seats. Between

flight deck and wing carry-through structure, on port side, are toilets, washing facilities, galley, equipment compartment, and two cabins for up to six relief crew, with table and facing bench seats convertible into bunks.

Aft of wing carry-through is passenger cabin for 88 persons. Hatches in upper deck provide access to wing and tail unit for maintenance when workstands not available. Flight deck and passenger cabin each accessible from cargo hold by hydraulically folding ladder, operated automatically with manual override. Rearward-sliding and jettisonable window each side of flight deck. Primary access to flight deck via airstair door, with ladder extension, forward of wing on port side. Smaller door forward of this and slightly higher. Door from main hold aft of wing on starboard side. Upper deck doors at rear of flight deck on starboard side and at rear of passenger cabin on each side. Emergency exit from upper deck aft of wing on each side.

Hydraulically operated visor-type upward-hinged nose takes 7 minutes to open fully, with simultaneous extension of folding nose loading ramp. When open, nose is steadied by reinforcing arms against wind gusts. No hydraulic, electrical or other system lines broken when nose is open. Radar wiring passes through tube in hinge. Hydraulically operated rear-loading doors take 3 minutes to open, with simultaneous extension of three-part folding ramp. This can be locked in intermediate position for direct loading from truck. Aft of ramp, centre panel of fuselage undersurface hinges upward; clamshell door to each side opens downward.

Completely unobstructed lower deck freight hold has titanium floor, attached 'mobilely' to lower fuselage structure to accommodate changes of temperature, with rollgangs and retractable attachments for cargo tiedowns. Load limits per tiedown fitting: 12,000 kgf (26,455 lbf) on main floor; 5,000 kgf (11,023 lbf) on rear ramp. Narrow catwalk along each sidewall facilitates access to, and mobility past, loaded freight. Payloads include largest CIS main battle tanks, complete missile systems, 12 standard ISO containers, oil well equipment and earth movers; HeavyLift/Volga-Dnepr aircraft previously transported Airbus wings in Europe. No personnel carried normally on lower deck in flight, because of low pressurisation, but can accommodate 360 troops (not including 88 on upper deck), plus two lavatories and oxygen bottles. Military aircraft equipped to airdrop up to 16 pallets, each of up to 4,500 kg (9,920 lb); or 268 paratroops in two passes. Medical evacuation capability is 288 stretchers and 28 attendants.

SYSTEMS: Automatic flight control system includes control loading for elevator and ailerons; stability augmentation for elevator and rudder; elevator trim and balance; elevator and rudder gear ratio system; and flight limit condition

Antonov An-124 heavy freight transport *(Yefim Gordon)*   0121102

Antonov An-124 flight deck *(Jane's/Paul Jackson)*   0051518

Antonov An-124 being loaded with a full-size mockup of the Baikal reusable space launcher   0121101

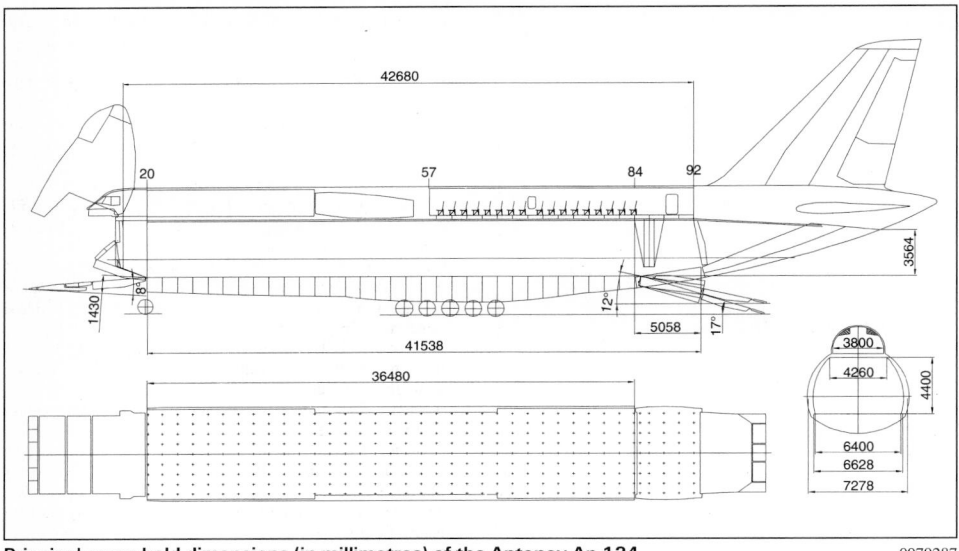

Principal cargo hold dimensions (in millimetres) of the Antonov An-124   0079287

restriction for elevator. Entire interior of aircraft is pressurised and air conditioned. Maximum pressure differential 0.55 bar (7.8 lb/sq in) on upper deck, 0.25 bar (3.55 lb/sq in) on lower deck. Four independent hydraulic systems. Quadruple redundant fly-by-wire flight control system, with mechanical emergency fifth channel to hydraulic control servos. Special secondary bus electrical system. Landing lights under nose and at front of each main landing gear fairing. APU in rear of each landing gear fairing for engine starting, can be operated in the air or on the ground to open loading doors for airdrop from rear or normal ground loading/unloading, as well as for supplying electrical, hydraulic and air conditioning systems. Bleed air anti-icing of wing leading-edges. Electro-impulse de-icing of fin and tailplane leading-edges.

AVIONICS: *Radar:* Two dielectric areas of nose visor enclose forward-looking weather radar and downward-looking ground-mapping/nav radar.

Flight: Hemispherical dielectric fairing above centre-fuselage for satellite nav receiver; quadruple INS; Loran and Omega.

Instrumentation: Conventional flight deck equipment, including automatic flight control system panel at top of glareshield, weather radar screen and moving map display forward of throttle and thrust reverse levers on centre console. No electronic flight displays. Dual attitude indicator/flight director and HSIs, and vertical tape engine instruments.

EQUIPMENT: Two electric travelling cranes in roof of hold, each with two lifting points, offer total lifting capacity of 20,000 kg (44,092 lb). First trial of 30,000 kg (66,139 lb) system in December 1999; development then launched of 40,000 kg (88,185 lb) lifting system. Two winches each pull a 3,000 kg (6,614 lb) load. Small two-face mirror, of V form, enables pilots to adjust their seating position until their eyes are reflected in the appropriate mirror, which ensures optimum field of view from flight deck.

DIMENSIONS, EXTERNAL:
Wing span	73.30 m (240 ft 5¾ in)
Wing aspect ratio	8.6
Length overall	69.10 m (226 ft 8½ in)
Fuselage max width	7.28 m (23 ft 10½ in)
Height overall	21.08 m (69 ft 2 in)
Wheel track	8.00 m (26 ft 3 in)

Wheelbase (centre row mainwheels)
	22.90 m (75 ft 1½ in)
Sill height: front door: normal	up to 2.79 m (9 ft 1¾ in)
kneeling	1.43 m (4 ft 8¼ in)
rear door, normal	up to 2.85 m (9 ft 4¼ in)

DIMENSIONS, INTERNAL:
Cargo hold: Length at floor:	
excl ramps	36.48 m (119 ft 8¼ in)
incl rear ramp	41.54 m (136 ft 3½ in)
Max length	42.68 m (140 ft 0¼ in)
Width: at floor	6.40 m (21 ft 0 in)
max	6.63 m (21 ft 9 in)
at ceiling	4.26 m (13 ft 11¾ in)
Max height	4.40 m (14 ft 5¼ in)
Area, incl ramp	2,650 m² (2,852 sq ft)
Volume, incl ramp	1,160 m³ (40,965 cu ft)
Passenger cabin: Width at floor	3.80 m (12 ft 5½ in)

AREAS:
Wings, gross	628.0 m² (6,760.0 sq ft)

WEIGHTS AND LOADINGS (A: basic An-124, B: An-124-100):
Operating weight empty: A	175,000 kg (385,800 lb)
Max payload: A	150,000 kg (330,700 lb)
B	120,000 kg (264,550 lb)
of which: rear ramp (A and B)	10,000 kg (22,046 lb)
Max fuel weight (An-124-210)	214,000 kg (471,790 lb)
Max T-O weight: A	405,000 kg (892,875 lb)
B	392,000 kg (864,200 lb)
Max ramp weight: B	398,000 kg (877,425 lb)
Max landing weight: B	330,000 kg (727,500 lb)
Max zero-fuel weight: A	325,000 kg (716,500 lb)
Max wing loading: A	644.9 kg/m² (132.09 lb/sq ft)
Max power loading: A	441 kg/kN (4.32 lb/lb st)

PERFORMANCE:
Max cruising speed: A	467 kt (865 km/h; 537 mph)
Normal cruising speed at 10,000-12,000 m (32,810-39,370 ft):	
A	432-459 kt (800-850 km/h; 497-528 mph)
Airdropping speed range	127-216 kt (235-400 km/h; 146-249 mph)
Approach speed:	
A	124-140 kt (230-260 km/h; 143-162 mph)
Max certified altitude	12,000 m (39,380 ft)
T-O balanced field length at max T-O weight:	
A	3,000 m (9,840 ft)

T-O run: A	2,520 m (8,270 ft)
Landing run at max landing weight: A	900 m (2,955 ft)
Range: with max payload:	
A	2,430 n miles (4,500 km; 2,795 miles)
B	2,591 n miles (4,800 km; 2,982 miles)
with 80,000 kg (176,375 lb) payload:	
A, B	4,535 n miles (8,400 km; 5,219 miles)
with 40,000 kg (88,184 lb) payload:	
A, B	6,479 n miles (12,000 km; 7,456 miles)
ferry, with max fuel:	
A, B	8,477 n miles (15,700 km; 9,755 miles)

OPERATIONAL NOISE LEVELS:
Stated to meet ICAO requirements

UPDATED

ANTONOV An-140

TYPE: Twin-turboprop airliner.

PROGRAMME: Announced 1993 as An-24 replacement; two prototypes with TV3-117 engines and static test airframe constructed at Kiev; rolled out 6 June 1997; first flight (UR-NTO) at Kiev-Svyatoshino (landing at Gostomel flight test centre, as base for certification trials) 17 September 1997; second airframe (0102) for static tests; second flying prototype (UR-NTP) rolled out 11 December 1997 and flew 26 December. Third (first production), with full systems, due to fly at Kharkov, early 1999, and undertake electromagnetic compatibility and climatic tests but maiden sortie delayed until 11 October 1999; this aircraft (UR-PWO) flew 41 sorties towards certification.

Certification to FAR/JAR 25/AP-25 and ICAO Part 3 Appendix 16 (FAR Pt 36/AP-36) standards intended; 940 hour certification programme began August 1998 and completed in late 1999; included cold trials at Arkhangelsk (by UR-NTO between 29 March and 1 May 1999, and by UR-NTP in Yakutia for 10 days in January 2000) and hot-and-high trials in Uzbekistan and Kyrgystan (by UR-NTP, concluding 3 September 1999). Following first series of flight testing, tailplanes of prototypes modified to obviate propwash-induced vibration, gaining 6° dihedral and shortened elevator horn balance. An-140's 1,000th hour flown (by UR-NTP) on 12 January 2000. Certification was achieved on 25 April 2000 coincident with that of TV3-117VMA-SBM1 engine and AV-140 propeller; some 1,000 sorties flown by April 2000.

Series production began 1999 at KhGAPP, Kharkov (where wings for prototypes were built) and at Aviacor, Samara, Russian Federation. Aviacor production intended to be 10 in 2000 (first in July) and to reach 30 per year by

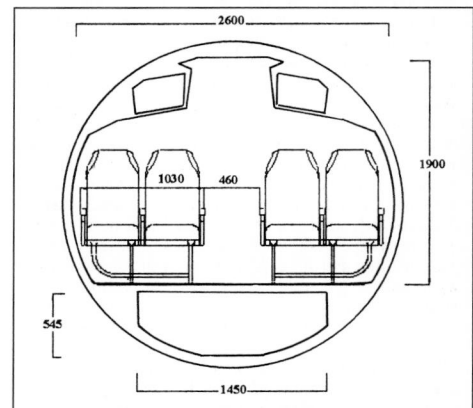

Cross-section of the An-140 cabin; dimensions in millimetres (*Jane's/Paul Jackson*)   0052938

First production Antonov An-140 twin-turboprop transport from Kharkov assembly (*Yefim Gordon*)   NEW/0525073

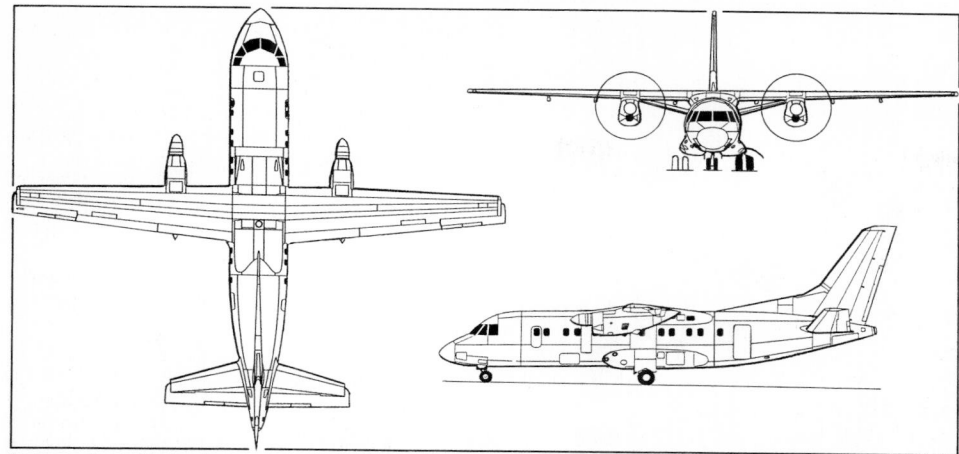

Antonov An-140 short-range transport (*Jane's/James Goulding*)   0079286

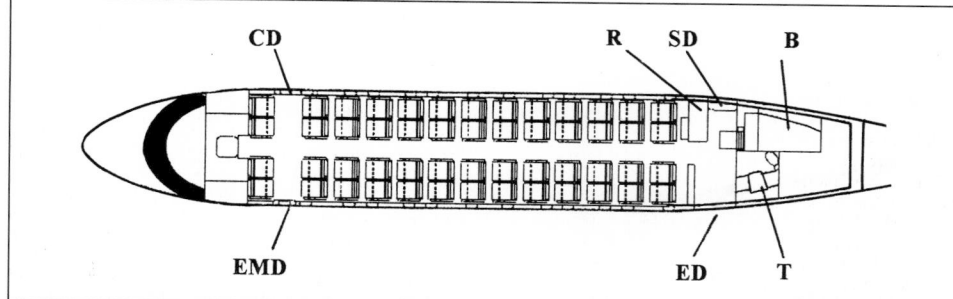

Antonov An-140 cabin seating for 52 passengers (*Jane's/Paul Jackson*)
B: baggage, CD: cargo door, ED: entrance door, EMD: emergency door, R: refreshments, SD: service door, T: lavatory
0051523

2001, but reduced to combined total of eight in 2000-01; five under assembly at Aviacor by May 2000, but none yet ordered and maiden flight of first slipped to late 2000. KhGAPP has capacity for 40 per year, initial batch comprising seven. First Kharkov production aircraft delivered March 2002 and made type's Western debut at Farnborough in July 2002.

Production share agreement of 1998 assigns empennage to Antonov; engine nacelles, wing and associated control surfaces to KhGAPP; landing gear to Aviagregat at Samara and Youzhmash at Dniepropetrovsk; and fuselage to Aviacor (incorporating Avio Interiors fittings from Italy).

Agreement signed February 1996 for assembly by HESA at rate of 12 a year from 1999, progressing to local parts manufacture, in new plant at Esfahan, Iran, with Ukrainian assistance (see AIO entry in Iranian section); first two kits shipped to Iran by late 2000; initial aircraft first flown 7 February 2001. In late 1999, Reims Aviation (which see in French section) was considering promotion of a Westernised version of Kiev-built An-140, but nothing more announced by mid-2002.

CURRENT VERSIONS: **An-140:** *As described.*

An-140A: Aeroflot version with PW127A engines; for regional services in western Russia.

An-140T: Proposed freighter with large door, port side, rear. Convertible **An-140TK** will be similar.

An-140-100: Proposed 68-seat version; fuselage stretched by 3.80 m (12 ft 5½ in); under development by 1997.

An-142: Under development for 2001 first flight; forward-retracting rear loading ramp similar to An-26. No further reports by mid-2002.

Iran 140: Licence-built by HESA (which see).

An-140 Military: Patrol, surveillance, photographic and similar variants proposed for military operators.

CUSTOMERS: Air Ukraine letter of intent for up to 40 by 2010 signed 17 September 1997 (first flight); initial four firm conversions made June 1999 (for manufacture at Kharkov), specifying deliveries from 2000; however, initial recipient from Ukrainian production is Ikar Airlines, with five on order from January 2000 contract, first being UR-PWO, delivered in 2001, following testing at Antonov. First order for 15 placed by republic of Sakha-Yakutia, Russian Federation, mid-1998, requiring 10 to be delivered to Sakha Airlines from Samara production in 2000-04. First Samara-built aircraft destined for Samara Airlines in first quarter of 2002. Aeroflot letter of intent for 50 PW127-powered An-140As signed mid-1999 (specifying Samara production). Tyumenaviatrans reportedly holds option on 25 An-140s, but this had been relegated to expression of interest (also from Polar Airlines and Mirny) when An-140 toured Siberia in early 2000. Odessa Airlines ordered five from Kharkov production, of which first (UR-14001) entered revenue-earning service on 29 March 2002, having been handed over to Ukrtransleasing on 4 March. Launch order for Iran 140 is 20 for Iran Asseman Airlines; Iran Air also will be operator; Iranian licence initially for 80 aircraft, but interest reported, late 1998, in

building further 160. Estimated sales of 645 by 2011, including 430 to Russia and 70 to Ukraine.

COSTS: US$7 million (2000).

DESIGN FEATURES: Light transport, designed to be capable of autonomous operation from airfields with unprepared runways at all altitudes and in all weathers, providing airline-standard comfort. International certification and various engine options to maximise sales prospects. Conventional high-wing monoplane; tapered wing with unswept leading-edge; sweptback fin and tailplane, latter with 6° dihedral; engines mounted underwing. Maintenance target of 6.5 mmh/fh. Service life 50,000 landings/50,000 hours/25 years.

FLYING CONTROLS: Conventional and manual. Control surfaces all horn balanced; ailerons with trim tab in each (two-section tab starboard); elevator with two-section tab in each; rudder with large single tab. Two-section flaps in each wing.

STRUCTURE: Largely of aluminium, with some titanium.

LANDING GEAR: Retractable tricycle type by Pivdennyi; twin Rubin wheels on each unit; nosewheels retract forward, mainwheels into fairings each side of lower fuselage. Mainwheels size 810×320-330; nosewheels 600×220-250. Rubin braking system. Able to operate from gravel or unpaved fields.

POWER PLANT: Two 1,839 kW (2,466 shp) AI-30 Series 1 turboprops (Klimov TV3-117VMA-SBM1 built under licence at Zaporozhye, Ukraine, by Motor-Sich), driving AV-140 propellers; optionally, two 1,864 kW (2,500 shp) Pratt & Whitney Canada PW127A turboprops, driving Hamilton Sundstrand 247F propellers. FED fuel-management system; Star engine control system.

ACCOMMODATION: Flight crew of two, plus cabin attendant; basic seating for 52 passengers, four-abreast with centre aisle, at 75 cm (30 in) pitch, or 48 at 81 cm (32 in) pitch.

Main passenger door with airstairs, at rear of cabin on port side, with service door opposite; emergency exit port side at front of cabin; cargo door starboard side, front. Coat stowage, galley and toilet at rear of cabin. Baggage/freight compartment at rear of cabin, plus forward underfloor freight hold, with door on port side. Cargo door on starboard side, forward part of cabin floor reinforced, and detachable equipment provided, enabling 1,900 to 3,650 kg (4,188 to 8,046 lb) of palletised cargo and 36 to 20 passengers to be carried with forward rows of seats removed. Overhead baggage lockers. Accommodation air conditioned and pressurised.

SYSTEMS: Motor-Sich AI-9-3B APU in rear fuselage. FED hydraulic system; Nauka air conditioning. Kommunar anti-icing and air conditioning control systems. Auxilec generators; Eros oxygen system.

Avionika SAU-140 digital flight control system offered for new installation, or retrofit for operation with existing analogue avionics.

AVIONICS: *Comms:* Satori ELT96 ELT. Television Engineering entertainment system.

Radar: Buran A-140 weather radar

Flight: Ukrainian Radio INS; Orizon Navigation SN-3301 GPS; VNIIRA-Navigator VND-94 VOR; Kurs-93M nav computer; other navigation aids by Russian Radio Equipment.

*DIMENSIONS, EXTERNAL:

Wing span	24.505 m (80 ft 4¾ in)
Length overall	22.605 m (74 ft 2 in)
Height overall	8.225 m (26 ft 11¾ in)
Wheel track (c/l shock-absorbers)	3.18 m (10 ft 5¼ in)
Wheelbase	8.01 m (26 ft 3¼ in)
Propeller diameter (AI-30)	3.60 m (11 ft 9¾ in)
Distance between propeller centres	8.20 m (26 ft 10¾ in)
Main cabin door: Height	1.605 m (5 ft 3¼ in)
Width	0.98 m (3 ft 3 in)
Emergency exit: Height	1.20 m (3 ft 11¼ in)
Width	0.515 m (1 ft 8¼ in)
Service door: Height	1.28 m (4 ft 2½ in)
Width	0.63 m (2 ft 0¾ in)
Cargo door: Height	1.34 m (4 ft 4¾ in)
Width	1.00 m (3 ft 3¼ in)
Underfloor freight hold door:	
Height	0.90 m (2 ft 11½ in)
Width	1.02 m (3 ft 4¼ in)

* *Data from Antonov design bureau; Kharkov figures mostly different*

DIMENSIONS, INTERNAL:

Cabin: Length:	
excl flight deck, galley/toilet area	10.50 m (34 ft 5¼ in)
incl galley/toilet and baggage	14.30 m (46 ft 11 in)
Max width	2.60 m (8 ft 6¼ in)
Max height	1.90 m (6 ft 2¾ in)
Volume	65.5 m³ (2,313 cu ft)
Aft baggage compartment: Volume	6.0 m³ (212 cu ft)
Underfloor freight hold: Length	3.98 m (13 ft 0¾ in)
Max width	1.45 m (4 ft 9 in)
Max height	0.545 m (1 ft 9¾ in)
Volume	3.0 m³ (106 cu ft)
Overhead baggage lockers:	
Volume (total)	2.4 m³ (85 cu ft)

WEIGHTS AND LOADINGS:

Baggage capacity	1,840 kg (4,057 lb)
Max payload	6,000 kg (13,227 lb)
Max fuel weight	4,370 kg (9,634 lb)
Max T-O weight	19,150 kg (42,218 lb)
Max landing weight	19,100 kg (42,108 lb)
Max zero-fuel weight	17,800 kg (39,242 lb)
Max power loading	5.21 kg/kW (8.56 lb/shp)

PERFORMANCE:

Max cruising speed at 7,200 m (23,620 ft)	
	310 kt (575 km/h; 357 mph)
Econ cruising speed at 7,200 m (23,620 ft)	
	280 kt (520 km/h; 323 mph)
Nominal cruising altitude	
	7,200-7,500 m (23,620-24,600 ft)

Odessa Airlines (Odeski Avialinii) began services with its first Antonov An-140 on 29 March 2002
(*Jane's/Paul Jackson*)
NEW/0525074

Balanced runway length   1,350 m (4,430 ft)
Design range at 280 kt (520 km/h; 323 mph) at 7,200 m
(23,620 ft), no reserves:
  with 6,000 kg (13,227 lb) payload
                             486 n miles (900 km; 559 miles)
  with 52 passengers:
    PW127   1,349 n miles (2,500 km; 1,553 miles)
    AI-30   1,133 n miles (2,100 km; 1,304 miles)
  with 46 passengers:
    PW127   1,646 n miles (3,050 km; 1,895 miles)
  with max fuel and 33 passengers:
    PW127   2,213 n miles (4,100 km; 2,548 miles)
    AI-30   1,997 n miles (3,700 km; 2,299 miles)
                                              UPDATED

Model of Antonov An-148 regional airliner (*Yefim Gordon*)   NEW/0525943

ANTONOV An-148

TYPE: Regional jet airliner.
PROGRAMME: In September 2001, Antonov revealed that it
was working on the design of an 80-seat, twin-jet (Progress
D-36-5AF), 1,349 n mile (2,500 km; 1,553 mile) range
airliner which could fly in 2004. Configuration is similar to
the An-74T-300; stretched (100-seat) and shrunk (55-seat)
versions are also planned.
   The first of three prototypes was stated to be under
construction by early 2002, with series manufacture
planned to be by KhGAPP at Kharkov and Ulan-Ude's
UUAP. Cost US$15 million to US$20 million (2002
estimate). Ukrainian participants include Motor Sich and
ZhEDB Progress.

                                              UPDATED

ANTONOV An-225 MRIYA

English name: Dream
NATO reporting name: Cossack
TYPE: Outsize freighter.
PROGRAMME: The world's largest aircraft was ordered in 1985
and first flew on 21 December 1988; it was last described in
the 1995-96 *Jane's*, the sole prototype having been in
storage at Gostomel since April 1994, with total flight time
of 671 hours in 339 sorties. In 1998, Antonov announced
that it was proceeding with assembly of the second
prototype, which had previously been abandoned in partly
complete condition, and was planning to return the first
aircraft to airworthy status; these projects were to cost
US$50 million and US$20 million, respectively. By July
2000, Antonov's own freight airline was concentrating on
refurbishing the prototype; this flew on 7 May 2001 and
was exhibited at Paris in the following month; retrofit
included collision-avoidance avionics and engine noise
suppression required for international service; crew
reduced to five; redesignated **An-225-100**. On 11
September 2001 lifted 253.82 tonne (559,575 lb) load to
2,000 m (6,560 ft) establishing new world record; six
further records established during same sortie: first
commercial charter flown 3 January 2002 from Stuttgart
(Germany) to Thumrait (Oman) with 188 tonne load. By
mid-2001, work on second aircraft (then 65 per cent
complete) was said to have been fully resumed but, by

Antonov An-225 outsize freighter with landing gear deployed (*Jane's/Paul Jackson*)   NEW/0525075

mid-2002, it emerged that none of the US$50 million to
US$100 million required for the project (to last 2½ years)
had been committed.
   Prompted by predictions of a growing market for ultra-
large air transports, including Volga-Dnepr requirement
for two or three, the Aviastar plant at Ulyanovsk (Russia)
has evaluated the prospects for placing the An-225 in series

production. Consideration also has been given to replacing
the aircraft's 230 kN (51,590 lb st) Progress D-18T
engines by 411 kN (91,300 lb st) Rolls-Royce Trent 892s
or 436 kN (98,000 lb st) Pratt & Whitney PW4098s.

                                              UPDATED

AVIAIMPEX

AVIAIMPEX JSC

Kiev
CHIEF DESIGNER: Mikhail Yu Kuchin
YANHOL PROJECT CHIEF: Sergey Tyurin

Aviation division of TVT Corporation formed to introduce
helicopter design and production to Ukraine. Initial liaison
with Aerokopter Ltd of Poltava (formed 14 December 1999);
however, on 3 May 2000 Aerokopter design bureau divided
and both elements became autonomous.
   Future plans include an eight-seat, turbine-powered
helicopter.

                                              NEW ENTRY

AVIAIMPEX YANHOL

English name: Angel.
TYPE: Three-seat helicopter.
PROGRAMME: Presented to the media on 31 August 2001,
when prototype structurally complete. First flight planned
for June 2002, followed by deliveries in 2003. By May
2002, Aviaimpex had revealed plans for gunship and UAV
versions.
CUSTOMERS: Options placed on 400 (mostly in two-seat
configuration) by early 2002, including Ukrainian Defence
Ministry (reportedly 100 for pilot training), Internal
Affairs Ministry (patrol), Georgian government and
several city administrations, including Moscow Mayor's
Office.
DESIGN FEATURES: Pod-and-boom configuration, with T tail
and landing skids. Three-blade main and two-blade tail
rotors.
COSTS: US$120,000 in Ukraine; US$150,000 export (2002).
Operating cost US$70 per hour (2002).

Aviaimpex Yanhol light helicopter (*Yefim Gordon*)   NEW/0525944

POWER PLANT: Two Rotax 912 ULS flat-four piston engines,
   each 73.5 kW (98.6 hp).
WEIGHTS AND LOADINGS:
   Max payload   350 kg (772 lb)
   Max T-O weight   870 kg (1,918 lb)

PERFORMANCE:
   Max cruising speed   89 kt (165 km/h; 103 mph)
   Range with max fuel   183 n miles (340 km; 211 miles)
                                              NEW ENTRY

AVIANT

AVIANT, KIEVSKY GOSUDARSTVENNY AVIATSIONNY ZAVOD (Aviant, Kiev State Aviation Plant)
prospekt Pobedy 100/1, 03062 Kiev
Tel: (+38 044) 441 52 01
Fax: (+38 044) 442 43 85
e-mail: aviant@carrier.kiev.ua
GENERAL DIRECTOR: Vasily K Tselykh
MARKETING DIRECTOR: Fedir I Hnatenko

The Kiev Aviation Plant has effectively terminated work on the Antonov An-32 twin-turboprop transport and An-124 heavy transport. However, it built two prototypes of the An-70 four-propfan transport and shares in production with Aviacor at Samara. The two firms will also manufacture the Tupolev Tu-334 twin-turbofan transport. Non-aviation activities include manufacture of trolley-buses, pressure chambers and customer goods. Recently contributed to manufacture of prototype Aviaimpex Yanhol helicopter prototype (which see).

Aviant previously built 3,320 An-2s, 1,028 An-24s, 1,402 An-26s, 123 An-30s, 361 An-32s and 17 An-124s.
UPDATED

Tupolev Tu-334 under construction at Aviant*(Yefim Gordon)* *NEW*/0525945

Uncompleted Antonov An-32Bs at the Aviant plant
(Yefim Gordon)
0121104

IKAR

FIRMA AEROKLUB 'IKAR' ('Ikar' Aero Club Firm)
ulitsa Vyborgskaya 99, 03115 Kiev
Tel/Fax: (+38 044) 458 48 16
e-mail: karpes@aviaikar.kiev.ua
Web: http://users.i.com.ua/~a-sergey

'Ikar' Aero Club was founded in 1995 to develop, build and provide engineering and maintenance support for light aircraft. In addition to the Ai-10 Lis, it has built the Ai-9 Lis of similar configuration; and Aist pod-and-boom ultralight. Projects include the Zhuravlik powered hang-glider; four-seat Ai-20 Dedal; Ai-30 Chirok four-seat amphibian; Ai-40 Lelica agricultural sprayer; tandem-seat Ai-50 Skvorets; and Ai-60 four-seat light twin.
NEW ENTRY

IKAR Ai-10 IKAR
TYPE: Side-by-side ultralight.
PROGRAMME: First flown in 1999.
COSTS: US$35,000 to US$39,000, according to equipment (2001).
DESIGN FEATURES: High-mounted, constant-chord, strut-braced wing. Sweptback fin and mid-mounted tailplane.
FLYING CONTROLS: Conventional and manual. Flaps.
STRUCTURE: Metal airframe. Fabric-covered wing.
LANDING GEAR: Tailwheel type; fixed.
POWER PLANT: One Rotax flat-four driving a three-blade, wooden propeller: 59.6 kW (79.9 hp) Rotax 912 UL; 73.5 kW (98.6 hp) 912 ULS; or 84.6 kW (113.4 hp) 914. Fuel capacity 88 litres (23.2 US gallons; 19.4 Imp gallons).

Ikar Ai-10 Ikar side-by-side ultralight *NEW*/0525358

DIMENSIONS, EXTERNAL:
Wing span	9.06 m (29 ft 8¾ in)
Length overall	6.21 m (20 ft 4½ in)
Height overall	1.70 m (5 ft 7 in)

AREAS:
Wings, gross	11.74 m² (126.4 sq ft)

WEIGHTS AND LOADINGS:
Weight empty	284 kg (626 lb)
Max T-O weight: Rotax 912 ULS	510 kg (1,124 lb)
Rotax 912 UL	450 kg (992 lb)

PERFORMANCE (Rotax 912 ULS):
Cruising speed	97 kt (180 km/h; 112 mph)
Stalling speed	27 kt (50 km/h; 32 mph)
Max rate of climb at S/L	360 m (1,181 ft)/min
T-O run	80 m (265 ft)
Landing run	60 m (200 ft)
Range with max fuel	540 n miles (1,000 km; 621 miles)
g limits	+6/-3

NEW ENTRY

KhGAPP

KHARKOVSKOYE GOSUDARSTVENNOYE AVIATSIONNOYE PROIZVODST-VENNOYE PREDPRIYATIE (Kharkov State Aviation Production Enterprise
ulitsa Sumskaya 134, 61023 Kharkov
Tel: (+38 0572) 43 19 85 and 43 08 32
Fax: (+38 0572) 47 52 81 and 47 80 01
e-mail: itl589@online.kharkov.ua
GENERAL DIRECTOR: Pavel O Naumenko

Established in 1926 as GAZ 135, Kharkov Aircraft Production Association has built military, passenger and freight aircraft, including the Su-2, Yak-18, MiG-15, Tu-104, Tu-124 and Tu-134. Export of its products began in 1964. ISO 9002 was gained in 1998.

Current production is centred on the Antonov An-74 in various versions, including the An-74-300, with underslung engines. Manufacture of the twin-turboprop An-140, described in Antonov entry, is now under way.

Also provides support services, including rectification under warranty, flying training, technical training, spare parts provision, ground equipment, test equipment and turnkey maintenance facilities.
UPDATED

MOTOR-SICH

OTKRITOYE AKTSIONERNOYE OBSHCHESTVO MOTOR-SICH (Motor-Sich JSC)
ulitsa 8 Marta 15, Zaporozh'e 69068
Tel: (+38 0612) 61 47 77

Fax: (+38 0612) 65 60 07
e-mail: motor@motorsich.com
Web: http://www.ukrainetrade.com/motorsich
GENERAL DIRECTOR: Vyacheslav Boguslaev

Long-established engine manufacturer Motor-Sich was confirmed in August 2000 as production centre for those

Kamov Ka-226 utility helicopters required for use in Ukraine. These to be powered by indigenous ZMKB AI-450 engines and employ components sourced entirely from Ukrainian manufacturers. Details of Ka-226 appear in Russian section.
VERIFIED

UNITED KINGDOM

BAE

BAE SYSTEMS plc

HEADQUARTERS: Warwick House, PO Box 87, Farnborough
Aerospace Centre, Farnborough, Hampshire GU14 6YU
Tel: (+44 1252) 37 32 32
Fax: (+44 1252) 38 30 00
Web: http://www.baesystems.com
CHAIRMAN: Sir Richard Evans
VICE-CHAIRMAN: Sir Charles Masefield
CEO: Mike Turner
CHIEF OPERATING OFFICERS: Chris Geoghegan (Operational,
Capital and Shared Services)

Ian King
Steve Mogford (Programmes)
Mark Ronald (North America)
MARKETING DIRECTOR: Mike Rouse
DIRECTOR, CORPORATE COMMUNICATIONS: Hugh Colver

On 19 January 1999, British Aerospace (BAe) announced
that it planned to merge with the Marconi Electronic Systems
business, then part of GEC-Marconi. Following regulatory
approval, BAE Systems was established on 30 November
1999. Sales in 2001 totalled £13.1 billion; order book at 31
December 2001 was £43.8 billion (including joint ventures).
Employees in 2001 totalled nearly 100,000 located in UK,

USA, Sweden, Saudi Arabia, France, Italy, Germany,
Australia and Canada.
Aircraft design, manufacturing and support divisions of
BAE Systems are as under:
Air Systems Group
Follows this entry
Aircraft Services Group
Follows Air Systems Group
Regional Aircraft
Follows Aircraft Services Group

UPDATED

AIR SYSTEMS GROUP

Group formed January 2002 to manage BAE Systems'
actvities in Hawk, Nimrod, Eurofighter Typhoon and
Lockheed Martin F-35 JSF programmes. Personnel total
10,000 at Warton (HQ), Samlesbury and Brough sites.
BROUGH: Brough, East Yorkshire HU15 1EQ
Tel: (+44 1482) 66 71 21
Fax: (+44 1482) 66 66 25

SAMLESBURY: Samlesbury, Balderstone, Lancashire BB2 7LF
Tel: (+44 1254) 481 23 71
Fax: (+44 1254) 481 36 23

WARTON: Warton Aerodrome, Preston, Lancashire PR4 1AX
Tel: (+44 1772) 63 33 33
Fax: (+44 1772) 63 47 24

UPDATED

BAE SYSTEMS HAWK 50, 60 and 100 SERIES

RAF/FAA designations: Hawk T. Mks 1 and 1A
US Navy designation: T-45A and T-45C Goshawk
Canadian Forces designation: CT-155
TYPE: Advanced jet trainer/light attack jet.
PROGRAMME: HS P1182 Hawk first flew 21 August 1974;
early history in 1989-90 *Jane's*; first-generation Hawk
remains available and is marketed with advanced 100
Series and single-seat 200 Series (detailed separately) to
meet customers' requirements; Hawk design leadership
transferred from Kingston to Brough 1988, and final
assembly and flight test from Dunsfold to Warton 1989.
Brough subsequently assumed responsibility for final
assembly, although first flights still at Warton.
Hawk 50 Series main exports made December 1980 to
October 1985; largely supplanted by 60 Series; Hawk 100
enhanced ground attack export model announced
mid-1982; first flight of 100 Series aerodynamic prototype
(G-HAWK/ZA101 converted as Mk 100 demonstrator) 21
October 1987; trials of wingtip Sidewinder rails started at
Warton in April 1990. Warton assembly line officially
opened 24 October 1991.

Hawk T. Mk 1A of No. 100 Squadron, RAF   NEW/0118414

CURRENT VERSIONS: **Hawk T. Mk 1:** Two-seater for RAF
advanced flying and weapon training; 23.1 kN (5,200 lb st)
Adour 151-01 (-02 in Red Arrows aircraft) non-
afterburning turbofan; two dry underwing hardpoints;
underbelly 30 mm gun pack; three-position flaps; simple
weapon sight in some aircraft of No. 4 FTS. Following
basic Tucano stage, future RAF fast-jet pilots undertake
100 hours of advanced flying, weapons and tactical
training with No. 4 FTS at Valley. Hawks introduced to
navigator training syllabus at No. 6 FTS, Finningley; first
delivery 10 September 1992; role to No. 4 FTS in 1995.
No. 100 Squadron received 15 Mks 1/1A from September
1991, replacing Canberras in target-towing role; initial
seven (subsequently increased to 15) Mk 1s loaned to
Royal Navy's Fleet Requirements and Air Direction Unit
(now Fleet Support Air Tasking Organisation) at Culdrose,
first arriving on 6 April 1994. In August 1996, Skyforce

Avionics Ltd Skymap II GPS moving map displays were
received for Hawks of the Red Arrows and No. 100
Squadron.
Hawk T. Mk 1A: Contract January 1983 to wire 89
Hawks (including Red Arrows) for AIM-9L Sidewinder on
each inboard wing pylon and optional activation of
previously unused outer wing hardpoints; last conversion
redelivered 30 May 1986; 72 (reduced to 50 by 1993
defence cuts) NATO-declared, for point defence and
participation in RAF's Mixed Fighter Force, to accompany
radar-equipped Tornado ADVs on medium-range air
defence sorties.
RAF Hawk re-wing programme began 1989; initial 85
wings completed by BAe in 1993; delivery of second batch
of 59 began November 1993 and completed in 1995.
Rebuild programme by BAe for 80 RAF Hawks authorised
27 December 1998 to replace centre and rear fuselage,
extending service life to 2010; work undertaken at DARA
St Athan and BAE Systems at Brough; first aircraft,
XX348, redelivered April 2000; last due in 2004. Avionics
upgrade plans being formulated in Staff Requirement (Air)
449; primary aim is 'glass cockpit'.
Hawk T. Mk 1W: Following re-winging, 24 RAF
Hawk T. Mk 1s gained the ability to carry stores on two
underwing pylons, although not the centreline gun pod.
The alternative designation **T. Mk 1FTS** is also used for
this modification.
Hawk 50 Series: Initial export version with 23.1 kN
(5,200 lb st) Adour 851 turbofan; maximum operating
weight increased by 30 per cent, disposable load by 70 per
cent, range by 30 per cent; revised tailcone shape to
improve directional stability at high speed; larger nose
equipment bay; four wing pylons, all configured for single
or twin store carriage; each pylon cleared for 515 kg
(1,135 lb) load; wet inboard pylons for 455 litre (122 US
gallon; 100 Imp gallon) fuel tanks; improved cockpit, with
angle of attack indication, fully aerobatic twin-gyro AHRS
and new weapon control panel; optional braking
parachute; suitable for day VMC ground attack and armed
reconnaissance with camera/sensor pod.
T-45A/C Goshawk: US Navy version (see Boeing/
BAE Systems in International section).
Hawk 60 Series: Development of 50 Series with
25.4 kN (5,700 lb st) Adour 861 turbofan; leading-edge
devices and four-position flaps to improve lift capability;
low-friction nose leg, strengthened wheels and tyres, and
adaptive anti-skid system; 591 litre (156 US gallon; 130
Imp gallon) drop tanks; provision for Sidewinder or Magic
AAMs; maximum operating weight increased by further
17 per cent over 50 Series, disposable load by 33 per cent
and range by 30 per cent; improved field performance,
acceleration, rate of climb and turn rate. Mk 67 is 'long-
nosed' version with nosewheel steering, supplied to South
Korea. Abu Dhabi upgraded 15 surviving Mk 63s to Mk

BAE Systems Hawk Mk 127 of Royal Australian Air Force   NEW/0533729

BAE Systems Hawk 100 with wingtip Sidewinders and additional side view (top) of Hawk 60 Series
(Jane's/Dennis Punnett)

Unconfirmed requirements reportedly include: Abu Dhabi, seven further Hawk Mk 63s; Brunei, six Hawk 100s and four Hawk 200s; India requested quotation in September 1999 for 92 Hawks, later amended to 66, of which eight kits and 42 complete aircraft to be assembled by HAL at Bangalore, and pricing negotiations were continuing in 2002,ᵇ following Anglo-Indian political reaffirmation of purchase plans on 12 December 2000; Kuwait, six attrition replacements; Philippines, commitment announced August 1991 for 12 Hawks, but abandoned.

COSTS: South African purchase of 24 Hawk LIFTs estimated at R4.7 billion (1998). Canadian contract for 20 valued at £400 million (1999). Hawk centre/rear fuselage replacement for 80 RAF aircraft valued at £100+ million (2000).

DESIGN FEATURES: Fully aerobatic two-seat advanced jet trainer, adaptable for ground attack and air defence; design capable of other optional roles, with wing improvements on developed Series to enhance combat efficiency. Low wing and mid-mounted, anhedral, sweptback tailplane; air intake on each side of fuselage, forward of wing leading-edge; single non-afterburning engine; elevated rear cockpit to enhance forward view; two strakes below rear fuselage; smurfs (refer Hawk 200 entry) on 100 Series; optional underwing hardpoints; wingtip AAM rails (100 Series).

Wing thickness/chord ratio 10.9 per cent at root, 9 per cent at tip; dihedral 2°; sweepback 26° on leading-edge, 21° 30′ at quarter-chord.

FLYING CONTROLS: Conventional and assisted. Ailerons and one-piece all-moving tailplane actuated hydraulically by tandem actuators; rudder manually actuated, with electric trim tab. Hydraulically actuated double-slotted flaps, outboard 300 mm (12 in) of flap vanes normally deleted; small fence on each wing leading-edge; 100 and 200 Series use special 'combat wing' with full-width flap vanes (refer Hawk 200 entry); large airbrake under rear fuselage, aft of wings. Hydraulic yaw damper on 100 Series rudder.

STRUCTURE: Aluminium alloy; one-piece wing, with machined torsion box of two main spars, auxiliary spar, ribs and skins with integral stringers; most of box forms integral fuel tank; honeycomb-filled ailerons; composites wing fences; frames and stringers fuselage. Wing attached to fuselage by six bolts.

LANDING GEAR: Wide-track, hydraulically retractable tricycle type, with single wheel on each unit. AP Precision Hydraulics oleos and jacks. Main units retract inward into wing, ahead of front spar; castoring (optionally power-steered) nosewheel retracts forward. Dunlop mainwheels, brakes and tyres size 6.50-10 (14 ply) tubeless, pressure 9.86 bar (143 lb/sq in). (Hawk Srs 60 and 100 mainwheel pressure 17.23 bar; 250 lb/sq in at 9,100 kg; 20,061 lb T-O weight.) Nosewheel and tyre size 16×4.4 (8 ply) tubeless, pressure 8.27 bar (120 lb/sq in). Tail bumper fairing under rear fuselage. Anti-skid wheel brakes. Tail braking parachute, diameter 2.64 m (8 ft 8 in), on Mks 52/53 and all 60 and 100 Series aircraft.

POWER PLANT: One Rolls-Royce Turbomeca Adour non-afterburning turbofan, as described under Current Versions. Adour Mk 861A for Switzerland assembled locally by Sulzer Brothers. Adour Mk 951, available for new-build or retrofitted Hawks rated at 31.1 kN (7,000 lb st) and has doubled TBO of 4,000 hours. Engine starting by Microturbo integral gas-turbine starter. Fuel in one fuselage bag tank of 832 litres (220 US gallons; 183 Imp gallons) capacity and integral wing tank of 823 litres (217 US gallons; 181 Imp gallons); total fuel capacity 1,655 litres (437 US gallons; 364 Imp gallons). Pressure refuelling point near front of port engine air intake trunk; gravity point on top of fuselage. Provision for carrying one 455 or 591 litre (120 or 156 US gallon; 100 or 130 Imp gallon) drop tank on each inboard underwing pylon, according to Series.

63A/B from 1991, incorporating Adour Mk 871 and new combat wing (four pylons and wingtip AAM rails); first two rebuilds at Brough; remainder at Al Dhafra. Surviving Kuwaiti Hawks were refurbished at Dunsfold in 1998-2000.

Description applies to Hawk 60 Series, except where otherwise specified.

Hawk 100 Series: Enhanced ground attack development of 60 Series, announced mid-1982, to exploit Hawk's stores carrying capability; two-seater, with perhaps pilot only on combat missions; 26.0 kN (5,845 lb st) Adour Mk 871 turbofan; new combat wing incorporating fixed leading-edge droop for increased lift and manoeuvrability from M0.3 to M0.7; full-width flap vanes; manually selected combat flaps; detail changes to wing dressing; structural provision for wingtip missile pylons; MIL-STD-1553B databus; advanced Smiths Industries HUDWAC and new air data sensor package with optional laser ranging and FLIR in extended nose; improved weapons management system allowing preselection in flight and display of weapon status; manual or automatic weapon release; passive radar warning; HOTAS controls; full-colour multipurpose CRT display in each cockpit; provision for ECM pod.

Demonstrator ZA101 (see Programme). Production prototype Mk 102D (ZJ100) flown 29 February 1992. Early orders from Abu Dhabi (placed 1989), Indonesia (signed June 1993), Malaysia (signed 10 December 1990) and Oman (signed 30 July 1990). FLIR, laser ranger and Sky Guardian RWR in Omani aircraft.

Selected by Australia in November 1996 as next-generation lead-in fighter trainer; order signed 24 June 1997 for 33 aircraft (MoU with ASTA March 1991 for licensed production if Hawk 100/200 chosen for RAAF trainer requirement); Australian Hawk Mk 127s have new, advanced instrumentation resembling that of the F-18 Hornet and including an integrated Smiths system of three 127 mm (5 in) square colour screens in each cockpit, HUD, upgraded mission computer, engine life computer and stores management system. First Australian Hawk (A27-01) flew (in UK, temporarily as ZJ632) on 16 December 1999. See BAE Systems entry in Australian section of 2002-03 and earlier *Jane's*.

BAE, Bombardier and partners supplying Raytheon T-6A Harvard II and Hawk Mk 115 as equipment of Canadian-based NATO Flying Training in Canada (NFTC) programme; contract agreed 17 November 1997 and formally issued on 12 May 1998 for 18, plus eight options, two of latter being taken up in 1999; further two stated to be required in early 2000, after Singapore joined programme; order total increased to 22 by early 2002. First aircraft, 155201, arrived in Canada 4 July 2000; officially received 6 July; instructor training began 12 July 2000.

Hawk LIFT: Variant of Series 100 as lead-in fighter trainer (LIFT). Features include 'combat wing', Adour Mk 871 or Mk 951 engine, new three MFD 'glass cockpit' with NVG compatibility, MIL-STD-1553B digital databus, revised mission planning system/data transfer unit, INS/GPS, OBOGS, APU, HUMS, new HUD and HOTAS controls. Provision for FLIR and aerial refuelling. Initial customer is South Africa (Mk 120). Offered to RAF (Mk 128) to meet MFTS (military flying training system) requirement for some 30 to enter service in 2007. Maximum weapon load 3,000 kg (6,614 lb).

Hawk 200 Series: Single-seat multirole version (described separately).

Hawk 100NDA: New Development Aircraft. Testbed for Adour Mk 951 turbofan first flew (ZJ951) at Warton 5 August 2002 (initially with interim standard engine). To South Africa for further trials associated with SAAF Hawk Mk 120.

CUSTOMERS: See table. BAE delivered 20 Hawks in 1996 and 22 in 1997; none in 1998; deliveries resumed in 1999 with 10 to Indonesia before supply of remaining six was suspended by UK government embargo of 11 September 1999; deliveries re-authorised in early 2000, but without certain US-sourced components. Deliveries in 2000 totalled 16 (eight Australian and eight Canadian), not including Australian assembly. In 2001, UK production line delivered 13 (three to Australia and 10 to Canada); 2002 production comprised two to Canada and one company trials aircraft.

Hawk LIFT selected for purchase by South Africa, as announced 18 November 1998; order for 12, plus 12 options, announced 15 September 1999 and signed 3 December 1999; these to be first Hawks with Adour Mk 951 engines. Bahrain announced selection of Hawk on 22 July 2002; version and quantity to be decided.

Hawk Mk 127 front cockpit   *NEW*/0533728

BAE Systems Hawk Mk 115 in Canada, where it is known as the CT-155   *NEW*/0533726

Hawk Mk 127 rear cockpit *NEW*/0533727

<table>
<tr><th colspan="6">BAE SYSTEMS HAWK CUSTOMERS</th></tr>
<tr><th>Customer</th><th>Qty</th><th>Mark</th><th>First aircraft</th><th>Deliveries</th><th>Squadrons</th></tr>
<tr><td>Abu Dhabi</td><td>16[8]</td><td>63</td><td>1001</td><td>Oct 1984 – May 1985</td><td></td></tr>
<tr><td></td><td>18*</td><td>102[1][6][10]</td><td>1051</td><td>Apr 1993 – Mar 1994</td><td>Khalif bin Zayed Air College</td></tr>
<tr><td></td><td>4*</td><td>63C</td><td>1017</td><td>Feb 1995 – Mar 1995</td><td></td></tr>
<tr><td>Australia</td><td>33[11]*</td><td>127</td><td>A27-01</td><td>Apr 2000 – Oct 2001</td><td>76, 79</td></tr>
<tr><td>Canada</td><td>22*</td><td>115/CT-155</td><td>155201</td><td>Jul 2000-</td><td>419</td></tr>
<tr><td>Dubai</td><td>8</td><td>61</td><td>501</td><td>Mar 1983 – Sep 1983</td><td>III Shaheen</td></tr>
<tr><td></td><td>1</td><td>61</td><td>509</td><td>Jun 1988</td><td>III Shaheen</td></tr>
<tr><td>Finland</td><td>50[2]</td><td>51</td><td>HW301</td><td>Dec 1980 – Oct 1985</td><td>3/11, 3/21, 3/31,</td></tr>
<tr><td></td><td>7*</td><td>51A</td><td>HW351</td><td>Nov 1993 – Sep 1994</td><td>Koulutuslentolaivue, 2/Tukilentolaivue</td></tr>
<tr><td>Indonesia</td><td>20[13]</td><td>53</td><td>LL-5301</td><td>Sep 1980 – Mar 1984</td><td>103</td></tr>
<tr><td></td><td>8*</td><td>109</td><td>TT-1201</td><td>May 1996 – Mar 1997</td><td>1, 12</td></tr>
<tr><td></td><td>16*</td><td>209</td><td>TT-1205</td><td>Feb 1996 – Mar 1997</td><td>12</td></tr>
<tr><td></td><td>16*</td><td>209</td><td>TT-0217</td><td>Apr 1999 –</td><td>1</td></tr>
<tr><td>Kenya</td><td>12</td><td>52</td><td>101</td><td>Apr 1980 – May 1982</td><td>(Laikipia AB)</td></tr>
<tr><td>Korea, South</td><td>20*</td><td>67</td><td>67-496</td><td>Sep 1992 – Aug 1993</td><td>216</td></tr>
<tr><td>Kuwait</td><td>12</td><td>64</td><td>140</td><td>Nov 1985 – Sep 1986</td><td>12</td></tr>
<tr><td>Malaysia</td><td>10*</td><td>108[1][6][10]</td><td>M40-01</td><td>Jan 1994 – Sep 1995</td><td>3 FTC (15 Sqdn)</td></tr>
<tr><td></td><td>18*</td><td>208[6][9][10]</td><td>M40-21</td><td>Aug 1994 – May 1995</td><td>6, 9, 3 FTC (15 Sqdn)</td></tr>
<tr><td>Oman</td><td>4*</td><td>103[1][6][10]</td><td>101</td><td>Dec 1993 – Jan 1994</td><td>6</td></tr>
<tr><td></td><td>12*</td><td>203[9][10]</td><td>121</td><td>Dec 1994 – May 1995</td><td>6</td></tr>
<tr><td>Saudi Arabia</td><td>30</td><td>65</td><td>2110</td><td>Aug 1987 – Oct 1988</td><td>21, 37</td></tr>
<tr><td></td><td>20</td><td>65A</td><td>7901</td><td>Mar 1997 – Dec 1997</td><td>79</td></tr>
<tr><td>South Africa</td><td>12</td><td>120/LIFT[1][10]</td><td>–</td><td>2005</td><td></td></tr>
<tr><td>Switzerland</td><td>20[3]</td><td>66</td><td>U-1251</td><td>Nov 1989 – Nov 1991</td><td>Pilotenschule</td></tr>
<tr><td>UK</td><td>176[7]</td><td>T. Mk 1</td><td>XX154</td><td>Nov 1976 – Feb 1982</td><td>4 FTS (19 & 208 Sqdns), Red Arrows, 100 Sqdn, FSATO, QinetiQ</td></tr>
<tr><td>USA</td><td>197[4]</td><td>T-45A/C</td><td>162787</td><td>Apr 1988 -</td><td>VT-7, VT-21, VT-22</td></tr>
<tr><td>Zimbabwe</td><td>8</td><td>60</td><td>600</td><td>Jul 1982 – Oct 1982</td><td>2</td></tr>
<tr><td></td><td>5*</td><td>60A</td><td>608</td><td>Jun 1992 – Sep 1992</td><td>2</td></tr>
<tr><td>Demonstrator</td><td>3[5]</td><td>60/102D/102NDA</td><td>G-HAWK</td><td></td><td>–</td></tr>
<tr><td></td><td>3[12]</td><td>200/200/200RDA</td><td>ZG200</td><td></td><td>–</td></tr>
<tr><td>**Total**</td><td>**781**</td><td></td><td></td><td></td><td></td></tr>
</table>

Notes

* Built at Brough, Hamble and Samlesbury, assembled at Warton or Brough and first flown at Warton; unless stated otherwise, remainder built at Kingston, assembled at Dunsfold (290) and Bitteswell (1)
[1] Laser nose
[2] 46 assembled by Valmet in Finland
[3] 19 assembled by F + W in Switzerland
[4] Production by Boeing in USA (see International section)
[5] Two first flown at Warton
[6] Wingtip Sidewinders
[7] 89 converted to T. Mk 1A
[8] 13 converted to 63A; two to 63B
[9] Removable refuelling probe
[10] Radar warning receiver
[11] First 12 assembled at Warton; remainder by BAe Australia at Williamtown, assisted by Hunter Aerospace, Hawker de Havilland and Qantas
[12] One first flown at Warton
[13] Five repurchased by BAE Systems in 1999

ACCOMMODATION: Crew of two in tandem under one-piece, fully transparent, acrylic canopy, opening sideways to starboard. Fixed front windscreen able to withstand a 0.9 kg (2 lb) bird at 454 kt (841 km/h; 523 mph). Improved front windscreen fitted retrospectively to RAF Hawks, able to withstand a 1 kg (2.2 lb) bird at 528 kt (978 km/h; 607 mph); this installed on all current export aircraft. Separate internal screen in front of rear cockpit. Rear seat elevated. Martin-Baker Mk 10LH zero/zero rocket-assisted ejection seats, with MDC (miniature detonating cord) system to break canopy before seats eject. MDC can also be operated from outside the cockpit for ground rescue. Dual controls standard. Entire accommodation pressurised, heated and air conditioned.

SYSTEMS: BAE Systems cockpit air conditioning and pressurisation systems, using engine bleed air. Two hydraulic systems; flow rate: System 1, 36.4 litres (9.6 US gallons; 8.0 Imp gallons)/min; System 2, 22.7 litres (6.0 US gallons; 5.0 Imp gallons)/min. Systems pressure 207 bar (3,000 lb/sq in). System 1 for actuation of control jacks, flaps, airbrake, landing gear and anti-skid wheel brakes. Compressed nitrogen accumulators provide emergency power for flaps and landing gear at pressure of 2.75 to 5.50 bar (40 to 80 lb/sq in). System 2 dedicated to powering flying controls. Hydraulic accumulator for emergency operation of wheel brakes. Pop-up Hamilton Sundstrand ram air turbine in upper rear fuselage provides emergency hydraulic power for flying controls in event of engine or No. 2 pump failure. No pneumatic system.

DC electrical power from single 12 kW 30 V DC brushless generator, with two 3 kVA 115/26 V 400 Hz three-phase inverters to provide AC power and two batteries for up to 20 minutes of standby power. Gaseous oxygen system for crew; optional Hamilton Sundstrand (HS Marston) OBOGS first installed in Australian Hawk Mk 127s from 2000.

AVIONICS: *Comms:* Mk 1/Srs 50 includes Sylvania UHF and VHF; Cossor 2720 Mk 10A IFF in Finnish aircraft; Srs 60 has Rockwell Collins UHF and VHF, Magnavox UHF and Raytheon 2720 IFF; Srs 100 has Rockwell Collins AN/ARC-182 U/VHF, Magnavox AN/ARC-164 UHF and Raytheon 4720 IFF. Australian Mk 127 has twin Rockwell Collins AN/ARC-210 UHF/VHF and Raytheon 4720 IFF.

Flight: Mk 1/Srs 50 with Raytheon CAT 7000 Tacan, Cossor ILS having CILS.75/76 localiser/glideslope receiver and marker receiver; Rockwell Collins VOR/ILS and ADF, Rockwell Collins Tacan, Smiths-Newmark 6000-05 AHRS and Smiths radar altimeter in Srs 60; Srs 100 has BAE Systems INS300 inertial platform, Rockwell Collins AN/ARC-118 Tacan, Rockwell Collins VIR-31A VOR/ILS and Smiths 0103-KTX-1 radar altimeter, all integrated via dual redundant MIL-STD-1553B databus. Optional Skyforce Skymap II GPS-driven moving map in some RAF Hawks.

Instrumentation: Smiths-Newmark compass, BAE Systems gyros and inverter, two Honeywell RAI-4 4 in (100 mm) remote attitude indicators and magnetic detector system in Mk 1/Srs 50; Smiths 1500 Series HUDWAC in Srs 100; BAE Systems F.195 weapon sight in approximately 90 RAF aircraft; BAE ISIS 195 sight in Srs 50 and Srs 60, except Saab RGS2 in Finnish Mk 51.

Mission: BAE Systems camera and recorder in F.195-equipped RAF aircraft (requirement announced December 1997 to re-equip 41 with video recorders by late 1999); Vinten camera and recorder in Srs 50 and Srs 60. Srs 100 has Smiths 3000 Series colour MFD, GEC data transfer system and Vinten colour video recording system; plus BAE Systems Type 105H laser range-finder; optional FLIR.

Self-defence (Series 100 only): Racal Prophet RWR in Mk 102s of Abu Dhabi; BAE Systems Sky Guardian in Indonesian, Malaysian and Omani aircraft, and retrofitted to Abu Dhabi's Mk 63s; optional chaff/flare dispenser at base of fin.

ARMAMENT: Underfuselage centreline-mounted 30 mm BAE Systems Aden Mk 4 cannon with 120 rounds (VKT 12.7 mm machine gun beneath Finnish aircraft), and two or four hardpoints under wing, according to Series. Provision for pylon in place of ventral gun pack. In RAF training roles, normal maximum external load is about 680 kg (1,500 lb), but the uprated Hawk 60 and 100 Series are cleared for an external load of 3,000 kg (6,614 lb), or 500 kg (1,102 lb) at 8 *g*. Typical weapon loadings on 60 Series include 30 mm or 12.7 mm centreline gun pod and four packs each containing eighteen 68 mm rockets; centreline reconnaissance pod and four packs each containing twelve 81 mm rockets; five 1,000 lb free-fall or retarded bombs; four launchers each containing four 100 mm rockets; nine 250 lb or 250 kg bombs; thirty-six 80 lb runway denial or tactical attack bombs; five 600 lb cluster bombs; four Sidewinder or two Magic air-to-air missiles; or four CBLS 100/200 carriers, each containing four practice bombs and four rockets. Vinten reconnaissance pod available for centre pylon. Similar options on 100 Series, plus wingtip air-to-air missiles. Mk 102s of Abu Dhabi/UAEAF can carry (but not designate for) two Alenia Marconi PGM-500 ASMs.

DIMENSIONS, EXTERNAL:

Wing span: Mk 1, Srs 50, Srs 60	9.39 m (30 ft 9¾ in)
Srs 100	9.08 m (29 ft 9½ in)
Srs 100 with Sidewinders	9.94 m (32 ft 7⅜ in)
Wing chord: at root	2.65 m (8 ft 8¼ in)
at tip	0.90 m (2 ft 11½ in)
Wing aspect ratio: Mk 1, Srs 50, Srs 60	5.3
Srs 100	4.9

BAE Systems' Hawk Mk 100 demonstrator (*Jane's/Paul Jackson*)
0110722

Length: fuselage (incl jetpipe):

Mk 1, Srs 50, Mks 60-66	10.775 m (35 ft 4¼ in)
Mk 67	11.375 m (37 ft 3¼ in)
Srs 100	11.40 m (37 ft 4¼ in)

fuselage and pitot:

Mk 1, Srs 50, Mks 60-66	11.455 m (37 ft 7 in)
Mk 67, Srs 100 (non combat)	12.035 m (39 ft 5¾ in)

Srs 100 (incl chaff/flare dispenser)

	12.095 m (39 ft 8 in)

overall:

Mk 1, Srs 50, Mks 60-66	11.845 m (38 ft 10¼ in)
Mk 67, Srs 100	12.425 m (40 ft 9¼ in)

Nose to pitot tip:

Mk 1, Srs 50, Mks 60-66	0.68 m (2 ft 2¾ in)
Mk 67	0.66 m (2 ft 2 in)
Srs 100	0.635 m (2 ft 1 in)
Tailplane overhang	0.39 m (1 ft 3½ in)
Height overall	3.98 m (13 ft 0¾ in)
Tailplane span	4.39 m (14 ft 4¾ in)
Wheel track	3.47 m (11 ft 5 in)
Wheelbase	4.50 m (14 ft 9 in)

AREAS:

Wings, gross	16.69 m² (179.6 sq ft)

Ailerons (total):

Mk 1, 50 and 60 Series	1.05 m² (11.30 sq ft)
100 Series	0.97 m² (10.44 sq ft)
Trailing-edge flaps (total)	2.50 m² (26.91 sq ft)
Airbrake	0.53 m² (5.70 sq ft)
Fin: Mk 1, 50 and 60 Series	2.51 m² (27.02 sq ft)
100 Series	2.61 m² (28.10 sq ft)
Rudder, incl tab	0.58 m² (6.24 sq ft)
Tailplane	4.33 m² (46.61 sq ft)

WEIGHTS AND LOADINGS:

Weight empty: 60 Series

Weight empty: 60 Series	4,012 kg (8,845 lb)
100 Series	4,400 kg (9,700 lb)
Max weapon load (60, 100 Series)	3,000 kg (6,614 lb)

Max fuel weight:

internal (usable)	1,277 kg (2,815 lb)
external (usable)	930 kg (2,050 lb)
Max T-O weight: T. Mk 1	5,700 kg (12,566 lb)
50 Series	7,350 kg (16,200 lb)
60, 100 Series	9,100 kg (20,061 lb)
Max landing weight: T. Mk 1	4,649 kg (10,250 lb)
60 Series	7,650 kg (16,865 lb)
Max wing loading: T. Mk 1	341.5 kg/m² (69.97 lb/sq ft)
50 Series	440.4 kg/m² (90.20 lb/sq ft)
60, 100 Series	545.4 kg/m² (111.70 lb/sq ft)
Max power loading: T. Mk 1	246 kg/kN (2.42 lb/lb st)
50 Series	318 kg/kN (3.12 lb/lb st)
60 Series	359 kg/kN (3.52 lb/lb st)
100 Series	350 kg/kN (3.43 lb/lb st)

PERFORMANCE:

Never-exceed speed (V$_{NE}$), clean:

at S/L	M0.87 (575 kt; 1,065 km/h; 661 mph EAS)

at and above 5,180 m (17,000 ft)

	M1.2 (575 kt; 1,065 km/h; 661 mph EAS)

Max level speed at S/L:

50 Series	535 kt (990 km/h; 615 mph)
60 Series	545 kt (1,010 km/h; 627 mph)
100 Series	540 kt (1,001 km/h; 622 mph)
Max level flight Mach No.	0.88
Stalling speed, flaps down	96 kt (177 km/h; 110 mph)
Max rate of climb at S/L	3,600 m (11,800 ft)/min

Time to 9,150 m (30,000 ft), clean: 60 Series   6 min 54 s

100 Series	7 min 30 s
Service ceiling: 60 Series	14,020 m (46,000 ft)
100 Series	13,565 m (44,500 ft)
T-O run (clean): 60 Series	710 m (2,330 ft)
100 Series	640 m (2,100 ft)
Landing run (10% fuel load): 60 Series	550 m (1,800 ft)
100 Series	605 m (1,980 ft)

Combat radius, Aden gun pod and two Sidewinder AAMs:

with four 500 lb bombs and two 130 Imp gallon tanks

	345 n miles (638 km; 397 miles)

with four 1,000 lb bombs

	125 n miles (231 km; 143 miles)

Ferry range:

60 Series, with two 591 litre (156 US gallon; 130 Imp gallon) drop tanks

	1,575 n miles (2,917 km; 1,812 miles)

100 Series, as above

	1,360 n miles (2,519 km; 1,565 miles)

Endurance, 100 n miles (185 km; 115 miles) from base:

60 Series	approx 2 h 42 min
100 Series	approx 2 h 6 min
g limits: full internal fuel, all	+8/−4

100 series, 60% fuel load:

1,360 kg (3,000 lb) external load	+8/−4
2,721 kg (6,000 lb) external load	+6/−3

UPDATED

BAE SYSTEMS HAWK 200 SERIES

TYPE: Multirole fighter.

PROGRAMME: Intention to build demonstrator (ZG200) announced 20 June 1984; first flight 19 May 1986 but lost 2 July 1986 in accident; replaced by first preproduction Hawk 200 (ZH200), first flown 24 April 1987; third demonstrator Series 200RDA (ZJ201) with full avionics and systems, including Northrop Grumman AN/APG-66H

BAE Systems Hawk Mk 203 of Royal Air Force of Oman   *NEW*/0533725

radar, flown 13 February 1992. BAe signed a collaborative production agreement with IPTN of Indonesia in June 1991, in anticipation of orders eventually totalling 144 Mks 100/200, but none ensued.

First production Hawk Srs 200 (Oman 121) flew 11 September 1993; first for Malaysia (Mk 208 M40-21) flew 4 April 1994; first Mk 209 for Indonesia followed in early 1996. Hawk 200 cleared to carry AGM-65 Maverick ASM by 1996.

CURRENT VERSIONS: Missions can include:

Airspace denial: Four Sidewinders, gun and two 591 litre (156 US gallon; 130 Imp gallon) drop tanks enabling 2 hour loiter on station at 100 n miles (185 km; 115 miles) from base.

Close air support: Typically four 1,000 lb bombs precision-delivered up to 115 n miles (213 km; 132 miles) from base in lo-lo-lo-lo mission with gun and wingtip Sidewinder missiles also carried.

Battlefield interdiction: Typically 907 kg (2,000 lb) load on hi-lo-lo-hi mission over 290 n mile (537 km; 340 mile) radius with gun and wingtip Sidewinder missiles also carried.

Long-range photo reconnaissance: 490 n mile (908 km; 564 mile) range with two external tanks, pod containing cameras and infra-red linescan and wingtip Sidewinder missiles for self-defence (rapid role change permits follow-on attack by the same aircraft).

Long-range deployment: 1,365 n mile (2,528 km; 1,571 mile) ferry range using two 591 litre (156 US gallon; 130 Imp gallon) external tanks, unrefuelled and with tanks retained (reserves allow 10 minutes over destination at 150 m; 495 ft).

Anti-shipping attack: Two rocket pods and two 591 litre (156 US gallon; 130 Imp gallon) external tanks plus wingtip Sidewinder missiles for self-defence, enabling ship attack and return with 10 per cent fuel reserves.

CUSTOMERS: See table. Oman (12 **Mk 203** ordered July 1990), Indonesia (16 **Mk 209** ordered June 1993; second 16 in June 1996), Malaysia (18 **Mk 208** ordered December 1990). Production of second Indonesian batch undertaken in 1998-99; first flight of initial aircraft (temporarily ZJ555) 22 October 1998; other customers' orders completed by March 1995; final six Indonesian aircraft built in UK, but embargoed by UK government on 11 September 1999; following lifting of embargo, were further delayed, pending arrival of US-sourced components. Total orders for, and production of, 200 Series totalled 62 by early 2000, excluding demonstrators;

none further by late 2002. Saudi Arabia signed MoU covering second batch of some 60 Hawks, reportedly including **Mk 205** with APG-66H radar, although none on firm order. All customers also ordered two-seat Hawks.

See Hawk 50, 60 and 100 Series entry for full description; principal differences given below.

DESIGN FEATURES: Hawk 200 virtually identical to current production Hawk two-seater aft of cockpit, giving 80 per cent airframe commonality. Significant changes from early two-seat versions include taller fin and 'combat wing' with fixed leading-edge droop to enhance lift and manoeuvrability at M0.3 to M0.7, manually selected combat flaps (less than quarter-flap setting) available below 350 kt (649 km/h; 403 mph) IAS to allow sustained 5 g+ at 300 kt (556 km/h; 345 mph) at sea level, full-width flap vanes reinstated and detail modifications to wing dressing; smurfs (strake ahead of each half of tailplane to restore control authority at high angles of attack). Intended to take advantage of new miniaturised, low-cost avionics and intelligent weapons; Hawk 100-type avionics include INS, DPMC, IFF, RWR and optional HOTAS controls; Lockheed Martin AN/APG-66H advanced multimode radar for all-weather target acquisition and navigation fixes; proposed alternative FLIR/laser range-finder nose no longer on offer; intended integral cannon also deleted; all four underwing pylons capable of 907 kg (2,000 lb) load, within maximum 3,493 kg (7,700 lb) external load; wingtip rails make possible four Sidewinders or similar AAMs (inboard pylons not cleared for these missiles).

LANDING GEAR: Mainwheel tyres size H22×6.5-11 (16 ply) tubeless, pressure 16.20 bar (235 lb/sq in). Nosewheel tyre size 18×5.5 (8 ply) tubeless, pressure 7.24 bar (105 lb/sq in).

POWER PLANT: One Rolls-Royce Turbomeca Adour Mk 871 non-afterburning turbofan, with uninstalled rating of 26.0 kN (5,845 lb st). Optional removable refuelling probe on starboard side of windscreen.

ACCOMMODATION: Pilot only, on Martin-Baker Mk 10LH zero/zero ejection seat, under starboard-hinged canopy.

SYSTEMS: 25 kVA generator with DC transformer-rectifier. Fairey Hydraulics yaw control system added, comprising rudder actuator and servo control system, incorporating an autostabiliser computer. Lucas Aerospace artificial feel system. 12 kVA APU for engine starting, ground running and emergency power.

AVIONICS: *Comms:* Rockwell Collins AN/ARC-182 U/VHF; Raytheon 4720 IFF; all integrated by MIL-STD-1553 databus.

The single-seat BAE Systems Hawk 200 Series with nose-mounted radar (*Jane's/Mike Keep*)

Radar: Northrop Grumman AN/APG-66H multimode radar.

Flight: Smiths DPMC. Rockwell Collins AN/ARN-153 Tacan and AN/ARN-147 VOR/ILS.

Instrumentation: BAE Systems 127 × 127 mm (5 × 5 in) multifunction display in cockpit; combined com/nav control and display unit allows control of all functions from one panel; HOTAS controls optional.

Self-defence: BAE Systems Sky Guardian 200 RWR in aircraft for Oman. Chaff/flare dispenser (Vinten Vicon 78 Srs 300 or equivalent) at base of fin.

ARMAMENT: None internally. All weapon pylons cleared for 8 *g* manoeuvres with 500 kg (1,102 lb) loads.

DIMENSIONS, EXTERNAL: As Hawk Series 100, except:

Wing span: 200RDA	9.39 m (30 ft 9¾ in)
Length: fuselage: 200	10.95 m (35 ft 11 in)
200 with chaff/flare dispenser	11.01 m (36 ft 1¼ in)
200RDA	10.99 m (36 ft 0¾ in)
200RDA with chaff/flare dispenser	
	11.05 m (36 ft 3 in)
overall: 200	11.34 m (37 ft 2½ in)
200RDA	11.38 m (37 ft 4 in)
Height overall: 200	4.130 m (13 ft 6¾ in)
200RDA	3.98 m (13 ft 0¾ in)
Wheelbase	3.56 m (11 ft 8 in)

WEIGHTS AND LOADINGS:

Basic weight empty	4,450 kg (9,810 lb)
Max fuel: internal (usable)	1,360 kg (3,000 lb)
external (usable)	932 kg (2,055 lb)
Max weapon load	3,000 kg (6,614 lb)
Max T-O weight	9,100 kg (20,061 lb)
Max wing loading	545.4 kg/m² (111.70 lb/sq ft)
Max power loading	350 kg/kN (3.43 lb/lb st)

PERFORMANCE (no external stores or role equipment unless stated):

Never-exceed speed (VNE):	
at S/L M0.87 (575 kt; 1,065 km/h; 661 mph EAS)	
at and above 5,180 m (17,000 ft)	
M1.2 (575 kt; 1,065 km/h; 661 mph EAS)	
Max level speed at S/L	540 kt (1,000 km/h; 621 mph)
Econ cruising speed at 12,500 m (41,000 ft)	
	430 kt (796 km/h; 495 mph)
Stalling speed, flaps down	
	96 kt (177 km/h; 110 mph) IAS
Max rate of climb at S/L	3,508 m (11,510 ft)/min
Time to 9,150 m (30,000 ft)	7 min 24 s
Service ceiling	13,715 m (45,000 ft)
Runway LCN: flexible pavement	15
rigid pavement	10
T-O run	630 m (2,070 ft)
Landing run	598 m (1,960 ft)
Ferry range (with two drop tanks)	
	1,365 n miles (2,528 km; 1,570 miles)
g limits	+8/−4

UPDATED

BAE SYSTEMS NIMROD MRA. Mk 4

TYPE: Maritime reconnaissance four-jet.

PROGRAMME: Original Hawker Siddeley HS 801 Nimrod first flew 23 May 1967 as adaptation of de Havilland D.H.106 Comet airliner. Total of two prototypes (converted Comet fuselages; now withdrawn), 46 Nimrod MR. Mk 1s and three Nimrod R. Mk 1 elint/sigint aircraft built for the RAF. Of these, 35 converted to MR. Mk 2 from 1979 onwards and 11 became Nimrod AEW. Mk 3s, but failed to

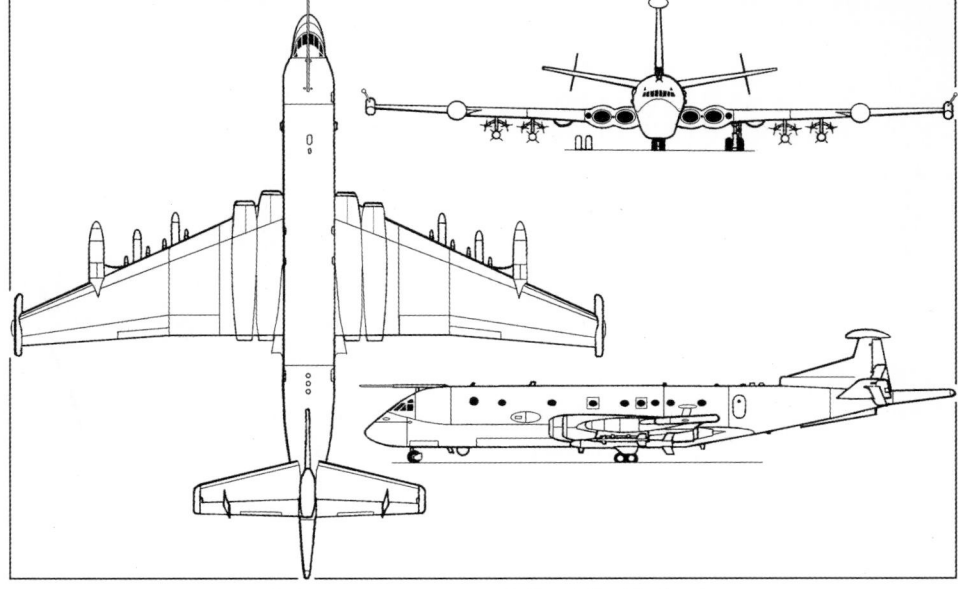

BAE Systems Nimrod MRA. Mk 4 maritime patrol aircraft *(Jane's/Paul Jackson)*   NEW/0533724

enter service and were scrapped. MR. Mk 2 fleet of Nos. 42(R), 120, 201 and 206 Squadrons at Kinloss in 1996 totalled 28 (including three in long-term storage), plus one under conversion to R. Mk 1, four scrapped or used for ground instruction and two lost in accidents. Two original R. Mk 1s remain with No. 51 Squadron at Waddington. These variants were last described in the 1987-88 *Jane's*.

Following RAF issue of Staff Requirement (Air) 420 for a Nimrod replacement, Dassault Atlantique 3, two versions of Lockheed Martin P-3 Orion (new production from manufacturer or refurbished P-3A/Bs from Loral) and upgraded Nimrod offered as potential solutions. UK official announcement on 25 July 1996 nominated last-mentioned proposal under commercial designation (since abandoned) Nimrod 2000. Although based on existing aircraft, the upgrade involves extensive (80 per cent) reconstruction of the airframe, plus incorporation of many new components, including engines, wings, landing gear and general systems, as well as new flight deck and detection systems. Contract MAR21a/100 awarded January 1997. Initial three development batch (DB) aircraft are conversions of stored Mk 2s, on which work began in early 1997; first delivery originally scheduled to RAF in 2001, followed by IOC in April 2003 on delivery of seventh aircraft; by early 1999, these dates had slipped 23 months due, in part, to excessive weight of new wing. New schedule called for first flight in June 2002; service deliveries in August 2004; IOC in March 2005; and final delivery in December 2008. Critical design review for air vehicle conducted late 1999 and finalised in early 2000, completing weapon system design. However, due to further slippage, prototype was given media debut on 16 August 2002 in expectation of maiden flight before year-end; delivery then stated to be late 2004.

By October 2002, first flight had been rescheduled for second half of 2003 due to discovery of "minor issues during airframe stress testing", although in-service date of March 2005 stated to be unchanged.

Ground testing and verification of engineering plans undertaken at Warton on Nimrod MR. Mk 1 prototype XV147. FR Aviation at Hurn, Bournemouth, selected for airframe relifing and upgrading as subcontractor to BAE, flying 'green' aircraft to Warton for mission avionics installation; first three (DB) aircraft (XV247/PA1, XV234/PA2 and XV242/PA3, for modification in that order) dismantled at RAF Kinloss and flown to Hurn by Antonov An-124 14 to 16 February 1997; redelivery to Warton originally due late 1999, January and February 2000. Fourth conversion subject, XV251/PA4, to Hurn under own power, 2 November 1998.

Programme revised in early November 1999, when FR Aviation contract cancelled and airframe work transferred to BAE at Woodford. PA1 completed at Hurn; PA2, PA3 and PA4 airfreighted to Woodford in An-124 in November/December 1999 for completion. Fifth subject, XV258/PA5, delivered directly ex-RAF to Woodford on 8 November 1999; further two followed in 2000 and one (a former ground instructional airframe) in October 2001. Wing for initial aircraft delivered from Chadderton to Woodford in September 2000; second aircraft similarly fitted in December 2000.

Short Brothers and various BAE plants supply airframe components and subassemblies. Ground training systems supplied by Thomson TSL.

Boeing supplying and integrating TCCS (tactical command and sensor system); 15-month laboratory test began at Warton in November 1998.

BAE offers the Nimrod MRA. Mk 4 for overseas sale, in which eventuality production would be relaunched. An agreement of June 1996 provided for McDonnell Douglas (now Boeing) to promote the Nimrod as a successor to the US Navy's Lockheed Martin P-3 Orion, with production at St Louis, if successful. This later expired, but the aircraft remained a contender in the US Navy's Multimission Maritime Aircraft (MMA) competition in early 2002, at which time Australia expressed an interest in joining the US programme. However BAE withdrew from MMA on 20 September 2002, citing lack of a US partner.

Static test wing to be mounted on Nimrod AEW. Mk 3 fuselage for 1,000 hour trial, beginning 2004.

CURRENT VERSIONS (specific): **PA1:** Air vehicle and general systems trials; planned 200 sorties, each between 2 and 4 hours.

PA2: Full mission system; airframe and systems tests. Assigned 160 sorties, each 6 hours, including weapons release and environmental tests.

PA3: Mission systems testbed. Assigned 140 sorties, each 6 hours, including integrated platform tests and actual submarine detection trials in Bahamas.

CUSTOMERS: Royal Air Force order was originally 21. In deference to comprehensiveness of rebuild, aircraft issued with new serial numbers ZJ514 to ZJ534. However, in March 2002 it was announced that the contract had been reduced to 18. Initial squadron to be No. 120 at Kinloss, operational from March 2005, using aircraft PA4 to PA11; no further deliveries until 2008.

COSTS: Ordered as £2 billion programme, including training system and initial logistic support, of which 75 per cent of work being placed with UK companies. Boeing TCCS mission system avionics contract valued at US$639 million (1996). Programme cost increased to £2.4 billion by March 1999, equivalent to 0.5 per cent above inflation. In mid-2000, BAE agreed to pay £46 million penalty because of 23-month programme slippage. BAE profits

Sextant seven-screen EFIS of the Nimrod MRA. Mk 4   0118205

Nimrod MRA. Mk 4 prototype at BAE Systems' Woodford plant   *NEW*/0521565

suffered £300 million charge in 2000 to offset expected Nimrod contract losses. Programme cost stated in 2002 to be £2.98 billion.

DESIGN FEATURES: World's first jet-powered, land-based maritime patrol and ASW aircraft on original 1969 service-entry; remains sole four-jet in this role. Optimised for anti-submarine warfare (ASW), anti-surface unit warfare (ASUW), search and rescue (SAR), maritime reconnaissance and provision of aid to civil authorities, including fisheries protection and operations against terrorism, drug smuggling and blockade running. Combines fast transit to operational area with low wing loading (and, hence, manoeuvrability) when on station. Can cruise on two or three engines to extend endurance. No special ground equipment required for off-base deployments.

Mid-mounted wing, with two engines in each root and auxiliary fuel tank in leading-edge at approximately two-thirds span; wingtip ESM pods; two hardpoints under each wing. Dihedral tailplane with auxiliary fins at one-third span to counter aerodynamic effects of electro-optical turret rotation; fin has large fillet and electrical equipment/decoy housing pod at tip. MAD tailboom. 'Double bubble' fuselage cross-section, with weapons bay below; latter restructured for Mk 4 and now in two, separated parts. Wing sweepback 20° at quarter-chord.

Nimrod Mk 4 upgrade extends operational lifetime by 25 years. Compared with original version, new power plant improves fuel consumption by over 20 per cent and thrust by 30 per cent. Larger wing and increased thrust restore performance despite 20 per cent increase in maximum take-off weight.

FLYING CONTROLS: Conventional control surfaces actuated hydromechanically. Plain flaps immediately outboard of engines.

STRUCTURE: All-metal, two-spar wing comprises centre-section, two stub-wings and two outer panels. Integrally machined stringers and machined spars. Fuselage is all-metal, semi-monocoque with pressurised upper component and unpressurised pannier including the weapons bay and nose radome. Pannier segments are free to move, so that structural loads are not transmitted to the pressure cell. Cantilever, all-metal tail surfaces surmounted by glass fibre fin-tip pod.

Nimrod MRA. Mk 4 introduces new wing centre box and wing inner panels of increased span designed by BAE Filton, built and assembled at Chadderton. Outer wings, to have been retained under BAE proposal, also new-build, as requested by RAF; designed by Filton, manufactured at Prestwick. Centre fuselage designed at Prestwick and built at Brough; lower (unpressurised) fuselage and front fuselage designed at Farnborough and built at Brough; rear fuselage designed by Dassault (France) and built at Brough; tailplane, elevators, rudder and weapons bay doors designed at Prestwick and Farnborough and manufactured at Prestwick and by FR Aviation. Flap

actuation system designed by Dowty. Other new items include wingtip pods, fin fillet, fin-tip pod and (enlarged) finlets. Retained structures, principally the fuselage pressure hull and empennage, are receiving new protective treatment, although all pressure bulkheads and floors are new. Some 80 per cent of the Mk 4 will be newly built.

LANDING GEAR: Retractable tricycle type; all-new Dowty unit in Nimrod MRA. Mk 4. Four-wheel tandem bogie main units; twin nosewheels. Dunlop tyres, wheels and brakes.

POWER PLANT: Four 'marinised', non-afterburning Rolls-Royce BR710 Mk 101 (BR710B3-40) turbofans with FADEC and EICAS, each rated at 68.9 kN (15,500 lb st). Additional 15 to 17 per cent of internal fuel compared with Nimrod MR. Mk 2.

ACCOMMODATION: Two-person flight crew; eight-person tactical team (Tacco 1, Tacco 2, radar, communications, ESM, two acoustics and one sonobuoy loader/general assistant); optional positions for two visual observers. Additional 13 seats for support personnel or replacement crew. Provision for further workstation. Crew areas, front to rear, are flight deck, lavatory, lateral observers' seats with bubble windows, tactical compartment, optional AEO compartment, galley and sonobuoy/general storage area, including crew door.

SYSTEMS: Smiths Aerospace flight management system (based on Boeing 737, but with vertical navigation capability). New Normalair-Garrett environmental control system (inlet at base of fin fillet) and hydraulic system. Hot air anti-icing system. Portable data storage system for fault analysis and rectification; compatible with Logistics Information Technology System. Honeywell APU in starboard wing trailing-edge. Lucas power-generation system; Normalair-Garrett oxygen system. Ram-air turbine scabbed to fuselage side, ahead of port wingroot.

AVIONICS: Mission avionics integration by BAE Systems, significant proportion of new equipment being obtained from Airbus and Eurofighter Typhoon programmes. Nimrod MRA. Mk 4 programmed with two million lines of computer software code.

Comms: Digital communications subsystem by Telephonics: five V/UHF, two HF.

Radar: Thales Avionics Searchwater 2000MR maritime surveillance radar. Modes include pulse Doppler (air-to-air), inverse SAR, weather and swath and spot SAR.

Flight: Smiths Industries navigation and flight management system. Rockwell Collins TCAS II with Thales IFF Mode S transponder; Thales (Sextant) AFCS with autothrottle and vertical hold/select. Two Litton LN-100G laser INS with embedded GPS; Avionics Specialities ground proximity warning and collision-avoidance. Triplex air data system. Rockwell Collins ADF; BAE Systems ILS/VOR/MLS; Flight Data Co MMS, Smiths NAV/FMS, Thales radar altimeter and Rockwell Collins Tacan.

Instrumentation: Thales (Sextant) seven-screen full-colour LCD EFIS, plus EICAS. Based on that of Airbus A330/A340.

Mission: 'Fourth-generation' sensor suite, including Boeing's tactical command sensor subsystem and Smiths Industries' armament control system conforming to MIL-STD-1760. Tactical display also capable of integrating weather radar, EO/FLIR imagery and defensive aids information. Seven identical, Boeing-supplied operator consoles integrated with Northrop Grumman (EOSDS) Nighthunter FLIR turret under forward fuselage. Retains CAE ASQ-504(V) AIMS MAD and Computing Devices Canada/Ultra Electronics AN/UYS-503 (AQS-970) processor from Nimrod MR. Mk 2, but MAD now fully integrated with tactical system. Ultra Link 11, Rockwell Collins Link 16 (JTIDS), MBDA SHF/satcom, Frederick TTY, Raytheon UHF/satcom. Full range of sonobuoys dispensed from two Normalair-Garrett rotary launchers (each holding 10 Size A buoys) in the rear fuselage; two single-barrel launchers for high-level release when fuselage pressurised. Internal racks for 180 buoys (or 360 in emergency), including new SR(SA) 903 active sonobuoy. Active search sonobuoy system (ASSS) ordered 2002 in form of Ultra Electronics CAMBS VI. Ultra acoustics system able to monitor up to 64 sonobuoys. Elta EL/L-8300UK ESM.

Self-defence: Integrated defensive aids subsystem (DASS) by BAE North America, including a Raytheon AN/ALE-50 towed decoy, Vicon 78, Thales (Vinten) chaff/flare dispensers, BAE Systems AN/AAR-57 missile approach warning and BAE AN/ALR-56M RWR, last mentioned being separate from ESM system. Potential for later addition of DIRCM, laser warning receiver and integrated countermeasures system.

ARMAMENT: Smiths armament control system capable of accepting all current and known future maritime patrol weapons. Four underwing hardpoints, each capable of carrying two side-mounted, self-protection Sidewinder AAMs in addition to an anti-ship missile. Weapons bay, with two pairs of doors, is able to carry up to six lateral rows of ASW weapons, including up to nine torpedoes as well as bombs. Maximum load of Harpoon ASMs is four underwing and one in each of two weapon bays.

DIMENSIONS, EXTERNAL:
Wing span over tip pods	38.71 m (127 ft 0 in)
Length overall (excl refuelling probe)	
	38.63 m (126 ft 9 in)
Height overall	9.45 m (31 ft 0 in)
Tailplane span	14.51 m (47 ft 7¼ in)

DIMENSIONS, INTERNAL:
Cabin (incl flight deck): Length	26.82 m (88 ft 0 in)
Max width	2.95 m (9 ft 8 in)
Max height	2.08 m (6 ft 10 in)
Volume	124.1 m³ (4,384 cu ft)

AREAS:
Wings, gross	235.80 m² (2,538.0 sq ft)
Ailerons (total)	5.63 m² (60.60 sq ft)
Tailplane	40.41 m² (435.00 sq ft)
Elevators, incl tabs	12.57 m² (135.30 sq ft)

WEIGHTS AND LOADINGS:
Weight empty	46,500 kg (102,515 lb)
Max weapon load	more than 5,443 kg (12,000 lb)
Max fuel weight	50,122 kg (110,500 lb)
Max T-O weight	104,420 kg (230,205 lb)
Max zero-fuel weight	58,287 kg (128,500 lb)
Max wing loading	442.8 kg/m² (90.70 lb/sq ft)
Max power loading	379 kg/kN (3.71 lb/lb st)

PERFORMANCE:
Max operating Mach No. (MMO)	0.77
Service ceiling	12,800 m (42,000 ft)
Range with max internal fuel	
more than 6,000 n miles	(11,112 km; 6,904 miles)
Endurance, unrefuelled	more than 15 h
g limit	+3

UPDATED

BAE SYSTEMS ADVANCED AIRCRAFT STUDIES

BAE involvement in the RAF's Future Offensive Air System studies was last described in the 2002-03 edition.

UPDATED

AIRCRAFT SERVICES GROUP

Woodford Aerodrome, Chester Road, Woodford, Cheshire SK7 1QR
Tel: (+44 161) 439 50 50
Fax: (+44 161) 955 30 08
Web: http://www.bae.regional.co.uk

SUBSIDIARY:
BAE Systems Aviation Services
PO Box 77, Filton, Bristol BS99 7AR
Tel: (+44 117) 936 41 69
Fax: (+44 117) 936 40 88

This group, formerly Regional Aircraft Group until 2001 merger with Filton-based Aircraft Services, has assembled the Avro RJ/RJX family of aircraft at Woodford; 185 ha (459

acre) site includes over 93,000 m² (1 million sq ft) of assembly hangars, not including flight test facilities. Also undertakes structural upgrade of Nimrod airframes (which see) as part of Mk 4 rebuild programme and aircraft conversions (including 37 Airbus A300B4 freighter conversions by April 2001).

UPDATED

REGIONAL AIRCRAFT

Prestwick International Airport, Ayrshire, Scotland KA9 2RW
Tel: (+44 1292) 47 98 88
Fax: (+44 1292) 47 97 03
MANAGING DIRECTOR: Alan Fraser

Division supports over 1,100 BAE-built regional aircraft remaining in service around the world. Divisions are Customer Support, Engineering and Asset Management; employs 1,150 staff at Prestwick, Hatfield and Weybridge.

NEW ENTRY

BAE SYSTEMS AVRO RJ and RJX

The final aircraft of this line, an RJ85 G-6-394/G-CBMH, made its first flight on 26 April 2002, being one of four unsold RJs held by BAE Systems. Production thus totalled 221 BAe 146s (including two prototype/demonstrator aircraft), 170 RJs and three RJXs, or 394 in all.

By individual versions, these were 35 BAe 146-100s, 116 BAe 146-200s (including 14 QT freighters and five QC quick-change versions), 70 BAe 146-300s (10 being QTs), 12 RJ70s, 87 RJ85s, 71 RJ100s, one RJX-85 and two RJX-100s. An illustrated description last appeared in the 2002-03 *Jane's*.

UPDATED

BNG

B-N GROUP LTD

Bembridge, Isle of Wight PO35 5PR
Tel: (+44 1983) 87 25 11
Fax: (+44 1983) 87 32 46
Web: http://www.britten-norman.com
CHAIRMAN: Alawi Zawawi
DEPUTY CHAIRMAN: Maurice Hynett
EXECUTIVE DIRECTOR: Robert Wilson
BOARD MEMBER:
  A Munim Zawawi
COMMUNICATIONS: Melanie Monk

Pilatus Aircraft Ltd of Switzerland, which acquired Britten-Norman (Bembridge) Ltd 1979, sold the company on 21 July 1998 to investment group Litchfield Continental Ltd, at which time the original name Britten-Norman Ltd was re-adopted. Company sold almost immediately to Global Spill Management (environmental protection) which reformed as Biofarm (pharmaceutical manufacturer) in Romania. Agreement to purchase Romaero (which see), long-term maker of Islander airframes, made January 1999, but not concluded. Proposed capital injection by UAE-based HSDP, early 2000, also failed to take place.

On 3 April 2000 Britten-Norman was placed in receivership; sale announced 26 April to Alawi Zawawi Enterprises of Oman, and renamed B-N Group Ltd (BNG). Order backlog at that time stood at nearly £12 million. Employment stood at approximately 110 in early 2002.

On 28 January 2002, B-N Group signed a 15-year agreement with Bembridge Airfield Ltd under which BNG will manage and develop Bembridge Airfield, thus securing the site for the manufacture and development of the Islander and Defender range.

UPDATED

BNG BN2B ISLANDER

TYPE: Light utility twin-prop transport.
PROGRAMME: Prototype (G-ATCT) first flight 13 June 1965 with two 157 kW (210 hp) Rolls-Royce Continental IO-360-B engines and 13.72 m (45 ft) span wings; subsequently re-engined with Textron Lycoming O-540s and flown 17 December 1965; wing span also increased by 1.22 m (4 ft) to initial production standard; production prototype BN2 (G-ATWU) flown 20 August 1966; domestic C of A received 10 August 1967; FAA type certificate 19 December 1967; Romanian manufacture (see Romaero entry) began 1969. In July 2002, further 24 Islanders ordered by BNG from Romaero with option to include two three-engined Trislanders.
CURRENT VERSIONS: **BN2 Islander:** Initial piston-engined production model (23 built: see earlier *Jane's*).
  BN2A Islander: Piston version built from 1 June 1969 until 1989 (see *Jane's Aircraft Upgrades*). Production totalled 890.
  BN2B Islander: Standard piston-engined version since 1979; higher maximum landing weight; improved interior design; available with two engine choices and optional wingtip fuel tanks as **BN2B-26** with O-540s and **BN2B-20** with IO-540s (BN2B-27 and -21 no longer available). Features include range of passenger seats and covers, more robust door locks, improved door seals and stainless steel sills, redesigned fresh air system to improve ventilation in hot and humid climates, smaller diameter propellers to decrease cabin noise, and redesigned flight deck and instrument panel. Total of 154 delivered by end December 1999.
  B-N, in conjunction with Hartzell Propeller Inc, has been testing a three-blade scimitar-shaped propeller developed as part of the NASA-sponsored AGATE/GAP programme, using a BN-2B operated by the North East Police Air Support Unit. The aim of the test programme is to minimise the aircraft's noise footprint through enhanced propeller aerofoil efficiency and reduced tip speeds.
  Detailed description applies to BN2B version.
    Series of modification kits available as standard or option for new production aircraft and can be fitted retrospectively to existing aircraft.
  BN2B Defender: Military version. Design features as detailed under Turbine Islander and Defender entry. Customers include Indian Navy (six recently converted to turbine power), Belgian Army, Botswana Defence Forces and Jamaican Defence Force.
    BN2T Turbine Islander: Described separately.
    BN2T-4R Defender: Described in 1997-98 and earlier editions.

BN2B-26 Islander of the Armed Forces of Malta (*Jane's/Paul Jackson*)   NEW/0121491

  BN2T-4S Defender 4000: Described separately.
  BN2A Mk III Trislander: Described separately.
CUSTOMERS: By December 2001, deliveries of Islanders and Defenders totalled 1,227. Recent customers have included the Botswana Defence Force (one attrition replacement), Hawker Pacific (Australia) and Tomen Aerospace (Japan). First to fly post-April 2000 purchase of company was G-BWNG exhibited at Farnborough Air Show, July 2000 having flown on 10 July 2000 and arrived at Bembridge on 19 July. First delivery under new ownership was a BN2B-26 delivered on 10 November 2000 to Aer Arann of Galway, Ireland, followed by another Islander delivered to Ryuku Air Commuter of Japan on 12 November 2000. Deliveries in 2001 included one BN2B-20 each to St Barth Commuter of St Barthelmy, Frisia Luftverkehr GmbH of Germany (four-blade propellers), Ryukyu Air Commuter and Kyokushin Air of Japan, and Calcasieu Parish Police Jury of the USA, last-named equipped with Micronair spraypods for mosquito control duties. See also BN2 status table.
COSTS: BN2B, about £500,000 (1997).
DESIGN FEATURES: Robust STOL light transport, suitable for operation from semi-prepared airstrips. High, unswept, constant-chord wings and tailplane; sweptback fin with small fillet; upswept rear fuselage. Three-blade propellers available on BN2B-20 and BN2B-26 giving quieter noise signature.
  NACA 23012 wing section; no dihedral; incidence 2°; no sweepback.
FLYING CONTROLS: Conventional and manual. Actuation by pushrods and cables. Slotted ailerons, with starboard ground-adjustable tab; mass-balanced elevator; trim tabs in rudder and elevator; single-slotted flaps operated electrically; fixed incidence tailplane. Dual controls standard.
STRUCTURE: L72 aluminium-clad aluminium alloys; two-spar wing torsion box in one piece; flared-up wingtips; integral fuel tanks in wingtips optional; four-longeron fuselage of pressed frames and stringers; two-spar tail unit with pressed ribs.
LANDING GEAR: Non-retractable tricycle type, with twin wheels on each main unit and single steerable nosewheel. Cantilever main legs mounted aft of rear spar. All three legs fitted with oleo-pneumatic shock-absorbers. All five wheels and tyres size 16×7-7, supplied by Goodyear. Tyre pressure: main 2.41 bar (35 lb/sq in); nose 2.00 bar (29 lb/sq in). Foot-operated air-cooled Cleveland hydraulic brakes on main units. Parking brake. Wheel/ski gear available optionally. Minimum ground turning radius 9.45 m (31 ft 0 in).
POWER PLANT: Two Textron Lycoming flat-six engines, each driving a Hartzell HC-C2YK-2B or -2C two-blade constant-speed feathering metal propeller; optional three-blade Hartzell HC-C3YR-2UF/FC8468-8R. Propeller synchronisers optional. Standard power plant 194 kW (260 hp) O-540-E4C5, but 224 kW (300 hp) IO-540-K1B5 fitted at customer's option.
  Integral fuel tank between spars in each wing, outboard of engine. Total fuel capacity (standard) 518 litres (137 US gallons; 114 Imp gallons). Usable fuel 492 litres (130 US gallons; 108 Imp gallons). With optional fuel tanks in wingtips, total capacity is increased to 814 litres (215 US gallons; 179 Imp gallons). Additional pylon-mounted underwing auxiliary tanks, each of 227 litres (60.0 US gallons; 50.0 Imp gallons) capacity, available optionally. Refuelling point in upper surface of wing above each internal tank. Total oil capacity 22.75 litres (6.0 US gallons; 5.0 Imp gallons).
ACCOMMODATION: Up to 10 persons, including pilot, on side-by-side front seats and four bench seats. No aisle. Seat backs fold forward. Access to all seats via three forward-opening doors, forward of wing and at rear of cabin on port side and forward of wing on starboard side. Baggage compartment at rear of cabin, with port-side loading door in standard versions. Exit in emergency by removing door windows. Special executive layouts available.
  Can be operated as freighter, carrying more than a ton of cargo; in this configuration passenger seats can be stored in rear baggage bay. In ambulance role, up to three stretchers and two attendants can be accommodated. Other layouts possible, including photographic and geophysical survey, parachutist transport or trainer (with accommodation for up to eight parachutists and dispatcher), firefighting, environmental protection and crop-spraying.
SYSTEMS: Janaero cabin heater standard. 45,000 BTU Janitrol combustion unit, with circulating fan, provides hot air for distribution at floor level outlets and at windscreen demisting slots. Fresh air, boosted by propeller slipstream, is ducted to each seating position for on-ground ventilation. Electrical DC power, for instruments, lighting and radio, from two engine-driven 24 V 70 A self-rectifying alternators and a controller to main busbar and circuit breaker assembly. Emergency busbar is supplied by a 24 V 25 Ah heavy-duty lead-acid battery in the event of a twin alternator failure. Ground power receptacle provided. Optional electric de-icing of propellers and windscreen, and pneumatic de-icing of wing and tail unit leading-edges. Oxygen system available optionally for all versions.
AVIONICS: *Comms:* Intercom, including second headset, and passenger address system are standard.
  Flight: Standard items include autopilot and wide range of VHF and HF communications and navigation equipment.
  Instrumentation: IFR standard.
DIMENSIONS, EXTERNAL:

Wing span	14.94 m (49 ft 0 in)
Wing chord, constant	2.03 m (6 ft 8 in)
Wing aspect ratio	7.4
Length overall	10.86 m (35 ft 7¾ in)
Fuselage: Max width	1.21 m (3 ft 11½ in)
Max depth	1.46 m (4 ft 9¾ in)
Height overall	4.18 m (13 ft 8¾ in)
Tailplane span	4.67 m (15 ft 4 in)
Wheel track (c/l of shock-absorbers)	3.61 m (11 ft 10 in)
Wheelbase	3.99 m (13 ft 1¼ in)
Propeller diameter	1.98 m (6 ft 6 in)
Cabin door (front, port): Height	1.10 m (3 ft 7½ in)
Width: top	0.64 m (2 ft 1¼ in)
Height to sill	0.59 m (1 ft 11¼ in)
Cabin door (front, starboard):	
Height	1.10 m (3 ft 7½ in)
Max width	0.86 m (2 ft 10 in)
Height to sill	0.57 m (1 ft 10½ in)
Cabin door (rear, port): Height	1.09 m (3 ft 7 in)
Width: top	0.635 m (2 ft 1 in)
bottom	1.19 m (3 ft 11 in)
Height to sill	0.52 m (1 ft 8½ in)
Baggage door (rear, port): Height	0.69 m (2 ft 3 in)

DIMENSIONS, INTERNAL:

Passenger cabin, aft of pilot's seat:	
Length	3.05 m (10 ft 0 in)
Max width	1.09 m (3 ft 7 in)

Max height	1.27 m (4 ft 2 in)
Floor area	2.97 m² (32.0 sq ft)
Volume	3.7 m³ (130 cu ft)
Baggage space aft of passenger cabin	1.4 m³ (49 cu ft)

Freight capacity:
aft of pilot's seat, incl rear cabin baggage space
4.7 m³ (166 cu ft)
with four bench seats folded into rear cabin baggage
space 3.7 m³ (130 cu ft)

AREAS:

Wings, gross	30.19 m² (325.0 sq ft)
Ailerons (total)	2.38 m² (25.60 sq ft)
Flaps (total)	3.62 m² (39.00 sq ft)
Fin	3.41 m² (36.64 sq ft)
Rudder, incl tab	1.60 m² (17.20 sq ft)
Tailplane	6.78 m² (73.00 sq ft)
Elevator, incl tabs	3.08 m² (33.16 sq ft)

WEIGHTS AND LOADINGS (A: 194 kW; 260 hp engines, B: 224 kW; 300 hp engines):
Weight empty, equipped (without avionics):

A	1,866 kg (4,114 lb)
B	1,925 kg (4,244 lb)
Max payload: A	929 kg (2,048 lb)
B	870 kg (1,918 lb)
Payload with max fuel: A	692 kg (1,526 lb)
B	633 kg (1,396 lb)
Max fuel weight: standard: A, B	354 kg (780 lb)
with optional tanks in wingtips: A, B	585 kg (1,290 lb)
Max T-O and landing weight: A, B	2,993 kg (6,600 lb)

Max zero-fuel weight (BCAR):

A, B	2,855 kg (6,300 lb)
Max floor loading, without cargo panels:	
A, B	586 kg/m² (120 lb/sq ft)
Max wing loading: A, B	99.2 kg/m² (20.31 lb/sq ft)
Max power loading: A	7.73 kg/kW (12.69 lb/hp)
B	6.70 kg/kW (11.00 lb/hp)

PERFORMANCE (A and B as above):
Never-exceed speed (VNE):

A, B	183 kt (339 km/h; 211 mph) IAS

Max level speed at S/L:

A	148 kt (274 km/h; 170 mph)
B	151 kt (280 km/h; 173 mph)

Max cruising speed at 75% power at 2,135 m (7,000 ft):

A	139 kt (257 km/h; 160 mph)
B	142 kt (264 km/h; 164 mph)

Cruising speed at 67% power at 2,745 m (9,000 ft):

A	134 kt (248 km/h; 154 mph)
B	137 kt (254 km/h; 158 mph)

Cruising speed at 59% power at 3,660 m (12,000 ft):

A	130 kt (241 km/h; 150 mph)
B	132 kt (245 km/h; 152 mph)

Stalling speed:

flaps up: A, B	50 kt (92 km/h; 57 mph) IAS
flaps down: A, B	40 kt (74 km/h; 46 mph) IAS
Max rate of climb at S/L: A	262 m (860 ft)/min
B	344 m (1,130 ft)/min
Rate of climb at S/L, OEI: A	44 m (145 ft)/min
B	60 m (198 ft)/min
Absolute ceiling: A	4,145 m (13,600 ft)
B	6,005 m (19,700 ft)
Service ceiling: A	3,445 m (11,300 ft)
B	5,240 m (17,200 ft)
Service ceiling, OEI: A	1,525 m (5,000 ft)
B	1,980 m (6,500 ft)

BNG BN2T Islander AL. Mk 1 of the UK Army Air Corps equipped for photographic surveillance (note partly open door with special insert, including camera window) *(Jane's/Paul Jackson)* 0064962

T-O run at S/L, zero wind, hard runway:

A	278 m (915 ft)
B	215 m (705 ft)
T-O run at 1,525 m (5,000 ft): A	396 m (1,300 ft)
B	372 m (1,225 ft)

T-O to 15 m (50 ft) at S/L, zero wind, hard runway:

A	371 m (1,220 ft)
B	355 m (1,165 ft)

T-O to 15 m (50 ft) at 1,525 m (5,000 ft):

A	528 m (1,735 ft)
B	496 m (1,630 ft)
Landing from 15 m (50 ft) at S/L, zero wind, hard runway: A, B	299 m (980 ft)

Landing from 15 m (50 ft) at 1,525 m (5,000 ft):

A, B	357 m (1,170 ft)
Landing run at 1,525 m (5,000 ft): A, B	171 m (560 ft)

Landing run at S/L, zero wind, hard runway:

A, B	140 m (460 ft)

Range, block-to-block, plus 10% plus 45 min reserves:
B, standard fuel:

IFR	503 n miles (931 km; 578 miles)
VFR	639 n miles (1,183 km; 735 miles)

B, with optional tanks:

IFR	896 n miles (1,659 km; 1,031 miles)
VFR	1,075 n miles (1,990 km; 1,237 miles)

UPDATED

BNG BN2T TURBINE ISLANDER AND DEFENDER

UK Army Air Corps designation: Islander AL. Mk 1
Royal Air Force designation: Islander CC. Mk 2/2A

TYPE: Utility turboprop twin.

PROGRAMME: Turboprop Islander; first flight of prototype (G-BPBN) 2 August 1980 with two Allison (now Rolls-Royce) 250-B17Cs; British CAA certification received end of May 1981; first production aircraft delivered December 1981; FAR Pt 23 US type approval 15 July 1982; full icing clearance to FAR Pt 25 gained 23 July 1984.

CUSTOMERS: Total 65 aircraft delivered by December 1999, mainly of the Defender variant. Customers include the Rhine Army Parachute Association. For turbine (BN2T) version, customers include Moroccan Ministry of Fisheries, Pakistan Maritime Security Agency, Belgian Gendarmerie, Netherlands Police, Royal Air Force, UK Army Air Corps and Mauritius Coast Guard.

DESIGN FEATURES: As for piston-engined version (previous entry); turboprops enable use of available low-cost jet fuel; low operating noise level; available for same range of applications as Islander. Defender has four underwing hardpoints for standard NATO pylons to attach fuel tanks, weapons and other stores; number of additional airframe options, including sliding door on rear port side which can be opened in flight, are also offered. Concept of Defender is to provide a low-cost airframe which can be fitted with best available sensors to meet operational needs of customers.

Description of BN2B Islander applies also to BN2T and Defender, except as follows:

POWER PLANT: Two 298 kW (400 shp) Rolls-Royce 250-B17C turboprops, flat rated at 238.5 kW (320 shp), and each driving a Hartzell three-blade constant-speed fully feathering metal propeller. Usable fuel 814 litres (215 US gallons; 179 Imp gallons). Pylon-mounted underwing tanks, each of 227 litres (60.0 US gallons; 50.0 Imp gallons) capacity, are available optionally for special purposes. Total oil capacity 5.7 litres (1.5 US gallons; 1.25 Imp gallons).

ACCOMMODATION: Generally as for BN2B. In ambulance role can accommodate, in addition to pilot, a single stretcher, one medical attendant and five seated occupants; or two stretchers, one attendant and three passengers; or three stretchers, two attendants and one passenger. Other possible layouts include photographic and geophysical survey; parachutist transport or trainer (with accommodation for up to eight parachutists and a dispatcher); and pest control or other agricultural spraying. Maritime Turbine Islander/Defender versions available for fishery protection, coast guard patrol, pollution survey, search and rescue, and similar applications. In-flight sliding parachute door optional.

AVIONICS: *Comms:* Optional maritime band and VHF transceivers.
Flight: VLF/Omega nav system; radar altimeter.
Self-defence (Military versions): Since at least 1994 Army Air Corps Islander AL. Mk 1s have had provision for AN/ALQ-144 IR jammer under fuselage. Lockheed Martin IRCM suite tested on AAC Islander in early 1998.

EQUIPMENT: According to mission, can include fixed tailboom or towed bird magnetometer, spectrometer, or electromagnetic detection/analysis equipment (geophysical survey); one or two cameras, navigation sights and appropriate avionics (photographic survey); 189 litre (50.0 US gallon; 41.5 Imp gallon) Micronair underwing spraypods complete with pump and rotary atomiser (pest control/agricultural spraying versions); dinghies, survival equipment and special crew accommodation (maritime versions). Optimal installations for British Army Islanders include photographic surveillance package of door-mounted Zeiss 610 camera, vertical Zeiss Trilens 80 mm and either F126 vertical,

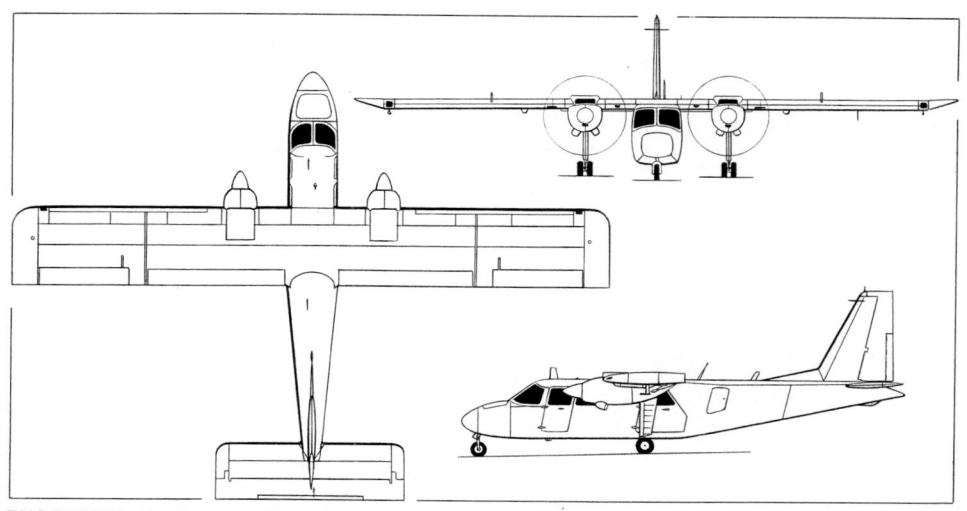

BNG BN2T Turbine Islander *(Jane's/Dennis Punnett)*

RMK vertical or two KS-153 vertical cameras; Vinten
F143 panoramic is also available.

DIMENSIONS, EXTERNAL: As for BN2B, except:

Length overall: standard nose	10.86 m (35 ft 7¾ in)
weather radar nose	11.07 m (36 ft 3¾ in)
Propeller diameter	2.03 m (6 ft 8 in)

WEIGHTS AND LOADINGS:

Weight empty, equipped	1,832 kg (4,040 lb)
Payload with max fuel	689 kg (1,519 lb)
Max T-O weight	3,175 kg (7,000 lb)
Max landing weight	3,084 kg (6,800 lb)
Max zero-fuel weight	2,994 kg (6,600 lb)
Max wing loading	105.2 kg/m² (21.54 lb/sq ft)
Max power loading	5.33 kg/kW (8.75 lb/shp)

PERFORMANCE (standard Turbine Islander/Defender):

Max cruising speed:	
at 3,050 m (10,000 ft)	170 kt (315 km/h; 196 mph)
at S/L	154 kt (285 km/h; 177 mph)
Cruising speed at 72% power:	
at 3,050 m (10,000 ft)	150 kt (278 km/h; 173 mph)
at 1,525 m (5,000 ft)	143 kt (265 km/h; 165 mph)
Stalling speed, power off:	
flaps up	52 kt (97 km/h; 60 mph) IAS
flaps down	45 kt (84 km/h; 52 mph) IAS
Max rate of climb at S/L	320 m (1,050 ft)/min
Rate of climb at S/L, OEI	66 m (215 ft)/min
Service ceiling	over 7,010 m (23,000 ft)
Absolute ceiling, OEI	over 3,050 m (10,000 ft)
T-O run	255 m (840 ft)
T-O to 15 m (50 ft)	381 m (1,250 ft)
Landing from 15 m (50 ft)	339 m (1,110 ft)
Landing run	228 m (750 ft)
Range (IFR) with max fuel, reserves for 45 min hold plus 10%	590 n miles (1,093 km; 679 miles)
Range (VFR) with max fuel, no reserves	728 n miles (1,348 km; 838 miles)

BN2 STATUS
(at 31 December 2000)

	Number delivered	Number remaining
BN2	23	3
BN2A	890	596
BN2B	155	149
BN2T	65	64
BN2T-4R	4	4
BN2T-4S	3	3
Trislander	82	52
Totals	**1,222**	**871**

Notes: Number remaining takes into account 11 BN-2s
converted to BN2As; eight BN2A-21s converted to
BN2B-21s; one BN2B converted to BN2T; and one BN2T
converted to BN2T-4R. Six BN2As/BN2Bs operated by the
Indian Navy converted to BN2Ts during 1996-98.

VERIFIED

BNG BN2T-4S DEFENDER 4000

TYPE: Multisensor surveillance twin-turboprop.

PROGRAMME: Marketing experience with BN2T Defenders
revealed that many military and government agencies
expressed need for greater payload; announced 1994;
prototype (G-SURV) made first flight 17 August 1994;
public launch at Farnborough Air Show September 1994;
CAA certification achieved 13 November 1995; CAA
transport (passenger) certification 4 April 1996.

Britten-Norman and Orenda Recip Inc of Canada
announced in September 1999 joint study to re-engine
Defender with Orenda OE600 V-8 piston engines,
resulting in improved performance and efficiency.
Projected specification includes maximum T-O weight
4,241 kg (9,350 lb), maximum cruising speed 176 kt
(326 km/h 202 mph) and maximum rate of climb 366 m
(1,200 ft)/min.

CUSTOMERS: Seven aircraft registered to manufacturer in
April 1996. Irish Ministry of Justice took delivery of one
on 14 August 1997; aircraft is operated by the Irish Air
Corps at Baldonnel, Dublin, on behalf of the Garda

Instrument panel of BN2T-4S Defender 4000
(Jane's/Paul Jackson) 0052936

First BNG Defender 4000 *NEW*/0134228

Siochana (National Police), and is equipped with a
comprehensive suite of law enforcement communications
equipment, thermal imager and observers' removable
consoles. Sabah Air of Malaysia took delivery of one in
November 1997; primarily equipped for aerial
photographic work, with quick-change facility to 12-
passenger transport. Police Aviation Services UK took
delivery of a single (ex-demonstrator) example in
November 1998, this later (2002) passed to Atlantic Air
Transport, Coventry. Hampshire Police Authority took
delivery of one (G-SCJH, named *Sir John Charles
Hoddinott*) on 26 February 2001 for service with its Lee-
on-Solent-based Air Support Unit, replacing a BN2B
Islander. Greater Manchester Police ordered one (G-
BWPU, c/n 4011) on 15 January 2002, for delivery by July
2002.

COSTS: £2.5 million (2002).

DESIGN FEATURES: As for Islander, but with enlarged wing,
based on that of Trislander, although retaining original
control surfaces; compared with BN2T, has tailplane and
elevator of increased span; modified fin with larger fillet;
fuselage stretched by means of 0.76 m (2 ft 6 in) plug
forward of wing to seat up to 16; more powerful engines for
greater sortie time and payload capacity; fuselage and
landing gear strengthened; flight deck windows deepened
for enhanced field of view; redesigned nose and tail unit;
rear-sliding door with blister window.

POWER PLANT: Two Rolls-Royce 250-B17F turboprops, each
flat rated at 298 kW (400 shp) and driving a three-blade
propeller. Internal fuel capacity 1,131 litres (299 US
gallons; 249 Imp gallons).

ACCOMMODATION: Flight crew of two on airline-type seats;
sliding seat rails permit seat and equipment positioning
anywhere within fuselage. Space for two or more consoles
and operators in tandem along one side of cabin. Up to 16
troops/passengers in tactical transport role. Certified to
carry pilot and 11 passengers, subject to local
airworthiness requirements.

SYSTEMS: Electrical system includes two 200 A engine-driven
generators.

AVIONICS: *Comms:* Full range of open or secure voice com
radios from UHF, VHF, HF and VHF-FM.

Radar: Modified nose can accommodate 68.5 cm
(27 in), 360° rotating antenna for maritime, simple search
or weather radar (BAE Seaspray 2000 in prototype
Defender 4000).

Flight: Fully integrated autopilot. GPS nav system,
integrated with Omega or INS.

Mission: Sensors can include thermal imagers and/or
hand-held or podded video or film cameras. Prototype on
debut fitted with Agema FLIR under fuselage on starboard
side; alternatives could include BAE pod or FLIR 2000.
Appropriate radars could include BAE Seaspray, Racal
Super Searcher, Thomson-CSF Detexis Ocean Master,
Telephonics 143 or Litton 504(V)5.

ARMAMENT: Two hardpoints under each wing, inboard pair
stressed for loads of up to 340 kg (750 lb) and outboard
pair for up to 159 kg (350 lb) each. Typical weapons on
inboard stations can include Sting Ray torpedo or Sea Skua
anti-ship missile.

DIMENSIONS, EXTERNAL: As for BN2B, except:

Wing span	16.15 m (53 ft 0 in)
Wing aspect ratio	8.0
Length overall	12.20 m (40 ft 0½ in)
Height overall	4.36 m (14 ft 3½ in)
Tailplane span	5.31 m (17 ft 5 in)
Wheelbase	4.76 m (15 ft 7½ in)

AREAS:

Wings, gross	32.61 m² (351.0 sq ft)

WEIGHTS AND LOADINGS:

Weight empty, equipped	2,223 kg (4,900 lb)
Max usable internal fuel	908 kg (2,002 lb)
Payload with max fuel	724 kg (1,598 lb)
Max T-O and landing weight	3,855 kg (8,500 lb)
Max zero-fuel weight	3,765 kg (8,300 lb)
Max wing loading	109.6 kg/m² (22.45 lb/sq ft)
Max power loading	6.46 kg/kW (10.63 lb/shp)

PERFORMANCE:

Max cruising speed: at S/L	154 kt (285 km/h; 177 mph)
at 3,050 m (10,000 ft)	176 kt (326 km/h; 202 mph)
Transit speed from base to patrol area	160 kt (296 km/h; 184 mph)
Cruising speed at 72% power:	
at 1,525 m (5,000 ft)	149 kt (276 km/h; 171 mph)
at 3,050 m (10,000 ft)	160 kt (296 km/h; 184 mph)
Stalling speed, power off:	
flaps up	53 kt (99 km/h; 61 mph) IAS
flaps down	47 kt (87 km/h; 54 mph) IAS
Max rate of climb at S/L	381 m (1,250 ft)/min
Rate of climb at S/L, OEI	61 m (200 ft)/min
Absolute ceiling	7,620 m (25,000 ft)
Absolute ceiling, OEI	3,660 m (12,000 ft)
T-O run	356 m (1,170 ft)
T-O to 15 m (50 ft)	565 m (1,855 ft)
Landing from 15 m (50 ft)	589 m (1,935 ft)
Landing run	308 m (1,015 ft)
Max range on internal fuel:	
with IFR reserves	861 n miles (1,594 km; 990 miles)
with VFR reserves	1,006 n miles (1,863 km; 1,157 miles)
Max endurance on internal fuel	8 h 30 min

UPDATED

BNG BN2A Mk III TRISLANDER

Planned reinstatement of Trislander production in 2000 was
not achieved, although the aircraft may be relaunched in the
near future, if demand warrants. Details last appeared in the
2001-02 *Jane's*.

UPDATED

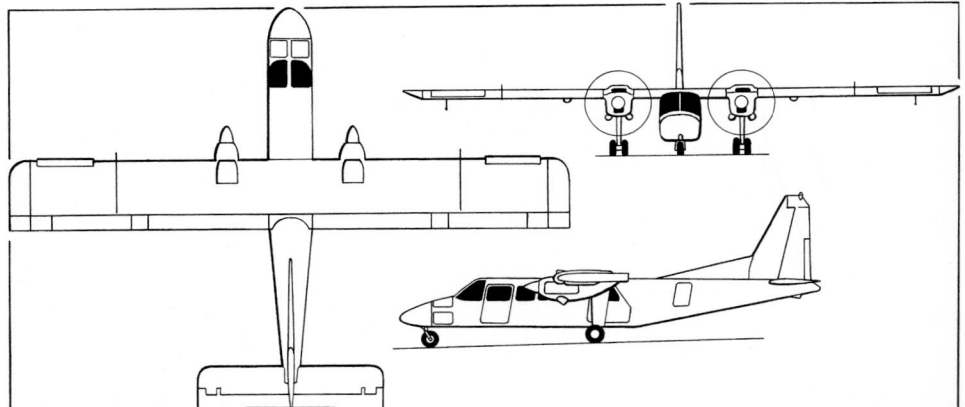

BNG BN2T-4S Defender 4000 *(Jane's/James Goulding)* 0051535

CFM

CFM AIRBORNE (UK) LTD

Unit 2D, Eastlands Industrial Estate, Leiston, Suffolk
IP16 4LL
Tel: (+44 1728) 83 23 53 and 83 21 00
Fax: (+44 1728) 83 29 44
e-mail: CFMAirborne@yahoo.com

Cook Flying Machines (trading as CFM MetalFax Ltd) went into receivership during 1996; assets were purchased from the liquidator by a small group of investors and the company was relaunched as CFM Aircraft Ltd, to manufacture and market the Shadow series of microlights and Streak/Star Streak light aircraft in kitbuilt and production versions as described below. By June 2001, 397 CFM ultralights had been delivered to customers in 40 countries. In 1999, CFM secured what was then the world's largest ultralight order when the Indian Air Force contracted for 24 Streak Shadows, of which 18 had been delivered by June 2001.

Workforce was 7 in 2001, with annual production capacity of 12 to 18 assembled aircraft, and a similar number of kits, at the company's 465 m² (5,000 sq ft) factory at Leiston. Flight testing undertaken at Parham (Framlingham), to which all company operations were scheduled to relocate in the summer of 2001.

However, in mid-2002 CFM went into receivership. Assets acquired by Airborne Innovations LLC (10 NASA Road One, Suite 209, Webster, Texas 77598), which had earlier purchased Laron Aviation Technologies, the former US distributor for CFM (see Airborne entry in 2000-01 edition). UK production was due to have restarted in early 2003.

UPDATED

CFM SHADOW SERIES D and E

TYPE: Tandem seat ultralight/kitbuilt.
PROGRAMME: Shadow first flew as prototype in 1983; in production since 1984; type approval to BCAR CAP 482 Section S gained May 1985. Certification of Series E was continuing in 2003.
CURRENT VERSIONS: **Series D:** Rotax 447 or 503 engine. Dual controls standard. Suitable for cropspraying with 64 litre (16.8 US gallon; 14.0 Imp gallon) chemical tank and multisensor surveillance for photography (Hasselblad survey camera), video recording, linescan or thermal imagery, or closed-circuit TV microwave transmission to a command vehicle/station. Certified to BCAR CAP 482 Section S in 1998.
Series E: As Series D, but with electric starter, long-range fuel tanks and optional 59.6 kW (79.9 hp) Rotax 912 UL four-stroke engine. *As described.*
CUSTOMERS: Total of 16 Series D and three Series E delivered by June 2001, including one fitted with hand controls for a disabled pilot.
COSTS: Series D: Assembled £19,950. Kit £12,950. Series E: Assembled £23,825. Kit £14,885. (All 2001, with Rotax 582 UL engine; add £2,000 to these prices for Rotax 912 UL).
DESIGN FEATURES: Design stated to have no 'defined stall' and to be unspinnable. Pod-and-boom layout; slightly tapered wings; three vertical stabilisers; sweptback tailplane with large rudder beneath; downturned wingtips. Options include agricultural spraygear and other specialised equipment. Quoted kit assembly time is 500 hours.
FLYING CONTROLS: Conventional and manual with cable actuation. Frise ailerons; full-span elevator with electrically actuated trim tab; fixed tab on rudder. Three-position flaps: 0, 15 and 30°.
STRUCTURE: Fuselage pod of Araldite-bonded Fibrelam. Wings comprise plywood formed leading-edge built on aluminium and plywood internal shearweb structure. Styrofoam formers provide both leading- and trailing-edge structures. Trailing surfaces covered with polyester cloth. Tailplane and rudder surfaces are polyester cloth-covered aluminium frames. Empennage connected to the monocoque by an aluminium boom.

CFM Shadow CD (Rotax 503 engine) *(Jane's/Paul Jackson)*   *NEW*/0526454

CFM Streak SA two-seat ultralight, fitted with a Rotax 582 *(Jane's/Paul Jackson)*   0109150

LANDING GEAR: Non-retractable tricycle type of aluminium and steel; castoring nosewheel; brakes. Options include wheel fairings and floats.
POWER PLANT: One 47.8 kW (64.1 hp) Rotax 582 UL two-stroke driving a three-blade Precision pusher propeller; or one 59.6 kW (79.9 hp) Rotax 912 UL four-stroke, driving a three-blade GSC propeller. Standard fuel capacity (582 UL) 62 litres (16.4 US gallons; 13.6 Imp gallons), (912 UL) 27 litres (7.1 US gallons; 5.9 Imp gallons). Options include a 30 litre (7.9 US gallon; 6.6 Imp gallon) auxiliary fuel tank.
EQUIPMENT: ASI, VSI, altimeter, tachometer, fuel and temperature gauges, slip ball indicator, electric elevator trim and front cockpit brakes indicator. Alloy wheels, rear cockpit brakes, intercom, GPS and ballistic parachute recovery system optional. Modified cockpit for disabled pilots comprises combined throttle/roll hand control on left console, sidestick controller for pitch/rudder on right console, each with hand-operated brake lever.
Data below refer to Shadow Series E with Rotax 582 UL engine.

DIMENSIONS, EXTERNAL:
Wing span	10.00 m (32 ft 9½ in)
Length overall	6.13 m (20 ft 1¼ in)
Height overall	1.75 m (5 ft 9 in)
Propeller diameter	1.32 m (4 ft 4 in)

AREAS:
Wings, gross	15.25 m² (164.1 sq ft)

WEIGHTS AND LOADINGS:
Weight empty	215 kg (474 lb)
Max T-O weight	425 kg (937 lb)

PERFORMANCE:
Cruising speed at 75% power	86 kt (159 km/h; 99 mph)
Stalling speed, flaps down	34 kt (63 km/h; 39 mph)
Max rate of climb at S/L, solo	366 m (1,200 ft)/min
T-O to 15 m (50 ft)	130 m (425 ft)
Landing from 15 m (50 ft)	183 m (600 ft)
Range with standard fuel	370 n miles (685 km; 426 miles)

UPDATED

CFM STREAK

TYPE: Tandem-seat ultralight/kitbuilt.
PROGRAMME: Construction of prototype started 1987 and this first flew June 1988; first flight of production aircraft (also known as Streak Shadow) October 1988. Also built under licence in South Africa, with Jabiru as engine option.
CURRENT VERSIONS: **Streak SA:** JAA Group A (Aeroplanes) certified version. *As described.*
Streak SLA: Microlight version. Differs from SA in having increased flap deflection and elevator gap seal to reduce minimum flying speed to comply with Microlight regulations.
CUSTOMERS: Total of 113 Streak SAs and seven Streak SLAs delivered by 11 June 2001. Recent customers include the Indian Air Force, which ordered 24 SAs in September 1999 for delivery over a 15-month period for bird control at airfields; total of 18 delivered by June 2001.
COSTS: SA/SLA with Rotax 582 UL: Assembled £22,975. Kit £14,465. With Rotax 912 UL: Assembled £25,105. Kit £16,940. (All excluding VAT, 2001). Factory-assembled aircraft available for export only.
DESIGN FEATURES: Derivative version of Shadow (which see); has new wing design of reduced span, light airframe weight and more powerful engine; no 'defined stall' and is spin-resistant.
FLYING CONTROLS: See Structure. Electric trim.
STRUCTURE: Similar to Shadow but with foam/glass fibre wings (CFM aerofoil section), and control surfaces with aluminium alloy ribs and polyester fabric covering (ailerons, flaps, rudder and elevators).
POWER PLANT: Engine options and fuel capacities as for Shadow Series D/E.

DIMENSIONS, EXTERNAL:
Wing span	8.53 m (28 ft 0 in)
Length overall	6.40 m (21 ft 0 in)
Height overall	1.75 m (5 ft 9 in)

AREAS:
Wings, gross	13.01 m² (140.0 sq ft)

WEIGHTS AND LOADINGS:
Weight empty	190 kg (419 lb)
Max T-O weight	408 kg (900 lb)

PERFORMANCE (Streak SA):
Max level speed	94 kt (174 km/h; 108 mph)
Stalling speed	34 kt (63 km/h; 39 mph)
Max rate of climb at S/L	366 m (1,200 ft)/min
T-O to 15 m (50 ft)	130 m (426 ft)
Landing from 15 m (50 ft)	183 m (600 ft)
Range	385 n miles (713 km; 443 miles)

UPDATED

CFM STAR STREAK

TYPE: Tandem-seat ultralight/kitbuilt.
PROGRAMME: Development of Streak with reduced wing chord, wider cockpit and uprated engine. Prototype flew in 1992; first deliveries February 1994. Total of nine delivered by 11 June 2001.
CURRENT VERSIONS: **Star Streak:** Standard version
Star Streak 912: Higher-powered version with 59.6 kW (79.9 hp) Rotax 912UL engine. *As described.*
COSTS: Star Streak: Assembled £23,350. Kit £14,650. Star Streak 912: Assembled £25,950. Kit £17,250. (All excluding VAT, 2001). Factory-assembled aircraft available for export only.
POWER PLANT: Star Streak has one 47.8 kW (64.1 hp) Rotax 582 UL two-stroke; Star Streak 912 has one 59.6 kW (79.9 hp) Rotax 912 UL four-stroke. Fuel capacity (both) 27 litres (7.1 US gallons; 5.9 Imp gallons).

DIMENSIONS, EXTERNAL:
Wing span	8.53 m (28 ft 0 in)
Length overall	6.05 m (19 ft 10 in)
Height overall	1.75 m (5 ft 9 in)

AREAS:
Wings, gross	11.00 m² (118.4 sq ft)

WEIGHTS AND LOADINGS:
Weight empty	205 kg (452 lb)
Max T-O weight	408 kg (900 lb)

PERFORMANCE:
Cruising speed at 75% power	94 kt (174 km/h; 108 mph)
Stalling speed, flaps down	38 kt (71 km/h; 44 mph)
Max rate of climb at S/L	427 m (1,400 ft)/min
T-O to 15 m (50 ft)	122 m (400 ft)
Landing from 15 m (50 ft)	183 m (600 ft)
Range with max fuel	210 n miles (388 km; 241 miles)

UPDATED

motor glider	358 kg (790 lb)
Baggage capacity	36 kg (80 lb)
Max T-O and landing weight: XS	621 kg (1,370 lb)
motor glider	635 kg (1,400 lb)
Payload with full fuel	197 kg (435 lb)
Max wing loading: XS	65.6 kg/m² (13.43 lb/sq ft)
motor glider	50.6 kg/m² (10.37 lb/sq ft)
Max power loading: XS	8.34 kg/kW (13.70 lb/hp)

PERFORMANCE, POWERED (XS and motor glider with Rotax 912 ULS engine, unless otherwise stated):
Max cruising speed at 75% power:
Rotax 912 ULS at 2,440 m (8,000 ft):
monowheel   140 kt (259 km/h; 161 mph)

tricycle	135 kt (250 km/h; 155 mph)
motor glider	130 kt (241 km/h; 150 mph)

Rotax 914 at 3,050 m (10,000 ft):
monowheel   174 kt (322 km/h; 200 mph)
tricycle   166 kt (307 km/h; 191 mph)
Stalling speed, power off:
flaps up   50 kt (92 km/h; 57 mph)
flaps down   45 kt (83 km/h; 51 mph)
Max rate of climb at S/L:
Rotax 912 ULS   305 m (1,000 ft)/min
motor glider   335 m (1,100 ft)/min
Rotax 914   396 m (1,300 ft)/min
T-O run: Rotax 912 ULS   180 m (590 ft)

motor glider   183 m (600 ft)
Rotax 914   153 m (500 ft)
Landing run: all   183 m (600 ft)
Range at econ cruising speed:
standard fuel   732 n miles (1,355 km; 842 miles)
optional fuel   1,093 n miles (2,024 km; 1,257 miles)
g limits (all): design   +3.8/−1.9
ultimate   +8.55/−4.27
PERFORMANCE, UNPOWERED (motor glider):
Best glide ratio   27
Min rate of sink at 47 kt (87 km/h; 52 mph)   0.90 m (2.94 ft)/s
UPDATED

Tri-gear and monowheel versions of Europa XS, each powered by a Rotax 914 flat-four *(Keith Wilson/Europa)*   NEW/0533398

FACL

FARNBOROUGH AIRCRAFT CORPORATION LTD

Cody Technology Park, Farnborough, Hampshire GU14 0LS
Tel: (+44 1753) 74 01 00
Fax: (+44 1753) 71 43 99
e-mail: information@farnborough-aircraft.com
Web: http://www.farnborough-aircraft.com
CHAIRMAN AND CEO: Andrew Taee
DIRECTOR: Geoffrey Galley
CHIEF DESIGNER: Afandi Darlington
COMPANY SECRETARY: Richard Blain

Former Farnborough-Aircraft.com (FAC: see 2002-03 *Jane's*) applied for insolvency protection in first half of 2002; this rejected, but 51 per cent of shareholders voted on 8 July 2002 for restructuring under new company title FACL, as above. First action by FACL is to renegotiate outstanding debts and attract up to US$100 million in additional investment capital (target figure of initial US$9 million by end of 2002).

UPDATED

FARNBOROUGH F1

TYPE: Business turboprop.
PROGRAMME: Launched 14 October 1999, at which time wind tunnel testing had been completed; first flight then scheduled for fourth quarter of 2002, FAR/JAR 23 certification in late 2004 and service entry in 2005, but delayed by financial uncertainties. Work continuing in late 2002; revised plan is for two flying prototypes, two for static/fatigue test, and possibly a fifth as demonstrator. Design and assembly to remain in UK, but manufacture by 'established companies' in Europe and/or USA. Funding permitting, first deliveries now targeted for 2007.

Artist's impression of Farnborough F1 turboprop business aircraft   NEW/0113314

CUSTOMERS: Intended for air taxi operators providing on-demand, Internet-booked service from a network of small airfields. Estimated total market for 24,000 aircraft from 2005 onwards. Launch orders for one each from separate British customers announced in September and October 2000.
COSTS: Development US$100 million; unit cost, equipped US$1.7 million (2002).
DESIGN FEATURES: Designed, via Internet communications, using computational fluid dynamics; design goals include ability to operate quietly from small airstrips carrying five

passengers at up to 330 kt (611 km/h; 380 mph) over a range of 1,000 n miles (1,852 km; 1,150 miles).
Wing aerofoil proprietary HLLF29140 high-lift, laminar flow section.
FLYING CONTROLS: Conventional and manual; 30 per cent chord Fowler flaps occupy approximately two-thirds span.
STRUCTURE: Primarily composites; fuselage of carbon-fibre skins with foam honeycomb core, moulded in two vertically split halves; carbon fibre/epoxy wing with aluminium alloy landing gear pick-ups, wing/fuselage fittings, flap tracks and aileron attachments.

LANDING GEAR: Retractable tricycle type; trailing-link main units retract inwards, nosewheel rearwards.

POWER PLANT: One Pratt & Whitney PT6A-60A turboprop, flat rated at 634 kW (850 shp).

ACCOMMODATION: Pilot and five passengers in pressurised cabin, maximum pressure differential 0.55 bar (8.0 lb/sq in) maintaining sea level cabin altitude to 5,790 m (19,000 ft). Four cabin windows on each side; door on port side aft of wing; overwing emergency exit on starboard side. Interior layout, designed by Coventry University, provides alternative configurations for business/executive, commuter, medevac and cargo applications.

AVIONICS: 'Glass cockpit' compatible with existing and planned ATC and GPS-based en route navigation and airfield approach systems. Weather radar in wingtip pod.

DIMENSIONS, EXTERNAL:

Wing span	12.41 m (40 ft 8½ in)
Wing aspect ratio	8.4
Length overall	10.70 m (35 ft 1¼ in)
Height overall	3.47 m (11 ft 4¾ in)

DIMENSIONS, INTERNAL:

Cabin: Length	3.44 m (11 ft 3½ in)
Max width	1.40 m (4 ft 7¼ in)
Max height	1.34 m (4 ft 4¾ in)

AREAS:

Wings, gross	18.26 m² (196.6 sq ft)

WEIGHTS AND LOADINGS:

Operating weight empty	1,805 kg (3,980 lb)
Max fuel weight	763 kg (1,682 lb)
Max payload	567 kg (1,250 lb)
Payload with max fuel	107 kg (236 lb)
Max T-O weight	2,675 kg (5,898 lb)
Max landing weight	2,541 kg (5,603 lb)
Max ramp weight	2,688 kg (5,927 lb)
Max zero-fuel weight	2,372 kg (5,230 lb)

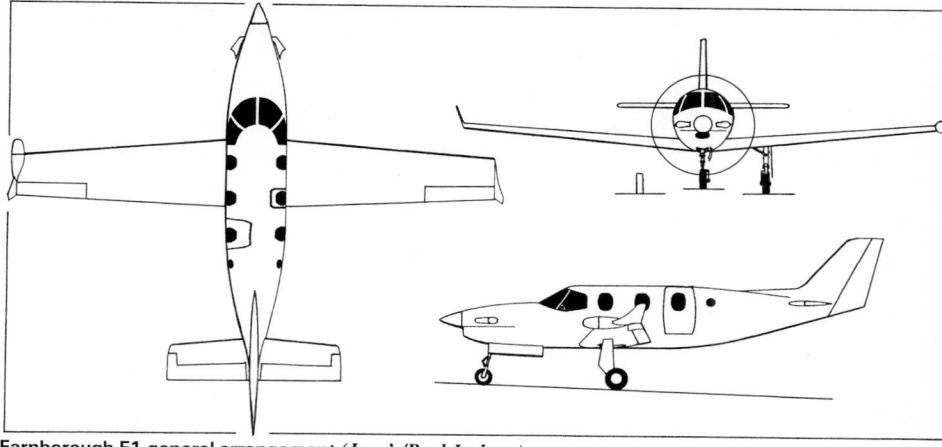

Farnborough F1 general arrangement *(Jane's/Paul Jackson)*   0110907

Max wing loading	146.5 kg/m² (30.00 lb/sq ft)		Time to 7,620 m (25,000 ft)	8 min
Max power loading	4.22 kg/kW (6.94 lb/shp)		Service ceiling	10,670 m (35,000 ft)

PERFORMANCE (estimated):

Maximum operating speed (V_{MO})
  295 kt (546 km/h; 339 mph)
High cruising speed at 9,145 m (30,000 ft)
  329 kt (609 km/h; 379 mph)
Long-range cruising speed at 9,145 m (30,000 ft)
  242 kt (448 km/h; 278 mph)
Stalling speed, flaps and landing gear down
  59 kt (110 km/h; 68 mph)

T-O run   466 m (1,530 ft)
Range with max fuel
  1,645 n miles (3,046 km; 1,893 miles)
Range with 363 kg (800 lb) payload, NBAA IFR reserves, 100 n mile (185 km; 115 mile) alternate
  975 n miles (1,805 km; 1,122 miles)
Ferry range   1,675 n miles (3,102 km; 1,927 miles)
UPDATED

RAF

ROYAL AIR FORCE

Main Building, Ministry of Defence, Whitehall, London SW1A 2HB
Tel: (+44 20) 72 18 90 00

Major RAF programmes in the formative stage now appear to be restricted to adaptations of existing aircraft.
UPDATED

FUTURE OFFENSIVE AIR SYSTEM (FOAS)

Staff Target (Air) 425 formulated as replacement for the RAF's Panavia Tornado GR. Mk 4; prefeasibility studies, 1993-95, under title of Future Offensive Aircraft (FOA). Programme launched on 16 December 1996 as FOAS, when £35 million assigned for allocation in 1997 to feasibility studies of broadest possible range of options, including variants of Eurofighter Typhoon, Lockheed Martin F-22 and JSF; new manned combat aircraft; uninhabited air vehicles, both combatant and non-combatant; and standoff missiles launched from transport aircraft.

Technologies considered included fly-by-light, stealth, virtual reality cockpits and integrated modular avionics. Study contracts awarded to several contractors, including the Defence Evaluation and Research Agency (QinetiQ);data sources included an Anglo-French technology demonstration programme permitting computer modelling of weapons systems. Studies into a crewed FOAS, with particular reference to LO technologies, have been undertaken by BAE Systems at Warton (which see). BAE Systems and partners received two-thirds of study funds, other contracts going to Logica and GKN Aerosystems.

Some 100 companies were working on FOAS feasibility by mid-1999. A third-phase study was launched that year; concept phase was then due to end in March 2001 with 'initial gate' decision to proceed to assessment phase. Force mix between different systems (if more than one solution chosen, as appeared likely in 1999) will be decided in time for selection of a solution in about 2010. By late 1999, French companies were working on joint studies with UK, while Germany was showing interest in joining.

During 2001, FOAS became a more modest programme, concentrating on technologies necessary to enhance existing aircraft. Tornado replacement now considered by advanced variants of Eurofighter and Lockheed Martin F-35, backed by unmanned air vehicles, new weapons and enhanced command and control.
UPDATED

FUTURE STRATEGIC TANKER AIRCRAFT (FSTA)

During 2001, FSTA became a direct competition between variants of the Airbus A330 and Boeing 767 (which see in International and US sections, respectively).
UPDATED

SUPPORT, AMPHIBIOUS AND BATTLEFIELD ROTORCRAFT (SABR)

This medium transport helicopter is expected to be chosen from a range of existing rotorcraft, including the Bell Boeing V-22 Osprey tiltrotor (promoted by BAE Systems), Boeing CH-47 Chinook, EHI EH 101, Eurocopter Cougar, NHI NH 90, Sikorsky S-92 Helibus and Sikorsky CH-53E Super Stallion. These are described in the current edition or in *Jane's Aircraft Upgrades.*
UPDATED

REALITY

REALITY AIRCRAFT LTD

Unit 1A, Amesbury Business Park, London Road, Amesbury, Wiltshire SP4 7LS
Tel: (+44 1980) 62 55 51
Fax: (+44 1980) 62 62 25
e-mail: info@realityaircraft.com
Web: http://www.realityaircraft.com
MANAGING DIRECTOR: Terry Francis
MARKETING DIRECTOR: Tom Lafferty
DIRECTOR: Kate Mather

Reality markets the Easy Raider, for which it has world rights, except in the USA.
NEW ENTRY

Jabiru-engined Reality Easy Raider *(Jane's/Paul Jackson)*   NEW/0137964

REALITY EASY RAIDER

TYPE: Tandem-seat ultralight kitbuilt.

PROGRAMME: Development of Flying K Sky Raider II (which see in US section), incorporating over 100 modifications. Prototype G-SRII (Rotax 503 engine) flew 2001; conforms to BCAR Section S. Kit fabrication by Just Aircraft of USA, with which Reality holds joint design rights.

CURRENT VERSIONS: **Easy Raider 503:** One 37.0 kW (49.6 hp) Rotax 503 DCDI-2V two-stroke engine driving a Powerfin FL370T three-blade, ground-adjustable pitch, composites propeller. Initial version.

Easy Raider J2.2: One 59.7 kW (80 hp) Jabiru 2200 engine driving a two-blade Powerfin ground-adjustable pitch or two-blade Newton wooden propeller. Prototype G-OESY (c/n 0002) built late 2001.

Easy Raider R100: BMW R100 four-stroke motorcycle engine driving a Powerfin two-blade, ground-adjustable pitch propeller. Prototype G-SLIP (c/n 0004) under construction in 2002.

CUSTOMERS: Three flying and five more under construction by July 2002.

COSTS: £12,563 (Rotax), £14,500 approx (BMW) or £16,298 (Jabiru), including engine, cowlings and propeller, excluding tax (2002).

DESIGN FEATURES: High wing with two tubular main, and two secondary bracing struts to each wing. Tailplane braced with twin wires above and twin rods below. Large fin fillet, compared with Sky Raider, enables sideslipping with cockpit doors in place. Wings fold for storage. Quoted build time 350 hours.

FLYING CONTROLS: Conventional and manual. Frise ailerons, cable actuated. Flight-adjustable tab in port tailplane. Horn-balanced tail surfaces. Slotted flaps, deflections 0, 15, 30 and 40°.

STRUCTURE: Fabric-covered chromoloy steel tube fuselage with composites engine cowling. Wing built on large diameter metal tube leading-edge/spar with wooden ribs, fabric covering and composites downturned wingtips.

LANDING GEAR: Tailwheel type; fixed. Rubber-in-tension shock-absorbers. Mainwheels 8.00-6 (4 ply); Matco steerable tailwheel. Optional floats or amphibious floats.

POWER PLANT: As under Current Versions. Two fuel tanks in wings, total capacity 37.9 litres (10.0 US gallons; 8.3 Imp gallons) standard; 72 litres (19.0 US gallons; 15.8 Imp gallons) optional.

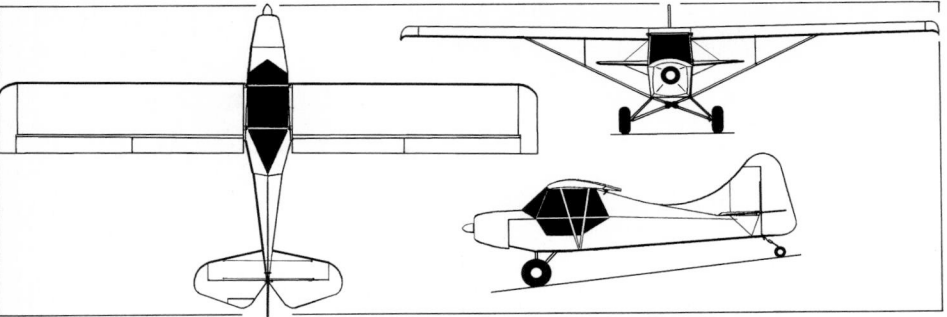

Reality Easy Raider (Rotax 503 engine) *(Jane's/Paul Jackson)*   **NEW**/0137970

DIMENSIONS, EXTERNAL:

Wing span	8.53 m (28 ft 0 in)	Length: overall	5.18 m (17 ft 0 in)
Width, wings folded	2.49 m (8 ft 2 in)	wings folded	6.10 m (20 ft 0 in)
		Height overall	1.75 m (5 ft 9 in)

DIMENSIONS, INTERNAL:

Cabin max width	0.67 m (2 ft 2½ in)

AREAS:

Wings, gross	9.75 m² (105.0 sq ft)

WEIGHTS AND LOADINGS (J: Jabiru 2200, R: Rotax 503):

Weight empty: J	225 kg (495 lb)
R	191 kg (420 lb)
Max T-O weight: J, R	450 kg (992 lb)

PERFORMANCE (J, R as above):

Max level speed: J	75 kt (139 km/h; 86 mph)
R	70 kt (129 km/h; 80 mph)
Normal cruising speed: J	60 kt (111 km/h; 69 mph)
R	55 kt (102 km/h; 63 mph)
Stalling speed, flaps down: J, R	34 kt (63 km/h; 40 mph)
Max rate of climb at S/L: J	213 m (700 ft)/min
R	122 m (400 ft)/min
T-O run: J	80 m (265 ft)
R	156 m (515 ft)
Landing run: J, R	139 m (460 ft)

NEW ENTRY

RN

ROYAL NAVY

Main Building, Ministry of Defence, Whitehall, London SW1A 2HB
Tel: (+44 20) 72 18 90 00

Fleet Air Arm requirements for a new class of 40,000 tonne aircraft carrier have been defined as the Lockheed Martin F-35 and an adaptation of an existing rotorcraft.

UPDATED

FUTURE JOINT COMBAT AIRCRAFT

Following the US Department of Defense's announcement in October 2001, the UK's FJCA is to be a variant of Lockheed Martin F-35 (which see).

UPDATED

FUTURE ORGANIC AIRBORNE EARLY WARNING (FOAEW)

A compound lift version of EH 101 Merlin or the Bell Boeing V-22 Osprey are considered likely candidates for this requirement. Details appear under EHI in the International section and Bell Boeing in the US section.

UPDATED

SLINGSBY

SLINGSBY AVIATION LIMITED

Ings Lane, Kirkbymoorside, York YO62 6EZ
Tel: (+44 1751) 43 24 74
Fax: (+44 1751) 43 11 73
e-mail: sales@slingsby.co.uk
Web: http://www.slingsby.co.uk
MANAGING DIRECTOR: Jeff Bevan
CHIEF DESIGNER: David Goddard
SALES AND MARKETING MANAGER: Steven Boyd

Formerly a subsidiary of ML Holdings, Slingsby was sold to Cobham plc in 1997 and is a component of Chelton Group within that holding. Company specialises in application of composites materials; was previously manufacturer of sailplanes but now concentrates on development and low-rate production of T67 Firefly series of aerobatic training aircraft, for which new orders received in 2001-02.

Other activities include design and manufacture of air cushion vehicles in composites materials; and design, development and manufacture of high-performance composites structures for marine and aerospace industries. Supplier to Lockheed Martin of C-130J cockpit interior trim panels in sculptured composites.

Work for UK MoD includes technical support for RAF Air Cadet gliders and Firefly aircraft.

UPDATED

SLINGSBY T67 FIREFLY

US Air Force designation: T-3A Firefly
TYPE: Primary prop trainer/sportplane.
PROGRAMME: Current composites constructed Firefly developed from wooden Slingsby T67A (licence-built version of French Fournier RF6B – see 1982-83 *Jane's*);

Slingsby T67M260 Firefly in the newly revised insignia of Babcock HCS *(Jane's/Paul Jackson)*   **NEW**/0533391

T67B gained CAA certification 18 September 1984; T67C was CAA certified 15 December 1987. Subsequent versions were T67M, T67M200 and T67M260.

Last-mentioned selected by USAF to meet Enhanced Flight Screener (EFS) requirement. Slingsby prime contractor for both acquisition contract and seven-year contractor logistic support (CLS) contract. Northrop Grumman subcontractor for final assembly at Hondo, Texas, and for operation of CLS activities at Hondo and USAF Academy at Colorado Springs.

Low-rate manufacture only, since 1997, but expanded in 2001 to cover order for 16 T67M260s placed by Royal Jordanian Air Force.

CURRENT VERSIONS: **T67C:** Basic version. See 1997-98 and previous editions for earlier C1, C2 and C3 subvariants.

One 119 kW (160 hp) Textron Lycoming O-320-D2A flat-four engine, driving a Sensenich M74DM6-O-64 two-blade, fixed-pitch, metal propeller. Fuel tank in each wing leading-edge, combined capacity 159 litres (42.0 US gallons; 35.0 Imp gallons). Nine T67Cs purchased by Netherlands government Civil Aviation Flying School for KLM and Royal Netherlands Navy pilot training; 12 T67Cs for Canadian Department of National Defence for military primary flying training; plus other T67C variants for UK schools. One (demonstrator) built in 2000.

T67M: Military variant. First flight of T67M Firefly 160 (G-BKAM) 5 December 1982; CAA certification 20 September 1983; designation changed to T67M Mk II as wing fuel tanks and two-piece canopy introduced. Powered by 119 kW (160 hp) Textron Lycoming AEIO-320-D1B flat-four engine, driving a Hoffmann HO-V72 two-blade constant-speed composites propeller.

Sold to Netherlands for military grading; Japan and UK for airline training; and Switzerland for aerobatic and general flying training. Used by RAF, RN and British Army for elementary flying training; Joint Elementary Flying Training School (JEFTS) formed at Topcliffe, July 1993, with first of eventual 18 (including some second-hand) T67M Mk IIs operated under contract by Hunting Aircraft Ltd (later Hunting Contract Services, and now Babcock HCS); transferred to Barkston Heath in April 1995.

T67M200: Development of T67M; 149 kW (200 hp) Textron Lycoming AEIO-360-A1E engine; Hoffmann HO-V123 three-blade, variable-pitch, composites propeller; wing fuel capacity as for T67C. First flight 16 May 1985; CAA certification 13 October 1985. First customer Turkish Aviation Institute, Ankara (16 delivered from 1985); others ordered by Dutch operator King Air (three T67M200s, plus one T67M Mk II) as screening trainers for prospective RNethAF pilots; Royal Hong Kong Auxiliary Air Force (now Government Flying Service) (four); and Norwegian government's Flying Academy (six).

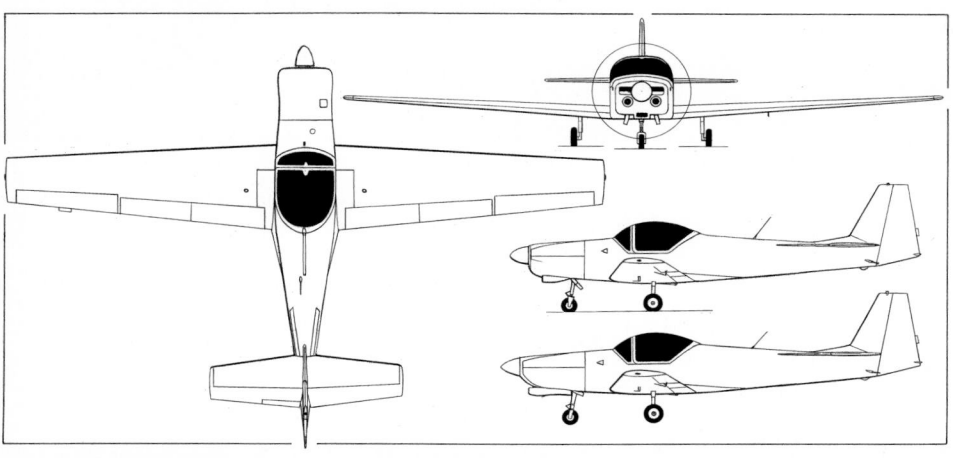

Slingsby T67M Mk II Firefly (Textron Lycoming AEIO-320-D1B engine) with additional side view (bottom) of T67M260 *(Jane's/Dennis Punnett)*

T67M260: Development of T67M; higher-powered Lycoming engine; electric elevator trim and electric flaps optional; cabin air conditioning system optional; higher maximum T-O and aerobatic weights, to allow 227 kg (500 lb) for two pilots and equipment plus full fuel load.

Prototype (G-BLUX) flown May 1991, evaluated at Wright-Patterson AFB; hot-and-high trials at USAF Academy in mid-1991. Preproduction aircraft (G-EFSM) flown September 1992. First flight of first USAF aircraft (92-0625/N7020D) 4 July 1993. T-3A type certificate awarded by CAA and FAA December 1993. Official USAF acceptance 25 February 1994; student pilot training began March 1994. Also acquired for UK military pilot training and by Bahrain and Jordan.

CUSTOMERS: Total of 257 civil/military T67s of all subtypes (including 10 T67As) delivered to customers in 13 countries by early 2001 (one built in 1998, two in 1999, two in 2000 and none in 2001). Manufacture of 19 for Bahrain (three) and Jordan (16) began in 2002. Production commitment 276.

US Air Force acquired 113 T-3As in three lots (38, 42 and 33) to replace Cessna T-41. USAF pilot conversion completed September 1993. Deliveries began January 1994 and were completed in November 1995. First operating unit (with 56 aircraft) was 1st Flight Screening Squadron of 12th Flying Training Wing based at Hondo, Texas; second and final operating unit (with 57 aircraft) is 557th FTS of USAF Academy at Colorado Springs; deliveries to this unit from August 1994; operational January 1995. All T-3As wore dual military/civilian identities. Remaining fleet of 110 grounded in July 1997 for fuel system modifications, and remained thus in early 2002, despite reassessment proving aircraft safe. In late 2000, USAF announced that it was contracting out future flight screening to civilian operators and in early 2002 was considering options for the T-3A fleet, including scrapping.

Hunting Contract Services (see T67M, above) took delivery of 25 T67M260s between June 1996 and March 1997 for extension to JEFTS tri-service training programme, followed by a further two in September 1999. JEFTS expanded to 18 M200s and 27 M260s, based at Barkston Heath (20), Middle Wallop, Cranwell (six) and Church Fenton (five). One each delivered to Belize Defence Force and a UK civilian customer in 1996. Royal Jordanian Air Force ordered 16 T67M260s in September 2001, for delivery between July and December 2002 to replace Bulldogs with No. 4 Squadron at Al Mafraq AFB. Three T67M260s ordered in March 2002 by Bahrain Amiri Air Force, for delivery (via BAE Systems) in first quarter of 2003.

DESIGN FEATURES: Conventional low-wing monoplane; design translated from wood to GFRP. Tapered wings and mid-mounted tailplane; sweptback fin.
Wing section NACA 23015 at root, 23013 at tip; dihedral 3° 30′; incidence 3°.

FLYING CONTROLS: Conventional and manual. Mass-balanced Frise ailerons, without tabs; mass-balanced elevators with manually operated port trim tab (electric trim optional); trailing-edge fixed hinge flaps; spin strakes forward of tailplane roots.

STRUCTURE: GFRP; single-spar wings with double skin (corrugated inner skin bonded to plain outer skin) and ribs in heavy load positions; frame and top-hat stringer fuselage; stainless steel firewall between cockpit and engine; fixed incidence tailplane of similar construction to wings (built-in VOR antenna); fin incorporates VHF antenna.

LANDING GEAR: Non-retractable tricycle type. Oleo-pneumatic shock-absorber in each unit. Steerable nosewheel. Mainwheel tyres size 6.00-6, pressure 1.38 bar (20 lb/sq in). Nosewheel tyre size 5.00-5, pressure 2.55 bar (37 lb/sq in). Hydraulic disc brakes. Parking brake.

Data follow for T67M260/T-3A; refer to 1999-2000 and earlier Jane's for other variants.

Slingsby T67M260 Firefly demonstrator posing in the colours of the Royal Jordanian Air Force   0121676

POWER PLANT: One 194 kW (260 hp) Textron Lycoming AEIO-540-D4A5 flat-six engine, driving a Hoffmann HO-V123K-KV/180DT three-blade constant-speed composites propeller. Fuel 159 litres (42.0 US gallons; 35.0 Imp gallons) in wing leading-edges; refuelling point in upper wing surface.

ACCOMMODATION: Two seats side by side, fixed windscreen and rearward-hinged upward-opening rear-canopy section. Optional raised canopy and lowered seats to accommodate crew wearing military-style helmets. Dual controls standard. Adjustable rudder pedals. Cockpit heated and ventilated. Baggage space aft of seats.

SYSTEMS: Hydraulic system for brakes only. Vacuum system for blind-flying instrumentation. Electrical power supplied by 28 V 70 A engine-driven alternator and 24 V 15 Ah battery.

AVIONICS: Optional avionics, available to customer requirements, include equipment by Bendix/King, up to full IFR standard.
Instrumentation: Standard avionics include artificial horizon and directional gyro, with vacuum system and vacuum gauge, electric turn and slip indicator, rate of climb indicator, recording tachometer, stall warning system, clock, outside air temperature gauge, accelerometer.

EQUIPMENT: Includes tiedown rings and towbar; cabin fire extinguisher, crash axe, heated pitot; instrument, landing, navigation and strobe lights. Optional equipment includes external power socket, and wingtip-mounted smoke system.

DIMENSIONS, EXTERNAL:
Wing span	10.59 m (34 ft 9 in)
Wing chord: at root	1.53 m (5 ft 0¼ in)
at tip	0.83 m (2 ft 8¾ in)
Wing aspect ratio	8.9
Length overall	7.57 m (24 ft 10 in)
Height overall	2.36 m (7 ft 9 in)
Tailplane span	3.40 m (11 ft 1¾ in)
Wheel track	2.44 m (8 ft 0 in)
Wheelbase	1.50 m (4 ft 11 in)
Propeller diameter	1.80 m (5 ft 10¾ in)

DIMENSIONS, INTERNAL:
Cockpit: Length	2.05 m (6 ft 8¾ in)
Max width	1.08 m (3 ft 6½ in)
Max height	1.08 m (3 ft 6½ in)

AREAS:
Wings, gross	12.63 m² (136.0 sq ft)
Ailerons (total)	1.24 m² (13.35 sq ft)
Trailing-edge flaps (total)	1.74 m² (18.73 sq ft)
Fin	0.80 m² (8.61 sq ft)
Rudder	0.82 m² (8.80 sq ft)
Tailplane	1.65 m² (17.76 sq ft)
Elevators (incl tab)	0.99 m² (10.66 sq ft)

WEIGHTS AND LOADINGS (with air conditioning installed):
Weight empty	794 kg (1,750 lb)
Max fuel weight	114 kg (252 lb)
Max T-O and landing weight:	
Utility/Aerobatic	1,157 kg (2,550 lb)
Max wing loading	91.5 kg/m² (18.75 lb/sq ft)
Max power loading	5.97 kg/kW (9.81 lb/hp)

PERFORMANCE:
Never-exceed speed (VNE)	195 kt (361 km/h; 224 mph)
Max level speed at S/L	152 kt (281 km/h; 175 mph)
Max cruising speed at 75% power at 2,590 m (8,500 ft)	
	140 kt (259 km/h; 161 mph)
Stalling speed, power off, flaps down	
	54 kt (100 km/h; 63 mph)
Max rate of climb at S/L	480 m (1,380 ft)/min
T-O run	334 m (1,095 ft)
T-O to 15 m (50 ft)	689 m (2,265 ft)
Landing from 15 m (50 ft)	984 m (3,228 ft)
Landing run	401 m (1,315 ft)
Range with max fuel at 65% power at 2,440 m (8,000 ft), allowances for T-O, climb and 30 min reserves	
	407 n miles (753 km; 468 miles)
Endurance at best econ setting at 2,440 m (8,000 ft), allowances as above	5 h 40 min
g limits	+6/−3

UPDATED

SPEEDTWIN

SPEEDTWIN DEVELOPMENTS LTD

DEVELOPMENT FACILITY:
Upper Cae Garw Farm, Trelleck, Monmouth, Gwent NP5 4PJ
Tel: (+44 1600) 86 90 46
Fax: (+44 1600) 86 91 27

MANAGING DIRECTOR: Malcolm Ducker

SALES OFFICE:
77 Culverden Down, Tunbridge Wells, Kent TN4 9SL
Tel: (+44 1892) 53 73 40
Fax: (+44 1892) 57 07 29
e-mail: speedtwin@ontel.net.uk

This company was formed to market the Speedtwin twin-engined aerobatic light sporting aircraft designed by the late Peter Phillips. It was bought by the present managing director in March 2000. A design contract was signed in May 2000 with RCS Aviation to complete a full stress analysis and to obtain design approval for the Mk II. No recent reports have been received. See 2001-02 edition for description.

UPDATED

WARRIOR

WARRIOR (AERO-MARINE) LTD

The Portway Centre, Old Sarum, Salisbury, Wiltshire SP4 6EB
Tel: (+44 1722) 42 18 60
Fax: (+44 1722) 42 18 61
e-mail: sales@centaurseaplane.com
Web: http://www.centaurseaplane.com
MANAGING DIRECTOR: James Labouchere

NORTH AMERICAN OFFICE:
Warrior (Aero-Marine) Inc, 23 Oceanwood Drive, Scarborough, Maine 04074, USA
Tel/Fax: (+1 207) 885 99 20
e-mail: dverrill@centaurseaplane.com
VICE-PRESIDENT AND DIRECTOR: David C Verrill

Company created in 1995 to develop and market Centaur six-seat amphibian; US subsidiary formed in January 2000 to promote aircraft in North American market. Funding acquired by March 2002 from private investors (more than

US$3 million), Maine Technology Institute (US$500,000) and UK Department of Trade and Industry (£450,000/US$640,000); further US$2 million then said to be required for marketing and 2004 US demonstration tour.

UPDATED

WARRIOR CENTAUR

TYPE: Six-seat amphibian.
PROGRAMME: First design studies 1992; Mk 1 hull design, and first flight by one-fifth scale model, in 1993; Mk 2 hull

configuration developed and trialled 1995; dynamically refined (and still current) one-fifth model first flown 1996; cockpit mockup completed 1997; initial funding by Royal Aeronautical Society Handley Page award. Revealed at Farnborough Air Show, September 1998, when development and planning continuing. Agreement in principle for Chichester-Miles Consultants Ltd (see CMC entry earlier in this section) to be the prototyping facility; funding for construction and testing of full-size prototype secured 1 September 2001. Prototype, manufactured by Warrior in association with Maine Composites Inc in USA and CMC Ltd in UK, began in December 2001 and was expected to fly in first half of 2003. Design changes since then include new contours on rear fuselage and outer wings, and deletion of wing bracing struts. Certification to JAR/FAR 23 anticipated 24 months after first flight. Production aircraft will be built in the UK and USA.

CUSTOMERS: Four or five orders reported by March 2002. Production target is 68 in first full year; 172 in second.

COSTS: Expected to be between US$525,000 and US$575,000 at 2002 prices and rates. Seat-mile cost estimated at 35 per cent less than comparable amphibians.

Following description applies to full-size aircraft:

DESIGN FEATURES: Concepts proven and refined using CATIA software. Objectives of design included low hydrodynamic drag and structural weight, use of composites materials, folding wing, comfortable operation in coastal wave conditions, competitive seat-mile costs and payload-range performance, safety, saline impermeability and access to all areas frequented by boats.

Slender boat hull without planing step improves rough-water ride and increases aerodynamic efficiency; parasol-mounted high-lift wing with centrally installed engine nacelle; stub-wing/sponsons with end-mounted floats; conventional tail surfaces. Main panels of wing fold back manually (optionally hydraulically) to within beamwidth of sponsons to permit docking in standard 12 m (40 ft) yacht berths.

FLYING CONTROLS: Conventional and manual. Electrohydraulically actuated, slotted trailing-edge flaps.

STRUCTURE: Composites construction (carbon and glass fibres). Top surfaces of stub-wings laminated with energy absorbent tread; door sills reinforced with stainless steel.

LANDING GEAR: Flying-boat or amphibian. Electrohydraulically retractable tricycle type in amphibian, nosewheel retracting rearward, mainwheels outward into sponson tip-floats. Stern-mounted 7 kW (9 hp) water-jet for surface manoeuvring.

POWER PLANT: One 224 kW (300 hp) Textron Lycoming IO-540 or 261 kW (350 hp) TIO-540-J2B flat-six engine; three-blade, constant-speed, variable-pitch propeller. One fuel tank in each sponson; see under Weights and Loadings for details.

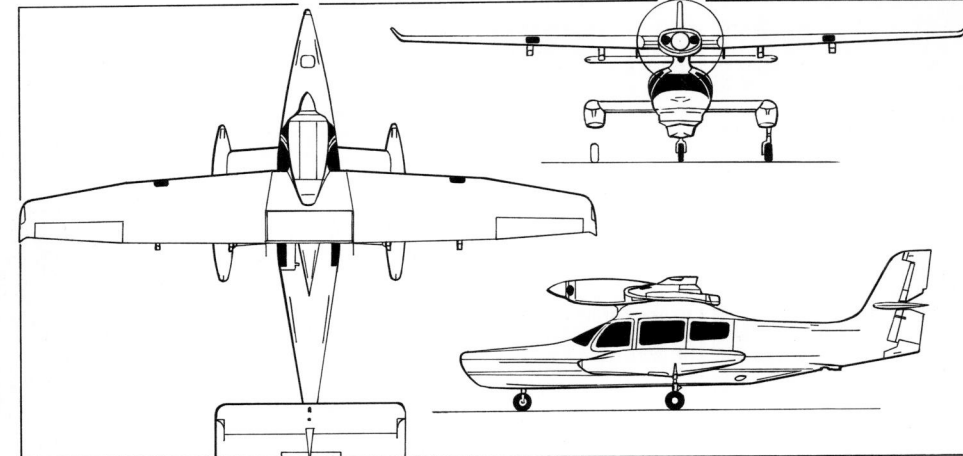

General arrangement of the Warrior Centaur amphibian (*Jane's/James Goulding*)   NEW/0526455

ACCOMMODATION: Pilot and one passenger in cockpit; five passenger seats in flying-boat cabin, four in amphibian. Baggage compartment behind seats. Three doors on port side; two on starboard.

SYSTEMS: Electrically driven hydraulic actuation of wing folding (optional), landing gear and flaps. Electrical systems powered by 28 V DC alternator and 24 V lead-acid batteries. Air conditioning optional.

AVIONICS: Honeywell APEX 1000 standard; marine HF radio, two-axis autopilot and bow-mounted weather radar.

DIMENSIONS, EXTERNAL:
Wing span	13.65 m (44 ft 9½ in)
Width over sponsons	4.54 m (14 ft 10¾ in)
Length overall	11.15 m (36 ft 7 in)
Height overall, on land	3.78 m (12 ft 4¾ in)
Wheel track	4.09 m (13 ft 5 in)

DIMENSIONS, INTERNAL:
Cabin, excl cockpit: Length	3.81 m (12 ft 6 in)
Max width	1.37 m (4 ft 6 in)
Max height	1.22 m (4 ft 0 in)
Volume	3.74 m³ (132 cu ft)
Baggage compartment volume	1.59 m³ (56 cu ft)

WEIGHTS AND LOADINGS:
Baggage capacity	90 kg (200 lb)
Fuel weight: max	354 kg (780 lb)
with max payload	68 kg (150 lb)

Payload: max: amphibian	641 kg (1,414 lb)
flying-boat	718 kg (1,583 lb)
with max fuel	264 kg (583 lb)

PERFORMANCE (amphibian with IO-540, estimated):
Cruising speed:
at max cruise power at 1,525 m (5,000 ft), 75%	
MTOW	127 kt (235 km/h; 146 mph)
at recommended cruise power at S/L	
	119 kt (220 km/h; 137 mph)

Stalling speed, power off:
flaps up	51 kt (95 km/h; 59 mph)
flaps down	46 kt (86 km/h; 53 mph)

Time to 3,050 m (10,000 ft) at max cruise power	
	13 min 24 s
T-O run: on water	173 m (569 ft)
T-O to 15 m (50 ft): on land	287 m (940 ft)
on water	363 m (1,192 ft)
Landing from 15 m (50 ft) at 1,700 kg (3,748 lb):	
on land	348 m (1,140 ft)

Range at max cruising speed, VFR reserves:
with max payload	150 n miles (278 km; 172 miles)
with 419 kg (923 lb) payload	
	600 n miles (1,111 km; 690 miles)
with 281 kg (620 lb) payload	
	1,200 n miles (2,222 km; 1,380 miles)*

* For flying-boat, add 72 kg (159 lb) to payload

UPDATED

WESTLAND

WESTLAND HELICOPTERS LIMITED (an AgustaWestland company)

Lysander Road, Yeovil, Somerset BA20 2YB
Tel: (+44 1935) 47 52 22
Fax: (+44 1935) 70 21 31
e-mail: info@gkn-whl.co.uk
Web: http://www.gkn-whl.co.uk
CEO: Richard Case

Westland Aircraft Ltd (later expanded to Westland Group plc) formed July 1935, taking over aircraft branch of Petters Ltd (known previously as Westland Aircraft Works) that had designed/built aircraft since 1915; entered helicopter industry having acquired licence to build US Sikorsky S-51 as Dragonfly 1947.

GKN shareholding in Westland Group progressively increased until overall control achieved on 18 April 1994. Merger of GKN's helicopter interests and those of Italy's Finmeccanica finalised on 26 July 2000 with announcement of **AgustaWestland**; this declared fully operational on 12 February 2001 (see International section).

Westland Helicopters Ltd includes GKN Westland Industrial Products Ltd at Weston-super-Mare, Somerset, and has 50 per cent interest in Aerosystems International Ltd at Yeovil (aerospace software; partnership with BAE Systems) and EH Industries Ltd. Last-mentioned was formed as partnership with Agusta of Italy for development and manufacture of EH101 Merlin helicopter (see EHI in International section). In addition to UK Merlins, Westland building 14 for Denmark and 12 for Portugal. Employment in helicopter and transmissions business was 4,800 in mid-2000; sales in 2000 exceeded £1,000 million, producing £116 million operating profit.

Under June 1989 agreement with former McDonnell Douglas, Westland obtained co-production rights for Boeing AH-64 Apache; this selected for British Army Air Corps, July 1995; Westland is prime contractor to UK MoD. Production now under way in newly built 5,200 m² (56,000 sq ft) assembly hall at Yeovil, opened 14 January 1999. On 8 January 2001, Westland acquired Boeing's aerospace fabrication business in St Louis, Missouri, becoming strategic supplier to Boeing.

First UK-assembled WAH-64D Apache, ZJ172   0106501

Aviation Training International (ATI) joint venture company formed with Boeing on 3 August 1998 to provide air, ground and maintenance crews for British Army Apaches under £650 million, 30 year contract; began operations 29 July 1999 at Sherborne HQ; is opening branches at Middle Wallop, Dishforth and Wattisham flying bases, plus engineering school at Arborfield.

Pending full production of EH 101 and Apache, Westland continued low-rate production of existing products; delivered four helicopters in 1993; none in 1994; three in 1995; and 11 in 1997, including the final Sea King. Production in 1999 comprised 10 Lynx and 13 EH 101s; increased in 2000 to 30 (nine WAH-64Ds, five Royal Navy Merlins, six RAF Merlins and 10 Lynx), plus three rebuilt Lynx and 14 kits for Lynx rebuilds.

Support of Sea King, Puma and Gazelle previously undertaken at Weston-super-Mare by Westland Industrial Products Ltd. Plant closure announced in January 2002.

UPDATED

WESTLAND WAH-64D APACHE LONGBOW

TYPE: Attack helicopter.

PROGRAMME: Westland selected, 13 July 1995, to build McDonnell Douglas (now Boeing) AH-64D Apache for Army Air Corps (AAC) and Royal Marines. Total 67 ordered, although original requirement was for 91; contract H12b/400 finalised 25 March 1996; programme involves some 240 UK companies (including 60 direct subcontractors) that will receive over 50 per cent of UK Apache work. Value increased to some £3,100 million by 1999, in part through addition of Marconi Avionics (now part of BAE Systems) to provide HIDAS self-protection. Apache description in US section; UK version similar, including Longbow radar, but powered by Rolls Royce Turbomeca RTM 322 turboshafts. Other differences from US Army version given below.

Engine integration trials conducted on leased sixth preproduction AH-64D (85-25408) at Boeing's Mesa plant; first engine run 6 April 1998; first flight 29 May 1998. Eight WAH-64Ds built and flown at Mesa; ZJ166/N9219G first flew 25 September 1998 and handed over to GKN Westland 28 September; initial trials and pilot training conducted in US, where four WAH-64s were retained by January 2000; further four supplied, complete, to Westland; remainder following in kit form from 30 August 1999; first delivery to AAC 15 March 2000; IOC on delivery of ninth aircraft, 14 December 2000; initial military aircraft release (IMAR) signed 22 December 2000; formal in-service date and release-to-service achieved 16 January 2001; initial weapons clearance planned for July 2001; HIDAS clearance December 2001; cannon and CRV-7 clearance August 2002, allowing full deployment; shipboard clearance November 2003; final delivery in second quarter of 2004; full environmental clearance in November 2004. First Apache with full suite of mission equipment (including HIDAS, HUMS and low-height warning system) delivered early 2003.

Shorts contracted in September 1996 to supply engine nacelles, stub-wings and horizontal stabiliser; heat exchangers manufactured by IMI Marston under Hughes-Treitler licence. UK and Netherlands agreed on 5 September 1996 on joint support of their Apache fleets. US and UK governments signed co-operation agreement 24 May 2000 for future Apache development, operation and support. UK contract also includes 68 Longbow radars, 980 Hellfire missiles and 204 launchers.

CURRENT VERSIONS (specific): **WAH1/ZJ166** (N9219G): US-built; first flight at Mesa 25 September 1998; handed over to Westland 28 September 1998 for RTM 322 engine trials in US.

WAH2/ZJ167 (N3266B): US-built; pilot training in US. Delivered 11 July 2000.

WAH3/ZJ168 (N3123T): US-built; airfreighted to Yeovilton 28 May 1999; thence Yeovil; first flight of WAH-64 in UK 26 August 1999; DERA Boscombe Down. Delivered June 2000.

WAH4/Z1169 (N3114H): US-built; pilot training in US. Delivered 11 July 2000.

WAH5/ZJ170 (N3065U): US-built; HIADS trials in USA until 2002.

WAH6/ZJ171 (N3266T): US-built; transferred to Yeovil 16 December 1999; instrumented aircraft; first army 'delivery' (actually roll-out) 15 March 2000, but not allocated to QinetiQ at Boscombe Down until April 2000.

WAH7/ZJ172: First UK kit; arrived Yeovil 30 August 1999. First flight 18 July 2000; delivered 31 July 2000. AHTU, Middle Wallop.

WAH8/ZJ173 (N3267A): US-built; transferred to Yeovil 30 December 1999. General trials and pilot training. To USA.

WAH9/ZJ174: Second UK kit; arrived Yeovil 16 December 1999. AHTU.

WAH10/ZJ175 (N3218V): US-built; arrived Yeovil 6 April 2000. Delivered June 2000. AHTU.

WAH11/ZJ176: Third UK kit. AHTU.

CUSTOMERS: British Army. Service allocation comprises 19 for operational evaluation, training (of which eight at Middle Wallop) and attrition replacement, plus 16 for each of three Army Air Corps regiments: No. 9 at Dishforth (656 and 657 Squadrons from third quarter of 2001) and

Manufactured image of a potential Future Lynx *NEW*/0533723

Nos. 3 (654 and 669 Squadrons from mid/late 2002) and 4 (662 and 663 Squadrons from early/mid-2003) at Wattisham. Units have additional responsibility for supporting Royal Marines with 'navalised' WAH-64s and integrated (Army/RM) crews. Re-equipment dates were not met because of crew shortage, and new aircraft delivered to storage at Fleetlands (beginning ZJ182 on 5 October 2001) and at Shawbury (beginning ZJ191 and ZJ192 on 2 July 2002). By mid-2002, Advanced Helicopter Trials Unit (AHTU) at Middle Wallop had some 12 Apaches for development work.

DESIGN FEATURES: Manual blade folding (windspeed limit 45 kt; 83 km/h; 52 mph).

FLYING CONTROLS: Manual back-up to FBW system.

POWER PLANT: Two 1,566 kW (2,100 shp) RRTI RTM 322-01/12 turboshafts.

SYSTEMS: Rotor blade de-icing system.

AVIONICS: Comms: Mode S IFF.
Instrumentation: Honeywell IHADSS helmet sight to be upgraded from about 2004.
Mission: Video recorder.
Self-defence: BAE Systems HIDAS helicopter integrated defensive aids system, including Sky Guardian 2000 RWR, Type 1223 LWR, Thales (Vinten) Vicon 78 Srs 455 chaff/flare dispenser and BAE Systems AN/AAR-57(V) common missile warning system (CMWS). Trial installation of three warning elements first flown (in USA) on UK Apache in July 2000. Standard US fit of AN/APR-48 RFI also included.

ARMAMENT: Additional provision for Bristol Aerospace CRV-7 70 mm rocket pods (reportedly cancelled in 2000) and Shorts Starstreak AAM.

UPDATED

WESTLAND LYNX

TYPE: Light utility helicopter.

PROGRAMME: Developed within Anglo-French helicopter agreement confirmed 2 April 1968; Westland given design leadership; first flight of first of 13 prototypes (XW835) 21 March 1971; first flight of fourth prototype (XW838) 9 March 1972, featuring production-type monobloc rotor head; first flights of British Army Lynx prototype (XX153) 12 April 1972, French Navy prototype (XX904) 6 July 1973; production Lynx (RN HAS. Mk 2 XZ229) 20 February 1976; first RN operational unit (No. 702

Squadron) formed on completion of intensive flight trials December 1977; AH. Mk 5 first flew (ZE375) 23 February 1985; other development details and records in 1975-76 and subsequent *Jane's*. Production shared 70 per cent Westland, 30 per cent Aerospatiale; for details of G-LYNX's 1986 world helicopter absolute speed record, and Lynx AH. Mk 7 XZ170's 1989 agility trials, see 1990-91 *Jane's*.

Second phase of development was Super Lynx (naval) and Battlefield Lynx, to which standards later UK military Lynx (Mks 8 and 9) were built; also exported. Battlefield Lynx mockup displayed at 1988 Farnborough Air Show (converted demonstrator G-LYNX), featuring wheeled landing gear, exhaust diffusers and provision for anti-helicopter missiles each side of fuselage; first flight of wheeled prototype (converted trials AH. Mk 7 XZ170) 29 November 1989; first flight of South Korean Super Lynx (90-0701, temporarily ZH219) 16 November 1989 (also first Lynx with Seaspray Mk 3 radar); first flight of Portuguese Super Lynx (9201, temporarily ZH580, ex-RN ZF559) 27 March 1992.

In September 1996, GKN Westland formally launched versions of Lynx powered by CTS800 turboshafts, with or without a six-screen EFIS. Current production versions were simultaneously redesignated, those now on offer being Srs 100, 200 or 300. Fuselages built from 1999 have potential for MTOW extension to 5,443 kg (12,000 lb) and beyond. First Super Lynx 300 (ZT800) flew 27 January 1999, although maiden flight with CTS800 engine was on 12 June 2001.

Potential life extension for Lynx possible following 30 January 2002 award to Westland of £20+ million, 18-month study contract expected to lead to upgrade of 80 British Army Mk 7 and Mk 9 Lynxes under Future Lynx Battlefield Light Utility Helicopter programme at total cost of £1 billion. This followed, on 22 July 2002, by announcement of assessment phase for Future Lynx to meet Royal Navy's Surface Combatant Maritime Rotorcraft requirement, requiring upgrade of 40 Lynx to enter service from 2008.

CURRENT VERSIONS: Early versions of Lynx for UK armed forces were last described in the 2002-03 *Jane's*. Those for overseas operators appeared in the 1990-01 and earlier editions.

Super Lynx Series 100: Introduced September 1996; export version of HMA. Mk 8; conventional cockpit instrumentation, Rolls-Royce Gem 42-1 turboshafts and 360° radar. Applies retrospectively to aircraft sold to Brazil (Mk 21A), Portugal (Mk 95) and South Korea (Mk 99) as well as to seven Mk 88As sold to Germany, also in September 1996.

Super Lynx Series 200: Conventional cockpit, but electronic power system displays and two LHTEC CTS800-4N turboshafts with FADEC. Original Lynx demonstrator G-LYNX fitted with two 1,007 kW (1,350 shp) T800 turboshafts as **Battlefield Lynx 800** private venture (LHTEC funding power plants and gearboxes, Westland provided airframe for full flight demonstration programme); first flight 25 September 1991; programme terminated early 1992 after 17 hours. CTS800 is derived from the T800 (as in the Boeing Sikorsky RAH-66 Comanche) and offers an additional 30 per cent of power to operators using the Lynx in hot climates. Version for production Lynx is CTS800-4N

Super Lynx Series 300: Prototype first flown (ZT800) 27 January 1999, initially with Gem 42 engines; flew with CTS800-4N engines 12 June 2001. Six-screen EFIS cockpit (including four 158 mm; 6¼ in square integrated display units); dual redundant MIL-STD-1553B and ARINC 429 databusses; new navigation system and AHRS; revised communications suite; CTS800 engines. Ordered by Malaysia (with Sea Skua anti-ship missiles); Thailand confirmed contract for two in 2001; Oman announced intention to purchase on 17 March 2001 and signed contract for 16 on 19 January 2002; further

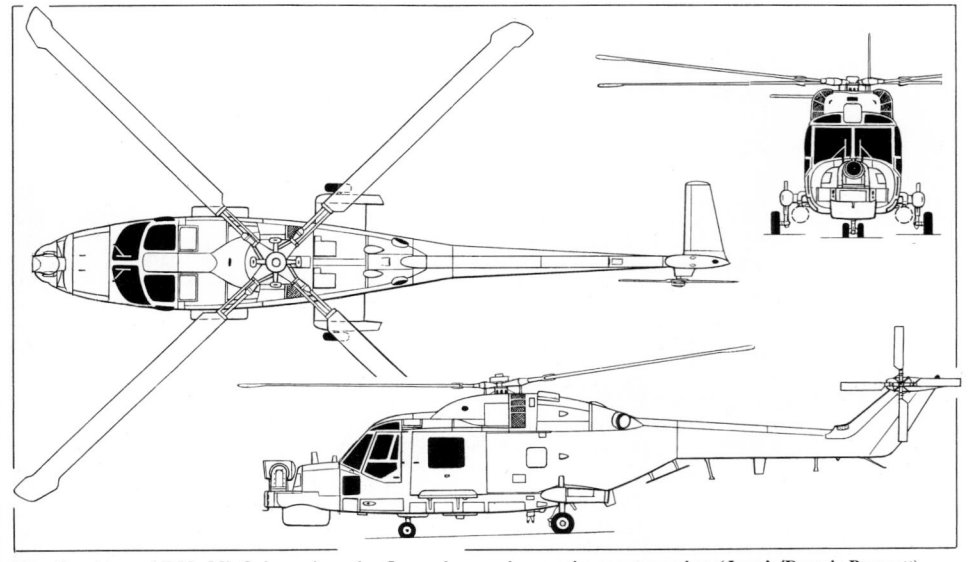

Westland Lynx HMA. Mk 8, based on the Super Lynx advanced export version *(Jane's/Dennis Punnett)*

customer expected to be South Africa, which announced intention to purchase on 18 November 1998, but later deferred expected contract for four to unspecified date; South African Srs 300s will be designated Mk 64 and equipped with Telephonics AN/APS-143 radar.

Future Lynx: Development of Series 300; offered to UK armed forces as retrofit. Growth potential to MTOW of 6,250 kg (13,778 lb).

CUSTOMERS: See table. Contracts for later versions include Super Lynx ordered by South Korea 1988 (12 **Mk 99** with Racal Avionics Doppler 71/TANS N nav system, Seaspray Mk 3 360° radar, AN/AQS-18 dipping sonar and Sea Skua), handed over between 26 July 1990 and May 1991 for 'Sumner' and 'Gearing' class destroyers; further 13 delivered 1999-2000 as Mk 99As to order confirmed in June 1997. Mk 99A has composites tailplane; first aircraft flown 3 July 1999 and departed Yeovil for surface delivery 1 September 1999. Portugal ordered five (first two ex-Royal Navy modified airframes) Super Lynx **Mk 95** 1990 (plus three options) with Racal RNS252 GPS-aided INS and Doppler 91 navigation systems and some US equipment including AN/AQS-18 sonar and Honeywell RDR 1500 radar; first two handed over 29 July 1993 for 'Vasco da Gama' class (MEKO 200) frigates; final two delivered 16 November 1993. Brazil ordered nine **Mk 21As** (five conversions from Mk 21). First for upgrade, N3027 delivered to Yeovil on 1 February 1995 and reflown as Mk 21A (N4010) on 22 December 1995; last two redelivered in late April 1998. First new-build helicopter (N4001) flew 12 June 1996; last delivered in August 1997. Mk 21A avionics include 360° Seaspray 3000 radar, RNS252 INS and Doppler 71 (but no FLIR or CTS); armament includes Sea Skua missiles. Mk 21A introduced 5,330 kg (11,750 lb) MTOW.

German Navy ordered seven Super Lynx **Mk 88As** in September 1996 and simultaneously took an option (confirmed on 25 June 1998) on upgrading (with new-build airframes) of its existing 17 (subsequently reduced by attrition to 15); new deliveries began with roll-out on 14 July 1999 (post first flight) of initial aircraft. Westland converted initial upgraded helicopter (8303), first flown 28 February 2001. First of 14 upgrades by Eurocopter at Donauwörth (of 8309) was redelivered on 7 March 2002, Mk 88A is a Srs 100 aircraft with 360° Seaspray Mk 3000 radar, FLIR turret (or nose fairing when not fitted), Rockwell Collins GPS and Racal Doppler 91 and RNS252; Sea Skua ASM armament.

Eight Danish aircraft being upgraded to **Mk 90B** with new airframes under contract announced 20 January 1998; completion due in 2004; all except one undertaken locally; initial conversion (S-191) by Westland and redelivered to Denmark on 1 November 2000. Super Lynx also offered to Australia and New Zealand (both unsuccessfully) and Malaysia, last-mentioned announcing order for six on 7 September 1999. By December 2000, production totalled 393 series-built aircraft, 14 prototypes and three demonstrators, or 415 in all; backlog then six complete aircraft and 10 kits for rebuilds.

COSTS: £100 million for seven Mk 88As (Germany), 1996. Proposed four Srs 300s for South Africa estimated to cost total of £80 million (1998). Two Thai Lynx 300s, plus logistic support and services, £25 million (2001). Six Malaysian Lynx 300s valued at RM700 million (US$184 million) (1999).

DESIGN FEATURES: Compact design suited to hunter-killer ASW and missile-armed anti-ship naval roles from frigates or larger ships (superseding ship-guided helicopters), armed/unarmed land roles with cabin large enough for squad, or other tasks; manually folding tail pylon on most (but not all) naval versions; single four-blade semi-rigid main rotor (foldable), each blade attached to main rotor hub by titanium root plates and flexible arm; rotor drives taken from front of engines into main gearbox mounted above cabin ahead of engines; in flight, accessory gears (at front of main gearbox) driven by one of two through shafts from first stage reduction gears; four-blade tail rotor, drive taken from main ring gear; single large window in each main cabin sliding door; provision for internally mounted armament, and for exterior universal flange mounting each side for other weapons/stores. Super Lynx has all-weather day/night capability; extended payload/range; advanced technology swept-tip (BERP) composites main rotor blades offering improved speed and aerodynamic efficiency and reduced vibration; and reversed direction tail rotor for improved control.

FLYING CONTROLS: Rotor head controls actuated by three identical tandem servojacks and powered by two independent hydraulic systems; control system incorporates simple stability augmentation system; each engine embodies independent control system providing full-authority rotor speed governing, pilot control being limited to selection of desired rotor speed range; in event of one engine failure, system restores power up to single-engine maximum contingency rating; main rotor can provide negative thrust to increase stability on deck after touchdown on naval versions; hydraulically operated rotor brake mounted on main gearbox; sweptback fin/tail rotor pylon, with starboard half-tailplane.

STRUCTURE: Conventional semi-monocoque pod and boom, mainly light alloy; glass fibre access panels, doors, fairings, pylon leading/trailing-edges, and bullet fairing

LYNX PRODUCTION

Variant	Customer	Qty	First aircraft	First flights	Operators
AH. Mk 1	Army Air Corps	113*	XZ170	11 Feb 1977-24 Jan 1984	Note A
HAS. Mk 2	Fleet Air Arm	60	XZ227	20 Feb 1976-26 May 1981	Converted to Mk 3
HAS. Mk 2(FN)	French Navy	26	260	4 May 1977-5 Sep 1979	31F, 34F, ERCE
HAS. Mk 3	Fleet Air Arm	31	ZD249	4 Jan 1982-21 Oct 1988	815, 702 Sqdns
HAS. Mk 4(FN)	French Navy	14	801	1 Apr 1982-26 Aug 1983	31F, 34F, ERCE
Mk 5	MoD(PE)	3*	ZD285	21 Nov 1984-23 Feb 1985	DERA
AH. Mk 6	Royal Marines	–		None built	–
AH. Mk 7	Army Air Corps	11*	ZE376	23 Apr 1985-5 Jun 1987	Note A
HMA. Mk 8	Fleet Air Arm	–		Conversions	815, 702 Sqdns
AH. Mk 9	Army Air Corps	16*†	ZG884	20 Jul 1990-21 Jun 1992	653, 659 Sqdns
Mk 21	Brazilian Navy	9	N3020	30 Sep 1977-14 Apr 1978	Five upgraded to 21A
Mk 21A	Brazilian Navy	9†	N4001	12 Jun 1996-24 Apr 1997	1° EHEAA
Mk 22	Egyptian Navy	–	–	None built	–
Mk 23	Argentine Navy	2	0734	17 May 1978-23 Jun 1978	Withdrawn from service
Mk 24	Iraqi Army	–	–	None built	–
Mk 25[1]	Netherlands Navy	6	260	23 Aug 1976-16 Sep 1977	7/860 Sqdns
Mk 26	Iraqi Army (armed)	–	–	None built	–
Mk 27[2]	Netherlands Navy	10	266	6 Oct 1978-12 Nov 1979	7/860 Sqdns
Mk 28	Qatari Police	3*	QP-31	2 Dec 1977-12 Apr 1978	Withdrawn from service[7]
Mk 64	South African Navy[10]	–	–	–	–
Mk 80[12]	Danish Navy	8[11]	S-134	3 Feb 1980-15 Sep 1981	See Mk 90
Mk 81[3]	Netherlands Navy	8	276	9 Jul 1980-24 Mar 1981	7/860 Sqdns
Mk 82	Egyptian Army	–	–	None built	–
Mk 83	Saudi Army	–	–	None built	–
Mk 84	Qatari Army	–	–	None built	–
Mk 85	UAE Army	–	–	None built	–
Mk 86	Norwegian Coast Guard	6	207	23 Jan 1981-11 Sep 1981	Skv 337
Mk 87	Argentine Navy	–	–	Embargoed	–
Mk 88	German Navy	19[9]	8301	26 May 1981-10 Dec 1988	3/MFG 3
Mk 88A	German Navy	7†	8320	30 Apr 1999 – Mar 2000	3/MFG 3
Mk 89	Nigerian Navy	3	01-F89	29 Sep 1983-14 Mar 1984	101 Sqdn
Mk 90[12]	Danish Navy	1[4]	S-256	19 Apr 1988[6]	Søværnets Flyvetjeneste
Mk 95	Portuguese Navy	3†[5]	9203	9 Jul 1993-1993	EHM[8]
Mk 99	South Korean Navy	12†	90-0701	16 Nov 1989-14 May 1991	627 Sqdn
Mk 99A	South Korean Navy	13†	99-0721	3 Jul 1999-Aug 2000	
Srs 300	Malaysian Navy	6	–	2002	499 Sqdn
	Thai Navy	2	–	2002	
	Omani Air Force	16	–	–	–
Subtotal		**417**			
Lynx 3		1	ZE477	14 Jun 1984	
Demonstrators		3	G-LYNX	18 May 1979-27 Jan 1999	
Prototypes		13	XW835	21 Mar 1971-5 Mar 1975	
Total		**434**			

Notes:
Note A: Army Air Corps 651, 652, 654, 655, 656, 657, 659, 661, 662, 663, 664, 665, 667, 669 and 671 Squadrons; 847 Squadron (Royal Marine Commando)
[1] Netherlands designation UH-14A; all to SH-14D
[2] Netherlands designation SH-14B; all to SH-14D
[3] Netherlands designation SH-14C; all to SH-14D
[4] Plus one conversion from demonstrator; these two being upgraded to Super Lynx Mk 90B with new airframes
[5] Plus two conversions from Mk 3
[6] Completion and first flight at Vaerløse, Denmark
[7] Sold to Royal Navy for spares recovery and use as ground instructional airframes
[8] Esquadrilha de Helicopteros de Marinha
[9] Surviving 15 being upgraded to Super Lynx with new airframes in 2001-2003
[10] Intends to order four, but programme delay announced in 1999
[11] Surviving six to be converted to Super Lynx Mk 90B with new airframes
[12] Converted during 1990s to Mk 80A/90A with Gem 42-1 engines
* Army version
† Super/Battlefield Lynx

First Westland Lynx Srs 300 for the Royal Malaysian Navy   *NEW*/0533722

Westland Super Lynx 300 demonstrator, incorporating CTS800 engines and 'glass cockpit'   *NEW*/0095707

rescue, with three crew, both versions can have a waterproof floor and a 272 kg (600 lb) capacity clip-on hydraulic or electric hoist on starboard side of cabin; cable length 30 m (98 ft).

ARMAMENT: For armed escort, anti-tank or air-to-surface strike missions, army version can be equipped with two 20 mm cannon mounted externally so as to permit fitment of pintle-mounted 7.62 mm machine gun inside cabin. External pylon can be fitted on each side of cabin for variety of stores, including two Minigun or other self-contained gun pods; two rocket pods; or up to eight HOT, Hellfire, TOW, or similar air-to-surface missiles. Additional six or eight reload missiles carried in cabin. For ASW role, armament includes two Mk 44, Mk 46, A244S or Sting Ray homing torpedoes, one each on an external pylon on each side of fuselage, and six marine markers; or two Mk 11 depth charges. Alternatively, up to four Sea Skua semi-active homing missiles; on French Navy Lynx, four AS.12 or similar wire-guided missiles.

Following data refer to Super Lynx.

DIMENSIONS, EXTERNAL:

Main rotor diameter	12.80 m (42 ft 0 in)
Tail rotor diameter	2.36 m (7 ft 9 in)
Distance between rotor centres	7.66 m (25 ft 1½ in)
Length of fuselage, tail rotor turning	13.34 m (43 ft 9¼ in)

Length overall:

main rotor turning	15.24 m (50 ft 0 in)
main rotor blades and tail folded	10.85 m (35 ft 7¼ in)

Width overall, main rotor blades folded

	2.94 m (9 ft 7¾ in)
Height overall: tail rotor turning	3.67 m (12 ft 0½ in)
main rotor blades and tail folded	3.25 m (10 ft 8 in)
Tailplane half-span	1.32 m (4 ft 4 in)
Wheel track	2.94 m (9 ft 7¾ in)
Wheelbase	3.02 m (9 ft 11 in)

DIMENSIONS, INTERNAL:

Cabin, from back of pilots' seats:

Min length	2.055 m (6 ft 9 in)
Max width	1.78 m (5 ft 10 in)
Max height	1.42 m (4 ft 8 in)
Floor area	3.45 m² (37.1 sq ft)
Volume	4.9 m³ (173 cu ft)

Cabin doorway: Width

Cabin doorway: Width	1.37 m (4 ft 6 in)
Height	1.19 m (3 ft 11 in)

AREAS:

Main rotor disc	128.71 m² (1,385.4 sq ft)
Tail rotor disc	4.37 m² (47.04 sq ft)

WEIGHTS AND LOADINGS:

Manufacturer's basic weight	3,291 kg (7,255 lb)

Operating weight empty (including crew and appropriate armament):

ASW (two torpedoes)	4,618 kg (10,181 lb)
ASV (four Sea Skuas)	4,373 kg (9,641 lb)
surveillance and targeting	3,708 kg (8,174 lb)
search and rescue	3,778 kg (8,329 lb)
Max underslung load	1,361 kg (3,000 lb)

Max fuel weight: normal

Max fuel weight: normal	787 kg (1,735 lb)
bench seat	275 kg (606 lb)
ferry tanks, total two	696 kg (1,534 lb)
Max T-O weight	5,330 kg (11,750 lb)
Max disc loading	41.4 kg/m² (8.48 lb/sq ft)

Transmission loading at max T-O weight and power

	3.89 kg/kW (6.39 lb/shp)

PERFORMANCE (CTS800 engines):

Max continuous cruising speed

	132 kt (244 km/h; 152 mph)
Hovering ceiling IGE, ISA +20°C	2,105 m (6,900 ft)
Hovering ceiling OGE, ISA +20°C	1,445 m (4,740 ft)

Radius of action:

range with auxiliary fuel

	540 n miles (1,000 km; 621 miles)

anti-submarine, 2 h on station, dipping sonar and one torpedo

	20 n miles (37 km; 23 miles)

point attack with four Sea Skuas

	125 n miles (232 km; 143 miles)

surveillance, 3 h 50 min on station

	75 n miles (139 km; 86 miles)
Max endurance, with auxiliary fuel	5 h 20 min

UPDATED

---

over tail rotor gearbox; composites main rotor blades; main rotor hub and inboard flexible arm portions built as complete unit, as titanium monobloc forging; tail rotor blades have light alloy spar, stainless steel leading-edge sheath and rear section as for main blades.

LANDING GEAR (general purpose military version): Non-retractable tubular skid type. Provision for a pair of adjustable ground handling wheels on each skid. Flotation gear optional. Battlefield Lynx and AH. Mk 9 equivalent have non-retractable tricycle gear with twin nosewheels.

LANDING GEAR (naval versions): Non-retractable oleo-pneumatic tricycle type. Single-wheel main units, carried on sponsons, fixed at 27° toe-out for deck landing; can be manually turned into line and locked fore and aft for movement of aircraft into and out of ship's hangar. Twin-wheel nose unit steered hydraulically through 90° by the pilot to facilitate independent take-off into wind. Sprag brakes (wheel locks) fitted to each wheel prevent rotation on landing or inadvertent deck roll. These locks disengaged hydraulically and re-engage automatically in event of hydraulic failure. Maximum vertical descent 1.83 m (6 ft)/s; with lateral drift 0.91 m (3 ft)/s for deck landing. Flotation gear, and hydraulically actuated harpoon deck lock securing system, optional.

POWER PLANT: Current option of two Rolls-Royce Gem 42-1 turboshafts, each rated at 746 kW (1,000 shp) for T-O or 664 kW (890 shp) max continuous; or two LHTEC CTS800-4Ns, each of 1,016 kW (1,362 shp) and 945 kW (1,267 shp), respectively. Transmission rating 1,372 kW (1,840 shp). Exhaust diffusers for IR suppression optional on Battlefield Lynx.

Fuel in five internal tanks; usable capacity 957 litres (253 US gallons; 210 Imp gallons) when gravity-refuelled; 985 litres (260 US gallons; 217 Imp gallons) when pressure-refuelled. Optional internal tank, replacing bench seat at rear of cabin, capacity 345 litres (91.0 US gallons; 75.9 Imp gallons). For ferrying, two tanks each of 441 litres (116 US gallons; 97.0 Imp gallons) in cabin, replacing bench tank. Maximum usable fuel 1,867 litres (493 US gallons; 411 Imp gallons). Engine oil tank capacity 6.8 litres (1.8 US gallons; 1.5 Imp gallons). Main rotor gearbox oil capacity 28 litres (7.4 US gallons; 6.2 Imp gallons).

ACCOMMODATION: Pilot and co-pilot or observer on side-by-side seats. Dual controls optional. Individual forward-hinged cockpit door and large rearward-sliding cabin door on each side; cockpit doors jettisonable; windows of cabin doors also jettisonable. Cockpit accessible from cabin area. Maximum high-density layout (military version) for one pilot and 10 armed troops or paratroops, on lightweight

bench seats in soundproofed cabin. Alternative VIP layouts for four to seven passengers, with additional cabin soundproofing. Seats can be removed quickly to permit carriage of up to 907 kg (2,000 lb) of freight internally. Tiedown rings provided. In casualty evacuation role, with a crew of two, Lynx can accommodate up to six Alphin stretchers and a medical attendant. Both basic versions have secondary capability for search and rescue (up to nine survivors) and other roles.

SYSTEMS: Two independent hydraulic systems, pressure 141 bar (2,050 lb/sq in). Third hydraulic system provided in naval version when sonar equipment installed. No pneumatic system. 28 V DC electrical power supplied by two 6.3 kW engine-driven starter/generators and an alternator. External power sockets. 24 V 23 Ah (optionally 40 Ah) Ni/Cd battery fitted for essential services and emergency engine starting. 200 V three-phase AC power available at 400 Hz from two 25 kVA transmission-driven alternators. Cabin heating and ventilation system. Optional supplementary cockpit air conditioning system. Electric anti-icing and demisting of windscreen, and electrically operated windscreen wipers, standard; windscreen washing system.

AVIONICS: *Comms:* Typically Rockwell Collins VOR/ILS; DME; Rockwell Collins AN/ARN-118 Tacan; I-band transponder (naval version only); BAE PTR446, Rockwell Collins APX-72, DaimlerChrysler STR 700/375 or Italtel APX-77 IFF.

Flight: BAE duplex three-axis automatic stabilisation equipment; BAE GM9 Gyrosyn compass system; Racal tactical air navigation system (TANS); Racal 71 Doppler, E2C standby compass. BAE Mk 34 AFCS. Additional units fitted in naval version, when sonar is installed, to provide automatic transition to hover and automatic Doppler hold in hover.

Latest versions have Racal Doppler 91 and RNS 252 navigation; Honeywell/Smiths AN/APN-198 radar altimeter; Rockwell Collins 206A ADF; Rockwell Collins VIR 31A VOR/ILS.

Radar: Optional Seaspray Mk 3000 or Honeywell RDR 1500 360° scan radar in chin fairing. BAE Sea Owl thermal imaging equipment optional above nose.

Mission: Optional Honeywell AN/AQS-18 or Thomson HS-312 sonars in naval variants. Detection of submarines by dipping sonar or magnetic anomaly detector. Dipping sonar operated by hydraulically powered winch and cable hover mode facilities within the AFCS.

EQUIPMENT: All versions equipped as standard with navigation, cabin and cockpit lights; adjustable landing light under nose; and anti-collision beacon. For search and

---

UNITED STATES OF AMERICA

AASI

ADVANCED AERODYNAMICS AND STRUCTURES INC

AASI, formed in 1990, succeeded Aerodynamics and Structures Inc (ASI), formed by former airline pilot Darius Sharifzadeh to develop Jetcruzer series of turboprop and turbofan business aircraft.

On 8 February 2002 AASI announced that it had purchased Congress Financial Corporation's position as senior creditor

for Mooney Aircraft Corporation. On 29 April 2002 AASI changed its name to Mooney Aerospace Group Ltd, under which title details of the Jetcruzer 500 and Mooney range will be found elsewhere in this section.

UPDATED

AASI JETCRUZER 500

Development of this single-engined business turboprop was suspended in February 2002 pending a design review, and

following AASI's acquisition of Mooney and reorganisation as Mooney Aerospace Group Ltd. A full description, photograph and three-view drawing last appeared in the 2002-03 *Jane's All the World's Aircraft*.

UPDATED

ADAM

ADAM AIRCRAFT INDUSTRIES

12876 East Jamison Circle, Englewood, Colorado 80112
Tel: (+1 303) 406 59 00 and (+1 800) 232 62 47
Fax: (+1 303) 406 59 50
e-mail: info@adamaircraft.com
Web: http://www.adamaircraft.com
CEO: George F 'Rick' Adam
COO: Cecil Miller
PRESIDENT: John C Knudsen
VICE-PRESIDENT, SALES AND MARKETING: Tom Wiesner
VICE-PRESIDENT, MANUFACTURING AND TEST ENGINEERING:
   Richard Boone
INFORMATION CONTACT: Mary Hammack

Adam Aircraft Industries formed in 1998 and is developing the CarbonAero centreline-thrust twin-engine business aircraft, which it will manufacture in a new 2,975 m² (32,000 sq ft) facility at Denver Centennial Airport, Colorado. By mid-2000 the workforce numbered 12 but had risen to 72 by October 2001 and 132 by July 2002. In October 2002 Adam announced that it was developing the A700 twin-engined business jet derivative.

Having designed its own seats for the A500, Adam has begun an auxiliary business as a seat producer to the aerospace industry.

UPDATED

ADAM A500 CARBONAERO

TYPE: Business twinprop.
PROGRAMME: Designed by Burt Rutan, with designation M-309 denoting his 309th aircraft design; development started September 1999; proof-of-concept aircraft (N309A), manufactured by Scaled Composites Inc, first flown at Mojave, California, 21 March 2000; formally rolled out 5 April 2000; total of 80 flight hours completed by the time of public debut at EAA AirVenture 2000 at Oshkosh in July 2000 and 155 hours by EAA AirVenture 2001; second aircraft, designated A500, and in proposed production configuration, first flew (N500AX) 11 July 2002 at Centennial Airport, Denver. First production fuselage, completed January 2002, used for promotional display; three conforming production aircraft to participate in FAA FAR Pt 23 IFR certification, using FAA's Certification Process Improvement programme including flight into known icing; no carry-forward certification from M-309 trials, and this aircraft had been withdrawn from use by mid-2002. Certification expected in second quarter 2003 in USA with deliveries starting immediately; certification in Europe will follow six months later. CarbonAero name chosen by competition during first six months of 2001.
CURRENT VERSIONS: **M-309:** Original proof-of-concept model, described in 2002-03 *Jane's.*
   A500: Production version with changes to wing span, length, flying controls and door positioning; *as described.*
   A family of aircraft, ultimately stretched to seat 19 passengers, is planned.
CUSTOMERS: Orders for 56 received by September 2002. Customers include a US Government agency, which has ordered two for surveillance at Kennedy Space Center.

Proof-of-concept prototype Adam M-309 CarbonAero (*Jane's/Paul Jackson*)   *NEW*/0525783

COSTS: US$935,000 (2002).
DESIGN FEATURES: Twin-boom configuration with swept fins and high-set tailplane; unswept low wings have dihedral on outboard panels. Tractor/pusher power plant simplifies handling with single engine failure.
FLYING CONTROLS: Conventional and manual. Flaperons of M-309 replaced by separate ailerons and Fowler flaps. Full-width elevator with tab and external mass balance.

Twin rudders have no tabs. No flaps on proof-of-concept aircraft; production aircraft will have separate flaps.
STRUCTURE: Single-cure graphite carbon composite with no secondary bonds or fasteners. Proof-of-concept aircraft has three-spar main wing spar, but production aircraft will have a single-piece spar, with wing mounted lower than on POC aircraft to minimise spar carry-through in cabin. Wing attached by four bolts to fuselage.

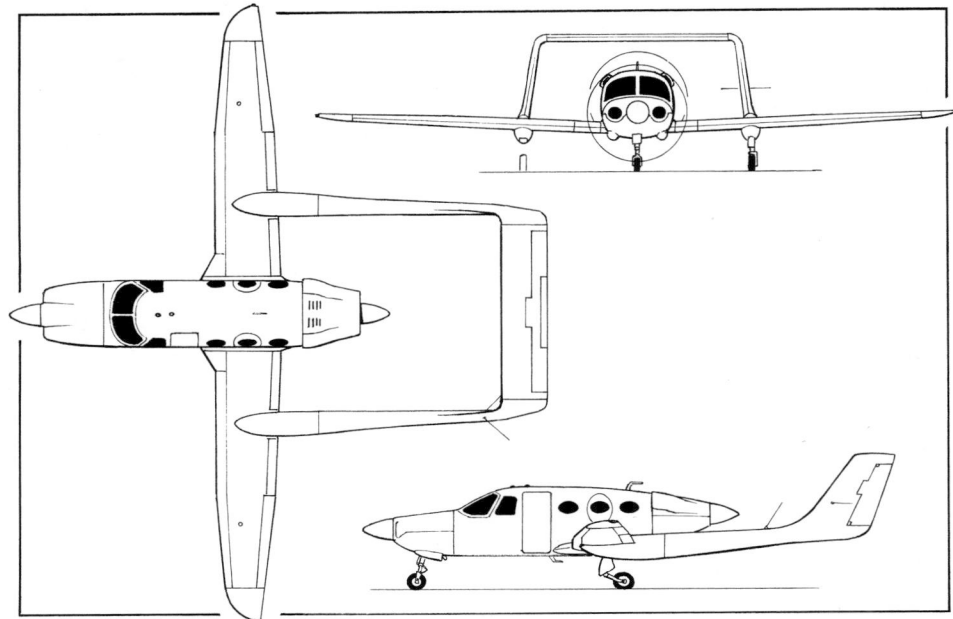

Adam A500 business twinprop (*Michael Badrocke*)   *NEW*/0536696

Adam A500 prototype on an early test flight   *NEW*/0536699

Instrument panel of Adam A500   *NEW*/0536700

LANDING GEAR: Retractable tricycle type, with single wheel on each unit; trailing link suspension on main units. All wheels retract rearwards; mainwheels 6.00-6, nosewheel 15×6.0-6. Hydraulic brakes.

POWER PLANT: Two 261 kW (350 hp) Teledyne Continental TSIO-550 flat-six piston engines, with FADEC, mounted in centreline thrust configuration and driving a three-blade Hartzell FC7663D-2R propeller at the front and a three-blade Hartzell FLC7663DF-2RX at rear. Future versions may be powered by diesel and turboprop engines. Fuel capacity 946 litres (250 US gallons; 208 Imp gallons) in outer wing tanks of which 871 litres (230 US gallons; 192 Imp gallons) are usable.

ACCOMMODATION: Pilot and five passengers in pressurised cabin on three pairs of seats. Door with integral airstair on port side, ahead of wing leading edge; emergency window on starboard side.

SYSTEMS: Pressurisation system maintains a 2,440 m (8,000 ft) cabin environment at 7,620 m (25,000 ft). Optional de-icing system.

AVIONICS: Meggitt Avionics MAGIC flat-panel display standard containing dual Garmin GNS 530 GPS/NAV/COM/ILS, Garmin 340 audio panel, Garmin 327 digital transponder and dual EFIS. Nexrad weather radar mounted in forward end of one tailboom. S-Tec System FiftyFiveX autopilot and VM1000 EMS. Options include Garmin GDL 49 WX Datalink transceiver, S-Tec ST-360 altitude selector/alerter and AirCell AGT02 communication system.

DIMENSIONS, EXTERNAL:

Wing span	13.41 m (44 ft 0 in)
Wing aspect ratio	10.9
Length overall	11.18 m (36 ft 8 in)
Height overall	2.90 m (9 ft 6 in)
Wheel track	3.57 m (11 ft 8½ in)
Wheelbase	3.08 m (10 ft 1¼ in)
Propeller diameter	1.93 m (6 ft 4 in)

DIMENSIONS, INTERNAL:

Cabin:

Length	4.15 m (13 ft 7¼ in)
Max width	1.37 m (4 ft 6 in)
Max height	1.29 m (4 ft 3 in)
Volume	7.4 m³ (262 cu ft)

AREAS:

Wings, gross	15.33 m² (165.0 sq ft)

WEIGHTS AND LOADINGS:

Weight empty	1,533 kg (3,380 lb)
Max fuel	499 kg (1,100 lb)
Max T-O weight	2,857 kg (6,300 lb)
Max wing loading	186.4 kg/m² (38.18 lb/sq ft)
Max power loading	5.48 kg/kW (9.00 lb/hp)

PERFORMANCE:

Max level speed at 6,100 m (20,000 ft)
  250 kt (463 km/h; 288 mph)
Cruising speed at 75% power at 6,100 m (20,000 ft)
  223 kt (413 km/h; 257 mph)
Econ cruising speed at 60% power at 6,100 m (20,000 ft)
  200 kt (370 km/h; 230 mph)
Stalling speed 70 kt (130 km/h; 81 mph)
Max rate of climb at S/L 549 m (1,800 ft)/min
Rate of climb at S/L, OEI 122 m (400 ft)/min
Service ceiling 7,620 m (25,000 ft)
Range with max fuel, IFR reserves:
  max cruise power
  1,020 n miles (1,887 km; 1,173 miles)
  economy power 1,150 n miles (2,129 km; 1,323 miles)

UPDATED

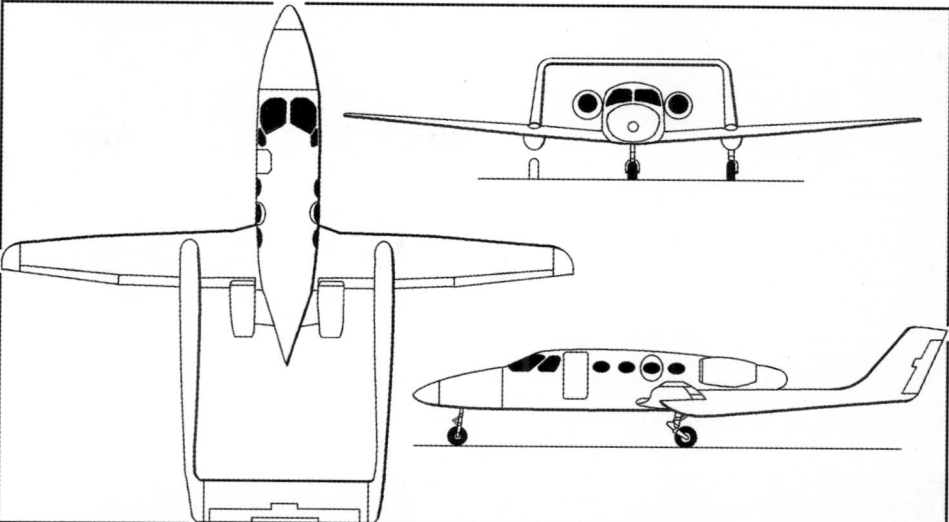

Adam A700 business jet *(Jane's/Paul Jackson)*   *NEW*/0536698

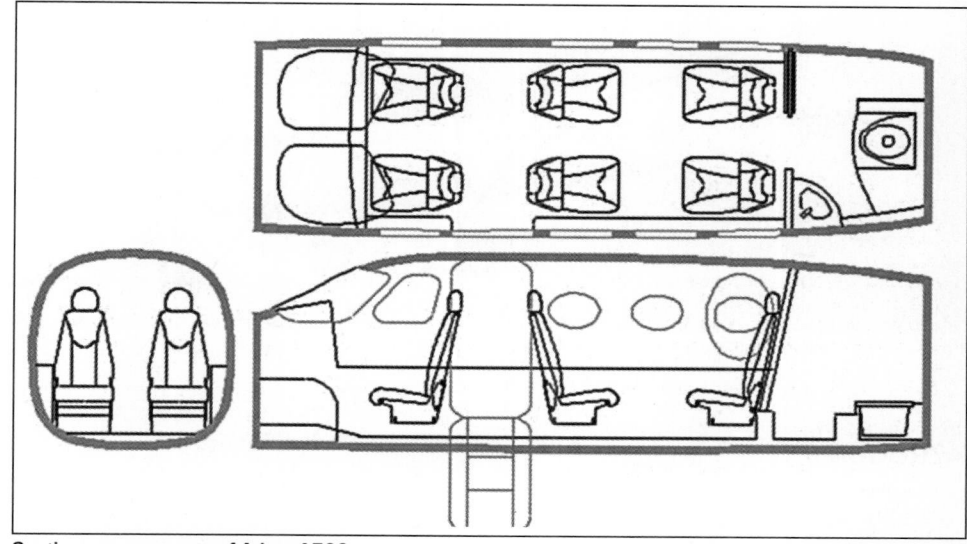

Seating arrangements of Adam A700   *NEW*/0536701

ADAM A700

TYPE: Light business jet.

PROGRAMME: Announced on 21 October 2002, at which time first flight of the prototype was anticipated during the second half of 2003, with customer deliveries from late 2004.

COSTS: Introductory price US$1,995,000 (2003).

DESIGN FEATURES: Design draws on A500 CarbonAero (which see). Engines pylon-mounted at rear of fuselage.

FLYING CONTROLS: As A500 CarbonAero. Twin tabs in elevator and on one rudder.

STRUCTURE: Monocoque fuselage of carbon fibre composites with honeycomb stiffening. Two-spar wing with integral fuel tank.

LANDING GEAR: As A500 CarbonAero; however nosewheel retracts forwards. Disc brakes on main wheels.

POWER PLANT: Two Williams International FJ33 turbofans, each rated at 5.34 kN (1,200 lb st).

ACCOMMODATION: One or two pilots in cockpit, plus four passengers in club seating in cabin, separated by divider and refreshment centre. Integral airstair on port side ahead of wing leading edge, with emergency exits on both sides of fuselage at centre of cabin. Lavatory fitted in aft of cabin. Baggage compartment in nose, forward of pressure bulkhead, accessible externally. Second baggage compartment at rear of cabin.

SYSTEMS: As A500 CarbonAero, but with de-icing protection for wing, horizontal tail, windscreen and engine inlets.

Dual starter-generators; dual bus electrical system with separate bus for emergency operation.

AVIONICS: 'Glass cockpit' includes dual VHF com, dual VHF NAV/LOC/GS, GPS, Mode C transponder, three-axis autopilot, attitude heading reference system and air data computer.

All data are provisional.

DIMENSIONS, EXTERNAL:

Wing span	13.41 m (44 ft 0 in)
Length overall	12.42 m (40 ft 9 in)
Height overall	2.92 m (9 ft 7 in)

DIMENSIONS, INTERNAL:

Cabin (aft of bulkhead): Length	4.88 m (16 ft 0 in)
Max width	1.37 m (4 ft 6 in)
Max height	1.31 m (4 ft 3½ in)
Volume	6.9 m³ (245 cu ft)
Baggage hold, volume: Front	0.71 m³ (25.0 cu ft)

WEIGHTS AND LOADINGS:

Max payload	329 kg (725 lb)

PERFORMANCE:

Max cruising speed at FL380
  340 kt (630 km/h; 391 mph) TAS
Service ceiling 12,500 m (41,000 ft)
T-O run 899 m (2,950 ft)
Range, 45 min reserves:
  IFR 1,100 n miles (2,037 km; 1,265 miles)
  VFR 1,400 n miles (2,592 km; 1,611 miles)

NEW ENTRY

ADI

ADI

5 Harris Court, Building S, Monterey, California 93940
Tel: (+1 831) 649 62 12
Fax: (+1 831) 649 57 38
e-mail: aircraft@mbay.net
Web: http://www.aircraftdesigns.com
PRESIDENT: Martin Hollmann

This company produces kits for the Hollmann-designed Stallion; sells plans for the Hollmann HA-2M Sportster and Hollmann Bumble Bee gyrocopters (see 1992-93 *Jane's*, Private Aircraft section); and is a major contributor to the Lancair series of kitbuilt aircraft (which see), as well as others including Seawind, Pulsar Sport 150, Prowler, Thunder Mustang, Kitfox and Condor. Covered plant area is 560 m² (6,025 sq ft).

UPDATED

ADI SUPER STALLION

TYPE: Six-seat kitbuilt.

PROGRAMME: Construction of prototype began in 1990; first flew (N408S) July 1994.

CUSTOMERS: Three production aircraft registered and 54 kits sold by July 2001; seven flying by early 2002.

COSTS: US$71,750 (2002) excluding engine, instruments and avionics.

DESIGN FEATURES: High-performance tourer for amateur construction. Designed to FAR Pt 23. Extensive use of composites. Employs Lancair IV wings and landing gear. Wing easily removed for storage and transportation. Winglets introduced from 2000. Quoted build time 1,500 hours.

Laminar flow, cantilever wing with 2° washout at tip. Wing section NACA 64-212 at tip, Jacosky RXM5-217 at root. Optional winglets.

FLYING CONTROLS: Conventional and manual. Flight-adjustable tab in port elevator; horn-balanced rudder with ground-adjustable tab. Fowler flaps, length 2.84 m (9 ft 4 in). Aileron length 1.78 m (5 ft 10 in).

STRUCTURE: Prebuilt centre-fuselage area of welded 4130 steel tubing; remainder of airframe primarily of graphite/Nomex honeycomb core/epoxy.

LANDING GEAR: Hydraulically retractable tricycle type; power is provided by an electric motor. Cleveland 6.00-6 mainwheels and tyres; nosewheel 5.00×5. Hydraulic disc brakes.

POWER PLANT: Choice of either 224 kW (300 hp) Teledyne Continental IO-550-G or 261 kW (350 hp) turbocharged Teledyne Continental TSIO-550-B six-cylinder engine driving a McCauley or Hartzell three-blade constant-speed propeller. Fuel capacity 681 litres (180 US gallons; 150 Imp gallons) in two wing tanks. Other engine options include 268 kW (360 hp) Textron Lycoming TIO-540-AE2A, 560 kW (751 shp) Walter M 601E and reconditioned 410 kW (550 shp) PT6A-20.

ACCOMMODATION: Two pilots and up to four passengers in three pairs of seats. Two rear pairs can be swivelled to provide club seating arrangement. Alternatively, centre row of seats can be quickly removed for carriage of cargo. Large (two-part, horizontally split) door on port side. Removable 1.88 m (6 ft 2 in) × 0.91 m (3 ft 0 in) panel on starboard side for loading. Baggage shelf.

AVIONICS: To customer's specification.

DIMENSIONS, EXTERNAL:

Wing span without winglets	10.67 m (35 ft 0 in)
Wing chord: at root	1.52 m (5 ft 0 in)
at tip	0.91 m (3 ft 0 in)
Wing aspect ratio	8.8
Length overall	7.62 m (25 ft 0 in)
Max width of fuselage	1.27 m (4 ft 2 in)
Height overall	2.90 m (9 ft 6 in)
Tailplane span	4.17 m (13 ft 8 in)
Tailplane chord: at root	0.91 m (3 ft 0 in)
at tip	0.56 m (1 ft 10 in)
Wheelbase	2.34 m (7 ft 8 in)
Wheel track	2.08 m (6 ft 10 in)

DIMENSIONS, INTERNAL:

Cabin: Length	3.30 m (10 ft 10 in)
Max width	1.24 m (4 ft 1 in)

AREAS:

Wings, gross	13.00 m² (140.0 sq ft)
Vertical tail surfaces (total)	1.78 m² (19.20 sq ft)
Tailplane	3.07 m² (33.00 sq ft)

WEIGHTS AND LOADINGS:

Weight empty	998 kg (2,200 lb)
Max payload	499 kg (1,100 lb)
Max T-O weight	1,724 kg (3,800 lb)
Max wing loading	132.5 kg/m² (27.14 lb/sq ft)
Max power loading: 300 hp	7.71 kg/kW (12.67 lb/hp)
350 hp	6.61 kg/kW (10.86 lb/hp)

PERFORMANCE:

Max operating speed (V_{MO}):	
300 hp	220 kt (407 km/h; 253 mph)
350 hp	266 kt (493 km/h; 306 mph)
Max cruising speed at 2,500 rpm:	
300 hp	200 kt (370 km/h; 230 mph)
350 hp	256 kt (474 km/h; 295 mph)
Stalling speed, power off, flaps down	
	62 kt (115 km/h; 71 mph)
Max rate of climb at S/L: 300 hp	488 m (1,600 ft)/min
350 hp	792 m (2,600 ft)/min
Service ceiling	9,750 m (32,000 ft)
T-O run	549 m (1,800 ft)
Landing run	213 m (700 ft)
Landing from 15 m (50 ft)	366 m (1,200 ft)
Range at cruising speed, no reserves	
	2,346 n miles (4,345 km; 2,700 miles)
	UPDATED

Prototype ADI Super Stallion, following retrofit with winglets (*Jane's/Paul Jackson*)   0110737

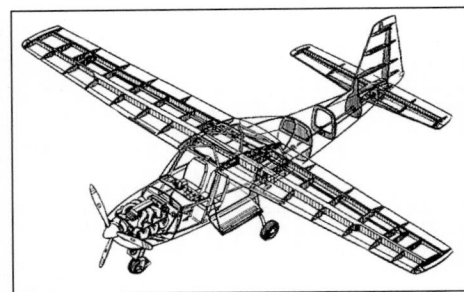

Structure of the ADI Super Stallion   0110976

AEROCOMP

AEROCOMP
800 Kemp Street, Merritt Island, Florida 32952
Tel/Fax: (+1 321) 453 66 41
e-mail: info@AerocompInc.com
Web: http://www.AerocompInc.com
OWNERS:
   Ron Lueck
   Stephen Young

Formed in 1993 and previously known for its production of floats, Aerocomp markets the Comp Air 3 and 4 kitbuilts, together with six-, seven-, eight- and 10-seat versions, known respectively as the Comp Air 6, 7, 8 and 10. Aerocomp acquired production rights for the Merlin GT and E-Z Flyer from Merlin Aircraft Inc, and has passed the latter to Blue Yonder Aviation of Canada (which see).

The company had 20 employees in 2001. In February 2002, Aerocomp announced that it was working on a jet-powered kitbuilt, the CA-J and its turbine-power comp Air 12 derivative.

UPDATED

AEROCOMP CA-J COMP AIR JET and COMP AIR 12

TYPE: Business monojet kitbuilt; business turboprop kitbuilt.

PROGRAMME: Announced 11 February 2002, and fuselage mockup shown at Sun 'n' Fun April 2002, when order book opened. Completion of first prototype was then set for August 2002. Kit deliveries due from first quarter 2003.

CURRENT VERSIONS: **Comp Air Jet:** *As described.*
   Comp Air 12: Turbine-engined version announced late 2002. Powered by a nose-mounted 746 kW (1,000 shp) Pratt & Whitney PT6.

COSTS: Kit cost US$349,000 including engine; estimated cost to finished standard US$600,000 (both 2002).

DESIGN FEATURES: Uses same structural technology as other aircraft in the Aerocomp product line, but company suggests builders use professional assistance in kit completion; quoted completion time six to eight months.

FLYING CONTROLS: Conventional and manual.

STRUCTURE: Low-wing monoplane with wing set towards rear of passenger cabin. Wing trailing-edge swept forward. Winglets. Tailplane set at base of fin, above fuselage. Engine inlets mounted at rear of fuselage behind cabin. Extensive use of carbon fibre hybrid sandwich throughout structure.

LANDING GEAR: Tricycle layout. Retractable.

POWER PLANT: One factory remanufactured ZMKB Progress AI-25 turbofan rated at 15.12 kN (3,400 lb st) for T-O and 11.12 kN (2,500 lb) during cruise. Design will also accommodate Pratt & Whitney JT12-8 or CJ610 engines. Fuel capacity 1,817 litres (480 US gallons; 400 Imp gallons).

ACCOMMODATION: Pilot and seven passengers. Main door on port side, immediately behind cockpit.

Aerocomp Comp Air Jet fuselage mockup displayed at Sun 'n' Fun, April 2002 (*Jane's/Paul Jackson*)
NEW/0525797

SYSTEMS: Pressurisation maintains 3,050 m (10,000 ft) cabin altitude at 9,115 m (29,900 ft) service ceiling.
All data are provisional.

DIMENSIONS, INTERNAL:

Cabin max height	1.78 m (5 ft 10 in)

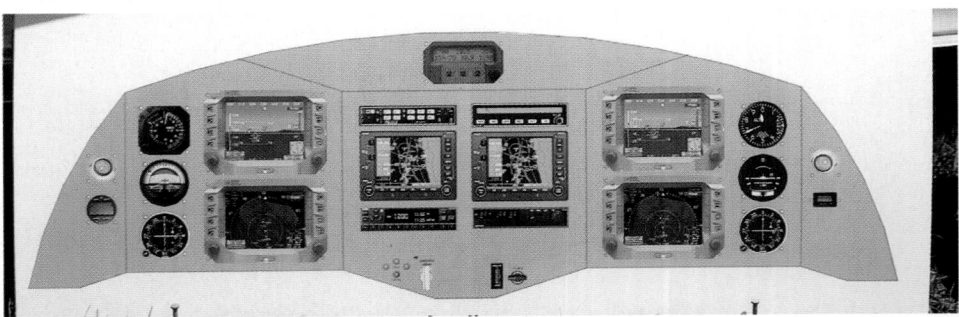

Aerocomp Comp Air Jet instrument panel preliminary layout (*Jane's/Paul Jackson*)   *NEW*/0525798

WEIGHTS AND LOADINGS:

Weight empty	2,223 kg (4,900 lb)
Max T-O weight	4,037 kg (8,900 lb)
Max power loading	267 kg/kN (2.62 lb/lb st)

PERFORMANCE (estimated):

Normal cruising speed	348 kt (644 km/h; 400 mph) TAS
Service ceiling	9,115 m (29,900 ft)
Range	955 n miles (1,770 km; 1,100 miles)

NEW ENTRY

AEROCOMP COMP AIR 3

TYPE: Three-seat kitbuilt.

PROGRAMME: Announced 1998; smaller version of Comp Air 4.

CUSTOMERS: At least one flying by early 2000.

COSTS: Kit US$25,995 (2002).

Data generally as for Comp Air 4, except as noted.

STRUCTURE: Wings removable for storage. Tapered wings, optional.

POWER PLANT: One 119 kW (160 hp) Textron Lycoming O-320 in prototype; engines in range 112 to 134 kW (150 to 180 hp) recommended. Fuel capacity: standard 170 litres (45.0 US gallons; 37.5 Imp gallons), optional 757 litres (200 US gallons; 167 Imp gallons).

ACCOMMODATION: Pilot and passenger side by side; third occupant, or cargo, in back of cockpit.

DIMENSIONS, EXTERNAL:

Wing span	10.52 m (34 ft 6 in)
Wing aspect ratio	6.6
Length overall	7.32 m (24 ft 0 in)
Max width of fuselage	0.91 m (3 ft 0 in)
Height overall	2.46 m (8 ft 1 in)

DIMENSIONS, INTERNAL:

Cabin: max width	1.09 m (3 ft 7 in)

AREAS:

Wings, gross	16.35 m² (176.0 sq ft)

WEIGHTS AND LOADINGS:

Weight empty	590 kg (1,300 lb)
Max T-O weight	1,111 kg (2,450 lb)
Max wing loading	68.0 kg/m² (13.92 lb/sq ft)

PERFORMANCE (119 kW; 160 hp Textron Lycoming engine):

Max operating speed	152 kt (281 km/h; 175 mph)
Normal cruising speed	126 kt (233 km/h; 145 mph)
Stalling speed	40 kt (73 km/h; 45 mph)
Max rate of climb at S/L	335 m (1,100 ft)/min
Service ceiling	4,724 m (15,500 ft)
T-O run	107 m (350 ft)
Landing run	183 m (600 ft)
Range	725 n miles (1,342 km; 834 miles)

UPDATED

AEROCOMP COMP AIR 4

TYPE: Four-seat kitbuilt.

PROGRAMME: First flew 3 April 1995. Public debut, then named Comp Monster, at Sun 'n' Fun 1995, when powered by 82 kW (110 hp) Hirth F 30 four-cylinder two-stroke engine.

CURRENT VERSIONS: **150G:** Powered by 112 kW (150 hp) Textron Lycoming O-320.

180G: 134 kW (180 hp) Textron Lycoming O-360; *as described.*

180SF: Float-equipped version of 180G.

Trainer: 'Two plus two' seat trainer version.

CUSTOMERS: Total of 60 flying by end 2001 (latest information).

COSTS: US$47,995 with 150 hp, US$46,995 with 180 hp Lycoming, US$26,995 basic kit without engine, propeller and instruments; Trainer kit US$22,900. Fast-build option US$1,795 (All 2002.)

DESIGN FEATURES: Easy-to-build composites aircraft; quoted build time 350 to 400 hours. High, braced wing; unswept, but with tapered trailing-edge.

Various construction options, including tricycle landing gear, tapered wing, enlarged flaps and 1,451 kg (3,200 lb) MTOW.

Modified Clark Y aerofoil wings with turned-down tips.

FLYING CONTROLS: Conventional and manual. Flaps of two optional sizes.

Aerocomp Comp Air 4 four-seat kitbuilt *(Jane's/Paul Jackson)*   NEW/0525786

Aerocomp Comp Air 6G six-seat kitbuilt *(Jane's/Paul Jackson)*   NEW/0525791

STRUCTURE: Composites construction with carbon and Kevlar reinforcement. Single-strut braced wings; braced tailplane. Optional tapered wing. Fuselage width upgrades available: 1.17 m (3 ft 10 in) and 1.21 m (3 ft 11½ in).

LANDING GEAR: Spring steel; available with nosewheel, tailwheel, floats or amphibious gear, last-mentioned using Matco wheels and hydraulic brakes with 5.00-5 in tyres. Non-retracting 0.15 m (6 in) Matco tailwheel acts as water rudder. Tailwheel version uses 6.00-6 in Matco wheels, tyres and brakes with a 0.20 m (8 in) spring-legged tailwheel.

POWER PLANT: One 134 kW (180 hp) Textron Lycoming O-360-A1A flat-four engine, driving a Sensenich 76 × 57 three-blade metal propeller. Options exist for engines ranging from 82 to 186 kW (110 to 250 hp); prototype employed 82 kW (110 hp) Hirth 95. Fuel capacity 197 litres (52.0 US gallons; 43.3 Imp gallons).

ACCOMMODATION: Pilot and three passengers in two side-by-side pairs. Glass-panelled door on each side of cabin for improved downwards visibility. Optional baggage pod below fuselage.

DIMENSIONS, EXTERNAL:

Wing span	11.46 m (37 ft 7 in)
Wing chord (constant version)	1.88 m (6 ft 2 in)
Length overall	7.92 m (26 ft 0 in)
Height overall	2.44 m (8 ft 0 in)
Tailplane span	3.05 m (10 ft 0 in)
Doors (each): Height	0.76 m (2 ft 6 in)
Width	0.91 m (3 ft 0 in)

DIMENSIONS, INTERNAL:

Cabin: Length	2.74 m (9 ft 0 in)
Max width	1.08 m (3 ft 6½ in)
Max height	1.30 m (4 ft 3 in)

AREAS:

Wings, gross	19.70 m² (212.0 sq ft)

WEIGHTS AND LOADINGS (G: 180G landplane, SF: 180SF floatplane):

Weight empty: G	630 kg (1,390 lb)
SF	753 kg (1,660 lb)
Baggage capacity	81.6 kg (180 lb)
Max T-O weight: both	1,292 kg (2,850 lb)
Max wing loading: both	65.6 kg/m² (13.44 lb/sq ft)
Max power loading: both	9.64 kg/kW (15.83 lb/hp)

PERFORMANCE:

Max operating speed: G	129 kt (239 km/h; 149 mph)
SF	110 kt (204 km/h; 127 mph)
Normal cruising speed at 70% power:	
G	115 kt (212 km/h; 130 mph)
SF	100 kt (185 km/h; 115 mph)
T-O speed	36 kt (68 km/h; 42 mph)
Stalling speed: G	34 kt (63 km/h; 39 mph)
SF	35 kt (65 km/h; 40 mph)
Max rate of climb at S/L: G	442 m (1,450 ft)/min
SF	366 m (1,200 ft)/min
Service ceiling	4,880 m (16,000 ft)
T-O run: G	91 m (300 ft)
SF	221 m (725 ft)
Landing run	213 m (700 ft)
Range	660 n miles (1,222 km; 759 miles)

UPDATED

AEROCOMP COMP AIR 6

TYPE: Six-seat kitbuilt.

PROGRAMME: Development of Comp Air 4; first flew January 1996.

CURRENT VERSIONS: **CA6G:** Normal landing gear, *as described.*

CA6SF: Float-equipped version.

CA6AF: Amphibian.

CA6TW: Tapered wing.

CA6TWHG: Tapered wing, high gross (1,451 kg; 3,200 lb MTOW).

CA6SHG: Super high gross (1,632 kg; 3,600 lb MTOW), wide-bodied version.

CUSTOMERS: At least 115 flying by early 2002.

COSTS: Kit US$29,995, without engine (2002).

DESIGN FEATURES: As Comp Air 4. Quoted build time 350 hours.

Data for CA6G generally as Comp Air 4, except those below.

STRUCTURE: Wings removable for storage. Optional wide-body fuselage, enlarged flap and tapered wing upgrades available.

LANDING GEAR: Choice of tricycle or tailwheel configuration; optional floats. Mainwheels 6.00-6 in. Optional 7075-T6 aluminium alloy spring landing gear available in place of composites version provided with kits.

POWER PLANT: Prototype had 164 kW (220 hp) Franklin engine; design compatible with engines in the 164 to 224 kW (220 to 300 hp) class. Automotive engine conversions are also possible. Fuel capacity 310 litres (82.0 US gallons; 68.3 Imp gallons).

ACCOMMODATION: Four adults plus full fuel and baggage; or four adults plus two children, with reduced fuel or baggage.

DIMENSIONS, EXTERNAL:

Wing span	10.52 m (34 ft 6 in)
Wing aspect ratio	5.6
Length overall	7.47 m (24 ft 6 in)
Height overall	2.44 m (8 ft 0 in)

DIMENSIONS, INTERNAL:

Cabin: Length	3.30 m (10 ft 10 in)
Max width	1.08 m (3 ft 6½ in)
Max height	1.30 m (4 ft 3 in)

AREAS:

Wings, gross	19.70 m² (212.0 sq ft)

WEIGHTS AND LOADINGS:

Weight empty	676 kg (1,490 lb)
Max T-O weight	1,293 kg (2,850 lb)
Max wing loading	65.6 kg/m² (13.44 lb/sq ft)
Max power loading	7.88 kg/kW (12.95 lb/hp)

PERFORMANCE (164 kW; 220 hp Franklin engine):

Max operating speed	145 kt (268 km/h; 167 mph)
Normal cruising speed	133 kt (246 km/h; 153 mph)
Stalling speed	35 kt (65 km/h; 40 mph)
Max rate of climb at S/L	366 m (1,200 ft)/min
Service ceiling	5,480 m (18,000 ft)
T-O run	107 m (350 ft)
Landing run	168 m (550 ft)
Range	695 n miles (1,287 km; 800 miles)

UPDATED

AEROCOMP COMP AIR 7

TYPE: Utility kitbuilt.

PROGRAMME: Development of Comp Air 6. First shown 1998.

CURRENT VERSIONS: **Comp Air 7P:** Piston-engine version.

Comp Air 7T: Turbine version.

Comp Air 7SL: Stretched version, introduced in 2001, with 30 cm (12 in) fuselage stretch and equal addition to wing chord; narrow window immediately to rear of front doors.

Comp Air 7SL T: Turbine version of 7SL.

CUSTOMERS: At least 22 flying or under construction by early 2002, including 14 turbine versions.

Comp Air 7T built by Brad Payne and J B Wilson (*Jane's/Paul Jackson*)   *NEW*/0525793

COSTS: Comp Air 7 US$39,995; Comp Air 7T US$49,995; Comp Air 7SL US$59,995; all minus engine (2002).

Data generally as for Comp Air 6, except those below.

DESIGN FEATURES: Suited for bush operations. Quoted build time 700 hours.

FLYING CONTROLS: Horn-balanced tail surfaces. Ground-adjustable tab on each aileron and on rudder. Trim tab in starboard elevator.

LANDING GEAR: Tailwheel type; fixed. Speed fairings optional. Mainwheels 8.00-6. Optional aluminium alloy landing gear available, as for Comp Air 6.

POWER PLANT: *Comp Air 7:* One Textron Lycoming TIO-540 rated between 194 and 261 kW (260 and 350 hp), driving a two-blade metal propeller. Fuel capacity 333 litres (88.0 US gallons; 73.3 Imp gallons); optional extra tank increases capacity to 455 litres (120 US gallons; 100 Imp gallons).

Comp Air 7T and *7SL:* One 490 kW (657 shp) Walter M 601D turboprop driving an Avia three-blade, constant-speed, feathering propeller; optionally a five-blade Avia Hamilton propeller can be used. Fuel capacity 568 litres (150 US gallons; 125 Imp gallons) in 7T, 719 litres (190 US gallons; 158 Imp gallons) in 7SL.

ACCOMMODATION: Pilot and up to six passengers in enclosed cabin. Third (rear passengers') door, starboard side.

DIMENSIONS, EXTERNAL (Comp Air 7 and Comp Air 7T):

Wing span: CA 7P, CA 7T	10.67 m (35 ft 0 in)
CA 7SL	10.06 m (33 ft 0 in)
Wing aspect ratio: CA 7P, CA 7T	6.9
CA 7SL	5.7
Length overall: CA 7P	8.08 m (26 ft 6 in)
CA 7T	8.99 m (29 ft 6 in)
CA 7SL	9.30 m (30 ft 6 in)
Height overall: CA 7P, CA 7T	2.44 m (8 ft 0 in)
CA 7SL	2.74 m (9 ft 0 in)

DIMENSIONS, INTERNAL:

Cabin max width: CA 7P standard	1.08 m (3 ft 6½ in)
CA 7P optional; CA 7T and CA 7SL standard	1.17 m (3 ft 10 in)
CA 7P and CA 7SL optional	1.21 m (3 ft 11½ in)

AREAS:

Wings, gross: CA 7P, CA 7T	16.54 m² (178.0 sq ft)
CA 7SL	17.74 m² (191.0 sq ft)

WEIGHTS AND LOADINGS:

Weight empty: CA 7P	953 kg (2,100 lb)
CA 7T	1,157 kg (2,550 lb)
CA 7SL: standard	1,179 kg (2,600 lb)
high gross upgrade	1,315 kg (2,900 lb)
Max T-O weight: CA 7P	1,678 kg (3,700 lb)
CA 7T: standard	1,710 kg (3,770 lb)
high gross upgrade	2,177 kg (4,800 lb)
CA 7SL: standard	2,086 kg (4,600 lb)
high gross upgrade	2,358 kg (5,200 lb)
Max landing weight:	
CA 7T high gross upgrade	2,087 kg (4,600 lb)
Max wing loading:	
CA 7P	101.5 kg/m² (20.79 lb/sq ft)
CA 7T: standard	103.4 kg/m² (21.18 lb/sq ft)
high gross upgrade	131.7 kg/m² (26.97 lb/sq ft)

CA 7SL: standard	117.6 kg/m² (24.08 lb/sq ft)
high gross upgrade	132.9 kg/m² (27.23 lb/sq ft)
Max power loading:	
CA 7T: standard	3.49 kg/kW (5.74 lb/shp)
high gross upgrade	4.45 kg/kW (7.31 lb/shp)
CA 7SL standard	4.26 kg/kW (7.00 lb/shp)
high gross upgrade	4.82 kg/kW (7.91 lb/shp)

PERFORMANCE:

Never-exceed speed (V$_{NE}$):	
CA 7P	191 kt (354 km/h; 220 mph)
CA 7T	223 kt (413 km/h; 257 mph) IAS
CA 7SL	206 kt (381 km/h; 237 mph) IAS
Max operating speed:	
CA 7P	178 kt (330 km/h; 205 mph)
Max cruising speed:	
CA 7T	239 kt (443 km/h; 275 mph)
CA 7SL	217 kt (402 km/h; 250 mph)
Stalling speed: CA 7	46 kt (86 km/h; 53 mph)
CA 7T	48 kt (89 km/h; 55 mph)
CA 7SL	53 kt (97 km/h; 60 mph)
Max rate of climb at S/L:	
CA 7P	457 m (1,500 ft)/min
CA 7T	1,219 m (4,000 ft)/min
CA 7SL	914 m (3,000 ft)/min
Service ceiling: CA 7	7,620 m (25,000 ft)
T-O run: CA 7P	145 m (475 ft)
CA 7T	122 m (400 ft)
CA 7SL	183 m (600 ft)
Landing run: CA 7	244 m (800 ft)
Range:	
CA 7P	1,100 n miles (2,037 km; 1,265 miles)
CA 7SL	900 n miles (1,666 km; 1.035 miles)
g limits: CA 7T	+6/−4

UPDATED

AEROCOMP COMP AIR 8

TYPE: Utility kitbuilt.

PROGRAMME: Prototype first flew mid-June 1999 and publicly displayed at Oshkosh in July 1999.

Aerocomp Comp Air 8T six-plus-two kitbuilt (*Jane's/Paul Jackson*)   *NEW*/0525795

Walter M 601D power plant fitted to Comp Air 7
(*Jane's/Susan Bushell*)   0110576

First customer-built Comp Air 7SL T stretched version, showing additional window to rear of door
(*Jane's/Paul Jackson*)   *NEW*/0525794

CUSTOMERS: Thirteen flying or under construction by early 2002. All believed to be CA 8T turboprop versions.

COSTS: Kit US$69,995 (2002), excluding engine.

DESIGN FEATURES: Generally similar to Comp Air 7. Quoted build time 800 hours.

FLYING CONTROLS: Mass-balanced controls for high-speed handling.

STRUCTURE: Generally as Comp Air 7, but with strengthened carbon fibre fuselage and tail unit. Tapered wings standard. Fuselage width upgrades available.

LANDING GEAR: Tailwheel type; fixed. Optional aluminium alloy landing gear available, as for Comp Air 6. Optional tricycle configuration; optional floats.

POWER PLANT: One 490 kW (657 shp) Walter M 601D turboprop driving an Avia three-blade, feathering, constant-speed propeller. Fuel capacity 681 litres (180 US gallons; 150 Imp gallons).

ACCOMMODATION: Pilot and five adults in three pairs of bucket seats, plus rear bench seat for two children.

DIMENSIONS, EXTERNAL:

Wing span	10.97 m (36 ft 0 in)
Wing aspect ratio	5.5
Length overall	9.60 m (31 ft 6 in)
Height overall	2.46 m (8 ft 1 in)

DIMENSIONS, INTERNAL:

Cabin max width: standard	1.17 m (3 ft 10 in)
wide option	1.21 m (3 ft 11½ in)

AREAS:

Wings, gross	22.02 m² (237.0 sq ft)

WEIGHTS AND LOADINGS:

Weight empty: standard	1,270 kg (2,800 lb)
high gross option	1,406 kg (3,100 lb)
Max T-O weight: standard	2,177 kg (4,800 lb)
intermediate	2,359 kg (5,200 lb)
high gross option	2,540 kg (5,600 lb)
Max wing loading: standard	98.9 kg/m² (20.25 lb/sq ft)
intermediate	107.1 kg/m² (21.94 lb/sq ft)
high gross option	115.4 kg/m² (23.63 lb/sq ft)
Max power loading: standard	4.45 kg/kW (7.31 lb/shp)
intermediate	4.82 kg/kW (7.91 lb/shp)
high gross option	5.19 kg/kW (8.52 lb/shp)

PERFORMANCE:

Never-exceed speed (V$_{NE}$)	199 kt (368 km/h; 229 mph) IAS
Max cruising speed at 6,400 m (21,000 ft)	217 kt (402 km/h; 250 mph) IAS
Normal cruising speed	182 kt (338 km/h; 210 mph)
Stalling speed, power off, flaps down	42 kt (78 km/h; 48 mph)
Max rate of climb at S/L	610 m (2,000 ft)/min
Service ceiling	8,230 m (27,000 ft)
T-O run	122 m (400 ft)
Landing run	183 m (600 ft)
Range	990 n miles (1,833 km; 1,139 miles)

UPDATED

AEROCOMP COMP AIR 10

TYPE: Utility kitbuilt.

PROGRAMME: Announced 1997; based on Comp Air 6.

CURRENT VERSIONS: **CA10:** Original version, no longer available; described in 2001-2002 *Jane's*.

CA10XL T: Turbine version. *As described.*

CUSTOMERS: Five flying by early 2002.
COSTS: US$79,995 turbopowered kit version (2002).
Data generally as for Comp Air 6, except that below.
DESIGN FEATURES: Twin outward-canted fins in addition to conventional horizontal tail surfaces. Quoted build time 800 hours. Optional HP (high performance) wing and two-stage increased MTOW options.
FLYING CONTROLS: Conventional and manual. Twin rudders operating in unison. Balance tab in starboard elevator.
STRUCTURE: Extensive use of composites throughout. Single-strut braced wings. Quoted build time 600 hours.
LANDING GEAR: Fixed tricycle type; sprung 7075-T6 aluminium alloy main legs. Mainwheel size 8.00-6 in; nosewheel 6.00-6 in. Floats and tailwheel configurations optional.
POWER PLANT: One 490 kW (657 shp) Walter M 601D driving an Avia three-blade, constant-speed propeller. Standard fuel capacity of 455 litres (120 US gallons; 100 Imp gallons); optionally 681 litres (180 US gallons; 150 Imp gallons). Oil capacity 5.7 litres (1.5 US gallons; 1.25 Imp gallons).
ACCOMMODATION: Pilot and up to nine passengers; passenger door, forward, each side; horizontally split, two-piece freight door, starboard, rear side; extra passenger door on port rear side. Optional luggage pannier.

DIMENSIONS, EXTERNAL:
Wing span	11.13 m (36 ft 6 in)
Length overall	9.45 m (31 ft 0 in)
Height overall	2.64 m (8 ft 8 in)
Wheel track	2.79 m (9 ft 2 in)

DIMENSIONS, INTERNAL:
Cabin: Length:	3.96 m (13 ft 0 in)
Max width	1.52 m (5 ft 0 in)
Max height	1.35 m (4 ft 5 in)

AREAS:
Wings, gross	22.48 m² (242.0 sq ft)

WEIGHTS AND LOADINGS:
Weight empty: standard	1,247 kg (2,750 lb)
high gross	1,429 kg (3,150 lb)
Baggage capacity: standard	136 kg (300 lb)
Max T-O and landing weight:	
standard	2,358 kg (5,200 lb)
intermediate	2,540 kg (5,600 lb)
high gross	2,721 kg (6,000 lb)
Max wing loading: standard	104.9 kg/m² (21.49 lb/sq ft)
intermediate	113.0 kg/m² (23.14 lb/sq ft)
high gross	121.1 kg/m² (24.79 lb/sq ft)
Max power loading: standard	4.82 kg/kW (7.91 lb/shp)
intermediate	5.19 kg/kW (8.52 lb/shp)
high gross	5.56 kg/kW (9.13 lb/shp)

Company demonstrator Comp Air CA10XL T (*Jane's/Susan Bushell*)   0110589

PERFORMANCE:
Max level speed	186 kt (346 km/h; 215 mph)
Max cruising speed at 75% power	
	167 kt (309 km/h; 192 mph)
Normal cruising speed at 65% power	
	152 kt (282 km/h; 175 mph)
Stalling speed: flaps up	51 kt (95 km/h; 59 mph)
flaps down	46 kt (86 km/h; 53 mph)
Max rate of climb at S/L	762 m (2,500 ft)/min
Service ceiling	7,620 m (25,000 ft)
T-O run	152 m (500 ft)
Landing run	183 m (600 ft)
Range with max fuel at 65% power	
	795 n miles (1,472 km; 915 miles)
Endurance at 65% power, 60 min reserves	3 h 42 min

UPDATED

AEROCOMP MERLIN GT
TYPE: Side-by-side kitbuilt.
PROGRAMME: Developed by Macair Aircraft in Canada; later Merlin Aircraft. Rotax 582-powered version first flown January 1985; 912-powered version in January 1993.
CURRENT VERSIONS: Available as floatplane, trainer, agricultural sprayer.
CUSTOMERS: At least 300 flying by mid-2002.
COSTS: Complete kits: Standard (with Rotax 582) US$24,995; Rotax 912-powered version US$29,995; Rotax 912S-powered version US$33,995 (2002).

DESIGN FEATURES: Strut-braced high wing; braced tailplane. No aerodynamic balances or tabs on empennage. Wings fold for storage. Quoted build time 400 hours.
FLYING CONTROLS: Manual. Full span Junkers flaperons; conventional rudder and elevator. Actuation by pushrods and cables.
STRUCTURE: Fuselage of welded 4130 chromoly steel tubing, with wooden floor and glass fibre engine cowling, otherwise fabric-covered. Metal wings with fabric covering.
LANDING GEAR: Non-retractable mainwheels and tailwheel. Metal tube mainwheel legs with bungee-bound shock-absorption; hydraulic brakes; steerable tailwheel. Full Lotus or Superfloats floats optional.
POWER PLANT: One 47.8 kW (64.1 hp) Rotax 582 engine, driving a two- or three-blade propeller; or a choice of 59.6 kW (79.9 hp) Rotax 912, 73.5 kW (98.6 hp) Rotax 912 ULS, 55.0 kW (73.8 hp) Rotax 618, 74.6 kW (100 hp) Canadian Automotive (CAM) 100, 59.7 kW (80 hp) Jabiru 2000 or 82.0 kW (110 hp) Formula Power Subaru EA81 engines. Fuel capacity 61 litres (16.0 US gallons; 13.3 Imp gallons).
SYSTEMS: 12 V electrical system, with battery.
ACCOMMODATION: Two persons, side by side with baggage stowage behind seats. Upward-opening door each side. Dual controls.

DIMENSIONS, EXTERNAL:
Wing span	9.75 m (32 ft 0 in)
Wing chord, constant	1.52 m (5 ft 0 in)
Wing aspect ratio	6.0
Length overall	6.10 m (20 ft 0 in)
Height overall	1.98 m (6 ft 6 in)
Tailplane span	2.15 m (7 ft 0½ in)
Wheel track	2.21 m (7 ft 3 in)
Wheelbase	6.10 m (20 ft 0 in)

DIMENSIONS, INTERNAL:
Cabin max width	1.04 m (3 ft 5 in)

AREAS:
Wings, gross	14.86 m² (160.0 sq ft)

WEIGHTS AND LOADINGS (Rotax 912):
Weight empty	279 kg (615 lb)
Max baggage capacity	45 kg (100 lb)
Max T-O weight: landplane	590 kg (1,300 lb)
floatplane	635 kg (1,400 lb)
Max wing loading: landplane	39.7 kg/m² (8.13 lb/sq ft)
floatplane	42.7 kg/m² (8.75 lb/sq ft)
Max power loading: landplane	9.89 kg/kW (16.25 lb/hp)
floatplane	10.65 kg/kW (17.5 lb/hp)

PERFORMANCE (Rotax 912):
Never-exceed speed (V$_{NE}$)	104 kt (193 km/h; 120 mph)
Cruising speed	81 kt (150 km/h; 93 mph)
Stalling speed	34 kt (63 km/h; 38 mph)
Max rate of climb at S/L	455 m (1,500 ft)/min
T-O distance	30 m (100 ft)
Landing distance	46 m (150 ft)
Max range	350 n miles (648 km; 402 miles)

UPDATED

Aerocomp Merlin GT, assembled in Malaysia by Langkawi Recreation Club and powered by a Rotax 912 ULS (*Jane's/Paul Jackson*)   0089492

AEROCOPTER

AEROCOPTER INC
300 Brickstone Square, Andover, Massachusetts 01810-1435
Tel: (+1 978) 749 99 99
Fax: (+1 978) 749 88 88
Web: http://www.aerocopter.com
CHAIRMAN: Rouzbeh Yassini
VICE-PRESIDENTS: George Syrovy (Engineering)
  Siamak Yassini (Business Development)

Founded in August 2000 by Messrs Syrovy and Siamak Yassini, the company is developing what is described as a "third generation" air transport with VTOL capability. Public debut was at the December 2001 NBAA Convention in New Orleans.

UPDATED

AEROCOPTER HUMMING
TYPE: Technology demonstrator.
PROGRAMME: Conceived as tiltdisc, VTOL technology demonstrator for commercial airliners (20 to 120 seats),

military transports, UAVs, cargo transport and business aircraft. Work began 2000; patent registered. Detail design being undertaken between 2002 and early 2005; certification and first deliveries in 2008.

	Hover mode	Transition Mode	Cruise Mode	Landing Mode
Flaperons		Active (pitch, roll, tilt)	Active (pitch, roll)	
Rudder		Active (yaw)	Active (yaw)	
Tail rotors Horizontal	Active (pitch, roll)	Active (pitch, roll, tilt)		Active (pitch, roll)
Vertical	Active (yaw)	Active (yaw)		Active (yaw)
Rotational Aerofoil	Active (stability)	Tilt control		Active (stability)

DESIGN FEATURES: Blended wing-body aircraft surrounded by rotational aerofoil comprising eight blades within concentric containing rings. Effect is similar to tiltrotor, but without requirement for synchronisation. Aerofoil rotates and generates thrust under power of up to four small jet engines (Williams EJ22, or similar) mounted tangentially on outer ring; complete unit pivots 90° about aircraft's lateral axis, lying in plane of the fixed wings for take-off and landing, and swivelling to provide thrust for forward flight, when conventional wings generate the required lift. Type of bearing between rotating aerofoil and fixed fuselage is not disclosed.

Advantages include reduction of airport congestion, inherent stability (thrust circularly distributed far from centre of gravity; gyroscopic damping), lightweight structure and size scalable between 20 and 120 passenger capacity.

FLYING CONTROLS: Fixed wing includes flaperons and rudder, plus two tail rotors for horizontal and vertical control. Interaction of controls is shown in the table.

Data follow for typical 50-passenger (all first class) Humming.

DIMENSIONS, EXTERNAL:
Rotor diameter: inner	18.3 m (60 ft)
outer	24.4 m (80 ft)
Fuselage: Length	17.1 m (56 ft)
Max width	13.4 m (44 ft)

WEIGHTS AND LOADINGS:
Max payload	5,900 kg (13,000 lb)
Max T-O weight	26,300 kg (58,000 lb)

PERFORMANCE:
Max cruising speed	261 kt (483 km/h; 300 mph)
Max certified altitude	9,145 m (30,000 ft)
Range	434-1,042 n miles (804-1,931 km; 500-1,200 miles)

UPDATED

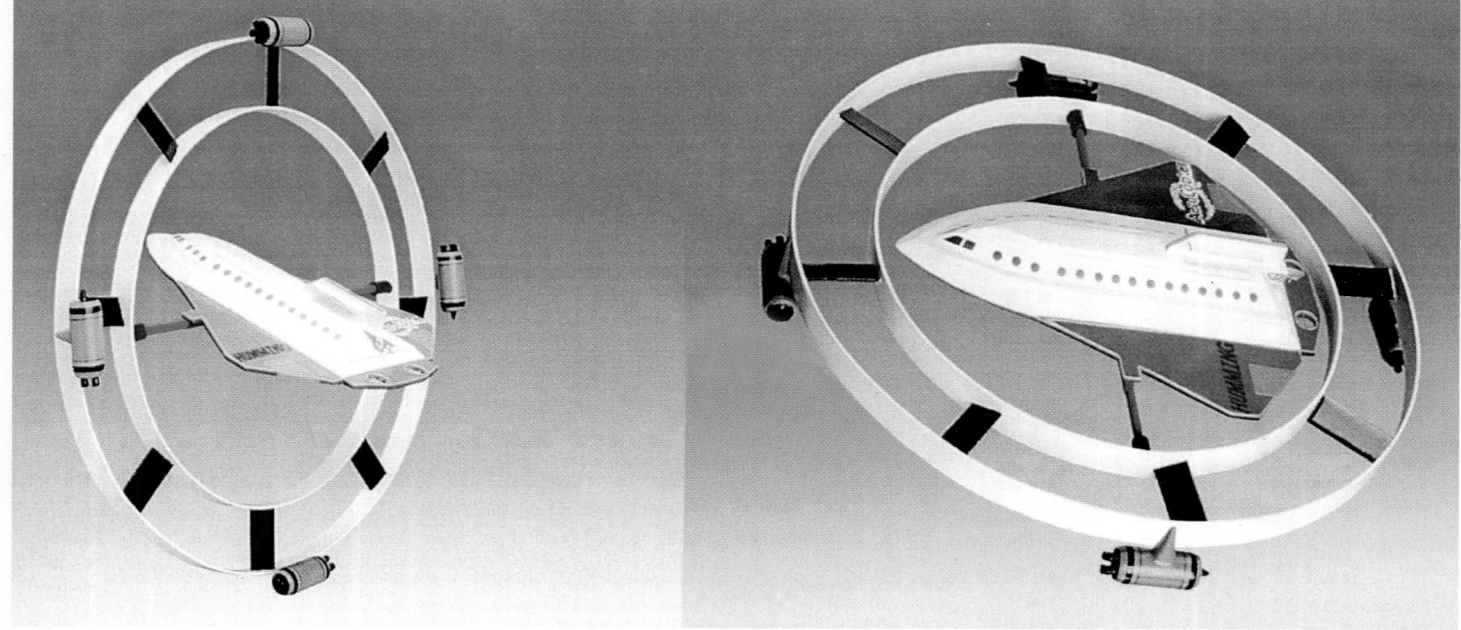

Model of AeroCopter Humming in forward (left) and vertical flight modes   NEW/0525799

AEROCOURIER

AEROCOURIER GROUP

10504 Southwest Indianola Road, Augusta, Kansas 67010
Tel: (+1 316) 733 52 38
e-mail: questions@aerocouriergroup.com
Web: http://www.aerocouriergroup.com
PRESIDENT: Paul Jackson
VICE-PRESIDENTS: John Guernsey (Engineering/Certification)
    Justin Ladner (Engineering/Operations)
    Mike Hahn (Operations)
MEDIA RELATIONS: Cindy Bagaus

AeroCourier was founded in April 2001 to design and produce a light freight aircraft in the Cessna 208 Caravan class. Detail design, and also production of up to 1,500 shipsets of fuselages and wings, has been subcontracted to Dirgantara of Indonesia (which see). Flight testing will be at Augusta Municipal Airport, Wichita, Kansas; assembly plant for production aircraft will be Wichita or Enid, Oklahoma.

UPDATED

AEROCOURIER AEROCOURIER

TYPE: Light utility turboprop.
PROGRAMME: Planning timetable began July 2001; announced 18 September 2001; initial order 12 December 2001; first flight due in June 2003; certification in July 2004. However, by late 2002, no further progress reports had been received.
CUSTOMERS: First 20 AeroCouriers ordered December 2001 for fractional ownership scheme managed by Airshares Elite of Atlanta, Georgia; deliveries begin 2005, primary use being executive transport.
COSTS: "Less than US$1 million" (2001).
DESIGN FEATURES: Configured for ease of loading new-specification LDX cargo containers. High wing and mid-mounted tailplane.
LANDING GEAR: Tricycle type; fixed.
POWER PLANT: One Pratt & Whitney Canada PT6A-114A turboprop, driving a three-blade propeller.
ACCOMMODATION: Pilot and optional co-pilot; six LDX containers or 10 passengers. Flight deck door each side; freight door port, rear, at truck-bed height.
WEIGHTS AND LOADINGS:
Max T-O weight	3,810 kg (8,400 lb)

UPDATED

Model of AeroCourier displayed at NBAA Convention, December 2001
(R W Simpson)   0130084

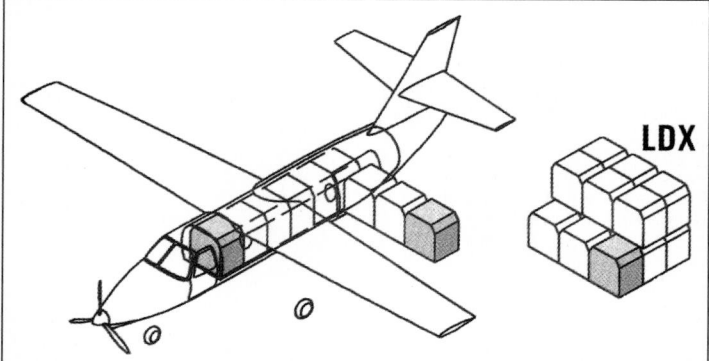

Conception of AeroCourier freighter being loaded with the new LDX containers   0126964

AEROLITES

AEROLITES INC

12104 David Road, Welsh, Louisiana 70591
Tel/Fax: (+1 337) 734 38 65
e-mail: aerolites@centurytel.net
Web: http://www.aerolites.com

VERIFIED

AEROLITES AEROMASTER AG

TYPE: Agricultural sprayer ultralight.
PROGRAMME: First flown 1993.
CUSTOMERS: Total of 30 sold and 24 flying by early 2001.
COSTS: Kit US$23,600 (2002).
DESIGN FEATURES: Strut-braced aluminium wing has ladder construction with internal diagonal bracing tested to +6/−4 g; aluminium ribs slot into pockets within fabric covering. Quoted build time 150 hours.
FLYING CONTROLS: Conventional and manual.
STRUCTURE: Fuselage of welded 4130 chromoly with fabric covering.
LANDING GEAR: Tailwheel type; fixed.
POWER PLANT: One 47.8 kW (64.1 hp) Rotax 582 two-cylinder engine driving Warp Drive carbon fibre three-blade propeller through C type 3:1 reduction gear. Fuel capacity 38 litres (10.0 US gallons; 8.3 Imp gallons).
EQUIPMENT: 12-nozzle SprayMiser CDA system fitted for agricultural operations; belly tank capacity 114 litres (30.0 US gallons; 25.0 Imp gallons). Optional ballistic parachute.

Normal cruising speed	56 kt (105 km/h; 65 mph)
Stalling speed	33 kt (62 km/h; 38 mph)
Max rate of climb at S/L	183 m (600 ft)/min
T-O run: water	168 m (550 ft)
land	134 m (440 ft)
Landing run: land	92 m (300 ft)
Range	190 n miles (351 km; 218 miles)

UPDATED

Aerolites Aeromaster agricultural ultralight (*Geoffrey P Jones*)   0092107

DIMENSIONS, EXTERNAL:
Wing span	8.74 m (28 ft 8 in)
Length overall	5.64 m (18 ft 6 in)
Height overall	2.11 m (6 ft 11 in)

AREAS:
| Wings, gross | 13.54 m² (145.7 sq ft) |

WEIGHTS AND LOADINGS::
| Weight empty | 193 kg (425 lb) |
| Max T-O weight | 453 kg (1,000 lb) |

PERFORMANCE (with SprayMiser system fitted):
Never-exceed speed (VNE)	95 kt (177 km/h; 110 mph)
Normal cruising speed	52 kt (97 km/h; 60 mph)
Spray speed	56 kt (105 km/h; 65 mph)
Stalling speed	28 kt (52 km/h; 32 mph)
Max rate of climb at S/L	244 m (800 ft)/min
T-O run	153 m (500 ft)
Landing run	46 m (150 ft)
Range with max fuel	150 n miles (277 km; 172 miles)
Swath width	9-30 m (30-100 ft)

UPDATED

AEROLITES AEROSKIFF

TYPE: Two-seat amphibian ultralight kitbuilt.
CUSTOMERS: Five flying by early 2002.

COSTS: Kit US$23,775 (2002).
DESIGN FEATURES: Pusher layout. Wings detach for storage. Quoted build time 150 hours.
FLYING CONTROLS: Conventional and manual. High-lift flaps.
STRUCTURE: Glass fibre fuselage and fabric-covered, strut-braced wings.
LANDING GEAR: Tailwheel type; retracts for water landings. Drum brakes. Floats mounted on tubular outriggers at wing mid-point.
POWER PLANT: One 47.8 kW (64.1 hp) Rotax 582 water-cooled engine is standard; optionally, 55.0 kW (73.8 hp) Rotax 618. Fuel capacity 45 litres (12.0 US gallons; 10.0 Imp gallons).
SYSTEMS: Optional bilge pump and electric starter.

DIMENSIONS, EXTERNAL:
Wing span	9.04 m (29 ft 8 in)
Length overall	6.81 m (22 ft 4 in)
Height overall	2.13 m (7 ft 0 in)

AREAS:
| Wings, gross | 14.49 m² (156.0 sq ft) |

WEIGHTS AND LOADINGS::
| Weight empty | 256 kg (565 lb) |
| Max T-O weight | 510 kg (1,125 lb) |

PERFORMANCE (Rotax 582):
| Never-exceed speed (VNE) | 82 kt (152 km/h; 95 mph) |

AEROLITES BEARCAT

TYPE: Single-seat ultralight kitbuilt.
CUSTOMERS: Total of 15 completed by early 2002.
COSTS: Kit US$14,125, including engine and propeller (2002).
DESIGN FEATURES: Replica Corben Baby Ace (1930s vintage). Wings are strut-braced and similar to AeroMaster (which see). Quoted build time 125 hours.
FLYING CONTROLS: Conventional and manual.
STRUCTURE: Fabric-covered 4130 chromoly fuselage and aluminium extended-wing spar.
LANDING GEAR: Tailwheel type; fixed. Bungee shock absorption on main legs; optional hydraulic disc brakes.
POWER PLANT: One 31.0 kW (41.6 hp) Rotax 447 in-line two-cylinder piston engine, driving a two-blade propeller. Optionally, 37.0 kW (49.6 hp) Rotax 503 UL-2V. Fuel capacity 19 litres (5.0 US gallons; 4.2 Imp gallons).

DIMENSIONS, EXTERNAL:
Wing span	9.35 m (30 ft 8 in)
Length overall	5.33 m (17 ft 6 in)
Height overall	1.98 m (6 ft 6 in)

AREAS:
| Wings, gross | 14.86 m² (160 sq ft) |

WEIGHTS AND LOADINGS:
| Weight empty | 136 kg (300 lb) |
| Max T-O weight | 340 kg (750 lb) |

PERFORMANCE:
Never-exceed speed (VNE)	95 kt (177 km/h; 110 mph)
Max operating speed	61 kt (113 km/h; 70 mph)
Normal cruising speed	48 kt (89 km/h; 55 mph)
Stalling speed	24 kt (44 km/h; 27 mph)
Max rate of climb at S/L	366 m (1,200 ft)/min
T-O and landing run	46 m (150 ft)
Range with max fuel	120 n miles (222 km; 138 miles)

UPDATED

AEROSTAR

AEROSTAR AIRCRAFT CORPORATION LLC

10555 Airport Drive, Coeur d'Alene Airport, Hayden Lake, Idaho 83835-9742
Tel: (+1 208) 762 03 38 or (+1 800) 442 42 42
Fax: (+1 208) 762 83 49
e-mail: info@aerostarjet.com
Web (1): http://www.aerostarjet.com
Web (2): http://www.aerostaraircraft.com
PRESIDENT: Steve Speer
VICE-PRESIDENT: Jim Christy
SENIOR PROJECT ENGINEER: William V Leeds

Aerostar Aircraft Corporation was formed in 1991 by former Ted Smith employees and acquired the rights to the Aerostar series of pressurised piston-engined twins from Piper in the same year. It offers upgrades, spares support and second-hand sales, and is planning to extend its 2,975 m² (32,000 sq ft) factory to 10,400 m² (112,000 sq ft) to accommodate production of a jet version.

A new company, Aerostar Jet LLC, was planned for launch in 2001 for the development of the FJ-100; in December 2001 it was seeking US$8.4 million as the first stage towards certification.

UPDATED

AEROSTAR FJ-100

TYPE: Business jet.
PROGRAMME: Jet derivative of former Piper/Ted Smith Aerostar 600/700 series, of which over 1,000 were built (see *Jane's* 1985-86); will be certified as a new-build design, but first flight-test example will be modified from an existing piston-engined aircraft. First flight originally scheduled for August 2001, with deliveries due to begin July 2002; but revised to first flight no earlier than December 2002, possibly preceded by development of turboprop conversion of the Aerostar, to be powered by a new, unspecified, turbine engine.
CURRENT VERSIONS: Family planned, ranging from four to eight seats.
CUSTOMERS: Total 25 deposits received by April 2001. First year (2003) production set at 17, rising to 34 in 2004 and 50 the following year; production expected to reach 70 per year.
COSTS: Development costs estimated at US$40 million. Unit price US$1.95 million (2001).
DESIGN FEATURES: Improved, jet-powered version of original piston-engined Aerostar; FAR Pt 23 compliant.
FLYING CONTROLS: Conventional and manual. Single-slotted Fowler flaps; maximum deflection 45°. Electrically operated trim tabs on port aileron, rudder and both elevators.

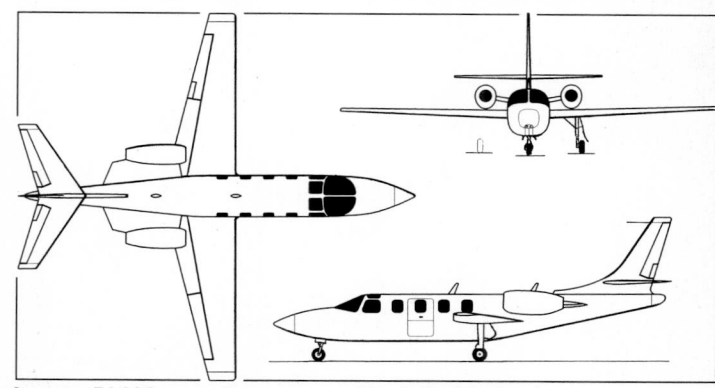

Aerostar FJ-100 general arrangement (*Jane's/Paul Jackson*)   0084553

Artist's impression of FJ-100 (two Williams FJ33-1 turbofans)
0084552

STRUCTURE: Metal throughout; semi-monocoque fuselage with flush-riveted skins; two-spar wing mounted mid-fuselage with 2° dihedral; contains integral fuel tanks. Mid-mounted tailplane; glass fibre dorsal fillet.

LANDING GEAR: Retractable, hydraulically actuated tricycle type; oleo-pneumatic suspension. Electrically actuated nosewheel steering. Dual calliper disc brakes on main wheels.

POWER PLANT: Two 5.35 kN (1,200 lb st) Williams FJ33-1 turbofans on elastomeric engine mounts on pylon on each side of rear fuselage. Fuel in integral wing tanks, plus bladder-type fuselage tank between rear of cabin and baggage compartment; all three tanks crossfeed for balanced fuel load.

ACCOMMODATION: Pilot and up to seven passengers in pairs of seats; standard layout is four club seats in cabin. Door on port side of fuselage behind cockpit. Baggage compartment to rear of cabin; optional lavatory on starboard side at rear of cabin; optional fold-out table.

SYSTEMS: 28 V DC battery for engine starting and as emergency/back-up source; DC external power point on aft fuselage underside. Cabin pressurised to 0.52 bar (7.6 lb/sq in) at 3,050 m (10,000 ft) and cooled using Freon 134 vapour-cycle system. De-icing system to be fitted to engine inlets, surface leading edges and windscreen; exact type to be determined.

AVIONICS: Cockpit modelled on original Aerostar; instrument panel canted to enhance readability by solo pilot.
Comms: Dual VHF nav/com; Dual Mode S transponders; audio control panel.
Radar: Colour weather radar.
Flight: GPS; DME; three-axis autopilot.

DIMENSIONS, EXTERNAL:	
Wing span	11.18 m (36 ft 8 in)
Wing aspect ratio	7.6
Length overall	11.76 m (38 ft 7 in)
Height overall	4.42 m (14 ft 6 in)
Tailplane span	4.37 m (14 ft 4 in)
Wheel track	4.37 m (14 ft 4 in)
Wheelbase	3.10 m (10 ft 2 in)
DIMENSIONS, INTERNAL:	
Cabin: Length	4.23 m (13 ft 10½ in)
Max width	1.16 m (3 ft 9½ in)
Max height	1.21 m (3 ft 11½ in)
Baggage compartment volume	0.68 m³ (24.0 cu ft)
Freight hold volume	0.85 m³ (30.0 cu ft)
AREAS:	
Wings, gross	16.54 m² (178.0 sq ft)
WEIGHTS AND LOADINGS:	
Weight empty	1,905 kg (4,200 lb)
Max fuel weight	1,131 kg (2,494 lb)
Max T-O and landing weight	3,311 kg (7,300 lb)

Max ramp weight	3,334 kg (7,350 lb)
Max zero-fuel weight	2,676 kg (5,900 lb)
Max wing loading	200.2 kg/m² (41.01 lb/sq ft)
Max power loading	310 kg/kN (3.04 lb/lb st)
PERFORMANCE:	
Max operating speed: S/L to 8,840 m (29,000 ft)	
	260 kt (482 km/h; 299 mph)
above 8,840 m (29,000 ft)	M0.71
Max cruising speed	415 kt (769 km/h; 478 mph)
Econ cruising speed	377 kt (698 km/h; 434 mph)
Manoeuvring speed	167 kt (309 km/h; 192 mph)
Stalling speed, landing configuration	
	78 kt (145 km/h; 90 mph)
Max rate of climb at S/L	975 m (3,200 ft)/min
Rate of climb at S/L, OEI	274 m (900 ft)/min
Service ceiling	12,495 m (41,000 ft)
T-O run	579 m (1,900 ft)
Landing run	488 m (1,600 ft)
Max range: VFR	1,750 n miles (3,241 km; 2,013 miles)
IFR with 100 n mile (185 km; 115 mile) alternate	
	1,550 n miles (2,870 km; 1,783 miles)
g limits	+3.6/−1.6
	UPDATED

AIR COMMAND

AIR COMMAND INTERNATIONAL INC

PO Box 1177, Caddo Mills Municipal Airport Building B, Caddo Mills, Texas 75135
Tel: (+1 903) 527 33 35
Fax: (+1 903) 527 38 05
e-mail: aircmd@aircommand.com
Web: http://www.aircommand.com

Company produces kits of autogyros. For details of earlier products, including the currently available 582, see the 1992-93 and 2001-02 editions of *Jane's All the World's Aircraft*.

Air Command also markets the Falcon 2000 derivative of the American Aircraft Falcon, described in the 1992-93 and earlier editions.

UPDATED

AIR COMMAND COMMANDER ELITE

TYPE: Two-seat autogyro kitbuilt.

PROGRAMME: Construction of tandem-seat F-30 and Mazda versions began in late 1999; both made first flights 1 March 2000; public debut at Sun 'n' Fun, April 2000. A side-by-side version of the F-30 has been introduced; a side-by-side Mazda model is under consideration.

CURRENT VERSIONS: **Elite F-30:** Basic version, with Hirth F-30 engine.

Elite F-30ES: As F-30, but side-by-side seating.

Elite Mazda: As F-30, but with more powerful engine.

CUSTOMERS: Over 200 kits sold by end 2001.

COSTS: F-30 US$20,000; F-30ES US$19,292; Mazda US$23,000, all including engine (2002).

DESIGN FEATURES: Open cockpit autogyro with wide range of options to enable individual requirements to be met; many subassemblies preconstructed by factory. Quoted build time 120 hours.

STRUCTURE: Pre-drilled and anodised 6061-T6 tubular airframe; 5 × 5 cm (2 × 2 in) tubing used for mast on F-30 models and 10 × 10 cm (4 × 4 in) on Mazda. Optional glass fibre enclosure. F-30 has aluminium rudder and composites horizontal stabiliser and tailfins; Mazda has all-composites tail surfaces.

LANDING GEAR: Fixed tricycle type with glass fibre speed fairings. Aluminium suspension system with Hager wheels and brakes on all wheels. Nosewheel is fully castoring with leg of 4130 steel tube and 6.00×6 Azusa tyres; 7.6 cm (3 in) tailwheel on F-30 models; 15 cm (6 in) tailwheel on Mazda. Mainwheel hydraulic brakes.

POWER PLANT: *Elite F-30 and F-30ES:* One 82.0 kW (110 hp) Hirth F-30 driving a four-blade Warp Drive propeller. Fuel capacity 37.9 litres (10.0 US gallons; 8.3 Imp gallons) in two seat tanks; optional 68.1 litre (18.0 US gallon; 15.0 Imp gallon) seat tank.

Elite Mazda: One 119 kW (160 hp) Mazda 13B with electric start, driving a four-blade Warp Drive propeller through a 1:2.5 ratio collinear planetary reduction drive. Fuel capacity 53.0 litres (14.0 US gallons; 11.7 Imp gallons) in two seat tanks.

ACCOMMODATION: Pilot and passenger in tandem on Elite F-30 and Elite Mazda; side by side on Elite F-30ES.

SYSTEMS: Basic engine instrumentation is provided with each kit. Options include hydraulic pre-rotator and clutch reduction drive on F-30; pre-rotator, C Box upgrade and electric start on F-30ES; hydraulic pre-rotator on Mazda. Optional New Horizons Components gyro recovery system rocket-powered ballistic parachute on F-30 and Mazda versions.

DIMENSIONS, EXTERNAL:	
Rotor diameter: F-30	8.53 m (28 ft 0 in)
F-30ES, Mazda	8.84 m (29 ft 0 in)
Rotor blade chord: F-30	0.19 m (7½ in)
Fuselage length: F-30	4.09 m (13 ft 5 in)
F-30ES	3.25 m (10 ft 8 in)
Mazda	4.27 m (14 ft 0 in)
Height to top of rotor head: F-30	2.74 m (9 ft 0 in)
F-30ES	2.13 m (7 ft 0 in)
Mazda	2.90 m (9 ft 6 in)
Width: F-30, Mazda	1.88 m (6 ft 2 in)
F-30ES	1.70 m (5 ft 7 in)
Propeller diameter: Mazda	1.73 m (5 ft 8 in)
AREAS:	
Rotor disc: F-30	57.2 m² (616.0 sq ft)
F-30ES, Mazda	61.4 m² (661.0 sq ft)
WEIGHTS AND LOADINGS:	
Weight empty: F-30	222 kg (490 lb)
F-30ES	141 kg (310 lb)
Mazda	318 kg (700 lb)
Max payload: F-30, F-30ES	191 kg (420 lb)
Mazda	352 kg (775 lb)
Max T-O weight: F-30	578 kg (1,275 lb)
F-30ES	412 kg (910 lb)
Mazda	680 kg (1,500 lb)
Max disc loading: F-30	10.1 kg/m² (2.07 lb/sq ft)
F-30ES	6.7 kg/m² (1.38 lb/sq ft)
Mazda	11.1 kg/m² (2.27 lb/sq ft)
Max power loading: F-30	7.05 kg/kW (11.59 lb/hp)
F-30ES	5.04 kg/kW (8.27 lb/hp)
Mazda	5.71 kg/kW (9.38 lb/hp)
PERFORMANCE:	
Max operating speed:	
F-30	95 kt (177 km/h; 110 mph)
F-30ES	73 kt (135 km/h; 84 mph)
Mazda	104 kt (193 km/h; 120 mph)
Cruising speed:	
F-30	65-70 kt (121-129 km/h; 75-80 mph)
F-30ES	56 kt (105 km/h; 65 mph)
Mazda	48-70 kt (89-129 kmh; 55-80 mph)
Min flying speed: F-30	18 kt (33 km/h; 20 mph)
F-30ES	20 kt (36 km/h; 22 mph)
Mazda	26 kt (49 km/h; 30 mph)
Max rate of climb at S/L: F-30	366 m (1,200 ft)/min
Mazda	396 m (1,300 ft)/min
Service ceiling: all	3,050 m (10,000 ft)
T-O run: F-30, F-30ES	31-76 m (100-250 ft)
Mazda	107-213 m (350-700 ft)
Landing run: all	0-6 m (0-20 ft)
	UPDATED

Air Command Commander Elite (*Jane's/Paul Jackson*)   NEW/0525800

AIR TRACTOR

AIR TRACTOR INC

PO Box 485, Municipal Airport, Olney, Texas 76374
Tel: (+1 940) 564 56 16
Fax: (+1 940) 564 23 48
e-mail: airmail@airtractor.com
Web: http://www.airtractor.com
PRESIDENT: Leland Snow
VICE-PRESIDENT FINANCE: David Ickert
DIRECTOR OF COMMUNICATIONS: Kristin Snow

Air Tractor agricultural aircraft based on 50-year experience of Leland Snow, who produced Snow S-2 series, which later became Rockwell S-2R (see earlier *Jane's*); AT-300/301/302 (587 built) no longer in production; 1,800th aircraft delivered in April 2000. Eight versions available, powered by various P&W PT6A and R-1340 engines. Company now has 160 employees and total of 13,660 m² (147,000 sq ft) of manufacturing space at three plants.

UPDATED

AIR TRACTOR AT-401 AIR TRACTOR

TYPE: Agricultural sprayer.
PROGRAMME: AT-401 developed 1986 from AT-301 (see *Jane's Aircraft Upgrades*), with increased wing span and larger hopper. AT-401A version with Polish PZL-3S radial engine abandoned, with just one aircraft produced (see 1992-93 *Jane's*). AT-401B version replaced the 401 and has Hoerner wingtips and increased wing span.
CURRENT VERSIONS: **AT-401B:** Standard version; *as described*.
  Increased T-O weight: Optional version with landing gear of AT-402A and 517 kg (1,140 lb) of additional disposable load.
  Orenda conversion: Retrofit available with 447 kW (600 hp) Orenda OE600 eight-cylinder diesel engine.
CUSTOMERS: By December 2001, 198 AT-401s, one AT-401A and 69 AT-401Bs delivered to Argentina, Australia, Brazil, Canada, Colombia, Mexico, Spain and USA. Production of AT-400/400A totalled 72 and 14, respectively.
COSTS: Standard AT-401B US$259,500; with customer-supplied engine US$219,900 (2002).
DESIGN FEATURES: Purpose-designed sprayer with cantilever low wing and hopper at CG. Constant-chord, unswept wings and tailplane; latter braced; moderately sweptback fin. High-mounted cockpit with protective reinforcement.
  Wing aerofoil NACA 4415; dihedral 3° 30′; incidence 2°.
FLYING CONTROLS: Conventional and manual. Balance tabs on ailerons, elevators and rudders; ailerons droop 10° when electrically operated Fowler flaps deflected to their maximum of 26°.
STRUCTURE: Two-spar wing structure of 2024-T3 light alloy, with alloy steel lower spar cap; bonded doubler inside wing leading-edge to resist impact damage; glass fibre wingroot fairings and skin overlaps sealed against chemical ingress; wing ribs and skins zinc chromated before assembly; flaps and ailerons of light alloy. Fuselage of 4130N steel tube, oven stress relieved and oiled internally, with skin panels of 2024-T3 light alloy attached by Camloc fasteners for quick removal; rear fuselage lightly pressurised to prevent chemical ingress; cantilever fin and strut-braced tailplane of light alloy, metal-skinned and sealed against chemical ingress.
LANDING GEAR: Tailwheel type; non-retractable. Cantilever heavy-duty E-4340 spring steel main gear, thickness 28.6 mm (1.125 in); flat spring suspension for castoring and lockable tailwheel. Cleveland mainwheels with tyre size 8.50-10 (8 ply), pressure 2.83 bar (41 lb/sq in). Tailwheel tyre size 5.00-5. Cleveland four-piston brakes with heavy-duty discs. Optional AT-402A-type landing gear.
POWER PLANT: One remanufactured 447 kW (600 hp) Pratt & Whitney R-1340 air-cooled radial engine with speed ring cowling or optional RamAir Scoopkit, driving a Hamilton Sundstrand 12D40/6101A-12 two-blade propeller; optional propellers include a Pacific Propeller 22D40/

Air Tractor AT-401 converted with Orenda OE600 diesel engine (*Jane's/Paul Jackson*)   0121709

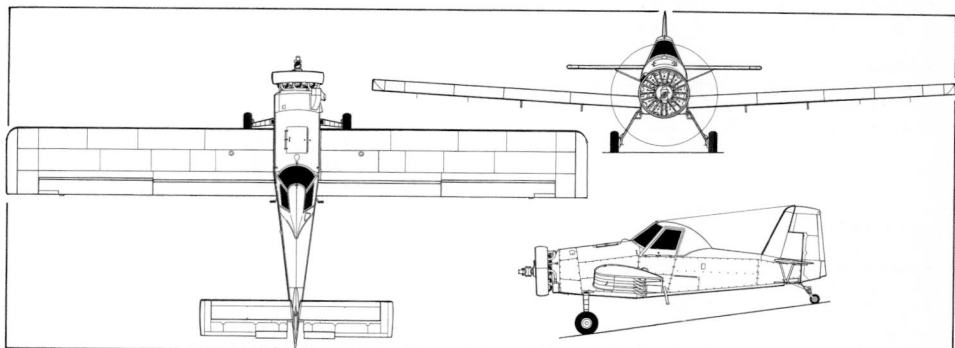

Air Tractor AT-401B (Pratt & Whitney R-1340 radial engine) (*Jane's/Paul Jackson*)   0051564

AG200-2 Hydromatic two-blade constant-speed metal and Hydromatic 23D40 three-blade propeller. Air Tractor has designed and is producing new FAA-approved replacement crankshaft for R-1340; other new replacement parts available include main and thrust bearings, and blower (impeller) bearings. Over 300 replacement crankshafts delivered by early 1997. 477 litre (126 US gallon; 105 Imp gallon) fuel tanks. Oil capacity 30 litres (8.0 US gallons; 6.7 Imp gallons). Orenda OE600 V-8 liquid-cooled piston engine of 447 kW (600 hp) tested in December 1999; development towards certification continued, with aim of deliveries during 2001.
ACCOMMODATION: Single seat with nylon mesh cover in enclosed cabin which is sealed to prevent chemical ingress. Downward-hinged window/door on each side. 'Line of sight' instrument layout, with swing-down lower instrument panel for ease of access for instrument maintenance. Baggage compartment in bottom of fuselage, aft of cabin, with door on port side. Cabin ventilation by 0.10 m (4 in) diameter airscoop.
SYSTEMS: 24 V electrical system, supplied by 35 A engine-driven alternator; 60 A alternator optional.
AVIONICS: Optional avionics include Bendix/King KX 155 nav/com/glideslope and ACK E-01 emergency locator transmitter.
EQUIPMENT: Agricultural dispersal system comprises a 1,514 litre (400 US gallon; 333 Imp gallon) Derakane vinylester resin/glass fibre hopper mounted in forward fuselage with hopper window and instrument panel-mounted hopper quantity gauge; 0.97 m (3 ft 2 in) wide Transland gatebox; Transland 5 cm (2 in) bottom loading valve; Agrinautics 6.4 cm (2½ in) spraypump with Transland on/off valve and five-blade variable-pitch plastics fan, and 38-nozzle stainless steel spray system with streamlined booms; swath width 24.39 m (80 ft 0 in). Ground start receptacle and hopper rinse system are standard.
  Optional equipment includes night flying package comprising strobe and navigation lights, night working lights, retractable 600 W landing light in port wingtip; and ferry fuel system. Alternative agricultural equipment

includes Transland 22358 extra high volume spreader, Transland 54401 NorCal Swathmaster, and 40 extra spray nozzles for high-volume spraying. Three-piece safety plate glass centre windshield and washer system optional, as is a new crew seat.

DIMENSIONS, EXTERNAL:
Wing span	15.57 m (51 ft 1¼ in)
Wing chord, constant	1.83 m (6 ft 0 in)
Wing aspect ratio	8.5
Length overall	8.23 m (27 ft 0 in)
Height overall	2.59 m (8 ft 6 in)
Propeller diameter: standard	2.77 m (9 ft 1 in)
optional	2.59 m (8 ft 6 in)

AREAS:
Wings, gross	28.43 m² (306.0 sq ft)
Ailerons (total)	3.55 m² (38.20 sq ft)
Trailing-edge flaps (total)	3.75 m² (40.40 sq ft)
Fin	0.90 m² (9.70 sq ft)
Rudder	1.30 m² (14.00 sq ft)
Tailplane	2.42 m² (26.00 sq ft)
Elevators, incl tabs	2.36 m² (25.40 sq ft)

WEIGHTS AND LOADINGS (S: standard, IGW: increased gross weight):
Weight empty, spray equipped	1,925 kg (4,244 lb)
Max T-O weight: S	3,565 kg (7,860 lb)
IGW	4,082 kg (9,000 lb)
Max landing weight: S	2,721 kg (6,000 lb)
IGW	3,175 kg (7,000 lb)
Max wing loading: S	125.4 kg/m² (25.69 lb/sq ft)
IGW	143.6 kg/m² (29.41 lb/sq ft)
Max power loading: S	7.97 kg/kW (13.10 lb/hp)
IGW	9.13 kg/kW (15.00 lb/hp)

PERFORMANCE (at standard max T-O weight, except where indicated):
Max cruising speed at S/L, hopper empty	135 kt (251 km/h; 156 mph)
Cruising speed at 1,220 m (4,000 ft)	124 kt (230 km/h; 143 mph)
Typical working speed	104-122 kt (193-225 km/h; 120-140 mph)
Stalling speed at 2,721 kg (6,000 lb):	
flaps up	64 kt (118 km/h; 73 mph)
flaps down	53 kt (99 km/h; 61 mph)
Stalling speed as usually landed	47 kt (87 km/h; 54 mph)
Max rate of climb at S/L:	
at max landing weight	335 m (1,100 ft)/min
at max T-O weight	158 m (520 ft)/min
T-O run	402 m (1,320 ft)
Range at econ cruising speed at 2,440 m (8,000 ft), no reserves	547 n miles (1,014 km; 630 miles)

UPDATED

AIR TRACTOR AT-402 TURBO AIR TRACTOR

TYPE: Agricultural sprayer.
PROGRAMME: Follow-on from AT-400 of 1980, AT-402 first flight August 1988; certified November 1988; first delivery late 1988. Current 402A and 402B have Hoerner wingtips and increased span.
CURRENT VERSIONS: **AT-402A:** Introduced in mid-1997, supplementing AT-402B, powered by a 410 kW (550 shp)

Air Tractor AT-402A Turbo Air Tractor, introduced in 1997   0102260

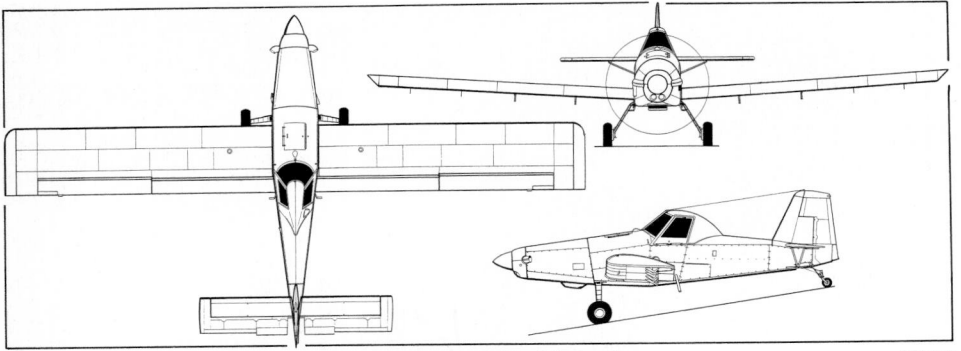

Air Tractor AT-402A/AT-402B Turbo Air Tractor (*Jane's/Paul Jackson*)   0079012

P&WC PT6A-11AG. Aimed at first-time turbine buyer and priced accordingly, complete with spray dispersal equipment, lights and air conditioning.

AT-402B: Combines fuselage, tail surfaces and landing gear of AT-400 with turboprop engine and wing of AT-401B.

Description refers to AT-402B, and is generally as for AT-401, except that below.

CUSTOMERS: Total of 68 AT-402s, 103 AT-402As and 31 AT-402Bs delivered by December 2001.

COSTS: Standard AT-402A with PT6A-11AG US$444,500; standard AT-402B US$536,500; or US$247,900 with customer-supplied PT6A-15AG, -27, -28 or -34AG engine (2002).

DESIGN FEATURES: Broadly as AT-401. All versions have steel alloy lower wing spar caps for long fatigue life and reinforced leading-edges to prevent birdstrike damage. Size 29×11.0-10 high-flotation tyres and wheels as standard.

POWER PLANT: One 507 kW (680 shp) P&WC PT6A-15AG, -27 or -28 turboprop, either new or customer-furnished, driving a Hartzell HCB3TN-3D/T1028N+4 three-blade constant-speed reversible-pitch propeller. Standard fuel capacity 644 litres (170 US gallons; 142 Imp gallons); optional fuel tankage 818 litres (216 US gallons; 180 Imp gallons) or 886 litres (234 US gallons; 195 Imp gallons).

SYSTEMS: 250 A starter/generator and two 24 V 21 Ah batteries.

EQUIPMENT: Hopper and gatebox size as for AT-401; optional equipment includes Transland extra high-volume dispersal system; engine-driven air conditioning system.

DIMENSIONS, EXTERNAL:
Length overall	9.32 m (30 ft 7 in)
Height overall	2.90 m (9 ft 6 in)
Tailplane span	5.23 m (17 ft 2 in)
Wheel track	2.62 m (8 ft 7 in)

WEIGHTS AND LOADINGS:
Weight empty, spray equipped	1,860 kg (4,100 lb)
Certified gross weight (FAR Pt 23)	3,175 kg (7,000 lb)
Typical operating weight (CAM 8):	
AT-402A	3,901 kg (8,600 lb)
AT-402B	4,159 kg (9,170 lb)
Max wing loading	137.2 kg/m² (28.10 lb/sq ft)
Max power loading	7.70 kg/kW (12.65 lb/shp)

PERFORMANCE:
Max level speed at S/L:	
clean	174 kt (322 km/h; 200 mph)
with dispersal equipment	160 kt (298 km/h; 185 mph)
Cruising speed at 283 kW (380 shp) at 2,440 m (8,000 ft)	142 kt (264 km/h; 164 mph)
Typical working speed	113-126 kt (209-233 km/h; 130-145 mph)
Stalling speed at 2,721 kg (6,000 lb) AUW:	
flaps up	64 kt (118 km/h; 73 mph)
flaps down	53 kt (99 km/h; 61 mph)
Stalling speed at 2,041 kg (4,500 lb) typical landing weight	46 kt (86 km/h; 53 mph)
Max rate of climb at S/L, dispersal equipment installed:	
AUW of 2,721 kg (6,000 lb)	495 m (1,625 ft)/min
AUW of 3,565 kg (7,860 lb)	320 m (1,050 ft)/min
T-O run at AUW of 3,565 kg (7,860 lb)	247 m (810 ft)
Landing run at 2,041 kg (4,500 lb)	122 m (400 ft)
Range at econ cruising speed at 2,400 m (8,000 ft), no reserves	573 n miles (1,062 km; 660 miles)
	UPDATED

AIR TRACTOR AT-502 TURBO

TYPE: Agricultural sprayer.

PROGRAMME: Developed as stretched AT-401. First flight of AT-502 April 1987; certified 23 June 1987.

CURRENT VERSIONS: **AT-502:** Original version.

Description refers to AT-502, and is generally as for AT-401, except that below.

AT-502A: Similar to AT-502B but with 820 kW (1,100 shp) PT6A-45R; slow turning (1,425 rpm) five-blade Hartzell (HCB5MP-3C/M10876AS) propeller; enlarged vertical tail surfaces. For operation in mountainous terrain or short strips. Prototype first flight February 1992; certified April 1992; 25 delivered by end 1999.

AT-502B: Stronger wing than AT-502, with 0.61 m (2 ft) longer span and Hoerner wingtips, which are stated to increase width of spray pattern by 0.91 m (3 ft).

AT-503A: Dual control, tandem-seat trainer; PT6A-34AG engine. Prototype designated AT-503. In low-rate production.

CUSTOMERS: Total of 526 AT-500 series delivered by December 2001, including out-of-production AT-501 (nine), AT-502 (208), AT-503 (one), plus AT-503A (three).

COSTS: Standard AT-502B US$585,900 with PT6A-15AG, or US$631,500 with-34AG; or US$292,500 with customer-supplied PT6A-15AG, -27, -28, -34AG (2002).

DESIGN FEATURES: See AT-401. 1,892 litre (500 US gallon; 416 Imp gallon) hopper handles low-density nitrogen-based fertilisers such as urea; safety glass centre windscreen with wiper. Alloy steel lower spar cap and bonded doubler on inside of wing leading-edge for increased resistance to impact damage, and glass fibre wingroot fairings.

LANDING GEAR: Non-retractable tailwheel type. Heavy-duty E-4340 spring steel main gear, thickness 37.2 mm (1.31 in); flat spring for castoring and lockable tailwheel. Cleveland mainwheels, tyre size 29×11.0-10, pressure 3.45 bar (50 lb/sq in); tailwheel tyre size 5.00-5; Cleveland six-piston brakes with heavy-duty discs.

POWER PLANT: One 507 kW (680 shp) Pratt & Whitney Canada PT6A-15AG, PT6A-27 or PT6A-28, or 559 kW (750 shp) PT6A-34 or PT6A-34AG turboprop, driving a Hartzell HCB3TN-3D/T10282+4 three-blade metal propeller. Standard fuel capacity 644 litres (170 US gallons; 142 Imp gallons). Optional capacities 818 litres (216 US gallons; 180 Imp gallons) and 886 litres (234 US gallons; 195 Imp gallons). AT-502A has 818 litre (216 US gallon; 180 Imp gallon) tanks as standard.

ACCOMMODATION: One or two persons (see Current Versions). Has quickly detachable instrument panel and removable fuselage skin panels for ease of maintenance.

SYSTEMS: Two 24 V 42 Ah batteries and 250 A starter/generator.

AVIONICS: *Comms:* Optional avionics include Bendix/King KX 155 nav/com/glideslope and KR 87 ADF, KY 196 VHF radio, KT 76A transponder and ACK E-01 emergency locator transmitter.

EQUIPMENT: Agricultural dispersal system comprises a 1,892 litre (500 US gallon; 416 Imp gallon) hopper

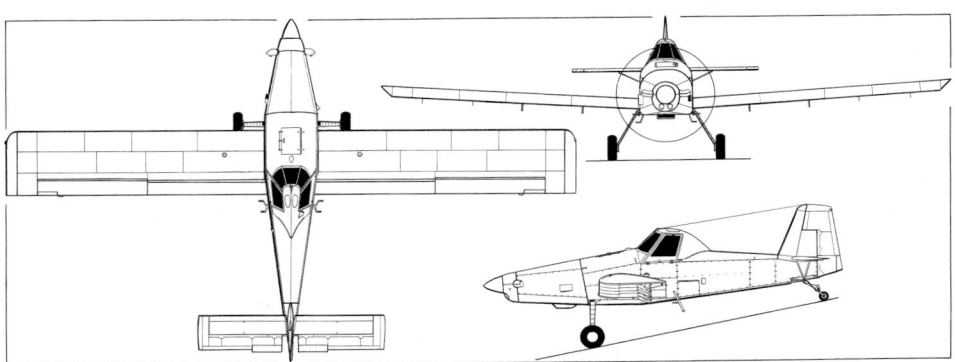

Air Tractor AT-502B agricultural aircraft (*Jane's/Paul Jackson*)   0079013

Air Tractor AT-502B   0121677

mounted in forward fuselage with hopper window and instrument panel-mounted hopper quantity gauge; 0.97 m (3 ft 2 in) wide Transland gatebox; Transland 6.4 cm (2½ in) bottom loading valve; Agrinautics 6.4 cm (2½ in) spraypump with Transland on/off valve and five-blade variable-pitch plastics fan and 38-nozzle stainless steel spray system with streamlined booms. Optional dispersal equipment includes 7.6 cm (3 in) spray system with 119 spray nozzles, and automatic flagman; spray swath 25.91 m (85 ft 0 in).

Standard equipment includes safety glass centre windscreen panel, ground start receptacle, three-colour polyurethane paint finish, strobe and navigation lights; windscreen washer and wiper; and twin nose-mounted landing/taxi lights. Optional equipment includes engine-driven air conditioning system, night flying package, comprising night working lights, retractable 600 W landing light in port wingtip; fuel flowmeter, fuel totaliser and ferry fuel system. Alternative agricultural equipment includes Transland 22356 extra high-volume spreader, Transland 54401 NorCal Swathmaster, 40 extra spray nozzles for high-volume spraying, and eight- or 10-unit Micronair Mini Atomiser unit; hopper rinse tank is standard. New crew seat is also optional.

DIMENSIONS, EXTERNAL:
Wing span: AT-502A/502B	15.85 m (52 ft 0 in)
AT-502	15.24 m (50 ft 0 in)
Wing chord, constant	1.83 m (6 ft 0 in)
Wing aspect ratio	8.7
Length overall	10.11 m (33 ft 2 in)
Height overall	3.12 m (10 ft 3 in)
Tailplane span	5.23 m (17 ft 2 in)
Wheel track	3.12 m (10 ft 3 in)
Wheelbase	6.64 m (21 ft 9½ in)
Propeller diameter: AT-502B	2.69 m (8 ft 10 in)

AREAS:
Wings, gross: AT-502A/502B	28.99 m² (312.0 sq ft)
AT-502	27.87 m² (300.0 sq ft)
Ailerons (total)	3.53 m² (38.0 sq ft)
Trailing-edge flaps (total)	3.75 m² (40.4 sq ft)
Fin	0.90 m² (9.7 sq ft)
Rudder	1.30 m² (14.0 sq ft)
Tailplane	2.41 m² (26.0 sq ft)
Elevators (total, incl tab)	2.44 m² (26.3 sq ft)

WEIGHTS AND LOADINGS:
Weight empty, spray equipped	1,949 kg (4,297 lb)
Max T-O weight	4,399 kg (9,700 lb)
Max landing weight	3,629 kg (8,000 lb)
Max wing loading:	
AT-502A/502B	151.8 kg/m² (31.09 lb/sq ft)
Max power loading: 502A	5.37 kg/kW (8.82 lb/shp)
502B (PT6A-34AG)	7.87 kg/kW (12.93 lb/shp)
502B (PT6A-15AG)	8.68 kg/kW (14.26 lb/shp)

PERFORMANCE (AT-502B with spray equipment installed):
Never-exceed speed (VNE) and max level speed at S/L, hopper empty	156 kt (290 km/h; 180 mph)
Cruising speed at 2,440 m (8,000 ft), 283 kW (380 shp)	136 kt (253 km/h; 157 mph)
Typical working speed	104-126 kt (193-233 km/h; 120-145 mph)
Stalling speed at 3,629 kg (8,000 lb):	
flaps up	72 kt (132 km/h; 82 mph)
flaps down	59 kt (110 km/h; 68 mph)
Stalling speed at 1,978 kg (4,360 lb) typical landing weight	46 kt (86 km/h; 53 mph)
Max rate of climb at S/L, AUW of 3,629 kg (8,000 lb):	
with PT6A-15AG	311 m (1,020 ft)/min
with PT6A-34AG	360 m (1,180 ft)/min
Max rate of climb at S/L, AUW of 4,309 kg (9,500 lb):	
with PT6A-15AG	232 m (760 ft)/min
with PT6A-34AG	282 m (925 ft)/min
T-O run at AUW of 3,629 kg (8,000 lb):	
with PT6A-15AG	236 m (775 ft)
with PT6A-34AG	222 m (730 ft)
T-O run at AUW of 4,309 kg (9,500 lb):	
with PT6A-15AG	356 m (1,170 ft)
with PT6A-34AG	302 m (990 ft)
Range with max fuel	538 n miles (998 km; 620 miles)

UPDATED

AIR TRACTOR AT-602

TYPE: Agricultural sprayer.

PROGRAMME: Prototype first flew 1 December 1995. Certification completed 6 June 1996 and deliveries began the following month. Aircraft fits into Air Tractor range between AT-502B and AT-802A, being AT-502 with increased wing and tail spans, plus taller fin.

CUSTOMERS: Total of 105 sold by early December 2001.

COSTS: Standard AT-602 with PT6A-60AG US$831,500; with customer-supplied engine US$358,500 (2001).

Description generally as for AT-401, except that following.

POWER PLANT: Choice of 783 kW (1,050 shp) Pratt & Whitney Canada PT6A-60AG turboprop, or PT6A-65AG of 966 kW (1,295 shp), driving a Hartzell five-blade constant-speed reversing propeller. Fuel capacity 818 litres (216 US gallons; 180 Imp gallons); optional fuel capacity 1,105 litres (292 US gallons; 243 Imp gallons).

SYSTEMS: Three batteries and 250 A starter/generator.

EQUIPMENT: Glass fibre hopper, capacity 2,385 litres (630 US gallons; 525 Imp gallons); Transland 0.96 m (3 ft 2 in)

AT-602, second-largest of the Air Tractor range   0102262

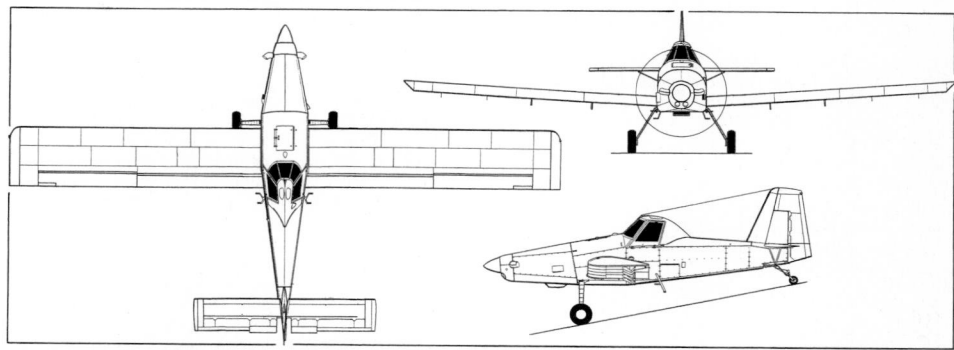

Air Tractor AT-602 (P&WC PT6A-60AG engine) *(Jane's/Paul Jackson)*   0079015

wide gatebox; five-blade ground-adjustable spraypump fan; pump shut-off valve. Optional crew seat.

DIMENSIONS, EXTERNAL:
Wing span	17.07 m (56 ft 0 in)
Wing chord, constant	1.83 m (6 ft 0 in)
Wing aspect ratio	9.3
Length overall	10.41 m (34 ft 2 in)
Height overall	3.38 m (11 ft 1 in)
Tailplane span	5.66 m (18 ft 7 in)
Wheel track	3.07 m (10 ft 1 in)
Wheelbase	6.71 m (22 ft 0 in)

AREAS:
Wings, gross	31.22 m² (336.0 sq ft)

WEIGHTS AND LOADINGS:
Weight empty, equipped	2,540 kg (5,600 lb)
Max T-O weight	5,670 kg (12,500 lb)
Max landing weight	5,443 kg (12,000 lb)
Max wing loading	181.6 kg/m² (37.20 lb/sq ft)
Max power loading	7.25 kg/kW (11.90 lb/shp)

PERFORMANCE:
Never-exceed speed (VNE)	189 kt (350 km/h; 217 mph)
Working speed	130 kt (241 km/h; 150 mph)
Stalling speed: flaps up	86 kt (160 km/h; 99 mph)
flaps down	75 kt (139 km/h; 87 mph)

UPDATED

AIR TRACTOR AT-802

TYPE: Agricultural sprayer.

PROGRAMME: Design started July 1989; optional configuration as firefighter; first flight of prototype (N802LS) 30 October 1990; second aircraft flew November 1991, with

PT6A-45R and configured as agricultural model with spraybooms, pump and Transland gatebox. Production deliveries started second quarter 1993.

CURRENT VERSIONS: **AT-802:** Tandem two-seater powered by P&WC PT6A-45R; certified for gross weight of 6,804 kg (15,000 lb) 27 April 1993. PT6A-65AG and -67AG versions certified April 1993 at maximum T-O weight of 7,257 kg (16,000 lb).

AT-802A: Single-seat version; third production aircraft in this configuration; can be powered by refurbished PT6A-65AG or -67AG engine. First flight 6 July 1992; FAA certification gained 17 December 1992; certified for gross weight of 6,804 kg (15,000 lb) with PT6A-45R on 27 April 1993. PT6A-65AG versions certified March 1993 at maximum T-O weight of 7,257 kg (16,000 lb).

AT-802F: Two-seat firefighting version; PT6A-67AG engine. FAA certification at 5,670 kg (12,500 lb) maximum T-O weight 17 December 1992; certified at 7,257 kg (16,000 lb) on 27 April 1993, giving useful load of 3,987 kg (8,790 lb).

Data apply to AT-802F version, except where indicated.

CUSTOMERS: Total 122 delivered by December 2001. Deliveries include 21 to Australia, 29 to Europe and eight to Canada plus two to Saudi Arabia for oil slick eradication. Total of 51 engaged in firefighting duties in six countries during 2001. Recent customers include Forest Protection Limited (FPL) of New Brunswick, Canada, which ordered six AT-802Fs, one purchased after lease during 2001, two for delivery in 2002 and three scheduled for delivery in 2003, to replace Grumman Avenger firebombers; and the US State Department's International Narcotics and Law Enforcement Division, which ordered

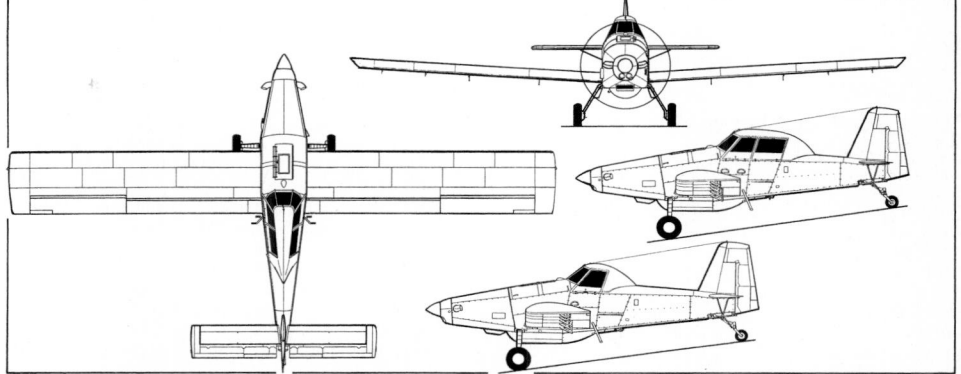

Air Tractor AT-802 two-seat agricultural aircraft with extra side view (lower) of AT-802A single-seater *(Jane's/Paul Jackson)*   0100424

Air Tractor AT-802 crop-sprayer (*Jane's/Paul Jackson*)   NEW/0525801

eight AT-802s for delivery from April 2002, for drug eradication duties in Colombia in a joint programme with the Colombian National Police. These aircraft are modified with self-sealing fuel tanks, armoured cockpits with bulletproof windscreens and engine fire-extinguishing systems.

COSTS: Standard AT-802A US$980,900 with new PT6A-65AG; with customer-supplied PT6A-65AG or -67AG US$449,900. Standard AT-802AF US$1,224,500 with PT6A-67AG; or US$606,900 with customer-supplied PT6A-67AG. Two-seat version with dual controls US$32,000 extra (2002).

DESIGN FEATURES: Generally as for AT-401. Largest aircraft built by company to date; full dual controls for training; also designed for firefighting; programmable logic computer with cockpit control panel and digital display enables pilot to select coverage level and opens hydraulically operated 'bomb bay' drop doors to prescribed width, closing them when selected amount of retardant released. Drop doors adjust automatically for changing head pressure and aircraft acceleration to provide a constant flow rate and even ground coverage.

Wing aerofoil section NACA 4415; dihedral 3° 30'; incidence 2°.

FLYING CONTROLS: Manually operated ailerons, elevators and rudder with balance tabs; electrically operated Fowler trailing-edge flaps deflect to maximum 30°.

STRUCTURE: Two-spar wing of 2024-T3 light alloy with alloy steel upper and lower spar caps and bonded doubler on inside of leading-edge for impact damage resistance; ribs and skins zinc chromated before assembly; glass fibre wingroot fairing and skin overlaps sealed against chemical ingress; flaps and ailerons of light alloy. Fuselage of welded 4130N steel tube, oven stress relieved and oiled internally, with skin panels of 2024-T3 light alloy attached by Camloc fasteners for quick removal; rear fuselage lightly pressurised to prevent chemical ingress. Cantilever fin and strut-braced tailplane of light alloy, metal skinned and sealed against chemical ingress.

LANDING GEAR: Non-retractable tailwheel type. Cantilever heavy-duty E-4340 spring steel main legs, thickness 44.5 mm (1.75 in); flat spring suspension for castoring and lockable tailwheel. Cleveland mainwheels with tyre size 11.00-12 (10 ply), pressure 4.14 bar (60 lb/sq in).

Tailwheel tyre size 6.00-6. Cleveland eight-piston brakes with heavy-duty discs.

POWER PLANT: One Pratt & Whitney Canada PT6A-65AG or -67AG turboprop, rated at 966 kW (1,295 shp) for -65AG and 1,007 kW (1,350 shp) for -67AG, both driving a Hartzell five-blade feathering and reversible-pitch constant-speed metal propeller. Fuel in two integral wing tanks, total usable capacity 961 litres (254 US gallons; 211 Imp gallons); optional tanks increase capacity to 1,438 litres (380 US gallons; 317 Imp gallons). Engine air is filtered through two large pleated paper industrial truck filters.

ACCOMMODATION: One or two seats in enclosed cabin, which is sealed to prevent chemical ingress and protected with overturn structure. Four downward-hinged doors, two on each side. Windscreen is safety-plate auto glass, with washer and wiper. Air conditioning system standard.

SYSTEMS: Hydraulic system, pressure 207 bar (3,000 lb/sq in).

AVIONICS: Advanced nav/com, including GPS.

EQUIPMENT: Two removable Derakane vinylester hoppers forward of cockpit and 227 litre (60 US gallon; 50 Imp gallon) gate tank in ventral bulge, for agricultural chemical, fire retardant or water; total capacity 3,104 litres (820 US gallons; 683 Imp gallons). Electronic spray monitoring system standard. Transland high-volume fertilising gate 0.96 × 0.25 m (3 ft 2 in × 10 in) optional.

DIMENSIONS, EXTERNAL:

Wing span	18.06 m (59 ft 3 in)
Wing chord, constant	2.07 m (6 ft 9½ in)
Wing aspect ratio	8.8
Length overall	10.95 m (35 ft 11 in)
Height overall	3.89 m (12 ft 9 in)
Tailplane span	6.03 m (19 ft 9½ in)
Wheel track	3.10 m (10 ft 2 in)
Wheelbase	7.25 m (23 ft 9½ in)
Propeller diameter (-65AG)	2.92 m (9 ft 7 in)

AREAS:

Wings, gross	37.25 m² (401.0 sq ft)
Ailerons (total)	4.61 m² (49.60 sq ft)
Trailing-edge flaps (total)	5.54 m² (59.60 sq ft)
Fin	1.24 m² (13.40 sq ft)
Rudder	1.57 m² (16.90 sq ft)
Tailplane	3.44 m² (37.00 sq ft)
Elevators (total, incl tab)	3.00 m² (32.30 sq ft)

WEIGHTS AND LOADINGS (AT-802A):

Weight empty, equipped: sprayer	2,951 kg (6,505 lb)
firefighter	3,270 kg (7,210 lb)
Max T-O and landing weight	7,257 kg (16,000 lb)
Max wing loading	194.8 kg/m² (39.90 lb/sq ft)
Max power loading (-67AG)	7.21 kg/kW (11.85 lb/shp)

PERFORMANCE:

Max level speed at S/L	182 kt (338 km/h; 210 mph)
Max cruising speed at 2,440 m (8,000 ft)	192 kt (356 km/h; 221 mph)
Stalling speed, power off, flaps down, at max landing weight	79 kt (147 km/h; 91 mph)
Max rate of climb at S/L	259 m (850 ft)/min
Service ceiling	3,965 m (13,000 ft)
T-O run	610 m (2,000 ft)
Range with max fuel	696 n miles (1,289 km; 800 miles)

UPDATED

Fire Boss firefighting version of AT-802 produced by float manufacturer Wipaire (*Wipaire*)   NEW/0536703

AIR TRACTOR S-22 SURVEYOR

TYPE: Utility turboprop.

PROGRAMME: Design started 2000; first delivery expected in mid-2004.

CUSTOMERS: US Department of the Interior has ordered one for delivery in 2004, with potential requirement for up to 15 for aerial survey of waterfowl.

DESIGN FEATURES: High wing; intended for passenger/freight carrying and aerial applications duties in remote areas; design goals include exceptional cockpit visibility and high cruising speed.

LANDING GEAR: Amphibious landing gear standard.

POWER PLANT: One Pratt & Whitney Canada PT6A turboprop.

ACCOMMODATION: Standard accommodation for up to 10 passengers.

NEW ENTRY

ALLIANCE

ALLIANCE AIRCRAFT CORPORATION

John D Rockefeller Industrial Park, Martinsburg, West Virginia

PRESIDENT AND CEO: Earl Robinson

VICE-PRESIDENT, CORPORATE DEVELOPMENT: Tim Ryan

CHIEF FINANCIAL OFFICER: Scott Gaul

CHIEF DESIGNER: Guido Pessotti

VICE-PRESIDENT, MARKETING: Michael E Cardellichio

Negotiations were under way in 2000 to construct a US$20 million, 37,160 m² (400,000 sq ft) factory at Pease International Tradeport, New Hampshire (the former Pease

AFB), where Alliance would assemble (from overseas-sourced components) the first of a family of jet airliners. The company has discussed provision of hardware and engineering systems by Sukhoi of Russia. Provisional agreement signed with newly formed Sukhoi Civil Aircraft Company on 5 June 2000; according to Russian reports, Sukhoi would "design and produce the aircraft". However, in August 2001, Sukhoi formally announced similar RRJ programme (which see), with inputs from Boeing, having earlier rejected Alliance's business plan.

In June 2001, Alliance announced it was looking to build new facilities at Martinsburg, West Virginia, and that it had entered into a deal with HAIG (Harbin) of China (which see) covering assembly of the SL 100/35 and purchase of up to 50

aircraft for local resale. First flight of Chinese-assembled aircraft set for late 2003, with deliveries 12 months later. Alliance's office equipment was auctioned in July 2001 to repay some of the company's debts. In 2002, Alliance was in discussions with the administrators of Fairchild Dornier (which see) with regard to funding continued production of Dornier 328 and manufacture of StarLiner in Germany.

UPDATED

ALLIANCE STARLINER

Details of this proposed regional jet airliner last appeared in the 2002-03 *Jane's All the World's Aircraft*.

UPDATED

AMD

AIRCRAFT MANUFACTURING & DEVELOPMENT

415 Airport Road, Heart of Georgia Regional Airport, PO Box 639, Eastman, Georgia 31023
Tel: (+1 912) 374 27 59
Fax: (+1 912) 374 27 93
e-mail: info@newplane.com
Web: http://www.NewPlane.com
PRESIDENT: Mathieu Heintz
MANUFACTURING DIRECTOR: Shawn Fallin
SALES MANAGER: Lisa Lewis

AMD was established in 1999 to produce and market the Zenair Zenith CH 2000 and CH2T. Its purpose-built, 2,600 m² (28,000 sq ft) factory became operational in December 1999; the first US-produced CH 2000 was completed in January 2000.

In January 2001, AMD became agent for the OMF Symphony (which see in German section), marketing and assembling aircraft for the North American market. On 1 March 2001, AMD first flew the CH 640 kitbuilt, which draws on the CH 2000.

UPDATED

AMD Zenith CH 2000 Alarus *(Jane's/Paul Jackson)*   NEW/0525804

AMD ZENITH CH 2000 ALARUS

TYPE: Two-seat lightplane.
PROGRAMME: Prototype C-FQCU first flown 26 June 1993; Canadian and US certification of prototypes 26 and 31 July 1994, respectively; first two production models (C-FRSK and C-FRSV) completed April 1994. First delivery September 1994; US type certification 25 July 1995. A test example fitted with 104.5 kW (140 hp) Lycoming O-320 was produced in 1995. Certified for IFR operations and spins on 2 October 1996. JAA certification received March 2000. Company revealed it would produce a lighter version during 2001, with 95 litre (25.0 US gallons; 20.8 Imp gallon) fuel capacity.
CURRENT VERSIONS: **CH 2000**: Standard production model, *as described.*
CH2T: Basic model intended for flight training market; avionics are not included.
CUSTOMERS: Total of 43 delivered by time production moved from Canada to US in June 1999. First US-built aircraft (N17KA, one of 11 CH2Ts for KeyFlite Academy at Utica, New York, and Nashua, New Hampshire) received C of A on 14 January 2000 and delivered following day. Additionally, Zenair built production aircraft 0063, '64 and '65. At least 74 (from both sources) built by August 2002.
COSTS: CH 2000 US$119,000 (2002). Provisional cost of lightweight version US$84,900 (2001).
DESIGN FEATURES: Side-by-side two-seat and smaller-span derivative of Tri-Z CH 300. Designed to conform to FAR Pt 23 (JAR-VLA equivalent level of safety). Conventional, low-wing configuration, with mid-mounted tailplane and sweptback fin.
Wing section LS (1) 0417 (mod.).
FLYING CONTROLS: Manual. All-moving tailplane deflects 12° up and 9° 30′ down with anti-balance tabs; all-moving fin/ rudder with horn balance deflects 21° 30′ each way. Electrically actuated split flaps lower to 50°; ailerons deflect ±15°.
STRUCTURE: Semi-monocoque fuselage of aluminium alloy, with stressed skins. Wings and tailplane aluminium skinned.
LANDING GEAR: Non-retractable tricycle type with steerable (±14°) nosewheel. Shock-cord absorption on nosewheel. Cleveland 5.00-5 in wheels with hydraulic disc brakes on mainwheels. Optional wheel fairings.
POWER PLANT: One 86.5 kW (116 hp) Textron Lycoming O-235-N2C, driving metal Sensenich 72-CK-0-46 or 72-

CK-0-48 propeller. Fuel capacity 106 litres (28.0 US gallons; 23.3 Imp gallons) standard of which 104 litres (27.5 US gallons; 22.9 Imp gallons) are usable. 129 litres (34.0 US gallons; 28.3 Imp gallons) optional. Oil capacity 5.7 litres (1.5 US gallons; 1.25 Imp gallons).
SYSTEMS: 12 V 60 A, heavy-duty battery.
ACCOMMODATION: Two persons, side by side, with dual controls; upward-hinged door each side; fixed windscreen; rear side windows. Baggage shelf behind seats.
AVIONICS: *Comms:* Based around Garmin GNS 430 COM/ VOR/GPS, GI 106A CDI, GMA 340 audio panel and GTX 327 transponder. Options include PS 100II intercom.

DIMENSIONS, EXTERNAL:
Wing span	8.79 m (28 ft 10 in)
Wing aspect ratio	5.4
Length overall	7.01 m (23 ft 0 in)
Height overall	2.08 m (6 ft 10 in)
Propeller diameter	1.83 m (6 ft 0 in)

DIMENSIONS, INTERNAL:
Cabin max width	1.17 m (3 ft 10 in)

AREAS:
Wings, gross	12.73 m² (137.0 sq ft)

WEIGHTS AND LOADINGS:
Weight empty	533 kg (1,175 lb)
Baggage weight	18 kg (40 lb)
Max T-O weight	728 kg (1,606 lb)
Max wing loading	57.2 kg/m² (11.72 lb/sq ft)
Max power loading	8.43 kg/kW (13.84 lb/hp)

PERFORMANCE:
Never-exceed speed (V$_{NE}$)	139 kt (257 km/h; 160 mph)
Econ cruising speed	100 kt (185 km/h; 115 mph)
Stalling speed	44 kt (81 km/h; 50 mph)
Max rate of climb at S/L	250 m (820 ft)/min
Service ceiling	3,660 m (12,000 ft)
T-O to 15 m (50 ft)	463 m (1,520 ft)
Landing run	183 m (600 ft)
Landing from 15 m (50 ft)	512 m (1,680 ft)
Range with max payload: standard fuel	
	434 n miles (804 km; 500 miles)
optional fuel	851 n miles (1,577 km; 980 miles)
g limits, Utility category	+4.4/–2.2

UPDATED

AMD ZODIAC CH 640

TYPE: Four-seat kitbuilt.
PROGRAMME: Prototype (N640Z) first flown 1 March 2001. Public debut at Sun 'n' Fun, April 2001.

COSTS: Standard kit US$24,800; quick-build kit introductory price US$35,190, both without engine, propeller, instruments, electrics and upholstery (2001). Can be purchased in smaller kit packages.
DESIGN FEATURES: Fuselage based on CH 2000 with cabin rear wall moved to allow extra seating. Influenced by other Zenith designs: area forward of firewall is based on similarly engined Zenith CH 801. However, wings are of new design, incorporating high-lift aerofoil. Quick-build kit includes almost-complete fuselage, wings, ailerons, flaps and tailplane; quoted build time 500 hours. Standard kit build time quoted as 1,500 hours. Meets FAA 51 per cent rule.
FLYING CONTROLS: Manual. Single-piece all-flying tailplane and rudder. Electrically operated split flaps. Anti-balance tab on both ailerons.
STRUCTURE: Stressed-skin semi-monocoque all-metal fuselage of 6061-T6 aluminium sheet riveted to aluminium extrusions. All-metal stressed-skin wings consist of two panels bolted to fuselage spar-box assembly; Hoerner wingtips.
LANDING GEAR: Tricycle type; fixed. Cantilever main legs comprise single piece of 19 mm (¾ in) aluminium. Steerable nosewheel with heavy-duty bungee shock absorption. Cleveland 5.00×5 wheels standard throughout, with option to fit 6.00×6. Cleveland hydraulic single-disc brakes on mainwheels. Optional wheel fairings. Tailskid protects empennage in hard landings.
POWER PLANT: One 134 kW (180 hp) Lycoming O-360 driving a Sensenich 76-EM8-0-63 two-blade, fixed-pitch propeller. Standard fuel capacity 144 litres (38.0 US gallons; 31.6 Imp gallons) in two wing tanks forward of wing spar; optional capacity 174 litres (46.0 US gallons; 38.3 Imp gallons).
ACCOMMODATION: Two pairs of seats, designed to withstand loads of 27 *g*. Dual controls. Gull-wing cabin door on each side of fuselage. Roll-over protection bar of 4130N tubing between two forward seat backs is bolted to top of wing main spar.
SYSTEMS: 12 V battery and 14 V 70 A alternator standard. Auxiliary electric fuel pump.
AVIONICS: To customer's choice.

DIMENSIONS, EXTERNAL:
Wing span	9.60 m (31 ft 6 in)
Wing chord at root	1.46 m (4 ft 9½ in)
Wing aspect ratio	6.6
Length overall	7.01 m (23 ft 0 in)
Height overall	2.24 m (7 ft 4 in)
Tailplane span	3.00 m (9 ft 10 in)

DIMENSIONS, INTERNAL:
Cabin: Length	1.88 m (6 ft 2 in)
Max width	1.17 m (3 ft 10 in)
Height, front seat to roof	0.96 m (3 ft 2 in)

AREAS:
Wings, gross	13.94 m² (150.0 sq ft)
Tailplane	2.79 m² (30.0 sq ft)

WEIGHTS AND LOADINGS:
Weight empty	520 kg (1,147 lb)
Max T-O weight	998 kg (2,200 lb)
Max wing loading	71.6 kg/m² (14.66 lb/sq ft)
Max power loading	7.44 kg/kW (12.22 lb/hp)

PERFORMANCE:
Never-exceed speed (V$_{NE}$)	139 kt (257 km/h; 160 mph)
Max operating speed	136 kt (253 km/h; 157 mph)
Normal cruising speed at 75% power at 2,135 m	
(7,000 ft)	130 kt (241 km/h; 150 mph)
Stalling speed: flaps up	51 kt (94 km/h; 58 mph)
flaps down	41 kt (76 km/h; 47 mph)
Max rate of climb at S/L	290 m (950 ft)/min
Service ceiling	3,900 m (12,800 ft)
T-O run	302 m (990 ft)
Landing run	351 m (1,150 ft)
Range with standard fuel	443 n miles (820 km; 510 miles)
Endurance	3 h 45 min

UPDATED

AMD Zodiac CH 640 prototype *(Jane's/Paul Jackson)*   NEW/0525805

AMERICAN CHAMPION

AMERICAN CHAMPION AIRCRAFT CORPORATION

PO Box 37, 32032 Washington Avenue, Highway D, Rochester, Wisconsin 53167
Tel: (+1 262) 534 63 15
Fax: (+1 262) 534 23 95
e-mail: aca-sales@mia.net
Web: http://www.amerchampionaircraft.com
PRESIDENT AND CEO: Jerry K Mehlhaff
GENERAL MANAGER: Dale Gauger

ACAC offers new-build Citabria (now called Aurora, Adventure and Explorer), Super Decathlon and Scout, all designed by Aeronca. Model 7 (based on Second World War L-3 Grasshopper liaison/observation machine) certified 18 October 1945. Series later marketed by Champion Aircraft Corporation (1954); from 1970 by Bellanca Aircraft Corporation (see 1979-80 *Jane's*); and then, in 1982, Champion Aircraft Company (see 1985-86 *Jane's*). Assets transferred to Jerry Mehlhaff's American Champion Aircraft Corporation, which restarted production in 1990. By early 1999, various manufacturers had built 14,750 of the Model 7 and Model 8 families. American Champion delivered a total of 96 aircraft in 2000, and on 26 April 2001 rolled out its 500th production aircraft, a Super Decathlon. In addition, ACAC offers its new metal spar wing for retrofit on all models, including the original 7AC Champion. A new model, the 7ACA, was displayed at Oshkosh in July 2000; the company is investigating ways of making the aircraft financially viable for production.

UPDATED

Prototype for proposed relaunch of American Champion 7ACA Champ (*Jane's/Paul Jackson*)   NEW/0525806

American Champion 7ECA Citabria Aurora (*Jane's/Susan Bushell*)   0110683

AMERICAN CHAMPION 7ACA CHAMP

TYPE: Two-seat lightplane.
PROGRAMME: Model 7AC Champion was initial production version (7,200 built with Continental A-65 engine) from 1945 onwards; Model 7ACA Champ certified 30 April 1971 with 44.7 kW (60 hp) Franklin engine, but only 71 built by Bellanca.

American Champion proposing relaunch with 59.7 kW (80 hp) Jabiru 2200 flat-four; new-build prototype (N82AC) displayed at Oshkosh in July 2000, shortly after first flight; will remain experimental while company explores demand. Target maximum take-off weight is 533 to 559 kg (1,220 to 1,232 lb) and maximum level speed 100 kt (185 km/h; 115 mph); cost estimated in region of US$60,000 (2000 prices), or US$45,000 for quick-build kit, if offered. At Sun 'n' Fun, in April 2002, the company confirmed that investigations were continuing, with emphasis then on propeller optimisation.

UPDATED

AMERICAN CHAMPION 7ECA CITABRIA AURORA

TYPE: Two-seat lightplane.
PROGRAMME: Original Aeronca Model 7 and successors built in 16 subvariants, with various letter suffixes. Currently the lowest powered of the American Champion range, the 7EC was certified on 30 November 1949 and 7ECA on 5 August 1964; 773 of former and 1,369 of latter had been built up to 1984; was reintroduced into range during 1995 as the cheapest aircraft. Current production aircraft has improved ventilation and heating; redesigned instrument panel and quick-jettison door; and dual controls and brakes.
CUSTOMERS: Total of 31 built by April 2001, including three in 2000.
COSTS: Standard aircraft US$70,900 (2002).
DESIGN FEATURES: High-wing cabin monoplane; wings braced by V-struts and two secondary struts each side; wire-braced fin and tailplane; constant-chord wings; sweptback fin with (current versions) squared tip.

Wing section NACA 4412; dihedral 1°.
FLYING CONTROLS: Conventional and manual. Horn-balanced elevators and rudder. Flight-adjustable trim tab in port

elevator. No flaps. Control surface movements: ailerons +27° 30'/–19°; elevators ±24°; rudder ±25°. Aileron suspended balance tabs (spades) optional.
STRUCTURE: Stainless steel exhaust system. Fuselage and empennage are welded chromoly steel tube with Dacron covering; two-spar wing has aluminium spars and ribs, Dacron covering, GFRP tips and steel tube V struts.
LANDING GEAR: Tailwheel type; non-retractable. Cantilever spring steel main gear; mainwheels 6.00-6 (4 ply); tailwheel 5.00, steerable. Hydraulic, toe-operated disc and parking brakes. Main wheel fairings optional. Floatplane option includes auxiliary fins above and below tailplane.
POWER PLANT: One 88 kW (118 hp) Textron Lycoming O-235-K2C flat-four piston engine driving a Sensenich 72 CKS8-52 fixed-pitch propeller. Fuel in two wing tanks, total capacity 136 litres (36.0 US gallons; 30.0 Imp gallons) of which 132 litres (35.0 US gallons; 29.1 Imp gallons) usable; overwing refuelling point for each tank. Oil capacity 5.7 litres (1.5 US gallons; 1.3 Imp gallons).
ACCOMMODATION: Pilot and passenger in tandem; five-point safety harness; quick jettison door on starboard side; pilot's port window opens. Space for baggage behind seats. Dual controls. Options include split door for photographic work, external baggage door, greenhouse roof and tinted cabin windows, wide rear seat and rear seat heater.
SYSTEMS: Electric starter; 60 A alternator; 12 V gel-cell battery; voltage regulator with protector. Optional lighting system.
AVIONICS: To customer's choice.

Comms: Optional Communication Group includes broadband transmitting antenna, twin headset jacks, cabin speaker and noise cancelling microphone. Intercom system, rear PTT switch and ELT also optional.

Flight: Full vacuum gyro package and electric attitude indicator and directional gyro set optional.

Instrumentation: ASI, sensitive altimeter, magnetic compass, fuel gauge, oil temperature and pressure gauges

and recording tachometer all standard. VSI, electric turn co-ordinator, OAT gauge, EGT gauge, CHT gauge, combination EGT/CHT eight-probe scanner and electric clock optional.
EQUIPMENT: Optional equipment includes Standard Operation Group comprising navigation lights, landing light, cabin light, dual wingtip strobe lights and bullet propeller spinner; deluxe Duraplush fabric/vinyl interior; landing gear leg steps; speed kit including wing strut fairings; CFP-2 corrosion protection; remote mounted oil filter; and oil quick drain.

DIMENSIONS, EXTERNAL:
Wing span	10.21 m (33 ft 6 in)
Wing aspect ratio	6.8
Length overall	6.73 m (22 ft 1 in)
Height overall	2.35 m (7 ft 8½ in)
Propeller diameter	1.83 m (6 ft 0 in)

DIMENSIONS, INTERNAL:
Length	2.69 m (8 ft 10 in)
Max width	0.76 m (2 ft 6 in)
Max height	1.19 m (3 ft 11 in)

AREAS:
Wings, gross	15.33 m² (165.0 sq ft)

WEIGHTS AND LOADINGS:
Weight empty, equipped	508 kg (1,120 lb)
Max baggage weight	45 kg (100 lb)
Max T-O weight	793 kg (1,750 lb)
Max wing loading	51.8 kg/m² (10.61 lb/sq ft)
Max power loading	9.03 kg/kW (14.83 lb/hp)

PERFORMANCE:
Never-exceed speed (VNE)	140 kt (260 km/h; 162 mph)
Max level speed at S/L	104 kt (193 km/h; 120 mph)
Cruising speed at 75% power	100 kt (185 km/h; 115 mph)
Stalling speed	46 kt (84 km/h; 52 mph)
Max rate of climb at S/L	226 m (740 ft)/min
Service ceiling	3,500 m (11,500 ft)
T-O run	138 m (450 ft)
T-O to 15 m (50 ft)	272 m (890 ft)
Landing from 15 m (50 ft)	233 m (765 ft)
Landing run	122 m (400 ft)

Range with max fuel, allowance for start, taxi, S/L T-O, climb and descent, no reserves:
at 75% power	556 n miles (1,030 km; 640 miles)
at 55% power	617 n miles (1,142 km; 710 miles)
g limits	+5/–2

UPDATED

AMERICAN CHAMPION 7GCAA CITABRIA ADVENTURE

TYPE: Two-seat lightplane.
PROGRAMME: American Champion 7GCAA Adventure relaunched 1997. Model 7GCA originally certified 17 November 1959 and 7GCAA 30 July 1965; these built up to 1980.
CUSTOMERS: First series totalled 396 built; 60 built by American Champion up to April 2001, including 23 in 2000.
COSTS: US$81,900 (2002).

American Champion 7GCAA Citabria Adventure (*Jane's/Paul Jackson*)   NEW/0525807

DESIGN FEATURES: As for Aurora.

FLYING CONTROLS: As for Aurora except elevator movements +28/−24°.

STRUCTURE: As for Aurora.

LANDING GEAR: As for Aurora.

POWER PLANT: One 119 kW (160 hp) Textron Lycoming O-320-B2B flat-four engine driving a two-blade fixed-pitch Sensenich 74DM6S8-1-56 propeller. Fuel capacity as for Aurora, oil capacity 7.6 litres (2.0 US gallons; 1.67 Imp gallons).

ACCOMMODATION: As for Aurora.

SYSTEMS: As for Aurora.

AVIONICS: As for Aurora.

EQUIPMENT: As for Aurora.

DIMENSIONS, EXTERNAL: As for Aurora, except:
Propeller diameter	1.88 m (6 ft 2 in)

DIMENSIONS, INTERNAL: As for Aurora

AREAS: As for Aurora

WEIGHTS AND LOADINGS: As for Aurora, except:
Weight empty	544 kg (1,200 lb)
Max power loading	6.66 kg/kW (10.94 lb/hp)

PERFORMANCE:
Never-exceed speed (V_{NE})	140 kt (260 km/h; 162 mph)
Max level speed	121 kt (225 km/h; 140 mph)
Max cruising speed at 75% power	117 kt (217 km/h; 135 mph)
Stalling speed, power off	46 kt (84 km/h; 52 mph)
Max rate of climb at S/L	356 m (1,167 ft)/min
Service ceiling	4,575 m (15,000 ft)
T-O run	108 m (355 ft)
T-O to 15 m (50 ft)	186 m (610 ft)
Landing from 15 m (50 ft)	230 m (755 ft)
Landing run	122 m (400 ft)
Range with max fuel:	
at 75% power, no reserves	482 n miles (893 km; 555 miles)
at 55% power	595 n miles (1,102 km; 685 miles)
g limits	+5/−2

UPDATED

AMERICAN CHAMPION 7GCBC CITABRIA EXPLORER

TYPE: Two-seat lightplane.

PROGRAMME: Formerly Citabria 150S, certified 3 December 1965.

CUSTOMERS: Total of 117 built since 1994. Overall production of 7GCBC totalled 1,314 by April 2001, including 22 in 2000.

COSTS: Standard aircraft US$84,900 (2002).

DESIGN FEATURES: As for Aurora. Specific 7GCBC features include optional wheel fairings. NACA 4412 aerofoil section. Dihedral 2°; incidence 1°.

FLYING CONTROLS: As for Aurora, except elevator movements +28/−24°. Flaps; maximum deflection 35°.

STRUCTURE: Aluminium alloy (front) and steel tube (aft) bracing struts.

American Champion 8GCBC Scout *(Jane's/Paul Jackson)* NEW/0525810

LANDING GEAR: As for Aurora.

POWER PLANT: As for Adventure. Fuel as for Aurora. Oil capacity 7.5 litres (2.0 US gallons; 1.7 Imp gallons).

ACCOMMODATION: As for Aurora.

SYSTEMS: As for Aurora.

AVIONICS: To customer's specification.

EQUIPMENT: As for Aurora.

DIMENSIONS, EXTERNAL:
Wing span	10.49 m (34 ft 5 in)
Wing chord, constant	1.52 m (5 ft 0 in)
Wing aspect ratio	6.9
Length overall	6.74 m (22 ft 1 in)
Height overall	2.35 m (7 ft 8½ in)
Wheel track	1.93 m (6 ft 4 in)
Wheelbase	4.90 m (16 ft 1 in)
Propeller diameter	1.85 m (6 ft 1 in)

DIMENSIONS, INTERNAL: As for Aurora

AREAS:
Wings, gross	15.97 m² (171.9 sq ft)
Ailerons (total)	1.53 m² (16.50 sq ft)
Fin	0.65 m² (7.02 sq ft)
Rudder	0.63 m² (6.83 sq ft)
Tailplane	1.14 m² (12.25 sq ft)
Elevators (total, incl tab)	1.35 m² (14.58 sq ft)

WEIGHTS AND LOADINGS:
Weight empty, equipped	567 kg (1,250 lb)
Baggage capacity	45 kg (100 lb)
Max T-O and landing weight	816 kg (1,800 lb)
Max wing loading	51.1 kg/m² (10.47 lb/sq ft)
Max power loading	6.85 kg/kW (11.25 lb/hp)

PERFORMANCE:
Never-exceed speed (V_{NE})	140 kt (261 km/h; 162 mph)
Max level speed at S/L	115 kt (212 km/h; 132 mph)
Cruising speed:	
at 75% power, at optimum height	111 kt (206 km/h; 128 mph)
at 65% power	107 kt (198 km/h; 123 mph)
Stalling speed, flaps up	45 kt (82 km/h; 51 mph)
flaps down	40 kt (74 km/h; 46 mph)
Max rate of climb at S/L	344 m (1,130 ft)/min
Service ceiling	4,725 m (15,500 ft)
T-O run	126 m (412 ft)
T-O to 15 m (50 ft)	200 m (656 ft)
Landing from 15 m (50 ft)	226 m (740 ft)
Landing run	110 m (360 ft)
Range with max fuel, allowance for start, taxi, S/L T-O, climb and descent, no reserves:	
at 75% power	431 n miles (799 km; 496 miles)
at 55% power	521 n miles (966 km; 600 miles)
g limits	+5/−2

UPDATED

AMERICAN CHAMPION 8GCBC SCOUT

TYPE: Two-seat lightplane.

PROGRAMME: Model 8 series of Champion developed by Bellanca as heavy-duty version of Model 7; second subvariant was 8GCBC, which received type approval on 30 April 1974.

CURRENT VERSIONS: **Scout:** Fixed-pitch propeller.

Super Scout: Constant-speed propeller.

CUSTOMERS: Total 428 built by April 2001, including 23 in 2000. Recent customers include the Department of Conservation and Land Management of Western Australia, which operates eight Scouts on fire spotting duties, replacing Piper Super Cubs.

COSTS: Scout: US$101,900; Scout CS: US$103,900 (both 2002).

DESIGN FEATURES: Upgraded version of Bellanca/Champion Scout, with new lighter and stronger metal spar wing with 300 per cent less deflection than previous wooden wings; circuit breakers; modern avionics; revised interior; and high-gloss weather-resistant exterior finish. Can operate from short fields while towing glider or banner; low speed assists pipeline, border patrol, forestry and wildlife management roles.

Wing section NACA 4412; dihedral 1°; incidence 1°.

FLYING CONTROLS: As for Explorer. Control surface movements: ailerons +27° 30'/−19°; elevators +29/−26°; rudder ±25°. Four-position trailing-edge flaps droop 7, 16, 21 and 27°.

STRUCTURE: As for Explorer, but metal belly skin optional.

LANDING GEAR: As for Aurora; main tyres 8.50-6 (4/6 ply); tail 5.00-5 (4 ply). Optional EDO 89-2000 floats or skis.

POWER PLANT: One 134 kW (180 hp) Textron Lycoming O-360-C1G flat-four engine, driving either a McCauley 1A200HFA80 fixed-pitch (in Scout) or Hartzell HC-C2YR-1BF/F7666A constant-speed (Super Scout) propeller; three-blade 1.83 m (6 ft 0 in) Hartzell HC-C3YR-1RF/F7282 constant-speed propeller optional for Super Scout. Standard fuel capacity as for Aurora; optional additional tank increases total to 265 litres (70.0 US gallons; 58.3 Imp gallons). Oil capacity 7.5 litres (2.0 US gallons; 1.7 Imp gallons).

ACCOMMODATION: As for Aurora. Heated and ventilated.

SYSTEMS: As for Aurora.

AVIONICS: As for Aurora.

EQUIPMENT: Options include cropduster package, long-range fuel tank and glider tow assembly.

DIMENSIONS, EXTERNAL (A: fixed-pitch propeller, B: constant-speed propeller):
Wing span	11.02 m (36 ft 2 in)
Wing chord, constant	1.52 m (5 ft 0 in)
Wing aspect ratio	7.3

General arrangement of the American Champion 7GCBC Citabria Explorer *(Jane's/James Goulding)* NEW/0525808

American Champion 7GCBC Citabria Explorer floatplane *(Jane's/Paul Jackson)* NEW/0525809

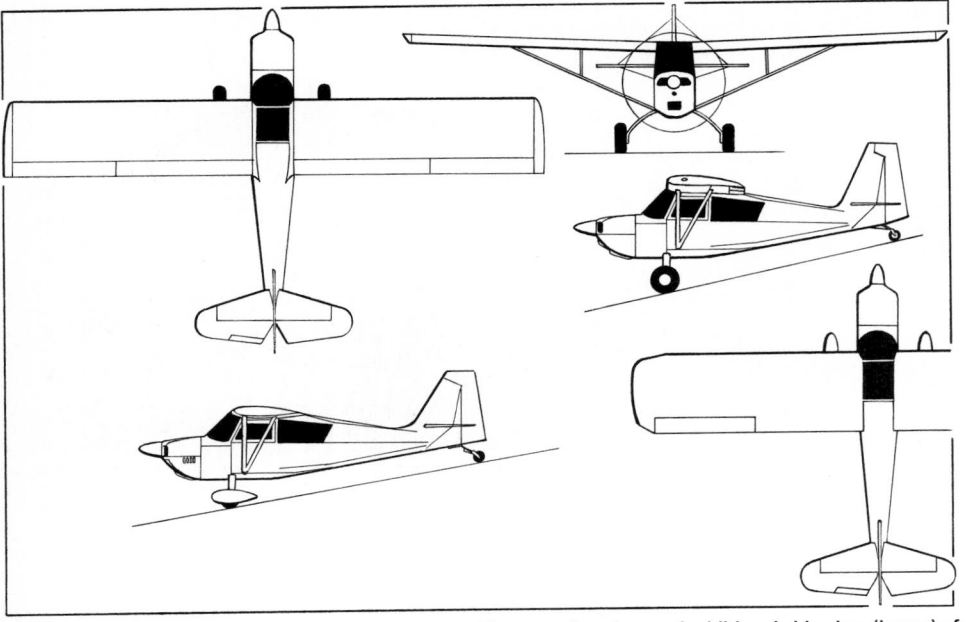

American Champion 8GCBC general arrangement, with scrap plan view and additional side view (lower) of 8KCAB (*Jane's/James Goulding*) NEW/0525811

Length overall: A	6.93 m (22 ft 9 in)
B	7.01 m (23 ft 0 in)
Height overall: A	2.65 m (8 ft 8½ in)
B	2.96 m (9 ft 8½ in)
Tailplane span	3.10 m (10 ft 2¼ in)
Propeller diameter: A	1.93 m (6 ft 4 in)
B	2.03 m (6 ft 8 in)

DIMENSIONS, INTERNAL: As for Aurora
AREAS:

Wings, gross	16.70 m² (180.0 sq ft)

WEIGHTS AND LOADINGS (A and B as above):

Weight empty: A	597 kg (1,315 lb)
B	635 kg (1,400 lb)
Baggage capacity	45 kg (100 lb)
Max T-O weight: A and B, Normal	975 kg (2,150 lb)
A, Restricted	1,179 kg (2,600 lb)
Max wing loading:	
A and B, Normal	58.3 kg/m² (11.94 lb/sq ft)
A, Restricted	70.5 kg/m² (14.44 lb/sq ft)
Max power loading:	
A and B, Normal	7.28 kg/kW (11.94 lb/hp)
A, Restricted	8.80 kg/kW (14.44 lb/hp)

PERFORMANCE (A and B as above):

Never-exceed speed (VNE)	140 kt (260 km/h; 162 mph)
Max level speed at S/L:	
A	117 kt (217 km/h; 135 mph)
B	121 kt (225 km/h; 140 mph)
Cruising speed at 75% power:	
A	106 kt (196 km/h; 122 mph)
B	113 kt (209 km/h; 130 mph)
Stalling speed, flaps up	47 kt (87 km/h; 54 mph)
flaps down: A	44 kt (81 km/h; 50 mph)
B	43 kt (79 km/h; 49 mph)
Max rate of climb at S/L: A	329 m (1,080 ft)/min
B	328 m (1,075 ft)/min
Service ceiling: A	4,420 m (14,500 ft)
B	5,180 m (17,000 ft)
T-O run: A	155 m (510 ft)
B	149 m (490 ft)
T-O to 15 m (50 ft)	312 m (1,025 ft)
Landing from 15 m (50 ft): A, B	376 m (1,235 ft)
Landing run: A, B	128 m (420 ft)
Range with standard tankage, no reserves:	
A, at 75% power	333 n miles (618 km; 384 miles)
A, at 75% power, optional fuel	
	680 n miles (1,260 km; 783 miles)

A, at 55% power, optional fuel	
	779 n miles (1,444 km; 897 miles)
B, at 75% power	360 n miles (668 km; 415 miles)
B, at 75% power, optional fuel	
	732 n miles (1,355 km; 842 miles)
B, at 55% power, optional fuel	
	838 n miles (1,551 km; 964 miles)

UPDATED

AMERICAN CHAMPION 8KCAB SUPER DECATHLON

TYPE: Two-seat lightplane.
PROGRAMME: Original Decathlon was powered by a 112 kW (150 hp) Lycoming IO-320 or AEIO-320: certified 16 October 1970 in Normal and Acrobatic categories. Super introduced by American Champion; first flight (N38AC) July 1990. The Fixed Pitch Decathlon (Sensenich propeller) is no longer marketed.
CUSTOMERS: Total of 240 built 1990 to April 2001 (883 overall).
COSTS: Standard price US$110,900 (2002).
DESIGN FEATURES: Generally as for Aurora. Cleared for limited inverted flight; constant-speed propeller. Wing

section NACA 1412 (modified); dihedral 1°; incidence 1° 30′.
FLYING CONTROLS: As for Aurora, but with balanced elevators. Control surface movements: ailerons ±30°; elevators ±20°; rudder ±30°.
STRUCTURE: As for Aurora. Optional metal wing spar.
LANDING GEAR: Tailwheel type; non-retractable. Cantilever spring steel main legs; mainwheel tyres 6.00-6 or 8.00-6 (4 ply), tail tyre 5.00-5 (4 ply). Cleveland disc brakes; optional wheel fairings.
POWER PLANT: One 134 kW (180 hp) Textron Lycoming AEIO-360-H1B flat-four engine, driving a Hartzell HC-C2YR-4CF/FC7666A-2 two-blade constant-speed propeller. Fuel capacity 151 litres (40.0 US gallons; 33.3 Imp gallons), of which 148 litres (39.0 US gallons; 32.5 Imp gallons) are usable. Oil capacity 9.5 litres (2.5 US gallons; 2.1 Imp gallons).
ACCOMMODATION: As for Aurora, but competition Hooker harness optional.
SYSTEMS: As for Aurora, plus accelerometer and electric fuel boost pump.
DIMENSIONS, EXTERNAL:

Wing span	9.75 m (32 ft 0 in)
Wing chord, constant	1.63 m (5 ft 4 in)
Wing aspect ratio	6.1
Length overall	6.98 m (22 ft 10¾ in)
Height overall	2.36 m (7 ft 9 in)
Tailplane span	3.10 m (10 ft 2¼ in)
Propeller diameter	1.88 m (6 ft 2 in)

DIMENSIONS, INTERNAL: As for Aurora
AREAS:

Wings, gross	15.71 m² (169.1 sq ft)

WEIGHTS AND LOADINGS:

Weight empty	608 kg (1,340 lb)
Baggage capacity	45 kg (100 lb)
Max T-O weight	816 kg (1,800 lb)
Max wing loading	52.0 kg/m² (10.64 lb/sq ft)
Max power loading	6.09 kg/kW (10.00 lb/hp)

PERFORMANCE:

Never-exceed speed	173 kt (321 km/h; 200 mph)
Max level speed at 2,135 m (7,000 ft)	
	135 kt (249 km/h; 155 mph)
Cruising speed:	
at 75% power	128 kt (237 km/h; 147 mph)
at 55% power	111 kt (206 km/h; 128 mph)
Stalling speed	46 kt (86 km/h; 53 mph)
Max rate of climb at S/L	390 m (1,280 ft)/min
Service ceiling	4,815 m (15,800 ft)
T-O run	151 m (495 ft)
T-O to 15 m (50 ft)	276 m (904 ft)
Landing from 15 m (50 ft)	321 m (1,051 ft)
Landing run	130 m (425 ft)
Range, no reserves:	
at 75% power	509 n miles (944 km; 587 miles)
at 55% power	542 n miles (1,006 km; 625 miles)
g limits	+6/−5

UPDATED

American Champion 8KCAB Super Decathlon (*Jane's/Paul Jackson*) NEW/0525812

AMERICAN HOMEBUILTS'

AMERICAN HOMEBUILTS' INC

10419 Vander Karr Road, Hebron, Illinois 60034
Tel: (+1 815) 648 46 17
Fax: (+1 815) 648 16 07
e-mail: info@AmericanHomebuilts.com
Web: http://www.AmericanHomebuilts.com
PRESIDENT: Steve Nusbaum

UPDATED

AMERICAN HOMEBUILTS' VAQUERO JOHN DOE

TYPE: Tandem-seat kitbuilt.
PROGRAMME: Design work started 1992 and construction of proof-of-concept aircraft (with Continental A-80-8 engine) began next year, latter first flown (N10CY) 1994; prototype (N20CY; designated John Doe Spirit and with Continental O-200 engine) first flown early 1997; first kit delivered October 2000; first production aircraft flown mid-2001.
CUSTOMERS: Three flying by mid-2001.
COSTS: Kit US$19,500 without coverings, instruments or power plant (2003); estimated cost to first flight US$35,000.
DESIGN FEATURES: Designed for operation from 'bush' airstrips; many repairs can be carried out 'off base'. Ribs replaceable without removing wing. Quoted build time 400 hours.
FLYING CONTROLS: Conventional and manual. Ailerons and flaps are interchangeable, as are elevators and rudders. Patented flap/aileron mechanism allows ailerons to droop progressively (8, 15 and 22°) as three-position flaps (15, 25 and 40°) are lowered. Slats and stall fences also fitted to reduce stall speeds further. Elevator incidence variable.
STRUCTURE: Fabric-covered welded 4130 steel tube fuselage and two-spar 6061-T6 aluminium tube wing; nose area metal panelled. High wing, with endplates, supported by two streamline steel V-struts.
LANDING GEAR: Tailwheel type; fixed. Bungee cord suspension. Cleveland wheels and disc brakes. Mainwheel tyre size 8.50-6. Maule tailwheel. Attachment points for floats fitted as standard.
POWER PLANT: One 93 kW (125 hp) Teledyne Continental IO-240B flat-four, driving a fixed-pitch IVO V-231 three-blade propeller; alternatively an 89.5 kW (120 hp) LOM 132A or 74.6 kW (100 hp) Teledyne Continental O-200 can be fitted; design will accept other engines in the 48.5 to 119 kW (65 to 160 hp) range. Fuel capacity 98 litres (26.0 US gallons; 21.7 Imp gallons) in two wing tanks.

ACCOMMODATION: Pilot and passenger in tandem; can be flown solo from rear seat. Upward-hinged Lexan doors fitted on both sides of fuselage.

DIMENSIONS, EXTERNAL:

Wing span	9.32 m (30 ft 7 in)
Wing chord, constant	1.30 m (4 ft 3 in)
Wing aspect ratio	7.2
Length overall	7.32 m (24 ft 0 in)
Height overall	2.06 m (6 ft 9 in)
Wheel track	1.68 m (5 ft 6 in)
Wheelbase	5.11 m (16 ft 9 in)
Propeller diameter	1.78 m (5 ft 10 in)
Propeller ground clearance	0.56 m (1 ft 10 in)

DIMENSIONS, INTERNAL:

Cabin: Length	2.18 m (7 ft 2 in)
Max width	0.69 m (2 ft 3 in)
Max height	1.19 m (3 ft 11 in)
Cargo area volume	0.51 m³ (18.0 cu ft)

AREAS:

Wings, gross	12.13 m² (130.55 sq ft)

WEIGHTS AND LOADINGS:

Weight empty	399 kg (880 lb)
Max T-O weight	680 kg (1,500 lb)
Max wing loading	56.1 kg/m² (11.49 lb/sq ft)
Max power loading	7.30 kg/kW (12.00 lb/hp)

PERFORMANCE:

Never-exceed speed (V_NE)	139 kt (257 km/h; 160 mph)
Max operating speed	122 kt (225 km/h; 140 mph)
Max cruising speed	113 kt (209 km/h; 130 mph)
Normal cruising speed	104 kt (193 km/h; 120 mph)

Stalling speed, flaps down	27 kt (49 km/h; 30 mph)
Max rate of climb at S/L	290 m (950 ft)/min
Service ceiling	3,960 m (13,000 ft)
T-O run	30 m (100 ft)
T-O to 15 m (50 ft)	137 m (450 ft)

Landing from 15 m (50 ft)	61 m (200 ft)
Landing run	38 m (125 ft)
Range with max fuel	391 n miles (724 km; 450 miles)
g limits	+4.4/−3

UPDATED

American Homebuilts' John Doe Spirit under conversion to LOM 132A engine *(Jane's/Paul Jackson)*
NEW/0533390

AQUASTAR

AQUASTAR INC

1335 Saratoga Street, Deland, Florida 32724
Tel: (+1 386) 736 43 21
Fax: (+1 386) 738 05 10
e-mail: info@seafireta16.com
Web: http://www.seafireta16.com
PRESIDENT: Doug Karlsen

At Sun 'n' Fun in April 2002, Aquastar relaunched promotion of the former Thurston Seafire with the intention of completing its FAA certification.

NEW ENTRY

AQUASTAR TA16 SEAFIRE

TYPE: Four-seat amphibian.
PROGRAMME: Designed by David B Thurston, noted for flying boat/amphibian aircraft including Colonial Skimmer (Lake Buccaneer) and TSC-1 Teal. Prototype (N16TA) first flew 10 December 1981; development passed from International Aeromarine to Thurston Aeromarine, but FAR Pt 23 certification was only 85 per cent completed; last appeared in 1996-97 *Jane's*. Prototype re-registered on 28 July 2000 to Aquastar Inc as Seafire Trojan TA16; refurbished and refitted with current avionics and test instrumentation by Seagull Aviation Parts of Wisconsin; certification trials in 2002.
CURRENT VERSIONS: Initial certification planned with SMA 230 turbo-diesel and Lycoming O-540; Continental O-520 to be later option.
CUSTOMERS: Thurston originally sold 70 sets of plans; by 2002, seven completed by private individuals as Experimental category homebuilts; two gained first and second places in Custom Plans-Built category at AirVenture 1998, Oshkosh. Marketed by Aquastar in complete form or as kit.
COSTS: US$300,000 complete; kit US$65,000, or US$130,000 including builder's assistance programme (within '51 per cent rule') at Deland lasting two weeks (2002).
DESIGN FEATURES: Originating requirement was 434 n mile (804 km; 500 mile) outward journey, collection of 227 kg (500 lb) load and return without refuelling or other support. Typical Thurston design featuring cabane-mounted single engine with (in this instance tractor) propeller; single-step flying-boat hull; and retractable landing gear for land operation. Mid-wing and high-mounted tailplane; upturned wingtips. Diesel option introduced to give 35 per cent increase in range. Steep 'V' hull and strakes improve handling on water; shear pins on floats protect wing structure from damage if floats strike waterborne debris.

Wing section NACA 64₂A215 laminar flow; dihedral 3°, incidence 4°.
FLYING CONTROLS: Thurston ailerons (US Patent 3,598,340) each with ground-adjustable trim tab; single-slotted trailing-edge flaps; two-piece (laterally divided) elevators with bungee trim system; conventional rudder with flight-adjustable tab. Dual controls.
STRUCTURE: All-metal constant-chord wings; all-metal single-step planing hull with retractable water rudder; cantilever all-metal tail surfaces. Wingtips and outboard floats of composites.
LANDING GEAR: Hydraulically retractable tricycle type. Main units retract inward, into wings; steerable, self-centring nosewheel retracts forward to close opening in hull, which needs no closure doors. Oleo-pneumatic shock-absorbers. Parker-Hannifin aluminium alloy wheels, all three with tyre size 6.00-6. Tyre pressure, mainwheels 2.76 bar (40 lb/sq in), nosewheel 2.07 bar (30 lb/sq in). Parker-Hannifin dual-pad disc hydraulic brakes. Toe brakes. Parking brake. Wheel landing gear designed to meet Canadian DoT snow-ski load conditions.
POWER PLANT: One 186 kW (250 hp) Textron Lycoming O-540-A4D5 flat-six engine driving MT-Propeller (recommended) or Hartzell two-blade, constant-speed (hydraulic), fully reversing, metal propeller. Alternatively 169 kW (227 hp) SMA 230 turbo-diesel or Teledyne Continental IO-520 of 224 kW (300 hp). Fuel tank in leading-edge of each wing, with combined capacity of 340 litres (90.0 US gallons; 75.0 Imp gallons). Refuelling point on upper surface of each wing. Oil capacity 11.5 litres (3.0 US gallons; 2.5 Imp gallons). Engine air intake incorporates filter and automatic inlet door which opens if main duct becomes blocked by ice or debris.

ACCOMMODATION: Pilot and three passengers in pairs in enclosed cabin, with two-section canopy. Forward section slides aft over rear canopy, and both can then be hinged and raised on either side of hull, or removed, to facilitate loading/unloading of bulky items. Space for baggage or freight at rear of cabin. Accommodation heated and ventilated.
SYSTEMS: Electrical system powered by 24 V 70 A engine-driven alternator; 24 V 37 Ah battery. Electrically driven hydraulic pump provides system pressure of 69 bar (1,000 lb/sq in) for actuation of landing gear.
AVIONICS: To customer's requirements.

DIMENSIONS, EXTERNAL:

Wing span	11.28 m (37 ft 0 in)
Wing chord, constant	1.52 m (5 ft 0 in)
Wing aspect ratio	7.4
Length: overall	8.27 m (27 ft 1½ in)
hull	7.42 m (24 ft 4 in)
Height overall	3.28 m (10 ft 9 in)
Tailplane span	3.05 m (10 ft 0 in)
Wheel track	4.01 m (13 ft 2 in)
Wheelbase	3.28 m (10 ft 9 in)
Propeller diameter	2.03 m (6 ft 8 in)

DIMENSIONS, INTERNAL:

Cabin: Length	2.44 m (8 ft 0 in)
Max width	1.07 m (3 ft 6 in)
Max height	1.12 m (3 ft 8 in)
Floor area	1.86 m² (20.0 sq ft)
Volume	2.3 m² (80 cu ft)

AREAS:

Wings, gross	17.00 m² (183.0 sq ft)
Ailerons	1.11 m² (12.00 sq ft)
Trailing-edge flaps (total)	2.51 m² (27.00 sq ft)
Fin	2.34 m² (25.20 sq ft)
Dorsal fin	0.12 m² (1.300 sq ft)
Rudder	0.67 m² (7.20 sq ft)
Tailplane	1.95 m² (21.00 sq ft)
Elevators (total)	1.45 m² (15.60 sq ft)

WEIGHTS AND LOADINGS:

Weight empty, equipped	885 kg (1,950 lb)
Baggage capacity	363 kg (800 lb)
Max fuel weight	245 kg (540 lb)
Max T-O and landing weight	1,451 kg (3,200 lb)
Max wing loading	85.2 kg/m² (17.49 lb/sq ft)
Max power loading	7.80 kg/kW (12.80 lb/hp)

PERFORMANCE:

Never-exceed speed (V_NE)	160 kt (298 km/h; 185 mph)
Max cruising speed at 75% power at S/L	152 kt (282 km/h; 175 mph)
Econ cruising speed at 67% power at FL80	145 kt (269 km/h; 167 mph)
Stalling speed:	
flaps up, power on	53 kt (99 km/h; 61 mph)
flaps down	48 kt (89 km/h; 56 mph)
Max rate of climb at S/L	323 m (1,060 ft)/min
Service ceiling	5,485 m (18,000 ft)
T-O run: land	198 m (650 ft)
water	259 m (850 ft)
T-O to 15 m (50 ft): land	305 m (1,000 ft)
water	366 m (1,200 ft)
Landing from 15 m (50 ft):	
land or water	366 m (1,200 ft)
Landing run: land	153 m (500 ft)
water	183 m (600 ft)
Range at 60% power	952 n miles (1,763 km; 1,095 miles)

NEW ENTRY

Thurston-built Aquastar TA16 Seafire prototype *(Jane's/Paul Jackson)*
NEW/0137381

ARCHEDYNE

ARCHEDYNE AEROSPACE

PO Box 568, Cape Canaveral, Florida 32920-0568
Tel: (+1 321) 453 24 43
Fax: (+1 321) 454 49 14
Web: http://www.archedyne.com
PRESIDENT AND CEO: G Leonard Gioia
COO: Thomas Marsh
BUSINESS DEVELOPMENT MANAGER: Denis Bonneaux

Archedyne, previously known as Amjet, planned to merge
with Lake Aircraft in 1999 in order to launch production of its
NauticAir 450 amphibian (last described in the 2002-03
edition). No agreement was reached and the merger was
abandoned; Archedyne subsequently initiated legal
proceedings which continued into 2002. In October 2000,
Archedyne disclosed it was in negotiations to purchase
another aerospace competitor, but by January 2002 there had
been no further announcements regarding work on the Nautic
Air 450 restarting.

UPDATED

Artist's impression of the Archedyne NauticAir 450   *NEW*/0533368

ARCHEDYNE NAUTICAIR 450

Design and development of this amphibious business jet has
been delayed because of legal proceedings, although it is
proposed to conduct trials of a radio-controlled model before
construction of a prototype begins in 2005-06.

UPDATED

ASII

AMERICAN SPORTSCOPTER INTERNATIONAL INC

PO Box 14608, Newport News, Virginia 23608
Tel: (+1 757) 872 87 78
Fax: (+1 757) 872 87 71
e-mail: asii@asiicopter.com
Web: http://www.ultrasport.rotor.com
FAR EAST OFFICE:
  Light's American Sportscopter Inc, 656 Jianshing Road,
  Bei Chiu, Taichung 404, Taiwan
Tel: (+886 4) 22 07 77 08
Fax: (+886 4) 22 08 77 88
e-mail: ultrasport@iname.com
Web: http://www.ultrasport.rotor.com

ASII was founded in 1990. North and South American
markets are covered by US office; rest of world by
Taiwanese office. An unmanned surveillance prototype
named Vigilante 496 (first flight February 1998) was
developed by SAIC (see *Jane's Unmanned Aerial Vehicles
and Targets*). In late 2001, ASII was looking at providing
armament, including 2.75 in rockets and Hellfire missiles.

UPDATED

ASII ULTRASPORT 254 and 331

TYPE: Single-seat ultralight helicopter kitbuilt.
PROGRAMME: Prototype (254 version) first flight 24 July 1993
and publicly displayed at Oshkosh that year. Two
prototypes built and tested. Prototype 331 (N331UV) first
flew December 1993. LBA certification was expected by
early 2002: version is only marketed in countries where
ultralight helicopters are regulated to company's
satisfaction. In late 2002, it was announced that Huzhou
Taixing Aviation Technology of Huzhou, China was
producing helicopters from kits, including examples of
Ultrasports.
CURRENT VERSIONS: Designations reflect empty weight in
pounds.
  Ultrasport 254: Single-seat, ultralight to FAR Pt 103.
  Ultrasport 331: Experimental category FAR Pt
  21.191(g); single-seat 'growth' model. Meets FAA 51 per
  cent amateur-built kit rules.
CUSTOMERS: Total of 70 single-seaters sold by end 2002, of
which more than 50 were flying.
COSTS: 254: US$34,900; 331: US$37,900 (2003).
DESIGN FEATURES: Design objective of 254 was basic weight
not to exceed 115 kg (254 lb) in order to comply with FAR
Pt 103. Two-blade composites construction main rotor
with tip weights for momentum conservation in event of
engine failure, 8° linear twist and infinite life; shielded
two-blade tail rotor; tailplane with fins at tip; tail rotor
drive carried in narrow streamlined tailboom. Centrifugal
sprag clutch for starting engages rotors at 2,000 engine rpm
and automatically disengages in the event of engine
failure. Quoted build time 80 hours. Broad (2.44 m; 8 ft
0 in) skid track helps to prevent rollovers.
FLYING CONTROLS: Conventional collective, cyclic and yaw
pedals. Floor-mounted cyclic option available since 1998;
early models had top-mounted stick.
STRUCTURE: Generally of epoxy resin, graphite fabric and
Nomex honeycomb; aluminium tailboom.
LANDING GEAR: Aluminium skids stressed for landings at up to
2.5 *g*; floats (weight 18 kg; 40 lb) optional.
POWER PLANT: *254:* One 41.0 kW (55 hp) Hirth 2703 dual-
carburettor two-stroke engine with pull starter and 12:1
planetary transmission. Normal fuel capacity 19 litres (5.0
US gallons; 4.1 Imp gallons).
  331: One 48.5 kW (65 hp) Hirth 2706 dual-carburettor
two-stroke engine with electric starter (also option for
254); 12:1 planetary transmission. Fuel capacity 38 litres
(10.0 US gallons; 8.3 Imp gallons).

ASII Ultrasport 331 single-seat ultralight helicopter with doors removed *(Jane's/Paul Jackson)*   0110714

ACCOMMODATION: Single seat partially enclosed (254
version); enclosed (331 version).
SYSTEMS: Electrical: 12 V battery; 14 V alternator fit
appropriate to engine.
AVIONICS: Customer choice.

DIMENSIONS, EXTERNAL:
Main rotor diameter	6.40 m (21 ft 0 in)
Main rotor blade chord	0.17 m (6¾ in)
Tail rotor diameter	0.76 m (2 ft 6 in)
Tail rotor blade chord	0.05 m (2 in)
Fuselage: Length, main rotor folded	5.84 m (19 ft 2 in)
Skid track	2.44 m (8 ft 0 in)
Height overall	2.39 m (7 ft 10 in)

DIMENSIONS, INTERNAL:
Cabin: Length	1.32 m (4 ft 4 in)
Max width	0.76 m (2 ft 6 in)
Max height	1.50 m (4 ft 11 in)

AREAS:
Main rotor disc	32.2 m² (346.4 sq ft)

WEIGHTS AND LOADINGS:
Weight empty: 254	114 kg (252 lb)
331	150 kg (330 lb)
Max T-O and landing weight: 254	238 kg (525 lb)
331	294 kg (650 lb)

PERFORMANCE:
Max level speed: 254	55 kt (101 km/h; 63 mph)
331	90 kt (167 km/h; 104 mph)
Cruising speed: 254, 331	55 kt (101 km/h; 63 mph)
Max rate of climb at S/L	305 m (1,000 ft)/min
Service ceiling	3,660 m (12,000 ft)
Hovering ceiling: IGE	3,290 m (10,800 ft)
OGE	2,135 m (7,000 ft)
Range, normal fuel, 55 kt:	
254	65 n miles (120 km; 75 miles)
331	130 n miles (241 km; 150 miles)

UPDATED

ASII ULTRASPORT 496

TYPE: Two-seat ultralight helicopter kitbuilt.
Details as for 254 and 331, except as follows:
PROGRAMME: Developed from the 331, the Ultrasport 496
(N496AS) first flew in July 1995. Deliveries began in April
1997.
CURRENT VERSIONS: **Ultrasport 496:** *As described.*
  Ultrasport 496RT: Announced 2002. Uses 84.6 kW
(113.4 hp) Rotax 914 turbocharged engine with electric
start and quadruple carburettors.

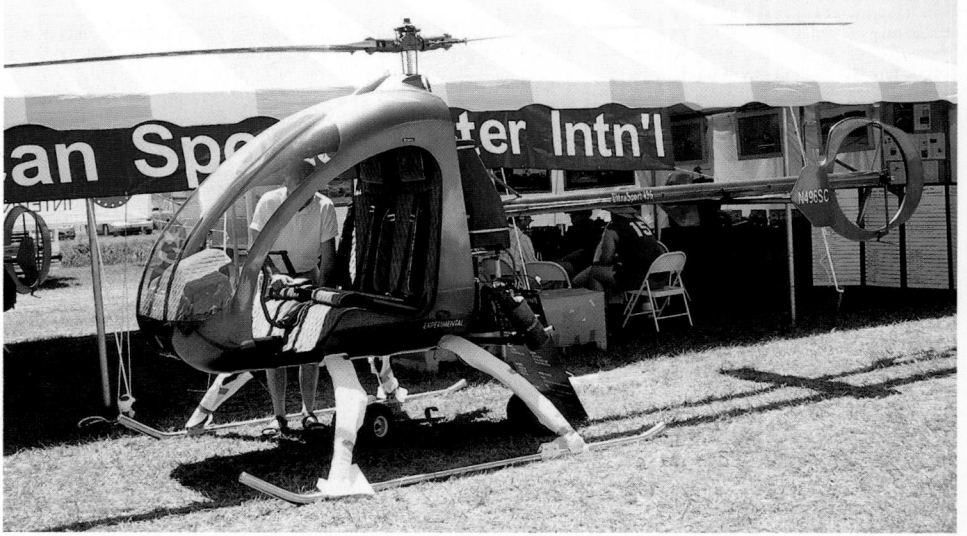

Two-seat American Sportscopter International Ultrasport 496 trainer *(Jane's/Susan Bushell)*   0110682

Sportscopter 600: New Zealand assembly; initial four registered in July 1999.

Vigilante: Remotely piloted version developed by Science Applications International Corporation (SAIC).

Vigilante 496 prototype (N496UV) evaluated as optionally piloted vehicle (OPV) by US Navy during first and second quarters of 1998. Programme for developed versions (Vigilante 500 and 600) appears to have lapsed.

CUSTOMERS: By end 2001, 135 had been delivered of which 35 were then flying.

COSTS: 496 US$62,900; 496RT US$77,900 (2003).

DESIGN FEATURES: Generally as for 254/331. Quick-build kit quoted build time 80 hours; dual controls standard.

STRUCTURE: Infinite-life composites rotor blades and fuselage. Shaft-driven tail rotor. Vertically mounted direct-drive Helical Spur Gears 11:1 two-stage main transmission.

LANDING GEAR: Aluminium skids; track as for 254/331. Floats optional.

POWER PLANT: One 85.8 kW (115 hp) Hirth F30 quad-carburettor engine with electric starter; 75.0 kW (101 hp) Hirth H30E version demonstrated at Sun 'n' Fun 2002. Fuel capacity 61 litres (16.0 US gallons; 13.3 Imp gallons).

EQUIPMENT: Ballistic parachute under consideration.

DIMENSIONS, EXTERNAL:
Main rotor diameter	7.01 m (23 ft 0 in)
Main rotor blade chord	0.17 m (6¾ in)
Length, blades folded	6.02 m (19 ft 9 in)
Tail rotor blade chord	0.05 m (2 in)
Height	2.39 m (7 ft 10 in)

DIMENSIONS, INTERNAL:
Cabin: Length	1.35 m (4 ft 5 in)
Max width	1.22 m (4 ft 0 in)
Max height	1.50 m (4 ft 11 in)

AREAS:
Main rotor disc	38.60 m² (415.5 sq ft)

WEIGHTS AND LOADINGS:
Weight empty: 496	245 kg (540 lb)
496RT	299 kg (660 lb)
Max T-O and landing weight:496	512 kg (1,130 lb)
496RT	499 kg (1,100 lb)

PERFORMANCE (at max T-O weight, ISA):
Never-exceed speed (VNE)	90 kt (167 km/h; 104 mph)
Max cruising speed	60 kt (111 km/h; 69 mph)
Max rate of climb at S/L	305 m (1,000 ft)/min
Hovering ceiling: IGE	3,290 m (10,800 ft)
OGE	2,135 m (7,000 ft)
Range	130 n miles (240 km; 149 miles)
Endurance	2 h 30 min

UPDATED

ATG

AVIATION TECHNOLOGY GROUP, INC

8001 South Interport Boulevard, Suite 310, Englewood, Colorado 80112-5951
Tel: (+1 303) 799 41 97
Fax: (+1 303) 799 48 13
e-mail: info@avtechgroup.com
Web: http://www.avtechgroup.com
PRESIDENT: George E Bye
VICE-PRESIDENT, ENGINEERING: Dr Robert S Wolf

Aviation Technology Group was formed in 1998 to develop the ATG Javelin two-seat jet; it was incorporated in June 2000. In June 2002 ATG signed an MoU with Luscombe Aircraft Corporation to consider the latter producing Javelins in its plant at Altus, Oklahoma.

UPDATED

ATG-1B Javelin revised design at Orlando in September 2002 (*Jane's/Paul Jackson*)   NEW/0533389

ATG-1 JAVELIN

TYPE: Two-seat jet sportplane.

PROGRAMME: Announced early 2001, at which time initial wind tunnel testing was being undertaken; this finished 28 February 2001. Construction of full-scale, non-flying mockup being undertaken during mid-2001. A second series of wind tunnel tests was scheduled for late 2001 and 2002, followed by a maiden flight during the first half of 2004. One non-conforming and two conforming prototypes to fly 1,400-hour test programme. Certification to FAR Pt 23 Amdt 52 (Aerobatic) intended late 2004 with deliveries in early 2005.

All data are provisional.

CURRENT VERSIONS: **ATG-1A:** Initial design. Mid-wing configuration and separate windscreen. Replaced by ATG-1B.

ATG-1B: Mockup unveiled 12 July 2002 and shown at NBAA Convention, Orlando, Florida, in September 2002. Low wing with leading-edge 'dog-tooth' and two-piece canopy.

Javelin HDI (Homeland Defense Interceptor): Single-seat version armed with wingtip-mounted AIM-9 Sidewinder or air-to-air Stinger missiles, was announced in early 2002.

UCAV: Unmanned combat version announced in July 2002.

CUSTOMERS: Production forecasts range from 711 to 1,638 units (2002 estimate). Deposits (US$25,000 each) taken on 26 aircraft by July 2002.

COSTS: US$2.20 million (2003). Development costs estimated at US$120 million up to certification. HDI version provisionally costed at US$4.5 million.

STRUCTURE: Use of aluminium provides optimum strength to weight ratio. Quarter-chord sweep: wings 33° 10'; tailplane 37° 0'; fins 41° 0'.

LANDING GEAR: Retractable tricycle type.

POWER PLANT: Two Williams FJ33-4 turbofan engines, each 6.7 kN (1,500 lb st). Fuel capacity 946 litres (250 US gallons; 208 Imp gallons).

ACCOMMODATION: Pilot and passenger in tandem. Dual controls. Max cabin pressure differential 0.48 bar (7.0 lb/sq in). Air conditioning standard.

SYSTEMS: Combination of de-icing/anti-icing systems is planned: electro-expulsive de-icing for wings, bleed air heat for engine inlets, electric de-icing for windscreen and electric anti-icing for probes.

AVIONICS: Full IFR capability with dual EFIS displays. Options include head-up display and airborne video camera systems.
Comms: VHF/UHF radios; Mode S transponder.
Radar: Colour weather radar.
Flight: E-TCAS, TAWS and GPS integrated FMS with waypoint map.
Instrumentation: Three-axis autopilot.

DIMENSIONS, EXTERNAL:
Wing span	6.10 m (20 ft 0 in)
Wing aspect ratio	3.8
Length overall	10.82 m (35 ft 6 in)
Height overall	3.20 m (10 ft 6 in)
Tailplane span	4.27 m (14 ft 0 in)

AREAS:
Wings, gross	9.75 m² (105.0 sq ft)
Horizontal tail	3.25 m² (35.00 sq ft)
Vertical tail (each)	3.96 m² (42.60 sq ft)

WEIGHTS AND LOADINGS:
Weight empty	1,139 kg (2,510 lb)
Baggage capacity	91 kg (200 lb)
Max T-O weight	2,090 kg (4,600 lb)
Max wing loading	213.9 kg/m² (43.81 lb/sq ft)
Max power loading	156 kg/kN (1.53 lb/lb st)

PERFORMANCE:
Never exceed speed (VNE)	M0.96
Max operating speed	425 kt (787 km/h; 489 mph) IAS
Max cruising speed	528 kt (978 km/h; 608 mph) TAS/M0.92
Stalling speed, power off; flaps down	95 kt (176 km/h; 110 mph) IAS
Max rate of climb at S/L	3,050 m (10,000 ft)/min
Service ceiling	1,050 m (10,000 ft)
Range with max fuel, 20 min reserve	1,250 n miles (2,315 km; 1,438 miles)
Endurance	3 h 22 min
g limits	+6/–3

UPDATED

ATLAS

ATLAS AEROSPACE INC

39520 Aviation Avenue, Zephyrhills Municipal Airport, Zephyrhills, Florida 33540
Tel: (+1 813) 783 33 61
Fax: (+1 813) 783 64 11
DIRECTORS: James D Stewart
David P Teichman

Company formed June 2001 to promote Liftmaster light freighter as replacement for Cessna 208 and stillborn Ayres Loadmaster as carrier of small parcels. In November 2001, received US$600,000 grant from Zephyrhills Council and similar amount from Florida State Transportation Board for infrastructure. However, no finance is known to have been secured towards construction of a prototype.

UPDATED

ATLAS LIFTMASTER

TYPE: Utility turboprop twin.

PROGRAMME: Announced September 2001; certification planned for late 2003; deliveries from July 2004 onwards. In 2002, company was still seeking US$5 million finance to build prototype and US$25 million to achieve certification.

Model of Atlas Liftmaster   NEW/0528445

COSTS: Initial US$3 million required to complete design, produce a prototype and initiate flight testing. Estimated US$2.1 million (2001) unit cost. Direct operating cost US$359/hour, US$1.55/n mile (fuel US$1.90/USg).

DESIGN FEATURES: Twin boom, high-wing transport with upward-hinged rear loading door; cargo capacity, three LD3 containers.

LANDING GEAR: Retractable tricycle type.

POWER PLANT: Two 507 kW (680 hp) Pratt & Whitney Canada PT6A-114A turboprops, each driving a five-blade, constant-speed, fully feathering and reversible propeller. Total fuel capacity 2,705 litres (714 US gallons; 595 Imp gallons).

ACCOMMODATION: Two pilots; seating for between 11 (FAR Pt 23) and 14 passengers.

DIMENSIONS, EXTERNAL:
Wing span	15.24 m (50 ft 0 in)
Length overall	12.27 m (40 ft 3 in)
Height overall	4.57 m (15 ft 0 in)

DIMENSIONS, INTERNAL:
Cabin (excl flight deck): Length	5.18 m (17 ft 0 in)
Max width	2.13 m (7 ft 0 in)
Max height	1.83 m (6 ft 0 in)
Volume	19.9 m³ (714 cu ft)

AREAS:
Wings, gross	29.22 m² (314.5 sq ft)

WEIGHTS AND LOADINGS:
Weight empty, equipped	2,195 kg (4,840 lb)
Max T-O and landing weight	5,420 kg (11,950 lb)
Max ramp weight	5,443 kg (12,000 lb)
Max wing loading	185.5 kg/m² (38.00 lb/sq ft)
Max power loading	5.35 kg/kW (8.79 lb/hp)

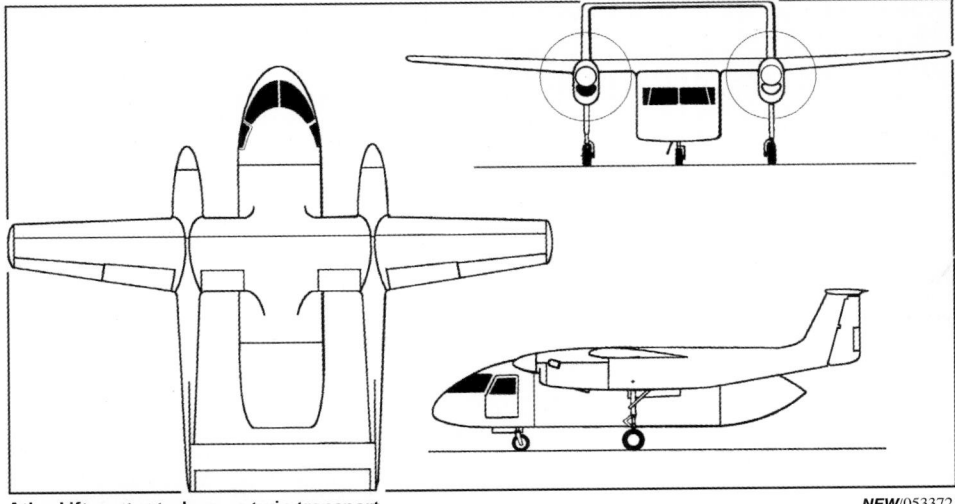

Atlas Liftmaster turboprop twin transport   *NEW*/0533721

PERFORMANCE:
Cruising speed at FL100	215 kt (398 km/h; 247 mph)
Stalling speed, landing configuration	58 kt (108 km/h; 67 mph)
Max rate of climb at S/L	651 m (2,136 ft)/min
T-O run	421 m (1,380 ft)
T-O to 15 m (50 ft)	756 m (2,480 ft)
Landing from 15 m (50 ft)	969 m (3,180 ft)
Landing run	476 m (1,560 ft)
Range	1,500 n miles (2,778 km; 1,726 miles)

UPDATED

AUC

AMERICAN UTILICRAFT CORPORATION

Suite B, 300 Petty Road NE, Lawrenceville, Georgia 30043
Tel: (+1 678) 376 08 98
Fax: (+1 678) 376 90 93
e-mail: info@utilicraft.com
Web: http://www.utilicraft.com
PRESIDENT AND CEO: John DuPont
EXECUTIVE VICE-PRESIDENT AND SENIOR VICE-PRESIDENT, MARKETING: James Carey

Company formed August 1990 to design and produce Freight Feeder cargo transport. Production will be undertaken at Gwinnett County, Atlanta, Georgia.

US$50 million of financing secured during June 2001 but economic slowdown in late 2001 affected progress with construction of prototype Freight Feeder; however on 25 June 2002 AUC signed an MoU with Averitt Express for lease of capacity of first 25 aircraft, and announced a more aggressive merger and acquisition strategy.

UPDATED

Impression of the mid-range Freight Feeder 200   0097154

AUC FF-1080 FREIGHT FEEDER

TYPE: Twin-turboprop freighter.

PROGRAMME: Design started 1990; patents filed 1991; original capacity for four LD3 containers increased to six in 1997. Two flying prototypes planned; first flight due 18 months after programme financing in place, followed by FAR Pt 25 certification. Mock up completed end 1999; pre-production prototype intended for completion early 2002. First flight was due before end 2002, with certification in 2003. FAA Pt 25 certification; this was revised following market downturn at end 2001. On 25 June 2002 company signed an MoU with transport provider Averitt Express for

an exclusive five year lease of first 25 aircraft, all of which are expected to be in service by end 2004.

CURRENT VERSIONS: **FF-1080-100:** Short fuselage version; four LD3 containers; PW121 engines. Not intended for construction at present; weights and loadings in 1999-2000 *Jane's*.

FF-1080-200: Standard version; *as described*.

FF-1080-500: Enlarged version; BR 715 engines.

CUSTOMERS: Company estimates market for 5,000 aircraft and aims to capture 10 per cent. Target market is Cessna 208 replacement/growth and F27 all-cargo market. Discussions in hand with several unidentified potential

customers; first production batch to comprise 48 aircraft. North Atlantic Industries of Netherlands had planned to take 50 aircraft as European distributor, but agreement was cancelled in 2001. Averitt Express (see above) is potential new launch customer.

COSTS: Development and certification to FAR Pt 25 estimated at US$75 million (1998). Standard flyaway version preliminary price US$7.0 million (2002).

DESIGN FEATURES: High-mounted wings, tapered outboard of overwing engine nacelles; box-section fuselage with ventral pannier; upswept rear fuselage with rear-loading door; angular tail surfaces with large dorsal fin fairing. Onboard freight management system.

FLYING CONTROLS: Conventional and manual. Upper surface blowing (USB) via overwing engine mounting.

STRUCTURE: All-aluminium construction. Final assembly by AUC using components provided by management partners under subcontract. Interest in component manufacture shown by companies in USA. Fuselage subassemblies constructed by Metalcraft Technologies Inc of Utah. Wing subassemblies designed by Aerostructures Corporation. Other risk-sharing partners include UPS Aviation Technologies, Goodrich AAR Cargo Systems, Castle Precision Industries, MPC Products, Auxilec Inc, Shaw Aero, HS Dynamic Controls, Securaplane Technologies, Lord Mounts, Hi-Temp Insulation Inc and General Electrodynamics Corp. Detailed engineering undertaken by Aircraft Design Services International.

LANDING GEAR: Non-retractable tricycle type; twin nosewheels and tandem pairs of mainwheels. Engineering undertaken by Castle Precision Industries.

POWER PLANT: *FF-1080-200:* Two 2,051 kW (2,750 shp) Pratt & Whitney Canada PW127F turboprops, each driving a Hamilton Sundstrand 568F six-blade propeller. FADEC fitted as standard. Fuel capacity 6,853 litres (1,810 US gallons; 1,507 Imp gallons), optional fuel 10,601 litres (2,800 US gallons; 2,332 Imp gallons).

FF-1080-100: PW121 engines each rated at 1,567 kW (2,100 shp).

FF-1080-500: Turboprop version of Rolls-Royce Deutschland BR 715 turbofan, (each 3,424 kW; 4,591 shp).

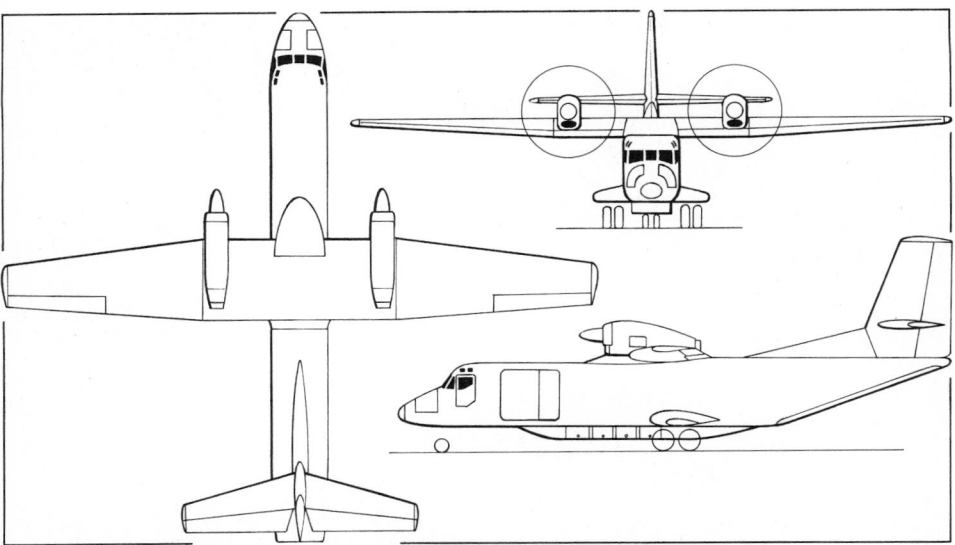

General appearance of the AUC FF-1080-200 Freight Feeder (*Jane's/James Goulding*)   0015552

ACCOMMODATION: Flight crew of two on IPECO seats. Main cargo hold accommodates up to six LD3 containers. Cargo roller floor from AAR Cargo Systems. Large cargo double door on port side, forward of wing; rear-loading door; crew airstair door on port side of flight deck. Flight deck pressurised; main cargo bay unpressurised. Additional capacity in cargo pannier and nose compartment.

SYSTEMS: Securaplane smoke detection system; HS Dynamic Controls anti-icing and de-icing systems; electrical system integration by Thales Avionics Electrical Systems. Shaw Aero Devices fuel management system.

AVIONICS: Flight and engine displays and autopilot by Meggitt Avionics of UK; GPS by UPS Aviation Technologies.

DIMENSIONS, EXTERNAL:

Wing span	27.10 m (88 ft 11 in)
Length overall	23.80 m (78 ft 1 in)
Height overall	9.04 m (29 ft 8 in)
Side cargo door: Height	2.13 m (7 ft 0 in)
Width	2.59 m (8 ft 6 in)
Rear cargo door: Height	2.13 m (7 ft 0 in)
Width	1.68 m (5 ft 6 in)

DIMENSIONS, INTERNAL:

Cargo bay: Length	13.31 m (43 ft 8 in)
Max width	1.93 m (6 ft 4 in)
Max height	2.13 m (7 ft 0 in)
Volume: cargo bay	54.8 m³ (1,936 cu ft)
pannier	5.61 m³ (198 cu ft)
nose compartment	5.47 m³ (193 cu ft)

WEIGHTS AND LOADINGS (estimated):

Weight empty: 200	8,117 kg (17,894 lb)
500	14,515 kg (32,000 lb)
Payload: for 400 n mile (740 km; 460 mile) range:	
200	6,619 kg (14,592 lb)
500	9,979 kg (22,000 lb)
for max range: 200	4,019 kg (8,860 lb)
500	7,348 kg (16,200 lb)
Max T-O weight: 200	17,236 kg (38,000 lb)
Max landing weight: 200	16,783 kg (37,000 lb)
Max zero-fuel weight: 200	12,247 kg (27,000 lb)
Max power loading: 200	4.21 kg/kW (6.91 lb/shp)

PERFORMANCE (estimated):

Cruising speed: 200	250 kt (463 km/h; 287 mph)
500	270 kt (500 km/h; 311 mph)

Stalling speed, power off, 200:

T-O flaps	85 kt (158 km/h; 98 mph)
flaps down	73 kt (136 km/h; 84 mph)
Minimum single-engine control speed (V_{MC}):	
200	82 kt (152 km/h; 95 mph)
Max rate of climb at S/L: 200	671 m (2,200 ft)/min
Rate of climb at S/L, OEI: 200	213 m (700 ft)/min
Service ceiling: 200	7,620 m (25,000 ft)
Service ceiling, OEI: 200	4,572 m (15,000 ft)
T-O run: 200	558 m (1,830 ft)
T-O to 15 m (50 ft): 200	856 m (2,810 ft)
Landing from 15 m (50 ft): 200	634 m (2,080 ft)
Landing run: 200	396 m (1,300 ft)
Range: with max fuel:	
200	1,500 n miles (2,778 km; 1,726 miles)
500	2,000 n miles (3,704 km; 2,301 miles)
with max payload:	
200	500 n miles (926 km; 575 miles)

UPDATED

AVIABELLANCA

AVIABELLANCA AIRCRAFT CORPORATION

PMB 47, 2315 B Forest Drive, Annapolis, Maryland 21401
Tel: (+1 410) 266 55 18
Fax: (+1 410) 266 86 97
e-mail: avbellanca@aol.com
Web: http://www.aviabellanca.com
PRESIDENT: August T Bellanca

The original Bellanca Aircraft Corporation was established in 1927 by Giuseppe Mario Bellanca and produced low-wing cabin monoplanes such as the Cruisair and Cruisemaster both before and after the Second World War. Bellanca left the original firm to set up Bellanca Development Company in 1954, which was renamed Bellanca Aircraft Engineering in 1963 and AviaBellanca in 1983. AviaBellanca's designer is August Bellanca, son of Giuseppe. New, all-composites August Bellanca-designed Skyrocket II introduced 1975 and set five NAA/FAI closed circuit speed records; improved Skyrocket III made public debut at Oshkosh 1997. Some 3,800 Bellanca-designed aircraft are registered in the USA.

VERIFIED

AviaBellanca Skyrocket II prototype (*Jim Koepnick/AviaBellanca*)   0062654

AVIABELLANCA SKYROCKET III

TYPE: Six-seat kitbuilt.
PROGRAMME: Design started 1969; prototype first flight (N14666) 1974; set five FAI international closed circuit speed records.
CURRENT VERSIONS: **Skyrocket II**: Original prototype. Currently used as flight test vehicle (re-registered N771AB).
Skyrocket III: Projected factory-manufactured version; available with or without pressurisation.
Description refers to Skyrocket III.
CUSTOMERS: By early 2000, one prototype was flying and 10 more were on order. Preproduction static test and flight test aircraft were under development.
DESIGN FEATURES: Designed for rapid, uncomplicated assembly. Intended to form the foundation for an entire family of aircraft from single-engined general aviation to twin-engined turboprop business aircraft. Aerodynamically clean, low-wing design with laminar flow wing section.
FLYING CONTROLS: Conventional and manual. Elevator with electric trim on entire stabiliser; actuation by cable and bellcrank; horn-balanced rudder and elevators; no trim tabs. Plain flaps.
STRUCTURE: All-composites. Skyrocket II wings comprise upper and lower covers and forward and rear spars; last-mentioned are spliced at centreline; all of sandwich construction, with glass fibre skins and aluminium honeycomb, plus glass fibre spar cap. Ailerons, flaps and

empennage similar, including left and right fin cover shells. Fuselage in left and right halves, with flat sandwich bulkheads and floor. Skyrocket III will have graphite fibre sandwich face skins with Nomex core.
LANDING GEAR: Tricycle hydraulically retractable type; mainwheels retract inboard; nosewheel aft. Oleo-pneumatic shock-absorption; hydraulic brakes.
POWER PLANT: One 324 kW (435 hp) Teledyne Continental GTSIO-520F flat-six, driving a Hartzell three-blade constant-speed propeller. Later option of 261 kW (350 hp) Teledyne Continental TSIO-550-B flat-six, 280 kW (375 hp) Teledyne Continental GTSIO-520-L 336 kW (450 hp) Textron Lycoming TIO-720 flat-eight and 540 kW (724 shp) Walter M 601E turboprop. Fuel capacity 666 litres (176 US gallons; 147 Imp gallons) in two wing tanks, of which 644 litres (170 US gallons; 142 Imp gallons) are usable. Gravity-fed refuelling points in wingtips.
ACCOMMODATION: Pilot and five passengers in three rows of side-by-side seats; pilot's door starboard side, above wing and passenger door on port side aft of wing. Optional club seating.
SYSTEMS: Air conditioning and pressurisation available on production version and on de luxe kitbuilt examples.
AVIONICS: To customer's specification; optional advanced IFR fit.
DIMENSIONS, EXTERNAL:

Wing span	10.67 m (35 ft 0 in)
Wing aspect ratio	6.7
Length	8.23 m (27 ft 0 in)
Height overall	2.74 m (9 ft 0 in)

DIMENSIONS, INTERNAL:

Cabin: Length	3.94 m (12 ft 11 in)
Max width	1.14 m (3 ft 9 in)
Max height	1.22 m (4 ft 0 in)

AREAS:

Wings, gross	16.96 m² (182.6 sq ft)

WEIGHTS AND LOADINGS:

Weight empty	1,129 kg (2,490 lb)
Baggage capacity	91 kg (200 lb)
Max T-O weight	1,905 kg (4,200 lb)
Max wing loading	112.3 kg/m² (23.00 lb/sq ft)
Max power loading	5.88 kg/kW (9.66 lb/hp)

PERFORMANCE:

Max operating speed (V_{MO})	295 kt (547 km/h; 340 mph)
Max cruising speed at 75% power at 7,620 m (25,000 ft)	284 kt (526 km/h; 327 mph)
Normal cruising speed at 65% power at 7,925 m (26,000 ft)	273 kt (505 km/h; 314 mph)
Stalling speed, flaps and landing gear down	59 kt (110 km/h; 68 mph)
Max rate of climb at S/L	634 m (2,080 ft)/min
Service ceiling	9,140 m (30,000 ft)
T-O run	207 m (680 ft)
T-O to 15 m (50 ft)	344 m (1,130 ft)
Landing from 15 m (50 ft)	546 m (1,790 ft)
Landing run	282 m (925 ft)
Range at 65% power, 45 min reserves:	
with max fuel	1,998 n miles (3,701 km; 2,300 miles)
with max payload	820 n miles (1,519 km; 944 miles)

UPDATED

AVIAT

AVIAT AIRCRAFT INC

672 South Washington, PO Box 1240, Afton, Wyoming 83110
Tel: (+1 307) 885 31 51
Fax: (+1 307) 885 96 74
e-mail: aviat@aviataircraft.com
Web: http://www.aviataircraft.com
CHAIRMAN AND PRESIDENT: Stuart Horn
DIRECTOR OF SALES AND MARKETING: Bob James
NATIONAL SALES MANAGER: Mark James

The Pitts Aerobatics company was acquired by Christen Industries in 1983 along with manufacturing and marketing rights for the Pitts Special aerobatic aircraft. In turn, Aviat

Aircraft Inc acquired Christen Industries in 1991 and production and type certificates for the Christen range. Aviat itself was bought by Stuart Horn in December 1995 and type certificates passed to Sky International Inc on 10 January 1996. The company has acquired manufacturing rights to the 1930s/1940s Globe Swift and Monocoupe 110. Aviat delivered a total of 83 aircraft in 1998, 85 in 1999, 91 in 2000 and 57 in 2001. Factory floor area 6,503 m² (70,000 sq ft). Workforce stood at 58 in 2002.

UPDATED

AVIAT HUSKY A-1

TYPE: Two-seat lightplane.
PROGRAMME: First flight (N6070H) 1986; FAA certification of original A-1 version under FAR Pt 23 achieved 1 May 1987, this having 816 kg (1,800 lb) MTOW.

CURRENT VERSIONS: **A-1** Original version, now superseded; 400 built.
A-1A Standard version from 395th aircraft (c/n 1395). Certified 28 January 1998, with MTOW increased by 41 kg (90 lb).
Description refers to A-1A unless otherwise stated.
A-1B Mission-specific version for government agencies. Has higher maximum take-off weight of 907 kg (2,000 lb). Certified 28 January 1998. Some 200 built by December 2001.
CUSTOMERS: Total 700 A-1s and A-1As delivered by September 2002, including two to US Department of Interior and four to US Department of Agriculture; also operated by US police agencies and in Kenya for wildlife protection patrols. Production for 2000 totalled 80 (four A-1A and 76 A-1B), and for 2001, 50 (all A-1Bs).

COSTS: A-1A US$127,500; A-1B US$135,328 (2002).

DESIGN FEATURES: Conventional high-wing monoplane; constant-chord wing with V-strut bracing and auxiliary struts; wire-braced empennage with elliptical surfaces.

Wing has modified Clark Y US 35B section; optional drooped Plane Booster wingtips and vortex generator kit.

FLYING CONTROLS: Conventional and manual. Symmetrical section ailerons with spade-type mass balance; trim tabs in elevators; slotted high-lift flaps. Fixed tailplane; trim by adjustable bungee. Control surface movements: ailerons ±20°; elevators +29/−15°; rudder ±25°; flaps 30°.

STRUCTURE: Tubular welded 4130 steel fuselage. Wing has two aluminium spars, metal ribs and metal leading-edge, Dacron covering overall. A-1B has 7075-T76 aluminium rear spar, replacing 6061-T6. Twin bracing struts each side of wings and wire- and strut-braced tail unit. Light alloy slotted flaps and ailerons, with Dacron covering. Fuselage and tail have chrome molybdenum steel tube frames, covered in Dacron except for metal skin to rear fuselage. Seven-coat corrosion protection.

LANDING GEAR: Non-retractable tailwheel type. Two faired side Vs and half-axles hinged to bottom of fuselage, with internal (under front seat) bungee cord shock-absorption. Cleveland mainwheels, tyres size 8.00-6 as standard; 6.00-6 or 8.50-6 tyres and 24×10-6 or 26×10.5-6 Tundra tyres optional. Cleveland mainwheel hydraulic disc brakes, toe-operated. Steerable leaf-spring tailwheel. Optional EDO 89-2000, Baumann BF-2100 and Wipline 2100 floats, Aero M1500, M1800, M2000 or M3000H wheel replacement skis, Aero wheel skis and Flairdyne hydraulic wheel-retracting skis.

POWER PLANT: One 134 kW (180 hp) Textron Lycoming O-360-A1P flat-four engine, driving a Hartzell HC-C2YK-1BF two-blade constant-speed metal propeller. Fuel in two metal tanks, one in each wing, total capacity 208 litres (52.0 US gallons; 45.75 Imp gallons), of which 189 litres (50.0 US gallons; 41.6 Imp gallons) are usable. Fuel filler point in upper surface of each wing, near root.

ACCOMMODATION: Enclosed cabin seating two in tandem, on ergonomic lumbar and side support seats, with dual controls. Custom leather seats optional. Five-point safety harness. Downward-hinged door on starboard side, with upward-hinged window above. Optional skylight window in roof. A-1B has luggage door on starboard side as standard; optional on A-1A. Rear compartment for extra 13.6 kg (30 lb) baggage standard from 2000; certified retrofitting kits available.

SYSTEMS: 12 V electrical system includes lights, 22 Ah battery, and 60 A alternator.

AVIONICS: VFR standard; Loran C transponder, GPS, nav/com, intercom, and IFR instrumentation, including Garmin GNS530 GPS/com and VM 1000 panel, optional.

Instrumentation: Sensitive altimeter, ASI, magnetic compass, digital tachometer, oil temperature/pressure gauge, manifold pressure gauge and CHT gauge all standard.

EQUIPMENT: Rear seat heater and windscreen defroster standard on A-1B.

Aviat Husky A-1B *(Jane's/Paul Jackson)*   NEW/0525813

DIMENSIONS, EXTERNAL:	
Wing span	10.82 m (35 ft 6 in)
Wing aspect ratio	6.9
Length overall	6.88 m (22 ft 7 in)
Height overall	2.01 m (6 ft 7 in)
Propeller diameter	1.93 m (6 ft 4 in)
DIMENSIONS, INTERNAL:	
Baggage hold volume	0.28 m³ (10 cu ft)
AREAS:	
Wings, gross	17.00 m² (183.0 sq ft)
Ailerons (total)	1.43 m² (15.40 sq ft)
Trailing-edge flaps (total)	2.09 m² (22.50 sq ft)
Fin	0.43 m² (4.66 sq ft)
Rudder	0.62 m² (6.76 sq ft)
Tailplane	1.48 m² (15.90 sq ft)
Elevators, incl tabs	1.31 m² (14.10 sq ft)
WEIGHTS AND LOADINGS (A-1B where different):	
Weight empty	540 kg (1,190 lb)
Baggage capacity	23 kg (50 lb)
Max T-O weight:	
landplane: A-1A	857 kg (1,890 lb)
A-1B	907 kg (2,000 lb)
floatplane: A-1A	943 kg (2,079 lb)
A-1B	998 kg (2,200 lb)
Max wing loading:	
landplane: A-1A	50.4 kg/m² (10.33 lb/sq ft)
A-1B	53.4 kg/m² (10.93 lb/sq ft)
floatplane: A-1A	55.5 kg/m² (11.36 lb/sq ft)
A-1B	58.7 kg/m² (12.02 lb/sq ft)
Max power loading:	
landplane: A-1A	6.39 kg/kW (10.50 lb/hp)
A-1B	6.76 kg/kW (11.11 lb/hp)
floatplane: A-1A	7.03 kg/kW (11.55 lb/hp)
A-1B	7.44 kg/kW (12.22 lb/hp)
PERFORMANCE (landplane; A-1B where different):	
Never-exceed speed (VNE)	132 kt (245 km/h; 152 mph)
Max level speed	126 kt (233 km/h; 145 mph)

Cruising speed: at 75% power at 1,220 m (4,000 ft)	
	122 kt (225 km/h; 140 mph)
at 55% power	113 kt (209 km/h; 130 mph)
Stalling speed, A-1B, flaps down:	
power on	37 kt (69 km/h; 43 mph)
power off	46 kt (86 km/h; 53 mph)
Landing speed: A-1A	42 kt (77 km/h; 48 mph)
A-1B	51 kt (94 km/h; 58 mph)
Max rate of climb at S/L	457 m (1,500 ft)/min
Service ceiling	6,100 m (20,000 ft)
T-O run: flaps down	61 m (200 ft)
T-O to 15 m (50 ft), flaps down	122 m (400 ft)
Landing from 15 m (50 ft), flaps down	244 m (800 ft)
Landing run, full flaps	107 m (350 ft)
Range: A-1A at 75% power, no reserves	
	608 n miles (1,126 km; 700 miles)
A-1B at 55% power, no reserves	
	695 n miles (1,287 km; 800 miles)
	UPDATED

AVIAT HUSKY PUP

TYPE: Two-seat lightplane.

PROGRAMME: Developed as simplified, cheaper version of Husky. Prototype (N180HY) first flown 7 March 2002; public debut at EAA Sun 'n' Fun at Lakeland, Florida, in April 2002.

COSTS: Approximately US$110,000 (2002).

DESIGN FEATURES: As Husky.

FLYING CONTROLS: Generally as Husky, but with flaperons instead of separate flaps and ailerons.

STRUCTURE: As Husky.

LANDING GEAR: Non-retractable tailwheel type; mainwheel size 6.00-6. Cleveland wheels and brakes.

POWER PLANT: One 119.3 kW (160 hp) Textron Lycoming O-320-D2A flat-four, driving a Sensenich two-blade fixed-pitch propeller. Fuel as for Husky.

ACCOMMODATION: Two in tandem.

AVIONICS: VFR panel standard.

Comms: Becker com, GTX 320 transponder, PM 10000 II intercom.

Flight: Garmin GPS III.

Instrumentation: ASI, altimeter, compass, digital tachometer, oil pressure/temperature gauge, EGT/CHT gauge and ammeter all standard.

EQUIPMENT: Two-colour paint scheme and polished wheel hub caps standard.

DIMENSIONS, EXTERNAL (As for Husky except):	
Propeller diameter	1.88 m (6 ft 2 in)
WEIGHTS AND LOADINGS:	
Weight empty	517 kg (1,140 lb)
Max T-O weight:	907 kg (2,000 lb)
Max wing loading:	53.4 kg/m² (10.93 lb/sq ft)
Max power loading:	7.61 kg/kW (12.50 lb/hp)
PERFORMANCE:	
Never-exceed speed (VNE)	133 kt (246 km/h; 153 mph)
Cruising speed	122 kt (225 km/h; 140 mph)
Stalling speed, flaps up	43 kt (79 km/h; 49 mph)
Service ceiling	5,180 m (17,000 ft)
Max rate of climb at S/L	241 m (790 ft)/min
T-O run	186 m (610 ft)
Landing run	181 m (595 ft)
Range, with reserves 869 n miles (1,609 km; 1,000 miles)	
Endurance	more than 7 h
	NEW ENTRY

AVIAT PITTS S-2C

TYPE: Aerobatic two-seat biplane.

PROGRAMME: Developed from Pitts S-2 series of two-seat biplanes using wing design from Curtis Pitts. Uses S-2B basic fuselage structure mated to new wings.

CUSTOMERS: 55 built by September 2002.

COSTS: Basic price US$188,635 (2002).

DESIGN FEATURES: Compact, positive-stagger, single-bay biplane; constant-chord wings, upper of which sweptback; wire-braced, elliptical tail surfaces. Optimised for aerobatic performance. Differs from S-2B in having symmetrical ailerons; squared fin-, tailplane- and wingtips; metal fuselage undersides removable for easier

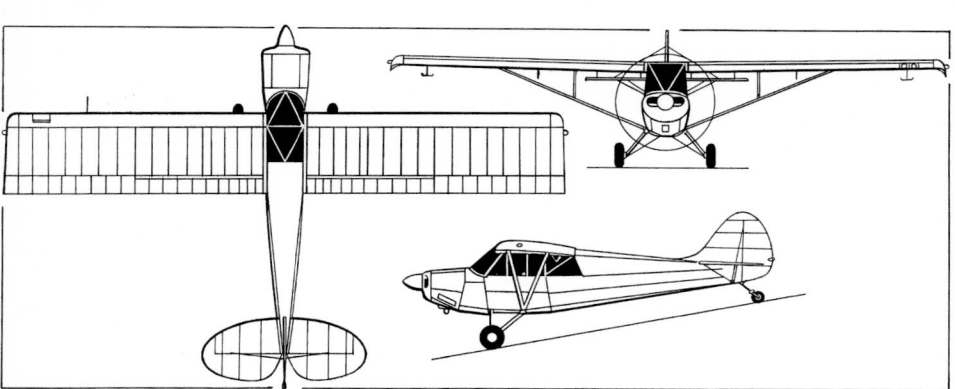

Aviat Husky A-1 (Textron Lycoming O-360 flat-four) *(Jane's/James Goulding)*   0015590

Aviat Pitts S-2C aerobatic biplane *(Jane's/Paul Jackson)*   NEW/0525814

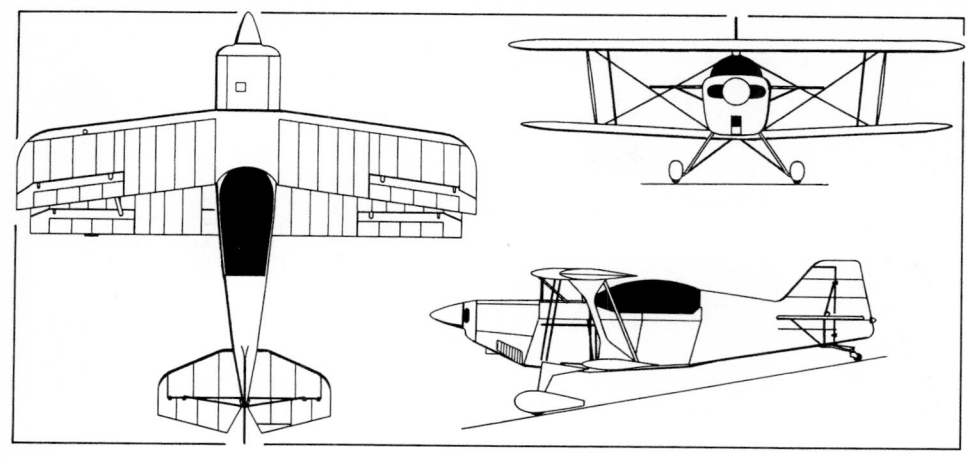

Aviat Pitts S-2C (Textron Lycoming flat-six)   *NEW*/0525815

maintenance. Stretch-formed compound-curved engine cowling. Rate of roll is greater than 300°/s.

Wing sections NACA 6400 series on upper wing, 00 series on lower wings. Both wings have 1° 30′ incidence and lower wing has 3° dihedral.

FLYING CONTROLS: Conventional and manual. Rudder and elevators aerodynamically balanced. Compared to S-2B, the S-2C only has a trim tab on starboard elevator.

STRUCTURE: Fuselage 4130 steel tube with wooden stringers, aluminium top decking and side panels; remainder Dacron covered. Steel tube, Dacron-covered fixed tail surfaces; Dacron-covered control surfaces.

LANDING GEAR: Tailwheel type; fixed. Bungee suspension in protective legs. Mainwheels 5.00-5; tyre pressure 2.34 bar (34 lb/sq in).

POWER PLANT: One 194 kW (260 hp) Textron Lycoming AEIO-540 flat-six driving a Hartzell 'Claw' three-blade aerobatic constant-speed composites propeller. Fuel capacity 110 litres (29.0 US gallons; 24.0 Imp gallons) of which 106 litres (28.0 US gallons; 23.3 Imp gallons) are usable; for aerobatic flight, the 19 litre (5.0 US gallon; 4.2 Imp gallon) wing tank is not used. Oil capacity 11.4 litres (3.0 US gallons; 2.5 Imp gallons).

ACCOMMODATION: Pilot and passenger in tandem under one-piece rearward-sliding canopy; single-piece windscreen. Optional replacement upper decking for single-seat operation, comprising front cockpit flush fairing and shortened canopy.

EQUIPMENT: Options include custom paint and exterior design, electronic gauges, custom parachutes, and canopy covers.

DIMENSIONS, EXTERNAL:
Wing span	6.10 m (20 ft 0 in)
Length overall	5.41 m (17 ft 9 in)
Height overall	1.96 m (6 ft 5 in)
Wheel track	1.54 m (5 ft 0¾ in)
Wheelbase	4.14 m (13 ft 7 in)
Propeller diameter	1.98 m (6 ft 6 in)

DIMENSIONS, INTERNAL:
Cabin: Length	2.11 m (6 ft 11 in)
Max width	0.71 m (2 ft 4 in)
Max height	1.19 m (3 ft 11 in)

AREAS:
Wings, gross	11.85 m² (127.5 sq ft)

WEIGHTS AND LOADINGS:
Weight empty	524 kg (1,155 lb)
Baggage capacity	9 kg (20 lb)
Max T-O and landing weight	771 kg (1,700 lb)
Max wing loading	65.1 kg/m² (13.33 lb/sq ft)
Max power loading	3.98 kg/kW (6.54 lb/hp)

PERFORMANCE:
Never-exceed speed (V_{NE})	185 kt (342 km/h; 212 mph)
Max level speed	169 kt (313 km/h; 194 mph)
Manoeuvring speed	134 kt (248 km/h; 154 mph) IAS
Cruising speed at 55% power	150 kt (278 km/h; 173 mph)
Stalling speed, power off, flaps up	56 kt (104 km/h; 65 mph)
Max rate of climb at S/L	884 m (2,900 ft)/min
Max roll rate	more than 300°/s
T-O run	169 m (555 ft)
T-O to 15 m (50 ft)	262 m (860 ft)

Aviat Eagle II in optional single-seat form (*Jane's/Paul Jackson*)   *NEW*/0525816

Landing from 15 m (50 ft)	366 m (1,200 ft)
Landing run	229 m (750 ft)
Max range at 55% power, 30 min reserves	300 n miles (555 km; 345 miles)
Endurance, 30 min reserves	1 h 36 min
g limits	+6/−5

UPDATED

AVIAT (CHRISTEN) EAGLE II

TYPE: Tandem-seat biplane kitbuilt.

PROGRAMME: Originated as rival to Pitts S-1, designed by Frank Christensen. First flown in February 1977. Kit production began October 1977, initially by Christen Industries.

CUSTOMERS: More than 350 completed and flown by mid-2001, including 320 in the United States.

COSTS: Kit US$85,000 (2002) less engine and propeller.

DESIGN FEATURES: Generally as for S-2B (see 2001-02 edition); similar to S-2C. Available in 24 individual subassemblies; quoted build time 1,800 hours.

FLYING CONTROLS: Conventional and manual. Ailerons on both sets of wings.

STRUCTURE: Fabric-covered wings with wooden spars and ribs plus metal leading- and trailing-edges. Welded 4130 steel tube fuselage, fabric covered aft of cockpit; metal panelled before.

LANDING GEAR: Tailwheel type; fixed. Steel cantilever main legs. Hydraulic brakes.

POWER PLANT: One 149 kW (200 hp) Textron Lycoming AEIO-360-A1D driving a Hartzell HC-C2YK-4/C7666A-2 constant-speed two-blade metal propeller. Fuel capacity 94.6 litres (25.0 US gallons; 20.8 Imp gallons), of which 90.8 litres (24.0 US gallons; 20.0 Imp gallons) are usable.

ACCOMMODATION: Pilot and passenger in tandem; flown solo from rear seat. Optional single-seat upper decking assembly comprises front cockpit flush fairing and shortened canopy.

SYSTEMS: Inverted oil system and manual fuel pump system as standard.

DIMENSIONS, EXTERNAL:
Wing span	6.07 m (19 ft 11 in)
Length overall	5.46 m (17 ft 11 in)
Height overall	1.98 m (6 ft 6 in)

AREAS:
Wings, gross	11.61 m² (125.0 sq ft)

WEIGHTS AND LOADINGS:
Weight empty	465 kg (1,025 lb)
Max T-O and landing weight	715 kg (1,578 lb)
Max baggage weight	13.6 kg (30 lb)
Max wing loading	61.6 kg/m² (12.62 lb/sq ft)
Max power loading	4.80 kg/kW (7.89 lb/hp)

PERFORMANCE:
Never-exceed speed (V_{NE})	182 kt (338 kt; 210 mph)
Max operating speed	159 kt (296 km/h; 184 mph)
Cruising speed	143 kt (266 km/h; 165 mph)
Stalling speed	51 kt (90 km/h; 58 mph)
Max rate of climb at S/L	640 m (2,100 ft)/min
Service ceiling	5,180 m (17,000 ft)
Max roll rate	187°/s
T-O run	244 m (800 ft)
Landing from 15 m (50 ft)	480 m (1,575 ft)
Range with 30 min reserves	380 n miles (703 km; 437 miles)
g limits	+9/−6

UPDATED

AVIAT MILLENNIUM SWIFT

Development of this redesigned version of the Globe GC-1B Swift has been terminated. A description, photograph and three-view last appeared in the 2002-03 edition.

UPDATED

AVIATION DEVELOPMENT

AVIATION DEVELOPMENT INTERNATIONAL LTD

110 Homestead Court, Hillsboro, Ohio 45133
Tel: (+1 937) 393 26 42
Fax: (+1 740) 335 89 96
Web: http://www.alaskanbushmaster.com
JOINT OWNERS:
  Rick Schneider
  Steve Sollars

Aviation Development International was formed in 1997 to build and market the Alaskan Bushmaster kitbuilt. The company has a 1,486 m² (16,000 sq ft) facility at Highland County Airport, Hillsboro.

UPDATED

AVIATION DEVELOPMENT ALASKAN BUSHMASTER

TYPE: Four-seat kitbuilt.

PROGRAMME: Developed during 1980s by Alaskan bush pilot Rick Schneider to replace his Piper Super Cub. Initial

Alaskan Bushmaster fitted with 119 kW (160 hp) LOM engine (*Jane's/Paul Jackson*)   *NEW*/0533388

design based on Merlin Explorer (see 1997-98 and earlier *Jane's*). Original production ceased in 1990 but restarted in 1997.

CUSTOMERS: 12 original examples and one new-build version flying by April 2001, at which time 10 further aircraft on order. One registered to Southland Holdings of Wilmington, Delaware, in July 2002.

COSTS: Kit US$22,500, excluding engine, propeller, instruments and upholstery.

DESIGN FEATURES: Piper Tri-Pacer family development. No welding, machining or heavy forming required during construction. Quoted build time 600 hours.

FLYING CONTROLS: Conventional and manual. Stainless steel control cables. Horn-balanced elevators and rudder; flight-adjustable tabs in both elevators.

STRUCTURE: Fabric-covered fuselage and tail of welded 4130 steel tube; fabric-covered extruded aluminium wings, flaps and ailerons. Wing braced with two extruded aluminium lift struts; tailplane and fin braced by wires.

LANDING GEAR: Tailwheel type; fixed. Spring aluminium main legs and tailwheel. Bungee cord suspension. Mainwheels 8.50-6. Options include Tundra tyres, Schneider Wheel Skis, skis and floats.

POWER PLANT: Prototype had 149 kW (200 hp) Textron Lycoming O-320 flat-four engine driving a two-blade fixed-pitch propeller; recommended engine power between 112 and 224 kW (150 and 300 hp), including engines from Teledyne Continental and LOM. Standard fuel capacity 114 litres (30.0 US gallons; 25.0 Imp gallons) per tank; option of two, four or six wing tanks, taking maximum possible capacity to 681 litres (180 US gallons; 150 Imp gallons).

ACCOMMODATION: Pilot and three passengers in two pairs of seats. Dual controls. Door each side, plus freight door to port, rear.

DIMENSIONS, EXTERNAL:

Wing span, without tips	11.89 m (39 ft 0 in)
Wing aspect ratio	7.2
Length overall: O-320 engine	7.09 m (23 ft 3 in)
LOM 322 engine	7.49 m (24 ft 7 in)
Height overall	2.08 m (6 ft 10 in)
Tailplane span	3.48 m (11 ft 5 in)
Wheel track	2.29 m (7 ft 6 in)

DIMENSIONS, INTERNAL:

Cabin max width	1.04 m (3 ft 5 in)

AREAS:

Wings, gross	19.51 m² (210.0 sq ft)
Ailerons (total)	1.77 m² (19.00 sq ft)
Flaps (total)	2.15 m² (23.10 sq ft)
Horizontal tail surfaces (total)	3.58 m² (38.50 sq ft)

WEIGHTS AND LOADINGS:

Weight empty, typical	680 kg (1,500 lb)
Baggage capacity	91 kg (200 lb)
Max T-O weight	1,361 kg (3,000 lb)
Max wing loading	69.7 kg/m² (14.29 lb/sq ft)
Max power loading, 119 kW (160 hp) engine	
	11.41 kg/kW (18.75 lb/hp)

PERFORMANCE (119 kW; 160 hp engine):

Never-exceed speed (V$_{NE}$)	130 kt (241 km/h; 150 mph)
Normal cruising speed	100 kt (185 km/h; 115 mph)
Stalling speed: flaps up	38 kt (70 km/h; 43 mph)
flaps down	31 kt (57 km/h; 35 mph)
Max rate of climb at S/L	305 m (1,000 ft)/min
T-O run	244 m (800 ft)
Landing run	91 m (300 ft)
Range with standard fuel	
	536 n miles (993 km; 617 miles)
	UPDATED

AVID

AVID AIRCRAFT INC

5057 Hwy 287 North, Ennis, Montana 59729
Tel: (+1 406) 682 56 15
Fax: (+1 406) 682 55 54
e-mail: avidair@avidair.com
Web: http://www.avidair.com
PRESIDENT: Jim Tomash
GENERAL MANAGER: Robert Stone

Avid Aircraft was formed in 1983 and currently markets the Avid Flyer Mark IV, Catalina (to special order only), Bandit, Magnum and Buzz. Avid Flyer also assembled in Europe by Yalo of Czech Republic (which see). In May 1999, the company was sold to a stock holder who restarted production in Ennis, Montana. The Avid Champion has been discontinued as the weight could not be kept low enough for FAR Pt 103.

UPDATED

Avid Aircraft Avid Flyer Speedwing in tailwheel configuration and with Rotax 582 engine
(Jane's/Paul Jackson) NEW/0533387

AVID FLYER MARK IV

TYPE: Side-by-side ultralight kitbuilt.

PROGRAMME: First flown 1983; available as single kit, or six separate kits to spread cost of purchase. Strongly influenced design of Indaer-Peru Chuspi light aircraft (see under Peru in 1992-93 *Jane's*), and the SkyStar series of kitbuilt aircraft. Avid **Bandit** identical, except for 454 kg (1,000 lb) maximum T-O weight.

CUSTOMERS: More than 1,600 Avid Flyer kits delivered, with over 900 flying.

COSTS: Kit: US$15,740 firewall back (2002).

DESIGN FEATURES: Strut-braced, high-wing ultralight of traditional configuration and construction. Two forms available (interchangeable), as original **High Gross STOL** with unique near full-span auxiliary aerofoil flaperons, and shorter span **Aerobatic Speedwing** using new wing section; cruising and stalling speeds raised with Aerobatic Speedwing fitted. Wings fold for storage. Quoted build time 650 hours.

FLYING CONTROLS: Manual. Junkers flaperons (see Design Features), elevators with adjustable trim tab, and rudder.

STRUCTURE: Aluminium wing spars and plywood ribs, covered with heat-shrunk Dacron. Welded steel tube fuselage, rudder, tailplane and elevators, Dacron covered except for fuselage nose which has premoulded GFRP cowlings. Fin integral with fuselage.

LANDING GEAR: Non-retractable tricycle or tailwheel landing gear, with Tundra tyres and brakes. Wheel rim size 12.7 cm (5 in). Steerable tailwheel. Mainwheels have bungee cord shock absorption. Optional Aqua 1500 floats, skis and wheel/skis. Wider landing gear available as option, introduced 2001.

POWER PLANT: One 47.8 kW (64.1 hp) Rotax 582 two-stroke engine, driving a two- or three-blade propeller; fixed-pitch wooden propeller for STOL, three-blade ground-adjustable for Speedwing. Fuel capacity 53 litres (14.0 US

Nosewheel version of Avid Flyer, fitted with rectangular rudder *(Jane's/Paul Jackson)*   0110716

gallons; 11.7 Imp gallons) for High Gross STOL and 68 litres (18.0 US gallons; 15.0 Imp gallons) for Aerobatic Speedwing; option to double capacity of each. Similar capacity fuel tanks may be added in port wing.

DIMENSIONS, EXTERNAL (A: STOL, B: Aerobatic Speedwing):

Wing span: A	9.11 m (29 ft 10½ in)
B	7.30 m (23 ft 11½ in)
Length overall	5.46 m (17 ft 11 in)
Height overall: tailwheel	1.80 m (5 ft 11 in)
tricycle	2.11 m (6 ft 11 in)

DIMENSIONS, INTERNAL:

Cabin max width	1.00 m (3 ft 3½ in)

AREAS (A and B as above):

Wings, gross: A	11.38 m² (122.5 sq ft)
B	9.04 m² (97.3 sq ft)

WEIGHTS AND LOADINGS (A and B as above):

Weight empty: A	200-231 kg (440-510 lb)
B	231 kg (510 lb)
Baggage capacity	16 kg (35 lb)
Max T-O weight: A	522 kg (1,150 lb)
B	476 kg (1,050 lb)

PERFORMANCE (A and B as above):
Max level speed at 1,525 m (5,000 ft):

A, B	117 kt (217 km/h; 135 mph)
Max cruising speed: A	78 kt (145 km/h; 90 mph)
B	100 kt (185 km/h; 115 mph)

Stalling speed:

A, flaps down, engine idling	28 kt (52 km/h; 32 mph)
B	37 kt (68 km/h; 42 mph)
Max rate of climb at S/L: A	518 m (1,700 ft)/min
B	366 m (1,200 ft)/min
Service ceiling: A, B	3,810 m (12,500 ft)
T-O run: A	27 m (90 ft)
B	38 m (125 ft)
Landing run: A	46 m (150 ft)
B	152 m (500 ft)
Range, no reserves: A	295 n miles (547 km; 340 miles)
B	491 n miles (910 km; 566 miles)
	UPDATED

AVID MAGNUM

TYPE: Side-by-side kitbuilt.

CUSTOMERS: About 200 under construction with 65 completed.

COSTS: Kit US$26,990 without engine (2002).

DESIGN FEATURES: High-wing braced monoplane, similar to Avid Flyer, but in certified category, using a Textron Lycoming engine. Recent upgrades include larger rudder and new external baggage door. Quoted build time 750 hours.

FLYING CONTROLS: Similar to Avid Flyer.

STRUCTURE: Similar to Avid Flyer, with Poly Fiber covering.

LANDING GEAR: Available as tricycle or tailwheel, with Cleveland wheels and brakes; wheel fairings. Spring aluminium landing gear standard. Optional floats.

POWER PLANT: One Textron Lycoming engine in 80.5 to 134 kW (108 to 180 hp) range, including O-235, O-320 and O-360. Tricycle version also accepts 119 kW (160 hp) Subaru EJ22 engine. Fuel capacity 110 litres (29.0 US gallons; 24.1 Imp gallons) of which 106 litres (28.0 US gallons; 23.3 Imp gallons) are usable. Optional 45 litre (12.0 US gallon; 10.0 Imp gallon) wingtip tanks available.

ACCOMMODATION: Two seats side by side, with dual controls, plus optional jump seat for small adult or two children in 0.81 m³ (28.5 cu ft) baggage area.

DIMENSIONS, EXTERNAL:

Wing span	10.06 m (33 ft 0 in)
Wing chord, constant	1.30 m (4 ft 3 in)
Wing aspect ratio	8.0
Width, wings folded	2.59 m (8 ft 6 in)
Length overall	6.40 m (21 ft 0 in)
Height overall	1.86 m (6 ft 1¼ in)

DIMENSIONS, INTERNAL:

Cabin max width	1.12 m (3 ft 8 in)

AREAS:

Wings, gross	12.63 m² (136.0 sq ft)

WEIGHTS AND LOADINGS:

Weight empty	442 kg (975 lb)
Baggage capacity	68 kg (150 lb)
Max T-O weight	794 kg (1,750 lb)
Max wing loading	62.8 kg/m² (12.87 lb/sq ft)
Power loading (119 kW; 160 hp)	
	6.66 kg/kW (10.94 lb/hp)

PERFORMANCE (119 kW; 160 hp O-320 engine):

Never-exceed speed (V$_{NE}$)	134 kt (249 km/h; 155 mph)
Cruising speed	113 kt (209 km/h; 130 mph)
Stalling speed	31 kt (58 km/h; 36 mph)
Max rate of climb at S/L	549 m (1,800 ft)/min
Service ceiling	more than 6,100 m (20,000 ft)
T-O run	approx 46 m (150 ft)
T-O to 15 m (50 ft)	69 m (225 ft)
Landing run	76 m (250 ft)
Range	450 n miles (833 km; 517 miles)
	UPDATED

AVID CHAMPION and BUZZ

TYPE: Single-seat ultralight kitbuilt.

CURRENT VERSIONS: **Champion:** Original version, first flown 5 June 1998, introduced in 1999; six flying by 2001. Now discontinued due to inability to meet weight criteria of

FAR Pt 103. Prototype converted to Buzz. Last described in 2002-03 edition.

Buzz: Stretched version introduced in place of Champion in 2002 at Sun 'n' Fun, when not yet flown; *as described.*

COSTS: Kit without engine US$10,995 (2002).

DESIGN FEATURES: Generally as for Flyer. Intended to conform to FAR Pt 103. Wings fold for storage. Flying controls provided as completed items. Quoted build time 160 hours.

FLYING CONTROLS: As for Flyer.

STRUCTURE: Fuselage 4130 chromoly steel tube covered with fabric; wing has aluminium spar and wooden ribs. Glass fibre cowling.

LANDING GEAR: Tailwheel type; fixed. Bungee cord suspension Maule tailwheel. Mainwheels 8.00-6; hydraulic brakes.

POWER PLANT: One 44.7 kW (60 hp) HKS 700E air-cooled four-stroke engine. Fuel capacity 34 litres (9.0 US gallons; 7.5 Imp gallons) in single wing tank.

DIMENSIONS, EXTERNAL:
Wing span	9.09 m (29 ft 10 in)
Length: overall	6.25 m (20 ft 6 in)

UPDATED

Prototype Avid Buzz single-seat kitbuilt making its public debut at Sun 'n' Fun 2002
(Jane's/Paul Jackson)
NEW/0533386

AVTEKAIR

AVTEKAIR INC

555 Airport Way, Suite A, Camarillo Airport, Camarillo, California 93010
Tel: (+1 805) 482 27 00
Fax: (+1 805) 987 00 68
e-mail: AvtekAir@aol.com
PRESIDENT: Quinten E Ward
SENIOR VICE-PRESIDENT, ENGINEERING: Niels Andersen
VICE-PRESIDENT, MARKETING: Robert D Honeycutt

Company founded 1982 as Avtek to develop Avtek 400, last described in 1995-96 *Jane's*. Planned preproduction aircraft failed to achieve 1995 first flight. However, in 1998 Avtek secured finance to certify a stretched, commuter version, the Avtek 419 Express and in 1999 changed its name to AvtekAir, renaming its product AvtekAir 9000T.

VERIFIED

AVTEKAIR 9000T

TYPE: Utility turboprop twin.

PROGRAMME: Design of Avtek 400 six/10-seat all-composites business aircraft started March 1981 and programme launched June 1982; construction of proof-of-concept aircraft N400AV (see 1985-86 *Jane's*) began January 1983; flew 17 September 1984; N400AV then fitted with P&WC PT6A-135M engines (PT6A-35s with counter-rotating gearboxes from PT6A-66); extensive changes made Spring 1985, including fuselage stretch 20 cm (8 in) forward and 76 cm (2 ft 6 in) aft of front pressure bulkhead, widened cabin, new outer wing and enlarged fuel tanks in forward-swept root extensions, foreplane with greater span and reduced chord ventral strakes (known as

delta fins), relocated main landing gear legs and specially developed P&WC PT6A-3s mounted closer to wings. Preproduction second prototype abandoned and late 1996 FAA certification failed to take place.

Programme relaunched in 1998, with priority to stretched 419 version for commuter airlines; renamed AvtekAir 9000T in 1999. No further information received.

CUSTOMERS: 40 on order in July 2002.

COSTS: Standard aircraft US$2,795,000; fully equipped IFR (2001).

DESIGN FEATURES: Twin-engine pusher configuration with foreplane. Initial design by Al W Mooney, founder of Mooney Aircraft; refined by Niels Anderson, Ford Johnston and Irvin Culver; computer analysis of configuration and wind tunnel testing by NASA; materials research by Dow Chemical and Dr Leo Windecker, Dow Chemical basic patents on Windecker Eagle (first all-composites aircraft to receive civil certification) licensed to Avtek.

Avtek 12 aerofoil section; anhedral 2° 30′ from roots; sweepback 50° inboard, 15° 30′ outboard; foreplane dihedral 1°; fuselage pressurised.

FLYING CONTROLS: Manual. Actuation by pushrods and cranks throughout; mass-balanced elevators on foreplane; mass-balanced rudder without trim tab; two-section ailerons, with inboard sections also electrically actuated as pitch-axis trim surfaces; no flaps.

STRUCTURE: All composites structure: 90 per cent graphite/carbon fibre, smaller quantitites of R-glass, S-glass, aluminium and nickel fibres; wire mesh incorporated to protect against lightning.

LANDING GEAR: Hydraulically retractable tricycle type, main units retracting inward and nosewheel forward. Emergency extension system. Oleo-pneumatic shock-absorber in each unit. Single wheel on each unit. Cleveland hydraulic disc brakes.

POWER PLANT: Two 582 kW (780 shp) Pratt & Whitney Canada PT6A-135 turboprops, one mounted within nacelle above each wing. Hartzell four-blade constant-speed fully feathering reversible-pitch pusher propellers.

ACCOMMODATION: Up to nine passengers. Unpressurised baggage compartment in nose with external door on port side. Accommodation pressurised, air conditioned, heated and ventilated.

SYSTEMS: AiResearch bleed air pressurisation system and air cycle air conditioning system. Electrically driven hydraulic pump provides pressure of 138 bar (2,000 lb/sq in) for landing gear actuation. Anti-icing of windscreen by electrical system, of propellers by engine efflux, electrically heated pitot. Choice of wing and foreplane pneumatic, alcohol, electric or engine bleed de-icing. Dual anti-icing inlets for each engine.

DIMENSIONS, EXTERNAL:
Wing span	10.67 m (35 ft 0 in)
Wing aspect ratio	8.5
Foreplane span	6.92 m (22 ft 8½ in)
Length overall	16.62 m (54 ft 5 in)
Passenger door (port): Height	1.17 m (3 ft 10 in)
Width	0.76 m (2 ft 6 in)
Height to sill	0.76 m (2 ft 6 in)
Baggage door (port, nose): Height	0.51 m (1 ft 8 in)
Width	0.56 m (1 ft 10 in)
Height to sill	0.79 m (2 ft 7 in)
Emergency exit (stbd): Height	0.51 m (1 ft 8 in)
Width	0.67 m (2 ft 2½ in)

DIMENSIONS, INTERNAL:
Cabin: Max width	1.40 m (4 ft 7 in)
Max height	1.37 m (4 ft 6 in)

AREAS:
Wings, gross	13.40 m² (144.2 sq ft)

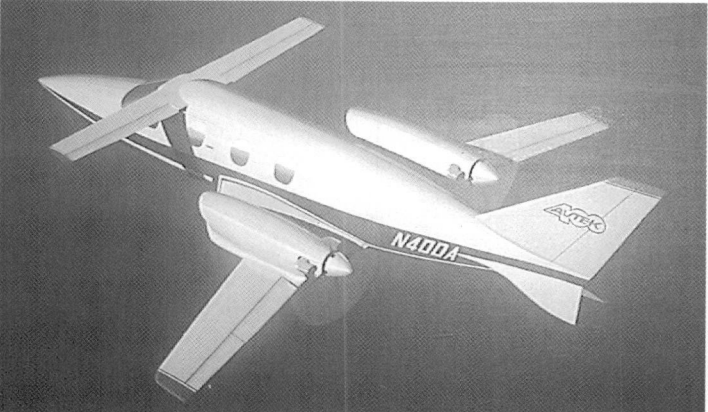

Artist's impression of Avtek 400A, precursor of the 9000T

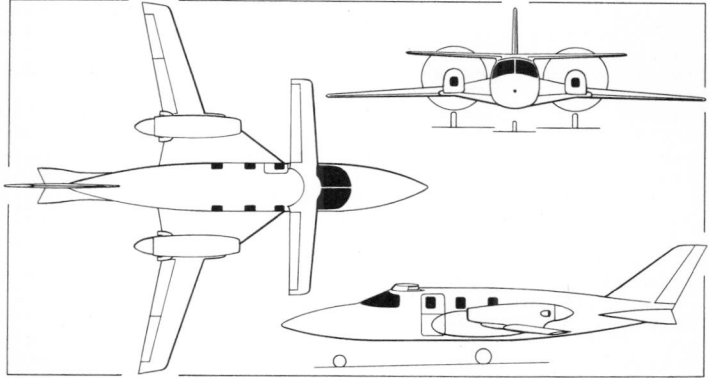

Avtek 400A six/10-seat twin-turboprop aircraft which is to be lengthened to become the AvtekAir 9000T *(Jane's/Dennis Punnett)*

For details of the latest updates to *Jane's All the World's Aircraft* online and to discover the additional information available exclusively to online subscribers please visit
jawa.janes.com

Foreplane, gross	4.52 m² (48.7 sq ft)	WEIGHTS AND LOADINGS:		PERFORMANCE (estimated):		
Elevators (total)	0.92 m² (9.9 sq ft)	Max T-O weight	5,670 kg (12,500 lb)	Cruising speed	364 kt (674 km/h; 419 mph)	
Ailerons (total)	0.60 m² (6.5 sq ft)	Max wing loading	423.2 kg/m² (88.69 lb/sq ft)	T-O run	463 m (1,520 ft)	
Fin	1.16 m² (12.5 sq ft)	Max power loading	4.88 kg/kW (8.01 lb/shp)	Range with 4 passengers		
Rudder	0.90 m² (9.7 sq ft)				1,911 n miles (3,540 km; 2,200 miles)	

UPDATED

AYRES

QUALITY AEROSPACE INC

PO Box 3050, 1 Ayres Way, Albany, Georgia 31707-3090
Tel: (+1 912) 883 14 40
Fax: (+1 912) 439 97 90
e-mail: mhumphries@qualityaero.com
Web: http://www.ayrescorp.com
PLANT MANAGER: Milt Humphries

Former Ayres Corporation bought Albany factory and manufacturing and world marketing rights to Thrush Commander-600 and -800 from Rockwell International General Aviation Division in November 1977. Acquired 93 per cent shareholding in Let Kunovice of Czech Republic (which see) in 1998, but stopped providing finance in July 2000 after Czech bank cancelled a debt repayment freeze; Let declared bankrupt in October 2000.

Ayres filed for Chapter 11 bankruptcy protection in late November 2000 and was seeking further investment of US$80 million; work restarted on the Loadmaster programme (see 2002-03 and earlier editions) after a financing agreement with GATX, as debtor-in-possession, was concluded. However, GATX foreclosed on the company on 7 August 2001, securing all assets in return for a debt reduction of US$10.3 million.

Quality Aerospace established, also on 7 August 2001, to take over almost all of Ayres's assets at 21,100 m² (227,000 sq ft) Albany plant and continues to produce aerostructures, support existing Thrush fleet and promote Turbo-Thrush.

UPDATED

AYRES TURBO-THRUSH S2R

TYPE: Agricultural sprayer.
PROGRAMME: Leland Snow designed S-1 prototype, first flown with radial engine on 17 August 1953; entered production as S-2; Type Certificate for original S2D issued 1 November 1965; SR2 followed 21 March 1968; over 1,800 built by Snow and (following 18 February 1970 transfer of Type Certificate) Rockwell. Type Certificate obtained by Ayres on 28 November 1977; Ayres introduced turboprop version, having previously specialised in turbine conversions. PT6-engined S2R-T34 certified 28 April 1977; Garrett TPE331 versions certified from 5 March 1992 (S2R-G6) onwards. Quality Aerospace obtained Type Certificate on 26 November 2001.
CURRENT VERSIONS: **S2R-T11:** 373 kW (500 shp) PT6A-11AG turboprop; standard 1,514 litre (400 US gallon; 333 Imp gallon) chemical hopper. Certified 26 October 1979.

S2R-T15: 507 kW (680 shp) P&WC PT6A-15AG turboprop; standard or optional 1,930 litre (510 US gallon; 425 Imp gallon) hopper. Certified 3 April 1979.

S2R-T34: 559 kW (750 shp) P&WC PT6A-34AG turboprop; standard or optional hoppers. High Gross (4,763 kg; 10,500 lb) MTOW version certified 5 November 1997.

S2R-T45: Powered by P&WC PT6R-45AG. Certified 23 July 1990.

S2R-T65 NEDS: Narcotics Eradication Delivery System; 1,026 kW (1,376 shp) P&WC PT6A-65AG turboprop and 2.82 m (9 ft 3 in) five-blade propeller; certified 3 September 1987; 19 delivered to US State Department (see Customers).

S2R-G1: Powered by 496 kW (665 shp) TPE331-1 turboprop; 1,514 litre (400 US gallon; 333 Imp gallon) hopper. Certified 29 August 1991.

S2R-G6: Introduced 1992; 559 kW (750 shp) TPE331-6 turboprop; standard or optional hopper; two 435 litre (115 US gallon; 95.75 Imp gallon) fuel tanks, with 863 litres (228 US gallons; 190 Imp gallons) usable; optional hopper capacity 1,930 litres (510 US gallons; 425 Imp gallons); pilot, with dual-cockpit option.

S2R-G10: First flown November 1992; certified 12 January 1993; CAM 8 certification; 701 kW (940 shp) TPE331-10 turboprop and four-blade propeller.

Ayres Turbo-Thrush S2R-T34 agricultural aircraft 0051558

V-1A Vigilante: Surveillance and close air support version; engineering designation S2RHG-T65; certified 8 June 1988; TPE331-14-GR turboprop; first flight (N3100A) 15 April 1989. Still carries hopper; has five-blade propeller for quiet operation; four hardpoints beneath wings; can be fitted with range of equipment including loudspeaker systems for policing duties. Fuel capacity 1,136 litres (300 US gallons; 250 Imp gallons) plus optional external tank containing 1,514 litres (400 US gallons; 333 Imp gallons).
CUSTOMERS: US State Department ordered 19 S2R-T65 NEDS during 1983-88 for use by International Narcotics Matters Bureau. Vigilante used in Africa for game and poacher surveillance. By July 2001, 15 G1s, 55 G6s, 68 G10s, 39 T15s, 272 T34s and 15 T45s had been produced and over 2,500 of all versions delivered to 80 countries. Production was continuing in 2001.
COSTS: About US$600,000, depending on engine and customer fit.
DESIGN FEATURES: Conventional crop-sprayer configuration, with hopper ahead of cockpit; cantilever low wing of constant chord; braced empennage with sweptback leading-edges; robust landing gear.

Advantages over piston-engined versions include much improved take-off and climb, 454 kg (1,000 lb) higher payload because of lower engine weight, operation on aviation turbine fuel or diesel, 3,500 hour TBO, quieter operation and ability to feather propeller without stopping engine while refuelling and reloading.
FLYING CONTROLS: Conventional and manual. Plain ailerons; servo tab in each elevator; electrically actuated flaps.
STRUCTURE: Two-spar light alloy wing with 4130 chrome molybdenum steel spar caps; welded chrome molybdenum steel tube fuselage structure covered with quickly removable light alloy skin panels; underfuselage skin of stainless steel; all-metal tail surfaces and strut-braced tailplane; metal ailerons and flaps. Extended wing, originally optional, is now standard, replacing earlier 13.54 m (44 ft 5 in), 30.34 m² (326.6 sq ft) version.
LANDING GEAR: Tailwheel type; fixed. Main units have rubber-in-compression shock-absorption and 8.50-10 (10 ply) tyres. Hydraulically operated Cleveland dual calliper disc brakes. Parking brakes. Wire cutters on main gear. Steerable, spring steel locking tailwheel, size 5.00-5.

POWER PLANT: One turboprop (see under Current Versions for engine and fuel details). All but NEDS, G10 and Vigilante have Hartzell three-blade, constant-speed, feathering and reversing propellers; usable fuel capacity 515 litres (136 US gallons; 113 Imp gallons) in 400 gallon hopper models and 863 litres (228 US gallons; 190 Imp gallons) in 510 gallon hopper models.
ACCOMMODATION: Single adjustable mesh seat in 'safety pod' sealed cockpit enclosure, with steel tube overturn structure. Tandem seating optional, with forward-facing second seat. Dual controls optional with forward-facing rear seat, for pilot training; these versions have 'DC' suffix to c/n; available since 1985. Adjustable rudder pedals. Downward-hinged door on each side. Tempered safety glass windscreen. Cockpit wire cutter. Dual inertia reel safety harness with optional second seat. Baggage compartment standard on single-seat aircraft.
SYSTEMS: Electrical system powered by a 24 V 50 A alternator. Dual lightweight 24 V 35 Ah batteries.
AVIONICS: To customer's requirements.
EQUIPMENT: GFRP hopper forward of cockpit can hold 1,514 litres (400 US gallons; 333 Imp gallons) of liquid or 1,487 kg (3,280 lb) of dry chemical, equivalent to 1.50 m³ (53.0 cu ft). Optional 1,931 litre (510 US gallon; 425 Imp gallon) hopper of 1.87 m³ (66.0 cu ft) can be installed. Hopper has a 0.33 m² (3.56 sq ft) lid, openable by two handles, and cockpit viewing window. Standard equipment includes Universal spray system with external 50 mm (2 in) stainless steel plumbing, 50 mm pump with wooden fan, Transland gate, 50 mm valve, quick-disconnect pump mount and strainer. Streamlined spraybooms with outlets for 68 nozzles. Micro-adjust valve control (spray) and calibrator (dry). A 51 mm (2 in) side-loading system is installed on the port side. Stainless steel rudder cables. Navigation lights, instrument lights and two strobe lights. Windscreen washer/wiper.

Optional equipment includes a rear cockpit to accommodate forward-facing seat for passenger, or flying instructor if optional dual controls installed; space can be used alternatively for cargo. Other optional items are a Transland high-volume spreader; agitator installation; 10-unit AU5000 Micronair installation in lieu of standard booms and nozzles; Transland gatebox with stiffener casting; quick-disconnect flange and kit; night working

Ayres Turbo-Thrush S2R with optional dual cockpit 0051557

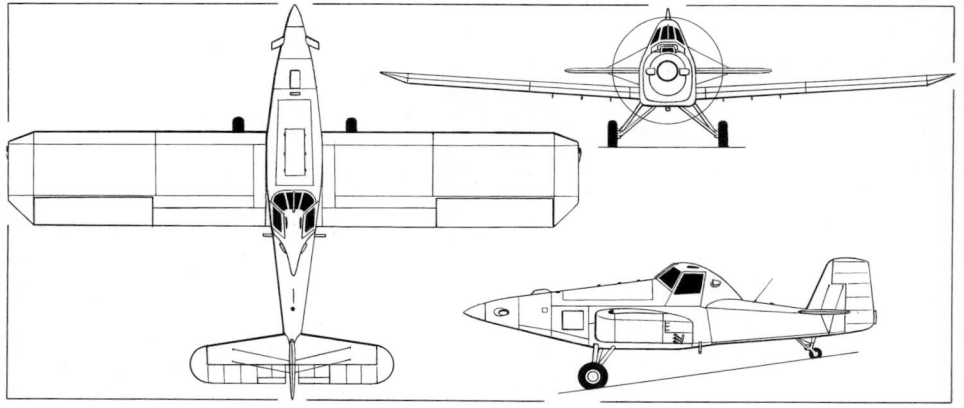
Ayres Turbo-Thrush S2R-T34 with optional 1,930 litre (510 US gallon; 425 Imp gallon) hopper and no longer available wing span option of 13.54 m (44 ft 5 in) *(Jane's/Dennis Punnett)*

lights including landing light and wingtip turn lights; cockpit fire extinguisher; and water-bomber configuration.

DIMENSIONS, EXTERNAL:
Wing span	14.48 m (47 ft 6 in)
Wing aspect ratio	6.4
Length overall	10.06 m (33 ft 0 in)
Height overall: except Vigilante	2.79 m (9 ft 2 in)
Vigilante	2.90 m (9 ft 6 in)
Wheel track	2.74 m (9 ft 0 in)

AREAS:
Wings, gross	32.52 m² (350.0 sq ft)

WEIGHTS AND LOADINGS: (A: standard hopper and PT6A, B: optional hopper and PT6A, C: with TPE331-6, D: with TPE331-10, E: Vigilante and High Gross option):
Weight empty: A	1,633 kg (3,600 lb)
B	1,769 kg (3,900 lb)
C	1,950 kg (4,300 lb)
D	2,041 kg (4,500 lb)
E	2,223 kg (4,900 lb)
Max T-O weight (CAR 3): A, B	2,721 kg (6,000 lb)
E	4,763 kg (10,500 lb)

Typical operating weight (CAM 8):
A	3,719 kg (8,200 lb)
B	3,856 kg (8,500 lb)
C	4,400 kg (9,700 lb)
D	4,082 kg (9,000 lb)

CAM 8 wing loading
A	114.4 kg/m² (23.43 lb/sq ft)
B	118.6 kg/m² (24.29 lb/sq ft)
C	135.3 kg/m² (27.71 lb/sq ft)
D	125.5 kg/m² (25.71 lb/sq ft)
Max wing loading: E	146.5 kg/m² (30.00 lb/sq ft)
Max power loading: A	7.30 kg/kW (12.00 lb/shp)
B	5.37 kg/kW (8.82 lb/shp)
E	6.80 kg/kW (11.17 lb/shp)

PERFORMANCE (A and B: with PT6A-34AG engine, C: with TPE331-6, D: with TPE331-10, E: Vigilante):

Never-exceed speed (V$_{NE}$):
D	138 kt (256 km/h; 159 mph)
E	191 kt (354 km/h; 220 mph)

Max level speed with spray equipment:
A, B, D	138 kt (256 km/h; 159 mph)
Max level speed: E	163 kt (302 km/h; 188 mph)

Cruising speed:
A, B and E at 50% power	130 kt (241 km/h; 150 mph)
C at 55% power	130 kt (241 km/h; 150 mph)

Working speed at 30-50% power
	78-130 kt (145-241 km/h; 90-150 mph)
Stalling speed: flaps up	61 kt (113 km/h; 70 mph)
flaps down	57 kt (106 km/h; 66 mph)

Stalling speed at normal landing weight:
flaps up	51 kt (95 km/h; 59 mph)
flaps down	50 kt (92 km/h; 57 mph)
Max rate of climb at S/L: A, B, C	530 m (1,740 ft)/min
D	762 m (2,500 ft)/min
E	1,067 m (3,500 ft)/min
Service ceiling: A, B, C, E	7,620 m (25,000 ft)
D	3,660 m (12,000 ft)
T-O run: A, B, D	183 m (600 ft)
C at 4,400 kg (9,700 lb)	366 m (1,200 ft)
Landing from 15 m (50 ft)	366 m (1,200 ft)
Landing run: A, B, C	183 m (600 ft)
D	244 m (800 ft)

Landing run with propeller reversal:
A, B, C	122 m (400 ft)
Range: D	500 n miles (926 km; 575 miles)

Ferry range at 45% power:
A, B, C	669 n miles (1,239 km; 770 miles)
E	900 n miles (1,667 km; 1,036 miles)

UPDATED

AYRES 660 TURBO-THRUST

TYPE: Agricultural sprayer.

PROGRAMME: Prototype produced in 1997 to verify new hopper and landing gear designs, but was otherwise as S2R. Certification achieved 13 March 2000 with first delivery same day.

CUSTOMERS: Nine delivered by October 2000; none further registered by mid-2002.

DESIGN FEATURES: Based on S2R (which see) but with enlarged hopper, made possible by change to cantilever main landing gear. Enlarged wing centre-section extends span.

Data as for Turbo-Thrust S2R, except as below.

FLYING CONTROLS: Aileron droop system, plus enlarged flaps.

LANDING GEAR: Tailwheel type; non-retractable. Cantilever mainwheel legs of spring steel, with heavy-duty 29 in tyres (31 in tyres optional) and Cleveland dual calliper hydraulic disc brakes. Wire cutters on main undercarriage legs. Tailwheel 6.00×6.

POWER PLANT: One 917 kW (1,230 shp) Pratt & Whitney Canada PT6A-65AG turboprop. Optionally, one 783 kW (1,050 shp) P&WC PT6A-60 turboprop. Hartzell HC-B5MP-3C five-blade, constant-speed feathering and reversing propeller. Fuel capacity 863 litres (228 US gallons; 190 Imp gallons); optional capacity 1,136 litres (300 US gallons; 250 Imp gallons) usable.

ACCOMMODATION: Optional air conditioning.

EQUIPMENT: GFRP hopper can hold 2,500 litres (660 US gallons; 550 Imp gallons) of liquid. Side loader size 76 mm

Prototype Ayres 660 Turbo-Thrust   0051559

(3 in) on port side. Standard equipment includes Universal spray system with external 76 mm (3 in) stainless steel plumbing, five-blade Weath-Aerofan, 76 mm Transland pump, 68-nozzle sprayboom extending over 70 per cent of span. Optional equipment as per Turbo-Thrust S2R.

DIMENSIONS, EXTERNAL (PT6A engine):
Wing span	16.46 m (54 ft 0 in)
Wing aspect ratio	7.3
Wing chord, constant	2.29 m (7 ft 6 in)
Length overall	10.21 m (33 ft 6 in)
Height overall	2.97 m (9 ft 9 in)
Tailplane: span	5.18 m (17 ft 0 in)
chord: at root	1.40 m (4 ft 7 in)
at tip	1.09 m (3 ft 7 in)
Wheel track	2.90 m (9 ft 6 in)
Wheelbase	6.74 m (22 ft 1¼ in)
Propeller diameter	2.82 m (9 ft 3 in)

DIMENSIONS, INTERNAL:
Hopper volume	2.5 m³ (88 cu ft)

AREAS:
Wings, gross	37.16 m² (400.0 sq ft)

WEIGHTS AND LOADINGS (PT6A-65AG):
Weight empty	2,880 kg (6,350 lb)
Max T-O weight	6,418 kg (14,150 lb)
Max landing weight	5,670 kg (12,500 lb)
Max wing loading	172.7 kg/m² (35.38 lb/sq ft)
Max power loading	7.00 kg/kW (11.50 lb/shp)

PERFORMANCE (PT6A-65AG):
Never-exceed speed (V$_{NE}$)	191 kt (354 km/h; 220 mph)
Cruising speed at 55% power	
	152 kt (282 km/h; 175 mph)
Working speed	86-152 kt (185-282 km/h; 100-175 mph)
Stalling speed, power off, flaps down	
	50 kt (92 km/h; 57 mph)
Max rate of climb at S/L	381 m (1,250 ft)/min
T-O run	457 m (1,500 ft)
Landing run	183 m (600 ft)
Ferry range at 50% power	
	521 n miles (966 km; 600 miles)

UPDATED

BARR

BARR AIRCRAFT

900 Airport Road, Montoursville, Pennsylvania 17754
Tel: (+1 570) 368 36 55
Fax: (+1 570) 368 20 95
e-mail: six@barraircraft.com
Web: http://www.barraircraft.com
MANAGING DIRECTOR: James L Barr

Barr Aircraft holds supplemental type certificates for the Piper PA-12, PA-14, PA-20 and PA-22 series of aircraft and is presently developing the BarrSix.

VERIFIED

BARR BARRSIX

TYPE: Six-seat kitbuilt.

PROGRAMME: Prototype (N83W) was due to fly late 1999, but delayed by hangar construction; new completion target was late 2000 but this has slipped into 2003; wing and tail mating undertaken in September 2001; final assembly taking place from September 2002.

CUSTOMERS: Launch orders being sought once prototype is flying.

COSTS: US$89,900 (2003).

DESIGN FEATURES: Designed to meet FAR Pt 23 and 51 per cent kitplane rules; extensive use of composites results in lightweight design. High wing with single bracing strut each side.
Dihedral 1° 44′; incidence 1° 30′ at wing root; −1° 30′ at tip.

FLYING CONTROLS: Conventional and manual. Frise-type ailerons; NACA single-slotted three-position (10, 20, 40°) flaps; horn-balanced rudder and elevators.

STRUCTURE: Primary structure is 7781 pre-preg 36-38 per cent resin with Nomex core and graphite rods. Fuselage has central keel fitted with load-bearing rods.

LANDING GEAR: Fixed tricycle type; speed fairings optional. Skis, floats and tailwheel layout optional.

POWER PLANT: One 298 kW (400 hp) Textron Lycoming IO-720-A1B flat-eight driving a Hartzell HC-E3YR-1RF/ F8468A-6R constant-speed propeller. Standard fuel capacity 341 litres (90.0 US gallons; 74.9 Imp gallons), optionally increased to 526 litres (140 US gallons; 116 Imp gallons) with auxiliary tank in place of rear cabin baggage area.

ACCOMMODATION: Pilot and up to five passengers; club seating for four in rear cabin. Passenger/cargo double door on starboard side; optional door in port side for pilot. Dual controls.

AVIONICS: Onboard computer will provide aeronautical charts and e-mail services.

DIMENSIONS, EXTERNAL:
Wing span	10.92 m (35 ft 10 in)
Wing chord: at root	1.63 m (5 ft 4 in)
at tip	1.07 m (3 ft 6 in)
Wing aspect ratio	7.4
Length overall	9.12 m (29 ft 11 in)
Height overall	2.18 m (7 ft 2 in)
Tailplane span	3.96 m (13 ft 0 in)
Propeller diameter	2.11 m (6 ft 11 in)
Propeller ground clearance	0.25 m (9¾ in)
Passenger door: Height	1.02 m (3 ft 4 in)
Width	0.94 m (3 ft 1 in)
Cargo double doors: Height	0.98 m (3 ft 2½ in)
Width	1.12 m (3 ft 8 in)

DIMENSIONS, INTERNAL:
Cabin: Length	3.89 m (12 ft 9 in)
Max width	1.35 m (4 ft 5 in)
Max height	1.24 m (4 ft 1 in)

AREAS:
Wings, gross	16.17 m² (174.0 sq ft)
Ailerons (total)	1.61 m² (17.32 sq ft)
Trailing-edge flaps (total)	2.69 m² (29.00 sq ft)

WEIGHTS AND LOADINGS:
Weight empty: One seat	1,086 kg (2,395 lb)
Fully IFR equipped	1,134 kg (2,501 lb)

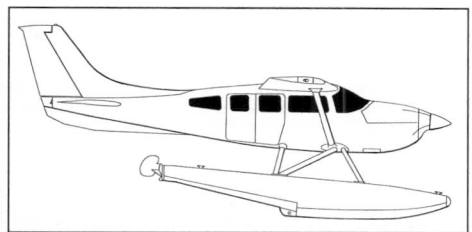
Floatplane BarrSix, showing underfin of this version and starboard freight door   0100429

Barr BarrSix six-seat tourer with optional pilot's door (*Jane's/James Goulding*)   0100430

Prototype BarrSix in late 2002   *NEW*/0526456

Baggage capacity	91 kg (200 lb)
Max T-O weight	2,041 kg (4,500 lb)
Max wing loading	124.8 kg/m² (25.86 lb/sq ft)
Max power loading	6.85 kg/kW (11.25 lb/hp)
PERFORMANCE:	
Never-exceed speed (VNE)	238 kg (441 km/h; 274 mph)
Max level speed	206 kt (381 km/h; 237 mph)
Max cruising speed	196 kt (362 km/h; 225 mph)
Normal cruising speed	180 kt (333 km/h; 207 mph)
Stalling speed, power off,	
flaps up	62 kt (115 km/h; 71 mph)
flaps down	54 kt (100 km/h; 62 mph)
Max rate of climb at S/L	274 m (900 ft)/min
Service ceiling	7,010 m (23,000 ft)
T-O run	274 m (900 ft)
T-O to 15 m (50 ft)	549 m (1,800 ft)
Landing from 15 m (50 ft)	427 m (1,400 ft)
Landing run	229 m (750 ft)
Range with max fuel	
	1,251 n miles (2,317 km; 1,440 miles)
Endurance	6 h 24 min
g limits	+3.86/−1.96

UPDATED

BEDE

BEDECORP LLC

6440 Norwalk Road, Unit C, Medina, Ohio 44256
Tel: (+1 330) 721 99 99
Fax: (+1 330) 721 99 98
e-mail: jim@bedecorp.com
Web (1): http://www.bedecorp.com
Web (2): http://www.bedeaerosport.com
PRESIDENT: James 'Jim' R Bede

Bedecorp specialises in design of high-performance jet aircraft for private ownership; civilian version of a previous product, BD-10, was transferred to Vortex Aircraft Company as the Phoenix. In June 2000, Bede announced it was working on the BD-17 Nugget, an all-metal low-wing single-seater and is actively marketing the BD-6 development of the BD-4; details of the BD-6 appeared in *Jane's* 1977-78 edition. The earlier BD-4 is also available in kit form from Tennessee Valley Aviation Products.

In May 2001, Jim Bede announced he was considering a further update of his 30-year-old BD-5 design.

UPDATED

BEDE BD-12

This single-seat sportplane kitbuilt was last described in the 2002-03 *Jane's*.

UPDATED

BEDE BD-17 NUGGET

TYPE: Single-seat sportplane kitbuilt.
PROGRAMME: Announced June 2000; prototype (N624BD) completed by end of that year and first flew 11 February 2001. Certification programme under way; four aircraft under construction by March 2001. Public debut at Sun 'n' Fun, April 2001.
COSTS: US$28,175 including engine (2003).
CUSTOMERS: Two aircraft registered by August 2002; three complete by January 2003
DESIGN FEATURES: Intended to be extremely easy to build, with approximately 110 parts. Optional folding wings. Constant-chord wings. Fin fillet.
FLYING CONTROLS: Conventional and manual. No flaps on prototype; kits will have flaps.
STRUCTURE: All-metal. Tubular spar and honeycomb fuselage covered with 5 mm (0.20 in) aluminium sheet; control surfaces of urethane foam.
LANDING GEAR: Fixed; choice of tricycle or tailwheel layout. Speed fairings optional. Scotch ply main legs. Differential braking. Prototype has 5.00-5 mainwheel tyres and 25 cm (10 in) nosewheel tyre.
POWER PLANT: Power plants in the range 33.6 to 59.7 kW (45 to 80 hp) being considered; prototype has 44.7 kW (60 hp) HKS-700 two-cylinder four-stroke engine. Fuel capacity 60.6 litres (16.0 US gallons; 13.3 Imp gallons).
ACCOMMODATION: Pilot only, rearward-sliding canopy.
All data provisional.

Prototype Bede BD-17 Nugget (*Jane's*/Paul Jackson)   *NEW*/0533385

Bede BD-17 Nugget single-seat kitbuilt (*Jane's*/James Goulding)   0121112

DIMENSIONS, EXTERNAL:
Wing span	6.55 m (21 ft 6 in)
Wing chord, constant	0.76 m (2 ft 6 in)
Wing aspect ratio	8.6
Length overall	5.33 m (17 ft 6 in)
Height overall	2.36 m (7 ft 9 in)

AREAS:
Wings, gross	4.97 m² (53.5 sq ft)

WEIGHTS AND LOADINGS:
Weight empty	191-204 kg (420-450 lb)
Max T-O weight	386 kg (850 lb)

PERFORMANCE (estimated):
Never-exceed speed (VNE)	169 kt (313 km/h; 195 mph)
Cruising speed	130 kt (241 km/h; 150 mph)
Stalling speed, flaps down	46 kt (84 km/h; 52 mph)
Max rate of climb at S/L	320 m (1,050 ft)/min
T-O run	250 m (820 ft)
Landing run	195 m (640 ft)
Range	610 n miles (1,129 km; 702 miles)

UPDATED

BEECH

RAYTHEON AIRCRAFT COMPANY

At the September 2002 NBAA Convention, held in Orlando, Florida, Raytheon Aircraft (which see for further details) announced its intention to revert to separate marketing of its Beech and Hawker lines of private and executive aircraft, reducing the emphasis on its corporate name.

Beech Aircraft Corporation founded in 1932 by Walter and Olive Ann Beech; trading name **Beechcraft**; became wholly owned subsidiary of Raytheon on 8 February 1980 and incorporated within Raytheon Aircraft Company upon its foundation on 15 September 1994.

NEW ENTRY

BEECH T-6A TEXAN II
Canadian Forces name: CT-156 Harvard II

TYPE: Basic turboprop trainer.
PROGRAMME: US version of Pilatus PC-9 Advanced Turbo Trainer (which see in Swiss section). For participation in USAF/USN Joint Primary Aircraft Training System (JPATS) competition, Beech and Pilatus reached agreement on joint approach in August 1990; Beech received two standard PC-9s from Pilatus, one of which (N26BA) converted as engineering development prototype and completed more than 260 hours' flight testing to reduce programme risk and develop engineering design baseline before return to Pilatus. Followed by two Beech-built production prototypes (Beech designation PD373; later, Beech 3000), first flights December 1992 (N8284M/ PT-2) and July 1993 (N209BA/PT-3); PT-2 used to complete flight test programme and evaluate systems

performance; PT-3 incorporated several improvements and was principal aircraft for USAF/USN flight evaluation; one prototype and four production aircraft completed more than 1,400 hours of flight testing to achieve FAA certification in July 1999. Promotional name for the purposes of competition was **Beech Mk II**.

US FORCES' T-6A TEXAN II PROCUREMENT

FY	Lot	Qty	First Aircraft	Order	Delivery
95	1	1[1]	95-3000	Jun 95	1999
96	2	3	95-3001	Jul 96	1999-2000
96	3	6	95-3004	Sep 96	2000
97	4	15	96-3010	Apr 97	2000
98	5	22	98-3025	Feb 98	2000-01
99	6	22	99-3547	May 99	2001-02
00	7	41[2]	00-3569	Jun 00	2002-03
01	8	58[3]		Apr 01	2003-04
02	9	47[4]		Jan 02	2004
02	10	35		Dec 02	2005
Total		**250**			

Notes: Serial number prefixes do not indicate year of order, as is usual in USAF. Early aircraft numbered 95-3000 to 3009, 96-3010 to 3013 and 97-3014 onwards
[1] Manufacturing development aircraft
[2] Includes 12 aircraft for US Navy
[3] Includes 24 aircraft for US Navy
[4] First batch of multiyear purchase involving 234 aircraft to be procured by USAF during FY02 to FY06; includes seven aircraft for US Navy

Selection as JPATS winner announced 22 June 1995; total requirement for 782 aircraft (454 USAF, 328 US Navy) by 2017; all to be built in USA; contract valued at US$4.7 billion awarded February 1996; allocated designation T-6 Texan II (in advance of T-4 and T-5) to honour North American AT-6 Texan of Second World War era; Canadian name reflects British Commonwealth name for AT-6.

First metal cut February 1997; roll-out of manufacturing development aircraft (N23262/95-3000/PT-4) 29 June 1998; first flight 15 July 1998; first delivery (95-3003/PT-7) to Randolph AFB, June 1999, for technical evaluation; further two were to have followed in third quarter of 1999 for six month multiservice operational test and evaluation programme (by first two Lot-3 aircraft), but were delayed by engineering fault with PT6 engine. First handover was 95-3004 on 1 March 2000; first squadron of 12th FTW, the 559th FTS, began equipping on 23 May 2000.

Initial production rate 12 aircraft per year, rising to maximum of 59 by 2004; Milestone III authorisation to proceed with full-rate production granted by USAF on 3 December 2001. Early aircraft to 12th FTW at Randolph AFB for instructor training; IOC with USAF at Moody AFB, Georgia, in June 2001, with student pilot training beginning on 10 October; IOC with Navy 2003; all USAF units equipped by 2011. FlightSafety selected on 21 April 1997 to provide associated ground training system at cost over 24 years of US$500 million.

US FORCES' T-6A TEXAN DELIVERIES

CY	USAF	USN
1999	1	
2000	49	
2001	47	

CURRENT VERSIONS: **T-6A Texan II**: USAF/USN version, *as described*.

T-6A-1/CT-156 Harvard II: Version for NATO Flying Training in Canada (NFTC) programme; generally similar to Texan II, but with blind-flying hood, dual VOR and ADF, and back-up VHF comm in place of UHF.

T-6B: Trainer and light attack (**AT-6B**) derivative of T-6A announced at Farnborough International in July 2002. To incorporate revised cockpit management systems, with avionics suite provided by Flight Visions; latter includes FV-4000 modular mission display processor (MMDP), SparrowHawk HUD, stores management system and associated multifunction displays (MFDs). Weaponry including guns, bombs and rockets will be available, permitting use in light attack/counter-insurgency role.

Beech/Pilatus PC-9 Mk II: Designation of Greek aircraft; first 25 are in USAF/US Navy configuration, remaining 20 in Greek-specified New Trainer Aircraft (NTA) configuration, with weapon sighting system plus provision for carriage of weaponry and auxiliary fuel tanks on three stores stations beneath each wing. Weapons trials, including tests with gun and rocket armament, conducted at Eglin AFB, Florida, in first half of 2002.
CUSTOMERS: USAF initial operator is 12th Flying Training Wing at Randolph AFB, Texas; second is 479th Flying

Beech T-6A Texan II of second training school to equip, 479th Flying Training Group at Moody AFB, Georgia
NEW/0530197

Training Group at Moody AFB, Georgia (49 aircraft during 2001-02); subsequent deliveries to 47th FTW, Laughlin AFB, Texas (96 aircraft during 2002-04); 71st FTW, Vance AFB, Oklahoma (91 aircraft during 2005-06), 14th FTW, Columbus AFB, Mississippi (89 aircraft during 2006-08) and 80th FTW, Sheppard AFB, Texas (69 aircraft during 2008-09). Initial Navy aircraft included in Lot 7; deliveries to Whiting Field, Florida, for TW-5 begin November 2002, with training to start in 2003; further deliveries to TW-4 at Corpus Christi, Texas, and TW-6 at Pensacola, Florida. US Navy aircraft allocated serial numbers 165958 to 166285.

Chilean Air Force signed a letter of intent in late 1996 for future purchase of 16 to 25 Beech Mk IIs, but has still to place firm order. Bombardier Services of Canada ordered 24 T-6A-1s in December 1997 for its NFTC programme for delivery to No. 2 FTS at Moose Jaw between April and December 2000; first delivery (156103) 29 February 2000; NFTC inaugurated 6 July 2000; all 24 aircraft handed over by end 2000. Further two aircraft ordered in mid-2002, with delivery expected by year-end. On 9 October 1998, Greek Air Force announced selection of T-6 and ordered 45, to be delivered between 2000 and 2003; option held on further five. First delivery (001) 17 July 2000 to 361 Squadron at Kalamata; total of 17 received by end of 2001, with training of student pilots having begun in September.

Raytheon holds marketing rights to PC-9 Mk II in all countries except Switzerland.
COSTS: Programme US$7 billion, of which US$362 million contracted by March 1998 for first 47 aircraft, comprising one manufacturing development aircraft and 46 production aircraft in five lots. Flyaway cost US$5.41 million over 740 aircraft (1998). Contracts for first 168 aircraft and associated ground training systems amount to US$852.6 million. Contract for two additional NFTC aircraft plus support valued at approximately US$11.6 million.
DESIGN FEATURES: Approximately 90 per cent redesign including strengthened fuselage; aerodynamic changes to rudder and elevator; improved birdproofing; stronger landing gear; more powerful engine with linear power response; trim aid device; increased fuel capacity and pressure refuelling; ergonomically improved pressurised

cockpits; numerous maintainability changes to decrease DOC including engine/systems diagnostics available through a central computer download connection; and new digital avionics. Certified to FAR Pt 23 in Aerobatic category.
FLYING CONTROLS: Conventional and manual. Selectable trim aid device (TAD) by Aeromach Labs automatically trims rudder in conjunction with throttle position to minimise effect of torque. Stall strip on starboard wing. Control deflections include rudder ±24°; elevators 18° up/16° down; ailerons 20° up/11° down; and flaps 23° for take-off, 50° for landing. Airbrake maximum deflection of 70°.
STRUCTURE: Generally all-metal; durability and damage-tolerant (DADT) structure includes wing and tailplane leading-edges reinforced for bird resistance; 18,000 hour service life. Avionics bay behind rear cockpit; forward-hinged door each side. Airframe built entirely in USA.
LANDING GEAR: Goodrich wheels and brakes. Capable of withstanding sink rates up to 4 m (13 ft)/s. Mainwheels 20×4.4 (14 ply (tubeless); nosewheel 16×4.4 (8 ply) tubeless. Otherwise generally as for PC-9.
POWER PLANT: One 1,274 kW (1,708 shp) Pratt & Whitney Canada PT6A-68 turboprop flat rated at 820 kW (1,100 shp), driving a four-blade, Hartzell propeller at a constant 2,000 rpm. Raytheon/P&WC power management unit provides jet-type linear throttle response. T-6A has integral 298 litre (78.5 US gallon; 65.5 Imp gallon) tank in each wing leading-edge, plus 26.5 litre (7.0 US gallon; 5.8 Imp gallon) collector tank below front cockpit; total fuel capacity 621 litres (164 US gallons; 137 Imp gallons). AT-6B fuel capacity 852 litres (225 US gallons; 187 Imp gallons). AT-6B has provision for two underwing auxiliary fuel tanks, each 265 litres (70.0 US gallons; 58.3 Imp gallons). Pressure refuelling/defuelling point in lower fuselage, adjacent to port wingroot; gravity refuelling points at three quarters' span in each wing upper surface.
ACCOMMODATION: Two pilots in stepped cockpits on Martin-Baker Mk 16LA zero/zero ejection seats. Limited HOTAS controls in T-6A; full multifunction HOTAS in AT-6B. Three-piece, Pilkington Aerospace acrylic starboard-hinged canopy; integral windscreen 19 mm (¾ in) thick for resistance to strike by 1.8 kg (4 lb) bird at 270 kt

Instrument panel of Beech T-6A Texan II
0131803

Beech CT-156 Harvard II in Canadian markings *(CAF)*   *NEW*/0527091

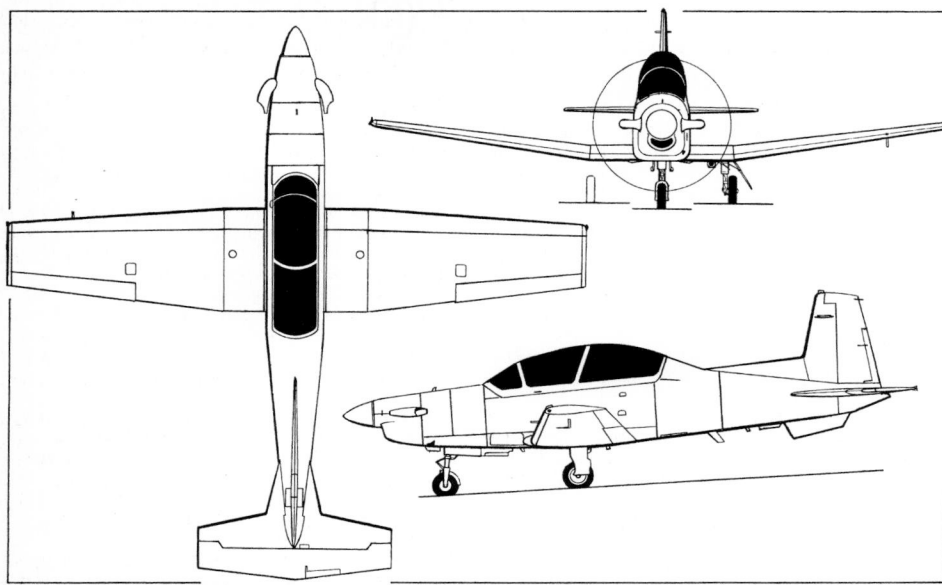

Beech T-6A Texan II primary trainer *(Jane's/Paul Jackson)*   0051810

(500 km/h; 311 mph); other panels 9 mm (⅓ in) thick. ATK canopy fracturing initiation unit and Teledyne McCormick fracturing system. Accommodation pressurised and air conditioned; warm air canopy defrosting/demisting. Baggage compartment in rear fuselage, behind avionics bay; upward-hinged door port side.

SYSTEMS: Generally as PC-9. Dowty Aerospace hydraulics; Enviro Systems air conditioning unit; Litton OBOGS. Cockpit pressurisation system has 0.24 bar (3.5 lb/sq in) max differential. Battery 28 V. Goodrich emergency power system.

AVIONICS: Honeywell primary contractor.

Comms: VHF/UHF transceiver, plus back-up VHF with separate antenna; intercom and interphone. MST 67A Mode S transponder. T-6A's radio management unit replaced in AT-6B by up-front control panel below HUD.

Flight: Dual VNS-41 navigation systems (including VN-411B receivers) for VOR, localiser, glideslope and marker beacon functions through common antenna. DFS-43A ADF (DF-431 receiver and AT-434 loop/sense antenna); KLN 90 GPS (combined INS-GPS in AT-6B); and Rockwell Collins 252-channel DM-441B DME. Litef fibre optic gyro for AHRS. Goodrich collision warning system. Flight data recorder. Integrated radar altimeter in AT-6B.

Instrumentation: Two independent EFIS 50 127 mm (5 in) square active matrix LCDs, plus control panel, in each cockpit (or one, plus 127 × 102 mm; 5 × 4 in MFD in AT-6B); attitude director indicator portion of EFIS provides primary attitude display, turn rate, mode selection annunciation, localiser/glideslope deviation as appropriate; horizontal situation indicator for primary heading display, primary navigation display, course select indication, navigation source annunciation, DME, localiser and glideslope deviation, remote map presentation and selected heading reference marker. Additional three 76 mm (3 in) square MFDs per cockpit provide engine and auxiliary instrument information, including fuel state and pressurisation (engine MFD being replaced in AT-6B by 152 × 203 mm; 6 × 8 in MFD). Standby instrumentation comprises airspeed, attitude, turn-and-slip and magnetic compass (electromechanical in T-6A, digital in T-6B). Korry Electronics warning panels. Flight Visions SparrowHawk HUD and second 152 × 203 mm (6 × 8 in) active matrix, liquid crystal MFD (with moving map provision) in T-6B version. AT-6B is compatible with Gen 4 NVGs.

Mission: Dual-mission computers in AT-6B. Provision for laser range-finder.

ARMAMENT: AT-6B version has three hardpoints under each wing and is able to carry 0.5 in gun pods, 2.75 in rocket pods and bombs in light attack role.

DIMENSIONS, EXTERNAL: As for PC-9M
AREAS:

Wings, gross	16.29 m² (175.3 sq ft)

WEIGHTS AND LOADINGS:

Weight empty	2,136 kg (4,709 lb)
Baggage capacity	36 kg (80 lb)
Max fuel weight	499 kg (1,100 lb)
Max underwing stores (AT-6B)	1,415 kg (3,120 lb)
Max T-O weight	2,948 kg (6,500 lb)
Max wing loading	181.0 kg/m² (37.08 lb/sq ft)
Max power loading	3.60 kg/kW (5.91 lb/shp)

PERFORMANCE:

Never-exceed speed (V_NE)	350 kt (648 km/h; 402 mph)
Max level speed: at altitude	316 kt (585 km/h; 364 mph)
at low level	270 kt (500 km/h; 311 mph) IAS
Max cruising speed at 2,285 m (7,500 ft)	
	230 kt (426 km/h; 265 mph)
Approach speed	100 kt (185 km/h; 115 mph) IAS
Stalling speed, power off:	
flaps up	82 kt (152 km/h; 95 mph)
flaps down	74 kt (137 km/h; 86 mph)
Max rate of climb at S/L	1,372 m (4,500 ft)/min
Service ceiling	9,450 m (31,000 ft)

T-O run	437 m (1,435 ft)
T-O to 15 m (50 ft)	654 m (2,145 ft)
Landing from 15 m (50 ft)	1,030 m (3,380 ft)
Landing run	739 m (2,425 ft)
Range at altitude	850 n miles (1,574 km; 978 miles)
Endurance at max cruising speed	3 h
g limits	+7/−3.5
	UPDATED

BEECH BONANZA A36

TYPE: Six-seat utility transport.

PROGRAMME: Descended from Model 35 of 1945 (10,403 built in V tail configuration), but now has little more than the name in common; further 1,911 late Model 33s built as Bonanzas, earlier version being Debonair. Model 36 certified 1 May 1968; developed from Bonanza E33A; A36 certified 24 October 1969 and introduced 1970; current subvariant announced 3 October 1983, succeeding model powered by 212.5 kW (285 hp) Continental IO-520-BB; certified in FAA Utility category.

CURRENT VERSIONS: **Bonanza A36:** Standard version. Features introduced for 2001 model year include dual Garmin GNS 430 nav/comm/GPS.
Description applies to A36, except where otherwise stated.
The **Bonanza A36AT (Airline Trainer)** and **Bonanza A36 Jaguar Special Edition** have both been discontinued. Descriptions in 2001-02 and previous editions.
Bonanza B36TC: Described separately.

CUSTOMERS: Total 3,488 Bonanza 36/A36s registered by November 2002, including 89 in 1995, 83 in 1996, 85 in 1997, 71 in 1998, 77 in 1999, 85 in 2000, 63 in 2001 and 51 in 2002.

COST: US$579,400 basic; US$596,000 typically equipped (2001).

DESIGN FEATURES: Conventional low-wing cabin monoplane. Tapered wings with leading-edge gloves; tapered tailplane and sweptback fin.
Beech modified NACA 23016.5 wing section at root, modified 23012 at tip; dihedral 6°; incidence 4° at root and 1° at tip.

FLYING CONTROLS: Conventional and manual. Electrically actuated trim tab in each elevator; ground-adjustable tabs in ailerons and rudder; single-slotted, three-position flaps.

STRUCTURE: Light alloy, with two-spar wing torsion box and stressed-skin tail surfaces. Conventional construction.

LANDING GEAR: Electrically retractable tricycle type, with steerable nosewheel. Mainwheels retract inward into wings, nosewheel rearward. Beech oleo-pneumatic shock-absorbers in all units. Cleveland mainwheels, size 6.00-6, and tyres, size 7.00-6 (6 ply), pressure 2.28 to 2.76 bar (33 to 40 lb/sq in). Cleveland nosewheel and tyre, size 5.00-5 (4 ply), pressure 2.76 bar (40 lb/sq in). Cleveland ring disc hydraulic brakes. Parking brake. Magic Hand landing gear system optional.

POWER PLANT: One 224 kW (300 hp) Teledyne Continental IO-550-B Raytheon Special Edition flat-six engine, driving a Hartzell three-blade constant-speed metal propeller. The engine is equipped with an altitude-compensating fuel pump which automatically makes the fuel/air mixture leaner and richer during climb and descent respectively. Usable fuel capacity 280 litres (74.0 US gallons; 61.6 Imp gallons).

ACCOMMODATION: Enclosed cabin seating four to six persons on individual seats. Pilot's seat is vertically adjustable. Dual controls standard. Two rear removable seats and two folding seats permit rapid conversion to utility configuration. Optional club seating with rear-facing third and fourth seats, executive writing desk, refreshment cabinet, headrests for third and fourth seats, reading lights and fresh air outlets for fifth and sixth seats. Double doors

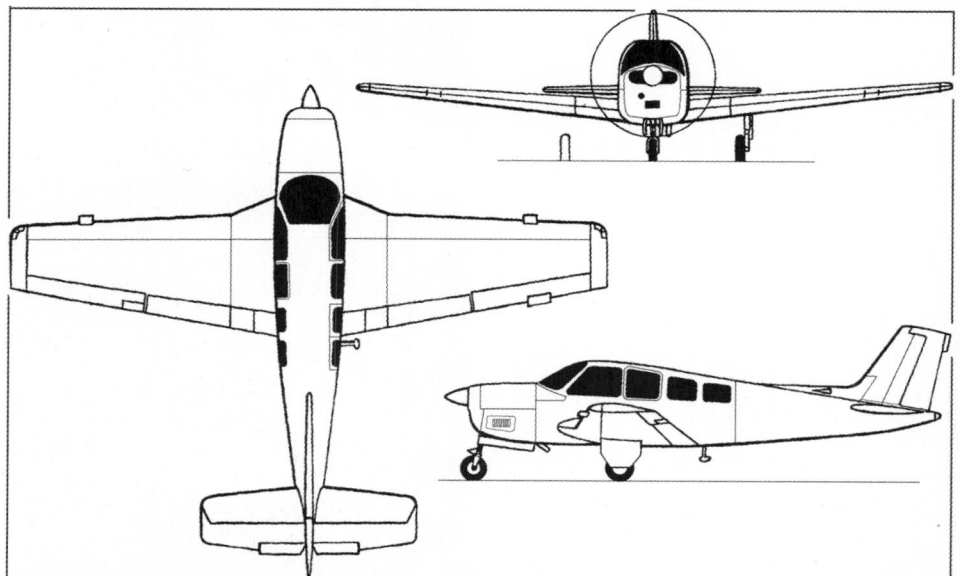

Beech Bonanza A36 four/six-seat utility aircraft *(Jane's/James Goulding)*   *NEW*/0528617

of bonded aluminium honeycomb construction on starboard side facilitate loading of cargo. As an air ambulance, one stretcher can be accommodated with ample room for a medical attendant and/or other passengers. Extra windows provide improved view for passengers.

SYSTEMS: Optional 12,000 BTU refrigeration-type air conditioning system comprises evaporator located beneath pilot's seat, condenser on lower fuselage, and engine-mounted compressor. Air outlets on centre console, with two-speed blower. Electrical system supplied by 28 V 100 A alternator, 24 V 15.5 Ah battery; optional standby generator. Hydraulic system for brakes only. Standby vacuum pump standard. Pneumatic system for instrument gyros optional. Oxygen system and electric propeller de-icing optional.

AVIONICS: *Comms:* Dual Garmin GNS 430 nav/comms with GI 106 indicators or one each GNS 430 and GNS 530; Bendix/King KT 76C transponder; PS Engineering PMA7000M-S audio control system.

Flight: Bendix/King encoding altimeter; KFC 225 AFCS; KI 256 flight command indicator/gyro horizon; KCS 55A compass; Goodrich WX-500 Stormscope; Shadin ADC 200+ fuel and air data computer.

EQUIPMENT: Standard equipment includes LCD digital chronometer, EGT and OAT gauges, rate of climb indicator, turn co-ordinator, 3 in horizon and directional gyros, four fore- and aft-adjustable and reclining seats, armrests, headrests, single diagonal strap shoulder harness with inertia reel for all occupants, pilot's storm window, ultraviolet-proof windscreen and windows, sun visors, large cargo door, emergency locator transmitter, stall warning device, alternate static source, heated pitot, rotating beacon, three-light strobe system, carpeted floor, super soundproofing, control wheel map lights, entrance door courtesy light, internally lit instruments, coat hooks, glove compartment, in-flight storage pockets, approach plate holder, utility shelf, cabin dome light, reading lights, instrument post lights, control wheel map light, electroluminescent subpanel lighting, landing light, taxying light, full-flow oil filter, three-colour polyurethane exterior paint, external power socket, static wicks and towbar.

Optional equipment includes dual controls, leather seats, co-pilot's wheel brakes, air conditioning, fifth passenger seat, fresh air vent blower and ground com switch.

DIMENSIONS, EXTERNAL:
Wing span	10.21 m (33 ft 6 in)
Wing chord: at root	2.13 m (7 ft 0 in)
at tip	1.07 m (3 ft 6 in)
Wing aspect ratio	6.2
Length overall	8.38 m (27 ft 6 in)
Height overall	2.62 m (8 ft 7 in)
Tailplane span	3.71 m (12 ft 2 in)
Wheel track	2.92 m (9 ft 7 in)
Wheelbase	2.39 m (7 ft 10¼ in)
Propeller diameter: A36	2.03 m (6 ft 8 in)
A36AT	1.93 m (6 ft 4 in)
Forward cockpit door: Height	0.91 m (3 ft 0 in)
Width	0.94 m (3 ft 1 in)
Baggage compartment door: Height	0.61 m (2 ft 0 in)
Width	0.99 m (3 ft 3 in)
Rear passenger/cargo door: Height	0.89 m (2 ft 11 in)
Width	1.14 m (3 ft 9 in)

Beech Bonanza A36 in latest, patriotic colour scheme (*Jane's/Paul Jackson*)   NEW/0526944

DIMENSIONS, INTERNAL:
Cabin (aft of firewall, incl extended baggage compartment):
Length	3.84 m (12 ft 7 in)
Max width	1.07 m (3 ft 6 in)
Max height	1.27 m (4 ft 2 in)
Volume	3.9 m³ (136 cu ft)

AREAS:
Wings, gross	16.80 m² (181.0 sq ft)
Ailerons (total)	1.06 m² (11.40 sq ft)
Trailing-edge flaps (total)	1.98 m² (21.30 sq ft)
Fin	0.93 m² (10.00 sq ft)
Rudder, incl tab	0.52 m² (5.60 sq ft)
Tailplane	1.75 m² (18.82 sq ft)
Elevators, incl tabs	1.67 m² (18.00 sq ft)

WEIGHTS AND LOADINGS:
Weight empty, standard:	
A36, A36AT	1,052 kg (2,320 lb)
Basic operating weight	1,148 kg (2,530 lb)
Baggage capacity	181 kg (400 lb)
Max T-O weight: A36	1,655 kg (3,650 lb)
A36AT	1,633 kg (3,600 lb)
Max ramp weight: A36	1,661 kg (3,663 lb)
A36AT	1,639 kg (3,613 lb)
Payload with max fuel	313 kg (689 lb)
Max wing loading: A36	98.5 kg/m² (20.17 lb/sq ft)
A36AT	97.1 kg/m² (19.89 lb/sq ft)
Max power loading: A36	7.40 kg/kW (12.17 lb/hp)
A36AT	7.59 kg/kW (12.46 lb/hp)

PERFORMANCE:
Max level speed (min weight)	184 kt (340 km/h; 212 mph)
Max cruising speed (mid-cruise weight):	
2,500 rpm at FL60	176 kt (326 km/h; 202 mph)
2,300 rpm at FL80	167 kt (309 km/h; 192 mph)
2,100 rpm at FL60	160 kt (296 km/h; 184 mph)
2,100 rpm at FL100	153 kt (283 km/h; 176 mph)
Stalling speed, power off:	
flaps up	68 kt (126 km/h; 78 mph) IAS
30° flap	59 kt (109 km/h; 68 mph) IAS
Max rate of climb at S/L	368 m (1,208 ft)/min
Service ceiling	5,640 m (18,500 ft)
T-O run: flaps up: A36	360 m (1,185 ft)
A36AT	450 m (1,475 ft)

12° flap: A36	296 m (970 ft)
A36AT	402 m (1,320 ft)
T-O to 15 m (50 ft): flaps up: A36	640 m (2,100 ft)
A36AT	662 m (2,170 ft)
12° flap: A36	583 m (1,913 ft)
A36AT	618 m (2,025 ft)
Landing from 15 m (50 ft)	442 m (1,450 ft)
Landing run	280 m (920 ft)
Range with max usable fuel, with allowances for engine start, taxi, T-O, climb and 45 min reserves at econ cruise power: 2,500 rpm at FL120	
	875 n miles (1,621 km; 1,008 miles)
2,300 rpm at FL120	
	903 n miles (1,672 km; 1,039 miles)
2,100 rpm at FL50	
	916 n miles (1,696 km; 1,054 miles)
Ferry range	930 n miles (1,722 km; 1,070 miles)

UPDATED

BEECH TURBO BONANZA B36TC

TYPE: Six-seat utility transport.

PROGRAMME: Derived from A36 (see previous entry). Certified as A36TC 7 December 1978; 271 A36TCs delivered; improved B36TC introduced 1982 following type certification on 15 January 1982.

CURRENT VERSIONS: **Bonanza B36TC:** Standard version, *as described.* New (2001) features as for A36.

Bonanza B36TC Jaguar Special Edition: Discontinued. Description in 2001-02 edition.

CUSTOMERS: Total of 695 A/B36TCs registered by November 2002, including 14 per year in 1993-97, 22 in 1998, 20 in 1999, 18 in 2000, 26 in 2001 and five in 2002.

COST: US$648,400 basic; US$665,000 typically equipped (2001).

DESIGN FEATURES: Compared with A36TC, B36TC has greater span; wing section NACA 23010.5 at tip; 0° incidence at tip; greater fuel capacity.

Data below summarise differences from A36TC:

POWER PLANT: One 223.7 kW (300 hp) Teledyne Continental TSIO-520-UB Raytheon Special Edition turbocharged flat-six engine, driving a three-blade constant-speed Hartzell metal propeller. Fixed engine cowl flaps. Two fuel tanks in each wing leading-edge, with total usable capacity of 386 litres (102 US gallons; 84.9 Imp gallons). Refuelling points above tanks. Oil capacity 11.5 litres (3.0 US gallons; 2.5 Imp gallons).

SYSTEMS: Air conditioning optional.

EQUIPMENT: EGT gauge not available. Turbine inlet temperature gauge is standard.

DIMENSIONS, EXTERNAL:
Wing span	11.53 m (37 ft 10 in)
Wing chord at tip	0.91 m (3 ft 0 in)
Wing aspect ratio	7.6
Propeller diameter	1.98 m (6 ft 6 in)

AREAS:
Wings, gross	17.47 m² (188.1 sq ft)

WEIGHTS AND LOADINGS:
Weight empty, standard	1,116 kg (2,460 lb)
Basic operating weight	1,241 kg (2,737 lb)
Max T-O and landing weight	1,746 kg (3,850 lb)
Max ramp weight	1,753 kg (3,866 lb)
Payload with max fuel	233 kg (514 lb)
Max wing loading	99.9 kg/m² (20.47 lb/sq ft)
Max power loading	7.81 kg/kW (12.83 lb/hp)

PERFORMANCE (speeds at mid-cruise weight):
Max level speed at FL220	213 kt (394 km/h; 245 mph)
Cruising speed at FL250	
at 79% power	200 kt (370 km/h; 230 mph)
at 75% power	195 kt (361 km/h; 224 mph)
at 69% power	188 kt (348 km/h; 216 mph)
at 56% power	173 kt (320 km/h; 199 mph)
Stalling speed, power off:	
flaps up	65 kt (120 km/h; 75 mph) IAS
30° flap	57 kt (106 km/h; 66 mph) IAS
Max rate of climb at S/L	321 m (1,053 ft)/min
Service ceiling	over 7,620 m (25,000 ft)
T-O run, 15° flap	311 m (1,020 ft)
T-O to 15 m (50 ft), 15° flap	649 m (2,130 ft)

Current configuration of Bonanza instrument panel   0110702

Landing from 15 m (50 ft)	516 m (1,692 ft)
Landing run	298 m (980 ft)

Range at FL250 with max fuel, allowances for engine start, taxi, T-O, cruise climb, descent, and 45 min reserves at 50% power:

at 79% power	956 n miles (1,770 km; 1,100 miles)
at 75% power	984 n miles (1,822 km; 1,132 miles)
at 69% power	1,022 n miles (1,893 km; 1,176 miles)
at 56% power at 6,100 m (20,000 ft)	
	1,087 n miles (2,013 km; 1,250 miles)
Ferry range	1,169 n miles (2,165 km; 1,345 mph)

UPDATED

Beech Baron 58 in 2002 colours   *NEW*/0526943

BEECH BARON 58

TYPE: Six-seat utility twin.

PROGRAMME: Prototype Beech 55 Baron flew 29 February 1960; total 3,728 built, not including 94 Model 56 Turbo Barons. Model 58 derived from E55 with 25 cm (10 in) fuselage stretch, plus internal rearrangement to lengthen cabin by 76 cm (30 in); certified in FAA Normal category 19 November 1969; marketing began in following month. Raytheon delivered 2,000th standard Model 58 (N558TH) in July 2001. Beech 58P Pressurised Baron and 58TC turbocharged version no longer produced.

CURRENT VERIONS: **Baron 58:** Standard version, *as described.* Features introduced for 2001 model year include dual Garmin GNS 530 nav/comm/GPS.

Baron 58 Jaguar Special Edition: Discontinued. Last described in 2001-02 edition.

CUSTOMERS: Operators include several commercial aviation flying schools. Total 2,690 Baron 58s delivered by late 2002, including 29 in 1995, 44 in 1996, 35 in 1997, 42 in 1998, 47 in 1999, 50 in 2000, 47 in 2001 and 27 in 2002. (Incorporates 497 of 58P and 151 of 58TC versions.)

COST: US$1,010,400 basic; US$1,058,030 typically equipped (2001).

DESIGN FEATURES: Conventional low-wing monoplane, ultimately derived from Beech Bonanza. Tapered wing with leading-edge gloves inboard of engines; sweptback fin.

Wing section NACA 23015.5 at root, 23010.5 at tip; dihedral 6°; incidence 4° at root and 0° at tip.

FLYING CONTROLS: Conventional and manual. Manually operated trim tabs in elevators, rudder and port aileron; electrically operated single-slotted flaps. Control surface movements: ailerons ±20°; elevator ±30/−15°; rudder ±25°; flap settings 15° for approach, 28° maximum.

STRUCTURE: Light alloy with two-spar wing box; elevators have smooth magnesium alloy skins.

LANDING GEAR: Electrically retractable tricycle type. Main units retract inward into wings, nosewheel aft. Beech oleo-pneumatic shock-absorbers in all units. Steerable nosewheel with shimmy damper. Cleveland wheels, with mainwheel tyres size 6.50-8 (8 ply) tubed, pressure 3.59 to 3.96 bar (52 to 56 lb/sq in) or 19.5×6.75-8 (10 ply) tubeless. Nosewheel tyre size 5.00-5 (6 ply) tubed, pressure 3.79 to 4.14 bar (55 to 60 lb/sq in) or 15×6.0-6 (4 ply) tubed. Cleveland ring disc hydraulic brakes. Heavy-duty brakes optional. Parking brake.

POWER PLANT: Two 224 kW (300 hp) Teledyne Continental IO-550-C Raytheon Special Edition flat-six engines, each driving a Hartzell three-blade constant-speed fully feathering metal propeller. The standard fuel system has a usable capacity of 628 litres (166 US gallons; 138 Imp gallons). Optional 'wet wingtip' installation also available, increasing usable capacity to 734 litres (194 US gallons; 161.5 Imp gallons).

ACCOMMODATION: Standard version has four individual seats in pairs in enclosed, soundproofed, heated and ventilated cabin, with door on starboard side. Single diagonal strap shoulder harness with inertia reel standard on all seats.

Pilot's vertically adjusting seat is standard. Co-pilot's vertically adjusting seat, folding fifth and sixth seats, or club seating comprising folding fifth and sixth seats and aft-facing third and fourth seats, are optional. Adjustable rudder pedals (retractable on starboard side). Executive desk available as option with club seating. Baggage compartment in nose. Double passenger/cargo doors on starboard side of cabin provide access to space for baggage or cargo behind the third and fourth seats. Pilot's storm window. Openable windows adjacent the third and fourth seats are used for ground ventilation and as emergency exits. Windscreen defrosting standard.

SYSTEMS: Cabin heated by Janitrol 50,000 BTU heater, which serves also for windscreen defrosting. Oxygen system of 1,389 litres (49 cu ft) or 1,814 litres (64 cu ft) capacity optional. Electrical includes two 28 V 100 A engine-driven alternators with alternator failure lights and two 12 V 25 Ah batteries. Two 100 A alternators optional. Hydraulic system for brakes only. Pneumatic pressure system for air-driven instruments, and optional wing and tail unit de-icing system. Oxygen system, cabin air conditioning and windscreen electric anti-icing optional.

AVIONICS: As for Beech A36 Bonanza, plus radar.

Radar: Honeywell RDR 2000 VP colour weather radar.

EQUIPMENT: Standard equipment includes ultraviolet-proof windscreen and cabin windows, super soundproofing, heated pitot head, instrument panel floodlights, navigation and position lights, steerable taxying light, dual landing lights, heated fuel vents, heated fuel and stall warning vanes and external power socket.

Options include alternate static source, internally illuminated instruments, strobe lights, electric windscreen anti-icing, wing ice detection light, static wicks, leather seat trim, cabin club seating, cockpit relief tube, cabin fire extinguisher and ventilation blower.

DIMENSIONS, EXTERNAL:

Wing span	11.53 m (37 ft 10 in)
Wing chord: at root	2.13 m (7 ft 0 in)
at tip	0.90 m (2 ft 11½ in)
Wing aspect ratio	7.2
Length overall	9.09 m (29 ft 10 in)
Height overall	2.97 m (9 ft 9 in)
Tailplane span	4.85 m (15 ft 11 in)
Wheel track	2.92 m (9 ft 7 in)
Wheelbase	2.72 m (8 ft 11 in)
Propeller diameter	1.96 m (6 ft 5 in)
Rear passenger/cargo doors:	
Max height	0.89 m (2 ft 11 in)
Width	1.14 m (3 ft 9 in)

Baggage door (fwd): Height	0.56 m (1 ft 10 in)
Width	0.64 m (2 ft 1 in)

DIMENSIONS, INTERNAL:

Cabin (incl rear baggage area):	
Length	3.84 m (12 ft 7 in)
Max width	1.07 m (3 ft 6 in)
Max height	1.27 m (4 ft 2 in)
Floor area	3.72 m² (40.0 sq ft)
Volume	3.9 m³ (136 cu ft)
Baggage compartment: nose	0.51 m³ (18.0 cu ft)

AREAS:

Wings, gross	18.51 m² (199.2 sq ft)
Ailerons (total)	1.06 m² (11.40 sq ft)
Trailing-edge flaps (total)	1.98 m² (21.30 sq ft)
Fin	1.46 m² (15.67 sq ft)
Rudder, incl tab	0.81 m² (8.75 sq ft)
Tailplane	4.95 m² (53.30 sq ft)
Elevators, incl tabs	1.84 m² (19.80 sq ft)

WEIGHTS AND LOADINGS:

Weight empty	1,633 kg (3,600 lb)
Basic operating weight	1,756 kg (3,871 lb)
Baggage capacity: cabin	181 kg (400 lb)
nose	136 kg (300 lb)
Max T-O weight	2,495 kg (5,500 lb)
Max ramp weight	2,506 kg (5,524 lb)
Payload with max fuel	213 kg (470 lb)
Max landing weight	2,449 kg (5,400 lb)
Max wing loading	134.8 kg/m² (27.61 lb/sq ft)
Max power loading	5.58 kg/kW (9.17 lb/hp)

PERFORMANCE (cruising speeds at average cruise weight):

Max level speed at S/L	208 kt (386 km/h; 239 mph)
Max cruising speed, 2,500 rpm at FL50	
	203 kt (376 km/h; 234 mph)
Cruising speed, 2,500 rpm at FL80	
	192 kt (356 km/h; 221 mph)
Econ cruising speed, 2,100 rpm at FL120	
	162 kt (300 km/h; 186 mph)
Stalling speed, power off:	
flaps up	84 kt (156 km/h; 97 mph) IAS
flaps down	75 kt (139 km/h; 86 mph) IAS
Max rate of climb at S/L	529 m (1,735 ft)/min
Rate of climb at S/L, OEI	119 m (390 ft)/min
Service ceiling	6,305 m (20,680 ft)
Service ceiling, OEI	2,220 m (7,280 ft)
T-O run	427 m (1,400 ft)
T-O to 15 m (50 ft)	701 m (2,300 ft)
Landing from 15 m (50 ft)	747 m (2,450 ft)
Landing run	434 m (1,425 ft)

Range with 628 litres (166 US gallons; 138 Imp gallons) usable fuel, power/altitude settings as above, with allowances for engine start, taxi, T-O, climb and 45 min reserves at econ cruise power:

at max cruising speed	
	860 n miles (1,593 km; 990 miles)
at cruising speed	
	974 n miles (1,804 km; 1,121 miles)
at econ cruising speed	
	1,313 n miles (2,432 km; 1,511 miles)
Ferry range	1,569 n miles (2,905 km; 1,805 miles)

UPDATED

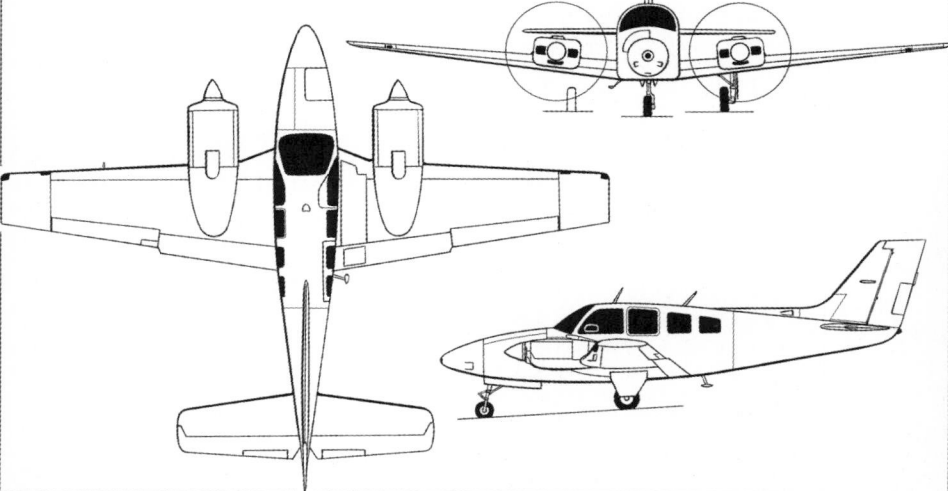

Beech Baron 58 (two Teledyne Continental IO-550-C piston engines) *(Jane's/Dennis Punnett)*   *NEW*/0527088

Instrument panel of Beech Baron   0110701

BEECH KING AIR C90B

TYPE: Business twin-turboprop.

PROGRAMME: King Air 90 first flew 21 November 1963. C90B announced at NBAA Convention, October 1991; superseded King Air 90, A90, B90, C90, C90-1, C90A; introduced four-blade McCauley propellers, special interior soundproofing, updated and redesigned interior, updated cockpit features and interior noise and vibration levels substantially reduced. A model delivered on 24 June 1996 to Jeld-Wen of Klamath Falls, Oregon, was the 5,000th King Air of all versions to be produced.

CURRENT VERSIONS: **C90B:** Standard version from 1991.

Detailed description applies to C90B.

King Air 200 and 350: Described in following entries.

CUSTOMERS: Total 1,670 commercial and 226 military King Air 90/A90/B90/C90/C90-1/C90A/C90B/C90SE delivered by October 2002, including 40 C90Bs and (now discontinued) C90SEs in 1995, 38 C90Bs in 1996, 41 C90Bs in 1997, 37 C90Bs in 1998, 41 C90Bs in 1999, 46 C90Bs in 2000, 41 C90Bs in 2001 and 21 C90Bs in 2002; 1,600th delivered to Larry V Plummer at Camarillo, California, September 2000.

COSTS: C90B US$2.867 million (2001).

DESIGN FEATURES: Conventional low-wing, twin-engined monoplane. Wing section NACA 23014.1 (modified) at root, 23016.22 (modified) at outer end of centre-section, 23012 at tip; dihedral 7°; incidence 4° 48′ at root, 0° at tip; tailplane 7° dihedral.

FLYING CONTROLS: Conventional and manual. Trim tabs on port aileron, in both elevators and rudders; single-slotted aluminium flaps.

STRUCTURE: Generally light alloy; magnesium ailerons. Internal corrosion-proofing.

LANDING GEAR: Hydraulically retractable tricycle type. Nosewheel retracts rearward, mainwheels forward into engine nacelles. Mainwheels protrude slightly beneath nacelles when retracted, for safety in a wheels-up emergency landing. Fully castoring steerable nosewheel with shimmy damper. Beech oleo-pneumatic shock-absorbers. Goodrich mainwheels with tyres size 8.50-10 (8 ply) tubeless, pressure 3.79 bar (55 lb/sq in). Goodrich nosewheel with tyre size 6.50-10 (6 ply) tubeless, pressure 3.59 bar (52 lb/sq in). Goodrich heat-sink and air-cooled multidisc hydraulic brakes. Parking brake. Minimum ground turning radius 10.82 m (35 ft 6 in).

POWER PLANT: Two 410 kW (550 shp) Pratt & Whitney Canada PT6A-21 turboprops, each driving a Hartzell four-blade constant-speed fully feathering propeller. Propeller auto ignition system, environmental fuel drain collection system, magnetic chip detector, automatic propeller feathering and propeller synchrophaser standard. Fuel in two tanks in engine nacelles, each with usable capacity of 231 litres (61.0 US gallons; 50.8 Imp gallons), and auxiliary bladder tanks in outer wings, each with capacity of 496 litres (131 US gallons; 109 Imp gallons). Total usable fuel capacity 1,454 litres (384 US gallons; 320 Imp gallons). Refuelling points in top of each engine nacelle and in wing leading-edge outboard of each nacelle. Oil capacity 13.2 litres (3.5 US gallons; 2.9 Imp gallons) per engine.

ACCOMMODATION: Two four-way adjustable seats side by side in cockpit with dual controls standard. Normally, four reclining seats in main cabin, in two pairs facing each other fore and aft. Standard furnishings include cabin forward partition, with fore and aft partition and coat rack, hinged nose baggage compartment door, seat belts and inertia reel shoulder harnesses for all seats. Standard accommodation for six passengers, one in club arrangement, one on sideways-facing seat adjacent door, and one on lavatory/seat in baggage area. Baggage racks at rear of cabin on starboard side, with lavatory on port side. Door on port side aft of wing, with built-in airstairs. Emergency exit on starboard side of cabin. Entire accommodation pressurised, heated and air conditioned. Electrically heated windscreen, windscreen defroster and windscreen wipers standard.

SYSTEMS: Pressurisation by dual engine bleed air system with pressure differential of 0.34 bar (5.0 lb/sq in). Cabin heated by 45,000 BTU dual engine bleed air system and auxiliary electrical heating system. Hydraulic system for landing gear actuation. Electrical system includes two 28 V 300 A starter/generators, 24 V 34 Ah air-cooled Ni/Cd battery with failure detector. Oxygen system, 623 litres (22 cu ft) or 1,814 litres (64 cu ft) capacity, optional. Vacuum system for flight instruments. Automatic pneumatic de-icing of wing/fin/tailplane leading-edges standard. Engine and propeller anti-icing systems standard. Engine fire detection and extinguishing system optional.

AVIONICS: Standard Rockwell Collins Pro Line II package with two-tube EFIS-84 in sectional instrument panel.

Comms: Dual VHF-22A transceivers with CTL-22 controls; dual TDR-64 transponders; dual DB Systems Model 415 audio systems; dual Lockheed Martin Fairchild CVR A1005; dual Flite-Tronics PC-250 inverters; edgelight radio panel; ELT; avionics master switch; ground clearance switch on com 1; control wheel push-to-talk switches; dual hand microphones and cockpit speakers; dual Telex headsets.

Radar: WXR-270 colour weather radar.

Beech King Air C90 business twin-turboprop (*Jane's/Paul Jackson*) NEW/0526942

Flight: Dual VIR-32 VOR/LOC/GLS/MKR receivers with CTL-32 controls; ADF-60A with CTL-62 control; DME-42 with IND-42 indicator; RMI-30; dual MCS-65 compass systems. Goodrich WX-1000+ Stormscope optional. APS-65H autopilot/flight director with EADI-84 and EHSI.

Instrumentation: EFIS-84; ALT-50A radio altimeter; pilot's encoding altimeter; dual 2 in electric turn and bank indicators; co-pilot's 75 mm (3 in) horizon indicator; co-pilot's HSI. Dual blind-flying instrumentation with dual instantaneous VSIs; standby magnetic compass; digital OAT gauge; LCD digital chronometer clock; vacuum gauge; de-icing pressure gauge; cabin rate of climb indicator; cabin altitude and differential pressure indicator; flight hour recorder; automatic solid-state warning and annunciator panel. Primary and secondary instrument lighting systems.

EQUIPMENT: Standard equipment includes fresh air outlets; oxygen outlets with overhead-mounted diluter demand masks with microphones; removable low-profile lavatory with shoulder harness and lap belt; two cabin tables; cabin fire extinguisher; wing ice lights; dual landing lights; nosewheel taxying light; flush position lights; dual rotating beacons; rheostat-controlled white cockpit lighting; wingtip recognition lights; wingtip and tail strobe lights; and vertical tail illumination lights.

Optional equipment includes electric flushing lavatory; engine fire detection system; oxygen bottle; cabinet with three drawers and stereo tape deck storage; and pilot-to-cabin paging with four stereo speakers.

DIMENSIONS, EXTERNAL:
Wing span	15.32 m (50 ft 3 in)
Wing chord: at root	2.15 m (7 ft 0½ in)
at tip	1.07 m (3 ft 6 in)
Wing aspect ratio	8.6
Length overall	10.82 m (35 ft 6 in)
Height overall	4.34 m (14 ft 3 in)
Tailplane span	5.26 m (17 ft 3 in)
Wheel track	3.89 m (12 ft 9 in)
Wheelbase	3.73 m (12 ft 3 in)
Propeller diameter	2.29 m (7 ft 6 in)
Propeller ground clearance	0.34 m (1 ft 1½ in)
Passenger door: Height	1.30 m (4 ft 3½ in)
Width	0.69 m (2 ft 3 in)
Height to sill	1.22 m (4 ft 0 in)

DIMENSIONS, INTERNAL:
Pressurised area: Length	5.43 m (17 ft 10 in)
Cabin: Length	3.78 m (12 ft 5 in)
Max width	1.37 m (4 ft 6 in)
Max height	1.45 m (4 ft 9 in)
Floor area	6.50 m² (70.0 sq ft)
Volume	6.4 m³ (227 cu ft)
Baggage compartment volume, rear	1.4 m³ (48 cu ft)

AREAS:
Wings, gross	27.31 m² (293.9 sq ft)
Ailerons (total)	1.29 m² (13.90 sq ft)
Trailing-edge flaps (total)	2.72 m² (29.30 sq ft)
Fin	2.20 m² (23.67 sq ft)

Rudder, incl tab	1.30 m² (14.00 sq ft)
Tailplane	4.39 m² (47.25 sq ft)
Elevators, incl tabs (total)	1.66 m² (17.87 sq ft)

WEIGHTS AND LOADINGS:
Weight empty	3,040 kg (6,702 lb)
Basic operating weight, typical	3,180 kg (7,010 lb)
Max fuel weight	1,167 kg (2,573 lb)
Max T-O weight	4,581 kg (10,100 lb)
Max ramp weight	4,608 kg (10,160 lb)
Max landing weight	4,354 kg (9,600 lb)
Max wing loading	167.8 kg/m² (34.36 lb/sq ft)
Max power loading	5.59 kg/kW (9.18 lb/shp)

PERFORMANCE:
Max level speed	249 kt (461 km; 286 mph)

Max cruising speed at AUW of 3,855 kg (8,500 lb):
at FL160	247 kt (457 km/h; 284 mph)
at FL210	243 kt (450 km/h; 280 mph)
at FL240	239 kt (442 km/h; 275 mph)
Approach speed	101 kt (187 km/h; 116 mph)

Stalling speed, power off:
wheels and flaps up	88 kt (163 km/h; 101 mph) IAS
flaps down	78 kt (144 km/h; 90 mph) IAS
Max rate of climb at S/L	610 m (2,003 ft)/min
Rate of climb at S/L, OEI	151 m (494 ft)/min
Service ceiling	9,150 m (30,000 ft)
Service ceiling, OEI	3,990 m (13,100 ft)
T-O run	620 m (2,035 ft)
T-O to 15 m (50 ft)	826 m (2,710 ft)
Accelerate/stop distance	1,113 m (3,650 ft)

Landing from 15 m (50 ft) at max landing weight, with propeller reversal 698 m (2,290 ft)

Landing run at max landing weight, with propeller reversal 384 m (1,260 ft)

Range with max fuel at max cruising speed, incl allowance for starting, taxi, take-off, climb, descent and 45 min reserves at max range power, ISA:
at FL160	930 n miles (1,722 km; 1,070 miles)
at FL210	1,080 n miles (2,000 km; 1,242 miles)
at FL240	1,165 n miles (2,157 km; 1,340 miles)

Max range at econ cruising power, allowances as above:
at FL160	1,115 n miles (2,065 km; 1,283 miles)
at FL210	1,280 n miles (2,370 km; 1,473 miles)
at FL240	1,340 n miles (2,481 km; 1,542 miles)
at FL290	1,400 n miles (2,592 km; 1,611 miles)

UPDATED

BEECH KING AIR B200

Israel Defence Force name: Tsofit (Thrush)
Swedish Air Force designation: Tp 101

TYPE: Utility turboprop twin.

PROGRAMME: Design of Super King Air 200 began October 1970; 'Super' prefix deleted from all 200, 300 and 350 series King Airs in 1996; first flight (c/n BB1) 27 October 1972; certified to FAR Pt 23 plus icing requirements of FAR Pt 25, 14 December 1973; design of B200 (prototype c/n BB343) began March 1980; production started May 1980; FAA certification 13 February 1981; on sale March 1981.

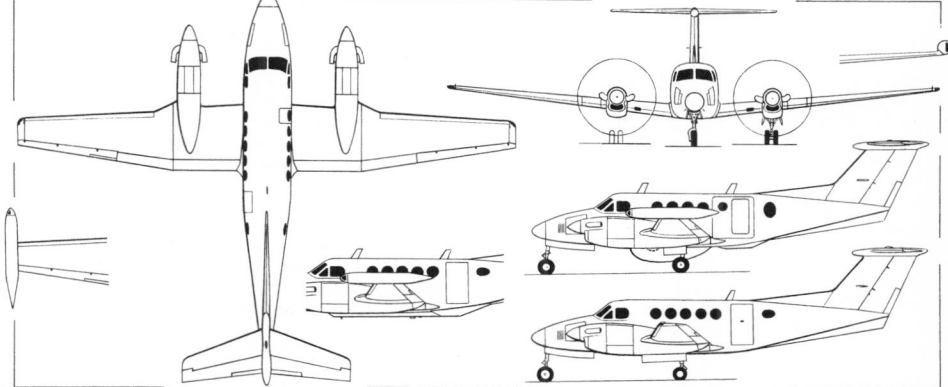

Beech King Air B200 twin-turboprop transport, with additional side view of Maritime Patrol B200T (centre right); scrap views of wingtip tanks and centre-fuselage of photo survey aircraft for IGN (*Jane's/Dennis Punnett*)

Beech King Air B200 used as an air ambulance in Australia ***(Jane's/Paul Jackson)*** ***NEW***/0526941

CURRENT VERSIONS: **King Air B200:** Baseline version. *Detailed description applies to B200.*

King Air B200C: As B200 but with 1.32 × 1.32 m (4 ft 4 in × 4 ft 4 in) cargo door. Two, identified as military C-12R/AP, ordered by US Army on behalf of Greece, January 2000; special missions fit includes cameras for geophysical survey, removable for VIP transport; delivered mid-2002.

King Air B200T: Standard provision for removable tip tanks, adding total 401 litres (106 US gallons; 88.25 Imp gallons), making total 2,460 litres (650 US gallons; 541 Imp gallons). Span without tip tanks 16.92 m (55 ft 6 in). Total 38 built up to 1993. Four ordered by US Army Missile Command in October 1998 for delivery by June 2000; however, these appear to be part of batch of five for Israel Defence Force, also completed in June 2000 and increasing production total to 43. Further three built in 2002 for undisclosed customer; total 46, all converted on production line from B200s.

King Air B200CT: Combines tip tanks and cargo door as standard. Four built by 1983 for Chile (one) and Peru (Navy, three). All converted on production line from King Air 200. Israel Defence Force/Air Force ordered five, of which first was delivered on 4 September 2002, with completion of deliveries scheduled for March 2003. All B200T/CTs converted on production line from King Air 200s.

Maritime patrol B200T: Described in 1998-99 and previous editions.

Beech 200 HISAR: Radar surveillance platform. Launched 1997; based on Beech 200T airframe; demonstrator (N4277E) undertook 16-country tour in 1997-98. First international sale to Traffic 2000 of Germany, November 1997, for North Sea environmental monitoring. Ventral radome for Hughes Integrated Synthetic Aperture Radar (HISAR) suitable for border surveillance, remote sensing, pollution monitoring, EEZ patrol and agricultural monitoring. Equipment operator's console in cabin.

C/RC/UC-12: Military versions; *described separately.*

King Air 300LW: Described in 1996-97 and earlier editions of *Jane's*.

King Air 350: *Described separately.*

CUSTOMERS: Deliveries by 30 September 2001 totalled 1,823 for commercial and private orders, plus 397 military versions (described in 2002-03 and previous *Jane's All the World's Aircraft*) to US armed forces and foreign customers. Total of 23 King Air B200s delivered in 1994, 28 in 1995, 33 in 1996, 45 in 1997, 45 in 1998, 55 in 1999, 59 in 2000, 46 in 2001 and 26 in 2002.

French Institut Géographique National has three B200Ts fitted with twin Wild RC-10 Superaviogon camera installations and Doppler navigation; maximum endurance 10 hours 20 minutes; high flotation landing gear; special French certification for maximum T-O weight 6,350 kg (14,000 lb) and maximum landing weight 6,123 kg (13,500 lb). Egyptian government acquired one King Air 200 in 1978 for water, uranium and other natural resources exploration over Sinai and Egyptian deserts as follow-up to satellite surveys; fitted with remote sensing gear, specialised avionics and special cameras. Navaid checking versions supplied to Taiwan government (one) and Malaysian government (two). Other special missions aircraft delivered to Taiwan Ministry of Interior May 1979; Royal Hong Kong Auxiliary Air Force (later Government Air Service) (two) 1986 and 1987; four King Air 200s operated by Swedish Air Force since 1988 as **Tp 101**.

COSTS: B200 US$3.839 million; B200C US$4.814 million (all 2001).

DESIGN FEATURES: Generally as for earlier versions. Wing aerofoil NACA 23018 to 23016.5 over inner wing, 23012 at tip; dihedral 6°; incidence 3° 48′ at root, −1° 7′ at tip; swept vertical and horizontal tail.

FLYING CONTROLS: Conventional and manual. Trim tabs in port aileron and both elevators; anti-servo tab in rudder; single-slotted trailing-edge flaps; fixed tailplane.

STRUCTURE: Two-spar light alloy wing; safe-life semi-monocoque fuselage.

LANDING GEAR: Hydraulically retractable tricycle type, with twin wheels on each main unit. Single wheel on steerable nose unit, with shimmy damper. Main units retract forward, nosewheel rearward. Beech oleo-pneumatic shock-absorbers. Goodrich mainwheels and tyres size 18×5.5 (10 ply) tubeless, pressure 7.25 bar (105 lb/sq in). Oversize and/or 10 ply mainwheel tyres optional. Goodrich nosewheel size 6.50×10 (8 ply) tubeless, with tyre size 22×6.75-10, pressure 3.93 bar (57 lb/ sq in). Goodrich hydraulic multiple-disc brakes. Parking brake.

POWER PLANT: Two 634 kW (850 shp) Pratt & Whitney Canada PT6A-42 turboprops, each driving a Hartzell four-blade, constant-speed, reversible-pitch, metal propeller with autofeathering and synchrophasing. Bladder fuel cells in each wing, with main system capacity of 1,461 litres (386 US gallons; 321.5 Imp gallons) and auxiliary system capacity of 598 litres (158 US gallons; 131.5 Imp gallons). Total usable fuel capacity 2,059 litres (544 US gallons; 453 Imp gallons). Two refuelling points in upper surface of each wing. Wingtip tanks optional, providing an additional 401 litres (106 US gallons; 88.3 Imp gallons) and raising maximum usable capacity to 2,460 litres (650 US gallons; 541 Imp gallons). Oil capacity 29.5 litres (7.8 US gallons; 6.5 Imp gallons).

ACCOMMODATION: Pilot only, or crew of two side by side, on flight deck, with full dual controls and instruments as standard. Seven cabin seats standard, each equipped with seat belts and inertia reel shoulder harness; optional seats in baggage compartment raise passenger capacity to nine. Partition with sliding door between cabin and flight deck, and partition at rear of cabin. Door at rear of cabin on port side, with integral airstair. Large cargo door optional. Inward-opening emergency exit on starboard side over wing. Lavatory and stowage for baggage in rear fuselage. Maintenance access door in rear fuselage; radio compartment access doors in nose. Cabin is air conditioned and pressurised, with electric heat panels to warm cabin before engine starting.

SYSTEMS: Cabin pressurisation by engine bleed air, with a maximum differential of 0.44 bar (6.5 lb/sq in). Cabin air conditioner of 32,000 BTU capacity. Auxiliary electric cabin heating. Oxygen system for flight deck, and 623 litre (22 cu ft) oxygen system for cabin, with automatic drop-down face masks; 2,182 litre (77 cu ft) system optional. Dual vacuum system for instruments. Hydraulic system for landing gear retraction and extension, pressurised to 171 to 191 bar (2,475 to 2,775 lb/sq in). Separate hydraulic system for brakes. Electrical system has two 300 A 28 V starter/generators and a 24 V 34 Ah air-cooled Ni/Cd battery with failure detector. AC power provided by dual 300 VA inverters. Engine fire detection system standard; engine fire extinguishing system optional. Pneumatic de-icing of wings and tailplane standard. Anti-icing of engine air intakes by hot air from engine exhaust, electrothermal anti-icing for propellers.

AVIONICS: Standard Rockwell Collins Pro Line II package.

Comms: Cockpit-to-cabin paging standard. Bendix/King KHF 950 transceiver, Fairchild A-100A cockpit voice recorder standard, Wulfsberg Flitefone optional.

Radar: Rockwell Collins WXR-270 colour weather radar standard, WXR-840 or WXR-850 turbulence detecting radar optional.

Flight: Collins APS-65H autopilot/flight director, Universal UNS-1D and UNS-1K navigation management systems, with GPS optional.

Instrumentation: Rockwell Collins EFIS-84/B-14, pilot's ALT-80A encoding altimeter; dual maximum allowable airspeed indicators, and flight director standard. Options include Rockwell Collins three-tube EFIS-85B with MFD.

EQUIPMENT: Standard/optional equipment generally as for King Air C90B except fluorescent cabin lighting, one-place couch with storage drawers, flushing chemical toilet, cabin electric heating, cockpit/cabin partition with sliding doors, and airstair door with hydraulic snubber and courtesy light, standard. FAR Pt 135 operational configuration includes cockpit fire extinguisher and 2.2 m^3 (77 cu ft) oxygen bottle with cockpit oxygen pressure indicator as standard. A range of optional cabin seating and cabinetry configurations is available, including quick-removable fold-up seats.

DIMENSIONS, EXTERNAL:	
Wing span	16.61 m (54 ft 6 in)
Wing chord: at root	2.18 m (7 ft 1¾ in)
at tip	0.90 m (2 ft 11½ in)
Wing aspect ratio	9.8
Length overall	13.36 m (43 ft 10 in)
Height overall	4.52 m (14 ft 10 in)
Tailplane span	5.61 m (18 ft 5 in)
Wheel track	5.23 m (17 ft 2 in)
Wheelbase	4.56 m (14 ft 11½ in)
Propeller diameter	2.39 m (7 ft 10 in)
Propeller ground clearance	0.43 m (1 ft 4¾ in)
Distance between propeller centres	5.23 m (17 ft 2 in)
Passenger door: Height	1.31 m (4 ft 3½ in)
Width	0.68 m (2 ft 2¾ in)
Height to sill	1.17 m (3 ft 10 in)
Cargo door (optional): Height	1.32 m (4 ft 4 in)
Width	1.24 m (4 ft 1 in)
Nose avionics service doors (port and stbd):	
Max height	0.57 m (1 ft 10½ in)
Width	0.63 m (2 ft 1 in)
Height to sill	1.37 m (4 ft 6 in)
Emergency exit (stbd): Height	0.66 m (2 ft 2 in)
Width	0.50 m (1 ft 7¾ in)
DIMENSIONS, INTERNAL:	
Cabin:	
Length from forward to rear pressure bulkhead	6.71 m (22 ft 0 in)
aft of cockpit divider	5.08 m (16 ft 8 in)
Max width	1.37 m (4 ft 6 in)
Max height	1.45 m (4 ft 9 in)
Floor area	7.80 m^2 (84 sq ft)
Volume	11.1 m^3 (392 cu ft)
Baggage hold, rear of cabin:	
Volume	1.5 m^3 (54 cu ft)
AREAS:	
Wings, gross	28.15 m^2 (303.0 sq ft)
Ailerons (total)	1.67 m^2 (18.00 sq ft)
Trailing-edge flaps (total)	4.17 m^2 (44.90 sq ft)
Fin	3.46 m^2 (37.20 sq ft)
Rudder, incl tab	1.40 m^2 (15.10 sq ft)
Tailplane	4.52 m^2 (48.70 sq ft)
Elevators, incl tabs (total)	1.79 m^2 (19.30 sq ft)
WEIGHTS AND LOADINGS:	
Weight empty	3,716 kg (8,192 lb)
Basic operating weight (one pilot)	3,910 kg (8,620 lb)
Baggage capacity	249 kg (550 lb)
Max fuel	1,653 kg (3,645 lb)
Max T-O and landing weight	5,670 kg (12,500 lb)
Max ramp weight	5,710 kg (12,590 lb)
Max zero-fuel weight	4,990 kg (11,000 lb)
Max wing loading	201.4 kg/m^2 (41.25 lb/sq ft)
Max power loading	4.48 kg/kW (7.35 lb/shp)
PERFORMANCE:	
Never-exceed speed (V_{NE})	259 kt (480 km/h; 298 mph) IAS
Max operating Mach No.	0.52
Max level speed at FL250, average cruise weight	292 kt (541 km/h; 336 mph)
Max cruising speed at FL280	288 kt (533 km/h; 331 mph)
at FL330	279 kt (517 km/h; 321 mph)
at FL350	273 kt (506 km/h; 314 mph)
Econ cruising speed at FL270, average cruise weight, normal cruise power	222 kt (411 km/h; 255 mph)
Stalling speed: flaps up	99 kt (183 km/h; 114 mph) IAS
flaps down	75 kt (139 km/h; 86 mph) IAS
Max rate of climb at S/L	747 m (2,450 ft)/min
Rate of climb at S/L, OEI	226 m (740 ft)/min
Service ceiling	over 10,670 m (35,000 ft)
Service ceiling, OEI	6,675 m (21,900 ft)
T-O run, 40% flap	567 m (1,860 ft)
T-O to 15 m (50 ft), 40% flap	786 m (2,580 ft)
Landing from 15 m (50 ft):	
without propeller reversal	867 m (2,845 ft)
with propeller reversal	632 m (2,075 ft)
Landing run	536 m (1,760 ft)
Range with max fuel, NBAA IFR reserves, 100 n mile (185 km; 115 mile) alternate:	
max cruise power:	
at FL330	1,673 n miles (3,098 km; 1,925 miles)
at FL350	1,772 n miles (3,281 km; 2,039 miles)
max range power:	
at FL330	1,807 n miles (3,346 km; 2,079 miles)
at FL350	1,825 n miles (3,379 km; 2,100 miles)

UPDATED

BEECH KING AIR 200/A200/B200 (MILITARY VERSIONS)

US basic military designation: C-12

Large-scale procurement of C-12s by the US armed forces has been completed. Details last appeared in the 2002-03 *Jane's All the World's Aircraft*.

UPDATED

BEECH KING AIR 350

US Army designation: C-12S

JGSDF designation: LR-2

TYPE: Utility turboprop twin.

NEW/0130523

Beech King Air 350, cutaway drawing key

1 Starboard navigation light (green) and strobe and recognition lights (white)
2 Flight deck glare shield
3 Composites wingtip fairing
4 Starboard winglet
5 Static dischargers
6 Overwing fuel filler
7 Outboard integral fuel tank
8 Outer leading-edge tank
9 Aileron hinge control, cable actuated
10 Fixed tab
11 Starboard aileron
12 Main wing panel centre fuel cell
13 Onboard leading-edge tank and filler
14 Main wing panel inboard fuel cell
15 Starboard outboard slotted flap segment
16 Outboard flap screw-jack and guide rail, cable driven via central electric motor
17 Nacelle fuel tank
18 Starboard engine nacelle hinged cowling panels
19 Exhaust stubs
20 Engine air intake
21 Spinner
22 Hartzell four-blade, constant-speed, fully feathering and reversible propeller
23 Glass fibre radome
24 Weather radar scanner
25 Scanner mounting bulkhead
26 Landing and taxying lights
27 Nosewheel steering link, hydraulically actuated
28 Torque scissor links
29 Aft-retracting nosewheel
30 Shock-absorber leg strut
31 Nosewheel doors
32 Breaker strut
33 Pitot head, port and starboard
34 Nosewheel hydraulic jack
35 Air conditioning system condenser, intake on starboard side
36 Condenser blower, ground running
37 Evaporator and fan on starboard side
38 Fuselage nose section frame structure
39 Avionics racks

40 Front pressure bulkhead
41 Access hatches, port and starboard
42 Marker beacon antenna
43 Rudder pedals
44 Control column handwheel
45 Instrument panel
46 Instrument panel shroud
47 Windscreen wipers
48 Electrically heated windscreen panels
49 Overhead switch panel
50 Cockpit bulkhead with sliding doors
51 Co-pilot's seat
52 Direct vision opening side window panel
53 Captain's seat
54 Portable fire extinguisher beneath seat
55 Cockpit bulkhead stowage units
56 Cabin conditioned air delivery duct
57 Refreshment cabinet
58 Aft-facing individual passenger seat
59 Wingroot leading-edge fairing
60 Centre-section main spar carry-through
61 Folding table, four places
62 Emergency exit hatches, port and starboard
63 Cabin roof frame structure
64 Cabin window panels
65 Stowage cabinet
66 Centre fuselage frame structure
67 Flap drive electric motor
68 Aileron cable quadrant
69 Main spar/fuselage attachment joint
70 Hydraulic reservoir and power pack, landing gear operation

71 Engine bleed-air pre-cooler via leading-edge ram air intake
72 Centre-section fuel cell
73 Inboard flap screw-jack and guide rail
74 Port inboard flap segment
75 Wingroot trailing-edge fillet
76 Underfloor rear cabin heater unit
77 Fuselage main longeron
78 Passenger seating, eight-seat club layout
79 Cabin trim panelling
80 Rear cabin bulkhead with sliding doors
81 Side-mounted toilet
82 VHF/UHF antenna
83 Baggage compartment window, port and starboard
84 Fin root fillet
85 ELT antenna
86 Fin two-spar torsion box structure
87 VOR antenna
88 Fin rib structure
89 Leading-edge bullet fairing
90 Fin/tailplane spar joint
91 Upper anti-collision beacon
92 Starboard tailplane structure
93 Elevator trim tab actuator
94 Starboard elevator
95 Elevator hinge control, cable actuated
96 Fin/tailplane trailing-edge fairing
97 Tail navigation light
98 Elevator trim tabs
99 Port elevator bonded structure

100 Elevator mass balance
101 Port tailplane two-spar torsion box structure
102 Logo light
103 Tailplane leading-edge de-icing boot
104 Rudder
105 Bonded rudder structure
106 Rudder trim tab
107 Detachable vented tailcone
108 Ventral fin
109 Rudder hinge control, cable actuated
110 Sloping fin spar attachment bulkheads
111 Pressurisation air vent duct
112 Oxygen bottle
113 Rear pressure bulkhead
114 Tailcone ventral access hatch
115 Baggage compartment heater air duct
116 Cabin pressurisation valves
117 Baggage compartment
118 Baggage restraint net
119 Door
120 Door balance strut
121 Door cables/handgrips
122 Airstairs
123 Optional Raisbeck extended nacelle baggage locker

124 Main wing panel bolted spar joint
125 Rear spar
126 Flap shroud ribs
127 Flap rib structure
128 Port outboard slotted flap segment
129 Aileron tab actuator
130 Port aileron hinge control, cable actuated
131 Aileron trim tab
132 Port aileron rib structure
133 Static dischargers
134 Port winglet
135 Port navigation (red) and strobe and navigation lights (white)
136 Outer wing panel dry bay
137 Wingtip joint rib
138 Port outboard integral fuel tank
139 Wing rib structure
140 Leading-edge de-icing boot
141 Stall warning vane
142 Leading-edge fuel tank
143 Wing main spar
144 Main wing panel mid and inboard fuel cells
145 Main landing gear leg pivot mounting
146 Twin mainwheels, forward retracting
147 Mainwheel doors, composites structure
148 Outer wing panel leading-edge root extension
149 Mainwheel leg drag/breaker strut
150 Hydraulic retraction jack
151 Port nacelle fuel tank
152 Nacelle structure
153 Firewall
154 Engine accessory equipment
155 Intake bypass air spill duct
156 Ventral oil cooler
157 Engine bearer struts
158 Annular engine air intake
159 Main engine anti-vibration mounting
160 Particle separator intake air duct
161 Exhaust stubs, port and starboard
162 Pratt & Whitney Canada PT6A-60A turboprop
163 Constant-speed propeller governor
164 Engine air intake, exhaust gas hot air de-icing
165 Propeller hub pitch change mechanism
166 Spinner
167 Port Hartzell four-blade propeller
168 Propeller blade root de-icing

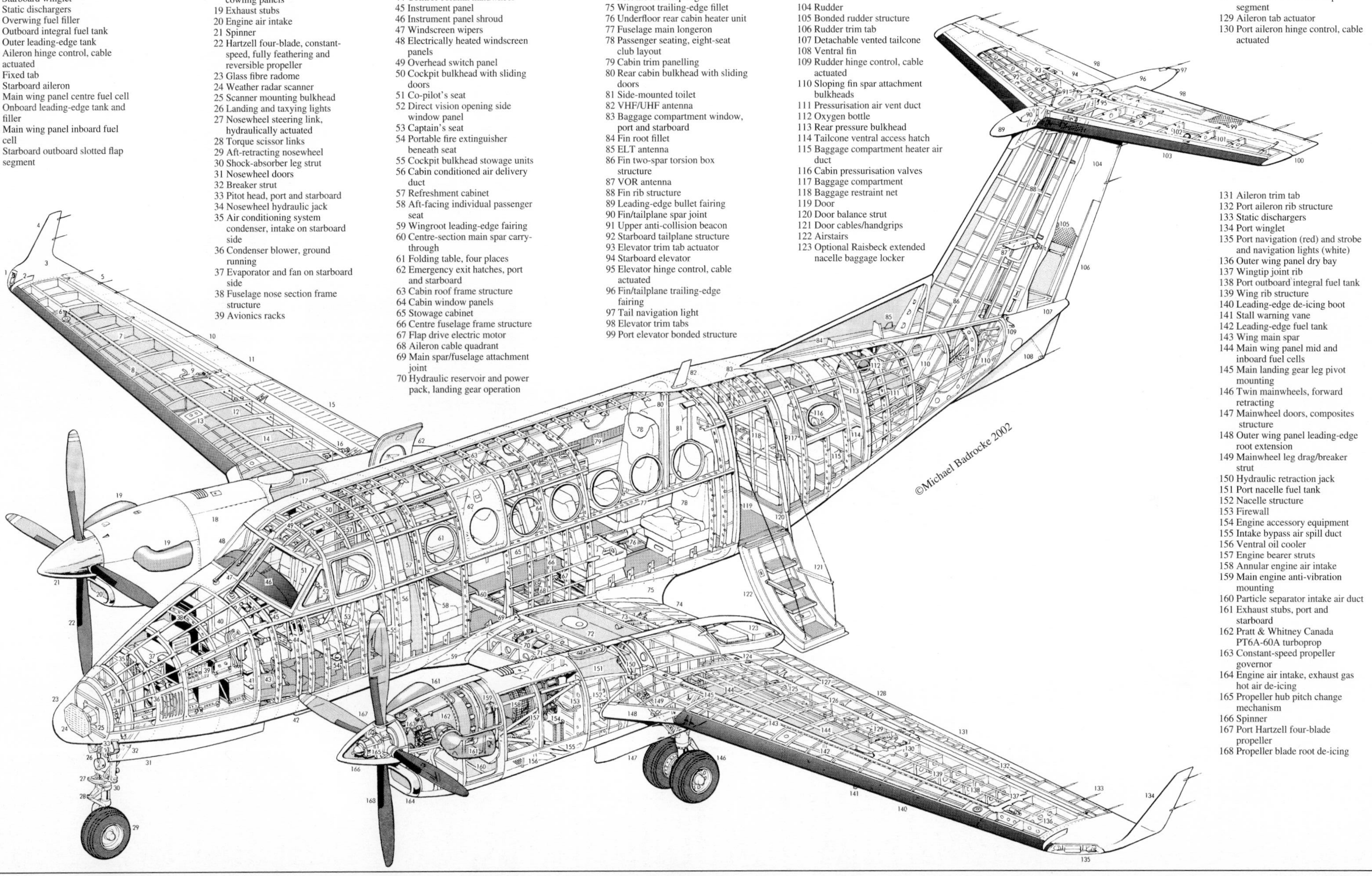

© Michael Badrocke 2002

PROGRAMME: Replaced King Air 300 (1991-92 *Jane's*); first flight (N120SK) September 1988; introduced at NBAA Convention 1989; certified to FAR Pt 23 (commuter category); first delivery 6 March 1990; Russian certification to AP 23 in November 1995; FAA approval for operation from unprepared runways granted during 1997.

CURRENT VERSIONS: **King Air 350**: Baseline version.
Detailed description applies to King Air 350.

King Air 350C: Has 132 × 132 cm (52 × 52 in) freight door with built-in airstair passenger door.

King Air 350 Special Missions: Versions available for aerial photography and airways and ground-based navaid checking. Single hardpoint under each wing for external stores; optional bubble windows and belly camera bay. Maritime patrol version has 7,257 kg (16,000 lb) maximum weight and a 1,292 kg (2,850 lb) mission payload, including ventral radar and FLIR; survival kits can be dispensed through an optional dropping hatch.

Australian Army received King Air 350 VH-HPJ in 1998 as follow-on (to Douglas C-47 Dakota) test platform for Ingara SAR/MTI radar in ventral pannier; installation by Hawker Pacific.

RC-350 Guardian: Elint version, converted from 350 prototype 1991 by Beech Aircraft Corporation; mission avionics include Raytheon AN/ALQ-142 ESM, Watkins-Johnson 9195C communications interceptor, Honeywell laser INS, GPS receiver and Cubic secure digital datalink; can loiter on station at 10,670 m (35,000 ft) for more than 6 hours; can locate/monitor radar emitters in 20 MHz to 18 GHz range, and intercept communications within 20 to 1,400 MHz bandwidths. Wingtip pods house AN/ALQ-142 antennas; underfuselage bulge contains antenna for comint system.

LR-2: Japan Ground Self-Defence Force funded two in FY97 for liaison and reconnaissance (undisclosed sensor in ventral radome). Total requirement for 20, of which third funded in 1999 and fourth in FY00. First delivery (23051) 22 January 1999; initial operator is the HQ Flight of 1st Helicopter Brigade at Kisarazu.

C-12S: US Army version with quick-change cargo capability and seating for up to 15 passengers. By late 2000, no aircraft of this type had been identified in service.

CUSTOMERS: Total 362 King Air 350s and 350Cs delivered by December 2002; first 350C delivery in 1990 to Rossing Uranium, Namibia; 15 King Air 350 deliveries 1995, 27 in 1996, 30 in 1997, 42 in 1998, 45 in 1999, 46 in 2000, 32 in 2001 and 24 in 2002. Recent customers include the Royal Australian Air Force, which signed a lease contract on 20 November 2002 for seven King Air 350s for delivery from March 2003, to be operated by 32 Squadron for the RAAF School of Navigation at East Sale, replacing H.S. 748s and King Air 200s.

COSTS: 350 US$5.404 million; 350C US$5.848 million (both 2001).

DESIGN FEATURES: Compared with King Air 300, fuselage stretched 0.86 m (2 ft 10 in) by plugs 0.37 m (1 ft 2½ in) forward of main spar and 0.49 m (1 ft 7½ in) aft; wing span increased by 0.46 m (1 ft 6 in) with NASA winglets 0.61 m (2 ft 0 in) high; two additional cabin windows each side. Can depart with full payload and full tanks. Raisbeck dual

aft body strakes (DABS), standard on production aircraft from c/n FL-312 (N3165M) in first quarter 2001, reduce drag, improve handling and stability and relax or eliminate restrictions on operations with inoperative yaw damper.

FLYING CONTROLS: Automatic cable tensioner in aileron circuit and larger elevator bobweight; larger rudder anti-servo tab; ailerons and rudder cleaned up.

STRUCTURE: As for B200.

LANDING GEAR: As for B200.

POWER PLANT: Two 783 kW (1,050 shp) Pratt & Whitney Canada PT6A-60A turboprops, each driving a Hartzell four-blade, constant-speed, fully feathering, reversible-pitch, metal propeller. Bladder cells and integral tanks in each wing, with usable capacity of 1,438 litres (380 US gallons; 316.5 Imp gallons); auxiliary tanks inboard of engine nacelles, capacity 601 litres (159 US gallons; 132.5 Imp gallons). Total fuel capacity 2,040 litres (539 US gallons; 449 Imp gallons). No provision for wingtip tanks. Oil capacity 30.2 litres (8.0 US gallons; 6.7 Imp gallons).

ACCOMMODATION: Double club seating for eight passengers; optionally two more seats in rear of cabin and one passenger on side-facing lavatory seat making maximum 11 passengers; certified for maximum 17 occupants including crew. Ultra Electronics UltraQuiet active noise control system installed as standard from 1998.

SYSTEMS: As for B200, except for automatic bleed air-type heating and 22,000 BTU cooling system with high-capacity ventilation system; 2,182 litre (77 cu ft) oxygen system standard; hydraulic landing gear retraction and extension; two 300 A 28 V starter/generators with triple bus electrical distribution system. Ultra Electronics Ltd UltraQuiet active noise control system introduced as standard from 1998, comprising 12 loudspeakers, 24 microphones and a high-speed digital processor which cancel propeller noise and reduce in-flight cabin sound level to less than 80 dB(A).

AVIONICS: Rockwell Collins Pro Line II as core system, with three- or five-tube EFIS, TWR-850 Doppler weather radar and Universal UNS-1K or UNS-1D FMS both with 12-channel GPS.

EQUIPMENT: Generally as for B200.

DIMENSIONS, EXTERNAL: As for B200 except:
Wing span over winglets	17.65 m (57 ft 11 in)
Wing aspect ratio	10.8
Length overall	14.22 m (46 ft 8 in)
Height overall	4.37 m (14 ft 4 in)
Propeller diameter	2.67 m (8 ft 9 in)
Propeller ground clearance	0.29 m (11½ in)
Emergency exit (each side of cabin, above wing):	
Height	0.66 m (2 ft 2 in)
Width	0.50 m (1 ft 7¾ in)

DIMENSIONS, INTERNAL:
Cabin (excl cockpit): Length	5.94 m (19 ft 6 in)
Max width	1.37 m (4 ft 6 in)
Height	1.45 m (4 ft 9 in)
Cabin volume	10.05 m³ (355 cu ft)
Baggage hold volume	1.6 m³ (55 cu ft)

AREAS:
Wings, gross	28.80 m² (310.0 sq ft)

WEIGHTS AND LOADINGS:
Weight empty	4,132 kg (9,110 lb)
Basic operating weight (one pilot)	4,317 kg (9,518 lb)
Max fuel weight	1,638 kg (3,611 lb)
Max T-O and landing weight	6,804 kg (15,000 lb)
Max ramp weight	6,849 kg (15,100 lb)
Max zero-fuel weight	5,670 kg (12,500 lb)
Max wing loading	236.2 kg/m² (48.39 lb/sq ft)
Max power loading	4.35 kg/kW (7.14 lb/shp)

PERFORMANCE:
Max level speed	315 kt (584 km/h; 363 mph)
Max cruising speed, AUW of 5,896 kg (13,000 lb):	
at FL280	308 kt (570 km/h; 354 mph)

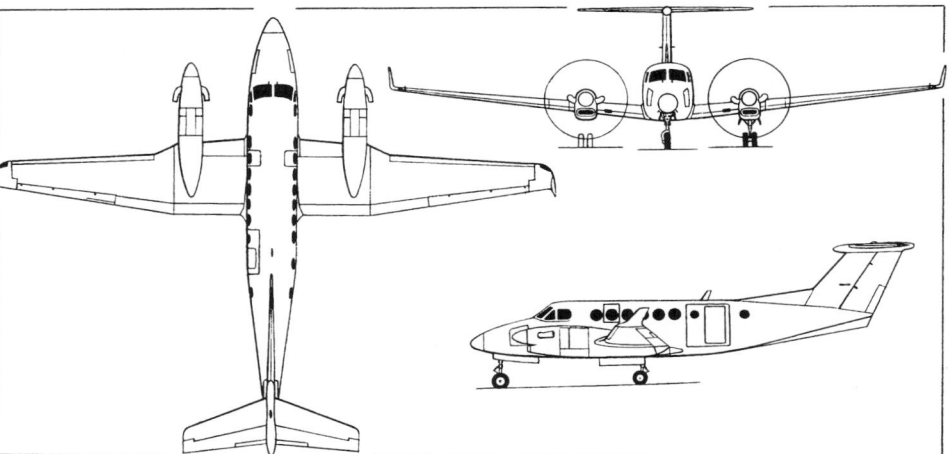

Beech King Air 350 (*Jane's/Dennis Punnett*)   0085295

Beech King Air 350 overflying the Grand Canyon

NEW/0527084

at FL330	297 kt (550 km/h; 342 mph)
at FL350	290 kt (537 km/h; 334 mph)

Cruising speed, normal cruising power, AUW of
5,896 kg (13,000 lb):

at FL240	301 kt (558 km/h; 347 mph)
at FL350	281 kt (521 km/h; 324 mph)

Cruising speed, max range power, AUW of 5,896 kg
(13,000 lb):

at FL180	210 kt (389 km/h; 242 mph)
at FL350	240 kt (445 km/h; 276 mph)

Stalling speed at max landing weight, flaps and wheels

down	81 kt (150 km/h; 94 mph)
Max rate of climb at S/L	832 m (2,731 ft)/min

Rate of climb at S/L, OEI, AUW of 6,350 kg (14,000 lb)

	236 m (775 ft)/min
Service ceiling	above 10,670 m (35,000 ft)
Service ceiling, OEI	6,555 m (21,500 ft)
T-O balanced field length	1,006 m (3,300 ft)
Landing from 15 m (50 ft)	821 m (2,695 ft)
Landing run	441 m (1,450 ft)

Range with 2,040 litres (539 US gallons; 449 Imp
gallons) usable fuel, allowances for start, T-O, climb
and descent plus 45 min reserves:
max cruising power:

at FL330	1,790 n miles (3,315 km; 2,060 miles)
at FL350	1,829 n miles (3,387 km; 2,105 miles)

max range power, allowances as above:

at FL330	1,463 n miles (2,709 km; 1,684 miles)
at FL350	1,554 n miles (2,878 km; 1,788 miles)

UPDATED

Beech 1900D (two P&WC PT6A turboprops) *(Jane's/Paul Jackson)*   *NEW*/0526940

BEECH 1900D

US Army designation: C-12J

TYPE: Twin-turboprop airliner.

PROGRAMME: Original Beech 1900 first flew on 3 September
1982; three prototypes followed by 74 1900Cs and 174
wet-wing 1900C-1s by late 1991. Current 1900D
announced at US Regional Airlines Association meeting
1989; development of 1900C (1991-92 *Jane's*); prototype
(converted from 1900C-1 N5584B) first flight 1 March
1990; certification to FAR Pt 23 Amendment 34 received
March 1991; full certification with supplements received,
and deliveries (to Mesa Airlines) began, November 1991;
contract signed February 1997 with Xian Aircraft
Company of People's Republic of China for supply to
Raytheon of 800 metal-bonded subassemblies for 1900D,
to be delivered between 1997 and 2001; 500th Model 1900
delivered in March 1997; 400th Model 1900D (N44640)
rolled out 21 March 2000. Beech also offers special
mission versions for signals intelligence, maritime patrol
and similar duties.

Manufactured to order from 2002, during which the
company expected to manufacture 11 1900Ds.

CURRENT VERSIONS: **1900D:** *As described.*

1900D Executive: Features custom-designed
executive interior ranging from twin double-club to
corporate shuttle configuration; refreshment bar,
entertainment system and flight phones optional.

C-12J: US Air Force ordered six Beech 1900C-1s (see
1991-92 *Jane's*) in 1986, of which four currently assigned
to 3rd, 46th and 51st (two) Wings for support, and two with
US Army (HQ Europe and 78th Aviation Battalion in
Japan). In March 1997, Army received a further Beech
1900D (96-0112) for Chemical and Biological Defense
Command at Aberdeen Proving Ground. C-12 designation
more properly belongs to King Air.

CUSTOMERS: Total 65 delivered in 1995, 69 in 1996, 42 in
1997, 45 in 1998, 24 in 1999, 54 in 2000, and 11 each in

2001 and 2002. Purchasers include the Algerian Air Force,
which ordered 12 in 2000 in sigint configuration, equipped
with Northrop Grumman Systems radar and FLIR; and Air
New Zealand, which ordered 16 in April 2001 for delivery
to its regional subsidiary Eagle Airways at the rate of one
per month from mid-2001.

COSTS: US$4.995 million (2001).

DESIGN FEATURES: Flat floor with stand-up headroom; cabin
volume increased by 28.5 per cent compared to 1900C;
winglets add better hot-and-high performance; tailplane
and fin swept; each tailplane carries small fin (tailet) on
underside near tip; auxiliary horizontal fixed tail surface
(stabilon) each side of rear fuselage improves centre of
gravity range; twin ventral strakes improve directional
stability and turbulence penetration; small horizontal
vortex generator on fuselage ahead of wingroots.

Wing aerofoil NACA 23018 (modified) at root, 23012
(modified) at tip; dihedral 6°; incidence 3° 29' at root,
−1° 4' at tip; no sweepback at quarter-chord.

FLYING CONTROLS: Conventional and manual. Automatic
cable tensioner in aileron circuit; trim tabs in elevators,
rudder and port aileron; primary and secondary controls
routed separately to improve protection from possible
engine-failure damage; single-slotted trailing-edge flaps in
two sections on each wing.

STRUCTURE: Generally of light alloy. Wing has continuous
main spar with fail-safe structure riveted and bonded;
fuselage pressurised and mainly bonded.

LANDING GEAR: Hydraulically retractable tricycle type; main
units retract forward and nose unit rearward; Beech oleo-
pneumatic shock-absorber in each unit. Twin Goodyear
wheels on each main unit, size 6.50×10, with Goodyear
tyres size 22×6.75-10 (10 ply) tubeless, pressure 6.69 bar
(97 lb/sq in); Goodyear steerable nosewheel size 6.50×8,
with Goodyear tyres size 19×6.75-8 (10 ply) tubeless,
pressure 4.14 bar (60 lb/sq in). Nosewheel power steering
optional. Goodyear multiple-disc hydraulic brakes.
Optional Beech Hydro-Aire anti-skid units, power steering
and brake de-icing. Ground turning radius based on
wingtip clearance 12.55 m (41 ft 2 in).

POWER PLANT: Two Pratt & Whitney Canada PT6A-67D
turboprops, each flat rated at 954 kW (1,279 shp) and
driving a Hartzell four-blade constant-speed fully
feathering reversible-pitch composites propeller. Wet
wing fuel storage (two integral tanks per wing) with a total
capacity of 2,528 litres (668 US gallons; 556 Imp gallons),
of which 2,519 litres (665 US gallons; 554 Imp gallons)
usable. Refuelling point in each wing leading-edge,
inboard of engine nacelle. Oil capacity (total) 29.5 litres
(7.8 US gallons; 6.5 Imp gallons).

ACCOMMODATION: Crew of one (FAR Pt 91) or two (FAR Pt
135) on flight deck, with standard accommodation in cabin
of commuter version for 19 passengers in single airline-
standard seats on each side of centre aisle. Forward
carry-on baggage lockers, underseat baggage stowage, rear
baggage compartment. Forward door, incorporating
airstairs, on port side. Upward-hinged rear cargo door, also
on port side. Three emergency exits over wing (two
starboard, one port). Accommodation air conditioned,
heated, ventilated and pressurised. Ultra Electronics
UltraQuiet active noise cancellation system optional.
Executive and corporate shuttle options seat between 10
and 18 passengers with options for forward and rear
compartments, combination lavatory/passenger seat and
two beverage bars at cabin compartment division. Club,
double club and triple club seating optional. Customised
interiors to customer's choice, including King Air 350
cockpit seats and passenger/cargo combi interiors. Oxygen
system, capacity 4,308 litres (152 cu ft) standard, with
outlets for all cabin occupants.

SYSTEMS: Bleed air cabin heating and pressurisation,
maximum differential 0.35 bar (5.1 lb/sq in), maintaining
sea level cabin environment to 3,350 m (11,000 ft), and
2,745 m (9,000 ft) cabin to 7,620 m (25,000 ft). Air cycle
and vapour cycle air conditioning. Hydraulic system,
pressure 207 bar (3,000 lb/sq in), for landing gear
actuation. 28 V electrical system includes two 300 A
engine-driven starter/generators, one 34 Ah Ni/Cd battery
and two 400 Hz solid-state invertors supplying 115 V DC
and 26 V AC power for avionics and instruments. Constant
flow oxygen system of 4,420 litre (156 cu ft) capacity
standard. Engine inlet screen anti-ice protection, exhaust
heated engine inlet lips, fuel vent heating, electric propeller
and windscreen de-icing systems standard. Brake de-icing
optional. Pneumatic de-icing boots on wings, tailplane,
tailets and stabilons. Self-monitoring continuous detection
loop and one-shot fire extinguisher in each engine nacelle.

AVIONICS: *Comms:* Rockwell Collins Pro Line II digital
technology radios; cabin briefer; cockpit voice recorder.

Flight: Dual flight directors; GPWS; provision for
TCAS1. GPS and Collins TWR-850 colour weather radar
optional.

Instrumentation: Rockwell Collins EFIS-84 four-tube
EFIS. Primary display consists of multicolour CRT panels,
remote display processor unit and system control units;
CRT displays provide conventional electronic attitude
director indicator (EADI) and electronic horizontal
situation indicator (EHSI) functions.

DIMENSIONS, EXTERNAL:

Wing span over winglets	17.67 m (57 ft 11¾ in)
Wing chord: at root	2.18 m (7 ft 1¾ in)
at tip	0.91 m (3 ft 0 in)
Wing aspect ratio	10.9
Length overall	17.63 m (57 ft 10 in)
Height overall	4.57 m (14 ft 11¾ in)
Tailplane span	5.63 m (18 ft 5¾ in)
Wheel track	5.23 m (17 ft 2 in)
Wheelbase	7.25 m (23 ft 9½ in)
Propeller diameter	2.78 m (9 ft 1½ in)
Propeller ground clearance	0.35 m (1 ft 1¾ in)
Distance between propeller centres	5.23 m (17 ft 2 in)
Passenger door: Height	1.63 m (5 ft 4¼ in)
Width	0.64 m (2 ft 1¼ in)
Cargo door: Height	1.45 m (4 ft 9 in)
Width	1.32 m (4 ft 4 in)
Emergency exits (each): Height	0.80 m (2 ft 7½ in)
Width	0.51 m (1 ft 8 in)
Turning radius	12.56 m (41 ft 2½ in)

DIMENSIONS, INTERNAL:
Cabin (incl flight deck and rear baggage compartment):

Length	12.03 m (39 ft 5½ in)
Max width	1.37 m (4 ft 6 in)
Max height	1.80 m (5 ft 10¾ in)
Floor area	15.3 m² (165 sq ft)
Pressurised volume	26.0 m³ (918 cu ft)
Volume of passenger cabin	18.1 m³ (640 cu ft)
Baggage volume, cabin: forward	0.34 m³ (12.0 cu ft)
underseat	0.91 m³ (32.0 cu ft)
rear	5.0 m³ (175 cu ft)

AREAS:

Wings, gross	28.80 m² (310.0 sq ft)
Ailerons (total)	1.67 m² (18.00 sq ft)

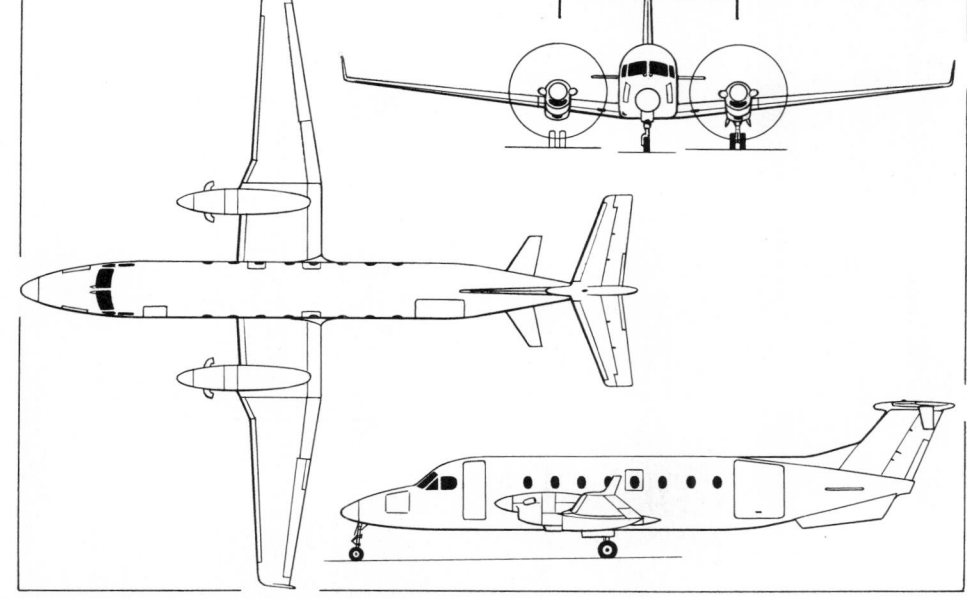

Beech 1900D regional transport *(Jane's/Dennis Punnett)*

Beechjet 400A twin-turbofan business aircraft   *NEW*/0527085

Trailing-edge flaps (total)	4.17 m² (44.90 sq ft)
Fin	4.86 m² (52.30 sq ft)
Rudder (incl tab)	1.40 m² (15.10 sq ft)
Tailets (total)	0.63 m² (6.80 sq ft)
Tailplane	6.32 m² (68.00 sq ft)
Elevator (incl tab)	1.79 m² (19.30 sq ft)
Stabilons (total)	1.44 m² (15.50 sq ft)

WEIGHTS AND LOADINGS:

Basic operating weight	4,831 kg (10,650 lb)
Max fuel (usable)	2,022 kg (4,458 lb)
Max baggage	939 kg (2,070 lb)
Max ramp weight	7,738 kg (17,060 lb)
Max T-O weight	7,688 kg (16,950 lb)
Max landing weight	7,530 kg (16,600 lb)
Max zero-fuel weight	6,804 kg (15,000 lb)
Payload with max fuel	947 kg (2,087 lb)
Max wing loading	267 kg/m² (54.68 lb/sq ft)
Max power loading	4.03 kg/kW (6.63 lb/shp)

PERFORMANCE:

Max cruising speed at AUW of 6,804 kg (15,000 lb):

at FL80	272 kt (504 km/h; 313 mph)
at FL160	284 kt (526 km/h; 327 mph)
at FL250	277 kt (513 km/h; 319 mph)

Unstick speed, T-O flap setting
105 kt (195 km/h; 121 mph) IAS

Approach speed at max landing weight
117 kt (217 km/h; 135 mph)

Stalling speed at max T-O weight:

wheels and flaps up	101 kt (187 km/h; 116 mph)
wheels down, T-O flap setting	90 kt (167 km/h; 104 mph)

Stalling speed at max landing weight, wheels and flaps

down	84 kt (156 km/h; 97 mph)
Max rate of climb at S/L	800 m (2,625 ft)/min
Rate of climb at S/L, OEI	206 m (676 ft)/min
Service ceiling	10,058 m (33,000 ft)
Max certified operating altitude	7,620 m (25,000 ft)
Service ceiling, OEI	5,334 m (17,500 ft)
T-O field length, T-O flap setting	1,139 m (3,737 ft)

Landing from 15 m (50 ft) at max landing weight
829 m (2,720 ft)

Max range with 19 passengers, at long-range cruise power, with allowances for starting, taxi, T-O, climb and descent, with reserves (45 min hold at 1,525 m; 5,000 ft)   708 n miles (1,311 km; 815 miles)

UPDATED

BEECH BEECHJET 400A
US Air Force designation: T-1A Jayhawk
JASDF designation: T-400

TYPE: Business jet.

PROGRAMME: Conceived as Mitsubishi MU-300 Diamond; first flight 29 August 1978; two prototypes; FAR Pt 25 certification awarded 6 November 1981; production aircraft fabricated in Japan and assembled at San Angelo, Texas; deliveries totalled 63 Diamond Is (JT15D-4 engines), 27 Diamond IAs (JT15D-4D) and one Diamond II (JT15D-5).

Beech acquired rights to Diamond II from Mitsubishi Heavy Industries and Mitsubishi Aircraft International, December 1985; made improvements to aircraft and renamed it Beechjet 400. First Beech-assembled Beechjet rolled out 19 May 1986; initial 64 used Japanese components. During 1989, Beech moved entire manufacturing operation to Wichita. Announced new Beechjet 400A November 1989, featuring certification to 13,715 m (45,000 ft), larger and more comfortable cabin, all Collins avionics with digital EFIS; customer deliveries began November 1990.

CURRENT VERSIONS: **Beechjet 400:** Initial production version (64 built; see earlier *Jane's*); superseded by 400A.

Beechjet 400A: Announced at 1989 NBAA show; production 400A first flight 22 September 1989; FAA certification received 20 June 1990; deliveries began November 1990. Also certified by July 1993 in Australia, Canada, France, Germany, Italy and UK; Brazilian and Pakistani type approval April 1994; Civil Aviation Authority of China certification achieved in second quarter of 1999.

Description applies to 400A, except where indicated.

Flight deck of Beechjet 400A
0130563

Beechjet 400A cabin
0130564

Beechjet T-1A Jayhawk: US Air Force selected McDonnell Douglas, Beech and Quintron to supply Tanker Transport Training System (TTTS) on 21 February 1990, including requirement for 180 Beechjet 400Ts, valued at US$755 million and designated T-1A Jayhawk; represents missionised version of 400A, sharing many components and characteristics with commercial counterpart; differences include cabin-mounted avionics, increased air conditioning capability, greater fuel capacity with single-point refuelling, and strengthened windscreen and leading-edges for low-level birdstrike protection. First production aircraft (90-0400) delivered 17 January 1992; deliveries at approximately three per month; final delivery 23 July 1997. By then total fleet time exceeded 182,000 flying hours, with 90 per cent operational availability, and more than 680 pilots had been trained on the Jayhawk.

IOC for USAF Jayhawks January 1993, for Air Education and Training Command Specialised Undergraduate Pilot Training (SUPT) programme at Reese AFB (52nd FTS/64th FTW) where establishment of 41 received by October 1993; Reese closed in 1997. Second recipient was 99th FTS/12th FTW at Randolph AFB, Texas, where 16 delivered for instructor training in 1993; third unit was 86th FTS/47th FTW at Laughlin AFB, Texas, from late 1993 with training courses beginning May 1994; fourth was 71st FTW at Vance AFB, Oklahoma (first aircraft December 1994); fifth was 14th FTW at Columbus AFB, Mississippi (early 1996). T-1A used for training crews for KC-10, KC-135, C-5 and C-17, with total fleet experience of more than 376,000 hours and more than 733,000 landings by October 1999.

In February 1997 Raytheon Aircraft and its subsidiary Raytheon Aerospace were awarded a contract, valued at US$6.2 million, to retrofit 62 Jayhawks with GPS; two further options to retrofit the entire fleet would bring the total value of the GPS upgrade to about US$25.3 million.

Beechjet 400T: JASDF T-400 version, featuring thrust reversers, long-range inertial navigation and direction-finding systems; interior changes. Meets TC-X trainer requirement; three, three, two and one ordered in 1992-95, plus one in 1998; first (41-5051) delivered 31 January 1994; 10th in 2000.

CUSTOMERS: The 500th Beechjet/Jayhawk (N500TH; 246th 400A) was delivered to Global Financial Services Group of Anderson, South Carolina, on 12 October 1999, during the NBAA Convention at Atlanta, Georgia. Total 30 delivered in 1995, 29 in 1996, 43 in both 1997 and 1998, 48 in 1999, 51 in 2000, 25 in 2001 and 10 in first six months of 2002. Recent customers include Hainan Airlines of China, which took delivery of one aircraft in February 1999. US Air Force 180 T-1As ordered, of which delivery completed 23 July 1997. JASDF had received 10 by December 1999 and ordered further two in 2000; operated by 41 Hikotai at Miho. By October 2002, 348 civil 400As had been registered, in addition to 64 Model 400s, 180 Jayhawks, 10 400Ts and 93 Diamonds, or 695 in all.

BEECHJET T-1A JAYHAWK PROCUREMENT

FY	Lot	Qty	First Aircraft	Delivery
89	–	1	89-0284	1991
90	1	14	90-0400	1992
91	2	28	91-0075	1992-93
92	3	34	92-0330	1993-94
93	4	36	93-0621	1994-95
94	5	35	94-0114	1995-96
95	6	32	95-0040	1996-97
Total		**180**		

COSTS: Jayhawk programme cost US$1.3 billion; Beech contracts for 180 aircraft, US$755 million. Civil 400A quoted at US$6.1 million (2001).

DESIGN FEATURES: Typical low-wing, T tail, rear-engined small business jet, with sweptback wings and empennage, plus small underfin. Compared to Diamond, Beechjet has increased payload and certified ceiling, greater cabin volume achieved by moving rear-fuselage fuel tank forward under floor (balanced by moving lavatory to rear of cabin), improved soundproofing, and emergency door moved one window forward to facilitate forward club seating.

Wing has computer-designed three-dimensional Mitsubishi MAC510 aerofoil; thickness/chord ratio 13.2 per cent at root, 11.3 per cent at tip; dihedral 2° 30′; incidence 3° at root, –3° 30′ at tip; sweepback 20° at quarter-chord.

T-1A Jayhawk features include student pilot in left seat, instructor on right and pupil/observer behind instructor; more bird-resistant windscreen and leading-edges; fewer cabin windows; strengthened wing carry-through structure and engine attachment points to meet low-level flight stresses; rails for four passenger seats in cabin for personnel transport; avionics relocated from nose to rack in cabin to facilitate nose installation of air conditioning; emergency door moved forward to position opposite main cabin door to allow straight-through egress; improved brakes; additional fuel tank; single-point pressure refuelling; Rockwell Collins five-tube EFIS; digital autopilot; weather radar; central diagnostic and maintenance system; Tacan with air-to-air capability.

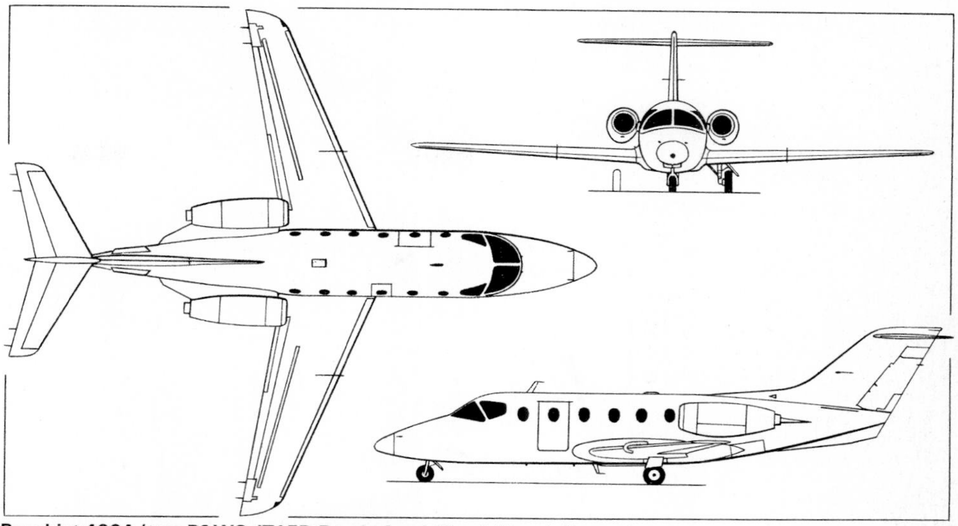

Beechjet 400A (two P&WC JT15D-5 turbofans) *(Jane's/Dennis Punnett)*

FLYING CONTROLS: Conventional and manual. Variable incidence tailplane and elevators for pitch axis; lateral control by small ailerons and almost full semi-span, narrow chord spoilers used also as airbrakes and lift dumpers; rudder with trim tab; narrow chord Fowler-type flaps, double-slotted inboard and single-slotted outboard, occupy most of trailing-edges and are hydraulically actuated; mid-span leading-edge fences on wing; small horizontal strakes on fuselage at base of fin; small ventral fin.

STRUCTURE: Wings include integrally machined metal upper and lower skins joined to two box spars forming integral fuel tank; tailplane and fin similar. Wing, fuselage and tail unit certified fail-safe for unlimited life (with periodic inspections and maintenance).

LANDING GEAR: Retractable tricycle type, with single wheel and oleo-pneumatic shock-absorber on each unit. Hydraulic actuation, controlled electrically. Emergency free-fall extension. Main tyres 24×7.7 (16 ply) tubeless; nose tyre 18×4.4 (10 ply) tubeless. Nosewheel, which is steerable by rudder pedals, retracts forward; mainwheels retract inward into fuselage. Goodyear wheels and tyres; Aircraft Braking Systems brakes.

POWER PLANT: Two Pratt & Whitney Canada JT15D-5 turbofans, each rated at 13.19 kN (2,965 lb st) for take-off. Nordam thrust reversers optional on 400A, but not fitted to T-1A. Total usable fuel capacity: 400A 2,775 litres (733 US gallons; 610 Imp gallons); 400T 2,998 litres (792 US gallons; 656 Imp gallons). One refuelling point in top of each wing, and one in rear fuselage for fuselage tank, capacity 1,158 litres (306 US gallons; 255 Imp gallons). (T-1A, single-point refuelling.) Oil capacity 7.7 litres (2.0 US gallons; 1.7 Imp gallons).

ACCOMMODATION: Crew of two on flight deck of 400A on vertically and horizontally adjustable reclining seats with five-point safety harnesses; T-1A has seats for trainee pilot, co-pilot/instructor and observer. Improved interior introduced 1996, featuring redesigned trim panels, enhanced acoustic panels and vibration-damping engine mounts.

Standard 'centre club' layout of 400A seats eight passengers in pressurised cabin. Of these, seven are on tracking, 360° swivelling, reclining seats: four in facing pairs, two forward-facing and one aft-facing; each with integral headrest, armrest and shoulder harness. Fold-out writing table between each pair of seats. Private flushing lavatory at rear with sliding doors and optional illuminated vanity unit and hot water supply. With seat belts, this compartment can serve as eighth passenger seat.

Interior options for up to nine passengers; these include substitution of carry-on baggage compartment for one of the forward centre seats, and hot and cold service refreshment centre with integral stereo entertainment system. Independent temperature control for flight deck and cabin heating systems standard. In-flight telephone optional. Tailcone baggage compartment with external access. Optional four passenger seats in main cabin of T-1A. The 400T has an aft club arrangement with swivel chairs.

SYSTEMS: Pressurisation system, with normal differential of 0.63 bar (9.1 lb/sq in) maintaining sea level cabin environment to 7,315 m (24,000 ft) and 2,286 m (7,500 ft) cabin environment to 13,715 m (45,000 ft). Back-up pressurisation system, using engine bleed air, for use in emergency. Hydraulic system, pressure 103.5 bar (1,500 lb/sq in), for actuation of flaps, landing gear and other services. Each variable volume output engine-driven pump has a maximum flow rate of 14.76 litres (3.9 US gallons; 3.25 Imp gallons)/min, and one pump can actuate all hydraulic systems. Reservoirs, capacity 4.16 litres (1.1 US gallons; 0.9 Imp gallon), pressurised by filtered engine bleed air at 1.03 bar (15 lb/sq in). All systems are, wherever possible, of modular conception: for example, entire hydraulic installation can be removed as a single unit. Stick shaker as back-up stall warning device.

AVIONICS: *Flight:* GPS retrofitted to some T-1As.

Instrumentation: Standard avionics include pilot's integrated Rockwell Collins Pro Line 4 EFIS featuring three-tube (optional four-tube) colour CRT primary flight display (PFD) and multifunction display (MFD) units mounted side by side, and control/display unit. PFD displays airspeed, altitude, vertical speed, flight director, attitude and horizontal situation information, while MFD displays navigation, radar, map, checklist and fault annunciation information. Smaller, single or dual CRTs mounted on central console function as independent navigation sensor displays or back-up displays for main CRTs. EFIS installation features strapdown attitude/heading referencing system, electronic map navigation display, airspeed trend information and V-speeds on Mach airspeed display, and solid-state Doppler turbulence detection radar.

DIMENSIONS, EXTERNAL:

Wing span	13.25 m (43 ft 6 in)
Wing aspect ratio	7.8
Length overall	14.75 m (48 ft 5 in)
Fuselage: Length	13.15 m (43 ft 2 in)
Max width	1.68 m (5 ft 6 in)
Max depth	1.85 m (6 ft 1 in)
Height overall	4.24 m (13 ft 11 in)
Tailplane span	5.00 m (16 ft 5 in)
Wheel track	2.84 m (9 ft 4 in)
Wheelbase	5.86 m (19 ft 3 in)
Crew/passenger door: Height	1.27 m (4 ft 2 in)
Width	0.71 m (2 ft 4 in)

DIMENSIONS, INTERNAL:

Cabin: Length: incl flight deck	6.32 m (20 ft 9 in)
excl flight deck	4.72 m (15 ft 6 in)
Max width	1.50 m (4 ft 11 in)
Max height	1.45 m (4 ft 9 in)
Volume: incl flight deck	11.3 m^3 (400 cu ft)
excl flight deck	8.6 m^3 (305 cu ft)
Baggage compartment volume	1.60 m^3 (56.4 cu ft)

AREAS:

Wings, net	22.43 m^2 (241.4 sq ft)
Trailing-edge flaps (total)	4.22 m^2 (45.40 sq ft)
Spoilers (total)	0.57 m^2 (6.20 sq ft)
Fin, incl dorsal fin	5.91 m^2 (63.60 sq ft)
Rudder, incl yaw damper	0.99 m^2 (10.70 sq ft)
Tailplane	5.25 m^2 (56.50 sq ft)
Elevators, incl tab	1.55 m^2 (16.70 sq ft)

WEIGHTS AND LOADINGS:

Basic operating weight, incl crew, avionics and interior fittings	4,921 kg (10,850 lb)
Baggage capacity: total	431 kg (950 lb)
tailcone	204 kg (450 lb)
Max fuel weight	2,228 kg (4,912 lb)
Max T-O weight	7,303 kg (16,100 lb)
Max ramp weight	7,393 kg (16,300 lb)
Max landing weight	7,121 kg (15,700 lb)
Max zero-fuel weight	5,896 kg (13,000 lb)
Max power loading	284 kg/kN (2.78 lb/lb st)

PERFORMANCE:

Max limiting Mach No.	0.78
Max level speed at FL270	468 kt (867 km/h; 539 mph)
Typical cruising speed at FL390	450 kt (834 km/h; 518 mph)
Long-range cruising speed at FL410	392 kt (726 km/h; 451 mph)
Stalling speed, flaps down, idling power	93 kt (173 km/h; 107 mph) CAS
Max rate of climb at S/L	1,149 m (3,770 ft)/min
Rate of climb at S/L, OEI	306 m (1,005 ft)/min
Max certified altitude	13,715 m (45,000 ft)
Service ceiling, OEI	6,279 m (20,600 ft)
FAA (FAR Pt 25) T-O field length at S/L, ISA	1,159 m (3,802 ft)

FAA landing distance at max landing weight
                                1,071 m (3,514 ft)
Range with max fuel, four passengers, incl 45 min
  reserves:
  VFR   1,669 n miles (3,091 km; 1,920 miles)
  IFR   1,457 n miles (2,698 km; 1,676 miles)
                                        UPDATED

BEECH 390 PREMIER I

TYPE: Light business jet.

PROGRAMME: Design started early 1994 as PD374 (later
PD390) and approved early 1995; originated in former
Beech design offices, but was first aircraft to carry only the
Raytheon name; brief details of 'new light business jet'
revealed June 1995; launched at National Business Aircraft
Association Convention in Las Vegas 26 September 1995
with full-scale fuselage/cabin mockup; wind-tunnel tests of
one-eighth model conducted early 1996 at Boeing, Boeing
V/STOL, NASA-Lewis and Wichita State University
facilities; to compete with Cessna CitationJet.

First forward fuselage completed in February 1997 and
mated to aft fuselage in April 1998; roll-out (N390RA, c/n
RB-1) 19 August 1998; first flight 22 December 1998.
Second aircraft (N704T) first flown 4 June 1999, followed
by third (N390TC), first with complete interior, on 17
September 1999; public debut (N390TC) at National
Business Aviation Association Convention at Atlanta,
Georgia, October 1999; more than 720 flight test hours
accumulated by 23 December 1999, at which time eight
production aircraft were in final assembly; static testing of
wing to 150 per cent of design load completed on 17
December 1999; four aircraft undertook 1,400 hour flight
test programme culminating in FAA FAR Pt 23
certification on 23 March 2001, followed by German
certification on 3 September 2001; certified in Bermuda,
Denmark, Mexico, Israel, South Africa and Switzerland in
2002. Deliveries began with three aircraft in third quarter
of 2001: RB-4, -6 and -7 to Tyrose Investments, Raytheon
and Town & Country Food Markets, respectively. Target
production rate 60 per year from 2003.

CUSTOMERS: More than 300 orders received by October 2001
from customers in 27 countries, of which some 100 were
from outside the USA and 51 from Europe, representing a
backlog until 2005. Total of 47 delivered 31 December
2002. Customers include Raytheon Travel Air, the
fractional ownership subsidiary of Raytheon Aircraft,
which has ordered 71 for delivery beginning 2001; the
Jordan Grand Prix racing team, which has ordered one; and
Aviation Leasing Group (ALG Transportation Inc) of
London, which ordered three in August 2000, two of which
will be used by the Civil Aviation Training Centre (CATC)
in Thailand for training student pilots for Thai Airways
International and other Pacific Rim carriers.

COSTS: US$5.3 million. Estimated direct operating cost
US$680 per hour (60th 2001).

DESIGN FEATURES: Conventional small business jet, developed
with assistance of CATIA programmes. Rear-mounted
engines, T tail and wing mounted below fuselage for
additional cabin space. Sweepback 20° at 25 per cent
chord; 2° 30′ dihedral; tailplane sweepback 25° at 25 per
cent chord.

FLYING CONTROLS: Conventional and manual. Activation via
pushrods and cables. Pitch trim via electrically actuated,
variable incidence tailplane and mechanically driven
geared tab on each elevator; electrically actuated trim tab
on each aileron; electrically actuated rudder trim tab.
Electrically signalled, hydraulically powered, three-
segment spoilers on upper surface of each wing augment
aileron roll control; outboard and middle panels provide
roll, airbrake and post-landing lift-dump functions;
inboard panel provides lift-dump function only. 75 per cent
span, four-segment, electrically controlled Fowler flaps,
deflections 0, 10, 20 and 30°. Rudder boost, for asymmetric
thrust and yaw damper, standard.

STRUCTURE: Fuselage of graphite/epoxy laminate and
honeycomb composites, formed by Cincinnati Milacron
Viper automatic fibre-placement machines over
aluminium mandrel, placing fibres at speeds up to 46 m
(150 ft) per minute, enabling entire fuselage to be
completed in one week; elimination of all internal frames,
and skin thickness of 20 mm (0.78 in), increase cabin
volume by 13 per cent and afford a weight saving of some
20 per cent over conventional alloy construction. Wing of
aluminium alloy with six-spar wing box, manufactured
using high-speed equipment capable of machining more
than 93 m² (1,000 sq ft) of material per minute, and
automatic riveting machines; with exception of three small
bays along trailing-edge, entire wing is used for fuel
storage. Ailerons and flaps of graphite/epoxy composites;
fin has aluminium alloy spars and ribs with graphite/epoxy
honeycomb skin; tailplane has one-piece, composites
forward-and-rear spar with alloy centre rib, composites
mid- and tip ribs and Nomex composites skin.

LANDING GEAR: Hydraulically actuated, retractable tricycle
type with free-fall emergency extension system; single
wheel on each unit. Mainwheel size 22×8.2 (12 ply);
nosewheel 18×4.4 (6 ply). Mainwheels retract inwards;
nosewheel forwards. Steerable nosewheel, maximum
deflection ±45°. Hydraulic disc brakes with electric anti-
skid system.

Beech Premier I cabin   0130569

POWER PLANT: Two pod-mounted Williams FJ44-2A
turbofans, each rated at 10.23 kN (2,300 lb st). Fuel
contained in integral wing tanks, total usable capacity
2,040 litres (539 US gallons; 449 Imp gallons), with
gravity filling point on each wing. Single-point pressure
refuelling/defuelling optional.

ACCOMMODATION: Crew of one or two, with dual controls
standard; six passengers in cabin, comprising four in
standard club seating arrangement with tracking,
swivelling and reclining capability and stowable writing
tables and two on fixed forward-facing seats to rear;
lavatory at rear, doubling as flight-accessible baggage
compartment, maximum capacity 64 kg (140 lb).
Refreshment/hang-up baggage cabinet on forward
starboard side of cabin. Airstair door on port side to rear of
flight deck; single plug-type emergency exit on starboard
side. Three cabin windows on each side. Accommodation
is air conditioned and pressurised. Externally accessible
main baggage compartment to rear of cabin, with upward-
opening door on port side, can accommodate large items
such as skis; heating optional; forward baggage
compartment in nose on port side with swing-up door.

SYSTEMS: Pressurisation system, maximum differential 0.58
bar (8.4 lb/sq in), maintains 2,440 m (8,000 ft) cabin
altitude to 12,500 m (41,000 ft). Vapour cycle, ozone-safe
R134a air conditioning system. Hydraulic system,
maximum pressure 207 bar (3,000 lb/sq in), for landing
gear, brakes, anti-skid and spoilers. Electrical system
comprises two 28 V 300 A engine-driven starter/
generators, 24 V 40 Ah lead/acid main battery, 24 V 5 Ah
standby battery and 28 V external power receptacle;
system is configured so that load-shedding is primarily
automatic in the event of failure of any or all main
electrical power sources. Oxygen system, capacity 1,134
litres (40 cu ft) standard, 2,182 litres (77 cu ft) optional,
with diluter-demand masks for crew and continuous flow

Beech Premier I flight deck   0126961

masks for passengers. Engine bleed-air anti-icing for wing leading-edges and nacelle inlets; electromagnetic expulsion de-icing (EMED) for tailplane leading-edges, automatically activated by dual nose-mounted, heated, ice detectors; electrically heated windscreens (with silicone coating for rain dispersal), pitot tubes and AoA probes.

AVIONICS: Rockwell Collins Pro Line 21 EFIS avionics suite as core system.

Comms: Dual Rockwell Collins VHF-422A transceivers, TDR-94 Mode S transponders and DB Model 438 audio systems; single CTL-23 nav/com tuning unit; four-speaker cabin paging unit.

Radar: Rockwell Collins RTA-800 colour weather radar.

Flight: Dual Rockwell Collins AHC-3000 AHRS, ADC-3000 air data computers, CDU-3000 control/display units and VIR-432 nav receivers; IAPS-3000 lightweight, integrated avionics processing system; FGC-3000 flight guidance system, FMS-3000 flight management system with database, ADF-462, DME-442, GPS-4000, ALT-4000 radio altimeter and MDC-3000 maintenance diagnostic computer.

Instrumentation: Rockwell Collins AFD-3010 integrated EFIS comprising two 254 × 203 mm (10 × 8 in) active matrix LCD adaptive flight displays providing PFD and MFD functions, with CRT HSI and back-up electromechanical rate/sensor/attitude instrument and ASI on right side; second PFD optional, but not mandatory for RVSM compliance.

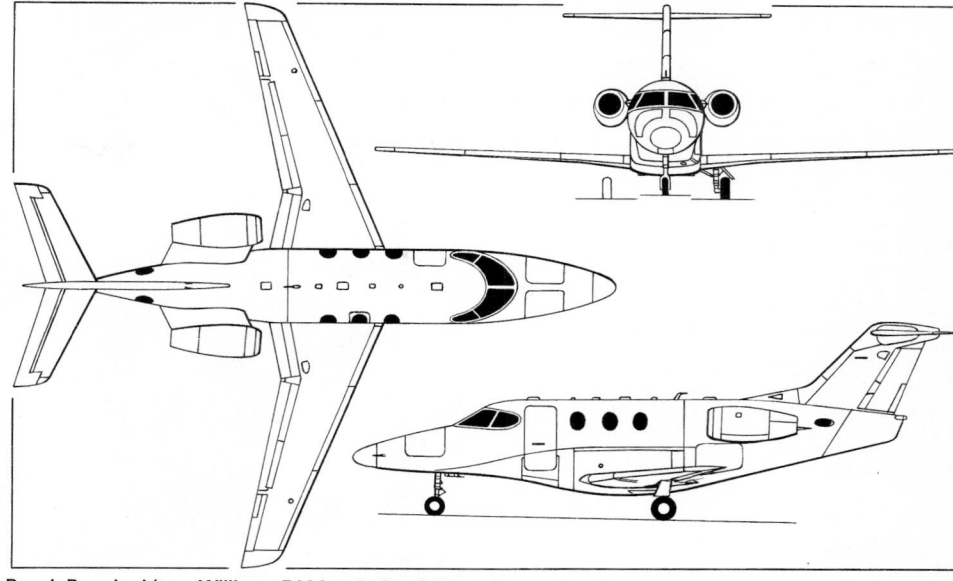
Beech Premier I (two Williams FJ44 turbofans) (*Jane's/James Goulding*) 0126960

DIMENSIONS, EXTERNAL:

Wing span	13.56 m (44 ft 6 in)
Wing aspect ratio	8.0
Length overall	14.02 m (46 ft 0 in)
Height overall	4.67 m (15 ft 4 in)
Tailplane span	4.90 m (16 ft 1 in)
Wheel track	2.79 m (9 ft 2 in)
Wheelbase	5.36 m (17 ft 7 in)
Crew/passenger door: Height	1.27 m (4 ft 2 in)
Width	0.64 m (2 ft 1½ in)

DIMENSIONS, INTERNAL:
Cabin:

Length: between pressure bulkheads	5.69 m (18 ft 8 in)
excl flight deck	4.11 m (13 ft 6 in)
Max width	1.68 m (5 ft 6 in)
Max height	1.65 m (5 ft 5 in)
Volume	9.4 m³ (331 cu ft)

External baggage compartment volume:	
main	1.27 m³ (45 cu ft)
forward	0.28 m³ (10.0 cu ft)

AREAS:

Wings, gross	22.95 m² (247.0 sq ft)
Horizontal tail surfaces (total)	4.65 m² (50.00 sq ft)

WEIGHTS AND LOADINGS:

Basic operating weight	3,627 kg (7,996 lb)
Baggage capacity: main	181 kg (400 lb)
forward	68 kg (150 lb)
internal	27 kg (60 lb)
Max fuel weight	1,638 kg (3,611 lb)
Max-T-O weight	5,670 kg (12,500 lb)
Max ramp weight	5,710 kg (12,590 lb)
Max landing weight	5,262 kg (11,600 lb)
Max zero-fuel weight	4,536 kg (10,000 lb)
Max wing loading	247.1 kg/m² (50.61 lb/sq ft)
Max power loading	277 kg/kN (2.72 lb/lb st)

PERFORMANCE:

Max operating speed:	
S/L to FL270	320 kt (593 km/h; 368 mph)
above FL270	M0.80
Max cruising speed at 10,060 m (33,000 ft)	461 kt (854 km/h; 530 mph)
Max operating altitude	12,500 m (41,000 ft)
T-O run	1,156 m (3,792 ft)
Landing run	966 m (3,170 ft)
Range with single pilot, four passengers, NBAA IFR reserves	1,430 n miles (2,648 km; 1,645 miles)
g limits	+3.2/−1.28

UPDATED

Beech Premier I light business jet *NEW*/0527092

BEL-AIRE

BEL-AIRE AVIATION INC
3891 Maxison Drive, Lyons, New York 14489
Tel: (+1 315) 923 77 45 and 946 50 58
PRESIDENT: Gerald Belcher

Having acquired and reverse-engineered a 1926 Travel Air Model B, Mr Belcher announced production of kits at the 2000 Sun 'n' Fun convention. However, by late 2002, no further completions were known.

UPDATED

BEL-AIRE 2000
No known production beyond the prototype. Last described and illustrated in 2002-03 *Jane's*.

UPDATED

BELL

BELL HELICOPTER TEXTRON INC
(Subsidiary of Textron Inc)
PO Box 482, Fort Worth, Texas 76101
Tel: (+1 817) 280 20 11
Fax: (+1 817) 280 23 21

Web: http://www.bellhelicopter.textron.com
CHAIRMAN AND CEO: John R Murphey
EXECUTIVE VICE-PRESIDENT: P D Shabay
SENIOR VICE-PRESIDENT, ACQUISITIONS AND INTERNATIONAL BUSINESS DEVELOPMENT: Fred N Hubbard
VICE-PRESIDENT, RESEARCH AND ENGINEERING: Art Lucas
DIRECTOR, PUBLIC AFFAIRS AND ADVERTISING: Carl L Harris

During 1970-81, Bell Helicopter Textron was unincorporated division of Textron Inc; became wholly owned subsidiary of Textron Inc from 3 January 1982. Bell Helicopter Canada (see Canadian section) formed at Montréal/Mirabel under contract with Canadian government October 1983; transfer to Mirabel, completed January 1987, of Bell 206B JetRanger and 206L LongRanger production. Production of Bell 212/

412 transferred mid-1988 and early 1989 respectively; Bell 230, 430, 427 and 407 programmes also undertaken in Canada, although some of these now terminated.

More than 34,000 Bell helicopters manufactured worldwide, including over 9,500 commercial models.

Bell and Boeing collaborate in design and manufacture of V-22 Osprey tiltrotor aircraft, as described in the following entry. New 41,800 m² (450,000 sq ft) factory at Amarillo International Airport, Texas, completed in 1999 as a Tiltrotor Assembly Centre (TAC) for the V-22 and the commercial BA609; latter tiltrotor, previously also a joint venture with Boeing, became a Bell programme on 1 March 1998 and is now (with the AB139) the core of a joint venture effort undertaken in conjunction with Agusta of Italy; an agreement establishing the Bell/Agusta Aerospace Company (see International section) was signed in November 1998.

Bell helicopters built in USA detailed here. Those currently built in Canada listed under Canada; other models built under licence by Dirgantara in Indonesia and Agusta in Italy (which see); KAI (which see) is co-producing Bell 427 as SB 427 in Republic of Korea; Bell Helicopter Asia (Pte) Ltd is wholly owned Singapore-based company for marketing and support in Southeast Asia.

UPDATED

BELL 449 SUPERCOBRA and KING COBRA
US Navy/Marine Corps designations: AH-1W and AH-1Z

TYPE: Attack helicopter.

PROGRAMME: Prototype Bell 209, derived from single-engined UH-1, first flew as tandem-seat combat aircraft on 7 September 1965. Built for US armed forces and export and under licence in Japan, as described in previous editions of *Jane's*. Universally known as HueyCobra.

First twin-engined Cobra was AH-1J SeaCobra, delivered from mid-1970; AH-1T Improved SeaCobra followed from 1977. All surviving US Marine Corps AH-1J SeaCobras withdrawn and 44 (including one ground-based trainer) AH-1T Improved SeaCobras converted to AH-1W to augment new production.

CURRENT VERSIONS: **AH-1W SuperCobra:** Bell flew AH-1T powered by two GE T700-GE-700; first flight of improved AH-1T+, including GE T700-GE-401 engines, 16 November 1983. USMC received 169 new-build examples as well as two composites maintenance trainers; 10 supplied to Turkey and 63 to Taiwan. Missions of AH-1W include anti-armour, escort, multiple-weapon fire support, armed reconnaissance, search and target acquisition.

AH-1W Upgrades: Following abandonment of the proposed Integrated Weapon System (IWS) project in July 1995 and the Marine Observation and Attack Aircraft programme which was intended to provide a replacement for both the AH-1W SuperCobra and the UH-1N Iroquois, the US Marine Corps has opted for a two-stage upgrade of the AH-1W, allowing it to be retained in the active inventory beyond 2020. Phase 1 concerned installation of a Night Targeting System (NTS), under which USMC AH-1Ws fitted with the Israeli Tamam laser NTS for dual TOW/Hellfire day, night and adverse weather capability.

Conversion of a prototype (162533) was authorised in December 1991, with an initial batch of 25 sets being built by Tamam for delivery from January 1993; joint production with Kollsman was approved in May 1994. A total of 250 sets was required by the USMC, with further sets produced for Turkey and Taiwan. Deliveries of modified aircraft to operational units of the USMC began in June 1994.

A further improvement programme, involving installation of an Embedded Global Positioning System/Inertial Navigation System (EGI), is being undertaken. Two prototype conversions (162532 and 163936) were delivered to test units for trials in November 1995 and March 1996, with EGI installed on new-build aircraft from Lot 9 onwards, as well as older AH-1Ws as a retrofit programme.

Phase 2 will involve installation of the Bell 680 four-blade rotor, offering a 70 per cent reduction in vibration; formerly designated **AH-1W (4BW)**, but now known as

Bell AH-1Z SuperCobra prototype   *NEW*/0137430

AH-1Z. Initial trials of the four-blade rotor system were undertaken with AH-1W 161022; bench testing of the new drive system began in second quarter of 1999 and was completed in first quarter of 2000. Bell also demonstrated 30-minute run-dry capability of new intermediate and tail rotor gearboxes in March 2000. The definitive aircraft will be fitted with a new four-blade, all-composites, hingeless/bearingless rotor system; four-blade composites tail rotor; a new transmission rated at 1,957 kW (2,625 shp); endplates on horizontal tail surfaces and new wing assemblies with the ability to carry twice the number of anti-armour missiles, 379 litres (100 US gallons; 83.0 Imp gallons) more fuel and additionally permitting concurrent carriage of two air-to-air self-defence missiles.

Lockheed Martin selected to develop and manufacture AN/AAQ-30 Hawkeye advanced target sighting system (TSS), with work on US$8 million, 54 month, engineering development and integration programme beginning in July 1998. TSS will feature imaging technology by Wescam of Canada and Lockheed Martin's Sniper third-generation FLIR, as well as colour TV camera, laser ranger, spot-tracker and designator.

Also provided in the second phase will be 'glass cockpits'; Northrop Grumman (formerly Litton Industries) has been selected as prime contractor for this aspect of the upgrade programme. Digital transfer of information on tactical situation, weaponry and flight data will enable crew interchangeability and allow AH-1Z to be flown from either front or rear seat. Major subcontractors include Rockwell Collins, which will supply active matrix liquid crystal displays (AMLCDs); Smiths Industries (fire-control system); Meggitt Avionics (standby air data and inertial sensing devices); and BAE Systems (air data computers). Other elements of the upgrade include new stores management system, onboard systems monitoring, mission data loader, HOTCC (hands on throttle, collective and cyclic) controls, airborne target handover system and a new EW suite.

A US$310 million cost-plus-fixed-fee contract was awarded to Bell in November 1996, for design, development, fabrication, installation, test and delivery of three engineering development AH-1W SuperCobra

Upgrade Aircraft. Assembly of first AH-1Z begun at Hurst, Texas, in April 1999, by which time 85 per cent of drawings had been released, with design work due for completion by end of 1999. Initial AH-1Z (162549, c/n 59001) completed final assembly in second quarter of 2000 and moved to Bell Flight Research Center at Arlington, Texas, for installation of instrumentation and functional testing that included restrained ground running which was completed in October 2000. Formal roll-out at Arlington on 20 November 2000, with first flight following on 7 December. Second development aircraft (162559, c/n 59002) was due to fly in 2001, but handling quality problems that emerged early in flight test programme necessitated redesign of horizontal stabiliser assembly and have caused delay; this and third AH-1Z were both due to fly in early 2002. Programme will include flight test and evaluation at Patuxent River, Maryland, to where first AH-1Z was airlifted by C-5 Galaxy on 31 March 2001. Weapons testing will take place at Yuma Proving Ground, Arizona with other trials at China Lake, California. OT-IIA operational assessment now due to take place in October 2002, with OT-IIB phase, including live-fire trials, set for second half of 2003, followed by operational evaluation starting in August 2004. Testing of full-scale AH-1Z fatigue and static test article at Arlington began in April 2000 and will include verification of predicted 10,000-hour service life. Finalisation of the cockpit upgrade design occurred in FY99, with first order for remanufacture due to be placed in FY04. IOC scheduled for 2007, with peak production rate requiring 24 AH-1Ws to be upgraded annually.

AH-1RO Dracula: Derivative of AH-1W for Romania, which intended to purchase initial batch of 96. Project abandoned by Bell in fourth quarter of 1999.

AH-1Z King Cobra: Version for Turkey, which plans to acquire 145 attack helicopters at cost of US$4 billion; bids for initial batch of 50 (including two prototypes) submitted by end 1997. Announcement of winning contender was expected at start of 1999 but deferred to mid-2000, following delays in flight evaluations of competing types. AH-1Z selected, with announcement made at Farnborough 2000 in late July, when revealed that initial batch of 50 to be purchased at approximate US$1.5 billion cost; contract signature was due in first quarter of 2001, but was delayed because of difficulties over indigenous production of key systems such as mission computer; these largely resolved by early 2002, when contract signature expected in June. Licensed production to be undertaken in Turkey by TAI at Ankara; current plan stipulates follow-on batches of 50 and 45 helicopters, with deliveries to begin in 2005 and continue until 2011.

ARH-1Z: Designation allocated to version unsuccessfully proposed for Australian Army Project Air 87 armed reconnaissance helicopter.

MH-1W: In April 1998, Bell revealed a reconnaissance, armed escort and fire support 'multimission' version of the SuperCobra under this designation. Evolved in response to a perceived need for armed helicopters to undertake anti-drug operations, marketing efforts principally aimed at Latin American countries, with presentations to Argentina, Brazil, Chile, Colombia and Venezuela. Configuration includes a nose-mounted sighting system, with a FLIRStar Safire FLIR sensor, laser range-finder, video recorder and automatic target tracker. Proposed weaponry includes a 20 mm cannon as well as 0.50 in gun pods and up to four

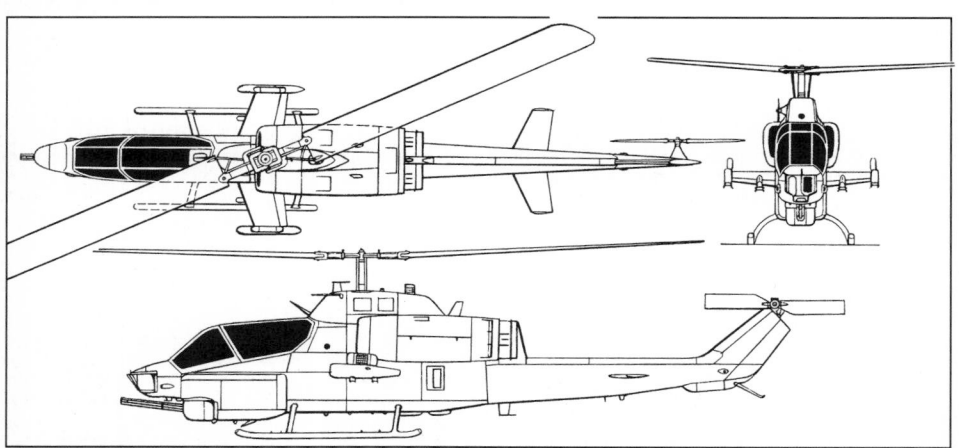

Bell AH-1W SuperCobra (*Jane's/Dennis Punnett*)

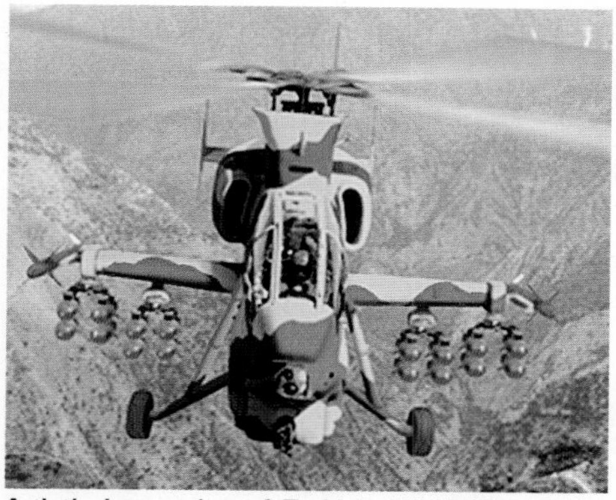

Artist's impression of Turkish AH-1Z King Cobra
NEW/0137431

70 mm rocket pods, but excludes anti-armour missiles and air-to-air missiles. Marketing estimates predict a need for up to 60 MH-1Ws in Latin America and initial deliveries could take place within 18 months of order being received.

CUSTOMERS: US Marine Corps (see under Current Versions); total of 179 AH-1W, including 10 diverted to Turkey.

Deliveries to USMC began on 27 March 1986 to Camp Pendleton, California, for HMLA-169, -267, -367 and -369, plus HMT-303 for training; further aircraft issued to USMC Reserve, beginning with HMA-775 (now HMLA-775) at Camp Pendleton from June 1992; followed by HMA-773 (now HMLA-773) at Atlanta, Georgia, and HMLA-767 (now HMLA-775 Detachment A) at New Orleans, Louisiana. Procurement augmented by remanufacture of 43 AH-1Ts to AH-1W for HMLA-167 and -269 at New River, North Carolina; last completed in 1993; 100th new/converted AH-1W delivered 8 August 1991. Last new-build aircraft for USMC delivered in fourth quarter of 1998.

Turkish Land Forces received five AH-1Ws in 1990 and five in 1993; all diverted from USMC contracts. Taiwan signed letter of offer and acceptance February 1992, for 42 over five years (plus one ground trainer); deliveries began 1993 for training with USMC; first aircraft 501 (ex-164913); in service with 1st Attack Helicopter Squadron at Lung Tan and 2nd AHS at Shinsur. Further 21 AH-1Ws subject of Taiwanese re-order announced in mid-1997; contract for first nine awarded in October 1997.

Remanufacture programme calls for 180 AH-1Ws to be produced, with procurement of first Low-Rate Initial Production (LRIP) batch of six helicopters expected in FY04, with delivery scheduled for 2006; another LRIP batch will follow in FY05, with full-rate production starting in FY06, assuming successful conclusion of operational evaluation.

COSTS: US$10.7 million (1996) projected unit cost. Total value of 1997 follow-on order for 21 AH-1Ws by Taiwan US$479 million. Unit cost of new-build AH-1Z estimated at US$14 million (Turkish proposal, 1998). In first quarter of 2002, total cost of remanufacturing 180 AH-1Zs and 100 UH-1Ys expected to be approximately US$3.7 billion.

DESIGN FEATURES: Essentially first-generation attack helicopter, with slightly stepped tandem seating and stub-wings for armament. Two-blade main rotor, similar to that of Bell 214, with strengthened rotor head incorporating Lord Kinematics Lastoflex elastomeric and Teflon-faced bearings. Blade aerofoil Wortmann FX-083 (modified); normal 311 rpm. Tail rotor also similar to that of Bell 214 with greater diameter and blade chord; normal 1,460 rpm. Rotor brake standard. Stub-wings have NACA 0030 section at root; NACA 0024 at tip; incidence 14°; sweepback 14.7°. AH-1Z will incorporate new four-blade rotor system and transmission (see Current Versions section for details).

STRUCTURE: Main rotor blades have aluminium spar and aluminium-faced honeycomb aft of spar; tail rotor has aluminium honeycomb with stainless steel skin and leading-edge. Airframe conventional all-metal semi-monocoque.

LANDING GEAR: Non-retractable tubular skid type on AH-1W; AH-1Z will incorporate new, lighter, design with rectangular cross tubes. Ground handling wheels optional.

POWER PLANT: Two General Electric T700-GE-401 turboshafts, each rated at 1,285 kW (1,723 shp). Transmission rating 1,515 kW (2,032 shp) for take-off; 1,286 kW (1,725 shp) continuous. Fuel (JP5) contained in two interconnected self-sealing rubber fuel cells in fuselage, with protection from damage by 0.50 in ballistic ammunition, total usable capacity 1,128 litres (298 US gallons; 248 Imp gallons). Gravity refuelling point in forward fuselage, pressure refuelling point in rear fuselage. Provision for carriage on underwing stores stations of two or four external fuel tanks each of 291 litres (77.0 US gallons; 64.1 Imp gallons) capacity; or two 378 litre (100 US gallon; 83.3 Imp gallon) tanks; or two 100 and two 77.0 US gallon tanks; large tanks on outboard pylons only. Oil capacity 19 litres (5.0 US gallons; 4.2 Imp gallons).

ACCOMMODATION: Crew of two in tandem, with co-pilot/gunner in front seat and pilot at rear in AH-1W; crew stations are interchangeable in AH-1Z. Cockpit is heated, ventilated and air conditioned. Dual controls; lighting compatible with night vision goggles, and armour protection standard. Forward crew door on port side and rear crew door on starboard side, both upward-opening. Inflatable body and head restraint system by Simula of Phoenix, Arizona, nearing end of development in mid-1995; retrofit provisions installed in 1996 production, with system incorporated in 1997 production.

SYSTEMS: Three independent hydraulic systems, pressure 207 bar (3,000 lb/sq in), for flight controls and other services. Electrical system comprises two 28 V 400 A DC generators, two 24 V 34.5 Ah batteries and three inverters: main 115 V AC, 1 kVA, single-phase at 400 Hz, standby 115 V AC, 750 VA, three-phase at 400 Hz and a dedicated 115 V AC 365 VA single-phase for AIM-9 missile system. AiResearch environmental control unit.

AVIONICS: *Comms:* Two AN/ARC-210(V) radios, KY-58 TSEC secure voice set; AN/APX-100(V) IFF.

Radar: None at present installed, but possibility exists that Longbow radar will eventually be fitted as so-called Cobra Radar System. Mockup of possible installation displayed at Asian Aerospace show in Singapore, February 2002, this depicting pod-mounted radar sited above wing assembly. Longbow radar and associated AGM-114L RF Hellfire missile being offered to Taiwan, as part of an upgrade package for existing AH-1W fleet.

Flight: AN/ASN-75 compass set, AN/ARN-89B ADF, AN/ARN-118 Tacan on AH-1W and AN/ARN-153(V-4) Tacan on AH-1Z, AN/APN-154(V) radar beacon set and Teledyne AN/APN-217 Doppler-based navigation system being replaced by Embedded Global Positioning System/Inertial Navigation System (EGI), which adds Honeywell CN-1689(V) EGI and Rockwell Collins AN/ARN-153(V) Tacan. Installation began with Lot 9 production in October 1996 and also involved retrofit of earlier aircraft. AN/APN-194 radar altimeter. Rockwell Collins CDU-800 control/display unit and dual Rockwell Collins ICU-800 processors.

Instrumentation: Kaiser HUD compatible with PNVS-5 and Elbit ANVIS-7 night vision goggles. Helmet-mounted sighting/aiming device selected June 2002 in form of Thales TopOwl. Rockwell Collins flat-panel colour AMLCD displays to be installed on AH-1Z.

Mission: Tamam/Kollsman Night Targeting System (NTSF-65) comprising FLIR, laser range-finder/designator, TV camera, day/night video tracker and full in-flight boresighting; being retrofitted (within M-65 sighting system for TOW missiles) from 1993 (see Design Features); first redelivery in June 1994; alternative Boeing Electronic Systems NightHawk system also offered for export SuperCobras, following 1992-93 flight testing. Lockheed Martin AN/AAQ-30 Hawkeye multisensor electro-optical/IR targeting system to be installed in AH-1Z; this features large-aperture mid-wave FLIR, colour TV camera, laser range-finder/designator, laser spot tracker and on-gimbal inertial measurement system.

Self-defence: AN/APR-39(V) pulse radar signal detecting set, AN/APR-44(V) CW radar warning system, and AN/ALQ-144(V) IR countermeasures set. Dual AN/ALE-39 chaff system with one MX-7721 dispenser mounted on top of each stub-wing. Improved countermeasures suite in AH-1Z replaces AN/APR-39 and AN/APR-44 by AN/APR-39A(XE2) radar warning receiver and adds AN/AVR-2 laser warner and AN/AAR-47 plume detecting set. AH-1Z to have ITT AN/ALQ-211 SIRFC as optional alternative for export aircraft.

ARMAMENT: Electrically operated General Electric undernose A/A49E-7(V4) turret housing an M197 three-barrel 20 mm gun. A 750-round ammunition container is located in the fuselage directly aft of the turret; firing rate is 675 rds/min; a 16-round burst limiter is incorporated in the firing switch. Either crew member can fire the gun, which can be slaved to a helmet-mounted sight/aiming device. Gun can be tracked 110° to each side, 18° upward, and 50° downward, but barrel length of 1.52 m (5 ft 0 in) makes it imperative that the M197 is centralised before wing stores are fired. Underwing attachments for up to four LAU-61A (19-tube), LAU-68A, LAU-68A/A, LAU-68B/A or LAU-69A (seven-tube) 2.75 in Hydra 70 rocket launcher pods; two CBU-55B fuel-air explosive weapons; four SUU-44/A flare dispensers; two M118 grenade dispensers; Mk 45 parachute flares; or two GPU-2A or SUU-11A/A Minigun pods.

Provision for carrying totals of up to eight TOW missiles, eight AGM-114 Hellfire missiles, two AIM-9L Sidewinder or AGM-122A Sidearm missiles, on outboard underwing stores stations. Canadian Marconi TOW/Hellfire control system enables AH-1W to fire both TOW and Hellfire missiles on same mission. AH-1Z expected to include FIM-92 Stinger AAM for self-defence.

Following data applicable to AH-1W except where stated.

DIMENSIONS, EXTERNAL:	
Main rotor diameter	14.63 m (48 ft 0 in)
Main rotor blade chord	0.84 m (2 ft 9 in)
Tail rotor diameter	2.97 m (9 ft 9 in)
Tail rotor blade chord	0.305 m (1 ft 0 in)
Distance between rotor centres	8.89 m (29 ft 2 in)
Wing span	3.28 m (10 ft 9 in)
Wing aspect ratio	3.7
Length: overall, rotors turning	17.68 m (58 ft 0 in)
fuselage	13.87 m (45 ft 6 in)
Width overall	3.28 m (10 ft 9 in)
Height: to top of rotor head	4.11 m (13 ft 6 in)
overall	4.44 m (14 ft 7 in)
Ground clearance, main rotor turning	2.74 m (9 ft 0 in)
Elevator span	2.11 m (6 ft 11 in)
Width over skids	2.24 m (7 ft 4 in)
AREAS:	
Main rotor blades (each)	6.13 m^2 (66.0 sq ft)
Tail rotor blades (each)	0.45 m^2 (4.835 sq ft)
Main rotor disc	168.11 m^2 (1,809.6 sq ft)
Tail rotor disc	6.94 m^2 (74.70 sq ft)
Vertical fin	2.01 m^2 (21.70 sq ft)
Horizontal tail surfaces	1.41 m^2 (15.20 sq ft)
WEIGHTS AND LOADINGS:	
Weight empty: AH-1W	4,953 kg (10,920 lb)
AH-1Z	5,579 kg (12,300 lb)
Mission fuel load (usable): AH-1W	946 kg (2,086 lb)
AH-1Z	1,256 kg (2,768 lb)
Max useful load (fuel and disposable ordnance):	
AH-1W	1,736 kg (3,828 lb)
AH-1Z	2,812 kg (6,200 lb)
Max T-O and landing weight:	
AH-1W	6,690 kg (14,750 lb)
AH-1Z	8,391 kg (18,500 lb)
Max disc loading: AH-1W	39.8 kg/m^2 (8.15 lb/sq ft)
AH-1Z	49.9 kg/m^2 (10.22 lb/sq ft)
Transmission loading at max T-O weight and power:	
AH-1W	4.42 kg/kW (7.26 lb/shp)
AH-1Z	5.54 kg/kW (9.10 lb/shp)
PERFORMANCE:	
Never-exceed speed (VNE):	
AH-1W	190 kt (352 km/h; 219 mph)
AH-1Z	222 kt (411 km/h; 255 mph)
Max level speed at S/L	152 kt (282 km/h; 175 mph)
Max cruising speed: AH-1W	150 kt (278 km/h; 173 mph)
AH-1Z	158 kt (293 km/h; 182 mph)
Rate of climb at S/L, OEI	244 m (800 ft)/min
Service ceiling	more than 4,270 m (14,000 ft)
Service ceiling, OEI	more than 3,660 m (12,000 ft)
Hovering ceiling: IGE	4,495 m (14,740 ft)
OGE	915 m (3,000 ft)
Range at S/L with standard fuel, no reserves:	
AH-1W	280 n miles (518 km; 322 miles)
AH-1Z	370 n miles (685 km; 426 miles)
Max endurance with standard fuel: AH-1W	2 h 48 min
AH-1Z	3 h 30 min
g limits: AH-1W	+2.5/–0.5
AH-1Z	+3.5/–0.5

UPDATED

BELL QUAD TILTROTOR

TYPE: Multimission tiltrotor.

PROGRAMME: Bell announced in early 1999 that it was studying a proposed Quad TiltRotor (QTR) to meet Future

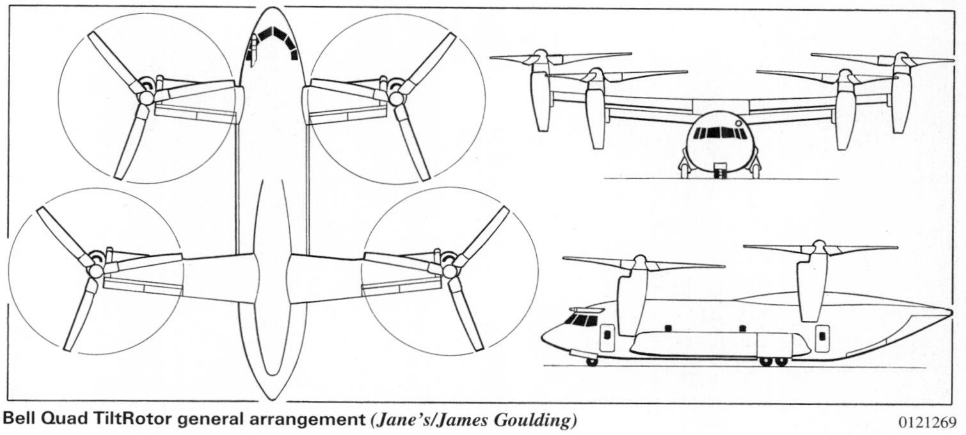

Bell Quad TiltRotor general arrangement *(Jane's/James Goulding)* 0121269

Transport Rotorcraft (FTR) requirements. As projected, the aircraft would feature a fuselage approximately the size of that of a Lockheed Martin C-130-30, mated to two sets of wings, engines and tiltrotors from the Bell Boeing V-22 Osprey, the rear units mounted on stub wings to extend span and ensure adequate fuselage clearance. Rear tiltrotors could fold in cruising flight, with their engines providing supplemental thrust. The Quad TiltRotor would be able to accommodate up to 90 passengers, or an AH-64, AH-1Z, RAH-66, UH-1Y or UH-60 helicopter, or three HMMWVs, or up to eight 463L pallets.

Provisional specifications include VTOL maximum T-O weight 45,360 kg (100,000 lb); STOL maximum T-O weight 63,505 kg (140,000 lb); and a payload up to 18,144 kg (40,000 lb). The Quad TiltRotor was in the conceptual design phase in mid-2000, with water tunnel testing of a 1/48th scale model to visualise complex airflow patterns around the tandem wings and four tiltrotors completed.

In mid-2000 DARPA awarded Bell a US$400,000 phase 1 contract as part of a three-phase, US$6 million cost-sharing programme to study the feasibility of the QTR concept. Phase 1 involved a detailed technology study. In phase 2, a 1/14th scale hovering model of the QTR was test-flown. Phase 3 comprised wind tunnel tests to determine loads and aerodynamic performance of the full-size aircraft. The feasibility study was scheduled for completion by the end of 2001.

Model of proposed Bell Quad TiltRotor (*Jane's/Paul Jackson*)   *NEW*/0525825

The first of two demonstrators could be flown by 2006, with production deliveries starting in 2010. Potential customers include the US Marine Corps (to replace Sikorsky CH-53E helicopters and KC-130 Hercules) and USAF (MH-53J combat SAR/special forces helicopters).

UPDATED

OTHER AIRCRAFT

Bell **TH-67 Creek** and **civilian helicopters** appear under Bell Helicopter Textron Canada in that country's section. **Bell/Agusta BA609** tiltrotor appears in the International section.

UPDATED

BELL BOEING
BELL HELICOPTER TEXTRON and THE BOEING COMPANY

Bell Boeing V-22 Joint Program Office, PO Box 70, Patuxent River, Maryland 20670-0070
Tel: (+1 301) 757 66 34
Web (1): http://www.boeing.com/rotorcraft/military/v22
Web (2): http://www.bellhelicopter.textron.com/products/tiltRotor/v22
BELL BOEING PROGRAMME DIRECTOR: Mike Tkach
V-22 PROGRAMME MANAGERS:
    Col Dan Schultz, USMC (Naval Air Systems Command)
    Col Dick Wohls, USAF (Deputy for CV-22)
COMMUNICATIONS MANAGERS:
    Gidge Dady (US Navy Public Affairs)
    Carl Harris (Bell Helicopter Textron)
    Douglas C Kinneard (Boeing)

UPDATED

BELL BOEING V-22 OSPREY
US Air Force designation: CV-22
US Navy designation: HV-22
US Marine Corps designation: MV-22
Manufacturer's model: 901
TYPE: Multimission tiltrotor.
PROGRAMME: Based on Bell/NASA XV-15 tiltrotor; initiated as US Department of Defense Joint Services Advanced

Vertical Lift Aircraft (JVX), run by US Army, FY82; programme transferred to US Navy January 1983; 24 month US Navy preliminary design contract 26 April 1983; aircraft named V-22 Osprey January 1985; seven year full-scale development (FSD) began 2 May 1986 with order for six prototypes (Nos. 1, 3 and 6 by Bell at Arlington, Texas; Nos. 2, 4 and 5 by Boeing at Wilmington, Delaware) plus static test airframes.

Prototype (163911) first flew 19 March 1989; joined by four further aircraft by June 1991; sixth not flown; all now retired (last flight 27 March 1997 by 163913); details of individual airframes and early development history appear in 1997-98 and previous editions of *Jane's*.

Osprey passed critical design review 13 December 1994; simultaneous defence review authorised V-22 production for both Marines and special forces, but latter version subsequently delayed, with decision to proceed with EMD phase not reached until January 1997. In meantime, contract for five low-rate initial production (LRIP) aircraft awarded June 1996.

All four EMD Ospreys flown in 1997-98. EMD Ospreys – see Current Versions (specific) – have significant changes from earlier aircraft, including substantial reduction in empty weight to approximately 14,800 kg (32,628 lb); aluminium cockpit cage, replacing titanium, but with smaller windows to preserve structural strength; upgraded flight controls; enhanced engine and drive system; improved tail unit construction (built by Aerostructures) including fibre placement aft fuselage; redesigned rotor system; absence of fin tuning weights; improved wing constructional techniques; redesigned wiring; and pyrotechnic escape hatches.

Manufacture of first LRIP MV-22B (165433) began 7 May 1997; splicing of three major fuselage sections took place in Philadelphia on 25 February 1998, with completed fuselage airlifted by C-17 to Arlington on 8 September 1998 for final assembly. Second LRIP aircraft also completed at Arlington, whereupon assembly transferred to new factory at Amarillo, Texas.

First flight of first LRIP MV-22B on 30 April 1999, followed by official roll-out and handover to US Marine Corps in first quarter 2000 and one aircraft temporarily sent to subsequently to Patuxent River for flight testing. Next major milestone was operational evaluation; seven-month opeval began 2 November 1999 with first two LRIP MV-22Bs, which joined by two more by January 2000. These four aircraft accumulated 805 flight hours in 522 sorties by end of opeval in July 2000.

Opeval included trials in USS *Essex* with four MV-22Bs in first quarter 2000 and one aircraft temporarily sent to Kirtland AFB, New Mexico, in March 2000 for trials with USAF 58th Special Operations Wing, before programme halted on 8 April 2000 when type grounded following fatal crash of fourth LRIP MV-22B (165436) at Avra Valley Airport, Arizona. Subsequent investigation established most likely cause as 'power settling', a condition in which it becomes difficult to stop descent because of recirculating air from rotor downwash. Clearance to return to flight given on 25 May 2000, with opeval resuming on 5 June. Final phase included trials at China Lake, California, and New River, North Carolina, before being concluded in late July 2000.

Further brief grounding order imposed on 11 aircraft (four EMD and seven LRIP machines) on 25 August 2000 in wake of problem encountered with retaining assembly for one of the interconnect driveshaft couplings; this came loose in flight, necessitating precautionary landing and inspection before flight operations resumed on 1 September 2000.

Subsequent events included final assessment by Multiservice Operational Test Team that MV-22 was operationally effective and suitable, with announcement on 8 November 2000 recommending full fleet introduction. Thereafter, decision on whether to authorise full-rate production was expected in December, but was indefinitely put on hold following loss of eighth LRIP MV-22B (165440) near New River on 11 December and imposition of grounding order. Cause of accident reported to be hydraulic system failure in first instance, compounded by error in software inputs to flight control system.

At time of grounding on 11 December 2000, V-22 total flight time was 3,884 hours, during which aircraft had demonstrated speed of 342 kt (633 km/h; 394 mph), 7,620 m (25,000 ft) height, 27,442 kg (60,500 lb) MTOW and 3.9 g load factor.

V-22 remained grounded throughout 2001, although production continued at low rate while various studies and reports prepared. In December 2001, it was decided to continue with the Osprey programme, subject to successful completion of further testing during 2002-04; in meantime, Bell Boeing directed to store total of 19 V-22s, comprising some completed examples and some to come from assembly line during 2002. On conclusion of test programme, these stored aircraft to be upgraded and fitted with latest safety features.

Flight testing resumed on 29 May 2002 at Patuxent River, with two EMD and three LRIP aircraft to participate in 18-month programme that will evaluate improved hydraulic and electrical systems, plus upgraded software; test objectives also to include exploration of aspects including vortex ring state boundaries, formation flying

V-22 Osprey production version flight deck   0093859

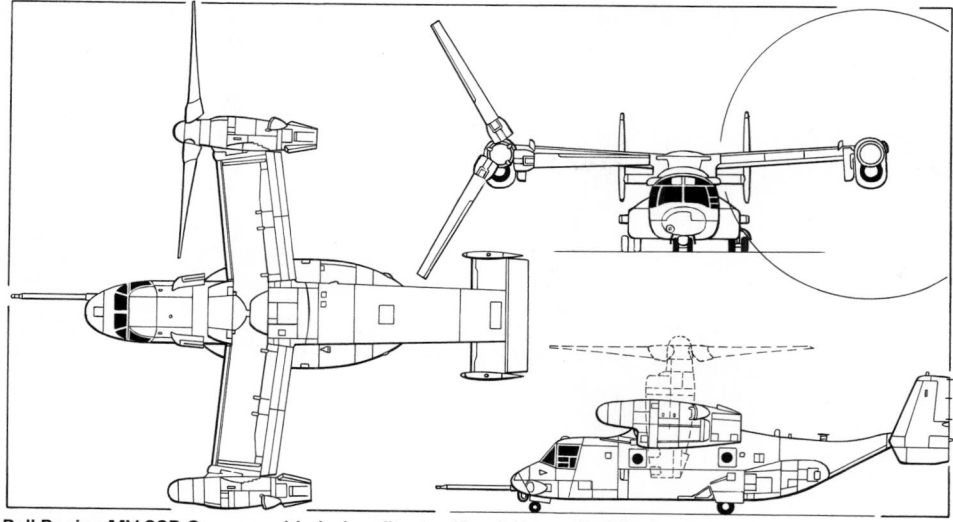

Bell Boeing MV-22B Osprey multimission tiltrotor *(Jane's/James Goulding)*   0093860

HV-22B: US Navy combat search and rescue (CSAR), special warfare and fleet logistics model. Requirement for 48 (originally 50); deliveries from FY10.

CV-22B: US Air Force long-range special missions aircraft to replace MH-53M helicopter and augment MC-130 (Hercules) with Air Force Special Operations Command (AFSOC). Original requirement for 80 reduced to 55 then 50, with first two production examples due for delivery in FY06; should carry 12 troops or 1,306 kg (2,880 lb) internal cargo over 520 n mile (964 km; 599 mile) radius at 250 kt (463 km/h; 288 mph), with ability to hover OGE at 1,220 m (4,000 ft) at 35°C (95°F). EMD go-ahead authorised in January 1997, with award of US$490 million contract; critical design review completed in mid-December 1998. Flight testing of remanufactured first and third EMD Ospreys at Edwards AFB, California.

Key feature of CV-22B will be ITT Avionics AN/ALQ-211 SIRFC (suite of integrated RF countermeasures); US$20.7 million LRIP contract awarded by Bell Boeing in January 2001 for four systems plus spares and engineering support. First SIRFC was due for delivery in 2002, with full-rate production decision dependent on successful outcome of flight testing.

V-22 also under consideration for USAF combat search and rescue (CSAR) mission as potential replacement for Sikorsky HH-60G; response to USAF request for information submitted on 3 February 2000, with analysis of alternatives due to be completed by March 2001 and procurement following from 2004. V-22 in competition with five helicopter types (HH-60, H-53, S-92, NH90 and EH 101).

US Army: Original requirement for 231 V-22s, based on USMC transport, withdrawn. Documented requirement remains for V-22 in medevac, special operations and combat assault support roles.

CUSTOMERS: Initial increment of production funding in FY96 budget, when US$48 million requested. First five production aircraft funded in FY97; further seven in FY98 and seven in FY99. Another 11 due to have been funded in FY00, to complete LRIP phase, although it appears that contract for these was delayed until first quarter of 2002; FY02 procurement anticipates acquisition of 10 MV-22s for US Marine Corps and two additional test specimens of the CV-22. Bell Boeing has proposed V-22 as suitable for UK's joint RAF/Navy Support Amphibious Battlefield Rotorcraft requirement, for which total of over 40 will be needed. In-service date currently expected to be 2008. Suitably modified V-22 could also satisfy Royal Navy's Future Organic Airborne Early Warning (FOAEW) requirement.

COSTS: Estimated cost (1991) to complete full-scale development, US$2,750 million. Unit cost US$32.3 million (November 1997); FY97 budget included US$1.385 billion for five MV-22s; FY98 budget included US$676.6 million for seven MV-22s; FY99 request included US$657.4 million for seven MV-22s; Bell Boeing

and low-speed hover and landing conditions. Test programme will involve 1,800 flight hours, with icing, cargo handling and radar warning systems also to be studied. Another two EMD aircraft that had previously been modified to CV-22 standard were expected to return to flight duty in July 2002; these will eventually be joined by two new-build test aircraft being purchased with FY02 funding.

US Marine Corps not expecting to make decision to move to full-rate production before 2006-07, but intends to continue purchasing Ospreys at rate of 10 to 12 per year; these will be to progressively improved standard, starting with Block A in 2003, incorporating re-routed and separated hydraulic lines and electrical wiring. USMC training unit VMMT-204 will receive total of 12 Block A aircraft and is expected to resume flight operations in November 2003, with IOC following by late 2004. Block B standard MV-22 will be available from 2005 on and will possess enhanced maintenance access, while the proposed Block C version will feature operational improvements, including gun turret, telescopic in-flight refuelling probe, revised environmental control system and other changes intended to improve payload and performance capabilities.

Block A enhancements will also be retrofitted to LRIP aircraft that have already been completed; these include total of eight examples at New River plus others at Amarillo production facility. This work is expected to begin in 2003, with all surviving LRIP aircraft having been upgraded to Block A standard by 2007.

CURRENT VERSIONS (specific): **No. 7/164939:** Assembly began 15 February 1995; fuselage sections mated at Boeing's Philadelphia plant on 1 August 1995; followed two months later by wing and nacelle mating at Fort Worth. Fuselage to Fort Worth for fitting of wings and engines. Ground vibration testing completed in late July 1996; tests of hydraulic lines concluded soon after, with integrated functional testing commencing in August. First flight in helicopter mode 5 February 1997 (at Fort Worth); first transition to conventional flight 6 March 1997; aircraft to V-22 Integrated Test Team at Patuxent River on 15 March 1997; ensuing objectives (structural, load and vibration tests) completed by mid-1998. Prepared at Arlington for trials of CV-22 version's auxiliary fuel tanks and terrain-following/terrain-avoidance radar system; modification work began 1 July 1999; flight status regained on 28 February 2000 and by 11 December had completed 498 hours and 244 sorties.

No. 8/164940: First flight 15 August 1997; to Patuxent River on 13 September 1997 and allocated to propulsion and systems testing and envelope expansion, including high-altitude trials and external loads demonstrations in 1998. During latter, it set new unofficial world record in August 1998 by carrying 4,536 kg (10,000 lb) external load at 220 kt (407 km/h; 253 mph). By 11 December 2000, had completed 557.5 hours of flight testing in 293 sorties.

No. 9/164941: First flight 17 July 1997; to Patuxent River 30 October 1997; allocated to validation of FLIR, navigation and other mission systems, as well as government technical evaluation. Had achieved total of 318 flight hours in 140 sorties, when arrived at Arlington on 7 June 1999 to be remanufactured as prototype for CV-22 special missions version, at cost saving of US$50 million; work completed in mid-2000, with aircraft joining

418th Flight Test Squadron at Edwards AFB, California, on 18 September 2000 for 18-month trials programme. By 11 December 2000, total flight time had risen to 339 hours in 152 sorties.

No. 10/164942: Wing and fuselage mating accomplished at Fort Worth in September 1996. To Patuxent River 15 February 1998, following first flight on 15 January; underwent modification in mid-1998, before participating (with No. 9) in operational evaluation during September and October 1998; performed sea trials in USS *Saipan* in January-February and USS *Tortuga* and *Saipan* in August-September 1999. By 11 December 2000, had completed 500 hours of flight testing in 213 sorties. First V-22 to resume flying after 17-month grounding, 29 May 2002.

CURRENT VERSIONS (general): **MV-22B:** Basic US Marine Corps transport; original requirement for 552 (now 360), to replace CH-46 Sea Knight and CH-53 Sea Stallion. First delivery (to Patuxent River via New River) in late May 1999 with three more handed over to test unit HMX-1 by January 2000. First unit is VMMT-204 at New River, North Carolina, which officially redesignated (from HMT-204) on 10 June 1999; received four opeval aircraft from HMX-1 at end of July 2000 and will function as fleet replacement squadron and also train USAF pilots; first deployable squadron expected to be VMM-264, also at New River. USMC plans called for 18 regular and four Reserve MV-22 squadrons (each with 12 aircraft) by 2013, plus training unit already mentioned.

An EMD Bell Boeing V-22 demonstrates carriage of a Humvee personnel carrier   *NEW*/0137432

received US$770 million contract in March 2002 to procure 11 MV-22s by 2005 for follow-on testing. Separate US$490 million contract modification awarded by USN in December 1996 for EMD of CV-22 special operations version; acquisition costs of 50 CV-22s estimated at US$3.72 billion based on FY01-08 procurement plan. Unit cost of MV-22B reported as US$57 million (2000).

DESIGN FEATURES: Unconventional design, with proprotors and engines mounted at tips of wings. Fuselage optimised for transport, featuring upswept rear, with loading ramp and twin fins of moderate sweepback. High-mounted, constant-chord wings with slight forward sweep; unswept tailplane; prominent landing gear sponsons.

During vertical take-off, wing begins to produce lift and ailerons, elevators and rudders become effective at between 40 and 80 kt (74 and 148 km/h; 46 and 92 mph). At this point, rotary-wing controls are gradually phased out by the flight control system. At approximately 100 to 120 kt (185 to 222 km/h; 115 to 138 mph), wing is fully effective and cyclic pitch control of proprotors is locked out.

In conversion from aeroplane flight to hover, fuselage and wing are free to remain in level attitude, eliminating tendency for wing to stall as speed decreases. Rotor lift fully compensates for decrease in wing lift. Because of great variability between aircraft and nacelle attitude, conversion corridor (range of permissible airspeeds for each angle of nacelle tilt) is very wide (about 100 kt; 185 km/h; 115 mph).

Engines are connected by shaft through wing. Under dual engine operations, shaft transmits very little power, but if one engine is lost, half remaining power is transferred to opposite proprotor. In event of double engine failure, can maintain proprotor rpm while descending without power. Pilot has options of making wingborne or rotorborne descent.

Engines, transmission and proprotors tilt through 97° 30′ between forward flight and steepest approach gradient or tail-down hover; cross-shaft keeps both proprotors turning after engine loss. Three-blade, contrarotating proprotors have special high-twist tapered format blades with elastomeric bearings and powered folding mechanisms; separate swashplates produce respectively yaw and fore-and-aft translation in hover and sideways flight in level attitude; tip speed 202 m (662 ft)/s.

Wing-fold sequence from helicopter mode involves power-folding of blades parallel to wing leading-edge, tilting engine nacelles down to horizontal and rotating entire wing/engine/proprotor group clockwise on stainless steel carousel to lie over fuselage; entire procedure for stowage takes about 90 seconds and MV-22B occupies same amount of deck space as Sikorsky CH-53E.

FLYING CONTROLS: Three-lane fly-by-wire (Moog actuators) with automatic stabilisation, full autopilot and formation-flying modes. Conventional aeroplane stick, rudder pedals and throttle automatically function as in helicopter cyclic stick, yaw pedals and collective control. Automatic control of configuration change during transition and of transfer of control from aerodynamic surfaces to rotor-blade pitch changing; flaperons and ailerons droop during hover to reduce download on wing. Pilot also controls nacelle angle. Failure of one nacelle actuator automatically results in both nacelles reverting to helicopter mode for a vertical or run-on emergency landing.

Bell Boeing V-22 Osprey   NEW/0137441

Rotors have separate cyclic control swashplates for sideways flight and fore-and-aft control (symmetrical for forward and rearward flight and differential for yaw) in hover. Lateral attitude controlled in hover by differential rotor thrust. Integrated electronic cockpit with six electronic display screens; helicopter-style control columns rather than aileron wheels, but left-handed power levers move forward for full power in opposite sense to helicopter collective lever. Using nacelle control provides additional method of manoeuvring, completely independent of fuselage attitude or cyclic pitch. Minimum time to accomplish a full conversion from hover to aeroplane flight mode is 12 seconds.

STRUCTURE: Approximately 43 per cent of airframe is composites; main composites are Hercules IM-6 graphite/epoxy in wing and AS4 in fuselage and tail; proprotor blades of graphite/glass fibre; nacelle cowlings and pylon supports are of GFRP. Cabin has composites floor panels and aluminium window frames. Crew seats of boron carbide/polyethylene laminate.

Different design approach adopted for EMD and subsequent production aircraft, whereby integrated product teams (IPTs) were tasked with using best available material to reduce cost and weight and simultaneously improve quality. In consequence, fuselage is now a hybrid structure, with mainly aluminium frames and composite skins. Benefits of IPT process are a 1,200 kg (2,645 lb) reduction in weight and a reduction in cost of over 22 per cent, with part count falling by 36 per cent and fastener count by 34 per cent.

High-strength wing, torsion box made up from one-piece upper and lower skins with moulded ribs and bonded stringers; two-segment graphite single-slotted flaperons

with titanium fittings; three-segment detachable leading-edge of aluminium alloy with Nomex honeycomb core. Wing locking and unlocking with Lucas Aerospace actuators; fuselage sponsons contain landing gear, air conditioning unit and fuel; tail unit of Hercules AS4 graphite/epoxy, built by Bell from first EMD aircraft onwards. Bell also contributes wing, nacelles, proprotor, ramp, overwing fairing and transmission systems and integrates engines. Boeing responsible for fuselage, landing gear, electric and hydraulic systems and integrates avionics.

LANDING GEAR: Dowty hydraulically actuated retractable tricycle type, with twin wheels and oleo-pneumatic shock-absorbers on each unit, Menasco Canada steerable nose unit. Main gear accommodated in sponsons on lower sides of centre-fuselage. Dowty Toronto two-stage shock-absorption in main gear is designed for landing impacts of up to 3.66 m (12 ft)/s normal, 4.48 m (14.7 ft)/s maximum, and has been drop tested to 7.32 m (24 ft)/s. Nosewheels retract rearward, mainwheels forward (was rearward on FSD aircraft only). Manual and nitrogen pressurised standby systems for emergency extension. Parker Bertea wheels and multidisc hydraulic carbon brakes. Main tyres 8.50-10 (12 ply) tubeless; nose tyres 18×5.7-8 (14 ply) tubeless.

POWER PLANT: Two Rolls-Royce T406-AD-400 (501-M80C) turboshafts, each with T-O and intermediate rating of 4,586 kW (6,150 shp) and maximum continuous rating of 4,392 kW (5,890 shp), installed in Bell-built hydraulically actuated tilting nacelles at wingtips and driving a three-blade proprotor incorporating substantial proportion of graphite/epoxy and glass fibre materials.

Engine output transferred to each rotor via proprotor gearbox which also drives tilt-axis gearbox, each of which absorbs 315 kW (423 shp) for electric and hydraulic pressure generation. Tilt-axis gearboxes connected by segmented shaft, driving (in fuselage) single mid-wing gearbox and APU. Transmission T-O rating 3,408 kW (4,570 shp) in MV-22; 3,706 kW (4,970 shp) in USN and USAF versions. OEI rating 4,415 kW (5,920 shp); rotor rpm remains constant. With total engine failure, aircraft generators can maintain essential systems on 67 per cent of rotor rpm. Transmission has 30 minute run dry capability.

Each nacelle has a Honeywell IR emission suppressor at rear. Air particle separator and Lucas inlet/spinner ice protection system for each engine. Lucas Aerospace FADEC for each engine, with analogue electronic back-up control. Pratt & Whitney originally named as second production source for engines, starting with production Lot 5.

Fuel capacity varies according to role; basic internal usable fuel (JP-5) in four crash-resistant, self-sealing (nitrogen pressurised) ILC Dover flexible fuel tanks: one 1,809 litre (478 US gallon; 398 Imp gallon) forward cell in each sponson and a 333 litre (88.0 US gallon; 73.3 Imp gallon) feed tank in each outer wing; basic fuel (MV-22) 4,285 litres (1,132 US gallons; 943 Imp gallons). Additional fuel for long range contained in 1,196 litre (316 US gallon; 263 Imp gallon) cell in rear of starboard sponson (MV-22 option 5,481 litres; 1,448 US gallons; 1,206 Imp gallons) and four 280 litre (74.0 US gallon; 61.6 Imp gallon) auxiliary cells in each wing leading-edge (CV-22 baseline 7,722 litres; 2,040 US gallons; 1,699 Imp gallons). Self deployment aided by up to four 2,328 litre (615 US gallon; 512 Imp gallon) auxiliary tanks in cabin. (Not all versions have all tanks.) Pressure refuelling point in starboard sponson leading-edge; gravity point in upper surface of each wing. Simmonds fuel management system. In-flight refuelling probe in lower starboard side of forward fuselage; standard item on CV-22B and available

A pair of Bell Boeing MV-22 Ospreys during trials   NEW/0137440

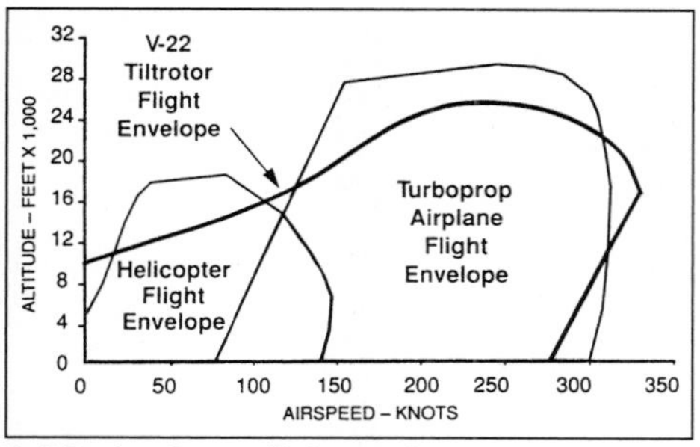

Bell Boeing Osprey flight envelope 0051561

in kit form for MV-22B. Fixed probe installed initially, but will be replaced by retractable type with effect from Lot 6; new probe likely to be provided by Flight Refuelling; will give wider field of fire for proposed gun turret and facilitate deck handling at sea.

ACCOMMODATION: Normal crew complement of pilot (in starboard seat), co-pilot and crew chief in USMC variant. USAF CV-22B will have third seat for flight engineer. Flight crew accommodated on Simula Inc crashworthy armoured seats capable of withstanding strikes from 0.30 in armour-piercing ammunition, 30 *g* forward and 14.5 *g* vertical decelerations. Flight deck has overhead and knee-level side transparencies in addition to large windscreen and main side windows, plus an overhead rearview mirror.

Main cabin can accommodate up to 24 combat-equipped troops, on IAI – Golan Industries inward-facing crash-worthy foldaway seats, plus two gunners; up to 12 litters plus medical attendants; or a 9,070 kg (20,000 lb) cargo load with energy absorbing tiedowns. Cargo handling provisions include a 907 kg (2,000 lb) capacity cargo winch and pulley system and removable roller rails. Main cabin door at front on starboard side, top portion of which opens upward and inward, lower portion (with built-in steps) downward and outward. Full-width rear-loading ramp/door in underside of rear fuselage, operated by Parker Bertea hydraulic actuators. Emergency exit windows on port side; escape hatch in fuselage roof aft of wing.

SYSTEMS: Environmental control system, utilising engine bleed air; control unit in rear of port main landing gear sponson. For NBC protection, ECS provides 0.062 bar (0.9 lb/sq in) flight deck pressure differential, and 0.048 bar (0.7 lb/sq in) in cabin. Three hydraulic systems (two independent main systems and one standby), all at operating pressure of 345 bar (5,000 lb/sq in), with Parker Bertea reservoirs.

Electrical power supplied by two Leland 40 kVA constant frequency AC generators, two 50/80 kVA variable frequency DC generators (one driven by APU), rectifiers, and a 24 Ah lead-acid battery. Latter provides 20 minutes' emergency flight power.

GE Aerospace triple redundant digital fly-by-wire flight control system, incorporating triple primary FCS (PFCS) and triple automatic FCS (AFCS) processors, and triple flight control computers (FCC) each linked to a MIL-STD-1553B databus; two PFCSs and one AFCS are fail-operational. FBW system signals hydraulic actuation of flaperons, elevator and rudders, controls aircraft transition between helicopter and aeroplane modes, and can be programmed for automatic management of airspeed, nacelle tilting and angle of attack. FCCs provide interfaces for swashplate, conversion actuator, flaperon, elevator, rudder and engine nacelle primary actuators, flight deck central drive, force feel, and nosewheel steering. Dual 1750A processors for PFCS and single 1750A for AFCS incorporated in each FCC. Non-redundant standby analogue computer (in development aircraft only) provides control of aircraft, including FADEC and pylon actuation, in the event of FBW system failure.

Hamilton Sundstrand Turbomach T-62T-46-2 224 kW (300 shp) APU, in rear portion of wing centre-section, provides power for mid-wing gearbox which, in turn, drives two electrical generators and an air compressor. Anti-icing of windscreens and engine air intakes; de-icing of proprotors and spinners. Clifton Precision combined oxygen (OBOGS) and nitrogen (OBIGGS) generating systems for cabin and fuel tank pressurisation respectively. Systron Donner pneumatic fire protection systems for engines, APU and wing dry bays.

AVIONICS: *Comms:* VHF/AM-FM, HF/SSB and (USAF only) UHF secure voice com; IFF. CV-22B will have four DCS2000 radios, combining AN/ARC-210 transceiver and KY-58 communications security encoder. Crash position indicator also to be installed on CV-22B.

Radar: USAF and USN only. Raytheon AN/APQ-174D terrain-following, terrain-avoidance multifunction radar in offset (to port) nose thimble for USN; advanced version, designated AN/APQ-186, with lower-altitude TF capability, in USAF aircraft.

Flight: AN/ARN-153(V) Tacan, AN/ARN-147 VOR/ILS, AHRS, AN/APN-194(V) radar altimeter, OA-8697/ARC UHF/VHF automatic direction-finder, lightweight inertial navigation system (LWINS), miniature airborne GPS receiver (MAGR) and digital map displays; Jet Inc ADI-350W standby attitude indicator; L-3 Communications data acquisition and storage system. Two Control Data AN/AYK-14 mission computers, with Boeing/IBM software. Low probability of intercept/detection radar altimeter may be added to CV-22B version at later date.

Instrumentation: Elbit/BAE Systems North America AN/AVS-7 pilots' night vision and integrated display system. Four full-colour. multifunction displays (MFDs) in cockpit provide pilots with primary flight symbology to control and navigate aircraft, plus video imagery such as FLIR and digital map data; these originally relied on CRTs, but replaced by EFW Inc flat-panel active matrix liquid crystal displays (AMLCDs) on US Marine Corps MV-22Bs starting in FY99, followed by USN HV-22s. AMLCDs to be installed on USAF CV-22B from outset. Lighting compatible with night vision goggles. Additional AMLCD control display unit and engine indication and crew alerting system (CDU/EICAS) in centre of console.

Mission: Raytheon AN/AAQ-27 FLIR incorporating laser range-finder in undernose fairing, with wide and narrow fields of view. Additional equipment expected to satisfy unique mission requirements of special operations forces and Navy combat search and rescue.

Self-defence: Honeywell AN/AAR-47 missile warning system; AN/APR-39A radar warning system; IR warning system; BAE Systems North America AN/ALE-47 countermeasures dispenser system (CMDS). CV-22B for USAF to utilise ITT Avionics AN/ALQ-211 integrated RF countermeasures suite, incorporating radar warning receiver, ESM radar location and jammers and will also have AN/AVR-2A laser detector system and a second, forward-firing, AN/ALE-47 dispenser system. Dedicated electronic warfare display (DEWD) unit also under development for CV-22B by Meggitt Avionics.

EQUIPMENT: Provision for internally stowed rescue hoist over forward (starboard) cabin door on CV-22B. MV-22B initially to have hoist in cabin, for deployment through side door, but now expected to position hoist near the rear ramp. USMC and USAF Ospreys have fast rope/rope ladders to facilitate insertion/extraction operations.

ARMAMENT: Provision stipulated in December 1994 for nose cannon; budget constraints resulted in this being deleted, but consideration again given in late 1998 to installing nose-mounted turreted defensive gun on MV-22B and this was evaluated in 2000, with General Dynamics Armament Systems 12.7 mm weapon selected in September 2000. Based on GAU-19/A, it consists of three-barrel weapon system weighing 209 kg (460 lb) and capable of firing 1,200 rounds per minute; magazine containing 750 rounds to be positioned under the floor and can be replenished in flight.

US Air Force also considering self-defence gun for CV-22B. USMC intended to request funding in 2001, followed by flight trials in 2004, with weapon to be installed on new-build MV-22Bs and retrofitted to earlier aircraft.

DIMENSIONS, EXTERNAL:

Rotor diameter, each	11.58 m (38 ft 0 in)
Rotor blade chord: at root	0.90 m (2 ft 11½ in)
at tip	0.56 m (1 ft 10 in)
Wing span: excl nacelles	14.02 m (46 ft 0 in)
incl nacelles	15.52 m (50 ft 11 in)
Width: rotors turning	25.40 m (83 ft 4 in)
stowed	5.61 m (18 ft 5 in)
Wing chord, constant	2.54 m (8 ft 4 in)
Wing aspect ratio, excl nacelles	5.5
Distance between proprotor centres	14.25 m (46 ft 9 in)
Length: fuselage, excl probe	17.47 m (57 ft 4 in)
overall, wings stowed/blades folded	19.08 m (62 ft 7 in)
Height: over tailfins	5.38 m (17 ft 7¾ in)
wings stowed/blades folded	5.51 m (18 ft 1 in)
overall, nacelles vertical	6.63 m (21 ft 9 in)
Tail span, over fins	5.61 m (18 ft 5 in)
Wheel track (c/l of outer mainwheels)	4.62 m (15 ft 2 in)
Wheelbase	6.59 m (21 ft 7½ in)
Nacelle ground clearance, nacelles vertical	1.31 m (4 ft 3¾ in)
Proprotor ground clearance, nacelles vertical	6.35 m (20 ft 10 in)
Dorsal escape hatch: Length	1.02 m (3 ft 4 in)
Width	0.74 m (2 ft 5 in)

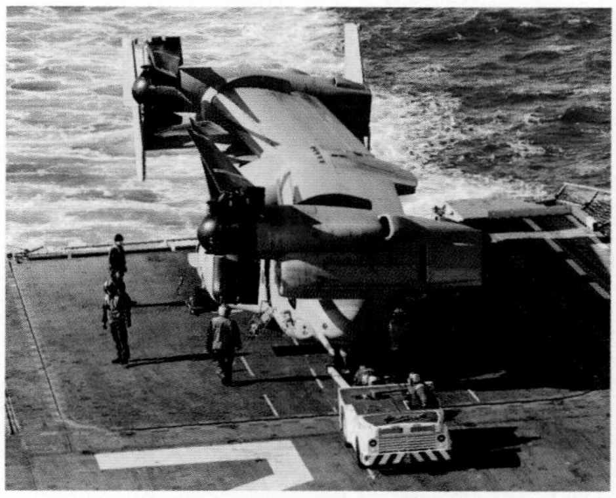

Bell Boeing V-22 Osprey folded for stowage 0093861

DIMENSIONS, INTERNAL:

Cabin: Length	7.37 m (24 ft 2 in)
Max width	1.80 m (5 ft 11 in)
Max height	1.83 m (6 ft 0 in)
Usable volume	24.3 m^3 (858 cu ft)

AREAS:

Rotor discs, each	105.36 m^2 (1,134.1 sq ft)
Rotor blades: each	4.05 m^2 (43.58 sq ft)
total	24.30 m^2 (261.5 sq ft)
Wing, total incl flaperons and fuselage centre-section	35.49 m^2 (382.0 sq ft)
Flaperons, total	8.25 m^2 (88.80 sq ft)
Fins, total	21.63 m^2 (232.80 sq ft)
Rudders, total	3.27 m^2 (35.20 sq ft)
Tailplane	8.22 m^2 (88.50 sq ft)
Elevators, total	4.79 m^2 (51.54 sq ft)

WEIGHTS AND LOADINGS:

Weight empty	15,032 kg (33,140 lb)
Max fuel weight: MV-22 baseline	3,493 kg (7,700 lb)
MV-22 option	4,468 kg (9,850 lb)
CV-22 baseline	6,282 kg (13,850 lb)
CV-22 with cabin tanks	11,970 kg (26,390 lb)
Max internal payload (cargo)	9,072 kg (20,000 lb)
Cargo hook capacity: single	4,536 kg (10,000 lb)
two hooks (combined weight)	6,804 kg (15,000 lb)
Rescue hoist capacity	272 kg (600 lb)
Normal mission T-O weight: VTO	21,545 kg (47,500 lb)
STO	24,947 kg (55,000 lb)
Max VTO weight	23,981 kg (52,870 lb)
Max STO weight for self-ferry	27,442 kg (60,500 lb)
Max floor loading (cargo)	1,464 kg/m^2 (300 lb/sq ft)
Max disc loading: VTO	113.8 kg/m^2 (23.31 lb/sq ft)
STO	130.2 kg/m^2 (26.67 lb/sq ft)
Transmission loading at max T-O weight and power	3.55 kg/kW (5.84 lb/shp)

PERFORMANCE:

Max level speed at S/L	275 kt (509 km/h; 316 mph)
Max cruising speed: at S/L, helicopter mode	100 kt (185 km/h; 115 mph)
Max forward speed with max slung load	214 kt (396 km/h; 246 mph)
Max rate of climb at S/L: vertical	332 m (1,090 ft)/min
inclined	707 m (2,320 ft)/min
Service ceiling	7,925 m (26,000 ft)
Service ceiling, OEI	3,441 m (11,300 ft)
Hovering ceiling OGE	4,331 m (14,200 ft)
T-O run at normal mission STO weight	less than 152 m (500 ft)
Range:	
amphibious assault	515 n miles (953 km; 592 miles)
VTO with 4,536 kg (10,000 lb) payload	350+ n miles (648+ km; 403+ miles)
VTO with 2,721 kg (6,000 lb) payload	700+ n miles (1,296+ km; 806+ miles)
STO with 4,536 kg (10,000 lb) payload	950+ n miles (1,759+ km; 1,093+ miles)
STO at 27,442 kg (60,500 lb) self-ferry gross weight, no payload	2,100 n miles (3,892 km; 2,418 miles)
g limits	+4/–1

UPDATED

BERKUT

BERKUT ENGINEERING INC

3025 Airport Avenue, Santa Monica, California 90405
Tel: (+1 310) 391 01 79
Fax: (+1 310) 391 86 45
e-mail: vickic@berkutengineering.com
Web: http://www.berkutengineering.com
PRESIDENT AND CEO: Vicki Cruse
CHIEF DESIGNER: Dave Ronneberg

Berkut Engineering purchased moulds and tooling for the Berkut series of kitbuilt aircraft from Renaissance Composites Inc in February 2001. Active promotion of the Berkut was discontinued in September 2001, although parts for the aircraft remain available to order and current builders are being supported. Berkut is looking for a company to purchase the tooling and resume production.

UPDATED

BERKUT BERKUT

Total of 63 kits sold up to September 2001, when manufacture was suspended. Only retractable landing gear versions will be available in future. Details last appeared in 2002-03 *Jane's*.

UPDATED

BOEING

THE BOEING COMPANY

100 North Riverside Plaza, Chicago, Illinois 60606
Tel: (+1 312) 544 20 02
Web: http://www.boeing.com
CHAIRMAN AND CEO: Philip M Condit
PRESIDENT AND COO: Harry C Stonecipher
CHIEF FINANCIAL OFFICER: Michael M Sears
VICE-PRESIDENT, COMMUNICATIONS: Judith Muhlberg

Company founded July 1916. Currently organised into five major business segments as detailed below, plus several smaller elements. Revenues for 2001 were US$58.2 billion. Unchallenged as leading US aerospace firm, but has recently experienced difficulties with airliner production. Deliveries of 563 commercial aircraft in 1998 increased to 620 in 1999, but fell to 489 in 2000.

Boeing and McDonnell Douglas merger under the Boeing name was completed on 4 August 1997, elevating Boeing to status of largest aerospace company in the world. Headquarters remained in Seattle, Washington, until September 2001, when moved to new location in Chicago, Illinois. Expansion of global interests occurred in 1997, with Boeing entering into alliances with AVIC and Taikoo Aircraft Engineering Co of China; with CSA Czech Airlines to establish joint venture company and acquire stake in Aero Vodochody; and with Instytut Lotnictwa of Poland for potential future collaboration on advanced technologies. In 1998, similar agreements and arrangements concluded with Israel Military Industries concerning military and civil programmes and with Hellenic Aerospace Industry to promote F-15H as new fighter for Greek Air Force, though latter failed to secure order. In January 1999, Boeing reached agreement with Netherlands-owned company MD Helicopters (which see in US section) on previously

announced disposal of light civil helicopter range. In November 2001, Boeing had 196,500 employees.

Following further reorganisation in latter half of 2002, operating components of The Boeing Company will comprise:

Integrated Defense Systems
Follows this entry
Boeing Commercial Airplanes
Follows Integrated Defense Systems
Phantom Works
Follows Boeing Commercial Airplanes
Shared Services Group
Follows Phantom Works

UPDATED

INTEGRATED DEFENSE SYSTEMS

PO Box 516, St Louis, Missouri 63166-0516
Tel: (+1 314) 232 02 32
Fax: (+1 314) 234 82 96
PRESIDENT AND CEO: Jim Albaugh
MEDIA CONTACTS:
   GENERAL MANAGER: Doug Kennett, *Tel:* (+1 314) 234 35 00
   DIRECTOR, MEDIA RELATIONS:
      Jo Anne Davis, *Tel:* (+1 314) 233 89 57
   MANAGER MEDIA/INTERNATIONAL PROGRAMMES:
      Mary Ann Brett, *Tel:* (+1 314) 234 71 11
   INTERNATIONAL PROGRAMMES:
      Jim Schlueter, *Tel:* (+1 314) 234 21 49

Created in latter half of 2002 through merger of Military Aircraft and Missile Systems with Space and Communications, Integrated Defense Systems has overall responsibility for military business activities. Management of aircraft and helicopter programmes as detailed below reflects the military customer base, but several subordinate organisations exist to oversee other key areas.

Other elements include **Aerospace Support** (service and logistics support, plus modifications).

Sales in 2001 were US$22.9 billion; workforce in mid-2002 numbered 78,600.

Major business units of Integrated Defense Systems are as follows:

US Navy Systems
VICE-PRESIDENT AND GENERAL MANAGER: John Lockard
(St Louis)

USAF Systems
VICE-PRESIDENT AND GENERAL MANAGER: George Muellner
(Long Beach)

US Army Systems
ACTING VICE-PRESIDENT AND GENERAL MANAGER: Jim Albaugh

Human Space Flight
VICE-PRESIDENT AND GENERAL MANAGER: Mike Mott (Houston)

Missile Defense Systems
VICE-PRESIDENT AND GENERAL MANAGER: Jim Evatt
(Washington, DC)

Homeland Defense
VICE-PRESIDENT AND GENERAL MANAGER: Rick Stephens
(Seal Beach)

Space and Intelligence Systems
VICE-PRESIDENT AND GENERAL MANAGER: Roger Roberts
(Seal Beach)

Commercial Space Systems
VICE-PRESIDENT AND GENERAL MANAGER: Bill Collopy
(Seal Beach)

Aerospace Support
VICE-PRESIDENT AND GENERAL MANAGER: David Spong
(St Louis)

UPDATED

BOEING F-15E EAGLE

Israel Defence Force name: Ra'am (Thunder)
TYPE: Multirole fighter.
PROGRAMME: F-15 initially employed as air superiority fighter, having first flown on 27 July 1972. Demonstration of industry-funded Strike Eagle prototype (71-0291) modified from F-15B, including accurate blind weapons delivery, completed at Edwards AFB and Eglin AFB in 1982; product improvements for the F-15E were tested on four Eagles, among which were the Strike Eagle prototype, an F-15C and an F-15D, between November 1982 and April 1983, including first take-off at 34,019 kg (75,000 lb), 3,175 kg (7,000 lb) more than F-15C with conformal tanks; new weight included conformal tanks,

Boeing F-15I Ra'am of 69 Squadron, Israel Defence Force (*Jane's/Paul Jackson*)   0121483

three other external tanks and eight 500 lb Mk 82 bombs; 16 different stores configurations tested, including 2,000 lb Mk 84 bombs, and BDU-38 and CBU-58 weapons delivered visually and by radar.

Full programme go-ahead announced 24 February 1984; first flight of first production F-15E (86-0183) 11 December 1986; first delivery to Luke AFB, Arizona, 12 April 1988; first delivery 29 December 1988 to 4th Wing at Seymour Johnson AFB, North Carolina.

In mid-2001, USAF and Boeing examining various options for mid-life upgrade of F-15E known as Block 6; among the possibilities are adoption of active electronically scanned array (AESA) radar, update of radar warning receiver equipment, re-engining with version of General Electric F110 turbofan, installation of new wing structure incorporating carbon fibre skin that would provide increased take-off weight and accommodate two extra hardpoints.
CURRENT VERSIONS: **F-15E:** Basic version, *as detailed.*
   F-15F: Proposed single-seat version, optimised for air combat; not built.
   F-15H: Proposed export version, lacking specialised air-to-ground capability; supplanted by F-15S. Designation subsequently allocated to proposed version for Greece, which eventually selected F-16 Fighting Falcon.
   F-15I Ra'am: Israeli export version of F-15E; selected November 1993; confirmed 27 January 1994; 21 ordered 12 May 1994; option on four more converted to firm order in November 1995. First flight of initial (unpainted) aircraft on 12 September 1997; this was subject of formal roll-out and handover ceremony at St Louis on 6 November 1997, with first pair of aircraft to be delivered leaving St Louis on 16 January 1998 and arriving in Israel three days later; total of 16 delivered during 1998, with final nine aircraft following in 1999. Operating unit No. 69 Squadron. Tactical electronic warfare system deleted; replaced by Israeli-built SPS-2100 integrated system including active jamming, radar and missile warning, and dispenser subsystems.
   Otherwise identical to USAF F-15E, with F100-PW-229 engines, LANTIRN pods, full capability AN/APG-70 radar, Kaiser holographic HUD, Litton ring laser INS and VHSIC central computer. Associated equipment includes four Sanders mission planning subsystems and one Sanders common mapping production system (CMPS) to assist ground planning, briefing and debriefing activities at total cost of US$6.2 million. F-15I includes significant number of co-produced components, such as airframe and wing subassemblies, heat exchangers and weapons pylons.
   F-15K: Proposed version to satisfy Republic of Korea's F-X fighter requirement; preliminary bids submitted in June 2000. Selection was expected in second half of 2001, but decision delayed, with formal announcement being made on 19 April 2002, when Korea revealed intent to purchase total of 40 aircraft for delivery during 2005-08. F-15K incorporates AN/APG-63(V)1 radar, expanded weapons capability (including JDAM, SLAM-ER and AIM-9X) and improved environmental control system,

plus Lockheed Martin AN/AAS-42 infra-red search and track system, Boeing Joint Helmet-Mounted Cueing System and revised cockpit with seven 152 × 152 mm (6 × 6 in) AMLCDs.
   F-15L: Proposed less costly version offered to Israel by Boeing in unsuccessful, last-minute, attempt to secure order for new combat aircraft; F-16I eventually selected.
   F-15S: Saudi Arabian export version of F-15E, lacking some air-to-air and air-to-ground capabilities; Saudi Arabian request for 72 aircraft approved by US government in December 1992; initially designated **F-15XP**; first funds assigned by US government on 23 December 1992; contract signature by Saudi government May 1993; planned delivery rate halved, early 1994, to one per month, beginning 1995. First F-15S flown 19 June 1995; official roll-out and handover 12 September 1995. First two examples delivered to Saudi Arabia in November 1995 were second and third built; total of 49 received by end of 1998 and 70 by end of 1999, with final two following in late July 2000.
   Saudi versions comprise 24 optimised for air-to-air missions and 48 optimised for air-to-ground; AN/APG-70 radar. Despite planned restrictions, aircraft delivered with full F-15E capability, plus conformal fuel tanks and tangential stores attachments. Armament includes AGM-65D/G Maverick, AIM-7 Sparrow, AIM-9M and AIM-9S Sidewinder missiles, CBU-87 submunitions dispenser and GBU-10/12 bombs. Saudi programme includes about 154 Pratt & Whitney F100-PW-229 engines.
   F-15SE: Proposed single-seat version of F-15E for USAF, for which pricing data supplied to Department of Defense as part of Quadrennial Defense Review. Not proceeded with.
   F-15U: Version conceived to satisfy United Arab Emirates requirement for long-range interdictor aircraft, in which it was competing against Lockheed Martin F-16 (selected), Dassault Rafale, Eurofighter Typhoon and Sukhoi Su-30MK. **F-15U Plus** proposal anticipated extended range, with additional 2,570 kg (5,665 lb) of fuel in thicker clipped-delta, 50° leading-edge sweep wing; more stores stations and internally situated IR navigation and targeting sensor suite in lieu of LANTIRN. Typical ordnance loads would have comprised nine 2,000 lb Mk 84 bombs or seven laser-guided GBU-24s.
   F-15MANX: Stealthy, tail-less proposal based on technology developed for ACTIVE thrust-vectoring programme.
CUSTOMERS: USAF funding for originally planned 392 reduced to 200; however, further nine funded in FY91 and FY92, comprising three Gulf War loss replacements and six with proceeds of sale to Saudi Arabia of 24 F-15C/Ds. Additional six in FY96 budget, despite not being requested by USAF. Further six aircraft in FY97 and five in FY98, raising total USAF buy to 226, with another five in FY00 and five in FY01, these final 10 for delivery between June 2002 and the end of 2004. Saudi Arabia 72 (F-15S); Israel 25 (F-15I). Latest customer is Republic of Korea, which is buying 40 F-15Ks to satisfy its F-X requirement.

USAF F-15E PROCUREMENT

FY	Batch	Qty	First aircraft
86	Lot 1	8	86-0183
87	Lot 2	42	87-0169
88	Lot 3	42	88-1667
89	Lot 4	36	89-0471
90	Lot 5	36	90-0227
91	Lot 6	36	91-0300
91	Lot 7	6	91-0600
92	Lot 8	3	92-0364
96	Lot 9	6	96-0200
97	Lot 10	6	97-0217
98	Lot 11	5	98-0131
00	Lot 12	5	00-3000
01	Lot 13	5	01-2000
Total		**236**	

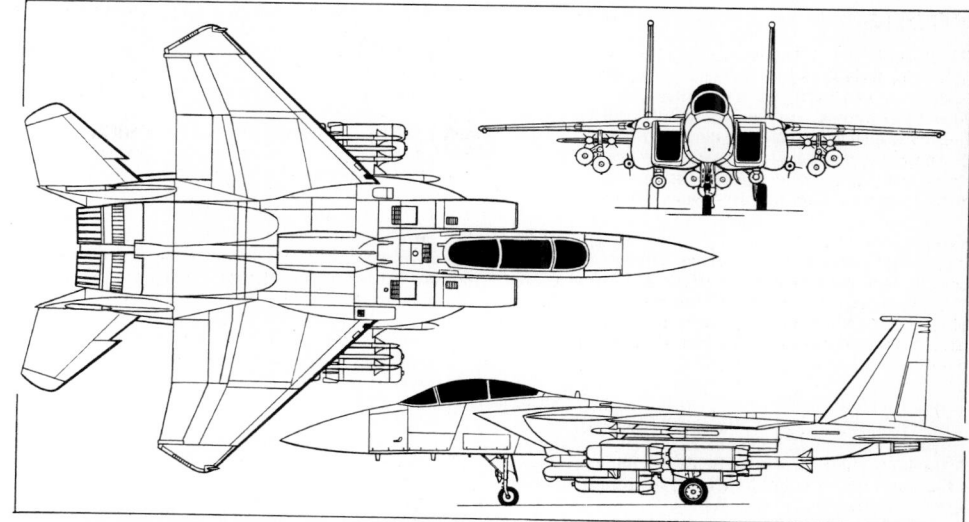

Boeing F-15E Eagle equipped for high ordnance payload air-to-ground mission *(Jane's/Dennis Punnett)*

Initial USAF combat-capable unit, 4th Wing, declared operational October 1989, currently with 333, 334, 335 and 336 Fighter Squadrons; and others with 57th Wing USAF Weapons School at Nellis AFB, Nevada; 90th FS of 3rd Wing at Elmendorf, Alaska, received first F-15E on 29 May 1991; 391st FS, sole F-15E squadron in multitype 366th Wing at Mountain Home AFB, Idaho, received first aircraft (reallocated from early production) 6 November 1991; 492nd and 494th FS of 48th FW at Lakenheath, UK, received first aircraft on 21 February 1992. Deliveries of Lot 7 began to 48th FW on 13 April 1994 and 209th USAF F-15E (92-0366) to 57th Wing on 11 July 1994. First aircraft from Lot 9 flown on 1 April 1999 and delivered to 57th Wing on 7 June; remaining five F-15Es from this batch all assigned to 48th Fighter Wing by November 1999; subsequent deliveries to 48th FW comprised all Lot 10 and Lot 11 aircraft, with final pair arriving on 26 August 2000.

Initial contract placed with McDonnell Douglas by US government on 18 December 1992 for 72 Saudi aircraft; project name, Peace Sun IX.

COSTS: US$35 million, flyaway; US$2,000 million for 21 F-15Is (1993), Israel; US$288.5 million appropriation for five F-15Es in FY00. Total cost of South Korean purchase expected to be US$4.1 billion.

DESIGN FEATURES: Conceived as air superiority fighter (F-15A to F-15D), but further developed for strike/attack mission, with provision for increased weight of ordnance. Typical 1970s fighter configuration, with twin fins positioned to receive vortex flow off wing and maintain directional stability at high angles of attack. Straight two-dimensional external compression engine air inlet each side of fuselage. High-mounted cockpit for all-round vision

Mission includes approach and attack at night and in all weathers; main systems of F-15E include high-resolution, synthetic aperture Raytheon AN/APG-70 radar, wide field of view FLIR, Lockheed Martin LANTIRN navigation (AN/AAQ-13) and targeting (AN/AAQ-14) pods beneath starboard and port air intakes respectively; air-to-air capability with AIM-7 Sparrow, AIM-9 Sidewinder and AIM-120 AMRAAM retained; rear cockpit has four multipurpose CRT displays for radar, weapon selection, and monitoring enemy tracking systems; front cockpit modifications include redesigned up-front controls, wide field of view HUD, colour CRT multifunction displays for navigation, weapon delivery, moving map, precision radar mapping and terrain-following. Engines have digital electronic control, engine trimming and monitoring; fuel tanks are foam-filled; more powerful generators; better environmental control.

NACA 64A aerofoil section with conical camber on leading-edge; sweepback 38° 42′ at quarter-chord;

thickness/chord ratio 6.6 per cent at root, 3 per cent at tip; anhedral 1°; incidence 0°.

FLYING CONTROLS: Powered. Plain ailerons and all-moving tailplane with dog-tooth extensions, both powered by National Water Lift hydraulic actuators; rudders have Ronson Hydraulic Units actuators; no spoilers or trim tabs; Moog boost and pitch compensator for control column; plain flaps; upward-opening airbrake panel in upper fuselage between fins and cockpit. Digital triple-redundant BAE Systems flight control system capable of automatic coupled terrain-following.

STRUCTURE: Wing based on torque box with integrally machined skins and ribs of light alloy and titanium; aluminium honeycomb wingtips, flaps and ailerons; airbrake panel of titanium, aluminium honeycomb and graphite/epoxy composites skin. F-15E version includes 60 per cent of earlier F-15 structure redesigned to allow 9 g and 16,000 hours fatigue life; superplastic forming/diffusion bonding used for upper rear fuselage, rear fuselage keel, main landing gear doors, and some fuselage fairings, plus engine bay structure.

LANDING GEAR: Hydraulically retractable tricycle type, with single wheel on each unit. All units retract forward. Cleveland nose and main units, each incorporating an oleo-pneumatic shock-absorber. Honeywell wheels and Michelin AIR X radial tyres on all units. Nosewheel tyre size 22×7.75-9 or 22×7.75R9 (26 ply) tubeless; mainwheel tyres size 36×11-18 or 36×11R18 (30 ply) tubeless; tyre pressure 21.03 bar (305 lb/sq in) on all units. Honeywell five-rotor carbon disc brakes.

POWER PLANT: Initially, two Pratt & Whitney F100-PW-220 turbofans, each rated for take-off at 104.3 kN (23,450 lb st), installed, with afterburning. USAF aircraft 135 onwards (90-0233), built from August 1991, have 129.4 kN (29,100 lb st) Pratt & Whitney F100-PW-229s, which also ordered for Saudi F-15S and Israeli F-15I. Air inlet controllers by Hamilton Sundstrand. Air inlet actuators by National Water Lift.

Internal fuel in foam-filled structural wing tanks and six Goodyear fuselage tanks, total capacity 7,643 litres (2,019 US gallons; 1,681 Imp gallons). Simmonds fuel gauge system. Optional conformal fuel tanks (CFT) attached to side of engine air intakes, beneath wing, each containing 2,737 litres (723 US gallons; 602 Imp gallons). Provision for up to three additional 2,309 litre (610 US gallon;

508 Imp gallon) external fuel tanks. Maximum total internal and external fuel capacity 20,044 litres (5,295 US gallons; 4,409 Imp gallons).

ACCOMMODATION: Two crew, pilot and weapon systems officer, in tandem on Boeing ACES II zero/zero ejection seats. Single-piece, upward-hinged, bird-resistant canopy; in-service modification undertaken by USAF replaced original windscreen with laminated polycarbonate material.

SYSTEMS: Lucas Aerospace generating system for electrical power, with Hamilton Sundstrand 60/75/90 kVA constant-speed drive units. Litton molecular sieve oxygen generating system (MSOGS) introduced in 1991 to replace liquid oxygen system. Honeywell air conditioning system. Three independent hydraulic systems (each 207 bar; 3,000 lb/sq in) powered by Abex engine-driven pumps; modular hydraulic packages by Hydraulic Research and Manufacturing Company. AlliedSignal APU for engine starting, and for provision of limited electrical or hydraulic power on the ground independently of main engines.

AVIONICS: *Comms:* Raytheon AN/ARC-164 UHF transceiver and UHF auxiliary transceiver with cryptographic capability; Honeywell AN/APX-101 IFF transponder; BAE Systems AN/APX-76 IFF interrogator with Litton reply evaluator. Some aircraft have AN/ARC-190 HF radio for very long-range communications.

Radar: Raytheon AN/APG-70 I-band pulse Doppler radar provides air-to-air capability equal to F-15C, plus high-resolution synthetic aperture mode for air-to-ground. Republic of Korea F-15K to have AN/APG-63(V)1 radar.

Flight: Triple redundant BAE Systems digital flight control system with automatic terrain-following standard. IBM CP-1075C very high-speed integrated circuit (VHSIC) central computer introduced in 1992 (replacing CP-1075). Honeywell AN/ASK-6 air data computer, Honeywell AN/ASN-108 AHRS, Honeywell CN-1655A/ASN ring laser gyro INS providing basic navigation data and serving as primary attitude reference system, Rockwell Collins AN/ARN-118 Tacan, Rockwell Collins HSI presenting aircraft navigation information on a symbolic pictorial display, Rockwell Collins AN/ARN-112 ILS receiver, Rockwell Collins ADF receiver, Dorne & Margolin glideslope localiser antenna and Teledyne Avionics angle of attack sensors. Rockwell Collins Miniaturised Airborne GPS Receiver installed from 1995.

Boeing F-15E Eagle of 48th FW at RAF Lakenheath, UK *(Jane's/Paul Jackson)*

Latest aircraft to join USAF (FY96 and subsequent production) feature new embedded GPS/INS.

In July 1998, Boeing began flight test of commercial computing technology in F-15E as part of Phantom Works Bold Stroke project, whereby commercial technology is being used to provide non-proprietary computer systems for installation on military aircraft. F-15E installation embodied advanced display core processor (ADCP) using PowerPC hardware to replace existing central computer and multipurpose display processor; initial trial demonstrated same capabilities as current military hardware.

Instrumentation: FLIR imagery displayed on Kaiser IR-2394/A wide field of view HUD; Honeywell vertical situation display set using CRT to present radar, electro-optical identification and attitude director indicator formats to pilot under all light conditions; moving map display by Honeywell RP-341/A remote map reader. Honeywell digital map system intended to replace remote map reader from 1996. F-15I has Elbit helmet-mounted sight and other cockpit displays. Final 10 aircraft for USAF will have Kaiser 125 × 125 mm (5 × 5 in) flat panel colour display (FPCD) instead of current CRT; this unit may also be retrofitted to earlier F-15Es. F-15K will feature seven 152 × 152 mm (6 × 6 in) AMLCDs.

Mission: Lockheed Martin LANTIRN externally mounted sensor package comprising AN/AAQ-13 navigation pod and AN/AAQ-14 targeting pod. Advanced Lockheed Martin Sniper targeting pod system to be acquired by Air Combat Command and will be fielded on F-15E.

Self-defence: Northrop Grumman Enhanced AN/ALQ-135(V) internal countermeasures set provides automatic jamming of enemy radar signals; Lockheed Martin AN/ALR-56C RWR, Raytheon AN/ALQ-128 EW warning set, (initially) BAE Systems AN/ALE-45 chaff dispenser. Unique SPS-2100 EW system developed by Elisra and Rokar for Israeli F-15I. Sanders and ITT Avionics AN/ALQ-214 Integrated Defensive Electronic CounterMeasures (IDECM) under development and will be installed on F-15E; IDECM package incorporates Sanders AN/ALE-55 fibre optic towed decoy (FOTD). Saab BOL chaff dispensers to be adopted universally, with first contract awarded to BAE Systems in fourth quarter of 2001.

ARMAMENT: 20 mm M61A1 six-barrel gun in starboard wingroot, with 512 rounds. General Electric lead computing gyro. Provision on underwing (one per wing) and centreline pylons for air-to-air and air-to-ground weapons and external fuel tanks. Wing pylons use standard rail and launchers for AIM-9 Sidewinder (Israeli F-15I also compatible with Rafael Python 4) and AIM-120 AMRAAM air-to-air missiles; AIM-7 Sparrow and AIM-120 AMRAAM can be carried on ejection launchers on the fuselage or on tangential stores carriers on CFTs. Maximum aircraft load (with or without CFTs) is four each AIM-7 and AIM-9, or up to eight AIM-120. Single or triple rail launchers for AGM-65 Maverick air-to-ground missiles can be fitted to wing stations only.

Tangential carriage on CFTs provides for up to six bomb racks on each tank, with provision for multiple ejector racks on wing and centreline stations. Edo BRU-46/A and BRU-47/A adaptors throughout, plus two LAU-106A/As each side of lower fuselage. F-15E can carry a wide variety and quantity of guided and unguided air-to-ground weapons, including Mk 20 Rockeye (26), Mk 82 (26), Mk 84 (seven), BSU-49 (26), BSU-50 (seven), GBU-10 (seven), GBU-12 (15), GBU-15 (two), GBU-24 (five), CBU-52 (25), CBU-58 (25), CBU-71 (25), CBU-87 (25) or CBU-89 (25) bombs; SUU-20 training weapons (three); A/A-37 U-33 tow target (one); and B57 and B61 series nuclear weapons (five). Is only USAF strike aircraft able to carry GBU-28. An AN/AXQ-14 datalink pod is used in conjunction with the GBU-15; LANTIRN pod illumination is used to designate targets for laser-guided bombs; AGM-130 powered standoff bomb integrated in 1993; AGM-88 HARM capability in 1996. Integration of JDAM, JSOW and WCMD weapons currently under way.

AN/AWG-27 armament control system; Programmable Armaments Control Set adopted by final 10 new-build USAF aircraft and likely to be retrofitted to most earlier aircraft.

Pneumatic weapon ejection system under development in early 2000; makes use of compressed air for weapons separation from aircraft.

DIMENSIONS, EXTERNAL:
Wing span	13.05 m (42 ft 9¾ in)
Wing aspect ratio	3.0
Length overall	19.43 m (63 ft 9 in)
Height overall	5.63 m (18 ft 5½ in)
Tailplane span	8.61 m (28 ft 3 in)
Wheel track	2.75 m (9 ft 0¼ in)
Wheelbase	5.42 m (17 ft 9½ in)

AREAS:
Wings, gross	56.49 m² (608.0 sq ft)
Ailerons (total)	2.46 m² (26.48 sq ft)
Flaps (total)	3.33 m² (35.84 sq ft)
Fins (total)	9.78 m² (105.28 sq ft)
Rudders (total)	1.85 m² (19.94 sq ft)
Tailplanes (total)	10.34 m² (111.36 sq ft)

WEIGHTS AND LOADINGS (F100-PW-220 engines):
Operating weight empty (no fuel, ammunition, pylons or external stores)	14,515 kg (32,000 lb)
Max weapon load	11,113 kg (24,500 lb)
Max fuel weight: internal (JP4)	5,952 kg (13,123 lb)
CFTs (two, total)	4,265 kg (9,402 lb)
external tanks (three, total)	5,396 kg (11,895 lb)
max internal and external	15,613 kg (34,420 lb)
Max T-O weight	36,741 kg (81,000 lb)
Max landing weight:	
unrestricted	20,094 kg (44,300 lb)
at reduced sink rates	36,741 kg (81,000 lb)
Max zero-fuel weight	28,440 kg (62,700 lb)
Max wing loading	650.5 kg/m² (133.22 lb/sq ft)
Max power loading	176 kg/kN (1.73 lb/lb st)

PERFORMANCE:
Max level speed at height	M2.5
Max combat radius	685 n miles (1,270 km; 790 miles)
Max range	2,400 n miles (4,445 km; 2,762 miles)

UPDATED

BOEING F/A-18E/F SUPER HORNET

TYPE: Multirole fighter.

PROGRAMME: Proposed 1991 as replacement for cancelled GD/MDC A-12 and follow-on for early F/A-18As and other USN/MC tactical aircraft as they phase out; based on earlier versions of Hornet (see 2001-02 edition, or *Jane's Aircraft Upgrades,* for full details); development funding approved by Congress for FY92; US$4.88 billion engineering and manufacturing development contract awarded June 1992, covering seven flight test aircraft (five Es; two Fs) and three ground test articles, plus associated 7½-year test programme; US$754 million award in 1992 to GE for F414 engine development.

Critical design review (CDR) undertaken 13 to 17 June 1994 at St Louis by team of independent government evaluators; successfully negotiated, with F/A-18E/F satisfying or surpassing all timescale, cost, technical, reliability and maintainability requirements.

Principal subcontractor Northrop Grumman launched assembly of first aircraft on 24 May 1994 with start of work on centre/aft fuselage section at Hawthorne, California. First forward fuselage section followed suit on new assembly line at St Louis, Missouri, 23 September 1994; completed 12 January 1995, except for wiring; mating with centre/aft section from Hawthorne effected 8 May 1995. Roll-out of prototype (E1/165164) on 18 September 1995; first flight 29 November 1995.

Prototype delivered to Naval Air Warfare Center at Patuxent River, Maryland, on 14 February 1996 for start of three year flight test programme involving seven aircraft; initial flight from Patuxent River made by E1 on 4 March 1996. Second F/A-18E prototype (E2/165165) made first flight on 26 December 1995 and first F/A-18F (F1/165166) on 1 April 1996. Supersonic speed exceeded for first time on 12 April 1996 when E1 achieved M1.1. Carrier suitability trials began mid-1996 and F1 completed three successful catapult launches at Patuxent River on 6 August that year.

By 1 February 1997, when the last development aircraft was delivered to Patuxent River, test fleet had flown 390 sorties, for a total of 631 flight hours. Subsequent milestones were passed on 13 May 1997 (1,000th flight hour); 9 September 1997 (1,500th flight hour); 24 September 1997 (1,000th sortie); 11 December 1997 (2,000th flight hour and 1,300th sortie); 1 July 1998 (3,000th hour) and 12 January 1999 (4,000th hour). First missile launch (an AIM-9 Sidewinder) accomplished by aircraft F2 on 5 April 1997; other weapons expended by mid-May 1997 included AIM-7 Sparrow, AIM-120 AMRAAM, SLAM/SLAM-ER, AGM-84 Harpoon, Mk 82 and Mk 83 bombs plus Rockeye CBUs. Successful deployment of AN/ALE-50 towed radar decoy also accomplished by mid-May 1997.

Flight test programme briefly halted in October 1997 for engine inspection after discovery of cracks in stator vanes. More intractable problem concerned wing-drop, with uncommanded departures from controlled flight evident as early as the seventh sortie in March 1996. Boeing and US Navy considered three solutions to eradicate this in January 1998, including fitting stall strips on upper surface; adding a chord-wise fence just inboard of the hinge fairing; and switching to a porous hinge fairing with slots that allow air to flow in both directions through wing fold fairing. Last-mentioned option selected in early 1998; subsequent flight testing confirmed efficacy of solution.

EMD phase of test programme completed end April 1999, by which time seven aircraft had accumulated 4,673 flight hours in 3,172 sorties; in the process, over 15,000 test points completed and 29 weapons configurations cleared for flight. By mid-2000, follow-on testing had increased totals to more than 5,500 flight hours in 3,800 flights.

Static testing began in August 1995 with airframe ST50; shock loading assessment from February 1996 with DT50; fatigue testing from 30 June 1997 with FT50, which

US F-15 PRODUCTION

	F-15A	F-15B	F-15C	F-15D	F-15DJ	F-15E	F-15I	F-15J	F-15K	F-15S	Total
Israel	19	2	18	13			25				77
Japan					12			10[1]			22
Saudi Arabia			55	19						72	146
South Korea									40[5]		40
USA	365[2]	59[3]	409	61		236[4]					1,120
Total	**384**	**61**	**482**	**93**	**12**	**236**	**25**	**10**	**40**	**72**	**1,415**

Note: Mitsubishi of Japan also produced 155 F-15Js and 36 F-15DJs to complete total JASDF procurement of 165 and 48, respectively
[1] Including eight kits to Mitsubishi
[2] Including 10 YF-15s
[3] Including two YF-15s, of which one converted to F-15E prototype
[4] Total includes five being purchased in FY00 and five in FY01
[5] On order for delivery during 2005-08.

Two-seat F/A-18F of training squadron VFA-122 (*Jane's/Paul Jackson*)

NEW/0527158

0100853

Boeing F/A-18E Super Hornet, cutaway drawing key

1 Composites radome
2 Radome open position for access
3 Raytheon AN/APG-73 multimode radar scanner
4 Radome hinge
5 Scanner tracking mechanism
6 Radar mounting bulkhead
7 Raytheon AN/APG-79 active electronically scanned array (AESA) radar for future integration
8 AN/ALR-67 low-band antenna
9 Radar equipment module
10 Electroluminescent formation lighting strip
11 M61A2 20 mm cannon barrels
12 Cannon port and blast-diffuser vents
13 Flight refuelling probe, extended
14 Probe actuating link
15 Upper combined interrogator IFF antenna
16 20 mm cannon
17 Cannon ammunition drum, 570 rounds
18 Incidence transmitter
19 Lower VHF/UHF D-band antenna
20 Pitot head
21 Gun gas vents
22 Cockpit front pressure bulkhead
23 Nosewheel door
24 Ground power socket
25 Avionics ground cooling air fan and ducting
26 Rudder pedals
27 Instrument panel, full-colour multifunction CRT displays
28 Instrument panel shroud
29 Frameless windscreen panel
30 Head-up display (HUD)
31 Upward-hinged cockpit canopy
32 Martin-Baker NACES zero/zero ejection seat
33 Starboard side console
34 Control column
35 Port console with engine throttle levers, full HOTAS controls
36 Sloping seat-mounting bulkhead
37 Boarding step
38 Forward fuselage lateral equipment bays, three per side
39 Nosewheel leg pivot mounting
40 Landing light
41 Deck approach signal lights
42 Nosewheel steering unit
43 Catapult engagement bar
44 Twin nosewheels, forward-retracting
45 Torque scissor links incorporating holdback fitting
46 Folding boarding ladder
47 Nosewheel retraction jack
48 AN/ALQ-165 EW transmitting antenna
49 Boarding ladder stowage
50 LEX equipment bay
51 Cockpit rear pressure bulkhead

52 Cockpit avionics equipment bay
53 Canopy rotary actuator
54 Starboard AN/ALQ-165 EW transmitting antenna
55 Canopy actuating strut
56 Canopy hinge point
57 No. 1 fuselage bag-type fuel tank
58 Sloping bulkhead, structural provision for two-seat F/A-18F
59 EW receiver
60 AN/ALR-67 low-band antenna
61 Port leading-edge extension (LEX) chine member
62 External fuel tank; centreline refuelling store as alternative
63 Port position light
64 Liquid cooling system equipment; reservoir, heat exchanger and ground running fan
65 Forward slinging point
66 Forward tank bay access panel
67 Starboard position light
68 Starboard LEX avionics equipment bay
69 Spoiler panel
70 LEX vent, operates in conjunction with leading-edge flap (initial production aircraft only)
71 Intake boundary layer spill duct
72 GPS antenna
73 No. 2 tank bay access panel
74 No. 2 bag-type fuel tank
75 Port spoiler
76 Spoiler hydraulic actuator
77 Boundary layer bleed air ducts
78 Bleed air spill duct
79 Port LEX vent
80 Perforated intake wall bleed air spill duct
81 Port fixed-geometry air intake
82 Mainwheel leg door
83 Main landing gear leg strut
84 Trailing axle suspension
85 Port mainwheel
86 Shock absorber strut
87 Mainwheel door

88 LAU-116 missile carrier/launch unit
89 Mainwheel leg pivot mounting
90 Hydraulic retraction jack
91 Intake duct framing
92 Wing panel attachment joints
93 Machined titanium fuselage main bulkheads
94 No. 3 bag-type fuel tank
95 No. 4 bag-type fuel tank
96 No. 3 tank access panel
97 IFF antenna
98 Dorsal fairing access panels
99 Upper VHF/UHF D-band antenna
100 Starboard wing panel bolted attachment joints
101 Starboard wing integral fuel tank
102 Leading-edge flap hydraulic drive unit and rotary actuator
103 Wing carbon fibre composites (CFC) skin panelling
104 Starboard stores pylons; wing pylons canted 4° inboard
105 Leading-edge dog-tooth

106 Wing-fold hinge fairing porous panel
107 Starboard leading-edge flap rotary actuator
108 Two-segment leading-edge flap
109 Outer wing panel dry bay
110 Wingtip position light
111 Formation light fairing
112 Wingtip missile installation
113 Starboard outer wing panel, folded position
114 Drooping aileron
115 Aileron hydraulic actuator
116 Wing-fold hydraulic jack
117 Aileron and flap opposed movement in airbrake mode
118 Starboard single-slotted trailing-edge flap
119 Hinged flap shroud

120 Flap hydraulic actuator
121 Dorsal equipment bay
122 No. 4 tank bay access panel
123 Ram air from intake duct to ECS (environmental control system)
124 Rear fuselage singeing points
125 ECS equipment bay
126 ECS hinged auxiliary intake doors
127 Fuselage fuel tank tanks, port and starboard
128 Primary (starboard) and secondary (port) heat exchangers

129 Heat exchanger exhaust ducts
130 Starboard all-moving tailplane
131 Starboard fin bolted attachment joints
132 Fin integral vent tank
133 Multispar fin structure
134 Leading-edge structure, CFC skin with honeycomb core
135 Fin CFC skin panelling
136 CFC fin-tip fairing
137 Rear position light
138 Aft AN/ALQ-165 EW receiving antenna
139 Starboard AN/ALR-67 RWR antenna
140 Fuel jettison
141 Starboard rudder, CFC skin with honeycomb core structure
142 Rudder hydraulic actuator
143 Starboard engine bay
144 Rear engine mounting support structure

145 Starboard all-moving tailplane
146 Flight data recorder
147 Fin formation lighting strip
148 Fuel venting ram air intake
149 Anti-collision beacon
150 AN/ALQ-165 high- and low-band transmitting antennas
151 Port AN/ALR-67 RWR antenna
152 Fuel jettison
153 Port rudder

154 Rudders move in opposing directions in airbrake mode
155 Variable area afterburner exhaust nozzles
156 Nozzle sealing flaps
157 Engine bay vent, above and below
158 Afterburner nozzle 'fueldraulic' actuator (three)
159 Afterburner duct
160 AN/ALE-50 towed radar decoy (TRD), three in ventral stowage; AN/ALE-55 for future integration
161 Port all-moving tailplane
162 CFC tailplane skin panel on aluminium honeycomb substrate
163 Tailplane pivot support structure
164 Pivot mounting

165 Tailplane hinge arm
166 Tailplane hydraulic actuator
167 Port fin-root attachment joints
168 Rear fuselage formation lighting strip
169 General Electric F414-GE-400 afterburning, low-bypass turbofan engine
170 Main engine mounting
171 Full authority digital engine controller (FADEC)
172 Deck arrester hook
173 Engine accessory equipment
174 Engine oil tank
175 Engine bay venting ram air intake
176 Compressor intake
177 Airframe-mounted accessory equipment gearbox, port and starboard, shaft-driven from engine
178 Generator
179 Stationary intake duct-mounted compressor radar-return shielding device
180 Trailing-edge flap root fairing

181 Central auxiliary power unit (APU)
182 Port mainwheel, stowed position
183 Wingroot attachment fittings
184 Port flap hydraulic actuator
185 Inboard flap hinge
186 Flap CFC rib and skin structure
187 Port hinged flap shroud
188 Wing panel multispar structure
189 Port wing integral fuel tank, fire suppressant foam-filled
190 Inboard 'wet' pylon hardpoints
191 Leading-edge flap rotary actuator
192 Hydraulic flap drive unit and torque shaft
193 Forward AN/ALR-67 and AN/ALQ-165 EW receiving antennas
194 Leading-edge flap CFC rib and skin structure
195 Inboard stores pylons
196 Outboard 'dry' pylon
197 Outboard pylon hardpoint
198 Wing-fold hinge joint
199 Wing-fold hydraulic jack
200 Outer wing panel hinge fitting
201 Port outer wing panel
202 Wingtip position light
203 Formation lighting strip
204 Wingtip missile launch rail
205 Port aileron
206 Aileron CFC skin on honeycomb core substrate
207 Aileron ventral hinge and actuator fairing
208 Supplementary position light
209 AIM-9M Sidewinder, close-range air-to-air missile
210 AIM-9X Advanced Sidewinder
211 AIM-7 Sparrow, intermediate-range air-to-air missile
212 AIM-120C AMRAAM
213 Twin missile carrier/launcher
214 AGM-84H SLAM-ER air-to-surface missile
215 Mk 83 1,000 lb bomb
216 Mk 82 500 lb HE bomb
217 AGM-84A Harpoon air-to-surface anti-ship missile
218 AGM-88C HARM air-to-surface anti-radar missile
219 GBU-16 1,000 lb laser-guided bomb
220 Advanced targeting forward-looking infra-red pod (ATFLIR)
221 ATFLIR mounting adaptor, port fuselage station
222 GBU-24 2,000 lb laser-guided bomb
223 Joint stand-off weapon (JSOW), submunitions or 500 lb HE unitary warhead
224 AGM-65 Maverick air-to-surface missile, television-homing, imaging IR or semi-active laser variants
225 GBU-31 2,000 lb joint direct attack munition (JDAM)

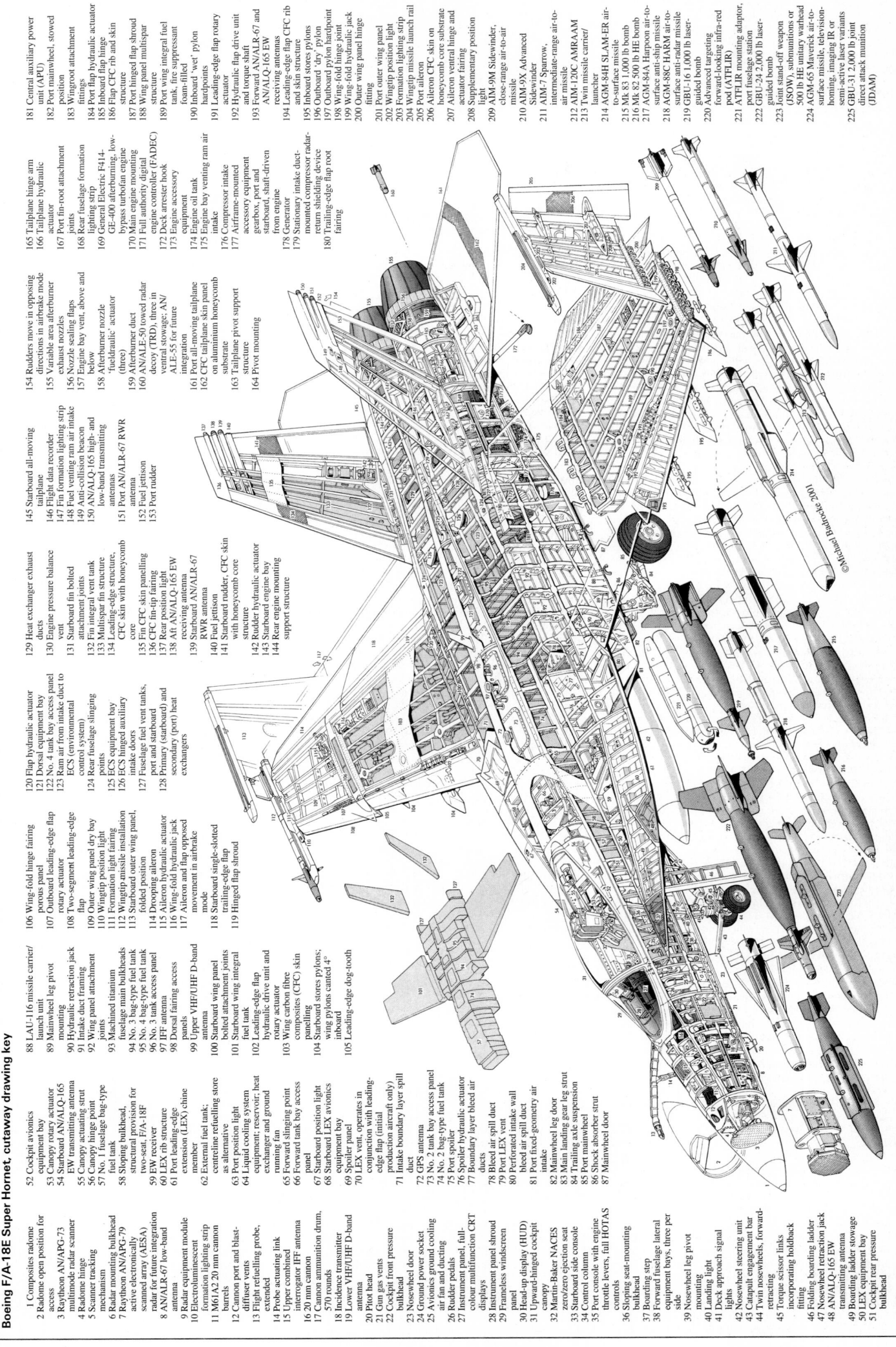

©Michael Badrocke 2001

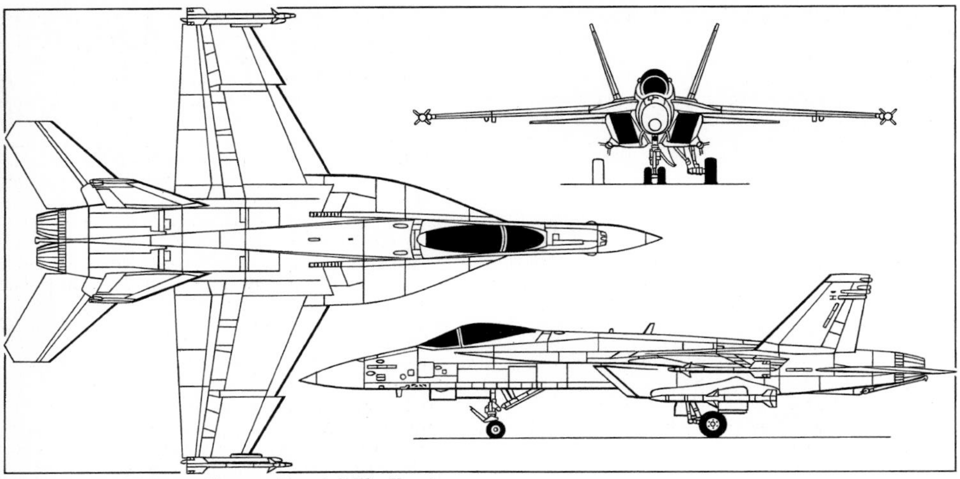

Boeing F/A-18E Super Hornet ***(Jane's/Mike Keep)***

completed first lifetime (6,000 hours) one month ahead of schedule, on 27 August 1998. ST50 transferred to Lakehurst, New Jersey, for series of six emergency barricade engagements; first successfully completed on 3 September 1997, but ST50 damaged during third test on 23 September when restraint cable failed. ST50 subsequently repaired and returned to test duty, for live-fire testing at China Lake, California; trials included firing large armour-piercing incendiary projectile through inlet duct into aft fuel tank, with resultant hole showing little evidence of leakage.

Approval for low-rate initial production (LRIP) of 62 aircraft in three lots given on 26 March 1997; first lot composed of eight F/A-18Es and four F/A-18Fs. Assembly of first LRIP F/A-18E (165533) started at Northrop Grumman in May 1997 and at Boeing in September 1997; final assembly began at St Louis on 19 June 1998, with mating of centre/aft and forward fuselage sections; first flight 6 November 1998, six weeks ahead of schedule; this aircraft officially accepted by US Navy on 18 December 1998 and flown to Patuxent River to join flight test programme before attachment to VX-9 squadron at China Lake, California, for operational evaluation. Latter programme began 27 May 1999 and involved total of 1,233 flight hours in 866 sorties by seven LRIP aircraft (three F/A-18Es and four F/A-18Fs) in six-month period, including testing of all mission capabilities in varying climates as well as operations at sea aboard USS *John C Stennis* and participation in 'Red Flag' exercise at Nellis AFB, Nevada. Results of operational evaluation announced 15 February 2000, with report stating aircraft to be "operationally effective and operationally suitable" and recommending fleet introduction with the US Navy.

First USN squadron is VFA-122, at Lemoore, California, as specialist FRS (Fleet Replacement Squadron), training pilots and ground crew; VFA-122 received first seven aircraft (of eventual 34) 17 November 1999. First operational squadron is VFA-115, which began transition from F/A-18C to F/A-18E in third quarter 2000; will make first deployment in USS *Abraham Lincoln* from June 2002. Next two squadrons are VFA-14 and VFA-41, for whom transition from F-14 Tomcat began in early 2001; VFA-14 has F/A-18E, while VFA-41 received F/A-18F, but both will deploy in May 2003 as part of Carrier Air Wing 11 (CVW-11) in USS *Nimitz*. VFA-97 also to get F/A-18E by FY04, with the F/A-18F to replace F-14 with VF-102 (FY03), VF-154 (FY07), VF-2 (FY08) and VF-32 (FY09). Second FRS to form at Oceana, Virginia, in second half of 2003.

CURRENT VERSIONS: **F/A-18E:** Single-seat. Initial aircraft (all LRIP examples plus those purchased in FY00 and FY01) to so-called Block 1 standard, with AN/APG-73 and introducing other items of equipment as they become available; subsequent aircraft (FY02 and later) to so-called Block 2 standard, incorporating revised forward fuselage as part of ECP 6038, with AN/APG-79, fibre optic data network and advanced crew station for NFO among other improvements.

F/A-18F: Two-seat. Also produced in Block 1 and Block 2 versions as detailed above.

F/A-18 C^{2}W: Original name of private venture development of F/A-18F as two-seat electronic warfare aircraft; announced 7 August 1995; following merger, Boeing is prime contractor, with Northrop Grumman as principal subcontractor with special responsibility for integration of electronic warfare suite; potential replacement for EA-6B Prowler; minimal structural changes; wideband receiver pods replace wingtip Sidewinder AAMs; other pods and antennas on weapon pylons; satcom receiver behind cockpit. US Navy formulated requirements for new EW aircraft to replace Prowler from 2008 onwards. Boeing anticipates using EA-6B equipment such as AN/ALQ-99 jamming system pods on Super Hornet. Future of this proposal, unofficially known as **EA-18** (previously **F/A-18G Growler**), still uncertain, following two-year Analysis of Alternatives study completed in December 2001. This tasked with establishing joint service needs in EW mission, but failed to recommend a single system to replace Prowler; various options available, including Growler/Super Hornet, UCAVs and larger aircraft of Boeing 757/767 size. Recent developments included initial flight demonstration of first EMD F/A-18F configured with three AN/ALQ-99 jamming pods and two fuel tanks in November 2001; further trials in first half of 2002 expanded performance envelope and provided data on noise and vibration characteristics.

CUSTOMERS: US Navy. Seven prototypes; approval for 12 LRIP aircraft in FY97 (Lot 1), with manufacture rising incrementally to peak rate of 48 per annum during FY98 to FY02. See table for details of three LRIP batches and first multiyear procurement (MYP) batch; contract for latter signed on 15 June 2000. All 62 LRIP aircraft delivered by end of third quarter of 2001, with first full-rate production Super Hornet (an F/A-18F) being handed over in late September; 100th Super Hornet (an F/A-18F) delivered 14 June 2002. Broad requirement originally identified for over 1,000 by FY15, but Quadrennial Defense Review reduced this to minimum of 548 and maximum of 785; defence guidance planning proposal in first quarter of 2002 advocates cutting procurement to 460. Decision awaited, with negotiations for follow-on multiyear contract in abeyance pending outcome. Boeing seeking export sales, following approval by US government; Malaysia is prime candidate to be first export customer, with Boeing offering 16 F/A-18F from 2005 in a deal that would involve taking back eight F/A-18Ds delivered in 1997. Latter aircraft could then be refurbished for Switzerland, which has indicated desire to obtain small number of first-generation Hornets for reconnaissance. Super Hornet will also be offered to Australia, which seeking new multirole fighter under Project Air 6000.

SUPER HORNET PROCUREMENT

FY	Batch	Total	Remarks
97	LRIP 1	12	Eight F/A-18E and four F/A-18F (Block 52)
98	LRIP 2	20	Eight F/A-18E and 12 F/A-18F (Block 53)
99	LRIP 3	30	14 F/A-18E and 16 F/A-18F (Block 54)
00	MYP1	36	
01	MYP1	42	
02	MYP1	48	
03	MYP1	48	
04	MYP1	48	
Total		**284**	

COSTS: Development estimated US$4.8 billion (1992); US$1,089 million in FY92 budget; US$943 million in FY93; approximately US$1,500 million in FY94 and US$1,348 million requested for FY95. US$2,600 million in provisional FY97 budget for first 12 aircraft; with US$2,100 million requested for FY98 procurement of 20 aircraft. US GAO estimated flyaway unit cost as being of the order of US$43.6 million in 1996, based on original planned procurement of 1,000. Most recent figures, for first MYP batch, put total cost at US$8.9 billion for 222 aircraft. Unit cost quoted as US$48 million in mid-2000, when Boeing pursuing measures to drive cost down to around US$40 million by 2005. Estimated cost of EA-18 approximately US$57 million to US$59 million.

DESIGN FEATURES: Generally as for first-generation Hornet. Stretched versions of F/A-18C/D; gross landing weight increased by 4,536 kg (10,000 lb); 0.86 m (2 ft 10 in) fuselage plug; wings photometrically increased in size to provide 9.29 m^2 (100.0 sq ft) extra area and 1.31 m (4 ft 3½ in) span increase; control surfaces disproportionately enlarged and dogtooth added to leading-edge for increased aileron authority. Wings 2.5 cm (1 in) deeper at root; larger horizontal tail surfaces; LEX size substantially increased in early 1993 (from total 5.8 m^2; 62.4 sq ft to 7.0 m^2; 75.3 sq ft, compared with 5.2 m^2; 56.0 sq ft on F/A-18C/D), ensuring full manoeuvre capability at beyond 40° AoA; also incorporates spoilers on upper surface of LEX as speedbrake and to increase nose-down control authority. Nevertheless, has 42 per cent fewer parts than immediate predecessor.

Additional 1,637 kg (3,600 lb) of internal and 1,406 kg (3,100 lb) of external fuel; 40 per cent extra range; further two (making 11) weapon hardpoints (stations 2 and 10, inboard of wingtips, for AAMs and ASMs of up to 520 kg; 1,146 lb); 'bring-back' weapons load increased to 4,082 kg (9,000 lb); additional survivability measures; air intakes redesigned and slewed to increase mass flow to more powerful F414-GE-400 engines and also changed to 'caret' shape to reduce radar signature. Incorporates several other 'affordable' stealth features to reduce radar cross-section, including saw-toothed doors and panels, realigned joints and edges and angled antennas.

FLYING CONTROLS: Full digital fly-by-wire controls using ailerons and tailerons for lateral control, plus flaps in flaperon form at low airspeeds; leading- and trailing-edge flaps scheduled automatically for high manoeuvrability, fast cruise and slow approach speed; horizontal stabilisers automatically assume neutral position if one is damaged, with pitch control then being passed to other surfaces; both rudders turned in at take-off and landing to provide extra nose-up trim effort; fly-by-wire returns towards 1 *g* flight if pilot releases controls; lateral and then directional control progressively washed out as angle of attack reaches extreme values; height, heading and airspeed holds provided in fly-by-wire system; aircraft can land automatically using carrier-based guidance system. Bertea hydraulic actuators for trailing-edge flaps; Hydraulic Research actuators for ailerons; National Water Lift actuators for tailerons.

STRUCTURE: Multispar wing mainly of light alloy, with graphite/epoxy inter-spar skin panels and trailing-edge flaps; tail surfaces mainly graphite/epoxy skins over aluminium honeycomb core; graphite/epoxy fuselage panels and doors; titanium engine firewall. Northrop Grumman responsible for rear and centre fuselages and delivered 100th shipset to St Louis at end of January 2002; assembly and test at St Louis factory; CASA produces horizontal tail surfaces, flaps, leading-edge extensions, speedbrakes, rudders and rear side panels for all F/A-18s.

LANDING GEAR: Messier-Dowty retractable tricycle type, with twin-wheel nose and single-wheel main units. Nose unit retracts forward, mainwheels rearward, turning 90° to stow horizontally inside the lower surface of the engine air ducts. Bendix wheels and brakes. Nosewheel tyres size 22×6.6-10 (20 ply) tubeless, pressure 24.13 bar (350 lb/sq in) for carrier operations, 10.34 bar (150 lb/sq in) for land operations. Mainwheel tyres size 30×11.5-14.5 (24/26 ply) tubeless, pressure 24.13 bar (350 lb/sq in) for carrier operations, 13.79 bar (200 lb/sq in) for land operations. Ozone nosewheel steering unit. Nose unit towbar for catapult launch. Arrester hook, for carrier landings, under rear fuselage.

POWER PLANT: Two General Electric F414-GE-400 turbofans, each rated at approximately 97.9 kN (22,000 lb st) with afterburning. Internal fuel capacity (JP-5 fuel) 8,063 litres (2,130 US gallons; 1,774 Imp gallons). Wing tanks protected from combat damage by low-density foam. Provision for five 1,250 litre (330 US gallon; 275 Imp gallon) or Lincoln Composites 1,817 litre (480 US gallon; 400 Imp gallon) external tanks, giving maximum fuel capacity of 17,148 litres (4,530 US gallons; 3,772 Imp gallons) and ability to operate in air refuelling role using hose drum unit on centreline station. Normal operational fit anticipated as three external tanks of either size.

ACCOMMODATION: Pilot only F/A-18E, on Martin-Baker SJU-5/6 NACES zero/zero ejection seat, in pressurised, heated and air conditioned cockpit. Upward-opening canopy, with separate windscreen, on all versions. Two pilots or pilot and Naval Flight Officer in F/A-18F.

SYSTEMS: High commonality with F/A-18C/D, incorporating two separate and independent hydraulic systems, each at 207 bar (3,000 lb/sq in), but with more powerful actuators to accommodate enlarged control surfaces with increased deflections. Leland Electrosystems power generating system; provides 60 per cent more electrical power than in F/A-18C. Hamilton Sundstrand air conditioning; Vickers hydraulic pumps; oxygen system; fire detection and extinguishing systems. Honeywell GTC36-200 APU.

AVIONICS: Over 90 per cent commonality with F/A-18C, but differences include following:

Comms: Rockwell Collins AN/ARC-210 secure UHF/VHF radio. Multifunction information distribution system (MIDS) datalink, Rockwell Collins digital communication system and Hazeltine combined interrogator transponder to be installed as part of upgrade package, before first operational deployment.

Radar: Raytheon AN/APG-73 multimode, digital air-to-air and air-to-ground radar as standard. Raytheon AN/APG-79 active electronically scanned array (AESA) radar under development for use on Super Hornet from 2006. Current plans anticipate procurement of at least 258 AESA radars, following trials in Boeing 737 avionics testbed from early 2002; low-rate initial production likely to involve 42 units in FY03-FY05, with full-rate production starting in FY06.

Flight: Litton embedded GPS; Smiths/Harris tactical aircraft moving map capability (TAMMAC); DRS Technologies deployable flight incident recorder set (DFIRS).

Instrumentation: Cockpit has 76 × 127 mm (3 × 5 in) touch-panel LCD upfront display and 159 mm (6¼ in) square colour LCD tactical situation display; also two 127 mm (5 in) square monochrome displays and will have monochrome programmable LCD in place of F/A-18C engine/fuel display; Kaiser AN/AVQ-28 HUD. Planar Advance/dpiX awarded joint development contract for Eagle-6 multifunction AMLCD in first half of 1998. Aft cockpit to be 'missionised' for strike operations, including 25 × 20 cm (10 × 8 in) AMLCD and two hand controllers for use by NFO.

Mission: Raytheon developing advanced targeting forward-looking infra-red (ATFLIR) targeting and navigation system, with award of EMD contract in March 1998. Low rate production authorised in March 2001 with IOC to follow in mid-2002 and initial operational deployment (with VFA-41) in 2003. Shared Reconnaissance Pod (SHARP) system, incorporating Recon/Optical cameras, under development and expected to be ready for use in second quarter of 2003. DRS Technologies WRR-818 cockpit video recording system.

Advanced mission computers based on commercial hardware and software and colour flat panel LCD displays are in development for incorporation in FY05, with DY4 Systems awarded US$1 million order to upgrade existing system which relies on Control Data Corporation AN/AYK-14 digital computers; upgrade will involve replacement of AN/AYK-14 by DMV-179 single board computers and PMC-642 fibre channel network interface module. Vision Systems International joint helmet-mounted cueing system (JHMCS) also to be installed by 2003 to target the AIM-9X high off-boresight missile; testing of JHMCS on Super Hornet began in first quarter of 2001, with delivery expected in FY03.

Self-defence: Management by BAE Systems (formerly Sanders) integrated defensive electronic countermeasures suite (IDECM); interfaces with Raytheon AN/ALR-67(V)3 radar warning receiver, BAE Systems (formerly Sanders) AN/ALE-55 fibre optic towed radar decoy (with triple dispenser stowed between jetpipes); and four BAE Systems AN/ALE-47 chaff/flare dispensers. However, delay and cost growth of IDECM means that first three operational squadrons will have aircraft fitted with less sophisticated Raytheon AN/ALE-50 towed decoy in conjunction with either AN/ALQ-165 ASPJ (first two squadrons) or BAE Systems (formerly Sanders) AN/ALQ-214(V)2 radio frequency countermeasures system (third squadron). Definitive AN/ALQ-214 and AN/ALE-55 combination to be deployed with effect from fourth squadron, by the end of 2003, while earlier aircraft will be updated.

ARMAMENT: See diagram. Full range of USN offensive and defensive ordnance. At least 29 weapons combinations cleared before service entry. M61A2 20 mm cannon with 400 rounds; will be compatible with forthcoming AIM-9X Advanced Sidewinder missile.

DIMENSIONS, EXTERNAL (approx):
Wing span over missiles	13.62 m (44 ft 8½ in)
Wing aspect ratio	4.0
Width, wings folded	9.94 m (32 ft 7¼ in)
Length overall	18.38 m (60 ft 3½ in)
Height overall	4.88 m (16 ft 0 in)

AREAS:
Wings, gross	46.45 m² (500.0 sq ft)

WEIGHTS AND LOADINGS:
Weight empty	14,009 kg (30,885 lb)
Max fuel weight: internal (JP-5)	6,781 kg (14,950 lb)
external (JP-5)	7,430 kg (16,380 lb)
Max external stores load: T-O	8,028 kg (17,700 lb)
landing	4,082 kg (9,000 lb)
T-O weight, attack mission	29,937 kg (66,000 lb)
Max wing loading	644.5 kg/m² (132.00 lb/sq ft)
Max power loading	153 kg/kN (1.50 lb/lb st)

PERFORMANCE (estimated):
Max level speed at altitude	more than M1.8
Approach speed	125 kt (232 km/h; 144 mph)
Combat ceiling	15,240 m (50,000 ft)

Min wind over deck:
launching	30 kt (56 km/h; 34.5 mph)
recovery	15 kt (28 km/h; 17.5 mph)

Combat radius: interdiction with two SLAM-ER, two AMRAAMs, two Sidewinders and three 1,817 litre (480 US gallon; 400 Imp gallon) external tanks, hi-hi-hi (including flight of SLAM-ER)
945 n miles (1,750 km; 1,087 miles)
fighter escort with four AMRAAMs, two Sidewinders and three 1,817 litre (480 US gallon; 400 Imp gallon) external tanks
795 n miles (1,472 km; 914 miles)
Combat endurance: maritime air superiority, six AAMs, three 1,818 litre (480 US gallon; 400 Imp gallon) external tanks, 150 n miles (278 km; 173 miles) from aircraft carrier
2 h 15 min
g limit +7.5

UPDATED

BOEING C-17A GLOBEMASTER III
TYPE: Strategic transport.

PROGRAMME: US Air Force selected McDonnell Douglas to develop C-X cargo aircraft 28 August 1981; full-scale development called off January 1982 and replaced on 26 July 1982 by slow-paced preliminary development order; development and three prototypes (one flying) ordered 31 December 1985; fabrication of first C-17A (T1/87-0025) began 2 November 1987; first production C-17A contract 20 January 1988; assembly started at Long Beach 24 August 1988; assembly of prototype completed 21 December 1990. Programme transferred from Douglas Aircraft Company to McDonnell Douglas Aerospace in 1992; to McDonnell Douglas Military Transport Aircraft in 1996 and to Boeing following merger of August 1997.

First flight (T1/87-0025) 15 September 1991 – also delivery to Edwards AFB; first flight of initial production aircraft (P1/88-0265) 18 May 1992; first delivery to operational unit 14 June 1993. First overseas service flight (P11/92-3291) to Mildenhall, UK, 25 May 1994.

C-17 was named Globemaster III on 5 February 1993; peak production target 15 per year (but could be increased to 18); assembly in 102,200 m² (1.1 million sq ft) facility at Long Beach, California.

Development flight testing completed 15 December 1994, by which time 16 production aircraft delivered and 22nd record set (further 13 world records set by 71st production aircraft during testing at Edwards in November 2001). Initial AMC Squadron (17th AS) received its 12th C-17A 22 December 1994; achieved IOC 17 January 1995 with acceptance of 13th (nominal spare) aircraft. Second squadron (14th AS) received its first aircraft 17 February 1995. Air Education and Training Command's 97th AMW at Altus AFB, Oklahoma, subsequently took delivery of eight aircraft between March 1996 and November 1997, all of which had previously been assigned to the 437th AW. Deliveries to 7th AS of 62nd AW at McChord AFB, Washington, began at end of July 1999. C-17 fleet passed 300,000 flying hours in 2001, excluding more than 1,000 hours accumulated by the test-dedicated prototype aircraft.

Further procurement of C-17 for USAF now almost certain, in light of statements by various defence officials on MRS05 (Mobility Requirements Study 2005) plan and BC-17X proposal; a decision was due to have been taken in 2002, with up to 60 additional aircraft in prospect. These are expected to be obtained at a rate of 15 per year during FY03-07, at a cost of more than US$9 billion; further purchases have not been ruled out and the USAF is looking at the possibility of buying another 42, raising total procurement to 222, excluding the original prototype.

CURRENT VERSIONS: **C-17A:** Basic standard.

Detailed description applies to C-17A.

C-17B: Not assigned. Used unofficially (along with C-17ER) to identify aircraft with extended-range fuel tanks (Lot 12 and upwards).

KC-17: Private venture tanker/transport project, offered unsuccessfully as replacement for USAF KC-135R/T Stratotanker.

BC-17X: Projected civil cargo version, known until 2000 as MD-17; *described separately.*

CUSTOMERS: US Air Force; original requirement 210; cut to 120 by 1991; capped at 40 in January 1994 for two year probationary period, during which contractor to achieve performance, cost and delivery targets; decision taken on 3 November 1995 to acquire the balance of 80 C-17s. Multiyear procurement contract signed 31 May 1996 for production through 2004; first aircraft from this contract (P41/97-0041) delivered 10 August 1998; last due on 30 November 2004.

Basing plans were revealed soon after the decision to procure 120 in all, when the USAF announced that the 437th AW at Charleston AFB, South Carolina, and the 62nd AW at McChord AFB, Washington, will each receive 48 Globemaster IIIs, with these respectively being supported by Air Force Reserve Command personnel of the 315th AW and 446th AW. A further eight C-17As are assigned to the 97th Air Mobility Wing at Altus AFB, Oklahoma, for training duties. Finally, six will be assigned from 2003 to the Air National Guard's 172nd AW at Jackson, Mississippi. The remaining 10 aircraft are to be

C-17 PROCUREMENT

FY	Lot	Qty	Cum	Line Numbers	First aircraft	Delivery
-	Proto	1	(1)	T1	87-0025	15 Sep 1991
88	1	2	2	P1-P2	88-0265	18 May 1992
89	2	4	6	P3-P6	89-1189	7 Sep 1992
90	3	4	10	P7-P10	90-0532	26 Aug 1993
91	-	-	-	-	-	-
92	4	4	14	P11-P14	92-3291	8 Apr 1994
93	5	6	20	P15-P20	93-0599	29 Sep 1994
94	6	6	26	P21-P26	94-0065	31 Jul 1995
95	7	6	32	P27-P32	95-0102	5 Nov 1996
96	8	8	40	P33-P40	96-0001	29 Aug 1997
97	9	8	48	P41-P48	97-0041	10 Aug 1998
98	10	9	57	P49-P57	98-0049	22 April 1999
99	11	13	70	P58-P70	99-0058	17 March 2000
00	12	15	85	P71-P85	00-0171	Feb/Mar 2001
	12	4 (RAF)	89	UK1-UK4	00-0201/ZZ171	May 2001
01	13	12	101	P86-P97	-	Aug 2002
02	14	15	116	P98-P112	-	Aug 2003
03	15	8	124	P113-P120	-	Aug 2004
Total		**124 + 1**				

Boeing C-17A Globemaster III strategic transport (*Jane's/Paul Jackson*)   *NEW*/0527160

Globemaster III of No. 99 Squadron, Royal Air Force (*Jane's/Paul Jackson*)   *NEW*/0527159

distributed among those units as reserves to cover scheduled depot level maintenance. In the event of further aircraft being ordered, McGuire AFB, New Jersey, is the preferred location for the next C-17 unit.

The first C-17A to be handed over to Air Mobility Command was the final aircraft from Lot 2 (P6/89-1192), which made its first flight on 8 May 1993 and was ferried to Charleston AFB for the 437th AW on 14 June 1993. Subsequent deliveries to Air Mobility Command comprise four Lot 3 aircraft between 26 August 1993 and 8 February 1994; four Lot 4 aircraft between 8 April and 20 August 1994; six Lot 5 aircraft between 28 September 1994 and 19 June 1995; six Lot 6 aircraft between 31 July 1995 and May 1996; six Lot 7 aircraft between August 1996 and June 1997; eight Lot 8 aircraft between August 1997 and May 1998; eight Lot 9 aircraft between August 1998 and February 1999; nine Lot 10 aircraft between April and December 1999, including the 50th C-17A, which was accepted by the USAF on 21 May 1999; and 13 Lot 11 aircraft during 2000. In 2001, USAF received total of 10 Lot 12 aircraft, with further four delivered to United Kingdom. By end of 2001, 80 C-17As had been delivered to the USAF; 15 more were due to be handed over during 2002.

In early 1997, UK Ministry of Defence reportedly considering acquisition of a small number of C-17s for service with RAF in strategic airlift role; no formal requirement then existed, but cost data were received from the USA. Invitation to submit bids for supply on lease basis of up to four 'C-17 equivalent' Short Term Strategic Airlift aircraft issued September 1998, but subsequent criticism by other potential bidders that requirement was weighted in favour of C-17 prompted UK MoD to waive some requirements. Submissions made in early 1999, but competition then suspended in August 1999, when none of competing bids proved acceptable. Restarted, late 1999; announced that C-17 selected on 16 May 2000, with total of four aircraft to be acquired for No. 99 Squadron. These taken from Lot 12 production; assembly of first officially begun 28 June 2000, although lease agreement with Chase Manhattan Bank not signed until 12 January 2001. Lease duration set at seven years, with options for two one-year extensions. Each aircraft valued at £197 million. First aircraft (UK1/ZZ171/00-0201) accepted by RAF on 17 May 2001 at Long Beach; ferried to UK via Charleston AFB and arrived at Brize Norton 23 May. Final aircraft (UK4/ZZ174/00-0204) accepted 24 August 2001 and at Brize Norton by end of August; first operational sortie accomplished in early June; fleet in-service date officially 30 September 2001, at which time UK MoD requested information about availability of further two aircraft and Boeing allocated three further airframes (with provisional US civil registrations) in anticipation of early decision which failed to materialise because of a shortage of funds.

Canada has requirement for new strategic transport and is believed to be examining several options, including lease or outright purchase of C-17; Australia has also begun studies of requirements for strategic airlift aircraft. Version of the C-17 to be offered for Japanese C-X requirement.

COSTS: Multiyear procurement contract for 80 aircraft signed in May 1996 worth US$14.2 billion; separate US$1.6 billion contract followed in December 1996 for 320 F117 engines. Unit price proposed in November 1995 was US$190 million; recent figures (early 2000) refer to unit cost of US$199 million. Boeing proposal for 60 plus C-17s submitted April 2001, at US$152 million unit cost (FY99 dollars); currently under review by USAF.

DESIGN FEATURES: Typical T tail, upswept rear fuselage, high-wing, podded-engine transport, deriving much detail, including externally blown flap system, from McDonnell Douglas YC-15 medium STOL transport prototypes, which involves extended flaps being in exhaust flow from engines during take-off and landing. Combines load-carrying capacity of Lockheed C-5 with STOL

performance of Lockheed Martin C-130; required to operate routinely from 915 m (3,000 ft) long and 27.45 m (90 ft) wide runways, complete 180° three-point turn in 25 m (82 ft) and back-taxi up 1 in 50 gradient when fully loaded using thrust reversers. Structure designed to survive battle damage and protect crew; rear-loading ramp.

Supercritical wing with 25° sweepback; 2.72 m (8.9 ft) high NASA winglets, angled at 15° from vertical and with 30° sweep.

FLYING CONTROLS: First military transport with all-digital FBW control system and two-crew cockpit with central stick controllers; outboard ailerons and four spoilers per wing; four elevator sections; two-surface rudder split into upper and lower segments; full-span leading-edge slats; two-slot, fixed-vane, simple hinged flaps over about two-thirds of trailing-edge; small strakes under tail. Quadruple-redundant BAE Systems digital fly-by-wire flight control system, with mechanical back-up.

STRUCTURE: Major subassemblies produced in factory at Macon, Georgia; some 50 subcontractors, of which 21 for airframe; subcontractors include Vought (composites ailerons, rudder, elevators, vertical and horizontal stabilisers, engine nacelles and thrust reversers), McCoak Metals (wing skins), Contour Aerospace (wing spars and stringers), Kaman Aerospace (wing ribs and bulkheads), Hitco Technologies (tailcone), Heath Tecna (wing-to-fuselage fillet), Aerostructures Hamble (composites flap hinge fairings and trailing-edge panels) and Northwest Composites (main landing gear pod panels). Raytheon was original supplier of composites winglets and landing gear doors, but replaced by Marion Composites with effect from 41st aircraft; Marion Composites also fabricates nose and tail radomes for last 80 aircraft of USAF order. C-17A structure is 69.3 per cent aluminium; 12.3 per cent steel alloys; 10.3 per cent titanium and 8.1 per cent composites. Wings of P1-P10 underwent local strengthening as consequence of load test of 1 October 1992; first rework to P8/90-0533 at McDonnell Douglas, Tulsa, between 3 January and 9 April 1994.

Proposal, early 1994, to design all-composites horizontal tail surfaces for weight-saving resulted in McDonnell Douglas (later Boeing) securing a US$40.7 million contract to build new unit; revised tail manufactured by Northrop Grumman (now Vought) using AS-4 carbon fibre for spars and skins and machined 7075 aluminium for ribs, resulting in assembly said to be 50 per

cent cheaper and 20 per cent lighter than existing tail. Incorporates 2,000 fewer components and 42,000 fewer fasteners. Following testing on prototype, completed April 1999, it was incorporated on production aircraft beginning with P51/98-0051.

Landing gear pod entirely redesigned in 1995, drastically reducing number of parts and fasteners and simplifying process of attachment to aircraft; introduced on first Lot 8 aircraft (P33/96-0001) and expected to save US$88 million over remainder of production programme for USAF. New, lighter and less costly engine nacelle, saving 113 kg (250 lb) per unit, flight tested at Edwards AFB between December 1997 and February 1998; first production aircraft with new nacelles (P41/97-0041) delivered to USAF on 10 August 1998.

LANDING GEAR: Hydraulically retractable tricycle type, with free-fall emergency extension; designed for sink rate of 3.81 m (12 ft 6 in)/s and suitable for operation from paved runways or unpaved strips. Mainwheel units, each consisting of two legs in tandem with three wheels on each leg, rotate 90° to retract into fairings on lower fuselage sides; tyre size 50×21.0-20 (30 ply) tubeless; pressure 9.52 bar (138 lb/sq in). Menasco (Goodrich from 41st aircraft) twin-wheel nose leg retracts forwards; tyre size 40×16-14 (26 ply) tubeless; pressure 10.69 bar (155 lb/sq in). Honeywell wheels and carbon brakes. Minimum ground turning radius at outside mainwheels 17.37 m (57 ft 0 in); minimum taxiway width for three-point turn 27.43 m (90 ft 0 in); wingtip/tailplane clearance 74.24 m (237 ft 0 in).

POWER PLANT: Four Pratt & Whitney F117-PW-100 (PW2040) turbofans, with maximum flat rating of 179.9 kN (40,440 lb st), pylon-mounted in individual underwing pods and each fitted with a directed-flow thrust reverser deployable both in flight and on the ground. With effect from the 20th production aircraft, an improved version of the F117 was adopted, this embodying single-crystal turbine blade technology, a supercharged compressor and enhanced thermal barrier coatings; benefits include a 20 per cent reduction in cost of maintenance, increased reliability and slightly better sfc. Provision for in-flight refuelling. Two outboard wing fuel tanks of 21,210 litres (5,603 US gallons; 4,665 Imp gallons) each; two inboard wing fuel tanks of 30,056 litres (7,940 US gallons; 6,611 Imp gallons) each; total capacity 102,532 litres (27,086 US gallons; 22,554 Imp gallons).

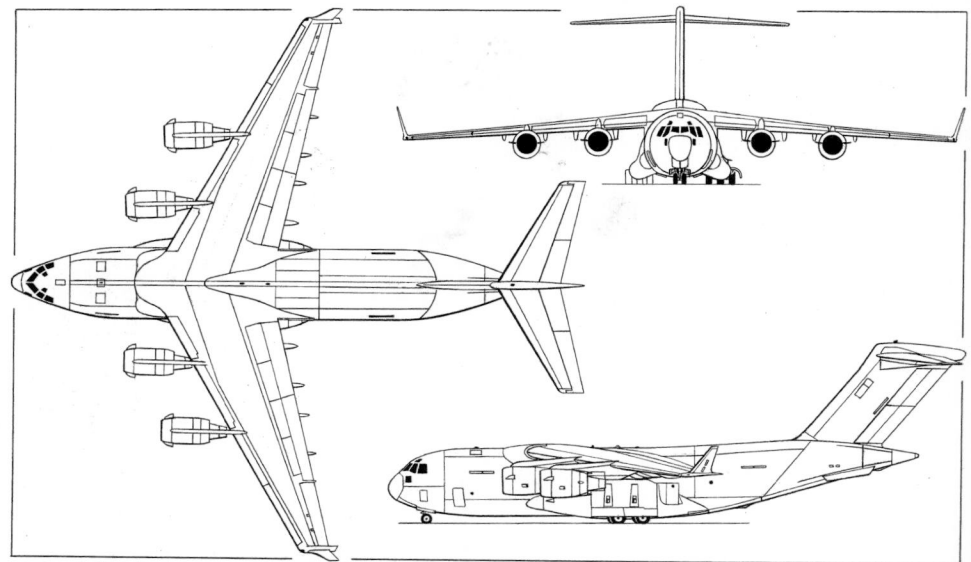

Boeing C-17A Globemaster III long-range heavy cargo transport (*Jane's/Dennis Punnett*)

BAE Systems fuel pumps. With effect from the 71st (first Lot 12) aircraft, an extended-range fuel tank containment system (ERFCS) was adopted; this converts wing dry bay into additional fuel tank containing approximately 36,339 litres (9,600 US gallons; 7,994 Imp gallons) of fuel, thus increasing range with 18,144 kg (40,000 lb) payload by 900 n miles (1,667 km; 1,036 miles). ERFCS also installed on four UK aircraft.

ACCOMMODATION: Normal flight crew of pilot and co-pilot side by side and two observer positions on flight deck, plus loadmaster station at forward end of main floor; access to flight deck via downward-opening airstair door on port side of lower forward fuselage. Bunks for crew immediately aft of flight deck area; crew comfort station at forward end of cargo hold.

Main cargo hold able to accommodate US Army wheeled and tracked vehicles up to M1 main battle tank, including 5 ton expandable vans in two rows, or up to three AH-64 Apache attack helicopters, with loading via hydraulically actuated rear-loading ramp which forms underside of rear fuselage when retracted. Aircraft fitted with 27 stowable tip-up seats along each sidewall and another 48 seats carried on board which can be erected along the centreline; optionally up to 36 litters for medical evacuation mission or up to 90 passengers on 10-passenger pallets in addition to 54 sidewall seats. Air delivery system capability for nine 463L pallets plus two on ramp in single row; or logistics handling system for 18 463L pallets in double row. Airdrop capability includes single platforms of up to 27,215 kg (60,000 lb), multiple platforms of up to 54,430 kg (120,000 lb), or up to 102 paratroops; aircraft originally configured for single-row airdrop, but dual-row capability operationally certified in 1998 and is standard feature of 51st and subsequent C-17s, as well as being retrofitted to earlier aircraft. Equipped for low-altitude parachute extraction system (LAPES) drops.

Cargo handling system includes rails for airdrops and rails/rollers for normal cargo handling. Each row of rails/rollers can be converted quickly by a single loadmaster from one configuration to the other. Total of 295 cargo tiedown rings, each stressed for 11,340 kg (25,000 lb), all over cargo floor forming grid averaging 74 cm (29 in) square. Three quick-erecting litter stanchions, each supporting three litters, permanently carried. Main access to cargo hold is via rear-loading ramp, which is itself stressed for 18,145 kg (40,000 lb) of cargo in flight. Underfuselage door aft of ramp moves upward inside fuselage to facilitate loading and unloading. Paratroop door at rear on each side; three overhead FEDS (flotation equipment deployment system) escape hatches and liferaft.

SYSTEMS: Include Honeywell computer-controlled integrated environmental control system and cabin pressure control system; 2,440 m (8,000 ft) equivalent cabin pressure up to 11,280 m (37,000 ft); quad-redundant flight control and four independent 276 bar (4,000 lb/sq in) hydraulic systems; independent fuel feed systems; electrical system; Honeywell GTCP331-250G APU (at front of starboard landing gear pod), provides auxiliary power for environmental control system, engine starting, and on-ground electronics requirements; onboard inert gas generating system (OBIGGS) for the explosion protection system, pressurised by engine bleed air at 4.14 bar (60 lb/sq in) to produce NEA (nitrogen-enriched air) and governed by a Parker-Gull system controller; fire suppression system; smoke detection systems. Separate OBIGGS for ERFCS-configured aircraft. All phases of cargo operation and configuration change capable of being handled by one loadmaster.

Electrical system includes single 90 kVA generator per engine and an APU, providing 115/200 V, three-phase, 400 Hz AC power; four 200 A transformer-rectifiers providing 28 V DC; single-phase, 1,000 VA inverter for ground refuelling and emergency AC; and two 40 Ah Ni/Cd batteries for APU starting and emergency DC. Aeromedical equipment provided with 60 Hz power.

AVIONICS: *Comms:* Telephonics Corporation IRMS integrated radio management system; Rockwell Collins AN/ARC-210 secure radio (from P50/98-0050, replacing Raytheon AN/ARC-187 plus Honeywell KY 58 encoder); satcom; VHF-AM/FM; HF; secure voice and jam-resistant UHF/VHF-FM intercom; IFF/SIF; en route army/marine UHF LOS and satcom hook-up; cockpit voice recorder; crash position indicator. Automatic communications processor and multiband radio in new-build aircraft starting with P50/98-0050 and retrofitted to earlier C-17As. In December 1998, Rockwell Collins selected by Boeing to provide SAT-2000 Aero-1 satcom systems and CMU-900 communication management units for future global air traffic management (GATM) environment; initial order for support and preproduction hardware, with options to install equipment on all 120 aircraft. Production line GATM introduction at P71/00-0171, involving TCAS, Mode S IFF, ADS-A (automatic dependent surveillance) and Inmarsat Aero I satcom, plus liquid crystal displays replacing all head-down CRTs. GATM augmented from P97 by HF datalink and RNP 4 (required navigation performance) measures. Under development in 1999 were two SOLL II (special operations low-level) roll-on/roll-off communications suites for use on specific missions; and carry-on radio suite for use on special forces flights, beginning in 2002.

Radar: Honeywell AN/APS-133(V) weather/mapping radar; this will be replaced by new unit identical to that selected for C-130 Hercules as part of the Avionics Modernization Program. New radar to be fitted from outset to all aircraft of planned follow-on order and will be retrospectively installed on older C-17s as their existing systems become inoperative.

Flight: Three Delco Electronics mission computers with MDC software and electronic control system on P1 to P40; beginning with P41/97-0041, two Lockheed Martin core integrated processors (CIPs) replaced Delco Electronics equipment and retrofitted on earlier aircraft by September 1998; additional memory added from P50/98-0050 onwards. Hamilton Sundstrand aircraft and propulsion data management computer; Honeywell dual air data computers; Litton warning and caution system; master warning system provides aural and voice alerts plus visual alerts on glareshields; General Dynamics automatic test equipment, and support equipment data acquisition and control system; GPS and four Honeywell inertial reference units, although new, embedded GPS INS introduced from P50/98-0050; VOR/DME; Tacan; ILS/marker beacon; UHF-DF; ground proximity warning system; radar altimeter; flight data recorder; flight plan entry manually or via laptop computer. BAE Systems terrain awareness warning system (TAWS) to be installed in conjunction with Lockheed Martin Control Systems video integrated processor (VIP), currently under development, from P86 onwards; TAWS underwent initial flight testing from Edwards AFB in 2001-02 and will complete further trials in first half of 2002.

Instrumentation: Advanced digital avionics and four Honeywell full-colour cathode ray tube (CRT) displays; new Honeywell 152 × 152 mm (6 × 6 in) AMLCD colour multifunction displays (MFDs) in 71st and subsequent production aircraft; two BAE Systems full flight regime foldable head-up displays.

Mission: Litton integrated mission and communications keyboards (MCKs) and displays (MCDs); Sierra Technologies AN/APN-243A(V) station-keeping equipment (SKE 2000) fitted as standard commencing with P33/96-0001, retrofitted to earlier aircraft and upgraded from P86 (first Lot 13) onwards to increase capacity from 17 to 100 aircraft. Loadmaster's laptop computer from P41/97-0041 onwards; this replaced from P58/99-0058 by Dolch aircrew data device.

Self-defence: Development of defensive electronic systems completed in 1994. ATK, Lockheed Martin or Honeywell AN/AAR-47 missile approach warning system and associated BAE Systems AN/ALE-47 automatic flare dispenser installed on five aircraft; USAF to adopt this equipment across the fleet. USAF seeking to counter IR homing missiles with laser jamming system and has plans to deploy these on C-17 from FY03 onward.

EQUIPMENT: Removable crew armour can be fitted around flight deck and loadmaster's area.

DIMENSIONS, EXTERNAL:

Wing span: wings only	50.29 m (165 ft 0 in)
at winglet tips	51.74 m (169 ft 9 in)
Wing aspect ratio	7.2
Length: overall	53.04 m (174 ft 0 in)
fuselage	48.49 m (159 ft 1¼ in)
Height: to flight deck roof	7.34 m (24 ft 1 in)
to winglet tips	6.93 m (22 ft 9 in)
overall	16.79 m (55 ft 1 in)
Fuselage diameter	6.85 m (22 ft 6 in)
Tailplane span	19.81 m (65 ft 0 in)
Wheel track	10.26 m (33 ft 8 in)
Wheelbase: to front mainwheel	17.60 m (57 ft 8¾ in)
to rear mainwheel	20.05 m (65 ft 9½ in)
Height to sill	1.63 m (5 ft 4 in)
Ground clearance under engine pods:	
inboard	2.71 m (8 ft 10½ in)
outboard	2.35 m (7 ft 8½ in)
Distance between fuselage centreline and engine	
centreline: inboard	7.44 m (24 ft 5 in)
outboard	13.94 m (45 ft 9 in)
Height of winglets	2.72 m (8 ft 11 in)

DIMENSIONS, INTERNAL:

Cargo compartment:	
Length, incl 6.05 m (19 ft 10 in) rear-loading ramp	26.82 m (88 ft 0 in)
Loadable width	5.49 m (18 ft 0 in)
Max height: under wing	3.96 m (13 ft 0 in)
aft of wing	4.50 m (14 ft 9 in)
Volume	591.8 m³ (20,900 cu ft)

AREAS:

Wings, gross	353.03 m² (3,800.0 sq ft)
Winglets (total)	3.33 m² (35.85 sq ft)
Ailerons (total)	11.83 m² (127.34 sq ft)
Tailplane	78.50 m² (845.00 sq ft)

WEIGHTS AND LOADINGS:

Operating weight empty:	
non-ERFCS aircraft	125,645 kg (277,000 lb)
ERFCS aircraft	127,685 kg (281,500 lb)
Max payload (2.5 *g* load factor):	
non-ERFCS aircraft	76,655 kg (169,000 lb)
ERFCS aircraft	75,250 kg (165,900 lb)
Max weight on rear-loading ramp	18,143 kg (40,000 lb)
Max usable fuel weight:	
non-ERFCS aircraft	82,125 kg (181,054 lb)
ERFCS aircraft	110,990 kg (244,688 lb)
Max T-O weight: both	265,350 kg (585,000 lb)
Max ramp weight: both	265,800 kg (586,000 lb)
Max wing loading	751.6 kg/m² (153.95 lb/sq ft)
Max power loading	366 kg/kN (3.59 lb/lb st)

PERFORMANCE:

Normal cruising speed at 8,535 m (28,000 ft) M0.74-0.77
Max cruising speed at low altitude
350 kt (648 km/h; 403 mph) CAS
Airdrop speed: at S/L
115-250 kt (213-463 km/h; 132-288 mph) CAS
at 7,620 m (25,000 ft)
130-250 kt (241-463 km/h; 150-288 mph) CAS
Approach speed with max payload
115 kt (213 km/h; 132 mph) CAS
Service ceiling 13,715 m (45,000 ft)
Runway LCN (paved surface) better than 49
T-O field length at MTOW 2,360 m (7,740 ft)
Landing field length with 72,575 kg (160,000 lb) payload, using thrust reversal 915 m (3,000 ft)
Range with payloads indicated, with no in-flight refuelling:
18,144 kg (40,000 lb) payload:
non-ERFCS aircraft
4,400 n miles (8,148 km; 5,063 miles)
ERFCS aircraft
5,300 n miles (9,815 km; 6,099 miles)
72,575 kg (160,000 lb), T-O in 2,286 m (7,500 ft), land in 915 m (3,000 ft), load factor of 2.25 *g*:
both: 2,400 n miles (4,444 km; 2,761 miles)
self-ferry (zero payload), T-O in 1,128 m (3,700 ft), land in 701 m (2,300 ft), load factor of 2.5 *g*:
non-ERFCS aircraft
4,700 n miles (8,704 km; 5,408 miles)
ERFCS aircraft
6,250 n miles (11,575 km; 7,192 miles)

UPDATED

BOEING 737 AEW&C

TYPE: Airborne early warning and control system.

PROGRAMME: Adaptation of Boeing Business Jet (BBJ, which combines 737-700 fuselage with strengthened wing and landing gear of 737-800). Additional features include extra fuel tanks in former baggage hold. Proposed for Australian Project Air 5077 Wedgetail by Boeing, Northrop Grumman Electronic Sensor Systems Division (ESSD) and BAE Australia. Competed against proposals from Lockheed Martin and Raytheon. Secured initial design activity contract worth about US$6 million in December 1997, paving way for submission of full tenders in early 1999. Selection of Boeing submission officially announced on 21 July 1999. RAAF initially planned to acquire seven aircraft for US$1.32 billion; cost concerns resulted in

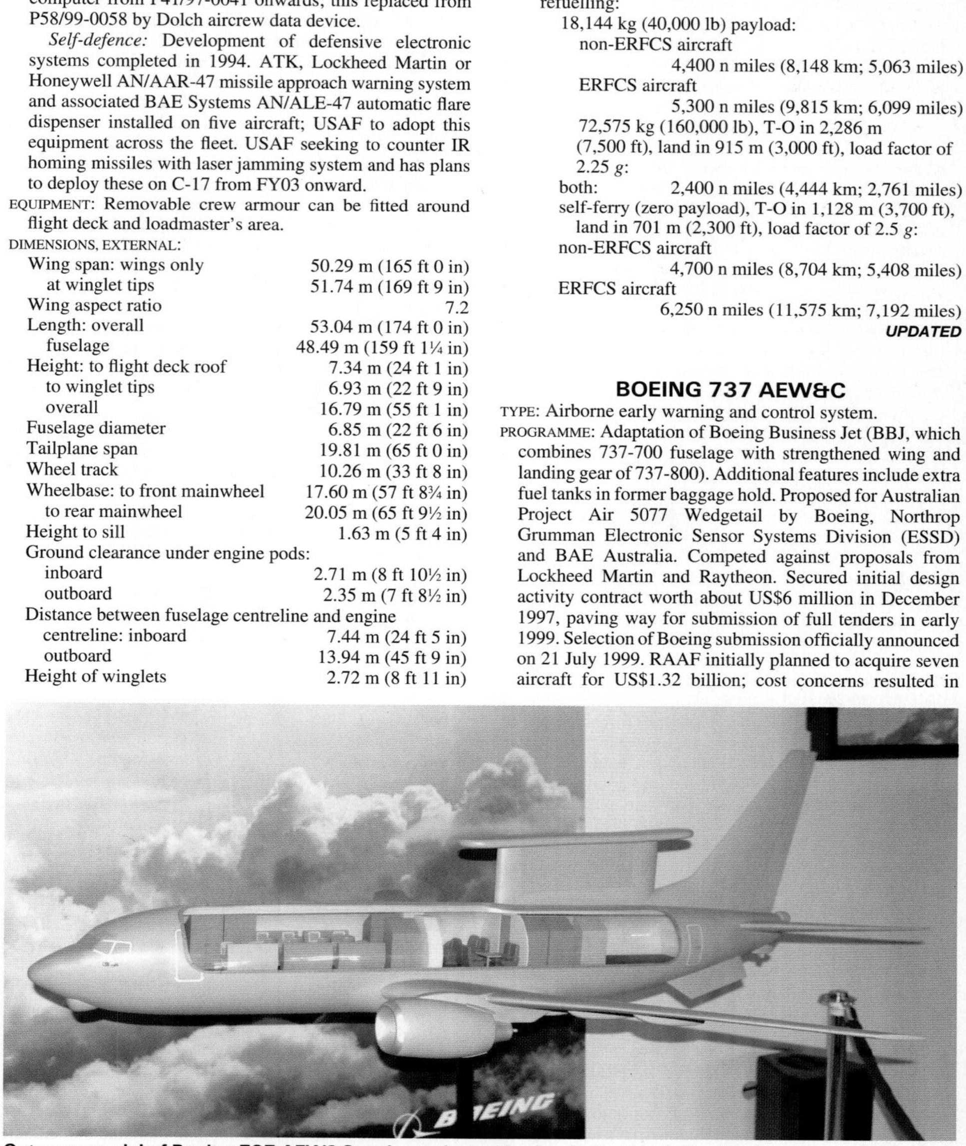

Cutaway model of Boeing 737 AEW&C variant adopted by Australia ***(Jane's/Paul Jackson)*** ***NEW***/0530173

reduction to six plus one option in May 2000, but when contract signed on 20 December, this had further reduced to four and three options, at reported cost of US$1.65 billion. Delivery of first two aircraft scheduled for the end of 2006, with operating unit to be No. 2 Squadron. Second pair to be handed over at end of 2007. Main base at Williamtown, New South Wales, with two aircraft permanently deployed to forward operating location at Tindal, Northern Territories; IOC to be achieved at end of 2008. Preliminary design review of radar and IFF systems successfully completed in September 2001; further reviews of navigation system, mission computing hardware and other airborne mission systems undertaken by January 2002, with ground-based elements not scheduled to undergo PDR until 2004.

ESSD L-band multirole electronically scanned array (MESA) radar mounted above rear fuselage ('top hat' configuration) providing 360° coverage from stationary antenna 10.7 m (35 ft) long and 2.35 m (7 ft 9 in) high. Operating modes will include acute long-range or broad short-range scanning and track-while-scan; maximum detection range said to exceed 216 n miles (400 km; 249 miles). Initial radar to be installed in Boeing 737 in 2003 for testing, with modification of this and remaining three aircraft to be undertaken in Wichita, Kansas. In late 1998, demonstrator BBJ temporarily fitted with full-scale replica of MESA radar, six operator consoles and equipment cabinets for inspection by Australian defence department officials; mockup also featured in-flight refuelling probe above cockpit and EW/ECM sensors. Definitive aircraft will, however, feature 10 operator consoles initially, with potential to add further two. Aerodynamic effects of MESA offset by two large strakes below rear fuselage. Electronic warfare self-protection (EWSP) system will include Northrop Grumman AN/AAQ-24(V) directed infra-red countermeasures system, plus chaff and flares; Elta providing advanced ESM/elint systems; these will be controlled by the ALR-2001 computer. Other mission equipment to include Link 11, JTIDS, Mode S IFF and satcom; flight deck tactical displays, three HF and eight VHF/UHF radios. Patrol endurance of 8 hours at 300 n miles (555 km; 345 miles) from base can be extended by airborne refuelling, with modification to include installation of flying boom receptacle and a removable probe. Maximum T-O weight 77,565 kg (171,000 lb); service ceiling 12,500 m (41,000 ft).

Republic of Korea interested in AEW-configured 737 as less costly solution to E-X requirement in lieu of Boeing 767, which considered too expensive; Boeing proposal in competition with rival offerings from Raytheon and Thales, both of which have Airbus A321 as platform. After studying proposals for AEW aircraft involving Airbus A310/Phalcon and Boeing 737/MESA combinations, in early December 2000 Turkey announced selection of latter and revealed intent to obtain total of six aircraft (with option on two more) at cost of US$1.5 billion; these will include some indigenous equipment. Contract signature was expected in early 2001, but negotiations continued throughout remainder of 2001; contract was finally signed by Turkish prime minister on 16 May 2002, at which time the number of aircraft to be purchased had been reduced to four (with two on option) because of concerns over the economy.

Boeing forecasting potential sales of up to 50 737 AEW&C aircraft, with other possible customers including Chile, Egypt, Italy, Malaysia, Singapore, Spain and Thailand.

UPDATED

BOEING 737 MMA

TYPE: Maritime surveillance twinjet.

PROGRAMME: Design study for 737 MMA (multimission maritime aircraft) revealed 18 April 2000 as potential replacement for Lockheed Martin P-3C Orion and EP-3E Aries II. Based on C-40A (see Boeing 737 entry under Boeing Commercial Airplanes), combining 737-800 wing and 737-700 fuselage, latter with internal weapons bay beneath forward section and former with hardpoints for air-to-surface missiles. Full range of maritime patrol equipment envisaged, including enlarged nose accommodating search radar, up to seven operators' consoles and rotary sonobuoy launcher, plus additional fuel in aft baggage hold.

Maximum T-O weight 77,565 kg (171,000 lb); maximum transit speed 490 kt (907 km/h; 564 mph); mission radius 2,000 n miles (3,704 km; 2,301 miles). US Navy awarded US$493,000 concept exploration study contract to Boeing in July 2000; one of four contracts awarded to industry, results of five month studies will be used by Center for Naval Analysis in selection of preferred concept. No recent news received.

UPDATED

BOEING 767 MILITARY VERSIONS

TYPE: Tanker-transport; airborne ground surveillance system.

PROGRAMME: Variants of Boeing 767 twin-turbofan airliner (which see under Boeing Commercial Airplanes for technical description).

Boeing 767 tanker-transport, as offered to the Royal Air Force by Tanker Transport Services Company *(TTSC)* ***NEW***/0528622

CURRENT VERSIONS: **767 AWACS:** Described in 2001-02 and previous editions.

KC-767 Global Tanker Transport Aircraft (GTTA): Tanker-transport version announced by Boeing, February 1995, in anticipation of Japanese order; Boeing in discussions with Kawasaki by mid-1996, concerning co-operative venture to offer KC-767 to Japan ASDF. Kawasaki involvement then expected to take form of post-production work (including fitment of boom refuelling gear, extra tanks and associated plumbing) on 767-300 derivative. JASDF intended to request funding from FY99, but economic downturn and reductions in defence budget caused delay. Purchase intention reaffirmed on 14 December 2001, by Japanese government, which announced plan to buy four aircraft for delivery from 2006.

Version of 767 with three hose drum units offered to UK Royal Air Force which has FSTA requirement for a new tanker aircraft to replace VC-10 and Tristar from about 2007 onwards. Total of 20 to 30 aircraft needed, with six consortia invited in September 1999 to tender submissions for a public-private finance initiative to satisfy requirement. Combinations had reduced this to four by late 1999, but only two consortia submitted proposals on 3 July 2001, of which one (Tanker & Transport Service Company) offering 767 solution based on use of ex-British Airways 767-300s. Selection of successful bid expected in mid-2002, with contract award following in 2003.

Tanker-transport proposals most recently based on 767-200ER (see Boeing Commercial Airplanes entry); fuel dispensed through 'flying boom' and two underwing pods; boom remotely controlled from cabin, assisted by CCTV; proximity trials at Patuxent River NAS in June 2000 showed 767 to be stable refuelling platform. Boeing offers first production aircraft 40 months from go-ahead. Fuel capacity 91,380 litres (24,140 US gallons; 20,101 Imp gallons), in standard (wing) tanks, plus 21,198 litres (5,600 US gallons; 4,663 Imp gallons) in supplementary underfloor tanks; total 112,578 litres (29,740 US gallons; 24,764 Imp gallons). As freighter (side cargo door and reinforced floor), can carry up to 18 463L pallets on main deck or 216 passengers, with additional cargo capacity on lower deck dependent upon auxiliary tank configuration. Version offered to RAF, with HDU in place of 'flying boom', has underfloor fuel of 29,942 litres (7,910 US gallons; 6,586 Imp gallons).

First Boeing 767 tankers will be flown by Italian Air Force, which revealed intention in July 2001 to purchase four new-build aircraft as replacements for current Boeing 707 tankers. Delivery will be accomplished during 2005-06 and total cost, including option on two additional aircraft, is about US$618 million. Three Smiths Aerospace hose drum units will be installed as standard, but consideration is being given to an optional centreline boom with a remote aerial refuelling operator station.

A derivative of the Boeing 767 is also a candidate to replace the USAF KC-135 Stratotanker. In 2001, proposal to lease initial tranche of 100 aircraft emerged, with this envisaging a 10-year period, after which the USAF could negotiate outright purchase; this won more support after September 2001 terrorist attack on USA, and USAF formally notified Congress in April 2002 of intention to begin negotiations. If deal is consummated, USAF anticipates receiving first aircraft in 2004-05 and is likely to accept about 20 per year; longer-term goal is to acquire up to 500 new tankers, allowing eventual retirement of veteran KC-135. Initial lease agreement cost projections are estimated to be US$26 billion.

Boeing formed 767 Tanker Programs office in March 2001 and subsequently selected its Wichita, Kansas facility as the centre for tanker modification work.

767 Multimission Command and Control Aircraft (MC2A): Boeing 767-400 version chosen by USAF in second quarter of 2002 to be platform for next-generation MC2A, which expected to replace RC-135 Rivet Joint, E-8 Joint STARS and E-3 Sentry AWACS from 2012 onwards. An initial batch of four operational MC2As is likely to be the forerunner of up to 50 systems that will incorporate Northrop Grumman Multi-Platform Radar Technology Insertion Program J-band radar, which is to merge moving target indicator over synthetic aperture radar to result in enhanced ground moving target indicator capability; subsequently, a dedicated air moving target indicator will be adopted, permitting retirement of the E-3 Sentry. Trials aimed at validating this new concept of intelligence gathering being undertaken with a Boeing 707 that flew for the first time in modified form on 18 April 2002; known as *Paul Revere*, this aircraft performed a series of operational test flights in mid-2002, including some as part of the joint expeditionary forces experiment at Nellis AFB, Nevada. Key objectives to be addressed by *Paul Revere* comprise directing activities of manned and unmanned air and space sensor systems used for surveillance and reconnaissance; integration and monitoring of all-source sensor data; location, identification, designation and tracking of targets using multisensor integration software techniques; undertake weapon-target matching; establishment of tactical priorities and objectives when engaging time-sensitive targets and undertake rapid and timely bomb damage assessment following an attack.

COSTS: Estimated US$216 million for Japanese 767 tanker (2001). Boeing price varies from US$150 million to US$225 million, according to quantity procured.

Data follow for Tanker-Transport.

WEIGHTS AND LOADINGS (estimated):

Operational weight, empty	90,720 kg (200,000 lb)
Max T-O weight	179,170 kg (395,000 lb)
Max ramp weight	179,625 kg (396,000 lb)
Max zero-fuel weight	117,934 kg (260,000 lb)

UPDATED

BOEING AL-1A

TYPE: Missile defence system.

PROGRAMME: Prototype YAL-1A (USAF serial number 00-0001) ordered on 12 November 1996; purchase of 747-400F airframe from Boeing confirmed on 30 January 1998. Formal authority to proceed received on 26 June 1998, after TRW successfully demonstrated laser firing and missile tracking. First metal cut on YAL-1A on 10 August 1999; rolled out at Everett on 12 December 1999; first flight 6 January 2000; delivered to Boeing Wichita on 21 January 2000 for outfitting with strengthened floor and modifications to take nose-mounted laser turret; subsequent critical design review completed in late April. On completion of 1.6 million man-hours of work at Wichita, aircraft was originally expected to fly in February 2002, then be ferried to Edwards AFB, California, for installation of six-module COIL laser from May 2002, followed by trials against various missiles, with PDRR phase culminating with demonstration against a ballistic missile fired from Vandenberg AFB, California, in September 2003; however, by early 2002, a combination of changed defence priorities and technical issues had combined to cause delay in programme and it had been confirmed that the ballistic missile shoot-down demonstration had been postponed to 2004; 30-month EMD phase expected to begin in early 2004, with IOC of first three aircraft set for late 2007 and full operational capability (seven aircraft) in 2009, programme now being one year late. Maiden flight eventually achieved on 18 July 2002. Funding for modification of second, EMD, aircraft being sought in FY03 budget.

CUSTOMERS: US Air Force (seven required).

COSTS: Programme cost (1997, estimated) US$5 billion for one concept prototype and one EMD (both eventually to be fully upgraded) and five production aircraft. Initial

Prototype YAL-1A taking off for the first time following equipment installation (*US DoD*)   NEW/0525360

programme definition and risk reduction (PDRR) contract of November 1996 valued at US$1.1 billion. US$598 million requested for FY03, including US$85 million for long-lead items for second aircraft.

DESIGN FEATURES: Based on airframe of Boeing 747-400F (which see under Boeing Commercial Airplanes). Equipped with TRW multihundred kW chemical oxygen iodine laser (COIL) and Lockheed Martin beam control/fire-control system, for target acquisition, plus aiming and firing of laser. Active Ranger System is housed in pod sited above forward fuselage section. Intended to destroy theatre ballistic missiles during their boost phase, but with additional capability against low-flying cruise missiles; other potential applications began emerging in 1998, these including protection of high-value aircraft such as AWACS and Joint STARS by destroying SAMs and AAMs; imaging and reconnaissance with aid of optical telescope; SEAD, by combining intelligence data to engage hostile missile sites and radar control systems; and command and control through search/detection of infrared signatures to aid cueing of other weapons. Laser able to fire 20 to 30 times per mission; titanium to be used instead of aluminium in certain areas to protect against heat damage to undersides caused by laser exhaust gases. Mission crew to comprise four specialists at individual consoles at rear of forward compartment, specifically the mission crew commander, airborne surveillance officer, weapon system operator and special equipment operator; standard flight deck crew of two could be augmented for long missions.

UPDATED

BOEING 114 and 414

US Army designations: CH-47 and MH-47 Chinook
Royal Air Force designations: Chinook HC. Mk 2, HC. Mk 2A and HC. Mk 3
Spanish Army designation: HT.17
JASDF and JGSDF designations: CH-47J and CH-47JA

TYPE: Medium-lift helicopter.

PROGRAMME: Design of all-weather medium transport helicopter for US Army began 1956; first flight of YCH-47A 21 September 1961. Performance increased in CH-47C by uprated transmissions and 2,796 kW (3,750 shp) T55-L-11A; internal fuel capacity increased; first flight 14 October 1967; delivered to US Army from 1968.

CURRENT VERSIONS: **CH-47D:** US Army contract to modify one each of CH-47A, B and C to prototype Ds placed 1976; first flight 11 May 1979; first production contract October 1980; first flight 26 February 1982; first delivery 31 March 1982; initial operational capability (IOC) achieved 28

February 1984 with 101st Airborne Division; first multiyear production contract for 240 CH-47Ds awarded 8 April 1985; second multiyear production contract for 144 CH-47Ds (including 11 MH-47Es) awarded 13 January 1989, bringing total CH-47D (and MH-47E) ordered to 472; two Gulf War attrition replacements authorised August 1992 (these new-build); seven ex-Australian rebuilds funded June 1993 for delivery January to November 1995. Additional new-build CH-47D ordered for US Army in June 1999 was delivered to Fort Hood, Texas, on 19 June 2002.

CH-47D update included strip down to bare airframe, repair and refurbish, fit Honeywell T55-L-712 turboshafts, uprated transmissions with integral lubrication and cooling, composite rotor blades, new flight deck compatible with night vision goggles (NVG), new redundant electrical system, modular hydraulic system, advanced automatic flight control system, improved avionics and survivability equipment, Solar T62-T-2B APU operating hydraulic and electrical systems through accessory gear drive, single-point pressure refuelling, and triple external cargo hooks. Principal external change is large, rectangular air intake in leading-edge of rear sail. Composites account for 10 to 15 per cent of structure. About 300 suppliers involved.

At maximum gross weight of 22,680 kg (50,000 lb), CH-47D has more than double useful load of CH-47A. Sample loads include M198 towed 155 mm howitzer, 32 rounds of ammunition and 11-man crew, making internal/external load of 9,980 kg (22,000 lb); D5 caterpillar bulldozer weighing 11,225 kg (24,750 lb) on centre cargo hook; US Army Milvan supply containers carried at up to 130 kt (256 km/h; 159 mph); up to seven 1,893 litre (500 US gallon, 416 Imp gallon), 1,587 kg (3,500 lb) rubber fuel blivets carried on three hooks.

MH-47D Special Operations Aircraft: Element of 160th Special Operations Aviation Regiment (at Hunter AAF, Georgia) equipped with 11 CH-47Ds modified to **MH-47D SOA** standard with refuelling probes (first refuelling July 1988), thermal imagers, Honeywell RDR-1300 weather radar, Rockwell Collins 'glass cockpits', improved communications and navigation equipment, and two pintle-mounted 7.62 mm six-barrel miniguns. Navigator/commander's station also fitted. It is intended to update all MH-47Ds to MH-47G configuration, incorporating technology improvements developed for the CH-47F.

GCH-47D: Limited number of Chinooks grounded for maintenance training at Fort Eustis, Virginia.

JCH-47D: At least two aircraft (84-24159 and 90-0180) modifed for temporary test tasks; both current in April 2001.

US ARMY CH-47D PROCUREMENT

FY	Qty	First aircraft	
81	9	81-23381	
82	19	82-23762	
83	24	83-24102	
84	36*	84-24152	
85	48	85-24322	
86	48	86-1635	
87	48	87-0069	MYP1
88	48†	88-0062	
89	48	89-0130	
90	48¹	90-0180	
91	48²	91-0230	MYP2
92	48³	92-0280	
	2⁴	92-0367	
93	7⁵	93-0928	
98	1⁶	98-2000	
Total	**482**		

* One crashed on test
† One to YMH-47E

Notes: All except one prototype issued with new serial numbers on remanufacture to CH-47D
MYP1 First multiyear procurement contract, 8 April 1985
MYP2 Second MYP, 13 January 1989
¹ This batch includes one aircraft subsequently further modified to MH-47E configuration under separate contract
² Total includes six which were updated to CH-47D standard and then further modified to MH-47E under separate contract
³ This batch includes two of original three prototypes (remaining prototype used as maintenance trainer at Fort Eustis, Virginia) and 11 from Italian production; total includes four which were updated to CH-47D standard and then further modified to MH-47E under separate contract
⁴ New-build Gulf War attrition replacements, authorised 2 August 1992
⁵ Former Australian Air Force CH-47Cs; contract 2 June 1993
⁶ New-build aircraft; delivered 19 June 2002

MH-47E: Special Forces variant; planned procurement 51, all originally to be deducted from total 472 CH-47D conversions, but only 25 initially received, including 14 not covered by multiyear production contract, but excluding one currently under conversion after use as VRTA testbed; prototype development contract 2 December 1987; long-lead items for next 11 helicopters authorised 14 July 1989; firm order for 11, plus option on next 14, awarded 30 June 1991; Lot 2 (14 helicopters) confirmed 23 June 1992. Prototype (88-0267) flew 1 June 1990; delivered 10 May 1991; initial production aircraft (90-0414) flown 1991; first 11 (of 24 intended) originally due to be delivered from November 1992 to 2 Battalion of 160th Special Operations Aviation Regiment at Fort Campbell, Kentucky. Following mission software problems, deliveries began January 1994 with 91-0498 to Fort Campbell; last of original 26 (including prototype) received at Fort Campbell April 1995. The 27th conversion (previously 86-1678) is expected to enter service by early 2003, this being a replacement for an earlier loss; at least three had been destroyed by April 2002, when it was revealed that US Special Operations Command was contemplating purchase of six CH-47Ds, from a Singaporean order, as interim attrition replacements.

Mission profile 5½ hour covert deep penetration over 300 n mile (560 km; 345 mile) radius in adverse weather, day or night, all terrain with 90 per cent success probability. Requirements include self-deployment to Europe in stages of up to 1,200 n miles (2,222 km; 1,381 miles), 44-troop capacity, powerful defensive weapons and ECM. Equipment includes IBM-Honeywell integrated avionics with four-screen NVG-compatible EFIS; dual MIL-STD-1553B digital databusses; AN/ASN-145 AHRS; jamming-resistant radios; Rockwell Collins CP1516-ASQ automatic target handoff system; inertial AN/ASN-137 Doppler, Rockwell Collins AN/ASN-149(V)2 GPS receiver and terrain-referenced positioning navigation systems; Rockwell Collins ADF-149; laser (Raytheon AN/AVR-2), radar (Lockheed Martin AN/APR-39A) and missile (Honeywell AN/AAR-47) warning systems; ITT AN/ALQ-136(V) pulse jammer and Northrop Grumman AN/ALQ-162 CW jammer; BAE Systems M-130 chaff/flare dispensers; Raytheon AN/APQ-174A radar with modes for terrain-following down to 30 m (100 ft), terrain-avoidance, air-to-ground ranging and ground-mapping; Raytheon AN/AAQ-16 FLIR in chin turret; digital moving map display; Elbit ANVIS-7 night vision goggles; uprated T55-L-714 turboshafts with FADEC; increased fuel capacity; additional troop seating (44 maximum); OBOGS; rotor brake; 272 kg (600 lb) rescue hoist with 61 m (200 ft) usable cable; two six-barrel miniguns in forward port and starboard positions (also on MH-47D); provisions for Stinger AAMs using FLIR for sighting. This system largely common with equivalent Sikorsky MH-60K (which see).

MH-47E has nose of Commercial Chinook to allow for weather radar, if needed; forward landing gear moved 1.02 m (3 ft 4 in) forward to allow for all-composites

Boeing MH-47E Chinook of 160th SOAR, detached to South Korea (*Peter R Foster*)   NEW/0525078

external fuel pods (also from Commercial Chinook) that double fuel capacity; Brooks & Perkins internal cargo handling system. Plans are under way to update 24 surviving MH-47Es to MH-47G configuration to improve performance and extend service life. See also Chinook HC. Mk 3, below.

US ARMY MH-47E PROCUREMENT

FY	Qty	First aircraft
90	1	90-0414
91	6	91-0496
92	4	92-0400
	14	92-0464
01	1	(86-1678)
Total	**26**	

Notes: First three batches included in MYP2 (see CH-47D Procurement); final batch is attrition replacement Excludes prototype (88-0267)

Chinook HC. Mk 2/2A: RAF version; Mk 1 designation CH47-352; all survivors of original 41 HC. Mk 1s upgraded to HC. Mk 1B; UK MoD authorised Boeing to update 33 (later reduced to 32) Mk 1Bs to Mk 2, equivalent to CH-47D, October 1989; changes include new automatic flight control system, updated modular hydraulics, T55-L-712F power plants with FADEC, stronger transmission, improved Solar 71 kW (95 shp) T62-T-2B APU, airframe reinforcements, low IR paint scheme, fuel system and standardisation of defensive aids package (IR jammers, chaff/flare dispensers, missile approach warning and machine gun mountings). Smiths Industries Health and Usage Monitoring System (HUMS) installed on fleet-wide basis at cost of about £100 million during 1997 to 2000. Requirement exists for FLIR.

Conversion continued from 1991 to July 1995. Chinook HC. Mk 1B ZA718 began flight testing Chandler Evans/ Hawker Siddeley dual-channel FADEC system for Mk 2 in October 1989. Same helicopter to Boeing, March 1991; rolled out as first Mk 2 19 January 1993; arrived RAF Odiham 20 May 1993; C(A) clearance November 1993. Initial deliveries pooled at Odiham by Nos 7 and 27 (Reserve) Squadrons – latter for training; first delivery to No 18 Squadron at Laarbruch, Germany, 1 February 1994; to No 78 Squadron, Falkland Islands, February 1994. Final Mk 1 withdrawn from service, May 1994, at which time 11 Mk 2s received. Further three new-build Mk 2s ordered 1993 (as Lot 2), for delivery from late 1995; decision to order further 14 Mks 2A/3 for delivery 1997-2000 at cost of US$365 million announced March 1995; these comprise Lots 3 and 4, respectively; Mk 2A has dynamically tuned fuselage (refer above). Latest purchase raised total RAF procurement to 58. First Lot 3 HC. Mk 2A (ZH891) handed over in USA 6 December 1997; shipped to UK and arrived Boscombe Down for clearance trials 18 December 1997. Rest of Lot 3 delivered by end 1998. Retrofits announced in 1996 comprise Smiths HUMS and Racal RA 800 secure communications control system. Current squadron dispositions are Nos. 7, 18 and 27 at Odiham; and No. 78 in Falkland Islands.

Chinook HC. Mk 3: Eight of 14 additional RAF Chinooks, announced March 1995, assigned to Special Forces; build standard as CH-47D, but with MH-47E's large fuel panniers, weather radar and refuelling probes. First flight of initial aircraft (ZH897, as N2045G) in mid-October 1998. Following flight test, all transferred to temporary store at Shreveport, Louisiana, for six to eight months, pending fitment of avionics (including Sky Guardian RWR) which form subject of separate contract; delivery was expected to begin in February 2000, but UK refused to accept initial aircraft, citing software problems as cause. First example eventually accepted in December 2000; initial deliveries to UK were ZH898 and ZH899 to Bristol by sea, arriving 15 July 2001; RAF service entry with Joint Helicopter Command expected in 2003.

HT.17 Chinook: Spanish Army version.

Boeing 234: Commercial version, currently out of production.

Boeing 414: Export military version. Superseded by CH-47D International Chinook and CH-47SD versions (see below).

CH-47D International Chinook: Boeing 414-100 first sold to Japan; Japan Defence Agency ordered two for JGSDF and one for JASDF in 1984; first flight (N7425H) January 1986 and, with second machine, delivered to Kawasaki April 1986 for fitting out; co-production arrangement (see under Japan: Kawasaki **CH-47J**). International Chinook was available in four versions with combinations of standard or long-range (MH-47E-type) fuel tanks and T55-L-712 SSB or T55-L-714.

CH-47SD 'Super D': Latest export variant, first flown on 25 August 1999; embodies some improvements first installed on the MH-47E for US special operations forces, including increased max T-O weight. Honeywell T55-GA-714A turboshaft with Coltec Industries FADEC chosen as standard power plant; single-point pressure refuelling and jettison capability on both sides of aircraft, with fuel contained in two ballistic and crash-resistant

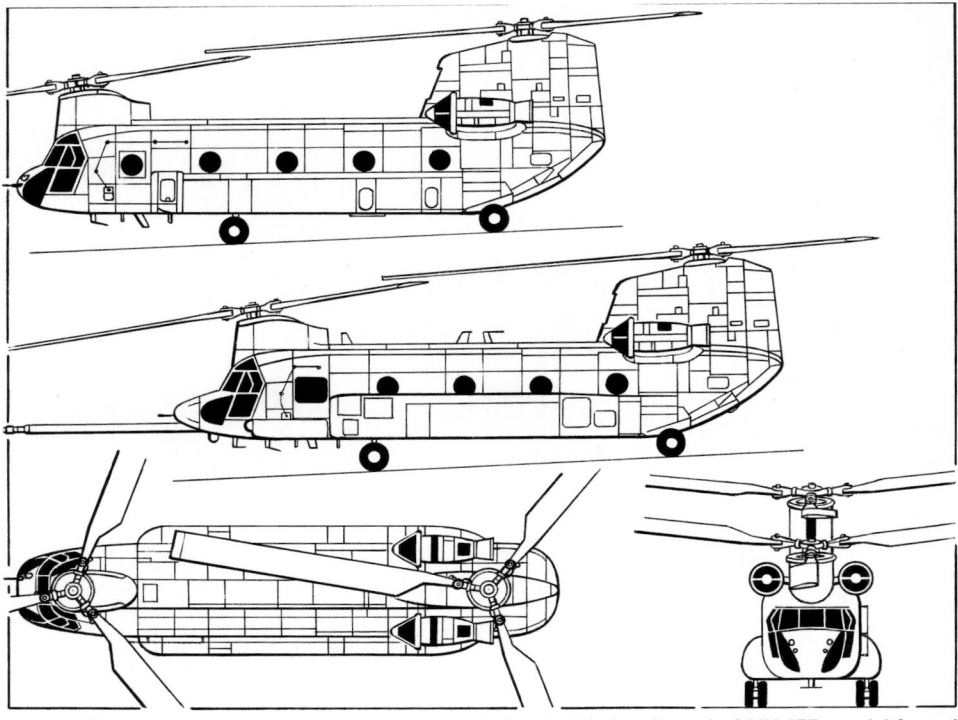

Boeing CH-47D military transport helicopter with additional side view (lower) of MH-47E special forces' variant (*Jane's/James Goulding*)

tanks; total usable capacity is 7,828 litres (2,068 US gallons; 1,722 Imp gallons); CH-47SD also has Smiths digital fuel quantity gauging system in place of Ragen analogue system.

Simplified structure offers benefits in maintainability and reliability. Specific changes include machined frames instead of standard frames.

CH-47SD also incorporates modernised NVG-compatible cockpit with avionics control management system (ACMS), utilising proven military and commercial off-the-shelf equipment on single console to reduce pilot workload and with provisions for growth. ACMS organised into three separate functional groups, specifically air vehicle group (including engines, drive system, fuel, electrics, flight controls and warning indicators); mission group (including communications, navigation and survivability equipment such as INS/GPS, weather radar and digital map display); and pilotage group (including EFIS). Radar nose (but not necessarily radar) fitted as standard.

Avionics suite is comparable to that of baseline CH-47D, but features two embedded INS/GPS units as well as AN/ARN-147 VOR/ILS and AN/ARN-149 ADF, plus space and power provisions for Tacan. AN/ARC-210 frequency-hopping radios to be installed on CH-47SD for Taiwan.

Initial order from Singapore received shortly before March 1998 announcement of formal SD launch, with first acceptance (in US) accomplished in early June 2001. Second customer is Taiwan which, in January 2000, announced purchase of nine CH-47SDs for use by Army; delivery accomplished by mid-2002.

CH-47F Improved Cargo Helicopter (ICH): Boeing programme for improved Chinook configuration for US Army involving the development and production of a new version that will remain operational and cost-effective until 2033, when last examples expected to be replaced by a new cargo helicopter to be developed between 2015 and 2020 under the current Army Aviation Modernization Plan. Programme will coincide with the existing Chinook beginning to reach planned life cycle limits in 2002. Benefits envisaged include greater airframe and systems reliability arising from lower levels of vibration; reduced operating and support costs; reduced pilot workload; and more efficient cargo handling. Weights unchanged, although US Army aircraft, like CH-47D, will have peacetime MTOW of 22,680 kg (50,000 lb). Key goal of upgrade is ability to airlift 7,257 kg (16,000 lb) load over 50 n miles (92 km; 56 miles) under 35°C (95°F) and 1,220 m (4,000 ft) altitude daytime conditions.

Studies indicated that further remanufacture, rather than new production, offered best degree of affordability, with a key factor being fleet-wide fitment of the T55-GA-714A, an improved variant of the T55-L-714 engine, as installed in the MH-47E and the International Chinook. Provision of Coltec Industries FADEC will result in lower specific fuel consumption and a modest reduction (about 3 per cent) in fuel burn. The T55-GA-714A results from a Honeywell (at Greer, South Carolina) kit upgrade programme begun December 1997 and affecting approximately 1,150 existing T55-L-712 engines in the Army inventory; upgrading brings power ratings up to those of the T55-L-714 and adds marinisation features.

Increased versatility also derived from replacement of the existing cargo handling system by integral flip-over roller panels in the cargo deck.

Cockpit modernisation by Rockwell Collins will include adding two dual-redundant MIL-STD-1553B databusses, four 127 mm (5 in) square MFD 255 EFIS LCDs, two 152 × 203 mm (6 × 8 in) MFD-268E1 situation awareness displays, two CDU 900 controller/processors, DR-200 data loader, moving digital map display, digital communications and electronic instrument displays. This allows for updated communications and navigation, enabling the Chinook to meet US Army Force XXI Battlefield requirements. Rockwell Collins selected by Boeing in April 1998 to provide integrated avionics suite, which will feature open architecture to facilitate future growth and updating; CH-47F will be first US Army helicopter to incorporate improved data modem (IDM), allowing it to link with digital battlefield and automatically send and receive information concerning targeting and aircraft position.

Under terms of Army development contract awarded in late 1997, two Chinooks (83-24107 and 83-24115) have been refurbished for the engineering and manufacturing development (EMD) phase, for which contract worth US$76 million was awarded to Boeing in May 1998. Both EMD aircraft arrived for conversion on 5 January 1999; first (98-0011/M8001; ex-83-24107) officially rolled out on 11 July 2001, having flown for the first time on 25 June; second (98-0012/M8002; ex-83-24115) completed upgrade in September 2001, with EMD phase planned to end in December 2002; second CH-47F was delivered to the US Army on 15 May 2002 and flown to Fort Campbell, Kentucky, for further test flights; is also to undergo testing at Fort Rucker, Alabama. Low-rate initial production (LRIP) scheduled to have begun in late 2002 with batch of seven aircraft; subsequent procurement of up to 26 per year until 2016 will give a total fleet of 300; delivery of first LRIP CH-47F due in third quarter of FY04, with first unit fully equipped by first quarter of FY05.

MH-47G: Upgraded version embodying improvements developed for the CH-47F. Plans are in hand to modernise all existing MH-47D and MH-47E helicopters starting in 2003, but specific details of extent of upgrade not yet available, although it will include a 'glass cockpit' and is expected to have enhanced self-protection equipment in the form of a radio frequency jammer and a directed-infra-red countermeasures system.

CH-47X: Proposed follow-on programme to CH-47F in response to diversion of funds from Future Transport Rotorcraft (FTR). MTOW would be in region of 31 tonnes (68,000 lb), requiring new engine in 4.5 MW (6,000 shp) class, as well as new dynamic components, including four-blade rotor system.

CUSTOMERS: US Army has received five YCH-47A, 349 CH-47A, 108 CH-47B, 270 CH-47C and three CH-47D as new-build aircraft from US production plus 11 CH-47C from Italian production; further 479 CH-47D/MH-47E obtained from conversion programme involving CH-47A/B/C models as detailed in table elsewhere. CH-47D, MH-47D and MH-47E inventory was 466 in September 1999 and had declined to just over 450 in mid-2002.

Exports include five CH-47Cs to Argentina (three Air Force; two Army); Australia 12 CH-47Cs with

CHINOOK PRODUCTION

Customer	Boeing	Boeing kits	Agusta	Kawasaki
Civilian	10*		6*	
Argentina	5			
Australia	14			
Canada	9			
Egypt	4		15	
Greece	7		· 10	
Iran		38	30	
Italy		2	35	
Japan	· 2	5		54
Libya			20	
Morocco			9	
Netherlands	6			
Singapore	16			
South Korea	30			
Spain	19			
Taiwan	12‡			
Thailand	6			
UK (RAF)	58			
US Army	735		11	
(Total 1,168)	**933**	**45**	**136**	**54**

Rebuilds to CH-47D

Customer	Boeing	Agusta
Australia	4	
Egypt		12
Greece	9	
Italy		23
Netherlands	7	
Spain	12	
UK (RAF)	32	
US Army	479	
(Total 578)	**543**	**35**

* Civilian standard
‡ Includes three to notional civilian standard and nine CH-47SD

COSTS: First US Army MYP (1985) US$1,200 million for 240 upgrades from CH-47C; second MYP (1989) US$773 million for 144 upgrades; US$67 million (1993) for 11 upgrades (four Australian Army; seven US Army); approximately US$23 million for two new-build CH-47Ds (1992). Unit cost of CH-47SD quoted as US$30-32 million (2002).

MH-47E: US$81.8 million (placed 1987) for development of MH-47E and conversion of one prototype; US$422 million (1989-91) for 11 MH-47Es and option on further 14. US$25.6 million allocated for conversion of single CH-47D to MH-47E (2000).

US Army estimates cost of CH-47F programme, including R&D, long-lead production and upgrade of 300 Chinooks, as approximately US$3.3 billion (1999).

DESIGN FEATURES: Two three-blade intermeshing contrarotating tandem rotors; front rotor turns anti-clockwise, viewed from above; rotor transmissions driven by connecting shafts from engine gearbox, which is driven by rear-mounted engines; normal rotor speed 225 rpm. Classic rotor heads with flapping and drag hinges; manually foldable blades, using Boeing VR7 and VR8 aerofoils with cambered leading-edges; blades can survive hits from 23 mm HEI and API rounds; rotor brake optional.

Development of new, low-maintenance elastomeric rotor hub begun at end August 1999, following signature of contract with US Army; development, testing and installation to be completed over four-year period, with new assembly to be installed on all US Army regular and reserve force Chinooks. Also under consideration for CH-47F and will be incorporated in CH-47SD; UK expected to participate in development programme, with view to adoption. Will have longer fatigue life (4,500 hours), 75 per cent fewer parts and offer 70 per cent reduction in requirement for special maintenance tools; will also retain same rotor flight dynamics and be fully interchangeable with existing hub.

Constant cross-section cabin with side door at front; rear-loading ramp that can be opened in flight; underfloor section sealed to give flotation after water landing; access to flight deck from cabin; main cargo hook mounting covered by removable floor panel so that load can be observed in flight.

FLYING CONTROLS: Differential fore and aft cyclic for pitch attitude control; differential lateral cyclic pitch (from rudder pedals) for directional control; automatic control to keep fuselage aligned with line of flight. Dual hydraulic rotor pitch-change actuators: secondary hydraulic actuators in control linkage behind flight deck for autopilot/autostabiliser input; autopilot provides stabilisation, attitude hold and outer-loop holds.

STRUCTURE: Blades based on D-shaped glass fibre spar, fairing assembly of Nomex honeycomb core and crossply glass fibre skin.

LANDING GEAR: Non-retractable quadricycle type, with twin wheels on each front unit and single wheels on each rear unit. Oleo-pneumatic shock-absorbers in all units. Rear units fully castoring; power steering on starboard rear unit.

Royal Australian Army Boeing CH-47D Chinook (*Jane's/Paul Jackson*)   0121038

crashworthy fuel system (four refurbished as CH-47D and redelivered from May 1995 to C Squadron of Army's 5 Aviation Regiment at Townsville; seven sold to US Army and converted to CH-47D for Hawaii National Guard; contract for further two signed 19 June 1998, with delivery to Townsville for further modification on 10 March 2000; these accepted by Army in first half of 2001); Canada nine CH-47Cs designated **CH-147**, delivered from September 1974 (details in 1985-86 *Jane's*), but withdrawn from use and seven sold to Netherlands in July 1993; Egypt finalised contract for four CH-47Ds on 20 August 1998, with delivery effected in mid-2000; Greece ordered seven CH-47Ds in October 1998, deliveries being effected between May 2001 and March 2002; Japan (two CH-47D International Chinooks, plus five kits to initiate licensed production, with further 54 co-produced by Kawasaki); Netherlands, six ordered in December 1993 for delivery in 1998-99 (first, D-101, on 29 May 1998), plus seven ex-Canadian CH-147s upgraded to CH-47D (including nose radar) by Boeing and redelivered from December 1995 for service with No. 298 Squadron at Soesterberg; Spanish Army 19, designated **HT.17**, of which 10 CH-47Cs (one lost), three Boeing 414-176s and six CH-47D International Chinooks delivered to 5th Helicopter Transport Battalion (Bheltra-V) at Colmenar Viejo, Madrid (last six have Honeywell RDR-1400 weather radar), nine survivors of original 10 upgraded by Boeing in USA to CH-47D between August 1990 and mid-1993, first redelivery to Spain on 27 September 1991, with three Boeing 414s also upgraded in 1997; South Korea, total 30 International

Chinooks, comprising 24 for Army (deliveries from 1988); and six for Air Force (ordered early 1990; delivered by February 1992); Singapore, six ordered April 1994 for delivery in 1996-97, (training operations conducted in 1996-97 at Grand Prairie, Texas, before transfer to Singapore base) followed by 10 CH-47SDs ordered in 1998 and delivered to 127 Squadron from December 2001 onwards; Taiwan (three, notionally civil Boeing 234MLR; finalised contract for nine CH-47SDs in January 2000; deliveries begun in February 2002); Thailand, six for Army (three ordered August 1988, three early 1990, deliveries 1990 to February 1992; all are International Chinook); UK (58, see entries for Chinook HC. Mk 2/2A/3); 16 civilian, of which few left operating in original oilfield support role; two for trials; 46 in kits, comprising 40 for assembly in Italy and six for Japan.

Agusta sold licence-built CH-47Cs to Egypt (15; 12 survivors to be upgraded to CH-47D standard); Greece (10, of which nine converted to CH-47D by Boeing between March 1992 and 1995; first redelivery in October 1993), Iran (68), Italy (37, including 10 CH-47C Plus, to which standard 23 earlier helicopters converted), Libya (20), Morocco (nine) and Pennsylvania US Army National Guard (11). Further six Italian helicopters to Civil Protection Agency. Kawasaki (which see) has firm orders for 61.

Total Chinook orders, including civil, 1,168, excluding additional order for unknown quantity of CH-47SD version from unspecified customer. Six CH-47Ds, ordered by China January 1989, embargoed by US government.

All wheels are size 24×7.7, with tyres size 8.50-10, pressure 6.07 bar (88 lb/sq in). Single-disc hydraulic brakes on all six wheels. Provision for fitting detachable wheel/skis.

POWER PLANT: Two Honeywell T55-L-712 turboshafts, pod-mounted on sides of the rear pylon, each with a standard power rating of 2,237 kW (3,000 shp) and maximum rating of 2,796 kW (3,750 shp). Honeywell T55-L-712 SSB engine has standard power rating of 2,339 kW (3,137 shp) and maximum of 3,217 kW (4,314 shp). Standard in MH-47E and International Chinook are two Honeywell T55-L-714 turboshafts, each with a standard power rating of 3,108 kW (4,168 shp) continuous and emergency rating of 3,629 kW (4,867 shp). CH-47SD (and Netherlands CH-47Ds) have T55-GA-714A turboshafts, with maximum continuous rating of 3,039 kW (4,075 shp). FADEC installed on late production CH-47Ds and CH-47SD. From January 1991, more than 100 CH-47Ds fitted with engine air particle separator (also available for RAF variant). Transmission capacity (CH-47D/SD and MH-47E) 5,617 kW (7,533 shp) on two engines and 3,430 kW (4,600 shp) OEI; rotor rpm 225.

Self-sealing pressure refuelled crashworthy fuel tanks in external fairings on sides of fuselage. Total usable fuel capacity 3,914 litres (1,034 US gallons; 861 Imp gallons) in CH-47D. Provision for up to three additional long-range tanks in cargo area, each of 3,028 litres (800 US gallons; 666 Imp gallons); maximum fuel capacity (fixed and auxiliary) 12,998 litres (3,434 US gallons; 2,859 Imp gallons). Oil capacity 14 litres (3.7 US gallons; 3.1 Imp gallons).

Normal fuel (MH-47E, International Chinook and CH-47SD) 7,828 litres (2,068 US gallons; 1,722 Imp gallons) but MH-47E can operate with three long-range tanks in cargo area, each containing 3,028 litres (800 US gallons; 666 Imp gallons), raising total fuel capacity to 16,913 litres (4,468 US gallons; 3,720 Imp gallons). MH-47D SOA and MH-47E have 8.97 m (29 ft 5 in) refuelling probe on starboard side of forward fuselage, which extends 5.41 m (17 ft 9 in) forward of the nose. Refuelling probe also an option for International Chinook.

ACCOMMODATION: Two pilots on flight deck, with dual controls. Lighting compatible with pilots' NVGs (Nite-Op in RAF variant). Jump seat for crew chief or combat commander. Jettisonable door on each side of flight deck. Depending on seating arrangement, 33 to 55 troops can be accommodated in main cabin, or 24 litters plus two attendants, or (see under Current Versions) vehicles and freight. Rear-loading ramp can be left completely or partially open, to permit transport of extra-long cargo and in-flight parachute or free-drop delivery of cargo and equipment.

Main cabin door, at front on starboard side, comprises upper hinged section which can be opened in flight, and lower section with integral steps. Upper section has a panel with window that is jettisonable. Triple external cargo hook system, with US Army aircraft having centre hook rated to carry maximum load of 11,793 kg (26,000 lb) and the forward and rear hooks 7,711 kg (17,000 lb) each, or 10,433 kg (23,000 lb) in unison, while International Chinook and CH-47SD ratings are 12,701 kg (28,000 lb) for centre hook and 9,072 kg (20,000 lb) each for forward and rear hooks, or 11,340 kg (25,000 lb) in unison. Options are available for a power-down ramp and water dam to permit ramp operation on water, internal ferry fuel tanks, external rescue hoist, and windscreen washers.

SYSTEMS: Hydraulic system contains a utility system, a No. 1 flight control system and a No. 2 flight control system. Each includes a separate, variable delivery pump and a reservoir cooler module. Utility system contains a pressure control module and a return control module. Both flight control systems contain a power control module, incorporating pressure and return in one module for each system. Each flight control system can be driven by the utility system for ground checkout through a power transfer unit without intermixing of hydraulic fluids. All hydraulic systems have a pressurised reservoir to prevent pump cavitation and are serviced by a common filter.

The CH-47D has a significant reduction in the number of hydraulic lines and fittings; approximately 200 tubes and hoses eliminated, thereby obviating approximately 700 leak points. Majority of all lines now swaged together for permanent joining; Rosan fluid adaptors for hardpoint connections. All systems capable of being monitored, both in flight and on the ground, for servicing prechecks and fault isolation. Subsystems designed for pressurisation on demand; power transfer units now used for flight control ground system checkout. Electrical system includes two 40 kVA oil-cooled alternators driven by transmission drive system. Solar T62-T-2B APU drives a 20 kVA generator and hydraulic motor pump, providing electrical and hydraulic power for main engine start and system operation on the ground.

AVIONICS: Baseline CH-47D. Specific MH-47E avionics listed under that heading. Avionics for RAF HC. Mk 1 listed in 1985-86 and earlier editions. Netherlands aircraft have Honeywell advanced cockpit management system (ACMS) and EFIS, latter adopted by Boeing as baseline for subsequent sales.

Comms: Honeywell AN/ARC-199 HF com radio, Rockwell Collins AN/ARC-186 VHF/FM-AM, Magnavox AN/ARC-164 UHF/AM com; C-6533 intercom (Netherlands aircraft have Telephonics Corp STARCOM); Honeywell AN/APX-100 IFF.

Flight: Honeywell AN/APN-209 radar altimeter; AN/ARN-89B ADF; BAE Systems AN/ASN-128 Doppler; AN/ARN-123 VOR/glideslope/marker beacon receiver; and AN/ASN-43 gyromagnetic compass. AFCS maintains helicopter stability, eliminating the need for constant small correction inputs by the pilot to maintain desired attitude. The AFCS is a redundant system using two identical control units and two sets of stabilisation actuators. RAF Chinooks have Racal RNS252 Super TANS INS including GPS.

Instrumentation: Flight instruments are standard for IFR, and include an AN/AQU-6A horizontal situation indicator. CH-47F will incorporate Integrated Engine Crew Advisory System (IECAS).

Mission: Chelton 19-400 satellite communications antenna on some RAF helicopters.

Self-defence: Provisions for Lockheed Martin AN/APR-39A RWR, Sanders AN/ALQ-156 missile warning equipment and BAE Systems M-130 chaff/flare dispensers. RAF Chinooks have BAE Systems M-206/M-1 chaff/flare dispensers, BAE Systems ARI 18228 Sky Guardian RWR and (from 1990) Lockheed Martin AN/ALQ-157 IR jammers and Honeywell AN/AAR-47 missile approach warning equipment; Elisra SPS-65V-3 integrated airborne self-protection system chosen for installation on HC. Mk 3 version. Northrop Grumman AN/AAR-54(V) passive missile approach warning system selected by Royal Netherlands Air Force, with installation to be accomplished during 2001-04.

EQUIPMENT: Hydraulically powered winch for rescue and cargo handling, rearview mirror, plus integral work stands and step for maintenance.

ARMAMENT: Provision for two machine guns or miniguns in crew door (starboard) and forward hold window (port).

DIMENSIONS, EXTERNAL:
Rotor diameter (each) 18.29 m (60 ft 0 in)
Rotor blade chord (each) 0.81 m (2 ft 8 in)
Distance between rotor centres:
CH-47SD 11.85 m (38 ft 10¾ in)
Length: overall, rotors turning:
MH-47E and CH-47SD 30.14 m (98 ft 10¾ in)
fuselage: MH-47E and CH-47SD 15.87 m (52 ft 1 in)
MH-47D/E, incl probe 21.08 m (69 ft 2 in)
Width, rotors removed: MH-47E 4.78 m (15 ft 8 in)
Height to top of rear rotor head:
CH-47SD 5.70 m (18 ft 8½ in)
MH-47E 5.59 m (18 ft 4 in)
Ground clearance, rotors turning:
rear approach 5.77 m (18 ft 11 in)
Ground clearance, static: rear approach:
MH-47E and CH-47SD 4.90 m (16 ft 0¾ in)
forward approach: CH-47SD 2.29 m (7 ft 6 in)
Wheelbase: MH-47E and CH-47SD 7.87 m (25 ft 10 in)
Passenger door (fwd, stbd): Height 1.68 m (5 ft 6 in)
Width 0.91 m (3 ft 0 in)
Height to sill 1.09 m (3 ft 7 in)
Rear-loading ramp entrance: Height 1.98 m (6 ft 6 in)
Width 2.31 m (7 ft 7 in)
Height to sill 0.79 m (2 ft 7 in)

DIMENSIONS, INTERNAL (MH-47E and CH-47SD):
Cabin, excl flight deck: Length 9.19 m (30 ft 2 in)
Width: mean 2.29 m (7 ft 6 in)
at floor 2.51 m (8 ft 3 in)
Height 1.98 m (6 ft 6 in)
Floor area 21.0 m^2 (226 sq ft)
Usable volume 41.7 m^3 (1,474 cu ft)

AREAS:
Rotor blades (each) 7.43 m^2 (80.0 sq ft)
Rotor discs (total):
MH-47E and CH-47SD 525.3 m^2 (5,655 sq ft)

WEIGHTS AND LOADINGS:
Weight empty: MH-47E 12,210 kg (26,918 lb)
CH-47SD 11,550 kg (25,463 lb)
Useful load: MH-47E 12,284 kg (27,082 lb)
CH-47SD 12,944 kg (28,537 lb)
Max underslung load: CH-47SD 12,700 kg (28,000 lb)
Max fuel weight:
MH-47E, CH-47SD 6,815 kg (15,025 lb)
Max T-O weight:
MH-47E, CH-47SD 24,494 kg (54,000 lb)
Transmission loading at max T-O weight and power:
MH-47E, CH-47SD 4.36 kg/kW (7.17 lb/shp)

PERFORMANCE (at 22,680 kg; 50,000 lb):
Max level speed: MH-47E 154 kt (285 km/h; 177 mph)
CH-47SD 155 kt (287 km/h; 178 mph)
Max cruising speed at S/L:
MH-47E, CH-47SD 140 kt (259 km/h; 161 mph)
Max rate of climb: MH-47E 561 m (1,840 ft)/min
CH-47SD 563 m (1,846 ft)/min
Service ceiling: MH-47E 3,090 m (10,140 ft)
CH-47SD 3,385 m (11,100 ft)
Hovering ceiling: IGE: MH-47E 2,985 m (9,800 ft)
CH-47SD 2,835 m (9,300 ft)
IGE, ISA + 20°C (68°F): MH-47E 2,410 m (7,900 ft)
CH-47SD 2,180 m (7,160 ft)
OGE: MH-47E, CH-47SD 1,675 m (5,500 ft)
OGE, ISA + 20°C (68°F):
MH-47E, CH-47SD 1,005 m (3,300 ft)
Radius of action, MH-47E, deploy special forces team (1,814 kg; 4,000 lb) at 1,220 m (4,000 ft), 35°C (95°F) ambient temperature 505 n miles (935 km; 581 miles)
Range, MH-47E, self-deployment at 24,494 kg (54,000 lb) T-O weight
1,260 n miles (2,333 km; 1,449 miles)
Range, CH-47SD, with 12,558 kg (27,686 lb) payload
651 n miles (1,207 km; 750 miles)

UPDATED

BOEING AH-64 APACHE

US Army designations: AH-64A and AH-64D
Israel Defence Force name: Pethen (Cobra)

TYPE: Attack helicopter.

PROGRAMME: Original Hughes Model 77 entered for US Army advanced attack helicopter (AAH) competition; first flights of two development prototype YAH-64s 30 September and 22 November 1975; selected by US Army December 1976; named Apache late 1981.

Deliveries started 26 January 1984; 900th delivered in October 1995, at which time US Army had ordered 821 AH-64As (excluding prototypes) with export contracts totalling 104; latter total increased to 242 (116 AH-64A; 126 AH-64D) by end of August 2001. Last of 821 AH-64As delivered to US Army on 30 April 1996; aircraft concerned was production vehicle 915, and manufacture of AH-64A variant terminated with completion of 937th example (for Egypt), in November 1996. Delivery of 1,000th Apache (including AH-64D rebuilds) effected on 30 March 1999.

Boeing awarded four year, US$15.9 million, contract on 27 May 1998 to design, manufacture and flight test new centre fuselage section incorporating advanced composites materials; Phantom Works responsible for design leadership, with support from Boeing facilities at Long Beach, Philadelphia and St Louis. If successful, new section from rear of aft cockpit to just behind engines will be simpler to manufacture as well as lighter and more durable than existing all-metal structure.

In separate development, on 5 October 1998, a modified AH-64D Apache Longbow prototype made initial flight with Rotorcraft Pilot's Associate (RPA) advanced cockpit management system. Also developed by Phantom Works over 60 month period, under terms of US$80 million

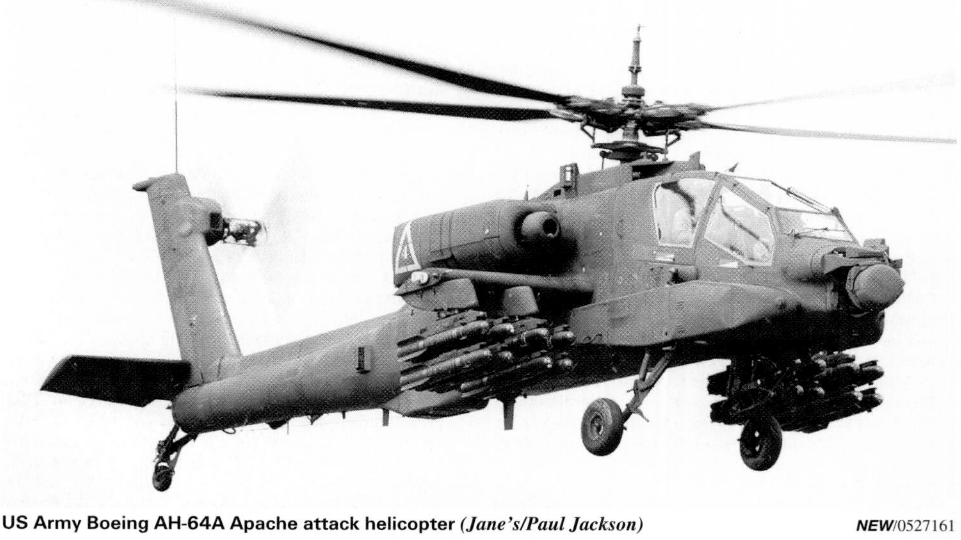

US Army Boeing AH-64A Apache attack helicopter ***(Jane's/Paul Jackson)*** ***NEW***/0527161

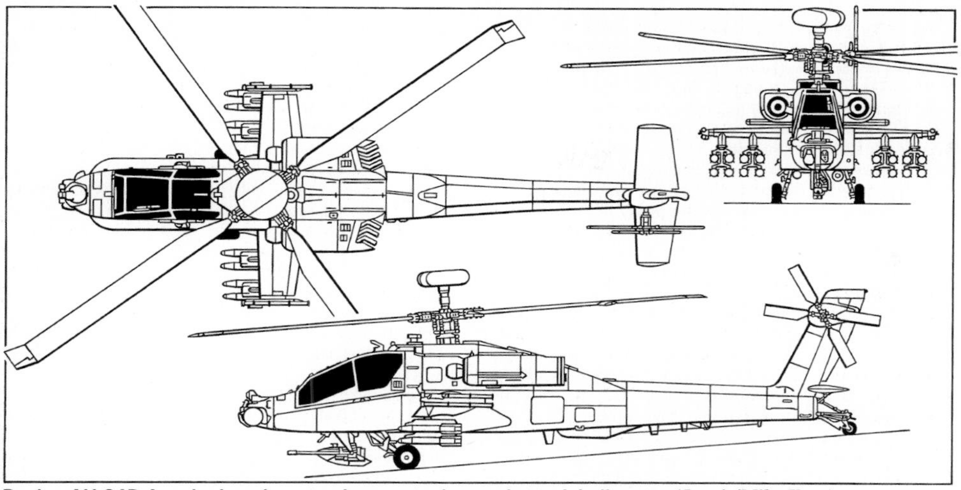

Boeing AH-64D Apache Longbow tandem-seat advanced attack helicopter ***(Jane's/Mike Keep)***

advanced technology demonstration contract, this featured updated controls and displays, including Boeing-developed four-axis, full authority, advanced digital flight control system. Flight and mission data presented to pilots on three large multipurpose colour displays; RPA also features advanced data fusion and an advanced pilotage system, as well as the ability to recognise and respond to verbal commands. Company flight trials continued throughout remainder of 1998, with test aircraft then visiting US Army's Yuma Proving Grounds for demonstration flights in January-February 1999.

In 2001, Boeing proposed series of upgrades designed to keep AH-64D in operational force until 2030; these included adoption of new split-face power plant gearbox, new main rotor assembly using five composite blades and titanium hub, composites fuselage structure elements for enhanced payload, and open architecture avionics with gradual introduction of Boeing RPA features such as digital navigation and communications equipment. Objective of upgrade package is to restore original performance, reduce purchase price by about 30 per cent and cut operating costs by 50 per cent, but US Army not enthusiastic and will instead direct available funding towards near-term reliability and sustainability concerns, such as adoption of Lockheed Martin Arrowhead targeting and navigation sensor system.

Some new features, including digital avionics incorporating commercial off-the-shelf (COTS) technology, being tested on preproduction AH-64D Apache Longbow first flown 12 July 2001; Boeing proposes introducing these enhancements during production of the second multiyear batch of helicopters.

CURRENT VERSIONS: **AH-64A:** Produced for US Army and export; 937 built. First 603 had two 1,265 kW (1,696 shp) T700-GE-701 engines. Total of 501 US Army examples currently programmed to be upgraded to AH-64D. Retrofit from 1993 with SINCGARS secure radios and GPS; first installed in Apaches of 5-501 Aviation Regiment on deployment to Camp Eagle, South Korea.

AH-64B: Cancelled in 1992. Was planned near-term upgrade of 254 AH-64As with improvements derived from operating experience in 1991 Gulf War, including GPS SINCGARS radios, target handover capability, better navigation, and improved reliability.

AH-64C: Previous designation for upgrade of AH-64As to near AH-64D standard, apart from omission of Longbow radar and retention of -701 engines; provisions for optional fitment of both; Army requested draft proposal, August 1991; funding for two prototype conversions awarded in September 1992. Designation abandoned late 1993.

AH-64D Apache Longbow: Current improvement programme based on Lockheed Martin and Northrop Grumman joint venture development of mast-mounted AN/APG-78 Longbow millimetre-wave radar and Hellfire missile with RF seeker; Northrop Grumman has lead on Longbow with Lockheed Martin taking principal role for Hellfire. Programme also includes more powerful engines, larger generators, MIL-STD-1553B databus allied to dual 1750A processors, and a vapour cycle cooling system for avionics; early user tests completed April 1990.

Detailed description applies to AH-64D.

Full-scale development programme, lasting 51 months, authorised by Defense Acquisition Board August 1990, but airframe work extended in December 1990 to 70 months to coincide with missile development; supporting modifications being incorporated progressively; first flight of AH-64A (82-23356) with dummy Longbow radome 11 March 1991; first (89-0192) of six AH-64D prototypes flown 15 April 1992; second (89-0228) flew 13 November 1992; fitted with radar in mid-1993 and flown 20 August 1993; No.3 (90-0324) flown 30 June 1993; No.4 (90-0423) on 4 October 1993; No.5 (85-25477; formerly AH-64C No.1) 19 January 1994 (first Apache with new Hamilton Sundstrand lightweight flight management computer); No.6 (85-25408) flown 4 March 1994; last two mentioned lack radar. Following termination of AH-64C in late 1993, original plan was to convert 748 AH-64As to AH-64D, although only 227 (original AH-64D total) to carry Longbow radar; subsequent review, revealed at start of 1999, proposed reducing total number of conversions to 530 and increasing purchase of Longbow radars to 500; these to equip regular Army units, with remaining 218 AH-64As of Army National Guard and Army Reserve undergoing service life extension programme. Most recent news indicates that US Army may eventually receive 600 AH-64Ds.

Under original programme, AH-64D to equip 26 battalions, although this may be reduced to 16 of regular Army; three companies, each with eight helicopters, per battalion. Longbow can track flying targets and see through rain, fog and smoke that defeat FLIR and TV. RF Hellfire, delivered to US Army from November 1996, can operate at shorter ranges; it can lock on before launch or launch on co-ordinates and lock on in flight; Longbow scans through 360° for aerial targets or scans over 270° in 90° sectors for ground targets; mast-mounted rotating antenna weighs 113 kg (250 lb). Longbow radar transmitter subject of redesign in late 1997 to overcome poor performance of some electrical components in low temperatures; eliminate lengthy and costly manual integration necessary to achieve required output; and avoid shortages of critical components which suppliers reported in 1995 that they would no longer provide. New transmitter meets or surpasses original specification and is fully interchangeable with original unit.

Further modifications include 'manprint' cockpit with large displays, air-to-air missiles, digital autostabiliser, integrated GPS/Doppler/INS/air data/laser/radar altimeter navigation system, digital communications, faster target handoff system, and enhanced fault detection with data transfer and recording. Cockpit displays initially monochrome, but these replaced by colour displays with effect from 27th production conversion (97-5027); first flight of AH-64D with colour displays on 12 September 1997. AH-64D No.1 made first Hellfire launch on 21 May 1993; first RF Hellfire launch 4 June 1994; first demonstration of digital air-to-ground data communications with Symetrics Industries improved data modem, 8 December 1993.

Advanced acquisition phase contract for remanufacture programme, worth US$279.6 million and covering 18 helicopters (later increased to 24), awarded on 14 December 1995; predated by arrival of first two AH-64As (85-25387 and 85-25394) at Mesa on 27 November 1995 for stripping to basic fuselage in readiness for start of remanufacture in early 1996. First fuselage moved to final assembly area on 15 August 1996; first flight (85-25387 with new identity 96-5001) 17 March 1997; formal roll-out at Mesa on 21 March.

Multiyear contract, worth US$1.87 billion, covering 232 AH-64Ds (retrospectively including advanced acquisition aircraft) over five year period signed 16 August 1996; further 269 conversions to be acquired in second multiyear contract covering FY01 to FY05, which signed 29 September 2000, at total cost of US$2.3 billion. US Army currently expects to complete AH-64D acquisition with final batch of 99, to give total buy of 600 for operational deployment. Initial contract also included 227 Longbow radars, 13,311 Hellfire missiles and 3,296 launchers. AH-64D deliveries to US Army began 31 March 1997 and total of 18 handed over by end of 1997. Delivery of 24th and last Lot 1 aircraft accomplished 4 March 1998, with US Army having accepted total of 55 by end of 1998, 102 by end of 1999 and 159 by end of 2000; 100th remanufactured AH-64D delivered to US Army on 9 December 1999; 200th followed on 23 August 2001. Final example of initial multiyear contract and first helicopter from second multiyear contract both delivered to US Army on 3 April 2002.

Initial AH-64D battalion (1-227 AvRgt) at Fort Hood, Texas fully equipped by end July 1998 and attained combat-ready status on 19 November 1998, after eight month training programme at company and battalion level which included four live fire exercises and more than 2,500 flight hours. Second unit is 2-101 AvRgt at Fort Campbell, Kentucky, which certified combat-ready on 28 October 1999; third is 1-3 AvRgt at Hunter AAF, Fort Stewart, Georgia, on 15 March 2001; fourth is 1-2 AvRgt, which deployed to South Korea in latter half of 2001; fifth is 1-101 AvRgt, also at Fort Campbell, which completed training on 6 December 2001; sixth is 6-6 Cav at Illesheim, Germany, which was declared combat-ready in May 2002 at Fort Hood and which introduced AH-64D to US Army Europe in mid-2002.

GAH-64A: At least 17 AH-64As grounded for technical instruction.

JAH-64A: Seven AH-64As for special testing, of which one reverted to standard.

WAH-64D: British Army version with Longbow radar and two Rolls-Royce/Turbomeca RTM 322 turboshafts; see Westland entry in the UK section.

CUSTOMERS: US Army 827 AH-64A (including six prototypes), of which last delivered 30 April 1996; see Programme and Current Versions for details; operational fleet in November 1999 totalled 743. Confirmed export orders and firm commitments totalled 242 in third quarter of 2001. Israel ordered 18 in March 1990; first two delivered 12 September 1990 to 113 Squadron; further 24 ex-US Army AH-64As supplied for 113 and 127 Squadrons at Ramon; deliveries to 190 Squadron at Ramat David, 1995. After rejecting new-build AH-64D Apache Longbow on cost grounds, Israel considered upgrade of existing AH-64A to AH-64D; however, subsequent change of plan involved abandoning upgrade proposal in favour of acquisition of new-build AH-64D; letter of offer and acceptance signed February 2001, covering nine helicopters (of which one will be a remanufactured AH-64A).

Deliveries began in April 1993 to Saudi Arabia (12) and in 1994 to Egypt (24). Further orders from Greece (20) and United Arab Emirates (20), both in December 1991; six handed over to UAE on 30 October 1993; 14 followed in 1994; all UAE examples are likely to be upgraded to AH-64D standard, incorporating improved Lockheed Martin Arrowhead piloting and weapon aiming system as well as enhanced EW suite. Greek deliveries from February 1995 for training in USA; initial six Apaches ferried to New Orleans, Louisiana, for shipment to Greece on 9 June 1995. Ten more ordered for UAE in June 1994; 12 more to Egypt approved early 1995, with Egypt revealing intent in July 2000 to upgrade 35 AH-64A to AH-64D; signature of contract followed in November 2001, with deliveries to begin in July 2003, although these will not have the Longbow radar.

Version of AH-64D Apache Longbow offered to British Army by consortium of McDonnell Douglas, Westland, Lockheed Martin, Northrop Grumman and Shorts; UK announced order for 67 on 13 July 1995; additional information under Westland entry in UK section. Netherlands signed contract on 24 May 1995 for 30 Apaches for Nos. 301 and 302 Squadrons at Gilze-Rijen; radar not required at outset, but formal request made in November 1999 and this will now be installed; US Army provided 12 AH-64As for use by No. 301 Squadron on lease basis for period 1996-99; all delivered 13 November 1996 and since returned. First AH-64Ds accepted by Netherlands at Mesa on 15 May 1998 and 21 had been delivered by late August 2001, with 30th and last handed over on 30 April 2002; No. 302 Squadron first unit to be equipped, with initial deliveries to Gilze-Rijen in mid-May 2000; No. 301 Squadron to become operational in 2003. Singapore selected AH-64D on 14 June 1999; total of eight initially to be acquired, with AGM-114K Hellfire 2 laser-guided missile, but lacking Longbow radar; option on further 12 taken up on 23 August 2001, at which time it was revealed that all 20 will have Longbow radar; first example accepted at Mesa on 17 May 2002. Initial eight, all of which were due to be delivered by August 2002, remain in USA (at Marana, Arizona) for pilot training. Kuwait was authorised in late 1997 to receive 16 AH-64Ds under FMS programme; letter of acceptance due for signature in last quarter of 1998, but debate continued over inclusion of Longbow radar and deal still to be finalised in mid-2002, although Congressional approval being sought for US$2.1 billion deal that includes eight Longbow radars as well as Arrowhead target acquisition and designation system plus spares, training, government and contractor support. AH-64D was offered to Australia, but was rejected, with Eurocopter Tiger being preferred. On 27 August 2001, Japan announced selection of AH-64D to satisfy its AH-X combat helicopter requirement; an initial batch of 10 helicopters is being acquired with FY02 funds, with the total buy likely to number 60, some of which will be fitted with Longbow radar. Future customers could include South Korea, which initially needs 18 combat helicopters, but could eventually acquire total of 36. Taiwan has indicated desire to obtain as many as 75 AH-64D Apache Longbow helicopters, while Norway, Spain and Sweden have expressed interest in the AH-64D.

APACHE REMANUFACTURE PROGRAMME

Lot	Year	Qty	First aircraft
-	Prototypes	6	89-0192
-	Preproduction	2	
1	FY96	24	96-5001
2	FY97	24	97-5025
3	FY98	44	98-5049
4	FY99	66	99-5093
5	FY00	74	00-5159
6-10	FY01-05	269	01-5233
	Total	**509**	

APACHE PROCUREMENT

	Qty	Remarks
US Army AH-64A		
FY73	3	Prototypes
FY79	3	Preproduction
FY82	11	
FY83	48	
FY84	112	
FY85	138	
FY86	116	
FY87	101	
FY88	77	
FY89	54	
FY90	132	
FY91	12	
FY92	10	
FY94	10	
Subtotal	**827[1]**	

Exports		First aircraft and date	
Israel (Lot 1)	18[2]	801	Sep 1990
Israel (Lot 2)	9[6]		
Saudi Arabia	12[3]	90-0291	Apr 1993
Egypt (Lot 1)	24[3]	3701	Feb 1994
Egypt (Lot 2)	12[3]	3725	1996
Greece	20[3]	E-1001	Feb 1995
UAE (Lot 1)	20[3]	050	Oct 1993
UAE (Lot 2)	10[3]	070	1996
Netherlands	30[4]	Q-01/98-0101	May 1998
UK	67[5]	ZJ166	Sep 1998
Singapore	20[6]		May 2002
Subtotal	**242**		
Total	**1,069**		

Notes:

[1] 501 programmed to be remanufactured as AH-64D, with further 99 in prospect, for total of 600

[2] AH-64A; further 24 obtained from US Army

[3] AH-64A

[4] Non-radar AH-64D; 12 US Army AH-64A operated during 1996-99 on lease pending delivery of AH-64D

[5] WAH-64D

[6] AH-64D

APACHE PRODUCTION

Lot	FY	Qty		First aircraft
		US	Export	
-	73	3[1]		73-22247
-	79	3[2]		79-23257
1	82	11		82-23355
2	83	48		83-23787
3	84	112		84-24200
4	85	138		85-25351
5	86	116		86-8940
6	87	101		87-0407
7	88	67		88-0197
		10		88-0275
8	89	54	18	89-0192
9	90	60	6	90-0280
10	90	72	36	90-0415
11	91/92	22	34	91-0112
12	94	10[3]	10	94-0328
13	95		12	95-0108
-	98 et seq		126[3]	98-0101
Subtotals		**827**	**242**	
Total		**1,069**		

Notes:

[1] Prototypes; 73-22247 was static test airframe

[2] Preproduction

[3] AH-64D/WAH-64D; all other aircraft are AH-64A version

COSTS: Programme cost (807 aircraft at 1991 values) US$1,169 million. Egyptian Lot 2 request costed at US$318 million for 12 Apaches plus four spare Hellfire launchers, thirty-four 70 mm (2.75 in) rocket launchers,

six spare T700 engines, one spare TADS/PNVS system and miscellaneous spares. Second multiyear procurement contract valued at US$2.3 billion, including training devices, spares, logistics and support services. Israeli AH-64D purchase of nine helicopters expected to cost almost US$500 million, including ordnance, spare parts, training and other support; remanufacture of 35 for Egypt valued at US$241.9 million.

DESIGN FEATURES: Modern, tandem-seat, armoured and damage-resistant combat helicopter; is required to continue flying for 30 minutes after being hit by 12.7 mm bullets coming from anywhere in the lower hemisphere plus 20°; also survives 23 mm hits in many parts; target acquisition and designation sight (TADS) and pilot night vision sensor (PNVS) sensors mounted in nose; low-airspeed sensor above main rotor hub; avionics in lateral containers; chin-mounted Chain Gun fed from ammunition bay in centre-fuselage; four weapon pylons on stub-wings (six when air-to-air capability is installed); engines widely separated, with integral particle separators and built-in exhaust cooling fittings; four-blade main rotor with lifting aerofoil blade section and swept tips; blades can be folded or easily removed; tail rotor consists of two teetering two-blade units crossed at 55° to reduce noise; airframe meets full crash-survival specifications. Two AH-64s will fit in C-141, six in C-5 and three in C-17A.

Main transmission, by Litton Precision Gear Division, can operate for 1 hour without oil; tail rotor drive, by Aircraft Gear Corporation, has grease-lubricated gearboxes with Bendix driveshafts and couplings; gearboxes and shafts can operate for 1 hour after ballistic damage; main rotor shaft runs within airframe-mounted sleeve, relieving transmission of flight loads and allowing removal of transmission without disturbing rotor; AH-64A has flown aerobatic manoeuvres and is capable of flying at 0.5 g.

FLYING CONTROLS: Fully powered controls with stabilisation and automatic flight control system; automatic hover hold; tailplane incidence automatically adjusted by Hamilton Sundstrand control to streamline with downwash during hover and to hold best fuselage attitude during climb, cruise, descent and transition.

STRUCTURE: Main rotor blades (by Tool Research and Engineering Corporation, Composite Structures Divisions) tolerant to 23 mm cannon shells, have five U-sections forming spars and skins bonded with structural glass fibre tubes, laminated stainless steel skin and composites rear section; blades attached to hub by stack of laminated steel straps with elastomeric bearings. Northrop Grumman produces all fuselages, wings, tail, engine cowlings, canopies and avionics containers.

LANDING GEAR: Menasco trailing arm type, with single mainwheels and fully castoring, self-centring and lockable tailwheel. Mainwheel tyres size 8.50-10 (10 ply) tubeless, tailwheel tyre size 5.00-4 (14 ply) tubeless. Hydraulic brakes on main units. Main gear is non-retractable, but legs fold rearward to reduce overall height for storage and transportation. Energy-absorbing main and tail gears are designed for normal descent rates of up to 3.05 m (10 ft)/s and heavy landings at up to 12.8 m (42 ft)/s. Take-offs and landings can be made at structural design gross weight on terrain slopes of up to 12° (head-on) and 10° (side-on).

POWER PLANT: Two General Electric T700-GE-701C turboshafts, each rated at 1,409 kW (1,890 shp) for 10 minutes, 1,342 kW (1,800 shp) for 30 minutes, 1,238 kW (1,660 shp) maximum continuous and 1,447 kW (1,940 shp) 2½ minutes OEI. Engines mounted one on each side of fuselage, above wings, with key components armour-protected. Upper cowlings let down to serve as maintenance platforms. General Electric and US Army to

undertake joint remanufacture programme of existing engines into T700-GE-701D standard from mid-2003; intended to alleviate poor levels of reliability, this will also increase power slightly and may be installed on more than 700 Apaches.

AH-64 modernisation may ultimately lead to installation of new 2,237 kW (3,000 shp) engine; US Army effort to prepare requirement for Common Engine Program (CEP) likely to result in adoption on H-60 series first in about 2007, but Army expects to draw up operational requirements document for AH-64 during 2001 or 2002. In near term, Boeing proposing adoption of new five-blade main rotor system allied to new drive system, offering improved performance, greater reliability and simplified maintenance.

Two crash-resistant fuel cells in fuselage, combined capacity 1,421 litres (375 US gallons; 312 Imp gallons). Modifications ordered September 1993 for carriage of four 871 litre (230 US gallon; 192 Imp gallon) Brunswick Corporation external tanks on 437 Apaches. Total internal and external fuel 4,910 litres (1,295 US gallons; 1,078 Imp gallons). New crashworthy, ballistically self-sealing internal auxiliary fuel tank entered evaluation phase in fourth quarter of 1997; tank holds 492 litres (130 US gallons; 108 Imp gallons) and is interchangeable with ammunition storage magazine, enabling all four weapons pylons to carry ordnance on long-range missions. Testing of preproduction system undertaken in 1998 in addition to formal test and qualification programme by US Army; total of 48 units ordered from Robertson Aviation in 1999; smaller 379 litre (100 US gallon; 83.3 Imp gallon) internal tank to be utilised by AH-64D, this permitting carriage of up to 300 rounds of 30 mm ammunition for Chain Gun. 'Black Hole' IR suppression system protects aircraft from heat-seeking missiles: this eliminates an engine bay cooling fan, by operating from engine exhaust gas through ejector nozzles to lower the gas plume and metal temperatures.

ACCOMMODATION: Crew of two in tandem: co-pilot/gunner (CPG) in front, pilot behind on 48 cm (19 in) elevated seat. Crew seats, by Simula Inc, are of lightweight Kevlar. Northrop Grumman canopy, with PPG transparencies and transparent acrylic blast barrier between cockpits, is designed to provide optimum field of view. Crew stations are protected by Ceradyne Inc lightweight boron armour shields in cockpit floor and sides, and between cockpits, offering protection against 12.7 mm armour-piercing rounds. Sierracin electric heating of windscreen. Seats and structure designed to give crew a 95 per cent chance of surviving ground impacts of up to 12.8 m (42 ft)/s. Simula of Phoenix, Arizona, under contract by US Army to develop cockpit-airbag system; development phase was due to be completed in early 1997, but cancelled in favour of modifications to gunner's sighting equipment to lessen risk of injury.

SYSTEMS: Honeywell totally integrated pneumatic system includes a shaft-driven compressor, air turbine starters, pneumatic valves, temperature control unit and environmental control unit. Fairchild Controls improved environmental control system comprises a distributed vapour-cycle cooling and heating unit, with two redundant systems incorporating dual-speed compressors, digital databus controllers and multiple heat exchangers, fans and control valves. Parker dual-hydraulic systems, operating at 207 bar (3,000 lb/sq in), with actuators ballistically tolerant to 12.7 mm direct hits. Redundant flight control system for both rotors. In the event of a flying control system failure, the system activates Honeywell secondary fly-by-wire control. Honeywell electrical power system, with two 45 kVA fully redundant engine-driven AC

Netherlands Air Force Boeing AH-64DN Apache (*Jane's/Paul Jackson*)   **NEW**/0132860

generators, two 300 A transformer-rectifiers, and URDC standby DC battery. Honeywell GTP 36-155(BH) 93 kW (125 shp) APU for engine starting and maintenance checking. DASA (TST) electric blade de-icing. Smiths Industries integrated electrical power management system (IEPMS) installed on AH-64D is currently being upgraded to incorporate multichannel remote interface unit (RIU) that will replace core electronics and wiring associated with conventional electrical control systems; improved IEPMS available from 2001 and is being installed on approximately 300 Apaches for US Army.

AVIONICS: *Comms:* AN/ARC-164 UHF, AN/ARC-222 SINCGARS secure UHF/VHF; AN/ARC-220 UHF to be retrofitted; KY-28/58/TSEC crypto secure voice, C-8157 secure voice control; AN/APX-100 IFF unit with KIT-1A secure encoding; C-10414 Tempest intercom.

Radar: Optional Lockheed Martin/Northrop Grumman AN/APG-78 Longbow mast-mounted 360° radar, presenting up to 256 targets on tactical situation display; detects air targets in air-to-ground mode; air-to-air mode for flying targets only.

Flight: BAE North America AN/ASN-157 lightweight Doppler navigation system, Litton LR-80 (AN/ASN-143) strapdown AHRS, AN/ARN-89B ADF, GPS, Honeywell digital automatic stabilisation equipment (DASE), Astronautics Corporation HSI, Pacer Systems omnidirectional, low-airspeed air data system, remote magnetic indicator, BITE fault detection and location. Doppler system, with AHRS, permits nap-of-the-earth navigation and provides data for storing target locations. BAE Systems air data system, comprising two omnidirectional airspeed and direction sensors (AADSs) mounted on engine cowlings and a high integration air data computer (HIADC) installed in avionics bay.

Instrumentation: Honeywell all-raster symbology generator processes TV data from IR and other sensors, superimposes symbology, and distributes the combination to CRT and helmet-mounted displays; Honeywell AN/APN-209 radar altimeter video display unit. 'Manprint' (manpower integration) instrumentation including Litton Canada upfront display and two Honeywell 152 × 152 mm (6 × 6 in) monochrome CRT displays in each cockpit of early aircraft; from 27th AH-64D, Honeywell flat-panel, colour, active matrix LCD multipurpose displays (MPDs) installed in both cockpits, as well as in aircraft for UK and Netherlands. AH-64A's 1,200 cockpit switches reduced to approximately 200 on AH-64D.

Mission: Lockheed Martin target acquisition and designation sight and AN/AAQ-11 pilot's night vision sensor (TADS/PNVS) comprises two independently functioning, fully integrated systems mounted on nose.

TADS consists of a rotating turret (±120° in azimuth, +30/–60° in elevation) housing sensor subsystems, optical relay tube (being replaced under 1998 contract by Planar Advance/dpiX flat panel display) in the CPG's cockpit, three electronic units in the avionics bay, and cockpit-mounted controls and displays; used principally for target search, detection and laser designation, with CPG as primary operator (can also provide back-up night vision to pilot in event of PNVS failure). Once acquired by TADS, targets can be tracked manually or automatically for autonomous attack with gun, rockets or Hellfire missiles. TADS daylight sensor consists of TV camera with narrow (0° 50′) and wide angle (4° 0′) fields of view; direct view optics (4° narrow and 18° wide angle); laser spot tracker; and International Laser Systems laser range-finder/designator. New switchable eyesafe laser range-finder designator (SELRD) currently being developed by Kollsman Inc under US$2.8 million contract awarded at start of 2000; to be installed in existing TADS turret on AH-64A and AH-64D and will have 80 per cent commonality with Kiowa Warrior SELRD. Night sensor, in starboard half of turret, incorporates FLIR sight with narrow, medium and wide angle (3° 6′, 10° 6′ and 50°) fields of view.

PNVS consists of FLIR sensor (30 × 40° field of view) in rotating turret (±90° in azimuth, +20/–45° in elevation) mounted above TADS; electronic unit in the avionics bay; and pilot's display and controls; provides pilot with thermal imaging for nap-of-the-earth flight to, from and within battle area at night or in adverse daytime weather, at altitudes low enough to avoid detection. Second-generation FLIR sensor to be installed on AH-64D; Lockheed Martin Arrowhead and Raytheon FIREsight systems conceived to satisfy this requirement, with Arrowhead selected in late October 2000 to progress to EMD phase. Production is currently expected to begin in late 2003, with operational deployment possibly to occur as early as 2004; Arrowhead also being offered to overseas operators. PNVS imagery displayed on monocle in front of one of pilot's eyes; flight information including airspeed, altitude and heading is superimposed on this imagery to simplify piloting. Monocle is part of Honeywell integrated helmet and display sighting system (HADSS) worn by both crew members. Symetrics Industries improved data modem for transmission of target data (and eventually real-time imagery) between helicopters, tactical jet, Joint STARS airborne command posts, HQs and ground units at 16,000 bits/s, plus radio frequency interferometer beneath radome for identification of hostile transmitters.

Self-defence: Aircraft survivability equipment (ASE) consists of Litton AN/APR-39 passive RWR, Sanders AN/ALQ-144 IR jammer, Raytheon AN/AVR-2 laser warning receiver, ITT AN/ALQ-136 radar jammer and chaff dispensers and Lockheed Martin AN/APR-48A radar frequency interferometer. Sanders AN/ALQ-212 Advanced Threat Infra-Red Countermeasures (ATIRCM) system and ITT AN/ALQ-211 suite of integrated RF countermeasures (SIRFC) system currently under development. ATIRCM combines next-generation directable IRCM system with Sanders AN/AAR-57 Common Missile Warning System (CMWS); contractor tests of SIRFC undertaken on Apache Longbow in latter half of 1999, followed by operational test and evaluation from early 2000. Elisra began flight test of passive airborne warning system in second half of 1999 and Israel plans to install this on its AH-64 fleet. Elisra-supplied EW equipment to be installed on IDF/AF AH-64D, which may also have LWS-20V-2 laser warning system. Upgraded Egyptian AH-64D to be fitted with Northrop Grumman AN/ALQ-162(V)6 'Shadowbox' high-band RF countermeasures system.

EQUIPMENT: Avpro of UK cleared Exint transport pod for use with AH-64 at start of 2000, but certification to carry personnel still required. In special forces insertion role, Apache can carry maximum of four pods, each able to accommodate 226 kg (500 lb) payload.

ARMAMENT: Boeing M230 Chain Gun 30 mm automatic cannon, located between the mainwheel legs in an underfuselage mounting with Smiths Industries electronic controls. Normal rate of fire is 625 rds/min of HE or HEDP (high-explosive dual-purpose) ammunition, which is interoperable with NATO Aden/DEFA 30 mm ammunition. Maximum ammunition load is 1,200 rounds. New 'Sideloader' system demonstrated June 1994 and now installed in starboard forward avionics bay; cuts normal loading time of 30 minutes by up to half and reduces number of personnel required from three to one. Gun mounting is designed to collapse into fuselage between pilots in the event of a crash landing.

New electric turret under development by Boeing, which received two year, US$5 million contract in first half of 1999; objective is to achieve accuracy of 0.5 mrads compared with current 3.0 mrads. Gun, mount and feed system to be retained in conjunction with redesigned mechanical system featuring electric rather than hydraulic drive as well as digital control; result should be at least 10 per cent lighter and require one instead of two electrical boxes. HR Textron responsible for controls, with Boeing providing the rest. Prototype delivery was scheduled for September 2001. Four underwing hardpoints, with Aircraft Hydro-Forming pylons and ejector units, on which can be carried up to 16 AGM-114 Hellfire anti-tank missiles or up to 76 2.75 in FFAR (folding fin aerial rockets) in their launchers or a combination of Hellfires and FFAR. Planned modification adds two extra hardpoints for four Stinger, four Mistral or two Sidewinder (including Sidearm anti-radiation variant) missiles; Shorts Starstreak high-velocity AAM system completed initial 17 month test programme on Apache in February 1997 and being promoted for use by US and UK, with funding allocated for follow-on two year test programme; second round of live fire tests completed October/November 1998. Hellfire remote electronics by Rockwell Collins; Honeywell aerial rocket control system; multiplex (MUX) system units by Honeywell. Co-pilot/gunner (CPG) has primary responsibility for firing gun and missiles, but pilot can override his controls to fire gun or launch missiles.

DIMENSIONS, EXTERNAL:

Main rotor diameter	14.63 m (48 ft 0 in)
Main rotor blade chord	0.53 m (1 ft 9 in)
Tail rotor diameter	2.79 m (9 ft 2 in)
Length overall: tail rotor turning	15.54 m (51 ft 0 in)
both rotors turning	17.76 m (58 ft 3¼ in)
Wing span: clean	5.23 m (17 ft 2 in)
over empty weapon racks	5.82 m (19 ft 1 in)
Height: over tailfin	3.55 m (11 ft 7½ in)
over tail rotor	4.30 m (14 ft 1¼ in)
to top of rotor head	3.84 m (12 ft 7 in)
overall (top of air data sensor)	4.66 m (15 ft 3½ in)
overall, Longbow radar	4.95 m (16 ft 3 in)
Main rotor ground clearance (turning)	3.59 m (11 ft 9¼ in)
Distance between c/l of pylons:	
inboard pair	3.20 m (10 ft 6 in)
outboard pair	4.72 m (15 ft 6 in)
Tailplane span	3.40 m (11 ft 2 in)
Wheel track	2.03 m (6 ft 8 in)
Wheelbase	10.59 m (34 ft 9 in)

AREAS:

Main rotor disc	168.11 m² (1,809.5 sq ft)
Tail rotor disc	6.13 m² (66.0 sq ft)

WEIGHTS AND LOADINGS:

Weight empty:	
without Longbow	approx 5,165 kg (11,387 lb)
with Longbow	5,352 kg (11,800 lb)
Max fuel weight: internal	1,108 kg (2,442 lb)
external (four Brunswick tanks)	2,712 kg (5,980 lb)
Primary mission gross weight	7,480 kg (16,491 lb)
Design mission gross weight	8,006 kg (17,650 lb)
Max T-O weight: -701 engines	9,525 kg (21,000 lb)
-701C engines, ferry mission, full fuel	10,432 kg (23,000 lb)
Max disc loading	62.1 kg/m² (12.71 lb/sq ft)

PERFORMANCE (A: with -701 engines, without Longbow at 6,552 kg; 14,445 lb AUW, L: Apache Longbow at 7,530 kg; 16,601 lb with -701C engines):

Never-exceed speed (VNE)	197 kt (365 km/h; 227 mph)
Max level and max cruising speed:	
A	158 kt (293 km/h; 182 mph)
L	143 kt (265 km/h; 165 mph)
Max rate of climb at S/L: L	736 m (2,415 ft)/min
Max vertical rate of climb at S/L:	
A	762 m (2,500 ft)/min
L	450 m (1,475 ft)/min
Service ceiling: A	6,400 m (21,000 ft)
L	5,915 m (19,400 ft)
Service ceiling, OEI: A	3,290 m (10,800 ft)
Hovering ceiling:	
IGE: A	4,570 m (15,000 ft)
L	4,170 m (13,690 ft)
OGE: A	3,505 m (11,500 ft)
L	2,890 m (9,480 ft)
Max range, internal fuel: 30 min reserves:	
A	260 n miles (482 km; 300 miles)
L	220 n miles (407 km; 253 miles)
no reserves: L	257 n miles (476 km; 295 miles)
Ferry range, max internal and external fuel, still air, 45 min reserves	1,024 n miles (1,899 km; 1,180 miles)
Endurance at 1,220 m (4,000 ft) at 35°C	1 h 50 min
Max endurance, L: internal fuel	2 h 44 min
internal and external fuel	8 h 0 min
g limits at low altitude and airspeeds up to 164 kt (304 km/h; 189 mph)	+3.5/–0.5

WEIGHTS FOR TYPICAL MISSION PERFORMANCE: (all without Longbow; A: anti-armour at 1,220 m/4,000 ft and 35°C, four Hellfire and 320 rounds of 30 mm ammunition; B: as A, but with 1,200 rounds; C: as A, but with six Hellfire and 540 rounds; D: anti-armour at 610 m/2,000 ft and 21°C, 16 Hellfire and 1,200 rounds; E: air cover at 1,220 m/4,000 ft and 35°C, four Hellfire and 1,200 rounds; F: as E but at 610 m/2,000 ft and 21°C, four Hellfire, 19 rockets, 1,200 rounds; G: escort at 1,220 m/4,000 ft and 35°C, 19 rockets and 1,200 rounds; H: escort at 610 m/2,000 ft and 21°C, 38 rockets and 1,200 rounds):

Mission fuel: A	727 kg (1,602 lb)
G	741 kg (1,633 lb)
E	745 kg (1,643 lb)
C	902 kg (1,989 lb)
B	1,029 kg (2,269 lb)
D	1,063 kg (2,344 lb)
H	1,077 kg (2,374 lb)
F	1,086 kg (2,394 lb)
Mission gross weight: A	6,552 kg (14,445 lb)
E	6,874 kg (15,154 lb)
G	6,932 kg (15,282 lb)
B, C	7,158 kg (15,780 lb)
D	7,728 kg (17,038 lb)
F	7,813 kg (17,225 lb)
H	7,867 kg (17,343 lb)

TYPICAL MISSION PERFORMANCE (A-H as above):

Cruising speed at intermediate rated power:	
C	147 kt (272 km/h; 169 mph)
D	148 kt (274 km/h; 170 mph)
F	150 kt (278 km/h; 173 mph)
B	151 kt (280 km/h; 174 mph)
E, H	153 kt (283 km/h; 176 mph)
A	154 kt (285 km/h; 177 mph)
G	155 kt (287 km/h; 178 mph)
Max vertical rate of climb at intermediate rated power:	
B, C	137 m (450 ft)/min
H	238 m (780 ft)/min
F, G	262 m (860 ft)/min
E	293 m (960 ft)/min
D	301 m (990 ft)/min
A	448 m (1,470 ft)/min
Mission endurance (no reserves): A, E, G	1 h 50 min
C	1 h 47 min
D, F, H	2 h 30 min
B	2 h 40 min

UPDATED

BOEING COMMERCIAL AIRPLANES

1901 Oakesdale Avenue Southwest, Renton, Washington 98055
Tel: (+1 206) 746 11 11
Fax: (+1 206) 746 29 49
Web: http://www.boeing.com
PRESIDENT: Alan R Mulally
EXECUTIVE VICE-PRESIDENTS:
 Airplane Programs: James M Jamieson
 Commercial Aviation Services: Michael Bair
VICE-PRESIDENT, COMMUNICATIONS: Tom Downey

Single-aisle aircraft are 717, 737 and 757; twin-aisle are 747, 767 and 777.
 Group revenue for 2001 was US$35.06 billion. Employment stood at 69,200 on 1 August 2002.

UPDATED

MCDONNELL DOUGLAS (BOEING) AIRLINER ORDERS AND DELIVERIES

With the exception of what is now the Boeing 717, there have been no recent deliveries of McDonnell Douglas jet airliners. A breakdown of the 3,483 such aircraft delivered up to 2001 appeared in the 2002-03 edition.

UPDATED

BOEING 737

US Navy designation: C-40A Clipper
US Air Force designations: C-40B and C-40C
TYPE: Twin-jet airliner.
PROGRAMME: Original Boeing 737 first flew 9 April 1967; -100 and -200 with Pratt & Whitney JT8D engines; -300 entered service with CFM56 engines, November 1984, followed by -400 and -500; 3,132nd and last of this generation delivered February 2000. These versions, including military T-43 and Surveiller, described in *Jane's Aircraft Upgrades*.
 Current 'Next-Generation' of family (initially 737X) originated in 1991, when Boeing asked more than 30 airlines to help define improved series; company board authorised offer for sale June 1993; Southwest Airlines ordered 63 737-700s (32 converted from options for 737-300s) plus 63 new options (all, and more, taken up) 18 November 1993; roll-out (737-700) 8 December 1996; first flight (N737X) 9 February 1997; certification 7 November 1997, immediately followed by first deliveries. CFM56-7B power plant first flew on Boeing 747 testbed on 16 January 1996. Flight test programme involved 10 aircraft: four 737-700s, three 737-800s and three 737-600s. FAA approval for 180-minute ETOPS granted in September 1999. By January 2000, B737s of all subtypes had flown over 100 million hours. 1,000th NG 737 (N418WN) first flew 1 November 2001 and delivered to Southwest Airlines on 13 November.
 Further enhancements, incorporated in 737-900 demonstrator, completed in March 2002; features include quiet climb system, GPS landing and synthetic vision system.
CURRENT VERSIONS: **737-600:** Smallest of current 737 family. Known as 737-500X until officially launched 15 March 1995; 110 two-class passengers; final assembly of prototype began at Renton 29 August 1997; roll-out December 1997; first flight 22 January 1998; FAA certification 18 August 1998; first delivery (SE-DNM to launch customer SAS) 18 September 1998.
 737-700: First to be ordered and manufactured; mid-size version of family, equivalent to previous ('Classic') 737-300, seating 126 passengers in two-class layout. First aircraft (N737X) rolled out 2 (officially 8) December 1996; first flight 9 February 1997, followed by second aircraft 27 February that year; second aircraft attained maximum certified altitude of 12,500 m (41,000 ft) for the first time on 19 March 1997; FAA certification 7 November 1997, with first delivery (fourth built, N700GS to Southwest Airlines) on 17 December 1997.
 737-700IGW: Formerly 737-700X; increased gross weight version, under study in 1997; based on Boeing Business Jet airframe; three versions under consideration: passenger, equipped with three underfloor auxiliary fuel tanks extending range beyond 3,496 n miles (6,475 km; 4,023 miles); quick-change -700QC version with cargo door aft of port forward and cabin door and cargo handling system ordered as **C-40A Clipper** to meet US Navy requirement for C-9B Skytrain II replacement; and all-cargo version. First C-40A flew 17 April 2000; first delivery to US Naval Reserve Fleet Logistics Support Squadron VR-59 at NAS/JRB Fort Worth, Texas 21 April 2001; three additional C-40As operated by VR-59, and two by VR-58 at NAS Jacksonville, Florida; sixth delivered 28 October 2002.
 US Air Force ordered first of up to seven C-40Bs in February 2001 to replace Air National Guard C-22s (Boeing 727s) from August 2003 onwards. Initial aircraft (01-0005) completed late 2002. However, at some time, USAF obtained two former civilian 737-700s (02-0201

BOEING AIRLINER FIVE-YEAR DELIVERIES

Year	717	737	747	757	767	777	MD-11	MD-80	MD-90	Total
1998		281	53	54	47	74	12	8	34	**563**
1999	12	320	47	67	44	83	8	26	13	**620**
2000	32	281	25	45	44	55	4		3	**489**
2001	49	299	31	45	40	61	2			**527**
2002	20	223	27	29	35	47				**381**

BOEING COMMERCIAL AIRPLANE GROUP ORDERS AND DELIVERIES
(at 1 January 2003)

	Orders Total	Deliveries Total	Orders in 2002	Deliveries in 2002
707/720	1,010	1,010	0	0
Total	**1,010**	**1,010**	**0**	**0**
717-200	153	113	16 (32)	20
Total	**153**	**113**	**16**	**20**
727	1,831	1,831	0	0
Total	**1,831**	**1,831**	**0**	**0**
737-100	30	30	0	0
737-200	1,114	1,114	0	0
737-300	1,113	1,113	0	0
737-400	486	486	0	0
737-500	389	389	0	0
737-600	81	47	0	5
737-700/C-40A	807	423	−26 (29)	73
737-700BBJ	71	64	6	9
737-800	1,028	677	135 (125)	126
737-800BBJ	8	7	0	2
737-900	50	29	4 (2)	8
Total	**5,177**	**4,379**	**119**	**223**
747-100	250	250	0	0
747-200	393	393	0	0
747-300	81	81	0	0
747-400	453	431	10	5
747-400D	19	19	0	0
747-400ER	6	3	0	3
747-400ERF	11	3	3 (1)	3
747-400F	97	78	6	15
747-400M	61	61	−2	1
Total	**1,371**	**1,319**	**17**	**27**
757-200	906	902	0	14
757-200M	1	1	0	0
757-200PF	80	80	0	0
757-300	63	39	0	15
Total	**1,050**	**1,022**	**0**	**29**
767-200	128	128	0	0
767-200ER	117	112	5	1
767-300	104	104	0	0
767-300ER	505	471	−4 (3)	20
767-300F	40	40	0	1
767-400ER	37	37	−3	13
Total	**931**	**892**	**−2**	**35**
777-200	86	81	−3	0
777-200ER	407	299	10 (15)	41
777-200LR	5	0	2	0
777-300	65	44	7 (4)	6
777-300ER	56	0	10 (11)	0
Total	**619**	**424**	**26**	**47**
Grand Totals	**12,142**	**10,990**	**176 (251)**	**381**

Note: Table does not include one each of 727-100, 747-100, 757-200, 767-200 and 777-200 owned by Boeing. Annual order total is net (after deduction of cancellations), some being negative quqntities; quantities in parentheses are 2002 gross announced orders (where different)

UPDATED

and -0202 as C-40Cs. Commercial launch customer for the 737-700QC was Saudi Aramco of Saudi Arabia, which took delivery of two in 2001.
 First civil equivalent, designated **737-700C**, was N743A, first flown 18 September 2001 and delivered to ARAMCO on 31 October.
 737-800: Known as 737-400X Stretch until launched 5 September 1994; seats 162 two-class passengers; roll-out (N737BX) 30 June 1997; first flight 31 July 1997; certified

by FAA 13 March and JAA 9 April 1998; first delivery (D-AHFC to launch customer Hapag-Lloyd) 22 April 1998. Hapag-Lloyd became first commercial courier to operate 737-800 fitted with optional winglets in May 2001.
 737-900: Formerly 737-900X; launched 10 November 1997 with an order for 10, plus 10 options, from Alaska Airlines; largest 737 variant to date, with (compared to -800) stretch by means of 1.57 m (5 ft 2 in) forward plug and 1.07 m (3 ft 6 in) aft plug and strengthened fuselage;

seating for 177 two-class passengers; deliveries from early 2001. One prototype/certification aircraft only; rolled out 23 July 2000; first flight (N737X) 3 August 2000; FAA certification achieved 17 April 2001 following 156 hours of ground testing and a two-aircraft flight test programme totalling 649 flight hours in 296 sorties; first delivery (Alaska Airlines) in April 2001.

737-900X: Increased capacity, long-range version under study in 2001 to compete with Airbus A321 in European charter market; would feature two additional Type I emergency exits aft of the wing, or enlarged forward doors, to increase certification-limited maximum capacity to more than 200 passengers; up to 4,536 kg (10,000 lb) increase in max take-off weight; max payload range 2,000 n miles to 2,400 n miles (3,704 km to 4,445 km; 2,301 miles to 2,762 miles).

Boeing Business Jet (BBJ): Corporate versions, *described separately.*

Special missions: Military versions, *described separately.*

CUSTOMERS: See table.

COSTS: List price (2002; all in millions): 737-600 US$41.0 to US$49.0; 737-700 US$47.0 to US$55.0; 737-800 US$57.5 to US$64.5; 737-900 US$60.5 to 68.5.

DESIGN FEATURES: Conventional, medium-size airliner with podded engines and sweptback wing and tail surfaces. Dihedral 6° at root; sweepback 25° at quarter-chord. Greater range and speed than previous 737s, with less noise and fewer emissions; wing area increased by some 25 per cent by means of 0.43 m (1 ft 5 in) increase in wing chord and about 4.83 m (15 ft 10 in) increase in wing span; new high-lift systems; larger tail surfaces; increased tankage gives US transcontinental range; new aircraft can use same runways, taxiways, ramps and gates as preceding variants; new variant of CFM56 turbofan derated from nominal thrust to suit smaller versions of the family. Noise on ground reduced by approximately 12 dB by new diffuser duct and cooling vent silencer on APU, new ECS fan and duct and new electrical/electronics cooling fan.

FLYING CONTROLS: Conventional and powered. All surfaces actuated by two independent hydraulic systems with manual reversion for ailerons and elevator; elevator servo tabs unlock on manual reversion; rudder has standby hydraulic actuator and system. Three outboard-powered overwing spoiler panels on each wing assist lateral control and also act as airbrakes. Variable incidence tailplane has two electric motors and manual standby.

Leading-edge Krueger flaps inboard and four sections of slats outboard of engines; two airbrake/lift dumper panels on each wing, inboard and outboard of engines; continuous-span, double-slotted trailing-edge flaps inboard and outboard of engines.

FAA Cat. II landing minima system standard using SP-300 dual digital integrated flight director/autopilot; Cat. IIIa capability optional.

STRUCTURE: Aluminium alloy dual-path fail-safe two-spar wing structure with corrosion-resistant 7055-T77 upper skin. Aluminium alloy two-spar tailplane. Graphite composites ailerons, elevators and rudder. Aluminium honeycomb spoiler/airbrake panels and trailing-edges of slats and flaps. Fuselage structure fail-safe aluminium. Elevators, rudder and ailerons contain graphite/Kevlar; other, unstressed, components in GFRP and CFRP include nosecone, wing/fuselage fairing, fin fillet, fintip and flap actuator fairings. Rears of engine nacelles are of graphite/Kevlar/glass fibre.

LANDING GEAR: Hydraulically retractable tricycle type, with Boeing oleo-pneumatic shock-absorbers; inward-retracting main units have no doors, wheels forming wheel well seal; nose unit retracts forward; free-fall emergency extension. Twin nosewheels have tyres size 27×7.75. Main units have heavy-duty twin wheels, H40×14.5-19 heavy-duty tyres, and Honeywell or Goodrich heavy-duty wheel brakes as standard. Mainwheel tyre pressure 13.45 to 14.00 bar (195 to 203 lb/sq in). Nosewheel tyre pressure 11.45 to 11.85 bar (166 to 172 lb/sq in).

POWER PLANT: *737-600:* Two CFM International CFM56-7B18 turbofans, each rated at 86.7 kN (19,500 lb st) standard, or two CFM56-7B22s, each rated at 101 kN (22,700 lb st) in high gross weight version.

737-700: Two CFM56-7B20s, each rated at 91.6 kN (20,600 lb st) standard, or two CFM56-7B24s, each rated at 101 kN (22,700 lb st) in high gross weight version.

737-800: Two CFM56-7B24s, each rated at 107.6 kN (24,200 lb st) standard, or two CFM56-7B27s, each rated at 121.4 kN (27,300 lb st) in high gross weight version.

737-900: Two CFM56-7B26s, each rated at 117 kN (26,300 lb st) standard, or two CFM56-7B27s, each rated at 121.4 kN (27,300 lb st) in high gross weight version.

Fuel capacity (all) 26,025 litres (6,875 US gallons; 5,725 Imp gallons).

ACCOMMODATION: *All:* Crew of two side by side on flight deck. One plug-type door at each corner of cabin, with passenger doors on port side and service doors on starboard side. Airstair for forward cabin door optional. Overwing emergency exit on each side. One or two galleys and one modular vacuum lavatory forward and one or two galleys and lavatories aft; all lavatories interconnected to single waste collection tank at port side rear. Lightweight interior, of crushed core materials, has movable class divider, overnight seating-pitch flexibility and modular passenger service unit (PSU) including fold-down video screen in

BOEING 737 NEXT-GENERATION ORDER BOOK
(at 1 January 2003)

Customer	Variant	First Order	600	700	800	900
Air Algérie	737-600	16 Jul 98	5			
	737-800	16 Jul 98			7	
Air Berlin	737-800	22 Dec 94			26	
Air China	737-700	31 Jul 01		6		
	737-800	30 Oct 97			11	
Air Pacific	737-700	-		1		
	737-800	-			2	
Alaska Airlines	737-700	10 Nov 97		17		
	737-900	10 Nov 97				14
American Airlines	737-800	21 Nov 96			118	
American Trans Air	737-800	30 Jun 00			20	
Ansett Worldwide	737-700	-		4		
Aramco Services Co	737-700	15 Dec 99		5		
Bavaria	737-700	30 Jan 95		6		
Bouillion Aviation	737-700	29 Jul 98		18		
	737-800	29 Jul 98			14	
Braathens	737-700	3 Feb 97		10		
Britannia Airways	737-800	18 Nov 98			4	
China Airlines	737-800	22 Dec 95			15	
China Eastern	737-700	2 Oct 01		4		
China Southern	737-800	2 Oct 01			20	
China Southwest	737-800	1 Jun 98			3	
China Yunnan	737-700	30 Oct 97		4		
CIT Leasing	737-700	14 Jun 99		19		
	737-800	14 Jun 99			11	
Continental	737-700	25 Jul 96		51		
	737-800	25 Jul 96			115	
	737-900	18 Mar 98				15
Copa Airlines	737-700	19 Jan 99		12		
	737-800	28 Jun 02			2	
Delta Air Lines	737-800	10 Jun 97			132	
Eastwind Airlines	737-700	20 Oct 97		2		
EasyJet	737-700	28 Jul 98		32		
El Al	737-700	1 Apr 98		2		
	737-800	1 Apr 98			3	
Ethiopian Airlines	737-700	28 Nov 02		3		
Garuda	737-700	15 Dec 99		18		
GATX	737-800	31 Jul 96			38	
GECAS	737-600	22 Jan 96	7			
	737-700	22 Jan 96		99		
	737-800	22 Jan 96			56	
Germania	737-700	28 Mar 95		12		
Hainan Airlines	737-800	30 Oct 97			11	
Hapag-Lloyd	737-800	18 Nov 94			27	
ILFC	737-600	25 Jul 95	31			
	737-700	25 Jul 95		109		
	737-800	25 Jul 95			74	
Itochu AirLease	737-800	11 Jul 96			10	
Jet Airways	737-700	14 Jun 99		4		
	737-800	11 Dec 96			11	
	737-900	14 Jun 99				1
Kenya Airways	737-700	7 Mar 00		2		
KLM	737-800	14 Mar 97			13	
	737-900	14 Mar 97				4
Korean Air	737-800	9 Jun 98			6	
	737-900	9 Jun 98				16
Lauda Air	737-600	31 Dec 98	2			
	737-700	8 Jan 99		2		
	737-800	8 Jan 99			4	
Linas Aereas Azteca	737-700	13 Feb 01		2		
LOT	737-800	9 Oct 96			2	
Maersk	737-700	16 Jun 94		12		
Midway Airlines	737-700	14 Jun 99		7		
Oman Air	737-700	30 Apr 01		1		
	737-800	30 Apr 01			1	
Pegasus Airlines	737-800				1	
Pembroke Capital Ltd	737-600	14 Mar 95	1			
	737-700	6 Mar 97		8		
Qantas	737-800	9 Nov 01			19	
Royal Air Maroc	737-700	30 Aug 96		5		
	737-800	30 Aug 96			14	
Ryanair	737-800	9 Mar 98			131	
SAS	737-600	14 Mar 95	28			
	737-700	14 Mar 95		6		
	737-800	14 Mar 95			23	
SA Victoria	737-800	11 Jul 96			2	
Shanghai Airlines	737-700	30 Oct 97		4		
	737-800	2 Oct 01			1	
Shenzhen Airlines	737-700	30 Oct 97		6		
South African Airways	737-800	31 May 00			5	
Southwest	737-700	17 Nov 93		242		
Sunrock	737-700	17 Jun 97		2		
	737-800	17 Jun 97			3	
Taiwan Air Force	737-800	30 Nov 98			1	
TAROM	737-700	17 Jun 99		4		
Tombo Aviation	737-700	31 Dec 96		7		
	737-800	31 Dec 96			8	
Transavia	737-700	15 Nov 01		4		
	737-800	6 Nov 95			14	
TunisAir	737-600	28 Oct 97	7			
Turkish Airlines	737-800	7 Oct 97			26	
Ukraine Airlines	737-700	15 Dec 99		1		
US Navy	737-700	3 Sep 97		6		
Westjet	737-700	23 Aug 00		30		
Wuhan Airlines	737-800	29 Jun 98			2	
Xiamen Airlines	737-700	30 Oct 97		6		
Various unidentified	737-700BBJ	22 Jan 96		71		
	737-800BBJ	19 Nov 97			8	
	737-700	18 Mar 02		12		
	737-800	21 Feb 01			22	
Subtotals			**81**	**878**	**1,036**	**50**
Total				**2,045**		

Note: For deliveries see Group Orders and Deliveries table

Germania Boeing 737-700 in the colours of a German tour operator (*Jane's/Paul Jackson*)   *NEW*/0526939

underside of baggage bin. Centreline stowage bins optional for emergency equipment and crew baggage.

Two underfloor baggage holds, forward and aft of wing. Rear hold has provision for post-delivery installation of telescopic baggage conveyor system (when additional fuel tanks not fitted). One baggage door in starboard side of each hold.

737-600: Alternative cabin layouts seat from 110 to 132 passengers. Typical arrangements offer eight first class seats four-abreast at 91 cm (36 in) pitch and 102 tourist class seats six-abreast at 81 cm (32 in) pitch in mixed class; and 132 all-tourist class at 76 cm (30 in) pitch. Total overhead baggage capacity of 6.1 m³ (216 cu ft), equivalent to 0.045 m³ (1.6 cu ft) per passenger.

737-700: Alternative cabin layouts seat from 126 to 149 passengers. Typical arrangements offer eight first class seats four-abreast at 91 cm (36 in) pitch and 118 tourist class seats six-abreast at 81 cm (32 in) pitch in mixed class; and 149 all-tourist class at 76 cm (30 in) pitch. Total overhead baggage capacity of 7.0 m³ (248 cu ft), equivalent to 0.05 m³ (1.8 cu ft) per passenger. *C-40A* options comprise 121 passengers, all-cargo (eight pallets) and combinations of 70 passengers and three pallets.

737-800: Alternative cabin layouts seat from 162 to 189 passengers. Typical arrangements offer 12 first class seats four-abreast at 91 cm (36 in) pitch and 150 tourist class seats six-abreast at 81 cm (32 in) pitch in mixed class; and 189 all-tourist class at 76 cm (30 in) pitch. Total overhead baggage capacity of 9.3 m³ (328 cu ft), equivalent to 0.05 m³ (1.7 cu ft) per passenger.

737-900: Alternative cabin layouts seat from 177 to 189 passengers. Typical arrangements offer 12 first class seats four-abreast at 91 cm (36 in) pitch and 165 tourist class seats six-abreast at 81 cm (32 in) pitch in mixed class; and 189 all-tourist class at 81 cm (32 in) pitch.

SYSTEMS: Honeywell 131-9(B) APU with air start capability to maximum certified altitude and 90 kVA electrical load capability to 11,278 m (37,000 ft). Three-wheel air cycle environmental control system with optional ozone converter and digital cabin pressure controls.

AVIONICS: *Flight:* Optional GPS, satcom and dual FMS (single standard) integrated with GPS.

Instrumentation: Honeywell Air Transport Systems common display system (CDS) with six-screen flat-panel liquid crystal display (LCD) technology and programmable software, enables operators to emulate previous 737 electronic flight instrument system (EFIS) and 747-400/777 primary flight display-navigation display (PFD-ND) flight deck formats. Optional HUD.

DIMENSIONS, EXTERNAL:
Wing span (all versions): standard	34.31 m (112 ft 7 in)
with winglets	35.79 m (117 ft 5 in)
Wing aspect ratio, standard	9.4
Length: overall: 600	31.24 m (102 ft 6 in)
700	33.63 m (110 ft 4 in)
800	39.47 m (129 ft 6 in)
900	42.11 m (138 ft 2 in)
fuselage: 700	32.18 m (105 ft 7 in)
800	38.02 m (124 ft 9 in)
900	40.67 m (133 ft 5 in)
Height overall: 600, 700	12.57 m (41 ft 3 in)
800, 900	12.55 m (41 ft 2 in)
Tailplane span: all	14.35 m (47 ft 1 in)
Wheel track (c/l shock-struts): all	5.71 m (18 ft 9 in)
Wheelbase: 700	12.60 m (41 ft 4 in)
800	15.60 m (57 ft 2 in)
900	17.17 m (56 ft 4 in)
Distance between engine centrelines	9.65 m (31 ft 8 in)

Main passenger door (port, fwd), all:
Height	1.83 m (6 ft 0 in)
Width	0.86 m (2 ft 10 in)
Height to sill: at OWE	2.74 m (9 ft 0 in)
at MTOW	2.59 m (8 ft 6 in)

Passenger door (port, rear):
Height: all	1.83 m (6 ft 0 in)
Width: all	0.76 m (2 ft 6 in)
Height to sill: 600, 700: at OWE	3.10 m (10 ft 2 in)
at MTOW	2.95 m (9 ft 8 in)
800, 900: at OWE	3.12 m (10 ft 2 in)
at MTOW	2.97 m (9 ft 9 in)

Emergency exits (overwing, port and stbd, each), all:
Height	0.96 m (3 ft 2 in)
Width	0.51 m (1 ft 8 in)

Service door (stbd, fwd), all:
Height	1.65 m (5 ft 5 in)
Width	0.76 m (2 ft 6 in)
Height to sill: at OWE	2.74 m (9 ft 0 in)
at MTOW	2.59 m (8 ft 6 in)

Service door (stbd, rear):
Height: all	1.65 m (5 ft 5 in)
Width: all	0.76 m (2 ft 6 in)
Height to sill: 600, 700: at OWE	3.10 m (10 ft 2 in)
at MTOW	2.97 m (9 ft 9 in)
800, 900: at OWE	3.12 m (10 ft 3 in)
at MTOW	2.97 m (9 ft 9 in)

Baggage hold door (stbd, fwd), all:
Height: door 1.30 m (4 ft 3 in) clear access	0.89 m (2 ft 11 in)
Width	1.22 m (4 ft 0 in)
Height to sill	1.45 m (4 ft 9 in)

Baggage hold door (stbd, rear), all:
Height: door 1.22 m (4 ft 0 in) clear access	0.84 m (2 ft 9 in)
Width	1.22 m (4 ft 0 in)
Height to sill: 700	1.78 m (5 ft 10 in)
800, 900	1.80 m (5 ft 11 in)

DIMENSIONS, INTERNAL::
Cabin, aft of flight deck to rear pressure bulkhead:
Length: 600	21.79 m (71 ft 6 in)
700	24.18 m (79 ft 4 in)
800	30.02 m (98 ft 6 in)
Max height: all	2.13 m (7 ft 0 in)
Floor area: 600	67.3 m² (725 sq ft)
700	75.1 m² (808 sq ft)
800	94.0 m² (1,012 sq ft)

Baggage hold:
Length: 700: front	4.67 m (15 ft 4 in)
rear	8.03 m (26 ft 4 in)
800: front	7.67 m (25 ft 2 in)
rear	10.87 m (35 ft 8 in)
900: front	9.25 m (30 ft 4 in)
rear	11.94 m (39 ft 2 in)
Max width, all: at roof	3.15 m (10 ft 4 in)
at floor	1.22 m (4 ft 0 in)
Min height, all	1.13 m (3 ft 8½ in)

Volume:
600: front	7.0 m³ (248 cu ft)
rear	13.4 m³ (472 cu ft)
700: front	11.5 m³ (406 cu ft)
rear	16.9 m³ (596 cu ft)
800: front	19.6 m³ (692 cu ft)
rear	25.5 m³ (899 cu ft)
900: front	23.8 m³ (840 cu ft)
rear	28.7 m³ (1,012 cu ft)

AREAS:
Wings, gross	125.00 m² (1,345.5 sq ft)
Vertical tail surfaces (total)	26.40 m² (284.2 sq ft)
Horizontal tail surfaces (total)	32.80 m² (353.1 sq ft)

WEIGHTS AND LOADINGS (600: A: CFM56-7B18 engines, B: CFM56-7B22s; 700: A: CFM56-7B20s, B: CFM56-7B24s; 800: A: CFM56-7B24s, B: CFM56-7B27s; 900: A: CFM56-7B26s, B: CFM56-7B27s):

Operating weight empty:
600, 110 passengers: A, B	37,104 kg (81,800 lb)
700, 126 passengers: A, B	38,147 kg (84,100 lb)
800, 162 passengers: A, B	41,145 kg (90,710 lb)
900, 177 passengers: A, B	42,493 kg (93,680 lb)

Max T-O weight:
600: A	56,245 kg (124,000 lb)
B	65,090 kg (143,500 lb)
700: A	60,330 kg (133,000 lb)
B	70,080 kg (154,500 lb)
800: A	70,535 kg (155,500 lb)
B	79,015 kg (174,200 lb)
900: A	74,840 kg (164,000 lb)
B	79,015 kg (174,200 lb)

Max ramp weight:
600: A	56,470 kg (124,500 lb)
B	65,315 kg (144,000 lb)
700: A	60,555 kg (133,500 lb)
B	70,305 kg (155,000 lb)
800: A	70,760 kg (156,000 lb)
B	79,245 kg (174,700 lb)
900: A	74,615 kg (164,500 lb)
B	79,245 kg (174,700 lb)

Max landing weight:
600 A, B	54,655 kg (120,500 lb)
700: A	58,060 kg (128,000 lb)
B	58,605 kg (129,200 lb)
800: A	65,315 kg (144,000 lb)
B	66,360 kg (146,300 lb)
900: A, B	66,360 kg (146,300 lb)

Max zero-fuel weight:
600: A, B	51,480 kg (113,500 lb)
700: A	54,655 kg (120,500 lb)
B	55,200 kg (121,700 lb)
800: A	61,690 kg (136,000 lb)
B	62,730 kg (138,300 lb)
900: A, B	62,730 kg (138,300 lb)

Max wing loading:
600: A	450.0 kg/m² (92.16 lb/sq ft)
B	520.7 kg/m² (106.65 lb/sq ft)
700: A	482.6 kg/m² (98.85 lb/sq ft)
B	560.6 kg/m² (114.83 lb/sq ft)
800: A	564.3 kg/m² (115.57 lb/sq ft)
B	632.1 kg/m² (129.47 lb/sq ft)
900: A	595.1 kg/m² (121.89 lb/sq ft)
B	632.1 kg/m² (129.47 lb/sq ft)

Max power loading:
600: A	324 kg/kN (3.18 lb/lb st)
B	322 kg/kN (3.16 lb/lb st)
700: A	329 kg/kN (3.23 lb/lb st)
B	347 kg/kN (3.40 lb/lb st)
800: A	327 kg/kN (3.21 lb/lb st)
B	325 kg/kN (3.19 lb/lb st)
900: A	318 kg/kN (3.12 lb/lb st)
B	325 kg/kN (3.19 lb/lb st)

PERFORMANCE (A, B as above):
Max operating Mach No. (MMO): all	0.82
Cruising speed: all	M0.785

Approach speed:
600: A, B	125 kt (232 km/h; 144 mph)
700: A	130 kt (241 km/h; 150 mph)
B	131 kt (243 km/h; 151 mph)
800: A	141 kt (261 km/h; 162 mph)
B	142 kt (263 km/h; 163 mph)
900: A, B	145 kt (268 km/h; 167 mph)
Max certified altitude: all	12,500 m (41,000 ft)

Initial cruising altitude, ISA +10°C:
600: A	12,500 m (41,000 ft)
B	12,220 m (40,100 ft)
700: A	12,500 m (41,000 ft)
B	10,710 m (35,140 ft)
800: A	11,675 m (38,300 ft)
B	10,955 m (35,940 ft)
900: A	11,310 m (37,100 ft)
B	10,975 m (36,000 ft)

T-O field length, S/L, 30°C:
600: A	1,616 m (5,300 ft)
B	1,796 m (5,890 ft)

Winglet-equipped Boeing 737-800 (*Jane's/Paul Jackson*)   *NEW*/0526938

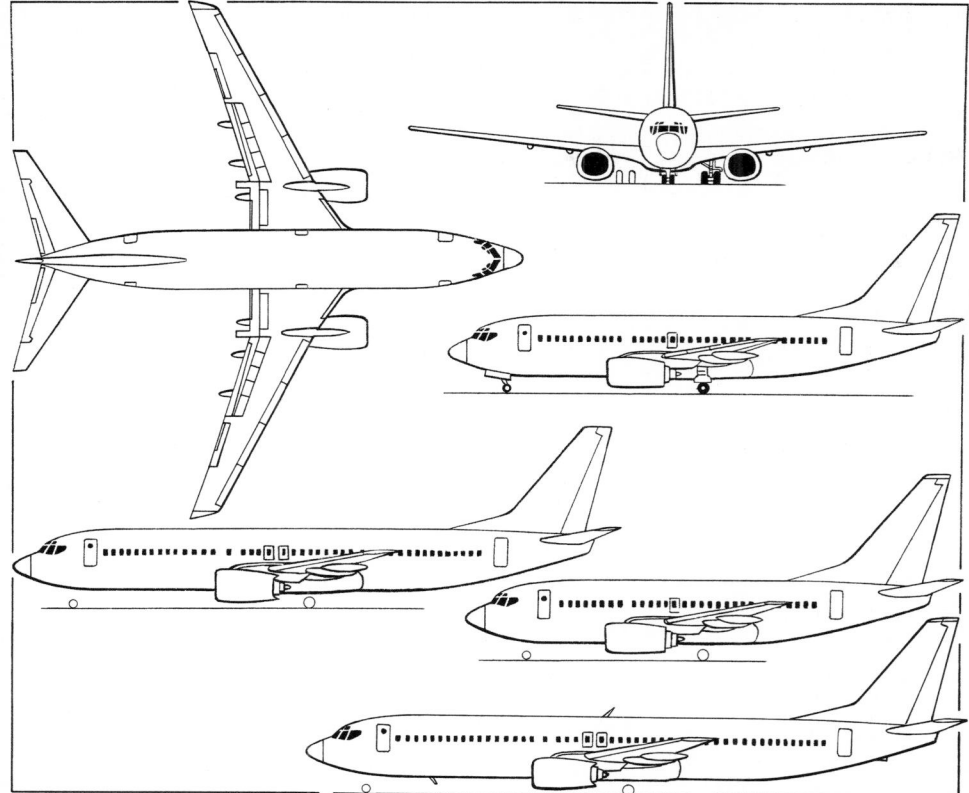

Boeing 737-700, with additional side views of stretched -800, shorter -600 and further stretched -900 (bottom) *(Jane's/James Goulding)*   0089520

700:	A	1,744 m (5,720 ft)
	B	1,921 m (6,300 ft)
800:	A	2,100 m (6,890 ft)
	B	2,308 m (7,570 ft)
900:	A	2,622 m (8,600 ft)
	B	2,500 m (8,200 ft)

Landing field length at max landing weight:

600:	A, B	1,342 m (4,400 ft)*
700:	A	1,418 m (4,650 ft)
	B	1,433 m (4,700 ft)*
800:	A	1,634 m (5,360 ft)
	B	1,659 m (5,440 ft)
900:	A, B	1,704 m (5,590 ft)

Design range:

600 with 110 passengers:

A	1,340 n miles (2,481 km; 1,542 miles)
B	3,050 n miles (5,648 km; 3,509 miles)

700 with 126 passengers:

A	1,540 n miles (2,852 km; 1,772 miles)
B	3,260 n miles (6,037 km; 3,751 miles)

800 with 162 passengers:

A	1,990 n miles (3,685 km; 2,290 miles)
B	2,940 n miles (5,444 km; 3,383 miles)

900 with 177 passengers:

A	2,060 n miles (3,815 km; 2,370 miles)
B	2,745 n miles (5,083 km; 3,158 miles)

* *Category D Honeywell brakes*

UPDATED

BOEING 747-400

TYPE: Wide-bodied airliner.

PROGRAMME (original): Announced 13 April 1966 (first ever wide-body jet airliner), with Pan American order for 25; official programme launch 25 July 1966; first flight 9 February 1969; FAA certification 30 December 1969; first delivery (to Pan Am) 12 December 1969; first route service New York-London flown 21 January 1970. 747-400 announced October 1985. In May 1990, Boeing decided to market only the -400; last -200 (a -200F Freighter for Nippon Cargo Air Lines) delivered 19 November 1991.

For all variants before 747-400, see *Jane's Aircraft Upgrades*. Production of earlier variants totalled 724 (205 -100, 45 SP, 393 -200 and 81 -300). Nineteen Pan American 747s modified as passenger/cargo **C-19A**s by Boeing Military Airplanes for Civil Reserve Air Fleet (see 1990-91 edition). Boeing board approved launch of 747-400IGW in December 1997. By 31 July 2002, 1,308 Boeing 747s (including 584 -400s) had been delivered, of which 1,100 remained in service. The worldwide fleet of 747s (all models) had flown more than 35 billion miles in 12 million flights, carrying 3.6 billion passengers, by January 2002.

PROGRAMME (current): Series 400 announced October 1985 as 747 development with extended capacity and range; design go-ahead July 1985; first order 22 October 1985; roll-out 26 January 1988; first flight 29 April 1988; certified with P&W PW4056 on 10 January 1989; first delivery 26 January 1989; entered service with Northwest Airlines 9

February 1989; certified with GE CF6-80C2B1F on 8 May 1989; R-R RB211-524G on 8 June 1989; R-R RB211-524H on 11 May 1990. Since May 1990, -400 is the only 747 marketed. 1,200th 747 delivered to British Airways on 17 February 1999.

CURRENT VERSIONS: **747-400:** Basic passenger version; standard and three optional gross weights (see below).

Detailed description applies to -400, except where indicated.

747-400M Combi: Passenger/freight version; initial order 9 April 1986; rolled out 23 March 1989; first flight 30 June 1989; certified 1 October 1989; first delivery 1 September 1989 to KLM. Maximum 266 three-class passengers with freight, 413 without; port-side rear freight door; main deck limit is seven pallets at 27,215 kg (60,000 lb); underfloor and fuel capacities as for passenger 747; 49 delivered by 31 December 1996. For all gross weights, maximum landing weight 285,763 kg (630,000 lb) and maximum zero-fuel weight 256,280 kg (565,000 lb). All three engine options available.

747-400F: All-freight version. See separate entry.

747-400 Domestic: Special high-density two-class 568-passenger version; first order 18 December 1988; rolled out 18 February 1991; first flown 18 March 1991; certified 10 October 1991 and delivered same day to Japan Air Lines (first of six) and later to All Nippon (six) and Japan Air System (one). Maximum T-O weight 272,155 kg (600,000 lb) but can be certified to 394,625 kg (870,000 lb). Structurally reinforced; no winglets; lower engine thrust; five more upper deck windows; revised avionics software and cabin pressure schedule; brake cooling fans; five pallets, 14 LD-1 containers and bulk cargo under floor; GE or P&W engines.

747-400 Performance Improvement Package (PIP): Announced April 1993, and first stage implemented in July 1993. Included gross weight increase of 2,268 kg (5,000 lb). Second stage, implemented in December 1993, included longer-chord dorsal fin made of CFRP, and wing

spoilers held down more tightly to reduce profile drag and leakage. These improvements were immediately applied to production aircraft and are retrofittable; PIP flight tested in leased United Airlines 747-400 May 1993.

747-400ER: Offered (as 747-400IGW) from December 1997 in response to Qantas requirement, for which the carrier has placed an order for six, with first delivery then scheduled for October 2002. One or two additional fuel tanks in hold. Range 7,500 n miles (13,890 km; 8,630 miles) with one additional tank; 7,700 n miles (14,260 km; 8,861 miles) with two. Structural strengthening around centrebody, wing/fuselage joint, flaps and landing gear.

Prototype N747ER (1,308th B747) rolled out 10 June 2002 (official ceremony on 17th); maiden flight 31 July 2002; became VH-OEE of Qantas.

747-400 ERF: See following entry.

747-400X QLR: Developed, Quiet Longer Range version, initially designated 747-400X. Based on 747-400 airframe, but with revised (B777-style) flight deck, crew rest/passenger sleeping area in upper aft fuselage, increased provision for carry-on baggage in cabin, 747-400F's thicker gauge outboard wing with B767-400ER-style raked wingtips (span 68.66 m; 225 ft 3 in) instead of winglets, MD-11-type trailing-edge wedges (of which flight tests began in October 1998), strengthened fuselage sections and landing gear, and modifications to cargo and fuel systems to permit installation of additional fuselage tank forward of centre wing tank, with second additional tank of same capacity optional. Max ramp weight 418,665 kg (923,000 lb); max T-O weight 417,760 kg (921,000 lb); operating weight empty 186,425 kg (411,000 lb), max structural payload 65,315 kg (144,000 lb); fuel load 248,714 litres (65,705 US gallons; 54,710 Imp gallons), giving max range, with 416 passengers in three classes, of 7,980 n miles (14,779 km; 9,183 miles); alternative 396,900 kg (875,000 lb) MTOW and 7,500 n mile (13,890 km; 8,630 mile) range to comply with QC2 noise regulations. Alternative seating up to 524 (including 42 first class). Cruising Mach No is 0.86. By late 2002, QLR had generated little airline interest and Boeing had developed the proposal with greater range and payload, provisionally designating it **747-800X**.

Initial engine is GE CF6-80C2B9F of 282 kN (63,300 lb st). New 'chevron' engine nacelles with serrated rear edges on core and fan nozzles promote mixing of bypass and core flows, and mixing of bypass flow and ambient air; combined with acoustic engine liners, these make QLR some 6 dB quieter (20 per cent on T-O; 40 per cent on approach), enabling it to meet QC2 noise standards.

QLR announced at Asian Aerospace, Singapore, 26 February 2002 (when 'chevron' design revealed); intended launch date was July 2002, permitting first deliveries in March 2004.

747-400XF QLR: Cargo version of Quiet Longer Range, with simplified and lightened handling system. Launched (then as Longer-Range 747-400 Freighter) April 2001 with order from ILFC. MTOW 417,760 kg (921,000 lb); range 5,150 n miles (9,537 km; 5,926 miles) with 112,810 kg (248,700 lb) payload. Cargo volumes (upper, lower and bulk) as for 747-400F. No extra fuel; capacity as for basic 747-400, GE-engined variant (203,325 litres; 53,765 US gallons; 44,769 Imp gallons). Typical cruising speed M0.845 on GE CF6-80C2B9F engines.

747X and 747X Stretch: Boeing cancelled development of its proposed 747X and 747X Stretch in March 2001.

747-800X: Under consideration by late 2002. Evolved from 747-400X QLR, with 1.98 m (6 ft 6 in) forward fuselage stretch; some 3,785 litres (1,000 US gallons; 833 Imp gallons) of additional fuel in tailplane tanks; between 20 and 40 more seats; and range increased to 8,000 n miles (14,816 km; 9,206 miles). Candidate engines in 276 to 285 kN (62,000 to 64,000 lb st) class.

CUSTOMERS: See table. Launch customer Northwest Orient Airlines ordered 10 -400s with PW4000s and 420-passenger interior October 1985; first delivery 26 January

Boeing 747-400 VH-OJH, name-ship of the Qantas *Longreach* class *(Jane's/Paul Jackson)*   0126850

1989. Total 1,338 of all 747 variants ordered (and 1,261 delivered) by 1 January 2001.

COSTS: US$186 million to US$211 million (2002); Combi version US$196 million to US$215 million (2002).

DESIGN FEATURES: Wide-bodied extrapolation of Boeing intercontinental jet configuration of low wing and four podded engines, optimised for greater passenger numbers and increased efficiency. Twin-deck forward fuselage; four mainwheel bogies for weight distribution.

According to engine type, fuel burn per seat over 3,000 n mile (5,556 km; 3,452 mile) sector varies between 135.8 kg (299.3 lb) and 138.5 kg (305.4 lb).

Sweepback at quarter-chord 37° 30′; thickness/chord ratio 13.44 per cent inboard, 7.8 per cent at mid-span, 8 per cent outboard; dihedral at rest 7°; incidence 2°; winglets, canted 22° outward and swept 60°, increase range by 3 per cent; upper deck extended rearward by 7.11 m (23 ft 4 in).

FLYING CONTROLS: Conventional and powered.

Elevators: Four elevator sections mechanically linked with breakable shear devices; each elevator has dual hydraulic-powered control units; control feel and three individual autopilot input servos mounted on central elevator quadrant; all surfaces have position transmitters; feel computer-operated by pitot pressure and tailplane angle.

Rudder: Upper rudder surface operated by three hydraulic actuators served by two hydraulic systems, lower surface by two actuators fed by remaining two hydraulic systems; no balance weights; each rudder has separate yaw damper module; left and right digital air data computers provide signals for controlling rudder ratio changer on each rudder surface according to air data and tailplane angle; combined feel actuator, rudder centring and trim actuator in rear servo area; mechanical cable linkage between rudder pedals and aft actuator area; rudder trim control switches on centre console. Maximum rudder deflection ±30°.

BOEING 747-400 ORDER BOOK
(at 1 January 2003)

Customer	Variant	First order	First delivery	Engine	400	400D	400ER	400ERF	400F	400M
Air Canada	400M	20 Jan 89	4 Jun 91	PW4056						3
Air China	400	31 May 90	20 Mar 92	PW4056	6					
	400M	16 May 86	13 Oct 89	PW4056						8
Air France	400	16 Dec 87	28 Feb 91	CF6-80	7					
	400ERF	24 Apr 01	none					2		
	400M	16 Dec 87	17 Sep 91	CF6-80						5
Air India	400	14 Aug 91	4 Aug 93	PW4056	6					
Air Namibia	400M	21 Apr 99	21 Oct 99	CF6-80						1
Air New Zealand	400	30 Jul 84	14 Dec 89	RB211-524	3					
	400	1 Mar 91	31 Oct 98	CF6-80	1					
All Nippon Airways	400	21 Oct 86	28 Aug 90	CF6-80	12					
	400D	21 Jan 86	13 Jan 92	CF6-80		11				
Amiri Flight	400	30 Nov 99	30 Nov 99	CF6-80	1					
Asiana Airlines	400	12 Jun 89	24 Jun 93	CF6-80	2					
	400F	3 Sep 90	4 Nov 94	CF6-80					6	
	400M	12 Jun 89	1 Nov 91	CF6-80						6
Atlas Air	400F	9 Jun 97	29 Jul 98	CF6-80					16	
British Airways	400	15 Aug 86	30 Jun 89	RB211-524	57					
Canadian Airlines Intnl	400	28 Jul 88	11 Dec 90	CF6-80	4					
Cargolux Airlines	400F	6 Dec 90	17 Nov 93	CF6-80					5	
	400F	13 Jun 95	8 Dec 98	RB211-524					7	
Cathay Pacific Airways	400	3 Jun 86	8 Jun 89	RB211-524	17					
	400F	28 Feb 90	1 Jun 94	RB211-524					5	
China Airlines	400	21 Jul 87	8 Feb 90	PW4056	19					
	400F	11 Aug 99	6 Jul 00	PW4056					17	
China Southern	400F	13 Feb 01	19 Jun 02	PW4062					2	
El Al	400	11 Dec 90	27 Apr 94	PW4056	4					
EVA Air	400	6 Oct 89	2 Nov 92	CF6-80	7					
	400F	6 May 99	20 Jul 00	CF6-80					3	
	400M	6 Oct 89	16 Sep 93	CF6-80						8
Garuda Indonesia	400	15 Nov 90	14 Jan 94	CF6-80	2					
GE Capital	400	22 Dec 95	18 Jun 99	CF6-80	1					
	400F	15 Dec 99	16 Oct 00	CF6-80					5	
ILFC	400	16 May 88	31 May 91	CF6-80	11					
	400	16 May 88	25 Sep 91	RB211-524	2					
	400F	30 Jan 90	14 Apr 99	CF6-80					1	
	400ERF	17 Apr 01	none	n/k				3		
Japan Airlines	400	21 Sep 87	25 Jan 90	CF6-80	37					
	400D	30 Jun 88	10 Oct 91	CF6-80		8				
	400F	7 Oct 02		CF6-80					2	
Japan Air SDF	400	23 Dec 87	17 Sep 91	CF6-80	2					
KLM	400	9 Apr 86	18 May 89	CF6-80	8					
	400M	9 Apr 86	1 Sep 89	CF6-80						17
Korean Air Lines	400	29 Aug 86	13 Jun 89	PW4056	27					
	400F	11 Jun 90	6 Sep 96	PW4056					10	
	400M	14 Apr 88	27 Jun 90	PW4056						1
Kuwait Airways	400M	12 Apr 92	29 Nov 94	CF6-80						1
Lufthansa	400	21 May 86	23 May 89	CF6-80	24					
	400M	21 May 86	19 Sep 89	CF6-80						7
Malaysia Airlines	400	19 Oct 88	27 May 90	CF6-80	2					
	400	12 Jan 89	27 Aug 92	PW4056	19					
	400M	30 Oct 87	6 Oct 89	CF6-80						2
Mandarin Airlines	400	15 Sep 94	14 Jun 95	PW4056	1					
Northwest Airlines	400	22 Oct 85	26 Jan 89	PW4056	16					
Omani Royal Flight	400	31 Jul 00	14 Dec 01	CF6-80	1					
Philippine Airlines	400	29 Oct 92	19 Nov 93	CF6-80	7					
	400M	18 Jan 96	29 Mar 96	CF6-80						1
Qantas Airways	400	2 Mar 87	11 Aug 89	RB211-524	21					
	400ER	19 Dec 00	31 Oct 02	CF6-80			6			
Saudia	400	18 Jun 95	24 Dec 97	CF6-80	5					
Singapore Airlines	400	27 Mar 86	18 Mar 89	PW4056	42					
	400F	16 Jan 90	5 Aug 94	PW4056					17	
South African Airways	400	6 May 89	19 Jan 91	RB211-524	6					
	400	30 Dec 98	30 Dec 98	CF6-80	2					
Thai Airways Intnl	400	16 Jun 87	21 Feb 90	CF6-80	16					
United Airlines	400	7 Nov 85	30 Jun 89	PW4056	44					
US Air Force (AL-1A)	400F	30 Jan 98	21 Jan 00	CF6-80					1	
UTA	400	3 Jul 86	22 Sep 90	CF6-80	1					
	400M	3 Jul 86	26 Jul 91	CF6-80						1
Virgin Atlantic Airways	400	20 Dec 96	17 Jun 97	CF6-80	9					
undisclosed	400	2 Apr 02	none	CF6-80	1					
	400ERF	31 Dec 01	none	n/k				6		
Subtotals					453	19	6	11	97	61
Total					647					

Note: For deliveries see Group Orders and Deliveries table

Tailplane: Tailplane angle set by hydraulic motor-driven shaft and ball screw with primary and secondary hydraulic brakes; flight control unit and air data computer signals sent to tailplane through dual stabiliser, trim and rudder ratio modules, which automatically apply Mach trim, and by dual-stabiliser control modules; tailplane trim limits computed according to flap positions.

Lateral control: Pilot and co-pilot aileron linkage can be physically separated if necessary; all four ailerons operate at low speeds; outboard ailerons are locked out at cruising speed; the inboard spoiler panel on each wing used on ground only; remainder have variable ratio response and spoiler mixer units; there are trim, centring and feel units.

Leading-edge and trailing-edge devices: Three-section Krueger flaps inboard of engines; variable camber slats between (five-section) and outboard (six-section) of engines lie flat when retracted and adopt camber curvature when extended. Two flap assemblies on each wing, one inboard of engines and the other between engines; three sections, fore flap, mid-flap and aft flap, move rearwards as single flat panel up to 5° deflection; thereafter, three sections separate progressively to form three slots, and camber angles relative to each other increase progressively.

Automatic flight control system: Combines autopilot, flight director and automatic tailplane trim and sends commands through triple independent flight control computers; system automates all flight phases except take-off; dual digital air data computers; pilots' primary flight and navigation displays are large-size cathode-ray tubes; two engine indicating and crew alerting screens, one on main panels, one on console; three multifunction control and display panels control flight management system, navigation and communications; flight control computers (autopilot) and inertial reference units are triplicated; new features include full-time autothrottle and dual-thrust management system included in flight management computer; integrated radio control panels and automatic start and shutdown of APU.

STRUCTURE: Wing and tail surfaces are aluminium alloy dual-path fail-safe structures; advanced aluminium alloys in wing torsion box save 2,721 kg (6,000 lb); advanced aluminium honeycomb spoiler panels; CFRP winglets and main deck floor panels; advanced graphite/phenolic and Kevlar/graphite in cabin fittings and engine nacelles; frame/stringer/stressed skin fuselage with some bonding. Improved corrosion protection and further coverage with compound introduced from 1993.

LANDING GEAR: Twin-wheel nose unit retracts forward; main gear consists of four four-wheel bogies; two, mounted side by side under fuselage at wing trailing-edge, retract forward; two, mounted under wings, retract inward; nosewheel steerable up to 70° left or right from tillers; full rudder pedal travel gives up to 7° for use at high speed; two centre main legs steer up to 13° when nosewheels are steered more than 20° and speed is less than 20 kt (37 km/h; 23 mph); carbon disc brakes on all mainwheels, with individually controlled digital anti-skid units; one of three brake pressure supplies automatically selected; main and nose tyres H49×19.0-20 or -22 (32 ply). Minimum ground turning radius, with body gear steering, is 48.46 m (159 ft 0 in) at wingtip and 27.73 m (91 ft 0 in) at nosewheels.

POWER PLANT: Four turbofans for baseline 747-400, these comprise: 252 kN (56,750 lb st) Pratt & Whitney PW4056; 258 kN (57,900 lb st) General Electric CF6-80C2B1F or 276 kN (62,100 lb st) CF6-80C2B5F; or 258 kN (58,000 lb st) Rolls-Royce RB211-524G or 270 kN (60,600 lb st) RB211-524H. Further optional engines (subject to certification) are 267 kN (60,000 lb st) PW4060, 276 kN (62,000 lb st) PW4062 and 274 kN (61,500 lb st) CF6-80C2B1F1. For 747-400ER, initial

Model of Boeing 747-400X QLR, showing extended wingtips and 'chevron' engine pod modifications *(Jane's/Paul Jackson)* 0131952

engine choices are CF6-80C2B5F, PW4062 and RB211-524H8T.

Fuel in four main tanks in wings can feed to any engine; in addition there are a centre-wing tank and reserve tanks in outer wing; optional tailplane tank; vent and surge tanks in outer wings and starboard tailplane; jettison pumps in inner main tanks; APU fed from port inner tank; automatic refuelling through two receptacles under each wing leading-edge between engines; automatic condensate scavenging and flame arresters in vent outlets.

Basic fuel capacity 204,355 litres (53,985 US gallons; 44,952 Imp gallons) with P&W and R-R engines; 203,523 litres (53,765 US gallons; 44,769 Imp gallons) with GE engines. At alternative higher T-O weights above 394,625 kg (870,000 lb) use of 12,492 litre (3,300 US gallon; 2,748 Imp gallon) tailplane/centre section tank is mandatory; fuel capacity including tailplane tank is therefore 216,846 litres (57,285 US gallons; 47,700 Imp gallons) with P&W and R-R engines and 216,013 litres (57,065 US gallons; 47,516 Imp gallons) with GE engines.

Usable fuel in 747-400ER comprises 239,389 litres (63,240 US gallons; 52,658 Imp gallons) with GE engines; 240,222 litres (63,460 US gallons; 52,841 Imp gallons) with P&W and R-R engines. Volumes include two tanks in cargo hold each of 11,583 litres (3,060 US gallons; 2,548 Imp gallons).

ACCOMMODATION: Two-crew flight deck, with seats for two observers; two-bunk crew rest cabin accessible from flight deck. Optional (but currently available on 90 per cent of B747-400 fleet) overhead cabin crew rest compartments above rear of main deck cabin (four bunks, four seats; eight bunks, two seats; two bunks, two seats, five sleeper seats). Typical 416-seat, three-class, long-range configuration accommodates 40 business class on upper deck; 23 first class in front cabin, 38 business class in middle cabin and 315 economy class in rear cabin on main deck. Maximum upper deck capacity 69 economy class. First class seating six abreast with two 86 cm (34 in) aisles, each twin-seat unit 1.45 m (4 ft 9 in) wide. Business passengers four abreast with 72 cm (28½ in) aisle and 1.37 m (4 ft 6 in) wide seat pairs on upper deck or two-three-two on lower deck with two 63 cm (24¾ in) aisles and 2.08 m (6 ft 10 in) triple seat. Economy seating three-four-three, with 49.5 cm (19½ in) aisles, two 1.51 m (4 ft 11½ in) triple seats and 2.07 m (6 ft 9½ in) quad seat. Five passenger doors on each side; upper deck emergency door each side.

747-400ER accommodates, typically, 500 in two-class arrangement (42 first, 458 economy or 416 three-class (as above).

Centre overhead stowage bins 0.16 m³ (5.7 cu ft) volume per 1.02 m (40 in) long bin; outboard bins 0.45 m³ (15.9 cu ft) volume per 1.52 m (60 in) long bin; 0.083 m³ (2.95 cu ft) bin volume per passenger (three-class). Two modular upper deck lavatories, 14 on main deck, relocatable (six upper deck optional locations; 33 on lower deck) and vacuum-drained into four waste tanks with combined volume of 1,136 litres (300 US gallons; 250 Imp gallons). Single-point drainage. Basic galley configuration, one on upper deck, seven centreline and two sidewall on main deck; lavatories and galleys can be quickly relocated if required fittings are installed; advanced integrated audio/video/announcement system.

Underfloor freight: forward compartment, five 2.44 m (96 in) × 3.18 m (125 in) pallets (totalling 58.8 m³, 2,075 cu ft) or 16 LD-1 containers (totalling 78.4 m³; 2,768 cu ft); aft compartment, 14 LD-1 containers (totalling 68.6 m³; 2,422 cu ft) or four pallets (totalling 47.0 m³; 1,660 cu ft); and bulk storage behind aft compartment 23.6 m³ (835 cu ft). Max capacity of 747-400 (16 LD-1s forward, 14 LD-1s aft and bulk storage in extreme rear) 170.6 m³ (6,025 cu ft). Door to each of three areas, all starboard side. Optional cargo door, port rear, on Combi version. 747-400ER lower deck accommodates 129 m³ (4,550 cu ft) with LD-1 containers, plus 22.3 m³ (789 cu ft) of bulk cargo, when two fuel tanks fitted. Capacity of 747-400X QLR reduced by up to six containers when both additional fuel tanks fitted; capacity 158.5 m³ (5,599 cu ft) with basic fuel; 137.0 m³ (4,837 cu ft) with max fuel.

SYSTEMS: Each engine drives a hydraulic pump feeding an independent system; services are connected to supplies in such a way that loss of one supply cannot disable one system; two hydraulic systems also have air-driven pumps to maintain pressure and two have electric pumps; one electric pump can be run to provide braking when the aircraft is being towed on the ground; all four hydraulic reservoirs can be filled from a single location in the port main landing gear bay.

Hot air bled from the low-pressure and high-pressure compressors of all four engines is precooled by fan exit air and fed via a manifold to the cabin pressurisation and air conditioning system and to provide de-icing of wing leading-edge and engine nose cowling and to pressurise hydraulic tanks. Three conditioning packs in wing/fuselage fairing provide cabin air; five cabin zones, each with digital temperature control.

Each engine drives an integrated drive generator supplying 90 kVA power to respective AC busses; three generators are a dispatch item, but one will supply essential loads; APU drives two further generators; automatic start-up, load transfers and load shedding reduce crew workload; power systems may be isolated from each other for triple-channel Cat. III autoland.

Completely self-contained 1,081 kW (1,450 shp) P&WC PW901A APU, mounted clear of all flight-critical structure and flight controls in the extreme tail, drives two 90 kVA generators that can supply electrical power for whole aircraft; also supplies compressed air to operate pneumatic components; can run at up to 6,100 m (20,000 ft) and supply compressed air below 4,575 m (15,000 ft). Capabilities include maintenance of 24°C (75°F) ground cabin temperature in 38°C (100°F) ambient conditions.

Forward underfloor cargo compartment heated to 5°C by hot air exhausted from flight deck cooling equipment and avionics in main equipment centre, boosted as necessary by two electrical heaters; rear underfloor hold heated to minimum 5°C or 18°C (selected by crew) by engine bleed.

Overheat detection and automatic extinguishing provided in all lavatories; APU automatically shut down and fire extinguisher bottles initiated on detection of fire;

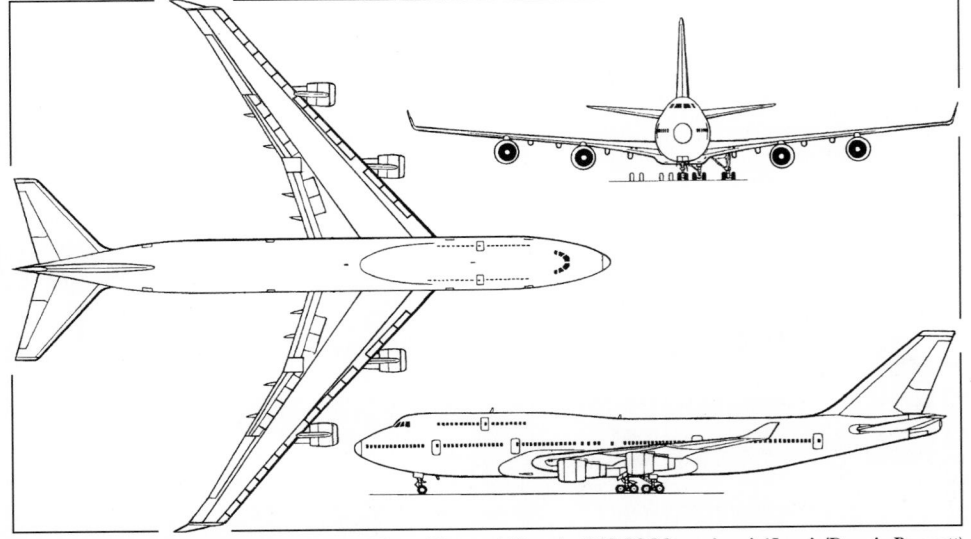

Boeing 747-400 advanced long-range airliner (General Electric CF6-80C2 engines) *(Jane's/Dennis Punnett)*
NEW/0526900

each engine has three dual fire detectors in series and a fourth detector for overheating. Underfloor freight compartments and upper deck hold of Combi have smoke detectors and extinguisher systems; wheel wells have overheat detectors.

AVIONICS: Boeing launched development of new Flight Management Computer software in January 1993 to match existing aircraft to international Future Air Navigation System (FANS-1) during 1995. Standard avionics fit as follows:

Comms: Dual VHF and HF transceivers with Selcal; dual transponders; flight intercom with air-to-ground facility, connectable also to satcom system; cabin entertainment and passenger address and service units.

Radar: Colour weather radar transmitting in I- and G-bands.

Flight: Dual VOR; triple ILS receivers with single marker beacon receiver; dual ADF; dual DME; all nav radios automatically tuned by flight management computer system (FMCS). Automatic flight control system (AFCS) integrates autopilot, flight director and automatic stabiliser trim functions; dual digital air data computers with dual selectable pressure sensors, angle of attack sensors and total air temperature probes; FMCS allows crew to preselect flight plan using standard air traffic control language; FMCS incorporates database, updated every 28 days, which includes data on waypoints, airports, standard instrument departures (SIDs), standard terminal arrival routes (STARs), airline routes and information on specific geographic areas; triple ring laser gyro inertial reference units provide navigation input on EFIS, flight management displays or radio magnetic indicators; other systems include ground proximity warning, triple low-range radio altimeters and TCAS.

Central maintenance computer monitors over 75 electrical and electromechanical systems, performs tests and centralises maintenance data; failures are indicated in EICAS displays and stored for future reference for in-flight use or line or hangar maintenance. Satcom datalink allows ground crews to interrogate system for additional information while aircraft in flight.

Instrumentation: Electronic flight instrument system (EFIS) comprising six (left/right inboard/outboard and central upper/lower) 20.3×20.3 cm (8×8 in) integrated display units (IDU), two each for primary flight display (PFD), navigation display (ND) and engine indicating and crew alerting (EICAS) functions; all IDUs receive data from all three EFIS/EICAS interface units (EIU), updated via software data loader; PFD and EICAS primary formats automatically switch to inboard and lower IDUs respectively, with facility for manual selection of formats on different IDUs as required. B747-400 has 181 switches, 171 lights and 13 gauges, compared with 284, 555 and 132 of earlier variants; total 365 is below average 450 for typical two-crew jet transport.

DIMENSIONS, EXTERNAL:

Wing span: normal	64.44 m (211 ft 5 in)
with winglets	64.92 m (213 ft 0 in)
Wing span, fully fuelled	64.92 m (213 ft 0 in)
Wing chord: at root	14.63 m (48 ft 0 in)
at tip	4.06 m (13 ft 4 in)
Winglet height	0.89 m (2 ft 11 in)
Wing aspect ratio	7.7
Length: overall	70.67 m (231 ft 10¼ in)
fuselage	68.63 m (225 ft 2 in)
Max width of fuselage	6.50 m (21 ft 4 in)
Height overall: at OWE: -400	19.51 m (64 ft 0 in)
-400ER	19.58 m (64 ft 3 in)
at MTOW: -400	18.77 m (61 ft 7 in)
-400ER	19.05 m (62 ft 6 in)
Tailplane span	22.17 m (72 ft 9 in)
Wheel track (c/l shock-struts):	
outer pair	11.00 m (36 ft 1 in)
inner pair	3.84 m (12 ft 7 in)
Wheelbase: mean	25.60 m (84 ft 0 in)
to forward main bogie	24.07 m (78 ft 11½ in)
to rear main bogie	27.14 m (89 ft 0½ in)
Distance between engine centrelines:	
outboard	41.66 m (136 ft 8 in)
inboard	23.37 m (76 ft 8 in)
Passenger doors (10, each):	
Height: door, clear access	1.93 m (6 ft 4 in)
Width: door	1.19 m (3 ft 11 in)
clear access	1.07 m (3 ft 6 in)
Height to sill: at OWE: front: -400	5.16 m (16 ft 11 in)
-400ER	5.21 m (17 ft 1 in)
rear: -400	5.31 m (17 ft 5 in)
-400ER	5.38 m (17 ft 8 in)
at MTOW: front: -400	4.72 m (15 ft 6 in)
-400ER	4.75 m (15 ft 7 in)
rear: 400	4.80 m (15 ft 9 in)
-400ER	4.98 m (16 ft 4 in)
Upper deck emergency door (two):	
Height: door	2.01 m (6 ft 7¼ in)
clear access	1.83 m (6 ft 0 in)
Width	1.07 m (3 ft 6 in)
Height to sill: at OWE: -400	7.90 m (25 ft 11 in)
-400ER	7.95 m (26 ft 1 in)
at MTOW: -400	7.52 m (24 ft 8 in)
-400ER	7.54 m (24 ft 9 in)
Baggage door (front hold):	
Height	1.68 m (5 ft 6 in)

Width	2.64 m (8 ft 8 in)
Height to sill: at OWE	3.10 m (10 ft 2 in)
at MTOW	2.69 m (8 ft 10 in)
Baggage door (rear hold):	
Height	1.68 m (5 ft 6 in)
Width	2.64 m (8 ft 8 in)
Height to sill: at OWE	3.17 m (10 ft 5 in)
at MTOW	2.82 m (9 ft 3 in)
Bulk loading door:	
Height: max (front)	1.42 m (4 ft 8 in)
min (rear)	1.24 m (4 ft 1 in)
Width	1.12 m (3 ft 8 in)
Mean height to sill: at OWE	3.40 m (11 ft 2 in)
at MTOW	3.00 m (9 ft 10 in)
Combi cargo door (port):	
Height (clear access)	3.05 m (10 ft 0 in)
Width	3.40 m (11 ft 2 in)
Height to sill: at OWE	5.26 m (17 ft 3 in)
at MTOW	4.87 m (16 ft 0 in)

747-400ER baggage door heights to sill: add 50 to 75 mm (2 to 3 in)

DIMENSIONS, INTERNAL:

Cabin (main): Max height	2.41 m (7 ft 11 in)
Passenger cabin volume	885.9 m³ (31,285 cu ft)

AREAS:

Wings, gross	541.16 m² (5,825.0 sq ft)
Ailerons (total)	20.90 m² (225.00 sq ft)
Trailing-edge flaps (total)	78.69 m² (847.00 sq ft)
Leading-edge flaps (total)	43.85 m² (472.00 sq ft)
Inboard spoilers (total)	12.78 m² (137.60 sq ft)
Outboard spoilers (total)	15.46 m² (166.40 sq ft)
Fin	77.11 m² (830.00 sq ft)
Rudder	21.37 m² (230.00 sq ft)
Tailplane	136.57 m² (1,470.00 sq ft)
Elevators (total, incl tabs)	30.38 m² (327.00 sq ft)

WEIGHTS AND LOADINGS (747-400: GB: CF6-80C2B1F, GM: CF6-80C2B5F, PB: PW4056, PM: PW4062, RB: RB211-524G, RM: 211-524M engines; all 416 passengers, five cargo pallets and 14 containers, 747-400ER: ER GM CF6-80C2B5F, ER PM: PW4062, ER RM: RB211-524H-8T with passengers/cargo as above):

Operating weight empty:

GB	180,485 kg (397,900 lb)
GM	180,895 kg (398,800 lb)
PB	180,845 kg (398,700 lb)
PM	181,255 kg (399,600 lb)
RB	181,435 kg (400,000 lb)
RM	181,845 kg (400,900 lb)
ER GM/PM/RM	184,565 kg (406,900 lb)

Baggage/freight capacity, all:

forward compartment	26,490 kg (58,400 lb)
aft compartment	22,938 kg (50,570 lb)
bulk compartment	6,749 kg (14,880 lb)

Max structural payload ER GM, PM, RM
67,175 kg (148,100 lb)

Max fuel weight:

GB	162,580 kg (358,425 lb)
GM	172,560 kg (380,425 lb)
PB, RB	163,250 kg (359,900 lb)
PM, RM	173,225 kg (381,900 lb)
ER, GM	192,190 kg (423,700 lb)
ER, PM, ER, RM	192,855 kg (425,175 lb)

Max T-O weight:

GB, PB, RB	362,875 kg (800,000 lb)
GM, PM, RM	396,895 kg (875,000 lb)
ER GM, PM, RM	412,770 kg (910,000 lb)

Max ramp weight:

GB, PB, RB	364,235 kg (803,000 lb)
GM, PM, RM	398,255 kg (878,000 lb)
ER, GM, PM, RM	414,130 kg (913,000 lb)

Max landing weight:

GB, PB, RB	260,360 kg (574,000 lb)
GM, PM, RM	295,745 kg (652,000 lb)
ER, GM, PM, RM	295,745 kg (652,000 lb)

Max zero-fuel weight:

GB, PB, RB	242,670 kg (535,000 lb)
GM, PM, RM	251,745 kg (555,000 lb)
ER, GM, PM, RM	251,745 kg (555,000 lb)

Max wing loading:

GB, PB, RB	670.5 kg/m² (137.34 lb/sq ft)
GM, PM, RM	733.4 kg/m² (150.21 lb/sq ft)

Max power loading:

GB, RB	352 kg/kN (3.45 lb/lb st)
GM, PB	359 kg/kN (3.52 lb/lb st)
PM	360 kg/kN (3.53 lb/lb st)
RM	368 kg/kN (3.61 lb/lb st)

PERFORMANCE (as above; landing at MLW):

Cruising Mach No.	0.85
Approach speed:	
GB, PB, RB	146 kt (270 km/h; 168 mph)
GM, PM, RM	157 kt (291 km/h; 181 mph)
Initial cruising altitude:	
GB, PB, RB	10,575 m (34,700 ft)
GM, PM, RM	10,000 m (32,800 ft)
T-O field length, 30°C (86°F):	
GB	2,820 m (9,250 ft)
GM	3,033 m (9,950 ft)
PB	2,820 m (9,250 ft)
PM	2,990 m (9,800 ft)
RB	2,850 m (9,350 ft)
RM	3,215 m (10,550 ft)
Landing field length:	
GB, PB, RB	1,905 m (6,250 ft)
GM, PM, RM	2,180 m (7,150 ft)
Design range:	
GB	6,185 n miles (11,454 km; 7,117 miles)
GM	7,260 n miles (13,445 km; 8,354 miles)*
PB	6,195 n miles (11,473 km; 7,129 miles)
PM	7,325 n miles (13,565 km; 8,429 miles)*
RB	6,040 n miles (11,186 km; 6,950 miles)
RM	7,170 n miles (13,278 km; 8,251 miles)*

Fuel volume limited

UPDATED

BOEING 747-400F
USAF designation: AL-1A

TYPE: Four-jet freighter.

PROGRAMME: Initial order 13 September 1989; rolled out 8 March 1993; first flight (N6005C) 4 May 1993; FAA certification 22 October 1993; JAR certification followed; first delivery (Cargolux) 17 November 1993. During 2000, Boeing began design of a freighter conversion of passenger 747-400s as an eventual partner to the first 747-300F conversion, begun in that year. However, this would differ from new-production -400F in several respects.

CURRENT VERSIONS: **747-400F:** *As described.*

747-400ERF: Freighter version of 747-400ER, described in previous entry; first order placed 30 April 2001. Deliveries to Air France (two), ILFC (three) and undisclosed customer (six).

Differences from 747-400ER include 302,090 kg (666,000 lb) max landing weight; 277,145 kg (611,000 lb) zero-fuel weight; 164,380 kg (362,400 lb) empty operating weight; 112,765 kg (248,600 lb) structural max payload; 530 m³ (18,720 cu ft) containerised volume on main deck; 159 m³ (5,600 cu ft) lower deck containerised volume and 14.7 m³ (520 cu ft) bulk cargo volume. Usable fuel is 203,523 litres (53,765 US gallons; 44,769 Imp gallons) for GE-engined version and 204,355 litres (53,985 US gallons; 44,952 Imp gallons) for P&W and R-R-engined versions.

747-400XF: Variant of 747-400X QLR (which see). ILFC to acquire five.

AL-1A: Anti-missile defence aircraft; details under Boeing Integrated and Defence Systems heading.

CUSTOMERS: See table with main 747-400 entry. First 747-400F delivered to Cargolux 17 November 1993; total of 34 delivered by October 1999. Production of 16 envisaged in 2000. Recent customers include China Airlines, which ordered 13 on 11 August 1999, for delivery between 2000 and 2007 (first on 6 July 2000); EVA Air, three for delivery from 2000 (first on 20 July); and Cathay Pacific Airways, which ordered two in October 1999 (first delivered on 12 September 2000), subsequently increasing its order to five.

COSTS: List price US$187.5 million to US$214.5 million (2002).

DESIGN FEATURES: 747-200F fuselage (short upper deck) with additional changes combined with stronger and larger 747-400 wing; strengthened floor of short upper deck, as offered for -200F, also integrated into 747-400F; further developed freight handling system; total cargo volume

Boeing 747-400F freighter of Singapore Airlines (*Jane's*/Paul Jackson)

NEW/0526931

777.8 m³ (27,467 cu ft), of which 604.5 m³ (21,347 cu ft) on main deck, 158.6 m³ (5,600 cu ft) in lower hold and 48.3 m³ (520 cu ft) available for bulk cargo. Compared to -200F, empty weight saving of 2,000 kg (4,409 lb) has raised maximum revenue freight load to about 113,000 kg (249,125 lb), at which range is 4,400 n miles (8,149 km; 5,063 miles); fuel consumption more than 15 per cent lower than 747-200F. Same gross weights as passenger 747-400; maximum landing weight at optional T-O weight, 302,090 kg (666,000 lb); maximum zero-fuel weight, 276,690 kg (610,000 lb), can be increased on condition T-O weight is decreased.

ACCOMMODATION: Two-pilot crew, as 747-400. Upward-opening nose cargo door and optional port-side rear cargo door; underfloor cargo doors fore and aft of wing and bulk cargo door aft of rear underfloor door; two crew doors to port. Capacity for 30 pallets on main deck and 32 LD-1 containers plus bulk cargo under floor.

UPDATED

BOEING 757-200
US Air Force designations: C-32A and C-32B

TYPE: Twin-jet airliner.

PROGRAMME: Announced early 1978; has 707/727/737 fuselage cross-section and two large turbofans; Eastern Air Lines and British Airways ordered 21 firm and 24 optioned and 19 + 18 respectively 13 August 1978; first flight (N757A) 19 February 1982 powered by 166.4 kN (37,400 lb st) Rolls-Royce RB535Cs and designated 757-200; first Boeing airliner launched with foreign engine.

FAA certification 21 December 1982; CAA certification 14 January 1983; revenue services began 1 January 1983 (EAL) and 9 February 1983 (BA). First flight of 757 powered by P&W PW2037s, 14 March 1984; certified October 1984 and delivered to Delta; first 757 with RB535E4s delivered to EAL 10 October 1984; first extended-range model delivered to Royal Brunei Airlines May 1986; 757 with RB535E4 engines approved FAA ETOPS December 1986 (extended to 180 minutes July 1990); 757 with PW2037/2040 ETOPS approved April 1990 (180 minutes for PW2037 April 1992); Boeing windshear guidance and detection system approved by FAA January 1987. Certified for operation in the Russian Federation and Associated States (CIS) September 1993. On 14 February 2002, 1,000th Boeing 757 was delivered, being 148th for American Airlines.

CURRENT VERSIONS: **757-200:** Initial production passenger airliner; extended range available.

Main description applies to -200 version, except where indicated.

757-200PF Package Freighter: Developed for United Parcel Service. Large freight door forward, single crew door and no windows; up to 15 standard 2.24 × 3.18 m (88 × 125 in) cargo pallets on main deck; operating weight empty 51,710 kg (114,000 lb), MTOW 115,665 kg (255,000 lb), MZFW 90,719 kg (200,000 lb), max structural payload 39,009 kg (86,000 lb). Cargo capacity 187 m³ (6,600 cu ft) on main deck plus 51.8 m³ (1,830 cu ft) on lower deck. UPS ordered 20 on 31 December 1985; deliveries began 17 September 1987.

757-200M Combi: Boeing's mixed cargo/passenger configuration with windows; upward-opening cargo door to port (forward) 3.40 × 2.18 m (134 × 86 in); carries up to three 2.24 × 2.74 m (88 × 108 in) cargo containers and 150 passengers; one delivered to Royal Nepal Airlines on 15 September 1988; no others by mid-2000.

757-200 Freighter: Developed by Pemco Aeroplex in 1992 as conversions of existing 757s; all-freight, combi and quick-change versions available; same weights as Boeing 757-200PF Package Freighter (see data below); choice of more powerful engines; large freight door forward on port side.

757SF: Under an agreement announced 5 October 1999, Boeing Airplane Services will purchase a total of 44 757-200s from British Airways and other operators and modify them, in conjunction with Israel Aircraft Industries and Singapore Technologies Aerospace, to 757SF Special Freighter configuration for lease to DHL Worldwide Express. The modification, which is also available to other customers, provides 226.5 m³ (8,000 cu ft) of cargo space with payload of 27,215 kg (60,000 lb) and range of over 2,000 n miles (3,704 km; 2,301 miles). Eventual DHL fleet will be 44, of which first two are 757-200PFs bought from Ansett on 15 November 1999. First converted 757SF, a former BA aircraft, entered service with DHL on 19 March 2001 and passed to EAT in Belgium on 28 March 2001 as OO-DLN. In 2002 a rival freighter conversion was offered by Precision Conversions of Goodyear, Arizona.

757-200 'Catfish': Boeing's own 757-200 prototype (N757A) fitted with radar nose in Lockheed Martin F-22A profile and representative F-22A swept wing section above flight deck containing conformal radar antennas for advanced radar trials; first flight in this configuration 11 March 1999. See also Lockheed Martin entry.

757-200ER: Projected extended-range version under study in 2001; would combine fuselage of 757-200 with strengthened wing structure of 757-300; up to four auxiliary fuel tanks in aft cargo hold; range 4,600 n miles; (8,519 km; 5,293 miles).

757-300: Stretched version. *Described separately.*

BOEING 757 ORDER BOOK
(at 1 January 2003)

Customer	First order	First Delivery	Variant	Engine	Qty
Air 2000	12 Dec 86	27 Apr 97	757-200	RB211	5
Air Europe	2 Jul 82	30 Mar 83	757-200	RB211	15
Air Holland	25 Jun 87	9 Mar 88	757-200	RB211	3
Airtours International	17 May 96	20 Mar 97	757-200	RB211	1
America West	6 Feb 87	10 Dec 87	757-200	RB211	4
American Airlines	25 May 88	17 Jul 89	757-200	RB211	126
American Trans Air	7 Sep 94	26 Sep 95	757-200	RB211	13
	30 Jun 00	4 Aug 01	757-300	RB211	12
Ansett Worldwide	7 Oct 87	15 Feb 89	757-200	RB211	19
	16 Nov 88	2 Dec 92	757-200	PW2037	2
	7 Oct 87	26 Jul 89	757-200PF	RB211	4
Arkia Israeli Airlines	31 Jul 98	31 Jan 00	757-300	RB211	2
Azerbaijan Airlines	30 Dec 97	27 Sep 00	757-200	RB211	2
Britannia Airways	20 Dec 91	10 Apr 92	757-200	RB211	11
British Airways	31 Aug 78	25 Jan 83	757-200	RB211	50
China Southern	1 Mar 88	22 Nov 88	757-200	RB211	19
China Southwest	31 May 90	5 Aug 92	757-200	RB211	12
China Xinjiang Airlines	30 Oct 97	30 Apr 98	757-200	RB211	6
CIT Leasing	6 Mar 01	none	757-300	n/k	1
Condor Flugdienst	29 Sep 88	19 Mar 90	757-200	PW2040	17
	9 Dec 96	10 Mar 99	757-300	PW2040	13
Continental Airlines	8 Oct 90	12 May 94	757-200	RB211	41
	2 Jan 01	20 Dec 01	757-300	RB211	15
Delta Air Lines	12 Nov 80	5 Nov 84	757-200	PW2037	110
	14 Nov 89	9 Jan 92	757-200	PW2040	6
Eastern Air Lines	31 Aug 78	22 Dec 82	757-200	RB211	25
El Al	1 Oct 86	25 Nov 87	757-200	RB211	7
Ethiopian Airlines	9 Jun 89	25 Feb 91	757-200	PW2040	4
	9 Jun 89	24 Aug 90	757-200PF	PW2040	1
Far Eastern Air Transport	8 Nov 94	8 Nov 94	757-200	PW2037	7
GATX Capital Corp	31 Jul 96	25 Apr 97	757-200	RB211	3
GE Capital	13 Jun 00	5 Feb 01	757-200	RB211	4
GPA Group Ltd	22 Apr 88	30 Aug 91	757-200	RB211	12
Iberia Airlines	25 May 90	7 Jun 93	757-200	RB211	24
Icelandair	19 Oct 88	4 Apr 90	757-200	RB211	7
	16 Jun 97	18 Mar 02	757-300	RB211	2
ILFC	18 Jun 86	1 Apr 87	757-200	RB211	43
	25 Apr 88	25 Feb 92	757-200	PW2037	13
	25 Apr 88	28 May 91	757-200	PW2040	26
JMC Airline	3 May 00	24 Apr 01	757-300	RB211	2
Kawasaki Leasing	25 Apr 88	13 May 92	757-200	RB211	2
LTE	25 Apr 88	15 Apr 92	757-200	RB211	1
LTU	25 Aug 83	25 May 84	757-200	RB211	12
Mexican Government	16 Nov 87	16 Nov 87	757-200	RB211	1
Mid East Jet	27 Jan 97	27 Jan 97	757-200	PW2040	1
Monarch Airlines	19 Feb 81	21 Mar 83	757-200	RB211	7
Northwest Airlines	29 Nov 83	28 Feb 85	757-200	PW2037	56
	16 Jan 01		757-300	PW2037	10
	16 Jan 01	20 Jul 02	757-300	PW2040	6
Republic Airlines	1 Oct 85	6 Dec 85	757-200	RB211	6
R-R Aircraft Mgmt	1 Aug 98	21 May 99	757-200	RB211	4
Royal Air Maroc	5 Feb 86	15 Jul 86	757-200	PW2037	2
Royal Brunei Airlines	30 May 85	6 May 86	757-200	RB211	3
Royal Nepal Airlines	17 Feb 86	9 Sep 87	757-200	RB211	1
	17 Feb 86	15 Sep 88	757-200M	RB211	1
Shanghai Airlines	14 Dec 88	4 Aug 89	757-200	PW2037	8
Singapore Airlines	31 May 83	12 Nov 84	757-200	PW2037	4
Starflite Corp	15 Dec 99	17 Dec 99	757-200	RB211	1
Sterling European	13 Jan 89	2 Aug 90	757-200	RB211	3
Sunrock Aircraft	25 Apr 88	19 Jun 91	757-200	RB211	2
Transavia Airlines	23 May 89	22 Feb 93	757-200	RB211	3
Trans World Airlines	5 Feb 96	10 Feb 97	757-200	PW2037	14
Turkmenistan Airlines	20 Oct 95	29 Aug 96	757-200	RB211	3
United Airlines	26 May 88	24 Aug 89	757-200	PW2037	47
	26 May 88	17 Aug 90	757-200	PW2040	51
UPS	6 Nov 90	2 Jul 94	757-200PF	RB211	40
	31 Dec 85	17 Sep 87	757-200PF	PW2040	35
US Airways	25 Apr 89	26 Feb 93	757-200	RB211	23
USAF	8 Aug 96	29 May 98	757-200	PW2040	4
Uzbekistan Airways	20 Oct 95	19 Oct 96	757-200	PW2037	3
Xiamen Airlines	27 Oct 89	12 Aug 92	757-200	RB211	7
Total					**1,050**

Note: For deliveries see Group Orders and Deliveries table

C-32A: Boeing 757-2G4. Four, with PW2040 engines, ordered 8 August 1996 as replacements for VC-137s of USAF's 89th Airlift Wing at Andrews AFB, Maryland; thus loosely described as 'VC-32s'. First aircraft (98-0001) flew 11 February 1998 and was delivered to 89th AW on 19 June. Further three followed on 23 June, 20 November and 25 November 1998. Post-production modifications, performed at Boeing's Wichita facility and completed on first aircraft on 2 April 1999, include installation of auxiliary fuel tanks, capacity 6,984 litres (1,845 US gallons; 1,536 Imp gallons) in forward and aft cargo holds, increasing range to 5,000 n miles (9,260 km; 5,753 miles); self-deploying forward airstair; crew ladder; satcom upgrade; and 378 litre (100 US gallon; 83.0 Imp gallon) potable water tank.

C-32B: One second-hand aircraft (86006) initially designated U-757, obtained for USAF's Foreign Emergency Support Team (FEST) by early 2000 at cost of US$45 million. Purchase of further FEST aircraft funded in FY02.

CUSTOMERS: See table.

COSTS: US$73.5 million to US$80.5 million 757-200; US$72.5 million to US$75.0 million 757-200F (2002).

DESIGN FEATURES: Low-wing, single-aisle airliner, with two podded turbofans below wings. Common design philosophy – including flight deck – with Boeing 767. Design goals also included multimarket performance, fuel efficiency, low noise and emissions, employment of advanced, digital avionics and 'dark' flight deck.

All flying surfaces sweptback and tapered; Boeing aerofoils; wing sweepback at quarter-chord 25°; dihedral 5°; incidence 3° 12'.

FLYING CONTROLS: Conventional and hydraulically powered. All-speed outboard ailerons assisted by five flight spoilers on each wing also acting variously as airbrakes and ground spoilers; one additional ground spoiler inboard on each wing; elevators and rudder; double-slotted trailing-edge flaps; full-span leading-edge slats, five sections each wing; variable incidence tailplane.

Rolls-Royce powered Boeing 757-200 of Britannia Airways (*Jane's/Paul Jackson*)   *NEW*/0526954

STRUCTURE: Aluminium alloy two-spar fail-safe wing box; centre-section continuous through fuselage; ailerons, flaps and spoilers extensively of honeycomb, graphite composites and laminates; tailplane has full-span light alloy torque boxes; fin has three-spar, dual-cell light alloy torque box; elevators and rudder have graphite/epoxy honeycomb skins supported by honeycomb and laminated spar and rib assemblies; CFRP wing/fuselage and flap track fairings. All landing gear doors of CFRP/Kevlar.

Subcontractors include Hawker de Havilland (wing in-spar ribs), Shorts (inboard flaps), CASA (outboard flaps), various Boeing divisions (leading-edge slats, main cabin sections, fixed leading-edges and flight deck), Northrop Grumman (fin and tailplane, extreme rear fuselage, overwing spoiler panels), Heath Tecna (wing/fuselage and flap track fairings), Schweizer (wingtips), Rohr Industries (engine support struts), IAI (dorsal fin) and Fleet Industries (APU access doors).

LANDING GEAR: Retractable tricycle type, with main and nose units manufactured by Menasco. Each main unit carries a four-wheel bogie, fitted with Dunlop or Goodrich wheels, carbon brakes and tyres. Twin-wheel nose unit, also with Dunlop or Goodrich tyres. Nose tyres H31×13.0-12 (20 ply); main tyres either H42×14.5-19 (24/26 ply) or, for higher weight options H42×16.0-19 (24 ply). Minimum ground turning radius 21.64 m (71 ft) at nosewheels, 29.87 m (98 ft) at wingtip.

POWER PLANT: Two 162.8 kN (36,600 lb st) Pratt & Whitney PW2037, 178.4 kN (40,100 lb st) PW2040, 178.8 kN (40,200 lb st) Rolls-Royce RB211-535E4, 189.5 kN (42,600 lb st) PW2043 or 193.5 kN (43,500 lb st) RB211-535E4-B turbofans, mounted in underwing pods. Standard fuel capacity 42,684 litres (11,276 US gallons; 9,389 Imp gallons), optional 43,489 litres (11,489 US gallons; 9,566 Imp gallons).

ACCOMMODATION: Crew of two on flight deck, with provision for an observer; common crew qualification with Boeing 767. Five to seven cabin attendants. Standard interior arrangements for 201 (12 first class/189 economy) or 195 (12 first class/183 economy) mixed-class passengers, or 224 or 231 all-economy passengers. First class seats are four-abreast, at 91 cm (36 in) pitch; business class five-abreast; economy seat pitch is 81 cm (32 in) in mixed class or 76 cm (30 in) in all-economy, mainly six-abreast. New cabin interior includes twin-door overhead luggage bins of 203 cm (80 in) width, replacing single-door, 152 cm

(60 in) type; optional ceiling-mounted stowage compartments and video screens; aesthetic improvements; and (typically, five) vacuum lavatories, as in 757-300.

Four doors each side of passenger deck: No. 1 (LH) passenger and No. 1 (RH) service doors immediately to rear of flight deck; Nos. 2 (LH) and 2 (RH) passenger doors; optional Nos. 3 (LH) and 3 (RH) emergency exit doors; and Nos. 4 (LH) passenger and 4 (RH) service doors; two overwing emergency exits each side if No. 3 doors not installed. Up to nine galleys at forward, mid-cabin and aft locations on starboard side; nine lavatory position options in typical three-door configuration, or up to 12 in four-door configuration. Movable class dividers. Cargo hold doors forward and aft on starboard side; optional bulk cargo hold door, starboard, extreme rear.

SYSTEMS: Honeywell ECS; General Electric engine thrust management system; Honeywell-Vickers engine-driven hydraulic pumps; four Abex electric hydraulic pumps. Hydraulic system maximum flow rate 140 litres (37.0 US gallons; 30.8 Imp gallons)/min at T-O power on engine-driven pumps; 25.4 to 34.8 litres (6.7 to 9.2 US gallons; 5.6 to 7.7 Imp gallons)/min on electric motor pumps; 42.8 litres (11.3 US gallons; 9.4 Imp gallons)/min on ram air turbine. Independent reservoirs, pressurised by air from pneumatic system, maximum pressure 207 bar (3,000 lb/sq in) on primary pumps. Hamilton Sundstrand electrical power generating system and ram air turbine; and Honeywell GTCP331-200 APU. Wing thermally anti-iced.

AVIONICS: *Flight:* Honeywell inertial reference system (IRS) (first commercial application of laser gyros); IRS provides position, velocity and attitude information to flight deck displays, and the flight management computer system (FMCS) and digital air data computer (DADC) supplied by Honeywell; FMCS provides automatic en route and terminal navigation capability, and also computes and commands both lateral and vertical flight profiles for optimum fuel efficiency, maximised by electronic linkage of the FMCS with automatic flight control and thrust management systems; CAT. IIIb instrument landing capability; Boeing windshear detection and guidance system is optional. Future Air Navigation System (FANS) FMS.

Instrumentation: Rockwell Collins EFIS-700 six-tube display with engine indication and crew alerting system (EICAS), EADI and EHSI functions; Rockwell Collins FCS-700 autopilot flight director system (AFDS).

DIMENSIONS, EXTERNAL:

Wing span	38.05 m (124 ft 10 in)
Wing chord: at root	8.20 m (26 ft 11 in)
at tip	1.73 m (5 ft 8 in)
Wing aspect ratio	7.8
Length: overall	47.32 m (155 ft 3 in)
fuselage	46.96 m (154 ft 10 in)
Fuselage max width	3.76 m (12 ft 4 in)
Height overall: at OWE	13.74 m (45 ft 1 in)
at MTOW	13.49 m (44 ft 3 in)
Tailplane span	15.21 m (49 ft 11 in)
Wheel track	7.32 m (24 ft 0 in)
Wheelbase	18.29 m (60 ft 0 in)
Distance between engine centrelines	12.95 m (42 ft 6 in)
Passenger doors (two, fwd, port and No. 2 starboard):	
Height	1.83 m (6 ft 0 in)
Width	0.84 m (2 ft 9 in)
Height to sill: at OWE	4.01 m (13 ft 2 in)
at MTOW	3.78 m (12 ft 5 in)
Passenger door (rear, port):	
Height	1.83 m (6 ft 0 in)
Width	0.76 m (2 ft 6 in)
Height to sill: at OWE	4.14 m (13 ft 7 in)
at MTOW	3.89 m (12 ft 9 in)
Service door (fwd, stbd): Height	1.65 m (5 ft 5 in)
Width	0.76 m (2 ft 6 in)
Service door (rear, stbd): Height	1.83 m (6 ft 0 in)
Width	0.76 m (2 ft 6 in)
Bulk cargo door (optional, starboard, extreme rear):	
Max height	0.81 m (2 ft 8 in)
Max width	1.22 m (4 ft 0 in)
Height to sill: at OWE	2.77 m (9 ft 1 in)
at MTOW	2.59 m (8 ft 6 in)
Emergency exits (four, overwing):	
Height	0.97 m (3 ft 2 in)
Width	0.51 m (1 ft 8 in)
Emergency exits, optional (two, aft of wings):	
Height	1.12 m (3 ft 8 in)
Width	0.61 m (2 ft 0 in)
Cargo door (fwd, port): Height	1.08 m (3 ft 6½ in)
Width	1.40 m (4 ft 7 in)
Height to sill: at OWE	2.67 m (8 ft 9 in)
at MTOW	2.46 m (8 ft 1 in)
Cargo door (aft, port): Height	1.12 m (3 ft 8 in)
Width	1.40 m (4 ft 7 in)
Height to sill: at OWE	2.51 m (8 ft 3 in)
at MTOW	2.36 m (7 ft 9 in)
Cargo door (-200PF):	
Max width	3.40 m (11 ft 2 in)
Max height	2.18 m (7 ft 2 in)
Height to sill: at OWE	4.01 m (13 ft 2 in)
at MTOW	3.81 m (12 ft 6 in)
Crew door (-200PF):	
Max width	0.56 m (1 ft 10 in)
Max height	1.22 m (4 ft 0 in)
Height to sill	as cargo door

DIMENSIONS, INTERNAL:

Cabin, aft of flight deck to rear pressure bulkhead:	
Length	36.09 m (118 ft 5 in)
Max width	3.53 m (11 ft 7 in)
Height: max	2.13 m (7 ft 0 in)
to ceiling video screen	1.85 m (6 ft 1 in)
to baggage lockers	1.66 m (5 ft 5½ in)
Floor area	116.0 m² (1,249 sq ft)
Passenger section volume	230.5 m³ (8,140 cu ft)
Underfloor baggage hold: Length:	
fwd	8.57 m (28 ft 1¼ in)
rear	11.17 m (36 ft 7¾ in)
Width at floor	1.26 m (4 ft 1¾ in)
Max height: fwd	1.12 m (3 ft 8 in)
rear	1.37 m (4 ft 6 in)
Volume (bulk loading): fwd	19.8 m³ (699 cu ft)
rear	27.5 m³ (971 cu ft)

AREAS:

Wings, gross	185.25 m² (1,994.0 sq ft)
Ailerons (total)	4.46 m² (48.00 sq ft)

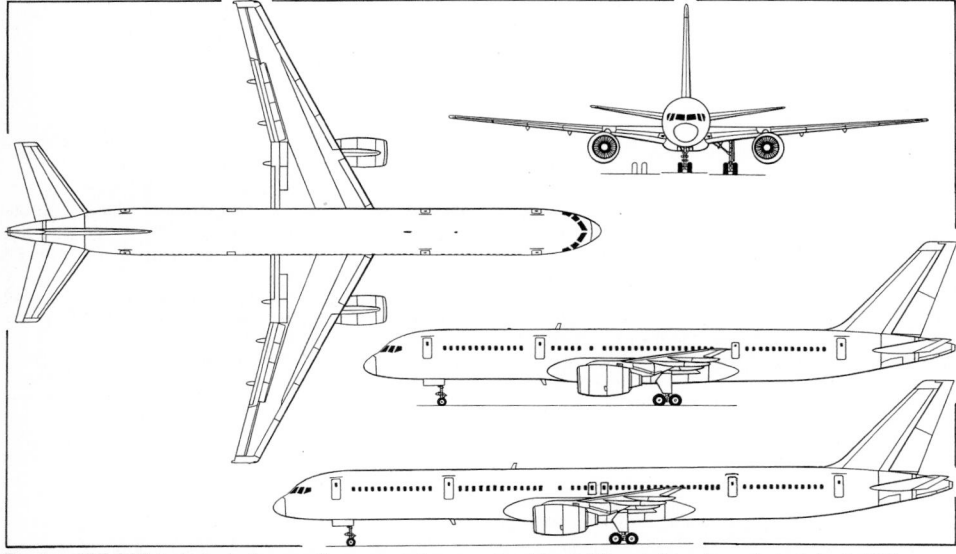

Boeing 757-200 twin-turbofan medium-range transport, with additional side view of stretched -300
(*Jane's/Dennis Punnett*)
0093862

Boeing 757-200 in Finnair colours (*Jane's/Paul Jackson*)   NEW/0526953

Trailing-edge flaps (total)	30.38 m² (327.00 sq ft)
Leading-edge slats (total)	18.39 m² (198.00 sq ft)
Flight spoilers (total)	10.96 m² (118.00 sq ft)
Ground spoilers (total)	12.82 m² (138.00 sq ft)
Fin	34.37 m² (370.00 sq ft)
Rudder	11.61 m² (125.00 sq ft)
Tailplane	50.35 m² (542.00 sq ft)
Elevators (total)	12.54 m² (135.00 sq ft)

WEIGHTS AND LOADINGS (with 201 passengers. A: PW2037 engines, B: PW2040s, C: RB211-535E4s, D: RB211-535E4-Bs):

Operating weight empty: A	58,325 kg (128,580 lb)
B	58,390 kg (128,730 lb)
C	58,550 kg (129,080 lb)
D	58,620 kg (129,230 lb)
Baggage capacity, underfloor:	
fwd hold	4,672 kg (10,300 lb)
aft hold	7,394 kg (16,300 lb)
Max structural payload: A, B	25,229 kg (55,620 lb)
C, D	26,712 kg (58,890 lb)
Max fuel weight	34,269 kg (75,550 lb)
Max T-O weight: A, C	99,790 kg (220,000 lb)
B, D	115,665 kg (255,000 lb)
Max landing weight: A, C	89,815 kg (198,000 lb)
B, D	95,255 kg (210,000 lb)
Max ramp weight: A, C	100,245 kg (221,000 lb)
B, D	116,120 kg (256,000 lb)
Max zero-fuel weight: A, C	83,460 kg (184,000 lb)
B	84,370 kg (186,000 lb)
D	85,275 kg (188,000 lb)
Max wing loading: A, C	538.7 kg/m² (110.33 lb/sq ft)
B, D	624.4 kg/m² (127.88 lb/sq ft)
Max power loading: A	306 kg/kN (3.01 lb/lb st)
B	324 kg/kN (3.18 lb/lb st)
C	279 kg/kN (2.74 lb/lb st)
D	299 kg/kN (2.93 lb/lb st)

PERFORMANCE (with 201 passengers; at max basic T-O weight except where indicated):

Max operating Mach No. (Mмo): A, B, C, D	0.86
Cruising speed: A, B, C, D	M0.80
Approach speed at S/L, flaps down, max landing weight:	
A, C	132 kt (245 km/h; 152 mph) EAS
B, D	137 kt (254 km/h; 158 mph)
Initial cruising height: A	11,675 m (38,300 ft)
B	10,790 m (35,400 ft)
C	11,795 m (38,700 ft)
D	10,880 m (35,700 ft)

Runway LCN at ramp weight of 100,244 kg (221,000 lb), optimum tyre pressure and subgrade C flexible pavement: H40×14.5-19.0 tyres   36

T-O field length (S/L, 29°C):

A	1,814 m (5,950 ft)
B	2,378 m (7,800 ft)
C	1,677 m (5,500 ft)
D	2,104 m (6,900 ft)
Landing field length at max landing weight:	
A	1,463 m (4,800 ft)
B	1,555 m (5,100 ft)
C	1,418 m (4,650 ft)
D	1,494 m (4,900 ft)
Range with 201 passengers:	
A	2,570 n miles (4,759 km; 2,957 miles)
B	3,930 n miles (7,278 km; 4,522 miles)*
C	2,375 n miles (4,398 km; 2,733 miles)
D	3,695 n miles (6,843 km; 4,252 miles)*

* Fuel volume limited

OPERATIONAL NOISE LEVELS (FAR Pt 36 Stage 3):

T-O, at max basic T-O weight, cutback power:	
A	82.2 EPNdB
B	86.2 EPNdB
C (estimated)	84.7 EPNdB
Approach at max landing weight, 30° flap:	
A	95.0 EPNdB
B, C	97.7 EPNdB
Sideline: A	93.3 EPNdB
B	94.0 EPNdB
C (estimated)	94.6 EPNdB

UPDATED

BOEING 757-300

TYPE: Twin-jet airliner.

PROGRAMME: Stretched 757-200. Launched 2 September 1996; major assembly began 9 August 1997; prototype (N757X; 804th B757; earmarked for Condor Flugdienst as D-ABOA) rolled out at Renton on 19 May 1998 (officially 31 May); first flight (transmitted live on the Internet) 2 August 1998. 912-hour, 356-sortie flight test programme undertaken by three aircraft. N757X/NU701, employed on initial airworthiness and basic controllability tests; D-ABOB/NU721, first flown 4 September 1998, for ground effect, autoland and fuel consumption tests; and D-ABOC/NU722, first flown 2 October 1998, initially for HERF testing, before installation of standard interior for a four-day period of simulated airline operations with launch customer Condor Flugdienst. FAA certification with 180-minute ETOPS approval 27 January 1999, followed by

first delivery (D-ABOE) to Condor on 10 March 1999 and first revenue-earning service 19 March 1999. Achieved 8.7 hours daily utilisation and 99.64 per cent reliability rate in first 12 months of operation. All early aircraft with Rolls-Royce engines; first with PW2040 power (N753JM) was flown 20 February 2002, before delivery to Northwest Airlines.

CUSTOMERS: See table accompanying Boeing 757-200 entry. Launch customer Condor Flugdienst ordered 12 firm, with 12 options, on 2 September 1996; Icelandair ordered two, with Rolls-Royce engines, in June 1997, for delivery in second quarter 2001 and second quarter 2002.

COSTS: US$82.0 million to US$89.5 million (2002). Seat-mile operating costs estimated at 10 per cent lower than those of 757-200.

Description generally as for 757-200, except that below.

DESIGN FEATURES: Fuselage stretched by 7.11 m (23 ft 4 in) – 4.06 m (13 ft 4 in) ahead of the wing and 3.05 m (10 ft 0 in) aft – to provide 20 per cent more passenger accommodation and nearly 50 per cent more cargo volume than 757-200; typical accommodation 240 passengers in two-class configuration or up to 289 in single class with pitch of between 71 and 74 cm (28 and 29 in). Other features include strengthened wings, engine pylons, high-lift devices and landing gear; strengthened overwing exit body section; new wheels, 26-ply tyres and brakes; nosewheel spray deflector; retractable tailskid, similar to those of 767-300 and 777-300, with 'body contact' indicator on flight deck; redesigned cabin interior based on 'Next-Generation' 737s, with upgraded environmental control system, larger precooler, more powerful fans, and vacuum lavatories with single-point drainage. Flight deck and operating systems generally as for 757-200, but with Pegasus FMS (including software options for GPS, FANS and satcom) and enhanced EICAS with improved BITE functions.

POWER PLANT: As for 757-200. Total fuel capacity 43,491 litres (11,490 US gallons; 9,567 Imp gallons) for all options.

ACCOMMODATION: Typical cabin arrangements provide for 243 passengers (12 first class, 231 economy) in mixed class configuration, four-abreast in first class, at 91 cm (36 in) seat pitch, six-abreast in economy at 81 cm (32 in) seat pitch; or 279 passengers in inclusive tour configuration, six-abreast at 71/74 cm (28/29 in) seat pitch.

DIMENSIONS, EXTERNAL: As for 757-200 except:

Length: overall	54.43 m (178 ft 7 in)
fuselage	54.08 m (177 ft 5 in)
Height overall: at OWE	13.64 m (44 ft 9 in)
at MTOW	13.56 m (44 ft 6 in)
Wheelbase	22.35 m (73 ft 4 in)
Passenger door, height to sill: No. 1:	
at OWE	4.01 m (13 ft 2 in)
at MTOW	3.78 m (12 ft 5 in)
No. 2: at OWE	as No. 1
No. 3 emergency exit: at OWE	4.06 m (13 ft 4 in)
at MTOW	3.94 m (12 ft 11 in)
No. 4: at OWE	4.06 m (13 ft 4 in)
at MTOW	3.96 m (13 ft 0 in)
Cargo door, height to sill: fwd: at OWE	2.67 m (8 ft 9 in)
at MTOW	2.44 m (8 ft 0 in)
rear: at OWE	2.39 m (7 ft 10 in)
at MTOW	2.29 m (7 ft 6 in)

DIMENSIONS, INTERNAL:

Cabin (rear of flight deck to rear pressure bulkhead):	
Length	43.21 m (141 ft 9 in)
Underfloor cargo hold: Length: fwd	12.80 m (42 ft 0 in)
rear	14.22 m (46 ft 7¾ in)
Width, height	as 757-200

Stretched Boeing 757-300 of launch customer, Condor (*Jane's/Paul Jackson*)   NEW/0526955

For details of the latest updates to *Jane's All the World's Aircraft* online and to discover the additional information available exclusively to online subscribers please visit

jawa.janes.com

Boeing 767-200 operated by Qantas (*Jane's/Paul Jackson*)   *NEW*/0526956

Volume (bulk loading): fwd   30.3 m³ (1,071 cu ft)
    rear   37.2 m³ (1,312 cu ft)
WEIGHTS AND LOADINGS (with 243 passengers: A: PW2040 engines, B: PW2043s, C: RB211-535E4s, D: RB211-535E4-Bs):
Operating weight empty: A, B   64,320 kg (141,800 lb)
    C, D   64,570 kg (142,350 lb)
Max structural payload: A, B   30,935 kg (68,200 lb)
    C, D   30,686 kg (67,650 lb)
Baggage capacity, underfloor:
    fwd hold   7,303 kg (16,100 lb)
    aft hold   8,845 kg (19,500 lb)
Max fuel weight: A, B, C, D   34,918 kg (76,980 lb)
Max T-O weight: A, B, C, D   122,470 kg (270,000 lb)
Max landing weight: A, B, C, D   101,605 kg (224,000 lb)
Max ramp weight: A, B, C, D   122,925 kg (271,000 lb)
Max zero-fuel weight: A, B, C, D 95,255 kg (210,000 lb)
Max wing loading: A, B, C, D
                                     661.1 kg/m² (135.41 lb/sq ft)
Max power loading: A   343 kg/kN (3.37 lb/lb st)
    B   323 kg/kN (3.17 lb/lb st)
    C   342 kg/kN (3.36 lb/lb st)
    D   316 kg/kN (3.10 lb/lb st)
PERFORMANCE (243 passengers):
Max operating Mach No. (M$_{MO}$): A, B, C, D   0.86
Cruising speed: A, B, C, D   M0.80
Approach speed, S/L, flaps down, at max landing weight:
    A, B, C, D   143 kt (265 km/h; 165 mph)
Initial cruising height: A   11,005 m (36,100 ft)
    B   10,400 m (34,120 ft)
    C   11,200 m (36,740 ft)
    D   10,420 m (34,180 ft)
T-O field length, S/L, 86°F: A   2,082 m (6,830 ft)
    B   2,625 m (8,610 ft)
    C   2,119 m (6,950 ft)
    D   2,607 m (8,550 ft)
Landing field length at max landing weight:
    A, B   1,750 m (5,740 ft)
    C, D   1,723 m (5,650 ft)
Range: A   2,235 n miles (4,139 km; 2,572 miles)
    B   3,465 n miles (6,417 km; 3,987 miles)*
    C   2,055 n miles (3,805 km; 2,364 miles)
    D   3,240 n miles (6,000 km ; 3,728 miles)*
* *Fuel volume limited*

UPDATED

BOEING 767

TYPE: Wide-bodied airliner.
PROGRAMME: Launched on receipt of United Air Lines order for 30 on 14 July 1978; construction of basic 220-passenger 767-200 began 6 July 1979; first flight (N767BA) 26 September 1981 with P&W JT9D turbofans; first flight fifth aircraft with GE CF6-80A 19 February 1982; 767 with JT9D-7R4D certified 30 July 1982; with CF6-80A 30 September 1982.
First delivery with JT9D (United Air Lines) 19 August 1982 (initial service 8 September); first delivery with CF6 (Delta) 25 October 1982. ETOPS approval for 767-200 with JT9D-7R4 or CF6-80A or -80A2 granted January 1987; ETOPS approval for 767-200 and -300 with PW4000 obtained April 1990; 180-minute ETOPS approval with PW4000 engines obtained August 1993. Joint 757/767 crew rating approved 22 July 1983. Boeing windshear detection and guidance system FAA approved for 767-200 and -300 February 1987.
Boeing 767-400 first flew 9 October 1999; FAA certification and 180-minute ETOPS granted 20 July 2000; JAA certification 24 July 2000; FAA common type rating with 767-200/300 and 757-200/300 issued 21 August 2000.

CURRENT VERSIONS: **767-200:** Basic model; no longer available. Medium-range variant (MTOW 136,080 kg; 300,000 lb) has reduced fuel; higher gross weight variant (142,880 kg; 315,000 lb) certified June 1983.
767-200ER: Extended-range version; announced January 1983; first flight 6 March 1984; basic -200ER with centre-section tankage and gross weight increased initially to 156,490 kg (345,000 lb) first delivered to El Al 26 March 1984; 175,540 kg (387,000 lb) certified 1988; current gross weights are 159,211 kg (351,000 lb), and 179,169 kg (395,000 lb).
767-300: Stretched 269-passenger version, with 3.07 m (10 ft 1 in) plug forward of wing and 3.35 m (11 ft) plug aft, and same gross weight as 767-200; strengthened landing gear and thicker metal in parts of fuselage and underwing skin; same flight deck and systems as other 767s; same engine options as 767-200ER; first ordered (by Japan Airlines) 29 September 1983. First flight with JT9D-7R4D engines 30 January 1986; certified with JT9D-7R4D and CF6-80A2 22 September 1986. First delivery (Japan Airlines) 25 September 1986. British Airways ordered 11 in August 1987, later increased to total 25, with Rolls-Royce RB211-524H engines; delivered from 8 February 1990. No longer available.
767-300ER: Extended-range, higher gross weight version; development began January 1985; optional gross weights 172,365 kg (380,000 lb) and, from 1992, 186,880 kg (412,000 lb); further increased centre-section tankage. Engine choice CF6-80C2, PW4000, RB211-524H; structural reinforcement; certified late 1987. Launch customer American Airlines (15), delivered from 19 February 1988. New interior introduced late 2000; based on Boeing 777; first recipient Lauda Air.
767-300ERX: Further range extension, under study from 1998; addition of tailplane fuel tank, capacity 7,571 litres (2,000 US gallons; 1,665 Imp gallons) would increase range to 6,695 n miles (12,400 km; 7,705 miles).
767-300X: Proposed early 2002. Main feature would be rapidly changeable 767-400-type wingtip, allowing airlines to customise aircraft to individual routes, according to distance, reverting to standard tips for long range.
767-300 General Market Freighter: See separate entry.
767-400ER: Stretched version with 10 to 15 per cent increase in passenger accommodation, seating 245 passengers in three-class configuration and 304 in two-class. Features include strengthened wing with thicker ribs, spars and skin; updated flight deck based on Boeing 777; fuselage lengthened 6.43 m (21 ft 1 in) by means of plugs forward (3.36 m; 11 ft 0¼ in) and aft (3.07 m; 10 ft 0¾ in) of centre-section; stringerless window belt with elliptical 777-type cabin windows; wing span increased by 4.42 m (14 ft 6 in) with highly sweptback (9° 50′) wingtips of composites construction which reduce take-off distance, increase climb rate and improve fuel consumption; redesigned interior; cargo volume 129.7 m³ (4,580 cu ft); new landing gear with 46 cm (18 in) longer main legs, Boeing 777 brakes and 127 cm (50 in) tyres and revised, hydraulically actuated tail skid; 120 kVA AC generators and more powerful Honeywell 331-400 APU also of 120 kVA. Engine choice 276 kN (62,100 lb st) CF6-80C2B7F1 or 282 kN (63,300 lb st) CF6-80C2B8, with PW4000 series as option; fuel capacity as currently offered on 767-300ER. Maximum T-O weight increase to 204,115 kg (450,000 lb) together with aerodynamic improvements to provide maximum range of approximately 5,600 n miles (10,371 km; 6,444 miles), enabling 767-400ER to operate most existing 767-300ER routes.

Offered from January 1997; launch customer Delta Air Lines announced intention to order 21 on 20 March 1997; confirmed 28 April 1997; Continental ordered 26 on 10 October 1997. Assembly of first aircraft began at Everett on 9 February 1999; roll-out 26 August 1999; first flight (N76400 No. 1) 9 October 1999. Four aircraft took part in test programme, comprising 1,150 flight hours and 1,200 ground testing hours: prototype used primarily to test and certify basic handling qualities; N76401 served as aerodynamics and avionics certification article; N87402, with full cabin interior used for systems development and certification; N47403 (first flown June 2000) for cabin entertainment and related evaluation. World tour (by N76400 No. 2) in July-August 2000. First delivery (N828MH), to launch customer Delta Airlines, 11 August 2000; deliveries to Continental Airlines began 30 August 2000 with refurbished second prototype (N76401/ N66051). Also on 30 August 2000, Delta received first 767-400ER with Rockwell Collins Large Format Display System, comprising six 203 × 203 mm (8 × 8 in) LCDs.
767-400ER Shrink: Under study in 1999 as alternative to 767-300ERX; would feature tailplane fuel tank, capacity 4,536 kg (10,000 lb); 302 kN (68,000 lb st) class engines providing a three to five per cent increase in cruising speed; and range up to 6,479 n miles (12,000 km; 7,456 miles).
Longer Range 767-400ER: Extended-range version; launched (as 767-400ERX) 13 September 2000. Features include strengthened wing and fuselage; strengthened main landing gear with 34-ply rated tyres; tailplane fuel tank capacity 7,949 litres (2,100 US gallons; 1,749 Imp gallons) bringing total fuel capacity to 98,891 litres (26,125 US gallons; 21,754 Imp gallons); 320 kN (72,000 lb st) class Engine Alliance GP7100 or Rolls-Royce Trent 600 engines with new nacelles and struts; advanced 777-type flight deck and cabin interior; front passenger doors moved forward 1.12 m (3 ft 8 in) to ensure adequate engine nacelle clearance for boarding bridges, passenger escape slides and galley service trucks; accommodation for 245 passengers in three-class configuration; maximum take-off weight 210,925 kg (465,000 lb); target take-off field length 2,895 m (9,500 ft); design maximum range: (GP7100), 6,245 n miles (11,565 km; 7,186 miles), (Trent 600) 6,230 n miles (11,538 km; 7,169 miles). Roll-out originally planned for September 2003; first flight at end of October 2004; FAA and JAA certification April 2004; first delivery, to launch customer Kenya Airways, which has ordered three, in May 2004. By early 2002, however, 767-400ER was being given lower priority than 767-300 improvements.
767 AWACS: See Boeing 767 AWACS earlier in Boeing Company entry. Also under consideration are a tanker version for boom and hose-reel refuelling systems and a carrier for Joint STARS radar.
767 AST: Boeing contracted by US Army on 11 October 1994 to supply company-owned 767 as Airborne Sensor Testbed for long wavelength infra-red surveillance system; operating contract was extended for 12 months in September 1998 at cost of US$4.1 million.
767 SF: Special Freighter conversion of 767-200 airliner; available from 2000; payload 39,010 kg (86,000 lb); freight door as 767-300F; strengthened floor, main landing gear and forward fuselage.
CUSTOMERS: See table for orders. Original prototype became 767 Airborne Surveillance Testbed (formerly AOA) for US Army. One reconfigured by E-Systems as medevac aircraft for Civil Reserve Air Fleet.
COSTS: US$101 million to US$112 million 767-200ER; US$115.5 million to US$127.5 million 767-300ER; US$126.5 million to US$138.5 million 767-400ER (all 2002).

DESIGN FEATURES: Low-wing, wide-bodied airliner with twin, podded turbofans underwing. Boeing aerofoils; quarter-chord sweepback 31° 30′; thickness/chord ratio 15.1 per cent at root, 10.3 per cent at tip; dihedral 6°; incidence 4° 15′.

FLYING CONTROLS: Conventional and hydraulically powered. Inboard, all-speed (between inner and outer flaps) and outboard low-speed ailerons supplemented by flight spoilers (four-section outboard; two-section inboard) also acting as airbrakes and lift dumpers; single-slotted, linkage-supported outboard trailing-edge flaps, double-slotted inboard; track-mounted leading-edge slats; variable incidence tailplane driven by hydraulic screwjack; two-piece elevators each side; no trim tabs; roll and yaw trim through spring feel system; triple digital flight control computers and EFIS; Boeing windshear detection and guidance system optional. Control surface deflections: outboard ailerons +30/−15°, inboard ailerons ±20°, inboard flaps 61° (first element 36°), outboard flaps 36°, spoilers +60°, elevators +28/−20°, rudder ±26°; tailplane incidence +2/−12°.

STRUCTURE: Fail-safe structure. Conventional aluminium structure augmented by graphite ailerons, spoilers, elevators, rudder and floor panels; advanced aluminium alloy keel beam chords and wing skins; composites engine cowlings, wing/fuselage fairing and rear wing panels; CFRP landing gear doors; and aramid flaps and engine pylon fairings.

Subcontractors include Boeing Military Aircraft (wing fixed leading-edges); Northrop Grumman (wing centre-section and adjacent lower fuselage section; fuselage bulkheads); Vought Aircraft (horizontal tail); Canadair (rear fuselage); Alenia (wing control surfaces, flaps and leading-edge slats, wingtips, elevators, fin and rudder, nose radome); Fuji (wing/body fairings and main landing gear doors); Kawasaki (forward and centre fuselage; exit hatches; wing in-spar ribs); Mitsubishi (rear fuselage body panels and rear fuselage doors).

LANDING GEAR: Hydraulically retractable tricycle type; Menasco twin-wheel nose unit retracts forward; Cleveland Pneumatic main gear, with two four-wheel bogies, retracts inward; oleo-pneumatic shock-absorbers; Honeywell wheels and brakes; mainwheel tyres of current production versions H46×18.0-20 (26/28 ply for -200/300; 32 ply for -200ER/300ER); nosewheel tyres size H37×14.0-15 (22/24 ply) for all; steel disc brakes on all mainwheels; electronically controlled anti-skid units. Nosewheel steerable ±16°; ±65° for towing.

POWER PLANT: Two high-bypass turbofans in pods, pylon-mounted on the wing leading-edges.

General Electric options: 225 kN (50,600 lb st) CF6-80C2B2F, 251 kN (56,500 lb st) CF6-80C2B4F, 268 kN (60,200 lb st) CF6-80C2B6F and 276 kN (62,100 lb st) CF6-80C2B7F.

Pratt & Whitney options: 233 kN (52,300 lb st) PW4052, 254 kN (57,100 lb st) PW4056, 268 kN (60,200 lb st) PW4060 and 282 kN (63,300 lb st) PW4062. P&W JT9D-7R4D of 213.5 kN (48,000 lb st) no longer offered.

Rolls-Royce options: 251 kN (56,400 lb st) RB211-524G4-T and 265 kN (59,500 lb st) RB211-524H2-T.

Fuel in one integral tank in each wing, and in centre tank, with total capacity of 63,216 litres (16,700 US gallons; 13,905 Imp gallons) in 200/300; 767-200ER and -300ER have additional 28,163 litres (7,440 US gallons; 6,195 Imp gallons) in second centre-section tank, raising total capacity to 91,379 litres (24,140 US gallons; 20,100 Imp gallons). Refuelling point in port outer wing.

ACCOMMODATION: Operating crew of two on flight deck; observer's seat and optional second observer's seat. Basic accommodation in -200 models for 224 passengers, made up of 18 first class passengers forward in six-abreast seating at 96.5 cm (38 in) pitch, and 206 tourist class in seven-abreast seating at 81 cm (32 in) pitch. Window or aisle seats comprise 86 per cent of total. Type A inward-opening plug doors provided at both front and rear of cabin on each side of fuselage, with options of Type A, I or III emergency exits at various mid-cabin locations on each side. Total of five lavatories installed, two centrally in main cabin, two aft in main cabin, and one forward in first class section. Galleys situated at forward and aft ends of cabin. Alternative single-class layouts provide for 255 tourist passengers seven-abreast (two-three-two) at 81 cm (32 in) pitch (one overwing exit each side) and maximum (requiring two additional overwing emergency exits) 290, mainly eight-abreast (two-four-two), at 76 cm (30 in) pitch. Three-class layout for 181 passengers: 15 first class (two-one-two) at 152 cm (60 in) pitch; 40 business class (two-two-two) at 91 cm (36 in); and 126 tourist class (two-three-two) at 81 cm (32 in).

Basic accommodation in -300 models for 269 passengers, made up of 24 first class passengers forward in six-abreast seating at 96.5 cm (38 in) pitch, 245 tourist class in seven-abreast at 78.7 cm (31 in) pitch, six lavatories and five galleys. Alternatives include 286 in two-three-two seating at 81 cm (32 in) pitch and 218 in three-class layout comprising 18 first, 46 business and 154 tourist class passengers arranged as in -200. Maximum seating capacity in -300 models is 350 passengers at 71 cm (28 in), six lavatories and four galleys; capacities from 291 upwards require standard -300 door configuration (each

BOEING 767 ORDER BOOK
(at 1 January 2003)

Customer	Variant	Engine	First order	First delivery	Qty
Aeromaritime	200ER	PW	17 Jan 89	26 Jul 90	2
	300ER	PW	17 Jan 89	22 Aug 91	1
Air Algérie	300ER	GE	1 May 89	28 Jun 90	3
Air Canada	200	PW	11 Jul 79	30 Oct 82	10
	200ER	PW	11 Jul 79	18 Oct 84	9
	300ER	PW	31 Aug 89	10 Aug 93	6
Air China	200ER	PW	23 May 85	9 Oct 85	6
	300	PW	31 May 90	20 May 92	4
Air France	300ER	PW	7 Feb 92	14 May 93	3
Air Mauritius	200ER	GE	19 Jan 87	5 Apr 88	2
Air New Zealand	200ER	GE	30 Jul 84	3 Sep 85	3
	300ER	GE	1 Mar 91	11 Aug 93	5
Airtours International	300ER	GE	17 Aug 93	16 Mar 94	3
Air Zimbabwe	200ER	PW	20 Jul 88	28 Nov 89	2
All Nippon Airways	200	GE	1 Oct 79	25 Apr 83	25
	300	GE	26 Dec 85	30 Jun 87	34
	300ER	GE	26 Dec 85	26 Jun 89	26
	300F		7 Aug 01	28 Aug 02	1
American Airlines	200	GE	15 Nov 78	4 Nov 82	13
	200ER	GE	15 Nov 78	18 Nov 85	17
	300ER	GE	3 Mar 87	19 Feb 88	58
Amiri Flight	300ER	GE	30 Nov 99	15 Dec 99	1
Ansett Australia	200	GE	17 Mar 80	7 Jun 83	5
Ansett Worldwide	200ER	GE	8 Sep 88	1 May 90	4
	300ER	PW	16 Nov 88	13 Dec 91	22
	300ER	GE	24 Jun 94	24 Jun 94	7
Asiana Airlines	300	GE	23 Dec 88	27 Sep 90	9
	300ER	GE	3 Sep 90	7 Nov 91	2
	300F	GE	3 Sep 90	23 Aug 96	1
Avianca	200ER	PW	24 Dec 80	26 Feb 90	2
BCC Leasing	300ER		25 Sep 01	24 Oct 02	2
Braathens	200	PW	28 Apr 80	23 Mar 94	2
Britannia Airways	200	PW	31 Mar 80	6 Feb 84	11
	300ER	GE	23 Nov 94	15 May 96	7
British Airways	300ER	RR	14 Aug 87	8 Feb 90	28
Canadian Airlines Int'l	300ER	GE	7 Apr 87	15 Apr 88	14
China Airlines	200	PW	20 Mar 80	20 Dec 82	2
China Yunnan	300ER	RR	10 Jan 95	26 Jul 96	3
Condor Flugdienst	300ER	PW	16 Nov 88	13 Jul 91	11
Continental Airlines	200ER	GE	23 Nov 98	9 Nov 00	10
	400ER	GE	10 Oct 97	30 Aug 00	16
Delta Airlines	200	GE	15 Nov 78	25 Oct 82	15
	300	GE	15 Nov 78	7 Nov 86	24
	300	PW	20 Dec 90	17 Jun 93	4
	300ER	PW	22 Sep 88	9 Jun 90	31
	300ER	GE	10 Jun 97	18 Jun 98	22
	400ER	GE	10 Jun 97	11 Aug 00	21
Egyptair	200ER	PW	12 Jan 84	20 Jul 84	3
	300ER	PW	25 Oct 88	15 Aug 89	2
El Al Israel Airlines	200	PW	18 Mar 81	12 Jul 83	2
	200ER	PW	18 Mar 81	26 Mar 84	2
Ethiopian Airlines	200ER	PW	16 Dec 82	23 May 84	3
	300ER	n/k	28 Nov 02		3
Eva Airways	200	GE	10 Jun 93	13 Jan 94	4
	300ER	GE	6 Oct 89	30 May 91	4
Flightlease	300ER	PW	15 Dec 99	4 Apr 00	4
GATX Capital Corp	200ER	GE	24 May 89	25 Jan 91	1
GECAS	300ER	GE	30 Aug 95	30 Aug 95	29
	300F	GE	15 Dec 99	28 Nov 01	1
GPA Ltd	200ER	PW	18 Apr 89	14 Jan 92	1
	300ER	GE	18 Apr 89	3 May 93	2
	300ER	PW	18 Apr 89	28 Feb 01	11
Gulf Air	300ER	GE	27 Apr 88	14 Jun 88	20
Hainan Airlines	300ER	PW	21 Aug 02	-	2
ILFC	200	GE	27 Aug 86	25 Aug 87	1
	200ER	GE	18 Feb 86	29 May 86	4
	300ER	GE	16 May 88	14 Jun 91	36
	300ER	PW	16 May 88	21 Mar 91	16
Itochu Air Lease Corp	300ER	GE	15 Apr 91	14 Aug 92	3
Itochu Corp	200ER	GE	4 Nov 93	1 Dec 94	4
Japan Airlines	200	PW	29 Sep 83	22 Jul 85	3
	300	PW	29 Sep 83	25 Sep 86	13
	300	GE	10 Jun 93	1 Aug 94	9
	300ER	GE	27 Nov 00	19 May 02	3
Kazakstan Airlines	200ER	GE	29 Dec 00	8 Feb 02	1
Kuwait Airlines	200ER	PW	18 Sep 84	20 Mar 86	3
LAM Mozambique	200ER	PW	1 Aug 90	24 Aug 93	1
LAN Chile	300ER	GE	28 May 97	29 Apr 98	3
	300F	GE	17 Nov 97	23 Sep 98	5
Lauda Air	300ER	PW	21 Apr 97	29 Apr 88	7
LOT Polish Airlines	200ER	GE	4 Nov 88	21 Apr 89	2
	300ER	GE	4 Nov 88	21 Aug 90	3
LTU	300ER	PW	21 Jan 88	2 Feb 89	5
Malev Hungarian Airlines	200ER	PW	21 Feb 91	30 Apr 93	2
Martinair Holland	300ER	PW	11 Mar 88	21 Sep 89	6
Mid East Jet	200ER	GE	4 Oct 96	4 Oct 96	1
Pacific Western	200	PW	21 Dec 78	4 Mar 83	2
Piedmont	200ER	GE	25 Jul 86	21 May 87	6
Qantas Airways	200ER	PW	7 Sep 83	3 Jul 85	7
	300ER	GE	24 Apr 87	30 Aug 88	22
Royal Brunei Airways	300ER	GE	8 Mar 96	8 Mar 96	2
SAS	200ER	PW	18 Jan 88	11 May 90	2
	300ER	PW	18 Jan 88	29 Mar 89	14
Shanghai Airlines	300	PW	14 May 93	22 Jul 94	4
Singapore Lease	300ER	GE	8 Aug 95	4 Aug 95	3
TACA Int'l Airlines	200	GE	22 May 86	22 May 86	1
	300ER	GE	27 Aug 91	26 Aug 91	2
Transbrasil	200	GE	30 Jul 81	23 Jun 83	3
TWA	200	PW	5 Dec 79	22 Nov 82	10
unidentified	200ER	n/k	15 Apr 02	none	1
	200ER	n/k	23 Apr 02	none	4
	300ER	n/k	21 Aug 01	none	3
	300ER	GE	15 Nov 01	none	5
United Airlines	200	PW	14 Jul 78	19 Aug 82	19
	300ER	PW	19 May 89	18 Apr 91	37
United Parcel Service	300F	GE	15 Jan 93	12 Oct 95	32
US Airways	200ER	GE	4 Apr 89	22 May 90	6
Uzbekistan	300ER	PW	20 Oct 89	27 Nov 96	2
Varig	200ER	GE	18 Mar 86	2 Jul 87	6
	300ER	GE	8 Sep 98	21 Dec 89	4
Total					**931**

Note: For deliveries see Group Orders and Deliveries table

Boeing 767-300 leased to Italy's Blue Panorama Airlines (*Jane's/Paul Jackson*)   *NEW*/0526957

side) of Type A front and rear and two Type Is overwing to be replaced by two Type As, plus third Type A ahead of wing and Type I adjacent to trailing-edge.

Underfloor cargo holds (forward and rear, combined) of -200 versions can accommodate, typically, up to 22 LD2 or 11 LD1 containers; 767-300 underfloor cargo holds can accommodate 30 LD2 or 15 LD1 containers. Starboard side forward and rear cargo doors of equal size on 767-200 and 767-300, but larger forward door standard on 767-200ER and 767-300ER and optional on 767-200 and 767-300. Bulk cargo door at rear on port side. Overhead stowage for carry-on baggage is 0.08 m³ (3.0 cu ft) per passenger. Cabin air conditioned, cargo holds heated.

SYSTEMS: Honeywell dual air cycle air conditioning system. Pressure differential 0.59 bar (8.6 lb/sq in). Electrical supply from two engine-driven 90 kVA three-phase 400 Hz constant frequency AC generators, 115/200 V output. 90 kVA generator mounted on APU for ground operation or for emergency use in flight. Three hydraulic systems at 207 bar (3,000 lb/sq in), for flight control and utility functions, supplied from engine-driven pumps and a Honeywell bleed air-powered hydraulic pump or from APU. Maximum generating capacity of port and starboard systems is 163 litres (43 US gallons; 35.8 Imp gallons)/min; centre system 185.5 litres (49.0 US gallons; 40.8 Imp gallons)/min, at 196.5 bar (2,850 lb/sq in). Reservoirs pressurised by engine bleed air via pressure regulation module. Reservoir relief valve pressure nominally 4.48 bar (65 lb/sq in). Additional hydraulic motor-driven generator, to provide essential functions for extended-range operations, standard on 767-200ER and 767-300ER and optional on 767-200 and 767-300. Nitrogen chlorate oxygen generators in passenger cabin, plus gaseous oxygen for flight crew. APU in tailcone to provide ground and in-flight electrical power and pressurisation. Anti-icing for outboard wing leading-edges (none on tail surfaces), engine air inlets, air data sensors and windscreen.

AVIONICS: *Radar:* Honeywell RDR-4A colour weather radar in aircraft for All-Nippon, Britannia and Transbrasil.

Flight: Standard ARINC 700 series equipment, including Honeywell VOR/ILS/marker beacon receivers, ADF, DME, RMI-743 radio magnetic indicator and radio altimeter. Honeywell IRS, FMCS and DADC, as described in Boeing 757 entry; dual digital flight management systems, and triple flight control computers, including FCS-700 flight control system; certified for Cat. IIIb landings; options include Boeing's windshear protection and guidance system.

Instrumentation: Honeywell EFIS-700 electronic flight instrument system.

DIMENSIONS, EXTERNAL:

Wing span: except 400ER	47.57 m (156 ft 1 in)
400ER	51.99 m (170 ft 7 in)
Wing chord: at root	8.57 m (28 ft 1¼ in)
at tip	2.29 m (7 ft 6 in)
Wing aspect ratio: except 400ER	8.0
400ER	9.3
Length: overall: 200/200ER	48.51 m (159 ft 2 in)
300/300ER	54.94 m (180 ft 3 in)
400ER	61.37 m (201 ft 4 in)
fuselage: 200/200ER	47.24 m (155 ft 0 in)
300/300ER	53.67 m (176 ft 1 in)
Fuselage: Max width	5.03 m (16 ft 6 in)
Height overall	15.85 m (52 ft 0 in)
Tailplane span	18.62 m (61 ft 1 in)
Wheel track	9.30 m (30 ft 6 in)
Wheelbase: 200/200ER	19.69 m (64 ft 7 in)
300/300ER	22.76 m (74 ft 8 in)
Passenger doors (two, fwd and rear, port):	
Height	1.88 m (6 ft 2 in)
Width	1.07 m (3 ft 6 in)
Galley service door (two, fwd and rear, stbd):	
Height	1.88 m (6 ft 2 in)
Width	1.07 m (3 ft 6 in)
Emergency exits (two, each): Height	0.97 m (3 ft 2 in)
Width	0.51 m (1 ft 8 in)
Cargo door (rear, stbd; fwd, stbd 200/300):	
Height	1.75 m (5 ft 9 in)
Width	1.78 m (5 ft 10 in)
Cargo door (fwd, stbd, 200ER/300ER):	

Height	1.75 m (5 ft 9 in)
Width	3.40 m (11 ft 2 in)
Bulk cargo door (port,rear): Height	1.22 m (4 ft 0 in)
Width	0.97 m (3 ft 2 in)

DIMENSIONS, INTERNAL:

Cabin, excl flight deck:	
Length: 200/200ER	33.93 m (111 ft 4 in)
300/300ER	40.36 m (132 ft 5 in)
Max width	4.72 m (15 ft 6 in)
Max height	2.87 m (9 ft 5 in)
Floor area: 200/200ER	154.9 m² (1,667 sq ft)
300/300ER	184.0 m² (1,981 sq ft)
Volume: 200/200ER	428.2 m³ (15,121 cu ft)
300/300ER	483.9 m³ (17,088 cu ft)
Volume, flight deck	13.5 m³ (478 cu ft)
Baggage holds (containerised), volume:	
200/200ER: fwd	40.8 m³ (1,440 cu ft)
rear	34.0 m³ (1,200 cu ft)
300/300ER: fwd	54.4 m³ (1,920 cu ft)
rear	47.6 m³ (1,680 cu ft)
Bulk cargo hold volume:	
all versions	12.2 m³ (430 cu ft)
Combined baggage hold/bulk cargo hold volume:	
200/200ER	87.0 m³ (3,070 cu ft)
300/300ER	114.1 m³ (4,030 cu ft)
Total cargo hold volume:	
200/200ER	111.3 m³ (3,930 cu ft)
300/300ER	147.0 m³ (5,190 cu ft)

AREAS:

Wings, gross: except 400ER	283.3 m² (3,050.0 sq ft)
400ER	290.70 m² (3,129.0 sq ft)
Ailerons (total)	11.58 m² (124.60 sq ft)
Trailing-edge flaps (total)	36.88 m² (397.00 sq ft)
Leading-edge slats (total)	28.30 m² (304.60 sq ft)
Spoilers (total)	15.83 m² (170.40 sq ft)
Fin	30.19 m² (325.00 sq ft)
Rudder	15.95 m² (171.70 sq ft)
Tailplane	59.88 m² (644.50 sq ft)
Elevators (total)	17.81 m² (191.70 sq ft)

WEIGHTS AND LOADINGS (2GB: 767-200ER basic with CF6-80C2B2F engines, 2GM: 767-200ER max optional T-O weight and CF6-80C2B7Fs, 2PB: 767-200ER basic

PW4052, 2PM: 767-200ER max PW4062, 3GB: 767-300ER basic CF6-80C2B4F, 3GM: 767-300ER max CF6-80C2B7F, 3PB: 767-300ER basic PW4056, 3PM: 767-300ER max PW4062, 3RB: 767-300ER basic RB211-524G4, 3RM: 767-300ER max RB211-524H; all -200 with 181 or 224 passengers, all -300ER with 218 or 269 passengers as indicated, where different):

Operating weight empty:	
2GB (181), 2GM (181)	84,685 kg (186,700 lb)
2GB (224), 2GM (224)	83,960 kg (185,100 lb)
2PB (181), 2PM (181)	84,730 kg (186,800 lb)
2PB (224), 2PM (224)	83,915 kg (185,000 lb)
3GB (218), 3GM (218)	90,535 kg (199,600 lb)
3GB (269), 3GM (269)	89,855 kg (198,100 lb)
3PB (218), 3PM (218)	90,585 kg (199,700 lb)
3PB (269), 3PM (269)	89,900 kg (198,200 lb)
3RB (218), 3RM (218)	91,580 kg (201,900 lb)
3RB (269), 3RM (269)	90,900 kg (200,400 lb)
Baggage capacity, underfloor:	
200ER: fwd: standard	9,798 kg (21,600 lb)
alternate	15,309 kg (33,750 lb)
rear: standard	8,165 kg (18,000 lb)
alternate	12,247 kg (27,000 lb)
300ER: fwd: standard	13,063 kg (28,800 lb)
alternate	20,412 kg (45,000 lb)
rear: standard	11,431 kg (25,200 lb)
alternate	17,574 kg (38,745 lb)
Bulk hold capacity (all versions)	2,926 kg (6,450 lb)
Max fuel weight:	
200, 300	51,130 kg (112,725 lb)
200ER, 300ER	73,635 kg (162,340 lb)
Max T-O weight: 2GB, 2PB	156,490 kg (345,000 lb)
2GM, 2PM	179,170 kg (395,000 lb)
3GB, 3PB, 3RB	172,365 kg (380,000 lb)
3GM, 3PM, 3RM	186,880 kg (412,000 lb)
Max ramp weight: 2GB, 2PB	156,945 kg (346,000 lb)
2GM, 2PM	179,625 kg (396,000 lb)
3GB, 3PB, 3RB	172,820 kg (381,000 lb)
3GM, 3PM, 3RM	187,335 kg (413,000 lb)
Max landing weight:	
2GB, 2PB	126,100 kg (278,000 lb)
2GM, 2PM	136,080 kg (300,000 lb)
3GB, 3GM, 3PB, 3PM, 3RB, 3RM	
	145,150 kg (320,000 lb)
Max zero-fuel weight:	
2GB, 2PB	114,755 kg (253,000 lb)
2GM, 2PM	117,935 kg (260,000 lb)
3GB, 3GM, 3PB, 3PM, 3RB, 3RM	
	133,810 kg (295,000 lb)
Max wing loading:	
2GB, 2PB	552.3 kg/m² (113.11 lb/sq ft)
2GM, 2PM	632.3 kg/m² (129.51 lb/sq ft)
3GB, 3PB, 3RB	608.3 kg/m² (124.59 lb/sq ft)
3GM, 3PM, 3RM	659.5 kg/m² (135.08 lb/sq ft)
Max power loading:	
2GB	348 kg/kN (3.41 lb/lb st)
2GM	324 kg/kN (3.18 lb/lb st)
2PB	336 kg/kN (3.30 lb/lb st)
2PM	318 kg/kN (3.12 lb/lb st)
3GB	343 kg/kN (3.36 lb/lb st)
3GM	338 kg/kN (3.32 lb/lb st)
3PB	339 kg/kN (3.33 lb/lb st)

Boeing 767-200 wide-bodied airliner, with additional side view of stretched -300 and further stretched -400, including last mentioned's revised wingtips (*Jane's/Dennis Punnett*)   0085775

3PM	332 kg/kN (3.25 lb/lb st)
3RB	344 kg/kN (3.37 lb/lb st)
3RM	353 kg/kN (3.46 lb/lb st)

PERFORMANCE:
Normal cruising speed, all versions   M0.80
Approach speed at MLW:

2GB, 2PB	137 kt (254 km/h; 158 mph)
2GM, 2PM	139 kt (258 km/h; 160 mph)
3GB, 3GM, 3PB, 3PM	145 kt (269 km/h; 167 mph)
3RB, 3RM	148 kt (275 km/h; 171 mph)

Initial cruising altitude at max T-O weight:

2GB	11,550 m (37,900 ft)
2GM, 2PM	10,670 m (35,000 ft)
2PB	11,310 m (37,100 ft)
3GB	10,700 m (35,100 ft)
3GM, 3PM	10,180 m (33,400 ft)
3PB	10,730 m (35,200 ft)
3RB	10,730 m (35,200 ft)
3RM	10,210 m (33,500 ft)

Service ceiling, OEI: 2GB   5,090 m (16,700 ft)

2GM	4,205 m (13,800 ft)
2PB	5,395 m (17,700 ft)
2PM	4,845 m (15,900 ft)
3GB	4,510 m (14,800 ft)
3GM	3,930 m (12,900 ft)
3PB	4,785 m (15,700 ft)
3PM	4,085 m (13,400 ft)
3RB	3,900 m (12,800 ft)
3RM	3,505 m (11,500 ft)

T-O field length S/L, 86°F: 2GB   2,301 m (7,550 ft)

2GM, 3PB	2,485 m (8,150 ft)
2PB	2,180 m (7,150 ft)
2PM	2,439 m (8,000 ft)
3GB	2,530 m (8,300 ft)
3GM	2,713 m (8,900 ft)
3PM	2,652 m (8,700 ft)
3RB	2,500 m (8,200 ft)
3RM	2,896 m (9,500 ft)

Landing field length at MLW: 2GB   1,524 m (5,000 ft)

2GM	1,555 m (5,100 ft)
2PB	1,509 m (4,950 ft)
2PM	1,540 m (5,050 ft)
3GB, 3GM, 3PB, 3PM	1,677 m (5,500 ft)
3RB, 3RM	1,707 m (5,600 ft)
3RM	2,896 m (9,500 ft)

Design range: 2GB (181)
5,125 n miles (9,491 km; 5,897 miles)

2GB (224)	4,830 n miles (8,945 km; 5,558 miles)
2GM (181)	6,655 n miles (12,325 km; 7,658 miles)
2GM (224)	6,545 n miles (12,121 km; 7,531 miles)
2PB (181)	5,030 n miles (9,315 km; 5,788 miles)
2PB (224)	4,740 n miles (8,778 km; 5,454 miles)
2PM (181)	6,555 n miles (12,139 km; 7,543 miles)
2PM (224)	6,450 n miles (11,945 km; 7,422 miles)
3GB (218)	5,230 n miles (9,686 km; 6,018 miles)
3GB (269)	4,890 n miles (9,056 km; 5,627 miles)
3GM (218)	6,150 n miles (11,389 km; 7,077 miles)
3GM (269)	5,990 n miles (11,093 km; 6,893 miles)
3PB (218)	5,155 n miles (9,547 km; 5,932 miles)
3PB (269)	4,820 n miles (8,926 km; 5,546 miles)
3PM (218)	6,075 n miles (11,250 km; 6,991 miles)
3PM (269)	5,915 n miles (10,954 km; 6,806 miles)
3RB (218)	4,880 n miles (9,037 km; 5,615 miles)
3RB (269)	4,555 n miles (8,435 km; 5,241 miles)
3RM (218)	5,850 n miles (10,834 km; 6,732 miles)
3RM (269)	5,620 n miles (10,408 km; 6,467 miles)

UPDATED

BOEING 767-300 GENERAL MARKET FREIGHTER

TYPE: Twin-jet freighter.
PROGRAMME: First 767 specialised package freighter launched January 1993 by United Parcel Service order; mockup completed early 1994; first flight (N301UP) 20 June 1995. The 767-300F for general operation was ordered by Asiana in November 1993 and differs from UPS version in having mechanical freight handling on main and lower decks, air conditioning for animals and perishables on main and forward lower decks and more elaborate crew facilities.
CUSTOMERS: UPS ordered 30 parcel freighters 15 January 1993; first delivery 12 October 1995; last on 9 September 1999; two of further 30 options taken up and delivered in October and November 2001. Asiana ordered two 767-300Fs in November 1993; contract reduced to one, which was delivered 23 August 1996. Five to LAN Chile between 23 September 1998 and 9 October 2001. One ordered by GECAS; one by undisclosed customer. Total 40.
COSTS: US$122.5 million to US$134.0 million (2002).
DESIGN FEATURES: Modifications include reinforced landing gear and internal wing structure; main deck floor strengthened to take 24 containers, each 2.24 × 3.18 m (88 × 125 in); no passenger windows; 2.67 × 3.40 m (8 ft 9 in × 11 ft 1¾ in) freight door forward to port; pilot-type rating and extensive component commonality with 757 Freighter.
POWER PLANT: Two General Electric CF6-80C2B6F/B7F, Pratt & Whitney PW4060/4062 or Rolls-Royce RB211-524/H turbofans; 267 kN (60,000 lb st) each.
DIMENSIONS, INTERNAL:
Cabin: Length   39.80 m (130 ft 7 in)

Main deck container capacity	339.5 m³ (11,990 cu ft)
Lower deck cargo capacity	92.9 m³ (3,282 cu ft)

WEIGHTS AND LOADINGS:

Max T-O weight: standard	185,065 kg (408,000 lb)
optional	186,880 kg (412,000 lb)
Max ramp weight	187,335 kg (413,000 lb)
Max zero-fuel weight	140,160 kg (309,000 lb)
Max landing weight	147,871 kg (326,000 lb)
Max wing loading	659.5 kg/m² (135.08 lb/sq ft)
Max power loading	350 kg/kW (3.43 lb/lb st)

PERFORMANCE:
Range: with 40,823 kg (90,000 lb) payload
4,000 n miles (7,408 km; 4,603 miles)
with 50,800 kg (112,000 lb) payload
3,000 n miles (5,556 km; 3,452 miles)

UPDATED

BOEING 777

TYPE: Wide-bodied airliner.
PROGRAMME: Formerly known as 767-X, initial variant now 777-200; Boeing board authorised firm offer for sale 8 December 1989; launch order by United Airlines 15 October 1990 (see Customers); Boeing launched production programme of initial market and Increased Gross Weight 777s and formed 777 Division 29 October 1990; configuration frozen March 1991; Boeing signed final agreement with Mitsubishi, Kawasaki and Fuji, making them risk-sharing programme partners for about 20 per cent of the 777 structure, on 21 May 1991; roll-out occurred 9 April 1994.
First flight (line no. WA001/N7771 with PW4084 engines), 12 June 1994, WA002 15 July, WA003 2 August, WA004 28 October, WA005 11 November; WA006/G-ZZZA (first with GE90, for British Airways) 2 February 1995, same day as FAA granted engine approval; PW-engined aircraft accumulated some 3,235 hours in 2,340 cycles in test programme, leading to joint FAA/JAA certification on 19 April 1995; FAA awarded 180-minute ETOPS approval 30 May 1995; first delivery, to United Airlines, on 15 May 1995; service entry with United on 7 June 1995 with inaugural revenue flight (by N777NA) from London to Washington, DC. Second GE90 aircraft for British Airways (WA010/G-ZZZB) joined WA006 in 1,750 hour, 1,260 cycle test programme; certification with GE engine 9 November 1995, with deliveries to BA commencing two days later.
Flight testing of Rolls-Royce Trent 800 on Boeing 747-100 testbed began late March 1995, with first flight of Trent-powered 777 (Boeing test aircraft) 26 May; certification and first delivery (to Thai International) on 3 April 1996 (formal hand-over 31 March), with ETOPS clearance following in October. The US National Aeronautic Association awarded Boeing and the 777 its 1995 Robert J. Collier Trophy for aeronautical achievement. On 2 April 1997, a Trent-powered 777 landed at Boeing Field having captured the eastbound world circumnavigation record in 41 hours 59 minutes with a single stop at Kuala Lumpur, Malaysia, its arrival at the latter also having secured the non-stop great circle distance record of 10,266.9 n miles (19,014.3 km; 11,814.9 miles).
Two new derivatives launched 29 February 2000 as 777-200LR and 777-300ER, both as extended-range

versions of current production types. Anticipated market of 500 for these derivatives, 45 per cent of which with Asian operators.
CURRENT VERSIONS: **777-100X**: Under study in 2000 to meet Singapore Airlines requirements for shortened, 250-seat version.
777-200: Initial production version. Maximum T-O weight 247,210 kg (545,000 lb); up to 440 passengers at maximum density.
777-200ER: Formerly -200-IGW (Increased Gross Weight). Maximum T-O weight initially 286,895 kg (632,500 lb); increased to 293,930 kg (648,000 lb) in March 1998 and 297,555 kg (656,000 lb) in January 1999; additional 53,828 litres (14,220 US gallons; 11,840 Imp gallons) of fuel in centre-section tank; same passenger capacity as basic aircraft; configuration frozen January 1994, including strengthened wing, fuselage, empennage, landing gear and engine pylons. First flew (G-VIIA, one of two for British Airways) 7 October 1996 (GE engines); FAA/JAA ETOPS certification 5 February 1997; delivery 6 February. First Trent-powered -200ER (A6-EMI, for Emirates) flew 21 November 1996. January 1999 MTOW increase resulted from fitting 777-300 main landing gear and restricting CG travel to 4 per cent, for increased taxiing weight. First flight with GE90-94B engines 12 June 2000; FAA certification 14 November 2000, followed immediately by first delivery (to Air France).
First 777 for private use is a -200ER (N777AS) first flown 3 November 1998 and delivered to Raytheon Systems at Waco, Texas, for outfitting; customer delivery to Middle East Jet due late 2000.
777-200LR: Ultra-long-range version (previously 777-200X) powered by two General Electric GE90-110B engines, each rated at 489 kN (110,000 lb/st); launched 29 February 2000. Maximum T-O weight 341,100 kg (752,000 lb); fuel capacity 195,283 litres (51,590 US gallons; 42,958 Imp gallons), with addition of tanks in rear cargo hold; range 8,860 n miles (16,408 km; 10,195 miles) with 301 passengers. Height 18.62 m (61 ft 1 in); see 777-300LR for further features. Firm launch order from EVA Airlines, 27 June 2000, for three. On 1 October 2001, Boeing announced suspension of development programme for up to 18 months; deliveries to Eva Air scheduled to begin in May 2006.
777-300: Initially known as 777 Stretch; revealed at Paris Air Show, 14 June 1995; launched by Boeing board 26 June 1995 when 36 commitments held; configuration frozen October 1995; major assembly started 7 April 1997; roll-out 8 September 1997; first flight (Rolls-Royce engines) 16 October 1997 (N5014K); first with P&W engines (HL-7534 of Korean Air) 4 February 1998; FAA certification achieved 4 May 1998 with 180-minute ETOPS approval; first delivery (B-HNH, 136th 777) to Cathay Pacific 22 May 1998.
Compared with first-generation 747s, 777-300 carries same number of passengers, but at two-thirds of fuel cost and with 40 per cent less maintenance. Features strengthened airframe, inboard wing and landing gear, ground-manoeuvring cameras on horizontal tail surfaces and wing/fuselage fairing; tailskid; Type A emergency door over each wing; and fuselage stretched by 19 frames, 10.13 m (33 ft 3 in) longer (5.33 m; 17 ft 6 in ahead of wing and 4.80 m; 15 ft 9 in aft of wing) than that of 777-200 to increase passenger capacity to 550 in single-class high-density configuration.

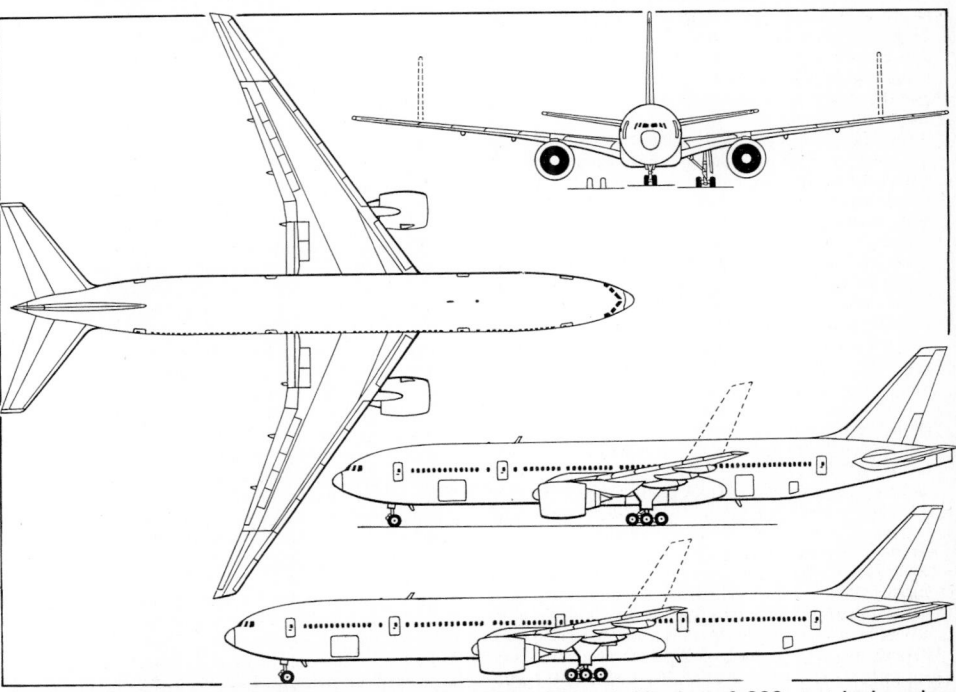

Boeing 777-200 twin-turbofan high-capacity airliner, with additional side view of -300 stretched version; optional folding wing shown by broken lines (*Jane's/Mike Keep*)   0015648

777-300ER: Stretched equivalent of 777-200R (previously 777-300X); launched 29 February 2000; assembly began 20 June 2002; rolled out 14 November 2002. GE90-115B engines each rated at 512 kN (115,000 lb st); height 18.57 m (60 ft 11 in). Maximum T-O weight as for 777-200LR; fuel capacity 181,283 litres (47,890 US gallons; 39,877 Imp gallons); range 7,200 n miles (13,334 km; 8,285 miles) with 359 passengers. First flight 24 February 2003; certification August 2003. Launch customer Japan Airlines ordered eight on 31 March 2000 for delivery from October 2003, but first recipient will now be ILFC/Air France in April 2004. Both -200LR and -300ER have strengthened horizontal and vertical tail surfaces, strengthened wings with new tips which increase span to 64.80 m (212 ft 7 in) and strengthened landing gear, including semi-levered main gear and extendable (up to 25 cm; 9¾ in) nosewheel leg on -300ER. Both will be certified for 207-minute ETOPS.

777-300ER Stretch: Under study in 1999 as possible replacement for 747-400 with Asian carrier on routes to Europe; fuselage stretch of about 7.9 m (26 ft) to accommodate an additional 60 passengers, but airframe otherwise unchanged; range as for 777-300ER. Also referred to as **777-400X.**

CUSTOMERS: See table.

COSTS: Estimated development cost US$4 billion (1990). US$153.5 million to US$171 million 777-200; US$162 million to US$182 million 777-200ER; US$188 million to US$213.5 million 777-200LR; US$178.5 million to US$203.5 million 777-300; US$203.5 million to US$231.5 million 777-300ER (all 2002). 777-300 estimated to burn 33 per cent less fuel and have 40 per cent lower maintenance cost than 747-100/200, resulting in direct operating cost savings of around 30 to 35 per cent.

DESIGN FEATURES: Objectives of design included replacement of McDonnell Douglas DC-10 Srs 10 and Lockheed TriStar in regional market, as well as DC-10 Srs 30 and Boeing 747SP in intercontinental service; also is replacement for early 747s. All features required for 180-minute ETOPS incorporated and tested in basic aircraft design. Cylindrical fuselage wider than 767 to allow twin aisle seating for from six- to 10-abreast; lavatories and overhead baggage bins designed to allow rapid change of layout.

Wing, of 31° 30′ sweepback at quarter-chord, incorporates new technology to allow minimum M0.83 cruise in combination with high thickness for economical structure and large internal volume, long span for improved take-off and payload/range and large area for high cruise altitude and low approach speed; no winglets. Outer 6.48 m (21 ft 3 in) of each wing can, as an option, be folded to vertical to reduce gate width requirement at airports.

FLYING CONTROLS: Fly-by-wire. Hydraulically actuated, with trim tab in rudder. Six-segment slats in each wing leading-edge. Single-slotted flaps mid-wing, double-slotted flaps inboard; flaperon between inboard and outboard flaps. Five-segment spoilers ahead of single-slotted flaps; two-segment spoilers ahead of double-slotted flaps.

Boeing's first airliner fly-by-wire system; fully powered control surface actuators (31 by Teijin Seiki America) electrically signalled from full FBW system; this signals slats, flaps, spoilers and control feel unit as well as inboard flaperons, outboard ailerons, elevators and rudder; system provides flight envelope protection as well as stabilisation and autopilot inputs, but the normal control columns and rudder pedals in cockpit are back-driven by the system to give the pilots direct appreciation of the activity of the automatic system.

In normal mode, flight guidance commands are generated by Rockwell Collins triple redundant digital autopilot/flight directors and the control laws and envelope protection commands are shaped by the Marconi Avionics triple digital primary flight computers; each of the three primary flight computers contains three 32-bit microprocessors (a Motorola 68040, an Intel 80486 and an AMD 29050), all three programmed in Ada to perform all FBW functions; with power supply and ARINC 629 modules, each microprocessor module constitutes a lane and the three lanes constitute a channel; each lane is compared with the others in its channel; the system not only has high fault tolerance, but allows deferred maintenance, by which failures can be carried over until the next scheduled maintenance.

Commands to the powered control units are produced by three BAE Systems and Teijin Seiki actuator control electronics units, which have a fourth analogue channel directly signalled from the sticks and pedals in the cockpit; normal operating mode is for the aircraft to be flown through autopilots, primary flight computers and actuator control electronics, which simultaneously back-drive the sticks and pedals in the cockpit; first degraded (secondary) mode is used if inertial units and standby attitude sensors all become disabled and the pilots take manual control through the digital primary flight computers; second degraded (direct) mode bypasses the main FBW system with the direct analogue link between cockpit and actuator control electronics; ultimate standby is mechanical control

of tailplane incidence for the pitch axis and two wing spoiler panels for lateral control. Some powered control units produced by Parker-Bertea and Moog; tailplane trim module and hydraulic brake by Raytheon Systems.

Pitch axis control law is C* U, effectively tending to make the aircraft hold an airspeed and to respond in pitch attitude to a departure from that airspeed; trim changes due to configuration changes are suppressed; the system returns the bank angle to 35° if that angle is exceeded by the pilots and the controls then released; the system prevents exceeding the limiting airspeed and stalling; asymmetric thrust is automatically countered; the variable feel system adjusts control forces to warn of approach to flight envelope limits in manual flight; the FBW system is linked to the ARINC 629 dual triplex digital databusses (see also aircraft information management system under Avionics heading).

STRUCTURE: Composites of carbon and toughened resin used in skins of tailplane and fin torsion boxes and cabin floor beams; CFRP used for rudder, elevators, ailerons, flaps, engine nacelles and landing gear doors; hybrid composites in wingroot fairing; GFRP in fixed-wing leading-edge, tailplane and fin fore and aft panels, wing aft panels, engine pylon fairings and radome. Toughened materials have high

damage resistance and allow simple low-temperature bolted repairs. Metal structure includes thick skins without need for tear straps; no bonding; single-piece fuselage frames; fuselage skin of advanced 2000-series aluminium alloy; wing, empennage and engine nacelle leading edges of 2000-series; wing top skin and stringers made in advanced 7055 aluminium alloy with greater compression strength; tailcone in standard 7000-series; 10 per cent of structure weight is composites.

Fully digital product definition with all parts created by Dassault/IBM CATIA CAD/CAM and communicated to manufacturing and publications; structure and systems integration, tube and cable run design completed before design release; 238 design/build teams have ensured that design, fabrication and test have proceeded concurrently for structure and systems. Whole aircraft defined in computer system; no mockup built.

Centre and rear fuselage barrel sections, tailcone, doors, wingroot fairing and landing gear doors made in Japan. Wing and tail leading-edges and moving wing parts, landing gear, floor beams, nose landing gear doors, wingtips, dorsal fin and nose radome made by Northrop Grumman, Kaman, Alenia (Italy), Embraer (Brazil), Short Brothers (UK), Singapore Aerospace Manufacturing,

BOEING 777 ORDER BOOK
(at 1 January 2003)

Customer	Variant	Engine	First Order	First Delivery	Qty
Air China	777-200	PW4000	24 Mar 97	26 Oct 98	10
Air France	777-200	GE90	20 Nov 96	27 Mar 98	18
	777-300ER	GE90	13 Nov 00	none	10
Alitalia	777-200ER	GE90	9 Nov 00	23 Aug 02	6
All Nippon Airways	777-200	PW4000	19 Dec 90	4 Oct 95	15
	777-200ER	PW4000	19 Dec 90	6 Oct 99	4
	777-300	PW4000	12 Sep 95	30 Jun 98	15
	777-300ER	GE90	18 Jul 00	none	6
American Airlines	777-200ER	Trent 800	21 Nov 96	21 Jan 99	47
Asiana Airlines	777-200ER	PW4000	20 Dec 96	none	6
	777-300	PW4000	20 Dec 96	none	1
British Airways	777-200	GE90	21 Aug 91	11 Nov 95	5
	777-200ER	GE90	21 Aug 91	6 Feb 97	24
	777-200ER	Trent 800	1 Aug 98	7 Jun 00	16
Cathay Pacific Airways	777-200	Trent 800	6 May 92	9 May 96	5
	777-300	Trent 800	6 May 92	22 May 98	10
China Southern Airlines	777-200	GE90	17 Dec 92	28 Dec 95	4
	777-200ER	GE90	17 Dec 92	28 Feb 97	2
Continental Airlines	777-200ER	GE90	12 May 93	28 Sep 98	16
Delta Air Lines	777-200ER	Trent 800	13 Nov 97	23 Mar 99	13
Egyptair	777-200ER	PW4090	23 Aug 95	23 May 97	5
El Al Israel Airlines	777-200ER	Trent 800	14 Dec 99	29 Jan 01	4
Emirates	777-200	Trent 800	4 Jun 92	5 Jun 96	3
	777-200ER	Trent 800	4 Jun 92	11 Apr 97	6
EVA Airways	777-200LR	GE90	27 Jun 00	none	3
	777-300ER	GE90	27 Jun 00	none	4
Garuda Indonesian	777-200ER	GE90	25 Jun 96	none	6
GECAS	777-200ER	GE90	26 Sep 00	none	5
	777-300	GE90	26 Sep 00	none	9
ILFC	777-200ER	GE90	15 Dec 92	8 Jan 98	21
	777-200ER	PW4000	15 Dec 92	16 Jul 98	7
	777-200ER	Trent 800	2 Sep 97	none	3
	777-200ER	n/k	28 Nov 00	none	20
	777-300	PW4000	6 Mar 96	none	1
	777-300	Trent 800	6 Mar 96	26 Sep 00	5
	777-300	n/k	28 Nov 00	none	2
	777-300ER	GE90	28 Nov 00	none	8
Japan Air System	777-200	PW4000	29 Jun 93	3 Dec 86	7
Japan Airlines	777-200	PW4000	24 Jan 92	15 Feb 96	7
	777-200ER	GE90	27 Nov 00	11 Jul 02	11
	777-300	PW4000	22 Dec 95	28 Jul 98	8
	777-300ER	GE90	31 Mar 00	none	8
Kenya Airways	777-200ER	n/k	22 Mar 02	none	3
KLM	777-200ER	n/k	26 Aug 02	none	4
Korean Air Lines	777-200ER	PW4000	16 Dec 93	21 Mar 97	9
	777-300	PW4000	16 Dec 93	12 Aug 99	4
Kuwait Airways	777-200	GE90	10 Jul 96	30 Mar 98	2
Lauda Air	777-200ER	GE90	13 Dec 91	24 Sep 97	4
Malaysia Airlines	777-200ER	Trent 800	8 Jan 96	23 Apr 97	15
Mid East Jet	777-200ER	GE90	1 Oct 97	24 Nov 98	1
Pakistan International Airlines	777-200ER	n/k	14 Nov 02	none	3
	777-200LR	n/k	14 Nov 02	none	2
	777-300ER	n/k	14 Nov 02	none	3
Saudia	777-200ER	GE90	18 Jun 95	26 Dec 97	23
Saudi Oger	777-200ER	GE90	30 Nov 98	22 Oct 99	1
Singapore Aircraft Leasing	777-200ER	Trent 800	22 Dec 95	20 Mar 01	2
	777-300	Trent 800	22 Dec 95	12 Nov 99	4
Singapore Airlines	777-200ER	Trent 800	22 Dec 95	5 May 97	50
	777-300	Trent 800	22 Dec 95	10 Dec 98	9
Thai Airways International	777-200	Trent 800	20 Jun 91	31 Mar 96	8
	777-300	Trent 800	22 Dec 95	23 Dec 98	6
unidentified	777-200ER	PW4000	31 Dec 99	none	7
	777-300ER	n/k	31 Dec 02	none	8
United Airlines	777-200	PW4000	15 Oct 90	15 May 95	22
	777-200ER	PW4000	15 Oct 90	7 Mar 97	39
Vietnam Airlines	777-200ER	PW4000	31 Jan 02	none	4
Total					**619**

Note: For delivery totals see Group Orders and Deliveries table

HDH and ASTA (Australia), Korean Air and other subcontractors. Boeing manufactures flight deck and forward cabin, basic wing and tail structures and engine nacelles; assembles and tests completed aircraft.

LANDING GEAR: Retractable tricycle type (Menasco/Messier-Bugatti joint design for main gear); two main legs carrying six-wheel bogies with steering rear axles automatically engaged by nose gear steering angle; six-wheel bogies avoid need for third leg in fuselage and simplify braking system; twin-wheel steerable nose gear; mainwheel tyres H49×19.0-22 or 50×20.0R22 (32 ply); nosewheel tyres 42×17.0R18 (26 ply). Honeywell Carbenix 4000 mainwheel brakes arranged so that initial toe-pedal pressure used during taxying applies brakes to alternate sets of three wheels to save brake wear; full toe-pedal pressure applies all six brakes together.

POWER PLANT: Two turbofans.

777-200: 343 kN (77,000 lb st) General Electric GE90-77B, 331 kN (74,500 lb st) Pratt & Whitney PW4074 or 343 kN (77,200 lb st) PW4077, or 327 kN (73,400 lb st) Rolls-Royce Trent 875 or 338 kN (76,000 lb st) Trent 877.

777-200ER: 377 kN (84,700 lb st) GE90-85B, 400 kN (90,000 lb st) GE90-90B, 417 kN (93,700 lb st) GE90-94B, 376 kN (84,600 lb st) PW4084, 401 kN (90,200 lb st) PW4090, 436 kN (98,000 lb st) PW4098, 372 kN (83,600 lb st) Trent 884, 400 kN (90,000 lb st) Trent 892 or 415 kN (93,400 lb st) Trent 895.

777-300: PW4090, PW4098 or Trent 892, as above.

All fuel of baseline version contained in integral tanks in wing torsion box, with reserve tank, surge tank and fuel vent and jettison pipes all inboard of wing fold; combined capacity of main, centre and reserve tanks in 777-200 is 117,348 litres (31,000 US gallons; 25,813 Imp gallons); 777-200ER and 777-300 fuel capacity increased by centre-section tank of 53,829 litres (14,220 US gallons; 11,841 Imp gallons) to maximum of 171,176 litres (45,220 US gallons; 37,653 Imp gallons).

ACCOMMODATION: Two-pilot crew; cabin cross-section, which is between that of 747 and 767, chosen to allow widest selection of twin-aisle class and seating layouts ranging from six- to 10-abreast; galleys and lavatories can be located at a selection of fixed points in front and rear cabins or freely positioned within large footprints in which they can be moved in 2.5 cm (1 in) increments and attached to prepositioned mounting, plumbing and electric fittings; overhead bins open downward and provide each passenger with 0.08 m³ (3 cu ft) volume; bins can be removed without disturbing ceiling panels, ducts or support structure; advanced cabin management system simplifies cabin management and includes digital sound system of hi-fi quality.

Typical configurations for 200/200ER include 261 passengers in three-class layout: 12 first class (two-two-two-abreast), 79 business class (two-three-two) and 170 economy class (three-three-three); 367 in two classes: 14 business (two-three-two) and 353 economy (three-three-three); 375 in two classes: 30 first class (two-two-two at 97 cm; 38 in pitch) and 345 economy (two-five-two at 79 cm; 31 in or 81 cm; 32 in pitch); 400 in two classes: 30 first class, as immediately previously and 370 economy (three-four-three, pitches as previously); and 440 passengers in single class (three-four-three).

For 777-300, options include 368 in three classes: 30 first (two-two-two at 152 cm; 60 in), 84 business (two-three-two at 97 cm; 38 in) and 254 economy (two-five-two at 79 cm; 31 in or 81 cm; 32 in); 386 in first (30), business (77) and economy (279) classes, as previously, except last-mentioned at three-four-three, with only four seats at minimum pitch; 451 passengers in two classes: 40 first, as above, and 411 economy (two-five-two, at usual pitches); 479, in first (44) and economy (435) classes, the former with 97 cm; 38 in pitch and latter three-four-three at usual pitches; and ultimate 550 single-class passengers, mostly two-four-two. Underfloor cargo compartments have mechanical handling system and can accommodate all LD formats and 88 in or 96 in width pallets; up to 32 LD-3 containers plus 17.0 m³ (600 cu ft) bulk cargo can be loaded in 777-200, or 44 LD-3s and some bulk cargo in 777-300.

Flight crew rest module on flight deck, port, contains two bunks. Optional underfloor crew rest module with same footprint as 96 in pallet contains six bunks, and stowage space and requires only electrical connection and hatch in passenger cabin floor, level with wing trailing-edge.

SYSTEMS: Honeywell air drive unit, using bleed air from engines, APU or ground supply, drives central hydraulic system; cabin air supply and pressure control by Honeywell; Hamilton Sundstrand variable-speed, constant frequency AC electrical power generating system, with two 120 kVA integrated drive generators, one APU-driven generator and Honeywell ram air turbine system. Honeywell GTCP331-500 APU; Hamilton Sundstrand air conditioning; Smiths Industries ultrasonic fuel quantity gauging system and electrical load management system; optional wingtip folding by Raytheon Montek Division and Frisby Airborne Hydraulics.

AVIONICS: *Radar:* Honeywell weather radar standard.

Flight: Main navigation system is Honeywell air data and inertial reference (ADIRS) containing the Hexad skewed axis arrangement of six ring laser gyros; standby system is the secondary attitude and air data reference unit (SAARU) containing interferometric fibre optic gyros (using light transmitted in two directions along fibre optic paths), which produces a secondary flight director attitude display, airspeed and altimeter; both are linked to the ARINC 629 digital databus (777 being first aircraft thus equipped); Honeywell TCAS; Honeywell/BAE Systems Canada global navigation satellite sensor with 12-channel receiver; Honeywell/Racal multichannel satcom system optional.

Emirates Boeing 777-200 taking off from Dubai *(Jane's/Paul Jackson)* ***NEW***/0526958

Dual Honeywell aircraft information management system (AIMS) contains the processing equipment required to collect, format and distribute onboard avionic information, including the flight management system (FMS), engine thrust control, digital communications management, operation of flight deck displays and monitoring of aircraft condition; both pilots and ground engineers can assess the condition of all onboard avionics systems.

Instrumentation: Based on five-screen EFIS using Honeywell 203 mm (18 in) ARINC D-size colour liquid crystal flat panel displays (two primary flight displays, two navigation displays and EICAS display); three multipurpose control and colour display units on centre console provide interface with integrated aircraft information management system, which handles flight management, thrust control and communications control as well as all systems information.

EQUIPMENT: Boeing 777-300 has Ground Maneuver Camera System with TV cameras in leading-edges of both horizontal stabilisers and underside of fuselage.

DIMENSIONS, EXTERNAL:

Wing span	60.93 m (199 ft 11 in)
Wing span with tips folded	47.32 m (155 ft 3 in)
Wing aspect ratio	8.7
Length overall: 200	63.73 m (209 ft 1 in)
300	73.86 m (242 ft 4 in)
Fuselage: Length	62.74 m (205 ft 10 in)
Max diameter	6.20 m (20 ft 4 in)
Max height: overall: 200	18.51 m (60 ft 9 in)
300	18.49 m (60 ft 8 in)
to tip of folded wing	14.22 m (46 ft 8 in)
Tailplane span	21.52 m (70 ft 7½ in)
Wheel track	10.97 m (36 ft 0 in)
Wheelbase	25.88 m (84 ft 11 in)
Passenger doors (four port, four stbd, each):	
Height	1.88 m (6 ft 2 in)
Width	1.07 m (3 ft 6 in)
Max height to sill	5.51 m (18 ft 1 in)
Forward cargo door, stbd: Height	1.70 m (5 ft 7 in)
Width	2.69 m (8 ft 10 in)
Max height to sill	3.05 m (10 ft 0 in)
Rear cargo door, stbd, standard:	
Height	1.70 m (5 ft 7 in)
Width	1.78 m (5 ft 10 in)
Max height to sill	3.40 m (11 ft 2 in)
Rear cargo door, stbd, optional width	2.69 m (8 ft 10 in)
Bulk cargo door, stbd: Height	0.91 m (3 ft 0 in)
Width	1.14 m (3 ft 9 in)
Max height to sill	3.48 m (11 ft 5 in)

DIMENSIONS, INTERNAL:

Cabin: Length	49.10 m (161 ft 1 in)
Max width	5.87 m (19 ft 3 in)
Floor area: 200, 200ER	279.1 m² (3,004 sq ft)
Max underfloor cargo hold volume:	
200, 200ER: forward	80.5 m³ (2,844 cu ft)
aft	62.6 m³ (2,212 cu ft)
bulk	17.0 m³ (600 cu ft)
total	160.2 m³ (5,656 cu ft)
300: forward	107.4 m³ (3,792 cu ft)
aft	89.5 m³ (3,160 cu ft)
bulk	17.0 m³ (600 cu ft)
total	213.8 m³ (7,552 cu ft)

AREAS:

Wings, projected	427.8 m² (4,605.0 sq ft)
Ailerons (total)	7.11 m² (76.50 sq ft)
Trailing-edge flaps (total)	67.13 m² (722.60 sq ft)
Slats (total)	36.84 m² (396.50 sq ft)
Inboard spoilers (total)	8.67 m² (93.30 sq ft)
Outboard spoilers (total)	14.34 m² (154.40 sq ft)
Flaperons	6.69 m² (72.00 sq ft)
Horizontal tail surfaces, projected	101.26 m² (1,090.0 sq ft)
Vertical tail surfaces, projected	53.23 m² (573.00 sq ft)
Elevators, incl tabs (total)	25.48 m² (274.30 sq ft)
Rudder, incl tab	18.16 m² (195.50 sq ft)

WEIGHTS AND LOADINGS (777-200 with 320 passengers, 24/61/235, 2GB: GE90-77B engines at basic MTOW, 2GM: GE90-77B maximum, 2PB: PW4074 basic, 2PM: PW4077 maximum, 2RB: Trent 875 basic, 2RM: Trent 877 maximum; 777-200ER, as above, 2ERGB: GE90-85B basic, 2ERGM: GE90-94 maximum, 2ERPB: PW4084 basic, 2ERPM: PW4090 maximum, 2ERRB: Trent 884 basic, 2ERRM: Trent 895 maximum; 777-300 with 386 passengers, 30/77/279, 3PB: PW4090 basic, 3PM: PW4098 maximum, 3RB: Trent 892 basic, 3RM: Trent 892 maximum):

Operating weight empty	
2GB	140,615 kg (310,000 lb)
2GM	141,115 kg (311,100 lb)
2PB	139,210 kg (306,900 lb)
2PM	139,345 kg (307,200 lb)
2RB	137,350 kg (302,800 lb)
2RM	137,485 kg (303,100 lb)
2ERGB	143,925 kg (317,300 lb)
2ERGM	144,105 kg (317,700 lb)
2ERPB	142,155 kg (313,400 lb)
2ERPM	142,925 kg (315,100 lb)
2ERRB	140,295 kg (309,300 lb)
2ERRM	140,480 kg (309,700 lb)
3PB	158,170 kg (348,700 lb)
3PM	158,620 kg (349,700 lb)
3RB, 3RM	155,675 kg (343,200 lb)
Max fuel weight: 200	94,210 kg (207,700 lb)
200ER/300	135,845 kg (299,490 lb)
Max T-O weight:	
2GB, 2PB, 2RB	229,575 kg (506,000 lb)
2GM, 2PM, 2RM	247,205 kg (545,000 lb)
2ERGB, 2ERPB, 2ERRB, 3PB, 3RB	263,080 kg (580,000 lb)
2ERGM, 2ERPM, 2ERRM	297,555 kg (656,000 lb)
3PM, 3RM	299,370 kg (660,000 lb)
Max ramp weight allowance:	
200, 200ERB, 300	907 kg (2,000 lb)
200ERM	454 kg (1,000 lb)
Max landing weight:	
200	201,845 kg (445,000 lb)
200ER	208,650 kg (460,000 lb)
300	237,680 kg (524,000 lb)
Max zero-fuel weight:	
200	190,510 kg (420,000 lb)
200ER	195,045 kg (430,000 lb)
300	224,525 kg (495,000 lb)
Max wing loading:	
2GB, 3PB, 2RB	536.5 kg/m² (109.88 lb/sq ft)
2GM, 3PM, 2RM	577.8 kg/m² (118.35 lb/sq ft)
2ERGB, 2ERPB, 2ERRB, 3PB, 3RB	614.9 kg/m² (125.95 lb/sq ft)
2ERGM, 2ERPM, 2ERRM	695.5 kg/m² (142.45 lb/sq ft)
3PM, 3RM	699.8 kg/m² (143.32 lb/sq ft)
Max power loading:	
2GB	335 kg/kN (3.29 lb/lb st)
2GM	361 kg/kN (3.54 lb/lb st)
2PB	346 kg/kN (3.40 lb/lb st)
2PM	360 kg/kN (3.53 lb/lb st)
2RB	351 kg/kN (3.45 lb/lb st)
2RM	366 kg/kN (3.59 lb/lb st)
2ERGB	349 kg/kN (3.42 lb/lb st)
2ERGM	357 kg/kN (3.50 lb/lb st)
2ERPB	350 kg/kN (3.43 lb/lb st)
2ERPM	371 kg/kN (3.64 lb/lb st)
2ERRB	354 kg/kN (3.47 lb/lb st)
2ERRM	358 kg/kN (3.51 lb/lb st)
3PB, 3RB	328 kg/kN (3.22 lb/lb st)
3PM	343 kg/kN (3.37 lb/lb st)
3RM	374 kg/kN (3.67 lb/lb st)

PERFORMANCE:

Cruising speed: all	M0.84
Approach speed: 200	136 kt (252 km/h; 157 mph)
200ER	138 kt (256 km/h; 159 mph)
300	149 kt (276 km/h; 171 mph)
Initial cruising altitude (ISA + 10°C):	
2GB	11,980 m (39,300 ft)
2GM, 2PB	11,550 m (37,900 ft)
2PM, 2ERGB	11,155 m (36,600 ft)

Thai Airways International Boeing 777-300 *(Jane's/Paul Jackson)*   0126871

2RB	11,645 m (38,200 ft)
2RM	11,370 m (37,300 ft)
2ERGM	10,605 m (34,800 ft)
2ERPB	10,820 m (35,500 ft)
2ERPM	10,270 m (33,700 ft)
2ERRB	11,005 m (36,100 ft)
2ERRM	10,455 m (34,300 ft)
3PB	10,975 m (36,000 ft)
3PM	10,485 m (34,400 ft)
3RB	11,245 m (36,900 ft)
3RM	10,395 m (34,100 ft)

Service ceiling, OEI (ISA +10°C):

2GB	5,515 m (18,100 ft)
2GM	4,755 m (15,600 ft)
2PB	4,940 m (16,200 ft)
2PM	4,905 m (16,100 ft)
2RB	4,815 m (15,800 ft)
2RM	5,365 m (17,600 ft)
2ERGB	3,995 m (13,100 ft)
2ERGM	3,720 m (12,200 ft)
2ERPB	4,235 m (13,900 ft)
2ERPM	3,660 m (12,000 ft)
2ERRB	4,785 m (15,700 ft)
23ERRM	3,749 m (12,300 ft)
3PB	4,480 m (14,700 ft)
3PM	3,535 m (11,600 ft)
3RB	4,725 m (15,500 ft)
3RM	3,415 m (11,200 ft)

T-O field length (30° C):

3GB	2,073 m (6,800 ft)
3GM	2,530 m (8,300 ft)
2PB, 2RB	2,164 m (7,100 ft)
2PM, 2RM	2,576 m (8,450 ft)
2ERGB	2,515 m (8,250 ft)
2ERGM	3,018 m (9,900 ft)
2ERPB	2,591 m (8,500 ft)
2ERPM	3,582 m (11,750 ft)
2ERRB	2,561 m (8,400 ft)
2ERRM	3,368 m (11,050 ft)
3PB	2,759 m (9,050 ft)
3PM	3,414 m (11,200 ft)
3RB	2,652 m (8,700 ft)
3RM	3,704 m (12,150 ft)

Landing field length:

2GB, 2GM	1,570 m (5,150 ft)
2PB, 2PM, 2RB, 2RM	1,555 m (5,100 ft)
2ERGB, 2ERGM	1,616 m (5,300 ft)
2ERPB, 2ERPM, 2ERRB, 2ERRM	1,601 m (5,250 ft)
300	1,844 m (6,050 ft)

Design range: 2GB 3,770 n miles (6,982 km; 4,338 miles)

2GM	4,920 n miles (9,111 km; 5,661 miles)
2PB	3,860 n miles (7,148 km; 4,442 miles)
2PM	4,990 n miles (9,241 km; 5,742 miles)
2RB	3,995 n miles (7,398 km; 4,597 miles)
2RM	5,065 n miles (9,380 km; 5,828 miles)
2ERGB	5,645 n miles (10,454 km; 6,496 miles)
2ERGM	7,625 n miles (14,121 km; 8,774 miles)
2ERPB	5,695 n miles (10,547 km; 6,553 miles)
2ERPM	7,505 n miles (13,899 km; 8,636 miles)
2ERRB	5,765 n miles (10,676 km; 6,634 miles)
2ERRM	7,595 n miles (14,065 km; 8,740 miles)
3PB	3,750 n miles (6,945 km; 4,315 miles)
3PM	5,515 n miles (10,213 km; 6,346 miles)
3RB	3,915 n miles (7,250 km; 4,505 miles)
3RM	5,820 n miles (10,778 km; 6,697 miles)

UPDATED

BOEING SONIC CRUISER

TYPE: New concept airliner.
PROGRAMME: Announced 29 March 2001, at concept stage, coincident with abandonment of plans to build Boeing 747X stretched version. Would have complemented existing Boeing airliners. Service entry possible in 2008, but project shelved in December 2002. Initial wind tunnel tests completed September 2001. Boeing also pursued parallel studies, including one with conventionally positioned wing and empennage.
DESIGN FEATURES: New class of airliner; capable of operating fractionally below M1.0 thereby offering reduced journey times without the attendant costs of supersonic flight. Typical saving would be 4 to 5 hours on London-Sydney journey, currently 23 hours. Operating height to allow more direct routing, above existing traffic. Required new turbofan engines, in which, by June 2001, world's three major manufacturers had shown interest in developing.

Canard configuration. Rear-mounted wing; 'double delta' layout, with steeply sweptback inboard leading-edges and moderately swept outboard panels ending in swept tips similar to those of Boeing 767-400. Podded engines, one each side, in trailing-edge and at approximately quarter span; long air intakes below wing. Inward-canted fins mounted at engine pod roots.
STRUCTURE: Wing all-composites. Airframe 60 per cent carbon fibre by weight; 17 per cent titanium; and 23 per cent aluminium, glass fibre and steel.
POWER PLANT: Two turbofans, each of some 400 kN (89,900 lb st).
ACCOMMODATION: Size undecided at time of initial announcement. Was expected to develop into family of aircraft, launched with lowest capacity version, seating some 225.
All data are provisional.
DIMENSIONS, EXTERNAL:

Wing span	36.5 m (120 ft)

PERFORMANCE (target):

Cruising speed	M0.95-0.98
Landing speed	145 kt (268 km/h; 167 mph)
Cruising altitude: initial	12,500 m (41,000 ft)
max	15,240 m (50,000 ft)
Required runway length	3,200 m (10,500 ft)
Range (developed versions)	
	9,000 n miles (16,668 km; 10,357 miles)

UPDATED

BOEING SUPER EFFICIENT AIRLINER

TYPE: New-concept airliner.
PROGRAMME: In parallel with its (now abandoned) Sonic Cruiser, Boeing was offering airlines a twin-engined aircraft of more conventional appearance which could enter service in 2008. Key features include turbofans with a bypass ratio in excess of 10:1; fuel burn 17 to 20 per cent less per passenger-mile than existing airliners; span 56.7 m (186 ft), including raked tips; and length 58.8 m (193 ft).

NEW ENTRY

BOEING ELECTRIC AIRPLANE

On 27 November 2001 Boeing Commercial Airplanes announced that it is developing a proof-of-concept electrically powered single-engined light aircraft employing fuel cell technology. Most of the development work will be carried out at the company's Research and Technology Center in Madrid, Spain. NASA, fuel cell manufacturers, the automotive industry and several European universities are supporting the project. Boeing plans to modify an as yet unidentified production lightplane by removing its engine and replacing it with fuel cells and an electric motor turning a conventional propeller. Test flights are expected to begin in 2004. No production of electrically powered light aircraft is planned. The ultimate goal of the project is to employ fuel cell technology to replace conventional APUs on transport category aircraft.

VERIFIED

DOUGLAS PRODUCTS DIVISION (Division of Boeing Commercial Airplane Group)

The Douglas Aircraft Company became Douglas Products Division when Boeing absorbed McDonnell Douglas in August 1997. Final MD-80 series, an MD-83, was delivered on 21 December 1999; MD-90 line closed at end of 2000; last MD-11 delivered 22 February 2001. Former MD-95 now marketed as Boeing 717. Having openly questioned the future of the 717, Boeing reconfirmed its commitment to continued production on 13 December 2001, albeit with lower production rate and revised delivery dates.

UPDATED

BOEING 717

TYPE: Twin-jet airliner.
PROGRAMME: Announced at Paris Air Show 1991 as MD-95; potential airline customers briefed and manufacturing partners announced in Berlin, November 1994; modification of former Eastern Airlines DC-9-30 into development prototype began late 1994; design acquired by Boeing in August 1997 and renamed Boeing 717 in January 1998. Three flight test aircraft (T1, T2 and T3); first nose section (built by MDC) delivered to Huntington Beach on 11 December 1996 for T1, which began final assembly in May 1997 and was rolled out on 10 June 1998. BR 715 engine certified 1 September 1998.
First flight of prototype (N717XA; T1) 2 September 1998; second prototype (N717XB; T2) first flew 26 October 1998; third (N717XC; T3) on 16 December 1998, at which time first two had flown 361 hours in 193 sorties. First production aircraft (N717XD; P1) rolled out 23 January 1999; JAA/FAA certification awarded 1 September 1999 at end of five aircraft, 2,000 hour, 1,900

BOEING 717-200 ORDERS AND DELIVERIES
(at 1 January 2003)

Customer	First Order	First Delivery	Qty
AeBal	16 May 00	22 Jun 00	3
AirTran	19 Oct 95	23 Sep 99	51
Bavaria International	4 May 98	29 Dec 99	5
Hawaiian Airlines	1 Mar 00	28 Feb 01	13
Impulse Airlines	29 Dec 00	29 Dec 00	3
Midwest Airlines	15 Apr 02	28 Feb 03	25
Pembroke Capital	31 Dec 98	17 Aug 00	26
TWA	31 Dec 98	18 Feb 00	24
Turkmenistan Airlines	25 Jul 00	1 Aug 01	3
Total			**153**

Note: Quantities are cumulative

sortie programme; first aircraft (N942AT; third production) delivered to AirTran on 23 September 1999, after demonstration flights; entered service on Atlanta-Washington, DC, route on 12 October 1999. Updated FMS certified 20 October 2000, adding GPS, fuel prediction, vertical guidance and automatic calculation of V-speeds. CIS certification awarded 20 February 2001. Delivery of 100th effected on 18 June 2002 as 37th for AirTran.
CURRENT VERSIONS: **Boeing 717-200:** Initial production version, *to which following description applies.* Available in basic (**BGW**) and high (**HGW**) gross weight versions.
Boeing 717-100: Proposed 86-seat version, formerly MD-95-20; studies will be four frames (1.9 m; 6 ft 3 in) shorter. Renamed **-100X**; wind tunnel tests began in early 2000; revised mid-2000 to eight-frame (3.86 m; 12 ft 8 in) shrink. Launch decision was deferred in December 2000 and again thereafter to an undisclosed date.

Boeing 717-100X Lite: Proposed 75-seat version, powered by Rolls-Royce Deutschland BR 710 turbofans; later abandoned.
Boeing 717-300: Proposed 130-seat version, formerly MD-95-50; studies suggest will be five frames (2.38 m; 7 ft 10 in) longer. Currently in abeyance, although AirTran has expressed interest in converting some -200 options to this model.
CUSTOMERS: Manufacturer foresees a market for over 3,000 aircraft in this class over the next 20 years. Launched 19 October 1995 with order for 50, plus 50 options, from ValuJet Airlines (later renamed AirTran), this increasing to 60 before being cut to 51 in 2002. Second customer was Bavaria International Leasing Company (five), followed by TWA (50, plus 50 options, later reduced to 30 orders). During 2001, production was set at five per month, Pembroke Capital (10, later increased to 25) and Hawaiian

(13, plus seven options). Boeing delivered 12 in 1999, 32 in 2000, 49 in 2001 and 20 in 2002. By 1 January 2003, orders stood at 153, of which 113 had been delivered.

COSTS: AirTran order for 50 aircraft worth US$1 billion. January 2001 list price US$35 million per aircraft.

DESIGN FEATURES: All major elements of airframe based on DC-9/MD-80 series; systems and avionics are blend of low cost and advanced technology. Conventional low-wing, rear-engine, T-tail configuration.

Wing from DC-9-34, but with 1° 34′ additional incidence; sweep 24° 30′ at quarter-chord; thickness/chord ratio 11.6.

FLYING CONTROLS: Conventional and partly assisted. Elevator and ailerons are manually actuated via cables; rudder powered hydraulically with manual reversion; fly-by-wire trimming of rudder, two-section spoilers and elevator; three-section, double-slotted flaps; full-span, two-position, five-section leading-edge slats.

STRUCTURE: All-metal, two-spar wing with riveted spanwise stringers; glass fibre trailing-edges on wings, ailerons, flaps, elevators and rudder. Variations from MD-80/MD-90 include thicker skins on tail surfaces; MD-87 fuselage lengthened forward of wing by three frames 1.45 m (4 ft 9 in) and fin tip by 250 mm (10 in); wing/fuselage fillet extended forwards by three frames, using composites structure. Composites also for fuselage tailcone, fintip, elevator and aileron tabs, radome and wing trailing-edge panels; otherwise of 2024T3 aluminium alloy.

Partners are: Alenia (fuselage sections), Korean Air Lines Aerospace Division (nose structure and main passenger door/entry area), KAI (Hyundai) (wings, in conjunction with Boeing Toronto Ltd, which built initial sets of wings for flight test aircraft and early production units), BAE Systems (wing join and underwing barrel) Rolls-Royce Deutschland (power plant), Goodrich (engine nacelles), ShinMaywa Industries Ltd (horizontal tail surfaces and engine pylons), Fischer Advanced Composite Components GmbH (cabin furnishings), Andalucia Aerospacial (slats, landing gear components, aft pressure bulkhead), Israel Aircraft Industries SHL Servo Systems (landing gear), Honeywell (environmental control system, wheels and brakes, flight guidance and avionics systems), Parker-Hannifin Corp (hydraulic and control systems), AIDC (empennage), Labinal (electric assemblies), Hamilton Sundstrand (electrical power generating system), and (in partnership with Hamilton Sundstrand) Auxiliary Power International Corporation (APU). Final assembly in Long Beach.

LANDING GEAR: Hydraulically retractable tricycle with steerable nosewheels; twin wheels on all legs. All-steel brakes; anti-skid units. Main tyres H41×15.0-19 (24 ply); nose tyres 26×6.6 (12 ply).

POWER PLANT: Two Rolls-Royce Deutschland BR 715 A1-30 turbofans, each 82.3 kN (18,500 lb st) at T-O at 30°C ambient, for 717-200 BGW; BR 715 C1-30 of 93.4 kN (21,000 lb st) for 717-200 HGW; optional 89.9 kN (20,000 lb st) available; switching between three thrust ratings does not involve engine hardware changes. Integrated drive generators system. Goodrich single-pivot door-type reversers for ground use only. Engine pylon based on MD-80, but thinner and without powered flap, although with extra frame for additional strength.

Standard (BGW) fuel capacity 13,904 litres (3,673 US gallons; 3,058 Imp gallons) in three tanks in wingroots and fuselage centre section. HGW has additional 1,628 litre (430 US gallon; 358 Imp gallon) tank in forward baggage hold and 1,022 litre (270 US gallon; 225 Imp gallon) tank in rear baggage hold for total volume of 16,667 litres (4,403 US gallons; 3,666 Imp gallons). Fuel recirculation system prevents wing upper surface ice accumulation and aids cooling.

ACCOMMODATION: Crew of two on advanced flight deck optimised for reduced parts count and high reliability.

Boeing 717 of Turkmenistan Airlines (*Jane's/Paul Jackson*)   NEW/0526959

Cabin cross-section as for MD-80. Main passenger door, fwd, port, with optional airstair stowage below; service door opposite; aft door on centreline with rearward-facing ramp stairs; two Type III emergency exits overwing on each side. Front and rear underfloor baggage hold doors, starboard. Typical two-class seating for 106 passengers: eight first-class in four-abreast seating with 0.91 m (36 in) pitch and 98 standard class with 0.81 m (32 in) pitch in five-abreast arrangement in new modern cabin, alternatively 117 single-class in 0.79/0.81 m (31/32 in) pitch. Seating designed by Avio Interiors and complies with 16 g impact regulations. Cabin designed with inputs from 500 airline executives, flight attendants and passengers; interior, manufactured by Fischer Advanced Composite Components of Austria, features wider and deeper overhead baggage bins, full-grip handrail throughout length of cabin. Two vacuum-operated lavatory units at rear of cabin, plus optional further unit at front for first-class. Galley units positioned at front of cabin. Integral cabin door, with optional airstairs, behind flight deck on port side; emergency exit on opposite side, plus two above each wing. Underfloor baggage and cargo hold; latter reduced in HGW version by auxiliary fuel tanks.

SYSTEMS: Honeywell dual air cycle air conditioning and pressurisation system with digital cabin air controllers, utilising engine bleed air, maximum differential 0.54 bar (7.77 lb/sq in). Three-wheel air cycle machine and modified water separator in rear fuselage. Two separate 207 bar (3,000 lb/sq in) hydraulic systems for operation of spoilers, flaps, slats, rudder, landing gear, nosewheel steering, brakes, thrust reversers and ventral stairway. Maximum flow rate 30.3 litres (8.0 US gallons; 6.7 Imp gallons)/min. Airless bootstrap-type reservoirs, output pressure 2.07 bar (30 lb/sq in). Pneumatic system, for air conditioning/pressurisation, engine starting and ice protection, utilises engine bleed air and/or APU. Electrical system includes two 35/40 kVA integrated drive generators, plus 60 kVA APU generator. Oxygen system of diluter demand type for crew on flight deck; continuous flow chemical canister type with automatic mask presentation for passengers. Anti-icing of wing, engine inlets and tailplane by engine bleed air. Electric windscreen de-icing. Thermal anti-icing of leading-edges. APIC APS 2100 APU.

AVIONICS: Honeywell Versatile Integrated Avionics VIA 2000 computer as core avionics management system; full Cat. IIIa capability will be upgraded later to IIIb with addition of radar altimeter, ILS receiver and inertial reference unit, SFE ARINC 700 avionics.

Flight: Honeywell flight management system (FMS), inertial reference system (IRS), digital flight guidance system (DFGS), digital air data computer and windshear detection system. To be certified for Cat. IIIb automatic landings. Flight computer system upgraded and certified 20 October 2000; changes include improved GPS.

Instrumentation: Six-tube EFIS with 203 × 203 mm (8 × 8 in) LCD screens providing navigation, flight management and systems data. Flight deck features include MD-11 AFCS with glareshield-mounted controls enabling crew to fly the aircraft automatically with only push-button and thumbwheel inputs. Simplified integrated flightcrew warning and alerting panel (IFWAP) overhead control panel with four LCDs replacing 13 gauges, meters and switch panels. Central fault display system for reduced maintenance time.

DIMENSIONS, EXTERNAL:

Wing span	28.45 m (93 ft 4 in)
Wing chord: at root	5.44 m (17 ft 10 in)
at tip	1.12 m (3 ft 8 in)
Wing aspect ratio	8.7
Length: overall	37.80 m (124 ft 0 in)
fuselage	34.34 m (112 ft 8 in)
Fuselage max width	3.34 m (10 ft 11½ in)
Height overall: at OWE	9.04 m (29 ft 8 in)
at MTOW	8.76 m (28 ft 9 in)
Tailplane span	11.23 m (36 ft 10 in)
Wheel track	4.88 m (16 ft 0 in)
Wheelbase	17.60 m (57 ft 8¾ in)
Distance between engine centrelines	6.60 m (21 ft 8 in)
Passenger door (fwd, port):	
Height	1.83 m (6 ft 0 in)
Width	0.86 m (2 ft 10 in)
Height to sill: at OWE	2.46 m (8 ft 1 in)
at MTOW	2.21 m (7 ft 3 in)
Aft door (centreline):	
Height	1.83 m (6 ft 0 in)
Width	0.70 m (2 ft 3¾ in)
Height to sill: at OWE	2.01 m (6 ft 7 in)
at MTOW	1.83 m (6 ft 0 in)
Service door (fwd, stbd):	
Height	1.22 m (4 ft 0 in)
Width	0.69 m (2 ft 3 in)
Height to sill: at OWE	2.46 m (8 ft 1 in)
at MTOW	2.21 m (7 ft 3 in)
Emergency exits (above wings, two per side):	
Height	0.91 m (3 ft 0 in)
Width	0.51 m (1 ft 8 in)
Height to sill: at OWE	2.87 m (9 ft 5 in)
at MTOW	2.77 m (9 ft 1 in)
Baggage door: forward:	
Height	1.27 m (4 ft 2 in)
Width	1.34 m (4 ft 4¾ in)
Height to sill: at OWE	1.30 m (4 ft 3 in)
at MTOW	1.09 m (3 ft 7 in)
rear:	
Height	1.27 m (4 ft 2 in)
Width	0.91 m (3 ft 0 in)
Height to sill: at OWE	1.35 m (4 ft 5 in)
at MTOW	1.17 m (3 ft 10 in)

DIMENSIONS, INTERNAL (BGW: standard, HGW: with extended-range tanks):

Cabin: Max width	3.12 m (10 ft 3 in)
Height at aisle	2.03 m (6 ft 8 in)
Underfloor baggage/freight hold: BGW:	
front: Length	11.04 m (36 ft 2½ in)
Height	0.99 m (3 ft 3 in)
Volume	18.3 m³ (646 cu ft)
rear: Length	5.21 m (17 ft 1 in)
Height	0.99 m (3 ft 3 in)
Volume	8.2 m³ (289 cu ft)
HGW: front: Volume	14.9 m³ (527 cu ft)
rear: Volume	5.7 m³ (203 cu ft)

AREAS:

Wings, gross	92.97 m² (1,000.7 sq ft)

WEIGHTS AND LOADINGS (BGW: standard, HGW: with extended-range tanks):

Operating weight empty: BGW	30,618 kg (67,500 lb)
HGW	31,071 kg (68,500 lb)
Max structural payload: BGW	12,020 kg (26,500 lb)
HGW	14,515 kg (32,000 lb)
Max fuel weight: BGW	11,162 kg (24,609 lb)
HGW	13,381 kg (29,500 lb)
Max T-O weight: BGW	49,895 kg (110,000 lb)
HGW	54,885 kg (121,000 lb)
Max ramp weight: BGW	50,349 kg (111,000 lb)
HGW	55,340 kg (122,000 lb)
Max landing weight: BGW	45,359 kg (100,000 lb)
HGW	49,895 kg (110,000 lb)
Max zero-fuel weight: BGW	42,638 kg (94,000 lb)
HGW	45,586 kg (100,500 lb)
Max wing loading: BGW	536.7 kg/m² (109.92 lb/sq ft)
HGW	590.4 kg/m² (120.92 lb/sq ft)

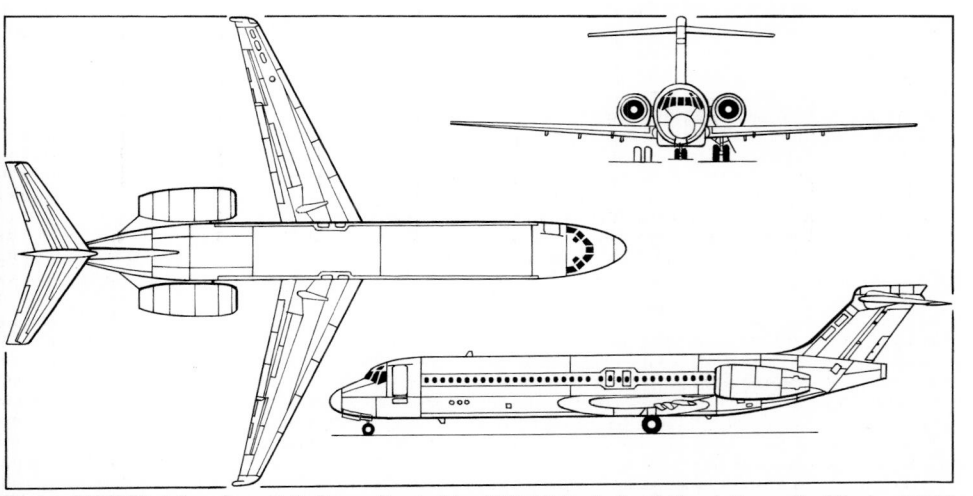

Boeing 717-200 airliner (two Rolls-Royce Deutschland BR 715 turbofans) (*Jane's/James Goulding*)   0126948

| Max power loading: BGW | 303 kg/kN (2.97 lb/lb st) |
| HGW | 294 kg/kN (2.88 lb/lb st) |

PERFORMANCE:
Max level speed:

BGW, HGW	438 kt (811 km/h; 504 mph) (M0.77)
Approach speed: BGW	132 kt (244 km/h; 152 mph)
HGW	139 kt (257 km/h; 160 mph)

Initial cruise altitude, MTOW, ISA +10°C:

BGW	10,424 m (34,200 ft)
HGW	9,815 m (32,200 ft)
Service ceiling	11,278 m (37,000 ft)

T-O field length at MTOW, S/L, ISA +10°C:

| BGW | 1,677 m (5,500 ft) |
| HGW | 1,753 m (5,750 ft) |

Landing field length at MLW: BGW 1,418 m (4,650 ft)

| HGW | 1,524 m (5,000 ft) |

Design range, domestic reserves, 106 passengers and baggage:

| BGW | 1,430 n miles (2,648 km; 1,645 miles) |
| HGW | 2,060 n miles (3,815 km; 2,370 miles) |

OPERATIONAL NOISE LEVELS (ICAO Annex 16 Ch 3):

T-O with cutback	81.4 dB
Approach	91.4 dB
Sideline	89.2 dB

UPDATED

PHANTOM WORKS (Division of Boeing Commercial Airplane Group)
PHANTOM WORKS
PRESIDENT: George Muellner
MANAGER, MEDIA RELATIONS: David Phillips, *Tel:* (+1 314) 232 13 72

Tasked with improving Boeing's competitive position through use of innovative technologies, improved processes and creation of new products, the Phantom Works was originally established by McDonnell Douglas at St Louis. Technologies and expertise have expanded in support of company-wide activities, including commercial transport aircraft; this division had workforce of 4,400 in November 2001, with personnel at every major Boeing facility.

Phantom Works activities are spread over six sites: Huntsville (operations analysis, advanced military aircraft and missiles, manufacturing technology & prototyping and advanced support concepts), Mesa (advanced rotorcraft), Philadelphia (advanced rotorcraft), Seattle (advanced commercial aircraft and information & engineering technology), Southern California (advanced tankers and transports; space and communications) and St Louis (computing technology).

In addition to the programmes described below in detail, Phantom Works is also involved with Naval Unmanned Combat Air Vehicle (UCAV-N), Solar Orbit Transfer Vehicle, X-31A Vector, X-37 Reusable Spaceplane, X-40 Space Maneuver Vehicle, X-43 Hyper-X and X-45 UCAV.

UPDATED

BOEING BIRD OF PREY
TYPE: Technology demonstrator.
PROGRAMME: Demonstrator for next generation of low-observables technology and proving project for rapid prototyping techniques. Developed by Boeing Phantom Works, beginning in 1992. Maiden flight in "Autumn of 1966"; flew over 40 hours in 36 sorties before retirement in 1999. Stored pending declassification and finally unveiled in public on 18 October 2002. Experience used in development of X-45A UCAV.
CUSTOMERS: One aircraft only.
COSTS: Total programme US$67 million; entirely funded by Boeing.
DESIGN FEATURES: Simple technology demonstrator, sufficiently sophisticated only to gather intended data; limited speed; no pressurisation. Heavily reliant on off-the-shelf components, including parts from F-15, and F/A-18.
Flattened, chined fuselage with spine intake for engine air. Wings, attached to rear fuselage, swept back at approximately 65°, have 9° dihedral changing to 55° anhedral at tips. No vertical tail surfaces, although a ventral strake was fitted for early flights. Flying characteristics reportedly docile, with 8° angle of approach and "T-38 typ" speed on finals.
FLYING CONTROLS: Gull wing configuration has rudderons in anhedral component and ailerons inboard.
STRUCTURE: Construction of aircraft proved techniques for manufacturing large, single-piece elements with complex shapes, combined with low-temperature curing of LTM-10 carbon composites and 3-D modelling to ensure accurate fit.
LANDING GEAR: Tricycle type; retractable. Single wheel on each unit. Nosewheel from North American F-100 Super Sabre.
POWER PLANT: One Pratt & Whitney JT15D turbofan rated at 14.2 kN (3,190 lb st).
ACCOMMODATION: Pilot only, on UPC/Stencel ejection seat. Single-piece canopy.
AVIONICS: Simple instrumentation only, plus GPS and radios.

BOEING BC-17X GLOBEMASTER
TYPE: Four-jet freighter.
PROGRAMME: Model displayed at Farnborough Air Show, September 1996. Civil version of C-17A, intially designated MD-17, lacking specific military equipment, such as refuelling receptacle, OBIGGS, ECM and paratroop doors, but otherwise identical. Standard version able to carry a 79,560 kg (175,400 lb) payload over 2,500 n miles (4,630 km; 2,877 miles); optional fuel tank in wing centre-section contains an additional 37,853 litres (10,000 US gallons; 8,327 Imp gallons). Cargo capacity eighteen 2.24 × 2.74 m (88 × 108 in) pallets, including four on ramp.

Formal launch awaits initial customer(s), although as few as between five and 10 commitments would be sufficient to initiate the programme; Boeing looking at fractional ownership scheme to encourage customers; civil certification would rely largely on military proving trials already accomplished; first delivery two years after launch. In third quarter of 1997, Boeing received Organizational Designated Airworthiness Representative delegation from FAA as part of progress towards type certification and announced it was building two 'white tail' aircraft which would be taken over by USAF if unsold, but these were not produced. Cost quoted as US$140 million at 2001 prices;

certification cost then quoted as US$300-400 million. Decision to launch was due in July 2001 but was rescheduled to an unspecified date; maximum production rate of four per year envisaged; Boeing and USAF have estimated market for at least 30 aircraft over period of 10 years; guaranteed DoD business offered with aircraft: from 50 per cent of utilisation in first year, gradually reducing to 13.7 per cent by 2015.

USAF support 'temporarily suspended' following increased production of C-17, plus concerns that semi-military status of BC-17X could inhibit export potential. However, business case for latter considered undiminished, and could be reactivated at short notice should circumstances change.

UPDATED

OTHER AIRCRAFT
See International section for description of **T-45 Goshawk** programme undertaken jointly with BAE Systems. Details of **V-22 Osprey** can be found under Bell Boeing entry in this section; **RAH-66 Comanche** appears under Boeing Sikorsky.

VERIFIED

Boeing Bird of Prey general arrangement (*Jane's/Paul Jackson*)   NEW/0527117

DIMENSIONS, EXTERNAL:

Wing span	6.91 m (22 ft 8 in)
Wing aspect ratio (estimated)	2.3
Length: fuselage	12.47 m (40 ft 11 in)
overall	14.22 m (46 ft 8 in)
Height overall	2.82 m (9 ft 3 in)

AREAS:

| Wings, gross, projected (estimated) | 20.4 m² (220 sq ft) |

WEIGHTS AND LOADINGS:

Max T-O weight	3,356 kg (7,400 lb)
Max wing loading (estimated)	164.2 kg/m² (33.64 lb/sq ft)
Max power loading	237 kg/kN (2.32 lb/lb st)

PERFORMANCE:

| Max level speed | 260 kt (482 km/h; 299 mph) |
| Max operating altitude | 6,095 m (20,000 ft) |

NEW ENTRY

BOEING BLENDED WING BODY LARGE COMMERCIAL TRANSPORT
TYPE: New-concept airliner.
PROGRAMME: In 1996, shortly before absorption by Boeing, McDonnell Douglas released details of the BWB-1-1 (blended wing body) study giving the most comprehensive analysis yet of the flying wing's advantages for very large transports. At least half-a-dozen derivatives under consideration at the beginning of 2002, ranging in passenger capacity from 180 to 570. An earlier incarnation, capable of carrying 800 passengers over a 7,000 n mile (12,964 km; 8,055 mile) range, gave benefits that included a 15.2 per cent reduction in T-O weight and 12.3 per cent less empty weight; 27.5 per cent reduction in fuel burn; 27.0 per cent reduction in installed thrust; and 20.6 per cent better lift:drag. These are achieved by increasing wing area by 27.9 per cent (reducing loading by 33.6 per cent) and having a 19.2 per cent greater span. Tests of a radio-controlled 5 m (17 ft) span scale BWB-1-1 were undertaken by Stanford University, California, at El Mirage dry lake in July 1997 as part of a US$2.3 million programme to evaluate flight control laws for a flying wing. Boeing expects to test a 14.2 per cent scale low-speed vehicle with a maximum T-O weight of 1,315 kg (2,900 lb) and a 10.70 m (35 ft 1¼ in) span during 2005-06 to study control and aerodynamic features. It will be

powered by three Williams FJ44 turbofans, each rated at 0.89 kN (200 lb st).

One potential application is a BWB tanker for multipoint aerial refuelling. Equipped with three 'smart' booms, two hose/drogue refuelling points and automated refuelling capabilities, BWB tanker would be able to accommodate simultaneous air-to-air refuelling of multiple conventional aircraft or UAVs. Fuel carried in wing tanks; maximum payload space would be available for up to 23 conventional pallets and 40 troops.

As a C² ISR (command, control, intelligence, surveillance, reconnaissance) platform, the BWB would provide increased loiter time, large interior space suitable for battle management control rooms, and ample exterior locations for conformal phased-array antennas for broadband communications with no increase in radar signature. These capabilities make the BWB additionally suitable as a long-range standoff weapons platform.

Several possible BWB derivatives have emerged as candidates for production, including the **BWB-250**, with accommodation for about 260 passengers, and the **BWB-450** with room for about 480 passengers. *Estimated data below refer to the 800-seat Boeing BWB-1-1 proposal powered by three 275 kN (61,900 lb st) turbofans.*

DIMENSIONS, EXTERNAL:

Wing span: excl winglets	85 m (280 ft)
incl winglets	88 m (289 ft)
Wing aspect ratio	5.1
Length overall	49 m (161 ft)
Height overall	15 m (50 ft)

AREAS:

| Wings: trap | 728 m² (7,840 sq ft) |
| gross | 1,423 m² (15,325 sq ft) |

WEIGHTS AND LOADINGS:

Weight empty	167,750 kg (369,800 lb)
Weight empty, equipped	186,900 kg (412,000 lb)
Max payload	104,800 kg (231,000 lb)
Max fuel weight	122,500 kg (270,000 lb)
Max T-O weight	373,300 kg (823,000 lb)
Max zero-fuel weight	291,650 kg (643,000 lb)
Fuel burn over 7,000 n miles with 800 passengers	96,820 kg (213,450 lb)
Max wing loading	513 kg/m² (105 lb/sq ft)
Max power loading	452 kg/kN (4.4 lb/lb st)

PERFORMANCE:
Normal cruising speed   M0.85
Max approach speed   150 kt (278 km/h; 173 mph) EAS
Initial cruising altitude   10,665 m (35,000 ft)
T-O field length   3,353 m (11,000 ft)
Range with 800 passengers
                     7,000 n miles (12,964 km; 8,055 miles)
                                                      UPDATED

BOEING X-50A DRAGONFLY CANARD ROTOR/WING

TYPE: New-concept rotorcraft.

PROGRAMME: The X-50A Canard Rotor/Wing (CR/W) began as a McDonnell Douglas Helicopters project for a VTOL reconnaissance and surveillance UAV in 1992; concept based on NASA-funded studies into high-speed rotorcraft and earlier (Hughes) rotor/wing studies. Design undertaken by Phantom Works personnel at Mesa, Arizona, under a March 1998 DARPA contract calling for the manufacture and testing of two prototypes. US Marine Corps is reported to have expressed interest in a manned version armed with missiles and other weaponry for use in the escort role.

As currently envisaged, the CR/W will possess VTOL attributes of the helicopter and the capability to fly like a conventional fixed-wing aircraft, with the addition of foreplanes being a key factor in bestowing this potential. Power will be provided by a single turbofan engine. In helicopter mode, exhaust gases will pass through channels in the two-blade rotor before exiting through vents in the blade tips. In aeroplane mode, as speed increases, exhaust efflux progressively transfers from driving the rotor to a nozzle at the rear, thus providing forward thrust.

With no tail rotor, directional control in helicopter mode will rely on what Boeing calls 'reaction drive', in which exhaust gases are diverted to nozzles on both sides of the tail; for aeroplane mode, conventional control surfaces will be provided, although consideration is being given to employing reaction jets for directional control in both modes of flight.

Following vertical take-off as a helicopter, the CR/W will accelerate to about 122 kt (225 km/h; 140 mph) before making the transition to aeroplane. This will be accomplished with the assistance of flap surfaces on the foreplanes and the aft-mounted wings. These will increase the amount of lift generated by the wings and reduce rotor loadings, whereupon the rotor will be slowed to a stop before being locked in position across the fuselage to act as an extra lifting surface. The flaps will then be retracted, with the three lifting surfaces sharing the load. For landing, the process will be reversed, enabling the CR/W to operate safely from confined areas, such as a carrier deck or helicopter platform. In conventional aeroplane mode, maximum speed is predicted to exceed 375 kt (695 km/h; 432 mph).

Initial research to include studies into ways and means of using exhaust gases to perform multiple functions, with US$21 million allocated jointly by Boeing and DARPA to pay for a three-year research effort. Boeing's Mesa facility is leading the project and had responsibility for assembly; further support provided by Boeing sites in Philadelphia and St Louis.

Initial technology demonstrator to be a subscale, unmanned aerial vehicle (UAV); first X-50A prototype was to have flown in 2001, but postponed; roll-out May 2002; maiden flight now unlikely to take place until December 2002, following discovery of high cyclic loads during ground testing that have necessitated modifying swashplate design and using steel instead of aluminium for

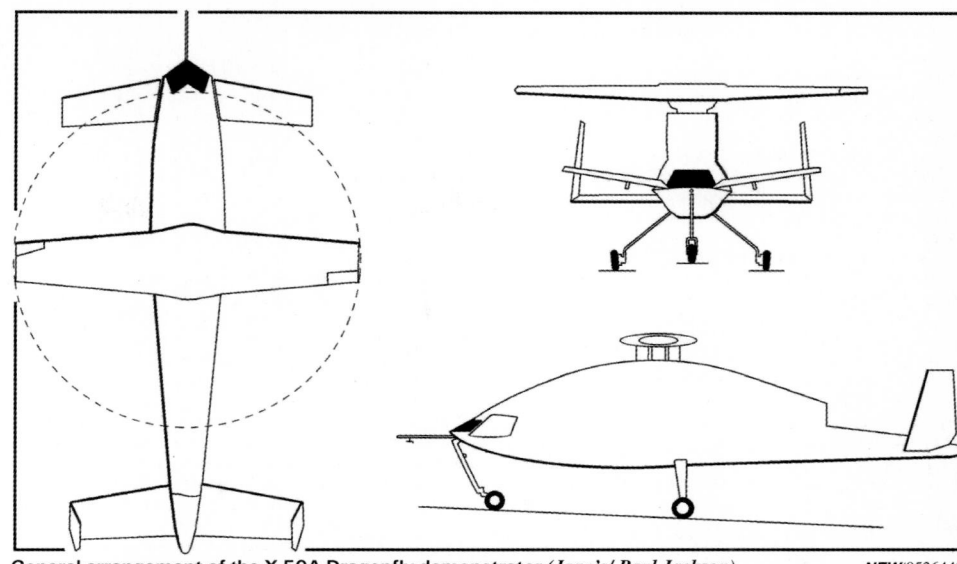

General arrangement of the X-50A Dragonfly demonstrator *(Jane's/ Paul Jackson)*   *NEW*/0526449

greater stiffness. Total of 11 flights planned, with test vehicle not to exceed 150 kt (278 km/h; 173 mph). X-50A will take-off vertically and operate as helicopter up to 60 kt (111 km/h; 69 mph), when canard and stabiliser begin to provide lift; transition to conventional wing-borne flight will occur at 130 kt (241 km/h; 150 mph). X-50A prototype is 5.38 m (17 ft 8 in) in length, 1.98 m (6 ft 6 in) in height to top of rotor mast and weighs 574 kg (1,265 lb) empty and 645 kg (1,423 lb) gross; it is powered by a 3.11 kN (700 lb st) Williams F112 turbofan and has a 3.65 m (12 ft) diameter rotor. Control systems based on those of the X-36, other spin-offs from this project including the engine and flight software and hardware in order to reduce development costs. Any subsequent production CR/W is likely to use advanced composites materials and be built in both manned and unmanned versions; Boeing studies include a 1,100 kg (2,425 lb) maritime UAV and a 10,000 to 11,000 kg (22,046 to 24,250 lb) manned aircraft that could function as an armed escort fighter in support of the MV-22B Osprey.
                                                      UPDATED

BOEING ADVANCED MEDIUM TRANSPORT

TYPE: Medium transport/multirole.

PROGRAMME: Design study revealed by McDonnell Douglas (now Boeing) in September 1996 for a potential replacement for the Lockheed Martin C-130 Hercules. Original study envisaged aircraft with four 8,950 kW (12,000 shp) turboprops fixed to wing, which tilted 15° for take-off and 45° for landing; potential in-service date of 2020. Those studies provided basis for No-Tail Advanced Theater Transport (NOTAIL ATT) Tilt-wing Super Short Takeoff and Landing (SSTOL) or 'Super Frog' as project was originally known. This also borrowed from BWB studies, as described under UHCA/VLCT entry in International section. Current Advanced Medium Transport (AMT) name adopted by mid-2002.

Concept revealed in September 1998, with 39 m (128 ft) span aircraft designed to carry up to four times the payload

of the C-130J Hercules and use runways as short as 183 m (600 ft). Baseline requirement is delivery of 27,215 kg (60,000 lb) load on to 229 m (750 ft) of rough airstrip, 1,220 m (4,000 ft) AMSL in 35°C (95°F) ambient temperature. Signature of co-operative research and development agreement (CRADA) between Boeing's Phantom Works and USAF Research Laboratory at Wright-Patterson AFB, Ohio, in December 1998 paved way for development and demonstration of 'enabling technologies' that could lead to production-configured aircraft; subsequently, in late 2000, Boeing also concluded development agreement with DARPA.

Fuselage interior width is 6.4 m (21 ft) and proposed AMT will be able to accommodate various loads up to a maximum of about 36,285 kg (80,000 lb), including one AH-64 Apache or two RAH-66 Comanche helicopters; or up to 10 cargo pallets; or 40 fully equipped soldiers. Wing features four widely separated podded turboprop engines, each driving eight-blade propeller. Wing originally pivoted about lateral axis near rear spar, with tilt used to increase lift during take-off and landing. Directional control of tail-less design uses similar system to B-2 bomber, which has split flaps. By early 2000, however, design had progressed to forward-swept (7 to 9°) wing, pivoted at leading-edge and tilting downwards at trailing-edge, with further refinements by second quarter of 2001 adding small horizontal tail surfaces and adopting C-17-style fuselage.

Initial wind tunnel testing of subscale model completed by end of 1998, followed by further trials with a 7 per cent subscale model during 1999; latter included tethered and untethered tests which began at Gray Butte, California, on 5 June. Further wind tunnel testing of latest configuration completed in January 2001, with Boeing planning to conduct manned simulations and powered wind tunnel tests with larger models in 2001-03. If project goes ahead, Boeing optimistic that production aircraft could be operational in 2015, with AMT a potential candidate for USAF MC-X special operations aircraft requirement.
                                                      UPDATED

SHARED SERVICES GROUP

2810 160th Avenue SE, Bellevue, Washington 98008
Tel: (+1 425) 865 51 66
Fax: (+1 425) 865 29 58
PRESIDENT: James F Palmer
PUBLIC RELATIONS DIRECTOR: Karen H Burt

Shared Services Group provides information management services and computing resources to Boeing operating divisions and government customers on a worldwide basis.
                                                      VERIFIED

BOEING BUSINESS JETS

BOEING BUSINESS JETS

PO Box 3707, Seattle, Washington 98124-2207
Tel: (+1 206) 655 98 00
Fax: (+1 206) 655 97 00
e-mail: business.jets@boeing.com
PRESIDENT: Lee Monson
COMMUNICATIONS DIRECTOR: Patricia York
MARKETING DIRECTOR: Charles Colburn
PARTICIPATING COMPANIES:
    Boeing Company: see this entry
    General Electric Company: see *Jane's Aero Engines*

In July 1996 The Boeing Company and The General Electric Company announced formation of a joint venture, Boeing Business Jets, to develop and market corporate versions of

the Next-Generation 737, deliveries of which began in 1998.

Company headquarters is at the corporate air services facility of Boeing Field's Flight Center.
                                                      UPDATED

BOEING BUSINESS JET (BBJ)

TYPE: Large business jet.

PROGRAMME: Launched July 1996. Aircraft are assembled at Boeing Commercial Airplane Group's Renton facility and supplied to Boeing Business Jets, which hands them over in 'green' condition to customers and delivers them to PATS Inc at Georgetown, Delaware, for installation of long-range fuel tanks before the aircraft are delivered to the customer's chosen completion centre for interior outfitting

and painting; designated completion centres are KC Aviation and Associated Air Center in Dallas, Texas; Greenpoint Aviation in Seattle, Washington; The Jet Center in Van Nuys, California; Ozark Aircraft Services in Bentonville, Arkansas; Raytheon in Waco, Texas; Lufthansa Technik in Hamburg, Germany; and Jet Aviation in Basle, Switzerland, but other completion centres may be used at the discretion of the customer. After completion, Boeing ferries the aircraft to the customer's base and carries out crew training.

First BBJ (101st N-G 737, N737BZ) rolled out 26 July 1998; first flight 4 September 1998; FAA and JAA certification achieved 29 October 1998; supplementary type certificate for long-range tanks awarded 20 May 1999, following demonstration non-stop flight of 6,252.5 n miles (11,580 km; 7,200.4 miles) in 13 hours 57 minutes 42 seconds.

CUSTOMERS: Total of 85 firm orders by December 2001, at which time 67 'green' airframes had been delivered and 43 completed aircraft were in customer service. By October 2002, confirmed order had been reduced to 67, with 59 delivered, according to Boeing Commercial Airplanes information, or 73 and 56, according to Boeing Business Jets. Launch customer General Electric ordered two in July 1996, of which first (N366G) flew 23 October 1993 and delivered 23 November 1998 for outfitting. First delivery of a completed aircraft to Dubai Air Wing Royal Flight 4 September 1999. See table of customers and deliveries.

BBJ/BBJ 2 ORDERS AND DELIVERIES

Year	BBJ orders	BBJ deliveries	BBJ 2 orders	BBJ 2 deliveries
1996	3			
1997	25		1	
1998	17	7		
1999		25		
2000	6	10	3	6
2001	12	13	4	
2002	4	4		1
Totals	**67**	**59**	**8**	**7**

BBJ/BBJ 2 CUSTOMERS

Registration	Owner
N367G	General Electric Company
N366G	General Electric Company
VP-BOC	TAG Group/Theberton USA Inc (France)
N50TC	Tracinda Corp
N500LS	Limited Inc
N737BZ	Boeing Business Jets
N737CC	Mid-East Jet (Saudi Arabia)
N737GG	Mid-East Jet (Saudi Arabia)
N21KR	Kjell Roekker (Norway)
N737WH	First Union Commercial Corp
TS-IOO	Government of Tunisia
HZ-TAA	HRH Talal bin AbdulAziz
N742PB	Chartwell Partners
N4AS	Air Shamrock
A6-HRS	Dubai Air Wing
A6-AIN	Government of Abu Dhabi
A6-SIR	Government of Abu Dhabi
N737SP	Raytheon Company
VP-BBJ	JABJ/Picton II (Switzerland)
9M-BBJ	Malaysian Airlines
VP-BWR	USAL Ltd (Switzerland)
N1011N	Aircraft Holdings
C6-TTB	Westmount Investments
P4-TBN	TBN Aviation (Saudi Arabia)
A6-LIW	Government of Abu Dhabi
A6-DAS	Government of Abu Dhabi
HB-IIO	Privat Air
HB-IIP	Privat Air
VP-BRM	Dobro Ltd
N371BJ	Boeing Business Jets
VP-BYA	Saudi Oger (Saudi Arabia)
VP-CEC	Bugshan Construction Co (Saudi Arabia)
N889NC	Newsflight Inc
VP-CZT	Sevel Argentina SA/Macair Jet SA
N127QS	EJI Inc/NetJets
N164RJ	Bausch & Lomb Inc
N129QS	EJI Inc/NetJets
N130QS	EJI Inc/NetJets
N180SM	Fun Air Corp
HZ-DG5	Dallah al Bakara
N171QS	EJI Inc/NetJets
P4-GJC	Tanzanite Investments (Switzerland)
HB-11Q	Privat Air
VP-BFE	Ford Motor Co (UK)
N79715	BBJ One Inc
VP-BFO	Ford Motor Co (UK)
N156QS	EJI Inc/NetJets
N315TS	Tutor-Saliba Corp
N515GM	Boeing Business Jets
N349BA	Boeing Aircraft Holding Inc
A36-001	Royal Australian Air Force
N191QS	EJI Inc/NetJets
A36-002	Royal Australian Air Force
N184QS	EJI Inc/NetJets
VP-BHN*	n/k
A6-MRM*	Government of Abu Dhabi
N374BJ*	Boeing Business Jets
N90R	Swiftlite Aircraft Corp
G-OBBJ*	Multi Flight Ltd
VP-CBB*	Bugshan Construction Co (Saudi Arabia)
ZS-RSA	South African Government
HL-7770	Samsung Aerospace Industries
N99ZL*	Lowa Ltd
00-0015	US Coast Guard
N103QS	EJI Inc/NetJets
N313P	Premier Executive Transport Services
N105QS	Boeing

Note: Some aircraft have subsequently been sold
* BBJ 2

Boeing Business Jet of the Dubai Air Wing (*Jane's/Paul Jackson*)   *NEW*/0526960

The BBJ has been proposed for USAF's Commander-in-Chief (CINC) support aircraft requirement.

COSTS: US$39 million 'green', estimated US$49 million to 54 million typically equipped (2002). Direct operating cost estimated at US$1,700 per hour based on operation within the USA and utilisation of 900 hours per year.

DESIGN FEATURES: Combines fuselage of 737-700, strengthened in aft section, with centre-section, wing and landing gear of 737-800. Aviation Partners Inc winglets standard, affording 5 to 7 per cent reduction in cruise drag, resulting in four to five per cent increase in range; winglets evaluated in mid-1998 by 737-800 (N737BX), first flown on BBJ prototype (N737BZ) on 20 February 1999, received FAA approval on 6 September 2000 and fitted as standard.

POWER PLANT: Two CFM International CFM56-7 turbofans, each rated at 121.4 kN (27,300 lb st). Standard N-G 737 fuel of 26,025 litres (6,875 US gallons; 5,725 Imp gallons) contained in wing, plus between three and nine belly tanks; maximum combined capacity 40,582 litres (10,721 US gallons; 8,927 Imp gallons).

ACCOMMODATION: To customer's choice; operating weights based on allowance of 5,624 kg (12,400 lb). Typical configuration includes forward lounge and private suite with double bed; mid-section conference room; 12 first class sleeper seats at 152 cm (60 in) pitch in two rows with centre aisle, and galley, lavatory and service area at rear, with crew rest area, galley and lavatory aft of flight deck. Alternative arrangements provide for exercise room/gymnasium, office, 24 first-class sleeper seats or high-density seating for up to 63 passengers, three abreast in two rows. Maximum 149 passengers in airline configuration.

AVIONICS: Rockwell Collins Series 90 as core system.
Comms: Triple VHF comm with 8.33 kHz channel spacing; dual HF comm; L-3 Communications 120-minute CVR and Coltech Selcal.
Flight: Dual Rockwell Collins multimode GPS/ILS/VOR/DME receivers; dual ADF; TCAS II; predictive windshear; dual Smiths Industries flight management computers; dual Honeywell ADIRU; Honeywell EGPWS; L-3 Communications FDR and CVR; Flight Dynamics HGS 4000 head-up guidance system; Teledyne airborne navigation data recorder, digital flight data acquisition unit and quick-access recorder; Teledyne Navlink, including two additional navigation computers and electronic standby artificial horizon.
Instrumentation: Honeywell flat-panel LCD displays.
Mission: Optional Teledyne Telelink.

DIMENSIONS, EXTERNAL AND AREAS: As for 737-700 except:
Wing span, incl winglets   35.79 m (117 ft 5 in)
DIMENSIONS, INTERNAL:
Cabin: Length   24.13 m (79 ft 2 in)
  Height   2.16 m (7 ft 1 in)
  Width   3.53 m (11 ft 7 in)

Floor area   75.0 m² (807 sq ft)
Volume   148.7 m³ (5,250 cu ft)
WEIGHTS AND LOADINGS:
Operating weight empty, typically equipped
    43,082 kg (94,980 lb)
Interior completion allowance   5,625 kg (12,400 lb)
Max fuel weight (incl supplementary tanks)
    32,825 kg (72,367 lb)
Max T-O weight   77,565 kg (171,000 lb)
Max ramp weight   77,790 kg (171,500 lb)
Max landing weight   60,780 kg (134,000 lb)
Max zero-fuel weight   57,155 kg (126,000 lb)
Max wing loading   620.5 kg/m² (127.09 lb/sq ft)
Max power loading   320 kg/kN (3.14 lb/lb st)
PERFORMANCE:
Max operating Mach No. (M_{MO})   0.82
Cruising speed: normal   M0.80
  long range   M0.79
Approach speed   132 kt (244 km/h; 152 mph)
Max rate of climb   980 m (3,215 ft)/min
Initial cruising altitude   11,580 m (38,000 ft)
Max certified altitude   12,500 m (41,000 ft)
Service ceiling, OEI: 7,070 m (23,200 ft)
T-O field length, S/L:
  fuel for range of 4,000 n miles (7,408 km; 4,603 miles)
    1,369 m (4,490 ft)
  fuel for range of 5,000 n miles (9,260 km; 5,754 miles)
    1,515 m (4,970 ft)
  fuel for range of 6,000 n miles (11,112 km;
    6,905 miles)   1,765 m (5,790 ft)
Landing run at typical landing weight   706 m (2,315 ft)
Range (nine belly tanks):
  with 8 passengers
    6,200 n miles (11,482 km; 7,134 miles)
  with 25 passengers
    5,935 n miles (10,991 km; 6,829 miles)
  with 50 passengers
    5,365 n miles (9,936 km; 6,173 miles)
OPERATIONAL NOISE LEVELS (FAR Pt 36 Stage 3):
T-O   85.6 EPNdB
Approach   95.9 EPNdB
Sideline   95.2 EPNdB
    UPDATED

BOEING BUSINESS JET 2 (BBJ 2)

TYPE: Large business jet.
PROGRAMME: Launched 11 October 1999; roll-out and first 'green' delivery 8 March 2001 to an undisclosed customer; first service entry was expected in early 2002; forecast production rate eight per year.
CUSTOMERS: BBJ 2 expected to account for some 25 per cent of all BBJ sales. Total of eight firm orders received by October 2002. See table in BBJ entry.

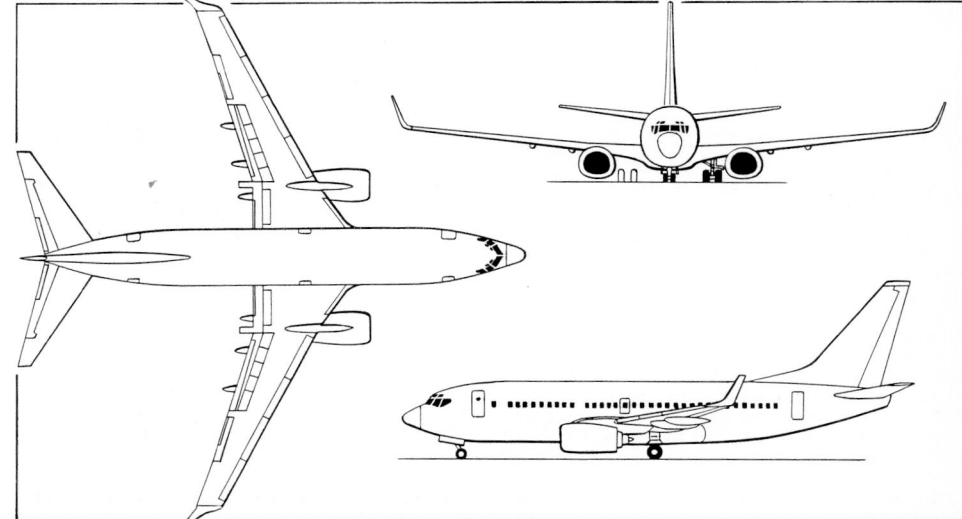

Boeing Business Jet corporate transport (*Jane's/James Goulding*)   0100433

COSTS: US$49 million, 'green'; estimated US$61 million to US$67 million, typically equipped (2002).

DESIGN FEATURES: Based on 737-800 airframe (which see), affording 25 per cent more cabin volume and twice the cargo volume of the BBJ; Aviation Partners Inc winglets standard.

POWER PLANT: Two CFM56-7 turbofans, each 121.4 kN (27,300 lb st). Standard 737 'Next-Generation' fuel of 26,025 litres (6,875 US gallons; 5,725 Imp gallons) contained in wing, plus between three and seven belly tanks; maximum combined capacity 39,466 litres (10,426 US gallons; 8,681 Imp gallons).

ACCOMMODATION: Up to 78 passengers, with executive lounge and private suite. Maximum 189 in airliner configuration. Operating weights based on completion allowance of 7,257 kg (16,000 lb).

DIMENSIONS, EXTERNAL: As for BBJ, except:
Length overall   39.47 m (129 ft 6 in)
DIMENSIONS, INTERNAL:
Cabin: Length   29.97 m (98 ft 4 in)
  Floor area   93.27 m² (1,004 sq ft)
  Max cargo volume   34.7 m³ (1,224 cu ft)

WEIGHTS AND LOADINGS:
Operating weight empty, typically equipped
                                          46,226 kg (101,910 lb)
Max fuel weight (incl supplementary tanks)
                                           31,922 kg (70,376 lb)
Max T-O weight   79,015 kg (174,200 lb)
Max ramp weight   79,245 kg (174,700 lb)
Max landing weight   66,360 kg (146,300 lb)
Max zero-fuel weight   62,730 kg (138,300 lb)
PERFORMANCE:
Max cruising speed   M0.82
Long-range cruising speed   M0.79
Max rate of climb at S/L   948 m (3,110 ft)/min
Initial cruising altitude   11,505 m (37,750 ft)
Max certified altitude   12,500 m (41,000 ft)
Service ceiling, OEI   6,090 m (22,600 ft)
T-O field length, S/L:
  fuel for range of 4,000 n miles (7,408 km; 4,603 miles)
                                              1,655 m (5,430 ft)
  fuel for range of 5,000 n miles (9,260 km; 5,753 miles)
                                              1,915 m (6,280 ft)

fuel for range of 5,650 n miles (10,464 km;
  6,502 miles)   2,118 m (6,950 ft)
Landing run at typical landing weight   758 m (2,485 ft)
Range (seven belly tanks):
  with 8 passengers
                        5,650 n miles (10,463 km; 6,501 miles)
  with 25 passengers
                        5,315 n miles (9,843 km; 6,116 miles)
  with 50 passengers
                        4,780 n miles (8,852 km; 5,500 miles)
OPERATIONAL NOISE LEVELS (FAR Pt 36 Stage 3):
T-O   86.0 EPNdB
Approach   96.3 EPNdB
Sideline   94.7 EPNdB
                                                        UPDATED

BOEING SIKORSKY

BOEING COMPANY and SIKORSKY AIRCRAFT

Boeing Sikorsky Comanche Joint Program Office, 5030 Bradford Drive NW, Building 48-99, Suite 100, Mailstop JE-01, Huntsville, Alabama 35805
Tel: (+1 256) 217 00 00
Fax: (+1 256) 217 00 50
Web: http://www.rah66comanche.com
VICE-PRESIDENT AND RAH-66 PROGRAMME DIRECTOR:
  Charles 'Chuck' Allen
ARMY PROGRAMME MANAGER: Colonel Robert Birmingham
PUBLIC AFFAIRS OFFICERS:
  John R Satterfield (Boeing)
  Ed Steadham (Sikorsky)

Boeing and Sikorsky began collaboration on what later became the RAH-66 in June 1985 and were awarded a contract for the demonstration/validation programme in 1991. In early 2002, Boeing Sikorsky announced selection of Bridgeport, Connecticut, as final assembly location for the production RAH-66; previously, in fourth quarter of 2001, it was revealed that the Joint Program Office was to move from Huntsville, Alabama, to Bridgeport during 2002.
                                                        UPDATED

BOEING SIKORSKY RAH-66 COMANCHE

TYPE: Attack helicopter.

PROGRAMME: Light Helicopter Experimental (LHX) design concepts requested by US Army 1981; numerous changes of programme; original plan for 5,000 to replace UH-1, AH-1, OH-58 and OH-6; reduced in 1987 to 2,096 scout/ attack only, replacing 3,000 existing helicopters; further cut to 1,292 in 1990 (with another 389 possible) and then to 1,096 in 1999, although subsequently raised to 1,213 (including eight for US Army operational evaluation) by mid-2000. Under latter procurement plan, initial production examples to so-called Block I standard for armed reconnaissance; definitive Block II standard was due to follow with heavy attack capability; Block III was to introduce added mission capabilities. Sixth and latest restructuring of programme has resulted in further cut in planned procurement to between 679 and 819 over 20 year period, according to defence acquisition review that was completed on 7 October 2002; Army still intends to field Comanche in Blocks incorporating incremental improvements with low-rate production to begin in FY06.

LHX request for proposals issued 21 June 1988; 23 month demonstration/validation contracts to Boeing Sikorsky and Bell/McDonnell Douglas. Boeing Sikorsky selected 5 April 1991; to build four YRAH-66 demonstration/validation prototypes in 78 month programme, plus static test article (STA) and propulsion system testbed (PSTB).

LHTEC T800 engine specified October 1988. LHX designation changed to LH early 1990, then US Army designation RAH-66 Comanche in April 1991. See *Jane's* 2001-02 and earlier editions for details of original programme timetable.

Prototype critical design review, completed in December 1993, authorised production of three YRAH-66 prototypes (first item for which manufactured in September 1993). At same time, however, further R&D economies under study; December 1994 decision reduced dem/val phase to two prototypes (lacking Longbow/ Hellfire capability).

Prototype construction began 29 November 1993 with forward fuselage at Sikorsky, Stratford; Boeing built aft fuselage in Philadelphia. STA airframe delivered to Stratford 1994, at which time PSTB under construction there. PSTB trials commenced in 1995 at West Palm Beach, with 100 per cent torque from both engines achieved during first 10 hours of running. PSTB subsequently suffered failure of left input bevel gear,

YRAH-66 Comanche with modified fin and added dummy radar housing   *NEW*/0110063

which disintegrated and punched hole in main gearbox housing during 110 per cent power test; resonance blamed for failure.

Front and rear sections of prototype joined at Stratford 25 January 1995; completed helicopter (94-0327) rolled out 25 May 1995. Following transfer to Sikorsky's Development Flight Test Center in West Palm Beach, Florida, during June 1995, first flight accomplished on 4 January 1996. First prototype was retired from flight test duty on 30 January 2002, by which time it had accumulated 387.1 flight hours in 318 sorties; details of test history appear in 2002-03 *Jane's*.

Aft fuselage section of second prototype (95-0001) delivered by Boeing to Stratford in early December 1996 for mating with forward fuselage; completed helicopter exhibited at Army Aviation Association's annual meeting in April 1998 and then to West Palm Beach. Made international debut when displayed statically at Farnborough Air Show in September 1998; flew for first time on 30 March 1999. Completed initial test schedule in April 1999, recording 4.9 hours in five sorties before temporary lay-up, due to funding constraints; also used for

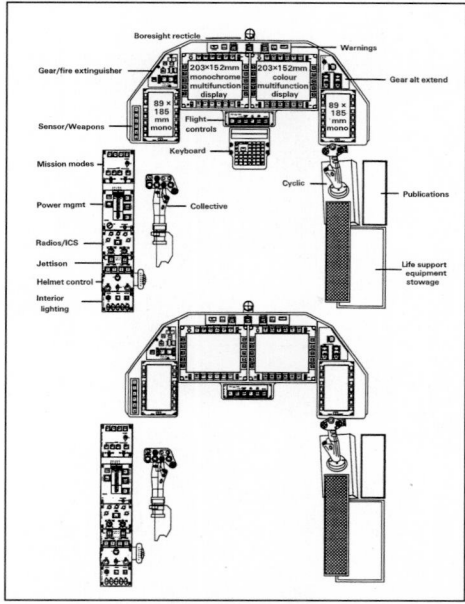

Boeing Sikorsky RAH-66 Comanche front and rear cockpit layout

vertical rate of climb demonstration later in year and will test integrated mission equipment package (MEP), including digital avionics, communications, navigation and target acquisition systems. By mid-December 2000, had logged almost 53 flight hours in 50 sorties; these figures had risen to 93 and 103.5 respectively in May 2001 when it was removed from flight status to be prepared for flight testing and validation of MEP. This phase of development began on 23 May 2002, when second prototype made first flight with MEP and new engines installed. Near-term objectives to be achieved by the second prototype include flight with the night vision pilotage system by October 2002, as well as completion of critical design review in April 2003 and delivery of the electro-optical sight system in June 2003.

Engineering and Manufacturing Development (EMD) officially began 1 June 2000, following RAH-66 meeting (on 4 April 2000) seven key Defense Acquisition Board Milestone 2 criteria, including a 107 m (350 ft)/min vertical climb rate, a specified detection range for the FLIR sensors, a radar cross-section specification, ballistic vulnerability and tolerance specifications and tower-testing of the selected FCR. Weight reduction effort under way in late 2000, to reduce from current level of about 4,310 kg (9,500 lb) to target weight of 4,218 kg (9,300 lb).

Under original plan, EMD expected to take six years and include production of five RAH-66s specifically for EMD testing, followed by further eight for initial operational test and evaluation (IOT&E) by the US Army. However, EMD contract and plan being restructured in mid-2002, at which time it seemed probable that number of EMD aircraft would be reduced to 11 (including those for IOT&E). This new plan expected to increase development cost and delay operational deployment until late 2009 but is considered to be more realistic and workable. The new plan was to include a start of low-rate initial production (LRIP) in late 2007, with full-rate production set at 96 per year; however, it now appears that the Army will receive no more than 819 helicopters instead of the anticipated 1,213. In meantime, second YRAH-66 to assume increasing burden of test duty.

First -801 growth version of T800 turboshaft began bench runs in March 1994; -801 preliminary design review completed May 1993; critical design review March 1995; prototypes originally fitted with less powerful T800-LHT-800 engines, but first flight with definitive -801 engine made on 1 June 2001 by first prototype. Same engine subsequently installed on second prototype in time for resumption of flight test duty in May 2002.

CUSTOMERS: US Army, two prototypes, plus still-to-be-determined number of EMD and initial operational test and evaluation (IOT&E) aircraft for test duties and operational trials. Planned procurement of between 679 and 819 production aircraft.

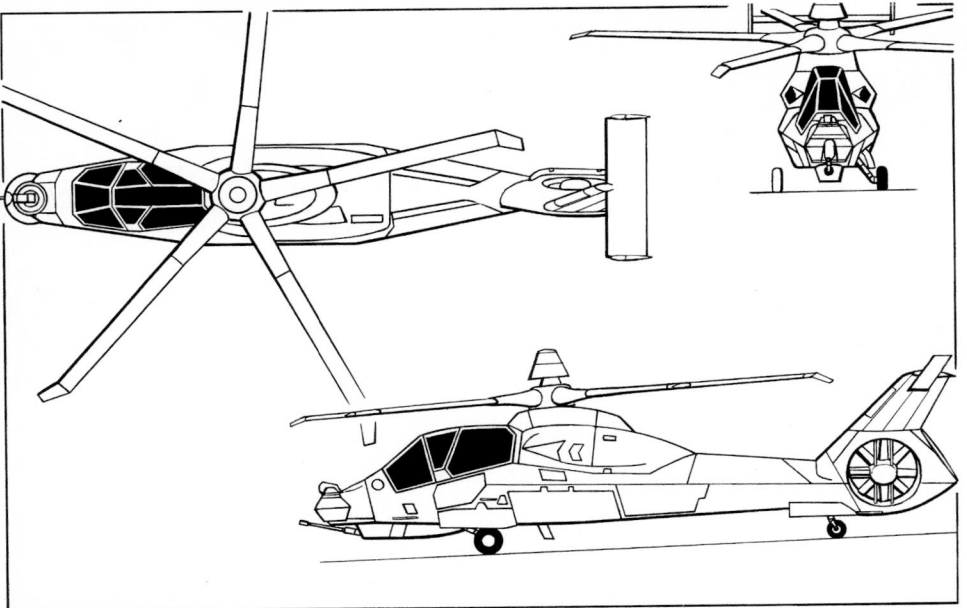

Boeing Sikorsky RAH-66 Comanche (*Jane's/James Goulding*)   NEW/0525828

COSTS: US$34,000 million programme, including US$1,960 million dem/val and US$900 million FSD but reduced to US$2,240 million dem/val/FSD between 1993 and 1997 by cancellation of three of six planned prototypes; US$8.9 million flyaway unit cost (1988 values), increased to US$13 million by early 1993. By 1993 (in 1994 dollars, estimated), procurement unit cost US$21 million, programme unit cost US$27 million. Appropriations for R&D in FY97-99 comprised US$338.6 million, US$282.0 million and US$367.8 million, with US$467 million in FY00 and US$614 million for FY01. EMD phase then projected to cost US$3.2 billion, with initial contract worth US$147.5 million awarded to Boeing Sikorsky on 1 June 2000. Under most recent programme plan (mid-2002), development costs expected to rise to about US$6.5 billion, some of which could be recouped from accelerated production.

DESIGN FEATURES: First combat helicopter designed from outset to have 'stealth' features and target acquisition radar. Embodies low-observables (LO) attributes and stated to have radar cross-section (RCS) lower than that of Hellfire missile; frontal RCS reportedly 360 times smaller than AH-64, 250 times smaller than OH-58D and 32 times smaller than OH-58D with mast-mounted sight. Also has quarter of AH-64D's IR emissions and is six times quieter, head-on.

RAH-66 is lighter, but only slightly smaller, than AH-64 Apache; specified empty weight of 3,402 kg (7,500 lb)

increased to 3,522 kg (7,765 lb) by early 1992, as result of Army add-ons, including allowance for Longbow radar; mission equipment package has maximum commonality with F-22A ATF technology. Design has faceted appearance for radar reflection; downward-angled engine exhausts; T tail with endplates; eight-blade fan-in-fin shrouded tail rotor; and five-blade all-composites bearingless main rotor system, with latter increased in diameter by 0.3 m (1 ft 0 in) and gaining noise-reducing anhedral tips on forthcoming EMD aircraft. New rotor incorporating anhedral tips flown for first time on first prototype on 20 July 2001. RAH-66 also features internal weapon stowage.

Split torque transmission, obviating need for planetary gearing. Upper part of T tail folds down for air transportation. Detachable stub-wings for additional weapon carriage and/or auxiliary fuel tanks (EFAMS: external fuel and armament management system). Radar, infra-red, acoustic and visual signature requirements set to defeat threats postulated by US Army. Eight deployable inside Lockheed C-5 Galaxy with only removal of main rotor; ready for flight 20 minutes after transport lands. Combat turnround time 13 minutes.

Subcontractors to Boeing Sikorsky include BAE Systems (flight control computer, controller grips), Boeing (flight control computer), Hamilton Sundstrand (environmental control system, air vehicle interface controller, air data system, electrical power generation,

control and distribution system), Harris Corporation (3-D digital map, controls and displays, super high-speed fibre optic databusses), Hughes-Link (operator training systems), ITT Industries (radar and laser warning receivers and point chemical detector), Kaiser Electronics (helmet integrated display sight system), Litton Guidance and Control Systems (PAGAN GPS navigation), Lockheed Martin Armament Systems (turreted gun system and ammunition feed system), Lockheed Martin Electronics and Missiles (target acquisition/designation system and night vision pilotage system), Moog (flight control actuators), Northrop Grumman (signal and data processors, aided target detection/classification systems), Smiths Industries (crash survivable memory unit), TRW Military Electronics and Avionics (communication, navigation and identification system and survivability systems) and Williams (subsystem power unit).

FLYING CONTROLS: Dual triplex fly-by-wire, with sidestick cyclic pitch controllers and normal collective levers. Main rotor blades removable without disconnecting control system.

STRUCTURE: Largely composites airframe and rotor system. Fuselage built around composites internal box-beam; non-load-bearing skin panels, more than half of which can be hinged or removed for access to interior (for example, weapons bay doors can double as maintenance work platforms). Eight-blade Fantail rear rotor operable with 12.7 mm calibre bullet hits; or for 30 minutes with one blade missing. Main rotor blades and tail section by Boeing, forward fuselage and final assembly by Sikorsky.

LANDING GEAR: Tailwheel type, retractable, with single wheel on each unit; main units retract aft, with tailwheel retracting forward. Main units can 'kneel' for air transportability.

POWER PLANT: Two LHTEC T800-LHT-801 turboshafts, each rated at 1,165 kW (1,563 shp). Transmission rating 1,639 kW (2,198 shp). Internal fuel capacity 1,142 litres (301.6 US gallons; 251.1 Imp gallons). Two external tanks totalling 3,407 litres (900 US gallons; 749.4 Imp gallons) for self-deployment; total fuel capacity 4,548 litres (1,201.6 US gallons; 1,000.5 Imp gallons). Additional fuel to be contained in two 424 litre (112 US gallon; 93.3 Imp gallon) tanks in side weapon bays, which in preliminary development in mid-1999. Main rotor tip speed (100 per cent N_r) 221 m (725 ft)/s; 355 rpm; normal operation from 95 per cent (quiet mode) to 107 per cent (load factor enhancement).

ACCOMMODATION: Pilot (in front) and WSO in identical stepped cockpits, pressurised for chemical/biological warfare protection. Crew seats resist 11.6 m (38 ft)/s vertical crash landing.

SYSTEMS: Williams International subsystem power unit (SPU), mounted aft of starboard engine, drives hydraulic pump for actuation of landing gear and weapons bay doors; further two hydraulic pumps provide pressure for flight control systems.

AVIONICS: Maximum commonality required with USAF Lockheed Martin F-22 Raptor programme.

Comms: Package known as Comanche Integrated Communications, Navigation and Identification (ICNIA)

Boeing Sikorsky YRAH-66 Comanche combat helicopter   NEW/0525826

systems comprises dual anti-jam VHF-FM and UHF-AM Have Quick tactical communications, VHF-AM, anti-jam HF-SSB; IFF.

Radar: Miniaturised version of Longbow fire-control radar to be installed in one-third of fleet although all to have carriage provision.

Flight: Litton PAGAN jam-resistant four-channel GPS and radar altimeter. Litton AHRS, comprising two fibre optic LN-210C and one LN-100C gyro platforms.

Instrumentation: Kaiser Electronics helmet-mounted display; NVG-compatible lighting; integrated cockpit, night vision pilotage systems, and digital map display. Two 152 × 203 mm (6 × 8 in) multifunction flat-screen LCDs in each cockpit (left is monochrome for FLIR/TV, right is colour for moving map, tactical situation and night operations), plus two 89 × 185 mm (3½ × 7¼ in) multipurpose display (MPD) flat-screen monochrome LCDs for fuel, armament and communications information in each cockpit. Microvision awarded contract by US Army to provide in 1998 a prototype helmet-mounted display incorporating virtual retinal-display technology as potential alternative to conventional helmet-mounted displays. Three redundant databusses: one low-speed (MIL-STD-1553B), one high-speed and one very high-speed (fibre optic-based) for signal data distribution.

Mission: Airborne target handover system. Laser range-finder/designator. Nose-mounted Lockheed Martin electro-optical target acquisition system with second-generation FLIR and TV camera. Northrop Grumman TASS (Target Acquisition System Software).

Self-defence: Goodrich Electro-Optical Systems AN/AVR-2A(V) Advanced Laser Warning Receiver, chemical warning and AN/ALQ-211(V)3 radar warning receivers; RF and IR jammers.

EQUIPMENT: Side armour for cockpit fitted as standard; optional armour kit available for floor of cockpit.

ARMAMENT: General Dynamics stowable XM-301 three-barrel 20 mm cannon in Giat undernose turret, with up to 500 rounds (320 rounds normal for primary mission) and either 750 or 1,500 rounds per minute firing rates. Aiming coverage of gun is +15 to −45° in elevation and ±120° in

Frontal aspect of YRAH-66 Comanche *NEW*/0525827

azimuth. Integrated retractable aircraft munitions system (IRAMS) features side-opening weapons bay door in each side of fuselage, on each of which can be mounted up to three Hellfire or six Stinger missiles or other weapons. Four more Hellfires or eight Stingers can be deployed from multiple carriers under tip of each optional stub-wing, or auxiliary fuel tank for self-deployment. Will be compatible with Starstreak and Mistral air-to-air missiles. Maximum of 56 Hydra 70 2.75 in FFARs or Sura D or Snora 81 mm equivalents. All weapons can be fired, and targets designated, from push-buttons on collective and sidestick controllers.

DIMENSIONS, EXTERNAL:
Main rotor diameter: YRAH-66   11.90 m (39 ft 0½ in)
  RAH-66   12.19 m (40 ft 0 in)
Fantail diameter   1.37 m (4 ft 6 in)
Fantail blade chord   0.17 m (6¾ in)

Length: overall, rotor turning   14.28 m (46 ft 10¼ in)
  fuselage (excl gun barrel)   13.20 m (43 ft 3¾ in)
Fuselage: Max width   2.04 m (6 ft 8¼ in)
Width over mainwheels   2.31 m (7 ft 7 in)
Height over tailplane   3.37 m (11 ft 0¾ in)
Tailplane span   2.82 m (9 ft 3 in)
AREAS:
Main rotor disc: YRAH-66   111.22 m² (1,197.1 sq ft)
Fantail disc   1.48 m² (15.90 sq ft)
WEIGHTS AND LOADINGS (estimated):
Weight empty   4,218 kg (9,300 lb)
Max useful load   2,296 kg (5,062 lb)
Max internal fuel weight   870 kg (1,918 lb)
T-O weight: primary mission   5,799 kg (12,784 lb)*
  max alternative   5,850 kg (12,897 lb)
  max (self-deployment)   7,896 kg (17,408 lb)
Max disc loading: YRAH-66   71.0 kg/m² (14.54 lb/sq ft)
Transmission loading at max T-O weight and power
    4.82 kg/kW (7.92 lb/shp)
with two crew, full internal fuel, 320 rds gun ammunition, radar, four Hellfires and two Stingers
PERFORMANCE (at 1,220 m; 4,000 ft and 35°C; 95°F, estimated):
Max level (dash) speed:
  without radar   175 kt (324 km/h; 201 mph)
  with radar   166 kt (307 km/h; 191 mph)
Cruising speed:
  without radar   160 kt (296 km/h; 184 mph)
  with radar   149 kt (276 km/h; 171 mph)
Vertical rate of climb: without radar   273 m (895 ft)/min
  with radar   152 m (500 ft)/min
Masking   1.6 s
180° hover turn to target   4.7 s
Snap turn to target at 80 kt (148 km/h; 92 mph)   4.5 s
Operational radius, internal fuel
    150 n miles (278 km; 173 miles)
Ferry range with external tanks
    1,260 n miles (2,334 km; 1,450 miles)
Endurance (standard fuel)   2 h 30 min
g limits   +3.5/−1
    UPDATED

BRANTLY

BRANTLY INTERNATIONAL INC

12399 Airport Drive, Wilbarger County Airport, Vernon, Texas 76384
Tel: (+1 940) 552 54 51
Fax: (+1 940) 552 27 03
e-mail: sales@brantly.com
Web: http://www.brantly.com
PRESIDENT AND CEO: Henry Yao
VICE-PRESIDENT: Cy A Russum
VICE-PRESIDENT, MARKETING AND SALES: George Stokes

On 23 December 1994, Brantly International obtained the type certificates for the Brantly B-2B and 305 helicopters from Japanese-American businessman James T Kimura's Brantly Helicopter Industries, which had acquired them in May 1989. In 2001, Brantly employed 40 in its 2,790 m² (30,000 sq ft) facility.

    VERIFIED

Brantly B-2B from the current production series *(Jane's/Susan Bushell)*   *NEW*/0525829

BRANTLY B-2B

TYPE: Two-seat helicopter.
PROGRAMME: Developed from coaxial twin-rotor B-1 by Newby O Brantly. First flight (B-2) 21 February 1953; FAA certification 27 April 1959. Total of 194 B-2s and 18 B-2As (with additional headroom) produced between 1960 and 1963. Improved Model B-2B with metal main blades and fuel-injected Lycoming IVO-360-A1A engine certified 1 July 1963; total of 165 built between 1963 and 1967 (company owned by Gates Learjet from 1966) and a further one (as H-2) in 1975 by Brantly-Hynes Helicopter.

Brantly Helicopter Industries (BHI) took over manufacturing and marketing rights and production facilities in 1989. First new-build B-2B (N25411 c/n 2001) flew 12 April 1991; three built under this name. Production continues under Brantly International, which received FAA production certificate on 19 July 1996.

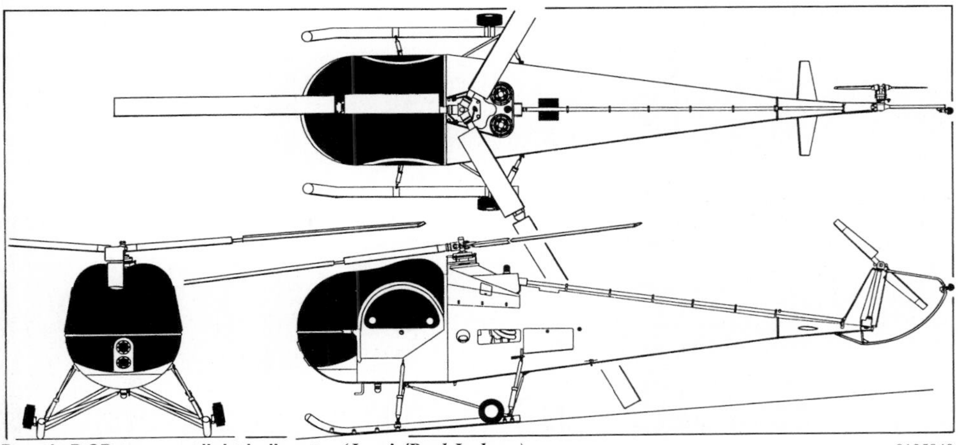

CUSTOMERS: Total of 23 of current series (including BHI) registered by mid-2002. Notified deliveries were two in 1998, none in 1999, six (to China) in 2000 and two in 2001. Total of 14 remained registered in USA at August 2002, and one in Australia. (Some 80 from earlier production remain registered in USA.)

COSTS: US$150,000 basic equipped (2001). Direct operating cost US$80 per hour (2000).
DESIGN FEATURES: Simple design, with blown main transparency and constant-taper fuselage. Double-articulated three-blade main rotor with pitch-change and flapping hinges close to hub and flap/lag hinges at 40 per cent blade span; symmetrical, rigid, inboard blade section with 29 per cent thickness/chord ratio, outboard section NACA 0012; outer blades quickly removable for compact storage; rotor brake standard; two-blade tail rotor mounted on starboard side, with guard. Transmission through automatic centrifugal clutch and planetary reduction gear. Bevel gear take-off from main transmission, with flexible coupling to tail rotor drive-shaft. Main rotor/engine rpm ratio 1:6.158; tail rotor ratio 1:1. Main rotor minimum speed 400 rpm; maximum 472 rpm.
FLYING CONTROLS: Conventional and manual; small fixed tailplanes on port and starboard sides of tailcone.
STRUCTURE: Semi-monocoque fuselage with alloy-stressed skin. Inboard rotor blades have stainless steel leading-edge spar; outboard blades have extruded aluminium spar; polyurethane core with bonded aluminium envelope riveted to spar. All-metal tail rotor blades.
LANDING GEAR: Fixed skid type with oleo-pneumatic shock-absorbers; small retractable ground handling wheels, size 10×3.5, pressure 4.12 bar (60 lb/sq in); fixed tailskid. Optional inflatable pontoons attach to standard skids for over-water operation.
POWER PLANT: One 134 kW (180 hp) Textron Lycoming IVO-360-A1A flat-four air-cooled piston engine, mounted vertically. Fuel contained in two interconnected bladder tanks behind cabin, total capacity 117 litres (31.0 US gallons; 25.8 Imp gallons) of which 115 litres (30.5 US gallons; 25.4 Imp gallons) are usable. Oil capacity 6.9 litres (1.83 US gallons; 1.52 Imp gallons).

Brantly B-2B two-seat light helicopter *(Jane's/Paul Jackson)*   0105040

ACCOMMODATION: Two, side by side in enclosed cabin; forward-hinged door on each side. Dual controls and cabin heater standard. Ground accessible baggage compartment, maximum capacity 22.7 kg (50 lb) in forward end of tailcone.

SYSTEMS: 60A alternator.

AVIONICS: To customer choice; GPS is standard.

EQUIPMENT: Twin landing lights in nose, navigation and anti-collision lights are standard.

DIMENSIONS, EXTERNAL:

Main rotor diameter	7.24 m (23 ft 9 in)
Main rotor blade chord: inboard	0.22 m (8¾ in)
Outboard	0.20 m (8 in)
Tail rotor diameter	1.30 m (4 ft 3 in)
Length: fuselage	6.63 m (21 ft 9 in)
overall, rotors turning	8.53 m (28 ft 0 in)
Height overall	2.11 m (6 ft 11 in)
Skid track	1.73 m (5 ft 8¼ in)
Passenger doors (each): Height	0.79 m (2 ft 7 in)
Width	0.86 m (2 ft 9¾ in)

Baggage door: Height	0.23 m (9 in)
Width	0.53 m (1 ft 9 in)

DIMENSIONS, INTERNAL:

Cabin: Length	1.83 m (6 ft 0 in)
Max width	1.19 m (3 ft 11 in)
Max height	0.99 m (3 ft 3 in)
Floor area	2.60 m² (28.0 sq ft)
Volume	2.78 m³ (98.0 cu ft)
Baggage compartment volume	0.17 m³ (6.0 cu ft)

AREAS:

Main rotor blades (each)	0.69 m² (7.42 sq ft)
Main rotor disc	41.16 m² (443.0 sq ft)
Tail rotor blades (each)	0.05 m² (0.50 sq ft)
Tail rotor disc	1.32 m² (14.19 sq ft)

WEIGHTS AND LOADINGS:

Weight empty: with skids	476 kg (1,050 lb)
with pontoons	494 kg (1,090 lb)
Baggage capacity	22.7 kg (50 lb)
Max fuel weight	82 kg (180 lb)

Max T-O weight	757 kg (1,670 lb)
Max disc loading	18.4 kg/m² (3.77 lb/sq ft)
Max power loading	5.65 kg/kW (9.28 lb/hp)

PERFORMANCE:

Never-exceed speed (VNE)	87 kt (161 km/h; 100 mph)
Max cruising speed at 75% power	
	78 kt (144 km/h; 90 mph)
Max rate of climb at S/L	427 m (1,400 ft)/min
Service ceiling	1,829 m (6,000 ft)
Hovering ceiling IGE	1,074 m (3,525 ft)
Range with max fuel, without reserves	
	191 n miles (353 km; 219 miles)
with reserves	173 n miles (321 km; 200 miles)

UPDATED

CAMERON

CAMERON & SONS AIRCRAFT

2772 Cessna Avenue, Coeur d'Alene Airport, Coeur d'Alene, Idaho 83814
Tel: (+1 208) 765 92 95
Fax: (+1 208) 765 94 15
e-mail: MustangMurdo@cs.com
Web: http://www.CameronAircraft.com
CEO: Murdo Cameron

Cameron Aircraft was formed in 1978. It's P-51 Mustang derivative is the company's first product; parts for the North American AT-6 Texan are also marketed. The 4,180 m² (45,000 sq ft) factory is situated on Coeur d'Alene Airport; in 2002 the workforce numbered 22

UPDATED

CAMERON P-51G

TYPE: Tandem-seat turboprop sportplane/kitbuilt.

PROGRAMME: Full-size, turbine-powered representation of North American P-51 Mustang. Design began 1988; publicly announced end 1997; first flight 1998 and displayed at Oshkosh in July 1998 as Grand 51; later renamed P-51G.

CUSTOMERS: Prototype flying; two production aircraft under construction. Factory capacity for 12 per year.

COSTS: Development costs US$4.5 million; kit price US$150,000; completed airframes US$615,000 (2003).

DESIGN FEATURES: Wing span matches racing P-51s; option to extend to 'normal' 11.38 m (37 ft 4 in) available; aircraft kit comprises 12 major components. Quoted build time 1,500 to 2,000 hours.

STRUCTURE: Airframe and flight control surfaces of carbon fibre epoxy; 85 per cent is composites. One-piece 'wet' wing.

FLYING CONTROLS: Conventional and manual. No flaps. Electric three-axis trim.

LANDING GEAR: Tailwheel configuration; hydraulically operated mainwheels retract inward; tailwheel rearward. Bridgestone 24×7.7 mainwheel tyres; 12.5 in tailwheel type. Cleveland wheels and brakes.

POWER PLANT: One 1,081 kW (1,450 shp) Textron Lycoming T53-L-701A turboprop, driving a Hamilton Sundstrand three-blade, fully feathering and reversible propeller. Rolls-Royce Merlin can be fitted. Fuel capacity 1,703 litres (450 US gallons; 375 Imp gallons) in prototype; kitbuilts have 946 litres (250 US gallons; 208 Imp gallons) with optional 757 litre (200 US gallon; 167 Imp gallon) auxiliary tank.

ACCOMMODATION: Pilot and passenger in tandem under one-piece hinged canopy. Optional P-51D canopy style available. Dual controls. Impact Dynamics leather seating. Dummy belly scoop provides baggage compartment with

starboard side access door.

AVIONICS: To customer's specification; IFR is standard.

DIMENSIONS, EXTERNAL:

Wing span	9.80 m (32 ft 2 in)
Wing chord at tip	1.52 m (5 ft 0 in)
Wing aspect ratio	4.7
Length overall	10.97 m (36 ft 0 in)
Height overall	3.28 m (10 ft 9 in)
Propeller diameter	3.05 m (10 ft 0 in)
Baggage door width	0.91 m (3 ft 0 in)

DIMENSIONS, INTERNAL:

Cockpit max width	0.86 m (2 ft 10 in)

AREAS:

Wings, gross	20.44 m² (220.0 sq ft)

WEIGHTS AND LOADINGS:

Weight empty	2,041 kg (4,500 lb)
Max T-O weight	3,628 kg (8,000 lb)
Max wing loading	177.5 kg/m² (36.36 lb/sq ft)
Max power loading	3.36 kg/kW (5.52 lb/shp)

PERFORMANCE:

Max operating speed	391 kt (724 km/h; 450 mph)
Normal cruising speed	313 kt (579 km/h; 360 mph)
Stalling speed	79 kt (145 km/h; 90 mph)
Max rate of climb at S/L	1,280 m (4,200 ft)/min
Service ceiling	9,144 m (30,000 ft)
T-O run	381 m (1,250 ft)
Landing run	533 m (1,750 ft)
Range:	
standard fuel	1,086 n miles (2,011 km; 1,250 miles)
auxiliary fuel	1,955 n miles (3,621 km; 2,250 miles)
g limits (prototype)	±8

UPDATED

Cameron P-51G (Textron Lycoming T53-L-701A)   NEW/0533396

Cameron P-51G instrument panel. Note four stick-top buttons   0121106

CARLSON

CARLSON AIRCRAFT INC

PO Box 88, 50643 SR 14, East Palestine, Ohio 44413-0088
Tel: (+1 330) 426 39 34
Fax: (+1 330) 426 11 44
e-mail: mlc@sky-tek.com
Web: http://www.sky-tek.com

In addition to the Skycycle and newly added Criquet, Carlson also markets the Sparrow series of ultralight aircraft (see 1992-93 *Jane's*) and a replacement wing for Baby Great Lakes and Piper Cub and Vagabond lightplanes. An ultralight version of the Cub was under construction in 2000.

Founder Ernest W Carlson was killed in the crash of the prototype Criquet on 24 May 2000.

UPDATED

CARLSON SKYCYCLE

There has been no known production of this ultralight kitplane beyond the prototype. Details last appeared in the 2002-03 *Jane's*.

UPDATED

CARLSON CA6 CRIQUET

TYPE: Tandem-seat kitbuilt.

PROGRAMME: Prototype N22CA under trial in 1999; lost in accident 24 May 2000. Construction of second aircraft was continuing in mid-2001.

COSTS: Kit US$25,070 excluding engine and instruments (2003).

DESIGN FEATURES: Three-quarter scale replica of Morane-Saulnier MS.500 Criquet, itself a copy of Fieseler Fi 156 Storch; uses Carlson wing, however.

POWER PLANT: One 103 kW (138 hp) LOM M332A four-cylinder piston engine driving a two-blade propeller. Fuel capacity 152 litres (40.0 US gallons; 33.3 Imp gallons) plus 7.6 litre (2.0 US gallon; 1.7 Imp gallon) header tank.

Optional 119 kW (160 hp) fuel-injected and supercharged M332B for higher gross weight version (998 kg; 2,200 lb) with 197 litre (52.0 US gallon; 43.3 Imp gallon) tankage.

DIMENSIONS, EXTERNAL:

Wing span	10.97 m (36 ft 0 in)
Wing chord, constant	1.63 m (5 ft 4 in)
Length: overall	7.49 m (24 ft 7 in)
Width, wings folded	3.00 m (9 ft 10 in)
Height overall	2.57 m (8 ft 5 in)

AREAS:

Wing, gross	17.84 m² (192.0 sq ft)

WEIGHTS AND LOADINGS:

Weight empty	476 kg (1,050 lb)
Max T-O weight	884 kg (1,950 lb)

PERFORMANCE:

Never-exceed speed (VNE)	117 kt (217 km/h; 135 mph)
Normal cruising speed	83 kt (153 km/h; 95 mph)
Stalling speed, power off	19 kt (34 km/h; 21 mph)
T-O and landing run	23 m (75 ft)

UPDATED

CARTER

CARTERCOPTERS LLC

5720 Seymour Highway, Wichita Falls, Texas 76310
Tel: (+1 940) 691 08 19
Fax: (+1 940) 691 59 77
e-mail: carter@wf.net
Web: http://www.cartercopters.com
CEO: Jay Carter
VICE-PRESIDENT, LICENSING: Claudius Klimt

Company founder Jay Carter previously worked on design of the Bell XV-15 tiltrotor. CarterCopters has produced a flying prototype of what it anticipates will be a family of convertiplanes, although its ultimate purpose is to market the design, not establish a manufacturing programme itself.
UPDATED

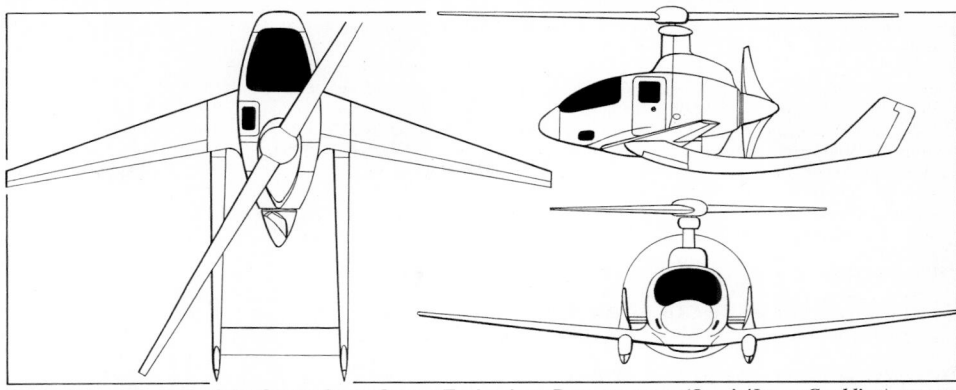

General arrangement of the Carter CarterCopter Technology Demonstrator (*Jane's/James Goulding*) 0087603

CARTER CARTERCOPTER CC1

TYPE: Convertiplane.
PROGRAMME: Developed with US$670,000 funding from NASA, the CarterCopter has been conceived as the first in a series of convertiplanes. First flight (N121CC) September 1998; development delayed following damage to prototype during aborted rolling take-off 16 December 1999; flight testing resumed late October 2000 and was interrupted by minor accidents in April 2001. Attempts are planned on rotary-wing world records. On 22 March 2002, the CCTD reached μ 0.87 (relationship between forward speed and blade speed); company hoped to attain μ 1 by early 2003.
CURRENT VERSIONS: **CC1, (Carter-Copter Technology Demonstrator):** also referred to as **CCTD :** One prototype version for proof-of-concept; *as described.*
   CC Heliplane Transport: See separate entry.
   CC-UCAR: Smaller, unmanned version of CC1 submitted for US Navy consideration in early 2000.
   CC Heliplane Trainer/Liaison/Utility (CCH-S): Passenger version for between five and nine occupants, with same fuselage width as CC1; intended to compete with medium-sized helicopters in the liaison and medical roles. Two TSX-1 turboshaft engines driving 2.44 m (8 ft 0 in) diameter wing-mounted pusher propellers; range estimated as 4,779 n miles (8,851 km; 5,500 miles) with maximum fuel or 608 n miles (1,126 km; 700 miles) with payload of 1,134 kg (2,500 lb).
   CC-AEH: Heavily armed escort version of CCH-S.
   CC-UBAV: Unmanned bomber version of CCH-S.
   CC-KC: Aerial tanker version of CCH-S.
   CC Heliplane Air-Express: Three-engined development planned for carrier on-board delivery role in competition with C-2 Greyhound and V-22 Osprey, Cruising speed 326 kt (604 km/h; 375 mph) at 9,144 m (30,000 ft); range 5,735 n miles (10,621 km; 6,600 miles) with maximum fuel and 1,303 n miles (2,414 km; 1,500 miles) with 9,072 kg (20,000 lb) payload.
   Kitplane: Two-seat kitbuilt version being marketed.
CUSTOMERS: Prototype only by end 2001.
COSTS: Production versions expected to sell for US$250,000 to US$300,000 (1997; latest data).
DESIGN FEATURES: Twin-boom configuration with high-aspect ratio sweptback wings. Pylon-mounted two-blade rotor is powered for take-off and landing, free-rotating in forward flight, when all power is transmitted to a pusher propeller.
   Maximum prerotation and flight speed 425 rpm; minimum in-flight rotor speed 80 rpm.

FLYING CONTROLS: All-moving elevator; twin rudders.
STRUCTURE: Fuselage of seamless moulded composites. Wing sections outboard of tailbooms removable. Tailbooms extend below propeller tips to protect latter in wheels-up landing.
   Bearingless rotor system. Rotor blades of carbon composites with twistable carbon spar; 29.5 kg (65 lb) weights in each blade tip.
LANDING GEAR: Retractable tricycle type. Mainwheels retract into tailbooms.
POWER PLANT: One Corvette LS1 piston engine, with two turbochargers in series, driving a two-blade carbon composites propeller with twistable carbon spar. Engine power 447 kW (600 hp) for short duration; 224 kW (300 hp) at high altitude. Total fuel capacity 606 litres (160 US gallons; 133 Imp gallons), comprising 227 litres (60.0 US gallons; 50.0 Imp gallons) in fuselage tank; 189 litres (50.0 US gallons; 41.6 Imp gallons) in centre wing (underfuselage) tank; and two 95 litre (25.0 US gallon; 20.8 Imp gallon) tanks in wing outboard sections.
ACCOMMODATION: Pilot and passenger in front, plus three further passengers on rear bench seat.
SYSTEMS: Cabin pressurised; maximum differential 0.69 bar (10.00 lb/sq in).

DIMENSIONS, EXTERNAL:

Rotor diameter	13.26 m (43 ft 6 in)
Rotor blade chord: at root	0.43 m (1 ft 5 in)
at tip	0.18 cm (7 in)
Wing span	9.75 m (32 ft 0 in)
Fuselage length	6.81 m (22 ft 4 in)
Height to top of rotor head	3.40 m (11 ft 2 in)
Propeller diameter	2.44 m (8 ft 0 in)

DIMENSIONS, INTERNAL:

Cabin max width	1.52 m (5 ft 0 in)

AREAS:

Rotor disc	138.10 m² (1,486.7 sq ft)
Wing area, gross	7.15 m² (77.0 sq ft)

WEIGHTS AND LOADINGS:

Weight empty	1,134 kg (2,500 lb)
Max T-O weight: STOL	2,268 kg (5,000 lb)
VTOL	1,905 kg (4,200 lb)

PERFORMANCE (two occupants):

Cruising speed: at S/L	191 kt (354 km/h; 220 mph)
at 13,715 m (45,000 ft)	304 kt (563 km/h; 350 mph)
Range, with max fuel, 45 min reserves	
	2,085 n miles (3,862 km; 2,400 miles)

UPDATED

CARTER HELIPLANE TRANSPORT

TYPE: Convertiplane.
PROGRAMME: Intended to be the largest of a family of convertiplanes; announced June 2000.
DESIGN FEATURES: Extensive use of composites throughout; fuselage constructed in one large piece to keep weight down. Composites thickness varies between 2.54 and 10.16 cm (1 and 4 in). Large stabiliser rotates downwards through 90° as collective is increased to reduce downwash in hover. Hydraulic cylinders at each side of rear ramp prevent aircraft tilting back and act as shock-absorbing tailskid during landings.
FLYING CONTROLS: Conventional and manual. No high-lift devices.
STRUCTURE: Fuselage of lightweight carbon composites. Carbon construction propellers counter rotor torque. Twistable spar eliminates pitch change spindle, bearings and housing.
   Wing of carbon composites stressed skins; single wing strut to keep weight of wing low. Outer wing sections (23.16 m; 76 ft 0 in) can be removed for aircraft stowage.
   Rotor tips each contain 408 kg (900 lb) tip weights to provide stored kinetic energy for high density-altitude zero-roll take-offs.
LANDING GEAR: Tricycle type; all wheels retract rearwards and can be retracted and extended while on ground to facilitate cargo handling. Mainwheels housed in sponsons; twin nosewheels. Mainwheels 1.83 m (6 ft 0 in) diameter and 0.76 m (2 ft 6 in) wide; nosewheels 0.76 m (2 ft 6 in) diameter and 0.305 m (1 ft 0 in) wide; all tyres pressure 4.137 bar (60 lb/sq in).
POWER PLANT: Three turbofan or turboprop engines driving a single gearbox, all located in fuselage with two-speed planetary drive to reverse-pitch propellers; plus second two-speed planetary drive unit to four-blade rotor, latter with overrunning clutch; computerised propeller controller. Airscoop at each side of engine pod; engines (KKBM NK-12MV is under consideration) provide 11,033 kW (14,795 shp) each, continuous; 3,740 kW (5,016 shp) at 9,900 m (32,500 ft).
ACCOMMODATION: As well as primary role of cargo carriage, up to 50 passengers can be carried in commercial airliner configuration. Wraparound windscreen enables pilot to see underslung loads. Large door on port side of fuselage between cockpit and wing.
SYSTEMS: Underslung hook and telescopic refuelling boom can be fitted.
All data are provisional.

DIMENSIONS, EXTERNAL:

Rotor diameter	45.72 m (150 ft 0 in)
Rotor blade chord: at root	1.22 m (4 ft 0 in)
at tip	0.76 m (2 ft 6 in)
Wing span	38.10 m (125 ft 0 in)
Fuselage length	32.31 m (106 ft 0 in)
Height to top of rotor head	13.11 m (43 ft 0 in)
Wheel track	9.14 m (30 ft 0 in)
Propeller diameter	7.32 m (24 ft 0 in)

DIMENSIONS, INTERNAL:

Cargo hold: Length: main area	19.81 m (65 ft 0 in)
over ramp	8.23 m (27 ft 0 in)
Max width	3.66 m (12 ft 0 in)
Max height	3.20 m (10 ft 6 in)

Prototype CarterCopter Technology Demonstrator *NEW*/0526459

Projected Heliplane Transport, based on the CarterCopter principle 0100517

Cargo hold volume: main area	312.4 m³ (11,033 cu ft)		Max T-O weight: VTOL	68,039 kg (150,000 lb)		Range, VTOL: with max fuel
over ramp	60.0 m³ (2,120 cu ft)		STOL	90,719 kg (200,000 lb)		4,344 n miles (8,046 km; 5,000 miles)
AREAS:			PERFORMANCE:			with 20,412 kg (45,000 lb) payload
Rotor blades (each)	20.16 m² (217.0 sq ft)		Cruising speed: at 4,570 m (15,000 ft)			695 n miles (1,287 km; 800 miles)
Rotor disc	1,642.4 m² (17,678.6 sq ft)			287 kt (531 km/h; 330 mph)		*VERIFIED*
Wings gross	90.02 m² (969.0 sq ft)		at 7,620 m (25,000 ft)	340 kt (629 km/h; 391 mph)		
WEIGHTS AND LOADINGS:			at 9,900 m (32,500 ft)	391 kt (724 km/h; 450 mph)		
Weight empty	40,823 kg (90,000 lb)		T-O run, STOL	91 m (300 ft)		

CENTURY

CENTURY AEROSPACE CORPORATION

3250 University Boulevard SE, Access Road B, Albuquerque, New Mexico 87106
Tel: (+1 505) 246 82 00
Fax: (+1 505) 246 83 00
e-mail: info@centuryaero.com
Web: http://www.centuryaero.com

PRESIDENT AND CEO: Bill Northrup
VICE-PRESIDENT AND GENERAL MANAGER: Roy Johnston
VICE-PRESIDENT, ENGINEERING: Dale Ruhmel
SALES DIRECTOR: Kevin Whited
MARKETING MANAGER: J Sheldon 'Torch' Lewis

The CA-100 Century Jet programme (see 2001-02 *Jane's*) became 'dormant' in August 2001 when a bank foreclosure ended plans to build the aircraft at Alliance Aerospace,

Macon, Georgia. Replacement funding of up to US$80 million is being sought to replace finance intended to have come from Taiwan. In mid-2002, Mooney Aerospace Group Ltd (which see in this section) was negotiating with Century Aerospace Corporation for manufacturing rights for the CA-100 Century Jet.

UPDATED

CESSNA

CESSNA AIRCRAFT COMPANY
(Subsidiary of Textron Inc)

PO Box 7706, Wichita, Kansas 67277-7706
Tel: (+1 316) 941 60 00
Fax: (+1 316) 941 78 12
Web: http://www.cessna.textron.com
CHAIRMAN: Russell W Meyer Jr
PRESIDENT AND COO: Charles B Johnson
SENIOR VICE-PRESIDENT, PRODUCT ENGINEERING: Jack Pelton
SENIOR VICE-PRESIDENT, SALES AND MARKETING: Roger Whyte
SENIOR VICE-PRESIDENT, FINANCE AND CHIEF FINANCIAL OFFICER:
Michael Shonka
VICE-PRESIDENT, MARKETING: Phil Michel
VICE-PRESIDENT, CITATION SALES: Mark Paolucci
VICE-PRESIDENT, CORPORATE COMMUNICATIONS:
Marilyn Richwine

SINGLE-ENGINE PISTON AIRCRAFT PRODUCTION PLANT:
One Cessna Boulevard, PO Box 1996, Independence, Kansas 67301
Tel: (+1 316) 332 00 00
Fax: (+1 316) 332 03 88
GENERAL MANAGER: Pat Boyarski

Founded by late Clyde V Cessna 1911; incorporated 7 September 1927; acquired by General Dynamics as wholly owned subsidiary 1985; acquisition by Textron announced February 1992.

Suspended manufacture of some piston-engined light aircraft types in mid-1980s; plans for resumed production of Cessna 172 Skyhawk, 182 Skylane, 206 Stationair and T206 Turbo Stationair announced 1994; 46,450 m² (500,000 sq ft) new factory at Independence Municipal Airport, Montgomery County, Kansas, opened 3 July 1996 to house final assembly, painting, flight test, engineering, finance, marketing and human resource operations for single-engined light aircraft range. First new model 172 rolled out 6 November 1996.

Owned subsidiaries include Cessna Finance Corporation in Wichita and Cessna-Columbus Division in Columbus, Georgia. Sold 49 per cent interest in Reims Aviation of France to Compagnie Française Chaufour Investissement (CFCI) February 1989; Reims continues manufacturing Cessna F406 Caravan II (which see in French section).

Total 183,666 aircraft produced by end of 2001. Total of 1,209 delivered in 2001, comprising 821 single-engined piston aircraft, 75 Caravans and 313 Citations. Total of 900 piston-engined aircraft was planned for production in 2002. Business jet production some 300 in 2002, falling to about 250 in 2003. 3,000th piston single from resumed production, a 182T Skylane, delivered in early February 2001; 3,000th Citation executive jet, a Citation X, delivered 19 November 1999. Workforce about 11,300 in December 2002.

UPDATED

CESSNA 172 SKYHAWK

TYPE: Four-seat lightplane.
PROGRAMME: Development of previous models described in 1985-86 and earlier editions of *Jane's*; total of 35,643 commercial aircraft in the earlier Model 172/Skyhawk series built between 1955 (first flight of prototype, 12 June) and 1985, when production was suspended. Prototype 'Restart 172' (N6786R), modified from 1978 Model 172N with IO-360 engine, made its first flight at Wichita 19 April 1995; certification flight testing began 10 January 1996; three preproduction 172Rs (N172KS, N172SE and N172NU) assembled at East Pawnee, Wichita, plant under Single-Engine Pilot Program, of which the first flew on 16 April 1996; FAA FAR Pt 23 certification achieved 21 June 1996 after 145 flight test hours; first production aircraft (c/n 80004; N172FN) rolled out at Independence 6 November 1996 and delivered 18 January 1997. First export delivery began on 30 January 1997 when second

Independence-built aircraft (c/n 80005; VH-NMN) left US for air ferry to Australia.
CURRENT VERSIONS: **172R Skyhawk:** Standard version. Compared with pre-1986 (172Q) versions, features fuel-injected engine; specially developed McCauley propeller optimised for reduced rpm operation; new Honeywell avionics; metal instrument panel with backlit, non-glare instruments; all-electric engine gauges, dual vacuum system, digital clock, EGT gauge, hour meter and centrally mounted annunciator panel; stainless steel control cables; epoxy corrosion-proofing; cabin improvements as detailed below; steps for visual checking of fuel tanks, which have quick-reference quantity tabs, and consolidation of all primary electrical components into a single junction box on the firewall for simplified servicing.

New standard features announced at the NBAA Convention at New Orleans on 9 October 2000, and introduced on 172R and 172S from 2001 model year, include redesigned wingtips, restyled and reclocked main landing gear fairings, improved nose gear fairing, low-drag wing strut fairings, streamlined refuelling and entrance steps, and improved interiors featuring sculptured composite side panels with integrated cupholders and cushioned armrests, removable overhead panels for simplified maintenance, Rosen sun visors, lightweight door handles and a 12 V electrical port for use with a portable GPS, CD player or laptop computer. New optional avionics include a 12.7 cm (5 in) Honeywell KMD 550 LCD colour MFD with Stormscope that interfaces with the previously available KLN 94 colour IFR GPS.
Detailed description applies to 172R.

172S Skyhawk SP: Special performance version, announced 19 March 1998; features Textron Lycoming IO-360-L2A engine rated at 134 kW (180 hp) driving a higher-performance McCauley fixed-pitch, 1.93 m (6 ft 4 in) diameter propeller; avionics and equipment as for baseline version except leather seats standard. Standard empty weight 746 kg (1,644 lb); maximum T-O and landing weight: normal 1,157 kg (2,550 lb), utility 998 kg (2,200 lb); maximum ramp weight: normal 1,160 kg (2,558 lb), utility 1,002 kg (2,208 lb); maximum level speed at S/L 126 kt (233 km/h; 145 mph); cruising speed, 75 per cent power at 2,590 m (8,500 ft) 124 kt (230 km/h; 143 mph); stalling speed, power off: flaps up 53 kt (99 km/h; 61 mph), flaps down 48 kt (89 km/h; 56 mph); maximum rate of climb at S/L 222 m (730 ft)/min; service ceiling 4,270 m (14,000 ft); T-O run 293 m (960 ft); T-O to 15 m (50 ft) 497 m (1,630 ft); landing from 15 m (50 ft) 407 m (1,335 ft); landing run 175 m (575 ft); range with maximum fuel: at 75 per cent power at 2,590 m (8,500 ft), 518 n miles (959 km; 596 miles); at 45 per cent power at 3,050 m (10,000 ft), 638 n miles (1,181 km; 734 miles); endurance 6 h 43 min. First delivery 31 July 1998. Further improvements introduced in October 2000, as noted above.

Millennium Edition: Special configuration limited edition Skyhawk SP, announced in February 2000 and

Cessna 172R Skyhawk *(Jane's/Paul Jackson)*   NEW/0533675

available from March 2000. Cessna announced on 9 April 2000 that limited edition would be extended, due to demand. Features redesigned cabin interior with leather seats; floor mats embossed with Cessna Millennium logo; new cup- and chart-holders; Rosen sun visors; polished spinner and cowling fasteners; Goodyear Flight Custom II tyres; avionics package plus Honeywell KLN 94 colour IFR GPS with moving map display; new exterior paint scheme; pilot's accessory collection.
CUSTOMERS: Total of 180 172Rs and 279 172Ss delivered in 1999, 150 and 340, respectively, in 2000 107 and 341, respectively in 2001 and 47 and 196 respectively in the first six months of 2002. The 1,000th 172R/SP was delivered in October 1999 to North Florida Medical. Recent customers include Embry-Riddle Aeronautical University (initial batch of 73 172Rs out of 300 piston-engined Cessnas of all models to be ordered over a 12 year period), TAM Training (10), Western Michigan University (38), US AFB Flight Training Centre (22 delivered from 30 March 1998), Civil Air Patrol (42 172Rs/SPs ordered by August 2000, Daniel Webster College of Nashua, New Hampshire (20 172Rs delivered by August 2000, Central Missouri State University (16 172Rs delivered by March 2000), Kansas State University (15 172Rs delivered), and Singapore Flying College at Jandakot, Perth, Australia (six 172Rs ordered in February 2000 in addition to five delivered in early 1998).
COSTS: 172R Skyhawk US$152,000 standard equipped; Skyhawk SP US$159,900 (2002). Price including NAV I US$161,700 (172R) and US$169,600 (172S); NAV II US$174,000 (172R) and US$181,900 (172S).
DESIGN FEATURES: Classic, all-metal, braced high-wing, touring lightplane. Tapered, mid-mounted tailplane and wing outer panels; sweptback fin.
NACA 2412 wing section. Dihedral 1° 44′. Incidence 1° 30′ at root, −1° 30′ at tip; fin sweepback 35° at quarter-chord.
FLYING CONTROLS: Conventional and manual. Actuation by stainless steel cables; modified Frise all-metal ailerons; electrically actuated single-slotted Para Lift flaps; trim tab on elevator.
STRUCTURE: Conventional light alloy, with composites for some non-structural components such as nose-bowl, wing-, tailplane- and fin-caps, and fairings. Epoxy corrosion proofing and polyurethane exterior paint standard.
LANDING GEAR: Fixed tricycle type with Cessna Land-O-Matic cantilever tapered steel tube main legs and oleo-pneumatic nose leg; steerable nosewheel, tyre size 5.00×5, mainwheel tyres size 6.00×6; hydraulic brakes. Wheel fairings optional, but included free when Nav I and II packages are purchased.
POWER PLANT: One 119 kW (160 hp) Textron Lycoming IO-360-L2A flat-four piston engine driving a two-blade, fixed-pitch McCauley metal propeller. Fuel contained in two integral wing tanks, combined capacity 212 litres (56.0

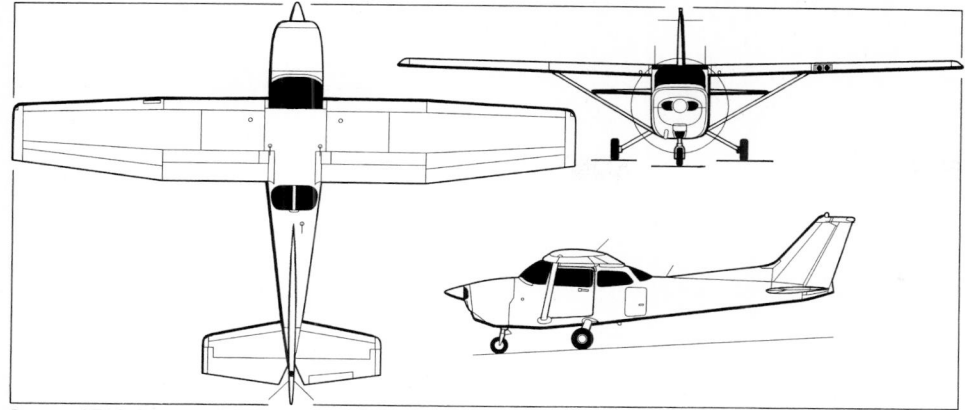

Cessna 172R Skyhawk *(Jane's/Paul Jackson)*   0015674

Cessna 172S Skyhawk SP *(Jane's/Paul Jackson)*   **NEW**/0533676

US gallons; 46.6 Imp gallons) of which 201 litres (53.0 US gallons; 44.1 Imp gallons) are usable. Refuelling points on upper surface of each wing. Oil capacity 7.6 litres (2.0 US gallons; 1.7 Imp gallons).

ACCOMMODATION: Four persons in two pairs on contoured, energy-absorbing seats dynamically tested to 26 *g*, those for pilot and front seat passenger being vertically adjustable, reclining and mounted on enlarged seat rails with dual locking pins; dual controls standard; rear bench seat with reclining back; door each side of cabin, hinged forward; inertia reel harnesses for all occupants; tinted windows with hinged side windows; ambient noise reduction soundproofing; composites headliner, and double-pin latching system for cabin doors. Baggage compartment to rear of cabin, with access door on port side. Cabin is heated and ventilated. Air conditioning optional.

SYSTEMS: Electrical system comprising 28 V 60 A alternator and 24 V 12.75 Ah battery.

AVIONICS: *Comms:* Bendix/King KX 155A nav/com; KMA 28 audio panel. KT 76C Mode C transponder all standard.
Flight: Bendix/King KI 208 VOR/LOC standard. Optional NAV I package comprises KLN 94 GPS, second KX 155A with glideslope, KI 209 VOR/LOC/GS and MD 41-231 GPS-Nav selector; optional NAV II package adds KAP 140 two-axis autopilot. Optional additions to NAV II include KCS 55 compass, KR 87 ADF and KMD 550 MFD.

DIMENSIONS, EXTERNAL:

Wing span	11.00 m (36 ft 1 in)
Wing chord: at root	1.63 m (5 ft 4 in)
at tip	1.12 m (3 ft 8½ in)
Wing aspect ratio	7.5
Length overall	8.28 m (27 ft 2 in)
Height overall	2.72 m (8 ft 11 in)
Wheel track	2.53 m (8 ft 3½ in)
Wheelbase	1.63 m (5 ft 4 in)
Propeller diameter	1.90 m (6 ft 3 in)
Cabin doors: Height	1.02 m (3 ft 4½ in)
Max width	0.94 m (3 ft 1 in)
Baggage door: Height	0.56 m (1 ft 10 in)
Max width	0.39 m (1 ft 3¼ in)

DIMENSIONS, INTERNAL:

Cabin (firewall to baggage area):	
Length	3.61 m (11 ft 10 in)
Max width	1.00 m (3 ft 3½ in)
Max height	1.22 m (4 ft 0 in)

AREAS:

Wings, gross	16.17 m² (174.0 sq ft)
Ailerons (total)	1.70 m² (18.3 sq ft)
Trailing-edge flaps (total)	1.98 m² (21.26 sq ft)
Fin	1.04 m² (11.24 sq ft)
Rudder, incl tab	0.69 m² (7.43 sq ft)
Tailplane	2.00 m² (21.56 sq ft)
Elevators, incl tab	1.35 m² (14.53 sq ft)

WEIGHTS AND LOADINGS:

Weight empty, standard	745 kg (1,642 lb)
Baggage capacity	54 kg (120 lb)

Max T-O and landing weight:

Normal	1,111 kg (2,450 lb)
Utility	952 kg (2,100 lb)
Max ramp weight: Normal	1,114 kg (2,457 lb)
Utility	956 kg (2,107 lb)
Max wing loading	68.7 kg/m² (14.08 lb/sq ft)
Max power loading	9.32 kg/kW (15.31 lb/hp)

PERFORMANCE:

Max level speed at S/L	123 kt (227 km/h; 141 mph)
Cruising speed at 80% power at 2,440 m (8,000 ft)	122 kt (226 km/h; 140 mph)
Stalling speed, power off:	
flaps up	51 kt (95 km/h; 59 mph)
flaps down	47 kt (87 km/h; 54 mph)
Max rate of climb at S/L	219 m (720 ft)/min
Service ceiling	4,115 m (13,500 ft)
T-O run	288 m (945 ft)
T-O to 15 m (50 ft)	514 m (1,685 ft)
Landing from 15 m (50 ft)	395 m (1,295 ft)
Landing run	168 m (550 ft)
Range with max fuel, 45 min reserves:	
at 80% power at 2,440 m (8,000 ft)	580 n miles (1,074 km; 667 miles)
at 60% power at 3,050 m (10,000 ft)	687 n miles (1,272 km; 790 miles)
Endurance	6 h 36 min

UPDATED

CESSNA 182 SKYLANE

TYPE: Four-seat lightplane.

PROGRAMME: Development of previous models described in 1985-86 and earlier editions of *Jane's*; total of 19,773 earlier Model 182/Skylanes were built between 1956 and 1985, when production was suspended. Manufacture resumed when three preproduction 182Ss assembled at East Pawnee, Wichita, plant under Single-Engine Pilot Program, of which the first (N182NU) flew on 16 July 1996; FAA FAR Pt 23 certification achieved 3 October

1996 after 164 flight test hours; assembly of first new production aircraft (N182FN) began at Independence 25 September 1996, with delivery 24 April 1997. Cessna 182S supplanted when delivery of improved 182T and T182T versions began in May 2001.

CURRENT VERSIONS: **182T Skylane:** Standard version from 2001 model year, announced at the NBAA Convention at New Orleans on 9 October 2000. Compared to 182S (last described in 2001-2002 and earlier editions) features redesigned wingtips, restyled and reclocked main landing gear fairings, improved nose gear fairing, low-drag wing strut fairings, streamlined refuelling and entrance steps, and improved interiors featuring sculptured composite side panels with integrated cupholders and cushioned armrests, removable overhead panels for simplified maintenance, Rosen sun visors, lightweight door handles, and a 12 V electrical port for use with a portable GPS, CD player or laptop computer. New optional avionics include a 12.7 cm (5 in) Honeywell KMD 550 LCD colour MFD with Stormscope that interfaces with the previously available KLN 94 colour IFR GPS. Standard empty weight 860 kg (1,897 lb); maximum level speed at S/L 149 kt (276 km/h; 171 mph); cruising speed at 80 per cent power at 1,829 m (6,000 ft), 144 kt (267 km/h; 166 mph); range with maximum fuel, 45 min reserves at 75 per cent power at 1,830 m (6,000 ft), 845 n miles (1,565 km; 972 miles).

T182T Turbo Skylane: Turbocharged model. Described separately.

Millennium Edition: Special configuration limited edition of 182S, available from April 2000. Description in 2001-02 edition.

CUSTOMERS: Recent customers include the Mexican Air Force, which had taken delivery of 76 Skylanes by June 2000, and the Louisiana State Department of Agriculture and Forestry, which ordered 18 for delivery in early 2003. Total of 248 delivered in 1999, 267 in 2000, 142 in 2001 and 72 in the first six months of 2002.

COSTS: US$253,300 standard equipped; with NAV I package, US$271,000; with NAV II package, US$290,000 (all 2002).

DESIGN FEATURES: Generally as for Cessna 172.
Wing section NACA 2412 (modified); incidence 0° 47′ at root, −2° 50′ at tip; dihedral 1° 44′.

FLYING CONTROLS: Conventional and manual. Actuation by stainless steel control cables; modified Frise ailerons; electrically actuated single-slotted Para Lift flaps; trim tabs on elevator and rudder.

STRUCTURE: Conventional light alloy, with composites for some non-structural components such as nose-bowl, wing-, tailplane- and fin-caps, and fairings. Epoxy corrosion proofing and polyurethane exterior paint standard.

LANDING GEAR: Fixed tricycle type with Cessna Land-O-Matic cantilever tapered steel tube main legs and oleo-pneumatic nose leg; steerable nosewheel, tyre size 5.00×5, mainwheel tyre size 6.00×6; hydraulic brakes. Wheel fairings optional.

POWER PLANT: One 172 kW (230 hp) Textron Lycoming IO-540-AB1A5 flat-six piston engine driving a three-blade, constant-speed McCauley metal propeller. Fuel contained in two integral wing tanks, combined capacity 348 litres (92.0 US gallons; 76.6 Imp gallons) of which 333 litres (88.0 US gallons; 73.3 Imp gallons) are usable. Refuelling points on upper surface of each wing. Oil capacity 8.5 litres (2.25 US gallons; 1.9 Imp gallons).

ACCOMMODATION: Four persons in two pairs on contoured, energy-absorbing seats, dynamically tested to 26 *g*, those for pilot and front seat passenger being vertically adjustable, reclining and mounted on enlarged seat rails with dual locking pins; dual controls standard; rear bench seat with reclining back; inertia reel harnesses for all occupants; tinted windows with hinged side windows; ambient noise reduction soundproofing; composites headliner, and double-pin latching system for cabin doors. Baggage compartment to rear of cabin, with access door on port side. Cabin is heated and ventilated. Air conditioning optional.

SYSTEMS: Electrical system comprising 28 V 60 A alternator and 24 V 12.75 Ah battery.

AVIONICS: *Comms:* Dual Bendix/King KX 155A nav/com glidesloop; KMA 28 audio panel, ELT and KT 76C Mode C transponder all standard.

Cessna 182T Skylane (Textron Lycoming IO-540 flat-six) *(Jane's/Paul Jackson)*   **NEW**/0533677

Flight: Bendix/King KAP 140 two-axis autopilot and KI 209A VOR/LOC/GS standard. NAV I package adds KI 209 VOR/LOC, KLN 94 moving map, KMD 550 MFD and MD 41-230 GPS-NAV selector; NAV II package adds KCS 55A compass and Goodrich WX 500 Stormscope and MD41-233 GPS-NAV. Options to NAV II include KR 87 ADF (also with NAV I) and KMH 880 hazard awareness system.

DIMENSIONS, EXTERNAL:

Wing span	10.97 m (36 ft 0 in)
Wing chord: at root	1.63 m (5 ft 4 in)
at tip	1.09 m (3 ft 7 in)
Wing aspect ratio	7.4
Length overall	8.84 m (29 ft 0 in)
Height overall	2.84 m (9 ft 4 in)
Wheel track	2.74 m (9 ft 0 in)
Wheelbase	1.69 m (5 ft 6½ in)
Propeller diameter	2.01 m (6 ft 7 in)
Cabin doors: Height	1.04 m (3 ft 5 in)
Max width	0.93 m (3 ft 0½ in)
Baggage door: Height	0.56 m (1 ft 10 in)
Width	0.40 m (1 ft 3¾ in)

DIMENSIONS, INTERNAL:

Cabin (firewall to baggage compartment):

Length	3.40 m (11 ft 2 in)
Max width	1.07 m (3 ft 6 in)
Max height	1.23 m (4 ft 0½ in)

AREAS:

Wings, gross	16.30 m² (175.50 sq ft)
Ailerons (total)	1.70 m² (18.3 sq ft)
Trailing-edge flaps (total)	1.97 m² (21.20 sq ft)
Fin	1.08 m² (11.62 sq ft)
Rudder	0.65 m² (6.95 sq ft)
Tailplane	2.13 m² (22.96 sq ft)
Elevators	1.54 m² (16.61 sq ft)

WEIGHTS AND LOADINGS:

Weight empty	873 kg (1,924 lb)
Baggage capacity	91 kg (200 lb)
Max T-O weight	1,406 kg (3,100 lb)
Max ramp weight	1,411 kg (3,110 lb)
Max landing weight	1,338 kg (2,950 lb)
Max wing loading	86.2 kg/m² (17.66 lb/sq ft)
Max power loading	8.20 kg/kW (13.48 lb/hp)

PERFORMANCE:

Max level speed at S/L	150 kt (278 km/h; 173 mph)
Cruising speed at 80% power at 2,135 m (7,000 ft)	145 kt (269 km/h; 167 mph)
Stalling speed, power off:	
flaps up	54 kt (100 km/h; 63 mph)
flaps down	49 kt (91 km/h; 57 mph)
Max rate of climb at S/L	282 m (924 ft)/min
Service ceiling	5,515 m (18,100 ft)
T-O run	242 m (795 ft)
T-O to 15 m (50 ft)	462 m (1,514 ft)
Landing from 15 m (50 ft)	412 m (1,350 ft)
Landing run	180 m (590 ft)
Range with max fuel, 45 min reserves:	
at 75% power at 1,981 m (6,500 ft)	820 n miles (1,518 km; 943 miles)
at 55% power at 3,050 m (10,000 ft)	968 n miles (1,792 km; 1,114 miles)

UPDATED

CESSNA T182T TURBO SKYLANE

TYPE: Four-seat lightplane.

PROGRAMME: Announced at NBAA Convention at New Orleans on 9 October 2000. Deliveries began in May 2001. *Description for Cessna 182S and 182T Skylanes applies also to the T182T Turbo Skylane, except as follows.*

CUSTOMERS: Total of 96 delivered in 2001, and 26 in the first six months of 2002.

COSTS: Standard equipped US$294,500; with NAV I package US$312,200; with NAV II package US$331,200 (all 2002).

POWER PLANT: One 175.2 kW (235 hp) Textron Lycoming TIO-540-AK1A turbocharged flat-six engine, driving a three-blade, constant-speed McCauley metal propeller.

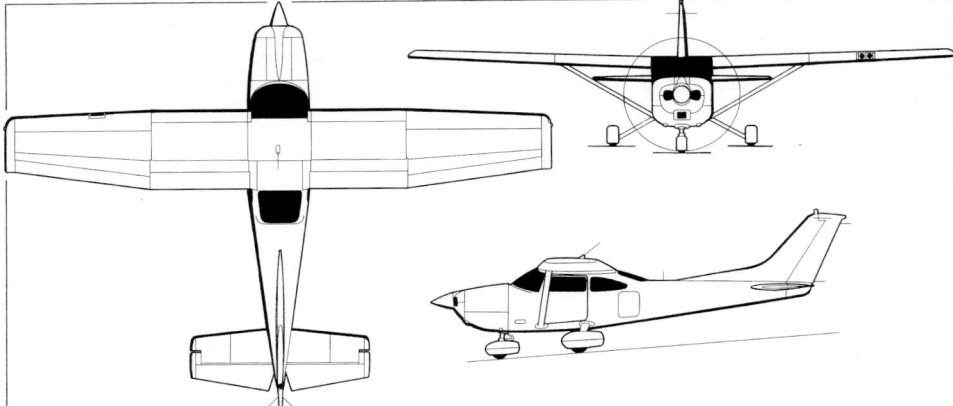

Cessna 182S Skylane (*Jane's/Paul Jackson*)   0015672

Cessna T206H turbocharged version of Stationair   *NEW*/0533720

AVIONICS: As for Skylane.

WEIGHTS AND LOADINGS:

Standard weight empty	918 kg (2,023 lb)
Max ramp weight	1,412 kg (3,112 lb)
Max power loading	8.03 kg/kN (13.19 lb/hp)

PERFORMANCE:

Max level speed at 6,100 m (20,000 ft)	176 kt (326 km/h; 203 mph)
Cruising speed at 88% power at 3,660 m (12,000 ft)	159 kt (294 km/h; 183 mph)
Max rate of climb at S/L	317 m (1,040 ft)/min
Max operating altitude	6,100 m (20,000 ft)
T-O run	236 m (775 ft)
T-O to 15 m (50 ft)	422 m (1,385 ft)
Max range	971 n miles (1,798 km; 1,117 miles)

UPDATED

Instrument panel of Cessna T182T Turbo Skylane with optional multifunction display   0100437

CESSNA 206 STATIONAIR and T206 TURBO STATIONAIR

TYPE: Six-seat utility transport.

PROGRAMME: Development of previous versions described in 1984-85 and earlier editions of *Jane's*; beginning in 1964, total of 7,556 earlier Model 206/Skywagons/Super Skylanes/Stationairs/Turbo Stationairs had been built when production was suspended in 1985. New model T206H prototype (N732CP) first flown from Wichita 28 August 1996; new model 206H prototype first flown 6 November 1996; assembly of first Pilot Program T206H began 3 December 1996.

FAA certification of 206H achieved 9 September 1998, followed by approval for T206H on 1 October 1998. First delivery, of 10 206Hs to Uruguayan Air Force, began 8 December 1998.

CURRENT VERSIONS: **206H Stationair** and **T206H Turbo Stationair (Stationair TC):** Standard versions. New features introduced on 206H and T206H comprise fuel-injected engines; new Honeywell avionics; metal instrument panel with backlit, non-glare instruments plus back-up cold fluorescent light system; all-electric engine gauges; stainless steel control cables; epoxy corrosion-proofing; cabin improvements as detailed below; strobe lights; handles and steps to check fuel levels; static wicks; GPU receptacle; and fire extinguisher.

New standard features announced at the NBAA Convention at New Orleans on 9 October 2000, and introduced on 206H and T206H from 2001 model year, include redesigned wingtips, restyled and reclocked main landing gear fairings, improved nose gear fairing, low-drag wing strut fairings, streamlined refuelling and entrance steps, and improved interiors featuring sculptured composite side panels with integrated cupholders and cushioned armrests, removable overhead panels for simplified maintenance, Rosen sun visors, lightweight door handles, and a 12 V electrical port for use with a portable GPS, CD player or laptop computer. New optional avionics include a 12.7 cm (5 in) Honeywell KMD 550 LCD colour MFD with Stormscope that interfaces with the previously available KLN 94 colour IFR GPS.

Description applies to the above version.

Millennium Edition: Special configuration limited edition, announced in February 2000 and available from April 2000. Features redesigned cabin interior with leather seats; floor mats embossed with Cessna Millennium logo; new cup- and chart-holders; Rosen sun visors; polished spinner and cowling fasteners; Goodyear Flight Custom II tyres; Honeywell KLN 94 IFR GPS; new exterior paint scheme; and a pilot's accessory collection.

CUSTOMERS: Total of 179 Stationairs and 365 Turbo Stationairs delivered by 30 June 2002, including 22 and 48 respectively in the first six months of 2002.

COSTS: Stationair US$336,900, Turbo Stationair US$381,200, both standard equipped (2002). NAV I

Cessna T182T Turbo Skylane (*Jane's/Paul Jackson*)   *NEW*/0533678

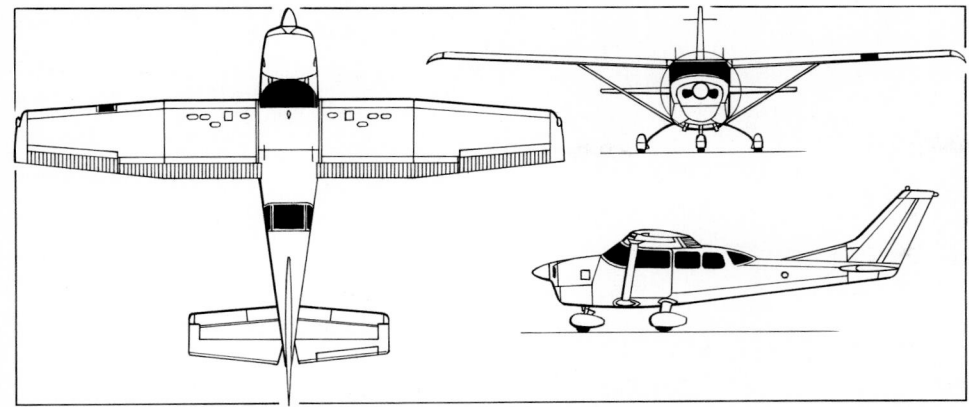

General arrangement of the Cessna 206H Stationair *(Jane's/James Goulding)*   0054031

aircraft US$355,000 and US$399,300, respectively; NAV II aircraft US$377,600 and US$421,900, respectively.

DESIGN FEATURES: Generally as for Cessna 172.
Wing section NACA 2412 (modified). Dihedral 2° 14'; incidence 1° 30' at root, −1° 30' at tip.

FLYING CONTROLS: As Cessna 182.

LANDING GEAR: As Cessna 182. Wipline 3450 floatplane conversion certified April 2000.

POWER PLANT: Stationair has one 224 kW (300 hp) Textron Lycoming IO-540-AC1A flat-six piston engine; Turbo Stationair has one 231 kW (310 hp) Textron Lycoming TIO-540-AJ1A turbocharged flat-six piston engine; each drives a three-blade, constant-speed McCauley metal propeller. Total fuel capacity 348 litres (92.0 US gallons; 76.6 Imp gallons), of which 333 litres (88.0 US gallons; 73.3 Imp gallons) are usable. Oil capacity 10.4 litres (2.75 US gallons; 2.3 Imp gallons).

ACCOMMODATION: Forward-hinged door, port side; double cargo doors starboard side, each opening away from centre. Six persons in three pairs on contoured, energy-absorbing seats, dynamically tested to 26 g, those for pilot and front seat passenger being vertically adjustable, reclining and mounted on enlarged seat rails with dual locking pins; other seats have reclining backs; inertia reel harnesses for all occupants; tinted windows; hinged front side windows; composites headliner; multilevel ventilation system; ambient noise reduction soundproofing and double-pin latching system on pilot's door, all standard. Utility interior optional.

AVIONICS: *Comms:* Bendix/King KX 155A nav/com/ glideslope, KMA 28 audio panel, KT 76C Mode C transponder and ELT.
Flight: Bendix/King KI 209 VOR/LOC/GS and KAP 140 two-axis autopilot. Optional NAV I package adds second KX 155A, second KI 209, plus KLN 94 moving map, KMD 550 MFD and MD41-231 GPS-NAV. NAV II adds Bendix/King KCS 55A HSI, Goodrich WX 500 Stormscope and replaces MD41-231 by MD41-233.

EQUIPMENT: Optional equipment includes electric propeller anti-icing, 0.45 m³ (16 cu ft) cargo pod, floatplane provisions kit and oversized rough terrain tyres.

DIMENSIONS, EXTERNAL:

Wing span	10.97 m (36 ft 0 in)
Wing chord: at root	1.63 m (5 ft 4 in)
at tip	1.13 m (3 ft 8½ in)
Wing aspect ratio	7.4
Length overall	8.61 m (28 ft 3 in)
Height overall	2.83 m (9 ft 3½ in)
Wheel track	2.46 m (8 ft 1 in)
Wheelbase	1.76 m (5 ft 9¼ in)
Pilot's door (port): Max height	1.04 m (3 ft 5 in)
Max width	0.94 m (3 ft 1 in)
Cargo double door (stbd):	
Max height	1.00 m (3 ft 3¼ in)
Max width	1.09 m (3 ft 7 in)
Height to sill	0.64 m (2 ft 1 in)

DIMENSIONS, INTERNAL:

Cabin: Length	3.68 m (12 ft 1 in)
Max width	1.12 m (3 ft 8 in)
Max height	1.26 m (4 ft 1½ in)
Volume available for payload	2.87 m³ (101.2 cu ft)

AREAS:

Wings, gross	16.30 m² (175.5 sq ft)
Ailerons (total)	1.61 m² (17.35 sq ft)
Trailing-edge flaps (total)	2.68 m² (28.85 sq ft)
Fin	1.08 m² (11.62 sq ft)
Rudder, incl tab	0.65 m² (6.95 sq ft)
Tailplane	2.31 m² (24.84 sq ft)
Elevators, incl tab	1.86 m² (20.08 sq ft)

WEIGHTS AND LOADINGS:

Weight empty, standard:	
206H	987 kg (2,176 lb)
T206H	1,034 kg (2,279 lb)
Max baggage	82 kg (180 lb)
Max fuel: 206H, T206H	239 kg (528 lb)
Max T-O and landing weight, 206H, T206H	
	1,632 kg (3,600 lb)
Max ramp weight: 206H	1,639 kg (3,614 lb)
T206H	1,641 kg (3,617 lb)
Max wing loading:	
206H, T206H	100.2 kg/m² (20.51 lb/sq ft)
Max power loading: 206H	7.30 kg/kW (12.00 lb/hp)
T206H	7.07 kg/kW (11.61 lb/hp)

PERFORMANCE:

Max level speed: 206H at S/L	
	151 kt (280 km/h; 174 mph)
T206H at 5,182 m (17,000 ft)	
	178 kt (330 km/h; 205 mph)
Cruising speed at 75% power:	
206H at 1,890 m (6,200 ft)	142 kt (263 km/h; 163 mph)
T206H at 6,100 m (20,000 ft)	
	164 kt (304 km/h; 189 mph)
Stalling speed:	
206H, T206H: flaps up	62 kt (115 km/h; 72 mph)
flaps down	54 kt (100 km/h; 63 mph)
Max rate of climb at S/L: 206H	301 m (988 ft)/min
T206H	320 m (1,050 ft)/min
Service ceiling: 206H	4,785 m (15,700 ft)
T206H	8,230 m (27,000 ft)
T-O run: 206H, T206H	277 m (910 ft)
T-O to 15 m (50 ft): 206H	567 m (1,860 ft)
T206H	530 m (1,740 ft)
Landing from 15 m (50 ft)	425 m (1,395 ft)
Landing run: 206H, T206H	224 m (735 ft)
Range with max fuel, 45 min reserves:	
206H: at 75% power at 1,890 m (6,200 ft)	
	605 n miles (1,120 km; 696 miles)
max range at 1,981 m (6,500 ft)	
	730 n miles (1,352 km; 840 miles)
T206H: at 75% power at 3,050 m (10,000 ft)	
	541 n miles (1,002 km; 622 miles)
at 75% power at 6,100 m (20,000 ft)	
	568 n miles (1,052 km; 653 miles)
max range at 6,100 m (20,000 ft)	
	692 n miles (1,281 km; 796 miles)

UPDATED

CESSNA 208 CARAVAN

Brazilian Air Force designation: C-98
US DoD designation: U-27A

TYPE: Light utility turboprop.

PROGRAMME: First flight of engineering prototype (N208LP) 9 December 1982; first production Caravan rolled out August 1984; FAA certification October 1984; full production started 1985; wheeled float version certified March 1986. First single-engined aircraft to achieve FAA certification for ILS in Cat. II conditions (Federal Express aircraft equipped); approval for IFR cargo operations 1989 made France and Ireland first European countries to allow single-engined public transport day/night IFR operation; also since approved in Canada, Denmark and Sweden. Russian certification of 208 and 208B achieved 4 September 1998.

The 1,000th of the 208B version was delivered on 12 December 2002 to Supap Puranitee of Thailand. 100th Caravan equipped with Wipaire 8000 floats delivered 16 May 2000. Total fleet operating time exceeded 6.1 million hours by December 2002, at which time Caravans were in service in 67 countries. Quoted dispatch reliability rate is 99.9 per cent. Basic utility model for passengers or cargo was Cessna 208. Engine flat rated at 447 kW (600 shp) to 3,800 m (12,500 ft). Federal Express Corporation version was 208A Cargomaster freighter with special features including T-O weight 3,629 kg (8,000 lb), Honeywell avionics, no cabin windows or starboard rear door, more cargo tiedowns, additional cargo net, underfuselage cargo pannier of composites materials, 15.2 cm (6 in) vertical extension of fin/rudder, jetpipe deflected to carry exhaust clear of pannier. These early versions now discontinued; military U-27A last described in 2001-02 edition.

CURRENT VERSIONS: **Caravan 675:** Combines airframe of 208 with fully rated engine of 208B. Announced at NBAA Convention in Dallas, Texas, September 1997; FAA certification achieved in April 1998 with first delivery (N900RG; c/n 00277) 15 April to Riversville Aviation Company of New York in amphibious configuration; replaces 208.

Detailed description applies to Caravan 675, except where indicated.

208B: Stretched version, lacking cabin windows, and with ventral cargo pod as standard, developed at request of Federal Express. Commissioned by FedEx as **Super Cargomaster;** first flight (N9767F) 3 March 1986; certified October 1986; first delivery to FedEx 31 October 1986. Features include 1.22 m (4 ft) fuselage plug aft of wing, payload of 1,587 kg (3,500 lb) and 12.7 m³ (450 cu ft) of cargo volume; 503 kW (675 shp) P&WC PT6A-114A from 1991.

Grand Caravan: Announced at NBAA Convention 1990; passenger version of 208B, with cabin windows, accommodating up to 14 in quick-change interior.

Soloy Pathfinder 21: Dual-engine conversion; described in *Jane's Aircraft Upgrades.*

CUSTOMERS: Federal Express Corporation received 40 208As and 260 208Bs. Recent customers include Ben Air Inc of Stauning, Denmark, which will take delivery of five Grand Caravans between 2000-2004 for onward sale to Scandinavian customers; Shandong Airlines of the People's Republic of China, which ordered one Grand Caravan and two Caravan 675s on amphibious floats for delivery in first quarter 2001, plus 37 options, of which two were exercised in 2002; China Northern Airlines of Shenyang, which took delivery of three Grand Caravans in 2002; Van Nuys Flight Center of California, which ordered 10 Grand Caravans for delivery between September 2001 and September 2005, four delivered to undisclosed

Float-equipped Cessna 206H Stationair Millennium Edition *(Jane's/Paul Jackson)*   0121054

For details of the latest updates to *Jane's All the World's Aircraft* online and to discover the additional information available exclusively to online subscribers please visit
jawa.janes.com

Cessna Grand Caravan Amphibian on Wipline floats (*Jane's/Paul Jackson*)   NEW/0533679

Russian customers in the first half of 2002, and four delivered to three cargo operators in Spain in the first half of 2002. Other customers include Royal Canadian Mounted Police (first amphibious version), Brazilian Air Force (eight), Chilean Army (three delivered from 7 February 1998 onwards), Liberian Army (one) and Malaysian Police (six delivered in 1994).

Total of 1,358 Caravans (all versions) delivered by 30 June 2002, including 102 in 1998; 87 in 1999, 92 in 2000, 75 in 2001 and 37 in the first six months of 2002.

COSTS: Caravan 675 US$1,465,000 (amphibian US$1,650,000); Super Cargomaster US$1,335,000; Grand Caravan US$1,485,000 (all 2002, typically equipped).

DESIGN FEATURES: Launched as first all-new single-engined turboprop general aviation aircraft; intention was to replace de Havilland Canada Beavers and Otters, Cessna 180s, 185s and 206s in worldwide utility role. Main qualities advertised are high speed with heavy load, compatibility with unprepared strips, economy and reliability with minimum maintenance; can also carry weather radar, air conditioning and oxygen systems; optional packs for firefighting, photography, spraying, ambulance/hearse, border patrol, parachuting and supply dropping, surveillance and government utility missions; optional wheel or float landing gear.

Braced, high-wing design; tapered wings and tailplane; sweptback fin with dorsal fillet; auxiliary fins on floatplane.

Wing aerofoil NACA 23017.424 at root, 23012 at tip; dihedral 3° from root; incidence 2° 37′ at root, −0° 36′ at tip.

FLYING CONTROLS: Conventional and manual. Lateral control by small ailerons and slot-lip spoilers ahead of outer section of flaps; aileron trim standard; all tail control surfaces horn balanced; fixed tailplane with upper surface vortex generators ahead of elevator; elevator trim tabs; electrically actuated single-slotted flaps occupy more than 70 per cent of trailing-edge and deflect to maximum 30°.

STRUCTURE: All-metal. Fail-safe two-spar wing; conventional fuselage.

LANDING GEAR: Non-retractable tricycle type, with single wheel on each unit. Tubular spring cantilever main units; oil-damped steerable nosewheel. Mainwheel tyres size 6.50-10 (8 ply); nosewheel 6.50-8 (8 ply). Oversize tyres, optional, mainwheels 8.50-10, nosewheel 22×8.00-8, and extended nosewheel fork, optional. Hydraulically actuated single-disc brake on each mainwheel. Certified with Wipline 8000 floats (with or without retractable land wheels).

POWER PLANT: One 503 kW (675 shp) P&WC PT6A-114A turboprop. McCauley three-blade constant-speed reversible-pitch and feathering metal propeller. Integral fuel tanks in wings, total capacity 1,268 litres (335 US gallons; 279 Imp gallons), of which 1,257 litres (332 US gallons; 276.5 Imp gallons) are usable.

ACCOMMODATION: Pilot and up to nine passengers or 1,360 kg (3,000 lb) of cargo. Maximum seating capacity with FAR Pt 23 waiver is 14. Cabin has a flat floor with cargo track attachments for a combination of two- and three-abreast seating, with an aisle between seats. Forward-hinged door for pilot, with direct vision window, on each side of forward fuselage. Airstair door for passengers at rear of cabin on starboard side. Cabin is heated and ventilated. Optional air conditioning. Two-section horizontally split cargo door at rear of cabin on port side, flush with floor at bottom and with square corners. Upper portion hinges upward, lower portion forward 180°. Optional electrically operated, flight openable tambour roll-up door with airflow deflecting spoiler. In a cargo role, cabin will accommodate, typically, two D-size cargo containers or up to ten 208 litre (55 US gallon; 45.8 Imp gallon) drums.

Upgraded interior with new colour-co-ordinated fabrics and leather, and new seat design introduced in February 2001.

SYSTEMS: Electrical system is powered by 28 V 200 A starter/generator and 24 V 45 Ah lead-acid battery (24 V 40 Ah Ni/Cd battery optional). Standby electrical system, with 75 A alternator, optional. Hydraulic system for brakes

only. Oxygen system, capacity 3,315 litres (117 cu ft), optional. Vacuum system standard. Cabin air conditioning system optional from c/n 208-00030 onwards. De-icing system, comprising electric propeller de-icing boots, pneumatic wing, wing strut and tail surface boots, electrically heated windscreen panel, heated pitot/static probe, ice detector light and standby electrical system, all optional.

AVIONICS: Customers' choice of Bendix/King and Garmin packages introduced in 2001; typical equipment listed below.

Comms: Dual Bendix/King KX 165 nav/com/glideslope, KT70 Mode S transponder, KHF 950 HF transceiver and KMA 24H audio control.

Radar: Honeywell RDR 200 weather radar with vertical profile.

Flight: Bendix/King KFC 150 flight director/autopilot, KLN 89B GPS, dual KR 87 ADF, KN 63 DME and KR 21 MKR. Garmin alternatives include GNS 340 com/nav/GPS; GNS 530 WAAS-upgradable IFR com/nav/GPS, with LOC and glideslope receiver, GMA 240 or GMA 340 audio panel and GTX 327 Mode C digital transponder with LCD display. Bendix/King KDR 510 flight information service and KMH 880 multi-hazard awareness system optional from 2003.

Instrumentation: Sensitive altimeter, electric clock, magnetic compass, attitude and directional gyros, true

airspeed indicator, turn and bank indicator, vertical speed indicator, ammeter/voltmeter, fuel flow indicator, ITT indicator, oil pressure and temperature indicator.

EQUIPMENT: Standard equipment includes windscreen defrost, ground service plug receptacle, variable intensity instrument post lighting, map light, overhead courtesy lights (three) and overhead floodlights (pilot and co-pilot), approach plate holder, cargo tiedowns, internal corrosion proofing, vinyl floor covering, emergency locator beacon, partial plumbing for oxygen system, pilot's and co-pilot's adjustable fore/aft/vertical/reclining seats with armrest and five-point restraint harness, tinted windows, control surface bonding straps, heated pitot and stall warning systems, retractable crew steps (port side), rudder gust lock, tiedowns and towbar.

Optional equipment includes passenger seats, stowable folding utility seats, digital clock, fuel totaliser, turn co-ordinator, flight hour recorder, fire extinguisher, dual controls, co-pilot flight instruments, floatplane kit (from c/n 208-00030 onwards), hoisting rings (for floatplane), inboard fuel filling provisions (included in floatplane kit), ice detection light, courtesy lights on wing underside, passenger reading lights, flashing beacon, retractable crew step for starboard side, oversized tyres, electric trim system, oil quick drain valve and fan-driven ventilation system.

DIMENSIONS, EXTERNAL (L: landplane, A: amphibian):

Wing span	15.88 m (52 ft 1 in)
Wing chord: at root	1.98 m (6 ft 6 in)
at tip	1.22 m (4 ft 0 in)
Wing aspect ratio	9.7
Length overall: L	11.46 m (37 ft 7 in)
Height overall: L	4.52 m (14 ft 10 in)
A (on land)	4.98 m (16 ft 4 in)
Tailplane span	6.25 m (20 ft 6 in)
Wheel track: L	3.56 m (11 ft 8 in)
A	3.25 m (10 ft 8 in)
Wheelbase: L	3.54 m (11 ft 7½ in)
A	4.44 m (14 ft 7 in)
Propeller diameter	2.69 m (8 ft 10 in)
Airstair door: Height	1.27 m (4 ft 2 in)
Width	0.61 m (2 ft 0 in)
Cargo door: Height	1.27 m (4 ft 2 in)
Width	1.24 m (4 ft 1 in)

DIMENSIONS, INTERNAL (208):

Cabin: Length, excl baggage area	4.57 m (15 ft 0 in)
Max width	1.57 m (5 ft 2 in)
Max height	1.30 m (4 ft 3 in)
Volume	9.7 m³ (341 cu ft)
Cargo pannier: Length	5.58 m (18 ft 3½ in)

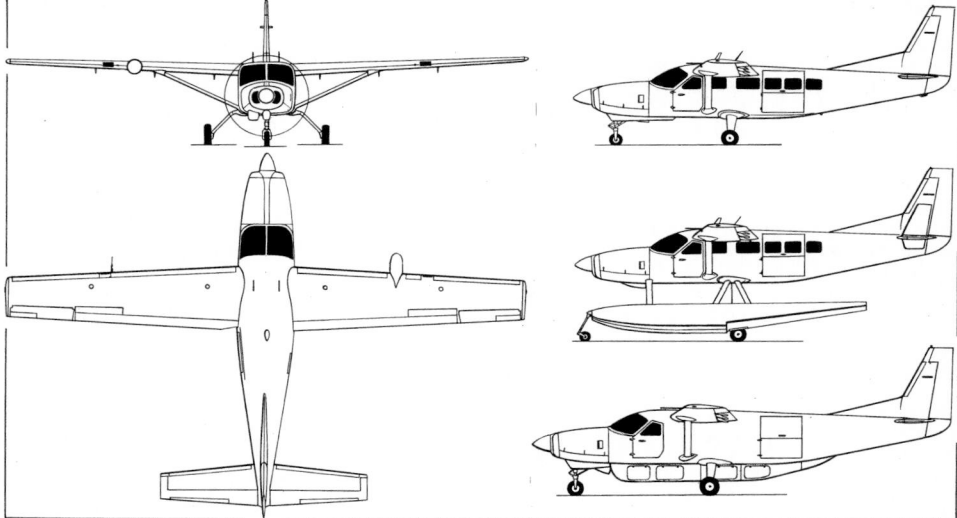

Cessna Caravan 675 with side views, top to bottom, of 208 Caravan 675, Caravan Amphibian and 208B Super Cargomaster (*Jane's/Dennis Punnett*)

Cessna Grand Caravan fitted with baggage pannier (*Jane's/Paul Jackson*)   NEW/0533680

Width	1.28 m (4 ft 2½ in)
Depth	0.50 m (1 ft 7½ in)
Volume	2.4 m³ (83 cu ft)

AREAS:

Wings, gross	25.96 m² (279.4 sq ft)
Vertical tail surfaces (total, incl dorsal fin)	
	3.57 m² (38.41 sq ft)
Horizontal tail surfaces (total)	6.51 m² (70.04 sq ft)

WEIGHTS AND LOADINGS (A: 208-675 Caravan; B: Super Cargomaster; C: Grand Caravan; D: Caravan Amphibian):

Weight empty: A	1,780 kg (3,925 lb)
D	2,220 kg (4,895 lb)
Baggage capacity	147 kg (325 lb)
Cargo pannier capacity	372 kg (820 lb)
Max fuel	1,009 kg (2,224 lb)
Max ramp weight: A, D	3,645 kg (8,035 lb)
B, C	3,985 kg (8,785 lb)
Max T-O weight: A, D	3,629 kg (8,000 lb)
B, C	3,969 kg (8,750 lb)
Max landing weight: A, D	3,538 kg (7,800 lb)
B, C	3,855 kg (8,500 lb)
Max wing loading: A, D	139.8 kg/m² (28.63 lb/sq ft)
B, C	152.9 kg/m² (31.32 lb/sq ft)
Max power loading: A, D	7.21 kg/kW (11.85 lb/shp)
B, C	7.89 kg/kW (12.96 lb/shp)

PERFORMANCE (A: 208-675 landplane; D: Caravan Amphibian):

Max operating speed (VMO)	
	175 kt (325 km/h; 202 mph) IAS
Max cruising speed at 3,050 m (10,000 ft):	
A	186 kt (344 km/h; 214 mph)
D	163 kt (302 km/h; 188 mph)
Stalling speed, power off:	
Flaps up: A	75 kt (139 km/h; 87 mph) CAS
D	74 kt (137 km/h; 86 mph) CAS
Flaps down: A	61 kt (113 km/h; 71 mph) CAS
D	59 kt (110 km/h; 68 mph) CAS
F, A, landing configuration	
	59 kt (109 km/h; 68 mph) CAS
Max rate of climb at S/L: A	376 m (1,234 ft)/min
D	338 m (1,110 ft)/min
Max certified altitude: A	7,620 m (25,000 ft)
D	6,100 m (20,000 ft)
T-O run: A	354 m (1,160 ft)
D: on land	335 m (1,100 ft)
on water	585 m (1,920 ft)
T-O to 15 m (50 ft): A	626 m (2,053 ft)
Landing from 15 m (50 ft) at S/L, without propeller reversal: A	505 m (1,655 ft)
D: on land	454 m (1,490 ft)
on water	590 m (1,935 ft)
Landing run at S/L, without propeller reversal:	
A	227 m (745 ft)
D: on land	224 m (735 ft)
on water	319 m (1,045 ft)

Range with max fuel, at max cruise power, allowances for start, taxi and reserves stated:

A at 3,050 m (10,000 ft), 45 min	
	932 n miles (1,726 km; 1,072 miles)
A at 6,100 m (20,000 ft), 45 min	
	1,220 n miles (2,259 km; 1,404 miles)
D at 3,050 m (10,000 ft), 30 min	
	990 n miles (1,833 km; 1,139 miles)

Range with max fuel at max range power, allowances as above: A at 3,050 m (10,000 ft)

	1,085 n miles (2,009 km; 1,248 miles)
A at 6,100 m (20,000 ft)	
	1,295 n miles (2,398 km; 1,490 miles)
g limits	+3.8/−1.52

PERFORMANCE (B: Super Cargomaster; C: Grand Caravan):

Max cruising speed:	
at 3,050 m (10,000 ft): B	175 kt (324 km/h; 201 mph)
C	184 kt (341 km/h; 212 mph)
Stalling speed, power off:	
flaps up: B, C	78 kt (145 km/h; 90 mph)
flaps down: B, C	61 kt (113 km/h; 71 mph)
Max rate of climb at S/L: B	282 m (925 ft)/min
C	297 m (975 ft)/min
Service ceiling: B	6,950 m (22,800 ft)
C	7,220 m (23,700 ft)

T-O run: B	428 m (1,405 ft)
C	416 m (1,365 ft)
T-O to 15 m (50 ft): B	762 m (2,500 ft)
C	738 m (2,420 ft)
Landing from 15 m (50 ft), without propeller reversal:	
B	530 m (1,740 ft)
C	547 m (1,795 ft)
Landing run, without propeller reversal:	
B	279 m (915 ft)
C	290 m (950 ft)

Range with max fuel, max cruise power, allowances for start, taxi, climb and descent, plus 45 min reserves: at 3,050 m (10,000 ft):

B	862 n miles (1,596 km; 992 miles)
C	907 n miles (1,679 km; 1,043 miles)
at 5,490 m (18,000 ft), conditions as above:	
B	1,044 n miles (1,933 km; 1,201 miles)
C	1,109 n miles (2,053 km; 1,276 miles)

Range with max fuel, max range power, allowances for start, taxi, climb and descent, plus 45 min reserves: at 3,050 m (10,000 ft):

B	963 n miles (1,783 km; 1,108 miles)
C	1,026 n miles (1,900 km; 1,180 miles)
at 5,490 m (18,000 ft), conditions as above:	
B	1,076 n miles (1,992 km; 1,238 miles)
C	1,163 n miles (2,153 km; 1,338 miles)

UPDATED

CESSNA CITATION MUSTANG

TYPE: Light business jet.

PROGRAMME: Initial design studies, then as turboprop, began in 1996-97; twin-jet configuration chosen in early 2001; announced at NBAA Convention, Orlando, Florida, 10 September 2002, when cabin mockup unveiled. New design but with wing scaled down from Citation Sovereign; smallest of Cessna jet family. First flight anticipated in May 2005; certification to FAR Pt 23 (day, night, VFR, IFR, single pilot, known icing and RVSM) expected by end of 2006. Target production rate 150 to 200 per year.

CUSTOMERS: More than 200 ordered by 12 September 2002. Announced customers include three-times Formula One World Drivers' Champion Nelson Piquet, who ordered one on 11 September 2002 and Delta Aero-Taxi of Florence, Italy, which ordered seven on 21 October 2002 for its fractional ownership programme.

COSTS: US$2.295 million (2002).

DESIGN FEATURES: Low-wing, T tail and twin, podded, rear-mounted turbofans in usual Cessna business jet configuration, but second type to employ new Sovereign wing planform.

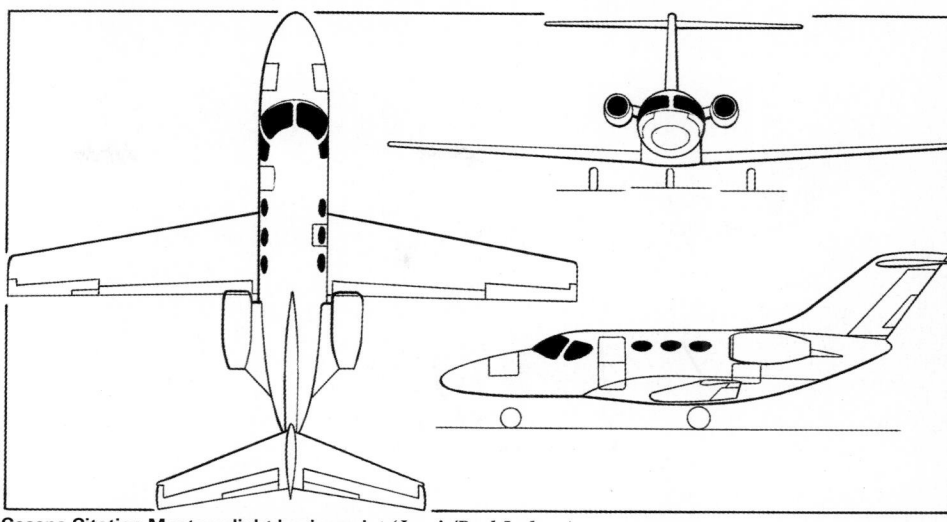

Cessna Citation Mustang light business jet *(Jane's/Paul Jackson)*   *NEW*/0526901

FLYING CONTROLS: Conventional and manual. Ailerons and elevator horn-balanced. Flight-adjustable tabs in port aileron, rudder and both elevators. Slotted flaps.

STRUCTURE: Generally of aluminium alloy; riveted and bonded joints, as required. Three-spar wing; centre-section, including fuel tank, mounted below fuselage to avoid penetration of pressure vessel. Semi-monocoque fuselage of frames and stringers with sheet metal skin; forward and aft pressure bulkheads.

LANDING GEAR: Nosewheel type; retractable. Single wheel on each leg. Trailing-link mainwheels retract inwards, nosewheel forwards; hydraulic actuation. Manually steerable nosewheel. Disc brakes on mainwheels.

POWER PLANT: Two pod-mounted Pratt & Whitney Canada PW615F turbofans, each rated at 6.00 kN (1,350 lb st) at ISA+15°C, with dual-channel FADEC.

ACCOMMODATION: Two pilots on flight deck; four passengers in club configuration in cabin. Door port side, ahead of wing; emergency exit, starboard, over wing. Chemical lavatory opposite door. Recessed central aisle; storage and refreshments cabinets behind each pilot's seat, facing cabin. Nose and aft baggage compartments both unpressurised, with external, port-side access only. Cabin pressuirised to 0.57 bar (8.3 lb/sq in), giving 2,440 m (8,000 ft) environment at 12,500 m (41,000 ft). Air conditioning vents in cabin sidewalls.

SYSTEMS: Electrical system supplied by two starter/generators via left and right busses; separate emergency bus. Hydraulic system, icing protection for wing, tailplane, windscreen and engine air intakes; certified for flight into known icing. Three-axis autopilot and yaw damper.

AVIONICS: *Comms:* Dual VHF, audio control, Mode S transponder.

Flight: Dual NAV/LOC/GS; GPS; marker; AHRS; air data computer. Traffic and weather awareness.

Instrumentation: Active matrix LCDs.

DIMENSIONS, EXTERNAL:

Wing span	12.88 m (42 ft 3 in)
Length overall	11.86 m (38 ft 11 in)
Height overall	4.19 m (13 ft 9 in)

DIMENSIONS, INTERNAL:

Cabin: Length between pressure bulkheads	
	4.37 m (14 ft 4 in)
Max width	1.40 m (4 ft 7 in)
Max height	1.37 m (4 ft 6 in)
Door width	0.61 m (2 ft 0 in)

WEIGHTS AND LOADINGS:

Basic operating weight	2,336 kg (5,150 lb)
Max fuel	1,170 kg (2,580 lb)
Payload with max fuel, one pilot	272 kg (600 lb)

PERFORMANCE (estimated):

Cruising speed at FL350	340 kt (630 km/h; 391 mph)
T-O balanced field length	950 m (3,120 ft)
Certified ceiling	12,500 m (41,000 ft)
Range, 45 min reserves	
	1,300 n miles (2,407 km; 1,496 miles)

NEW ENTRY

REIMS-CESSNA F406/CARAVAN II

See Reims Aviation in French section.

CESSNA 525 CITATION CJ1

TYPE: Light business jet.

PROGRAMME: Original CitationJet announced at NBAA Convention 1989 as replacement for Citation 500 and I (production of which stopped 1985); first flight of FJ44 turbofans in Citation 500 April 1990; first flight of CitationJet (N525CJ) 29 April 1991; first flight of second (preproduction) prototype 20 November 1991; FAA certification for single-pilot operation received 16 October 1992; first customer delivery 31 March 1993. Russian

Cessna 208B Super Cargomaster used for carrying light freight between the Canary Islands *(Jane's/Paul Jackson)*   *NEW*/0533681

Cessna CJ1 six/seven-seat light business jet *(Jane's/Paul Jackson)*   NEW/0526961

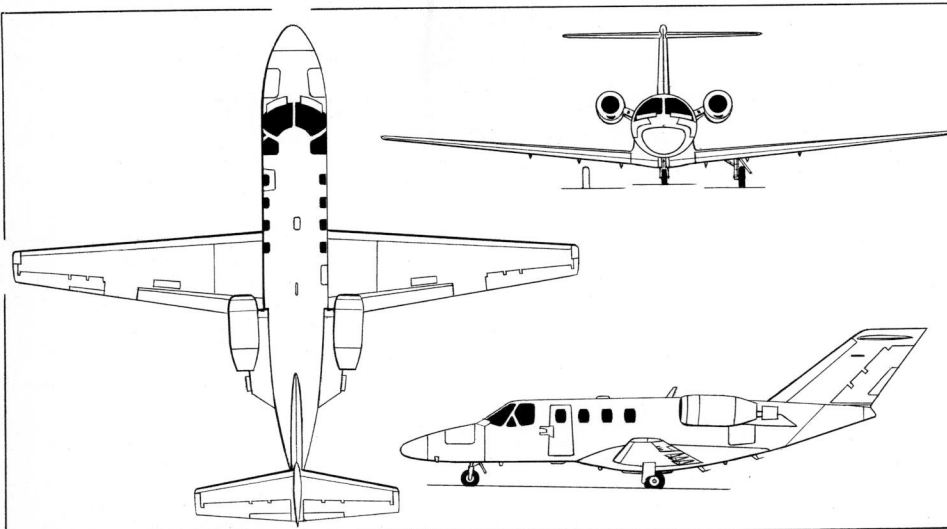

Cessna 525 Citation CJ1 (two Williams FJ44 turbofans) *(Jane's/Dennis Punnett)*

Crew/passenger door: Height	1.29 m (4 ft 2¾ in)
Width	0.60 m (1 ft 11½ in)

DIMENSIONS, INTERNAL:
Cabin: Length between pressure bulkheads

	4.85 m (15 ft 11 in)
Max width	1.47 m (4 ft 10 in)
Max height	1.45 m (4 ft 9 in)

Baggage compartment volume:

nose	0.69 m³ (24.4 cu ft)
cabin	0.11 m³ (4.0 cu ft)
tailcone	0.86 m³ (30.2 cu ft)
total	1.66 m³ (58.6 cu ft)

AREAS:

Wings, gross	22.30 m² (240.0 sq ft)
Vertical tail surfaces (total, incl tab)	4.35 m² (46.8 sq ft)
Horizontal tail surfaces (total, incl tabs)	
	5.64 m² (60.7 sq ft)

WEIGHTS AND LOADINGS:

Weight empty, typically equipped	3,016 kg (6,650 lb)
Max usable fuel weight	1,460 kg (3,220 lb)
Payload with max fuel, single pilot	306 kg (675 lb)
Max T-O weight	4,808 kg (10,600 lb)
Max landing weight	4,445 kg (9,800 lb)
Max ramp weight	4,853 kg (10,700 lb)
Max wing loading	215.6 kg/m² (44.17 lb/sq ft)
Max power loading	284 kg/kN (2.79 lb/lb st)

PERFORMANCE:

Max operating speed (V_{MO}):	
S/L to FL305	263 kt (487 km/h; 302 mph) IAS
above FL305	M0.70
Max cruising speed at 3,992 kg (8,800 lb) AUW at	
FL350	381 kt (706 km/h; 438 mph)
Stalling speed, landing configuration at max landing	
weight	82 kt (152 km/h; 94 mph) CAS
Max rate of climb at S/L	985 m (3,230 ft)/min
Rate of climb at S/L, OEI	259 m (850 ft)/min
Max certified altitude	12,500 m (41,000 ft)
Service ceiling, OEI	6,462 m (21,200 ft)
FAR Pt 25 T-O field length	1,000 m (3,280 ft)
Landing run	841 m (2,760 ft)
IFR range with 100 n mile (185 km; 115 mile) alternate	
and 45 min reserves	
	1,247 n miles (2,309 km; 1,435 miles)

OPERATIONAL NOISE LEVELS:

T-O	73.6 EPNdB
Approach	89.7 EPNdB
Sideline	83.6 EPNdB

UPDATED

certification achieved 4 September 1998. RVSM Group approval granted by FAA in January 2000.

Replacement Citation CJ1 announced at NBAA Convention 18 October 1998; first aircraft (N31CJ) completed late 1999; FAA certification 16 February 2000; first customer delivery 31 March 2000 to the Commercial Envelope Company of Deer Park, New York; 500th Cessna 525 series, a CJ1, rolled out 7 June 2002. By June 2002 a total of 493 Model 525s were in service, at which time their flight time totalled more than 550,000 hours.

CUSTOMERS: Total 359 CitationJets, plus three prototypes, delivered up to early 2000, when manufacture turned to CJ1. Total of 159 CJ1s delivered by 31 December 2002, including 30 delivered in 2002. Recent customers include Avemex SA of Toluca, Mexico (one); Taxi Marilia (TAM) of Brazil (three, including c/n 0500); Atlas Air Service GmbH of Germany (four), and Cessna's TAG Aviation USA's fractional ownership operation, CitationShares, which has received one, and the Chilean Air Force, which took delivery of three in November 2001.

COSTS: US$3.986 million (2002).

DESIGN FEATURES: Small business jet of conventional appearance; wings attached below fuselage; T tail; tapered wings and tailplane; podded engines mounted clear of upper rear fuselage. Natural Laminar Flow (NLF) aerofoil section. CJ1 is generally identical to CitationJet; increased maximum T-O, ramp and landing weights to provide improved range/payload; Pro Line 21 avionics.

FLYING CONTROLS: Conventional and manual. Ailerons and elevators horn balanced; trim tab on port aileron; trim tabs on rudder and both elevators; hydraulically actuated single-slotted flaps, deflections 15, 35 and 60°, last-named for ground use only, and selectable to any intermediate position between 0 and 35°. Speed brake in upper and lower surface of each wing.

STRUCTURE: All-metal. Three-spar wing.

LANDING GEAR: Hydraulically retractable tricycle type, with single wheel on each unit. Trailing-link main units retract inward into wing; nose gear retracts forward. Main tyres 22×7.75-10 (8 ply) tubeless; nose 18×4.4 (6 ply) tubeless.

POWER PLANT: Two 8.45 kN (1,900 lb st) Williams FJ44-1A turbofans pod-mounted, with thrust attenuators. Two independent fuel systems, with fuel contained in integral wing tanks, and single filler port in each wing; see under Weights and Loadings.

ACCOMMODATION: Crew of two on flight deck. Main cabin with standard seating for five passengers, one on sideways-facing seat at front of cabin, with four in club arrangement; tray tables which stow in elbow rails, and individual reading lights and ventilation ducts, standard.

SYSTEMS: Digitally controlled pressurisation system, maximum differential 0.58 bar (8.5 lb/sq in).

AVIONICS: Rockwell Collins Pro Line 21 as core system.

Comms: Honeywell CN1500 radios and radio altimeter.

Radar: Rockwell Collins RTA 800 colour weather radar.

Flight: Honeywell KLN-900 FMS. Dual VHF nav receivers, ADF, DME, RMI, Honeywell KLN-90B GPS.

Instrumentation: Two-tube EFIS with 203 × 253 mm (8 × 10 in) adaptive flight displays (third optional), LCD attitude director indicator (ADI) and horizontal situation indicator (HSI) for co-pilot.

DIMENSIONS, EXTERNAL:

Wing span	14.26 m (46 ft 9½ in)
Wing aspect ratio	9.1
Length overall	12.98 m (42 ft 7 in)
Height overall	4.20 m (13 ft 9¼ in)
Tailplane span	5.73 m (18 ft 9½ in)
Wheel track	3.96 m (13 ft 0 in)
Wheelbase	4.67 m (15 ft 4 in)

CESSNA 525A CITATION CJ2

TYPE: Light business jet.

PROGRAMME: Design began 1 May 1998; construction of prototype started August 1998; announced at NBAA Convention at Las Vegas, Nevada, 18 October 1998; first flight of prototype (N2CJ, a rebuilt CitationJet) 27 April 1999, followed by first preproduction aircraft (N525AZ) on 15 October and second (N765CT) in mid-December 1999; total of 1,008 flight test hours accumulated by three aircraft by 9 April 2000; first production aircraft c/n 525A-0003/(N132CJ) rolled out 14 January 2000; FAA FAR Pt 23 certification achieved 21 June 2000; public debut at EAA AirVenture Oshkosh in July 2000; first customer delivery, November 2000, was N200KP to King Pharmaceuticals of Tennessee (second production aircraft). By June 2002 the 75 CJ2s then in service with operators in seven countries had logged a total of 15,500 flight hours.

CUSTOMERS: Total of 76 firm orders held at time of launch; by December 2000 order backlog accounted for planned production until first quarter of 2004. Recent customers include WD40 Something LLC (second preproduction aircraft, re-registered N400WD), Avemex SA of Toluca, Mexico (one); Taxi Aereo Marilia (TAM) of Brazil (five); and Atlas Air Service GmbH of Germany (three). First delivery to overseas customer was D-IBBB, which arrived at Bremen for Atlas on 21 June 2001. 100th production CJ2 (N170TM) delivered to TitleMax Aviation Inc of Savannah, Georgia, on 27 August 2002. Total of 86 delivered in 2002.

COSTS: US$4.879 million (2002).

DESIGN FEATURES: Development of CitationJet/CJ1 (which see) with fuselage stretched by 1.30 m (4 ft 3 in) in cabin

Flight deck of Cessna CitationJet   0015668

The 97th production Cessna 525A Citation CJ2 *(Jane's/Paul Jackson)*   NEW/0526962

Standard six-seat cabin of Cessna 525A Citation CJ2   0054034

Flight deck of Cessna Citation CJ2   0075948

and tailcone areas (extra two cabin windows), wing span increased by 0.84 m (2 ft 9 in), swept tailplane of greater span, Williams FJ44-2C turbofans and Rockwell Collins Pro Line 21 avionics. Design goals included improvements in cabin room and comfort, speed and range over CitationJet. Common type rating with CJ1.

Description for the Citation CJ1 applies also to the Citation CJ2, except as follows:

FLYING CONTROLS: Spoilers, 60° flap deflection and engine nacelle thrust attenuators provide lift dump function during landing roll.

STRUCTURE: Primarily metal, with composites in non-critical areas such as fairings, wing and tailplane tips and exhaust nozzles. Three-spar wing.

POWER PLANT: Two pod-mounted Williams FJ44-2C turbofans, each flat rated at 10.68 kN (2,400 lb st) at 22°C (72°F), with thrust attenuators. Integral fuel tank in each wing, total capacity 2,245 litres (593 US gallons; 494 Imp gallons); overwing gravity refuelling.

ACCOMMODATION: One or two crew on flight deck; six passengers in standard centre club layout with refreshment centre on starboard side; customised interiors to customer's choice. Flight accessible baggage area at rear of cabin; other baggage compartments in nose and tailcone. Cabin is pressurised, heated and air conditioned. Six cabin windows per side. Main door on port side forward of wing; emergency exit starboard aft.

SYSTEMS: Pressurisation system, maximum pressure differential 0.61 bar (8.9 lb/sq in). Open-centre hydraulic system for operation of landing gear, flaps, speedbrakes and thrust attenuators. Vapour cycle air conditioning system. Electrical system supplied by battery and two engine-driven starter/generators. Oxygen system, capacity 623 litres (22.0 cu ft) standard, 1,417 litres (50.0 cu ft) optional.

AVIONICS: Rockwell Collins Pro Line 21 suite as core system.
Comms: Honeywell CNI-5000 panel-mounted radios.
Radar: Rockwell Collins RTA-800 solid-state colour weather radar.
Flight: Honeywell KLN 900 GPS FMS, VOR, DME, ADF, Rockwell Collins digital autopilot, digital air data computer and AHRS standard.
Instrumentation: Two-tube EFIS with 203 × 254 mm (8 × 10 in) PFD and MFD active matrix LCDs.

DIMENSIONS, EXTERNAL:
Wing span	15.09 m (49 ft 6 in)
Wing aspect ratio	9.3
Length overall	14.30 m (46 ft 11 in)
Height overall	4.24 m (13 ft 10¾ in)
Tailplane span	6.34 m (20 ft 9½ in)
Wheel track	4.88 m (16 ft 0 in)
Wheelbase	5.59 m (18 ft 4 in)
Passenger door: Height	1.29 m (4 ft 2¾ in)
Width	0.60 m (1 ft 11½ in)

DIMENSIONS, INTERNAL:
Cabin: Length: overall	5.77 m (18 ft 11 in)
excl cockpit	4.19 m (13 ft 9 in)
Max width	1.47 m (4 ft 10 in)
Max height	1.45 m (4 ft 9 in)
Baggage volume: nose	0.58 m³ (20.4 cu ft)
cabin	0.11 m³ (4.0 cu ft)
tailcone	1.42 m³ (50.0 cu ft)
total	2.11 m³ (74.4 cu ft)

AREAS:
Wings, gross	24.53 m² (264.0 sq ft)
Vertical tail surfaces	4.35 m² (46.8 sq ft)
Horizontal tail surfaces	6.50 m² (70.0 sq ft)

WEIGHTS AND LOADINGS:
Weight empty, typically equipped	3,402 kg (7,500 lb)
Max fuel	1,783 kg (3,930 lb)
Payload with max fuel	363 kg (800 lb)
Baggage capacity: nose	181 kg (400 lb)
cabin	45 kg (100 lb)

tailcone	272 kg (600 lb)
total	499 kg (1,100 lb)
Max T-O weight	5,613 kg (12,375 lb)
Max landing weight	5,216 kg (11,500 lb)
Max ramp weight	5,670 kg (12,500 lb)
Max zero-fuel weight	4,218 kg (9,300 lb)
Max wing loading	228.9 kg/m² (46.88 lb/sq ft)
Max power loading	263 kg/kN (2.58 lb/lb st)

PERFORMANCE (estimated):
Max operating Mach No. (MMO)	0.72
Max level speed (VMO):	
S/L to FL80	260 kt (481 km/h; 299 mph)
S/L to FL293	275 kt (509 km/h; 316 mph)
above FL293	M0.72
Max cruising speed at FL330	410 kt (759 km/h; 472 mph)
Econ cruising speed at FL410	332 kt (615 km/h; 382 mph)
Stalling speed in landing configuration at MLW	86 kt (160 km/h; 99 mph)
Max rate of climb at S/L	1,180 m (3,870 ft)/min
Rate of climb at S/L, OEI	360 m (1,180 ft)/min
Time to: FL370	17 min
FL430	36 min
Max certified altitude	13,715 m (45,000 ft)
FAR Pt 25 T-O balanced field length	1,042 m (3,420 ft)
Landing field length at max landing weight	908 m (2,980 ft)
Range, 100 n mile (185 km; 115 mile) alternate and 45 min reserves	1,546 n miles (2,863 km; 1,779 miles)

OPERATIONAL NOISE LEVELS:
T-O	74.5 EPNdB
Approach	91.4 EPNdB
Sideline	88.8 EPNdB

UPDATED

CESSNA 525B CITATION CJ3

TYPE: Light business jet.

PROGRAMME: Announced on eve of NBAA Convention, Orlando, Florida, 9 September 2002; fuselage mockup shown following day. First flight scheduled for second quarter of 2003; certification second quarter of 2004; deliveries from third quarter of 2004.

DESIGN FEATURES: Further development of CJ1/CJ2 family. Fuselage stretched by 1.00 m (3 ft 3½ in) (seven windows each side), giving 0.61 m (2 ft 0 in) extra in cabin, and wing span extended by 1.03 m (3 ft 4¼ in). Uprated engines with 14 per cent more T-O thrust and 12 per cent more in cruise. Improved range and passenger comfort.

Description for Citation CJ1/CJ2 applies to CJ3, except as follows:

CUSTOMERS: Orders for more than 120 held by mid-September 2002.

COSTS: US$5,795,000 (2002).

DESIGN FEATURES: Wing dihedral 5° 0′; sweepback 0° at 31 per cent chord; taper ratio 0.30. Tailplane sweepback 20° 0′ at 25 per cent chord; no dihedral; taper ratio 0.43. Fin sweepback 49° 0′ at 25 per cent chord; taper ratio 0.56.

FLYING CONTROLS: Flaps fitted (deflection 55° on ground).

POWER PLANT: Two 12.37 kN (2,780 lb st) Williams FJ44-3A turbofans with FADEC.

ACCOMMODATION: As CJ2, but chemical lavatory starboard, rear, opposite baggage area.

SYSTEMS: Hydraulic system with two engine-driven pumps, pressure 103.5 bar (1,500 lb/sq in). Secondary hydraulic system for mainwheel brakes. Cabin pressurisation as CJ2. Emergency oxygen system reservoir 1,134 litres (40.0 cu ft). Electrical system 28 V DC includes two engine-driven generators and one 44 Ah Ni/Cd battery.

Engine bleed air for anti-icing of engine air intakes and windscreen, and inflation of de-icing boots on horizontal leading-deges. Backup windscreen alcohol de-icing.

AVIONICS: Rockwell Collins Pro Line 21 suite as core system.
Comms: Dual VHF-4000 transceivers. Dual TDR-94 Mode S transponders. Optional CVR. Artex C-406-2 ELT.
Radar: As CJ2.
Flight: Collins NAV-4000/NAV-4500 dual VOR, LOC, glideslope and MKR; single (optional second) ADF. DME-4000 DME. AHC-3000 AHRS, ADC-3000 ADC, FMS-3000 FMS. GPS-4000A GPS, ALT-4000 radio altimeter, Goodrich Skywatch TCAS, Goodrich TAWS 8000 Landmark terrain-avoidance. Maintenance diagnostic system.

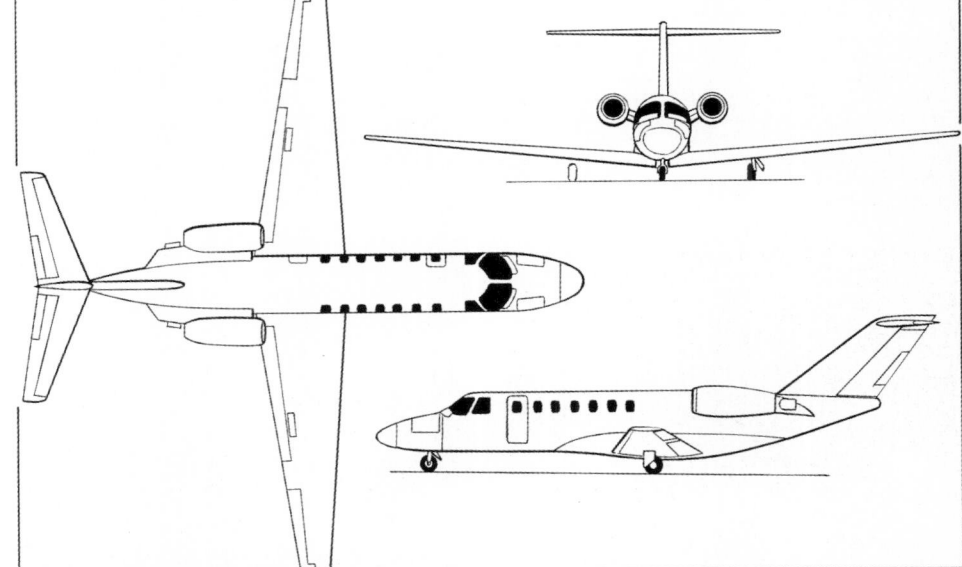

Cessna 525B Citation CJ3 *(Jane's/James Goulding)*   *NEW*/0526904

Instrumentation: Three-tube EFIS with 203 × 254 mm (8 × 10 in) PFD and MFD active matrix LCDs. Goodrich GH-3000 standby instruments; Smiths standby EHSI.

DIMENSIONS, EXTERNAL:

Wing span	16.11 m (52 ft 10¼ in)
Wing mean chord	1.90 m (6 ft 2¾ in)
Wing aspect ratio	9.5
Length: overall	15.30 m (50 ft 2½ in)
fuselage	13.79 m (45 ft 2¾ in)
Height overall	4.60 m (15 ft 1¼ in)
Tailplane span	as CJ2
Tailplane mean chord	1.09 m (3 ft 7 in)
Wheel track	4.88 m (16 ft 0 in)
Wheelbase	6.10 m (20 ft 0¼ in)
Door	as CJ2
Emergency exit width	0.51 m (1 ft 8 in)

DIMENSIONS, INTERNAL:

Cabin: Length between pressure bulkheads	
	6.35 m (20 ft 10 in)
Max width	1.45 m (4 ft 9¼ in)
Max height	1.45 m (4 ft 9 in)
Baggage volume	as CJ2

AREAS:

Wings, gross	27.32 m² (294.1 sq ft)
Vertical tail surfaces (total)	5.23 m² (56.30 sq ft)
Horizontal tail surfaces (total)	6.57 m² (70.68 sq ft)

WEIGHTS AND LOADINGS:

Weight empty	3,701 kg (8,160 lb)
Max fuel	2,136 kg (4,710 lb)
Payload with max fuel, two pilots	(800 lb)
Baggage capacity	as CJ2
Max ramp weight	6,382 kg (14,070 lb)
Max T-O weight	6,291 kg (13,870 lb)
Max landing weight	5,783 kg (12,750 lb)
Max zero-fuel weight	4,767 kg (10,510 lb)
Max wing loading	230.3 kg/m² (47.16 lb/sq ft)
Max power loading	254 kg/kN (2.49 lb/lb st)

PERFORMANCE (estimated):

Max operating speed (V$_{MO}$):	
S/L to FL80	260 kt (482 km/h; 299 mph) CAS
FL80 to FL293	275 kt (509 km/h; 316 mph) CAS
Max cruising speed, at mid-cruise weight: at FL350	
	414 kt (767 km/h; 476 mph)
at FL390	408 kt (756 km/h; 496 mph)
Stalling speed, flaps and gear down	
	87 kt (162 km/h; 101 mph) CAS
Time to: FL410	22 min
FL450	35 min
Max certified altitude	13,715 m (45,000 ft)
T-O balanced field length	1,052 m (3,450 ft)
Landing runway length	936 m (3,070 ft)
Range: with max fuel	
	1,900 n miles (3,518 km; 2,186 miles)
with four passengers, IFR reserves	
	1,664 n miles (3,081 km; 1,914 miles)

NEW ENTRY

CESSNA 550 CITATION BRAVO

TYPE: Business jet.

PROGRAMME: Announced at Farnborough Air Show September 1994; replaced Citation II; prototype (N550BB, c/n 0734) first flight 19 April 1995; initial production aircraft (N801BB, c/n 0801) flew mid-1996; FAA certification in January 1997; first delivery, to Firebond Corporation of Minden, Louisiana, on 25 February 1997. Russian certification achieved 4 September 1998. Certification for operation into London City Airport achieved in May 2002.

CUSTOMERS: First 18 months' production had been sold by time of initial delivery. Earlier production in this series accounted for 621 Model 550 Citation IIs, 97 Model 551 Citation II SPs and 15 Model 552 T-47A Citations, or 733

Cessna Citation Bravo twin-turbofan business jet (*Jane's/Paul Jackson*)   NEW/0526963

in all. Total of 217 Bravos delivered by 30 June 2002, comprising 28 in 1997, 34 in 1998, 36 in 1999, 54 in 2000, 48 in 2001, and 41 in 2002. Recent customers include Avemex SA of Toluca, Mexico (two, for delivery from May 2002); Taxi Aereo Marilia (TAM) of Brazil (three, for delivery from April 2002); Flying Partners CV of Antwerp, Belgium, which has ordered two for its shared ownership programme, for delivery from January 2002; Cessna's and TAG Aviation USA's fractional ownership operation, CitationShares, which placed an initial order for six, for delivery before the end of 2000; and The Company Jet of Grand Rapids, Michigan, which has ordered five for delivery in 2003 for its fractional ownership operation. By September 2002 some 228 Bravos were in service, and total fleet time stood at more than 210,000 hours.

COSTS: US$5.446 million, typically equipped (2002).

DESIGN FEATURES: Small/mid-size business jet. Based on Citation II airframe; tapered wing with sweptback inboard leading-edge; tapered, mid-set tailplane; sweptback fin; podded engines mounted clear of upper rear fuselage; wide track landing gear.

Wing aerofoil NACA 23014 (modified) at centreline, NACA 23012 at wing station 247.95; dihedral 4°; tailplane dihedral 9°.

FLYING CONTROLS: Conventional and manual. Trim tab on port aileron is manually operated; manual rudder trim; electric elevator trim tab with manual standby; electrically actuated single-slotted flaps; hydraulically actuated airbrake.

STRUCTURE: Two primary, one auxiliary metal wing spars; three fuselage attachment points; conventional ribs and stringers. All-metal pressurised fuselage with fail-safe design providing multiple load paths.

LANDING GEAR: Hydraulically retractable tricycle type with single wheel on each unit. Trailing-link main units retract inward into the wing, nose gear forward. Free-fall and pneumatic emergency extension systems. Mainwheel tyres H22.0×8.25-10 (14 ply); steerable nosewheel with tyre size 18×4.4DD (10 ply); all tubeless. Hydraulic brakes. Parking brake and pneumatic emergency brake system. Anti-skid system optional.

POWER PLANT: Two 12.84 kN (2,887 lb st) Pratt & Whitney Canada PW530A turbofans; Nordam target-type thrust reversers standard. Integral fuel tank in each wing, with combined usable capacity of 2,725 litres (720 US gallons; 600 Imp gallons).

ACCOMMODATION: Crew of two on separate flight deck, on fully adjustable seats, with seat belts and inertia reel shoulder harness. Sun visors standard. Fully carpeted main cabin equipped with seats for seven to 10 passengers (seven standard, on pedestal-mounted seats). Main baggage areas in nose and tailcone; flight accessible baggage area at rear of cabin. Refreshment centre standard.

Cabin is pressurised, heated and air conditioned. Individual reading lights and air inlets for each passenger. Dropout constant-flow oxygen system for emergency use. Plug-type door with integral airstair at front on port side and one emergency exit on starboard side. Doors on each side of nose baggage compartment. Tinted windows, each with curtains. Pilot's storm window, birdproof windscreen with de-fog system, anti-icing, standby alcohol anti-icing and bleed air rain removal system.

SYSTEMS: Pressurisation system supplied with engine bleed air, maximum pressure differential 0.61 bar (8.9 lb/sq in), maintaining a sea level cabin altitude to 6,720 m (22,040 ft), or a 2,440 m (8,000 ft) cabin altitude to 13,110 m (43,000 ft). Hydraulic system, pressure 103.5 bar (1,500 lb/sq in), with two pumps to operate landing gear and speed brakes. Separate hydraulic system for wheel brakes. Electrical system supplied by two 28 V 400 A DC starter/generators, with two 350 VA inverters and 24 V 40 Ah Ni/Cd battery. Oxygen system of 1,814 litre (64 cu ft) capacity includes two crew demand masks and five dropout constant-flow masks for passengers. Engine fire detection and extinguishing systems. Wing leading-edges electrically de-iced ahead of engines; pneumatic de-icing boots on outer wings.

AVIONICS: Honeywell Primus 1000 with dual digital flight directors and IC-600 integrated avionics computer as core system.

Comms: Dual Honeywell CN15000 transceivers; dual Mode S transponders; Lockheed Martin cockpit voice recorder.

Radar: Honeywell P-650 colour weather radar.

Flight: Dual Honeywell CN15000 receivers, DME, ADF and radio altimeter; Global GNS-XLS flight management system with GPS and Vnav.

Instrumentation: EFIS with two 178 × 203 mm (7 × 8 in) primary flight display (PFD) screens combining attitude and HSI formats with additional air data; additional multifunction display (MFD) screen of same size with map/plan/checklist capability.

DIMENSIONS, EXTERNAL:

Wing span	15.90 m (52 ft 2 in)
Wing aspect ratio	8.4
Length overall	14.39 m (47 ft 2½ in)
Height overall	4.57 m (15 ft 0 in)
Tailplane span	5.79 m (19 ft 0 in)
Wheel track	4.06 m (13 ft 4 in)
Wheelbase	5.64 m (18 ft 6 in)
Crew/passenger door: Height	1.29 m (4 ft 2¾ in)
Width	0.60 m (1 ft 11½ in)

DIMENSIONS, INTERNAL:

Cabin: Length (between pressure bulkheads)	
	6.37 m (20 ft 11 in)
Max width	1.48 m (4 ft 10¼ in)
Max height	1.43 m (4 ft 8¼ in)

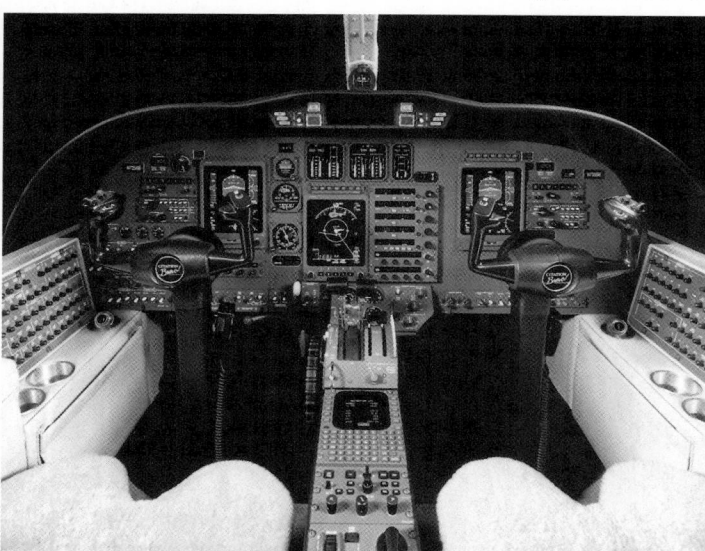

Flight deck of Cessna Citation Bravo   0015667

Standard cabin interior of Cessna Citation Bravo, facing rear   0015666

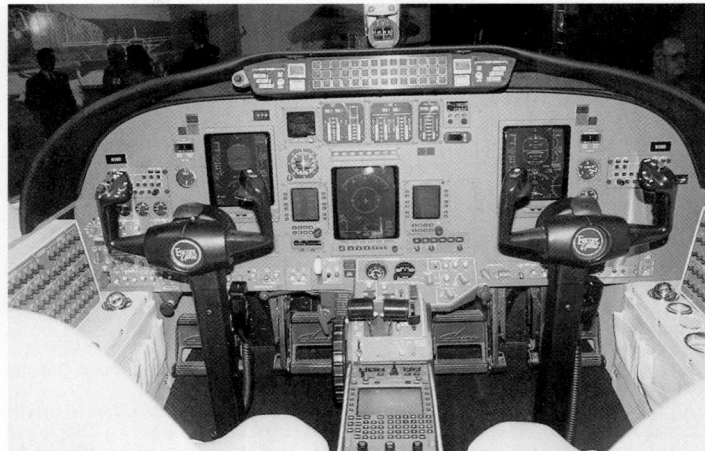

Flight deck of Cessna 560 Citation Encore *(Jane's/Paul Jackson)*   0075949

Cabin interior of Cessna 560 Citation Encore
0054037

Baggage volume: nose	0.50 m³ (17.6 cu ft)
cabin	0.78 m³ (27.7 cu ft)
tailcone	0.80 m³ (28.2 cu ft)

AREAS:

Wings, gross	30.00 m² (322.9 sq ft)
Vertical tail surfaces (total)	4.73 m² (50.9 sq ft)
Horizontal tail surfaces (total, incl tab)	
	6.48 m² (69.8 sq ft)

WEIGHTS AND LOADINGS:

Weight empty, typically equipped	4,060 kg (8,950 lb)
Max fuel weight (usable)	2,188 kg (4,824 lb)
Max T-O weight	6,713 kg (14,800 lb)
Max ramp weight	6,804 kg (15,000 lb)
Max landing weight	6,123 kg (13,500 lb)
Max zero-fuel weight	5,126 kg (11,300 lb)
Max wing loading	223.8 kg/m² (45.83 lb/sq ft)
Max power loading	262 kg/kN (2.56 lb/lb st)

PERFORMANCE:

Max operating speed (V$_{MO}$):	
S/L to FL80	260 kt (481 km/h; 299 mph)
FL80 to FL279	275 kt (509 km/h; 316 mph)
at FL279 and above	M0.70
Max cruising speed at FL310 at average cruise weight of	
5,670 kg (12,500 lb)	403 kt (746 km/h; 464 mph)
Stalling speed in landing configuration at max landing	
weight	86 kt (160 km/h; 99 mph) CAS
Max rate of climb at S/L	974 m (3,195 ft)/min
Rate of climb at S/L, OEI	345 m (1,133 ft)/min
Max certified altitude	13,715 m (45,000 ft)
Service ceiling, OEI	8,458 m (27,750 ft)
T-O balanced field length (FAR Pt 25)	1,098 m (3,600 ft)
FAR Pt 25 landing field length at max landing weight	
	970 m (3,180 ft)
IFR range, 100 n mile (185 km; 115 mile) alternate, plus	
45 min reserves	1,779 n miles (3,295 km; 2,047 miles)

OPERATIONAL NOISE LEVELS:

T-O	73.7 EPNdB
Approach	91.2 EPNdB
Sideline	85.2 EPNdB

UPDATED

CESSNA 560 CITATION ENCORE
US Army designation: UC-35B
US Marine Corps designation: UC-35D

TYPE: Business jet.

PROGRAMME: Prototype, based on rebuilt Citation Ultra, first flown (N560VU) 9 July 1998; announced at NBAA Convention at Las Vegas, Nevada 18 October 1998; first production aircraft (c/n 560-0539) rolled out in early March 2000. FAA certification achieved 26 April 2000; first delivery (N539CE/N5108G) on 29 September 2000 to J R Simplot Company of Boise, Idaho.

CURRENT VERSIONS: **Citation Encore:** Civilian business jet; *as described.*

UC-35B: In January 1996, US Army selected Ultra for its C-XX medium-range transport aircraft programme, for which 35 aircraft are required over a five year period; equipped with 0.90 × 1.15 m (2 ft 11½ in × 3 ft 9 in) upward-opening clamshell cargo door on port forward fuselage; first order was for two UC-35As, of which one (95-0123) was delivered in last quarter of 1996; first two serve with 207 AvnCo at Heidelberg, Germany. Subsequent orders for five in each of FY96, FY97, FY98 and FY99; (last two aircraft of this batch are first UC-35Bs), three in 2000 and one in 2001 (delivered December 2001). Total six UC-35Bs, plus 20 earlier UC-35As.

UC-35D: Equivalent version to UC-35B for US Marine Corps.

CUSTOMERS: Deliveries of Ultra began July 1994; deliveries comprised 24 in 1994, 56 in 1995, 52 in 1996, 47 in 1997, 41 in 1998, 32 in 1999, six in 2000, 37 in 2001 and 36 in 2002. Announced customers for Encore include J R Simplot Co of Boise, Idaho, which has ordered two; Taxi Aereo Marilia (TAM) of Brazil (one); Flying Partner CV of Antwerp, Belgium (one). By September 2002 more than 70 Encores were in service and total fleet time exceeded 25,800 flight hours.

COSTS: US$7.559 million, typically equipped (2002).

DESIGN FEATURES: Stretched version of Citation S/II for full eight-seat cabin and with fully enclosed toilet/vanity area; seventh cabin window each side; two baggage compartments outside main cabin. Rear-engined executive jet with low-mounted wing of tapered planform with root gloves; tapered horizontal tail with sweptback fin and fillet.

Encore generally as Citation Ultra and Citation Bravo (which see). Compared to Ultra, has increased wing span; bleed air anti-icing for wing leading-edges; boundary layer energisers and stall fences to improve stall characteristics; trailing-link main landing gear with narrower track; increased fuel capacity; Pratt & Whitney Canada PW535 turbofans offering 10 per cent increase in thrust and 15 per cent improvement in specific fuel consumption; reduced fuel capacity; fuel heaters obviating the need for additives; digital pressurisation system; improved braking system; single level access electrical junction box; and redesigned interior with increased headroom, new passenger service units, pin-mounted seats for easy removal/replacement and increased serviceability.

Description for Citation Ultra applies also to Citation Encore except as follows:

FLYING CONTROLS: Conventional and manual. Trim tab on port aileron manually actuated; manual rudder trim; electric elevator trim tab with manual standby; hydraulically actuated Fowler flaps; hydraulically actuated airbrake.

STRUCTURE: Two primary, one auxiliary metal wing spars; four fuselage attachment points, conventional ribs and stringers. All-metal pressurised fuselage with fail-safe design providing multiple load paths.

LANDING GEAR: Hydraulically retractable trailing-link tricycle type with single wheel on each unit. Main units retract inward into the wing, nose gear forward. Free-fall and pneumatic emergency extension systems. Goodyear mainwheels with tubeless tyres size 22×8.0-10 (12 ply), pressure 6.90 bar (100 lb/sq in). Steerable nosewheel (±20°) with Goodyear wheel and tyre size 18×4.4DD (10

Cessna 560 Citation Encore (two Pratt & Whitney Canada JT15D-5D turbofans) *(Jane's/Dennis Punnett)*

Cessna 560 Citation Encore   *NEW*/0527096

ply), pressure 8.27 bar (120 lb/sq in). Goodyear hydraulic brakes. Parking brake and pneumatic emergency brake system. Anti-skid system optional. Minimum ground turning radius about nosewheel 8.38 m (27 ft 6 in).

POWER PLANT: Two Pratt & Whitney Canada PW535A turbofans, each rated at 15.12 kN (3,400 lb st) at 27°C (80°F). Integral fuel tank in each wing, combined usable capacity 3,047 litres (805 US gallons; 670 Imp gallons).

ACCOMMODATION: Standard seating for seven passengers in three forward-facing and four in club arrangement, or eight passengers in double-club arrangement, on swivelling and fore/aft/inboard-tracking pedestal seats; refreshment centre in forward cabin area; lavatory/vanity centre with sliding doors to rear, metallic plating on cabin fittings, veneer overlay on armrests, optional pleated window shades and cabin divider mirror standard; space in aft section of cabin for 272 kg (600 lb) of baggage, in addition to baggage compartments in nose and rear fuselage.

SYSTEMS: Pressurisation system supplied with engine bleed air, maximum pressure differential 0.61 bar (8.8 lb/sq in), maintaining a sea level cabin altitude to 6,720 m (22,040 ft), or a 2,440 m (8,000 ft) cabin altitude to 12,495 m (41,000 ft). Hydraulic system, pressure 103.5 bar (1,500 lb/sq in), with two pumps to operate landing gear and speed brakes. Separate hydraulic system for wheel brakes. Electrical system supplied by two 28 V 300 A DC starter/generators, with two 350 VA inverters and 24 V 40 Ah Ni/Cd battery. Oxygen system of 0.62 m³ (22 cu ft) capacity includes two crew demand masks and five dropout constant flow masks for passengers. High-capacity oxygen system optional. Engine fire detection and extinguishing system. Bleed air anti-icing system for wing leading-edges and engine inlets; pneumatic de-icing boots on tailplane leading-edges.

AVIONICS: Standard avionics package based on Honeywell Primus 1000 digital flight control system with integrated avionics computer.
Comms: Dual Honeywell Primus II transceivers, dual TDR-94 transponders, dual altitude reporting systems and Lockheed Martin A200S cockpit voice recorder standard.
Radar: Honeywell Primus 660 colour weather radar standard.
Flight: Dual Honeywell Primus II, single Honeywell ADF, ALT-55 radio altimeter, coupled vertical navigation system, Global GNS-XL with GPS, expanded keyboard and colour CDU display standard. Universal Avionics UNS-1C*sp* FMS optional.
Instrumentation: Three-tube EFIS with 203 × 178 mm (8 × 7 in) CRTs comprising pilot's and co-pilot's primary flight displays (PFDs) and centrally mounted multifunction display (MFD); PFDs integrate functions of five flight instruments and several sources of navigation data, and provide trend data for airspeed, altitude and rate of climb.

DIMENSIONS, EXTERNAL:
Wing span	16.48 m (54 ft 1 in)
Wing aspect ratio	7.2
Length overall	14.90 m (48 ft 10¾ in)
Height overall	4.57 m (15 ft 0 in)
Tailplane span	6.55 m (21 ft 6 in)
Wheelbase	6.06 m (19 ft 10¾ in)
Wheel track	4.04 m (13 ft 3 in)

DIMENSIONS, INTERNAL:
Cabin:
Length: between pressure bulkheads	
	6.89 m (22 ft 7¼ in)
excl cockpit	5.28 m (17 ft 4 in)
Max width	1.49 m (4 ft 10¾ in)
Max height	1.40 m (4 ft 7 in)
Baggage compartment volume	1.95 m³ (69 cu ft)

AREAS:
Wings, gross	29.94 m² (322.3 sq ft)

Early production Cessna Citation Excel (*Jane's/Paul Jackson*)   *NEW*/0526964

Vertical tail surfaces (total, incl tab)	4.73 m² (50.90 sq ft)
Horizontal tail surfaces (total)	7.88 m² (84.80 sq ft)

WEIGHTS AND LOADINGS:
Weight empty, typically equipped	4,593 kg (10,125 lb)
Max fuel	2,468 kg (5,440 lb)
Max T-O weight	7,543 kg (16,630 lb)
Max ramp weight	7,634 kg (16,830 lb)
Max landing weight	6,895 kg (15,200 lb)
Max zero-fuel weight	5,715 kg (12,600 lb)
Max wing loading	251.9 kg/m² (51.60 lb/sq ft)
Max power loading	249 kg/kN (2.45 lb/lb st)

PERFORMANCE:
Max cruising speed at FL350	
	429 kt (795 km/h; 494 mph)
Stalling speed, flaps down	83 kt (154 km/h; 96 mph)
Max rate of climb at S/L	1,445 m (4,740 ft)/min
Time to: FL350	12 min
FL450	28 min
Rate of climb at S/L, OEI	439 m (1,440 ft)/min
Max certified altitude	13,715 m (45,000 ft)
T-O balanced field length (FAR Pt 25)	1,064 m (3,490 ft)
Landing from 15 m (50 ft) at max landing weight	
	844 m (2,770 ft)

OPERATIONAL NOISE LEVELS:
T-O	70.0 EPNdB
Approach	90.5 EPNdB
Sideline	89.8 EPNdB

UPDATED

CESSNA 560XL CITATION EXCEL

TYPE: Business jet.

PROGRAMME: Announced at National Business Aircraft Association Convention in New Orleans, October 1994. Construction of prototype (N560XL) began February 1995; first flight 29 February 1996; public debut at NBAA convention at Orlando, Florida, November 1996, by which time prototype and preproduction aircraft (N561XL) had completed 350 hours of flight testing in 400 sorties; first production aircraft rolled out 21 November 1997; FAA certification achieved 22 April 1998; first delivery 2 July 1998 to Swift Transportation Inc of Phoenix, Arizona. First export, September 1998, to Automobilvertriebs AG, Austria; 100th production Excel rolled out 17 April 2000 and delivered in August 2000.

CUSTOMERS: Total market expected to exceed 1,000. Total of 15 delivered in 1998, 39 in 1999, 79 in 2000, 85 in 2001, and 81 in 2002. Recent customers include Avemex SA of Toluca, Mexico (three), Taxi Aereo Marilia (TAM) of Brazil (one) and the Swiss Federal Office for Civil Aviation, which took delivery of one on 28 August 2002. By September 2002 more than 260 Excels were in service and had flown a total of more than 170,000 hours.

COSTS: US$9.451 million (2002).

DESIGN FEATURES: Combines systems and wing and tail surfaces of Citation Ultra (Encore) with shortened version of Citation X's fuselage, providing 10-seat cabin with stand-up headroom; dual ventral strakes.

FLYING CONTROLS: As Citation Encore.

LANDING GEAR: Hydraulically retractable tricycle type with single wheel on each unit; trailing-link suspension on main legs. Mainwheel tyre size 22×8.0-10 (12 ply) tubeless, pressure 10.48 bar (152 lb/sq in); mechanically steerable nosewheel (±20°) with chined tyre, size 18×4.4DD (10 ply) tubeless, pressure 9.65 bar (140 lb/sq in). Hydraulic multiple disc carbon brakes with anti-skid and pneumatic emergency system.

POWER PLANT: Two 16.92 kN (3,800 lb st) Pratt & Whitney Canada PW545A turbofans. Nordam clamshell-type thrust reversers standard. Integral fuel tank in each wing, total capacity 3,804 litres (1,005 US gallons; 837 Imp gallons) of which 3,668 litres (969 US gallons; 807 Imp gallons) usable; single-point pressure refuelling.

ACCOMMODATION: Choice of four standard seating configurations for up to 10 passengers in various layouts; seats recline, swivel and track forward and aft and laterally; forward refreshment centre and cupboard; aft lavatory and centreline cupboard. Airstair door on port side aft of flight deck. Baggage compartment in rear fuselage, capacity 317.5 kg (700 lb), with external access door incorporating integral step.

SYSTEMS: Honeywell RE-100(XL) APU optional from c/n 5021 (mid-1999) onwards, and retrofittable to earlier aircraft. Pressurisation system maximum differential 0.64 bar (9.3 lb/sq in), maintaining a sea level cabin altitude to 7,700 m (25,200 ft) or a 2,070 m (6,800 ft) cabin altitude to 13,715 m (45,000 ft). Hydraulic system, pressure 103.5 bar (1,500 lb/sq in), with two engine-driven pumps to operate landing gear, flaps, horizontal stabiliser, speed brakes and thrust reversers; separate hydraulic system for wheel brakes and anti-skid. Electrical system supplied by two 28 V 300 A DC starter/generators, with 24 V 44 Ah Ni/Cd battery. Vapour cycle air conditioning system. Oxygen system, capacity 1,417 litres (50 cu ft) with pressure demand masks for crew and dropout constant-flow masks for passengers. Engine inlets and wing leading-edges supplied by engine bleed air for anti-icing; tailplane has de-icer boots; fin unprotected; electrically heated windscreen and cockpit side windows with PPG SurfaceSeal coating for rain dispersal, with electric blower assistance.

AVIONICS: Standard Honeywell Primus 1000 integrated digital avionics suite with IC-600 avionics computer as core system. Rockwell Collins system optional.
Comms: Dual Honeywell TR-850 transceivers, dual XS-852B Mode S transponders, Artex 110-4 ELT,

Eight-seat cabin interior of Cessna Citation Excel   0054039

Cockpit of Cessna Citation Excel showing Honeywell Primus 1000 avionics with three-tube EFIS   0054038

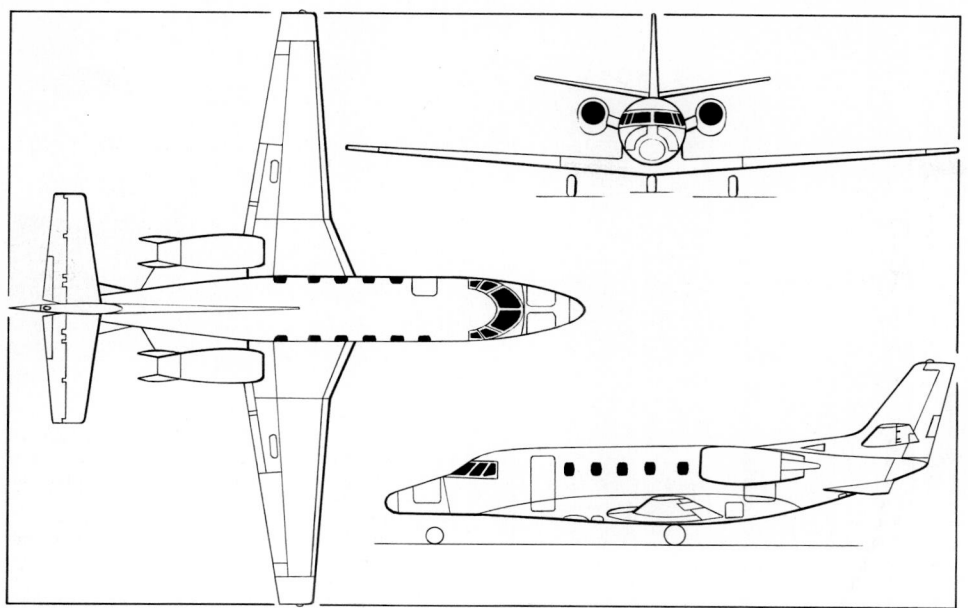

Cessna Citation Excel (PW545 turbofans) *(Jane's/James Goulding)* 0126686

Lockheed Martin A200S cockpit voice recorder, airborne telephone system.

Radar: Honeywell Primus 880 colour weather radar.

Flight: Dual Honeywell NV-850, dual DM-850 DME, single DF-850 ADF, long-range navigation management system incorporating GPS, and Rockwell Collins ALT-55 radio altimeter.

Instrumentation: Three-tube EFIS with 178 × 203 mm (7 × 8 in) CRT screens comprising dual primary flight displays (PFDs) showing attitude/heading and all air data information, and single multifunction display (MFD) for map/plan, weather and checklist data.

DIMENSIONS, EXTERNAL:

Wing span	16.98 m (55 ft 8½ in)
Wing aspect ratio	8.4
Length: overall	15.79 m (51 ft 9½ in)
Height overall	5.24 m (17 ft 2½ in)
Tailplane span	6.55 m (21 ft 6 in)
Wheelbase	6.67 m (21 ft 10¾ in)
Wheel track	4.54 m (14 ft 10¾ in)

DIMENSIONS, INTERNAL:

Cabin: Length:

between pressure bulkheads	7.01 m (23 ft 0 in)
excl cockpit	5.69 m (18 ft 8 in)
Max width	1.70 m (5 ft 7 in)
Max height	1.73 m (5 ft 8 in)
Baggage capacity (aft)	2.55 m³ (90 cu ft)

AREAS:

Wings, gross	34.35 m² (369.7 sq ft)
Vertical tail surfaces (total, incl tab)	4.73 m² (50.9 sq ft)
Horizontal tail surfaces (total, incl tab)	7.88 m² (84.8 sq ft)

WEIGHTS AND LOADINGS:

Weight empty, typically equipped	5,579 kg (12,300 lb)
Max fuel weight (usable)	3,057 kg (6,740 lb)
Max T-O weight	9,071 kg (20,000 lb)
Max ramp weight	9,163 kg (20,200 lb)
Max landing weight	8,482 kg (18,700 lb)
Max zero-fuel weight	6,804 kg (15,000 lb)
Max wing loading	264.1 kg/m² (54.10 lb/sq ft)
Max power loading	269 kg/kN (2.64 lb/lb st)

PERFORMANCE:

Max operating speed (V$_{MO}$):

S/L to FL80	260 kt (481 km/h; 299 mph)
FL80 to FL265	305 kt (564 km/h; 351 mph) CAS
above FL265	M0.75
Max cruising speed at FL350	429 kt (795 km/h; 494 mph)

Stalling speed in landing configuration, at max landing weight 90 kt (167 km/h; 104 mph)

Max rate of climb at S/L	1,155 m (3,790 ft)/min
Rate of climb at S/L, OEI	259 m (850 ft)/min
Max certified altitude	13,715 m (45,000 ft)
Service ceiling, OEI	8,717 m (28,600 ft)
T-O balanced field length (FAR Pt 25)	1,095 m (3,590 ft)

FAR Pt 25 landing field length at max weight 969 m (3,180 ft)

IFR range 2,165 n miles (4,009 km; 2,491 miles)

OPERATIONAL NOISE LEVELS:

T-O	72.4 EPNdB
Approach	93.1 EPNdB
Sideline	85.3 EPNdB

UPDATED

CESSNA 680 CITATION SOVEREIGN

TYPE: Business jet.

PROGRAMME: Design started mid-1998; announced at NBAA Convention at Las Vegas, Nevada, 18 October 1998; critical design review completed in late 1999; structural testing of fatigue test fuselage began in late 1999; official launch 3 January 2000; construction of cyclic fatigue test airframe started October 2000; manufacture of prototype started in 2001; first flight (prototype c/n 000P/N680CS) 27 February 2002, followed by first preproduction aircraft (c/n 0001/N681CS) on 27 June 2002. Public debut (N681CS) at NBAA Convention in Orlando, Florida, 10 September 2002. These aircraft, plus c/n 0002 which flew in fourth quarter 2002, will undertake a 2,000-hour, 19-month test programme. By 20 January 2003 the three aircraft had logged more than 670 flight hours in 370 sorties. Two static test airframes have also been built, one of which will undertake a 36,000-cycle fatigue test programme. After completion of initial flight tests and envelope expansion in June 2002, 000P was dedicated to stability and control checks, high altitude stall tests, cruise performance validation, systems and APU testing and airfield performance; c/n 0001 is the systems verification and certification article; and c/n 0002/N682CS is dedicated to avionics and autopilot testing. FAA certification to FAR Pt 25 Amendment 92 and JAR 25 Change 15, both including TVSM compliance, is scheduled for late 2003; c/n 0002 will also perform a 150-hour function and reliability programme and, with c/n 0001, will undertake 300 hours of post-certification in-service testing before first customer delivery in January 2004; both pre-production aircraft will then be refurbished and brought up to final production configuration before sale to customers. Delivery targets are 21 in 2004, 40 in 2005 and 58 in 2006.

CUSTOMERS: Launch customer Swift Air of Phoenix, Arizona, ordered six on day of launch announcement; Executive Jet Aviation ordered 50, with 50 options, on 20 October for its NetJets fractional ownership scheme. Other announced customers include Atlas Air Service GmbH of Germany, which has ordered one. Firm orders totalled more than 100 by January 2003.

COSTS: Basic US$12.695 million, typically equipped US$13.257 million (2001).

DESIGN FEATURES: Design goals included large cabin, good short-field performance and US coast-to-coast range. Low wing with sweptback leading-edge; mid-mounted tailplane with leading-edge sweepback; podded engines on rear fuselage shoulders.

New wing design; sweepback 16° 18′ at leading-edge, −12° 42′ at quarter-chord, dihedral 3°. Mid-mounted tailplane, sweepback 17° 36′ at leading-edge, dihedral 5°.

FLYING CONTROLS: Conventional. Five hydraulically actuated spoiler panels per wing; three innermost panels function as roll spoilers, variable position speedbrakes and ground spoilers; inner and outboard panels function as variable speedbrakes and ground spoilers. Trim tabs in ailerons and rudder; trimmable tailplane; yaw damper.

STRUCTURE: Primarily metal. Fokker Aerostructures of the Netherlands selected in 2000 to manufacture the tail surfaces of production aircraft.

LANDING GEAR: Hydraulically retractable tricycle type with twin wheels on trailing-link main units and nosewheel; main units retract inboard, nosewheel forwards. Carbon brakes; anti-skid standard; hydraulically boosted nosewheel steering.

POWER PLANT: Two Pratt & Whitney Canada PW306C turbofans with FADEC, each developing 25.3 kN (5,686 lb st) flat rated to ISA + 15°C; target-type thrust reversers. Integral fuel tanks in wings; single-point pressure fuelling.

ACCOMMODATION: Crew of two on flight deck and up to 11 passengers in cabin; standard accommodation for eight passengers in double club arrangement with forward galley and aft lavatory. Fully dimmable LED cabin lighting. Baggage compartment in tailcone with external access. Cabin is pressurised, heated and air conditioned. Airstair door at front on port side; two emergency exits aft, above wings.

SYSTEMS: Pressurisation system, maximum differential 0.64 bar (9.3 lb/sq in), and air-conditioning system supplied by engine bleed air; pressurisation system maintains sea level cabin to 7,690 m (25,230 ft); closed-centre hydraulic

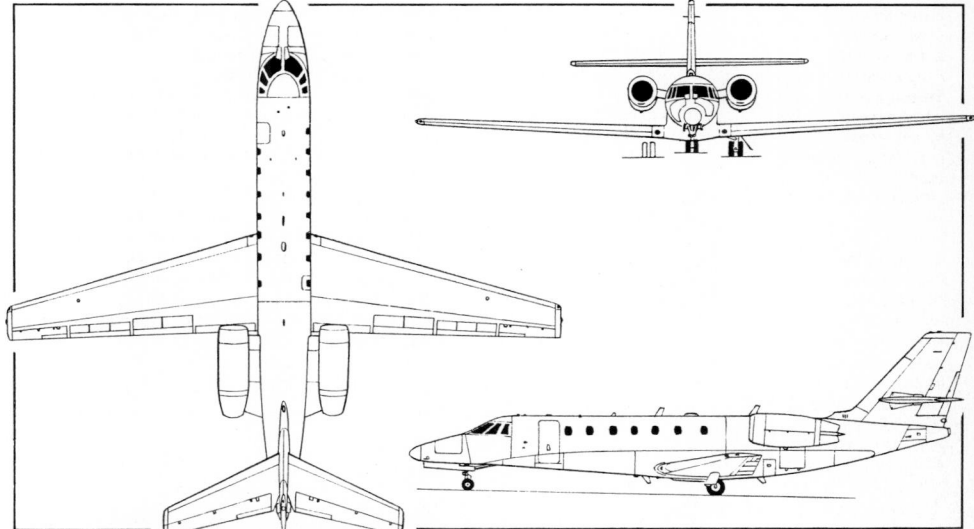

Citation Sovereign eight/11-passenger business jet *(Michael Badrocke)* NEW/0526865

First preproduction Cessna 680 Citation Sovereign making its public debut at NBAA, Orlando, September 2002 *(Jane's/Paul Jackson)* NEW/0526965

Eight-seat double club cabin of Cessna 680 Citation Sovereign   0054041

Cessna 750 Citation X flight deck   0015654

system, pressure 207 bar (3,000 lb/sq in), for operation of landing gear, nosewheel steering, braking system, spoilers and thrust reversers. 28 V DC electrical system, supplied by two 300 A starter/generators and two AC alternators. Honeywell RE100 APU certified for in-flight operation up to 915 m (3,000 ft); and Honeywell environmental control and cabin pressure control systems. Oxygen system standard. Wing and tailplane leading-edges and engine inlets anti-iced by engine bleed air; electrically anti-iced windscreens and air data probes.

AVIONICS: Honeywell Primus Epic as core system.
  Radar: Honeywell Primus 880 colour weather radar.
  Flight: VOR, ILS, ADF, GPS, dual NZ-2000 FMS and three-axis autopilot standard.
  Instrumentation: Four-tube EFIS with 203 × 254 mm (8 × 10 in) active matrix quartz PFD, MFD and EICAS displays. Max-Viz EVS2000 enhanced visibility system optional.

DIMENSIONS, EXTERNAL:
Wing span	19.24 m (63 ft 1½ in)
Wing aspect ratio	7.7
Length overall	19.35 m (63 ft 6 in)
Height overall	6.07 m (19 ft 11 in)
Tailplane span	8.38 m (27 ft 6 in)
Wheel track	3.11 m (10 ft 2½ in)
Wheelbase	8.51 m (27 ft 11 in)

DIMENSIONS, INTERNAL:
Cabin: Length	7.38 m (24 ft 2½ in)
Max width	1.70 m (5 ft 7 in)
Max height	1.73 m (5 ft 8 in)
Volume	19.25 m³ (680 cu ft)
Baggage compartment volume	2.83 m³ (100 cu ft)

AREAS:
Wings, gross	47.94 m² (516.0 sq ft)
Vertical tail surfaces	8.85 m² (95.3 sq ft)
Horizontal tail surfaces	12.87 m² (138.5 sq ft)

WEIGHTS AND LOADINGS:
Basic operating weight, two crew	7,961 kg (17,550 lb)
Max fuel	4,863 kg (10,720 lb)
Max ramp weight	13,721 kg (30,250 lb)
Max T-O weight	13,607 kg (30,000 lb)
Max landing weight	12,292 kg (27,100 lb)
Max zero-fuel weight	9,208 kg (20,300 ft)
Max wing loading	283.9 kg/m² (58.14 lb/sq ft)
Max power loading	269 kg/kN (2.64 lb/lb st)

PERFORMANCE:
Max operating speed (V$_{MO}$)	M0.80
Max cruising speed	446 kt (826 km/h; 513 mph)
Time to FL370	14 min
Max certified altitude	14,325 m (47,000 ft)
Service ceiling	13,105 m (43,000 ft)
T-O balanced field length	1,126 m (3,695 ft)
Landing run	959 m (3,145 ft)
IFR range, 100 n mile (185 km; 115 mile) alternate,	
45 min reserves	2,686 n miles (4,974 km; 3,091 miles)

UPDATED

CESSNA 750 CITATION X

TYPE: Business jet.
PROGRAMME: Announced at NBAA Convention in New Orleans in October 1990; engine flew on Citation VII testbed (N650) 21 August 1992; first flight (N750CX) 21 December 1993; two preproduction aircraft to aid integration of production systems; first of these (N751CX) flown 27 September 1994, second (N752CX) flown 11 January 1995; FAA FAR Pt 25, Amendment 74 certification 3 June 1996 after flight test programme totalling more than 3,000 hours; JAA certification achieved during 1999. First customer delivery July 1996. Citation X design team awarded the National Aeronautic Association's Robert J. Collier Trophy in February 1997.

Cessna delivered its 3,000th Citation, a Citation X, on 19 November 1999.

In October 2000, Cessna announced improvements to the Citation X aimed at boosting range/payload performance and enabling the aircraft to operate from shorter runways. Improvements, to be incorporated on all aircraft delivered after 1 January 2002, beginning with c/n 0173 include uprated 30.01 kN (6,764 lb st) Rolls-Royce AE 3007C-1 turbofans; 181 kg (400 lb) increase in maximum take-off weight to 16,374 kg (36,100 lb), enabling a typically equipped aircraft to carry seven passengers with maximum fuel; and take-off balanced field length at MTOW of 1,585 m (5,200 ft). Several currently optional items of avionics will become standard, including Honeywell TCAS II and EGPWS, CVR, satcom, VHF/AFIS, provisions for an FDR, and second HF transceiver, plus Teledyne angle of attack indicator/indexer, tail floodlights, red strobe light, pulse lights, Litton ELT, and 2,154 litre (76 cu ft) oxygen bottle. First delivery of an upgraded citation X took place on 5 February 2002 to golfer Arnold Palmer (c/n 0176/N1AP).
CUSTOMERS: First delivery (0003/N1AP) to Arnold Palmer July 1996; 100th Citation X delivered 23 December 1999 to Townsend Engineering of Des Moines, Iowa and 200th delivered 14 October 2002 to NetJets Inc; total 203 by 31 December 2002, comprising seven, 28, 30, 36 in 1996-99, 37 in 2000, 34 in 2001 and 31 in 2002; most to US operators, but others exported to Canada, Finland, Germany, Mexico, South Africa and UK. Executive Jet Aviation (EJA) ordered 31 for delivery beginning in 1997 and extending beyond 2000, for its NetJets fractional ownership operation. Other early recipients included General Motors (five), Honeywell (two) and Williams Companies (three). Recent customers include former World Motor Racing Champions Nigel Mansell, who took delivery of one in February 2002, and Nelson Piquet, who ordered one at the NBAA Convention at Orlando, Florida, in September 2002, and the Air Traffic Management Bureau of the Civil Aviation Authority of China, which ordered one on 1 October 2001. By September 2002 more than 160 Citation Xs were in service, and had accumulated 180,377 flight hours and 120,059 landings.
COSTS: Basic US$18.795 million, typically equipped US$18.995 million (2001).
DESIGN FEATURES: Optimised for high maximum operating Mach number; US transcontinental and transatlantic range. Design generally as for Citation VII (which see), but with greater angles of sweepback on all flying surfaces and of increased size and weight.
  Wing sweepback 37°; dihedral 3°.
FLYING CONTROLS: Dual hydraulically powered controls with manual reversion. One-piece all-moving tailplane; two-piece rudder, lower portion hydraulically powered, upper portion electrically powered; speed brakes/spoilers with manual back-up. Five spoiler panels per wing, operating in combination as aileron augmentors, airbrakes and lift dumpers.
STRUCTURE: Thick wing skins, milled from solid; all control surfaces, spoilers, speedbrakes and flaps are of composites construction.
LANDING GEAR: Trailing-link main units, each with twin wheels; powered anti-skid carbon brakes; hydraulically steerable nose unit with twin wheels. Main tyres 26×6.6R14 (14 ply) tubeless; nose tyres 16×4.4D (6 or 10 ply) tubeless.
POWER PLANT: Two Rolls-Royce AE 3007C-1 turbofans, each rated at 30.01 kN (6,764 lb st) for take-off, pod-mounted on sides of rear fuselage; FADEC. Hydraulically operated target-type thrust reversers standard. Fuel contained in three separate tanks, one in each outer wing and one in centre-section/forward fairing, combined usable capacity 7,291 litres (1,926 US gallons; 1,604 Imp gallons). Two independent fuel supply systems; fuel is fed from centre tank to wing tanks; single point refuelling.
ACCOMMODATION: Crew of two on separate flight deck, and up to 12 passengers; interior custom designed; cabin is pressurised, heated and air conditioned; heated and pressurised baggage compartment in rear fuselage with external door. Windscreen electrically heated and demisted.
SYSTEMS: Pressurisation system, maximum pressure differential 0.64 bar (9.3 lb/sq in), maintains 2,440 m (8,000 ft) cabin altitude at 15,545 m (51,000 ft). Dual isolated hydraulic systems, pressure 207 bar (3,000 lb/sq in), maintained by pressure-compensated pumps. Electrical system is powered by two engine-driven 400 A DC generators, with two 24 V 44 Ah Ni/Cd batteries; wiring designed to minimise susceptibility of critical systems to HIRF interference. Wing and tail leading-edges and engine inlets heated by engine bleed air for ice protection; wing cuffs, pitot/static system, AoA system and windscreen electrically heated.

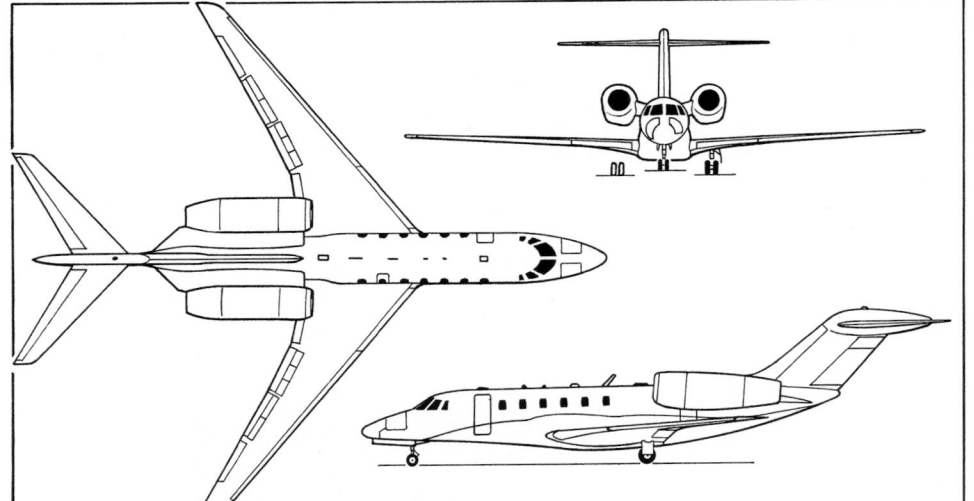

Cessna 750 Citation X (*Jane's/Dennis Punnett*)   0075950

NEW/0526923

Cessna Citation Sovereign, cutaway drawing key

1 Hinged composites radome
2 Weather radar antenna
3 ILS glideslope antenna
4 Radar mounting bulkhead
5 Pitot head
6 Nose compartment hinged access doors, port and starboard
7 Windscreen blower
8 Nose avionics bay
9 Position of oxygen cylinder on starboard side
10 Nosewheel doors
11 Temperature probe
12 Taxying lights
13 Twin nosewheels with chined tyres, forward retracting
14 Nose landing gear leg strut and torque links
15 Hydraulic jack
16 Canted front pressure bulkhead
17 Rudder pedals
18 Control column
19 Instrument panel, four full-colour LCD displays
20 Windscreen rain dispersal air ducts
21 Instrument panel shroud
22 Electrically heated windscreen panels
23 Cockpit roof frames
24 Curtained flight deck bulkhead
25 Warning horn
26 First officer's seat
27 Captain's seat
28 Direct vision opening side window panel
29 Side console panel with nosewheel steering tiller
30 Adjustable seat rails
31 Incidence vane
32 Door closure panel
33 Door with integral airstairs
34 Door internal latch
35 Door struts and balance cables
36 Folding handgrip
37 Doorway
38 Forward wardrobe
39 Starboard side refreshment cabinet
40 GPS 1 and 2 antennae
41 Cockpit section joint frame
42 Single side-facing seat to starboard
43 TCAS antenna
44 ATC 1 and 2 antennae
45 Fold-out table, four positions
46 Individual reclining seats with adjustable armrests
47 Cabin window panels
48 Sidewall storage pockets
49 Cabin sidewall trim panelling
50 Fuselage frame and stringer structure, frames riveted to bonded circumferential stiffeners

51 Wing inspection light
52 Forward ventral fairing
53 Tailplane control cables, outside fuselage pressure shell
54 Cabin insulation blankets
55 Individual window blinds
56 Wing spar/fuselage drag fitting
57 Wing centre section, continuous beneath fuselage pressure shell
58 Dropped aisle cabin floor
59 Fold-out table stowage
60 Cup holders and individual LCD screen terminals
61 VHF 1 antenna
62 Starboard wing integral fuel tank
63 Structural fuel venting channels
64 Overwing fuel filler
65 Thermally de-iced leading-edge
66 Dual starboard navigation lights
67 De-icing air venting louvres
68 Wingtip strobe light
69 Starboard aileron
70 Aileron tab
71 Aileron actuator, cable operated
72 Outboard three-segment spoiler/ speedbrake panels
73 Three-segment Fowler flaps, extended
74 Inboard two-segment spoiler/ speedbrake panels
75 Spoiler hydraulic jacks
76 Flap tracks and guide rails
77 Flap operating screw-jacks
78 ADF antenna
79 Starboard side emergency escape hatch
80 Lavatory
81 Cabin bulkhead with sliding doors
82 Composites floor panels
83 Wing main and rear spar/ fuselage attachment links
84 Rear spar/centre section yaw fittings
85 Wing spar attachment double fuselage frames
86 Dual cabin pressurisation valves

87 Wash basin
88 Rear wardrobe
89 Rear pressure bulkhead
90 Baggage compartment
91 Electrical equipment racks
92 Rear avionics equipment shelf
93 VHF 2 antenna
94 Transverse engine mounting beams
95 Engine pylon
96 Starboard engine installation
97 Target-type thrust reverser, deployed
98 Reverser door actuator
99 Cabin air conditioning pack
100 Heat exchanger ram air intake
101 Aft fuselage frame and stringer structure
102 Canted fin spar mounting bulkhead
103 Tailplane de-icing air ducts
104 Variable incidence tailplane control jack
105 Tailplane pivot mounting
106 Elevator actuating levers, cable operated
107 Tailplane sealing plate
108 Two-spar and rib fin torsion box structure
109 Composites fin leading edge
110 Starboard tailplane
111 Starboard elevator
112 Static dischargers
113 VOR localiser antenna

114 Anti-collision beacon
115 Rudder rib structure
116 Rudder tab
117 Tab actuator, cable operated
118 Elevator tab
119 Tail navigation lights
120 Port elevator rib structure
121 Three-spar and rib tailplane torsion box structure
122 Tailplane leading-edge thermal de-icing
123 APU exhaust
124 Auxiliary Power Unit (APU)
125 Rudder control quadrant, cable operated
126 Fin rear spar mounting canted bulkhead
127 Engine fire suppression bottles
128 Port engine thrust reverser cowlings
129 Rear engine support links
130 Pylon-mounted engine bleed-air pre-cooler
131 Main engine mounting
132 Baggage bay hatch
133 Hinged cowling panels
134 Engine drain mast
135 Pratt & Whitney Canada PW306C turbofan
136 Engine accessory equipment gearbox
137 FADEC controller
138 Baggage bay door with integral steps
139 Generator cooling air duct
140 Intake lip bleed-air de-icing
141 Acoustically lined intake duct
142 Battery bay, port and starboard
143 Cabin conditioned air delivery ducting
144 Flap shroud ribs
145 Graphite composites flaps multispar structure
146 Port three-segment Fowler flaps
147 Spoiler panel graphite composites construction
148 Two-segment inboard spoiler/ speedbrake panels
149 Outboard three-segment spoiler/ speedbrake panels

150 Aileron tab
151 Port aileron
152 Wingtip strobe light
153 Port dual navigation lights
154 Wing panel three-spar and rib torsion box structure
155 Wing bonded bottom skin/ stringer panel with access hatches
156 Overwing fuel filler
157 Port wing integral fuel tank
158 Leading-edge double skin de-icing air ducting
159 Wing rib structure
160 De-icing air piccolo tube
161 Twin mainwheels
162 Levered suspension axle beam
163 Shock-absorber strut
164 Main landing gear leg strut pivot mounting
165 Hydraulic jack
166 Mainwheel well
167 Centre-section fuel collector with boost pump, port and starboard
168 Front spar/fuselage attachment link
169 De-icing air delivery ducting
170 Landing light, port and starboard
171 Position of pressure refuelling connection on starboard side

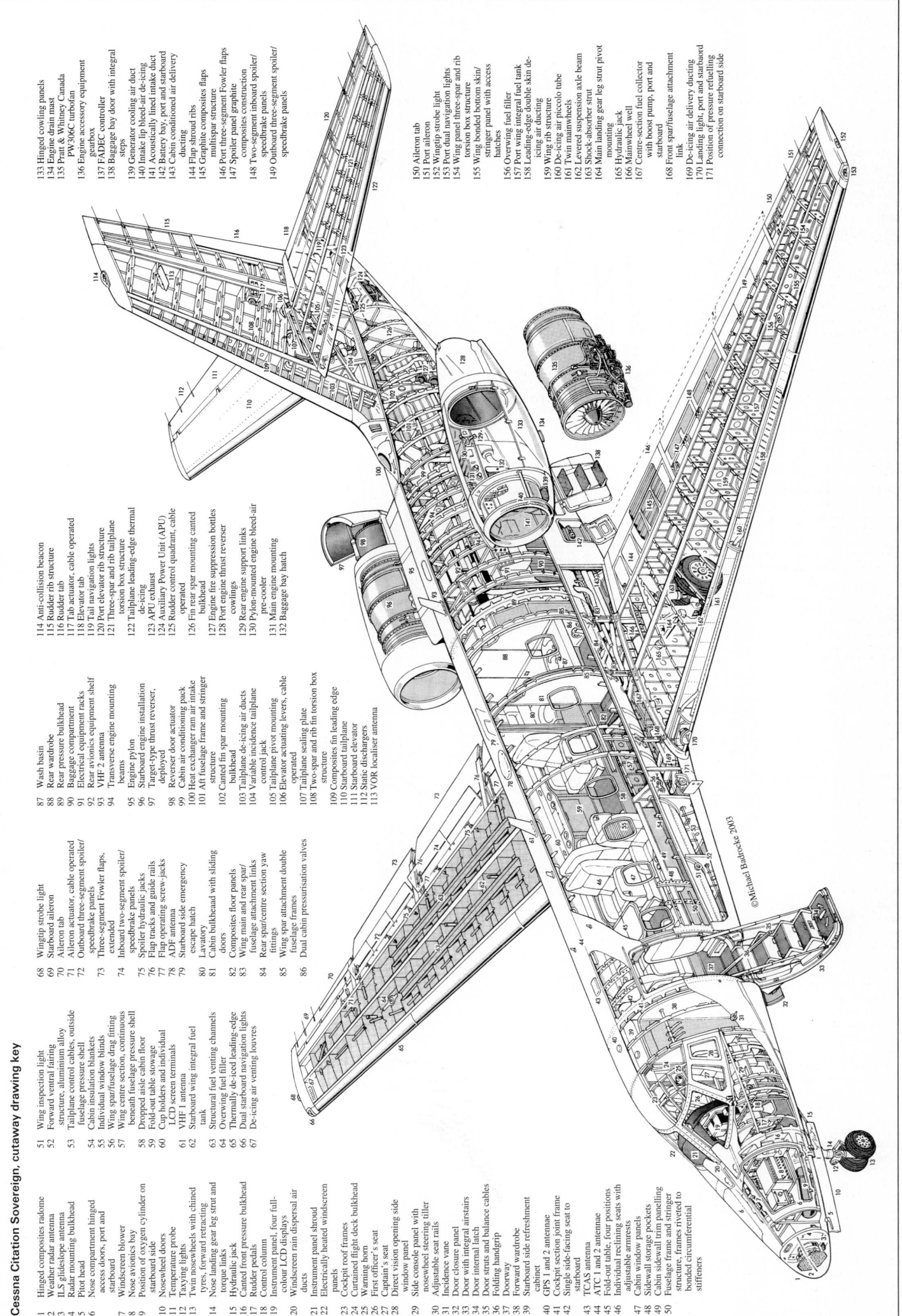

©Michael Badrocke 2003

AVIONICS: Honeywell Primus 2000 autopilot/flight director as core system.

Flight: Dual flight management systems (FMS), dual Honeywell Laseref IV IRS; dual attitude and heading reference systems, and Honeywell GPS standard.

Instrumentation: Five-tube EFIS with 178 × 203 mm (7 × 8 in) primary flight display (PFD) and multifunction display (MFD) for pilot and co-pilot; similarly sized engine instrument and crew alerting system (EICAS) display in centre. Max-Viz EVS 2000 enhanced visibility system optional.

DIMENSIONS, EXTERNAL:

Wing span	19.38 m (63 ft 7 in)
Wing aspect ratio	7.8
Length overall	22.05 m (72 ft 4 in)
Height overall	5.84 m (19 ft 2 in)
Tailplane span	7.95 m (26 ft 1 in)
Wheel track	3.23 m (10 ft 7 in)
Wheelbase	8.74 m (28 ft 8 in)

DIMENSIONS, INTERNAL:

Cabin (front to mid-pressure bulkhead):

Length: incl flight deck	8.89 m (29 ft 2 in)
excl flight deck	7.16 m (23 ft 6 in)
Max width	1.70 m (5 ft 7 in)
Max height	1.73 m (5 ft 8 in)
Baggage compartment volume (aft, including ski compartment)	2.32 m³ (82 cu ft)

AREAS:

Wings, gross	48.96 m² (527.0 sq ft)
Vertical tail surfaces (total)	10.31 m² (111.0 sq ft)
Horizontal tail surfaces (total)	11.15 m² (120.0 sq ft)

WEIGHTS AND LOADINGS:

Weight empty, typically equipped	9,798 kg (21,600 lb)
Max fuel weight	5,865 kg (12,931 lb)
Max T-O weight	16,374 kg (36,100 lb)
Max ramp weight	16,511 kg (36,400 lb)
Max landing weight	14,424 kg (31,800 lb)
Max zero-fuel weight	11,068 kg (24,400 lb)
Max wing loading	334.5 kg/m² (68.50 lb/sq ft)
Max power loading	272 kg/kN (2.67 lb/lb st)

PERFORMANCE:

Max operating Mach No. (M_{MO})	0.92

Max operating speed (V_{MO}):

S/L to FL800	270 kt (500 km/h; 310 mph)
FL800 to FL306	350 kt (648 km/h; 403 mph)
Max cruising speed, mid-cruise weight at FL370	M0.91

Max cruising speed at FL350

	525 kt (972 km/h; 604 mph)
Max rate of climb at S/L	1,113 m (3,650 ft)/min
Max certified altitude	15,545 m (51,000 ft)
T-O balanced field length (FAR Pt 25)	1,567 m (5,140 ft)
FAR Pt 25 landing field length	1,036 m (3,400 ft)

IFR range with two crew, M0.82 at FL490, 100 n mile (185 km; 115 mile alternate and 45 min reserves

	3,216 n miles (5,956 km; 3,700 miles)

OPERATIONAL NOISE LEVELS (FAR Pt 36 Amendment 20):

T-O	72.3 EPNdB
Approach	90.2 EPNdB
Sideline	83.0 EPNdB

UPDATED

Cessna 750 Citation X business jet (*Jane's/Paul Jackson*)   NEW/0526966

CIRRUS

CIRRUS DESIGN CORPORATION

4515 Taylor Circle, Duluth International Airport, Duluth, Minnesota 55811
Tel: (+1 218) 727 27 37
Fax: (+1 218) 727 21 48
Web: http://www.cirrusdesign.com
PRESIDENT: Alan Klapmeier
EXECUTIVE VICE-PRESIDENT: Dale Klapmeier
EXECUTIVE VICE-PRESIDENT AND CFO: Peter McDermott
EXECUTIVE VICE-PRESIDENT, OPERATIONS: David Coleal
VICE-PRESIDENT, ENGINEERING: Patrick Waddick
VICE-PRESIDENT, RESEARCH AND TECHNOLOGY: Dean Vogel
VICE-PRESIDENT, SALES AND MARKETING: Thomas Shea

Founded 1984 by Klapmeier brothers. Previously engaged in production of kits, Cirrus is concentrating on fully certified factory-built aircraft; first such product was ST-50 (see under Israviation in Israeli section of 1998-99 and earlier editions). Support for the VK-30 (see 1996-97 *Jane's*) continues.

Current products are SR20 and SR22. Cirrus purchased 20 per cent of parachute systems company BRS in September 1999. Workforce was 595 in January 2002, based at Duluth and Hibbing, Minnesota, and Grand Forks, North Dakota. In August 2001, Crescent Capital acquired 58 per cent share in Cirrus for US$100 million. The 250th SR-series aircraft was delivered in early November 2001. By late 2002, 600 had been delivered and a further 350 were on order. Production rate was two per working day in April 2002, scheduled to rise to three per working day in 2003.

UPDATED

CIRRUS SR20

TYPE: Four-seat lightplane.
PROGRAMME: Development began 1990; mockup revealed at Oshkosh 1994; first flight (N200SR) 31 March 1995; second prototype (N202CD) flown November 1995; FAR Pt 23 certification aircraft (N203FT), designated C-1, made first flight 28 January 1998 after completion of wing redesign to lower stall speed and improve lateral control. By end 1997, the two prototypes had accumulated 1,500 hours of test flying. A second production-standard aircraft (N204CD; C-2) joined the flight test programme on 3 June 1998, committed to trials of fuel and electrical systems and avionics. Recovery parachute trials involved eight deployments, including three for FAA. Late 1998 start of deliveries delayed by decision of avionics supplier, Trimble, to withdraw from general aviation. FAR Pt 23 certification Amdt 47 received 23 October 1998; Transport Canada certification granted in March 2002; first aircraft manufactured to Canadian certification standard delivered 2 April 2002 to Tim Harpell of Toronto.

Initial production aircraft first flew (N115CD) 22 March 1999, but lost on following day. First delivery (c/n 1005, N415WM) 20 July 1999, at which time 325 on order (although many since upgraded to SR22 orders). Production certificate, to enable Cirrus to carry out its own inspections, awarded 12 June 2000.

CURRENT VERSIONS: **SR20:** Total 133 delivered to customers by April 2001, when production suspended. MTOW 1,315 kg (2,900 lb).

SR20A: In production from July 2001. Additional 45 kg (100 lb) MTOW; landing light repositioned lower on engine cowling; and optional Goodrich Skywatch and

Cirrus Design SR20 four-seat lightplane (*Jane's/Paul Jackson*)   NEW/0533384

Sandel SN3308 EHSI for traffic information. Upgrades (including MTOW) retrofittable. *As described.*

SR20 *Version 2.0, Version 2.1 and Version 2.2:* Upgraded versions available from 2003; launched at EAA Sun 'n' Fun in April 2002. Differ from SR20A principally in having no vacuum instruments and featuring electrical system and avionics packages based on those of the SR22. *Version 2.0* has a single-alternator, dual-battery, dual-bus electrical system and features Avidyne FlightMax MFD, single Garmin GNS 430 GPS/com/nav and GNC 250XL GPS/com and S-Tec System 20 autopilot as standard. *Version 2.1* adds a dual-alternator electrical system and GNS 430/420 package. *Version 2.2* has dual GNS 430s, S-Tec Fifty FiveX autopilot with altitude preselect and Sandel EHSI.

CUSTOMERS: Orders for 552 received by July 2001. Total of 53 delivered in 2001, and 55 in first six months of 2002. By December 2002, 267 had been registered.

COSTS: US$207,800 with baseline/Configuration A avionics (2002).

DESIGN FEATURES: Low-wing, composites construction monoplane with upturned wingtips; mid-set tailplane with horn-balanced elevators; horn-balanced rudder; fixed tab on starboard aileron. First certified aircraft with ballistic parachute as standard equipment (Cirrus Airplane Parachute System or CAPS), fitted just aft of baggage compartment. Wing dihedral 4° 30′.

STRUCTURE: Composites monocoque; one-piece main spar attached to fuselage on two locations under front seats; spar carry-through attached to fuselage side-walls. Wing upper and lower surfaces bonded to spars and ribs, forming torsion box. All flying control surfaces are flush-riveted aluminium. Fin integral with fuselage; one-piece tailplane. Cabin incorporates composites roll-cage; acrylic windscreen and windows.

FLYING CONTROLS: Conventional and manual. Actuation by pushrods, cables and bellcranks. Electric trim tabs for roll and pitch activated through springed centring devices; ground-adjustable rudder tab. Three-position flaps.

LANDING GEAR: Fixed, tricycle; composites cantilever main legs; single wheels and speed fairings throughout. Main tyres 15×6.00-6, nosewheel tyre 5.00-5. Hydraulic calliper disk brakes on mainwheels. Castoring nosewheel.

POWER PLANT: One 149 kW (200 hp) Teledyne Continental IO-360-ES flat-six engine driving a two-blade Hartzell BHC-J2YF-1BF/F7694 propeller (three-blade 1.88 m; 6 ft 2 in PHC-J3YF-1MF/F77392-1 propeller optional). Single fuel tank in each wing, each of 113.5 litres (30.0 US gallons; 25.0 Imp gallons) capacity, of which 212 litres (56.0 US gallons; 46.7 Imp gallons) usable. Oil capacity 7.6 litres (2.0 US gallons; 1.7 Imp gallons).

ACCOMMODATION: Four 26 g energy attenuating seats in pairs; two forward-hinged passenger doors. Dual controls.

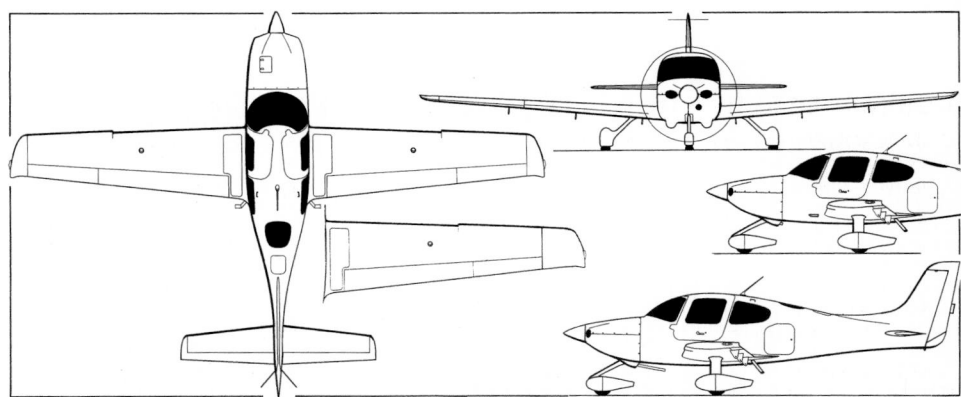

Cirrus Design SR20, with scrap views of SR22 wing and forward fuselage (*Jane's/Paul Jackson*)   0110908

Baggage compartment door aft of cabin on port side of fuselage. Interior reflects modern motorcar design.

SYSTEMS: Single-axis S-Tec System 20 autopilot with GPS; dual-axis S-Tec System 30 and Fifty FiveX optional; 24 V DC power system with 28 V 75 A alternator and 24 V 10 Ah battery. 12 V power port.

AVIONICS: Garmin integrated package.

Comms: GNS 430 colour GPS/com/nav IFR approach-certified GPS plus back-up, GNS 250XL GPS/com, GMA 340 audio panel and GTX 327 transponder. Options include GNS 420 colour GPS/com and/or additional GNS 430 (back-up IFR GPS).

Flight: Garmin GI 106 GPS/VOR/LOC/GSI. S-Tec System FortyX autopilot. Options include Goodrich Stormscope WX-500, S-Tec System FiftyX or Fifty FiveX, Goodrich Skywatch system, GI 102 or second GI 106.

Instrumentation: Avidyne EX3000C MFD standard, EX5000C optional.

EQUIPMENT: Integral ballistic recovery parachute (Cirrus Airplane Parachute System); vertical descent rate 7.3 m (24 ft)/s at S/L. Navigation, landing and strobe lights standard.

DIMENSIONS, EXTERNAL:

Wing span	10.82 m (35 ft 6 in)
Wing aspect ratio	9.1
Length overall	7.92 m (26 ft 0 in)
Height overall	2.80 m (9 ft 2¼ in)
Tailplane span	3.93 m (12 ft 10¾ in)
Wheel track	3.38 m (11 ft 1 in)
Wheelbase	2.29 m (7 ft 6 in)
Propeller diameter (two blade)	1.93 m (6 ft 4 in)
Cabin doors (each): Height	0.94 m (3 ft 1 in)
Width	0.86 m (2 ft 10 in)

DIMENSIONS, INTERNAL:

Cabin: Length	3.30 m (10 ft 10 in)
Max width	1.25 m (4 ft 1¼ in)
Max height	1.27 m (4 ft 2 in)

AREAS:

Wings, gross	12.56 m² (135.2 sq ft)
Horizontal tail	3.48 m² (37.50 sq ft)

WEIGHTS AND LOADINGS:

Weight empty	930 kg (2,050 lb)
Max fuel weight	163 kg (360 lb)
Max T-O weight	1,360 kg (3,000 lb)
Max wing loading	108.3 kg/m² (22.19 lb/sq ft)
Max power loading	9.12 kg/kW (15.00 lb/hp)

PERFORMANCE:

Never-exceed speed (VNE)	200 kt (370 km/h; 230 mph)

Cruising speed at 75% power

	160 kt (296 km/h; 184 mph)
Manoeuvring speed	135 kt (250 km/h; 155 mph) IAS
Stalling speed: flaps up	64 kt (119 km/h; 74 mph) IAS
flaps down	56 kt (104 km/h; 65 mph) IAS
Max rate of climb at S/L	268 m (880 ft)/min
Service ceiling	5,335 m (17,500 ft)
T-O run	435 m (1,430 ft)
T-O and landing from 15 m (50 ft)	622 m (2,040 ft)
Landing run	309 m (1,014 ft)

Range with reserves: at 75% power

	686 n miles (1,270 km; 789 miles)
at econ cruise	831 n miles (1,539 km; 956 miles)
g limits	+3.8/−1.9

UPDATED

CIRRUS SR22

TYPE: Four-seat lightplane.

PROGRAMME: Modifications to the SR20 (which see) have resulted in the SR22, which comprises an SR20 airframe with modified wing, more powerful engine and all-electric instrumentation (no vacuum system); prototype (N140CD) displayed at AOPA convention in October 2000; FAA certification awarded 30 November 2000. Delivery of first production aircraft (N415PJ) 6 February 2001.

CURRENT VERSIONS: **SR22:** *As described.*

SR21 tdi: Announced at Friedrichshafen in April 2001; comprises SR22 airframe powered by 169 kW (227 hp) SMA SR-305 turbocharged diesel engine; aimed initially at European market. Will offer lower fuel consumption. Initial deliveries expected during 2003.

CUSTOMERS: Orders for 234 received by June 2001. Launch of SR22 saw approximately 20 per cent of SR20 customers upgrade to new model. Total of 124 delivered in 2001, and 113 in first six months of 2002. By December 2002, 399 had been registered. Recent customers include AirShares Elite of Atlanta, Georgia, which ordered 25, plus 25 options, on 8 April 2002, for its fractional ownership programme.

COSTS: Configuration A: US$289,400; configuration B: US$307,500 (both 2002). Fractional (one-eighth) ownership averages US$147 per hour for 75 hours annual usage; initial cost US$50,000, less resale value after four years.

DESIGN FEATURES: As for SR20, but with fuselage-mounted vortex generator ahead of wingroots and wingtip extensions.

STRUCTURE: As for SR20.

FLYING CONTROLS: As for SR20, but flight-adjustable rudder tab and ground-adjustable tab on port elevator.

LANDING GEAR: As for SR20.

POWER PLANT: One 231 kW (310 hp) Teledyne Continental IO-550-N flat-six engine driving a three-blade Hartzell PHC-J3YF-1MF/F77392-1 constant-speed propeller. Single fuel tank in each wing, total capacity 318 litres (84.0 US gallons; 69.9 Imp gallons) of which 307 litres (81.0 US gallons; 67.4 Imp gallons) are usable. Oil capacity as SR20.

ACCOMMODATION: As for SR20.

SYSTEMS: Dual 24 V DC power system with 28 V 75 A alternator and 24 V 10 Ah battery. 12 V power port.

AVIONICS: *Comms:* Configuration A package includes single Garmin GNS 430 colour GPS/com/nav, GNS 420 GPS/com, GMA 340 audio panel and GTX 327 transponder as standard. Configuration B package substitutes second GNS 430 for GNS 420.

Flight: S-Tec Thirty autopilot standard in package A, replaced by S-Tec Fifty FiveX with altitude preselect in package B, which also adds Sandel SN3308 EHSI. Goodrich WX-500 Stormscope and Skywatch traffic information system optional.

Instrumentation: Avidyne FlightMax EX5000C 10.4 in MFD standard.

EQUIPMENT: Cirrus Airframe Parachute System standard.

DIMENSIONS, EXTERNAL:

Wing span	11.73 m (38 ft 6 in)
Wing aspect ratio	10.2
Length overall	7.92 m (26 ft 0 in)
Height overall	2.80 m (9 ft 2¼ in)
Tailplane span	3.93 m (12 ft 10¾ in)
Wheel track	3.23 m (10 ft 7¼ in)
Wheelbase	2.20 m (7 ft 2½ in)
Propeller diameter	1.98 m (6 ft 6 in)
Cabin doors (each): Height	0.99 m (3 ft 3 in)
Width	0.86 m (2 ft 10 in)

DIMENSIONS, INTERNAL:

Cabin: Length	3.30 m (10 ft 10 in)
Max width	1.25 m (4 ft 1¼ in)
Max height	1.27 m (4 ft 2 in)
Baggage compartment volume	0.91 m³ (32.0 cu ft)

AREAS:

Wings, gross	13.46 m² (144.9 sq ft)
Horizontal tail	3.48 m² (37.50 sq ft)

WEIGHTS AND LOADINGS:

Weight empty	1,021 kg (2,250 lb)
Baggage weight	59 kg (130 lb)
Max fuel weight	218 kg (480 lb)
Max T-O weight	1,542 kg (3,400 lb)
Max wing loading	114.6 kg/m² (23.46 lb/sq ft)
Max power loading	6.68 kg/kW (10.97 lb/hp)

PERFORMANCE:

Never-exceed speed (VNE)	204 kt (377 km/h; 234 mph)

Cruising speed at 75% power

	181 kt (335 km/h; 208 mph)
Manoeuvring speed	142 kt (263 km/h; 163 mph) IAS
Stalling speed: flaps up	70 kt (130 km/h; 81 mph) IAS
flaps down	59 kt (110 km/h; 68 mph) IAS
Max rate of climb at S/L	427 m (1,400 ft)/min
Service ceiling	5,335 m (17,500 ft)
T-O run	311 m (1,020 ft)
T-O to 15 m (50 ft)	480 m (1,575 ft)
Landing from 15 m (50 ft)	709 m (2,325 ft)
Landing run	348 m (1,140 ft)

Range with reserves:

at 75% power	744 n miles (1,377 km; 856 miles)
at econ cruise	more than 1,000 n miles (1,852 km; 1,150 miles)
g limits	+3.8/−1.9

UPDATED

Cirrus SR22 four-seat tourer (*Jane's/Paul Jackson*)   NEW/0533383

CLASSIC

CLASSIC FIGHTER INDUSTRIES INCORPORATED

Renamed Me 262 Project (which see).

UPDATED

COMMANDER

COMMANDER AIRCRAFT COMPANY

Wiley Post Airport, 7200 North-West 63rd Street, Bethany, Oklahoma 73008

Tel: (+1 405) 495 80 80
Fax: (+1 405) 495 83 83
e-mail: cacsales@telepath.com
Web: http://www.commanderair.com
CEO: Wirt Walker
PRESIDENT AND COO: Mat Goodman
EXECUTIVE VICE-PRESIDENT: Carl Gull

Company (a division of Aviation General Inc since 1998) acquired manufacturing, marketing and support rights for Rockwell Commander 112 and 114 from Gulfstream Aerospace Corporation in 1988; spares and support services for existing aircraft and manufacturing based in Oklahoma; 120 employees in 1999. Commander 114 remains in production, including TC version introduced in 1995. Deliveries in 1997-2002 comprised 14, 13, 13, 21, 10 and seven (none in final quarter). In March 2000, Commander announced the introduction of the 115, an updated and refined replacement for the 114.

Factory floor area 9,638 m² (103,742 sq ft). Commander has over 140 authorised service centres worldwide.

Commander filed for Chapter 11 bankruptcy protection on 27 December 2002, citing US$1 million cash flow shortfall and US$3.7 million total net indebtedness.

UPDATED

COMMANDER 115

TYPE: Four-seat lightplane.

PROGRAMME: Original Rockwell 112 first flew (N112AC) 4 December 1970; 149 kW (200 hp) Lycoming IO-360; production deliveries began August 1972 and ended in February 1978; more powerful Rockwell 114 subsequently added to range; all production terminated in September 1979. Following formation of Commander Aircraft, 114B recertified 5 May 1992. Commander 114 series was withdrawn in early 2000 in favour of the 115; see 2000-01 edition for details. New version uses type certificate for 114 and continues to be registered as such.

CURRENT VERSIONS: **Commander 115:** Basic version, announced in early 2000. Improvements over 114B include increased fuel capacity; wing-mounted landing

cords. Goodyear tyres: mainwheels 10.00-6, pressure 1.86 bar (27 lb/sq in); tailwheel 8.00-4, pressure 3.10 bar (45 lb/sq in). Cleveland brakes. Minimum ground turning radius 7.92 m (26 ft 0 in).

POWER PLANT: One 265 kW (355 hp) VOKBM M-14PT radial engine driving a constant-speed two-blade wooden propeller. Main fuel tank in fuselage contains 121 litres (32.0 US gallons; 26.6 Imp gallons); wing tank contains 38 litres (10.0 US gallons; 8.3 Imp gallons) and ferry tank a further 151 litres (40.0 US gallons; 33.3 Imp gallons); total usable fuel 299 litres (79.0 US gallons; 65.8 Imp gallons). Oil capacity 15 litres (4.0 US gallons; 3.3 Imp gallons).

ACCOMMODATION: Pilot and either one or two passengers, depending on model, in open cockpit. Baggage compartment behind headrest.

SYSTEMS: 10 A electrical system for radio and lights; pneumatic engine starting system.

AVIONICS: *Comms:* Honeywell radio, transponder and GPS.

DIMENSIONS, EXTERNAL:
Wing span: upper	7.32 m (24 ft 0 in)
lower	7.01 m (23 ft 0 in)
Wing chord, constant	1.07 m (3 ft 6 in)
Length overall	6.40 m (21 ft 0 in)
Max diameter of fuselage	1.05 m (3 ft 5½ in)
Height overall	2.44 m (8 ft 0 in)
Tailplane span	2.29 m (7 ft 6 in)
Wheel track	1.98 m (6 ft 6 in)
Wheelbase	5.18 m (17 ft 0 in)
Propeller diameter	2.39 m (7 ft 10 in)
Propeller ground clearance	0.91 m (3 ft 0 in)

DIMENSIONS, INTERNAL:
Cabin: Length	2.44 m (8 ft 0 in)
Max width	0.81 m (2 ft 8 in)
Max height	0.81 m (2 ft 8 in)
Floor area	0.56 m² (6.0 sq ft)
Baggage hold volume	0.11 m³ (4.0 cu ft)

AREAS:
Wings, gross	14.96 m² (161.0 sq ft)
Ailerons (total)	2.60 m² (28.00 sq ft)
Rudder, incl tab	1.02 m² (11.00 sq ft)
Tailplane	2.97 m² (32.00 sq ft)
Elevators, incl tab	1.49 m² (16.00 sq ft)

WEIGHTS AND LOADINGS:
Weight empty	635 kg (1,400 lb)
Max fuel weight	218 kg (480 lb)
Max T-O and landing weight	1,043 kg (2,300 lb)
Max zero-fuel weight	848 kg (1,870 lb)
Max wing loading	69.7 kg/m² (14.29 lb/sq ft)
Max power loading	3.94 kg/kW (6.48 lb/hp)

PERFORMANCE:
Never-exceed speed (V$_{NE}$) at S/L	
	208 kt (386 km/h; 240 mph)
Max operating speed (V$_{MO}$)	191 kt (354 km/h; 220 mph)
Econ cruising speed at 2,590 m (8,500 ft)	
	148 kt (274 km/h; 170 mph)
Stalling speed, power off	59 kt (108 km/h; 67 mph)
Max rate of climb at S/L	1,371 m (4,500 ft)/min
Service ceiling	3,660 m (12,000 ft)
T-O run	91 m (300 ft)

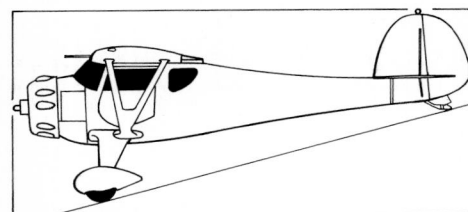

Culp's MonoCulp *(Jane's/Paul Jackson)*   0092144

T-O to 15 m (50 ft)	122 m (400 ft)
Landing from 15 m (50 ft)	213 m (700 ft)
Landing run	244 m (800 ft)
Range: with max fuel at econ cruising speed	
	955 n miles (1,770 km; 1,100 miles)
with max payload	487 n miles (902 km; 561 miles)
g limits	±10

UPDATED

CULP'S MONOCULP

TYPE: Four-seat kitbuilt.

PROGRAMME: Announced at Sun 'n' Fun, April 2000, when prototype under construction.

CUSTOMERS: Two kits and three sets of plans sold by May 2000.

COSTS: Kit price US$58,000 with wings completed (2002).

DESIGN FEATURES: Scaled-up (125 per cent) version of 1930s Monocoupe 90A. Quoted build time 2,500 hours.

FLYING CONTROLS: Conventional and manual.

STRUCTURE: Strut-braced high wing; wire-braced tailplane.

LANDING GEAR: Tailwheel type; fixed. Speed fairings on mainwheels.

POWER PLANT: One 294 kW (394 hp) VOKBM M-14PF nine-cylinder radial engine. Fuel capacity 318 litres (84.0 US gallons; 70.0 Imp gallons).

ACCOMMODATION: Pilot and three passengers in two pairs of seats; cabin door on port side.

DIMENSIONS, EXTERNAL:
Wing span	9.75 m (32 ft 0 in)
Length overall	7.47 m (24 ft 6 in)
Height overall	2.54 m (8 ft 4 in)

DIMENSIONS, INTERNAL:
Cabin max width	1.22 m (4 ft 0 in)

AREAS:
Wings, gross	17.84 m² (192.0 sq ft)

WEIGHTS AND LOADINGS:
Weight empty	771 kg (1,700 lb)
Max T-O weight	1,315 kg (2,900 lb)
Max wing loading	73.7 kg/m² (15.10 lb/sq ft)
Max power loading	4.48 kg/kW (7.36 lb/hp)

PERFORMANCE:
Never-exceed speed (V$_{NE}$)	234 kt (434 km/h; 270 mph)
Normal cruising speed	148 kt (274 km/h; 170 mph)
Stalling speed, power off	54 kt (100 km/h; 62 mph)
Max rate of climb at S/L	914 m (3,000 ft)/min
Service ceiling	3,660 m (12,000 ft)
T-O run	122 m (400 ft)

Original Mono Aircraft Monocoupe 90, built in 1930, on which the MonoCulp is based *(Jane's/Paul Jackson)*
NEW/0533382

Landing run	274 m (900 ft)
Range with max fuel	827 kt (1,532 km; 952 miles)
Endurance	4 h 0 min

UPDATED

CULP'S SOPWITH PUP

TYPE: Tandem-seat biplane kitbuilt.

PROGRAMME: Announced at Sun 'n' Fun, April 2000, prototype then under construction.

CUSTOMERS: Orders for 12 kits and 18 sets of plans received by May 2000. Three completed by January 2003.

COSTS: Plans US$200; kit US$36,252 (2003).

DESIGN FEATURES: Based on First World War Sopwith Pup fighter, but using all-new materials. Quoted build time 2,500 hours.

STRUCTURE: Wood, fabric and tube construction. Biplane wings with interplane struts.

LANDING GEAR: Tailwheel type; fixed.

POWER PLANT: One VOKBM M-14P nine-cylinder radial derated to 179 kW (240 hp); engines in range 164 to 268 kW (220 to 360 hp) can be fitted. Fuel capacity 189 litres (50.0 US gallons; 41.7 Imp gallons).

ACCOMMODATION: Pilot and passenger in tandem in open cockpit.

DIMENSIONS, EXTERNAL:
Wing span	8.08 m (26 ft 6 in)
Length overall	5.49 m (18 ft 0 in)
Height overall	2.74 m (9 ft 0 in)

DIMENSIONS, INTERNAL:
Cabin max width	0.71 m (2 ft 4 in)

AREAS:
Wings, gross	24.62 m² (265.0 sq ft)

WEIGHTS AND LOADINGS:
Weight empty	590 kg (1,300 lb)
Max T-O weight	1,088 kg (2,400 lb)
Max wing loading	44.2 kg/m² (9.06 lb/sq ft)
Max power loading	6.09 kg/kW (10.00 lb/hp)

PERFORMANCE:
Never-exceed speed (V$_{NE}$)	191 kt (354 km/h; 220 mph)
Normal cruising speed	130 kt (241 km/h; 150 mph)
Stalling speed	33 kt (62 km/h; 38 mph)
Max rate of climb at S/L	1,372 m (4,500 ft)/min
Service ceiling	3,660 m (12,000 ft)
T-O run	76 m (250 ft)
Landing run	152 m (500 ft)
Range with max fuel	521 n miles (965 km; 600 miles)

UPDATED

DERRINGER

DERRINGER AIRCRAFT COMPANY LLC

1246 Sabovich Street, Mojave, California 93501
Tel: (+1 661) 824 22 22
Fax: (+1 661) 824 22 02
e-mail: derringer@derringeraircraft.com
Web: http://www.derringeraircraft.com
MARKETING MANAGER: Roger Lewis

SALES DIVISION:
558 S Broad Street, Mobile, Alabama 36603
Tel: (+1 888) 909 82 22
Fax: (+1 334) 433 33 64

Derringer continues to promote the D-1 New Derringer utility twin, but no production is known. A description last appeared in the 2001-02 *Jane's*.

UPDATED

Prototype D-1 New Derringer receiving attention at Mojave in 2001 *(Peter J Cooper)*
NEW/0525831

DFL

DFL HOLDINGS INC

Team Tango Division, 3001 Northeast 20th Way, Gainesville, Florida 32609
Tel: (+1 800) 340 40 35 or (+1 352) 373 41 07
Fax: (+1 352) 373 53 07
e-mail: ttango@fdn.com
Web: http://www.teamtango.com

Current projects of DFL are the two-seat Tango-2 and a four-seat development, the Foxtrot-4. The company has two Builders' Centres located in Gainesville, Florida, and Saint Louis, Missouri.

VERIFIED

DFL FOXTROT-4

TYPE: Four-seat kitbuilt.
PROGRAMME: Development began in 1999; first engine runs of prototype (N747F) completed 27 August 2001; first flight was due before end of 2001; no further information received by late 2002, however.
COSTS: Basic kit US$29,995, plus optional US$11,500 assisted assembly.
CUSTOMERS: Two under construction by August 2002.
DESIGN FEATURES: Conventional design similar in appearance to Tango-2. Vinylester resin used in place of epoxy to reduce cure time of composites.
FLYING CONTROLS: Conventional and manual. Electric flaps.
STRUCTURE: Extensive use of composites throughout.
LANDING GEAR: Tricycle type; fixed. Speed fairings on all wheels.
POWER PLANT: Prototype fitted with 224 kW (300 hp) Textron Lycoming IO-540; engines in the range 149 to 224 kW (200 to 300 hp) are recommended. Standard fuel capacity 265 litres (70.0 US gallons; 58.3 Imp gallons); optional capacity 409 litres (108 US gallons; 90.0 Imp gallons).
ACCOMMODATION: Pilot and three passengers in two pairs of seats. Upward-opening, centreline-hinged door each side; fixed windscreen. Baggage stowage behind seats.

DIMENSIONS, EXTERNAL:
Wing span	9.75 m (32 ft 0 in)
Wing aspect ratio	8
Length overall	7.56 m (24 ft 9½ in)
Height overall	2.21 m (7 ft 3 in)

DIMENSIONS, INTERNAL:
Cabin: Max width	1.17 m (3 ft 10 in)
Max height	1.09 m (3 ft 7 in)
Baggage volume	0.40 m³ (14.0 cu ft)

AREAS:
Wings, gross	11.89 m² (128.0 sq ft)

Second DFL Tango-2 kitbuilt tourer, built by Dr Sid Hood, which first flew in May 1999 (*Jane's/Paul Jackson*)
0103550

WEIGHTS AND LOADINGS (194 kW; 260 hp engine):
Weight empty	635 kg (1,400 lb)
Baggage capacity	68 kg (150 lb)
Max payload	250 kg (552 lb)
Max T-O weight	1,179 kg (2,600 lb)
Max wing loading	99.2 kg/m² (20.31 lb/sq ft)
Max power loading	6.09 kg/kW (10.00 lb/hp)

PERFORMANCE (estimated, 194 kW; 260 hp engine):
Never-exceed speed (V_NE)	212 kt (394 km/h; 245 mph)
Max operating speed	200 kt (370 km/h; 230 mph)
Normal cruising speed	189 kt (351 km/h; 218 mph)
Stalling speed, flaps down	51 kt (94 km/h; 58 mph)
Max rate of climb at S/L	914 m (3,000 ft)/min
Service ceiling	7,315 m (24,000 ft)
Range with max fuel	1,564 n miles (2,896 km; 1,800 miles)
g limits	+6/–4

UPDATED

DFL TANGO-2

TYPE: Side-by-side kitbuilt.
PROGRAMME: Developed from the Aero Mirage TC-2 of 1983. Prototype flew November 1996. First two production aircraft displayed at Oshkosh in July 1999; third flew March 2000.
CUSTOMERS: Five flying and 13 under construction by mid-2002.
COSTS: Basic kit US$20,895, plus US$10,500 assisted assembly, US$36,500 O-360 engine kit and US$11,995 VFR avionics kit (2001).

DESIGN FEATURES: Conventional design optimised for amateur assembly; low-mounted, constant-chord wings and mid-positioned tailplane. Steeply raked windscreen. Quoted build time 570 hours for quick-build kit version; company has building facility to enable purchasers to complete in 30 to 45 days.
FLYING CONTROLS: Conventional and manual. Flaps deflect 10/20/35°.
STRUCTURE: Generally of composites. Two-spar wing has modified NACA64415 aerofoil and 3° dihedral.
POWER PLANT: One 134 kW (180 hp) Textron Lycoming O-360 flat-four driving a two-blade Hartzell HCC2YR-7666A constant-speed propeller. Fuel capacity 152 litres (40.0 US gallons; 33.3 Imp gallons) in two wing tanks; two optional extra 38 litre (10.0 US gallon; 8.3 Imp gallon) tanks can be fitted. Optionally, 149 kW (200 hp) IO-360 or 194 kW (260 hp) IO-540.
LANDING GEAR: Tricycle type; fixed. Cantilever mainwheel legs are fibre glass; nose leg is 4130 steel tube; speed fairings on all wheels. Main units have Cleveland wheels and brakes with Goodyear 5.00-5 tyres; nosewheel is steerable and has 4.50-5 Lamb tyre.
ACCOMMODATION: Two, side by side. Upward-opening, centreline hinged door each side; fixed windscreen. Baggage stowage behind seats.
AVIONICS: To customer's specification. Standard VFR option includes Honeywell KLX 135 GPS/com, KT 76A transponder and Sigtronics intercom; IFR version adds KX 155 nav/com and KI 209 VOR/LOC.

DIMENSIONS, EXTERNAL:
Wing span	7.62 m (25 ft 0 in)
Wing aspect ratio	8.3
Length overall	6.31 m (20 ft 8½ in)
Height overall	2.21 m (7 ft 3 in)
Wheel track	2.54 m (8 ft 4 in)
Propeller diameter	1.93 m (6 ft 4 in)

AREAS:
Wings, gross	6.97 m² (75.0 sq ft)

DIMENSIONS, INTERNAL:
Cockpit Max width	1.12 m (3 ft 8 in)
Max height	0.95 m (3 ft 1½ in)
Baggage compartment volume	0.34 m³ (12.0 cu ft)

WEIGHTS AND LOADINGS:
Weight empty	488 kg (1,075 lb)
Baggage capacity	45 kg (100 lb)
Max T-O weight	839 kg (1,850 lb)
Max wing loading	120.4 kg/m² (24.67 lb/sq ft)
Max power loading	6.26 kg/kW (10.28 lb/hp)

AREAS:
Wings, gross	6.97 m² (75.0 sq ft)

PERFORMANCE:
Never-exceed speed (V_NE)	212 kt (394 km/h; 245 mph)
Max level speed	191 kt (354 km/h; 220 mph)
Max cruising speed	174 kt (322 km/h; 200 mph)
Stalling speed	55 kt (102 km/h; 63 mph) IAS
Max rate of roll	120°/s
Max rate of climb at S/L	549 m (1,800 ft)/min
Service ceiling	7,315 m (24,000 ft)
T-O run	183 m (600 ft)
Landing run	244 m (800 ft)
Max range	869 n miles (1,609 km; 1,000 miles)
g limits	+6/–4

UPDATED

DFL Foxtrot-4 prototype displayed incomplete at Sun 'n' Fun, 2001 (*Jane's/Susan Bushell*)
0121673

EAGLE

EAGLE R&D LTD CO

2512 Caldwell Boulevard, Nampa, Idaho 83651-1518
Tel: (+1 208) 461 25 67
Fax: (+1 208) 454 37 52
Web: http://www.helicycle.com
CEO: Buford J Schramm

Company founded by B J Schramm, originator of the RotorWay kit helicopter programme. Eagle's sales policy for its Helicycle includes mandatory test flying of customer-completed aircraft by a factory pilot before handover of final key components.

NEW ENTRY

EAGLE HELICYCLE

TYPE: Single-seat helicopter kitbuilt.
PROGRAMME: Prototype (N3275Q) registered in April 1995. First customer-built aircraft (N9019U) assembled by Douglas Swochert of Burlington, Wisconsin; initially intended with two-stroke Rotax, but completed with T62 turboshaft and registered as a Swochert A/W 95.
CUSTOMERS: At least 30 kits sold by 2002. Mr Swochert registered a second in his name in April 2002.
DESIGN FEATURES: Embodies best features from Mr Schramm's experience of rotary-wing design; comparable in capabilities to industry-standard Robinson R22, but single-seat. Conventional appearance; small streamlined, enclosed cabin with power plant, gearbox and fuel

immediately to rear; tail rotor and empennage mounted on open boom; dorsal and ventral fins, latter incorporating tail bumper. Two-blade main and tail rotors.
Handling assisted by fully harmonised rotor; modulated collective system; elastomeric thrust bearings; generous flapping angle for low g and slope landing conditions; control friction devices; and electronic throttle control.
STRUCTURE: Metal tube structure with composites cabin.
LANDING GEAR: Twin-skid type.
POWER PLANT: One piston or turbine engine. Two-stroke Rotax initially proposed, but support changed in 2002 to Hirth 3701 rated at 64.9 kW (87 hp) at 5,000 rpm. Alternative 59.6 kW (79.9 hp) Rotax 912 flat-four installation under development in 2002.

First production helicopter flown with Solar T62 turbine ground power plant derated to 78.3 kW (105 hp); modified T62 installation offered by Eagle. Fuel capacity 79.0 litres (21.0 US gallons; 17.5 Imp gallons).

Data for T62 engine installation:

DIMENSIONS, EXTERNAL:

Main rotor diameter	6.35 m (20 ft 10 in)
Tail rotor diameter	1.12 m (3 ft 8 in)
Length overall	6.05 m (19 ft 10 in)
Height overall	2.18 m (7 ft 2 in)

WEIGHTS AND LOADINGS:

Weight empty	229 kg (505 lb)
Max T-O weight	372 kg (820 lb)

PERFORMANCE:

Max cruising speed	87 kt (161 km/h; 100 mph)
Max rate of climb at S/L	274 m (900 ft)/min
Hovering ceiling IGE	2,895 m (9,500 ft)
Max range	156 n miles (289 km; 180 miles)

NEW ENTRY

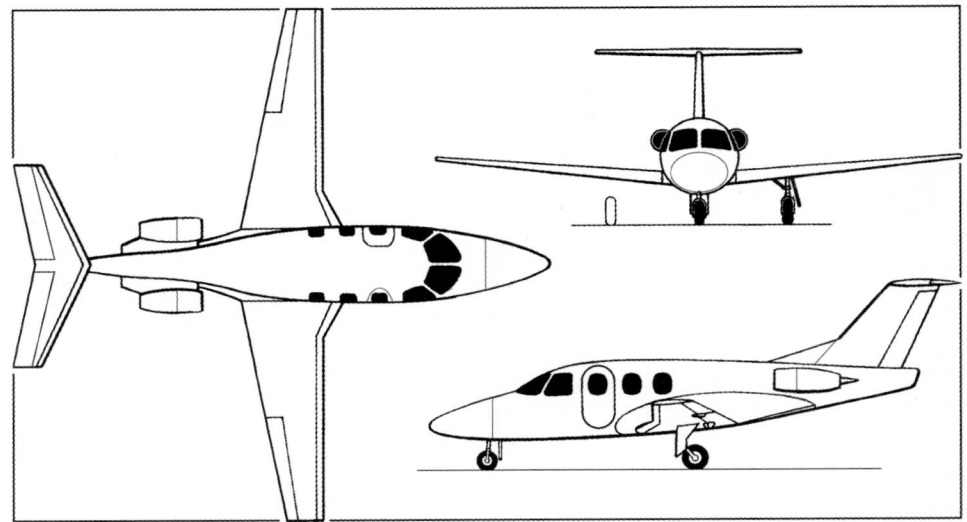

Prototype Eagle Helicycle with Rotax engine   *NEW*/0527118

ECLIPSE

ECLIPSE AVIATION CORPORATION

2503 Clark Carr Loop SE, Albuquerque, New Mexico 87106
Tel: (+1 505) 245 75 55
Fax: (+1 505) 245 78 88
e-mail: info@eclipseaviation.com
Web: http://www.eclipseaviation.com
CHAIRMAN: Harold A Poling
PRESIDENT AND CEO: Vern Raburn
VICE-PRESIDENT, AVIONICS AND ELECTRONICS: Don Burtis
VICE-PRESIDENT, BUSINESS AFFAIRS: Jack Harrington
VICE-PRESIDENT, CUSTOMER AND PRODUCT SUPPORT:
    Michael McConnell
VICE-PRESIDENT, ENGINEERING: Oliver Masefield
VICE-PRESIDENT, MANUFACTURING: Larry Jones
VICE-PRESIDENT, SUPPLY MANAGEMENT: Chris Herzog
VICE-PRESIDENT, SALES AND MARKETING: Dottie Hall
VICE-PRESIDENT, TRAINING AND FLIGHT OPERATIONS: Don Taylor
VICE-PRESIDENT, FINANCE AND ADMINISTRATION/CFO:
    Peter C Reed
DIRECTOR OF PUBLIC RELATIONS: Cory Canada

Eclipse Aviation formed May 1998 (as Pronto Aircraft) to market the Eclipse 500, which supersedes the Williams V-Jet II (1999-2000 *Jane's*). Eclipse had funding of US$238 million in place by July 2002. At that time the company had more than 200 employees working on the project.
UPDATED

ECLIPSE ECLIPSE 500

TYPE: Light business jet.
PROGRAMME: Development began June 1999 and announced March 2000; 22 per cent scale model wind tunnel testing undertaken March to April 2000; preliminary design review completed September 2000; full-size mockup displayed at EAA AirVenture 2000 and NBAA 2000. Third round of wind tunnel tests completed January 2001 and critical design review began May 2001; transfer of engineering and certification responsibility from Williams completed July 2001. Designated 3.43 kN (770 lb st) Williams EJ-22 engine first flown in Rockwell Sabreliner 60 testbed (N15HF) 30 May 2002. Prototype (c/n 500-100/N500EA) rolled out 13 July 2002; first flight 26 August 2002, but no further flights reported by mid-November when Eclipse announced abandonment of Williams engine; selection of Pratt & Whitney Canada announced 19 February 2003. Eight test airframes will be used in the

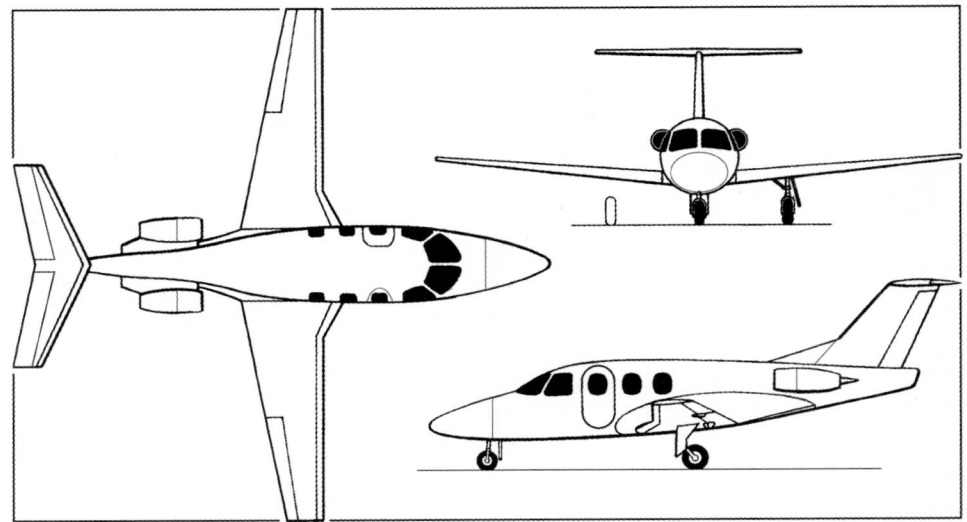

Eclipse 500 in original Williams-powered configuration (*Jane's/James Goulding*)   *NEW*/0527132

development and certification programme, dedicated as follows: c/n 500-100: low-speed envelope handling qualities including airspeed calibration, stall characteristics, longitudinal stability and control, plus trim, flap and landing gear operation; c/n 500-101: flight controls, flight handling characteristics, flutter, flight loads, ice protection, windows and windscreens; c/n 500-102: standard oxygen system, fuel system, landing gear, engine controls, power plant installation, lighting, pressurisation, environmental control system, electrical system; c/n 500-103: Pt 135 oxygen system, fire protection, air data system, avionics, noise emissions; c/ns 500-104 and 500-105; static structural and fatigue test airframes; c/ns 500-106 and 500-107: 'beta' airframes, completed to full production configuration, each to be employed on a 1,000-hour performance and durability trial before first customer deliveries, and will continue to be operated by the manufacturer on intensive post-service entry evaluation and problem-solving trials. Pending availability of Pratt & Whitney Canada PW610F, the prototype will resume flight testing during 2003 with two 4.48 kN (1,000 lb st) Teledyne CAE F408-CA-400

turbofans. First flight with PW610Fs is expected before the end of 2004. The first four aircraft will accumulate some 1,000 flight hours during the test programme, leading to FAR Pt 23 plus take-off and landing to Pt 25 standard, with target date of first quarter 2006.
CUSTOMERS: Total of 1,357 firm orders and 715 options held by 9 September 2002, of which 530 were from individuals, from customers in the USA, Europe and South America. Announced customers include Aviace AG of Switzerland, which ordered 112 on 28 May 2002 for its private jet club fractional ownership scheme. A provisional order for 1,000 aircraft placed by Nimbus Group on 31 August 2001 subsequently lapsed in July 2002.
COSTS: Development programme US$350 million (estimated). Unit cost US$950,000-US$975,000 for pre-2003 orders, US$1.175 million thereafter (2003).
DESIGN FEATURES: Designed to fulfil goals of NASA's Small Air Transportation System programme. Friction stir welding to be used for first time on high-volume production, obviating need for large numbers of rivets and reducing weight of finished article. Unswept, low-mounted wing and T tail. Redesign work undertaken during 2001

Eclipse 500 business jet prototype on its maiden flight   *NEW*/0527134

included increasing the size of the horizontal engine pylons, modifying the rear fuselage to improve engine intake airflow and modifications to wingroot shape, wing placement (moved aft 9 cm (3½ in)) and redesigned tail. Airframe modifications that will result from installation of the PW610Fs will include moving the engine mounts forward by 0.24 m (9 ½ in), strengthening the wing spar, moving the vapour-cycle machine from the wingroot fairing to the nose, modifying the flaps to increase Fowler effect, and addition of small tip tanks to accommodate the additional fuel needed to meet range guarantees.

FLYING CONTROLS: Conventional and manual. Sidestick controller. Electrically actuated Fowler flaps and split tailcone speedbrakes.

STRUCTURE: Principally of aluminium; sharply tapered rear fuselage means distance between engine centres will be 1.04 m (3 ft 5 in).

LANDING GEAR: Retractable tricycle type with trailing-link main gear designed by Cal-Draulics. Cleveland wheels and brakes with Michelin radial tyres fitted for improved rough-field operations.

POWER PLANT: Two pod-mounted Pratt & Whitney Canada PW610F turbofans, each rated at 4.00 kN (900 lb st) at ISA+10°C, with FADEC.

ACCOMMODATION: Pilot and up to five passengers; dual sidestick controllers for pilot(s); cabin laid out in club format. Polarised windows throughout; door on port side of cabin behind flight deck. Optional lavatory. Horizontally split door on port side, behind pilots.

SYSTEMS: Pressurisation system maintained sea level cabin environment to 6,553 m (21,500 ft) and 2,440 m (8,000 ft) cabin at 12,500 m (41,000 ft) altitude. Pneumatic de-icing system for wing leading-edges, bleed anti-icing for engine inlet, and electrically heated windscreen and air system probes. Lighting system by DeVore Aviation.

AVIONICS: Avio intelligent flight system, developed in conjunction with Avidyne, BAE Systems and General Dynamics; features all-glass, three-screen, fully IFR cockpit with two PFDs and one MFD. Avionics suite includes dual three-axis autopilots; FMS; autothrottle; aircraft performance computer; colour weather radar; traffic information services; ADS-B; TAWS-B; dual VHF nav/com/LOC/GS, Mode S transponders, GPS and AHRS with air data computer and pitot static systems; active route moving map display; flight path predictor; VNAV; LNAV; ELT; data loader, health monitoring, and RVSM capability. *All data are provisional and apply to prototype configuration with EJ-22 engine.*

DIMENSIONS, EXTERNAL:
Wing span	10.97 m (36 ft 0 in)
Length overall	10.08 m (33 ft 1 in)
Height overall	3.35 m (11 ft 0 in)

DIMENSIONS, INTERNAL:
Cabin: Length	3.76 m (12 ft 4 in)
Max width	1.42 m (4 ft 8 in)
Max height	1.27 m (4 ft 2 in)
Volume	5.4 m³ (191 cu ft)
Baggage compartment volume	1.20 m³ (42.5 cu ft)

WEIGHTS AND LOADINGS:
Weight empty: standard	1,225 kg (2,700 lb)
with extended-range tip tanks	1,256 kg (2,770 lb)
Max fuel weight: standard	603 kg (1,330 lb)
Max T-O weight	2,131 kg (4,700 lb)
Max ramp weight	2,143 kg (4,725 lb)
Max landing weight	2,086 kg (4,600 lb)
Max power loading	311 kg/kN (3.05 lb/lb st)

PERFORMANCE:
Max operating speed (MMO/VMO)	M0.68 (285 kt; 528 km/h; 328 mph)
Max cruising speed	355 kt (657 km/h; 409 mph)
Stalling speed	62 kt (115 km/h; 72 mph)
Max rate of climb at S/L	817 m (2,680 ft)/min
Rate of climb at S/L, OEI	256 m (840 ft)/min
Time to climb to 10,670 m (35,000 ft)	20 min
Service ceiling	12,495 m (41,000 ft)
Service ceiling, OEI	8,535 m (28,000 ft)
T-O run	628 m (2,060 ft)
Landing run	625 m (2,050 ft)
Range: with max fuel and tip tanks	1,825 n miles (3,380 km; 2,100 miles)
without tip tanks	1,600 n miles (2,963 km; 1,841 miles)
with four occupants, NBAA IFR alternate	1,300 n miles (2,407 km; 1,496 miles)

UPDATED

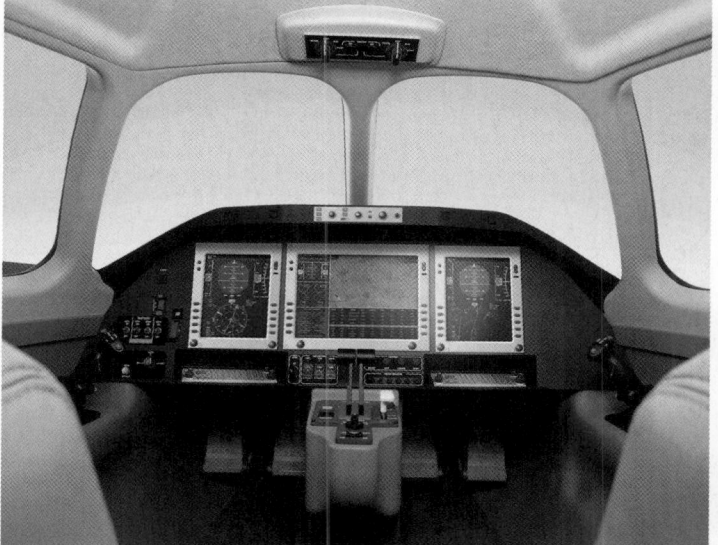

Eclipse 500 instrument panel *NEW*/0527133

Five-seat 'semi-club' seating arrangement option for the Eclipse 500 *NEW*/0527131

ENSTROM

THE ENSTROM HELICOPTER CORPORATION

PO Box 490, 2209 North 22nd Street, Twin County Airport, Menominee, Michigan 49858-0490
Tel: (+1 906) 863 12 00
Fax: (+1 906) 863 68 21
e-mail: enstrom@cybrzn.com
Web: http://www.enstromhelicopter.com
CHAIRMAN: Robert M Tuttle
CHIEF EXECUTIVE OFFICER: Steve Daniels
VICE-PRESIDENT, ENGINEERING: Robert L Jenny
DIRECTOR, SALES: P Bayard duPont
DIRECTOR, MANUFACTURING: James Stutson

Enstrom Helicopter Corporation was established in 1959 to develop and produce light helicopters, based on Rudy Enstrom's initial design work. Extensive development and certification efforts led to start of production of the F28 helicopter in April 1965; development of turbine-powered model was initiated in 1988, culminating in certification of the TH28 in September 1992 and the 480, to FAR Pt 27 standards, in December 1994.

Initially publicly owned, the company has passed through a period as a subsidiary of a large corporation and is now privately owned. By the beginning of 2000, Enstrom had produced over 1,000 helicopters of which over 700 remain in operation; deliveries have comprised eight in 1999, seven in 2000 and eight in 2001. The fleet has amassed over three million flying hours. Negotiations continue with the Wuhan Helicopter Company regarding possible licensed production in China; see China section.

UPDATED

ENSTROM F28 and 280

TYPE: Three-seat helicopter.
PROGRAMME: Prototype F28 flew 12 November 1960. Basic F28A and 280 described in 1978-79 *Jane's*; replaced by

Enstrom 280FX light helicopter *(Jane's/Paul Jackson)* *NEW*/0533381

turbocharged F28C and 280C, certified by FAA 8 December 1975 and last described in 1984-85 *Jane's*; production of these models ceased November 1981; succeeded by F28F and 280F Shark, described in 1985-86 *Jane's*, and 280FX.

CURRENT VERSIONS: **F28F Falcon:** Basic model certified to FAR Pt 6 on 31 December 1980. The two major changes incorporated into the F28F and 280F/280FX over the earlier F28C/280C series were an increase in power from 153 kW (205 shp) to 168 kW (225 shp) and the addition of a throttle correlator to reduce pilot workload. Most recent changes to the F28F and 280FX are the redesigned main gearbox with a heavy wall main rotor shaft (standard equipment on all new aircraft and retrofittable to all existing F models); optional lightweight exhaust silencer,

which reduces noise in the hover by 40 per cent and gives a 30 per cent reduction when flying at 152 m (500 ft) (can also be retrofitted to F28F, 280F and 280FX); and a lightweight starter motor. No recent production.

F28F-P Sentinel: Dedicated police patrol version; first delivery October 1986; can be fitted with a PA and siren system, Spectrolab SX-5 or Carter searchlight, specialised police equipment, FLIR and datalink. F28F-P has same specification and performance as F28F. No recent production.

280FX Shark: Certified to FAR Pt 6 on 14 January 1985. Prototype N280FX. Improvements over previous 280 series (see also F28F entry) include new seats with lumbar support and energy-absorbing NASA foam, new tailplane with endplate fins, tail rotor guard, covered tail

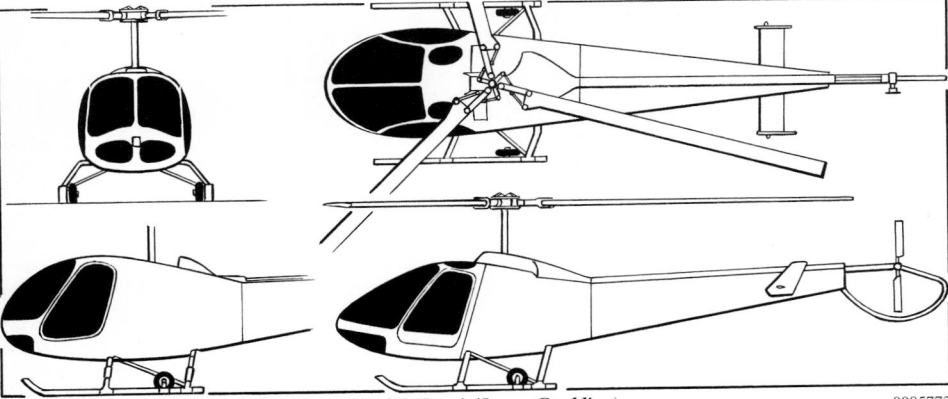

Enstrom 280FX, with scrap side view of F28F *(Jane's/James Goulding)*   0085773

rotor drive shaft, redesigned air inlet system, and completely faired landing gear. Optional internal auxiliary fuel tank extends range to 339 n miles (627 km; 390 miles).

CUSTOMERS: Total of 735 of earlier versions (14 F28, 315 F28A, 121 F28C, 56 F28C-2, 21 280, 206 280C and two 280L). Last of 135 F28Fs delivered 1999 to Fresno Police Department, California. Total of 93 280FXs built up to mid-2002, including four in 2001, increasing Enstrom piston-engined helicopter production to 963. Chilean Army operates 15 280FXs for primary and instrument training; Peruvian Army has 10 F28Fs for flight training; Colombian Air Force operates 12 F28Fs for primary and instrument training. Numerous US police departments operate F28F-P for patrol and surveillance missions. Most recent military customer is the Venezuelan National Guard, which ordered four for delivery in August 2001.

COSTS: Basic 2000 price US$275,000 for F28F Falcon; US$287,000 for 280FX Shark (2001).

DESIGN FEATURES: Conventional light helicopter with skid landing gear and tubular metal tail rotor protector; horizontal stabiliser with fins at tips. High inertia, three-blade fully articulated rotor head with blades attached by retention pin and drag link; control rods pass inside tubular rotor shaft to swashplate inside fuselage; no rotor brake; blade section NACA 0013.5; blades do not fold; two-blade teetering tail rotor. Goodyear 30-groove belt drive from horizontally mounted engine to transmission.

FLYING CONTROLS: Conventional and manual. Trim system absorbs feedback from rotor and repositions stick datum as required by pilot.

STRUCTURE: Bonded light alloy blades. Fuselage has glass fibre and light alloy cabin section, steel tube centre-section frame, and stressed skin aluminium tailboom.

LANDING GEAR: Skids carried on Enstrom oleo-pneumatic shock-absorbers. Air Cruiser inflatable floats available optionally.

POWER PLANT: One 168 kW (225 hp) Textron Lycoming HIO-360-F1AD flat-four engine with Rotomaster 3BT5EE10J2 turbocharger. Two fuel tanks, each of 79.5 litres (21.0 US gallons; 17.5 Imp gallons). Total standard fuel capacity 159 litres (42.0 US gallons; 35.0 Imp gallons), of which 151 litres (40.0 US gallons; 33.3 Imp gallons) are usable. Auxiliary tank, capacity 49 litres (13.0 US gallons; 10.8 Imp gallons), can be installed in the baggage compartment. Oil capacity 9.5 litres (2.5 US gallons; 2.1 Imp gallons).

ACCOMMODATION: Pilot and two passengers, side by side on bench seat; centre place removable. Removable door on each side of cabin. Baggage space aft of engine compartment, with external door. Cabin heated and ventilated.

SYSTEMS: Electrical power on F28F provided by 12 V 70 A engine-driven alternator; 24 V 70 A system optional on F28F, standard on 280FX. No hydraulic system.

AVIONICS: Variety of fits from Honeywell and other avionics suppliers.

Instrumentation: Standard equipment includes airspeed indicator, sensitive altimeter, compass, outside air temperature gauge, turn and bank indicator, rotor/engine tachometer, manifold pressure/fuel flow gauge, EGT gauge, oil pressure gauge, gearbox and oil temperature gauge, ammeter, cylinder head temperature gauge, and fuel quantity gauge. Eight-light annunciator panel consisting of low rotor rpm, chip detectors (main and tail rotor transmissions), overboost, clutch not fully engaged, low fuel pressure, starter and low-voltage warning lights.

EQUIPMENT: Shoulder harnesses for three seats. Night lighting is optional for F28F and standard on 280FX. Night lighting includes instrument lighting with dimmer control, position light on each horizontal stabiliser tip, anti-collision light and nose-mounted landing light. Optional equipment for both F28F and 280FX includes fixed float kit, wet or dry agricultural spray kit and cargo hook for utility missions. Wide instrument panel available for IFR training.

DIMENSIONS, EXTERNAL (F28 and 280, except where specified):

Main rotor diameter	9.75 m (32 ft 0 in)
Tail rotor diameter	1.42 m (4 ft 8 in)
Distance between rotor centres	5.56 m (18 ft 3 in)
Main rotor blade chord	0.24 m (9½ in)
Tail rotor blade chord	0.11 m (4½ in)
Length overall, rotors stationary	8.92 m (29 ft 3 in)
Fuselage: Length: F28	8.56 m (28 ft 1 in)
280	8.75 m (28 ft 8½ in)
Max width: F28	1.55 m (5 ft 1 in)
280	1.52 m (5 ft 0 in)
Height: to top of cabin: F28	1.85 m (6 ft 1 in)
280	1.83 m (6 ft 0 in)
to top of rotor head	2.74 m (9 ft 0 in)
Tailplane span	1.60 m (5 ft 3 in)
Skid track	2.21 m (7 ft 3 in)
Cabin doors (each): Height	1.04 m (3 ft 5 in)
Width	0.84 m (2 ft 9 in)
Height to sill	0.64 m (2 ft 1 in)
Baggage door: Height	0.55 m (1 ft 9½ in)
Width	0.39 m (1 ft 3½ in)
Height to sill	0.86 m (2 ft 10 in)

DIMENSIONS, INTERNAL:

Cabin max width: F28F	1.55 m (5 ft 1 in)
280FX	1.50 m (4 ft 11 in)
Baggage compartment volume	0.18 m³ (6.3 cu ft)

AREAS:

Main rotor disc	74.72 m² (804.25 sq ft)
Tail rotor disc	1.66 m² (17.88 sq ft)

WEIGHTS AND LOADINGS (F28F Normal category):

Weight empty, equipped: F28F	712 kg (1,570 lb)
280FX	719 kg (1,585 lb)
Baggage capacity: tailboom	49 kg (108 lb)
Max T-O weight: F28F, 280FX	1,179 kg (2,600 lb)
Max disc loading:	
F28F, 280FX	15.8 kg/m² (3.23 lb/sq ft)

PERFORMANCE (both versions at AUW of 1,066 kg; 2,350 lb, except where indicated):

Never-exceed speed (VNE):	
F28F	97 kt (180 km/h; 112 mph)
280FX	102 kt (189 km/h; 117 mph)
Max level speed, S/L to 915 m (3,000 ft):	
F28F	97 kt (180 km/h; 112 mph) IAS
280FX	102 kt (189 km/h; 117 mph) IAS
Econ cruising speed: F28F	89 kt (165 km/h; 102 mph)
280FX	93 kt (172 km/h; 107 mph)
Max rate of climb at S/L:	
at AUW of 1,066 kg (2,350 lb)	442 m (1,450 ft)/min
at AUW of 1,179 kg (2,600 lb)	350 m (1,150 ft)/min
Max certified altitude	3,660 m (12,000 ft)
Hovering ceiling:	
IGE: at 1,066 kg (2,350 lb)	4,020 m (13,200 ft)
at 1,179 kg (2,600 lb)	2,345 m (7,700 ft)
OGE: at 1,066 kg (2,350 lb)	2,650 m (8,700 ft)
at 1,179 kg (2,600 lb)	1,707 m (5,600 ft)
Max range, standard fuel, no reserves:	
F28F	229 n miles (423 km; 263 miles)
280FX	260 n miles (483 km; 300 miles)
Max endurance	3 h 30 min

UPDATED

ENSTROM 480 and TH-28

TYPE: Light utility helicopter.

PROGRAMME: Proof-of-concept 280FX, powered by Allison 250-C20W turbine engine, flown December 1988; first flight of wide-cabin 480/TH-28 prototype (N8631E) in October 1989; second 480/TH-28 (production prototype) flew December 1991; TH-28 FAA certified September 1992 (three-seat), 480 FAA certified in June 1993 and recertified to FAR Pt 27 in December 1994.

CURRENT VERSIONS: **480:** Five-seat (two plus three) configuration with staggered seating for forward visibility from all seats. Easily reconfigured to three seats for training or executive transport; or pilot's seat only for

Enstrom 480 five-seat turbine-powered light helicopter *(Jane's/Paul Jackson)*   NEW/0533380

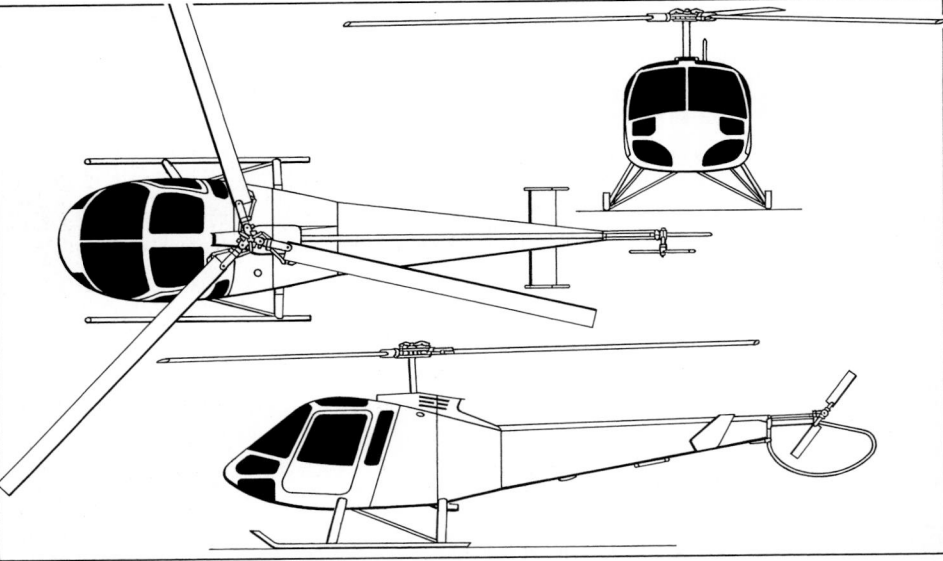

Enstrom 480 light helicopter (Rolls-Royce 250-C20W turboshaft) *(Jane's/Mike Keep)*   0016199

transporting light cargo. Instrument panel centrally located above avionics console.

480B: Certification received 9 February 2001; has increased gross weight and improved transmission, incorporates cyclic control and airframe vibration damping, providing significant improvement in comfort. Demonstrator, N480EN (43rd production 480) exhibited at NBAA Convention, New Orleans, October 2000.

TH-28: Prototype (N8631E) first flew 7 October 1989. Specially configured as a training/light patrol helicopter; unique features include three crashworthy seats, crashworthy fuel system, and a large instrument panel which will accommodate dual instrumentation for either VFR or IFR training. Seating configuration allows training of two students simultaneously.

CUSTOMERS: Production began in 1994 with initial deliveries to Europe. By June 2002, 48 480s and four TH-28s had been built. Final TH-28 produced in 1993 and delivered to China in 1997. Four 480s delivered in 2001. Now certified and operating in 13 countries: Belgium, Brazil, Canada, China, France, Germany, Japan, South Africa, Sweden, Switzerland, Thailand, UK and USA. In addition, 480s are also operating in Burkina Faso and Russian Federation.

COSTS: Initial basic price US$580,000; later increased to US$598,000 (2001) for 480B; TH-28 price on application. *Description generally as for F28/280, except that below (for 480B).*

DESIGN FEATURES: High-inertia, three-blade, fully articulated main rotor system with upgraded and sturdier main rotor gearbox than piston models; larger tail rotor; unboosted flight controls and high skid gear are standard equipment. Rotor speed 300 rpm nominal; 334 rpm maximum permissible; anti-nodal vibrating cantilever beam to damp vibration at low speeds. Extensive crashworthiness and safety features incorporated into basic design. The 480 cabin layout can be quickly reconfigured to seat one, three or five persons.

POWER PLANT: One 313 kW (420 shp) Rolls-Royce 250-C20W turboshaft engine derated to 227 kW (305 shp) for take-off and 207 kW (277 shp) for continuous operation. Engine air inlet particle separator is standard. Fuel capacity 340 litres (90.0 US gallons; 74.9 Imp gallons) in two interconnected tanks.

ACCOMMODATION: See Current Versions. Options include air conditioning (two or three evaporator system). Baggage compartment in tailboom.

EQUIPMENT: Latest additions to optional equipment include emergency pop-out floats, air conditionig and cargo hook rated for external loads up to 454 kg (1,000 lb); 480 can also be fitted with police patrol equipment including special police radios, searchlight, siren/PA system, FLIR and data recording facility.

DIMENSIONS, EXTERNAL:
Main rotor diameter	9.75 m (32 ft 0 in)
Tail rotor diameter	1.54 m (5 ft 0½ in)
Distance between rotor centres	5.64 m (18 ft 6 in)
Fuselage: Length	9.09 m (29 ft 10 in)
Max width	1.79 m (5 ft 10½ in)
Height: to top of cabin	2.11 m (6 ft 11¼ in)
to top of rotor head	3.00 m (9 ft 10 in)
Skid track	2.50 m (8 ft 2½ in)

DIMENSIONS, INTERNAL:
Cabin max width	1.80 m (5 ft 10¾ in)

AREAS:
Main rotor disc	74.72 m² (804.25 sq ft)
Tail rotor disc	1.85 m² (19.96 sq ft)

WEIGHTS AND LOADINGS:
Weight empty	769 kg (1,695 lb)
Baggage capacity: cabin	23 kg (50 lb)
tailboom	70 kg (155 lb)
Max T-O weight	1,360 kg (3,000 lb)
Max disc loading	18.2 kg/m² (3.73 lb/sq ft)

PERFORMANCE (A: at 1,360 kg; 3,000 lb AUW, B: at 1,134 kg; 2,500 lb):
Never-exceed speed (VNE)	125 kt (231 km/h; 144 mph)
Cruising speed: A	109 kt (202 km/h; 125 mph)
B	115 kt (213 km/h; 132 mph)
Max rate of climb at S/L: A	360 m (1,180 ft)/min
B	457 m (1,500 ft)/min
Service ceiling	3,965 m (13,000 ft)
Hovering ceiling:	
IGE: A	3,350 m (11,000 ft)
B	4,265 m (14,000 ft)
OGE: A	1,370 m (4,500 ft)
B	3,655 m (12,000 ft)
Max range (no reserves) at 914 m (3,000 ft):	
A	350 n miles (648 km; 402 miles)
B	435 n miles (806 km; 501 miles)
Endurance (no reserves): A	4 h 24 min
B	4 h 36 min

UPDATED

EXPLORER

EXPLORER AIRCRAFT INC

One Great Place, Jasper, Texas 75951
Tel: (+1 409) 489 15 00
Fax: (+1 409) 489 17 00
e-mail: info@exploreaircraft.com
Web: http://www.exploreaircraft.com
CEO: Graham Swannell
PRESIDENT: Donald G Joseph
DIRECTOR, OPERATIONS, RESEARCH AND DEVELOPMENT:
Geoffrey P Danes

Explorer Aircraft Inc was established in Colorado in 1998 to develop and market the Australian-designed Explorer range of single-engined utility aircraft. The company moved to Texas in July 2001, and has established a new 1,300 m² (14,000 sq ft) office and hangar facility there.

VERIFIED

EXPLORER EXPLORER

TYPE: Light utility turboprop.
PROGRAMME: Designed by Aeronautical Engineers Australia (AEA), beginning in 1993, with assistance from US aerodynamicist John Roncz; company renamed Explorer Aircraft Corporation. Proof-of-concept (POC) prototype (VH-OKA), designated Explorer 350R, made first flight 23 January 1998 and had flown more than 70 hours up to award of Developmental C of A on 27 April 1998. Prototype (re-registered VH-ONA in May 1998) and given Experimental C of A on 27 April 1999; IFR certification 6 May 1999; departed for USA 20 May 1999.

Re-engined with 447 kW (600 shp) Pratt & Whitney Canada PT6-135B turboprop as prototype Explorer 500T, in which form it first flew on 9 June 2000; simultaneously fitted with six-screen EFIS; public debut at EAA AirVenture Oshkosh in August 2000; will later be re-engined with Orenda OE600A V-8 as prototype 500R; manufacture on North American production line and certification test programme expected to begin in 2002; certification and first customer deliveries of launch model 500T in 2004, with deliveries of 500R following six months later and 750T in 2006.

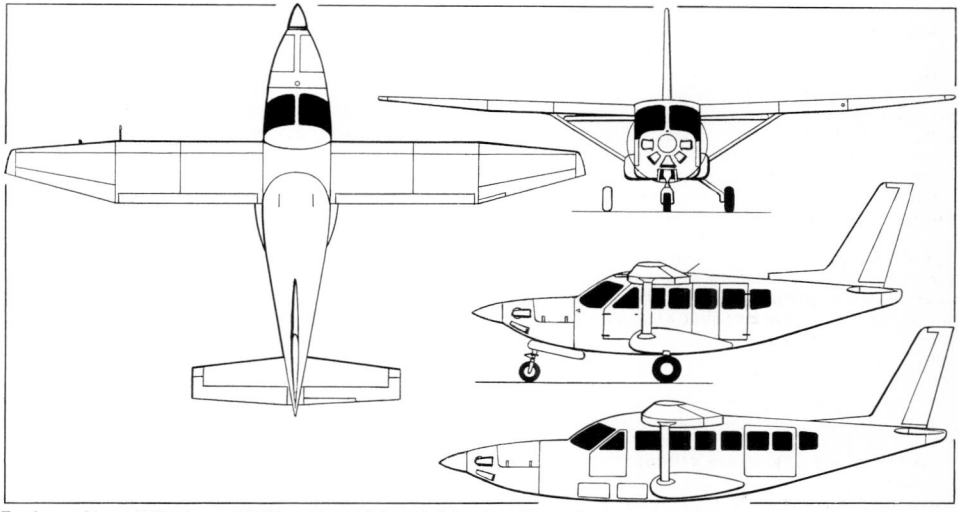

Explorer Aircraft Explorer 500T, with additional side view (lower) of stretched Explorer 750T *(Jane's/James Goulding)*   0105701

CURRENT VERSIONS: **350R:** Proof-of-concept prototype, powered by one 261 kW (350 hp) Teledyne Continental TSIO-55-E3B piston engine. No production decision taken by mid-2000. Described in 2000-01 edition of *Jane's*.

500T: Initial production version employing airframe similar to that of 350R but with 447 kW (600 shp) Pratt & Whitney Canada PT6-135B turboprop. *As described.*

500R: Airframe, payload and performance as for 500T, but with 447 kW (600 hp) Orenda OE600 liquid-cooled V-8 piston engine.

750T: Stretched version with plugs fore and aft of wing; up to 16 passengers; PT6-60A engine flat rated to 597 kW (800 shp).

COSTS: 500T US$1.035 million; 500R US$820,000; 750T US$1.16 million (2000).

DESIGN FEATURES: High-mounted, strut-braced wing with Roncz advanced aerofoil profile and constant chord, thick centre-section; outer panels (approximately 40 per cent of each half-span) sharply tapered; tall, sweptback fin and rudder. Constant cross-section cabin. 'Productionisation' to be accelerated by use of state-of-the-art computer modelling.

Wing thickness/chord ratio 19.6; washout 2°.

FLYING CONTROLS: Conventional and manual. Fowler flaps with three settings for take-off, approach and landing, plus neutral; three-axis trim via cable/screw-jack actuated tabs in starboard elevator, starboard aileron and rudder.

STRUCTURE: Carbon fibre fuselage shell over metal frame; remainder of primary structure is of high-grade aluminium alloy, with composites flaps, elevators, ailerons and rudder.

LANDING GEAR: Fully retractable tricycle type, with single wheel on each unit. Mainwheel tyre size 22×7.75-10, nosewheel 6.00-6. Steerable nose oleo unit retracts rearward; glass fibre mainwheel legs retract by first extending downward and then crossing over and upward so that port wheel is housed in streamline fairing on starboard side, and vice versa. Mainwheels do not intrude into cabin. Hydraulic brakes. Floats may be offered as future option.

Explorer 500T multirole utility transport with freight door open *(Jane's/Paul Jackson)*   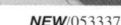NEW/0533379

Instrument panel of Explorer 500T
(Jane's/Paul Jackson)   0100506

POWER PLANT: *500T:* One Pratt & Whitney Canada PT6-135B turboprop, flat rated at 447 kW (600 shp), driving a four-blade, constant-speed, fully reversing Hartzell D9511FK-2 propeller. Fuel tanks in wing centre-section, total capacity 1,049 litres (277 US gallons; 231 Imp gallons).

500R: One 447 kW (600 hp) Orenda V-8 OE600A piston engine driving a three-blade constant-speed propeller. Fuel capacity 795 litres (210 US gallons; 175 Imp gallons).

750T: One Pratt & Whitney Canada PT6-60A turboprop, flat rated at 597 kW (800 shp), driving a four-blade constant-speed Hartzell reversing propeller. Fuel capacity 1,514 litres (400 US gallons; 333 Imp gallons).

All versions have refuelling port on top of each wing.

ACCOMMODATION: *500T, 500R:* 10 seats, including pilot(s); *750T:* 16 seats, including pilot(s). Forward-opening crew door on each side of forward fuselage. Large cargo door at rear of cabin on port side. Cabin is extensively glazed with five (500T, 500R) or seven (750T) large, square windows on each side. Fuselage hardpoints for tiedowns and cargo nets. Air conditioning standard.

SYSTEMS: 28 V DC electrical system. Engine-driven hydraulic pump. De-icing system, comprising leading-edge boots and heated propeller.

AVIONICS: *Comms:* Dual Garmin GNS 430 GPS/com/nav; GTX 327 transponder; GMA 340 audio panel.
Flight: Garmin GI 106A VOR/LOC/GS; Meggitt System 55 programmer/computer; remote annunciator; altitude selector/alerter; 6446 flux gate; Meggitt autopilot.

Instrumentation: Dual Meggitt Avionics 84-132-1 EDU; 84-133-1 PFD; 84-134-1 ND.

DIMENSIONS, EXTERNAL (A: 500T, B: 500R, C: 750T):

Wing span: A, B	14.43 m (47 ft 4 in)
C	17.68 m (58 ft 0 in)
Wing aspect ratio: A, B	11.3
Length overall: A, B	9.68 m (31 ft 9 in)
C	12.34 m (40 ft 6 in)
Height overall: A, B, C	4.72 m (15 ft 6 in)
Passenger door: A, B: Height	1.24 m (4 ft 1 in)
Width	1.27 m (4 ft 2 in)
Height to sill	0.96 m (3 ft 2 in)
Baggage door: A, B: Height	1.24 m (4 ft 1 in)
Width	1.27 m (4 ft 2 in)
Height to sill	0.96 m (3 ft 2 in)

DIMENSIONS, INTERNAL:

Cabin: Length: A, B	3.35 m (11 ft 0 in)
C	5.11 m (16 ft 9 in)
Max width: A, B, C	1.55 m (5 ft 1 in)
Max height: A, B, C	1.35 m (4 ft 5 in)
Volume: A, B	7.1 m³ (250 cu ft)
C	11.3 m³ (400 cu ft)
Cargo pod volume: A, B	1.42 m³ (50 cu ft)
C	2.265 m³ (80 cu ft)

AREAS:

Wings, gross: A, B	18.36 m² (197.6 sq ft)
Horizontal tail surfaces (total): A, B	5.29 m² (56.90 sq ft)
Vertical tail surfaces (total): A, B	2.87 m² (30.90 sq ft)

WEIGHTS AND LOADINGS:

Weight empty: A, B	1,724 kg (3,800 lb)
C	2,268 kg (5,000 lb)
Max T-O weight: A, B	2,812 kg (6,200 lb)
C	4,082 kg (9,000 lb)
Max wing loading: A, B	153.2 kg/m² (31.38 lb/sq ft)
Max power loading: A	6.29 kg/kW (10.33 lb/lb shp)
B	6.29 kg/kW (10.33 lb/lb shp)
C	6.85 kg/kW (11.25 lb/lb shp)

PERFORMANCE (estimated):

Cruising speed: A at max power	205 kt (380 km/h; 236 mph)
Cruising speed:	
A at 65% power	180 kt (333 km/h; 207 mph)
C	190 kt (352 km/h; 219 mph)
Stalling speed, flaps down: A, B, C	61 kt (113 km/h; 71 mph)
Service ceiling: A, B, C	7,620 m (25,000 ft)
Max rate of climb at S/L: A, B	305 m (1,000 ft)/min
C	366 m (1,200 ft)/min
T-O run: A, B, C	366 m (1,200 ft)
Max range: A, B	950 n miles (1,759 km; 1,093 miles)
C	900 n miles (1,666 km; 1,035 miles)

UPDATED

EXPRESS

EXPRESS AIRCRAFT COMPANY

PO Box 236, Olympia, Washington 98507-0236
Tel: (+1 360) 352 05 60
Fax: (+1 360) 352 05 53
e-mail: information@express-aircraft.com
Web: http://www.express-aircraft.com
PRESIDENT: Larry Olson
EXECUTIVE DIRECTOR: Paul Fagerstrom
GENERAL MANAGER: Chris Michalak

Assets of former Wheeler Technology Inc (see 1992-93 *Jane's*), which became bankrupt in late 1990, were acquired by Express Design Inc; EDI resumed production and customer support for former Wheeler Express kitbuilt aircraft. Factory floor area 1,950 m² (21,000 sq ft). EDI sold in June 1997 and renamed Express Aircraft Company. Also holds rights to Series 90, based on the Express, of which seven were flying by early 1999. A development of the Express, the **Auriga**, was shown at Sun 'n' Fun in April 1999.

UPDATED

EXPRESS EXPRESS

TYPE: Four-seat kitbuilt.

PROGRAMME: Designed by Wheeler Technology Inc as high-speed cross-country kitplane. Three prototypes built from kits of premoulded parts; first flight 28 July 1987. First, larger, production line Express demonstrator made first flight May 1990, powered by Teledyne Continental engine. Delivered kits incomplete at time of Wheeler bankruptcy; deficiencies made good by EDI. Complete kits, available from Express, incorporate some modifications which can also be retrofitted to existing aircraft.

CURRENT VERSIONS: **Express FT:** Original version. Unusual seating arrangement of one forward- and one aft-facing seat in rear, behind two side-by-side front seats with dual controls. Has 157 kW (210 hp) Teledyne Continental IO-360-ES1.

Express CT: Replaced FT.

Express 2000: Originally introduced as S-90 in 1992; became 2000 during 1999; four seats; higher-powered version. First aircraft completed at factory by November 2000, using builder-assist programme.

Express S300RG (formerly 2000RG): Available from late 2001; new wing with additional area provided by rearward extended trailing-edge (23 cm; 9 in at root, 2.5 cm; 1 in at tip); retractable landing gear. Planned increase in VNE to 290 kt (537 km/h; 333 mph) IAS; cruising speed to 265 kt (143 km/h; 305 mph).

Loadmaster 3200: Six seats; 194 kW (260 hp) engine; optional ventral pannier; no longer available.

CUSTOMERS: Over 300 kits of all versions sold, of which around 70 had been completed by mid-2002. Quoted build time 1,600 hours, reducing to 400 hours with builder-assist programme.

COSTS: 2000 Airframe kit US$48,900; S 300RG US$62,500, excluding engine, propeller, upholstery, paint and instruments (2003).

DESIGN FEATURES: Conforms to FAR Pt 23. Streamlined low-wing monoplane; tapered wing with downturned tips; sweptback fin and tailplane. Optional Precise Flight Schempp-Hirth-type spoilers in upper wing surface, electrically actuated.
Wing section NASA NFL-1 0215-F (laminar flow); dihedral 5°.

FLYING CONTROLS: Conventional and manual. Horn-balanced rudder and elevators; flight-adjustable tabs in port elevator, rudder (2000 only) port aileron. Plain flaps. Retractable spoilers in wing upper surface.

STRUCTURE: Constructed of composites sandwich material, comprising polyurethane foam core, glass fibre, unidirectional glass fibre tape and vinylester resin. CT and FT have cruciform tail; 2000 has larger, conventional tail unit.

LANDING GEAR: Non-retractable tricycle type standard; speed fairings. Mainwheels 15×6.00-6. Differential Cleveland brakes; castoring nosewheel. Retractable gear 2000RG available from late 2001; mainwheels retract inboard, nosewheel rearward; hydraulic actuation, with electric alternative.

POWER PLANT: *CT:* One 186 kW (250 hp) Textron Lycoming IO-540-C4B5 recommended.
2000: 224 kW (300 hp) Textron Lycoming IO-540-K1G5, or 231 kW (310 hp) Teledyne Continental IO-550-N1B both driving Hartzell three-blade constant-speed propeller.
S 300RG: One 261 kW (350 hp) Teledyne Continental TSIO-550-E, giving cruise speed of 225 kt (417 km/h; 259 mph).
Fuel capacity of CT and 2000 is 310 litres (82.0 US gallons; 68.3 Imp gallons); that of S 300RG is 341 litres (140 US gallons; 74.9 Imp gallons). Oil capacity 6.62 litres (1.75 US gallons; 1.46 Imp gallons).

ACCOMMODATION: Four forward-facing seats. Two additional seats in enlarged versions. Upward-hinged door each side; baggage door on port side.

DIMENSIONS, EXTERNAL:

Wing span: 2000	9.60 m (31 ft 6 in)
Wing chord: at root	1.52 m (5 ft 0 in)
at tip	0.95 m (3 ft 1½ in)
Wing aspect ratio	7.6
Length overall: CT	7.92 m (26 ft 0 in)
2000, S 300RG	7.62 m (25 ft 0 in)
Height overall: CT	2.13 m (7 ft 0 in)
-2000, S 300RG	2.44 m (8 ft 0 in)
Tailplane span	3.05 m (10 ft 0 in)
Wheel track	3.43 m (11 ft 3 in)
Wheelbase	1.97 m (6 ft 5½ in)

DIMENSIONS, INTERNAL:

Cabin max width	1.17 m (3 ft 10 in)

AREAS:

Wings, gross: CT, 2000	12.08 m² (130.0 sq ft)
S 300RG	13.01 m (140.0 sq ft)
Rudder	0.40 m² (4.33 sq ft)

WEIGHTS AND LOADINGS:

Weight empty: CT	830 kg (1,830 lb)
2000	839 kg (1,850 lb)
S 300RG	873 kg (1,925 lb)
Baggage capacity, with four 77 kg (170 lb) occupants	29 kg (64 lb)
Max T-O weight: CT, 2000	1,542 kg (3,400 lb)
S 300RG	1,723 kg (3,800 lb)
Max wing loading: CT, 2000	127.7 kg/m² (26.15 lb/sq ft)
S 300RG	132.5 kg/m² (27.14 lb/sq ft)
Max power loading (IO-540): 2000	6.90 kg/kW (11.33 lb/hp)
S 300RG	6.61 kg/kW (10.86 lb/hp)

PERFORMANCE (CT with IO-540, 2000 with IO-550, where different):

Never-exceed speed (VNE)	230 kt (426 km/h; 264 mph)
Cruising speed at 75% power at 2,290 m (7,500 ft):	
CT	185 kt (343 km/h; 213 mph)
2000	190 kt (352 km/h; 219 mph)
Stalling speed: flaps up	58 kt (108 km/h; 67 mph)
flaps down: except S 300RG	53 kt (102 km/h; 64 mph)
S 300RG	50 kt (93 km/h; 58 mph)
Demonstrated crosswind speed for T-O and landing	29 kt (54 km/h; 33 mph)
Max rate of climb at S/L: CT	457 m (1,500 ft)/min
2000	366 m (1,200 ft)/min
S 300RG	427 m (1,400 ft)/min
Service ceiling	6,700 m (22,000 ft)
T-O run	275 m (900 ft)
Landing run	244 m (800 ft)
Range: at 55% power, standard fuel, no reserves:	
CT	1,110 n miles (2,056 km; 1,227 miles)
2000	1,120 n miles (2,074 km; 1,289 miles)
with auxiliary fuel tank	1,688 n miles (3,125 km; 1,942 miles)
S 300RG	1,300 n miles (2,407 km; 1,496 miles)
g limits	+4.4/−2.2
	+8.8/−4.4 ultimate

UPDATED

Express Express 2000 kit aircraft (*Jane's/Paul Jackson*)   NEW/0533378

FH-1100

FH-1100 MANUFACTURING CORPORATION

6080 Industrial Boulevard, Century, Florida 32535
Tel: (+1 850) 256 00 26
Fax: (+1 850) 256 50 90
e-mail: info@fh1100.com
Web: http://www.fh1100.com
PRESIDENT: Georges R van Nevel
VICE-PRESIDENT: Remy Y van Nevel

The company was formed to return to production the Fairchild Hiller FH-1100 light helicopter.

NEW ENTRY

FH-1100 FHOENIX

TYPE: Light utility helicopter.
PROGRAMME: Hiller 1100 designed to meet US Army requirement for light observation helicopter; first flew 26 January 1963; three evaluated as HO-5s (later OH-5As), but not ordered in quantity. Type certificate awarded 22 May 1964; became Fairchild Hiller FH-1100 in September 1964; total 254 (including prototypes) built up to 1974.

Hiller Aviation formed January 1973 and acquired FH-1100 rights; built one; renamed Rogerson Hiller and produced five examples of RH-1100B Pegasus between 1983 and 1986; converted one in 1985 to RH-1100M Hornet military version. See 1983-84 *Jane's*. Promotion continued and further variants proposed, including seven-seat RH-1100S, but none built; last full description in 1992-93 edition. Rights eventually passed to Siam Hiller Holdings.

Type certificate acquired by current owners on 28 February 2000; two FH-1100s (N372F and Pegasus N4035G) obtained simultaneously as demonstrators; promotion of newly named FHoenix began at Heli-Expo, Anaheim, California, in February 2002.
DESIGN FEATURES: Conventional helicopter of early 1960s design; extensively glazed nose; low-set tailboom.

Two-blade, semi-rigid main rotor; aerofoil section NACA 63$_2$015; each blade attached to rotor head by single main retention bolt and drag link. Main rotor rpm 368. Main blades fold; rotor brake optional. Rotor drive through single-stage bevel and two-stage planetary main transmission, with intermediate and tail rotor gearboxes. Rotor/engine rpm ratio 1:16.30 main, 1:2.47 tail. Tubular guard for tail rotor.
FLYING CONTROLS: Electrically controlled trim.
STRUCTURE: All-metal main rotor with rolled stainless steel leading-edge spar bolted to an aluminium trailing section

with a honeycomb core. Tail rotor of stainless steel and honeycomb. Aluminium alloy, semi-monocoque pod-and-boom fuselage; aluminium alloy and honeycomb construction of vertical and horizontal tail surfaces.
LANDING GEAR: Skid type; torsion tube suspension.
POWER PLANT: One 313 kW (420 shp) Rolls-Royce 250-C20B turboshaft engine, derated to 204 kW (274 shp) for T-O; 174 kW (233 shp) max continuous. Single bladder fuel tank in bottom of centre-fuselage with usable capacity of 259 litres (68.5 US gallons; 57.0 Imp gallons). Refuelling point on starboard side of rear fuselage. Oil capacity 2.6 litres (0.7 US gallons; 0.6 Imp gallons).
ACCOMMODATION: Pilot and co-pilot side by side with three passengers to rear, or pilot and four passengers. Four forward-hinged doors, two on each side of cabin. Dual internal stretcher kit optional. Baggage compartment to rear of cabin, capacity 0.30 m³ (10.5 cu ft). Accommodation ventilated. Cabin heater and windscreen defroster optional.
SYSTEMS: Hydraulic system for cyclic and collective pitch controls; operating pressure 65.5 bar (950 lb/sq in). Electrical system includes a 28 V 60 A DC starter/generator and Ni-Cd battery.

DIMENSIONS, EXTERNAL:
Main rotor diameter	11.05 m (36 ft 3 in)
Tail rotor diameter	1.83 m (6 ft 0 in)
Distance between rotor centres	6.29 m (20 ft 7½ in)
Main rotor blade chord	0.33 m (1 ft 1 in)
Length overall, rotors turning	12.57 m (41 ft 3 in)
Length of fuselage	9.08 m (29 ft 9½ in)
Fuselage max width	1.32 m (4 ft 4 in)
Height overall	2.80 m (9 ft 2¼ in)
Skid track	2.20 m (7 ft 2¾ in)

AREAS:
Main rotor disc	95.89 m² (1,032.1 sq ft)
Tail rotor disc	2.63 m² (28.3 sq ft)

WEIGHTS AND LOADINGS:
Weight empty	687 kg (1,515 lb)
Payload: max	463 kg (1,020 lb)
with max fuel	328 kg (723 lb)
Max fuel weight	200 kg (442 lb)
T-O weight: normal	1,292 kg (2,850 lb)*
FAR Pt 133	1,451 kg (3,200 lb)
Max disc loading: normal	13.48 kg/m² (2.76 lb/sq ft)
FAR Pt 133	15.14 kg/m² (3.10 lb/sq ft)

* Awaiting certification; currently 1,247 kg (2,750 lb)
PERFORMANCE:
Never-exceed speed (VNE) at S/L	110 kt (204 km/h; 127 mph)
Max cruising speed	105 kt (194 km/h; 121 mph)
Max rate of climb at S/L	488 m (1,600 ft)/min
Vertical rate of climb at S/L	244 m (800 ft)/min
Service ceiling	5,275 m (17,300 ft)
Hovering ceiling: IGE	5,180 m (17,000 ft)
OGE	3,660 m (12,000 ft)
Range at FL50	330n miles (611 km; 379 miles)
Endurance	3 h

NEW ENTRY

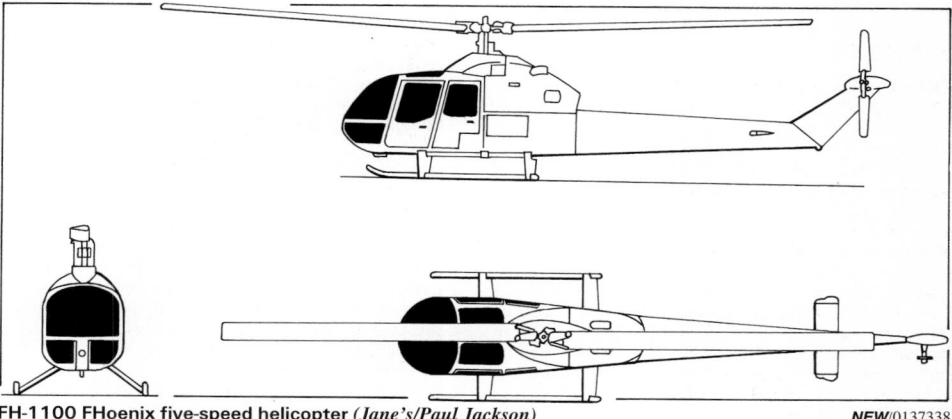

FH-1100 FHoenix five-speed helicopter (*Jane's/Paul Jackson*)   NEW/0137338

FISHER

FISHER FLYING PRODUCTS

PO Box 468, Industrial Park, Edgeley, North Dakota 58433
Tel: (+1 701) 493 22 86
Fax: (+1 701) 493 25 39
e-mail: ffpjac@daktel.com
Web: http://www.fisherflying.com
PRESIDENT: Darlene Jackson-Hanson
VICE-PRESIDENT: Gene Hanson

Fisher Flying Products markets a range of 14 homebuilt light aircraft. These comprise the FP-202 Koala, FP-303 microlight, FP-404, FP-505 Skeeter, FP-606 Sky Baby microlight, Avenger and Youngster single-seat aircraft; and Super Koala, Classic, Dakota Hawk, Tiger Moth, Celebrity and Horizon two-seaters. For information on the FP-202 Koala, Super Koala, FP-404 and Classic, see the Private Aircraft section of *Jane's* 1992-93. In 1998, Fisher took over Fisher Aero of Portsmouth, Ohio, which added the Celebrity, Horizon 1 and Horizon 2 (see *Jane's* 1992-93) to the product line. Workforce in late 2002 was nine.

UPDATED

FISHER AVENGER

TYPE: Single-seat kitbuilt.
PROGRAMME: First flown June 1994.
CURRENT VERSIONS: **Avenger:** Baseline version.
Avenger V: Powered by one 37.3 kW (50 hp) Volkswagen four-cylinder engine.
CUSTOMERS: Total of 160 sold by 2002, of which around 60 were flying.
COSTS: Avenger US$4,350 less engine (2002); Avenger V US$4,750 less engine (2002).
DESIGN FEATURES: Low-wing monoplane; wings braced by V-struts to mainwheel hubs; mutually braced tail surfaces. Wire-braced tailplane. Quoted build time 300 hours.
FLYING CONTROLS: Conventional and manual.
STRUCTURE: Wood and fabric.
LANDING GEAR: Tailwheel type; fixed.
POWER PLANT: One 37.0 kW (49.6 hp) two-cylinder Rotax 503 UL-2V; other engines from 20.9 to 48.5 kW (28 to 65 hp) are optional. Fuel capacity 19 litres (5.0 US gallons; 4.2 Imp gallons).

Fisher Avenger single-seat kitbuilt   0110843

DIMENSIONS, EXTERNAL:
Wing span	8.23 m (27 ft 0 in)
Length overall	4.95 m (16 ft 3 in)
Height overall	1.52 m (5 ft 0 in)

AREAS:
Wings, gross	11.24 m² (121.0 sq ft)

WEIGHTS AND LOADINGS:
Weight empty: Avenger	113 kg (250 lb)
Avenger V	159 kg (350 lb)
Max T-O weight: Avenger	272 kg (600 lb)
Avenger V	294 kg (650 lb)

PERFORMANCE:
Max level speed: Avenger	82 kt (152 km/h; 95 mph)
Avenger V	86 kt (160 km/h; 100 mph)
Normal cruising speed	52 kt (97 km/h; 60 mph)
Stalling speed: Avenger	23 kt (42 km/h; 26 mph)
Avenger V	28 kt (52 km/h; 32 mph)
Max rate of climb at S/L: Avenger	274 m (900 ft)/min
Avenger V	259 m (850 ft)/min
T-O run: Avenger	30 m (100 ft)
Avenger V	61 m (200 ft)
Landing run: Avenger	61 m (200 ft)
Avenger V	76 m (250 ft)
Range: Avenger	150 n miles (277 km; 172 miles)
Avenger V	250 n miles (463 km; 287 miles)

UPDATED

FISHER DAKOTA HAWK

TYPE: Side-by-side ultralight kitbuilt.
PROGRAMME: First flight September 1994.
CUSTOMERS: Total of 70 kits sold by mid-2002. Ten flying by end 2002.

Fisher Dakota Hawk two-seat kitbuilt *(Jane's/Paul Jackson)*   0110963

COSTS: Airframe kit US$10,500; quick-build airframe kit US$12,500; both excluding engine (2002).

DESIGN FEATURES: Geodetic wing structure built upon I-beam spar and braced with V struts. Wire-braced tailplane. Quoted build time 500 hours. Wings fold for storage.

FLYING CONTROLS: Conventional and manual.

STRUCTURE: Precut, shaped and slotted wooden pieces; wooden truss fuselage with plywood skin; glass fibre cowling.

LANDING GEAR: Tailwheel type; fixed. Matco wheels and hydraulic brakes. Steerable tailwheel; shock-absorbers and Matco speed fairings on main legs.

POWER PLANT: One 59.6 kW (79.9 hp) Rotax 912 UL four-stroke engine driving a two-blade propeller. Optional alternatives include Teledyne Continental engines between 48.4 and 63 kW (65 and 85 hp) or 74.6 kW (100 hp) Subaru. Fuel capacity 45.4 litres (12.0 US gallons; 10.0 Imp gallons).

DIMENSIONS, EXTERNAL:
Wing span	8.69 m (28 ft 6 in)
Length overall	6.02 m (19 ft 9 in)
Height overall	1.85 m (6 ft 1 in)

AREAS:
Wings, gross	11.89 m² (128.0 sq ft)

WEIGHTS AND LOADINGS:
Weight empty	272 kg (600 lb)
Max T-O weight	522 kg (1,150 lb)

PERFORMANCE:
Max cruising speed	87 kt (161 km/h; 100 mph)
Normal cruising speed	78 kt (145 km/h; 90 mph)
Stalling speed	31 kt (57 km/h; 35 mph)
Max rate of climb at S/L	244 m (800 ft)/min
T-O run	107 m (350 ft)
Landing run	122 m (400 ft)
Range with max fuel	217 n miles (402 km; 250 miles)

UPDATED

FISHER YOUNGSTER and YOUNGSTER V

TYPE: Single-seat ultralight biplane kitbuilt.

PROGRAMME: First flown June 1994.

CURRENT VERSIONS: **Youngster:** *As described.*

Youngster V: Uprated version with 48.5 kW (65 hp) Volkswagen motorcar engine.

CUSTOMERS: Total of 20 flying by mid-2002.

COSTS: Youngster kit US$5,700 less engine; Youngster V US$5,950 less engine (2002).

DESIGN FEATURES: Open cockpit, strut-braced biplane with detachable cockpit transparency for winter operations. Constant-chord wings of geodetic construction. Wire-braced low-mounted tailplane. Stainless steel control wires. Additional central strut between nose and upper wing forward of cockpit. Aerofoil type 2315. Quoted build time 400 hours.

FLYING CONTROLS: Conventional and manual.

STRUCTURE: Wood and fabric.

LANDING GEAR: Tailwheel type; fixed. Tyres 8.00-6. Bungee cord shock absorption.

POWER PLANT: One 37.3 kW (50 hp) Great Plains four-cylinder engine; Rotax engines up to 47.8 kW (64.1 hp) are optional. Fuel capacity 30.3 litres (8.0 US gallons; 6.7 Imp gallons).

DIMENSIONS, EXTERNAL:
Wing span	5.49 m (18 ft 0 in)
Length overall	4.72 m (15 ft 6 in)
Height overall	1.85 m (6 ft 1 in)

AREAS:
Wings, gross	11.71 m² (126.0 sq ft)

WEIGHTS AND LOADINGS:
Weight empty	181 kg (400 lb)
Max T-O weight	294 kg (650 lb)

PERFORMANCE:
Max level speed	95 kt (177 km/h; 110 mph)
Normal cruising speed	74 kt (137 km/h; 85 mph)
Stalling speed	30 kt (55 km/h; 34 mph)
Max rate of climb at S/L	244 m (800 ft)/min
T-O run	61 m (200 ft)
Landing run	76 m (250 ft)
Range	225 n miles (416 km; 259 miles)

UPDATED

FISHER R-80 and RS-80 TIGER MOTH

TYPE: Tandem-seat ultralight biplane kitbuilt.

PROGRAMME: Scale (80 per cent) replica of de Havilland D.H.82A Tiger Moth. First flight June 1994.

CURRENT VERSIONS: **R-80 Tiger Moth:** *As described.*

RS-80 Tiger Moth: Introduced 1999; has steel tube fuselage and 89.5 kW (120 hp) LOM M132 engine.

CUSTOMERS: Total of 60 kits sold by mid-2002, of which nine were then registered. Four under construction in Zimbabwe for use on aerial safaris.

COSTS: R.80 standard airframe kit US$11,250; quick-build airframe kit US$14,250; both excluding engine (2001); RS-80 standard airframe kit US$16,500; quick-build airframe kit US$18,500; both excluding engine (2001).

DESIGN FEATURES: Can be built using the contents of a basic tool kit. Quoted build time 550 hours.

Single-bay, staggered biplane with N-type interplane struts and slight sweepback to constant-chord wings.

FLYING CONTROLS: Conventional and manual, with horn-balanced rudder.

STRUCTURE: Wood construction using epoxy adhesives on preformed components.

LANDING GEAR: Tailwheel type; fixed. Tubular steel mainwheel legs; Matco wheels with hydraulic brakes.

POWER PLANT: Prototype has one 70.8 kW (95 hp) Mid West AE 100 rotary, but any engine between 55.9 and 89 kW (75 and 120 hp) can be fitted, including Rotax 912, Jabiru 3300, Subaru EA81-100, Norton rotary and LOM M132. One 72 litre (19.0 US gallon; 15.8 Imp gallon) centre-section fuel tank.

DIMENSIONS, EXTERNAL:
Wing span	7.01 m (23 ft 0 in)
Length overall	5.79 m (19 ft 0 in)
Height overall	2.24 m (7 ft 4 in)

DIMENSIONS, INTERNAL:
Cockpit max width	0.66 m (2 ft 2 in)

AREAS:
Wings, gross	15.79 m² (170.0 sq ft)

WEIGHTS AND LOADINGS:
Weight empty: R-80	254 kg (560 lb)
RS-80	363 kg (800 lb)
Max T-O weight: R-80	522 kg (1,150 lb)
RS-80	612 kg (1,350 lb)

PERFORMANCE:
Max level speed: R-80	95 kt (177 km/h; 110 mph)
RS-80	104 kt (193 km/h; 120 mph)
Cruising speed: R-80	69 kt (129 km/h; 80 mph)
RS-80	78 kt (145 km/h; 90 mph)
Stalling speed: R-80	31 kt (57 km/h; 35 mph)
RS-80	35 kt (65 km/h; 40 mph)
Max rate of climb: both	244 m (800 ft)/min
T-O run: R-80	91 m (300 ft)
RS-80	76 m (250 ft)
Landing run: R-80	122 m (400 ft)
RS-80	91 m (300 ft)
Range with max fuel: both	250 n miles (463 km; 287 miles)

UPDATED

Replica Tiger Moth produced by Fisher Flying Products *(Jane's/Paul Jackson)*   0110964

Fisher Youngster (37.3 kW; 50 hp Great Plains)   0110842

FLIGHTSTAR

FLIGHTSTAR SPORTSPLANES

PO Box 760, Ellington, Connecticut 06029
Tel: (+1 860) 875 81 85
Fax: (+1 860) 870 54 99
e-mail: fstar@mail2.nai.net
Web: http://www.fly-flightstar.com
PRESIDENT: Thomas A Peghiny
TECHNICAL DIRECTOR: Mark Lamontagne

Pioneer International formed as a division of Pioneer Parachute Company in 1981 to produce the initial Flightstar; design was sold to Pampa's Bull of Argentina as the Aviastar I (see 1992-93 *Jane's*) when Pioneer closed in 1985. Pioneer Flightstar (later Flightstar) formed in the 1990s initially to import the design, and has since developed it into the current range of ultralights. It became part of Leza Lockwood (which see) in 2001, but trades as separate company. The initial Flightstar Classic is no longer available.

UPDATED

Flightstar Spyder single-seat ultralight 0110699

FLIGHTSTAR IISL and IISC

TYPE: Side-by-side ultralight kitbuilt.
PROGRAMME: Originally designed by Pioneer International; construction of prototype Flightstar started 1992; first flight 1993. Certified to German BFU-95 standards 1997; also certified in Australia. SL version introduced in 1995.
CURRENT VERSIONS: **IISL:** (SL = Sport Light) Standard version, *as described.*
  IISC: (SC = Sport Cabin) Introduced 1999; cabin version with doors for cold-weather operations.
CUSTOMERS: Over 1,000 of original Flightstar II built. Current version orders total more than 400, of which 375 were flying by mid-2001.
COSTS: IISL US$12,495; IISC US$13,695, both without engines (2001).
DESIGN FEATURES: Wings fold for storage and transportation. Quoted build time 100 hours for IISL; 130 hours for IISC.
  Flightstar FS-II-2 aerofoil section; dihedral 2° 48′; incidence 4°, twist 2°.
FLYING CONTROLS: Conventional and manual. Push-pull cables; optional flaps on IISL.
STRUCTURE: 6061-T6 aluminium tube spar, strut-braced wing with Dacron covering; Five-ply Mylar covering optional. Composites-covered chromoly 4130 steel fuselage cage with stainless steel brackets; engine attached to heavy-duty aluminium tube running length of aircraft; all metal parts anodised to obviate painting. Strut-braced, low-mounted tailplane.
LANDING GEAR: Tricycle type; fixed; steerable nosewheel. Azusa wheels and optional drum brakes. Tyres size 13 in nose; 15 in main. Shock cord absorption. Optional Full Lotus floats.

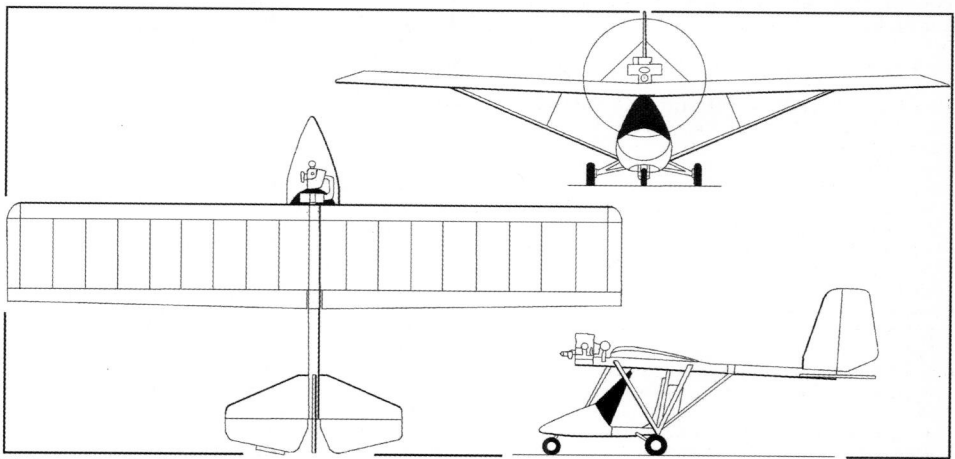

Flightstar Spyder (Rotax 447 two-cylinder engine) (*Jane's/Paul Jackson*)   **NEW**/0533719

POWER PLANT: One 37.0 kW (49.6 hp) Rotax 503 UL-2V driving three-blade Powerfin propeller; optional engines include 47.8 kW (64.1 hp) Rotax 582 UL and 44.7 kW (60.0 hp) four-stroke HKS 700E; latter is standard on IISC and has 3.47:1 gearbox reduction. Fuel capacity 37.9 litres (10.0 US gallons; 8.3 Imp gallons), of which 36.0 litres (9.5 US gallons; 7.9 Imp gallons) usable.

Two-seat Flightstar IISC (*Jane's/Paul Jackson*)   **NEW**/0533682

ACCOMMODATION: Fastback fabric rear fairing fitted to IISL. Cabin doors closed by zips.
EQUIPMENT: Optional BRS parachute.
DIMENSIONS, EXTERNAL:

Wing span	9.75 m (32 ft 0 in)
Length overall	5.79 m (19 ft 0 in)
Height overall	2.39 m (7 ft 10 in)

AREAS:

Wings, gross	14.59 m² (157.0 sq ft)

WEIGHTS AND LOADINGS:

Weight empty	166 kg (365 lb)
Max T-O weight	431 kg (950 lb)

PERFORMANCE (with 37.0 kW; 49.6 hp engine):

Never-exceed speed (VNE)	83 kt (154 km/h; 96 mph)
Max operating speed	74 kt; (137 km/h; 85 mph)
Normal cruising speed	56 kt; (105 km/h; 65 mph)
Stalling speed	32 kt (58 km/h: 36 mph)
Max rate of climb at S/L	213 m (700 ft)/min
T-O run	61 m (200 ft)
Landing run	91 m (300 ft)
Range with max fuel	156 n miles (289 km; 180 miles)

UPDATED

FLIGHTSTAR SPYDER and FORMULA

TYPE: Single-seat ultralight kitbuilt.
PROGRAMME: Derived from Pioneer Flightstar II. Can conform to FAR Pt 103.
CURRENT VERSIONS: **Spyder:** Standard version, *as described.*
  Formula: High-performance upgraded model with wider windscreen and cockpit doors for cold weather flying.
CUSTOMERS: Total of 112 kits sold, and 110 flying, by mid-2001.
COSTS: Spyder US$10,495; Formula US$11,395, both without engines (2001).
DESIGN FEATURES: Flightstar FS-I-2 aerofoil. Wings fold for storage and transportation. Optional BRS parachute recovery system.
FLYING CONTROLS: As Flightstar IISL.
STRUCTURE: As Flightstar IISL.
LANDING GEAR: As Flightstar IISL.
POWER PLANT: One 31.0 kW (41.6 hp) Rotax 447 UL-2V driving Tennessee two-blade or Ivo three-blade propeller. Options include 37.0 kW (49.6 hp) Rotax 503 UL-2V and 47.8 kW (64.1 hp) Rotax 582 UL. Standard fuel capacity of Spyder 18.9 litres (5.0 US gallons; 4.2 Imp gallons); optionally 37.9 litres (10.0 US gallons; 8.3 Imp gallons); Formula fuel capacity 37.9 litres (10.0 US gallons; 8.3 Imp gallons).
DIMENSIONS, EXTERNAL:

Wing span	9.14 m (30 ft 0 in)

Open cockpit Flightstar IISL (*Jane's/Paul Jackson*)   0105029

Length overall: Spyder	5.03 m (16 ft 6 in)	Max T-O weight: Spyder	294 kg (650 lb)	T-O run: both	62 m (205 ft)
Formula	5.94 m (19 ft 6 in)	Formula	431 kg (950 lb)	Landing run: both	61 m (200 ft)
Height overall	2.39 m (7 ft 10 in)	PERFORMANCE:		Range with standard fuel: Spyder	
AREAS:		Never-exceed speed (VNE):	83 kt (154 km/h; 96 mph)		86 n miles (161 km; 100 miles)
Wings, gross	13.38 m² (144.0 sq ft)	Normal cruising speed: Spyder	48 kt; (89 km/h; 55 mph)	Formula	191 n miles (354 km; 220 miles)
WEIGHTS AND LOADINGS:		Formula	56 kt (105 km/h; 65 mph)		*UPDATED*
Weight empty: Spyder	127 kg (280 lb)	Stalling speed: both	28 kt (52 km/h: 32 mph)		
Formula	181 kg (400 lb)	Max rate of climb at S/L: both	201 m (660 ft)/min		

FLYING K

FLYING K ENTERPRISES, INC

416 Caldwell Boulevard, Nampa, Idaho 83651
Tel: (+1 208) 465 71 16
Fax: (+1 208) 465 79 37
e-mail: flyingk@mindspring.com
Web: http://www.skyraider.net

Flying K Enterprises was established by former SkyStar employee, the late Kenny Schrader, and is run by his widow, Kyna Schrader. A version is produced in the UK as Reality Easy Raider (which see).

UPDATED

Flying K Sky Raider with HKS engine and wings folded *(Jane's/Paul Jackson)*   NEW/0533684

FLYING K SKY RAIDER and SKY RAIDER II

TYPE: Single-seat ultralight kitbuilt; tandem-seat ultralight kitbuilt.

PROGRAMME: First flown in 1996.

CURRENT VERSIONS: **Sky Raider:** Single-seat model, available as certified aeroplane or under FAR Pt 103 ultralight regulations.

Sky Raider II: Tandem-seat version; introduced in 1999.

CUSTOMERS: Total of 162 SkyRaiders and 26 SkyRaider IIs sold by 4 June 2001.

COSTS: Sky Raider kit US$8,500; Sky Raider II kit US$9,950 (2000).

DESIGN FEATURES: High wing with two tubular main, and two secondary bracing struts to each wing. Mutually wire-braced fin and tailplane. Wings foldable for storage without disconnection of control runs. Quoted build time 300 hours for Sky Raider and 350 hours for Sky Raider II.

FLYING CONTROLS: Conventional and manual. Frise ailerons, cable actuated. Slotted flaps.

STRUCTURE: Fabric-covered steel tube. Optional wingtip extensions to Sky Raider add 55 cm (21½ in) to span; choice of rounded or square fin on Sky Raider. Heavier wing spar (6 cm; 2½ in diameter) on Sky Raider II. Horn-balanced rudder on Sky Raider II.

LANDING GEAR: Tailwheel type: fixed. Mainwheels 8.00-6. Optional floats and skis.

POWER PLANT: *Sky Raider:* One 20.1 kW (27 hp) Rotax 277 two-cylinder two-stroke driving a two-blade propeller for Pt 103; 31.0 kW (41.6 hp) Rotax 447 for normal operations. Optionally, engines up to 37.3 kW (50 hp) can be fitted. Fuel capacity 18.9 litres (5.0 US gallons; 4.2 Imp gallons).

Sky Raider II: One 44.7 kW (60 hp) HKS flat-two, four-stroke engine driving a GSC ground-adjustable three-blade or Powerfin carbon fibre propeller. Fuel capacity 37.9 litres (10.0 US gallons; 8.3 Imp gallons) due to extra wing fuel tank.

ACCOMMODATION: *Sky Raider:* Pilot only, in open-sided cabin.

Sky Raider II: Two seats in tandem within fully enclosed cabin.

DIMENSIONS, EXTERNAL (A: Sky Raider, B: Skyraider II):
Wing span: A	7.99 m (26 ft 2½ in)
B	8.53 m (28 ft 0 in)
Length overall: A, B	5.18 m (17 ft 0 in)
Height overall: A	1.57 m (5 ft 2 in)
B	1.73 m (5 ft 8 in)

AREAS:
Wings, gross, A, B:	
without extensions	9.10 m² (98.0 sq ft)
wth extensions	9.94 m² (107.0 sq ft)

WEIGHTS AND LOADINGS:
Weight empty: A with Rotax 447	115 kg (253 lb)
B with HKS	184 kg (405 lb)
Max T-O weight: A	249 kg (550 lb)
B	431 kg (950 lb)

PERFORMANCE (A with Rotax 447; B with HKS):
Never-exceed speed (VNE):	
A, B	86 kt (160 km/h; 100 mph)
Normal cruising speed: A	56 kt (105 km/h; 65 mph)
B	65 kt (121 km/h; 75 mph)
Stalling speed, flaps down: A	20 kt (37 km/h; 23 mph)
B	26 kt (49 km/h; 30 mph)
Max rate of climb at S/L: A	320 m (1,050 ft)/min
B	396 m (1,300 ft)/min
T-O and landing run: both	23 m (75 ft)
Range: A	120 n miles (222 km; 138 miles)
B	250 n miles (463 km; 287 miles)

UPDATED

Flying K Sky Raider *(Jane's/Paul Jackson)*   NEW/0533683

FOUR WINDS

FOUR WINDS AIRCRAFT

1501 Airway Crescent, New Smyrna Beach, Florida 32168
Tel: (+1 386) 426 77 95
Fax: (+1 386) 426 53 39
Web: http://www.fourwindsaircraft.com
PRESIDENT: Jeff Rahm

The company's 3,900 m² (42,000 sq ft) plant employed 20 persons in early 2002. During 1990s, completed 22 Lancairs, two GlaStars, two Stallions, two Glass Gooses and one BD-5 on behalf of private owners.

NEW ENTRY

FOUR WINDS 192, 210 and 250T

TYPE: Four-seat kitbuilt.

PROGRAMME: Stretched version of GlaStar (which see); partly completed prototype exhibited at Sun 'n' Fun, Lakeland, Florida, April 2002; first kit deliveries then scheduled for September 2002.

CURRENT VERSIONS: See Power Plant.

CUSTOMERS: Total 15 kits sold by April 2002.

COSTS: Kit US$38,000 (2002); estimated US$100,000 to flyable, equipped configuration, including builder assistance programme of 14 days in-works support.

DESIGN FEATURES: Four-seat version of GlaStar, but with cabin reinforcement cage similar to OMF Symphony (which see in German section); however, wing of composites, with no bracing struts.

Quoted build time 1,000 hours, or 500 hours with builder assistance programme.

Unswept wing and tailplane; fin sweepback 48° 50′.

FLYING CONTROLS: Generally as for GlaStar.

STRUCTURE: Generally of composites but 4130 steel (3.18 cm; 1¼ in diameter tube) structural cage surround to cabin and 7075 aluminium main landing gear legs.

LANDING GEAR: Tricycle type, fixed; single wheel with speed fairing, on each unit. Mainwheels 15×6.00-6; nosewheel 11×4.00-5. Optional twin-float and tailwheel versions.

POWER PLANT: Choice of three engines to drive single propeller, as under.
192: One 149 kW (200 hp) Textron Lycoming IO-360 flat-six.
210: One 164 kW (220 hp) PZL-Franklin 220 flat-six.
250T: One 313 kW (420 shp) Rolls-Royce 250 turboprop. Fuel capacity 227 litres (60.0 US gallons; 50.0 Imp

gallons) standard in all versions. Optional extended-range tanks.

ACCOMMODATION: Four persons, including one or two pilots, in two side-by-side pairs. Four forward-hinged doors. Baggage space behind rear seats; latter tip up, or rapidly removable, for additional stowage.

DIMENSIONS, EXTERNAL:
Wing span	11.19 m (36 ft 8½ in)
Wing chord, mean	1.14 m (3 ft 9 in)
Wing aspect ratio	9.8
Length overall	7.39 m (24 ft 3 in)

Tailplane span	3.73 m (12 ft 3 in)
Tailplane chord, constant	0.91 m (3 ft 0 in)

DIMENSIONS, INTERNAL:
Cabin: Length	2.76 m (9 ft 0½ in)
Max width	1.12 m (3 ft 8 in)
Max height	1.30 m (4 ft 3¼ in)

AREAS:
Wings, gross	12.79 m² (137.7 sq ft)
Vertical tail surfaces (total)	1.77 m² (19.03 sq ft)
Horizontal tail surfaces (total)	3.41 m² (36.75 sq ft)

WEIGHTS AND LOADINGS:
Weight empty: 192	562 kg (1,240 lb)
210	517 kg (1,140 lb)
250T	472 kg (1,040 lb)
Baggage capacity: 192	23 kg (50 lb)
210	68 kg (150 lb)
250T	113 kg (250 lb)
Max T-O weight: all	1,124 kg (2,480 lb)
Max wing loading: all	87.9 kg/m² (18.01 lb/sq ft)
Max power loading: 192	7.55 kg/kW (12.40 lb/hp)
210	6.86 kg/kW (11.27 lb/hp)
250T	3.59 kg/kW (5.90 lb/shp)

PERFORMANCE (estimated):
Cruising speed at 70% power:	
192 at FL60	174 kt (322 km/h; 200 mph)
210 at FL60	182 kt (338 km/h; 210 mph)
250T at FL120	209 kt (386 km/h; 240 mph)
Stalling speed: all	53 kt (99 km/h; 61 mph)
Max range:	
192, 210	1,042 n miles (1,931 km; 1,200 miles)
250T	869 n miles (1,609 km; 1,000 miles)
g limits: all	+3.8/−1.6

Four Winds 192 four-seat kitbuilt (*Jane's/James Goulding*)   NEW/0143321

NEW ENTRY

GLASAIR

NEW GLASAIR LLC

18701 58th Avenue Northeast, Arlington, Washington 98223
Tel: (+1 360) 435 85 33
Fax: (+1 360) 435 95 25
e-mail: info@newglasair.com
Web: http://www.newglasair.com
PRESIDENT: Thomas W Wathen
COO: Mikael Vai
VICE-PRESIDENT, MARKETING: Ted Setzer

Stoddard-Hamilton company formed in 1979. Glasair I, II and II-S were superseded by Super II and III. Details of Glasair I in 1989-90 *Jane's.* Two-seat GlaStar added to product range in 1995; float production rights sold to Aero Marine Inc in 1996. Company had delivered more than 2,400 kits by April 1999; by 2002 over 800 Glasairs and GlaStars were flying.

Work began on a six-seater, named Aurora, in early 2000, but Stoddard-Hamilton ceased trading in May 2000 and filed for Chapter 11 bankruptcy on 17 July. At a bankruptcy hearing on 18 August 2000, the bulk of the assets was sold (subject to final agreement on some matters) to Anduril Private Investment Fund, run by Strider Capital Management. However, at a court hearing on 27 November 2000 this agreement was set aside and bidding for the company was re-opened.

Stoddard-Hamilton then received offer from Phoenix Composites for rights to the Glasair which again failed to complete before rights to both aircraft were purchased from Aglington Advanced Development by Thomas Wathen on 15 June 2001. Separate marketing companies were established by Mr Wathen for the two main product lines, namely New Glasair (this entry) and New GlaStar (which see).

VERIFIED

GLASAIR SUPER II

TYPE: Side-by-side kitbuilt.

PROGRAMME: Derived from Glasair TD, first flown in 1979. Jump-start kit options, to reduce quoted build time by around 750 hours, offered from 1999 onwards.

CURRENT VERSIONS: Glasair Super II (stretched) available in **RG** (retractable landing gear), **FT** (non-retractable tricycle gear) and **TD** (taildragger) forms introduced in 1992 to supersede earlier Glasair series; TD no longer marketed in 2002, but remains available on a special order basis.

CUSTOMERS: 430 kits sold by 2002.

COSTS: Kits: US$29,655 for RG; US$24,705 for FT; US$22,280 for TD; both without engine, propeller, instruments and avionics; engine US$28,833 for 180 hp version; fixed-pitch propeller US$2,122, constant-speed propeller US$6,400 (2003).

DESIGN FEATURES: Low-wing monoplane with moderately tapered wing and sweptback tail surfaces; optimised for high performance consistent with amateur assembly. Quoted build time 2,000 hours.
Wing section NASA LS(I)-0413; dihedral 3°; incidence 2° 20′. Upswept Hoerner style wingtips. Optional 61 cm (2 ft 0 in) extension on each wingtip.

FLYING CONTROLS: Conventional and manual. Horn-balanced rudder and elevators; ailerons with trim tab; plain flaps (slotted flaps optional).

Glasair Super II-RG two-seat tourer (*Jane's/Paul Jackson*)   NEW/0533685

STRUCTURE: Glass fibre and foam composites construction. One-piece wing spar.

LANDING GEAR: Three types available, as detailed above. RG version has inward-retracting mainwheels and rearward-retracting, free-castoring nosewheel. Brakes fitted. Wheel size 5.00-5 in. TD has full-swivel, locking tailwheel.

POWER PLANT: One 112 to 149 kW (150 to 200 hp) Textron Lycoming O-320 or IO-360 flat-four engine driving a Sensenich constant-speed metal propeller. Fuel capacity 151 litres (40.0 US gallons; 33.3 Imp gallons) in main wing tanks plus 30 litres (8.0 US gallons; 6.7 Imp gallons) in header tank. Auxiliary tanks of 41.6 litres (11.0 US gallons; 9.2 Imp gallons) capacity in optional wingtip extensions.

ACCOMMODATION: Two persons, side by side. Enclosed cockpit with gull-wing doors. NACA air vents provide cabin ventilation.

DIMENSIONS, EXTERNAL (all versions):
Wing span: standard	6.81 m (22 ft 4 in)
optional	8.33 m (27 ft 4 in)
Wing aspect ratio: standard	6.1
optional	8.2
Length overall	6.31 m (20 ft 8½ in)
Height overall	2.07 m (6 ft 9½ in)
Propeller diameter	1.78 m (5 ft 10 in)

DIMENSIONS, INTERNAL:
Cabin max width	1.07 m (3 ft 6 in)
Baggage volume	0.34 m³ (12.0 cu ft)

AREAS:
Wings, gross: standard	7.55 m² (81.3 sq ft)
optional	8.50 m² (91.5 sq ft)

WEIGHTS AND LOADINGS:
Weight empty: RG	601 kg (1,325 lb)
FT	567 kg (1,250 lb)
TD	544 kg (1,200 lb)
Baggage capacity: all versions	45 kg (100 lb)
Max T-O weight: standard wing:	
RG, FT	953 kg (2,100 lb)
TD	907 kg (2,000 lb)
extended wing: RG, FT	998 kg (2,200 lb)
TD	952 kg (2,100 lb)
Max wing loading: standard wing:	
RG, FT	126.1 kg/m² (25.83 lb/sq ft)
TD	120.1 kg/m² (24.60 lb/sq ft)
extended wing: RG, FT	117.4 kg/m² (24.04 lb/sq ft)
TD	112.1 kg/m² (22.95 lb/sq ft)
Max power loading: standard wing, 119 kW (160 hp)	
engine: FT	7.99 kg/kW (13.13 lb/hp)
extended wing: FT	8.37 kg/kW (13.75 lb/hp)
standard wing, 134 kW (180 hp) engine:	
RG	7.10 kg/kW (11.67 lb/hp)
TD	6.76 kg/kW (11.11 lb/hp)
extended wing: RG	7.44 kg/kW (12.22 lb/hp)
TD	7.10 kg/kW (11.67 lb/hp)

PERFORMANCE (180 hp engine):
Never-exceed speed (VNE)	226 kt (418 km/h; 260 mph)
Max level speed at S/L:	
RG	207 kt (383 km/h; 238 mph)
FT, TD	198 kt (367 km/h; 228 mph)
Econ cruising speed at 75% at 2,440 m (8,000 ft):	
RG	192 kt (356 km/h; 221 mph)
FT, TD	182 kt (338 km/h; 210 mph)
Stalling speed, power off: plain flaps down:	
RG	57 kt (105 km/h; 65 mph)
FT, TD	55 kt (102 km/h; 63 mph)
slotted flaps down:	
RG	52 kt (95 km/h; 59 mph)
FT, TD	50 kt (92 km/h; 57 mph)
Max rate of climb at S/L: all	823 m (2,700 ft)/min
Max rate of roll: standard wing:	
FT, TD	130°/s
RG	140°/s
extended wing: FT, TD	85°/s
RG	90°/s
Service ceiling: all	approx 5,790 m (19,000 ft)
T-O and landing run: TD, FT	213 m (700 ft)
RG	244 m (800 ft)
Range: RG, standard fuel	
	938 n miles (1,738 km; 1,080 miles)
RG, with auxiliary fuel	
	1,155 n miles (2,140 km; 1,330 miles)
FT, TD, standard fuel	
	903 n miles (1,673 km; 1,040 miles)
FT, TD, with auxiliary fuel	
	1,110 n miles (2,056 km; 1,278 miles)
g limits at AUW of 708 kg (1,560 lb): all	+6/−4
	(+9/−6 ultimate)

UPDATED

GLASAIR III

TYPE: Side-by-side kitbuilt.
PROGRAMME: First flight July 1986.
CURRENT VERSIONS: **Glasair III:** *As described.*

Glasair III Turbo: Powered by 224 kW (300 hp) Textron Lycoming TIO-540. Maximum level speed 284 kt (526 km/h; 327 mph); maximum rate of climb 1,143 m (3,750 ft)/min; withdrawn in 1998 and superseded by Glasair Super III.

Glasair Super III: Prototype (N401KT) flying by third quarter 1998, when demonstrated 322 kt (596 km/h; 280 mph) at 10,060 m (33,000 ft), powered by IO-540-AA1B5 uprated with Honeywell turbocharger to 261 kW (350 hp) at 11,275 m (37,000 ft) and with Hartzell four-blade propeller. Predicted maximum speed of 350 kt (648 km/h; 403 mph). Has enlarged rudder, extra fuel capacity and highly modified cowling. Public debut at Sun 'n' Fun 1999.

CUSTOMERS: 220 flying by early 2002.
COSTS: Kit: US$35,982 without engine, propeller, instruments and avionics; engine US$43,713; constant-speed propeller US$6,580 (2003). Jumpstart options for wing (US$8,350) and fuselage (US$8,450) offered.
DESIGN FEATURES: Similar configuration to earlier Glasairs, but designed to offer better performance, constructional simplicity and economical kit price. Larger fuselage for increased baggage space, payload capacity and comfort, also improving the longitudinal and directional stability for better cross-country and IFR performance; windows to rear of cockpit doors. Thicker windscreen to improve protection against bird strikes at higher speeds, and additional glass fibre laminates, integral longerons, and lay-up schedule which provides structurally stronger and torsionally stiffer fuselage. Quoted build time 2,500 working hours.

Wing section LS(I)-0413; wing incidence 2° 18′, dihedral 3°; strengthened; carries more fuel than previous models. Optional wingtip extensions for 8.32 m (27 ft 3½ in) or 7.86 m (25 ft 9½ in) spans.
FLYING CONTROLS: As Glasair Super II. Slotted flaps optional.

Glasair III *(Jane's/Paul Jackson)*   *NEW*/0533686

LANDING GEAR: Retractable tricycle only, otherwise as Super II-RG. Mainwheels size 5.00-5; nosewheel 4.00-5. Differential Cleveland disc brakes on mainwheels.
POWER PLANT: One 224 kW (300 hp) Textron Lycoming IO-540-K1H5 flat-six piston engine driving a two-blade constant-speed propeller. Fuel capacity in wings 201 litres (53.0 US gallons; 44.1 Imp gallons). Fuselage header tank, capacity 30 litres (8.0 US gallons; 6.7 Imp gallons). Optional tanks in wingtip extensions, combined capacity 41.6 litres (11.0 US gallons; 9.2 Imp gallons).
ACCOMMODATION: As for Super II.
Data below refer to standard wing span Glasair III.

DIMENSIONS, EXTERNAL:

Wing span: standard	7.11 m (23 ft 4 in)
optional	8.33 m (27 ft 4 in)
Wing chord: at root	1.28 m (4 ft 2½ in)
at tip	0.84 m (2 ft 9 in)
Wing aspect ratio: standard	6.7
optional	8.2
Length overall	6.52 m (21 ft 4 in)

Height overall	2.29 m (7 ft 6 in)
Tailplane span	2.64 m (8 ft 8 in)
Wheel track	3.10 m (10 ft 2 in)

DIMENSIONS, INTERNAL: As for Glasair Super II

AREAS:

Wings, gross: standard	7.55 m² (81.3 sq ft)
optional	8.50 m² (91.5 sq ft)
Fin	1.06 m² (11.40 sq ft)
Rudder	0.57 m² (6.10 sq ft)
Tailplane	1.51 m² (16.30 sq ft)
Elevator	0.53 m² (5.70 sq ft)

WEIGHTS AND LOADINGS:

Weight empty	737 kg (1,625 lb)
Baggage capacity	45 kg (100 lb)
Max T-O weight: without wingtip extensions	1,089 kg (2,400 lb)
with wingtip extensions	1,134 kg (2,500 lb)
Max wing loading: standard	144.1 kg/m² (29.52 lb/sq ft)
optional	133.4 kg/m² (27.32 lb/sq ft)
Max power loading: standard	4.86 kg/kW (8.00 lb/hp)
optional	5.07 kg/kW (8.33 lb/hp)

PERFORMANCE (standard wings):

Never-exceed speed (VNE)	291 kt (539 km/h; 335 mph)
Max level speed at S/L	234 kt (433 km/h; 269 mph)
Cruising speed: at 75% power at 2,440 m (8,000 ft)	224 kt (415 km/h; 258 mph)
at 65% power at 2,440 m (8,000 ft)	216 kt (400 km/h; 249 mph)
Stalling speed: pilot only, plain flaps down	68 kt (126 km/h; 79 mph)
at max T-O weight, slotted flaps down	63 kt (117 km/h; 73 mph)
Max rate of climb at S/L	1,143 m (3,750 ft)/min
Max rate of roll: standard wing	140°/s
extended wing	90°/s
Service ceiling: Glasair III	approx 7,315 m (24,000 ft)
Glasair Super III	9,144 m (30,000 ft)
T-O run	274 m (900 ft)
Landing run	305 m (1,000 ft)

Range at 55% power:

standard fuel	955 n miles (1,770 km; 1,100 miles)
with tip tanks	1,129 n miles (2,092 km; 1,300 miles)
g limits at AUW of 962 kg (2,120 lb): all	+6/−4
	+9/−6 ultimate

UPDATED

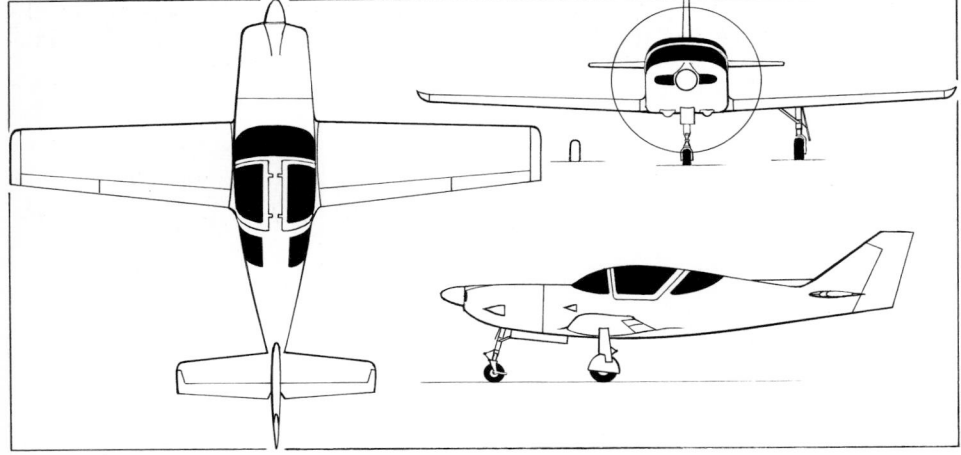

General arrangement of the Glasair III *(Jane's/James Goulding)*   0093599

GLASTAR

NEW GLASTAR LLC
Web: http://www.newglastar.com

A partial successor to Stoddard-Hamilton, New GlaStar shares accommodation, contact details and executives with New Glasair (which see for further information).

VERIFIED

GLASTAR GLASTAR

TYPE: Side-by-side kitbuilt.
PROGRAMME: Designed by Arlington Aircraft Developments and produced until 2000 by Stoddard-Hamilton. Prototype (N824G) first flew 29 November 1994; deliveries of kits started 1995. Over 1,400 hours flown by December 1998. Prototype converted to tailwheel configuration and then back to tricycle. Chosen by EAA as flagship aircraft for Young Eagle junior pilot scheme; first customer-built aircraft completed in July 1996. Early examples had a 93 kW (125 hp) Teledyne Continental IO-240.

Re-engineered version being built in Germany as OMF Symphony (which see). Four-seat, Four Winds versions produced in US (which see).
CUSTOMERS: Over 750 kits sold in 16 countries up to April 1999; 150 flying by February 2002. Total of 205 registered in USA alone by December 2001 increasing to 229 in October 2002.
COSTS: US$24,705 per kit, excluding engine, instruments, upholstery and paint (2003).

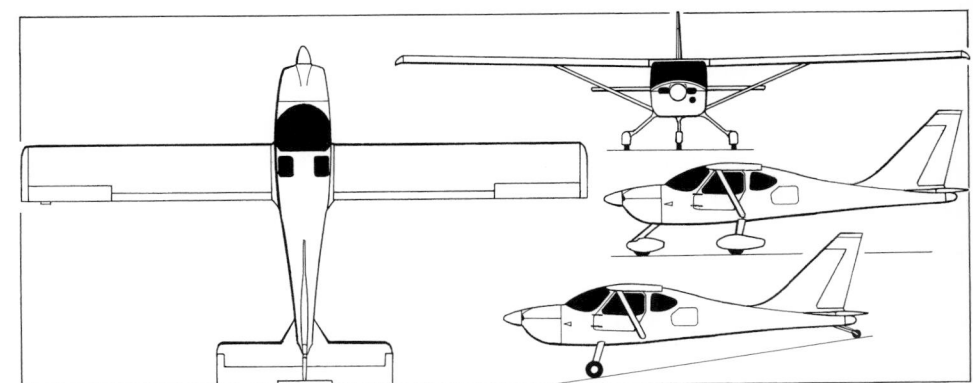

Tricycle version of GlaStar two-seat kitplane, with additional side view of tailwheel version *(Jane's/Paul Jackson)*   0093597

DESIGN FEATURES: Conceived as inexpensive two-seat personal lightplane with wide performance envelope. Convertible landing gear, float compatible. Wings fold and horizontal tail surfaces removable for compact hangarage or for trailer mounting. Quoted build time 1,000-1,500 hours. 'Jump-start' options introduced in November 1998 to reduce time by half; wing kits for this option are assembled in the Czech Republic.

Braced, high wing of constant chord, unswept. Tailplane has root extensions; tall, sweptback fin. Wing section LS(I)-0413.

FLYING CONTROLS: Conventional and manual. Frise ailerons; Fowler flaps. Elevator tab for pitch trim. Ground-adjustable tab on ailerons. Three-position flaps: 0, 20 and 40°. Vortex generators added to wings in front of ailerons and at wingroots, and root strakes to horizontal stabiliser. Optional electric elevator trim.
STRUCTURE: Fuselage construction of GFRP with 4130 steel tube frame surrounding cockpit section. Wings and tail surfaces of 2024-T3 aluminium. Two-spar wings have six full ribs per side, plus stiffeners.

LANDING GEAR: Fixed nosewheel type with spats; steerable tailwheel or Aerocet 2200 float versions; no differences in airframe among landing gear alternatives. Tailwheel (15 cm; 6 in tyre) conversion effected in 3 hours; choice of three mainwheel sizes: 5.00-5 (with speed fairings), 6.00-6 and 8.00-6. Optional Scott 3200 20 cm (8 in) tailwheel with larger mainwheels.

POWER PLANT: One 119 kW (160 hp) Textron Lycoming O-320-D1F flat-four, driving a Sensenich fixed-pitch or Hartzell constant-speed two-blade propeller; power plants in the range 93-168 kW (125-225 hp) are in use. Usable fuel capacity 104 litres (27.6 US gallons; 23.0 Imp gallons). Optional 66 litre (17.5 US gallon; 14.6 Imp gallon) auxiliary tank.

ACCOMMODATION: Two persons side by side; seats adjust for pilot heights between 1.52 and 1.98 m (5 ft 0 in and 6 ft 6 in). Door each side. Baggage compartment behind crew with separate baggage door on port side.

SYSTEMS: Electrical system: 12 V 30 Ah battery; alternator fit appropriate to engine.

AVIONICS: To customer's specification.

DIMENSIONS, EXTERNAL:
Wing span	10.67 m (35 ft 0 in)
Wing chord, constant	1.12 m (3 ft 8 in)

Wing aspect ratio	9.6
Length: fuselage	6.78 m (22 ft 3 in)
overall, Lycoming O-320	6.95 m (22 ft 9½ in)
wings folded	7.47 m (24 ft 6 in)
Height overall: tricycle	2.77 m (9 ft 1 in)
tailwheel	2.11 m (6 ft 11 in)
Tailplane span	3.28 m (10 ft 9 in)
Propeller diameter	1.83 m (6 ft 0 in)
Doors (each): Height	0.80 m (2 ft 7½ in)
Width	0.94 m (3 ft 1 in)
Height to sill	0.84 m (2 ft 9 in)

DIMENSIONS, INTERNAL:
Cabin max width	1.17 m (3 ft 10 in)
Baggage compartment volume	0.91 m³ (32.0 cu ft)

AREAS:
Wings, gross	11.89 m² (128.0 sq ft)
Ailerons (total)	0.37 m² (3.97 sq ft)
Trailing-edge flaps (total)	0.65 m² (6.95 sq ft)
Fin	1.29 m² (13.92 sq ft)
Rudder	0.47 m² (5.08 sq ft)
Tailplane	1.11 m² (12.00 sq ft)
Elevators (total)	0.99 m² (10.67 sq ft)

WEIGHTS AND LOADINGS:
Weight empty	544 kg (1,200 lb)

Fuel weight	92 kg (203 lb)
Baggage capacity	113 kg (250 lb)
Max T-O weight: landplane	889 kg (1,960 lb)
floatplane	952 kg (2,100 lb)
Max wing loading: landplane	74.8 kg/m² (15.31 lb/sq ft)
floatplane	80.1 kg/m² (16.41 lb/sq ft)
Max power loading: landplane	7.46 kg/kW (12.25 lb/hp)
floatplane	7.99 kg/kW (13.13 lb/hp)

PERFORMANCE (119 kW, 160 hp, engine, constant-speed propeller):
Never-exceed speed (VNE)	162 kt (300 km/h; 186 mph)
Max level speed	145 kt (269 km/h; 167 mph)
Max cruising speed at 75% power	
	140 kt (259 km/h; 161 mph)
Econ cruising speed at 65% power	
	133 kt (246 kmh; 153 mph)
Stalling speed, power off:	
flaps up	49 kt (91 km/h; 57 mph)
flaps down	43 kt (79 km/h; 49 mph)
Max rate of climb at S/L	632 m (2,075 ft)/min
Service ceiling	6,096 m (20,000 ft)
T-O run	88 m (290 ft)
T-O to 15 m (50 ft)	155 m (510 ft)
Landing from 15 m (50 ft)	168 m (550 ft)
Landing run	79 m (260 ft)
Range: with standard fuel	
	481 n miles (890 km; 553 miles)
with max fuel	828 n miles (1,533 km; 953 miles)
Endurance with max fuel	6 h
g limits	+3.8/−1.5

UPDATED

GlaStar GlaStar assembled in the USA (*Jane's/Paul Jackson*)   NEW/0533687

Wipline 2100 floats fitted to a GlaStar
(*Jane's/Paul Jackson*)   NEW/0533688

GRIFFON

GRIFFON AEROSPACE INC

301C Nick Fitchard Road, Huntsville, Alabama 35806
Tel: (+1 256) 859 38 80
Fax: (+1 256) 859 34 97
e-mail: info@griffon-aerospace.com
Web: http://www.griffon-aerospace.com
PRESIDENT: Larry French

Griffon's first product, the Lionheart, is stated to be the first kitplane designed (by Larry French) entirely on 3-D CAD.
VERIFIED

GRIFFON LIONHEART

TYPE: Utility biplane kitbuilt.

PROGRAMME: Representation of 1930s Beech 17 'Staggerwing'. Design began 1993; prototype (N985L) first flown July 1997 and publicly displayed at Sun 'n' Fun 1998.

CUSTOMERS: Nine kits sold by April 2001; four had flown by end 2001, first being N985CC (c/n 3), built by Chuck Cianchette in 1999 and second, N223TC (c/n 2), built by Tom Constantino; however, two had been lost in accidents by 2002 and there were no further known completions up to late 2002.

COSTS: US$99,900, excluding engine (2002); 'express build' kits available at extra US$18,000.

DESIGN FEATURES: Classic, equal-span negative stagger biplane. While retaining superficial Beech 17 appearance, is entirely new, computer-aided design. Changes from original include additional cabin glazing, addition of elevator horn balance and enlargement of rudder horn, unbraced wings lacking elliptical tips, and replacement of full span upper ailerons and lower flaps by separate flaps and ailerons on each wing. Quoted build time 2,500 hours. Aerobatics permissible below 1,905 kg (4,200 lb).

Aerofoil modified NACA 64-215 at root, tapering to 64-212 at tip. Laminar flow maintained over 45 per cent of chord. Upper wing dihedral 0° 54′; lower wing dihedral 0° 24′. Washout 1°.

FLYING CONTROLS: Conventional and manual. Ailerons and elevator actuated by pushrods; cable-operated rudder. Flaps and ailerons on all wings. Flight-adjustable tab in port elevator.

STRUCTURE: Extensive use of composites; mainly Nomex honeycomb core with carbon skins and 7781 glass fibre. Cantilever wings; fuselage contains four longerons and underfloor keel beam.

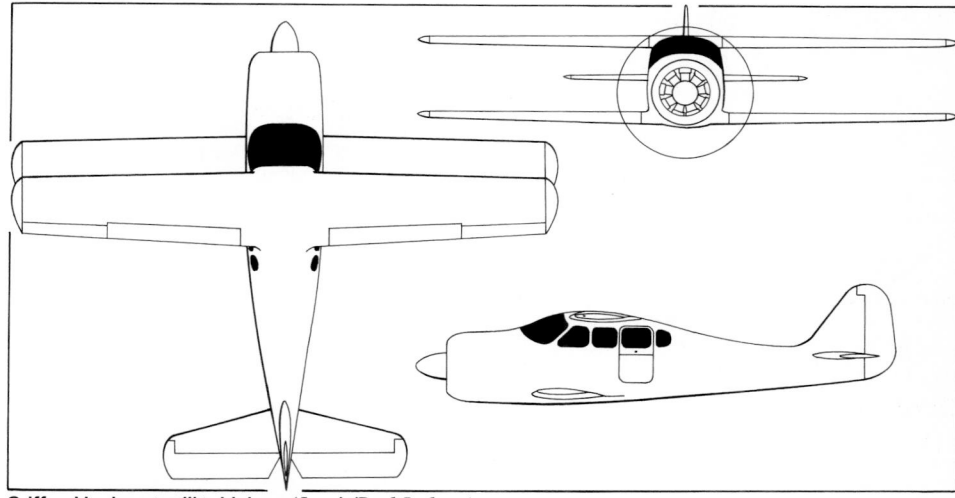

Griffon Lionheart utility biplane (*Jane's/Paul Jackson*)   0099164

Prototype Griffon Lionheart (*Jane's/Paul Jackson*)   NEW/0533689

LANDING GEAR: Tailwheel type; all three units electrohydraulically retractable. Mainwheels 6.50-8. Pressurised gas damping on main legs; gas-charged air/oil cylinder on tail units.

POWER PLANT: One 336 kW (450 hp) Pratt & Whitney R-985 Wasp Junior nine-cylinder radial engine (as in Beech D17S), driving a constant-speed two-blade Hamilton Sundstrand or three-blade Hartzell R10152-5.5 or MT propeller. Alternative engines in the range 261 to 447 kW (350 to 600 hp). Fuel capacity 704 litres (186 US gallons; 155 Imp gallons) in two upper wing tanks, each holding 208 litres (55.0 US gallons; 45.8 Imp gallons) and two lower wing tanks each of 144 litres (38.0 US gallons; 31.7 Imp gallons); total usable fuel 681 litres (180 US gallons; 150 Imp gallons). Oil capacity 28.4 litres (7.5 US gallons; 6.25 Imp gallons).

ACCOMMODATION: Pilot and up to six passengers in enclosed cabin in either forward-facing or club seating. Dual controls. Two-piece, horizontally split door behind wing on port side.

SYSTEMS: Vision Microsystems VM1000 engine monitoring system.

DIMENSIONS, EXTERNAL:	
Wing span	9.45 m (31 ft 0 in)
Wing chord, mean	1.19 m (3 ft 10¾ in)
Wing aspect ratio (effective)	5.2
Negative stagger	0.63 m (2 ft 1 in)
Length overall	8.31 m (27 ft 3 in)
Height overall	2.54 m (8 ft 4 in)
Wheel track	5.30 m (17 ft 4¾ in)
Wheelbase	2.90 m (9 ft 6 in)
Propeller diameter	2.44 m (8 ft 0 in)
DIMENSIONS, INTERNAL:	
Cabin: Length	3.91 m (12 ft 10 in)
Max width	1.32 m (4 ft 4 in)
Max height	1.32 m (4 ft 4 in)
AREAS:	
Wings, gross (each)	10.96 m² (118.0 sq ft)
WEIGHTS AND LOADINGS:	
Weight empty	1,412 kg (3,114 lb)
Max T-O weight	2,358 kg (5,200 lb)
Max wing loading	107.6 kg/m² (22.03 lb/sq ft)
Max power loading	7.03 kg/kW (11.56 lb/hp)

PERFORMANCE:	
Never-exceed speed (VNE)	278 kt (514 km/h; 320 mph)
Max level speed	200 kt (370 km/h; 230 mph)
Max cruising speed	182 kt (338 km/h; 210 mph)
Normal cruising speed at 60% power at 2,895 m (9,500 ft)	170 kt (315 km/h; 196 mph)
Stalling speed, power off:	
flaps up	61 kt (113 km/h; 71 mph)
flaps down	56 kt (104 km/h; 65 mph)
Max rate of climb at S/L	533 m (1,750 ft)/min
Service ceiling	6,095 m (20,000 ft)
T-O run	320 m (1,050 ft)
T-O to 15 m (50 ft)	488 m (1,600 ft)
Landing from 15 m (50 ft)	640 m (2,100 ft)
Landing run	396 m (1,300 ft)
Range with max fuel at 75% power, 45 min reserves	1,450 n miles (2,685 km; 1,668 miles)
Endurance	7 h 12 min
g limits	+6/−3
	UPDATED

GROEN

GROEN BROTHERS AVIATION INC

2640 West California Avenue, Suite A, Salt Lake City, Utah 84104
Tel: (+1 801) 973 01 77
Fax: (+1 801) 973 40 27
e-mail: mktg@gbagyros.com
Web: http://www.gbagyros.com
PRESIDENT AND CEO: David Groen
CHAIRMAN: Jay Groen
VICE-PRESIDENT AND COO: James P Mayfield III
CHIEF FINANCIAL OFFICER: Robin H H Wilson
DIRECTOR OF MARKETING: Scott Muir

MARKETING:
18000 Pacific Highway South, Suite 408, Seattle, Washington 98188
Tel: (+1 206) 241 81 16
Fax: (+1 206) 241 81 17
MARKETING MANAGER: Cyndi Upthegrove

SUBSIDIARY COMPANY:
Nanjing Groen Aviation Industrial Ltd, 140 Guangzhou Road, Suite 22D, Nanjing, Jiangsu 210024, People's Republic of China
Tel: (+86 25) 360 50 73
Fax: (+86 25) 360 49 53

Groen Brothers has been working on its Hawk Gyroplane family since 1986. Prototypes leading up to the Hawk 4, currently in the process of FAA type certification, include the Hawk I and H2X, as described in previous editions of *Jane's*. The Hawk 4 Gyroplane was the first in a series of VTOL-capable aircraft. Development and FAA certification of the Hawk 6 (six place) and Hawk 8 (eight place) will follow the Hawk 4, and preliminary designs of larger aircraft have been completed. The RevCon 6G uses the fuselage of a Cessna 337 as the basis of its fuselage. The company has a flight test facility at Buckeye, Arizona, where, on 12 July 2000, the prototype Jet Hawk 4T/Hawk 4 made its initial flight.

In July 2001, Groen announced plans to move to a new 18,580 m² (200,000 sq ft) facility at Phoenix, Arizona. The plant, which will cost US$14 million and employ 425, was intended to become operational by end 2002 and have the capacity to produce four aircraft per day.

In August 2001, the company concluded a joint venture with Al-Obayya Corporation to produce and market gyroplanes in Saudi Arabia. However, economic downturn of late 2001 resulted in 85 of 130-strong workforce being laid off.

UPDATED

GROEN HAWK 4

TYPE: Four-seat autogyro.

PROGRAMME: Design started April 1996 and prototype two-seat H2X first flew 4 February 1997; later converted to three-seat Hawk III standard. First deliveries had been due June 1998, but design changes to H2X and later Hawk III in October 1998 resulted in Hawk 4. Initial aircraft (N402GB) first flew 29 September 1999, powered by a Continental piston engine, and made first vertical take-off on 9 December 1999; had flown 120 hours in 200 sorties by early April 2000. In September 2000 company switched certification effort to Jet Hawk 4T, which was renamed Hawk 4 at this time; the following October Groen changed its focus to seek government contracts for Hawk 4, slowing certification process for both piston- and turbine-powered versions until it sees market upturn.

On 28 December 2001, Groen announced contract with Utah Olympic Public Safety Command for lease of Hawk 4, beginning 20 January 2002, for security patrols at Salt Lake International Airport, equipped with video downlink system, Spectrolab SX-5 searchlight and additional radios; aircraft flew 75 hours in 67 operations.

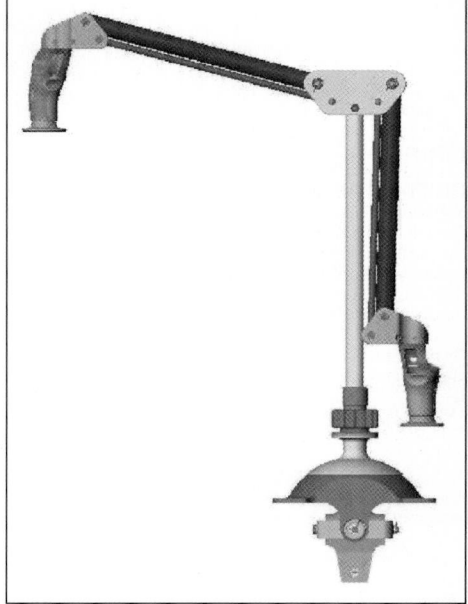

Groen Hawk 4 four-seat autogyro in its original configuration (*Jane's/Paul Jackson*)   0084613

CURRENT VERSIONS: **Hawk 4:** Currently powered by 313 kW (420 shp) Rolls-Royce 250-B17C turboprop; first flew (N403GB) 12 July 2000; originally intended for certification late 2001. Other changes include addition of underfins and taller landing gear. Two further prototypes under construction.

Applications include airborne law enforcement, electronic news gathering, aerial surveillance, utility/passenger transport and aerial application.

Jet Hawk 4T: Original designation for turbine-powered version. Renamed Hawk 4 by 2002.

CUSTOMERS: By May 2002, deposits on 148 aircraft had been taken, via 10 dealerships. Fractional ownership programme announced July 2001.

COSTS: Around US$749,000. Fractional ownership US$99,000 plus US$900 per month and US$226 per flight hour for one-eighth share (2003).

DESIGN FEATURES: Twin tailbooms supported by stub-wings which also house main landing gear; large twin rudders; fixed horizontal tail surface. Two-blade, semi-rigid aluminium teetering rotor with swashplate. Rotor speed 270 rpm.

FLYING CONTROLS: Patented collective pitch-controlled rotor head allows smooth vertical take-off (zero ground roll) and enhanced flight performance. Rotor brake is standard. Fixed horizontal stabiliser; Actuation by pushrods. Patented dual-control stack cyclic flight controls.

STRUCTURE: Steel mast and engine mounts; stressed skin aluminium semi-monocoque fuselage, tail unit, hub structure and propeller; composites nose, engine cowling and wingtips; acrylic windscreen and doors; glass fibre nosecone and engine cowling.

LANDING GEAR: Initially fixed tricycle type; with electrohydraulically operated retractable gear to be offered

Groen Hawk 4 cyclic control provides for one set of controls to be folded away when not in use
NEW/0134514

Groen Hawk 4 used for security work at the 2002 Winter Olympics and Paralympics, Salt Lake City
NEW/0134511

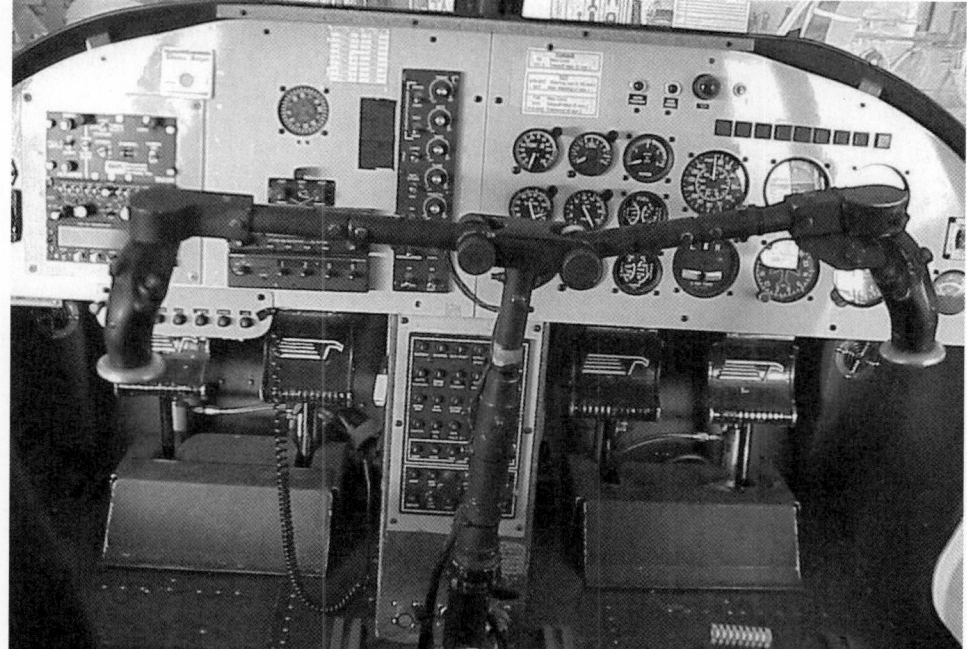
Dual cyclic control fitted to Hawk 4   *NEW*/0134513

PERFORMANCE:

Never-exceed speed (VNE)	128 kt (238 km/h; 148 mph)
Cruising speed at 75% power	
	115 kt (212 km/h; 132 mph)
Max rate of climb at S/L	457 m (1,500 ft)/min
Service ceiling	4,875 m (16,000 ft)
Take-off run	8 m (25 ft)
T-O to 15 m (50 ft)	76 m (250 ft)
Landing from 15 m (50 ft)	46 m (150 ft)
Range with max fuel at 75% power	
	315 n miles (584 km; 363 miles)
	UPDATED

GROEN HAWK 6 and HAWK 8

TYPE: Utility autogyro.

PROGRAMME: Projected developments of Hawk 4 (see previous entry); will benefit from Hawk 4 certification. Preliminary design work has begun on both.

CURRENT VERSIONS: **Hawk 6:** Six-seat version.

RevCon 6G: Announced late 2000 as Hawk 6G with an expected first flight (prototype N9112A) in January 2001; first flight eventually made 22 September. Based around Cessna 337 fuselage with single 335 kW (420 shp) Rolls-Royce 250-B17F2 turboprop in nose for propulsion, driving FC9684C-6RX three-blade propeller; Rolls-Royce 250-C18 above fuselage for rotation; inverted Cessna 337 tailplanes and shortened wing with spoilers and two-blade rotor system from Hawk 4. Estimated cruising speed 113 to 130 kt (209 to 241 km/h; 130 to 150 mph); useful load 907 kg (2,000 lb).

Hawk 8: Eight-seat version.

COSTS: RevCon 6G approximately US$950,000 (2002).

DESIGN FEATURES: Basically similar to Hawk 4, but Hawk 8 will have four rotor blades.

FLYING CONTROLS: Broadly similar to Hawk 4.

STRUCTURE: As Hawk 4.

POWER PLANT: Each powered by a 559 kW (750 shp) turboprop engine. Fuel capacity 416 litres (110 US gallons; 91.6 Imp gallons) in Hawk 6 and 606 litres (160 US gallons; 133 Imp gallons) in Hawk 8. External tanks are modified Piper PA-24 Comanche wingtip units.

AVIONICS: *Instrumentation:* Full IFR.

DIMENSIONS, EXTERNAL:

Main rotor diameter: RevCon 6G	13.41 m (44 ft 0 in)
Hawk 8	12.80 m (42 ft 0 in)
Fuselage length: RevCon 6G	9.07 m (29 ft 9 in)
Height: RevCon 6G	4.11 m (13 ft 6 in)

AREAS:

Main rotor disc: RevCon 6G	168.11 m² (1,809.6 sq ft)
Hawk 8	128.71 m² (1,385.4 sq ft)

WEIGHTS AND LOADINGS (estimated):

Weight empty: RevCon 6G	964 kg (2,125 lb)
Hawk 8	1,406 kg (3,100 lb)
Max T-O weight: RevCon 6G	2,041 kg (4,500 lb)
Hawk 8	2,495 kg (5,500 lb)
Max disc loading: RevCon 6G	12.14 kg/m² (2.49 lb/sq ft)
Hawk 8	19.38 kg/m² (3.97 lb/sq ft)
Max power loading: RevCon 6G	
	6.52 kg/kW (10.71 lb/shp)
Hawk 8	4.46 kg/kW (7.33 lb/shp)

PERFORMANCE (estimated):

Never-exceed speed (VNE)	147 kt (273 km/h; 170 mph)
Cruising speed	104 kt (193 km/h; 120 mph)
Service ceiling	4,875 m (16,000 ft)
Range:	
RevCon 6G at 50% power	
	360 n miles (667 km; 415 miles)
Hawk 8 at 75% power	
	347 n miles (643 km; 400 miles)
Endurance: RevCon 6G	2 h 58 min
Hawk 8	2 h 53 min
	UPDATED

later. Mainwheel tyres 6.00×6; nosewheel 5.00×5; Cleveland hydraulic brakes. Twin safety wheels at rear of tailbooms.

POWER PLANT: Production prototype for Hawk seriesX was powered by a 134 kW (180 hp) Textron Lycoming O-360-A4M flat-four. Hawk H2X had one 335 kW (450 hp) Geschwinder V-8 aluminium liquid-cooled engine, derated to 261 kW (350 hp) at 2,500 rpm, driving a Hartzell three-blade constant-speed propeller.

Hawk 4 piston-powered version has air-cooled, six-cylinder Teledyne Continental TSIO-550 rated at 261 kW (350 hp) at 2,700 rpm; prototype had four-blade MTV propeller but production models will have Hartzell three-blade constant-speed propeller. Engine provides power to rotor for prerotation to provide for short and vertical take-off capability; power to rotor system never engaged during flight. Hawk 4 turbine-powered version has one 313 kW (420 shp) gas turbine engine driving three-blade constant-speed propeller.

Fuel capacity 284 litres (75.0 US gallons; 62.5 Imp gallons) in single tank at rear of fuselage; refuelling point at top of fuselage. Oil capacity 11.4 litres (3.0 US gallons; 2.5 Imp gallons).

ACCOMMODATION: Pilot and up to three passengers in enclosed cabin in two pairs of seats; rear seats fold to provide baggage space.

SYSTEMS: Electrical system 28 V DC.

AVIONICS: *Comms:* Honeywell KLX 135A GPS/COM, KT 76A transponder with Mode C, PM 3000 intercom.

Flight: United Instruments suite.

DIMENSIONS, EXTERNAL:

Rotor diameter	12.80 m (42 ft 0 in)
Rotor blade chord	0.34 m (1 ft 1½ in)
Propeller diameter	1.93 m (6 ft 4 in)
Length of fuselage	7.31 m (24 ft 0 in)
Height: overall	4.11 m (13 ft 6 in)
to top of fuselage	3.35 m (11 ft 0 in)
Wheel track	2.72 m (8 ft 11 in)
Wheelbase	2.47 m (8 ft 1¼ in)
Passenger doors: Width	0.94 m (3 ft 1 in)
Height to sill	0.89 m (2 ft 11 in)

DIMENSIONS, INTERNAL:

Cabin: Length	1.91 m (6 ft 3 in)
Max width	1.37 m (4 ft 6 in)

AREAS:

Main rotor disc	128.71 m² (1,385.4 sq ft)
Vertical tails	1.88 m² (20.2 sq ft)
Horizontal stabiliser	1.49 m² (16.0 sq ft)

WEIGHTS AND LOADINGS:

Weight empty	835 kg (1,840 lb)
Max T-O weight	1,587 kg (3,500 lb)
Max disc loading	12.33 kg/m² (2.53 lb/sq ft)
Max power loading	6.09 kg/kW (10.00 lb/hp)

Proof of concept Groen RevCon 6G (*Jane's/Paul Jackson*)   *NEW*/0533690

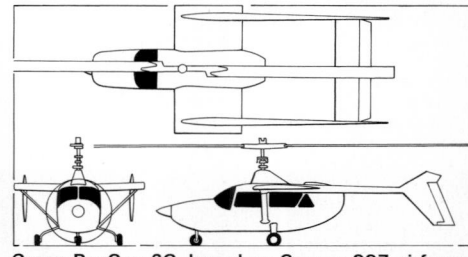
Groen RevCon 6G, based on Cessna 337 airframe (*Jane's/Paul Jackson*)   0105705

GROSSO

GROSSO AIRCRAFT INC

229 Yarrow Hill Drive, Cottage Grove, Wisconsin 53527
Tel: (+1 608) 345 04 06
e-mail: grossoaircraft@msn.com
PRESIDENT: Ron Grosso

Grosso Aircraft was set up to market the Easy Eagle ultralight.
UPDATED

GROSSO EASY EAGLE

TYPE: Tandem-seat ultralight biplane kitbuilt.

PROGRAMME: Construction began 1996; Easy Eagle first displayed at Oshkosh 1998. Two-seat Easy Eagle II introduced in 1999.

CURRENT VERSIONS: **Easy Eagle:** Single-seat version.
Easy Eagle II: Two-seat version.

CUSTOMERS: Three Easy Eagles and one Easy Eagle II flying by mid-2002.

COSTS: Kit US$9,900 (Easy Eagle) or US$13,900 (Easy Eagle II) (2002).

DESIGN FEATURES: Classic open-cockpit, single-bay biplane with steel tube cabane and flying struts and steel landing wires. Modified Clark Y aerofoil. No aerofoil on tailplane. Quoted build time 300 to 500 hours for Easy Eagle and 500 to 700 hours for Easy Eagle II.

FLYING CONTROLS: Conventional and manual. Large ailerons on lower wings only.

STRUCTURE: Fabric-covered 4130 steel tube fuselage with metal-covered fuselage forward of cockpit and two-spar wooden wings with aluminium leading-edge.

LANDING GEAR: Tailwheel configuration; one-piece sprung aluminium mainwheel legs. Azusa wheels and cable brakes; tyres 5.00-5 on Easy Eagle and 6.00-6 Easy Eagle II. Solid tailwheel.

POWER PLANT: *Easy Eagle:* One 1,914 cc Volkswagen four-cylinder air-cooled motorcar engine; design accepts engines to 63.4 kW (85.0 hp). Wooden propeller. Fuel capacity 45.4 litres (12.0 US gallons; 10.0 Imp gallons).
Easy Eagle II: One 83.5 kW (112 hp) Textron Lycoming O-235 flat-four. Fuel capacity 68.1 litres (18.0 US gallons; 15.0 Imp gallons).

DIMENSIONS, EXTERNAL:

Wing span: Easy Eagle	5.49 m (18 ft 0 in)
Easy Eagle II	6.20 m (20 ft 4 in)
Length overall: Easy Eagle	4.37 m (14 ft 4 in)
Easy Eagle II	5.49 m (18 ft 0 in)
Height overall: Easy Eagle	1.78 m (5 ft 10 in)
Easy Eagle II	1.98 m (6 ft 6 in)

AREAS:

Wings, gross: Easy Eagle	9.57 m² (103.0 sq ft)
Easy Eagle II	13.01 m² (140.0 sq ft)

WEIGHTS AND LOADINGS:

Weight empty: Easy Eagle	193 kg (425 lb)
Easy Eagle II	328 kg (725 lb)
Max T-O weight: Easy Eagle	328 kg (725 lb)
Easy Eagle II	521 kg (1,150 lb)

PERFORMANCE:

Max operating speed: both	95 kt (177 km/h; 110 mph)
Normal cruising speed: both	78 kt (145 km/h; 90 mph)
Stalling speed: both	40 kt (73 km/h; 45 mph)
Max rate of climb at S/L: both	274 m (900 ft)/min
T-O run: Easy Eagle	92 m (300 ft)
Easy Eagle II	122 m (400 ft)
Landing run: both	92 m (300 ft)
Range with max fuel: Easy Eagle	260 n miles (482 km; 300 miles)
Easy Eagle II	173 n miles (321 km; 200 miles)

UPDATED

Newly completed Grosso Easy Eagle I *(Jane's/Paul Jackson)*   0126879

GULFSTREAM

GULFSTREAM AEROSPACE CORPORATION

500 Gulfstream Road, Savannah, Georgia 31408
Tel: (+1 912) 965 30 00
Fax: (+1 912) 965 37 75/41 71
Web: http://www.gulfstream.com
VICE-CHAIRMAN: Bryan T Moss
PRESIDENT: William (Bill) W Boisture Jr
CHIEF OPERATING OFFICER: Joe Lombardo
SENIOR VICE-PRESIDENT, MARKETING AND SALES: Raynor Reavis
SENIOR VICE-PRESIDENT, PRODUCT SUPPORT: Larry R Flynn
SENIOR VICE-PRESIDENT, PROGRAMS, ENGINEERING AND TEST: Preston Henne
SENIOR VICE-PRESIDENT, GOVERNMENT SALES AND MARKETING: Buddy Sams
SENIOR VICE-PRESIDENT, FINANCE AND PLANNING: Dan Clare
SENIOR VICE-PRESIDENT, ADMINISTRATION: Ira Berman
VICE-PRESIDENT, MARKETING AND COMMUNICATIONS: Stephanie Snyder
DIRECTOR OF CORPORATE COMMUNICATIONS: Robert Baugniet

Originally an offshoot of Grumman Corporation; passed through several owners before purchase by General Dynamics for US$4.8 billion in May 1999; deal completed 31 July 1999.

Corporate HQ and production centre at Savannah, Georgia, has 4,430 employees. Subassembly and support locations are Oklahoma City, Oklahoma (350) and Mexicali, Mexico (600); however, the company announced that its Oklahoma City subsidiary would close by end 2002. Completions and service undertaken at Long Beach, California (865), Dallas, Texas (705), Appleton, Wisconsin (400) and Brunswick, Georgia (215). Service and refurbishment centre at Westfield, Massachusetts (150); a 4,985 m (53,650 sq ft) refurbishment support facility was added at Savannah in April 2000, at which time the total direct workforce was 8,200. In early 2001, Gulfstream exchanged its engine overhaul service in Dallas with units of BBA Aviation, acquiring four regional maintenance centres in Dallas, Las Vegas, Minneapolis and West Palm Beach; the

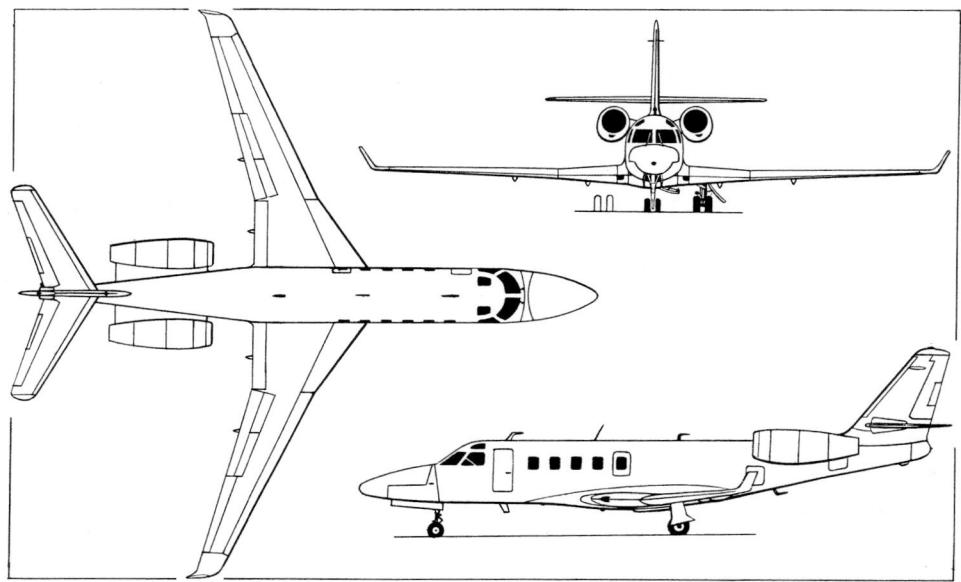

Gulfstream G100 business transport (two Honeywell TFE731-40R-200G turbofans) *(Jane's/Dennis Punnett)*

service operation was rebranded as General Dynamics Aviation Services as a result.

The 1,000th Gulfstream, a Gulfstream V, was delivered to the company's Long Beach completion centre in September 1997. During 1998, Gulfstream and Lockheed Martin's Skunk Works were revealed to be undertaking preliminary studies into the potential of a supersonic business jet; however Skunk Works withdrew from the project in 2000, following financial difficulties.

Gulfstream delivered 51 aircraft (22 Gulfstream IV-SPs and 29 Vs) in 1997 and 61 (32 Gulfstream IV-SPs and 29 Vs) in 1998. Gulfstream simultaneously handed over the 400th Gulfstream IV and 100th Gulfstream V at Savannah,

Georgia, on 25 April 2000. Sales for 2000 were 82, comprising 27 new-generation V-SPs, 35 IV-SPs and 20 Vs; deliveries for same period were 70; and for 2001 36 IV-SPs and 35 Vs, plus five Gulfstream 100s and 24 200s.

Following the purchase of Galaxy Aerospace of Israel by General Dynamics on 5 June 2001, the former's Astra and Galaxy business jets were incorporated into the Gulfstream product line as the Gulfstream 100 and 200 respectively. Production continues at Tel Aviv with 'green' aircraft flown to Lincoln, Nebraska (100) and Alliance Airport, Texas (200) for outfitting by Gulfstream; the Alliance facility closed in mid-2002 as a completion centre and became a sales and design centre.

UPDATED

GULFSTREAM G100
US Air Force designation: C-38A
TYPE: Business jet.
PROGRAMME: Descendant of US Aero Commander 1121 Jet Commander (first flight 27 January 1963), acquired by IAI 1967 and developed successively as 1121 Commodore Jet, 1123 and 1124 Westwind and 1124A Westwind 2; sales of 1121/23/24/24A models by Aero Commander and IAI totalled 441 by end of 1987; 1125 model launched at NBAA Convention October 1979, renamed Astra 1981.

Construction of two prototypes and one static/fatigue test aircraft started April 1982; roll-out 1 September 1983; first flight (4X-WIN, c/n 001) 19 March 1984; first flight 4X-WIA (c/n 002) August 1984; first flight production Astra (4X-CUA) 20 March 1985; FAR Pts 25 and 36 certification 29 August 1985; first delivery 30 June 1986. Astra SP introduced at NBAA Convention October 1989; first delivery of this version (N60AJ) late 1990.

Gulfstream G100 twin-turbofan business transport   *NEW*/0530174

Interior of the Gulfstream G100 with executive and cargo accommodation
NEW/0527120

Flight deck of Gulfstream G100 with Pro Line 4 avionics
0100871

Astra SPX announced at NBAA Convention 1994 and certified by FAA 8 January 1996; type certificate acquired by Gulfstream Aerospace in June 2001, renamed Gulfstream 100 and formally unveiled as such at Paris on 17 June 2001; Gulfstream G100 name adopted from September 2002. Production rate 12 to 15 per year.

CURRENT VERSIONS: **Astra:** Initial version; 32 production aircraft; see 1992-93 and earlier *Jane's All the World's Aircraft*.

Astra SP: Introduced 1989; superseded by SPX; 36 built; details in 1997-98 *Jane's*

Gulfstream G100: First flight (as Astra SPX, 4X-WIX) 18 August 1994; production deliveries from early 1996. Winglets and Rockwell Collins Pro Line 4 avionics. Change to more powerful TFE731-40R-200G turbofans with FADEC, hydromechanical fuel control back-up and Dee Howard thrust reversers; increased weights and payload/range performance and shorter T-O run.

Detailed description applies to Gulfstream 100 from c/n 138 onwards.

C-38A: Two Astra SPXs, outfitted by Tracor Flight Systems for transport and medevac duties; delivered as C-38As to 201st Airlift Squadron of the US Air National Guard at Andrews AFB, Maryland, in April and May 1998, replacing C-21A Learjets. Contract contained options for an additional two.

Gulfstream G150: Wide-cabin version; *described separately.*

Gulfstream G200: Improved and redesigned version; *described separately.*

CUSTOMERS: Combined production of Astra and Astra SP totalled 68 (plus two prototypes), of which approximately 80 per cent to US customers.

Total of 52 SPXs and 18 G100s produced by late 2002, including customer deliveries of 14 in 1998, 12 in 1999, 11 in 2000 and five in 2001.

DESIGN FEATURES: Typical swept wing, rear-engined business jet configuration. Meets FAR Pt 36 Stage 3 (noise), SFAR Pt 27 (fuel venting and exhaust emission) and RVSM requirements.

Wing section high-efficiency IAI Sigma 2; leading-edge sweep 34° inboard, 25° outboard; dihedral 2°; trailing-edge sweep on outer panels; winglets.

FLYING CONTROLS: Conventional, with hydraulically powered ailerons, manually actuated elevators and rudder. Control surfaces operated by pushrods; tailplane incidence controlled by three motors running together to protect against runaway or elevator disconnect; ailerons can be separated in case of jam; spoiler/lift dumper panels ahead of Fowler flaps; elevators and horn-balanced rudder each have geared tab; flaps interconnected with outboard leading-edge slats, both electrically actuated.

STRUCTURE: One-piece, two-spar wing with machined ribs and skin panels, attached by four main and five secondary frames; wing/fuselage fairings, elevator and fin tips and tailcone of GFRP; ailerons, spoilers, inboard leading-edges and wingtips of Kevlar and Nomex honeycomb; nose avionics bay door and nosewheel doors of Kevlar; Kevlar-reinforced nacelle doors and panels; chemically milled fuselage skins; some titanium fittings; heated windscreens of laminated polycarbonate with external glass layer to resist scratching.

LANDING GEAR: SHL hydraulically retractable tricycle type, with oleo-pneumatic shock-absorber and twin wheels on each unit. Trailing-link main units retract inward, nosewheels forward. Tyre sizes 23×7.00-12 (10 ply) (main), 16×4.4-10 (6 ply) deflector type (nose), pressures 8.55 bar; 124 lb/sq in and 6.90 bar; 100 lb/sq in respectively. Hydraulic extension, retraction and nosewheel steering (±58°); hydraulic multidisc anti-skid mainwheel brakes. Compressed nitrogen cylinder provides additional power source for emergency extension.

POWER PLANT: Two 18.90 kN (4,250 lb st) Honeywell TFE731-40R-200G turbofans, with Dee Howard hydraulically actuated target-type thrust reversers and FADEC, pylon-mounted in nacelle on each side of rear fuselage. Standard fuel in integral tank in wing centre-section, two outer-wing tanks, and upper and lower tanks in centre-fuselage (combined usable capacity 4,910 litres; 1,297 US gallons; 1,080 Imp gallons). Additional fuel can be carried in 378.5 litre (100 US gallon; 83.3 Imp gallon) removable auxiliary tank in forward area of baggage compartment. Single pressure refuelling point in lower starboard side of fuselage aft of wing, or single gravity point in upper fuselage, allow refuelling of all tanks from one position. Fuel sequencing automatic.

ACCOMMODATION: Crew of two on flight deck. Dual controls standard. Sliding door between flight deck and cabin. Standard accommodation in pressurised cabin for six persons, two in forward-facing seats at front and four in club layout; galley (port or starboard) at front of cabin, coat closet forward (starboard), lavatory at rear. All six seats individually adjustable fore and aft, laterally, and can be swivelled or reclined; all fitted with armrests and headrests. Two wall-mounted foldaway tables between club seat pairs. Coat closet houses stereo tape deck. Maximum accommodation for nine passengers.

New cabin interior (first redesign since 1986), introduced in September 1999, has less intrusive (curved) cabinetry and galley units; restyled headliner and sidewall panels; new seats, tables and lighting. Forward club seating maximises recline angle and personal space for each occupant; lavatory has brighter lighting; more storage space for coats and carry-on baggage.

Plug-type airstair door at front on port side; emergency exit over wing on each side. Heated baggage compartment aft of passenger cabin, with external access and service ladder. Service compartment in rear fuselage houses aircraft batteries (or optional APU), electrical relay boxes, inverters and miscellaneous equipment.

SYSTEMS: Honeywell environmental control system, using engine bleed air, with normal pressure differential of 0.61 bar (8.8 lb/sq in). Honeywell 36-150(W) APU available optionally. Two independent hydraulic systems, each at pressure of 207 bar (3,000 lb/sq in). Primary system operated by two engine-driven pumps for actuation of anti-skid brakes, landing gear, nosewheel steering, spoilers/lift dumpers and primary control surfaces. Back-up system, operated by electrically driven pump, provides power for emergency/parking brake, primary control surfaces and thrust reversers.

Electrical system comprises two Lucas Aerospace 300 A 30 V DC engine-driven starter/generators, with two 1 kVA single-phase solid-state inverters operating in unison to supply single-phase 115 V AC power at 400 Hz and 26 V AC power for aircraft instruments. Two 24 V 24 Ah Ni/Cd batteries for engine starting and to permit operation of essential flight instruments and emergency equipment. 28 V DC external power receptacle standard.

Pneumatic de-icing of wing leading-edge slats and tailplane leading-edges; thermal anti-icing of engine intakes. Oxygen system for crew (pressure demand) and passengers (drop-down masks) supplied by 2.18 m³ (77 cu ft) cylinder, with second cylinder of same capacity optional. Two-bottle Freon-type engine fire extinguishing system standard.

AVIONICS: Rockwell Collins Pro Line 4 suite standard.

Comms: Dual VHF-422C radios, RTU-4220 radio tuners, TDR-94D transponders, Baker B12135 audio

Gulfstream C-38A of US Air National Guard (*Jane's/Paul Jackson*)
NEW/0526998

For details of the latest updates to *Jane's All the World's Aircraft* online and to discover the additional information available exclusively to online subscribers please visit

jawa.janes.com

systems and Magnastar Flightphones, Rockwell Collins HF 9000 HF, Motorola NA-1335 Selcal, Artex ELT and Universal CVR-30B CVR.

Radar: TWR-850 colour weather radar and WXP-4220 control panel.

Flight: Dual Universal UNS-1C FMS with embedded GPS, Rockwell Collins FCC-4005 autopilots, AHC-85E AHRS, ADC-850C air data systems, VIR-432 nav/GS/ markers and DME-442; single ADF-462, ALT-55B radio altimeter, TCAS-94 and Honeywell EGPWS.

Instrumentation: Rockwell Collins EFD-4077 EFIS displays all flight information on four 18.4 cm (7¼ in) screens; dual Grimes engine displays and Davtron M850A digital clocks; single Flight Line 8047-10 standby altimeter, 8059-2B standby ASI, Jet AI-804CE standby AI, Precision PAI-700-02 standby compass and Hobbs 15007 hour meter.

EQUIPMENT: Standard equipment includes electric windscreen wipers, electric (warm air) windscreen demisting, wing ice inspection lights, landing light in each wingroot, taxying light inboard of each mainwheel door, navigation and Precise Flight pulse lights at wingtips and tailcone, rotating beacons under fuselage and on top of fin, provision for DeVore logo light and wing/tailplane static wicks.

DIMENSIONS, EXTERNAL:
Wing span over winglets	16.64 m (54 ft 7 in)
Wing mean aerodynamic chord	2.19 m (7 ft 2¼ in)
Wing aspect ratio	8.8
Length overall	16.94 m (55 ft 7 in)
Fuselage: Max width	1.57 m (5 ft 2 in)
Max depth	1.91 m (6 ft 3 in)
Height overall	5.54 m (18 ft 2 in)
Tailplane span	6.40 m (21 ft 0 in)
Wheel track (c/l of shock-struts)	2.77 m (9 ft 1 in)
Wheelbase	7.34 m (24 ft 1 in)
Passenger door (fwd, port): Height	1.37 m (4 ft 6 in)
Width	0.66 m (2 ft 2 in)
Overwing emergency exits (each):	
Height	0.69 m (2 ft 3 in)
Width	0.48 m (1 ft 7 in)

DIMENSIONS, INTERNAL:
Cabin: Length: incl flight deck	6.86 m (22 ft 6 in)
excl flight deck	5.21 m (17 ft 1 in)
Max width	1.45 m (4 ft 9 in)
Max height	1.70 m (5 ft 7 in)
Volume, incl flight deck	12.03 m³ (425 cu ft)
Baggage compartment volume (A: with, B: without fuel tank extension):	
A	1.2 m³ (42 cu ft)
B	1.6 m³ (55 cu ft)

AREAS:
Wings, gross (excl winglets)	29.41 m² (316.6 sq ft)
Ailerons (total)	1.20 m² (12.92 sq ft)
Trailing-edge flaps (total)	4.20 m² (45.18 sq ft)
Leading-edge slats (total)	2.39 m² (25.76 sq ft)
Winglets (total)	0.52 m² (5.57 sq ft)
Fin	3.34 m² (35.97 sq ft)
Rudder (incl tab)	1.17 m² (12.63 sq ft)
Tailplane	4.86 m² (52.28 sq ft)
Elevators (total, incl tabs)	2.01 m² (21.66 sq ft)

WEIGHTS AND LOADINGS:
Basic operating weight empty	6,214 kg (13,700 lb)
Max usable fuel: standard	3,942 kg (8,692 lb)
with long-range tank	4,248 kg (9,365 lb)
Max payload	1,073 kg (2,365 lb)
Payload with max fuel	363 kg (800 lb)
Baggage compartment capacity (A: with, B: without fuel tank extension):	
A	168 kg (370 lb)
B	499 kg (1,100 lb)
Max ramp weight	11,249 kg (24,800 lb)
Max T-O weight	11,181 kg (24,650 lb)
Max landing weight	9,389 kg (20,700 lb)
Max zero-fuel weight	7,711 kg (17,000 lb)
Max wing loading	380.3 kg/m² (77.89 lb/sq ft)
Max power loading	296 kg/kN (2.90 lb/lb st)

PERFORMANCE:
Max operating Mach No. (MMO)	0.875
Max operating speed (VMO), S/L to FL270	353 kt (653 km/h; 406 mph) IAS
Max cruising speed at FL330, 8,618 kg (19,000 lb) mid-cruise weight	484 kt (896 km/h; 557 mph)
Normal cruising speed	M0.80 (459 kt; 850 km/h; 528 mph)
Long-range cruising speed	430 kt (796 km/h; 495 mph)
Max rate of climb at S/L	1,160 m (3,805 ft)/min
Rate of climb at S/L, OEI	411 m (1,348 ft)/min
Initial cruise altitude	12,497 m (41,000 ft)
Max certified altitude	13,715 m (45,000 ft)
FAR Pt 25 T-O balanced field length	1,645 m (5,395 ft)
FAR Pt 25 landing distance at MLW	890 m (2,920 ft)
Range: with eight passengers	2,286 n miles (4,235 km; 2,631 miles)
with four passengers, max fuel and NBAA IFR reserves	2,949 n miles (5,461 km; 3,393 miles)
with max fuel, VFR reserves	3,256 n miles (6,030 km; 3,747 miles)

OPERATING NOISE LEVELS (FAR Pt 36, Stage 3):
T-O	79.1 EPNdB
Sideline	89.5 EPNdB
Approach	91.9 EPNdB

UPDATED

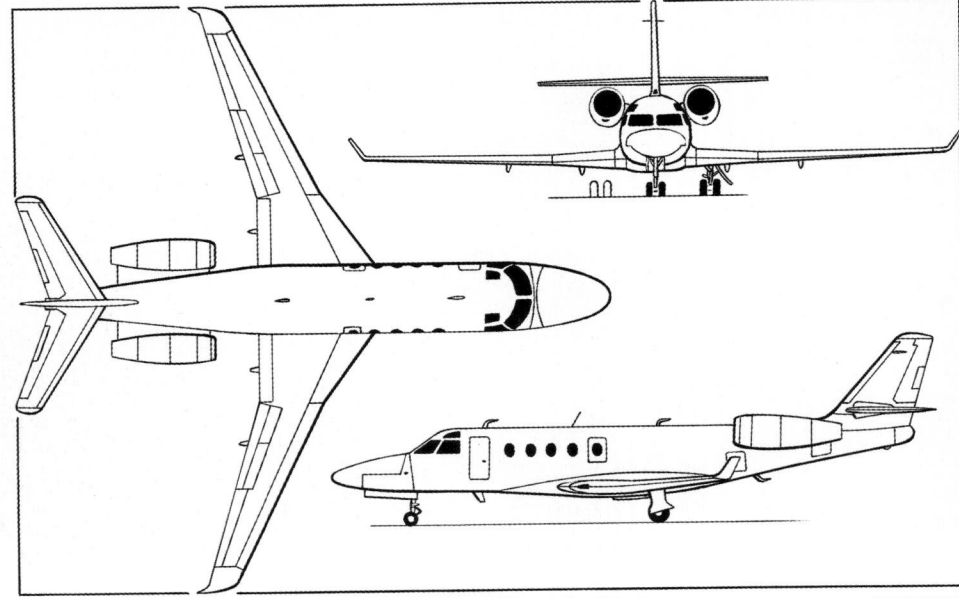

Gulfstream G150, widened derivative of G100 (*Jane's/James Goulding*)   *NEW*/0526905

GULFSTREAM G150

TYPE: Business jet.

PROGRAMME: Announced on eve of NBAA Convention at Orlando, Florida, 8 September 2002. Deliveries to launch customer NetJets scheduled to begin in second quarter of 2005.

CUSTOMERS: Launch customer NetJets Inc ordered 50, plus 50 options, on 9 September 2002, for delivery between 2005 and 2010.

COSTS: US$12.5 million (2002).

DESIGN FEATURES: Wide-cabin version of G100; max internal dimensions increased by 31 cm (12¼ in) in width and 5 cm (2 in) in height. Uprated engines, increased weights, but same range and cruising speed. G200-style windscreen and cabin windows, latter reduced in number from six to five.

Differences from G100 summarised below:

POWER PLANT: Two 19.6 kN (4,400 lb st) Honeywell TFE731-40R turbofans.

ACCOMMODATION: Two pilots and between six and eight passengers in three choices of interior.

DIMENSIONS, EXTERNAL:
Wing span	16.94 m (55 ft 7 in)
Length overall	17.25 m (56 ft 7¼ in)
Height overall	5.54 m (18 ft 2 in)

DIMENSIONS, INTERNAL:
Cabin: Length	5.41 m (17 ft 9 in)
Max width	1.75 m (5 ft 9 in)
Max height	1.75 m (5 ft 9 in)
Volume	13.17 m³ (465 cu ft)
Baggage volume:	
main compartment	1.56 m³ (55 cu ft)
overhead bins	0.25 m³ (8.8 cu ft)

WEIGHTS AND LOADINGS:
Basic operating weight (including two pilots)	6,849 kg (15,100 lb)
Payload: max	1,089 kg (2,400 lb)
with max fuel	363 kg (800 lb)
Max fuel weight	4,581 kg (10,100 lb)
Max T-O weight	11,725 kg (25,850 lb)
Max landing weight	9,843 kg (21,700 lb)
Max zero-fuel weight	7,938 kg (17,500 lb)
Max power loading	300 kg/kN (2.94 lb/lb st)

PERFORMANCE:
Max operating Mach No. (MMO)	0.85
Normal cruising speed	459 kt; (850 km/h; 528 mph)
Long-range cruising speed	430 kt (796 km/h; 495 mph)
FAR Pt 25 T-O balanced field length	1,777 m (5,830 ft)
FAR Pt 25 landing distance at MLW	1,052 m (3,450 ft)

NEW ENTRY

GULFSTREAM G200

TYPE: Business jet.

PROGRAMME: Initiated as IAI 1126 derivative of Astra SP; design (then called Astra IV) finalised late 1992 in anticipation of 1993 launch; co-production with Yakovlev of Russia discussed during early part of 1993; formal announcement of launch as Galaxy, with minor design changes, announced 20 September 1993 just before NBAA Convention in Atlanta, Georgia, USA, followed next day by news that Yakovlev to be risk-sharing partner; other partners to be Rockwell Collins (avionics supplier) and eventual engine manufacturer (P&WC since selected). Replacement for Yakovlev sought in 1995, and Sogerma (France) contracted August 1996 to build production aircraft fuselages and tail units, but latter contract terminated mid-2001.

Four prototypes (003 and 004 flying plus 001 static and 002 fatigue test); first flight rescheduled for fourth quarter 1996 (later changed to second quarter 1997 and later still to fourth quarter 1997); 003 rolled out 4 September 1997 and made first flight (4X-IGA) 25 December 1997. Static testing of 001 completed September 1998. Second flight test aircraft (4X-IGO, c/n 004) made first flight 21 May 1998; 005 (company demonstrator 4X-IGB/N505GA) first flew on 23 September 1998; combined total of 750 hours in 260 flights by three aircraft by December 1998. Israeli CAA and US FAA certification to FAR Pt 25 Amendment 25 and FAR Pts 34 and 36 awarded 16 December 1998; European JAA certification process started mid-1999. First customer delivery, to TTI of Fort Worth, Texas, made in January 2000. Production rate two per month in mid-2000. Type certificate acquired by Gulfstream Aerospace in June 2001, aircraft renamed Gulfstream 200 and formally unveiled as such at Paris on 17 June 2001; name Gulfstream G200 adopted in September 2002. Civil Aviation Administration of China certification achieved 10 September 2002.

CURRENT VERSIONS: Seen as four/eight-passenger business/ executive (standard model), with option of alternative

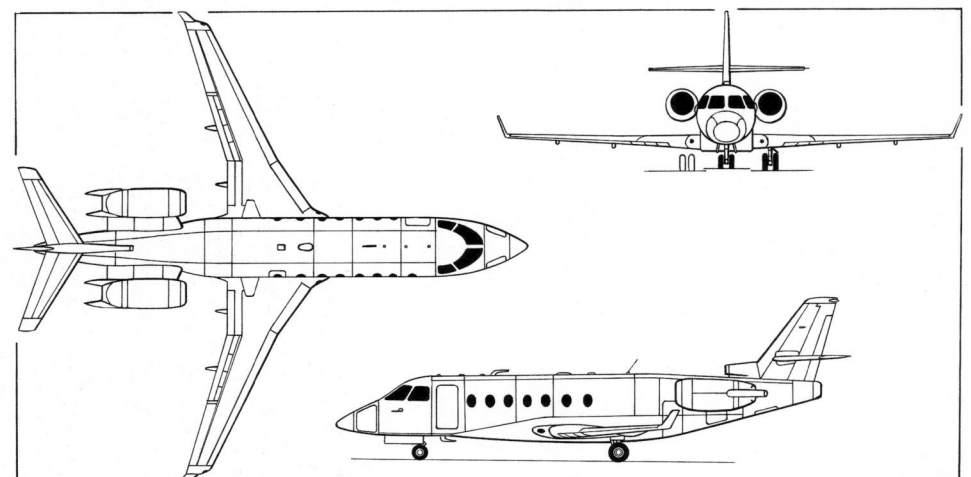

Gulfstream G200 twin-turbofan business and commuter transport (*Jane's/James Goulding*)   0062974

Typical Gulfstream G200 business jet interior 0099570

Flight deck of Gulfstream G200 with Rockwell Collins Pro Line 4 avionics 0100870

interior seating up to 18 passengers for regional transport operation.

CUSTOMERS: Between 45 and 50 commitments by October 1999, of which about 35 per cent from non-US operators in Canada, Europe, Israel, Mexico and South America. First bulk order, for seven, plus option on 15 more, placed mid-1999 by charter operator ILI Aviation of Zurich, Switzerland. IAI estimates break even at 100 sales and potential market for 200. Totals of one delivered in 1999, six in 2000, 24 in 2001, and 12 in the first six months of 2002. Orders by September 2002 included 50 firm and 50 options by Executive Jet Inc, three by Hainan Airlines of Beijing, China, and one by MetroJet of Hong Kong. Total 71 supplied to completion centres or customers by October 2002.

COSTS: Approximately US$18 million flyaway (July 2000); total development programme cost forecast at approximately US$152 million (1993).

Detailed description applies to Gulfstream G200 from c/n 40 onwards.

DESIGN FEATURES: Designed for transatlantic range (non-stop Paris to New York). Essentially same wing as Gulfstream 100, except for 34° 30′ inboard leading-edge sweep and addition of Krueger flaps; new wide-body fuselage is longer and has more headroom.

FLYING CONTROLS: All-hydraulic dual actuation except for rudder (manual); aileron movement 10° up/15° down, elevators 27° up/20° down, rudder 20° left/right; tailplane and rudder tab have electric trim. Wings fitted with outboard leading-edge slats (25° fully out), inboard Krueger leading-edge flaps (110°), four-segment upper surface airbrakes/lift dumpers (45° fully up) and Fowler single-slotted inboard and outboard trailing-edge flaps (settings 0, 12, 20 and 40°). Ailerons and elevators can be operated manually in event of hydraulic failure.

STRUCTURE: Generally similar to that described for G100. Sogerma (France) produced tail units, doors, access panels, tailcones and wing/fuselage fairings for prototypes. Flight Environments cabin flame protection and sound attenuation system.

LANDING GEAR: Hydraulically retractable tricycle type; twin wheels and oleo-pneumatic shock-absorbers on each unit. Nose unit has electrohydraulic steering (±60°) and retracts forward. Trailing-link mainwheel units retract inward and are equipped with Honeywell multidisc anti-skid carbon brakes.

POWER PLANT: Two FADEC-equipped Pratt & Whitney Canada PW306A turbofans, each flat rated at 26.9 kN (6,040 lb st), pylon-mounted on sides of rear fuselage. Nordam nacelles and thrust reversers. Usable fuel capacity 8,472 litres (2,238 US gallons; 1,864 Imp gallons).

ACCOMMODATION: One or two pilots; provision for jump-seat for third crew member. Standard club-type seating for four to eight persons in business/executive version; legroom between facing seats allows enough space for full reclining and berthing; large galley with room for refrigerator, microwave oven, coffee maker and storage. Alternative 10-passenger layout has four club seats at front, with a four-place conference group and two on a divan to the rear, plus optional lavatory and additional baggage space aft. Three-abreast seating for up to 18 passengers, with single aisle, can be provided in corporate shuttle configuration. Generous baggage compartment in rear fuselage, accessed by external airstair door, can accommodate baggage for all 18 passengers. Entire accommodation, including baggage compartment, is pressurised. Airshow in-flight cabin entertainment system optional.

SYSTEMS: Dual hydraulic systems, each 207 bar (3,000 lb/sq in). Electrical system comprises three (including one on APU) Lucas Aerospace 28 V 400 Ah engine-driven starter/generators, two 24 V 43 Ah Ni/Cd batteries and a Honeywell GTCP36-150 APU. Third (24 V 27 Ah) battery for back-up powering of essential flight instruments and

emergency systems. Pneumatic system for emergency extension of landing gear, actuation of wheel brakes and thrust reversers, and de-icing of wing leading-edges. Cabin pressurisation and air conditioning system, differential 0.61 bar (8.8 lb/sq in). One (optionally two) oxygen bottles.

AVIONICS: Rockwell Collins Pro Line 4 suite standard.

Comms: Dual VHF-422C radios, RTU-4220 radio tuners, TDR-94D transponders, Bendix/King KHF 950 HF and Baker B1045-F512 audio systems; triple Magnastar Flightphones; single Avtech Selcal, Artex ELT and Universal CVR-30B CVR.

Radar: TWR-850 colour weather radar with turbulence detection and WXP-4220 control panel.

Flight: Dual Universal UNS-1C FMS with embedded GPS, Rockwell Collins FCC-4005 autopilots, AHS-3000 AHRS, ADC-850C air data systems, VIR-432 VOR/ILS/GS/markers and DME-442; single ADF-462, ALT-4000 radio altimeter, TCAS-4000, Honeywell Laseref IV IRS, EGPWS Mk V and FDR.

Instrumentation: Rockwell Collins EFD-4077 EFIS displays all flight and EICAS information on five 18.4 cm (7¼ in) screens; dual Davtron M850A digital clocks; Flight Line 8047-10 standby altimeter, 8059-2B standby ASI, Jet AI-804CE standby AI, Precision PAI-700-04 standby compass and Hobbs 15007 hour meter.

DIMENSIONS, EXTERNAL:
Wing span	17.70 m (58 ft 1 in)
Length overall	18.97 m (62 ft 3 in)
Wing aspect ratio	9.1
Height overall	6.53 m (21 ft 5 in)
Tailplane span	6.86 m (22 ft 6 in)
Wheel track	3.30 m (10 ft 10 in)
Wheelbase	7.39 m (24 ft 3 in)
Passenger door: Height	1.83 m (6 ft 0 in)
Width	0.84 m (2 ft 9 in)
Baggage compartment door:	
Height	1.14 m (3 ft 9 in)
Width	0.89 m (2 ft 11 in)

DIMENSIONS, INTERNAL:
Cabin: Length: incl flight deck	9.30 m (30 ft 6 in)
excl flight deck	7.44 m (24 ft 5 in)
Max width	2.18 m (7 ft 2 in)
Max height	1.91 m (6 ft 3 in)
Volume, excl flight deck	24.6 m³ (868 cu ft)
Baggage compartment volume	3.5 m³ (125 cu ft)

AREAS:
Wings, gross	34.28 m² (369.0 sq ft)

WEIGHTS AND LOADINGS:
Basic operating weight empty	8,981 kg (19,800 lb)
Max usable fuel weight	6,804 kg (15,000 lb)
Max payload	1,905 kg (4,200 lb)
Baggage compartment capacity	898 kg (1,980 lb)
Payload with max fuel	363 kg (800 lb)
Max ramp weight	16,148 kg (35,600 lb)
Max T-O weight	16,079 kg (35,450 lb)
Max landing weight	13,608 kg (30,000 lb)
Max zero-fuel weight	10,886 kg (24,000 lb)
Max wing loading	469.1 kg/m² (96.07 lb/sq ft)
Max power loading	299 kg/kN (2.93 lb/lb st)

PERFORMANCE (estimated):
Max operating Mach No. (M_{MO})	0.85
Max operating speed (V_{MO}):	
S/L to 3,050 m (10,000 ft)	310 kt (574 km/h; 356 mph) IAS
3,050-6,100 m (10,000-20,000 ft)	310-330 kt (574-611 km/h; 356-379 mph) IAS
6,100-7,620 m (20,000-25,000 ft)	360 kt (667 km/h; 414 mph) IAS
Max cruising speed (V_{MC}) at 9,450 m (31,000 ft), mid-cruise weight of 12,247 kg (27,000 lb)	494 kt (915 km/h; 568 mph)
Typical cruising speed at 11,880 m (39,000 ft)	475 kt (880 km/h; 547 mph)
Long-range cruising speed	430 kt (797 km/h; 495 mph)
Stalling speed, flaps and gear down, at MLW	112 kt (208 km/h; 129 mph) IAS
Max certified altitude	13,715 m (45,000 ft)

Gulfstream G200 (formerly Galaxy) business jet (two PW306A turbofans) 0130549

T-O distance (S/L, ISA) 1,855 m (6,080 ft)
Landing distance at MLW (S/L, ISA) 1,001 m (3,285 ft)
Range with four passengers, NBAA IFR reserves
3,600 n miles (6,667 km; 4,142 miles)
g limits, flaps and gear up +2.63/−1.0
OPERATING NOISE LEVELS (FAR Pt 36 Stage 3):
T-O 81.4 EPNdB
Sideline 85.8 EPNdB
Approach 90.9 EPNdB
UPDATED

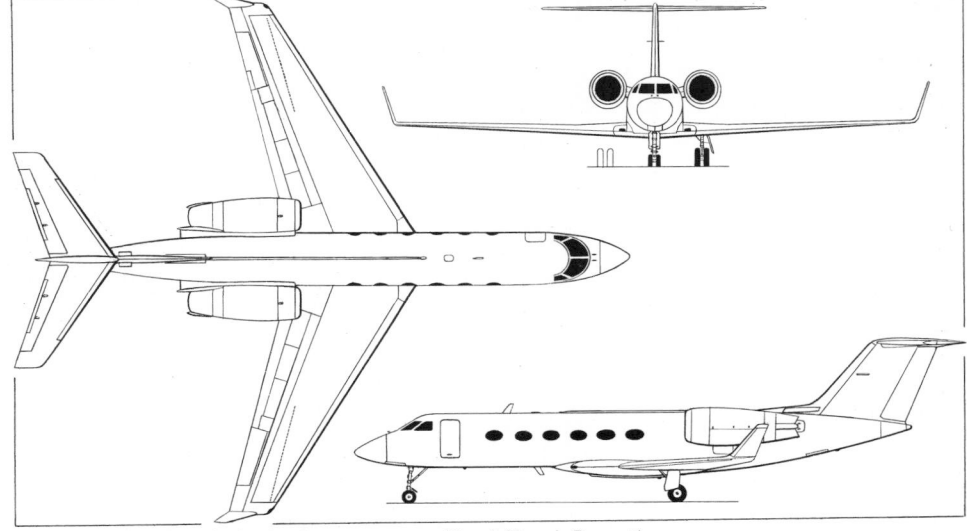

Gulfstream IV twin-turbofan business transport (*Jane's/Dennis Punnett*)

GULFSTREAM IV, G300 and G400
Swedish Air Force designations: S 102B Korpen and Tp 102
US military designations: C-20F/G/H
JASDF designation: U-4

TYPE: Long-range business jet.

PROGRAMME: Design of Gulfstream IV started March 1983; manufacture of four production prototypes (one for static testing) began 1985; first aircraft (N404GA) rolled out 11 September 1985; first flight 19 September 1985; first flight of second prototype 11 June 1986 and third prototype August 1986; FAA certification 22 April 1987 after 1,412 hours' flight testing; certificate is extension of original Gulfsteam II, first approved on 19 October 1967. Westbound round-the-world flight from Le Bourget Airport, Paris, on 12 June 1987, covering 19,887.9 n miles (36,832.44 km; 22,886.6 miles), took 45 hours 25 minutes at average speed of 437.86 kt (811.44 km/h; 504.2 mph) and set 22 world records; eastbound round-the-world flight in N400GA from Houston, Texas, on 26 and 27 February 1988 covered 20,028.68 n miles (37,093.1 km; 23,048.6 miles) in 36 hours 8 minutes 34 seconds at average speed of 554.15 kt (1,026.29 km/h; 637.71 mph), setting 11 records.

In March 1993, Gulfstream IV-SP N485GA set new world speed and distance records in class, at 503.57 kt (933.21 km/h; 579.87 mph) and 5,139 n miles (9,524 km; 5,918 miles) respectively, on routine business flight from Tokyo, Japan, to Albuquerque, USA. Russian Federation and Associated States (CIS) certification achieved in June 1996. FAA approval for RVSM operation of Gulfstream IV series granted 11 August 1997. European JAA validation of FAA certification of Gulfstream IV and IV-SP achieved 16 October 2001. Gulfstream IV-SP holds 75 flight records; by October 2001 fleet had accumulated more than 1.5 million flying hours and 770,000 sorties, with a 99.6 per cent despatch reliability rate.

G300/G400 introduced from January 2003 on common production line; G400 completed to full specification; G300 as baseline aircraft, with only those avionics and fittings specified by customer and lavatory location only at rear. G400 assembly includes 360 items installed on production line which previously were part of outfitting.

CURRENT VERSIONS: **Gulfstream IV:** Built until 1992. MTOW 33,203 kg (73,200 lb).

Gulfstream IV-SP: Improved (Special Performance), higher-weight version announced at NBAA Convention, Houston, in October 1991. Prototype (N476GA, converted from standard) first flown 24 June 1992; designation applied to all new IVs sold after 6 September 1992 (c/n 1214 and upwards); maximum payload increased by 1,134 kg (2,500 lb) and maximum landing weight increased by 3,402 kg (7,500 lb), with no increase in guaranteed manufacturer's empty weight. Payload/range envelope extended; expanded capability Honeywell SPZ-8400 flight guidance and control system. Production ended 3 December 2002 when 500th and final IV/IV-SP (N499GA: c/n 1499) rolled out.

Detailed description applies to Gulfstream IV-SP, except where otherwise indicated.

Gulfstream IV-MPA: Multipurpose Aircraft, announced September 1994; derived from US Navy C-20G Operational Support Aircraft (which see, below) to provide commercial operators with quick-change interior for up to 26 passengers in high-density shuttle layout, low-density executive configuration, 2,177 kg (4,800 lb) cargo capacity, or combination; large cargo door and larger/additional emergency exits standard.

Gulfstream IV-B: Improved, longer-range version, announced September 1994 as subject of development study by Gulfstream Aerospace and Texas Aerostructures; would feature increased wing span incorporating winglet design from Gulfstream V and additional fuel capacity, providing 400 n mile (741 km; 460 mile) increase in range; project indefinitely postponed, June 1995.

Gulfstream G300: Mid-range (basic specification) version, announced on the eve of the NBAA Convention at Orlando, Florida, 8 September 2002; service entry in 2003. Generally as IV-SP except: basic operating weight (three crew) 19,504 kg (43,000 lb); payload with max fuel 1,134 kg (2,500 lb); max fuel weight 12,202 kg (26,900 lb); T-O distance 1,554 m (5,100 ft); range with three crew, eight passengers, NBAA IFR reserves 3,600 n miles (6,667 km; 4,142 miles). First aircraft N520GA (c/n 1500).

Gulfstream G400: New (full specification) version, announced on the eve of the NBAA Convention at Orlando, Florida, 8 September 2002; service entry in 2003. Generally as IV-SP except HUD standard and: basic operating weight (three crew) 19,913 kg (43,900 lb); max payload 2,313 kg (5,100 lb); payload with max fuel 726 kg (1,600 lb); range with three crew, eight passengers, NBAA IFR reserves 4,100 n miles (7,593 km; 4,718 miles). First aircraft N401SR (c/n 4001).

CUSTOMERS: Total 500 Gulfstream IV and IV-SPs built by December 2002 (c/n 1000 to 1499), when production switched to G300/G400. Deliveries totalled 22 in 1997, 32 in 1998, 39 in 1999, 37 in 2000, 36 in 2001 and 17 in the first six months of 2002.

COSTS: US$24.652 million 'green' (1996).

DESIGN FEATURES: Rear-engined, T-tail configuration, with all flying surfaces sweptback; wing mounted below cabin. Differences from Gulfstream III (1987-88 and earlier *Jane's All the World's Aircraft*) include aerodynamically redesigned wing, with winglets, contributing to lower cruise drag; wing also structurally redesigned with 30 per cent fewer parts, 395 kg (870 lb) lighter and carrying 544 kg (1,200 lb) more fuel; increased tailplane span; fuselage 1.37 m (4 ft 6 in) longer, with sixth window each side; Rolls-Royce Tay turbofans; flight deck with electronic displays; digital avionics; and fully integrated flight management and autothrottle systems.

Advanced sonic rooftop aerofoil; sweepback at quarter-chord 27° 40′; thickness/chord ratio 10 per cent at wing station 50, 8.6 per cent at station 414; dihedral 3°; incidence 3° 30′ at root, −2° at tip; NASA (Whitcomb) winglets.

FLYING CONTROLS: Conventional. Hydraulically powered flying controls with manual reversion; trim tab in port aileron and both elevators; two spoilers on each wing act differentially to assist aileron and, with third spoiler each side, act collectively as airbrakes and lift dumpers; single-slotted Fowler flaps; four vortillons and a single 'tripper' strip under leading-edge of each wing ensure inboard part of wing stalls before outboard section; variable incidence tailplane.

STRUCTURE: Light alloy airframe except for carbon composites ailerons, spoilers, rudder and elevators, some tailplane parts, some cabin floor structure, and parts of flight deck; winglets of aluminium honeycomb. Wing box manufactured by Aerostructures Corporation.

LANDING GEAR: Retractable tricycle type with twin wheels on each unit. Main units retract inward, steerable nose unit forward. Mainwheel tyres size 34×9.25-16 (18 ply) tubeless; pressure 12.07 bar (175 lb/sq in). Nosewheel tyres size 21×7.25-10 (10 ply) tubeless, pressure 7.93 bar (115 lb/sq in); maximum steering angle ±82°. Dunlop air-cooled carbon brakes; Aircraft Braking Systems anti-skid units and digital electronic brake-by-wire system. Dowty electronic steer-by-wire system. Turning circle about wingtip 14.43 m (47 ft 4 in); about nosewheel 12.04 m (39 ft 6 in).

Gulfstream IV-SP (Special Performance)

0130573

POWER PLANT: Two Rolls-Royce Tay Mk 611-8 turbofans, each flat rated at 61.6 kN (13,850 lb st) to ISA +15°C. Target-type thrust reversers. Fuel in two integral wing tanks, with total capacity of 16,542 litres (4,370 US gallons; 3,639 Imp gallons). Single pressure fuelling point in leading-edge of starboard wing. In 1999 Gulfstream considered re-engining of IV-SP as part of product improvement policy, possibilities including Rolls-Royce Deutschland BR710, and General Electric CF34; in May 2000 it placed an order for US$1.4 billion with Rolls-Royce for improved Tays.

ACCOMMODATION: Crew of two plus cabin attendant. Standard seating for up to 19 passengers (typically 12 to 14 in corporate configuration) in pressurised and air conditioned cabin. 'Quick Change' cargo/passenger version, certified for up to 26 passengers, announced 22 December 1993. Galley, lavatory and large baggage compartment, capacity 907 kg (2,000 lb), at rear of cabin. Integral airstair door at front of cabin on port side. Baggage compartment door on port side. Electrically heated wraparound windscreen. Six cabin windows, including two overwing emergency exits, on each side.

SYSTEMS: Cabin pressurisation system maximum differential 0.65 bar (9.45 lb/sq in) maintains 1,980 m (6,500 ft) cabin altitude at 13,715 m (45,000 ft); dual air conditioning systems. Two independent hydraulic systems, each 207 bar (3,000 lb/sq in). Maximum flow rate 83.3 litres (22 US gallons; 18.3 Imp gallons)/min. Two bootstrap-type hydraulic reservoirs, pressurised to 4.14 bar (60 lb/sq in). Honeywell GTCP36-100G APU in tail compartment, flight rated to 12,500 m (41,000 ft) since s/n 1156. Electrical system includes two 36 kVA alternators with two solid-state 30 kVA converters to provide 23 kVA 115/200 V 400 Hz AC power and 250 A of regulated 28 V DC power; two 24 V 40 Ah Ni/Cd storage batteries and external power socket. Wing leading-edges and engine inlets anti-iced.

AVIONICS: *Comms:* Dual VHF/HF transceivers, transponders and cockpit audio systems; cockpit voice recorder; Calquest CD-400 satellite communications equipment optional.

Radar: Digital colour weather radar.

Nav: Dual VOR/LOC/GS with marker beacon receivers; dual DME; dual ADF; dual radio altimeters; optional MLS, GPS and VLF Omega. Optional Northstar Technologies CT-1000 flight deck organiser.

Flight: Honeywell SPZ-8400 digital AFCS; Honeywell SPZ-8000 flight management system (FMS); dual fail-operational flight guidance systems including autothrottles; dual air data systems; dual flight guidance and performance computers; dual laser IRS; AHRS; VNAV; flight data recorder. System integration is accomplished through a Honeywell avionics standard communications bus (ASCB). Optional TCAS and Honeywell EGPWS.

Instrumentation: Six 203 × 203 mm (8 × 8 in) colour CRT EFIS screens, two each for primary flight display (PFD), navigation display (ND) and engine instrument and crew alerting system (EICAS). Honeywell/BAE HUD 2020 head-up display received FAA approval for Cat. II operations in early 1997, with first installation in Gulfstream IV-SP completed in May 1997. Northstar Technologies CT-1000G flight deck organiser system optional. EVS system used on Gulfstream V received FAA certification for installation on IV-SP on 19 December 2002.

Self-defence: Optional BAE Systems AN/ALQ-204 Matador IRCM available at cost of US$3.5 million.

DIMENSIONS, EXTERNAL:

Wing span over winglets	23.72 m (77 ft 10 in)
Wing chord: at root (fuselage c/l)	5.94 m (19 ft 5¾ in)
at tip	1.85 m (6 ft 0¾ in)
Wing aspect ratio	6.4
Length overall	26.92 m (88 ft 4 in)
Fuselage: Length	24.03 m (78 ft 10 in)
Max diameter	2.39 m (7 ft 10 in)
Height overall	7.44 m (24 ft 5 in)
Tailplane span	9.75 m (32 ft 0 in)
Wheel track	4.17 m (13 ft 8 in)
Wheelbase	11.61 m (38 ft 1¼ in)
Passenger door (fwd, port): Height	1.57 m (5 ft 2 in)
Width	0.91 m (3 ft 0 in)
Baggage door (rear): Height	0.90 m (2 ft 11¾ in)
Width	0.72 m (2 ft 4½ in)

DIMENSIONS, INTERNAL:

Cabin:

Length, incl galley, lavatory and baggage compartment	13.74 m (45 ft 1 in)
Max width	2.24 m (7 ft 4 in)
Max height	1.88 m (6 ft 2 in)
Floor area	22.9 m² (247 sq ft)
Volume	43.2 m³ (1,525 cu ft)
Flight deck volume	3.5 m³ (124 cu ft)
Rear baggage compartment volume	4.8 m³ (169 cu ft)

AREAS:

Wings, gross	88.29 m² (950.4 sq ft)
Ailerons (total, incl tab)	2.68 m² (28.86 sq ft)
Trailing-edge flaps (total)	11.97 m² (128.84 sq ft)
Spoilers (total)	7.46 m² (80.27 sq ft)
Winglets (total)	2.38 m² (25.60 sq ft)
Fin	10.92 m² (117.53 sq ft)

Rudder, incl tab	4.16 m² (44.75 sq ft)
Horizontal tail surfaces (total)	18.83 m² (202.67 sq ft)
Elevators (total, incl tabs)	5.22 m² (56.22 sq ft)

WEIGHTS AND LOADINGS:

Manufacturer's weight empty	16,102 kg (35,500 lb)
Allowance for outfitting	3,175 kg (7,000 lb)
Typical operating weight empty	19,278 kg (42,500 lb)
Max payload	2,948 kg (6,500 lb)
Payload with max fuel	1,361 kg (3,000 lb)
Max usable fuel	13,381 kg (29,500 lb)
Max T-O weight	33,838 kg (74,600 lb)
Max ramp weight	34,019 kg (75,000 lb)
Max landing weight	29,937 kg (66,000 lb)
Max zero-fuel weight	22,226 kg (49,000 lb)
Max wing loading	383.2 kg/m² (78.49 lb/sq ft)
Max power loading	275 kg/kN (2.69 lb/lb st)

PERFORMANCE:

Max operating speed (V$_{MO}$/M$_{MO}$)	340 kt (629 km/h; 391 mph) CAS or M0.88
Max cruising speed at FL310	505 kt (936 km/h; 582 mph) or M0.85
Normal cruising speed at FL450	M0.80 (459 kt; 850 km/h; 528 mph)
Approach speed at max landing weight	149 kt (276 km/h; 172 mph)

Stalling speed at max landing weight:

wheels and flaps up	130 kt (241 km/h; 150 mph)
wheels and flaps down	115 kt (213 km/h; 133 mph)
Max rate of climb at S/L	1,256 m (4,122 ft)/min
Rate of climb at S/L, OEI	314 m (1,030 ft)/min
Initial cruising altitude	12,500 m (41,000 ft)
Max certified altitude	13,715 m (45,000 ft)
Runway PCN	25
FAA balanced T-O field length at S/L	1,662 m (5,450 ft)
Landing run	973 m (3,190 ft)

Range:

with max payload, normal cruising speed and NBAA IFR reserves   3,338 n miles (6,182 km; 3,841 miles)

with max fuel, eight passengers, at M0.80 and with NBAA IFR reserves
   4,220 n miles (7,815 km; 4,856 miles)

OPERATIONAL NOISE LEVELS (FAR Pt 36):

T-O	77.5 EPNdB
Approach	92.0 EPNdB
Sideline	86.6 EPNdB

UPDATED

GULFSTREAM V, G500 and G550
US military designation: C-37

TYPE: Long-range business jet.

PROGRAMME: Study announced at NBAA Convention, Houston, in October 1991. Go-ahead commitment and engine selection (BR710) announced at Farnborough Air Show in September 1992. Risk-sharing agreement with wing designers/manufacturers (Vought and ShinMaywa) announced at Paris Air Show in June 1993, and with tail (and later floor panel) manufacturer (Fokker) at NBAA Convention in September 1993.

Prototype (N501GV) rolled out 22 September 1995; first flight 28 November 1995; second aircraft, c/n 502 (N502GV) was structural test article before completion as company demonstrator; c/n 503 (N503GV), first flown 10 March 1996, used for systems testing; c/n 504 (N504GV), first flown May 1996, for engine, flight loads and environmental trials and JAA certification testing; c/n 505 (N505GV), first flown August 1996, outfitted in standard production configuration for operational testing and HIRF evaluation. Public debut (N502GV) at NBAA Convention at Orlando, Florida, in November 1996.

Provisional FAA certification achieved 16 December 1996 after more than 1,100 hours of flight testing in 550 sorties. Full FAA type certification (extension of Gulfstream II) granted on 11 April 1997, and FAA production certificate awarded on 11 June. First fully completed aircraft for a customer (c/n 507) delivered to Walter Annenberg, former US Ambassador to UK, on 1 July 1997. RVSM approval January 2000. JAA certification eventually granted on 31 October 2002.

From 2003, original versions discontinued and replaced by Gulfstream G500 and G550. Latter to full specification; former is baseline version with customer-specified additions only.

By 17 November 1997 company demonstrators had achieved 39 world records in time to climb and maximum altitude with payload, and city pair categories, including the first ever non-stop flights by a business jet between Los Angeles and London, London and Hong Kong, Tokyo and New York, and Washington, DC and Dubai; last-named 6,330 n mile (11,723 km; 7,284 mile) journey completed on 13 November 1997 in 12 hours 40 minutes 48 seconds with four crew and seven passengers. On 19 February 1998, N502GV became first business jet to fly non-stop from New York to Hawaii. On 4 March 2002 a Gulfstream V broke a 44-year-old record for speed over a recognised course, flying from Tokyo to Washington, DC, in 11 hours 54 minutes covering the 5,543 n miles (10,845 km; 6,739 miles) at an average speed of 492 kt (911 km/h; 556 mph). This record was previously held by a USAF KC-135 Stratotanker, which flew Tokyo to Washington, DC, in 13 hours 46 minutes in April 1958. By March 2002, the Gulfstream V held 70 world and national records. By that time the in-service fleet of 140 aircraft had flown more than 120,000 hours, with 99 per cent despatch reliability.

CURRENT VERSIONS: **Gulfstream V:** *As described.*

Gulfstream V-SP: Announced on eve of NBAA Convention, 8 October 2000; larger cabin, enhanced performance and better range by reason of aerodynamic refinement of existing V. First flight of GV-SP test article (N5SP; modified from 132nd GV) 31 August 2001, followed by first production GV-SP (c/n 5001, also N5SP) rolled out 19 June 2002 and first flown 18 July 2002, at which time the test article airframe had completed 100 flights totalling 185 hours. Redesigned as G550 (which see).

Gulfstream G500: Reduced-range version of Gulfstream V, announced on the eve of the NBAA Convention at Orlando, Florida, 8 September 2002; provisional type certificate granted by FAA 11 December 2002; service entry in 2003. Powered by 68.4 kN (15,385 lb st) BR710 turbofans; HUD standard. Data generally as for Gulfstream V except: basic operating weight (four crew) 21,682 kg (47,800 lb); max payload 3,039 kg (6,700 lb); payload with max fuel 1,134 kg (2,500 lb); max fuel weight 15,966 kg (35,200 lb); max T-O weight 38,600 kg (85,100 lb); T-O run 1,570 m (5,150 ft); landing distance 844 m (2,770 ft); initial cruising altitude 13,106 m (43,000 ft); range with eight passengers, three crew, NBAA IFR reserves 5,800 n miles (10,741 km; 6,674 miles).

Gulfstream G550: New version of Gulfstream V-SP, announced on the eve of the NBAA Convention at Orlando, Florida, 8 September 2002; service entry in fourth quarter of 2003. Power plant as for G500; other data generally as for Gulfstream V except: basic operating weight (four crew) 21,909 kg (48,300 lb); max payload 2,858 kg (6,300 lb); payload with max fuel 816 kg (1,800 lb); max T-O weight 41,277 kg (91,000 lb); T-O run 1,801 m (5,910 ft); landing distance 844 m (2,770 ft); range with eight passengers, three crew, NBAA IFR reserves 6,750 n miles (12,501 km; 7,767 miles).

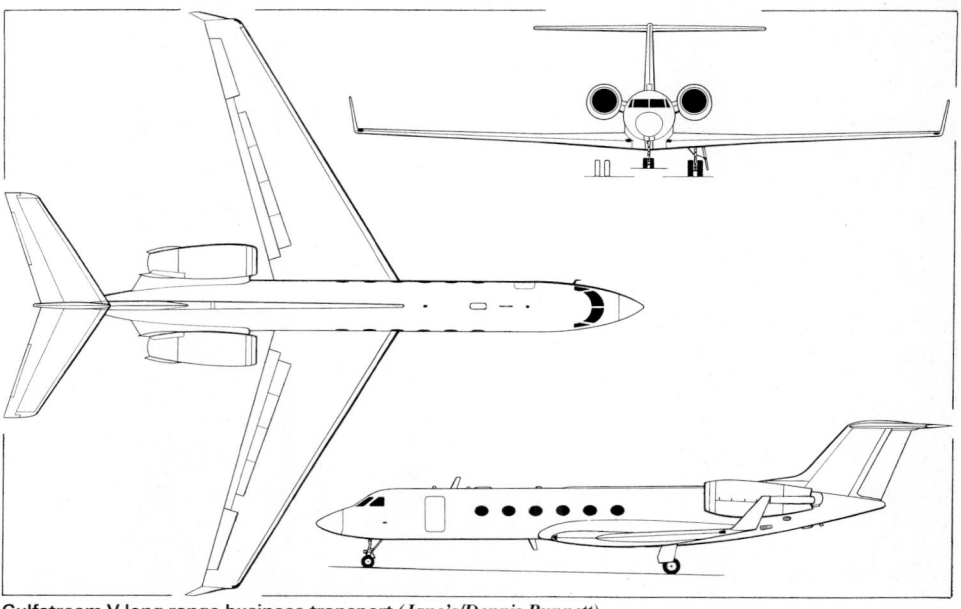

Gulfstream V long-range business transport (*Jane's/Dennis Punnett*)

Flight deck uses Honeywell Primus Epic suite with Gulfstream PlaneView cockpit, comprising four 360 mm (14 in) LCDs, including HUD and EVS, latter was certified in August 2001 and first became operational, on a USAF C-37A, on 15 May 2002. First customer is Executive Jets, which ordered 20 for delivery up to 2008, at cost of US$800 million; unit cost with most popular options US$24.9 million. Other customers include GATX Leasing (two).

Special Missions: Lockheed Martin, teamed with (then) GEC-Marconi Defence, Logica, Marshall Aerospace, MSI, Racal and CAE Electronics Montréal , chose the Gulfstream V as its platform for the UK's Airborne Stand-Off Radar (ASTOR) requirement, involving a Racal dual-mode, electronically scanned surveillance radar in a ventral fairing, with satcom antennas above and below the fuselage for real-time data transfer to ground stations, and provision for in-flight refuelling.

Additionally, when an upgraded version of the Joint STARS system was readmitted to the ASTOR competition in January 1998, the Gulfstream V was selected by Northrop Grumman as platform in preference to Boeing E-8 (707). Neither application was successful. Currently competing for NATO Ground Surveillance System programme. Three purchased by Israel in December 2001 for special electronic missions to be delivered from 2003, with requirement for a further six for AEW missions.

C-37A: Military version; two (plus four options, one of which converted to firm order in 1998, another in April 1999 and a third subsequently, with fourth converted in 2001) ordered by USAF as part of VCX requirement to replace Boeing VC-137s of 89th Airlift Wing, Andrews AFB, Maryland. Announced 5 May 1997. First (97-0400) delivered 14 October 1998; second in January 1999 and third (for C-in-C USAF) 21 February 2000. US Army's Priority Air Transport Squadron operates one example. Five C-37As leased by US Air Force for delivery between August 2001 and September 2003, first aircraft handed over 28 August 2001. Further contract awarded 13 March 2002 for up to 20 for delivery between 2002 and 2012; first firm order under this contract is for one C-37A for the USAF. US Navy requires five VC-37As as replacements for VP-3A Orions; deliveries began 2002 and extend to 2009. In December 2000 US Coast Guard ordered a C-37A which was delivered in May 2002.

EC-37A: Proposed airborne combat support aircraft; Gulfstream announced at RIAT 2000, Cottesmore, UK, that it was looking for funding during late 2000 and partner to develop systems. Aimed at two roles: standoff electronic jammer and intelligence gathering, using interchangeable underwing and underfuselage pods and in-flight workstation reconfiguration.

CUSTOMERS: Three produced in 1996, 29 in 1997, 25 in 1998, 31 in 1999, 34 in 2000, 35 in 2001 and 16 in the first six months of 2002; of these (including prototypes) the 10th was delivered to BMW on 7 January 1998. Executive Jet International took options on two aircraft in January 1995, for delivery to its Gulfstream Shares fleet in 1998-99. Recent customers include Nigerian government; Kuwait Airways, (three, first aircraft delivered November 1999); Brunei government (one); Time Warner (two); Chrysler Corporation (two) and Executive Jet International (10, for delivery by 2004, plus 12 options); Saudi Arabian Ministry of Defence and Aviation (first of two in medevac role, handed over 15 May 2000), and the US National Center for Atmospheric Research, which ordered one in January 2002 for service entry in 2005 as a high-performance instrumented airborne platform for environmental research

(HIAPER). Japanese coast guard ordered two Gulfstream Vs on 14 November 2001 at cost of approximately US$100 million.

Gulfstream delivered 100th aircraft in April 2000. Manufacture of Gulfstream V completed in 2003 with 193 aircraft (c/n 501 to 665, 667 to 693, and 699); G500/G550 followed immediately.

COSTS: US$29.5 million (fixed price) for first 24 aircraft, then US$30.5 million (fixed price) up to 39th aircraft; typically equipped price after outfitting, US$35 million; aircraft sold on Internet during December 1999 in deal valued at US$40 million. Cost of EVS system estimated at US$1 million.

Contract for one C-37A valued at US$43.4 million (2002).

DESIGN FEATURES: Gulfstream IV fuselage re-engineered to increase length by 2.13 m (7 ft 0 in); larger wing of same basic shape and interior structure, but 10 per cent more efficient than Gulfstream IV's; larger vertical and horizontal tail surfaces; flight deck volume increased by moving bulkhead 0.30 m (1 ft) aft to provide more space for pilots and to accommodate full-size jump seat; cockpit layout and instrumentation generally similar to Gulfstream IV-SP, but redesigned to incorporate human engineering changes in system control functions; airstair door moved aft by 1.52 m (5 ft); avionics bay relocated. Computational fluid dynamics and CATIA design system used extensively in development.

FLYING CONTROLS: As Gulfstream IV.

LANDING GEAR: As Gulfstream IV. Turning circle about wingtip 17.07 m (56 ft 0 in); about nosewheel 14.15 m (46 ft 5 in).

Gulfstream V-SP PlaneView flight deck   0130571

POWER PLANT: Two 65.6 kN (14,750 lb st) Rolls-Royce Deutschland BR710-48 turbofans with FADEC. Fuel capacity 23,417 litres (6,186 US gallons; 5,151 Imp gallons) in integral wing tanks, of which 22,993 litres 6,074 US gallons; 5,058 Imp gallons) are usable.

ACCOMMODATION: Crew of two/three plus cabin attendant. Standard seating for 15 to 19 passengers in pressurised and air conditioned cabin. Rear windows, each side, are emergency exits. Customised interiors according to requirements.

SYSTEMS: Digitally controlled automatic cabin pressurisation system; maximum pressure differential 0.703 bar (10.2 lb/sq in). Hamilton Sundstrand electrical power generating system profile integrated with the flight management system (FMS), and will maintain equivalent of 1,830 m (6,000 ft). Honeywell RE220 APU, designed specifically for Gulfstream V, provides engine-starting capability up to 13,110 m (43,000 ft), 40 kVA of electrical power for ground and flight use up to 13,715 m (45,000 ft), and ground air conditioning, with almost twice the cooling airflow rate of the Gulfstream IV's APU.

AVIONICS: Honeywell SPZ-8500 as core system.

Flight: Three independent IRS integrated into FMS; GPS. Honeywell enhanced ground proximity warning system (EGPWS); TCAS; turbulence-detecting Doppler radar; Hamilton Sundstrand maintenance data acquisition unit; optional Northstar Technologies CT-1000 flight deck organiser system.

Instrumentation: Honeywell SPZ-8500 digital AFCS/FMS with six 20.3 × 20.3 cm (8 × 8 in) colour LCD EFIS displays with EICAS; Honeywell/BAE Model 2020 HUD. Kollsman All Weather Window IR sensor flight testing began September 1999 to provide enhanced vision system (EVS) in conjunction with Honeywell 2020 HUD, facilitating operations in Cat. III weather on Cat. I runway, with decision height of 30 m (100 ft) and RVR of 220 m (722 ft). IR sensor flight is mounted beneath radome. Certification was due early 2001.

Mission: IBM satellite-based international communications system, including airborne voice, data, networking, fax and teleconferencing facilities, provides Gulfstream V with 'office in the sky' capability; optional Sanders AN/ALQ-204 Matador IRCM system.

DIMENSIONS, EXTERNAL:

Wing span: basic	27.69 m (90 ft 10 in)
over winglets	28.50 m (93 ft 6 in)
Length overall	29.39 m (96 ft 5 in)
Height overall	7.87 m (25 ft 10 in)
Tailplane span	10.72 m (35 ft 2 in)
Wheel track (c/l shock-absorbers)	4.37 m (14 ft 4 in)
Wheelbase	13.72 m (45 ft 0 in)

DIMENSIONS, INTERNAL:

Cabin: Length, aft of flight deck	15.57 m (50 ft 1 in)
Max width	2.24 m (7 ft 4 in)
Max height	1.88 m (6 ft 2 in)
Volume	47.3 m³ (1,669 cu ft)
Baggage compartment volume	6.4 m³ (226 cu ft)

AREAS:

Wings, gross	105.63 m² (1,137.0 sq ft)

WEIGHTS AND LOADINGS:

Weight empty, 'green'	17,917 kg (39,500 lb)
Operating weight empty	21,772 kg (48,000 lb)
Allowance for outfitting	3,856 kg (8,500 lb)

Prototype Gulfstream V-SP, now known as G550   *NEW*/0530194

Baggage capacity	1,134 kg (2,500 lb)
Max payload	2,948 kg (6,500 lb)
Max fuel	18,733 kg (41,300 lb)
Payload with max fuel	726 kg (1,600 lb)
Max T-O weight	41,050 kg (90,500 lb)
Max ramp weight	41,231 kg (90,900 lb)
Max landing weight	34,155 kg (75,300 lb)
Max zero-fuel weight	24,720 kg (54,500 lb)
Max wing loading	382.2 kg/m² (78.28 lb/sq ft)
Max power loading	308 kg/kN (3.02 lb/lb st)

PERFORMANCE (at max T-O weight, except where indicated):

Max operating Mach No. (V$_{MO}$)	M0.885
Cruising speed: max	499 kt (924 km/h; 574 mph)
normal	488 kt (904 km/h; 562 mph) or M0.85
long-range	459 kt (850 km/h; 528 mph) or M0.80
Max rate of climb at S/L	1,276 m (4,188 ft)/min
Initial cruising altitude	12,500 m (41,000 ft)
Max certified altitude	15,545 m (51,000 ft)
T-O balanced field length	1,862 m (6,110 ft)
Landing distance, S/L, max landing weight	841 m (2,760 ft)

Max range, eight passengers and four crew, M0.80,
NBAA IFR reserves

6,500 n miles (12,038 km; 7,480 miles)

UPDATED

GULFSTREAM SBJ

TYPE: Supersonic business jet.

PROGRAMME: Gulfstream Aerospace and Lockheed Martin (which see) revealed at Farnborough on 7 September 1998 that they were jointly conducting an 18 to 24 month feasibility study into an SBJ. As currently envisaged, the aircraft, similar in size to the Gulfstream II, would feature a stand-up-headroom cabin accommodating eight passengers, and would cruise at M1.6 to M2.0 over a range

of more than 4,000 n miles (7,408 km; 4,603 miles). Engines in the class of P&W F119 and GE F120 could be considered. Key design goals are the ability to operate out of existing business aviation airfields; take-off noise compatible with anticipated future emissions regulations; fuel-efficient operation at subsonic speeds; and an initial cruising altitude above that of subsonic traffic.

The feasibility phase of the project, completed in mid-2000, is being followed by a further two years of wind-tunnel testing and project definition before a launch decision is taken; studies carried out by the company confirm that there is a market, but sonic boom suppression, engine emissions and noise present technological challenges. Gulfstream and Lockheed Martin, through its Advanced Development Projects division, in defining the SBJ, will have access to data from NASA's (now discontinued) High-Speed Research (HSR) programme for a future supersonic airliner. The partners lobbied the US government for support via the Defense Advanced Research Projects Agency which received US$15 million for its quiet supersonic aircraft technology (QSAT) budget. However, Lockheed Martin has since withdrawn from the programme. A model of a twin-engined aircraft, displayed at the time of the project's announcement at the Farnborough Air Show in September 1998, was only 'notional', according to the two companies, and does not necessarily reflect their current thinking on configuration; at 1999 NBAA Convention the companies agreed the model was no longer accurate; in 2000 patents were filed with different layouts including tailplane mounted at tail fintip and drooping to attach to engine nacelles; further patents included one with large delta wing shape plus nose canard; in October Gulfstream displayed, at NBAA, the Quiet Supersonic Jet (QSJ) which has swept wing and wingtip-mounted swept tail.

Estimated unit cost is in the region of US$70 million to US$80 million (2002), based on 200 produced. The

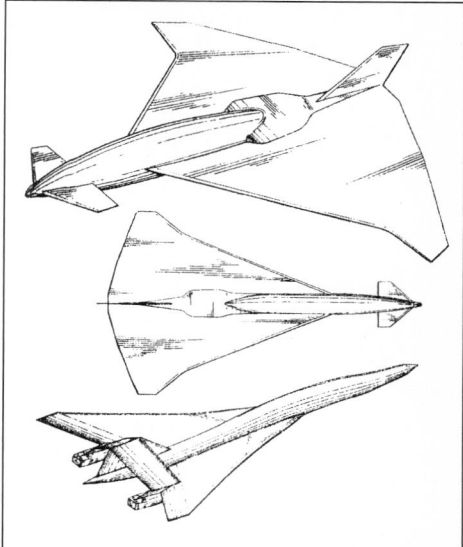

Potential configurations for a Gulfstream SBJ, as revealed by patent applications   0100448

proposed aircraft is not related to the abortive Gulfstream-Sukhoi SSBJ project from which Gulfstream withdrew in 1992. Executive Jet Aviation, which operates the NetJets fractional ownership programme, is reportedly interested in the concept of supersonic business jets. During 2002, there were no significant announcements, however.

UPDATED

GUT WORKS

GUT WORKS LLC

1617, St Andrews Drive, Lawrence, Kansas 66047
Tel: (+1 785) 312 98 80

MARKETING DIRECTOR: Travis Atwood

Plans to produce the Thunder Mustang were abandoned in July 2002 (see TBG entry this section).

UPDATED

HAWKER

RAYTHEON AIRCRAFT COMPANY

At the September 2002 NBAA Convention, held in Orlando, Florida, Raytheon Aircraft (which see for further details) announced its intention to revert to separate marketing of its Beech and Hawker lines of private and executive aircraft, reducing the emphasis on its corporate name.

Raytheon acquired British Aerospace's Corporate Jets division for US$372 million on 6 August 1993; founded Raytheon Corporate Jets Inc at Little Rock, Arkansas, with responsibility for design, development, production and support of renamed Hawker family of corporate jets. Hawker name derived by Raytheon from Hawker Siddeley, parent company of de Havilland at conception of DH/HS/BAe 125 twin-jet, which forms basis of the current Hawker line. Raytheon Corporate Jets included in Raytheon Aircraft Company upon its foundation on 15 September 1994.

NEW ENTRY

HAWKER 800

JASDF designation: U-125

TYPE: Business jet.

PROGRAMME: Derived from the de Havilland/Hawker Siddeley/British Aerospace 125, which was built in the UK from 1962 onwards, progressing through Srs 1 to 3 and 400 to 700. Prototype Srs 800 first flew (G-BKTF) 26 May 1983; type certificate gained 4 May 1984 and Public Transport Category C of A on 30 May 1984; FAA certification 7 June 1984; Russian certification May 1993; Canadian certification (800XP) awarded in third quarter of 1997. Adopted Hawker nomenclature when programme purchased by Raytheon in 1993, at which time some 850 aircraft had been sold in 44 countries.

Final assembly of Hawker 800XP gradually transferred to Wichita; first US-assembled aircraft flew on 5 November 1996, being N297XP, the 297th Series 800; second (N1105Z; No. 301) followed on 24 November 1996; transition complete with flight of last UK-assembled aircraft, No. 337, 29 April 1997; FAA production certificate awarded to Raytheon in May 1997. One thousandth 125/Hawker series aircraft, an 800XP, was delivered as the 'Millennium Hawker' to Gainey Corporation of Grand Rapids, Michigan (N984GC), October 1998. Winglets available as option on new aircraft, or as retrofit, from 2003.

CURRENT VERSIONS: **800:** Original version superseded by 800XP in late 1995; 275 built, of which last delivered December 1995.

800XP (Extended Performance): Announced March 1995, when prototype (G-BVYW, modified from 800) completed; this and preproduction 800XP used in development programme, culminating in CAA and FAA certification in July 1995; first delivery (to Green Tree Financial of St Paul, Minnesota) October 1995 after public debut at National Business Aviation Association Convention in Las Vegas during previous month. First seven delivered in 1995, followed by 26, 33, 48 and 55 in 1996-99.

800SP: Winglet modification of 800/800XP devised by Aviation Partners Inc and shown at NBAA Convention, Orlando, Florida, September 2002; available from early 2003. Height 107 cm (3 ft 6 in); 7 per cent consequential drag reduction translates into M0.03 speed increase (18 kt; 33 km/h; 21 mph) and range extension of 180 n miles (333 km; 207 miles).

Detailed description applies to this version.

800FI, SM, RA and SIG: Special missions versions; described in 2002-03 and earlier editions.

U-125A: Refer to KAC entry in Japanese section.

CUSTOMERS: By June 2002 596 Series 800/800XPs built including 55 delivered in 2001 and 46 in 2002. Largest contract for 125/Hawker placed in May 1997 when Executive Jet Inc ordered 20 800XPs, followed in September 1998 by order for a further 20, plus 16 options for NetJets fractional ownership scheme; deliveries between 1997 and 2004. Recent customers include

National Air Service (NAS) of Jeddah, Saudi Arabia, which ordered 14 in November 1999 for its NetJets Middle East fractional ownership programme, five of these being delivered in 2000, including the first (HZ-KSRA) at the Farnborough International Air Show on 24 July 2000, and three per year thereafter until 2003; and Hainan Airlines of China, which took delivery of one aircraft on 19 July 1999.

COSTS: 800XP: US$12.49 million (2001).

DESIGN FEATURES: Classic small business jet; sweptback wing mounted below cabin floor; podded engines on rear fuselage sides; and high tailplane.

Improvements of baseline Series 800, compared with earlier 700 variant, include curved windscreen, sequenced nosewheel doors, extended fin leading-edge, larger ventral fuel tank, and increased wing span which reduces induced drag, enhances aerodynamic efficiency and carries extra fuel; outboard 3.05 m (10 ft) of each wing redesigned.

In XP version, TFE731-5BR-1H turbofans boost performance, including 14 kt (26 km/h; 16 mph) increase in cruising speed at 11,800 kg (26,015 lb) at 11,890 m (39,000 ft); 225 kg (496 lb) payload increase with eight passengers; 15 to 23 per cent reduction in time to cruising altitude, to reach 11,280 m (37,000 ft) in 23 minutes at maximum take-off weight in ISA + 10°C conditions; and enhanced take-off performance. Other improvements include installation of vortillons in place of wing fences, permitting lower V-speeds and reducing drag; enhanced TKS de-icing system with increased fluid capacity;

Winglet-equipped Hawker 800SP (*Jane's/Paul Jackson*)   NEW/0526967

Hawker Horizon cabin   0130567

Hawker Horizon flight deck   0130566

operation, braking, thrust reversers, spoilers, rudder, nosewheel steering and emergency electrical generation. Two-bottle oxygen system, total capacity 4,080 litres (144 cu ft), with quick-donning diluter-demand masks for crew and auto-deploy constant flow masks in overhead boxes for passengers. Tailcone-mounted Honeywell AE-36-150(HH) APU approved for in-flight operation from sea level to 10,670 m (35,000 ft) and for main engine starting from sea level to 7,925 m (26,000 ft).

AVIONICS: Honeywell Primus Epic lightweight modular avionics system based on Virtual Backplane Network architecture, which combines the cabinet-based modular capabilities of Honeywell's 777 AIMS system with the aircraft-wide network capabilities of its Primus 2000 system and built-in maintenance recording with portable access terminal. The system has 'point and click' capability via two Cursor Control Devices (CCDs) which include touchpad, numeric control and on-screen 'soft keys'. Voice Command will be a future option for some Epic functions.

The Horizon's avionics system provides functions which Raytheon states have never previously been offered in a super-mid-size business jet, including digital circuit breaker control, automatic refuelling control, and full-authority autothrottle. All avionics boxes are installed in a cabinet behind the co-pilot's seat, in an environmentally controlled, pressurised area with easy access for maintenance.

Comms: Dual VHF comms; single Collins HF-9000 HF with Selcal; dual Mode S Diversity transponders; dual audiophone/interphone/PA systems; airborne telephone.

Radar: Primus 880 colour weather radar.

Flight: Primus Epic AFCS and FMS with integrated performance computer; dual VHF nav VOR/LOC/GS/Markers; dual DME; dual IRU; dual GPS; EGPWS; TCAS II; solid-state FDR and CVR.

Instrumentation: Five 203 × 254 mm (8 × 10 in) colour, active, flat panel LCD screens comprising two PFDs, two MFDs and one EICAS, plus two smaller multifunction control and display units (MCDUs).

DIMENSIONS, EXTERNAL:

Wing span	18.82 m (61 ft 9 in)
Mean aerodynamic chord	2.92 m (9 ft 7 in)
Wing aspect ratio	7.2
Length overall	21.11 m (69 ft 3 in)
Height overall	5.97 m (19 ft 7 in)
Tailplane span	7.90 m (25 ft 11 in)
Wheel track	2.79 m (9 ft 2 in)
Wheelbase	8.46 m (27 ft 9 in)
Passenger door: height	1.58 m (5 ft 6 in)
width	0.76 m (2 ft 6 in)

DIMENSIONS, INTERNAL:

Cabin:	
Length, excl flight deck	7.62 m (25 ft 0 in)
Max width	1.97 m (6 ft 5½ in)
Width at floor	1.27 m (4 ft 2 in)
Max height	1.83 m (6 ft 0 in)
Volume	9.37 m³ (331 cu ft)
Baggage compartment volume	2.83 m³ (100 cu ft)

AREAS:

Wings, gross	49.3 m² (531.0 sq ft)
Horizontal tail surfaces	13.01 m² (140.00 sq ft)
Vertical tail surfaces	10.26 m² (110.40 sq ft)

Ailerons (total)	1.35 m² (14.52 sq ft)
Horizontal surfaces (total)	13.01 m² (140.0 sq ft)
Fin, excl dorsal fairing	10.26 m² (110.40 sq ft)
Rudder	2.40 m² (25.87 sq ft)

WEIGHTS AND LOADINGS:

Basic operating weight (incl crew)	9,494 kg (20,930 lb)
Max fuel weight	6,350 kg (14,000 lb)
Max payload	1,619 kg (3,570 lb)
Payload with max fuel	544 kg (1,200 lb)
Max zero fuel weight	11,113 kg (24,500 lb)
Max T-O weight	16,329 kg (36,000 lb)
Max ramp weight	16,420 kg (36,200 lb)
Max wing loading	331.0 kg/m² (67.80 lb/sq ft)
Max power loading	282 kg/kN (2.77 lb/lb st)

PERFORMANCE (estimated):

Max operating speed	M0.84
Max certified altitude	13,715 m (45,000 ft)
Cruising speed: S/L to FL80	280 kt (519 km/h; 322 mph)
FL80 to FL264	350 kt (648 km/h; 403 mph)
T-O field length	1,600 m (5,250 ft)
Landing run	713 m (2,340 ft)
Max range	3,400 n miles (6,297 km; 3,912 miles)
Range at cruising speed of M0.82, six passengers, NBAA	
IFR reserves	3,100 n miles (5,741 km; 3,567 miles)

UPDATED

HAWKER 450

TYPE: Business jet.

PROGRAMME: Announced at NBAA Convention in New Orleans, 9 October 2000; Hawker 450 is interim designation pending announcement of name; full-scale development began in late 2000 but suspended in second quarter of 2002 pending decision on whether to postpone or cancel the programme.

1.80 m
(5 ft 11 in)

1.83 m (6 ft 0 in)

Hawker 450 cabin cross-section   0130550

CUSTOMERS: Launch customer Raytheon Travel Air ordered 50, with 25 options, on 10 October 2000; total of 105 orders held by mid-October 2000. Tulip Air of Rotterdam, Netherlands, ordered one in June 2001, then scheduled for delivery in 2006.

COSTS: US$8.4 million to US$9.6 million (2001).

DESIGN FEATURES: Conventional swept-wing, T-tail design.

STRUCTURE: Composites fuselage formed over aluminium mandrel by automatic fibre-placement machines. Aluminium wing and tail surfaces, last named with composites skins.

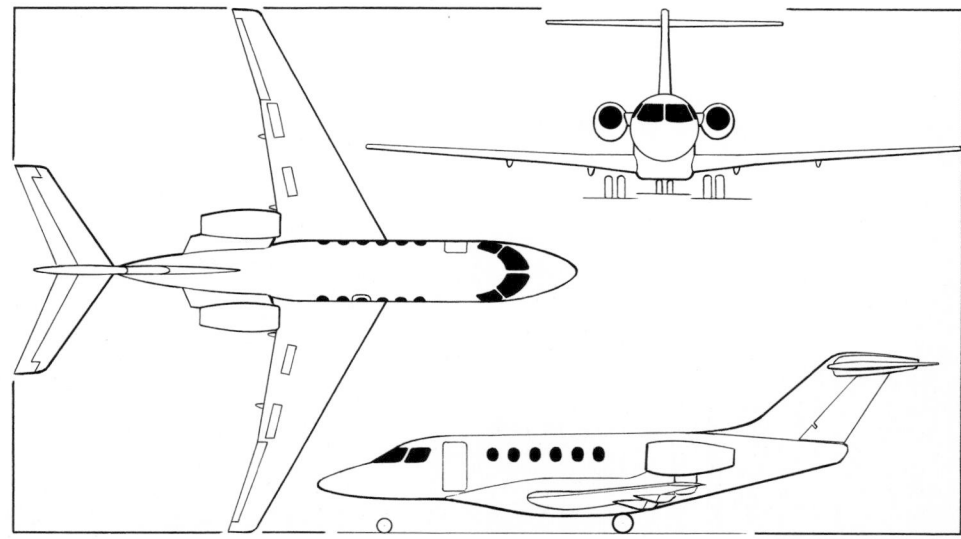

Hawker 450 business jet (*Jane's/James Goulding*)   0131843

LANDING GEAR: Retractable tricycle type, with twin wheels on each unit.

POWER PLANT: Two pod-mounted Honeywell TFE731-40 turbofans, each flat-rated at 18.1 kN (4,070 lb st) at ISA +13°C.

ACCOMMODATION: Crew of two, standard accommodation for eight passengers in 'centre club' arrangement; 'double club four' layout optional; baggage area and lavatory at rear of cabin. Door on port side to rear of flight deck; six cabin windows on each side. Internally accessed baggage compartment, volume 0.57 m³ (20 cu ft) and externally accessed baggage compartment, volume 1.98 m³ (70 cu ft).

SYSTEMS: Cabin pressurisation system, maximum differential 0.64 bar (9.25 lb/sq in).

AVIONICS: Honeywell Primus Epic suite as core system.
    Flight: EICAS, TCAS II and EGPWS standard.
    Instrumentation: EFIS panel with four 203 × 254 mm (8 × 10 in) active matrix LCD screens for PFD and MFD functions.

DIMENSIONS, EXTERNAL:

Wing span	16.79 m (55 ft 1 in)
Length overall	18.41 m (60 ft 5 in)
Height overall	5.64 m (18 ft 6 in)
Tailplane span	6.68 m (21 ft 11 in)

DIMENSIONS, INTERNAL:

Cabin (excl flight deck): Length	5.89 m (19 ft 4 in)
Max width	1.83 m (6 ft 0 in)
Max height	1.80 m (5 ft 11 in)

WEIGHTS AND LOADINGS:

Basic operating weight empty	6,441 kg (14,200 lb)
Max payload	1,338 kg (2,950 lb)
Max fuel weight	3,765 kg (8,300 lb)
Max T-O weight	10,795 kg (23,800 lb)
Max landing weight	10,115 kg (22,300 lb)

Computer-generated image of Hawker 450 mid-size business jet   *NEW*/0130565

Max ramp weight	10,863 kg (23,950 lb)	Service ceiling	13,715 m (45,000 ft)
Max zero-fuel weight	7,779 kg (17,150 lb)	T-O balanced field length	1,433 m (4,700 ft)
Max power loading	298 kg/kN (2.92 lb/lb st)	Range with max fuel, 658 kg (1,450 lb) payload,	
PERFORMANCE (estimated):		NBAA IFR reserves, 200 n mile (370 km; 230 mile)	
Max level speed at FL350	Mach 0.82	alternate	2,045 n miles (3,787 km; 2,353 miles)
Max cruising speed at FL350			*UPDATED*
	472 kt (874 km/h; 543 mph)		

HENDERSON

HENDERSON AERO SPECIALTIES INC

36 Taramino Place, Lewes, Delaware 19958-1668
e-mail: Litbear@aol.com
CEO: David O Henderson

Mr Henderson, owner of a 1946 Piper J3C-65 Cub, has produced replicas of this aircraft in two forms. There has been no known recent production. A description last appeared in the 2002-03 edition.

UPDATED

HENSLEY

HENSLEY AIRCRAFT INC

888 Arbor Drive, Chaska, Minnesota 55318
Tel: (+1 952) 442 85 67
Fax: (+1 603) 737 09 63
e-mail: joe@hensleyaircraft.com
Web: http://www.hensleyaircraft.com
CEO: Robert Hensley
VICE-PRESIDENT: Joe Penaz

The H-1 Wolf is Hensley's initial offering.

UPDATED

HENSLEY H-1 WOLF

TYPE: Four-seat kitbuilt twin.

PROGRAMME: Design began in 1991, initially as part-time venture. Construction of prototype started in 1998; exhibited, unflown, at AirVenture, Oshkosh, July 2001; maiden flight then scheduled for December 2001 and kit deliveries for second quarter of 2002.

CURRENT VERSIONS: **Wolf:** Baseline version.
    Wolf GT: Higher-powered, long-range version, with two cabin doors and increased MTOW.

CUSTOMERS: First aircraft sold October 2001.

COSTS: Wolf US$124,900 including engines (2003).

DESIGN FEATURES: Intended as manoeuvrable twin, able to operate into remote, short airstrips; affordable; easy to build and maintain.
    Low-drag fuselage; constant-chord, low-mounted wings, with winglets; pusher engines located close inboard and above wings.
    Wing section VH-1 Twin Turbo, purpose developed by Jeff Viken for high lift, low drag laminar flow.

FLYING CONTROLS: Conventional and manual. Fowler flaps immediately outboard of engines. Horn-balanced rudder and elevator.

STRUCTURE: Composites throughout.

LANDING GEAR: Tricycle type; fixed. Mainwheels, 6.00-6, on aluminium sprung legs; trailing-link nosewheel, 11×4.00-5, with rubber-in-compression suspension. Hydraulic brakes on mainwheels. Speed fairings on all wheels.

POWER PLANT: Two Quantum Powermills engines (based on Honda VTEC) with pusher propellers and dual fuel system. Baseline aircraft has 101 kW (135 hp) Q-135s, Sensenich wooden propellers and 189 litres (50.0 US gallons; 41.6 Imp gallons) of fuel; Wolf GT has 123 kW (165 hp) Q-165Ts, MT constant-speed propellers and 662 litres (175 US gallons; 146 Imp gallons).

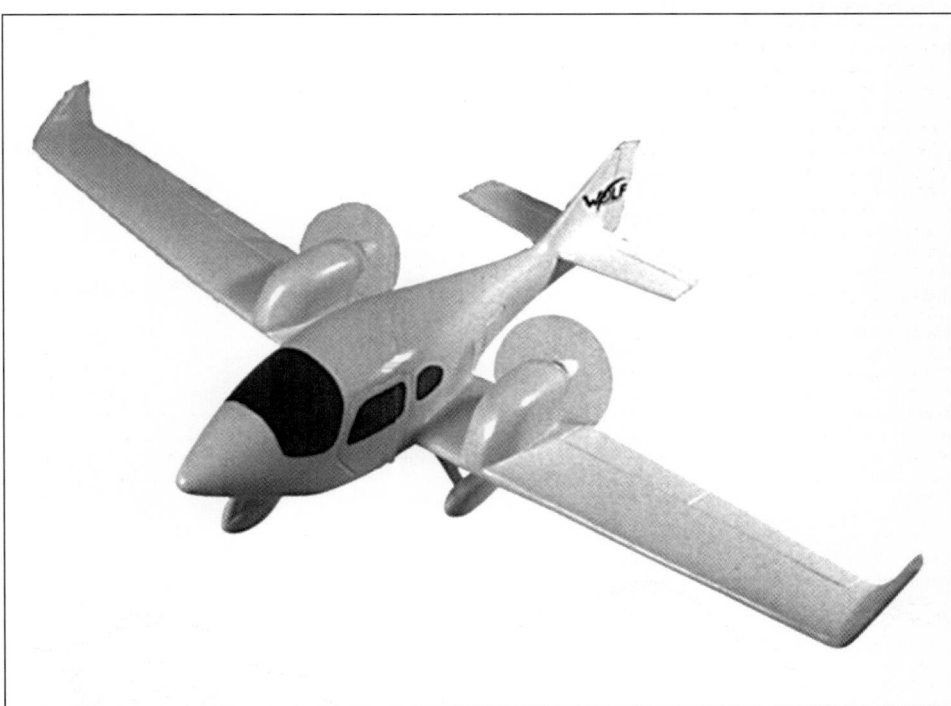

Model of Hensley H-1 Wolf   0110977

ACCOMMODATION: Four persons in back-to-back pairs. Door, port side. Dual sidestick controls. Two doors on GT version.

SYSTEMS: Dual electrical system.

DIMENSIONS, EXTERNAL:

Wing span	10.36 m (34 ft 0 in)
Wing aspect ratio	9.6
Length overall	6.71 m (22 ft 0 in)

DIMENSIONS, INTERNAL:

Cabin max width	1.22 m (4 ft 0 in)

AREAS:

Wings, gross	11.15 m² (120.0 sq ft)

WEIGHTS AND LOADINGS (A: Wolf, B: Wolf GT):

Weight empty: A	907 kg (2,000 lb)
B	953 kg (2,100 lb)

Max T-O weight: A	1,360 kg (3,000 lb)
B	1,587 kg (3,500 lb)
Wing loading: A	122.1 kg/m² (25.00 lb/sq ft)
B	142.4 kg/m² (29.17 lb/sq ft)
Power loading: A	6.76 kg/kW (11.11 lb/hp)
B	6.46 kg/kW (10.61 lb/hp)

PERFORMANCE (A, B as above):

Max cruising speed: A	182 kt (338 km/h; 210 mph)
B	196 kt (362 km/h; 225 mph)
Stalling speed: A	57 kt (105 km/h; 65 mph)
B	66 kt (121 km/h; 75 mph)
Range: A	651 n miles (1,207 km; 750 miles)
B	2,281 n miles (4,224 km; 2,625 miles)
g limits: A	+6.0/−3.0
B	+5.0/−2.5

UPDATED

HILLBERG

HILLBERG HELICOPTERS
PO Box 8974, Fountain Valley, California 92728-8974
Tel: (+1 909) 279 56 78
e-mail: Rotermouse@earthlink.com
PRESIDENT: Don Hillberg

Company founded 1990 to provide maintenance and
modification services to helicopter operators, and moved into
kit and experimental fields to help builders.

Hillberg has also developed the four-seat 1-04, which
awaits funding for production. Produces a retrofit kit for the
RotorWay Exec (which see), upgrading it to turbine power
with a Solar T62-T32 of 112 kW (150 shp) and reducing
empty weight to 340 kg (750 lb).

UPDATED

Hillberg EH 1-01 RotorMouse (one Honeywell 36-55-C)

HILLBERG EH 1-01 ROTORMOUSE and EH 1-02 TANDEMMOUSE

TYPE: Single-seat helicopter kitbuilt; two-seat helicopter
kitbuilt.
PROGRAMME: Design and construction of prototype began in
August 1990; first flew (N10TE) August 1992.
Construction of first production aircraft began in
September 1996. Type is currently certified as
experimental by FAA.
CURRENT VERSIONS: **EH 1-01 RotorMouse:** Single-seat
version.
Description applies to EH 1-01 except where indicated.
  EH 1-02 TandemMouse: Tandem two-seat version;
under development; was due to fly 1999; this date slipped,
but this version is believed to have flown by early 2001.
Powered by 112 kW (150 shp) Solar T62-T32 turbine
engine. Provisional figures appear below.
  EH-1-02A: Development of EH 1-02; has 313 kW (420
shp) Rolls-Royce 250-C20B with 681 litre (180 US gallon;
150 Imp gallon) fuel tank.
  CombatMouse: Proposed military version.
CUSTOMERS: By June 2002, 31 had been ordered, of which at
least four had been built.
COSTS: Kit: EH-1-01 US$68,000 (2003) ex-works
US$100,000; EH 1-02 US$120,000 (2002).
DESIGN FEATURES: Forward fuselage resembles Bell AH-1
HueyCobra. Stub-wing behind cockpit for carriage of
external loads. Main rotor speed 510 rpm; tail rotor 3,250
rpm.
FLYING CONTROLS: Auto-governed hydromechanical flying
controls. Optional rotor brake.
STRUCTURE: Aluminium alloy (2024-T3) monocoque
fuselage, bulkheads and keel; stainless steel engine deck
and firewall. Rotor has Gyrodyne QH-50 modified blades
of NACA 1200 aerofoil section with 8° twist and block tip
caps. Modified Robinson R22 hub and transmission.
Conventional tail rotor using aluminium-skinned blades.
LANDING GEAR: Fixed steel cross-tube skid type; floats
optional.

POWER PLANT: One 108 kW (145 shp) Honeywell 36-55-C
turboshaft engine driving a non-folding, two-blade, semi-
rigid wood-laminated main rotor. Engines in range 108-
186 kW (145-250 hp) are suitable. Single 125 litre (32.9
US gallon; 27.4 Imp gallon) crash-resistant bladder fuel
tank around transmission; refuelling point behind main
mast. Optional 76 litre (20.0 US gallon; 16.7 Imp gallon)
fuel tank on stub-wing. EH 1-02 has 170 litre (45.0 US
gallon; 37.5 Imp gallon) fuel capacity.
ACCOMMODATION: Single pilot in enclosed cockpit. Entry
through upward-hinged port and starboard side windows.
Baggage compartment behind seat.
SYSTEMS: 24 V 60 A electrical system; Electrocraft alternator.
EQUIPMENT: Optional cargo hook.

DIMENSIONS, EXTERNAL:
Main rotor diameter: EH 1-01		6.10 m (20 ft 0 in)
EH 1-02		7.62 m (25 ft 0 in)
Tail rotor diameter		0.91 m (3 ft 0 in)
Length: overall, rotors turning:		
EH 1-01		7.32 m (24 ft 0 in)
fuselage: EH 1-01		6.35 m (20 ft 10 in)
EH1-02		8.53 m (28 ft 0 in)
Height overall		2.26 m (7 ft 5 in)

DIMENSIONS, INTERNAL:
Cabin max width: EH 1-01	0.61 m (2 ft 0 in)
EH 1-02	0.89 m (2 ft 11 in)

AREAS:
Main rotor disc: EH 1-01	29.20 m² (314.3 sq ft)
EH 1-02	45.62 m² (491.1 sq ft)

WEIGHTS AND LOADINGS:
Weight empty: EH 1-01	295 kg (650 lb)
EH 1-02	408 kg (900 lb)
EH 1-02A	more than 499 kg (1,100 lb)

Max T-O and landing weight:
EH 1-01	621 kg (1,370 lb)
EH 1-02	816 kg (1,800 lb)
EH 1-02A	1,451 kg (3,200 lb)

PERFORMANCE:
Never-exceed speed (VNE)	184 kt (341 km/h; 212 mph)
Max level speed: EH 1-01	139 kt (257 km/h; 160 mph)
EH 1-02	148 kt (274 mph; 170 mph)
Cruising speed: both	113 kt (209 km/h; 130 mph)
Max rate of climb at S/L:	
EH 1-01	1,433 m (4,700 ft)/min
Vertical rate of climb at S/L:	
EH 1-01	823 m (2,700 ft)/min
EH 1-02	457 m (1,500 ft)/min
Service ceiling:	
EH 1-01	4,495 m (14,750 ft)
EH 1-02	4,115 m (13,500 ft)
Range, 20 min reserves: EH 1-01:	
with max payload	300 n miles (555 km; 345 miles)
with max internal and external fuel	
	695 n miles (1,287 km; 800 miles)
EH 1-02: with max payload	
	390 n miles (722 km; 448 miles)

UPDATED

HILLBERG 1-04 BABY HUEY
TYPE: Four-seat helicopter kitbuilt.
PROGRAMME: Work is continuing on a proposed development
of RotorMouse/TandemMouse with four seats and
external cabin styling resembling Bell UH-1 'Huey'.
MTOW 1,134 kg (2,500 lb).

UPDATED

JAG

JAG HELICOPTER GROUP LLC
15866 Sturgeon Street, Roseville, Michigan 48066-6623
Tel: (+1 810) 775 11 84
Fax: (+1 810) 775 13 38
e-mail (1): radial@bignet.net
e-mail (2): info@jaghelicopter.com
Web (1): http://www.rinke-aerospace.com
Web (2): http://jaghelicopter.com

Jag is promoter of the world's first turbine-powered, multi-
bladed kit helicopter. Company manufacturing plant at
Roseville covers 5,110 m² (55,000 sq ft) and employed 32
people in September 2002.

UPDATED

JAG JAG
TYPE: Two-seat helicopter kitbuilt.
PROGRAMME: Mockup exhibited at Sun 'n' Fun, April 2001
and EAA AirVenture, Oshkosh, July 2001. Prototype
(unflown, but later registered N255JH) shown at Heli
Expo, February 2002.
CURRENT VERSIONS: **JAG 235:** Baseline version; two seats,
three main rotor blades and three tail rotor blades.
  JAG 255: Two seats, five main blades and five tail rotor
blades. Otherwise generally as for JAG 235.
  JAG 285: Two seats, eight main and five tail rotor
blades. Otherwise generally as for JAG 235.
CUSTOMERS: Total 30 sold by July 2001.
COSTS: US$162,000 per kit for first 100 JAG 235s;
approximately US$175,000 thereafter. JAG 255
US$184,000; JAG 285 US$223,000 (2003).
DESIGN FEATURES: Multiblade helicopter with shrouded,
otherwise conventional, tail rotor. Three, four or five
blades standard; six, seven or eight blade versions
available to special order. Streamlined design with track
main landing gear mounted on large sponsons.

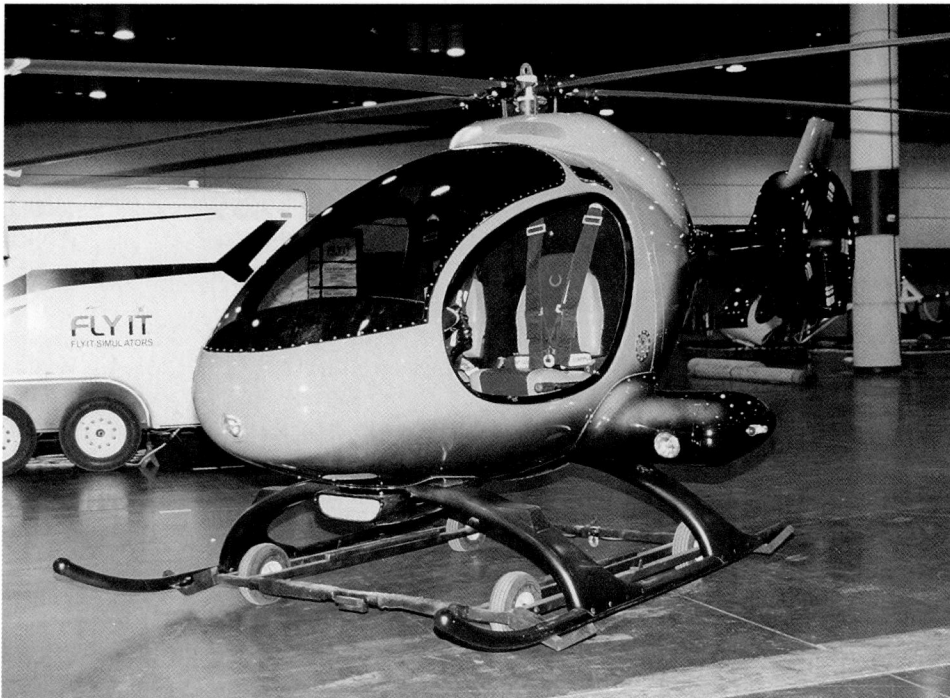

JAG 255 at Heli Expo in 2002 (*Falcon Aviation/Peter J Cooper*)   NEW/0533639

FLYING CONTROLS: Conventional and manual.
STRUCTURE: Composites.
LANDING GEAR: Wheeled tricycle type; fixed.

POWER PLANT: One 236 kW (317 shp) derated Rolls-Royce
250-C18 turboshaft. Fuel capacity 265 litres (70.0 US
gallons; 58.3 Imp gallons).

ACCOMMODATION: Two persons, side by side. Cockpit open at each side.

DIMENSIONS, EXTERNAL:

Main rotor diameter	6.30 m (20 ft 8 in)
Tail rotor diameter	0.91 m (3 ft 0 in)
Length overall, rotors turning	7.21 m (23 ft 8 in)
Height overall	2.44 m (8 ft 0 in)
Width: overall	2.82 m (9 ft 3 in)
fuselage	2.54 m (8 ft 4 in)

DIMENSIONS, INTERNAL:

Cockpit: Max width	1.37 m (4 ft 6 in)
Max height	1.23 m (4 ft 0½ in)

AREAS:

Main rotor disc	31.57 m² (339.8 sq ft)
Tail rotor disc	2.63 m² (28.27 sq ft)

WEIGHTS AND LOADINGS (JAG 255, estimated):

Weight empty, equipped	499 kg (1,100 lb)
Max T-O weight	1,451 kg (3,200 lb)
Max disc loading	46.0 kg/m² (9.42 lb/sq ft)

PERFORMANCE (JAG 255, estimated):

Never-exceed speed (V_{NE})	154 kt (286 km/h; 178 mph)
Max level speed at S/L	126 kt (233 km/h; 145 mph)
Max rate of climb at S/L	549 m (1,800 ft)/min
Service ceiling	4,420 m (14,500 ft)

Hovering ceiling: IGE	2,590 m (8,500 ft)
OGE	1,829 m (6,000 ft)
Range	346 n miles (642 km; 399 miles)
Endurance	2 h 45 min

UPDATED

JOPLIN

JOPLIN LIGHT AIRCRAFT

PO Box 3805, 523 North Schifferdecker, Joplin, Missouri 64801
Tel: (+1 417) 623 29 50
Fax: (+1 417) 623 76 06
e-mail: jla@compnmore.com
Web: http://www.compnmore/jla
SALES MANAGER: John Lukey

Joplin Light Aircraft manufactures the Tundra and ½ TUN previously produced by Laron Aviation Technologies; it also markets the Suzi-Air engine.

UPDATED

JOPLIN TUNDRA

TYPE: Tandem-seat ultralight kitbuilt.
PROGRAMME: Originally produced by Laron; meets Transport Canada TP 10141 requirements.
CUSTOMERS: Over 30 sold and 25 flying by mid-2002.
COSTS: US$9,500 (2001).
DESIGN FEATURES: High-wing monoplane with high-life wing and pusher engine; cockpit enclosure optional. Quoted build time 250 hours.
FLYING CONTROLS: Manual. Full-span Junkers ailerons.
STRUCTURE: Welded metal tube fuselage. Preformed glass fibre nosecone.
LANDING GEAR: Fixed, wide-track tricycle type; ski and float options. Steerable nosewheel.
POWER PLANT: One 47.8 kW (64.1 hp) Rotax 582 two-cylinder two-stroke engine; options include 34.0 kW (45.6 hp) Rotax 503 and 47.7 kW (64 hp) Suzi Air. Fuel capacity 38 litres (10.0 US gallons; 8.3 Imp gallons).

DIMENSIONS, EXTERNAL:

Wing span	9.75 m (32 ft 0 in)
Length overall	6.40 m (21 ft 0 in)
Height overall	1.98 m (6 ft 6 in)

AREAS:

Wings, gross	15.79 m² (170.0 sq ft)

WEIGHTS AND LOADINGS:

Weight empty	181 kg (400 lb)
Max T-O weight	408 kg (900 lb)

PERFORMANCE:

Max level speed	87 kt (161 km/h; 100 mph)
Normal cruising speed at 80% power	74 kt (137 km/h; 85 mph)
Stalling speed	31 kt (57 km/h; 35 mph)
Max rate of climb at S/L	366 m (1,200 ft)/min
T-O run	30 m (100 ft)

Tundra kitbuilt, previously produced by Laron

Landing run	61 m (200 ft)
Range at max level speed	202 n miles (374 km; 232 miles)

UPDATED

JOPLIN ½ TUN

TYPE: Single-seat ultralight kitbuilt.
CUSTOMERS: Three flying by mid-2002.
COSTS: US$5,995 (2001).
DESIGN FEATURES: As Tundra, but meets FAR Pt 103 weight rules. Quoted build time 90 hours.
FLYING CONTROLS: As Tundra.
STRUCTURE: As Tundra.
LANDING GEAR: Tailwheel type.
POWER PLANT: One 29.5 kW (39.6 hp) two-cylinder, liquid-cooled Rotax 447; options include 2 SI 460 F35 and Suzi Air. Fuel capacity 18.9 litres (5.0 US gallons; 4.2 Imp gallons).

DIMENSIONS, EXTERNAL:

Wing span	7.92 m (26 ft 0 in)
Length overall	6.40 m (21 ft 0 in)
Height overall	1.68 m (5 ft 6 in)

AREAS:

Wings, gross	10.87 m² (117.0 sq ft)

WEIGHTS AND LOADINGS:

Weight empty	113 kg (250 lb)
Max T-O weight	249 kg (550 lb)

PERFORMANCE:

Max level speed	54 kt (101 km/h; 63 mph)
Normal cruising speed at 80% power	48 kt (89 km/h; 55 mph)
Stalling speed	22 kt (41 km/h; 25 mph)
Max rate of climb at S/L	305 m (1,000 ft)/min
T-O run	46 m (150 ft)
Landing run	31 m (100 ft)

UPDATED

KAMAN

KAMAN AEROSPACE CORPORATION
(Subsidiary of Kaman Corporation)

Old Windsor Road, PO Box No. 2, Bloomfield, Connecticut 06002
Tel: (+1 860) 243 71 00 or 242 44 61
Fax: (+1 860) 243 75 14
e-mail: kmax-kac@kaman.com
Web: http://www.kamanaero.com
CHAIRMAN, PRESIDENT AND CEO, KAMAN CORPORATION:
  Paul R Kuhn
CHAIRMAN EMERITUS: Charles Kaman
VICE-CHAIRMAN: C William Kaman II
PRESIDENT, KAMAN AEROSPACE: Joseph H Lubenstein
VICE-PRESIDENT, ENGINEERING: Michael Bowes
VICE-PRESIDENT, MILITARY HELICOPTERS:
  M Terry Higginbotham
DIRECTOR, BUSINESS DEVELOPMENT: William D Brown
PUBLIC RELATIONS MANAGER: David M Long

Founded 1945 by Charles H Kaman. Developed servo-flap control of helicopter main rotor, initially in contrarotating two-blade main rotors, and still used in SH-2G Super Seasprite four-blade main rotor and on the K-MAX. R&D programmes sponsored by US Army, Air Force, Navy and NASA include advanced design of helicopter rotor systems, blades and rotor control concepts, component fatigue life determination and structural dynamic analysis and testing. In late 2001, company was said to be in early conceptual stages

for a multirole helicopter of similar size to the NH 90, with a cabin wide enough to admit stretchers stowed widthways. Kaman has also undertaken helicopter drone programmes since 1953; is continuing advanced research in rotary-wing unmanned aerial vehicles (UAVs).

Kaman is major subcontractor on many aircraft and space programmes, including design, tooling and fabrication of components in metal, metal honeycomb, bonded and composites construction, using techniques such as filament winding, braiding and RTM. Participates in programmes including Northrop Grumman E-2C, Bell Boeing V-22, Boeing 737, 747, 757, 767, 777 wing trailing-edge, KC-135 and C-17, Boeing Sikorsky RAH-66, Sikorsky UH-60 and SH-60, and NASA Space Shuttle Orbiter main fuel tank. It also designs and produces fuzes for a range of weaponry.

Kaman designed and, since 1977, has been producing all-composites rotor blades for Bell AH-1 Cobras for US and foreign forces. Appointed in April 2000 to build fuselages for MD Helicopters' single-engine line (MDH 500, 520, 530 and 600) under 10-year contract; a further contract to build MDH Explorer rotor blade systems was awarded in July 2000.

KAC's activities were reorganised in early 2002 into three new business units, of which Helicopter Programs has responsibility for the Super Seasprite and K-MAX described in the following entries. Aerostructures Subcontracting in Moosup, Connecticut, focuses on non-rotary-wing aircraft programmes including commercial airliner wing structures and components; major structural assemblies for military transports; aircraft thrust reversers; and business jet subassembly components. Helicopter Subcontracting, with

operations in Bloomfield, Connecticut and Jacksonville, Florida, undertakes work on helicopter airframes, composites rotor blades and components, including manufacture of MD 500 and MD 600 airframes for MD Helicopters (which see).

Kaman Aerospace International Corporation was established in 1995 as the international affiliate of Kaman Aerospace with offices in Canberra and Nowra, Australia; Kuala Lumpur, Malaysia; Cairo, Egypt; and Bloomfield, Connecticut.

Net sales of Kaman's Aerospace business were US$382.7 million in 1998, US$371.8 million in 1999 and US$381.9 million in 2000; net sales for 2001 were US$301.6 million. The company employs 1,400 people.

UPDATED

KAMAN (K894) SH-2G SUPER SEASPRITE

TYPE: Naval combat helicopter.
CURRENT VERSIONS: **SH-2G Super Seasprite:** Available as either SH-2F remanufactured airframe or as new-build helicopter. For full specification and detailed description see *Jane's All the World's Aircraft 1994-95* and the current edition of *Jane's Aircraft Upgrades.*
CUSTOMERS: Most recent customers include Egypt, for which a US$150 million contract was awarded on 22 February 1995; first of 10 **SH-2G(E)s**, remanufactured from SH-2Fs, was delivered 21 October 1997; deliveries were completed November 1998. Possible further four to eight, for search and rescue role, being negotiated in 2002.

Royal Australian Navy ordered 11 **SH-2G(A)**s remanufactured from SH-2Fs for delivery from early 2001 in a US$600 million contract for use on its eight new 'ANZAC' class frigates; the first (N351KA, ex-163210) made its first flight 6 January 2000 and was delivered in early March 2001. However, entire programme has become victim of time and cost overruns, partly due to concerns over age of airframes to be remanufactured (seven from 1963 and four from 1985 production), but primarily because of development problems with Integrated Tactical Avionics System (ITAS) originated by Litton but now taken over by Northrop Grumman. Eight aircraft delivered by mid-2002, with two more then in production, but these capable only of basic utility missions pending installation of fully functional ITAS. This due for critical design review (CDR) in March 2003; if passed, delivery of fully functional SH-2G(A)s should be achieved by end of 2004 and IOC by late 2006 – some three and a half years later than intended.

In 1997 Royal New Zealand Navy ordered four new-build **SH-2G(NZ)**s (later increased to five: NZ3601 to 3605) for delivery from 2001 as replacements for Westland Wasp, also for use on 'ANZAC' class frigates as well as older 'Leander' class, equipped with Raytheon AGM-65 Maverick missiles. In the interim it used four SH-2F Seasprites for training. Initial RNZAF SH-2G(NZ) (N352KA) made first flight 2 August 2000. First two aircraft officially handed over 18 August 2001; two further deliveries had been made by mid-2002, with fifth due in early 2003; operating unit is No. 3 Squadron at Whenuapai.

Kaman received a USN contract in February 2002 to reactivate four stored SH-2Gs for the Polish Navy, for service aboard its frigates *Pulawski* and *Kosciuszko*. The first two were expected to be delivered to Poland in the second half of 2002. A similar sale of four to the Mexican Navy was under discussion in the early part of the year.

Following details refer specifically to SH-2G(NZ):
STRUCTURE: Composites main rotor blades.
LANDING GEAR: Emergency flotation system.
ACCOMMODATION: Crew of two: pilot and co-pilot/TACCO; no instructor's seat. Three-person bench seat and two single seats in cabin, plus one stretcher.
SYSTEMS: Hydraulic system rated at 207 bar (3,000 lb/sq in); maximum flow rate 30.3 litres (8.0 US gallons; 6.7 Imp gallons)/min. Electrical system powered by two 30 kVA AC generators, two 100 A DC converters and an Ni/Cd battery. Honeywell GTCP 36-150(BH) APU.
AVIONICS: *Comms:* Rockwell Collins AN/ARC-210 V/UHF (AM/FM) and HF-9000D radios; ITT KY 100 for HF (embedded in ARC-210 for V/UHF); Bendix/King AN/APX-100 IFF with KIT 1C encoder; Telephonics Starcom intercom. Three-frequency (121.5/243.0/406 MHz) ELT.
Radar: Telephonics AN/APS-143(V)3 surveillance radar.
Flight: Rockwell Collins AN/ARN-147(V) VOR/ILS and marker beacon receiver, DME-42 and DF-301E VHF/UHF ADF; Trimble Tasman GPS/INS; Litton AN/APN-217(V)6 Doppler velocity sensor and AN/ASN-150 tactical data system; AN/APN-194 radar altimeter; Rosemount 542CBAY8 air data computer; AN/APQ-107 RAWS; AN/ASN-50 AHRS.
Instrumentation: Standby instruments and magnetic compass for pilot and co-pilot.
Mission: FLIR Systems AN/AAQ-27 FLIR; AN/ASN-150 tactical datalink; UYS-503 acoustic data processor; AN/APN-194 radar altimeter; ARR-84 sonobuoy receiver; AQS-81 MAD; AKT-22 datalink; ASQ-188 torpedo presetter; Panasonic AG-750 video cassette recorder.

Kaman SH-2G(A) maritime helicopter in Australian markings conducting dummy deck landings
NEW/0526927

Self-defence: Elisra AES-210/E ESM suite; BAE Systems AN/ALE-47 chaff/flare dispenser. RWR.
EQUIPMENT: Two external launchers each for four Mk 25 marine markers; provision for LUUIIB flares on outboard stores stations. Fairey Hydraulics harpoon and Scott three-wire deck securing and handling system. NVG-compatible internal and external lighting. Pilot's and TACCO's personal survival packs. Hoist and cargo hook as for standard SH-2G.
ARMAMENT: Two AGM-65D Maverick ASMs; two Mk 46 torpedoes; two Mk 11 depth charges; door-mounted 7.62 mm MAG 58 machine gun.

UPDATED

KAMAN K-1200 K-MAX
TYPE: Light lift helicopter.
PROGRAMME: First flight (N3182T) 23 December 1991; first public showing 22 March 1992; first flights of second prototype (N131KA) 18 September 1993 and first production aircraft (N132KA) 12 January 1994; third prototype is static and drop test aircraft to prove 20 year life at 1,000 hours per year and with 30 return logging sorties per hour. Certification to FAR Pts 27 and 133 achieved 30 August 1994 after 800 hour/32 month programme. N133KA of Scott Paper (now Kimberly-Clark) achieved 1,000 hours in eight months, 22 June 1995, as first K-MAX to reach this total. Canadian certification awarded 23 November 1994. Uses include logging, firefighting, agricultural spraying, constructing and surveying.

A US$690,000 contract to demonstrate Vertrep (Vertical Replenishment) to US Navy awarded August 1995: two month demonstration period saw two K-MAXs (c/n 0010 and 0013) lift 453,592 kg (1,000,000 lb) with

follow-on assessment in Guam during May 1996 during which 142 hours were flown and 2,449,400 kg (5,400,000 lb) was lifted. Two helicopters then began US$5.7 million six-month deployment in Arabian Gulf on 3 June 1996 aboard USS *Niagara Falls*. During late August 1998, the prototype K-MAX was involved in trials for Magic Lantern mine detection systems. FAA Pt 27 IFR certification received 14 May 1999 as part of bid for US Navy Vertrep contract. In November 2001, on behalf of the US Naval Undersea Warfare Center at Newport, Rhode Island, the K-MAX recovered more than two dozen torpedoes used during exercises in local waters. On 14 July 1999 USMC placed contract worth US$4.2 million for remote piloting package and another worth US$2.7 million to May 2000: for further details see *Jane's Unmanned Aerial Vehicles and Targets*.

Kaman received certification for an external seat in June 1999; up to two may be fitted.
CUSTOMERS: Kaman moved from its original helicopter lease programme to a sales programme, but reinstated leasing arrangements in 2002, with Eagle Helicopters (Switzerland) and Superior Helicopters (Oregon) becoming lease customers shortly afterwards. Current operators include Mountain West, Woody Contracting, Superior Helicopter (two), Petroleum Helicopters, Rainier Helicopter Logging and US State Department in the USA; Midwest Helicopters and Cariboo Chilcotin in Canada; Helog of Switzerland, HeliAir Zagel of Germany, Japan Royal Helicopters (two) of Japan, Rotex Helicopters AG of Liechtenstein and Wucher Helikopter of Austria. A K-MAX operated by Woody Contracting was the first to exceed 10,000 flying hours. Peruvian National Police indicated interest in purchasing five aircraft in early 2000 and in December 2000 the US State Department ordered these five (c/n 28 to 32, the last of which was lost in a pre-delivery crash on 14 November 2001) at a contract cost of US$21 million for anti-drug operations in Peru. Three sold to US and European operators in first three quarters of 2001. Total of 34 had been registered by May 2002; five had been written off in non-fatal accidents and a further two substantially damaged by that time.
COSTS: US$4,056 million (2002).
DESIGN FEATURES: Kaman intermeshing, contrarotating rotors ensure all engine power produces lift; rotor disc loading is very low and airspeed is limited to reduce rotor stresses; all internal spaces painted white and provided with lights to ease night-time servicing; freight compartment beneath transmission casing allows carriage of special tools or parts; no hydraulics; transmission and engine fluid lines located on opposite sides of helicopter to avoid servicing errors; fuselage is narrow to give good downward view; panels in domed side windows can be opened to give direct vision or doors can be removed altogether. Electronic engine instruments and hook load measuring sensor record engine cycles and any overloads, but also give the pilot an immediate record of operations for billing purposes, avoiding paperwork for pilot. Inter-blade friction damper stops can be removed so that blades swing sideways to align for parking in narrow spaces.

Normal rotor rpm range between 250 and 270 unloaded, 260 and 270 with external load, giving maximum blade tip speed of 200 m (658 ft)/min; translational lift is attained at 12 kt (22 km/h; 14 mph); rpm reduced to 200 for autorotation and airspeed of 50 kt (92 km/h; 57 mph) then gives a power-off descent rate of 366 to 427 m (1,200 to 1,400 ft)/min.

Northrop Grumman Integrated Tactical Avionics System 'glass cockpit' in the SH-2G(A)　*NEW*/0526926

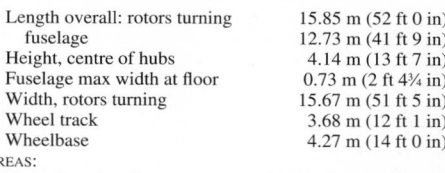

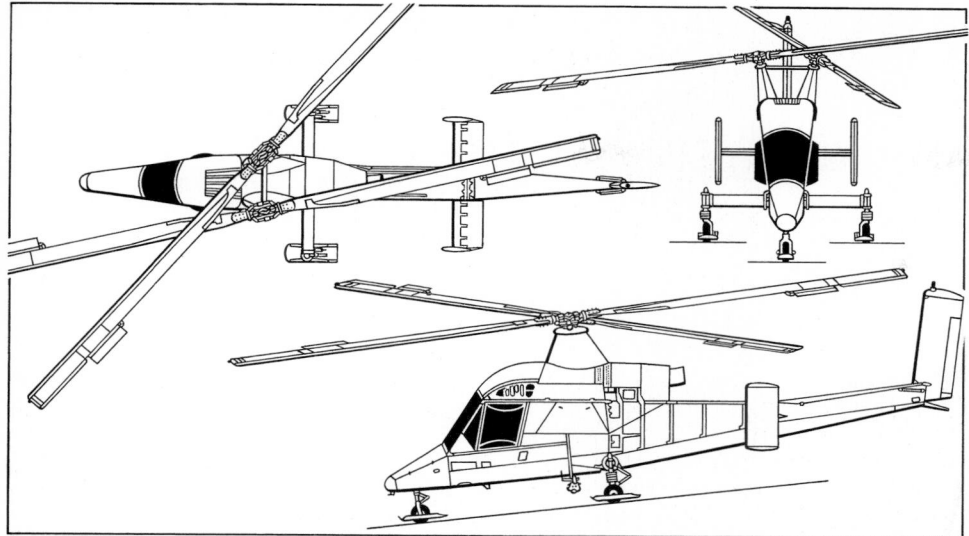

Production version of Kaman K-MAX intermeshing rotor helicopter (*Jane's/Mike Keep*)

Length overall: rotors turning	15.85 m (52 ft 0 in)
fuselage	12.73 m (41 ft 9 in)
Height, centre of hubs	4.14 m (13 ft 7 in)
Fuselage max width at floor	0.73 m (2 ft 4¾ in)
Width, rotors turning	15.67 m (51 ft 5 in)
Wheel track	3.68 m (12 ft 1 in)
Wheelbase	4.27 m (14 ft 0 in)

AREAS:

Rotor discs (total)	340.9 m² (3,669.0 sq ft)

WEIGHTS AND LOADINGS:

Operating weight empty	2,334 kg (5,145 lb)
Max hook capacity	2,721 kg (6,000 lb)
Max payload, ISA +15°C: at S/L	2,721 kg (6,000 lb)
at 1,525 m (5,000 ft)	2,568 kg (5,663 lb)
at 3,050 m (10,000 ft)	2,341 kg (5,163 lb)
at 3,810 m (12,100 ft)	2,273 kg (5,013 lb)
at 4,920 m (15,000 ft)	1,956 kg (4,313 lb)
Max fuel weight	705 kg (1,554 lb)
Max T-O weight:	
without jettisonable load	2,948 kg (6,500 lb)
with jettisonable load	5,443 kg (12,000 lb)
Max disc loading	17.20 kg/m² (3.52 lb/sq ft)
Max transmission power loading (T-O power)	
	5.83 kg/kW (9.57 lb/shp)

PERFORMANCE:

Never-exceed speed (VNE):	
clean	100 kt (185 km/h; 115 mph)
with external load	80 kt (148 km/h; 92 mph)
Max rate of climb at S/L, normal flat-rated torque	
	762 m (2,500 ft)/min
Service ceiling	4,572 m (15,000 ft)
Hovering ceiling at AUW of 2,721 kg (6,000 lb) ISA:	
IGE	8,020 m (26,300 ft)
OGE	8,875 m (29,120 ft)
Range with max fuel:	
without external load	267 n miles (494 km; 307 miles)
with external load	214 n miles (396 km; 246 miles)
Endurance	2 h 41 min
OPERATIONAL NOISE LEVELS (FAR Pt 36), limit 87 dB(A):	
Flyover	Average of 82 dB(A)

UPDATED

FLYING CONTROLS: Blade angle of attack controlled by trailing-edge flaps and light control linkage, avoiding need for hydraulic power; an electric actuator in each flap control run is operated by pilot to track the blades on ground or in flight.

In normal powered flight, turns at or near the hover are effected by applying differential collective pitch to the rotors by means of differential collective pitch commanded from the foot pedals; a small fore-and-aft cyclic pitch change also occurs; at low powers near autorotation, this would produce directional control reversal, because differential collective pitch change would cause drag on the unwanted side; so a non-linear cam between collective lever and rotor blade controls phases out differential collective from a 'dwell zone' at about 25 per cent collective demand and replaces it progressively with differential cyclic.

Intermeshing rotors cause pronounced pitch attitude change in response to collective pitch change; the K-MAX tailplane is connected to collective to alleviate this, to reduce blade stresses and produce touchdown and lift-off in level attitude; fixed fins on tailplane are in rotor downwash and rear fin is outside rotor disc; rudder is connected to foot pedals to help balance turns.

STRUCTURE: Semi-monocoque fuselage constructed from 2024T-3 and T-4 aluminium; GFRP and CFRP rotor blades and flaps. Tail assembly weighs 36.3 kg (80 lb) and can be removed quickly by two people. Rotor blades fold to fore and aft.

LANDING GEAR: Fixed, tricycle; nosewheel is out of pilot's field of view; impact sustaining suspension with transverse mounting tube for mainwheels; rubber-in-compression suspension for mainwheels; oleo for nosewheel; bear paw plate round each wheel for operation from soft ground and snow; nosewheel swivels and locks; mainwheels have individual foot-powered brakes and parking brake.

POWER PLANT: One 1,119 kW (1,500 shp) Honeywell T53-17A-1 turboshaft (civil equivalent of military T53-L-703), with particle separator; flat rated at 1,007 kW (1,350 shp) for take-off up to approximately 8,875 m (29,120 ft). Transmission designed for 1,119 kW (1,500 shp), but operated at 1,007 kW (1,350 hp) with a 10,000 hour life with 1,800 hour overhaul intervals, raised to 2,500 hours in September 1998. Fuel capacity 865 litres (228.5 US gallons; 190.0 Imp gallons) located at aircraft CG of which 831 litres (219.5 US gallons; 183 Imp gallons) usable. Hot refuelling capable; dual electric fuel pumps. Oil capacity 12.1 litres (3.2 US gallons; 2.7 Imp gallons) in both engine and transmission tanks.

ACCOMMODATION: Pilot only, in Simula crash impact-absorbing seat with five-point harness; external seat for one passenger can be attached rapidly to either side, immediately ahead of the mainwheel legs. Pilot's seat and rudder pedals adjustable; heater and windscreen demister; doors removable for operation in hot weather; curved windscreen in production version. Tool/cargo compartment 0.74 m³ (26 cu ft) fitted with 2,268 kg (5,000 lb) stress tiedown rings.

SYSTEMS: DC electrical system with starter/generator; no hydraulics.

EQUIPMENT: Pilot-controlled swivelling landing light; standard configuration is for lifting slung loads, but fittings provided for 2,650 litre (700 US gallon; 583 Imp gallon) Bambi firefighting bucket, Loadcell and long-line hook gear; also 2,000 litre (528 US gallon; 440 Imp gallon) capacity, twin-barrel IFEX 3000 water cannon, which can fire 18 or 25 litres (4.8 or 6.6 US gallons; 4.0 or 5.5 Imp gallons) per shot and can traverse 90 to 180°. Kits planned for patrol, firefighting tank and snorkel system, agricultural applications and similar missions. Optional rear-view mirror, port side of nose.

DIMENSIONS, EXTERNAL:

Rotor diameter, each	14.73 m (48 ft 4 in)

Kaman K-MAX sold to US State Department for use in Peru   0121679

KIMBALL

JIM KIMBALL ENTERPRISES INC

PO Box 849, 5354 Cemetary Road, Zellwood, Florida 32798-0849
Tel: (+1 407) 889 34 51
Fax: (+1 407) 889 71 68
e-mail: jwkimball@aol.com
Web: http://www.jimkimballenterprises.com

PRESIDENT: Jim Kimball
VICE-PRESIDENT: Kevin Kimball

Kimball Enterprises offers kits of the Pitts Model 12 Super Stinker after purchasing design rights from Mid-America Aircraft. Plans are marketed through Model 12 Inc at same address.

Recent additions to the Kimball range are the Raptor and McCullocoupe.

VERIFIED

KIMBALL McCULLOCOUPE

TYPE: Two-seat sportplane kitbuilt.

PROGRAMME: Design initiated late 2000. Construction of prototype began in early 2001, with intended completion date of end 2002. By July 2002 the fuselage was complete and wings and tail had been trial-mated.

All data are provisional.

DESIGN FEATURES: Similar to clip-wing Monocoupe designs of 1930s, incorporating some Kimball (Pitts) Model 12 items, including engine cowling and landing gear legs.

FLYING CONTROLS: Conventional and manual.
STRUCTURE: Strut-braced, two-piece wood-covered wing. Compared to original Monocoupe 110 Special, has larger tail surfaces to improve handling. Strut-braced tailplane.
LANDING GEAR: Identical to that used on Kimball (Pitts) Model 12 with fairings covering legs and wheels.
POWER PLANT: One 294 kW (394 hp) VOKBM M-14PF nine-cylinder radial engine driving a three-blade MT propeller. Fuel capacity 265 litres (70.0 US gallons; 58.3 Imp gallons) in two wing tanks.
ACCOMMODATION: Pilot and passenger side by side.
DIMENSIONS, EXTERNAL:

Wing span	7.92 m (26 ft 0 in)
Wing chord at root	1.52 m (5 ft 0 in)
Wing aspect ratio	5.4
Length overall	6.86 m (22 ft 6 in)
Height to top of wing	2.44 m (8 ft 0 in)

AREAS:

Wings, gross	11.57 m² (124.6 sq ft)

WEIGHTS AND LOADINGS:

Weight empty	635 kg (1,400 lb)
Baggage capacity	36 kg (80 lb)
Max T-O weight	1,043 kg (2,300 lb)
Max wing loading	90.1 kg/m² (18.45 lb/sq ft)
Max power loading	3.55 kg/kW (5.84 lb/hp)

PERFORMANCE:

Never-exceed speed (V_{NE})	234 kt (434 km/h; 270 mph)
Normal cruising speed	182 kt (338 km/h; 210 mph)
Stalling speed	57 kt (105 km/h; 65 mph)
Endurance	6 h 0 min
	UPDATED

KIMBALL MODEL 15 RAPTOR A S

TYPE: Tandem-seat sportplane kitbuilt.
PROGRAMME: Work began late 1999 and project details revealed early 2001. Development is continuing slowly while Mullocoupe is completed; expected construction start mid-2003.
All data are provisional.
DESIGN FEATURES: Based on the Kimball (Pitts) Model 12, using tail, landing gear, firewall forward and canopy. Rate of roll in excess of 360°/s.
FLYING CONTROLS: As Kimball (Pitts) Model 12. Horn-balanced rudder.
STRUCTURE: Generally as Kimball (Pitts) Model 12 with new aerobatic wing; sports wing will be available later, if market demands.
LANDING GEAR: As Kimball (Pitts) Model 12.
POWER PLANT: Choice of 265 kW (355 hp) VOKBM M-14P or 294 kW (394 hp) M-14PF nine-cylinder radial engine driving a three-blade MT propeller; option to use Textron Lycoming power plant will be added later. Standard fuel capacity as Kimball (Pitts) Model 12; optional capacity 280 litres (74.0 US gallons; 61.7 Imp gallons).
ACCOMMODATION: As Kimball (Pitts) Model 12.
DIMENSIONS, EXTERNAL:

Wing span	7.92 m (26 ft 0 in)
Wing aspect ratio	5.8
Length overall	5.99 m (19 ft 8 in)
Height overall	2.36 m (7 ft 9 in)

AREAS:

Wings, gross	10.82 m² (116.5 sq ft)

WEIGHTS AND LOADINGS:

Weight empty	637 kg (1,405 lb)
Max T-O weight	1,043 kg (2,300 lb)
Max wing loading	96.4 kg/m² (19.74 lb/sq ft)
Max power loading: M-14P	3.94 kg/kW (6.48 lb/hp)
M-14PT	3.56 kg/kW (5.84 lb/hp)

PERFORMANCE:

Normal cruising speed	161 kt (298 km/h; 185 mph)
Max rate of climb at S/L	1,219 m (4,000 ft)/min
Endurance: with standard fuel	3 h 30 min
with max fuel	5 h 0 min
	UPDATED

KIMBALL (PITTS) MODEL 12

TYPE: Tandem-seat biplane/kitbuilt.
PROGRAMME: Designed by Curtis Pitts and originally named Macho Stinker and, later, Pitts Monster. Design started in 1994; construction of prototype began 1996; first flight March 1996; first delivery April 1997.
CUSTOMERS: Total of 75 under construction in nine countries: Australia, Canada, Finland, Germany, Iceland, Lithuania, South Africa, Sweden and the USA; plus 198 sets of plans sold in a further six countries. Total of 15 flying by September 2002, including one in South Africa.
COSTS: Kit US$47,995; complete aircraft US$185,000 (2003).
DESIGN FEATURES: Traditional Pitts design, optimised for aerobatics; contours dictated by M-14P radial engine. Wire- and strut-braced wings of unequal span; upper wing mounted on cabane; single interplane strut each side; wire-braced tailplane and fin. Roll rate over 300°/s.
FLYING CONTROLS: Conventional and manual. Rod-connected ailerons on upper and lower wings. Flight-adjustable trim tab in each elevator.
STRUCTURE: 4130 steel tube fuselage; wooden spars and ribs; fabric covering.
LANDING GEAR: Tailwheel type; non-retractable. Optional mainwheel fairings. Aluminium sprung main legs; mainwheels 6.00-6 in; Cleveland disc brakes.

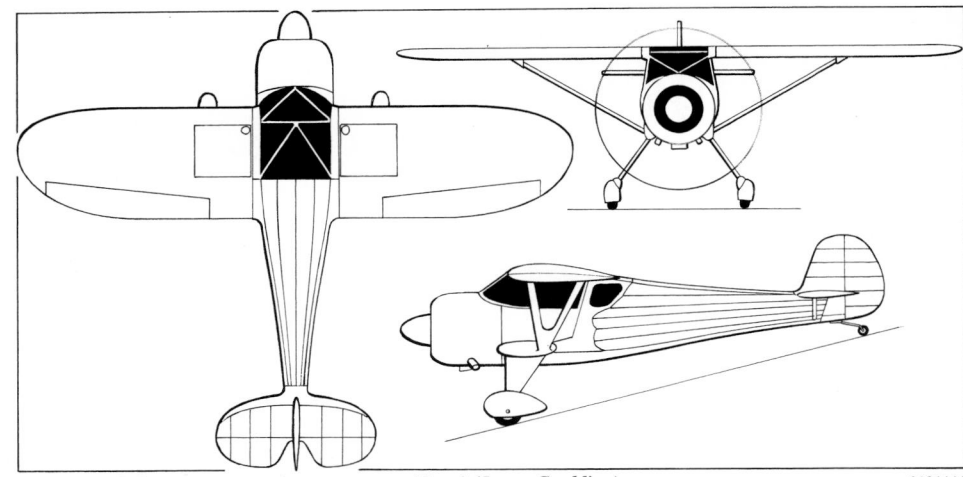

Kimball McCullocoupe general arrangement (*Jane's/James Goulding*)

0121111

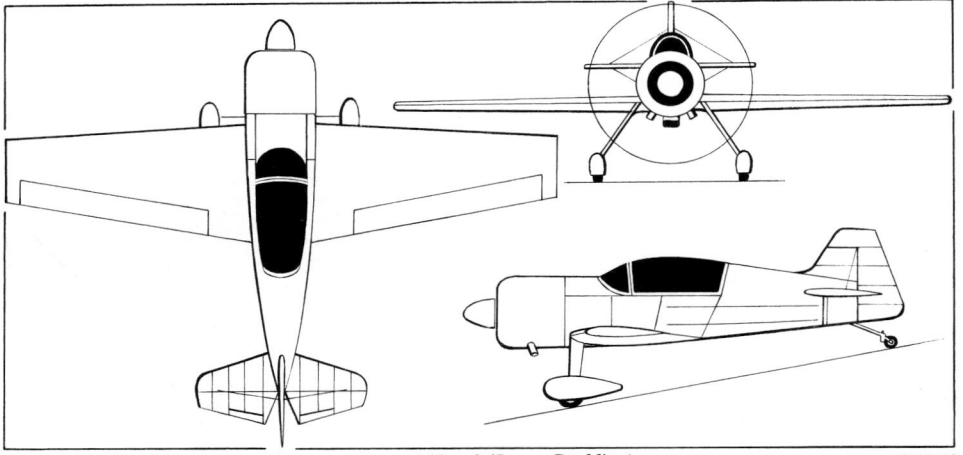

Kimball Model 15 Raptor general arrangement (*Jane's/James Goulding*)

0121110

Kimball (Pitts) Model 12 (*Jane's/Paul Jackson*)

NEW/0533691

General arrangement of the Kimball (Pitts) Model 12 (VOKBM M-14P radial engine) (*Jane's/James Goulding*)

0079018

POWER PLANT: One 265 kW (355 hp) VOKBM M-14P or 294 kW (394 hp) P.14PF nine-cylinder radial engine driving an MT three-blade constant-speed propeller. Fuel capacity 204 litres (54.0 US gallons; 45.0 Imp gallons) in 68 litre (18.0 US gallon; 15.0 Imp gallon) wing tank and 136 litre (36.0 US gallon; 30.0 Imp gallon) fuselage tank. Oil capacity 15 litres (4.0 US gallons; 3.3 Imp gallons).

ACCOMMODATION: Pilot and passenger in enclosed cockpit under one-piece Perspex canopy hinged on right side. Fixed windscreen.

SYSTEMS: 14 V or 28 V battery.

DIMENSIONS, EXTERNAL:

Wing span: upper	6.71 m (22 ft 0 in)
lower	6.40 m (21 ft 0 in)
Wing chord at tip	1.07 m (3 ft 6 in)
Length overall	5.99 m (19 ft 8 in)
Max width of fuselage	1.07 m (3 ft 6 in)

Height: overall	2.97 m (9 ft 9 in)
to top of upper wing	2.51 m (8 ft 3 in)
Tailplane span	2.44 m (8 ft 0 in)
Wheel track	2.29 m (7 ft 6 in)
Wheelbase	4.42 m (14 ft 6 in)
Propeller diameter	2.49 m (8 ft 2 in)
Propeller ground clearance	0.36 m (1 ft 2 in)

AREAS:

Wings, gross	13.94 m² (150.1 sq ft)
Ailerons (total)	1.88 m² (20.21 sq ft)
Fin	0.34 m² (3.69 sq ft)
Rudder, incl tab	0.63 m² (6.82 sq ft)
Elevators, incl tab	0.68 m² (7.30 sq ft)

WEIGHTS AND LOADINGS:

Weight empty	658 kg (1,450 lb)
Baggage capacity	18 kg (40 lb)
Max T-O weight	1,043 kg (2,300 lb)

Max wing loading	73.2 kg/m² (14.99 lb/sq ft)
Max power loading	3.86 kg/kW (6.34 lb/hp)

PERFORMANCE:

Never-exceed and max operating speed (VNE, VMO)	207 kt (384 km/h; 239 mph)
Max cruising speed	178 kt (330 km/h; 205 mph)
Econ cruising speed	148 kt (274 km/h; 170 mph)
Stalling speed, power off	56 kt (103 km/h; 64 mph)
Roll rate	300°/s
Max rate of climb at S/L	more than 914 m (3,000 ft)/min
T-O run	76 m (250 ft)
T-O to 15 m (50 ft)	91 m (300 ft)
Landing from 15 m (50 ft)	274 m (900 ft)
Range with max fuel	525 n miles (972 km; 604 miles)

UPDATED

KINETIC

KINETIC AVIATION

7481 Northland Avenue, San Ramon, California 94583-3736
Tel: (+1 925) 373 93 96
e-mail: mtgoatacft@aol.com
Web: http://members.aol.com/mtgoatacft
PRESIDENT: Bill Montagne

PRODUCTION PLANT:
Goatworks Inc
74 North Oakland Avenue, Sharon, Pennsylvania 16146-2384

Kinetic Aviation has selected Wasilla in Alaska as the production site for the Mountain Goat and its proposed Big Horn development. Few details of the latter have been revealed.

UPDATED

KINETIC MOUNTAIN GOAT

TYPE: Two-seat lightplane.

PROGRAMME: Prototype registered N101MG in February 1997; certification awaits raising of finance.

CUSTOMERS: Orders being taken; first production aircraft is N93GW, built by Goat Works Inc and registered in September 2001. Company intends construction of 50 to 100 aircraft in first full year of production, rising to 500 per year in fifth.

COSTS: US$125,000, equipped (2001), US$150,000 with titanium frame.

DESIGN FEATURES: Similar in appearance to Piper PA-18 Super Cub. High, strut-braced wings; empennage strut-braced. Sweptback fin. Heavy-duty construction and refined aerodynamics for inhospitable terrain.

FLYING CONTROLS: Conventional and manual. Actuation bellcrank and pushrod. Fowler flaps and flaperons. Horn-balanced rudder and elevators; flight-adjustable trim tab on port elevator; ground-adjustable tab on rudder.

STRUCTURE: Fuselage and empennage of fabric-covered 4130 steel tube. All-metal flush riveted high wing has twin bracing struts. Strut-braced tailplane. Optional titanium replacing steel.

LANDING GEAR: Tailwheel type; fixed. Bungee cord or spring steel suspension. Mainwheels have Goodyear 26×10.5-6 tyres and toe-operated dual hydraulic brakes; dual-piston brakes optional. Scott 3200 tailwheel with heavy-duty tail spring. Float mountings standard; skis optional.

Kinetic Mountain Goat prototype with appropriate registration and decoration   *NEW*/0533718

POWER PLANT: One 134 kW (180 hp) Textron Lycoming IO-360-B2E flat-four driving a Kinetic Aviation two-blade, ground-adjustable pitch propeller; McCauley, Hartzell and MT propellers can be fitted. Fuel capacity 250 litres (66.0 US gallons; 55.0 Imp gallons) of which 235 litres (62.0 US gallons; 51.6 Imp gallons) usable.

ACCOMMODATION: Two seats in tandem with optional jump seat. One-piece full Plexiglas door on each side of fuselage. Dual controls. Four-point seat belts. Baggage space behind rear seat.

SYSTEMS: Skytec lightweight starter; B&C 40 A alternator; electric auxiliary fuel pump.

AVIONICS: *Comms:* Garmin GPS/COM 250 XL, Mode C transponder. Vision Microsystems engine management system.

EQUIPMENT: Wingtip strobe lighting; landing light and anti-collision beacon.

DIMENSIONS, EXTERNAL:

Wing span	10.82 m (35 ft 6 in)
Wing aspect ratio	6.7
Length overall	7.42 m (24 ft 4 in)
Height overall	2.06 m (6 ft 9 in)
Propeller diameter	2.03 m (6 ft 8 in)

DIMENSIONS, INTERNAL:

Baggage hold volume	0.685 m³ (24.2 cu ft)

AREAS:

Wings, gross	17.47 m² (188.0 sq ft)

WEIGHTS AND LOADINGS:

Weight empty: standard	556 kg (1,225 lb)
titanium version	510 kg (1,125 lb)
Baggage capacity	159 kg (350 lb)
Max T-O weight	1,122 kg (2,475 lb)
Max zero-fuel weight	946 kg (2,085 lb)
Max wing loading	64.3 kg/m² (13.16 lb/sq ft)
Max power loading	8.37 kg/kW (13.75 lb/hp)

PERFORMANCE:

Never-exceed speed (VNE)	185 kt (342 km/h; 213 mph)
Max cruising speed at 75% power	138 kt (256 km/h; 159 mph)
Econ cruising speed at 55% power	111 kt (206 km/h; 128 mph)
Stalling speed, power off: flaps up	44 kt (81 km/h; 50 mph)
flaps down	36 kt (66 km/h; 41 mph)
Max rate of climb at S/L	518 m (1,700 ft)/min
T-O run	92 m (300 ft)
Landing run	84 m (275 ft)
Range with max fuel, 45 min reserves	659 n miles (1,287 km; 800 miles)

UPDATED

KNR

KNR AIRCRAFT CO INC

6831 Oxford Street, St Louis Park, Minnesota 55426-4412
Tel: (+1 877) 304 19 68 and (+1 612) 920 98 11
Fax: (+1 612) 924 11 11

KNR Aircraft proposed to produce the G10 Aeropony, an updated version of the 1939 Bücker Bü 181 Bestmann, manufactured by the Kader factory at Nasser City, Egypt. No sales had been reported by late 2002.

UPDATED

KNR BESTMANN and AEROPONY

No known production. A description last appeared in the 2002-03 edition.

UPDATED

KOLB

NEW KOLB AIRCRAFT COMPANY INC

8375 Russell Dyche Highway, London, Kentucky 40741
Tel: (+1 606) 862 96 92
Fax: (+1 606) 862 96 22
e-mail: customersupport@tnkolbaircraft.com
Web: http://www.kolbaircraft.com
PLANT MANAGER: Ray Brown

The New Kolb Aircraft Company purchased design rights to Kolb series of light aircraft in 1999. For details of the FireStar and TwinStar, see 1992-93 *Jane's*; TwinStar has since been renamed Kolb Mark III. Kolbra launched at EAA convention, Oshkosh, July 2000. Total sales of all products exceed 1,500. In its 1,347 m² (14,500 sq ft) factory, Kolb has 18 employees

and also produces assemblies for other kit manufacturers, including Rihn Aircraft.

In April 2002 Kolb announced that it had been appointed US assembly centre and distributor for the Ultravia Pelican Sport (which see) which is marketed as the Kolb Sport 600.

UPDATED

KOLB FIREFLY

TYPE: Single-seat kitbuilt ultralight.

PROGRAMME: Introduced at Sun 'n' Fun 1995.

CUSTOMERS: Over 100 flying by mid-2002.

COSTS: US$9,927 including engine (2003).

DESIGN FEATURES: High, constant-chord wing and boom-mounted empennage, based on FireStar; meets FAR Pt 103.7. Folding wings and tail for easy storage. Quoted build time 400 hours.

Modified NACA 4412 series aerofoil.

FLYING CONTROLS: Full-span flaperons.

STRUCTURE: Welded 4130 chromoly steel fuselage cage; 127 mm (5 in) rear fuselage tube and wing spar. Glass fibre nosecone; aluminium tube, fabric-covered wings.

LANDING GEAR: Tailwheel type; brakes optional.

POWER PLANT: One 29.5 kW (39.6 hp) Rotax 447 UL two-cylinder piston engine with Rotax 2.58:1 reduction unit, driving IVO two-blade wooden propeller (three-blade IVO propeller optional). Fuel capacity 18.9 litres (5.0 US gallons; 4.2 Imp gallons).

ACCOMMODATION: Optional full enclosure cockpit.

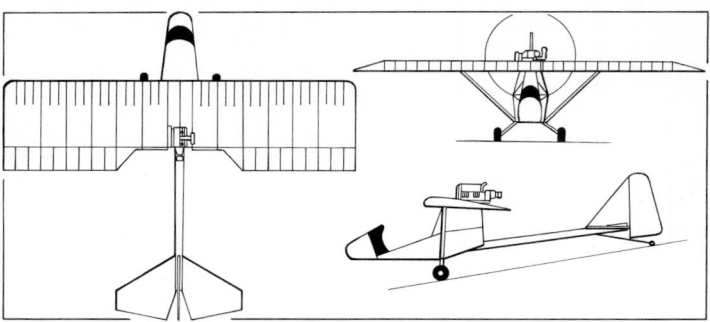

Kolb FireFly single-seat ultralight (*Jane's/James Goulding*)   0045105

Kolb FireFly (29.5 kW; 39.6 hp Rotax 447 engine) (*Jane's/Susan Bushell*)
0059976

DIMENSIONS, EXTERNAL:
Wing span	6.71 m (22 ft 0 in)
Width, wings folded	1.75 m (5 ft 9 in)
Length overall	5.94 m (19 ft 6 in)
Height overall	1.75 m (5 ft 9 in)

AREAS:
Wings, gross	10.87 m² (117.0 sq ft)

WEIGHTS AND LOADINGS:
Weight empty	115 kg (253 lb)
Max T-O weight	226 kg (500 lb)

PERFORMANCE:
Max operating speed	60 kt (112 km/h; 70 mph)
Stalling speed	23 kt (42 km/h; 26 mph) CAS
Max rate of climb at S/L	305 m (1,000 ft)/min
T-O and landing run	46 m (150 ft)
g limits	+4/−2

UPDATED

KOLB KOLBRA

TYPE: Tandem-seat ultralight kitbuilt.

PROGRAMME: First flown 7 May 2000 and publicly announced July 2000. Debut at Oshkosh that month. Prototype N718KA.

CUSTOMERS: Over 50 kits sold and 12 aircraft flying by mid-2002.

COSTS: Kolbra kit US$18,835 including engine and covering (2003).

DESIGN FEATURES: Comprises Mark III wings and tail, FireStar nose and Sling Shot-style landing gear; resulting hybrid then optimised for training role.

FLYING CONTROLS: Manual. Full-span flaperons.

STRUCTURE: Welded 4130 chromoly steel fuselage cage; 152 mm (6 in) diameter rear fuselage tube and wing spar. Glass fibre nosecone; aluminium tube wings, fabric-covered.

LANDING GEAR: Tailwheel type; fixed. Matco wheels and tyres. Chromoly construction for high strength. High flotation tyres. Optional hydraulic disc brakes; Tundra tyres can be fitted.

POWER PLANT: One 47.8 kW (64.1 hp) Rotax 582 UL two-cylinder piston engine with 'B' type 2.58:1 ratio gearbox, or 'C' or 'E' type 3.47:1 ratio gearbox, driving a two- or three-blade IVO or three-blade Warp Drive propellers. Alternatively, a 59.6 kW (79.9 hp) Rotax 912 UL or 73.5 kW (98.6 hp) Rotax 912 ULS can be fitted driving 1.83 m (6 ft 0 in) two- or 1.63 m (5 ft 4 in) three-blade IVO or Warp Drive propeller.

Fuel capacity (both) 37.9 litres (10.0 US gallons; 8.3 Imp gallons).

SYSTEMS: Optional electric starter and dual Magnum strobes.

EQUIPMENT: Optional BRS ballistic parachute.

DIMENSIONS, EXTERNAL:
Wing span	8.64 m (28 ft 4 in)
Length overall	7.32 m (24 ft 0 in)
Height overall, three-blade propeller	2.03 m (6 ft 8 in)

AREAS:
Wings, gross	14.49 m² (156.0 sq ft)

WEIGHTS AND LOADINGS (A: Kolbra Trainer with Rotax 582, B: King Kolbra with Jabiru):
Weight empty: A	225 kg (496 lb)
B	249 kg (550 lb)
Max T-O weight: A, B	454 kg (1,000 lb)

PERFORMANCE:
Never-exceed speed (VNE):		
A, B		95 kt (177 km/h; 110 mph)
Normal cruising speed: A		65 kt (121 km/h; 75 mph)
B		74 kt (137 km/h; 85 mph)
Stalling speed, flaps down: A		31 kt (57 km/h; 35 mph)
B		40 kt (73 km/h; 45 mph)
Max rate of climb at S/L: A, B		335 m (1,100 ft)/min
T-O and landing run: A, B		76 m (250 ft)
Range: A		160 n miles (296 km; 184 miles)
B		250 n miles (463 km; 287 miles)

UPDATED

KOLB LASER

The Kolb Laser has not been put into production; details last appeared in the 2002-03 edition.

UPDATED

KOLB SLING SHOT

TYPE: Tandem-seat ultralight.

PROGRAMME: First-flown July 1996.

CUSTOMERS: Over 30 flying by July 2002.

COSTS: US$17,407 with Rotax 582 engine; US$22,966 with Rotax 912 (2000).

DESIGN FEATURES: Generally as for FireFly. Quoted build time 400 hours.

FLYING CONTROLS: Full-span flaperons.

STRUCTURE: Welded 4130 chromoly steel fuselage cage; 127 mm (5 in) rear fuselage tube and wing spar. Glass fibre

Kolb Sling Shot (*Jane's/Paul Jackson*)   0089501

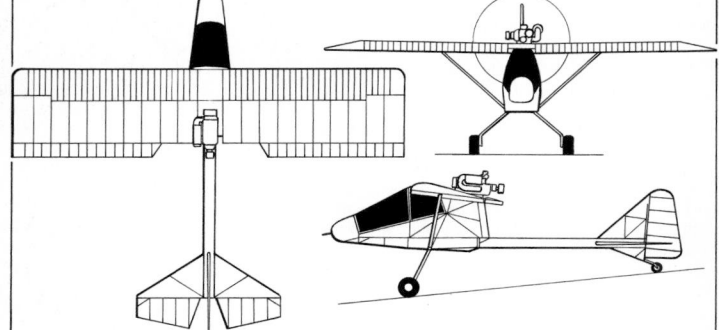

Kolb Sling Shot two-seat ultralight (*Jane's/James Goulding*)   0045107

Kolb Kolbra tandem-seat ultralight (*Jane's/Paul Jackson*)   *NEW*/0533692

nosecone and steel-reinforced, aluminium tube, fabric-covered wings.

LANDING GEAR: Tailwheel type; fixed; hydraulic disc brakes optional.

POWER PLANT: One 47.8 kW (64.1 hp) Rotax 582 using 2.58:1 or 3.47:1 reduction unit driving IVO two- or three-blade wooden propeller (metal propeller optional); further options are 37.0 kW (49.6 hp) Rotax 503 with 2.58:1 reduction unit; 59.6 kW (79.9 hp) Rotax 912 and 59.7 kW (80 hp) Jabiru 2200. Fuel capacity 37.0 litres (10.0 US gallons; 8.3 Imp gallons). Electric starter optional.

EQUIPMENT: Optional ballistic parachute and strobe lighting.

DIMENSIONS, EXTERNAL:
Wing span	6.71 m (22 ft 0 in)
Width, wings folded	1.91 m (6 ft 3 in)
Length overall	5.79 m (19 ft 0 in)
Height overall	2.08 m (6 ft 10 in)

AREAS:
Wings, gross	10.22 m² (110.0 sq ft)

WEIGHTS AND LOADINGS:
Weight empty	156 kg (345 lb)
Max T-O weight	385 kg (850 lb)

PERFORMANCE (Rotax 582 engine):
Max operating speed	95 kt (176 km/h; 109 mph)
Normal cruising speed	76 kt (140 km/h; 87 mph)
Stalling speed, power off	36 kt (66 km/h; 41 mph)
Max rate of climb at S/L	396 m (1,300 ft)/min
T-O run	53 m (175 ft)
Landing run	61 m (200 ft)
g limits	+4/−2

UPDATED

LANCAIR

LANCAIR GROUP INC

Formed in 1984, the Lancair Group comprises Lancair International, producing kit aircraft; and Lancair Company, marketing factory-built aircraft.

Lancair is also US dealer for the HeliSport CH-7 Kompress (see Italian section).

UPDATED

LANCAIR INTERNATIONAL

2244 Airport Way, Redmond, Oregon 97756
Tel: (+1 541) 923 22 33
Fax: (+1 541) 923 22 55
e-mail: lancair@lancair.com
Web: http://www.lancair.com
CEO: Lance Neibauer
PRESIDENT: Bing Lantis
GENERAL MANAGER: Bob Fair
SALES MANAGERS: Mike Schrader
    Orin Riddell

More than 2,000 Lancairs (including out-of-production Model 235) sold to customers in 40 countries; around 800 are currently flying.

In July 1999 Lancair International announced the Legacy 2000, deliveries of which began in August 2000, superseding the earlier 320/360 series, which last appeared in the 1999-2000 *Jane's*.

UPDATED

LANCAIR LANCAIR IV, IV-P and SENTRY

TYPE: Four-seat kitbuilt.

PROGRAMME: Kit deliveries began 1990. On 20 February 1991, prototype set NAA world speed record between San Francisco and Denver of 314.7 kt (583.2 km/h; 362.4 mph). First kit completed July 1991.

CURRENT VERSIONS: **Lancair IV:** Standard version.
Description mainly applies to Lancair IV.

Lancair IV-P: Pressurised version; first flight 1 November 1993. Provides 0.34 bar (5.0 lb/sq in) differential; equipment from Dukes Research of California.

Lancair Propjet IV-P: Powered by 560 kW (751 shp) Walter M 601E turboprop; prototype (N750TL, registered as a IV-TP) first flown 9 July 2001. Available as option to standard IV-P at cost of US$7,500; completed aircraft can be retrospectively converted; main structural alteration is repositioning of firewall. Cruise speed 348 kt (644 km/h; 400 mph); max rate of climb 1,158 m (3,800 ft)/min; range with optional 462 litre (122 US gallon; 101.5 Imp gallon) fuel capacity 1,000 n miles (1,852 km; 1,150 miles). Max T-O weight 1,610 kg (3,550 lb).

On 11 September 2002, a Lancair IV-P Propjet flown by Wesley E Behel Jr claimed a Class C-1c Group 2 speed record over a 3 km course at restricted altitude by flying at 307 kt (568 km/h; 352 mph).

Lancair Sentry: Tandem-seat version based on Lancair IV and retaining wing and lower fuselage, with redesigned upper fuselage and vertical tail. Prototype (N806EY) first flown 7 September 2002. Power supplied by same engine as IV-P Propjet, but with larger fuel tank. Intended for military training and civil air racing markets.

Lancair IV-P performance kitbuilt, with optional winglets (*Jane's/Paul Jackson*)   NEW/0533694

Prototype Lancair Sentry turboprop   NEW/0533717

CUSTOMERS: Total of 500 kits sold; more than 160 flying by mid-2001. The Mexican Navy is to assess the **Sentry**, to augment its Aermacchi Redigo fleet.

COSTS: Fastbuild US$79,900. Pressurised version: Fastbuild US$104,900. Firewall fastbuild options and two-week builder workshop programmes available. Turbine IV-P US$115,900; Walter engine US$84,000. Sentry US$119,900 (2003).

DESIGN FEATURES: Conventional seating for four persons in airframe of new design, but following Lancair formula for high cruising speeds. Optional winglets. Quoted fastbuild construction time 1,000 hours.

FLYING CONTROLS: Conventional and manual. Fowler flaps. Sidestick controllers for both pilots. Trim tabs in port aileron and port elevator. Optional speed brakes.

STRUCTURE: Carbon fibre/epoxy airframe, with Nomex honeycomb cores.

LANDING GEAR: Tricycle layout. Mainwheels 15×6.00-6; nosewheel is 5.00-5. Cleveland dual-piston caliper brakes. Hydraulic retraction system.

POWER PLANT: One 261 kW (350 hp) Teledyne Continental TSIO-550-E1B, -E2B or -E3B twin-turbocharged flat-six engine driving a three- or four-bladed MT or Hartzell constant-speed propeller. Walter M 601E turboprop can also be fitted; as described above. Two versions of PT-6A turboprop tested during 1999, but apparently not proceeded with. Lancair IV-P N540LM flew September 2000 as testbed for 485 kW (650 hp) Eagle 540 piston engine. Fuel in integral wing tanks; total standard usable capacity 341 litres (90.0 US gallons; 74.9 Imp gallons); long-range tanks, total capacity 416 litres (110 US gallons; 91.6 Imp gallons). Oil capacity 11.4 litres (3.0 US gallons; 2.5 Imp gallons).

SYSTEMS: Thermoelectric de-icing system tested during 2001.

DIMENSIONS, EXTERNAL:
Wing span: plain tips	9.19 m (30 ft 2 in)
winglets	9.93 m (32 ft 7 in)
Wing chord: at root	1.18 m (3 ft 10½ in)
at tip	0.77 m (2 ft 6¼ in)

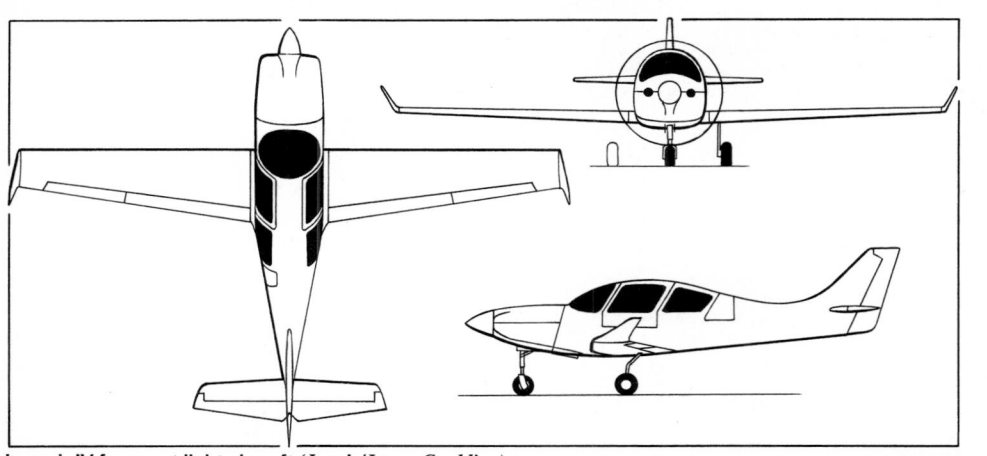

Lancair IV four-seat light aircraft (*Jane's/James Goulding*)   0106496

Prototype turboprop-powered Lancair IV-P Propjet (*Jane's/Paul Jackson*)   *NEW*/0533695

Wing aspect ratio	9.3
Width, wings removed	2.41 m (7 ft 11 in)
Length overall	7.62 m (25 ft 0 in)
Height overall	2.44 m (8 ft 0 in)
Tailplane span	3.35 m (11 ft 0 in)
Wheel track	2.13 m (7 ft 0 in)
Wheelbase	1.88 m (6 ft 2 in)
Propeller diameter	1.93 m (6 ft 4 in)

DIMENSIONS, INTERNAL:
Cabin: Length	3.20 m (10 ft 6 in)
Max width	1.17 m (3 ft 10 in)
Max height	1.22 m (4 ft 0 in)

AREAS:
Wings, gross: plain tips	9.10 m² (98.0 sq ft)
winglets	10.03 m² (108.0 sq ft)
Winglets, total	1.30 m² (14.00 sq ft)
Flaps, total	1.12 m² (12.07 sq ft)

WEIGHTS AND LOADINGS:
Weight empty: IV	907 kg (2,000 lb)
IV-P	998 kg (2,200 lb)
Baggage capacity	68 kg (150 lb)
Max T-O weight	1,451 kg (3,200 lb)
Max landing weight	1,360 kg (3,000 lb)
Max wing loading, plain tips	159.4 kg/m² (32.65 lb/sq ft)
Max power loading	5.56 kg/kW (9.14 lb/hp)

PERFORMANCE:
Never-exceed speed (VNE)	274 kt (507 km/h; 315 mph)
Cruising speed at 75% power at 7,315 m (24,000 ft)	291 kt (539 km/h; 335 mph)
Stalling speed: standard wing	66 kt (121 km/h; 75 mph)
with winglets	61 kt (113 km/h; 70 mph)
Max rate of climb at S/L	792 m (2,600 ft)/min
Service ceiling	8,840 m (29,000 ft)
T-O run	457 m (1,500 ft)
T-O to 15 m (50 ft)	549 m (1,800 ft)
Landing run	518 m (1,700 ft)
Range, with max fuel, no reserves	1,347 n miles (2,494 km; 1,550 miles)
Endurance	4 h 48 min
g limits	+4.4/−2.3 (allowable), +8/−4 ultimate

UPDATED

LANCAIR LANCAIR ES and SUPER ES

TYPE: Four-seat kitbuilt.

PROGRAMME: Prototype unveiled at Oshkosh in 1992; flight testing completed 1995. Certification of Fastbuild kit granted early 1996 under simplified version of FAR Pt 23 governing light aircraft. Further improvements to Fastbuild kit introduced in early 2002.

CURRENT VERSIONS: **ES:** Baseline version.
Description applies mainly to this version.
   Super ES: Higher-powered version.

CUSTOMERS: Total of 163 kits sold by early 2001, of which 30 were then flying. Production continues.

COSTS: Standard Fastbuild US$58,900; New Fastbuild kit US$68,900 (2003).

DESIGN FEATURES: Generally as for earlier Lancairs; hardtop cabin with paired side-by-side seats. Sidestick controls. Quoted Fastbuild kit construction time 1,800 hours; with new 2002 improvements reduced to quoted time of 700 hours.

FLYING CONTROLS: Generally as for Lancair IV.

STRUCTURE: All-composites.

LANDING GEAR: Non-retractable tricycle type. Main tyres 6.00-15×6; nose 5.00-5. Nosewheel steerable through 90°.

POWER PLANT: *ES:* One 149 kW (200 hp) Teledyne Continental IO-360-ES flat-six engine; two-blade McCauley constant-speed propeller.
   Super ES: One 209 kW (280 hp) Teledyne Continental IO-550-G or 231 kW (310 hp) IO-550-N with three-blade Hartzell constant-speed propeller. Standard fuel capacity 288 litres (76.0 US gallons; 63.3 Imp gallons), of which 284 litres (75.0 US gallons; 62.5 Imp gallons) are usable, in two wing tanks. Long-range fuel capacity 546 litres (100 US gallons; 120 Imp gallons) usable. Oil capacity 7.6 litres (2.0 US gallons; 1.7 Imp gallons).

ACCOMMODATION: Pilot and three passengers in two pairs of seats.

DIMENSIONS, EXTERNAL:
Wing span	10.82 m (35 ft 6 in)
Wing chord: at root	1.50 m (4 ft 11 in)
at tip	0.91 m (3 ft 0 in)
Wing aspect ratio	9.0
Width, wings removed	2.41 m (7 ft 11 in)
Length overall	7.62 m (25 ft 0 in)
Height overall	2.34 m (7 ft 8 in)
Tailplane span	4.01 m (13 ft 2 in)
Wheel track	2.36 m (7 ft 9 in)
Wheelbase	1.88 m (6 ft 2 in)
Propeller diameter	1.93 m (6 ft 4 in)

DIMENSIONS, INTERNAL:
Cabin: as Lancair IV	
Baggage hold volume	1.0 m³ (35 cu ft)

AREAS:
Wings, gross	13.01 m² (140.0 sq ft)

WEIGHTS AND LOADINGS:
Weight empty:	
ES, Super ES with IO-550-G	862 kg (1,900 lb)
Super ES with IO-550N	907 kg (2,000 lb)
Baggage capacity	79 kg (175 lb)
Max T-O weight: Normal	1,451 kg (3,200 lb)
Utility	1,361 kg (3,000 lb)
Max wing loading: Normal	111.60 kg/m² (22.86 lb/sq ft)
Utility	104.6 kg/m² (21.43 lb/sq ft)
Max power loading: Normal:	
ES	9.74 kg/kW (16.00 lb/hp)
Super ES	6.96 kg/kW (11.43 lb/hp)
Utility: ES	9.13 kg/kW (15.00 lb/hp)
Super ES	6.52 kg/kW (10.71 lb/hp)

PERFORMANCE:
Never-exceed speed (VNE)	220 kt (467 km/h; 253 mph)
Cruising speed at 75% power at 3,050 m (10,000 ft):	
ES	174 kt (332 km/h; 200 mph)
Super ES	196 kt (362 km/h; 225 mph)
Stalling speed: ES	50 kt (92 km/h; 57 mph)
Super ES	57 kt (105 km/h; 65 mph)
Rate of climb at S/L: ES	381 m (1,250 ft)/min
Super ES	610 m (2,000 ft)/min

Service ceiling	5,490 m (18,000 ft)
T-O run	183 m (600 ft)
Landing run	244 m (800 ft)
Range, max fuel, no reserves:	
ES	1,260 n miles (2,333 km; 1,450 miles)
Super ES	1,364 n miles (2,526 km; 1,570 miles)
Endurance: ES	7 h
Super ES	6 h
g limits: Normal	+4.4/−2.3
Utility	+9.0/−4.5

UPDATED

LANCAIR LEGACY

TYPE: Side-by-side kitbuilt.

PROGRAMME: Prototype (N199L) first flown 16 July 1999, publicly displayed at Oshkosh 28 July 1999. First kitbuilt example (N540L) first flown 15 May 2001. Early aircraft were marked "Legacy 2000".
   At Reno, Nevada, on 13 September 2002, a Legacy flown by Frederick E Schrameck claimed a Class C-1c Group 1 record for speed, over a 3 km course at restricted altitude, of 297 kt (550 km/h; 342 mph).

CUSTOMERS: Ten sold during Oshkosh debut; 16 by end of 1999, 150 by May 2001; by September 2002, over 100 had been delivered.

COSTS: US$44,900 (2003).

DESIGN FEATURES: Improved version of Lancair series of two-seat kitbuilts, offering increased cabin area, new wing and better performance; only part common to Lancair 360 is horizontal stabiliser. Quoted build time under 1,000 hours.
   Streamlined design. Sweptforward wing trailing-edge, tapered tailplane and sweptback fin; low wing and mid-mounted horizontal tail.
   Wing section NLF(1)-0215F; incidence 1°; dihedral 3°; twist 1° 30′; thickness/chord ratio 15 per cent.

FLYING CONTROLS: Conventional and manual. Fowler flaps. Aileron and elevator electric trim system included. Optional rudder trim and speedbrakes.

STRUCTURE: Carbon fibre airframe.

LANDING GEAR: Tricycle type. Mainwheels retract inwards; nosewheel rearwards. Cleveland 5.00-5 in wheels and brakes. Hydraulic retraction system.

POWER PLANT: One 149 kW (200 hp) Textron Lycoming IO-360-C1D6 flat-four driving a two-blade Hartzell or three-blade MTV-12B constant-speed propeller. Hartzell introduced 7694-4T two-blade propeller specifically for Legacy in February 2002. Optional engines include 134 kW (180 hp) Lycoming IO-360-B1F and -M1A, 194 kW (260 hp) Lycoming IO-540 and 231 kW (310 hp) Teledyne Continental IO-550-N. Standard fuel capacity 189 litres (50.0 US gallons; 41.6 Imp gallons) in two tanks forward of main spar; can be expanded to 250 litres (66.0 US gallons; 55.0 Imp gallons) for long-range operations.
   Experimental example with supercharger displayed at Sun 'n' Fun 2002, offering 295 kt (547 km/h; 340 mph) at 4,570 m (15,000 ft).

ACCOMMODATION: Pilot and passenger in 25° reclined side-by-side seats beneath forward-hinging, one-piece canopy.

DIMENSIONS, EXTERNAL:
Wing span	7.77 m (25 ft 6 in)

Lancair ES fixed-gear four-seater (*Jane's/Paul Jackson*)   *NEW*/0533696

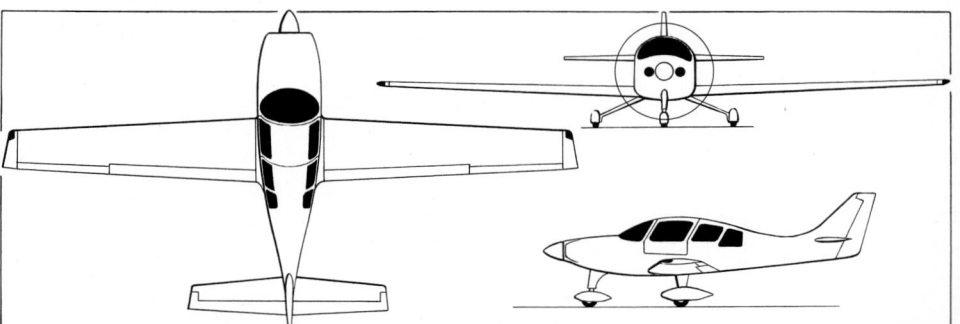

Lancair Super ES general arrangement (*Jane's/James Goulding*)   0126951

Wing aspect ratio	7.9
Length overall	6.71 m (22 ft 0 in)
Height overall	2.54 m (8 ft 4 in)
Tailplane span	2.54 m (8 ft 4 in)
Wheelbase	1.37 m (4 ft 6 in)

DIMENSIONS, INTERNAL:

Cabin: Length	1.60 m (5 ft 3 in)
Max width	1.10 m (3 ft 7½ in)
Max height	1.17 m (3 ft 10 in)
Baggage volume	1.49 m³ (16.0 cu ft)

AREAS:

Wings, gross	7.66 m² (82.5 sq ft)

WEIGHTS AND LOADINGS (IO-360-C1D6 engine, unless stated otherwise):

Weight empty: IO-360	635 kg (1,400 lb)
IO-550	680 kg (1,500 lb)
Baggage capacity	41 kg (90 lb)
Max T-O weight: Normal category	997 kg (2,200 lb)
Utility category	861 kg (1,900 lb)

Max wing loading:

Normal category	130.2 kg/m² (26.67 lb/sq ft)
Utility category	112.4 kg/m² (23.03 lb/sq ft)

Max power loading:

Normal category	6.70 kg/kW (11.00 lb/hp)
Utility category	5.78 kg/kW (9.50 lb/hp)

PERFORMANCE (IO-550N engine, unless stated otherwise):

Max cruising speed: at 5,485 m (18,000 ft)

	243 kt (451 km/h; 280 mph)
at sea level (IO-550)	261 kt (483 km/h; 300 mph)

Normal cruising speed: at 2,440 m (8,000 ft)

	240 kt (444 km/h; 276 mph)
at 3,660 m (12,000 ft)	217 kt (402 km/h; 250 mph)

Stalling speed, flaps down:

IO-360	53 kt (97 km/h; 60 mph)
IO-550	59 kt (108 km/h; 67 mph)
Max rate of climb at S/L: IO-360	594 m (1,950 ft)/min
IO-550	914 m (3,000 ft)/min
Service ceiling: IO-360	7,315 m (24,000 ft)
IO-550	5,485 m (18,000 ft)
T-O run: IO-360	290 m (950 ft)
IO-550	244 m (800 ft)
Landing run: IO-360, IO-550	275 m (900 ft)

Range with max fuel, no reserves: IO-360

	1,129 n miles (2,092 km; 1,300 miles)

IO-550 with reserves

	1,042 n miles (1,931 km; 1,200 miles)
g limits: Normal	+3.8/−1.7
Utility	+4.4/−2.2

UPDATED

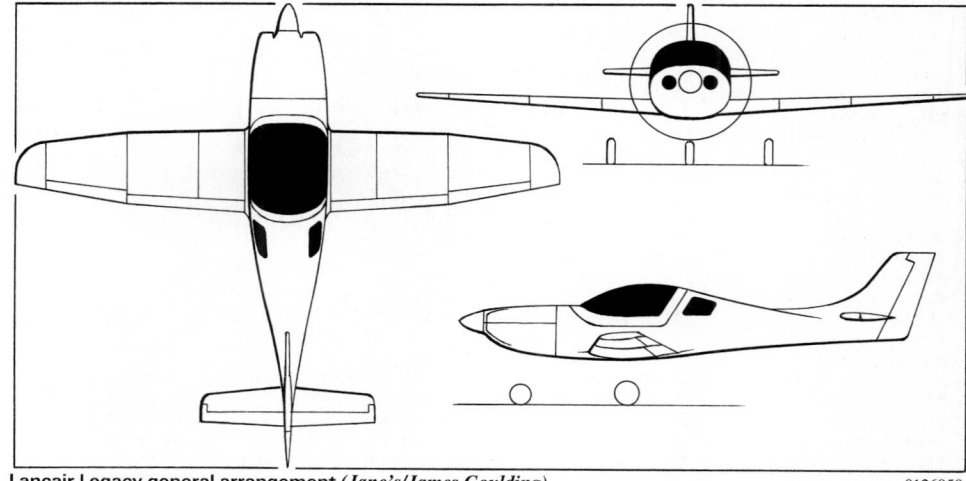

Lancair Legacy two-seat kitplane prototype   *NEW*/0101602

Lancair Legacy general arrangement (*Jane's/James Goulding*)   0126950

LANCAIR COMPANY

22550 Nelson Road, Bend, Oregon 97701
Tel: (+1 541) 318 11 44
Fax: (+1 541) 318 11 77
e-mail: info@lancair.com/certified
Web: http://www.lancair.com/certified
CEO: Lance Neibauer
PRESIDENT: Bing Lantis
GENERAL MANAGER: Bob Fair
SALES MANAGER: Mike Schrader

EUROPEAN AGENT:
Hanseatische Luftwerft GmbH, Emoeweg 16, NL-8218 PC Lelystad Airport, Netherlands

Augmenting the manufacture of kits, Lancair moved into the certified aircraft market with the four-seat Columbia 300, which was unveiled at Oshkosh in August 1997 and assembled at a new 13,100 m² (141,000 sq ft) facility at Bend Municipal Airport, Redmond, Oregon. In July 2002 Lancair's certified company began to lay off part of its workforce, reducing from 320 to 43 by September 2002, following the company's failure to achieve a US$25 million finance deal. On 23 September 2002 the company was offered for sale and on 12 December 2002 announced it had reached agreement with a Malaysian company; the workforce was increased by 60 to 70 by the end of 2002.

UPDATED

LANCAIR COLUMBIA

TYPE: Four-seat lightplane.
PROGRAMME: Announced 1996 as LC-40; aerodynamic prototype flown July 1996; first flight of certification prototype (N140LC), early 1997. Second prototype (N141LC) first flew 14 April 1997; public debut at Oshkosh August 1997; modified for FAA testing to confirm spin-resistance. Certification completed 18 September 1998; RLD certification completed March 2001.

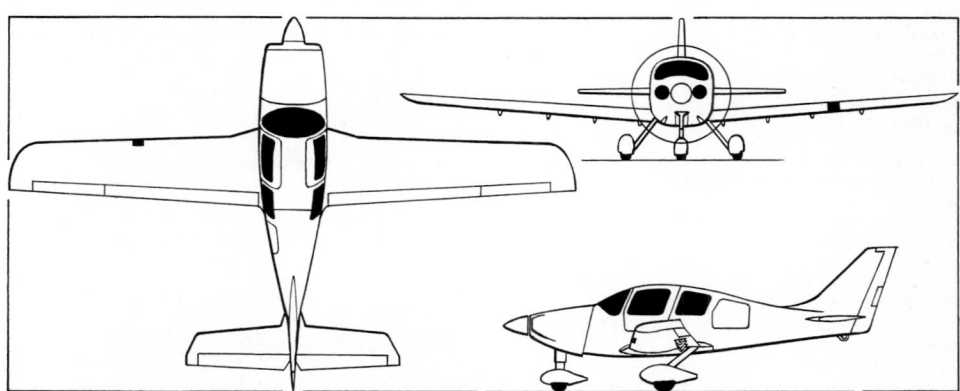

Lancair Columbia 300 (Teledyne Continental IO-550 flat-six) (*Jane's/James Goulding*)   0089517

First production aircraft (N424CH) delivered 24 February 2000.
CURRENT VERSIONS: **Columbia 300:** *As described.* Engineering designation **LC-40-550FG.**
Trainer: Lower-powered version with 116 kW (155 hp) engine, planned for flight training market, rivalling Cessna 172 and Cirrus SR20.
Columbia IV: Proposed version with pressurisation and retractable landing gear.
Columbia 350: Announced and publicly displayed at Sun 'n' Fun 2002; prototype (N70090) first flown early 2002 with deliveries due to start late 2002. All-electric systems replacing Columbia 300's vacuum pumps. Powered by 231 kW (310 hp) Teledyne Continental IO-550-N driving Hartzell three-blade propeller. Empty weight 1,043 kg (2,300 lb); performance data generally similar to Columbia 300 except range in economy cruise 1,320 n miles (2,444 km; 1,519 miles).

Columbia Turbo 400: Engineering designation **LC-41-T550FG.** Turbocharged Columbia 300 with 231 kW (310 hp) Continental TSIO-550E and (when available) FADEC. Announced 9 April 2000; prototype (N143LC converted from Columbia 300) first flew June 2000; first production aircraft (N166PD) registered September 2001; certification had been due in 2002 but was delayed; April 2003 was estimated. Provisional empty weight 1,066 kg (2,350 lb), normal cruising speed 220 kt (407 km/h; 253 mph), max rate of climb 366 m (1,200 ft)/min and service ceiling 7,315 m (24,000 ft). Other data as Columbia 300. Cockpit will eventually include electronic primary flight display projection system; prototype equipped with Highways in the Sky (HITS) instrumentation.
CUSTOMERS: Deposits for 295 aircraft taken by end 1998; firm orders stood at 240 by late 2002. May be offered in fractional ownership schemes. Total 18 built (and nine

registered in US) during 2000 and 27 sold in 2001; by September 2002, 72 had been registered and some 57 built. One example delivered to NASA at Langley 10 January 2001 for use as testbed for general aviation-based AGATE and SATS programmes; designated **Columbia 300X**.

COSTS: US$299,500 fully IFR-equipped (2001). Columbia Turbo 400 US$349,500.

DESIGN FEATURES: Low-wing monoplane of typical Lancair appearance and design. Integral roll-over cage to protect cockpit area.

FLYING CONTROLS: Conventional and manual. Fowler flaps droop to 40° maximum. Aileron travel limits 22° up and 18° down; elevator travel up 13°, down 12°; rudder travel ±17°. Optional Precise Flight speedbrake.

STRUCTURE: Extensive composites construction throughout. One-piece dual-spar wing.

LANDING GEAR: Fixed tubular steel tricycle type. Mainwheels 6.00-6, nosewheel 5.00-5. Cleveland brakes. Speed fairings on all wheels. Retractable version planned. Minimum ground turning radius 4.72 m (15 ft 6 in); nosewheel steerable ±60°.

POWER PLANT: One 224 kW (300 hp) Teledyne Continental IO-550-N2B flat-six engine, driving three-blade constant-speed Hartzell PHC-J3YF-1RF/F7691D-1 propeller. FADEC. Total fuel capacity 402 litres (106 US gallons; 88.3 Imp gallons) in two integral wing tanks; usable fuel 371 litres (98.0 US gallons; 81.6 Imp gallons). Oil capacity 7.6 litres (2.0 US gallons; 1.7 Imp gallons).

ACCOMMODATION: Pilot and three passengers in two pairs of seats in enclosed cabin. Seating certified to 26 g. Double gull-wing doors hinge upwards. Emergency exit through large baggage door behind rear passenger seats.

SYSTEMS: Single 12 V battery; 14 V 60 A alternator; air conditioning.

AVIONICS: *Comms:* Apollo SL30 nav/com, SL 70 transponder.
Flight: Apollo nav indicator, S-Tec 30 or 55 autopilot, Apollo GX60 GPS and Goodrich Stormscope WX-950 lightning sensor. Options include UPS Aviation Technologies dual MX-20 colour moving map displays, GX50 GPS, Honeywell KR 87 ADF, KI 227 remote indicator, KCS 55A HSI and KI 256 flight director.
Instrumentation: AvroTec 265 mm (10½ in) flat panel display optional.

EQUIPMENT: Fire extinguisher standard.

DIMENSIONS, EXTERNAL:
Wing span	10.97 m (36 ft 0 in)
Wing chord: at root	1.40 m (4 ft 7 in)
at tip	1.02 m (3 ft 4 in)
Wing aspect ratio	9.2
Length overall	7.67 m (25 ft 2 in)
Fuselage max width	1.27 m (4 ft 2 in)
Height overall	2.74 m (9 ft 0 in)
Tailplane span	4.17 m (13 ft 8 in)
Wheel track	2.24 m (7 ft 4 in)
Wheelbase	2.04 m (6 ft 8¼ in)
Propeller diameter	1.96 m (6 ft 5 in)

DIMENSIONS, INTERNAL:
Cabin: Length	3.54 m (11 ft 7½ in)
Max width	1.24 m (4 ft 1 in)
Max height	1.30 m (4 ft 3 in)
Baggage compartment: volume	0.76 m³ (27.0 cu ft)

AREAS:
Wings, gross	13.12 m² (141.2 sq ft)

WEIGHTS AND LOADINGS:
Weight empty	1,021 kg (2,250 lb)
Baggage capacity	54 kg (120 lb)
Max fuel weight	288 kg (636 lb)
Max T-O weight	1,542 kg (3,400 lb)
Max landing weight	1,465 kg (3,230 lb)
Max wing loading	117.6 kg/m² (24.08 lb/sq ft)
Max power loading	6.90 kg/kW (11.33 lb/hp)

PERFORMANCE:
Never-exceed speed (VNE)	235 kt (435 km/h; 270 mph)
Normal cruising speed at 75% power	
	190 kt (352 km/h; 219 mph)
Stalling speed: flaps up	71 kt (132 km/h; 82 mph)
flaps down	57 kt (106 km/h; 66 mph)

First production Columbia 300 N142LC, which Erik Lindbergh flew from New York to Paris on 1-2 May 2002, emulating his grandfather's 1927 flight in a Ryan NYP *(Jane's/Paul Jackson)*   NEW/0533697

Prototype Lancair Columbia 350 at its first public showing, April 2002 *(Jane's/Paul Jackson)*   NEW/0533698

Lancair Turbo Columbia 400 prototype *(Jane's/Paul Jackson)*   NEW/0533699

Max rate of climb at S/L	427 m (1,400 ft)/min	Landing run	472 m (1,550 ft)
Service ceiling	5,480 m (18,000 ft)	Range with max fuel, economy cruise	
T-O run	213 m (700 ft)		1,385 n miles (2,565 km; 1,593 miles)
T-O to 15 m (50 ft)	381 m (1,250 ft)	g limits	+4.4/−1.76
Landing from 15 m (50 ft)	716 m (2,350 ft)		*UPDATED*

LEARJET

BOMBARDIER AEROSPACE LEARJET
(Subsidiary of Bombardier Inc)

One Learjet Way, PO Box 7707, Wichita, Kansas 67277
Tel: (+1 316) 946 20 00
Fax: (+1 316) 946 22 20
VICE-PRESIDENT AND GENERAL MANAGER: Jim Ziegler
VICE-PRESIDENT, OPERATIONS: Ghislain Bourque
VICE-PRESIDENT, FINANCE: Chris Chrashaw
VICE-PRESIDENT, ENGINEERING: Keith Miller
DIRECTOR, PUBLIC RELATIONS: Dave Franson

Company originally founded 1960 by Bill Lear Snr as Swiss American Aviation Corporation (SAAC); transferred to Kansas 1962 and renamed Lear Jet Corporation; prototype Lear Jet 23 (N801L) first flew 7 October 1963; Gates Rubber Company bought about 60 per cent share in 1967 and renamed it Gates Learjet Corporation, moving much of manufacturing to Tucson, Arizona; 64.8 per cent acquired by

Integrated Acquisition Inc September 1987 and renamed Learjet Corporation; all manufacturing moved from Tucson to Wichita during 1988, leaving customer service completion and modification centre in Tucson.

Acquisition of Learjet by Canada's Bombardier announced April 1990 and concluded 22 June 1990 for US$75 million; name changed to Learjet Inc; Bombardier assumed responsibility for Learjet's line of credit; now part of Bombardier Inc (which see under Canada). Learjet 31A, 45 and 60 assembled in Wichita and flown to Tucson for completion and delivery.

Over 2,200 Learjets built, including 105 LJ23s, 259 LJ24s, 368 LJ25s, five LJ28s, four LJ29s, 184 LJ31s, 673 LJ35s, 62 LJ36s, 50 LJ45s, 147 LJ55s and 162 LJ60s. 2,000th Learjet, Model 45 N158PH, delivered to Parker Hannifin Corporation in August 1999.

Learjet bought manufacturing and marketing rights and tooling of Aeronca thrust reversers, for application to Learjet and other aircraft, March 1989.

Total of 36 deliveries in 1994; 43 in 1995; 34 in 1996; 45 in 1997; 61 in 1998; 99 in 1999, 134 in 2000; 182 in 2001, and 60 in 2002.

UPDATED

LEARJET 31

TYPE: Business jet.

PROGRAMME: Learjet 31 introduced September 1987 following first flight of aerodynamic prototype (modified Learjet 35A) on 11 May 1987; first production aircraft (N311DF) used as systems testbed; FAA certification 12 August 1988. Learjet 31A and 31A/ER announced October 1990 to replace Learjet 31. Certified in 21 countries, including Argentina, Australia, Bermuda, Brazil, Canada, Czech Republic, Denmark, Germany, Grand Cayman, Guatemala, Indonesia, Italy, Japan, Luxembourg, Mexico, Namibia, Pakistan, Philippines, South Africa, Switzerland and USA. Total fleet time 400,000 hours by 1 October 2000. Will be superseded by Learjet 40 from 2004.

Learjet 31A (two Honeywell TFE731-2 turbofans) *(Jane's/Paul Jackson)*   NEW/0526981

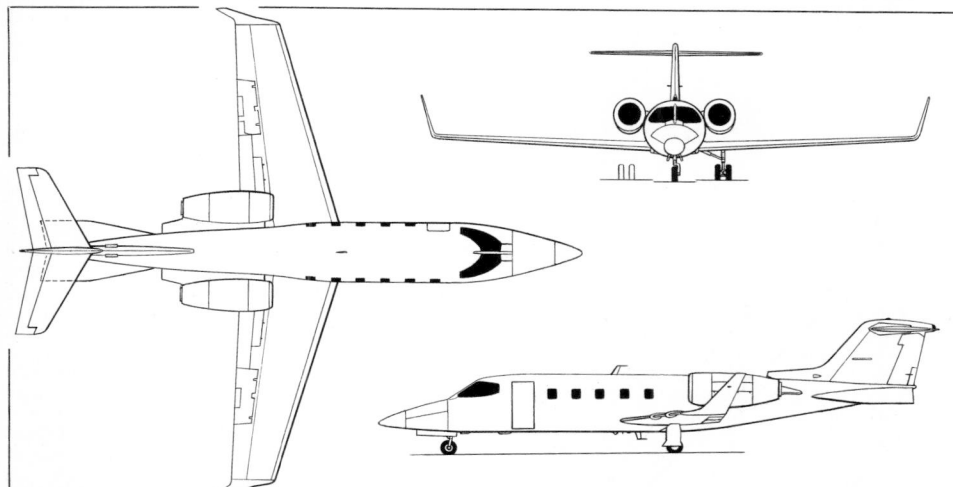

Learjet 31A business aircraft *(Jane's/Dennis Punnett)*

CURRENT VERSIONS: **Learjet 31:** Original version. No longer produced.

Learjet 31A: Current version.

Description applies to Learjet 31A, except where indicated.

Learjet 31A/ER: Optional extended-range version with 2,627 litres (694 US gallons; 578 Imp gallons) of fuel.

CUSTOMERS: Total of 205 built by June 2002, including 36 of original Learjet 31; 14 Learjet 31As delivered in 1994, 19 in 1995, 13 in 1996, 21 in 1997, 22 in 1998, 24 in 1999, 28 in 2000, 17 in 2001 and nine in 2002. Twenty-one were in the Bombardier FlexJets fleet in early 1999. 200th Learjet 31A delivered 3 October 2000, to Falcon Air Services of Phoenix, Arizona.

COSTS: US$6.419 million, equipped (2000).

DESIGN FEATURES: Small, rear-engined business jet; low wing with leading-edge sweepback; T tail and two large strakes below rear fuselage.

Original Learjet 31 combined fuselage/cabin and power plant of Learjet 35A/36A with wing of Learjet 55; delta fins added to eliminate Dutch roll, stabilise aircraft at high airspeeds, induce docile stall and reduce approach speeds and field lengths; stick pusher/puller and dual yaw dampers no longer required; stick shaker and single yaw damper retained for comfort.

Additional features of Learjet 31A include cruise Mach number up to 13,105 m (43,000 ft) increased 4 per cent to 0.81 and V$_{MO}$ increased from 300 kt (556 km/h; 345 mph) to 325 kt (602 km/h; 374 mph) IAS. Increases mainly benefit descent from high altitudes. Learjet 31A also features integrated digital avionics package.

Improvements announced at the NBAA Convention in Atlanta in October 1999, and incorporated from c/n 31A-194, include: increased maximum T-O weight of 7,711 kg (17,000 lb) and maximum landing weight of 7,257 kg (16,000 lb), latter available for retrofit on earlier aircraft; revised winglet design, based on that of Learjet 60, for enhanced high-speed/high-altitude performance and handling; thrust reversers standard; Universal UNS-1C FMS standard; lighter and more reliable AHRS; dual R134a air conditioning systems to provide increased capacity and redundancy, with one system dedicated to cockpit; and digital engine control for increased reliability and provision of trend monitoring.

FLYING CONTROLS: Conventional and manual. Ailerons have brush seals and geared tabs; electrically actuated trim tab in port aileron. Electrically actuated tailplane incidence control has separate motors for pilot and co-pilot and single-fault survival protection; aircraft can be manually controlled following tailplane runaway and landed with reduced flap. Rudder has electric trim tab; automatic electric rudder assist servo operates automatically if rudder pedal loads exceed 22.6 kg (50 lb). Full-chord fences bracket the ailerons. Single spoiler panel in each wing used as airbrake and lift dumper. Hydraulically actuated flaps extend to 40°. Optional drag parachute mounted on inside of baggage hatch under tail.

STRUCTURE: Semi-monocoque fuselage with eight-spar internal structure; multispar wing with machined skins; winglets have full-depth honeycomb core bonded to skin.

LANDING GEAR: Retractable tricycle gear; main legs retract inward, nose leg forward; twin-wheel main units with anti-skid disc brakes; nosewheel has full-time digital steer-by-wire. Tyres 17.5×5.75-8 (12 ply) main; 18×4.4 (10 ply) nose. Ground turning radius about nosewheel 8.08 m (26 ft 6 in).

POWER PLANT: Two 15.56 kN (3,500 lb st) Honeywell TFE731-2-3B turbofans with N$_1$ digital electronic engine controller giving engine trend monitoring, automatic retention of power settings above 4,575 m (15,000 ft) and special idling control for descent from 15,545 m (51,000 ft). Engine synchroniser fitted. Dee Howard 4000 thrust reverser system standard. One integral fuel tank in each wing standard or ER tank in fuselage (see under Weights and Loadings for quantities); fuselage fuel transferred by gravity or pump; single-point pressure refuelling standard.

ACCOMMODATION: Cabin furnishings include a three-seat divan, four Erda 10-way adjustable individual seats in club seating arrangement, side-facing seat with lavatory, two folding tables, baggage compartment, overhead panels with reading lights, indirect lighting, air vents and oxygen masks. Revised interior offering more headroom and three cabin configurations announced October 1994 and introduced on 100th aircraft May 1995. Aft lavatory option introduced in 1996, providing an additional 0.28 m³ (10.0 cu ft) of baggage space. Optional external baggage locker increases total baggage volume to 1.76 m³ (62.0 cu ft).

SYSTEMS: Hydraulic system operates flaps, landing gear, airbrakes, wheelbrakes and thrust reversers; system pressure 69 to 120.6 bar (1,000 to 1,750 lb/sq in); pneumatic standby for gear extension and wheelbrakes. Normal cabin pressure differential 0.64 bar (9.4 lb/sq in) with automatic flood engine bleed if cabin altitude exceeds 2,820 m (9,250 ft); pop-out emergency oxygen for passengers and masks for crew. Electrical system based on two starter/generators, two lead-acid batteries and two inverters; both busses can run from one engine; electrics operate tailplane incidence, rudder assister and nosewheel steering. Anti-icing by bleed air for wing, engine intakes and windscreen; tailplane electrically heated; fin not protected. Alcohol spray for radome to stop shed ice entering engines; controls prevent internal ice and condensation during long descents.

AVIONICS: Honeywell integrated digital avionics package with five-tube EFIS 50, Universal UNS-1M flight management system (FMS), dual Series III nav/comms, and dual KFC 3100 autopilots/flight directors.

EQUIPMENT: Throttle-mounted landing gear warning mute and go-around switches, nacelle heat annunciator, engine synchroniser and synchroscope, recognition light, wing ice light, emergency press override switches, transponder ident switch in pilot's control wheel, flap preselect, crew

lifejackets, cockpit dome lights, cockpit speakers, crew oxygen masks and fire extinguisher are standard.

DIMENSIONS, EXTERNAL:
Wing span	13.36 m (43 ft 10 in)
Wing aspect ratio	7.2
Length overall	14.83 m (48 ft 8 in)
Max diameter of fuselage	1.63 m (5 ft 4 in)
Height overall	3.76 m (12 ft 4 in)
Tailplane span	4.48 m (14 ft 8½ in)

DIMENSIONS, INTERNAL:
Cabin: Length:	
incl flight deck: 31A	6.63 m (21 ft 9 in)
31A/ER	6.27 m (20 ft 7 in)
excl flight deck: 31A	5.21 m (17 ft 1 in)
31A/ER	4.85 m (15 ft 11 in)
Max width: 31A, 31A/ER	1.50 m (4 ft 11 in)
Max height: 31A, 31A/ER	1.32 m (4 ft 4 in)
Floor area, excl flight deck	3.60 m² (38.75 sq ft)
Volume, cockpit divider to rear pressure bulkhead	7.7 m³ (271 cu ft)
Baggage compartment volume: 31A	1.1 m³ (40 cu ft)
31A/ER	0.76 m³ (26.7 cu ft)
Optional external baggage locker	0.34 m³ (12.0 cu ft)

AREAS:
Wings, basic	24.57 m² (264.5 sq ft)
Horizontal tail surfaces (total)	5.02 m² (54.0 sq ft)
Vertical tail surfaces (total)	3.57 m² (38.4 sq ft)

WEIGHTS AND LOADINGS:
Weight empty	4,651 kg (10,253 lb)
Basic operating weight empty, typical	5,087 kg (11,214 lb)
Max payload	1,070 kg (2,360 lb)
Payload with max fuel: standard	878 kg (1,936 lb)
optional	865 kg (1,907 lb)
Max fuel weight:	
wing tanks: 31A	1,272 kg (2,804 lb)
31A/ER	1,282 kg (2,826 lb)
fuselage tank: 31A	599 kg (1,320 lb)
31A/ER	829 kg (1,827 lb)
total: 31A	1,871 kg (4,124 lb)
31A/ER	2,111 kg (4,653 lb)
Max T-O weight: 31A (standard)	7,711 kg (17,000 lb)
31A (optional)	8,028 kg (17,700 lb)
Max ramp weight: 31A (standard)	7,802 kg (17,200 lb)
31A (optional)	8,119 kg (17,900 lb)
Max landing weight	7,257 kg (16,000 lb)
Max zero-fuel weight	6,124 kg (13,500 lb)
Payload with max fuel: 31A (standard)	896 kg (1,976 lb)
31A (optional)	1,089 kg (2,400 lb)
Max wing loading:	
31A (standard)	313.8 kg/m² (64.27 lb/sq ft)
31A (optional)	326.7 kg/m² (66.92 lb/sq ft)
Max power loading:	
31A (standard)	248 kg/kN (2.43 lb/lb st)
31A (optional)	258 kg/kN (2.53 lb/lb st)

PERFORMANCE (A: standard max T-O weight, B: optional max T-O weight):
Max operating Mach No. (M$_{MO}$):	
FL275-FL430	M0.81
FL430-FL470	M0.81-0.79
above FL470	0.79
Max cruising speed	463 kt (857 km/h; 533 mph)
Typical cruising speeds at mid-cruise weight:	
at FL410	462 kt (856 km/h; 532 mph)
at FL430	456 kt (845 km/h; 525 mph)
at FL 450	447 kt (828 km/h; 514 mph)
at FL 470	435 kt (806 km/h; 501 mph)
Long-range cruising speed	420 kt (778 km/h; 483 mph)
Max speed with landing gear extended	260 kt (481 km/h; 299 mph)
Tyre limiting speed	182 kt (337 km/h; 209 mph)
Approach speed	131 kt (243 km/h; 151 mph) IAS
Stalling speed, flaps and landing gear down	96 kt (178 km/h; 111 mph) IAS
Max rate of climb at S/L: A	1,554 m (5,100 ft)/min
B	1,490 m (4,890 ft)/min
Rate of climb at S/L, OEI: A	494 m (1,620 ft)/min
B	462 m (1,515 ft)/min
Time to FL430	19 min 48 s
Max certified altitude	15,545 m (51,000 ft)
Service ceiling, OEI: A	7,986 m (26,200 ft)
B	7,620 m (25,000 ft)
T-O field length: A	1,064 m (3,490 ft)
B	1,158 m (3,800 ft)

Learjet 31A instrument panel   0016192

Learjet 45, twin-turbofan business jet   *NEW*/0526918

FAR Pt 91 landing distance: A, B   874 m (2,866 ft)
Range with two crew and four passengers:
   VFR: 31A   1,654 n miles (3,063 km; 1,903 miles)
       31A/ER   1,882 n miles (3,485 km; 2,166 miles)
   IFR reserves:
       31A   1,259 n miles (2,331 km; 1,448 miles)
       31A/ER   1,488 n miles (2,755 km; 1,712 miles)
OPERATIONAL NOISE LEVELS (FAR Pt 36):
T-O   81.9 EPNdB
Approach   92.8 EPNdB
Sideline   86.9 EPNdB
                                     UPDATED

LEARJET 45 and 45�Xᴿ

TYPE: Business jet.

PROGRAMME: Design started September 1992; unveiled at NBAA Convention 20 September 1992. Other members of Bombardier group are involved; Learjet is responsible for project co-ordination, final assembly, testing and certification.

Engine testing began January 1995 with TFE731-20 installed in one nacelle of Learjet 31A testbed. Wing and fuselage of first production aircraft (N45XL) mated at Wichita 4 November 1994; first flight 7 October 1995; second prototype (N452LJ) first flown 6 April 1996, assigned to flutter testing; third aircraft (N453LJ), first flown 24 April 1996, assigned to avionics testing; fourth aircraft assigned to HIRF and lightning-strike testing, engine testing and fuel system operation; fifth aircraft fitted with production interior and assigned to function and reliability testing, including interior noise measurement.

Initial FAA certification granted 22 September 1997 followed by full approval in May 1998; JAA certification achieved 3 July 1998; FAA RVSM approval granted 25 July 2000. First customer delivery (N903HC) 28 July 1998, to Hytrol Conveyor of Jonesboro, Arkansas. First delivery to Europe (N459LJ) 8 September 1998 for Eifel Holdings Ltd of Jersey. More than 50 delivered by February 2000, including 2,000th Learjet, N158PH, to Parker Hannifin Corporation in August 1999. 100th Learjet 45 delivered in October 2000, initially as Bombardier company demonstrator, but destined for a Philippine operator; 200th delivered in April 2002.

CURRENT VERSIONS: **Learjet 45**: Current version, *as described.*

   Learjet 45�Xᴿ: Improved version offering enhanced payload/range capability. Announced at Farnborough International Air Show, 21 July 2002 and formally launched at NBAA Convention, Orlando, Florida, 10 September 2002; certification and service entry scheduled for mid-2003. Features include a 454 kg (1,000 lb) increase in maximum T-O weight; 15.57 kN (3,500 lb st) Honeywell TFE731-20-BR turbofans flat rated at ISA +25°C (104°F) with 0.67 kN (150 lb st) automatic power reserve at T-O; and redesigned cabin interior with seats that are 5 cm (2 in) wider and increase legroom by 15 cm (6 in), and an LED lighting system. Learjet 45�Xᴿ maximum T-O weight increase and engine upgrade will be retrofittable to existing Learjet 45s.

CUSTOMERS: Total of 200 delivered by April 2002; deliveries began with seven in 1998, followed by 43 in 1999, 71 in 2000, 63 in 2001 and 33 in 2002. Major orders include 40, with 10 options, by JetSolutions for its FlexJets fractional ownership scheme. Recent customers include Singapore Airlines, which ordered two in April 2001, supplementing four aircraft delivered in 1998, Cathay Pacific Airways, which has ordered two for crew training, operated by BAE Systems Flight Training (Australia) Pty Ltd at Adelaide, South Australia, of which the first was delivered on 24 August 2001 with the second scheduled for delivery in 2002, and Gold Air International which ordered five 45�Xᴿs on 5 February 2003.

COSTS: Learjet 45 US\$9.125 million, Learjet 45�Xᴿ US\$9.7 million (both 2002).

DESIGN FEATURES: Generally as Model 31; Learjet 45 designed to combine docile handling characteristics of 31/31A and 60 with exceptional fuel efficiency and good overall performance, and offer increased maintainability and reliability; new larger fuselage, wing and tail unit; increased head and shoulder room; wing carry-through spar recessed beneath floor; latest technology systems. Wing designed with NASA; winglets; sweepback 13° 25′ at 25 per cent chord.

Performance enhancement package announced at the Paris Air Show in June 1999 includes maximum T-O weight increased by 136 kg (300 lb); reductions in T-O speeds; improved nosewheel steering and removal of 40 kt (74 km/h; 46 mph) steering system limitation; improved brake-by-wire effectiveness; reconfiguration of flap selector module to permit use of 8° flap setting, instead of 20°, for approach climb in the event of a go-around; improved climb performance with bleed air anti-icing systems operating; and updated Honeywell avionics software.

FLYING CONTROLS: Conventional and manual. Two spoiler/lift dumper panels in each wing. Horn-balanced elevators. Trim tab in rudder; two in each aileron. Flaps.

Further improvements introduced in late 2000 include restyled seats providing greater freedom of movement in the cabin, a 10 to 12 dB reduction in cabin noise levels and improvements in the cabin air distribution system.

STRUCTURE: Unigraphics, CATIA and Computervision digital design systems adopted by Learjet, de Havilland of Canada and Shorts for engineering design. Short Brothers of UK manufactures the fuselage and de Havilland Canada the wings.

LANDING GEAR: Retractable tricycle type; semi-articulated trailing-link main legs retract inward, nose leg forward; twin-wheel main units, size 22×5.75-12 (10 ply) with brake-by-wire anti-skid carbon multidisc brakes; nosewheel has dual-chine tyre, size 18×4.4 (10 ply), with steer-by-wire.

POWER PLANT: Two Honeywell TFE731-20-AR turbofans, each flat rated at 15.57 kN (3,500 lb st) at ISA +16°C (61° F); Dee Howard target-type thrust reversers; digital electronic engine control. Total fuel capacity 3,426 litres (905 US gallons; 754 Imp gallons).

ACCOMMODATION: Two crew and up to nine passengers; eight-passenger cabin typically with PMP fully adjustable swivelling and reclining seats in double club arrangement, galley/refreshment centre and storage cabinet at front of cabin; lavatory, doubling as optional ninth seat, at rear; cabin pressurised, maximum differential 0.65 bar (9.43 lb/ sq in); clamshell door with integral steps on port side at front of cabin, upper part serves as emergency exit; eight cabin windows per side, one forward of starboard wing leading-edge serving as emergency exit; externally accessible heated and lined baggage compartment, capacity 227 kg (500 lb), in aft fuselage accessed via door on port side beneath engine nacelle.

Learjet 45 flight deck with Honeywell Primus 1000 integrated avionics system
0016184

Learjet 45 cabin
0016185

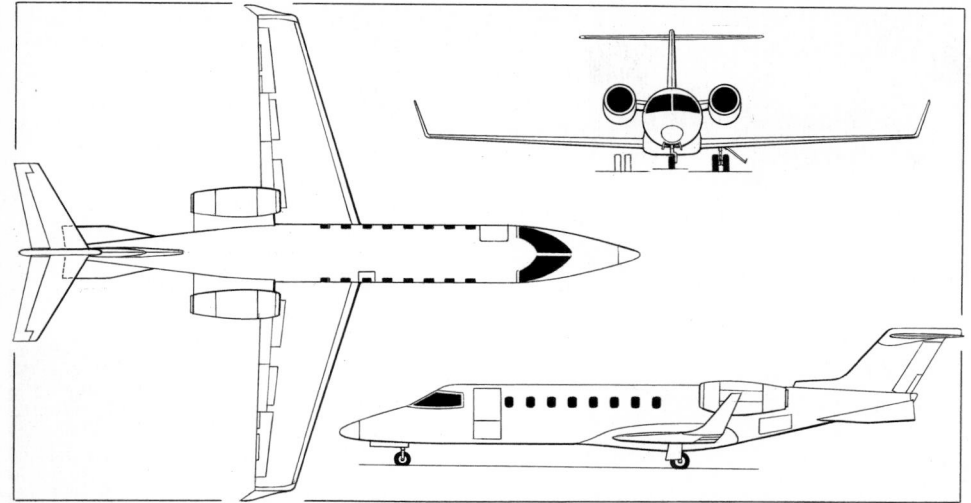

Learjet 45 business jet (two Honeywell TFE731-20-AR turbofans) *(Jane's/Dennis Punnett)*

SYSTEMS: Honeywell air conditioning and pressurisation system, maximum differential 0.65 bar (9.4 lb/sq in), with dual independent digital control system and pneumatic redundancy; dual-zone automatic temperature control. Gaseous oxygen system, pressure 127.6 bar (1,850 lb/sq in). Main and auxiliary back-up hydraulic systems, pressure 207 bar (3,000 lb/sq in). Dual independent anti-icing and de-icing systems comprising bleed air anti-icing on wing and tailplane leading-edges and engine inlets, electric de-icing on pitot-static probes and electric anti-icing and de-fogging on windscreen. Honeywell RE 100 APU optional.

AVIONICS: Honeywell Primus 1000 integrated avionics system.

Comms: Dual Primus II nav/ident radios.

Radar: Primus 660 weather radar standard.

Flight: Dual Primus II nav radios. Primus 1000 digital autopilot/flight director standard; Honeywell traffic-alert and collision-avoidance (TCAS II) optional.

Instrumentation: Primus 1000 with EICAS; dual PFDs and MFDs; flight and navigation information displayed on four 203 × 178 mm (8 × 7 in) EFIS screeens; heart of system is IC-600 integrated avionics computer, which combines EFIS and EICAS processor.

DIMENSIONS, EXTERNAL:
Wing span	14.56 m (47 ft 9¼ in)
Wing aspect ratio	7.3
Length overall	17.56 m (57 ft 7¼ in)
Max diameter of fuselage	1.75 m (5 ft 9 in)
Height overall	4.31 m (14 ft 1½ in)
Wheel track	2.84 m (9 ft 4 in)
Wheelbase	7.87 m (25 ft 9¾ in)

DIMENSIONS, INTERNAL:
Cabin:
Length: incl flight deck	7.52 m (24 ft 8¼ in)
excl flight deck	6.02 m (19 ft 9 in)
Max width	1.55 m (5 ft 1 in)
Max height	1.50 m (4 ft 11 in)
Floor area, excl flight deck	5.76 m² (62.0 sq ft)
Volume, excl flight deck	11.6 m³ (410 cu ft)
Baggage compartment volume	1.4 m³ (50 cu ft)

AREAS:
Wings, basic	28.95 m² (311.6 sq ft)

WEIGHTS AND LOADINGS (A: Learjet 45, B: Learjet 45XR):
Basic operating weight, typical: A	6,212 kg (13,695 lb)
B	6,230 kg (13,734 lb)
Max payload: A	1,046 kg (2,305 lb)
B	1,028 kg (2,266 lb)
Payload with max fuel: A	450 kg (993 lb)
B	886 kg (1,954 lb)
Max fuel weight: A, B	2,750 kg (6,062 lb)
Max T-O weight: A	9,298 kg (20,500 lb)
B	9,752 kg (21,500 lb)
Max landing weight: A, B	8,709 kg (19,200 lb)
Max ramp weight: A	9,412 kg (20,750 lb)
B	9,866 kg (21,750 lb)
Max zero-fuel weight: A, B	7,257 kg (16,000 lb)
Max wing loading: A	321.2 kg/m² (65.79 lb/sq ft)
B	336.9 kg/m² (69.00 lb/sq ft)
Max power loading: A	299 kg/kN (2.93 lb/lb st)
B	313 kg/kN (3.07 lb/lb st)

PERFORMANCE:
Max operating speed (VMO)	330 kt (611 km/h; 379 mph)
Max operating Mach No. (MMO)	0.81
High cruising speed: A, B	464 kt (859 km/h; 534 mph)
Normal cruising speed: A, B	457 kt (846 km/h; 526 mph)
Long-range cruising speed: A, B	430 kt (796 km/h; 495 mph)
Time to FL430 after MTOW departure: A	23 min 6 s
B	24 min 54 s
Max certified altitude: A, B	15,545 m (51,000 ft)
T-O field length: A	1,326 m (4,350 ft)
B	1,542 m (5,060 ft)
FAR Pt 91 landing distance: A, B	811 m (2,660 ft)

Range with two crew and four passengers, IFR reserves:
A	2,102 n miles (3,892 km; 2,419 miles)
B	2,098 n miles (3,885 km; 2,414 miles)

Range with two crew and eight passengers, IFR reserves:
B	2,007 n miles (3,717 km; 2,309 miles)

OPERATIONAL NOISE LEVELS:
T-O: A	74.4 EPNdB
B	75.5 EPNdB
Sideline: A	85.2 EPNdB
B	85.1 EPNdB
Approach: A, B	93.4 EPNdB

UPDATED

LEARJET 40

TYPE: Business jet.

PROGRAMME: Announced at Farnborough International Air Show 21 July 2002 as replacement for Learjet 31A. Prototype (N40LX, converted from Learjet 45 prototype) first flight 31 August 2002, immediately followed by first production aircraft (45-2001/N40LJ) 5 September. Public debut (N40LJ) and formal launch at NBAA Convention, Orlando, Florida, 10 September 2002; prototype and first production aircraft undertaking flight test programme, leading to FAA and Transport Canada certification in third quarter 2003, followed by JAA certification and service entry in first quarter 2004.

COSTS: Approximately US$7.5 million (2002).

DESIGN FEATURES: Generally as for Learjet 45, but fuselage shortened by 0.62 m (2 ft 0½ in) forward of wing, three cabin windows removed and capacity of fuselage fuel tank reduced.

POWER PLANT: As Learjet 45.

ACCOMMODATION: Two crew and up to seven passengers in forward club arrangement with galley on starboard side at front of cabin and full-width lavatory at rear. Compared with Learjet 45, cabin features redesigned seats that are 5 cm (2 in) wider and increase legroom by 15 cm (6 in), and an LED lighting system. The first 40 aircraft will feature an optional racing car-inspired red and black custom interior reflecting a sponsorship agreement between Bombardier and Indianapolis Motor Speedway Corporation.

DIMENSIONS, EXTERNAL:
Length overall	16.92 m (55 ft 6 in)

DIMENSIONS, INTERNAL:
Cabin: Length: incl flight deck	6.92 m (22 ft 8¼ in)
excl flight deck	5.41 m (17 ft 9 in)
Floor area, excl flight deck	5.17 m² (55.7 sq ft)
Volume, excl flight deck	10.3 m³ (363 cu ft)

WEIGHTS AND LOADINGS:
Basic operating weight	6,184 kg (13,633 lb)
Payload: max	1,074 kg (2,367 lb)
with max fuel	756 kg (1,667 lb)

Learjet 40 cabin interior *(Bombardier Aerospace)*   *NEW*/0526919

First production Learjet 40 *(Jane's/Paul Jackson)*   *NEW*/0526980

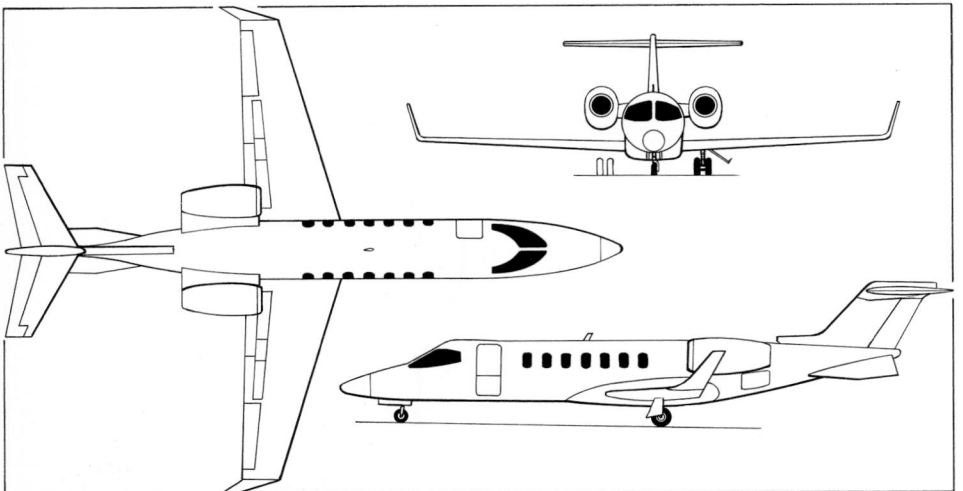

Learjet 40 business jet (*Jane's/James Goulding*)   *NEW*/0526908

Fuel weight: max	2,404 kg (5,300 lb)
with max payload	2,087 kg (4,600 lb)
Max ramp weight	9,344 kg (20,600 lb)
Max T-O weight	9,230 kg (20,350 lb)
Max landing weight	as Learjet 45
Max zero-fuel weight	as Learjet 45
Max wing loading	318.9 kg/m² (65.31 lb/sq ft)
Max power loading	296 kg/kN (2.91 lb/lb st)

PERFORMANCE:

Max operating Mach No	as Learjet 45
Cruising speed: high	as Learjet 45
normal	as Learjet 45
long-range	430 kt (796 km/h; 495 mph)
Time to FL430 at MTOW	22 min 42 s
Max certified altitude	as Learjet 45
T-O balanced field length	1,306 m (4,285 ft)
FAR Pt 91 landing distance	as Learjet 45
Range with two crew and four passengers, IFR reserves:	
	1,803 n miles (3,339 km; 2,074 miles)

OPERATIONAL NOISE LEVELS: As Learjet 45.

NEW ENTRY

LEARJET 60

TYPE: Business jet.

PROGRAMME: Announced 3 October 1990 as Learjet 55C successor; first flight of proof-of-concept aircraft with one PW305 turbofan 18 October 1990; flight testing resumed 13 June 1991 with two PW305s and stretched fuselage (more than 300 hours flown by May 1992); first production aircraft first flight (N601LJ) 15 June 1992; certification awarded 15 January 1993; deliveries started immediately. Certified in Argentina, Austria, Bermuda, Brazil, Canada, China, Denmark, Germany, Grand Cayman, Italy, Luxembourg, Malaysia, Mexico, the Philippines, South Africa, Switzerland, Turkey, the United Arab Emirates and USA. Production increased from two to three per month in 2000.

CUSTOMERS: Total of 250 in service by June 2002, including 22 delivered in 1994, 24 in 1995, 23 in 1996, 24 in 1997, 32 in 1998, 32 in 1999, 35 in 2000, 29 in 2001 and 18 in 2002. Avolar made commitment to purchase 10, and placed 12 options, in December 2001.

COSTS: US$11.584 million (2000).

DESIGN FEATURES: Largest Learjet; otherwise generally as for earlier versions.

FLYING CONTROLS: As Learjet 31A. Spoilers can be partially extended to adjust descent rates.

LANDING GEAR: Retractable tricycle type, hydraulically actuated and electrically controlled; single wheel on nose unit and twin wheels on main units; nosewheel retracts forward, mainwheels inward; steerable nosewheel (±60°). Mainwheel tyre size 17.5×5.75-8, pressure 14.76 bar (214 lb/sq in); nosewheel tyre size 18×4.4 (10 ply), pressure 7.24 bar (105 lb/sq in).

POWER PLANT: Two Pratt & Whitney Canada PW305A turbofans with FADEC, each flat rated at 20.46 kN (4,600 lb st) at up to 27°C (80°F). Fuel details under Weights and Loadings below. Single-point pressure refuelling system standard; gravity-feed fuel filler ports in each wing.

ACCOMMODATION: Two crew and up to 10 passengers; gross pressure cabin volume 15.57 m³ (550 cu ft); compared with 55C, main cabin is 0.71 m (2 ft 4 in) longer and rear baggage hold section 0.38 m (1 ft 3 in) longer; full-across aft lavatory has flat floor, large mirror, wardrobe and external servicing; total 1.67 m³ (59 cu ft) baggage capacity divided between an externally accessible hold (larger than that of Learjet 55C) and internal pressurised, heated compartment that is accessible in flight; galley cabinet has warming oven, cold liquid dispensers, ice compartment and storage for dinnerware; entertainment centre; 10-way adjusting seating is standard. New interior

introduced in 1998 (from c/n 115), featuring redesigned passenger service unit, wider headliner, revised beverage cabinet, optional television monitor and wider choice of upholstery fabrics and leather trim.

SYSTEMS: Environmental control system uses engine bleed air for air conditioning and pressurisation, maximum differential 0.65 bar (9.4 lb/sq in), maintaining sea level cabin altitude to 7,835 m (25,700 ft) and 2,440 m (8,000 ft) cabin altitude to 15,545 m (51,000 ft). Hydraulic system, pressure 103.4 bar (1,500 lb/sq in), provided by two engine-driven, variable volume pumps, with electrically driven auxiliary pump. Electrical system comprises two 30 V DC 400 A engine-driven starter-generators and two 24 V DC batteries. Oxygen system, capacity 2.18 m³ (77.0 cu ft), with demand-type masks for crew and drop-down constant-flow masks for passengers. Bleed air anti-icing on wing leading-edges, engine inlets, engine spinners and inlet guide vanes, electric anti-icing on windscreen. Hamilton Sundstrand T-20G-10C3A APU.

AVIONICS: Standard fully integrated all-digital Rockwell Collins Pro Line 4.

Radar: Rockwell Collins WXR840 colour weather radar.

Flight: Four-tube (152 × 178 mm; 6 × 7 in) EFIS, dual digital air data computers, dual navigation and communications radios, UNS-1C FMS, dual automatic

AHRS, Rockwell Collins AMS-850 avionics management system, advanced autopilot and long-range navaid as standard; circuit breaker and control panels redistributed, as in Learjet 31A.

DIMENSIONS, EXTERNAL:

Wing span	13.34 m (43 ft 9 in)
Wing chord: at root	2.74 m (9 ft 0 in)
at tip	1.12 m (3 ft 8 in)
Wing aspect ratio	7.2
Length: overall	17.89 m (58 ft 8¼ in)
fuselage	17.02 m (55 ft 10 in)
Fuselage max diameter	1.96 m (6 ft 5 in)
Height overall	4.44 m (14 ft 6¾ in)
Tailplane span	4.48 m (14 ft 8½ in)
Wheel track	2.51 m (8 ft 3 in)
Wheelbase	7.73 m (25 ft 4½ in)
Cabin door: Width	0.64 m (2 ft 1 in)
Height to sill	0.69 m (2 ft 3 in)

DIMENSIONS, INTERNAL:

Cabin: Length: incl flight deck	7.07 m (23 ft 2½ in)
excl flight deck	5.38 m (17 ft 8 in)
Max width	1.80 m (5 ft 11 in)
Max height	1.73 m (5 ft 8 in)
Floor area, excl flight deck	6.40 m² (68.9 sq ft)
Volume, excl flight deck	12.8 m³ (453 cu ft)

AREAS:

Wings, gross	24.57 m² (264.5 sq ft)
Horizontal tail surfaces (total)	5.02 m² (54.00 sq ft)
Vertical tail surfaces (total)	4.79 m² (51.53 sq ft)

WEIGHTS AND LOADINGS:

Weight empty	6,364 kg (14,030 lb)
Basic operating weight empty, typical	
	6,689 kg (14,746 lb)
Max payload	1,022 kg (2,254 lb)
Payload with max fuel	496 kg (1,094 lb)
Max usable fuel weight	3,588 kg (7,910 lb)
Max T-O weight	10,659 kg (23,500 lb)
Max ramp weight	10,773 kg (23,750 lb)
Max landing weight	8,845 kg (19,500 lb)
Max wing loading	433.8 kg/m² (88.85 lb/sq ft)
Max power loading	260 kg/kN (2.55 lb/lb st)

PERFORMANCE:

Max operating speed (V$_{MO}$):	
S/L-2,440 m (8,000 ft)	
	300 kt (555 km/h; 345 mph) IAS
2,440-6,100 m (8,000-20,000 ft)	
	340 kt (629 km/h; 391 mph) IAS
6,100-7,010 m (20,000-23,000 ft)	
	340-330 kt (630-611 km/h; 391-378 mph) IAS
7,010-8,155 m (23,000-26,750 ft)	
	330 kt (611 km/h; 378 mph) IAS
Max operating Mach No. (M$_{MO}$):	
FL265-FL370	0.81

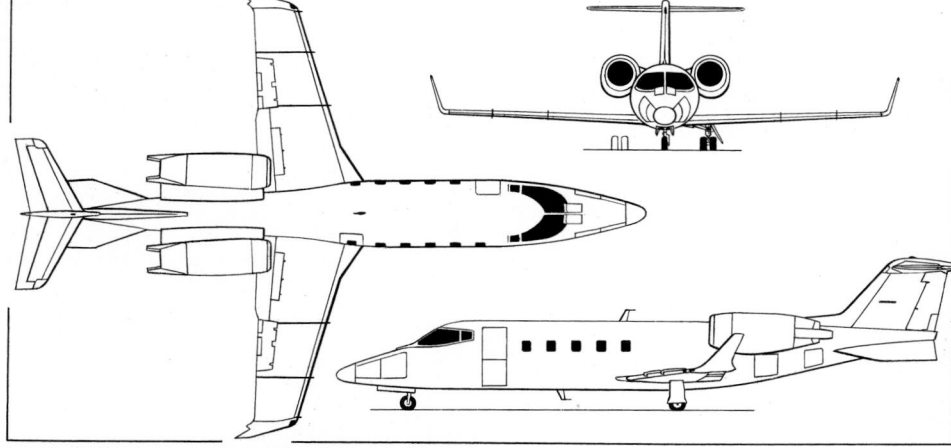

Learjet 60 business transport (*Jane's/Dennis Punnett*)

Learjet 60 (Pratt & Whitney PW305 turbofans) (*Jane's/Paul Jackson*)   *NEW*/0526979

FL370-FL430	0.81-0.78		
above FL430	0.78		
Cruising speed: high	464 kt (859 km/h; 534 mph)		
normal	457 kt (846 km/h; 526 mph)		
long-range	422 kt (782 km/h; 486 mph)		

Stalling speed, flaps and landing gear down
106 kt (197 km/h; 122 mph) CAS
Approach speed   139 kt (257 km/h; 160 mph) IAS
Max rate of climb at S/L   1,371 m (4,500 ft)/min

Rate of climb at S/L, OEI   378 m (1,240 ft)/min
Time to FL410 after MTOW departure   18 min 30 s
Max certified altitude   15,545 m (51,000 ft)
Service ceiling, OEI   7,195 m (23,600 ft)
T-O balanced field length   1,661 m (5,450 ft)
FAR Pt 91 landing distance   1,043 m (3,420 ft)
Range with two crew and four passengers:
VFR   2,685 n miles (4,972 km; 3,089 miles)
NBAA IFR   2,496 n miles (4,622 km; 2,872 miles)

OPERATIONAL NOISE LEVELS:
T-O   78.9 EPNdB
Sideline   83.2 EPNdB
Approach   87.7 EPNdB
UPDATED

Learjet 60 cabin   0051813

Learjet 60 flight deck
0016183

LEGEND

LEGEND AIRCRAFT INC

PO Box 11, Winnsboro, Louisiana 71295
Tel: (+1 318) 435 44 01
e-mail: lrundell@3g.quik.com
Web: http://www.turbinelegend.com
PRESIDENT: Lanny Rundell

Lanny Rundell of Southern Air purchased the assets of Performance Aircraft in February 2002.
NEW ENTRY

LEGEND LEGEND and LEGEND TURBINE LEGEND

TYPE: Tandem-seat sportplane kitbuilt/tandem-seat turboprop sportplane kitbuilt.
PROGRAMME: Prototype (N620L) first flown by Performance Aircraft Inc in 1996 with Chevrolet V-8 engine; subsequently converted to Turbine Legend with Walter M 601E turboprop; public debut in turboprop form at EAA Sun 'n' Fun at Lakeland, Florida, in April 1999.
CURRENT VERSIONS: **Legend:** Piston-engined version, powered by one 429 kW (575 hp) liquid-cooled Chevrolet V-8 driving a three-blade Hartzell propeller via a Geschwender 2:1 reduction gearbox. Wing span 8.22 m (27 ft 0 in). Ventral (P-51-type) air scoop.
Turbine Legend: Turboprop version, *as described*.
JC 100: Designation of Turbine Legend built in 2000 by Toys 4 Boys at Deland, Florida.
CUSTOMERS: At least eight under construction, of which seven flying by December 2002.
COSTS: Fast-build kit, less engine US$119,700 (2003). Engine (new) US$104,000 (1999).
DESIGN FEATURES: Highly streamlined, low-wing monoplane with sweptback tail surfaces and steeply raked windscreen. Tapered wings and mid-mounted tailplane. Optional winglets. Quoted build time: 2,500 hours, standard; 1,900 hours fast-build.
FLYING CONTROLS: Conventional and mechanical; horn-balanced elevators and rudder; three-axis electric trim; electrically actuated slotted flaps, maximum deflection 38°.
STRUCTURE: Mostly CFRP.
LANDING GEAR: Retractable tricycle type; mainwheel size 15×6.0-6, nosewheel size 5.00×5; mainwheels retract

Performance Turbine Legend built by Toys 4 Boys at Deland, Florida, and designated JC 100
(Jane's/Paul Jackson)   ***NEW***/0533702

inwards, nosewheel rearwards; steering by differential braking; dual disc brakes standard.
POWER PLANT: One Walter M 601 turboprop, rated at 540 kW (724 shp) for take-off and 490 kW (657 shp) maximum continuous, driving an Avia V 508E/84 three-blade, constant-speed, reversible-pitch feathering propeller. Fuel contained in integral tanks in outer wings 379 litres (100 US gallons; 83.3 Imp gallons); optional tip tanks, combined capacity 95 litres (25.0 US gallons; 20.8 Imp gallons).
ACCOMMODATION: Two persons in tandem under rear-hinged, upward-opening canopy with fixed single-piece windscreen. Dual controls standard.

DIMENSIONS, EXTERNAL:
Wing span   8.69 m (28 ft 6 in)
Wing aspect ratio   8.0
Length overall   7.84 m (25 ft 8½ in)
Height overall   2.86 m (9 ft 4¼ in)
Propeller diameter   2.13 m (7 ft 0 in)
DIMENSIONS, INTERNAL:
Cabin: max width   0.75 m (2 ft 5½ in)
AREAS:
Wings, gross   9.38 m² (101.0 sq ft)
WEIGHTS AND LOADINGS:
Weight empty   930 kg (2,050 lb)
Baggage capacity   54 kg (120 lb)

Max T-O weight   1,496 kg (3,300 lb)
Max wing loading   159.5 kg/m² (32.67 lb/sq ft)
Max power loading   3.06 kg/kW (5.02 lb/shp)
PERFORMANCE:
Never-exceed speed (VNE)   347 kt (643 km/h; 400 mph)
Max level speed   309 kt (573 km/h; 356 mph)
Max cruising speed at 7,620 m (25,000 ft)
290 kt (537 km/h; 334 mph)
Econ cruising speed at 7,620 m (25,000 ft)
245 kt (454 km/h; 282 mph)
Manoeuvring speed   261 kt (482 km/h; 300 mph)
Stalling speed: flaps up   73 kt (136 km/h; 84 mph)
landing configuration   66 kt (123 km/h; 76 mph)
Max rate of climb at S/L   1,981 m (6,500 ft)/min
Service ceiling   10,670 m (35,000 ft)
T-O and landing run   244 m (800 ft)
Range: with standard fuel and reserves:
at max cruising speed
821 n miles (1,520 km; 944 miles)
at econ cruising speed
993 n miles (1,839 km; 1,142 miles)
with optional fuel
1,207 n miles (2,237 km; 1,390 miles)
g limits   +6/–4
UPDATED

LEZA

LEZA AIRCAM CORPORATION

1 Leza Drive, Sebring, Florida 33870
Tel: (+1 863) 655 42 42
Fax: (+1 863) 655 03 10
e-mail: aircam@ct.net
Web: http://www.lezaaircam.com
CEO AND PRESIDENT: Antonio Leza

Leza AirCam has a 3,900 m² (42,000 sq ft) factory at Sebring Airport and also produces the Drifter and Super Drifter series of single- and two-seat ultralights (for details see Maxair Drifter, in *Jane's* 1992-93).

UPDATED

LEZA AIRCAM

TYPE: Two-seat kitbuilt twin.
PROGRAMME: Initially designed as camera platform for *National Geographic* magazine and first flew 1994. Redesigned as a kit and made public debut at Sun 'n' Fun 1996. First float-equipped version made initial flight 9 June 1999.

Leza AirCam with camera-carrying passenger (*Jane's/Paul Jackson*)   NEW/0533700

CUSTOMERS: More than 120 kits sold by mid-2002; over 40 flying by early 2001. Export destinations include Australia, Namibia and Zimbabwe.

COSTS: US$52,435 including Rotax 582 engines; US$69,950 with Rotax 912 S engines (2001).

DESIGN FEATURES: Configured for slow flight. Strut- and wire-braced wings and mid-mounted tailplane wire-braced to tall, narrow-chord fin.

FLYING CONTROLS: Conventional and manual. Full-width in-flight-adjustable trim tab on elevator, Pushrod-actuated ailerons taper at tips. Electric flaps.

STRUCTURE: All-metal monocoque fuselage with glass fibre cockpit enclosure. Constant chord, swept-tip wings of 6061-T6 aluminium tubing with aluminium spars and ribs, and fabric covering. Low-drag wing struts. Tailfin has glass fibre fillet and fabric-covered rudder and elevator with metal trim tab.

LANDING GEAR: Tailwheel type; fixed. Spring steel mainwheel legs; differential hydraulic disc brakes and 6.00-6 in tyres. Maule steerable tailwheel. Full Lotus floats optional.

POWER PLANT: Two 47.8 kW (64.1 hp) Rotax 582s or two 73.5 kW (98.6 hp) Rotax 912 ULSs or mounted above wings, driving three-blade ground-adjustable Warp Drive pusher propellers. Optionally two 85.8 kW (115 hp) turbocharged Rotax 914s can be fitted. Fuel in two wing-mounted aluminium tanks, total capacity 106 litres (28.0 US gallons; 23.3 Imp gallons).

ACCOMMODATION: Pilot and passenger/observer in tandem seats within open cockpit. Lexan windshield. Dual controls. Open baggage compartment behind second seat.

AVIONICS: To customer's specification. Options include Terra T X 760D COM, TRT 250D transponder.

EQUIPMENT: Fire extinguisher standard.

DIMENSIONS, EXTERNAL:
Wing span	11.07 m (36 ft 4 in)
Wing aspect ratio	6.5
Length overall	8.23 m (27 ft 0 in)
Height overall	2.54 m (8 ft 4 in)
Tailplane span	3.96 m (13 ft 0 in)
Wheel track	2.59 m (8 ft 6 in)
Min distance between propeller blade tips	0.15 m (6 in)

DIMENSIONS, INTERNAL:
Baggage compartment: Length	1.07 m (3 ft 6 in)
Max width	0.63 m (2 ft 0¾ in)

AREAS:
Wings, gross	18.95 m² (204.0 sq ft)

WEIGHTS AND LOADINGS (Rotax 912):
Weight empty	472 kg (1,040 lb)
Max T-O weight	762 kg (1,680 lb)
Max wing loading	40.21 kg/m² (8.24 lb/sq ft)
Max power loading	5.19 kg/kW (8.52 lb/hp)

PERFORMANCE (Rotax 912):
Never-exceed speed (VNE)	95 kt (177 km/h; 110 mph)
Max operating speed	87 kt (161 km/h; 100 mph)
Normal cruising speed	43-87 kt (80-161 km/h; 50-100 mph)
Stalling speed, power off, flaps down	34 kt (63 km/h; 39 mph)
Max rate of climb at S/L	610 m (2,000 ft)/min
Rate of climb at S/L, OEI	91 m (300 ft)/min
Service ceiling	5,485 m (18,000 ft)
T-O run	less than 61 m (200 ft)
Landing run	91 m (200 ft)
Range at 57 kt (106 km/h; 70 mph)	295 n miles (547 km; 340 miles)
Endurance	6 h
g limits	±6

UPDATED

LIBERTY

LIBERTY AEROSPACE INC

1500 East Oak Grove, Montrose, Colorado 81401
Tel: (+1 970) 626 57 80
Fax: (+1 970) 249 40 97
e-mail: sales@libertyaircraft.com
Web: http://www.libertyaircraft.com
PRESIDENT AND CEO: Anthony Tiarks
VICE-PRESIDENT, SALES AND MARKETING: Ivan Shaw

NORTH AMERICAN AND REST-OF-WORLD SALES OFFICE
Liberty Aerospace Inc, 801 International Parkway, 5th Floor, Orlando, Florida 32746
Tel: (+1 407) 562 19 66
Fax: (+1 407) 562 17 66

EUROPEAN SALES OFFICE
Liberty Aerospace PLC, Kirby Mills Industrial Estate, Kirkbymoorside, North Yorkshire YO62 6NR, United Kingdom
Tel: (+44 1751) 43 17 73
Fax: (+44 1751) 43 17 06

AUSTRALASIA AND ASIA SALES OFFICE
Liberty Aircraft Company Pty Ltd, 2 Third Street, Moorabbin Airport, PO Box 2750, Cheltenham, Victoria 3192, Australia
Tel: (+61 3) 95 80 76 67
Fax: (+61 3) 95 87 65 24

UPDATED

LIBERTY XL-2

TYPE: Two-seat lightplane.

PROGRAMME: Design, by the team that created the Europa kitbuilt (which see in UK section), began in 1997. Announced 26 May 2000; mockup displayed at NBAA Convention in New Orleans and AOPA-USA Convention in Long Beach during October 2000, when Scaled Technology Works announced as manufacturing partner. 'Official' first flight of prototype (N202XL) 2 April 2001, followed by public debut at EAA Sun 'n' Fun at Lakeland, Florida later that month; second prototype (N203XL) appeared at 2002 Sun 'n' Fun fitted with a redesigned engine cowling. FAR Pt 23 certification was anticipated in September 2002, (but had not been notified up to late December), followed shortly thereafter by JAR-VLA certification and first customer deliveries in early 2004. Will be certified for IFR. Target production rate up to 400 aircraft per year.

CURRENT VERSIONS: **XL-2:** Standard version, adoption of Teledyne Continental IOF-240 engine announced 21 July 2001. *As described.*

XL-2R: Proposed future version with Rotax 912 S flat-four.

CUSTOMERS: Launch customer Civil Flying School at Moorabbin Airport, Melbourne, Australia, ordered two in July 2000; Bill Crokaris of Melbourne has ordered one.

Second prototype Liberty XL-2 two-seat light trainer flight testing a new design of engine cowling in mid-2002   NEW/0533716

First 50 aircraft (for Founders Club members, at special price of US$85,000); total of 70 sold by April 2002.

COSTS: US$116,500 with Continental engine (2002).

DESIGN FEATURES: Based on Europa, but with modifications to optimise airframe for mass production; cabin is 10 cm (4 in) wider than that of Europa. Design goals included low price, efficient high-speed cruise with STOL performance, economy and ease of maintenance, advanced structural materials offering strength and durability, excellent handling with positive stability, and ability to store at home on a purpose-designed transporter.

FLYING CONTROLS: Conventional and manual, via pushrods. Dual controls and adjustable rudder pedals standard. All-moving tailplane with electrically actuated 1.3:1 geared anti-balance/trim tab; mass-balanced ailerons with differential action, maximum deflection +24/–20°; electrically actuated slotted flaps occupying 70 per cent span and 30 per cent chord, maximum deflection 27°; rudder maximum deflection ±30°.

STRUCTURE: Welded 4130 steel tube forward fuselage and centre-section to carry engine, landing gear and wing attachment loads, with modular carbon fibre skin; flying surfaces comprise riveted subassembly with bonded aluminium skin; single-spar wings have optional folding facility, quick connect attachment to centre-section and self-connecting flap and aileron controls; de-riggable tailplane optional. Scaled Technology Works will undertake manufacturing, final assembly and flight testing at Montrose, Colorado.

LANDING GEAR: Non-retractable tricycle type; aluminium main legs; tubular steel nose leg; steering via castoring nosewheel and differential brakes. Cleveland wheels and hydraulic disc brakes. Tyre size 5.00×5.

POWER PLANT: One 93 kW (125 hp) Teledyne Continental IOF-240 flat-four with Aerosance PowerLink FADEC driving a Sensenich two-blade metal propeller, in standard version. Alternatively, one 73.5 kW (98.6 hp) Rotax 912 S flat-four, driving a purpose-designed two-blade, fixed-pitch carbon fibre Dowty propeller via 2.43:1 reduction gearing. Fuel capacity 106 litres (28 US gallons; 23.3 Imp gallons).

ACCOMMODATION: Two persons, side by side. Baggage compartment, with cargo net, behind seats, maximum capacity 45 kg (100 lb). Upward-hinged door on port side. Cabin heating.

SYSTEMS: Electrical system powered by 14 V 60 A engine-driven alternator. 14 V battery and 14 V auxiliary power outlet standard.

AVIONICS: Optional Garmin and UPS (Apollo) packages.

Basic VFR: Garmin GNC 250XL VFR GPS/com and GTX 327 transponder/altitude encoder; or UPS Apollo GX 65 GPS/com and SL 70 transponder/altitude encoder.

Basic IFR: Garmin GNS 430 GPS/com/nav/map, GNC 300XL IFR GPS/com, GTX 327 transponder/altitude encoder, GMA 340 audio panel, and GI 102A and GI 106A indicators; or UPS (Apollo) GX 60 GPS/com, SL 30 nav/com/GPS, SL 70 transponder/altitude encoder, SL 15M audio panel, ACU-14V annunciator and Mid-Continent MD200-306 indicator.

Deluxe IFR: as Basic IFR with the addition of second Garmin GNS 430 and substitution of GI 105A indicator for GI 102A; or with addition of UPS (Apollo) MX20 MFD. ELT standard on aircraft for US market, optional elsewhere.

Instrumentation: Vision Microsystems VM1000FX digital engine information display. Sensitive altimeter, magnetic compass, airspeed indicator, vertical speed indicator, electric tachometer, trim position indicator, flap position indicator, electric stall warning, annunciator panel, fuel contents gauge, fuel pressure gauge, manifold pressure gauge, percentage power gauge, OAT gauge, CHT gauge, EGT gauge, oil temperature and pressure gauges, voltmeter, ammeter and quartz clock.

EQUIPMENT: Standard equipment includes tinted windows, windscreen defrost, four-point safety harnesses, cloth upholstery, map/storage pockets, map shelf, baggage restraint net, static port and alternate static port, wingtip navigation lights, anti-collision beacon, tiedown points, fuel quick-drain and white paint finish. Optional equipment includes electric gyro package comprising artificial horizon, direction indicator and turn co-ordinator; pitot heat; speed kit comprising nose- and mainwheel fairings and flap hinge fairings; landing/taxi/instrument/ cabin flights; deluxe leather interior; pearlescent exterior paint; choice of two exterior decal designs; and towbar.

DIMENSIONS, EXTERNAL:

Wing span	8.53 m (28 ft 0 in)
Wing aspect ratio	7.0
Length overall	6.25 m (20 ft 6 in)
Height overall	2.24 m (7 ft 4 in)
Propeller diameter	1.60 m (5 ft 3 in)

DIMENSIONS, INTERNAL:

Cabin max width	1.22 m (4 ft 0 in)
Baggage volume	0.68 m³ (24.0 cu ft)

AREAS:

Wings, gross	10.41 m² (112.0 sq ft)

WEIGHTS AND LOADINGS (Continental engine):

Weight empty, VFR	476 kg (1,050 lb)
Baggage capacity	45 kg (100 lb)
Max T-O weight	749 kg (1,653 lb)
Max wing loading	72.1 kg/m² (14.76 lb/sq ft)
Max power loading	8.05 kg/kW (13.22 lb/hp)

PERFORMANCE:

Never-exceed speed (V$_{NE}$)	180 kt (333 km/h; 207 mph)
Max cruising speed	132 kt (244 km/h; 152 mph)
Econ cruising speed, 55% power	120 kt (222 km/h; 138 mph)
Manoeuvring speed	102 kt (189 km/h; 117 mph)
Stalling speed: flaps up	52 kt (97 km/h; 60 mph)
flaps down	45 kt (84 km/h; 52 mph)
Max rate of climb at S/L	350 m (1,150 ft)/min
T-O to 15 m (50 ft)	381 m (1,250 ft)
T-O and landing run	229 m (750 ft)
Range with max fuel and 55% power with reserves	500 n miles (926 km; 575 miles)

UPDATED

LITTLE WING

LITTLE WING AUTOGYROS INC
746 Highway 89 North, Mayflower, Arkansas 72106
Tel: (+1 501) 470 74 44
Fax: (+1 501) 470 74 87
e-mail: rotopup@aol.com
Web: http://www.flygyro.com/littlewing
PRESIDENT: Ronald Herron

The company is developing a series of autogyros; the original Roto-Pup will be joined by a new design, the tricycle landing gear, open cockpit **Skylark**.

UPDATED

LITTLE WING ROTO-PUP

TYPE: Single-seat autogyro kitbuilt; two-seat autogyro kitbuilt.

PROGRAMME: Initial design work began in 1980 and construction of LW-1 prototype (N45LW), using a Piper PA-11 fuselage, started December 1990; this made its first flight 21 October 1994. LW-2, scaled-down version of LW-1, first flew April 1995; LW-3, controlled by full universal tilting of rotor head, first flew (N46LW) 22 September 1996. Certified by FAA in Experimental category.

CURRENT VERSIONS: **Roto-Pup LW-2:** Controlled by elevator, rudder and laterally tilting rotor head.

Roto-Pup LW-3: Controlled by rudder and universally tilting rotor head. Single seat.

Roto-Pup LW-3+2: Tandem two-seat version. Two flying by end 2000.

Description applies to LW-3, unless otherwise stated.

Roto-Pup LW-5: Tandem-seat, shorter version. Landing gear moved 24 cm (9½ in) forward to compensate for 61 cm (2 ft 0 in) reduction in length. Flown solo from rear seat.

Roto-Pup Ultralight: First flew April 1995. McCulloch O-100 engine; fuselage uncovered to remain within 115 kg (254 lb) ultralight limit. Controlled by elevator, lateral tilting rotor head and rudder.

CUSTOMERS: Four prototypes undertook flight test programme. Total of 130 sets of plans also sold. Two customer kit-built examples have also flown, including one

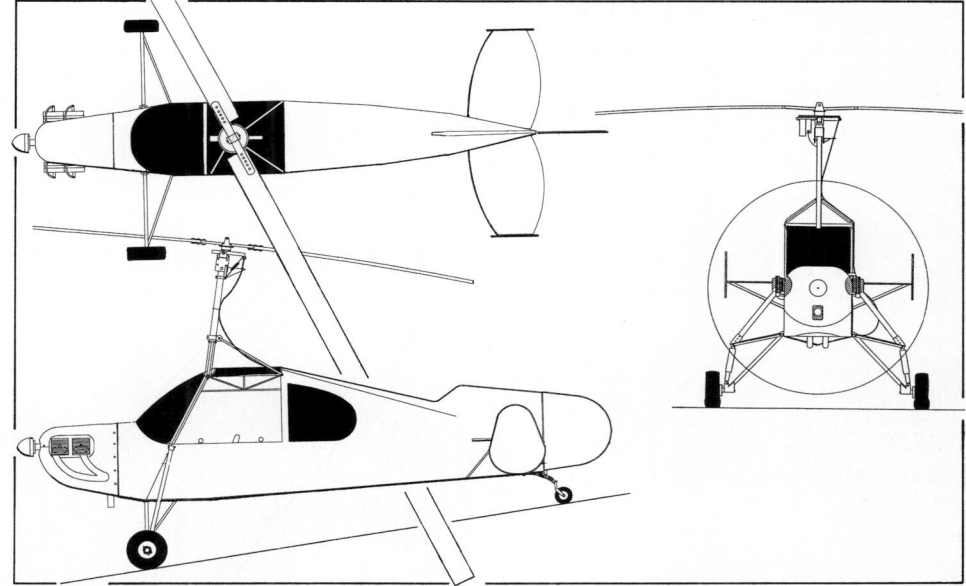
Little Wing LW-5 general arrangement   0059969

Rotax 914-engined LW-5 which achieved 1,609 km 869 n mile; 1,000 mile cross-country flight in June 2002. Others nearing completion in Belgium, France and Japan, as well as USA.

COSTS: LW-3 kit approximately US$10,000; LW-3+2 US$15,000; Ultralight US$8,995 including engine, rotor system and instruments (1999).

DESIGN FEATURES: Fuselage design based on Piper PA-11 Cub Special, giving classic autogyro appearance, but scaled-down and using Pratt trusses in place of Cub's Warren trusses. Two-blade Rotor Flight Dynamics rotor; each blade is aluminium bonded and has positive twist. Rotor speed 390 rpm, blade tip speed 140 m (459 ft)/s. (On LW-2, pitch link to airframe is isolated by elastomeric

dampers, determining rotor axis inclination relative to airframe). Rotor mast dampened by large rubber bushes at aft pylon braces. Rotor brake optional, but recommended on all versions.

FLYING CONTROLS: Floor-mounted control stick; cable and pulley actuated elevator in LW-2; LW-3 has no elevators, but floor-mounted control column, pushrods and universally tilting rotor head for pitch and roll, plus cable-actuated rudder. Ground-adjustable horizontal stabiliser for airframe pitch; tip shields at ends of horizontal stabilisers. Push-pull cables used on two-seat versions.

STRUCTURE: Full chromoly 4130 steel tube welded fuselage and rotor pylon; fuselage covered with Dacron fabric and polyurethane finish. Hub structure of 2024-T3 aluminium bar with bolt attachments, pivoted at rotor attachment point for ground adjustment of blade pitch.

LANDING GEAR: Non-retractable type with tailwheel. Mainwheels Hegar 6.50-6 in with 1.03 bar (15 lb/sq in) tyre pressure. Matco 3.00-2 steerable tailwheel of solid

Little Wing LW-2 Roto-Pup   0106522

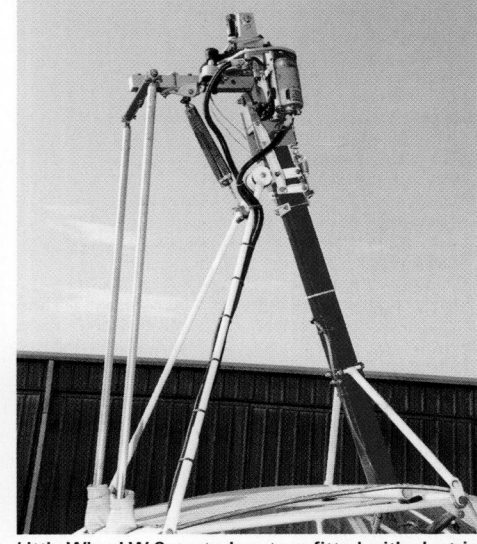
Little Wing LW-3 control system, fitted with electric pre-rotator and centrifugal teeter stops   0059972

rubber. Internal expanding go-kart style brakes by Leaf. Optional shock-absorbing gear. Floats optional.

POWER PLANT: *LW-3:* One 52.2 kW (70 hp) TEC converted Volkswagen four-stroke engine with dual ignition; options include 2si 52.2 kW (70 hp) two-stroke water-cooled engine and various Rotax, Subaru and Hirth engines. Fuel capacity 38 litres (10.0 US gallons; 8.3 Imp gallons) in seat tank. Refuelling position on starboard side of cabin. Oil capacity 2.4 litres (5.0 US pints; 4.2 Imp pints).

LW-3+2: One 82.0 kW (110 hp) Hirth F30 or Rotax 912; or 84.6 kW (113.4 hp) Rotax 914 optional power plants in the range 67.1 to 112 kW (90 to 150 hp) can be fitted. Fuel capacity 49.2 litres (13.0 US gallons; 10.8 Imp gallons).

ACCOMMODATION: One or two occupants, according to version, in enclosed cockpit; forward-opening door on starboard side. Baggage compartment behind single seat. Lexan polycarbonate windscreen and side windows.

AVIONICS: Micronair 760 nav/com and EIS engine electronic monitoring system are recommended.

SYSTEMS: 1,000 A 12 V battery for ignition back-up and rotor pre-spin. Hydraulic pre-spin optional.

DIMENSIONS, EXTERNAL:

Rotor diameter: LW-2/3	7.62 m (25 ft 0 in)
LW-3+2	8.53 m (28 ft 0 in)
Rotor blade chord	0.18 m (7 in)
Fuselage length: standard	5.49 m (18 ft 0 in)
LW-5	4.88 m (16 ft 0 in)
Height to top of rotor head	2.59 m (8 ft 6 in)
Tail unit span	2.13 m (7 ft 0 in)
Wheel track	2.13 m (7 ft 0 in)
Wheelbase: standard	3.96 m (13 ft 0 in)
LW-5	3.51 m (11 ft 6 in)
Propeller diameter: VW engine	1.57 m (5 ft 2 in)
2si engine	1.73 m (5 ft 8 in)

DIMENSIONS, INTERNAL:

Cabin: Length: standard	1.24 m (4 ft 1 in)
LW-5	1.90 m (6 ft 3 in)
Max width: standard	0.57 m (1 ft 10½ in)
LW-5	0.66 m (2 ft 2 in)
Max height	1.07 m (3 ft 6 in)

AREAS:

Rotor blades (each)	0.54 m² (5.83 sq ft)
Rotor disc: LW-2/3	45.62 m² (491.1 sq ft)
LW-3+2	57.21 m² (615.8 sq ft)
Dorsal fin	0.42 m² (4.52 sq ft)
Auxiliary tip plates (each)	0.15 m² (1.60 sq ft)
Rudder	0.28 m² (3.03 sq ft)
Tailplane	0.81 m² (8.80 sq ft)
Elevators (total)	0.84 m² (9.00 sq ft)

WEIGHTS AND LOADINGS:

Weight empty: LW-2	160 kg (352 lb)
LW-3+2	215 kg (475 lb)
LW-3	204 kg (450 lb)
Baggage capacity	11 kg (25 lb)
Max fuel weight	24 kg (54 lb)
Max T-O weight: LW-2	340 kg (750 lb)
LW-3+2	499 kg (1,100 lb)
Max disc loading: LW-3+2	8.72 kg/m² (1.79 lb/sq ft)

PERFORMANCE:

Never-exceed speed (VNE)	86 kt (160 km/h; 100 mph)
Max operating speed at S/L:	
LW-3	70 kt (129 km/h; 80 mph)
LW-3+2	78 kt (145 km/h; 90 mph)
Econ cruising speed at 305 m (1,000 ft):	
LW-3	52 kt (97 km/h; 60 mph)
LW-3+2	65 kt (121 km/h; 75 mph)
Touchdown speed for power-off landing	
	9-13 kt (16-24 km/h; 10-15 mph)
Max rate of climb at S/L: LW-3	305 m (1,000 ft)/min
LW-3+2	183 m (600 ft)/min
Service ceiling: standard	3,050 m (10,000 ft)
with Rotax 914	7,315 m (24,000 ft)
T-O run: LW-3	61 m (200 ft)
LW-3+2	92 m (300 ft)
Landing run: LW-3, LW-3+2	3-6 m (10-20 ft)
Range: LW-3	100 n miles (185 km; 115 miles)
LW-3+2	150 n miles (277 km; 172 miles)

UPDATED

Little Wing LW-5 tandem-seat kitbuilt   *NEW*/0059974

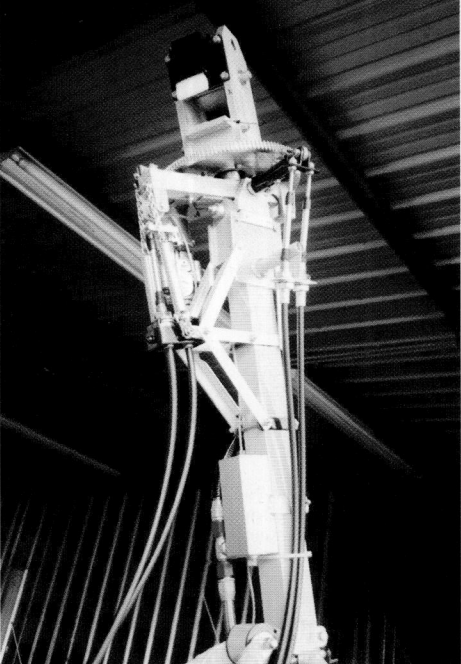

Little Wing LW-3+2 control system using redundant push-pull cables, hydraulic pre-rotation, centrifugal teeter stops and spring-loaded lateral pivot which absorbs vibration   0059973

LOCKHEED MARTIN

LOCKHEED MARTIN CORPORATION

6801 Rockledge Drive, Bethesda, Maryland 20817
Tel: (+1 301) 897 60 00
Fax: (+1 301) 897 60 28
Web: http://www.lmco.com
CHAIRMAN AND CEO: Vance D Coffman
PRESIDENT AND COO: Robert J Stevens
VICE-PRESIDENT AND CHIEF FINANCIAL OFFICER:
   Christopher E Kubasik
EXECUTIVE VICE-PRESIDENT, AERONAUTICS COMPANY:
   Dain M Hancock
EXECUTIVE VICE-PRESIDENT, SPACE SYSTEMS:
   Albert E Smith
EXECUTIVE VICE-PRESIDENT, SYSTEMS INTEGRATION:
   Robert B Coutts
EXECUTIVE VICE-PRESIDENT, TECHNOLOGY SERVICES:
   Michael F Camardo
SENIOR VICE-PRESIDENT, CORPORATE COMMUNICATIONS:
   Dennis Boxx

Former Lockheed Aircraft Corporation renamed Lockheed Corporation in September 1977. Merger with Martin Marietta announced 30 August 1994 and completed 15

March 1995. Further expansion resulted from the acquisition of Loral Corporation's defence electronics and systems integration businesses in April 1996 for approximately US$9.1 billion and completed purchase of Comsat in August 2000 for US$2.6 billion. Workforce total of about 125,000 at end of 2001, with sales of US$24.0 billion. Activities include design and production of aircraft, electronics, satellites, space systems, missiles, ocean systems, information systems, and systems for strategic defence and for command, control, communications and intelligence.

Following major strategic and organisational review, revised corporate structure came into being in early 2000; under this, four major business groups of Lockheed Martin are:

Aeronautics Company (see next entry)
Space Systems, comprising:
   Lockheed Martin Space Systems-Astronautics Operations
   Lockheed Martin Space Systems-Michoud Operations
   Lockheed Martin Space Systems-Missiles & Space Operations
   Lockheed Martin Commercial Space Systems
   Lockheed Martin Management & Data Systems
   International Launch Services
Systems Integration, comprising:
   Lockheed Martin Air Traffic Management

   Lockheed Martin Systems Integration-Owego
   Lockheed Martin Canada
   Lockheed Martin Distribution Technologies
   Lockheed Martin Information Systems
   Lockheed Martin Tactical Systems
   Lockheed Martin Missiles and Fire Control
   Lockheed Martin Mission Systems
   Lockheed Martin Naval Electronics & Surveillance Systems-Akron
   Lockheed Martin Naval Electronics & Surveillance Systems-Surface Systems
   Lockheed Martin Naval Electronics & Surveillance Systems-Undersea Systems
   Lockheed Martin Naval Electronics & Surveillance Systems-Syracuse
   Lockheed Martin UK
Technology Services, comprising
   Knolls Atomic Power Laboratory
   Lockheed Martin Aircraft & Logistics Centers
   Lockheed Martin Information Support Services
   Lockheed Martin Space Operations
   Lockheed Martin Technical Operations
   Sandia National Laboratories
   Technology Ventures

UPDATED

LOCKHEED MARTIN AERONAUTICS COMPANY (LM Aero)

1 Lockheed Boulevard, Fort Worth, Texas 76108
Tel: (+1 817) 777 20 00
Fax: (+1 817) 777 21 15
PRESIDENT AND COO: Dain M Hancock
EXECUTIVE VICE-PRESIDENT FOR CUSTOMER REQUIREMENTS:
   Tom Burbage

EXECUTIVE VICE-PRESIDENT FOR PROGRAMS: Robert T Elrod
EXECUTIVE VICE-PRESIDENT FOR FINANCE: John C McCarthy
EXECUTIVE VICE-PRESIDENT FOR OPERATIONS: Ralph D Heath

LM Aero includes the following operating units:
Lockheed Martin Aeronautics Company-Marietta
   Next entry

Lockheed Martin Aeronautics Company-Palmdale
   Follows Lockheed Martin Aeronautics Company-Marietta
Lockheed Martin Aeronautics Company-Fort Worth
   Follows Lockheed Martin Aeronautics Company-Palmdale
Before merger that culminated in creation of Lockheed Martin Aeronautics Sector (subsequently renamed Lockheed

Martin Aeronautical Systems Division and, since January 2000, Lockheed Martin Aeronautics Company), Lockheed aircraft manufacturing activity consolidated at Marietta, Georgia in 1991. Sales in 2001 were valued at US$5.4 billion.

Teaming arrangement agreed with Northrop Grumman to collaborate in development and marketing of AEW aircraft;

also with KAI (formerly Samsung) to develop and co-produce the T/A-50 Golden Eagle advanced trainer/light combat aircraft (see entry in Korea, South, section), in which it has 13 per cent share of programme investment and responsibility for about 20 per cent of production work.

On 31 October 2001, Lockheed Martin and AgustaWestland announced a joint venture promoting the EHI EH101 helicopter (which see) for US military and coast guard requirements under the new designation, US101.

UPDATED

LOCKHEED MARTIN AERONAUTICS COMPANY – MARIETTA

86 South Cobb Drive, Marietta, Georgia 30063-0264
Tel: (+1 770) 494 44 11
Fax: (+1 770) 494 76 56
Web: http://www.lockheedmartin.com
MANAGER, MEDIA RELATIONS: Sam Grizzle

In April 1991, Lockheed won competition to produce F-22 (now F/A-22) with General Dynamics (now Lockheed Martin Tactical Aircraft Systems) and (then) Boeing Military Airplanes. Lockheed Martin Aeronautics Company-Marietta also involved in studies into advanced mobility aircraft for 21st century strategic transport; was subcontractor to Boeing on NASA High-Speed Civil Transport (HSCT) programme; working with NASA on advanced subsonic technology transport programme; and engaged in other, classified projects.

Further long-term activities at Marietta include production of C-130J Hercules and F/A-22 Raptor aircraft. LMAS signed joint venture agreement with Alenia of Italy in September 1996 concerning development and marketing of the C-27J tactical transport (see entry in Italian section) that incorporates systems developed for the C-130J Hercules.

Lockheed Martin Aeronautics Company-Marietta working to develop advanced avionics suite for future versions (upgrades) of P-3 Orion and AEW aircraft; open architecture design for maximum flexibility and system growth; elements include AN/APS-145 radar, GPS and communications/navigation system. Also interest from several countries on possible new-build aircraft.

Aeronautical Systems Support, with headquarters in Smyrna, Georgia, is subordinate element, with responsibility for after sales support, including provision of spare parts and technical back-up for variety of aircraft types.

UPDATED

LOCKHEED MARTIN (645) F/A-22 RAPTOR

TYPE: Multirole fighter.
PROGRAMME: US Air Force Advanced Tactical Fighter (ATF) requirement called for 750 McDonnell Douglas F-15 Eagle replacements incorporating low observables technology and supercruise (supersonic cruise without afterburning); parallel assessment of two new power plants; request for information issued 1981; concept definition studies awarded September 1983 to Boeing, General Dynamics, Grumman, McDonnell Douglas, Northrop and Rockwell; requests for proposals issued September 1985; submissions received by 28 July 1986; USAF selection announced 31 October 1986 of demonstration/validation phase contractors: Lockheed YF-22 and Northrop YF-23 (see 1991-92 *Jane's All the World's Aircraft*); each produced two prototypes and ground-based avionics testbed. Competing engine demonstration/validation programmes launched September 1983; ground testing began 1986-87; flight-capable Pratt & Whitney YF119s and General Electric YF120s ordered early 1988; all four aircraft/engine combinations flown.

Decision of 11 October 1989 extended evaluation phase by six months; draft request for engineering and manufacturing development (EMD) proposals issued

F/A-22 cockpit concept demonstrator at Marietta
0105079

April 1990; first artists' impressions released May 1990; Lockheed teamed with General Dynamics (Fort Worth) and Boeing Military Airplanes to produce two YF-22 prototypes, which first flew on 29 September and 30 October 1990. Details of flight testing, configuration and dimensions appeared in the 1996-97 and previous editions. Second aircraft placed in USAF Museum, Wright Patterson AFB, 31 March 1998.

Acquisition of former General Dynamics gave Lockheed Martin control of 67.5 per cent of programme; this involves 1,150 suppliers in more than 40 US states. Final engineering and manufacturing development (EMD) requests issued for both weapon system and engine 1 November 1990; proposals submitted 2 January 1991; F-22 and F119 power plant announced by USAF as winning combination, 23 April 1991; EMD contract given 2 August 1991 for 11 (later reduced to nine) flying prototypes, plus one static test article and one fatigue test airframe; design underwent several detail refinements through early 1990s, immediately previous layout being Configuration 644.

Combat roles reassessment of May 1993 added air-to-ground attack with precision-guided munitions (PGMs) to F-22's roles. Under US$6.5 million contract addition on 25 May 1993, main weapon bay and avionics adapted for delivery of AIM-9X missile and 454 kg (1,000 lb) GBU-32 Joint Direct Attack Munition (JDAM). Addition of ground attack capability eventually resulted in redesignation as the **F/A-22**, which was announced by USAF Chief of Staff General Jumper on 17 September 2002.

Target first flight date delayed some 12 months, until May 1997, due to three consecutive fiscal year budget cuts; preliminary design review, covering all aspects of the design, completed 30 April 1993; critical design review completed February 1995; preproduction verification (PPV) batch of four aircraft was scheduled to be ordered 1997 but these deleted from overall programme at start of that year; long lead items for first production batch (two aircraft), subsequently reclassified as Production Representative Test Vehicles (PRTV), ordered in 1998, with full funding in FY99. Minor design changes for production aircraft announced July 1991. Suggested name of SuperStar rejected in 1991 and it remained unnamed

until occasion of roll-out in April 1997 when it was announced that the name Raptor had been chosen.

Fabrication of first component for first EMD aircraft (91-4001, c/n 4001) began 8 December 1993 at Boeing's facility in Kent, Washington; assembly of forward fuselage launched at Marietta on 2 November 1995 with start of work on nose landing gear well; assembly work also begun at Fort Worth in mid-1995 with mating of three assemblies that comprise the mid-fuselage of first EMD aircraft taking place in early 1996, followed by road transfer of entire section to Marietta in September 1996 for start of final assembly. Delivery by Boeing to Lockheed Martin at Marietta of subassemblies, including wing and aft fuselage, for first EMD aircraft took place on schedule in third quarter of 1996. Pratt & Whitney also delivered first F119 flight test power plant in September. Prototype rolled out 9 April 1997; planned May 1997 first flight delayed by fuel leaks and hardware-related anomalies. First flight accomplished 7 September 1997, with aircraft airborne for 58 minutes during which initial handling evaluation undertaken before landing gear was retracted and further handling assessment performed at speeds of up to 250 kt (463 km/h; 288 mph); maiden flight also included simulated powered approach at medium altitude before landing at Marietta. Second sortie of 35 minutes took place on 14 September 1997, after which aircraft underwent minor structural modifications and was then placed in structural test fixture for load ground tests and strain gauge calibration.

Low-rate initial production (LRIP) decision originally dependent upon accumulating 183 hours of flight testing; this milestone passed on 23 November 1998 and cleared way for release of US$195.5 million in late December 1998 for advance procurement (long lead items) for six Lot 1 LRIP aircraft, which subsequently reclassified as PRTV 2 aircraft after wrangle over funding in mid-1999. Earlier, in December 1998, Lockheed Martin received contract worth US$503 million for two PRTVs and associated programme support.

Wind tunnel testing occupied 19,195 hours up to YF-22 stage and a further 16,930 hours up to mid-1995, when Configuration 645 was finalised; six major categories were investigated (aerodynamic loads and weapons bay acoustics, inlet and engine compatibility, mission/manoeuvre performance, inlet icing, stability and control flying qualities, weapons and stores separation), with last-named comprising majority of 900 hours that remained to be done in 1995-97.

Avionics trials from November 1997 in a Boeing 757 (N757A) Flying Test Bed (FTB) with AN/APG-77 radar in F-22-type nosecone; block 1 software, permitting basic radar operation including simultaneous search-and-track modes, delivered to Boeing, April 1998 and subsequently tested on 757; block 2 software delivered on 7 December 1998, with block 3S beginning flight testing on 24 April 2000; this is early version of block 3.0 software, which delivered to Seattle on 11 August 2000 in readiness for airborne trials programme that began in mid-September. Communications/navigation/identification (CNI) system and EW suite being tower-tested on full-scale model of forward fuselage at Fort Worth, Texas, during 1998-99; subsequently tested on Boeing 757 FTB, which is fitted with three common integrated processors (CIPs) for 1,400 hour software flight test programme. Further modification

Second (first to fly) production representative Lockheed Martin F/A-22 during its initial sortie on 16 September 2002   *NEW*/0522852

Raptor 10, the first production representative F/A-22 pictured on delivery to Lockheed Martin at Palmdale on 30 October 2002 for pre-delivery modifications *(USAF/Kevin Robertson)*   NEW/0527095

of Boeing 757 to mount representative wing section above forward fuselage completed in late 1998, and flight testing of conformal antennas began on 11 March 1999; by December 2000, 757 FTB had accumulated 641.9 flight test hours in 126 sorties. Radar testing also accomplished using T-39 Sabreliner as target aircraft and has involved a Lockheed T-33 for calibrated airborne trials since December 1998.

Milestones in 1999 included delivery of first AN/ALR-94 EW system on 15 February by Sanders to the Avionics Integration Laboratory in Seattle; 100th sortie in May; successful completion of design load limit testing of article 3999 on 25 September; and compliance with all five major 1999 flight test objectives (including flight at altitude of 50,000 ft; opening of side and main weapon bay doors in flight; supercruise demonstration; and flutter envelope expansion) by 24 September.

Flight test programme interrupted at least twice during 2000, with most serious occurrence arising in May after discovery of hairline cracks in canopies of first two aircraft; grounding order lifted on 5 June, when second EMD F-22 resumed testing with some restrictions while awaiting replacement unit. Major event of year was expected to be Pentagon Defense Acquisition Board review to culminate in award of contract for initial batch of LRIP aircraft; this was scheduled for 21 December, but slipped to 3 January 2001 and was further delayed by poor weather that prevented three of 11 critical test objectives being achieved; funding release dependent upon compliance with several 'exit criteria', including first flight of F-22 with block 3.0 software, first AMRAAM launch and first flights of aircraft 4004, 4005 and 4006. First AIM-120C AMRAAM launch (of 60 planned) achieved 24 October 2000; static tests completed 28 December 2000. Remaining three objectives satisfied by 6 February 2001, with first flight of aircraft 4006. However, decision to proceed with LRIP phase not taken until 15 August 2001, when initial batch of 10 aircraft approved with FY01 funds, to be followed by 13 more in FY02 and 23 in FY03; at same time, it was revealed that LRIP will continue until FY05, with full-rate production starting in FY06 and running until at least FY13.

Assembly of first LRIP aircraft (4018) began at Fort Worth on 19 March 2001; first flight and delivery to Tyndall AFB, Florida, expected in 2003. Notable milestone passed on 18 April 2001, when 1,000th flight test hour recorded with 2,000th hour completed on 7 June 2002. Nevertheless, delays in production and delivery of test aircraft meant programme was running behind schedule, with knock-on effect on start of dedicated initial operational test and evaluation (DIOT&E); this was scheduled to begin in August 2002, but is now to start in third quarter of 2003.

Block 3.1 software package delivered 5 February 2002 and flown for first time (on 91-4006) on 25 April at Edwards AFB.

CURRENT VERSIONS: (specific): Test programme involves total of nine EMD aircraft, plus two non-flying test articles. Airframe 3999 for static loads testing built between second and third flying F-22s, and airframe 4000 for fatigue testing built between third and fourth flying F-22s. Final assembly of 3999 began in July 1998; following completion in January 1999, it was transferred to the structural test facility at Marietta and began load testing in March 1999. All planned static testing completed by mid-May 2002, with 'first service-life' fatigue testing satisfactorily concluded on 17 May 2002, at which time second cycle of full lifetime testing began.

Clear division of test assignments resulted in three aircraft (4001-03) being allocated to airframe structure evaluations, with remaining six concentrating on avionics test taskings. Use of separate instrumentation configurations for airframe and avionics-dedicated test articles provides back-up for almost every aircraft and offers potential to switch missions between test fleet if necessary. Tasks allocated to EMD aircraft are:

4001/91-4001: 'Spirit of America'. Rolled out on 9 April 1997 at Lockheed Martin's Marietta, Georgia, facility and made first flight on 7 September 1997. Following completion of structural ground tests, disassembled and airlifted to Edwards AFB on 5 February 1998; used for evaluation of flying qualities, flutter and loads characteristics. Flown to Wright-Patterson AFB, Ohio, on 2 November 2000 and formally retired from flight test duty; 175 flights and 372.7 hours; stripped of useful components and used for live-fire testing, involving exploding shells and missile fragments, during 2001. Further details of this aircraft can be found in *Jane's All the World's Aircraft* 2001-02.

4002/91-4002: 'Old Reliable'. Rolled out 10 February 1998; maiden flight on 29 June 1998 at Marietta; flew to Edwards AFB on 26 August and by end October 1998 had made 27 flights (66.1 hours), expanding flutter and handling qualities envelope, including 26° AoA. By 8 September 2001, had completed 253 flights (including nine from Marietta and ferry flight to Edwards) and accumulated 538.8 hours. Used to launch first AIM-9M Sidewinder on 25 July 2000, followed by first AIM-120C AMRAAM on 24 October and had previously been employed for fit checks and captive-carry trials with AGM-88 HARM in April 1999. Tasks include performance assessment (propulsion; high AoA) plus stores separation and jettison as well as some electronic warfare and IR signature evaluations.

4003/91-4003: First Block 2 aircraft, with internal structure fully representative of production version; rolled out at Marietta on 25 May 1999; first engine runs in October 1999; taxi trials completed at beginning of March 2000, with first flight following on 6 March and delivery to Edwards AFB (fourth flight) on 15 March; regained flight status on 19 September 2000, being previously engaged on planned ground testing and upgrade programme. By 8 September 2001, had completed 68 flights and accumulated 135.2 hours. Earmarked for envelope expansion, replacing 4001, including loads testing, crosswind landings, validation of arrester hook and weapons bay environment work. Mid-fuselage section used for fit checks of Sidewinder and AMRAAM weapons at Fort Worth in July 1998 shortly before delivery to Marietta for final assembly. Accomplished first AIM-120 AMRAAM launch at supersonic speed on 21 August 2002, when single missile fired while flying at M1.2 at 3,650 m (12,000 ft). Will also test M61A2 cannon and JDAM integration.

4004/91-4004: First EMD aircraft with Hughes CIP software, including AN/APG-77 radar and ILS. Allocated to avionics development, but also to be used for low observables evaluation including radar cross-section and IR signature assessments; and comms/nav/ident (CNI) testing. Mid-fuselage section delivered from Fort Worth to Marietta on 28 December 1998. Block 1.1 avionics installed at Marietta, with electrical power applied for first time on 31 August 1999; Block 1.2 avionics then incorporated to support taxi trials and initial flight testing at Dobbins Air Reserve Base, from where first flight took place on 15 November 2000. Aircraft ferried to Edwards AFB at end January 2001. By 8 September 2001, had completed 48 flights and accumulated 119.0 hours.

4005/91-4005: Radar, CNI and armament development tasks. Assigned to initial testing of Block 3.0 software. Primary fire-control evaluation aircraft, with first

Third EMD F/A-22A launches an AIM-120 AMRAAM   NEW/0527093

flight made on 5 January 2001; by 8 September 2001, had completed 41 flights and accumulated 103.6 hours; made first guided launch of AIM-120C against target drone over Point Mugu test range on 21 September 2001.

4006/91-4006: Avionics development tasks. Primarily for integrated avionics testing, RCS testing and, eventually, for systems effectiveness/military utility evaluation. Expected to fly for first time in December 2000, but did not do so until 5 February 2001; by 8 September 2001, it had completed 16 flights and accumulated 33.0 hours.

4007/91-4007: Was to have been initial two-seat F-22B but completed as single-seat F-22A and assigned to integrated avionics testing. First flight made 15 October 2001, thereafter joining test fleet at Edwards AFB on 5 January 2002.

4008/91-4008: Allocated to avionics development tasks and also destined to validate observability specification data. Subsequently to be used for DIOT&E with two other EMD aircraft (4007 and 4009) and two PRTV aircraft (4010 and 4011). Flown for first time on 8 February 2002 and ferried to Edwards AFB on 31 May 2002.

4009/91-4009: Was to have been the second F-22B but completed as F-22A. Avionics development and observability trials. First flight originally planned for 1 June 2001, but retained for ground testing at Marietta, where formally delivered for maintenance trials 15 April 2002; first flight had not occurred by end October 2002; at that time, it was expected to fly in December 2002.

4010/98-4010 and **4011/98-4011:** First PRTV aircraft; to participate in DIOT&E and will also be used for additional service testing at Nellis AFB, Nevada from the third quarter of 2003. First flight of a PRTV aircraft (98-4011, the second example) took place at Marietta on 16 September 2002; was formally accepted by USAF on 26 November 2002, completing DIOT&E fleet. First PRTV aircraft (98-4010) made maiden flight on 12 October 2002; was officially delivered to the USAF at Marietta on 23 October; and transferred to Palmdale on 30 October 2002 for modifications before joining Detachment 6 at Edwards AFB. Both aircraft assigned to DIOT&E programme at Edwards AFB from mid-2003.

Trials by 4001 to 4009 were originally to occupy 4,337 hours in 2,409 sorties. Of these totals, 2,110 hours and 1,200 sorties dedicated to airframe and systems testing, with balance allocated to mission avionics testing; however, as part of an effort to make up for delays in test programme, amount of time allocated to avionics flight testing was cut to 1,530 hours in third quarter of 2001.

CURRENT VERSIONS (general): **F/A-22A:** Originally designated F-22A. Single-seat production version for USAF. Planned in-service date is December 2005.

F-22B: Projected two-seat version for USAF; development terminated 10 July 1996 to reduce overall programme costs.

FB-22: Proposed long-range strike-dedicated derivative under study by Lockheed Martin at USAF request in 2002; is larger than F/A-22, with greater combat radius and significantly bigger payload of 30 small diameter bombs rather than eight as on F/A-22. If goes ahead, likely to have 90 per cent avionics and 30 per cent structural commonality with F/A-22, but may feature different power plant, with General Electric F110 and Pratt & Whitney F135 also under consideration. Other changes include modified delta wing planform, insertion of fuselage plug to increase weapons bay size and carriage characteristics and compromised air-to-air combat capability, with lower *g* limit of 5 rather than 9 on F/A-22.

NATF: Projected US Navy variant to replace Grumman F-14 Tomcat; development abandoned.

CUSTOMERS: US Air Force: two YF-22 demonstrators; nine EMD aircraft plus one static and one fatigue test airframes;

original 648 production aircraft programme reduced to 442 in January 1994; latter originally to be funded from 1997 (long lead), beginning with four preproduction verification (PPV) aircraft in FY98, followed by series production of 438 (but PPV aircraft cancelled in early 1997). Quadrennial Defense Review (QDR) report in May 1997 resulted in planned procurement falling to 339 (including eight PRTV aircraft), although state of uncertainty still exists over total number to be acquired, with some projections calling for as few as 180 aircraft. Current plan anticipates funding for 295 aircraft, with additional F/A-22s up to 339 to be purchased if economies can be made. Contracts for two PRTV 1 aircraft and long lead items for batch of six PRTV 2 aircraft signed December 1998. Decision to authorise LRIP expected in December 2000 but delayed until 15 August 2001. Pilot training to be accomplished at Tyndall AFB, Florida, by 325th Fighter Wing from 2003, with IOC of first squadron following in December 2005, this being element of 1st Fighter Wing at Langley AFB, Virginia.

F/A-22 PROCUREMENT
(at 1 November 2002)

FY	Lot	Quantity
99	PRTV 1	2
00	PRTV 2	6
01	LRIP 1	10
02	LRIP 2	13
03	LRIP 3	20
04	LRIP 4	22
05	LRIP 5	24
Total		**97**

Note: Excludes two prototypes and nine EMD aircraft. FY04 and FY05 totals subject to approval and could change.

COSTS: US$818 million contracts to both ATF teams, October 1986, for 54-month studies; each airframe team investing own funds (Lockheed/Boeing/GD team investment totalled US$675 million in addition to DoD funding); each engine contractor, about US$50 million; total US$3,800 million spent by USAF on both ATFs up to April 1991; programme cost for 648 aircraft was US$13,000 million for development (1991 base year) and US$52,500 million for production (1994 then-year); programme acquisition cost US$162 million (1994); flyaway cost US$61.2 million at 1991 prices. EMD contract, 2 August 1991, comprised US$9,550 million for 11 (subsequently nine) airframes, plus US$1,375 million to P&W for 33 (later amended to 27) engines. FY94 US Congressional appropriation of US$2,100 million was US$163 million below expectations, resulting in slippage of critical design review and first flight. Similarly, FY95 appropriation of US$2,300 million was US$110 million below expected figure, leading to further delay in maiden flight; and on 9 December 1994, Defense Secretary William Perry announced 10 per cent cut (approximately US$210 million) in FY96 budget, necessitating a third restructuring of the programme. December 1996 announcement by USAF revealed that additional US$1.5 billion would be allocated for development phase and that total procurement cost of 438 aircraft by FY10 would be US$87 billion. Flyaway cost per aircraft stated to be US$92.6 million at same time.

Most recent cost figures (April 1999), in then-year dollars, adjusted for inflation, value total programme at US$62.6 billion, comprising US$22.7 billion for RDT & E (including Dem/Val and EMD), US$39.7 billion for production of 339 aircraft and US$0.2 billion for related construction projects at military bases. At that time, unit

flyaway cost of one production aircraft, including engines but excluding fuel and consumables, quoted as US$84.7 million; latter figure rises to US$184.0 million if RDT&E expenditures taken into account.

On 30 December 1999, Lockheed Martin awarded contract for approximately US$1.3 billion, covering procurement of six PRTV 2 aircraft, augmenting earlier US$195.5 million contract for long-lead items; at same time, Pratt & Whitney received separate US$180 million award for 12 F119 engines. Additional appropriation of US$277.1 million allocated to Lockheed Martin for long-lead items for 10 Lot 1 LRIP aircraft; further US$862 million contract awarded 19 September 2001 towards remaining cost of producing these 10 aircraft. More than US$770 million 'bridging funds' allocated during early 2001, pending authorisation to proceed with LRIP, with another US$441 million appropriation being made on 29 June for first two LRIP batches. Total cost of LRIP 2 batch of 13 aircraft quoted as US$3.03 billion, with US$4.47 billion appropriation for 20 LRIP 3 aircraft in FY03. Congressional cap on production budget currently US$37.6 billion although, in third quarter of 2001, US DoD was considering asking Congress to increase this to US$45 billion.

DESIGN FEATURES: Low-observables configuration and construction; stealth/agility trade-off decided by design team. Antennas located in leading- or trailing-edges of wings and fins, or flush with surfaces, to minimise radar signature. Target thrust/weight ratio 1.4 (achieved ratio 1.2 at T-O weight); greatly improved reliability and maintainability for high sortie-generation rates, including under 20 minutes combat turnround time; enhanced survivability through 'first-look, first-shot, first-kill' capability; short T-O and landing distances; supersonic cruise and manoeuvring (supercruise) in region of M1.5 without afterburning; internal weapons storage and generous internal fuel; conformal sensors.

Highly integrated avionics for single-pilot operation and rapid reaction. Radar, RWR and comms/ident managed by single system presenting relevant data only, and with emissions controlled (passive to fully active) in stages, according to tactical situation. CIP handles all avionics functions, including self-protection and radio, and automatically reconfigures to compensate for faults and failures. Two CIPs, with space for third, linked by 400 Mbits/s fibre optic network (see Avionics).

Wing and horizontal tail leading-edge sweep 42°; trailing-edge 17° forward, increased to 42° outboard of ailerons; all-moving five-edged horizontal tail. Vertical tail surfaces canted outwards at 28°; leading- and trailing-edge sweep 22.9°; biconvex aerofoil. Wing taper ratio 0.169; leading-edge anhedral 3.25°; root twist 0.5°; tip twist −3.1°; thickness/chord ratio 5.92 per cent at root, 4.29 per cent at tip; custom-designed aerofoil. Horizontal tails have no dihedral or twist.

Diamond-shaped cheek air intakes with highly contoured air ducts; single-axis thrust vectoring included on F119, but most specified performance achievable without.

Production aircraft to be coated with new, Boeing-developed, stealthy paint intended to enhance low-visibility attributes; this first applied to second EMD aircraft in March 2000.

FLYING CONTROLS: Triplex, digital, fly-by-wire system with GEC sidestick control, using line-replaceable electronic modules to enhance maintainability; thrust vectoring utilised to augment aerodynamic pitch control power and provide firm control even at low speeds and high angles of attack. Technology and control concepts demonstrated throughout the flight envelope during prototype air vehicle test programme, including flight at AoA greater than 60°; wind tunnel testing with models of production aircraft successfully attained AoAs greater than 85°.

Fifth prototype Lockheed Martin F/A-22A Raptor undertaking a test flight near Edwards AFB *(USAF)*   *NEW*/0527094

Ailerons and flaperons occupy almost entire wing trailing-edge; full-span leading-edge flaps; conventional rudders in vertical tail surfaces; slab taileron surfaces; airbrake not included (differential rudder and wing trailing-edge surfaces for speed control). Control surface authorities: leading-edge flaps 0° up/35° down (2°/37° overtravel); trailing-edge flaperons 20° up/35° down; ailerons ±25°; horizontal tail leading-edges 30° up/25° down; rudders ±30°; speedbrake (rudder) 30° out.

STRUCTURE: Lockheed Martin at Marietta constructs forward fuselage, including cockpit (with avionics architecture, displays, controls, air data system), apertures, edges, tail assembly, landing gear and environmental control system and undertakes final assembly. Lockheed Martin's Fort Worth plant builds the centre-fuselage; Palmdale, the radome. Boeing responsible for wings, fuselage aft sections, power plant installation, auxiliary power generation system, radar, arresting gear system and avionics integration laboratory.

Total airframe weight comprises approximately 36 per cent titanium 64, three per cent titanium 62222, 24 per cent thermoset composites (both epoxy resin and bismaleimide), one per cent thermoplastic composites, 16 per cent aluminium, 6 per cent steel and 14 per cent other materials. Forward fuselage substructure is aluminium and composites; centre-fuselage includes titanium, aluminium and composites; both forward and centre-fuselage skins primarily graphite bismaleimide. Four main mid-fuselage section bulkheads are single-piece, closed-die, titanium forgings. Rear fuselage approximately 67 per cent titanium, 22 per cent aluminium and 11 per cent composites. Engine bay doors titanium honeycomb, produced by liquid-interface diffusion bonding. Tailbooms assembled by electron beam welding. Titanium vent screens (to reduce radar reflectivity) contain thousands of precisely shaped holes in special alignment, each cut by abrasive water-jet. Thermoplastics used in areas requiring high tolerance to damage, examples including doors for landing gear and weapon bay.

Wing skins are monolithic graphite bismaleimide. Main (front) wing spars are machined titanium forgings; intermediate spars are mix of resin transfer moulded (RTM) sine-wave composites and titanium for strength to meet vulnerability requirements in wing fuel tank; rear spars composites and titanium. Wingroot and control surface actuator attachment fittings are titanium HIP castings. Horizontal stabiliser incorporates 'tow placed' composites pivot shaft in addition to aluminium honeycomb core and graphite bismaleimide skins; vertical stabilisers use solid graphite bismaleimide skins over graphite epoxy RTM spars. Wing control surfaces are combination of co-cured composites skins/substructure and non-metallic honeycomb core construction.

LANDING GEAR: Menasco retractable tricycle type, stressed for no-flare landings of up to 3.05 m (10 ft)/s. Honeywell wheels, brakes and anti-skid system. Nosewheel tyre Goodyear or Michelin 23.5×7.5-10 (22 ply) tubeless; mainwheel tyres Goodyear or Michelin 37×11.50-18 (30 ply) tubeless. Kaiser airfield arrester hook in enclosed fairing between engines.

POWER PLANT: Two 156 kN (35,000 lb st) class Pratt & Whitney F119-PW-100 advanced technology reheated turbofans, each fitted with a two-dimensional, convergent/divergent thrust vectoring (±20° in vertical plane) exhaust nozzle for enhanced performance and manoeuvrability.

Fuel in eight tanks located in forward fuselage, mid-fuselage, wings and tailbooms; provision for later addition of fuel in saddle and fin tanks. Additionally, up to four external fuel tanks, each 2,271 litres (600 US gallons; 500 Imp gallons), on underwing hardpoints. Dorsal Xar Industries aerial refuelling receptacle covered by doors, except when required. Fuel grade JP-8.

ACCOMMODATION: Pilot only, on zero/zero modified Boeing ACES II ejection seat and wearing tactical life support system with improved *g*-suits, pressure breathing and arm restraint. Pilot's view over nose is −15°. Canopy manufactured by Sierracin Sylmar Corp as single-piece unit, hinged at rear.

SYSTEMS: Twin 276 bar (4,000 lb/sq in) hydraulic systems with four pumps (each 273 litres; 72.0 US gallons; 60.0 Imp gallons per minute), but single Parker Bertea actuator on each control surface to save weight and cost. Curtiss Wright actuators on leading-edge flaps. Two 65 kW engine-driven generators. Honeywell G250 335 kW (450 hp) APU driving a Hamilton Sundstrand 27 kW generator and 100 litre (26.5 US gallon; 22.0 Imp gallon)/min pump. Smiths 270 V DC electrical distribution system. Honeywell environmental control system for life support system and flight avionics (open loop air cycle), mission avionics (closed loop vapour cycle) and fuel cooling (thermal management). OBIGGS and Normalair-Garrett OBOGS.

AVIONICS: Final integration, as well as integration of entire suite with non-avionics systems, undertaken at Avionics Integration Laboratory, Seattle, Washington; airborne integration supported by Boeing 757 flying testbed and the Air Vehicle Integration Facility (including coherent RF stimulation) at Marietta.

Comms: TRW AN/ASQ-220 communication/navigation/identification (CNI) system includes Mk 12 IFF

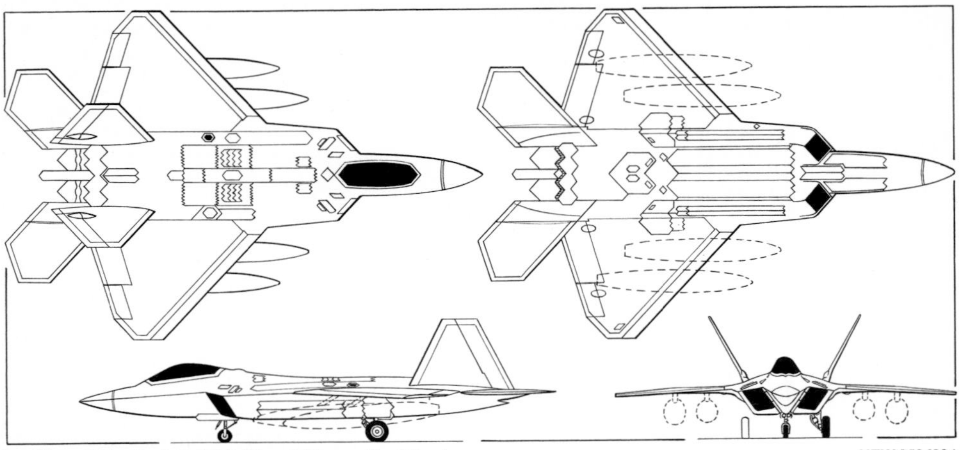

Lockheed Martin F/A-22A *(Jane's/James Goulding)* ***NEW***/0526924

and UHF/VHF. CNI system uses modules contained in two integrated CNI racks, CIP assets and integrated display/control panels. Inter/Intra-Flight Data Link (IFDL) allows all Raptors in a flight to share target and system data without voice communication. Supplier team, responsible for electronics and software, also includes Rockwell Collins, ITT and GEC.

Radar: Northrop Grumman/Raytheon AN/APG-77 multimode radar incorporates active electronically scanned array (AESA), capable of interleaving air-to-air search and multitarget track functions. Also has weather mapping mode and provisions for air-to-ground modes and side arrays. Radar reported to be capable of detecting 1 m² (10.76 sq ft) target at range of approximately 109 n miles (201 km; 125 miles). AN/APG-77 to be replaced starting with Lot 5 production in about 2008 by new AESA radar that will eventually incorporate advanced air-to-ground computer software.

Flight: Vehicle management system (VMS) combines flight and propulsion controls; integrated vehicle subsystem control (IVSC) operates utilities via digital databus. Total of 18 Raytheon 1750A common processor modules used for VMS, IVSC and stores management system. F/A-22 is first fighter with triplex digital flight control computers and no electrical or mechanical back-up; control reconfiguration modes provide safe flying after actuator or hydraulic failures. VMS controls 14 surfaces (horizontal tail, ailerons, flaperons, rudders, leading-edge flaps and inlet bleed and bypass doors). No AoA limitation, but overstressing made impossible by restrictions to roll rate and load factor, according to fuel state, stores carriage and flight condition. Lear Astronics VMS integrated with Rosemount low-observable air data system, including two AoA probes and four sideslip plates on nose. CNI system includes GPS, Tacan and ILS. Twin Litton LN-100F INS. Throttle and stick contain 20 controls with 63 functions.

Software Block 0 for initial flight tests. Block 1.1 installed on aircraft 4004 in 1999, is primarily for radar, but includes more than half of the avionics suite's source lines of code. Block 2, installed in the 757 FTB in October 1999, begins sensor fusion, including radio frequency co-ordination and some electronic warfare functions. Block 3S brings CNI and ECCM; Block 3.0, tested in the 757 FTB in fourth quarter of 2000 and flown for first time in F/A-22 on 5 January 2001, provides full sensor fusion and weapon delivery function; Block 3.1 adds GBU-32 JDAM, JTIDS receive and GPS and will be available to IOC aircraft. Block 4 will incorporate helmet cueing, AIM-9X and JTIDS send. Work is under way on a Block 5 upgrade that should provide enhanced air-ground capability from around 2006, including compatibility with forthcoming Small Diameter Bomb (SDB).

Instrumentation: Fused situational awareness information is displayed to pilot via four Lockheed Martin colour liquid crystal multifunction displays (MFD); MFD bezel buttons provide pilot format control. Centre screen measures 203 × 203 mm (8 × 8 in) and typically will function as situation display; right and left screens measure 152 × 152 mm (6 × 6 in) and function as attack and defensive displays respectively; fourth screen (directly below situation display) can be used to provide global side-view depiction of tactical situation and may also present other data such as fuel status, engine parameters, stores data, BIT reports and electronic checklists. Additionally, 76 × 102 mm (3 × 4 in) upfront display screens are each side of the integrated control panel, immediately below the BAE Systems HUD. Illumination is fully NVG-compatible.

Mission: Two Hughes common integrated processors (CIP); CIP also contains mission software that uses tailorable mission planning data for sensor emitter management and multisensor fusion; mission-specific information delivered to system through Fairchild data transfer equipment/mass memory (DTE/MM) system that contains mass storage for default data and air vehicle operational flight programme; stores management system. General purpose processing capacity of CIP is rated at more than 700 million instructions per second (Mips) with growth to 2,000 Mips; signal processing capacity greater than 20 billion operations per second (Bops) with expansion capability to 50 Bops; CIP contains more than 300 Mbytes of memory with growth potential to 650 Mbytes. Of 132 slots available in CIPs 1 and 2, 41 are initially vacant and thus available for growth. Intra-flight datalink automatically shares tactical information between two or more F/A-22s. CNI system includes JTIDS (receive-only terminal). Lockheed Martin airborne video recorder. Lockheed Martin stores management system. Airframe contains provisions for IRST and side-mounted phased-array radar.

Self-defence: BAE Systems AN/ALR-94 electronic warfare (RF warning and countermeasures and missile launch detection functions) subsystem. AN/ALE-52 flare dispenser.

ARMAMENT: Internal long-barrel General Dynamics M61A2 20 mm cannon with hinged muzzle cover and 480-round magazine capacity (production aircraft). Three internal bays for AIM-9M Sidewinder or next-generation AIM-9X (one in each side bay on Hughes LAU-141/A trapeze-type launcher) and six AIM-120C AMRAAM AAMs and/or 460 kg (1,015 lb) GBU-32 JDAM PGMs on Edo LAU-142/A hydraulic ejection launchers in main weapons bay. Four underwing stores stations at 317 mm (125 in) and 442 mm (174 in) from centreline of fuselage capable of carrying 2,268 kg (5,000 lb) each.

Typical weapon loads include six AIM-120s and two AIM-9s carried internally for air combat; two JDAMs, two AIM-120s and two AIM-9s carried internally for air-to-ground attack; and six AIM-120s and two AIM-9s internally, plus two external fuel tanks and further four AAMs (AIM-120 or AIM-9) underwing for long-range air combat.

Other weaponry envisaged for use by the F/A-22 includes the BLU-109 Penetrator, the wind-corrected munitions dispenser (WCMD), AGM-88 HARM, GBU-22 Paveway 3 guidance unit (with 500 lb bomb), new Small Diameter Bomb (SDB) and the low-cost autonomous attack system (LOCAAS) submunitions dispenser package.

DIMENSIONS, EXTERNAL:

Wing span	13.56 m (44 ft 6 in)
Wing chord: at root (theoretical)	9.85 m (32 ft 3½ in)
at tip (reference)	1.66 m (5 ft 5½ in)
at tip (actual)	1.14 m (3 ft 9 in)
Wing aspect ratio	2.4
Length overall	18.92 m (62 ft 1 in)
Height overall	5.08 m (16 ft 8 in)
Tail span: horizontal surfaces	8.84 m (29 ft 0 in)
vertical surfaces	5.97 m (19 ft 7 in)
Wheelbase	6.04 m (19 ft 9¾ in)
Weapon bay ground clearance	0.94 m (3 ft 1 in)

AREAS:

Wings, gross	78.0 m² (840.0 sq ft)
Leading-edge flaps (total)	4.76 m² (51.20 sq ft)
Flaperons (total)	5.10 m² (55.00 sq ft)
Ailerons (total)	1.98 m² (21.40 sq ft)
Vertical tails (total)	16.54 m² (178.00 sq ft)
Rudders/speedbrakes (total)	5.09 m² (54.80 sq ft)
Stabilators (total)	12.63 m² (136.00 sq ft)

WEIGHTS AND LOADINGS (estimated):

Weight empty (target)	14,365 kg (31,670 lb)
Max T-O weight	almost 27,216 kg (60,000 lb)
Max wing loading	348.7 kg/m² (71.43 lb/sq ft)
Max power loading	87 kg/kN (0.86 lb/lb st)

PERFORMANCE (YF-22, demonstrated):

Max level speed: supercruise	M1.58
with afterburning	M1.7 at 9,150 m (30,000 ft)
Ceiling	15,240 m (50,000 ft)
g limit	+7.9

PERFORMANCE (F/A-22A, design target, estimated):

Max level speed at S/L	800 kt (1,482 km/h; 921 mph)
g limit	+9

UPDATED

LOCKHEED MARTIN 382U/V HERCULES

US Air Force designations: C-, CC-, EC-, WC-130J
US Coast Guard designation: HC-130J
US Marine Corps designation: KC-130J
RAF designations: Hercules C. Mk 4 (C-130J-30);
 Hercules C. Mk 5 (C-130J)

TYPE: Medium transport/multirole.

PROGRAMME: US Air Force specification issued 1951, leading to first-generation Hercules (Allison T56 turboprops); first production contract for C-130A to Lockheed September 1952; first flight 23 August 1954; two YC-130 prototypes, 231 C-130As, 230 C-130Bs, 491 C-130Es, 1,089 C-130Hs and 113 L-100s manufactured before introduction of C-130J and commercial L-100J equivalent (details in earlier *Jane's* and current *Jane's Aircraft Upgrades*). Official total of 2,156 (including prototypes) delivered by January 1998, when final C-130H handed over.

Privately funded development and flight test programme for next-generation version began in 1991 as Hercules II, but now known as C-130J Hercules, or L-100J in equivalent civilian form. Early development detailed in 2000-01 and previous editions of *Jane's*.

Initial British delivery accomplished on 24 August 1998; aircraft involved was ZH865, which arrived at Boscombe Down on 26 August for start of clearance trials by UK Defence Evaluation Research Agency (DERA). First aircraft ferried to RAF transport force at Lyneham was ZH878 on 21 November 1999. Final RAF aircraft flown from Marietta to Cambridge in late May 2000; final aircraft delivered to RAF on 21 June 2001.

First 'operational' mission accomplished by Lockheed Martin test crew in late November 1998, when C-130J completed three sorties from Marietta and airlifted 37,650 kg (83,000 lb) of hurricane relief supplies to Tegucigalpa, Honduras. Subsequently, 1999 witnessed start of deliveries to US Air Force Reserve Command and Air National Guard, as well as to operating units in Australia and the UK. Entire fleet had accumulated 30,000 flight hours by February 2002.

Block 5.3 software configuration available from third quarter 2001 is standard equipment for latest deliveries and will be retrofitted to existing aircraft by end of 2002; Block 5.3 offers ability to fly integrated precision radar approaches, gives enhanced navigation capabilities and permits fully automatic formation flying using co-ordinated aircraft positioning system (CAPS).

CURRENT VERSIONS: **C-130J:** Baseline version. Dimensionally similar to preceding C-130H, but incorporating new equipment and features as subsequently described. Subject of initial order for two from USAF in FY94, with subsequent contract for two in FY96; these initially earmarked for trials and eventually to Air Force Reserve Command (AFRC), while further eight funded in FY97 and FY98 assigned to Air National Guard's 135th Airlift Squadron at Warfield ANGB, Martin State Airport, Baltimore, Maryland, which fully equipped by mid-July 2000. Initial aircraft delivered to AFRC 403rd Wing at Keesler AFB, Mississippi, on 31 March 1999 for training.
Description applies mainly to baseline C-130J except where indicated.

C-130J-30: Stretched version of current production C-130; fuselage lengthened by 4.57 m (15 ft 0 in), offering increases in capability of between 31 and 50 per cent,

US Air Force/ANG Lockheed Martin C-130J Hercules   NEW/0127025

dependent upon mission and configuration (see accompanying diagram). Orders received from Australia, Italy, UK and USA. Last named could ultimately purchase as many as 168 to replace C-130E version with regular USAF units.

CC-130J: Designation allocated to C-130J-30 aircraft in USAF service. First three funded in FY99, with two more in FY00 and five in FY02. Initial deliveries to ANG units in Rhode Island and California; FY02 aircraft to be assigned to 146th AW (two aircraft), 143rd AW (one), 403rd Wing (one) and a new USAF C-130J training unit that will be established at Little Rock AFB, Arkansas (one); all five to be delivered in 2004. USAF in early 2002 was considering multiyear procurement of 40 CC-130Js, commencing with four aircraft in FY04, followed by nine each year from FY05 up to and including FY08.

EC-130J: 'Commando Solo' psychological warfare version; first two funded in FY98 budget, with first (99-1933) handed over on 17 October 1999; after flight testing, it moved to Palmdale, California, in July 2000 for fitting out before being delivered to the 193rd Special Operations Squadron, Air National Guard, at Harrisburg IAP, Pennsylvania. Additional procurement comprises third example in FY99 budget, fourth in FY00 and fifth in FY01.

HC-130J: Replacement for earlier US Coast Guard HC-130s; funding for initial six contained in FY01 budget.

KC-130J: Tanker/transport version for US Marine Corps, which has requirement for 79 to replace KC-130F, KC-130R and KC-130T variants. Fitted with two Flight Refuelling Mk32B-901E hose-and-drogue wing-mounted refuelling pods; three aircraft funded in FY97 budget, two in FY98, two in FY99, one in FY00, three in FY01 and two in FY02. USMC contemplating multiyear purchase of 24 KC-130Js at rate of four per year, starting in FY03 and running to FY08. First contract for five USMC conversions (to be accomplished by end of 2000) announced 21 July

1998. Final assembly of first KC-130J (165735) began 22 March 1999, with first flight on 9 June 2000; total of three aircraft assigned to test programme at Patuxent River in latter half of 2000; first drogue engagement accomplished by Navy F/A-18 Hornet on 30 August 2000; initial trials revealed that aft fairing of pod was unsatisfactory, with cracks appearing in hose/drogue coupling; redesign of fairing in 2000-01 cleared way for further testing in second quarter of 2001, which confirmed much improved performance in areas of flying quality and durability. On completion of trials, all three test aircraft assigned to USMC tanker/transport squadron VMGR-252 at MCAS Cherry Point, North Carolina, which took delivery at beginning of September 2001; total of seven handed over by end 2001. Pod refuelling rate (each) 1,136 litres (300 US gallons; 250 Imp gallons) per minute; total offload capability is 32,005 litres (8,455 US gallons; 7,040 Imp gallons) using only wing and external tanks. Provisions for installation of refuelling probe incorporated in basic aircraft. Additional fuel in underwing and cargo hold tanks; see Power Plant.

WC-130J: Weather reconnaissance version to be equipped with aerial reconnaissance weather officer console, dropsonde system operator console and dropsonde launch tube; initial batch of four included in FY96 budget, with three more in FY97, two in FY98 and one in FY99; for 53rd WRS, Air Force Reserve Command at Keesler AFB, Mississippi. US$46.9 million contract for modification of first six, with option for further four, signed 18 September 1998; deliveries began on 30 September 1999; formal acceptance 12 October 1999.

CUSTOMERS: See tables. Total of 118 ordered and 85 delivered by beginning of May 2002. First customer was RAF, which ordered 25 in December 1994; of these, 15 are stretched C-130J-30, designated C. Mk 4, with final 10 as standard C-130Js, designated Hercules C. Mk 5. Delivery of first

Lockheed Martin CC-130J of Rhode Island Air National Guard   NEW/0127027

Flight deck of the C-130J, showing HUDs and EFIS   0075954

example to the trials unit at Boscombe Down was due to take place in November 1996, but was delayed until 26 August 1998. First service recipient was J Conversion Flight of No. 57 (Reserve) Squadron, followed by No. 24 and then No. 30 Squadrons at RAF Lyneham. Deliveries to operating base at Lyneham began on 21 November 1999, when C. Mk 4 ZH878 arrived for duty as temporary ground procedures trainer; two days later, on 23 November, C. Mk 4 ZH875 was formally handed over in official ceremony at Lyneham. Completion of deliveries against original order occurred on 21 June 2001; operational service with RAF began 14 November 2000, when No.24 Squadron flew scheduled mission to Puerto Rico. No.30 Squadron attained operational status in June 2002, at which time tactical release to service was expected imminently. In mid-2002, it also appeared likely that some RAF C. Mk 5 aircraft would be fitted with in-flight refuelling systems identical to those of the USMC KC-130J.

Second order covered two C-130Js for evaluation by the USAF and was finalised on 13 October 1995, one week before the first was rolled out at Marietta. Further two funded in FY96, four in FY97 and four in FY98 for Air Force Reserve Command and Air National Guard units; first ANG squadron is 135th AS at Martin State Airport, Baltimore. Further USAF orders began FY99 with contract for three CC-130J aircraft, with FY01 procurement including two more CC-130Js for USAF and initial batch of six C-130Js for long-range SAR duties with US Coast Guard; five more CC-130Js in FY02 budget. First USAF CC-130J rolled out 25 January 2001; following testing, this delivered to 143rd AW, Rhode Island ANG, on 2 December 2001. Two more CC-130Js to 143rd AW by end 2001, with next two to 146th AW, California ANG, which received its first example on 2 June 2002.

One further order received in 1995 was from Australia, for 12 C-130J-30s to replace C-130Es of No. 37 Squadron at Richmond, at total cost of US$660 million. Order placed on 21 December 1995 and included options for an additional 24 aircraft, plus eight options for New Zealand, which was guaranteed pricing based on Australia's larger order. First Australian C-130J (A97-440/N130JQ, c/n 5440) flew on 15 February 1997; delivery began 7 September 1999, when A97-464 arrived at Richmond; last of initial batch handed over on 1 June 2000, with operational status achieved in December 2001.

Next firm orders, placed in January 2000 and March 2000, were for batches of two additional aircraft from Italy, which had previously contracted for 18 C-130Js; first aircraft formally rolled out at Marietta on 11 July 2000. Italian military certification also awarded in July 2000, with first delivery (actually second aircraft, MM 62176, c/n 5497) to 2° Gruppo, 46° Aerobrigata at Pisa departing USA on 16 August; formal acceptance 21 September 2000. All will be tanker-capable and configured to receive fuel, although only six likely to be operated as such at any one time. Newest orders specified C-130J-30, which also subject of option for up to two further aircraft; C-130J-30 version assigned to 50° Gruppo, 46° Aerobrigata, from early 2002. At same time as placing latest firm order, Italy also revised overall procurement plan and will acquire 12 C-130J and 10 C-130J-30 versions. Denmark ordered three C-130J-30s, with option on fourth, in December 2000.

Other potential customers reported to be in discussion with Lockheed Martin include Bahrain, Canada, Israel,

Kuwait and Portugal. Israel considering purchase of 12, but early order appears unlikely. Norway has also expressed interest in obtaining six as replacements for its current fleet of six C-130Hs, while Saudi Arabia reported to have requirement for up to 24 aircraft. Extensive world promotional tour undertaken by C-130J between February and June 1998, visiting 32 countries and completing 130 demonstration flights.

MILITARY/GOVERNMENT C-130J SALES
(To July 2002, excluding USA)

Country	J	J-30	Total
Australia		12	12
Denmark		3	3
Italy	12	10	22
UK	10	15	25
Total	**22**	**40**	**62**

Note: Options not included

COSTS: US$55 million programme unit cost (Australia) (1995). Italian order of November 1997 valued at US$1.2 billion. Purchase of two USAF CC-130J aircraft in FY01 budget costed at total of US$149 million. Five USAF CC-130J procured in FY02 at total cost of US$355 million. Baseline price of C-130J-30 quoted as US$67 million in early 2002.

DESIGN FEATURES: Archetypal tactical transport: upswept rear fuselage for ramp access; high wing for propeller ground clearance despite low floor height; latter provided by pannier-mounted landing gear, which obviates long mainwheel legs stowed in wings. Can deliver loads and parachutists over open ramp and parachutists through side doors; cargo hold pressurised.

Significant changes introduced with C-130J, which optimised for economical operation, justifying customers' substitution for earlier C-130s on 30 year lifetime savings alone. Entirely revised flight deck reduces LRUs by half and wire assemblies by 53 per cent, with wire terminations cut by 81 per cent; has four MFDs, plus HUD for both pilots; lighting compatible with NVGs. Most systems have digital interfaces with the main mission computer to include unmodified mechanical systems like the

hydraulics, which are largely unaltered from those of C-130H; provision for integrated self-defence suite (RWR, MAW, chaff/flare dispensers and IR jammers). Propulsion system provides 29 per cent more take-off thrust and is 15 per cent more efficient; fuel efficiencies obviate requirement for external tanks on most types of mission; propeller has 50 per cent fewer parts and weighs 15 per cent less. Manpower requirements of typical 16-aircraft squadron cut by 38 per cent compared with earlier versions of C-130, as result of reduced flight crew and 50 per cent better maintainability. Comprehensive computerised maintenance system employs a hand-held data module as interface between aircraft's BITE and operating base's central technical records.

Wing section NACA 64A318 at root and NACA 64A412 at tip; dihedral 2° 30'; incidence 3° at root, 0° at tip.

FLYING CONTROLS: All flying controls integrated with digital autopilot/flight director and comprise control surfaces boosted by dual hydraulic units; trim tabs on ailerons, both elevators and rudder; elevator tabs have AC main supply and DC standby; Lockheed-Fowler composites trailing-edge flaps.

STRUCTURE: All-metal two-spar wing with integrally stiffened taper-machined skin panels up to 14.63 m (48 ft 0 in) long. Incorporates carbon fibre composites materials for flaps and 32 graphite-epoxy trailing-edge panels.

LANDING GEAR: Hydraulically retractable tricycle type. Each main unit has two wheels in tandem, retracting into fairing built on to fuselage side. Nose unit has twin wheels and is steerable ±60°. Mainwheels 20.00-20 (26 ply) tubeless; nose 12.50-16 (12 ply) tubed or tubeless. Oleo shock-absorbers. Minimum ground turning radius: C-130J, 11.28 m (37 ft) about nosewheel and 25.91 m (85 ft) about wingtip; C-130J-30 14.33 m (47 ft) about nosewheel and 27.43 m (90 ft) about wingtip.

POWER PLANT: Four Rolls-Royce AE 2100D3 turboprops, flat rated to 3,424 kW (4,591 shp) (manufacturer's rating 3,458 kW; 4,637 shp at ISA + 25°C), fitted with Dowty Aerospace R391 six-blade composites propellers and Lucas Aerospace FADEC. Automatic thrust control system (ATCs) and autofeather systems, plus engine monitoring system (EMS) which is incorporated into aircraft's integrated diagnostic system (IDS).

Total internal fuel capacity of 25,552 litres (6,750 US gallons; 5,621 Imp gallons) without foam and 24,363 litres (6,436 US gallons; 5,359 Imp gallons) with foam. Provisions only for two optional underwing pylon tanks, each with capacity of 5,220 litres (1,379 US gallons; 1,148 Imp gallons) without foam and 4,883 litres (1,290 US gallons; 1,074 Imp gallons) with foam. Total fuel capacity 35,992 litres (9,508 US gallons; 7,917 Imp gallons) without foam and 34,129 litres (9,016 US gallons; 7,507 Imp gallons) with foam. Tanker versions can carry cargo hold tank with additional 13,578 litres (3,587 US gallons; 2,987 Imp gallons). Single pressure refuelling point and overwing gravity fuelling. In-flight refuelling probe fitted as standard on port side of RAF aircraft; optional for all others, with Italian aircraft currently unique in being configured as receiver/tankers.

ACCOMMODATION: Crew of two on flight deck, comprising pilot and co-pilot, with provisions for optional third workstation. Ergonomic problems suffered by short pilots necessitated a number of alterations to cockpit, including redesign of seat and seat track, HUD and main control yoke, with throttle quadrant also modified. Two crew bunks, with lower incorporating three additional seats and harnesses for relief crew/flight deck passengers. Galley. Separate loadmaster's station in cargo hold, including folding desk, is standard equipment on RAF aircraft; optional for all others. Flight deck and main cabin pressurised and air conditioned.

Standard complements for C-130J are as follows: 92 troops, 64 paratroopers, 74 litter patients plus two attendants, 54 passengers on palletised airline seating. Corresponding data for C-130J-30 are 128 troops, 92 paratroopers, 97 litter patients plus four attendants and 79 passengers on palletised airline seating. Airdrop loads comparable to C-130H/H-30 and include light armoured vehicles. Light and medium towed artillery pieces, wheeled and tracked vehicles and 463L palletised loads (five in C-130J and seven in C-130J-30, plus one on ramp in each model) are transportable. Hydraulically operated (with dual actuators) main loading door and ramp at rear of

US C-130J PROCUREMENT
(To July 2002)

FY	C-130J	CC-130J	EC-130J	HC-130J	KC-130J	WC-130J	Total
94	2						2
96	2					4	6
97	4				3	3	10
98	4		2		2	2	10
99		3	1		2	1	7
00			1		1		2
01		2	1	6	3		12
02		5			2		7
Total	**12**	**10**	**5**	**6**	**13**	**10**	**56**

Lockheed Martin C-130J-30 in Royal Australian Air Force service (*Jane's/Paul Jackson*) 0121714

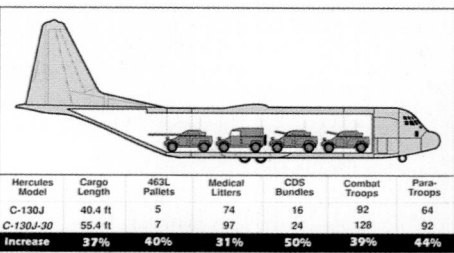

Hercules Model	Cargo Length	463L Pallets	Medical Litters	CDS Bundles	Combat Troops	Para-Troops
C-130J	40.4 ft	5	74	16	92	64
C-130J-30	55.4 ft	7	97	24	128	92
Increase	**37%**	**40%**	**31%**	**50%**	**39%**	**44%**

C-130J Hercules accommodation; standard and (illustrated) stretched versions compared 0105080

hold; can be opened in flight at up to 250 kt (463 km/h; 288 mph). Crew door on port forward fuselage side. Paratroop door on each side aft of landing gear fairing. Two emergency exit doors standard. Optional cargo handling system ordered by the USAF includes flush-mounted winch; in-ramp towplate for airdrop operations; low-profile rails and electric locks; flip-over rollers with covers; container delivery system centre vertical restraint rails. Capable of automatic preprogrammed cargo drops.

SYSTEMS: Lucas generator. Environmental control system in starboard undercarriage fairing similar to that of the C-130H, incorporating dual air cycle machines, but with 30 per cent greater cooling capacity and a digital electronic control system. Honeywell GTCP85-185L(A) auxiliary power unit in port undercarriage fairing furnishes ground electrical power and bleed air for environmental control system. MIL-STD-1553B digital databus architecture. Integrated diagnostic system (IDS) incorporating fault detection/isolation subsystem with BIT (built-in test) facility. Goodrich pneumatic fin anti-icing system.

AVIONICS: *Comms:* Honeywell com/nav/ident management system, with Intel 80960 processor. AN/ARC-222 VHF, HF/UHF radio, intercom and IFF, with provisions for satcom system. All communication radios have secure features.

Radar: Northrop Grumman AN/APN-241 low-power colour radar incorporates Doppler beam-sharpening ground mapping mode, air-to-air skin paint mode and protective windshear mode as well as conventional colour weather radar.

Flight: HG-9550 radar altimeter, AN/ARN-153(V) Tacan, digital autopilot/flight director (DA/FD), dual Honeywell laser INS with embedded GPS receivers, Doppler velocity sensor, VOR, ILS, marker beacon receiver, UHF/VHF DF, ADF, E-TCAS, ground collision avoidance system (GCAS), global digital map display units and provision for microwave landing system.

Instrumentation: 'Dark cockpit' concept. Flight Dynamics HUD as certified primary flight instrument at pilot and co-pilot positions, four 152 × 203 mm (6 × 8 in) Avionics Display Corporation active matrix liquid-crystal display (AMLCD) colour multipurpose display systems (CMDSs) which are NVG-compatible for flight instrumentation, navigation and engine information and five Avionics Display Corporation 58 × 76 mm (2.3 × 3 in) monochrome AMLCDs for digital selector panels.

Mission: Sierra Technologies AN/APN-243(V) station-keeping equipment. Provision for secure voice communication system.

Self-defence: Provisions for Lockheed Martin AN/AAR-47 missile warning system, Sanders AN/ALQ-157 IR countermeasures system, BAE Systems AN/ALE-47 chaff/flare dispensing systems and AN/ALR-56M radar warning receiver or AN/ALR-69 enhanced radar warning system. RAF aircraft originally ordered with US-supplied systems, but RWR and countermeasures dispensing systems now subject to bidding from UK defence sector, with formal contest beginning early 1998.

EQUIPMENT: USMC KC-130Js and Italian C-130Js have Flight Refuelling Mk 32B-901E hose pods underwing. Latest CC-130Js for Air National Guard's 146th AW are equipped with Airborne Fire Fighting System (AFFS), which allows them to drop up to 15,142 litres (4,000 US gallons; 3,331 Imp gallons) of fire retardant in a single pass.

DIMENSIONS, EXTERNAL:
Wing span	40.41 m (132 ft 7 in)
Wing aspect ratio	10.1
Length overall: C-130J	29.79 m (97 ft 9 in)
C-130J-30	34.37 m (112 ft 9 in)
Height overall: C-130J	11.84 m (38 ft 10 in)
C-130J-30	11.81 m (38 ft 9 in)
Tailplane span	16.05 m (52 ft 8 in)
Wheel track	4.34 m (14 ft 3 in)
Propeller diameter	4.11 m (13 ft 6 in)

Main cargo door (rear of cabin):
Height	2.77 m (9 ft 1 in)
Width	3.12 m (10 ft 3 in)
Height to sill	1.03 m (3 ft 5 in)

Paratroop doors (each): Height	1.83 m (6 ft 0 in)
Width	0.91 m (3 ft 0 in)
Height to sill	1.03 m (3 ft 5 in)
Emergency exits (each): Height	1.22 m (4 ft 0 in)
Width	0.71 m (2 ft 4 in)

DIMENSIONS, INTERNAL:
Cabin, excl flight deck:
Length excl ramp: C-130J	12.19 m (40 ft 0 in)
C-130J-30	16.76 m (55 ft 0 in)
Length incl ramp: C-130J	15.32 m (50 ft 3 in)
C-130J-30	19.89 m (65 ft 3 in)
Max width	3.12 m (10 ft 3 in)
Max height	2.74 m (9 ft 0 in)
Total usable volume: C-130J	128.9 m³ (4,551 cu ft)
C-130J-30	170.5 m³ (6,022 cu ft)

AREAS:
Wings, gross	162.12 m² (1,745.0 sq ft)
Ailerons (total)	10.22 m² (110.00 sq ft)
Trailing-edge flaps (total)	31.77 m² (342.00 sq ft)
Fin	20.90 m² (225.00 sq ft)
Rudder, incl tab	6.97 m² (75.00 sq ft)
Tailplane	35.40 m² (381.00 sq ft)
Elevators, incl tabs	14.40 m² (155.00 sq ft)

WEIGHTS AND LOADINGS (internal fuel only, except where specified):
Operating weight empty:
C-130J	34,274 kg (75,562 lb)
C-130J-30	35,966 kg (79,291 lb)
Max fuel weight: internal	20,819 kg (45,900 lb)
external (optional)	8,506 kg (18,754 lb)
Max payload, 2.5 g: C-130J	18,955 kg (41,790 lb)
C-130J-30	17,264 kg (38,061 lb)
Max normal T-O weight	70,305 kg (155,000 lb)

Max overload T-O weight:
C-130J, C. Mk 4	79,380 kg (175,000 lb)
Max normal landing weight	58,965 kg (130,000 lb)
Max overload landing weight	70,305 kg (155,000 lb)
Max zero-fuel weight, 2.5 g	53,230 kg (117,350 lb)
Max wing loading (normal)	433.7 kg/m² (88.83 lb/sq ft)
Max power loading (normal)	5.14 kg/kW (8.44 lb/shp)

PERFORMANCE (C-130J except where indicated):
Never-exceed speed (VNE) at 3,050 m (10,000 ft)
C. Mk 4	378 kt (700 km/h; 435 mph)

Max cruising speed:
C-130J	348 kt (645 km/h; 400 mph)
C. Mk 4 at 7,620 m (25,000 ft)	356 kt (659 km/h; 410 mph)

Econ cruising speed	339 kt (628 km/h; 390 mph)
Stalling speed	100 kt (185 km/h; 115 mph)
Max rate of climb at S/L	640 m (2,100 ft)/min
Time to 6,100 m (20,000 ft)	14 min
Cruising altitude	8,535 m (28,000 ft)
Service ceiling at 66,680 kg (147,000 lb) AUW	9,315 m (30,560 ft)
Service ceiling, OEI, at 66,680 kg (147,000 lb) AUW	6,955 m (22,820 ft)
T-O run	930 m (3,050 ft)
T-O to 15 m (50 ft)	1,433 m (4,700 ft)
T-O run using max effort procedures	549 m (1,800 ft)
Landing from 15 m (50 ft) at 58,967 kg (130,000 lb) AUW	777 m (2,550 ft)
Landing run at 58,967 kg (130,000 lb) AUW	427 m (1,400 ft)
Runway LCN: asphalt	37
concrete	42

Range with 18,144 kg (40,000 lb) payload and Mil-C-5011A reserves
2,835 n miles (5,250 km; 3,262 miles)
UPDATED

LOCKHEED MARTIN ADVANCED MOBILITY AIRCRAFT (AMA)

TYPE: Medium transport/multirole.

PROGRAMME: Studies under way since early 1995; initially known as 'World Airlifter' and New Strategic Aircraft (NSA). Lockheed Martin's primary objective was to develop a replacement for Boeing KC-135 Stratotanker in-flight refuelling aircraft, Lockheed C-141 StarLifter strategic transport and tanker/transport types such as the Lockheed L-1011 TriStar and McDonnell Douglas KC-10 Extender, although it is also intended to offer commercial freighter versions as well as special mission aircraft configured for AEW and battlefield surveillance tasks.

Lockheed Martin is proposing to develop the AMA as a private venture and is seeking two or three risk-sharing international partners to form a consortium; discussions have taken place with several potential partners, including Aerospatiale Matra, DaimlerChrysler Aerospace and BAE Systems.

Recent studies envisage a twin-engined design with high bypass turbofans in the 267 to 311 kN (60,000 to 70,000 lb st) class, an M0.85 cruise speed, with 30 per cent greater fuel offload than the KC-135R/T Stratotanker. In the cargo role, AMA will carry a payload of 45,360 to 54,430 kg (100,000 to 120,000 lb) over 4,000 n miles (7,408 km; 4,603 miles).

Over 40 advanced aircraft designs examined, leading to further study of four basic concepts, all of which feature modular design using common basic structure and systems

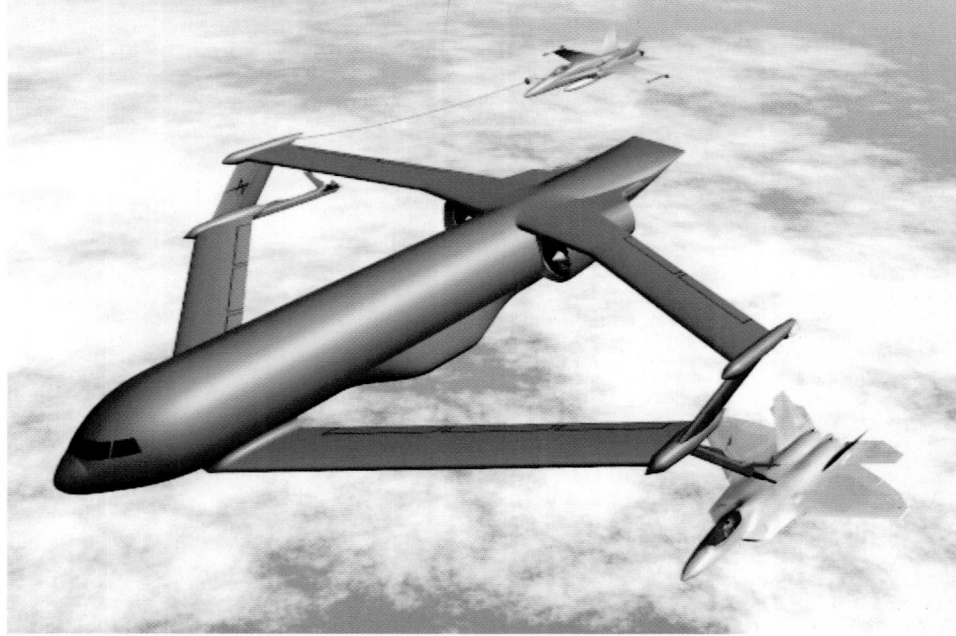

Computer-generated image of Lockheed Martin box-wing tanker/transport design 0121681

to reduce initial manufacturing costs and facilitate airframe upgrades during service life. Modular systems and avionics bus architecture will easily accommodate mission-orientated equipment for specific roles. Configurations studied include a conventional high-wing aircraft; a blended wing/body aircraft; a box-wing aircraft with two refuelling booms and two hose-and-drogue assemblies (see accompanying illustration); and a global transport with an unrefuelled range of 12,000 n miles (22,220 km; 13,810 miles).

Design effort directed to the box-wing aircraft concept during 1997-2000, by virtue of aerodynamic and structural efficiency, combined with greatly reduced aircraft size. As recently envisaged, it will have two flight deck crew, plus advanced refuelling and loadmaster workstations; incorporate roll-on/roll-off cargo handling capability and be compatible with 20 and 40 ft ISO containers; and embody fly-by-light/power-by-wire flight control systems. Testing of a radio-controlled scale model began on 7 March 1997, this exhibiting excellent flight characteristics

and meeting, or surpassing, test objectives during its 18-sortie programme.

Current planning expects AMA development effort to reach a peak during 2004-13, with resultant production aircraft ready for delivery to USAF and other customers from 2013; however, USAF purchase seems less likely in view of early 2002 decision to acquire tanker version of Boeing 767 on lease. World market estimated at 700 to 1,000 by 2030.

UPDATED

LOCKHEED MARTIN AERONAUTICS COMPANY – PALMDALE

1011 Lockheed Way, Palmdale, California 93599-3740
Tel: (+1 661) 572 41 55
Fax: (+1 661) 572 41 63
VICE-PRESIDENT AND SITE GENERAL MANAGER: Richard S Baker
VICE-PRESIDENT, COMMUNICATIONS: Mary Jo Polidore

Nickname, Skunk Works became official title when former Lockheed Advanced Development Company renamed in 1995, but renamed Palmdale Operations in January 2000, before adopting current title. Functions autonomously at Palmdale; has specialised in 'black' or covert development

programmes, including U-2, SR-71 and F-117A and retains responsibility for support of U-2 and F-117 aircraft. Participant in Joint Strike Fighter (JSF) project and responsible for the production of both X-35 prototypes.

Organisation expanded in latter half of 1995 through assumption of responsibility for Lockheed Martin Aircraft Services at Ontario, California. Former separate division has designed, fabricated and installed major aircraft structural modifications and integrated complete avionic systems for US military, foreign governments and commercial customers.

UPDATED

LOCKHEED MARTIN AEROCRAFT

TYPE: Outsize freighter.
PROGRAMME: Revealed February 1999 as part of Revolutionary Concepts (RevCon) project. Original 'pure' airship abandoned in favour of partially buoyant hybrid (and thus heavier than air) design. Projected in both rigid and non-rigid forms. In August 1999, NASA announced plans for US$10 million funding for a 7.8 per cent scale AeroCraft testbed to fly in 2001 at its Dryden base. However, project subsequently abandoned because of inability to meet NASA schedule. See 2002-03 *Jane's* for additional details.

UPDATED

LOCKHEED MARTIN AERONAUTICS COMPANY – FORT WORTH

1 Lockheed Boulevard, Fort Worth, Texas 76108
Tel: (+1 817) 777 20 00
Fax: (+1 817) 763 47 97
Web: http://www.lmtas.com
VICE-PRESIDENT, F-16 PROGRAMMES: John L Bean
EXECUTIVE VICE-PRESIDENT, JSF PROGRAMME:
    C T 'Tom' Burbage

General Dynamics' Fort Worth Division sold to Lockheed; became Lockheed Fort Worth Company on 1 March 1993; renamed LMTAS following merger between Lockheed and Martin Marietta in 1995. Adopted current title in January 2000 with consolidation of all Lockheed Martin aeronautical operations into single company with headquarters in Fort Worth, Texas.

Activities include production, development and support of F-16 Fighting Falcon; one-third share of F/A-22 Raptor development and initial production; and leadership of F-35 Joint Strike Fighter development team (with Northrop Grumman and BAE Systems). Is principal subcontractor to Mitsubishi for production of F-2 in Japan and to KAI for development of T-50 in South Korea.

UPDATED

LOCKHEED MARTIN F-35 JOINT STRIKE FIGHTER

TYPE: Multirole fighter.
PROGRAMME: Origins of Joint Strike Fighter (JSF) programme vested in separate USAF/USN Joint Advanced Strike Technology (JAST) and Defense Advanced Research Project Agency (DARPA) Common Affordable Lightweight Fighter (CALF) projects of early 1990s; designation X-32 then assigned to planned CALF demonstrator (see US Navy entry in 1997-98 and previous editions of *Jane's* for more details of JAST and CALF programmes).

Projects merged in November 1994, as JAST, after Congressional directive in mid-1994; programme renamed JSF in latter half of 1995. Previously, formal request for

Article 301, the first Lockheed Martin X-35 to fly, pictured in X-35A guise   0098624

proposals (RFP) for preliminary research contracts released on 2 September 1994, stipulating industry response by 4 November and issue of contract awards by 16 December.

Some elements of US industry joined forces to win JAST/JSF work, with international collaboration in evidence. McDonnell Douglas led one team after signing October 1994 Memorandum of Understanding (MoU) with Northrop Grumman and British Aerospace; each company submitted individual bids, but all three would participate in event of securing contract. Boeing allied with Dassault of France on aspects of subsystem design effort.

Subsequent research contracts worth US$99.8 million were distributed between four companies: Boeing (US$27.6 million), Lockheed Martin (US$19.9 million), McDonnell Douglas (US$28.2 million) and Northrop Grumman (US$24.1 million). Further US$28 million allocated for associated avionics, propulsion systems, structures and materials, and modelling and simulation.

Merger of JAST and CALF resulted in expanded flight test programme, involving two finalists; each to build two demonstrators, one with ASTOVL capability and the other to use conventional take-off and landing (CTOL).

Draft RFP issued December 1995, with USA and UK signing MoU on 20 December 1995, which committed UK to participate in four year weapons system concept demonstration (WSCD) phase. MoU also stipulated that UK must contribute some 10 per cent (approximately US$200 million) of demonstration phase costs as full collaborative partner.

Formal release of the final RFP for JSF was expected on 7 March 1996, but was delayed to June 1996, with contract award date in November 1996. X-32 and X-35 designations allocated to demonstration phase, requiring successful teams to build two aircraft each, with CTOL version to fly first. STOVL versions to fly second and participate in assessment and demonstration of hover and transition qualities.

All three WSCD contenders chose Pratt & Whitney's F119 engine for their proposals, although a General Electric/Allison/Rolls-Royce team secured a US$7 million contract in March 1996 to examine alternative power plants. These were based on the General Electric F110 and YF120 engines, with the latter being chosen in May 1996 following Congressional directive aimed at fostering competition and also overcoming possible impact of

X-35B in transition mode with fan doors open and landing gear deployed   *NEW*/0527090

Lockheed Martin X-35C US Navy version *NEW*/0528624

developmental or operational problems with the F119. Further US$96 million multiyear contract awarded in February 1997 to cover technology maturation and core engine development of YF120-FX version over four year period; this likely to result in follow-on development programme, culminating in full-scale EMD from 2004.

On 16 November 1996, US Secretary of Defense, William J Perry announced that Boeing and Lockheed Martin had been chosen to participate in WSCD and that the team headed by McDonnell Douglas had been eliminated. Simultaneously, Boeing was awarded a US$661.8 million contract, while Lockheed Martin received US$718.8 million; in addition, Pratt & Whitney secured a contract worth US$804 million for the associated Engine Ground and Flight Demonstration Program. Subsequently, Northrop Grumman and British Aerospace joined Lockheed Martin team.

Australia, Canada, France, Germany, Greece, Israel, Singapore, Spain and Sweden were all briefed on JSF programme. For System Development and Demonstration (SDD) phase, four partnership options available. Most costly is Level 1, with responsibility for 10 per cent of cost; UK is only partner at this level. Italy and the Netherlands are Level 2 partners, each contributing about 5 per cent of cost. Level 3 involves payment of 1 to 2 per cent, with Denmark and Norway having teamed up to share burden, while Australia, Canada and Turkey meet the cost alone. Finally, Security Co-operation Participant Level involves smaller contribution of US$75 million, but no subscribers by late 2002, although Israel and Singapore expected to sign up, with Greece, Poland and South Korea potential future additions.

JSF acquisition planning little affected by Quadrennial Defense Review of May 1997 and production run of successful design for US and UK expected to number at least 3,000, with export sales likely to account for at least 600 more in short-term.

Lockheed Martin development included production of 91 per cent scale powered model of JAST demonstrator for wind-tunnel tests. Model JAST, with F100 engine, began trials at Pratt & Whitney's West Palm Beach, Florida, facility in February 1995; subsequently to outdoor hover test rig at NASA Ames and then installed in 24 × 36 m wind tunnel at Mountain View, California, for series of powered hover and transition tests which ran from December 1995 to 5 March 1996; total of 196 hours accumulated, representative of approximately 2,400 take-offs and landings with the vertical lift system. Midway through outdoor hover tests, design was reconfigured to eliminate canards.

Lockheed Martin design, development, construction and flight testing of full-scale demonstrator aircraft, required one initially to be flown as CTOL version (X-35A) to demonstrate land-based USAF model, before being reconfigured to serve as STOVL version (X-35B) for US Marine Corps, Royal Air Force and Royal Navy; other aircraft representative of carrier-capable US Navy model (X-35C). Although Fort Worth is team leader, both X-35 aircraft built at Palmdale, California, using rapid prototyping techniques.

Design of X-35 frozen as **Configuration 220** 13 June 1997 after Initial Design Review, at which time 11,000 hours of model testing accumulated. Development team joined by Northrop Grumman on 8 May 1997 and BAe (now BAE Systems) on 18 June 1997.

Release of engineering drawings in early September 1997 heralded start of parts production at Palmdale for demonstrator aircraft; by end of 1997, Lockheed Martin had completed about 70 per cent of required tooling and had conducted second interim programme review. X-35 final design review completed in September 1998 and coincided with roll-out of full-scale mockup at Fort Worth.

Assembly of first aircraft began in April 1998, with main wing carry-through bulkhead installed in early August 1998, by which time manufacturing of large composites skins for upper and lower wing surfaces also complete; aircraft moved from assembly testing to factory floor on 18 September 1999, in readiness for installation of flight control surfaces and landing gear as well as systems checks.

Flight control system software tested on NF-16D VISTA in 1998 as part of integrated subsystem technology demonstration. Further trials using AFTI/F-16 in 1999-2000 involved all-electric flight control system and modular electric power system planned for JSF. JSF avionics also tested on Northrop Grumman's BAe One-Eleven Co-operative Avionics Test Bed (CATB), which was fitted with sensors, processors and software in 1999; trials begun of Northrop Grumman radar and distributed infra-red sensor system, Kaiser helmet-mounted display and Lockheed Martin core processor in first quarter 2000.

More than 100 hours of flight time accumulated by early September 2000 using CATB, when avionics development and integration process completed; successful demonstrations included automatic target cueing (ATC), whereby sensors acquire targets rapidly and automatically; electro-optical targeting system (EOTS); electronic warfare suite and electronically scanned radar. Subsequent testing included co-operative engagement between CATB and Northrop Grumman E-8C Joint STARS, demonstrating all-weather precision targeting and combat identification techniques for fixed and moving targets.

Design of JSF continued to be refined after selection of Configuration 230-1. By 1998, third version of **Configuration 230 (230-3)** had reduced area of USAF/ USMC JSF wings by seven per cent, but increased USN variant's wing area by 11 per cent. Further redesign occurred in 1999, culminating in September with Configuration 230-5, which has enlarged wing to satisfy sustained turn performance requirement and strengthened to meet 9 g stress requirement for the CTOL variant; further main change involved redesign of lift-fan nozzle from D-shape extendible box to venetian blind-type box, offering dual benefits of simpler design and reduced weight for the STOVL configuration.

Redesign process continued into 2000, with Configuration 230-5 adding further refinements aimed at reducing weight and increasing payload bringback capability; resultant structural modification entailed weight reduction in several areas such as weapon bay and

landing gear door assemblies. Configuration 230-5 finally submitted as Preferred Weapons System Concept design in mid-2000.

Significant programme events in latter part of 1998 and 1999 included first test run of basic Pratt & Whitney JSF119 engine (designated FX661) at West Palm Beach, Florida, facility on 11 June 1998; over 330 hours of developmental testing performed by end August 1999, validating complete X-35 flight envelope. Final altitude flight qualification testing undertaken with another engine (designated FX663) in fourth quarter of 1999. National Aerospace Laboratory low-speed wind tunnel in the Netherlands used for testing of scale models of JSF starting in June 1998; initial trials of lift fan system's clutch, fan and gearbox rigs at Indianapolis, Indiana, in May and June 1998; and installation of engine inlet duct in assembly tool at Palmdale at beginning of July 1998. Subsequently, also in July, over 50 hours of preliminary engine testing were completed at West Palm Beach, including vibration surveys, fan and core running-in, operating performance calibration and engine control stability assessment; next key phase of engine test programme involved altitude testing at Arnold Engine Development Center, Tennessee.

Testing of first STOVL engine (designated FX662) undertaken initially at West Palm Beach; first run with lift fan engaged accomplished on 10 November 1998, with operation to 100 per cent speed following on 22 November; first series of tests included stress surveys and performance calibration and occupied about six weeks. Validation of STOVL propulsion system achieved in late August 1999, with high-power clutch engagement of shaft-driven lift fan, simulating seamless conversion from conventional wing-borne flight configuration to jet-borne flight configuration for STOVL approach ending with vertical landing.

Following this, on 9 December 1999, flight engine YF001 successfully installed on the X-35A demonstrator at Palmdale, with integration tests, including plumbing connections, data communications and electrical checks, beginning immediately thereafter.

In UK, DERA vectored-thrust advanced aircraft control (VAAC) Harrier T. Mk 4 completed 20 hour, 36 sortie, flight test programme in November 1998, during which a sidestick control column was evaluated by civilian and military pilots. Two-phase programme began with initial calibration to validate STOVL control laws and stick characteristics; subsequent evaluation included pattern work, approach and transition to hover and precision and

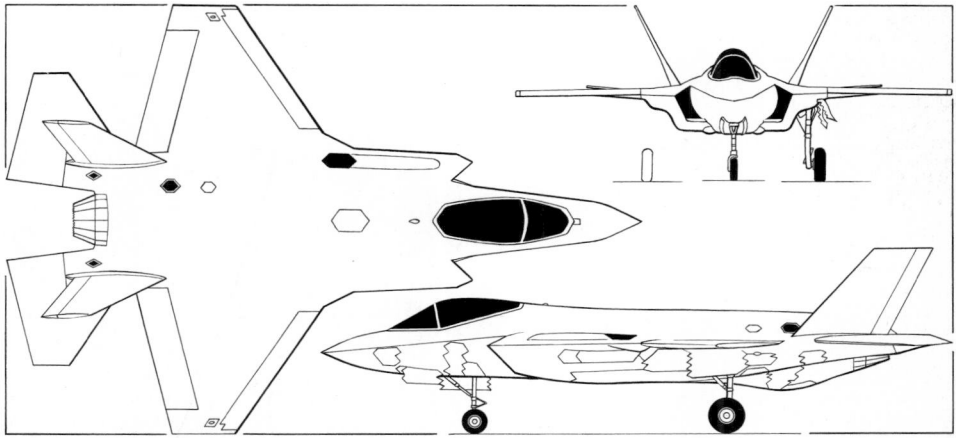

Lockheed Martin F-35A production configuration (*Jane's/James Goulding*) 0131841

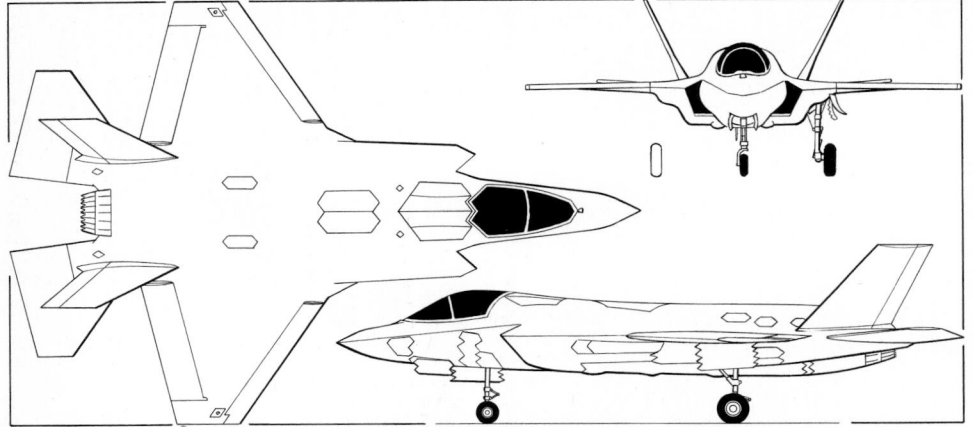

F-35B STOVL version for US Marine Corps and United Kingdom (*Jane's/James Goulding*)   NEW/0526907

aggressive hover tasks, resulting in confirmation that side-stick provides satisfactory control of STOVL aircraft at low speeds.

Lockheed Martin also reached preliminary agreement with partners over work-sharing arrangements to be implemented for large-scale production. Lockheed Martin has responsibility for forward fuselage, cockpit, edges and final assembly; Northrop Grumman to fabricate mid-fuselage and wing box, with BAE Systems producing aft fuselage section.

Radar signature testing of full-scale pole model began in late 1999 at Helendale, California; this included measurement of radar cross-section, assessment of antenna performance and demonstration of the robustness of supportable low-observable materials.

Significant events in 2000 included X-35A flight readiness review in March. Subsequently, in April, testing associated with development and flight qualification of the JSF119-PW-611 engine for the X-35A and X-35C was completed after 193 hours of operating time. Final assembly and painting of X-35A in late May was followed by lengthy series of ground tests, including systems checkout, engine running at full military power and with full afterburner augmentation, and low- and medium-speed taxi tests. These extended into October and culminated in a successful first flight by X-35A Article 301 from Palmdale to Edwards on 24 October; by 6 November, further four flights had been made, including first by USAF pilot, and envelope had been expanded to 390 kt (722 km/h; 449 mph); first aerial refuelling (from KC-135) on 7 November; maiden supersonic flight 21 November, when 25 hours had been flown in 25 sorties. X-35A test programme completed on 22 November, whereupon Article 301 returned to Palmdale for conversion to X-35B. X-35C (Article 300) first flight took place on 16 December 2000 from Palmdale to Edwards AFB. Initial X-35C testing at Edwards was completed in early February 2001, with aircraft making transcontinental ferry flight via Fort Worth, Texas, to Navy test centre at Patuxent River, Maryland on 9-10 February for specialised trials.

Following installation of shaft-driven lift fan in late December 2000, X-35B STOVL version began series of hover-pit trials on 22 February 2001; these concluded 16 March and included 26 lift-fan clutch engagements from CTOL to STOVL mode at high rpm settings. Accelerated mission testing followed and was concluded on 6 April, with X-35B then being readied for flight trials; this process included installation of flight-ready lift fan. Taxi tests began 12 June 2001, followed by first brief vertical take-off and landing on 23 June, with sustained hover accomplished on next day. Initial testing at Palmdale completed successfully by 3 July, when X-35B flown to Edwards AFB for remainder of trials programme; notable events included first airborne transition from STOVL to conventional mode on 9 July; first vertical landing from wingborne flight on 16 July and 'Mission X' demonstrations involving short take-off, level supersonic dash and vertical landing on 20 and 26 July, before STOVL testing concluded on 30 July.

X-35 FLIGHT TESTS

	X-35A	X-35B	X-35C
First flight	24 Oct 00	23 Jun 01	16 Dec 00
Last flight	22 Nov 00	6 Aug 01	10 Mar 01
Sorties	27	39	73
Hours	27.4	21.5	58.0
Max g	5.0	5.0	4.8
Max speed	M1.05	M1.2	M1.22
Max AoA	20°	20°	20°
Max alt (ft)	34,000	34,000	34,000
Dummy deck landings			252
VTOs		17	
STOs		14	
Short landings		6	
Vertical landings		27	

After study of test results and company proposals, the US Department of Defense announced on 26 October 2001 that Lockheed Martin would be awarded a US$19 billion contract to cover SDD of what now became known as the F-35. SDD began immediately and is scheduled to last 126 months. Simultaneously, it was revealed that Pratt & Whitney would receive a contract worth US$4 billion for development and production of the associated F135 engine. As a major partner, the UK will contribute US$2 billion towards total SDD expenditures of approximately US$25 billion.

Key events during the next few years include a preliminary design review (PDR) at the end of FY02, with a critical design review (CDR) following about nine months later, in mid-2003. A total of 14 fully instrumented flying aircraft will be built at Fort Worth and assigned to SDD. Five will emerge as **F-35As** (USAF CTOL version); five will be **F-35Cs** (US Navy CV version); and the remaining four will be **F-35Bs** (US Marine Corps/UK STOVL version). Further eight non-flying airframes: two of each version as static test articles, plus another F-35C for drop testing and one pole-test airframe for radar signature evaluation. F-35A first flight scheduled for October 2005, 48 months into SDD, although Lockheed Martin is aiming for maiden flight on 28 August 2005. F-35B to fly after 53 months (early 2006); and F-35C after 62 months (late 2006). Flight trials will be undertaken by joint industry/service teams at Edwards AFB, California (USAF) and NAS Patuxent River, Maryland (US Navy). Programme originally expected to involve almost 15,000 flight hours, although this reduced to 10,185 (about 5,700 sorties) according to most recent estimates.

Full funding for Low-Rate Initial Production (LRIP) is scheduled to begin in about mid-2006 (see table), with the first delivery of an operational aircraft (an F-35B for the US Marine Corps) due in late 2008, leading to IOC in last quarter of FY10. Respective IOC dates for the F-35A and F-35C are final quarter of FY11 and final quarter of FY12, with the UK also expected to achieve IOC in FY12. Initial aircraft will be to Block 1 standard, with basic warfighting capability (including compatibility with JDAM and AIM-120 AMRAAM); additional weaponry will be added on Block 2 standard aircraft, while Block 3 will introduce SEAD and deep strike capability. Lockheed Martin anticipates production rate of 17 aircraft per month from 2011, but would like to

achieve rate of 20 per month and could go as high as 30 to 35 if required to do so, although latter could only be accomplished with use of multiple subassembly production centres.

CURRENT VERSIONS: Three variants of basic design, optimised for the mission requirements of different armed services. As currently envisaged, basic company designation for operational JSF is Configuration 230-5, with following variants expected to enter service with US armed forces.

F-35A: Land-based **CTOL** derivative for USAF. Evaluated by X-35A between 24 October and 22 November 2000.

F-35B: STOVL version for US Marine Corps, Royal Air Force and Royal Navy. Engine-driven lifting fan behind cockpit replaces some fuel. Wing folding originally specified on Royal Navy version only, but requirement since dropped. Broader and higher spine behind cockpit to accommodate lifting fan and air intake; shortened cockpit canopy replaces blister-type glazing of F-35A and B. Evaluated by X-35B between 23 June and 30 July 2001; final flight on 6 August 2001 was from Edwards AFB to Lockheed Martin facility at Palmdale.

F-35C: Carrier-based CTOL (CV). Wing, fin and elevator areas increased by chord extension; ailerons in addition to flaperons on wing; enlarged control surfaces and modified control system; strengthened landing gear with catapult launch bar on twin-wheel nose leg; arrester hook; and folding wing. Evaluated by X-35C between 16 December 2000 and 10 March 2001 from test centres at Edwards AFB and Patuxent River, Maryland (from 10 February 2001).

CUSTOMERS: Two X-35 demonstrators; potential for some 3,000 F-35s for USA and UK (see table). UK confirmed on 30 September 2002 that F-35B (STOVL) version will be procured for both RAF and Royal Navy. Orders also expected from Australia, Canada, Denmark, Italy, the Netherlands, Norway and Turkey.

US and UK F-35 REQUIREMENTS

Service	Qty	Remarks
US Air Force	1,763	Replaces A-10 and F-16; complements F-22. IOC FY11
US Navy	480	Complements F/A-18E/F. IOC FY12
US Marine Corps	609	Replaces AV-8B and F/A-18. IOC FY10
Royal Navy (UK)	60	Replaces Sea Harrier. IOC FY12
Royal Air Force (UK)	90	Replaces Harrier
Total	**3,002**	

Note: In early 2003, it appeared likely that procurement for the US Navy and Marine Corps could be significantly reduced.

COSTS: At time of down select for SDD phase, unit cost of USAF F-35A quoted as US$40 million, with CV/STOVL versions costing under US$50 million each. US$3.41 billion SDD funding for FY03, following US$1.52 billion appropriation in FY02; requests for FY04 and FY05 are US$4.37 billion and US$4.47 billion respectively. Fly away costs mentioned in connection with Italian involvement quoted as US$36.6 million for CTOL version, US$47.4 million for CV version and US$45.3 million for STOVL version.

Cutaway view of F-35C, showing lifting fan and vectored engine thrust   0116922

F-35 INITIAL PRODUCTION

Batch	Total	F-35A	F-35B	F-35C	RAF/RN	Last Delivery
LRIP 1	10	6		4		1Q2009
LRIP 2	22	14		8		4Q2009
LRIP 3	54	20	9	20	5	4Q2010
LRIP 4	91	30	20	32	9	4Q2011
LRIP 5	120	44	32	32	12	4Q2012
LRIP 6	168	72	48	36	12	4Q2013
Totals	**465**	**186**	**109**	**132**	**38**	

DESIGN FEATURES: Trapezoidal mid-wing configuration, optimised for low observability from frontal aspect. Twin tailfins; internal weapon bays. Wing and tailplane leading-edges swept back approximately 33°; trailing-edges swept forward approximately 14°; fins swept back approximately 42° and canted outward at tips by approximately 25°. Twin 'divertless' fibre-placed graphite/epoxy composites engine air intakes with no moving parts produced by ATK. All-electric flight control system.

STOVL version employs a lifting fan behind the cockpit, driven by a shaft from the single engine; inlet and outlet are covered by doors, except when in use. The resultant cold air barrier prevents hot air from being reingested when on or near the ground.

FLYING CONTROLS: All-electric flight control system for movement of primary flight control surfaces (flaps, tailerons and rudder) incorporating Parker Hannifin electric actuators. Moog leading-edge flap drive system. Flight control computer originally to be supplied by Honeywell, but replaced by advanced Lockheed Martin unit in third quarter of 1998 to eliminate anticipated throughput problems arising from growth in flight control software.

LANDING GEAR: Retractable tricycle type; mainwheels retract inwards; nosewheel(s) forward. Single wheels on each unit, except twin nosewheels and catapult towbar on F-35C, which also has reinforced gear for deck landings. Tyres by Goodyear will feature embedded transponder including integrated circuit and capacitive pressure sensor to facilitate monitoring of pressure and condition.

POWER PLANT: One 177 kN (40,000 lb) class (111 kN; 16,000 lb St dry thrust) Pratt & Whitney F135 (formerly JSF119-PW-611; F119 derivative) turbofan. Axisymmetric (thrust-vectoring) exhaust nozzle on USAF and US Navy -611C versions. Rolls-Royce three-bearing swivel-duct nozzle on -611S version to deflect thrust downwards for STOVL, plus a Rolls-Royce engine-driven fan behind cockpit and a bleed air reaction control valve in each wingroot to provide stability at low speeds. For F-35B, total vertical lift of 173 kN (39,000 lb st) comprises some 40 per cent from main nozzle, 48 per cent from fan and 12 per cent from reaction control valves. F-35A has in-flight refuelling receptacle on spine; US Navy specifies retractable probe on starboard side.

ACCOMMODATION: Pilot only; canopy by Sierracin; STOVL versions have canopy of reduced length. Martin-Baker Mk 16E lightweight ejection seat in X-35s; F-35 expected to have new seat resulting from Joint Ejection Seat Program (JESP).

SYSTEMS: Hamilton Sundstrand 80 kW engine-driven switched-reluctance starter/generator providing two independent channels of 270 V DC electrical power; electric distribution units, power panels, power distribution centres, batteries and battery charger equipment provided by Smiths Industries; Honeywell thermal- and energy-management module (T/EMM) combining functions of auxiliary and emergency power units and environmental control system; 270 V DC lithium ion emergency battery. Weapons bay door drive system by TRW Aeronautical Systems.

AVIONICS: *Radar:* Northrop Grumman MIRFS/MFA (multifunction integrated RF system/multifunction nose array) active electronically scanned array combining radar, electronic warfare and communications functions planned for production F-35.

Flight: Lockheed Martin Tactical Defense Systems Integrated Core Processor (ICP) incorporating open system architecture.

Instrumentation: Kaiser 200 × 500 mm (8 × 20 in) single flat panel MFD. Meggitt secondary flight display system. Kaiser/Vision Systems International selected in third quarter of 1999 to provide advanced integrated helmet-mounted display system.

Mission: Northrop Grumman conformal array multifunction imaging IR sensor system under development for JSF; distributed architecture IR system (DAIRS) functions include air-to-air search-and-track, target cueing and missile warning, and air-to-ground surface-target tracking; uses six conformal compact lightweight imaging IR sensors around airframe, with combined data providing all-aspect multifunction imaging to pilot via wide-angle helmet-mounted display, overlaid with target and threat data information. Lockheed Martin internal Electro-Optical Targeting System (EOTS) can be used when engaging targets on the ground and in the air.

Self-defence: EW capability incorporated into MIRFS/MFA. BAE Systems is prime contractor for EW equipment and systems integration; Litton Amecom to supply low-cost ESM equipment.

ARMAMENT: Internal cannon (in port engine air intake trunk upper surface) specified by USAF and under consideration by US Navy; Boeing/Mauser BK 27 (27 mm) weapon selected in second quarter of 1999; US Marine Corps and UK variants to have 'missionised' cannon in low-observables pod. Internal weapon bays, incorporating pneumatic weapon suspension and release equipment by EDO; preferred USAF weapons fit is two AIM-120 AMRAAMs and two 1,000 lb JDAM bombs; US Navy

specifies 2,000 lb JDAM (which optional on all variants). Optional external stores on four hardpoints.
Data are provisional.

DIMENSIONS, EXTERNAL (estimated):

Wing span: F-35A, F-35B	10.70 m (35 ft 1¼ in)
F-35C	13.26 m (43 ft 6 in)
Length overall: F-35A, F-35B	15.39 m (50 ft 6 in)
F-35C	15.47 m (50 ft 9 in)
Height overall: F-35A, F-35B	4.57 m (15 ft 0 in)
F-35C	4.72 m (15 ft 6 in)
Tailplane span: F-35A, F-35B	7.24 m (23 ft 9 in)
F-35C	8.69 m (28 ft 6 in)
Wheel track	4.42 m (14 ft 6 in)
Wheelbase: F-35A, F-35B	6.10 m (20 ft 0 in)
F-35C	5.87 m (19 ft 3 in)

AREAS:

Wings, gross: F-35A, F-35B	42.7 m² (460 sq ft)
F-35C	57.6 m² (620 sq ft)

WEIGHTS AND LOADINGS (approx):

Weight empty: F-35A	12,020 kg (26,500 lb)
F-35B, F-35C	13,608 kg (30,000 lb)
Max weapon load:	
F-35A, F-35B	more than 5,897 kg (13,000 lb)
F-35C	more than 7,711 kg (17,000 lb)
Max internal fuel weight:	
F-35A	more than 8,165 kg (18,000 lb)
F-35B	more than 5,897 kg (13,000 lb)
F-35C	more than 8,618 kg (19,000 lb)
Max T-O weight: all	approx 27,215 kg (60,000 lb)
Max wing loading:	
F-35A, F-35B	530.7 kg/m² (108.70 lb/sq ft)
F-35C	393.7 kg/m² (80.65 lb/sq ft)

PERFORMANCE (approx):

Max level speed: all	M1.6
Combat radius: F-35A, F-35C	
	more than 600 n miles (1,111 km; 690 miles)
F-35B	more than 450 n miles (833 km; 517 miles)

UPDATED

LOCKHEED MARTIN F-16 FIGHTING FALCON

Israel Defence Force names: F-16C Barak (Lightning), F-16D Brakeet (Thunderbolt) and F-16I Suefa (Storm)

Turkish Air Force name: Savaşan Şahin (Fighting Falcon)

TYPE: Multirole fighter.

PROGRAMME: Emerged from YF-16 of US Air Force Lightweight Fighter prototype programme 1972 (details under General Dynamics in 1977-78 and 1978-79 *Jane's*); first flight of prototype YF-16 (72-01567) 2 February 1974; first flight of second prototype (72-01568) 9 May 1974; selected for full-scale development (FSD) 13 January 1975; day fighter requirement extended to add air-to-ground capability with radar and all-weather navigation; production of FSD aircraft began July 1975; first flight of FSD aircraft 8 December 1976. F-16 achieved 5 millionth flying hour late in 1993; 7 millionth flying hour passed in early 1997; 3,900th aircraft delivered 26 March 1999, coincidentally 3,035th from Fort Worth factory and 2,202nd for USAF. 4,000th aircraft delivered (to Egyptian Air Force) on 28 April 2000; total of 4,046 delivered at end of 2001.

Total of 234 F-16s ordered in 2000, increasing sales to 4,285 as of 1 January 2001; Israeli and Greek decisions to exercise options on additional F-16s raised total sales to 4,347 at end of 2001. Further orders placed by Chile (10 aircraft) and Oman (12) in 2002, increasing sales to 4,369 by end of third quarter.

CURRENT VERSIONS: **F-16A:** First version for air-to-air and air-to-ground missions; not currently in production, but still available for export customers. See 2002-03 and previous *Jane's* and *Jane's Aircraft Upgrades* for details.

F-16B: Standard tandem two-seat version of F-16A; fully operational both cockpits; fuselage length unaltered; reduced fuel. See 2002-03 and previous *Jane's* and *Jane's Aircraft Upgrades* for details.

F-16C/D: Single-seat and two-seat USAF Multinational Staged Improvement Program (**MSIP**) aircraft respectively, implemented February 1980. MSIP expands growth capability to allow for ground attack and BVR missiles, and all-weather, night and day missions; **Stage I** applied to Block 15 F-16A/Bs delivered from November 1981; **Stage II** applied to Block 25 F-16C/Ds from July 1984 includes core avionics, cockpit and airframe changes. Block 25 aircraft originally delivered with F100-PW-200 engine, but all surviving examples now have F100-PW-220E following retrofit, as well as other improvements as detailed under Stage III of MSIP. **Stage III** involved installation of systems as they became

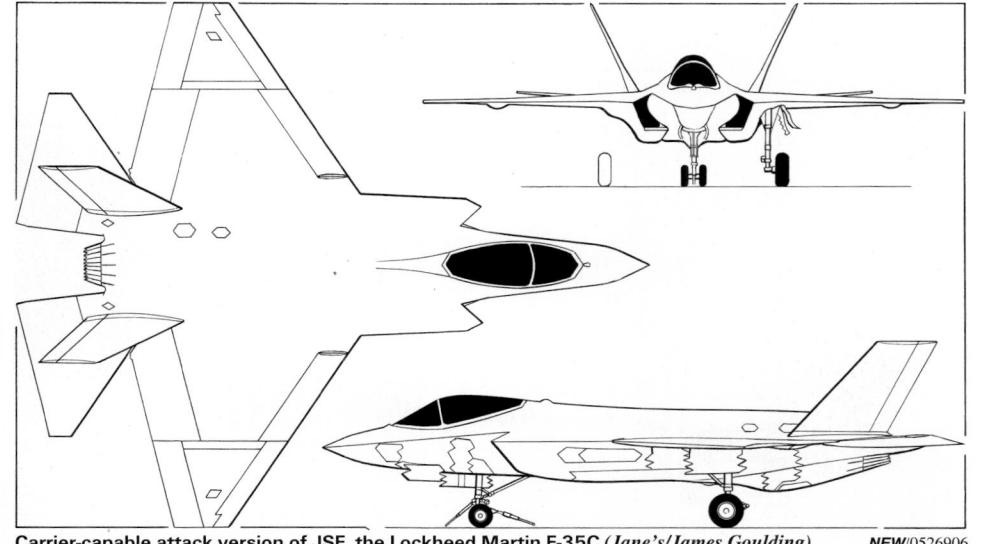

Carrier-capable attack version of JSF, the Lockheed Martin F-35C (*Jane's/James Goulding*)   NEW/0526906

For details of the latest updates to *Jane's All the World's Aircraft* online and to discover the additional information available exclusively to online subscribers please visit

jawa.janes.com

Prototype IAI ACE upgrade (an F-16B) first flew on 29 May 2001   0095639

available, beginning 1988 and extending to Block 50/52, including selected retrofits back to Block 25. Changes include Northrop Grumman AN/APG-68 multimode radar with improved range, resolution, more operating modes and better ECCM than AN/APG-66; advanced cockpit with multifunction displays and upfront controls, BAE Systems wide-angle HUD, Fairchild mission data transfer equipment and radar altimeter; expanded base of fin giving space for proposed later fitment of AN/ALQ-165 Airborne Self-Protection Jamming system (since cancelled, though now being installed in Korean aircraft); increased electrical power and cooling capacity; structural provision for increased take-off weight and manoeuvring limits; and smart weapons such as AIM-120A AMRAAM and AGM-65D IR Maverick.

Common engine bay introduced at **Block 30/32** (deliveries from July 1986) to allow fitting of either P&W F100-PW-220 (Block 32) or GE F110-GE-100 (Block 30) Alternate Fighter Engine. Other changes include computer memory expansion and seal-bonded fuselage fuel tanks. First USAF wing to use F-16C/Ds with F110 engines was 86th TFW at Ramstein AB, Germany, from October 1986. Additions in 1987 included voice message unit, doubled chaff/flare capacity, repositioning of RWR antennas to provide better coverage in forward hemisphere, Shrike anti-radiation missiles (from August 1997), crash survivable flight data recorder and modular common inlet duct allowing full thrust from F110 at low airspeeds.

Software upgraded for full Level IV multitarget compatibility with AMRAAM early 1988. Industry-sponsored development of radar missile capability for several European air forces resulted in firing of AIM-7F and AIM-7M missiles from F-16C in May 1988; capability introduced mid-1991; missiles guided using pulse Doppler illumination while tracking targets in a high PRF mode of the AN/APG-68 radar.

Block 40/42 Night Falcon (deliveries from December 1988) upgrades include AN/APG-68(V) radar allowing 100 hour operation before maintenance, full compatibility with Lockheed Martin low-altitude navigation and targeting infra-red for night (LANTIRN) pods, four-channel digital flight control system, expanded capacity core computers, diffractive optics HUD, enhanced envelope gunsight, GPS, improved leading-edge flap drive system, improved cockpit ergonomics, high gross weight landing gear, structural strengthening, increased performance battery and provision for improved EW equipment, including advanced interference blanker. LANTIRN targeting pod gives day/night standoff target identification, automatic target handoff for multiple launch of Mavericks, autonomous laser-guided bomb delivery and precision air-to-ground laser ranging. LANTIRN navigation pod provides real-world IR view through HUD for night flight plus automatic/manual terrain following with dedicated radar sensor. Combat Edge pressure breathing system installed 1991 for higher pilot *g* tolerance.

A total of 39 Block 40 F-16C/Ds of the 31st FW was involved in a quick response capability (QRC) modification effort, known as 'Sure Strike', to install Improved Data Modem (IDM) equipment. Work was undertaken by a joint USAF/LMTAS team at Aviano AB, Italy, and was completed in December 1995, these being the first Block 40 aircraft to receive the IDM which is standard equipment on current production Block 50 F-16s. In mid-1998, a demonstration programme was conducted to adapt existing IDM with Lockheed Martin kit to provide 'Gold Strike' system capable of two-way transmission of digitised video imagery of targets and thus enhance pilot's situational awareness.

First Block 40 F-16C/Ds issued in late 1990 to 363rd FW (Shaw AFB, South Carolina); first LANTIRN pods to 36th FS/51st FW at Osan, South Korea, in 1992. USAF Block 40/42 F-16Cs unofficially designated **F-16CG**. Following retirement of F-111F, Block 40/42 F-16Cs and F-16Ds with LANTIRN comprise more than 50 per cent of USAF night/precision strike force.

Block 50/52 (deliveries began with F-16C 90-0801 in October 1991 for operational testing) upgrades include F110-GE-129 and F100-PW-229 increased performance

engines (IPE), AN/APG-68(V)5 radar with advanced programmable signal processor employing VHSIC technology, Have Quick IIA UHF radio and AN/ALR-56M advanced RWR. Changes initiated at **Block 50D/52D** in 1993 include full integration of HARM anti-radiation missiles via HARM aircraft launcher interface computer (ALIC), improved data modem (IDM), upgraded programmable display generator with growth potential for colour and map, expanded data transfer cartridge, ring laser INS (Honeywell and Litton units both used) and improved VHF/FM antenna. AN/ALE-213 advanced chaff/flare dispenser fitted to all FMS Block 50 aircraft delivered after mid-1996 and incorporated as standard on USAF aircraft with effect from FY97 purchase; also retrofitted to earlier USAF machines.

First Block 50D/52D (91-0360) delivered to USAF on 7 May 1993; optimised for defence suppression missions, having software for horizontal situation display on existing two MFDs and provision for one of 112 HARM (AGM-88 High-Speed Anti-Radiation Missile) targeting systems ordered by USAF; sensor in pod on starboard side of engine inlet; AN/ASQ-213 HARM targeting system (HTS) has capability similar to F-4G 'Wild Weasel' which it replaced in SEAD role. USAF has a current programme

to raise HARM targeting system inventory to 150 pods and is incorporating an upgrade that features software and hardware improvements enabling more targets to be tracked with enhanced ambiguity resolution and speedier reaction time.

Deliveries of Block 50/52 began to 4th FS of 388th FW at Hill AFB, Utah, from October 1992; others to 52nd FW at Spangdahlem, Germany, replacing Block 30 aircraft from (first delivery) 20 February 1993. Block 50D/52D aircraft initially to 309th FS (now 79th FS) of 363rd (now 20th) FW at Shaw AFB, South Carolina; followed by 23rd FS/52nd FW at Spangdahlem, Germany, from 14 January 1994, then squadrons at Mountain Home AFB, Idaho, and Misawa, Japan. Production for USAF was due to terminate with FY94 batch, but six additional F-16Cs funded in each of FY96 and FY97, plus three in FY98, one in FY99, 10 in FY00 and four in FY01; FY96 aircraft to Block 50 standard. FY97 and subsequent aircraft delivered from mid-2000 are to improved configuration, incorporating modular mission computer by Raytheon (replacing three core avionics processors) that was developed for F-16A/B MLU programme, Honeywell colour liquid crystal multifunction displays (replacing monochrome CRT MFDs), Honeywell colour programmable display generator, Teac colour airborne videotape recorder, colour cockpit TV sensor and Litton onboard oxygen generating system (OBOGS). FY00 and subsequent aircraft incorporate AN/APX-113 and associated avionics improvements. FY00 and FY01 aircraft delivered to USAF between April and December 2002. USAF Block 50/52 F-16Cs unofficially designated **F-16CJ**.

First Samsung-assembled F-16C Block 52D rolled out in South Korea on 7 November 1995. First flight of initial TAI-built F-16C Block 50D in late May 1996, with delivery to Turkish Air Force following on 29 July; last TAI-produced F-16 delivered October 1999. First Block 50D delivered to Greece 29 January 1997. First Block 52D delivered to Singapore 30 January 1998. First production lease (PL) Block 52D aircraft for Singapore was delivered 28 May 1998; this was under terms of commercial contract, with delivery achieved in less than 24 months from placing of order.

Block 50+/52+: Improved versions: first customer was Greece, which revealed intention on 30 April 1999 to buy as many as 70 and subsequently placed an order for 50

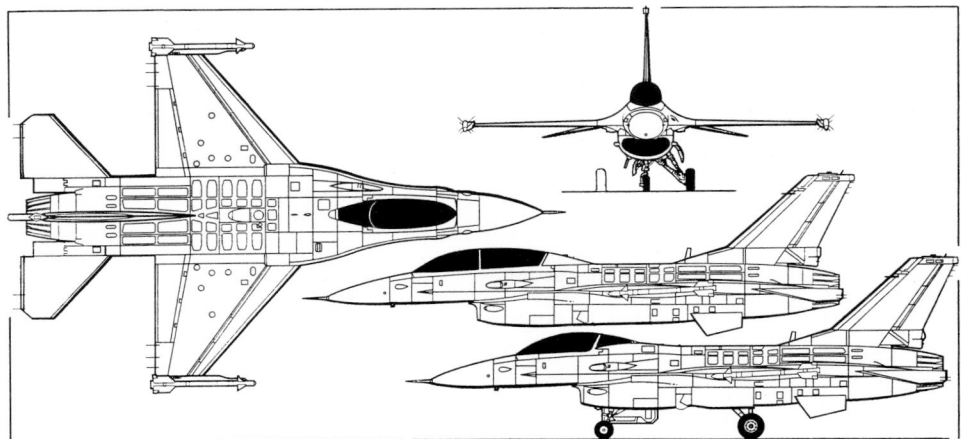

F-16C (GE F110 turbofan) with extra side view (top) of two-seat F-16D (P&W F100 turbofan) *(Jane's/Paul Jackson)*   0016166

Artist's impression of F-16I ordered by Israel for delivery from 2003   0083057

Block 52+ aircraft; at same time, a fixed price option was taken for up to 10 additional aircraft and this was converted to a firm order on 14 September 2001. Block 50+/52+ features upgraded AN/APG-68(V)9 radar with greater detection range and synthetic aperture radar (SAR) mode for high-resolution ground mapping, improved modular mission computer, two 102 mm (4 in) colour cockpit displays, helmet-mounted cueing system and a digital terrain system. The airframe structure is enhanced to meet design requirements for a full 8,000 hour service life, with (or without) the conformal 2,271 litre (600 US gallon; 500 Imp gallon) fuel tanks that can be carried by both single- and two-seat versions. Two-seat aircraft incorporate a dorsal equipment compartment. Raytheon Advanced Self-Protection Integrated Suite (ASPIS) EW equipment will be installed on Greek aircraft and Greece is also contemplating procurement of a high off-boresight missile system. Aircraft deliveries were due to begin in October 2002.

'GF-16C': Unofficial designation allocated to non-flying aircraft in use at Sheppard AFB for instructional purposes from 1993.

USAF F-16C/D Retrofit Programmes: All USAF F-16C/D aircraft are undergoing structural upgrade in a programme called 'Falcon UP'. This intended to ensure an 8,000 hour service life without depot inspection; late Block 50/52s receiving this during production, with earlier aircraft retrofitted concurrently with other depot-level upgrades. Follow-up effort called 'Falcon Star', in preparation in mid-1999, is expected to involve Block 40/42 and earlier aircraft delivered during 1984-91. 'Falcon Star' will be a structural enhancement to sustain operational life; USAF planning anticipates the modification programme to take six years to complete at a cost of between US$250 million and US$500 million. No start date yet determined, but work could begin in FY03.

ANG/AFRC Block 25/30/32 F-16C/Ds currently subject to combat upgrade plan integration details (CUPID); scheduled for completion by mid-2003, CUPID will bring about 620 older F-16Cs to a standard close to that of the Block 50/52 aircraft. Among the improvements being incorporated are situation awareness datalink (SADL), improved airborne videotape recorder, colour camera, initial NVG-compatible cockpit lighting, LANTIRN and Rafael Litening II FLIR targeting pod capability, AN/ALQ-213 countermeasures system and provisions for GPS/laser INS. Future improvements include expanded central computer, joint helmet-mounted cueing system, AIM-9X missile, follow-on NVIS capability, PIDS-3 pylon upgrade for smart weapon compatibility, ACES II ejection seat improvements, enhanced main battery and software upgrades.

Block 40/42 F-16C/Ds are currently being upgraded to include NVG compatibility, MD-1295/A improved data modem, digital terrain system, and smart weapons compatibility (including the GBU-24/27, JDAM, JSOW and WCMD). Block 50/52 F-16C/Ds already have (or will receive) these capabilities.

In June 1998, USAF launched an upgrade effort known as common configuration implementation program (CCIP), which is intended to provide common hardware and software capability to 650 Block 40/42/50/52 aircraft. CCIP is a multiphase effort and is being implemented in stages, based on availability of subsystems. Work is being undertaken at the Ogden Air Logistics Center, Hill AFB, Utah, to where the first eight modification kits were shipped on 29 June 2001; these kits include the modular mission computer and colour MFDs applicable to the Block 50/52 versions. The first aircraft was completed ahead of schedule and delivered to the 20th FW on 11 January 2002.

Phase two Block 50/52 kits include the AN/APX-113 combined electronic interrogator/transponder which gives

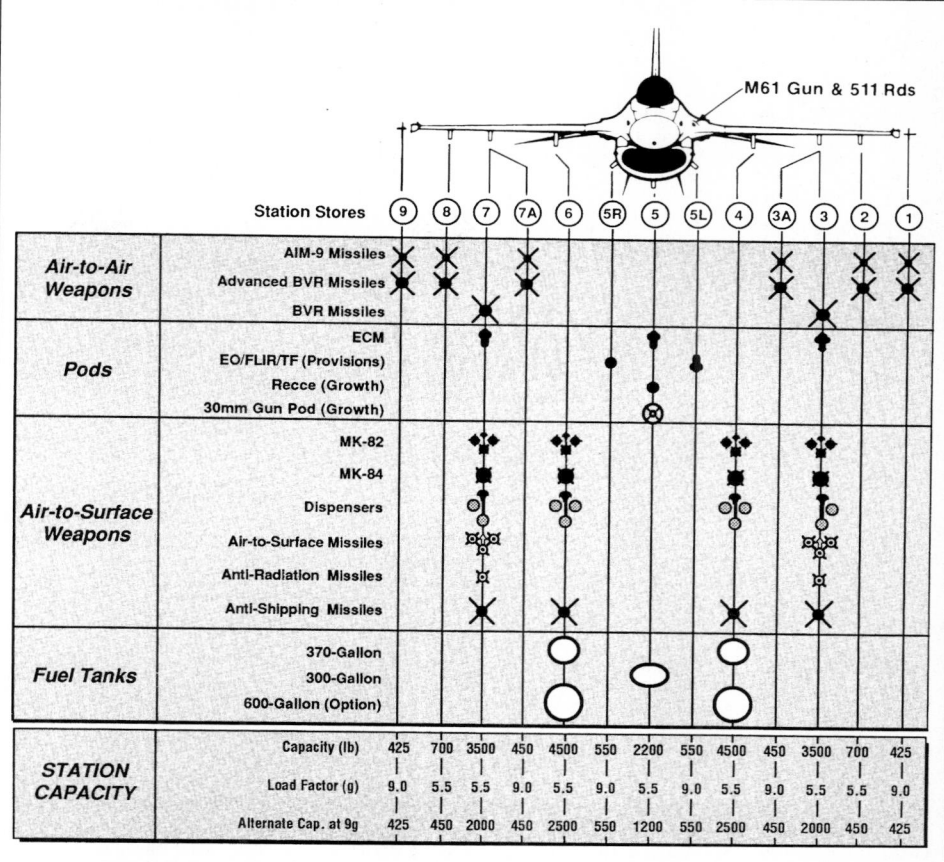

Lockheed Martin F-16C Fighting Falcon weapon options

autonomous BVR intercept capability. These aircraft also capable of alternative carriage of an advanced Lockheed Martin Sniper XR FLIR targeting pod in addition to the HARM targeting system pod. The first aircraft with this capability was delivered in September 2002.

The next phase, to begin in July 2003, adds Link 16 multifunctional information distribution system (MIDS) datalink, the Vision Systems International joint helmet-mounted cueing system (JHMCS) and an electronic horizontal situation indicator to Block 50/52 aircraft. Starting in 2005, Block 40/42 Fighting Falcons will receive the full package of modifications detailed above.

Block 60: On 12 May 1998, the United Arab Emirates announced selection of the Block 60 F-16, although signature of contract delayed by UAE decision to reopen competition for EW system; further announcement, on 5 March 2000, revealed that contractual agreement had been signed by UAE and Lockheed Martin. Total of 80 **Desert Falcon** aircraft will be delivered between 2004 and 2007. First three two-seat aircraft will be used for flight testing, with maiden flight expected in last quarter of 2003; unofficial designation **F-16F** adopted by trainers to comply with FAA regulations. Major differences from the Block 50/52 F-16 include Northrop Grumman AN/APG-80 active electronically scanned array (AESA) radar, Northrop Grumman AN/AAQ-32 integrated FLIR targeting system, advanced internal electronic countermeasures system, enhanced performance General

Electric F110-GE-132 engine in the 144.6 kN (32,500 lb st) class, conformal tanks carrying an additional 1,666 litres (440 US gallons; 366 Imp gallons) of fuel, an advanced cockpit (including three 127 × 178 mm (5 × 7 in) liquid crystal displays) and Thales secure radio and datalink equipment. Initial deliveries to UAE will be to so-called Lot 1 configuration which marries Block 50 capability to new EW suite, advanced cockpit and new radar. Lot 2 in early 2006 will feature advanced EW modes as well as terrain-following radar capability and additional weaponry; definitive Lot 3 version in 2007 will include BAE Systems TERPROM digital terrain avoidance and navigation system and, possibly, a helmet-mounted cueing system.

NF-16D: Variable stability in-flight simulator test aircraft (**VISTA**) modified from Block 30 F-16D (86-0048) ordered December 1988 to replace NT-33A testbed. Features include Calspan variable stability flight control system, fully programmable cockpit controls and displays, additional computer suite, permanent flight test data recording system, variable feel centrestick and sidestick in front cockpit, with latter also available for use in F-16 mode; safety pilot in rear cockpit. Internal gun, RWR and chaff/flare equipment removed, providing space for Phase II and III growth including additional computer, reprogrammable display generator and customer hardware allowance. Dorsal avionics compartment in bulged spine. Aircraft transferred to USAF Test Pilot School at Edwards AFB in 2001 to support TPS missions and various research and development activities. See 2002-03 *Jane's* for more details of aircraft history.

F-16I: Two-seat version (to F-16D-52+ standard) for Israel, featuring Pratt & Whitney F100-PW-229 engine, conformal fuel tanks and SAR. Will incorporate Northrop Grumman AN/APG-68(V)9 radar and significant amount of Israeli avionics, including EW suite, cockpit displays and helmet-mounted sight and advanced self-protection system (ASPS). EW suite, by Elisra, will include radar and missile approach warning systems plus jammers, while Elbit Systems to provide central mission computer, advanced display processor, DASH IV display and sight helmet system and stores management systems. RADA teamed with Smiths Aerospace to supply data acquisition system; Elop to provide HUD unit. Aircraft also to utilise Rafael Litening II targeting pod. Total of 50 initially purchased in US$2.5 billion deal, with delivery to begin in 2003; further 60 aircraft on option, of which 52 subsequently converted to firm order on 19 December 2001.

F-16N: US Navy supersonic adversary aircraft (SAA) modified from F-16C/D Block 30. Four of 26 were two-seat **TF-16N**. Entire fleet retired during 1994-95; last described in *Jane's* 2001-02.

FS-X and TFS-X: F-16 derivatives selected by Japan Defence Agency for its FS-X (now F-2) requirement 19 October 1987; details under Mitsubishi in Japanese section.

F-16C Fighting Falcons of the 555th 'Triple Nickel' Fighter Squadron, 31st Fighter Wing, Aviano, Italy
NEW/0533365

F-16 TABLES

Table 1 provides a rapid reference to customers and quantities; more detailed information on production block numbers appears in Tables 2 and 3.

TABLE 1: F-16 CUSTOMERS

Operator	Total	Single-seat	Qty	Two-seat	Qty	Power plant	First aircraft A/C	B/D	First delivery	Squadrons (or base)
Bahrain	22	F-16C-40	18	F-16D-40	4	F110-GE-100	101	150	March 1990	1, 2
Belgium	160[1]	F-16A-10	55	F-16B-10	12	F100-PW-200	FA01	FB01	January 1979	
		F-16A-15	41	F-16B-15	8	F100-PW-200	FA56	FB13	October 1982	1, 2, 23, 31, 349, 350
		F-16A-15OCU	40	F-16B-15OCU	4	F100-PW-220	FA97	FB21	January 1988	
Denmark	70	F-16A-10	30[1]	F-16B-10	8[1]	F100-PW-200	E-174	ET-204	January 1980	723, 727, 730
		F-16A-15	16[1]	F-16B-15	4[1]	F100-PW-200	E-596	ET-613	May 1982	723, 727, 730
		F-16A-15OCU	8[3]	F-16B-15OCU	4[3]	F100-PW-220	E-004	ET-197	December 1987	726
Egypt	220	F-16A-15	34	F-16B-15	8[2]	F100-PW-200	9301	9201	March 1982	72, 74
		F-16C-32	34	F-16D-32	6	F100-PW-220	9501	9401	August 1986	68, 70
		F-16C-40	34	F-16D-40	7	F110-GE-100	9901	9801	October 1991	60, 64
		F-16C-40	1	F-16D-40	5	F110-GE-100	9935	9808	1994	
		F-16C-40	34[7]	F-16D-40	12[7]	F110-GE-100	9951	9851	March 1994	75, 77
		F-16C-40	21			F110-GE-100B	(96-0086)	—	May 1999	71, 73
		F-16C-40	12	F-16D-40	12	F110-GE-100B	(99-0105)	(99-0117)	June 2001	
Greece	140	F-16CG-30	34	F-16DG-30	6	F110-GE-100	110	144	November 1988	330, 346
		F-16CG-50	32	F-16DG-50	8	F110-GE-129	045	077	January 1997	341, 347
		F-16CG-52+	40	F-16DG-52+	20	F100-PW-229			Late 2002	
Indonesia	12	F-16A-15OCU	8	F-16B-15OCU	4	F100-PW-220	S-1605	S-1601	December 1989	3
Israel	260	F-16A-10	67	F-16B-10	8	F100-PW-200	100	001	January 1980	140, 147, 253
		F-16C-30	51	F-16D-30	24	F110-GE-100A	301	020	December 1986	101, 105, 109, 110, 117
		F-16C-40	30	F-16D-40	30	F110-GE-100A	502	601	July 1991	101, 105, 109, 110, 117
				F-16D-52+ (F-16I)	102	F110-PW-229			2003	
Korea, South	180	F-16C-32	30	F-16D-32	10	F100-PW-220	85-574	84-370	March 1986	161, 162, 155
		F-16C-52	95[8]	F-16D-52	45[8]	F100-PW-229	92-000	92-028	December 1994	
Lockheed Martin	12	F-16C-52	4	F-16D-52	8	F100-PW-229	96-5033	96-5025	June 1998	Singapore (lease in USA)
Netherlands	213[3]	F-16A-10	46	F-16B-10	13	F100-PW-200	J-212	J-259	June 1979	306, 311, 312, 313,
		F-16A-15	84	F-16B-15	18	F100-PW-200	J-258	J-649	May 1982	315, 322, 323
		F-16A-15OCU	47	F-16B-15OCU	5	F100-PW-220	J-141	J-065	1988	
Norway	74	F-16A-10	28[3]	F-16B-10	7[3]	F100-PW-200	272	301	January 1980	332, 338
		F-16A-15	32[3]	F-16B-15	5[3]	F100-PW-200	300	690	June 1982	331, 334
				F-16B-15OCU	2	F100-PW-220	—	711	July 1989	331
Pakistan	68	F-16A-15	28	F-16B-15	12	F100-PW-200	82-701	82-601	January 1983	9, 11, 14
		F-16A-15OCU	13	F-16B-15OCU	15	F100-PW-220	91-729	91-613	—	See note 10
Portugal	20	F-16A-15OCU	17	F-16B-15OCU	3	F100-PW-220E	15101	15118	February 1994	201
Singapore	58	F-16A-15OCU	4	F-16B-15OCU	4	F100-PW-220	880	884	February 1988	140
		F-16C-52	18	F-16D-52	12	F100-PW-229	608	624	January 1998	
		F-16C-52	20[11]			F110-PW-229			Late 2003	
Taiwan	150	F-16A-20	120	F-16B-20	30	F100-PW-220	6601	6801	July 1996	12, 14, 17, 21, 22, 23, 26, 27
Thailand	36	F-16A-15OCU	14	F-16B-15OCU	4	F100-PW-220	10305	10301	June 1988	103
		F-16A-15OCU	12	F-16B-15OCU	6	F100-PW-220	07020	07032	September 1995	403
Turkey	240[4]	F-16C-30	34	F-16D-30	9	F110-GE-100	86-0066	86-0191	May 1987	142, Oncel Flight
		F-16C-40	102	F-16D-40	15	F110-GE-100	88-0033	88-0014	July 1990	141, 161, 162, 191, 192, 181, 182
		F-16C-50	60	F-16D-50	20	F110-GE-129	93-0657	93-0691	July 1996	141, 151, 152
UAE	80	F-16C-60	55	F-16D-60	25	F110-GE-132			2004	
USAF	2,230[9]	F-16A-10	255	F-16B-10	74	F100-PW-200	78-0001	78-0077	August 1978	See note A
		F-16A-15	409[5]	F-16B-15	46[6]	F100-PW-200	80-0541	80-0635	September 1981	See note A
		F-16C-25	209	F-16D-25	35	F100-PW-200	83-1118	83-1174	July 1984	See note B
		F-16C-30/32	360/56	F-16D-30/32	48/5	both (100/220)	85-1398	85-1509	July 1986	See note B
		F-16C-40/42	234/150	F-16D-40/42	31/47	both (100/220)	87-0350	87-0391	December 1988	See note B
		F-16C-50/52	189/42	F-16D-50/52	28/12	both (129/229)	90-0801	90-0834	October 1991	See note B
US Navy	26	F-16N-30	22	TF-16N-30	4	F110-GE-100	163268	163278	June 1987	Stored
Venezuela	24	F-16A-15	18	F-16B-15	6	F100-PW-200	1041	1715	September 1983	161, 162
Totals	**4,347**		**3,447**		**900**					

Notes:

[1] Built by Sabca (Belgium); 222nd and last Sabca F-16 (BAF FA-136) delivered 22 October 1991

[2] One built by Fokker (Netherlands)

[3] Built by Fokker; 300th and last Fokker F-16 (RNethAF J-021) delivered 27 February 1992

[4] Two F-16Cs and six F-16Ds built by GD; remainder by TAI

[5] Two built by Fokker

[6] Four built by Sabca

[7] TAI production

[8] 12 built by GD, 36 CKD kits and 92 produced locally by Samsung Aerospace

[9] Table excludes full-scale development (FSD) aircraft (six F-16A and two F-16B)

[10] Embargoed. Pakistan offered compromise batch of 38 aircraft. Offer rejected and 28 completed aircraft of follow-on order for 71 placed in storage at Davis-Monthan AFB, Arizona. To be distributed between US Navy (10 F-16A and four F-16B) for aggressor training and US Air Force (three F-16A and 11 F-16B) for test support duties

[11] Includes unspecified number of F-16D-52 aircraft

Note A: Currently operated by US Air Force Flight Test Center/412th Test Wing, Edwards AFB, California; US Air Force Air Armament Center/46th Test Wing, Eglin AFB, Florida (39, 40 FITSs); 56th FW, Luke AFB, Arizona (21 FS/Taiwan AF training squadron); 148, 178, 179, 184, 186 and 195 FSs, Air National Guard.

Note B: Currently operated by US Air Force Flight Test Center/412th TW, Edwards AFB, California; US Air Force Air Armament Center/46th Test Wing, Eglin AFB, Florida (39, 40 FITSs); 8th FW, Kunsan AB, South Korea (35, 80 FSs); 20th FW, Shaw AFB, South Carolina (55, 77, 78, 79 FSs); 27th FW, Cannon AFB, New Mexico (522, 523, 524 FSs and 428 FS/Singapore AF training squadron); 31st FW, Aviano AB, Italy (510, 555 FSs); 35th FW, Misawa AB, Japan (13, 14 FSs); 51st FW, Osan AB, South Korea (36 FS); 52nd FW, Spangdahlem AB, Germany (22, 23 FSs); 53rd W, Eglin AFB, Florida (422 Test and Evaluation Squadron, at Nellis AFB, Nevada); 56th FW, Luke AFB, Arizona (61, 62, 63, 308, 309, 310, 425 FSs); 57th W, Nellis AFB, Nevada (F-16 Fighter Weapons School, 414 Combat Training Sqdn, Thunderbirds Air Demonstration Sqdn); 354th FW, Eielson AFB, Alaska (18 FS); 366th W, Mountain Home AFB, Idaho (389 FS); 388th FW, Hill AFB, Utah (4, 34, 421 FSs); and also 93, 301, 302, 457 and 466 FSs, Air Force Reserve Command; 107, 111, 112, 113, 119, 120, 121, 124, 125, 134, 138, 149, 152, 157, 160, 162, 163, 170, 174, 175, 176, 182, 184, 188 and 194 FSs, Air National Guard.

Blocks 1 and 5 retrofitted to Block 10 standard 1982-84; Block 15 retrofitted to Block 15OCU (avionics standard only) from 1987. New-build F-16As are Block 15OCU from November 1987.

Export programme codenames are: Bahrain – Peace Crown I-II; Egypt – Peace Vector I, II, III, IIIA and IV; Greece – Peace Xenia I-II; Indonesia – Peace Bima-Sena; Israel – Peace Marble I-V; Jordan – Peace Falcon; South Korea – Peace Bridge I-II; Pakistan – Peace Gate I-IV; Portugal – Peace Atlantis I-II; Singapore – Peace Carvin I-IV; Taiwan – Peace Fenghuang; Thailand – Peace Naresuan I-II; Turkey – Peace Onyx I-II (and CAE-Link simulator, Peace Onyx III); and Venezuela – Peace Delta.

Recent orders: Contracts agreed in 2000 with Greece (34 F-16C, 16 F-16D); Israel (50 F-16I); South Korea (15 F-16C, 5 F-16D); Singapore (20 F-16C/D) and the United Arab Emirates (55 F-16C, 25 F-16D). Greek option for 10 aircraft (six F-16C and four F-16D) converted to firm order in 2001; Israel also sought to convert option in 2001, when further 52 F-16I added to outstanding order. USAF resumed procurement with batch of six aircraft in FY96 plus six in FY97, three in FY98, one in FY99, 10 in FY00 and four in FY01; up to 20 more may be purchased by 2010. Chile expected to be next customer, with order for 10 F-16C/D Block 50+ aircraft due to be signed by end of 2001; Oman has also requested 12 new aircraft.

Transfers: Israel received 36 F-16As and 14 F-16Bs from surplus USAF stocks August 1994 to March 1995 for 140, 144 and 253 Squadrons. Denmark received three ex-USAF F-16As in July 1994, plus three F-16As and one F-16B in 1997 and plans to obtain another three in the post-2000 period. **Jordan** took delivery of an initial batch of ex-USAF ADF aircraft (12 F-16As and four F-16Bs) on a lease basis in December 1997/April 1998 and may eventually receive 36 to 48 examples; Portugal to receive 25 second-hand aircraft (21 F-16As and four F-16Bs, of which five F-16As to be broken-down for spares); **Thailand** request for 16 surplus USAF F-16A/B approved by US Congress at beginning of 2000; **Italy** to acquire 30 F-16A and four F-16B in 2003 on five-year lease arrangement; and Austria, Bulgaria, Croatia, Czech Republic, Oman, Philippines, Poland, Romania, Slovenia and Venezuela may obtain F-16s on purchase/lease from this source.

F-16 Recce: Following trials with a prototype system in mid-1995, an LMTAS-designed reconnaissance pod for the US Air National Guard was flown for the first time on 29 September 1995 and subsequently tested on an F-16C (86-227) of the 149th Fighter Squadron, Virginia ANG.

Four pods were built and then deployed with the 149th FS to Aviano AB, Italy, in May 1996 for operational validation in support of NATO missions over Bosnia. This successful trial culminated in decision to procure a total of 20 podded systems for service with F-16C Block 30

aircraft of five ANG squadrons, each of which will have four pods and one ground exploitation system. The core of the programme, known as the Theater Airborne Reconnaissance System (TARS), is the Per Udsen MRP with USAF-supplied KS-87 cameras incorporating E-O

TABLE 3: F-16C/D PRODUCTION

Block	USAF		USN		Bahrain		Egypt		Greece		Israel		Korea		Singapore		Turkey		United Arab Emirates		Lockheed Martin	
	F-16C	F-16D	F-16N	TF-16N	C	D	C	D	C	D	C	D	C	D	C	D	C	D	C	D	C	D
25	7	4																				
25A	16	3																				
25B	25	5																				
25C	35	5																				
25D	40	4																				
25E	47	4																				
25F	39	10																				
Subtotal	**209**	**35**																				
30	16	2									4	–										
32							1	4					2	6								
30A	40	2									15	1										
32A							13	2					4	–								
30B	47	3	4	–							15	–					17	6				
32B							20	–					4	–								
30C	35	2	8	–							16	–										
32C	20	3											4	–								
30D	42	4	8	–							1	4										
32D	13	–											4	–								
30E	55	5	2	4							–	10					17	3				
32E													4	–								
30F	51	6									–	9										
32F	2	1											4	–								
30H	34	12							2	4												
32H	11	1											4	–								
30J	25	8							12	2												
32J	10	–																				
30K	15	3							16	–												
30L									4													
32Q													–	4								
30VISTA	–	1																				
Subtotal	**625**	**88**	**22**	**4**			**34**	**6**	**34**	**6**	**51**	**24**	**30**	**10**			**34**	**9**				
40	2	–																				
40A	6	3															17	3				
42A	5	3																				
40B	22	–																				
42B	7	13																				
40C	38	3																				
42C	17	2																				
40D	38	2			2	4											17	3				
42D	17	3																				
40E	37	–			6	–																
42E	18	5																				
40F	33	4															17	3				
42F	15	8																				
40G	31	5					2	–														
42G	21	3																				
40H	14	5					–	4			6	4										
42H	27	6																				
40J	7	5					9	2			8	8					16	3				
42J	21	4																				
40K	6	4					15	1			8	8										
42K	2	–																				
40L							8				8	10					17	3				
40M							1	1														
40N							–	8														
40P																	18	–				
40Q							3	3														
40R							12	5														
40S							19															
40T							5															
40U							8															
40V							8															
40W					10																	
40							12	12														
Subtotal	**1,009**	**166**	**22**	**4**	**18**	**4**	**136**	**42**			**81**	**54**					**136**	**24**				
50	4	–																				
50A	7	7																				
52A	1	1																				
50B	24	8																				
50C	21	4																				
50D	133	9							32	8							60	20				
52D	41	11											95	45	38	12					4	8
52+							40	20				102										
Subtotal	**1,240**	**206**	**22**	**4**	**18**	**4**	**136**	**42**	**106**	**34**	**81**	**156**	**125**	**55**	**38**	**12**	**196**	**44**			**4**	**8**
60																			55	25		
Totals	**1,240**	**206**	**22**	**4**	**18**	**4**	**136**	**42**	**106**	**34**	**81**	**156**	**125**	**55**	**38**	**12**	**196**	**44**	**55**	**25**	**4**	**8**
Grand total											**2,611**											

video back instead of wet film; the Lockheed Martin Fairchild Systems medium-altitude E-O (MAEO) camera; an Ampex DCRsi-240 digital data recorder and a TERMA Elektronik cockpit control device. The TARS pods were delivered in 1999 and certified for operational use by the F-16 in first half of 2000.

Elta EL/M-2060P pod also cleared for use by F-16 in 1999. System is contained in standard 300 US gallon drop tank and comprises autonomous, all-weather, day and night high-resolution reconnaissance synthetic aperture radar sensor with ability to transmit imagery to ground station via bi-directional datalink.

Most F-16s delivered since the early 1990s have provisions for a reconnaissance pod.

F-16ES: Enhanced Strategic two-seat, long-range interdictor F-16 proposal; now defunct (see 1998-99 *Jane's* for details), but provided basis for Block 50+/52+ and Israeli F-16I.

F-16U: Proposed two-seat version unsuccessfully offered to United Arab Emirates. See 1995-96 *Jane's* for more details.

F-16 'Falcon 2000': Private venture design of early 1990s, similar to F-16U, which Lockheed Martin aimed at USAF as a follow-on to the F-16 before introduction of JSF. Also known as the **F-16X**; now defunct (see 1998-99 *Jane's* for details).

CUSTOMERS: See tables. Total 4,369 production aircraft ordered or requested by beginning of October 2002, including planned USAF procurement of 2,230 and 28 embargoed Pakistan Air Force examples that are to be distributed equally between the USAF and US Navy. Backlog of 301 aircraft at end of 2001.

COSTS: Approximately US$24 million, flyaway, USAF version (1999). UAE purchase of 80 aircraft and associated equipment valued at US$7.9 billion.

DESIGN FEATURES: Conceived as 'lo' complement to Boeing F-15 Eagle in hi-lo fighter mix; optimised for high agility in air combat. Cropped delta wings blended with fuselage, with highly swept vortex control strakes along fuselage forebody and joining wings to increase lift and improve directional stability at high angles of attack; wing section NACA 64A-204; leading-edge sweepback 40°; relaxed stability (rearward CG) to increase manoeuvrability; deep wingroots increase rigidity, save 113 kg (250 lb) structure weight and increase fuel volume; fixed geometry engine intake; pilot's ejection seat inclined 30° rearwards; single-piece birdproof forward canopy section; two ventral fins below wing trailing-edge.

Baseline F-16 airframe life planned as 8,000 hours with average usage of 55.5 per cent in air combat training, 20 per cent ground attack and 24.5 per cent general flying; structural strengthening programme for pre-Block 50 aircraft was required during 1990s.

FLYING CONTROLS: Four-channel digital fly-by-wire (analogue in earlier variants); pitch/lateral control by pivoting monobloc taperons and wing-mounted flaperons; maximum rate of flaperon movement 52°/s; automatic wing leading-edge manoeuvring flaps programmed for Mach number and angle of attack; flaperons and tailerons interchangeable left and right; sidestick control column with force feel replacing almost all stick movement.

STRUCTURE: Wing, mainly of light alloy, has 11 spars, five ribs and single-piece upper and lower skins; attached to fuse-lage by machined aluminium fittings; leading-edge flaps are one-piece bonded aluminium honeycomb and driven by rotary actuators; fin is multispar, multirib with graphite epoxy skins; brake parachute or ECM housed in fairing aft of fin root; tailerons have graphite epoxy laminate skins, attached to corrugated aluminium pivot shaft and removable full-depth aluminium honeycomb leading-edge; ventral fins have aluminium honeycomb and skins; split speedbrakes in fuselage extensions inboard of tailerons open to 60°. Nose radome by Brunswick Corporation.

LANDING GEAR: Menasco hydraulically retractable type, nose unit retracting rearward and main units forward into fuse-lage. Nosewheel is located aft of intake to reduce the risk of foreign objects being thrown into the engine during ground operation, and rotates 90° during retraction to lie horizontally under engine air intake duct. Oleo-pneumatic struts in all units. Aircraft Braking Systems mainwheels and brakes; Goodyear or Goodrich tubeless mainwheel tyres, size 27.75×8.75-14.5 (or 27.75×8.75R14.5) (24 ply), pressure 14.48 to 15.17 bar (210 to 220 lb/sq in) at T-O weights less than 13,608 kg (30,000 lb). Steerable nosewheel with Goodyear, Goodrich or Dunlop tubeless tyre, size 18×5.7-8 (18 ply), pressure 20.68 to 21.37 bar (300 to 310 lb/sq in) at T-O weights less than 13,608 kg (30,000 lb). All but two main unit components interchangeable. Brake-by-wire system on main gear, with Aircraft Braking Systems anti-skid units. Runway arresting hook under rear fuselage; Irvin 7.01 m (23 ft 0 in) diameter braking parachute fitted in Greek and Turkish (Block 30/40 only) F-16s. Israeli (F-16C/D models only) and Singaporean (F-16Ds) aircraft have braking parachute compartment configured for electronic equipment. Landing/taxying lights on nose landing gear door.

POWER PLANT: One 131.6 kN (29,588 lb st) General Electric F110-GE-129, or one 129.4 kN (29,100 lb st) Pratt & Whitney F100-PW-229 afterburning turbofan as

Conformal fuel tanks trials with an Edwards AFB F-16C   *NEW*/0533364

alternative standard. These Increased Performance Engines (IPE) installed from late 1991 in Block 50 and Block 52 aircraft. Pratt & Whitney has proposed F100-PW-229A version, with new fan module among other radical improvements that will raise airflow by more than 10 per cent, lower turbine temperatures by almost 50°C (122°F) and permit inspection intervals to rise from 4,300 cycles to 6,000. New version offers potential to increase maximum augmented thrust rating to about 142 kN (31,860 lb st), although this would require larger inlet on F-16. General Electric also engaged in improvement efforts, using company funding to begin development of F110-GE-129 EFE (Enhanced Fighter Engine) in October 1997; EFE initially to be rated at up to 151.0 kN (33,945 lb st), with further growth potential to 160.0 kN (35,970 lb st); alternatively, improved thrust levels can be sacrificed for up to a 50 per cent increase in TBO and servicing intervals. Production derivative known as F110-GE-132 rated at 144.6 kN (32,500 lb st) installed in F-16 Block 60 aircraft for UAE. Immediately prior standard was 128.9 kN (28,984 lb st) F110-GE-100 or 105.7 kN (23,770 lb st) F100-PW-220 in Blocks 40/42.

Of 1,446 F-16Cs and F-16Ds ordered by USAF, 556 with F100 and 890 with F110. Fixed geometry intake, with boundary layer splitter plate, beneath fuselage. Apart from first few, F110-powered aircraft have intake widened by 30 cm (1 ft 0 in) from 368th F-16C (86-0262); Israeli second-batch F-16C-30s have power plants locally modified by Bet-Shemesh Engines to F110-GE-110A with provision for up to 50 per cent emergency thrust at low level.

Standard fuel contained in wing and five seal-bonded fuselage cells which function as two tanks; 3,986 litres (1,053 US gallons; 876 Imp gallons) in single-seat aircraft; 3,297 litres (871 US gallons; 726 Imp gallons) in two-seat aircraft. Halon inerting system. In-flight refuelling receptacle in top of centre-fuselage, aft of cockpit. Auxiliary fuel can be carried in drop tanks: one 1,136 litre (300 US gallon; 250 Imp gallon) under fuselage; 1,402 litre (370 US gallon; 308 Imp gallon) under each wing. Optional Israel Military Industries 2,271 litre (600 US gallon; 500 Imp gallon) underwing tanks initially adopted only by Israel, but have since been selected by one or two

other operators; also adopted for F-16 Block 60 version, which will have conformal fuel tanks (CFTs) with a combined capacity of 1,666 litres (440 US gallons; 366 Imp gallons). CFTs to be fitted as standard on Block 50+/52+ and F-16I.

ACCOMMODATION: Pilot only in F-16C, in pressurised and air conditioned cockpit. Boeing (formerly McDonnell Douglas) ACES II zero/zero ejection seat. Bubble canopy made of polycarbonate advanced plastics material. Inside of USAF F-16C/D canopy coated with gold film to dissipate radar energy. In conjunction with radar-absorbing materials in air intake, this reduces frontal radar signature by 40 per cent. Windscreen and forward canopy are an integral unit without a forward bow frame, and are separated from the aft canopy by a simple support structure which serves also as the breakpoint where the forward section pivots upward and aft to give access to the cockpit. A redundant safety lock feature prevents canopy loss. Windscreen/canopy design provides 360° all-round view, 195° fore and aft, 40° down over the side, and 15° down over the nose.

To enable the pilot to sustain high *g* forces, and for pilot comfort, the seat is inclined 30° aft and the heel line is raised. In normal operation the canopy is pivoted upward and aft by electrical power; the pilot is also able to unlatch the canopy manually and open it with a back-up handcrank. Emergency jettison is provided by explosive unlatching devices and two rockets. A limited displacement, force-sensing control stick is provided on the right-hand console, with a suitable armrest, to provide precise control inputs during combat manoeuvres.

The F-16D has two cockpits in tandem, equipped with all controls, displays, instruments, avionics and life support systems required to perform both training and combat missions. The layout of the F-16D second station is similar to the F-16C, and is fully systems-operational. A single-enclosure polycarbonate transparency, made in two pieces and spliced aft of the forward seat with a metal bow frame and lateral support member, provides outstanding view from both cockpits.

F-16Ds supplied to Israel and Singapore are configured with weapon system operator station in rear cockpit, plus a large dorsal equipment compartment extending from the

0130929

Lockheed Martin F-16 Fighting Falcon (two-seat Block 60 equivalent), cutaway drawing key

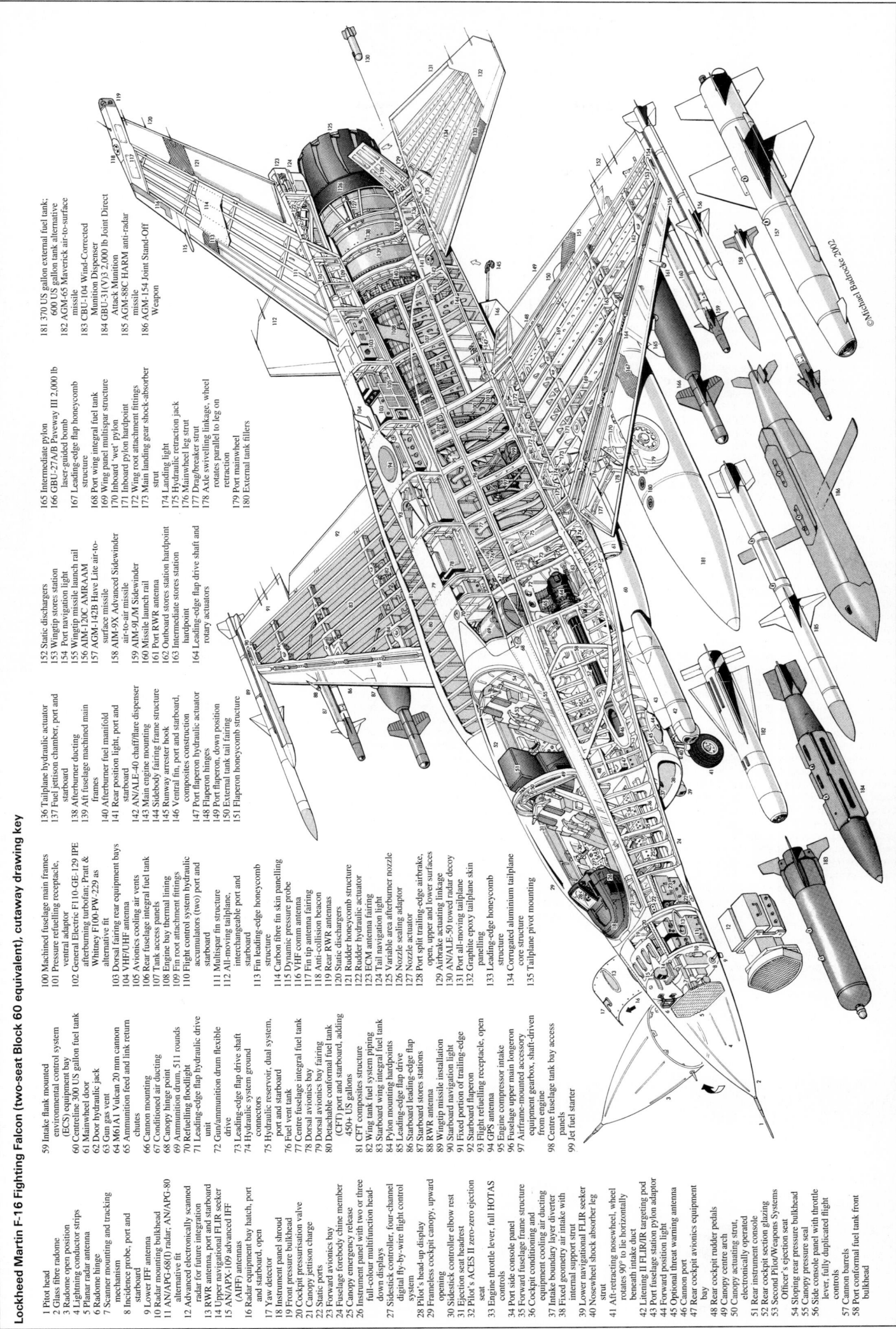

©Michael Badrocke 2002

1 Pitot head
2 Glass fibre radome
3 Radome open position
4 Lightning conductor strips
5 Planar radar antenna
6 Radome hinge
7 Scanner mounting and tracking mechanism
8 Incidence probe, port and starboard
9 Lower IFF antenna
10 Radar mounting bulkhead
11 AN/APG-68(I) radar; AN/APG-80 alternative fit
12 Advanced electronically scanned radar for future integration
13 RWR antenna, port and starboard
14 Upper navigational FLIR seeker
15 AN/APX-109 advanced IFF (AIFF) antennas
16 Radar equipment bay hatch, port and starboard, open
17 Yaw detector
18 Instrument panel shroud
19 Front pressure bulkhead
20 Cockpit pressurisation valve
21 Canopy jettison charge
22 Static ports
23 Forward avionics bay
24 Fuselage forebody chine member
25 Canopy emergency release
26 Instrument panel with two or three full-colour multifunction head-down displays
27 Sidestick controller, four-channel digital fly-by-wire flight control system
28 Pilot's head-up display
29 Frameless cockpit canopy, upward opening
30 Sidestick controller elbow rest
31 Ejection seat headrest
32 Pilot's ACES II zero-zero ejection seat
33 Engine throttle lever, full HOTAS controls
34 Port side console panel
35 Forward fuselage frame structure
36 Cockpit conditioning and equipment cooling air ducting
37 Intake boundary layer diverter
38 Fixed geometry air intake with internal support strut
39 Lower navigational FLIR seeker
40 Nosewheel shock absorber leg strut
41 Aft-retracting nosewheel, wheel rotates 90° to lie horizontally beneath intake duct
42 Litening II FLIR/IR targeting pod
43 Port fuselage station pylon adaptor
44 Forward position light
45 Optional threat warning antenna
46 Cannon port
47 Rear cockpit avionics equipment bay
48 Rear cockpit rudder pedals
49 Canopy centre arch
50 Canopy actuating strut, electronically operated
51 Rear instrument console
52 Rear cockpit section glazing
53 Second Pilot/Weapons Systems Officer's ejection seat
54 Sloping rear pressure bulkhead
55 Canopy pressure seal
56 Side console panel with throttle lever, fully duplicated flight controls
57 Cannon barrels
58 Port conformal fuel tank front bulkhead

59 Intake flank mounted environmental control system (ECS) equipment bay
60 Centreline 300 US gallon fuel tank
61 Mainwheel door
62 Door hydraulic jack
63 Gun gas vent
64 M61A1 Vulcan 20 mm cannon
65 Ammunition feed and link return chutes
66 Cannon mounting
67 Conditioned air ducting
68 Canopy hinge point
69 Ammunition drum, 511 rounds
70 Refuelling floodlight
71 Leading-edge flap hydraulic drive unit
72 Gun/ammunition drum flexible drive
73 Leading-edge flap drive shaft
74 Hydraulic system ground connectors
75 Hydraulic reservoir, dual system, port and starboard
76 Fuel vent tank
77 Centre fuselage integral fuel tank
78 Dorsal avionics bay
79 Dorsal avionics bay fairing
80 Detachable conformal fuel tank (CFT) port and starboard, adding 450+ US gallons
81 CFT composites structure
82 Wing tank fuel system piping
83 Starboard wing integral fuel tank
84 Pylon mounting hardpoints
85 Starboard leading-edge flap
86 Starboard flaperon
87 Starboard stores stations
88 RWR antenna
89 Wingtip missile installation
90 Starboard navigation light
91 Fixed portion of trailing-edge
92 Starboard flaperon
93 Flight refuelling receptacle, open
94 GPS antenna
95 Engine compressor intake
96 Fuselage upper main longeron
97 Airframe-mounted accessory equipment gearbox, shaft-driven from engine
98 Centre fuselage fuel tank bay access panels
99 Jet fuel starter

100 Machined fuselage main frames
101 Pressure refuelling receptacle, ventral adaptor
102 General Electric F110-GE-129 IPE afterburning turbofan; Pratt & Whitney F100-PW-229 as alternative fit
103 Dorsal fairing rear equipment bays
104 VHF/UHF antenna
105 Avionics cooling air vents
106 Rear fuselage integral fuel tank
107 Tank access panels
108 Engine bay thermal lining
109 Fin root attachment fittings
110 Flight control system hydraulic accumulators (two) port and starboard
111 Multispar fin structure
112 All-moving tailplane, interchangeable port and starboard
113 Fin leading-edge honeycomb structure
114 Carbon fibre fin skin panelling
115 Dynamic pressure probe
116 VHF comm antenna
117 Fin tip antenna fairing
118 Anti-collision beacon
119 Rear RWR antennas
120 Static dischargers
121 Rudder honeycomb structure
122 Rudder hydraulic actuator
123 ECM antenna fairing
124 Tail navigation light
125 Variable area afterburner nozzle
126 Nozzle sealing adaptor
127 Nozzle actuator
128 Port split trailing-edge airbrake, open, upper and lower surfaces
129 Airbrake actuating linkage
130 AN/ALE-50 towed radar decoy
131 Port all-moving tailplane
132 Graphite epoxy tailplane skin panelling
133 Leading-edge honeycomb structure
134 Corrugated aluminium tailplane core structure
135 Tailplane pivot mounting

136 Tailplane hydraulic actuator
137 Fuel jettison chamber, port and starboard
138 Afterburner ducting
139 Aft fuselage machined main frames
140 Afterburner fuel manifold
141 Rear position light, port and starboard
142 AN/ALE-40 chaff/flare dispenser
143 Main engine mounting
144 Sidebody fairing frame structure
145 Runway arrester hook
146 Ventral fin, port and starboard, composites construction
147 Port flaperon hydraulic actuator
148 Flaperon hinges
149 Port flaperon, down position
150 External tank tail fairing
151 Flaperon honeycomb structure
152 Static dischargers
153 Wingtip stores station
154 Port navigation light
155 Wingtip missile launch rail
156 AIM-120C AMRAAM
157 AGM-142B Have Lite air-to-surface missile
158 AIM-9X Advanced Sidewinder air-to-air missile
159 AIM-9L/M Sidewinder
160 Missile launch rail
161 Port RWR antenna
162 Outboard stores station hardpoint
163 Intermediate stores station hardpoint
164 Leading-edge flap drive shaft and rotary actuators
165 Intermediate pylon
166 GBU-27A/B Paveway III 2,000 lb laser-guided bomb
167 Leading-edge flap honeycomb structure
168 Port wing integral fuel tank
169 Wing panel multispar structure
170 Inboard 'wet' pylon
171 Inboard pylon hardpoint
172 Wing root attachment fittings
173 Main landing gear shock-absorber strut
174 Landing light
175 Hydraulic retraction jack
176 Mainwheel leg strut
177 Drag/breaker strut
178 Axle swivelling linkage, wheel rotates parallel to leg on retraction
179 Port mainwheel
180 External tank fillers
181 370 US gallon external fuel tank; 600 US gallon tank alternative
182 AGM-65 Maverick air-to-surface missile
183 CBU-104 Wind-Corrected Munition Dispenser
184 GBU-31(V)3 2,000 lb Joint Direct Attack Munition
185 AGM-88C HARM anti-radar missile
186 AGM-154 Joint Stand-Off Weapon

rear of the canopy to the leading edge of the fin; compartment houses avionics unique to each operator, plus additional chaff/flare dispensers and an in-flight refuelling receptacle.

SYSTEMS: Regenerative 12 kW environmental control system, with digital electronic control, uses engine bleed air for pressurisation and cooling of crew station and avionics compartments. Two separate and independent hydraulic systems supply power for operation of the primary flight control surfaces and the utility functions. System pressure (each) 207 bar (3,000 lb/sq in), rated at 161 litres (42.5 US gallons; 35.4 Imp gallons)/min. Bootstrap-type reservoirs, rated at 5.79 bar (84 lb/sq in).

Electrical system powered by engine-driven 60 kVA main generator and 10 kVA standby generator (including ground annunciator panel for total electrical system fault reporting), with Hamilton Sundstrand constant speed drive and powered by a Hamilton Sundstrand accessory drive gearbox. 17 Ah battery. Four dedicated, sealed cell batteries provide transient electrical power protection for the fly-by-wire flight control system.

An onboard Hamilton Sundstrand/Solar jet fuel starter is provided for engine self-start capability. Simmonds fuel measuring system. AlliedSignal emergency power unit automatically drives a 5 kVA emergency generator and emergency pump to provide uninterrupted electrical and hydraulic power for control in the event of the engine or primary power systems becoming inoperative.

AVIONICS: *Comms:* Magnavox AN/ARC-164 UHF transceiver (AN/URC-126 Have Quick IIA in Block 50/52); provision for Magnavox KY-58 secure voice system; Rockwell Collins AN/ARC-186 VHF AM/FM transceiver, ARC-190 HF radio, government-furnished AN/AIC-18/25 intercom and SCI advanced interference blanker, Teledyne Electronics AN/APX-101 IFF transponder with government-furnished IFF control, government-furnished National Security Agency KIT-1A/TSEC cryptographic equipment. F-16C/D Block 52 aircraft of Singapore and South Korea have Litton AN/APX-109+ advanced interrogator/transponder. AN/APX-113 advanced interrogator/transponder installed in Greek Block 50, Taiwanese Block 20 and Turkish Block 50 aircraft and is standard equipment on USAF aircraft procured in FY00 and FY01; is also being retrofitted to earlier USAF aircraft as part of CCIP upgrade package.

Radar: Northrop Grumman AN/APG-68(V) pulse Doppler range and angle track radar, with mechanically scanned planar array in nose. Provides air-to-air modes for range-while-search, uplook search, velocity search with ranging, air combat, track-while-scan (10 targets), raid cluster resolution, single target track and pulse Doppler track to provide target illumination for AIM-7 missiles, plus air-to-surface modes for ground-mapping, Doppler beam-sharpening, ground moving target, sea target, fixed target track, target freeze after pop-up, beacon, and air-to-ground ranging. Improved AN/APG-68(V)9 radar derivative installed on new Block 50+/52+ aircraft from 2002 and Northrop Grumman plans to offer an upgrade kit enabling existing radars to be brought to latest standard, which has synthetic aperture radar (SAR) mapping and terrain following (TF) modes, plus interleaving of all modes; if internal FLIR targeting system is selected, this could share processor with ABR. Elta EL/2032 multimode pulse Doppler radar could be installed in future aircraft for Israel. Block 60 aircraft for UAE to have Northrop Grumman AN/APG-80 active electronically scanned array (AESA) radar.

Flight: Litton LN-39 standard inertial navigation system (ring laser Litton LN-93 or Honeywell H-423 in Block 50/52: LN-93 for Egypt, Indonesia, Israel, South Korea, Pakistan, Portugal and Taiwan, plus Netherlands retrofit and Greek second batch); Rockwell Collins AN/ARN-108 ILS, Rockwell Collins AN/ARN-118 Tacan, Rockwell Collins GPS, Honeywell central air data computer, Elbit Fort Worth enhanced stores management computer, Gould AN/APN-232 radar altimeter. Fairchild digital terrain system (incorporating BAE Systems Terprom algorithms) to be installed in all new USAF F-16s and USAF Reserve F-16C/Ds. Optional equipment includes Rockwell Collins VIR-130 VOR/ILS.

Instrumentation: Marconi wide-angle holographic electronic HUD with raster video capability (for LANTIRN) and integrated keyboard; data entry/cockpit interface and dedicated fault display by Litton Canada and Elbit Fort Worth; Astronautics cockpit/TV set. Total of 43 aircraft (F-16C/D Block 40) of 31st FW at Aviano AB, Italy, modified between September 1997 and March 1998 as part of quick response capability (QRC) programme to provide night vision information system (NVIS) compatible lighting and facilitate use of NVGs; it is eventually planned to fit NVIS lighting on all USAF F-16Cs and F-16Ds.

Mission: Honeywell multifunction displays. Lockheed Martin LANTIRN package comprises AN/AAQ-13 (navigation) and AN/AAQ-14 (targeting) pods. Turkish aircraft (150+ modified by 1996) to share 60 LANTIRN pod systems; LANTIRN also purchased by Greece, South Korea and Singapore, although Singapore now seeking a replacement system. Sharpshooter pod (down-rated export version of AAQ-14 LANTIRN targeting system) acquired by Bahrain and Israel, but latter obtained indigenous Rafael Litening IR targeting and navigation pod as replacement. Total of 168 Litening II navigation/targeting pods ordered from Rafael and Northrop Grumman to equip Block 25/30/32/40/42 F-16C/Ds of the Air National Guard (136) and Air Force Reserve Command (32); programme called Precision Attack Targeting System, with first AFRC unit (457th FS at Fort Worth, Texas) receiving initial batch of four pods in February 2000; this and three more AFRC squadrons expected to be fully equipped by fourth quarter of 2000. ANG also accepted first pods during 2000; goal is to allocate eight pods to each 15-aircraft squadron. Under CCIP modification programme, Block 40/42/50/52 aircraft of USAF to adopt new Lockheed Martin Sniper XR advanced FLIR targeting pod system from late 2002 onwards.

Raytheon AN/ASQ-213 HARM Targeting System (HTS) pod introduced on Block 50D/52D aircraft and subsequently retrofitted to entire Block 50/52 fleet. Entered service 1994 and currently deployed by USAF units in USA, Japan and Germany. New modular mission computer and colour displays for retrofit as part of CCIP and new production in Block 50/52 F-16C/Ds.

Self-defence: Dalmo Victor AN/ALR-69 radar warning system replaced in USAF Block 50/52 by BAE Systems AN/ALR-56M advanced RWR, which also ordered for USAF Block 40/42 retrofit and (first export) Korean Block 52s. Korean aircraft being retrofitted with the ITT Avionics/Northrop Grumman AN/ALQ-165 Airborne Self-Protection Jammer (ASPJ) from mid-2000. Provision for Northrop Grumman AN/ALQ-131 or Raytheon AN/ALQ-184 jamming pods. AN/ALQ-131 supplied to Bahrain and Egypt. Israeli Air Force F-16s extensively modified with locally designed and manufactured equipment, as well as optional US equipment to tailor them to the IAF defence role. This includes Elisra SPS 3000 self-protection jamming equipment in enlarged spines of F-16D-30s and Elta EL/L-8240 ECM in third batch of F-16C/Ds, replacing AN/ALQ-178(V)1 Rapport ECM in Israeli F-16As. Chilean aircraft will have ITT Industries Advanced Integrated Defensive Electronic Warfare Suite (AIDEWS), incorporating radar warning and RF countermeasures.

BAE Systems AN/ALQ-178(V)3 Rapport III integral self-protection system in Turkish F-16C/Ds will almost certainly be replaced by improved AN/ALQ-178(V)5 system. In March 1993, Greece ordered Raytheon ASPIS (Advanced Self-Protection Integrated Suite) self-defence system, comprising AN/ALR-93 RWR, Tracor chaff/flare dispensers and (initially on only 35 aircraft) Raytheon AN/ALQ-187 I-DIAS jammer.

USAF Air National Guard procured Terma PIDS wing weapon pylon with additional chaff/flare dispensers.

Tracor AN/ALE-40(V)-4 chaff/flare dispensers (AN/ALE-47 in FY97 Block 50, FMS Block 20/50 since mid-1996 and for retrofit to Block 40/42 and 50/52 of USAF). Raytheon AN/ALE-50(V)2 towed decoy installed in AMRAAM missile pylons and adopted by USAF for all Block 40/42/50/52 aircraft was widely fielded in 1999; USAF plans to buy total of 961. USAF Air National Guard participating in wing weapon pylon upgrade programme (to be fielded in 2004) that adds MIL-STD-1760 interface to existing PIDS pylons. This programme also includes provisions for future incorporation of a passive missile approach warning system.

ARMAMENT: General Dynamics M61A1 20 mm multibarrel cannon in the port side wing/body fairing, equipped with a General Dynamics ammunition handling system and an enhanced envelope gunsight (part of the head-up display system) and 511 rounds of ammunition. There is a mounting for an air-to-air missile at each wingtip, one underfuselage centreline hardpoint, and six underwing hardpoints for additional stores. For manoeuvring flight at 5.5 *g* the underfuselage station is stressed for a load of up to 1,000 kg (2,200 lb), the two inboard underwing stations for 2,041 kg (4,500 lb) each, the two centre underwing stations for 1,587 kg (3,500 lb) each, the two outboard underwing stations for 318 kg (700 lb) each, and the two wingtip stations for 193 kg (425 lb) each. For manoeuvring flight at 9 *g* the underfuselage station is stressed for a load of up to 544 kg (1,200 lb), the two inboard underwing stations for 1,134 kg (2,500 lb) each, the two centre underwing stations for 907 kg (2,000 lb) each, the two outboard underwing stations for 204 kg (450 lb) each, and the two wingtip stations for 193 kg (425 lb) each. There are mounting provisions on each side of the inlet shoulder for the specific carriage of sensor pods (electro-optical, FLIR and so on); each of these stations is stressed for 408 kg (900 lb) at 5.5 *g*, and 250 kg (550 lb) at 9 *g*.

Typical stores loads can include two wingtip-mounted AIM-9L/M/P Sidewinders, with up to four more on the outer underwing stations; Rafael Python 3 on Israeli F-16s from early 1991 and Python 4 from mid-1997; centreline GPU-5/A 30 mm cannon; drop tanks on the inboard underwing and underfuselage stations; HARM targeting system pod along the starboard side of the nacelle; and bombs, air-to-surface missiles or flare pods on the four inner underwing stations. Stores can be launched from Aircraft Hydro-Forming MAU-12C/A bomb ejector racks, Hughes LAU-88 launchers, Orgen triple or multiple ejector racks and Lucas Aerospace Flight Structures Twin Store Carrier (TSC). New BRU-57 bomb racks to be fitted to Block 50/52 aircraft of USAF in fourth quarter of 2001; to be installed at mid-span hardpoint, each BRU-57 will be able to carry two smart munitions such as JSOW, JDAM and WCMD.

Weapons launched successfully from F-16s, in addition to AIM-9 Sidewinder and AIM-120A AMRAAM, include radar-guided AIM-7 Sparrow, Rafael Derby and Sky Flash BVR air-to-air missiles, AIM-132 ASRAAM and Magic 2 IR homing air-to-air missiles, AGM-65A/B/D/G Maverick air-to-surface missiles, AGM-88 HARM and AGM-45 Shrike anti-radiation missiles, AGM-84 Harpoon anti-ship missiles (clearance trials 1993-94) and, in Royal Norwegian Air Force service, the Penguin Mk 3 anti-ship missile. LGBs include GBU-10, GBU-12, GBU-22, GBU-24 and GBU-27; F-16 can also deliver GBU-15 glide bomb, which used in conjunction with datalink pod. Israeli IMI STAR-1 anti-radiation weapon has also begun carriage trials on F-16D, although full-scale development is dependent upon receipt of a firm order; IMI runway attack munition (RAM) introduced into IDF/AF service in about 2000. CMS Defense Systems Autonomous Free-flight Dispenser System (AFDS) was tested at Eglin AFB, Florida, during 1992-93 and can be loaded with a variety of submunitions, including cratering bombs, shaped charge bomblets, anti-tank mines, area denial submunitions and general purpose bomblets.

Newest capability, introduced on Block 50/52 aircraft of USAF, incorporates 50T5 software upgrade, allowing F-16 to carry and deliver latest family of precision munitions; release to service occurred in mid-2000. New weapons comprise GBU-31 Joint Direct Attack Munition (JDAM), AGM-154 Joint StandOff Weapon (JSOW) and CBU-103, CBU-104 and CBU-105 wind-corrected munitions dispensers (WCMDs). First operational unit with JSOW and WCMD is 20th FW at Shaw AFB, South Carolina. AGM-158 Joint Air-to-Surface Standoff Missile (JASSM) currently being tested on F-16 and expected eventually to be deployed operationally.

Lockheed Martin F-16C of 35th Fighter Wing at Misawa, Japan ***NEW***/0533366

DIMENSIONS, EXTERNAL (F-16C, D):

Wing span: over missile launchers	9.45 m (31 ft 0 in)
over missiles	10.00 m (32 ft 9¾ in)
Wing aspect ratio	3.2
Length overall	15.03 m (49 ft 4 in)
Height overall	5.09 m (16 ft 8½ in)
Tailplane span	5.58 m (18 ft 3¾ in)
Wheel track	2.36 m (7 ft 9 in)
Wheelbase	4.00 m (13 ft 1½ in)

AREAS (F-16C, D):

Wings, gross	27.87 m² (300.0 sq ft)
Flaperons (total)	2.91 m² (31.32 sq ft)
Leading-edge flaps (total)	3.41 m² (36.72 sq ft)
Fin, incl dorsal fin	4.00 m² (43.10 sq ft)
Rudder	1.08 m² (11.65 sq ft)
Horizontal tail surfaces (total)	5.92 m² (63.70 sq ft)

WEIGHTS AND LOADINGS:

Weight empty:	
F-16C: F100-PW-229	8,433 kg (18,591 lb)
F110-GE-129	8,581 kg (18,917 lb)

F-16D: F100-PW-229	8,645 kg (19,059 lb)
F110-GE-129	8,809 kg (19,421 lb)
Max internal fuel (JP-8): F-16C	3,249 kg (7,162 lb)
F-16C Block 60	3,084 kg (6,800 lb)
F-16D	2,687 kg (5,924 lb)
F-16D Block 60	2,521 kg (5,558 lb)
Max external fuel (JP-8), F-16C/D non Block 60:	
normal	3,208 kg (7,072 lb)
optional	4,627 kg (10,200 lb)
F-16C/D Block 60	5,987 kg (13,200 lb)
Max external load (full internal fuel):	
F-16C: F100-PW-229	7,226 kg (15,930 lb)
F110-GE-129	7,072 kg (15,591 lb)
Typical combat weight (two AAMs, 50% fuel):	
F-16C: F100-PW-229	10,659 kg (23,498 lb)
F110-GE-129	10,812 kg (23,837 lb)
Max T-O weight: with two AAMs, no tanks:	
F-16C: F100-PW-229	12,138 kg (26,760 lb)
F110-GE-129	12,292 kg (27,099 lb)

with full external load:

| F-16C/D Block 50/52 | 19,187 kg (42,300 lb) |
| F-16C/D Block 60 | 22,679 kg (50,000 lb) |

Wing loading: at 12,927 kg (28,500 lb) AUW
463.8 kg/m² (95.00 lb/sq ft)
at 19,187 kg (42,300 lb) AUW
688.4 kg/m² (141.00 lb/sq ft)
Thrust/weight ratio (clean) 1.1 to 1
PERFORMANCE:
Max level speed at 12,200 m (40,000 ft) above M2.0
Service ceiling more than 15,240 m (50,000 ft)
Radius of action:
F-16C Block 50, two 907 kg (2,000 lb) bombs, two Sidewinders, 3,940 litres (1,040 US gallons; 867 Imp gallons) external fuel, tanks retained, hi-lo-lo-hi
676 n miles (1,252 km; 780 miles)
F-16C Block 50, armament as above, 5,678 litres (1,500 US gallons; 1,249 Imp gallons) external fuel, tanks retained, hi-lo-lo-hi

802 n miles (1,485 km; 923 miles)
F-16C Block 50, two BVR missiles, two Sidewinders, 3,940 litres (1,040 US gallons; 867 Imp gallons) external fuel, tanks dropped when empty, combat air patrol mission 866 n miles (1,604 km; 997 miles)
Ferry range:
F-16C Block 50, with 3,940 litres (1,040 US gallons; 867 Imp gallons) external fuel
1,961 n miles (3,632 km; 2,257 miles)
F-16C Block 50, with 5,678 litres (1,500 US gallons; 1,249 Imp gallons) external fuel
2,276 n miles (4,215 km; 2,619 miles)
Symmetrical g limit with full internal fuel +9
UPDATED

LOGO

LOPRESTI GORDON VTOL INC

2620 Airport North Drive, Vero Beach, Florida 32960
Tel: (+1 561) 562 47 57
Fax: (+1 561) 563 04 46 and (+1 800) 859 47 57
Web (1): http://www.LoGoVTOL.com
Web (2): http://www.TurboHawk.com
PRESIDENT AND BUSINESS MANAGER: Larry Gordon

LoGo was formed in early 2001 to develop a family of VTOL aircraft platforms to meet the specific needs of military Special Forces groups, when operating in complex urban environments. The family at present comprises two such vehicles: the **Guardian**, designed by the late Roy LoPresti, and (in a joint venture with UrbanAero of Israel, which see) the **TurboHawk** designed by Dr Rafi Yoeli. A primary design feature of both vehicles is the ability to dock with high-rise buildings and transfer people, thus widening their civil appeal for such missions as fire rescue, medevac and chemical/biological decontamination. Versions have been defined to carry up to eight people in addition to the pilot, perform amphibious operations and then depart a danger zone at speeds of up to 348 kt (644 km/h; 400 mph).
UPDATED

Artist's impression of LoGo Guardian effecting a rescue from a tall, burning building
0131873

LOPRESTI

LOPRESTI INC

2620 North Airport Drive, Vero Beach, Florida 32960
Tel: (+1 561) 562 47 57 or (+1 800) 859 47 57
Fax: (+1 561) 563 04 46
e-mail: info@speedmods.com
Web: http://www.speedmods.com

Company founder LeRoy (Roy) LoPresti died on 7 August 2002. Before that, however, LoPresti Inc had decided to abandon its plans to build the LP-1 Fury. Refer also to the LoGo entry in this section.
UPDATED

LOPRESTI LP-1 FURY

Despite its success in a legal dispute, LoPresti had abandoned plans to build this two-seat lightplane by early 2002.
UPDATED

LUSCOMBE

LUSCOMBE AIRCRAFT CORPORATION

PO Box 80062, Las Vegas, Nevada 89180-0062
Tel: (+1 702) 434 67 22 and 521 92 04
Fax: (+1 702) 434 10 04
e-mail: sales@luscombeaircraft.com
Web: http://www.luscombeaircraft.com
WORKS: 5333 North Main Street, Altus, Oklahoma 73521
SALES & MARKETING: 3301 Bayshore Boulevard, Suite 1102, Tampa, Florida 33629
FOUNDER AND CEO: Bill McKown
PRESIDENT: Billy McCoy
SENIOR VICE-PRESIDENT: Charles Gibson Jr

Original Luscombe company formed in 1934, specialising in all-metal private light aircraft; ownership passed to Temco in 1949, Silvaire Aircraft in 1955, and to Luscombe Aircraft Corporation on termination of aircraft production in 1960. Type certificate of Model 11 transferred to Classic Air of Lansing, Michigan; relaunch of modified version undertaken from 1998 in newly built 11,150 m² (120,000 sq ft) factory at Altus. Company also holds type certificates for Luscombe 4 and Luscombe Phantom 1/1S.
A second design of the original Luscombe company, the Model 8, is marketed in the US by Renaissance (which see).
UPDATED

LUSCOMBE 185-11E

TYPE: Four-seat lightplane.
PROGRAMME: Model 11A Sedan first flew (NX74202) 11 September 1946; total of 198 built with tailwheel and

Luscombe 185-11E preproduction aircraft after re-registration as NX11E *(Jane's/Paul Jackson)* NEW/0533377

123 kW (165 hp) Continental E-165 engine; production ended in 1949; 41 remained registered in US during 2002. Planned 1970 relaunch as Alpha Aviation Alpha IID failed to materialise. Considerably redesigned Luscombe 185-11E being produced at Altus, Oklahoma, following conversion of Sedan N1674B as proof-of-concept followed by construction of preproduction prototype (N747BM, later NX11E) for certification. Initially marketed under name of Spartan, which had been dropped by 2002.

Structural test fuselage completed December 1997; production of new aircraft due to have started April 2000, but this has slipped. Certification for sales and marketing purposes awarded November 1998, with full type certification expected by mid-2002.
CUSTOMERS: By late 1998, orders believed to exceed 300; two (including prototype) registered by October 2001. Announced customers include the US Civil Air Patrol, which has ordered eight.
COSTS: US$158,900 basic (2002).

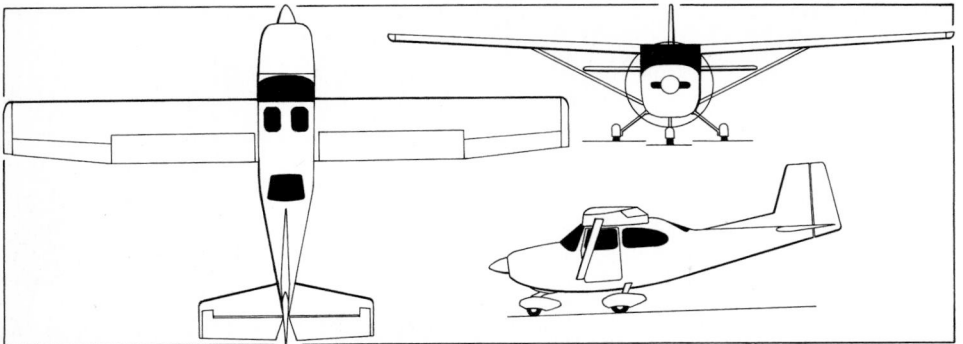

General arrangement of the Luscombe 185-11E *(Jane's/Paul Jackson)*   0016820

Wing aspect ratio	8.9
Length overall	7.32 m (24 ft 0 in)
Height overall	2.69 m (8 ft 10 in)
Tailplane span	3.58 m (11 ft 9 in)
Wheel track	2.64 m (8 ft 8 in)
Wheelbase	2.24 m (7 ft 4 in)
Propeller diameter	1.93 m (6 ft 4 in)
Passenger door: Max height	1.19 m (3 ft 11 in)
Max width	0.81 m (2 ft 8 in)

DIMENSIONS, INTERNAL:

Cabin: Length	2.70 m (8 ft 10¼ in)
Max width	1.16 m (3 ft 9½ in)
Max height	1.30 m (4 ft 3¼ in)

AREAS:

Wings, gross	15.51 m² (167.0 sq ft)

WEIGHTS AND LOADINGS:

Weight empty	658 kg (1,450 lb)
Baggage capacity	45 kg (100 lb)
Max T-O and landing weight	1,034 kg (2,280 lb)
Max ramp weight	1,037 kg (2,286 lb)
Max wing loading	66.7 kg/m² (13.65 lb/sq ft)
Max power loading	7.50 kg/kW (12.32 lb/hp)

PERFORMANCE:

Never-exceed speed (V_{NE})	157 kt (290 km/h; 180 mph)
Max level speed	130 kt (241 km/h; 150 mph)

Cruising speed at 2,286 m (7,500 ft):

70% power	117 kt (217 km/h; 135 mph)
60% power	110 kt (204 km/h; 127 mph)
T-O speed	60 kt (111 km/h; 69 mph)

Stalling speed, power off:

flaps up	47 kt (87 km/h; 54 mph)
flaps down	43 kt (79 km/h; 49 mph)
Max rate of climb at S/L	267 m (876 ft)/min
Service ceiling	4,877 m (16,000 ft)
T-O run	274 m (900 ft)
T-O to 15 m (50 ft)	419 m (1,375 ft)
Landing from 15 m (50 ft)	579 m (1,900 ft)
Landing run	264 m (866 ft)

Range with reserves:

at 70% power cruising speed	500 n miles (926 km; 575 miles)
at 60% power cruising speed	550 n miles (1,018 km; 632 miles)
Endurance	5 h 42 min
g limits	+3.8/−1.52
Noise level	72.4 dB(A)
	UPDATED

DESIGN FEATURES: High-wing braced monoplane; original 1946 design modified for nosewheel and 0.30 m (1 ft 0 in) increase in constant-chord inner wing, with sweptforward outer trailing-edge and Hoemer-style wingtips; tapered tailplane; slightly sweptback fin with ventral fillet. Wing section NACA 43012A.

FLYING CONTROLS: Conventional and manual. Stainless steel control cables. All control surfaces aluminium skinned with external stiffeners. Horn-balanced elevators with flight-adjustable trim tab on port side; ground-adjustable tab on starboard aileron. Three-position electric flaps; maximum deflection 25°.

STRUCTURE: Steel tube semi-monocoque corrosion-proofed fuselage and strut-braced two-spar wings, all covered with aluminium skins. Dorsal tail fillet.

LANDING GEAR: Alloy spring leaf main legs with 6.00-6 tyres, pressure 2.4 bar (35 lb/sq in). Air/oil shock-absorbing strut nose gear with 5.00-5 tyre, pressure 1.7 bar (25 lb/sq in), steerable ±10°. Hydraulic disc brakes and parking brake on mainwheels; turning radius approximately 8.23 m (27 ft 0 in).

POWER PLANT: One 157 kW (210 hp) Teledyne Continental IO-360-ES4 six-cylinder fuel-injected piston engine derated to 138 kW (185 hp) and driving a Sensenich 76EC8S10 two-blade fixed-pitch metal propeller. Fuel capacity 152 litres (40.0 US gallons; 33.3 Imp gallons) in two bladder wing tanks, of which 148 litres (39.0 US gallons; 32.5 Imp gallons) are usable. Filler port in each wing. Oil capacity 7.5 litres (2.0 US gallons; 1.6 Imp gallons). Proof-of-concept aircraft has three-blade propeller, which is optional on production airframes.

ACCOMMODATION: Pilot and three passengers in side-by-side pairs; door each side; front seats fully adjustable; dual controls. Inertia reel harnesses for all seats. Cabin windows are tinted, as are two vision ports in cabin roof.

SYSTEMS: 28 V 70 A alternator; 24 V 12.75 Ah battery; dual vacuum pumps.

AVIONICS: Garmin and Bendix/King options. Representative (but not exclusive) equipment listed below.

Comms: Bendix/King KX 155A nav/com/glideslope; KI 209A VOR/LOC/GS; KT 76C transponder with ACK-30 Mode C encoder; KI 227 ADF; KN 64 DME; emergency locator transmitter; PS Engineering 6000C audio panel.

Flight: Altimeter, compass, ASI, OAT, VSI, turn co-ordinator; Garmin GX60 GPS with moving map display, S-Tec autopilot.

Instrumentation: Ammeter, vacuum gauge, fuel capacity, tachometer, EGT, oil pressure.

EQUIPMENT: Strobe position lights, white paint with accent stripe.

Data apply to both versions, except when otherwise specified.

DIMENSIONS, EXTERNAL:

Wing span	11.73 m (38 ft 6 in)

MAULE

MAULE AIR INC

2099 GA Highway 133 South, Lake Maule, Moultrie, Georgia 31768
Tel: (+1 912) 985 20 45
Fax: (+1 912) 985 96 28
e-mail: bdmaule@alltel.net
Web: http://members.surfsouth.com/~mauleair
CHAIRMAN: June D Maule
DIRECTOR: Ray Maule
VICE-PRESIDENT: David Maule
ENGINEER: Don Richie
SALES MANAGER: Brent Maule

Original Maule Aircraft Corporation formed by the late Belford D Maule to manufacture M-4, a four-seat extrapolation of Piper Cub; transferred to Moultrie, Georgia, 1968; production ceased 1975; Maule Air Inc formed 1984 to produce uprated M-5 Lunar Rocket and M-7 Super Rocket; Lunar Rocket discontinued, but variants listed below currently available. Delivered 63 aircraft in 1998, and 69 in 1999, 57 in 2000 (including one each M-6-235 and MX-7-180AC, and 57 in 2001.

UPDATED

MAULE M-7 SERIES

TYPE: Five-seat lightplane/turboprop.

PROGRAMME: Introduced 1984 as M-7-235 Super Rocket (prototype N5656A: based on M-6-235 but with extended cabin and additional windows). In May 2000, Maule received first of an initial batch of 10 SMA SR305 turbocharged diesel engines, rated at 169 kW (227 hp), for installation in MX-9s.

CURRENT VERSIONS: Produced with short (M-5), mid-length (M-6) or long (A Model) wing span and engine/landing gear options as detailed below; short option deleted on 1999 versions, while piston-engined and turboprop/nosewheel aircraft have only M-6 wing from that year, the turboprop/tailwheel variants employing A Model wings. Designation prefixes are T = tricycle and X = short span.

MX-7-160 Sportplane: Four-seater; 119 kW (160 hp) Textron Lycoming O-320-B2D engine, Sensenich two-blade fixed-pitch metal propeller, and four-position flaps. One delivered in 1999, since when no further deliveries reported.

MXT-7-160 Comet: As MX-7-160, but tricycle landing gear; two seats standard, four seats optional. Previously known as Maule Trainer. No recent deliveries reported.

MX-7-160C: As M-7-160, but spring aluminium landing gear. Available from 2000, but no reported deliveries by 30 June 2002.

MX-7-180A Sportplane: As MX-7-160, but 134 kW (180 hp) Textron Lycoming O-360-C4F engine. No recent deliveries reported.

MXT-7-180A Comet: As MX-7-180A, but tricycle landing gear and four seats standard. Introduced and first production aircraft (N1002N) flown 1996; optimised for flight training schools. Total of 18 delivered in 1999, six in 2000, one in 2001 and two in the first six months of 2002.

MX-7-180B Star Rocket: 134 kW (180 hp) Textron Lycoming O-360-C1F engine, Hartzell two-blade constant-speed metal propeller and five-position flaps. Two delivered in first six months of 2002.

MX-7-180C Millenium: As MX-7-180B, but with spring aluminium main landing gear. Two delivered in 1999, two in 2000, one in 2001 and one in the first six months of 2002.

MXT-7-180 Star Rocket: As MX-7-180B, but with tricycle landing gear. Previously known as Star Craft. One delivered in 1999, seven in 2000, and two in 2001; no reported deliveries in first six months of 2002.

MX-7 Rocket: 153 kW (205 hp) PZL-Franklin 6A-350-C1R engine, McCauley two-blade propeller. Prototype was being test flown by second quarter of 1999. Engine may be offered as an option on Textron Lycoming O-360-powered models. No reported deliveries by 30 June 2002.

Maule MT-7-235 Super Rocket of the Civil Air Patrol *(Jane's/Paul Jackson)*   NEW/0533376

M-7-235B Super Rocket: Five-seater; choice of 175 kW (235 hp) carburetted Textron Lycoming O-540-J1A5, low-compression Mogas approved O-540-B4B5 or fuel-injected IO-540-W1A5 engines; McCauley constant-speed propeller, five-position flaps; fuselage raised 7.6 cm (3 in) at trailing-edge of wing and baggage area moved aft 12.7 cm (5 in) to accommodate fifth seat; recommended for high gross weight short-field operation, and for floatplane operation. Eight delivered in 1999, seven in 2000, five in 2001 and six in the first six months of 2002.

M-7-235C Orion: As for M-7-235B but with spring aluminium main landing gear. Sixteen delivered in 1999, 18 in 2000, 14 in 2001 and eight in the first six months of 2002.

MT-7-235 Super Rocket: As M-7-235B, but tricycle landing gear, four-position flaps and IO-540-W1A5 engine only. Four delivered in 1999, five in 2000, 16 in 2001 and 11 in the first six months of 2002. Civil Air Patrol approved purchase of 15 in glider towing configuration mid-2000.

M-7-260: As M-7-235B, but 194 kW (260 hp) Textron Lycoming IO-540-V4A5 engine; McCauley two-blade constant-speed propeller standard, Hartzell and MT propellers optional. Seven delivered in 1999, one in 2000, and four in 2001.

M-7-260C: As M-7-235C, but 194 kW (260 hp) Textron Lycoming IO-540-V4A5 engine; McCauley two-blade constant-speed propeller standard, Hartzell and MT propellers optional. Eight delivered in 1999, seven each in 2000 and 2001.

MT-7-260: As MT-7-235, but 194 kW (260 hp) Textron Lycoming IO-540-V4A5 engine; McCauley two-blade constant-speed propeller standard, Hartzell and MT propellers optional. Three delivered in 1999, two in 2000, and four in 2001.

M-7-420AC: M-7 fuselage with long-span wings, 313 kW (420 shp) Rolls-Royce 250-B17C turboprop, spring aluminium tailwheel landing gear. One delivered in 1999, none in 2000, and one in 2001.

MT-7-420: As M-7-420AC, but tricycle gear. One delivered in 2001.

MX-9-230: M-7 with 169 kW (227 hp) SMA SR305 diesel engine with cowling designed by LoPresti Speed Merchants; five seats; max take-off weight 1,270 kg (2,800 lb); fuel capacity 322 litres (85.0 US gallons; 70.8 Imp gallons); range 869 n miles (1,609 km; 1,000 miles). Prototype reported to have been test flown in early 2002, but unconfirmed. Estimated cost US$200,000 (2001).

CUSTOMERS: Total of 885 produced by mid-2001, plus 908 of earlier M-5 and M-6 series. At 17 April 2001 Maule's order backlog stood at 100 aircraft, of which 10 were for SMA SR305-powered M-7s.

COSTS: MX-7-160 US$99,069; MX-7-180A US$104,499; MXT-7-160 US$108,570; MXT-7-180A US$113,999; MX-7-180B US$116,435; MX-7-180C US$121,435; MXT-7-180 US$126,888; M-7-235B (O-540 engine) US$129,868; M-7-235B (IO-540 engine) US$137,568; M-7-235C (O-540 engine) US$135,706; M-7-235C (IO-540 engine) US$143,406; MT-7-235 US$149,278; M-7-260 US$147,568; M-7-260C US$153,988; MT-7-260 US$159,278; M-7-420AC US$450,000. All 1998; standard equipment.

DESIGN FEATURES: Rugged, STOL utility aircraft. High, constant chord, wing braced by V-struts; mid-mounted tailplane; large sweptback fin.
USA 35B (modified) wing section; dihedral 1°; incidence 0° 30'; cambered wingtips standard.

FLYING CONTROLS: Conventional and manual. Ailerons linked to rudder servo tab to reduce adverse yaw; trim tab in port elevator; rudder trim by spring to starboard rudder pedal; flap deflection 40° down for slow flight (further setting of 48° down on all except -420), 24° down, 0 and 7° up for improved cruise performance; underfin on floatplane and amphibious versions.

STRUCTURE: All-metal two-spar wing with dual struts and glass fibre tips; fuselage frame of welded 4130 steel tube with Ceconite covering aft of cabin and metal doors and skin forward of cabin; glass fibre engine cowling.

LANDING GEAR: Non-retractable tailwheel or nosewheel type. Maule oleo-pneumatic shock-absorbers in narrow track main units on MX-7-160, MX-7-180A, MX-7-180B, M-7-235B and M-7-260 models; wide chord main units standard. Maule steerable tailwheel. Cleveland mainwheels with Goodyear or McCreary tyres size 7.00-6, pressure 1.79 bar (26 lb/sq in). Tailwheel tyre size 8×3.5-4, pressure 1.03 to 1.38 bar (15 to 20 lb/sq in). Cleveland hydraulic disc brakes. Parking brake. Oversize tyres, size 20×8.5-6 (pressure 1.24 bar; 18 lb/sq in), and fairings aft of mainwheels optional.
Provisions for fitting optional Aqua 2400 Baumann 2790, Edo 2400B or Edo 797-2500 amphibious or Wipline 2350 or Wipline 3000 amphibious floats (also available on some tricycle models). Float option available for MX-7-160, MX-7-180A/B/C, M-7-235B/C, M-7-260 and M-7-420AC. Ski option available for M-7-235B.

POWER PLANT: One flat-four or flat-six engine as described under Current Versions, driving a Sensenich two-blade fixed-pitch or Hartzell two-blade constant-speed propeller (three-blade McCauley propeller optional on 175 kW; 235 hp models); or Rolls-Royce 250-B17C turboprop with three-blade, Hartzell constant-speed propeller in -420 models. Piston versions have two fuel tanks in wings with total usable capacity of 163 litres (43.0 US gallons; 35.8 Imp gallons). Auxiliary fuel tanks in outer wings (standard on all 1999 models), to provide total capacity of 276 litres (73.0 US gallons; 60.8 Imp gallons). Turboprop versions have total fuel capacity of 322 litres (85.0 US gallons; 70.8 Imp gallons). Refuelling points on wing upper surface.

ACCOMMODATION: Four or five seats according to model, as described under Current Versions; individual, adjustable front seats; rear bench seat. Dual controls standard. Baggage compartment, capacity 113 kg (250 lb), aft of seats; cargo capacity with passenger seats removed 349 kg (770 lb). One front-hinged door on port side; three doors on starboard side, forward and centre doors hinged at front edge, rear baggage door hinged at rear edge to form double cargo door providing an opening 1.30 m (4 ft 3 in) wide to facilitate loading of bulky cargo; aircraft may be flown with doors removed. Accommodation heated and ventilated.

SYSTEMS: Hydraulic system for brakes only; electrical system powered by 60 A engine-driven alternator; 12 V battery (24 V battery on turboprops).

AVIONICS: *Comms:* Single Bendix/King KX 125-01 nav/com or KLX 135A GPS/com, KT 76A-00 transponder, Narco AR-850 remote altitude encoder, PS Engineering PM1000 intercom, ELT, broadband antenna, omni antenna, avionics master switch, cabin overhead speaker, microphone and microphone/headset jacks front and rear, all standard. Options on most models include KX 155-42,

Maule M-7-420AC floatplane *(Jane's/Paul Jackson)*   *NEW*/0533375

KX 165-21, KI 206, VOR indicator, KT 76C transponder, Garmin GNC 300 GPS and PS Engineering PMA-6000 audio panel.
Flight: Option of KI 206-04 or KI 209-01 VOR/LOC/GS indicators, KN 62A-01 or KN 34-00 DME, KR 87-16 digital ADF and KA 33-00 avionics cooling blower.
Instrumentation: Standard (MX-7, MXT-7 and M-7 models, but others generally similar) includes full gyro panel/vacuum system, carburettor air temperature gauge, acoustic stall warner, vertical speed indicator, altimeter, compass tachometer, electric turn co-ordinator, electric fuel gauge, fuel pressure gauge, cylinder head temperature, manifold pressure and oil temperature/pressure gauges, OAT gauge, ammeter, clock, and stall warning light.

EQUIPMENT (MX-7, MXT-7, and M-7; others similar): Includes instrument and dome lights, auxiliary cabin heater, auxiliary fuel pump, auxiliary power plug, heated pitot tube, cabin soundproofing, cloth velour upholstery, cabin steps, cargo tiedowns, landing light in port wing, navigation lights, wingtip strobe lights, tinted windscreen, pilot's swing-out window, windscreen defroster, airframe powder coating, wing corrosion proofing and standard external paint scheme in Insignia White base colour with two-tone trim. Optional equipment includes dual-calliper brakes, dynamically balanced propeller, McCauley three-blade constant-speed metal propeller (175 kW; 235 hp models only), Hartzell three-blade constant-speed propeller (turboprop models), wing corrosion proofing, Alcor EGT gauge (one-, four- or six-cylinder), Aviall hour meter, jump seat in baggage compartment, cabin door pockets, Plexiglas door, observation window, skylight, co-pilot's swing-out window, shoulder harnesses, landing light in starboard wing, fuselage grab handles, fire extinguisher, float reinforcement, lift rings and fairing, Schweizer glider tow/release kit and aircraft towbar.

DIMENSIONS, EXTERNAL:
Wing span: mid-length (M-6)	10.01 m (32 ft 10 in)
long (A Model)	10.31 m (33 ft 10 in)
Wing chord, constant: all	1.60 m (5 ft 3 in)
Wing aspect ratio: M-6	6.5
A Model	6.7
Length overall:	
piston-engined versions	7.16 m (23 ft 6 in)
turboprop versions	7.32 m (24 ft 0 in)
Height overall: tailwheel versions	1.93 m (6 ft 4 in)
tricycle versions	2.54 m (8 ft 4 in)

floatplane	3.05 m (10 ft 0 in)
amphibian	3.20 m (10 ft 6 in)
Wheel track:	
MX-7-160, MX-7-180A, MX-7-180B, M-7-235B, M-7-260	1.83 m (6 ft 0 in)
other versions	2.39 m (7 ft 10 in)
Propeller diameter:	
MX-7-160, MXT-160	1.88 m (6 ft 2 in)
MX-7-180A, MX-7-180C, MXT-7-180A, MX-7-180B, MXT-7-180	1.93 m (6 ft 4 in)
M-7-235, MT-7-235, MX-7-235	2.06 m (6 ft 9 in)
M-7-235, MT-7-235, MX-7-235 three-blade:	
option 1	2.03 m (6 ft 8 in)
option 2	1.98 m (6 ft 6 in)
M-7-420, MT-7-420	2.03 m (6 ft 8 in)

DIMENSIONS, INTERNAL:
Cabin: max width	1.07 m (3 ft 6 in)

AREAS:
Wings, gross: M-6	15.38 m² (165.6 sq ft)
A Model	15.72 m² (169.2 sq ft)

WEIGHTS AND LOADINGS:
Weight empty: MX-7-160	603 kg (1,330 lb)
MXT-7-160	649 kg (1,430 lb)
MX-7-180A	612 kg (1,350 lb)
MXT-7-180A	658 kg (1,450 lb)
MX-7-180B	633 kg (1,395 lb)
MXT-7-180	640 kg (1,410 lb)
MX-7-180C landplane	680 kg (1,500 lb)
M-7-235B	694 kg (1,530 lb)
M-7-235C	726 kg (1,600 lb)
M-7-235B floatplane	826 kg (1,821 lb)
MX-7-420	626 kg (1,380 lb)
Max T-O weight:	
MX-7-160, MXT-7-160	998 kg (2,200 lb)
MX-7-180A, MXT-7-180A	1,089 kg (2,400 lb)
all other landplanes	1,134 kg (2,500 lb)
M-7-235B floatplane	1,247 kg (2,750 lb)
Max wing loading:	
MX-7-180B, MX-7-180C, M-7-235B, M-7-235C	73.7 kg/m² (15.10 lb/sq ft)
M-7-235B floatplane	81.1 kg/m² (16.61 lb/sq ft)
Max power loading:	
MX-7-160, MXT-7-160	8.37 kg/kW (13.75 lb/hp)
MX-7-180A, MXT-7-180A	8.12 kg/kW (13.33 lb/hp)
MX-7-180B, MXT-7-180	8.45 kg/kW (13.89 lb/hp)
M-7-235B, MX-7-235	6.47 kg/kW (10.64 lb/hp)

MAULE M-7 OPTIONS

Version	Power Plant							Gear		
	B2D	C4F	C1F	B4B5	W1A5	V4A5	B17	TW/S	TW/A	TR/A
MX-7-160	•							•		
MXT-7-160	•									•
MX-7-160C	•								•	
MX-7-180A		•						•		
MX-7-180B			•					•		
MX-7-180C			•						•	
MXT-7-180			•							•
MXT-7-180A		•								•
M-7-235B				•				•		
M-7-235C				•					•	
MT-7-235					•				•	
M-7-260						•		•		
M-7-260C						•			•	
MT-7-260						•			•	
M-7-420AC							•	•		
MT-7-420							•			•

Notes: Power plants are Lycoming O-320-B2D, O-360-C4F, O-360-C1F, O-540-B4B5, IO-540-W1A5, IO-540-V4A5 (or earlier equivalents) and Rolls-Royce (Allison) 250-B17-C.
Landing gears are tailwheel/oleo main gear, tailwheel/aluminium and tricycle/aluminium.
'X' designations indicate shorter wing span..

MAULE M-7 PRODUCTION

Version	1983	1984	1985	1986	1987	1988	1989	1990	1991	1992	1993	1994	1995	1996	1997	1998	1999	2000	2001	Totals
																				(Annual Deliveries)
MX-7-160											15	16	9	3			1			**44**
MXT-7-160											1	1	1			5				**8**
MX-7-180			20	12	8	17	6	7	12	9	4	1								**96**
MX-7-180A											9	17	22	5	7	1				**61**
MX-7-180B												1	4	8		4				**17**
MX-7-180C														1	2	6	2	2	1	**14**
MXT-7-180								3	21	7	16	4	4	18	3	4	1	7	2	**90**
MXT-7-180A											1	1	3	6	22	24	18	6	1	**82**
M-7-235	2	17	13	11	16	15	13	9	13	3	13	7								**132**
M-7-235B												5	17	11	7	6	8	7	5	**66**
M-7-235C														7	11	5	16	18	14	**71**
MT-7-235										2	6	9	6	4	2	6	4	5	16	**60**
MX-7-235		2	23	25	18	14	5	3	16	9	2		1							**118**
M-7-260																1	7	1	4	**13**
M-7-260C																1	8	7	7	**23**
MT-7-260																	3	2	4	**9**
M-7-420																	1		2	**3**
M-7-420AC																			1	**1**
MX-7-420								2	1		1									**4**
Yearly totals	2	19	56	48	42	46	24	24	63	30	67	63	67	63	54	63	69	55	57	**912**

M-7-235B floatplane   7.12 kg/kW (11.70 lb/hp)
MX-7-420   3.62 kg/kW (5.95 lb/shp)

PERFORMANCE:
Max level speed:
  M-7-235B floatplane   130 kt (241 km/h; 150 mph)
  MX-7-420   174 kt (322 km/h; 200 mph)
Cruising speed at 75% power at optimum altitude:
  MXT-7-160   113 kt (209 km/h; 130 mph)
  MX-7-160, MXT-7-180A   117 kt (217 km/h; 135 mph)
  MXT-7-180, MX-7-180A   122 kt (225 km/h; 140 mph)
  MX-7-180, MX-7-180C   126 kt (233 km/h; 145 mph)
  M-7-235B, M-7-235C   139 kt (257 km/h; 160 mph)
  M-7-235 floatplane   125 kt (232 km/h; 144 mph)
  MX-7-420: at 50% power   156 kt (290 km/h; 180 mph)
    at 75% power   169 kt (314 km/h; 195 mph)
Stalling speed, flaps down, power off:
  MX-7-160, MXT-7-160, MX-7-180A, MX-7-180C,
    MXT-7-180A, MX-7-180B, MXT-7-180, MX-7-
    235, M-7-235C   35 kt (64 km/h; 40 mph)
  M-7-235B   31 kt (57 km/h; 35 mph)
  M-7-235B floatplane   47 kt (87 km/h; 54 mph)
  MX-7-420   44 kt (81 km/h; 50 mph)
Max rate of climb at S/L:
  MX-7-160, MXT-7-160   251 m (825 ft)/min
  MX-7-180A, MXT-7-180A   280 m (920 ft)/min
  MX-7-180B, MX-7-180C, MXT-7-180   365 m (1,200 ft)/min

  M-7-235B, M-7-235C, MT-7-235   609 m (2,000 ft)/min
  M-7-235B floatplane   411 m (1,350 ft)/min
  MX-7-420   1,432 m (4,700 ft)/min
Service ceiling:
  MX-7-160, MXT-7-160   3,965 m (13,000 ft)
  MX-7-180A, MX-7-180C, MXT-7-180A, MX-7-180B, MXT-7-180   4,575 m (15,000 ft)
  M-7-235B, MT-7-235, M-7-235C, MX-7-420   6,100 m (20,000 ft)
  M-7-235B floatplane   5,180 m (17,000 ft)
T-O run (solo, half fuel):
  MX-7-160, MXT-7-160   183 m (600 ft)
  MX-7-180C, MX-7-180B   92 m (300 ft)
  MXT-7-180, MX-7-420   61 m (200 ft)
  MX-7-180A, MXT-7-180A   168 m (550 ft)
  M-7-235C   77 m (250 ft)
  M-7-235B, M-7-235B floatplane   229 m (750 ft)
T-O to 15 m (50 ft):
  MX-7-160, MXT-7-160   360 m (1,180 ft)
  MX-7-180A, MXT-7-180A   350 m (1,150 ft)
  MX-7-180B, MX-7-180C, MXT-7-180, M-7-235B, M-7-235C   183 m (600 ft)
  M-7-235B floatplane   381 m (1,250 ft)
Landing from 15 m (50 ft):
  all landplane models   152 m (500 ft)
  M-7-235B floatplane   305 m (1,000 ft)

Landing run: MX-7-420   91 m (300 ft)
Range with main tank fuel only, optimum altitude, 30 min reserves:
  MX-7-160, MXT-7-160   469 n miles (869 km; 540 miles)
  MX-7-180A, MXT-7-180A   434 n miles (804 km; 500 miles)
  MXT-7-180, all tanks   825 n miles (1,528 km; 950 miles)
Range with max fuel, no reserves:
  MX-7-180C   951 n miles (1,762 km; 1,095 miles)
  M-7-235B:
    O-540 engine   792 n miles (1,467 km; 912 miles)
    IO-540 engine   829 n miles (1,537 km; 955 miles)
  M-7-235C:
    O-540 engine   792 n miles (1,467 km; 912 miles)
    IO-540 engine   864 n miles (1,601 km; 995 miles)
  M-7-235B floatplane:
    O-540 engine   708 n miles (1,311 km; 815 miles)
    IO-540 engine   772 n miles (1,430 km; 889 miles)
  MX-7-420:
    at 75% power   551 n miles (1,022 km; 635 miles)
    at 50% power   782 n miles (1,448 km; 900 miles)
UPDATED

MAVERICK

MAVERICK JETS INC

1371 General Aviation Drive, Melbourne, Florida 32935
Tel: (+1 321) 752 41 11
Fax: (+1 321) 752 44 55
e-mail: info@maverickjets.com
Web (1): http://www.twinjet.com
Web (2): http://www.maverickjets.com
PRESIDENT AND CEO: Jim McCotter
SENIOR VICE-PRESIDENT: David Hall
VICE-PRESIDENT, FLIGHT OPERATIONS AND CUSTOMER SERVICE: Jack Reed
DIRECTOR OF ENGINEERING: Alex Portalatin

The company moved into new premises in May 1998 and immediately began manufacture of the first Twinjet. In early 2001, Colorado businessman Jim McCotter acquired a majority share in Maverick Air Inc, renaming it Maverick Jets, and announced plans to establish a new company, known as McCotter Aviation, to develop a six-seat, factory-built, FAA-certified version of the Twinjet, designated MC2400.

UPDATED

MAVERICK LEADER

TYPE: Light business jet/kitbuilt.
PROGRAMME: Launched at EAA Convention at Oshkosh, Wisconsin, July 1997; then known as Twinjet 1200; construction of proof-of-concept prototype, undertaken by Composites Unlimited of Scappoose, Oregon, started September 1999; first flight (N750TJ) 4 August 1999; designated TJ-1500 Twinjet 1500; public debut at EAA AirVenture, Oshkosh, August 2000. Became Maverick Leader in 2002.

Prototype Maverick Twinjet 1500, now known as Maverick Leader (*Jane's/Paul Jackson*)   NEW/0533374

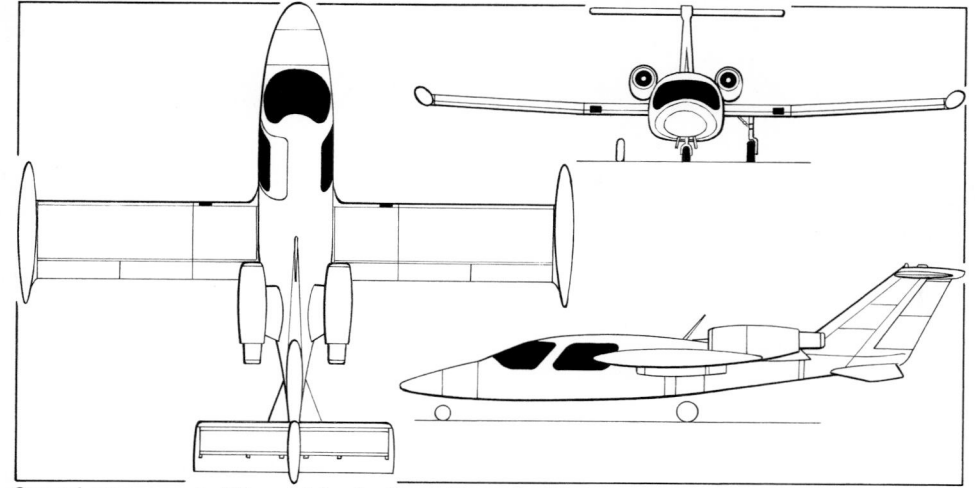

General arrangement of Maverick Leader (*Jane's/James Goulding*)   0105706

CURRENT VERSIONS: **Maverick Leader:** *As described.*

MC2400: Proposed six-seat, European-built, certified derivative, announced August 2001. Wind tunnel testing due to have begun in October 2001, followed by first flight in July 2002 and JAR-23 certification in July 2003.

Features include longer, taller cabin accommodating six people, enlarged wing, more powerful engines, 400 kt (741 km/h; 460 mph) cruising speed and 1,500 n mile (2,778 km; 1,726 mile) range. Cost US$1.5 million (2001).

CUSTOMERS: Orders for 14 Twinjets received by October 1999, for which five kits delivered by early 2001. Orders in September 2002 stated to be ''more than 12''; two aircraft flying by that time.

COSTS: Kit, less engines, US$219,000; engines (two), US$110,000; estimated cost to completion US$750,000 (2002). Factory-built US$1.2 million to US$1.4 million (2001). Projected operating cost US$170 per hour or 55 cents per mile (2001).

DESIGN FEATURES: Designed using advanced CAD and three-dimensional modelling techniques; design goals include comfort and performance of a traditional business jet in a light piston twin configuration and good short-field performance; unswept, mid-mounted, constant-chord wing with optional tip tanks; T tail; engines pod-mounted on rear fuselage shoulders.

FLYING CONTROLS: Conventional and manual. Flaps, approximately two-thirds span, standard; speedbrakes in upper wing optional; single control column between front seats.

STRUCTURE: Composites, employing prepreg glass fibre with carbon fibre reinforcement in high-stress areas. Supplied as kit, with fast-build options; quoted build time less than 2,000 hours.

LANDING GEAR: Retractable tricycle type with single wheel on each unit; main units retract inwards; nosewheel, forwards; Cleveland brakes.

POWER PLANT: Kitbuilt version has two reconditioned General Electric T58 turbojets, converted from turboshafts used in Bell UH-1 helicopters, each derated to 3.34 kN (750 lb st) and designated Maverick MC-750. Fuel in two wing tanks and one fuselage tank, combined capacity 1,249 litres (330 US gallons; 275 Imp gallons); gravity refuelling point on each wing. Factory-built version will be powered by two turbofans in the 4.45 to 5.34 kN (1,000 to 1,200 lb st) class

from Agilis, Pratt & Whitney Canada, Teledyne Continental or Williams International.

ACCOMMODATION: Four or five persons in pressurised cabin; single upward-opening gullwing door on port side; baggage compartments in nose and to rear of cabin.

SYSTEMS: Pressurisation system, maximum differential 0.39 bar (5.6 lb/sq in) maintains 3,050 m (10,000 ft) cabin altitude to 9,150 m (30,000 ft). Electrical system supplied by two 24 V batteries and two alternators. Airframe de-icing and oil-heated engine inlets standard.

AVIONICS: Sierra Flight Systems EFIS-2000 as core system; Garmin 430 GPS.

DIMENSIONS, EXTERNAL:

Wing span: standard	10.13 m (33 ft 3 in)
with tip tanks	10.52 m (34 ft 6 in)
Wing aspect ratio	7.6
Length overall	8.69 m (28 ft 6 in)
Height overall	2.74 m (9 ft 0 in)

DIMENSIONS, INTERNAL:

Cabin: Length	2.64 m (8 ft 8 in)
Max width	1.32 m (4 ft 4 in)
Max height	1.09 m (3 ft 7 in)
Baggage volume	0.42 m³ (15.0 cu ft)

AREAS:

Wings, gross	13.43 m² (144.6 sq ft)

WEIGHTS AND LOADINGS:

Weight empty	1,315 kg (2,900 lb)
Max T-O weight	2,630 kg (5,800 lb)
Max wing loading	195.8 kg/m² (40.11 lb/sq ft)
Max power loading	394 kg/kN (3.87 lb/lb st)

PERFORMANCE (estimated):

Never-exceed speed (VNE) and max level speed	
	391 kt (724 km/h; 450 mph)
Max cruising speed	350 kt (648 km/h; 403 mph)
Econ cruising speed	330 kt (611 km/h; 380 mph)
Cruising speed, OEI	200 kt (370 km/h; 230 mph)
Stalling speed, flaps down	78 kt (145 km/h; 90 mph)
Max rate of climb at S/L	914 m (3,000 ft)/min
Rate of climb at S/L, OEI	305 m (1,000 ft)/min
Max certified altitude	9,450 m (31,000 ft)
Service ceiling, OEI	6,095 m (20,000 ft)
T-O run	671 m (2,200 ft)
Landing run	607 m (1,990 ft)
Range with max internal fuel	
	1,000 n miles (1,852 km; 1,150 miles)
with tip tanks	1,300 n miles (2,407 km; 1,496 miles)

UPDATED

MD

MD HELICOPTERS INC

4555 East McDowell Road, Mesa, Arizona 85215-9797
Tel: (+1 480) 346 63 00
Fax: (+1 480) 346 68 07
e-mail: helicoptersales@mdhelicopters.com
Web: http://www.mdhelicopters.com
CHAIRMAN AND CEO: Henk Schaeken
COO: Albert Halder
VICE-PRESIDENT OF SALES AND MARKETING: Colin Whicher
MARKETING MANAGER: Mike McNabb

McDonnell Douglas' line of light helicopters, derived from the Hughes company's products, was acquired by Boeing as part of its purchase of the former company in August 1997. Having no part in the Boeing business strategy, all except the AH-64 Apache were offered for sale and in January 1999 it was announced that Netherlands-based holding company MD Helicopters, owned by the Rotterdam Dockyard Company, had been successful in its bid.

MD now owns all production jigs and tooling for the MD 500, 530F, 520N, 600N and Explorer, as well as a licence to employ the NOTAR system in future helicopters (the technology remaining in Boeing ownership) and in August 2000 acquired the former Boeing facility at Falcon Field, Mesa, Arizona, and began a 6,975 m² (75,000 sq ft) expansion to include a spares warehouse, completion and delivery centre and customer training facility.

In April 2000, Kaman (which see) was contracted to build fuselages for MD's single-engine product range at its Moosup and Jacksonville plants, deliveries beginning June 2000 and, in July 2000, was contracted also to supply rotor blades for the MD Explorer. Composite Solutions of Auburn, Washington, supplies tailbooms for the NOTAR range of helicopters. TAI (which see in Turkish section) is building Explorer fuselages at Akinci.

Hongdu MD Helicopters established in early 2003 as 60:40 venture between Chinese manufacturer and MD's parent (RDM Holdings); RDM investing US$10 million for Chinese assembly line for MD 500 and MD 600 at Nanchang.

MD Helicopters delivered 33 aircraft in 1999 (five MD 500Es, six MD 530Fs, five MD 520Ns, six MD 600Ns and 11 Explorers), 41 in 2000 (11 MD 500Es, three MD 530Fs, four MD 520Ns, seven MD 600Ns and 16 Explorers), 28 in 2001 (four MD 500Es, two MD 520Ns, two MD 600Ns and 20 Explorers). Had expected to deliver 58 in 2002 (10 MD 500Es, five MD 530Fs, seven MD 520Ns, 14 MD 600Ns and 22 Explorers), but actual total was 15 (five MD 500Es, four

MD 520Ns, two MD 600Ns and four Explorers). Order book at January 2003 was 35 and target production, 44. Revenues for 2001 totalled US$133 million.

UPDATED

MD 500 and MD 530

TYPE: Light utility helicopter.

PROGRAMME: Derived from Hughes OH-6A/civil Model 500 first flown in 1963; see McDonnell Douglas Helicopters entries in earlier *Jane's All the World's Aircraft* for MD 500D and previous versions. First flight MD 500E (N5294A) 28 January 1982; first flight MD 530F, 22 October 1982.

CURRENT VERSIONS: **MD 500E:** Replaced MD 500D in production 1982; deliveries started following issue of Type Certificate on 15 December 1982; Rolls-Royce 250-C20R became optional replacement for standard 250-C20B in late 1988; window area of forward canopy increased in 1991 model. MD 500E introduced many cabin improvements including more space for front and rear seat occupants, lower bulkhead between front and rear seats, T tail and optional four-blade Quiet Knight tail rotor.

MD 530F Lifter: Powered by Rolls-Royce 250-C30; transmission rating increased from 280 kW (375 shp) to 317 kW (425 shp) from 11 July 1985; diameter of main

rotor increased by 0.3 m (1 ft 0 in) and of tail rotor by 5 cm (2 in); cargo hook kit for 907 kg (2,000 lb) external load available; certified 29 July 1983; first delivery 20 January 1984. Upgraded drive system in **369FF**, certified 11 July 1985.

Detailed description applies to MD 500E and 530F, except where indicated.

MD 500/530 Defender: Described in 2001-02 and earlier editions.

MD 500N: See MD 520N.

CUSTOMERS: Some 4,663 OH-6/MD 500/530 series (excluding licensed manufacture) produced by late 2002. Recent customers for the MD 500E include the San Diego County, California, Sheriff's Department (one); Columbus, Ohio, Police Department (one); and an unnamed US operator (two), all delivered in the first half of 2000. The San Diego County Sheriff's Department has two MD 530Fs. Eleven MD 500s and four MD 530s delivered in 2000, four MD 500Es in 2001 and five in 2002. The Police Departments of Mesa, Arizona, Columbus, Ohio, Wichita, Kansas, and Atlanta, Georgia; and the Xiongying Aero Club of Zhongshan, China, each took delivery of an MD500E during 2002; the Metropolitan Police Department of Las Vegas, Nevada, has three MD 530Fs. Mesa fleet rose to three in December 200, while Oklahoma City's fleet simultaneously increased to two.

MD 500E registered in Belgium as OO-SOO (*Jane's/Paul Jackson*)   *NEW*/0526978

COSTS: Typical MD 500E US$860,000; MD 530F US$1.1 million (2002). Direct operating cost: MD 500E US$211, MD 530F US$237 per hour (both 2002).

DESIGN FEATURES: Fully articulated five-blade main rotor with blades retained by stack of laminated steel straps; blade aerofoil section NACA 015; blades can be folded after removing retention pins; two-blade tail rotor with optional X-pattern four-blade Quiet Knight tail rotor to reduce external noise; optional high-skid landing gear to protect tail rotor in rough country; protective skid on base of lower fin; narrow chord fin with high-set tailplane and endplate fins introduced with MD 500D. Main rotor rpm (500E/530F) 492/477 normal; main rotor tip speed 207 to 208 m (680 to 684 ft)/s; tail rotor rpm, 2,933/2,848.

FLYING CONTROLS: Plain mechanical without hydraulic boost.

STRUCTURE: Airframe based on two A frames from rotor head to landing gear legs, enclosing rear-seat occupants; front-seat occupants protected within straight line joining rotor hub and forward tips of landing skids; engine mounted inclined in rear of fuselage pod, with access through clamshell doors; main rotor blades have extruded aluminium spar hot-bonded to wraparound aluminium skin; tail rotor blades have swaged tubular spar and metal skin. Thicker fuselage skins, to reduce surface rippling, introduced during 2001.

LANDING GEAR: Tubular skids carried on oleo-pneumatic shock-absorbers. Utility floats, snow skis and emergency inflatable floats optional.

POWER PLANT: *MD 500E:* One 313 kW (420 shp) Rolls-Royce 250-C20B or 335.5 kW (450 shp) 250-C20R turboshaft, derated in both cases to 280 kW (375 shp) for T-O (5 minutes); maximum continuous rating 261 kW (350 shp).

MD 530F: One 485 kW (650 shp) Rolls-Royce 250-C30 turboshaft, derated to 317 kW (425 shp) up to 50 kt (92 km/h; 57 mph) and 280 kW (375 shp) above 50 kt (92 km/h; 57 mph) and for maximum continuous power (MCP).

MCP transmission rating 261 kW (350 shp); improved, heavy-duty transmission, rating 447 kW (600 shp), derated to 280 kW (375 shp) optional on production aircraft or for retrofit from June 1995.

Two interconnected bladder fuel tanks with combined usable capacity of 242 litres (64.0 US gallons; 53.3 Imp gallons). Self-sealing fuel tank optional. Refuelling point on starboard side of fuselage. Auxiliary fuel system, with 79.5 litre (21.0 US gallon; 17.5 Imp gallon) internal tank, available optionally. Oil capacity 5.7 litres (1.5 US gallons; 1.2 Imp gallons). Greater capacity internal fuel tanks also available.

ACCOMMODATION: Forward bench seat for pilot and two passengers, with two or four passengers, or two litter patients and one medical attendant, in rear portion of cabin. Pilot sits on left instead of normal right-hand seating. Low-back front seats and individual rear seats, with fabric or leather upholstery, optional. Baggage space, capacity 0.31 m³ (11 cu ft), under and behind rear seat in five-seat form. Clear space for 1.2 m³ (42 cu ft) of cargo or baggage with only three front seats in place. Two doors on each side. Interior soundproofing optional.

SYSTEMS: Aero Engineering Corporation air conditioning system or Fargo pod-mounted air conditioner optional.

AVIONICS (MD 500E): Optional avionics listed below.

Comms: Dual Bendix/King KY 195 or Rockwell Collins VHF-251 transceivers; Bendix/King KT 76 or Rockwell Collins TDR-950 transponder; intercom system, headsets, microphones; and optional public address system.

Radar: Optional installation.

Flight: Dual Bendix/King KX 175 or Rockwell Collins VHF-251/231 nav receivers, latter with IND-350 nav indicator; Bendix/King KR 85 or Rockwell Collins ADF-650 ADF.

Instrumentation: Basic VFR instruments and night flying lighting, attitude and directional gyros and rate of climb indicator.

Mission: Optional, FLIR and 30 Mcd Spectrolab SX-16 Nightsun searchlight.

EQUIPMENT: Optional equipment includes shatterproof glass, heating/demisting system, nylon mesh seats, dual controls, cargo hook, cargo racks, underfuselage cargo pod, heated pitot tube, extended landing gear, blade storage rack, litter kit, emergency inflatable floats and inflated utility floats.

DIMENSIONS, EXTERNAL:

Main rotor diameter: 500E	8.05 m (26 ft 5 in)
530F	8.33 m (27 ft 4 in)
Main rotor blade chord	0.171 m (6¾ in)
Tail rotor diameter: 500E	1.37 m (4 ft 6 in)
530F	1.42 m (4 ft 8 in)
Distance between rotor centres:	
500E	4.67 m (15 ft 4 in)
530F	4.88 m (16 ft 0 in)
Length overall, rotors turning:	
500E	8.61 m (28 ft 3 in)
530F	9.94 m (32 ft 7¼ in)
Length of fuselage	7.49 m (24 ft 7 in)
Height to top of rotor head:	
standard skids	2.67 m (8 ft 9 in)
extended skids	2.97 m (9 ft 8¾ in)
Tailplane span	1.65 m (5 ft 5 in)
Skid track (standard)	1.91 m (6 ft 3 in)
Cabin doors (each): Height	1.13 m (3 ft 8½ in)
Max width	0.76 m (2 ft 6 in)

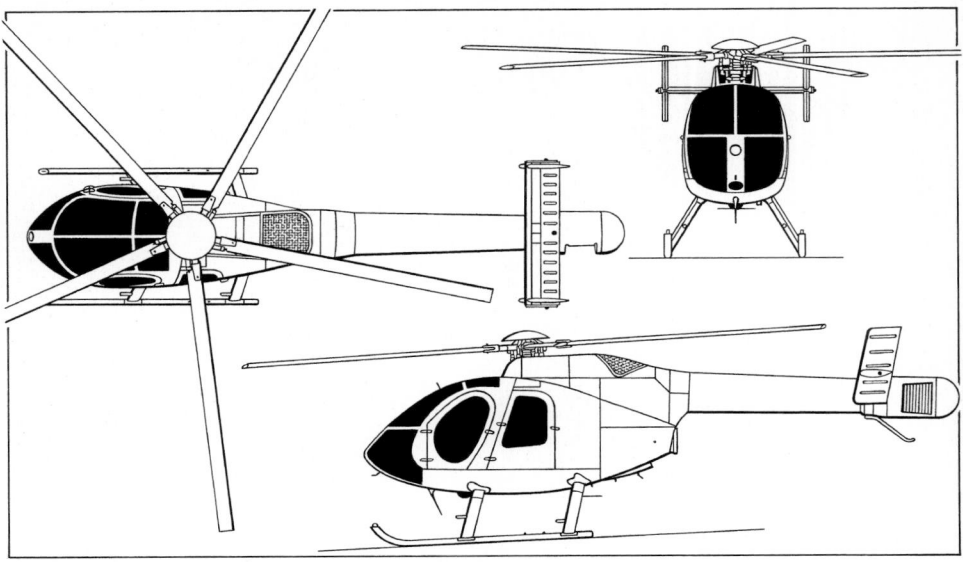

MD 520N five-seat NOTAR helicopter *(Jane's/Mike Keep)*

Height to sill: 500E	0.79 m (2 ft 7 in)
530F	0.76 m (2 ft 6 in)
Cargo compartment doors (each):	
Height	1.12 m (3 ft 8¼ in)
Width	0.88 m (2 ft 10½ in)
Height to sill: 500E	0.71 m (2 ft 4 in)
530F	0.66 m (2 ft 2 in)

DIMENSIONS, INTERNAL:

Cabin: Length	2.44 m (8 ft 0 in)
Max width	1.31 m (4 ft 3½ in)
Max height	1.52 m (5 ft 0 in)

AREAS:

Main rotor blades (each): 500E	0.62 m² (6.67 sq ft)
530F	0.65 m² (6.96 sq ft)
Tail rotor blades (each): 500E	0.063 m² (0.675 sq ft)
530F	0.066 m² (0.711 sq ft)
Main rotor disc: 500E	50.89 m² (547.8 sq ft)
530F	54.58 m² (587.5 sq ft)
Tail rotor disc: 500E	1.53 m² (16.47 sq ft)
530F	1.65 m² (17.72 sq ft)
Fin	0.56 m² (6.05 sq ft)
Tailplane	0.76 m² (8.18 sq ft)

WEIGHTS AND LOADINGS:

Weight empty: 500E	672 kg (1,481 lb)
530F	722 kg (1,591 lb)
Max normal T-O weight: 500E	1,361 kg (3,000 lb)
530F	1,406 kg (3,100 lb)
Max overload T-O weight:	
500E, 530F	1,610 kg (3,550 lb)
Max T-O weight, external load:	
530F	1,701 kg (3,750 lb)
Max hook capacity: 530F	907 kg (2,000 lb)
Disc loading at max normal T-O weight:	
500E	26.8 kg/m² (5.48 lb/sq ft)
530F	25.8 kg/m² (5.28 lb/sq ft)
Transmission loading at max T-O weight and power:	
500E standard	5.22 kg/kW (8.57 lb/shp)
500E optional	4.87 kg/kW (8.00 lb/shp)
530F standard	5.39 kg/kW (8.86 lb/shp)
530F optional	5.03 kg/kW (8.27 lb/shp)

PERFORMANCE (at max normal T-O weight, except where indicated):

Never-exceed speed (VNE) at S/L:	
500E, 530F	152 kt (282 km/h; 175 mph)
Max cruising speed at S/L:	
500E	134 kt (248 km/h; 154 mph)
530F	133 kt (247 km/h; 154 mph)
Max cruising speed at 1,525 m (5,000 ft):	
500E	132 kt (245 km/h; 152 mph)
530F	135 kt (249 km/h; 155 mph)
Econ cruising speed at S/L:	
500E	129 kt (239 km/h; 149 mph)
530F	131 kt (243 km/h; 151 mph)
Econ cruising speed at 1,525 m (5,000 ft):	
500E, 530F	123 kt (228 km/h; 142 mph)
Max rate of climb at S/L, ISA:	
500E	536 m (1,760 ft)/min
530F	631 m (2,070 ft)/min
Max rate of climb at S/L, ISA+20°C:	
530F	628 m (2,061 ft)/min
Vertical rate of climb at S/L: 500E	248 m (813 ft)/min
530F	446 m (1,462 ft)/min
Service ceiling: 500E	4,575 m (15,000 ft)
530F	5,700 m (18,700 ft)
Hovering ceiling IGE: ISA: 500E	2,590 m (8,500 ft)
530F	4,875 m (16,000 ft)
ISA+20°C: 500E	1,830 m (6,000 ft)
530F	4,358 m (14,300 ft)
Hovering ceiling OGE: ISA: 500E	1,830 m (6,000 ft)
530F	4,389 m (14,400 ft)

ISA+20°C: 500E	975 m (3,200 ft)
530F	3,535 m (11,600 ft)
Range, 2 min warm-up, standard fuel, no reserves:	
500E at S/L	233 n miles (431 km; 268 miles)
530F at S/L	200 n miles (371 km; 231 miles)
500E at 1,525 m (5,000 ft)	
	264 n miles (488 km; 303 miles)
530F at 1,525 m (5,000 ft)	
	232 n miles (429 km; 267 miles)
Endurance, 530F at S/L	2 h 0 min

UPDATED

MD 520N

TYPE: Light utility helicopter.

PROGRAMME: First flight OH-6A NOTAR (no tail rotor) testbed 17 December 1981; extensive modifications during 1985 with second blowing slot, new fan, 250-C20B engine and MD 500E nose; flight testing resumed 12 March 1986 and completed in June; retired to US Army Aviation Museum, Fort Rucker, Alabama, October 1990. Commercial MD 520N and uprated (485 kW; 650 shp Rolls Royce 250-C30) MD 530N NOTAR helicopters announced February 1988 and officially launched January 1989; first flight of MD 530N (N530NT) 29 December 1989, but this variant not pursued; first flight 520N (N520NT) 1 May 1990 and first production 520N 28 June 1991; 520N certified 12 September 1991; first production 520N (N521FB) delivered to Phoenix Police Department 31 October 1991. MD 520N set new Paris to London speed record in September 1992, at 1 hour 22 minutes 29 seconds. Now certified in 21 countries.

CURRENT VERSIONS: **MD 520N:** NOTAR version of MD 500, offering more power, higher operating altitude and greater maximum T-O weight than MD 500E. Engineering designation is **MD 500N**.

Description applies to 520N.

MD 600N: Stretched version, described separately.

MD 520N Defender: Military variant, being developed.

CUSTOMERS: Total of 99 registered by mid-2002. Four delivered in 2000, two in 2001 and four in 2002. Law enforcement agencies flying MD 520Ns include Phoenix, Arizona (first operator; two more delivered in early 2002); Burbank, Glendale, Huntington Beach, Los Angeles, Ontario and San Jose, California; Hernando County, Florida; Orange County, Florida; Jefferson County, Kentucky, which took delivery of one in March 2002; Prince George County, Clinton, Maryland, which took delivery of two on 4 October 2000; Hamilton County, Ohio, which has seven 520Ns; El Salvador, whose Policia Nacional Civil took delivery of two in 1996; San Juan, Puerto Rico; Honolulu, Hawaii; and Calgary, Alberta, Canada. Other operators include the Tata Group of Mumbai, India, Weetabix Ltd, UK, and Belgian Gendarmerie (two).

COSTS: US$980,000 (2002). Direct operating cost US$224.71 per hour (2002).

DESIGN FEATURES: NOTAR system provides anti-torque and steering control without an external tail rotor, thus eliminating the danger of tail strikes; air emerging through Coanda slots and steering louvres is cool and at low velocity. Believed to be currently the quietest turbine helicopter, based on FAA certification test noise figures. Main rotor rpm 477; main rotor tip speed 208 m (684 ft)/s; NOTAR system fan rpm 5,388. Emergency floats among options. Redesigned diffuser in NOTAR tailboom and revised fan rigging introduced from early 2000 and available for retrofit; combined with uprated Rolls-Royce 250-C20R+ engine (see below); these improvements increase the aircraft's payload capability.

MD 520N operated by Prince George's County (Maryland) Police Department　　　0126956

FLYING CONTROLS: Unboosted mechanical, as in earlier versions. Rotor downwash over tailboom deflected to port by two Coanda-type slots fed with low-pressure air from engine-driven variable-pitch fan in root of tailboom; this counters normal rotor torque; some fan air is also vented at tail through variable-aperture louvres controlled by pilot's foot pedals, giving steering control in hover and forward flight. Port moving fin on tailplane connected to foot pedals, primarily to increase directional control during autorotation and allow touchdown at under 20 kt (37 km/h; 23 mph); starboard fin operated independently by yaw damper.

STRUCTURE: Same as for MD 500E/530F, except graphite composites tailboom; metal tailplane and fins; new high-efficiency fan with composites blades fitted in production aircraft. NOTAR system components now have twice the lifespan of conventional tail rotor system assemblies. During 1993, NOTAR system components' warranty increased from two to three years. Thicker fuselage skins, to reduce surface rippling, will be introduced during 2001.

POWER PLANT: One Rolls-Royce 250-C20R turboshaft, derated to 317 kW (425 shp) for T-O (5 minutes) and 280 kW (375 shp) maximum continuous. Improved, heavy-duty transmission, rating 447 kW (600 shp), derated to 317 kW (425 shp) for T-O, 280 kW (375 hp) maximum continuous, on production aircraft from June 1995. Rolls-Royce 250-C20R+ engine, providing improved hot weather performance and 3 to 5 per cent more power, introduced from early 2000. Fuel capacity 235 litres (62.0 US gallons; 51.6 Imp gallons).

SYSTEMS: Electrical system includes 87 Ah starter-generator. Aero Aire air conditioning system optional.

EQUIPMENT: Phoenix Police Department pioneered use of MD 520Ns for firefighting, using 341 litre (90.0 US gallon; 75.0 Imp gallon) 'Bambi' buckets to drop 151,416 litres (40,000 US gallons; 33,307 Imp gallons) of water on fires in remote desert and mountain areas.

DIMENSIONS, EXTERNAL:

Rotor diameter	8.33 m (27 ft 4 in)
Length: overall, rotor turning	9.78 m (32 ft 1¼ in)
fuselage	7.77 m (25 ft 6 in)
Height to top of rotor head:	
standard skids	2.74 m (9 ft 0 in)
extended skids	3.01 m (9 ft 10¾ in)
Height to top of fins	2.83 m (9 ft 3½ in)
Tailplane span	2.01 m (6 ft 7¼ in)
Skid track	1.92 m (6 ft 3½ in)

AREAS:

Rotor disc	54.51 m² (586.8 sq ft)

WEIGHTS AND LOADINGS:

Weight empty: standard	719 kg (1,586 lb)
Max fuel weight	183 kg (404 lb)
Max hook capacity	1,004 kg (2,214 lb)
Max T-O weight: normal	1,519 kg (3,350 lb)
with external load	1,746 kg (3,850 lb)
Max normal disc loading	27.9 kg/m² (5.71 lb/sq ft)
Transmission loading at max T-O weight and power	
	4.80 kg/kW (7.88 lb/shp)

PERFORMANCE (at normal max T-O weight, ISA, except where indicated):

Never-exceed speed (VNE)	152 kt (281 km/h; 175 mph)
Max cruising speed at S/L	135 kt (249 km/h; 155 mph)
Max rate of climb at S/L: ISA	564 m (1,850 ft)/min
ISA + 20°C	480 m (1,575 ft)/min
Service ceiling	4,320 m (14,175 ft)
Hovering ceiling IGE	3,414 m (11,200 ft)
Hovering ceiling OGE: ISA	1,830 m (6,000 ft)
ISA +20°C	1,292 m (4,240 ft)
Range at S/L	229 n miles (424 km; 263 miles)
Endurance at S/L	2 h 24 min

OPERATIONAL NOISE LEVELS:

T-O	85.4 EPNdB
Approach	87.9 EPNdB
Flyover	80.2 EPNdB

UPDATED

MD 600N

TYPE: Light utility helicopter.

PROGRAMME: Stretched version of MD 520N. Announced as 'concept', 8 November 1994; prototype, then known as MD 630N (N630N, converted from MD 530F demonstrator), first flight 22 November 1994, 138 days after project approval; public debut at Heli-Expo in Las Vegas January 1995; production go-head 28 March 1995, at which time designation changed to MD 600N; prototype first flown with production standard engine and rotor system 6 November 1995; production prototype (N600RN) first flown 15 December 1995, and became certification test vehicle leading to FAR Pt 27 certification, but was destroyed in ground fire 28 May 1996, following emergency landing after rotor/tailboom strike during abrupt control reversal tests. This resulted in changes to tailboom/rotor clearance; third prototype (N605AS) flown 9 August 1996; further accidents to N630N on 4 and 21 November 1996 and on 18 January 1997, all during autorotational descents, culminated in total loss and delayed certification and first delivery, originally scheduled for 18 December 1996, to 15 May and 6 June 1997, respectively.

In July 1998, Boeing completed a year-long envelope expansion programme for the MD 600N leading to FAA approval for operation at a density altitude of 2,135 m (7,000 ft) at a T-O weight of 1,746 kg (3,850 lb) and at a density altitude of 1,220 m (4,000 ft) at a T-O weight of 1,860 kg (4,100 lb). Other performance enhancements approved by the FAA included provision for doors-off operation at speeds up to 115 kt (213 km/h; 132 mph), operation at temperatures –40°C/52°C (–40°F/126°F), lifting up to 968 kg (2,134 lb) on the external cargo hook, making slope landings up to 10° in any direction, operation with emergency floats and for installation of a movable landing light and additional wire strike protection on the fuselage. The MD 600N also completed HIRF trials at NAS Patuxent River, Maryland. Yaw-stability augmentation system (Y-SAS) under development during 2000, aimed at reducing pilot workload during extended flights and in turbulent conditions; Y-SAS was expected to be available from March 2001, following FAA certification, and will be field-installable.

CUSTOMERS: Total of 66 registered by October 2002. Launch customer AirStar Helicopter of Arizona (two, of which first delivered 6 June 1997); Saab Helikopter AB of Sweden and Rotair Limited of Hong Kong ordered one each in June 1995; other customers include Guangdong General Aviation Company (GGAC) of the People's Republic of China, which took delivery of one MD 600N in November 2000, during Airshow China 2000 in Zhuhai, Los Angeles County Sheriff's Department Aero Bureau (three), Orange County, California, Sheriff's Department, Indianapolis, Indiana, Police Department (one), Presta Services of France (one), Turkish National Police, which ordered 10 in December 2000 for delivery during 2002 (subsequently postponed to 2003), UND (University of North Dakota) Aerospace (one), West Virginia State Police (one) and the US Border Patrol (45, of which 11 delivered by end of 1998, when procurement halted pending evaluation of UAVs for border patrol role). Deliveries totalled 15 in 1997, 21 in 1998, six in 1999, seven in 2000, two in 2001 and two in 2002.

COSTS: US$1.315 million (2002). Direct operating cost US$269 per hour (2002).

DESIGN FEATURES: Stretched MD 520N airframe (less than 1 per cent new parts) by means of 0.76 m (2 ft 6 in) plug aft of cockpit/cabin bulkhead and 0.71 m (2 ft 4 in) plug in tailboom, combined with more powerful engine, uprated transmission and six-blade main rotor. Cabin has flat floor to assist cargo handling, and will feature quick-change interior configurations to suit multiple-use operators. Intended for civil, utility, offshore, executive transport,

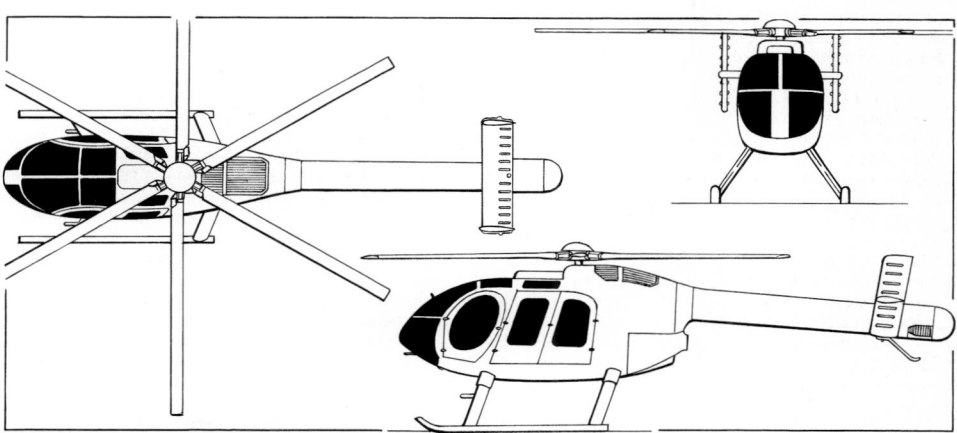

MD 600N eight-seat NOTAR helicopter (*Jane's/James Goulding*)　　　0051783

MD 600N operated by the US Border Patrol (*Jane's/Paul Jackson*)　　　*NEW*/0526977

medevac, aerial news gathering, touring, law enforcement and other noise-sensitive operations; also adaptable for armed scout, utility and other military missions.

Description generally as for MD 520N except as follows:

POWER PLANT: One 603 kW (808 shp) Rolls-Royce 250-C47M turboshaft, derated to 447 kW (600 shp) for T-O (5 minutes) and 395 kW (530 shp) maximum continuous, with FADEC. Transmission manufactured from WE43A magnesium alloy for lower weight, greater strength and enhanced corrosion resistance, rating 447 kW (600 shp). Fuel contained in two crashworthy bladder tanks in lower fuselage, total capacity 440 litres (116 US gallons; 96.8 Imp gallons), of which 434 litres (115 US gallons; 95.5 Imp gallons) are usable.

ACCOMMODATION: Standard seating for eight, including pilot, in 3+3+2 configuration with centre bench; alternative six-seat club and coach layouts; utility with main cabin seats removed; EMS with accommodation for pilot, one stretcher case and two medical attendants; or electronic news gathering, with accommodation for pilot and passenger in cockpit and operator with cabin and monitor in cabin. All facilitated by quick-release seat mechanisms. Two removable centre-opening doors on each side of cabin; door to cockpit on each side. Bubble 'comfort' windows and sliding windows for cabin, custom soundproofing and Integrated Flight Systems Inc air conditioning optional.

SYSTEMS: Electrical system comprises 28 V 200 Ah starter-generator and 28 V 17 Ah Ni/Cd battery. 24 V auxiliary power receptacle inside starboard cockpit door standard.

AVIONICS: To customer's choice, including VHF/UHF/FM/AM transceivers, GPS, transponder, ELT, intercom, stereo tape system and PA system.

EQUIPMENT: Optional equipment includes dual controls, particle separator, heated pitot tube, rotor brake, cargo hook, wire strike kit and emergency floats.

DIMENSIONS, EXTERNAL:
Rotor diameter	8.38 m (27 ft 6 in)
Rotor ground clearance	2.65 m (8 ft 8½ in)
Length: overall, rotor turning	10.79 m (35 ft 4¾ in)
fuselage	8.99 m (29 ft 6 in)
Fuselage width at cabin	1.40 m (4 ft 7 in)
Height to top of rotor head:	
standard skids	2.65 m (8 ft 8½ in)
extended skids	2.74 m (9 ft 0 in)
Height to top of fins:	
standard skids	2.68 m (8 ft 9½ in)
extended skids	2.83 m (9 ft 3½ in)
Fuselage ground clearance, extended skids	
	0.58 m (1 ft 10¾ in)
Skid track	2.47 m (8 ft 1¼ in)
Tailplane span, over fins	2.01 m (6 ft 7¼ in)

DIMENSIONS, INTERNAL:
Cabin, excl cockpit:	
Length	1.83 m (6 ft 0 in)
Width	1.22 m (4 ft 0 in)
Height	1.22 m (4 ft 0 in)

AREAS:
Rotor disc	55.18 m² (594.0 sq ft)

WEIGHTS AND LOADINGS:
Weight empty, standard	952 kg (2,100 lb)
Useful load: internal	907 kg (2,000 lb)
external	1,179 kg (2,600 lb)
Max hook capacity	1,361 kg (3,000 lb)
Max T-O weight: internal load	1,859 kg (4,100 lb)
slung load	2,131 kg (4,700 lb)
Max disc loading:	
internal load	33.7 kg/m² (6.90 lb/sq ft)
slung load	38.6 kg/m² (7.91 lb/sq ft)
Transmission loading at max T-O weight and power:	
internal load	4.16 kg/kW (6.83 lb/shp)
slung load	4.77 kg/kW (7.83 lb/shp)

PERFORMANCE:
Never-exceed speed (VNE)	135 kt (250 km/h; 155 mph)
Max cruising speed, S/L to 1,525 m (5,000 ft), ISA	
	134 kt (248 km/h; 154 mph)
Max rate of climb at S/L, ISA	411 m (1,350 ft)/min
Max operating altitude	6,100 m (20,000 ft)
Hovering ceiling, ISA: IGE	3,383 m (11,100 ft)
OGE	1,829 m (6,000 ft)
Max range:	
at S/L	342 n miles (633 km; 393 miles)
at 1,525 m (5,000 ft), ISA	
	382 n miles (707 km; 439 miles)
Endurance: at S/L	3 h 36 min
at 1,525 m (5,000 ft)	3 h 54 min

UPDATED

MD EXPLORER

US Coast Guard designation: MH-90 Enforcer

TYPE: Light utility helicopter.

PROGRAMME: Initially known as MDX, then MD 900 (proposed MD 901 with Turbomeca engines was not pursued); McDonnell Douglas design; announced February 1988; launched January 1989; Hawker de Havilland of Australia designed and manufactures airframe; Canadian Marconi tested initial version of integrated instrumentation display system (IIDS) early 1992; Kawasaki completed 50 hour test of transmission early 1992. Other partners include Aim Aviation (interior),

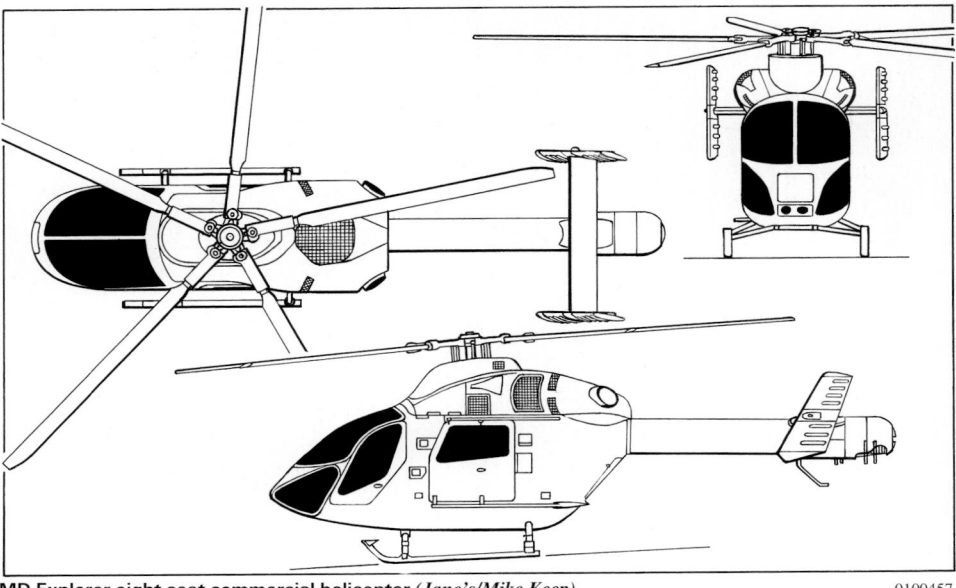

MD Explorer eight-seat commercial helicopter *(Jane's/Mike Keep)* 0100457

IAI (cowling and seats) and Lucas Aerospace (actuators). Ten prototypes and trials aircraft, of which seven (Nos. 1, 3-7 and 9) for static tests; first flight (No.2/N900MD) 18 December 1992, followed by No.8/N900MH 17 September 1993 and No.10/N9208V 16 December 1993; first production/demonstrator Explorer (No.11/N92011) flown 3 August 1994. FAA certification 2 December 1994; first delivery 16 December 1994; JAA certification July 1996; FAA certification for single-pilot IFR operation achieved January 1997. Type certificate transferred to MDHI on 18 February 1999.

FAA certification of uprated PW207E engine achieved in July 2000, providing 11 per cent more power for take-off and 610 m (2,000 ft) increase in hovering capability OEI in hot-and-high conditions; first delivery of PW207E-engined Explorer to Police Aviation Services, UK, 27 September 2000. "100th production" Explorer (actually 89th overall, including prototypes) delivered 1 March 2002 to Tomen Aerospace Corporation of Japan for ENG operations by Aero Asahi of Hiroshima. Total fleet time stood at more than 120,000 hours by December 2002.

CURRENT VERSIONS: **MD Explorer:** Initial civilian utility version, as described.

Details apply to civilian version except where indicated.

MD Enhanced Explorer: Improved version, announced September 1996; originally MD 902, but now known as "902 Configuration". Main features include Pratt & Whitney Canada PW206E engines with increased OEI ratings; transmission approved for dry running for 30 minutes at 50 per cent power; improved engine air inlets, NOTAR inlet design and engine fire suppression system, and more powerful stabiliser control system, resulting in 7 per cent increase in range, 4 per cent increase in endurance and 113 kg (250 lb) increase in payload over Explorer. First flight (N9224U; c/n 900-0051, 41st Explorer) 5 September 1997. FAA certification to Category A performance standards (including continued take-off with one failed engine) and single-pilot IFR operation achieved 11 February 1998; JAA certification for Category A performance achieved July 1998. Retrofit kits to convert Explorers to Category A standard. First Enhanced Explorer delivery in May 1998 to Tomen Aerospace of Japan. PW206E replaced by PW207E from late 2000, beginning at c/n 900-0077, allowing further MTOW increase to 2,948 kg (6,500 lb).

MH-90 Enforcer: Beginning March 1999, under a programme code-named Operation New Frontier, the US Coast Guard used two leased MD 900 Explorers for shipboard anti-drug smuggling operations. Armed with a pintle-mounted M240 7.62 mm minigun at the door station. In September 1999 the MD900s were exchanged for two leased MD 902 Enhanced Explorers. These subsequently replaced by Agusta A 109s. Six delivered to Mexican Navy at Acapulco (two each respectively in May and December 1999 and April 2000) for anti-drug operations, equipped with 0.50 in General Dynamics GAU-19/A Gatling guns, and 70 mm rocket pods; further four in process of delivery. Weapons qualification trials were completed at Fort Bliss, Texas in November 2000.

Combat Explorer: Displayed at Paris Air Show, June 1995; demonstrator N9015P (No.15), an MD 900 variant. Can be configured for utility, medevac or combat missions; armament and mission equipment may include seven- or 19-tube 70 mm rocket pods, 0.50 calibre machine gun pods, chin-mounted FLIR night pilotage system and roof-mounted NightHawk surveillance and targeting systems. Combat weight 3,130 kg (6,900 lb); two P&WC PW206A engines. No customers announced by January 2000, but N9015P became one of initial two MH-90s (with third prototype, N9208V).

CUSTOMERS: Market estimated at 800 to 1,000 in first decade; 100th Explorer registered in 2002 (to become seventh for Netherlands police); total of 108 manufactured by December 2002; first delivery 16 December 1994 to Petroleum Helicopters Inc (PHI) which ordered five; second delivery (N901CF) December 1994 to Rocky Mountain Helicopters for EMS duties with affiliate Care Flight unit of Regional Emergency Medical Services Authority (REMSA) in Reno, Nevada. Total of two delivered in 1994, 12 in 1995, 15 in 1996, one in 1997, four in 1998, 11 in 1999, 16 in 2000, 20 in 2001 and four in 2002; initial (MD 900) series comprised 40 aircraft, including three flying prototypes; PW207E engine from 64th production (67th overall) aircraft.

Other disclosed customers include Aero Asahi of Japan (15, of which the first, JA6757, was delivered in July 1995 and seven in service by October 2000), Airfast Indonesia, which ordered two in 2001 for offshore oil support in the Java Sea, Allegheny General Hospital, Pittsburgh, Pennsylvania (four, delivered mid-1996), Belgian

MD Explorer light utility helicopter *(Jane's/Paul Jackson)* NEW/0526976

Dual-pilot instrument panels of MD Explorer   0022147

government (two delivered in 1996, in law enforcement configuration for Rijkswacht/Gendarmerie), Netherlands National Police (eight, plus two options, originally for delivery beginning June 2002 and scheduled for completion by end of 2002 but all postponed to 2003); German State Police of Lower Saxony operates three, based at Hanover, and State Police of Baden-Württemberg at Stuttgart has ordered five, of which delivery rescheduled from 2001 to 2003. Guangdong General Aviation Company of Shanghai, China, which ordered one for delivery in 2002; IBCOL Group of Germany (one, delivered in early 1996, operated by Air Lloyd of Bonn), Idaho Helicopters Inc/Life Flight (one, for EMS service at St Alphonsus Regional Medical Center, Boise, Idaho), Japan Digital Laboratory (one), Luxembourg Air Rescue (one, in EMS configuration), New York Suffolk County Police, Police Aviation Services (PAS) UK (10, of which initial delivery was made in June 1998; Tata Group of India (one, delivered in fourth quarter 2000); Televisa of Mexico (one); Japanese distributor Tomen (two); UND (University of North Dakota) Aerospace (one); and Virgin HEMS (London) Ltd (one, delivered in October 2000). By early 2003, UK Police Aviation Services fleet totalled three, while further six Explorers were owned by forces of Dorset, Greater Manchester, Humberside, Sussex, West Midlands and West Yorkshire.

COSTS: US$2.285 million (2002); direct operating cost US$408.11 (2002) per hour.

DESIGN FEATURES: NOTAR anti-torque system; all-composites five-blade rotor of tapered thickness with parabolic swept outer tip with bearingless flexbeam retention and pitch case; tuned fixed rotor mast and mounting truss for vibration reduction; replaceable rotor tips; maximum rotor speed 392 rpm; modified A-frame construction from rotor mounting to landing skids protects passenger cabin; energy-absorbing seats absorb 20 *g* vertically and 16 *g* fore and aft; onboard health monitoring, exceedance recording and blade track/balance.

FLYING CONTROLS: NOTAR tailboom (see details under MD 520N); mechanical engine control from collective pitch lever is back-up for electronic FADEC. Automatic stabilisation and autopilot available for IFR operation.

STRUCTURE: Cockpit, cabin and tail largely carbon fibre; top fairings Kevlar composites; no magnesium; lightning strike protection embedded in composites skin. Transmission overhaul life 5,000 hours; glass fibre blades have titanium leading-edge abrasion strip and are attached to bearingless hub by carbon fibre encased glass fibre flexbeams; rotor blades and hub on condition.

LANDING GEAR: Fixed skids with replaceable abrasion pads; emergency floats optional.

POWER PLANT: Baseline MD 900 powered by two Pratt & Whitney Canada PW206E turboshafts with FADEC, each rated at 463 kW (621 shp) for 5 minutes for T-O, 489 kW (656 shp) for 2½ minutes OEI and 410 kW (550 shp) maximum continuous. Transmission rating 820 kW (1,100 shp) for T-O, 746 kW (1,000 shp) maximum continuous, 507 kW (680 shp) for 2½ minutes OEI and 462 kW (620 shp) maximum continuous OEI.

Fuel contained in single tank under passenger cabin, capacity 564 litres (149 US gallons; 124 Imp gallons), of which 553 litres (146 US gallons; 122 Imp gallons) are usable. Single-point refuelling; self-sealing fuel lines.

ACCOMMODATION: Two pilots or pilot/passenger in front on energy-absorbing adjustable crew seats with five-point shoulder harnesses/seat belts; six passengers in club-type energy-absorbing seating with three-point restraints; rear baggage compartment accessible through rear door; cabin can accept long loads reaching from flight deck to rear door; hinged, jettisonable door to cockpit on each side; sliding door to cabin on each side.

SYSTEMS: Hydraulic system, operating pressure 34.475 bar (500 lb/sq in).

AVIONICS: Full IFR capability for single- or two-pilot operation.

Comms: Two headsets standard. Bendix/King Silver Crown VHF transceiver, audio control panel, ELT, cockpit voice recorder and Wulfsberg Flexcomm II optional.

Radar: Honeywell RDR 2000 vertical profile colour weather radar optional with IFR package.

Flight: Optional equipment includes Bendix/King Silver Crown ADF, HSI, ADF, DME, marker beacon receiver, radar altimeter, Loran C and KLN 90B GPS. Coupled three-axis autopilot optional.

Instrumentation: Single- or two-pilot instrument panels incorporate Canadian Marconi integrated instrumentation display system (IIDS) with high-resolution sunlight-readable LCD screen displaying engine and system

information including engine condition trend monitoring, exceedance recording, caution annunciators, onboard track and balance of rotor and fan, weight on cargo hook, outside air temperature, digital clock, running time meter and RS-232 download modem interface for personal computer. Other standard instrumentation includes airspeed indicator, encoding altimeter, vertical speed indicator, turn and slip indicator, wet compass and clock. EFIS 40 electronic flight information system and parallel IIDS monitor for long-line hook operations from left seat optional.

Mission: Law enforcement panel with space for FLIR screen available. Avionics and equipment options for law enforcement conversions are described in *Jane's Aircraft Upgrades.*

EQUIPMENT: Standard equipment includes magnetic chip detectors on engines, tiedown fittings, flush-mounted cargo tiedowns, rotor blade tiedowns, right side passenger step, utility beige colour carpet, trim, wall and ceiling panels, soundproofing, tinted windows, map case, recessed hover and approach light, wander and white dome lights in cockpit, white dome light in cabin, utility light in baggage compartment, single 28 V DC power outlet each in cockpit and cabin, single colour external paint with two-colour accent stripes, FOD covers, pitot tube cover and cockpit fire extinguisher.

Optional equipment includes dual controls, heated pitot head, rotor brake, pilot-activated engine fire extinguisher, engine air particle separator, maintenance hand pump for hydraulics, external cargo hook with 1,361 kg (3,000 lb) capacity 272 kg (600 lb) personnel hoist, wire strike kit, emergency floats for skids, retractable landing light, port side cabin step, landing gear and rotor fairing canopy cover, heater/defogger, vapour-cycle air conditioner, upgraded soundproofing, passenger service unit with air gaspers and reading lights, window reveal panels, matching close-out panel for aft baggage area, upgraded passenger seats, smoke detector in baggage compartment, jack-point fittings and ground handling wheels.

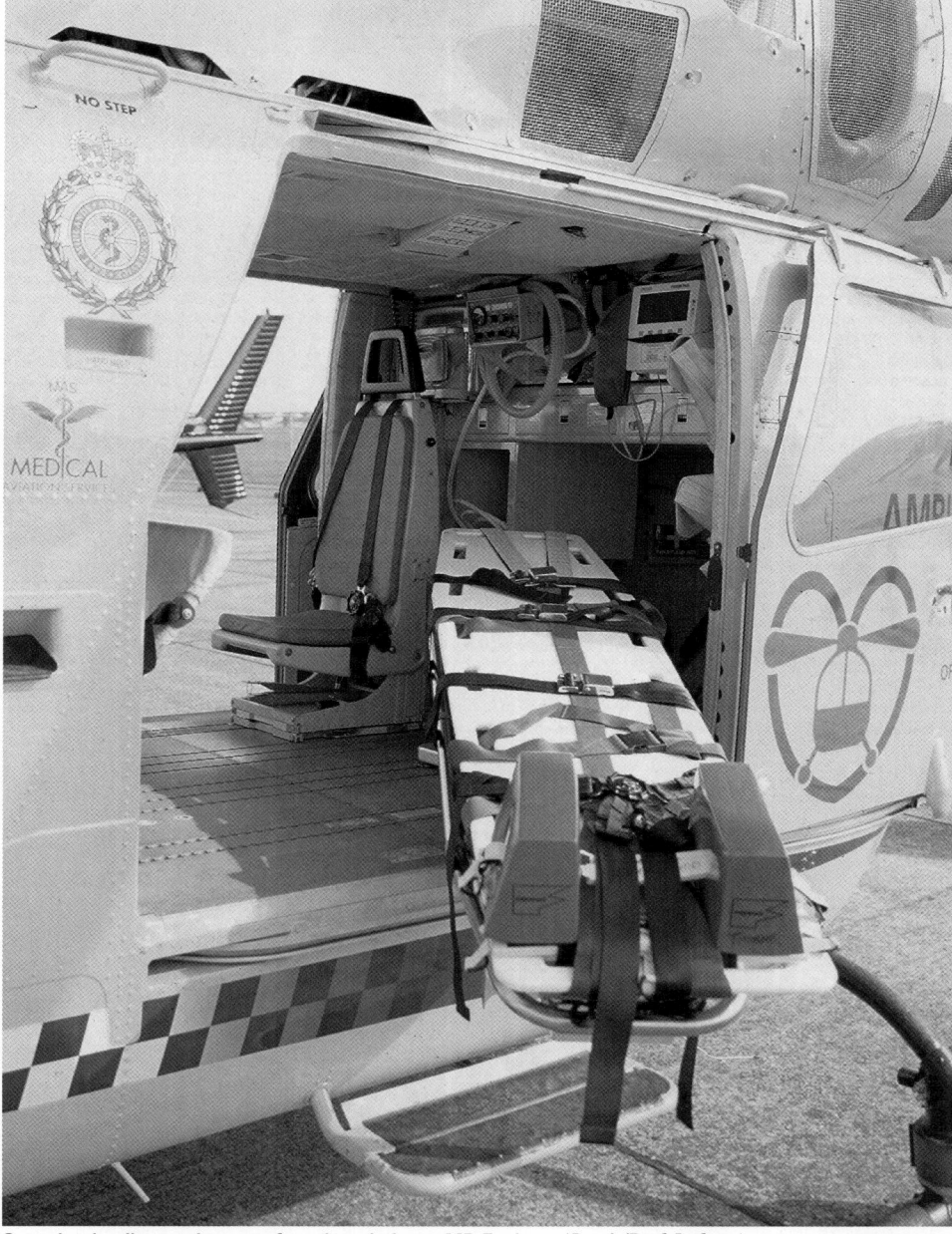
Stretcher-loading equipment of an air ambulance MD Explorer (*Jane's/Paul Jackson*)   NEW/0526974

DIMENSIONS, EXTERNAL:
Rotor diameter	10.31 m (33 ft 10 in)
Length: overall, rotor turning	11.83 m (38 ft 10 in)
fuselage	9.85 m (32 ft 4 in)
Fuselage width at cabin	1.63 m (5 ft 4 in)
Height: to top of rotor head	3.66 m (12 ft 0 in)
to top of fins	2.79 m (9 ft 2 in)
Min fuselage ground clearance	0.38 m (1 ft 3 in)
Tailplane span	2.84 m (9 ft 4 in)
Skid track	2.24 m (7 ft 4 in)
Cabin door width	1.27 m (4 ft 2 in)

DIMENSIONS, INTERNAL:
Cabin: Length overall, incl baggage compartment	3.93 m (12 ft 10¾ in)
Length, passenger compartment only	1.91 m (6 ft 3 in)
Max height	1.24 m (4 ft 1 in)
Max width	1.45 m (4 ft 9 in)
Volume	4.9 m³ (173 cu ft)
Baggage (if closed off)	1.39 m³ (49 cu ft)

AREAS:
Rotor disc	83.52 m² (899.0 sq ft)

WEIGHTS AND LOADINGS:
Weight empty, standard configuration	1,531 kg (3,375 lb)
Standard fuel load	489 kg (1,078 lb)
Max internal payload	1,292 kg (2,848 lb)
Max slung load	1,361 kg (3,000 lb)
Max T-O weight: internal load	2,835 kg (6,250 lb)
external load	3,129 kg (6,900 lb)
Max disc loading:	
internal load	33.9 kg/m² (6.95 lb/sq ft)
external load	37.5 kg/m² (7.68 lb/sq ft)
Transmission loading at max T-O weight and power:	
internal load	3.80 kg/kW (6.25 lb/shp)
external load	3.82 kg/kW (6.27 lb/shp)

PERFORMANCE (at max internal load T-O weight, ISA, except where indicated):
Never-exceed speed (VNE) at S/L, ISA	140 kt (259 km/h; 161 mph)

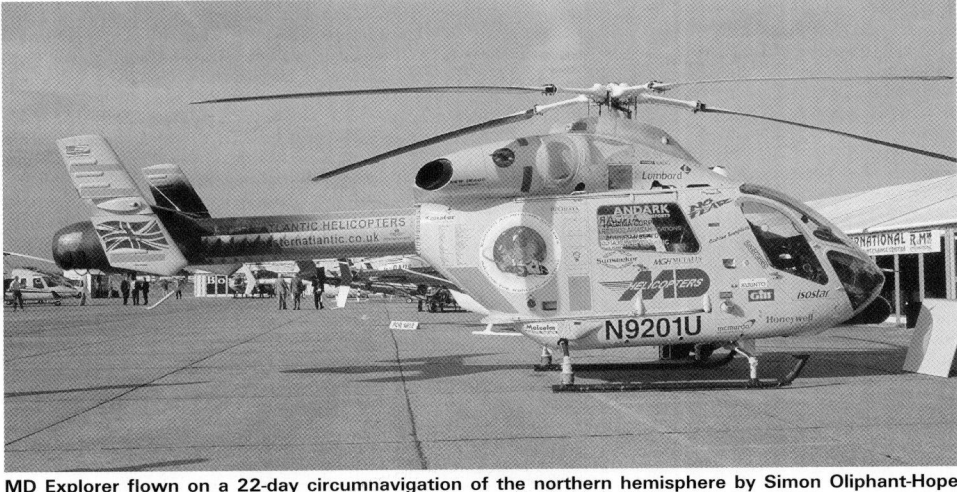
MD Explorer flown on a 22-day circumnavigation of the northern hemisphere by Simon Oliphant-Hope (*Jane's/Paul Jackson*)   NEW/0526975

Max cruising speed at S/L, 38°C (100°F)	134 kt (248 km/h; 154 mph)
Max rate of climb at S/L	579 m (1,900 ft)/min
Vertical rate of climb at S/L	411 m (1,350 ft)/min
Rate of climb at S/L, OEI	305 m (1,000 ft)/min
Service ceiling: both engines	5,335 m (17,500 ft)
OEI	3,200 m (10,500 ft)
Hovering ceiling:	
IGE, ISA	3,353 m (11,000 ft)
IGE, ISA + 20°C	2,042 m (6,700 ft)
OGE, ISA	2,743 m (9,000 ft)
OGE, ISA + 20°C	1,433 m (4,700 ft)
Hovering ceiling, OEI:	
IGE, ISA at 87% max T-O weight	1,220 m (4,000 ft)
Max range: at S/L	257 n miles (476 km; 295 miles)
at 1,525 m (5,000 ft), ISA	293 n miles (542 km; 337 miles)
Max endurance: at S/L	2 h 54 min
at 1,525 m (5,000 ft), ISA	3 h 12 min

OPERATIONAL NOISE LEVELS:
T-O	84.1 EPNdB
Approach	88.9 EPNdB
Flyover	83.1 EPNdB

UPDATED

Me 262 PROJECT

Me 262 PROJECT

Paine Field, Building 221, 10727 36th Avenue West, Everett, Washington 98204
Tel: (+1 425) 290 78 78
Fax: (+1 425) 355 13 23
e-mail: me262project@juno.com
Web: http://www.stormbirds.com
PROJECT MANAGER: Bob Hammer

Founded by the late Stephen L Snyder, Classic launched a programme in 1993 to build five exact flying replicas of a Second World War German jet fighter, the Messerschmitt Me 262. Work was originally contracted to the Texas Airplane Factory, but halted during 1997 because of a legal dispute. The production line was relocated to Paine Field (Snohomish County Airport) in early 1999; first aircraft flew in late 2002.

NEW ENTRY

MESSERSCHMITT Me 262

TYPE: Two-seat jet sportplane
PROGRAMME: Original prototype first flew 25 March 1942; entered operational service June 1944; flying terminated May 1945, except for limited number assigned to Allied technical evaluation or flown by Czechoslovak Air Force.
  Classic programme based on reverse-engineering of WkNr 110639, former US Me 262B-1a evaluation aircraft, refurbished by Classic from 1993 onwards and redelivered to US Navy at Willow Grove, Pennsylvania, on 8 September 2000. Completion of first new-build aircraft rescheduled to 2002, due to company restructuring. First

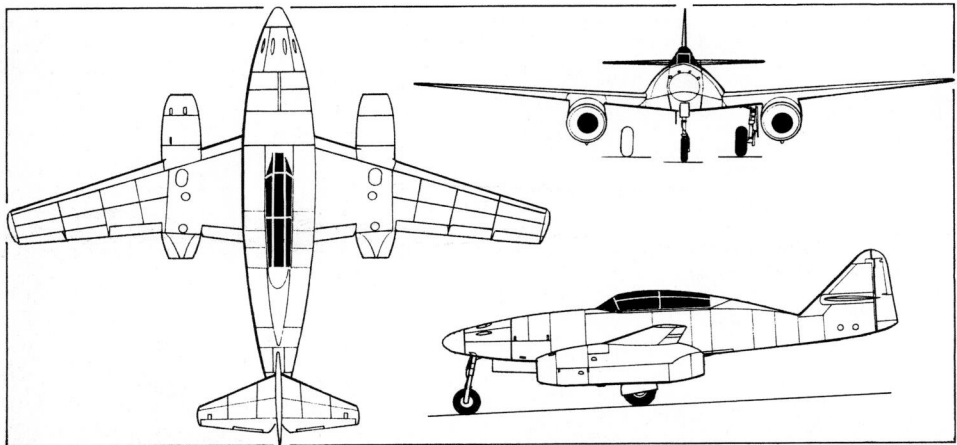
Me 262 Project Messerschmitt Me 262B-1c two-seat trainer version of Second World War jet fighter (*Jane's/Paul Jackson*)   0103513

taxi testing undertaken 25 June 2002, following engine tests conducted from February 2002. Maiden flight (following earlier short 'hops') (N262AZ) 20 December 2002, but damaged in landing accident on 18 January 2003.
CURRENT VERSIONS: **Me 262A-1c:** Single-seat version; one (501245/'white 3') under construction by Classic; 55 per cent complete in mid-2001; but suspended, pending customer interest.
  Me 262B-1c: Two-seat trainer; two (501241/N262AZ/ 'White 1' and 501243/'White ') under construction; first airframe for private owner in US; second 55 per cent complete, but suspended.

'Me 262A/B-1c': Further two aircraft (501242/'White 8' and 501244/N262MS/'Red 13'); can be converted to either of above variants; second machine earmarked for private owner in Germany and was 72 per cent complete in mid-2001; will be second aircraft to fly.
  Current production aircraft assigned 'c' type suffix by Messerschmitt Foundation to indicate employment of CJ610 engine; maker's serial numbers follow on from German 1945 production.
Descriptions of Me 262 appear in 1945-46 Jane's *and in various historical references. Differences from original are noted below.*
COSTS: Around US$2 million, excluding engines and avionics.
DESIGN FEATURES: Modifications minimal, except where required to enhance safety. Replacement of Jumo 004 engine involves smaller, lighter and more powerful CJ610 being located in rear of original-sized nacelles; interior air duct and false Jumo engine exterior shape below access panels restore weight to original for aerodynamic and balance purposes. Landing gear mountings strengthened to obviate original weakness; modern disc brakes within mainwheel hubs; nosewheel brake deleted.
POWER PLANT: Two 12.7 kN (2,850 lb st) General Electric CJ610 turbojets; soft throttle stop at 8.0 kN (1,800 lb st) to emulate Jumo engine max output; full estimated installed thrust rating of up to 11.1 kN (2,500 lb st) employed for initiating (only) take-off roll and may be re-engaged above 226 kt (418 km/h; 260 mph) asymmetric safety speed.
PERFORMANCE:
Never-exceed speed (VNE)	434 kt (804 km/h; 500 mph)*
Max level speed (estimated)	539 kt (998 km/h; 620 mph)
Range with max fuel	more than 869 n miles (1,609 km; 1,000 miles)

**voluntary limitation*

UPDATED

MICCO

MICCO AIRCRAFT COMPANY

3100 Airman's Drive, Fort Pierce, Florida 34946
Tel: (+1 561) 465 99 96
Fax: (+1 561) 465 99 97
e-mail: micco_aircraft@msn.com
Web: http://www.miccoair.net
PRESIDENT: F DeWitt Beckett
MARKETING DIRECTOR: Decki J Beckett

Micco is owned by the Seminole tribe of native Americans and is descended from the Meyers company, formed in 1936, which produced the OTW (102 built), MAC-125/145 (22 built), Meyers 200 (46 built) and Meyers 400 Interceptor turboprop; after purchase by Rockwell in July 1965, the Meyers 200 became the Aero Commander 200 (88 built), until production ceased in 1967. Type certificate transferred to Interceptor Corporation in 1969; to Prop-Jets Inc 1977; Nydia Myers Trust 1981; Ralph Haven 1987; New Myers 21 March 1994; and to Estumketda Ltd on 18 January 1995.

New Meyers renamed Micco (Seminole for chief) in 1997, remaining under Estumketda ownership. The Micco plant has 3,720 m² (40,000 sq ft) of factory space and 465 m² (5,000 sq ft) of engineering and administrative offices and employed 73 people in 2000. Production of the SP20 and its derivative, the SP26, ceased in November 2001 because of poor order book; outstanding deposits returned to customers; staff numbers reduced to 25, pending sale of the company and anticipated resumption of manufacturing in 2002, although no news of this by year's end.

UPDATED

MICCO SP20 and SP26

TYPE: Aerobatic two-seat lightplane.
PROGRAMME: Based on original 1947 Meyers MAC-145A, certified 2 November 1948, but with increased power. Redesign started March 1994, using converted MAC-145 (N520SP); construction started June 1995 and prototype (N720SP) first flown 17 December 1997. First production aircraft (N820SP) construction started January 1999. Certification to FAR Pt 23 issued 5 January 2000, with deliveries soon after. Second production aircraft shown at Sun 'n' Fun, Lakeland, Florida, April 2000, before delivery to TE Aero, Texas; third production aircraft to Clipper Aviation of Oneonta, New York, June 2000. Aircraft registered using MAC-145A (SP20) and MAC-145B (SP26) type certificates.
CURRENT VERSIONS: **SP20:** Baseline version.
  SP26: Updated and uprated aerobatic version. Six-cylinder engine in longer cowling, but firewall moved rearwards by identical distance (30 cm; 12 in), resulting in no overall change of fuselage length. Prominent strakes ahead of tailplane on prototype have now been removed. Prototype converted from N820SP; debut at Sun 'n' Fun, April 2000. Certification in Utility category granted on 19 October 2000 for 113 kg (250 lb) extension of MTOW over SP20. First production aircraft (N201MA) flew 15 November 2000 and delivered on 26 December 2000 to Andrew Spaulding of Cumberland, Maine. Aerobatic certification achieved 4 March 2002.
CUSTOMERS: Over 50 on order by end 2000, including 24 SP26; deliveries in 2000 totalled five SP20s and one SP26, and in 2001, 10 SP26s. When production suspended in October 2001, status was eight SP20s completed (two prototypes, including one to SP26 prototype; one static test; and single aircraft to SoCal Aviation, P T Esposito, Sportsman Air Park, Douglas Taylor and Dean Maggio), 11 complete production SP26s (Andrew Spaulding, Robert A Csognor, Grant Air Servicing, SoCal Aviation, Jim B Sayers, Cordes Air, Sportsman Air Park, Westminster

Micco SP26 aerobatic two-seater *(Jane's/Paul Jackson)*   *NEW*/0533373

College, Robert and Brittan William, Robert G Boston and Micco), and 11 SP26s registered but incomplete. Further 16 original MAC-145s currently registered in USA.
COSTS: SP20: VFR equipped US$165,000; IFR equipped US$190,000; SP26: VFR equipped US$199,500; IFR equipped US$225,000 (2001).
DESIGN FEATURES: Low-wing, high-performance, metal sport aircraft; low development cost achieved by employment of existing design. Major changes from MAC-145A include replacement of the 108 kW (145 hp) Continental engine; substitution of a Meyers 200 fin and rudder; wings and flaps from the Meyers 400 Interceptor; and provision of a sliding canopy in place of the fixed canopy and side doors.
FLYING CONTROLS: Conventional and manual. Horn-balanced cable-operated rudder with ground-adjustable trim tab; horn-balanced push/pull tube-controlled elevators with flight-adjustable tabs in each. Control movements: elevator +28/−20°; ailerons +22/−12°; rudder ±22°. Electrically operated Fowler flaps, max deflection 30°.
STRUCTURE: 4130 steel tube centre section and all-aluminium wings.
LANDING GEAR: Tailwheel type; main units retract inward into wings, tailwheel rearward; mainwheel doors cover legs only when retracted, wheel wells remaining exposed. Electrohydraulic Parker Hannifin hydraulic brakes and leaf spring shock-absorbers. 6.00-6 tyres on mainwheels; 10 cm (4 in) solid tailwheel. Ground turning radius about wingtip 6.07 m (19 ft 11 in).
POWER PLANT: *SP20:* One 149 kW (200 hp) Textron Lycoming IO-360-C1E6 flat-four driving a three-blade, constant-speed McCauley QZP (B3D36C424/74SA-0) propeller.
  SP26: One 194 kW (260 hp) Textron Lycoming IO-540-T4B5 flat-six driving a three-blade, constant-speed Hartzell HC-C3YR-1RF/F7693F Compact propeller.
  Fuel (both versions) in integral wing tanks, combined capacity 273 litres (72.0 US gallons; 60.0 Imp gallons), of which 257 litres (68.0 US gallons; 56.6 Imp gallons) are usable. Three drain points. Oil capacity 7.6 litres (2.0 US gallons; 1.67 Imp gallons in SP20, 9.5 litres (2.5 US gallons; 2.1 Imp gallons) in SP26. Aerobatic versions have two additional 57 litre (15.0 US gallon; 12.5 Imp gallon) tanks located in the centre-section outboard of cabin area.
ACCOMMODATION: Two, side by side, under rearward-sliding canopy. Baggage area behind seats.
SYSTEMS: 28 V 70 A alternator and battery; landing lights.
AVIONICS: VFR or IFR standards available, recommended Trimble Terra suite; S-Tec 30 two-axis autopilot and Apollo MX20 multifunction display optional.
EQUIPMENT: Landing/taxying light in each wing leading-edge.
Data apply to both versions, except where stated.

DIMENSIONS, EXTERNAL:	
Wing span	9.25 m (30 ft 4 in)
Wing chord: at root	2.29 m (7 ft 6 in)
at tip	1.02 m (3 ft 4 in)
Wing aspect ratio	5.9
Length overall	7.34 m (24 ft 1 in)
Max diameter of fuselage	1.22 m (4 ft 0 in)
Height overall	1.98 m (6 ft 6 in)
Tailplane span	3.76 m (12 ft 4 in)
Wheel track	3.15 m (10 ft 4 in)
Wheelbase	4.02 m (13 ft 2¼ in)
Propeller diameter: SP20	1.88 m (6 ft 2 in)
SP26	1.98 m (6 ft 6 in)
Propeller ground clearance: SP20	0.76 m (2 ft 6 in)

DIMENSIONS, INTERNAL:	
Cabin: Length	1.65 m (5 ft 5 in)
Max width	1.09 m (3 ft 7 in)
Max height	0.99 m (3 ft 3 in)
Baggage volume	0.40 m³ (14.0 cu ft)

AREAS:	
Wings, gross	14.55 m² (156.6 sq ft)

WEIGHTS AND LOADINGS:	
Weight empty: SP20	839 kg (1,850 lb)
SP26	925 kg (2,040 lb)
Baggage capacity	45 kg (100 lb)
Max aerobatic weight: SP26	1,202 kg (2,650 lb)
Max T-O weight: SP20	1,179 kg (2,600 lb)
SP26	1,292 kg (2,850 lb)
Max landing weight: SP20	1,130 kg (2,492 lb)
SP26	1,243 kg (2,742 lb)
Max ramp weight: SP20	1,193 kg (2,630 lb)
Max wing loading: SP20	81.1 kg/m² (16.60 lb/sq ft)
SP26	88.9 kg/m² (18.20 lb/sq ft)
Max power loading: SP20	7.91 kg/kW (13.00 lb/hp)
SP26	6.67 kg/kW (10.96 lb/hp)

PERFORMANCE:	
Never-exceed speed (VNE):	
SP20	180 kt (333 km/h; 207 mph)
SP26	195 kt (361 km/h; 224 mph)
Max level speed: SP20	145 kt (269 km/h; 167 mph)
SP26	165 kt (306 km/h; 190 mph)
Max cruising speed at 75% power:	
SP20	135 kt (250 km/h; 155 mph)
SP26	155 kt (287 km/h; 178 mph)
Normal cruising speed at 65% power:	
SP20	128 kt (237 km/h; 147 mph)
SP26	148 kt (274 km/h; 170 mph)
Econ cruising speed at 55% power:	
SP20	120 kt (222 km/h; 138 mph)
SP26	135 kt (250 km/h; 155 mph)
Stalling speed, power off:	
flaps and landing gear up	55 kt (102 km/h; 64 mph)
flaps and landing gear down	49 kt (91 km/h; 57 mph)
Max rate of climb at S/L: SP20	290 m (950 ft)/min
SP26	458 m (1,500 ft)/min
Service ceiling: SP20	3,660 m (12,000 ft)
SP26	5,486 m (18,000 ft)
T-O run: SP20	460 m (1,510 ft)
SP26	244 m (800 ft)
T-O to 15 m (50 ft): SP20	618 m (2,025 ft)
SP26	335 m (1,100 ft)
Landing from 15 m (50 ft): SP20, SP26	610 m (2,000 ft)
Landing run: SP20, SP26	213 m (700 ft)
Range with 30 min reserves: at max level speed:	
SP20, SP26	750 n miles (1,389 km; 863 miles)
at max cruising speed:	
SP20	850 n miles (1,574 km; 978 miles)
SP26	800 n miles (1,481 km; 920 miles)
at econ cruising speed:	
SP20	1,004 n miles (1,859 km; 1,155 miles)
SP26	1,040 n miles (1,926 km; 1,196 miles)
g limits (SP26)	+6/−3

UPDATED

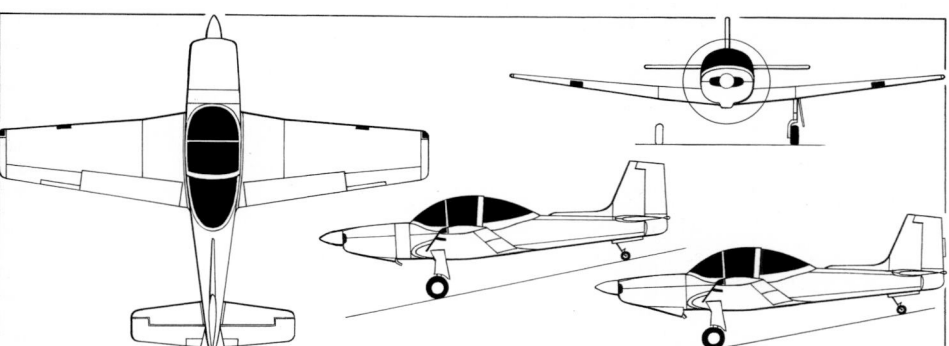

Micco SP20 two-seat light aircraft, with additional side view (lower) of SP26 *(Jane's/James Goulding)*   0059968

MILLENNIUM JET

MILLENNIUM JET INC

Renamed Trek Aerospace Inc

UPDATED

MINI-IMP

MINI-IMP AIRCRAFT COMPANY

PO Box 2011, Weatherford, Texas 76086-2011
Tel: (+1 817) 596 32 78
e-mail: info@mini-imp.com
Web: http://www.mini-imp.com
PRESIDENT: Gary S James

Company formed to produce kit version of Mini-IMP high-performance lightplane, designed and marketed in 1970s by Moulton B 'Molt' Taylor, who gained earlier prominence with the FAA-certified (1956) Aerocar flying car. Also holds rights to Micro-IMP.

NEW ENTRY

MINI-IMP MINI-IMP

TYPE: Single-seat sportplane kitbuilt.
PROGRAMME: Single-seat version of four-place Aerocar IMP ('Independently Made Plane'); first flown in mid-1970s (ahead of Imp); fitted in early 1976 with folding wing. Marketed as minimal kit or plans-only homebuilt by Aerocar Inc. Last described in 1992-93 *Jane's.*
 Relaunched as partial kit by current owner in April 2002.
CUSTOMERS: Some 25 of original series sold as minimal kits, of which between six and 10 have flown, including one in Brazil.
COSTS: Kit US$12,500 (2002), excluding engine.
DESIGN FEATURES: Compact, high-speed lightplane of high aerodynamic efficiency; mid-mounted, engine and propeller at extreme rear to achieve low cross-sectional area and minimal airflow interruption. 'Inverted butterfly' tail surfaces. Drive-shaft resonance obviated by Flexodyne coupling.
 High, constant-chord wing, with downturned tips; wings foldable for storage in 10 minutes; flaperons deflect +10° in cruising flight to unload tail surfaces and increase speed. Aerofoil GA-PC(1).

Mini-IMP of original series (*Jane's/Paul Jackson*)   *NEW*/0137384

FLYING CONTROLS: Manual. Almost full-span flaperons. Inverted V ruddervators with 40° anhedral.
STRUCTURE: Aluminium overall, with glass fibre for some non-structural parts.
LANDING GEAR: Tricycle type. Electric retraction. Mainwheels hinge upwards, legs stowed in fuselage side, wheels in wing; nosewheel retracts forward on separate electrical circuit. Optional spring aluminium mainwheel legs. Typical tyre sizes 5.00-5 main; 10×3.50-4 nose.
POWER PLANT: Mid-mounted power plant of 60 to 75 kW (80 to 100 hp), including Teledyne Continental O-200, Rotax 912 or Jabiru 3300. Internal cooling fan.
 Centre-section fuel tank, capacity 45 litres (12.0 US gallons; 10.0 Imp gallons). Optional external wingtip tanks, each 34 litres (9.0 US gallons; 7.5 Imp gallons).
DIMENSIONS, EXTERNAL:
Wing span   7.57 m (24 ft 10 in)
Wing chord, constant   0.91 m (3 ft 0 in)

Length overall: normal   4.88 m (16 ft 0 in)
 elongated nose   5.18 m (17 ft 0 in)
Height: over fuselage   1.22 m (4 ft 0 in)
 propeller turning   1.89 m (6 ft 2½ in)
WEIGHTS AND LOADINGS:
Weight empty: normal   236 kg (520 lb)
 elongated nose   354 kg (780 lb)
Max T-O weight: normal   362 kg (800 lb)
 elongated nose   453 kg (1,000 lb)
PERFORMANCE (75 kW; 100 hp engine):
Max level speed   more than 174 kt (322 km/h; 200 mph)
Cruising speed   more than 156 kt (290 km/h; 180 mph)
Stalling speed: 44 kt (81 km/h; 50 mph)
Max rate of climb at S/L   457 m (1,500 ft)/min
T-O run   244 m (800 ft)
g limits   +6/−4

NEW ENTRY

MOONEY

MOONEY AIRPLANE COMPANY (MAC)

Louis Schreiner Field, Kerrville, Texas 78028
Tel: (+1 830) 896 60 00 or 792 29 10
Fax: (+1 830) 896 81 80
e-mail: sales@mooney.com
Web: http://www.mooney.com
CEO: Nelson Happy
VICE-PRESIDENT, SALES AND MARKETING: Nicholas Chabbert

Original Mooney Aircraft Corporation formed in Wichita, Kansas, 1948; produced single-seat M-18 Mite until 1952 (total 282); M20 family in production since then. Alexandre Couvelaire, President of Euralair/Avialair Paris, France, and Michel Seydoux, President of MSC, jointly acquired Mooney in 1985; Mooney and Aerospatiale (Socata) announced joint development of TBM 700 June 1987 (see under Socata in French section), but Mooney withdrew in 1991. Controlled by AVAQ Group Inc since 1998. By early 2001, Mooney had produced 10,320 aircraft, including 9,940 of M20 family. New Mooney Encore announced October 1996; Eagle in 1998. Company decided in 1999 not to proceed with a turboprop, but was considering reintroducing a pressurised aircraft with 240 kt (444 km/h; 276 mph) cruising speed. Mooney delivered 100 aircraft in 2000, compared with 97 in 1999, 93 in 1998, 86 in 1997 and 73 in 1996. Personnel totalled 220 in 2001, including those engaged on Lockheed Martin F-16 and C-130, Boeing, Air Tractor, Aviat Pitts Special and Commander subcontracts; by early February 2002, this had been reduced to 11.
 Mooney Aircraft Corporation filed for Chapter 11 bankruptcy protection on 27 July 2001, continuing production while restructuring the company; delivered 29 aircraft in 2001 against year's target of 70. Announced on 16 November 2001 that extended financing had been received to continue operations, including spares support, pending sale of the company. On 8 February 2002 AASI announced it had purchased Congress Financial Corporation's position as senior creditor and intended restarting production as well as moving its own production facility to Kerrville.
 AASI subsequently reorganised as Mooney Aerospace Group, Ltd, with Mooney Airplane Company as wholly owned subsidiary. Plans announced in February 2002 for redesign of AASI Jetcruzer 500, but by late 2002 further development of this and related designs thought unlikely as company concentrates on production and support of Mooney range of single-engined piston aircraft. MAC received FAA production certificate on 26 June 2002 for Eagle 2, Ovation 2 and Bravo 2. Employment at Kerrville totalled 130 in July 2002. Total of 10 aircraft delivered in 2002, all in final quarter.

UPDATED

MOONEY M20M BRAVO

TYPE: Four-seat lightplane.
PROGRAMME: Wooden M20 prototype first flew 10 August 1953; some 700 of early series built before design translated into metal with M20B, introduced in 1961. Progressed through several further subvariants, of which M20M announced 2 February 1989 as TLS (Turbo Lycoming Sabre, with 201 kW; 270 hp TIO-540-AF1A); Bravo introduced in 1996.
CUSTOMERS: Total of 315 TLS/Bravos delivered by June 2001 halt of production. Deliveries resumed in July 2002. Da Kine, Hawaii Inc of Hood River, Oregon, became the first customer for a Mooney Airplane Company-built aircraft (a Bravo), with an order announced on 25 July 2002.
COSTS: Standard IFR equipped US$399,950 (2002).
DESIGN FEATURES: High-efficiency touring aircraft, originally designed by Mooney brothers. Low wing and mid-mounted tailplane; flying surfaces of characteristic Mooney 'reversed' configuration with unswept leading-edges and sharply forward-swept trailing-edges.
 Wing section NACA 63_2182-215 at root, 64_1181-412 at tip; dihedral 5° 30'; incidence 2° 30' at root, 1° at tip; wing swept forward 2° 29' at quarter chord.

MOONEY M20 PRODUCTION

Model	Name	Eng desig	Built	Certification
M20			200	24 Aug 55
M20A			501	13 Feb 58
M20B	Mk 21		224	14 Dec 60
M20C	Mk 21		2,222	20 Oct 61
	Ranger			
	Aerostar 200			
M20D	Master		160	15 Oct 62
M20E	Super 21	21	1,485	4 Sep 63
	Chapparal	21		
	Aerostar 201	21		
M20F	Executive	22	1,251	25 Jul 65
	Aerostar 220	22		
M20G	Statesman		190	13 Nov 67
M20J	201	24	2,135	27 Sep 76
	201LM			
	201SE			
	205			
	ATS			
	MSE			
	Allegro			
MT20	TX-1		2	nil
M20K	231	25	1,151	16 Nov 78
	231SE	25		
	252 TSE	25		
	Encore	25		
M20L	PFM	26	41	25 Feb 88
M20M	TLS	27	315*	28 Jun 89
	Bravo	27		
M20R	Ovation	29	280*	30 Jun 94
	Ovation 2	29		
M20S	Eagle	30	62*	7 Feb 99
M20T	EFS	28	1	nil
Total			**10,220**	

*Currently in production. Total produced up to February 2002
Notes: Individual models have had several marketing names. Engineering designation forms part of aircraft's serial number. Certification date is for FAA.

Rudder	0.58 m² (6.23 sq ft)
Tailplane	1.99 m² (21.45 sq ft)
Elevators (total)	1.11 m² (12.05 sq ft)

WEIGHTS AND LOADINGS:
Weight empty	1,029 kg (2,268 lb)
Max T-O weight	1,528 kg (3,368 lb)
Max landing weight	1,452 kg (3,200 lb)
Max wing loading	94.0 kg/m² (19.25 lb/sq ft)
Max power loading	7.59 kg/kW (12.47 lb/hp)

PERFORMANCE:
Max cruising speed:	
at FL130	195 kt (361 km/h; 224 mph)
at FL 250	220 kt (407 km/h; 253 mph)
Stalling speed:	
flaps and wheels up	66 kt (122 km/h; 76 mph)
flaps and wheels down	59 kt (109 km/h; 68 mph)
Max rate of climb at S/L	344 m (1,130 ft)/min
Max certified altitude	7,620 m (25,000 ft)
Range with reserves, at FL130	1,070 n miles (1,982 km; 1,231 miles)
Endurance	6 h 42 min

UPDATED

Mooney Bravo four-seat turbocharged light aircraft *(Jane's/Paul Jackson)* NEW/0526973

Mooney Ovation2 Platinum Edition *(Jane's/Paul Jackson)* NEW/0526972

MOONEY M20R OVATION2 and M20S EAGLE2

TYPE: Four-seat lightplane.

PROGRAMME: Prototype (N20XR) rolled out April 1994; first flight May 1994; FAA certification June 1994 (Ovation) and February 1999 (Eagle).

CURRENT VERSIONS: **Mooney M20R Ovation2:** Baseline version, superseded Ovation; unveiled at the AOPA Convention on 22 October 1999. A limited edition Mooney Ovation2 Platinum Edition, with advanced avionics and instrumentation, plus airframe and cabin upgrades, has been introduced. An Ovation2 which made its first flight on 18 June 2002 was the first new production Mooney following the formation of Mooney Aerospace Group, Ltd.

Description applies to M20R Ovation2 unless otherwise stated.

Mooney M20S Eagle2: Entry-level version, using basic M20R fuselage and power plant, derated to 182 kW (244 hp), two-blade McCauley 2A34C239/90DMC-15 1.91 m (6 ft 3 in) diameter propeller and lower-cost avionics. Usable fuel capacity 284 litres (75.0 US gallons; 62.5 Imp gallons). Cruising speed 180 kt (333 km/h; 207 mph); maximum climb rate 350 m (1,150 ft)/min; range 1,200 n miles (2,222 km; 1,380 miles) at 3,048 m (10,000 ft); service ceiling 5,640 m (18,500 ft); max T-O weight 1,451 kg (3,200 lb). First Eagle certified and delivered February 1999.

CUSTOMERS: Most sold in USA, but others exported to Brazil, France, Germany, Israel, Italy, Paraguay, South Africa, Switzerland, Thailand and UK; orders include three for Western Michigan University. Total of 280 Ovations and 62 Eagles had been registered before Mooney reconstituted in February 2002. Ten delivered in 2002, comprising eight Ovations and two Eagles.

COSTS: Ovation2 US$349,950, Eagle2 US$299,950 (both 2002).

FLYING CONTROLS: Conventional and manual. Sealed gap, differential ailerons; fin and tailplane integral so that both tilt, varying tailplane incidence for trimming; no trim tabs; electrically actuated single-slotted flaps; electric rudder trim; Precise Flight speed brakes. Control surface movements: ailerons +14° 30'/−8°, elevator ±22°, rudder ±24°, flaps −10° T-O and −33° landing.

STRUCTURE: Single-spar wing with auxiliary spar out to mid-position of flaps; wing and tail surfaces covered with stretch-formed wraparound skins. Steel tube cabin section covered with light alloy skin; semi-monocoque rear fuselage with extruded stringers and sheet metal frames.

LANDING GEAR: Electrically retractable levered suspension tricycle type with airspeed safety switch bypass. Nosewheel retracts rearward, main units inward into wings. Rubber disc shock-absorbers in main units. Cleveland mainwheels, size 6.00-6 (6-ply), and steerable nosewheel, size 5.00-5 (6-ply). Tyre pressure, mainwheels 2.07 bar (30 lb/sq in), nosewheel 3.38 bar (49 lb/sq in). Cleveland hydraulic single-disc (double optional) brakes on mainwheels. Parking brake.

POWER PLANT: One 201 kW (270 hp) turbocharged Textron Lycoming TIO-540-AF1B flat-six engine, driving a McCauley B3D32C417/82NRD-7 three-blade, metal propeller. Two integral fuel tanks in inboard wing leading-edges, with a combined capacity of 363 litres (96.0 US gallons; 79.9 Imp gallons), of which 337 litres (89.0 US gallons; 74.1 Imp gallons) usable. Two-piece nose cowling is of glass fibre/graphite composites construction.

ACCOMMODATION: Cabin accommodates four persons in pairs on individual vertically adjusting seats with reclining back, armrests, lumbar support and headrests; side armrests removable in rear seats. Front seats have inertia reel shoulder harnesses. Dual controls standard. Overhead ventilation system. Cabin heating and cooling system, with adjustable outlets and illuminated control. One-piece wraparound windscreen. Tinted Plexiglas windows. Rear seats removable for freight stowage. Rear seats fold forward for carrying cargo. Single door on starboard side. Compartment for 54 kg (120 lb) baggage behind cabin, with access from cabin or through door on starboard side.

SYSTEMS: Dual 70 Ah 28 V alternators, dual 24 V 10 Ah batteries, voltage regulator and warning lights, together with protective circuit breakers. Oxygen system, capacity 3,259 litres (115 cu ft), with masks and overhead outlets, standard. TKS known ice protection system and air conditioning optional.

AVIONICS: Standard equipment comprises Garmin GNS 530 nav/com/GPS, GNS 430 with GI 106A indicator, transponder/encoder, two-axis autopilot, and slaved HSI. Goodrich WX-500 Stormscope and SkyWatch optional.

EQUIPMENT: Standard equipment includes attitude indicator, IFR directional gyro, two electric fuel quantity gauges, electric OAT gauge, CHT and EGT gauges, Precise Flight speed brakes, push-pull tube-actuated flight control system, annunciator panel; navigation lights, wing-mounted dual landing/taxying lights, internally lit instruments with rheostat; sound-damping composites interior, vertically adjusting front seats, wing jackpoints and external tiedowns, fuel tank quick drains, auxiliary power plug, heated pitot tube, polished propeller spinner, epoxy polyimide and conversion anti-corrosion treatment, external polyurethane paint finish, dual static ports with drains, and alternate static air source.

DIMENSIONS, EXTERNAL:
Wing span	11.00 m (36 ft 1 in)
Wing chord, mean	1.50 m (4 ft 11¼ in)
Wing aspect ratio	7.4
Length overall	8.15 m (26 ft 9 in)
Height overall	2.54 m (8 ft 4 in)
Tailplane span	3.58 m (11 ft 9 in)
Wheel track	2.79 m (9 ft 2 in)
Wheelbase	2.02 m (6 ft 7½ in)
Propeller diameter	1.91 m (6 ft 3 in)
Baggage door: Width	0.53 m (1 ft 9 in)
Height	0.43 m (1 ft 5 in)

DIMENSIONS, INTERNAL:
Cabin: Length	3.20 m (10 ft 6 in)
Max width	1.10 m (3 ft 7½ in)
Max height	1.13 m (3 ft 8½ in)
Floor area	3.53 m² (38.0 sq ft)
Volume	3.9 m³ (137 cu ft)
Baggage compartment volume	0.64 m³ (22.6 cu ft)

AREAS:
Wings, gross	16.26 m² (175.0 sq ft)
Ailerons (total)	1.06 m² (11.40 sq ft)
Trailing-edge flaps (total)	1.66 m² (17.90 sq ft)
Fin	0.73 m² (7.92 sq ft)

Mooney M20S Eagle *(Jane's/Paul Jackson)* NEW/0526971

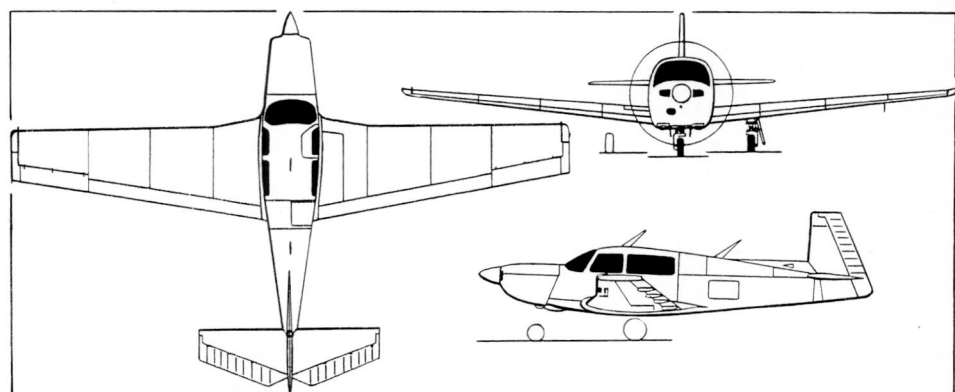

Mooney M20R Ovation2 *(Jane's/James Goulding)* 0092656

DESIGN FEATURES: Combines airframe of M20M Bravo (which see) with normally aspirated flat-six engine. Completely restyled instrument panel, seats and cabin interior with sandwich-core rigid trim/soundproofing panels and leather and wool fabric upholstery.

POWER PLANT: M20R Ovation2 has one 224 kW (300 hp) Teledyne Continental IO-550-G5B flat-six engine, derated to 209 kW (280 hp), driving a McCauley 3A32C418-G/82NRC-9 three-blade, constant-speed metal propeller; Ovation2 has redesigned propeller giving slight performance improvement. Usable fuel 337 litres (89.0 US gallons; 74.1 Imp gallons).

AVIONICS: Generally as for Mooney M20M Bravo but Eagle2 has single GNS 430.

EQUIPMENT: Generally as for M20M Bravo, except oxygen system optional on both models, and polished spinner and Precise Flight speed brakes optional on Eagle2.

DIMENSIONS, EXTERNAL:
Propeller diameter	1.85 m (6 ft 1 in)

WEIGHTS AND LOADINGS:
Weight empty	1,009 kg (2,225 lb)
Baggage capacity	54 kg (120 lb)
Max usable fuel	242 kg (534 lb)
Max T-O weight	1,528 kg (3,368 lb)
Max landing weight	1,451 kg (3,200 lb)
Max wing loading	94.0 kg/m² (19.25 lb/sq ft)
Max power loading	7.32 kg/kW (12.03 lb/hp)

PERFORMANCE (at max T-O weight, ISA, except where indicated):
Max cruising speed	192 kt (356 km/h; 221 mph)
Stalling speed:	
flaps and wheels up	66 kt (123 km/h; 76 mph)
flaps and wheels down	59 kt (110 km/h; 68 mph)
Max rate of climb at S/L	381 m (1,250 ft)/min
Service ceiling	6,100 m (20,000 ft)
Max range, economy cruise, with reserves, at 2,745 m	
(9,000 ft)	1,280 n miles (2,370 km; 1,473 miles)

UPDATED

MORROW

MORROW AIRCRAFT CORPORATION

3280 25th Street NE, PO Box 12665, Salem, Oregon 97389
Tel: (+1 503) 365 02 00
Fax: (+1 503) 365 03 00
e-mail: comments@morrowaircraft.com
Web: http://www.morrowaircraft.com
CHAIRMAN: Ray Morrow
VICE-PRESIDENT, PRODUCT MANAGEMENT: Dale Johnson
CHIEF MARKETING OFFICER: Dan Waldron

Morrow Aircraft Corporation formed 1996 to develop general aviation aircraft using advanced composites

materials and state-of-the-art technology. Ray Morrow is co-founder of American Blimp Corporation (which see) and established II Morrow avionics company (now UPS Aviation Technology). The company's first product is a development of the Rutan Boomerang, which it also intends to use for its own SkyTaxi programme. In mid-2002, Morrow employed 14 at its 465 m² (5,000 sq ft) factory.

UPDATED

MORROW MB-300

PROGRAMME: Development of Rutan 202-11 Boomerang (see 1994-95 *Jane's* for details); Morrow MB-300 is 15 per cent longer than original with approximately 20 per cent greater

payload and 30 per cent greater cabin volume. Proof-of-concept studies undertaken with N24BT owned by Burt Rutan; 250 hours accumulated by mid-2002. In October 1998, Scaled Composites contracted to build one prototype and two preproduction examples for certification; prototype one-third complete by early 2002; company froze development in mid-2002 to concentrate on its SkyTaxi fleet of Cessna aircraft with intention of re-starting development at a later date. MB-300 provisionally due to enter service during 2005.

CUSTOMERS: Company estimates market for up to 1,500 aircraft for its own SkyTaxi programme by 2011.

A description last appeared in the 2002-03 Jane's.

UPDATED

NASA

NATIONAL AERONAUTICS AND SPACE ADMINISTRATION
(Office of Aeronautics)

600 Independence Avenue SW, Washington, DC 20546
Tel: (+1 202) 453 26 93
Fax: (+1 202) 426 42 56
Web: http://www.nasa.gov
ADMINISTRATOR: Sean O'Keefe

CHIEF ENGINEER: Daniel R Mulville
ASSOCIATE ADMINISTRATOR, AEROSPACE TECHNOLOGY:
Spence M Armstrong
CHIEF FINANCIAL OFFICER: Arnold G Holz

NASA is looking to increase expenditure on research and development allied to aerospace technology during FY03, after a period in which manned space flight activity took the greater share of the annual budget. For FY03, total of US$15 billion requested. Personnel totalled approximately 17,800 in

2000. Prime NASA concerns are R&D, rather than operationally driven, with about US$800 million earmarked for research into hypersonic flight during next five years. In addition to selected programmes described below, NASA funds are also allocated to the Boeing Blended Wing/Body programmes described elsewhere in this edition. NASA activities in space are described in *Jane's Space Directory*.

UPDATED

NASA HYPERSONIC FLIGHT PROGRAMMES

Since the days of NACA and the Bell X-1, NASA has been involved in experiments to increase the speed of atmospheric and stratospheric flight. Latest projects include that described below.

UPDATED

NASA X-43 HYPER-X

Second of NASA's 'Waverider' projects (see 2001-02 and earlier *Jane's*), X-43 has objective of allowing testing of subscale hydrogen-fuelled GASL Inc scramjet engine at speeds of M7 to M10. Forebody is shaped to generate shock waves which compress air entering scramjet intake; hydrogen-based fuel combusts spontaneously on contact with oxygen contained in the compressed air.

Request for proposals was issued on 13 October 1996, with contract, worth US$33.4 million over 55 month period, following in March 1997 to MicroCraft Inc of Tullahoma, Tennessee, although total project cost, including flight testing, has since risen to around US$170 million. MicroCraft heads a consortium that includes Accurate Automation, Boeing North American and GASL, and has produced three (originally to have been four) test vehicles, each approximately 3.7 m (12 ft) long, with a span of about 1.5 m (5 ft) and gross weight of 1,270 kg (2,800 lb). Preliminary design review completed in June 1997, with fabrication commencing in third quarter of 1997. First flight vehicle delivered to NASA's Dryden Flight Research Center in October 1999.

Three flights are scheduled, with first originally due to take place in May 2000, after captive-carry flight on NASA's NB-52B launch platform; however, development delays resulted in postponement to 2001. Initial 'flight' was a captive-carry test on 28 April 2001. The first launch was planned to reach M7, followed by second flight, in May or June 2003, to M7 and final flight, in 2004, to M10. After drop launch at altitude of 6,100 m (20,000 ft), an Orbital Sciences' Pegasus booster rocket accelerates test vehicle to specified speed at approximately 30,500 m (100,000 ft), whereupon X-43 separates to fly under own power. Flight path is entirely over water and it is not planned to recover test vehicles from

First **X-43A** Hyper-X and its Orbital Sciences booster beneath the wing of NASA's Boeing NB-52   0109155

Pacific Ocean. First flight, on 2 June 2001, was terminated prematurely when X-43A suffered structural failure and deviated from planned flight path following booster ignition; investigations identified loss of control of Pegasus booster as cause of failure.

Three more versions of X-43 now planned, though not all are currently funded. **X-43B** is scaled-up version of X-43A with objective of testing reusable combined-cycle propulsion system; NASA engaged in development of rocket-based and turbine-based engine options. X-43B not currently funded, but could fly in 2010. **X-43C** expected to be next version to

fly, in 2007-08 and will be powered by a HyTech hydrocarbon-fuelled dual-mode scramjet that is now under development for the USAF by Pratt & Whitney. Maximum speeds of the order of M7 are expected to be achieved by the X-43C, which is a 2,270 kg (5,000 lb), 5.03 m (16 ft 6 in) long lifting-body design. Finally, there is the **X-43D**, a potential successor to the X-43A that would be powered by a cooled hydrogen-fuelled dual-mode scramjet capable of propelling the vehicle to M15. The X-43D is currently in the study phase.

UPDATED

NASA THRUST-VECTORING CONTROL PROGRAMME

A new series of X-craft was proposed in 1999 to continue from earlier studies by NASA and the aerospace industry.

NASA X-44A

TYPE: Trials platform.
PROGRAMME: Possible new design for research aircraft. Currently moribund, this project anticipated building on

experience gained with F-15B assigned to ACTIVE project; feasibility study completed in 1999, involving NASA, Lockheed Martin, Pratt & Whitney and the USAF. Notional X-44A MANTA (Multi-Axis No-Tail Aircraft) design resembled F/A-22A Raptor, but without horizontal and vertical tail surfaces and also lacking moving control surfaces on wings; thrust-vectoring would have been only method of aerodynamic control.

UPDATED

NASA AIRCRAFT TRIALS PROGRAMMES

Of some 100 aircraft operated by NASA, most are standard types assigned to specific research programmes. The fleet comprises about 35 dedicated to research and 65 to programme support; most carry civilian registrations between N400NA and N999NA, assigned according to base. Main operating bases of NASA's Aerospace Technology division are:

Ames Research Center, Moffett Federal Airfield, California (N700-799A)

Dryden Flight Research Center, Edwards AFB, California (N800-899NA)

Glenn Research Center, Cleveland-Hopkins IAP, Ohio (N600-699NA)

Langley Research Center, Langley AFB, Hampton, Virginia (N500-599NA)

Other significant aircraft operations bases are:

Johnson Space Center, Ellington ANGB, Houston, Texas (N900-999NA)

Wallops Flight Facility, Wallops Island, Virginia (N400-499NA)

Details of some significant research projects involving aircraft appear below.

Boeing F-15B: Original F-15B prototype (NASA837/71-0290) modified for ACTIVE (Advanced Control Technology for Integrated Vehicles) project involving thrust vectoring nozzles developed by Pratt & Whitney. Details of initial testing appear in *Jane's* 2001-02, while most recent series of trials, in non-reversioning mode, undertaken in 1999, included take-offs and landings using vectoring only as well as flight to unlimited angles of attack and speeds of M2. New nozzles can turn up to 20° in any direction, permitting thrust control in pitch and yaw. ACTIVE programme is a joint venture involving NASA, the USAF's Wright Laboratory, Boeing (formerly McDonnell Douglas) and Pratt & Whitney.

ACTIVE F-15B also used in early 1999 to evaluate 'neural network'-based controller as opening part of intelligent flight control system (IFCS) programme; total of 15 flights accomplished during March and April 1999.

A second F-15B (NASA836/74-0141), known as the Aerodynamic Flight Facility and equipped with a flight test fixture (airborne test-bench) on the centreline pylon, flew seven sorties between 26 February and 1 April 1996 to test materials for the X-33's thermal protection system; used during 1999 for investigation of supersonic natural laminar flow of advanced wing design, carrying 0.91 by 1.22 m (3 by

Active Aeroelastic Wing F/A-18B Hornet *(NASA/Tony Landis)* *NEW*/0533363

4 ft) test article on centreline pylon during four test flights at speeds approaching M2 and altitudes of up to 13,715 m (45,000 ft). This aircraft recently involved in experiments into gust monitoring and aeroelasticity, with 0.45 m (1 ft 6 in) long, flexible, test wing mounted on the flight test fixture. Future projects include trials of Pulse Detonation Engine; these will involve an experimental engine suspended from a new centreline Propulsion Flight Test Fixture (PFTF) that was flown for the first time in late 2001.

Boeing F/A-18: Includes F/A-18A High Angle of Attack Research Vehicle (HARV) (NASA840/161251) with thrust vectoring system; first flight 16 January 1991 at Dryden; first flight with vectoring 15 July 1991; reached 79° AoA in sustained flight, 1992; initial series of Phase 2 trials completed in January 1993. Aircraft then further modified by January 1994, initially with inlet-mounted pressure sensor and new programmable Ada software for quadruplex digital FCS; addition of vortex-flow nose strakes, late 1994; strakes 1.22 m (4 ft 0 in) long and 0.15 m (6 in) wide, hinged on lower edges and rotated outwards in flight to control nose vortices. After second series of Phase 2 testing between January and June 1994, HARV was fitted with strakes and commenced Phase 3 trials, involving approximately 65 flights. These began in April 1995 and concluded with final sortie on 29 May 1996, by which time HARV had undertaken total of 383 research flights over nine-year period; final test objectives included evaluation of nose strakes for yaw control at high AoA.

F/A-18B NASA853 refurbished to serve as testbed for Active Aeroelastic Wing program (AAW) which will examine use of 'flexible wings' to research airflow over the wing and assess how this affects controllability. Following modification, aircraft was rolled out at Dryden on 27 March 2002 and made ready for initial flight testing in mid-2002. Second series of trials scheduled for 2003.

Two NASA F/A-18Bs used for demonstration of autonomous close formation flight in 2001. Achieved automatic 61 m (200 ft) horizontal and 15 m (50 ft) lateral separation on 21 February 2001 on seventh of planned 12 sorties. This test effort also included study of wingtip vortex effect, confirming that close formation flying could offer significant fuel economies; in one test, one F/A-18 flew in wingtip vortex of another F/A-18, with subsequent analysis revealing that trailing aircraft consumed 270 kg (600 lb) less fuel than a third F/A-18 which was not part of the formation. This equates to about a 12 per cent reduction in fuel burn.

Boeing 747SP: Engineering work under way to build Stratospheric Observatory For Infra-red Astronomy (SOFIA) with 2.69 m (8 ft 10 in) telescope housed in unpressurised cavity in aft fuselage and looking to port; partner is German Ministry of Science (BMFT), which delivered telescope to USA on 4 September 2002; cruise altitude 12,500 to 13,700 m (41,000 to 45,000 ft); to fly up to 160 sorties (8 hours each) per year for 20 years. US$484 million, 10 year contract secured in December 1996 by a team led by Universities Space Research Association (USRA) and including Raytheon Systems and United Airlines to design, assemble, test and operate SOFIA. Former United Airlines Boeing 747SP (N145UA) chosen from several in storage at Las Vegas, Nevada, and ferried to San Francisco in February 1997 for attention by United before being handed over to NASA; following acceptance, 747SP delivered to Waco, Texas, in May 1997 for modification and fitting out by Raytheon. Conversion now under way, with 747SP due to undergo a series of flight tests in 2003-04, before being delivered to NASA at Ames Research Center in May or June 2004.

Boeing 757: The so-called ARIES (Airborne Research Integrated Experiments System) aircraft (N557NA) undertook research into turbulence associated with convective storms in early 2002, using an experimental radar designed to detect atmospheric disturbance. Total of 13 missions undertaken from Langley with object of encountering turbulence and comparing it with what radar predicted.

Boeing NB-52B Stratofortress: Used for aerial launch of test vehicles, with most recent activity involving X-43A Hyper-X and X-38 CRVprojects. NB-52B currently in use (NASA008/52-0008) to be replaced in 2003-04 by B-52H (61-0025) which was delivered on 30 July 2001 and then flown to Tinker AFB, Oklahoma for programmed depot maintenance, returning on 17 April 2002; it will be fitted with a universal pylon able to accept different adaptors and initially accommodate loads of up to 11,340 kg (25,000 lb), although it is eventually planned to increase load-carrying capability to 31,750 kg (70,000 lb).

McDonnell Douglas DC-8-72: Joined Airborne Science Program in April 1998 at Dryden, following outfitting with Jet Propulsion Laboratory (Pasadena, California) advanced airborne synthetic aperture radar and other sensor systems including imaging spectroradiometer and thermal emission and reflection radiometer. Used for convection and moisture experiments, Earth science studies and also supporting NASA investigation into ozone depletion. In first quarter of 2002, used for ice and snow studies, employing synthetic aperture radar and other sensors to gather data that will give scientists a better understanding of how the snow process works. Previously, the DC-8 had been deployed to the Pacific Rim to conduct studies of volcanic activity and gather data to assist in planning for, and responding to, natural disasters.

UPDATED

NASA's new Mother Ship is this white Boeing B-52H Stratofortress replacing a B-52B *(NASA/Jim Ross)* *NEW*/0533362

Still in the NASA fleet, albeit in flyable storage awaiting a new mission, is a rare Schweizer YO-3A observation and monitoring powered sailplane which saw operational service in the Vietnam War *(NASA/Tony Landis)* *NEW*/0533361

NORTHROP GRUMMAN

NORTHROP GRUMMAN CORPORATION

1840 Century Park East, Los Angeles, California 90067
Tel: (+1 310) 553 62 62
Fax: (+1 310) 201 30 23 and (+1 310) 553 20 76
Web: http://www.northropgrumman.com
CHAIRMAN AND CEO: Kent Kresa
PRESIDENT AND COO: Ronald D Sugar
CORPORATE VICE-PRESIDENT AND CHIEF FINANCIAL OFFICER:
   Richard B Waugh Jr
CORPORATE VICE-PRESIDENT, COMMUNICATIONS:
   Rosanne O'Brien

Northrop company formed 1939 to produce military aircraft; activities extended to missiles, target drones, electronics, space technology, communications, support services and commercial products. Grumman Aircraft Engineering Corporation incorporated 6 December 1929; became major supplier of carrierborne aircraft to US Navy.

Acquisition of Grumman by Northrop completed 1 May 1994 and new corporation formed 18 May. Northrop Grumman subsequently completed acquisition of Vought Aircraft Company, with US$130 million purchase of Carlyle Group's 51 per cent interest, in August 1994. Further expansion announced on 3 January 1996, when Northrop Grumman revealed agreement to acquire the defence and electronic systems businesses of the Westinghouse Electric Corporation for US$3 billion. Finalisation of this purchase in

March 1996 resulted in significant increase in products and technologies offered by Northrop Grumman. A major addition, finalised in July 1999, was that of Teledyne Ryan Aeronautical at cost of approximately US$140 million. Subsequently, in early April 2001, Northrop Grumman completed acquisition of Litton Industries Inc; value of this acquisition estimated at US$5.1 billion (including US$1.3 billion net debt). Another acquisition, announced on 20 April 2001, involved purchase of Electronic and Information Systems Group of Aerojet-General Corporation for US$315 million. Further growth followed on 30 November 2001, with acquisition of Newport News Shipbuilding, manufacturer of nuclear-powered submarines and aircraft carriers. In first quarter of 2002, Northrop Grumman announced intention of merging with TRW Inc; deal was finalised by 1 July, when Northrop Grumman chairman Kent Kresa revealed details of US$7.8 billion purchase. Workforce totalled about 39,000 in November 2000, rising to about 80,000 after purchase of Litton and further increasing to approximately 100,000 with addition of Newport News. Net sales of US$13,558 million were achieved in 2001, compared with US$7,618 million in 2000.

Collaboration with Lockheed Martin announced in September 1996 concerning development and marketing of airborne early warning and control system aircraft. Further co-operation occurred in May 1997, when Northrop Grumman joined Lockheed Martin Joint Strike Fighter team, having previously been associated with unsuccessful McDonnell Douglas/British Aerospace JSF submission.

Closer collaboration with DASA on surveillance and reconnaissance systems to result from MoU signed on 19 April 2000. In late 2000, Northrop Grumman received a Phase I study contract from US Defense Advanced Research Projects Agency for the Quiet Supersonic Platform (QSP) programme, intended to develop technology required for any future long-range, supersonic, military and/or civil aircraft; follow-on contracts were awarded in 2002 for continued development.

Northrop Grumman now organised into six major business units as follows:

Integrated Systems
   Follows this entry.
Electronic Systems
   Follows Integrated Systems.
Information Technology
   Based in Herndon, Virginia, this wholly owned subsidiary, now including much of Litton, provides technical engineering, project management and support resource services for information technology systems, plus technical and professional services in operations and maintenance.
Newport News Shipbuilding
   Production and assembly of nuclear-powered submarines, warships and aircraft carriers.
Ship Systems
   At Pascagoula, Mississippi.
Component Technologies
   At Iselin, New Jersey.

UPDATED

INTEGRATED SYSTEMS

225 East John W Carpenter Freeway, Suite 1500, Irving, Texas 75062
Tel: (+1 469) 420 00 00
Web: http://www.iss.northropgrumman.com
CORPORATE VICE-PRESIDENT AND PRESIDENT, INTEGRATED
   SYSTEMS: Scott J Seymour
DIRECTOR, COMMUNICATIONS: Patrick Mullaney

Having divested itself of the aerostructures business, Integrated Systems now functions as prime contractor for the B-2A Spirit stealth bomber and the E-8C Joint STARS airborne targeting and battlefield management system, while continuing production of the E-2C Hawkeye for the US Navy and export customers. Serves as principal subcontractor for the F/A-18 (see Boeing) and is teamed with Lockheed Martin on the F-35 JSF programme; also received study contract valued at US$500,000 in July 2000 for concept exploration work in connection with US Navy's Multi-Mission Maritime Aircraft (MMMA) programme and the associated Broad Area Maritime Surveillance (BAMS) project, whereby a mix of manned and unmanned platforms may eventually replace existing Lockheed Martin P-3/EP-3 aircraft. Other work includes modification and support of the EA-6B Prowler, F-5 Tiger II, F-14 Tomcat and Fairchild A-10 Thunderbolt II. Also engaged in production and support of UAVs for reconnaissance, surveillance and deception as well as aerial target systems.

Headquartered in Texas since 1999, Integrated Systems was due to have moved to One Hornet Way, El Segundo, California 90245-2804 at the end of 2002.

Operating elements of Integrated Systems are now organised into business areas, with responsibilities as detailed below; first mentioned includes RQ-8A Fire Scout unmanned version of Schweizer 333 helicopter (which see).

Air Combat Systems

One Hornet Way, El Segundo, California 90245-2804
Tel: (+1 310) 331 36 16
CORPORATE VICE-PRESIDENT AND GENERAL MANAGER:
   William H Lawler
MANAGER, COMMUNICATIONS: James F Hart

B-2 Spirit, F/A-18 Hornet/Super Hornet, F-35 Joint Strike Fighter, F-5 Tiger II/T-38 Talon and UAV systems. Has sites at Palmdale and San Diego, California; New Town, North Dakota; Whiteman AFB, Missouri; Tinker AFB, Oklahoma; and Hill AFB, Utah.

Airborne Early Warning and Electronic Warfare Systems

South Oyster Bay Road, Bethpage, New York 11714
Tel: (+1 516) 575 51 19
MANAGER, COMMUNICATIONS: John Vosilla

Production of E-2C Hawkeye 2000, plus modification, repair and support of other types, including EA-6B Prowler, A-10 Thunderbolt II, C-2 Greyhound, F-5 Tiger II, F-14 Tomcat and S-2T Turbo-Tracker. Has sites at St Augustine and Cecil Field, Florida; Point Mugu, California, plus field support services at many other locations.

Airborne Ground Surveillance and Battle Management Systems

2000 NASA Boulevard, Melbourne, Florida 32902
Tel: (+1 516) 575 51 19
MANAGER, COMMUNICATIONS: John Vosilla

Northrop Grumman design for a quieter supersonic aircraft   *NEW*/0533360

Production of E-8C Joint STARS, plus associated advanced surveillance and battle management systems; currently developing new surveillance systems as part of Multi-Platform Radar Technology Insertion Program (MP-RTIP) for USAF. Has additional site at Lake Charles, Louisiana.

UPDATED

NORTHROP GRUMMAN QSP

In 2002, the Defense Advanced Research Projects Agency awarded Northrop Grumman two contracts for continuation of

conceptual studies for a future Quiet Supersonic Platform (QSP). The first contract, valued at US$2.7 million, will permit Northrop Grumman to validate earlier studies and undertake wind-tunnel testing of a scale model of its preferred QSP design. At least a dozen different concepts were considered after work began in 2000, culminating in selection of the so-called 'dual relevant' approach, whereby the 'joined wing' or 'strut-braced wing' configuration could be adapted into either a strike/attack platform or an executive jet.

In its current guise, the QSP utilises a slender fuselage that can accommodate side-by-side weapons bays some 8.3 m

Artist's impression of a modified Northrop Grumman F-5E and F-111 (lower) both in transonic flight
NEW/0533359

(27 ft) in length or a 6.9 m (23 ft) long cabin. The main wing is shoulder-mounted and has a sharply swept cranked arrow planform with a straight trailing-edge; inboard leading-edge sweep is 84°, reducing to 71.5° outboard. The QSP is approximately 51.9 m (170 ft) long with a span of 17.6 m (57 ft 9 in); max T-O weight will be about 45,360 kg (100,000 lb), including payload of approximately 9,072 kg (20,000 lb). Project objectives include a very low sonic boom signature, range of 6,000 n miles (11,112 km; 6,904 miles) and maximum cruise speed of M2.4. Two General Electric GE449 engines are located at the rear of the aircraft, with a single vertical tail protruding above the engine nacelles.

The second contract received by Northrop Grumman was worth US$3.4 million and scheduled to involve flight trials of a modified Northrop F-5E Tiger II in the latter half of 2002. This aircraft was to be given a revised forward fuselage that is predicted to result in a shaped sonic boom, which will be significantly quieter.

NEW ENTRY

Hawkeye 2000 demonstrator   *NEW*/0533358

NORTHROP GRUMMAN E-2C HAWKEYE

TYPE: Airborne early warning and control system.
PROGRAMME: First flight of first of three prototypes 21 October 1960; total 59 production E-2As, of which 51 updated to E-2B by end 1971 apart from two TE-2A trainers and two converted to E-2C prototypes (see earlier *Jane's*); first flight of E-2C prototype 20 January 1971; production started mid-1971; first flight production aircraft 23 September 1972; 211 of all E-2C versions ordered, of which 189 had been delivered by end of 2001.
CURRENT VERSIONS: **E-2C:** Current service and production version (*as detailed*). Baseline aircraft (Group '0') had AN/APS-120, but remaining examples to be phased out of US Navy inventory by 2004.

AN/APS-139 and Allison T56-427 engines form **Group I** update; first operational aircraft (163538) delivered to VAW-112 on 8 August 1989; 18 built; AN/APS-139 can detect cruise missiles at ranges exceeding 100 n miles (185 km; 115 miles); also monitors maritime traffic; radar coverage extended by AN/ALR-73 passive detection system (PDS), detecting electronic emitters at twice radar detection range; more detailed description in 1979-80 *Jane's*; AN/APS-145 in **Group II** aircraft from December 1991; other enhancements give Group II 96 per cent expansion in radar volume, 400 per cent extra target tracking capability, 40 per cent more radar and identification range and 960 per cent increase in numbers of targets displayed. Group II added JTIDS in 1993-94; also has GPS. Final Group II aircraft for US Navy delivered in mid-2001. Retrofit with eight-blade propellers was due to start in 2001, with first operational example then planned to join fleet in fourth quarter; however, vibration problems encountered in testing have caused delay and new propeller not now expected to enter service until late 2003. Group II aircraft expected to begin phase-out in 2002-03, when improved Hawkeye 2000 joins fleet, although US Navy has examined possibility of upgrading Group II aircraft to Hawkeye 2000 standard, but may ultimately opt for more sophisticated Advanced Hawkeye derivative (see later).
TE-2C: Training model, based on E-2C; two conversions (158639, 158648) originally undertaken, of which one (158639) was later assigned to JTIDS development with Northrop Grumman. At least three further conversions (159105, 163029, 163848) subsequently done for training purposes with VAW-120 at Norfolk, Virginia.

E-2T: Originally reported to be conversion of E-2B for Taiwan, but new-build aircraft actually supplied; AN/APS-138 radar and electronic warfare upgrades. Delivery of initial four aircraft began in 1995, with further two to be obtained following agreement in July 1999; these will be to Hawkeye 2000E standard, with AN/APS-145 radar, but lacking CEC and satcom equipment.
Hawkeye 2000: In December 1994, company received US$155 million contract to redefine E-2C as the Hawkeye 2000. Key element is mission computer upgrade (MCU), with new equipment based on Raytheon's Model 940.

Initial trials of upgraded mission computer installed on second Group II aircraft (164109) began with first flight on 24 January 1997 and were completed in July 1997, at which time authorisation given for low-rate initial production of new mission computer. However, early flight trials revealed software problems that delayed production of new mission computer by about a year. More ambitious technical and operational evaluations undertaken with five modified aircraft in 1999-2001. All were Group II aircraft fitted with MCU and ACIS (see below) elements of proposed Hawkeye 2000; first two delivered to Patuxent River for initial evaluation by May 1999. At least four to Point Mugu from August 1999, joining VAW-117 for operational evaluation from October, with latter phase including deployed duty aboard a carrier for full battlegroup operations. In meantime, another E-2C (163849) used as a testbed for satcom, vapour-cycle cooling upgrade and Navy's USG-3 co-operative engagement capability (CEC) package, following first flight in April 1998; latter results in addition of 1.37 m (4 ft 6 in) antenna dish under belly containing omnidirectional transceiver that connects with command centres on parent aircraft carrier and surface combatant warships; provision of satcom evident through addition of 'inverted cone' fairing on top of rotodome.

New mission computer is less than half the weight of current L-304, one-third of volume, and offers 15 times the processing power; other improvements for Hawkeye 2000 include government-furnished advanced control indicator set (ACIS), satellite-based voice and data communications capability, a new Honeywell vapour-cycle cooling system, air-to-air refuelling capability (if required) and inclusion of equipment and systems that form part of Navy CEC package. MCU and ACIS make use of commercial off-the-shelf technology incorporating open architecture.

In April 1999, contract awarded for 24 aircraft for US Navy (21 Hawkeye 2000s) as five-year procurement package, Taiwan (two Hawkeye 2000Es) and France (one Group II); in early 2002, it appeared likely that US Navy would buy further five Hawkeye 2000s in 2004-05 in order to keep production line open until Advanced Hawkeye (see below) becomes available. IOC with US Navy will occur in 2004. Export-configured Hawkeye 2000s lack CEC and satcom facilities.
Advanced Hawkeye: Currently in proposal stage and unlikely to be operationally available until at least 2010 if project goes ahead. As now envisaged, Advanced Hawkeye will incorporate Lockheed Martin's AN/APS-XX solid-state, electronically steered UHF radar, surveillance infra-red search and track (SIRST) system, modular communications equipment, multisensor integration and tactical (glass) cockpit, with latter enabling co-pilot to augment system operators by performing some tactical functions. New systems will offer improved detection capability against theatre air and missile threats, including overland cruise missiles. US Navy expected to be prime customer, but other potential buyer is UK Royal Navy, which has emerging requirement for new AEW platform to operate from future aircraft carrier.

Northrop Grumman received US$49 million Pre-Systems Development and Demonstration contract for the E-2C Radar Modernization Program (RMP) in late 2001; work to be done under this 12-month contract concerns definition of physical architecture of future mission system. Northrop Grumman also to conduct initial flight testing of new radar's transmitter, receiver, antenna and rotary coupler on an NC-130H Hercules testbed; this was set to begin in third or fourth quarter of 2002 and to last six months.
CUSTOMERS: US Navy orders for E-2C totalled 139 by FY92, of which all delivered by March 1994. Further procurement authorised December 1994, when Northrop Grumman awarded US$122.5 million contract for start-up of new assembly line at St Augustine, Florida. Initial orders for seven Group II aircraft (four with FY95 funds, three with FY96 funds); first of these new aircraft (165293) was rolled out on 24 February 1997, flew on 22 March, and was delivered to VAW-120 at Norfolk, Virginia. Further contract in December 1996 to cover advance acquisition costs of four Group II aircraft with FY97 funds; four more ordered before April 1999 award of multiyear contract for 24 aircraft for US Navy, France and Taiwan. Delivery of first new-build Hawkeye 2000 to US Navy accomplished in October 2001, following handover of single upgraded aircraft converted from existing E-2C under US$17.8 million contract awarded in April 2000; four Hawkeye 2000s to be delivered in 2002, then five per year in 2003-05, with last in February 2006. First operational deployment of Hawkeye 2000 will be by VAW-117 in USS *Nimitz* in late 2002. Sole Group II example for France ordered by US Navy on 28 March 2001, for delivery in final quarter of 2003. Current US Navy programme status in adjacent table:

Variant	Qty	First aircraft
E-2C 'Group 0'	100	158638
E-2C Group I	18[1]	163029
E-2C Group II	36	164108
Hawkeye 2000	21	165648
Total	**175**	

[1] Two aircraft later converted to TE-2C and 16 upgraded to Group II standard

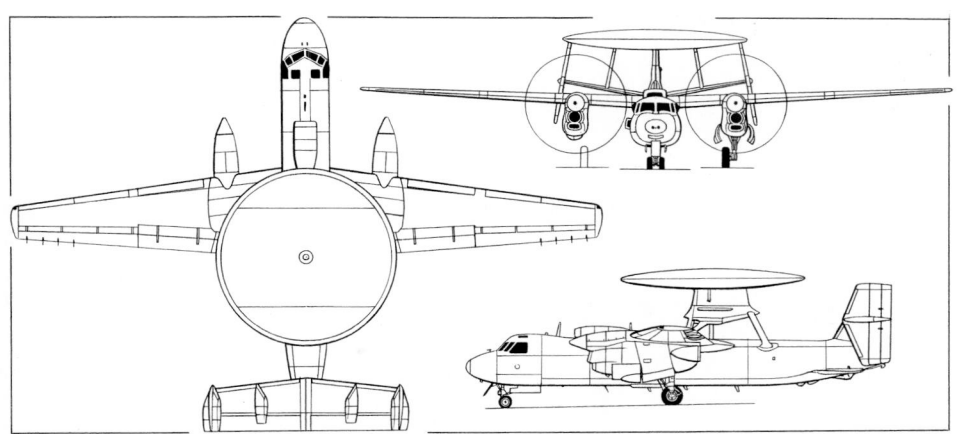

Northrop Grumman E-2C Hawkeye (Allison T56 turboprop engines) (*Jane's/Dennis Punnett*)

Major retrofit programme for USN and FMS aircraft was planned, including upgrade of all 18 Group I and minimum of 36 older ('Group 0') aircraft to Group II standard from FY95. Instead, 1994 defence review established new E-2Cs more economical than retrofit; however, 16 Group I E-2Cs have been upgraded to Group II at St Augustine, Florida; first returned to service on 21 December 1995. First operational squadron with upgraded Group II E-2C was VAW-123 which accepted its fourth and final aircraft on 29 April 1996; VAW-121 followed later in year. Total cost of upgrade programme originally to be US$135 million; however, an amendment to contract in third quarter of 1997 added US$8.5 million to cover conversion of an additional aircraft by May 1999.

E-2C entered service with VAW-123 at NAS Norfolk, Virginia, November 1973 and went to sea on board USS *Saratoga* late 1974; E-2C issued to 19 other squadrons, including three of Naval Reserve; current squadrons are VAW-112, 113, 116 and 117 at Point Mugu, California; VAW-115 at Atsugi, Japan; VAW-120, 121, 123 to 126 with VAW-78 of Reserves at Norfolk, Virginia, plus VAW-77 of Reserves at Atlanta, Georgia, for anti-drug smuggling surveillance duty. VAW-120 is training unit. Miramar was base of first Group II squadron, VAW-113, June 1992. VAW-113 first operational evaluation cruise in USS *Carl Vinson*, 1993. New Build Group II aircraft also issued to VAW-110 (disbanded September 1994), 112, 116 and 117.

See table for exports. Singaporean aircraft with AN/APS-138 radar; Taiwan received E-2T with AN/APS-138 radar. Israeli aircraft in storage by 1994, with three being sold to Mexico in 2002. France signed Letter of Offer and Acceptance (LOA) in June 1995 for two aircraft; first flight of first Aéronavale aircraft on 12 March 1998, with formal roll-out ceremony at St Augustine, Florida, on 28 April 1998. Both aircraft initially retained in USA for crew training, with first E-2C (actually No 2/165456) delivered to Lann-Bihoué, France, on 14 December 1998; second in April 1999. Operating unit, 4 Flottille, formed at Lann-Bihoué January 1999.

French Navy given authorisation in 1998 to purchase third Hawkeye, which, according to Northrop Grumman, will be a Group II aircraft. Italy a potential customer in longer term, with requirement for at least two AEW aircraft by 2007. Egypt and Singapore said to be considering acquisition of surplus US Navy aircraft in mid-2000, with former subsequently signing contract for single example as attrition replacement.

United Arab Emirates may obtain five refurbished aircraft from surplus US Navy stocks; possibility of sale at estimated cost of US$400 million notified to Congress on 4 September 2002.

COSTS: Procurement of 21 new Hawkeye 2000s for US Navy will cost US$1.47 billion. Upgrading of 13 Japanese aircraft to Hawkeye 2000 standard expected to cost approximately US$400 million, with cost of Egyptian upgrade of five aircraft estimated at US$210 million.

DESIGN FEATURES: E-2C can cover naval task force in all weathers flying at 9,150 m (30,000 ft) and can detect and assess approaching aircraft at in excess of 300 n miles (556 km; 345 miles); AN/APS-145 has total radiation aperture control antenna (TRAC-A) to reduce sidelobes to offset jamming; radar sweeps 6 million cu mile envelope and simultaneously monitors surface ships; long-range, automatic target track initiation and high-speed processing enable each E-2C to track more than 2,000 targets simultaneously and automatically, and control more than 40 intercepts; Randtron Systems AN/APA-171 antenna housed in 7.32 m (24 ft) diameter radome, rotating at 5 to 6 rpm above rear fuselage; antenna arrays in rotodome provide radar sum and difference signals and IFF.

High-mounted tapered wing; moderately sweptback tailplane with pronounced dihedral; twin fins and twin auxiliary fins, all with sweptback leading-edges and at right angles to tailplane. Rotating, circular radar housing mounted on cabane above rear fuselage. Nose-tow catapult attachment, arrester hook and tail bumper; parts of tail made of composites to reduce radar reflection; wings fold hydraulically on skewed hinges to lie parallel to fuselage. Wing incidence 4° at root, 1° at tip.

FLYING CONTROLS: Conventional and fully powered, with artificial feel; tailplane has 11° dihedral; four fins and three double-hinged rudders; long-span ailerons droop automatically when hydraulically operated Fowler flaps are extended; autopilot provides autostabilisation or full flight control. Empennage (including horizontal stabilisers, elevators, rudders, vertical fins and tabs) produced by Potez Aéronautique of France, following award of contract in mid-1997 for initial batch of five assemblies; first completed assembly delivered in second quarter of 1999.

STRUCTURE: Wing centre-section has three beams, ribs and machined skins; hinged leading-edge provides access to flying and engine controls. Fuselage conventional light metal. Composites used in parts of tail.

LANDING GEAR: Hydraulically retractable tricycle type. Pneumatic emergency extension. Steerable nosewheel unit retracts rearward. Mainwheels retract forward and rotate to lie flat in bottom of nacelles. Twin wheels on nose unit only. Oleo-pneumatic shock-absorbers. Mainwheel tyres size 36×11 (24 ply) tubeless, pressure 17.93 bar (260 lb/sq in) on ship, 14.48 bar (210 lb/sq in) ashore. Nosewheel tyres 20×5.5 (12/14 ply) tubeless. Hydraulic brakes. Hydraulically operated retractable tailskid. A-frame arrester hook under tail.

POWER PLANT: Two 3,803 kW (5,100 ehp) Allison T56-A-427 turboprops, driving Hamilton Sundstrand Type 54460-1 four-blade fully feathering reversible-pitch constant-speed propellers. These have foam-filled blades which have a steel spar and glass fibre shell. T56-A-427 engines provide 15 per cent improvement in efficiency, compared with -425 installed before 1989. Two Israeli Hawkeyes fitted locally with fixed in-flight refuelling probes; first aircraft (941) seen 1993. Hamilton Sundstrand NP2000 eight-blade, all-composites propeller flight tested on E-2C 163535 starting on 19 April 2001, following successful ground running on T56 and completion of critical design review in September 1998; US Navy has selected NP2000 for E-2C and C-2A Greyhound and placed US$44.5 million contract for 187 propellers, with option on further

54. New propellers will eventually be installed on all Hawkeye 2000s as well as retrofitted to surviving E-2C Group II aircraft. In 2002, US Navy requested Northrop Grumman to undertake studies into possibility of adopting a new engine; if one is chosen, it is unlikely to be introduced until the Advanced Hawkeye but could be retrofitted to earlier aircraft.

ACCOMMODATION: Normal crew of five, consisting of pilot and co-pilot on flight deck, plus ATDS Combat Information Center (CIC) staff of combat information centre officer, air control officer and radar operator. Downward-hinged door, with built-in steps, on port side of centre-fuselage and three overhead escape hatches.

SYSTEMS: Pneumatic boot de-icing on wings, tailplane and fins. Spinners and blades incorporate electric anti-icing.

AVIONICS: *Comms:* AN/AIC-14A intercom. AN/ARC-210 wideband/narrowband radio to be installed on Hawkeye 2000.

Radar: Lockheed Martin AN/APS-145 advanced radar processing system (ARPS) with fully automatic overland/overwater detection capability, Randtron AN/APA-171 rotodome (radar and IFF antennas). New Lockheed Martin radar now being developed for installation on Advanced Hawkeye.

Flight: Lockheed Martin AN/ASN-92 CAINS carrier aircraft inertial navigation system, GPS, AN/ASN-50 heading and attitude reference system, Rockwell Collins AN/ARA-50 UHF ADF, AN/ASW-25B ACLS, BAE Systems standard central air data computer, ASM-400 in-flight performance monitor, Honeywell AN/APN-171(V) radar altimeter.

Mission: BAE Systems/Hazeltine AN/APA-172 control indicator group with Lockheed Martin enhanced (colour) main display units (EMDU), which being replaced by L-3 Communications flat panel display screen during 2000-07; Litton OL-77/ASQ computer programmer (L-304) with Lockheed Martin enhanced high-speed processor, BAE Systems/Hazeltine OL-483/AP airborne interrogator system, Litton AN/ALR-73 passive detection system, AN/ARC-158 UHF datalink, AN/ARQ-34 HF datalink and JTIDS Class 2 HP terminal. Barco Display Systems to supply graphics controllers with radar display capability and colour flat panel displays for Hawkeye 2000 upgrade. Hawkeye 2000 has Lockheed Martin AN/ALQ-217 ESM and Multi-Mission Advanced Tactical Terminal (MATT) for data communications.

E-2C EXPORTS

Customer	Qty	Group	First aircraft	Delivery	Unit (or base)
Egypt[4]	6	0	162791	1987-88 (5)	(Cairo West)
		0	164626	1993 (1)	
France[5]	3	II	1(165455)	1998	4 Flottille
Israel[1]	4	0	160771	1978	192 Sqdn
Japan[2]	13	0	34-3451	1982 (4)	601 Sqdn
		0	54-3455	1984 (4)	
		0	34-3459	1992-93 (3)	
		0	44-3462	1993 (2)	
Singapore	4	0	162793	1987	111 Sqdn
Taiwan[3]	6		2501	1995	78 Sqdn
Total	**36**				

Notes:

[1] Israeli aircraft withdrawn from use in 1994 and stored awaiting disposal; three sold to Mexico in 2002 to be refurbished by IAI before delivery

[2] Japanese aircraft being upgraded to Hawkeye 2000 standard

[3] Taiwanese aircraft known as E-2Ts; total includes two on order as E-2T Hawkeye 2000E

[4] Five Egyptian aircraft to be upgraded to Hawkeye 2000 standard; a sixth, former US Navy aircraft, being acquired for upgrading (not included in above table)

[5] Third French aircraft for delivery in 2003

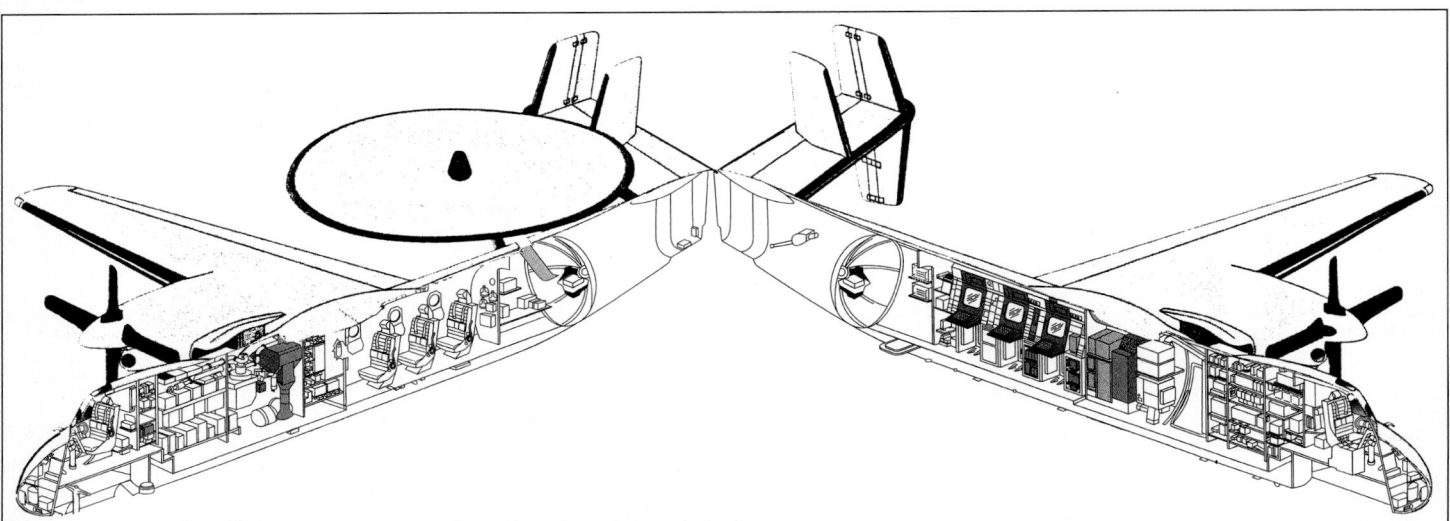

Interior layout of E-2C Hawkeye, showing three combat staff positions directly beneath the rotodome

0051794

DIMENSIONS, EXTERNAL:

Wing span	24.56 m (80 ft 7 in)
Wing chord: at root	3.96 m (13 ft 0 in)
at tip	1.32 m (4 ft 4 in)
Wing aspect ratio	9.3
Width, wings folded	8.94 m (29 ft 4 in)
Length overall	17.60 m (57 ft 8¾ in)
Height overall	5.58 m (18 ft 3¾ in)
Diameter of rotodome	7.32 m (24 ft 0 in)
Tailplane span	7.99 m (26 ft 2½ in)
Wheel track	5.93 m (19 ft 5¾ in)
Wheelbase	7.06 m (23 ft 2 in)
Propeller diameter	4.11 m (13 ft 6 in)

AREAS:

Wings, gross	65.03 m² (700.0 sq ft)
Ailerons (total)	5.76 m² (62.00 sq ft)
Trailing-edge flaps (total)	11.03 m² (118.75 sq ft)
Fins, incl rudders and tabs: outboard (total)	10.25 m² (110.36 sq ft)
inboard (total)	4.76 m² (51.26 sq ft)
Tailplane	11.62 m² (125.07 sq ft)
Elevators (total)	3.72 m² (40.06 sq ft)

WEIGHTS AND LOADINGS:

Weight empty	18,363 kg (40,484 lb)
Max fuel (internal, usable)	5,624 kg (12,400 lb)
Max T-O weight	24,687 kg (54,426 lb)
Max wing loading	379.6 kg/m² (77.75 lb/sq ft)
Max power loading	3.25 kg/kW (5.34 lb/ehp)

PERFORMANCE:

Max level speed	338 kt (626 km/h; 389 mph)
Max cruising speed	325 kt (602 km/h; 374 mph)
Cruising speed (ferry)	259 kt (480 km/h; 298 mph)
Approach speed	103 kt (191 km/h; 119 mph)
Stalling speed (landing configuration)	75 kt (138 km/h; 86 mph)
Service ceiling	11,275 m (37,000 ft)
Min T-O run	564 m (1,850 ft)
T-O to 15 m (50 ft)	793 m (2,600 ft)
Min landing run	439 m (1,440 ft)
Ferry range	1,541 n miles (2,854 km; 1,773 miles)
Time on station, 175 n miles (320 km; 200 miles) from base	4 h 24 min
Endurance with max fuel	6 h 15 min

UPDATED

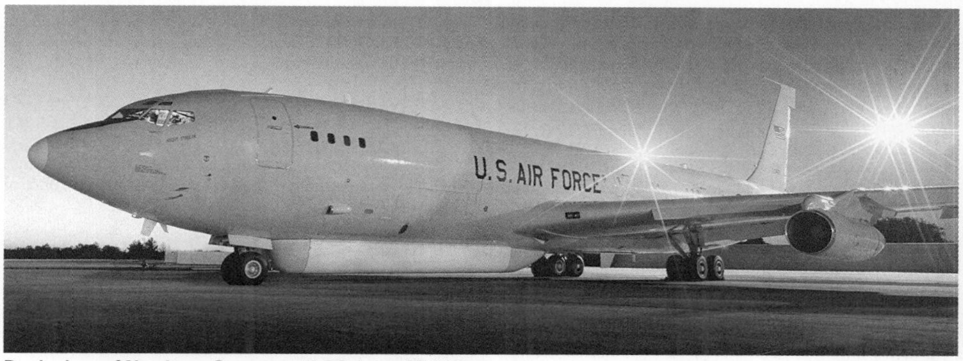

Dusk view of Northrop Grumman E-8C Joint STARS 0109157

E-8 JOINT STARS PROCUREMENT

Lot	FY Advance Acquisition	FY Full Funding	Type	Identity	Delivered
FSD	85[1]		E-8A	86-0416/T1	
FSD	85[1]		E-8A	86-0417/T2	
FSD	90[2]		E-8C	90-0175/T3	
1	92	93	E-8C	92-3289/P1	4 March 1996
1	92	93	E-8C	92-3290/P2	11 December 1996
2	93	94	E-8C	93-0597/P3	25 November 1997
2	93	94	E-8C	93-1097/P4	18 August 1998
3	94	95	E-8C	94-0284/P5	13 August 1999
3	94	95	E-8C	94-0285/P6	27 November 2000
4	95	96	E-8C	95-0122/P7	1 December 1999
4	95	96	E-8C	95-0121/P8	30 March 2001
5	96	97	E-8C	96-0042/P9	6 March 2000
5	96	97	E-8C	96-0043/P10	27 July 2000
6	97	98	E-8C	97-0100/P11	6 August 2001
7	98	99	E-8C	97-0200/P12	5 November 2001
7	98	99	E-8C	97-0201/P13	25 April 2002
8	99	00	E-8C	99-0006/P14	19 August 2002
9	00	01	E-8C	00-2000/P15	(31 March 2003[3])
10	01	02	E-8C	01-2005/P16	(31 March 2004[3])
11	02	03	E-8C	02-0911/P17	(31 March 2005[3])

Notes: [1] Year of full-scale development (FSD) contract award; may eventually be upgraded to E-8C standard and assigned to operational fleet
[2] Year of follow-on FSD contract award; aircraft serves as permanent testbed

NORTHROP GRUMMAN E-8 JOINT STARS

TYPE: Airborne ground surveillance system.

PROGRAMME: Full-scale development contract for Joint Surveillance Target Attack Radar System (Joint STARS) programme awarded to Grumman (later Northrop Grumman), 27 September 1985; two Boeing 707-328Cs used as EC-18C testbeds, later redesignated E-8A. See 2002-03 and earlier *Jane's* for details of initial development and operational deployment

Airframe modifications done at company's Lake Charles, Louisiana, plant following Boeing's withdrawal from programme; avionics installation at Melbourne.

Northrop Grumman awarded US$132 million (two contracts) in June 1997 for computer replacement programme taking advantage of commercial off-the-shelf (COTS) technology and intended to reduce costs; will result in integration of new, more powerful central computers significantly faster than those originally installed. It is also intended to replace programmable signal processors and substitute high-capacity switch and fibre-optic cable for existing copper-wired workstation network. This expected to produce savings of about US$19.3 million per aircraft, with original five central computers (three operating and two 'hot' spares) giving way to just two Compaq Clippers (one operating and one 'hot' spare). Testing of the new computers was completed in early August 2000; installation in production aircraft began in 2001; earlier E-8Cs are to be retrofitted during routine depot maintenance.

Major radar upgrade programme begun in December 1998, with award of US$14.5 million contract to Northrop Grumman for pre-Engineering and Manufacturing Development (EMD) phase of radar technology insertion program (RTIP) to replace existing AN/APY-3 with new 2-D electronically scanned array radar, offering upgraded signal processing and operation and control performance. Scope of programme subsequently broadened and renamed Multi-Platform Radar Technology Insertion Programme (MP-RTIP), with Northrop Grumman securing US$303 million contract at end of 2000; resultant system destined for installation in new Multi-sensor Command and Control Aircraft (MC2A). This will be a modified Boeing 767-400ER; USAF at present planning to buy five such platforms to supplement E-8C fleet. MP-RTIP EMD will begin in fourth quarter of 2003 with full-rate production decision due in third quarter of 2007. Potential applications include installation in RQ-4A Global Hawk UAV in addition to manned platform, with latter expected to attain operational capability in about 2012; some E-8Cs may also be retrofitted with MP-RTIP as the E-8D. Research and development effort likely to be divided equally between Northrop Grumman (prime contractor) and Raytheon (subcontractor), with former taking overall responsibility for radar and latter working on antenna development. New radar expected to offer a five-fold increase in resolution in MTI mode, a 10-fold acuity increase in SAR mode, a faster scan rate and additional operating modes to improve ground clutter elimination and resistance to jamming. Also likely to feature smaller antenna with resolution improved from current 3.66 m (12 ft) to 0.30 to 0.91 m (1 to 3 ft).

In short-term future, several upgrades and improvements are planned for implementation by 2006 that will take E-8C from initial **Block 0** (essentially, test and evaluation) to Block 50 configuration with full battle management capabilities, as described under Current Versions.

CURRENT VERSIONS: **E-8A**: Two development aircraft, as above. Fitted with consoles for 10 operators; Pratt & Whitney JT3D-3B engines. Returned to Grumman after participation in Gulf War for completion of performance testing in 100 sorties and several thousand hours of ground testing. Both accepted by USAF, December 1993, for trials at Edwards AFB. May eventually be upgraded as E-8Cs, but first currently serving as **TE-8A** crew trainer with 93rd Wing.

E-8B: Originally proposed production version, based on new-built airframe; F108 turbofan engines; 15 operator consoles. One prototype YE-8B, 88-0322, flown 12 June 1990 in 'green' state; avionics not installed; overtaken by decision to use remanufactured Boeing 707 airframes; delivered to USAF 3 October 1991 for storage at Davis-Monthan AFB, Arizona; bartered with Omega Air, 1993, for five used 707s.

E-8C: Production version; 18 operator consoles (17 operations, one navigation/self-defence). Initial two aircraft to Lake Charles for conversion, May-June 1992. First E-8C (T3/90-0175) is permanent testbed, not for delivery to USAF; is currently expected to serve as Universal Testbed (UTB) for Multi-Platform Radar Technology Insertion Program (MP-RTIP). First operational aircraft (P1/92-3289) commenced flight testing on 17 August 1995 and was delivered to USAF in March 1996, making operational debut with mission from Frankfurt over former Yugoslavia on 15 November 1996.

Following description applies to E-8C, except where otherwise indicated.

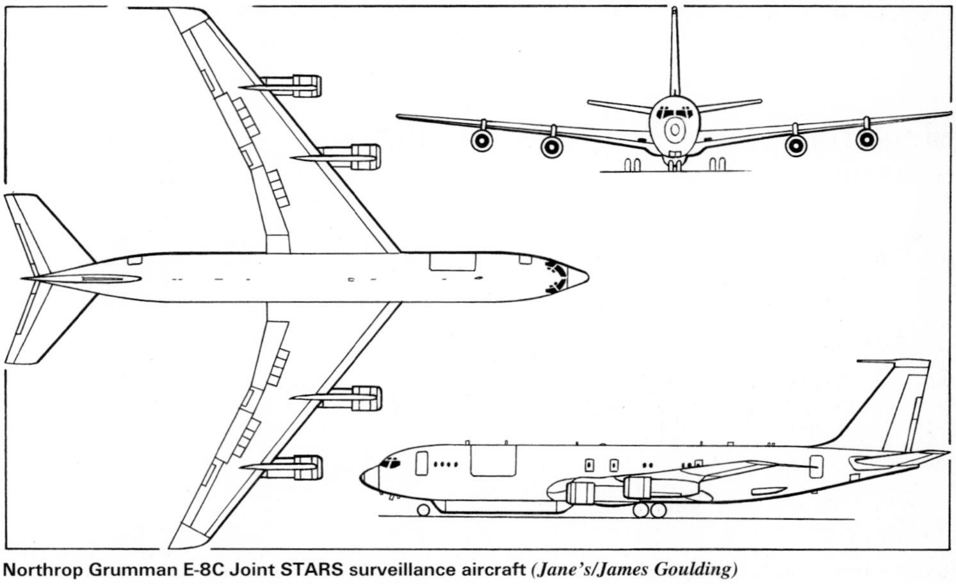
Northrop Grumman E-8C Joint STARS surveillance aircraft *(Jane's/James Goulding)*

E-8C Block 10: Baseline USAF aircraft, featuring interoperability with other US and allied aircraft; specific enhancements include TADIL-J secure communications, a digital autopilot and Y2K computer compliance. First example was P5/94-0284, in service 1999; same standard also applicable to P6-P10. These, plus initial four 'Block 0' aircraft, to be upgraded to Block 20 configuration; first aircraft to receive upgrade was P7/95-0122, which was redelivered on 11 February 2002; second is P9/96-0042, due for redelivery on 5 December 2002, with P6/94-0285 and P8/95-0121 to follow in 2003.

E-8C Block 20: Modified to accept COTS hardware, with computer replacement programme tied to networked communications; introduces US Army improved data modem, basic SINCGARS radio and radar advanced signal processing. Planned for 2000, but first aircraft to this configuration was P11/97-0100, which was delivered in August 2001. Four more delivered by August 2002, including first of 10 aircraft due to be upgraded by 2006.

E-8C Block 30: Baseline attack control platform with UHF satcom, broadcast intelligence, TADIL-J attack support upgrade (ASU) and phase one of global air traffic management (GATM). P16 will be first aircraft delivered with satcom.

E-8C Block 40: Baseline information dominance platform featuring enhanced and inverse SAR, improved MTI and EHF/SHF satcom.

E-8C Block 50: Baseline battle management platform, initially planned to incorporate RTIP, elint capability, combat identification, helicopter detection/tracking, maritime detection, automatic target recognition and Kinematic Auto Tracker capability. Some of these features may not now be adopted.

CUSTOMERS: US Air Force and US Army (aircraft operated by USAF). See table for details of USAF procurement; total 20 originally expected, including two upgraded E-8As and one E-8C permanent testbed, but proposed fleet reduced to 13 by 1997 Quadrennial Defense Review. Recent budget requests have increased fleet to 16 (with long-lead funding for a 17th requested in FY02).

Operating unit has been 93rd Air Control Wing which activated at Robins AFB, Georgia, in January 1996, although some (possibly all) aircraft to be reassigned to 116th Air Control Wing of Georgia ANG at same base in fourth quarter of 2002. IOC achieved with delivery of third aircraft (93-0597) to Robins AFB on 18 December 1997.

COSTS: E-8A development costs US$1,006.2 million; E-8C development costs US$868.7 million (FY91). Procurement of Lot 1 to Lot 4 (eight aircraft) totalled US$2,141 million; advance acquisition costs (long lead) for two aircraft in FY96 US$104.6 million. Total of US$221.8 million appropriated for 16th aircraft in FY01-02.

DESIGN FEATURES: Boeing 707-320C airliner converted with 7.32 m (24 ft) antenna covered by a 12.19 m (40 ft) 'canoe' radome and fairing under forward fuselage, for phased-array SLAR.

FLYING CONTROLS, STRUCTURE, LANDING GEAR: As, or similar to, commercial Boeing 707, last described in 1980-81 *Jane's* and detailed in the current *Jane's Aircraft Upgrades*.

POWER PLANT: Four 80.1 kN (18,000 lb st) Pratt & Whitney TF33-P-102B turbofans initially installed; now being upgraded to 85.4 kN (19,200 lb st) TF33-P-102C standard following award of US$10.5 million contract in December 1998 to United Technologies for 42 modification kits. See Weights and Loadings for fuel. In late June 2002, it was announced that re-engining with Pratt & Whitney JT8D-219 turbofans is to be undertaken so as to reduce need for aerial refuelling, increase time on station and raise operating ceiling to 12,800 m (42,000 ft) to increase radar coverage significantly. It is hoped to secure initial funding in FY03 or FY04.

ACCOMMODATION: Standard mission crew of 21, comprising pilot, co-pilot, flight engineer, navigator/self-defence suite operator and 17 Air Force/Army operators; on longer missions replacement flight deck crew and system operators can be carried up to maximum of 34.

SYSTEMS: As Boeing 707; additional electrical generating capacity. World Auxiliary Power Company APU.

AVIONICS: *Comms:* Telephonics multiple intercom net control system; Raytheon UHF communications system.

Radar: Northrop Grumman (Norden Systems) AN/APY-3 multimode side-looking phased-array I-band radar, scanned electronically in azimuth and steered mechanically in elevation from either side of aircraft to provide 120° field of view. Synthetic aperture (SAR) mode used to detect stationary objects, such as parked tanks. Can interleaf Doppler mode to detect moving targets in moving target indicator/wide-area search (MTI/WAS), which is primary operating mode. Coverage is approximately 50,000 km² (19,305 sq miles) per minute, cruising at 9,150 to 12,200 m (30,000 to 40,000 ft), with ability to detect targets at range of 50 km (31 miles) to 250 km (155 miles) from aircraft.

Flight: Litton INS, Rockwell Collins flight management system.

Mission: Five Raytheon Model 920/866 supermini computers per aircraft (being replaced by two Compaq Clippers from 2001), Computing Devices International programmable signal processors (three), Interstate Electronics graphic displays, 18 Raytheon Model 920 workstations, Orbit International workstation keyboards, Miltope message printers (replaced by Data Metrics printers on fifth and subsequent aircraft), Cubic Defense Systems surveillance and control datalink, JTIDS for TADIL-J generation and processing. Satellite communications link, Magnavox encrypted UHF radios (12), two encrypted HF radios, three encrypted VHF radios with single channel ground and airborne radio system (SINCGARS) provision. Radar data can be transmitted instantaneously to ground stations; or attacks by aircraft and ground forces directed via JTIDS datalink.

Self-defence: Unspecified, but includes chaff/flare dispensers and probably also incorporates threat-sensing equipment.

ARMAMENT: Nil.

DIMENSIONS, EXTERNAL (abbreviated):

Wing span	44.42 m (145 ft 9 in)
Length overall	46.61 m (152 ft 11 in)
Height overall	12.95 m (42 ft 6 in)

WEIGHTS AND LOADINGS:

Weight empty	77,564 kg (171,000 lb)
Max fuel weight	70,307 kg (155,000 lb)
Max T-O weight	152,407 kg (336,000 lb)

PERFORMANCE:

Max operating Mach No. (M$_{MO}$)	0.84
Service ceiling	12,800 m (42,000 ft)
Endurance: internal fuel	11 h
with one in-flight refuelling	20 h

UPDATED

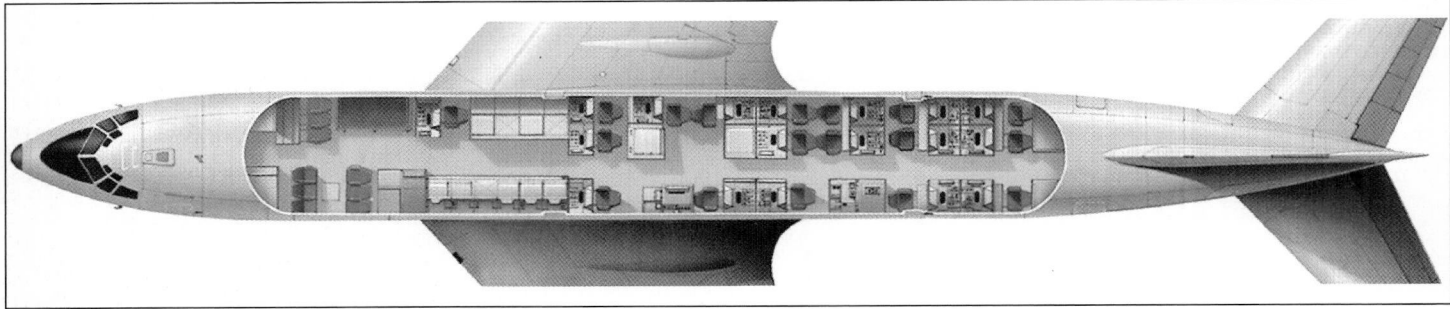

Inboard view of E-8C Joint STARS, showing operators' positions

ELECTRONIC SYSTEMS

1580-A West Nursery Road, Linthicum, Baltimore, Maryland 21090
Tel: (+1 410) 993 24 63
Fax: (+1 410) 993 23 94
Web: http://sensor.northgrum.com
CORPORATE VICE-PRESIDENT AND PRESIDENT, ELECTRONIC SYSTEMS: Robert P Iorizzo
COMMUNICATIONS OFFICER: Jack Martin Jr

Electronic Systems responsibilities include the development and production of radar and electronic systems for installation in the F-16 Fighting Falcon, F-22 Raptor, B-1B Lancer, AH-64D Apache Longbow, C-130 Hercules, E-3B/C Sentry and Boeing 737 AWACS and E-8C Joint STARS. Teamed with Rafael of Israel for sale and production of Litening II target-designation and navigation pod.

Development, integration and manufacturing expertise also embraces military airborne, space and undersea radar, surveillance and reconnaissance systems, air defence systems, tactical communications equipment, anti-submarine warfare sensors and systems, submersibles, mine countermeasures equipment, marine systems and shipboard instrumentation. Radar systems are produced for civil air traffic control agencies in the USA, Europe, Africa, the Middle and Far East, Asia and Latin America.

California Microwave Systems was acquired in April 1999, retaining its original name.

UPDATED

NUVENTURE

NUVENTURE AIRCRAFT

4184 West Kelley, Fresno, California 93722-9798
Tel: (+1 559) 447 11 12
e-mail: alantolle@cs.com
Web: http://www.nuventureaircraft.com
PRESIDENT: Alan E Tolle

NuVenture was established to maintain availability of components for the former Questair Venture. All jigs and stocks of spares were bought from Henry Bouley of North Windham, Connecticut by Richard Clawson and transferred to Visalia, California. In February 2001, ownership of NuVenture was purchased by Alan Tolle, manufacture remaining at Visalia.

NEW ENTRY

NUVENTURE VENTURE

TYPE: Side-by-side kitbuilt.
PROGRAMME: Questair M20 Venture designed by Jim Griswald and associates. Prototype (N62V) first flew 1 July 1987; made public debut at Oshkosh in same month; several improvements introduced in following year. Marketed in kit form; some 100 sold, of which about half completed during early 1990s and 36 remained on US civil register in 2002. Further five of fixed landing gear version, known as Spirit, also produced, first having flown in February 1991; last described in 1992-93 *Jane's*. Venture established nine speed and climb records in FAI Class C1b during 1989 and 1990; inaugurated Sport Class in Reno air races.

Kit production resumed by NuVenture; Spirit also available.

COSTS: US$49,900 in kit form, excluding engine (2002).

DESIGN FEATURES: Optimised for high-speed flight, but with comfortably large cockpit in spite of overall small size. 'Egg shape' fuselage; high-aspect ratio wings; narrow-chord fin and tapered tailplane.

Wing section NACA 23017 at root; 23010 at tip; twist 3°. Leading-edge cuff on each wing, from 60 per cent semi-span to tip, adds droop; leading-edge slots at cuff end and 40 per cent span generate vortices at high AoA to retard stalled airflow spreading from root to tip.

FLYING CONTROLS: Conventional and manual. No flaps, but ailerons droop when landing gear extended. Electrically operated trim tabs on ailerons and elevators.

STRUCTURE: All-aluminium.

LANDING GEAR: Retractable tricycle type. Single, small wheel on each leg. Electric retraction; separate systems for mainwheels and nosewheel. Steerable nosewheel. All wheels retract rearwards. Hydraulic brakes.

POWER PLANT: One 231 kW (310 hp) Teledyne Continental IO-550-N6B flat-six, derated to 209 kW (280 hp), driving a McCauley two-blade, constant-speed propeller. Fuel capacity 212 litres (56.0 US gallons; 46.6 Imp gallons).

DIMENSIONS, EXTERNAL:

Wing span	8.38 m (27 ft 6 in)
Length overall	4.95 m (16 ft 3 in)
Height overall	2.35 m (7 ft 8½ in)
Propeller diameter	1.73 m (5 ft 8 in)

AREAS:

Wings, gross	6.74 m² (72.50 sq ft)

WEIGHTS:

Weight empty	553 kg (1,220 lb)
Max T-O weight	907 kg (2,000 lb)

Max wing loading	134.7 kg/m² (27.59 lb/sq ft)
Max power loading	4.35 kg/kW (7.14 lb/hp)

PERFORMANCE:

Never-exceed speed (V$_{NE}$)	300 kt (555 km/h; 345 mph)
Max level speed	265 kt (491 km/h; 305 mph)
Cruising speed at FL120	240 kt (144 km/h; 276 mph)
Stalling speed:	
flaps up	68 kt (126 km/h; 79 mph)
flaps down	61 kt (113 km/h; 71 mph)
Max rate of climb at S/L	762 m (2,500 ft)/min
T-O to 15 m (50 ft)	305 m (1,000 ft)
Landing from 15 m (50 ft)	488 m (1,600 ft)
Service ceiling	8,840 m (29,000 ft)
Range, cruising at FL120:	
VFR reserves	1,000 n miles (1,853 km; 1,151 miles)
IFR reserves	833 n miles (1,542 km; 958 miles)
Endurance	4 h
g limits: solo	+6/−3
at max T-O weight	+5/−2.5

NEW ENTRY

Alan Tolle's own Questair M20 Venture, pattern for the relaunched version (*Jane's/Paul Jackson*) *NEW*/0137385

PAM

PERFORMANCE AVIATION MANUFACTURING GROUP

PO Box 80, Williamsburg, Virginia 23187
Tel: (+1 757) 229 03 67
Fax: (+1 757) 229 73 80
e-mail: pamgroup@erols.com
Web: http://www.flying-platform.com
CEO: Clement Makowski
PRESIDENT: Robert J Pegg

PAM Group was founded in 1998 and is developing an individual lifting vehicle.

UPDATED

PAM 100B INDIVIDUAL LIFTING VEHICLE

TYPE: Lifting vehicle.
PROGRAMME: Project initiated October 1989; construction of prototype started January 1993; first flight (N6172N) June 1994, powered by single engine; rebuilt with two Hirth engines and reflown May 1998; FAA certification April 1999. A total of 50 hours (tethered and free) flown between 1994 and mid-1999; over 120 hours had been accumulated by July 2002.

CURRENT VERSIONS: **PAM 100B:** *As described.*
 PAM 200: Kitbuilt version of PAM 100B.
 PAM 100T: Turbine-powered (JFS-100-13) version of PAM 100B with 37.9 litre (10.0 US gallon; 8.3 Imp gallon) fuel capacity. Empty weight 263 kg (580 lb).
 UAV: Unmanned military version, powered by two JFS-100-13 turbines; design started in December 2001.
COSTS: Kit US$50,500, including engines.
DESIGN FEATURES: Two co-axial counter-rotating two-blade, ground-adjustable pitch rotors mounted below main open-frame structure. Rotor section NACA 0012, constant; no twist; end caps. Rotor speed 1,150 rpm. Tip speed 168 m (550 ft)/s. Two small propellers at ends of cross-tubes provide directional control.
FLYING CONTROLS: Kinesthetic control for pitch and roll functions; manual throttle on engine for vertical control; two small propellers for directional control.
STRUCTURE: Tubular main structure. Rotors of extruded 6063T6 aluminium; aluminium flexure rotor hub.
LANDING GEAR: Fixed skid type.
POWER PLANT: Two 78.3 kW (105 hp) Hirth F30A flat-four, two-stroke engines with two-spur 1:2.64 reduction gear on each engine; over-running and centrifugal clutches on transmission. Fuel in two tanks, total capacity 26.5 litres (7.0 US gallons; 5.8 Imp gallons).

ACCOMMODATION: One person, standing within protective tubular framework.
DIMENSIONS, EXTERNAL:

Rotor diameter (each)	2.79 m (9 ft 2 in)
Rotor blade chord	0.20 m (0 ft 8 in)
Height overall	2.59 m (8 ft 6 in)
Skid track	3.05 m (10 ft 0 in)

AREAS:

Rotor blades (each)	0.20 m² (2.11 sq ft)
Rotor discs (total)	12.26 m² (132.0 sq ft)

WEIGHTS AND LOADINGS:

Weight empty	300 kg (660 lb)
Max fuel	20 kg (45 lb)
Max T-O weight	431 kg (950 lb)
Max disc loading	35.1 kg/m² (7.20 lb/sq ft)
Max power loading	2.75 kg/kW (4.52 lb/hp)

PERFORMANCE:

Max level speed	52 kt (97 km/h; 60 mph)
Max cruising speed	39 kt (72 km/h; 45 mph)
Service ceiling	1,830 m (6,000 ft)
Range with max fuel	21 n miles (40 km; 25 miles)

UPDATED

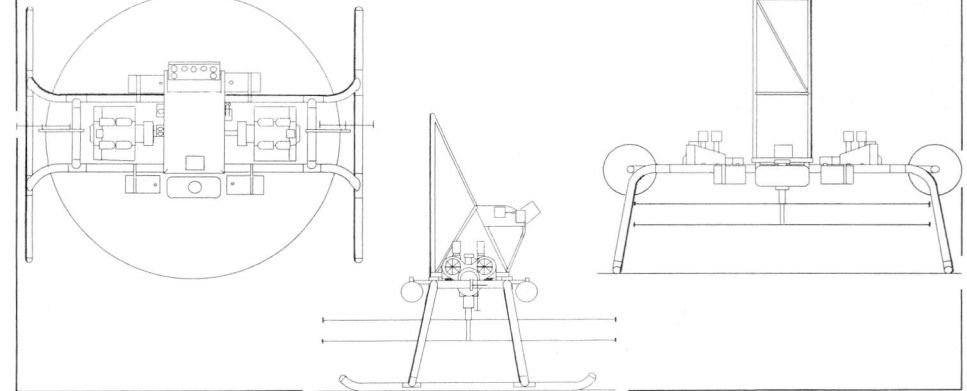

PAM 100B Individual Lifting Vehicle 0064492

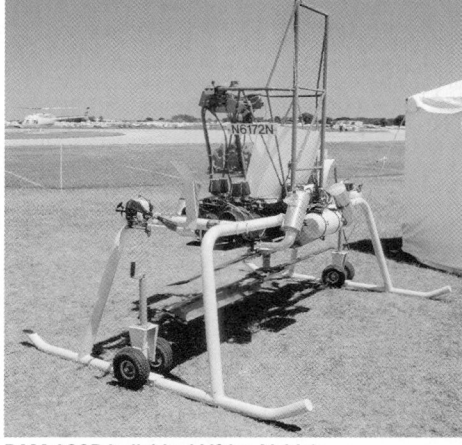

PAM 100B Individual Lifting Vehicle
(Jane's/Paul Jackson) *NEW*/0533701

PERFORMANCE

PERFORMANCE AIRCRAFT INC

12897 West 151 Street, Suite C, Olathe, Kansas 66062
Tel: (+1 913) 780 91 40

Fax: (+1 913) 782 10 92
e-mail: perfair@msn.com
Web: http://www.performanceaircraft.com
PRESIDENT: Jeff Ackland

The Performance Legend has been sold to Legend Aircraft (which see).

UPDATED

PIASECKI

PIASECKI AIRCRAFT CORPORATION

Second Street West, Essington, Pennsylvania 19029-0360
Tel: (+1 610) 521 57 00
Fax: (+1 610) 521 59 35
e-mail: piac@concentric.net
PRESIDENT: Frank N Piasecki
VICE-PRESIDENTS:
 Frederick W Piasecki (Technology)
 John W Piasecki (Contracts)
DIRECTOR, MILITARY REQUIREMENTS: Jimmy R Hayes
DIRECTOR, PROGRAM REQUIREMENTS: Joseph P Cosgrove

Frank Piasecki flew the USA's second successful helicopter on 11 April 1943; created Piasecki Helicopter Corporation in 1946,

this forming subsidiary Piasecki Aircraft Corporation in 1955, following which the original company was sold to Boeing.

Piasecki assisted PZL Swidnik of Poland (which see) in FAA certification of the W-3A helicopter, which was awarded in May 1993. Piasecki has exclusive sales agreement for W-3A in the Americas and Pacific Rim countries. W-3A Sokół complies with FAR Pt 29, is certified for full IFR operations and has US instrumentation.

UPDATED

PIASECKI VECTORED THRUST DUCTED PROPELLER (VTDP)

TYPE: New-concept rotorcraft.
PROGRAMME: Began with US Army contract to develop a compound helicopter incorporating the Piasecki VTDP

concept for the AH-64 Apache and AH-1W SuperCobra. Programme objectives were met or exceeded by both the AH-64 VTCAD (Vectored Thrust Combat Agility Demonstrator) and AH-1W VTCAD configurations, resulting in increased maximum level flight speed; 50 per cent improvement in longitudinal acceleration and deceleration capability in level flight; 50 per cent decrease in turn and pull-up radii at speeds in excess of 95 kt (176 km/h; 109 mph); and handling qualities at least as good as those of the baseline AH-64A and AH-1W. In addition, tactical simulations confirmed superiority of the VTCADs over the standard Apache and AH-1W SuperCobra.

A separate US Navy contract, awarded in April 1997 and valued at US$16.1 million, involved investigation into application of VTDP technology to the AH-1W(4BW)

Model of HH-60G/VTDP compound helicopter proposal   0120255

four-blade rotor configuration. The Navy contract, recently completed, included ground testing of the full-scale VTDP and additional flight controls simulation and testing of the 4BW/VTDP configuration.

Piasecki then proposed flight demonstration of this technology to the Navy on an AH-1W(4BW), but instead

was awarded a four-year US$26.1 million contract on 28 September 2000 for integration, testing and flight demonstration of VTDP on a modified Sikorsky YSH-60F. This flight test programme, to begin in 2003, will be conducted jointly by Piasecki and the US Navy following modification of the YSH-60F at the US Navy Naval Air

Warfare Center at NAS Patuxent River, Maryland. The VTDP concept is being investigated by the DoD as an affordable means of upgrading the capabilities extending the service life of existing single main rotor helicopters such as the UH-1, UH/SH-60 and AH-64, until the follow-on Joint Replacement Aircraft is fielded some time after 2025. Most recently, the US Air Force selected the H-60/VTDP concept as one of a number of alternatives being considered as an upgrade or replacement, to be fielded as early as 2007, for its ageing HH-60G combat search and rescue helicopters.

By mid-2000, Piasecki had constructed a 2.44 m (8 ft 0 in) diameter ring tail unit and integrated its five-blade ducted propeller. In addition to installation of this unit, modifications to the YSH-60F will include addition of lifting wings (from an Aerostar business twin) and integration of VTDP controls into the helicopter's existing mechanical control system, all of which is expected to add some 726 kg (1,600 lb) to the YSH's empty weight.

DESIGN FEATURES: The VTDP comprises a ducted propeller with integral vanes and spherical sectors to vector propeller thrust. It provides lateral thrust for anti-torque and lateral control, in lieu of a tail rotor, as well as forward thrust for auxiliary propulsion. Combined with a lifting wing, it provides for increasing speed to more than 200 kt (370 km/h; 230 mph); greater manoeuvrability; reduced vibration and fatigue loads; and consequent improvement in component lives and reduction in maintenance requirements and life cycle costs.

Piasecki estimated that a VTDP-based H-60 compound helicopter would have an unrefuelled combat radius of 762 n miles (1,411 km; 876 miles) following a rolling take-off, and 520 n miles (963 km; 598 miles) after a vertical take-off, representing a three-fold increase over the standard HH-60.

UPDATED

PIPER

THE NEW PIPER AIRCRAFT INC

2926 Piper Drive, Vero Beach, Florida 32960
Tel: (+1 561) 567 43 61
Fax: (+1 561) 778 21 44
Web: http://www.newpiper.com
PRESIDENT AND CEO: Charles (Chuck) M Suma
VICE-PRESIDENT, PRODUCT SERVICES: Werner Hartlieb
MARKETING AND SALES DIRECTOR: Larry Bardon

In July 1995, after Piper had stabilised its financial position during four years of Chapter 11 protection, the US Bankruptcy Court approved a new reorganisation plan under which Piper's assets were bought for US$95 million by Newco Pac Inc, a new company jointly owned by Philadelphia-based investment firm Dimeling, Schreiber and Park, Teledyne Continental Motors (which was Piper's largest creditor), and the remaining creditors. Under this ownership, The New Piper Aircraft Inc was established. Earlier history of Piper last appeared in the 1995-96 *Jane's*.

During 1999, Piper delivered 329 aircraft, 395 in 2000; 441 in 2001, and expected to deliver 331 in 2002. Workforce 1,100 by February 2002. Sales revenue totalled US$243 million in 2001, up from US$157 million in 2000.

Kits and parts for classic Pipers are available from other outlets. These include:

J-3 Cub: See Wag-Aero Sport trainer. Replacement fuselages (standard and with additional 10 cm; 4 in cockpit width) from Airframes Inc, Big Lake, Alaska 99652; *tel* (+1 907) 892 82 44; http://www.supercubs.com.

PA-12 Super Cruiser: Fuselages manufactured by Dakota Air Frame Inc, 1372 Cleveland Avenue, Larchwood, Iowa 51241; *tel/fax* (+1 712) 477 20 84.

PA-14 Family Cruiser: See Wag-Aero Sportsman.

PA-15 Vagabond: See Wag-Aero Wag-A-Bond.

PA-18 Super Cub: Fuselages from Airframes Inc (see J-3 Cub); complete aircraft from Cub Crafters.

UPDATED

Piper PA-28-161 Warrior III *(Jane's/Paul Jackson)*   *NEW*/0533704

panels; dihedral 7°, incidence 2° at root, −1° at tip; sweepback at quarter-chord 5°.

FLYING CONTROLS: Manual. Conventional ailerons and rudder, plus all-moving tailplane with combined anti-servo and trim tab. Four-position manually operated flaps.

STRUCTURE: Conventional light alloy, with semi-monocoque fuselage, single-spar wings and ribbed light alloy skins on fin and tailplane; glass fibre nose cowl and wing/tailplane/fin-tips.

LANDING GEAR: Non-retractable tricycle type. Steerable nosewheel. Piper oleo-pneumatic shock-absorbers; single wheel on each unit. Cleveland wheels with tyres size 6.00-6 (4 ply) on main units, pressure 1.65 bar (24 lb/sq in). Cleveland nosewheel and tyre size 5.00-5 (4 ply), pressure 2.06 bar (30 lb/sq in). Cleveland disc brakes. Parking brake. Glass fibre speed fairings standard.

POWER PLANT: One 119 kW (160 hp) Textron Lycoming O-320-D3G flat-four engine, driving a Sensenich 74DM6-0-60 two-blade fixed-pitch metal propeller. Fuel in two wing tanks, with total capacity of 189 litres (50.0 US gallons; 41.6 Imp gallons), of which 181.5 litres (48.0 US gallons; 40.0 Imp gallons) are usable. Refuelling point on upper surface of each wing. Oil capacity 7.5 litres (2.0 US gallons; 1.7 Imp gallons).

ACCOMMODATION: Four persons in pairs in enclosed cabin. Individual horizontally adjustable front seats with seat belts and shoulder harnesses; bench-type rear seat with seat belts and shoulder harnesses. Dual controls standard. Large door on starboard side. Baggage compartment at rear of cabin, with external baggage door on starboard side. Heating, ventilation and windscreen defrosting standard.

PIPER PA-28-161 WARRIOR III

TYPE: Four-seat lightplane.

PROGRAMME: As replacement for Cherokee 140 series, Piper redesigned airframe with several refinements, including fuselage stretch and new wing; first flight of prototype PA-28-151 17 October 1972; FAA certification 9 August 1973; PA-28-161 Warrior II first flown 27 August 1976; two/four-seat Cadet trainer version introduced April 1988, but no longer in production; Warrior III introduced late 1994.

CUSTOMERS: Five delivered in 1996, seven in 1997, 20 in 1998, 25 in 1999, 43 in 2000, 32 in 2001 and 19 in the first six months of 2002. Piper has built some 30,000 of PA-28 series since prototype Cherokee flew on 10 January 1960.

COSTS: Standard equipped price US$152,500 (2000).

DESIGN FEATURES: Classic tourer and trainer. Low, moderately tapered wing (with root glove) and low-set tailplane; sweptback fin.

NACA 65_2-415 wing section on inboard panels, Mod No. 5 of NACA TN 2228 on leading-edge of outboard

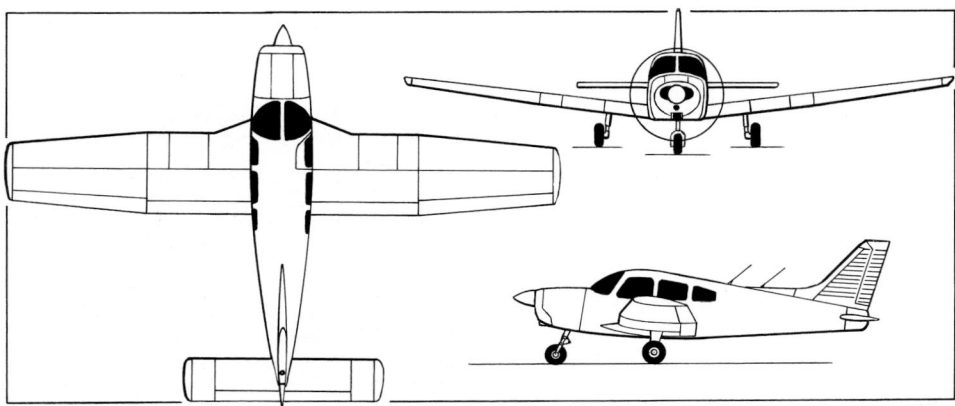

Piper PA-28-161 Warrior III (Lycoming O-320 flat-four) *(Jane's/James Goulding)*   0100458

SYSTEMS: Hydraulic system for brakes only. Electrical system powered by 28 V 60 A engine-driven alternator. 24 V 10 Ah battery standard.

AVIONICS: Standard and optional Advanced Training Group packages by Garmin/Meggitt, as described.

Comms: Standard: Garmin GNS-430 com/nav/IFR GPS; GMA-340 audio panel with marker beacon lights and four-position intercom; GTX-320 transponder; ELT; avionics master switch; cabin speaker; Telex 100T microphone and Airman 760 headset. Advanced Training Group: as above but with dual GNS-430.

Flight: Standard: GI-106A VOR/LOC/GS/GPS indicator; Narco AR-850 altitude reporter. Advanced Training Group: as above, plus dual GI-106A and RCR 650A ADF with IND 650A indicator.

Instrumentation: Piper true airspeed indicator, magnetic compass, sensitive altimeter, gyro horizon, directional gyro, rate of turn indicator, rate of climb indicator, OAT gauge, digital ammeter, electric clock, annunciator panel with push-to-test, recording tachometer, three-way oil temperature/pressure and fuel-pressure gauge, dual fuel quantity gauges and vacuum gauge.

EQUIPMENT: Standard equipment includes engine hour recorder, alternate static source, heated pitot head, colour co-ordinated control wheels, internally lit rocker switches, external power receptacle, electrical engine primer system, instrument panel lighting package, cabin dome light, navigation lights, landing/taxying light, wingtip Comet strobe lights, avionics dimming switch, pilot's and co-pilot's vertically adjustable seats in fabric and vinyl with magazine storage pockets on backs, rear bench seat, four floor-mounted cabin fresh-air vents, vinyl cabin side panels, wall-to-wall carpet and headliner, crew armrests, pilot's storm window, two sun visors, two map pockets, 'Quietised' soundproofing, carpeted baggage compartment with security straps, halon fire extinguisher, tiedown points, jack pads, static discharge wicks, and Du Pont Imron polyurethane exterior paint in base colour with two contrasting trim stripes. Meggitt electric trim, carburettor ice detector, second altimeter, two-tone base colour exterior paint and foreign certification gross weight kit optional.

DIMENSIONS, EXTERNAL:

Wing span	10.67 m (35 ft 0 in)
Wing chord: at root, excl glove	1.60 m (5 ft 3 in)
at tip	1.07 m (3 ft 6¼ in)
Wing aspect ratio	7.2
Length overall	7.25 m (23 ft 9½ in)
Height overall	2.22 m (7 ft 3½ in)
Tailplane span	3.96 m (12 ft 11¾ in)
Wheel track	3.05 m (10 ft 0 in)
Wheelbase	2.03 m (6 ft 8 in)
Propeller diameter	1.88 m (6 ft 2 in)
Propeller ground clearance	0.21 m (8¼ in)
Cabin door: Height	0.89 m (2 ft 11 in)
Width	0.91 m (3 ft 0 in)
Baggage door: Height	0.51 m (1 ft 8 in)
Max width	0.56 m (1 ft 10 in)
Height to sill	0.71 m (2 ft 4 in)

DIMENSIONS, INTERNAL:

Cabin (instrument panel to rear bulkhead):

Length	2.49 m (8 ft 2 in)
Max width	1.05 m (3 ft 5¼ in)
Max height	1.14 m (3 ft 8¾ in)
Floor area	2.28 m² (24.5 sq ft)
Volume (incl baggage)	3.0 m³ (106 cu ft)
Baggage compartment volume	0.68 m³ (24.0 cu ft)

AREAS:

Wings, gross	15.79 m² (170.0 sq ft)
Ailerons (total)	1.23 m² (13.20 sq ft)
Trailing-edge flaps (total)	1.36 m² (14.60 sq ft)
Fin	0.69 m² (7.40 sq ft)
Rudder	0.38 m² (4.10 sq ft)
Tailplane, incl tab	2.76 m² (29.70 sq ft)

WEIGHTS AND LOADINGS:

Weight empty, equipped	695 kg (1,533 lb)
Baggage capacity	91 kg (200 lb)
Max T-O weight	1,106 kg (2,440 lb)
Max wing loading	70.1 kg/m² (14.35 lb/sq ft)
Max power loading	9.28 kg/kW (15.25 lb/hp)

PERFORMANCE:

Max level speed	117 kt (216 km/h; 134 mph)

Normal cruising speed at 75% power

	115 kt (213 km/h; 132 mph)
Stalling speed: flaps up	50 kt (93 km/h; 58 mph)
flaps down	44 kt (82 km/h; 51 mph)
Max rate of climb at S/L	196 m (644 ft)/min
Service ceiling	3,355 m (11,000 ft)
T-O to 15 m (50 ft)	494 m (1,620 ft)
Landing from 15 m (50 ft)	354 m (1,160 ft)

Range, 45 min reserves:

at 75% power	426 n miles (789 km; 490 miles)
at 55% power at 3,050 m (10,000 ft)	
	513 n miles (950 km; 590 miles)

UPDATED

PIPER PA-28-181 ARCHER III

TYPE: Four-seat lightplane.

PROGRAMME: Introduced (as Cherokee Challenger 180) 9 October 1972 as successor to Cherokee 180; Archer 180,

Piper PA-28-181 Archer III *(Jane's/Paul Jackson)*   *NEW*/0533705

featuring minor changes, available from 1974; PA-28-181 Archer II launched 1976, and since 1978 has featured the tapered wings of Warrior II; Archer III from 1994, with axisymmetric engine inlets, redesigned windscreen and cabin side windows, plus interior restyling and improvements.

CUSTOMERS: 37 Archer IIIs delivered in 1995, and 45 each in 1996 and 1997; 91 in 1998; 100 in 1999, 102 in 2000, 88 in 2001 and 16 in the first six months of 2002.

Description of the Warrior III applies also to Archer III except as follows:

COSTS: Standard, equipped US$179,900 (2000).

LANDING GEAR: Tricycle type. Tyres size 6.00-6 (4 ply), on all three wheels. Mainwheel tyre pressure 1.65 bar (24 lb/sq in), nosewheel 1.24 bar (18 lb/sq in). Cleveland high-capacity disc brakes. Parking brake. Wheel speed fairings standard.

POWER PLANT: One 134 kW (180 hp) Textron Lycoming O-360-A4M flat-four engine, driving a Sensenich 76EM8S14-O-62 two-blade fixed-pitch metal propeller. Fuel in two tanks in wing leading-edges, with total capacity of 189 litres (50.0 US gallons; 41.6 Imp gallons), of which 181.5 litres (48.0 US gallons; 40.0 Imp gallons) are usable. Oil capacity 7.5 litres (2.0 US gallons; 1.7 Imp gallons).

ACCOMMODATION: Four persons in pairs in enclosed cabin. Individual adjustable front seats, with dual controls; individual rear seats. Large door on starboard side. Baggage compartment at rear of cabin; door on starboard side. Rear seats removable to provide 1.3 m³ (44 cu ft) cargo space. Accommodation heated and ventilated. Windscreen defrosting.

SYSTEMS: 28 V 70 A alternator and 24 V 10 Ah battery.

AVIONICS: Standard Garmin International and Meggitt package, Optional Premium Select package adds second GNS-430 and GI-106A VOR/LOC/GS/GPS indicator, Meggitt System 55 dual-axis autopilot with automatic electric trim, DG with heading bug and trim indicator and Meggitt ST-360 altitude preselect.

Comms: Garmin GMA-340 audio panel with marker beacon receiver and stereo intercom; GTX-320 solid-state transponder; Telex 100T microphone and Airman 760 headset.

Flight: Integrated VOR/LOC/GS/COM and GPS.

Instrumentation: Standard GNS-430 high-resolution colour LCD MFD with GI-106A VOR/LOC/GS/GPS indicator.

EQUIPMENT: Standard equipment generally as for Warrior III except: polished propeller spinner, metal instrument panel, illuminated side-mounted OAT gauge, overhead switch panel, EGT gauge, flush locking fuel caps, overhead vent fan system, restyled windscreen and window lines with tinted transparencies, full chemical corrosion protection, Aztec Silver interior decor, DuPont Imron 6000 clearcoat/basecoat and exterior paint, wool carpeting and Hobnail side panels, passenger armrests and headrests. Additional options include Piper Aire air conditioning, carburettor ice detector, stainless steel cowling fasteners; Meggitt electric trim low-noise exhaust system and leather seats in choice of three colours.

DIMENSIONS, EXTERNAL: As for Warrior III except:

Length overall	7.32 m (24 ft 0 in)
Tailplane span	3.92 m (12 ft 10½ in)

Instrument panel of baseline-equipped Piper PA-28-181 Archer III *(Jane's/Paul Jackson)*.   0110707

Wheelbase	2.00 m (6 ft 7 in)
Propeller diameter	1.93 m (6 ft 4 in)

DIMENSIONS, INTERNAL: As for Warrior III except:

Cabin: Width	1.06 m (3 ft 5¾ in)
Height	1.14 m (3 ft 9 in)

Baggage compartment volume, incl hatshelf

	0.74 m³ (26.0 cu ft)

AREAS: As for Warrior III

WEIGHTS AND LOADINGS:

Weight empty, equipped	766 kg (1,689 lb)
Baggage capacity	90 kg (200 lb)
Max T-O weight	1,156 kg (2,550 lb)
Max ramp weight	1,160 kg (2,558 lb)
Max wing loading	73.2 kg/m² (15.00 lb/sq ft)
Max power loading	8.62 kg/kW (14.17 lb/hp)

PERFORMANCE:

Max level speed	133 kt (246 km/h; 153 mph)

Normal cruising speed at 75% power at 2,410 m

(7,900 ft)	128 kt (237 km/h; 147 mph)
Stalling speed, flaps down	45 kt (84 km/h; 52 mph)
Max rate of climb at S/L	211 m (692 ft)/min
Service ceiling	4,300 m (14,100 ft)
T-O run	346 m (1,135 ft)
T-O to 15 m (50 ft)	490 m (1,608 ft)
Landing from 15 m (50 ft)	427 m (1,400 ft)
Landing run	280 m (920 ft)

Range at S/L, allowances for taxi, T-O, climb and descent and 45 min reserves, at 1,830 m (6,000 ft):

at 75% power	444 n miles (822 km; 511 miles)
at 65% power	487 n miles (901 km; 560 miles)
at 55% power	522 n miles (966 km; 600 miles)

UPDATED

PIPER PA-28R-201 ARROW

TYPE: Four-seat lightplane.

PROGRAMME: Derived from Cherokee Archer II, but with retractable landing gear, more powerful engine and untapered wings of 1975 PA-28-180 Archer; replaced original retractable gear Cherokee, the PA-28R Arrow 180, which had first flown 1 February 1967; PA-28R-201 Arrow III with increased span, tapered wings first flown 16 September 1975; first production aircraft 7 January 1977; new Arrow IV and Turbo Arrow IV models introduced 1979 with all-moving T tails; these subsequently discontinued and earlier low-tail Arrow III model restored to production.

CUSTOMERS: Seven delivered in 1996, two in 1997, two in 1998, six in 1999, 18 in 2000, 23 in 2001 and 14 in the first six months of 2002.

Description of Archer III applies also to Arrow except as follows:

COSTS: Standard, equipped US$228,700 (2000).

LANDING GEAR: Tricycle type, retracted hydraulically with an electrically operated pump supplying the hydraulic pressure. Main units retract inward into wings, nose unit rearward. All units fitted with oleo-pneumatic shock-absorbers. Mainwheels and tyres size 6.00-6 (6 ply), pressure 2.07 bar (30 lb/sq in). Nosewheel and tyre size 5.00-5 (4 ply), pressure 1.86 bar (27 lb/sq in). High-capacity dual hydraulic disc brakes and parking brake.

POWER PLANT: One 149 kW (200 hp) Textron Lycoming IO-360-C1C6 flat-four engine, driving a McCauley two-blade constant-speed metal propeller. Fuel tanks in wing leading-edges with total capacity of 291 litres (77.0 US gallons; 64.1 Imp gallons), of which 273 litres (72.0 US gallons; 60.0 Imp gallons) are usable. Oil capacity 7.5 litres (2.0 US gallons; 1.7 Imp gallons).

SYSTEMS: Generally as for Archer III and Warrior III except for electrohydraulic system for landing gear actuation.

AVIONICS: Standard and optional Advanced Training Group (IFR) packages by Bendix/King, as described.

Comms: Standard: Bendix/King KX 155 nav/com/glideslope with audio amplifier and broadband and VOR antennas; KT 76A transponder; ELT; avionics master switch; cabin speaker; Telex 100T microphone; Telex headset; Telex ProCom four-position intercom; microphone and microphone and headset jacks. Advanced Training Group: As above but with second KX 155 and PS Engineering PM6000M audio selector panel.

Flight: Standard: KI 203 VOR/LOC indicator; Narco AR-850 altitude reporter. Advanced Training Group: As

Piper PA-28R-201 Arrow *(Jane's/Paul Jackson)*   NEW/0533706

above, plus KI 204-02 VOR/LOC/GS indicator; KLN 89B GPS, KR 87 ADF and KI 227-00 and KCS 55A slaved compass system.

DIMENSIONS, EXTERNAL:

Wing span	10.80 m (35 ft 5 in)
Wing chord: at root, excl glove	1.60 m (5 ft 3 in)
at tip	1.07 m (3 ft 6¼ in)
Wing aspect ratio	7.4
Length overall	7.52 m (24 ft 8¼ in)
Height overall	2.39 m (7 ft 10¼ in)
Tailplane span	3.92 m (12 ft 10½ in)
Wheel track	3.19 m (10 ft 5½ in)
Wheelbase	2.39 m (7 ft 10¼ in)

AREAS:

Wings, gross	15.79 m² (170.0 sq ft)

WEIGHTS AND LOADINGS:

Weight empty, equipped	812 kg (1,790 lb)
Max T-O weight	1,247 kg (2,750 lb)
Max wing loading	79.0 kg/m² (16.18 lb/sq ft)
Max power loading	8.37 kg/kW (13.75 lb/hp)

PERFORMANCE:

Max level speed	145 kt (268 km/h; 166 mph)
Normal cruising speed	137 kt (253 km/h; 157 mph)
Stalling speed: flaps up	60 kt (112 km/h; 69 mph)
flaps down	55 kt (102 km/h; 64 mph)
Max rate of climb at S/L	253 m (831 ft)/min
Service ceiling	4,935 m (16,200 ft)
T-O to 15 m (50 ft)	488 m (1,600 ft)
Landing from 15 m (50 ft)	463 m (1,520 ft)
Cruising range at 55% power, at 2,743 m (9,000 ft), 45 min reserves	880 n miles (1,630 km; 1,013 miles)

UPDATED

PIPER PA-32R-301 SARATOGA II HP

TYPE: Six-seat utility transport.

PROGRAMME: Saratoga family of fixed and retractable landing gear light aircraft announced 17 December 1979 to replace earlier PA-32 Cherokee SIX 300 and T-tail Lance series; fixed-gear models now out of production; Saratoga II HP introduced in 1993, featuring axisymmetric engine inlets, aerodynamic clean-up and interior improvements and restyling.

CURRENT VERSIONS: **Saratoga II HP:** Standard version, *as described.*

Saratoga II TC: Turbocharged version with Textron Lycoming TIO-540 engine; *described separately.*

CUSTOMERS: Total of 38 HPs delivered in 1997, 28 in 1998, 30 in 1999, 28 in 2000, 22 in 2001, and one in the first six months of 2002.

Piper has built over 7,200 of PA-32 family since prototype first flew on 6 December 1963.

COSTS: Standard, equipped US$409,800 (2000).

DESIGN FEATURES: Larger counterpart to PA-28 series, featuring stretched fuselage. Current versions have 'semi-tapered' wing, with root glove, unswept inboard panel and tapered outer panel with small fence inboard of tip.

Wing section NACA 66₂-415, dihedral 7°, thickness/chord ratio 15 per cent, twist 2°, incidence 2°, washout 0°.

FLYING CONTROLS: Conventional. Ailerons, rudder and all-moving tailplane with combined anti-servo and trim tab. Rudder trim; four-position electrically actuated flaps with preselect.

STRUCTURE: Conventional light alloy, with semi-monocoque fuselage, single-spar wings and ribbed light alloy skins on fin and tailplane; glass fibre nose cowl and wing/tailplane/fintips.

LANDING GEAR: Hydraulically retractable tricycle type with single wheel on each unit. Main units retract inward, nosewheel aft. Emergency free-fall extension. Piper oleo-pneumatic shock-absorbers. Steerable nosewheel. Mainwheels and tyres size 6.00-6 (8 ply), pressure 2.62 bar (38 lb/sq in). Nosewheel and tyre size 5.00-5 (6 ply), pressure 2.41 bar (35 lb/sq in).

POWER PLANT: One 224 kW (300 hp) Textron Lycoming IO-540-K1G5 flat-six engine, driving a Hartzell three-blade constant-speed metal propeller. Two fuel tanks in each wing with combined capacity of 405 litres (107 US gallons; 89.1 Imp gallons), of which 386 litres (102 US gallons; 84.9 Imp gallons) are usable. Refuelling points on wing upper surface. Oil capacity 11.5 litres (3.0 US gallons; 2.5 Imp gallons).

ACCOMMODATION: Enclosed cabin seating six people, rear four in club arrangement; dual controls standard. Two forward-hinged doors, one on starboard side forward, overwing, and one on port side at rear end of cabin. Space for 45 kg (100 lb) baggage at rear of cabin, with external, lockable baggage/utility door on port side. Additional baggage space, capacity 45 kg (100 lb), between engine fireproof bulkhead and instrument panel, with external door on starboard side. Pilot's storm window. Accommodation heated and ventilated. Piper Aire air conditioning system optional. Windscreen defroster standard.

SYSTEMS: Electrically driven hydraulic pump for landing gear actuation. Electrical system includes 28 V 90 A engine-driven alternator and 24 V 10 Ah battery. Standby vacuum system standard. Optional oxygen system and Piper Aire air conditioning.

AVIONICS: Standard Garmin International and Meggitt package.

Comms: Garmin GMA-340 audio panel with marker beacon receiver and stereo intercom; GTX-320 solid-state transponder.

Flight: Dual Garmin GNS-430 integrated VOR/LOC/GS/COM and GPS, GI-106A VOR/LOC/GD/GPS indicator, Meggitt ST-180 HSI, DME 450 and Meggitt 55 two-axis autopilot with automatic electric trim, DG with heading bug and trim indicator, ST-361 ADI with flight director features.

Instrumentation: Dual GNS-430 high-resolution colour LCD MFD; Piper true airspeed indicator, illuminated magnetic compass, sensitive altimeter (with second optional) ADI, HSI, rate of turn indicator, rate of climb indicator, OAT gauge, digital ammeter, annunciator panel with push-to-test, recording tachometer, manifold pressure/fuel flow gauge, oil pressure gauge, oil temperature gauge, dual fuel quantity gauges, fuel quantity sight gauges, CHT gauge, EGT gauge and vacuum gauge. Co-pilot's 76 mm (3 in) instrument panel with electric attitude indicator optional.

EQUIPMENT: Standard equipment includes engine hour recorder, alternate static source, electric clock, internally lit rocker switches, heated pitot head, resettable circuit breakers in CB panel, standby electric vacuum pump, electric emergency fuel pump, external power receptacle, instrument panel lighting package, avionics dimming, map lights, navigation lights, landing/taxying light, wingtip strobe lights, pulsating recognition lights, pilot's and co-pilot's vertically adjustable all-leather seats, pilot's vent window, sun visors, four all-leather passenger seats with headrests, shoulder harnesses, seat belts, and quick-release facility, fold-down armrests (fifth and sixth seats), refreshment console, executive writing table, Hobnail fabric side panels, super 'Quietised' soundproofing, utility door for rear baggage access and cargo loading, static discharge wicks, and Du Pont Imron 6000 clearcoat/basecoat exterior paint. Optional equipment includes entertainment/executive console; provision for AM/FM radio and CD player, multimedia entertainment system and laptop computer workstation; and metallic paint.

DIMENSIONS, EXTERNAL:

Wing span	11.02 m (36 ft 2 in)
Wing aspect ratio	7.3
Length overall	8.43 m (27 ft 8 in)
Height overall	2.59 m (8 ft 6 in)
Tailplane span	3.94 m (12 ft 11 in)
Wheel track	3.38 m (11 ft 1 in)
Wheelbase	2.41 m (7 ft 11 in)
Cabin door (fwd, stbd): Height	0.89 m (2 ft 11 in)
Width	0.91 m (3 ft 0 in)
Cabin door (rear, port): Height	0.72 m (2 ft 4½ in)
Width	0.71 m (2 ft 4 in)
Baggage door (fwd): Height	0.41 m (1 ft 4 in)
Width	0.56 m (1 ft 10 in)
Baggage/utility door (fwd): Height	0.52 m (1 ft 8½ in)
Width	0.66 m (2 ft 2 in)

DIMENSIONS, INTERNAL:

Cabin: Length (instrument panel to rear bulkhead)	3.16 m (10 ft 4¼ in)
Max width	1.24 m (4 ft 0¾ in)
Max height	1.07 m (3 ft 6 in)
Volume (incl rear baggage area)	5.5 m³ (195 cu ft)

Baggage compartment volume:

forward	0.20 m³ (7.0 cu ft)
rear	0.49 m³ (17.3 cu ft)

AREAS:

Wings, gross	16.56 m² (178.3 sq ft)
Ailerons (total)	0.98 m² (10.60 sq ft)
Trailing-edge flaps (total)	1.36 m² (14.60 sq ft)
Fin	0.70 m² (7.50 sq ft)
Rudder, incl tab	0.40 m² (4.30 sq ft)
Horizontal tail surfaces (total)	2.94 m² (31.60 sq ft)

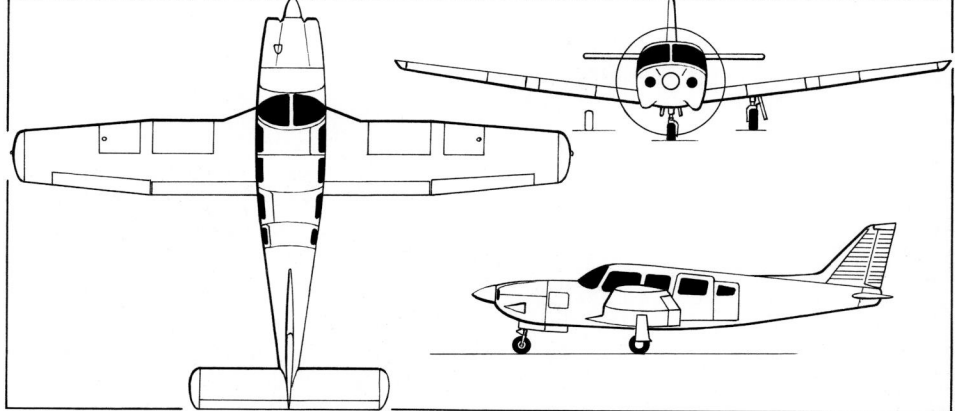

Piper PA-32R-301 Saratoga II HP *(Jane's/Paul Jackson)*   NEW/0533707

General arrangement of the Piper PA-32R-301 Saratoga II HP *(Jane's/James Goulding)*   0100459

WEIGHTS AND LOADINGS:
Weight empty, equipped	1,087 kg (2,396 lb)
Max T-O weight	1,633 kg (3,600 lb)
Max ramp weight	1,639 kg (3,615 lb)
Max wing loading	98.6 kg/m² (20.19 lb/sq ft)
Max power loading	7.30 kg/kW (12.00 lb/hp)

PERFORMANCE:
Max level speed	175 kt (324 km/h; 201 mph)
Normal cruising speed	166 kt (307 km/h; 191 mph)
Stalling speed: flaps up	67 kt (124 km/h; 78 mph) CAS
flaps down	63 kt (117 km/h; 73 mph)
Max rate of climb at S/L	398 m (1,305 ft)/min
Service ceiling	5,029 m (16,500 ft)
T-O run	366 m (1,200 ft)
T-O to 15 m (50 ft)	540 m (1,770 ft)
Landing from 15 m (50 ft)	464 m (1,520 ft)
Landing run	195 m (640 ft)

Range, long-range cruise power, allowances for start, taxi, T-O, climb and descent, and 45 min reserves
859 n miles (1,590 km; 988 miles)

UPDATED

PIPER PA-32R-301T SARATOGA II TC

TYPE: Six-seat utility transport.

PROGRAMME: Development of Saratoga II HP. Rolled out 15 July 1997; FAA certification 21 July 1997; first customer deliveries August 1997.

CUSTOMERS: Total of 26 delivered in 1997, 48 in 1998, 46 in 1999, 70 in 2000, 68 in 2001 and 24 in the first six months of 2002.

COSTS: Standard, equipped US$439,600 (2000).

The description for the Saratoga II HP applies also to the Saratoga II TC except:

DESIGN FEATURES: New interior and relocated battery and ground power receptacle.

POWER PLANT: One 224 kW (300 hp) Textron Lycoming TIO-540-AH1A turbocharged flat-six engine, with automatic wastegate, driving a Hartzell three-blade constant-speed metal propeller. Fuel as for Saratoga II HP.

WEIGHTS AND LOADINGS:
Weight empty, equipped	1,118 kg (2,465 lb)

PERFORMANCE:
Max level speed	187 kt (346 km/h; 215 mph)

Cruising speed:
at 2,440 m (8,000 ft)	172 kt (318 km/h; 198 mph)
at 3,050 m (10,000 ft)	175 kt (324 km/h; 201 mph)
at 4,575 m (15,000 ft)	185 kt (343 km/h; 213 mph)
Max certified altitude	6,100 m (20,000 ft)
T-O run	338 m (1,110 ft)
T-O to 15 m (50 ft)	552 m (1,810 ft)
Landing from 15 m (50 ft)	519 m (1,700 ft)
Landing run	269 m (880 ft)

Range, 45 min reserves and normal allowances, normal (N) or long-range (LR) cruise power:
at 2,440 m (8,000 ft):		
N		842 n miles (1,559 km; 969 miles)
LR		950 n miles (1,759 km; 1,093 miles)
at 3,050 m (10,000 ft):		
N		844 n miles (1,563 km; 971 miles)
LR		948 n miles (1,756 km; 1,091 miles)
at 4,575 m (15,000 ft):		
N		844 n miles (1,563 km; 971 miles)
LR		945 n miles (1,750 km; 1,087 miles)

UPDATED

PIPER PA-34-220T SENECA V

TYPE: Six-seat utility twin.

PROGRAMME: Original PA-34 Seneca certified 7 May 1971 and announced 23 September 1971, having been preceded by prototypes of fixed-gear PA-34-180 Twin Six (flown 25 April 1967) and retractable gear version in following year; 149 kW (200 hp) engines; redesignated Seneca II with increased MTOW, certified 18 July 1974 and introduced for 1975; improved Seneca III with more powerful counter-rotating (C/R) engines certified 17 December 1980 and introduced 15 February 1981; Seneca IV with axisymmetric engine inlets, aerodynamic refinements and interior improvements and restyling certified 17 November 1993 and introduced in 1994. Seneca V with

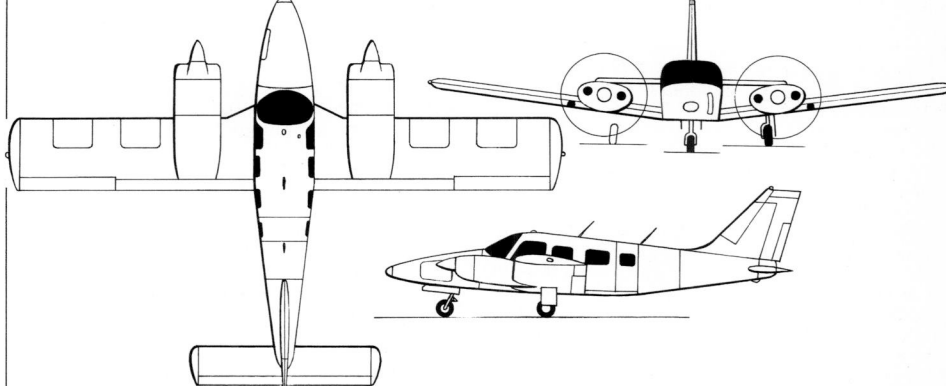

Piper PA-34-220T Seneca V *(Jane's/Paul Jackson)*   *NEW*/0533708

Piper PA-34-220T Seneca V six-seat twin *(Jane's/James Goulding)*   0016212

L/TSIO-360-RB engines unveiled 16 January 1997 (having been secretly certified on 11 December 1996); engines, equipped with intercoolers and density control system (automatic wastegate), maintain S/L rated power to 5,945 m (19,500 ft). Other new features include redesigned instrument panel with digital display monitoring panel (DDMP), overhead switch/dome light/ speaker panel, entertainment/executive console with extendable work table and in-flight phone/fax option replacing the second row seat on the starboard side. Senecas have also been built in Brazil and Poland.

CUSTOMERS: Total of 28 Seneca IVs delivered in 1995, and 18 in 1996; 39 Seneca Vs delivered in 1997, 55 each in 1998 and 1999, 42 in 2000, 38 in 2001 and 21 in the first six months of 2002. Widely used as twin-conversion trainer; recent customers include Sabena Airlines, which took delivery of two in 1999 for its training centre at Scottsdale, Arizona.

Production of the PA-34 series in the USA totals some 4,700; the prototype first flew on 30 August 1968.

COSTS: Standard equipped price US$19,900 (2000).

DESIGN FEATURES: Twin-engined version of Cherokee SIX/ Saratoga. Low-mounted, constant-chord wings and tailplane; leading-edge gloves inboard of engines; constant-chord tailplane.

Wing section NACA 65₂-415; dihedral 7°; thickness/ chord ratio 15 per cent; washout 0° 41′; twist 2° 41′.

FLYING CONTROLS: Manual. Frise ailerons, rudder and one-piece all-moving tailplane with combined anti-balance and trim tab; anti-servo tab in rudder; wide-span electrically operated slotted flaps. Control surface movements: ailerons +35/−20°; tailplane +12° 3′/−7° 30′; rudder ±35°. Maximum flap deflection 40°.

STRUCTURE: Conventional light alloy, with semi-monocoque fuselage; single-spar wings; glass fibre wingtips.

LANDING GEAR: Hydraulically retractable tricycle type. Main units retract inward, nose unit forward. Oleo-pneumatic shock-absorbers. Steerable nosewheel. Emergency free-fall extension system. Mainwheels and tyres size 6.00-6

(8 ply), pressure 3.79 bar (55 lb/sq in); nosewheel and tyre size 6.00-6 (6 ply), pressure 2.76 bar (40 lb/sq in). Nosewheel safety mirror. High-capacity disc brakes. Parking brake.

POWER PLANT: One 164 kW (220 hp) Teledyne Continental TSIO-360-RB (port) and one 164 kW (220 hp) LTSIO-360-RB (starboard) flat-six turbocharged counter-rotating engine, each driving a Hartzell BHC-C2YF-2/ FC8459(B)-8R (port) or /FJC8459(B)-8R (starboard) two-blade, constant-speed, fully feathering metal propeller; McCauley 82NJA-6 and L82NJA-6 three-blade propellers optional, but mandatory when de-icing equipment installed. Propeller synchrophasers standard. Fuel in four tanks in wings, with a total capacity of 485 litres (128 US gallons; 107 Imp gallons), of which 462 litres (122 US gallons; 102 Imp gallons) are usable. Oil capacity 7.5 litres (2.0 US gallons; 1.7 Imp gallons). Glass fibre engine cowlings.

ACCOMMODATION: Enclosed cabin, seating five people on individual seats with 0.25 m (10 in) centre aisle; sixth seat (second row, starboard side) optional, replacing entertainment/executive console. Dual controls standard. Pilot's storm window. Two forward-hinged doors, one on starboard side at front, the other on port side at rear. Large utility door adjacent rear cabin door provides an extra-wide opening for loading bulky items. Passenger seats easily removable to provide different seating/baggage/cargo combinations. Space for 39 kg (85 lb) baggage at rear of cabin, and for 45 kg (100 lb) in nose compartment with external door on port side. Cabin heated and ventilated. Windscreen defrosters standard.

SYSTEMS: Electrohydraulic system for landing gear actuation. Electrical system powered by two 28 V 85 A alternators; 24 V 19 Ah battery. Optional pneumatic de-icing boots on wing, tailplane and fin leading-edges. Oxygen system optional.

AVIONICS: *Comms:* Dual Garmin GNS-430 com/nav/GS/ GPS; GMA-340 audio panel with marker beacon receiver and intercom; GTX-320 transponder; Telex 100T microphone and Airman 760 headset.

Radar: Provision in nose and on instrument panel for weather radar installation.

Flight: Dual Garmin GNS-430; Meggitt ST-361 ADI; ST-180 slaved HIS; dual GI-106A VOR/LOC/GS/GPS indicators; AR-850 altitude encoder; Meggitt System 55 dual-axis FCS.

Instrumentation: Piper true airspeed indicator, magnetic compass, sensitive altimeter, ADI, HIS, rate of turn indicator, rate of climb indicator, illuminated OAT gauge, digital ammeter, annunciator panel with push-to-test, recording tachometer, dual manifold pressure gauges, three-way oil pressure, temperature and CHT gauge, dual fuel quantity gauges and vacuum gauge with warning indicator. Co-pilot's 75 mm (3 in) instrument and United 5035-P40 encoding altimeter panel optional.

EQUIPMENT: Standard equipment as listed for Saratoga II HP, plus emergency landing gear extension system, cabin dome light, pilot's storm window, sun visors with power setting table and checklist, chemical corrosion protection,

Piper PA-32R-301T Saratoga II TC *(Jane's/Paul Jackson)*   0100490

flush fuel caps and nose gear safety mirror. Piper Aire air conditioning system, built-in oxygen system and foreign certification gross weight kit optional.

DIMENSIONS, EXTERNAL:

Wing span	11.85 m (38 ft 10¾ in)
Wing chord, constant	1.60 m (5 ft 3 in)
Wing aspect ratio	7.3
Length overall	8.72 m (28 ft 7½ in)
Height overall	3.02 m (9 ft 10¾ in)
Tailplane span	4.14 m (13 ft 6¾ in)
Wheel track	3.38 m (11 ft 1 in)
Wheelbase	2.13 m (7 ft 0 in)
Propeller diameter	1.93 m (6 ft 4 in)
Distance between propeller centres	3.80 m (12 ft 5½ in)
Cabin door (stbd, fwd): Height	0.89 m (2 ft 11 in)
Width	0.91 m (3 ft 0 in)
Cabin door (port, rear): Height	0.72 m (2 ft 4½ in)
Width	0.71 m (2 ft 4 in)
Baggage door (stbd, rear): Height	0.52 m (1 ft 8½ in)
Width	0.66 m (2 ft 2 in)
Baggage door (port, fwd): Height	0.53 m (1 ft 9 in)
Width	0.61 m (2 ft 0 in)

DIMENSIONS, INTERNAL:

Cabin (incl flight deck): Length	3.15 m (10 ft 4¼ in)
Max width	1.24 m (4 ft 0¾ in)
Max height	1.07 m (3 ft 6 in)
Volume	5.5 m³ (195 cu ft)
Forward baggage compartment	0.43 m³ (15.3 cu ft)
Rear baggage compartment	0.49 m³ (17.3 cu ft)

AREAS:

Wings, gross	19.39 m² (208.7 sq ft)
Ailerons, incl tab (total)	1.17 m² (12.60 sq ft)
Trailing-edge flaps (total)	1.94 m² (20.84 sq ft)
Fin	1.14 m² (12.32 sq ft)
Rudder, incl tab	0.71 m² (7.62 sq ft)
Horizontal tail surfaces (total)	3.60 m² (38.74 sq ft)

WEIGHTS AND LOADINGS:

Weight empty, equipped	1,548 kg (3,413 lb)
Baggage capacity	83 kg (185 lb)
Max T-O weight	2,154 kg (4,750 lb)
Max ramp weight	2,165 kg (4,773 lb)
Max zero-fuel weight	2,031 kg (4,479 lb)
Max landing weight	2,047 kg (4,513 lb)
Max wing loading	111.1 kg/m² (22.76 lb/sq ft)
Max power loading	6.57 kg/kW (10.80 lb/hp)

PERFORMANCE:

Max level speed	217 kt (402 km/h; 250 mph)
Cruising speed: at 3,050 m (10,000 ft):	
at max cruise power	182 kt (337 km/h; 209 mph)
at normal cruise power	176 kt (326 km/h; 202 mph)
at 5,640 m (18,500 ft):	
at max cruise power	201 kt (372 km/h; 231 mph)
at normal cruise power	190 kt (352 km/h; 219 mph)
Stalling speed, flaps down	61 kt (113 km/h; 71 mph)
Max rate of climb at S/L	446 m (1,462 ft)/min
Rate of climb at S/L, OEI	76 m (250 ft)/min
Max certified altitude	7,620 m (25,000 ft)
Service ceiling, OEI	5,030 m (16,500 ft)
T-O run	349 m (1,145 ft)
T-O to 15 m (50 ft)	521 m (1,710 ft)
Landing from 15 m (50 ft)	665 m (2,180 ft)
Landing run	427 m (1,400 ft)
Range at long-range cruise power, incl allowance for taxi, T-O, climb and 45 min reserves:	
at 3,050 m (10,000 ft)	
	812 n miles (1,503 km; 934 miles)
at 4,575 m (15,000 ft)	
	826 n miles (1,529 km; 950 miles)
at 5,640 m (18,500 ft)	
	819 n miles (1,516 km; 942 miles)

UPDATED

PIPER PA-44-180 SEMINOLE

TYPE: Four-seat utility twin.

PROGRAMME: Prototype first flown May 1976; production version announced 21 February 1978; FAA certification 10 March 1978; two versions, Seminole, and Turbo Seminole (latter certified 29 November 1979), produced until 1982; normally aspirated Seminole restored to production in 1988, suspended in 1990 and restored again in 1995.

CUSTOMERS: Four delivered in 1995, eight in 1996, six in 1997, four in 1998, six in 1999, 11 in 2000, 62 in 2001 and 39 in the first six months of 2002.

Instrument panel of Piper Seminole
(Jane's/Paul Jackson) 0075915

Piper PA-44-180 Seminole *(Jane's/Paul Jackson)* *NEW*/0533709

Manufacture of all PA-44 variants reached 600 in 2002. Many used in twin-conversion and IFR training.

COSTS: Standard, equipped US$356,600 (2000).

DESIGN FEATURES: Twin-engined derivation of PA-28 Arrow (which see). Constant-chord low wing and T tail; leading-edge gloves inboard of engines; sweptback fin with small fillet.

Wing section NACA 65₂-415; thickness/chord ratio 15 per cent; dihedral 7°; incidence 2°; twist 3°; washout −1°.

FLYING CONTROLS: Manual. Ailerons, rudder and all-moving tailplane with full-span anti-servo tab; rudder tab; four-position manually operated flaps. Control surface movements: ailerons +23/−17°; tailplane +15/−3°; rudder ±37°; maximum flap deflection 40°.

STRUCTURE: Conventional light alloy, with semi-monocoque fuselage; single-spar wings.

LANDING GEAR: Hydraulically retractable tricycle type. Free-fall emergency extension system. Piper oleo-pneumatic shock-absorbers. Mainwheels and tyres size 6.00-6 (8 ply), with tubes. Steerable nosewheel with tyre size 5.00-5 (6 ply), with tube. Dual toe-operated high-capacity disc brakes. Heavy-duty brakes and tyres optional.

POWER PLANT: Two 134 kW (180 hp) Textron Lycoming flat-four counter-rotating engines (one O-360-A1H6 to port and one LO-360-A1H6), each driving a Hartzell two-blade constant-speed fully feathering metal propeller: HC-C2Y (K,R)-2CEUF/FC7666A-2R and /FJC7666A-2R, respectively. One bladder-type fuel tank in each engine nacelle, with total capacity of 416 litres (110 US gallons; 91.6 Imp gallons), of which 409 litres (108 US gallons; 89.9 Imp gallons) are usable. Refuelling point on upper surface of each nacelle. Oil capacity 11.5 litres (3.0 US gallons; 2.5 Imp gallons).

ACCOMMODATION: Cabin seats four in two pairs of individual seats. Dual controls standard. Emergency exit on port side. Pilot's storm window. Baggage compartment at rear of cabin, capacity 91 kg (200 lb). Accommodation heated and ventilated. Windscreen defrosters.

SYSTEMS: Electrohydraulic system for landing gear actuation and brakes. Electrical system includes two engine-driven 14 V 60 A alternators and 12 V 35 Ah battery. Janitrol combustion heater of 45,000 BTU capacity. Dual vacuum systems standard.

AVIONICS: Standard and optional Advanced Training Group (IFR) packages by Garmin, as described.

Comms: Standard: Garmin GNS-430 com/nav/IFR GPS; GTX-320 transponder; GMA-340 audio panel with marker beacon receiver and four-position intercom; Telex 100T noise-cancelling microphone and Airman 760 headset.

Flight: Standard: Garmin GNS-430 com/nav/IFR GPS; Narco AR-850 altitude reporter. Advanced IFR training group: as above, plus second GNS 430; Meggitt ST-180 slaved compass system; and RCR-650A ADF with IND-650A indicator. Optional equipment includes ST-361 ADI; ST-360 altitude select/alerter; DME-450; Meggitt System 55 dual-axis autopilot; and Meggitt electric trim (standard with autopilot).

Instrumentation: Garmin GI-106A VOR/LOC/GPS indicator. Advanced IFR training group adds second GI-106A.

EQUIPMENT: Metal instrument panel, engine hour recorders, alternate static source, heated pitot head, instrument panel lights and overhead blue lighting, avionics dimming, dome light, navigation lights, landing/taxying light, wingtip strobe lights, pilot's and co-pilot's vertically adjustable seats in fabric and vinyl, with optional lumbar support, two reclining rear passenger seats, vinyl cabin side panels, crew armrests, tinted windscreen and windows, pilot's storm window, 'Quietised' soundproofing, tiedown points, jack pads, nose gear safety mirror, external power receptacle, static discharge wicks, and Du Pont Imron polyurethane exterior paint in white base colour with two contrasting trim stripes. Metallic paint optional.

DIMENSIONS, EXTERNAL:

Wing span	11.75 m (38 ft 6½ in)
Wing aspect ratio	8.1
Length overall	8.41 m (27 ft 7¼ in)
Height overall	2.59 m (8 ft 6 in)
Tailplane span	3.05 m (10 ft 0 in)
Wheel track	3.21 m (10 ft 6½ in)
Wheelbase	2.56 m (8 ft 4¾ in)
Propeller diameter	1.88 m (6 ft 2 in)
Distance between propeller centres	3.86 m (12 ft 7¼ in)
Cabin door (stbd): Height	0.89 m (2 ft 11 in)
Width	0.91 m (3 ft 0 in)
Baggage door: Height	0.51 m (1 ft 8 in)
Width	0.56 m (1 ft 10 in)

DIMENSIONS, INTERNAL:

Cabin (instrument panel to rear bulkhead):	
Length	2.46 m (8 ft 1 in)
Max width	1.05 m (3 ft 5½ in)
Max height	1.25 m (4 ft 1 in)
Volume	3.0 m³ (106 cu ft)
Baggage compartment volume	0.74 m³ (26.0 cu ft)

AREAS:

Wings, gross	17.08 m² (183.8 sq ft)
Ailerons (total)	1.13 m² (12.12 sq ft)
Trailing-edge flaps (total)	1.36 m² (14.60 sq ft)
Fin	1.26 m² (13.56 sq ft)
Rudder, incl tab	0.76 m² (8.21 sq ft)
Horizontal tail surfaces (total)	2.27 m² (24.40 sq ft)

WEIGHTS AND LOADINGS:

Weight empty, equipped	1,179 kg (2,600 lb)
Max T-O weight	1,723 kg (3,800 lb)
Max wing loading	100.9 kg/m² (20.67 lb/sq ft)
Max power loading	6.42 kg/kW (10.55 lb/hp)

PERFORMANCE:

Max level speed	168 kt (311 km/h; 193 mph)
Cruising speed:	
at 75% power	162 kt (300 km/h; 186 mph)
at 65% power	157 kt (291 km/h; 181 mph)
Stalling speed: flaps up	57 kt (106 km/h; 66 mph) IAS
flaps down	55 kt (102 km/h; 63 mph) IAS
Max rate of climb at S/L	408 m (1,340 ft)/min
Rate of climb at S/L, OEI	65 m (212 ft)/min

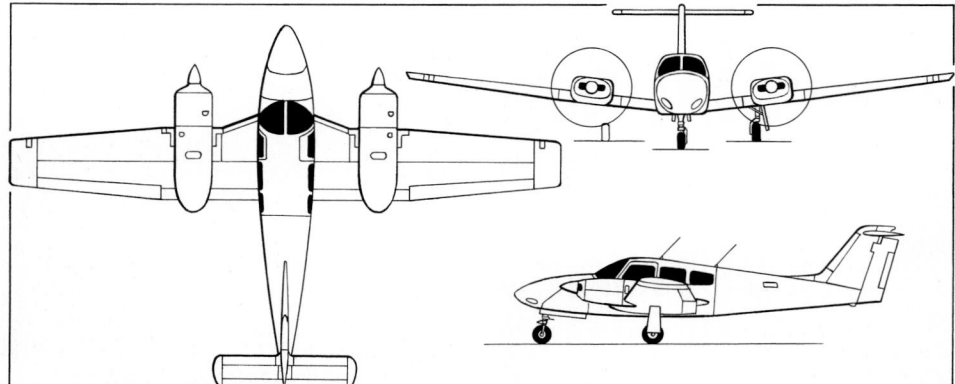

Piper PA-44-180 Seminole four-seat light twin *(Jane's/James Goulding)* 0100460

Piper PA-46-350P Malibu Mirage *(Jane's/Paul Jackson)*   *NEW*/0533710

Service ceiling	4,575 m (15,000 ft)
Service ceiling, OEI	1,155 m (3,800 ft)
T-O to 15 m (50 ft)	671 m (2,200 ft)
Landing from 15 m (50 ft)	454 m (1,490 ft)
Cruising range at 55% power, 45 min reserves	
	770 n miles (1,426 km; 886 miles)

UPDATED

PIPER PA-46-350P MALIBU MIRAGE

TYPE: Business prop.

PROGRAMME: Prototype Malibu first flew 30 November 1979; FAA certification of original PA-46-310P Malibu (TSIO-520 engine and two-blade propeller) received 27 September 1983; production deliveries began December 1983; 404 built before replaced by PA-46-350P Malibu Mirage October 1988, FAA certification having been received on 30 August 1988. Production temporarily suspended in 2000 to enable the company to concentrate on launching the turboprop Malibu Meridian; resumption then planned for late 2001.

CURRENT VERSIONS: **Malibu Mirage:** *As described.* Improvements introduced on 1995 model include pilot's heated glass windscreen, inflatable lumbar support on pilot/co-pilot's seats, colour co-ordinated control wheels and restyled interior trim, cabinetry and seats. Strengthened wing structure of Malibu Meridian (which see) introduced as standard from 1999, affording 18 kg (40 lb) increase in maximum take-off weight.

Malibu Meridian: Described separately.

JetPROP DLX: Supplemental Type Certificate awarded in 1998 to JetPROP LLC and Rocket Engineering, both of Spokane, Washington, for conversion of Malibu and Malibu Mirage with Pratt & Whitney PT6A-35 turboprop and Hartzell four-blade propeller.

CUSTOMERS: Total of 40 delivered in 1995, 57 in 1996, 54 in 1997, 55 in 1998, 61 in 1999, 63 in 2000, 10 in 2001 and 12 in the first six months of 2002, making overall total of some 534.

COSTS: Standard, equipped US$869,800 (2000).

DESIGN FEATURES: High-speed, long-range Piper single, with streamlined appearance and moderately high-aspect ratio wing, which is low mounted and tapered; mid-tailplane, also tapered; sweptback fin; wide track landing gear.

Wing section NASA 23016 at root, 23009 at tip; dihedral 4° 30′; thickness/chord ratio 16 per cent; twist 2° 57′; incidence 3° 38′.

FLYING CONTROLS: Conventional and manual. Horn-balanced elevators and rudder; stainless steel control cables. Electronic trim tab in elevator; electrically operated trailing-edge flaps. Precise Flight speed brakes standard. Control surface movements: ailerons ±18°; elevator ±23° 30′/−14° 30′; rudder 26° port/30° starboard; elevators +19/−14° 30′; maximum flap deflection 35°.

STRUCTURE: Cantilever high-aspect ratio all-metal wings; light alloy fuselage, fail-safe construction in pressurised area; light alloy tail surfaces.

LANDING GEAR: Hydraulically retractable tricycle type with single wheel on each unit; main units retract inward into wingroots, nosewheel rearward, rotating 90° to lie flat under baggage compartment. Mainwheel size 6.00-6 (8 ply), pressure 3.80 bar (55 lb/sq in); nosewheel 5.00-6 (6 ply) pressure 3.45 bar (50 lb/sq in). Toe-operated brakes.

POWER PLANT: One 261 kW (350 hp) Textron Lycoming TIO-540-AE2A turbocharged and intercooled flat-six engine, driving a Hartzell HC-13YR-1E/7890K three-blade constant-speed propeller with polished spinner. Composites (Kevlar) propeller replaced metal from 1998. Fuel system capacity 462 litres (122 US gallons; 102 Imp gallons), of which 454 litres (120 US gallons; 100 Imp gallons) are usable. Oil capacity 11.5 litres (3.0 US gallons; 2.5 Imp gallons).

ACCOMMODATION: Pilot and five passengers in pressurised, heated and air conditioned cabin; dual controls standard; front two occupants have vertical, fore-and-aft adjusting and reclining leather seats with inflatable lumbar supports, stowaway armrests, inertia reel shoulder harnesses and map-holders. Leather reclining passenger seats in club arrangement with stowaway armrests and inertia reel shoulder harnesses; unpressurised baggage compartment in nose, and pressurised space at rear of cabin. Door with integral steps on port side aft of wing.

SYSTEMS: Pressurisation, maximum differential 0.38 bar (5.5 lb/sq in), to provide a cabin altitude of 2,400 m (7,900 ft) to a height of 7,620 m (25,000 ft). Hydraulic system pressure 107 bar (1,550 lb/sq in). Dual engine-driven vacuum pumps standard. Split bus electrical system has two 28 V/70 A alternators; 24 V 10 Ah battery; full icing protection standard. Optional electrically heated pilot's windscreen and fuel management system.

AVIONICS: Standard Garmin New Generation IFR avionics package, as detailed.

Comms: Dual GNS-430 com/nav/IFR GPS with MFDs; GTX-320 transponder (second optional); GMA-340 audio panel with marker beacon receiver and intercom; Telex 100T noise-cancelling microphone and Airman 760 headset.

Radar: RDR-2000 vertical profile colour weather radar in wing-mounted pod. Goodrich WS-1000+ Stormscope optional.

Flight: Garmin GNS-430 com/nav/IFR GPS; ST-361 ADI; ST-180 HSI with slaved compass system; ST-360 altitude select/alerter; United 5035P-P40 altitude encoder; Meggitt System 55 three-axis autopilot with electric trim, turn indicator and yaw damper; co-pilot's longitudinal electric trim button. Options include single or dual RCR-650A ADF with IND-650A indicators; DME 450;

Honeywell KI 229 RMI (exchange for IND-650A): and KRA-10A radar altimeter.

Instrumentation: Dual MFDs with GNS-430 installation; dual GI-106A VOR/LOC/GS/GPS indicators.

EQUIPMENT: Standard equipment includes heated lift detector; stall warning computer and horn; digital ammeter/voltmeter; six-channel CHT monitoring system with selectable cylinder readout; gyro air filter; alternate static source; heated pitot head; nosewheel light; wingtip taxying light; navigation lights, wingtip strobe lights; cockpit dome lights; solid-state dimming landing gear position and instrument panel lights; seven cabin overhead lights; PPG pilot's heated windscreen; windscreen defrosters; pilot's opening storm window; pilot's relief tube; supplemental electric heater; emergency oxygen system; leather sidewalls; stowaway executive writing table; forward refreshment/entertainment centres; super 'Quietised' soundproofing with inner passenger windows; halon fire extinguisher; Truax static discharge wicks; chemical corrosion protection; and Du Pont Imron polyurethane exterior paint in single- or two-tone base colour with graphics striping in choice of three trim colours. Optional equipment includes 'Infinity' paint scheme and stainless steel cowling fasteners.

DIMENSIONS, EXTERNAL:

Wing span	13.11 m (43 ft 0 in)
Wing aspect ratio	10.6
Length overall	8.81 m (28 ft 10¾ in)
Height overall	3.44 m (11 ft 3½ in)
Tailplane span	4.42 m (14 ft 6 in)
Wheel track	3.75 m (12 ft 3½ in)
Wheelbase	2.44 m (8 ft 0 in)
Propeller diameter	2.03 m (6 ft 8 in)
Passenger door (port, rear): Height	1.17 m (3 ft 10 in)
Width	0.61 m (2 ft 0 in)
Baggage door (port, nose): Height	0.58 m (1 ft 11 in)
Width	0.48 m (1 ft 7 in)

DIMENSIONS, INTERNAL:

Cabin (instrument panel to rear pressure bulkhead):	
Length	3.76 m (12 ft 4 in)
Max width	1.26 m (4 ft 1½ in)
Max height	1.19 m (3 ft 11 in)
Baggage compartment volume:	
nose	0.37 m³ (13.0 cu ft)
rear cabin	0.57 m³ (20.0 cu ft)

AREAS:

Wings, gross	16.26 m² (175.0 sq ft)

WEIGHTS AND LOADINGS:

Weight empty, equipped	1,416 kg (3,121 lb)
Baggage capacity: nose	45 kg (100 lb)
rear cabin	45 kg (100 lb)
Max T-O weight	1,968 kg (4,340 lb)
Max ramp weight	1,976 kg (4,358 lb)
Max landing and max zero-fuel weight	
	1,870 kg (4,123 lb)

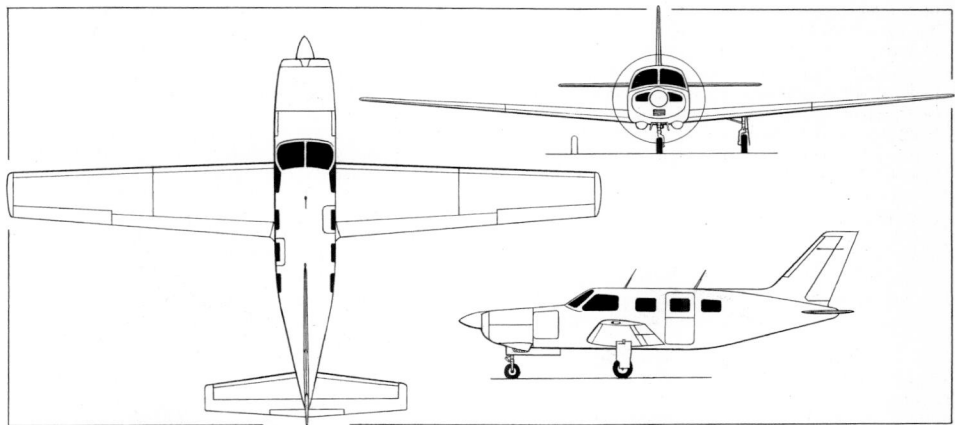

Piper PA-46-350P Malibu Mirage (Textron Lycoming TIO-540-AE2A) *(Jane's/Dennis Punnett)*

JetPROP DLX turboprop conversion of Piper Malibu Mirage *(Jane's/Paul Jackson)*   *NEW*/0533711

Piper Malibu Meridian turboprop (*Jane's/Paul Jackson*)   *NEW*/0533712

Piper Malibu Meridian EFIS instrument panel 0100472

Max wing loading	121.1 kg/m² (24.80 lb/sq ft)
Max power loading	7.55 kg/kW (12.40 lb/hp)

PERFORMANCE:
Max level speed at mid-cruise weight
    220 kt (407 km/h; 253 mph)
Cruising speed at optimum altitude, mid-cruise weight,
    high-speed cruise power   213 kt (394 km/h; 245 mph)
Stalling speed, flaps and wheels down
    58 kt (108 km/h; 67 mph)

Max rate of climb at S/L	372 m (1,220 ft)/min
Max certified altitude	7,620 m (25,000 ft)
T-O run	332 m (1,090 ft)
T-O to 15 m (50 ft)	637 m (2,090 ft)
Landing from 15 m (50 ft)	599 m (1,965 ft)
Landing run	309 m (1,015 ft)

Range with max fuel, allowances for start, T-O, climb
    and descent, plus 45 min reserves, at optimum altitude,
    normal cruise power
    1,065 n miles (1,972 km; 1,225 miles)
    UPDATED

PIPER PA-46-500TP MALIBU MERIDIAN

TYPE: Business turboprop.

PROGRAMME: Launched at 1997 National Business Aviation Association Convention in Dallas, Texas, where a full-scale fuselage mockup was displayed; prototype (N400PT, converted from second Malibu Mirage) rolled out 13 August 1998; first flight 21 August 1998. Static test airframe and three further certification flight test aircraft (N403MM, first flown in July 1999, N402MM first flown 27 August 1999 and N401MM first flown in September 1999) built on production tooling. N401MM was dedicated to stability and autopilot testing and ice shape testing; N402MM, first with production standard interior and exterior paint finish, conducted performance and avionics testing and certification for flight into known icing, as well as serving as marketing demonstrator, N403MM was responsible for systems and power plant testing, high-speed flight testing and flutter testing. Public debut (N402MM) at the NBAA Convention in Atlanta, Georgia, in October 1999. First flight of production aircraft (c/n 003/N375RD) 30 June 2000. FAR Pt 23 certification achieved 27 September 2000, followed by UK CAA approval on 21 June 2001.

Increase in maximum take-off weight to 2,310 kg (5,092 lb), resulting in 15 per cent increase in useful load, scheduled for certification by end of 2002; modifications include strengthening of airframe, modified stall strips and addition of vortex generators to the wings and undersurface of tailplane.

CUSTOMERS: Total of 239 orders held by October 2000. First delivery (N375RD) to Richard Dumais of Richardson,

Piper Malibu Meridian cabin with optional entertainment centre 0100470

Texas, in November 2000. Total of 98 delivered in 2001, and 35 projected for 2002, of which 11 had been delivered by 30 June.

COSTS: US$1.6 million typically equipped (2002).

Description generally as for Malibu Mirage, except that below.

DESIGN FEATURES: Strengthened wing, incorporating wingroot leading-edge gloves to increase area and reduce stalling speed; strengthened tail surfaces; and 37 per cent increase in area of horizontal stabiliser.

FLYING CONTROLS: Deflections as for Malibu Mirage except flaps 36°. Flight-adjustable trim tab in rudder. Ground-adjustable tab on starboard aileron.

POWER PLANT: One Pratt & Whitney Canada PT6A-42A turboprop, thermodynamic rating 901 kW (1,209 shp), flat rated at 373 kW (500 shp) maximum continuous, driving a Hartzell HC-E4N-3Q/E8501B-3.5 four-blade constant-speed reversible propeller. Fuel capacity 655 litres (173 US gallons; 144 Imp gallons), of which 644 litres (170 US gallons; 141.5 Imp gallons) are usable. Oil capacity 11 litres (3.0 US gallons; 2.5 Imp gallons).

ACCOMMODATION: Pilot and five passengers in standard club layout; passenger capacity reduced to four with optional entertainment centre comprising beverage cooler, storage cabinets, AM/FM/CD stereo and provision for VCR with flat-panel monitor. Fully automatic bleed-air conditioning and temperature control systems.

SYSTEMS: Upgraded electrical system, with 200 A starter/generator and 130 A standby alternator.

AVIONICS: Standard Garmin International and Meggitt package with dual Garmin GNS 530 nav/com/GPS as core system.
    Comms: Garmin transceiver with 8.33 kHz spacing; GMA-340 audio panel; GTX-327 transponder.
    Radar: Honeywell RDR 2000 colour weather radar.
    Flight: Meggitt System 550 autopilot/flight director, fully coupled to GNS 530; air data, attitude, heading and reference system (ADAHRS) provides digital readout of pitch and roll attitudes, roll and yaw rates, altitude, rate of altitude change, airspeed and heading.
    Instrumentation: Meggitt Avionics engine instrument display system (EIDS) standard. Optional dual MAGIC EFDS fit comprising pilot's and co-pilot's colour flat panel

LCD primary flight display and navigation display driven by dual air data attitude heading reference systems (ADAHRS).

DIMENSIONS, EXTERNAL: As for Malibu Mirage except:
Length overall	9.02 m (29 ft 7¼ in)
Height overall	3.45 m (11 ft 4 in)
Propeller diameter	2.095 m (6 ft 10½ in)

DIMENSIONS, INTERNAL: As for Malibu Mirage

AREAS:
Wings, gross	17.00 m² (183.0 sq ft)

WEIGHTS AND LOADINGS:
Weight empty, equipped	1,471 kg (3,243 lb)
Baggage capacity	45 kg (100 lb)
Max T-O weight: current	2,200 kg (4,850 lb)
projected	2,310 kg (5,092 lb)
Max ramp weight	2,219 kg (4,892 lb)
Max wing loading: current	129.4 kg/m² (26.50 lb/sq ft)
projected	135.8 kg/m² (27.82 lb/sq ft)
Max power loading: current	5.90 kg/kW (9.7 lb/shp)
projected	6.20 kg/kW (10.2 lb/shp)

PERFORMANCE:
Max cruising speed at 9,150 m (30,000 ft) at mid-cruise
    weight   262 kt (485 km/h; 302 mph)
Max rate of climb at S/L	530 m (1,739 ft)/min
Max certified altitude	9,150 m (30,000 ft)
T-O run	518 m (1,698 ft)
T-O to 15 m (50 ft)	776 m (2,545 ft)
Landing from 15 m (50 ft)	644 m (2,114 ft)
Landing run	308 m (1,009 ft)

Max range at 9,150 m (30,000 ft), max cruise power,
    mid-cruise weight, 45 min reserves
    1,018 n miles (1,885 km; 1,171 miles)
Endurance   4 h 48 min
    UPDATED

PIPER NEW PROJECTS

At the NBAA Convention at New Orleans in October 2000, New Piper Aircraft CEO Chuck Suma announced that the company is "actively developing new platforms at the lighter end of our product line" that will "embrace advanced propulsion technologies, aerodynamic refinements and cutting edge avionics capabilities". Among these are expected to be FADEC-equipped piston engines across the company's range and a new model to fill the market gap between the Seneca and Malibu Mirage and Meridian. New Piper would also consider a joint venture to develop a new turboprop aircraft bigger than the Meridian, Suma said. No further details had been revealed by September 2002.
    UPDATED

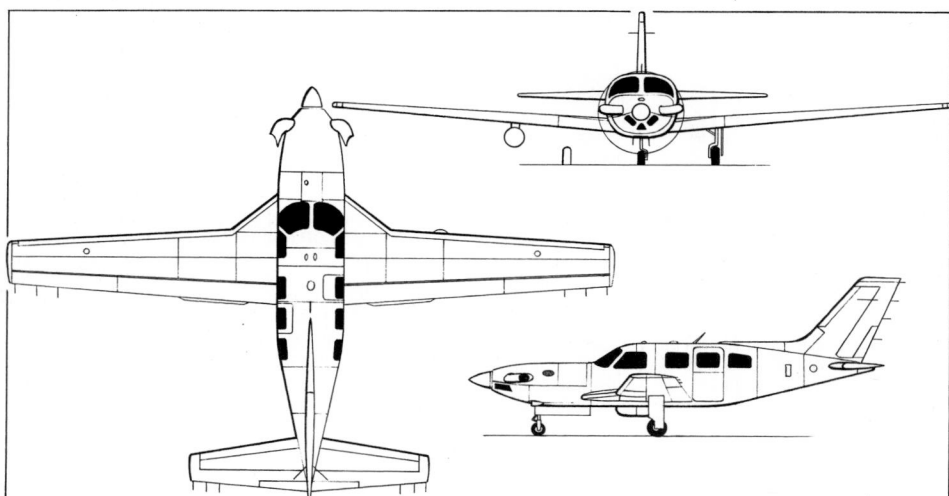

Piper Malibu Meridian (Pratt & Whitney Canada PT6A turboprop) (*Jane's/James Goulding*)   0075961

PRECISION TECH

PRECISION TECH AIRCRAFT INC

155E Hangar 33, Highway 61 SE, Cartersville, Georgia 30120

Tel: (+1 770) 607 40 09
Fax: (+1 770) 386 63 55
e-mail: info@fergy.net
Web: http://www.fergy.net
PRESIDENT: Lance McAfee

Precision Tech took over manufacture and distribution rights for the Ferguson F-II from its designer during 1999.
    VERIFIED

PRECISION TECH F-IIB FERGY

TYPE: Side-by-side ultralight kitbuilt.

PROGRAMME: Designed by Bill Ferguson; awarded Best New Design at Sun 'n' Fun 91.

CURRENT VERSIONS: **Ferguson F-II**: Initial version, no longer produced.

F-IIB Fergy: Current production model, *as described*.

CUSTOMERS: At least 40 sold by mid-2001; 19 flying by mid-2002.

COSTS: Kit US$11,500 (2003).

DESIGN FEATURES: Pod-and-boom configuration; high wing and low tailplane. Wings and tail fold for transportation. Quoted build time 300-350 hours.

FLYING CONTROLS: Conventional and manual. Dual controls. Full-span flaperons deflect to −15°. Deflections: aileron +27/−23°, rudder ±35°.

STRUCTURE: Fuselage of 4130 chromoly steel tubing with glass fibre cowling; 6063-T6 aluminium tube tailboom and wing spars; wing ribs and tail 6061-T6 aluminium tubing with chromoly steel brackets. Fabric covering; aluminium wing struts. Wing dihedral 0°45′.

LANDING GEAR: Tailwheel type; fixed. Brakes on mainwheels; optional speed fairings. Main wheels 6.00-6; larger tyres can be fitted. Options include skis and straight or amphibious floats.

POWER PLANT: One 37.0 kW (49.6 hp) Rotax 503UL-2V engine mounted on rear of wing driving two-blade (optional three-blade) Warp Drive propeller mounted above wing; engines in the range 33.6 to 48.5 kW (45 to 65 hp) recommended. Fuel capacity 47.3 litres (12.5 US gallons; 10.4 Imp gallons).

DIMENSIONS, EXTERNAL:

Wing span	8.99 m (29 ft 6 in)
Length overall	6.71 m (22 ft 0 in)
Height overall	1.73 m (5 ft 8 in)

AREAS:

Wings, gross	13.01 m² (140.0 sq ft)

WEIGHTS AND LOADINGS:

Weight empty	181 kg (400 lb)
Max T-O weight	453 kg (1,000 lb)

PERFORMANCE (48.5 kW; 65 hp engine):

Never-exceed speed (VNE):	86 kt (161 km/h; 100 mph)
Normal cruising speed range	
	43-70 kt (80-129 km/h; 50-80 mph)
Stalling speed, power off:	
flaperons up	30 kt (55 km/h; 34 mph)
flaperons down	25 kt (45 km/h; 28 mph)
Max rate of climb at S/L	366 m (1,200 ft)/min
T-O run	53 m (175 ft)
Landing run	61 m (200 ft)
Range	191 n miles (354 km; 220 miles)

UPDATED

Precision Tech F-IIB Fergy II kitplane (*Jane's/Paul Jackson*)   0089508

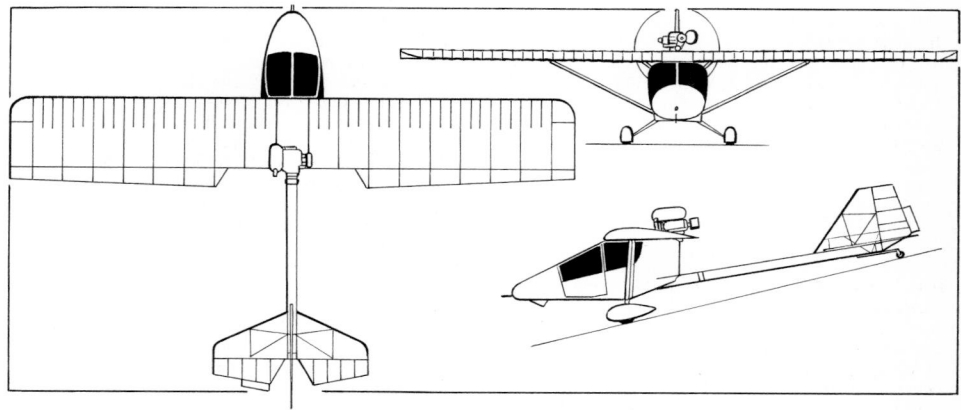

General arrangement of Precision Tech Fergy (*Jane's/James Goulding*)   0110973

PROSPORT

PROSPORT AVIATION

PO Box 3428, Tega Cay, South Carolina 29708
Tel: (+1 803) 802 01 98 and (+1 800) 850 37 08
Fax: (+1 803) 802 01 96
e-mail: eagle@flashlink.net
Web: http://www.sportlite103.com

Prosport has acquired the rights to the Freebird from Freebird, and also markets the Sportlite 103 ultralight.

VERIFIED

PROSPORT FREEBIRD CLASSIC

TYPE: Side-by-side kitbuilt.

PROGRAMME: Launched 1993 by Freebird. Meets FAR Pt 103 requirements.

CUSTOMERS: Over 100 under construction and over 50 flying by mid-2001.

COSTS: US$9,995 (2003) excluding engine; ready to fly models from US$16,000.

DESIGN FEATURES: Wings removable for storage. Quoted build time 125 hours.

FLYING CONTROLS: Conventional and manual; flaps. Large horn balance on rudder.

STRUCTURE: Fabric-covered aluminium tubing and composites nose skin. Strut-braced wings and tailplane. Wing-to-tailplane bracing struts augment slim rear fuselage.

LANDING GEAR: Tricycle type; fixed. Semi-recessed nosewheel.

POWER PLANT: One 47.8 kW (64.1 hp) Rotax 582 UL-2V two-cylinder engine driving a wooden fixed-pitch pusher propeller. Fuel capacity 37.8 litres (10.0 US gallons; 8.3 Imp gallons).

Prosport Freebird II (*Geoffrey P Jones*)   0051787

ACCOMMODATION: Optional enclosed cockpit with provision for heater.

SYSTEMS: Optional electric starter.

DIMENSIONS, EXTERNAL:

Wing span	8.84 m (29 ft 0 in)
Length overall	5.33 m (17 ft 6 in)
Height overall	1.57 m (5 ft 2 in)

AREAS:

Wings, gross	12.26 m² (132.0 sq ft)

WEIGHTS AND LOADINGS:

Weight empty	175 kg (385 lb)
Max T-O weight	394 kg (870 lb)

PERFORMANCE (enclosed cockpit):

Max level speed	70 kt (129 km/h; 80 mph)
Normal cruising speed at 75% power	
	61 kt (113 km/h; 70 mph)
Stalling speed, flaps down	28 kt (52 km/h; 32 mph)
Max rate of climb at S/L	244 m (800 ft)/min
T-O run	38 m (125 ft)
Landing run	46 m (150 ft)
Range at 75% power	173 n miles (321 km; 200 miles)
g limits	+6/−3

UPDATED

PULSAR

PULSAR AIRCRAFT CORPORATION

4233 North Santa Anita Avenue, Suite 5, El Monte Airport, El Monte, California 91731

Tel: (+1 626) 443 10 19
Fax: (+1 626) 443 13 11
e-mail: info@pulsaraircraft.com
Web: http://www.pulsaraircraft.com
PRESIDENT: Solly Melyon

Pulsar Aircraft Corporation acquired rights to Pulsar series of light aircraft from SkyStar in July 1999 and announced the Super Pulsar 100, based on the Pulsar III, at that time. It has also obtained rights to the former Tri-R KIS and KIS Cruiser, marketing of which began in 2001.

VERIFIED

PULSAR SUPER PULSAR 100

TYPE: Side-by-side kitbuilt.

PROGRAMME: Designed by Mark Brown. Prototype first flew 3 April 1988 as two-seat version of Aero Designs Star-Lite (itself first flown in 1983). Originally with 47.8 kW (64.1 hp) Rotax 582 engine. Until suspended in 1999, in favour of Super Pulsar, parallel main production versions were Pulsar II, with 59.6 kW (79.9 hp) Rotax 912 UL and Pulsar III with 84.6 kW (113.4 hp) turbocharged Rotax 914, giving improved performance. Larger canopy and redesigned cowling with chin intake; 100 per cent mass-balanced controls. Nose- and tailwheel versions of both were available. Programme acquired by SkyStar in late 1996 and by Pulsar in 1999.

CURRENT VERSIONS: **Super Pulsar 100:** Refined version of Pulsar III with repositioned tail unit for improved stability and higher, wider cabin. Prototype (N601SP) first flew August 2000 but damaged 20 August 2000; flight testing resumed 2 April 2001 and aircraft made public debut at Sun 'n' Fun 13 April 2001; by April 2002, 228 hours had been flown by prototype. Tailwheel version under consideration.

Description applies to Super Pulsar 100.

Wega: Version of Pulsar II-914 Turbo, certified to JAR-VLA with 600 kg (1,323 lb) maximum T-O weight, marketed by HK Aircraft Technology in Germany, which see.

Pulsar XP: Returned to production in 2001 as kit; cost US$19,950; described in 1997-98 *Jane's*

CUSTOMERS: More than 250 of all Pulsar variants flying by mid-2002, with a further 300 then under construction.

COSTS: Basic Super Pulsar kit (no engine) US$26,950 (2003); fast-build options add US$7,500.

DESIGN FEATURES: Highly streamlined low-wing configuration, with sweptback fin. Designed for ease of home construction with premoulded components where possible. Wings removable in about 15 minutes for mounting on trailer. Quoted build time 1,000 hours, or 600 hours for 'Super' kit.

FLYING CONTROLS: Conventional and manual; 60 per cent mass-balanced surfaces operated by rods and cables.

STRUCTURE: GFRP with carbon fibre and foam cores. Wing skins of composites.

LANDING GEAR: Tricycle type; fixed. Hydraulic brakes.

POWER PLANT: One 88.0 kW (118 hp) Pulsar Aeromaxx flat-four piston engine recommended; alternatively 93 kW (125 hp) Teledyne Continental IO-240, 59.7 kW (80 hp) Jabiru 2200 or 89 kW (120 hp) Jabiru 3300. Fuel capacity 106 litres (28.0 US gallons; 23.3 Imp gallons).

DIMENSIONS, EXTERNAL:

Wing span	7.62 m (25 ft 0 in)
Wing aspect ratio	7.8
Length overall	6.10 m (20 ft 0 in)
Height overall	1.80 m (5 ft 11 in)

DIMENSIONS, INTERNAL:

Cabin: Max width	1.09 m (3 ft 7 in)
Max height	1.14 m (3 ft 9 in)

AREAS:

Wings, gross	7.43 m² (80.0 sq ft)

WEIGHTS AND LOADINGS (Aeromaxx):

Weight empty, equipped	340 kg (750 lb)
Baggage capacity	27 kg (60 lb)
Max T-O weight	635 kg (1,400 lb)
Max wing loading	85.4 kg/m² (17.50 lb/sq ft)
Max power loading	7.22 kg/kW (11.86 lb/hp)

PERFORMANCE (Aeromaxx, estimated):

Never-exceed speed (VNE)	191 kt (354 km/h; 220 mph)
Max level speed at S/L	165 kt (305 km/h; 190 mph)
Stalling speed, power off	46 kt (84 km/h; 52 mph)
Max rate of climb at S/L, solo	610 m (2,000 ft)/min
T-O run	152 m (500 ft)
Landing run	244 m (800 ft)
Range	782 n miles (1,448 km; 900 miles)
g limits	+6/−4

UPDATED

PULSAR SPORT 150

TYPE: Side-by-side kitbuilt.

PROGRAMME: First flown 1991 and initially known as Tri-R KIS; name derived from 'keep it simple'. Programme sold to Pulsar in 1999; new owner modifying design with revised canopy (rear-hinged or sliding) and empennage.

CURRENT VERSIONS: **TR-1:** Tricycle landing gear version, *as described.*

TD: Tailwheel version, introduced 1992; powered by 88.0 kW (118 hp) Textron Lycoming O-235-C1B; seven flying by end of 1996.

CUSTOMERS: At least 31 (24 tricycle and seven tailwheel) flying by 2002.

COSTS: Basic kit US$22,950, without engine, avionics, propeller, spinner, upholstery, battery, instruments and paint (2001).

DESIGN FEATURES: Conventional low-wing monoplane of composites construction. Upturned wingtips and sharply tapered fin. Quoted build time 1,000 hours.

FLYING CONTROLS: Conventional and manual. Horn-balanced rudder and elevators, latter with trim tab.

STRUCTURE: Constructed of high-temperature epoxy pre-impregnated GFRP/CFRP premoulded components, with either Divinycell or honeycomb core. All metal

Aero Designs Pulsar assembled from a kit by a Danish builder *(Jane's/Paul Jackson)*   NEW/0533713

Prototype Super Pulsar *(Jane's/Paul Jackson)*   NEW/0533714

components prewelded or premachined. Aerofoil NACA 63215 slightly modified.

LANDING GEAR: Non-retractable, with single wheels and cantilever main legs; optional speed fairings. Tricycle or tailwheel versions available. Matco wheels and brakes; McCreary 5.00-5 tyres. Tricycle version has steerable nosewheel.

POWER PLANT: One piston engine driving a two- or three-blade propeller: typically 88 kW (118 hp) Textron Lycoming O-235-C1B; 93 kW (125 hp) Teledyne Continental IO-240; and 119 kW (160 hp) Textron Lycoming O-320. Alternatives include 59.7 kW (80 hp) Limbach L 2000 and 74.6 kW (100 hp) Fire Wall Forward CAM 100 four-cylinder converted Honda motorcar engine. Fuel capacity 76 litres (20.0 US gallons; 16.7 Imp gallons); optionally 129 litres (34.0 US gallons; 28.3 Imp gallons) for Lycoming- and Continental-powered versions.

ACCOMMODATION: Two, side by side. Upward-hinged door each side.

DIMENSIONS, EXTERNAL:

Wing span	7.01 m (23 ft 0 in)
Wing chord	1.17 m (3 ft 10 in)
Wing aspect ratio	6.0
Length overall	6.71 m (22 ft 0 in)
Height overall: TR-1	1.83 m (6 ft 0 in)
TD	1.89 m (6 ft 2½ in)
Propeller diameter	1.42 m (4 ft 8 in)

DIMENSIONS, INTERNAL:

Cabin: Length	1.65 m (5 ft 5 in)
Max width	1.07 m (3 ft 6 in)
Max height	1.02 m (3 ft 4 in)

AREAS:

Wings, gross	8.18 m² (88.0 sq ft)

WEIGHTS AND LOADINGS (L: Lycoming O-320 engine, C: Continental engine):

Weight empty: L	386 kg (850 lb)
C	367 kg (810 lb)
Max T-O weight: L, C	658 kg (1,450 lb)

Max wing loading: L, C	80.4 kg/m² (16.48 lb/sq ft)
Max power loading: L	5.52 kg/kW (9.06 lb/hp)
C	7.06 kg/kW (11.60 lb/hp)

PERFORMANCE (L, C, as above):

Never-exceed speed (VNE)	191 kt (354 km/h; 220 mph)
Max level speed: L	182 kt (338 km/h; 210 mph)
C	165 kt (306 km/h; 190 mph)
Cruising speed at 75% power:	
L	161 kt (298 km/h; 185 mph)
C	148 kt (274 km/h; 170 mph)
Stalling speed: L, C	51 kt (94 km/h; 58 mph)
Max rate of climb at S/L: L	457 m (1,500 ft)/min
C	366 m (1,200 ft)/min
Service ceiling: C	5,180 m (17,000 ft)
T-O run: L, C	229 m (750 ft)
Landing run: C	396 m (1,300 ft)
Range at cruising speed, standard tankage:	
L	521 n miles (965 km; 600 miles)
C	695 n miles (1,287 km; 800 miles)
g limits	+4.4/−2.2

UPDATED

PULSAR CRUISER and SUPER CRUISER

TYPE: Four-seat kitbuilt.

PROGRAMME: First flown 1994, when known as Tri-R Cruiser TR-4. Design purchased by Pulsar in 1999.

CURRENT VERSIONS: **Cruiser:** Baseline model; being replaced by Super Cruiser.

Super Cruiser: Higher-powered version with minor redesign to dorsal fin to give larger area. First aircraft to this standard (N98WG) flew 8 March 1999.

CUSTOMERS: More than 60 kits sold and 18 flying by mid-2002.

COSTS: Kit US$34,500 (2003); fast-build options add US$7,000.

DESIGN FEATURES: Larger version of the KIS TR-1. Quoted build time 1,500 hours.

Description generally as for Sport 150 except following.

Tri-R KIS TD (now known as Pulsar Sport 150) two-seat kitplane *(Jane's/Paul Jackson)*   NEW/0533641

POWER PLANT: *Cruiser:* One 134 kW (180 hp) Textron Lycoming O-360 flat-six, driving a two-blade, fixed-pitch propeller.

Super Cruiser: One 157 kW (210 hp) six-cylinder Teledyne Continental IO-360, driving a two-blade, constant-speed propeller.

Usable fuel capacity 189 litres (50.0 US gallons; 41.6 Imp gallons).

DIMENSIONS, EXTERNAL:
Wing span	8.83 m (29 ft 0 in)
Length overall	7.77 m (25 ft 6 in)
Height overall	2.29 m (7 ft 6 in)

DIMENSIONS, INTERNAL:
Cabin: Length	1.98 m (6 ft 6 in)
Max width	1.12 m (3 ft 8 in)
Max height	1.17 m (3 ft 10 in)

AREAS:
Wings, gross	1.25 m² (135.0 sq ft)

WEIGHTS AND LOADINGS:
Weight empty: Cruiser	544 kg (1,200 lb)
Super Cruiser	590 kg (1,300 lb)
Baggage capacity	29 kg (65 lb)
Max T-O weight:	
Cruiser, Super Cruiser	1,088 kg (2,400 lb)
Max wing loading:	
Cruiser, Super Cruiser	86.8 kg/m² (17.78 lb/sq ft)
Max power loading:	
Cruiser	8.12 kg/kW (13.33 lb/hp)
Super Cruiser	6.96 kg/kW (11.43 lb/hp)

Pulsar Super Cruiser four-seat tourer (*Jane's/Paul Jackson*) NEW/0533670

PERFORMANCE:
Never-exceed speed (VNE)	186 kt (346 km/h; 215 mph)
Cruising speed: at 75% power:	
Cruiser	161 kt (298 km/h; 185 mph)
Super Cruiser	174 kt (322 km/h; 200 mph)
at 60% power: Cruiser	152 kt (282 km/h; 175 mph)
Super Cruiser	156 kt (290 km/h; 180 mph)
Stalling speed	48 kt (89 km/h; 55 mph)
Max rate of climb at S/L: Cruiser	305 m (1,000 ft)/min
Super Cruiser	457 m (1,500 ft)/min
Service ceiling	5,180 m (17,000 ft)
T-O run: Cruiser	305 m (1,000 ft)
Super Cruiser	274 m (900 ft)
T-O to 15 m (50 ft): Cruiser	366 m (1,200 ft)
Super Cruiser	335 m (1,100 ft)
Landing run: Cruiser	395 m (1,300 ft)
Super Cruiser	411 m (1,350 ft)
Range:	
Cruiser	more than 782 n miles (1,448 km; 900 miles)
Super Cruiser	more than 695 n miles (1,287 km; 799 miles)

UPDATED

QUIKKIT

RAINBOW FLYERS INC, QUIKKIT DIVISION
9002 Summer Glen, Dallas, Texas 75243-7445
Tel/Fax: (+1 214) 349 04 62
e-mail: quikkit@glassgoose.com
Web: http://www.glassgoose.com
PRESIDENT: Thomas W Scott
PRODUCTION MANAGER: Robert Giddens

WORKS:
Lakeview Airport, Lake Dallas, Texas 75065

Quikkit was initially involved in the production of parts and modification kits for the Aero Composites Sea Hawker (see 1992-93 *Jane's*); the Glass Goose, while visually similar, is a new design.

VERIFIED

QUIKKIT GLASS GOOSE
TYPE: Two-seat amphibian kitbuilt.
PROGRAMME: Extensively redesigned Aero Composite Sea Hawker (see 1992-93 *Jane's*), obviating deficiencies of that discontinued aircraft. Modifications initially applied to Tom Scott's Sea Hawker and later marketed for incorporation by other Sea Hawker operators, although no longer available.

Prototype new-build Glass Goose flew 1996, but lost to flaperon flutter. Second prototype, in definitive configuration, first flew (N96GG) 20 March 1999. Production undertaken on Sea Hawker tools and jigs, acquired by Quikkit in 1993.
CUSTOMERS: Total 42 under construction, of which five flying, by April 2002. At least six others extant by 2000, reflecting Sea Hawker conversions.
COSTS: US$29,500, excluding engine, avionics and equipment (2002).
DESIGN FEATURES: Pusher, cantilever biplane, with single-step hull, sponsons and underwing pods for retractable main landing gear. Dished upper rear fuselage; mid-mounted tailplane. Wing positive stagger of 20 cm (8 in); upswept lower wingtips; downswept upper wingtips.

Modifications to Sea Hawker design include extended (91 cm; 3 ft 0 in) upper wing span; sponsons of aerofoil

Second prototype Quikkit Glass Goose (*Jane's/Paul Jackson*) 0110733

section (replacing wingtip floats); and extended pylon chord, with associated fillets, vortex generators and augmentation slots in reshaped cowling to eliminate propeller vibration.
FLYING CONTROLS: Conventional and manual, with drooping ailerons on all wings, max deflection +10/−40°. Actuation by pushrods and cables.
STRUCTURE: Glass fibre, with Kevlar reinforcement. Forward wing spars of glass fibre beam type, with carbon caps, at 27 per cent chord; glass fibre C-type rear spar at 85 per cent chord; glass fibre/foam core sandwich ribs. Pre-moulded flaperons aid construction. Intersecting fuselage bulkheads for additional strength. Quoted build time 1,000 hours.
LANDING GEAR: Retractable tricycle type for land operation. Two, side-by-side Cheng Shin 11×4.00-5 tyres on each main unit, underside of landing gear stowage pod also being forward-hinged leg/door. Mainwheel brakes.

Rearward retracting, stainless steel, fully castoring nose leg, with 4.00-5 tyre, extends against spring-loaded single door. Electrohydraulic actuation. Ventral rudder, cable operated, for water manoeuvring.

POWER PLANT: One 119 kW (160 hp) Textron Lycoming O-320-B2B flat-four driving a Warp Drive, four-blade, ground-adjustable pitch propeller. Four equal-size, integral fuel tanks in wings (upper pair main; lower pair for long range), total capacity 265 litres (70.0 US gallons; 58.3 Imp gallons). No crossfeed. Optional provision for total of 455 litres (120 US gallons; 100 Imp gallons) involves fuselage tank.
ACCOMMODATION: Two, side by side, with baggage stowage areas behind seats and between lower wings.

DIMENSIONS, EXTERNAL:
Wing span: upper	8.23 m (27 ft 0 in)
lower	7.62 m (25 ft 0 in)
Wing chord: at root	0.84 m (2 ft 9 in)
at tip	0.65 m (2 ft 1½ in)
Length overall	5.94 m (19 ft 6 in)
Height overall	2.29 m (7 ft 6 in)
Tailplane span	2.44 m (8 ft 0 in)

DIMENSIONS, INTERNAL:
Cockpit max width	1.07 m (3 ft 6 in)
Baggage volume: cockpit	0.24 m³ (8.5 cu ft)
between wings	0.45 m³ (16.0 cu ft)

AREAS:
Wings, gross	12.17 m² (131.0 sq ft)
Sponsons (lifting)	1.44 m² (15.50 sq ft)
Flaperons (total)	1.86 m² (20.0 sq ft)

WEIGHTS AND LOADINGS:
Weight empty	454 kg (1,000 lb)
Max T-O weight	816 kg (1,800 lb)
Max wing loading	67.1 kg/m² (13.74 lb/sq ft)
Max power loading	6.85 kg/kW (11.25 lb/hp)

PERFORMANCE:
Max level speed	139 kt (257 km/h; 160 mph)
Max cruising speed at 75% power	122 kt (225 km/h; 140 mph)
Stalling speed: flaps up	44 kt (81 km/h; 50 mph)
flaps down	37 kt (68 km/h; 42 mph)
Max rate of climb at S/L	366 m (1,200 ft)/min
Service ceiling	3,660 m (12,000 ft)
T-O run	244 m (800 ft)
Landing run	274 m (900 ft)
Max range, 30 min reserves	955 n miles (1,770 km; 1,100 miles)
g limits	±9

UPDATED

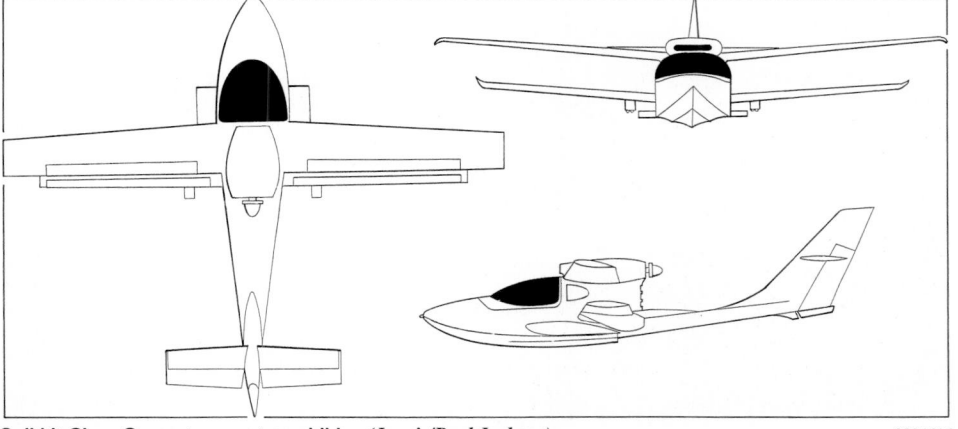

Quikkit Glass Goose two-seat amphibian (*Jane's/Paul Jackson*) 0084616

RANS

RANS INC

4600 Highway 183 Alternate, Hays, Kansas 67601
Tel: (+1 785) 625 63 46
Fax: (+1 785) 625 27 95
e-mail: rans@media-net.net
Web: http://www.rans.com
PRESIDENT AND CEO: Randy J Schlitter
VICE-PRESIDENT: Paula Schlitter

Earlier designs include S-4 Coyote, S-6 Coyote II, S-9 Chaos, S-10 Sakota, S-11 Pursuit and S-14 Airaile. Rans delivered its 3,000th kitplane during 1998 and in 1999 achieved FAA '51 per cent rule' certification of the S-7, S-12 and S-16 kits. Latest products are the related S-17 and S-18 single- and two-seat ultralights.

UPDATED

Rans S-7C Courier tandem-seat kitbuilt *(Jane's/Paul Jackson)*   NEW/0526970

RANS S-7 COURIER

TYPE: Tandem-seat lightplane/kitbuilt.
PROGRAMME: Prototype first flew November 1985; in production from 1986.
CURRENT VERSIONS: **S-7:** Kitbuilt version, *as described.*

S-7C: Factory-built version; changes include additional fuselage side stringers, smaller ailerons, larger flaps, taller (15 cm; 6 in) fin and rudder, extended dorsal strake, control surface gap seals, improved low-friction control cable runs, tab for pitch trim, tapered spring steel main landing gear legs with improved fairings, dual calliper brakes, larger baggage compartment and corrosion-proofing as standard. First flight (N11632; second airframe) 20 December 1996. Further aircraft, N2506A for trials, built 2000. Type certificate awarded 14 September 2001, permitting start to deliveries.
CUSTOMERS: Total of 296 kits sold, of which 267 completed, by July 2002.
COSTS: S-7 kit, with engine: Rotax 912UL, US$27,745; Rotax 912ULS, US$29,550 (2003). Production version cost US$55,000 for first 50 examples.
DESIGN FEATURES: High-wing, mid-tailplane configuration with sweptback fin, constant-chord wings and tapered tailplane. Wings braced by A-struts; empennage mutually wire-braced. Wings fold rearwards and tailplanes upwards for stowage; usual rigging/de-rigging time 30 to 45 minutes. Quoted build time for kit version is 500 to 700 hours.
   NACA 2412 wing section, thickness/chord ratio 14 per cent; dihedral 1°; twist 25°.
FLYING CONTROLS: Conventional and manual. Cable-operated rudder and ailerons, the last-named spade-assisted, and torque tube-operated elevators; trim tab on starboard elevator; cable-operated four-position flaps (0, 8, 16 and 26°).

STRUCTURE: Wings have two tubular 6061-T6 alloy spars, preformed aluminium tube ribs riveted and clipped to spars and sheet alloy leading-edges; fuselage and tail surfaces of welded 4130 steel; glass fibre forward decking and engine cowling. Airframe is fabric-covered.
LANDING GEAR: Non-retractable tailwheel type with tubular spring steel main legs; steerable, full-swivelling tailwheel; mainwheel size 16×5 (6.00-6 on S-7C), maximum pressure 2.41 bar (35 lb/sq in); tailwheel with solid tyre, size 8×2; Matco/Cleveland wheels with hydraulic brakes. Optional speed fairings on main legs increase cruising speed by 2 to 4 kt (5 to 8 km/h; 3 to 5 mph). Tundra tyres, skis, floats and Stoddart-Hamilton Aerocet amphibious floats optional.
POWER PLANT: One 59.6 kW (79.9 hp) Rotax 912 UL or one 73.5 kW (98.6 hp) Rotax 912 ULS four-cylinder piston engine, each with 2.27:1 reduction gear, driving a two-blade, ground-adjustable propeller (Tennessee on 912 UL and Sensenich on 912 ULS). S-7C has Rotax 912 ULS driving Sensenich two-blade propeller. Fuel contained in two wing tanks, combined capacity 68 litres (18.0 US gallons; 15.0 Imp gallons), of which 58 litres (15.2 US gallons; 12.7 Imp gallons) are usable. Oil capacity 2.9 litres (0.75 US gallon; 0.6 Imp gallon).
ACCOMMODATION: Pilot and passenger in tandem on fore- and aft-adjustable seats; lap and shoulder harness standard; optional deluxe seats. Dual controls standard. Cabin is heated and ventilated. Upward-hinged door on each side is flight-openable with uplocks; quick-release hinge pins optional. Baggage compartment at rear of cabin.
AVIONICS: To customer's choice; optional GPS, VHF com and intercom on S-7C.
DIMENSIONS, EXTERNAL: (all versions, except: A: S-7 with Rotax 912 UL, B: S-7 with Rotax 912 ULS, C: S-7C):

Wing span	8.92 m (29 ft 3 in)
Wing chord, constant	1.52 m (5 ft 0 in)
Wing aspect ratio	5.8
Length overall: A, B	6.96 m (22 ft 10 in)
C	7.09 m (23 ft 3 in)
Length overall, wings folded: A, B	7.11 m (23 ft 4 in)
Height overall: A, B	1.90 m (6 ft 3 in)
C	2.01 m (6 ft 7 in)
Tailplane span	2.44 m (8 ft 0 in)
Wheel track	1.93 m (6 ft 4 in)
Wheelbase	5.18 m (17 ft 0 in)
Propeller diameter:	
A, B	1.73 m to 1.83 m (5 ft 8 in to 6 ft 0 in)
C	1.83 m (6 ft 0 in)
Cabin door: Max height	0.97 m (3 ft 2 in)
Max width	1.52 m (5 ft 0 in)
Height to sill	0.81 m (2 ft 8 in)

DIMENSIONS, INTERNAL (all versions):

Cabin: Length	1.93 m (6 ft 4 in)
Max width	0.76 m (2 ft 6 in)
Max height	1.07 m (3 ft 6 in)
Floor area	1.11 m² (12.00 sq ft)
Volume	1.4 m³ (48 cu ft)
Baggage compartment:	
Volume	0.28 m³ (10.0 cu ft)

AREAS (A, B and C as above):

Wings, gross: all	13.67 m² (147.1 sq ft)
Ailerons (total): A, B	1.11 m² (12.00 sq ft)
Trailing-edge flaps (total): A, B	1.11 m² (12.00 sq ft)
Fin: A, B	0.74 m² (8.00 sq ft)
Rudder: A, B	0.56 m² (6.00 sq ft)
Tailplane: A, B	1.11 m² (12.00 sq ft)
Elevators, incl tabs (total): A, B	0.93 m² (10.00 sq ft)

WEIGHTS AND LOADINGS (A, B and C as above):

Weight empty: A, B	306 kg (675 lb)
C	318 kg (700 lb)
Max T-O weight: all	544 kg (1,200 lb)
Baggage capacity	23 kg (50 lb)
Max wing loading: all	39.8 kg/m² (8.16 lb/sq ft)
Max power loading: A	9.14 kg/kW (15.02 lb/hp)
B, C	7.41 kg/kW (12.17 lb/hp)

PERFORMANCE (A, B and C as above):

Never-exceed speed (VNE)	113 kt (209 km/h; 130 mph)
Cruising speed: A	96 kt (177 km/h; 110 mph)
B	103 kt (190 km/h; 118 mph)
Stalling speed:	
Flaps up	40 kt (74 km/h; 46 mph)
Flaps down	36 kt (66 km/h; 41 mph)
Max rate of climb at S/L: A	244 m (800 ft)/min
B	335 m (1,100 ft)/min
C	305 m (1,000 ft)/min
Service ceiling: A	4,270 m (14,000 ft)
B, C	4,420 m (14,500 ft)
T-O run: A, C	72 m (235 ft)
B	69 m (225 ft)
Landing run	99 m (325 ft)
Range: A	420 n miles (779 km; 484 miles)
B	410 n miles (759 km; 472 miles)
C	417 n miles (772 km; 480 miles)
Endurance: A	4 h 24 min
B, C	4 h 0 min
g limits	+4/−2

UPDATED

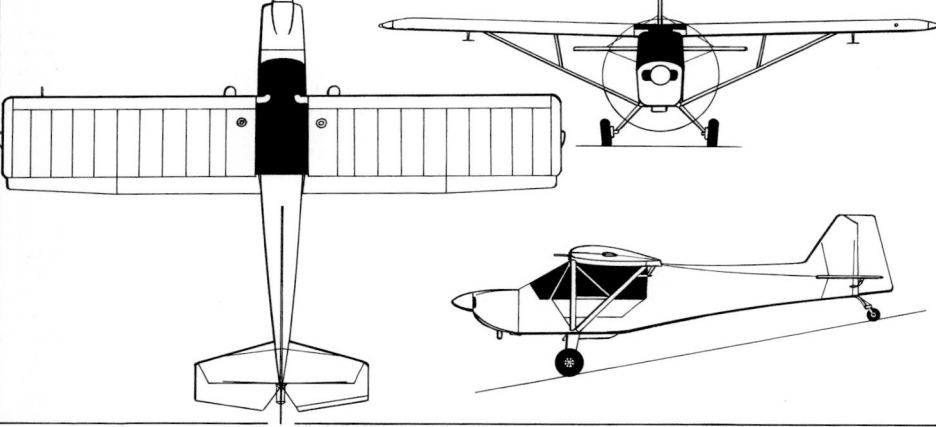

Rans S-7C Courier two-seat factory-built light aircraft *(Jane's/James Goulding)*   0089516

Rans S-12S Super Airaile side-by-side ultralight   0109159

RANS S-12XL AIRAILE and S-12S SUPER AIRAILE

TYPE: Side-by-side ultralight kitbuilt.
PROGRAMME: Improved, production versions of S-12 Airaile kitbuilt. R&D prototype of S-12XL (N8045X) made its debut at Sun 'n' Fun in Lakeland, Florida, April 1995.
CURRENT VERSIONS: **S-12XL Airaile:** Standard version.
Description applies to S-12XL, except where indicated.
S-12S Super Airaile: Improved version, first flown (N80887; modification of an earlier S-12XL) early 1999. 59.6 kW (79.9 hp) Rotax 912 or 73.5 kW (98.6 hp) Rotax 912 ULS engines, with thrust line lowered 25 mm (1 in) and driving ground-adjustable, composites Warp Drive propeller; new wing with improved aerofoil section, identical in rib construction to that of S-7C Courier; stamped aluminium ribs in tail surfaces; reduced tailplane incidence; cabin height raised 50 mm (2 in), with more sharply raked windscreen incorporating a new flow fence;

increased range of seat tilt to improve headroom and comfort; hydraulic brakes standard; pre-sewn Dacron skins replaced by traditional doped fabric secured by pop rivets.

Performance improvements include 9 kt (16 km/h; 10 mph) increase in cruising speed. Marketed in parallel with S-12XL.

CUSTOMERS: Over 1,000 S-12s (all models) ordered by early 2002, of which 805 S-12XLs and seven S-12Ss flying.

COSTS: S-12XL kits, with engine: Rotax 503SC, US$15,020; Rotax 503DC, US$15,150; Rotax 582, US$16,840; Rotax 912, US$24,450; Rotax 912S, US$26,355 (2003).

S-12S kits, with engine: Rotax 912, US$28,750; Rotax 912S, US$30,455 (2003).

DESIGN FEATURES: Pod and boom fuselage; V-strut-braced wings; wire-braced tail surfaces; wings fold rearwards and tailplanes upwards for storage. Quoted build time 175 to 225 hours; or 325 to 500 hours with full enclosure cockpit. (Extra 200 hours for S-12S.)

FLYING CONTROLS: Conventional and manual. Cable-operated; ground-adjustable flaps primarily used for setting best trim, but can also be lowered for landing; tailplane incidence is ground-adjustable. Dual controls standard.

STRUCTURE: Welded steel tube cockpit cage; aluminium alloy tubular tailboom; wings have two tubular alloy spars and preformed ribs; ailerons, flaps and tail surfaces of alloy tube with stamped alloy or moulded ABS ribs; fabric covering.

LANDING GEAR: Fixed tricycle type with small tailwheel/bumper; tubular spring steel main legs, spring-loaded telescoping nose leg; pedal-operated hydraulic brakes on mainwheels; steerable nosewheel. All wheels 15 cm (6 in) aluminium on S-12XL and 13 cm (5 in) on S-12S. Optional spats, tundra tyres and Puddle Jumper floats.

POWER PLANT: One air-cooled 34.0 kW (45.6 hp) Rotax 503 or liquid-cooled 47.8 kW (64.1 hp) Rotax 582 two-cylinder engine with 2.58:1 reduction gear, or 59.6 kW (79.9 hp) Rotax 912 or 73.5 kW (98.6 hp) Rotax 912 ULS flat-fours with 2.27:1 reduction gear, driving a two-blade Tennessee wooden propeller; three-blade Warp Drive propeller optional; electric start optional on 503 and 582, standard on 912. Version with 48.5 kW (65 hp) Hirth 2706 two-cylinder two-stroke is available in Germany. Fuel in single wing tank, maximum capacity 34 litres (9.0 US gallons; 7.5 Imp gallons); dual tanks optional, (standard in S-12S) combined capacity 68 litres (18.0 US gallons; 15.0 Imp gallons).

ACCOMMODATION: Pilot and passenger side by side in open cockpit; transparent mini pod fairing standard; optional partial- and full-enclosure cockpits provide increased leg, shoulder and head room and lower noise levels and, combined with flow fences attached to fuselage pod and boom, raise cruising speed.

DIMENSIONS, EXTERNAL:
Wing span	9.45 m (31 ft 0 in)
Length overall	6.25 m (20 ft 6 in)
Height overall	2.36 m (7 ft 9 in)
Propeller diameter:	
S-12XL	1.73 to 1.83 m (5 ft 8 in to 6 ft 0 in)
S-12S	1.73 m (5 ft 8 in)

AREAS:
Wings, gross	14.12 m² (152.0 sq ft)

WEIGHTS AND LOADINGS (A: S-12XL with Rotax 912 UL, B: S-12S with Rotax 912 ULS):
Weight empty: A	261 kg (575 lb)
B	295 kg (650 lb)
Max T-O weight: A	499 kg (1,100 lb)
B	521 kg (1,150 lb)

PERFORMANCE (A and B as above):
Cruising speed: A	74 kt (137 km/h; 85 mph)
B	87 kt (161 km/h; 100 mph)*
Stalling speed, flaps down: A, B	31 kt (57 km/h; 35 mph)
Max rate of climb at S/L: A	274 m (900 ft)/min
B	366 m (1,200 ft)/min

Rans S-16 Shekari with nosewheel landing gear (*Jane's/Paul Jackson*)   NEW/0526969

Rans S-16 Shekari, with additional side view of optional nosewheel version (*Jane's/Paul Jackson*)   0064496

T-O run: A	87 m (285 ft)
B	81 m (265 ft)
Landing run: A, B	61 m (200 ft)
Range: A, B	325 n miles (601 km; 374 miles)
Endurance: A	4 h 24 min
B	3 h 54 min

* With cockpit and wheel fairings

UPDATED

RANS S-16 SHEKARI

TYPE: Side-by-side kitbuilt.

PROGRAMME: Prototype (N8072U), powered by 59.6 kW (79.9 hp) Rotax 912 UL engine, first flew 25 March 1997 in tricycle landing gear configuration, and won its class in the 1997 Sun 'n' Fun Race at Lakeland, Florida, averaging 113 kt (209 km/h; 130 mph) over a 52 n mile (97 km; 60 mile) course; re-engined with Teledyne Continental IO-240B and flown April 1998; subsequently converted to tailwheel configuration, although this configuration not marketed. Further prototype, N8073U, (plus one static test airframe) completed for trial installation of 119.3 kW (160 hp) Textron Lycoming O-320; FAA approval under homebuilt rules achieved 16 December 1998; first kit delivered to customer two days later; production capacity 50 kits per year.

CUSTOMERS: Total 22 sold and seven flying by December 2001.

COSTS: Quick-build kit US$25,000, not including engine (2003).

DESIGN FEATURES: Conventional configuration low-wing monoplane with tailwheel or nosewheel landing gear. Quoted build time 500 to 1,000 hours.

FLYING CONTROLS: Conventional and manual. Pushrod-operated elevators, cable-operated rudder; half-span, five-position flaps. Triangular trim tab at base of rudder.

STRUCTURE: Composites fuselage, made up of four moulded shells, with powder-coated welded chromoly steel cockpit cage; aluminium wings, elevators and rudder with composites tips.

LANDING GEAR: Non-retractable nosewheel configuration; nosewheel steerable; hydraulic toe brakes standard; leg and wheel speed fairings on all units.

POWER PLANT: One 93.2 kW (125 hp) Teledyne Continental IO-240B flat-four or one 119.3 kW (160 hp) Textron Lycoming O-320 flat-four, driving a three-blade ground-adjustable Warp Drive wooden propeller. Alternative engines include Subaru EA81. Fuel in two fuselage tanks, total capacity 121 litres (32.0 US gallons; 26.6 Imp gallons); oil capacity 4.7 litres (1.25 US gallons; 1.04 Imp gallons).

ACCOMMODATION: Two persons, side by side under one-piece forward-hinged canopy supported by gas struts; dual controls, central throttle, shoulder harnesses, lap belts and roll-over protection standard.

DIMENSIONS, EXTERNAL:
Wing span	7.32 m (24 ft 0 in)
Wing aspect ratio	6.6
Wing chord, mean	1.12 m (3 ft 8 in)
Length overall	5.89 m (19 ft 4 in)
Height overall	2.24 m (7 ft 4 in)
Tailplane span	2.54 m (8 ft 4 in)
Propeller diameter	1.78 m (5 ft 10 in)

DIMENSIONS, INTERNAL:
Cabin max width	1.02 m (3 ft 4 in)
Baggage volume	0.20 m³ (7.0 cu ft)

AREAS:
Wings, gross	8.08 m² (87.0 sq ft)

WEIGHTS AND LOADINGS (IO-240 engine):
Weight empty	422 kg (930 lb)
Baggage capacity	23 kg (50 lb)
Max T-O weight: Normal	657 kg (1,450 lb)
Aerobatic	590 kg (1,300 lb)
Max wing loading	81.4 kg/m² (16.67 lb/sq ft)
Max power loading	7.06 kg/kW (11.60 lb/hp)

PERFORMANCE (IO-240 engine):
Never-exceed speed (VNE)	191 kt (354 km/h; 220 mph)
Max cruising speed	139 kt (257 km/h; 160 mph)
Stalling speed: flaps up	54 kt (100 km/h; 62 mph)
flaps down	51 kt (94 km/h; 58 mph)

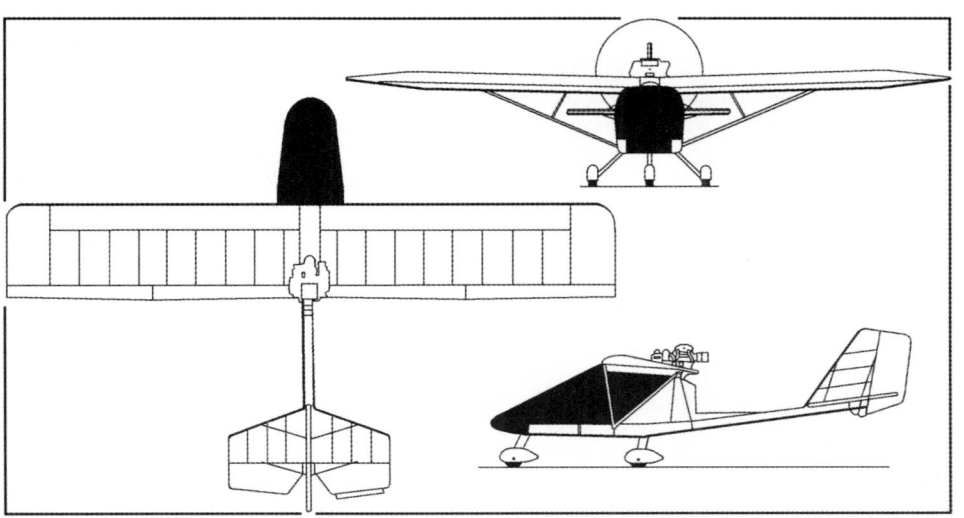

Rans S-12S Super Airaile general arrangement   NEW/0527089

Rans S-17 Stinger (*Jane's/Paul Jackson*)   0089510

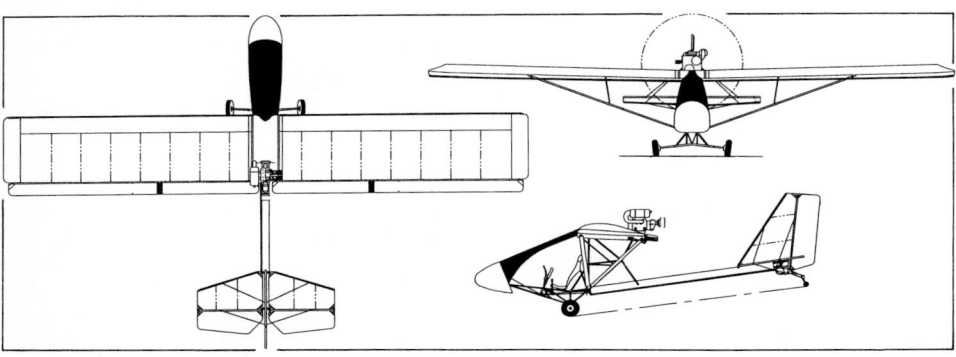

Rans S-17 general arrangement   0089513

Max rate of climb at S/L	305 m (1,000 ft)/min		Cruising speed	52 kt (97 km/h; 60 mph)
Service ceiling	4,420 m (14,500 ft)		Stalling speed, power off, flaps up	
T-O run	152 m (500 ft)			25 kt (45 km/h; 28 mph)
Landing run	160 m (525 ft)		Max rate of climb at S/L	335 m (1,100 ft)/min
Range	741 n miles (1,372 km; 853 miles)		Service ceiling	3,965 m (13,000 ft)
Endurance	5 h 18 min		T-O run	24 m (80 ft)
g limits	+4/−4		Landing run	30 m (100 ft)
	UPDATED		Range	52 n miles (97 km; 60 miles)
			g limits	+4/−2
				UPDATED

RANS S-17 STINGER

TYPE: Single-seat ultralight/kitbuilt.

PROGRAMME: First flight April 1996; development continuing towards FAA FAR Pt 103 certification. Had undergone fundamental redesign by 2000, changing from tractor to pusher propeller and relocating tailboom to low position, complete with tailwheel; by 2003, certification still continuing.

CURRENT VERSIONS: **S-17:** *As described.*
S-18: Two-seat; described separately.

CUSTOMERS: 22 completed by early 2002.

COSTS: Kit, with engine: Rotax 447, US$10,840; Rotax 503 SC, US$11,485; Rotax 503 DC, US$11,640 (all 2003). Flyaway US$3,000 to US$4,000 extra.

DESIGN FEATURES: Pod and boom fuselage; V-strut-braced wings; quoted building time 150 hours.

FLYING CONTROLS: Conventional and manual.

STRUCTURE: Light alloy fuselage, pre-assembled with all controls in situ; wing has tubular spar. Fuselage is 12.7 cm (5 in) aluminium tube with 4130 welded steel cockpit cage; optional composites nosecone. Dacron skin; spring steel landing gear.

LANDING GEAR: Fixed tailwheel type; tapered spring steel main legs; hand-lever operated brakes on mainwheels.

POWER PLANT: Prototype has one 34.0 kW (45.6 hp) Rotax 503 two-cylinder piston engine driving a three-blade IVO propeller via 2.58:1 reduction gearing. Alternatives include 31.0 kW (41.6 hp) Rotax 447 UL-2V. Fuel capacity 19 litres (5.0 US gallons; 4.2 Imp gallons) standard; 34 litres (9.0 US gallons; 7.5 Imp gallons) optional.

ACCOMMODATION: One person in open cockpit with Lexan windscreen/nosecone; four-point Hooker harness.

DIMENSIONS, EXTERNAL:
Wing span	8.99 m (29 ft 6 in)
Length overall	5.28 m (17 ft 4 in)
Height overall	2.13 m (7 ft 0 in)
Propeller diameter	1.73 m (5 ft 8 in)

AREAS:
Wings, gross	11.80 m² (127.0 sq ft)

WEIGHTS AND LOADINGS (Rotax 503):
Weight empty	118 kg (261 lb)
Max T-O weight	243 kg (536 lb)

PERFORMANCE (Rotax 503):
Never-exceed speed (VNE)	82 kt (152 km/h; 95 mph)

RANS S-18 STINGER II

TYPE: Tandem-seat ultralight kitbuilt.

PROGRAMME: Developed from single-seat S-17 Stinger. First flight of prototype S-18 (N24052) 9 September 2000; production from December 2000; first kit deliveries scheduled for March 2001.

CUSTOMERS: Three flying by early 2002.

COSTS: Kit, including engine and exhaust: Rotax 503 DC US$16,630; Rotax 582 US$18,320; Rotax 912 US$25,440 (2003).

DESIGN FEATURES: Pod-and-boom fuselage; V-strut-braced wings are identical to those of S-12XL Airaile and fold rearwards for transport or storage; wire-braced tail surfaces. Quoted build time under 200 hours.

FLYING CONTROLS: Conventional and manual. Flaps.

STRUCTURE: Welded 4130 steel tube fuselage cage and 15 cm (6 in) light alloy tailboom, pre-assembled and painted; wing has tubular spar; all parts prefabricated. Wing and empennage covered by fabric; control gaps sealed by Velcro.

LANDING GEAR: Tailwheel type; fixed. One-piece tapered aluminium cantilever main legs; main tyre size 8.00-6; manually operated brakes on mainwheels.

POWER PLANT: Prototype has one 37.0 kW (49.6 hp) Rotax 503 DC air-cooled, two-cylinder piston engine with 1:2.58 reduction gearing, driving a three-blade Ivo pusher propeller; alternatives are 47.8 kW (64.1 hp) Rotax 582 and 73.5 kW (98.6 hp) Rotax 912 liquid-cooled engines with angled radiator on wing upper surface, cooling airflow optimised by louvres. Fuel contained in single tank, capacity 34 litres (9.0 US gallons, 7.5 Imp gallons); second tank of similar capacity optional.

ACCOMMODATION: Two persons in tandem. Options of open cockpit with Lexan windscreen/nosecone; full windscreen; and fully enclosed cabin. Baggage in optional 'saddlebags' each side of cabane on open versions.

DIMENSIONS, EXTERNAL:
Wing span	9.45 m (31 ft 0 in)
Width, wings folded	1.75 m (5 ft 9 in)
Length overall	6.88 m (22 ft 7 in)
Height overall	2.06 m (6 ft 9 in)
Propeller diameter	1.68 m (5 ft 6 in)

AREAS:
Wings, gross	14.12 m² (152.0 sq ft)

WEIGHTS AND LOADINGS:
Weight empty	217 kg (478 lb)
Max T-O weight	417 kg (920 lb)

PERFORMANCE (Rotax 503 DC):
Never-exceed speed (VNE)	78 kt (144 km/h; 90 mph)
Cruising speed	52 kt (97 km/h; 60 mph)
Stalling speed, power off	32 kt (58 km/h; 36 mph)
Max rate of climb at S/L	152 m (500 ft)/min
T-O run	101 m (330 ft)
Landing run	61 m (200 ft)
Endurance with max fuel, with reserves	3 h 0 in
	UPDATED

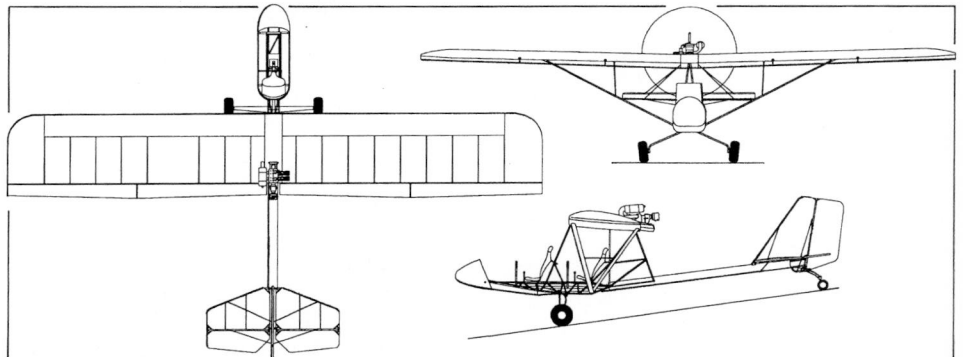

Fully enclosed version of Rans S-18 Stinger II ultralight kitbuilt (*Jane's/Paul Jackson*)   NEW/0526968

Two-seat Rans S-18 Stinger II open cockpit version (*Jane's/Paul Jackson*)   0105043

RAYTHEON

RAYTHEON AIRCRAFT COMPANY
(Subsidiary of Raytheon Company)
9709 East Central, Wichita, Kansas 67201-0085
Tel: (+1 316) 676 71 11
Fax: (+1 316) 676 82 86
Web: http://www.raytheonaircraft.com
BRANCH DIVISIONS: Salina, Kansas and Little Rock, Arkansas.
CHAIRMAN AND CEO: James E Schuster
COO: Bob Horowitz
SENIOR VICE-PRESIDENT, AIRCRAFT BUSINESS:
   Bradley A Hatt
VICE-PRESIDENTS:
   Douglas Debrecht (Chief Information Officer)
   Edward P Dolanski (Customer Support)
   Paul R Dudek (Chief Financial Officer)
   Robert Feazell (Quality Assurance)
   Jackson L Hulsey (Development)
   Ray Lapointe (Jet Programmes)
   David H Riemer (Government Business)
   Peter S Ross (Supply Chain Management)
   Thomas J Sarama (Engineering)
   Michael J Scheidt (Commuter Airline Programme)
   Paul R Schumacher (Operations)
   Wayne W Wallace (General Counsel)
DIRECTOR, GOVERNMENT CONTRACTS: Howard E Stewart
DIRECTOR, HORIZON PROGRAMME: C Dwayne Johnston
DIRECTOR, PREMIER PROGRAMME: Bill Patterson
DIRECTOR, CORPORATE AFFAIRS AND COMMUNICATIONS:
   James M Gregory

Refer to **Beech** and **Hawker** entries earlier in this section for aircraft descriptions.

Raytheon Aircraft Company (RAC) formed 15 September 1994, combining Raytheon Company subsidiaries Beech Aircraft Corporation and Raytheon Corporate Jets Inc. Raytheon acquired British Aerospace's Corporate Jets

BEECH/HAWKER FIVE-YEAR DELIVERIES

Type	1998	1999	2000	2001	2002
T-6A	–	1	49	47	48
Bonanza	93	96	103	89	56
Baron	42	49	50	47	27
King Air	124[1]	128[2]	151[3]	119[4]	71[5]
Beech 1900	45	24	54	11	11
Beechjet (civil)	43	48	51	25	19
Premier				18	29
Hawker 800	48	55	67	55	46
Annual totals	**395**	**401**	**525**	**411**	**307**

Notes:
[1] 37 King Air C90Bs, 45 King Air B200s and 42 King Air 350s
[2] 41 King Air C90Bs; 44 King Air B200s and 43 King Air 350s
[3] 46 King Air C90Bs; 59 King Air B200s and 46 King Air 350s
[4] 41 King Air C90Bs, 46 King Air B200s and 32 King Air 350s
[5] 21 King Air C90Bs, 26 King Air B200s and 24 King Air 350s

division 6 August 1993. RAC builds civil and military aircraft, and components; Salina division supplies wings for most models, non-metallic interior components, ventral fins, nosecones and tailcones.

Final assembly of Hawkers transferred from UK to new 41,800 m² (450,000 sq ft) plant at Wichita beginning last quarter of 1995 and finalised by April 1997, when last UK-built Hawker was completed; 17,745 m² (191,000 sq ft) Training Systems Division plant added for Texan II assembly line. Little Rock continues to provide custom interiors, avionics, sales and customer support services for Hawker models and aircraft painting services for Hawkers. Beech and Hawker product names are retained and separate Beechcraft and Hawker divisions were established in mid-2002 for marketing purposes, Raytheon title having been dropped from aircraft designations.

Wholly owned subsidiaries include Raytheon Aircraft Credit Corporation Inc (business aircraft retail financing and leasing); Travel Air Insurance Company Ltd (aircraft liability insurance); Raytheon Aircraft Services network of fixed-base operations; Raytheon Travel Air, formed in 1997, which offers fractional ownership; and Raytheon Aircraft Charter and Management, which offers aircraft charter and full-service management.

In November 2001 RAC had 11,700 employees worldwide, of whom 8,800 were in Kansas, and occupied 380,900 m² (4,100,000 sq ft) of plant area in Wichita.

Total production by RAC more than 53,867 by 31 December 2002.

UPDATED

RENAISSANCE

RENAISSANCE AIRCRAFT LLC
PO Box 596, Cape Girardeau, Missouri 63702
Tel: (+1 573) 651 39 33
Fax: (+1 573) 651 39 23
Web: http://www.renaissanceaircraft.com

Renaissance Aircraft has relaunched production of the classic Luscombe 8F light aircraft, assembling airframes, manufactured on original tooling, by subcontractors in the Czech Republic at its 4,459 m² (48,000 sq ft) factory in Missouri.

UPDATED

RENAISSANCE 8F
TYPE: Two-seat lightplane.
PROGRAMME: Luscombe Airplane Corp Model 8 Silvaire first flew 18 December 1938; progressed through several subvariants, built by Luscombe, Temco Engineering and Silvaire Aircraft until 1960; total of some 5,867 produced. In 1999, Renaissance refurbished an original Luscombe 8F (N999RA) as prototype for relaunched production. First deliveries were due late 2000, but not effected. However, promotion was continuing in mid-2001, and assembly of first aircraft (N722J)began late July 2001.
CUSTOMERS: More than 10 deposits placed by July 2001.
COSTS: US$74,500, basic VFR; US$83,000 Mode C VFR; US$94,400 IFR (all 2002).
DESIGN FEATURES: High wing, all-metal touring lightplane of late 1930s technology, but incorporating modifications and some modern features. Changes include considerable increase in power from original 67.1 kW (90 hp) Continental C-90; 60 per cent more baggage area or optional seating for two children; additional fuel; and new cockpit interior.
FLYING CONTROLS: Conventional and manual. Flight-adjustable tab in port elevator; ground-adjustable tabs on both ailerons. No horn balances. Plain flaps; deflections 0, 10, 20 and 30°.
STRUCTURE: Alclad aluminium monocoque fuselage; oval section duralumin bulkhead stampings and riveted Alclad skin. Wings constructed on two I-type spars of extruded aluminium and ribs of riveted T-section extrusions with Alclad skin. Ailerons covered with beaded Alclad sheet riveted to a single duralumin spar. Wings attached to upper sides of cabin and braced by single streamlined strut each side. Wing struts and landing gear attached to aluminium forgings riveted to forward section of metal seat bottom on each side. Duralumin/Alclad tailplane bolted to fuselage. Steel tube main landing gear legs hinged at fuselage sides and with single oleo-spring unit within fuselage.
LANDING GEAR: Tailwheel type; fixed. Mainwheel tyres 6.00-6; toe-operated hydraulic brakes; optional speed fairings. Steerable tailwheel, size 2.30/2.50-4. Optional skis, floats or tundra tyres.
POWER PLANT: One 112 kW (150 hp) Textron Lycoming O-320 flat-four driving a Sensenich 74-62, two-blade,

Prototype Renaissance 8F *(Jane's/Paul Jackson)*
                                          0084574

fixed-pitch aluminium propeller with spinner. Optionally, 103 kW (138 hp) LOM M332A straight-four. Fuel tank in each wing, total 117 litres (30.0 US gallons; 25.0 Imp gallons).
ACCOMMODATION: Two, on individual leather seats, with dual controls. Forward-hinged door each side. Baggage compartment behind seats may be replaced by two child seats.
SYSTEMS: Electrical system with starter, ammeter and circuit breakers
AVIONICS: *Comms:* ELT standard. Optional Mode C VFR package includes Bendix/King KX 155 nav/com/glideslope with KI 209 VOR/LOC/GS head; KT 76C-10 transponder; KMA 12801 audio panel with intercom and push-to-talk switches on control columns, and Ameriking AK 350 encoder. Optional IFR package includes Mode C items as above and adds a second KX 155 with KI 1209 VOR/LOC/GS; KT 76C-10X transponder in place of KT 76C-10, and KA 33-00 cooling blower.
   Flight: Optional Mode C VFR and IFR packages each include Bendix/King KMD 150-05 GPS with colour moving map display.
   Instrumentation: Optional full gyro package includes engine-driven vacuum system and gauge; vacuum-driven directional gyro and attitude indicator; electric turn-and-bank indicator; ASI, OAT gauge and precision electric clock.
EQUIPMENT: Navigation, landing and wingtip strobe lights; vinyl and fabric cabin upholstery, and carved wood control column grips and fuel escutcheons, all standard. Combination EGT/CHT gauge, custom exterior paint; leather interior; thermal acoustic cabin insulation; custom wood instrument panel overlays with integral lighting, and

interior and panel lights, all optional. Heated pitot tube included in optional IFR avionics package.
DIMENSIONS, EXTERNAL:
Wing span	6.71 m (35 ft 0 in)
Wing aspect ratio	8.8
Length overall	6.70 m (22 ft 0 in)
Height overall	2.13 m (7 ft 0 in)
Propeller diameter	1.88 m (6 ft 2 in)

DIMENSIONS, INTERNAL:
Cabin max width	0.99 m (3 ft 3 in)

AREAS:
Wings, gross	13.01 m² (140.0 sq ft)

WEIGHTS AND LOADINGS:
Weight empty	449 kg (990 lb)
Max T-O weight: interim	635 kg (1,400 lb)
planned	793 kg (1,750 lb)
Max wing loading: interim	48.8 kg/m² (10.00 lb/sq ft)
planned	61.0 kg/m² (12.50 lb/sq ft)
Max power loading: interim	5.68 kg/kW (9.33 lb/hp)
planned	7.10 kg/kW (11.67 lb/hp)

PERFORMANCE:
Max level speed	130 kt (241 km/h; 150 mph)
Max cruising speed at 75% power:	
at S/L	122 kt (225 km/h; 140 mph)
at 2,440 m (8,000 ft)	126 kt (233 km/h; 145 mph)
Econ cruising speed at 65% power	
	119 kt (220 km/h; 137 mph)
Stalling speed: flaps up	41 kt (76 km/h; 47 mph)
flaps down	38 kt (70 km/h; 43 mph)
Max rate of climb at S/L	457 m (1,500 ft)/min
Service ceiling	6,400 m (21,000 ft)
g limits	+4.6/–2.2

UPDATED

RFW

REFLEX FIBERGLASS WORKS INC

538 Rentz Road, PO Box 497, Walterboro, South Carolina 29488-0494
Tel: (+1 843) 538 66 82
Fax: (+1 843) 538 66 83
PRESIDENT: Howell C Jones Jr

RFW took over production of the Lightning Bug and White Lightning kitbuilt sport aircraft in early 1996 and was housed in a 1,400 m² (15,000 sq ft) purpose-built factory. By 2002 the company's rights, tooling and mouldings had been acquired by Jackson Fields Mathews LLC, which intends to relaunch the White Lightning and sell on the rights to the Lightning Bug.

UPDATED

RFW LIGHTNING BUG

TYPE: Single-seat sportplane kitbuilt.
CUSTOMERS: At least nine completed, of which four current in September 2002; two confirmed lost in accidents. Most recent completions, in 1999, by Roger Pearce of Easley, South Carolina, and John Gibbs of Monterey, California.
COSTS: US$25,000 (2001) for kit which includes engine and propeller. US$1,000 deposit on ordering.
DESIGN FEATURES: Compact, mid-wing monoplane; conforms to FAR Pt 23 Utility category at 430 kg (950 lb); Aerobatic category at 340 kg (750 lb). Quoted build time 400 hours.
FLYING CONTROLS: Conventional and manual. Elevator tab for pitch trim; ground-adjustable tabs on ailerons and rudder.
STRUCTURE: General construction of composites, stainless steel and aluminium 2024-T3 or 7075-T6. Moulded parts use E glass, epoxy resin, PVC foam and Baltek mat.
LANDING GEAR: Fixed mainwheels with spats; option of (retractable) nosewheel, or tailwheel. Same wheels and tyres as Long-EZ.
POWER PLANT: One 74.6 kW (100 hp) AMW 808 FIAGD three-cylinder liquid-cooled two-stroke engine, with fuel injection and dual ignition, driving fixed-pitch propeller

Lightning Bug high-performance monoplane (AMW 808 two-stroke engine)

through reduction gear; variable-pitch propeller under investigation. Fuel capacity 87 litres (23.0 US gallons, 19.2 Imp gallons); grades 100 LL or Mogas.
ACCOMMODATION: Single seat accommodates pilot with maximum weight of 127 kg (280 lb) and maximum height 1.98 m (6 ft 6 in).
SYSTEMS: Electrical system includes 12 V 30 Ah battery.
AVIONICS: Customer specified.
DIMENSIONS, EXTERNAL:

Wing span	5.44 m (17 ft 10 in)
Length overall	5.31 m (17 ft 5 in)
Fuselage max width	0.63 m (2 ft 1 in)
Height overall	1.55 m (5 ft 1 in)

AREAS:

Wings, gross	3.72 m² (40.0 sq ft)

WEIGHTS AND LOADINGS:

Weight empty	215 kg (475 lb)
Max T-O weight: Aerobatic	340 kg (750 lb)
Utility	430 kg (950 lb)

PERFORMANCE:

Max level speed	217 kt (402 km/h; 250 mph)
Econ cruising speed at 75% power	
	196 kt (362 km/h; 225 mph)
Stalling speed, power off:	
flaps up	74 kt (137 km/h; 85 mph)
flaps down	54 kt (100 km/h; 62 mph)
Max rate of climb at S/L at 139 kt (257 km/h; 160 mph)	
	366 m (1,200 ft)/min
T-O run	244 m (800 ft)
Landing run	305 m (1,000 ft)
Range with max fuel at 75% power	
	764 n miles (1,416 km; 880 miles)

UPDATED

RFW WLAC-1 WHITE LIGHTNING

TYPE: Four-seat kitbuilt.
PROGRAMME: Previously produced by the White Lightning Aircraft Corporation. Prototype first flew 8 March 1986. In same year it established several world speed records in Classes C1b and C1c.
CUSTOMERS: 32 kits sold, 15 aircraft flying by mid-1999. By September 2002, 11 remained on US civil register; most recent completions in 1998 by Leonard Moore of Hurst, Texas.
COSTS: Kit: US$45,000 (2001) including 40-hour builders' course. Many optional subkits, including electrical for US$1,110.
DESIGN FEATURES: Highly streamlined mid-wing monoplane. Options include electrical, various engine installations, strobe light and interior subkits, plus other items. Quoted build time 1,500 hours.
Wing section NACA 66₂-215.

FLYING CONTROLS: Conventional and manual. Horn-balanced elevators. Ground-adjustable tab on starboard aileron and rudder, trim tab on port elevator.
STRUCTURE: Wings have graphite tubular 'wet' main spar, and glass fibre/epoxy front and rear spars and ribs, all precast in the lower glass fibre/epoxy skin. Glass fibre/epoxy fuselage moulded in upper and lower halves. Spars and ribs of fin and tailplane precast into one skin of each.
LANDING GEAR: Retractable tricycle type, with brakes. Main and nosewheels retract rearwards into fuselage. Single wheel on each unit.
POWER PLANT: One 156.6 kW (210 hp) Teledyne Continental IO-360-CB flat-six engine. Fuel capacity 265 litres (70.0 US gallons; 58.0 Imp gallons) of which 257 litres (68.0 US gallons; 56.6 Imp gallons) are usable.
ACCOMMODATION: Four persons in two side-by-side pairs. Door port side, rear.
DIMENSIONS, EXTERNAL:

Wing span	8.43 m (27 ft 8 in)
Wing aspect ratio	8.6
Length overall	7.11 m (23 ft 4 in)
Height overall	2.18 m (7 ft 2 in)
Tailplane span	2.74 m (9 ft 0 in)
Propeller diameter	1.85 m (6 ft 1 in)

AREAS:

Wings, gross	8.29 m² (89.2 sq ft)
Tailplane	1.81 m² (19.50 sq ft)

WEIGHTS AND LOADINGS:

Weight empty	635 kg (1,400 lb)
Max baggage capacity	154 kg (340 lb)
Max T-O weight	1,224 kg (2,700 lb)
Max wing loading	147.8 kg/m² (30.27 lb/sq ft)
Max power loading	7.83 kg/kW (12.86 lb/hp)

PERFORMANCE:

Max level speed at S/L	243 kt (450 km/h; 280 mph)
Max cruising speed at 2,440 m (8,000 ft)	
	230 kt (426 km/h; 265 mph)
Econ cruising speed at 3,200 m (10,500 ft)	
	217 kt (402 km/h; 250 mph)
Stalling speed: flaps up	80 kt (148 km/h; 92 mph)
flaps down	53 kt (99 km/h; 61 mph)
Max rate of climb at S/L	610 m (2,000 ft)/min
Service ceiling	6,100 m (20,000 ft)
T-O and landing run, full flap	400 m (1,300 ft)
Range with max fuel:	
at 75% power	1,423 n miles (2,636 km; 1,638 miles)
at 55% power	1,738 n miles (3,218 km; 2,000 miles)
Endurance	6 h 10 min
g limits	+4.4/−2 utility

UPDATED

RFW WLAC-1 White Lightning four-seat composites kit aircraft
(Geoffrey P Jones)   0062652

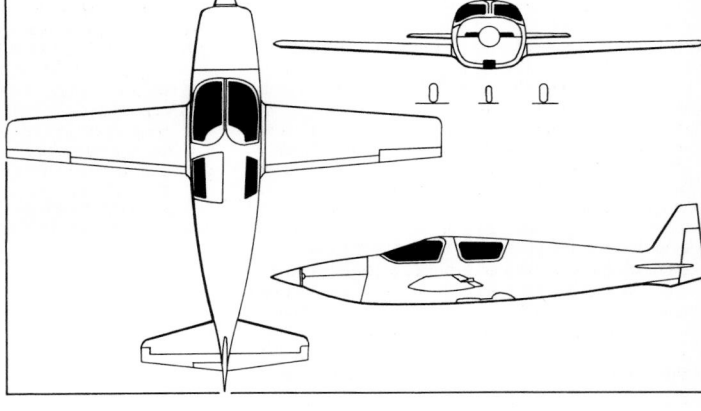

RFW WLAC-1 White Lightning *(Jane's/Paul Jackson)*

ROBINSON

ROBINSON HELICOPTER COMPANY

2901 Airport Drive, Torrance, California 90505
Tel: (+1 310) 539 05 08
Fax: (+1 310) 539 51 98
Web: http://www.robinsonheli.com
PRESIDENT: Franklin D Robinson
VICE-PRESIDENT, MARKETING: Wayne Walden
MARKETING DIRECTOR: Tim Goetz
PUBLIC RELATIONS: Barbara K Robinson

By February 2002, Robinson had produced 5,400 helicopters, including 390 in 2000 and 328 in 2001. Factory floor area 24,150 m² (260,000 sq ft) Workforce totals 780. Company is ISO 9001 certified.

UPDATED

ROBINSON R22

TYPE: Two-seat helicopter.
PROGRAMME: Design began 1973; first flight 28 August 1975; first flight of second R22 early 1977; FAA certification 16

March 1979; UK certification June 1981; deliveries began October 1979; early versions were R22 (O-320-A2B engine; 199 built) and R22HP (O-320-B2C; 151 built), both with 621 kg (1,300 lb) max T-O weight. R22 Alpha (O-320-B2C and 621 kg; 1,370 lb MTOW) certified October 1983 (further 151 built); R22 Beta certified 5 August 1985. Current Beta II introduced in 1995.
CURRENT VERSIONS

R22 Beta II: More powerful Textron Lycoming O-360 engine provides better high-level hover performance and allows take-off power to be sustained up to 2,285 m (7,500 ft). Previously optional, tinted windscreen and door windows fitted as standard. Production began at c/n 2571 in 1995.
Main description applies to Beta II except where indicated.

R22 Mariner II: Float-equipped **Beta II**; introduced in 1995, replacing original Mariner introduced 1985. Total of 228 R22 Mariners of both versions produced by December 2000.

R22 Police: Special communications and other equipment including removable port-side controls, 70 A alternator, searchlight, loudspeaker, siren and transponder.

R22 IFR: Equipped with flight instruments and radio to allow training for helicopter IFR flying; includes Honeywell KCS 55A HSI, KR 87 ADF, KX 165 nav/com, KT 76A transponder; KR 22 marker, plus optional KN 63 DME and KLN 89B GPS (replacing KR 87). Four delivered in 1997, six in 1998, six in 1999 and eight in 2000.

External load R22: Hook kit certified for 181 kg (400 lb) produced by Classic Helicopter Corporation, Boeing Field, Seattle, Washington; weighs 2.3 kg (5 lb); used for slung load training.

R22 Agricultural: Equipped with Apollo Helicopter Services DTM-3 spray system; FAA approved December 1991. Low-drag belly tank contains 151 litres (40.0 US gallons; 33.3 Imp gallons); boom length 7.31 m (24 ft 0 in); tank frame attached to landing gear mounting points with four bolts and wing nuts; installation of entire system requires no tools and can be completed by one person in approximately 5 minutes.
CUSTOMERS: Total production of 3,273 by 1 September 2001. 3,000th handed over 15 October 1999; deliveries in 1999 totalled 128, followed by 126 in 2000 and 134 in 2001.

Robinson R22 Mariner 0110909

Instrument panel of Robinson R22 0075972

COSTS: US$164,000 for basic R22 Beta II (2002). Direct operating cost US$27.30 per hour; total cost (500 hours annually) US$75.68 per hour (2002). Mariner II basic cost US$174,000 (2001).

DESIGN FEATURES: Simple, pod-and-boom light helicopter; horizontal stabiliser, starboard side only; vertical stabiliser above and below boom; offset to starboard; tall rotor mast. Horizontally mounted piston engine drives transmission through multiple V belts and sprag-type overrunning clutch; main and tail gearboxes use spiral bevel gears; maintenance-free flexible couplings of proprietary manufacture used in both main and tail rotor drives. Two-blade semi-articulated main rotor, with tri-hinged underslung rotor head to reduce blade flexing, rotor vibration and control force feedback, and an elastic teeter hinge stop to prevent blade-boom contact when starting or stopping rotor in high winds; blade section NACA 63-015 (modified); two-blade tail rotor on port side; rotor brake standard.

FLYING CONTROLS: Manual. Cyclic stick mounted between pilots with handgrips on swing arm for comfortable access from either seat.

STRUCTURE: All-metal bonded blades with stainless steel spar and leading-edge, light alloy skin and light alloy honeycomb filling; cabin section of steel tube with metal and plastics skinning; full monocoque tailboom.

LANDING GEAR: Welded steel tube and light alloy skid landing gear, with energy-absorbing crosstubes. Twin float/skid gear on Mariner with additional tailplane surface on lower tip of fin.

POWER PLANT: One Textron Lycoming O-360 flat-four engine, derated to 97.5 kW (131 hp), mounted in the lower rear section of the main fuselage, with cooling fan. Light alloy main fuel tank in upper rear section of the fuselage on port side, usable capacity 72.5 litres (19.2 US gallons; 16.0 Imp gallons). Optional auxiliary fuel tank, capacity 39.75 litres (10.5 US gallons; 8.7 Imp gallons). Oil capacity 5.7 litres (1.5 US gallons; 1.25 Imp gallons). No transmission limit; transmission overhaul interval 2,200 hours or 12 years.

ACCOMMODATION: Two seats side by side in enclosed cabin, with inertia reel shoulder harness. Curved two-panel, tinted windscreen. Removable door, with tinted window, on each side. Police version has observation doors with bubble windows, which are also available as options on other models. Baggage space beneath each seat. Cabin heated and ventilated.

SYSTEMS: Electrical system, powered by 12 V DC alternator, includes navigation, panel and map lights, dual landing lights, anti-collision light and battery. Second battery optional.

AVIONICS: *Comms:* Bendix/King KY197A VHF com radio or Garmin 430 GPS/COM/VOR/LOC/GS; KT 76C transponder and remote altitude encoder.

Flight: Optional Garmin 150XL/250XL/400. IFR trainer avionics listed under Current Versions above.

Instrumentation: AIM 305-1 AL DVF artificial horizon, Bendix/King KEA 129 encoding altimeter, Astronautics DC turn indicator, rate of climb indicator, sensitive altimeter, quartz clock, hour meter, low rotor rpm warning horn, temperature and chip warning lights for main gearbox and chip warning light for tail gearbox.

EQUIPMENT: Landing and anti-collision lights.

DIMENSIONS, EXTERNAL:

Main rotor diameter	7.67 m (25 ft 2 in)
Tail rotor diameter	1.07 m (3 ft 6 in)
Main rotor blade chord	0.18 m (7¼ in)
Distance between rotor centres	4.39 m (14 ft 5 in)
Length overall, rotors turning	8.76 m (28 ft 9 in)
Fuselage: Length	6.30 m (20 ft 8 in)
Max width	1.12 m (3 ft 8 in)
Height: overall	2.72 m (8 ft 11 in)
to top of cabin	1.75 m (5 ft 9 in)
Skid track	1.93 m (6 ft 4 in)

DIMENSIONS, INTERNAL:

Cabin max width	1.12 m (3 ft 8 in)

AREAS:

Main rotor blades (each)	0.70 m² (7.55 sq ft)
Tail rotor blades (each)	0.04 m² (0.40 sq ft)
Main rotor disc	46.21 m² (497.4 sq ft)
Tail rotor disc	0.89 m² (9.63 sq ft)
Fin	0.21 m² (2.28 sq ft)
Stabiliser	0.14 m² (1.53 sq ft)

WEIGHTS AND LOADINGS:

Weight empty (without auxiliary fuel tank):	
Beta II	388 kg (855 lb)
Mariner II	405 kg (892 lb)
Fuel weight: standard	52 kg (115 lb)
auxiliary	28.6 kg (63 lb)
Max T-O and landing weight	621 kg (1,370 lb)
Max zero-fuel weight	569 kg (1,255 lb)
Max disc loading	13.4 kg/m² (2.75 lb/sq ft)
Max power loading at T-O	6.37 kg/kW (10.46 lb/hp)

PERFORMANCE (Beta II):

Never-exceed speed (VNE):	
without sling load	102 kt (190 km/h; 118 mph)
with sling load	75 kt (139 km/h; 86 mph)
Max level speed	97 kt (180 km/h; 112 mph)
Cruising speed at 70% power at 2,440 m (8,000 ft)	96 kt (177 km/h; 110 mph)
Econ cruising speed	82 kt (153 km/h; 95 mph)
Max rate of climb at S/L	more than 305 m (1,000 ft)/min
Rate of climb at 3,050 m (10,000 ft)	more than 183 m (600 ft)/min
Service ceiling	4,265 m (14,000 ft)
Hovering ceiling IGE	2,865 m (9,400 ft)
Range with max payload, no reserves:	
normal fuel	more than 173 n miles (321 km; 200 miles)
max fuel	more than 260 n miles (482 km; 300 miles)
Endurance at 65% power, auxiliary fuel, no reserves	3 h 20 min

UPDATED

ROBINSON R44

TYPE: Four-seat helicopter.

PROGRAMME: Development began 1986; first flight (N44RH) 31 March 1990; first flight of third R44 March 1992; marketing began 22 March 1992; certification completed 10 December 1992. In August 1997 an R44 became first piston helicopter flown around the world. In June 2002 an R44 Raven (G-NUDE) became the first piston-engined helicopter to land at the North Pole.

CURRENT VERSIONS: **R44 Astro** and **Raven:** Versions available for passenger transport, water bombing, law enforcement and logging support. Raven, introduced in April 2000, has hydraulic flight control system as standard, plus elastomeric tail rotor bearings and adjustable pedals for pilot.

R44 Clipper: Float-equipped version; certified 17 July 1996. Standard landing gear set extra US$6,000. Optional avionics include ICOM IC-M59 marine radio, IC-271 2 m radio and Garmin GPS MAP 130. Pop-out floats certified June 1999. Total of 141 Clippers sold by 1 June 2001.

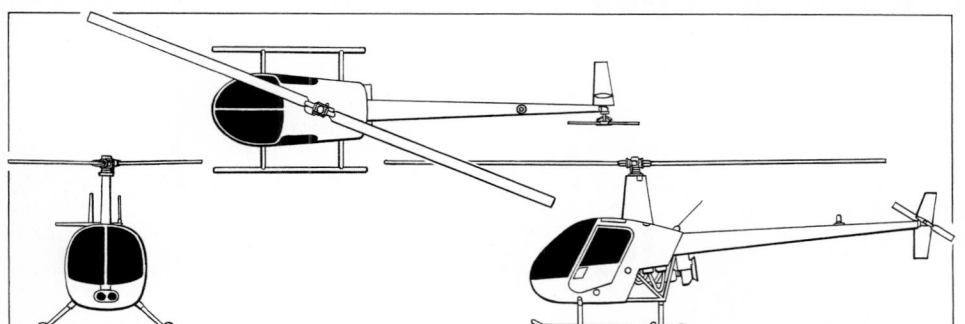

Robinson R22 Beta two-seat helicopter (*Jane's/James Goulding*) 0051888

For details of the latest updates to *Jane's All the World's Aircraft* online and to discover the additional information available exclusively to online subscribers please visit

jawa.janes.com

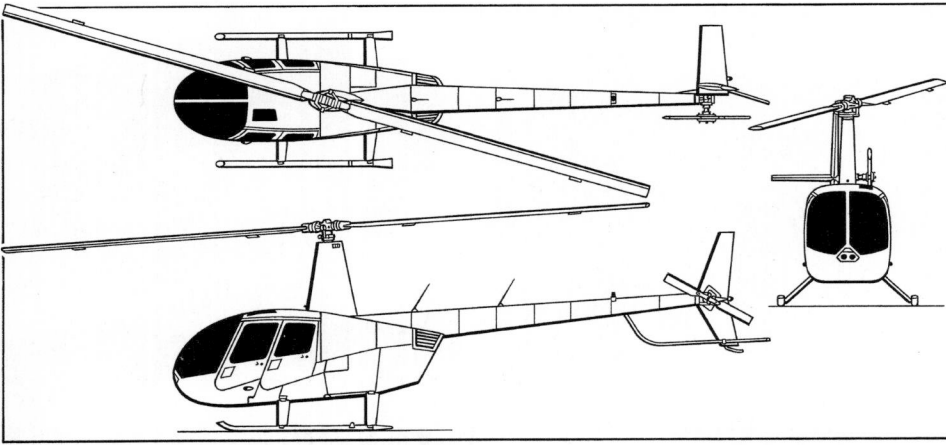

Robinson R44 light helicopter *(Jane's/Mike Keep)*

R44 Police: Specially modified for law enforcement, with Inframetrics 445G MkII IR sensor, TV camera and ×7 zoom lens mounted in gyrostabilised nose turret; video monitor; Spectrolab SX-5E searchlight; FM radio package; bubble door windows; Bendix/King KFM 985 dual-band transceiver, KY 196A VHF, II Morrow SL-60 standby comms, KT 76C transponder, 28 V electrical system and extended landing gear struts. Empty weight, equipped, approx 715 kg (1,570 lb). Certified July 1997; first customer El Monte Police Agency of Los Angeles. Recent customers include the Estonian Air Force, which ordered two in February 2002 for delivery commencing May 2002, and China's Zheng Zhou Public Security Bureau, which took delivery of one in June 2002 following its appearance at the China Expo 2002 exhibition..

R44 Newscopter: Electronic news gathering (ENG) version intended for media companies and fitted with FLIR Systems UltraMedia RS gyrostabilised video-camera and microwave transmitter; four video monitors, two micro cameras, KY 196 VHF, KT 76C transponder (Mode C) standard, two FM radios and bubble window in port door. Empty weight, equipped, 725 kg (1,598 lb). Deliveries began January 1998; 29 sold by 1 June 2002. Digital ENG version introduced in 2002 featuring Ikegami HL-59WNA digital camera with ×21 lens and 360° continuous rotation with five-axis gyro stabilisation; first R44 so equipped was delivered in June 2002 to Metro Networks, and was Metro's ninth R44 and the 29th ENG version delivered.

R44 IFR: Equipped for IFR helicopter training; generally as for R22 IFR. 18 Sold by 1 June 2001.

R44 Raven II: Upgraded version, introduced in June 2002. FAA certification and first deliveries in November 2002. Features a Textron Lycoming IO-540 engine with take-off rating increased to 183 kW (245 hp) for five

minutes, 153 kW (205 hp) maximum continuous, and 28 V 70 A electrical system. Provisional data include: empty weight 680 kg (1,500 lb); maximum take-off weight 1,134 kg (2,500 lb); hovering ceiling IGE: 2,835 m (9,300 ft) at ISA; 2,469 m (8,100 ft) at ISA + 20°C; hovering ceiling OGE: 1,859 m (6,100 ft) at ISA; 1,463 m (4,800 ft) at ISA + 20°C. More than 50 ordered by November 2002, including two ENG variants.

CUSTOMERS: Total of 1,125 delivered by 1 September 2001. Deliveries totalled 150 in 1999, 264 in 2000 and 194 in 2001. 1,000th delivery took place in February 2001.

COSTS: Standard price US$307,000 (2002). Clipper US$323,000 with utility floats, US$330,000 with pop-out floats (2002). Direct operating cost US$43.30 per hour; total (500 hours annually) US$124.96 per hour (2002). R44 Police US$485,000; Newscopter, typically US$525,000 (both 2001). Raven II US$335,000 (2002)

DESIGN FEATURES: New design, incorporating general configuration (but larger cabin) and some proven concepts of R22; designed to requirements of FAR Pt 27; features for comfort and safety include electronic throttle governor to reduce pilot workload by controlling rotor and engine rpm during normal operations, rotor brake, advanced warning devices, automatic clutch engagement to simplify and reduce start-up procedure and reduce chance of overspeed, T-bar cyclic control, and crashworthy features including energy-absorbing landing gear and lap/shoulder strap restraints designed for high forward *g* loads. High reliability and low maintenance; patented rotor design with tri-hinge (see R22); low noise levels.

FLYING CONTROLS: Conventional, with Robinson central cyclic stick; rpm governor; rotor brake standard; left-hand collective lever and pedals removable. Customised hydraulic flight control system certified July 1999 and

offered from September 1999 as option at US$13,000 (2000).

LANDING GEAR: Fixed skids or, on R44 Clipper, twin utility floats or pop-out helium floats which inflate in 2 seconds.

POWER PLANT: One 194 kW (260 hp) Textron Lycoming O-540 flat-six engine, derated to 165 kW (225 hp) at T-O, 153 kW (205 hp) continuous. Standard fuel capacity 116 litres (30.6 US gallons; 25.5 Imp gallons) in tank on port side; optional extra 69 litres (18.3 US gallons; 15.2 Imp gallons) in starboard tank. No transmission limit; transmission overhaul 2,200 hours or 12 years.

ACCOMMODATION: Four persons seated 2 + 2. Concealed baggage compartments. Dual controls. Cabin heated and ventilated. Tinted windscreen and door windows.

AVIONICS: Selection of optional avionics and instruments.
   Comms: Bendix/King KY 197A VHF radio; Inter-VOX AA-80 intercom for all four occupants; KT 76C transponder with remote altitude encoder. Optional Pointer 3000-10 or 4000-10 ELT.
   Flight: Optional Garmin 150XL/250XL/400 GPS.

EQUIPMENT: Standard equipment includes auxiliary fuel system and heater and rotor brake. Options include belly hardpoint.

DIMENSIONS, EXTERNAL:

Main rotor diameter	10.06 m (33 ft 0 in)
Tail rotor diameter	1.47 m (4 ft 10 in)
Length overall, rotors turning	11.76 m (38 ft 7 in)
Fuselage: Length	9.07 m (29 ft 9 in)
Max width	1.28 m (4 ft 2½ in)
Height: overall	3.28 m (10 ft 9 in)
to top of cabin	1.83 m (6 ft 0 in)
Skid track	2.18 m (7 ft 2 in)

DIMENSIONS, INTERNAL:

Cabin max width	1.19 m (3 ft 11 in)

AREAS:

Main rotor disc	79.46 m² (855.3 sq ft)
Tail rotor disc	1.70 m² (18.35 sq ft)

WEIGHTS AND LOADINGS:

Weight empty: standard	654 kg (1,442 lb)
Clipper with utility floats	676 kg (1,490 lb)
with pop-out floats	684 kg (1,509 lb)
Fuel weight: standard	83 kg (184 lb)
auxiliary	50 kg (110 lb)
Max T-O and landing weight	1,088 kg (2,400 lb)
Max disc loading	13.7 kg/m² (2.81 lb/sq ft)
Max power loading at T-O	6.49 kg/kW (10.67 lb/hp)

PERFORMANCE:

Cruising speed at 75% power	113 kt (209 km/h; 130 mph)
Max rate of climb at S/L	over 305 m (1,000 ft)/min
Service ceiling	4,270 m (14,000 ft)
Hovering ceiling:	
IGE at 1,088 kg (2,400 lb)	1,950 m (6,400 ft)
OGE at 998 kg (2,200 lb)	1,555 m (5,100 ft)
Max range, no reserves	approx 347 n miles (643 km; 400 miles)

UPDATED

Robinson R44 Raven II

NEW/0533357

ROTORWAY

ROTORWAY INTERNATIONAL

4140 West Mercury Way, Stellar Air Park, Chandler, Arizona 85226
Tel: (+1 480) 961 10 01
Fax: (+1 480) 961 15 14
e-mail: rotorway@rotorway.com
Web: http://www.rotorway.com
VICE-PRESIDENT, PRODUCTION: Rusty Mueller
VICE-PRESIDENT, SALES: Brent Marshall
MARKETING DIRECTOR: Susie Bell

Initial RotorWay Company set up by B J Schramm in early 1960s; successively produced the Javelin, Scorpion I and Scorpion II before launching Exec in 1980. Company purchased and re-established on 1 June 1990 by the late John Netherwood; assets acquired in 1996 by company's employees, who are current owners. Original product was Exec 90, the former Exec with some 23 modifications, which was certified in UK in 1990 and in Poland in 1993. In August 1994, Exec 162F introduced and replaced the Exec 90; 162F improvements can be retrofitted to Exec 90.

RotorWay currently manufactures Exec 162F helicopter kits for final assembly by amateur builders and operators. The Exec is powered by the company's own engine, which has been in production for more than 20 years and has recently been upgraded to improve performance at altitude.

Some foreign distributors also sell helicopters; approximately 50 per cent of sales are outside the USA. RotorWay, which had 40 employees in mid-2002, offers flight orientation and maintenance training, and customer service programme; 3,440 m² (37,000 sq ft) facility near Phoenix, Arizona, incorporates all departments, including manufacturing, sales and flight school.

UPDATED

ROTORWAY EXEC 162F

TYPE: Two-seat helicopter kitbuilt.
PROGRAMME: Development of first Rotorway helicopter (Scorpion) began in 1967; followed by Scorpion Too in 1971 and Exec in 1980. Kits, which lack only avionics and paint, are currently marketed in USA and 50 other countries; 50 per cent of production is shipped abroad. French certification received in October 2001 following approval of modifications to meet DGAC requirements, including improved fire protection, repositioned instrumentation and foot controls.
CURRENT VERSIONS: **Exec 162F:** *As described.*
  Exec 162 Pro: Not approved by Rotorway. Optional improvement package from Pro-Drive Inc, comprising modified main rotor drive system, 'E Z Start' clutch and carbon fibre tail rotor.
  JetExec: Not approved by Rotorway. Turbine conversion offered by Kiss Aviation of Perris, California;

JetExec turbine-powered version of Rotorway Exec 162, a modification not approved by the manufacturer (*Jane's/Paul Jackson*)   **NEW**/0526997

Rotorway Exec 162F   **NEW**/0522804

powered by 112 kW (150 hp) Solar T-62T-32 Titan. Fuel capacity 151 litres (40.0 US gallons; 33.3 Imp gallons); max rate of climb 548 m (1,800 ft)/min; range 260 n miles (482 km; 300 miles).
CUSTOMERS: Including earlier Exec 90 models, over 850 kits sold and 600 flying; 500th 162F delivered to Mexican Navy June 2000 and 600th to Heli Diffusion of France in June 2002. Russian version (50 on order) designated **162E**; first example (FLARF-02133) completed and flown in June 2000. International uses include police surveillance, forestry observation, powerline inspection, ranch work, crop-spraying, recreation and flight training.
COSTS: Kit: US$64,350 complete (2001). ACIS US$4,250 extra. Also available in four batches of parts to spread cost.

DESIGN FEATURES: Asymmetrical Rotorway-designed aerofoil section two-blade main rotor. All-metal aluminium alloy blades attached to aluminium alloy teetering rotor hub by retention straps. Teetering tail rotor, with two blades each comprising steel spar and aluminium alloy skin. Elastomeric bearing rotor hub system with dual push/pull cable-controlled swashplate for cyclic pitch control. Quoted build time 300 to 400 hours. Company offers on-line diagnostic testing by modem.
STRUCTURE: Blades as detailed under Design Features. Basic 4130 steel tube airframe structure, with wraparound glass fibre fuselage/cabin enclosure. Aluminium alloy monocoque tailboom.
LANDING GEAR: Twin-skid type. Floats optional (from outside source).
POWER PLANT: One 113 kW (152 hp) RotorWay International RI 162F 2.66 litre (162 cu in) liquid-cooled engine with FADEC. Optional Altitude Compensation Induction System (ACIS) supercharger supplies pressurised air to engine at altitude, obviating need for turbocharging and prolonging engine life. Standard fuel capacity 64.4 litres (17.0 US gallons; 14.2 Imp gallons).
ACCOMMODATION: Two persons side by side; optional external HeliPak baggage pod fits between skids, capacity 0.21 m³ (7.4 cu ft).
AVIONICS: Extra navigational equipment optional on US market helicopters.
EQUIPMENT: Optional crop-spraying kit, including two 42 litre (11.0 US gallon; 9.2 Imp gallon) tanks.

DIMENSIONS, EXTERNAL:
Main rotor diameter	7.62 m (25 ft 0 in)
Tail rotor diameter	1.28 m (4 ft 2½ in)
Length of fuselage	6.71 m (22 ft 0 in)
Length overall, rotors turning	8.99 m (29 ft 6 in)
Height to top of main rotor	2.44 m (8 ft 0 in)
Skid track	1.65 m (5 ft 5 in)

DIMENSIONS, INTERNAL:
Cabin max width	1.12 m (3 ft 8 in)

AREAS:
Main rotor disc	45.60 m² (490.9 sq ft)
Tail rotor disc	1.17 m² (12.57 sq ft)

WEIGHTS AND LOADINGS:
Weight empty	442 kg (975 lb)
Crew weight	193 kg (425 lb)
Max T-O weight	680 kg (1,500 lb)
Max disc loading	14.9 kg/m² (3.06 lb/sq ft)
Max power loading at T-O	6.01 kg/kW (9.87 lb/hp)

PERFORMANCE:
Never-exceed (VNE) and max level speed	100 kt (185 km/h; 115 mph)
Normal cruising speed	82 kt (153 km/h; 95 mph)
Max rate of climb at S/L	305 m (1,000 ft)/min
Service ceiling	3,050 m (10,000 ft)

Hovering ceiling, with two persons:
IGE	2,135 m (7,000 ft)
OGE	1,525 m (5,000 ft)
Range with max fuel at optimum cruising power	
	156 n miles (289 km; 180 miles)
Endurance with max fuel at optimum cruising power	2 h

UPDATED

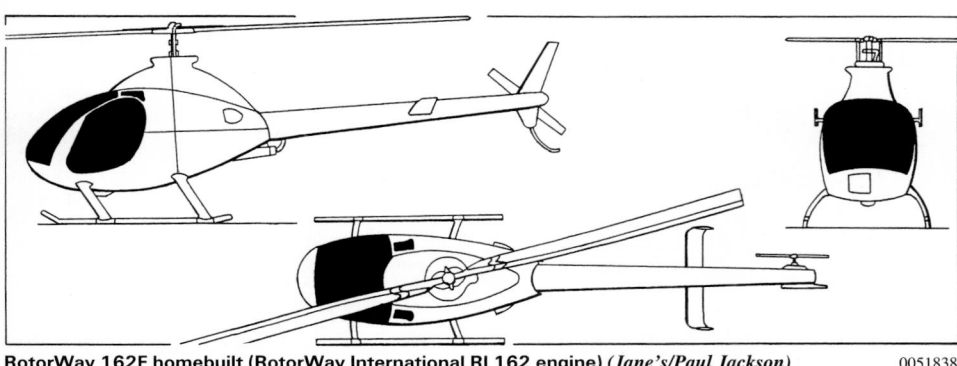

RotorWay 162F homebuilt (RotorWay International RI 162 engine) (*Jane's/Paul Jackson*)   0051838

Rotorway Exec 162F kit with completed aircraft in background   **NEW**/0522805

SAFIRE

SAFIRE AIRCRAFT COMPANY

400 Clematis Street, Suite 207, West Palm Beach, Florida
33401
Tel: (+1 561) 650 08 30 and (+1 800) 862 86 80
Fax: (+1 561) 659 63 19
e-mail: info@safireaircraft.com
Web: http://www.safireaircraft.com
CHAIRMAN: Dimitri Margaritoff
PRESIDENT: Robert M Kuhn
EXECUTIVE VICE-PRESIDENT: David Humphries
DIRECTOR, SALES AND MARKETING: Bruce Hamilton
MANUFACTURING ENGINEERING MANAGER: Joe DiVito

Safire, formed in September 1998, is developing the S-26 six-
seat business jet. Later projects will include the two-seat S-12
and four-seat S-14.

UPDATED

SAFIRE S-26

TYPE: Light business jet.
PROGRAMME: Preliminary details released early 1999. By
2002, S-26 was being envisaged with increased engine
power, weight and speed. Wind-tunnel testing of a one-
fifth scale model was scheduled for completion in June
2002 to be followed by selection of key components and
hardware. First flight planned for February 2004 and FAR
Pt 23 certification for single pilot operation a year later.
Target production rate four per day.
CUSTOMERS: Total of 900 delivery positions reserved by
January 2003.
COSTS: Launch flyaway cost US$869,000 (2000), increased to
US$919,000 in September 2001 for all subsequent orders.
Estimated direct operating cost US$280 per hour (2002).
DESIGN FEATURES: Conventional low-wing monoplane with
rear-mounted engines and sweptback cruciform tail.
Supercritical laminar-flow wing aerofoil.
FLYING CONTROLS: Conventional and manual. Control
surfaces aerodynamically and mass balanced. Ailerons
coupled with outboard spoilers to provide roll control.
Inboard spoilers function as airbrakes. Trim tabs in port
aileron, starboard elevator and rudder. Yaw damper
planned. Fowler flaps.
STRUCTURE: Combined aluminium and carbon fibre
composites. Wing is three-spar design with widely spaced
ribs. Fuselage uses keel beam and longerons with few
frames.
LANDING GEAR: Hydraulically retractable tricycle type, with
single wheel on each unit. Oleo-pneumatic shock-
absorbers and simple disc brakes with composites pads.
Trailing-link main units retract inward into wing and wing
fuselage fairing; nose gear retracts forward. Nosewheel
size 6.00-6 with 2.90 bar (42.0 lb/sq in) tyre pressure;
mainwheels size 6.50-8 with 5.17 bar (75.0 lb/sq in) tyres.
POWER PLANT: Original choice was two Williams FJX-2
turbofans, each approximately 3.11 kN (700 lb st); this

Artist's impression of Safire S-26   *NEW*/0533715

was changed to two Agilis TF-800 turbofans each of
4.00 kN (900 lb st) and subsequently to two Agilis
TF-1000 turbofans, each rated at 4.45 kN (1,000 lb st), but
relationship with Agilis terminated in late 2002 when an
alternative power plant was being sought. Usable fuel
capacity 852 litres (225 US gallons; 187 Imp gallons) in
integral wing tanks. Single refuelling point in port fuselage
side.
ACCOMMODATION: Pilot and co-pilot or passenger in cockpit.
Seating for four in club arrangement in main cabin.
Clamshell door with integral steps on port side of cabin;
emergency exit on starboard side. Lavatory and in-flight
accessible baggage compartment aft of main cabin.
SYSTEMS: Autopilot standard. Pressurisation system supplied
with engine bleed air, maximum pressure differential 0.54
bar (7.8 lb/sq in), maintaining 2,440 m (8,000 ft) cabin
altitude to 11,280 m (37,000 ft) certified ceiling. Vapour
cycle air conditioning system. Hydraulic system, pressure
103.4 bar (1,500 lb/sq in), operates landing gear only.
Separate hydraulic system for brakes. Electrical system
supplied by two 42 V 300 A DC starter/generators. Oxygen
system includes two demand masks for crew and four
drop-down constant-flow masks for passengers. Engine
fire detection and extinguishing system. Engine inlet lip
anti-iced with engine bleed air; wings and empennage anti-
iced with TKS fluid system.
AVIONICS: Fully digital avionics suite designed to support
NASA/FAA Highways-in-the-Sky and synthetic vision
programmes as technology becomes available.
Instrumentation: Full EFIS with three 260 mm (10.4 in)
flat panel displays. Left and right screens present attitude,
airspeed, altitude, and short-range navigation (ILS, TCAS,

flight director, terrain avoidance) information. Centre
screen provides navigation (moving map), systems
management (engine data, fuel management, electrical
system status), radio frequency management (nav/com),
transponder (Mode S), master caution and warning
information and checklists (normal and emergency).
Following data are provisional:

DIMENSIONS, EXTERNAL:

Wing span	11.58 m (38 ft 0 in)
Length overall	10.97 m (36 ft 0 in)
Fuselage max diameter	1.56 m (5 ft 1½ in)
Height overall	4.39 m (14 ft 5 in)
Tailplane span	4.44 m (14 ft 7 in)
Wheel track	2.49 m (8 ft 2 in)
Wheelbase	4.05 m (13 ft 3½ in)
Crew/passenger door (port):	
Height	1.19 m (3 ft 11 in)
Width	0.61 m (2 ft 0 in)
Height to sill	0.94 m (3 ft 1 in)
Emergency exit (starboard):	
Height	0.66 m (2 ft 2 in)
Width	0.61 m (2 ft 0 in)
Height to sill	1.48 m (4 ft 10¼ in)

DIMENSIONS, INTERNAL:

Cabin (between pressure bulkheads)	
Length:	4.11 m (13 ft 6 in)
Max width	1.45 m (4 ft 9 in)
Max height	1.35 m (4 ft 5 in)
Baggage volume	0.40 m³ (14.0 cu ft)

AREAS:

Ailerons, including tab (total)	0.38 m² (4.06 sq ft)
Trailing-edge flaps (total)	2.75 m² (29.58 sq ft)
Roll-control spoilers (total)	0.36 m² (3.85 sq ft)
Airbrakes (total)	0.36 m² (3.83 sq ft)
Fin	1.98 m² (21.31 sq ft)
Rudder, incl tab	0.74 m² (8.01 sq ft)
Tailplane	2.61 m² (28.05 sq ft)
Elevators (total) incl tab	0.63 m² (6.78 sq ft)

WEIGHTS AND LOADINGS:

Weight empty, equipped	1,654 kg (3,646 lb)
Max payload	635 kg (1,400 lb)
Max take-off weight	2,676 kg (5,900 lb)
Max power loading	301 kg/kN (2.95 lb/lb st)

PERFORMANCE:

Max operating Mach number (M_{MO})	0.61
Max cruising speed	340 kt (630 km/h; 391 mph) TAS
Econ cruising speed at 11,280 m (37,000 ft)	
	260 kt (482 km/h; 299 mph) TAS
Approach speed	90 kt (167 km/h; 104 mph)
Stalling speed, landing configuration	
	69 kt (128 km/h; 80 mph) CAS
Max rate of climb at S/L	884 m (2,900 ft)/min
Rate of climb at S/L, OEI	244 m (800 ft)/min
Max certified altitude	11,280 m (37,000 ft)
T-O to, and landing from, 15 m (50 ft)	
	less than 762 m (2,500 ft)
Max range	1,400 n miles (2,592 km; 1,611 miles)
Range with 363 kg (800 lb) payload, NBAA IFR reserves	
	1,020 n miles (1,889 km; 1,173 miles)

UPDATED

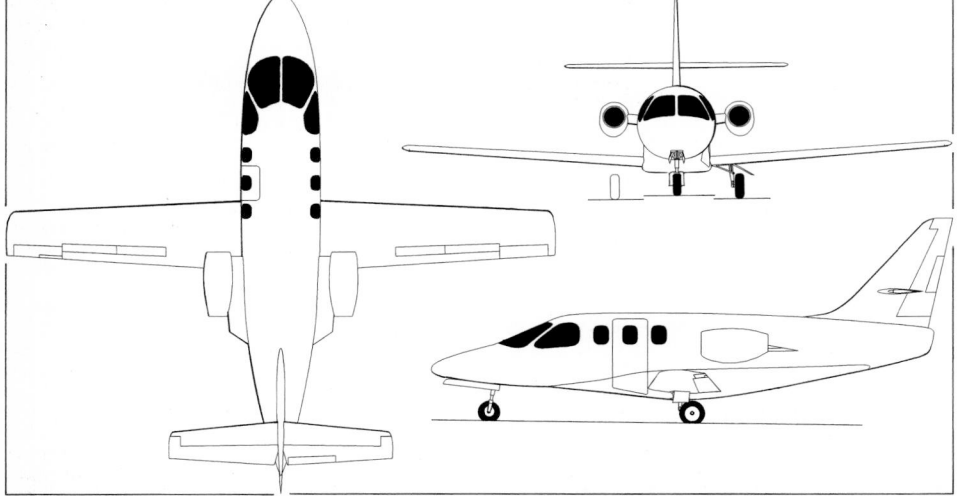

Safire S-26 general arrangement *(Jane's/Paul Jackson)*   0126966

SCALED

SCALED COMPOSITES LLC

1624 Flight Line, Mojave, California 93501-1663
Tel: (+1 661) 824 45 41
Fax: (+1 661) 824 41 74
e-mail: info@scaled.com
Web: http://www.scaled.com
PRESIDENT: Burt (Elbert L) Rutan
VICE-PRESIDENT AND GENERAL MANAGER: Michael W Melvill

Scaled Composites founded 1982; bought by Beech Aircraft
Corporation (now Raytheon) June 1985; sold back to Burt
Rutan November 1988 and integrated in joint venture with
Wyman-Gordon Company of North Grafton, Massachusetts;
is now owned by private investors following buy-out by Burt
Rutan and 10 other investors; continues to provide R&D
facilities to individuals and companies; several projects
developed for Beech retained by Scaled.
Scaled Composites is responsible for research and
development and prototype construction; company also

active in UAV field. Current programmes include proprietary
products.
Scaled produced the prototype and undertook early test
flights of the Visionaire Vantage business jet. Similarly,
Scaled was involved in the Adam M-309 (now A500, which
see), which it designed, built and flight tested.
The company's V-Jet II testbed for the Williams FJX-2
turbofan has been replaced by the Eclipse Eclipse 500.
Company currently has 145 employees and annual revenue
of around US$18 million.

UPDATED

SCALED 281 PROTEUS

TYPE: High-altitude platform.

PROGRAMME: Announced 24 October 1997. First flight (N281PR) 26 July 1998. In 1998, Angel Technologies of St Louis teamed with Raytheon to develop air and ground elements for HALO fixed and mobile radio, data and video services, and signed agreement with Scaled owners Wyman-Gordon for 100 Proteus-type production aircraft. By late 2000, this project was understood to have lapsed, and prototype remains only example so far built.

During 1999, participated in NASA's Environmental Research Aircraft and Sensor Technology (ERAST) programme, carrying (from 2 February onwards) ventral pod containing HyperSpectral Sciences Inc Airborne Real-Time Imaging System (ARTIS) camera. NASA's Dryden facility also assisted with development of station-keeping autopilot and satcom-based uplink/downlink data system.

On 25 and 27 October 2000, the Proteus set three world altitude records – peak altitude 19,277 m (63,245 ft), sustained altitude in horizontal flight 19,015 m (62,385 ft) and peak altitude of 17,067 m (55,994 ft) with 1,000 kg (2,205 lb) payload.

During 2000-01, the aircraft successfully completed four atmospheric sounding science deployments in support of NASA/NOAA/DoD Integrated Program Office, including a flight around the Pacific rim and over the North Pole.

In January 2001 aircraft was sold to Northrop Grumman as a sensor and mission platform and continues to serve as a systems and atmospheric research testbed, operated on behalf of Northrop Grumman by Scaled. Deployments in 2002 included Wallops Island, Virginia (upper atmosphere research); Oklahoma City (study of thunderstorm development); Key West, Florida (atmospheric research); and Ponca City, Oklahoma. A NASA deployment to Sweden was expected in 2003.

By August 2002 the Proteus had accumulated more than 850 flight hours, including about 160 flights above 15,240 m (50,000 ft).

CUSTOMERS: There are currently no plans to build more aircraft.

CURRENT VERSIONS: **Telecommunications:** *As described.*

Atmospheric Research: Single pilot operation; two scientists in cabin; able to loiter for 18 hours at 500 n mile (926 km; 575 mile) radius or 6 hours at 3,000 n mile (5,556 km; 3,452 mile) radius. Service ceiling 19,810 m (65,000 ft) at 3,175 kg (7,000 lb) weight.

Reconnaissance/surveillance: Single pilot plus one crew member; carries integrated aircraft/payload thermal management system. Max T-O weight 7,167 kg (15,800 lb) including payload of 2,268 kg (5,000 lb). Service ceiling 19,510 m (64,000 ft) at 2,438 kg (8,000 lb); loiter time 22 hours at 500 n mile (926 km; 575 mile) radius or 12 hours at 2,000 n mile (3,704 km; 2,301 mile) radius. Fuel weight up to 4,082 kg (9,000 lb) can be carried using fuselage tank. Can also be operated as a UAV (refer *Jane's Unmanned Aerial Vehicles and Targets*).

Micro Satellite Launcher: Capable of carrying micro satellites up to 29 kg (65 lb) using two-stage rocket, including reusable booster. Rocket would be launched at 10,973 m (36,000 ft) and speed of 180 kt (333 km/h; 207 mph) up to 70°γ. Satellite launch cost US$600,000. Crew of pilot plus launch officer. Gross weight 6,350 kg (14,000 lb); payload 2,948 kg (6,500 lb). Port aft wing and canard wingtips installed for asymmetric loads.

DESIGN FEATURES: Optimised for loiter at 18,000 to 20,000 m (59,050 to 65,620 ft) carrying underslung antenna, but most flights to date made between 16,760 and 17,680 m (55,000 and 58,000 ft). Highly unconventional configuration, incorporating large foreplane; twin, unbraced tailbooms carrying vertical fins; and twin, podded turbofans mounted on rear fuselage shoulders. Anhedral on outboard wing panels. Provision for extended wing- and foreplane tips. Fuselage centre-section interchangeable depending on mission.

FLYING CONTROLS: Manual elevators in foreplanes with servo tab on port elevator; ailerons in centre-section of each wing, outboard of booms; twin rudders with trim tab to port.

STRUCTURE: Composites throughout. Participating companies include Hickok Inc (engine instruments), Puroflow Inc (fuel filters), Concorde Battery Corporation (batteries), Hamilton Sundstrand (engine and throttle controls), Whittaker Corporation, Dodson International and Arens Controls.

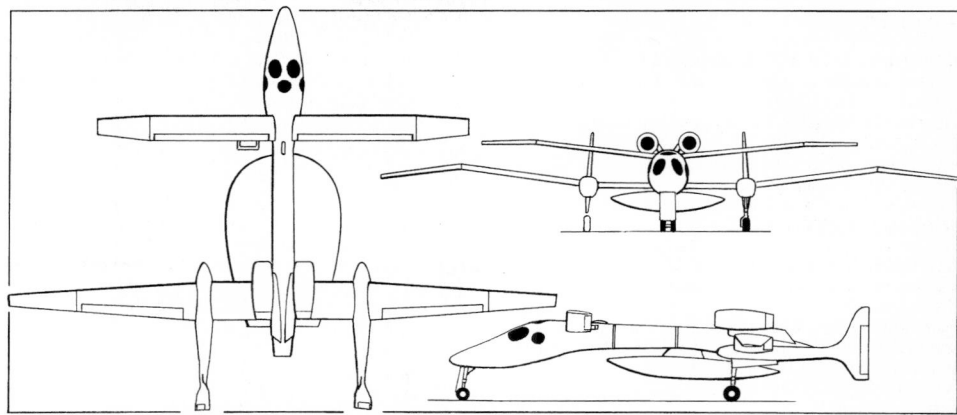

Scaled Proteus high-altitude research aircraft (*Jane's/James Goulding*)   0084579

The sole Scaled Proteus (*Jane's/Paul Jackson*)   *NEW*/0526996

LANDING GEAR: Retractable tricycle type. Twin steerable nosewheels (size 6.00-6) retract rearwards; single mainwheels (6.50-10) retract rearwards into tailbooms. Hydraulic disc brakes on mainwheels. Wide wheelbase improves crosswind take-off capability.

POWER PLANT: Two 10.2 kN (2,293 lb st) Williams FJ44-2 turbofans. Fuel capacity 2,722 kg (6,000 lb) in 10 wing and fuselage sump tanks. Additional auxiliary tanks can be fitted.

ACCOMMODATION: Pilot and relief pilot in cabin pressurised by engine bleed air system to 0.69 bar (10.0 lb/sq in) differential, providing 2,440 m (8,000 ft) equivalent altitude at cruising height. Crew door starboard side.

SYSTEMS: Electrical system 28 V. 30 kW electrical power available for payload. Primary coolant loop and power conditioning unit housed in fuselage.

EQUIPMENT: A range of ventral pods can be fitted to carry a variety of scientific equipment.

DIMENSIONS, EXTERNAL:
Wing span: normal	23.65 m (77 ft 7 in)
extended	27.99 m (91 ft 10 in)
Foreplane span: normal	16.66 m (54 ft 8 in)
extended	19.71 m (64 ft 8 in)
Length overall	17.17 m (56 ft 4 in)
Height overall	5.38 m (17 ft 8 in)

DIMENSIONS, INTERNAL:
Cabin: Length	2.74 m (9 ft 0 in)
Max diameter	1.55 m (5 ft 1 in)

AREAS:
Wings, gross, normal	27.92 m² (300.5 sq ft)
Foreplanes, gross, normal	16.60 m² (178.7 sq ft)

WEIGHTS AND LOADINGS:
Weight empty	2,658 kg (5,860 lb)
Mission payload	1,043 kg (2,300 lb)
Max fuel weight	3,311 kg (7,300 lb)
Max T-O weight: FAR Pt 23	5,670 kg (12,500 lb)
military	7,166 kg (15,800 lb)
Max landing weight	4,900 kg (11,000 lb)
Max wing/foreplane loading, normal span, military	
	161.0 kg/m² (32.97 lb/sq ft)
Max power loading, military	351 kg/kN (3.45 lb/lb st)

PERFORMANCE (FAR Pt 23, relay role):
Max level speed	272 kt (504 km/h; 313 mph)
Econ cruising speed:	
at FL200	190 kt (352 km/h; 219 mph)
at FL400	280 kt (519 km/h; 322 mph)
Unstick speed	71 kt (130 km/h; 81 mph)
Max rate of climb at S/L	1,036 m (3,400 ft)/min
Cruising altitude, mid-mission	18,590 m (61,000 ft)
T-O run: at 3,629 kg (8,000 lb)	223 m (730 ft)
at FAR Pt 23 MTOW	433 m (1,420 ft)
Endurance on station, 1,000 n miles (1,852 km;	
1,150 miles) from base	14 h
g limits	+3.2/–1.8

UPDATED

SCALED (TOYOTA) TAA-1

TYPE: Four-seat lightplane.

PROGRAMME: Outcome of four-year study sponsored by Toyota Motor Corporation and Toyota Motor Sales (US arms of Japanese motorcar manufacturer), with objective of entering US lightplane market. TAA-1 (Toyota Advanced Aircraft) POC prototype (N72TA) made maiden flight at Mojave Airport, California, on 31 May 2002, flown by Scaled Composites' test pilot Jon Karkow. No plans to produce TAA-1 in quantity; Toyota spokesperson quoted as saying that prototype has no connection with "the new aircraft Toyota is considering now", which would have both fixed- and retractable-gear versions and advanced avionics.

COSTS: Reported investment of some US$50 million for conception, design and prototype manufacture; unit cost objective of US$150,000 (2002).

DESIGN FEATURES: Conventional lightplane configuration with low-mounted, tapered wings and sweptback vertical tail; hardtop cabin.

Toyota proprietary laminar flow wing section.

STRUCTURE: Mainly composites; fuselage is one-piece, co-cured (single-moulded) shell of carbon fibre and resin. Prototype built by Scaled Composites.

LANDING GEAR: Non-retractable tricycle type.

POWER PLANT: One 149 kW (200 hp) Textron Lycoming IO-360 flat-four engine; two-blade propeller.

NEW ENTRY

SCALED 318

TYPE: High-altitude platform.

PROGRAMME: Revealed following first flight of prototype (N318SL) on 1 August 2002, but no other details divulged. Three further flights made by end of that month, including one lasting 'for hours'.

TAA-1 testbed built for Toyota by Scaled Composites   *NEW*/0530178

Scaled 318 during test flying at Mojave   *NEW*/0527146

DESIGN FEATURES: Twin-boom configuration with separate T tails and short fuselage pod. High-aspect ratio wings have marked dihedral inboard of booms and approximately equal dihedral outboard, raising fuselage high above ground. Podded jet engine pylon-mounted to fuselage above each wingroot. Tailbooms sharply tapered to narrow point at front. Forward section of fuselage pod bulged, in similar fashion to Proteus but with many more circular windows (at least 14) and forward-opening circular crew door, sugesting possible extremely high-altitude role. Very wide-track landing gear.

LANDING GEAR: Four-wheel type. Rearward-retracting mainwheel in each tailboom in line with wing anhedral/dihedral junction; small, non-retractable, 'spatted' nosewheel at front of each boom.

NEW ENTRY

SCHWEIZER

SCHWEIZER AIRCRAFT CORPORATION

PO Box 147, Elmira, New York 14902
Tel: (+1 607) 739 38 21
Fax: (+1 607) 796 24 88
e-mail: marketing@sacusa.com
Web: http://www.schweizer-aircraft.com
PRESIDENT: W Stuart Schweizer
EXECUTIVE VICE-PRESIDENTS:
  Leslie E Schweizer
  Paul H Schweizer
VICE-PRESIDENT: Michael D Oakley
MARKETING CO-ORDINATOR: Peter S Schweizer
MARKETING DIRECTORS:
  David K Savage
  Barbara J Tweedt

Established 1939 by Ernest Schweizer (1912-2000) to produce sailplanes; from mid-1957 to 1995 made Grumman (later Gulfstream American) Ag-Cat, initially under subcontract, but as Schweizer-owned product from January 1981; manufacturing rights sold to Ag-Cat Corporation of Malden, Missouri, but production terminated soon thereafter.

Schweizer acquired rights for sole US manufacture of Hughes 300 light helicopter 13 July 1983; continues to support earlier Hughes 300s; first Elmira-built 300C completed June 1984; Schweizer purchased US rights for whole 300C programme from McDonnell Douglas Helicopter Company (formerly Hughes Helicopters) 21 November 1986. Deliveries totalled 56 helicopters in 1996, 44 in 1997, 41 in 1998, 43 in 1999, 36 in 2000 and 33 in 2001, by which time Schweizer had delivered nearly 850 helicopters.

Schweizer also manufactures the Model 300C, 300CB, 300CBi and 333 helicopters and the 2-37 surveillance aircraft. An unmanned version of the 300CB, known as the RoboCopter 300, has been developed in collaboration with Kawada Industries Inc of Japan and is described in *Jane's Unmanned Aerial Vehicles and Targets*.

At the Heli-Expo show at Orlando, Florida, in February 2002 Schweizer, Sikorsky Aircraft and Shanghai Little Eagle Science and Technology (SLEC) of the People's Republic of China announced an agreement whereby a new joint-venture company, Shanghai Sikorsky Helicopter Co, in which SLEC and Sikorsky hold 51 per cent and 49 per cent shares respectively, will produce Schweizer Model 300 series helicopters under licence. Initially, Schweizer 300C/CBs will be assembled from Schweizer-supplied kits, leading to manufacture of detail parts and eventually to full manufacture in China. First three kits arrived in Shanghai on 26 August 2002, followed by two more in September.

Subcontracts include aircraft structural components and assemblies for Northrop Grumman, Sikorsky and Boeing. The company employed 460 in 2001 in its 18,580 m² (200,000 sq ft) facility.

UPDATED

SCHWEIZER SA 2-37

US military designation: RG-8A

TYPE: Multisensor surveillance lightplane.
PROGRAMME: First flight 1985; prototype fitted with Hughes AN/AAQ-16 and other manufacturers' thermal imaging systems. Full description in 1991-92 and earlier *Jane's*.
CURRENT VERSIONS: **SA 2-37A:** Baseline version, *as described.*
  SA 2-37B: Powered by 186 kW (250 hp) Textron Lycoming TIO-540 engines. First aircraft (N6150U, c/n 015) registered February 2001; all subsequent production is to this standard.
CUSTOMERS: Total 16 built by late 2001, including demonstrator. Three for US Army, of which one lost in accident; remaining two transferred in 1987 to US Coast Guard at Opa Locka for anti-narcotics operations; one converted to SA 2-38A (see next entry).
  Fifth, sixth and seventh 2-37As built for Central Intelligence Agency late 1989/early 1990; others to Colombian and Mexican air forces, possibly to assist US drugs interdiction efforts; one (No. 010) possibly used as testbed by Lockheed Martin. See table of production.
  CIA aircraft employed as airborne communications relay platforms for long-range reconnaissance UAVs; believed used over former Yugoslavia in 1994, supporting General Atomics Gnat 750 vehicles; also fitted with surface surveillance equipment to identify sites for NATO air strikes in same operational theatre. Participated in hostage rescue at Japanese embassy in Peru, 1997.
DESIGN FEATURES: Modification of Schweizer SGM 2-37 motor glider, but with slightly greater wing span, drooped leading-edge and leading-edge fences on outer wing panels

to improve stall, much more powerful engine with large exhaust silencers on fuselage sides, three-blade quiet propeller, fuselage modified to accept bulged canopy and larger engine, streamlined fairings and hydraulic parking brake on mainwheels; more than treble standard fuel capacity, but optional extra tank also available; removable underfuselage skin and hatches give access to 1.84 m³ (65 cu ft) payload bay behind cockpit, in which pallets holding LLLTV, FLIR or camera payloads can be quickly removed and installed; other engines and larger payloads available for surveillance, basic and advanced training, operator training, glider and banner towing, and priority cargo delivery. Certified to FAR Pt 23 for day and night IFR; inaudible when overflying at 'quiet mode' speed at about 610 m (2,000 ft) using 38.8 kW (52 hp) from its 175 kW (235 hp) Textron Lycoming IO-540 engine.
  Wing section Wortmann FX-61-163 at root and FX-60-126 (modified) at tip; outer wing panels and horizontal tail can be removed for transport.
FLYING CONTROLS: Manual. All-moving tailplane with anti-balance tab; externally mass-balanced ailerons; rudder.
LANDING GEAR: Tailwheel type; fixed. Cleveland disc brakes.
POWER PLANT: Initially one 175 kW (235 hp) Textron Lycoming IO-540-W3A5D flat-six engine, driving a McCauley three-blade constant-speed propeller. Reportedly re-engined in service with 186 kW (250 hp) turbocharged TIO-540. Standard fuel capacity of 196.8 litres (52.0 US gallons; 43.3 Imp gallons), increasable optionally to 253.6 litres (67.0 US gallons; 55.8 Imp gallons). Shadin fuel flow system.
ACCOMMODATION: Seats for two persons side by side under two-piece upward-opening canopy, hinged on centreline. Dual controls, seat belts and inertia reel harnesses standard. Compartment aft of seats enlarged to accommodate pallet containing up to 340 kg (750 lb) of sensors or other equipment.

DIMENSIONS, EXTERNAL:
Wing span	18.745 m (61 ft 6 in)
Wing aspect ratio	19
Fuselage length	8.46 m (27 ft 9 in)
Height overall (tail down)	2.36 m (7 ft 9 in)
Wheel track	2.79 m (9 ft 2 in)
Wheelbase	5.99 m (19 ft 8 in)
Propeller diameter	2.18 m (7 ft 2 in)

AREAS:
Wings, gross	18.52 m² (199.4 sq ft)

SCHWEIZER SA 2-37 PRODUCTION

c/n	Reg/serial	Date	Operator
1	N3623C	1985	Schweizer
	85-0047	1986	US Army; to
	8101	1987	USCG; to SA 2-38A
2	N3623F	1985	Schweizer
	85-0048	1986	US Army; lost
3	N9237A		Schweizer; cancelled Apr 94
4	86-0404	1987	US Army; to
	8102	1987	USCG; lost 1996
5	N7508U	Jul 89	CIA; to
	N701AN	Dec 96	CIA
6	N7508W	Nov 89	CIA; to
	N87260	Feb 97	CIA
7	N7508Y	Jan 90	CIA; to
	N4122L	Dec 96	CIA
8	N7508Z	Feb 91	Schweizer; to
	FAC 2046		Colombian AF; to
	N12139	Mar 97	CIA
9	N7510U	May 94	Schweizer; to
	OES-2252	Jul 94	Mexican AF; to
	EAOE-2252		Mexican AF
010	N6147U	Oct 97	LAIRD International
011	N61474	Nov 97	Crashed 19 Feb 98
012	N61499	Oct 98	Schweizer; to
	FAC 5751		Colombian AF
014	N20086	May 99	Schweizer
015	N6150U	Jun 00	Schweizer; to
	N2061L	Feb 01	Schweizer; to
	n/k	May 02	Colombian AF
016	N40623	Sep 01	Schweizer
017	N10673	Nov 01	Schweizer
018	n/k		
019	N30745	Feb 02	Schweizer
020	N2067F	Apr 02	Schweizer

Notes: CIA aircraft registered to Vantage Leasing Inc of Dover, Delaware.

Nos. 1 and 2 remain currently registered (2002) as N3623C and N3623F, apparently due to administrative oversight.

No. 010 owner at Marietta, Georgia.

No. 013: This serial number not allocated by Schweizer company.

Nos. 015 and upwards are SA 2-37Bs.

Schweizer RG-8A of the Colombian Air Force

0085838

WEIGHTS AND LOADINGS:

Weight empty	918 kg (2,025 lb)
Max mission payload: SA 2-37A	340 kg (750 lb)
SA 2-37B	231 kg (510 lb)
Max T-O weight	1,587 kg (3,500 lb)
Max wing loading	85.65 kg/m² (17.55 lb/sq ft)
Max power loading	9.06 kg/kW (14.89 lb/hp)

PERFORMANCE (IO-540):

Max permissible diving speed (VD)	176 kt (326 km/h; 202 mph) CAS
Cruising speed at 1,525 m (5,000 ft):	
75% power	138 kt (256 km/h; 159 mph)
65% power	129 kt (239 km/h; 148 mph)
Approach speed	117 kt (217 km/h; 135 mph) CAS
Quiet mode speed	70-80 kt (130-148 km/h; 80-92 mph)
Optimum climbing speed	77 kt (142 km/h; 88 mph) CAS
Stalling speed:	
airbrakes open	71 kt (132 km/h; 82 mph)
airbrakes closed	67 kt (124 km/h; 77 mph)
Max rate of climb at S/L	292 m (960 ft)/min
Service ceiling: IO-540	5,490 m (18,000 ft)
TIO-540	7,315 m (24,000 ft)
T-O run (S/L, ISA): hard surface	387 m (1,270 ft)
grass	533 m (1,750 ft)
T-O to 15 m (50 ft) (S/L, ISA):	
hard surface	612 m (2,010 ft)
grass	759 m (2,490 ft)
Landing from 15 m (50 ft) (S/L, ISA):	
hard surface	680 m (2,230 ft)
grass	732 m (2,400 ft)
Best glide ratio	20
Endurance: SA 2-37B	up to 12 h
g limits	+6.6/−3.3

UPDATED

Schweizer 300C three-seat helicopter (*Jane's/Paul Jackson*)   NEW/0533671

SCHWEIZER 300C
Engineering designation: Model 269

TYPE: Three-seat helicopter.

PROGRAMME: Hughes 269 first flew 2 October 1956; production deliveries began October 1961; marketed later as Model 300. Developed 300C first flew August 1969; first flight Hughes production model December 1969; FAA certification May 1970 (basic Hughes 300 described in 1976-77 *Jane's*). Production of 300C transferred from Hughes Helicopters to Schweizer July 1983; first flight Schweizer 300C (1,166th 300C) June 1984; Schweizer bought entire programme November 1986. Deliveries of 300CB began in August 1995 with initial three of 10 ordered by launch customer Helicopter Adventures, of Concord, California. By late 2001, some 3,600 Hughes/Schweizer 300/330s had been produced.

CURRENT VERSIONS: **300C:** Standard civil version.
Main description applies to 300C, unless otherwise indicated.

300C Sky Knight: Special police version; options include safety mesh seats with inertia reel shoulder harnesses, public address/siren system, searchlight, integrated communications, infra-red sensor, heavy-duty 28 V 100 A electrical system, cabin heater, night lights with strobe beacons, cabin utility light, fire extinguisher, first aid kit and map case.

TH-300C: Military training version.

300CB: 'Basic' version (otherwise designated 269C-1) for training role; 134 kW (180 hp) Textron Lycoming HO-360-C1A engine, 132 litre (35 US gallon; 29.1 Imp gallon) fuel tank and 2,000 hour TBO; stated to be cheaper to operate than 300C. First flight (N6002X) 28 May 1993; type certificate awarded August 1995. Changes from standard model include one-piece upper pulley hub; improved air filtration system; lightweight exhaust; and improved oil filter. 200,000 hours flown by October 2002, at which time 11 had been involved in accidents without fatalities.

300CBi: Enhanced version of 300CB, which it was scheduled to replace from 2002; introduced at Heli-Expo 2002 at Orlando, Florida, in February 2002. Features fuel-injected engine, new splined (replacing bolted) main rotor

driveshaft and hub with 4,000-hour life, and electronic automatic engagement system (AES). First delivery (G-OCBI, c/n 0139) to CSE Aviation at Oxford, UK, in August 2002. UK certification delivered to Great Slave Hellicopters at Yellowknife, NWT, October 2002.

RoboCopter: Unmanned version developed by Kawada Industries in Japan for cropspraying, pipeline patrol, aerial crane and reconnaissance; first flight October 1996; Schweizer teamed with Northrop Grumman to offer helicopter for US Navy VTOL UAV requirement (subsequently met by RQ-8A Fire Scout version of Schweizer 333, which see); RoboCopter still being promoted in late 2001 under US marketing name **Argus**; further details in *Jane's Unmanned Aerial Vehicles and Targets*.

CUSTOMERS: Hughes produced 2,775 of all versions, including TH-55A Osage for US Army, before transfer to Schweizer. Royal Thai Army received 58 TH-300C helicopters between March 1986 and mid-1989; 500th Schweizer-manufactured 300C delivered at Heli Expo 1994. Recent Sky Knight customers include police departments of Baltimore, Maryland; Lakewood, California; and Kansas City, Missouri. Recent customers include the Civil Aviation Flying College in Sichuan Province, China, which ordered two 300CBs and one 300C, and the Yang Jiang Aviation Company in Guangdong Province, which ordered two 300CBs, all for delivery in the second quarter of 2002. By September 2002, Schweizer had built some 670 300Cs and 140 300CBs; during 2000, 13 300Cs and 17 300CBs were delivered, followed by 17 and 12 in 2001. Argentine Coast Guard took delivery of two 300Cs in August 2001 to supplement two supplied in 1996.

COSTS: 300C: US$235,650 (2001); 300CB US$204,750 (2001).

DESIGN FEATURES: Simple, pod-and-boom configuration, with exposed power plant and fuel tanks. Aerofoil-section stabiliser, starboard side, rear, at approx 45° dihedral; triangular underfin. Fully articulated three-blade main rotor, turning at 471 rpm; fully interchangeable blades; blade section NACA 0015; elastomeric dampers; two-blade teetering tail rotor; limited blade folding; no rotor brake; multiple V-belt and pulley reduction gear/drive system between horizontally mounted engine and transmission, with electrically controlled belt-tensioning system instead of clutch; braced tubular tailboom; separate dihedral tailplane and fin.

FLYING CONTROLS: Manual. Electric two-axis cyclic trim, cyclic and collective friction control.

STRUCTURE: Main rotor blades bonded with constant-section extruded aluminium spar, wraparound skin and trailing-

edge; tail rotor blades have steel tube spar and glass fibre skin; steel tube cabin section with light alloy and stainless steel skin and Plexiglas transparencies.

LANDING GEAR: Tubular skids carried on booted oleo-pneumatic shock-absorbers. Replaceable heavy-duty skid shoes. Two ground handling wheels with 0.25 m (10 in) balloon tyres, pressure 4.14 to 5.17 bar (60 to 75 lb/sq in). Available optionally on floats made of polyurethane-coated nylon fabric, 4.70 m (15 ft 5 in) long and with a total installed weight of 27.2 kg (60 lb). Heavy-duty skid plates optional.

POWER PLANT: One 168 kW (225 hp) Textron Lycoming HIO-360-D1A flat-four engine, derated to 142 kW (190 hp), mounted horizontally aft of seats. Normal fuel capacity of 300C and 300CB is 123 litres (32.5 US gallons; 27.1 Imp gallons) of which 114 litres (30.0 US gallons; 25.0 Imp gallons) are usable; auxiliary tank of 133 litres (35.0 US gallons; 29.2 Imp gallons) can be fitted to 300C and tank of 114 litres (30.0 US gallons; 25.0 Imp gallons) to 300CB. Oil capacity 9.5 litres (2.5 US gallons; 2.1 Imp gallons).

ACCOMMODATION: Three persons side by side on sculptured and cushioned bench seat, with shoulder harness, in Plexiglas enclosed cabin. Accommodation ventilated. Carpet and tinted canopy standard. Forward-hinged, removable door on each side. Dual controls optional in 300C, standard in 300CB; captain sits on left in 300C and on right in 300CB. Offset anti-torque pedals. Exhaust muff heating kit available.

SYSTEMS: Standard electrical system includes 24 V 70 A alternator, 24 V battery, lightweight high-torque starter and external power socket. Night lights with strobe beacons standard on 300CB. Automatic engagement system (AES) introduced as an option from February 2002 incorporates a start-up overspeed governor, computer-controlled rotor engagement device and low rotor rpm warning.

AVIONICS: *Comms:* Avionics include Honeywell KY 96A com transceiver and headsets, KR 86 ADF and GTX 320 transponders and ACK A-30 blind encoder.

EQUIPMENT: Standard for all models are fire extinguisher, first aid kit and engine hour meter. Model 300C has external power plug and fuel pressure indicator; Model 300CB has night lights with strobes. Optional equipment includes stretcher kits, cargo racks with combined capacity of 91 kg (200 lb), external load sling of 408 kg (900 lb) capacity, Simplex Model 5200 agricultural spray or dry powder dispersal kits, instrument training package, throttle governor, start-up overspeed control unit, night flying kit, single or dual exhaust mufflers, door lock and dual oil coolers.

DIMENSIONS, EXTERNAL:

Main rotor diameter	8.18 m (26 ft 10 in)
Main rotor blade chord	0.17 m (6¾ in)
Tail rotor diameter	1.30 m (4 ft 3 in)
Distance between rotor centres	4.66 m (15 ft 3½ in)
Length: fuselage	6.77 m (22 ft 2½ in)
overall, rotors turning	9.40 m (30 ft 10 in)
Height: to top of rotor head	2.66 m (8 ft 8¾ in)
to top of cabin	2.19 m (7 ft 2 in)
Width: rotor partially folded	2.44 m (8 ft 0 in)
cabin	1.30 m (4 ft 3 in)
Skid track	1.99 m (6 ft 6½ in)
Length of skids	2.51 m (8 ft 3 in)
Passenger doors (each): Height	1.09 m (3 ft 7 in)
Width	0.97 m (3 ft 2 in)
Height to sill	0.91 m (3 ft 0 in)

DIMENSIONS, INTERNAL:

Cabin: Length	1.50 m (4 ft 11 in)
Max width	1.45 m (4 ft 9 in)

AREAS:

Main rotor blades (each)	0.70 m² (7.55 sq ft)
Tail rotor blades (each)	0.08 m² (0.86 sq ft)

General arrangement of the Schweizer 300 (*Jane's/Paul Jackson*)   0064928

Latest 300CB*i* version of Schweizer helicopter   *NEW*/0527163

Main rotor disc	52.54 m² (565.5 sq ft)
Tail rotor disc	1.32 m² (14.20 sq ft)
Fin	0.23 m² (2.50 sq ft)
Horizontal stabiliser	0.246 m² (2.65 sq ft)

WEIGHTS AND LOADINGS:

Weight empty: 300C	499 kg (1,100 lb)
300CB	494 kg (1,088 lb)
Baggage capacity	45 kg (100 lb)

Max T-O weight:

Normal category: 300C	930 kg (2,050 lb)
300CB	794 kg (1,750 lb)
external load	975 kg (2,150 lb)

Max disc loading:

Normal category: 300C	17.7 kg/m² (3.63 lb/sq ft)
300CB	15.1 kg/m² (3.09 lb/sq ft)
external load	18.6 kg/m² (3.80 lb/sq ft)

Max power loading at T-O:

Normal category: 300C	6.57 kg/kW (10.79 lb/hp)
300CB	5.61 kg/kW (9.21 lb/hp)
external load	6.89 kg/kW (11.32 lb/hp)

PERFORMANCE (at max Normal T-O weight, ISA):

Never-exceed speed (V~NE~) at S/L:

300C	95 kt (176 km/h; 109 mph) IAS
300CB	94 kt (174 km/h; 108 mph) IAS
Max cruising speed: 300C	86 kt (159 km/h; 99 mph)
300CB	80 kt (148 km/h; 92 mph)

Speed for max range, at 1,220 m (4,000 ft)

	67 kt (124 km/h; 77 mph)
Max rate of climb at S/L: 300C	229 m (750 ft)/min
300CB	381 m (1,250 ft)/min

Hovering ceiling: IGE: 300C

	3,290 m (10,800 ft)
300CB	2,135 m (7,000 ft)
OGE: 300C	2,620 m (8,600 ft)
300CB	1,465 m (4,800 ft)

Range at 1,220 m (4,000 ft), 2 min warm-up, max normal

fuel, no reserves	208 n miles (386 km; 240 miles)

Max endurance at S/L: 300C

	3 h 48 min
300CB	3 h 18 min

UPDATED

SCHWEIZER 330 and 333

TYPE: Four-seat helicopter.

PROGRAMME: Announced 1987; first flight in public (N330TT; converted from a 300C) 14 June 1988; FAA certification September 1992. Italian certification April 2002. Deliveries started mid-1993. Three sets of controls available for student training. Model 330SP is marketing designation; type is 269D.

CURRENT VERSIONS: **330:** Original version delivered from 1993.

330SP: Announced May 1997, incorporates larger main rotor hub, increased blade chord and raised landing gear. Modifications can be retrofitted to existing 330s.

333: Announced early 2000 as an upgraded version of the 330SP; lead customer was San Antonio Police Department (two aircraft). FAA certification awarded 28 September 2000, at which time first three already delivered. 330SP can be upgraded to 333 standard. Differences include new cambered wider chord aerofoil main rotor blades, larger diameter rotor system, increased height landing gear and increased transmission rating.

Details apply to 333 unless otherwise stated.

RQ-8A Fire Scout: Schweizer 333 selected as airframe basis for Northrop Grumman Model 379 VTOL tactical UAV under development for US Navy (see *Jane's Unmanned Aerial Vehicles and Targets*) for which a US$93.7 million contract was awarded on 10 February 2000.

CUSTOMERS: Include Fuchs Helikopter of Switzerland, Houston Police Department (two) and West Palm Beach Police Department (four including first 333 conversion).

Three aircraft delivered in 1993; five in 1994, plus one prototype and one demonstrator; five in 1996, five in 1997, three in 1998, one in 1999, six in 2000 and four in 2001 for a total of 36, including two Fire Scout prototypes.

COSTS: US$607,500 for 333 (2001).

Schweizer 333 turbine-powered helicopter   *NEW*/0098867

DESIGN FEATURES: Developed from Model 300C, with turbine power, enclosed engine and tapered tailboom. Civil roles include law enforcement, scout/observation, aerial photography, light utility, agricultural spraying and personal transport; uses turbine fuel rather than scarcer Avgas; extremely low flat rating of engine for good hot-and-high performance; streamlined external envelope and large tailplane with endplate fins added during development; rotor rpm 471.

FLYING CONTROLS: Similar to 300C; see Design Features.

STRUCTURE: Generally similar to 300C.

POWER PLANT: *330/330SP:* One 313.2 kW (420 shp) Rolls-Royce 250-C20W turboshaft derated to 175 kW (235 shp). Transmission rating 175 kW (235 hp) for T-O; 164 kW (220 hp) maximum continuous. Maximum usable fuel capacity 276 litres (73.0 US gallons; 60.8 Imp gallons).

333: One 313.2 kW (420 shp) Rolls-Royce 250-C20W turboshaft derated to 173 kW (232 shp). Transmission rating 189 kW (253 shp) for T-O; 173 kW (232 shp) maximum continuous. Fuel capacity as 330SP.

ACCOMMODATION: Three persons side by side (central occupant slightly forward), with up to three sets of controls; four point inertia reel seat belts. Fourth occupant in 'high density' configuration (central seat replaced by double seat). Cabin ventilated.

SYSTEMS: Engine inlet anti-ice system; 24 V lead-acid battery; 150 A starter/generator; external APU plug; circuit breaker protection.

AVIONICS: *Instrumentation:* Tachometer, engine torquemeter, engine oil temperature and pressure indicators; electric digital OAT, turbine outlet temperature indicator; annunciator warning panel.

Mission: Law enforcement options include FLIR, video recorder, Spectrolab SX-5 Starburst searchlight, Pronet and Litton computer system.

DIMENSIONS, EXTERNAL:

Main rotor diameter	8.38 m (27 ft 6 in)
Tail rotor diameter	1.30 m (4 ft 3 in)
Length overall, rotors turning	9.50 m (31 ft 2 in)
Fuselage: Length	6.82 m (22 ft 4½ in)
Max width	1.72 m (5 ft 7¾ in)

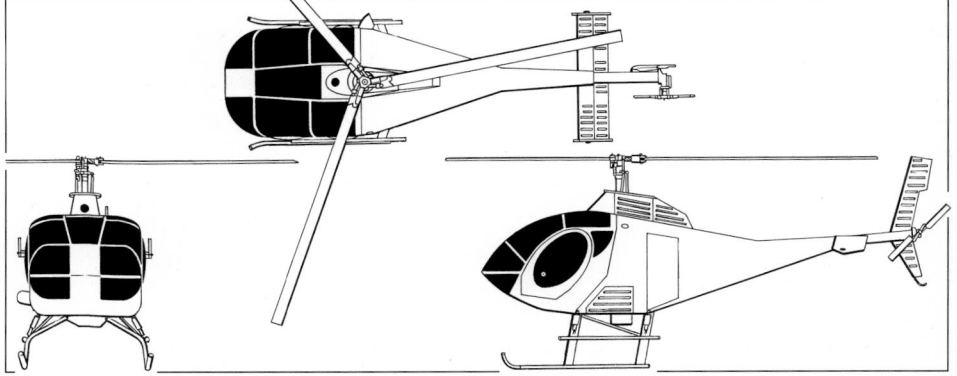

Schweizer 330SP three/four-seat light helicopter (*Jane's/Mike Keep*)

0130925

Height overall	3.35 m (11 ft 0 in)
Width overall	2.13 m (7 ft 0 in)
Tailplane span	2.04 m (6 ft 8½ in)
Skid track	1.92 m (6 ft 3½ in)
DIMENSIONS, INTERNAL:	
Cabin: Max width	1.84 m (6 ft 0½ in)
Length	1.98 m (6 ft 6 in)
Height at seat	1.35 m (4 ft 5¼ in)
AREAS:	
Main rotor disc area	55.20 m² (594.2 sq ft)
Tail rotor disc area	1.32 m² (14.20 sq ft)

WEIGHTS AND LOADINGS:	
Weight empty	567 kg (1,250 lb)
Max T-O weight	1,156 kg (2,550 lb)
Max disc loading	20.95 kg/m² (4.29 lb/sq ft)
Max power loading	6.16 kg/kW (10.12 lb/shp)
Transmission loading at max T-O weight and power	
	6.16 kg/kW (10.12 lb/shp)
PERFORMANCE:	
Never-exceed speed (Vne)	120 kt (222 km/h; 138 mph)
Normal cruising speed	105 kt (194 km/h; 121 mph)
Econ cruising speed	94 kt (174 km/h; 108 mph)

Max rate of climb at S/L	420 m (1,380 ft)/min
Hovering ceiling at MTOW: IGE	2,651 m (8,700 ft)
OGE	1,554 m (5,100 ft)
Max range at 1,220 m (4,000 ft), no reserves	
	319 n miles (590 km; 367 miles)
Max endurance, no reserves	4 h 10 min
	UPDATED

SEASTAR

SEASTAR AIRCRAFT INC

1500 Beville Road, Suite 606-314, Daytona Beach, Florida 32114-5644
Tel: (+1 904) 822 97 40
e-mail: info@seastarplane.com
Web: http://www.seastarplane.com
PRESIDENT: János Dósa
DESIGNER: Craig Easter
WORKS
Box 107, Szigetszentmiklós, H-2311, Hungary
Tel: (+36 24) 45 10 00
Fax: (+36 24) 45 20 00

Seastar has undertaken a fundamental redesign of the Seawind amphibian which it now offers complete or as a kit. The next product will be the Adventurer turbo-powered version; other versions planned include an armed version for patrol work and a search and rescue equipped model.

The company is establishing a 5,000 m² (53,820 sq ft) facility in Hungary where certified aircraft will be built before being transferred to the USA for completion.

UPDATED

SEASTAR SEASTAR

TYPE: Six-seat amphibian kitbuilt.
PROGRAMME: Production prototype Seawind first flew 23 August 1982. Progressed through several versions, last of which built by Seawind International Inc of Canada.

Compared to Seawind, the completely new design SeaStar offers a changed aerofoil, single-piece Fowler flaps, gull-wing cabin doors and stainless steel landing gear. Adapted for optional pressurisation and alternative diesel and turboprop engines. Prototype (N601SS) exhibited, partly complete, at Sun 'n' Fun, April 2000; first flown 12 November 2000; by April 2002, 111 hours had been flown and certification programme begun. Certification expected 2005-06.
CURRENT VERSIONS: **SeaStar 7T:** piston-engined version, *as described.*

SeaStar 7P: Intended pressurised version, envisaged also for law enforcement duties. Will have two pop-in doors and flight controls for pilot only. Prototype due for construction 'in near future'.
COSTS: Basic kit, less engine and avionics, US$109,000, pressurised US$119,000; estimated cost by completion US$250,000 to US$270,000. Factory-built, M 601 turboprop-powered law enforcement version, but without avionics, US$590,000 (2003).
CUSTOMERS: Several kits under construction by end 2001; company assisting with completion of first ten; over 30 deposits taken by April 2002.
DESIGN FEATURES: Unconventional configuration; shoulder wing with stabilising floats at tips; long-chord fin mounts engine and tailplane at mid-position, shielding propeller from water spray.
FLYING CONTROLS: Conventional and manual. Full-span elevator with two flight-adjustable tabs; ground-adjustable tab on rudder; flight-adjustable tab in port aileron. Electrically operated two-section slotted flaps, max deflection 40°.
STRUCTURE: Carbon fibre throughout; stainless steel landing gear. Wing aerofoil GA37A315.
POWER PLANT: *Piston version:* One 261 kW (350 hp) Textron Lycoming TIO-540-J2B flat-six driving a four-blade, reversible pitch MT propeller or three-blade Avia propeller, with standard fuel capacity of 379 litres (100 US gallons; 83.3 Imp gallons). Auxiliary fuel tank optional, capacity 151 litres (40.0 US gallons; 33.3 Imp gallons).
Turbine version: One 490 kW (657 shp) derated Walter M 601D turboprop driving a three-blade, reversible pitch Avia propeller. Standard fuel capacity 530 litres (140 US gallons; 117 Imp gallons); auxiliary fuel tank capacity 227 litres (60.0 US gallons; 50.0 Imp gallons).
LANDING GEAR: Tricycle type; electrically retractable; spring steel legs. Mainwheels are 7.00-6, have hydraulic brakes and retract upwards into wing; steerable 6.00-6 nosewheel retracts forwards.
ACCOMMODATION: Two front seats and bench for three passengers in rear. Four centreline-hinged, upward-opening doors in piston-engined version.
ARMAMENT: Military version has provision for total of four underwing pylons.

DIMENSIONS, EXTERNAL:	
Wing span	11.58 m (38 ft 0 in)
Length overall	9.19 m (30 ft 2 in)
Height overall	3.35 m (11 ft 0 in)
Baggage door: width	0.76 m (2 ft 6 in)
height	0.69 m (2 ft 3 in)
DIMENSIONS, INTERNAL:	
Cabin: Length	3.25 m (10 ft 8 in)
Width: at front	1.35 m (4 ft 5 in)
at rear	1.52 m (5 ft 0 in)
Height	1.08 m (3 ft 6½ in)
Baggage compartment: Length	1.83 m (6 ft 0 in)
Volume	0.76 m³ (27.0 cu ft)
AREAS:	
Wings, gross	15.89 m² (171.0 sq ft)
WEIGHTS AND LOADINGS:	
Weight empty	1,134 kg (2,500 lb)
Max T-O weight: piston	1,814 kg (4,000 lb)
turboprop	1,886 kg (4,600 lb)
Max wing loading: piston	114.2 kg/m² (23.39 lb/sq ft)
turboprop	131.3 kg/m² (26.90 lb/sq ft)
Max power loading: Lycoming	6.96 kg/kW (11.43 lb/hp)
Walter	4.26 kg/kW (7.00 lb/shp)
PERFORMANCE:	
Max level speed: Lycoming	181 kt (335 km/h; 208 mph)
Walter	217 kt (402 km/h; 250 mph)
Max cruising speed: Lycoming at 75% power:	
at 3,050 m (10,000 ft)	175 kt (323 km/h; 201 mph)
at 6,100 m (20,000 ft)	185 kt (343 km/h; 213 mph)
Walter at 90% power:	
at 3,050 m (10,000 ft)	216 kt (400 km/h; 249 mph)
at 6,100 m (20,000 ft)	237 kt (439 km/h; 273 mph)
Stalling speed	52 kt (97 km/h; 60 mph)
Service ceiling	7,010 m (23,000 ft)
Max rate of climb at S/L: Lycoming	427 m (1,400 ft)/min
Walter	610 m (2,000 ft)/min
T-O run on land: Lycoming	274 m (900 ft)
Walter	235 m (770 ft)
Max range, no reserves:	
Lycoming	1,215 n miles (2,251 km; 1,399 miles)
Walter	1,203 n miles (2,228 km; 1,385 miles)
	UPDATED

SeaStar SeaStar multipurpose amphibian (*Jane's/Susan Bushell*)   0121672

SEQUOIA

SEQUOIA AIRCRAFT CORPORATION

2000 Tomlynn Street, PO Box 6861, Richmond, Virginia 23230
Tel: (+1 804) 353 17 13
Fax: (+1 804) 359 26 18
e-mail: seqair@aol.com
Web: http://www.SeqAir.com
PRESIDENT: Alfred P Scott

Sequoia has marketed the F.8L Falco since 1979.

UPDATED

SEQUOIA FALCO F.8L

TYPE: Side-by-side lightplane/kitbuilt.
PROGRAMME: Sequoia Aircraft markets plans and kits to build improved version of Falco F.8L high-performance monoplane, designed in Italy by Ing Stelio Frati of General Avia (which see) and first flown on 15 June 1955. Italian production totalled 77; kit version first flown 1974.
CUSTOMERS: Total of 72 flown by mid-2002, in Australia, Brazil, Canada, Chile, France, Germany, Italy, Netherlands, New Zealand, Norway, South Africa, UK and USA.
COSTS: Kit: US$96,000 (2001) including propeller but excluding engine. Plans: US$400. Extra costs to complete estimated by manufacturer as US$13,000.
DESIGN FEATURES: Streamlined low-wing, retractable-gear monoplane. Quoted build time 2,000 hours.
NACA 64₂212.5 wing section at root, NACA 64₂210 at tip. Wing dihedral 5°, washout 3°. Tailplane section NACA 65009.
FLYING CONTROLS: Conventional and manual.
STRUCTURE: Entire airframe has plywood-covered wood structure, with overall fabric covering and glass fibre wing fillets. Optional metal control surfaces.
LANDING GEAR: Retractable tricycle type. Steerable nosewheel. Cleveland single calliper disc brakes. 5.00-5 mainwheels and 4.00-5 nosewheel.
POWER PLANT: One 119 kW (160 hp) Textron Lycoming IO-320-B1A is standard for kitbuilt aircraft, driving Hartzell HCC-2YL-1BF/F7663-4 two-blade constant-speed metal propeller. Optional engines for which Sequoia offers installation kits are the 112 kW (150 hp) IO-320-A1A and 134 kW (180 hp) IO-360-B1E. Fuel capacity 151 litres (40.0 US gallons; 33.3 Imp gallons). Optional 11.4 litre (3.0 US gallon; 2.5 Imp gallon) header tank to permit inverted flight.
ACCOMMODATION: Two seats, side by side; child's seat or baggage area to rear. Dual controls.

DIMENSIONS, EXTERNAL:	
Wing span	8.00 m (26 ft 3 in)
Wing chord: at root	1.65 m (5 ft 5 in)
at tip	0.83 m (2 ft 8½ in)
Wing aspect ratio	6.4
Length overall	6.63 m (21 ft 9 in)
Height overall	2.29 m (7 ft 6 in)
Tailplane span	2.71 m (8 ft 10½ in)
Wheel track	2.08 m (6 ft 10 in)
Wheelbase	1.50 m (4 ft 11 in)
Propeller diameter	1.83 m (6 ft 0 in)
DIMENSIONS, INTERNAL:	
Cabin max width	1.07 m (3 ft 6 in)
AREAS:	
Wings, gross	10.00 m² (107.6 sq ft)
Rudder	0.48 m² (5.20 sq ft)

Tailplane	2.17 m² (23.40 sq ft)
Elevator	0.83 m² (9.00 sq ft)

WEIGHTS AND LOADINGS (119 kW; 160 hp engine):

Weight empty	550 kg (1,212 lb)
Payload with max fuel	194 kg (428 lb)
Baggage capacity	41 kg (90 lb)
Max aerobatic weight	748 kg (1,650 lb)
Max T-O weight	852 kg (1,880 lb)
Max wing loading	85.3 kg/m² (17.47 lb/sq ft)
Max power loading	7.15 kg/kW (11.75 lb/hp)

PERFORMANCE (119 kW; 160 hp engine):

Never-exceed speed (VNE)	208 kt (386 km/h; 240 mph)
Max level speed at S/L	184 kt (341 km/h; 212 mph)
Cruising speed at 75% power	
	165 kt (306 km/h; 190 mph)
Stalling speed: clean	66 kt (121 km/h; 75 mph)
flaps and wheels down	54 kt (100 km/h; 62 mph)
Max rate of climb at S/L	347 m (1,140 ft)/min
Service ceiling	5,790 m (19,000 ft)
T-O run	174 m (570 ft)
T-O to 15 m (50 ft)	351 m (1,150 ft)
Landing from 15 m (50 ft)	351 m (1,150 ft)
Landing run	229 m (750 ft)
Range at econ cruising speed	
	869 n miles (1,609 km; 1,000 miles)
Endurance	4 h 24 min
g limits	+6/−3

UPDATED

Sequoia F.8L two-seat kitbuilt (*Jane's/Paul Jackson*)

0121041

SHERPA

SHERPA AIRCRAFT MANUFACTURING CO

17350 SW Shaw Street, Aloha, Oregon 97007
Tel: (+1 503) 649 85 58
Fax: (+1 503) 591 73 73
e-mail: sherpaworldwide@att.net
Web: http://www.sherpaaircraft.com
DIRECTORS:
  Byron Root
  Glen Gordon

Sherpa Aircraft is in the process of certifying the Sherpa 300 derivative of its earlier six-seat aircraft version. In 2001, the company announced that it had entered into a joint venture agreement with Khrunichev (of Russia) which would result in parts being manufactured in Moscow and Sherpa marketing the T-411 Aist and Aviotechnica SL-90. Sherpa is expanding its manufacturing facility at Scappoose, Oregon, at present 1,400 m² (15,000 sq ft), to accommodate increased production.

UPDATED

SHERPA SHERPA

TYPE: Light utility transport.
PROGRAMME: Development of the Sherpa series began in 1991, with first prototype (N1415B) making debut at Oshkosh in July 1994. Second prototype (N711SA) first flown 1995; both prototypes five-seat versions. Third prototype (N712SA), shown at AirVenture 2001, is the eight-seat Sherpa C-400 version. All versions use same airframe, except for modified wings on K-200.
CURRENT VERSIONS: **Sherpa:** Original six-place version; kit production under consideration.
  Sherpa K-200: Five/six-seat version; under development. Reduced wing span and area.
  Sherpa K-300: Six-seat version; main differences from K-200 are in cabin and CG envelope.
  Sherpa K-300T: Turbine-powered eight-seater; same airframe as C-400.
  Sherpa C-400: Eight-seat version; *as described unless otherwise stated.*
  Sherpa C-500T: Proposed certified eight/10-seat version.
CUSTOMERS: Three flying by end 2001.

COSTS: Sherpa K-200: US$135,000; Sherpa K-300: US$150,000, Sherpa C-400 US$425,000, Sherpa C-500T US$850,000 (2002).
DESIGN FEATURES: Optimised for use in inhospitable terrain, with STOL performance, strengthened landing gear, option for large tyres and easy maintenance access.
  Constant-chord, high-mounted wing with V bracing struts and auxiliary struts; wire-braced empennage. Tail surfaces have rounded tips and horn balances. Fully glazed flight deck door.
FLYING CONTROLS: Conventional and manual. Two-track Fowler flaps increase wing area by 15 per cent when fully deployed at 40°. Inboard trim tab each side above flap. Tabs on rudder and starboard elevator adjustable in flight, as is tailplane incidence.
STRUCTURE: Steel 4130 fuselage frame has fabric covering aft and metal sheet covering forward of cabin. Fabric-covered 4130 steel tube wing has metal leading-edges and control surfaces. Fabric-covered steel tube empennage. Graphite composites nose cowlings. Single-piece 6 mm (¼ in) thick wrap-round windscreen.
LANDING GEAR: Tailwheel type; fixed. Standard tyre size 29×11.0-10; optional 1.07 m (42 in) Tundra tyres can be fitted. Cleveland hydraulic brakes. Optional floats, wheeled floats and skis; conversion time quoted as 90 minutes.
POWER PLANT: *Sherpa K-200:* One 265 kW (355 hp) VOKBM M-14X nine-cylinder radial. Fuel capacity 530 litres (140 US gallons; 117 Imp gallons).
  Sherpa K-300T and *C-500T:* One 560 kW (751 shp) Walter M 601 turboprop driving a three-blade propeller. Fuel capacity 852 litres (225 US gallons; 187 Imp gallons).
  Sherpa K-300 and *C-400:* One 347 kW (465 hp) Textron Lycoming TIO-720-X152 flat-eight driving a three-blade Hartzell F8483 propeller. Normal fuel capacity 568 litres (150 US gallons; 125 Imp gallons) in six individually removable tanks.
ACCOMMODATION: Pilot and up to seven passengers on individual seats with central aisle; seats can be removed individually to allow carriage of freight (up to 10 empty 208 litre (55.0 US gallon; 45.8 Imp gallon) drums) or two casualties plus two attendants. Double door on starboard side of fuselage for cabin; individual doors for crew positions. Additional baggage door on starboard side aft of cabin area.
SYSTEMS: 12 V electrical system.

AVIONICS: Honeywell avionics suite plus S-Tek autopilot.
DIMENSIONS, EXTERNAL (all, except where indicated):

Wing span: except K-200	13.69 m (44 ft 11 in)
K-200	13.41 m (44 ft 0 in)
Wing chord, constant	1.83 m (6 ft 0 in)
Wing aspect ratio	7.3
Length overall	9.96 m (32 ft 8 in)
Height overall	2.87 m (9 ft 5 in)
Tailplane span	5.03 m (16 ft 6 in)
Wheel track	2.69 m (8 ft 10 in)
Propeller diameter	2.54 m (8 ft 4 in)
Propeller ground clearance	0.69 m (2 ft 3 in)
Cabin door: Height	1.12 m (3 ft 8 in)
Width	1.55 m (5 ft 1 in)
Baggage door: Height	0.51 m (1 ft 8 in)
Width	0.76 m (2 ft 6 in)

DIMENSIONS, INTERNAL (all):

Cabin volume	4.42 m³ (156 cu ft)
Baggage hold	0.36 m³ (12.7 cu ft)

AREAS (all, except where indicated):

Wings, gross: except K-200	25.83 m² (278.0 sq ft)
K-200	24.71 m² (266.0 sq ft)
Ailerons (total)	1.83 m² (19.7 sq ft)
Trailing-edge flaps (total)	4.51 m² (48.5 sq ft)
Spoilerons (total)	0.39 m² (4.2 sq ft)
Fin	1.24 m² (13.4 sq ft)
Rudder, incl tab	1.37 m² (14.7 sq ft)
Tailplane	2.81 m² (30.3 sq ft)
Elevators, incl tab	2.39 m² (25.7 sq ft)

WEIGHTS AND LOADINGS:

Weight empty: K-200	1,139 kg (2,510 lb)
K-300 landplane	1,390 kg (3,065 lb)
K-300T	1,393 kg (3,070 lb)
C-400	1,445 kg (3,185 lb)
C-500T	1,427 kg (3,146 lb)
Max T-O weight: K-200	1,846 kg (4,070 lb)
K-300: landplane	2,268 kg (5,000 lb)
floatplane	2,404 kg (5,300 lb)
K-300T	2,494 kg (5,500 lb)
C-500T	2,844 kg (6,270 lb)
Max wing loading: K-200	74.7 kg/m² (15.30 lb/sq ft)
K-300: landplane	87.8 kg/m² (17.99 lb/sq ft)
floatplane	93.1 kg/m² (19.06 lb/sq ft)
K-300T	96.6 kg/m² (19.78 lb/sq ft)
C-500T	110.1 kg/m² (22.55 lb/sq ft)

Prototypes of the Sherpa C-400 (left) and K-200 (in floatplane configuration) (*Jane's/Paul Jackson*)

NEW/0526995

Max power loading: K-200	6.98 kg/kW (11.46 lb/hp)		K-300T	146 kt (270 km/h; 168 mph)	K-300	159 m (522 ft)
K-300: landplane	6.54 kg/kW (10.75 lb/hp)		C-400	129 kt (238 km/h; 148 mph)	K-300T	108 m (355 ft)
floatplane	6.94 kg/kW (11.40 lb/hp)		Stalling speed: flaps up: K-200	51 kt (94 km/h; 58 mph)	C-400	151 m (494 ft)
K-300T	4.46 kg/kW (7.33 lb/hp)		K-300, K-300T, C-400	54 kt (100 km/h; 62 mph)	Landing run: K-200	104 m (340 ft)
C-500T	5.09 kg/kW (8.36 lb/hp)		flaps down: K-200	35 kt (65 km/h; 40 mph)	K-300	122 m (400 ft)

PERFORMANCE:

Max operating speed: K-200 123 kt (229 km/h; 142 mph)
K-300T 182 kt (338 km/h; 210 mph)
Max cruising speed: K-300 132 kt (245 km/h; 152 mph)
K-300T 162 kt (301 km/h; 187 mph)
C-400 140 kt (259 km/h; 161 mph)
Normal cruising speed at 75% power: K-200
105 kt (195 km/h; 121 mph)
C-400 144 kt (267 km/h; 166 mph)
C-500T 162 kt (301 km/h; 187 mph)
Econ cruising speed: K-300 120 kt (222 km/h; 138 mph)

K-300 40 kt (73 km/h; 45 mph)
K-300T, C-400 42 kt (78 km/h; 48 mph)
C-500T 46 kt (84 km/h; 52 mph)
Max rate of climb at S/L: K-200, C-400
300 m (985 ft)/min
K-300 274 m (900 ft)/min
K-300T 509 m (1,670 ft)/min
Service ceiling: K-200 5,030 m (16,500 ft)
K-300 5,640 m (18,500 ft)
K-300T, C-400 7,620 m (25,000 ft)
T-O run: K-200 119 m (390 ft)

K-300T 81 m (265 ft)
C-400 86 m (283 ft)
Range with max cruising speed:
K-300A 938 n miles (1,738 km; 1,080 miles)
K-300T 1,147 n miles (2,124 km; 1,320 miles)
C-400 833 n miles (1,543 km; 959 miles)
Range at econ cruising speed:
K-300T 1,364 n miles (2,526 km; 1,570 miles)
C-400 943 n miles (1,747 km; 1,086 miles)
UPDATED

SIKORSKY

SIKORSKY AIRCRAFT
(Subsidiary of United Technologies Corporation)
6900 Main Street, Stratford, Connecticut 06615-9129
Tel: (+1 203) 386 40 00
Fax: (+1 203) 386 73 00
Web: http://www.sikorsky.com
OTHER WORKS: Troy, Alabama; South Avenue, Bridgeport, Connecticut; Shelton, Connecticut; West Haven, Connecticut; Development Flight Test Center, West Palm Beach, Florida
PRESIDENT AND CEO: Dean C Borgman
SENIOR VICE-PRESIDENT:
    Mark J Evans (Production Operations)
VICE-PRESIDENT, PROGRAMS: Kenneth J Kelly
VICE-PRESIDENT, CIVIL PROGRAMS: Tommy H Thomason
VICE-PRESIDENT, DEVELOPMENT ENGINEERING AND ADVANCED PROGRAMS: Dr Kenneth M Rosen
VICE-PRESIDENT, PRODUCTION ENGINEERING: Donald P Gover
VICE-PRESIDENT, RESEARCH AND ENGINEERING: Paul Martin
VICE-PRESIDENT, FINANCE AND CFO: Jay Haberland
VICE-PRESIDENT, GOVERNMENT BUSINESS DEVELOPMENT:
    Rene I Beauchamp
COMMUNICATIONS MANAGER: Ed Steadham
MANAGER OF PUBLIC RELATIONS: William S Tuttle

Founded as Sikorsky Aero Engineering Corporation in 1923 by late Igor I Sikorsky; has been division of United Technologies Corporation since 1929, but established as a subsidiary with effect 1 January 1995; began helicopter production in 1940s.

Headquarters and main plant at Stratford, Connecticut; also has manufacturing facilities elsewhere in Connecticut; other smaller facilities in Alabama and Florida. Workforce in October 2002 was about 8,000 worldwide. Main current programmes include UH-60 Black Hawk and derivatives, S-76 series and, in co-operation with international partners, development of S-92 Helibus. Sikorsky delivered total of 105 helicopters in 1996, 101 in 1997, 87 in 1998, 65 in 1999, 70 in

Sikorsky MH-60S KnightHawk   *NEW*/0526920

2000 and 86 in 2001 (comprising 62 Black Hawk versions, 16 MH-60Ss and eight S-76s). Sikorsky and Boeing Helicopters won US Army RAH-66 Comanche light helicopter demonstration/validation order on 5 April 1991 (is joint Boeing Sikorsky product).

Sikorsky licensees include Agusta of Italy, Eurocopter of France and Germany, Korean Air Lines of South Korea, Mitsubishi of Japan, Pratt & Whitney Canada Ltd and UK element of AgustaWestland. Sikorsky and Embraer of Brazil signed agreement in mid-1983 to transfer technology covering design and manufacture of composites components. Sikorsky and CASA of Spain signed MoU in June 1984 covering long-term helicopter industrial co-operation programme; CASA builds tail rotor pylon, tailcone and stabiliser components for H-60 and S-70, with first CASA S-70 components delivered to Sikorsky January 1986. Most recent overseas venture, in collaboration with the Alpata Group of Turkey, concerns creation of Alp Aviation, which manufactures high-technology, precision-machined aerospace and defence components in 6,500 m² (70,000 sq ft) facility in Eskisehir; announcement of venture made at Paris Air Show in June 1999.

In October 1998, Sikorsky purchased Helicopter Support Inc, so as to offer enhanced support to helicopter operators on worldwide basis. In latter half of 1998, Sikorsky also secured US$150 million deal with US Navy covering contractor maintenance of jet trainer aircraft as well as HH-1N Iroquois and UH-3H Sea King helicopters. Overall responsibility for latter allocated to Sikorsky Support Services Inc, with the work undertaken at Meridian, Mississippi; Pensacola, Florida; and Corpus Christi, Texas. Joint venture Shanghai Sikorsky Helicopter Company markets Schweizer 300 light helicopter in China.
UPDATED

SIKORSKY S-70A
US Army designations: UH-60A, UH-60C, UH-60L, UH-60M and UH-60Q Black Hawk, AH-60L, EH-60A, HH-60L, HH-60M, MH-60K, MH-60L and MH-60M
US Air Force designation: HH-60G Pave Hawk
US Navy designation: MH-60S KnightHawk
US Marine Corps designation: VH-60N
Israel Defence Force name: Yanshuf (Owl)
Japan Self-Defence Forces designations: UH-60J and UH-60JA
South Korean Army designation: UH-60P
TYPE: Multirole medium helicopter.
PROGRAMME: UH-60A declared winner of US Army Utility Tactical Transport Aircraft System (UTTAS) competition against Boeing Vertol YUH-61A 23 December 1976; first flight of first of three YUH-60A competitive prototypes 17

October 1974; early development history recorded in 1982-83 and earlier *Jane's All the World's Aircraft*; 2,000th H-60 delivered May 1994; 2,500th followed at end of 2001. For retrofitted improvements, see UH-60L below.

AgustaWestland in UK, Korean Air Lines of South Korea and Mitsubishi of Japan have licences to build the S-70 series, although UK programme has not been activated.

CURRENT VERSIONS: **UH-60A Black Hawk:** Initial production version, designed to carry crew of three and 11 troops; also can be used without modification for medevac, reconnaissance, command and control, and troop supply; cargo hook capacity 4,082 kg (9,000 lb); one UH-60A can be carried in C-130, two in C-141 and six in C-5.

Medevac kits delivered from 1981; missile qualification completed June 1987, with day and night firing of Hellfire in various flight conditions; airborne target handover system (ATHS) qualified; cockpit lighting suitable for night vision goggles fitted to production UH-60s since November 1985 and retrofitted to those built earlier; US Army began testing Alliant Techsystems Volcano mine dispensing system July 1987; modular Volcano container is disposable and dispenses 960 Gator anti-tank and anti-personnel mines; deployment of 2,940 kg (6,482 lb) system began in FY95; usage monitor to measure certain rotor loads installed in 30 UH-60As; wire strike protection added to UH-60s and EH-60s during 1987; accident data recorders also fitted. Total 1,049 built for US Army (including 66 conversions to EH-60A) before production change to UH-60L in 1989; more than 600 UH-60Ls since produced for US Army, with procurement continuing.
Detailed description applies to UH-60A/L except where indicated.

Enhanced Black Hawk: Incorporates active and passive self-defence systems, retrofitted by Corpus Christi Army Depot, Texas, to new-build UH-60A/Ls; first 15 delivered to US Army in South Korea November 1989. Equipment includes BAE Systems AN/ARN-148 Omega navigation receiver, Motorola AN/LST-5B satellite UHF communications transceiver, Honeywell AN/ARC-199 HF-SSB, and AEL AN/APR-44(V)3 specific threat RWR complementing existing AN/APR-39 general threat RWR; M134 Minigun can be fitted on each of two pintle mounts, replacing M60 machine gun.

JUH-60A: At least seven used temporarily for trials.

GUH-60A: At least 20 grounded airframes for technical training.

HH-60D Night Hawk: One prototype (82-23718) completed for abandoned USAF combat rescue variant; subsequently became an HH-60G.

EH-60A: (Designation **EH-60C** reserved, but not adopted by US Army.) Prototype YEH-60A (79-23301) ordered in October 1980 to carry 816 kg (1,800 lb) Quick

Sikorsky Firehawk firefighting helicopter   0110604

One of four Sikorsky S-70As of the Royal Brunei Air Force (*Jane's/Paul Jackson*)   0099726

Fix IIB battlefield ECM detection and jamming system. TRW Electronic Systems Laboratories was prime contractor for AN/ALQ-151(V)2 ECM kit, with installation by Tracor Aerospace; four dipole antennas on fuselage and deployable whip antenna; hover IR suppressor system (HIRSS) standard. YEH-60A first flight 24 September 1981; order for Tracor Aerospace to modify 40 UH-60As to EH-60A standard under US$51 million contract placed October 1984; first delivery July 1987 as part of US Army Special Electronics Mission Aircraft (SEMA) programme; 66 funded by FY87 excluding prototype; programme completed 1989. Under current Army planning, EH-60A to be phased out in 2005.

Intercepts/locates AM, FM, CW and SSB radio emissions from upper HF to mid-VHF ranges over bandwidths of 8, 30 or 50 kHz; jams VHF communications. Protective systems of UH-60A/L (M-130 chaff/flare and AN/ALQ-144 IR jammer) augmented by Sanders AN/ALQ-156(V) missile approach warning system, ITT AN/ALQ-136(V) pulsed transmitter, Northrop Grumman AN/ALQ-162(V) CW transmitter and Litton AN/APR-39(V) RWR.

AN/ALQ-151(V)3 Advanced Quick Fix mission system tasked with providing ESM capability to forward ground units at division level; at least six EH-60As (84-24027, 85-24468, 85-24473, 87-24657, 87-24662 and 87-24670) adapted to **EH-60L** configuration. As fielded for Task Force XXI trials (1997), the EH-60L utilised the following equipment: sensor subsystem comprised Sanders TACJAM-A ESM for detection, direction-finding, identification and tracking of communications signals in HF, UHF, VHF and SHF frequency bands; Lockheed Martin Federal Systems communications high-accuracy locating system – exploitable (CHALS-X) for direction-finding in HF, UHF and SHF frequency bands; and signal location subsystem (SILO) for direction-finding in VHF frequency band. Navigation and timing subsystem featured INS, GPS and control display unit in cockpit. Mission control and interface subsystem comprised control, navigation, workstation and graphics processors, mass storage unit, keyboards with trackball and 483 × 483 mm (19 × 19 in) colour monitors. Communications subsystem comprised modified PRC-118 wideband control datalink, two AN/ARC-201A SINCGARS radios, AN/ARC-164 tasking and reporting datalink, intercom and AN/UYH-15 digital temporary storage recorder/reproducer set. Antenna subsystem. Airborne survivability subsystem comprised AN/ALQ-144(V)1 IR jammer, AN/ALQ-156(V)2 missile approach warning system (MAWS), AN/ALQ-162(V)2 CW transmitter and AN/APR-39(V) RWR.

Improvements to airframe include installation of UH-60L engines and gearbox for increase in maximum weight from 7,845 kg (17,295 lb) to 10,206 kg (22,500 lb).

MH-60A: About 30 modified for Army 160th Special Operations Aviation Regiment (SOAR); fitted with Raytheon Systems AN/AAQ-16 FLIR, BAE Systems AN/ARN-148 Omega/VLF navigation, M-130 chaff/flare dispensers, AN/ALQ-144 IR jammer, night vision equipment, multifunction displays, auxiliary fuel tanks and door-mounted Minigun; fitted with -701C engines; interim equipment, pending MH-60K. Replaced by MH-60L in late 1990 and reverted to UH-60A configuration.

UH-60C: Designation for command and control (C2) version; was under development at the US Naval Research Laboratory on behalf of the US Army. This version subsequently abandoned but UH-60C project continues, with Raytheon as prime contractor for C2 system that is to be installed on 34 UH-60Ls. Contract worth US$110 million for development and production awarded in late 2001. Original plan was to obtain 207 helicopters, although this reduced to 119 by start of 2001. Rockwell Collins AN/ASC-15B/C consoles have been fitted to more than 50 US Army aircraft as **UH-60A(C)**. Communications suite to be

finalised, but expected to include Have Quick II, SINCGARS/SIP and JTIDS; other capabilities also to be incorporated, including FLIR, NVG compatibility, digital map display, mission planner facility, improved ECCM and storage space for ground power generators and antennas.

HH-60G Pave Hawk: Replaced US Air Force HH-60D Night Hawk rescue helicopters, which were not funded (see 1987-88 *Jane's All the World's Aircraft*); converted from UH-60A/L, including 10 originally delivered to 55th Aerospace Rescue and Recovery Squadron (later Special Operations Squadron) at Eglin AFB, Florida, in 1982-83, initially remaining as UH-60As; all progressively fitted by Sikorsky Support Services at Troy, Alabama, with aerial refuelling probe, 443 litre (117 US gallon; 97.5 Imp gallon) internal auxiliary fuel tank and fuel management panel; then to Pensacola NAD for mission avionics and modified instrument panel; some retrofitted with replacement internal tank of 700 litres (185 US gallons; 154 Imp gallons) capacity; -701C engines fitted to 10 special operations examples and later production aircraft (FY89 onwards); recent retrofit programme, begun in November 1999, entailed installation of -701C engine on 11 HH-60Gs of California and New York ANG by April 2001, with remaining 22 aircraft likely to follow suit, if funding permits.

Further procurement began with batch of nine in FY87, followed by purchases of 16, 18, 22, 15 and 13 in FY88-92; eight more funded in FY97 and delivered in 1998. All designated MH-60G until 1 January 1992, when 82 in combat rescue role redesignated HH-60G, with balance of 16 remaining as MH-60G for special operations units; by fourth quarter of 1998, only nine still in MH-60G configuration and at start of 2000, all 102 in inventory were using HH-60G designation. All have rescue hoist, Doppler/INS, electronic map display, Tacan, Honeywell AN/APN-239 lightweight weather/ground-mapping radar, secure HF, and satcom; MH-60G had ESSS (see Armament paragraph) for weapons and additional fuel carrying capability, plus door-mounted 0.50 in machine guns and Raytheon AN/AAQ-16 Pave Low III FLIR.

Upgraded version of HH-60G, known as **Block 152**, made debut at Stratford, Connecticut, on 29 April 1999, when first of 49 planned aircraft rolled out in Upgraded Communication, Navigation/Integrated Electronic Warfare (UCN/IEW) configuration; new features include

enhanced com/nav system and EW suite integrated into MIL-STD-1553 databus to reduce crew workload. Contractor trials in May and June 1999, after which modified HH-60G (possibly 92-26460) delivered to Nellis AFB, Nevada, for operational test and evaluation with 422nd TES. Retrofit also includes installation of revised, externally mounted, armament system with 0.50 in machine guns, expendable chaff/flare defensive countermeasures and repositioned nose radar.

Further extensive upgrade under consideration by USAF for HH-60G fleet from about 2003, with various options being examined; these include remanufacture to new **Block 162** standard with 'glass cockpit', new defensive aids and other changes; and less ambitious structural life extension programme.

MH-60K: US Army special operations aircraft (SOA); prototype (89-26194) ordered in January 1988; first flight 10 August 1990. US Army funded two batches of 11 with options for another 38, which not taken up; first production aircraft (91-26368) completed, February 1992; trials at Patuxent River and Edwards AFB before intended first deliveries in June 1992 to 160th Special Operations Aviation Group (part of 160 SOA Regiment). Deliveries delayed by software problems with special operations equipment; first 10 accepted in 1992 in non-operational state; remaining 12 initially stored, then delivered with new software installed, October to December 1993, to permit start of training by 160 SOA Group, February 1994.

Features include provision for additional 3,141 litres (829 US gallons; 691 Imp gallons) of internal and external fuel (see Power Plant), plus flight refuelling capability, integrated avionics system with electronic displays, Raytheon AN/AAQ-16 FLIR and AN/APQ-174B terrain-following, ground-mapping and air-to-ground ranging radar, T700-GE-701C engines and uprated transmission, external hoist, wire-strike protection, rotor brake, tiedown points, folding tailplane, AFCS similar to that of SH-60B, strengthened pintle mounts for 0.50 in machine guns, provision for Stinger missiles, missile warning receiver, pulse radio frequency jammer, CW radio jammer, laser detector, chaff/flare dispensers, and IR jammer.

SH-60B Seahawk: US Navy ASW/ASST helicopter, *described separately.*

SH-60F Seahawk: US Navy carrierborne inner-zone ASW helicopter to replace SH-3D Sea King. See Seahawk entry.

HH-60H and HH-60J Jayhawk: Search and rescue/special warfare helicopters; see Seahawk entry.

UH-60J: Designation of Japanese-built S-70A-12 for Air and Maritime Self-Defence Forces; procurement details under Mitsubishi in Japanese section.

UH-60JA: Japanese Ground Self-Defence Force version; requirement for 50 to 70; procurement began in 1995; refer to Mitsubishi in Japanese section.

UH-60L: Replaced UH-60A in production for US Army from October 1989 (aircraft 89-26179 onwards); prototype (84-23953) first flight 22 March 1988; two pre-series aircraft (89-26149 and 26154); first delivery 7 November 1989 to Texas ArNG. Powered by T700-GE-701C engines with uprated 2,535 kW (3,400 shp) transmission. Current production aircraft fitted with hover IR suppression system (HIRSS) to cool exhaust in hover as well as forward flight; older UH-60s retrofitted. Composites wide-chord main rotor blades of improved design flight tested at West Palm Beach, beginning 8 December 1993; new blade 16 per cent wider than current titanium rotor and has anhedral tip angled down at 20°; testing reveals much lower vibration plus anticipated benefits in payload, speed and manoeuvrability; projected retrofit from 1997 has still to be implemented. Underslung load capability increased to 4,082 kg (9,000 lb).

HH-60L: Alternative designation for new-build medical evacuation version based on UH-60L airframe but incorporating UH-60Q specialised mission equipment.

UH-60A transferred from US Army to Customs Service, but retaining its military serial number (*Jane's/Paul Jackson*)   NEW/0526994

Australian Army Sikorsky S-70A-9 Black Hawk (*Jane's/Paul Jackson*)   *NEW*/0526993

Sikorsky awarded US$11 million contract on 22 February 2000 for design definition and conversion of four UH-60Ls to this standard. Initial delivery expected in 2000, but delayed until March 2001, when three helicopters assigned to 507th Medical Company (Air Ambulance) at Fort Hood, Texas. At least five more acquired, following award of contract in May 2001, which stipulated delivery by March 2002.

MH-60L: Similar to MH-60A; for 160th SOAR, US Army; further modified as below.

AH-60L: 'Direct Action Penetrator'. Upgrade of MH-60L in 1990 with FLIR, radar and standard UH-60 external stores support system; two Black Hawk companies of 160th SOAR each have MH-60K platoon and AH-60L platoon. Armament includes multiple 30 mm Chain Gun, racks of four Hellfires and 2.75 in rocket pods, 40 mm grenade launcher or trainable 7.62 mm Gatling guns.

HH-60M: Medical evacuation version based on upgraded UH-60M; as many as 357 could be acquired for service with first- and second-line medical evacuation units.

MH-60M: Special forces derivative, intended to replace existing MH-60 fleet with 160th SOAR. Few details available, but total of 96 required by US Army, which hopes to complete deployment in 2010.

UH-60M: First use of designation was for proposed enhanced version for US Army. Cancelled early 1989 in favour of UH-60L. Designation subsequently re-used in 2000; see immediately below for details.

UH-60M: Improved Black Hawk version (originally known as the **UH-60L+**) for service with US Army; involves avionics and power plant modernisation effort to extend operational life by 25 to 30 years, with benefits in terms of payload (up to 907 kg; 2,000 lb advantage over UH-60A) and performance (up to 15 kt; 28 km/h; 17 mph faster). Sikorsky awarded US$7.45 million contract in August 2000 for preliminary risk reduction effort. US Army envisages initial procurement via major upgrade programme, whereby approximately 906 UH-60As will be brought to UH-60M standard, while another 311 UH-60Ls are probably to be similarly upgraded. New-build examples also to be produced from 2007, with US Army having requirement for up to 300.

Improvements for the UH-60M include a wide-chord, composite-spar main rotor, a digitised 'glass cockpit' based on the MIL-STD-1553 databus and new avionics, Stormscope weather mapping system, an advanced flight control computer, new diagnostic monitoring systems, a strengthened centre fuselage, larger fuel tanks and advanced infra-red suppression.

Upgrade programme will also allow entire US Army Black Hawk fleet to standardise on General Electric T700-GE-701D engine and improved durability main gearbox. General Electric and US Army announced joint effort to remanufacture existing engines into more reliable -701D form at start of 2001, benefits including longer engine life and lower operating costs, as well as offering about 4 per cent more power.

On 30 March 2001, Defense Acquisition Board approved Milestone B system development and demonstration phase; this followed in May by US$219.7 million contract for research, development, test and evaluation. Total of four UH-60M prototypes planned, comprising three conversions of existing Black Hawks and one new-build helicopter. Upgrade programme began in November 2001, with arrival of three UH-60s at Troy, Alabama, for dismantling and evaluation. Some structural reconditioning will be undertaken at Troy, before components and assemblies are shipped to Stratford for

reconditioning of dynamic components and reassembly. First aircraft to be upgraded is UH-60A 85-24432, which is due to fly in 2003. Second will be conversion of UH-60L 89-26217; third will be conversion of UH-60A 77-22716 to **HH-60M** medical evacuation configuration. Development phase to take four years, leading to low-rate initial production of 10 aircraft in FY04, 15 in FY05 and 36 in FY06, eventually rising to rate of 90 helicopters per year by FY09. Total cost of upgrade estimated to be about US$11.6 billion.

VH-60N: Nine for US Marine Corps Executive Flight Detachment of squadron HMX-1 at Quantico, Virginia, to replace UH-1Ns; deliveries started November 1988; known as VH-60A until redesignated 3 November 1989. Name **White Hawk** adopted by Marine Corps.

Additional equipment includes more durable gearbox, weather radar, SH-60B-type flight control system and ASI, -401 engines as in SH-60B, cabin soundproofing, VIP interior, cabin radio operator station, EMP hardening, 473 litre (125 US gallon; 104 Imp gallon) internal fuel tank and extensive avionics upgrading. SPAR (Special Progressive Aircraft Rework) undertaken on VH-60N fleet from 1998.

UH-60P: South Korean Army version of UH-60L (S-70A-18) with minor avionics modifications to meet local requirements; first (KA-1602) of three UH-60Ls delivered by Sikorsky 10 December 1990; balance of 81 UH-60Ps on initial contract assembled locally by Korean Air Lines with increasing indigenous content, in US$500 million, five year programme. Deliveries from follow-on batch of 57 since completed, but acquisition of third batch, to replace remaining UH-1 Iroquois, has not taken place. South Korea has requirement for about 12 medical evacuation helicopters and could obtain UH-60Q or equivalent.

UH-60Q: 'Dustoff' (Dedicated Unhesitating Service To Our Fighting Forces) medical evacuation/search and rescue version for US Army. Development began in early 1990s, after Gulf War, when it was realised that a requirement existed for a longer-range medevac helicopter to replace the UH-1V Iroquois. A proof-of-principle conversion of a UH-60A (86-24560) was undertaken by Serv-Air Inc of Richmond, Kentucky, and flown for the first time as the **YUH-60A(Q)** on 31 January 1993. This aircraft was subsequently delivered to the Tennessee Army National Guard at Lovell Field, Chattanooga, on 12 March 1993, where it underwent a 12-month evaluation programme beginning in September 1993, with an organisation known as CECAT (Combat Enhancing Capable Aviation Team).

Sikorsky eventually selected as prime integrator for production and was awarded an initial contract on 9 February 1996 for two Phase 2 conversions in FY96, with second contract for further two in FY97; YUH-60A(Q) designation also applied to these aircraft, which participated in two-year qualification programme, with initial flight tests successfully completed in second quarter of 1997; formal operational test followed at Fort Campbell, Kentucky, between July and September 1998, using three YUH-60A(Q)s, with fourth delivered to CECAT in early 1999. Major subcontractors are Air Methods (medical interiors); Breeze-Eastern (HS-29900 external electric rescue hoist); BAE Systems Canada (mission management system); FLIR Systems Inc (SAFIRE thermal imaging system); Litton (LITOX onboard oxygen generating system); Simula (medical attendant seats) and Telephonics (intercom).

Definitive UH-60Q configuration includes a medical interior able to accommodate six stretcher patients, with integrated suction and oxygen systems plus defibrillation, ventilation and intubation equipment, as well as apparatus for monitoring of vital signs. It also has a 'glass cockpit'

incorporating Litton smart multifunction displays (SMFDs); Doppler 128C with embedded GPS; NVG-compatible lighting; AN/ARS-6(V)2 personnel locator system; HIRSS; chaff/flare dispensers; ESSS; -701C engines; plus an improved data modem and SINCGARS radios, which allow it to transmit and receive digital data.

MH-60S KnightHawk: Shipboard transport helicopter. Key element in 1996 US Navy Helicopter Master Plan, entailing retirement of CH/HH-46Ds, HH-60Hs and SH-3s by FY12 and their replacement by navalised MH-60S (previously designated CH-60S until 6 February 2001). Design is fundamentally baseline UH-60L Black Hawk with T700-GE-401C engines and dynamics of SH-60 Seahawk, plus automatic rotor blade folding system, folding tail pylon, improved durability gearbox, rotor brake, automatic flight control system (AFCS), HIFR capability and rescue hoist for SAR/CSAR missions. Also has 'glass cockpit', with active matrix liquid crystal displays (AMLCDs); common cockpit on MH-60S and MH-60R, with Lockheed Martin securing US$61 million contract in August 1998 to develop and produce two prototypes for flight testing from late 1999. Equipment includes two integrated inertial navigation/GPS units, mass memory unit, mission computer, flight management computer, operational software and four Litton flat-panel displays replacing all but standby instruments; MH-60S cockpit is 'scaled-down' version, with potential for upgrading if combat SAR mission is added at later date. MH-60S also has provision for external stores support system (ESSS), allowing carriage of additional fuel and forward-firing weapons.

Design includes a convertible cargo handling system; when configured for pure cargo operations, MH-60S can carry two 1.02 × 1.22 × 1.02 m (40 × 48 × 40 in) tri-wall pallets with total weight of 1,588 to 1,814 kg (3,500 to 4,000 lb); as a personnel transport, is able to accommodate a crew of four, plus 13 passengers. Underslung loads up to 4,082 kg (9,000 lb) may also be carried, with total payload capacity about 4,536 kg (10,000 lb).

MH-60S potential demonstrated by modified UH-60L in June 1995, when internal and external cargo-carrying capability studied by Sikorsky and the US Navy. Detail design began October 1996, with award in April 1997 of US$5.75 million contract to Sikorsky for demonstrator. This was hybrid vehicle, based around UH-60L (96-26673) borrowed from the US Army, married to components from SH-60F furnished by US Navy. Resulting YCH-60S (Navy identity 966673) made first flight on 6 October 1997 and was used for joint Navy/Sikorsky 35 hour flight test programme that ended on 10 January 1998. First shipboard demonstration on 19 November 1997, with YCH-60S completing 17 landings aboard combat store ship USS *Saturn*; initial trial included 12 vertical replenishment lifts with 680 kg (1,500 lb) slung load and three hot refuellings.

YCH-60S evaluated in 1999-2000 as potential airborne mine countermeasures (AMCM) platform to replace MH-53E; existing minesweeping sleds used by MH-53E are too heavy for MH-60S, but new lightweight towed systems and laser imaging detection and ranging equipment are in prospect and could bestow adequate minehunting capability. Initial trials at Stratford in third quarter of 1999, followed by transfer to Patuxent River in fourth quarter for tow tests, plus carriage, winch deployment and recovery of AN/AQS-20/X mine detection sonar. Thicker frames for helicopter's rear-cabin structure will permit cable loads up to 2,722 kgf (6,000 lbf). Replacement of MH-53E by MH-60S expected to begin in 2005, with ultimate total of 66 MH-60Ss planned for AMCM mission.

Features not embodied in demonstrator include fuel dump vents, flotation gear, HIFR and navalised T700 engines. Avionics also unrepresentative, although featuring a databus and four 127 × 127 mm (5 × 5 in) Rockwell Collins AMLCDs for pilot and co-pilot vertical and horizontal situation data, plus two control display/navigation units and an LN-100G embedded GPS/INS.

Decision to proceed with MH-60S low-rate initial production (LRIP) taken in early 1998, although firm fixed-price contract for first lot not awarded to Sikorsky until September 1999; valued at US$67.4 million, this was for initial five aircraft, plus option for one more (taken up in November 1999) and associated engineering and logistic services. Maiden flight of initial production MH-60S (165742) at Stratford, Connecticut on 27 January 2000; delivered to Patuxent River, Maryland, on 15 May to begin US Navy development testing and operational evaluation. At least four helicopters involved in trials, with initial technical evaluation completed by November 2000; three-month Opeval began November 2001 and yielded disappointing result, with MH-60S considered neither operationally "effective" nor "suitable". Changes to software and operational requirements ensued, allowing approval to be given for start of full-rate production at end of August 2002.

Navy has requirement for 237 MH-60Ss. First example funded in FY98, with five more (Lot 1) in FY99, 14 (Lot 2) in FY00 and 15 (Lot 3) for FY01. Navy procurement in sixth multiyear purchase (FY02 to FY06) expected to total 82. First squadron is HC-3 at North Island, California, which received initial example (165745) for use as ground maintenance trainer in early 2001 and which assumed responsibility for flight training in February 2002; second squadron is HC-5 at Andersen AFB, Guam, which began converting from H-46 in mid-2002. HC-6 at Norfolk, Virginia, also accepted its first MH-60S in second quarter of 2002. Turkey also planning to acquire initial batch of four to six MH-60Ss.

UH-60X: Designation allocated to potential follow-on to UH-60M, but now known as Future Utility Rotorcraft (FUR), for which UH-60M may provide basis. Total of 256 required for service with 'first-to-fight' units, but currently unfunded and unlikely to enter inventory until 2025.

Firehawk: Trials of specialist firefighting version began in July 1998, using modified UH-60L (96-26728) with extended landing gear and removable 3,785 litre (1,000 US gallon; 833 Imp gallon) ventral watertank manufactured by Aero Union of Chico, California. Replenishment of tank can be accomplished in two ways: by landing next to water source for water to be pumped into tank via side connector, or by hovering over source and using snorkel hose and pump assembly to suck up water. Subsequent three month demonstration of firefighting capability by Los Angeles County Fire Department proved validity of system and first customer was involved in negotiations with Sikorsky during fourth quarter of 1998. Identity of potential buyer not confirmed, but thought to be Sultan of Brunei. Demonstrator subsequently returned to US Army for further testing, before delivery to Oregon Army National Guard in 1999. Two examples delivered to Los Angeles County Fire Department in 2001, with Army National Guard units in California and Florida due to receive single examples during 2002-03.

Maple Hawk: Unsuccessful contender in contest to supply Canadian Forces with new SAR helicopter; offer reportedly priced at C$300 million for 15 helicopters, but EHI EH 101 Cormorant selected in December 1997.

Battle Hawk: Offered to Australian Army for Project Air 87 armed reconnaissance requirement; based on MH-60K; two -701C engines and 2,535 kW (3,400 shp) transmission; up to 680 kg (1,500 lb) extra payload; improved electromagnetic and corrosion protection; belly turret with 20 mm Giat THL 20 cannon slaved to Elbit Toplite II FLIR/day TV/laser designator-ranger targeting sensor or Elbit Midash helmet sight; and full 'glass cockpit' with Rockwell Collins active matrix colour LCDs and HOCAS controls. Eliminated from bidding, but competition reopened in December 2000 and Battle Hawk reinstated; however, competition won in 2001 by Eurocopter Tiger. Australian Army also requires further 12 troop-lift helicopters to augment existing S-70s.

Exports comprise: S-70A-1: FMS deal for Royal Saudi Land Forces Army Aviation Command; 12 delivered January to April 1990 to squadron based at King Khaled Military City; modified to **Desert Hawk** and one (delivered December 1990) fitted with VIP interior; Desert Hawk has 15 troop seats, blade erosion protection using polyurethane tape and spray-on coating, Racal Jaguar 5 frequency-hopping radio, provision for searchlights, internal auxiliary fuel tanks, and external hoist.

S-70A-1L: Medical evacuation version for Saudi Arabia; IR filtered searchlight, rescue hoist, improved AN/ARC-217 HF com, AN/ARN-147 VOR/ILS, AN/ARN-149 ADF, air conditioning and provision for six stretchers; eight delivered from December 1991; further eight required.

S-70A-5: Two for Philippine Air Force, delivered March 1984.

S-70A-9: Royal Australian Air Force; 39 replaced Bell UH-1s; deliveries from October 1987 to 1 February 1991; first completed by Sikorsky, remainder assembled by Hawker de Havilland in Australia; aircraft transferred to Australian Army in February 1989, but RAAF continues to maintain them.

Israel Defence Force Sikorsky UH-60A Yanshuf *(IDF)* 0095401

S-70A-11: Three to Jordan in 1986-87.

S-70A-12: Japan Self-Defence Forces acquiring **UH-60J/JA** versions of Mitsubishi SH-60J for search and rescue. Sikorsky-built prototype (N7267D), plus two CKD kits, delivered late 1990. Further production by Mitsubishi (which see).

S-70A-16: Reserved for Westland Helicopters (see UK section of 1995-96 and earlier *Jane's All the World's Aircraft*).

S-70A-17: Turkish Jandarma ordered six in September 1988; deliveries completed December 1988; further six (including two VIP) delivered from late 1990 to Turkish National Police. See also S-70A-28.

S-70A-18: Korea (see UH-60P).

S-70A-19: Reserved for GKN Westland of UK.

S-70A-21: Two VIP versions to Egypt, 1990. Two VIP-configured UH-60L ordered by Egypt in third quarter of 1999; delivery due by end of 2003. Further two VIP-configured UH-60Ls requested in September 2002.

S-70A-22: Korean VIP version. Three aircraft built by Sikorsky.

S-70A-24: Two UH-60Ls for Mexico. Delivered 1991. Further four received in about 1996.

S-70A-25/26: Moroccan Gendarmerie ordered two Black Hawks with different seating arrangements in 1991; delivered October 1992; began operations 11 November 1992; fitted with colour weather radar.

S-70A-27: Hong Kong. Two delivered 16 December 1992 to Royal Hong Kong Auxiliary Air Force; unit became Government Flying Service 1 April 1993. Fitted with FLIR and searchlight. Requirement for further four reportedly existed in 1995, but only one additional aircraft delivered.

S-70A-28: Turkish follow-on batch; 90 ordered 8 December 1992, of which first five to Jandarma on 4 January 1993, followed by 40 to armed forces during 1993-94; balance of 45 to have been co-produced in Turkey by TAI (which see) but programme suspended. However, fresh negotiations for 50 additional Black Hawks concluded in latter half of 1998, with first five airlifted to Turkey by An-124 in mid-June 1999 and deliveries completed in 2001. Final 30 produced with 'glass cockpit', which to be retrofitted to earlier machines; first flight of **S-70A-28D** in this guise on 29 March 2000. Installation includes four Rockwell Collins MFDs, dual flight management system and LN-100G INS/GPS; four to incorporate Tadiran Spectralink ASR-700 airborne search and rescue system (ASARS) for use by Turkish Army Special Forces in combat search and rescue and covert operations.

S-70A-30: One VIP transport ordered for Argentine Air Force, January 1994; delivered 4 September 1994.

S-70A-33: Four ordered by Brunei in 1995; delivered 1997-98. Equipment includes radar, AN/AAQ-21 FLIR and external stores support system.

S-70A-34: Malaysia ordered two S-70A Black Hawks in 1996 as replacements for AS 332 Super Puma in VIP transport role; first of pair delivered by end of 1997, with second following in February 1998.

S-70A-36: Brazil received four S-70As in August 1997 for use in Peru/Ecuador peacekeeping support mission; equipment includes GPS, HF radio, internal rescue hoist and weather radar.

S-70A-37: Version of Firehawk; two to Sultan of Brunei, 2000, replacing S-70Cs.

S-70A-39: Chilean order for one, announced in March 1998; delivered in July 1998, with further purchases expected to replace UH-1H Iroquois over next few years.

S-70A-41: Colombia. Believed to refer to 22 aircraft delivered during 1994 (two), 1995 (two), 1997 (seven) and 1998 (11); these are UH-60L derivative, unlike initial delivery of 1988-89 which was baseline UH-60A. Further helicopters requested in November 1999; two contracts received by Sikorsky in mid-December 2000 cover purchase of 21 helicopters for Colombian Army, plus seven for Air Force and two for National Police. Ultimately, as many as 60 in prospect as part of US drive to combat increased narcotics trafficking in this region.

S-70A-42: Austria signed contract in December 2000 for nine aircraft to begin replacement of Agusta-Bell 204/212; total cost of deal put at US$184 million. First example handed over in USA on 10 June 2002.

S-70A-50: Israel. Request for 15 UH-60Ls revealed by US DoD in April 1997; first so-called 'Peace Hawk' handed over at Stratford, Connecticut, on 23 March 1998, being airlifted (with four others) to Israel by C-5 on 27 May 1998. Deliveries completed by end of 1998. Additional 35 requested in September 2000 at estimated cost of US$525 million, but subsequent contract received on 31 January 2001 covered supply of 24 aircraft at total cost of US$211.8 million. Delivery of these was accomplished in second half of 2002.

Other recent customers include Thailand, which ordered first two of planned total of 33 Black Hawks in 2000; total of three delivered by early 2002, with procurement set to continue until 2009. Los Angeles County Fire Department accepted two S-70A Firehawks in the first half of 2001.

Direct transfers include one UH-60L to Bahrain, early 1991; five UH-60As delivered to Colombian Air Force in July 1988 for anti-narcotics operations; five more sold February 1989. Israel received 10 former US Army UH-60As in August 1994, for 124 Squadron at Palmachim, under local name of Yanshuf (Owl).

Taiwan requires new utility helicopter as replacement for existing UH-1H Iroquois; S-70A in contention and reportedly preferred option over competing Bell 412; up to 80 of chosen type to be obtained, with local manufacture expected to be a key factor in selection. Decision originally anticipated in 1998 but has been deferred, with initial contract for 20 to 25 helicopters likely.

Black Hawk also a contender for Greek Army requirement for up to 66 utility/combat search and rescue helicopters; comparative evaluation undertaken in July 2000, with Mil Mi-17 and Eurocopter AS 532 Cougar Mk II among candidates for co-production agreement. Follow-on purchase of up to 12 Black Hawks by Australia is also possible, following decision to establish another Army troop lift squadron by 2007; funding request was expected in May 2001 budget request, but has been deferred.

S-70C: Commercial version, described separately and under SH-60B Seahawk.

CUSTOMERS: By 1999, over 2,400 H-60s of all variants had flown more than 3,600,000 hours. US Army total includes EH-60As and diversions to USAF, Bahrain, Colombia, Egypt and Saudi Arabia; 1,000th of S-70 series accepted 17 October 1988 and 2,000th May 1994; US Army Black Hawks in service in Germany, Hawaii and South Korea and with Army National Guard and Army Reserve.

US Army UH-60As loaned to the US Drug Enforcement Agency, augmenting five bought direct from Sikorsky. Fleet in 1999 totalled 13 (all with military serial numbers), plus three in storage and further two lost in accidents.

US BLACK HAWK PROCUREMENT

FY	Lot	Army	USAF	USN	FMS	First aircraft*	Remarks
73	-	6				73-21650	RDT&E
77	1	15				77-22714	
78	2	56				78-22960	Includes YEH-60B (78-23013)
79	3	92				79-23265	
80	4	94				80-23416	
81	5	80	5			81-23547	
82	6	96	6			82-23660 ⎤	
83	7	96				83-23837 ⎬ MYP1, 18 December 1981	
84	8	96				84-23933 ⎦	
85	9	104				85-24387 ⎤	
86	10	96				86-24483 ⎬ MYP2, 31 October 1984	
87	11	102	9			87-24579 ⎦	
88	12	72	16		20[2]	88-26015 ⎤	
89	13	72	18		5[3]	89-26123 ⎬ MYP3, 11 January 1988	
90	14	76	22		1[4]	90-26218 ⎮	
91	15	67[1]	15		8[5]	91-26318 ⎦	
92	16	52	13			92-26408 ⎤	
93	17	60				93-26473 ⎮	
94	18	63				94-26533 ⎬ MYP4, 28 April 1992	
95	19	68				95-26596 ⎮	
96	20	65			9[9]	96-26664 ⎦	
97	21	34	8		15[8]	97-26738 ⎤	
98	22	34		1[10]		98-26795 ⎬ MYP5, 18 July 1997[6]	
99	23	29		5[10]		99-26829 ⎮	
00	24	27		14[10]		00-26847 ⎦	
01	25	24		15			
Totals		**1,676**	**112**	**35**	**58**		

Notes: [1] Includes 22 MH-60K for special operations
[2] For Saudi Arabia (13), Colombia (5) and Egypt (2)
[3] For Colombia
[4] For Bahrain
[5] For Saudi Arabia
[6] MYP5 (FY97-01) originally to consist of 108 aircraft (58 Army, 42 Navy and 8 USAF), with 36 in first year and 18 per year thereafter; however, additional aircraft have been purchased in each year. MYP6 (FY02-06) is currently expected to include at least 162 aircraft (80 UH-60L for US Army and 82 MH-60S for US Navy)
[7] For Egypt
[8] For Israel
[9] Includes two for Colombia
[10] Serial numbers not included in US Army sequence
* Serial numbers not always consecutive; batches include transfers to USAF and Foreign Military Sales programme

See Current Versions for export models and details and also entry for S-70C version. Sikorsky exported 31 S-70s on commercial terms in 2000, following 22, 32 and 12 in 1997-99.

COSTS: UH-60L US$8.6 million (1997) US Army unit cost; MH-60G US$10.2 million. Two VIP aircraft for Egypt cost about US$47 million (2002), with spare engines, spare parts, tools and support equipment and other logistical support. Two utility aircraft for Brunei, with firefighting kits, believed to cost US$25-30 million (1999); two standard aircraft for Thailand will cost about US$20 million. MYP6 purchase of 80 UH-60L and 82 MH-60S estimated to be worth US$1.5 billion during FY02-06.

DESIGN FEATURES: Represented new generation in technology for performance, survivability and ease of operation when introduced to replace UH-1 as US Army's main squad-carrying helicopter; adapted to wide variety of other roles, including several maritime applications. Four-blade main rotor; one-piece forged titanium rotor head with elastomeric blade retention bearings providing all movement and requiring no lubrication; hydraulic drag dampers; bifilar self-tuning vibration absorber above head; blades have 18° twist, and tips swept at 20°; thickness and camber vary over the length of blades, based on Sikorsky SC-1095 aerofoil; blades tolerant up to 23 mm hits and spar tubes pressurised with gauges to indicate loss of pressure following structural degradation.

Two pairs of tail rotor blades fastened in cross-beam arrangement, mounted to starboard; tail rotor pylon tilted to port to produce lift as well as anti-torque thrust and to extend permissible CG range; fixed fin large enough to allow controlled run-on landing following loss of tail rotor.

FLYING CONTROLS: Rotor pitch control powered by two independent hydraulic systems; Hamilton Sundstrand AFCS with digital three-axis autopilot provides speed and height control and coupled modes. Full-time autostabilisation includes feet-off heading hold cancelling torque-induced yaw at all airspeeds and during hover; positive fuselage attitude control provided by electrically driven variable incidence tailplane moving from +34° in hover to −6° during autorotation; angle is controlled by combined sensing of airspeed, collective-lever position, pitch attitude rate and lateral acceleration.

STRUCTURE: Main blade spar is formed and welded into oval titanium tube, with Nomex core, graphite trailing-edge, and covered by glass fibre/epoxy skin; titanium leading-edge abrasion strip and Kevlar tip. New main blades, with modified tips and 16 per cent increase in chord, under development for UH-60L; available for retrofit from 1997. Cross-beam composites tail rotor, eliminating all rotor head bearings. Light alloy airframe designed to retain 85 per cent of its flight deck and passenger space intact after vertical impact at 11.5 m (38 ft)/s, lateral impact at 9.1 m

(30 ft)/s, and longitudinal impact at 12.2 m (40 ft)/s; also withstands simultaneous 20 g forward and 10 g downward impact; glass fibre and Kevlar used for cockpit doors, canopy, fairings and engine cowlings; glass fibre/Nomex floors; tailboom folds to starboard and main rotor mast can be lowered for transport/storage.

LANDING GEAR: Non-retractable tailwheel type with single wheel on each unit. Energy-absorbing main gear with a tailwheel which gives protection for the tail rotor in taxying over rough terrain or during a high-flare landing. Axle assembly and main gear oleo shock-absorbers by General Mechatronics. Mainwheel tyres size 26×10.00-11, pressure 8.96 to 9.65 bar (130 to 140 lb/sq in); tailwheel tyre size 15×6.00-6, pressure 6.21 to 6.55 bar (90 to 95 lb/sq in). Alaskan-based H-60s have Airglass Engineering ski landing gear.

POWER PLANT: Two 1,210 kW (1,622 shp) intermediate rating General Electric T700-GE-700 turboshafts initially. From late 1989 (UH-60L), two T700-GE-701C engines, each developing intermediate 1,342 kW (1,800 shp). (T700-GE-701A engines with maximum T-O rating of 1,285 kW;

1,723 shp optional in export models.) Transmission rating 2,109 kW (2,828 shp) in UH-60A, uprated to 2,535 kW (3,400 shp) in models with T700-GE-701C engines.

Two crashworthy, bulletproof fuel cells, with combined usable capacity of 1,361 litres (360 US gallons; 300 Imp gallons), aft of cabin. Single-point pressure refuelling via point on each tank. Auxiliary fuel can be carried internally in one of several optional arrangements, or externally by the ESSS system. Two external tanks each of 871 litres (230 US gallons; 192 Imp gallons); up to two internal tanks, each of 700 litres (185 US gallons; 154 Imp gallons).

ACCOMMODATION: Two-man flight deck, with pilot and co-pilot on armour-protected seats. A third crew member is stationed in the cabin at the gunner's position adjacent forward cabin windows. Forward-hinged jettisonable door on each side for access to flight deck area. Main cabin open to cockpit to provide good communication with flight crew and forward view for squad commander. Accommodation for 11 fully equipped troops, or 14 in high-density configuration; 20 minimally armed personnel in optional configuration. Eight troop seats can be removed and replaced by four litters for medevac missions, or to make room for internal cargo. An optional layout is available to accommodate a maximum of six litter patients. Executive interiors for seven to 12 passengers available for the S-70A. Cabin heated and ventilated. Simula Safety Systems Inc received US$7.1 million contract in April 1999 covering supply of 290 cockpit airbag systems for installation in US Army UH-60A/L; this is low-rate initial production phase and total comprises 275 aircraft units and 15 spares.

External cargo hook, having a 3,630 kg (8,000 lb) lift capability, enables UH-60A to transport a 105 mm howitzer, its crew of five and 50 rounds of ammunition. Rescue hoist of 272 kg (600 lb) capacity optional. Large rearward-sliding door on each side of fuselage for rapid entry and exit.

SYSTEMS: Solar 67 kW (90 hp) T-62T-40-1, Honeywell and Hamilton Sundstrand APU. An optional winterisation kit provides a second hydraulic accumulator installed in parallel with the APU hydraulic start accumulator, maintaining engine start capability at low ambient temperatures; Honeywell 30 to 40 kVA and 20 to 30 kVA electrical power generators; 17 Ah Ni/Cd battery. Engine fire extinguishing system. Rotor blade de-icing system standard on US Army aircraft, optional for export. Electric windscreen de-icing.

AVIONICS: Configurations vary between aircraft. Additional avionics and self-protection equipment installed in Enhanced Black Hawk, as described under Current Versions. Improvement options offered from 1996 for new-build and retrofit on S-70 series include 'glass cockpit' and digital avionics; equipment available includes EFIS and digital automated flight computer system (AFCS).

Comms: Raytheon AN/ARC-186 VHF-FM, Raytheon AN/ARC-115 VHF-AM, Raytheon AN/ARC-164 UHF-AM, Rockwell Collins AN/ARC-186(V) VHF-AM/FM, Honeywell AN/APX-100 IFF transponder, Raytheon TSEC/KT-28 voice security set, and intercom. HH-60G has AN/URC-108 satcom and is being upgraded with Rockwell Collins AN/ARC-210 integrated communications system; Rockwell Collins AN/ARC-220 nap of the earth digital radio and AN/ARC-222 installed on Block 152 Upgrade HH-60G. UH-60M will have Telephonics secure digital intercom system.

Radar: MH-60K has Raytheon AN/APQ-147A terrain-following/terrain-avoidance radar, HH-60G has

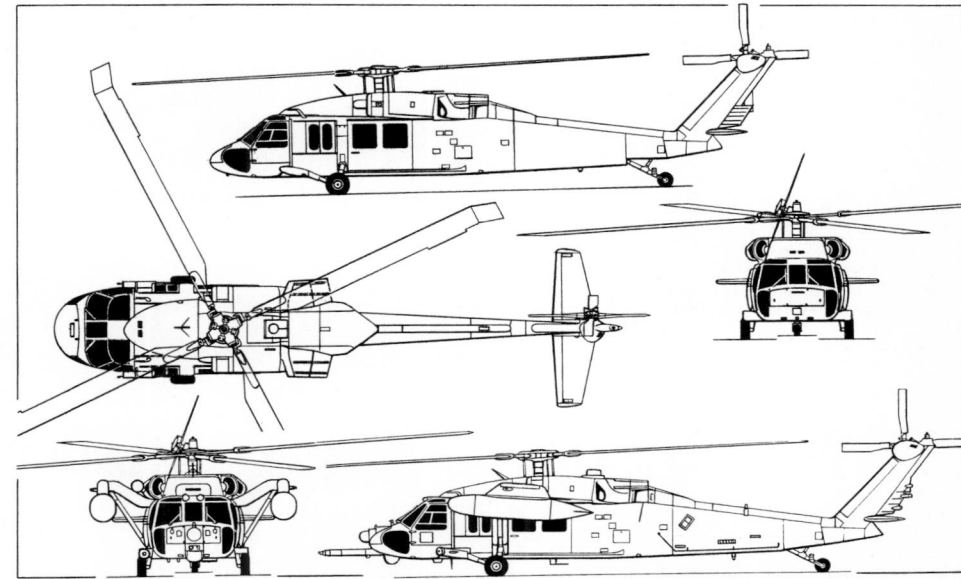

Sikorsky UH-60L Black Hawk combat assault helicopter, with additional lower side view and lower front view of MH-60K special operations variant (*Jane's/Dennis Punnett*)

Honeywell AN/APN-239 (RDR-1400C) radar. AH-60L and some export S-70s also equipped with radar.

Flight: Hamilton Sundstrand AFCS with digital three-axis autopilot, Honeywell AN/ARN-123(V)1 VOR/marker beacon/glideslope receiver, Emerson AN/ARN-89 ADF, BAE Systems AN/ASN-128 Doppler, AN/ASN-43 gyrocompass, Honeywell AN/APN-209(V)2 radar altimeter. HH-60G has BAE Systems AN/ASN-137 Doppler, Rockwell Collins AN/ASN-149 GPS and Litton ring laser gyro INS (replacing Carousel IV).

Instrumentation: HH-60G has Teldix KG-10 map display. US Army MH-60 special operations versions to receive Elbit ANVIS 7 NVG/HUD system as retrofit; system already fitted in UH-60A and L. Elbit ANVIS/HUD system ordered by Australia in August 1999; 12 units acquired for installation by Raytheon Australia on S-70A-9. NVG-compatible version of Lockheed Martin GH-3000 electronic standby instrument system to be installed on MH-60K. UH-60M will have four Rockwell Collins 152 × 203 mm (6 × 8 in) landscape colour AMLCDs.

Mission: HH-60G has Raytheon AN/AAQ-16 FLIR. UH-60Q to have FLIR Systems Inc AN/AAQ-21 SAFIRE thermal imaging system. Trials of Northrop Grumman Airborne Standoff Minefield Detection System (ASTAMIDS) undertaken over Bosnia in latter half of 1997; intended to be precursor of procurement of at least 85 production systems but poor test results, especially in areas of dense undergrowth, combined with budget cuts to force US Army to slow pace of development and delay deployment of this system.

Self-defence: Baseline UH-60 Black Hawk has Raytheon AN/APR-39(V)1 RWR, Sanders AN/ALQ-144 IR countermeasures set and BAE Systems M-130 chaff/flare dispenser. MH-60K has BAE Systems AN/AAR-47 missile warning system, Northrop Grumman AN/ALQ-136 pulse radio frequency jammer, Northrop Grumman AN/ALQ-162 CW radio jammer, Raytheon AN/APR-39A and AN/APR-44 pulse/CW warning receivers, Raytheon AN/AVR-2 laser detector, BAE Systems M-130 chaff/flare dispenser and Sanders AN/ALQ-144 IR countermeasures set. HH-60G has chaff/flare dispenser (BAE Systems M-130 being replaced by AN/ALE-47 since 1998) and Sanders AN/ALQ-144 IR countermeasures set. Development testing of Sanders AN/ALQ-212 Advanced Threat IR Countermeasures (ATIRCM) system on EH/MH-60s of US Army to begin in 1999. AN/AAR-47 missile warning system, AN/ALR-69(V) RWR, AN/ALE-47 chaff/flare dispensers and AN/ALQ-213 EW management system installed on Block 152 Upgrade HH-60G.

EQUIPMENT: HH-60G has Lucas Aerospace internal rescue hoist with 76 m (250 ft) of cable. Lucas awarded contract in mid-2000 to supply initial batch of 21 hoists for installation on MH-60S and other S-70 variants, with potential options for additional 70 in 2001 and 2002 plus 34 in 2003. UH-60Q has external Breeze-Eastern HS-29900 electric rescue hoist.

ARMAMENT: New production UH-60As and Ls from c/n 431 onward incorporate hardpoints for an external stores support system (ESSS). This consists of a combination of fixed provisions built into the airframe and four removable external pylons from which fuel tanks and a variety of weapons can be suspended. Able to carry more than 2,268 kg (5,000 lb) on each side of the helicopter, the ESSS can accommodate two 871 litre (230 US gallon; 192 Imp gallon) fuel tanks outboard, and two 1,703 litre (450 US gallon; 375 Imp gallon) tanks inboard. This allows the UH-60A to self-deploy 1,200 n miles (2,222 km; 1,381 miles) without refuelling. The ESSS also enables the Black Hawk to carry Hellfire laser-guided anti-armour missiles, gun or M56 mine dispensing pods, FIM-92 Stinger AAMs, ECM packs, rockets and motorcycles. Up to 16 Hellfires can be carried externally on the ESSS, with another 16 in the cabin to provide capability to land and reload. Two pintle mounts in cabin can each accommodate a 0.50 in calibre General Electric GECAL 50 or 7.62 mm six-barrel Minigun.

DIMENSIONS, EXTERNAL:

Main rotor diameter	16.36 m (53 ft 8 in)
Main rotor blade chord	0.53 m (1 ft 8¾ in)
Tail rotor diameter	3.35 m (11 ft 0 in)
Length overall: rotors turning	19.76 m (64 ft 10 in)
rotors and tail pylon folded	12.60 m (41 ft 4 in)
Length of fuselage:	
UH-60A/HH-60G, excl flight refuelling probe	15.26 m (50 ft 0¾ in)
HH-60G, incl retracted refuelling probe	17.38 m (57 ft 0¼ in)
Fuselage max width: UH-60A	2.36 m (7 ft 9 in)
Max depth of fuselage	1.75 m (5 ft 9 in)
Height: overall, tail rotor turning	5.13 m (16 ft 10 in)
to top of rotor head	3.76 m (12 ft 4 in)
in air-transportable configuration	2.67 m (8 ft 9 in)
Tailplane span	4.38 m (14 ft 4½ in)
Tailplane chord	0.88 m (2 ft 10½ in)
Wheel track	2.705 m (8 ft 10½ in)
Wheelbase	8.83 m (28 ft 11¾ in)
Tail rotor ground clearance	1.98 m (6 ft 6 in)
Cabin doors (each): Height	1.37 m (4 ft 6 in)
Width	1.75 m (5 ft 9 in)

DIMENSIONS, INTERNAL:

Cabin: Volume	11.6 m^3 (410 cu ft)

AREAS:

Main rotor blades (each)	4.34 m^2 (46.70 sq ft)
Tail rotor blades (each)	0.41 m^2 (4.45 sq ft)
Main rotor disc	210.15 m^2 (2,262.0 sq ft)
Tail rotor disc	8.83 m^2 (95.03 sq ft)
Fin	3.00 m^2 (32.30 sq ft)
Tailplane	4.18 m^2 (45.00 sq ft)

WEIGHTS AND LOADINGS:

Weight empty: UH-60A	5,118 kg (11,284 lb)
UH-60L	5,224 kg (11,516 lb)
Payload: internal, UH-60A/L	1,197 kg (2,640 lb)
underslung, UH-60A	3,629 kg (8,000 lb)
underslung, UH-60L/Q and MH-60S	4,082 kg (9,000 lb)
Mission T-O weight: UH-60A	7,708 kg (16,994 lb)
UH-60L	7,907 kg (17,432 lb)
HH-60G	8,119 kg (17,900 lb)
MH-60K	11,113 kg (24,500 lb)
Max alternative T-O weight:	
UH-60A	9,185 kg (20,250 lb)
UH-60L	11,113 kg (24,500 lb)
Max disc loading:	
UH-60L at mission T-O weight	36.7 kg/m^2 (7.52 lb/sq ft)
UH-60L at max alternative T-O weight	52.9 kg/m^2 (10.83 lb/sq ft)
Transmission loading:	
UH-60L at mission T-O weight and max power	3.04 kg/kW (5.00 lb/shp)
UH-60L at max alternative T-O weight and power	4.20 kg/kW (6.91 lb/shp)

PERFORMANCE (UH-60A at mission T-O weight, except where indicated):

Never-exceed speed (VNE):	
UH-60L/Q	195 kt (361 km/h; 224 mph)
MH-60S	180 kt (333 km/h; 207 mph)
Max level speed at S/L	160 kt (296 km/h; 184 mph)
Max level speed at max T-O weight	158 kt (293 km/h; 182 mph)
Max cruising speed:	
UH-60A	139 kt (257 km/h; 160 mph)
UH-60L	159 kt (294 km/h; 183 mph)
MH-60S	142 kt (263 km/h; 163 mph)
Single-engine cruising speed at 1,220 m (4,000 ft) and 35°C (95°F)	105 kt (195 km/h; 121 mph)
Vertical rate of climb at 1,220 m (4,000 ft) and 35°C (95°F): UH-60A	119 m (390 ft)/min
UH-60L	472 m (1,550 ft)/min
Service ceiling: UH-60A	5,700 m (18,700 ft)
UH-60L	5,835 m (19,150 ft)
Hovering ceiling: IGE at 35°C	2,895 m (9,500 ft)
OGE, ISA	3,170 m (10,400 ft)
OGE at 35°C: UH-60A	1,645 m (5,400 ft)
UH-60L	2,330 m (7,650 ft)
Range with max internal fuel at max T-O weight, 30 min reserves: UH-60A	319 n miles (592 km; 368 miles)
UH-60L	315 n miles (584 km; 363 miles)
Range with external fuel tanks on ESSS pylons:	
with two 870 litre (230 US gallon; 191.5 Imp gallon) tanks	880 n miles (1,630 km; 1,012 miles)
with two 870 litre (230 US gallon; 191.5 Imp gallon) and two 1,703 litre (450 US gallon; 375 Imp gallon) tanks	1,200 n miles (2,222 km; 1,381 miles)
Endurance: UH-60A	2 h 18 min
UH-60L	2 h 6 min

UPDATED

SIKORSKY S-70B

US Navy designations: SH-60B and MH-60R Seahawk, SH-60F and HH-60H

US Coast Guard designation: HH-60J Jayhawk

Japan Maritime Self-Defence Force designation: SH-60J

Spanish Navy designation: HS.23

Republic of China Navy designation: S-70C(M)-1 and S-70C(M)-2 Thunderhawk

TYPE: Naval combat helicopter.

PROGRAMME: Naval development of Sikorsky UTTAS (UH-60A Black Hawk) utility helicopter; won US Navy LAMPS Mk III competition for shipboard helicopter in 1977; first flight of first of five YSH-60B prototypes (161169) 12 December 1979; development details in 1982-83 *Jane's*; first 18 SH-60Bs authorised FY82. Changed USN planning in 1993 resulted in premature end to SH-60B/F production; original intent was to remanufacture SH-60B/Fs and HH-60Hs as SH-60R (redesignated MH-60R in mid-2001), but acquisition strategy changed in 2001 and most MH-60Rs will be new-build helicopters.

Most recent development, revealed in model form at Asian Aerospace, Singapore, February 2000, involves incorporation of external stores support system (ESSS) of UH-60/S-70A series, as well as MH-60R sensor suite. According to Sikorsky, this proposed version evolved in response to requests from potential customers in and around Pacific Rim, such as Indonesia, Malaysia and Singapore. If proceeded with, new version would have maximum take-off weight of around 11,340 kg (25,000 lb).

CURRENT VERSIONS: **SH-60B**: Initial production version for ASW/ASST; 181 built, excluding prototypes.

Detailed description applies to SH-60B, unless otherwise stated.

NSH-60B: Designation applied to two SH-60Bs (162337 and 162974) assigned to permanent test duties at Patuxent River, Maryland.

SH-60F: CV Inner Zone ASW helicopter, known as CV-Helo, for close-in ASW protection of aircraft carrier groups; US$50.9 million initial US Navy contract for full-scale development and production options placed 6 March 1985; replacing SH-3H Sea King; Seahawk prototype modified as SH-60F test aircraft; first flight 19 March 1987; initial fleet deployment by HS-2 aboard USS *Nimitz* in 1991. Currently assigned to 10 deployable squadrons (HS-2 to HS-8, HS-11, HS-14 and HS-15) plus one training unit (HS-10) and one Reserve Force squadron (HS-75). Production terminated with delivery of 82nd example 1 December 1994.

SH-60F has all LAMPS Mk III avionics, fairings and equipment removed, including cargo hook and RAST system main and tail probes, but installation provisions retained. Replaced by integrated ASW mission avionics including Honeywell AN/AQS-13F dipping sonar, MIL-STD-1553B databus, dual Litton AN/ASN-150 tactical navigation computers and AN/ASM-614 avionics support equipment, automatic flight control system with quicker automatic transition and both cable and Doppler autohover, tactical datalink with other aircraft, communications control system, multifunction keypads and displays for each of four crew members; internal/external fuel system and extra weapon station to port allowing carriage of three Mk 50 homing torpedoes; provision for surface search radar, FLIR, night vision equipment, passive ECM, MAD, air-to-surface missile capability, sonobuoy datalink, chaff/sonobuoy dispenser,

First of four test Sikorsky MH-60Rs on its maiden flight, 19 July 2001 0114483

Initial aircraft of Spain's second S-70B order, delivered in 2002   *NEW*/0530172

attitude and heading reference system (AHRS), Navstar GPS, fatigue monitoring system and increase of maximum T-O weight to 10,659 kg (23,500 lb); secondary missions include SAR and plane guard.

YSH-60F: Designation applied to second production SH-60F (163283) which serves as 'prototype' on test duties at Patuxent River, Maryland. To be fitted with vectored thrust ducted propeller ('ring tail') by Piasecki Aircraft Corporation for trials project at Patuxent River during 2003-04 as part of advanced technology demonstration programme. Refer to Piasecki entry in US section for more details.

HH-60H: US Navy procurement of 42 completed in 1996; used for strike-rescue/special warfare support (HCS); designated HH-60H in September 1986; first flight (163783) 17 August 1988; accepted by USN 30 March 1989; in service with HCS-4 at Norfolk, Virginia, January 1990; initial procurement ended with 18th delivery July 1991, completing HCS-5 at Point Mugu, California; both squadrons are part of Navy Reserve. Regular SH-60F squadrons later added pairs of HH-60Hs for deployed duty when embarked aboard aircraft carriers; missions are to recover four-man crew at 250 n miles (463 km; 288 miles) from launch point or fly 200 n miles (371 km; 230 miles) and drop eight SEALs from 915 m (3,000 ft).

Close derivative of SH-60F, with T700-GE-401C engines and HIRSS as SH-60B/F; equipment includes Litton AN/APR-39A(XE)2 RWR, Raytheon AN/AVR-2A (V) laser warning receiver, Honeywell AN/AAR-47 missile plume detector, Lockheed Martin AN/ALE-47 chaff/flare dispenser, Sanders AN/ALQ-144 IR jammer, Elbit ANVIS 7 NVG/HUD system and two cabin-mounted M60D 7.62 mm machine guns; provision for weapon pylons; required to operate from decks of FFG-7, DD-963, CG-47 and larger vessels, as well as unprepared sites. Cubic AN/ARS-6 personnel locator system installed from FY91. Some equipped with Indal RAST (recovery assist, secure and traverse) equipment. Armament development authorised October 1991 for installation of Hellfire ASM, 70 mm (2.75 in) rockets and forward-firing guns. Some HH-60H now fitted with nose-mounted AN/AAS-44(V) FLIR/laser designator system for use with Hellfire missile.

HH-60J Jayhawk: Ordered in parallel with HH-60H; adapted for US Coast Guard medium-range recovery (MRR) role; last of 42 delivered in 1996. First flight (USCG 6001) 8 August 1989; first delivery to USCG (6002 at Elizabeth City CGAS) 16 June 1990; subsequently to Mobile, Traverse City, San Diego, Astoria, San Francisco, Cape Cod, Sitka, Kodiak and Clearwater CGAS. When carrying three 455 litre (120 US gallon; 100 Imp gallon) external tanks, HH-60J can fly out 300 n miles (556 km; 345 miles) and return with six survivors in addition to four-man crew, or loiter for 1 hour 30 minutes when investigating possible smugglers; other duties include law enforcement, drug interdiction, logistics, aids to navigation, environmental protection and military readiness; compatible with decks of 'Hamilton' and 'Bear' class USCG cutters. Equipment includes Honeywell RDR-1300C search/weather radar, AN/ARN-147 VOR/ILS, KDF 806 direction-finder, GPS, Tacan, VHF/UHF-DF, TacNav, dual U/VHF-FM radios, HF radio, IFF, V/U/HF IFF crypto computers, NVG-compatible cockpit, rescue hoist and external cargo hook.

XSH-60J: Japan Maritime Self-Defence Force (JMSDF) placed US$27 million order for two **S-70B-3s**

for installation of Japanese avionics and mission equipment; first flights 31 August and early October 1987; 1,007 hour test programme by Japan Defence Agency Technical Research and Development Institute between 1 June 1989 and 7 April 1991 to evaluate largely Japanese avionics for SH-60J, but AN/APS-124 radar.

SH-60J: Mitsubishi (which see) is manufacturing SH-60J Seahawk for JMSDF.

MH-60R Strikehawk: Originally designated SH-60R and also known as LAMPS Block II; combines SH-60B capabilities with dipping sonar of SH-60F; original plan was for rebuild of existing fleet; first two conversions to be funded in FY98; 15 in FY99 and more thereafter; however, concerns over cost led to one year delay in launch of remanufacture programme, which began in FY00 with batch of four helicopters for test duties (ordered 25 April 2000) and was followed by five low-rate initial production (LRIP) helicopters in FY01, also converted from existing airframes, to be delivered in 2002. First conversion was due for delivery to USN in 2001, but unforeseen avionics and sensor integration problems resulted in delay and fleet introduction deferred to latter half of 2002; training unit HSL-41 at North Island, California, will be first squadron.

US Navy Helicopter Master Plan called for 170 SH-60Bs and 18 SH-60Fs to be converted by 2011; thereafter, it was intended that remaining 59 SH-60Fs and 42 HH-60Hs would be included in the remanufacture programme. Development delays and cost concerns emerging in 2000-01 prompted Navy to restructure programme with mostly new-build MH-60Rs, which will

be purchased for US$1 million to US$3 million more than remanufactured examples and also allow the Navy to implement measures to improve power to weight performance. Of total 243 MH-60Rs required, only two prototypes and first nine 'production' aircraft will be remanufactured airframes.

MH-60R systems orientated towards littoral warfare operations, with ability to process and prosecute large number of air and sea contacts in a comparatively confined space, the latter in relatively shallow water. New systems added to enhance countermeasures and passive and active detection capability. Initial upgrade package abandoned on grounds of high cost in 1998, when less costly programme, making extensive use of COTS technology, was adopted.

Lockheed Martin secured US$61 million contract in third quarter of 1998 for development of common cockpit prototype applicable to MH-60R and MH-60S variants. Under terms of two-year contract, Lockheed Martin providing flight instrument displays, two MFDs, two operator keysets and digital communications suite as well as Litton integrated INS/GPS, mass memory unit, mission and flight management computers and applicable operational software for both versions. New 'glass cockpit' centred around Lockheed Martin-developed computer systems, but using commercial PowerPC processors, with data presented to pilots via electronic flight instrument display and multifunction mission display.

Other changes to be made on MH-60R include deletion of MAD, addition of AGM-114 Hellfire anti-armour missile and two additional stores stations, databus, Telephonics AN/APS-147 multimode radar, Raytheon/Thomson-Marconi Sonar AN/AQS-22 (FLASH) advanced airborne low-frequency sonar, AN/AYK-14 mission processor and AN/UYS-2A enhanced modular signal processor, Lockheed Martin AN/ALQ-210 ESM, Raytheon AN/AAS-44 FLIR/laser ranger and NVG compatibility. MTOW expected to rise to 10,659 kg (23,500 lb). May eventually be fitted with new, more powerful engine.

Two SH-60Bs (162976 and 162977) selected to serve as prototypes; conversion undertaken by Lockheed Martin Systems Integration at Owego, New York, where first (then designated SH-60R) was rolled out on 5 August 1999. First flight scheduled for October 1999, following electronic systems functional test and checkout on the ground, but delayed until 11 December; first prototype half analogue/half 'glass cockpit' for initial testing, with full 'glass cockpit' installed after about three months. After initial trials at Owego, first prototype delivered to Patuxent River, Maryland, in early May 2000 for start of two-year Navy/contractor developmental test programme.

Initial test aircraft (166402), remanufactured by Sikorsky, first flew 19 July 2001 and formally accepted by Navy (still in manufacturer's hands) later in same month. Subsequently to Patuxent River, Maryland, on 10 August 2001 for installation of flight test instrumentation and then to Lockheed Martin at Owego for fitting of new mission

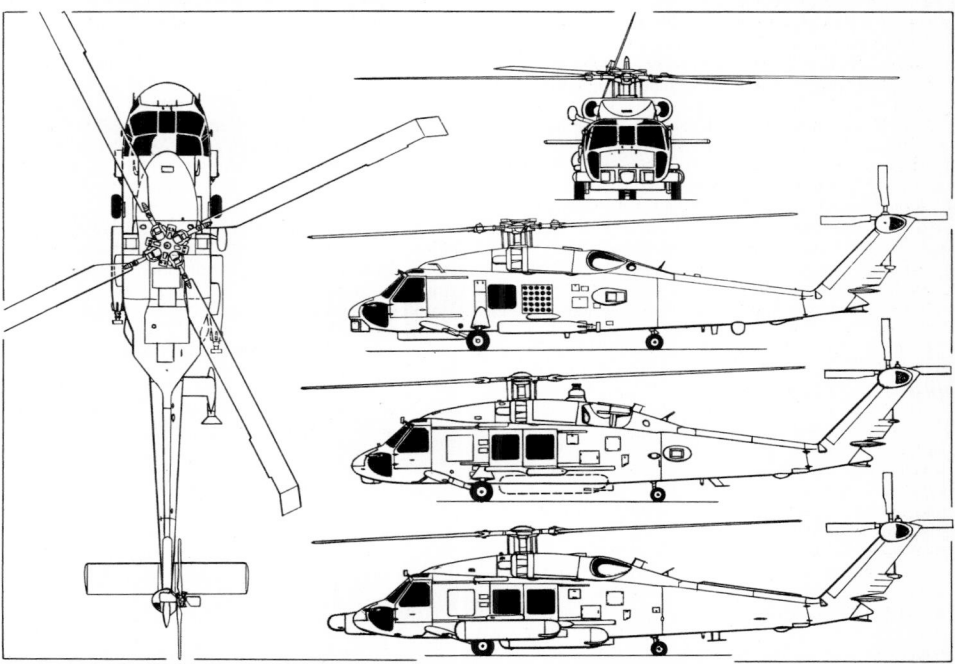

Sikorsky SH-60B Seahawk twin-turbine ASW/ASST helicopter, with additional side views of HH-60H (centre) and HH-60J Jayhawk (bottom) (*Jane's/Dennis Punnett*)

For details of the latest updates to *Jane's All the World's Aircraft* online and to discover the additional information available exclusively to online subscribers please visit
jawa.janes.com

Sikorsky HH-60H fitted with nose-mounted AN/AAS-44(V) FLIR and laser designator (*Jane's/Patrick Allen*)   *NEW*/0527144

systems. First flight with 'total weapon system' made on 4 April 2002.

All four test aircraft (166402 to' 405) had been delivered to the Navy by the beginning of February 2002, with first of five LRIP MH-60Rs (166406 to '410) flying on 9 July and being delivered by end of that month. Further two LRIP machines were delivered in third quarter of 2002 and then allocated to Lockheed Martin at Owego for fitment of mission systems. Final hurdle before start of full-rate production is US Navy OpEval (operational evaluation), currently set for 2003. A second LRIP batch (of six new-build helicopters) will be funded in 2004, with Milestone III approval anticipated in 2005.

Exports comprise: S-70B-1: Spanish Navy received six from December 1988 (designated **HS.23**) for operation from four FFG-7 frigates by Escuadrilla 010 at Rota; similar to USN SH-60B, but with Honeywell AN/AQS-13F dipping sonar. Spanish government approval to order additional six granted in December 1998, with order placed in third quarter 2000, for delivery by end of 2004. Cost of deal is US$77.4 million, which includes funds to upgrade original six to same standard, including armament kits and compatibility with AGM-114 Hellfire missile. First to be upgraded to Lockheed Martin Systems Integration at Owego, New York in April 2001; this example redelivered to Spanish Navy by end of 2001, with remaining five subjected to upgrade process by end of 2002.

S-70B-2: Royal Australian Navy (RAN) selected Seahawk for role adaptable weapon system (RAWS) full-spectrum ASW helicopter with autonomous operating capability; order for eight confirmed 9 October 1984; eight more ordered May 1986. S-70B-2 has substantially different avionics from USN version: Racal Super Searcher radar (capable of tracking 32 surface targets) and Rockwell Collins advanced integrated avionics including cockpit controls and displays, navigation receivers, communications radios, airborne target handoff datalink and tactical data system (TDS). Upgrade of Australian Seahawks, known as Project Sea 1405, to include installation of Raytheon AN/AAQ-27 FLIR and an electronic warfare support measures package based on Elisra's AES-210 system; also entails installation of Smiths NVG-compatible aircraft standby attitude indicators and Northrop Grumman AN/AAR-54(V) passive MAWS. Total of 16 Seahawks to be upgraded by third quarter of 2003; first helicopter handed over to Tenix Defence Systems in first quarter 2000. Total cost of upgrade project expected to be US$63 million. Mid-life upgrade (MLU) expected to follow in due course, with project definition study to begin in 2003-04; MLU is likely to involve provision of dipping sonar and integration of ASM, with Penguin Mk2 anti-ship missile a strong possibility as this already purchased for use by RAN Seasprite helicopters.

S-70B-6: Hybrid SH-60B/F for Greece, unofficially known as **Aegean Hawk**; selected December 1991 and initial quantity of five ordered 17 August 1992 for MEKO 200 frigates, with option for three more subsequently converted to firm order and contract for further two signed on 12 June 2000. Armament includes NFT Penguin Mk2 ASMs; avionics include AN/AQS-18(V)-3 dipping sonar, AN/APS-143(V3) radar and AN/ALR-66(V)-2 ESM; towed MAD and sonobuoy launcher omitted. First two delivered fourth quarter of 1994, with three more in 1995, one in 1997 and two in 1998.

S-70B-7: Six Seahawks ordered by Royal Thai Navy in October 1993; equipped for coastal surveillance, maritime patrol and SAR from aircraft carrier HTMS *Chakri Naruebet*; first handed over at Stratford on 6 March 1997, with all six delivered by June.

S-70B-28: Initial batch of four ordered by Turkish Navy on 14 February 1997, with option on another four subsequently converted to firm order; the first example made its maiden flight on 18 January 2001 and all eight were delivered in 2002 for service aboard Oliver Hazard Perry and MEKO 200 frigates in ASW and surveillance roles. They are first export Seahawks with a Rockwell Collins 'glass cockpit' and also have L-3 Communications Ocean Systems HELRAS long-range active dipping sonar and Telephonics AN/APS-143(V) radar installed. Original order includes supply of AGM-114 Hellfire II ASM. Turkey has ultimate requirement for up to 28 S-70Bs, of which further eight ordered in 2002.

S-70C(M)-1/2 Thunderhawk: Delivery began July 1990 to Taiwanese Navy of 10 SH-60B Seahawks; given S-70C(M)-1 designation; operated from Hualien by 701 Squadron; shipboard deployment from 1993 aboard six FFG-7 frigates. Equipment includes Honeywell AN/AQS-18(V) dipping sonar, Telephonics AN/APS-128PC radar and Litton AN/ALR-606(V)-2 ESM integrated with radar antenna; no MAD. At least two of original batch modified for sigint gathering; large antenna array positioned on top of main rotor assembly. Sigint system used to intercept HF and VHF radio communications and may also possess jamming capability. Second order, for 11 S-70C(M)-2 version, placed on 26 June 1997; first three of this batch delivered in mid-2000, with remainder handed over by end of that year.

CUSTOMERS: Total US Navy requirement originally 260 SH-60Bs; 186 on order, including five prototypes, when procurement prematurely terminated in FY94. First flight production Seahawk 11 February 1983; last SH-60B delivered to US Navy on 25 September 1996; first squadron was HSL-41 at NAS North Island, San Diego, California; operational deployment began 1984; 10 US Navy squadrons operating by March 1991 (HSLs 41, 43, 45, 47 and 49 at NAS North Island; 40, 42, 44, 46 and 48 at NAS Mayport, Florida); subsequently HSL-51 formed at Atsugi, Japan, 1 October 1991, and HSL-37 at NAS Barbers Point, Hawaii, began converting from SH-2Fs on 6 February 1992; most recent unit to equip is HSL-60 of the Reserve Force, also at Mayport. SH-60Bs deployed in 'Oliver Hazard Perry' (FFG-7) class frigates, 'Spruance'

SEAHAWK DELIVERIES

Year	SH-60B series							SH-60F series	HH-60H	HH-60J
	USN	Australia	Greece	Japan	Spain	Taiwan	Thailand	USN		
1983	9									
1984	27									
1985	24			2						
1986	24									
1987	22							2		
1988	16	(14)			6					
1989	8	4						10	8	
1990	6	6				2		19	4	6
1991	6	6				8		18	6	10
1992	6							17		12
1993	9							9		7
1994	14		2					7	2	4
1995	8		3						14	2
1996	2								8	1
1997			1				6			
1998			2							
2000						11				
Totals	**181**	**16**	**8**	**2**	**6**	**21**	**6**	**82**	**42**	**42**

Notes: Kits in parentheses; NOT to be included in totals (appear also under year of completion). No deliveries in 1999 and 2001

class and Aegis-equipped destroyers and 'Ticonderoga' class guided missile cruisers. US Navy originally required 150 SH-60Fs; total 82 completed, comprising seven pre-series plus 18 each in FY88, 89 and 91, 12 in FY92 and nine in FY93; procurement then prematurely halted; two used for operational evaluation; in West Coast service with HS-2, 4, 6, 8, 10 and 14 squadrons at NAS North Island, California; HS-3 at Jacksonville, Florida, equipped from 27 August 1991 as first East Coast squadron, followed by HS-1, 5, 7, 11 and 15, of which training squadron HS-1 since disestablished, leaving HS-10 of Pacific Fleet to conduct all US Navy SH-60F instruction. Reserve Force unit HS-75 at Jacksonville now also has SH-60F.

Exported to Australia, Greece, Japan, Spain, Taiwan, Thailand and Turkey (see Current Versions). S-70B-4 and -5 are derivatives of SH-60F and HH-60H, respectively; not taken up.

COSTS: US$20.25 million (1992) USN programme unit cost. Flyaway cost of final USN SH-60B about US$16 million; total MH-60R development programme costs expected to be around US$400 million, with unit flyaway cost US$16 million to US$18 million (FY96 dollars) for remanufactured aircraft and slightly more for new-build examples.

DESIGN FEATURES: SH-60B Seahawk designed to provide all-weather detection, classification, localisation and interdiction of surface ships and submarines, either controlled through datalink from parent ship or operated independently; secondary missions include SAR, vertical replenishment, medevac, fleet support and communications relay.

Revised features, compared with UH-60A, include more powerful navalised GE T700-GE-401 engines, additional fuel, sensor operator's station, port-side internal launchers for 25 sonobuoys, pylon on starboard side of tailboom for MAD bird, lateral pylons for two torpedoes or external tanks, chin-mounted ESM pods, sliding cabin door, rescue hoist, electrically actuated blade folding, rotor brake, folding tail, short-wheelbase tailwheel landing gear with twin tailwheels stressed for lower crash impact, DAF Indal RAST recovery assist, secure and traversing for haul-down landings on small decks and moving into hangar, hovering in-flight refuelling system, and emergency flotation system; pilots' seats not armoured. SH-60B gives 57 minutes' more listening time on station and 45 minutes' more ship surveillance and targeting time than SH-2F Seasprite LAMPS Mk I.

Initial testing of new Fairey Hydraulics Decklock landing system for S-70B was completed in mid-1995; ensuing one year development programme was expected to lead to manufacture of prototype unit for operational trials. Decklock consists of a pair of steel jaws attached to a two-stage actuator which extends during approach to landing platform; jaws then automatically secure helicopter to deck-installed grid on landing, permitting operation without assistance of deck crew during storm-force weather conditions.

For operation in Gulf during mid-1980s Iran-Iraq war, 25 SH-60Bs fitted with upper and lower Sanders AN/ALQ-144 IR jammers, BAE Systems AN/ALE-39 chaff/flare dispensers, Honeywell AN/AAR-47 electro-optical missile warning, and a single 7.62 mm machine gun in door, for a weight penalty of 169 kg (369.5 lb); seven Seahawks fitted with Raytheon AN/AAS-38 FLIR on root weapon pylon with instantaneous relay to parent ship.

First Block I SH-60B update, introduced in production Lot 9, delivered from October 1991, includes provision for NFT AGM-119 Penguin anti-ship missile, Mk 50 advanced lightweight torpedo, Flightline AN/ARR-84 99-channel sonobuoy receiver (replacing ARR-75), Rockwell Collins AN/ARC-182 V/UHF FM radio and Rockwell Collins Class 3A Navstar GPS; before production cutbacks, 115 Penguin-capable Seahawks to come from retrofitting back to Lot 5, but only 28 launch kits (delivered 1997) so far ordered.

FLYING CONTROLS: As for UH-60.

STRUCTURE: Basically as for UH-60 plus marine corrosion protection; single cabin door, starboard side, narrower than on UH-60.

LANDING GEAR: Generally as for UH-60, but with twin tailwheel positioned further forward to facilitate operation from landing platforms on warships.

POWER PLANT: Two 1,260 kW (1,690 shp) intermediate rating General Electric T700-GE-401 turboshafts in early aircraft; 1,342 kW (1,800 shp) T700-GE-401C turboshafts introduced in 1988 and on HH-60H/J. Transmission rating 2,535 kW (3,400 shp). Internal fuel capacity 2,233 litres (590 US gallons; 491 Imp gallons). Hovering in-flight refuelling capability. Two 455 litre (120 US gallon; 100 Imp gallon) auxiliary fuel tanks on fuselage pylons optional (three on HH-60J). Hover IR suppressor subsystem (HIRSS) exhaust cowling fitted to HH-60H.

ACCOMMODATION: Pilot and airborne tactical officer/back-up pilot in cockpit, sensor operator in specially equipped station in cabin. Dual controls standard. Sliding door with jettisonable window on starboard side. Accommodation heated, ventilated and air conditioned.

SYSTEMS: Generally as for UH-60A.

AVIONICS: Refer also to Current Versions for variants other than SH-60B.

Comms: Rockwell Collins AN/ARC-159(V)2 UHF, Rockwell Collins AN/ARC-174(V)2 HF, Hazeltine AN/APX-76A(V) and Honeywell AN/APX-100(V)1 IFF transponders, TSEC/KG-45(E-1) communications security set, TSEC/KY-75 voice security set, Telephonics OK-374/ASC communications system control group. Satellite communications planned for MH-60R.

Radar: Raytheon AN/APS-124 search radar on SH-60B and Telephonics AN/APS-147 on MH-60R (Racal Super Searcher for Australia; Telephonics AN/APS-128PC for Taiwan; Telephonics AN/APS-143(V) for Turkey).

Flight: Rockwell Collins AN/ARN-118(V) Tacan, Northrop Grumman AN/APN-127 Doppler, Rockwell Collins AN/ARA-50 UHF DF, Honeywell AN/APN-194(V) radar altimeter. US Navy began testing version of Raytheon Joint Precision Approach and Landing System (JPALS) on an SH-60 in 2002 and this system is expected to begin replacing existing radar-based shipboard precision approach system from 2005-06.

Mission: Sikorsky sonobuoy launcher, Flightline AN/ARR-75 and R-1651/ARA sonobuoy receiving sets (AN/ARR-84 receiver in Australian Seahawks and for USN Block 1 upgrade), Raytheon AN/ASQ-81(V)2 towed MAD (CAE AN/ASQ-504(V) internal MAD in Australian Seahawks), Raymond MU-670/ASQ magnetic tape memory unit, Astronautics IO-2177/ASQ altitude indicator, Fairchild AN/ASQ-164 control indicator set, Fairchild AN/ASQ-165 armament control indicator set, IBM AN/UYS-1(V)2 Proteus acoustic processor (Computing Devices UYS-503 for Australia) and CV-3252/A converter display, GD Information AN/AYK-14 (XN-1A) digital computer, Raytheon AN/ALQ-142 ESM, Sierra Research AN/ARQ-44 datalink and telemetry (Rockwell Collins DHS-901 in Australian Seahawks). SH-60F has Honeywell AN/AQS-13F dipping sonar (AN/AQS-18 in Taiwanese S-70s). During 1991 Gulf War, pod-mounted Hughes (now Raytheon) AN/AAQ-16 FLIR fitted to five SH-60Bs and Raytheon AN/AAQ-17 FLIR deployed on one SH-60B; BAE Systems Sea Owl IR turret evaluated later in 1991. Raytheon AN/AAS-44 FLIR/laser ranger on MH-60R. Australian examples also acquired AN/AAQ-16 FLIR for 1991 Gulf War. Raytheon AN/AAQ-27 FLIR system adopted as part of upgrade of Australian Seahawks to be undertaken by Tenix Defence Systems. US Navy to acquire airborne laser mine detection system for SH-60B and MH-60R; Northrop Grumman system selected in mid-2000, with 36 month engineering and manufacturing development (EMD) from 2001, followed by production decision before the end of 2004.

Self-defence: ESM systems include Raytheon AN/APR-39 RWR on HH-60H; none on SH-60F. MH-60R has Lockheed Martin AN/ALQ-210 ESM. Australian Seahawks fitted with AN/ALE-47 chaff/flare dispensers and AN/AAR-47 missile detectors for 1991 Gulf War. Upgrade of Australian Seahawks to include ESM based on Elisra AES-210 system.

EQUIPMENT: External cargo hook (capacity 2,722 kg; 6,000 lb) and rescue hoist (272 kg; 600 lb) standard. SH-60B/F and MH-60R have provision for eight sonobuoys.

ARMAMENT: US Navy armament includes up to three Mk 46 torpedoes and (IOC 1993) NFT AGM-119B Penguin Mk 2 Mod 7 anti-shipping missiles. Block I upgrade integrated Penguin and Honeywell Mk 50 Advanced Lightweight Torpedo from 1993. HH-60H has two pintle-mounted M240G 7.62 mm machine guns and cleared in 1996 to operate with AGM-114 Hellfire ASM. HH-60H can operate with 70 mm (2.75 in) rocket pods and GAU-17/A 7.62 mm forward-firing guns. Hellfire to be included in SH-60B and MH-60R armament for attacking small ships. In early 2001, US Navy contemplating undertaking a risk reduction flight test programme for the Israel Military Industries Light Defender standoff loiter weapon system using an SH-60B; at present, US Navy has no definite plans to use this weapon, but it could be deployed for soft targets.

DIMENSIONS, EXTERNAL: As UH-60A except:

Length overall, rotors and tail pylon folded:	
SH-60B	12.47 m (40 ft 11 in)
HH-60H	12.51 m (41 ft 0⅝ in)
HH-60J	13.13 m (43 ft 0⅞ in)
Length of fuselage: HH-60J	15.87 m (52 ft 1 in)
Width, rotors folded	3.26 m (10 ft 8½ in)
Height:	
overall, tail rotor turning	5.18 m (17 ft 0 in)
overall, pylon folded	4.04 m (13 ft 3¼ in)
to top of rotor head	3.79 m (12 ft 5⅜ in)
Wheelbase	4.83 m (15 ft 10 in)
Tail rotor ground clearance	1.83 m (6 ft 0 in)
Main/tail rotor clearance	6.6 cm (2⅝ in)

AREAS: As UH-60A

WEIGHTS AND LOADINGS:

Weight empty: SH-60B ASW	6,191 kg (13,648 lb)
HH-60H	6,114 kg (13,480 lb)
HH-60J	6,086 kg (13,417 lb)
Useful load: HH-60J	3,551 kg (7,829 lb)
Internal payload: HH-60H	1,860 kg (4,100 lb)
Mission gross weight:	
SH-60B	9,575 kg (21,110 lb)
Max T-O weight:	
SH-60B Utility, HH-60H	9,926 kg (21,884 lb)
SH-60F, MH-60R	10, 659 kg (23,500 lb)
HH-60J	9,637 kg (21,246 lb)
Max disc loading:	
SH-60B Utility, HH-60H	47.2 kg/m² (9.67 lb/sq ft)
SH-60F, MH-60R	50.7 kg/m² (10.39 lb/sq ft)
HH-60J	45.9 kg/m² (9.39 lb/sq ft)
Transmission loading at max T-O weight and power:	
SH-60B Utility, HH-60H	3.92 kg/kW (6.44 lb/shp)
SH-60F, MH-60R	4.21 kg/kW (6.91 lb/shp)
HH-60J	3.80 kg/kW (6.25 lb/shp)

PERFORMANCE:

Cruising speed at S/L:	
HH-60H	147 kt (272 km/h; 169 mph)
HH-60J	146 kt (271 km/h; 168 mph)
Dash speed at 1,525 m (5,000 ft), tropical day:	
SH-60B	126 kt (234 km/h; 145 mph)
Vertical rate of climb at S/L, 32.2°C (90°F):	
SH-60B	213 m (700 ft)/min
Vertical rate of climb at S/L, 32.2°C (90°F), OEI:	
SH-60B	137 m (450 ft)/min

UPDATED

SIKORSKY S-70C

Republic of China Air Force name: Super Blue Hawk

TYPE: Multirole medium helicopter.

PROGRAMME: H-60 version for customers not qualifying for FMS purchases.

CUSTOMERS: Delivery of 24, designated S-70C-2, with nose radar to People's Republic of China completed December 1985 but offered for sale in 1992 (apparently without success) due to spares embargo; 14 to Republic of China Air Force, Taiwan in 1985-86; one each for GKN Westland (designated **WS-70L**) and Rolls-Royce in UK; R-R aircraft used as testbed for R-R Turbomeca RTM 322 turboshafts; Westland demonstrator sold to Brainerd Helicopters of Florida in 1995 and used in agricultural role; two VIP-configured S-70C-14s for Brunei delivered 1986 but replaced by S-70A-37s in 2000 and one repurchased by Sikorsky, also certified for agricultural use; all based on Black Hawk. **S-70C(M)-1/2** designation and name Thunderhawk assigned to 10 Seahawks for Taiwan (see SH-60B entry) which ordered second Thunderhawk batch of 11 helicopters in 1997; first three of follow-on batch delivered in mid-2000.

Further four S-70C-6 Super Blue Hawks handed over to Taiwan at Stratford, Connecticut, on 7 April 1998; for use in SAR role by Rescue Squadron of 4th Tactical Fighter Wing, Republic of China Air Force.

DESIGN FEATURES: Certified to FAR Pt 21.25; roles include utility transport, external lift, maritime and environmental survey, forestry and conservation, mineral exploration and heavy construction support; powered by General Electric CT7-2C or -2D engines; options include de-icing kit for main and tail rotors, cargo hook for 3,630 kg (8,000 lb) loads, rescue hoist, aeromedical evacuation kit and winterisation kit.

FLYING CONTROLS: As for UH-60.

STRUCTURE: As for UH-60.

POWER PLANT: Two 1,212 kW (1,625 shp) General Electric CT7-2C or 1,285 kW (1,723 shp) CT7-2D turboshafts, or equivalent military T700s. Combined transmission rating (continuous) 2,334 kW (3,130 shp). Maximum fuel capacity 1,370 litres (362 US gallons; 301 Imp gallons).

Former Brunei VVIP Sikorsky S-70C after repurchase by manufacturer ***(Jane's/Paul Jackson)*** 0110649

ACCOMMODATION: Flight deck crew of two, with provision for 12 passengers in standard cabin configuration or up to 19 passengers in high-density layout. Forward-hinged door on each side of flight deck for access to cockpit area. Large rearward-sliding door on each side of main cabin.

DIMENSIONS, INTERNAL:

Cabin: Length	3.84 m (12 ft 7 in)
Max width	2.34 m (7 ft 8 in)
Max height	1.37 m (4 ft 6 in)
Floor area	8.18 m² (88 sq ft)
Volume	11.0 m³ (387 cu ft)
Baggage compartment volume	0.52 m³ (18.5 cu ft)

WEIGHTS AND LOADINGS:

Weight empty	4,607 kg (10,158 lb)
Max external load	3,630 kg (8,000 lb)
Max T-O weight	9,185 kg (20,250 lb)
Max disc loading	43.7 kg/m² (8.96 lb/sq ft)
Transmission loading at max T-O weight and power	
	3.94 kg/kW (6.47 lb/shp)

PERFORMANCE:

Never-exceed speed (V_{NE})	195 kt (361 km/h; 224 mph)
Max level speed at S/L	157 kt (290 km/h; 180 mph)
Cruising speed at S/L	145 kt (268 km/h; 167 mph)
Max rate of climb at S/L	615 m (2,020 ft)/min
Service ceiling	4,360 m (14,300 ft)
Service ceiling, OEI	1,095 m (3,600 ft)
Hovering ceiling OGE	1,204 m (3,950 ft)
Range at 135 kt (250 km/h; 155 mph) at 915 m (3,000 ft) with max standard fuel, 30 min reserves	
	255 n miles (473 km; 294 miles)
Range, max fuel, no reserves	
	297 n miles (550 km; 342 miles)

UPDATED

SIKORSKY S-76

TYPE: Multirole medium helicopter.

PROGRAMME: S-76A announced 19 January 1975; first flight (N762SA) 13 March 1977; certification to FAR Pt 29, Category B, on 21 November 1978; Category A on 9 January 1979; deliveries started early 1979; delivery of Mk II began 1 March 1982; S-76B programme initiated October 1983; first flight (N3123U) 22 June 1984; certified in category B on 31 October 1985 and in Category A on 3 February 1987; final S-76B delivered December 1997; see 1997-98 and previous *Jane's All the World's Aircraft* for details of these early models.

S-76C announced June 1989; replaced S-76B's P&WC PT6B turboshafts with Arriel engines; first flight 18 May 1990; FAA certification in Category B on 15 March 1991 and in Category A on 12 April 1991, deliveries beginning immediately thereafter.

Following manufacture of c/n 760514 in mid-2000, production of S-76C+ fuselages progressively transferred to Aero Vodochody of the Czech Republic under a US$200 million contract, beginning with Sikorsky-built c/n 760515, which was shipped to the Czech Republic for final assembly and scheduled to return to the USA in January 2001. Three further shipsets to be sent from Sikorsky, with first wholly Czech-built fuselage expected to be completed in late 2001. Fiescher Advanced Composites of Austria supplies composites components to Aero Vodochody. Keystone Helicopter Corporation of West Chester, Pennsylvania, selected in April 2000 as principal completion centre for S-76C+s, following flight testing and certification at Sikorsky's Stratford, Connecticut, facility. Production rate 10 per year in 2000 and 2001, rising to 15 per year from 2002.

Improvements, announced at Heli-Expo 2001 in February 2001 for introduction in 2004, include new low-noise tail rotor and transmission, 6 per cent increase in take-off power, a fully integrated, 'all-glass cockpit', possibly based on the Honeywell Primus Epic system; Goodrich HUMS; a Differential GPS approach and landing system and passive and active noise/vibration control systems for the cabin.

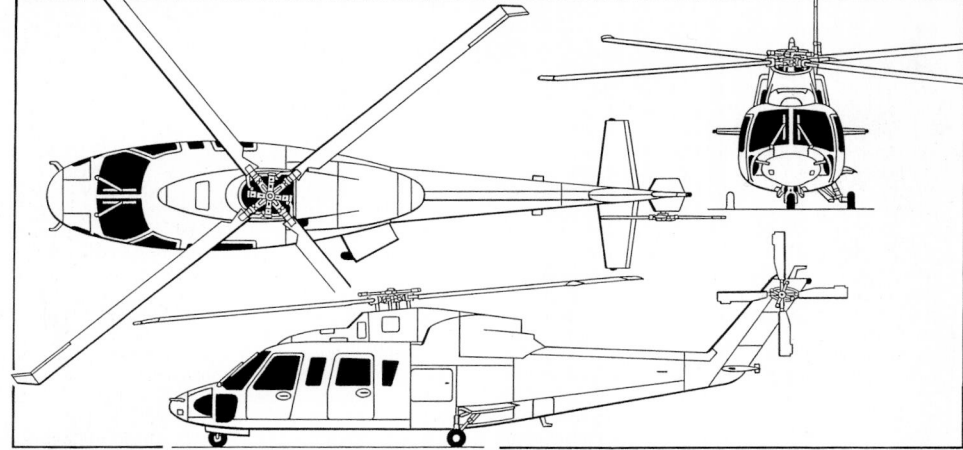

Sikorsky S-76C+ commercial transport helicopter (*Jane's/Mike Keep*)   0016738

By September 2001, more than 500 S-76s of all models were in service with 192 operators in 44 countries, and had accumulated some 2.7 million flight hours.

CURRENT VERSIONS: **S-76C:** Initial production version with Arriel 1S1 engines; now superseded by S-76C+.

S-76C+: FAA and CAA certification and first delivery, mid-1996; features Arriel 2S1 turboshafts with FADEC for improved single-engine performance and fuel efficiency. *As described.*

H-76 Eagle and S-76N Eagle: Military and naval developments of S-76B; last described in 1997-98 and 1998-99 editions, respectively.

Shadow: One-off modification (N765SA) of fourth (flying) prototype S-76; Sikorsky Helicopter Advanced Demonstrator and Operator Workload fitted with nose-mounted single-seat cockpit for US Army Rotorcraft Technology Integration (ARTI) programme; first flight 24 June 1985; used 1990 in Boeing Sikorsky LH First Team programme to flight test night vision pilotage system (NVPS) developed by Martin Marietta for RAH-66 Comanche; details in 1997-98 and earlier *Jane's All the world's Aircraft.*

S-76D: Proposed utility variant to be developed jointly with Mil Helicopters of Russia. New features would include fixed landing gear, Mil-developed composites main rotor blades with electrothermal de-icing, X-configuration quiet tail rotor, aimed at reducing manufacturing costs and increasing payload and performance. New rotors would be offered as retrofit to existing S-76 models. If proceeded with, the S-76D would be produced in Russia for domestic and international markets.

CUSTOMERS: See table. The 513th S-76 was registered in June 2002. Recent customers include Palm Beach County Health Care Service, which has taken delivery of two in 'Trauma Hawk' EMS configuration; Copterline of Finland, which took delivery of two S-76C+s in March 2000 for scheduled services between Helsinki and Tallinn, Estonia; Toho Air Service of Japan, which ordered one S-76C+ for delivery in second quarter 2001 for operations around six island destinations from its base on Hachijo Island; and Chinese Ministry of Communications, which ordered two SAR helicopters in December 2000 to be based at Shanghai. One S-76C+ is used by the British Royal Family.

DESIGN FEATURES: Meets FAR Pt 29 with Category A IFR; intended for offshore support, business transport, medical evacuation and general utility use; technology and aerodynamics based on those of UH-60 Black Hawk.

Four-blade main rotor with high twist and varying section and camber based on Sikorsky SC-1095; tapered blade tip has 30° leading-edge sweep; fully articulated

rotor head with single elastomeric bearings; hydraulic drag dampers; dual bifilar vibration absorber assemblies above rotor head; four-blade cross-beam tail rotor on port side; rotor brake optional. Max rotor speed, power on, 316 rpm (108 per cent). Main rotor blade life of 28,000 hours; tail rotor life unlimited. Optional manual blade folding.

Emergency medical service installation includes multiple-position pivoting primary patient litter, a second litter, and track-mounted seats for four attendants, forward and rear oxygen systems, and dual access to external power on ground; cabin volume 4.0 m³ (141 cu ft).

FLYING CONTROLS: Dual powered hydraulic controls with autostabilisation and autopilot; releasable spring-centring trim system for both cyclic and collective controls.

STRUCTURE: Main rotor blades have formed and welded titanium oval-section tubular main spar with Nomex honeycomb aerofoil core, glass fibre composites outer skin and titanium/nickel leading-edge abrasion strips. Fuselage has bonded aluminium/honeycomb skin panels to permit flush riveting; extensive use made of carbon fibre (replacing Kevlar and glass fibre in earlier S-76s). Aerospace Industrial Development Corporation (AIDC) of Taiwan signed five-year contract in December 1998 for supply of composites crew and passenger doors.

LANDING GEAR: Hydraulically retractable tricycle type, with single wheel on each unit. Nosewheel retracts rearward, main units inward into rear fuselage; all three wheels are enclosed by doors when retracted. Mainwheel tyres size 14.5×5.5-6 (14 ply) tubeless, pressure 11.38 bar (165 lb/sq in); nosewheel tyre size 5.00-4 (12 ply) tubeless, pressure 9.31 bar (135 lb/sq in). Hydraulic brakes; hydraulic mainwheel parking brake. Non-retractable tricycle gear, with low-pressure tyres and skid landing gear both optional.

POWER PLANT: Two Turbomeca Arriel 2S1 turboshafts with FADEC, each rated at 638 kW (856 shp) for take-off, 731 kW (980 shp) OEI for 30 seconds and 592 kW (794 shp) maximum continuous; transmission rating 1,193 kW (1,600 shp) for take-off and maximum continuous. Standard fuel capacity 1,083 litres (286 US gallons; 238 Imp gallons) of which 1,064 litres (281 US gallons; 234 Imp gallons) are usable; optional auxiliary tank capacity 405 litres (107 US gallons; 89.1 Imp gallons). Self-sealing tanks optional.

ACCOMMODATION: One or two pilots and 12 to 13 passengers. Three four-abreast rows of seats, floor-mounted at a pitch of 79 cm (31 in). A number of executive layouts are available, including a four-passenger 'office in the sky' configuration. Executive versions have luxurious interior trim, full carpeting, special soundproofing, radiotelephone and co-ordinated furniture. Dual controls optional. Two large doors on each side of fuselage, hinged at forward edges; sliding doors available optionally. Baggage hold aft of cabin, with external door each side of fuselage. Cabin heated and ventilated; air conditioning optional. Windscreen demisting and dual windscreen wipers. Windscreen heating and external cargo hook optional.

SYSTEMS: Hydraulic system at pressure of 207 bar (3,000 lb/sq in) supplied by two pumps driven from main gearbox. Hydraulic system maximum flow rate 15.9 litres (4.2 US gallons; 3.5 Imp gallons)/min. Bootstrap reservoir. Pump head pressure 3.45 bar (50 lb/sq in). In VFR configuration, electrical system comprises two 200 A DC starter/generators and a 24 V 17 Ah Ni/Cd battery. In IFR configuration, system comprises gearbox-driven 7.5 kVA generator, and a 115 V 600 VA 400 Hz static inverter for AC power. 34 Ah battery optional. Engine fire detection and extinguishing system.

AVIONICS: *Comms:* Honeywell Primus II integrated radios.

Radar: Honeywell Primus 440 colour weather radar.

Flight: Honeywell SPZ-7600 dual digital AFCS. Optional Universal UNS-1D navigation management system, L-3 Com/Sextant TCAS and Argus moving map display.

Instrumentation: Four-tube Honeywell colour EFIS; Parker Hannifin Gull Electronics active colour matrix LCD three-screen integrated instrument display system (IIDS).

Sikorsky S-76C+ twin-turbine helicopter   NEW/0527143

VIP interior of Sikorsky S-76   NEW/0527142

Sikorsky S-76 instrument panel
NEW/0527145

EQUIPMENT: Standard equipment includes cabin fire extinguishers; cockpit, cabin, instrument, navigation and anti-collision lights; landing light; external power socket; and utility soundproofing. Optional equipment includes 1,497 kg (3,000 lb) capacity cargo hook, 272 kg (600 lb) capacity rescue hoist, emergency flotation gear, engine air particle separators and litter installation.

DIMENSIONS, EXTERNAL:
Main rotor diameter	13.41 m (44 ft 0 in)
Main rotor blade chord	0.39 m (1 ft 3½ in)
Tail rotor diameter	2.44 m (8 ft 0 in)
Tail rotor blade chord	0.16 m (6½ in)
Length: overall, rotors turning	16.00 m (52 ft 6 in)
fuselage	13.21 m (43 ft 4 in)
Height overall, tail rotor turning	4.42 m (14 ft 6 in)
Tailplane span	3.05 m (10 ft 0 in)
Width of fuselage	2.13 m (7 ft 0 in)
Wheel track	2.44 m (8 ft 0 in)
Wheelbase	5.00 m (16 ft 5 in)
Tail rotor ground clearance	1.97 m (6 ft 5¾ in)

DIMENSIONS, INTERNAL:
Passenger cabin: Length	2.46 m (8 ft 1 in)
Max width	1.93 m (6 ft 4 in)
Max height	1.35 m (4 ft 5 in)
Floor area	4.18 m² (45.0 sq ft)
Volume	5.8 m³ (204 cu ft)
Baggage compartment volume	1.1 m³ (38 cu ft)

AREAS:
Main rotor disc	141.26 m² (1,520.5 sq ft)
Tail rotor disc	4.67 m² (50.27 sq ft)
Tailplane	2.00 m² (21.5 sq ft)

WEIGHTS AND LOADINGS:
Operating weight empty	2,545 kg (5,611 lb)
Weight empty, executive configuration	
	3,691 kg (8,138 lb)
Max baggage	272 kg (600 lb)
Useful load	1,616 kg (3,562 lb)
Max fuel	3,030 kg (6,680 lb)
Max T-O weight	5,307 kg (11,700 lb)
Max disc loading	37.6 kg/m² (7.69 lb/sq ft)
Transmission loading at max T-O weight and power	
	4.45 kg/kW (7.31 lb/shp)

PERFORMANCE:
Never-exceed speed (VNE) and max level speed	
	155 kt (287 km/h; 178 mph)
Long-range cruising speed	140 kt (259 km/h; 161 mph)
Max rate of climb at S/L	495 m (1,625 ft)/min
Service ceiling	3,871 m (12,700 ft)
Service ceiling, OEI	975 m (3,200 ft)
Hovering ceiling: IGE	1,722 m (5,650 ft)
OGE	549 m (1,800 ft)
Range at 140 kt (259 km/h; 161 mph) at 1,220 m	
(4,000 ft) with standard fuel:	
no reserves	439 n miles (813 km; 505 miles)
30 min reserves	385 n miles (713 km; 443 miles)

UPDATED

SIKORSKY S/H-92

TYPE: Medium transport helicopter.

PROGRAMME: Announced March 1992; originally envisaged as S-92C 'Growth Hawk' development of S-70; market evaluation co-ordinated with Mitsubishi Corporation and Mitsubishi Heavy Industries. Launched at Paris Air Show, June 1995 as S-92A Helibus and military S-92IU (international utility). Risk-sharing partners Mitsubishi Heavy Industries (7.5 per cent), Jingdezhen Helicopter Group of China (2 per cent) and Gamesa of Spain (7 per cent), with Taiwan Aerospace (6.5 per cent) and Embraer (4 per cent) as additional fixed-price supplier/partners; Russia's Mil is associated with programme, but not yet a full partner; other suppliers include Aerazur (fuel cells), Goodrich (health and usage monitoring system), Dunlop

S-76 DELIVERIES

Year	S-76A	S-76A+	S-76B	S-76C	S-76C+	Total	Cum
1979	35					35	35
1980	86					86	121
1981	50					50	171
1982	29					29	200
1983	36					36	236
1984	30					30	266
1985	8		11			19	285
1986	3		13			16	301
1987	3		7			10	311
1988	2		9			11	322
1989		3	14			17	339
1990	1	13	5			19	358
1991		1	8	10		19	377
1992			4	8		12	389
1993	1		3	7		11	400
1994			6	7		13	413
1995			6	10		16	429
1996			11	1	7	19	448
1997			4		14	18	466
1998					16	16	482
1999					7	7	489
2000					7	7	496
2001					8	8	504
Totals	**284**	**17**	**101**	**43**	**59**	**504**	

Notes: Some S-76As retrofitted to A+. Excludes seven (six flying) prototypes and pre-series

(engine inlet), Eaton (emergency flotation bag electro-optical sensors); Hamilton Sundstrand (automatic flight control system and active vibration computers), Honeywell (radar and APU), Lucas-Western (rescue hoist), Martin-Baker (crew seats), Messier-Bugatti (wheels and brakes), Moog (active vibration controls), Parker Bertea (servos), Rockwell Collins (multifunction displays and nav/com suite); Universal (FMS) and Vickers (hydraulic pumps). Programme rationalisation resulted in discontinuation of S-92IU and adoption of standard civil/military configuration with new designation S/H-92 applied in 2002.

Design changes announced in July 2000 in response to customer requests include a 0.41 m (1 ft 4 in) increase in cabin length to permit installation of a 1.27 m (4 ft 2 in) wide cabin door to improve hoisting capability and to accommodate a Stokes litter during SAR operations; reduction in height of tail rotor pylon by about 1.02 m (3 ft 4 in) to offset additional weight of cabin extension; and relocation of the horizontal stabiliser. These changes will provide additional benefits in creating an improved fold configuration for shipboard operations, increased birdstrike protection deriving from the relocation of tail rotor drive shaft and controls aft of the tail spar, and a flatter hover attitude arising from a forward shift of the helicopter's centre of gravity, improving visibility for confined space and shipboard landing, and increasing aft fuselage ground clearance. The revised configuration has been incorporated from the third prototype and on all production aircraft.

Test programme comprises one ground test vehicle (PA1/GTV; first airframe built, first ground run 4 September 1998 and completed 200-hour FAA certification endurance run in September 1999) and four flying prototypes, of which first (PA2/N292SA) completed maiden flight at Sikorsky's Development Flight Center in Florida on 23 December 1998 and was subsequently modified to the revised configuration described above. GTV and first flying prototype have 1,305 kW (1,750 shp) CT7-6D engines; remaining three flying prototypes have

production-standard CT7-8s and APUs. PA3/N392SA (second flying) joined the test programme in May 1999, flew in October, and is devoted principally to engine and AFCS development; subsequently upgraded to production configuration with extended main rotor shaft; new main rotor servos; redesigned electrical system to maintain rotor ice protection after a generator failure; new tail rotor pylon; and state-of-the-art cockpit displays. Flown again in this configuration on 22 March 2002. PA4/N492SA delayed during assembly to incorporate airframe changes described above and final avionics updates and first flew in definitive production configuration on 5 October 2001. PA5/N592SA, exhibited unflown at the Paris Air Show in June 1999, is in utility configuration with MIL-STD-1553 databus, rear loading ramp, folding seats for 22 troops, sliding windows, cargo hook and provision for 7.62 mm machine gun pintle mounts; was later modified to revised configuration as described above, and made its first flight on 8 February 2001.

By February 2002, PA2 had flown more than 380 hours, and PA3 221 hours, including flights at maximum take-off weight of 13,998 kg (30,860 lb), demonstrating a range of 800 n miles (1,482 km; 921 miles) with mid-mission hover; PA4 had flown 60 hours; and PA5 more than 260 hours. PA2 was withdrawn from use in May 2002. FAA FAR Pt 29 certification achieved in December 2002 after 1,570-hour flight test programme; JAA certification expected in late 2003; first customer deliveries in first quarter of 2004.

CUSTOMERS: Nine deposits (and 18 further prospects or commitments) held by February 2003. Launch customer Cougar Helicopter of Eastern Canada announced intent to order up to five for delivery in 2002; following programme delays, reaffirmed requirement for three at February 2003 Heli-Expo, Dallas, Texas. Other announced customers include Helijet of Vancouver, Aircontactgruppen AS of Norway (six for delivery between 2004 and 2007), Copterline of Finland, Norsk Helikopter (two for Stratoil contract, agreed February 2003), Era Aviation Inc, which ordered three in August 2001 for offshore operations in the

Fourth (nearest), third and second S/H-92 prototypes   *NEW*/0527170

in civil version. Main door, starboard side, front in three options: (a) split horizontally in upward- and downward-hinged sections, latter including steps; (b) lower airstair section with inward/backward-sliding upper portion (allowing installation of external SAR winch) and (c) full-height sliding door incorporating Type IV emergency exit. Further three FAA/JAA Type IV cabin emergency exits, plus pop-out windows at each seat row, for rapid emergency evacuation. Total of 22 combat-ready troops in military version; both versions have rear-loading ramp. Military version has sliding cabin windows and weapons mounts. Martin-Baker crew and passenger seats. Accommodation is heated and ventilated with independent cockpit and cabin zones; air conditioning optional. Active noise suppression system under consideration.

SYSTEMS: Goodrich IMD-HUMS health and usage monitoring system standard, with cockpit displays and downloading facility to enable ground crew to access system via hand-held diagnostic equipment; active noise control system will reduce cabin noise by 3 to 4 dB; active vibration control may be employed in key airframe areas. BAE Systems CVR/FDR.

Honeywell 36-150 (S92) APU with in-flight start and continuous run capability. Electrical system similar to that of H-60 series, comprising two 45 kVA main gearbox-driven AC generators and one 35 kVA APU-driven back-up generator with improved generator control unit.

Three hydraulic systems, supplied by main gearbox-driven pumps, two serving main and tail rotor and stability augmentation system, while third serves utilities and acts as back-up. Anti-icing system for engine inlets, windscreen and pitot head; main and tail rotor de-icing optional. Automatically deployed emergency flotation system meets or exceeds FAR/JAR stability requirements and has demonstrated Sea State 5 capability; externally mounted 14-person liferaft in forward end of each sponson, with 50 per cent overload capability.

AVIONICS: Open architecture avionics system accommodating ARINC 429 and MIL-STD-1553 interfaces with Collins Pro Line 21 as core system.

Comms: Honeywell Primus II VHF.

Radar: Provision for radar in nose compartment.

Flight: UNS-1C FMS; Hamilton Sundstrand automatic flight control system featuring three-axis stability augmentation system and fully coupled dual digital autopilots with automatic approach-to-hover option.

Instrumentation: Rockwell Collins EFIS cockpit with four 203 × 152 mm (8 × 6 in) MFD-268EP active matrix liquid crystal displays (AMLCD) for PFD, EICAS, health monitoring, navigational and weather radar functions; fifth AMLCD, centre-mounted, optional for displaying sensor information such as moving map or FLIR data.

All avionics housed in removable mission equipment rack behind co-pilot station with wiring routed through conduits in fuselage frames for added protection.

EQUIPMENT: Optional hydraulically powered, electrically controlled SAR rescue hoist, maximum capacity 272 kg (600 lb), with cable viewing window and spotlight. Optional cargo handling system, which is compatible with 1.07 × 1.22 m (3 ft 6 in × 4 ft 0 in) pallets, includes 1,814 kg (4,000 lb) capacity centreline winch with 28 V DC winch motor, easy on/off cabin rollers, and cargo tiedowns at all major frames.

DIMENSIONS, EXTERNAL:

Main rotor diameter	17.71 m (56 ft 4 in)
Tail rotor diameter	3.35 m (11 ft 0 in)
Length overall, rotors turning	20.88 m (68 ft 6 in)
Fuselage: Length: airframe	17.12 m (56 ft 2 in)
incl tail rotor	18.47 m (60 ft 7 in)
Max width: over sponsons	3.89 m (12 ft 9 in)
incl horizontal stabiliser	5.26 m (17 ft 3 in)
Height: overall, rotors turning	5.46 m (17 ft 11 in)
to top of main rotor head	4.70 m (15 ft 5 in)
Wheel track (c/l shock-absorbers)	3.17 m (10 ft 5 in)
Wheelbase	6.20 m (20 ft 4 in)
Forward door (stbd): Height	1.85 m (5 ft 11 in)
Width	1.02 m (3 ft 4 in)

DIMENSIONS, INTERNAL:

Cabin: Length	6.10 m (20 ft 0 in)
Max width	2.01 m (6 ft 7 in)
Max height	1.83 m (6 ft 0 in)
Volume	16.9 m³ (596 cu ft)
Baggage volume (civil)	3.97 m³ (140 cu ft)

AREAS:

Main rotor disc	231.55 m² (2,492.4 sq ft)
Tail rotor disc	8.83 m² (95.03 sq ft)

WEIGHTS AND LOADINGS:

Weight empty: military utility	6,895 kg (15,200 lb)
offshore oil	7,121 kg (15,700 lb)
SAR	7,257 kg (16,000 lb)
Max internal fuel	2,143 kg (4,725 lb)
Max hook capacity	4,536 kg (10,000 lb)
Max T-O weight:	
internal load	11,861 kg (26,150 lb)
external load	12,836 kg (28,300 lb)
Max disc loading:	
internal load	51.2 kg/m² (10.49 lb/sq ft)
external load	55.4 kg/m² (11.35 lb/sq ft)
Transmission loading at max T-O weight and power:	
internal load	3.81 kg/kW (6.26 lb/shp)
external load	4.12 kg/kW (6.77 lb/shp)

PERFORMANCE (at max internal load T-O weight):

Never-exceed speed (VNE)	165 kt (305 km/h; 189 mph)
Max cruising speed	151 kt (280 km/h; 174 mph)
Econ cruising speed	137 kt (254 km/h; 158 mph)
Service ceiling	4,575 m (15,000 ft)
Hovering ceiling, IGE	3,261 m (10,700 ft)
OGE	1,942 m (6,370 ft)
Range, normal fuel, 19 passengers, no reserve	
	538 n miles (996 km; 619 miles)
Range with internal auxiliary fuel (1,401 litres, 370 US gallons; 308 Imp gallons)	
	670 n miles (1,240 km; 771 miles)

UPDATED

VIP interior of S/H-92   *NEW*/0527135

OTHER AIRCRAFT
Description of **RAH-66 Comanche** attack helicopter appears under Boeing Sikorsky in this section.

SINO SWEARINGEN

SINO SWEARINGEN AIRCRAFT COMPANY

1770 Skyplace Boulevard, San Antonio, Texas 78216
Tel: (+1 210) 258 86 99
Fax: (+1 210) 258 39 73
e-mail: ssacsj30@aol.com
Web: http://www.sj30jet.com
CHAIRMAN: Chuen-Huei Tsai
PRESIDENT AND CEO: Dr Carl Chen
VICE-CHAIRMAN: Ed Swearingen
SENIOR VICE-PRESIDENT, SALES AND MARKETING: Gene Comfort
SENIOR VICE-PRESIDENT, OPERATIONS: Ronald D Neal
VICE-PRESIDENTS:
    John W Dieker (Manufacturing)
    Robert E Homan (Quality Assurance)
    Chester J Schickling (Sales and Marketing)
    T H Tsiang (Operations)
    Roger Wilson (Engineering)
    Jim L Wooley (Human Resources)
    Carol Zuniga (Finance)
DIRECTORS:
    Chris Danner (Product Support)
    William R Moreu (Flight Operations)
    Mike Potts (Corporate Communications)
    Donnie Rex (Management Information Systems)

Swearingen SJ30-2 prototype   *NEW*/0526911

Ed Swearingen is well known for designing Fairchild Merlin and Metro commuter and business aircraft, and for engineering such aircraft as Piper Twin Comanche and Lockheed JetStar II. Current project is SJ30-2 business jet.

Swearingen Aircraft Inc re-formed on 1 April 1995 as joint venture company with Sino Aerospace of Taiwan, which provides financial support; will certify and produce SJ30-2 at new 8,130 m² (87,500 sq ft) facility at Martinsburg, West Virginia, where the workforce totalled 85 in September 2002, and is expected to rise to about 350 by 2006. San Antonio workforce stood at 400 in late 2002.

UPDATED

First production-conforming Sino Swearingen SJ30-2 at NBAA Convention, Orlando, September 2002 *(Jane's/Paul Jackson)*
NEW/0526992

SINO SWEARINGEN SJ30-2

TYPE: Light business jet.
PROGRAMME: Announced 30 October 1986 as SA-30 Fanjet; Gulfstream Aerospace, Williams International and Rolls-Royce announced they were joining programme in October 1988 and aircraft renamed Gulfstream SA-30 Gulfjet; Gulfstream withdrew from programme 1 September 1989; place taken by Jaffe Group of San Antonio, Texas, and aircraft renamed Swearingen/Jaffe SJ30; now a Sino Swearingen project. First flight of prototype (N30SJ) 13 February 1991 (with FJ44-1 engines).

Certification originally intended in 1995 but delayed pending development of increased performance SJ30-2, *as described*, which features fuselage stretched by 1.32 m (4 ft 4 in), wing span increased by 1.83 m (6 ft 0 in), increased wing dihedral, revised wing/fuselage fairing, increased fuel capacity and more powerful FJ44-2A engines. Prototype (N30SJ, modified from SJ30 prototype) first flown (original engines) 8 November 1996.

FJ44-2A engines installed and flown for first time on 4 September 1997; total of 270 flight hours accumulated with these engines (or 370 hours overall in SJ30-2 configuration) by late April 1999, including operation at speeds up to M0.83 and maximum altitude of 13,105 m (43,000 ft), 1.8 *g* turns at 12,495 m (41,000 ft), verification of stall and engine-out minimum control speeds, stability and control tests, engine operating tests throughout the flight envelope, simulated icing tests, yaw/damper/rudder bias system testing and handling trials with simulated flight control failures including trim runaways and asymmetric

speed brake deployment. N30SJ withdrawn from use in mid-1999.

Static and fatigue tests on a representative section of primary wing joint completed February 1999, to 150 per cent of limit load in 230,000 loading cycles. Risk-sharing partner Gamesa Aeronautica of Vitoria, Spain, manufactured fuselages and wings of five certification test airframes; first of three production conforming flying prototypes (c/n 002/N138BF) rolled out at San Antonio 17 July 2000 and first flown 30 November 2000; public debut at the NBAA Convention in Orlando, Florida, 9 September 2002, at which time it had recorded some 250 flight hours of a planned 700; 003/N30SJ (No. 2) and 004/N709JB will also be flying prototypes, scheduled to fly in 2003, and will each complete 400 flight hours during the test and certification programme; c/n TF-2 and TF-3 are static test and fatigue airframes respectively, last-named being the final airframe to be assembled in Texas; TF-2 transferred to Sino Swearingen static test facility at San Antonio 8 March 2001, with static tests scheduled for completion by mid-2002; assembly of first production airframe (c/n 005) began in March 2002. The 1,500 hour flight test programme is scheduled to culminate in FAA certification in FAR Pt 23 Commuter category and first customer deliveries in early 2004; total of 30 scheduled for production in 2004, with production rate rising to one per week by 2005 and two per week in 2006.
CUSTOMERS: More than 150 orders placed by customers and distributors in Brazil, Canada, France, Germany, Mexico, South Africa, Sweden, United Kingdom and USA by September 2002, representing planned production to mid-2005.

COSTS: US$4.869 million (2000). Estimated direct operating cost, based on 1,000 n mile (1,852 km; 1,150 mile) stage length, US$601.15 per hour (2002).
DESIGN FEATURES: High-speed and high-altitude cruise with long-range capabilities, plus good slow speed handling; highly efficient swept wing planform.

Proprietary, computer-designed aerofoil sections. Wing sweep 32° 36', dihedral 2° 30'.
FLYING CONTROLS: Conventional and manual. Electrically actuated variable incidence tailplane, trim tab on rudder, aileron force bias spring for trimming in roll axis; single ventral fin under tail for yaw damping; slotted Fowler trailing-edge flaps actuated electrically; hydraulically actuated full-span leading-edge slats; single electrohydraulically actuated spoiler/lift dumper panel on each wing ahead of flap.
STRUCTURE: All-metal with chemically milled skins on fuselage. Aerostructures Corporation of Nashville, Tennessee, will build primary structure, for final assembly by Sino Swearingen.
LANDING GEAR: Retractable tricycle type, manufactured by Integrated Aerospace of Santa Ana, California, with twin wheels on each unit. Trailing-link oleo-pneumatic suspension on main units. Hydraulic actuation, main units retracting inward and rearward into fuselage, nose unit forward. Hydraulically steerable (±60°) nose unit. All wheels 41 cm (16 in) diameter. Main unit tyre pressure 10.76 bar (155 lb/sq in), nosewheel tyre pressure 2.14 bar (31 lb/sq in). Standard air-cooled power braking; dual ABS anti-skid.
POWER PLANT: Two 10.23 kN (2,300 lb st) Williams FJ44-2A turbofans flat rated to 72°C, pod-mounted on pylons on sides of rear fuselage. Six fuel tanks: one in each outer wing (total 1,055 litres; 278.8 US gallons; 232.1 Imp gallons); two wing hopper tanks (total 103 litres; 27.2 US gallons; 22.6 Imp gallons); centre wing tank (199 litres; 52.7 US gallons; 43.9 Imp gallons); and fuselage (torsion box) tank (1,440 litres; 380.3 US gallons; 316.7 Imp gallons) for total of 2,797 litres (739 US gallons; 615.3 Imp gallons) under FAR Pt 23 certification, or 2,487 litres (657 US gallons; 547 Imp gallons) under JAR 23 certification. Fuel is burned sequentially to maintain aircraft centre of gravity. Single refuelling point in fuselage tank. Oil capacity 3.4 litres (0.9 US gallon; 0.75 Imp gallon) per engine.
ACCOMMODATION: Pilot and one passenger (or co-pilot) on flight deck. Main cabin separated by a bulkhead; six passengers seats comprising club seats and two forward-facing seats; aft-facing seats recline and have lateral tracking capability; forward-facing seats have full berthing capability; lavatory, separated from main cabin by a half-cabin height partition with fan close-out, has belted, side-facing seat that is approved for occupation during take-off and landing. Full-height refreshment centre and foldaway tables, dual cupholders and cabin environment controls standard; recessed storage area aft of lavatory, with foldaway coat hooks, optional. Airstair passenger door with light at front on port side. Emergency exit over starboard wing. Baggage compartment aft of main cabin, with external access via port side door under engine

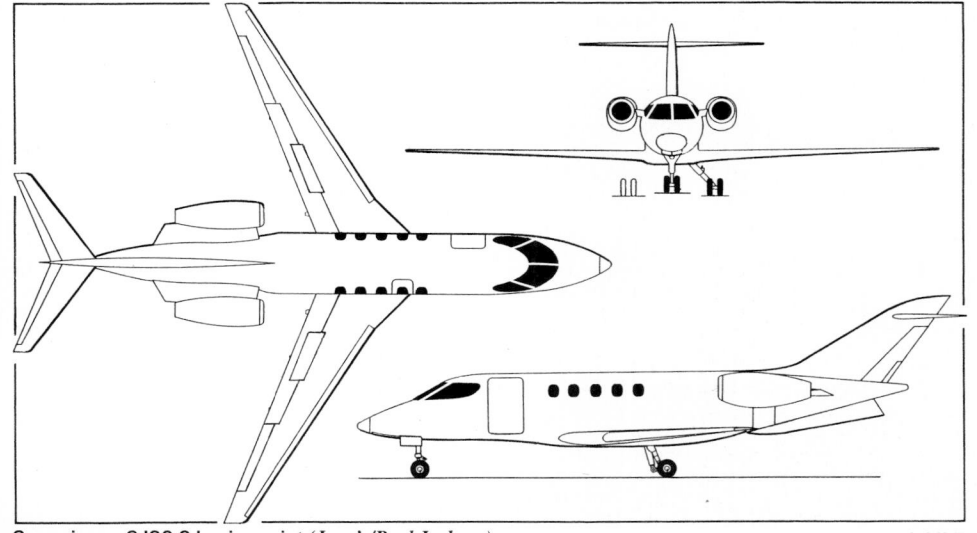

Swearingen SJ30-2 business jet *(Jane's/Paul Jackson)*   0126967

nacelle. Two-piece birdproof electrically heated wraparound windscreen.

SYSTEMS: Cabin pressurised to 0.83 bar (12.0 lb/sq in), maintaining sea level pressure to 12,500 m (41,000 ft); cabin heated by engine bleed air; cooled by a vapour-cycle system. Hydraulic system (207 bar; 3,000 lb/sq in) for actuation of leading-edge slats, airbrake/lift dumpers and landing gear extension/retraction. Two 300 A 28 V DC engine-driven starter/generators and static inverters. Redundant frequency-wild alternators provide power for windscreen heating. Wing and engine inlets anti-iced by engine bleed air; tailplane has pneumatic de-icer boots, de-icing system. Oxygen system has 646 litre (22.8 cu ft) capacity.

AVIONICS: Honeywell Primus Epic CDS as core system.

Comms: Honeywell Primus II radio system; optional 8.33 kHz spacing.

Radar: Honeywell Primus 331 colour weather radar.

Flight: Dual IC-615 integrated avionics computers with combined flight director, autopilot and FMS/GPS functions; dual RVSM-compatible Honeywell AZ-850 micro air data computers; ADF, DME. Options include TCAS 2000, second FMS/GPS integrated with IC-615, second ADF, second DME and lightning sensor system.

Instrumentation: EFIS cockpit comprising three Honeywell DU-180 flat panel 203 × 254 mm (8 × 10 in) active matrix colour LCDs, including two PFDs and one MFD.

DIMENSIONS, EXTERNAL:

Wing span	12.90 m (42 ft 4 in)
Wing chord: at root	2.41 m (7 ft 11 in)
at tip	0.50 m (1 ft 7½ in)
Wing aspect ratio	9.4
Length overall	14.26 m (46 ft 9½ in)
Fuselage: Length	12.79 m (41 ft 11½ in)
Max diameter	1.52 m (5 ft 0 in)
Height overall	4.34 m (14 ft 3 in)
Tailplane span	4.44 m (14 ft 7 in)
Wheel track (mean)	2.09 m (6 ft 10¼ in)
Wheelbase	5.70 m (18 ft 8½ in)
Passenger door: Height	1.42 m (4 ft 8 in)
Width	0.81 m (2 ft 8 in)
Baggage door: Height	0.65 m (2 ft 1½ in)
Width	0.47 m (1 ft 6½ in)
Emergency door: Height	0.56 m (1 ft 10 in)
Width	0.48 m (1 ft 7 in)

DIMENSIONS, INTERNAL:

Cabin: Length:	
between pressure bulkheads	5.36 m (17 ft 7 in)
passenger section	3.81 m (12 ft 6 in)
Max width	1.43 m (4 ft 8½ in)
Max height	1.31 m (4 ft 3½ in)
Floor area	5.46 m² (58.8 sq ft)
Volume	5.4 m³ (190.6 cu ft)
Baggage compartment volume	1.50 m³ (53 cu ft)

AREAS:

Wings, gross	17.72 m² (190.7 sq ft)
Ailerons (total)	0.76 m² (8.20 sq ft)
Trailing-edge flaps (total)	3.31 m² (35.63 sq ft)
Leading-edge flaps (total)	2.83 m² (30.48 sq ft)
Fin	1.06 m² (11.40 sq ft)
Rudder, incl tab	0.69 m² (7.40 sq ft)
Tailplane	3.41 m² (36.70 sq ft)
Elevator, incl tabs	0.79 m² (8.53 sq ft)

WEIGHTS AND LOADINGS: (A: FAR Pt 23 commuter certification; B: JAR23 certification):

Weight empty, equipped: A, B	3,493 kg (7,700 lb)
Operating weight empty: A, B	3,629 kg (8,000 lb)
Max baggage weight: A, B	227 kg (500 lb)
Max payload: A, B	635 kg (1,400 lb)
Max fuel weight: A	2,245 kg (4,950 lb)
B	1,996 kg (4,400 lb)
Max T-O weight: A	6,123 kg (13,500 lb)
B	5,670 kg (12,500 lb)
Max ramp weight: A	6,169 kg (13,600 lb)
B	5,715 kg (12,600 lb)
Max landing weight: A	5,688 kg (12,540 lb)
B	5,670 kg (12,500 lb)
Max zero-fuel weight: A, B	4,536 kg (10,000 lb)
Max wing loading: A	345.6 kg/m² (70.79 lb/sq ft)
B	320.0 kg/m² (65.55 lb/sq ft)
Max power loading: A	299 kg/kN (2.93 lb/lb st)
B	277 kg/kN (2.72 lb/lb st)

Instrument panel of the SJ30-2   *NEW*/0526912

SJ30-2 cabin with fan-type divider open and closed   *NEW*/0526915/0526916

PERFORMANCE (estimated):

Max operating Mach No. (MMO)	0.83
High-speed cruising speed	M0.80 (459 kt; 850 km/h; 528 mph)
Long-range cruising speed	M0.78 (447 kt; 828 km/h; 514 mph)
Stalling speed, flaps down, flight idle power	91 kt (169 km/h; 105 mph)
Max rate of climb at S/L: A	1,189 m (3,900 ft)/min
B	1,250 m (4,100 ft)/min
Rate of climb at S/L, OEI: A	305 m (1,000 ft)/min
B	335 m (1,100 ft)/min
Service ceiling	14,935 m (49,000 ft)
Service ceiling, OEI: A	6,126 m (20,100 ft)
B	6,340 m (20,800 ft)
FAA T-O balanced field length: A	1,217 m (3,993 ft)
B	1,103 m (3,620 ft)

FAA landing distance: at max landing weight: A, B	1,042 m (3,420 ft)
at typical mission completed weight, pilot and three passengers: A, B	847 m (2,780 ft)

Range, pilot and three passengers, NBAA IFR reserves, 100 n mile (185 km; 115 mile) alternate, at M0.80:

A, B	2,363 n miles (4,376 km; 2,719 miles)

Range, pilot and three passengers, NBAA IFR reserves, 100 n mile (185 km; 115 mile) alternate at M0.78

A	2,500 n miles (4,630 km; 2,877 miles)
B	1,840 n miles (3,407 km; 2,117 miles)

OPERATIONAL NOISE LEVELS (estimated):

T-O	74.7 EPNdB
Approach	93.6 EPNdB
Sideline	85.6 EPNdB

UPDATED

SKYCRAFT

SKYCRAFT INTERNATIONAL INC

5000 West Airport Drive, Canton/Akron Regional Airport, North Canton, Ohio 44720
Tel: (+1 330) 966 69 66
Fax: (+1 330) 862 62 62
e-mail: arvsuper2@aol.com
Web: http://www.arv-super2.com
PRESIDENT: John Hall

SkyCraft purchased rights to Super2 from Highlander Aircraft Corporation of Golden Valley, Minnesota, in 1999. Previously, Highlander had obtained the design from Aviation Scotland in UK and became sole source when co-licensee, ASL Hagfors Aero of Sweden, ceased trading in September 1995.

UPDATED

SKYCRAFT ARV SUPER2

TYPE: Side-by-side ultralight/kitbuilt.

PROGRAMME: Initial production in UK from assembly of Scottish-built kits. Original prototype Island Aircraft ARV-1 Super2 (G-OARV) first flew 11 March 1985; see 2000-01 and earlier *Jane's* for other non-US history.

Design rights acquired by Highlander in USA; company started construction of US prototype in September 1995 and this (N280KT) flew for first time in February 1996. By August 1996, Highlander had sold six kits, two of which were flying by year end. Programme transferred to SkyCraft in 1999. No known production by last-mentioned.

Description applies to US-built version, unless otherwise stated.

CUSTOMERS: Total of 32 Super2s built by early 1994 (seven by Aviation Scotland). UK production then discontinued. Aircraft c/n 033 to Sweden in October 1994 as SE-KYP for demonstrations. Total of 37 aircraft believed flying by early 2001.

COSTS: Quick-build kit US$18,950; factory-built version US$130,000 (2002).

DESIGN FEATURES: Design objective low initial cost and low maintenance cost; engine with aluminium radiator in recessed duct in rear fuselage; superplastically formed aluminium alloy pressings to save weight; shoulder location of wing with forward sweep, low panel line and close cowling of small engine combine to provide optimum view from cabin, particularly in turns. Designed to BCAR Section K and FAR Pt 21.91; intended to be suitable for home building in kit form; quoted build time 900 hours.

Wing section NACA 2415 (modified); 5° forward sweep, 2° dihedral and 3° incidence.

FLYING CONTROLS: Conventional and manual. Mass-balanced ailerons operated by torque tube through leading-edge of manually operated, three-position, plain flaps; mass-balanced rudder and elevator with trim tab.

STRUCTURE: Single-spar aluminium alloy wing, cold bonded and flush riveted. Single uncompromised main bulkhead carries wings, bracing struts, controls, seats, fuel tank and main landing gear; conventional double beam forward of main bulkhead carries firewall, nose landing gear and engine; front fuselage skinning in four panels; rear fuselage of aluminium alloy with single curvature skinning. Cowlings, fairings and wingtips of glass fibre. Tail unit conventional aluminium alloy structure with three-spar fixed surfaces; single-spar rudder and elevator. Airframe corrosion-treated.

LANDING GEAR: Non-retractable tricycle type. Cantilever main legs of tapered steel leaf spring. All three wheels size 3.50×6.0, with tyres size 13×4.00-6, pressure 1.72 bar (25 lb/sq in). Hydraulic single disc brakes. Steerable nose leg has rubber shock rings and gas-filled damper.

POWER PLANT: One 59.6 kW (79.9 hp) Rotax 912 UL flat-four engine driving two-blade propeller. Mid-West AE110 rotary engine reinstated as option in 1998, along with 2SI three-cylinder two-stroke and Jabiru 2200 four-cylinder four-stroke engines. Fuel capacity 76 litres (20.0 US gallons; 16.7 Imp gallons) in single fuselage tank. Oil capacity 1.9 litres (0.5 US gallon; 0.4 Imp gallon).

ACCOMMODATION: Rearward-hinged one-piece Perspex canopy. Dual controls standard.

AVIONICS: To customers' requirements.
Comms: Optional Honeywell or Becker nav/com, Narco transponder.
Instrumentation: Three basic flight instruments plus engine instruments. Optional vacuum instruments driven by dual venturis mounted under front fuselage.

DIMENSIONS, EXTERNAL:
Wing span	8.99 m (29 ft 6 in)
Length overall	5.49 m (18 ft 0 in)
Height overall	2.29 m (7 ft 6 in)
Propeller diameter	1.07 m (3 ft 6 in)

DIMENSIONS, INTERNAL:
Cabin: Length	1.27 m (4 ft 2 in)

Max width	0.99 m (3 ft 3 in)
Max height	1.09 m (3 ft 7 in)

AREAS:
Wings, gross	8.59 m² (92.5 sq ft)

WEIGHTS AND LOADINGS:
Weight empty, equipped	313 kg (690 lb)
Max T-O and landing weight	529 kg (1,168 lb)
Max wing loading	61.7 kg/m² (12.63 lb/sq ft)
Max power loading	8.89 kg/kW (14.60 lb/hp)

PERFORMANCE (at max T-O weight):
Never-exceed speed (VNE)	131 kt (243 km/h; 151 mph)
Max level speed	119 kt (220 km/h; 137 mph)
Cruising speed at 75% power	100 kt (185 km/h; 115 mph)
Stalling speed, power off, 40° flap	40 kt (73 km/h; 45 mph)
Max rate of climb at S/L	244 m (800 ft)/min
Service ceiling	3,050 m (10,000 ft)
T-O and landing run	138 m (450 ft)
T-O to 15 m (50 ft)	423 m (1,387 ft)
Landing from 15 m (50 ft)	442 m (1,450 ft)
Range	499 n miles (925 km; 575 miles)
g limits	+3.8/−2.0

UPDATED

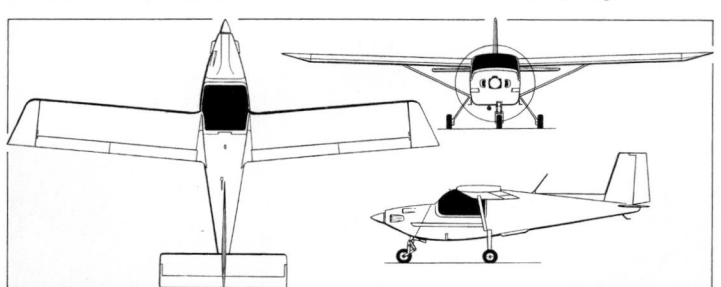

Skycraft ARV Super2 two-seat monoplane (*Jane's/Dennis Punnett*)

ARV Super2 from the original production series (*Jane's/Paul Jackson*)
NEW/0526991

SKYSTAR

SKYSTAR AIRCRAFT CORPORATION
3901 Aviation Way, Caldwell, Idaho 83605
Tel: (+1 800) 554 83 69 and (+1 208) 454 24 44
Fax: (+1 800) 454 64 64
e-mail: info@skystar.com
Web: http://www.skystar.com
PRESIDENT AND CEO: Edward S Downs
VICE-PRESIDENT, ENGINEERING: Frank Miller

SkyStar Aircraft Corporation took over Kitfox programme from former Denney Aerocraft Company in 1993. Kitfox production continues. By 2000, more than 4,000 Kitfox kits had been sold. In 1998, the company added the Kitfox Lite and in the following year streamlined its range. The Series 6 Kitfox and Kitfox Lite² were added in January 2000; the Series 7 was introduced in early 2002 to replace the Series 6.

Rights to Aero Designs Pulsar purchased in late 1996, but in July 1999 the design was sold to the Pulsar Aircraft Corporation (which see).

In early 2001, Skystar opened a spare parts shop at Caldwell and launched an Internet retailing operation.
UPDATED

SKYSTAR KITFOX
TYPE: Side-by-side kitbuilt.
PROGRAMME: Prototype Kitfox first flown 7 May 1984. Initial variants, now discontinued, had MTOW below US 544 kg (1,200 lb) ultralight limit; kit production totalled 257 Model 1s, 491 Model 2s, and 467 Model 3s. Series IV (retrospectively renamed Classic IV) introduced 1991; 323 kits of 476 kg (1,050 lb) MTOW version produced before introduction of current version. First certified version was Series 5, introduced April 1994, but replaced by Series 6 in January 2000; itself replaced by Series 7 in early 2002.
CURRENT VERSIONS: **Classic IV:** Ultralight version, introduced in 1992. Engines up to 74.6 kW (100 hp) can be fitted under helmeted engine cowling. MTOW 544 kg (1,200 lb). **Speedster** option includes short-span wings and improved streamlining for up to 17 kt (32 km/h; 20 mph) additional speed. Still available.
Kitfox 4B: Biplane version. One only built (N95FX).
Kitfox 6 Convertible: Introduced 2000; available in tailwheel and nosewheel subvariants. Considerably refined design, with only 5 per cent parts commonality with Kitfox V. Now discontinued.
Kitfox 7: Public debut at Sun 'n' Fun, April 2002; modifications include different struts and redesigned cowling, plus wing technology developed for Kitfox Lite.

Detailed description applies to above version except where indicated.
Kitfox Sport: Intended to meet Sport Pilot category; will have empty weight of 318 kg (700 lb). In abeyance pending finalisation of legislation.
Kitfox Lite: *Described separately.*
Kitfox Lite²: *Described with Kitfox Lite.*
CUSTOMERS: Some 4,000 Kitfox aircraft ordered, and over 2,500 built.
COSTS: Classic IV US$14,995; Series 7 US$20,995, excluding engine and avionics; Sport US$29,995 including engine and instruments (2003).
DESIGN FEATURES: Designed to have good short-field performance. Constant-chord, high wing with V strut bracing each side; braced tailplane, mid-mounted. With Kitfox IV, introduced in 1991, new wing with laminar flow section and all-metal hinge brackets for full-span flaperons became standard; windscreen material thickened and given increased slope and, with other changes, resulted in increase in cruising speed; stalling and landing speeds decreased, former due to use of flaperons; wing folding standard. Optional agricultural spray pods. Quoted build time 700 hours.

FLYING CONTROLS: Manual. Junkers-type differential flaperons have been redesigned for Series 7; rudder and elevators with mass balance. Variable incidence tailplane.
STRUCTURE: Wings of aluminium alloy, plywood ribs and glass fibre tips, with fabric covering overall. Steel tube fuselage and tail unit, with fabric covering.
LANDING GEAR: Non-retractable type, with hydraulic disc brakes; tailwheel or tricycle configuration; can be converted after completion; optional speed fairings. Sprung aluminium main legs. Optional floats, amphibious floats and skis.
POWER PLANT: Options include 73.5 kW (98.6 hp) Rotax 912 ULS or 84.6 kW (113.4 hp) Rotax 914 driving a Hoffmann HO-V352F two-blade constant-speed propeller; 74.6 kW (100 hp) Teledyne Continental O-200, 93.2 kW (125 hp) Teledyne Continental IO-240B and 86.5 kW (116 hp) Textron Lycoming O-235. Fuel in two wing tanks, total capacity 102 litres (27.0 US gallons; 22.5 Imp gallons) of which 98 litres (26.0 US gallons; 21.7 Imp gallons) are usable.
ACCOMMODATION: Two, side by side, can have 0.24 m³ (8.5 cu ft) storage space behind seat. Optional drop tank style cargo pod.

Continental/Lycoming-powered SkyStar Kitfox Series 6, with additional profiles of Rotax version with helmeted cowling and (bottom left) Series 5 Voyager, final examples of which are still being completed (*Jane's/James Goulding*)
0092109

DIMENSIONS, EXTERNAL:

Wing span: normal:	9.75 m (32 ft 0 in)
optional (Speedster and Sport)	8.84 m (29 ft 0 in)
Wing aspect ratio, normal	7.8
Width, wings folded	2.44 m (8 ft 0 in)
Length: overall	5.99 m (19 ft 8 in)
wings folded: Series 7	6.88 m (22 ft 7 in)
Sport	6.71 m (22 ft 0 in)
Max width of fuselage	1.08 m (3 ft 6½ in)
Height overall: tailwheel	1.73 m (5 ft 8 in)
nosewheel	2.44 m (8 ft 0 in)
Tailplane span	2.41 m (7 ft 11 in)
Wheel track	2.03 m (6 ft 8 in)
Propeller diameter	1.83 m (6 ft 0 in)

DIMENSIONS, INTERNAL:

Cabin max width	1.09 m (3 ft 7 in)

AREAS:

Wings, gross (normal): Series 7	12.26 m² (132.0 sq ft)
Sport	11.89 m² (128.0 sq ft)

WEIGHTS AND LOADINGS (Rotax 912 ULS):

Weight empty: Series 7	363 kg (800 lb)
Sport	318 kg (700 lb)
Baggage capacity	68 kg (150 lb)
Max T-O weight: Series 7	703 kg (1,550 lb)
Sport	558 kg (1,232 lb)
Max wing loading: Series 7	57.3 kg/m² (11.74 lb/sq ft)
Sport	47.0 kg/m² (9.63 lb/sq ft)
Max power loading: Series 7	9.57 kg/kW (15.72 lb/hp)
Sport	7.60 kg/kW (12.49 lb/hp)

PERFORMANCE (two crew, Rotax 912 ULS, unless stated otherwise):

Max level speed	111 kt (205 km/h; 125 mph)
Cruising speed	104 kt (193 km/h; 120 mph)
Stalling speed	33 kt (60 km/h; 37 mph)
Max rate of climb at S/L	366 m (1,200 ft)/min
T-O and landing run	92 m (300 ft)
Range, with reserves: Series 7	
	664 n miles (1,231 km; 765 miles)
Sport	573 n miles (1,062 km; 660 miles)
g limits, pilot only	+3.8/−1.52

UPDATED

SkyStar Kitfox Series 7 prototype *(Jane's/Paul Jackson)*   NEW/0526990

Single-seat Kitfox Lite *(Jane's/Paul Jackson)*   NEW/0526989

Skystar Kitfox Lite² in tricycle configuration *(Jane's/Paul Jackson)*   0110650

SKYSTAR KITFOX LITE and KITFOX LITE²

TYPE: Single-seat ultralight/kitbuilt; side-by-side ultralight/kitbuilt.

PROGRAMME: First displayed at Sun 'n' Fun in April 1998. Planned to be certified in Europe under BCAR standards.

CURRENT VERSIONS: **Kitfox Lite:** Single-seat version.
Description applies to the above.
Kitfox Lite²: Two-seat training version introduced in 2000. Has 37.0 kW (49.6 hp) Rotax 503 UL-2V engine. Tricycle version first flown 15 February 2001.

CUSTOMERS: 200 kits and 10 factory-built aircraft delivered by January 2001.

COSTS: Kitfox Lite kit US$13,995 including engine, factory built US$21,995 (2002). Kitfox Lite² kit US$21,995 including engine, factory built US$28,995 (2002).

DESIGN FEATURES: Generally as for Kitfox. Meets FAR Pt 103 standards. Wings fold for storage. Quoted build time 200 hours; wing pre-built by factory.

FLYING CONTROLS: Manual, including flaperons.

STRUCTURE: Welded 4130 steel fuselage; aluminium wing spar; aluminium and wood wing, all fabric-covered.

LANDING GEAR: Tricycle or tailwheel configuration; free-castoring 6 cm (2½ in) tailwheel. 4.00-6 main tyres. Individual wheel brakes. Options include floats and skis.

POWER PLANT: One 26.1 kW (35 hp) Skystar/2SI Series 460 engine driving Tennessee two-bladed propeller through 2.5:1 reduction gear. Fuel capacity 19 litres (5.0 US gallons; 4.2 Imp gallons) in Lite; 38 litres (10.0 US gallons; 8.3 Imp gallons) in Lite².

EQUIPMENT: Kitfox Lite option includes BRS parachute (standard on factory-built examples).

DIMENSIONS, EXTERNAL:

Wing span: Lite	7.67 m (25 ft 2 in)
Lite²	8.89 m (29 ft 2 in)
Length overall: Lite	5.00 m (16 ft 5 in)
Lite²	5.61 m (18 ft 5 in)
Length: wings folded: Lite	5.36 m (17 ft 7 in)
Height overall: Lite	1.57 m (5 ft 2 in)
Lite²	1.88 m (6 ft 2 in)

AREAS:

Wings, gross: Lite	9.38 m² (101.0 sq ft)
Lite²	11.61 m² (125.0 sq ft)

WEIGHTS AND LOADINGS:

Weight empty: Lite	115 kg (254 lb)
Lite²	223 kg (492 lb)
Max T-O weight: Lite	249 kg (550 lb)
Lite²	476 kg (1,050 lb)

PERFORMANCE:

Max level speed: Lite	55 kt (101 km/h; 63 mph)
Lite²	72 kt (133 km/h; 83 mph)
Normal cruising speed: Lite	48 kt (89 km/h; 55 mph)
Lite²	65 kt (121 km/h; 75 mph)
Stalling speed: Lite	24 kt (44 km/h; 27 mph)
Lite²	32 kt (58 km/h; 36 mph)
Max rate of climb at S/L	229 m (750 ft)/min
T-O run: Lite	31 m (100 ft)
Lite²	77 m (250 ft)
Landing run: Lite	23 m (75 ft)
Lite²	67 m (220 ft)
Range: Lite	95 n miles (177 km; 110 miles)
Lite²	199 n miles (370 km; 230 miles)
Endurance: Lite	2 h 0 min
Lite²	3 h 30 min

UPDATED

SLIPSTREAM

SLIPSTREAM INDUSTRIES INC

W 8407 Cottonville Drive, Wautoma, Wisconsin 54982
Tel: (+1 920) 787 58 86 or (+1 800) 464 36 64
Fax: (+1 920) 787 57 15
e-mail: mail@slipstreamind.com
Web: http://www.slipstreamind.com
PRESIDENT: Michael Puhl

SlipStream produces a range of ultralight aircraft, and also offers the Dragonfly, originally produced by Viking Aircraft.

The company also manufactures the Shark range of amphibious floats at its 1,394 m² (15,000 sq ft) factory.

UPDATED

SLIPSTREAM DRAGONFLY

TYPE: Side-by-side ultralight kitbuilt.

PROGRAMME: Prototype first flight 16 June 1980, designed by Bob Walters. Available as kit or as plans.

CURRENT VERSIONS: **Dragonfly Mark I:** Original configuration, with non-retractable mainwheels at tips of foreplane. EAA Oshkosh Outstanding Design, 1980.

Dragonfly Mark II: In parallel production, for operation from unprepared strips and narrow taxiways. Main landing gear in short non-retractable cantilever units under wings, with individual hydraulic toe brakes, and increased foreplane and elevator areas; wheel track 2.44 m (8 ft).

Dragonfly Mark III: Flight tested in 1985; non-retractable, narrow-track tricycle landing gear with steerable nosewheel and cantilever mainwheel legs attached to fuselage. **Mark III Millenium** displayed at EAA AirVenture, Oshkosh, July 2001 (N204AS, then as yet unflown) as conversion of earlier aircraft.

Tricycle gear SlipStream Dragonfly Mk III Millenium (*Jane's/Paul Jackson*)   NEW/0526988

Recommended engines: 59.7 kW (80 hp) Jabiru 2200 flat-four or 89 kW (120 hp) Jabiru 3300 flat-six.

CUSTOMERS: More than 100 kits delivered, together with 2,000 sets of plans; some 70 kitbuilt and more than 430 plans-built aircraft flying.

COSTS: Standard kit, including Jabiru 2200 engine, US$38,500 (2001).

DESIGN FEATURES: Unconventional shoulder-wing monoplane with equal-span foreplane; latter also mounting landing gear in one of two alternative locations.

Quoted build time 1,000 hours, or 500 hours for quick-build kit.

Wing section Eppler 1213. Foreplane GU25 section. Thickness/chord ratio 15 per cent; dihedral 3°; no incidence or sweepback. Foreplane t/c 17 per cent; anhedral 3° on Mk I or dihedral on Mk II; incidence −1°; no sweepback.

FLYING CONTROLS: Manual. Ailerons on inboard trailing-edge of wing; near full-span elevators on foreplane; horn-balanced rudder. Flettner tab on each foreplane.

STRUCTURE: Composites wing; foreplane and tail unit structures of styrene foam, glass fibre, carbon fibre and epoxy. Semi-monocoque fuselage, formed (not carved) from 12.5 mm (½ in) thick urethane foam, with strips of 18 mm (¾ in) foam bonded along edges to allow large-radius external corners. Fuselage covered with glass fibre inside and out.

LANDING GEAR: See Current Versions. Brakes fitted.

POWER PLANT: Mks I and II have one 44.5 kW (60 hp) 1,835 cc modified Volkswagen motorcar engine; 1,600 cc engine, rated at 33.5 kW (45 hp), optional. Fuel capacity 56.8 litres (15.0 US gallons; 12.5 Imp gallons).

DIMENSIONS, EXTERNAL:
Wing span	6.71 m (22 ft 0 in)
Foreplane span	6.71 m (22 ft 0 in)
Length overall	5.79 m (19 ft 0 in)
Height overall: Mk I	1.22 m (4 ft 0 in)
Mk III	1.68 m (5 ft 6 in)

DIMENSIONS, INTERNAL:
Cabin: Max width	1.09 m (3 ft 7 in)
Height	0.79 m (2 ft 7 in)

AREAS:
Wings, gross	4.24 m² (45.6 sq ft)
Foreplane, net	4.33 m² (46.6 sq ft)

WEIGHTS AND LOADINGS (Millenium, Jabiru 2200):
Weight empty	272 kg (600 lb)
Max T-O weight	567 kg (1,250 lb)

PERFORMANCE (Millenium, Jabiru 2200):
Never-exceed speed (VNE)	156 kt (290 km/h; 180 mph)
Cruising speed	113 kt (209 km/h; 130 mph)
Stalling speed	48 kt (89 km/h; 55 mph)
Max rate of climb at S/L	259 m (850 ft)/min
T-O to 50 ft (15 m)	366 m (1,200 ft)
Landing from 50 ft (15 m)	610 m (2,000 ft)
Range	434 n miles (804 km; 500 miles)
g limits	+4.4/−2.0

UPDATED

SLIPSTREAM GENESIS and REVELATION

TYPE: Side-by-side ultralight kitbuilt.

PROGRAMME: Prototype Genesis first flew 1992; Revelation developed as a lower-powered 'recreational' version.

CURRENT VERSIONS: **Genesis:** Baseline version.

Revelation: Lower-powered version; can be upgraded to Genesis standard by replacement only of windscreen and engine.

CUSTOMERS: Total of 101 Genesis flying, by end 2001; 100 Revelations sold, and 75 flying, by same date.

COSTS: Kit including engine (2002): Genesis US$16,491-US$28,494; Revelation US$13,295-US$26,798, depending on engine.

DESIGN FEATURES: Quoted build time for Genesis 200 hours; for Revelation 150 hours (or 90 when opting for sailcloth-covered wing).

FLYING CONTROLS: Conventional and manual. Horn-balanced elevator.

STRUCTURE: Preformed glass fibre nose; fabric-covered strut-braced wings; tubular steel rear fuselage.

LANDING GEAR: Fixed tricycle layout with mechanical brakes; hydraulic brakes optional. Optional floats and speed fairings.

POWER PLANT: *Genesis:* One 47.8 kW (64.1 hp) Rotax 582 driving Tennessee two-blade wooden propeller. Fuel capacity 76 litres (20.0 US gallons; 16.7 Imp gallons). Other engines including Rotax 912 UL, 912 ULS and 914 can be fitted.

Revelation: One 39.3 kW (52.7 hp) Rotax 503B. Fuel capacity 38 litres (10.0 US gallons; 8.3 Imp gallons). Same engine options as Genesis.

Data apply to both versions, except where otherwise stated.

DIMENSIONS, EXTERNAL:
Wing span	9.35 m (30 ft 8 in)
Length overall	5.89 m (19 ft 4 in)
Height overall	1.92 m (6 ft 3½ in)

AREAS:
Wings, gross	16.62 m² (179.0 sq ft)

WEIGHTS AND LOADINGS (A: Genesis, B: Revelation):
Weight empty: A	295 kg (650 lb)
B	211 kg (465 lb)
Max T-O weight: A	589 kg (1,300 lb)
B	499 kg (1,100 lb)

PERFORMANCE (A and B as above):
Never-exceed speed (VNE):	
A	104 kt (193 km/h; 120 mph)
Normal cruising speed: A	70 kt (129 km/h; 80 mph)
B	57 kt (106 km/h; 66 mph)

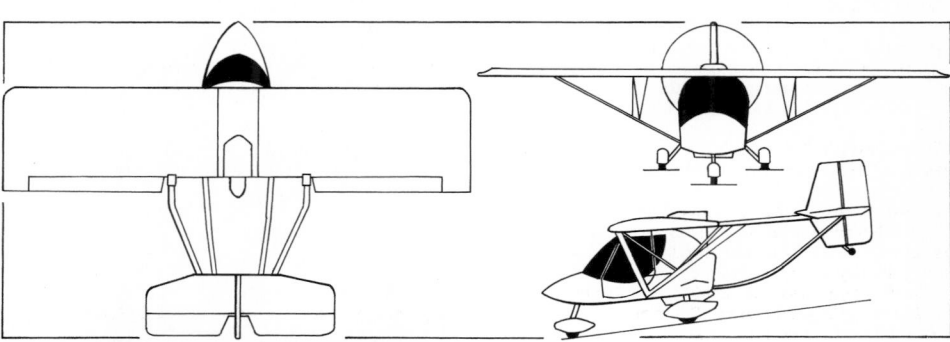

SlipStream Revelation two-seat ultralight (*Jane's/Paul Jackson*)   0100466

SlipStream Genesis two-seat kitbuilt (*Jane's/Paul Jackson*)   NEW/0526987

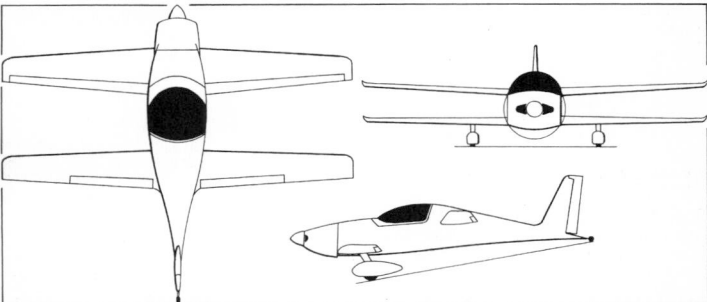

SlipStream Dragonfly Mark II (*Jane's/Paul Jackson*)   0016522

SlipStream Dragonfly Mark II (*Jane's/Paul Jackson*)
0110731

SlipStream Revelation ultralight derivative of Genesis named 'Pink Panther' (*Jane's/Paul Jackson*)
NEW/0526986

Stalling speed: A	39 kt (72 km/h; 45 mph)
B	43 kt (79 km/h; 49 mph)
Max rate of climb at S/L: A	305 m (1,000 ft)/min
B	244 m (800 ft)/min
T-O run: A, B	122 m (400 ft)
Landing run: A	77 m (250 ft)
B	46 m (150 ft)
Range with max fuel: A 195 n miles (362 km; 225 miles)	
B	130 n miles (241 km; 150 miles)

UPDATED

SLIPSTREAM SCEPTER

TYPE: Single-seat ultralight kitbuilt.
PROGRAMME: Development of Genesis, introduced in 1999.
CUSTOMERS: Three flying by mid-2002.
COSTS: US$10,995 to US$14,699 (2002) including engine.
DESIGN FEATURES: Quoted build time 120 hours.
Data as Genesis, except that below.
POWER PLANT: One 39.3 kW (52.7 hp) Rotax 503B, 29.5 kW (39.6 hp) Rotax 447 UL-1V or 47.7 kW (64.0 hp) Zanzottera MZ 202R. Fuel capacity 19 litres (5.0 US gallons; 4.2 Imp gallons).
WEIGHTS AND LOADINGS (Rotax 503):

Weight empty	199 kg (438 lb)
Max T-O weight	499 kg (1,100 lb)

PERFORMANCE (Rotax 447):

Normal cruising speed	61 kt (113 km/h; 70 mph)
Stalling speed	24 kt (44 km/h; 27 mph)
Max rate of climb at S/L	305 m (1,000 ft)/min
T-O and landing run	23 m (75 ft)
Range with max fuel	60 n miles (112 km; 70 miles)

UPDATED

SLIPSTREAM SKYBLASTER

TYPE: Two-seat kitbuilt twin.
PROGRAMME: Introduced in 2000; based on Genesis/Revelation with 98 per cent commonality of parts.
CUSTOMERS: Four flying by mid-2002.
COSTS: US$22,995 with Rotax 503E engines (2002).
DESIGN FEATURES: Two in-line engines, with counter-rotating propellers, mounted on front and back of wing; layout simplifies single-engine handling. Quoted build time 325 hours.
Data as Genesis, except that below.
POWER PLANT: Two 39.3 kW (52.7 hp) Rotax 503E engines in push/pull layout. Fuel capacity 151 litres (40.0 US gallons; 33.3 Imp gallons).
DIMENSIONS EXTERNAL:

Height overall	2.42 m (7 ft 11¼ in)

WEIGHTS AND LOADINGS:

Weight empty	363 kg (800 lb)
Max T-O weight	635 kg (1,400 lb)

PERFORMANCE:

Normal cruising speed	70 kt (129 km/h; 80 mph)
Stalling speed	44 kt (81 km/h; 50 mph)
Max rate of climb at S/L	366 m (1,200 ft)/min
T-O run	23 m (75 ft)

UPDATED

SLIPSTREAM SKYQUEST

TYPE: Two-seat kitbuilt twin.
PROGRAMME: Introduced at Sun'n'Fun 2001.
CUSTOMERS: One flying by mid-2002.

STRUCTURE: Tube fuselage; glass fibre cockpit enclosure. Aluminium wings and Dacron-covered control surfaces.
LANDING GEAR: Tricycle type; fixed. Mechanical (optionally hydraulic) brakes. Floats, skis or speed fairings can be fitted.
POWER PLANT: Two 39.3 kW (52.7 hp) Rotax 503E engines, each driving a Powerfin three-blade F composite ground-adjustable propeller; options include two 47.7 kW (64 hp) Zanzottera MZ 202s. Fuel capacity 151 litres (40.0 US gallons; 33.3 Imp gallons) in two wing tanks.
ACCOMMODATION: Pilot and passenger side by side.
AVIONICS: To customer's requirements; MicroAir 760 radio, Ameri-King AK-450 ELT and Lowrance Airmap 300 GPS are recommended options.
DIMENSIONS EXTERNAL:

Wing span	9.35 m (30 ft 8 in)
Wing chord, constant	1.85 m (6 ft 1 in)
Wing aspect ratio	5.3
Length overall	5.89 m (19 ft 4 in)
Height overall	2.30 m (7 ft 6½ in)
Tailplane span	2.44 m (8 ft 0 in)
Wheel track	2.13 m (7 ft 0 in)

AREAS:

Wings, gross	16.63 m² (179.0 sq ft)

WEIGHTS AND LOADINGS (Rotax 503s):

Weight empty	374 kg (825 lb)
Max T-O weight	635 kg (1,400 lb)
Max wing loading	38.2 kg/m² (7.82 lb/sq ft)
Max power loading	8.08 kW/kg (13.28 lb/hp)

PERFORMANCE:

Never-exceed speed	104 kt (193 km/h; 120 mph)
Normal cruising speed	74 kt (137 km/h; 85 mph)
Stalling speed	44 kt (81 km/h; 50 mph)
Max rate of climb at S/L	366 m (1,200 ft)/min
Service ceiling	3,660 m (12,000 ft)
T-O run	61 m (200 ft)

UPDATED

SlipStream Scepter single-seat ultralight (*Jane's/Paul Jackson*)
NEW/0526985

COSTS: US$23,995 including Rotax 503E engines and propellers (2001). Also available in four subcomponent kits.
DESIGN FEATURES: Based on company's Genesis two-seat ultralight/kitbuilt with 95 per cent commonality of parts. Redesigned rear fuselage allows two pusher engines to be placed close together, reducing asymmetric thrust in single-engine operation. Quoted build time 325 hours.
FLYING CONTROLS: Conventional and manual. Optional flaps. Optional electric trim.

SlipStream SkyQuest twin-engined kitbuilt (*Jane's/Susan Bushell*)
0110685

SOLARIS

SOLARIS AVIATION INC
Palm Beach North County Airport, 11250-4 Aviation Boulevard, West Palm Beach, Florida 33412
Tel: (+1 954) 757 54 80
Fax: (+1 954) 757 51 82
Web: http://www.solarisaviation.com
PRESIDENT: Nick Schneider
VICE-PRESIDENTS: George Rodgers
    J Cook
CHAIRMAN: David Schuldt

Solaris Aviation was established in 1999 and purchased the assets of Ruschmeyer Luftfahrttechnik GmbH, including Type Certificates, moulds and tooling. A German company (see 1997-98 and earlier *Jane's*), Ruschmeyer had filed for bankruptcy in January 1996, having built a small number of R 90 four-seat touring aircraft. Solaris has also acquired a

Ruschmeyer R 90-230 RG displayed at Sun 'n' Fun in April 2002 to promote Solaris Sigma
(Jane's/Paul Jackson)   *NEW*/0533672

10,220 m² (110,000 sq ft) manufacturing facility in Bielsko-Biała, Poland, where labour-intensive parts of the relaunched aircraft are produced before being shipped to Florida for final assembly. The Sigma, as the R 90 is now known, was first shown publicly at AirVenture, Oshkosh, in July 2001, the project having been formally announced on 5 July 2001.
VERIFIED

SOLARIS SIGMA and ALPHA

TYPE: Four-seat lightplane.
PROGRAMME: Development of original Ruschmeyer MF-85 started in Germany in 1985; first prototype MF-85P-RG (V0001) flew with Porsche engine; changed to R 90-230 RG using flat-rated Textron Lycoming IO-540 after cessation of Porsche production in March 1990; first flight of prototype V001 (D-EEHE) 8 August 1988; first flight of V002 (D-EERO) 25 September 1990; V003 (D-EERH) first flight 12 February 1992; LBA certification June 1992; FAA 24 June 1994. Main production variant was R 90-230 RG, of which 25 built; Ruschmeyer also produced prototype fixed landing gear R 90-230 FG; lower and higher-powered versions were overtaken by company's bankruptcy. Programme transferred to Solaris in USA, 1999; German-built R 90-230 RG registered N230S in June 2001 as demonstrator. Production deliveries due to begin in third quarter of 2002, but no production reported by August 2002 and Type Certificate remained in Ruschmeyer ownership in late 2002.
CURRENT VERSIONS: **Sigma 230:** One Textron Lycoming IO-540-C4D5 flat-six, derated to 172 kW (230 hp). Retractable landing gear; intended for European market.
Alpha 230: As above. Fixed gear.
Sigma 250: IO-540-C4D5 engine, fully rated at 186 kW (250 hp). Retractable gear. Baseline US version.
Alpha 250: As immediately above. Fixed gear.
Sigma 310: Teledyne Continental IO-550N flat-six, rated at 231 kW (310 hp). Retractable gear. Uprated US version.
Alpha 310: As immediately above. Fixed gear.
COSTS: Sigma 250 US$244,900; Sigma 310 US$279,900 (2002).
DESIGN FEATURES: Objectives were FAR Pt 23 compliance; high performance by low-drag all-composites airframe, use of derated engine for European version, requiring less cooling, reduced noise by lower engine rpm, special silencer and, originally, matched four-blade constant-speed propeller. Consequently, reduced fuel consumption and emissions; noise level under German regulations demonstrated 66 dBA, 8 dBA below limit; also 10.2 dBA below ICAO Chapter 10 limit. Laminar flow aerofoil. Certified 18,000 hour life.
FLYING CONTROLS: Conventional and manual; all push/pull rod operated to minimise friction; dual controls with sticks standard; fixed tailplane; combined trim and trim tab on port elevator; stall strips on inboard wing leading-edge; three-position Fowler flaps operated by electrohydraulic system are linked to elevator to avoid trim changes; max deflection 30°. Aileron deflections +16/−11°; elevator +18/−15°; rudder starboard 240 mm (9.45 in), port 220 mm (8.66 in).
STRUCTURE: All-composites moulded airframe with 70 per cent fewer parts than equivalent metal structure. Ruschmeyer qualified BASF Palatal A430 resin fibre composites material for aviation; advantages are durability at up to 72°C ambient, improved strength, field repair

without special tools, negligible allergic risks during manufacture, and material is recyclable. GFRP with Rohacell foam cores and CFRP reinforcement round doors; steel tube running from firewall, over roof and into tail supports doors and protects against roll-over; components earth bonded to German and FAA standards. Meets 1.75 safety factor for composites airframes.
LANDING GEAR: Retractable tricycle type; electrohydraulic actuation; trailing-link levered mainwheel suspension retracts inward and covered by mechanically linked doors; foot-powered hydraulic disc brakes; steerable nosewheel retracts rearward, also covered by doors.
POWER PLANT: As for Current Versions. Hartzell three-blade, constant-speed propeller. Adjustable cowl flaps. Standard Sigma 250/Alpha 250 fuel 311 litres (82.0 US gallons; 68.3 Imp gallons) in integral tanks in inner wings; Sigma 310/Alpha 310 349 litres (92.0 US gallons; 76.7 Imp gallons). Oil capacity 11.4 litres (3.0 US gallons; 2.5 Imp gallons) for Sigma 250/Alpha 250; 12 litres (3.25 US gallons; 2.7 Imp gallons) for Sigma 310/Alpha 310.
ACCOMMODATION: Four-seat interior with ergonomically designed seats; three-point harness; upward-opening gull wing doors; heating and ventilation system with electric blower; baggage compartment accessible from inside and outside.
SYSTEMS: S-Tec 40 (optional 55X) single-axis autopilot. Electrohydraulic actuation for flaps; pitot heater and night lighting. Electrical system 28 V DC; 70 Ah alternator; 24 V 18 Ah battery.
AVIONICS: *Comms:* Garmin GNS-430 transceiver; GTX-327 transponder; GMA-340 audio system; ELT. Optional GNS-530 nav/com with GI-106 moving map indicator.
Flight: Garmin GNS-420 GPS. Optional second GNS-430, S-Tec ST-180 slaved compass, ST-361 flight director, ST-360 altitude selector, Goodrich WX-500 Stormscope and Shadin fuel/air data computer.
DIMENSIONS, EXTERNAL:
Wing span   9.50 m (31 ft 2 in)

Wing aspect ratio	7.0
Length overall	7.98 m (26 ft 2 in)
Height overall	2.74 m (9 ft 0 in)
Wheelbase	1.93 m (6 ft 4 in)
Propeller diameter	1.88 m (6 ft 2 in)

DIMENSIONS, INTERNAL:
Cabin: Length	2.86 m (9 ft 4½ in)
Width	1.14 m (3 ft 9 in)
Height	1.24 m (4 ft 0¼ in)
Baggage compartment volume	0.80 m³ (28.3 cu ft)

AREAS:
Wings, gross	12.94 m² (139.3 sq ft)

WEIGHTS AND LOADINGS (Sigma 230, 250 and 310):
Weight empty: S230	850 kg (1,874 lb)
S250	880 kg (1,940 lb)
S310	1,051 kg (2,317 lb)
Baggage capacity: all	50 kg (110 lb)
Max T-O weight: S230	1,352 kg (2,980 lb)
S250	1,372 kg (3,025 lb)
S310	1,587 kg (3,500 lb)
Max wing loading: S230	104.4 kg/m² (21.39 lb/sq ft)
S250	106.0 kg/m² (21.72 lb/sq ft)
S310	122.7 kg/m² (25.13 lb/sq ft)
Max power loading: S230	7.89 kg/kW (12.96 lb/hp)
S250	7.36 kg/kW (12.10 lb/hp)
S310	6.87 kg/kW (11.29 lb/hp)

PERFORMANCE (Sigma 230, 250 and 310):
Max level speed: S230	175 kt (324 km/h; 201 mph)
S250	184 kt (340 km/h; 211 mph)
S310	215 kt (398 km/h; 247 mph)
Cruising speed: S250	173 kt (320 km/h; 199 mph)
S310	205 kt (380 km/h; 236 mph)
Econ cruising speed: S250	125 kt (232 km/h; 144 mph)
S310	150 kt (278 km/h; 173 mph)
Stalling speed: flaps up:	
S230, S250	67 kt (125 km/h; 78 mph)
S310	71 kt (132 km/h; 82 mph)
flaps down: S230, S250	57 kt (106 km/h; 66 mph)
S310	60 kt (112 km/h; 69 mph)
Max rate of climb at S/L: S250	399 m (1,310 ft)/min
S310	408 m (1,340 ft)/min
Service ceiling: S250, S310	6,100 m (20,000 ft)
T-O run: S230	261 m (855 ft)
S250	249 m (815 ft)
S310	250 m (820 ft)
T-O to 15 m (50 ft): S250	427 m (1,400 ft)
S310	503 m (1,650 ft)
Landing from 15 m (50 ft): S250	412 m (1,350 ft)
S310	488 m (1,600 ft)
Landing run: S230, S250	328 m (1,075 ft)
S310	406 m (1,330 ft)
Range: normal cruise:	
S250	840 n miles (1,555 km; 966 miles)
S310	1,150 n miles (2,129 km; 1,323 miles)
long-range cruise:	
S230	870 n miles (1,611 km; 1,001 miles)
S250	1,160 n miles (2,148 km; 1,334 miles)
S310	1,488 n miles (2,755 km; 1,712 miles)

UPDATED

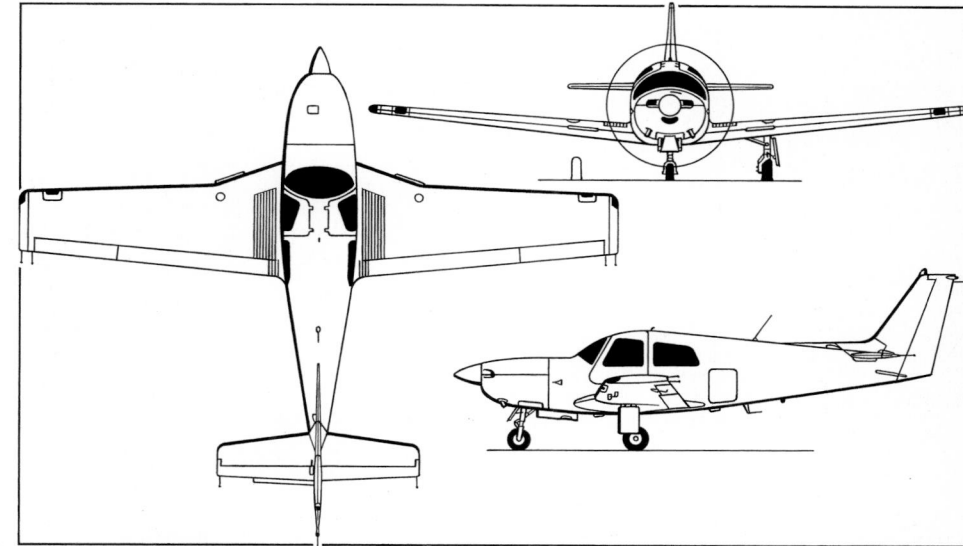

Solaris Sigma four-seat tourer *(Jane's/Mike Keep)*

SONEX

SONEX LTD
PO Box 2521, Oshkosh, Wisconsin 54903-2521
Tel: (+1 920) 231 82 97
Fax: (+1 920) 426 83 33
e-mail: info@sonex-ltd.com
Web: http://www.sonex-ltd.com

PRESIDENT: John Monnett
GENERAL MANAGER: Jeremy Monnett

EUROPEAN AGENT
ARES srl, Via Papiria 54a, I-00175 Roma, Italy
Tel: (+39 06) 76 90 01 32
Fax: (+39 06) 76 90 00 89
e-mail: arescomp@tin.it
Web: http://www.rce.it/sonex

Company formed by John Monnett, designer of the Sonerai, Monerai and Moni light aircraft, to produce kits of his latest design, the Sonex and its motor glider derivative, the Xenos. In late 2002 the company had a workforce of four in its 968 m² (10,424 sq ft) factory.
UPDATED

SONEX SONEX

TYPE: Side-by-side ultralight kitbuilt.

PROGRAMME: Announced March 1996 in response to request for ultralight to meet Italian requirements; first flight of prototype (N12SX) with Jabiru 2200 engine, 28 February 1998. Five prototypes. First customer-built aircraft flown 12 June 2001.

CURRENT VERSIONS: **Sonex:** Kitbuilt, *as described.*

Xenos: Motor glider, based around Sonex front fuselage; *described separately.*

CUSTOMERS: Over 500 sets of plans and kits sold and 36 flying by October 2002.

COSTS: Airframe kit US$11,985 tailwheel, US$12,265 tricycle; plans US$600 (including two-day workshop) (2002).

DESIGN FEATURES: Simple to assemble; 198 parts to easy-build kit. Wings removable for transport and storage. Quoted build time 700 hours.

NACA 64-415 aerofoil.

FLYING CONTROLS: Conventional and manual. Flaps.

STRUCTURE: Of 6061-T6 aluminium throughout.

LANDING GEAR: Choice of tailwheel or tricycle; both fixed. Titanium mainwheel legs and glass fibre speed fairings. Azusa wheels and brakes. Mainwheels 5.00×5, tailwheel (where fitted) 4×1 in.

POWER PLANT: Choice of power plants between 59.7 and 89.5 kW (80 and 120 hp), including Volkswagen 2180 and Jabiru 2200 and 3300. Two-blade, wooden, fixed-pitch Sensenich or Sterba propeller. Fuel capacity 61 litres (16.0 US gallons; 13.3 Imp gallons) in single fuel tank.

DIMENSIONS, EXTERNAL:

Wing span	6.71 m (22 ft 0 in)
Length overall	5.36 m (17 ft 7 in)
Height overall: tailwheel	1.63 m (5 ft 4 in)
nosewheel	1.93 m (6 ft 4 in)

AREAS:

Wings, gross	9.10 m² (98.0 sq ft)

WEIGHTS AND LOADINGS (59.7 kW; 80 hp Jabiru 2200):

Weight empty	249 kg (550 lb)
Baggage capacity	18 kg (40 lb)
Max T-O weight	499 kg (1,100 lb)

PERFORMANCE (engine as above):

Never-exceed speed (V$_{NE}$)	171 kt (317 km/h; 197 mph)
Max level speed	130 kt (241 km/h; 150 mph)
Normal cruising speed	113 kt (209 km/h; 130 mph)
Stalling speed: flaps up	40 kt (74 km/h; 46 mph)
flaps down	35 kt (65 km/h; 40 mph)
Max rate of climb at S/L	305 m (1,000 ft)/min
T-O run	91 m (300 ft)
Landing run	153 m (500 ft)
Range with max fuel	478 n miles (885 km; 550 miles)
g limits	+4.4/−2.2

UPDATED

SONEX XENOS

TYPE: Side-by-side ultralight motor glider kitbuilt.

PROGRAMME: Announced in 2000; wing static load testing completed March 2001; prototype under construction during late 2002.

DESIGN FEATURES: Based on Sonex fuselage (which see).

FLYING CONTROLS: Conventional and manual. Speed brakes.

STRUCTURE: Primarily of 6061 aluminium. Wing has Wortmann FX-61 aerofoil. V tail similar to designer's earlier Moni motor glider.

Third prototype Sonex Sonex in nosewheel configuration (*Jane's/Paul Jackson*)   NEW/0526984

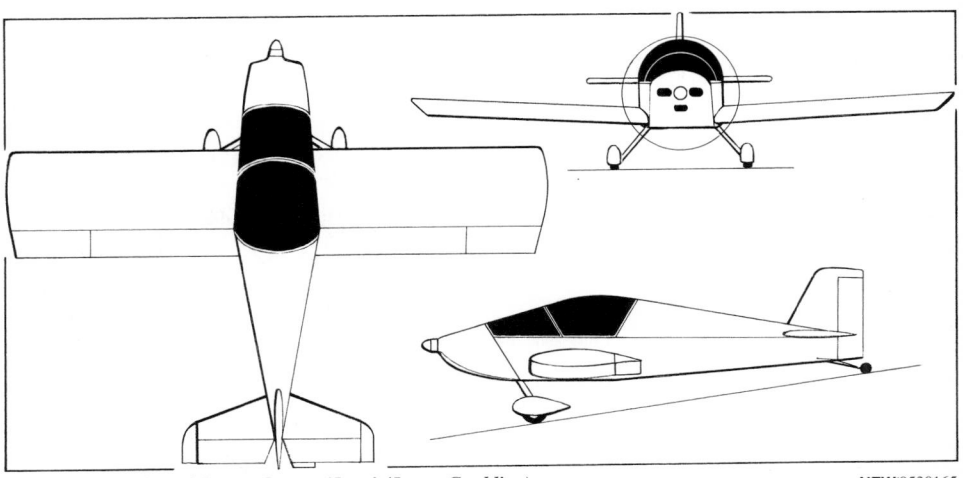

Production version of Sonex Sonex (*Jane's/James Goulding*)   NEW/0530165

LANDING GEAR: Tailwheel type; fixed.

POWER PLANT: Choice of engines including 2,180 cc Aero-Vee 2002 VW conversion and 2200 and 3300 Jabiru. Fuel capacity 61 litres (16.0 US gallons; 13.3 Imp gallons) in single fuel tank.

All data are provisional.

DIMENSIONS, EXTERNAL:

Wing span: Utility	13.46 m (44 ft 2 in)
Aerobatic	11.99 m (39 ft 4 in)
Length overall	6.02 m (19 ft 9 in)

AREAS:

Wings, gross: Utility	14.68 m² (158.0 sq ft)
Aerobatic	13.38 m² (144.0 sq ft)

WEIGHTS AND LOADINGS:

Weight empty: Utility	306 kg (675 lb)
Aerobatic	304 kg (670 lb)
Max T-O weight: Utility	544 kg (1,200 lb)
Aerobatic	430 kg (950 lb)

PERFORMANCE, POWERED:

Never-exceed speed (V$_{NE}$)	130 kt (241 km/h; 150 mph)
Stalling speed	33 kt (60 km/h; 37 mph)

PERFORMANCE, UNPOWERED:

Best glide ratio: Utility	23
Aerobatic	21

NEW ENTRY

STAUDACHER

STAUDACHER AIRCRAFT INC

2648 East Beaver Road, Kawkawlin, Michigan 58631-9402

DESIGNER: Jon A Staudacher

PRESIDENT: Greg Panzl

Since 1990, some 30 aerobatic aircraft in the S-300 family have been built by Staudacher and have gained high placings in competitions in the US and abroad. The company was purchased in 1999 by Greg Panzl, brother of S-300 aerobatic pilot Chris Panzl, and has launched an improved version under that name. Staudachers have proved popular subjects for scale radio-controlled aeromodellers.

UPDATED

STAUDACHER S-300

TYPE: Aerobatic two-seat sportplane.

PROGRAMME: First aircraft built for Mike Goulian in 1990; design progressively refined throughout 1990s. Seven Staudachers participated in 1993 US national championships.

CURRENT VERSIONS: **S-300:** Baseline version; *as described.* Several subvariants, up to S-300E.

S-300RG: Retractable landing gear; cantilever tailplane; IO-540-K engine and two-blade Hartzell propeller. One only (N126RG), first flown 22 February 1998.

S-600: Tandem two-seat version; cockpit enlarged forward. Four built.

Panzl S-330 P: Prototype (N330LS) delivered to Loren Smith 24 July 2000; second built 2001; third registered in

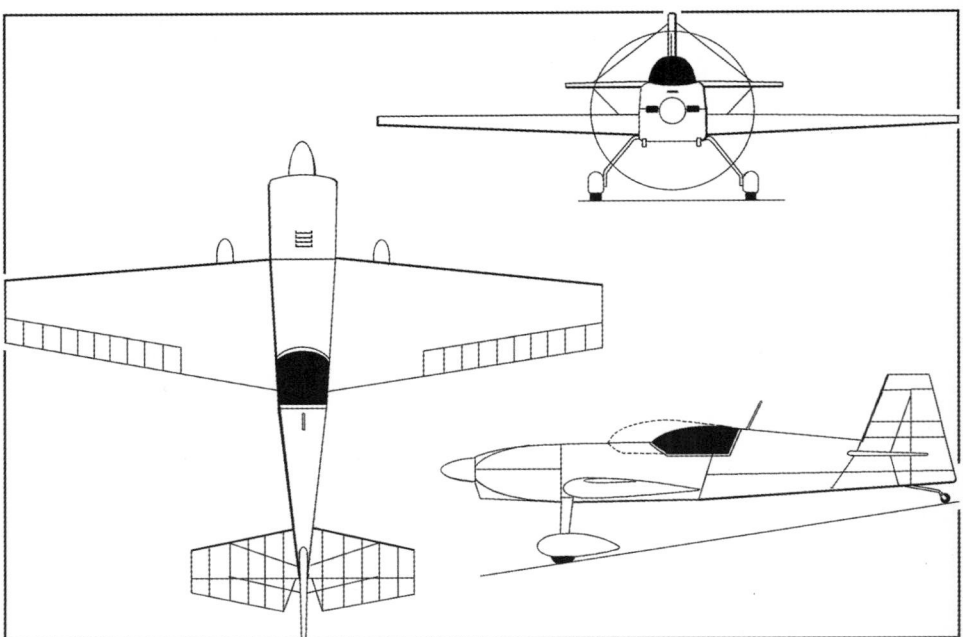

Staudacher S-300 single-seat version, with two-seat canopy shown by broken line (*Jane's/Paul Jackson*)
NEW/0527168

July 2002 (N330TA of Wingtip-to-Wingtip Inc, Onstead, Michigan). Derived from S-300, but with improved canopy and carbon fibre panels throughout, resulting in weight saving. Empty weight, equipped, 585 kg (1,289 lb); 246 kW (330 hp) Textron Lycoming IO-540-G1D5 and MTV-9-B-C propeller. Aerobatic fuel 87 litres (23.0 US gallons; 19.2 Imp gallons); ferry tank in each wingroot, combined capacity 159 litres (42.0 US gallons; 35.0 Imp gallons).

Additionally, several custom versions of Staudacher, comprising **S-341** (N341SA), built 1992 with Textron Lycoming GO-480; **S-1000** (N6195V), built 1996 for Jon Staudacher; **DC 11** (N327MS), with Textron Lycoming O-320 built 1992; and **GU 11** (N11GU), with Teledyne Continental O-200 built for Jon Staudacher in 2000.

CUSTOMERS: Total 31 aircraft of Staudacher design built by mid-2001.

DESIGN FEATURES: Created to rival Sukhoi and Extra Unlimited class aerobatic aircraft; roll rate in excess of 420°/s. Low-wing monoplane with wire-braced tail surfaces. Moderately tapered wing; ailerons occupy two-thirds of trailing-edges; all subvariants have same aerofoil.

FLYING CONTROLS: Conventional and manual. Spade-type aileron servos. Trim tab in each elevator. Twin arrowhead-shape horn balance at fin tip.

STRUCTURE: Sitka spruce (wooden) wing reinforced with carbon fibre and covered with plywood. Metal tube fuselage and empennage; aluminium turtledeck. Fabric-covered ailerons.

LANDING GEAR: Tailwheel type; fixed. Aluminium cantilever mainwheel legs.

POWER PLANT: One 186 kW (250 hp) Textron Lycoming IO-540-B1A5 or 224 kW (300 hp) IO-540-K1B5 flat-six

Staudacher S-300D flown by Chris Panzl, brother of company president, Greg Panzl (*Jane's/Paul Jackson*)
NEW/0526983

driving an MTV-9-B-C/C200-15 three-blade, constant-speed propeller. Fuel tank ahead of pilot, capacity 129 litres (34.0 US gallons; 28.3 Imp gallons).

ACCOMMODATION: One or two (in tandem) pilots, according to version.

DIMENSIONS, EXTERNAL:
Wing span	7.72 m (25 ft 4 in)
Length overall	6.64 m (21 ft 9½ in)
Height overall	1.80 m (5 ft 11 in)

AREAS:
Wings, gross	10.29 m² (110.8 sq ft)

WEIGHTS AND LOADINGS:
Weight empty	567 kg (1,250 lb)
Max T-O weight (Aerobatic)	725 kg (1,600 lb)
Max wing loading	70.3 kg/m² (14.44 lb/sq ft)
Max power loading (IO-540-K)	3.25 kg/kW (5.33 lb/hp)

PERFORMANCE:
Never-exceed speed (VNE)	217 kt (402 km/h; 250 mph)
Stalling speed	52 kt (95 km/h; 59 mph)
Max rate of climb at S/L, S-330	more than 1,219 m (4,000 ft)/min

UPDATED

TBG

THUNDER BUILDERS' GROUP LLC

Tel: (+1 719) 591 93 73
Fax: (+1 509) 472 61 08
e-mail: kwmacs@earthlink.net
MANAGING MEMBER: John McCartney

Scale replica of the classic P-51 Mustang fighter was originally produced by Papa 51 company which ceased trading on 1 December 2000. Group of 25 kitbuilders (six of them with almost complete aircraft) then formed Thunder Builder's Group LLC and acquired company's intellectual rights and production moulds with a view to sustaining their own projects and resuming manufacture. This seemed to have been achieved on 13 September 2001, when GUT Works reached agreement with Group; however, the two partners parted on 25 July 2002 when TBG again put the assets up for sale.

NEW ENTRY

TBG THUNDER MUSTANG

TYPE: Two-seat sportplane kitbuilt.

PROGRAMME: Design, by Dan Denney, started 1993; first flight of prototype (N151TM) 16 November 1996; prototype destroyed in crash 30 May 1998. First customer aircraft (N7TR) owned by Tommy Rose of Hickory, Mississippi, first flown from Nampa factory, where assembly took place, 11 January 1999. In September 1999 this aircraft qualified in the Sport Class of the US National Championship Air Races at Reno, Nevada, at a speed of 269 kt (499 km/h; 310 mph). Programme sustained by Thunder Builder Group after closure of original kit manufacturer, resulting in three new aircraft flying during August 2001.

CUSTOMERS: Total of 30 aircraft under construction by May 1999, including at least two in South Africa. Kit production

TBG Thunder Mustang (*Jane's/Paul Jackson*)
NEW/0533640

was due to resume in 2002. Total 10 registered in USA by November 2001, of which four were flying by July 2002.

COSTS: Complete kit, including engine, but not including instruments, avionics, upholstery or paint, US$285,000 (2000).

DESIGN FEATURES: Scale replica of Second World War North American P-51D Mustang fighter; faithful to the original in outline although fuselage is 62 per cent and wing 75 per cent. Low-wing, mid-tailplane configuration with sweptback fin and fillet, plus prominent belly air intake for radiator. Baseline kit quoted build time is 5,000 hours. Three fast-build options (4,000, 2,000 and 1,500 hours) planned.

Wing section NACA 65-series laminar flow with thickness/chord ratio 15 per cent at root, 12 per cent at tip, 1 per cent camber and 1.25 per cent leading-edge droop.

FLYING CONTROLS: Conventional and manual. Electrically actuated plain flaps; trim tab in rudder and on starboard aileron.

STRUCTURE: Primary structure of carbon/graphite composites, supplied in kit of over 200 ready-formed components; assembled with two-part epoxy adhesive; quoted build time approximately 1,000 hours.

LANDING GEAR: Tailwheel type; main units retract inwards into wings, tailwheel rearwards. Mainwheel tyre size 6.50-8; tailwheel tyre size 4.10-4. Hydraulic brakes.

POWER PLANT: One 477 kW (640 hp) Ryan Falconer V-12 with 2.8:1 reduction gearbox, driving a four-blade constant-speed wood/composites MT propeller. Supercharged 895 kW (1,200 hp) version of Falconer engine under development during 1999 for planned attempts on several class world records. Fuel contained in integral wing tanks, combined capacity 386 litres (102 US gallons; 84.9 Imp gallons). External wing tanks, each of 114 litres (30.0 US gallons; 25.0 Imp gallons) capacity, optional.

ACCOMMODATION: Two in tandem under rearward-sliding bubble canopy with fixed three-piece windscreen. Baggage compartment at rear of cockpit. Custom upholstery optional.

SYSTEMS: 60A alternator and regulator. Optional oxygen system.

AVIONICS: Instrument and avionics packages and autopilot optional.

EQUIPMENT: Optional gas-operated dummy cannon.

DIMENSIONS, EXTERNAL:
Wing span	7.24 m (23 ft 9 in)
Wing aspect ratio	5.4
Length overall	7.37 m (24 ft 2¼ in)
Height overall	2.77 m (9 ft 1¼ in)
Tailplane span	2.70 m (8 ft 10¼ in)
Propeller diameter	2.39 m (7 ft 10 in)

AREAS:
Wings, gross	9.66 m² (104.0 sq ft)

WEIGHTS AND LOADINGS:
Weight empty	998 kg (2,200 lb)
Baggage capacity	23 kg (50 lb)
Payload with max fuel	181 kg (400 lb)

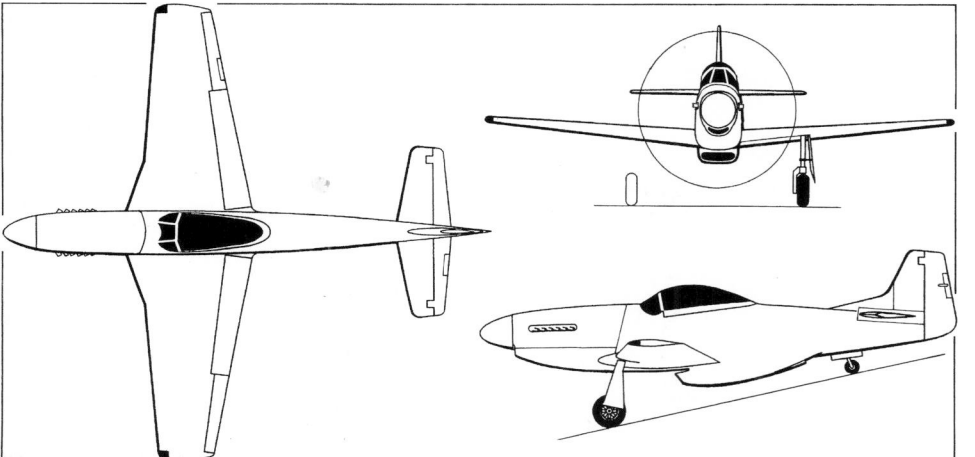
Precautionary landing by a Thunder Mustang scale kitbuilt replica of the P-51D following loss of canopy (*Jane's/Paul Jackson*)
NEW/0533703

jawa.janes.com

Jane's All the World's Aircraft 2003-2004

Max T-O weight	1,451 kg (3,200 lb)		Cruising speed at 75% power		Max rate of climb at S/L	1,585 m (5,200 ft)/min
Max wing loading	150.2 kg/m² (30.77 lb/sq ft)			300 kt (556 km/h; 345 mph)	Service ceiling	7,620 m (25,000 ft)
Max power loading	3.04 kg/kW (5.00 lb/hp)		Manoeuvring speed	222 kt (411 km/h; 255 mph)	Range at econ cruising speed	

PERFORMANCE (provisional):
Never-exceed speed (VNE) 439 kt (813 km/h; 505 mph)
Max level speed at S/L 326 kt (604 km/h; 375 mph)

Stalling speed, power off:
flaps and landing gear up 77 kt (143 km/h; 89 mph)
flaps and landing gear down 68 kt (126 km/h; 79 mph)

approx 1,600 n miles (2,963 km; 1,841 miles)
g limits: at 1,179 kg (2,600 lb) +9/–6
at 1,451 kg (3,200 lb) +7.3/–4.9
UPDATED

TEAM ROCKET

TEAM ROCKET
80 CR 406, Taylor, Texas 76574
Tel: (+1 512) 365 81 31
Fax: (+1 512) 352 50 80
e-mail: mark@teamrocketaircraft.com
Web: http://www.teamrocketaircraft.com
SALES MANAGER: Mark Frederick

The F1 kitbuilt produced by Team Rocket was developed by International High Performance Aircraft of Lobeček 732, CZ-728 01 Kralupy nad Vlatava, Czech Republic (http://www.hpai.cz), and Venice, Florida (http://www.ihpaircraft.com).

VERIFIED

Team Rocket F1 sportplane making its debut at EAA AirVenture, Oshkosh, July 2001 *(Jane's/Paul Jackson)*
0110735

TEAM ROCKET F1 and F2

TYPE: Tandem-seat sportplane kitbuilt; side-by-side sportplane kitbuilt.

PROGRAMME: Developed from Van's RV-4 by way of Harmon Rocket. First flight (in USA) November 2000 (N121JC). Second flew in Australia, July 2001.

CURRENT VERSIONS: **Team Rocket F1**: Tandem-seat version. *Description applies to this version, unless otherwise indicated.*

Team Rocket F2: Side-by-side two-seat version, announced at Oshkosh in July 2002. Based on major components from the F1, including wings, tail and landing gear. Quick-build kits available.

CUSTOMERS: Sales totalled 92 by April 2002, of which 84 had been delivered and seven were flying.

COSTS: Kit, excluding engine, propeller and avionics, US$29,999 fast-build, or US$19,999 unassembled (2003). F2: US$24,990; fast build kit US$33,595 (2003).

DESIGN FEATURES: Intended to combine high performance and low price. Low-wing monoplane with closely cowled engine. Constant-chord wings, tapered tailplane and moderately sweptback fin.
Meets '51 per cent' rule. Quoted build time 2,000 hours (1,200 for fast-build kit).

FLYING CONTROL: Conventional and manual. Flaps. Horn-balanced rudder and elevators.

STRUCTURE: Principally of aluminium, primed with epoxy chromate for corrosion resistance. Titanium mainwheel legs; glass fibre engine cowling and wingtips.

LANDING GEAR: Tailwheel type; fixed. Spring titanium mainwheel legs. Mainwheels 5.00-5.

POWER PLANT: One flat-six piston engine rated between 175 and 224 kW (235 and 300 hp), driving a two- or three-blade propeller recommended for both versions; typically prototypes have Textron Lycoming IO-540 and three-blade, constant-speed MTV-9-B-C/C193-109 propeller. Fuel capacity 197 litres (52.0 US gallons; 43.3 Imp gallons).

ACCOMMODATION: Two persons in tandem, beneath rearwards-sliding canopy. Fixed windscreen.

DIMENSIONS, EXTERNAL:
Wing span: F1 6.56 m (22 ft 6 in)
F2 7.62 m (25 ft 0 in)
Wing aspect ratio: F1 4.8
F2 5.5
Length overall: F1, F2 6.40 m (21 ft 0 in)

DIMENSION, INTERNAL:
Cockpit max width: F2 1.16 m (3 ft 9¾ in)

AREAS:
Wings, gross: F1 9.85 m² (106.0 sq ft)
F2 10.59 m² (114.0 sq ft)

WEIGHTS AND LOADINGS:
Weight empty: F1 544 kg (1,200 lb)
F2 590 kg (1,300 lb)
Max T-O weight: F1 907 kg (2,000 lb)
F2 952 kg (2,100 lb)
Max wing loading: F1 92.1 kg/m² (18.87 lb/sq ft)
F2 89.9 kg/m² (18.42 lb/sq ft)
Max power loading: F1 4.06 kg/kW (6.67 lb/hp)
F2 4.26 kg/kW (7.00 lb/hp)

PERFORMANCE (224 kW; 300 hp engine):
Max level speed 220 kt (407 km/h; 253 mph)
Cruising speed at 75% power
200 kt (370 km/h; 230 mph)
Stalling speed 47 kt (87 km/h; 54 mph)
Max rate of climb at S/L 1,067 m (3,500 ft)/min
T-O run 92 m (300 ft)
Landing run 214 m (700 ft)
Range at 55% power
1,000 n miles (1,850 km; 1,150 miles)
UPDATED

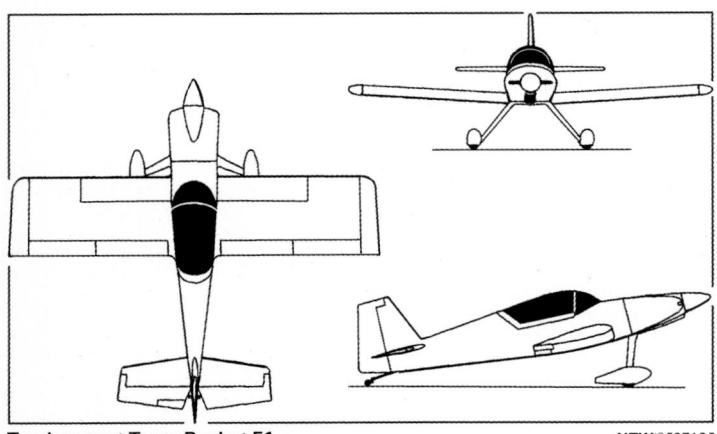

Tandem-seat Team Rocket F1   *NEW*/0527125   Side-by-side Team Rocket F2   *NEW*/0527124

TIGER

TIGER AIRCRAFT LLC
226 Pilot Way, Martinsburg, West Virginia 25401
Tel: (+1 304) 267 10 00
Fax: (+1 304) 262 00 69
e-mail: billcrum@tigeraircraft.com
Web: http://www.tigeraircraft.com
CEO AND CHAIRMAN: Bob Crowley
VICE-CHAIRMAN: Peter Lo
CFO: Ron Shade
MARKETING: Bill Crum

On 12 May 2000, Tiger Aircraft acquired the Type Certificate for the AG-5B Tiger, last employed by American General Aircraft Corporation. It also holds the AA-1 Type Certificate and intends to produce new designs. The company employed 63 at the end of 2001 with plans to increase workforce to 85 over following 12 months. In late 2002, it was owned jointly by Teleflex (30 per cent) and Taiwanese investors.

Production certification for Tiger granted by FAA on 3 September 2002.

UPDATED

TIGER AG-5B TIGER

TYPE: Four-seat lightplane.

PROGRAMME: Original two-seat BD-1 designed by Jim Bede; prototype first flown 11 July 1963; developed into AA-1 by American Aviation (later bought by Grumman and then by Gulfstream American), which built 461 AA-1s, 470 AA-1As, 680 AA-1Bs and 211 AA-1Cs. Four-seat AA-5 certified 12 November 1971 and 3,054 built before designs sold to American General Aircraft in 1989. Updated AG-5B Tiger first flown 21 April 1990; certified 21 September 1990; company sold to Teleflex in early 1992 but later filed for bankruptcy in January 1994 after building 89 examples.

First Tiger Aircraft AG-5B (N999TE) made maiden flight 9 July 2001. Symbolically delivered to new owner,

Herb Hortman, 26 July 2001 at AirVenture 2001. Certification completed 3 December 2001, when first three aircraft handed over. In late 2002 Tiger Aircraft was negotiating with Chengdu Aircraft Industrial Corporation (CAC) of China with a view to licensed assembly of the AG-5B for Chinese commercial pilot training schools.

CUSTOMERS: 41 orders by July 2001. Company planned to build 30 in 2002, of which 15 (in total) registered by October 2002, of which nine still held by Tiger Aircraft; 90 planned in 2003.

COSTS: US$219,500 (2002), including pilot training (to IFR standard if already holder of PPL) and first annual inspection.

DESIGN FEATURES: Conventional low-wing monoplane with constant-chord wings and tailplane and moderately sweptback fin with fillet. Extensive use of honeycomb and adhesive bonding throughout the structure.

FLYING CONTROLS: Conventional and manual with adjustable tab on each aileron and rudder. Electrically operated flaps, maximum deflection 45°.

STRUCTURE: Semi-monocoque fuselage of aluminium honeycomb. Cantilever low wing with tubular single spar, section NACA 64₂415 (modified), clad in aluminium skin. Dihedral 5°, incidence 1° 25′.

LANDING GEAR: Tricycle type; fixed. Mainwheels 6.00-6, tyre pressure 2.34 bar (34 lb/sq in); nosewheel 5.00-5, tyre pressure 1.72 bar (25 lb/sq in). Cleveland hydraulic brakes.

POWER PLANT: One 134 kW (180 hp) Textron Lycoming O-360-A4K flat-four, driving a two-blade fixed-pitch Sensenich 76EM8S-10-0 propeller. Fuel in two wing tanks, combined capacity 199 litres (52.6 US gallons; 43.8 Imp gallons), of which 193 litres (51.0 US gallons; 42.5 Imp gallons) are usable. Oil capacity 7.6 litres (2.0 US gallons; 1.7 Imp gallons).

ACCOMMODATION: Pilot and three passengers in individual pairs of leather seats; baggage area behind rear seats. Fixed wraparound windscreen; entry to cockpit via sliding canopy.

SYSTEMS: 24 V electrical system.

AVIONICS: Full Garmin IFR panel standard.
 Comms: Dual GNS 430 com/nav with VOR/LOC/GPS/GPS-LOC; GMA 340 audio panel with marker beacon and four-place intercom system; GTX 327 digital transponder, and ground clearance system
 Flight: Dual Garmin 106A glideslope indicators; S-Tec 30 two-axis autopilot with altitude hold.
 Instrumentation: Sigma/Tek electronic instrument cluster comprising vacuum, load amp, oil pressure, oil temperature, fuel pressure and CHT gauges; Davtron M800 chronometer and Datcom Hobbs meter.

EQUIPMENT: Standard equipment includes leather interior; leather-wrapped control yokes, machined aluminium instrument panel with backlit instruments and panel/

Fifth Tiger Aircraft AG-5B Tiger *(Jane's/Paul Jackson)*   NEW/0533673

instrument light dimmers, cabin dome light, map light, and flush-mounted anti-collision beacons and landing lights.

DIMENSIONS, EXTERNAL:
Wing span	9.60 m (31 ft 6 in)
Wing chord, constant	1.35 m (4 ft 5¼ in)
Wing aspect ratio	7.1
Length overall	6.71 m (22 ft 0 in)
Height overall	2.44 m (8 ft 0 in)
Tailplane span	2.65 m (8 ft 8½ in)
Wheel track	2.51 m (8 ft 3 in)
Wheelbase	1.64 m (5 ft 4½ in)
Propeller diameter	1.98 m (6 ft 6 in)

DIMENSIONS, INTERNAL:
Cabin: Length	1.98 m (6 ft 6 in)
Max width	1.04 m (3 ft 5 in)
Max height	1.23 m (4 ft 0¼ in)
Floor area	2.18 m² (23.5 sq ft)
Baggage hold volume	0.34 m³ (12.0 cu ft)

AREAS:
Wings, gross	13.02 m² (140.1 sq ft)

WEIGHTS AND LOADINGS:
Weight empty	680 kg (1,500 lb)
Baggage capacity: four-seat	54 kg (120 lb)
two-seat	154 kg (340 lb)
Max T-O weight	1,089 kg (2,400 lb)
Max wing loading	83.6 kg/m² (17.13 lb/sq ft)
Max power loading	8.12 kg/kW (13.33 lb/hp)

PERFORMANCE:
Never-exceed speed (V_NE)	173 kt (93 km/h; 200 mph)
Max operating speed (V_MO)	148 kt (274 km/h; 170 mph)
Max cruising speed at 75% power at 2,590 m (8,500 ft)	143 kt (265 km/h; 165 mph)
Stalling speed: flaps up	56 kt (104 km/h; 65 mph)
flaps down	53 kt (99 km/h; 61 mph)
Max rate of climb at S/L	259 m (850 ft)/min
Service ceiling	4,205 m (13,800 ft)
T-O run	260 m (854 ft)
T-O to 15 m (50 ft)	472 m (1,550 ft)
Landing from 15 m (50 ft)	341 m (1,120 ft)
Landing run	125 m (410 ft)
Range: no reserves	681 n miles (1,261 km; 783 miles)
with 45 min reserves	572 n miles (1,059 km; 658 miles)
Endurance, no reserves	4 h 42 min

UPDATED

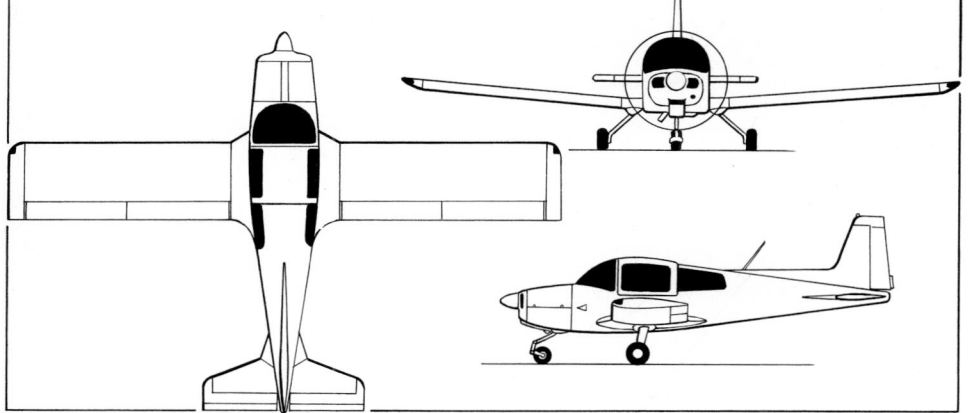

Tiger Aircraft AG-5B Tiger general arrangement *(Jane's/James Goulding)*   0131346

TKEF

THE KING'S ENGINEERING FELLOWSHIP
Municipal Airport, Orange City, Iowa 51041
Tel: (+1 712) 737 44 44
e-mail: tkef@angelaircraft.com
Web: http://www.angelaircraft.com
PRESIDENT: Carl A Mortenson
CHIEF ENGINEER: Ed Mortenson

MANUFACTURING:
Angel Aircraft Corporation
1410 Arizona Place SW,
Orange City, Iowa 5141
Tel: (+1 712) 737 33 44
Fax: (+1 712) 737 33 99

MARKETING:
Kansair Inc, 1533 20th Park Place, Emporia, Kansas 66801
Tel: (+1 620) 342 35 00
e-mail: kansair@osprey.net

Angel, developed by The King's Engineering Fellowship (TKEF) through donations, and designed by Carl Mortenson, follows earlier Evangel. TKEF holds Type Certificate.

NEW ENTRY

TKEF 44 ANGEL
TYPE: Light utility twin-prop transport.

PROGRAMME: Design started November 1972; prototype construction began January 1977; prototype (N44KE) built on production tooling; first flight 13 January 1984; structural testing began 1990; certification to FAR Pt 23 (Normal category) for day, night, VFR and IFR conditions gained 20 October 1992. Production began December 1993; second aircraft completed July 1994 but held in abeyance, pending acquisition of larger production premises. Second Angel eventually registered in August 2002 to Kansair Inc as N442KB, with 2002 build date quoted.

DESIGN FEATURES: Designed partly, but not exclusively, for missionary aviation; other commercial uses include air

taxi, air ambulance, air observation/patrol, fishery/pipeline/border inspection, tracking, mining/oil/rubber/forestry operations, ranching and firefighting control. Design goals include STOL capability, operation from soft and rough fields, easy repair in field, and easy loading of bulky cargo.

Vortex generators on wing outboard panels for enhanced lift. Crashworthiness built and tested in key structures; seats dynamically tested to absorb 20 *g* vertically and 26 *g* horizontally; cabin structure includes areas of double-wall and tested for overturn impact survivability.

Wing section NACA 23018-23010 with modified leading-edge; sweepback 15° 36′ at leading-edge, 11° at quarter-chord; dihedral 5° 24′; incidence 3° at root, –0° 22′ at tip.

FLYING CONTROLS: Manual. Actuation by cables. Almost full-span, hydraulically actuated semi-Fowler flaps deflecting to maximum 37°. Lateral control by multiple small-plate spoilers immediately forward of flaps; spoilers outboard of flaps to assist during single-engined flight; elevators and rudder, both with trim tabs.

STRUCTURE: Riveted aluminium alloy and welded 4130/4340 steel tube; wing has built-up capstrip spars and 19 die-formed ribs each side; broad-chord fin and rudder with large dorsal fin; CFRP spinners; GFRP/epoxy in some areas.

LANDING GEAR: Retractable tricycle type. Electrohydraulic retraction, mainwheels inward into wingroots, nosewheel rearward. Emergency extension by handpump or gravity. Cleveland wheels; McCreary mainwheel tyres size 8.50-10 (pressure 2.41 bar; 35 lb/sq in), McCreary or Goodyear nosewheel tyre size 8.50-6 (pressure 1.03 bar; 15 lb/sq in). Cleveland disc brakes. Min ground turning radius 5.11 m (16 ft 9 in).

POWER PLANT: Two 224 kW (300 hp) Textron Lycoming IO-540-M1C5 flat-six engines, each driving a Hartzell HC-E3YR-2ATLF/FCL7468 three-blade, constant-speed (hydraulic), feathering pusher propeller. Two wing fuel

tanks, gravity flow cross-fed, with total capacity 849 litres (224 US gallons; 187 Imp gallons). Refuelling point in top of each wing. Oil capacity 22.7 litres (6.0 US gallons; 5.0 Imp gallons).

ACCOMMODATION: Enclosed cabin seating up to eight persons, including pilot. Five seats can be removed for carrying cargo, including four 208 litre (55 US gallon; 45.8 Imp gallon) drums. Rearmost bench seat is fixed. Four large windows and one smaller circular window on each side of cabin. Horizontally divided clamshell door on port side at front of cabin: emergency exit on starboard side. Heating and window air vents standard. Compartment for baggage at rear of fuselage, with door on port side.

SYSTEMS: Hydraulic system, with electric pump, for landing gear and flap actuation. Electrical system with two alternators includes 12 V DC battery in nose.

AVIONICS: IFR. Bendix/King and Garmin avionics, including twin com/nav transceivers, glideslope, ADF, transponder, ELT and GPS Weather radar, Loran C and HF com optional.

DIMENSIONS, EXTERNAL:
Wing span	12.16 m (39 ft 10¾ in)
Wing aspect ratio	7.06
Wing chord: at root	2.59 m (8 ft 6 in)
at tip	1.00 m (3 ft 3½ in)
Length overall	10.21 m (33 ft 6 in)
Height overall	3.51 m (11 ft 6 in)
Tailplane span	5.30 m (17 ft 4½ in)
Wheel track	3.95 m (12 ft 11¾ in)
Wheelbase	4.67 m (15 ft 4 in)
Propeller diameter	1.93 m (6 ft 4 in)
Propeller ground clearance at T-O	0.23 m (9.01 in)
Cabin door (port, fwd): Height	0.91 m (3 ft 0 in)
Width	1.04 m (3 ft 5 in)
Height to sill	0.91 m (3 ft 0 in)
Baggage door: Height	0.41 m (1 ft 4 in)
Width	0.61 m (2 ft 0 in)
Height to sill	1.17 m (3 ft 10 in)

Emergency exit door (stbd, fwd):	
Height	0.48 m (1 ft 7 in)
Width	1.04 m (3 ft 5 in)
Max height	1.32 m (4 ft 4 in)

DIMENSIONS, INTERNAL:
Cabin, incl flight deck:

Length	3.51 m (11 ft 6 in)
Max width	1.07 m (3 ft 6 in)
Max height	1.14 m (3 ft 9 in)
Volume	2.38 m³ (84.0 cu ft)
Baggage compartment volume (aft)	0.28 m³ (10 cu ft)

AREAS:

Wings, gross	20.94 m² (225.4 sq ft)
Ailerons (total)	0.39 m² (4.20 sq ft)
Trailing-edge flaps (total)	4.05 m² (43.60 sq ft)
Spoilers (total)	0.37 m² (4.00 sq ft)
Rudder, incl tab	1.41 m² (15.15 sq ft)
Tailplane (total)	6.23 m² (67.02 sq ft)
Elevators (total, incl tabs)	2.76 m² (29.67 sq ft)

WEIGHTS AND LOADINGS:

Weight empty	1,760 kg (3,880 lb)
Baggage capacity (rear)	91 kg (200 lb)
Max fuel weight	612 kg (1,350 lb)
Max T-O and landing weight	2,631 kg (5,800 lb)
Max wing loading	125.6 kg/m² (25.73 lb/sq ft)
Max power loading	5.87 kg/kW (9.67 lb/hp)

PERFORMANCE:

Never-exceed speed (V_{NE})	209 kt (387 km/h; 240 mph)
Max level speed at S/L	180 kt (333 km/h; 207 mph)
Cruising speed at 65% power at FL115	169 kt (313 km/h; 195 mph)
Stalling speed, flaps and gear down:	
power off	57 kt (106 km/h; 66 mph)
power on	51 kt (95 km/h; 59 mph)
Max rate of climb at S/L	410 m (1,345 ft)/min
Rate of climb, OEI	60 m (196 ft)/min
Service ceiling	6,265 m (20,560 ft
Service ceiling, OEI	1,580 m (5,180 ft)

T-O run	201 m (660 ft)
T-O to 15 m (50 ft)	428 m (1,405 ft)
Landing from 15 m (50 ft)	320 m (1,050 ft)
Landing run	178 m (585 ft)

Range with max fuel, VFR reserves:

65% power	1,263 n miles (2,339 km; 1,453 miles)
45% power	1,605 n miles (2,972 km; 1,847 miles)
35% power	1,720 n miles (3,185 km; 1,979 miles)
Endurance, VFR reserves: with eight occupants	3 h

with max fuel and three occupants:

65% power	7 h 54 min
45% power	11 h 18 min
35% power	13 h 6 min

NEW ENTRY

TREK

TREK AEROSPACE INC

931 Benicia Avenue, Sunnyvale, California 94085
Tel: (+1 408) 732 17 00
Fax: (+1 408) 730 45 88
e-mail: info@solotrek.com
Web: http://www.solotrek.com
CEO: Michael W Moshier
VICE-PRESIDENT AND CFO: Harry Falk
CHIEF ENGINEER: Robert Bulaga
PUBLIC RELATIONS OFFICER: Michelle Garrett

Company was founded in 1996 by Michael Moshier to produce a 'strap-on' personal flying vehicle. Known as Millennium Jet Inc until 2002. Once development of a military version of the initial SoloTrek has been completed, Trek will begin work on the two-seat, enclosed DuoTrek and related Mule UAV, both of which employ the same ducted fan technology.

However, following the crash of the second prototype and failure to meet a DARPA deadline, the company laid-off its staff on 31 December 2002.

NEW ENTRY

TREK SOLOTREK

TYPE: Exoskelitor flying vehicle.
PROGRAMME: Begun in 1996 as a private venture; awarded US Defense Advanced Research Projects Agency (DARPA) 36-month, US$5 million development contract in December 2000; NASA Ames assisting with development of stability augmentation system. Prototype registered N102MJ in September 1999; first tethered flight 25 November 2001 preceded by 36 hours of static tests; first controlled, manned (still tethered) flight 18 December 2001; initial untethered flight 15 March 2002. Retired from flight testing in mid-2002 and offered for sale in deliberately non-flyable condition in January 2003.

Pre-series vehicle (002), in production configuration, was due to fly in mid-2002, but eventually did so in December 2002. However, it was seriously damaged almost immediately afterwards when the rotors ingested the safety tether and the craft fell from a height of 2 m (5 ft).

COSTS: Target cost of possible civil version will be equivalent to a top-of-the-range sports car.

DESIGN FEATURES: Exoskelitor flying vehicle (XFV), or strap-on personal aircraft; initial variant optimised for military and law-enforcement users. Twin, counter-rotating, fixed-pitch ducted fans provide 40 per cent more lift than equivalent unducted propellers and are tilted in conjunction with power variation to achieve vertical take-off, climb, forward flight and vertical landing; maximum fan forward tilt angle 35°. Time between overhauls 1,000 hours.

Central structure supports pilot, engine, fans and tail-leg with fin for stability in forward flight. Engine located behind pilot; power transmission by short driveshaft with universal joints to self-lubricating main gearbox above and behind pilot's head; twin driveshafts to fans, each with satellite gearboxes. Fuel tank below engine. Small cooling fan for main gearbox.

FLYING CONTROLS: No conventional aerodynamic surfaces. Ducted fans tilted by electromechanical actuators; small vanes, electromechanically actuated, behind each fan for precise control at slow speed and hover. Three-axis trim.

STRUCTURE: Welded chromoly steel tube with non-structural composites fairings and composites ducted fan blades; limited use of carbon fibre, Kevlar and titanium, including carbon-Kevlar driveshafts.

LANDING GEAR: Twin wheels and rear skid (tripod) on prototype; pre-series aircraft with two-point (bipod) landing gear of main support plate and rear wheel.

POWER PLANT: Prototype has 70.8 kW (95 hp) Hirth F30. Production version assigned undisclosed 89 kW (120 hp) rotary engine driving two five-blade, fixed-pitch fans; candidates include 93 kW (125 shp) Williams WTS-125. Fuel capacity 57 litres (15.0 US gallons; 12.5 Imp gallons) of kerosene.

ACCOMMODATION: One person in standing position, supported by safety harness. Two hand-levers for engine and flight control: engine/vertical axis, left; longitudinal and lateral axes, plus trim, right.

SYSTEMS: Fully FBW stability control system is programmed with pilot's weight before each flight. Electrical system, 24 V DC.

AVIONICS: *Instrumentation:* Helmet-mounted display for all necessary data.

EQUIPMENT: Pilot's zero/zero parachute extraction system for emergency recovery.

DIMENSIONS, EXTERNAL:

Length (height) overall	1.575 m (5 ft 2 in)
Width over fan ducts	2.64 m (8 ft 8 in)

Prototype Trek SoloTrek personal air vehicle
NEW/0131752

WEIGHTS AND LOADINGS:

Weight empty	147 kg (325 lb)
Max fuel weight	44 kg (98 lb)
Max T-O weight	317 kg (700 lb)

PERFORMANCE (estimated):

Max level speed	70 kt (130 km/h; 81 mph)
Normal cruising speed	50-60 kt (93-111 km/h; 58-69 mph)
Hovering ceiling, OGE	2,440 m (8,000 ft)
Range	120 n miles (222 km; 138 miles)
Endurance	more than 2 h

NEW ENTRY

TURBINE DESIGN

TURBINE DESIGN INC

1335 Saratoga Street, Deland, Florida 32724
Tel: (+1 386) 736 82 62
Fax: (+1 386) 738 05 10
Web: http://www.turbinedesign.com

Company activities include US distribution of Khrunichev T-411 Aist (which see in Russian section); sales of a kit version of Eurospace F-15-F Excalibur (see 1998-99 *Jane's*, Italian section) using 10 uncompleted airframes obtained from Sokol of Russia (which see), of which a turboprop Rolls-Royce 250-engined F-15-F is designated **TD-4**; and Maxima conversion of Piper Malibu and Mirage with Walter M 601 turboprop.

NEW ENTRY

TURBINE DESIGN TD-2 TEMPEST

TYPE: Tandem-seat turboprop sportplane kitbuilt.
PROGRAMME: First aircraft almost complete by April 2002. Further example made public debut at Sun 'n' Fun in April 2002 in disassembled state.
DESIGN FEATURES: Optimised for private owner-builder, requiring high performance; suitable for aerobatics or

Incomplete TD-2 Tempest displayed by Turbine Design at Lakeland, Florida, April 2002 (*Jane's/Paul Jackson*)
NEW/0137386

touring. Low wing with winglets; deep centre fuselage; mid-mounted tailplane. Wing sweep at quarter-chord 0°; taper ratio 0.54; fin sweepback 36° 35′.

FLYING CONTROLS: Conventional and manual. Fowler flaps.

STRUCTURE: Carbon fibre throughout; epoxy bonding. Two-spar wing; monocoque fuselage with oval reinforcement frame behind cockpit.

LANDING GEAR: Tricycle type, retractable.

POWER PLANT: One 560 kW (751 shp) Walter M 601 Turboprop driving a five-blade, variable-pitch Avia propeller. Fuel in wing tanks, normal capacity 613 litres (162 US gallons; 135 Imp gallons); optional capacity 719 litres (190 US gallons; 158 Imp gallons).

ACCOMMODATION: Two persons in tandem, rear seat raised 15 cm (6 in), beneath single-piece canopy. Cockpit pressurised.

DIMENSIONS, EXTERNAL:

Wing span	9.63 m (31 ft 7 in)
Wing mean chord	1.37 m (4 ft 6 in)
Wing aspect ratio	8.0
Length overall	8.36 m (27 ft 5 in)
Fuselage max width	0.81 m (2 ft 8 in)
Tailplane span	3.43 m (11 ft 3 in)

AREAS:

Wings, gross	11.61 m² (125.0 sq ft)
Vertical tail surfaces (total)	1.66 m² (17.90 sq ft)
Horizontal tail surfaces (total)	2.42 m² (26.00 sq ft)

WEIGHTS AND LOADINGS:

Weight empty	1,026 kg (2,262 lb)
Max T-O weight	1,769 kg (3,900 lb)
Max wing loading	152.3 kg/m² (31.20 lb/sq ft)
Max power loading	3.16 kg/kW (5.19 lb/hp)

PERFORMANCE (estimated):

Speed at FL250: max level	323 kt (598 km/h; 372 mph)
70% power	279 kt (517 km/h; 321 mph)
Stalling speed: flaps up	85 kt (158 km/h; 98 mph)
flaps down	69 kt (128 km/h; 80 mph)
Max rate of climb at S/L	1,646 m (5,400 ft)/min
Cruising altitude	7,620 m (25,000 ft)
T-O run	464 m (1,520 ft)
Landing run	441 m (1,445 ft)
Range: at max level speed	1,380 n miles (2,557 km; 1,589 miles)
at 70% power	1,477 n miles (2,735 km; 1,700 miles)
g limits (ultimate)	±10

NEW ENTRY

ULLMANN

ULLMANN AIRCRAFT COMPANY

5132 Willow Point, Wichita, Kansas 67220
Tel/Fax: (+1 877) 227 19 85
e-mail: brian.ullmann@ullmannaircraft.com
Web: http://www.ullmannaircraft.com
PRESIDENT AND CEO: Bill Ullmann
CHIEF ENGINEER: Brian Ullmann

The Ullmann Aircraft Company's first aircraft is the Panther.

UPDATED

ULLMANN PANTHER

TYPE: Four-seat kitbuilt.

PROGRAMME: Design began 15 April 1997 and construction of first prototype on 2 June 1998. Formally announced at AirVenture '99; mockup displayed at Sun 'n' Fun 2000. Part-completed prototype displayed at Sun 'n' Fun 2001 and again at AirVenture 01, with first flight then expected before end 2001; this later slipped to June 2002 but had not taken place by November 2002, when it was said to be expected in the following month. Kit deliveries planned to begin July 2003.

COSTS: Development costs estimated at US$500,000 in 1997. Kit US$51,000, excluding engine, propeller, avionics, instruments, interior or paint; four sub-kits are available (2003 prices).

DESIGN FEATURES: Cantilever high-wing monoplane of rectangular fuselage cross-section. Constant-chord wings and tailplane; sweptback fin with fillet. High-performance kitbuilt meeting FAR Pt 23 certification and 51 per cent homebuilt rule. Many parts supplied pre-formed, pre-drilled or pre-assembled. Quoted build time 1,200 to 1,600 hours.

Aerofoil section NACA 64₂A215, dihedral 1° 30′; incidence 1°; no twist.

FLYING CONTROLS: Conventional and manual. Horn-balanced elevator and rudder, both with electric trim. Single-slotted Fowler flaps. Dorsal fillet and drooped outboard cuff assists spin resistance.

STRUCTURE: All-metal wing with square tips. Fuselage of steel tube covered with aluminium sheet.

LANDING GEAR: Tricycle type; fixed, Wittmann sprung steel main legs. Cleveland wheels; disc brakes on mainwheels. Nosewheel 5.00-5, steerable ±60°. Mainwheels 6.00-6, all with 2.07 bar (30.0 lb/sq in) tyre pressure. Pressure Recovery wheel fairings.

POWER PLANT: Prototype has one 224 kW (300 hp) Teledyne Continental IO-550-L driving a three-blade Hartzell constant-speed propeller; engine mount also compatible with IO-520-L. Mounts and cowlings for 194 kW (260 hp) Textron Lycoming IO-540 and 224 kW (300 hp) Zehrbach engines under consideration. Fuel in two integral wing

Ullmann Panther prototype nearing completion *NEW*/0526917

leading-edge tanks, combined capacity 277 litres (73.0 US gallons; 60.8 Imp gallons); or alternative long-range tanks combined capacity 390 litres (103 US gallons; 85.8 Imp gallons), of which 386 litres (102 US gallons; 84.9 Imp gallons) are usable. Gravity refuelling point on wing. Oil capacity 7.6 litres (2.0 US gallons; 1.7 Imp gallons).

ACCOMMODATION: Pilot and three passengers in two pairs of seats. Split doors on each side of fuselage; single baggage door at rear of cabin.

SYSTEMS: 12 V 24 Ah Concorde RG-25XC battery. Electric fuel pump.

AVIONICS: Prototype has two Bendix/King KX 155A com radios, Bendix/King KLN 94 VOR/DME and GPS.

DIMENSIONS, EXTERNAL:

Wing span	10.43 m (34 ft 2½ in)
Wing chord, constant	1.16 m (3 ft 9½ in)
Wing aspect ratio	9.0
Length overall	7.85 m (25 ft 9 in)
Fuselage max width	1.29 m (4 ft 2¼ in)
Height overall	3.15 m (10 ft 4 in)
Tailplane span	3.43 m (11 ft 3 in)
Wheel track	2.84 m (9 ft 4 in)
Wheelbase	2.34 m (7 ft 8 in)
Propeller diameter	1.98 m (6 ft 6 in)
Propeller ground clearance	0.43 m (1 ft 5 in)
Passenger doors (each): Height	1.07 m (3 ft 6 in)
Width	1.05 m (3 ft 5¼ in)
Height to sill	0.88 m (2 ft 10½ in)
Baggage door: Height	0.45 m (1 ft 5¾ in)
Width	0.54 m (1 ft 9¼ in)
Height to sill	0.89 m (2 ft 11 in)

DIMENSIONS, INTERNAL:

Cabin: Length	2.32 m (7 ft 7½ in)

Max width	1.27 m (4 ft 2 in)
Max height	1.22 m (4 ft 0 in)
Volume	2.97 m³ (105.0 cu ft)
Baggage hold volume	0.61 m³ (21.5 cu ft)

AREAS:

Wings, gross: flaps in	12.08 m² (130.0 sq ft)
flaps out	13.19 m² (142.0 sq ft)
Ailerons (total)	1.48 m² (15.88 sq ft)
Trailing-edge flaps (total)	2.37 m² (25.52 sq ft)
Fin	1.32 m² (14.25 sq ft)
Rudder, incl tab	0.47 m² (5.08 sq ft)
Tailplane	2.35 m² (25.31 sq ft)
Elevators, (total incl tab)	0.84 m² (9.00 sq ft)

WEIGHTS AND LOADINGS (estimated):

Weight empty	798 kg (1,760 lb)
Baggage capacity	54 kg (120 lb)
Max payload	363 kg (800 lb)
Max fuel weight	281 kg (620 lb)
Max T-O and landing weight	1,360 kg (3,000 lb)
Max wing loading	112.7 kg/m² (23.08 lb/sq ft)
Max power loading	6.09 kg/kW (10.00 lb/hp)

PERFORMANCE (estimated):

Never-exceed speed (VNE)	250 kt (463 km/h; 287 mph)
Max operating speed	200 kt (370 km/h; 230 mph)
Stalling speed, power off, flaps down	58 kt (108 km/h; 67 mph)
Range, 45 min reserves: with max fuel	1,227 n miles (2,272 km; 1,412 miles)
with max payload	823 n miles (1,524 km; 947 miles)
Endurance	6 h 8 min
g limits	+3.8/−1.52

UPDATED

UNITED STATES AIR FORCE

AIR FORCE MATERIEL COMMAND

3753 Chidlaw Road, Wright-Patterson AFB, Ohio 45433-5006
Tel: (+1 937) 257 63 06
Fax: (+1 937) 257 25 58
e-mail: HQAFMC.PA@wpafb.af.mil
Web: http://www.afmc.wpafb.af.mil
COMMANDER, AFMC: General Lester L Lyles
DIRECTOR, PUBLIC AFFAIRS: Col Donna Pastor

Formed 1 July 1992, incorporating former Air Force Logistics Command and Air Force Systems Command. Resulting Air Force Materiel Command (AFMC) is composed of four product centres, each responsible for developing and acquiring specific weapon systems with the assistance of the Air Force Research Laboratory at Wright-Patterson AFB, Ohio and the Air Force Office of Scientific Research at Bolling AFB, Washington, DC. These comprise the Aeronautical Systems Center, Wright-Patterson AFB, Ohio; the Air Armament Center, Eglin AFB, Florida; the Electronic Systems Center, Hanscom AFB, Massachusetts;

and the Space and Missile Systems Center, Los Angeles AFB, California.

AFMC also has two test establishments, namely Air Force Flight Test Center at Edwards AFB, California; and Arnold Engineering Development Center at Arnold AFB, Tennessee. These are responsible for evaluation of systems under development and in the field.

Air Logistics Centers (Ogden ALC at Hill AFB, Utah; Oklahoma City ALC at Tinker AFB, Oklahoma and Warner Robins ALC at Robins AFB, Georgia) provide weapon system sustainment and maintenance services for just under

50 types of aircraft, with a total inventory of about 6,200 airframes and 20,000 engines. Finally, two specialist organisations (Aerospace Maintenance and Regeneration Center at Davis-Monthan AFB, Arizona, and Air Force Security Assistance Center at Wright-Patterson AFB, Ohio) provide for storage, regeneration, reclamation and disposal of aircraft and associated aerospace equipment.

VERIFIED

REPLACEMENT INTERDICTOR AIRCRAFT (RIA)

Exploratory studies, for an aircraft to replace the Lockheed Martin F-117A Nighthawk and Boeing F-15E Eagle in the interdiction role and also augment the bomber force, have been launched by the US Air Force, with Boeing, Lockheed Martin and Northrop Grumman undertaking studies into a potential RIA. Service entry approximately 2015-20. No recent news received.

UPDATED

NEW TANKER PROGRAMME

Acquisition of a new tanker aircraft to replace USAF Boeing KC-135 Stratotanker, could be eventual outcome of a preliminary tanker replacement study for 2005 (TRS-05) undertaken by Air Mobility Command during 2000; this expected to lead to follow-on analysis of alternatives (AOA) study lasting 18 to 24 months. Various options likely to be put forward by Boeing, Lockheed Martin and Raytheon (in collaboration with EADS), with earliest IOC anticipated in 2013-14, when retirement of KC-135 expected to begin. No recent news received.

UPDATED

UNITED STATES NAVY

NAVAL AIR SYSTEMS COMMAND

47123 Buse Road, Unit Moffett Building, Patuxent River, Maryland 20650-1547
Tel: (+1 301) 757 78 25
Web: http://www.navair.navy.mil

NAVAIR has over 31,500 military and civilian employees and provides total life cycle support of all naval aviation weapon systems. Subordinate elements include test organisations at China Lake, California; Patuxent River, Maryland and Point Mugu, California, as well as aircraft repair and overhaul depots at Cherry Point, North Carolina; Jacksonville, Florida; and North Island, California. It is prime authority for the USAF/USN Lockheed Martin F-35 Joint Strike Fighter, via the Program Executive Office (PEO); other PEOs have responsibility for tactical aircraft programmes; air ASW, assault and special mission programmes; and strike weapons and unmanned aviation.

UPDATED

MULTIMISSION MARITIME AIRCRAFT (MMA)

In late 1997, US Navy initiated requirements study, lasting two years, into replacement for Lockheed Martin P-3C Orion maritime aircraft, as well as intelligence-gathering EP-3E, submarine-communications Boeing E-6 Mercury and tanker/transport Lockheed Martin KC-130 Hercules of US Marine Corps. Latter type being replaced by new KC-130J aircraft, but maritime requirement remains. Further exploratory study contracts awarded in July 2000, when it was anticipated that service entry of MMA could occur as early as 2010 and be no later than 2015.

Options considered by US Navy included new development aircraft, off-the-shelf purchase of commercial airliners such as Boeing 737, or executive jets like Gulfstream V, acquisition of new-build Lockheed Martin P-3 Orion 21s with new engines, propellers and avionics, and remanufacturing existing P-3 Orions. Latter idea seems increasingly unlikely, following award on 10 September 2002 of separate US$7 million contracts to Boeing and Lockheed Martin. These cover five-month Phase 1 of MMA Component Advanced Development (CAD) programme that is intended to provide US Navy with variety of concepts from which two or three will be selected to progress to 13-month second phase, followed by choice of single type for full-scale development in early 2004. Boeing studies centre around derivative of 737-700, while Lockheed Martin is concentrating on the Orion 21. Entry into service is now anticipated in 2012, with 150 to 200 MMAs expected to be acquired in total.

UPDATED

US LIGHT AIRCRAFT

US LIGHT AIRCRAFT CORPORATION

27080 Rancho Ballena Lane, Ramona, California 92065
Tel/Fax: (+1 760) 789 86 07
e-mail: j.millett@gte.net or sales@flyhornet.com
Web: http://www.flyhornet.com
PRESIDENT: James Millet

US Light Aircraft formed from partnership between Millett Engineering and Forespar Company in 1993.

VERIFIED

US LIGHT AIRCRAFT HORNET

TYPE: Tandem-seat ultralight kitbuilt.
PROGRAMME: Design began in 1991 and construction of prototype in 1992; first flight June 1993. Winner of 15 awards at Copperstate, Oshkosh and Sun 'n' Fun between 1993 and 1999. Qualifies under FAR Pt 21.191 as 51 per cent experimental kit.
CUSTOMERS: About 40 sold and 25 flying by mid-2000 (latest data supplied).
COSTS: Kit US$20,400, including engine; US$13,750 without engine (2003).
DESIGN FEATURES: 400 parts in kits include 7,500 factory-installed rivets. Quoted build time 300 hours.
FLYING CONTROLS: Conventional and manual. Electrically operated flaps with maximum deflection of 30°. Horizontal stabiliser trimmed electrically. Mass-balanced elevator.
STRUCTURE: Fabric-covered fuselage of 3 cm (1.25 in) 6063-T8 aluminium tube with tapered-base tubular longitudinal structure. Fabric-covered dual-spar wings with single bracing struts.
LANDING GEAR: Fixed tricycle type with 15 cm (6 in) Goodyear Airspring pneumatic suspension on mainwheels

US Light Aircraft Corporation Hornet *(Geoffrey P Jones)*   0100462

and 7.5 cm (3 in) Airspring on nosewheel; hydraulic brakes on mainwheels with aluminium Hegar wheels and 15×6-6 Cheng Shin tyres; nose tyre 13×5-6.
POWER PLANT: One 41.0 kW (55 hp) Hirth 2703 engine with electric start is standard, driving Tennessee two-blade, wooden pusher propeller of diameter up to 1.52 m (5 ft 0 in); 48.5 kW (65 hp) Hirth 2706 optional. Fuel capacity 68 litres (18.0 US gallons; 15.0 Imp gallons) in two tanks between spars.
EQUIPMENT: Optional BRS VLS 1050 ballistic parachute.
DIMENSIONS, EXTERNAL:

Wing span	8.38 m (27 ft 6 in)
Length overall	6.10 m (20 ft 0 in)
AREAS:	
Wings, gross	12.82 m² (138.0 sq ft)

WEIGHTS AND LOADINGS:

Weight empty	215 kg (475 lb)
Max T-O weight	454 kg (1,000 lb)
PERFORMANCE (41.0 kW; 55 hp engine):	
Never-exceed speed (V_{NE})	108 kt (201 km/h; 125 mph)
Max level speed	104 kt (193 km/h; 120 mph)
Normal cruising speed at 75% power	61 kt (113 km/h 70 mph)
Stalling speed, flaps down	35 kt (65 km/h; 40 mph)
Max rate of climb at S/L	305 m (1,000 ft)/min
T-O run	114 m (375 ft)
Landing run	76 m (250 ft)
Range, 10% reserves	304 n miles (563 km; 350 miles)

UPDATED

VAN'S

VAN'S AIRCRAFT INC

14401 NE Keil Road, Aurora, Oregon 97002
Tel: (+1 503) 678 65 45
e-mail: info@vansaircraft.com
Web: http://www.vansaircraft.com
PRESIDENT: Richard VanGrunsven

Company derives its name from its president's nickname and began in 1973 with the sale of plans (and, later, kits) for the RV-3 sporting homebuilt. More than 10,000 Van's kits sold by November 2002, of which 3,075 were then flying. The four-seat RV-10 was announced in October 2001.

The company constructed a new 5,110 m² (55,000 sq ft) facility at Aurora, Oregon, to which it moved during 2000; by June 2002 the workforce numbered 60. All Van's kits conform to the FAA's 51 per cent rule.

UPDATED

Van's RV-4 kitbuilt upgraded to Harmon Rocket configuration *(Jane's/Paul Jackson)*   NEW/0527012

VAN'S RV-4

TYPE: Tandem-seat kitbuilt.

PROGRAMME: First flight of prototype 21 August 1979. Plans and kits available to homebuilders.

CURRENT VERSIONS: **RV-4**: *As described.*

CUSTOMERS: Over 4,600 sets of plans and kits sold, with about 1,800 aircraft under construction and 1,033 RV-4s flying by 14 November 2002.

COSTS: Kit: US$12,805 (2003). Information pack: US$8.

DESIGN FEATURES: Cantilever low-wing monoplane with constant-chord, low aspect ratio wings, tapered horizontal tail surfaces and sweptback fin.

Wing section Van's Aircraft 135; dihedral 3° 30′; incidence 1°.

FLYING CONTROLS: Conventional and manual. Horn-balanced elevators; Frise ailerons; plain flaps. No tabs.

STRUCTURE: Generally of light alloy. I-beam main wing spar and pressed ribs; moulded glass fibre wingtips and engine cowling.

LANDING GEAR: Tailwheel type; fixed cantilever tapered steel spring main gear struts. Glass fibre speed fairings on mainwheels, tyre size 5.00-5; steerable tailwheel, tyre size 6 in.

POWER PLANT: One 112 kW (150 hp) Textron Lycoming O-320-E1F flat-four engine; two-blade fixed-pitch propeller. Options for engine up to 134 kW (180 hp). Fuel capacity 121 litres (32.0 US gallons; 26.6 Imp gallons).

ACCOMMODATION: Two persons in tandem under starboard-hinged canopy. Baggage compartments forward of instrument panel and aft of rear seat.

DIMENSIONS, EXTERNAL:
Wing span	7.01 m (23 ft 0 in)
Wing aspect ratio	4.8
Length overall	6.20 m (20 ft 4 in)
Height overall	1.65 m (5 ft 5 in)
Wheel track	1.88 m (6 ft 2 in)
Propeller diameter	1.73 m (5 ft 8 in)

DIMENSIONS, INTERNAL:
Cabin: Length	1.91 m (6 ft 3 in)
Max width	0.72 m (2 ft 4 in)
Max height	1.02 m (3 ft 4 in)
Baggage volume	0.21 m³ (7.3 cu ft)

AREAS:
Wings, gross	10.22 m² (110.0 sq ft)

WEIGHTS AND LOADINGS:
Weight empty	411 kg (905 lb)
Baggage capacity	23 kg (50 lb)
Max T-O weight	680 kg (1,500 lb)
Max wing loading	66.6 kg/m² (13.64 lb/sq ft)
Max power loading	6.09 kg/kW (10.00 lb/hp)

PERFORMANCE (112 kW; 150 hp engine; MTOW):
Max level speed at S/L	174 kt (322 km/h; 200 mph)
Cruising speed at 2,440 m (8,000 ft):	
at 75% power	163 kt (303 km/h; 188 mph)
at 55% power	148 kt (274 km/h; 170 mph)
Stalling speed	47 kt (87 km/h; 54 mph)
Max rate of climb at S/L	457 m (1,500 ft)/min
Service ceiling	5,485 m (18,000 ft)
T-O run	145 m (475 ft)
Landing run	130 m (425 ft)
Range with max fuel at 55% power	
	686 n miles (1,271 km; 790 miles)

UPDATED

VAN'S RV-6 and RV-6A

No longer available as complete kit; component kits still available, however. Some 1,560 flying by November 2002. Details last appeared in 2002-03 *Jane's All the World's Aircraft*.

UPDATED

VAN'S RV-7 and RV-7A

TYPE: Side-by-side kitbuilt.

PROGRAMME: Replacement for the RV-6 and RV-6A. Design work undertaken during 2000; prototype (N137RV) first flown 19 March 2001. Public debut at Sun 'n' Fun in April 2001. RV-6/6A kits currently under construction can be completed to RV-7/7A standard by using optional parts.

Van's RV-4 N91685 modified in 2002 as testbed for ATP turboprop   *NEW*/0527011

CURRENT VERSIONS: **RV-7**: Tailwheel version.

RV-7A: Nosewheel version.

CUSTOMERS: Total of 935 kits sold by July 2002, of which six were flying by 14 November 2002.

COSTS: RV-7 kit US$16,140, quick-build kit US$23,865; RV-7A kit US$16,775, quick-build kit US$24,500 (2003).

DESIGN FEATURES: Shares commonality of parts with the RV-8/8A and RV-9A. Quoted build time reduced to 1,400 hours compared to RV-6 family due to improvements made in kit production.

FLYING CONTROLS: Conventional and manual. Horn-balanced rudder and elevators; flight-adjustable tab in port tailplane. Plain electric flaps.

STRUCTURE: Fuselage and wings of metal, with composites wingtips, tailplane tip, fin-tip, fin-root fairing, cowling, landing gear legs, fairings and spats.

LANDING GEAR: Fixed; choice of tricycle or tailwheel; Wittman leaf-spring main legs. Speed fairing on main legs. Solid tailwheel.

POWER PLANT: Prototype has 149 kW (200 hp) Textron Lycoming IO-360 driving a two-blade Hartzell constant-speed propeller; the design accepts engines in the 112 to 149 kW (150 to 200 hp) range. Fuel capacity 159 litres (42.0 US gallons; 35.0 Imp gallons) in two wing tanks.

ACCOMMODATION: Pilot and passenger side by side under forward-hinged one-piece canopy; optional rearward-sliding canopy with fixed windscreen. Compared to RV-6, cabin room is increased.

DIMENSIONS, EXTERNAL:
Wing span	7.62 m (25 ft 0 in)
Wing aspect ratio	5.2
Length overall	6.20 m (20 ft 4 in)
Height overall: RV-7	1.78 m (5 ft 10 in)
RV-7A	2.39 m (7 ft 10 in)

DIMENSIONS, INTERNAL:
Cabin max width	1.09 m (3 ft 7 in)
Baggage volume	0.34 m³ (12.0 cu ft)

AREAS:
Wings gross	11.15 m² (120.0 sq ft)

WEIGHTS AND LOADINGS (149 kW; 200 hp engine):
Weight empty	505 kg (1,114 lb)
Baggage capacity	45 kg (100 lb)

Van's RV-7 at Sun 'n' Fun, April 2002 *(Jane's/Paul Jackson)*   *NEW*/0527010

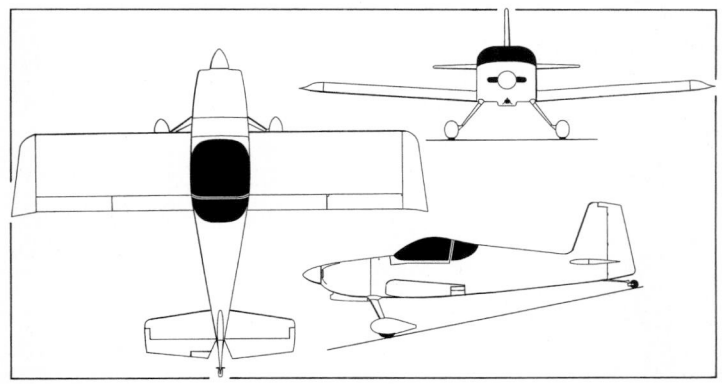

Van's RV-7 general arrangement *(Jane's/Paul Jackson)*   0110689

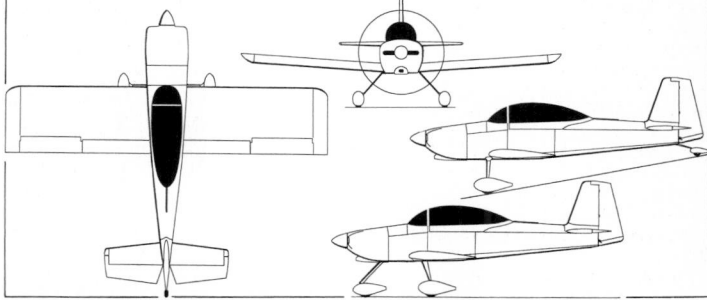

Van's RV-8 two-seat light aircraft, with additional side view of RV-8A *(Jane's/Paul Jackson)*   0051856

Van's RV-8 tailwheel version (*Jane's/Paul Jackson*)   **NEW**/0527009

Max T-O weight	816 kg (1,800 lb)
Max wing loading	73.2 kg/m² (15.00 lb/sq ft)
Max power loading	5.48 kg/kW (9.00 lb/hp)

PERFORMANCE (149 kW; 200 hp engine):
Never-exceed speed (VNE)	199 kt (370 km/h; 230 mph)
Max operating speed: RV-7	188 kt (348 km/h; 216 mph)
RV-7A	185 kt (343 km/h; 213 mph)
Cruising speed: at 75% power:	
RV-7	179 kt (332 km/h; 206 mph)
RV-7A	177 kt (328 km/h; 204 mph)
at 55% power: RV-7	162 kt (299 km/h; 186 mph)
RV-7A	159 kt (295 km/h; 183 mph)
Stalling speed, flaps down: RV-7, RV-7A	
	51 kt (94 km/h; 58 mph)
Max rate of climb at S/L: RV-7	579 m (1,900 ft)/min
RV-7A	564 m (1,850 ft)/min
Service ceiling: RV-7	6,860 m (22,500 ft)
RV-7A	6,555 m (21,500 ft)
T-O and landing run	152 m (500 ft)
Range with max fuel, 55% power	
RV-7	812 n miles (1,504 km; 935 miles)
RV-7A	803 n miles (1,488 km; 925 miles)

UPDATED

VAN'S RV-8 and RV-8A

TYPE: Tandem-seat kitbuilt.

PROGRAMME: First flown (N118RV) 22 July 1995; launched at Oshkosh five days later. Second prototype (N58RV) flew early 1997 and accumulated 325 hours in seven months.

CURRENT VERSIONS: **RV-8**: Tailwheel version.

RV-8A: Tricycle version; first flown (N58VA) April 1998; public debut at Sun 'n' Fun, same month.

CUSTOMERS: By mid-2002, 1,300 sets of plans and kits had been sold. By 14 November 2002, 251 were flying.

COSTS: RV-8 kit US$16,495, quick-build kit US$24,220; RV-8A kit US$16,865, quick-build kit US$24,580 (2003 prices).

DESIGN FEATURES: Similar in appearance to the RV-4, but with wider, longer fuselage; rounded engine cowling; non-sweptback main landing gear legs; larger canopy; however, very little parts commonality with RV-4.

Description generally as for RV-4/RV-6.

LANDING GEAR: Wittman leaf-spring main legs; nose- or tailwheel. Speed fairings on mainwheels and (when fitted) nosewheel.

POWER PLANT: One 149 kW (200 hp) Textron Lycoming IO-360-A16D four-cylinder piston engine driving two-blade Hartzell HC-C2YK-BF/F7666A4 constant-speed propeller. Fuel capacity 159 litres (42.0 US gallons; 35.0 Imp gallons). Oil capacity 7.6 litres (2.0 US gallons; 1.7 Imp gallons).

ACCOMMODATION: Tandem seating for pilot and passenger under sliding canopy. Separate windscreen.

Specification for RV-8/RV-8A, except where noted.

DIMENSIONS, EXTERNAL:
Wing span	7.32 m (24 ft 0 in)
Wing chord, constant	1.47 m (4 ft 10 in)
Wing aspect ratio	4.8
Length overall: RV-8	6.40 m (21 ft 0 in)
RV-8A	6.35 m (20 ft 10 in)
Height overall: RV-8	1.70 m (5 ft 7 in)
RV-8A	2.24 m (7 ft 4 in)
Wheel track	1.90 m (6 ft 3 in)
Propeller diameter	1.83 m (6 ft 0 in)

DIMENSIONS, INTERNAL:
Cockpit max width	0.91 m (3 ft 0 in)
Baggage volume	0.21 m³ (7.5 cu ft)

AREAS:
Wings, gross	10.78 m² (116.0 sq ft)

WEIGHTS AND LOADINGS:
Weight empty	508 kg (1,120 lb)
Baggage capacity	57 kg (125 lb)
Max T-O weight	816 kg (1,800 lb)
Max wing loading	79.9 kg/m² (16.36 lb/sq ft)
Max power loading	5.48 kg/kW (9.00 lb/hp)

PERFORMANCE (two persons, 149 kW; 200 hp engine):
Max operating speed: RV-8	191 kt (354 km/h; 220 mph)
RV-8A	189 kt (351 km/h; 218 mph)
Cruising speed at 55% power at 2,440 m (8,000 ft):	
RV-8	162 kt (301 km/h; 187 mph)
RV-8A	161 kt (298 km/h; 185 mph)

Stalling speed, power off:	
one occupant: both	45 kt (83 km/h; 51 mph)
two occupants: both	51 kt (94 km/h; 58 mph)
Max rate of climb at S/L: RV-8	579 m (1,900 ft)/min
RV-8A	549 m (1,800 ft)/min
Service ceiling: RV-8	6,860 m (22,500 ft)
RV-8A	6,550 m (21,500 ft)
T-O and landing run: both	153 m (500 ft)
Range at 55% power:	
RV-8	816 n miles (1,512 km; 940 miles)
RV-8A	808 n miles (1,496 km; 930 miles)

UPDATED

VAN'S RV-9 and RV-9A

TYPE: Side-by-side kitbuilt.

PROGRAMME: Prototype RV-9A (N96VA) first flown late 1997; was hybrid proof-of-concept aircraft using unique RV-6T fuselage plus new wings; had flown more than 200 hours when lost in accident 2 April 2000; second prototype, and first 'true' RV-9A (N129RV), flown 15 June 2000. Empennage kits became available in late 1999, followed by wing and fuselage/finishing kits in mid-2000 and quick-build kits in late 2000; first customer-built aircraft flown 17 June 2001. RV-9 prototype (N179RV) first flown 4 March 2002.

CURRENT VERSIONS: **RV-9**: Tailwheel version.

RV-9A: Tricycle version.

CUSTOMERS: 400 kits sold by April 2001; 20 flying by 14 November 2002.

COSTS: Basic kit RV-9 US$16,135, RV-9A US$16,770; quick-built RV-9 US$24,065, RV-9A US$24,495 (2003).

DESIGN FEATURES: Based upon, and similar in appearance to, RV-6/6A, but with redesigned wing (Roncz aerofoil and sheared wingtips) and enlarged, constant-chord horizontal tail surfaces.

FLYING CONTROLS: Conventional and manual. Horn-balanced rudder. Slotted electric flaps extend approximately two-thirds of wing span, compared with one-half on other RV designs.

LANDING GEAR: Fixed tricycle type on RV-9A; tailwheel on RV-9. Brakes.

POWER PLANT: First prototype flown with 88.0 kW (118 hp) Textron Lycoming O-235-L2C flat-four driving two-blade Sensenich 72 × 66 fixed-pitch metal propeller. Second

Van's RV-9A kitbuilt (*Jane's/Paul Jackson*)   **NEW**/0527008

Van's RV-9A general arrangement (*Jane's/Paul Jackson*)   0110690

Prototype RV-9 tailwheel version at its public debut, Sun 'n' Fun, April 2002 (*Jane's/Paul Jackson*) **NEW**/0527007

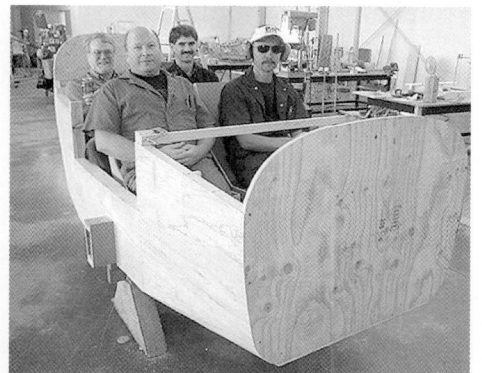
Wooden mockup of the RV-10 cabin   0126970

prototype has 119 kW (160 hp) Textron Lycoming O-320-D3G driving three-blade MT propeller. Engines between 80.5 and 119 kW (108 and 160 hp) can be fitted. Fuel capacity 136 litres (36.0 US gallons; 30.0 Imp gallons) in two wing tanks. Wilksch developing conversion during

2002, to enable use of its 90 kW (120 hp) WAM-120 diesel engine.

ACCOMMODATION: Pilot and passenger side by side under sliding canopy. Dual controls.

DIMENSIONS, EXTERNAL:

Wing span: both	8.53 m (28 ft 0 in)
Length overall: both	6.22 m (20 ft 5 in)
Height overall: RV-9	1.83 m (6 ft 0 in)
RV-9A	2.39 m (7 ft 10 in)
Propeller diameter	1.83 m (6 ft 0 in)

DIMENSIONS, INTERNAL: As RV-7.

AREAS:

Wings, gross	11.52 m² (124.0 sq ft)

WEIGHTS AND LOADINGS (119 kW; 160 hp engine):

Weight empty	476 kg (1,050 lb)
Baggage capacity	45 kg (100 lb)
Max T-O weight	793 kg (1,750 lb)
Max wing loading	68.9 kg/m² (14.11 lb/sq ft)
Max power loading	6.66 kg/kW (10.94 lb/hp)

PERFORMANCE (119 kW; 160 hp engine):

Max operating speed: RV-9	170 kt (315 km/h; 196 mph)
RV-9A	169 kt (312 km/h; 194 mph)
Cruising speed at 75% power: RV-9	
	163 kt (303 km/h; 188 mph)
RV-9A	162 kt (299 km/h; 186 mph)

Stalling speed, flaps down: both	44 kt (81 km/h; 50 mph)
Max rate of climb at S/L: both	427 m (1,400 ft)/min
Service ceiling: RV-9	5,790 m (19,000 ft)
RV-9A	5,640 m (18,500 ft)
T-O run: both	145 m (475 ft)
Landing run: both	137 m (450 ft)
Range at 75% power: RV-9	
	617 n miles (1,142 km; 710 miles)
RV-9A	608 n miles (1,126 km; 700 miles)
	UPDATED

VAN'S RV-10

In late 2001, Van's built a wooden mockup of the cabin of its first four-seat project, the RV-10, the prototype of which was expected to make its first flight during the second quarter of 2003; by November 2002 the tailcone and cabin sections of the fuselage had been joined. The aircraft has preliminary design criteria including a 9.60 m (31 ft 6 in) span, 13.65 m² (147 sq ft) wing area, slotted flaps and an empty weight of 726 kg (1,600 lb). Power plant will be in the range 149 to 194 kW (200 to 260 hp), with fuel capacity of 227 litres (60.0 US gallons; 50.0 Imp gallons).

UPDATED

VAT

VERTICAL AVIATION TECHNOLOGIES INC

1609 Hangar Road, Sanford, Florida 32772-2527
Tel: (+1 407) 322 94 88
Fax: (+1 407) 330 26 47
e-mail: sales@vertical-aviation.com
Web: http://www.vertical-aviation.com
PRESIDENT: Bradley G Clark

Vertical Aviation Technologies Inc is FAA-approved repair facility for various Sikorsky helicopters. The company remanufactures Sikorsky S-55/H-19s and S-58/H-34s, as described in *Jane's Aircraft Upgrades*. In 1984, began to modify four-seat Sikorsky S-52-3 former production helicopter (of which 95 originally produced, mainly for US Marine Corps as HO5S) into kit for assembly by individuals, corporations or military; all tooling and fixtures completed by 1988. Promotion and manufacture was interrupted by S-55 upgrade effort, but resumed at Sun 'n' Fun in April 2002, when upgraded S-52 was first shown.

UPDATED

VAT HUMMINGBIRD

TYPE: Four-seat helicopter kitbuilt.

PROGRAMME: Based on previously type-certified Sikorsky S-52-3 and first flown 1990. (S-52 prototype flew 12 February 1947; total 95 built). First batch of sales between 1989 and 1993; promotion lapsed between 1993 and 2001, when VAT efforts directed towards S-55QT Whisper Jet conversion. Modified Hummingbird shown at Sun 'n' Fun in April 2002; twin anhedral tail strakes replaced by more conventional horizontal stabilisers with endplates; Lycoming engine replacing 194 kW (260 hp) Ford Taurus V-8; doors and engine cowling of composites.

CUSTOMERS: Fourteen Hummingbird kits sold by mid-1990s, of which 12 flying. Customers in China, Hungary, Slovak Republic, USA and UK.

COSTS: Kit US$140,500 (2002), including engine.

DESIGN FEATURES: Newly manufactured airframes and tailcones identical to S-52-3, except for nose section. Quoted build time 800 hours; kit includes original Sikorsky components.

FLYING CONTROLS: Conventional and manual. Two anhedral stabilisers of original S-52 replaced from 2002 by horizontal unit with endplates.

STRUCTURE: Mainly aluminium, with GFRP nose (from Bell JetRanger), doors, engine cowling, tailplane endplates and tailcone. Monocoque cabin and tailboom; metal tube centre-section.

LANDING GEAR: Four-wheel type, fixed; oleo-pneumatic shock-absorbes on each leg; rear wheels 5.00-5; forward wheels 4.10/3.50. Hydraulic brakes on rear.

POWER PLANT: One 194 kW (260 hp) Textron Lycoming VO-435-A1F driving three-blade fully articulated main rotor and two-blade tail rotor. Unlimited transmission. Fuel capacity 216 litres (57.0 US gallons; 47.5 Imp gallons). Oil capacity 11 litres (3.0 US gallons; 2.5 Imp gallons).

ACCOMMODATION: Four persons, including pilot(s); door each side at front; rear door port only.

DIMENSIONS, EXTERNAL:

Main rotor diameter	10.06 m (33 ft 0 in)
Tail rotor diameter	1.75 m (5 ft 9 in)
Length: overall, rotors turning	12.11 m (39 ft 9 in)
fuselage	9.30 m (30 ft 6 in)
Distance between rotor centres	6.07 m (19 ft 11 in)
Height: to top of rotor head	2.62 m (8 ft 7 in)
overall	2.95 m (9 ft 8 in)
Fuselage max width	1.52 m (5 ft 0 in)

Current version of Vertical Aviation Technologies Hummingbird (*Jane's/Paul Jackson*)   *NEW*/0137387

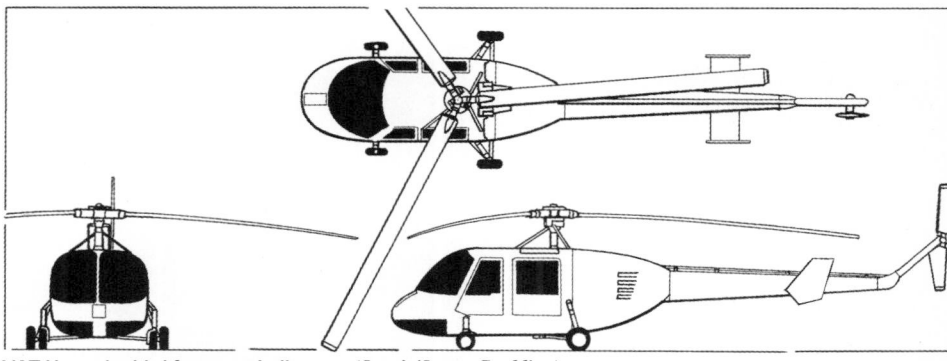
VAT Hummingbird four-seat helicopter (*Jane's/James Goulding*)   *NEW*/0137335

Wheel track: forward	1.83 m (6 ft 0 in)
rear	2.49 m (8 ft 2 in)
Wheelbase	1.88 m (6 ft 2 in)

AREAS:

Main rotor disc	79.46 m² (855.3 sq ft)
Tail rotor disc	2.41 m² (25.97 sq ft)

WEIGHTS AND LOADINGS:

Weight empty	816 kg (1,800 lb)
Max T-O weight	1,225 kg (2,700 lb)
Max disc loading	15.4 kg/m² (3.16 lb/sq ft)
Max power loading	6.32 kg/kW (10.38 lb/hp)

PERFORMANCE (three-blade rotor):

Never-exceed speed (V_NE)	95 kt (177 km/h; 110 mph)
Cruising speed	78 kt (145 km/h; 90 mph)
Max rate of climb at S/L at 1,043 m (3,420 lb)	
	381 m (1,250 ft)/min
Hovering ceiling: IGE	2,745 m (9,000 ft)
OGE	1,220 m (4,000 ft)
Service ceiling	3,660 m (12,000 ft)
Range	286 n miles (531 km; 330 miles)
	UPDATED

VENTURE

VENTURE LIGHT AIRCRAFT RESOURCES LLC

Air Center West, Ryan Field, HCR 2, Box 270, Tucson, Arizona 85735
Tel/Fax: (+1 520) 883 82 95
e-mail: l.rebling@gte.net
Web: http://www.venture-thorpt-211.com
PRESIDENT: Larry A Rebling

Venture builds kits of the Thorp T-211 and by July 2001 had supplied three complete aircraft and six kits to AD Aerospace in the UK (see 2001-02 and previous editions).

UPDATED

First UK-assembled Thorp T-211 experimentally fitted with a Wilksch WAM-120 diesel engine *(Jane's/Paul Jackson)*   *NEW*/0527006

VENTURE THORP T-211

TYPE: Side-by-side kitbuilt.
PROGRAMME: Originated in T-11 Sky Skooter, designed by John Thorp and first flown 15 August 1946. See 1992-93 and earlier editions for previous development history. T-211 certified 20 April 1964; production rights passed to Phoenix Aircraft at Mesa, Arizona, in 1992, but that company declared bankrupt in mid-1994. Acquired by AD Aircraft in the UK; market launch (and first sale, to Ken Fowler) of kits at PFA International Air Rally & Exhibition at Cranfield, Bedfordshire, in July 1998. AD sold six homebuild kits produced under licence in the USA by Venture Light Aircraft Resources (the previous owner); Venture production began with c/n 104. UK production of complete aircraft planned, but postponed in 1999. Type certificate owned by First Kentucky Bank.
CURRENT VERSIONS: **T-211**: Factory-built version.
   T-211A: Continental-engined kitbuilt version, supplied complete, or in eight sub-kits for stage-by-stage assembly. *Data refer to Continental-powered version.*
   T-211B: Jabiru-engined kitbuilt. Initial aircraft is first UK-built T-211, G-TZII.
CUSTOMERS: Five kits sold during PFA International Air Rally in July 1999, increasing total to six. (At least nine built in USA by Tubular Aircraft, one; Aircraft Engineering Associates, one; and Venture.)
COSTS: Basic kit US$10,995, finishing kit US$3,200; wiring harness US$595, excluding engine and avionics (2003).
FLYING CONTROLS: Manual. Plain ailerons; all-moving tailplane with anti-balance tab occupying 90 per cent span; wide-span manually operated three-position trailing-edge flaps with maximum deflection of 20°.
STRUCTURE: Light alloy fuselage; wing with main spar, false spar and externally stiffened skin; ailerons, flaps, tailplane and rudder have externally stiffened skin.
LANDING GEAR: Non-retractable tricycle type; oleo-pneumatic shock-absorber in each leg; all three wheels size 5.00-5; steerable nosewheel; Cleveland brakes on main units.
POWER PLANT: One 74.6 kW (100 hp) Teledyne Continental O-200A flat-four, driving a Sensenich 69CK two-blade fixed-pitch propeller. Fuel tank aft of cabin, capacity 91 litres (24.0 US gallons; 20.0 Imp gallons), of which 79.5 litres (21.0 US gallons; 17.5 Imp gallons) are usable; oil capacity 5.7 litres (1.5 US gallons; 1.25 Imp gallons).

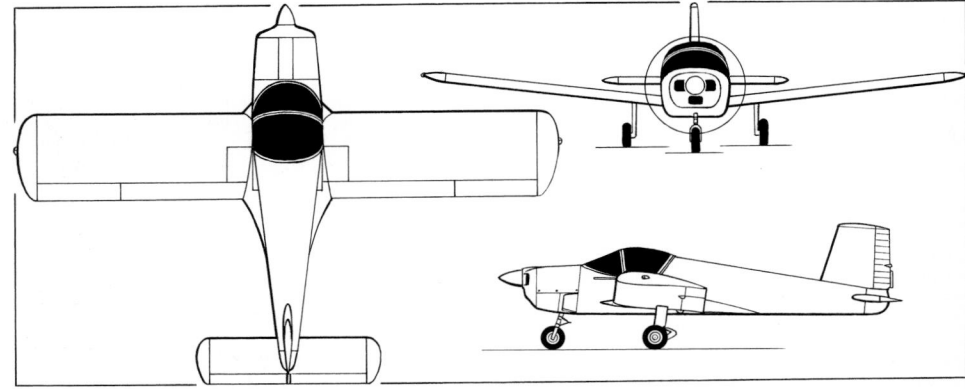

Venture Thorp T-211 two-seat kitbuilt *(Jane's/James Goulding)*   0084388

Jabiru 3300 six-cylinder four-stroke, rated at 89 kW (120 hp), available as an option from 1999.
ACCOMMODATION: Two seats side by side beneath rearward-sliding transparent canopy; fixed windscreen. Baggage compartment behind seats, capacity 8.1 kg (17.8 lb).

DIMENSIONS, EXTERNAL:
Wing span	7.62 m (25 ft 0 in)
Wing aspect ratio	6.0
Length overall	5.50 m (18 ft 0½ in)
Height overall	1.90 m (6 ft 2¾ in)
Propeller diameter	1.70 m (5 ft 7 in)

DIMENSIONS, INTERNAL:
Cabin max width	0.94 m (3 ft 1 in)

AREAS:
Wings, gross	9.72 m² (104.6 sq ft)

WEIGHTS AND LOADINGS:
Weight empty	340 kg (750 lb)
Baggage capacity	36 kg (80 lb)
Max T-O weight	576 kg (1,270 lb)
Max wing loading	59.3 kg/m² (12.14 lb/sq ft)
Max power loading	7.73 kg/kW (12.70 lb/hp)

PERFORMANCE:
Never-exceed speed (V_{NE})	138 kt (255 km/h; 158 mph)
Max level speed	109 kt (202 km/h; 125 mph)
Cruising speed at 75% power	104 kt (193 km/h; 120 mph)
Stalling speed, flaps down, power on	39 kt (73 km/h; 45 mph)
Max rate of climb at S/L	more than 229 m (750 ft)/min
Service ceiling	3,810 m (12,500 ft)
T-O run	91 m (300 ft)
Landing run	122 m (400 ft)
Max range	413 n miles (764 km; 475 miles)

UPDATED

VIPER

VIPER AIRCRAFT CORPORATION

3908 Stearman Avenue, Pasco, Washington 99301
Tel: (+1 509) 543 35 70
Fax: (+1 509) 547 31 42
e-mail: info@viper-aircraft.com
Web: http://www.viper-aircraft.com
PRESIDENT: Scott Hanchette
VICE-PRESIDENT: Dan Hanchette

Viper Aircraft's first product is the ViperJet. In May 2001 it was announced that the ViperJet would be marketed by Ascendair International and assembled in Georgia by its sister company, Tbilisi Aerospace Manufacturing.

UPDATED

VIPER VIPERJET

TYPE: Two-seat jet sportplane kitbuilt.
PROGRAMME: Development began in 1988; engineering by AirBoss Aerospace; tooling began in 1996; prototype was initially designed around a Teledyne Continental TSIO-520 and named ViperFan; in early 1999 a switch to jet propulsion was revealed and aircraft renamed ViperJet; development of ViperFan will continue as a lower priority. Prototype (N520VF), officially registered as Hanchette ViperJet) first flown 22 October 1999 and displayed publicly at Oshkosh in July 2000; underwent further modifications, but seriously damaged in crash at St George, Utah, on 13 March 2002. Figures quoted below are for ViperJet Mk II production aircraft, supplanting data released earlier.
CURRENT VERSIONS: **ViperFan**: Initial piston-engined version.
   ViperJet: Jet-powered prototype.
   ViperJet Mk II: Production version, *as described.*

Viper ViperJet prototype, following repainting and removal of winglets   *NEW*/0530171

CUSTOMERS: Ten sold and one flying by end 2002.

COSTS: Kit US$139,900; engine extra (2003).

DESIGN FEATURES: High-performance personal jet at self-build price. Low wing with sweptback leading-edge and initially with winglets; sweptback tail surfaces with fin fillet and ventral strakes.

Wing aerofoil section NACA 65-218 at root, NACA 65-212 at tip; horizontal tail NACA 66-012; vertical tail NACA 63-015.

Quoted build time 2,000 hours.

FLYING CONTROLS: Conventional and manual. Electric double-slotted flaps, span 1.65 m (5 ft 5 in) deflecting between 0 and 40°; aileron span 1.22 m (4 ft 0 in), deflection 27° up and 18° down; rudder deflection ±25°; elevator deflection ±30°. Dual controls.

STRUCTURE: Extensive use of composites; much of this provided by Unlimited Composites of Scappoose, Oregon. Preformed skins and wing spars.

LANDING GEAR: Retractable tricycle type; nosewheel retracts rearwards, mainwheels inwards. Production aircraft will not use landing gear doors.

POWER PLANT: Prototype initially powered by one refurbished Turbomeca Marboré 2 turbojet rated at 3.91 kN (880 lb st);

later changed to converted General Electric T58-8F rated at 4.44 kN (1,000 lb st). Other engine options include versions of Marboré up to 4.89 kN (1,100 lb st). Also recommended is 13.12 kN (2,950 lb st) General Electric J85/CJ-610. Fuel capacity 511 litres (135 US gallons; 112 Imp gallons) in prototype; 1,136 litres (300 US gallons; 250 Imp gallons).

ACCOMMODATION: Pilot and passenger in tandem in unpressurised cockpit.

DIMENSIONS, EXTERNAL:

Wing span	7.87 m (25 ft 10 in)
Wing chord: at root	1.78 m (5 ft 10 in)
at tip	0.86 m (2 ft 10 in)
Wing aspect ratio	7.7
Length overall	7.57 m (24 ft 10 in)
Height overall	3.25 m (10 ft 8 in)
Tailplane span	4.06 m (13 ft 4 in)

DIMENSIONS, INTERNAL:

Cockpit: Length	2.92 m (9 ft 7 in)
Max width	0.76 m (2 ft 6 in)
Max height	1.24 m (4 ft 1 in)

AREAS:

Wings, gross	10.50 m² (113.0 sq ft)

WEIGHTS AND LOADINGS (J85 engine):

Weight empty	998 kg (2,200 lb)
Max T-O weight	1,950 kg (4,300 lb)
Baggage capacity	45 kg (100 lb)
Max wing loading	185.8 kg/m² (38.05 lb/sq ft)
Max power loading	149 kg/kN (1.46 lb/lb st)

PERFORMANCE (estimated):

Never-exceed speed (VNE)	434 kt (805 km/h; 500 mph)
Max level speed at 7,620 m (25,000 ft)	
	391 kt (724 km/h; 450 mph)
Stalling speed: flaps down	73 kt (136 km/h; 84 mph)
Max rate of climb at S/L	3,050 m (10,000 ft)/min
Service ceiling	7,620 m (25,000 ft)
T-O run	366 m (1,200 ft)
Landing run	610 m (2,000 ft)
Range with max fuel, 45 min reserve	
	869 n miles (1,609 km/h; 1,000 miles)
Endurance	2 h 0 min
g limits	±12

UPDATED

VISIONAIRE

VISIONAIRE CORPORATION

Spirit of St Louis Airport, 595 Bell Avenue, Chesterfield, Missouri 63005

Tel: (+1 636) 530 10 07 and 530 64 00
Fax: (+1 636) 530 00 05
e-mail: info@visionaire.com
Web: http://www.visionaire.com

CHAIRMAN AND CEO: James O Rice Jr
PRESIDENT AND COO: Daniel D Mickelson
SENIOR VICE-PRESIDENT: Tom Stark
EXECUTIVE VICE-PRESIDENT: Glen Monigold
VICE-PRESIDENT, MARKETING AND SALES: Jan G Gilbert
VICE-PRESIDENT, FINANCE: Gary Pluth
VICE-PRESIDENT, ENGINEERING: Joe Furnish
DIRECTOR, COMMUNICATIONS: Mark A Jones
DIRECTOR, SALES: Mike Page
VANTAGE PROJECT MANAGER: Del Potts

VisionAire Corporation, formed in 1988 by James O Rice Jr, produces the Vantage all-composites business jet. Intended for mid-sized companies, the Vantage is specifically designed to offer cabin-class jet transportation at about half the cost of twin-jets currently on the market. Assembly of additional test aircraft was due to begin during first quarter of 1999 at 10,775 m² (116,000 sq ft) Ames, Iowa factory, opened 26 May 1998; however, this has been delayed. An agreement of October 1997 provides for SimCom of Orlando, Florida, to provide pilot and maintenance training on behalf of Visionaire as part of the Vantage's purchase price. Visionaire employed 130 at the end of 1998, but laid off half of these in January 1999, pending a restructuring of the Vantage programme, which was completed in June 1999; by October 2000 workforce numbered 20.

FutureWorks, VisionAire's new product development division, introduced a full-size mockup, at EAA Oshkosh in 1998, of a two-seat tandem personal sport jet to be powered by a single turbofan. Designated VA-12, this is the first in a proposed line of aircraft to be marketed under the Spirit name.

In mid-1998, VisionAire was in negotiation with Israeli government with a view to forming VisionArad company to build the Vantage at Arad, Israel; in 2001 it was announced that IAI (which see) was undertaking design verification and generally overseeing technical aspects of the programme. Scaled Technology Works is designing and certifying airframe parts for the Vantage.

The company has received a Chapter 7 bankruptcy motion by five creditors, but in August 2002 this was converted to Chapter 11 voluntary bankruptcy, allowing time for VisionAire to seek further financial support; in September 2002, this was said to be in prospect from a European syndicate, but no firm decision had been reached by December 2002, when re-imposition of Chapter 7 was imminent.

UPDATED

VISIONAIRE VA-10 VANTAGE

TYPE: Business mono-jet.

PROGRAMME: Refinement of earlier design began on 15 January 1993; tested by two scale models; programme announced September 1995; design and construction of proof of concept (POC) aircraft started January 1996; built by Scaled Composites Inc under US$3.1 million contract; rolled out 8 November 1996; first flight (N247VA) 16 November 1996.

By 100th flight (180th hour), in February 1998, prototype had received 10 cm (4 in) stretch ahead of wings to accommodate aircraft systems; extra 2° of wing dihedral; further 27 per cent of horizontal tail area; and reduced rudder area. Changes then identified for following aircraft included redesigned landing gear; independently

Proof of concept VisionAire VA-10 Vantage (N247VA)   *NEW*/0533739

actuated mainwheel doors operating as brakes; reduced size windscreen posts; and a 1° reduction in jetpipe angle. Total of 193 sorties flown by August 2000; original plans for seven development aircraft: one POC, two flight development (VT1 and VT2), two preproduction and two static test.

In December 1998, work on Vantage was temporarily halted for a design review and restructuring as a consequence of control and weight problems, the aircraft being at least 363 kg (800 lb) over target. Modifications announced in June 1999 include revised, thinner wing for improved low-speed handling and lower drag; reduced forward sweep and whole wing moved aft; vertical tail increased in size and horizontal tail moved aft; engine moved back 51 cm (20 in) and up 13 cm (5 in) to reduce inlet curvature and decrease susceptibility to icing; main landing gear legs moved to wings and redesigned as trailing link units, as well as being lowered 15 cm (6 in); baggage compartment added in rear fuselage. Changes (to what is now known as VA-10A, or configuration 1A5) result in 181 kg (400 lb) MTOW increase.

Cost reduction plans announced in April 2000 to reduce required investment from US$130 million to US$90 million; involve modification of original prototype to exterior shape of final version and reduction in flight development and preproduction fleet from four to three. In late 2002, investment funds of US$100 million to US$150 million were being sought to see the project through to certification and production.

First flight of VT-1 (initial development aircraft) due third quarter 2002, followed by VT-2, and two preproduction aircraft (PT-1 and PT-2) at regular intervals. FAA certification to FAR Pt 23 Amdt 51 expected early 2003 with deliveries from mid-2003. These deadlines not being met, but first flight of VT1 expected within 18 months of refinancing of company, and certification 10 months thereafter.

CUSTOMERS: Having reached 150, firm orders had declined to some 60 by December 2002.

COSTS: US$2.195 million (2001). Direct operating cost (excluding finance and annual fixed expense; 456 hours per year utilisation) US$262 per hour.

DESIGN FEATURES: Low-cost high-performance corporate jet, meeting FAR Pt 23 Amendment 51 (single pilot operation) and making use of conventional aerodynamic techniques. Sweptforward wing minimises drag on mid-wing design and allows main spar to pass behind cabin. Bifurcated engine air intakes on fuselage shoulders, level with wing leading-edge.

Wing leading-edge sweep −6°; dihedral 5° 54′.

FLYING CONTROLS: Conventional and manual. Aerodynamically balanced ailerons, elevators and rudder; three-axis electrical trim; yaw damper; autopilot. Electrically actuated Fowler flaps; two-element spoilers on each wing.

STRUCTURE: Prepreg carbon-graphite construction. Monocoque fuselage structure with composites sandwich panels.

LANDING GEAR: Retractable tricycle type with hydraulic actuation. Single-wheel main units retract inward; single-wheel nose gear retracts forward. ABSC wheels and hydraulic brakes.

POWER PLANT: One 13.55 kN (3,045 lb st) Pratt & Whitney Canada JT15D-5D turbofan. Two integral wing fuel tanks. Gravity refuelling point over each wing. Fuel capacity 908 litres (240 US gallons; 200 Imp gallons) in prototype; up to 1,261 litres (333 US gallons; 277 Imp gallons) in production version.

ACCOMMODATION: Pilot and co-pilot or passenger in cockpit, with two pairs of seats in cabin; crew-seating by Fischer + Entwicklungen, crashworthy to 26 g. Forward baggage area; screened lavatory and baggage area at rear of cabin. Main door on port side, immediately behind cockpit. Cabin pressurised to equivalent altitude of 2,440 m (8,000 ft).

SYSTEMS: 28 V dual electrical power. Enviro systems electric vapour-cycle air conditioning. Cox mechanical de-icing. Available options permit use as single-engined charter aircraft under FAR Pt 135.

AVIONICS: *Radar:* Honeywell RDR 2000 weather radar.

Flight: Integrated suite, comprising components from Garmin and Meggitt, includes Meggitt MAGIC integrated display and sensor system; Garmin GNS 530 comprising integrated IFR GPS/comm with VOR, LOC and glideslope, GTX327 digital transponder and GMA340 audio panels; and Meggitt autopilot.

Instrumentation: Dual IFR system with EFIS displays.

DIMENSIONS, EXTERNAL:

Wing span	14.48 m (47 ft 6 in)
Wing chord: at root	1.90 m (6 ft 3 in)
at tip	0.99 m (3 ft 3 in)
Wing aspect ratio	10.2
Length overall: prototype	12.53 m (41 ft 1½ in)
production	12.42 m (40 ft 9 in)
Fuselage: Length	11.32 m (37 ft 1½ in)
Max diameter	1.71 m (5 ft 7¼ in)
Height	4.37 m (14 ft 4 in)
Wheel track	3.61 m (11 ft 10 in)
Wheelbase	5.18 m (17 ft 0 in)
Passenger door: Width	0.75 m (2 ft 5½ in)

DIMENSIONS, INTERNAL:

Cabin (excl flight deck): Length	3.30 m (10 ft 10 in)
Max width	1.57 m (5 ft 2 in)
Max height	1.52 m (5 ft 0 in)
Floor area	5.48 m² (59 sq ft)
Volume	8.8 m³ (310 cu ft)
Baggage compartment volume:	
internal	0.90 m³ (31.8 cu ft)
external	0.42 m³ (15.0 cu ft)

AREAS:

extended sleeping deck (conversion from aircraft to camper interior taking about 2 minutes and accommodating two persons), extended baggage area, engine of up to 85.7 kW (115 hp), and provision for full electrical system.

Wing section USA 35B.

FLYING CONTROLS: Conventional and manual. Horn-balanced rudder. Tab on port elevator.

STRUCTURE: All-wood wing and aluminium alloy aileron structures. Welded 4130 steel tube and flat plate fuselage structure, and steel tube tail unit. Complete airframe is fabric covered.

LANDING GEAR: Non-retractable, tailwheel type; as Sport Trainer; mainwheel tyres 7.00-6; Cleveland brakes. Optional skis.

POWER PLANT: Traveler can be powered by a Textron Lycoming engine of 80.5 to 85.7 kW (108 to 115 hp). Classic can be powered by a Teledyne Continental engine of 48.5 to 74.5 kW (65 to 100 hp). Fuel capacity: Traveler 98.4 litres (26.0 US gallons; 21.6 Imp gallons), Classic 45.4 litres (12.0 US gallons; 10.0 Imp gallons).

ACCOMMODATION: Two persons side by side on adjustable bench seat; cabin includes baggage stowage. Forward-opening cabin door on each side of fuselage.

SYSTEMS: Optional electrical system in Traveler.

DIMENSIONS, EXTERNAL:

Wing span	8.32 m (29 ft 3½ in)
Wing chord, constant	1.60 m (5 ft 3 in)
Wing aspect ratio	5.0
Length overall	5.70 m (18 ft 8½ in)
Height overall	1.83 m (6 ft 0 in)
Wheel track	1.83 m (6 ft 0 in)

DIMENSIONS, INTERNAL:

Cabin max width	1.02 m (3 ft 4 in)

AREAS:

Wings, gross	13.70 m² (147.5 sq ft)

WEIGHTS AND LOADINGS:

Weight empty: Classic	318 kg (700 lb)
Traveler	363 kg (800 lb)
Baggage capacity: Classic	18 kg (40 lb)
Traveler	27 kg (60 lb)
Max T-O weight: Classic	567 kg (1,250 lb)
Traveler	658 kg (1,450 lb)
Max wing loading: Classic	41.4 kg/m² (8.47 lb/sq ft)
Traveler	48.0 kg/m² (9.83 lb/sq ft)
Max power loading: Classic	7.61 kg/kW (12.50 lb/hp)
Traveler	7.67 kg/kW (12.61 lb/hp)

PERFORMANCE:

Max level speed: Classic	91 kt (169 km/h; 105 mph)
Traveler	118 kt (219 km/h; 136 mph)
Cruising speed: Classic	83 kt (153 km/h; 95 mph)
Traveler	104 kt (193 km/h; 120 mph)
Stalling speed:	
Classic, Traveler	39 kt (71 km/h; 44 mph)
Max rate of climb at S/L: Classic	190 m (625 ft)/min
Traveler	259 m (850 ft)/min
Service ceiling: both	4,260 m (14,000 ft)
T-O run: both	119 m (390 ft)
Landing run: both	232 m (760 ft)
Range: both	620 n miles (1,148 km; 713 miles)

UPDATED

WAG-AERO 2+2 SPORTSMAN

TYPE: Four-seat kitbuilt.

PROGRAMME: Version of the Piper PA-14 Family Cruiser. Plans and material kits available. Conforms to FAA 51 per cent rule.

CUSTOMERS: 1,971 sets of plans and kits sold by May 2001.

COSTS: Plans: US$89; kit US$15,000 (2001).

Wag-Aero Sportsman *(Jane's/Paul Jackson)*   0093689

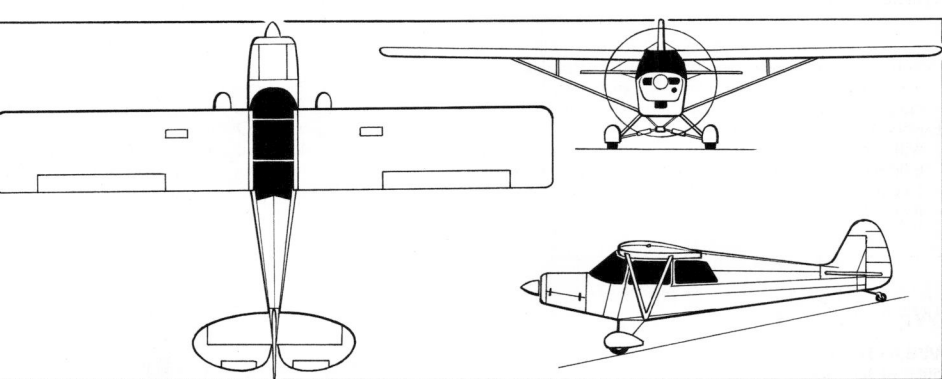

Wag-Aero 2+2 Sportsman *(Jane's/James Goulding)*   *NEW*/0526903

DESIGN FEATURES: True four-seater, with option of hinged rear fuselage decking to provide access to baggage and rear seat areas. Conventional, high-wing configuration with V-type bracing struts, wire-braced empennage, mid-mounted tailplane and classic fin and rudder. Wing section USA 35B.

FLYING CONTROLS: Conventional and manual. Horn-balanced rudder. Upper and lower spoilers; trim tabs on both elevators. Aileron limits ±17°; elevator limits 34° up, 29° down.

STRUCTURE: Similar construction to Wag-A-Bond, with glass fibre wingtips. Alternatively, drawings and materials provided to modify standard PA-12, PA-14 or PA-18 wings. Prewelded fuselage structure also available.

LANDING GEAR: As Wag-A-Bond; main units use 8.00-6 or 7.00-6 tyres. Optional floats.

POWER PLANT: Engine of 93 to 149 kW (125 to 200 hp). Usable fuel capacity 148 litres (39.0 US gallons; 32.5 Imp gallons).

ACCOMMODATION: Four persons; two separate front seats, plus rear bench seat which is removable for stretcher or cargo carrying. Baggage area to rear of bench seat. Forward-opening cabin door on each side.

DIMENSIONS, EXTERNAL:

Wing span	10.90 m (35 ft 9 in)
Wing chord, constant	1.60 m (5 ft 3 in)
Wing aspect ratio	7.3
Length overall	7.12 m (23 ft 4½ in)
Height overall	2.02 m (6 ft 7½ in)
Wheel track	1.98 m (6 ft 6 in)

DIMENSIONS, INTERNAL:

Cabin max width	0.99 m (3 ft 3 in)

AREAS:

Wings, gross	16.18 m² (174.1 sq ft)

WEIGHTS AND LOADINGS:

Weight empty	490 kg (1,080 lb)
Max T-O weight	998 kg (2,200 lb)
Max wing loading	61.7 kg/m² (12.63 lb/sq ft)
Max power loading	6.70 kg/kW (11.00 lb/hp)

PERFORMANCE (typical; actual data depend on engine fitted):

Max level speed	112 kt (207 km/h; 129 mph)
Cruising speed	108 kt (200 km/h; 124 mph)
Stalling speed	33 kt (62 km/h; 38 mph)
Max rate of climb at S/L	244 m (800 ft)/min
Service ceiling	4,510 m (14,800 ft)
T-O run	70 m (230 ft)
Landing run	104 m (340 ft)
Range at cruising speed	
	670 n miles (1,240 km; 771 miles)

UPDATED

WARNER

WARNER AEROCRAFT COMPANY

9415 Laura Court, Seminole, Florida 33776-1625
Tel: (+1 727) 595 23 82
Fax: (+1 617) 598 10 14
e-mail: info@warnerair.com
Web: http://www.warnerair.com
PRESIDENT: Dana Axelrod
SALES MANAGER: Mike Dwyer

Warner Aerocraft markets the Warner Space Walker (see *Jane's* 1990-91), Revolution and Sportster.

VERIFIED

WARNER SPORTSTER

TYPE: Tandem-seat sportplane kitbuilt.

PROGRAMME: Development of Country Air (later Warner) Space Walker, of which 46 built. Two-seat versions of Space Walker developed as Revolution I (48.5 kW; 65 hp Teledyne Continental) and Revolution II (74.6 kW; 100 hp Textron Lycoming); further power increase produced Sportster; prototype (N269U) first flown 18 April 1998. New version with revised canopy introduced 2002 and has replaced original model aircraft; length reduced to 6.20 m (20 ft 4 in). For details of accommodation, see below. Prototype (N69DA) unflown by April 2002.

Single-seat, enclosed cockpit option for Warner Sportster *(Jane's/Paul Jackson)*   *NEW*/0527005

CUSTOMERS: 125 Space Walkers and Revolutions built, plus six of original Sportster.

COSTS: Basic kit US$14,495; fast-build kit US$22,995 (2002).

DESIGN FEATURES: Quoted build time 700 hours. Roll rate 45°/s.

FLYING CONTROLS: Conventional and manual. Horn-balanced rudder and elevators; trim tab in port elevator. Dual controls.

STRUCTURE: 4130 steel tube and fabric fuselage with wooden turtle deck; fabric-covered two-spar wooden cantilever wings. Glass fibre cowling and wingtips. Tail unit identical

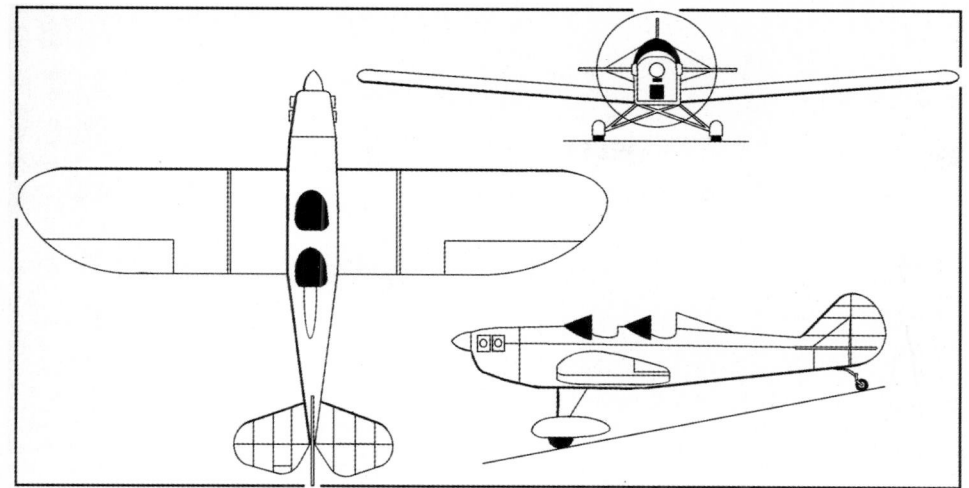

Warner Sportster tandem-seat, open-cockpit version *(Jane's/Paul Jackson)*   **NEW**/0527126

Compared to early Sportster, new version has slightly more curved lips to accommodate canopy curvature, and instrument panel moved forward.

DIMENSIONS, EXTERNAL:	
Wing span	7.92 m (26 ft 0 in)
Wing chord, constant	1.37 m (4 ft 6 in)
Wing aspect ratio	6.0
Length overall	6.20 m (20 ft 4 in)
Max diameter of fuselage	0.76 m (2 ft 6 in)
Height overall	1.68 m (5 ft 7 in)
Propeller ground clearance	0.46 m (1 ft 6 in)
Tailplane span	2.41 m (7 ft 11 in)
Wheel track	1.83 m (6 ft 0 in)
Wheelbase	4.41 m (14 ft 5½ in)
DIMENSIONS, INTERNAL:	
Cockpit max width	0.76 m (2 ft 6 in)
AREAS:	
Wings, gross	10.41 m² (115.0 sq ft)
WEIGHTS AND LOADINGS (O-290 engine):	
Weight empty	386 kg (850 lb)
Baggage capacity	11 kg (25 lb)
Max T-O weight	680 kg (1,400 lb)
Max wing loading	65.4 kg/m² (13.39 lb/sq ft)
Max power loading	7.94 kg/kW (13.04 lb/hp)
PERFORMANCE (O-290 engine):	
Never-exceed speed (VNE)	130 kt (241 km/h; 140 mph)
Max level speed	115 kt (212 km/h; 132 mph)
Normal cruising speed	104 kt (193 km/h; 120 mph)
Stalling speed	40 kt (73 km/h; 43 mph)
Max rate of climb at S/L	358 m (1,200 ft)/min
Service ceiling	3,660 m (12,000 ft)
T-O and landing run	130 m (350 ft)
Range with max fuel	356 n miles (659 km; 410 miles)
g limits	+6/−4

UPDATED

to Revolution/Space Walker. Compared to early version, new Sportster has lower bottom longeron and lowered bulkheads forward of cockpit.

LANDING GEAR: Tailwheel type; fixed. Tubular welded V legs with tubular bracing and spring shock-absorbers. 6.00-6 tyres, Cleveland wheels with disc brakes. Steerable 15 cm (6 in) Maule tailwheel. Speed fairings on mainwheels.

POWER PLANT: Engines from 63.4 to 119 kW (85 to 160 hp) can be fitted; prototype has 85.6 kW (115 hp) Textron Lycoming O-290-D. Fuel capacity 76 litres (20.0 US gallons; 16.7 Imp gallons).

ACCOMMODATION: Pilot and passenger in tandem in individual open cockpits; choice of three styles of windscreen; alternatively, rear cockpit can be covered with a bubble canopy and front canopy temporarily faired over.

WEATHERLY

WEATHERLY AIRCRAFT COMPANY

5000 Bailey Loop, McClellan, California 95652
Tel: (+1 916) 640 01 20
PRESIDENT: Hal Weatherly

Weatherly developed conversion of Fairchild M-62 in the early 1960s for agricultural role, and later designed and built the Model 201 (see *Jane's* 1979-80 for details). Further refinement of the design resulted in the Weatherly 620, production of which continues at low rate. The company is in process of moving to McClellan Park, California, the former McClellan AFB, which closed in July 2001.

UPDATED

WEATHERLY 620

TYPE: Agricultural sprayer.

PROGRAMME: Developed from Weatherly 201; prototype (N9256W) first flown 1979.

CURRENT VERSIONS: **620:** Initial version; no longer available. See *Jane's* 1982-83 for details.

620B: Current piston-engined version, powered by Pratt & Whitney radial engine. Length 8.15 m (26 ft 9 in); height 2.44 m (8 ft 0 in). Slightly different spray equipment but same sized hopper.

620TP: Initial turboprop version, with P&WC PT6A-11AG; no longer available. See *Jane's* 1982-83 for details.

620BTG: Current turboprop version; engine can be retrofitted to earlier versions.

Description applies to Weatherly 620BTG.

CUSTOMERS: By 1999, 155 Weatherly 620s of all versions had been built. The 156th was registered in 2001.

COSTS: US$185,000, less engine (1998).

DESIGN FEATURES: High lift:drag ratio wings, with tip vanes for vortex diffusion and increase of swath width; vanes fold downwards for ground handling. Hinged wing leading-edges aid inspection.

FLYING CONTROLS: Conventional and manual. Stainless steel rudder cables.

STRUCTURE: All-metal monocoque fuselage; metal wings and control surfaces; glass fibre tail fairings. Rudder deflects ±20°; elevator 27° up and 15° down; ailerons 29° up and 11° 30′ down.

LANDING GEAR: Tailwheel type; fixed. Mainwheel oleos connected to spar to disperse stresses. Spring tailwheel assembly with fully castoring and locking wheel. Three-piston Cleveland brakes. Mainwheel tyre size 8.5-10; tailwheel 12.5-4.5.

Weatherly 620B (one Pratt & Whitney R-985) *(W J Taylor)*   0084556

POWER PLANT: *620B:* One Pratt & Whitney 336 kW (450 hp) R-985-AN radial engine driving a Hartzell HC-B3R30-4 or -4B three-blade propeller. Fuel capacity 369 litres (97.5 US gallons; 81.1 Imp gallons), of which 341 litres (90.1 US gallons; 75.1 Imp gallons) is usable. Oil capacity 25.3 litres (6.7 US gallons; 5.6 Imp gallons).

620BTG: One Honeywell TPE331-1 turboprop, derated to 373 kW (500 shp), driving a McCauley 3GFR34C602/100LA-2 three-blade propeller. Fuel capacity 492 litres (131 US gallons; 108 Imp gallons) in seven tanks (two in each wing and three in wing centre-section) of which 439 litres (116 US gallons; 96.5 Imp gallons) usable. Oil capacity 0.6 litre (1.6 US gallons; 1.3 Imp gallons).

ACCOMMODATION: Pilot only. Four-way adjustable mesh seat. Baggage compartment behind seat.

SYSTEMS: 24 V electrical system; 250 A starter/generator. Optional air conditioning and windshield washer/wiper.

EQUIPMENT: Agricultural dispersal system comprises a 1,344 litre (355 US gallon; 296 Imp gallon) hopper in forward fuselage, size 1.35 m³ (47.5 cu ft); Weath-Aero fan; 5 cm (2 in) SS bottom load pump; 5 cm (2 in) Agrinautics spray pump with three-way valve; 6.10 m (20 ft) pylon-mounted drop boom with nozzles every 15 cm (6 in); 63.5 cm (2 ft 1 in) Transland gate box. Main gear and cockpit wire deflectors, plus deflector cable from cabin to fin. Rinse-out system optional, as are stainless steel spray system and flow meters.

DIMENSIONS, EXTERNAL:	
Wing span (with vanes extended)	14.22 m (46 ft 8 in)
Wing aspect ratio	7.9
Length overall	9.04 m (29 ft 8 in)
Height overall	2.90 m (9 ft 6 in)
Wheel track	3.20 m (10 ft 6 in)
Propeller diameter: 620B	2.43 m (7 ft 11½ in)
620BTG	2.49 m (8 ft 2 in)
AREAS:	
Wings, gross	25.73 m² (277.0 sq ft)
WEIGHTS AND LOADINGS:	
Weight empty	1,374 kg (3,030 lb)
Max T-O weight under CAM 8: 620B	2,721 kg (6,000 lb)
620BTG	2,858 kg (6,300 lb)
Baggage shelf	11 kg (25 lb)
Max wing loading: 620B	105.8 kg/m² (21.66 lb/sq ft)
620BTG	111.0 kg/m² (22.74 lb/sq ft)
Max power loading: 620B	8.12 kg/kW (13.33 lb/shp)
620BTG	7.67 kg/kW (12.60 lb/shp)
PERFORMANCE:	
Never-exceed speed (VNE)	153 kt (283 km/h; 176 mph)
Normal operating speed	122 kt (226 km/h; 140 mph)
Stalling speed	62 kt (115 km/h; 72 mph)
Max rate of climb at S/L	427 m (1,400 ft)/min
Service ceiling	4,572 m (15,000 ft)
T-O run	287 m (940 ft)

UPDATED

ZENITH

ZENITH AIRCRAFT COMPANY

Mexico Memorial Airport, PO Box 650, Mexico, Missouri 65265-0650
Tel: (+1 573) 581 90 00
Fax: (+1 573) 581 00 11

e-mail: info@zenithair.com
Web: http://www.zenithair.com
PRESIDENT: Sebastien C Heintz
MANAGER, OFFICE: Linda Tucker
MANAGER, PRODUCTION: Nicholas M Heintz

Zenith Aircraft Company has acquired all rights to manufacture and sell the Canadian Zenair range of light aircraft kits, detailed below. For details of the CH 2000, see AMD in this section. For details of the CH620, see 2001-02 *Jane's All the World's Aircraft*. By the end of 2001, over 2,400 kits had been produced.

In 1997, Zenith announced a joint venture with Czech Aircraft Works to assemble the CH 601 and CH 701 for the European market; see CZAW entry in Czech Republic section. Additionally these Zenith aircraft are marketed both as complete aircraft and in kit form in Germany and Austria by Roland Aircraft; they are distinguished by the suffix 'D' in model numbers.

UPDATED

ZENITH ZODIAC CH 601HD

TYPE: Side-by-side ultralight kitbuilt.

PROGRAMME: First flight of CH 600 prototype June 1984. Plans and kits of parts (45 per cent premanufactured) for improved model followed. New **CH 601UL** developed to meet TP 10141 advanced ultralight (AULA) category in Canada. Construction of prototype CH 601 (59.7 kW; 80 hp Rotax 912) began September 1990 (first flight October that year). Kit production started January 1991. UL version (2001 cost US$12,380) remains available and was last described in the 1997-98 *Jane's*.

CH 601HD offers wider cockpit and extra power for increased performance and true cross-country capability; **CH 601HDS** variant (first flown August 1991) has shorter, tapered 'speed' wing of 7.01 m (23 ft 0 in) span. HD and UL are available as kits and as 85 per cent pre-assembled kits requiring only quoted 120 working hours to complete, compared to quoted 400 hours for unassembled kits.

Description applies to CH 601HD.

CUSTOMERS: Over 300 of all versions flying by end 2000 (latest data supplied).

COSTS: Plans: US$315. Kit: US$12,620. Information pack: US$15 (2001). Materials and component parts are available.

DESIGN FEATURES: Conventional low-wing monoplane with all-moving vertical tail. Conforms to Canadian TP 10141 airworthiness standards.
Wing section NACA 65018 (modified); dihedral 6°, incidence 8°.

FLYING CONTROLS: Manual. Long-span ailerons; elevators; and Zenith all-moving vertical fin.

STRUCTURE: All-metal monocoque. Single-piece, starboard-hinged canopy.

LANDING GEAR: Non-retractable tricycle or tailwheel type. Skis and floats optional.

POWER PLANT: One engine of 47.7 to 85.75 kW (64 to 115 hp), including 59.6 kW (79.9 hp) Rotax 912 UL, with reduction gear, driving three-blade ground adjustable propeller, 47.8 kW (64.1 hp) Rotax 582 UL, 48.5 to 74.5 kW (65 to 100 hp) Teledyne Continental or 52 kW (70 hp) Volkswagen. 74.6 kW (100 hp) four-cylinder four-stroke liquid-cooled Subaru EA81 is available from Stratos for the CH 601HD. Fuel capacity 60 litres (16.0 US gallons; 13.3 Imp gallons); wing baggage lockers may be fitted with two 26.5 litre (7.0 US gallon; 5.8 Imp gallon) auxiliary tanks.

Zenith CH 601 XL *(Jane's/Paul Jackson)*   NEW/0527003

DIMENSIONS, EXTERNAL:

Wing span	8.23 m (27 ft 0 in)
Wing chord (constant)	1.24 m (4 ft 1 in)
Wing aspect ratio	5.6
Length overall	5.79 m (19 ft 0 in)
Height overall: tailwheel	1.90 m (6 ft 2¾ in)
nosewheel	1.98 m (6 ft 6 in)
Tailplane span	2.31 m (7 ft 7 in)
Wheel track	1.88 m (6 ft 2 in)
Wheelbase	1.30 m (4 ft 3¼ in)
Propeller diameter	1.73 m (5 ft 8 in)

DIMENSIONS, INTERNAL:

Cockpit max width	1.12 m (3 ft 8 in)

AREAS:

Wings, gross	12.08 m² (130.0 sq ft)

WEIGHTS AND LOADINGS (Rotax 912 for power loading):

Weight empty	290 kg (640 lb)
Baggage capacity	36 kg (80 lb)
Max T-O weight	544 kg (1,200 lb)
Max wing loading	45.1 kg/m² (9.23 lb/sq ft)
Max power loading	9.11 kg/kW (15.00 lb/hp)

PERFORMANCE (Rotax 912):

Never-exceed speed (VNE)	130 kt (241 km/h; 150 mph)
Max level speed at S/L	117 kt (217 km/h; 135 mph)
Econ cruising speed	104 kt (193 km/h; 120 mph)
Stalling speed	41 kt (76 km/h; 47 mph)
Max rate of climb at S/L	351 m (1,150 ft)/min
Service ceiling	4,875 m (16,000 ft)
T-O run	131 m (430 ft)
Landing run	159 m (520 ft)
Range with max fuel	677 n miles (1,255 km; 780 miles)
Endurance	4 h
g limits	±6

UPDATED

ZENITH SUPER ZODIAC CH 601XL

TYPE: Side-by-side kitbuilt.

PROGRAMME: Certifiable version of CH 601 (which see), with new wing planform and 589 kg (1,300 lb) MTOW. Introduced at Oshkosh in July 1998.

Description generally as for CH 601, except that below.

CURRENT VERSIONS: **CH 601XL:** US version.
CH 601DX: German designation.

COSTS: Kit US$15,890 (2003).

DESIGN FEATURES: Wing section NACA 23018 at root, NACA 23015 at tip.

FLYING CONTROLS: Flaps added.

POWER PLANT: One 89 kW (120 hp) Jabiru 3300 flat-six or 86.5 kW (116 hp) Textron Lycoming O-235. Options include 73.5 kW (98.6 hp) Rotax 912 ULS and conversions of Subaru motorcar engines. Fuel capacity 91 litres (24.0 US gallons; 20.0 Imp gallons) in two wing tanks.

DIMENSIONS, EXTERNAL:

Wing span	8.23 m (27 ft 0 in)
Wing chord: at main panel root	1.60 m (5 ft 3 in)
at tip	1.33 m (4 ft 4¼ in)
Wing aspect ratio	5.5

AREAS:

Wings, gross	12.26 m² (132.0 sq ft)

WEIGHTS AND LOADINGS (O-235):

Weight empty	363 kg (800 lb)
Baggage capacity	18 kg (40 lb)
Max T-O weight	589 kg (1,300 lb)
Max wing loading	47.5 kg/m² (9.85 lb/sq ft)
Max power loading	6.82 kg/kW (11.21 lb/hp)

PERFORMANCE (O-235):

Never-exceed speed (VNE)	156 kt (289 km/h; 180 mph)
Max level speed	129 kt (238 km/h; 148 mph)
Normal cruising speed	120 kt (222 km/h; 138 mph)
Stalling speed, flaps down	39 kt (71 km/h; 44 mph)
Max rate of climb at S/L	283 m (930 ft)/min
Service ceiling	4,267 m (14,000 ft)
T-O and landing run	152 m (500 ft)
Max range	521 n miles (965 km; 600 miles)
Endurance, no reserves	4 h 30 min
g limits	±6

UPDATED

ZENITH STOL CH 701

TYPE: Side-by-side ultralight kitbuilt.

PROGRAMME: Experimental category prototype made first flight mid-1986. Plans, 49 or 85 per cent kits available. Meets Canadian TP 10141 standards for advanced ultralight trainers. German promotion as **CH-701D**. Also available in Czech Republic as **Kappa-1**, with slight performance differences. Ekoflug of Slovak Republic builds version designated MXP-740 (which see). CH-701-AG agricultural version ceased US production after 55 completed, but revived by Czech Aircraft Works (CZAW, which see). Marketed in Italy as **ICP Savannah** and previously in France as **Aerotrophy TT 2000**.

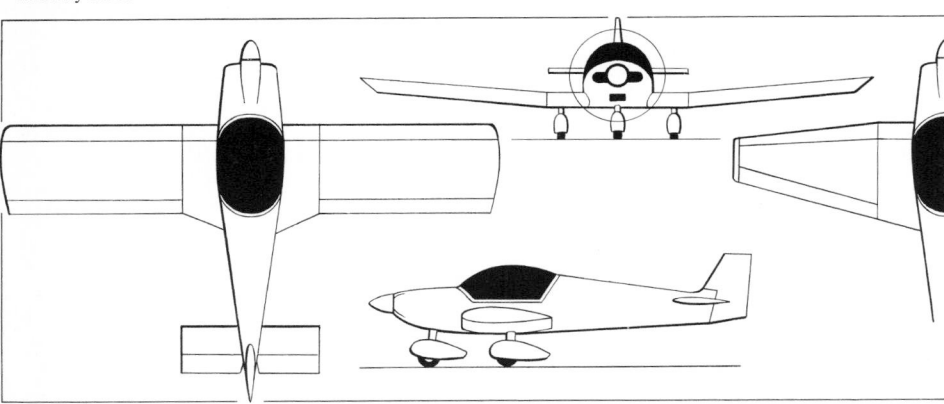

Zenith Zodiac CH 601HD, with half-plan view of tapered CH 601HDS wing *(Jane's/James Goulding)*   0016514

Zenith CH 601 HD *(Jane's/Paul Jackson)*   NEW/0527004

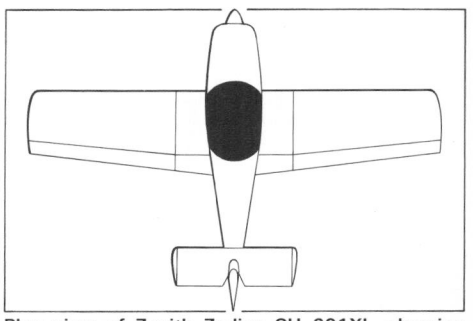

Plan view of Zenith Zodiac CH 601XL, showing revised wing shape of this version *(Jane's/James Goulding)*   0062649

In July 2001 design was upgraded with redesigned wing spar to accommodate increased MTOW, increased fuel capacity, top-hinged cabin doors and redesigned windscreen. Aircraft from serial number 7-4512 are affected.

CUSTOMERS: Over 6,300 flying by mid-2002. Indian National Cadet Corps, part of the Indian Air Force, ordered 85 plus 48 options in 2000; these are being assembled by CZAW.

COSTS: Plans: US$320. Kit: US$12,980. Information pack: US$35. Amphibian kit: US$17,730 (2003). Materials and component parts available.

DESIGN FEATURES: Strut-braced, high-wing, STOL cabin monoplane with foldable wings and fixed leading-edge slats. In most countries, CH 701 can be registered either as advanced ultralight (AUL) or as experimental homebuilt. Quoted build time of standard CH 701 400 hours.

FLYING CONTROLS: Manual. Junkers-type flaperons; elevators; and Zenith all-moving fin. Full-span leading-edge slats for STOL performance.

STRUCTURE: All-metal semi-monocoque.

LANDING GEAR: Tricycle or tailwheel gear; optional floats, amphibious or ski gears.

POWER PLANT: One 37.0 kW (49.6 hp) Rotax 503 UL-2V engine in prototype, but 59.6 kW (79.9 hp) Rotax 912 UL, 47.8 kW (64.1 hp) Rotax 582 UL, 44.7 to 52.2 kW (60 to 70 hp) VW or 48.5 to 67.1 kW (65 to 90 hp) Teledyne Continental engine optional. Two- or three-blade propeller. Standard fuel capacity 75.7 litres (20.0 US gallons; 16.7 Imp gallons) in two wing tanks.

DIMENSIONS, EXTERNAL:
Wing span	8.23 m (27 ft 0 in)
Length overall	6.38 m (20 ft 11 in)
Width, wings folded	2.06 m (6 ft 9 in)
Height overall	2.62 m (8 ft 7 in)

DIMENSIONS, INTERNAL:
Cabin max width	1.07 m (3 ft 6 in)

AREAS:
Wings, gross	11.33 m² (122.0 sq ft)

WEIGHTS AND LOADINGS:
Weight empty	218 kg (480 lb)
Max T-O weight	499 kg (1,100 lb)

PERFORMANCE (47.8 kW; 64.1 hp engine):
Max level speed at S/L	82 kt (152 km/h; 95 mph)
Max cruising speed	70 kt (129 km/h; 80 mph)
Stalling speed	27 kt (49 km/h; 30 mph)
Max rate of climb at S/L	427 m (1,400 ft)/min
T-O run	28 m (90 ft)
Landing run	24 m (80 ft)
Range: standard fuel	182 n miles (338 km; 210 miles)
with wing tanks	360 n miles (667 km; 415 miles)

UPDATED

ZENITH SUPER STOL CH 801

TYPE: Four-seat kitbuilt.

PROGRAMME: Based on STOL CH 701 design; work started in 1988 for an undisclosed customer, but was suspended when order cancelled. Subsequently revived; prototype built by Flypass Ltd in Ontario, Canada and first flown March 1998. Launched at Oshkosh, July 1998, with kit deliveries scheduled for late 1998.

CUSTOMERS: 150 kits sold and 20 flying by end 2002. Marketing by CZAW (which see in Czech section) which offers builder's assistance programme.

COSTS: Kit excluding engine US$20,950 (2003).

DESIGN FEATURES: High-wing, strut-braced monoplane. Although visually similar to the CH 701, no commonality of parts. Quoted build time 700 hours.
Wing aerofoil NACA65015.

FLYING CONTROLS: Manual. Full-span Junkers flaperons with outboard section offset 2° higher; deflection to 35° elevator; all-moving, horn-balanced fin. Fixed full-span leading-edge wing slats.

STRUCTURE: All-metal semi-monocoque construction, making extensive use of blind rivets. Single-spar 6061-T6 aluminium wing. Fuselage welded 4130 tubular steel frame.

LANDING GEAR: Tricycle type; cantilever mainwheel legs optimised for rough-field operations with Matco or Cleveland wheels and brakes. Nosewheel steerable ±15°. Tundra tyres on all wheels. Optional flotation gear.

POWER PLANT: One 134 kW (180 hp) Textron Lycoming O-360-A driving two-blade, fixed-pitch Sensenich 76-EMB-O-54 metal propeller in production prototype; engines in the range 112 to 179 kW (150 to 240 hp) can be fitted, including four-stroke LOM M337B of 173 kW (232 hp). Standard fuel capacity 151 litres (40.0 US gallons; 33.3 Imp gallons); optional fuel capacity 303 litres (80.0 US gallons; 66.6 Imp gallons).

ACCOMMODATION: Pilot and three passengers in enclosed cabin. Single door on port side. Underfuselage cargo pod optional.

DIMENSIONS, EXTERNAL:
Wing span	9.45 m (31 ft 0 in)
Wing chord, constant	1.60 m (5 ft 3 in)
Wing aspect ratio	5.8
Length overall	7.47 m (24 ft 6 in)
Height overall	2.67 m (8 ft 9 in)
Tailplane span	2.54 m (8 ft 4 in)
Tailplane chord	0.91 m (3 ft 0 in)
Wheel track	2.03 m (6 ft 8 in)

Zenith STOL CH 701 resident in Germany (*Jane's/Paul Jackson*)   *NEW*/0527002

Zenith CH 701 STOL floatplane (*Jane's/Paul Jackson*)   *NEW*/0527001

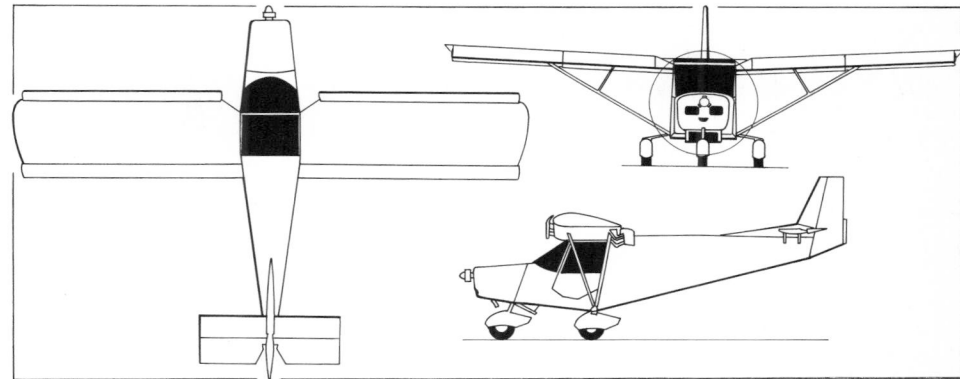

Zenith STOL CH 701 light aircraft (*Jane's/Paul Jackson*)   0051860

Zenith Super STOL CH 801 (Textron Lycoming O-360A flat-four)   *NEW*/0527136

Wheelbase	1.65 m (5 ft 5 in)
Passenger door: Height	0.97 m (3 ft 2 in)
Width	0.91 m (3 ft 0 in)
DIMENSIONS, INTERNAL:	
Cabin: Length	1.98 m (6 ft 6 in)
Max width	1.12 m (3 ft 8 in)
Max height	1.17 m (3 ft 10 in)
AREAS:	
Wings, gross	15.51 m² (167.0 sq ft)
Tailplane	2.32 m² (25.00 sq ft)
WEIGHTS AND LOADINGS:	
Weight empty	522 kg (1,150 lb)
Max T-O weight	975 kg (2,150 lb)
Max wing loading	62.9 kg/m² (12.87 lb/sq ft)
Max power loading	7.27 kg/kW (11.94 lb/hp)
PERFORMANCE:	
Never-exceed speed (V_{NE})	130 kt (241 km/h; 150 mph)
Max level speed	96 kt (177 km/h; 110 mph)
Max cruising speed at 75% power	
	91 kt (169 km/h; 105 mph)
Stalling speed: flaps up	48 kt (89 km/h; 56 mph)
flaps down	34 kt (63 km/h; 39 mph)
Max rate of climb at S/L	219 m (720 ft)/min
Service ceiling	4,267 m (14,000 ft)
T-O run	119 m (390 ft)
T-O to 15 m (50 ft)	122 m (400 ft)
Landing from 15 m (50 ft)	91 m (300 ft)
Landing run	46 m (150 ft)
Range:	
with standard fuel	273 n miles (507 km; 315 miles)
with optional fuel	547 n miles (1,013 km; 630 miles)
Endurance: with standard fuel	3 h 0 min
with optional fuel	6 h 0 min
g limits	+6/−3

UPDATED

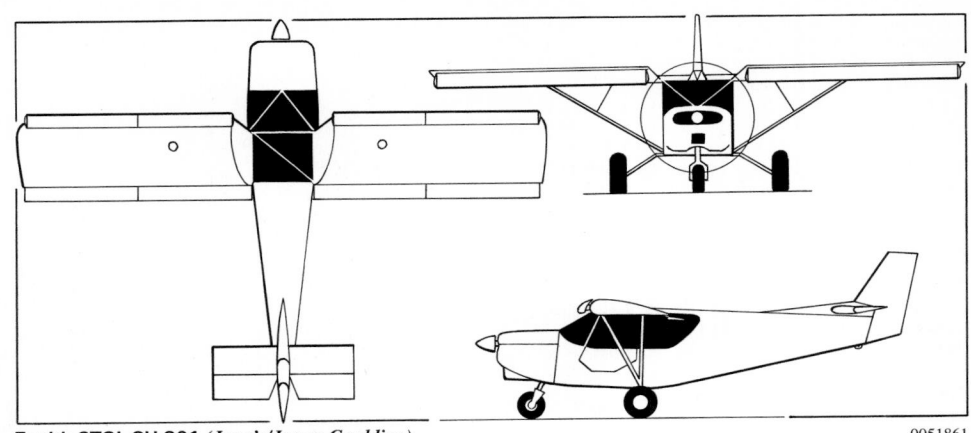

Zenith STOL CH 801 (*Jane's/James Goulding*)   0051861

Demonstrator Zenith Super STOL CH 801 advertising the builder's assistance programme offered by CZAW in conjunction with Sky Shop of the USA (*Jane's/Paul Jackson*)
NEW/0527000

ZIVKO

ZIVKO AERONAUTICS INC

502 Airport Road, Building 11, Guthrie, Oklahoma 73044
Tel: (+1 405) 282 13 30
Fax: (+1 405) 282 13 39
Web: http://www.zivko.com
PRESIDENT: Bill Zivko

Founded in 1987, Zivko manufactures the Edge 540 series of aerobatic monoplanes from a 1,858 m² (20,000 sq ft) factory on the airport at Guthrie and also provides wings for retrofitting to other aerobatic aircraft. In August 2000, it received FAA certification under Pt 21.191(g). The improved Edge 540A was introduced in April 2001. In mid-2002 the workforce numbered 20.

UPDATED

ZIVKO EDGE 540

TYPE: Aerobatic single-seat sportplane; aerobatic two-seat sportplane.
PROGRAMME: Initial version was Zivko Edge 360. Design work on 540 began in 1992 with first flight in 1993; 540T followed in 1999 with first flight 2000.
CURRENT VERSIONS: **540:** Single-seat version.
   540A: Improved version introduced at Sun 'n' Fun, April 2001; has larger rudder and elevator for improved handling.
   540T: Two-seat model introduced March 2000; differences listed below.
CUSTOMERS: 31 540s and four 540Ts built by July 2001, including examples to Mexico, South Africa and Australia.

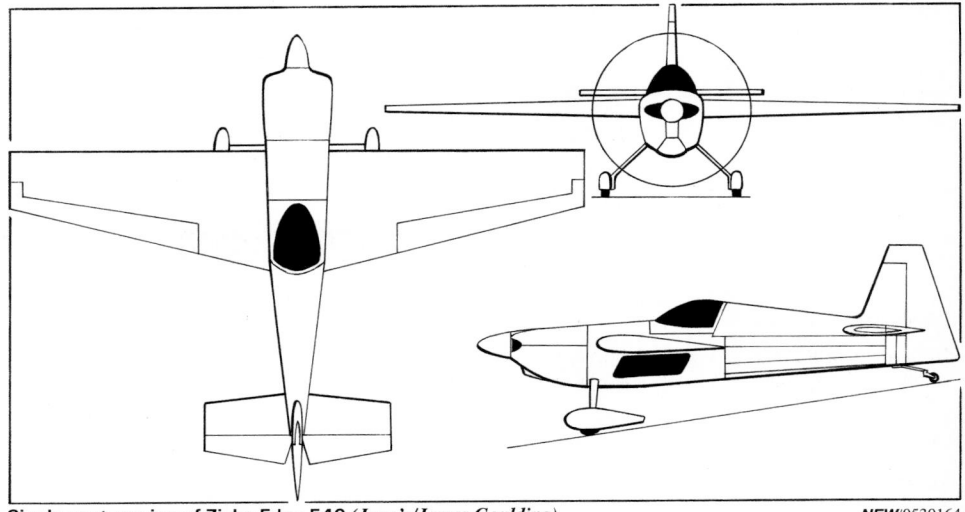

Single-seat version of Zivko Edge 540 (*Jane's/James Goulding*)   *NEW*/0530164

COSTS: Programme cost US$1.1 million by 2001. Edge 540 US$224,908; Edge 540T US$238,477 (2003).
DESIGN FEATURES: Optimised for aerobatics; customised John Roncz aerodynamic design; rate of roll 420°/s. Mid-positioned, tapered wing with straight leading-edge and large, horn-balanced ailerons and unique aerofoil. Transparencies in lower sides and floor of cockpit.

FLYING CONTROLS: Conventional and manual. Ailerons and elevator actuated by pushrods; horn-balanced rudder by cables. Flight adjustable tabs in port elevator. Suspended, spade-type tab on each aileron; electric trim.
STRUCTURE: 4130 Steel tube fuselage; upper fuselage and fairings of composites; lower aft fuselage fabric covered, stressed to ultimate ±15 g; all-composites fully cantilevered two-spar wing. Fully composites empennage with no wire bracing.
LANDING GEAR: Tailwheel type; fixed. Main legs aluminium sprung steel; Cleveland 5.00-5 wheels and hydraulic brakes; optional speed fairings. Tyre pressure 3.45 bar (50.0 lb/sq in). Spring steel tail gear, with chrome-plated, solid 4.00-4 tailwheel, choice of fixed or steerable.
POWER PLANT: One Textron Lycoming IO-540 modified to produce 254 kW (340 hp), driving Hartzell HC-C3YR-4AX or optional Hartzell HC-C3YR-1AX1 three-blade, composites propeller. Edge 540 has single 72 litre (19.0 US gallon; 15.8 Imp gallon) forward fuselage fuel tank, plus optional total of 159 litres (42.0 US gallons; 35.0 Imp gallons) in wing tanks; Edge 540T has single 76 litre (20.0 US gallon; 16.6 Imp gallon) fuselage tank and two optional 83 litre (22.0 US gallon; 18.3 Imp gallon) wing tanks. Oil capacity 11.4 litres (3.0 US gallons; 2.5 Imp gallons). Inverted oil (Christen) and fuel (ZAI) systems.
ACCOMMODATION: Pilot (pilot and passenger in tandem in 540T) under one-piece perspex canopy hinged on starboard side; carbon fibre contoured seats. Baggage compartment forward of front seat.

Zivko Edge 540A N24KC, flown by Kirby Chambliss, made its debut at Sun 'n' Fun, April 2001; note earlier braced tailplane (*Jane's/Susan Bushell*)   0110967

SYSTEMS: 12 V DC system with light weight B&C 8 A alternator and regulator. Bendix fuel injection system.

AVIONICS: *Comms:* Becker com and Mode C transponder.

Flight: Optional Sierra Flight Systems EFIS.

DIMENSIONS, EXTERNAL:

Wing span: 540	7.42 m (24 ft 4 in)
540T	7.87 m (25 ft 10 in)
Wing chord: at root: 540	1.57 m (5 ft 2 in)
540T	1.68 m (5 ft 6 in)
at tip: 540	0.74 m (2 ft 5 in)
540T	0.84 m (2 ft 9 in)
Length overall: 540	6.27 m (20 ft 7 in)
540T	7.01 m (23 ft 0 in)
Height overall	2.36 m (7 ft 9 in)
Tailplane span	2.46 m (8 ft 1 in)

AREAS:

Wings, gross: 540	9.10 m² (98.0 sq ft)
540T	9.87 m² (106.2 sq ft)

WEIGHTS AND LOADINGS:

Weight empty: 540	531 kg (1,170 lb)
540T	572 kg (1,261 lb)
Baggage capacity	5 kg (10 lb)
Max T-O weight: 540: Aerobatic	703 kg (1,550 lb)
Utility	816 kg (1,800 lb)
540T: Aerobatic	726 kg (1,600 lb)
Utility	885 kg (1,950 lb)
Max wing loading: 540:	
Aerobatic	77.2 kg/m² (15.82 lb/sq ft)
Utility	89.7 kg/m² (18.37 lb/sq ft)
540T: Aerobatic	73.6 kg/m² (15.07 lb/sq ft)
Utility	89.6 kg/m² (18.36 lb/sq ft)
Max power loading: 540	3.22 kg/kW (5.29 lb/hp)
540T: Aerobatic	2.86 kg/kW (4.71 lb/hp)
Utility	3.49 kg/kW (5.74 lb/hp)

PERFORMANCE:

Never-exceed speed (V$_{NE}$)	230 kt (426 km/h; 265 mph)
Max cruising speed at 75% power	195 kt (361 km/h; 224 mph)
Stalling speed: 540: Aerobatic	53 kt (98 km/h; 61 mph)
Utility	61 kt (113 km/h; 71 mph)
540T: Aerobatic	53 kt (99 km/h; 61 mph)
Utility	61 kt (113 km/h; 71 mph)
Max rate of climb at S/L: 540	1,128 m (3,700 ft)/min
540T	1,067 m (3,500 ft)/min
Service ceiling	3,810 m (12,500 ft)
T-O run	61 m (199 ft)
T-O to 15 m (50 ft)	90 m (293 ft)
Landing from 15 m (50 ft)	300 m (982 ft)
Range with max fuel	456 n miles (844 km; 525 miles)
g limits: Aerobatic	±10
Utility	±8

UPDATED

Zivko Edge 540T two-seat aerobatic lightplane with earlier braced tailplane (*Jane's/Paul Jackson*)

NEW/0526999

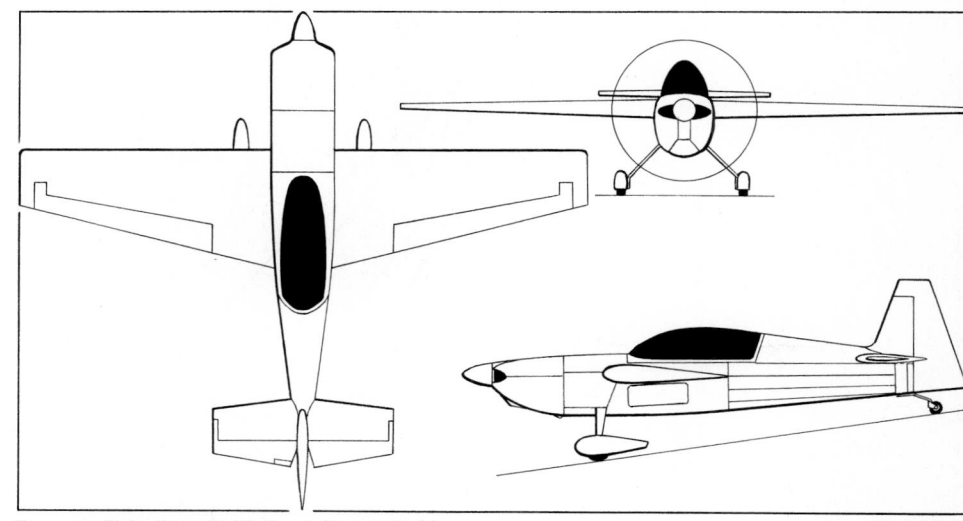

Two-seat Zivko Edge 540T (*Jane's/James Goulding*)

NEW/0530163

UZBEKISTAN

TAPO

TASHKENTSKOYE AVIATSIONNOYE PROIZDSTVENNOYE OBEDINENIE IMENI V P CHKALOVA (Tashkent Aircraft Production Enterprise named for Valery P Chkalov)

ulitsa Elbek 61, 700016 Tashkent

Tel: (+810 99871) 136 11 67

Fax: (+810 99871) 268 03 18

e-mail: oms@tapo.tashkent.ru

GENERAL DIRECTOR: Vladim P Kucherov

DEPUTY GENERAL DIRECTOR: Nabil A Artykov

TECHNICAL DIRECTOR: Nikolai A Kuznetsov

MARKETING DIRECTOR: Zafar M Islamov

Tashkent plant established 20 November 1941 on arrival of first elements of GAZ 84 (formed 1932 and incorporating Polikarpov from 1936) evacuated from Khimki, Moscow. Enterprise, dating from 1972 in current form, is said to be the largest aircraft manufacturing centre on the Asian continent. It comprises five plants at Tashkent, Andizhan and Fergana. Total of over 7,000 aircraft built, including I-15s, I-16s, I-153s, Li-2s (total 2,258), Il-14s, An-8s, Ka-22s, An-12s, An-22s and Il-76s. It has manufactured wing centre-sections for the An-124 and An-225 transports and two An-70 prototypes, and has repaired MiG and Sukhoi fighters and Tu-22M bombers. It developed the Il-78 tanker version and collaborated with Beriev on the A-50 AEW derivative of the Il-76. Low-rate production is under way of the Ilyushin Il-114 twin-turboprop transport and Il-76, for which TAPO is the sole source, although airframes are stored partly complete until orders are received. Il-114s retrofitted with US engines, as required by demand.

TAPO was registered as a joint stock company in May 1996; the state holds 82.7 per cent, National Bank of Foreign Economic Activity 6.7 per cent and employees 10.6 per cent. An attempt in 1997 to sell 14 per cent of the company was unsuccessful; by mid-1998, offers were being invited for a 25 per cent share, but no acceptances have been reported. Company began airbrake manufacture for Avro RJ in 1997; this now discontinued. Following April 2000 foundation of Ilyushin International company, TAPO expected to join forces with design bureau responsible for its current production aircraft.

Subsidiary, TAPO-Avia operates commercial freight charter flights with fleet of four An-12s, one An-24 and five Il-76MDs. TAPO aircraft are marketed via Uzavialeasing, established December 1997 in collaboration with local government and Russian and Uzbek banks.

UPDATED

LIGHTER THAN AIR

CANADA

21ST CENTURY

21ST CENTURY AIRSHIPS INC

1188B Gorham Street, Newmarket, Ontario L3Y 7V1
Tel: (+1 905) 898 62 74
Fax: (+1 905) 898 72 45
e-mail: ballof@rogers.com
Web: http://www.21stCenturyAirships.com
CHAIRMAN AND CEO: Hokan Colting

Formed 1988 to focus on R&D aimed at improving airship low-speed manoeuvrability, a problem its chairman identified and has studied since 1978, particularly with spherical airships. He moved from Sweden to Canada 1981; has extensive experience in manufacturing and piloting lighter than air craft; achieved all objectives of his original 1978 research programme. Company has built and flight tested nine spherical airship manned prototypes since 1991.

Details of the early prototypes can be found in the 1995-96 *Jane's*. Descriptions of latest known programmes follow. A new product, the SPAS-R1, was produced in 2001, registered in October of that year as C-GJKI; it flew for the first time on 30 June 2002.

UPDATED

21ST CENTURY SPAS-13

TYPE: Helium non-rigid special shape.
PROGRAMME: Demonstrator prototype: first flight (C-FRLM) 10 June 1994. Made 112 flights as a semi-rigid before being deflated on 20 November 1994. Reinflated in 1995 after a number of improvements was incorporated; 40 more flights completed; flew again in early 1998 as full-scale platform for a drag-reducing device which enables engine horsepower to be decreased while maintaining same speed. This invention planned to be incorporated in all the company's spherical airships. Remained registered in 2001.
DESIGN FEATURES: Spherical shape, with outer load-bearing envelope and separately sealed inner helium envelope. Only two ground crew required.
FLYING CONTROLS: Steering and altitude controlled through varied and deflected thrust technology developed and patented by 21st Century Airships Inc.
POWER PLANT: Two 37.0 kW (49.6 hp) Rotax 503 UL-2V piston engines.
ACCOMMODATION: Flight deck and passenger compartment inside load-bearing envelope.
DIMENSIONS, EXTERNAL:
Envelope diameter 13.72 m (45 ft 0 in)
DIMENSIONS, INTERNAL:
Envelope volume 1,167 m³ (41,200 cu ft)

WEIGHTS AND LOADINGS:
Weight empty 772 kg (1,702 lb)
Max T-O weight 1,100 kg (2,425 lb)
PERFORMANCE:
Max level speed 30 kt (56 km/h; 35 mph)
Max rate of climb at S/L 366 m (1,200 ft)/min
Max tested pressure altitude 1,400 m (4,600 ft)
Endurance, 30 min reserves 1 h 30 min
VERIFIED

21ST CENTURY SPAS-70

TYPE: Helium non-rigid special shape.
PROGRAMME: Prototype (C-FYOK) made first flight 5 August 1997 with four 74.6 kW (100 hp) piston engines; used for certification trials, which were delayed to incorporate drag reducer (see SPAS-13 entry) and change to twin diesel engines. Latter installed in 1999, but funding problems apparently still delaying flight testing in this form by late 2001. Modified 1999-2000 to permit take-offs and landings on water. Remained registered in 2001, in which year work started on pre-production example with 18.29 m (60 ft) diameter envelope.
DESIGN FEATURES: As described for SPAS-13.
FLYING CONTROLS: All controls are fly-by-wire; otherwise as described for SPAS-13.
POWER PLANT: Two 67.1 kW (90 hp) turbo-diesel engines planned for eventual production version.
ACCOMMODATION: Four seats inside load-bearing envelope; only two fitted for certification flights; positioning of other two optional within large internal floor area.
AVIONICS: *Comms:* VHF transceivers; transponder.
  Flight: GPS nav receiver and radar altimeter. Goodrich Stormscope adverse weather warning system.
  Instrumentation: Regular engine monitoring, plus internal pressure gauges, outside/inside and helium temperature gauges, and fire warning lights for each engine and APU. All engine instruments and controls colour-coded to simplify management of four engines.
DIMENSIONS, EXTERNAL:
Envelope diameter 17.07 m (56 ft 0 in)
Propeller diameter 3.96 m (13 ft 0 in)
DIMENSIONS, INTERNAL:
Envelope volume 2,605 m³ (92,000 cu ft)
Usable floor space inside load-bearing envelope
65.0 m² (700 sq ft)
WEIGHTS AND LOADINGS:
Weight empty 1,775 kg (3,913 lb)
Max T-O weight 2,520 kg (5,555 lb)
PERFORMANCE:
Max level speed 35 kt (64 km/h; 40 mph)
Max rate of climb at S/L 335 m (1,100 ft)/min

SPAS-70 flight deck, with colour-coded engine instruments 0105831

Max tested pressure altitude 3,155 m (10,350 ft)
Endurance, 30 min reserves 11 h
VERIFIED

21ST CENTURY SPAS-R1

TYPE: Helium non-rigid special shape.
PROGRAMME: Prototype completed in 2001; flew for first time (C-GJKI) 30 June 2002, reaching approximately 5,480 m (18,000 ft). Forerunner of planned range of manned/unmanned stratosphere balloons for telecommunications and environmental monitoring. Typical operational

SPAS-13 in current non-rigid form, painted to resemble a baseball 0105832

The four-seat, four-engined SPAS-70 0051874

version could have 79.25 m (260 ft) diameter and 260,000 m³ (9.2 million cu ft).

DESIGN FEATURES: Spherical shape, similar to others in company's range. Designed for flight in stratosphere (18,280 to 20,720 m; 60,000 to 68,000 ft) gathering data for surveillance, communications and environmental monitoring by airships at these altitudes.

POWER PLANT: Two turbo-diesel engines with large-diameter propellers, plus electric motor.

ACCOMMODATION: Four persons, including pilot.

DIMENSIONS, EXTERNAL:
Envelope diameter   18.29 m (60 ft 0 in)

DIMENSIONS, INTERNAL:
Envelope volume   approx 3,001.6 m³ (106,000 cu ft)
UPDATED

21st Century's new SPAS-R1 stratospheric airship
***NEW**/0525379*

PAN ATLANTIC

PAN ATLANTIC AEROSPACE CORPORATION (Subsidiary of Av-Intel Corporation)

PO Box 599, Station B, Ottawa, Ontario K1P 5P7
Tel: (+1 561) 776 13 51
Fax: (+1 561) 776 23 70
e-mail: ltadrone@aol.com
CHAIRMAN: Fredrick D Ferguson
VICE-PRESIDENT: Ian Stewart
INFORMATION OFFICER: Nick Baumberg

Pan Atlantic has been involved in lighter than air design and systems integration since 1978, including the Magnus Aerospace rotating sphere development project. Many of its designs have been awarded patents. Company's recent LTA-based projects have included a high-altitude, satellite-like platform in partnership with Skysat Communications Networks Corporation of New York; the Cargo Airship System (CAS); the LEAP surveillance drone; and the Magnus advertising vehicle. There have been no recent reports of the Cargo Airship System.
UPDATED

PAN ATLANTIC CARGO AIRSHIP SYSTEM (CAS)

Last described in 2002-03 *Jane's*. Latest known information suggests that the company is still searching for outside investor(s) to contribute US$20 million to build a demonstrator.
UPDATED

CHINA, PEOPLE'S REPUBLIC

SVAM

SHANGHAI VANTAGE AIRSHIP MANUFACTURING COMPANY (Shanghai Vantage Feiting Zhizao Gongsi)

Tel: (+86 21) 39 51 01 81
Fax: (+86 21) 62 46 30 43
e-mail: svam@sh163.com
Web: http://www.vantageship.com

Company registered in May 2000; workforce mainly former military personnel, many (including three pilots) wth helium airship experience; has CAAC approval. Completed its first prototype airship (single-place CA-50) in September 1999; second prototype followed September 2001.

Latest product is medium-sized, twin-engined CA-80 for advertising, aerial photography and surveillance. Long-term plans include 60 tonne payload CA-800.

NEW ENTRY

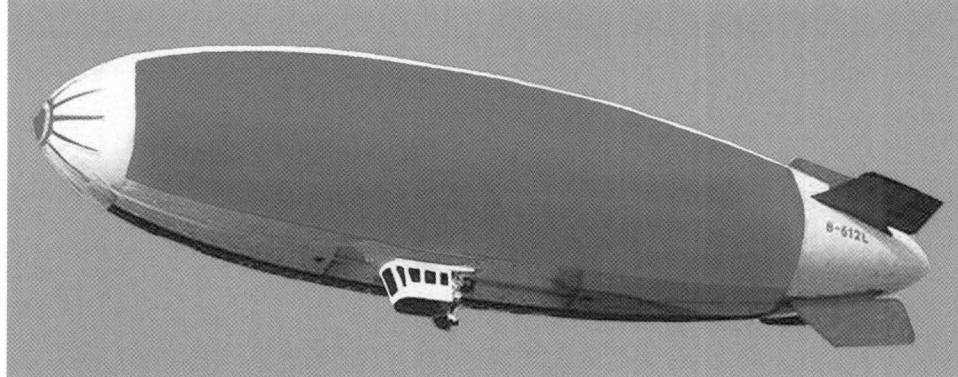

The twin-engined SVAM CA-80 helium semi-rigid   ***NEW**/0525362*

CZECH REPUBLIC

KUBICEK

BALONY KUBICEK Spol. s.r.o.

Francouzská 81, CR-602 00 Brno
Tel: (+420 5) 45 21 19 17 and 45 21 47 55
Fax: (+420 5) 45 21 30 74
e-mail: sales@kubicekballoons.cz
Web: http://www.kubicekballoons.cz
MANAGING DIRECTOR: Dipl Ing Ales Kubíček
SALES MANAGER: Michael Suchý

First Kubíček hot-air balloon was manufactured in 1983 and was followed by about 50 more of Kubíček design by various state-owned companies such as Aviatik and Aerotechnik. Present company was formed as private enterprise in 1991 and began new range of hot-air balloons under BB series designations (total of over 180 produced by early 2002). Currently markets standard, ready-to-fly balloons in nine envelope sizes from 1,200 m³ (42,380 cu ft) to 6,000 m³ (211,900 cu ft); also special-shape balloons to order.

Workforce of 70 in mid-2001, in factory area of 2,400 m² (25,833 sq ft).

Kubíček's first hot-air airship design was the AV-1. A new design, the AV-2, made its maiden flight in 1999. Both continue to be marketed in Russia by Avgur and RosAeroSystems (which see).
UPDATED

KUBICEK AV-2

TYPE: Hot-air non-rigid.
PROGRAMME: First flight 26 May 1999; public debut (c/n 106/OK-8036) June 1999; then sold to ADS Kosice in Slovak Republic as OM-ADS.
COSTS: €75,882 with Rotax 912 (2000).
LANDING GEAR: Tricycle type; fixed.
POWER PLANT: One 34.0 kW (45.6 hp) Rotax 503 UL-2V engine and Avia V-230C two-blade wooden propeller standard; Rotax 912 optional. Fuel tank beneath centre seat, capacity 30 litres (7.9 US gallons; 6.6 Imp gallons).

Kubicek AV-2 three-place hot-air airship   0121731

ACCOMMODATION: Pilot and two passengers in all-metal (welded chromoly steel tube frame, laminated plastics skin) gondola.

AVIONICS: *Comms:* Icom A3E 760-channel VHF radio.
   Flight: Garmin 12XL nav receiver.
   Instrumentation: Flytec 3040 instrument pack.
DIMENSIONS, EXTERNAL:
   Envelope: Length   38.00 m (124 ft 8 in)
      Max diameter   14.20 m (46 ft 7 in)

Gondola: Length	2.90 m (9 ft 6¼ in)
Max width	1.70 m (5 ft 7 in)
Max height	2.00 m (6 ft 6¾ in)
DIMENSIONS, INTERNAL:	
Envelope volume	3,500 m³ (123,601 cu ft)
WEIGHTS AND LOADINGS:	
Weight empty	500 kg (1,102 lb)

Fuel weight	20 kg (44 lb)
Max T-O weight	910 kg (2,006 lb)
PERFORMANCE (Rotax 503):	
Max level speed	10.5 kt (20 km/h; 12 mph)
Pressure ceiling (estimated)	2,440 m (8,000 ft)
Endurance	1 h 30 min to 2 h
	UPDATED

FRANCE

AERALL

ASSOCIATION POUR L'ETUDE ET LA RECHERCHE SUR LES AERONEFS ALLEGES

2 bis, avenue Odette, F-94130 Nogent-sur-Marne
Tel: (+33 1) 49 00 15 14
Fax: (+33 1) 47 75 09 56
PRESIDENT: Jean René Fontaine

VERIFIED

AERALL AVEA

TYPE: Helium non-rigid.
PROGRAMME: Originated in mid-1990s in sounding balloon department of CNES space agency; has French Ministry of Transport, regional and private funding. Phase 1 (preliminary design evaluation) undertaken early 2000; Phase 2 (feasibility and market studies) was undertaken in late 2000 by Office National d'Etudes et de Recherche

Aérospatiale (ONERA). Single MAP (see below) test-flown in September 2000; manufacture of full-size AVEA projected to start in 2004, with first flight 2006 and service entry in 2008.
DESIGN FEATURES: AVEA (*Aile Volante Epaisse Aérostatique:* thick aerostatic flying wing) is an unorthodox, ovoid design comprising a number of tall, cylindrical, controllable modules (*modules aérostatiques pilotables,* or MAPs), attached in tandem and linked across their tops and bottoms by a trellis-like framework. Each MAP is provisionally 20 m (65.6 ft) in diameter and 100 m (328.1 ft) tall and comprises an inner and outer envelope separated by an overpressure layer of air, the latter serving both to add rigidity and act as ballast. The number of MAPs can be varied according to the payload capacity required, envisaged as ranging from 70 to 500 tonnes (154,325 to 1,102,300 lb). An **Lv-001** prototype is envisaged as a 200 m (656 ft) long, 1.5 million m³ (53 million cu ft) craft having a range of up to 5,400 n miles (10,000 km; 6,215 miles).

UPDATED

AERALL AVEA Lv-001 *(Jane's/James Goulding)*
0131832

AMSAT

ADVANCED MECHANICAL SYSTEMS AND ALTERNATIVE TECHNOLOGY

CHIEF DESIGNER: Jacques Lantini

As an alternative to trans-Europe road transportation of large Airbus A380 subassemblies from Germany, Spain and UK to Toulouse, a number of outsize airship projects are currently being studied. One such design is the Mantha.

NEW ENTRY

AMSAT MANTHA

English name: Manta
TYPE: Helium semi-rigid.
PROGRAMME: Design study in progress in 2002.
DESIGN FEATURES: Flying wing configuration, with symmetrical aerofoil section; ventral crew gondola. Ballonets for pressure control, maintained in place by composites trellis framework.
LANDING GEAR: Non-retractable.
POWER PLANT: Two turboprop/turboshafts, mounted on top of envelope, generating energy to drive five ducted fans; two tandem pairs in centre of envelope for vertical lift and one, with vectored thrust, at rear to provide forward/upward propulsion.
ACCOMMODATION: Crew of three (pilot, co-pilot and loadmaster).
Following data are provisional.
DIMENSIONS, INTERNAL:
   Envelope volume   50,000 m³ (1.77 million cu ft)
PERFORMANCE (estimated):
   Cruising speed   54 kt (100 km/h; 62 mph)
   Cruising altitude   1,500-2,000 m (4,920-6,560 ft)

NEW ENTRY

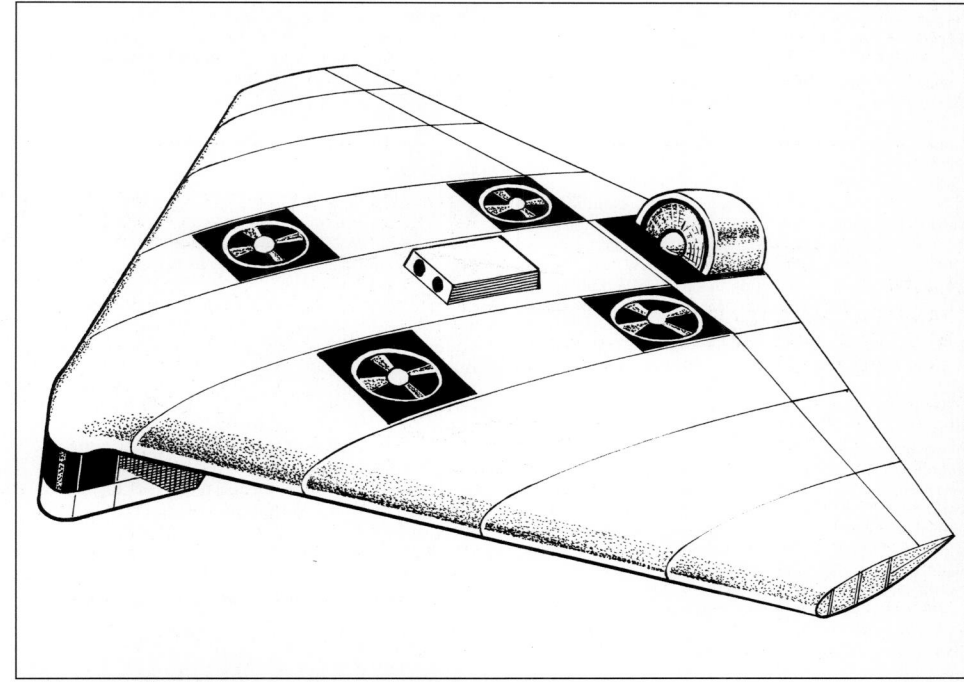

Artist's impression of AMSAT Mantha *(Jane's/James Goulding)*   *NEW*/0526044

EADS FRANCE

EADS FRANCE

37 boulevard Montmorency, F-75781 Paris Cedex 16
Tel: (+33 1) 42 24 24 24
Fax: (+33 1) 42 24 26 19
Web: http://www.eads.net

The Délégation Générale des Armements (DGA) is studying a number of domestic and foreign airship proposals for future strategic and tactical transport use by the French Army. EADS is involved in two of these, the other being the Aerall AVEA (which see).

NEW ENTRY

EADS FIRST 1A

TYPE: Helium semi-rigid.
PROGRAMME: FIRST (*Force pour l'Innovation, la Recherche Scientifique et Technique*) is a current design study for an outsize transport airship.
DESIGN FEATURES: Not yet finalised (2002); ballonet pressure control.
STRUCTURE: Composites.
Following data are provisional.
POWER PLANT: Six 4,500 kW (6,035 shp) turboprops, each driving a 10-blade propeller.
DIMENSIONS, EXTERNAL:
   Length overall   300 m (984.3 ft)
   Envelope max diameter   73 m (239.5 ft)

DIMENSIONS, INTERNAL:	
Envelope volume	600,000 m³ (21.2 million cu ft)
Cargo bay volume	5,760 m³ (203,400 cu ft)
WEIGHTS AND LOADINGS:	
Max payload	250,000 kg (551,155 lb)
PERFORMANCE (estimated):	
Cruising speed	86 kt (160 km/h; 99 mph)
Range	3,239 n miles (6,000 km; 3,728 miles)
	NEW ENTRY

VOLIRIS

VOLIRIS SARL

10 Chemin des Heuleux, F-78240 Chambourcy
Tel: (+33 1) 39 79 39 88
Fax: (+33 1) 30 65 88 48

e-mail: info@voliris.com
Web: http://www.voliris.com

Voliris was formed in 1999. It has already flown a 9.5 m (31.2 ft) long, 25 m³ (883 cu ft) radio-controlled test model,

and hoped to complete the prototype Voliris 900 by mid-2002.

NEW ENTRY

VOLIRIS 900

TYPE: Helium semi-rigid.

PROGRAMME: Prototype planned to make maiden flight in mid-2002.

POWER PLANT: One (unspecified) 89.5 kW (120 hp) piston engine, driving pusher propeller at rear of gondola.

ACCOMMODATION: Pilot and one passenger.

DIMENSIONS, EXTERNAL:
Length overall	31.00 m (101 ft 8½ in)
Envelope max diameter	8.00 m (26 ft 3 in)

DIMENSIONS, INTERNAL:
Envelope volume	1,000 m³ (35,315 cu ft)
Helium volume	900 m³ (31,783 cu ft)

AREAS:
Envelope surface area	615.0 m² (6,620 sq ft)

PERFORMANCE (estimated):
Max level speed	45-48 kt (85-90 km/h; 52-55 mph)
Cruising speed	35-38 kt (65-70 km/h; 40-43 mph)
Endurance	5 h

NEW ENTRY

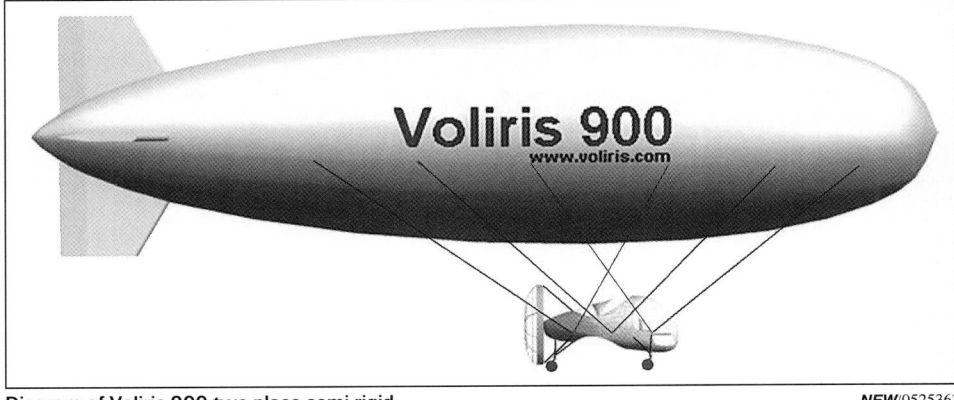

Diagram of Voliris 900 two-place semi-rigid

NEW/0525363

GERMANY

CARGOLIFTER

CARGOLIFTER AG

Grüneburgweg 102, D-60323 Frankfurt-am-Main
Tel: (+49 69) 15 05 70
Fax: (+49 69) 150 58 18
e-mail: info@cargolifter.com
Web (1): http://www.cargolifter.de
Web (2): http://www.cargolifter-world.de
MANAGEMENT BOARD:
   Dr Wolfgang Schneider (CEO)
   Karl Bangert
MARKETING COMMUNICATIONS: Silke Rösser

Company formed on 1 September 1996 to continue development of CargoLifter large multipurpose airship for transportation of heavy and outsize cargoes. Nominal start-up capital of DM15.5 million; by June 2001, some 27 million shares sold in Germany and internationally and raised capital €40.5 million (DM79.2 million). Huge, 360 × 210 × 107 m (1,181 × 689 × 351 ft), 5.5 million m³ (194.2 million cu ft) hangar at Brand, south of Berlin, began construction March 1999 and completed in November 2000; can house two CL 160s. Second hangar was planned at new site in North Carolina, USA, for completion in about 2005. Full-size CL 160 preceded by subscale 'Joey' as proof-of-concept prototype and CL 75 AirCrane transport balloon; Skyship 600B, known as 'Charly', acquired in July 2000 for crew training, certified by LBA August 2001 and now registered D-LCLA.

Additional shares offered in November 2001 in attempt to increase capital to a potential €50.6 million, but insufficient investment received. US office closed as economy measure in second quarter 2002, and on 9 May company offered convertible bonds to general public in renewed effort to raise working capital. However, this also failed to meet expectations and, on 28 May, CargoLifter admitted inability to pay creditors or employees. Applications for insolvency filed, on 31 May and 7 June 2002 respectively, by CargoLifter Development GmbH (wholly owned subsidiary) and CargoLifter AG; Professor Dr Rolf-Dieter Mönning appointed administrator for both companies. Number of shareholders at latter date was 72,000. CargoLifter AG restructured under two-man management board in order to develop new programme concept which would focus first on sales and production of CL 75 AirCrane, while maintaining core competencies for eventual realisation of CL 160. However, plans suffered further setback on 11 July 2002 when envelope of sole CL 75 AirCrane was ruptured during a storm.

Letter of intent signed with Boeing Phantom Works on 2 May 2002, under which each to investigate business opportunities for CL 160 "beyond CargoLifter's current focus"; Boeing interest said to include military surveillance possibilities; was reaffirmed in 30 July 2002 announcement of contract signature between CargoLifter AG and Boeing Unmanned Systems for joint exploration of stratospheric airship concepts.

Insolvency proceedings began on 1 August 2002; on 9 August, application for public sector assistance was rejected by German Ministry of Economic Affairs. Workforce was

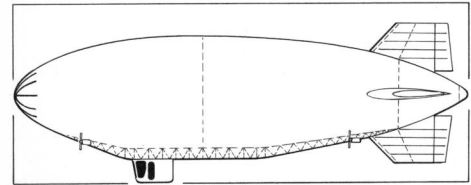

CargoLifter Joey one-eighth scale prototype for the CL 160 (*Jane's/James Goulding*) 0051886

The Joey on its mooring mast at Brand 0130557

491 in early June 2002, of whom 284 involved in airship development at Brand, but most of these suspended by insolvency administrator, leaving a staff of only 30 by mid-August 2002.

UPDATED

CARGOLIFTER JOEY

TYPE: Helium semi-rigid.

PROGRAMME: One-eighth scale prototype to undertake materials, load suspension, flight behaviour and certification trials on behalf of CL 160; name Joey (as in baby kangaroo) chosen because dimensionally it could fit

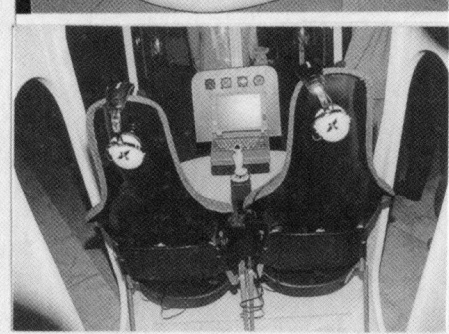

Crew gondola of the Joey prototype 0051878

inside the loading bay 'pouch' of the CL 160. Designed by engineers from Stuttgart University; exhibited at Leipzig Trade Fair in May 1998; first flight (D-LJOE) 18 October 1999.

STRUCTURE: Envelope manufactured by Lindstrand Balloons (see under UK).

POWER PLANT: Two 17.2 kW (23 hp) Solo piston engines, each driving an MT 140R 60 propeller. Fuel capacity 40 litres (10.6 US gallons; 8.8 Imp gallons).

ACCOMMODATION: Small gondola for pilot and test-flight engineer.

DIMENSIONS, EXTERNAL:
Length overall	32.00 m (104 ft 11¾ in)
Max diameter	8.00 m (26 ft 3 in)
Fineness ratio	4.0

DIMENSIONS, INTERNAL:
Envelope volume	1,050 m³ (37,109 cu ft)

WEIGHTS AND LOADINGS:
Max payload	220 kg (485 lb)
Max T-O weight	830 kg (1,829 lb)

PERFORMANCE (estimated):
Max level speed	33 kt (61 km/h; 38 mph)
Cruising speed	20 kt (37 km/h; 23 mph)
Pressure ceiling	300 m (985 ft)
Endurance at cruising speed	4 h

VERIFIED

CARGOLIFTER CL 75 AIRCRANE

TYPE: Helium balloon.

PROGRAMME: AirCrane introduced as transportation balloon, initially as materials, function and loads testbed for CL 160 programme; assembly began January 2001. Used 30/31 July 2001 to validate inflation procedure for CL 160, and is largest aircraft yet to be filled with helium, having greater volume than pre-war *Graf Zeppelin* airship. Ballast exchange procedure tested on 5 October 2001; taken out of hangar for first time seven days later.

CL 75 originally intended for commercial operation after completing commitments to CL 160 development, but became prime focus of current company activity following 2002 insolvency. Japanese shipping company Nippon Yusen Kaishe (NYK) appointed in November 2001 to market CL 75 in Japan and other Asian countries.

CUSTOMERS: First sale announced 16 March 2002, to Heavy Lift Canada Ltd, which has option on a further 25. First delivery was then scheduled for December 2002, but compromised by storm damage to first AirCrane on 11 July 2002.

World's largest helium-filled aircraft: the envelope of the CL 75 AirCrane   0130555

COSTS: US$10 million (2002).
DESIGN FEATURES: Intended for transporting loads of up to 75 tonnes (73.8 UK tons) in weight and 13 × 6 × 6 m (42.7 × 19.7 × 19.7 ft) in size, with range (towed) of approximately 135 n miles (250 km; 155 miles); can be towed by ground vehicle, ship or helicopter. Balloon has a suspended load frame on which cargo is mounted.
DIMENSIONS, EXTERNAL (approx):
  Diameter   61.00 m (200 ft)
  Height overall, incl load frame   85.00 m (279 ft)
DIMENSIONS, INTERNAL (approx):
  Envelope volume   110,000 m³ (3.885 million cu ft)
WEIGHTS AND LOADINGS:
  Max payload   75,000 kg (165,346 lb)
PERFORMANCE:
  Transportation speed (depending on towing mode)
                             up to 38 kt (70 km/h; 43 mph)
                                                 UPDATED

CARGOLIFTER CL 160

TYPE: Helium semi-rigid.
PROGRAMME: Began 1994; outcome of research project funded by German VDMA engineering plant construction organisation, indicating possibility of lower operating costs than conventional aircraft, higher speeds than a ship, and minimal need for ground handling infrastructures. Project had the support of Brandenburg state government. Design frozen 5 November 1998. Construction plans were for two prototypes and two pre-series airships; first flight rescheduled for 2005 prior to June 2002 insolvency.

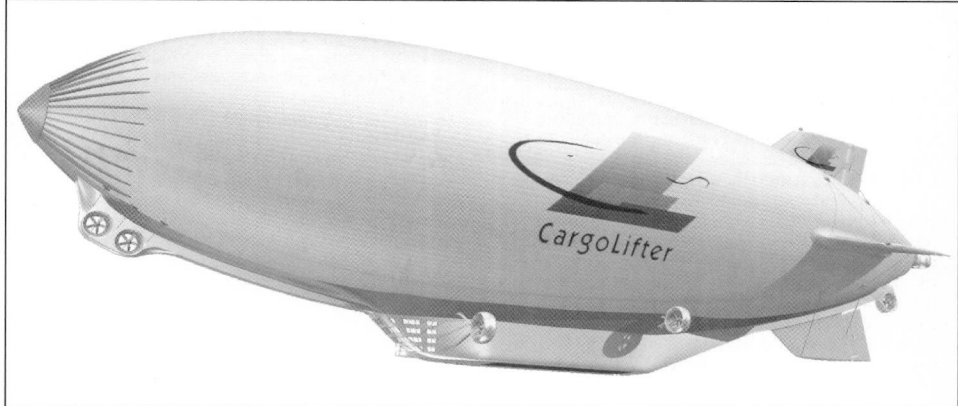

Computer-generated impression of the projected CargoLifter CL 160   0130558

Developments during 2001 included completion (July) of lightning protection tests by Brandenburg Technical University, using a one-twenty-fifth scale helium-filled model known as **Scala**; contract with BAE Systems Controls of Santa Monica, California, USA signed 16 July to co-develop avionics suite; CL 75 AirCrane used (30/31 July) to validate CL 160 inflation procedure; contract with GKN-Engage of UK (1 August) to support integration of propulsion system; contract with Eaton Fluid Power of USA (10 August) to develop and supply hydraulic systems for platform winches, ballonet blowers and front engine swivel arms); and official launch of prototype manufacture on 27 September. Contraves Space AG of Switzerland named as nosecone supplier in January 2002; other participants announced by early 2002 included Hamilton Sundstrand (electrical power system), Denel Aviation (mid-keel structure) and MT-Propeller (propellers).
  Scala testbed (10 m; 32.8 ft long, 35 m³; 1,236 cu ft volume) made maiden flight inside hangar at Brand on 9 July 2002.
CUSTOMERS: Sixteen letters of intent (including one from United Nations) reportedly received by mid-2000.
COSTS: Development costs reassessed in early 2002 at US$633 million, with further US$370 million then required for production start.
DESIGN FEATURES: Semi-rigid structure with keel and catenary curtain; heart-shaped cross-section; large central cargo bay houses a specially designed 'multibox' container which can be loaded and unloaded from all sides; in-flight access to all main parts of the airship.
FLYING CONTROLS: Digital fly-by-wire control of aerodynamic control surfaces, all engines, and load exchange system.
STRUCTURE: Full-length carbon fibre structural keel supports cargo, load exchange installations, power plant and auxiliary systems; 20 cell envelope with central catenary curtain; internal ballonets. Nose and tail fuel/water tanks provide ballast in event of any gas cell failure (water replaces fuel as flight progresses). Subcontractors include

MAN and Dornier Luftfahrt (keel), Contraves Space (composites nosecone) and Liebherr (loading crane).
POWER PLANT: Eight 1,864 kW (2,500 shp) General Electric CT7-8 turboshaft engines, four providing forward main propulsion, plus two bow and two stern thrusters for lateral manoeuvring.
ACCOMMODATION: Crew of 12. Payload is carried in an internal ventral cargo bay on a suspended load platform (multibox) that is winched up and down hydraulically for loading and unloading. Multibox can accommodate 36 standard 12.2 × 2.44 × 2.44 m (40 × 8 × 8 ft) freight containers. Winches are coupled electronically to keep load platform level regardless of airship movement.
Following data are provisional:
DIMENSIONS, EXTERNAL:
  Envelope: Length, excl nose boom   265 m (869 ft)
    Max diameter   65 m (213 ft)
  Height overall   82 m (269 ft)
  Tail unit span   74 m (243 ft)
  Propeller diameter (main engines)   6.5 m (21.3 ft)
DIMENSIONS, INTERNAL:
  Envelope volume   550,000 m³ (19.4 million cu ft)
  Cargo multibox: Length   50 m (164 ft)
    Width   8 m (26 ft)
    Depth   8 m (26 ft)
    Volume   3,200 m³ (113,000 cu ft)
WEIGHTS AND LOADINGS:
  Weight empty (approx)   260,000 kg (573,200 lb)
  Max payload   120,000-160,000 kg (264,555-352,740 lb)
  Max T-O weight   400,000 kg (881,850 lb)
PERFORMANCE (estimated):
  Max level speed   72 kt (135 km/h; 83 mph)
  Typical cruising speed at 2,000 m (6,560 ft)
                                49 kt (90 km/h; 56 mph)
  Max flight altitude   2,000 m (6,560 ft)
  Max range   5,400 n miles (10,000 km; 6,215 miles)
                                                 UPDATED

GEFA-FLUG

GESELLSCHAFT ZUR ENTWICKLUNG UND FÖRDERUNG AEROSTATISCHER FLUGSYSTEME GmbH (Aerostatic Flight Systems Development and Promotion Company)

Weststrasse 24C, D-52074 Aachen
Tel: (+49 241) 88 90 40
Fax: (+49 241) 889 04 20

Latest design of four-seat gondola for AS 105 GD non-rigid (*Jane's/Paul Jackson*)   0121717

e-mail: j.leisten@gefa-flug.de
Web: http://www.gefa-flug.de
MANAGING DIRECTOR: Karl Ludwig Busemeyer
SALES AND MARKETING DIRECTOR: Jürgen Leisten

GEFA-Flug was established 1975 to operate advertising and passenger-carrying hot-air balloons; has R&D department to develop airships for manned and unmanned civil and environmental research applications such as aerial photogrammetry and pollution monitoring.
  Company currently operates more than six hot-air balloons, hot-air airships and a number of remotely controlled aerostats, and has sold about 100 systems (mostly unmanned and remotely controlled) worldwide. Its 20th hot-air airship was sold in mid-2000. Fleet includes three AS

105GD hot-air airships, including one four-seater; production of this model continues. Future market seen as survey projects for environmental research; GEFA-Flug also developing scientific measuring equipment in partnership with German Mining Technology, a partly government-funded body.
                                                 VERIFIED

GEFA-FLUG AS 105 GD

TYPE: Hot-air non-rigid.
PROGRAMME: Modified version of AS 105 Mk II first flown 1990 and built until 1994 by Thunder & Colt in UK; German development began 1993. Certification (by UK CAA) and first production AS 105 GD completed in 1996; also certified by German LBA.

GEFA-Flug AS 105 GD   0121682

CUSTOMERS: Total of 21 built in Germany by early 2002, of which at least six have German-certified four-seat gondolas. First sale to UK effected in early 2002.

COSTS: Development programme DM1.4 million; standard airship approximately DM300,000 (2000).

DESIGN FEATURES: Longer and slimmer envelope than UK-built version; cruciform tail surfaces with rudder on each vertical fin; twin catenary load suspension system distributes loads of gondola weight and power plant forces evenly into envelope.

FLYING CONTROLS: Burners located inside envelope and operated electrically, with manual override; pilot lights fitted with electric spark and piezoelectric ignition, specially modified to operate directly underneath inflation fan without blowing out; steering by rudder on each vertical fin, operated by cables connecting them to gondola.

STRUCTURE: Envelope made from high-tenacity and high temperature-resistant fabric; engine drives generator supplying electric fan used to pressurise envelope;

additional air for envelope pressurisation via scoop located in propeller slipstream. Gondola is a chromoly aircraft-grade steel tube spaceframe with Macrolon skin panels; windscreen, of polycarbonate sheet, forms partially enclosed cockpit.

LANDING GEAR: Non-retractable; two front and two rear wheels; spring shock-absorbers.

POWER PLANT: One 47.8 kW (64.1 hp) Rotax 582 two-cylinder water-cooled engine, driving a Helix four-blade carbon fibre pusher propeller. Fuel capacity 28 litres (7.4 US gallons; 6.2 Imp gallons).

ACCOMMODATION: Pilot, co-pilot and two passengers in pairs.

AVIONICS: *Comms:* Dittel FSG 71 720-channel VHF transceiver; transponder optional.
Instrumentation: Standard VFR.

DIMENSIONS, EXTERNAL:

Envelope: Length overall	41.00 m (134 ft 6¼ in)
Max diameter	12.80 m (42 ft 0 in)
Fineness ratio	3.2

Tail unit span	18.00 m (59 ft 0¾ in)
Propeller diameter	1.60 m (5 ft 3 in)
Propeller ground clearance	0.10 m (4 in)
DIMENSIONS, INTERNAL:	
Envelope volume	2,973 m³ (105,000 cu ft)
Gondola cabin: Length	4.00 m (13 ft 1½ in)
Max width	1.60 m (5 ft 3 in)
Max height	1.80 m (5 ft 10¾ in)
WEIGHTS AND LOADINGS:	
Weight empty	approx 450 kg (992 lb)
Max T-O weight: two-seat	850 kg (1,873 lb)
four-seat	900 kg (1,984 lb)
PERFORMANCE:	
Max level speed	22 kt (40 km/h; 25 mph)
Max cruising speed	20 kt (37 km/h; 23 mph)
Econ cruising speed	15 kt (28 km/h; 17 mph)
Max rate of climb at S/L	183 m (600 ft)/min
Endurance with max payload	1 h

UPDATED

WDL

WDL LUFTSCHIFFGESELLSCHAFT mbH

Flughafen Essen-Mülheim, D-45470 Mülheim/Ruhr
Tel: (+49 208) 37 80 80
Fax: (+49 208) 378 08 33 (Management)
    (+49 208) 378 08 41 (Operations)
e-mail: salesdirector@wdl-luftschiff.de
Web: http://www.wdl-luftschiff.de

OWNER: Theodor Wüllenkemper
MARKETING, SALES AND OPERATIONS MANAGER: Arnold D Beier

WDL resumed airship construction in 1987 with first of new design known as WDL 1B; three of these built to date; described and illustrated in 1997-98 *Jane's*; further examples manufactured to firm order only.

Company plans include US$8 million project, launched in March 1997, for new factory and base at Ormond Beach

Business Park and Airpark, Florida, USA. Contract for lease of 4.9 ha (12 acre) site signed in March 1998; intended to build there, eventually, both the WDL 1B and a new, smaller airship. Construction was due to begin in second half of 2001; however, Florida Department of Transportation declined to renew its US$2 million grant in September 2001, placing this project in jeopardy.

VERIFIED

ZEPPELIN

ZEPPELIN LUFTSCHIFFTECHNIK GmbH

Allmannsweilerstrasse 132, D-88046 Friedrichshafen
Tel: (+49 7541) 20 25 33
Fax: (+49 7541) 20 24 99
e-mail: info@zepplin-nt.com
Web: http://www.zeppelin-nt.com
CHAIRMAN AND CEO: Dr Bernd Sträter
COO: Günter Schwenk
VICE-PRESIDENT, ENGINEERING: Mathias Mandel
SALES AND MARKETING MANAGER: Dietmar Blasius
HEAD OF PUBLIC RELATIONS: Dr Jeannine Meighörner

Zeppelin Luftschifftechnik (ZLT) was formed in September 1993 for the development and production of new technology (NT) airships. Luftschiffbau Zeppelin GmbH, of which ZLT is a subsidiary, was founded by Count Ferdinand von Zeppelin in 1908 and still exists under the auspices of the Zeppelin Foundation. It had a mid-2001 workforce of 75 full-time and 25 freelance personnel.

Corporate members (percentage shareholdings in parentheses) are ZF Friedrichshafen AG (44.8), Luftschiffbau Zeppelin GmbH (38.9), Zeppelin GmbH (11.5) and ZF-Lemförder Metallwaren AG (4.8). Deutsche Zeppelin Reederei (DZR) formed in January 2001 as 100 per cent subsidiary of Zeppelin Luftschifftechnik; will be developed as operator, pilot training school and service centre for latter's airships; first commercial service flown 15 August 2001. (First-generation Zeppelin undertook initial passenger service on 24 June 1910.)

UPDATED

ZEPPELIN N 07

TYPE: Helium semi-rigid.

PROGRAMME: Study group for revival of modern semi-rigid airships formed 1989; concluded that combination of vectored thrust and new constructional approach offers best solution, resulting in NT programme; 10 m (32 ft 10 in) long, remotely controlled proof-of-concept model tested extensively in 1991. Design definition of N 07 series (originally designated N 05) completed late 1992; construction started October 1995, leading to completion in July 1997 and first flight (D-LZFN) on 18 September 1997 (69th anniversary of maiden flight (1928) of LZ 127 *Graf Zeppelin*) following public debut (April) at Friedrichshafen Air Show. Test flights undertaken from

new airship base at Friedrichshafen Airport; damaged by gale in February 1999, but returned to test flying by following April; made 100th flight in August 1999; D-LZFN named *Friedrichshafen* 2 July 2000; flight test programme completed in December 2000 (more than 1,000 hours in approximately 250 flights); certified by LBA 26 April 2001. Being brought up to production standard in 2002; then to undertake mineseeking operations in Croatia; scheduled for traffic monitoring duties during 2004 Olympic Games in Athens.

First production airship (D-LZZR, c/n 02) made maiden flight 19 May 2001; second (D-LION) entered construction in third quarter of 2000 and was due for completion in 2002. First revenue-earning flight (by 02) 15 August 2001; had carried 3,400 fare-paying passengers by end of 2001.

CUSTOMERS: Five production airships ordered by four pleasure cruise operators by early 1997 (see 2001-02 *Jane's*); however, subsequently decided to operate first three airships via DZR subsidiary initially, to gain experience.

COSTS: €7 million (DM14.5 million), including ground infrastructure (2001).

DESIGN FEATURES: Revives original design principles of Count Ferdinand von Zeppelin; first airship in world since 1940 with load-bearing internal structure. Intended as precursor of product line of commercial airships for variety of applications including scientific research, environmental monitoring, TV missions and tourism. Internal 'prism' primary structure modified and optimised from Stuttgart University proposal of 1975 and British patent of 1924; single helium cell; placement of vectored thrust units near CG and also at rear enhances manoeuvrability and control at low speeds, further improving safety and reducing ground crew requirements to three persons, thereby also reducing operating costs. Tailfins in inverted Y configuration.

FLYING CONTROLS: Computer-assisted flight control, propulsion and landing. Dual fly-by-wire flight control system; no mechanical back-up. Elevator or rudder on each tail surface. Engine thrust control (see under Power Plant) in pitch and yaw.

STRUCTURE: Internal primary structure consists of three rows of aluminium tubular longerons, interconnected with triangular carbon fibre frames and cross-braced with Kevlar cables, providing internal prism-shaped structure accounting for only 12 per cent of gross weight; this frame and pressurised envelope are both load-carrying structures to enhance safety factor (airship remains fully manoeuvrable even if internal pressure drops). Envelope, attached continuously to all longerons, is a high-strength, multilayer laminate of polyester and Tedlar; this material has low gas permeability and allows airship to be moored permanently in the open. Air ballonets in lower part of envelope afford protection to cabin and airframe in the event of a hard landing.

LANDING GEAR: Twin landing/ground handling wheels in tandem under gondola and rear of envelope. Airship can be moored to a fixed or mobile mast.

POWER PLANT: Three 149 kW (200 hp) Textron Lycoming IO-360 flat-four engines: one each side of hull above gondola in vectored-thrust propulsion unit, each driving a three-blade, variable-pitch, swivelling (up to 120°) tractor propeller; engine in tailcone drives, via a belt-driven gearbox, a lateral fan (for precise yaw control during low-speed and hovering manoeuvres) and a vectored propeller used as a pusher in forward flight and for precise pitch control in landing configuration. Combination of thrust vectoring and fuel trim tanks virtually obviates need for ballast management.

ACCOMMODATION: Crew of two, on ergonomically designed flight deck; gondola main cabin accommodates up to 12 passengers on individual seats, with galley, lavatory and wardrobe provision, or equivalent cargo or other payload.

Zeppelin N 07 cabin interior   0130553

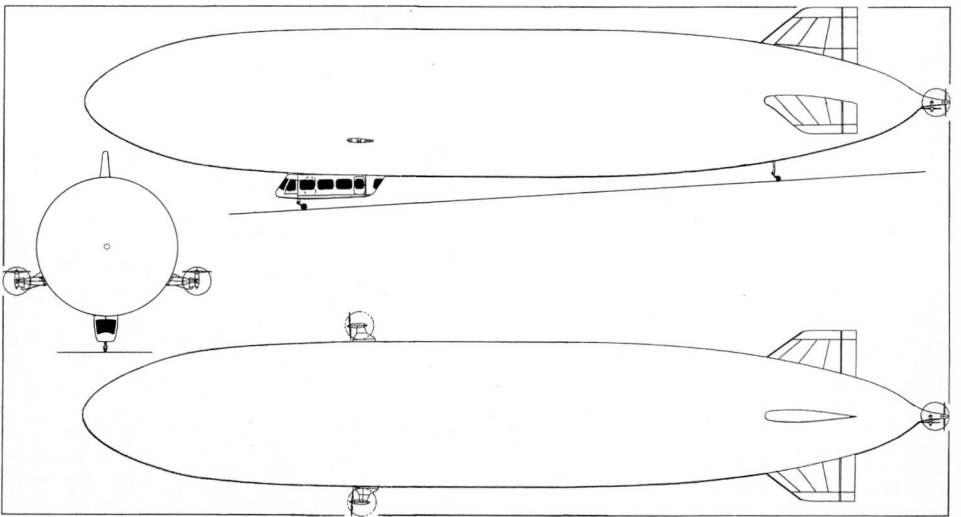

Zeppelin N 07 airship *(Jane's/Paul Jackson)*   NEW/0143738

AVIONICS: *Radar:* Weather radar.

Flight: Dual digital fly-by-wire aerodynamic steering; triple thrust vector steering; modern LCD flight deck avionics, including EFIS, EICAS, radio nav and GPS.

DIMENSIONS, EXTERNAL:
Envelope: Length overall   75.00 m (246 ft 0¾ in)
   Max diameter   14.16 m (46 ft 5½ in)
Fineness ratio   5.3
Max width over propeller ducts   19.50 m (63 ft 11¾ in)
Height overall   17.40 m (57 ft 1 in)

DIMENSIONS, INTERNAL:
Envelope volume   8,225 m³ (290,450 cu ft)
Ballonet volume (max)   2,200 m³ (77,700 cu ft)
Gondola: Length   10.70 m (35 ft 1¼ in)
   Max width   2.25 m (7 ft 4½ in)
   Volume   26.0 m³ (918.2 cu ft)
WEIGHTS AND LOADINGS:
Gondola: structural weight   440 kg (970 lb)
   incl equipment and furniture   1,200 kg (2,646 lb)
Max payload   1,900 kg (4,189 lb)
Max T-O weight   10,690 kg (23,567 lb)

PERFORMANCE (estimated):
Max level speed   67 kt (125 km/h; 77 mph)
Cruising speed   62 kt (115 km/h; 71 mph)
Cruising altitude   1,000 m (3,280 ft)
Pressure ceiling   2,600 m (8,530 ft)
Range   486 n miles (900 km; 559 miles)
Endurance at 38 kt (70 km/h; 43 mph):
   with max payload   14 h
   with reduced payload   24 h
UPDATED

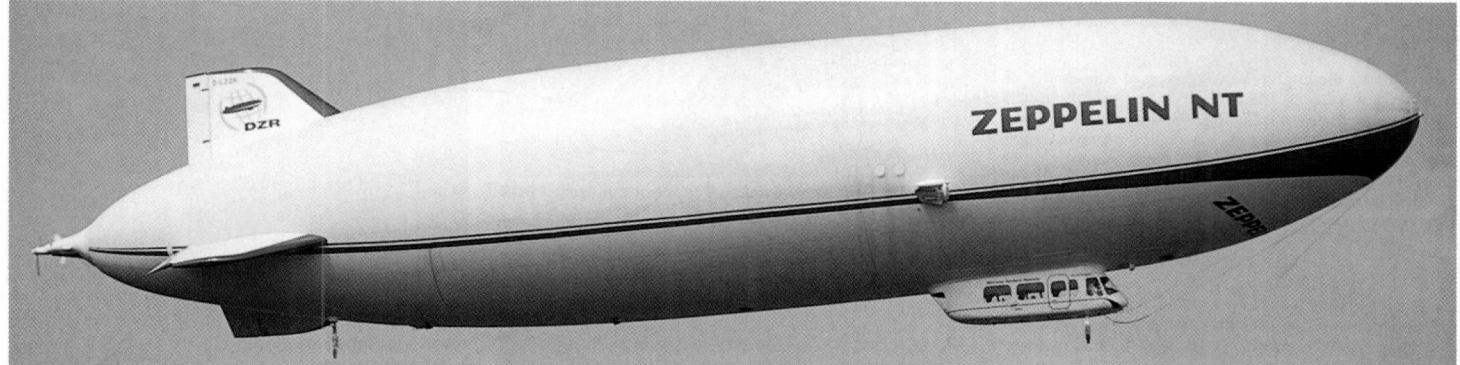

First production Zeppelin N 07 (*Jane's/Paul Jackson*)   NEW/0143726

INDIA

IITB

INDIAN INSTITUTE OF TECHNOLOGY BOMBAY
Department of Aerospace Engineering, Mumbai 400 076
Tel: (+91 22) 576 71 01 and 576 71 02
Fax: (+91 22) 572 26 02
e-mail: rkpant@aero.iitb.ac.in
Web: http://www.aero.iitb.ac.in/~airships
PROGRAMME DIRECTOR, PADD: Professor Rajkumar S Pant

The IITB is heading a team of Indian national aerospace organisations, sponsored by a department of the Ministry of Science and Technology, to develop airship designs for various domestic applications under the generic title PADD (Programme on Airship Design and Development). As proposed, PADD will proceed through three phases (conceptual design, preliminary design, and detailed design and development, of six, 12 and 18 months duration respectively.

Initial Phase 1 activity has included a small unmanned prototype, the **PADD Micro**, which made its first flight on 23 January 2002; this craft is 5 m (16.4 ft) long and has an envelope volume of 6.8 m³ (240 cu ft). It is to be followed by a larger version (**PADD Demo**) with 15 to 20 kg (33 to 44 lb) payload capacity. Future plans include a 100 kg (220 lb) payload airship and, eventually, one with the ability to carry 15 passengers or 1,500 kg (3,307 lb) of cargo.
NEW ENTRY

IITB students with their Micro remotely controlled prototype   NEW/0525364

KOREA, SOUTH

KARI

KOREA AEROSPACE RESEARCH INSTITUTE
PO Box 113, Yu-Sung, 305600 Taejon
Tel: (+42 860) 23 52
Fax: (+42 860) 20 06
e-mail: hanch@viva.kari.re.kr

Web: http://www.kari.re.kr
INFORMATION: Jong-Won Lee

The Aircraft Division of KARI is undertaking a feasibility study for a stratospheric airship. Details were originally expected to become available in mid-1998 on completion of the basic design, but none have been released so far. However, at the Seoul Air Show in October 2001 it was

confirmed that some US$38 million (US$28.5 million from government and US$9.5 million from the private sector) was to be invested in the project by August 2007. A subscale (50 m; 164 ft long) version, to have a 3 hour endurance at 3,000 m (9,840 ft), will be built first, to be followed by a full-size (200 m; 656 ft) airship. The project is supported by Ministry of Commerce, Industry and Energy (MOCIE).
VERIFIED

NETHERLANDS

RIGIDAIR

RIGID AIRSHIP DESIGN NV
Franse Kampweg 7, NL-1243 JC 's-Graveland
Tel: (+31 35) 656 48 77
Fax: (+31 35) 656 30 07
e-mail: rigidair@worldonline.nl
MANAGING DIRECTOR: Nico Nanninga
NL-1 PROJECT LEADER: Ian Alexander
DESIGN: Arnout De Jong

OPERATIONS: Andrew Horler
AIRWORTHINESS: Koos Tromp
MARKETING: Jos Williams

Rigid Airship Design consortium formed on 26 May 1998 for the design, development and commercial manufacture of rigid airships, based on ideas put forward by Scottish airship expert Ian Alexander. Shareholders were RDM Aerospace NV (42.3 per cent); Airshot International NV (36.3 per cent); Stork NV (10.0 per cent); M Caransa BV (7.0 per cent); and

Greenfield Capital Partners (4.4 per cent). A US subsidiary, Rigid Airship USA Inc, was established in late 1998.

Initial (NL-1) project was for a 180 m (590 ft) long airship, provisionally designated RA-180 (see 2002-03 edition for description). Subject to RLD design approval, Rigidair had hoped to have the prototype ready for flight test by the end of 2001, but the company was reported bankrupt in August of that year. No buyer had emerged by mid-2002.
UPDATED

trusses form cross-sectional contours; lower keel serves as platform for main equipment, fuel and crew compartments; central part of envelope forms spacious cargo hold.

POWER PLANT: Eight ducted propellers: two forward, two amidships and four (in line with tailfins) at rear.

Following data are provisional.

DIMENSIONS, EXTERNAL:

Envelope: Length	268 m (880 ft)
Max diameter	54 m (177 ft)

DIMENSIONS, INTERNAL:

Envelope volume	500,000 m³ (17.66 million cu ft)

WEIGHTS AND LOADINGS:

Lifting capacity	180,000 kg (397,000 lb)

PERFORMANCE:

Max level speed	91 kt (170 km/h; 105 mph)
Cruising speed	65 kt (120 km/h; 75 mph)
Range at cruising speed	
	8,100 n miles (15,000 km; 9,320 miles)

UPDATED

2001 model of the MD-900 (*Jane's/Paul Jackson*)   0126886

ROSAEROSYSTEMS MD-900

TYPE: Helium semi-rigid.

PROGRAMME: Details first made public August 1995; preliminary design by 1997 and still being refined in 2001; company seeking risk-sharing partners.

DESIGN FEATURES: *Modulnyi dirizhabl*: modular airship. Intended for variety of missions (see Equipment paragraph below); can be configured to transport passengers, cargo or specialised, removable mission modules. Conventional shape envelope with ventral gondola; four tailfins, each with rudder/elevator, indexed in X configuration; auxiliary fin between lower pair serves as mount for stern thruster engine.

FLYING CONTROLS: Rudders and elevators controlled hydraulically; stern thruster for yaw control. Two internal air ballonets for pressure control.

STRUCTURE: Envelope of domestic aerostatic fabric with an outer layer of Lavsan treated with titanium dioxide; fin truss of aluminium alloy; reinforced nose with mooring attachment. Semi-monocoque crew compartment.

LANDING GEAR: Non-retractable tricycle type.

POWER PLANT: Two MAN-D2866 thrust-vectoring turbo-diesel cruise engines (each 280 kW; 375 hp) plus a smaller turbo-diesel (37.3 kW; 50 hp) at rear for directional control.

ACCOMMODATION: Flight crew of three; dismountable payload module for up to 18 passengers or 24 casualties/medical staff. Ground crew six persons.

SYSTEMS: Single hydraulic system for tail control surface actuation.

AVIONICS: Flight envelope computer-controlled to reduce pilot workload; com/nav avionics standard.

EQUIPMENT: Standard gondola module is 9.1 × 2.4 × 2.4 m (29.9 × 7.9 × 7.9 ft), and can be outfitted as 12- to 18-seat passenger transport or flying hospital, or with equipment for radar surveillance, ecological monitoring, electricity generation or other missions.

DIMENSIONS, EXTERNAL:

Envelope: Length	60.00 m (196 ft 10¼ in)
Max diameter	17.10 m (56 ft 1¼ in)
Fineness ratio	3.5
Height overall	22.00 m (72 ft 2¼ in)
Payload module: Length	10.00 m (32 ft 9¾ in)
Max width	3.00 m (9 ft 10 in)
Max height	2.50 m (8 ft 2½ in)

DIMENSIONS, INTERNAL:

Envelope volume	9,050 m³ (319,600 cu ft)
Ballonet volume (two, total)	1,850 m³ (65,325 cu ft)

AREAS:

Tail surface area (total)	130.0 m² (1,399.3 sq ft)

WEIGHTS AND LOADINGS:

Weight empty	4,680 kg (10,318 lb)
Max payload	3,170 kg (6,989 lb)
Max dynamic lift	785 kg (1,730 lb)
Max T-O weight	8,630 kg (19,026 lb)

PERFORMANCE (estimated):

Max level speed	70 kt (130 km/h; 80 mph)
Cruising speed at 1,000 m (3,280 ft)	
	54 kt (100 km/h; 62 mph)
Typical operating altitude	up to 1,000 m (3,280 ft)
Pressure ceiling	2,000 m (6,560 ft)
Range at cruising speed above:	
standard fuel	1,620 n miles (3,000 km; 1,864 miles)
with auxiliary fuel tank	
	2,294 n miles (4,250 km; 2,640 miles)
Endurance: at max speed	20 h
at cruising speed above	50 h

VERIFIED

ROSAEROSYSTEMS PD-300

TYPE: Helium semi-rigid.

PROGRAMME: Development of PD-220 design (1999-2000 *Jane's*), which it apparently supersedes (*patrulnyi dirizhabl*: patrol airship). Revealed 1998 as PD-300, later apparently redesignated PD-3000 but now quoted with original designation. Reported by Novosti news agency in

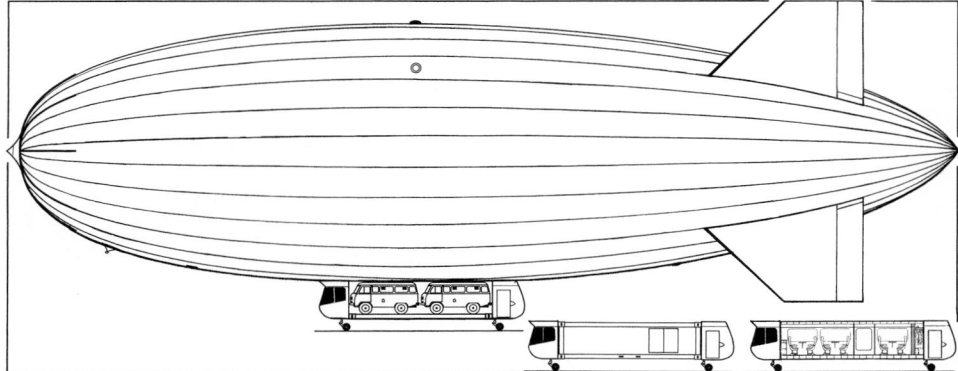

MD-900 fitted with open-sided cargo module, and showing alternative freight and passenger modules (*Jane's/Paul Jackson*)   0051882

November 2000 as "built and tested" at Avialine Scientific and Technical Design Bureau in Moscow, but photographic confirmation not yet seen (mid-2002).

DESIGN FEATURES: Designed for patrol, inspection, surveillance, people/payload transportation, search and rescue, photographic reconnaissance, agricultural roles, and other civil or paramilitary applications. Conventional shape envelope; cruciform tailfins, each with elevator or rudder. Compliant with Russian Federation and US FAA airship design criteria.

FLYING CONTROLS: Vectored thrust (+30/−105°) from cruise engines for fore-and-aft control; yaw control by stern thruster mounted on port side of lower vertical fin. Two internal air ballonets for pressure control; up to 200 litres (52.8 US gallons; 44.0 Imp gallons) water ballast for trim control.

STRUCTURE: Envelope of high-strength polyurethane-coated nylon with titaniaum dioxide UV protection; all-composites gondola.

LANDING GEAR: Single, non-retractable and self-centring wheel under rear of gondola.

POWER PLANT: Two 64.9 kW (87 hp) Subaru 1.8 cruise engines, with similar cascade-type thrust vectoring to Au-12; 5 kW electric motor (stern thruster) for directional control. Fuel capacity 285 litres (75.3 US gallons; 62.7 Imp gallons).

ACCOMMODATION: Pilot and co-pilot, with dual controls; detachable bulkhead, aft of which are seats for four passengers. Wardrobe, washstand and lavatory at rear. Access via let-down rear door with built-in steps. Accommodation heated and ventilated. Ground crew of four to six.

SYSTEMS: Include envelope anti-icing and APU.

AVIONICS: *Comms*: Dual nav/com radios; transponder.

Flight: Computer-aided flight controls; ADF; marker beacon receiver. Autopilot optional.

Instrumentation: IFR standard.

EQUIPMENT: Could include video cameras, loudhailers and other appropriate mission equipment.

DIMENSIONS, EXTERNAL:

Envelope: Length	45.40 m (148 ft 11½ in)
Max diameter	11.30 m (37 ft 1 in)
Fineness ratio	4.0
Gondola length	7.80 m (25 ft 7 in)

DIMENSIONS, INTERNAL:

Envelope volume	3,001 m³ (105,979 cu ft)
Ballonet volume (two, total)	750.0 m³ (26,486 cu ft)

AREAS:

Tail surface area (total)	65.90 m² (709.3 sq ft)

WEIGHTS AND LOADINGS:

Weight empty	1,500 kg (3,307 lb)
Max water ballast	200 kg (441 lb)
Max payload	1,200 kg (2,646 lb)
Max dynamic lift	400 kg (882 lb)
Max T-O weight	3,350 kg (7,385 lb)

PERFORMANCE (estimated):

Max level speed	59 kt (110 km/h; 68 mph)
Max cruising speed	49 kt (90 km/h; 56 mph)
Cruising altitude	1,000 m (3,280 ft)
Pressure ceiling	2,500 m (8,200 ft)
Range: at 38 kt (70 km/h; 43 mph) cruising speed	
	540 n miles (1,000 km; 621 miles)
ferry	810 n miles (1,500 km; 932 miles)
Endurance: at max speed	5 h
at cruising speed above	15 h

UPDATED

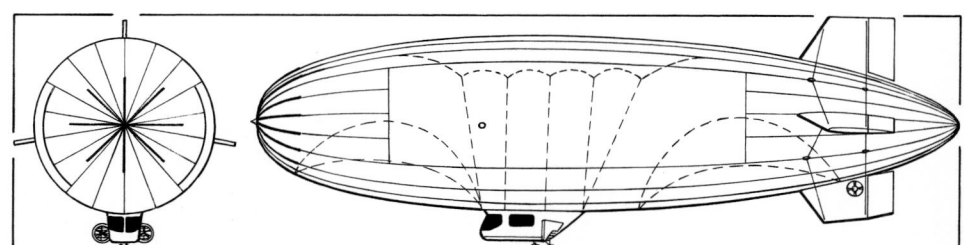

General arrangement of the PD-300 patrol airship (*Jane's/James Goulding*)   0085413

RVO

RUSSKOE VOZDUKHOPLAVATELNOE OBSHCHESTVO (Russian Aeronautical Society)

Web: http://www.pbo.ru

According to an announcement by the ITAR-TASS news agency in September 2000, the RVO is involved in the design of the large, heavy-lift DTs-N1. Details of this airship can be found under the RosAeroSystems heading in this section.

UPDATED

THERMOPLANE

THERMOPLANE DESIGN BUREAU, MOSCOW AVIATION INSTITUTE

bolshaya Volokolamskoe 4, 125871 Moskva
Tel: (+7 095) 158 41 27
Fax: (+7 095) 158 29 77
e-mail: globex@orc.ru
Web: http://www.mai.ru
CHIEF DIRECTOR: Yuri G Ishkov

According to reports in early 1999, Thermoplane has submitted designs for an international patent for a modified version of the ALA-40 'two-gas' airship (see 1996-97 *Jane's*) capable of use as a shuttle-type air and space vehicle. Side-mounted jet/ramjet engines would accelerate the vehicle to Mach 5 before being jettisoned at 40 km (25 miles) altitude, when a liquid-propellant rocket engine would be ignited to propel the ALA-40 to orbital insertion speed.

According to reports in June 2000, additional funding for the original Thermoplane project has been forthcoming from

an Egypt-based, US-backed agricultural concern interested in a system to transport produce from Egypt to markets in Europe. A free-flying second prototype was then said to be nearing completion, with a modified third example (different propulsion and guidance systems) some 40 per cent complete. No further news of either project had emerged by mid-2002.

UPDATED

VOZDUKh

VOZDUKh AERONAUTICAL ENTERPRISE

Moskva
CHIEF DESIGNER: Yuri S Boyko

Although first reported in early 2000, no further details of this company's B-150 airship had emerged by late 2000.

UPDATED

VOZDUKh B-150

TYPE: Helium semi-rigid hybrid.
PROGRAMME: Project first reported in early 2000; programme status not established at time of going to press in October 2002.
DESIGN FEATURES: Envelope elliptical in profile, with heart-shaped cross-section; dorsal tailfin with inset rudder; anhedral tailplanes and elevators; large underfin with inset shrouded propeller. Powered by same engines as Mi-26 helicopter (which see).

Combines buoyant lift of helium airship with large-diameter, outrigged rotors which have reversible-pitch blades to provide assistance in ascent and descent and controlled acceleration during take-off and landing.
FLYING CONTROLS: Electromechanical and hydraulic.
POWER PLANT: Five 8,500 kW (11,399 shp) ZMKB Progress D-136 turboshafts, two driving each large-diameter rotor; rotors interconnected by synchronising shaft to maintain

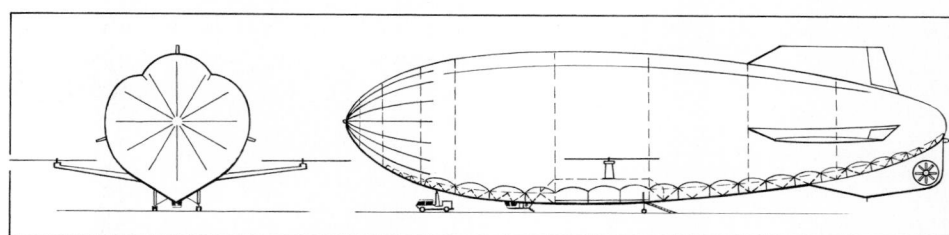

General arrangement of the semi-rigid B-150 *(Jane's/James Goulding)*   0105834

safe flight if both engines on one side fail. Fifth engine drives contrarotating pusher propellers aft of tailcone and lateral control shrouded propeller in ventral fin.
ACCOMMODATION: Crew of six.
Following data all provisional:
DIMENSIONS, EXTERNAL:

Length overall	187 m (614 ft)
Rotor diameter (each)	30 m (98 ft)
Width overall, incl rotors	100 m (328 ft)
Height overall	46 m (151 ft)
Propeller diameter: pusher pair, each	18 m (59.1 ft)
lateral controller	8 m (26.2 ft)

DIMENSIONS, INTERNAL:

Envelope volume	200,000 m³ (7,062,900 cu ft)
Payload compartment: Length	40 m (131.2 ft)
Width	6 m (19.7 ft)
Height	5 m (16.4 ft)

WEIGHTS AND LOADINGS:

Payload: typical	100,000 kg (220,460 lb)
max	150,000 kg (330,695 lb)

PERFORMANCE (design):

Max level speed	81 kt (150 km/h; 93 mph)
Cruising speed	70 kt (130 km/h; 81 mph)
Pressure ceiling	1,500 m (4,920 ft)

Range:
with 100,000 kg (220,460 lb) payload
1,349 n miles (2,500 km; 1,553 miles)
with 150,000 kg (330,695 lb) payload
270 n miles (500 km; 310 miles)
Endurance with 100,000 kg (220,460 lb) payload   20 h

UPDATED

UNITED KINGDOM

ATG

ADVANCED TECHNOLOGIES GROUP

No 2 Hangar, Cardington Field, Shortstown, Bedford MK42 0TJ
Tel: (+44 1234) 74 35 81
Fax: (+44 1234) 74 35 82
e-mail: airshiptech@compuserve.com
Web: http://www.airship.com
PRESIDENT: Air Marshal Sir John Walker
EXECUTIVE CHAIRMAN AND TECHNICAL DIRECTOR: Roger Munk
MANAGING DIRECTOR: Peter Ward
SALES AND MARKETING DIRECTOR: Gordon Taylor
PUBLIC RELATIONS MANAGER: Joanna Amis

US SUBSIDIARY:
SkyCat Technologies Inc, 3201 Airpark Drive, Suite 101, Santa Maria, California 93455
Tel: (+1 805) 614 08 24
Fax: (+1 805) 614 09 24
e-mail: skycattech@bigfoot.com
Web: http://www.skycattech.com

ATG is the trading name, from June 2000, of Airship Technologies Services Ltd, which was created in February 1996 to revive the activities of the former Airship Industries company (1991-92 *Jane's*), responsible for the British-designed Skyship 500/600 series.

ATG's design and production team draws on experience in designing and operating more transport category, passenger certified airships than any other world operator. Current work is directed towards two totally new concepts in lighter-than-air design: SkyCat, a hybrid air vehicle, and StratSat, a stratospheric telecommunications platform. SkyCat has a fully amphibious design incorporating a sophisticated lifting body, ellipsoidal cross-section hull and a catamaran hover cushion landing system which equips it to land on a grass field, a swamp, sand, snow or open sea.

In March 2002, ATG announced the first flight of its AT-10 airship, intended for advertising work, as a camera platform,

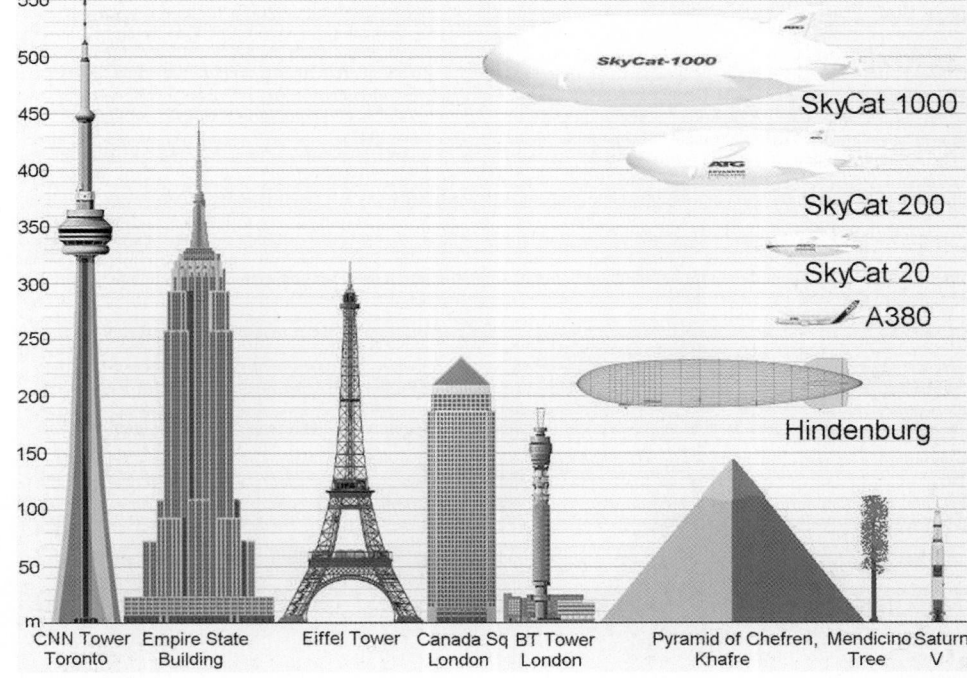

Size comparisons of ATG products   *NEW*/0525365

and for surveillance and pilot training applications. The AT-10, SkyCat 20 and StratSat will all be powered by ATG diesel engines ranging from 74.6 to 447 kW (100 to 600 hp).

UPDATED

ATG AT-10

TYPE: Helium non-rigid.

PROGRAMME: First flight expected in early 2002; prototype registered G-OATG in November 2001; made first flight 28 March 2002.

CUSTOMERS: First sale announced in May 2001, to Chinese operator for advertising and police surveillance duties in Shanghai area. This airship reported as partially completed by September 2002.

COSTS: US$3 million (2001).

DESIGN FEATURES: Commercial airship with capability for advertising and camera platform use and for other applications such as pilot training. X configuration tailfins.

FLYING CONTROLS: Split-channel, optically signalled flight control system incorporating sidestick controllers, powered flight actuators and an Ada control system interface; provision for autopilot. Single ballonet centrally located over gondola for envelope pressure control.

STRUCTURE: Envelope of laminated translucent fabric with external catenary collar system supporting gondola. Provision for internal mercury illumination system.

LANDING GEAR: Non-retractable bicycle type, coil-sprung with oleo damping; twin wheels on each unit.

POWER PLANT: Two 74.6 kW (100 hp) (59.7 kW; 80 hp maximum continuous) ATG A-Tech 100 horizontally opposed, two-stroke, direct injection diesel engines with vectorable-thrust ducted propellers. Supercharged induction system.

ACCOMMODATION: One or two pilot stations with sidestick controls; four or five passenger seats on multiple tracks; large flight deck and passenger cabin transparencies; large passenger door on port side. Rear of gondola contains lavatory, ballonet air system, pneumatics, electrical services and cabin heater.

SYSTEMS: Low-pressure pneumatic system to power flight control actuators and provide low susceptibility to lightning strike and electromagnetic interference. Electrical power (28 V DC) from 6 kVA alternator.

DIMENSIONS, EXTERNAL:
Length overall	41.40 m (135 ft 10 in)
Max diameter	10.70 m (35 ft 1¼ in)
Height overall	13.60 m (44 ft 7½ in)

DIMENSIONS, INTERNAL:
Envelope volume	2,500 m³ (88,290 cu ft)
Gondola: Length	4.40 m (14 ft 5¼ in)
Max width	1.52 m (4 ft 11¾ in)
Width at floor	0.83 m (2 ft 8¾ in)
Min height	1.86 m (6 ft 1¼ in)

WEIGHTS AND LOADINGS:
Max payload	740 kg (1,631 lb)

PERFORMANCE (design):
Max level speed	60 kt (111 km/h; 69 mph)
Cruising speed	50 kt (93 km/h; 58 mph)
Pressure ceiling	2,745 m (9,000 ft)
Max range at 30 kt (56 km/h; 35 mph)	
	1,020 n miles (1,889 km; 1,173 miles)
Endurance at 30 kt (56 km/h; 35 mph)	34 h

UPDATED

ATG SKYCAT

TYPE: Hybrid non-rigid air vehicle.

PROGRAMME: Launched 29 June 2000, with three versions announced; name derived from 'sky catamaran'; numerals in designations indicate payload capacity in tonnes. Preceded by SkyKitten I, a 12.2 m (40 ft) long proof-of-concept prototype; first flight (retrospectively allocated, but did not carry, UK 'B Conditions' registration G-86-01 which eventually applied instead to one-tenth scale StratSat prototype) 28 June 2000; more than 30 short flights, from land and water, by early 2001; bow thruster added. SkyKitten II (13.5 m; 44.3 ft long, envelope volume 145 m³; 5,120 cu ft) under construction as one-seventh scale representation of SkyCat 20; more powerful engines; retractable hover skirts; envelope shape identical to SkyCat 20. Shown statically at Paris, June 2001, registered G-86-02.

Prototype ATG AT-10 four/five-passenger airship, first flown 28 March 2002   NEW/0525366

SkyKitten II G-86-02 on display at Paris in June 2001 (Jane's/Paul Jackson)   0126884

CURRENT VERSIONS: **SkyCat 20**: Initial operational version. Manufacture under way by mid-2000; first flight targeted for second quarter of 2002.

SkyCat 200: Second, and larger, operational version. Work already under way on payload module. Industrial partner IAR Brasov (which see under Romania) was due to begin deliveries of major subassemblies in 2001, but not confirmed by mid-2002.

SkyCat 1000: Largest in current range; targeted to fly in late 2008. Subject in 2001 of US Army study contract to assess ultra-heavy-lift potential.

CUSTOMERS: SkyCat 20 ordered by World Airships (UK) for delivery in January 2003.

COSTS: SkyCat 20 approximately US$25 million to US$28 million (2001).

DESIGN FEATURES: Brings together lighter than air and hovercraft technologies, enabling airship to land on virtually any flat land or water surface without need for landing infrastructure or ground crew. Larger versions seen as heavy-lift military or cargo-carrying vehicles (SkyCat 1000 has payload capacity of 10 Boeing 747s) and can be constructed in the open, eliminating need for costly hangars.

Elliptical cross-section created by twin hulls, allied to cambered longitudinal shape, provides up to 40 per cent of lift.

FLYING CONTROLS: Flight capability derived from combination of aerodynamic lift and helium buoyancy. All versions have dual-channel, optically signalled flight control system, plus pressure control by multiple ballonets fore and aft in each of the outer hulls.

STRUCTURE: Envelope of laminated fabric with internal catenary system supporting payload module. Internal diaphragms supporting envelope shape allow for limited compartmentalisation, further enhancing vehicle's fail-safe properties. Payload module located on centreline, with roll-on/roll-off freight capability in SkyCat 200 and 1000.

LANDING GEAR: Air-cushion landing system. Hover skirts on underside of outer hulls provide amphibious capability, as well as enhancing ground handling compared with conventional airships. Hover skirts are 'sucked in' for clean in-flight profile and enhanced all-round field of view. System shares use of ballonet fans with hull pressure system.

POWER PLANT: *SkyCat 20:* Four 447 kW (600 hp) (358 kW; 480 hp maximum continuous) ATG A-Tech 600 horizontally opposed, four-cylinder two-stroke, direct injection diesel engines, two mounted forward on hull and two on hull stern for cruise operation; all four in ducts with blown vanes to vector thrust for T-O, landing and ground handling. Supercharged induction system.

SkyCat 200: Four 5,966 kW (8,000 shp) (4,474 kW; 6,000 shp maximum continuous) turboshafts, in thrust-vectoring ducts positioned as for SkyCat 20.

Impression of SkyCat 20 as ordered by World Airships   0103490

Gondola of the AT-10 under construction (Arnold Nayler)   0105838

SkyCat 1000: Six 11,185 kW (15,000 shp) (8,948 kW; 12,000 shp maximum continuous) turboshafts, two within each stern duct each driving a single propeller and one each side of hull forward.

ACCOMMODATION: Flight deck has side-by-side pilot stations, with sidestick controls and large transparencies for wide field of view; on larger models, also provides space (18.6 m²; 200 sq ft in SkyCat 200, 83.6 m²; 900 sq ft in 1000) for off-duty crew members.

SkyCat 20: Four passenger seats on individual tracks in payload compartment; large door each side with built-in steps; full-width door at rear for cargo use.

SkyCat 200 and *1000:* Main deck provides clear load space for cargo/freight on a military-rated floor structure. Mezzanine decking can be provided to give multiple lower load area. Ramps (at rear on 200, fore and aft on 1000) provide roll-on/roll-off access to cargo area. Above door aperture is further accommodation space (37.2 m²; 400 sq ft in 200, 195.1 m²; 2,100 sq ft in 1000).

SYSTEMS: All versions have low-pressure pneumatic system as described for AT-10.

DIMENSIONS, EXTERNAL:
Length overall: 20		81.30 m (266 ft 8¾ in)
200		185.00 m (606 ft 11½ in)
1000		307.00 m (1,007 ft 2½ in)
Max width: 20		41.00 m (134 ft 6¼ in)
200		77.00 m (252 ft 7½ in)
1000		136.00 m (446 ft 2¼ in)
Height overall: 20		23.90 m (78 ft 5 in)
200		47.00 m (154 ft 2½ in)
1000		77.00 m (252 ft 7½ in)

DIMENSIONS, INTERNAL:
Envelope volume: 20	32,000 m³ (1.13 million cu ft)
200	457,500 m³ (16.16 million cu ft)
1000	2 million m³ (70.63 million cu ft)
Payload deck:	
Length: 20	22.30 m (73 ft 2 in)
200	49.40 m (162 ft 1 in)
1000	80.80 m (265 ft 1 in)
Width: 20	3.00 m (9 ft 10 in)
200	7.50 m (24 ft 7¼ in)
1000	12.20 m (40 ft 0¼ in)

Height: 20	2.00 m (6 ft 6¾ in)
200	5.00 m (16 ft 4¾ in)
1000	8.00 m (26 ft 3 in)

WEIGHTS AND LOADINGS:
Max payload: 20	20,000 kg (44,092 lb)
200	200,000 kg (440,925 lb)
1000	1,000,000 kg (2,204,625 lb)

PERFORMANCE (estimated):
Max level speed: 20	80 kt (148 km/h; 92 mph)
200	90 kt (166 km/h; 103 mph)
1000	110 kt (203 km/h; 126 mph)
Cruising speed: 20	70 kt (130 km/h; 81 mph)
20	75 kt (139 km/h; 86 mph)
1000	100 kt (185 km/h; 115 mph)
Pressure ceiling: 20, 200, 1000	2,745 m (9,000 ft)*
Range:	
20 (max payload)	1,225 n miles (2,268 km; 1,409 miles)
20 (ferry)	4,000 n miles (7,408 km; 4,603 miles)
200	3,225 n miles (5,972 km; 3,711 miles)
1000	4,000 n miles (7,408 km; 4,603 miles)

* 5,485 m (18,000 ft) for SkyCat 20 high-altitude version

UPDATED

CAMERON

CAMERON BALLOONS LTD

St John Street, Bedminster, Bristol BS3 4NH
Tel: (+44 117) 963 72 16
Fax: (+44 117) 966 11 68
e-mail: hcameron@cameronballoons.co.uk
Web (1): http://www.cameronballoons.co.uk
Web (2): http://www.thunderandcolt.co.uk
MANAGING DIRECTOR: Don Cameron
SALES DIRECTOR: Nick Purvis
MARKETING DIRECTOR: Alan Noble
TECHNICAL DIRECTOR: Tom Sage
OPERATIONS AND PRODUCTION DIRECTOR: Jim Howard
PUBLIC RELATIONS: Hannah Cameron

Cameron Balloons, formed in 1970 and now part of the Cameron Holdings Group, holds CAA, FAA, French CNT and German Musterzulassungsschein type certificates for its hot-air balloons. It is the world's largest manufacturer of conventional and special-shape balloons and, by mid-2002, had produced nearly 5,000 from its main factory in Bristol, of which 352 or more (built or on order) were special shapes. A sister company in Dexter, Michigan, USA, has produced more than 1,200.

Details of Cameron's conventional-shape balloon range are currently available on the company website.

Cameron also designs and produces hot-air airships, being the first company to develop a craft of this type. Current production hot-air airships are predominantly the DP-80, DP-90, AS-105 (1996-97 *Jane's*), the slightly more streamlined AS 105GD and the AS-120. Production of these had totalled 70 by June 2001, comprising three D-38 type, 25 DP type, 10 D-50 type, 20 D-96 type and 12 AS type, in addition to six earlier DG type helium airships. No further sales had been reported by mid-2002.

Shortly after the 11 September 2001 terrorist attacks in the USA, Cameron proposed a revival of the barrage balloon concept as a comparatively inexpensive means of defence against such actions. Later that month, it indicated that it was designing a helium balloon for this purpose, approximately 15.24 m (50 ft) long and 6.10 m (20 ft) in diameter. It would use modern materials for both envelope and cable, making it more durable than its Second World War predecessors.
UPDATED

CAMERON DP SERIES

TYPE: Hot-air non-rigid.
PROGRAMME: DP prototype (G-BMEZ) made first flight April 1986; DP series first/second/third in World Hot-Air Airship Championships 1990; DP 70 first in British and European Championships 1991; DP 80 holds current world endurance record and in 1994 received CAA approval for

Cameron DP 70 airship (*Phil Dunnington/Cameron Balloons*)   **NEW**/0525079

night flying. Cameron still averaging one or two a year, and series remains available to order; figures in designations indicate volume in thousands of cubic feet. Full description in 1996-97 and previous editions; shortened version follows.

CURRENT VERSIONS: **DP 60:** Single/two-place; details in 2000-01 and earlier editions.

DP 70: As DP 60, but larger envelope; two-seater.

DP 80: Two-seater, with better performance in hot/high conditions.

DP 90: Larger-envelope version of DP 80.

CUSTOMERS: By mid-2001 (latest figure received) Cameron had completed, in addition to the DP 50 prototype, 25 DP series airships; customer countries comprise Australia, Belgium, Brazil, Chile, Czech Republic, Germany, Hungary, Luxembourg, Switzerland, UK, USA and Venezuela.

POWER PLANT: Standard engine for DP 80/90 is 47.8 kW (64.1 hp) Rotax 582 UL driving a three-blade shrouded pusher propeller. Fuel capacity 22.7 litres (6.0 US gallons; 5.0 Imp gallons) of 40:1 two-stroke mixture.

ACCOMMODATION: Gondola seats two persons side by side with full-harness seat belts. Full-height polycarbonate windscreen.

DIMENSIONS, EXTERNAL:
Envelope: Length overall: DP 80	33.83 m (111 ft 0 in)
DP 90	35.05 m (115 ft 0 in)
Height: DP 80	15.24 m (50 ft 0 in)
DP 90	15.54 m (51 ft 0 in)
Max width: DP 80	12.19 m (40 ft 0 in)
DP 90	12.80 m (42 ft 0 in)

DIMENSIONS, INTERNAL:
Envelope volume: DP 80	2,265 m³ (80,000 cu ft)
DP 90	2,549 m³ (90,000 cu ft)

AREAS:
Display area (each side): DP-80	92.0 m² (990 sq ft)
DP-90	99.0 m² (1,066 sq ft)

WEIGHTS AND LOADINGS:
Gondola (all models)	195 kg (430 lb)
Useful passenger load (S/L, 15°C):	
DP 80	285 kg (628 lb)
DP 90	359 kg (791 lb)

PERFORMANCE:
No details provided

UPDATED

CAMERON ROZIERE BALLOONS

Rozières are combination helium/hot-air balloons in which a small propane burner maintains helium efficiency at night when solar heating is absent, providing greatly extended flight duration without expending helium or sand ballast. Cameron manufactures several sizes, from the R-15 to the R-900. Some earlier achievements of this series have been detailed in previous editions. Most recent of these was the R-650 *Spirit of Freedom 'Bud Light'*, which circumnavigated the southern hemisphere in June/July 2002. US pilot Steve Fosset, flying in an unpressurised gondola, took off from Northam, Australia, on 19 June and landed at Birdsville, Australia, on 4 July, a journey of 18,343.7 n miles (33,972.6 km; 21,109.6 miles).

UPDATED

LINDSTRAND

LINDSTRAND BALLOONS LTD

Maesbury Road, Oswestry, Shropshire SY10 8ZZ
Tel: (+44 1691) 67 17 17
Fax: (+44 1691) 67 11 22
e-mail: info@lindstrand.co.uk

Web: http://www.lindstrand.co.uk
MANAGING DIRECTOR: Per Lindstrand

Formed November 1991. Workforce more than 70 in 2002. Company is now part of Cameron (which see) Holdings Group, but Per Lindstrand retains a 35 per cent shareholding. Product range includes hot-air and gas balloons and airships.

Its round-the-world *ICO Global Challenger*, with a volume of more than 36,000 m³ (1,271,330 cu ft), is the largest superpressure balloon so far built. For full description of HS 110 hot-air airship, see 2001-02 *Jane's*. Last delivery was made in 2000. Other products include tethered aerostats supplied to the UK Ministry of Defence and for civilian passenger-carrying.

UPDATED

For details of the latest updates to *Jane's All the World's Aircraft* online and to discover the additional information available exclusively to online subscribers please visit

jawa.janes.com

QINETIQ

QINETIQ

A1 Cody Technology Park, Ively Road, Farnborough, Hampshire GU14 0LX
Tel: (+44 1252) 39 46 11
Fax: (+44 1252) 39 33 99
Web: http://www.qinetiq.com
MARKETING DIRECTOR: Brenda Jones
HEAD OF MEDIA RELATIONS: Joanna Sale

QinetiQ (pronounced 'kinetic') was created in July 2001 from the major portion of the former Defence Evaluation and Research Agency (DERA), itself derived from the earlier Royal Aircraft Establishment at Farnborough and elsewhere. One of Europe's largest science and technology organisations, its activities and facilities include indoor and outdoor ranges for air-, land- and sea-launched weapons trials; wind tunnels; underwater target ranges and marine testing facilities; automotive test tracks; and climatic testing laboratories. Currently structured as a public-private partnership (PPP), it is eventually to be found a strategic investment partner as a prelude to flotation as a fully commercial company. Carlyle Group named as preferred bidder in September 2002.

QinetiQ is sponsoring the huge QinetiQ 1 experimental balloon, which was planned to be launched during the third quarter of 2002.

NEW ENTRY

QINETIQ QINETIQ 1

TYPE: Helium balloon.
PROGRAMME: Announced in November 2001 as attempt on balloon world altitude record (first since present record of 34,668 m; 113,740 ft was established by two US Navy fliers on 4 May 1961) and for scientific study of cosmic rays and solar radiation in the stratosphere; planned to reach 40,235 m (132,000 ft). Launch window July to September 2002; actual launch date to be governed by optimum weather expectations following 72 hour countdown; however, launch not achieved during stated window; rescheduled for second quarter of 2003.
STRUCTURE: Polyethylene envelope, reinforced with Kevlar straps. Carbon fibre crew platform.

ACCOMMODATION: Crew of two (Andy Elson and Colin Prescot, current holders of world balloon endurance record), on non-pressurised open flight deck platform.
EQUIPMENT: Cameras and real-time video downlink. Zvezda space suits, pressurised to 10,660 m (35,000 ft) equivalent altitude, and oxygen cylinders for crew.
Following data are approximate.
DIMENSIONS, EXTERNAL:
Height overall at launch   387 m (1,270 ft)
DIMENSIONS, INTERNAL:
Envelope volume at 40,235 m (132,000 ft)
        more than 1.13 million m³ (40 million cu ft)
WEIGHTS AND LOADINGS:
Weight empty, uninflated   less than 3.05 tonnes (3 tons)
PERFORMANCE (estimated):
Max rate of ascent/descent   335 m (1,100 ft)/min
Time to 40,235 m (132,000 ft)   3-5 h
NEW ENTRY

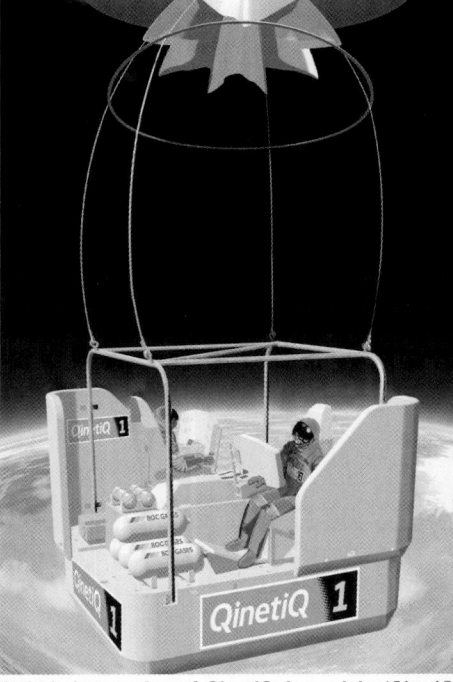

Artist's impression of QinetiQ 1 gondola *(QinetiQ)*
NEW/0525947

When launched, QinetiQ 1 will be the world's largest manned aircraft *(QinetiQ)*   NEW/0525367

UNITED STATES OF AMERICA

ABC

AMERICAN BLIMP CORPORATION

1900 North-East 25th Avenue, Suite 5, Hillsboro, Oregon 97124
Tel: (+1 503) 693 16 11
Fax: (+1 503) 681 09 06
Web (1): http://www.americanblimp.com
Web (2): http://www.lightships.com
PRESIDENT AND CEO: James Thiele
EXECUTIVE VICE-PRESIDENT: Charles Ehrler
CHIEF ENGINEER: Rudy Bartel

AIRBORNE SURVEILLANCE GROUP:
302 Ritchie Highway, Severna Park, Maryland 21146-1910
Tel: (+1 410) 544 65 08
Fax: (+1 410) 544 65 09
e-mail: Jud4Blimps@aol.com
VICE-PRESIDENT, MARKETING: E Judson Brandreth Jr

In 1995 ABC announced the design of a series of airships larger than the A-60, designated A-1-50 (marketing designation A-150). These are a refinement of the earlier A-120 conceptual design with a new technology envelope. The civil version incorporates the ABC internal illumination system for advertising and is designated a Lightship; the non-illuminating surveillance version is designated SPECTOR. Lightship envelopes remain translucent for internal illumination, whereas SPECTOR envelopes are opaque. Each is offered in three sizes to optimise costs with mission requirements. Other structural elements are identical.

The same envelope technology was introduced as an option for production on existing A-60+ Lightships from 1996 (first contract, for seven envelopes to be supplied by ILC Dover, announced 18 March 1996).

Addition to the range of the new, diesel-powered A-100 Lightship/SPECTOR 31 was announced in June 1998. Company is currently developing airship designs for medium-altitude (2,000 to 3,000 m; 6,560 to 9,840 ft) operation with military and commercial payloads for surveillance and communications relay. It has also, on behalf of the Jet Propulsion Laboratory, been studying the use of a helium airship as a possible vehicle for unmanned exploration of Titan, the largest of Saturn's moons. Named Aerover, the projected airship would be approximately 10 m (33 ft) long, 2.44 m (8 ft) in diameter and weigh about 100 kg (220 lb).

In 1999, ABC increased its manufacturing area by 30 per cent with a 410 m² (4,400 sq ft) extension at Hillsboro, which now covers almost 2,325 m² (25,000 sq ft). At the same time, a 280 m² (3,000 sq ft) fabrication shop was being readied at the Tillamook flight test centre. On 12 February 2000, ABC's combined fleet of 17 A-60+/A-1-50 airships achieved a cumulative total of 100,000 flight hours.

ABC purchased, on 21 November 2000, the type certificate (originally issued on 17 November 1992) for the Colt GA 42 non-rigid helium airship, last described in the 1994-95 *Jane's*, UK section, under Thunder & Colt. It is believed that three airships of this type were included in the transfer.

UPDATED

ABC LIGHTSHIP A-60+

TYPE: Helium non-rigid.
PROGRAMME: A-50 prototype N5132A (see 1990-91 *Jane's*) registered 31 March 1988 and made first flight 9 April 1988; first flight of production A-60 (1991-92 *Jane's*) June 1989; A-60 certified by FAA 18 May 1990 for day/night and VFR/IFR flight; first delivery (to Virgin Lightships Ltd) November 1990; A-60+ certified third quarter 1991 by FAA; APU upgrade certified by FAA, and FLIR Systems Inc gyrostabilised IR camera approved by FAA, in 1993; all A-60s converted to A-60+ in 1993. Certified also in Argentina, Australia, Brazil, Canada, Germany, Italy and South Africa; approved for operation in China, Japan and Turkey. Airship 05 reached 10,000 flight hours on 28 October 1999, and two others had achieved this total

Wescam surveillance equipment on an ABC Lightship A-60+   NEW/0525368

by February 2000. A Lightship Group two-man British crew in 016/N606LG set a new FAI Class BA absolute world speed record for gas airships on 19 January 2000, reaching 50.11 kt (92.80 km/h; 57.66 mph) over a 1 km course.

One A-60+ loaned by The Lightship Group was used in January 2000 for UK-based flight trials fitted with a DERA (now QinetiQ) Mineseeker ultra-wideband, ground-penetrating radar for detection of undersurface metal and plastics mines. Subsequently, after the addition of a Wescam 16DS-M dual-sensor (daylight TV and FLIR) payload, this airship conducted a six-week programme (4 October to 14 November 2000) from bases in Italy and Kosovo, surveying 30 sites in the latter locality for mines and other unexploded ordnance on behalf of the United Nations Mine Action Co-ordination Centre (UN MACC). As a result of the success of this operation, a Mineseeker

Lightship A-60+ promoting another aircraft manufacturer (*Jane's/Paul Jackson*)   *NEW*/0525080

Inside the A-1-50's 10-seat gondola, looking forward   0016727

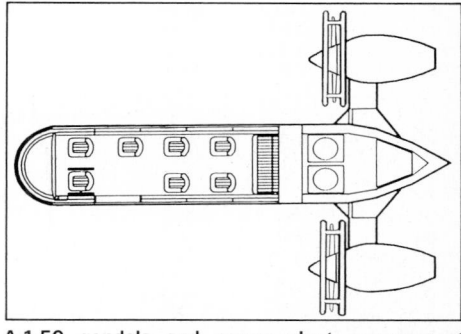

A-1-50 gondola and power plant arrangement   0016729

A-60+ OPERATORS
(at August 2002)

c/n[1]	Identity	Operator	Contractor	Country (base)
01	N2012P *Snoopy 1*	Icarus Aircraft	Met Life	USA
03	N2017A/I-TIRE *Spirit of Europe 1/G-TLEL*	Lightship Group	Goodyear	UK
04	N2022B/ZS-ODY/N460LG	Lightship Group		
05	N560VL *Skeeter*	Virgin Lightships	Monster.com	USA
06	N660VL *Van Heusen*	Virgin Lightships	CDW	USA
07	N760VL *Snoopy 2*	Icarus Aircraft	Met Life	USA
08	N860AB	Lightship Group	out of service	
09	N960AB/B–	Beijing Orient	Beijing Orient	China
010	N3119W/VH-ZIC *Spirit of the South Pacific*	Lightship Group	Goodyear	Australia
011	N11ZP	Lightship Group	NatWest Bank	UK
012	N12ZP *Spirit of Europe 2*	Virgin Lightships	Goodyear	UK
014	N604LG/PT-MKJ *Spirit of the Americas*	Lightship Group	Goodyear	Brazil
015	N605LG	Lightship Group	Premiere World	UK
016	N606LG/G-OLEL	Lightship Group	Mazda	UK
017	N607LG *Horizon*	Lightship Group	Hood Dairies	USA
018	N618LG	Lightship Group	not known	Singapore
019-023	N619LG-N623LG[2]	Lightship Group	no allocation	USA

[1] 02 and 13 not built
[2] Reservations only; none completed by mid-2002

Foundation was set up as a joint venture with DERA/QinetiQ which, in April 2001, announced plans to raise £10 million to purchase five similarly equipped airships.

CURRENT VERSIONS: **A-60:** Initial production version, described in 1991-92 *Jane's*. All since converted to A-60+.

A-60+: Current version from 1991; larger envelope; payload increased by 181 kg (400 lb) compared with A-60; Lightsign message board option from 1992. Also offered in **SPECTOR 19** form (see later entry).

Description applies to A-60+.

CUSTOMERS: Total of 16 registered at January 2001: see table; 018 made its first flight 4 August 2000; 019 nearing completion at end of that year; 01 badly damaged by storm on 7 September 2001 and may not be repairable.

DESIGN FEATURES: Conventional-shape envelope; cruciform tail surfaces; gondola suspended by 12 catenary cables each attached to external patches, eliminating need for internal cables, gastight fittings and bellow sleeves. Can be inflated without a net by attaching ballasted gondola before adding helium; gondola can be removed from inflated Lightship without a net by ballasting catenary cables.

FLYING CONTROLS: Single internal ballonet; rudder or elevator on each of four tailfins; primary flight controls for left-hand front seat only.

STRUCTURE: Outer envelope skin of Dacron/Mylar with separate urethane film inner gastight bladder and single ballonet. All structural attachments to sewn outer bag, such as nose mooring, fin base and guy wires, car catenary and handling lines, made with webbing reinforcements sewn directly to envelope.

LANDING GEAR: Single twin-wheel unit beneath gondola; small tailwheel at base of lower vertical fin. Operating area diameter required: 229 m (750 ft)

POWER PLANT: Twin 59.7 kW (80 hp) Limbach L 2000 engines, pusher-mounted to enhance propulsive efficiency and reduce noise. Standard capacity of rear-mounted fuel tank is 280 litres (74.0 US gallons; 61.6 Imp gallons); can be refuelled in the field without mooring.

ACCOMMODATION: Two single seats and rear bench seat in gondola (pilot plus three adult passengers or two adults and two small children.) Ground crew of 13.

SYSTEMS: 28 V 70 A electrical system with two 28 V 35 A alternators. APU capable of delivering 2.5 kW at 110 V certified December 1993 and now standard. Used mainly to power internal illumination lights, but can also be used for other airborne equipment.

EQUIPMENT: Two 1 kW bulbs for internal illumination system. Certified for gyrostabilised TV or IR camera mount, complete with microwave downlink for live broadcast; can be operated simultaneously with APU.

DIMENSIONS, EXTERNAL:
Overall: Length	39.01 m (128 ft 0 in)
Width	10.97 m (36 ft 0 in)
Height	13.41 m (44 ft 0 in)

Envelope: Length	39.01 m (128 ft 0 in)
Max diameter	10.01 m (32 ft 10 in)
Fineness ratio	3.9
Advertising banner areas:	
Each side: Length	27.43 m (90 ft 0 in)
Max depth	6.10 m (20 ft 0 in)
Underside, forward:	
Length	6.10 m (20 ft 0 in)
Width	7.62 m (25 ft 0 in)
Gondola: Length	3.96 m (13 ft 0 in)
Width	1.52 m (5 ft 0 in)
Height (incl landing gear)	2.90 m (9 ft 6 in)
Propeller diameter	1.52 m (5 ft 0 in)

DIMENSIONS, INTERNAL:
Envelope volume	1,926 m³ (68,000 cu ft)
Gondola: Cabin length	2.68 m (8 ft 9½ in)
Cabin height	1.93 m (6 ft 4 in)

AREAS:
Tailfins (four, total)	42.74 m² (460.0 sq ft)

WEIGHTS AND LOADINGS:
Total weight empty	1,393 kg (3,070 lb)
Max buoyancy	1,814 kg (4,000 lb)
Max dynamic lift	113 kg (250 lb)
Max useful lift, ISA at 660 m (2,000 ft)	680 kg (1,500 lb)
Max gross weight	1,993 kg (4,394 lb)

PERFORMANCE:
Max level speed	47 kt (88 km/h; 55 mph)
Cruising speed	28 kt (51 km/h; 32 mph)
Max rate of ascent	488 m (1,600 ft)/min
Service ceiling	2,225 m (7,300 ft)
Max rate of descent	396 m (1,300 ft)/min
Min T-O distance	112 m (366 ft)
Max range at 28 kt (51 km/h; 32 mph)	425 n miles (787 km; 489 miles)
Max endurance at 28 kt (51 km/h; 32 mph)	15 h

UPDATED

ABC LIGHTSHIP A-100

TYPE: Helium non-rigid.

PROGRAMME: Announced 26 June 1998; first flight then planned for 1999, but delayed awaiting certification of SR305 diesel engine, which was still unavailable by mid-2002. Prototype had not been completed or registered by then. Description in 2002-03 edition.

UPDATED

ABC LIGHTSHIP/SPECTOR SERIES

TYPE: Helium non-rigid.

PROGRAMME: Three envelope sizes offered for both civil and surveillance versions. First to be built, a Lightship A-1-50 (N5132A, c/n 001), made its first flight on 8 January 1997. FAA type certificate issued 3 October 1997, at which time a total of 683 hours had been flown. Marketing designations are A-150 and A-170; designation on Type Certificate and registration documents of current version is A-1-50.

CURRENT VERSIONS: **Lightship A-60+/SPECTOR 19.** See A60+ entry.

Lightship A-1-50/SPECTOR 42. Only version so far built (mid-2002).

Lightship A-170/SPECTOR 50.

CUSTOMERS: Six A-1-50s delivered by January 2001; see table. Further three (N157LG to N159LG) then under construction, but had not been completed by mid-2002.

COSTS: SPECTOR 42 approximately US$3.5 million (2002).

DESIGN FEATURES: As described for A-60+; larger gondola; also seat tracks for quick reconfiguration of cabin; water ballast trim system.

ABC Lightship A-1-50 with 10-place gondola   0023307

FLYING CONTROLS: Standard as for A-60+; dual controls and autopilot optional. Trimming with ballonet accomplished by transferring water between the nose and a tank in the rear portion of the gondola. Certified for day/night, VFR/IFR, single-pilot operation (A-1-50).

STRUCTURE: Single-walled ILC Dover envelope with single (26 per cent of volume) ballonet; Tedlar outer film plus inner ripstop fabric/helium barrier film laminated material with heat-sealed seams. Envelopes translucent on Lightship versions, opaque for SPECTOR; those for SPECTOR series have carbon black in the material for ease of maintenance.

LANDING GEAR: Non-retractable tricycle type. Spring-damped main units each with single wheel and tyre; slightly raised nose unit with twin wheels/tyres; small tailwheel at base of lower vertical fin. Required operating area diameter: 305 m (1,000 ft).

POWER PLANT: Two 134 kW (180 hp) Textron Lycoming IO-360-B1G6 flat-four engines, each driving an MTV-25-B-C five-blade constant-speed, reversible-pitch tractor propeller. Rear-mounted fuel tank, standard capacity 560 litres (148 US gallons; 123 Imp gallons). Airship can be refuelled in the field without mooring.

ACCOMMODATION: Single seats for pilot and one passenger in cockpit; five single seats and a three-person bench seat in main cabin. Lavatory and galley optional. Ground crew of 16.

SYSTEMS: Electrical system is 28 V DC, with two 90 A alternators; additional 2.5 kW/110 V at 60 Hz available from removable APU.

EQUIPMENT: Two 1 kW bulbs for internal illumination system. Gyrostabilised camera mount, as in A-60+. Gondola hardpoint provisions for external installation of a variety of sensor equipment. Standard seat tracks in main cabin facilitate sensor/electronics rack installations.

DIMENSIONS, EXTERNAL (A: A-1-50/SPECTOR 42, B: A-170/SPECTOR 50):

Overall: Length: A	50.29 m (165 ft 0 in)
B	52.42 m (172 ft 0 in)
Width: A	14.02 m (46 ft 0 in)
B	14.63 m (48 ft 0 in)
Height: A	16.76 m (55 ft 0 in)
B	17.37 m (57 ft 0 in)
Envelope: Length: A	49.38 m (162 ft 0 in)
B	51.51 m (169 ft 0 in)
Max diameter: A	13.11 m (43 ft 0 in)
B	13.72 m (45 ft 0 in)
Fineness ratio: A, B	3.8
Advertising banner areas:	
Each side: Length	35.97 m (118 ft 0 in)
Max depth	8.53 m (28 ft 0 in)
Underside, forward:	
Length	8.53 m (28 ft 0 in)
Width	9.75 m (32 ft 0 in)
Gondola: Length: A, B	8.08 m (26 ft 6 in)
Max width: A, B	6.25 m (20 ft 6 in)
Max height: A, B	3.35 m (11 ft 0 in)
Propeller diameter: A, B	1.68 m (5 ft 6 in)

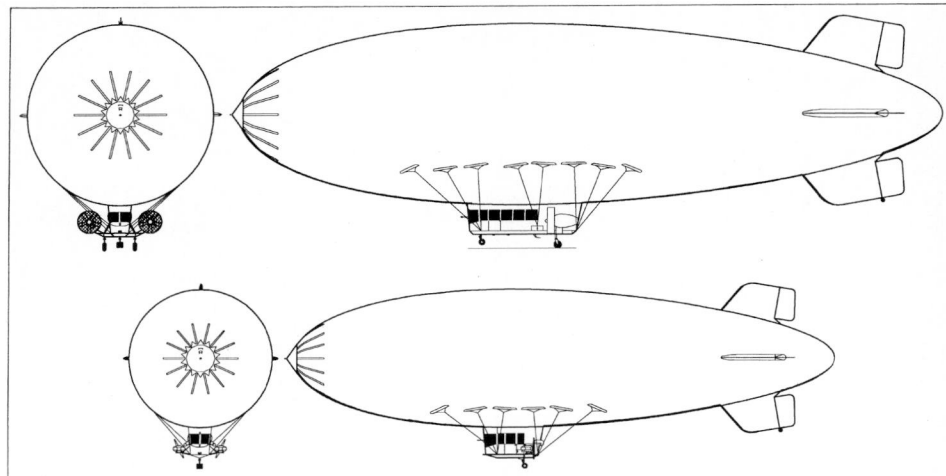

Comparative views, to scale, of the Lightship A-1-50 (top) and A-60+ *(Jane's/Paul Jackson)*

A-1-50 OPERATORS
(at mid-2002)

c/n	Identity	Operator	Contractor	Country (base)
101	N151AB	Lightship Group	Sanyo	USA
102	N152LG/N151LG *Bud 1*	Lightship Group	Budweiser	USA
103	N153LG/TC-AVK	KOC Holdings	Out of service	
104	N154ZP	Lightship Group	Vegas.com	USA
105	N155LG *Trumpasaurus*	Lightship Group	Monster.com	USA
106	N156LG	Lightship Group	QinetiQ	UK

DIMENSIONS, INTERNAL:

Envelope volume: A	4,248 m³ (150,000 cu ft)
B	4,814 m³ (170,000 cu ft)
Gondola: Cabin overall length:	
A, B	4.79 m (15 ft 8½ in)
Main cabin length: A, B	4.27 m (14 ft 0 in)
Average cabin width: A, B	1.52 m (5 ft 0 in)
Cabin height: A, B	1.93 m (6 ft 4 in)

WEIGHTS AND LOADINGS:

Total weight empty: A	2,860 kg (6,305 lb)
B	2,925 kg (6,449 lb)
Max buoyancy: A	4,055 kg (8,939 lb)
B	4,595 kg (10,131 lb)
Max dynamic lift: A	340 kg (750 lb)
B	386 kg (850 lb)
Max useful lift at 656 m (2,000 ft), ISA:	
A	1,633 kg (3,600 lb)
B	2,253 kg (4,967 lb)

Max T-O weight: A	4,394 kg (9,689 lb)
B	4,981 kg (10,981 lb)

PERFORMANCE (B: estimated):

Max level speed: A	55 kt (101 km/h; 63 mph)
B	50 kt (92 km/h; 57 mph)
Cruising speed: A	40 kt (74 km/h; 46 mph)
Max single-engine speed: A	33 kt (61 km/h; 38 mph)
B	31 kt (57 km/h; 35 mph)
Max rate of ascent: A	427 m (1,400 ft)/min
Service ceiling: A, B	3,050 m (10,000 ft)
Max rate of descent: A	488 m (1,600 ft)/min
Max range at 40 kt (74 km/h; 46 mph):	
A	493 n miles (913 km; 567 miles)
B	504 n miles (933 km; 580 miles)
Max endurance at 40 kt (74 km/h; 46 mph):	
A	12 h 20 min
B	14 h

UPDATED

AEROS

WORLDWIDE AEROS CORPORATION

8411 Canoga Avenue, Canoga Park, California 91304-2607
Tel: (+1 818) 993 55 33
Fax: (+1 818) 993 94 35
e-mail: aeros-us@worldnet.att.net
Web: http://www.aeros-airships.com
PRESIDENT: Igor Pasternak
SALES AND MARKETING DIRECTOR: Daniel Weissman
TECHNICAL DIRECTOR: Andrei Rybkin
OPERATIONS DIRECTOR: Anatoliy Pasternak
PUBLIC RELATIONS DIRECTOR: Fred Edworthy

Formed in Ukraine in 1988, Aeros transferred its activities to the USA and was incorporated in Delaware in 1992, relocating to California in October 1993. Its current operation combines manufacturing and flight test facilities, R&D, management and marketing. Aeros Airship Company subsidiary formed September 1999 to lease and operate Aeros airships; operations began, with an Aeros 40B, in November 1999. Aeros' technical capabilities are based on more than 20 years of research (including that initiated before 1988 by former Soviet design bureaux) into LTA technologies, especially those relating to cargo airships. It has also produced a wide range of manned and unmanned civil and military tethered aerostats for commercial, advertising, surveillance and scientific applications.

The company's current airship products are the 40A, 40B and 40C family of passenger non-rigids, all of which are designed to meet US FAA standards; 12 had been registered by the end of 2000. Latest known design is the large ML passenger airship, brief details of which follow those for the Aeros 40 family.

UPDATED

AEROS 40A

TYPE: Helium non-rigid.

PROGRAMME: Two sold to China Sky Corporation for government advertising in 1997. No further sales by March 2001, but remains available.

Description generally as for Aeros 40B except as follows:

POWER PLANT: Two 50.7 kW (68 hp) Limbach L 1700 flat-four engines. Fuel capacity 83.3 litres (22.0 US gallons; 18.3 Imp gallons).

ACCOMMODATION: Pilot and one passenger. Single cockpit door.

DIMENSIONS, EXTERNAL:

Length overall	37.25 m (122 ft 2½ in)
Envelope: Length	36.58 m (120 ft 0 in)
Max diameter	9.50 m (31 ft 2 in)
Fineness ratio	3.9
Height overall	11.86 m (38 ft 11 in)

DIMENSIONS, INTERNAL:

Envelope volume	1,700 m³ (60,035 cu ft)
Ballonet volume (two, total)	476 m³ (16,810 cu ft)
Gondola cabin: Length	1.68 m (5 ft 6 in)
Max width	1.60 m (5 ft 3 in)
Max height	1.55 m (5 ft 1 in)

AREAS:

Banner area (each side)	260.13 m² (2,800.0 sq ft)

WEIGHTS AND LOADINGS:

Weight empty	363 kg (800 lb)
Useful load	181.5 kg (400 lb)

The 1,700 m³ Aeros 40A two-seat airship
NEW/0525369

Aeros 40B Sky Dragon twin-engined airship in flight   *NEW*/0525370

PERFORMANCE: (As 40B except):

Max level speed	36 kt (67 km/h; 42 mph)
Pressure ceiling	3,050 m (10,000 ft)
Max endurance	24 h
	UPDATED

AEROS 40B SKY DRAGON

TYPE: Helium non-rigid.

PROGRAMME: Prototype completed; first flight (N818AC) 11 September 1998; exported to China (Thakral Media Corporation) in November 1998; operating in Hong Kong 1999. Three more (one for Argos Medien AG of Germany and two more for China) under construction in 1999-2001, of which N819AC for China substantially damaged in accident on 9 January 2001 and not delivered by early 2002; N820AC for Argos Medien registered in same month. FAA certification awarded 21 June 2000. Second of Chinese pair still not completed by early 2002.

DESIGN FEATURES: Göttingen 409 root rib profile; X configuration tail unit, each with ruddervator. One ellipsoid ballonet forward and one aft (together, 22 per cent of total volume). Internal illumination by three 400 W lamps: one above gondola and one in each ballonet.

FLYING CONTROLS: Fly-by-wire electronically controlled, pneumatically actuated ruddervators, with manual back-up.

STRUCTURE: Envelope made up of 1,020 heat-sealed panels of a transparent, nylon-based, 7 ply polyurethane coated fabric; tailfins aluminium alloy with fabric covering. Aluminium alloy semi-monocoque gondola, with polycarbonate sheet transparencies.

LANDING GEAR: Two mainwheels, with shock-absorbers.

POWER PLANT: Two 93 kW (125 hp) Teledyne Continental IO-240-B8 flat-four engines; MTV-7-D/LD170-12 variable- and reversible-pitch, three-blade pusher propellers. Fuel capacity 287 litres (75.9 US gallons; 63.2 Imp gallons), of which 267 litres (70.5 US gallons; 58.7 Imp gallons) are usable.

ACCOMMODATION: Pilot and four passengers (pilot and co-pilot only in prototype). Front two seats fully articulating and reclining; three bucket seats to rear. Full-depth windscreen; door, with upper and lower windows, in each side of gondola.

SYSTEMS: Pressure maintained by two air valves in each ballonet, each pneumatically actuated by an electrical blower and having mechanical return springs. Three emergency rip panels for fast deflation. Two ballast compartments in gondola. Electrical systems 28 V DC and 100 V AC. Nosecone mast mooring; ground handling crew of 12.

AVIONICS: Digital 'glass cockpit' package includes audio panel, voice annunciator, colour LEDs, GPS/com, second com, and transponder/encoder; readings displayed in both digital and analogue form. Provisions for full IFR package.

EQUIPMENT: Halon fire extinguisher and first aid kit. Navigation lights (bow, stern, port and starboard); anti-collision beacons (one each port, starboard and beneath gondola); two propeller lights; two red ceiling lights in gondola cabin. Three optional lamps (see Design Features) for internal envelope illumination.

DIMENSIONS, EXTERNAL:

Length overall	43.59 m (143 ft 0 in)
Envelope: Length	42.61 m (139 ft 9½ in)
Max diameter	10.60 m (34 ft 9½ in)
Fineness ratio	4.0
Height overall	13.35 m (43 ft 9½ in)
Banner (each side): Length	22.86 m (75 ft 0 in)
Height	7.92 m (26 ft 0 in)
Propeller diameter	1.70 m (5 ft 7 in)

DIMENSIONS, INTERNAL:

Envelope volume	2,508 m³ (88,570 cu ft)
Ballonet volume (two, total)	549.2 m³ (19,396 cu ft)
Gondola cabin: Length	1.96 m (6 ft 5 in)
Max width	1.57 m (5 ft 2 in)
Max height	1.73 m (5 ft 8 in)

AREAS:

Banner area (each side)	185.34 m² (1,995.0 sq ft)

WEIGHTS AND LOADINGS:

Weight empty	1,914 kg (4,220 lb)
Useful load	676 kg (1,490 lb)

PERFORMANCE:

Max level speed	44 kt (82 km/h; 51 mph) IAS
Max rate of climb	610 m (2,000 ft)/min
Max rate of descent	610 m (2,000 ft)/min
Max certified altitude	2,885 m (7,500 ft)
Pressure ceiling	3,000 m (9,840 ft)
Endurance: at 45% power	6 h 42 min
max	24 h
	UPDATED

AEROS 40C

TYPE: Helium non-rigid.

PROGRAMME: Projected for sightseeing tours to Grand Canyon; funding of US$5 million for two airships sought in 1998. Prototype under construction 1999, but apparently not completed by mid-2002.

Description generally as for Aeros 40B except as follows:

POWER PLANT: Two 224 kW (300 hp) Textron Lycoming AEIO-540-K2A5 flat-six engines. Hoffmann HOV 155 five-blade variable-pitch propellers. Fuel capacity 927 litres (245 US gallons; 204 Imp gallons).

ACCOMMODATION: Pilot and up to 13 passengers. Door on each side.

DIMENSIONS, EXTERNAL:

Length overall	60.00 m (196 ft 10¼ in)
Envelope: Length	58.70 m (192 ft 7 in)
Max diameter	14.90 m (48 ft 10½ in)
Fineness ratio	4.0
Height overall	18.70 m (61 ft 4¼ in)
Propeller diameter	1.37 m (4 ft 6 in)

Banner (each side): Length	32.31 m (106 ft 0 in)
Height	11.28 m (37 ft 0 in)

DIMENSIONS, INTERNAL:

Envelope volume	7,000 m³ (247,205 cu ft)
Ballonet volume (two, total)	2,450 m³ (86,520 cu ft)
Gondola cabin: Length	6.10 m (20 ft 0 in)
Max width	2.59 m (8 ft 6 in)
Max height	2.74 m (9 ft 0 in)

AREAS:

Banner area (each side)	364.37 m² (3,922.0 sq ft)

WEIGHTS AND LOADINGS:

Weight empty	3,143 kg (6,929 lb)
Useful load	4,250 kg (9,370 lb)

PERFORMANCE (estimated):

Max level speed	48 kt (90 km/h; 56 mph)
Pressure ceiling	2,890 m (9,480 ft)
Max endurance	24 h
	UPDATED

AEROS ML

TYPE: Helium rigid.

PROGRAMME: Current incarnation of former D-1 project, now envisaged as mass passenger transport with cabin comfort comparable with cruise liners. No launch date yet announced.

DESIGN FEATURES: Gondola/cabin integrated within hull structure; 'glass cockpit' controls; turboprop engines; air cushion landing gear system to eliminate need for ground handling crew.

FLYING CONTROLS: Digital flight control system; five tailfins, each with electrically actuated control surface; twin canards at nose. Pressurisation control by internal ballonets (12), helium valves and water ballast.

STRUCTURE: Hull of tubular trusses and bonded composites panels; aluminium alloy and composites cabin; aluminium alloy semi-monocoque tailfins and canards.

LANDING GEAR: Air cushion system for assisted T-O and vacuum mooring.

POWER PLANT: Two 1,044 kW (1,400 shp), thrust-vectoring Pratt & Whitney Canada PT6A-series turboprops; plus two electric motors to power bow thrusters and tail control surfaces.

ACCOMMODATION: Passenger cabin inside hull, seating maximum of 120 people. Polycarbonate and acrylic windows in hull sides.

EQUIPMENT: Provision for hull-side illuminations.

DIMENSIONS, EXTERNAL (approx):

Hull: Length overall	81.08 m (266 ft)
Max diameter	35.36 m (116 ft)
Propeller diameter (PT6A)	7.92 m (26 ft)

DIMENSIONS, INTERNAL (approx):

Hull max volume	25,000 m³ (882,875 cu ft)
Ballonet volume (12, total)	5,000 m³ (176,575 cu ft)
Cabin: Length	38.10 m (125 ft)
Max width	24.99 m (82 ft)
Max height	2.50 m (8.2 ft)

WEIGHTS AND LOADINGS:

Weight empty	20,340 kg (44,842 lb)
Useful load	12,712 kg (28,025 lb)

PERFORMANCE (estimated):

Max level speed	150 kt (277 km/h; 172 mph)
Max rate of ascent	610 m (2,000 ft)/min
Pressure ceiling	2,440 m (8,000 ft)
Max rate of descent	610 m (2,000 ft)/min
Max range	2,400 n miles (4,444 km; 2,761 miles)
Max endurance	24 h
	NEW ENTRY

Computer graphic depicting the projected Aeros ML passenger-carrying rigid   *NEW*/0525373

AHA

ADVANCED HYBRID AIRCRAFT INC

PO Box 144, Eugene, Oregon 97440
Tel/Fax: (+1 604) 541 80 64
e-mail: bnb@ahausa.com
Web: http://www.ahausa.com
MANAGER AND DIRECTOR: Bruce N Blake

Formerly based in Isle of Man, UK (see 1993-94 *Jane's*); relocated to Australia in 1993; to USA in 1996; and to British Columbia, Canada, in 1998 (while retaining a presence in Oregon, USA). A facility at Boundary Bay Airport, Vancouver, was being established in 2002. Venture capital for airship development, expected in early 1997, did not materialise, but reported that AHA signed letter of intent on 6 October 1998 with Xiao Bang Group (XBG) of China initiating 10 year joint venture to build and market new buoyant aircraft in China. Intention is to manufacture first Light Utility and a two-seat LV2 Hornet in USA, then to establish assembly plant in Zhuhai after training Chinese workforce. A formal agreement was signed in February 1998 with the Blimp Group of Buenos Aires, Argentina, to promote AHA buoyant aircraft. In March 2001, a similar agreement was signed with Denebola (DSA) SA de CV of Ensenada, Mexico.

Numerous letters of interest for the commercial Hornet LV2 were received during 2001, and several serious enquiries were being negotiated. AHA was then also seeking new investment to revive the projected Advanced Non-Rigid (ANR) airship last described in the 1992-93 *Jane's*, but funding for this had not been located by mid-2002.

Details of AHA's Hornet Hybrid RPV version and the Wasp remotely piloted blimp can be found in *Jane's Unmanned Aerial Vehicles and Targets.*

UPDATED

AHA HORNET

TYPE: Helium non-rigid hybrid.
PROGRAMME: Design of Hornet LV (leisure variant) initiated October 1991; funding for development and manufacture, which will be to FAA P-8110-2 airship design criteria standards, was still being sought in mid-2001.
CURRENT VERSIONS: **Hornet LV2:** Two-seat LV, available also as single-seat LV kit. Meets FAA '49 per cent rule' to qualify for Experimental category operation. To be used for promotions at major world air shows and elsewhere.
  Light Utility: More robust, four-seat version; described separately.
  Hornet AW: Aerial work version, planned to follow if sales of LV2 are successful; intended specifically to meet needs of electronic news-gathering organisations. Expected to provide capabilities similar to those of current twin-engined helicopters but at about half the acquisition and operating cost.
COSTS: LV2, US$162,100 (uncertified) or US$289,400 (FAA certified); AW (basic configuration), US$1.46 million (2001).
DESIGN FEATURES: Two-seat gondola/cabin for LV2, with fixed tricycle landing gear; can be taxied like an ultralight, mast-moored by one person (mainly an automated operation), and remain on ground when vacated by pilot.
  Designed to operate at heaviness of about 50 per cent, compared with conventional types with which operation at or near equilibrium is essential; one half of total lift is provided aerostatically by the buoyancy of the helium-filled envelope, the other half being generated

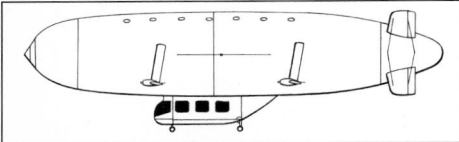

Hornet AW projected version
(Jane's/James Goulding)

aerodynamically by the stub-wings attached to the sides of the envelope. Hybrid buoyant aircraft do not require ballast and have flexibility to carry greater or lesser load, thus providing greater productivity.
FLYING CONTROLS: Fly-by-wire; automatic envelope pressure control.
STRUCTURE: Major subcontractors expected to be ILC Dover (envelope) and Scaled Composites (gondola).
LANDING GEAR: Non-retractable tricycle type.
POWER PLANT: Two 67.1 kW (90 hp) MWE AE 100R rotary engines in LV2, mounted on stub-wings that can be tilted upward to vector thrust 30° downward for climb. Hornet AW would have four 74.6 kW (100 hp) AE 100Rs and 45° thrust vectoring.
ACCOMMODATION: Two seats in LV2, one in LV.
DIMENSIONS, EXTERNAL:

Envelope: Length overall: LV2		19.51 m (64 ft 0 in)
	AW	30.48 m (100 ft 0 in)
Max diameter: LV2		4.88 m (16 ft 0 in)
	AW	6.10 m (20 ft 0 in)
Fineness ratio: LV2		4.0
	AW	5.0
Wing span: LV2		9.14 m (30 ft 0 in)
	AW	9.21 m (30 ft 2½ in)
Wing chord, constant: LV2		1.52 m (5 ft 0 in)
	AW	1.25 m (4 ft 1¼ in)

DIMENSIONS, INTERNAL:

Envelope volume: LV2	250.0 m³ (8,830 cu ft)
AW	623.0 m³ (22,000 cu ft)

Max ballonet volume: LV2		37.5 m³ (1,325 cu ft)
	AW	124.6 m³ (4,400 cu ft)
WEIGHTS AND LOADINGS:		
Disposable load: LV2		209 kg (460 lb)
	AW	427 kg (941 lb)
Buoyancy: LV2		249 kg (550 lb)
	AW	560 kg (1,235 lb)
Max heaviness: LV2		249 kg (550 lb)
	AW	720 kg (1,587 lb)
Max T-O weight: LV2		499 kg (1,100 lb)
	AW	1,280 kg (2,822 lb)
PERFORMANCE (estimated):		
Max level speed: LV2		61 kt (113 km/h; 70 mph)
	AW	81 kt (150 km/h; 93 mph)
Ceiling: LV2		1,525 m (5,000 ft)
	AW	2,135 m (7,000 ft)
Range: LV2, AW		200 n miles (370 km; 230 miles)

UPDATED

AHA LIGHT UTILITY

TYPE: Helium non-rigid hybrid.
PROGRAMME: Development of Hornet.
CURRENT VERSIONS: **Light Utility:** For similar applications to Hornet LV2, but more robust and intended to operate from rough fields and away from built-up areas.
  Light Utility.XR Patroller: Long-range version, designed for coastal patrol and drug interdiction duties. Presented to Mexican Navy in June 2001.

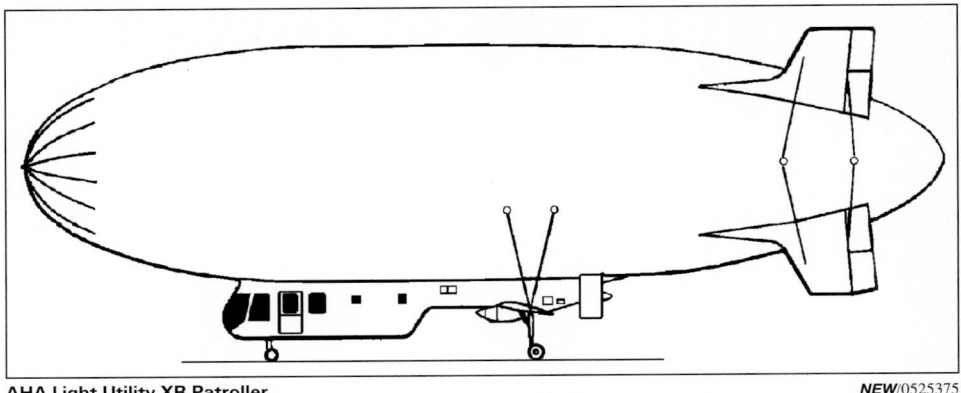

AHA Light Utility.XR Patroller *NEW*/0525375

Technology demonstration model of the AHA Light Utility airship 0016730

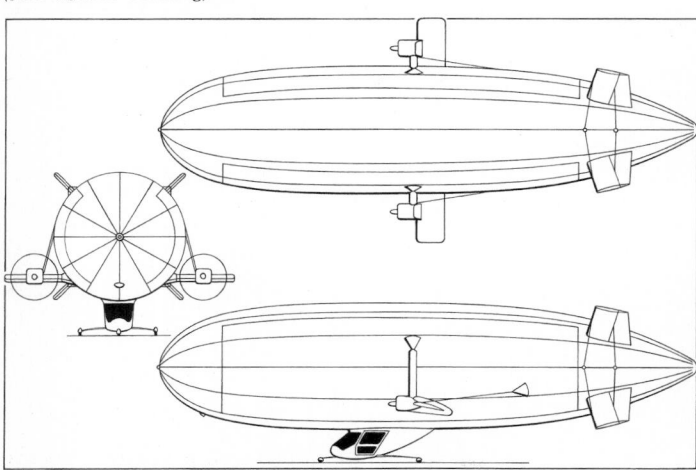

General arrangement of the AHA Hornet LV2 *(Jane's/Mike Keep)*

Scale model of the AHA Hornet LV2

COSTS: Light Utility (four- to six-seat) US$972,000 (2001); XR prices on application.

DESIGN FEATURES: Generally as for Hornet.

FLYING CONTROLS: As for Hornet.

LANDING GEAR: Non-retractable tricycle type.

POWER PLANT: *Light Utility:* Two non-vectoring 74.6 kW (100 hp) MWE AE 100R rotary engines, mounted on stub-wings at rear of gondola; plus stern thruster.
Light Utility. XR: Two 97 kW (130 hp) DAIR 100 two-cylinder diesel engines.

ACCOMMODATION: Four seats, including pilot(s), standard; six or eight seats optional. Four crew plus bunks, galley and lavatory in XR.

DIMENSIONS, EXTERNAL (A: Light Utility, B: Light Utility. XR):

Envelope: Length overall: A	30.48 m (100 ft 0 in)
B	38.10 m (125 ft 0 in)
Max diameter: A	7.49 m (24 ft 7 in)
B	10.24 m (33 ft 7 in)
Fineness ratio: A	4.1
B	3.7
Wing span: A	5.92 m (19 ft 5 in)
Wing chord, constant: A	1.44 m (4 ft 8½ in)
Height overall: B	13.89 m (45 ft 7 in)

DIMENSIONS, INTERNAL:

Envelope volume: A	1,200.6 m³ (42,400 cu ft)
B	2,406.9 m³ (85,000 cu ft)
Max ballonet volume: A	155.7 m³ (5,500 cu ft)

WEIGHTS AND LOADINGS (B):

Disposable load	431 kg (950 lb)
Buoyancy	1,080 kg (2,380 lb)
Max heaviness	359 kg (791 lb)
Max T-O weight	1,436 kg (3,166 lb)

PERFORMANCE (estimated. A and B as above):

Max level speed: A	51 kt (94 km/h; 58 mph)
B, all three engines	75 kt (139 km/h; 86 mph)
Max cruising speed, two engines, at 75% power:	
B	52 kt (96 km/h; 60 mph)
Econ cruising speed, one engine, at 75% power:	
B	41 kt (76 km/h; 47 mph)
Ceiling: A	2,745 m (9,000 ft)
Range: A	300 n miles (555 km; 345 miles)
B (two engines, 50 kt; 93 km/h; 58 mph cruise)	
	1,000 n miles (1,852 km; 1,150 miles)
B (one engine, 30 kt; 56 km/h; 35 mph cruise)	
	3,000 n miles (5,556 km; 3,452 miles)
Endurance, no reserves, conditions as above	
B (two engines)	24 h
B (one engine)	100 h

UPDATED

GSI

GLOBAL SKYSHIP INDUSTRIES INC

1001 Armstrong Boulevard, Unit A, Kissimmee, Florida 34741
Tel: (+1 407) 932 37 79
Fax: (+1 407) 932 29 16
e-mail: information@airshipoperations.com
Web (1): http://www.globalskyships.com
Web (2): http://www.airshipoperations.com
TECHNICAL DIRECTOR: Gary Burns
PUBLIC RELATIONS: Mary Kenny

Global Skyship Industries is the US subsidiary of Aviation Support Group Ltd of the UK which, in December 1996, acquired the assets of the former Westinghouse Airships Inc (WAI). As a result, GSI became holder of the type certificates for the UK-designed Skyship 500 HL and Skyship 600 and the WAI Sentinel series. In late 1997, GSI announced that it had re-opened the production line and had these airships available for immediate delivery. One Skyship 600B was delivered to CargoLifter GmbH of Germany in July 2000.

See 1999-2000 *Jane's* for details of Sentinel 1000 and 1240, which remain available; no orders by mid-2002.

UPDATED

GSI SKYSHIP 500HL

TYPE: Helium non-rigid.

PROGRAMME: First flight (G-SKSB) (converted Skyship 500) 30 July 1987.

CUSTOMERS: Three conversions completed in UK; one (N501LP) in service in USA with Airship Operations Inc in 1998, but destroyed by severe storm on 7 September 1998. Type remains available from GSI which, in mid-2001, still had three (including 07/N503LP and 09/N504LP) ready for assembly on receipt of orders.

DESIGN FEATURES: Combines modified Skyship 500 gondola with larger envelope of Skyship 600, providing ability to operate with greater payload in hotter climates and at higher altitudes than Skyship 500.

POWER PLANT: Two 153 kW (205 hp) Porsche 930/01 non-turbocharged engines; fuel capacity 545 litres (144 US gallons; 120 Imp gallons), of which 530 litres (140 US gallons; 117 Imp gallons are usable).

ACCOMMODATION: Pilot, co-pilot and up to nine passengers. Ground crew of 18.

DIMENSIONS, EXTERNAL: As for Skyship 600 except:

Length overall	59.13 m (194 ft 0 in)
Gondola: Length	9.75 m (32 ft 0 in)
Max width	2.13 m (7 ft 0 in)
Max height	1.93 m (6 ft 4 in)

GSI Skyship 600B with ground crew and support vehicles   0121730

DIMENSIONS, INTERNAL: As for Skyship 600 except:

Envelope volume	6,666 m³ (235,400 cu ft)
Gondola cabin: Length	4.20 m (13 ft 9½ in)
Height	1.96 m (6 ft 5 in)

WEIGHTS AND LOADINGS:

Max usable fuel	382 kg (842 lb)
Gross disposable load	2,190 kg (4,829 lb)

PERFORMANCE:

Max level speed	53 kt (99 km/h; 62 mph)
Cruising speed	30 kt (56 km/h; 35 mph)
Pressure ceiling	2,135 m (7,000 ft)
Max range	347 n miles (643 km; 400 miles)
Endurance at 35 kt (65 km/h; 40 mph)	17 h

UPDATED

GSI SKYSHIP 600

TYPE: Helium non-rigid.

PROGRAMME: First flight 6 March 1984; special category C of A awarded by UK CAA 1 September 1984; aerial work certification received second quarter 1986; full passenger-carrying C of A 8 January 1987, initiating Skycruise aerial sightseeing service over London, San Francisco, Munich and Sydney in 1987, and over Paris in 1988. First US FAA type certificate awarded to an airship for civil use was issued to Skyship 600 on 9 May 1989. Certificate passed to Slingsby Aviation on 24 September 1990; Westinghouse Airships Inc on 24 February 1994; and Global Skyship

Industries on 7 August 1997. During 1990 Farnborough Air Show a Skyship 600 (G-SKSC) with two 227 litre (60 US gallon; 50 Imp gallon) long-endurance fuel tanks made unrefuelled flight of 50 hours 15 minutes; sufficient fuel for a further 20 hours remained at end of flight.

CURRENT VERSIONS: **Skyship 600:** Initial (UK-built) version (nine completed).

Description applies to Skyship 600 except where indicated.

Skyship 600B: With larger envelope, using Tedlar fabric, manufactured by TCOM LP. First example, converted from Skyship 600 c/n 07 (N602SK) and operated for Fuji Film by Airship Management Services, flew for first time 25 July 1998. Second example (c/n 10, N610SK) completed for Airship Operations Inc (Hilfiger contract) to replace destroyed Skyship 500HL.

CUSTOMERS: Total of nine built in UK, of which 01 (G-SKSC) purchased by Global Skyships from UK MoD and 07 (N602SK) converted to 600B in 1998; four others then still in operation with Interport Marine Agencies (03/G-SKSG), Airship International (02/N502LP and 04) and Wilmington Trust (06/N602SA). Two more 600Bs began construction in 1999, of which 02/B-04 (N602CL) was delivered to CargoLifter, Germany, in May 2000, certified by LBA 3 August 2001 and re-registered as D-LCLA *Charly;* for use as a crew trainer. In 2001, 08/N606SA had its Porsche engines replaced by Lycoming IO-540s, making its maiden flight in this form on 18 January 2002 before being redelivered to Airship Management Services Europe in Paris in the second quarter of 2002, bearing the name *Santos-Dumont.*

COSTS: Skyship 600B approximately US$6 million (2000).

DESIGN FEATURES: Non-rigid envelope, with one ballonet forward and one aft (together 26 per cent of total volume); ballonet air intake aft of each propulsor unit. Cruciform tail unit.

FLYING CONTROLS: Vectored thrust propulsion (see Power Plant); differential inflation of ballonets for static fore and aft trim; cable-operated rudders and elevators, each with spring tab; ballast in box below crew seats; disposable water ballast in tanks at rear.

STRUCTURE: Envelope manufactured by Aerazur (France) from single ply polyester fabric, coated with titanium dioxide-loaded polyurethane to reduce ultraviolet degradation; polyvinylidene chloride film bonded on to inner coating of polyurethane on inside of envelope minimises loss of helium gas. Four parabolic arch load curtains, carrying multiple Kevlar 29 gondola suspension cables. Nose structure is domed disc, moulded from GFRP and carrying fitting by which airship is moored to its mast. Each tail surface attached to envelope at root and braced by wires on each side; all four surfaces constructed from interlocking ribs and spars of Fibrelam with GFRP skins.

One-piece moulded gondola of Kevlar-reinforced plastics, with flooring and bulkheads of Fibrelam panels; those forming engine compartment at rear are faced with titanium for fire protection.

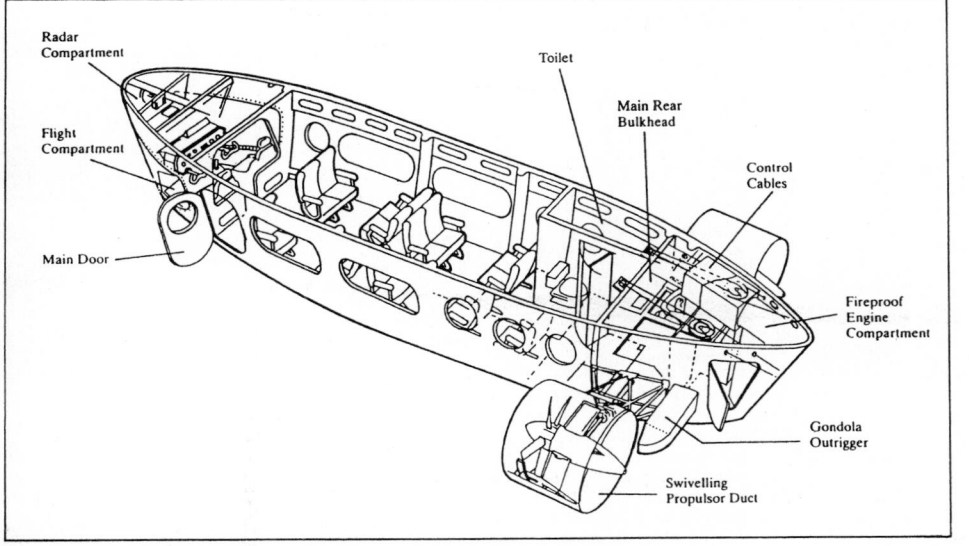

Skyship 600 13-passenger gondola (*Airship Management Services*)   0051869

LANDING GEAR: Single two-wheel assembly with double tyres, mounted beneath rear gondola.

POWER PLANT: Two 190 kW (255 hp) Porsche 930/67/AI/3 six-cylinder air-cooled and turbocharged piston engines mounted in rear of gondola. Each drives a ducted propulsor consisting of a Hoffmann HO-V155A-R/137 five-blade reversible-pitch propeller within an annular duct of carbon fibre-reinforced GFRP. Each propulsor can be rotated about its pylon attachment to gondola through an arc of 210°, 90° upward and 120° downward, vectored thrust thus providing both V/STOL and in-flight hovering ability.

Fuel tank, capacity 682 litres (180 US gallons; 150 Imp gallons), at rear of engine compartment; 668 litres (176.5 US gallons; 147 Imp gallons) usable. Auxiliary fuel tanks optional. Engine modifications include provision of automatic mixture control, fuel injection and electronic ignition.

ACCOMMODATION: Seats for pilot and co-pilot, with dual controls. Maximum capacity 12 passengers in addition to pilots. Ground crew of 18.

SYSTEMS: 28 V electrical system, supplied by engine-driven alternators.

AVIONICS: Include Honeywell Silver Crown series dual nav/com, ADF, Omega, VOR/ILS and weather radar.

EQUIPMENT: Night signs (30.48 m; 100 ft long and 9.14 m; 30 ft high) are full-colour aerial boards comprising 8,400 primary-coloured lamps computerised to enable the display of signs, animation and graphic logos in 16 colours.

DIMENSIONS, EXTERNAL:

Envelope: Length overall: 600	59.00 m (193 ft 7 in)	
600B	60.96 m (200 ft 0 in)	
Max diameter	15.20 m (49 ft 10½ in)	
Fineness ratio: 600	3.88	
600B	3.91	
Height overall, incl gondola and landing gear		
	20.30 m (66 ft 7¼ in)	

Vectored-thrust Porsche power plant of the Skyship 600/600B   0054440

Width over tailfins	19.20 m (63 ft 0 in)
Gondola: Length overall	11.67 m (38 ft 3½ in)
Max width	2.56 m (8 ft 4¾ in)
Max height	1.93 m (6 ft 4 in)
Propeller diameter	1.37 m (4 ft 6 in)

DIMENSIONS, INTERNAL:

Envelope volume: 600	6,666 m³ (235,400 cu ft)
600B	7,192.5 m³ (254,000 cu ft)
Ballonet volume (two, total)	1,733 m³ (61,200 cu ft)
Gondola cabin: Length	6.89 m (22 ft 7¼ in)
Height	1.92 m (6 ft 3½ in)
Floor area (usable)	12.00 m² (130.0 sq ft)

WEIGHTS AND LOADINGS:

Max usable fuel	484 kg (1,067 lb)
Gross disposable load	2,343 kg (5,165 lb)

PERFORMANCE:

Max level speed	56 kt (104 km/h; 65 mph)

Skyship 600/600B first class seating arrangement   0121683

Cruising speed	35 kt (64 km/h; 40 mph)
Pressure ceiling	2,135 m (7,000 ft)
Max range (standard fuel)	
	347 n miles (643 km; 400 miles)

UPDATED

OHIO

OHIO AIRSHIPS INC

11177 Bowen Road, Mantua, Ohio 44255
Tel: (+1 330) 823 30 90
Web: http://www.ohioairships.com
PRESIDENT: Robert Rist
VICE-PRESIDENT AND CORPORATION STRATEGIST: Brian Martin
VICE-PRESIDENT AND CHIEF FINANCIAL OFFICER:
  Scott Gindlesberg
ENGINEERING ADVISER: Don VanFossen

This company has a hybrid semi-rigid in the development stage. Design assistance is being given by Conceptual Research Corporation of Sylmar, California.
UPDATED

OHIO DYNALIFTER

TYPE: Hybrid rigid air vehicle.

PROGRAMME: Announced in 2000, taxi tests of subscale prototype having begun in March of that year.

DESIGN FEATURES: Heavy-lift winged airship, combining static (helium) and dynamic (wing) lift, advantage of which is to enable Dynalifter to be smaller than traditional airship for given weight of payload; also offers capability for higher speeds (87 to 130 kt; 161 to 241 km/h; 100 to 150 mph cruise) and operating altitude (up to 3,050 m; 10,000 ft), STOL runway performance and ability to use existing mid- to large-size airports. Cargo in detachable ventral bay some two-thirds of overall envelope length. Full-size

General appearance of the Dynalifter hybrid   *NEW*/0525376

Dynalifter could have envelope volume of up to 198,220 m³ (7 million cu ft), with cargo container of 13,592 m³ (480,000 cu ft), and an estimated range of 5,000 n miles (9,260 km; 5,754 miles) with 101,605 kg (224,000 lb) of payload.

CURRENT VERSIONS (projected): **PSC-1:** Up to 453,590 kg (1 million lb) MTOW, including 158,760 kg (350,000 lb) of cargo payload.

PSC-2: Smaller version, with up to 20,411 kg (45,000 lb) of payload and MTOW of 90,718 kg (200,000 lb).

Following details refer to subscale prototype:

STRUCTURE: Semi-buoyant, aerofoil-shaped envelope with internal framework and five gas cells.

POWER PLANT: Two wing-mounted glow-plug engines.

DIMENSIONS, EXTERNAL:

Wing span	5.79 m (19 ft 0 in)
Length overall	6.40 m (21 ft 0 in)

DIMENSIONS, INTERNAL:

Envelope volume	14.16 m³ (500 cu ft)

UPDATED

SKYMEDIA

SKYMEDIA AIRSHIPS INC

1497 Poinsettia Avenue, Suite 156, Vista, California 92083-8542
Tel: (+1 760) 295 13 66
Fax: (+1 760) 295 13 67
e-mail: info@skymediaairships.com
Web: http://www.skymediaairships.com
PRESIDENT: Paul K Cohen
CHAIRMAN: Paul M Thomas

SkyMedia Airships was formed in 2001 with equity holding, management and marketing assistance from VentureNet Capital Group of Temecula, California. Its first venture is a low-cost hot-air airship for advertising purposes.
NEW ENTRY

SKYMEDIA SMA

TYPE: Hot-air non-rigid.

PROGRAMME: Small-scale (19.8 m; 65 ft long) prototype (N9156K) made maiden flight 14 September 2001; full-size version planned to fly in 2002.

DESIGN FEATURES: Envelope can be sealed and pressurised, hot air then being utilised in form of jets to provide main propulsion. Internal digital system projects advertising displays directly on to envelope, obviating need for

Computer-generated image of the full-size SkyMedia SMA   *NEW*/0525377

external screens or banners. Conventional envelope shape; triple tailfins, each with movable control surface. Quoted time of 20 minutes for deflation, pack-up and storage in ground transport vehicle.

ACCOMMODATION: Ventral gondola for pilot only in scale prototype; two persons in full-size SMA. Ground crew of three.

DIMENSIONS, EXTERNAL:

Length overall	42.06 m (138 ft 0 in)

PERFORMANCE (estimated):

Max level speed	43 kt (80 km/h; 50 mph)
Pressure ceiling	1,525 m (5,000 ft)

NEW ENTRY

UPship

UPship CORPORATION
5198 Highway 84, Elba, Alabama 36323
Tel/Fax: (+1 334) 897 61 32
e-mail: airship@alaweb.com
PRESIDENT AND TECHNICAL DIRECTOR: Jesse Blenn

According to the last information received, UPship's first airship to be built will be the 36 m, three-seat version described below. Design and cost projections for other, different sized versions have been made for craft with one, six, 16 and 50 seats; last-mentioned would be fitted with bow and stern thrusters for heavy lifting and precision hover.

Other design work was under way in 2001 on 50 and 90 tonne payload cargo airships (lengths 180 and 216 m; 590.6 and 708.7 ft, respectively) for a Latin American client, based on features of the small prototype. Brazilian Army known to be interested in airship for supplies transport in Amazonia.

UPDATED

UPship 001
TYPE: Helium semi-rigid.
PROGRAMME: Design started 1989; original 750-001 (see 1992-93 *Jane's*) later enlarged as two-seat, 32 m 001 proof-of-concept vehicle; further enlarged in 1999 to 36 m, lower-drag envelope profile and three seats (or two occupants and heavier commercial or scientific payload). Progress by late 1998 included land acquisitions, detail design work, granting of two US patents, and completion of residence and design studio at the operations site. Refinement of design continuing in 2001, but funding efforts were then still slower than hoped. Prototype construction and certification estimated to take two years once a start is made.
CUSTOMERS: Markets foreseen in tourism, advertising, scientific research, cargo, resource management, and minimum-impact logging operations, where UPships can be built and operated at competitive price.
COSTS: Up to US$2 million for set-up, construction and FAA certification of prototype.

DESIGN FEATURES: Shape of minimum resistance, with propulsion engines in inverted V tailfins; nose-mounted thruster for control at all speeds. Designed for ease of maintenance, with reduced dependence on internal pressure. Greatly improved control, multiple helium cells and smoother ride increase usefulness and safety while minimising number of ground crew.
FLYING CONTROLS: Joystick regulates pitch and yaw by ruddervators in propeller slipstream, with aerodynamic assist; automatic pitch trim; foot pedals control bow thruster for enhanced and low-speed control (up/down, right/left and reverse) and for heavy lifting.
STRUCTURE: Aluminium and steel tubing hard structures; some carbon composites as appropriate; envelope of proprietary ripstop construction with internal divisions and three helium cells; tailfins deflect under excessive ground or air loads.
POWER PLANT: Three 17.9 kW (24 hp) König two-stroke radial piston engines with electric starting, two operating in tailfin openings and one as bow thruster. Fuel capacity 75.7 litres (20.0 US gallons; 16.7 Imp gallons).
ACCOMMODATION: Pilot and two passengers, seated in line; rear seat reclinable. Space aft of seats for rescuee(s), equipment or rest area. Cabin has side door and hatch for camera or rescue hoist; is electrically heated; noise- and vibration-free due to distance from engines.
SYSTEMS: Two alternators provide electrical power for night illumination system and cabin heating.

DIMENSIONS, EXTERNAL:
Envelope: Length overall	36.00 m (118 ft 1¼ in)
Max diameter	7.20 m (23 ft 7½ in)
Height overall	8.70 m (28 ft 6½ in)
Tail unit span	10.00 m (32 ft 9¾ in)
Propeller diameter	1.50 m (4 ft 11 in)

DIMENSIONS, INTERNAL:
Envelope volume	894.0 m³ (31,571 cu ft)
Ballonet volume (30%)	235.0 m³ (8,300 cu ft)
Helium fill	785.0 m³ (27,722 cu ft)
Gondola: Length	5.00 m (16 ft 4¾ in)
Max width	1.00 m (3 ft 3¼ in)

WEIGHTS AND LOADINGS (approx):
Weight empty	470 kg (1,036 lb)

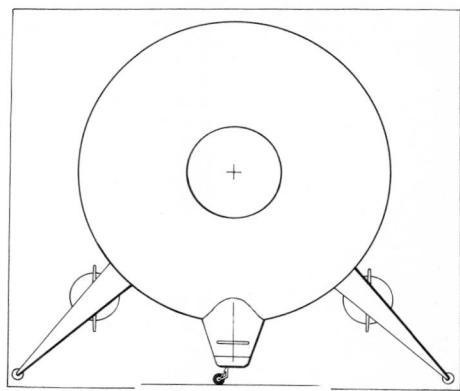

Nose view of UPship 001 as currently envisaged
(Jane's/James Goulding)   0085294

Useful lift	315 kg (694 lb)
Total static lift	785 kg (1,731 lb)
Additional thruster lift	50 kg (110 lb)

PERFORMANCE (estimated):
Max level speed	52 kt (98 km/h; 60 mph)
Cruising speed: at 50% power	42 kt (78 km/h; 48 mph)
at 25% power	34 kt (63 km/h; 39 mph)
Ceiling above T-O: normal	1,000 m (3,280 ft)
max	3,000 m (9,840 ft)

Range, two engines, 20% reserves:
at 50% power cruising speed
                216 n miles (400 km; 248 miles)
at 25% power cruising speed
                345 n miles (640 km; 397 miles)
Endurance, two engines, 20% reserves:
at 50% power cruising speed	5 h
at 25% power cruising speed	10 h

VERIFIED

AIR-LAUNCHED MISSILES

Raytheon AIM-120 AMRAAM being launched from a Lockheed Martin F-16C Fighting Falcon *NEW*/0527137

The following pages support the Armament paragraphs of aircraft descriptions in this book by explaining in brief how the potential of an individual aircraft is enhanced by its missile armament. Coverage is restricted to missiles carried by, or applicable to, aircraft in the current edition; for this reason, some older missiles are excluded, as are future projects still in the early stages of definition. Contents do include certain anti-tank and shoulder-launched anti-aircraft missiles which have airborne applications, mostly on helicopters.

To expedite retrieval of data, missiles are listed in alphabetical order of name or designation, with full cross-references to alternative epithets. In many instances, the 'manufacturer' of Chinese and Russian missiles is actually the export sales agency. More detailed information is contained in *Jane's Air-Launched Weapons*.

KEY

Roles		Guidance					
AAM	Air-to-air missile	AL	Active laser	LR	Laser radar		
ARM	Anti-radiation missile	AP	Autopilot	MMW	Millimetric-wave		
ASM	Air-to-surface missile	A/P	Active/passive radar	PR	Passive radar		
AshM	Anti-ship missile	ARH	Active radar homing	RC	Radio command		
ATM	Anti-tank missile	GPS	Global Positioning System	RF	Radio frequency		
LGB	Laser-guided bomb	I	Inertial	SAL	Semi-active laser		
SOM	Standoff missile	Im	Imaging	SARH	Semi-active radar homing		
		IR	Infra-red	T	Terrain reference		
		L	Laser	TV	Television		

Name/designation	Role	Manufacturer/country	Length m (in)	Diameter m (in)	Weight kg (lb)		Guidance	Range n miles (km)	
3M55 see 'Yakhont'									
3M60 see 'Kayak'									
3M80 see Moskit									
9M14 see 'Sagger'									
9M17 see 'Swatter'									
9M32 see 'Grail'									
9M36 see 'Gremlin'									
9M39 see 'Grouse'									
9M114 see 'Spiral'									
9M120 see 'Spiral' and AT-X-16									
9M121 see AT-X-16									
9M313 see 'Gimlet'									
AA-6 see 'Acrid'									
AA-7 see 'Apex'									
AA-8 see 'Aphid'									
AA-9 see 'Amos'									
AA-10 see 'Alamo'									
AA-11 see 'Archer'									
AA-12 see 'Adder'									
AA-X-13 see R-37									
AAM-3, Type 90	AAM	Mitsubishi/Japan	3.00 (118)	0.13 (5)	91	(201)	IR	4.3	(8)
AAM-4	AAM	Mitsubishi/Japan					ARH	medium	
AAM-5	AAM	not assigned/Japan					ImIR		
AAM-L, KS-172	AAM	Novator/Russia	7.40 (291)		750	(1,653)	I/ARH	216	(400)
AARGM	ARM	SAT & NAWC/USA					GPS/MMW	100	(185)
ACM, AGM-129	ASM	Raytheon/USA	6.35 (250)	0.70 (28)	1,250	(2,756)	I/LR	1,620	(3,000)
'Acrid', AA-6/R-46TD	AAM	Vympel/Russia	6.20 (244)	0.36 (14)	467	(1,030)	I/IR	27	(50)

Name/designation	Role	Manufacturer/country	Length m (in)	Diameter m (in)	Weight kg (lb)		Guidance	Range n miles (km)	
A-Darter, V3E	AAM	Denel/South Africa	2.98 (117)	0.17 (7)	89	(196)	AL	11	(20)
'Adder', AA-12/R-77	AAM	Vympel/Russia	3.60 (142)	2.00 (79)	175	(386)	I/ARH	40	(75)
'Adder', R-77M-PD	AAM	Vympel/Russia	3.60 (142)	2.00 (79)	225	(496)	I/ARH	81	(150)
AFDS (unpowered)	SOM	LFK/Germany	3.47 (137)	0.63 (25)	660	(1,455)	GPS/I		
AGM-65 see Maverick									
AGM-78 see Standard									
AGM-84 see Harpoon/SLAM									
AGM-88 see HARM									
AGM-114 see Hellfire									
AGM-119 see Penguin									
AGM-122 see Sidearm									
AGM-123 see Skipper									
AGM-129 see ACM									
AGM-130A	LGB	Boeing/USA	3.94 (155)	0.46 (18)	1,323	(2,917)	TV, ImIR	24	(45)
AGM-130C	LGB	Boeing/USA	3.95 (156)	0.46 (18)	1,353	(2,983)	TV/IIR	24	(45)
AGM-142 see Popeye									
AGM-154C JSOW	SOM	Raytheon/USA	4.26 (168)		475 (1,047)/ 680 (1,499)		I/GPS	32	(60)
AGM-158 see JASSM									
AIM-7 see Sparrow									
AIM-9 see Sidewinder									
AIM-120 see AMRAAM									
AIM-132 see ASRAAM									
'Alamo', AA-10/R-27AE	AAM	Vympel/Russia	4.78 (188)	0.26 (10)	350	(772)	I/ARH	43	(80)
'Alamo', AA-10/R-27EM	AAM	Vympel/Russia	4.78 (188)	0.26 (10)	350	(772)	I/SARH	59	(110)
'Alamo', AA-10/R-27ER	AAM	Vympel/Russia	4.70 (185)	0.26 (10)	350	(772)	I/SARH	40	(75)
'Alamo', AA-10/R-27ET	AAM	Vympel/Russia	4.50 (177)	0.26 (10)	343	(756)	I/IR	38	(70)
'Alamo', AA-10/R-27R	AAM	Vympel/Russia	4.00 (157)	0.23 (9)	253	(558)	I/SARH	27	(50)
'Alamo', AA-10/R-27T	AAM	Vympel/Russia	3.70 (146)	0.23 (9)	254	(560)	I/IR	22	(40)

AA-12 'Adder' AAM (*Jane's/Paul Jackson*)   NEW/0121705

MBDA ALARM anti-radar weapon (*Jane's/Paul Jackson*)   NEW/0121684

Name/designation	Role	Manufacturer/country	Length m (in)	Diameter m (in)	Weight kg (lb)		Guidance	Range n miles (km)	
ALARM	ARM	MBDA/International	4.30 (169)	0.22 (9)	265	(584)	PR	24	(45)
AM 39 see Exocet									
'Amos' AA-9/R-33	AAM	Vympel/Russia	4.15 (163)	0.38 (15)	490	(1,080)	I/SARH	65	(120)
AMRAAM, AIM-120A/B/C	AAM	Raytheon/USA	3.65 (144)	0.18 (7)	157	(346)	I/ARH	27	(50)
APACHE AP	SOM	MBDA/International	5.10 (201)	0.63 (25)	1,230	(2,712)	I/ARH	76	(140)
'Apex', AA-7/R-24R	AAM	Vympel/Russia	4.46 (176)	0.20 (8)	235	(518)	SARH	27	(50)
'Apex', AA-7/R-24T	AAM	Vympel/Russia	4.16 (164)	0.20 (8)	235	(518)	IR	27	(50)
'Aphid', AA-8/R-60	AAM	Vympel/Russia	2.08 (82)	0.13 (5)	65	(143)	IR	1.6	(3)
'Aphid', AA-8/R-60M	AAM	Vympel/Russia	2.08 (82)	0.13 (5)	65	(143)	AL	2.7	(5)
'Archer', AA-11/R-73M1	AAM	Vympel/Russia	2.90 (114)	0.17 (7)	105	(231)	I/IR	11	(20)
'Archer', AA-11/R-73M2	AAM	Vympel/Russia	2.90 (114)	0.17 (7)	110	(243)	I/IR	16	(30)
ARMAT	ARM	MBDA/International	4.15 (163)	0.40 (16)	550	(1,213)	I/PR	49	(90)
Armiger	ARM	BGT/Germany	3.90 (153)	0.20 (8)	230	(507)	PR/ImIR	108	(200)
AS-7 see 'Kerry'									
AS-9 see 'Kyle'									
AS-10 see 'Karen'									
AS-11 see 'Kilter'									
AS-12 see 'Kegler'									
AS-13 see 'Kingbolt'									
AS-14 see 'Kedge'									
AS-15 see 'Kent'									
AS 15TT	AShM	MBDA/International	2.30 (91)	0.18 (7)	96	(212)	Radio	8.1	(15)
AS-16 see 'Kickback'									
AS-17 see 'Krypton'									
AS-18 see 'Kazoo'									
AS-19 see 'Koala'									
AS-20 see 'Kayak'									
AS-30L	ASM	MBDA/International	3.65 (144)	0.34 (13)	520	(1,146)	I/SAL	5.4	(10)
ASM-1, Type 80	AShM	Mitsubishi/Japan	4.00 (157)	0.34 (13)	600	(1,323)	I/ARH	27	(50)
ASM-1C, Type 91	AShM	Mitsubishi/Japan	4.00 (157)	0.35 (14)	510	(1,124)	I/ARH	35	(65)
ASM-2, Type 93	AShM	Mitsubishi/Japan	4.10 (161)	0.35 (14)	520	(1,146)	I/ImIR	54	(100)
ASMP	SOM	MBDA/International	5.38 (212)	0.38 (15)	860	(1,896)	I/T	135	(250)
ASMP Plus (ASMP-A)	SOM	MBDA/International						270	(500)
Aspide 1	AAS	MBDA/International	3.70 (146)	0.20 (8)	220	(485)	SARH	19	(35)
ASRAAM (AIM-132)	AAM	MBDA/International	2.90 (114)	0.17 (7)	87	(192)	ImIR	11	(20)
AT-2 see 'Swatter'									
AT-3 see 'Sagger'									
AT-6 see 'Spiral'									
AT-9 see 'Spiral 2'									

Name/designation	Role	Manufacturer/country	Length m (in)	Diameter m (in)	Weight kg (lb)	Guidance	Range n miles (km)
AT-12 see 'Swinger'							
AT-16, 9M120M/9M121 Vikhr M	ATM	Shipunov/Russia	2.80 (110)	0.13 (5)	45 (99)	SAL	5.4 (10)
Ataka see 'Swinger'							
ATAM see Mistral							
ATASK see Helstreak							
BGM-71 see TOW							
Black Shaheen see SCALP EG							
Brimstone	ASM	MBDA/International	1.63 (64)	0.18 (7)	50 (110)	MMW/I	4.3 (8)
Burya see 'Kitchen'							
C-101, YJ-16	AShM	CPMIEC/China	7.50 (295)	0.54 (21)	1,850 (4,079)	I/ARH	24 (45)
C-201, HY-4	AShM	CPMIEC/China	7.36 (290)	0.76 (30)	1,740 (3,836)	AP/ARH	73 (135)
C-601, CAS-1 'Kraken'/YJ-6	AShM	CPMIEC/China	7.36 (290)	0.76 (30)	2,440 (5,379)	AP/ARH	54 (100)
C-701	AShM	CPMIEC/China	2.51 (99)	0.54 (21)	100 (220)	I/TV	8.1 (15)
C-801, YJ-1	AShM	CPMIEC/China	4.65 (183)	0.36 (14)	655 (1,444)	I/ARH	27 (50)
C-802, YJ-2	AShM	CPMIEC/China	6.40 (252)	0.36 (14)	715 (1,576)	I/ARH	70 (130)
CAS-1 see C-601							
Dandy see NT-D							
Darter, V-3C (see U-Darter)	AAM	Denel/South Africa	2.75 (108)	0.16 (6)	90 (198)	IR	2.7 (5)
Derby	AAM	Rafael/Israel	3.62 (143)	0.16 (6)	118 (260)	ARH	63 (34)
DWS24/DWS39 (unpowered)	SOM	LFK/Germany	3.50 (138)		600 (1,323)	I	5.4 (10)
Exocet, AM 39	AShM	MBDA/International	4.70 (185)	0.35 (14)	670 (1,477)	I/ARH	27 (50)
FIM-92 see Stinger							
Gabriel 3AS	AShM	IAI/Israel	3.85 (152)	0.34 (13)	560 (1,235)	I/ARH	19 (35)
Gabriel 4LR	AShM	IAI/Israel	4.70 (185)	0.44 (17)	960 (2,116)	I/ARH	108 (200)
'Gimlet', SA-16/9M313 Igla 1	AAM	Kolomna/Russia	1.69 (67)	0.07 (3)	11 (24)	IR	2.7 (5)
'Grail' SA-7/9M32 Strela 2	AAM	Turopov/Russia	1.22 (48)	0.07 (3)	10 (22)	IR	2.7 (5)
'Gremlin', SA-14/9M36 Strela 3	AAM	Turopov/Russia	1.47 (58)	0.07 (3)	11 (24)	IR	2.7 (5)
'Grouse', SA-18/9M39 Igla	AAM	Kolomna/Russia	1.69 (67)	0.07 (3)	11 (24)	IR	2.7 (5)

SA-18 Grouse AAM (Jane's/Paul Jackson)   NEW/0105006

BGT IRIS-T AAM   NEW/0527138

Hakeem see PGM							
HARM, AGM-88A/B/B+/C/D	ARM	Raytheon/USA	4.17 (164)	0.25 (10)	361 (796)	PR	43 (80)
Harpoon, AGM-84A	AShM	Boeing/USA	3.90 (154)	0.34 (13)	530 (1,168)	I/ARH	65 (120)
Have Lite see Popeye 2							
Have Nap see Popeye 1							
Hellfire, AGM-114A	ATM	Hellfire Systems LLC/ USA	1.63 (64)	0.18 (7)	46 (101)	SAL	4.3 (8)
Hellfire, AGM-114B/C	ATM	Hellfire Systems LLC/ USA	1.73 (68)	0.18 (7)	48 (106)	SAL, ImIR or RF+IR	4.3 (8)
Hellfire, AGM-114F	ATM	Hellfire Systems LLC/ USA	1.80 (71)	0.18 (7)	49 (108)	SAL	4.3 (8)
Hellfire 2, AGM-114K	ATM	Hellfire Systems LLC/ USA	1.63 (64)	0.18 (7)	46 (101)	SAL	4.9 (9)
Hellfire 2, AGM-114 Longbow	ATM	Hellfire Systems LLC/ USA	1.78 (70)	0.18 (7)	50 (110)	MMW/I	4.3 (8)
Helstreak/ATASK (Starstreak)	AAM	Shorts/UK	1.40 (55)	0.13 (5)	16 (35)	RC	3.2 (6)
HJ-8A	ATM	Norinco/China	0.88 (35)	0.12 (5)	11 (24)	Wire	1.6 (3)
HJ-8B	ATM	Norinco/China	1.00 (39)	0.12 (5)	13 (29)	Wire	2.2 (4)
HOT 1	ATM	MBDA/International	1.27 (50)	0.14 (6)	24 (53)	Wire	2.2 (4)
HOT 2	ATM	MBDA/International	1.30 (51)	0.15 (6)	24 (53)	Wire	2.2 (4)
HOT 3	ATM	MBDA/International	1.30 (51)	0.15 (6)	33 (73)	Wire	2.2 (4)
Hsiung Feng 2	AShM	Chung Shan/Taiwan	3.90 (154)	0.34 (13)	520 (1,146)	I/ARH+ImIR	43 (80)
HY-4 see C-201							
Igla see 'Grouse' and 'Gimlet'							
Ingwe, ZT35	ATM	Denel/South Africa	1.75 (69)	0.40 (16)	29 (64)	L	2.7 (5)
IRIS-T	AAM	BGT/Germany	3.00 (118)	0.13 (5)	87 (192)	ImIR	6.5 (12)
JASSM, AGM-158	ASM	Lockheed Martin/USA	4.26 (168)	0.55 (22)	1,024 (2,257)	ImIR/GPS/INS	200 (370)
JSOW see AGM-154							
'Karen', AS-10/Kh-25MR	ASM	Zvezda/Russia	4.04 (159)	0.28 (11)	300 (661)	RC	5.4 (10)
'Karen', AS-10/Kh-25ML	ASM	Zvezda/Russia	4.04 (159)	0.28 (11)	300 (661)	SAL	11 (20)

Name/designation	Role	Manufacturer/country	Length m (in)	Diameter m (in)	Weight kg (lb)	Guidance	Range n miles (km)
'Kayak', AS-20/Kh-35/3M60 Uran	AShM	Zvezda/Russia	3.75 (148)	0.42 (17)	480 (1,058)	I/ARH	70 (130)
'Kazoo', AS-18/Kh-59M Ovod M	ASM	Raduga/Russia	5.37 (211)	0.38 (15)	930 (2,050)	I/TV	62 (115)
'Kedge', AS-14/Kh-29L	ASM	Vympel/Russia	3.87 (152)	0.38 (15)	657 (1,448)	SAL	5.4 (10)
'Kedge', AS-14/Kh-29T	ASM	Vympel/Russia	3.87 (152)	0.38 (15)	670 (1,477)	TV	6.5 (12)
'Kedge', AS-14/Kh-29TE	ASM	Vympel/Russia	3.87 (152)	0.38 (15)	700 (1,543)	TV	16 (30)
'Kegler', AS-12/Kh-25MP/Kh-27	ASM	Zvezda/Russia	4.36 (172)	0.28 (11)	310 (683)	I/PR	22 (40)
'Kent', AS-15A/Kh-55/RKV-500	SOM	Raduga/Russia	6.04 (238)	0.51 (20)	1,400 (3,086)	I/T	1,296 (2,400)
'Kent', AS-15B/Kh-55SM/ RKV-500M	SOM	Raduga/Russia	7.10 (280)	0.77 (30)	1,700 (3,748)	I/T	1,620 (3,000)
KEPD 150 (PDWS 200)	SOM	Taurus Systems/ International	4.50 (177)		1,060 (2,337)	I/GPS/T/ImIR	81 (150)
KEPD 350 (Taurus)	SOM	Taurus Systems/ International	5.00 (197)		1,400 (3,086)	I/GPS/T/ImIR	189 (350)
'Kerry', AS-7/Kh-23	ASM	Zvezda/Russia	3.53 (139)	0.28 (11)	287 (633)	SAL or RC	2.7 (5)
Kh-15 see 'Kickback'							
Kh-22 see 'Kitchen'							
Kh-23 see 'Kerry'							
Kh-25 see 'Karen' and 'Kegler'							
Kh-27 see 'Kegler'							
Kh-28 see 'Kyle'							
Kh-29 see 'Kedge'							
Kh-31 see 'Krypton'							
Kh-35 see 'Kayak'							
Kh-41 see Moskit							
Kh-55/65 see 'Kent'							
Kh-58 see 'Kilter'							
Kh-59 see 'Kingbolt'							
Kh-101	SOM	Raduga/Russia	7.45 (293)		2,400 (5,291)	I/Im	2,700 (5,000)
'Kickback', AS-16/Kh-15/RKV-500B	ASM	Raduga/Russia	4.78 (188)	0.46 (18)	1,200 (2,646)	I/PR or I/ARH	81 (150)
'Kilter' AS-11/Kh-58E	ARM	Raduga/Russia	5.00 (197)	0.38 (15)	650 (1,433)	I/PR	86 (160)
'Kingbolt', AS-13/Kh-59 Ovod	ASM	Raduga/Russia	5.40 (213)	0.38 (15)	850 (1,874)	TV	86 (160)
'Koala', AS-19 (Kh-90/BL-10)	ASM	reportedly terminated					
Kokon see 'Spiral'							
'Kraken' see C-601							

Members of the Maverick family   NEW/0527129

Raytheon AGM-154 JSOW
NEW/0527130

'Krypton', AS-17/Kh-31A-1	ASM	Zvezda/Russia	4.70 (185)	0.36 (14)	600 (1,323)	I/ARH	27 (50)
'Krypton', AS-17/Kh-31A-2	ASM	Zvezda/Russia	5.23 (206)	0.36 (14)	600 (1,323)	I/ARH	38 (70)
'Krypton', AS-17/Kh-31P-1	ASM	Zvezda/Russia	4.70 (185)	0.36 (14)	600 (1,323)	I/PR	54 (100)
'Krypton', AS-17/Kh-31P-2	ASM	Zvezda/Russia	5.23 (206)	0.36 (14)	600 (1,323)	I/PR	108 (200)
Kukri, V-3B	AAM	Denel/South Africa	2.94 (116)	0.13 (5)	73 (161)	IR	2.2 (4)
'Kyle', AS-9/Kh-28	ARM	Zvezda/Russia	6.00 (236)	0.43 (17)	715 (1,576)	PR	49 (90)
LY-60 see PL-11								
MAA-1 Piranha/Mol	AAM	Orbita/Brazil	2.82 (111)	0.15 (6)	90 (198)	IR	2.7 (5)
Magic 1, R 550	AAM	MBDA/International	2.72 (107)	0.16 (6)	89 (196)	IR	1.6 (3)
Magic 2, R 550	AAM	MBDA/International	2.75 (108)	0.16 (6)	89 (196)	IR	11 (20)
Marte 2	AShM	AOSM/Italy	4.80 (189)	0.32 (13)	340 (750)	I/ARH	11 (20)
Marte 2A	AShM	AOSM/Italy	3.90 (154)	0.32 (13)	269 (593)	I/AR	16 (30)
Marte 2B	ARM	AOSM/Italy	3.90 (154)	0.32 (13)	269 (593)	PR	32 (60)
Maverick, AGM-65A	ASM	Raytheon/USA	2.49 (98)	0.31 (12)	210 (463)	TV	1.6 (3)
Maverick, AGM-65B	ASM	Raytheon/USA	2.49 (98)	0.31 (12)	210 (463)	TV	4.3 (8)
Maverick, AGM-65D	ASM	Raytheon/USA	2.49 (98)	0.31 (12)	220 (485)	ImIR	11 (20)
Maverick, AGM-65E	ASM	Raytheon/USA	2.49 (98)	0.31 (12)	293 (646)	SAL	11 (20)
Maverick, AGM-65F/G/J/K	ASM	Raytheon/USA	2.49 (98)	0.31 (12)	307 (677)	ImIR	13 (25)

Name/designation	Role	Manufacturer/country	Length m (in)	Diameter m (in)	Weight kg (lb)		Guidance	Range n miles (km)	
Maverick, AGM-65H	ASM	Raytheon/USA	2.60 (102)	0.31 (12)	305	(672)	ARH	13	(25)
Meteor	AAM	MBDA/International	3.65 (144)		185	(408)	ARH	54	(100)
MICA	AAM	MBDA/International	3.10 (122)	0.16 (6)	110	(243)	I/ARH or IR	27	(50)

Maverick (centre), flanked by AIM-9L Sidewinder and LAU 5002 rocket launcher (*Jane's/Paul Jackson*) NEW/0121703

Mokopa anti-tank missile (*Jane's/Paul Jackson*) NEW/0121685

Name/designation	Role	Manufacturer/country	Length m (in)	Diameter m (in)	Weight kg (lb)		Guidance	Range n miles (km)	
Mistral, ATAM	AAM	MBDA/International	1.80 (71)	0.09 (4)	18	(40)	IR	2.7	(5)
Mokopa, ZT6	ATM	Denel/South Africa	1.80 (71)	0.18 (7)	52	(115)	MMW or SAL	4.3	(8)
Moskit, Kh-41/3M80	ASM	Raduga/Russia	9.74 (383)	0.76 (30)	4,500	(9,921)	I/ARH or I/PR	135	(250)
Nag	ATM	DRDO/India			43	(95)	RC + ImIR or MMW	2.2	(4)
Nimrod	ASM	IAI/Israel	2.84 (112)	0.21 (8)	100	(220)	I/SAL	13	(25)
NTD Dandy	ATM	Rafael/Israel	1.20 (47)	0.14 (6)	11	(24)	I/TV/IR	2.2	(4)
NT-S Spike	ATM	Rafael/Israel	1.90 (75)	0.14 (6)	27	(60)	I//TV/IR	6.2	(10)
Ovod see 'Kazoo' and 'Kingbolt'									
Penguin 2, AGM-119B	AShM	Kongsberg/Norway	2.96 (117)	0.28 (11)	385	(849)	I/IR	19	(35)
Penguin 3, AGM-119A	AShM	Kongsberg/Norway	3.18 (125)	0.28 (11)	370	(816)	I/IR	30	(55)
PGM-A (1A, 2A, 3A) Hakeem	ASM	MBDA/International	3.60 (142)	0.35 (14)	300	(661)	I+SAL/IR/TV	11	(20)
PGM-B (1B, 2B, 3B) Hakeem	ASM	MBDA/International	4.00 (157)	0.43 (17)	1,115	(2,458)	I+SAL/IR/TV	11	(20)
Piranha see MAA-1									
PL-2/PL-3	AAM	CATIC/China	2.99 (118)	0.13 (5)	76	(168)	IR	1.6	(3)
PL-5	AAM	CATIC/China	2.89 (114)	0.13 (5)	85	(187)	IR	1.6	(3)
PL-7	AAM	CATIC/China	2.75 (108)	0.16 (6)	90	(198)	IR	1.6	(3)
PL-8	AAM	CATIC/China	3.00 (118)	0.16 (6)	120	(265)	IR	2.7	(5)
PL-9	AAM	Luoyang/China	2.99 (118)	0.16 (6)	115	(254)	IR	2.7	(5)
PL-10	AAM	CATIC/China	3.99 (157)	0.29 (11)	300	(661)	SARH	8.1	(15)
PL-11, LY-60	AAM	CATIC/China	3.89 (153)	0.20 (8)	220	(485)	SARH	13	(25)
Popeye 1, AGM-142A Have Nap	ASM	Rafael/Israel	4.82 (190)	0.53 (21)	1,360	(2,998)	I/TV or ImIR	43	(80)
Popeye 2, AGM-142B Have Lite	ASM	Rafael/Israel	4.00 (157)	0.53 (21)	1,115	(2,458)	I/TV or ImIR	40	(75)
Popeye, AGM-142C	ASM	Rafael/Israel		0.53 (21)			I/TV		
Popeye, AGM-142D	ASM	Rafael/Israel		0.53 (21)			I/ImIR		

Popeye 1/AGM-142A ASM NEW/0105052

Israel's Python 4 AAM NEW/0073040

Name/designation	Role	Manufacturer/country	Length m (in)	Diameter m (in)	Weight kg (lb)		Guidance	Range n miles (km)	
Python 3	AAM	Rafael/Israel	3.00 (118)	0.16 (6)	120	(265)	IR	8.1	(15)
Python 4	AAM	Rafael/Israel	3.00 (118)	0.16 (6)	105	(231)	IR	8.1	(15)
QW-1 Vanguard	AAM	CPMIEC/China	1.53 (60)	0.07 (3)	16.5	(36)	IR	2.7	(5)
QW-2 Vanguard	AAM	CPMIEC/China	1.59 (63)	0.07 (3)	11.3	(25)	AL	3.2	(6)
R-24 see 'Apex'									
R-27 see 'Alamo'									
R-33 see 'Amos'									
R-37, AA-X-13	AAM	Vympel/Russia	4.20 (165)	0.38 (15)	600	(1,323)	I/ARH	81	(150)

Name/designation	Role	Manufacturer/country	Length m (in)	Diameter m (in)	Weight kg (lb)	Guidance	Range n miles (km)
R-46 see 'Acrid'							
R-60 see 'Aphid'							
R-73 see 'Archer'							
R-77 see 'Adder'							
R 530 see Super 530							
R 550 see Magic							
RB 04	AShM	Saab/Sweden	4.25 (167)	0.50 (20)	600 (1,323)	I/AR	16 (30)
RB 05	ASM	Saab/Sweden	3.60 (142)	0.30 (12)	305 (672)	RC	4.3 (8)
RB 15F Mk 1	AShM	Saab/Sweden	4.35 (171)	0.50 (20)	598 (1,318)	I/AR	49 (90)
RB 15F Mk 2/3	AShM	Saab/Sweden	4.33 (170)	0.50 (20)	630 (1,389)	I/AR	81 (150)/108 (200)
RB 24J Swedish AIM-9J							
RB 71 Swedish Sky Flash							
RB 74 Swedish AIM-9L							
RB 75 Swedish Maverick							
R-Darter, V4	AAM	Denel/South Arfica	2.87 (113)	0.13 (5)	90 (198)	IR/AL	4.3 (8)
S-25LD	ATM	PMED/Russia	4.10 (161)	0.34 (13)	400 (882)	SAL	6.2 (10)
SA-7 see 'Grail'							
SA-14 see 'Gremlin'							
SA-16 see 'Gimlet'							
SA-18 see 'Grouse'							
'Sagger', AT-3/9M14 Malyutka	ATM	Kolomna/Russia	0.86 (34)	0.13 (5)	113 (249)	Wire	1.6 (3)
SCALP EG (Storm Shadow)	SOM	MBDA/International	5.10 (201)	0.63 (25)	1,300 (2,866)	I/MMW/ImIR	135 (250)
Sea Eagle	AShM	MBDA/International	4.14 (163)	0.40 (16)	600 (1,323)	I/ARH	59 (110)
Sea Skua	AShM	MBDA/International	2.50 (98)	0.25 (10)	147 (324)	SARH	8.1 (15)
Shafrir 2	AAM	Rafael/Israel	2.60 (102)	0.16 (6)	95 (209)	IR	1.6 (3)
Shturm see 'Spiral 2'							
Sidearm, AGM-122	ARM	Motorola/USA	3.00 (118)	0.13 (5)	91 (200)	PR	4.3 (8)
Sidewinder, AIM-9L/M/S	AAM	several/USA and Europe	2.87 (113)	0.13 (5)	87 (192)	IR	4.3 (8)
Sidewinder, AIM-9P	AAM	several/USA	3.07 (121)	0.13 (5)	87 (192)	IR	4.3 (8)
Sidewinder, AIM-9R	AAM	several/USA	2.87 (113)	0.13 (5)	87 (192)	Visual band CCD	4.3 (8)
Sidewinder, AIM-9S	AAM	Raytheon/USA	2.87 (113)	0.13 (5)	86 (190)	IR	4.3 (8)
Sidewinder, AIM-9X	AAM	Raytheon/USA	2.90 (114)	0.13 (5)	85 (187)	ImIR	5.4 (10)
Skipper, AGM-123	LGB	ESC/USA	4.33 (170)	0.36 (14)	582 (1,283)	SAL	3.8 (7)
Sky Flash	AAM	MBDA/International	3.66 (144)	0.20 (8)	195 (430)	SARH	22 (40)
Skyshark (unpowered)	ASM	CASMU/Italy	4.76 (187)		1,050 (2,315)	I	6.5 (12)
Skyshark (powered)	ASM	CASMU/Italy	4.76 (187)		1,170 (2,579)	I	13 (25)
Sky Sword (Tien Chien) 1	AAM	Chung Shan/Taiwan	2.87 (113)	0.13 (5)	90 (198)	IR	2.7 (5)
Sky Sword (Tien Chien) 2	AAM	Chung Shan/Taiwan	3.60 (142)	0.20 (8)	190 (419)	SARH	22 (40)
SLAM, AGM-84E	ASM	Boeing/USA	4.50 (177)	0.34 (13)	630 (1,389)	I/GPS/ImIR	51 (95)
SLAM-ER, AGM-84H	ASM	Boeing/USA	4.37 (172)	0.34 (13)	726 (1,600)	I/GPS/ImIR	151 (280)
Sparrow, AIM-7F	AAM	Raytheon/USA	3.66 (144)	0.20 (8)	227 (500)	SARH	22 (40)
Sparrow, AIM-7M	AAM	Raytheon/USA	3.66 (144)	0.20 (8)	230 (507)	SARH	24 (45)
Sparrow, AIM-7P	AAM	Raytheon/USA	3.66 (144)	0.20 (8)	230 (507)	RC/SARH	24 (45)
Sparrow, AIM-7R	AAM	Raytheon/USA	3.66 (144)	0.20 (8)	230 (507)	RC/SARH/IR	24 (45)
Spike see NT-S							
'Spiral', AT-6/9M114 Kokon	ATM	Kolomna/Russia	1.83 (72)	0.13 (5)	34 (75)	RC	3.2 (6)
'Spiral 2', AT-9/9M/14 Shturm	ATM	Kolomna/Russia	1.83 (72)	0.13 (5)	40 (88)	RC	4.3 (8)
Standard, AGM-78	ARM	Raytheon/USA	4.57 (180)	0.34 (13)	615 (1,356)	PR	30 (55)
Starstreak see Helstreak							
Stinger, FIM-92	AAM	Raytheon/USA	1.52 (60)	0.07 (3)	16 (35)	IR	1.6 (3)

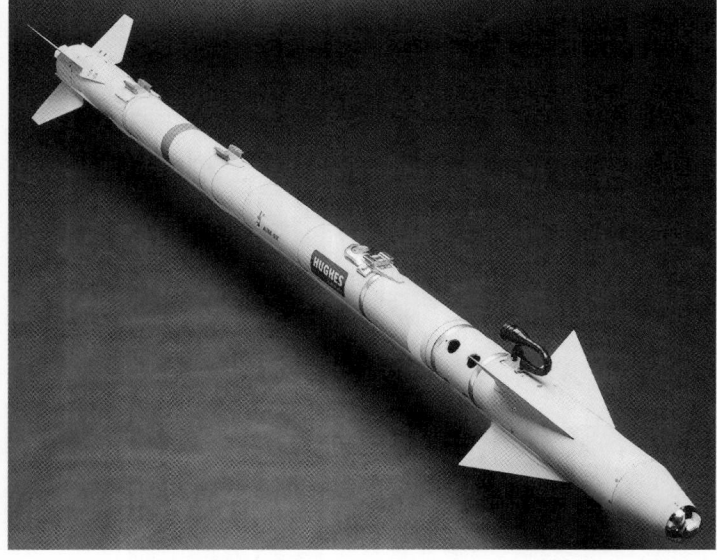

Raytheon AIM-9X Sidewinder
NEW/0022869

MBDA Storm Shadow   *NEW*/0527139

Name/designation	Role	Manufacturer/country	Length m (in)	Diameter m (in)	Weight kg (lb)	Guidance	Range n miles (km)
Storm Shadow see SCALP EG							
Strela see 'Grail' and 'Gremlin'							
Super 530D	AAM	MBDA/International	3.80 (150)	0.26 (10)	270 (595)	SARH	22 (40)
Super 530F1	AAM	MBDA/International	3.54 (139)	0.26 (10)	245 (540)	SARH	13 (25)
'Swatter', AT-2C/9M17 Skorpion	ATM	Nudelman/Russia	1.16 (46)	0.13 (5)	30 (66)	RC	2.2 (4)
Swift, ZT3	ASM	Denel/South Africa	1.35 (53)/	0.13 (5)	19 (42)	L	2.2 (4)
'Swinger', AT-12/9M120 Vikhr/Ataka	ATM	Shipunov/Russia	1.70 (67)	0.13 (5)	43 (95)	RC	4.3 (8)
Taurus see KEPD 350							
Tien Chien see Sky Sword							
Torgos	SOM	Kentron/South Africa	4.86 (191)		980 (2,161)	I/GPS/ImIR	162 (300)
TOW, BGM-71A/B	ATM	Raytheon/USA	1.17 (46)	0.15 (6)	19 (42)	Wire	2.2 (4)
TOW, BGM-71C I-TOW	ATM	Raytheon/USA	1.45 (57)	0.15 (6)	19 (42)	Wire	2.2 (4)

Name/designation	Role	Manufacturer/country	Length m (in)	Diameter m (in)	Weight kg (lb)	Guidance	Range n miles (km)
TOW, BGM-71D/E TOW 2/2A	ATM	Raytheon/USA	1.55 (61)	0.15 (6)	22 (49)	Wire	2.2 (4)
TOW, BGM-71F TOW 2B	ATM	Raytheon/USA	1.17 (46)	0.15 (6)	23 (51)	Wire	2.2 (4)
TRIGAT, ATGW-3LR	ATM	consortium/Europe	1.57 (62)	0.15 (6)	49 (108)	ImIR	2.7 (5)
Type 80 see ASM-1							
Type 88 see ASM-2							
Type 90 see AAM-3							
U-Darter (see Darter)	AAM	Denel/South Africa	2.75 (108)	0.16 (6)	96 (212)	IR	4.3 (8)
Uran see 'Kayak'							
V3B see Kukri							
V3C see Darter							
V3E see A-Darter							
V4 see R-Darter							
Vikhr see 'Swinger' and AT-16							
X- see Kh-							
Yakhont (3M55)	AShM	Strela/Russia	8.30 (327)	0.67 (26)	2,550 (5,622)		
YJ-1 see C-801							
YJ-2 see C-802							
YJ-6 see C-601							
YJ-16 see C-101							
ZT3 see Swift							
ZT35 see Ingwe							
ZT6 see Mokopa							

MBDA Magic AAM being launched from Dassault Rafale   *NEW*/0527140

AERO-ENGINES

Introduction

The following pages summarise the vital statistics of power plants mentioned in the main body of this book. They are divided into piston engines, turboprops/turboshafts and jet engines, and listed in alphabetical order of the manufacturer's name. Readers requiring further data on the two last-mentioned categories are referred to *Jane's Aero-Engines*. Some very small engines, employed by UAVs (and ultralights), are described in *Jane's Unmanned Aerial Vehicles and Targets*; those for large missiles and spacecraft, in *Jane's Space Directory*; and turboshafts, in *Jane's Helicopter Markets and Systems*. Space precludes an entry on each subvariant of more widely produced power plants, and therefore the aero-engine or helicopter publications should be consulted for these data.

Note that engine power ratings given below have been supplied, principally, by their manufacturers and are usually uninstalled output; data may thus vary from that quoted in the Aircraft section of this book, in which information is mainly supplied by the producer of the airframe. In most cases, the engine manufacturer's data reflect take-off power, generally under ISA sea-level conditions. Many gas-turbine engines, especially those for helicopters, are cleared to higher powers for brief periods in emergency.

UPDATED

Diesel engines are now available in production light aircraft, as evidenced by this combination of Thielert TAE 125 and Diamond Star
NEW/0527155

Piston engines

Notes:
Arrangement: 4-O (2) means an engine with four cylinders, horizontally opposed, two-stroke; 8-IV (4) means eight cylinders in inverted-vee form, four-stroke; 4-X (2D) means four cylinders in X configuration, two-stroke diesel; 9-R is a nine-cylinder radial; L indicates in line. Wankel-type engines (W) are alternatively known as rotary (or, more accurately, rotating-piston) engines; 2 × 2-O is two horizontally opposed pistons within single cylinder.
Cooling: A = air cooling, L = liquid
In many cases the engines exist in numerous variants, for example with a geared drive or a supercharger (mechanically driven or turbo). Horsepower ratings to one place of decimals are based on metric (CV/PS) original data.

Engine type	Arrangement	Cooling	Bore	Stroke	Capacity	Weight, dry	Max power
Aero Prag (Czech Republic)							
AP-45							64.1 kW (86 hp)
Alturair (USA)							
A-650	1-rotor (W)	L	-	-	650 cc (39.54 cu in)	62.1 kg (137 lb)	74.6 kW (100 hp)
American Eagle (USA)							
540	8-V (4)	L	-	-	8,827 cc (540 cu in)	316 kg (697 lb)	484.9 kW (650 hp)
Arrow (Italy)							
AE 530AC	2-O (2)	A	74.6 mm (2.94 in)	61.0 mm (2.40 in)	533 cc (32.53 cu in)	50.0 kg (110 lb)	50.7 kW (68 hp)
AE 1070AC	4-O (2)	A	74.6 mm (2.94 in)	61.0 mm (2.40 in)	1,066 cc (65.1 cu in)	65.0 kg (143 lb)	90 kW (120 hp)
GP 1000	4-O (2)	A	74.6 mm (2.94 in)	57.0 mm (2.24 in)	996 cc (60.78 cu in)	65.0 kg (143 lb)	90 kW (120 hp)
GP 1500	6-O (2)	A	74.6 mm (2.94 in)	57.0 mm (2.24 in)	1,495 cc (91.2 cu in)	87.5 kg (193 lb)	134 kW (180 hp)
ATG (UK)							
A-Tech 100	2-O (2D)	L	-	-	1,810 cc (110.5 cu in)	100 kg (220 lb)	74.6 kW (100 hp)
A-Tech 600	4-O (2D)	L	-	-	9,162 cc (559 cu in)	225 kg (496 lb)	447 kW (600 hp)
CAM (Canada)							
100	4-L (4)	L	74.0 mm (2.91 in)	86.5 mm (3.41 in)	1,488 cc (90.7 cu in)	92.1 kg (203 lb)	74.6 kW (100 hp)
Continental - see Teledyne Continental							
CRM (Italy)							
18D/SS	18-W (4D)	L	150.0 mm (5.91 in)	180.0 mm (7.09 in)	57,260 cc (3,495 cu in)	1,700 kg (3,745 lb)	1,380 kW (1,850 hp)
DaimlerChrysler (Germany)							
Suprex ('Smart')	3-L(4)	L	-	-	599 cc (36.55 cu in)	66 kg (146 lb)	40.0 kW (54 hp)
Diesel Air (Germany)							
D 280	2 × 2-O	A	74.0 mm (2.91 in)	2 × 78.0 mm (3.07 in)	1,400 cc (85.43 cu in)	85 kg (187.4 lb)	55 kW (73.7 hp)
Dair 100	2 × 2-O	L	-	-	1,800 cc (109.8 cu in)	90 kg (198 lb)	74.6 kW (100 hp)
FAM (France)							
200	6-V (4)	L	-	-	3,000 cc (183.0 cu in)	179 kg (395 lb)	136 kW (182 hp)
HCI (USA)							
R180	5-R (4)	A	85.5 mm (3.365 in)	87.5 mm (3.445 in)	2,520 cc (154 cu in)	55.3 kg (122 lb)	55.9 kW (75 hp)
Hirth (Germany)							
F23A	2-O (2)	A	72.0 mm (2.835 in)	64.0 mm (2.52 in)	521 cc (31.79 cu in)	24.0 kg (52.9 lb)	29.8 kW (40 hp)
F30	4-O (2)	A	72.0 mm (2.835 in)	64.0 mm (2.52 in)	1,042 cc (63.6 cu in)	36.0 kg (79.4 lb)	70.8 kW (95 hp)

Diesel Air 280 (*Jane's/Paul Jackson*)   NEW/0526933

DaimlerChrysler Suprex ('Smart')
(*Jane's/Paul Jackson*)   NEW/0126831

Hirth F33   NEW/0527154

Engine type	Arrangement	Cooling	Cylinders			Weight, dry	Max power
			Bore	Stroke	Capacity		
F30A	4-O (2)	A	72.0 mm (2.835 in)	64.0 mm (2.52 in)	1,042 cc (63.6 cu in)	39.0 kg (86.0 lb)	77.2 kW (104 hp)
F30A36	4-O (2)	A	72.0 mm (2.835 in)	64.0 mm (2.52 in)	1,042 cc (63.6 cu in)	39.0 kg (86.0 lb)	88.0 kW (118 hp)
F30E	4-O (2)	A	72.0 mm (2.835 in)	64.0 mm (2.52 in)	1,042 cc (63.6 cu in)	42.0 kg (92.6 lb)	61.0 kW (82 hp)
F30ES	4-O (2)	A	72.0 mm (2.835 in)	64.0 mm (2.52 in)	1,042 cc (63.6 cu in)	42.0 kg (92.6 lb)	75.0 kW (101 hp)
F31	2-L (2)	A	76.0 mm (2.99 in)	69.0 mm (2.72 in)	625 cc (38.13 cu in)	26.5 kg (58.4 lb)	29.1 kW (39 hp)
F33A	1 (2)	A	76.0 mm (2.99 in)	69.0 mm (2.72 in)	313 cc (19.1 cu in)	12.7 kg (28.0 lb)	18.1 kW (24 hp)
F33B	1 (2)	A	76.0 mm (2.99 in)	69.0 mm (2.72 in)	313 cc (19.1 cu in)	13.0 kg (28.7 lb)	18.1 kW (24 hp)
2701	2-L (2)	A	70.0 mm (2.755 in)	64.0 mm (2.52 in)	493 cc (30.08 cu in)	32.8 kg (72.5 lb)	32.1 kW (43 hp)
2703	2-L (2)	A	72.0 mm (2.835 in)	64.0 mm (2.52 in)	521 cc (31.79 cu in)	32.8 kg (72.5 lb)	40.4 kW (55 hp)
2704	2-L (2)	A	76.0 mm (2.99 in)	69.0 mm (2.72 in)	625 cc (38.14 cu in)	31.0 kg (68.3 lb)	39.0 kW (52.3 hp)
2706	2-L (2)	A	76.0 mm (2.99 in)	69.0 mm (2.72 in)	625 cc (38.13 cu in)	31 kg (68.3 lb)	48.5 kW (64.1 hp)
3701	3-L (2)	A	76.0 mm (2.99 in)	69.0 mm (2.72 in)	939 cc (57.3 cu in)	45.0 kg (99.2 lb)	59.0 kW (79 hp)
3701ES	3-L (2)	A	76.0 mm (2.99 in)	69.0 mm (2.72 in)	939 cc (57.3 cu in)	45.0 kg (99.2 lb)	74.0 kW (99 hp)
HKS (Japan)							
700E	2-O (4)	A	85.0 mm (3.346 in)	60.0 mm (2.36 in)	680 cc (41.5 cu in)	55.0 kg (121.25 lb)	44.0 kW (60.0 hp)
HPower (USA)							
HKS 700E	(4)						44.7 kW (60 hp)
Jabiru (Australia)							
1600	4-O (4)	A	88.0 mm (3.465 in)	66.0 mm (2.60 in)	1,606 cc (98.0 cu in)	54.0 kg (119 lb)	44.7 kW (60 hp)
2200	4-O (4)	A	97.5 mm (3.84 in)	74.0 mm (2.91 in)	2,200 cc (134.3 cu in)	55.8 kg (123 lb)	59.7 kW (80 hp)
3300	6-O (4)	A	97.5 mm (3.84 in)	740 mm (2.91 in)	3,300 cc (201.4 cu in)	73.0 kg (161 lb)	89 kW (120 hp)
6000	8-O (4)	A	97.5 mm (3.84 in)	100.0 mm (3.94 in)	5,973 cc (364.5 cu in)	108.4 kg (239.0 lb)	149.2 kW (200 hp)
JPX (France)							
4T60/A	4-O (4)	A	93.0 mm (3.66 in)	75.4 mm (2.97 in)	2,050 cc (125.0 cu in)	73.0 kg (161 lb)	47.8 kW (65 hp)
4TX75/A	4-O (4)	A	95.0 mm (3.74 in)	82.0 mm (3.23 in)	2,325 cc (141.9 cu in)	78.0 kg (172 lb)	59.5 kW (79.8 hp)
4TX75/M	4-O (4)	A	95.0 mm (3.74 in)	82.0 mm (3.23 in)	2,325 cc (141.9 cu in)	71.0 kg (156.5 lb)	59.5 kW (79.8 hp)
König (Germany)							
SD 750	4-R (2)	A	66.0 mm (2.60 in)	42.0 mm (1.655 in)	570 cc (34.78 cu in)	18.5 kg (41.0 lb)	20.8 kW (28 hp)
SF 930	4-R (2)	A	70.0 mm (2.755 in)	60.0 mm (2.36 in)	930 cc (56.75 cu in)	36.0 kg (79.4 lb)	35.8 kW (48 hp)
Limbach (Germany)							
L 550	4-O (2)	A	66.0 mm (2.60 in)	40.0 mm (1.57 in)	548 cc (33.44 cu in)	15.5 kg (34.0 lb)	32.0-33.6 kW (43-45 hp)
SL 1700	4-O (4)	A	88.0 mm (3.46 in)	69.0 mm (2.72 in)	1,680 cc (102.5 cu in)	73.0 kg (161 lb)	50.7 kW (68 hp)
L 1800	4-O (4)	A	90.0 mm (3.54 in)	69.0 mm (2.72 in)	1,756 cc (107.0 cu in)	70.0 kg (154 lb)	49.2 kW (66 hp)
L 2000	4-O (4)	A	90.0 mm (3.54 in)	78.4 mm (3.09 in)	1,994 cc (120.3 cu in)	70.0 kg (154 lb)	59.7 kW (80 hp)
L 2400EB	4-O (4)	A	97.0 mm (3.82 in)	82.0 mm (3.23 in)	2,424 cc (147.9 cu in)	82.0 kg (181 lb)	64.9 kW (87 hp)
L 2400EF	4-O (4)	A	97.0 mm (3.82 in)	82.0 mm (3.23 in)	2,424 cc (147.9 cu in)	82.0 kg (181 lb)	73.5 kW (99 hp)
L 2400EFI	4-O (4)	A+L	97.0 mm (3.82 in)	82.0 mm (3.23 in)	2,424 cc (147.9 cu in)	77.0 kg (170 lb)	74.6 kW (100 hp)
L 2400DWFIG	4-O (4)	A	97.0 mm (3.82 in)	82.0 mm (3.23 in)	2,424 cc (147.9 cu in)	100 kg (220 lb)	96 kW (128 hp)
L 2400EFI turbo	4-O (4)	A+L	97.0 mm (3.82 in)	82.0 mm (3.23 in)	2,424 cc (147.9 cu in)	105 kg (231 lb)	118 kW (158 hp)
LOM (Czech Republic)							
M132A, AK, AR	4-L (4)	A	105.0 mm (4.13 in)	115.0 mm (4.53 in)	3,980 cc (242.9 cu in)	102 kg (225 lb)	90 kW (121 hp)
M132B	4-L (4)	A	105.0 mm (4.13 in)	115.0 mm (4.53 in)	3,980 cc (242.9 cu in)	105 kg (231.5 lb)	97.0 kW (130 hp)
M137A, AZ, AR	6-L (4)	A	105.0 mm (4.13 in)	115.0 mm (4.53 in)	5,970 cc (364.3 cu in)	141 kg (311 lb)	134 kW (180 hp)
M137B	6-L (4)	A	105.0 mm (4.13 in)	115.0 mm (4.53 in)	5,970 cc (364.3 cu in)	141 kg (311 lb)	143 kW (192 hp)
M332A	4-L (4)	A	105.0 mm (4.13 in)	115.0 mm (4.53 in)	3,980 cc (242.9 cu in)	102 kg (225 lb)	103 kW (138 hp)
M332B	4-L (4)	A	105.0 mm (4.13 in)	115.0 mm (4.53 in)	3,980 cc (242.9 cu in)	113 kg (249 lb)	118 kW (158 hp)
M337A	6-L (4)	A	105.0 mm (4.13 in)	115.0 mm (4.53 in)	5,970 cc (364.3 cu in)	153 kg (337 lb)	154 kW (207 hp)
M337B	6-L (4)	A	105.0 mm (4.13 in)	115.0 mm (4.53 in)	5,970 cc (364.3 cu in)	153 kg (337 lb)	173 kW (232 hp)
LPE (USA)							
IVG-600	8-IV (4)	L	-	-	9,832 cc (600.0 cu in)	255 kg (562 lb)	448 kW (600 hp)
Lycoming - see Textron Lycoming							
Mid-West (Germany)							
AE 50 Harrier	1-rotor (W)	A+L	-	-	294 cc (17.94 cu in)	33.0 kg (72.75 lb)	37.3 kW (50 hp)
AE75	3-L (2)	L	-	-	748 cc (45.50 cu in)	50.0 kg (110 lb)	57.4 kW (77 hp)
AE 100	2-rotor (W)	A+L	-	-	558 cc (35.90 cu in)	52.0 kg (115 lb)	70.8 kW (95 hp)
AE 110 Hawk	2-rotor (W)	A+L	-	-	588 cc (35.88 cu in)	53.0 kg (116.9 lb)	78.3 kW (105 hp)
Morane Renault – see Société de Motorisations Aéronautiques							
Nelson (USA)							
H-63CP	4-O (2)	A	68.3 mm (2.69 in)	70.0 mm (2.75 in)	1,030 cc (63.00 cu in)	30.8 kg (68 lb)	35.8 kW (48 hp)

Jabiru 3300 (*Jane's/Paul Jackson*)   NEW/0126832

Orenda OE600 (*Jane's/Paul Jackson*)   NEW/0126833

PZL-F4A   NEW/0526914

Teledyne Continental O-470-R *(Jane's/Paul Jackson)*
NEW/0526934

Thielert TAE 125 *(Jane's/Paul Jackson)* NEW/0126834

Wilksch WAM 160 *(Jane's/Paul Jackson)* NEW/0526935

Engine type	Arrangement	Cooling	Cylinders			Weight, dry	Max power
			Bore	Stroke	Capacity		
Novikov (RKBM) (Russian Federation)							
DN-200	3 × 2-O (2D)	L	72.0 mm (2.835 in)	72.0 mm (2.835 in)	4,440 cc (270.9 cu in)	105 kg (231 lb)	110 kW (148 hp)
Orenda (Canada)							
OE600	8-V(4)	L	112.6 mm (4.433 in)	101.6 mm (4.00 in)	8,112 cc (495 cu in)	340 kg (750 lb)	447 kW (600 hp)
OE600 Turbo	8-V (4)	L	112.6 mm (4.433 in)	101.6 mm (4.00 in)	8,112 cc (495 cu in)		559 kW (750 hp)
Parma-Technik (ex-Walter) (Czech Republic)							
Mikron IIIAE	4-L (4)	A	90.0 mm (3.54 in)	96.0 mm (3.78 in)	2,440 cc (149.0 cu in)	70.0 kg (154 lb)	48.5 kW (65 hp)
Mikron IIIB	4-L (4)	A	90.0 mm (3.54 in)	96.0 mm (3.78 in)	2,440 cc (148.89 cu in)	69.0 kg (152.1 lb)	55.0 kW (74 hp)
Pulsar Aircraft (USA)							
Aeromaxx	4-O (4)	A	-	-	1,998 cc (121.9 cu in)	74.8 kg (165 lb)	88.0 kW (118 hp)
PZL (Poland)							
PZL-3S	7-R (4)	A	155.5 mm (6.12 in)	155.0 mm (6.10 in)	20,600 cc (1,265 cu in)	411 kg (906 lb)	442 kW (592 hp)
PZL AI-14RA	9-R (4)	A	105.0 mm (4.125 in)	130.0 mm (5.118 in)	10,160 cc (620 cu in)	200 kg (441 lb)	191 kW (256 hp)
PZL ASz-62	9-R (4)	A	155.0 mm (6.10 in)	174.0 mm (6.85 in)	29,870 cc (1,823 cu in)	580 kg (1,279 lb)	735 kW (985 hp)
PZL K-9	9-R (4)	A	155.0 mm (6.10 in)	174.0 mm (6.85 in)	29,870 cc (1,823 cu in)	580 kg (1,279 lb)	860 kW (1,170 hp)
PZL-F 2A-120-C1	2-O (4)	A	117.48 mm (4.625 in)	88.9 mm (3.50 in)	1,916 cc (117.0 cu in)	58.5 kg (129 lb)	44.7 kW (60 hp)
PZL-F 4A-235-B31	4-O (4)	A	117.48 mm (4.625 in)	88.9 mm (3.50 in)	3,850 cc (235.0 cu in)	103 kg (226 lb)	86.5 kW (116 hp)
PZL-F 6A6350-C1	6-O (4)	A	117.48 mm (4.625 in)	88.9 mm (3.50 in)	5,735 cc (350.0 cu in)	150 kg (330 lb)	153 kW (205 hp)
Rotax (Austria)							
447 UL-1V	2-L (2)	A	67.5 mm (2.66 in)	61.0 mm (2.40 in)	436.5 cc (26.64 cu in)	26.8 kg (59.1 lb)	29.5 kW (39.6 hp)
447 UL-2V	2-L (2)	A	67.5 mm (2.66 in)	61.0 mm (2.40 in)	436.5 cc (26.64 cu in)	26.8 kg (59.1 lb)	31.0 kW (41.6 hp)
462	2-L (2)		-	-	-	-	38.8 kW (52 hp)
503 UL-1V	2-L (2)	A	72.0 mm (2.835 in)	61.0 mm (2.40 in)	496.7 cc (30.31 cu in)	31.4 kg (69.2 lb)	34.0 kW (45.6 hp)
503 UL-2V	2-L (2)	A	72.0 mm (2.835 in)	61.0 mm (2.40 in)	496.7 cc (30.31 cu in)	31.4 kg (69.2 lb)	37.0 kW (49.6 hp)
582 UL	2-L (2)	L	76.0 mm (2.99 in)	64.0 mm (2.52 in)	580.7 cc (35.44 cu in)	27.4 kg (60.5 lb)	48.0 kW (64.4 hp)
618 UL-2V	2-L (2)	L	76.0 mm (2.99 in)	68.0 mm (2.68 in)	617.0 cc (37.65 cu in)	31.0 kg (68.3 lb)	55.0 kW (73.8 hp)
912 UL, 912A, 912F	4-O (4)	A+L	79.5 mm (3.13 in)	61.0 mm (2.40 in)	1,211.2 cc (73.9 cu in)	59.0 kg (130 lb)	59.6 kW (79.9 hp)
912 ULS, 912S	4-O (4)	A+L	84.0 mm (3.31 in)	61.0 mm (2.40 in)	1,352 cc (82.50 cu in)	56.6 kg (125 lb)	73.5 kW (98.6 hp)
914 UL DCDI, 914F	4-O (4)	A+L	79.5 mm (3.13 in)	61.0 mm (2.40 in)	1,211.2 cc (73.9 cu in)	64.0 kg (141 lb)	84.5 kW (113.3 hp)
936	6-V	-	-	-	-	-	162 kW (217 hp)
RotorWay (USA)							
RI 162	4-O (4)	L	-	-	2,660 cc (162.0 cu in)	77.1 kg (170 lb)	113 kW (152 hp)
SAEC (China) HS5: ASh-62IR made under licence, see PZL ASz-62 (Poland)							
Sauer (Germany)							
UL 1800	4-O (4)	A	90.0 mm (3.54 in)	69.0 mm (2.72 in)	1,745 cc (106.49 cu in)	56.5 kg (124.56 lb)	35.0 kW (46.9 hp)
UL 2100	4-O (4)	A	90.0 mm (3.54 in)	84.0 mm (3.31 in)	2,135 cc (130.3 cu in)	60.5 kg (133.4 lb)	62.0 kW (83.1 hp)
SE 1800	4-O (4)	A	90.0 mm (3.54 in)	69.0 mm (2.72 in)	1,745 cc (106.49 cu in)	63.3 kg (139.55 lb)	40.0 kW (53.6 hp)
SA 2100	4-O (4)	A	90.0 mm (3.54 in)	84.0 mm (3.31 in)	2,135 cc (130.3 cu in)	69.0 kg (152.1 lb)	59.0 kW (79.1 hp)
SD 2500	4-O (4)	A	97.0 mm (3.82 in)	84.0 mm (3.31 in)	2,481 cc (151.4 cu in)	79.0 kg (174.2 lb)	68.0 kW (91.2 hp)
SM 2700	4-O (4)	A	97.0 mm (3.82 in)	90 mm (3.54 in)	2,660 cc (162.3 cu in)	82.0 kg (180.8 lb)	75.0 kW (100.6 hp)
Société de Motorisations Aéronautiques (France)							
SR305	4-O (4D)	A	undisclosed	undisclosed	4,988 cc (305.0 cu in)	192 kg (423.3 lb)	169 kW (227 hp)
Stol Techno (Japan)							
HKS 700E	2-O (4)	A+L	85.0 mm (3.35 in)	60.0 mm (2.36 in)	680 cc (41.5 cu in)	47.0 kg (103.6 lb)	44.7 kW (60.0 hp)
Subaru (Japan)							
EJ22	4-L (4)	L	96.9 mm (3.81 in)	75.0 mm (2.95 in)	2,212 cc (136.0 cu in)	119 kg (264 lb)	119 kW (160 hp)
EA81-100	4-O (4)	L	92.0 mm (3.62 in)	67.0 mm (2.64 in)	1,781 cc (108.7 cu in)	97.1 kg (214 lb)	74.6 kW (100 hp)
EA81-140	4-O (4)	L	92.0 mm (3.62 in)	67.0 mm (2.64 in)	1,781 cc (108.7 cu in)	100 kg (222 lb)	104 kW (140 hp)
Teledyne Continental Motors (USA)							
O-200-A & B	4-O (4)	A	103.1 mm (4.06 in)	98.4 mm (3.875 in)	3,280 cc (201.0 cu in)	77.2 kg (170.18 lb)	74.6 kW (100 hp)
IO-240-A & B	4-O (4)	A	112.7 mm (4.44 in)	98.4 mm (3.875 in)	3,940 cc (240.0 cu in)	93 kg (205.00 lb)	93.0 kW (125 hp)
IOF-240	4-O (4)	A	112.7 mm (4.44 in)	98.4 mm (3.875 in)	3,940 cc (240.0 cu in)		
O-300-A, C & D	6-O (4)	A	103.1 mm (4.06 in)	98.4 mm (3.875 in)	4,916 cc (300 cu in)	112.8 kg (248.70 lb)	108.0 kW (145 hp)
IO-300-CCB, D.DB, G.GB, H, HB	6-O (4)	A	103.1 mm (4.06 in)	98.4 mm (3.875 in)	4,916 cc (300 cu in)	148.4 kg (327.25 lb)	156.7 kW (210 hp)
IO-300-ES	6-O (4)	A	103.1 mm (4.06 in)	98.4 mm (3.875 in)	4,916 cc (300 cu in)	138.3 kg (305.00 lb)	156.7 kW (210 hp)
IO-300-J, JB, K, KB	6-O (4)	A	103.1 mm (4.06 in)	98.4 mm (3.875 in)	4,916 cc (300 cu in)	148.4 kg (327.25 lb)	145.5 kW (195 hp)
TSIO-300-A, AB	6-O (4)	A	103.1 mm (4.06 in)	98.4 mm (3.875 in)	4,916 cc (300 cu in)	128.7 kg (283.81 lb)	156.7 kW (210 hp)
TSIO-300-C, CB	6-O (4)	A	103.1 mm (4.06 in)	98.4 mm (3.875 in)	4,916 cc (300 cu in)	136.5 kg (301.00 lb)	167.8 kW (225 hp)
TSIO-300-D, DB	6-O (4)	A	103.1 mm (4.06 in)	98.4 mm (3.875 in)	4,916 cc (300 cu in)	128.7 kg (283.81 lb)	167.8 kW (225 hp)
L/TSIO-360-E, EB, F, -FB	6-O (4)	A	112.7 mm (4.44 in)	98.4 mm (3.875 in)	5,900 cc (360.0 cu in)	145.8 kg (321.35 lb)	149.1 kW (200 hp)
TSIO-360-H, -HB	6-O (4)	A	112.7 mm (4.44 in)	98.4 mm (3.875 in)	5,900 cc (360.0 cu in)	134.6 kg (295.75 lb)	156.7 kW (210 hp)
TSIO-360-JB	6-O (4)	A	112.7 mm (4.44 in)	98.4 mm (3.875 in)	5,900 cc (360.0 cu in)	134.6 kg (295.75 lb)	167.8 kW (225 hp)
L/TSIO-360-KB	6-O (4)	A	112.7 mm (4.44 in)	98.4 mm (3.875 in)	5,900 cc (360.0 cu in)	148.9 kg (328.35 lb)	164.1 kW (220 hp)
TSIO-360-LB	6-O (4)	A	112.7 mm (4.44 in)	98.4 mm (3.875 in)	5,900 cc (360.0 cu in)	155.7 kg (343.35 lb)	156.7 kW (210 hp)
TSIO-360-MB	6-O (4)	A	112.7 mm (4.44 in)	98.4 mm (3.875 in)	5,900 cc (360.0 cu in)	148.6 kg (327.50 lb)	156.7 kW (210 hp)
L/TSIO-360-RB	6-O (4)	A	112.7 mm (4.44 in)	98.4 mm (3.875 in)	5,900 cc (360.0 cu in)	148.6 kg (327.50 lb)	164.1 kW (220 hp)
TSIO-360-SB	6-O (4)	A	112.7 mm (4.44 in)	98.4 mm (3.875 in)	5,900 cc (360.0 cu in)	148.9 kg (328.35 lb)	164.1 kW (220 hp)
O-470-GCI	6-O (4)	A	127.0 mm (5.0 in)	101.6 mm (4.0 in)	7,702 cc (470 cu in)	195.9 kg (431.60 lb)	179.0 kW (240 hp)

For details of the latest updates to *Jane's All the World's Aircraft* online and to discover the additional information available exclusively to online subscribers please visit

jawa.janes.com

Engine type	Arrangement	Cooling	Cylinders			Weight, dry	Max power
			Bore	Stroke	Capacity		
O-470-J	6-O (4)	A	127.0 mm (5.0 in)	101.6 mm (4.0 in)	7,702 cc (470 cu in)	160.6 kg (354.15 lb)	167.8 kW (225 hp)
O-470-K, L & M	6-O (4)	A	127.0 mm (5.0 in)	101.6 mm (4.0 in)	7,702 cc (470 cu in)	172.4 kg (380.00 lb)	171.5 kW (230 hp)
O-470-R & S	6-O (4)	A	127.0 mm (5.0 in)	101.6 mm (4.0 in)	7,702 cc (470 cu in)	172.2 kg (379.66 lb)	171.5 kW (230 hp)
O-470-U	6-O (4)	A	127.0 mm (5.0 in)	101.6 mm (4.0 in)	7,702 cc (470 cu in)	176.4 kg (388.93 lb)	171.5 kW (230 hp)
IO-470-C	6-O (4)	A	127.0 mm (5.0 in)	101.6 mm (4.0 in)	7,702 cc (470 cu in)	182.6 kg (402.48 lb)	186.4 kW (250 hp)
IO-470-D & E	6-O (4)	A	127.0 mm (5.0 in)	101.6 mm (4.0 in)	7,702 cc (470 cu in)	193.3 kg (426.06 lb)	193.9 kW (260 hp)
IO-470-F	6-O (4)	A	127.0 mm (5.0 in)	101.6 mm (4.0 in)	7,702 cc (470 cu in)	181.0 kg (399.06 lb)	193.9 kW (260 hp)
IO-470-H	6-O (4)	A	127.0 mm (5.0 in)	101.6 mm (4.0 in)	7,702 cc (470 cu in)	182.6 kg (402.48 lb)	193.9 kW (260 hp)
IO-470-J & K	6-O (4)	A	127.0 mm (5.0 in)	101.6 mm (4.0 in)	7,702 cc (470 cu in)	182.3 kg (401.90 lb)	167.8 kW (225 hp)
IO-470-L & M	6-O (4)	A	127.0 mm (5.0 in)	101.6 mm (4.0 in)	7,702 cc (470 cu in)	186.6 kg (411.28 lb)	193.9 kW (260 hp)
IO-470-N	6-O (4)	A	127.0 mm (5.0 in)	101.6 mm (4.0 in)	7,702 cc (470 cu in)	188.0 kg (414.44 lb)	193.9 kW (260 hp)
IO-470-S	6-O (4)	A	127.0 mm (5.0 in)	101.6 mm (4.0 in)	7,702 cc (470 cu in)	185.7 kg (409.37 lb)	193.9 kW (260 hp)
IO-470-U, V & VO	6-O (4)	A	127.0 mm (5.0 in)	101.6 mm (4.0 in)	7,702 cc (470 cu in)	184.2 kg (406.00 lb)	193.9 kW (260 hp)
IO-520-A & J	6-O (4)	A	133.0 mm (5.25 in)	101.6 mm (4.0 in)	8,500 cc (520.0 cu in)	187.1 kg (412.43 lb)	212.5 kW (285 hp)
IO-520-B, BA, BB	6-O (4)	A	133.0 mm (5.25 in)	101.6 mm (4.0 in)	8,500 cc (520.0 cu in)	184.5 kg (406.65 lb)	212.5 kW (285 hp)
IO-520-C, CB	6-O (4)	A	133.0 mm (5.25 in)	101.6 mm (4.0 in)	8,500 cc (520.0 cu in)	180.9 kg (398.72 lb)	212.5 kW (285 hp)
IO-520-D & E	6-O (4)	A	133.0 mm (5.25 in)	101.6 mm (4.0 in)	8,500 cc (520.0 cu in)	186.7 kg (411.43 lb)	212.5 kW (285 hp)
IO-520-F	6-O (4)	A	133.0 mm (5.25 in)	101.6 mm (4.0 in)	8,500 cc (520.0 cu in)	186.7 kg (411.43 lb)	212.5 kW (285 hp)
IO-520-K	6-O (4)	A	133.0 mm (5.25 in)	101.6 mm (4.0 in)	8,500 cc (520.0 cu in)	186.7 kg (411.43 lb)	212.5 kW (285 hp)
IO-520-L	6-O (4)	A	133.0 mm (5.25 in)	101.6 mm (4.0 in)	8,500 cc (520.0 cu in)	187.2 kg (412.80 lb)	212.5 kW (285 hp)
IO-520-M, MB	6-O (4)	A	133.0 mm (5.25 in)	101.6 mm (4.0 in)	8,500 cc (520.0 cu in)	186.5 kg (411.20 lb)	212.5 kW (285 hp)
L/IO-520-P	6-O (4)	A	133.0 mm (5.25 in)	101.6 mm (4.0 in)	8,500 cc (520.0 cu in)	174.4 kg (384.44 lb)	186.4 kW (250 hp)
L/TSIO-520-AE	6-O (4)	A	133.0 mm (5.25 in)	101.6 mm (4.0 in)	8,500 cc (520.0 cu in)	165.5 kg (364.80 lb)	186.4 kW (250 hp)
TSIO-520-AF	6-O (4)	A	133.0 mm (5.25 in)	101.6 mm (4.0 in)	8,500 cc (520.0 cu in)	197.5 kg (435.47 lb)	212.5 kW (285 hp)
TSIO-520-B, BB	6-O (4)	A	133.0 mm (5.25 in)	101.6 mm (4.0 in)	8,500 cc (520.0 cu in)	184.7 kg (407.17 lb)	212.5 kW (285 hp)
TSIO-520-BE	6-O (4)	A	133.0 mm (5.25 in)	101.6 mm (4.0 in)	8,500 cc (520.0 cu in)	200.5 kg (442.10 lb)	231.2 kW (310 hp)
TSIO-520-C & H	6-O (4)	A	133.0 mm (5.25 in)	101.6 mm (4.0 in)	8,500 cc (520.0 cu in)	193.4 kg (426.30 lb)	212.5 kW (285 hp)
TSIO-520-CE	6-O (4)	A	133.0 mm (5.25 in)	101.6 mm (4.0 in)	8,500 cc (520.0 cu in)	197.5 kg (435.47 lb)	242.4 kW (325 hp)
TSIO-520-D, DB	6-O (4)	A	133.0 mm (5.25 in)	101.6 mm (4.0 in)	8,500 cc (520.0 cu in)	184.5 kg (406.72 lb)	212.5 kW (285 hp)
TSIO-520-E, EB	6-O (4)	A	133.0 mm (5.25 in)	101.6 mm (4.0 in)	8,500 cc (520.0 cu in)	184.0 kg (405.57 lb)	223.7 kW (300 hp)
TSIO-520-G	6-O (4)	A	133.0 mm (5.25 in)	101.6 mm (4.0 in)	8,500 cc (520.0 cu in)	188.2 kg (415.00 lb)	212.5 kW (285 hp)
TSIO-520-J, JB, N, NB	6-O (4)	A	133.0 mm (5.25 in)	101.6 mm (4.0 in)	8,500 cc (520.0 cu in)	179.8 kg (395.35 lb)	231.2 kW (310 hp)
TSIO-520-K, KB	6-O (4)	A	133.0 mm (5.25 in)	101.6 mm (4.0 in)	8,500 cc (520.0 cu in)	179.8 kg (395.35 lb)	212.5 kW (285 hp)
TSIO-520-L, LB	6-O (4)	A	133.0 mm (5.25 in)	101.6 mm (4.0 in)	8,500 cc (520.0 cu in)	188.7 kg (416.10 lb)	231.2 kW (310 hp)
TSIO-520-(M) P & R	6-O (4)	A	133.0 mm (5.25 in)	101.6 mm (4.0 in)	8,500 cc (520.0 cu in)	189.4 kg (417.52 lb)	212.5 kW (285 hp)
TSIO-520-T	6-O (4)	A	133.0 mm (5.25 in)	101.6 mm (4.0 in)	8,500 cc (520.0 cu in)	193.4 kg (426.30 lb)	231.2 kW (310 hp)
TSIO-520-U & UB	6-O (4)	A	133.0 mm (5.25 in)	101.6 mm (4.0 in)	8,500 cc (520.0 cu in)	191.5 kg (422.27 lb)	223.7 kW (300 hp)
TSIO-520-VB	6-O (4)	A	133.0 mm (5.25 in)	101.6 mm (4.0 in)	8,500 cc (520.0 cu in)	184.4 kg (406.50 lb)	242.4 kW (325 hp)
TSIO-520-WB	6-O (4)	A	133.0 mm (5.25 in)	101.6 mm (4.0 in)	8,500 cc (520.0 cu in)	216.0 kg (416.10 lb)	242.4 kW (325 hp)
GTSIO-520-D & H	6-O (4)	A	133.0 mm (5.25 in)	101.6 mm (4.0 in)	8,500 cc (520.0 cu in)	219.5 kg (484.00 lb)	275.9 kW (375 hp)
GTSIO-520-L	6-O (4)	A	133.0 mm (5.25 in)	101.6 mm (4.0 in)	8,500 cc (520.0 cu in)	220.0 kg (485.00 lb)	275.9 kW (375 hp)
GTSIO-520-M	6-O (4)	A	133.0 mm (5.25 in)	101.6 mm (4.0 in)	8,500 cc (520.0 cu in)	219.1 kg (483.00 lb)	275.9 kW (375 hp)
GTSIO-520-N	6-O (4)	A	133.0 mm (5.25 in)	101.6 mm (4.0 in)	8,500 cc (520.0 cu in)	220.4 kg (486.00 lb)	275.9 kW (375 hp)
IO-550-A	6-0 (4)	A	133.0 mm (5.25 in)	108.0 mm (4.25 in)	9,000 cc (550.0 cu in)	188.0 kg (414.40 lb)	223.7 kW (300 hp)
IO-550-B	6-0 (4)	A	133.0 mm (5.25 in)	108.0 mm (4.25 in)	9,000 cc (550.0 cu in)	184.5 kg (406.65 lb)	223.7 kW (300 hp)
IO-550-C	6-0 (4)	A	133.0 mm (5.25 in)	108.0 mm (4.25 in)	9,000 cc (550.0 cu in)	188.3 kg (415.10 lb)	223.7 kW (300 hp)
IO-550-D, E, F & (L)	6-0 (4)	A	133.0 mm (5.25 in)	108.0 mm (4.25 in)	9,000 cc (550.0 cu in)	192.8 kg (425.00 lb)	223.7 kW (300 hp)
IO-550-G	6-0 (4)	A	133.0 mm (5.25 in)	108.0 mm (4.25 in)	9,000 cc (550.0 cu in)	186.9 kg (412.00 lb)	208.8 kW (280 hp)
IO-550-N	6-0 (4)	A	133.0 mm (5.25 in)	108.0 mm (4.25 in)	9,000 cc (550.0 cu in)	186.9 kg (412.00 lb)	231.2 kW (310 hp)
TSIO-550-B & E	6-0 (4)	A	133.0 mm (5.25 in)	108.0 mm (4.25 in)	9,000 cc (550.0 cu in)	200.5 kg (442.00 lb)	261.0 kW (350 hp)
TSIO-550-C	6-0 (4)	A	133.0 mm (5.25 in)	108.0 mm (4.25 in)	9,000 cc (550.0 cu in)	200.5 kg (442.00 lb)	231.2 kW (310 hp)
TSIOL-550-A	6-0 (4)	L	133.0 mm (5.25 in)	108.0 mm (4.25 in)	9,000 cc (550.0 cu in)	182.1 kg (401.50 lb)	261.0 kW (350 hp)
TSIOL-550-B	6-0 (4)	L	133.0 mm (5.25 in)	108.0 mm (4.25 in)	9,000 cc (550.0 cu in)	188.4 kg (415.40 lb)	242.4 kW (325 hp)
TSIOL-550-C	6-0 (4)	L	133.0 mm (5.25 in)	108.0 mm (4.25 in)	9,000 cc (550.0 cu in)	188.4 kg (415.40 lb)	261.0 kW (350 hp)
Textron Lycoming (USA)							
O-235-C	4-O (4)	A	111.0 mm (4.375 in)	98.4 mm (3.875 in)	3,850 cc (235.0 cu in)	96.6 kg (213 lb)	85.8 kW (115 hp)
O-235-I, M	4-O (4)	A	111.0 mm (4.375 in)	98.4 mm (3.875 in)	3,850 cc (235.0 cu in)	98.0 kg (218 lb)	88.0 kW (118 hp)
O-235N, P	4-O (4)	A	111.0 mm (4.375 in)	98.4 mm (3.875 in)	3,850 cc (235.0 cu in)	98.0 kg (218 lb)	86.5 kW (116 hp)
O-320-A, E	4-O (4)	A	130.0 mm (5.118 in)	98.4 mm (3.875 in)	5,200 cc (319.8 cu in)	110.7 kg (244 lb)	112 kW (150 hp)
(H)O-320-B2C	4-O (4)	A	130.0 mm (5.118 in)	98.4 mm (3.875 in)	5,200 cc (319.8 cu in)	112 kg (246 lb)	119 kW (160 hp)
O-320-D	4-O (4)	A	130.0 mm (5.118 in)	98.4 mm (3.875 in)	5,200 cc (319.8 cu in)	114 kg (253 lb)	119 kW (160 hp)
AEIO-320-D	4-O (4)	A	130.0 mm (5.118 in)	98.4 mm (3.875 in)	5,200 cc (319.8 cu in)	123 kg (271 lb)	119 kW (160 hp)
AEIO-320-E	4-O (4)	A	130.0 mm (5.118 in)	98.4 mm (3.875 in)	5,200 cc (319.8 cu in)	117 kg (258 lb)	112 kW (150 hp)
(L)IO-320-B, C	4-O (4)	A	130.0 mm (5.118 in)	98.4 mm (3.875 in)	5,200 cc (319.8 cu in)	117.5 kg (259 lb)	119 kW (160 hp)
(L)O-360-A	4-O (4)	A	130.0 mm (5.118 in)	111.0 mm (4.375 in)	5,920 cc (361.0 cu in)	120 kg (265 lb)	134 kW (180 hp)
O-360-F	4-O (4)	A	130.0 mm (5.118 in)	111.0 mm (4.375 in)	5,920 cc (361.0 cu in)	122 kg (269 lb)	134 kW (180 hp)
TO-360-C, F	4-O (4)	A	130.0 mm (5.118 in)	111.0 mm (4.375 in)	5,920 cc (361.0 cu in)	154 kg (343 lb)	157 kW (210 hp)

Zoche ZO 01 (*Jane's/Paul Jackson*)   NEW/0126835

Zoche ZO 02 (*Jane's/Paul Jackson*)   NEW/0526936

Engine type	Arrangement	Cooling	Cylinders				Weight, dry	Max power
			Bore	Stroke		Capacity		
IO-360-A, C	4-O (4)	A	130.0 mm (5.118 in)	111.0 mm (4.375 in)		5,920 cc (361.0 cu in)	133 kg (293 lb)	149 kW (200 hp)
IO-360-B	4-O (4)	A	130.0 mm (5.118 in)	111.0 mm (4.375 in)		5,920 cc (361.0 cu in)	122.5 kg (270 lb)	134 kW (180 hp)
LIO-360-C	4-O (4)	A	130.0 mm (5.118 in)	111.0 mm (4.375 in)		5,920 cc (361.0 cu in)	139 kg (306 lb)	149 kW (200 hp)
TIO-360-C	4-O (4)	A	130.0 mm (5.118 in)	111.0 mm (4.375 in)		5,920 cc (361.0 cu in)	158 kg (348 lb)	157 kW (210 hp)
HIO-360-B1A	4-O (4)	A	130.0 mm (5.118 in)	111.0 mm (4.375 in)		5,920 cc (361.0 cu in)	118.4 kg (261 lb)	134.3 kW (180 hp)
HO-360-C1A	4-O (4)	A	130.0 mm (5.118 in)	111.0 mm (4.375 in)		5,920 cc (361.0 cu in)	116.6 kg (257 lb)	134.3 kW (180 hp)
HIO-360-D1A	4-O (4)	A	130.0 mm (5.118 in)	111.0 mm (4.375 in)		5,920 cc (361.0 cu in)	132 kg (290 lb)	142 kW (190 hp)
(L)HIO-360-F1AD	4-O (4)	A	130.0 mm (5.118 in)	111.0 mm (4.375 in)		5,920 cc (361.0 cu in)	133 kg (293 lb)	142 kW (190 hp)
AEIO-360-A	4-O (4)	A	130.0 mm (5.118 in)	111.0 mm (4.375 in)		5,920 cc (361.0 cu in)	139 kg (307 lb)	149 kW (200 hp)
AEIO-360-B	4-O (4)	A	130.0 mm (5.118 in)	111.0 mm (4.375 in)		5,920 cc (361.0 cu in)	125 kg (277 lb)	134 kW (180 hp)
O-540-A	6-O (4)	A	130.0 mm (5.118 in)	111.0 mm (4.375 in)		8,860 cc (541.5 cu in)	170 kg (374 lb)	186 kW (250 hp)
IO-540-A1A5	6-O (4)	A	130.0 mm (5.118 in)	111.0 mm (4.375 in)		8,860 cc (541.5 cu in)	203 kg (448 lb)	201 kW (270 hp)
O-540-B	6-O (4)	A	130.0 mm (5.118 in)	111.0 mm (4.375 in)		8,860 cc (541.5 cu in)	166 kg (366 lb)	175 kW (235 hp)
IO-540-C	6-O (4)	A	130.0 mm (5.118 in)	111.0 mm (4.375 in)		8,860 cc (541.5 cu in)	170 kg (375 lb)	186 kW (250 hp)
IO-540-D	6-O (4)	A	130.0 mm (5.118 in)	111.0 mm (4.375 in)		8,860 cc (541.5 cu in)	173 kg (381 lb)	194 kW (260 hp)
O-540-E	6-O (4)	A	130.0 mm (5.118 in)	111.0 mm (4.375 in)		8,860 cc (541.5 cu in)	167 kg (368 lb)	194 kW (260 hp)
O-540-F1B5	6-O (4)	A	130.0 mm (5.118 in)	111.0 mm (4.375 in)		8,860 cc (541.5 cu in)	167 kg (369 lb)	194 kW (260 hp)
O-540-J	6-O (4)	A	130.0 mm (5.118 in)	111.0 mm (4.375 in)		8,860 cc (541.5 cu in)	165 kg (364 lb)	175 kW (235 hp)
IO-540-K	6-O (4)	A	130.0 mm (5.118 in)	111.0 mm (4.375 in)		8,860 cc (541.5 cu in)	199 kg (438 lb)	224 kW (300 hp)
IO-540-S	6-O (4)	A	130.0 mm (5.118 in)	111.0 mm (4.375 in)		8,860 cc (541.5 cu in)	200 kg (441 lb)	224 kW (300 hp)
IO-540-T4B5	6-O (4)	A	130.0 mm (5.118 in)	111.0 mm (4.375 in)		8,860 cc (541.5 cu in)	176 kg (387 lb)	196 kW (260 hp)
IO-540-W1A5	6-O (4)	A	130.0 mm (5.118 in)	111.0 mm (4.375 in)		8,860 cc (541.5 cu in)	167 kg (369 lb)	175 kW (235 hp)
AEIO-540-D	6-O (4)	A	130.0 mm (5.118 in)	111.0 mm (4.375 in)		8,860 cc (541.5 cu in)	174 kg (386 lb)	194 kW (260 hp)
AEIO-540-L	6-O (4)	A	130.0 mm (5.118 in)	111.0 mm (4.375 in)		8,860 cc (541.5 cu in)	204 kg (449 lb)	224 kW (300 hp)
TIO-540-A	6-O (4)	A	130.0 mm (5.118 in)	111.0 mm (4.375 in)		8,860 cc (541.5 cu in)	231.8 kg (511.0 lb)	231.2 kW (310 hp)
TIO-540-AB1AD	6-O (4)	A	130.0 mm (5.118 in)	111.0 mm (4.375 in)		8,860 cc (541.5 cu in)	205 kg (456 lb)	186 kW (250 hp)
TIO-540-AF1A, AF1B	6-O (4)	A	130.0 mm (5.118 in)	111.0 mm (4.375 in)		8,860 cc (541.5 cu in)	211 kg (466 lb)	201 kW (270 hp)
TIO-540-AE2A	6-O (4)	A	130.0 mm (5.118 in)	111.0 mm (4.375 in)		8,860 cc (541.5 cu in)	246 kg (542 lb)	261 kW (350 hp)
TIO-540-C	6-O (4)	A	130.0 mm (5.118 in)	111.0 mm (4.375 in)		8,860 cc (541.5 cu in)	205 kg (456 lb)	186 kW (250 hp)
(L)TIO-540-F	6-O (4)	A	130.0 mm (5.118 in)	111.0 mm (4.375 in)		8,860 cc (541.5 cu in)	233 kg (514 lb)	242 kW (325 hp)
(L)TIO-540-J	6-O (4)	A	130.0 mm (5.118 in)	111.0 mm (4.375 in)		8,860 cc (541.5 cu in)	239 kg (527 lb)	261 kW (350 hp)
TIO-540-S	6-O (4)	A	130.0 mm (5.118 in)	111.0 mm (4.375 in)		8,860 cc (541.5 cu in)	228 kg (502 lb)	224 kW (300 hp)
(L)TIO-540-U	6-O (4)	A	130.0 mm (5.118 in)	111.0 mm (4.375 in)		8,860 cc (541.5 cu in)	248 kg (547 lb)	261 kW (350 hp)
(L)TIO-540-V	6-O (4)	A	130.0 mm (5.118 in)	111.0 mm (4.375 in)		8,860 cc (541.5 cu in)	248 kg (547 lb)	269 kW (360 hp)
(L)TIO-540-W	6-O (4)	A	130.0 mm (5.118 in)	111.0 mm (4.375 in)		8,860 cc (541.5 cu in)	243 kg (536 lb)	261 kN (350 hp)
TIO-541-E	6-O (4)	A	130.0 mm (5.118 in)	111.0 mm (4.375 in)		8,860 cc (541.5 cu in)	270 kg (596 lb)	283 kW (380 hp)
TIGO-541-E	6-O (4)	A	130.0 mm (5.118 in)	111.0 mm (4.375 in)		8,860 cc (541.5 cu in)	319 kg (704 lb)	317 kW (425 hp)
AEIO-580	6-O (4)	A						246 kW (330 hp)
IO-720-A, B, D	8-O (4)	A	130.0 mm (5.118 in)	111.0 mm (4.375 in)		11,840 cc (722.0 cu in)	258 kg (568 lb)	298 kW (400 hp)
Thielert (Germany)								
TAE 125	4-L(4D)	L	-	-		1,700 cc (103.74 cu in)	134 kg (295.4 lb)	99.3 kW (133 hp)
UFA (Russian Federation)								
Turbo-Diesel								331 kW (444 hp)
VAZ (Russian Federation)								
VAZ-416	2-rotor (W)	L	-	-		1,308 cc (79.8 cu in)	125 kg (275.6 lb)	134 kW (180 hp)
VAZ-4161	2-rotor (W)	L	-	-		1,308 cc (79.8 cu in)	125 kg (275.6 lb)	132 kW (178 hp)
VAZ-4162	2-rotor (W)	L	-	-		1,308 cc (79.8 cu in)	125 kg (275.6 lb)	147 kW (197 hp)
VAZ-426	3-rotor (W)	L	-	-		1,962 cc (119.7 cu in)	155 kg (341.7 lb)	201 kW (270 hp)
VAZ-4261	3-rotor (W)	L	-	-		1,962 cc (119.7 cu in)	155 kg (341.7 lb)	177 kW (237 hp)
VAZ-4262	3-rotor (W)	L	-	-		1,962 cc (119.7 cu in)	155 kg (341.7 lb)	198 kW (266 hp)
VAZ-4263	3-rotor (W)	L	-	-		1,962 cc (119.7 cu in)	145 kg (319.7 lb)	221 kW (296 hp)
VAZ-4265	3-rotor (W)	L	-	-		1,962 cc (119.7 cu in)	130 kg (287 lb)	201 kW (270 hp)
VAZ-526	4-rotor (W)	L	-	-		2,616 cc (159.6 cu in)	175 kg (386 lb)	298 kW (400 hp)
Verner (Germany)								
SVS 1400	2-O (4)	A	94.0 mm (3.70 in)	100.0 mm (3.94 in)		1,400 cc (85.4 cu in)	70.0 kg (154 lb)	58.8 kW (78.9 hp)
133 M	2-O (4)	A	97.0 mm (3.82 in)	90.0 mm (3.54 in)		1,329 cc (81.0 cu in)	64.0 kg (141.0 lb)	58.2 kW (78.0 hp)
VM (Italy)								
1304HF	4-O (4D)	A	130.0 mm (5.118 in)	110.0 mm (4.33 in)		5,840 cc (356.4 cu in)	185 kg (408 lb)	154 kW (206 hp)
1306HF	6-O (4D)	A	130.0 mm (5.118 in)	110.0 mm (4.33 in)		8,760 cc (534.6 cu in)	243 kg (536 lb)	235 kW (315 hp)
1308HF	8-O (4D)	A	130.0 mm (5.118 in)	110.0 mm (4.33 in)		11,680 cc (713 cu in)	298 kg (657 lb)	316 kW (424 hp)
VOKBM (Russian Federation)								
M-3	3-R (4)	A	105.0 mm (4.125 in)	130.0 mm (5.118 in)		3,387 cc (206.5 cu in)	119 kg (262 lb)	77.2 kW (104 hp)
M-5	5-R (4)	A	105.0 mm (4.125 in)	130.0 mm (5.118 in)		5,644 cc (344.4 cu in)	115 kg (254 lb)	118 kW (158 hp)
M-7	7-R (4)	A	105.0 mm (4.125 in)	130.0 mm (5.118 in)		7,902 cc (482.2 cu in)	156 kg (343.9 lb)	191 kW (256 hp)
M-9F	9-R (4)	A	105.0 mm (4.125 in)	130.0 mm (5.118 in)		10,160 cc (620 cu in)	214 kg (472 lb)	294 kW (394 hp)
M-14NTK	9-R (4)	A	105.0 mm (4.125 in)	130.0 mm (5.118 in)		10,160 cc (620 cu in)		316 kW (424 hp)
M-14PF	9-R (4)	A	105.0 mm (4.125 in)	130.0 mm (5.118 in)		10,160 cc (620 cu in)	214 kg (472 lb)	294 kW (394 hp)
M-14PT	9-R (4)	A	105.0 mm (4.125 in)	130.0 mm (5.118 in)		10,160 cc (620 cu in)	217 kg (478 lb)	265 kW (355 hp)
Walter (Czech Republic)								
M202	2-O (2)	A	82.0 mm (3.23 in)	64.0 mm (2.52 in)		676 cc (41.25 cu in)	36.0 kg (79.4 lb)	48.5 kW (65 hp)
Wankel Rotary (Germany)								
LCR-407 SGti	1-rotor (W, D)	L	-	-		407 cc (24.8 cu in)	25.0 kg (55.1 lb)	27.2 kW (36.4 hp)
LOCR-407 SD	1-rotor (W, D)	L	-	-		407 cc (24.8 cu in)	38.0 kg (83.8 lb)	33.1 kW (44.4 hp)
LCR-814 TG ti	2-rotor (W, D)	L	-	-		814 cc (49.7 cu in)	35.0 kg (77.2 lb)	55.1 kW (73.9 hp)
LOCR-814 TD	2-rotor (W, D)	L	-	-		814 cc (49.7 cu in)	50.0 kg (110 lb)	66.2 kW (88.8 hp)
Twinpack	4-rotor (W, D)	L	-	-		-	119 kg (262 lb)	110 kW (148 hp)
Wilksch (UK)								
WAM 120	3-L(2D)	L	-	-		-	100 kg (220.5 lb)	89.5 kW (120 hp)
WAM 160	4-L(2D)	L	-	-		-	120 kg (264.6 lb)	119.3 kW (160 hp)
Zoche (Germany)								
ZO 01A	4-X (2D)	A	95.0 mm (3.74 in)	94.0 mm (3.70 in)		2,665 cc (162.6 cu in)	84.0 kg (185 lb)	110 kW (147.5 hp)
ZO 02A	8-X (2D)	A	95.0 mm (3.74 in)	94.0 mm (3.70 in)		5,330 cc (325.3 cu in)	118 kg (259 lb)	220 kW (295 hp)
ZO 03A	2-IV (2D)	A	95.0 mm (3.74 in)	94.0 mm (3.70 in)		1,332.5 cc (81.3 cu in)	55.0 kg (121 lb)	51.0 kW (68.4 hp)
Zöllner (Germany)								
DZ	4-O (4)	A	-	-		1,998 cc (121.9 cu in)	74.0 kg (163 lb)	86.8 kW (116 hp)

UPDATED

Turboprop engines

Notes:
Arrangement: A = axial stages, C = centrifugal stages, C+C= two centrifugal stages on the same shaft, C, C two stages on different shafts.
Prop drive: FT = free turbine, SS = single shaft, 2S = two-shaft engine but no independent power turbine
For turboshaft engines, T-O rating is the maximum except contingency, usually a 30 minute, or maximum continuous power.

Engine type	Arrangement	Air flow	Prop drive	Length	Width	Dry weight	T-O rating
AlliedSignal - see Honeywell							
Allison – see Rolls-Royce (USA)							
DEMC (China)							
WJ5A I: AI-24A made under licence, details as AI-24T under Progress (ZMKB) (Ukraine)							2,080 kW (2,790 shp)
WJ5E	10A	14.6 kg (32.2 lb)/s	SS	2,381 mm (93.7 in)	770 mm (30.3 in)	720 kg (1,587 lb)	2,130 kW (2,856 shp)
							720 kg (1,587 lb)*
General Electric (USA)							
CT7-5A	5A+C	4.41 kg (10.0 lb)/s	FT	2,438 mm (96 in)	737 mm (29 in)	355 kg (783 lb)	1,294 kW (1,735 shp)
CT7-9	5A+C	5.20 kg (11.5 lb)/s	FT	2,438 mm (96 in)	737 mm (29 in)	365 kg (805 lb)	<1,447 kW (1,940 shp)
T64-P4D	14A	12.2 kg (27.0 lb)/s	FT	2,793 mm (110 in)	683 mm (26.9 in)	538 kg (1,188 lb)	2,535 kW (3,400 shp)
Honeywell (USA)							
TPE331-3	C+C	3.54 kg (7.80 lb)/s	SS	1,092 mm (43 in)	533 mm (21 in)	161 kg (355 lb)	626 kW (840 shp)
TPE331-12	C+C	3.49 kg (7.7 lb)/s	SS	1,168 mm (46 in)	533 mm (21 in)	181 kg (400 lb)	834 kW (1,100 shp)
TPE331-14GR	C+C	5.26 kg (11.60 lb)/s	SS	1,333 mm (52.5 in)	533 mm (21 in)	281 kg (620 lb)	1,462 kW (1,960 shp)
LTP 101-700	A+C	2.31 kg (5.10 lb)/s	FT	949 mm (37.4 in)	592 mm (23.3 in)	147 kg (325 lb)	522 kW (700 shp)
KKBM (Russian Federation)							
NK-12MV	14A	65.0 kg (143 lb)/s	FT	4,785 mm (188 in)	1,150 mm (45.3 in)	2,900 kg (6,393 lb)	11,033 kW (14,795 shp)
Klimov (Russian Federation)							
TV3-117VMA-SB2	12A	9.0 kg (19.84 lb)/s	FT	2,055 mm (80.9 in)	650 mm (25.6 in)	560 kg (1,235 lb)	1,864 kW (2,500 shp)
TV7-117S	5A+C	7.95 kg (17.5 lb)/s	FT	2,143 mm (84.4 in)	886 mm (34.9 in)	520 kg (1,146 lb)	1,839 kW (2,466 shp)
LHTEC (USA)							
CTP800-4T Twin	2×(C+C)	2×3.27 kg (7.20 lb)/s	2×FT	1,769 mm (69.6 in)	1,252 mm (49.28 in)	517 kg (1,140 lb)	2,013 kW (2,700 shp)
OMKB (Russian Federation)							
TVD-20	7A+C	5.40 kg (11.9 lb)/s	FT	1,770 mm (69.7 in)	850 mm (33.5 in)	240 kg (529 lb)	1,081 kW (1,450 shp)
Pratt & Whitney Canada (Canada)							
PT6A-27	3A+C	3.08 kg (6.80 lb)/s	FT	1,575 mm (62 in)	483 mm (19 in)	149 kg (328 lb)	507 kW (680 shp)
PT6A-41	3A+C	3.40 kg (7.50 lb)/s	FT	1,701 mm (67 in)	483 mm (19 in)	183 kg (403 lb)	634 kW (850 shp)
PT6A-65R	4A+C	4.31 kg (9.50 lb)/s	FT	1,905 mm (75 in)	483 mm (19 in)	218 kg (481 lb)	1,026 kW (1,376 shp)
PT6A-68	4A+C	4.5 kg (9.9 lb)/s	FT	1,930 mm (76 in)	483 mm (19 in)	259.5 kg (572 lb)	1,193 kW (1,600 shp)
PW120	C, C	—	FT	2,134 mm (84 in)	635 mm (25 in)	418 kg (921 lb)	1,491 kW (2,100 shp)
PW121	C, C	—	FT	2,134 mm (84 in)	635 mm (25 in)	425 kg (936 lb)	1,603 kW (2,150 shp)
PW123 C/D	C, C	—	FT	2,134 mm (84 in)	635 mm (25 in)	450 kg (992 lb)	1,603 kW (2,150 shp)
PW127	C, C	—	FT	2,134 mm (84 in)	660 mm (26 in)	481 kg (1,060 lb)	2,051 kW (2,750 shp)
PW150A	3A+C	—	FT	2,423 mm (95.4 in)	767 mm (30.2 in)	690 kg (1,522 lb)	3,781 kW (5,071 shp)
Progress (ZMKB) (Ukraine)							
AI-20M	10A	20.7 kg (45.6 lb)/s	SS	3,096 mm (121.9 in)	842 mm (33.15 in)	1,040 kg (2,293 lb)	2,940 kW (3,943 shp)
AI-24T	10A	14.4 kg (31.7 lb)/s	SS	2,346 mm (92.4 in)	677 mm (26.65 in)	600 kg (1,323 lb)	1,880 kW (2,520 shp)
PZL (Poland)							
TWD-10B	6A+C	4.58 kg (10.1 lb)/s	FT	2,060 mm (81.1 in)	555 mm (21.9 in)	230 kg (507 lb)	754 kW (1,011 shp)
RKBM (Russian Federation)							
TVD-1500S	3A+C	4.0 kg (8.8 lb)/s	FT	1,965 mm (77.4 in)	620 mm (24.4 in)	240 kg (529 lb)	1,044 kW (1,400 shp)

General Electric CT7 turboprop   0105053

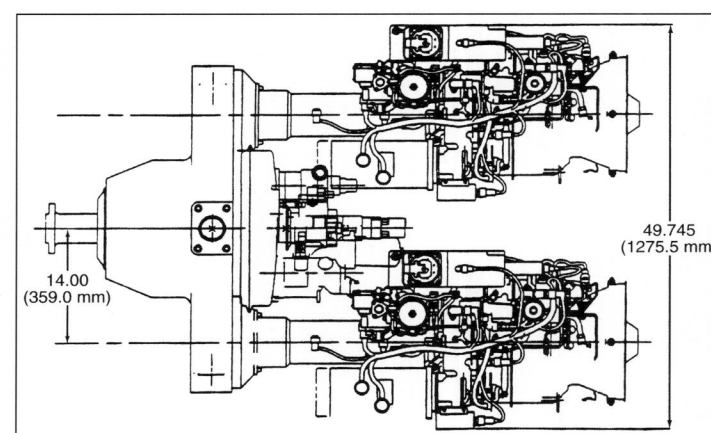

14.00 (359.0 mm)   49.745 (1275.5 mm)

LHTEC CTP800-4T coupled turboprop   0105056

Walter M 601B turboprop (*Jane's/Paul Jackson*)   NEW/0126836

Experimental ATP turboprop for light aircraft, first flown in USA in 2002 (*Jane's/Paul Jackson*)   NEW/0526932

Engine type	Arrangement	Air flow	Prop drive	Length	Width	Dry weight	T-O rating
Rolls-Royce (UK)							
Dart Mk 536	C+C	10.7 kg (23.5 lb)/s	SS	2,480 mm (97.6 in)	963 mm (37.9 in)	569 kg (1,257 lb)	1,700 kW (2,280 shp)
Tyne 21	6A, 9A	21.1 kg (46.5 lb)/s	2S	2,762 mm (108.7 in)	1,400 mm (55 in)	1,085 kg (2,391 lb)	4,226 kW (5,665 shp)
Rolls-Royce (USA)							
250-B17	6A+C	1.56 kg (3.45 lb)/s	FT	1,143 mm (45 in)	483 mm (19 in)	88.4 kg (195 lb)	313 kW (420 shp)
T56-15	14A	14.7 kg (32.4 lb)/s	SS	3,708 mm (146 in)	686 mm (27 in)	828 kg (1,825 lb)	3,424 kW (4,591 shp)
T56-427	14A	15.2 kg (33.5 lb)/s	SS	3,711 mm (146.1 in)	686 mm (27.0 in)	880 kg (1,940 lb)	3,910 kW (5,250 shp)
AE 2100A	14A	16.3 kg (36.0 lb)/s	FT	—	—	715.8 kg (1,578 lb)	3,095 kW (4,150 shp)
AE 2100C	14A	—	FT	—	—	715.8 kg (1,578 lb)	2,685 kW (3,600 shp)
AE 2100D2, D3	14A	16.3 kg (36.0 lb)/s	FT	2,743 mm (108 in)	1,151 mm (45.3 in)	702 kg (1,548 lb)	3,424 kW (4,591 shp)
SAEC (China)							
WJ6: AI-20M made under licence; details as under Progress (ZMKB) (Ukraine)							
Saturn (Russian Federation)							
AL-34-1	C	—	FT	1,700 mm (66.9 in)	640 mm (25.2 in)	178 kg (392 lb)	809 kW (1,085 shp)
Turbomeca (France)							
Arrius 1D	C	not disclosed	FT	826 mm (32.5 in)	476 mm (18.74 in)	111 kg (245 lb)	313 kW (420 shp)
Arrius 2F	C	not disclosed	FT	945 mm (37.5 in)	459 mm (18.07 in)	103 kg (227 lb)	376 kW (504 shp)
Walter (Czech Republic)							
M 601E	2A+C	3.60 kg (7.94 lb)/s	FT	1,675 mm (65.9 in)	590 mm (23.23 in)	200 kg (441 lb)	560 kW (751 shp)
M 601F	2A+C	3.60 kg (7.94 lb)/s	FT	1,675 mm (65.9 in)	590 mm (23.23 in)	202 kg (445 lb)	580 kW (778 shp)
M 602	C, C	7.33 kg (16.2 lb)/s	FT	2,669 mm (105 in)	753 mm (29.65 in)	570 kg (1,257 lb)	1,360 kW (1,824 shp)

*Residual thrust

UPDATED

P&W Canada PT6A-67B (*Jane's/Paul Jackson*)   NEW/0126837

Klimov TV3-117VMA-SB2 (*Jane's/Paul Jackson*)   0085458

Turboshaft engines

Notes:
See Turboprop engines

Engine type	Arrangement	Air flow	Drive	Length	Width	Dry weight	T-O rating
AlliedSignal – see Honeywell							
Allison – see Rolls-Royce							
Aviadvigatel (Russian Federation)							
D-25V	9A	26.2 kg (57.8 lb)/s	FT	2,737 mm (107.75 in)	1,086 mm (42.8 in)	1,325 kg (2,921 lb)	4,050 kW (5,430 shp)
CLXMW (China)							
WZ6		Turbomeca Turmo IIIC made under licence					1,145 kW (1,536 shp)
General Electric (USA)							
CT7-2	5A+C	4.5 kg (10.0 lb)/s	FT	1,194 mm (47.0 in)	635 mm (25.0 in)	212 kg (466 lb)	1,212 kW (1,625 shp)
CT7-6	5A+C	5.9 kg (13.0 lb)/s	FT	1,194 mm (47.0 in)	660 mm (26.0 in)	220 kg (485 lb)	1,491 kW (2,000 shp)
CT7-6D	5A+C	6.1 kg (13.5 lb)/s	FT	1,194 mm (47.0 in)	660 mm (26.0 in)	229 kg (504 lb)	1,514 kW (2,030 shp)
CT7-8	5A+C	5.9 kg (13.0 lb)/s	FT	1,194 mm (47.0 in)	660 mm (26.0 in)	—	1,879 kW (2,520 shp)
T64-419	14A	13.3 kg (29.4 lb)/s	FT	2,006 mm (79.0 in)	660 mm (26.0 in)	343 kg (755 lb)	3,542 kW (4,750 shp)
T700-401	5A+C	4.5 kg (10.0 lb)/s	FT	1,168 mm (46.0 in)	635 mm (25.0 in)	197 kg (434 lb)	1,260 kW (1,690 shp)
T700-700	5A+C	4.5 kg (10.0 lb)/s	FT	1,168 mm (46.0 in)	635 mm (25.0 in)	198 kg (437 lb)	1,210 kW (1,622 shp)
T700/T6	5A+C	5.9 kg (13.0 lb)/s	FT	1,205 mm (47.4 in)	660 mm (26.0 in)	220 kg (485 lb)	1,652 kW (2,215 shp)
Honeywell (USA) (formerly AlliedSignal)							
LTS 101-600A	1A+C	2.31 kg (5.10 lb)/s	FT	785 mm (30.9 in)	599 mm (23.6 in)	115 kg (253 lb)	459 kW (615 shp)
LTS 101-750B	1A+C	2.30 kg (5.10 lb)/s	FT	795 mm (31.3 in)	470 mm (18.5 in)	123 kg (271 lb)	515 kW (690 shp)
T53-703	5A+C	5.90 kg (13.0 lb)/s	FT	1,209 mm (47.6 in)	584 mm (23.0 in)	247 kg (545 lb)	1,343 kW (1,800 shp)
T5317A	5A+C	5.53 kg (12.2 lb)/s	FT	1,209 mm (47.6 in)	584 mm (23.0 in)	256 kg (564 lb)	1,119 kW (1,500 shp)
T55-714	7A+C	13.19 kg (29.1 lb)/s	FT	1,181 mm (46.5 in)	610 mm (24.0 in)	377 kg (832 lb)	3,630 kW (4,868 shp)
Klimov (Russian Federation)							
GTD-350 see under PZL (Poland)							
TV2-117	10A	8.4 kg (18.5 lb)/s	FT	2,842 mm (111.9 in)	547 mm (21.7 in)	338 kg (745 lb)	1,118 kW (1,500 shp)
TV3-117V	12A	8.7 kg (19.18 lb)/s	FT	2,085 mm (82.1 in)	640 mm (25.2 in)	285 kg (628 lb)	<1,633 kW (2,190 shp)
TV3-117VMA-SB3	12A	9.0 kg (19.84 lb)/s	FT	2,055 mm (80.9 in)	650 mm (25.6 in)	310 kg (683 lb)	1,864 kW (2,500 shp)
VK-2500	12A	9.3 kg (20.5 lb)/s	FT	2,055 mm (80.91 in)	660 mm (26.0 in)	295 kg (650 lb)	1,790 kW (2,400 shp)
LHTEC (USA)							
T800-800	C+C	3.27 kg (7.20 lb)/s	FT	843.3 mm (33.2 in)	550.1 mm (21.7 in)	143 kg (315 lb)	995 kW (1,334 shp)
T800-801	C+C	4.43 kg (9.8 lb)/s	FT	843.3 mm (33.2 in)	550.1 mm (21.7 in)	149.7 kg (330 lb)	1,165 kW (1,563 shp)
CTS800-1G	C+C	4.01 kg (8.84 lb)/s	FT	856.0 mm (33.7 in)	550 mm (21.66 in)	149.7 kg (330 lb)	1,312 kW (1,760 shp)
CTS800-4	C+C	3.54 kg (7.8 lb)/s	FT	1,047.2 mm (41.23 in)	561.6 mm (22.11 in)	173.7 kg (383 lb)	1,016 kW (1,362 shp)
Mitsubishi (Japan)							
MG5-100	—	—	FT	—	—	—	597 kW (800 shp)
XTS1-10	—	—	FT	—	—	—	659 kW (884 shp)
MTR (International)							
390	C+C	3.20 kg (7.05 lb)/s	FT	1,078 mm (42.4 in)	442 mm (17.4 in)	169 kg (373 lb)	958 kW (1,285 shp)

OMKB (Russian Federation) GTD-3 see under Poland as PZL-10W

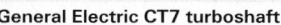

General Electric CT7 turboshaft 0105054

RKBM RD-600V *NEW*/0126840

Turbomeca Arrius 2K turboshaft 0075984

LHTEC T800 turboshaft
NEW/0130078

Engine type	Arrangement	Air flow	Drive	Length	Width	Dry weight	T-O rating
Pratt & Whitney Canada (Canada)							
PT6B-36	3A+C	3.08 kg (6.80 lb)/s	FT	1,504 mm (59.2 in)	495 mm (19.5 in)	169 kg (373 lb)	732 kW (981 shp)
PT6T-3B Twin-Pac	2×(3A+C)	2×3.08 kg (6.80 lb)/s	2×FT	1,702 mm (67.0 in)	1,118 mm (44.0 in)	299 kg (660 lb)	1,342 kW (1,800 shp)
PT6C-67	4A+C	4.3 kg (9.5 lb)/s	FT	1,930 mm (76.0 in)	483 mm (19.0 in)	—	1,252 kW (1,679 shp)
PW206A	C	—	FT	912 mm (35.9 in)	500 mm (19.7 in)	108 kg (237 lb)	477 kW (640 shp)
PW207D	C	—	FT	—	—	—	529 kW (710 shp)
Progress (ZMKB) (Ukraine)							
AI-450	C	1.72 kg (3.79 lb)/s	FT	1,085 mm (42.72in)	515 mm (20.28 in)	110 kg (242.5 lb)	345 kW (462 hp)
D-136	6A, 7A	35.6 kg (78.4 lb)/s	FT	3,715 mm (146.3 in)	1,382 mm (4.4 in)	1,077 kg (2,374 lb)	7,457 kW (10,000 shp)
D-127	5A, 2A+C	27.4 kg (60.4 lb)/s	FT	3,950 mm (155.5 in)	1,400 mm (55.1 in)	—	10,700 kW (14,350 shp)
PZL (Poland)							
PZL-10W	6A+C	4.6 kg (10.1 lb)/s	FT	1,875 mm (73.8 in)	740 mm (29.0 in)	141 kg (310 lb)	662 kW (888 shp)
PZL GTD-350	7A+C	2.19 kg (4.83 lb)/s	FT	1,385 mm (54.53 in)	626 mm (24.9 in)	140 kg (307 lb)	294 kW (394 shp)
RKBM (Russian Federation)							
TVD-1500V (RD-600V)	3A+C	4.0 kg (8.8 lb)/s	FT	1,250 mm (49.2 in)	620 mm (24.4 in)	220 kg (485 lb)	969 kW (1,300 shp)
Rolls-Royce (UK)							
Gem 42	4A, C	3.41 kg (7.50 lb)/s	FT	1,099 mm (43.2 in)	575 mm (22.6 in)	183 kg (404 lb)	746 kW (1,000 shp)
Gnome	10A	6.3 kg (13.8 lb)/s	FT	1,392 mm (54.8 in)	577 mm (22.7 in)	148 kg (326 lb)	1,145 kW (1,535 shp)
Rolls-Royce (USA) (formerly Allison)							
250-C20B	6A+C	1.56 kg (3.45 lb)/s	FT	985 mm (38.8 in)	483 mm (19.0 in)	71.5 kg (158 lb)	313 kW (420 shp)
250-C28	C	2.02 kg (4.45 lb)/s	FT	1,021 mm (40.2 in)	557 mm (21.9 in)	106 kg (233 lb)	373 kW (500 shp)
250-C30	C	2.54 kg (5.60 lb)/s	FT	1,041 mm (41.0 in)	557 mm (21.9 in)	114 kg (251 lb)	485 kW (650 shp)
250-C40B	C	2.77 kg (6.10 lb)/s	FT	1,041 mm (41 in)	557 mm (21.9 in)	127 kg (280 lb)	533 kW (715 shp)
T406	14A	16.1 kg (35.5 lb)/s	FT	1,958 mm (77.0 in)	671 mm (26.4 in)	440 kg (971 lb)	4,586 kW (6,150 shp)
RRTI (International)							
RTM 322-01/8	3A+C	5.75 kg (12.68 lb)/s	FT	1,171 mm (46.1 in)	604 mm (23.8 in)	240 kg (538 lb)	1,566 kW (2,100 shp)
RTM 322-01/9	3A+C	5.9 kg (13.0 lb)/s	FT	1,171 mm (46.1 in)	659 mm (25.9 in)	233 kg (512 lb)	1,785 kW (2,393 shp)
RTM322-01/12	3A+C	5.75 kg (12.68 lb)/s	FT	1,171 mm (46.1 in)	736 mm (28.98 in)	249 kg (549.0 lb)	1,566 kW (2,101 shp)
RTM322-02/8	3A+C	5.8 kg (12.79 lb)/s	FT	1,171 mm (46.1 in)	736 mm (28.98 in)	248 kg (546.7 lb)	1,670 kW (2,241 shp)
Soyuz (Russia)							
TV-O-100	2A+C	2.66 kg (5.86 lb)/s	FT	1,275 mm (50.2 in)	780 mm (30.7 in)	125 kg (276 lb)	537 kW (720 shp)
Turbomeca (France)							
Ardiden (Shakti)	C+C	not disclosed	FT	—	—	198 kg (437 lb)	1,165 kW (1,562 shp)
Arriel 1	1A+C	not disclosed	SS	1,090 mm (42.9 in)	430 mm (16.9 in)	120 kg (265 lb)	551 kW (739 shp)
Arriel 2	1A+C	not disclosed	SS	—	—	—	632 kW (848 shp)
Arrius 1A	C	not disclosed	FT	782 mm (30.8 in)	360 mm (14.2 in)	87.0 kg (192 lb)	380 kW (509 shp)
Arrius 1A/1M	C	not disclosed	FT	793 mm (31.2 in)	367 mm (14.45 in)	101 kg (223 lb)	357 kW (479 shp)
Arrius 2B	C	not disclosed	FT	782 mm (30.8 in)	360 mm (14.2 in)	87.0 kg (192 lb)	519 kW (696 shp)
Arrius 2B1	C	not disclosed	FT	1,158 mm (45.6 in)	518 mm (20.4 in)	114 kg (251 lb)	559 kW (750 shp)
Arrius 2K1	C	not disclosed	—	968 mm (38.1 in)	470 mm (18.5 in)	115 kg (253.5 lb)	500 kW (670 shp)
Artouste III	A+C	4.31 kg (9.50 lb)/s	SS	1,815 mm (71.5 in)	507 mm (20.0 in)	178 kg (392 lb)	440 kW (590 shp)
Astazou XIVM	2A+C	2.49 kg (5.50 lb)/s	SS	1,470 mm (57.9 in)	460 mm (18.1 in)	160 kg (353 lb)	640 kW (858 shp)
Makila 1A1	3A+C	5.5 kg (12.1 lb)/s	FT	2,103 mm (82.8 in)	528 mm (20.8 in)	241 kg (531 lb)	1,357 kW (1,820 shp)
Makila 1A2	3A+C	5.5 kg (12.1 lb)/s	FT	2,117 mm (83.3 in)	498 mm (19.6 in)	247 kg (545 lb)	1,376 kW (1,845 shp)
Makila 1A4	3A+C	5.7 kg (12.6 lb)/s	FT	1,836 mm (71.6 in)	498 mm (19.4 in)	247 kg (545 lb)	1,567 kW (2,101 shp)
TM 333	2A+C	—	FT	1,045 mm (41.1 in)	454 mm (17.9 in)	156 kg (345 lb)	747 kW (1,001 shp)
Turmo IIIC	1A+C	5.9 kg (13.0 lb)/s	FT	2,184 mm (85.5 in)	637 mm (25.1 in)	225 kg (496 lb)	1,104 kW (1,480 shp)
ZEF (China)							
WZ8: Turbomeca Arriel 1C made under licence							522 kW (700 shp)

UPDATED

MTR 390 on Eurocopter Tiger *(MTU)*

NEW/0526913

Jet engines

Notes:
Arrangement: A = axial stages, C = centrifugal stages, F = fan stages, a/b = afterburner or augmentor
Air flow: Flow is through fan, where applicable, not core; BPR = bypass ratio; PF = propfan
<: Various ratings up to this maximum
[1] without reverser (other NK-8 family include it)
[2] measured over propfan blades
[3] has vectored nozzle(s)
W: Afterburning or augmented ('wet') thrust
D: Unaugmented ('dry') thrust of afterburning engine

Engine type	Arrangement	Air flow	BPR	Length	Diameter	Dry weight	T-O rating
Aerosud-Marvol (South Africa/Russia)							
SMR-95	4F, 9A, a/b	75.5 kg (166 lb)/s	0.49	4,229 mm (166.5 in)	1,080 mm (42.52 in)	1,235 kg (2,701 lb)	81.4 kN (18,300 lb) W 49.4 kN (11,110 lb) D
Agilis (USA)							
turbofan	1F, 1A	—	5.0	1,420 mm (55.9 in)	580 mm (22.83 in)	104 kg (229.3 lb)	4.45 kN (1,000 lb)
Aviadvigatel (Russian Federation)							
D-20P	3F, 8A	113 kg (249 lb)/s	1.0	3,304 mm (130.0 in)	976 mm (38.3 in)	1,468 kg (3,236 lb)	53.0 kN (11,905 lb)
D-21A1	5F, 10A	153 kg (337 lb)/s	0.83	4,837 mm (190.4 in)	1,020 mm (40.2 in)	2,100 kg (4,630 lb)	52.3 kN (11,750 lb)
D-30 III	5F, 10A	127 kg (280 lb)/s	1.0	3,983 mm (156.8 in)	1,050 mm (41.3 in)	1,550 kg (3,417 lb)	66.7 kN (14,990 lb)
D-30F6	5F, 10A, a/b	150 kg (331 lb)/s	0.57	7,040 mm (277.2 in)	1,020 mm (40.2 in)	2,416 kg (5,326 lb)	186.1 kN (41,843 lb) W 93.2 kN (20,944 lb) D
D-30KU	3F, 11A	269 kg (593 lb)/s	2.42	5,700 mm (224.0 in)	1,560 mm (61.4 in)	2,668 kg (5,882 lb)	107.9 kN (24,250 lb)
D-30KU-90	3F, 13A	265 kg (584 lb)/s	2.44	5,700 mm (224.4 in)	1,560 mm (61.4 in)	2,400 kg (5,291 lb)	117.7 kN (26,455 lb)
PS-7	1F+2A, 5A	—	—	—	—	—	68.7 kN (15,432 lb)
PS-9	1F+3A, 5A	—	—	—	—	—	100.0 kN (22,487 lb)
PS-14	1F+3A, 5A	—	—	—	—	—	137.3 kN (30,864 lb)
PS-18R	1F+4A, 5A	—	—	—	—	—	176.5 kN (30,683 lb)
PS-90A	2F+2A, 13A	470 kg (1,036 lb)/s	4.6	4,964 mm (195.4 in)	1,900 mm (74.8 in)	2,950 kg (6,503 lb)	156.9 kN (35,275 lb)
PS-90A10	1F, 12A	264 kg (582 lb)/s	3.76	4,280 mm (168.5 in)	1,400 mm (55.1 in)	1,900 kg (4,180 lb)	90.2 kN (20,283 lb)
PS-90A12	1F, 13A	370 kg (816 lb)/s	5.05	4,795 mm (188.8 in)	1,670 mm (65.8 in)	2,300 kg (5,071 lb)	117.7 kN (26,455 lb)
CFE (USA)							
CFE-738	1F, 5A+C	95.3 kg (210 lb)/s	5.3	2,514 mm (99.0 in)	1,092 mm (43.0 in)	601 kg (1,325 lb)	26.5 kN (5,957 lb)
CFM (International)							
CFM56-2B	1F+3A, 9A	370 kg (817 lb)/s	6.0	2,430 mm (95.7 in)	1,735 mm (68.3 in)	2,119 kg (4,671 lb)	97.9 kN (22,000 lb)
CFM56-3C	1F+3A, 9A	312 kg (688 lb)/s	5.0	2,360 mm (93.0 in)	1,524 mm (60.0 in)	1,951 kg (4,301 lb)	<104.5 kN (23,500 lb)
CFM56-5A	1F+3A, 9A	386 kg (852 lb)/s	6.0	2,422 mm (95.4 in)	1,735 mm (68.3 in)	2,257 kg (4,975 lb)	<117.9 kN (26,500 lb)
CFM56-5C	1F+4A, 9A	466 kg (1,027 lb)/s	6.6	2,616 mm (103.0 in)	1,836 mm (72.3 in)	2,644 kg (5,830 lb)	<151.3 kN (34,000 lb)
CFM56-7B	1F+4A, 9A	<355 kg (783 lb)/s	5.1	2,507 mm (98.7 in)	1,549 mm (61.0 in)	2,384 kg (5,256 lb)	<117.4 kN (26,400 lb)
CFM56-9	1F+2A, 9A	<282 kg (621 lb)/s	5.08	2,329 mm (91.7 in)	1,422 mm (56.0 in)	—	>82.3 kN (18,500 lb)
Engine Alliance (USA)							
GP7167	1F+3A, 9A	c907 kg (2,000 lb)/s	7	4,293 mm (169.0 in)	2,565 mm (101.0 in)	c5,216 kg (11,500 lb)	298 kN (67,000 lb)
GP7275	IF+5A, 9A	1,361 kg (3,000 lb)/s	8.7	4,750 mm (187 in)	3,150 mm (124 in)	c6,350 kg (14,000 lb)	333.46 kN (75,000 lb)
Eurojet (International)							
EJ200	3F, 5A, a/b	<77.1 kg (170 lb)/s	c0.4	4,000 mm (157.0 in)	740 mm (29.0 in)	990 kg (2,183 lb)	c89.0 kN (20,000 lb) W c60 kN (13,500 lb) D
General Electric (USA)							
GE90-76B	1F+3A, 10A	1,361 kg (3,000 lb)/s	9.0	5,182 mm (204.0 in)	3,404 mm (134.0 in)	7,559 kg (16,664 lb)	340 kN (76,400 lb)
GE90-85B	1F+3A, 10A	1,415 kg (3,120 lb)/s	8.7	5,182 mm (204.0 in)	3,404 mm (134.0 in)	7,559 kg (16,664 lb)	377 kN (84,700 lb)
GE90-90B	1F+3A, 10A	1,449 kg (3,195 lb)/s	8.4	5,182 mm (204.0 in)	3,404 mm (134.0 in)	7,559 kg (16,664 lb)	401 kN (90,000 lb)
GE90-92B	1F+3A, 9A	1,461 kg (3,221 lb)/s	8.3	5,182 mm (204.0 in)	3,404 mm (134.0 in)	7,559 kg (16,664 lb)	409 kN (92,000 lb)
GE90-115B	1F+4A, 9A	1,641 kg (3,641 lb)/s	8.9	c5,400 mm (212.5 in)	3,556 mm (140 in)	c8,618 kg (19,000 lb)	511.6 kN (115,000 lb)
CF6-50C2/E2	1F+3A, 14A	658 kg (1,450 lb)/s	4.3	4,394 mm (173.0 in)	2,195 mm (86.4 in)	3,956 kg (8,721 lb)	234 kN (52,500 lb)
CF6-80A2/A3	1F+3A, 14A	663 kg (1,460 lb)/s	4.6	3,998 mm (157.4 in)	2,195 mm (86.4 in)	3,819 kg (8,420 lb)	222.4 kN (50,000 lb)
CF6-80C2	1F+3A, 14A	802 kg (1,769 lb)/s	5.05	4,267 mm (168.0 in)	2,692 mm (106.0 in)	4,309 kg (9,499 lb)	<270 kN (60,690 lb)
CF6-80E1	1F+3A, 14A	874 kg (1,926 lb)/s	5.3	4,405 mm (173.5 in)	2,794 mm (110.0 in)	5,075 kg (11,189 lb)	<298 kN (66,870 lb)
CF34-3A	1F, 14A	151 kg (332 lb)/s	6.3	2,616 mm (103.0 in)	1,245 mm (49.0 in)	739 kg (1,625 lb)	<41.0 kN (9,220 lb)
CF34-8C1	1F, 10A	c181 kg (400 lb)/s	—	3,264 mm (128.5 in)	1,321 mm (52.0 in)	1,059 kg (2,335 lb)	56.38 kN (12,670 lb)
CF34-8D3	1F, 10A	c181 kg (400 lb)/s	—	3,264 mm (128.5 in)	1,321 mm (52.0 in)	1,120 kg (2,470 lb)	59.17 kN (13,300 lb)
CF34-8E	1F, 10A	—	—	—	—	1,120 kg (2,470 lb)	62.28 kN (14,000 lb)
CF34-10E	1F, 10A	—	5.3	—	—	—	82.26 kN (18,500 lb)
F101-102	2F, 9A, a/b	159.7 kg (352 lb)/s	2.01	4,590 mm (180.7 in)	1,402 mm (55.2 in)	2,023 kg (4,460 lb)	136.9 kN (30,780 lb) W
F110-100	3F, 9A, a/b	113 kg (250 lb)/s	0.87	4,630 mm (182.3 in)	1,181 mm (46.5 in)	1,492 kg (3,289 lb)	124.6 kN (28,000 lb) W 78.0 kN (17,530 lb) D
F110-129	3F, 9A, a/b	122 kg (270 lb)/s	0.76	4,620 mm (181.9 in)	1,181 mm (46.5 in)	1,791 kg (3,940 lb)	129.0 kN (29,000 lb) W 75.7 kN (17,000 lb) D
F110-400	3F, 9A, a/b	113 kg (250 lb)/s	0.87	4,620 mm (181.9 in)	1,181 mm (46.5 in)	1,599 kg (3,525 lb)	119.2 kN (26,080 lb) W 71.6 kN (16,800 lb) D
F118-100	3F, 9A	130+ kg (287+ lb)/s		2,553 mm (100.5 in)	1,181 mm (46.5 in)	1,452 kg (3,200 lb)	84.5 kN (19,000 lb)
F136	—	—	—	—	—	—	180+ kN (40,490+lb)W
F404-400	3F, 7A, a/b	64.4 kg (142 lb)/s	0.34	4,030 mm (158.8 in)	880 mm (34.8 in)	989 kg (2,180 lb)	71.2 kN (16,000 lb) W 48.9 kN (11,000 lb) D

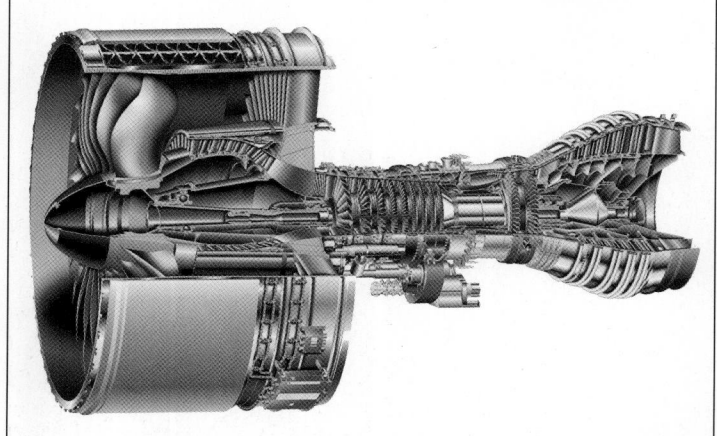

Cutaway of Engine Alliance GP7200 *NEW*/0126841

GTRE Kaveri (*Jane's/Craig Hoyle*) *NEW*/0130524

Engine type	Arrangement	Air flow	BPR	Length	Diameter	Dry weight	T-O rating
F404-402	3F, 7A, a/b	66.2 kg (146 lb)/s	0.27	4,030 mm (158.8 in)	880 mm (34.8 in)	1,035 kg (2,282 lb)	78.7 kN (17,700 lb) W
							53.2 kN (11,950 lb) D
F404-F1D2	3F, 7A	65 kg (143 lb)/s	—	—	880 mm (34.8 in)	—	46.8 kN (10,540 lb)
F414-400	3F, 7A, a/b	76.8 kg (169 lb)/s	—	—		—	97.9 kN (22,000 lb) W
							55.6 kN (12,500 lb) D
GTRE (India)							
Kaveri	3F, 6A, a/b	—	—	—	—	—	81.0 kN (18,210 lb) W
							52.0 kN (11,687 lb) D
Honeywell (USA)							
ALF502R-6	1F, 7A+C	87.0 kg (192 lb)/s	5.6	1,487 mm (58.6 in)	1,059 mm (41.7 in)	624 kg (1,375 lb)	33.4 kN (7,500 lb)
AS907	1F, 4A+C	—	—	2,290 mm (90.0 in)	868.7 mm (34.2 in)	619 kg (1,364 lb)	28.9 kN (6,500 lb st)
ATF3-6A	1F, 5A, C	73.5 kg (162 lb)/s	2.8	2,591 mm (102.0 in)	853 mm (33.6 in)	510 kg (1,125 lb)	24.2 kN (5,440 lb)
LF507	1F, 7A+C	87.0 kg (192 lb)/s	5.7	1,487 mm (58.6 in)	1,059 mm (41.7 in)	624 kg (1,375 lb)	31.1 kN (7,000 lb)
TFE731-2	1F, 4A+C	51.3 kg (113 lb)/s	2.66	1,520 mm (59.8 in)	716 mm (28.2 in)	337 kg (743 lb)	15.57 kN (3,500 lb)
TFE731-3	1F, 4A+C	53.7 kg (118 lb)/s	2.8	1,520 mm (59.8 in)	716 mm (28.2 in)	342 kg (754 lb)	16.46 kN (3,700 lb)
TFE731-5B	1F, 4A+C	64.9 kg (143 lb)/s	3.48	1,665 mm (65.5 in)	754 mm (29.7 in)	408 kg (899 lb)	21.13 kN (4,750 lb)
TFE731-20	1F, 4A+C	66.2 kg (146 lb)/s	3.1	1,547 mm (60.9 in)	716 mm (28.2 in)	406 kg (895 lb)	15.57 kN (3,500 lb)
TFE731-40	1F, 4A+C	65.8 kg (145 lb)/s	2.9	1,547 mm (60.9 in)	716 mm (28.2 in)	406 kg (895 lb)	18.91 kN (4,250 lb)
TFE731-60	1F, 4A+C	84.8 kg (187 lb)/s	3.9	2.083 mm (82.0 in)	781 mm (30.7 in)	448 kg (988 lb)	22.24 kN (5,000 lb)
TFE1042 (F124)	3F, 4A+C	42.7 kg (94.1 lb)/s	0.4	1,925 mm (75.8 in)	591 mm (23.25 in)	499 kg (1,100 lb)	28.0 kN (6,300 lb)
TFE1042-70 (F125)	3F, 4A+C, a/b	43.3 kg (95.4 lb)/s	0.4	3,561 mm (140.2 in)	591 mm (23.25 in)	617 kg (1,360 lb)	41.1 kN (9,250 lb) W
							26.8 kN (6,025 lb) D
IAE (International)							
V2522-A5	1F+4A, 10A	384 kg (848 lb)/s	4.9	3,200 mm (126.0 in)	1,600 mm (63.0 in)	2,359 kg (5,200 lb)	97.9 kN (22,000 lb)
V2527-A5	1F+4A, 10A	384 kg (848 lb)/s	4.75	3,200 mm (126.0 in)	1,600 mm (63.0 in)	2,500 kg (5,511 lb)	117.9 kN (26,500 lb)
V2528-D5	1F+4A, 10A	384 kg (848 lb)/s	4.8	3,200 mm (126.0 in)	1,600 mm (63.0 in)	2,540 kg (5,600 lb)	124.6 kN (28,000 lb)
V2533-A5	1F+4A, 10A	384 kg (848 lb)/s	4.4	3,200 mm (126.0 in)	1,600 mm (63.0 in)	2,500 kg (5,511 lb)	146.8 kN (33,000 lb)
IHI (Japan)							
F3-30	2F, 5A	34.0 kg (75.0 lb)/s	0.9	1,340 mm (52.76 in)	560 mm (22.0 in)	340 kg (750 lb)	16.37 kN (3,680 lb)
IL (Poland)							
D-18A	2F, 5A	38.4 kg (84.66 lb)	0.7	1,940 mm (76.37 in)	750 mm (29.5 in)	380 kg (837.7 lb)	17.65 kN (3,968 lb)
K-15	6A	23.5 kg (51.8 lb)/s	0	1,560 mm (61.42 in)	725 mm (28.5 in)	320 kg (705.5 lb)	14.71 kN (3,307 lb)
SO-3	7A	—	0	2,151 mm (84.7 in)	707 mm (27.8 in)	321 kg (708 lb)	10.79 kN (2,425 lb)
KKBM (Russian Federation)							
NK-8-2U	2F+2A, 6A	228 kg (503 lb)/s	1.049	5,288 mm (206.2 in)	1,442 mm (56.8 in)	2,350 kg (5,180 lb)	103.0 kN (23,150 lb)
NK-8-4	2F+2A, 6A	222 kg (489 lb)/s	1.042	5,101 mm (201.0 in)	1,442 mm (56.8 in)	2,440 kg (5,379 lb)	103.0 kN (23,150 lb)
NK-86	2F+2A, 6A	292 kg (644 lb)/s	1.6	3,638 mm (143.0 in)[1]	1,600 mm (63.0 in)	2,450 kg (5,401 lb)	127.5 kN (28,660 lb)
Klimov (Russian Federation)							
RD-33	4F, 9A, a/b	75.5 kg (166 lb)/s	0.49	4,229 mm (166.5 in)	1,040 mm (40.95 in)	1,055 kg (2,326 lb)	81.4 kN (18,300 lb) W
RD-35 – see PS/ZMK DV-2							49.4 kN (11,110 lb) D
LM (China)							
WP6	9A, a/b	46.3 kg (102 lb)/s	0	5,483 mm (215.9 in)	668 mm (26.3 in)	725 kg (1,598 lb)	39.72 kN (8,929 lb) W
							29.42 kN (6,614 lb) D
WP6A	9A, a/b	46.3 kg (102 lb)/s	0	5,483 mm (215.9 in)	668 mm (26.3 in)	725 kg (1,598 lb)	36.78 kN (8,267 lb) W
WS6	3F, 11A, a/b	155 kg (342 lb)/s	1.0	4,654 mm (183.2 in)	1,370 mm (53.94 in)	2,100 kg (4,630 lb)	122.1 kN (27,445 lb) W
							71.1 kN (15,991 lb) D
LMC (China)							
WS6 uprated for JH-7							138.3 kN (31,085 lb) W
							71.3 kN (16,027 lb) D
WP7B(BM)	3A+5A, a/b	65.0 kg (143 lb)/s	0	4,600 mm (181.1 in)	825 mm (32.5 in)	1,053 kg (2,321 lb)	59.8 kN (13,448 lb) W
							43.2 kN (9,700 lb) D
WP7F	3A+5A, a/b	—	0	—	—	—	63.7 kN (14,330 lb) W
							44.1 kN (9,921 lb) D
WP13	3A+5A, a/b	65.6 kg (145 lb)/s	0	4,600 mm (181.1 in)	907 mm (35.7 in)	1,211 kg (2,670 lb)	64.7 kN (14,550 lb) W
							40.2 kN (9,039 lb) D
WP13A II	3A+5A, a/b	65.6 kg (145 lb)/s	0	5,150 mm (202.8 in)	907 mm (35.7 in)	1,201 kg (2,648 lb)	65.9 kN (14,815 lb) W
							47.1 kN (10,582 lb) D
WP13B	3A+5A, a/b	—	0	—	—	—	68.7 kN (15,432 lb) W
							47.1 kN (10,582 lb) D
WP13F	3A+5A, a/b	—	0	—	—	—	64.7 kN (14,550 lb) W
							44.1 kN (9,921 lb) D
Pratt & Whitney (USA)							
F100-220	3F, 10A, a/b	—	0.7	5,280 mm (208.0 in)	1,181 mm (46.5 in)	1,451 kg (3,200 lb)	106.0 kN (23,830 lb) W
							65.3 kN (14,670 lb) D
F100-220P	3F, 10A, a/b	—	0.6	5,280 mm (208.0 in)	1,181 mm (46.5 in)	1,526 kg (3,365 lb)	120.1 kN (27,000 lb) W
							74.3 kN (16,700 lb) D
F100-229	3F, 10A, a/b	—	0.36	4,855 mm (191.15 in)	1,181 mm (46.5 in)	1,681 kg (3,705 lb)	129.5 kN (29,100 lb) W
							79.2 kN (17,800 lb) D
F117-100	1F+4A, 12A	608 kg (1,340 lb)/s	6.0	3,729 mm (146.8 in)	2,154 mm (84.8 in)	3,220 kg (7,100 lb)	181.0 kN (40,700 lb)
F119-100	3F, 6A, a/b		0.45				c155.6 kN (35,000 lb) W
F135	3F, 9A, a/b	—	—	—	—	—	c177.9 kN (40,000 lb) W
J52-408	5A, 7A	64.9 kg (143 lb)/s	0	3,020 mm (118.9 in)	814 mm (32.1 in)	1,052 kg (2,318 lb)	49.8 kN (11,200 lb)

Experimental thrust vectoring nozzle on Eurojet EJ2000 (*Jane's/Paul Jackson*)
NEW/0126839

Saturn AL-55 (*Jane's/Paul Jackson*)
0105011

Engine type	Arrangement	Air flow	BPR	Length	Diameter	Dry weight	T-O rating
JT8D-219	1F+6A, 7A	221 kg (488 lb)/s	1.77	3,911 mm (154.0 in)	1,250 mm (49.2 in)	2,092 kg (4,612 lb)	93.4 kN (21,000 lb)
JT9D-7R4H	1F+4A, 11A	769 kg (1,695 lb)/s	4.8	3,371 mm (132.7 in)	2,463 mm (97.0 in)	4,029 kg (8,885 lb)	249 kN (56,000 lb)
PW2040	1F+4A, 12A	608 kg (1,340 lb)/s	6.0	3,729 mm (146.8 in)	2,154 mm (84.8 in)	3,311 kg (7,300 lb)	185.5 kN (41,700 lb)
PW4050	1F+4A, 11A	798 kg (1,759 lb)/s	5.1	3,371 mm (132.7 in)	2,463 mm (97.0 in)	4,179 kg (9,213 lb)	222 kN (50,000 lb)
PW4060	1F+4A, 11A	816 kg (1,800 lb)/s	4.8	3,371 mm (132.7 in)	2,463 mm (97.0 in)	4,179 kg (9,213 lb)	267 kN (60,000 lb)
PW4168	1F+5A, 11A	903 kg (1,990 lb)/s	5.1	4,143 mm (163.1 in)	2,715 mm (106.9 in)	6,396 kg (14,100 lb)	305 kN (68,600 lb)
PW4084	1F+6A, 11A	1,089 kg (2,400 lb)/s	6.4	4,869 mm (191.7 in)	3,010 mm (118.5 in)	6,768 kg (14,920 lb)	386 kN (86,760 lb)
PW4098	1F+7A, 11A	1,293 kg (2,850 lb)/s	5.8	4,945 mm (194.7 in)	3,035 mm (119.5 in)	7,484 kg (16,500 lb)	436 kN (98,000 lb)
PW6000	1F+4A, 5A	—	4.9	2,743 mm (108 in)	1,524 mm (60.0 in)	2,313 kg (5,100 lb)	<102.3 kN (23,000 lb)
PW8000	1F+3A, 5A	—	10.0	3,150 mm (124 in)	1,930 mm (76.0 in)	3,629 kg (8,000 lb)	<155.7 kN (35,000 lb)
PW7000 (XTE-66)	—	—	—	—	—	—	<155.7 kN (35,000 lb)
Pratt & Whitney Canada (Canada)							
JT15D-1B	1F+C	34.0 kg (75.0 lb)/s	3.3	1,506 mm (59.3 in)	691 mm (27.2 in)	235 kg (519 lb)	9.79 kN (2,200 lb)
JT15D-5	1F+1A, C	42.0 kg (92.6 lb)/s	—	1,549 mm (61.0 in)	711 mm (28.0 in)	288 kg (635 lb)	12.9 kN (2,900 lb)
PW305	1F, 4A+C	—	4.3	2,070 mm (81.5 in)	970 mm (38.2 in)	450 kg (993 lb)	23.4 kN (5,266 lb)
PW306	1F, 4A+C	—	4.5	1,920 mm (75.6 in)	970 mm (38.2 in)	473 kg (1,043 lb)	25.4 kN (5,700 lb)
PW308	1F, 4A+C	—	4.5	—	—	—	35.6 kN (8,000 lb)
PW308A	1F, 4A+C	—	3.8	—	—	—	28.9 kN (6,500 lb)
PW530A	1F, 2A+C	—	3.9	1,524 mm (60.0 in)	701 mm (27.6 in)	278 kg (613 lb)	13.34 kN (3,000 lb)
PW535A	1F, 2A+C	—	3.9	1,524 mm (60.0 in)	701 mm (27.6 in)	—	14.94 kN (3,360 lb)
PW545A	1F, 2A+C	—	4.0	1,727 mm (68.0 in)	813 mm (32.0 in)	347 kg (765 lb)	17.24 kN (3,876 lb)
Progress (ZMKB) (Ukraine)							
AI-222-25F	2F, 8A	50.3 kg (110.9 lb)/s	1.19	3,010 mm (118.5 in)	810 mm (31.89 in)	540 kg (1,190.5 lb)	41.1 kN (9,259 lb) W 24.5 kN (5,511 lb) D
AI-222-25KFK	2F, 8A	50.3 kg (110.9 lb)/s	1.19	2,460 mm (96.85 in)	810 mm (31.89 in)	520 kg (1,146 lb)	30.07 kN (6,768 lb) W 24.5 kN (5,511 lb) D
AI-25TL	3F, 7A	46.8 kg (103 lb)/s	2.0	1,993 mm (78.46 in)	820 mm (32.3 in)	350 kg (772 lb)	16.87 kN (3,792 lb)
D-18T	1F, 7A, 7A	765 kg (1,687 lb)/s	5.6	5,400 mm (212.6 in)	2,330 mm (91.7 in)	4,100 kg (9,039 lb)	230 kN (51,660 lb)
D-18TM	1F, 7A, 7A	770 kg (1,698 lb)/s	5.5	5,700 mm (224.5 in)	2,330 mm (91.7 in)	4,750 kg (10,472 lb)	248 kN (55,777 lb)
D-27 propfan	2PF, 5A, 2A+C	27.4 kg (60.4 lb)/s	29.95	—	—	1,650 kg (3,638 lb)	109.8 kW (24,690 lb)
D-36	1F, 6A, 7A	255 kg (562 lb)/s	5.6	3,470 mm (136.6 in)	1,373 mm (54.1 in)	1,109 kg (2,445 lb)	63.7 kN (14,330 lb)
D-436T	1F, 6A, 7A	265 kg (584 lb)/s	4.98	3,470 mm (136.6 in)	1,373 mm (54.1 in)	1,250 kg (2,756 lb)	73.5 kN (16,535 lb)
D-436T2	1F+A, 6A, 7A	265 kg (584 lb)/s	4.9	4,157 mm (163.7 in)	1,655 mm (65.2 in)	1,450 kg (3,197 lb)	80.4 kN (18,078 lb)
PS/ZMK (International)							
DV-2	1F+2A, 7A	49.5 kg (109 lb)/s	1.46	1,721 mm (67.75 in)	994 mm (39.1 in)	440 kg (970 lb)	21.58 kN (4,852 lb)
PW/MTU (International)							
MTFE	1F+2A, 6A	c370 kg (c815 lb)/s	5.2	2,515 mm (99.0 in)	1,397 mm (55.0 in)	1,769 kg (3,900 lb)	89.0 kN (20,000 lb)
RKBM (Russian Federation)							
RD-41	7A	53.5 kg (118 lb)/s	0	1,594 mm (62.75 in)	635 mm (25.0 in)	290 kg (639 lb)	40.2 kN (9,040 lb)
Rolls-Royce (Germany)							
BR710	1F, 10A	197 kg (435 lb)/s	4.2	3,409 mm (134.0 in)	1,321 mm (52.0 in)	1,633 kg (3,600 lb)	<68.9 kN (15,500 lb)
BR715	1F+2A, 10A	288 kg (636 lb)/s	4.5	3,599 mm (142.0 in)	1,575 mm (62.0 in)	2,114 kg (4,660 lb)	<102.3 kN (23,000 lb)
Rolls-Royce (UK)							
535C	1F, 6A, 6A	518 kg (1,142 lb)/s	4.4	3,010 mm (118.5 in)	1,877 mm (73.9 in)	3,309 kg (7,294 lb)	166.4 kN (37,400 lb)
535E4B	1F, 6A, 6A	522 kg (1,150 lb)/s	4.3	2,995 mm (117.9 in)	1,892 mm (74.5 in)	3,261 kg (7,189 lb)	<191.7 kN (43,100 lb)
Pegasus 11-61	3F, 8A³	208 kg (459 lb)/s	1.2	3,485 mm (137.2 in)	1,222 mm (48.1 in)	1,932 kg (4,260 lb)	105.9 kN (23,800 lb)
RB211-22B	1F, 7A, 6A	626 kg (1,380 lb)/s	5.0	3,033 mm (119.4 in)	2,154 mm (84.8 in)	4,171 kg (9,195 lb)	186.8 kN (42,000 lb)
RB211-524B	1F, 7A, 6A	671 kg (1,480 lb)/s	4.4	3,106 mm (122.3 in)	2,180 mm (85.8 in)	4,452 kg (9,814 lb)	222 kN (50,000 lb)
RB211-524G/H	1F, 7A, 6A	728 kg (1,604 lb)/s	4.3	3,175 mm (125.0 in)	2,192 mm (86.3 in)	4,387 kg (9,671 lb)	273 kN (60,600 lb)
Spey 512	5A, 12A	94.3 kg (208 lb)/s	0.71	2,911 mm (114.6 in)	942 mm (37.1 in)	1,168 kg (2,574 lb)	55.8 kN (12,550 lb)
Spey 807	5A, 12A	91.6 kg (202 lb)/s	0.96	2,456 mm (96.7 in)	825 mm (32.5 in)	1,096 kg (2,417 lb)	49.1 kN (11,030 lb)
Tay 611, 620	1F+3A, 12A	176 kg (388 lb)/s	3.18	2,405 mm (94.7 in)	1,118 mm (44.0 in)	1,422 kg (3,135 lb)	61.6 kN (13,850 lb)
Tay 651	1F+3A, 12A	193 kg (425.5 lb)/s	3.10	2,405 mm (94.7 in)	1,138 mm (44.8 in)	1,533 kg (3,380 lb)	68.5 kN (15,400 lb)
Trent 556	1F, 8A, 6A	858 kg (1,892 lb)/s	8.5	3,912 mm (154.0 in)	2,474 mm (97.4 in)	4,719 kg (10,400 lb)	252 kN (56,000 lb)
Trent 600	1F, 8A, 6A	922 kg (2,032 lb)/s	8.0	3,912 mm (154.0 in)	2,474 mm (97.4 in)	4,719 kg (10,400 lb)	306 kN (68,000 lb)
Trent 772	1F, 8A, 6A	897 kg (1,978 lb)/s	4.9	3,912 mm (154.0 in)	2,474 mm (97.4 in)	4,785 kg (10,550 lb)	320 kN (71,100 lb)
Trent 892	1F, 8A, 6A	1,200 kg (2,645 lb)/s	5.7	4,369 mm (172.0 in)	2,794 mm (110.0 in)	5,957 kg (13,133 lb)	411 kN (91,300 lb)
Trent 895	1F, 8A, 6A	1,217 kg (2,684 lb)/s	5.79	4,369 mm (172 in)	2,794 mm (110.0 in)	5,981 kg (13,186 lb)	425 kN (95,500 lb)
Trent 976	1F, 8A, 6A	1,179 kg (2,600 lb)/s	8.0	4,369 mm (172.0 in)	2,794 mm (110.0 in)	5,806 kg (12,800 lb)	338 kN (76,000 lb)
Viper 535	8A	23.9 kg (52.7 lb)/s	0	1,806 mm (71.1 in)	740 mm (29.1 in)	358 kg (790 lb)	14.94 kN (3,360 lb)
Viper 632	8A	26.5 kg (58.4 lb)/s	0	1,806 mm (71.1 in)	740 mm (29.1 in)	376.5 kg (830 lb)	17.66 kN (3,970 lb)
Viper 680	8A	27.2 kg (60.0 lb)/s	0	1,806 mm (71.1 in)	740 mm (29.1 in)	379 kg (836 lb)	19.39 kN (4,360 lb)
Rolls-Royce (USA)							
AE 3007A	1F, 14A	96.6 kg (213 lb)/s	5.0	2,705 mm (106.5 in)	1,105 mm (43.5 in)	717.1 kg (1,581 lb)	33.7 kN (7,580 lb)
AE 3007H	1F, 14A	118 kg (260 lb)/s	5.0	2,705 mm (106.5 in)	1,105 mm (43.5 in)	717.1 kg (1,581 lb)	36.9 kN (8,290 lb)
RRTI (International)							
Adour 811/815	2F, 5A, a/b	43.1 kg (95.0 lb)/s	0.75	2,970 mm (116.9 in)	559 mm (22.0 in)	738 kg (1,627 lb)	37.4 kN (8,400 lb) W 24.6 kN (5,520 lb) D
Adour 871	2F, 5A	44.0 kg (97.0 lb)/s	0.80	1,956 mm (77.0 in)	559 mm (22.0 in)	603 kg (1,330 lb)	26.6 kN (5,990 lb)
Samara (Russian Federation)							
NK-25	3F, 9A	339 kg (747 lb)/s	0.9	5,200 mm (205 in)	1,348 mm (53.1 in)	2,850 kg (6,283 lb)	245 kN (55,115 lb)
NK-44	—	—	6.0	—	3,100 mm (122.0 in)	—	392 kN (88,180 lb)
NK-93	2PF, 7A, 8A	1,000 kg (2,205 lb)/s	17.0	5,972 mm (235.1 in)	2,900 mm (114.2 in)²	3,650 kg (8,047 lb)	176.5 kN (39,683 lb)
NK-321	3F, 5A, 7A, a/b	365 kg (805 lb)/s	1.4	c6,000 mm (236.0 in)	1,460 mm (57.5 in)	3,442 kg (7,588 lb)	245 kN (55,077 lb) W 137.2 kN (30,843 lb) D

Williams FJ-33 small turbofan *(Jane's/Paul Jackson)*   NEW/0126838

Williams FJ44-3 *(Jane's/Paul Jackson)*   NEW/0526937

Engine type	Arrangement	Air flow	BPR	Length	Diameter	Dry weight	T-O rating
Saturn (Russian Federation)							
AL-21F3	14A, a/b	104 kg (229 lb)/s	—	5,340 mm (210.0 in)	1,030 mm (40.6 in)	1,800 kg (3,968 lb)	110.5 kN (24,800 lb) W 76.5 kN (17,200 lb) D
AL-31FM	4F, 9A, a/b	112 kg (247 lb)/s	0.57	4,950 mm (195.0 in)	1,277 mm (50.3 in)	1,488 kg (3,280 lb)	122.6 kN (27,560 lb) W 79.3 kN (17,857 lb) D
AL-37FU	4F, 9A, a/b	—	0.65	c5,000 mm (c197 in)	1,277 mm (50.3 in)	1,660 kg (3,660 lb)	142.2 kN (31,967 lb) W 83.4 kN (18,740 lb) D
AL-41 (SAT-41)	2F, 6A, a/b	c150 kg (331 lb)/s	—	—	—	c1,850 kg (c4,078 lb)	c196.1 kN (c44,090 lb) W c113.8 kN (25,600 lb) D
AL-55	4F, 6A	29.8 kg (65.7 lb)/s	0.6	1,210 mm (47.6 in)	590 mm (23.2 in)	315 kg (694 lb)	21.58 kN (4,850 lb)
AL-55F	4F, 6A, a/b	29.8 kg (65.7 lb)/s	0.6	2,520 mm (99.2 in)	590 mm (23.2 in)	385 kg (849 lb)	34.34 kN (7,716 lb)
SNECMA (France)							
Atar 9K50	9A, a/b	73.0 kg (161 lb)/s	0	5,944 mm (234.0 in)	1,020 mm (40.2 in)	1,582 kg (3,487 lb)	70.6 kN (15,870 lb) W 49.2 kN (11,055 lb) D
M53 P2	3F, 5A, a/b	86.0 kg (190 lb)/s	0.35	5,070 mm (199.6 in)	1,055 mm (41.5 in)	1,500 kg (3,307 lb)	95.0 kN (21,355 lb) W 64.3 kN (14,455 lb) D
M88-2	3F, 6A, a/b	67.0 kg (148 lb)/s	0.25	3,540 mm (139.0 in)	660 mm (26.0 in)	909 kg (2,004 lb)	75.0 kN (16,872 lb) W 50.0 kN (11,250 lb) D
Soyuz (Russian Federation)							
R-195	3A	67.0 kg (148 lb)/s	0	3,300 mm (130.0 in)	914 mm (36.0 in)	990 kg (2,183 lb)	44.13 kN (9,921 lb)
R-29-300	5A, 6A, a/b	110 kg (243 lb)/s	0	4,960 mm (195.0 in)	912 mm (35.9 in)	1,880 kg (4,145 lb)	122.3 kN (27,500 lb) W 78.5 kN (17,635 lb) D
R-35	5A, 6A, a/b	110 kg (243 lb)/s	0	4,950 mm (194.9 in)	912 mm (35.9 in)	1,765 kg (3,891 lb)	127.5 kN (28,660 lb) W 83.9 kN (18,850 lb) D
R-79	5F, 6A, a/b[3]	120 kg (265 lb)/s	1.0	5,229 mm (205.9 in)	1,100 mm (43.3 in)	2,750 kg (6,063 lb)	152.0 kN (34,170 lb) W 107.6 kN (24,200 lb) D
RD-1700	2F, 4A	30 kg (66.1 lb)/s	0.78	—	—	297.58 kg (656 lb)	19.61 kN (4,409 lb) W 16.67 kN (3,748 lb)D
Turbomeca-SNECMA (France)							
Larzac 04-20	2F, 4A	28.6 kg (63.0 lb)/s	1.04	1,179 mm (46.4 in)	602 mm (23.7 in)	302 kg (666 lb)	14.12 kN (3,175 lb)
Turbo-Union (International)							
RB199 Mk 105	3F, 3A, 6A, a/b	74.6 kg (164 lb)/s	c1	3,302 mm (130.0 in)	752 mm (29.6 in)	980 kg (2,160 lb)	74.7 kN (16,800 lb) W 42.5 kN (9,550 lb) D
Volvo/GE (Sweden)							
RM12	3F, 7A, a/b	68.0 kg (150 lb)/s	0.28	4,100 mm (161.4 in)	880 mm (34.8 in)	1,050 kg (2,315 lb)	80.5 kN (18,100 lb) W 54.0 kN (12,140 lb) D
Williams and Williams Rolls (USA)							
FJ33-1	1F+2A, C	—	—	1,216 mm (47.9 in)	439.4 mm (17.3 in)	<136 kg (300 lb)	5.34 kN (1,200 lb)
FJ44-1A	1F+1A, C	29.575 kg (65.2 lb)/s	3.28	1,356 mm (53.4 in)	530.9 mm (20.9 in)	205 kg (452 lb)	8.45 kN (1,900 lb)
FJ44-1C	1F+1A, C	26.49 kg (58.4 lb)/s	3.28	1,356 mm (53.4 in)	530.9 mm (20.9 in)	208 kg (459 lb)	6.67 kN (1,500 lb)
FJ44-2A	1F+3A, C	35.8 kg (79.0 lb)/s	4.1	1,507 mm (59.4 in)	553.7 mm (21.8 in)	—	10.23 kN (2,300 lb)
FJ44-2C	1F+3A, C	—	—	1,507 mm (59.4 in)	553.7 mm (21.8 in)	—	10.68 kN (2,400 lb)
FJX/EJ-22	1F+1A, C	—	—	1,041 mm (41.0 in)	368.3 mm (14.5 in)	39.5 kg (85 lb)	3.42 kN (770 lb)
XAE (China)							
WS9: Rolls-Royce Spey Mk 202 made under licence; data presumed identical							
WP8	8A	150 kg (331 lb)/s	—	5,380 mm (211.8 in)	1,400 mm (55.1 in)	3,133 kg (6,906 lb)	93.2 kN (20,944 lb) W

UPDATED

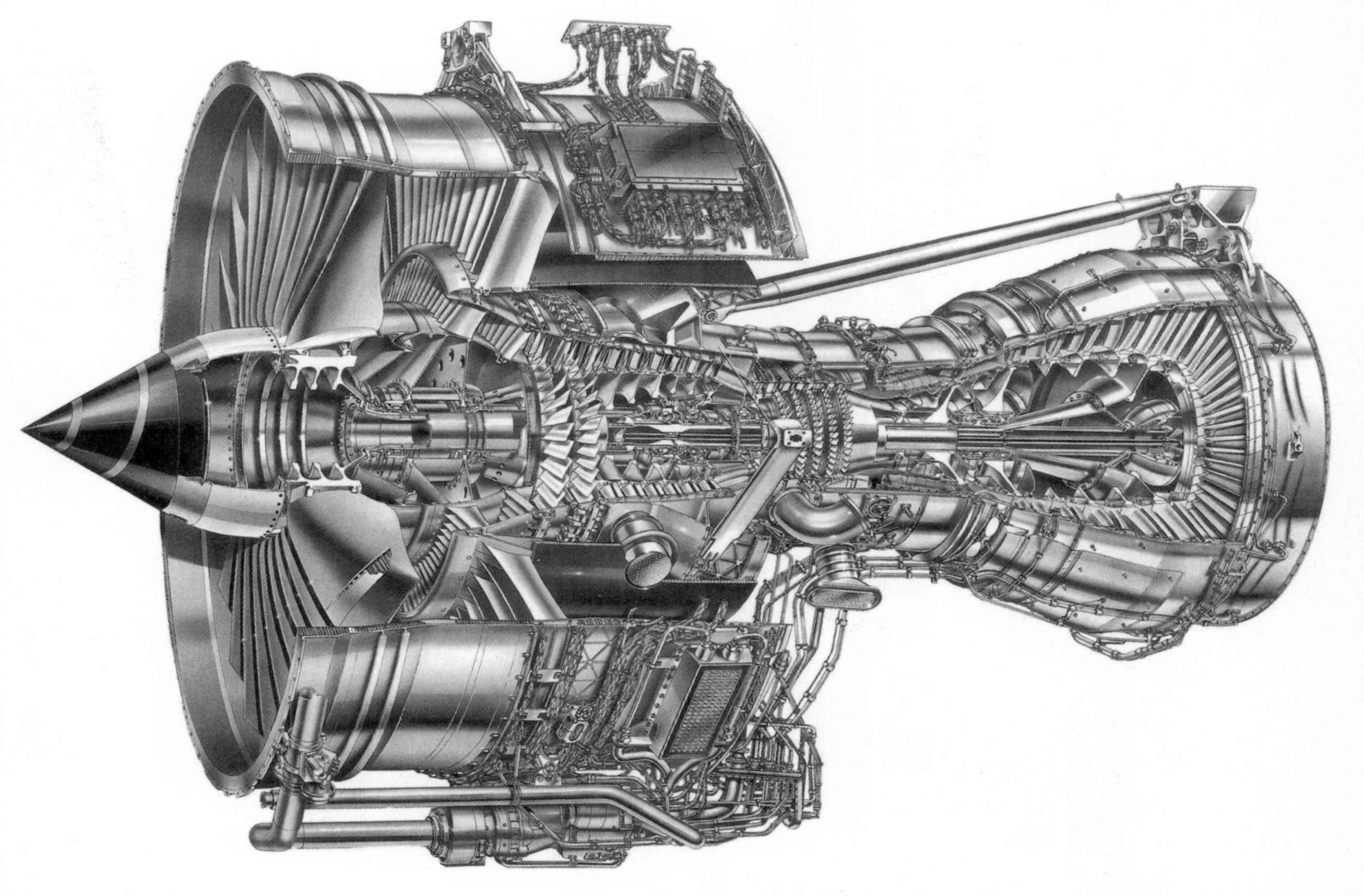

Rolls-Royce Trent 900 turbofan

NEW/0527153

Auxiliary power units (APUs)

APU type	Start-up ceiling	Dry weight	S/L compression	Ground power	Rating
Aerosila (Russia)					
TA-4FE	<2,500 m (8,200 ft)	180 kg (397 lb)	2.4 bar (35 lb/sq in)	18 kVA	124 kW
TA-6A	<3,000 m (9,840 ft)	245 kg (540 lb)	4.4 bar (64 lb/sq in)	40 kVA	235 kW
TA-6V	<3,000 m (9,840 ft)	255 kg (562 lb)	4.4 bar (64 lb/sq in)	40 kVA	235 kW
TA-8	<3,000 m (9,840 ft)	165 kg (364 lb)	3.2 bar (46 lb/sq in)	—	107 kW
TA-8V	<5,000 m (16,400 ft)	185 kg (408 lb)	3.2 bar (46 lb/sq in)	40 kVA	107 kW
TA-8K	<3,500 m (11,480 ft)	178 kg (392 lb)	3.1 bar (45 lb/sq in)	60 kVA	55 kW
TA-12	<7,000 m (22,960 ft)	290 kg (639 lb)	4.8 bar (70 lb/sq in)	40 kVA	287 kW
TA-12-60	<7,000 m (22,960 ft)	335 kg (739 lb)	4.8 bar (70 lb/sq in)	60 kVA	287 kW
TA-14	<8,000 m (26,250 ft)	50 kg (110 lb)	3.6 bar (52 lb/sq in)	20 kVA	79 kW
TA-18-100	<11,000 m (36,090 ft)	140 kg (309 lb)	4.4 bar (64 lb/sq in)	60 kVA	214 kW
Honeywell (USA)					
36-100	6,095 m (20,000 ft)	53 kg (117 lb)	3.52 bar (51 lb/sq in)	30 kVA	127 kW
36-150W	6,095 m (20,000 ft)	52-55 kg (115-121 lb)	3.24 bar (47 lb/sq in)	10 kVA	142 kW
36-150CX/IAI	6,095 m (20,000 ft)	57 kg (126 lb)	3.38 bar (49 lb/sq in)	10 kVA	142 kW
36-150F2	6,095 m (20,000 ft)	55 kg (121 lb)	3.17 bar (46 lb/sq in)	10 kVA	142 kW
36-150F	6,095 m (20,000 ft)	53 kg (117 lb)	3.24 bar (47 lb/sq in)	10 kVA	142 kW
36-150DD	6,095 m (20,000 ft)	53 kg (117 lb)	3.24 bar (47 lb/sq in)	10 kVA	145 kW
36-150 M	6,095 m (20,000 ft)	65 kg (143 lb)	3.59 bar (52 lb/sq in)	23 kVA	145 kW
36-150RJ	11,280 m (37,000 ft)	73 kg (160 lb)	3.11 bar (45 lb/sq in)	30 kVA	179 kW
RE100	6,095 m (20,000 ft)	40 kg (88 lb)	3.45 bar (50 lb/sq in)	10 kVA	101 kW
RE220	13,105 m (43,000 ft)	109 kg (240 lb)	3.86 bar (56 lb/sq in)	45 kVA	261 kW
36-280	10,670 m (35,000 ft)	159 kg (350 lb)	3.24 bar (47 lb/sq in)	90 kVA	291 kW
36-300	10,670 m (35,000 ft)	136 kg (300 lb)	3.24 bar (47 lb/sq in)	90 kVA	291 kW
131-9A	12,495 m (41,000 ft)	159 kg (350 lb)	3.52 bar (51 lb/sq in)	90 kVA	343 kW
131-9B	12,495 m (41,000 ft)	186 kg (410 lb)	3.52 bar (51 lb/sq in)	96 kVA	90 kW
331-200	13,105 m (43,000 ft)	227 kg (500 lb)	3.45 bar (50 lb/sq in)	90 kVA	432 kW
331-250	13,105 m (43,000 ft)	227 kg (500 lb)	3.04 bar (44 lb/sq in)	90 kVA	447 kW
331-350	12,495 m (41,000 ft)	254 kg (560 lb)	3.45 bar (50 lb sq in)	115 kVA	618 kW
331-500	13,105 m (43,000 ft)	331 kg (730 lb)	3.65 bar (53 lb/sq in)	120 kVA	969 kW
700	7,620 m (25,000 ft)	295 kg (650 lb)	3.05 bar (44 lb/sq in)	100 kVA	648 kW
Microturbo (France)					
Saphir 20	6,095 m (20,000 ft)	38 kg (84 lb)		10 kVA	70 kW
Omsk (Russian Federation)					
VSU-10					
Pratt & Whitney (Canada)					
PW901	7,620 m (25,000 ft)	379 kg (835 lb)	3.72 bar (54 lb/sq in)	180 kVA	1,145 kW
ZMKB Progress (design), **Motor Sich** (manufacture) (Ukraine)					
AI-8	7,620 m (25,000 ft)	145 kg (320 lb)	—	60 kVA	60 kW
AI-9	—	45 kg (99 lb)	2.35 bar (34 lb/sq in)	—	
AI-9K	—	52 kg (115 lb)	2.35 bar (34 lb/sq in)		
AI-9V	—	70 kg (154 lb)	2.89 bar (41 lb/sq in)		
AI-9-3B	—	128 kg (282 lb)	3.92 bar (57 lb/sq in)	—	29 kW
Stupino (Russian Federation)					
TA-6					
TA-12					
Sundstrand (USA)					
APS500	<9,145 m (<30,000 ft)	46-52 kg (100-115 lb)	c3.6 bar (c52 lb/sq in)	19 kVA	200 kW
APS1000	<7,620 m (<25,000 ft)	70-82 kg (155-180 lb)	4.14 bar (60 lb/sq in)	45 kVA	261 kW
APS2000	11,280 m (37,000 ft)	132 kg (290 lb)	3.65 bar (53 lb/sq in)	60 kVA	376 kW
APS2100	11,280 m (37,000 ft)	122 kg (270 lb)	3.65 bar (53 lb/sq in)	60 kVA	376 kW
APS3200	11,885 m (39,000 ft)	138 kg (305 lb)	4.14 bar (60 lb/sq in)	90 kVA	450 kW
Ufa (Russian Federation)					
VD-100					

VERIFIED

ZMKB Progress/Motor Sich AI-8

NEW/0130075

PROPELLERS

Introduction

Although a vital component of the majority of aircraft built, the airscrew has received little attention until now. In the table will be found explanations of various manufacturers' designation systems, to be read in conjunction with data supplied in the 'Engines' paragraph of aircraft descriptions.

NEW ENTRY

The offset thrust line of the Pilatus PC-21's five-blade propeller is indicated by the angle of its Hartzell E881KX spinner
(*Jane's/Paul Jackson*)
NEW/0528615

Manufacturer	Example designation	Explanation
Aerosila		
Aerosila NPP OAO	AV-17	Three-blade, variable pitch; diameter 3.60 m
ulitsa Metallistov 6	AV-24AN	Three-blade, variable pitch, Antonov An-28; diameter 2.80 m
Stupino	SV-27	8 + 8 contrarotating propfan; diameter 4.50 m
142800 Moscovskoya oblast	SV-34	Six-blade, constant-speed; diameter 3.60 m
Russian Federation	AV-36	Six-blade, constant-speed; diameter 2.65 m
Tel: (+7 095) 333 22 72	SV-92	8 + 10 ducted fan; diameter 2.90 m
Fax: (+7 095) 333 51 47	AV-101	Two/three/four-blade, variable pitch; diameter 1.20 m
e-mail: vint@aerosila.ru	AV-110	Four-blade, variable pitch; diameter 1.40 m
Web: http://www.aerosila.ru	AV-140	Six-blade, constant-speed; diameter 3.72 m
Avia		
Avia Propeller sro	AV 723-5-B	Avia; variable pitch (G = ground adjustable, F = fixed pitch); 72 mm blade shank diameter; 3 blades;
Beranových 666		constant-speed, non-feathering, non-counterweighted (1 = constant speed, non-feathering,
CZ-199 02 Prague 9		counterweighted, 2 = CS, feathering, 3 = CS, feathering, reversing, 4 = constant speed, reversing, 6 =
Czech Republic		autonomous); SAE 2 flange with ½ in bolts (A = special flange, 80 mm bolts, C = SAE 2 ⁷⁄₁₆ in bolts, D
Tel: (+420 266) 11 20 30		= ARP 502 flange, E = ARP 880 flange, F = SAE 1 flange, G = Walter/LOM flange, H = PW115
Fax: (+420 266) 11 20 31		flange)
e-mail: brunclik@aviapropeller.cz	V-500A	2,000 mm diameter (/1690 = 1,690 mm diameter, /1905 = 1,905 mm diameter). Two-blade
Web: http://www.aviapropeller.com		
	V 508D	2,500 mm diameter. Three-blade. Walter M601D engine. (508Z = Walter M601Z, 508E/84 = 84 in
		diameter, 508E-AG/99B = 99 in diameter, agricultural.)
	V 510	2,300 mm diameter. Five-blade. Walter M601E-F. (510T = Walter M601F-T, 510AG agricultural.)
	V 520	2,700 mm diameter. Two-blade
Avtek		
Avtek srl		Carbon fibre, two- or three-blade, flight-adjustable
Via A Gramsci 18		
I-20050 Zoccorino (MI)		
Italy		
Tel: (+39 0362) 60 20 77		
Fax: (+39 0362) 80 26 20		
e-mail: lucianomerati@tin.it		
Baoding		
Baoding Propeller Plant	J17-G13	Four-blade, constant-speed, fully feathering, metal; diameter 4.50 m
Box 609, PO Box 818		
Baoding 072152		
Hebei		
People's Republic of China		
Tel: (+86 312) 705 15 82		
Fax: (+86 312) 705 14 54		

Manufacturer	Example designation	Explanation
Bolly The Propeller Company Pty Ltd Unit 8 100 Hewittson Road Elizabeth West, SA 5113 Australia *Tel:* (+61 08) 82 55 96 88 *Fax:* (+61 08) 82 55 96 66 *e-mail:* bolly@bolly.com.au *Web:* http://www.bolly.com.au	BOS1-54-24-N-L BOS2-58-33-S-R BOS3-62-52-S-R BOS4-54-22-N-L	Bolly Optima Series 1; 54 in diameter (60 = 60 in) ; 24 in nominal pitch (22 = 22 in); narrow aspect ratio blade (S = Standard, W = wide); left-hand (anticlockwise) rotation viewed from behind (R = clockwise). Composites, fixed pitch; two-blade Series 2; 58 (55½, 59½, 62) in diameter; 33 (31, 34, 36, 38, 39, 42) in pitch; standard aspect ratio; clockwise rotation. Composites; individual, bolted, fixed pitch blades; two to five blades Series 3; 62 (68, 72) in diameter; 52 (58, 60) in pitch; S, R as above. Composites, ground-adjustable pitch; two-, three- or four-blade Series 4 (folding blades); 54 in diameter; 22 in pitch; N, L as above. Composites, ground-adjustable pitch; two-blade
Dowty Dowty Propellers Anson Business Park Cheltenham Road East Staverton Gloucester GL2 9QN United Kingdom *Tel:* (+44 1452) 71 60 00 *Fax:* (+44 1452) 71 61 01 *Web:* http://www.dowty.com	R352/6-123-F/1 R381 R391 R408 R414	Rotol type 352; six-blade; blade shank size 123; engine flange mounted; first subvariant Six-blade, constant-speed, feathering/reversing; Saab 2000 diameter 12 ft 6 in Six-blade, constant-speed, feathering/reversing; composites; diameter 13 ft 6 in; Lockheed C-130J, Alenia C-27J Six-blade, constant-speed, feathering/reversing; composites; diameter 13 ft 6 in; Dash 8 Q400 Six-blade, constant-speed, feathering/reversing; composites; ShinMaywa US-1A Kai
EVRA EVRA 4 avenue de la Forêt d'Halatte F-60100 Creil France *Tel:* (+33 3) 44 25 50 31 *Fax:* (+33 3) 44 25 90 72	160.150.136 D9.28 D11.29 6006 Merville DH.5220 180.170 Evolution	Wood or composites; various specific designations according to aircraft type
GT GT Propellers Viale del Commercio I-47838 Riccione (RN) Italy *Tel:* (+39 0541) 69 33 99 *Fax:* (+39 0541) 69 33 31 *e-mail:* info@gt-propellers.com *Web:* http://www.gt-propellers.com		Wood, two- and three-blade, fixed pitch
Hamilton Sundstrand Hamilton Sundstrand Propulsion Systems One Hamilton Road Windsor Lock Connecticut 06096-1010, USA *Tel:* (+1 860) 654 27 72 *e-mail:* props.info@hs.utc.com *Web:* http://www.hamiltonsundstrandcorp.com	14SF-5 247F-1 568F-1	Four-blade, 13 ft diameter; aluminium spar with glass fibre blade shell; hydromechanical constant-speed; ATR 42 (7 = Dash 8 Q100, 11 = ATR 72, 11E = ATR 72 electronic pitch control, 15 = Dash 8 Q100/Q200/Q300, 17 = Canadair 215T, 19 = Canadair 215T/415, 23 = Dash 8 Q100/Q200/Q300) Four-blade, 13 ft diameter, graphite spar and Kevlar shell; hydromechanical constant-speed; ATR 72 Srs 211/212 (1E = ATR 72 Srs 211/212 electronic pitch control; 3 = XAC MA-60) Six-blade, 13 ft diameter, graphite spar and Kevlar shell; electromechanical constant-speed; ATR 42 Srs 400/500 (5 = CASA C-295, 7 = Ilyushin Il-114, 11 = previously assigned to Ayres Loadmaster)
Hartzell Hartzell Propeller Inc One Propeller Place Piqua USA Ohio 45356-2634 USA *Tel:* (+1 937) 778 42 00 *Fax:* (+1 937) 778 43 91 *Web:* http://www.hartzellprop.com	BHC-J2YF-1BF/F7694	Hartzell hub series B (blank = A, C = C, D = D); two-blade (3 = three, 4 = four); /; blades diameter 76 in; subtype 94 (7068 = 70 in, 7280 = 72/72½ in, 7663 = 76 in, 7666 = 76 in, 7681 = 76 in, 7692 = 76 in, 8052 = 80 in, 8459 = 84 in)
Hoffmann Hoffmann Propeller GmbH Küpferlingstrasse 9 D-83022 Rosenheim Germany *Tel:* (+49 8031) 187 80 *Fax:* (+49 8031) 18 78 78 *e-mail:* info@hoffmann-prop.com *Web:* http://www.hoffmann-prop.com	HO-V343K-V/S180GY-B+10 HO27HM-180/138	Hoffmann; Variable pitch; Type 34; three blades; AS 127D flange, SAE 2, ½ in bolts (F = ARP 502 Type 1, L= AS 127D SAE 2, 7/16 in bolts, B = AS 127D SAE 1, G = 3/8 in bolts); hydraulic pitch change, oil pressure to decrease/counterweights to increase (blank = oil pressure increase/no counterweights, F = oil pressure increase/counterweights, S = oil pressure decrease/counterweights to increase up to feathering); /; feathering blades (blank = right hand, tractor, D = right hand, pusher, L = left hand, tractor, LD = left hand, pusher, V = pitch change pin position, altered to allow oil pressure to decrease pitch); 1,800 mm diameter; blade design GY; electrical de-icing; 10 cm over basic size Fixed pitch Hoffmann, 1,800 mm diameter
Junkers Profly/Woodcomp Junkers Profly GmbH Blaich 6b D-95326 Kulmbach Germany *Tel:* (+49 9221) 54 44 *Fax:* (+49 9221) 87 93 13 *e-mail:* Junkers_Profly_GmbH@t-online.de *Web:* http://www.junkers-profly.de	Junkers JU 160 Junkers Vario 165 Kremen SR 116 Kremen SR 200 A Kremen SR 2000 XA SportProp Winglet 136 SportProp Varia 165/2R	Carbon fibre; 1,600 mm diameter (165 = 1,650 mm; 170 = 1,700 mm); ground-adjustable pitch Carbon fibre; 1,650 mm diameter (170 = 1,700 mm); mechanical flight-adjustable pitch, two-blade Wood; 1,600 mm diameter (117 = 1,450 mm); ground-adjustable pitch Wood; 1,450 mm diameter (B/C = 1,680 mm) Wood; electrically variable pitch design; 1,700 mm diameter (XB = 1,500 mm) Glass fibre; ground-adjustable pitch (Winglet 165, Speedy 165, Klassisch 160, Speedy 175, Para 110, Para 137) Glass fibre; mechanical flight-adjustable pitch (Varia 170); two-blade
LOM Praha LOM Praha SP Černokostelcká 270 CZ-100 38 Praha 10 Czech Republic *Tel:* (+420 270) 36 38 *Fax:* (+420 270) 65 23 *e-mail:* lompraha@mbox.vol.cz *Web:* http://www.lom.cz	V-231 V-341 V-532 V-541 V-546	Two-blade, wood, fixed-pitch, 1,650 to 1,800 mm diameter Two-blade, metal, ground-adjustable pitch, 1,600 to 1,900 mm diameter Three-blade, wood, hydraulically variable pitch, 1,850 mm diameter Two-blade, metal, hydraulically constant-speed, 1,750 to 1,900 mm diameter Three-blade, metal, hydraulically constant-speed, reversing/feathering, 1,750 to 1,900 mm diameter
McCauley McCauley Propeller Systems PO Box 5053 Vandalia Ohio 45377-5053, USA *Fax:* (+1 937) 890 60 01 *Web:* http://www.mccauley.textron.com	D3AF32C72-A/S-82NC-0 1A100/MCM6950	Dowel (indexing hole) location 90° and 270° (blank = 60°/240°, B = 0°/180°, C = 30°/210°); three-blade; 4 in flange (D = SAE 2 flange, F = 4½ in flange); feathering; shank size 32; constant-speed; version 72; minor change; / ; blade minor change; 82 in diameter; blade design characteristics; zero reduction from basic diameter Basic design (fixed pitch); / ; SAE 1 flange (SLM = SAE 2); 69 in diameter; 50 in pitch at ¾ radius

Manufacturer	Example designation	Explanation
MT-Propeller MT-Propeller Entwicklung GmbH Flugplatzstrasse 1 D-94348 Atting Germany *Tel:* (+49 9429) 940 90 *Fax:* (+49 9429) 84 32 *e-mail:* sales@mt-propeller.com *Web:* http://www.mt-propeller.com	MTV-9-B-C/CL160-03 MT186R140-3D	MT-Propeller; variable pitch; Type 9; SAE 2½ in bolts (A-F as for Avia propellers); counterweight moment towards high pitch position (blank = none or small counterweights with low pitch moment); blade pitch change pin moment towards high pitch (blank = low pitch); left hand rotation, tractor (blank = right tractor; RD = right, pusher, LD = left, pusher); 160 cm diameter; Type 3 MT-Propeller; two-blade, fixed pitch; 1,860 mm diameter
Neuform Neuform Composites GmbH Ameke 63 D-48317 Dreinsteinfurt Germany *Tel:* (+49 2387) 940 26 *Fax:* (+49 2387) 940 28 *e-mail:* info@neuform-propeller.de *Web:* http://www.neuform-propeller.de	T2 C3-V	Composites, T-type blade (C, TX), 1,730 to 1,750 mm diameter, fixed pitch, two-blade (3 = three blade, 4 = four blade) Composites, C-type blade (T, TX), 1,730 to 1,750 mm diameter, variable pitch (2 = two blade, 4 = four blade)
Powerfin Powerfin Inc 4700, 188th Street NE Arlington Washington 98223, USA *Tel:* (+1 360) 403 06 35 *Fax:* (+1 360) 403 05 99 *e-mail:* info@powerfin.com *Web:* http://www.powerfin.com	Model A Model B Model C Model D Model E Model F	64 to 72 in diameter, three-blade, composites 60 to 64 in diameter, three-blade, composites 64 to 69 in diameter, three-blade, composites 69 to 72 in diameter, three-blade, composites 51 in diameter, three-blade, composites 66 to 70 in diameter, three-blade, composites
PZL Warszawa PZL Warszawa-Okecie SA aleja Krakowska 110/114 PL-00-971 Warszawa, Poland *Tel:* (+48 22) 846 61 52 *Fax:* (+48 22) 846 27 01 *e-mail:* wasiucionek@pzl-okecie.pl *Web:* http://www.pzl-okecie.com.pl	AW-2-30	Four-blade, constant-speed; metal; Antonov An-2; 30 cm reduction from full AW-2 diameter of 2.60 m (M-18 Dromader)
Sensenich Sensenich Wood Propeller Co Inc 2008 Wood Court Plant City Florida 33567 USA *Tel:* (+1 813) 752 37 11 *Fax:* (+1 813) 742 28 18 *Web:* http://www.sensenichprop.com	W70L-43 W72CK-42	Wood; 70 in diameter; Lycoming engine (A = Aeronca, C = Continental, F = Franklin, J = Jacobs, R = Ranger, W = Warner); pitch angle Wood; 72 in diameter; Continental engine (AC/BR/BE/BS/BRCK/DK/EY/FK/GK/JR/RK Continental, AF/CF = Franklin, JA = Jacobs, AA/FE/FM/JB/LY/RM = Lycoming, BB/BS/RB = Ranger, CA/CB = Warner); pitch angle
Sensenich Propeller Manufacturing Co 14 Citation Lane Lititz Pennsylvania 17543 USA *Tel:* (+1 717) 569 04 35 *Fax:* (+1 717) 560 37 25 *e-mail:* denisen@sensenich.com *Web:* http://www.sensenich.com	74DM6S5-2-60L	Metal; 74 in diameter; blade design D; hub design M6 for SAE 2 flange, ⅜ in bolts (C = ARP 502 flange, K = SAE 1, M7 = SAE 2, 7⁄16 in bolts, M8 = SAE 2, ½ in bolts, R = SAE 3); integral dowled spacer 5 × ¼ in; 2 in reduction from basic diameter; blade pitch 60 in at 75% radius; design change
VPERED ooo VPERED pr Entuziastov 15 111024 Moskva Russian Federation *Tel:* (+7 095) 273 49 28 *Fax:* (+7 095) 273 49 28 *e-mail:* vpered@online.ru	VM-3	Wood, two-blade, fixed pitch 1,850 mm diameter
VZLU VZLU Beranových 130 Prague-Letňany CZ-199 05 Czech Republic *Tel:* (+420 272) 11 53 04 *Fax:* (+420 286) 92 05 18 *e-mail:* vrtule@vzlu.cz *Web:* http://www.vzlu.cz	V230B V234 V237 V331 V534AD	Wood, two-blade, fixed pitch; 1,625 mm diameter; Rotax 447/503 (C = Rotax 912/582/618, E = Rotax 912F) two-blade Wood, two-blade, fixed pitch; 1,800 mm diameter Wood, two-blade, fixed pitch; 1,600 mm diameter (237A 1,600 mm; 237DF 1,450 mm) Wood, three-blade, ground-adjustable pitch; 1,650 mm diameter Wood, two-blade, hydraulically controlled constant-speed; 1,613 mm diameter (BD = 1,650 mm)
Warp Drive Warp Drive Inc 1207 Highway 18 East Ventura Iowa 50482 USA *Tel:* (+1 641) 357 60 00 *Fax:* (+1 641) 357 75 92		Carbon fibre, ground-adjustable pitch; two-, three-, four-, five-blade

NEW ENTRY

INDEXES

AIRCRAFT

To help users of this title evaluate the published data, *Jane's Information Group* has divided entries into three categories.

N   NEW ENTRY   Information on new equipment and/or systems appearing for the first time in the title.

V   VERIFIED   The editor has made a detailed examination of the entry's content and checked its relevance and accuracy for publication in the new edition to the best of his ability.

U   UPDATED   During the verification process, significant changes to content have been made to reflect the latest position known to *Jane's* at the time of publication.

AIRCRAFT APPEARING IN THIS EDITION ONLY. For those appearing in previous 10 editions, please see following index on tinted paper.

Index begins with aircraft generally identified only by numbers. Most were previously listed as 'Model 707' and so on under M

AIRCRAFT–PREVIOUS TEN EDITIONS

Index begins with aircraft generally identified only by numbers. Most were previously listed as 'Model 707' and so on under M

LIGHTER THAN AIR

To help users of this title evaluate the published data, *Jane's Information Group* has divided entries into three categories.

N   NEW ENTRY   Information on new equipment and/or systems appearing for the first time in the title.

V   VERIFIED   The editor has made a detailed examination of the entry's content and checked its relevance and accuracy for publication in the new edition to the best of his ability.

U   UPDATED   During the verification process, significant changes to content have been made to reflect the latest position known to *Jane's* at the time of publication. Items in italics refer to entries which have been deleted from this edition with the relevant page numbers from last year.

(AIRSHIPS APPEARING IN THIS EDITION ONLY. For those in the previous ten editions, please see the following index on tinted paper)

LIGHTER THAN AIR–PREVIOUS TEN EDITIONS

Listing includes recreational balloons, which have not appeared after the 1992-93 edition

AERO-ENGINES–PREVIOUS TEN EDITIONS

Detailed descriptions of aero-engines last appeared in the 1995-96 edition. For current data refer to table in this edition, or to full details in Jane's Aero-Engines

A

AE 301X (Allison) .... (1995-96) 756
AE 530 AC (Arrow) .... (1995-96) 715
AE 1070 AC (Arrow) .... (1995-96) 715
AE 1107 (Allison) .... (1995-96) 754
AE 2100 (Allison) .... (1995-96) 755
AE 3007 (Allison) .... (1995-96) 755
AE 3012 (Allison) .... (1995-96) 756
AI-14R (WSK-PZL Kalisz) .... (1995-96) 718
AI-20 (Progress) .... (1995-96) 739
AI-24 (Progress) .... (1995-96) 739
AI-25 (Progress) .... (1995-96) 740
AI-25TL (Progress) .... (1995-96) 740
AL-7 (Saturn) .... (1993-94) 642
AL-21 (Saturn) .... (1995-96) 732
AL-31 (Saturn) .... (1995-96) 732
AL-34 (Saturn) .... (1995-96) 733
AL-35 (Saturn) .... (1995-96) 733
ALF 502 (AlliedSignal) .... (1995-96) 751
ASSAD (Russia) .... (1995-96) 720
ASTOVL (General Electric) .... (1995-96) 757
ASz-621R (WSK-PZL Kalisz) .... (1995-96) 718
ATF-3 (AlliedSignal) .... (1995-96) 749
AVIC (see Aviation Industries of China) .... (1995-96) 695

Adour (Rolls-Royce Turbomeca) .... (1995-96) 712
Advanced STOVL (Rolls-Royce) .... (1995-96) 748
AEROSTAR SA (Romania) .... (1995-96) 720
AEROTECHNIK (Czech Republic) .. (1995-96) 700
ALFA ROMEO AVIO SpA (Italy) .... (1995-96) 714
ALLIEDSIGNAL INC (USA) .... (1995-96) 749
ALLISON ENGINE COMPANY ...... (1995-96) 753
ALVIS UAV ENGINES (UK) .... (1994-95) 731
AOI ENGINE FACTORY (EF) (Egypt) .... (1995-96) 702
Arriel (Turbomeca) .... (1995-96) 704
Arrius (Turbomeca) .... (1995-96) 704
Arrius 1D (Turbomeca) .... (1995-96) 704
ARROW ENGINEERING SRL (Italy) .... (1995-96) 715
Artouste III (Turbomeca) .... (1995-96) 705
Astazou Turboshaft (Turbomeca) .. (1995-96) 704
Atar 9K50 (SNECMA) .... (1995-96) 702
ATLAS AVIATION (South Africa) .. (1995-96) 738
AVIATION INDUSTRIES OF CHINA
 (China) .... (1995-96) 695
AVIO AR.318 (Alfa Romeo) .... (1995-96) 714

B

BAZ-21083 (Volszhsky) .... (1995-96) 737
BR 700 (BMW-RR) .... (1994-95) 694
BR 710 (BMW-RR) .... (1995-96) 706
BR 715 (BMW-RR) .... (1995-96) 706
BSEL (see Bet-Shemesh Engines) .. (1995-96) 714
BUAA (see Beijing University of Aeronautics and
 Astronautics) .... (1995-96) 695

BEIJING UNIVERSITY OF AERONAUTICS AND
 ASTRONAUTICS
 (China) .... (1995-96) 695
BET-SHEMESH ENGINES LTD (Israel) ...... (1995-96) 714
BMW ROLLS-ROYCE GmbH .... (1995-96) 706
BOMBARDIER-ROTAX GmbH (Austria) ..... (1995-96) 690

C

CAM (see Canadian Airmotive Inc) .... (1995-96) 691
CAM 100 (CAM) .... (1995-96) 691
CAREC (see China National Aero-Engine
 Corporation) .... (1995-96) 696
CEC (see Chengdu Engine Company) .... (1995-96) 696
CF 077A (Emdair) .... (1994-95) 731
CF 077B (Emdair) .... (1994-95) 731
CF 092A (Emdair) .... (1994-95) 731
CF 112 (Emdair) .... (1994-95) 731
CF6 (General Electric) .... (1995-96) 758
CF6-80C2 (General Electric) .... (1995-96) 759
CF6-80E1 (General Electric) .... (1995-96) 759
CF34 (General Electric) .... (1995-96) 757
CFE738 (CFE) .... (1995-96) 756
CFM56 (CFM International) .... (1995-96) 708
CFM88 (CFM International) .... (1995-96) 710
CLXMW (see Changzhou Lan Xiang Machinery
 Works) .... (1995-96) 696
CRM 18D/SS (CRM) .... (1995-96) 715
CT7 (General Electric) .... (1995-96) 761

CANADIAN AIRMOTIVE INC (Canada) ..... (1995-96) 691
CELMA-CIA ELECTROMECANICA
 (Brazil) .... (1995-96) 691
CFE COMPANY (USA) .... (1995-96) 756

CFM INTERNATIONAL SA
 (International) .... (1995-96) 708
CHANGZHOU LAN XIANG MACHINERY WORKS
 (China) .... (1995-96) 696
CHENGDU ENGINE COMPANY (China) ... (1995-96) 696
CHERNYSHOV (Russia) .... (1995-96) 724
CHINA NATIONAL AERO-ENGINE CORPORATION
 (China) .... (1995-96) 696
CRM (Italy) .... (1995-96) 715

D

D-18A (IL) .... (1995-96) 717
D-18T (Progress) .... (1995-96) 742
D-18TI (Progress) .... (1995-96) 742
D-18TM (Progress) .... (1995-96) 742
D-18TR (Progress) .... (1995-96) 742
D-21A1 (Aviadvigatel) .... (1995-96) 723
D-25V (JSC Aviadvigatel) .... (1994-95) 708
D-27 (Progress) .... (1995-96) 743
D-30 (JSC Aviadvigatel) .... (1995-96) 721
D-30F6 (JSC Aviadvigatel) .... (1995-96) 723
D-30K (JSC Aviadvigatel) .... (1995-96) 721
D-36 (Progress) .... (1995-96) 740
D-100/110/112 (JSC Aviadvigatel) .... (1995-96) 724
D-127 (Progress) .... (1995-96) 743
D-136 (Progress) .... (1995-96) 741
D-200 (JSC Aviadvigatel) .... (1995-96) 724
D-200/VAZ-430 (VAZ) .... (1995-96) 736
D-227 (ZMKB Progress) .... (1994-95) 730
D-236 (Progress) .... (1995-96) 741
D-436 (Progress) .... (1995-96) 741
D-727 (Progress) .... (1995-96) 743
DEMC (see Dongan Engine Manufacturing
 Company) .... (1995-96) 696
DN-200 (RKBM) .... (1995-96) 730
DV-2 (Progress) .... (1995-96) 740

DONGAN ENGINE MANUFACTURING COMPANY
 (China) .... (1995-96) 696
Dual Pac (Soloy) .... (1995-96) 767
DVADESETPRIVI MAJ BEOGRAD
 (Yugoslavia) .... (1995-96) 772

E

EJ-200 (Eurojet) .... (1995-96) 710

EMDAIR LTD (UK) .... (1994-95) 731
EUROJET TURBO GmbH (International) ... (1995-96) 710

F

F3 (IHI) .... (1995-96) 716
F23A (Hirth) .... (1995-96) 706
F30 (Hirth) .... (1995-96) 707
F100 (Pratt & Whitney) .... (1995-96) 766
F103-GE-100 (General Electric) .... (1995-96) 758
F104-GA-100 (AlliedSignal) .... (1995-96) 749
F108 (CFM International) .... (1995-96) 708
F109-GA-100 (AlliedSignal) .... (1995-96) 749
F110 (General Electric) .... (1995-96) 759
F117 (Pratt & Whitney) .... (1995-96) 765
F118 (General Electric) .... (1995-96) 760
F119 (Pratt & Whitney) .... (1995-96) 766
F124-GA-100 (AlliedSignal) .... (1995-96) 750
F125-GA-100 (AlliedSignal) .... (1995-96) 750
F129 (Williams) .... (1995-96) 772
F402, Pegasus (Rolls-Royce) .... (1995-96) 748
F404 (General Electric) .... (1995-96) 756
F405 (Rolls-Royce Turbomeca) .... (1995-96) 712
F414 (General Electric) .... (1995-96) 757
FAM (see France Aéro Moteurs) .... (1995-96) 702
FJ44 (Williams) .... (1995-96) 772

FIATAVIO (Italy) .... (1995-96) 715
FLYGMOTOR (Sweden) .... (1994-95) 725
FRANCE AERO MOTEURS SARL
 (France) .... (1995-96) 702

G

GADRI (see Guizhou Aero-Engine Design
 and Research Institute) .... (1995-96) 696
GE90 (General Electric) .... (1995-96) 757
GLC 38 (General Electric) .... (1994-95) 746
GP 1000 (Arrow) .... (1995-96) 715
GP 1500 (Arrow) .... (1995-96) 715

GTD-3 (Mars) .... (1995-96) 728
GTD-350 (WSK-PZL-Rzeszów) .... (1995-96) 719
GTE-400 (Soyuz) .... (1995-96) 736
GTRE (see Gas Turbine Research
 Establishment) .... (1995-96) 708
GTX (GTRE) .... (1995-96) 708
Galigusov (Russia) .... (1995-96) 724
GARRETT (USA) .... (1994-95) 740
GAS TURBINE RESEARCH ESTABLISHMENT
 (India) .... (1995-96) 708
Gavrilov (Russia) .... (1995-96) 724
GE AIRCRAFT ENGINES (USA) .... (1995-96) 756
Gem (Rolls-Royce) .... (1995-96) 748
GENERAL ELECTRIC (USA) .... (1995-96) 756
Gnome (Rolls-Royce) .... (1995-96) 749
GOBLER-HIRTHMOTOREN GmbH
 (Germany) .... (1995-96) 706
GROUPEMENT TURBOMECA-SNECMA
 (GRTS) (France) .... (1995-96) 706
GUIZHOU AERO-ENGINE DESIGN AND
 RESEARCH INSTITUTE (China) .... (1995-96) 696

H

H-63 (Nelson) .... (1995-96) 762
HAL (see Hindustan Aeronautics Ltd.) .... (1995-96) 708
HDHV (see Hawker de Havilland Victoria
 Ltd) .... (1995-96) 690
HIRTH (see Gobler-Hirthmotoren) .... (1995-96) 706
HIRTH 2701 R 03 (Hirth) .... (1995-96) 706
HIRTH 2702 R 03 (Hirth) .... (1995-96) 706
HIRTH 2703 (Hirth) .... (1995-96) 706
HIRTH 2704 (Hirth) .... (1995-96) 707
HS7 (DEMC) .... (1995-96) 696
HS8 (DEMC) .... (1995-96) 696

HAWKER DE HAVILLAND VICTORIA
 LTD (Australia) .... (1995-96) 690
HINDUSTAN AERONAUTICS LTD
 (India) .... (1995-96) 708

I

IAE (see International Aero Engines AG) ..... (1995-96) 711
IAI (see Israel Aircraft Industries Ltd) .... (1995-96) 714
IHI (see Ishikawajima-Harima Jukogyo
 Kabushiki Kaisha) .... (1995-96) 716
IL (see Instytut Lotnictwa) .... (1995-96) 717
IO-240 (TCM) .... (1995-96) 767
IO-320 Series (Textron Lycoming) .... (1995-96) 770
IO-360 Series (Textron Lycoming) .... (1995-96) 770
IO-520 Series (TCM) .... (1995-96) 768
IO-540 Series (Textron Lycoming) .... (1995-96) 770
IO-550 (TCM) .... (1995-96) 768
IO-720 Series (Textron Lycoming) .... (1995-96) 770
ITP (see Industria de Turbo Propulsores) .... (1995-96) 738

INDUSTRIA DE TURBO PROPULSORES
 (Spain) .... (1995-96) 738
INSTYTUT LOTNICTWA (Poland) .... (1995-96) 717
INTERNATIONAL AERO ENGINES AG
 (International) .... (1995-96) 711
INTREPRINDEREA TURBOMECANICA
 BUCURESTI (Romania) .... (1995-96) 720
ISHIKAWAJIMA-HARIMA JUKOGYO
 KABUSHIKI KAISHA (Japan) .... (1995-96) 716
ISRAEL AIRCRAFT INDUSTRIES LTD
 (Israel) .... (1995-96) 714
IVCHENKO PROGRESS ZMKB
 (Ukraine) .... (1995-96) 739

J

J52 (Pratt & Whitney) .... (1995-96) 763
J69 (Teledyne CAE) .... (1994-95) 756
JPX 4T60/A (JPX) .... (1995-96) 702
JPX 4TX 75 (JPX) .... (1995-96) 702
JT8 (Pratt & Whitney) .... (1995-96) 763
JT8D (Pratt & Whitney) .... (1994-95) 749
JT8D-200 Series (Pratt & Whitney) .... (1995-96) 763
JT9D (Pratt & Whitney) .... (1995-96) 764
JT15D (P&WC) .... (1995-96) 691
Jabiru 1600 (Jabiru Aircraft) .... (1995-96) 690
JABIRU AIRCRAFT PTY (Australia) .... (1995-96) 690
JPX, SARL (France) .... (1995-96) 702
JSC AVIADVIGATEL (Russia) ...... (1995-96) 721

NOTES

NOTES

NOTES